Encyclopedia of Pain

Encyclopedia of Pain

Gerald F. Gebhart • Robert F. Schmidt
Editors

Encyclopedia of Pain

Second Edition

Volume 1

A–Ch

With 795 Figures and 217 Tables

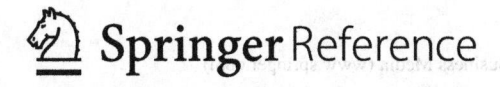

Springer Reference

Editors
Gerald F. Gebhart
Department of Anesthesiology
Center for Pain Research
University of Pittsburgh
Pittsburgh, PA, USA

Robert F. Schmidt
Physiologisches Institut der
Universität Würzburg
Würzburg, Germany

ISBN 978-3-642-28752-7   ISBN 978-3-642-28753-4 (eBook)
ISBN 978-3-642-28754-1 (print and electronic bundle)
DOI 10.1007/978-3-642-28753-4
Springer Heidelberg New York Dordrecht London

Library of Congress Control Number: 2013951551

Preface to the Second Edition

As was pointed out in the Preface to the first edition, pain is the principal reason individuals seek medical and dental attention. Fortunately, acute tissue insults associated with pain – and typically inflammation – readily resolve with appropriate treatment and care. Chronic pain states, in contrast, are notoriously difficult to manage and are commonly associated with comorbidities (e.g., depression, cross-organ sensitization) and reduced quality of life. Indeed, chronic pain has come to be considered by some as a disease in its own right. This edition of the *Encyclopedia of Pain* updates the content of the 1st edition and introduces new topics consistent with advances in our knowledge of underlying mechanisms of pain. Thus, new content on channelopathies, expanded and updated material on the role of glia and immune-competent cell interactions with nociceptive neurons, and advances in imaging and changes in brain structure in chronic pain states are included in this edition.

The cardinal objective of research on pain is the translation of knowledge between the laboratory and the clinic. The present prevailing refrain is "bench to bedside," but in fact many ideas for research in pain derive from clinical observations and the absence of knowledge to explain clinical pain states. In the recent past, basic and clinical science research has seen the application of imaging techniques to visualize areas of the brain that are involved in both the sensory and affective aspects of pain, cloning of pain-related channels and receptors, and the application of elegant genetic manipulations that selectively "label" functionally distinct dorsal root ganglion sensory and spinal neurons, two-photon imaging, and optogenetic approaches.

Accordingly, much is new in this second edition of the *Encyclopedia of Pain*. The number of authors who have contributed to this edition is close to 800. In addition to a print edition, the publisher, Springer-Verlag, will produce an electronic version that will make comprehensive searches easy to manage. The electronic version, to be made available on the online publishing platform SpringerReference, can be updated regularly to ensure that the content remains current.

October 2013
Gerald F. Gebhart   Robert F. Schmidt
Pittsburgh, PA, USA   Würzburg, Germany

Preface to the First Edition

As all medical students know, pain is the most common reason for a person to consult a physician. Under ordinary circumstances, acute pain has a useful, protective function. It discourages the individual from activities that aggravate the pain, allowing faster recovery from tissue damage. The physician can often tell from the nature of the pain what its source is. In most cases, treatment of the underlying condition resolves the pain. By contrast, children born with congenital insensitivity to pain suffer repeated physical damage and die young (see Sweet WH (1981) Pain 10:275).

Pain resulting from difficult to treat or untreatable conditions can become persistent. Chronic pain "never has a biologic function but is a malefic force that often imposes severe emotional, physical, economic, and social stresses on the patient and on the family...." (Bonica JJ (1990) The Management of Pain, vol 1, 2nd edn. Lea & Febiger, Philadelphia, p 19). Chronic pain can be considered a disease in its own right.

Pain is a complex phenomenon. It has been defined by the Taxonomy Committee of the International Association for the Study of Pain as "An unpleasant emotional and sensory experience associated with actual or potential tissue damage, or described in terms of such damage" (Merskey H and Bogduk N (1994) Classification of Chronic Pain, 2nd edn. IASP Press, Seattle). It is often ongoing, but in some cases it may be evoked by stimuli. Hyperalgesia occurs when there is an increase in pain intensity in response to stimuli that are normally painful. Allodynia is pain that is evoked by stimuli that are normally non-painful.

Acute pain is generally attributable to the activation of primary efferent neurons called nociceptors (Sherrington CS (1906) The Integrative Action of the Nervous System. Yale University Press, New Haven; 2nd edn, 1947). These sensory nerve fibers have high thresholds and respond to strong stimuli that threaten or cause injury to tissues of the body. Chronic pain may result from continuous or repeated activation of nociceptors, as in some forms of cancer or in chronic inflammatory states, such as arthritis.

However, chronic pain can also be produced by damage to nervous tissue. If peripheral nerves are injured, peripheral neuropathic pain may develop. Damage to certain parts of the central nervous system may result in central neuropathic pain. Examples of conditions that can cause central neuropathic pain include spinal cord injury, cerebrovascular accidents, and multiple sclerosis.

Research on pain in humans has been an important clinical topic for many years. Basic science studies were relatively few in number until experimental work on pain accelerated following detailed descriptions of peripheral nociceptors and central nociceptive neurons that were made in the 1960s and 1970s, by the discovery of the endogenous opioid compounds and the descending pain control systems in the 1970s and the application of modern imaging techniques to visualize areas of the brain that are affected by pain in the 1990s. Accompanying these advances has been the development of a number of animal models of human pain states, with the goal of using these to examine pain mechanisms and also to test analgesic drugs or non-pharmacologic interventions that might prove useful for the treatment of pain in humans. Basic research on pain now emphasizes multidisciplinary approaches, including behavioral testing, electrophysiology, and the application of many of the techniques of modern cell and molecular biology, including the use of transgenic animals.

The *Encyclopedia of Pain* is meant to provide a source of information that spans contemporary basic and clinical research on pain and pain therapy. It should be useful not only to researchers in these fields but also to practicing physicians and other health care professionals and to health care educators and administrators. The work is subdivided into 35 fields, and the Field Editor of each of these describes the areas covered in each in a brief review entry. The topics included in a field are the subject of a series of short essays, accompanied by key words, definitions, illustrations, and a list of significant references. The number of authors who have contributed to the encyclopedia exceeds 550. The plan of the publisher, Springer-Verlag, is to produce both print and electronic versions of this encyclopedia. Numerous links within the electronic version should make comprehensive searches easy to manage. The electronic version will be updated at sufficiently short intervals to ensure that the content remains current.

The editors thank the staff at Springer-Verlag who have provided oversight for this project, including Rolf Lange, Thomas Mager, Claudia Lange, Natasja Sheriff, and Michaela Bilic. Working with these outstanding individuals has been a pleasure.

July 2006
ROBERT F. SCHMIDT   WILLIAM D. WILLIS
Würzburg, Germany   Galveston, Texas, USA

About the Editors

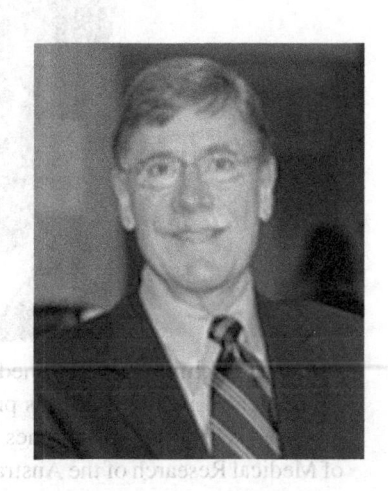

Gerald F. **Gebhart** received his PhD from The University of Iowa at Iowa City, Iowa, USA, after which he completed a 2-year fellowship at the Université de Montréal, Montréal, Canada, under the direction of Herbert H. Jasper. He began his academic career in the Department of Pharmacology at The University of Iowa in 1973, where he developed a productive research program and led a Pain Interest Group. His career includes a sabbatical year in Heidelberg, Germany (1981–1982), and leadership as head of the Department of Pharmacology from 1996 to 2006. In 2006, Dr. Gebhart was named director of the Center for Pain Research, Department of Anesthesiology, the University of Pittsburgh at Pittsburgh, Pennsylvania, USA. Dr. Gebhart has served the pain community throughout his career. He was elected to the Board of Directors of the American Pain Society (1991–1994) and subsequently elected president of the APS (1997). In 2000, he became the founding editor of *The Journal of Pain*, the official journal of the APS, serving as editor-in-chief until 2010. Dr. Gebhart also served on the Council and later as president-elect, president, and past president of the International Association for the Study of Pain (2005–2012).

Dr. Gebhart is best known for his research contributions in descending modulation of pain and mechanisms of visceral pain, and was designated in 2004 by Thomson ISI as a Highly Cited Researcher (Neuroscience). His scientific contributions have been recognized with several awards, including a 5-year Bristol Myers-Squibb Award for Excellence in Pain Research (1989–1994), a 10-year MERIT Award from the National Institutes of Health (1993–2003),

the Frederick W.L. Kerr Award from the American Pain Society (1994), the Purdue Pharma Prize for Pain Research (2004), the Janssen Award in Gastroenterology (2005), the Founders Award from the American Academy of Pain Medicine (2006), and the Patrick D. Wall Award from the British Pain Society (2012).

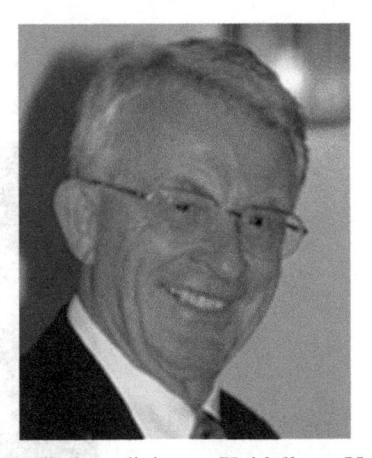

Robert F. Schmidt studied medicine at Heidelberg University (Germany). He graduated in 1959 and was promoted to Dr.med. in the same year. After residencies in clinical disciplines, Dr. Schmidt went to the John Curtin School of Medical Research of the Australian National University in Canberra, ACT, where he worked under the head of the Department of Neurophysiology, Sir John C. Eccles. He received his Ph.D. in 1963, the same year Sir John was awarded the Nobel Prize for Physiology and Medicine.

Returning to Heidelberg Medical School, Dr. Schmidt continued his academic career. He completed his habilitation in 1964, becoming a tenured lecturer (Wissenschaftlicher Rat) in 1966 and an associate professor in 1970. After an 8-month sabbatical with Sir John at the State University of New York, Dr. Schmidt became professor and director of the Department of Physiology at the University of Kiel, Germany. In 1982, he moved to the University of Würzburg to take the same position in the Department of Physiology. In 2000, Dr. Schmidt became professor emeritus at this department.

The research work of Dr. Schmidt and his associates in both Kiel and Würzburg focused on the neurophysiology of fine sensory afferents and their connections and functions in the spinal cord. In Würzburg, he concentrated on the processes leading to the changes in response characteristics of nociceptive afferents (i.e., peripheral sensitization) during inflammation of joints, discovering in the process silent nociceptors, which have subsequently been described in all vertebrates studied, including humans. In addition, Dr. Schmidt has edited, coedited, authored, and coauthored numerous textbooks and monographs, most of them having several editions and translations.

In recognition of his research contributions, Dr. Schmidt received many prizes, awards, and honors, including a D.Sc. honoris causa from the University of New South Wales, Sydney, Australia, and an Honorary Professorship from the University of Tübingen, Germany. He was elected to membership in the Akademie der Wissenschaften und der Literatur, Mainz, Germany. Dr. Schmidt is an Honorary Member of the Colombian Association for the Study of Pain, the Japan Physiological Society, the German Society for the Study of Pain, the German Pain Association, the International Association for the Study of Pain (IASP), and the German Physiological Society (in which he served as president in 1979). In 2000, he was awarded the Bundesverdienstkreuz 1st Class (Order of Merit of the German Federal Republic).

Section Editors

Ebtesam Ahmed College of Pharmacy and Health Sciences, St. John's University, Department of Pain Medicine and Palliative Care, Beth Israel Medical Center, New York, NY, USA

A. Vania Apkarian Department of Physiology, Northwestern University Feinberg School of Medicine, Chicago, IL, USA

Lars Arendt-Nielsen Center for Sensory-Motor Interaction, Department of Health Science and Technology, School of Medicine, Aalborg University, Aalborg, Denmark

Carlos Belmonte Instituto de Neurociencias, Universidad Miguel Hernandez-CSIC, San Juan de Alicante, Spain

Niels Birbaumer Institute of Medical Psychology and Behavioral Neurobiology, University of Tübingen, Tübingen, Germany

Nikolai Bogduk University of Newcastle, Newcastle Bone & Joint Institute,
Royal Newcastle Centre, Newcastle, NSW, Australia

Jin Mo Chung Department of Neuroscience and Cell Biology, University Chair in Neuroscience, University of Texas Medical Branch, Galveston, TX, USA

Joyce DeLeo Dartmouth Hitchcock Medical Center, Dartmouth Medical School, Hanover, NH, USA

Marshall Devor Department of Cell & Developmental Biology, Institute of Life Sciences and Center for Research on Pain, The Hebrew University of Jerusalem, Jerusalem, Givat Ram, Israel

Hans-Christoph Diener Department of Neurology, University Hospital Essen, University Duisburg-Essen, Essen, Germany

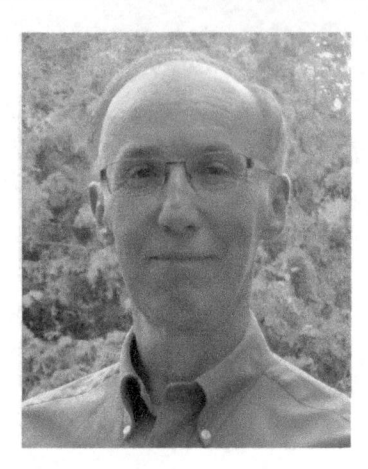

Jonathan O. Dostrovsky Department of Physiology, University of Toronto, Toronto, ON, Canada

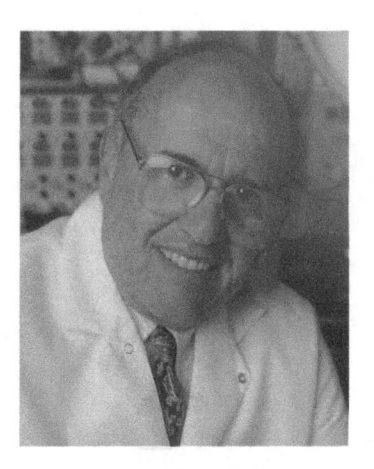

Ronald Dubner Department of Neural and Pain Sciences, University of Maryland Dental School, Baltimore, MD, USA

Michel Y. Dubois NYU Medical School, NYU Langone Medical Centers, Center for Research & Treatment of Pain, NYU Ambulatory Care Center, New York, NY, USA

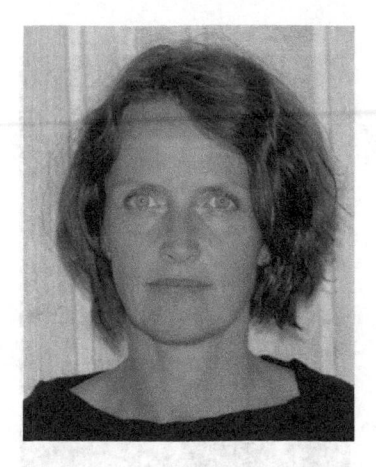

Nanna Brix Finnerup Aarhus University Hospital, Danish Pain Research Center, Aarhus, Denmark

Herta Flor Department of Cognitive and Clinical Neuroscience, Central Institute of Mental Health, Medical Faculty Mannheim, Heidelberg University, Mannheim, Germany

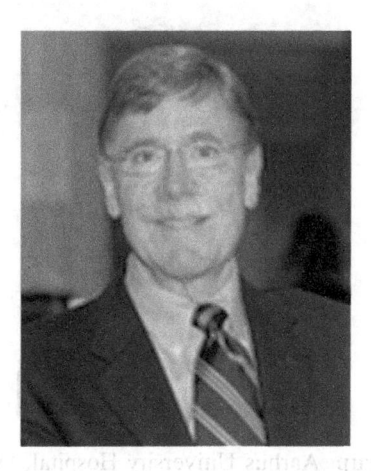

Gerald F. Gebhart Department of Anesthesiology, Center for Pain Research, University of Pittsburgh, Pittsburgh, PA, USA

Maria Adele Giamberardino Pathophysiology of Pain Laboratory and Centre for Fibromyalgia and Musculoskeletal Pain, Department of Medicine and Science of Aging, University of Chieti, Chieti, Italy

Glenn J. Giesler Department of Neuroscience, University of Minnesota, Minneapolis, MN, USA

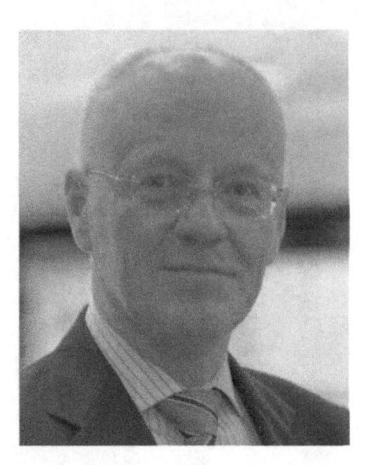

Peter Goadsby Headache Center, Department of Neurology, University of California, San Francisco, San Francisco, CA, USA

Michael S. Gold Department of Anesthesiology, Center for Pain Research, University of Pittsburgh, Pittsburgh, PA, USA

Hermann O. Handwerker Department of Physiology and Pathophysiology, University of Erlangen/Nürnberg, Erlangen, Germany

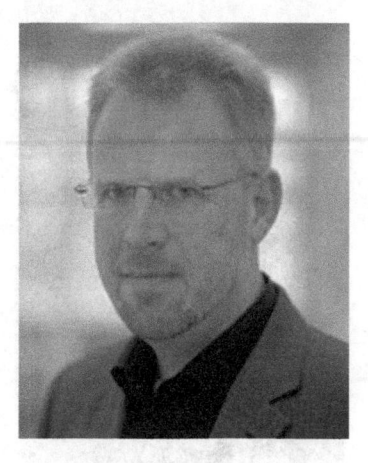

Burkhard Hinz Institute of Toxicology and Pharmacology, University of Rostock, Rostock, Germany

Wilfrid Jaenig Physiologisches Institut, Christian-Albrechts-Universität zu Kiel, Kiel, Germany

Michaela Kress Department für Physiologie und Medizinische Physik, Medizinische Universität Innsbruck, Innsbruck, Austria

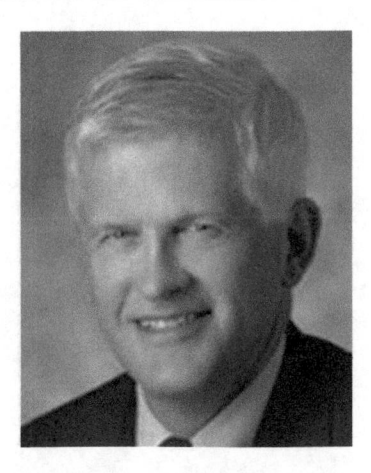

Frederick A. Lenz Department of Neurosurgery, Johns Hopkins Hospital, Baltimore, MD, USA

Tonya M. Palermo Seattle Children's Hospital Research Institute, CW8-6 - Child Health, Behavior and Development, University of Washington, Seattle, WA, USA

Frank Porreca Department of Pharmacology, University of Arizona Health
Sciences Center, Tucson, AZ, USA

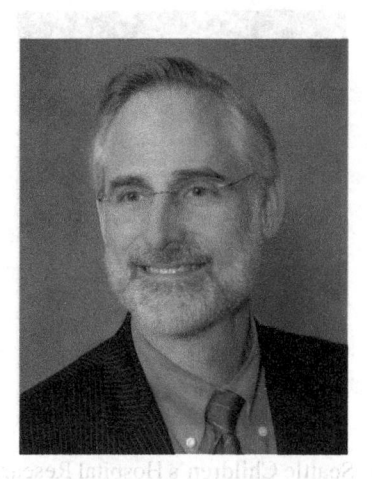

Russell K. Portenoy Department of Pain Medicine and Palliative Care, Beth
Israel Medical Center, New York, NY, USA

James P. Robinson Department of Anesthesiology & Pain Medicine, University of Washington, Seattle, WA, USA

Hans-Georg Schaible Department of Physiology - Neurophysiology, Friedrich-Schiller-University of Jena, Jena, Germany

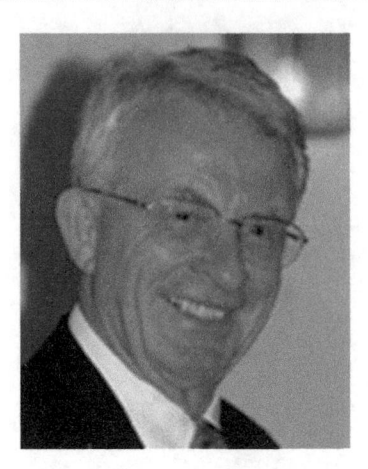

Robert F. Schmidt Physiologisches Institut der Universität Würzburg, Würzburg, Germany

Stephan A. Schug School of Medicine and Pharmacology, Royal Perth Hospital, Faculty of Medicine, Dentistry & Health Sciences, The University of Western Australia, UWA Anaesthesiology, Perth, Australia

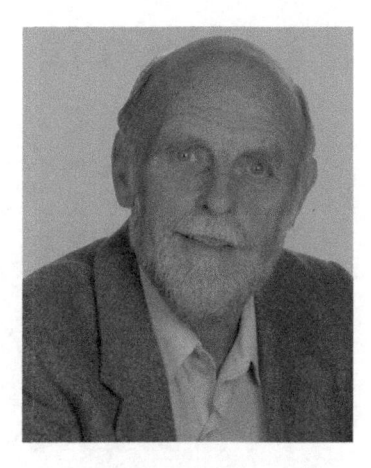

Barry J. Sessle Faculty of Dentistry, University of Toronto, Toronto, ON, Canada

Bengt H. Sjölund Department of Community Medicine and Rehabilitation Medicine, Umea University, Umea, Sweden

Mark D. Sullivan Psychiatry and Behavioral Sciences, University of Washington, Seattle, WA, USA

Irene Tracey Pain Imaging Neuroscience (PaIN) Group, Department of Physiology, Anatomy and Genetics, Oxford University, Oxford, UK

Dennis C. Turk Department of Anesthesiology & Pain Research, University of Washington, Seattle, WA, USA

Félix Viana Instituto de Neurociencias, Universidad Miguel Hernandez-CSIC, San Juan de Alicante, Alicante, Spain

Katy Vincent Nuffield Department of Obstetrics and Gynaecology, The Women's Centre, John Radcliffe Hospital, University of Oxford, Oxford, UK

Gary A. Walco Department of Anesthesiology and Pain Medicine, University of Washington School of Medicine, Seattle Children's Hospital, Seattle, WA, USA

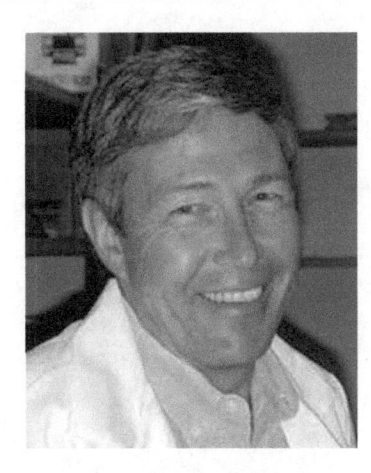

George L. Wilcox Department of Neuroscience, Pharmacology and Dermatology, University of Minnesota Medical School, Minneapolis, MN, USA

Katja Wiech Nuffield Department of Clinical Neurosciences, John Radcliffe Hospital, Oxford, UK

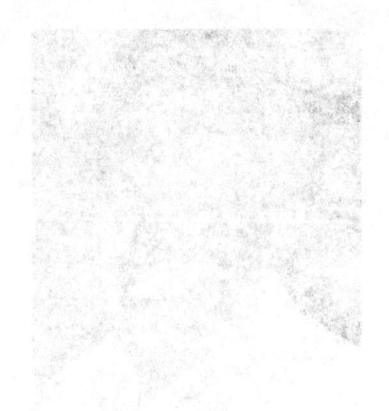

George L. Wilcox Department of Neuroscience, Pharmacology, and Toxicology, University of Minnesota Medical School, Minneapolis, MN, USA

Katja Wiech Nuffield Department of Clinical Neurosciences, John Radcliffe Hospital, Oxford, UK

Contributors

Tipu Aamir The Auckland Regional Pain Service, Auckland Hospital, Auckland, New Zealand

Catherine Abbadie Department of Pharmacology, Merck Research Laboratories, Rahway, NJ, USA

Frances V. Abbott Departments of Psychiatry and Psychology, McGill University, Montreal, QC, Canada

Heather Adams Department of Surgery, McGill University Health Centre, Montreal, QC, Canada

Mary O. Adedoyin Department of Anatomy, University of California San Francisco, San Francisco, CA, USA

Rolf H. Adler University of Berne Medical School, Kehrsatz, Switzerland

Claire D. Advokat Department of Psychology, Louisiana State University, Baton Rouge, LA, USA

Reto M. Agosti Headache Center Hirlsanden, Zurich, Switzerland

Ebtesam Ahmed College of Pharmacy and Health Sciences, St. John's University, Department of Pain Medicine and Palliative Care, Beth Israel Medical Center, New York, NY, USA

Marie-Claire Albanese Department of Psychology, McGill University, Montreal, QC, Canada

Kathryn M. Albers Department of Neurobiology, University of Pittsburgh School of Medicine, Pittsburgh, PA, USA

Phillip J. Albrecht Center for Neuropharmacology and Neuroscience, Albany Medical College, Albany, NY, USA

Elie D. Al-Chaer Center for Pain Research, University of Arkansas for Medical Sciences, Little Rock, AR, USA

Terry Altilio Department of Pain Medicine and Palliative Care, Beth Israel Medical Center, New York, USA

Rainer Amann Medical University Graz, Graz, Austria

Ron Amir Institute of Life Sciences, The Hebrew University of Jerusalem, Jerusalem, Israel

Praveen Anand Peripheral Neuropathy Unit, Imperial College London, London, UK

Karen O. Anderson The University of Texas MD Anderson Cancer Center, Houston, TX, USA

Ole K. Andersen Department of Health Science & Technology, Aalborg University, Aalborg, Denmark

Stanley W. Anderson Department of Neurosurgery, Johns Hopkins Hospital, Johns Hopkins Medical Institutions, Johns Hopkins University, Baltimore, MD, USA

Trevor Anderson Fairview Pain & Palliative Care, University of Minnesota Medical Center, Minneapolis, MN, USA

Karl-Erik Andersson Wake Forest Institute for Regenerative Medicine, Wake Forest University School of Medicine, Winston Salem, USA

Frank Andrasik Department of Psychology, Applied Psychophysiology and Biofeedback, The University of Memphis, Memphis, TN, USA

A. Vania Apkarian Department of Physiology, Northwestern University Feinberg School of Medicine, Chicago, IL, USA

Lars Arendt-Nielsen Center for Sensory-Motor Interaction, Department of Health Science and Technology, School of Medicine, Aalborg University, Aalborg, Denmark

Kirsten Arndt CNS Pharmacology, Boehringer Ingelheim Pharma GmbH & Co. KG, Biberach, Germany

Charles-Daniel Arreto Université Paris Descartes, INSERM UMR 894, France

Erik Askenasy Neurobiology and Anatomy, Health Science Center, University of Texas, Houston, TX, USA

J. H. Atkinson Psychiatry Service, VA San Diego Healthcare System and Department of Psychiatry, University of California, San Diego, CA, USA

Nadine Attal Centre d'Evaluation et de Traitement de la Douleur, Hôpital Ambroise Paré, INSERM U 987 APHP, Boulogne-Billancourt, France

Christoph Aufenberg Functional Neurosurgery, Neurosurgical Clinic University Hospital, Zürich, Switzerland

Beate Averbeck Sanofi-Aventis Deutschland GmbH, Frankfurt, Germany

Qasim Aziz Section of GI Sciences, Hope Hospital University of Manchester, Salford, UK

Cathrine Baastrup Danish Pain Research Center, Aarhus University Aarhus University Hospital, Aarhus, Denmark

F. W. Bach Danish Pain Research Center, Department of Neurology, Aarhus University Hospital, Aarhus, Denmark

S. K. Back Neuroscience Research Institute and Department of Physiology, Korea University, College of Medicine, Seoul, Republic of Korea

Misha-Miroslav Backonja Department of Neurology, University of Wisconsin-Madison, Madison, WI, USA

Banani Banerjee Division of Gastroenterology and Hepatology and Pediatric Gastroenterology, Medical College of Wisconsin, Milwaukee, WI, USA

John D. Banja Department of Rehabilitation Medicine, Center for Ethics Emory University, Atlanta, GA, USA

Carsten Bantel Department of Surgery & Cancer? Anaesthetics Section, Imperial College of Science and Medicine Chelsea and Westminster Campus, London, UK

Andrew Paul Baranowski The Pain Management Centre, National Hospital for Neurology and Neurosurgery, UCL Hospitals NHS Foundation Trust, London, UK

Alex Barling Department of Neurology, Birmingham Muscle and Nerve Centre Queen Elizabeth Hospital, Birmingham, UK

Ralf Baron Department of Neurological Pain Research and Therapy, Christian Albrechts University Kiel, Kiel, Germany

Gordon A. Barr Department of Anesthesiology and Critical Care Medicine, Children's Hospital of Philadelphia University of Pennsylvania, Philadelphia, PA, USA

Emily J. Bartley Department of Psychology, The University of Tulsa, Tulsa, OK, USA

Jeffrey R. Basford Department of Physical Medicine and Rehabilitation, Mayo Clinic and Mayo Foundation, Rochester, NY, USA

Michele C. Battie Department of Physical Therapy, University of Alberta, Edmonton, AB, Canada

Ulf Baumgartner Chair of Neurophysiology, Center for Biomedicine and Medical Technology Mannheim, Heidelberg University, Mannheim, Germany

Nicola Beale Nuffield Department of Anaesthetics, Oxford University Hospitals NHS Trust, UK

Alain Beaudet Montreal Neurological Institute, McGill University, Montreal, QC, Canada

Lino Becerra P.A.I.N. Group, Department of Anesthesiology, Perioperative and Pain Medicine, Boston Children's Hospital Department of Radiology Massachusetts General Hospital Center for Pain and the Brain, Harvard Medicals, Boston, MA, USA

Yaakov Beilin Department of Anesthesiology, and Obstetrics, Mount Sinai School of Medicine, New York, NY, USA

Alvin J. Beitz Department of Veterinary and Biomedical Sciences, University of Minnesota, St. Paul, MN, USA

Miles Belgrade Fairview Pain Management Center, University of Minnesota Medical Center, Minneapolis, MN, USA

Carlo Valerio Bellieni Neonatal Intensive Care Unit, University Hospital, Siena, Italy

Carlos Belmonte Instituto de Neurociencias, Universidad Miguel Hernandez-CSIC, San Juan de Alicante, Spain

Allan J. Belzberg Department of Neurosurgery, Johns Hopkins School of Medicine, Baltimore, MD, USA

Matt Bender Department of Neurosurgery, Johns Hopkins Hospital, Baltimore, MA, USA

Fabrizio Benedetti Department of Neuroscience, Clinical and Applied Physiology Program, University of Turin Medical School, Turin, Italy

David L. H. Bennett Wolfson CARD, Kings College London, London, UK

Nuffield Department of Clinical Neurosciences, John Radclife Hospital, The University of Oxford, Oxford, UK

Robert Bennett Department of Medicine, Oregon Health and Science University, Portland, USA

Edward C. Benzel Department of Neurosurgery, Center for Spine Health Neurological Institute Cleveland Clinic, Cleveland, OH, USA

David A. Bereiter Department of Diagnostic and Biological Sciences, University of Minnesota School of Dentistry, Minneapolis, MN, USA

Elmar Berendes Klinik für Anästhesiologie, Operative Intensivmedizin und Schmerztherapie, Klinikum Krefeld, Krefeld, Germany

Karen J. Berkley Program in Neuroscience, Florida State University, Tallahassee, FL, USA

Peter Berlit Department of Neurology, Alfried Krupp Hospital, Essen, Germany

Jean-François Bernard INSERM UMR-894, UPMC Site Pitié-Salpêtrière, Paris, France

Anne K. Bertelsen Neurology, Aleris Hospital, Bergen, Norway

Jennifer Bickel Integrative Pain Management, Children's Mercy Hospitals and Clinics, University of Missouri-Kansas City School of Medicine, Kansas City, MO, USA

Marcelo E. Bigal Department of Neurology and Montefiore Headache Unit, Albert Einstein College of Medicine, Bronx, NY, USA

Yitzchak M. Binik McGill University, Montreal, QC, Canada

Niels Birbaumer Institute of Medical Psychology and Behavioral Neurobiology, University of Tübingen, Tübingen, Germany

Frank Birklein Department of Neurology, University of Mainz, Mainz, Germany

Kathryn A. Birnie Department of Psychology, Dalhousie University, Halifax, NS, Canada

Thomas Bishop Centre for Neuroscience, King's College London, London, UK

Henning Bliddal The Parker Institute, Frederiksberg Hospital, Copenhagen, Denmark

Nikolai Bogduk University of Newcastle, Newcastle Bone & Joint Institute, Royal Newcastle Centre, Newcastle, NSW, Australia

Jorgen Boivie Department of Neurology, University Hospital, Linköping, Sweden

Michael R. Bond University of Glasgow, Glasgow, UK

Richard L. Boortz-Marx Division of Pain Medicine, Department of Anesthesiology, University of North Carolina, Chapel Hill, North Carolina, USA

James M. Borowczyk Department of Orthopaedics and Musculoskeletal Medicine, University of Otago, Christchurch, Christchurch, New Zealand

David Borsook P.A.I.N. Group, Department of Anesthesiology, Perioperative and Pain Medicine, Boston Children's Hospital Department of Radiology Massachusetts General Hospital Center for Pain and the Brain, Harvard Medicals, Boston, MA, USA

George S. Borszcz Department of Psychology, Wayne State University, Detroit, USA

Jasenka Borzan Department of Anesthesiology and Critical Care Medicine, Johns Hopkins University, Baltimore, USA

Regina M. Botting London, UK

Didier Bouhassira Centre d'Evaluation et de Traitement de la Douleur, Hôpital Ambroise Paré, INSERM U 987 APHP, Boulogne-Billancourt, France

Laurence Bourgeais INSERM UMR 894, Centre de Psychiatrie et Neurosciences, Paris, France

David Bowsher Pain Research Institute, University Hospital Aintree, Liverpool, UK

Emma L. Brandon Anaesthesiology and Pain Medicine, University of Western Australia Royal Perth Hospital, Perth, WA, Australia

Jennifer Brawn P.A.I.N. Group, Center for Pain and the Brain, Boston Children's Hospital, Harvard Medical School, Waltham, MA, USA

Harald Breivik Department of Pain Management and Research, Oslo University Hospital, Rikshospitalet and University of Oslo, Oslo, Norway

Kori L. Brewer Department of Emergency Medicine, Brody School of Medicine East Carolina University, Greenville, NC, USA

Christopher R. Brigham Brigham and Associates, Inc., Portland, USA

Abigail Broderick Nuffield Department of Obstetrics and Gynaecology, University of Oxford, Oxford, UK

Jeffrey A. Brown Department of Neurological Surgery, Wayne State University School of Medicine, Detroit, MI, USA

Stephen C. Brown Department of Anesthesia, The Hospital for Sick Children, Toronto, ON, Canada

Eduardo Bruera Department of Palliative Care and Rehabilitation Medicine, The University of Texas M. D. Anderson Cancer Center, Houston, TX, USA

Pablo Brumovsky Faculty of Biomedical Sciences, Austral University, Buenos Aires, Argentina

Consejo Nacional de Investigaciones Científicas y Técnicas (CONICET), Buenos Aires, Argentina

Kay Brune Department of Experimental and Clinical Pharmacology and Toxicology, Friedrich Alexander University Erlangen-Nürnberg, Erlangen, Germany

Maria Burian Department of General, Visceral and Transplantation Surgery, University of Heidelberg, Heidelberg, Germany

Dan Buskila Rheumatic Disease Unit, Soroka Medical Center, Beer Sheva, Israel

Catherine M. Cahill Anesthesiology & Perioperative Care, University of California, Irvine, Irvine, CA, USA

Brian E. Cairns Faculty of Pharmaceutical Sciences, University of British Columbia, Vancouver, BC, Canada

Nigel A. Calcutt Department of Pathology, University of California San Diego, La Jolla, CA, USA

Margarita Calvo Wolfson CARD, Kings College London, London, UK

James N. Campbell Johns Hopkins University School of Medicine, Baltimore, MD, USA

Lauren Campbell Department of Psychology, York University, Toronto, ON, Canada

Ling Cao Department of Biomedical Sciences, College of Osteopathic Medicine, University of New England, ME, USA

Adam Carinci Department of Anesthesia, Critical Care and Pain Medicine, Massachusetts General Hospital Harvard Medical School, Boston, MA, USA

Giancarlo Carli Department of Physiology, University of Siena, Siena, Italy

Christer P. O. Carlsson Rehabilitation Clinic, Lunds University Hospital, Lund, Sweden

Susan M. Carlton Neuroscience and Cell Biology, University of Texas Medical Branch, Galveston, TX, USA

Richard Carr Clinic for Anaesthesiology, Heidelberg University, Mannheim, Baden-Wuerttemberg, Germany

Benjamin S. Carson Department of Neurosurgery, Johns Hopkins Hospital, Baltimore, MA, USA

Brian Carter University of Missouri at Kansas City Children's Mercy Hospital and Bioethics Center, Kansas City, MO, USA

Kenneth L. Casey Department of Neurology, University of Michigan Medical Center, Ann Arbor, MI, USA

Christine T. Chambers Departments of Pediatrics and Psychology, Dalhousie University, Halifax, NS, Canada

Sandra Chaplan Janssen Research and Development, LLC, San Diego, CA, USA

C. Richard Chapman Department of Anesthesiology, University of Utah School of Medicine, Salt Lake City, UT, USA

Santosh K. Chaturvedi Psychiatry, National Institute of Mental Health & Neurosciences, Bangalore, India

Andrew C. N. Chen Human Brain Mapping and Cortical Imaging Laboratory, Aalborg University, Aalborg, Denmark

Hsinlin Thomas Cheng Department of Neurology, University of Michigan, Ann Arbor, MI, USA

Pavel Cherkas Faculty of Dentistry, University of Toronto, Toronto, ON, Canada

Andrea Cheville Department of Physical Medicine and Rehabilitation, Mayo Clinic, Rochester, MN, USA

Chen Yu Chiang Faculty of Dentistry, University of Toronto, Toronto, ON, Canada

N. L. Chillingworth School of Physiology and Pharmacology, University of Bristol, Bristol, UK

Edward Chow Department of Radiation Oncology, Sunnybrook Health Sciences Centre Odette Cancer Centre (T2-182), Toronto, ON, Canada

Julie A. Christianson Department of Anatomy and Cell Biology, University of Kansas Medical Center, Kansas City, KS, USA

MacDonald J. Christie Discipline of Pharmacology, University of Sydney, Sydney, Australia

Jin Mo Chung Department of Neuroscience and Cell Biology, University Chair in Neuroscience, University of Texas Medical Branch, Galveston, TX, USA

Kyungsoon Chung Department of Neuroscience and Cell Biology, University of Texas Medical Branch, Galveston, TX, USA

Anna K. Clark Wolfson Centre for Age Related Diseases, King's College London, London, UK

W. Crawford Clark College of Physicians and Surgeons, Columbia University, New York, USA

James F. Cleary Pain & Policy Studies Group, UW Carbone Cancer Center, Madison, WI, USA

John De Clercq Department of Rehabilitation Sciences and Physiotherapy, Ghent University Hospital, Ghent, Belgium

Jacqui Clinch Royal National Hospital for Rheumatic Diseases NHS Foundation Trust, Bath, UK

Terence J. Coderre Departments of Anesthesia, Neurology & Neurosurgery and Psychology, McGill University, Montreal, QC, Canada

Robert Coghill Department of Neurobiology and Anatomy, Wake Forest University School of Medicine, Winston-Salem, NC, USA

Lindsey Cohen Psychology, Georgia State University, Atlanta, GA, USA

Alan L. Colledge Utah Labor Commission, International Association of Industrial Accident Boards and Commissions Central Utah Medical Clinic, South Salt Lake City, UT, USA

Beverly J. Collett University Hospitals of Leicester, Leicester, UK

Lesley A. Colvin Department of Anaesthesia, Critical Care and Pain Medicine Western General Hospital, Edinburgh, UK

Kevin C. Conlon Trinity College Dublin, University of Dublin, Dublin, Ireland

Mark Connelly Integrative Pain Management, Children's Mercy Hospitals and Clinics, University of Missouri-Kansas City School of Medicine, Kansas City, MO, USA

Brian Y. Cooper Department of Oral and Maxillofacial Surgery, University of Florida, Gainesville, FL, USA

Christine Cordle University Hospitals of Leicester, Leicester, UK

Laura A. Cousins Psychology, Georgia State University, Atlanta, GA, USA

Michael J. Cousins Department of Anesthesia and Pain Management, Royal North Shore Hospital University of Sydney, Sydney, NSW, Australia

Verne C. Cox Department of Psychology, University of Texas at Arlington, Arlington, TX, USA

George A. Crabill Department of Neurosurgery, Johns Hopkins Hospital, Baltimore, MA, USA

Rebecca M. Craft Psychology, Washington State University, Pullman, WA, USA

Kenneth D. Craig Department of Psychology, University of British Columbia, Vancouver, BC, Canada

Richard Crevenna Medizinische Universität Wien, Wien, Austria

Geert Crombez Department of Experimental Clinical and Health Psychology, Ghent University, Ghent, Belgium

Joan Crook School of Nursing, Faculty of Health Sciences, McMaster University, Hamilton, ON, Canada

Giorgio Cruccu Department of Neurological Sciences, La Sapienza University, Rome, Italy

Michele Curatolo Department of Anesthesiology, Division of Pain Therapy, University Hospital of Bern Inselspital, Bern, Switzerland

Nachum Dafny Neurobiology and Anatomy, Health Science Center, University of Texas, Houston, TX, USA

Jorgen B. Dahl Department of Anaesthesiology, Glostrup University Hospital, Herlev, Denmark

Yi Dai Department of Anatomy and Neuroscience, Hyogo College of Medicine, Hyogo, Japan

Stefaan Van Damme Department of Experimental Clinical and Health Psychology, Ghent University, Ghent, Belgium

Alexandre F. M. DaSilva Biologic & Materials Sciences, University of Michigan Dental School, MI, USA

Steve Davidson Pain Center & Department of Anesthesiology, Washington University School of Medicine, MO, USA

Brian Davis Department of Neurobiology, University of Pittsburgh School of Medicine, Pittsburgh, PA, USA

Karen D. Davis Brain, Imaging and Behaviour – Systems Neuroscience, Toronto Western Research Institute, Toronto, ON, Canada

Department of Surgery, University of Toronto, Toronto, ON, Canada

Richard O. Day Department of Clinical Pharmacology and Toxicology, St Vincent's Hospital, Sydney, NSW, Australia

Isabelle Decosterd Pain Center, University Hospital Center and University of Lausanne, Lausanne, Switzerland

Ruth Defrin Physical Therapy, Faculty of Medicine, Tel-Aviv University, Tel-Aviv, Israel

Antoon F. C. De Laat Department of Oral/Maxillofacial Surgery, Catholic University of Leuven, Leuven, Belgium

Joyce DeLeo Dartmouth Hitchcock Medical Center, Dartmouth Medical School, Hanover, NH, USA

A. Lee Dellon Department of Plastic Surgery and Neurosurgery, Johns Hopkins University, Baltimore, USA

Dellon Institutes for Peripheral Nerve Surgery, Baltimore, MD, USA

Marshall Devor Department of Cell & Developmental Biology, Institute of Life Sciences and Center for Research on Pain, The Hebrew University of Jerusalem, Jerusalem, Givat Ram, Israel

Perry Dhaliwal Department of Neurosurgery, Cleveland Clinic Spine Institute, Cleveland, OH, USA

David Diamant Neurological and Spinal Surgery, Lincoln, NE, USA

Anthony H. Dickenson Department of Pharmacology, University College London, London, UK

Hans-Christoph Diener Department of Neurology, University Hospital Essen, University Duisburg-Essen, Essen, Germany

Natalia Dmitrieva Program in Neuroscience, Florida State University, Tallahassee, FL, USA

L. F. Donaldson School of Life Sciences, University of Nottingham, Nottingham, UK

Henri Doods CNS Pharmacology, Boehringer Ingelheim Pharma GmbH & Co. KG, Biberach, Germany

Victoria Dorf Social Security Administration, Baltimore, MD, USA

Michael J. Dorsi Department of Neurosurgery, Johns Hopkins School of Medicine, Baltimore, MD, USA

Jonathan O. Dostrovsky Department of Physiology, University of Toronto, Toronto, ON, Canada
Faculty of Dentistry, University of Toronto, Toronto, ON, Canada

Robyn Drach Fairview Pain Management Center, Minneapolis, MN, USA

Ron Dressler Utah Labor Commission, International Association of Industrial Accident Boards and Commissions Central Utah Medical Clinic, South Salt Lake City, UT, USA

Paul Dreyfuss Department of Rehabilitation Medicine, University of Washington, Seattle, USA

Peter D. Drummond School of Psychology, Murdoch University, Perth, Australia

Ronald Dubner Department of Neural and Pain Sciences, University of Maryland Dental School, Baltimore, MD, USA

Anne Ducros Headache Emergency Department, INSERM, Paris, France

Gary H. Duncan Département de stomatologie, Faculté de médecine dentaire, Université de Montréal, Montreal, QC, Canada

Department of Neurology and Neurosurgery, McGill University, Montreal, QC, Canada

Mary J. Eaton Research Service (151), Miami VA Health Care System, Miami VAMC, Miami, FL, USA

Andrea Ebersberger Department of Physiology, Friedrich Schiller University of Jena, Jena, Germany

Christopher Eccleston Pain Management Unit, University of Bath, Bath, UK

John Edmeads Sunnybrook and Woman's College Health Sciences Centre, University of Toronto, Toronto, Canada

Edzard Ernst Complementary Medicine, Peninsula Medical School Universities of Exeter and Plymouth, Exeter, UK

Reuben Escorpizo Swiss Paraplegic Research, Nottwil, Switzerland

Department of Health Sciences and Health Policy, University of Lucerne, Lucerne, Switzerland

ICF Research Branch in cooperation with the World Health Organization (WHO) Collaborating Centre for the Family of International Classifications in Germany, Nottwil, Switzerland

Schonstein Eva RehabCo, Fyshwick, ACT, Australia

Joshua C. Eyer Department of Psychology, University of Alabama, Tuscaloosa, AL, USA

Marie Faaborg-Andersen McGill University, Montreal, QC, Canada

Keri L. Fakata College of Pharmacy and Pain Management Center, University of Utah, Salt Lake City, USA

Marie Fallon Department of Oncology, Edinburgh Cancer Center University of Edinburgh, Edinburgh, UK

Melissa Farmer Department of Physiology / Feinberg School of Medicine, Northwestern University, Chicago, IL, USA

Samantha R. Fashler Department of Psychology, University of British Columbia, Vancouver, BC, Canada

Jean-Baptiste Fassier Department of Occupational Health and Medicine, UMRESTTE-Université Claude Bernard Lyon 1, Lyon, France

Charlotte Feinmann Eastman Dental Institute London, University College London, London, UK

Elizabeth Roy Felix The Miami Project to Cure Paralysis, University of Miami Miller School of Medicine, Miami, FL, USA

Bin Feng Center for Pain Research, Department of Anesthesiology, School of Medicine, University of Pittsburgh, Pittsburgh, Pennsylvania, USA

Yi Feng Department of Anesthesiology, Peking University People's Hospital, Peking, China

Andres Fernandez Department of Neurosurgery, Johns Hopkins Hospital, Baltimore, MD, USA

Antonio Vicente Ferrer-Montiel Institute of Molecular and Cell Biology, University Miguel Hernández, Elche, Alicante, Spain

Perry G. Fine Pain Research Center, School of Medicine, University of Utah, Salt Lake City, UT, USA

David J. Fink Department of Neurology, University of Michigan and VA Ann Arbor Healthcare System, Ann Arbor, MI, USA

Nanna Brix Finnerup Aarhus University Hospital, Danish Pain Research Center, Aarhus, Denmark

David A. Fishbain Department of Psychiatry, University of Miami, Miami, FL, USA

Maria Fitzgerald Neuroscience, Physiology & Pharmacology, University College London, London, UK

Per Flisberg Department of Anaesthesia and Intensive Care, Lund University Hospital, Lund, Sweden

Herta Flor Department of Cognitive and Clinical Neuroscience, Central Institute of Mental Health, Medical Faculty Mannheim, Heidelberg University, Mannheim, Germany

Christopher M. Flores Drug Discovery, Johnson and Johnson Pharmaceutical Research and Development, Spring House, PA, USA

Wilbert E. Fordyce Department of Rehabilitation, University of Washington School of Medicine, Seattle, WA, USA

Robert D. Foreman Department of Physiology, College of Medicine, University of Oklahoma Health Sciences Center, Oklahoma City, OK, USA

Gary M. Franklin Department of Environmental and Occupational Health Sciences, University of Washington, Seattle, WA, USA

Department of Health Services, University of Washington, Seattle, WA, USA

Jan Fridén Department of Hand Surgery, Sahlgrenska University Hospital, Gothenborg, Sweden

Justin Friedlander The Arthur Smith Institute for Urology, North Shore-LIJ Healthcare System, Hofstra North Shore LIJ School of Medicine, New Hyde Park, NY, USA

Allan H. Friedman Division of Neurosurgery, Duke University Medical Center, Durham, NC, USA

Perry N. Fuchs Department of Psychology, University of Texas at Arlington, Arlington, TX, USA

Deborah Fulton-Kehoe Department of Environmental and Occupational Health Sciences, University of Washington, Seattle, WA, USA

Andrea D. Furlan Institute for Work & Health, Toronto, ON, Canada

Vasco Galhardo Departamento de Biologia Experimental, Faculdade de Medicina, Universidade do Porto, Portugal

Andreas R. Gantenbein RehaClinic Bad Zurzach, Bad Zurzach, Switzerland

I. Garonzik Department of Neurosurgery, Johns Hopkins Medical Institutions, Baltimore, MD, USA

Robert Gassin Musculoskeletal Medicine Clinic, Frankston, VIC, Australia

Robert J. Gatchel Department of Psychology, The University of Texas, Arlington, TX, USA

Gerald F. Gebhart Department of Anesthesiology, Center for Pain Research, University of Pittsburgh, Pittsburgh, PA, USA

Gerd Geisslinger Department of Clinical Pharmacology, University Frankfurt am Main, Frankfurt, Germany

Robert D. Gerwin Department of Neurology, Johns Hopkins University, Bethesda, MD, USA

Maria Adele Giamberardino Pathophysiology of Pain Laboratory and Centre for Fibromyalgia and Musculoskeletal Pain, Department of Medicine and Science of Aging, University of Chieti, Chieti, Italy

Stephen Gibson National Ageing Research Institute Department of Medicine, University of Melbourne, Parkville, VIC, Australia

Glenn J. Giesler Department of Neuroscience, University of Minnesota, Minneapolis, MN, USA

Philip L. Gildenberg Baylor Medical College, Houston, TX, USA

Myra Glajchen Department of Pain Medicine and Palliative Care, Beth Israel Medical Center Albert Einstein College of Medicine, New York, NY, USA

Hartmut Göbel Kiel Migraine and Headache Center, Kiel Pain Center, Kiel, Germany

Peter Goadsby Headache Center, Department of Neurology, University of California, San Francisco, San Francisco, CA, USA

Philippe Goffaux Faculty of Medicine, University of Sherbrooke, Sherbrooke, QC, Canada

Ana Gomis Instituto de Neurociencias de Alicante, Universidad Miguel Hernández-CSIC, San Juan de Alicante, Spain

Gilbert R. Gonzales Department of Neurology, Memorial Sloan-Kettering Cancer Center, New York, USA

Allan Gordon Wasser Pain Management Centre, Mount Sinai Hospital University of Toronto, Toronto, ON, Canada

Liesbet Goubert Department of Experimental Clinical and Health Psychology, Ghent University, Ghent, Belgium

Jayantilal Govind Pain Clinic, Department of Anaesthesia, University of New South Wales, Sydney, NSW, Australia

Richard H. Gracely Departments of Medicine-Rheumatology and Neurology, University of Michigan Health System, Ann Arbor, MI, USA

Garry G. Graham Department of Pharmacology, School of Medical Sciences, University of New South Wales, Sydney, NSW, Australia

David K. Grandy Department of Physiology and Pharmacology, Oregon Health and Science University, Portland, OR, USA

Thomas Graven-Nielsen Laboratory for Musculoskeletal Pain and Motor Control, Center for Sensory-Motor Interaction, Aalborg University, Aalborg, Denmark

Barry Green Pierce Laboratory, Yale School of Medicine, New Haven, CT, USA

Paul G. Green Oral & Maxillofacial Surgery, University of California San Francisco, San Francisco, CA, USA

Joel D. Greenspan Department of Neural and Pain Sciences, University of Maryland School of Dentistry, Baltimore, MD, USA

John W. Griffin Department of Neurology, Johns Hopkins University School of Medicine, Baltimore, MD, USA

Cornelius Groenwald Department of Anesthesiology and Pain Medicine, Seattle Children's Hospital, University of Washington School of Medicine, Seattle, WA, USA

Douglas P. Gross Department of Physical Therapy, University of Alberta WCB-Alberta Millard Health, Edmonton, AB, Canada

Sabine Grösch Institute of Clinical Pharmacology, Clinical Center of the Goethe University, Frankfurt, Germany

Blair D. Grubb Department of Cell Physiology and Pharmacology, University of Leicester, Leicester, UK

Ruth Eckstein Grunau Pediatrics Dept, University of British Columbia Developmental Neurosciences & Child Health, Child & Family Research Institute, Vancouver, BC, Canada

Christoph Gutenbrunner Department of Physical Medicine and Rehabilitation, Hanover Medical School, Hanover, Germany

Maija Haanpää Department of Neurosurgery, Helsinki University Central Hospital, Helsinki, Finland

Sherri Haas Fairview Pain Management Center, Minneapolis, MN, USA

Ute Habel Department of Psychiatry and Psychotherapy, RWTH Aachen University, Aachen, Germany

Thomas Hachenberg Clinic for Anaesthesiology and Intensive Therapy, Otto-von-Guericke University, Magdeburg, Germany

Nouchine Hadjikhani Martinos Center for Biomedical Imaging, Massachusetts General Hospital Harvard Medical School, Charlestown, MA, USA

Ole Haegg Department of Orthopedic Surgery, Sahlgren University Hospital, Goteborg, Sweden

Bryan C. Hains Department of Neurology, Yale University School of Medicine, West Haven, CT, USA

Else K. B. Hals University of Oslo, Faculty of Dentistry, Oslo, Norway

Darryl T. Hamamoto Diagnostic and Biological Sciences, University of Minnesota, Minneapolis, MN, USA

Donna L. Hammond Department of Anesthesia, University of Iowa, Iowa City, IA, USA

Menachem Hanani Hadassah-Hebrew University Medical Center Hebrew University Medical School, Jerusalem, Israel

Hermann O. Handwerker Department of Physiology and Pathophysiology, University of Erlangen/Nürnberg, Erlangen, Germany

George M. Hanna Department of Anesthesia, Critical Care and Pain Medicine, Massachusetts General Hospital Harvard Medical School, Boston, MA, USA

Jing-Xia Hao Department of Physiology and Pharmacology, Karolinska Institutet, Stockholm, Sweden

Eleni G. Hapidou Department of Psychiatry and Behavioural Neurosciences, McMaster University and Hamilton Health Sciences Chronic Pain Management Unit, Chedoke Hospital, Hamilton, ON, Canada

Geoffrey Harding Sandgate Spinal Medicine Clinic, Sandgate, QLD, Australia

Louise Harding Hospitals NHS Foundation Trust, University College London, London, UK

Kenneth M. Hargreaves Department of Endodontics, University of Texas Health Science Center at San Antonio, San Antonio, USA

D. Paul Harries Pain Treatment Center, Samaritan Hospital Lexington, Lexington, KY, USA

Denise Harrison Nursing, Children's Hospital of Eastern Ontario (CHEO), University of Ottawa, Ottawa, ON, Canada

Steven Harte Anesthesiology and Internal Medicine/Rheumatology, University of Michigan VA Ann Arbor Healthcare System, Ann Arbor, MI, USA

Barbara A. Hastie Community Dentistry & Behavioral Science, University of Florida College of Dentistry, Gainesville, FL, USA

Jennifer A. Haythornthwaite Psychiatry & Behavioral Sciences, Johns Hopkins University, Baltimore, MD, USA

Heinz-Joachim Häbler FH Bonn-Rhein-Sieg, Rheinbach, Germany

John H. Healey Orthopaedic Service, Department of Surgery, Memorial Sloan-Kettering Cancer Center Weill Medical College of Cornell University, New York, NY, USA

Alicia A. Heapy VA Connecticut Healthcare System, Westhaven, CT, USA

Anne M. Heffernan The Adelaide and Meath Hospital Incorporating The National Children's Hospital, Dublin, Ireland

Geert Van der Heijden Julius Center, Department of Epidemiology, University Medical Center Utrecht, Utrecht, The Netherlands

Mary M. Heinricher Departments of Neurological Surgery and Behavioral Neuroscience, Oregon Health & Science University, Portland, USA

Robert D. Helme Department of Medicine, Royal Melbourne Hospital, University of Melbourne, Melbourne, Australia

Kim Helsen Research Group on Health Psychology, KU Leuven Katholieke Universiteit Leuven, Leuven, Belgium

Department of Experimental–Clinical and Health Psychology, Ghent University, Ghent, Belgium

Amin S. Herati The Arthur Smith Institute for Urology, North Shore-LIJ Healthcare System, Hofstra North Shore LIJ School of Medicine, New Hyde Park, NY, USA

Robert D. Herbert Neuroscience Research Australia, Sydney, Australia

Christiane Hermann Department of Clinical Psychology, Justus-Liebig University Giessen, Giessen, Germany

Juan F. Herrero Department of Physiology, Faculty of Medicine University of Alcala, Alcalá de Henares Madrid, Spain

Aki J. Hietaharju Department of Neurology and Rehabilitation, Pain Clinic Tampere University Hospital, Tampere, Finland

Karin Westlund High Department of Physiology, University of Kentucky, Lexington, KY, USA

Marita Hilliges Dalarna University, Halmstad, Sweden

Max J. Hilz University of Erlangen-Nuremberg, Erlangen, Germany

Burkhard Hinz Institute of Toxicology and Pharmacology, University of Rostock, Rostock, Germany

Anthony R. Hobson Section of GI Sciences, Hope Hospital University of Manchester, Salford, UK

Edward Högestätt Department of Laboratory Medicine, Lund University, Division of Clinical Chemistry and Pharmacology, Lund, Sweden

Tomas Hökfelt Department of Neuroscience, Karolinska Institute, Stockholm, Sweden

Sarah V. Holdridge Department of Pharmacology and Toxicology, Queen's University, Kinston, Canada

Kjell Hole University of Bergen, Bergen, Norway

Graham R. Holland Department of Cariology, University of Michigan, Ann Arbor, MI, USA

Chang-Zern Hong Department of Physical Medicine and Rehabilitation, University of California, Irvine, CA, USA

Minoru Hoshiyama Department of Integrative Physiology, National Institute for Physiological Sciences, Okazaki, Japan

Fred M. Howard University of Rochester, Rochester, NY, USA

Sung-Tsang Hsieh Department of Anatomy and Cell Biology, Department of Neurology, National Taiwan University Hospital, National Taiwan University, Taipei, Taiwan

James W. Hu Faculty of Dentistry, University of Toronto, Toronto, Canada

Claire E. Hulsebosch Spinal Cord Injury Research, University of Texas Medical Branch, Galveston, TX, USA

Thomas Hummel Department of Otorhinolaryngology, Technical University of Dresden Medical School, Dresden, Germany

Mark R. Hutchinson School of Medical Sciences, University of Adelaide, Adelaide, SA, Australia

Charles E. Inturrisi Department of Pharmacology, Weill Cornell Medical College, New York, USA

Koji Inui Department of Integrative Physiology, National Institute for Physiological Sciences, Okazaki, Japan

Koichi Iwata Department of Physiology, Nihon University School of Dentistry, Chiyoda-ku, Japan

Suhayl J. Jabbur Anatomy, Cell Biology and Physiology, Faculty of Medicine, American University of Beirut, Beirut, Lebanon

Neuroscience Program, Faculty of Medicine, American University of Beirut, Beirut, Lebanon

Nora Janjan University of Texas M. D. Anderson Cancer Center, Houston, TX, USA

Luc Jasmin Department of Anatomy, University of California San Francisco, San Francisco, CA, USA

Anne Jaumees Royal North Shore Hospital, St Leonards, NSW, Australia

Wilfrid Jaenig Physiologisches Institut, Christian-Albrechts-Universität zu Kiel, Kiel, Germany

Daniel Jeanmonod Functional Neurosurgery, The Center of Ultrasound Functional Neurosurgery, Solothurn, Switzerland

Andrew Jeffreys Director Department of Anaesthesia & Pain Medicine, Western Health, Melbourne, VIC, Australia

Mark P. Jensen Department of Rehabilitation Medicine, University of Washington, Seattle, WA, USA

Troels Staehelin Jensen Danish Pain Research Center, Department of Neurology, Aarhus University Hospital, Aarhus, Denmark

Kurt L. Johnson Department of Rehabilitation Medicine, University of Washington, Seattle, WA, USA

Mark I. Johnson Faculty of Health and Social Sciences, Leeds Pallium Research Group, Leeds Metropolitan University, Leeds, UK

Mark Johnston Hibiscus Coast Highway, Orewa, New Zealand

Edward Jones Center for Neuroscience, University of California, Davies, CA, USA

Jeroen R. De Jong Department of Rehabilitation, University Hospital Maastricht, Maastricht, The Netherlands

Sven-Eric Jordt Department of Pharmacology, Yale University School of Medicine, New Haven, CT, USA

Ellen Jørum The Laboratory of Clinical Neurophysiology, Rikshospitalet University, Oslo, Norway

Stefan Just CNS Pharmacology, Bochringer Ingelheim Pharma GmbH & Co. KG, Biberach, Germany

Timothy K. Y. Kaan Department of Anatomy, University of California San Francisco, San Francisco, CA, USA

Noufissa Kabli Department of Pharmacology and Toxicology, Queen's University, Kinston, Canada

Ryusuke Kakigi Department of Integrative Physiology, National Institute for Physiological Sciences, Okazaki, Japan

Peter C. A. Kam Anaesthesia, Central Clinical School, The University of Sydney, Sydney, Australia

Giresh Kanji The Sports and Pain Clinic, Wellington, New Zealand

Joel Katz Department of Psychology, McGill University, Montreal, QC, Canada

Valerie Kayser NSERM U677 91Bd de l'Hôpital, Paris, France

Francis J. Keefe Pain Prevention and Treatment Research, Department of Psychiatry and Behavioral Sciences, Duke University Medical Center, Durham, NC, USA

Lois J. Kehl Department of Anesthesiology, University of Minnesota, Minneapolis, MN, USA

Maureen Kelley Department of Pediatrics, Bioethics Division, University of Washington School of Medicine, WA, USA

Seattle Children's Hospital, Seattle, WA, USA

Stephen Kennedy Nuffield Department of Obstetrics and Gynaecology, University of Oxford, Oxford, UK

Edmund Keogh Department of Psychology, University of Bath, Bath, UK

Robert D. Kerns VA Connecticut Healthcare System, Westhaven, CT, USA

Sanjay Khanna Department of Physiology, National University of Singapore, Singapore

Kok Eng Khor Department of Pain Management, Prince of Wales Hospital University of New South Wales, Randwick, NSW, Australia

Mina Khoshnejad Department of Neuroscience, University of Montreal, Montreal, QC, Canada

H. J. Kim Department of Life Science, Yonsei University, Wonju, Republic of Korea

Sara King Faculty of Education (School Psychology), Mount Saint Vincent University, Halifax, NS, Canada

Wade King Department of Clinical Research, Royal Newcastle Centre University of Newcastle, Newcastle, NSW, Australia

K. L. Kirsh Behavioral Medicine, The Pain Treatment Center of the Blue grass, Lexington, KY, USA

Henriette Klit Danish Pain Research Center, Aarhus University Hospital, Aarhus C, Denmark

Ricardo J. Komotar Department of Neurological Surgery, Neurological Institute Columbia University, New York, NY, USA

Sungtae Koo Division of Meridian and Structural Medicine, School of Korean Medicine, Pusan National University, Yangsan, Korea

Rhonda K. Kotarinos Oakbrook Terrace, IL, USA

Katalin J. Kovacs Department of Veterinary and Biomedical Sciences, University of Minnesota, St. Paul, MN, USA

Malgorzata Krajnik Chair of Palliative Care, Nicolaus Copernicus University in Torun, Collegium Medicum in Bydgoszcz, Bydgoszcz, Poland

Elliot Krane Anesthesia and Pediatrics, Stanford University School of Medicine Lucile Packard Children's Hospital at Stanford, Stanford, CA, USA

Ajit Krishnaney Department of Neurosurgery, Cleveland Clinic, Cleveland, OH, USA

Greg Krohm International Association of Industrial Accident Boards and Commissions (IAIABC), Utah Valley University, USA

Lawrence Kruger UCLA Geffen School of Medicine, Los Angeles, CA, USA

M. M. ter Kuile Department of Gynecology, Leiden University Medical Center, Leiden, The Netherlands

Promil Kukreja Department of Anesthesiology, University of Alabama at Birmingham, Birmingham, AL, USA

Alice Kvåle Department of Public Health and Primary Health Care, Faculty of Medicine and Dentistry University of Bergen, Bergen, Norway

Gunnvald Kvarstein Department of Pain Management and Pain Research, Oslo University Hospital, Rikshospitalet, Oslo, Norway

Jean-Jacques Labat Neurology and Rehabilitation, Clinique Urologique, Nantes, France

Gaston Labrecque Faculty of Pharmacy, University of Laval, Quebec City, QC, Canada

Marco Lacerenza Pain Medicine Center, Casa di Cura S. Pio X, Opera San Camillo Foundation, Milan, Italy

Uri Ladabaum Gastroenterology & Hepatology, Stanford Cancer Institute, Redwood City, CA, USA

Marie Andree Lahaie McGill University, Montreal, QC, Canada

Miguel J. A. Lainez-Andres Department of Neurology, University of Valencia, Valencia, Spain

Robert H. LaMotte Department of Anesthesiology, Yale School of Medicine, New Haven, CT, USA

James W. Lance Institute of Neurological Sciences, University of New South Wales, Sydney, Australia

Robert G. Large The Auckland Regional Pain Service, Auckland Hospital, Auckland, New Zealand

Alice A. Larson Department of Veterinary and Biomedical Sciences, University of Minnesota, St. Paul, MN, USA

Roberta Lattanzi Department of Physiology and Pharmacology, Sapienza University of Rome, Rome, Italy

Peter Lau Department of Clinical Research, Royal Newcastle Hospital, University of Newcastle, Newcastle, NSW, Australia

Stefan A. Laufer Institute of Pharmacy, Eberhard-Karls University of Tuebingen, Tuebingen, Germany

Stefan Lautenbacher Physiologische Psychologie, Otto-Friedrich-Universität, Bamberg, Germany

Gilles J. Lavigne Facultés de Médecine Dentaire et de Médecine, Université de Montréal, Hopital Sacre Coeur de Montreal. Surgery-trauma dept, Montréal, QC, Canada

Emily Law Center for Child Health, Behavior, and Development, Seattle Children's Research Institute, Seattle, WA, USA

Peter G. Lawlor Our Lady's Hospice, Medical Department, Dublin, Ireland

Michel Lazdunski Institut de Pharmacologie Moléculaire et Cellulaire (IPMC) CNRS/UNS UMR7275, Valbonne-Sophia Antipolis, France

Vincenzo Di Lazzaro Institute of Neurology, Campus Biomedico University, Rome, Italy

Alyssa Lebel Boston Children's Hospital, Boston, MA, USA

Frank Lee Johns Hopkins University, Baltimore, MD, USA

Maaike Leeuw Department of Medical, Clinical and Experimental Psychology, Maastricht University, Maastricht, The Netherlands

Siri Graff Leknes Research Fellow, Psykologisk institutt, Oslo, Norway

Frederick A. Lenz Department of Neurosurgery, Johns Hopkins Hospital, Baltimore, MD, USA

Guillaume Léonard Faculty of Medicine, University of Sherbrooke, Sherbrooke, QC, Canada

Pauline Lesage Department of Pain Medicine and Palliative Care, Beth Israel Medical Center, New York, USA

Isobel Lever Neural Plasticity Unit, Institute of Child Health University College London, London, UK

Khan W. Li Department of Neurosurgery, Johns Hopkins Hospital, Baltimore, MD, USA

Alan R. Light Department of Anesthesiology, University of Utah, Salt Lake City, UT, USA

Michael Lim Department of Neurosurgery and Department of Neurology, Johns Hopkins Hospital, Baltimore, MA, USA

Deolinda Lima Department of Experimental Biology, Faculdade de Medicina da Universidade do Porto, Porto, Portugal

Volker Limmroth Department of Neurology, University of Essen, Essen, Germany

Eric Lingueglia Institut de Pharmacologie Moléculaire et Cellulaire (IPMC) CNRS/UNS UMR7275, Valbonne-Sophia Antipolis, France

Steven J. Linton Department of Occupational and Environmental Medicine, Orebro Medical Center, Orebro, Sweden

Arthur G. Lipman Departments of Pharmscotherpay and Anesthesiology and Pain Management Center, University of Utah, Salt Lake City, UT, USA

Richard B. Lipton Departments of Neurology, Epidemiology and Population Health, Albert Einstein College of Medicine, Bronx, NY, USA

Diana Lisi Department of Psychology, York University, Toronto, ON, Canada

Kenneth M. Little Division of Neurosurgery, Duke University Medical Center, Durham, NC, USA

Edward Littleton Queen Elizabeth Hospital, Birmingham, UK

Spencer S. Liu Virginia Mason Medical Center, Seattle, WA, USA

Anne Elisabeth Ljunggren Section for Physiotherapy Science, Department of Public Health and Primary Health, Faculty of Medicine, University of Bergen, Bergen, Norway

John D. Loeser Department of Neurological Surgery, University of Washington, Seattle, WA, USA

Joern Loetsch Institute for Clinical Pharmacology, Pharmaceutical Center Frankfurt, Johann Wolfgang Goethe University, Frankfurt, Germany

Patrick Loisel Disability Prevention Research and Training Centre, Université de Sherbrooke, Longueuil, Canada

Elisa Lopez-Dolado Servicio de Rehabilitacion, National Hospital of Paraplèjicos SESCAM, Toledo, Spain

Jürgen Lorenz Faculty of Life Science, Competence Center Gesundheit, University of Applied Science Hamburg, Hamburg, Germany

Dayna R. Loyd US Army Institute of Surgical Research, Fort Sam Houston, TX, USA

Daniel Lubelski Cleveland Clinic Lerner College of Medicine, Cleveland, OH, USA

Stephen Lutz Blanchard Valley Regional Cancer Center, Findlay, OH, USA

Bruce Lynn Department of Physiology, University College London, London, UK

Ross MacPherson Department of Anesthesia and Pain Management, Royal North Shore Hospital University of Sydney, Sydney, NSW, Australia

Christophe Maes Manager Innovatie AZ Alma, Belgium

Michel Magnin INSERM, Neurological Hospital, Lyon, France

Bryan Mahony Department of Anesthesiology, Ohio State University Medical Center, Mount Sinai Hospital, New York, NY, USA

Steven F. Maier Department of Psychology and Neuroscience, and The Center for Neuroscience, University of Colorado at Boulder, Boulder, CO, USA

Thorsten J. Maier Institute of Pharmaceutical Chemistry, Goethe University, Frankfurt, Germany

Chris J. Main Keele University, Keele, Staffordshire, UK

Marzia Malcangio Wolfson Centre for Age Related Diseases, King's College London, London, UK

Paolo L. Manfredi Department of Neurology, Memorial Sloan-Kettering Cancer Center, New York, USA

Kaisa Mannerkorpi Department of Rheumatology and Inflammation Research, Institute of Medicine, University of Gothenburg, Göteborg, Sweden

Barton H. Manning Amgen Inc., Thousand Oaks, CA, USA

Patrick W. Mantyh Department of Pharmacology, The University of Arizona, AZ, USA

Jianren Mao Department of Anesthesia and Critical Care, Harvard Medical School, Boston, MA, USA

Serge Marchand Faculty of Medicine, University of Sherbrooke, Sherbrooke, QC, Canada

Paolo Marchettini Pain Medicine Center, Scientific Institute San Raffaele, Milan, Italy

Sarah R. Martin Psychology, Georgia State University, Atlanta, GA, USA

Peggy Mason Department of Neurobiology, Pharmacology and Physiology, University of Chicago, Chicago, IL, USA

Scott Masters Caloundra Spinal and Sports Medicine Centre, Caloundra, QLD, Australia

Laurence E. Mather Department of Anaesthesia and Pain Management, University of Sydney at Royal North Shore Hospital, St. Leonards, Sydney, NSW, Australia

Bill McCarberg Pain Services, Kaiser Permanente, San Diego, USA

Lance McCracken Health Psychology/Psychology Department, King's College London, London, UK

Kathleen McGinn Anesthesiology & Pain Medicine, Seattle Children's Hospital, Seattle, WA, USA

Patricia A. McGrath The Hospital for Sick Children Pain Research Center, Toronto, ON, Canada

Patrick McGrath Departments of Psychology, Pediatrics and Psychiatry, Dalhousie University, Canada

Greg McIntosh Canadian Back Institute, Toronto, ON, Canada

Peter McIntyre Department of Pharmacology, University of Melbourne, Melbourne, Australia

Elspeth M. McLachlan Neuroscience Research Australia, University of New South Wales, Randwick, NSW, Australia

University of Glasgow, Glasgow, UK

Peter A. McNaughton Department of Pharmacology, University of Cambridge, Cambridge, UK

J. Mark Melhorn Section of Orthopaedics, Department of Surgery, University of Kansas School of Medicine, Wichita, KS, USA

Ronald Melzack Department of Psychology, McGill University, Montreal, QC, Canada

Lorne M. Mendell Department of Neurobiology and Behavior, Stony Brook University, Stony Brook, NY, USA

George Mendelson Department of Psychological Medicine, Monash University, Caulfield, VIC, Australia

Siegfried Mense Neuroanatomy, Heidelberg University, Medical Faculty Mannheim, CBTM, Mannheim, Germany

Daniel Menétrey Biomédicale, Paris, France

Sebastiano Mercadante Pain Relief & Palliative Care, La Maddalena Cancer Center University of Palermo, Palermo, Italy

Susan Mercer School of Biomedical Sciences, The University of Queensland, Brisbane, Australia

Karl Messlinger University of Erlangen-Nürnberg, Erlangen, Germany

Richard A. Meyer Department of Neurosurgergy, School of Medicine Johns Hopkins University, Baltimore, MD, USA

Walter J. Meyer III Shriners Hospitals for Children, Shriners Burns Hospital, Galveston, TX, USA

Department of Psychiatry, University of Texas Medical Branch, Galveston, TX, USA

Björn A. Meyerson Department of Clinical Neuroscience, Section of Neurosurgery, Karolinska Institute/University Hospital, Stockholm, Sweden

Martin Michaelis Sanofi-Aventis Deutschland GmbH, Frankurt am Main, Germany

Joseph Mignogna Michael E. DeBakey VA Medical Center, Houston VA HSR&D Center of Excellence (152), Houston, TX, USA

Judith A. Mikacich Department of Obstetrics and Gynecology, University of California Los Angeles Medical Center, Los Angeles, CA, USA

Kensaku Miki Department of Integrative Physiology, National Institute for Physiological Sciences, Okazaki, Japan

Vjekoslav Miletic School of Medicine and Public Health, University of Wisconsin, Madison, WI, USA

Scott R. Miller Memorial Sloan-Kettering Cancer Center, New York, USA

Erin D. Milligan Department of Neurosciences, University of New Mexico, Albuquerque, NM, USA

Michael S. Minett Molecular Nociception Group, Wolfson Institute for Biomedical Research, University College London, London, WC, UK

J. Mocco Department of Neurological Surgery, Neurological Institute Columbia University, New York, NY, USA

Jeffrey S. Mogil Department of Psychology, McGill University, Montreal, QC, Canada

Harvey Moldofsky Sleep Disorders Clinic of the Centre for Sleep and Chronobiology, Toronto, ON, Canada

Robert M. Moldwin The Arthur Smith Institute for Urology, North Shore-LIJ Healthcare System, Hofstra North Shore LIJ School of Medicine, New Hyde Park, NY, USA

Derek C. Molliver Department of Neurobiology, University of Pittsburgh School of Medicine, Pittsburgh, PA, USA

Robert D. Mootz Washington State Department of Labor and Industries, Olympia, WA, USA

Laura Mordillo-Mateos FENNSI Group, Hospital Nacional de Parapléjicos, Toledo, Spain

Anne Morel Laboratory for Functional Neurosurgery, Neurosurgical Clinic University Hospital Zürich, Zürich, Switzerland

Anne Morinville Montreal Neurological Institute, McGill University, Montreal, QC, Canada

Stephen Morley Academic Unit of Psychiatry and Behavioral Sciences, University of Leeds, Leeds, UK

David B. Morris University of Virginia, Charlottesville, VA, USA

G. Lorimer Moseley Sansom Institute for Health Research, University of South Australia Neuroscience Research Australia, Adelaide, Australia

Joachim Mossner Medical Clinic and Policlinic II, University Clinical Center Leipzig AöR, Leipzig, Germany

Eric A. Moulton P.A.I.N. Group, Anesthesiology, Boston Children's Hospital, Harvard Medical School, Waltham, MA, USA

Francis Githae Muriithi Department of Obstetrics and Gynaecology, Aga Khan University, Nairobi, Kenya

Anne Z. Murphy Neuroscience Institute, Georgia State University, Atlanta, GA, USA

Paul M. Murphy Department of Anaesthesia and Pain Management, Royal North Shore Hospital, St. Leonards, NSW, Australia

Alison Murray Foothills Medical Center, Calgary, AB, Canada

H. S. Na Neuroscience Research Institute and Department of Physiology, Korea University, College of Medicine, Seoul, Republic of Korea

Daisuke Naka Department of Integrative Physiology, National Institute for Physiological Sciences, Okazaki, Japan

Yoshio Nakamura Pain Research Center, Department of Anesthesiology University of Utah School of Medicine, Salt Lake City, UT, USA

Barbara Namer Department of Physiology & Pathophysiology, FAU, Erlangen/Nuremberg, Bavaria, Germany

Matti V. O. Närhi Department of Biomedicine/Physiology, Department of Dentistry, University of Eastern Finland, Institute of Medicine, Kuopio, Finland

H. J. W. Nauta University of Texas Medical Branch, Galveston, TX, USA

Lucia Negri Department of Physiology and Pharmacology, Sapienza University of Rome, Rome, Italy

Gary Nesbit Department of Radiology, Oregon Health and Science University, Portland, OR, USA

Timothy Ness Department of Anesthesiology, University of Alabama at Birmingham, Birmingham, AL, USA

Volker Neugebauer Department of Neuroscience & Cell Biology, University of Texas Medical Branch, Galveston, TX, USA

Lawrence C. Newman Albert Einstein College of Medicine, New York, NY, USA

Michael K. Nicholas Royal North Shore Hospital, University of Sydney, St. Leonards, NSW, Australia

Ellen Niederberger Pharmazentrum frankfurt, Institute of Clinical Pharmacology, Goethe-University Hospital, Frankfurt, Germany

Lone Nikolajsen Department of Anaesthesiology, Danish Pain Research Center Aarhus University Hospital, Aarhus, Denmark

Katie Nixdorf Comprehensive Pain Center, Oregon Health and Science University, Portland, OR, USA

Donald R. Nixdorf Department of Diagnostic and Biological Sciences, School of Dentistry University of Minnesota, Minneapolis, MN, USA

Carl E. Noe University of Texas Southwestern Medical Center, Lubbock, TX, USA

Koichi Noguchi Department of Anatomy and Neuroscience, Hyogo College of Medicine, Hyogo, Japan

Richard B. North The Johns Hopkins University, Baltimore, MD, USA

Turo J. Nurmikko Department of Neurological Science, University of Liverpool, and Pain Research Institute, The Walton Centre NHS Foundation Trust, Liverpool, UK

Anne Louise Oaklander Nerve Injury Unit, Departments of Neurology and Pathology, Massachusetts General Hospital Harvard Medical School, Boston, MA, USA

Koichi Obata Department of Anatomy and Neuroscience, Hyogo College of Medicine, Hyogo, Japan

Tim F. Oberlander Division of Developmental Pediatrics, University of British Columbia, Vancouver, BC, Canada

Bruno Oertel Institute for Clinical Pharmacology, Pharmaceutical Center Frankfurt, Johann Wolfgang Goethe University, Frankfurt, Germany

Andrea C. Ogbonna Wolfson Centre for Age Related Diseases, King's College London, London, UK

Alfred T. Ogden Department of Neurological Surgery, Neurological Institute Columbia University, New York, NY, USA

S. Ohara Department of Neurosurgery, Johns Hopkins Hospital, Johns Hopkins Medical Institutions, Johns Hopkins University, Baltimore, MD, USA

Peter T. Ohara Department of Anatomy, University of California San Francisco, San Francisco, CA, USA

Akiko Okifuji Department of Anesthesiology, University of Utah, Salt Lake City, UT, USA

Jean-Louis Oliveras Inserm, Paris, France

Antonio Oliviero FENNSI Group, National Hospital of Paraplejicos, Toledo, Spain

Sean O'Mahony Palliative medicine, Rush University Medical Center, Chicago, IL, USA

Peregrine B. Osborne Dept Anatomy & Neuroscience, University of Melbourne, Melbourne, VIC, Australia

Michael H. Ossipov Department of Pharmacology, University of Arizona College of Medicine, Tucson, AZ, USA

Tonya M. Palermo Seattle Children's Hospital Research Institute, CW8-6 - Child Health, Behavior and Development, University of Washington, Seattle, WA, USA

Shreela Palit Department of Psychology, The University of Tulsa, Tulsa, OK, USA

Juan A. Pareja Department of Neurology, Fundación Hospital Alcorcón, Madrid, Spain

Steven D. Passik Psychiatry and Anesthesiology, Vanderbilt University, Nashville, TN, USA

Gavril W. Pasternak Laboratory of Molecular Neuropharmacology, Memorial Sloan-Kettering Cancer Center, New York, USA

S. H. Patel Department of Neurosurgery, Johns Hopkins Hospital, Baltimore, MD, USA

Gavin Pattullo Department of Anesthesia and Pain Management, Royal North Shore Hospital, University of Sydney, St. Leonards, NSW, Australia

Kevin Pauza Tyler Spine and Joint Hospital, Tyler, TX, USA

Eric M. Pearlman Pediatrics, Mercer University School of Medicine, The Children's Hospital at Memorial University Medical Center, Savannah, USA

Elvira De la Pena Instituto de Neurociencias, Universidad Miguel Hernández-CSIC, Alicante, Spain

Yuan Bo Peng Department of Psychology, University of Texas at Arlington, Arlington, TX, USA

Jose Pereira Foothills Medical Center, Calgary, AB, Canada

Edward Perl Department of Cell and Molecular Physiology, School of Medicine University of North Carolina, Chapel Hill, NC, USA

Tiffany Grace Perry Center for Spine Health, The Cleveland Clinic Foundation, Cleveland, OH, USA

Marie Pertin Pain Center, University Hospital Center and University of Lausanne, Lausanne, Switzerland

Antti Pertovaara Department of Physiology, Institute of Biomedicine University of Helsinki, Helsinki, Finland

Peter Peter Department of Neurobiology and Anatomy, University of Rochester Medical Center, Rochester, NY, USA

Lisa M. Peters Department of Anesthesiology and Pain Medicine, Seattle Children's Hospital, Seattle, WA, USA

Pramit Phal Department of Radiology, Oregon Health and Science University, Portland, OR, USA

Rebecca Pillai Riddell Department of Psychology, York University, Toronto, ON, Canada

Department of Psychiatry, Hospital for Sick Children, Toronto, ON, Canada

Issy Pilowsky University of Adelaide, Adelaide, Australia

Leon Plaghki Institute of Neuroscience (IoNS), Université Catholique de Louvain, Brussels, Belgium

Markus Ploner Department of Neurology, Technische Universität München, Munich, Germany

Esther M. Pogatzki-Zahn Department of Anesthesiology and Intensive Care Medicine, University Muenster, Muenster, Germany

Michael Polydefkis Department of Neurology, Johns Hopkins University School of Medicine, Baltimore, MD, USA

James D. Pomonis Algos Preclinical Services, Roseville, MN, USA

Dieter Pongratz Friedrich Baur Institute, Ludwig Maximilians University, Munich, Germany

Hayley Poole Hadley House, Leicester General Hospital, Leicester, UK

Frank Porreca Department of Pharmacology, University of Arizona Health Sciences Center, Tucson, AZ, USA

Russell K. Portenoy Department of Pain Medicine and Palliative Care, Beth Israel Medical Center, New York, NY, USA

Ian Power Critical Care and Pain Medicine, The University of Edinburgh, Edinburgh, UK

Donald D. Price Oral Surgery and Neuroscience, McKnight Brain Institute University of Florida, Gainesville, FL, USA

Daryl Pullman Medical Ethics, Memorial University of Newfoundland, St. John's, NL, Canada

Yunhai Qiu Department of Integrative Physiology, National Institute for Physiological Sciences, Okazaki, Japan

Michael Quittan Department of Physical Medicine and Rehabilitation, Kaiser-Franz-Joseph Hospital, Vienna, Austria

Raymond M. Quock Department of Pharmaceutical Sciences, Washington State University College of Pharmacy, Pullman, WA, USA

Jennifer Rabbitts Department of Anesthesiology and Pain Medicine, Seattle Children's Hospital, University of Washington School of Medicine, Seattle, WA, USA

Nicole Racine Psychology, York University, Toronto, ON, Canada

Gabor B. Racz Texas Tech University Health Sciences Center, Lubbock, TX, USA

Lukas Radbruch Department of Palliative Medicine, University of Bonn, Bonn, Germany

Robert B. Raffa Department of Pharmaceutical Sciences, Temple University School of Pharmacy, Philadelphia, PA, USA

Vasudeva Raghavendra Algos Therapeutics Inc., Saint Paul, MN, USA

Pierre Rainville Department of Stomatology, Faculty of Dental Medicine, University of Montreal, Montreal, QC, Canada

Srinivasa N. Raja Department of Anesthesiology and Critical Care Medicine, Johns Hopkins University, Baltimore, USA

Alan Randich University of Alabama, Birmingham, AL, USA

Andrea J. Rapkin Department of Obstetrics and Gynecology, University of California Los Angeles Medical Center, Los Angeles, CA, USA

Z. Harry Rappaport Department of Neurosurgery, Rabin Medical Center Tel-Aviv University, Petah Tiqva, Israel

Matthew N. Rasband Department of Molecular and Cellular Biology, Baylor College of Medicine, Farmington, Houston TX, USA

University of Connecticut Health Center, Farmington, CT, USA

Douglas Rasmusson Department of Physiology and Biophysics, Dalhousie University, Halifax, Canada

James P. Rathmell Department of Anesthesiology, University of Vermont College of Medicine, Burlington, USA

K. Ravishankar The Headache and Migraine Clinic, Jaslok Hospital and Research Centre Lilavati Hospital and Research Centre, Mumbai, India

Ke Ren Department of Neural and Pain Sciences, University of Maryland Dental School, Baltimore, MD, USA

Seth Resnick Silver Hill Hospital, New York University School of Medicine, USA

Cielito Reyes-Gibby The University of Texas MD Anderson Cancer Center, Houston, TX, USA

Jamie L. Rhudy Department of Psychology, The University of Tulsa, Tulsa, OK, USA

Alfredo Ribeiro-da-Silva Department of Pharmacology and Therapeutics, McGill University, Montreal, QC, Canada

Frank L. Rice Center for Neuropharmacology and Neuroscience, Albany Medical College, Albany, NY, USA

Matthias Ringkamp Department of Neurosurgergy, School of Medicine Johns Hopkins University, Baltimore, MD, USA

Pierre J. M. Rivière Ferring Research Institute, San Diego, CA, USA

Claude Robert Université Paris Descartes, INSERM UMR 894, France

Roger Robert Neurosurgery, Hotel Dieu, Nantes, CHU, France

James P. Robinson Department of Anesthesiology & Pain Medicine, University of Washington, Seattle, WA, USA

John Robinson Pain Management Centre, Burwood Hospital, Christchurch, New Zealand

Alfonso Romero-Sandoval Department of Pharmaceuticals and Administrative Sciences, Presbyterian College School of Pharmacy, Clinton, SC, USA

David Roselt Aberdovy Clinic, Bundaberg, QLD, Australia

Jackie Ross Early Pregnancy Unit, King's College Hospital Early Pregnancy Unit, London, UK

Shlomo Rotshenker Department of Medical Neurobiology, IMRIC, Faculty of Medicine Hebrew University of Jerusalem, Jerusalem, Israel

Paul C. Rousseau Medical University of South Carolina, Charleston, SC, USA

Ranjan Roy Department of Clinical Health Psychology, Faculty of Medicine, University of Manitoba, Winnipeg, MB, Canada

Carolina Roza Biología de Sistemas (Unidad Fisiología), Universidad de Alcalá, Alcalá de Henares, Madrid, Spain

Todd D. Rozen Department of Neurology, Geisinger Health System, Wilkes-Barre, PA, USA

Roman Rukwied Department of Anaesthesiology and Intensive Care Medicine, Faculty of Clinical Medicine Mannheim University of Heidelberg, Heidelberg, Germany

Irwin Jon Russell Division of Clinical Immunology and Rheumatology, Department of Medicine, The University of Texas Health Science Center, San Antonio, TX, USA

Nayef E. Saadé Anatomy, Cell Biology and Physiology, Faculty of Medicine, American University of Beirut, Beirut, Lebanon

Mary Ann C. Sabino Oral and Maxillofacial Surgery, Hennepin County Medical Center University of Minnesota, Minneapolis, MN, USA

Thomas E. Salt Institute of Ophthalmology, University College London, London, UK

A. Samdani Department of Neurosurgery, Johns Hopkins Medical Institutions, Baltimore, MD, USA

Yasmin Sana King's College Hospital, London, UK

Margarita Sanchez-del-Rio Neurology Department, Hospital Ruber Internacional, Madrid, Spain

Jürgen Sandkühler Department of Neurophysiology, Center for Brain Research Medical University of Vienna, Vienna, Austria

Peter S. Sándor RehaClinic, Cantonal Hospital Baden, Baden, Switzerland

Johannes Sarnthein Functional Neurosurgery, Neurosurgical Clinic University Hospital, Zürich, Switzerland

Susanne K. Sauer Department of Physiology and Pathophysiology, University of Erlangen, Erlangen, Germany

Steven Savvas Biomedical Division, National Ageing Research Institute, Melbourne, VIC, Australia

Jana Sawynok Department of Pharmacology, Dalhousie University, Halifax, Canada

John W. Scadding The National Hospital for Neurology and Neurosurgery, London, UK

Frederieke Geraldine Schaafsma Department of Public and Occupational Health, EMGO+ Institute/VU Medical Center, Amsterdam, MB, The Netherlands

Hans-Georg Schaible Department of Physiology - Neurophysiology, Friedrich-Schiller-University of Jena, Jena, Germany

Steven S. Scherer Department of Neurology, The Perelman School of Medicine at the University of Pennsylvania, Philadelphia, PA, USA

Michael Schäfer Department of Anesthesiology and Intensive Care Medicine, Charité University, Berlin, Germany

Martin Schmelz Department of Anesthesiology in Mannheim, University of Heidelberg, Mannheim, Germany

Robert F. Schmidt Physiologisches Institut der Universität Würzburg, Würzburg, Germany

Suzan Schneeweiss The Hospital for Sick Children, Toronto, ON, Canada

Frank Schneider Department of Psychiatry and Psychotherapy, RWTH Aachen University, Aachen, Germany

Alfons Schnitzler Department of Neurology, Heinrich Heine University, Düsseldorf, Germany

Jean Schoenen Department of Neurology, University of Liège, CHR Citadelle, Belgium

Department of Neuroanatomy, University of Liège, CHR Citadelle, Belgium

Jens Schouenborg Neuronano Research Center, Lund University, Lund, Sweden

Stephan A. Schug School of Medicine and Pharmacology, Royal Perth Hospital, Faculty of Medicine, Dentistry & Health Sciences, The University of Western Australia, UWA Anaesthesiology, Perth, Australia

Patrick Schuss Department of Neurosurgery, University-Hospital Bonn, Bonn, Germany

Anthony C. Schwarzer Faculty of Medicine and Health Sciences, The University of Newcastle, Newcastle, Australia

Whitney Scott Department of Psychology, McGill University, Montreal, QC, Canada

Laura C. Seidman Pediatric Pain Program, Department of Pediatrics David Geffen School of Medicine at UCLA, Los Angeles, CA, USA

Ze'ev Seltzer Faculty of Dentistry, University of Toronto Centre for the Study of Pain, Toronto, ON, Canada

Peter Selwyn Albert Einstein College of Medicine, Bronx, NY, USA

Patrick B. Senatus Department of Neurological Surgery, Neurological Institute Columbia University, New York, NY, USA

Jyoti N. Sengupta Division of Gastroenterology and Hepatology and Pediatric Gastroenterology, Medical College of Wisconsin, Milwaukee, WI, USA

Mariano Serrao Department of Neurology and Otorinolaringoiatry, University of Rome La Sapienza, Rome, Italy

Barry J. Sessle Faculty of Dentistry, University of Toronto, Toronto, ON, Canada

Faculty of Dentistry, University of Toronto, Toronto, ON, Canada

Virginia S. Seybold Department of Neuroscience, University of Minnesota, Minneapolis, MN, USA

T. Shah Royal Perth Hospital and University of Western Australia, Crawley, WA, Australia

Yair Sharav Department of Oral Medicine, School of Dental Medicine, Hebrew University-Hadassah, Jerusalem, Israel

S. Murray Sherman Department of Neurobiology, University of Chicago, Chicago, IL, USA

Yoshio Shigenaga Department of Oral Anatomy and Neurobiology, Osaka University, Osaka, Japan

Susan Shin Neurology, Mount Sinai School of Medicine, New York, NY, USA

Edward A. Shipton Department of Anaesthesia, Christchurch School of Medicine and Health Sciences, University of Otago, Christchurch, New Zealand

Yoram Shir The Alan Edwards Pain Management Unit, McGill University Health Centre, Montreal, QC, Canada

Peter Shrager Department of Neurobiology and Anatomy, University of Rochester Medical Center, Rochester, NY, USA

Marc J. Shulman VA Connecticut Healthcare System, Yale University, New Haven, CT, USA

David M. Sibell Comprehensive Pain Center/Spine Center, Oregon Health and Science University, Portland, OR, USA

Philip J. Siddall Department of Pain Management, Greenwich Hospital, University of Sydney, Sydney, NSW, Australia

Stephen D. Silberstein Jefferson Medical College, Thomas Jefferson University and Jefferson Headache Center, Thomas Jefferson University Hospital, Philadelphia, PA, USA

Donald A. Simone Diagnostic and Biological Sciences, University of Minnesota, Minneapolis, MN, USA

David G. Simons Department of Rehabilitation Medicine, Emory University, Atlanta, GA, USA

David Simpson Clinical Neurophysiology Laboratories, Mount Sinai Medical Center, New York, NY, USA

Marc Sindou Department of Neurosurgery, Hopital Neurologique Pierre Wertheimer, University of Lyon 1, Lyon, France

Soeren H. Sindrup Department of Neurology, Odense University Hospital, Odense, Denmark

Bengt H. Sjölund Department of Community Medicine and Rehabilitation Medicine, Umea University, Umea, Sweden

Carsten Skarke Institute for Translational Medicine and Therapeutics (ITMAT), University of Pennsylvania Perelman School of Medicine, Philadelphia, PA, USA

Paul A. Sloan Department of Anesthesiology, University of Kentucky, Lexington, KY, USA

Kathleen A. Sluka Physical Therapy and Rehabilitation Science Graduate Program, University of Iowa, Iowa City, IA, USA

Seymour Solomon Montefiore Medical Center, Albert Einstein College of Medicine, Bronx, NY, USA

Claudia Sommer Department of Neurology, University of Würzburg, Würzburg, Germany

Linda S. Sorkin Occupational Therapy and Rehabilitation, Anesthesia Research Laboratories, University of California, San Diego, CA, USA

Ingrid Söderback Uppsala University, Uppsala, Sweden

Kari Sørensen Department of Pain Management and Research, Oslo University Hospital, Rikshospitalet, Oslo, Norway

Michael Spaeth Friedrich-Baur-Institute, University of Munich, Munich, Germany

Peregrin O. Spielholz Washington State Department of Labor and Industries, SHARP Program, Olympia, WA, USA

Peter S. Staats Division of Pain Medicine, The Johns Hopkins University, Baltimore, MD, USA

Brett R. Stacey Comprehensive Pain Center, Oregon Health and Science University, Portland, OR, USA

Wendy M. Stein San Diego Hospice and Palliative Care, UCSD, San Diego, CA, USA

Christoph Stein Department of Anesthesiology and Intensive Care Medicine, Freie Universitaet Berlin Charité Campus Benjamin Franklin, Berlin, Germany

Michael P. Steinmetz Department of Neurosurgery, MetroHealth Medical Center, Case Western Reserve University School of Medicine, Cleveland, OH, USA

Jair Stern Functional Neurosurgery, Neurosurgical Clinic University Hospital, Zürich, Switzerland

Bonnie J. Stevens University of Toronto and The Hospital for Sick Children, Toronto, ON, Canada

Carl-Olav Stiller Division of Clinical Pharmacology and Department of Medicine, Karolinska University Hospital, Stockholm, Sweden

Jennifer N. Stinson The Hospital for Sick Children, Toronto, ON, Canada

Henry Stockbridge University of Washington School of Public Health and Community Medicine, Seattle, WA, USA

Christian S. Stohler Baltimore College of Dental Surgery, University of Maryland, Baltimore, MD, USA

Laura Stone Faculty of Dentistry, Alan Edwards Centre for Research on Pain, Montreal, QC, Canada

William Stones Department of Obstetrics and Gynaecology, Aga Khan University, Nairobi, Kenya

Andreas Straube Department of Neurology, Ludwig Maximilians University, Munich, Germany

Jenny Strong University of Queensland, Brisbane, QLD, Australia

Audun Stubhaug Department of Pain Management and Research, Oslo University Hospital, Rikshospitalet and University of Oslo, Oslo, Norway

Gerold Stucki Swiss Paraplegic Research, Nottwil, Switzerland

Department of Health Sciences and Health Policy, University of Lucerne, Lucerne, Switzerland

ICF Research Branch in cooperation with the World Health Organization (WHO) Collaborating Centre for the Family of International Classifications in Germany, Nottwil, Switzerland

Cheryl L. Stucky Department of Cell Biology Neurobiology and Anatomy, Medical College of Wisconsin, Milwaukee, WI, USA

Mathias Sturzenegger Department of Neurology, University Hospital, Inselspital, University of Bern, Bern, Switzerland

Yasuo Sugiura School of Human Care Study, Nagoya University of Arts and Sciences, Nissin City, Aichi, Japan

Mark D. Sullivan Psychiatry and Behavioral Sciences, University of Washington, Seattle, WA, USA

Michael J. L. Sullivan Department of Psychology, McGill University, Montreal, QC, Canada

Rie Suzuki Department of Pharmacology, University College London, London, UK

Kristina B. Svendsen Danish Pain Research Center, Aarhus University Hospital, Aarhus, Denmark

Peter Svensson Clinical Oral Physiology, Aarhus University, Aarhus, Denmark

Peter Szucs Spinal Neuronal Networks, Instituto de Biologia Molecular e Celular - IBMC, Porto, Portugal

A. Taghva Department of Neurosurgery, Johns Hopkins Hospital, Baltimore, MD, USA

Kimberson Cochien Tanco Department of Palliative Care and Rehabilitation Medicine, The University of Texas M. D. Anderson Cancer Center, Houston, TX, USA

Kersi Taraporewalla Royal Brisbane and Womens' Hospital, University of Queensland, Brisbane, QLD, Australia

Irmgard Tegeder Pharmazentrum frankfurt, Institute of Clinical Pharmacology, Goethe-University Hospital, Frankfurt, Germany

Astrid Juhl Terkelsen Danish Pain Research Center, Department of Neurology, Aarhus University Hospital, Aarhus, Denmark

Gregory W. Terman Department of Anesthesiology and the Graduate Program in Neurobiology and Behavior, University of Washington, Seattle, WA, USA

Alison Thomas Lismore, NSW, Australia

Benjamin Thomas Chelsea & Westminster Hospital Foundation Trust, London, UK

Jacob Thomas School of Medical Sciences, University of Adelaide, Adelaide, SA, Australia

Miles Thompson Goldsmiths, University of London, London, UK

Beverly E. Thorn Department of Psychology, University of Alabama, Tuscaloosa, AL, USA

Cindy L. Thurston-Stanfield Department of Biomedical Sciences, University of South Alabama, Mobile, AL, USA

Arne Tjølsen Department of Bomedicine, University of Bergen, Haukeland University Hospital, Bergen, Norway

Andrew J. Todd Institute of Neuroscience and Psychology, College of Medical, Veterinary and Life Sciences, University of Glasgow, Glasgow, UK

Makoto Tominaga Department of Cellular and Molecular Physiology, Mie University School of Medicine, Tsu Mie, Japan

Victor Tortorici Instituto Venezolano de Investigaciones Cientificas (IVIC), Caracas, Venezuela

Irene Tracey Pain Imaging Neuroscience (PaIN) Group, Department of Physiology, Anatomy and Genetics, Oxford University, Oxford, UK

Diep Tuan Tran Department of Integrative Physiology, National Institute for Physiological Sciences, Okazaki, Japan

Rolf-Detlef Treede Institute of Physiology and Pathophysiology, Johannes Gutenberg University, Mainz, Germany

Jennie C. I. Tsao Pediatric Pain Program, Department of Pediatrics David Geffen School of Medicine at UCLA, Los Angeles, CA, USA

Lawrence Tsen Department of Anesthesiology, Perioperative and Pain Medicine, Brigham and Women's Hospital Harvard Medical School, Boston, MA, USA

Jeanna Tsenter Department of Physical Medicine and Rehabilitation, Hadassah University Hospital, Jerusalem, Israel

Maurits Van Tulder Institute for Research in Extramural Medicine, Free University Amsterdam, Amsterdam, The Netherlands

Eldon Tunks Hamilton Health Sciences Corporation, Hamilton, ON, Canada

Susan Tupper Pediatrics, University of Saskatchewan, Saskatoon, SK, Canada

Dennis C. Turk Department of Anesthesiology & Pain Research, University of Washington, Seattle, WA, USA

Judith A. Turner Department of Psychiatry and Behavioral Sciences, University of Washington, Seattle, WA, USA

Department of Rehabilitation Medicine, University of Washington, Seattle, WA, USA

Stephen P. Tyrer Royal Victoria Infirmary, University of Newcastle upon Tyne, Newcastle-on-Tyne, UK

Alexander Tzabazis Department of Anesthesia, Stanford University, Stanford, CA, USA

Kene Ugokwe Department of Neurosurgery, St Elizabeth Health Center, Youngstown, OH, USA

Lindsay S. Uman Department of Psychology, IWK Health Centre, Halifax, NS, Canada

Christiane Vahle-Hinz Department of Neurophysiology and Pathophysiology, University Medical Center Hamburg-Eppendorf, Hamburg, Germany

Muthukumar Vaidyaraman Division of Pain Medicine, The Johns Hopkins University, Baltimore, MD, USA

Todd W. Vanderah Department of Pharmacology, College of Medicine University of Arizona, Tucson, AZ, USA

Guy Vanderstraeten Department of Rehabilitation Sciences and Physiotherapy, Ghent University Hospital, Ghent, Belgium

Horacio Vanegas Instituto Venezolano de Investigaciones Cientificas (IVIC), Caracas, Venezuela

Jean-Jacques Vatine Outpatient and Research Division, Reuth Medical Center, Sackler Faculty of Medicine, Tel Aviv University, Tel Aviv, Israel

Leigh M. Vaughan Medical University of South Carolina, Charleston, SC, USA

Linda K. Vaughn Department of Biomedical Sciences, Marquette University, Milwaukee, WI, USA

Enrique Vazquez Instituto Venezolano de Investigaciones Cientificas (IVIC), Caracas, Venezuela

Louis Vera-Portocarrero Department of Pharmacology, College of Medicine University of Arizona, Tucson, AZ, USA

Paul Verrills Metropolitan Spinal Clinic, Prahran, VIC, Australia

Tine Vervoort Department of Experimental Clinical and Health Psychology, Ghent University, Ghent, Belgium

Félix Viana Instituto de Neurociencias, Universidad Miguel Hernández-CSIC, San Juan de Alicante, Alicante, Spain

Charles J. Vierck Jr. Department of Neuroscience and McKnight Brain Institute, University of Florida College of Medicine, Gainesville, FL, USA

Luis Villanueva INSERM UMR 894, Centre de Psychiatrie et Neurosciences, Paris, France

Marcelo Villar Faculty of Biomedical Sciences, Austral University, Buenos Aires, Argentina

Katy Vincent Nuffield Department of Obstetrics and Gynaecology, The Women's Centre, John Radcliffe Hospital, University of Oxford, Oxford, UK

David Vivian Metro Spinal Clinic, South Caulfield, VIC, Australia

Johan W. S. Vlaeyen Research Group Health Psychology, Catholic University of Leuven, Leuven, Belgium

Department of Clinical Psychological Science, Maastricht University, Maastricht, The Netherlands

Stéphanie Volders Research Group on Health Psychology, Leuven, Belgium

Ernest Volinn Pain Research Center, University of Utah, Salt Lake City, UT, USA

Lucy Vulchanova Department of Veterinary and Biomedical Sciences, University of Minnesota, St. Paul, MN, USA

Paul W. Wacnik Neuromodulation Research, Medtronic Inc., Minneapolis, MN, USA

Gordon Waddell University of Glasgow, Glasgow, UK

Gary A. Walco Department of Anesthesiology and Pain Medicine, University of Washington School of Medicine, Seattle Children's Hospital, Seattle, WA, USA

Suellen M. Walker University of Sydney Pain Management and Research Centre, Royal North Shore Hospital, NSW, Australia

Victoria C. J. Wallace Department of Anaesthetics, Pain Medicine and Intensive Care Chelsea and Westminster Hospital Campus, London, UK

Xiaohong Wang Department of Integrative Physiology, National Institute for Physiological Sciences, Okazaki, Japan

Gunnar Wasner Department of Neurological Pain Research and Therapy, Neurological Clinic, Kiel, Germany

Shoko Watanabe Department of Integrative Physiology, National Institute for Physiological Sciences, Okazaki, Japan

Linda R. Watkins Department of Psychology and Neuroscience, and The Center for Neuroscience, University of Colorado at Boulder, Boulder, CO, USA

Peter N. Watson Department of Medicine, University of Toronto, Toronto, ON, Canada

James Watt North Shore Hospital, Auckland, New Zealand

Stephen G. Waxman Department of Neurology, Yale School of Medicine, New Haven, CT, USA

Inge Wegner Julius Center, Department of Epidemiology, University Medical Center Utrecht, Utrecht, The Netherlands

Feng Wei Department of Neural and Pain Sciences, University of Maryland Dental School, Baltimore, MD, USA

Christian Weidner Department of Physiology and Experimental Pathophysiology, University Erlangen, Erlangen, Germany

P. T. M. Weijenborg Department of Gynecology, Leiden University Medical Center, Leiden, The Netherlands

Robin Weir School of Nursing, Faculty of Health Sciences, McMaster University, Hamilton, ON, Canada

Nirit Weiss Department of Neurosurgery, Mount Sinai Medical Center, New York, NY, USA

Ursula Wesselmann Departments of Neurology and Anesthesiology / Division of Pain Management, University of Alabama at Birmingham, Birmingham, AL, USA

Dagmar Westerling Department of Anesthesiology, Central Hospital of Kristianstad, Kristianstad, Sweden

Karin N. Westlund Department of Physiology, University of Kentucky, Lexington, KY, USA

Thomas M. Wickizer Health Services Management and Policy, Ohio State University, Colombus, OH, USA

Eva Widerström-Noga The Miami Project to Cure Paralysis, Department of Neurological Surgery, University of Miami Miller, School of Medicine, Miami, FL, USA

Julie Wieseler Department of Psychology and Neuroscience, and The Center for Neuroscience, University of Colorado at Boulder, Boulder, CO, USA

Zsuzsanna Wiesenfeld-Hallin Karolinska Institute, Stockholm, Sweden

Ann Wigham McCoull Clinic, Prudhoe Hospital, Northumberland, UK

George L. Wilcox Department of Neuroscience, Pharmacology and Dermatology, University of Minnesota Medical School, Minneapolis, MN, USA

Oliver H. G. Wilder-Smith Pain Centre, Department of Anaesthesiology, Radboud University Nijmegen Medical Centre, Nijmegen, The Netherlands

Kjersti Wilhelmsen Section for Physiotherapy Science, Department of Public Health and Primary Health, Faculty of Medicine, University of Bergen, Bergen, Norway

Victor Wilk Brighton Spinal Group, Brighton, VIC, Australia

W. A. Williams Royal Perth Hospital and University of Western Australia, Perth, WA, Australia

Sara E. Williams Division of Pediatric Gastroenterology, Medical College of Wisconsin, Milwaukee, WI, USA

William D. Willis Department of Neuroscience and Cell Biology, University of Texas Medical Branch, Galveston, TX, USA

John B. Winer Department of Neurology, Birmingham Muscle and Nerve Centre Queen Elizabeth Hospital, Birmingham, UK

Christopher J. Winfree Department of Neurological Surgery, Neurological Institute Columbia University, New York, NY, USA

Janet Winter Novartis Institute for Medical Science, London, UK

Alain Woda Faculté de Chirurgie Dentaire, University Clermont-Ferrand, Clermont-Ferrand, France

John N. Wood Molecular Nociception Group, Wolfson Institute for Biomedical Research, University College London, London, WC, UK

Lee Woodson Shriners Hospitals for Children, Shriners Burns Hospital, Galveston, TX, USA

Department of Anesthesia, University of Texas Medical Branch, Galveston, TX, USA

Anava A. Wren Pain Prevention and Treatment Research, Department of Psychiatry and Behavioral Sciences, Duke University Medical Center, Durham, NC, USA

Xiao-Jun Xu Department of Physiology and Pharmacology, Section of Integrative Pain Rsearch Karolinska Institutet, Stockholm, Sweden

Hiroshi Yamasaki Department of Integrative Physiology, National Institute for Physiological Sciences, Okazaki, Japan

Michael Yelland School of Medicine, Logan campus, Griffith University, Logan, Australia

Sriram Yennurajalingam Department of Palliative Care and Rehabilitation Medicine, The University of Texas M. D. Anderson Cancer Center, Houston, TX, USA

David C. Yeomans Department of Anesthesia, Stanford University, Stanford, CA, USA

Robert P. Yezierski Comprehensive Center for Pain Research and Department of Neuroscience, The McKnight Brain Institute University of Florida, Gainesville, FL, USA

Way Yin Department of Anesthesiology, University of Washington, Seattle, WA, USA

Atsushi Yoshida Department of Oral Anatomy and Neurobiology, Osaka University, Osaka, Japan

Joanna M. Zakrzewska Eastman Dental Hospital UCLH NHS Foundation Trust, London, UK

Jenna E. Zauk St. George's University, School of Medicine, Grenada, West Indies

Hanns Ulrich Zeilhofer Institute for Pharmacology and Toxicology, University of Zurich, and Swiss Federal Institute of Technology (ETH) Zurich, Zürich, Switzerland

Lonnie K. Zeltzer Pediatric Pain Program, Department of Pediatrics David Geffen School of Medicine at UCLA, Los Angeles, CA, USA

Giovambattista Zeppetella St. Clare Hospice, Hastingwood, UK

Xijing Zhang Department of Neuroscience, University of Minnesota, Minneapolis, MN, USA

Min Zhuo Department of Physiology, University of Toronto, Toronto, ON, Canada

C. Zöllner Department of Anesthesiology and Intensive Care Medicine, Charité University, Berlin, Germany

Krina Zondervan Genetic and Genomic Epidemiology Unit, Nuffield Dept of Obstetrics and Gynaecology, Wellcome Trust Centre for Human Genetics, Oxford, UK

Peter Zygmunt Department of Laboratory Medicine, Lund University, Division of Clinical Chemistry and Pharmacology, Lund, Sweden

Zbigniew Zylicz Hildegard Hospiz, Basel Switzerland, Basel, Switzerland

David C. Yeomans Department of Anesthesia, Stanford University, Stanford, CA, USA

Robert P. Yezierski Comprehensive Center for Pain Research and Department of Neuroscience, The McKnight Brain Institute, University of Florida, Gainesville, FL, USA

Wei Yin Department of Anesthesiology, University of Washington, Seattle, WA, USA

Atsushi Yoshida Department of Oral Anatomy and Neurobiology, Osaka University, Osaka, Japan

Joanna M. Zakrzewska Eastman Dental Hospital UCLH NHS Foundation Trust, London, UK

Jonna E. Zaus St. George's University School of Medicine, Grenada, West Indies

Hanns Ulrich Zeilhofer Institute for Pharmacology and Toxicology, University of Zürich, and Swiss Federal Institute of Technology (ETH), Zürich, Switzerland

Lonnie K. Zeltzer Pediatric Pain Program, Department of Pediatrics, David Geffen School of Medicine at UCLA, Los Angeles, CA, USA

Shravanthini Zeppetella St. Clad Hospice, Hertfordshire, UK

Sijing Zhang Department of Neuroscience, University of Minnesota, Minneapolis, MN, USA

Min Zhuo Department of Physiology, University of Toronto, Toronto, ON, Canada

C. Zöllner Department of Anaesthesiology and Intensive Care Medicine, Charité University, Berlin, Germany

Irina Zondervan Chronic and Genetic of Pain Epidemiology Unit, Nuffield Dept. of Obstetrics and Gynaecology, Wellcome Trust Centre for Human Genetics, Oxford, UK

Peter Zygmunt Department of Laboratory Medicine, Lund University, Division of Clinical Chemistry and Pharmacology, Lund, Sweden

Grigorios Nylou Hildgard Hospital, Basel Switzerland, Basel, Switzerland

A

α$_{2A}$ or α$_{2C}$-Adrenoceptor Agonists

▶ Alpha(α) 2-Adrenergic Agonists in Pain Treatment

α$_2$-Adrenergic Agonists

▶ Alpha(α) 2-Adrenergic Agonists in Pain Treatment

α-Agonists

▶ Alpha(α) 2-Adrenergic Agonists in Pain Treatment

Aδ-mechanoreceptor

▶ Mechanonociceptors

A Afferent Fibers (Neurons)

Definition

These are types of sensory afferent nerve fibers that are myelinated (encased in a myelin sheath)

and are classified according to their conduction velocity and sensory modality.

Aβ fibers are medium diameter afferent fibers with conduction velocities of 30–80 ms, and encode signals from non-noxious stimuli such as touch.

Aδ fibers are smaller caliber afferent fibers with conduction velocities of 5–30 ms, and principally encode signals from noxious stimuli. They are commonly thought to be responsible for the rapid sensation of "first pain" following injury.

It is often difficult to precisely identify the different classes of A fibers during development, as growth in fiber diameter and myelination occur slowly, so the eventual fate of fibers is not necessarily obvious at earlier stages of development.

Cross-References

▶ Infant Pain Mechanisms
▶ Insular Cortex, Neurophysiology, and Functional Imaging of Nociceptive Processing
▶ Magnetoencephalography in Assessment of Pain in Humans
▶ Nociceptor, Categorization
▶ Spinothalamic Tract Neurons, in Deep Dorsal Horn

A Delta(δ)-mechanoheat Receptor

▶ Polymodal Nociceptors, Heat Transduction

G.F. Gebhart, R.F. Schmidt (eds.), *Encyclopedia of Pain*, DOI 10.1007/978-3-642-28753-4,

A Fibers (A-Fibers)

Definition

The terminology refers to compound action potential deflections; A fibers are the most rapidly conducting category representing activity of myelinated fibers. Most A fibers are afferent nerve fibers that carry non-noxious somatosensory information.

Cross-References

▶ A Afferent Fibers (Neurons)
▶ Opiates During Development

AAV

▶ Adenoassociated Virus Vectors

Abacterial Meningitis

▶ Headache in Aseptic Meningitis

Abdominal Skin Reflex

Definition

A polysynaptic reflex triggered by stroking of the abdomen around the umbilicus. Excitation of low- and high-threshold mechanosensors induces a contraction of the abdominal muscles. The umbilicus moves toward the source of the stimulation. This reflex can be readily evoked in children but may have been lost in adult persons. The reflex action is tested in neurological examinations, e.g., for testing the integrity of spinal nerves. The interpretation as a "protective" reflex is plausible, but the reflex is not purely nocifensive.

Cross-References

▶ Mechanosensory Nociceptors
▶ Nocifensive

Abduction

Definition

It is the movement of a body part away from the midline of the body.

Cross-References

▶ Cancer Pain Management, Orthopedic Surgery

Aberrant Drug-Related Behaviors

Definition

Use of a prescription medication in a manner that violates expectations for responsible drug use. May be applied to verbal responses or actions. Occur on a continuum from relatively mild (e.g., unsanctioned dose escalation on one or two occasions) to severe (e.g., injecting oral formulations). Must be assessed to determine appropriate diagnosis (e.g., addiction, pseudoaddiction, and other psychiatric disorder).

Cross-References

▶ Cancer Pain, Evaluation of Relevant Comorbidities and Impact

Ablation

Definition

The basic definition of ablation is "elimination or removal." Medically, it is a procedure involving destruction of brain tissue to decrease the activity of a brain structure or interrupt information transmitted along a specific tract.

Cross-References

▶ Facet Joint Pain
▶ Pain Treatment, Intracranial Ablative Procedures

Abnormal Illness Affirming States

Definition

A group of psychiatric disorders (conversion disorder, hypochondriasis, somatization, pain disorder, factitious disorder, and malingering), where primary and/or secondary gain is believed to be important to the production of some or all of the patient's symptoms. Primary gain refers to intrapersonal (e.g., anxiety reducing) benefits and secondary gain refers to interpersonal benefits. It is to be noted that for malingering, secondary gain is thought to operate on a conscious level, but at an unconscious level for the other illness affirming states.

Cross-References

▶ Abnormal Illness Behavior
▶ Malingering, Primary and Secondary Gain

Abnormal Illness Behavior

Definition

Refers to an inappropriate or maladaptive mode of perceiving and acting in relation to one's own state of health, despite the fact that the doctor has offered an accurate and lucid explanation about the illness based on an adequate assessment of all biological, psychological, social, and cultural factors.

Cross-References

▶ Abnormal Illness Affirming States
▶ Pain as a Cause of Psychiatric Illness
▶ Psychiatric Aspects of the Management of Cancer Pain

Abnormal Illness Behavior of the Unconsciously Motivated, Somatically Focussed Type

▶ Hypochondriasis, Somatoform Disorders, and Abnormal Illness Behavior

Abnormal Sensation

▶ Dysesthesia, Assessment

Abnormal Temporal Summation

Synonyms

Wind-up

Definition

Abnormal Temporal Summation ("wind-up"): Repeated stimuli (e.g., touch) delivered in rapid succession to the skin are normally felt individually, or as vibration. Patients with neuropathic pain sometimes report that light touch stimuli repeated about once per second cause a sensation that builds up abnormally into an intensely painful crescendo.

Cross-References

▶ Diagnosis and Assessment of Clinical Characteristics of Central Pain

Abnormal Ureteric Peristalsis in Stone Rats

Definition

A marked increase in amplitude of phasic contractions (such that the intraureter pressure reaches levels likely to be sufficient to activate ureteric nociceptors) associated with a decrease in rate of contractions, and a reduced basal tone compared to peristalsis seen in normal rats.

Cross-References

▶ Visceral Pain Model, Kidney Stone Pain

Abscess

Definition

An abscess is a circumscribed area of injury and inflammation in which considerable necrosis has occurred and a fluid containing dead tissue and bacteria has collected. It may drain and be relatively comfortable, but if closed, tissue distension results in pain.

Cross-References

▶ Dental Pain, Etiology, Pathogenesis, and
  Management

Absorption

Definition

The absorption of a drug begins with transport or diffusion from one compartment (e.g., the stomach) into another (e.g., the blood). Alternatively, after a drug dissolved in a vehicle is administered intravenously, absorption can refer to transport or diffusion from the blood into the site of action, for example, the brain parenchyma.

Cross-References

▶ NSAIDs, Pharmacokinetics

ACC

▶ Anterior Cingulate Cortex

Accelerated Recovery Programs

▶ Postoperative   Pain,   Importance   of
  Mobilization

Acceleration-deceleration Injury

▶ Whiplash

Accelerometer

Definition

It is an instrument for measuring acceleration or change of velocity with respect to time.
www.ncbi.nlm.nih.gov/snp

Cross-References

▶ Assessment of Pain Behaviors

Acceptance

Lance McCracken[1] and Miles Thompson[2]
[1]Health Psychology/Psychology Department, King's College London, London, UK
[2]Goldsmiths, University of London, London, UK

Synonyms

Willingness

Definition

An ongoing quality of behavior that reflects a willingness to experience unwanted psychological experiences, such as pain, without defense, in the present moment.

Introduction

The word acceptance has philosophical and religious roots that span millennia. In more contemporary times, it appears within many varied approaches to psychological therapy and has become increasingly prominent (Williams and Lynn 2010–2011). Within the field of chronic pain, however, the term most frequently appears in association with the form of cognitive behavioral therapy called Acceptance and Commitment Therapy (ACT) where acceptance is one of six interconnected processes (Hayes et al. 2006; Hayes et al. 1999; Ruiz 2010). ACT is a contextual cognitive behavioral therapy and, in common with other approaches such as Mindfulness (Kabat-Zinn 1990) and Dialectical Behavior Therapy (Linehan 1993), it emphasizes behavior change and qualities of daily functioning rather than symptom reduction (Hayes et al. 2011). A key distinction in ACT and these other approaches is the difference between the form and the function of experienced events. From a "functional" perspective, psychological experiences, such as pain, are not seen as being problematic in their form alone but rather in the processes by which they influence an individual's behavior and, thus, their participation in life.

Pragmatic, functional, and contextual approaches, such as ACT, can be especially useful in working with long-term health conditions where consistent symptom reduction is often unobtainable. In such situations, moments spent wrestling for control over symptoms often have an ironic quality. In these moments, the influence of symptoms on behavior is actually increased and not decreased as intended, and a quality of connection with other goals or engagement in "normal life" can be lost. Acceptance can neutralize these ironic processes.

Characteristics

Expanded Definition

Similar to everyone, people experiencing chronic pain engage in behavior patterns that are natural and automatic given their circumstances. Many of these behaviors aim to avoid, control, reduce, or allow escape from unwanted experiences. This kind of behavior can take place in relation to physical sensations associated with pain as well as other psychological experiences such as thoughts, feelings, and memories that are painful. While entirely understandable, behavior of this kind is often unsuccessful in controlling experiences. In addition, such behavior often incurs additional costs. The dominance of avoiding unwanted experiences can narrow the range of behavior options, blocking other potentially more successful behavior and reducing quality of life. In fact, experimental and clinical evidence indicate that attempts to control the frequency or form of psychological experiences are difficult, *and* it can actually result in paradoxical increases in their occurrence and their impact (Hayes and Gifford 1997; Wenzlaff and Wegner 2000). Acceptance is at first an unusual and even counterintuitive approach to take toward pain and other distressing psychological experiences, and at the same time, it provides an alternative to the potential problems that emerge from the more avoidant and control-focused behavior patterns highlighted above.

Acceptance is defined as an active, ongoing, purposeful, quality of behavior that involves a willingness to experience psychological experiences, in the present moment, and without defense. The term is notoriously easy to misunderstand, and a few distinctions can help. Acceptance is not a mental act. It does not reside in thoughts or beliefs. It is not a onetime act of saying, "I accept this." It is an act of the whole person. It involves ongoing moment-by-moment acts of openness to experiences that might otherwise invite resistance, defense, or refusal. Acceptance is typically found in the coordination of patterns of behavior around unwanted experiences. If what is coordinated is avoidance or struggling for control of these experiences, this is not acceptance. If what is coordinated is engagement in goals and normal life activities at the same time that unwanted experiences are being contacted, this is acceptance. In fact, it appears that the best way to consider acceptance is as having two components: one of these is *engaging* in activities while pain is present, and the other is *refraining* from attempts at avoidance (McCracken 2010).

Evidence for Acceptance

Over recent years, evidence for the utility of acceptance within chronic pain management has grown steadily in experimental research, clinical studies, and treatment outcome data. More detailed information on this body of work can be found in recent review papers and chapters (e.g., McCracken and Vowles 2006; Thompson and McCracken 2011; Vowles and Thompson 2011).

Experimental, laboratory-based, research in this area often involves healthy subjects without chronic pain. Here, participants may receive transient pain stimulation, for example, during cold-pressor tasks. In research of this kind, different experimental groups are given different instructions or brief training experiences while pain tolerance or task persistence is assessed. In some conditions, participants might be asked to distract themselves or to attempt to suppress or control the pain, while others are encouraged to respond with willingness to experience painful sensations. The growing body of work in this area suggests that acceptance-oriented instructions are more successful than distraction, suppression, or control techniques, particularly with respect to pain tolerance measures or persistence with experimental tasks.

In clinical settings, many studies have examined the relationships between measures of acceptance and reports of daily functioning in patients seeking treatment with chronic pain. Overall, results from these studies suggest that higher levels of acceptance are associated with lower levels of pain intensity, depression, pain-related anxiety, physical and psychosocial disability, higher levels of daily activity, and overall well-being. Recent research has also compared the utility of acceptance with other more traditional psychological variables thought to be important in this area such as attention, anxiety sensitivity, coping, and catastrophizing. Preliminary results suggest that levels of acceptance are better than most of these more traditional variables at predicting patient functioning. This suggests that research in these others areas might be enhanced by a closer understanding of the developing evidence base surrounding acceptance.

There is increasingly robust evidence for efficacy of acceptance-based treatment in the area of chronic pain. The first randomized trial was published in 2004 (Dahl et al. 2004). Results of treatment often include reductions in measures of disability, depression, pain-related anxiety, and distress along with improvements in measures of physical functioning and performance and reduced health care use. For example, one recent study suggested that 75.4 % of patients demonstrated reliable change in either disability, pain-related anxiety, or depression, while 61.4 % showed reliable change in two or more of these areas (Vowles and McCracken 2008). Notably, these treatments do not target acceptance alone but also the other processes from the ACT model, such as values, committed action, and others (see McCracken 2005 for more details). Research has been carried out in both adult and adolescent populations. While much of this research has involved group-based interdisciplinary treatment with highly disabled individuals, there are also trials of treatment delivered individually and in the form of a self-help book with limited therapist support (Johnston et al. 2010). The status of this evidence base is now such that ACT is regarded as having "strong research support" by the Society of Clinical Psychology within the American Psychological Association (APA, Division 12; "Society of Clinical Psychology," 2011).

Wider Psychological Model

Acceptance is only one of six interrelated processes within the psychological model proposed by ACT. The other five processes are cognitive defusion, contact with the present moment, self as context, values, and committed action (see Fig. 1). Together, all six processes entail "psychological flexibility": an ability to be fully connected to the present moment and from there to be able to either maintain or change one's behavior to pursue one's goals and values according to what the situation directly affords. It is likely to be most useful to see acceptance within this wider context and not in isolation. Acceptance is not the complete answer to most behavior problems. It is also not an end in itself. It does, however, provide, in combination with other processes of behavior change, a pragmatic means for individuals with chronic pain to move toward the things which matter most to them in their lives. As mentioned, evidence for the wider set of processes, in combination with acceptance and by themselves, is growing (Thompson and McCracken 2011).

Measures of Acceptance

CPAQ: The Chronic Pain Acceptance Questionnaire is the measure most frequently used in the assessment of acceptance within chronic pain populations (McCracken et al. 2004). It is a 20-item scale consisting of two subscales: activity engagement and pain willingness. Activity engagement measures participation in activities with continuing pain (e.g., "I lead a full life even though I have chronic pain"), while pain willingness assesses an individual's capacity to have pain without attempts to avoid

Acceptance, Fig. 1 The relationship between acceptance, psychological flexibility, and related processes

or control it (e.g., "Before I make any serious plans, I have to get some control over my pain"). A recent systematic review confirmed the status of the measures' internal consistency, construct validity, and reliability (Reneman et al. 2010). The CPAQ has been translated into a number of languages including Cantonese, German, Spanish, and Swedish. An adapted version of the measure for adolescent populations has also been developed (CPAQ-A; McCracken et al. 2010).

AAQ-II: The Acceptance and Action Questionnaire-II (Bond et al. 2011) is a broader measure of acceptance that explores an individual's relationship with thoughts, feelings, and other physical symptoms generally, not just those related to chronic pain. The AAQ-II contains seven items (e.g., "I'm afraid of my feelings") which relate to one factor. Higher scores indicate lower levels of acceptance and psychological flexibility. Research indicates that in studies involving both the CPAQ and the AAQ-II, the AAQ-II explains unique variance independent to that explained by the CPAQ (McCracken and Zhao-O'Brien 2010). Importantly, this suggests that it is both, willingness to experience psychological experiences not directly associated with chronic pain and willingness to experience pain, that facilitate healthy vital functioning.

References

Bond, F. W., Hayes, S. C., Baer, R. A., Carpenter, K. M., Guenole, N., Orcutt, H. K., et al. (2011). Preliminary psychometric properties of the acceptance and action questionnaire – II: A revised measure of psychological inflexibility and experiential avoidance. *Behavior Therapy, 42*(4), 676–688.

Dahl, J., Wilson, K. G., & Nilsson, A. (2004). Acceptance and commitment therapy and the treatment of persons at risk for long-term disability resulting from stress and pain symptoms: A preliminary randomized trial. *Behavior Therapy, 35*, 785–801.

Hayes, S. C., & Gifford, E. V. (1997). The trouble with language: Experiential avoidance, rules, and the nature of verbal events. *Psychological Science, 8*, 170–173.

Hayes, S. C., Strosahl, K., & Wilson, K. G. (1999). *Acceptance and commitment therapy: An experiential approach to behavior change.* New York: Guilford Press.

Hayes, S. C., Luoma, J., Bond, F. W., Masuda, A., & Lillis, J. (2006). Acceptance and commitment therapy: Model, processes, and outcomes. *Behaviour Research and Therapy, 44*, 1–25.

Hayes, S. C., Villatte, M., Levin, M., & Hildebrandt, M. (2011). Open, aware, and active: Contextual approaches as an emerging trend in the behavioral and cognitive therapies. *Annual Review of Clinical Psychology, 7*, 141–168.

Johnston, M., Foster, M., Shennan, J., Starkey, N. J., & Johnson, A. (2010). The effectiveness of an acceptance and commitment therapy self-help intervention for chronic pain. *The Clinical Journal of Pain, 26*, 393–402.

Kabat-Zinn, J. (1990). *Full catastrophe living: Using the wisdom of your body and mind to face stress, pain, and illness.* New York: Delacorte Press.

Linehan, M. M. (1993). *Cognitive-behavioral treatment of borderline personality disorder.* New York: Guilford.

McCracken, L. M. (2005). *Contextual cognitive-behavioral therapy for chronic pain.* Seattle: IASP Press.

McCracken, L. M. (2010). Toward understanding acceptance and psychological flexibility in chronic pain. *Pain, 149*, 420–421.

McCracken, L. M., & Vowles, K. E. (2006). Acceptance of chronic pain. *Current Pain and Headache Reports, 10*, 90–94.

McCracken, L. M., & Zhao-O'Brien, J. (2010). General psychological acceptance and chronic pain: There is more to accept than the pain itself. *European Journal of Pain, 14*, 170–175.

McCracken, L. M., Vowles, K. E., & Eccleston, C. (2004). Acceptance of chronic pain: Component analysis and a revised assessment method. *Pain, 107*, 159–166.

McCracken, L. M., Gauntlett-Gilbert, J., & Eccleston, C. (2010). Acceptance of pain in adolescents with chronic pain: Validation of an adapted assessment instrument and preliminary correlation analyses. *European Journal of Pain, 14*, 316–320.

Reneman, M. F., Dijkstra, A., Geertzen, J. H., & Dijkstra, P. U. (2010). Psychometric properties of chronic pain acceptance questionnaires: A systematic review. *European Journal of Pain, 14*, 457–465.

Ruiz, F. J. (2010). A review of acceptance and commitment therapy (ACT) empirical evidence: Correlational, experimental psychopathology, component and outcome studies. *International Journal of Psychology and Psychological Therapy, 10*, 125–162.

Society of Clinical Psychology. (2011). Acceptance and Commitment Therapy for chronic pain. Retrieved December 11, 2011, from http://www.div12.org/PsychologicalTreatments/treatments/chronicpain_act.html

Thompson, M., & McCracken, L. M. (2011). Acceptance and related processes in adjustment to chronic pain. *Current Pain and Headache Reports, 15*, 144–151.

Vowles, K. E., & McCracken, L. M. (2008). Acceptance and values-based action in chronic pain: A study of treatment effectiveness and process. *Journal of Consulting and Clinical Psychology, 76*, 397–407.

Vowles, K. E., & Thompson, M. (2011). Acceptance and commitment therapy for chronic pain. In L. L. M. McCracken (Ed.), *Mindfulness and acceptance in behavioral medicine: Current theory and practice* (pp. 31–60). Oakland: New Harbinger.

Wenzlaff, R. M., & Wegner, D. M. (2000). Thought suppression. *Annual Review of Psychology, 51*, 59–91.

Williams, J. C., & Lynn, S. J. (2010–2011). Acceptance: An historical and conceptual review. *Imagination, Cognition and Personality, 30*, 5–56.

Acceptance-based Treatment

▶ Psychological Treatment of Pain in Children

Accommodation (of a Nerve Fiber)

Definition

Changes in the threshold for initiation of an action potential or in conduction velocity of a nerve fiber; for example, imposing a slow depolarizing stimulus on an axon will move the action potential threshold to a more depolarized potential, making it more difficult to initiate an action potential.

Cross-References

▶ Mechano-insensitive C-Fibers, Biophysics

Accuracy and Reliability of Memory

Definition

Reliability of memory of pain concerns the correlation for a group of patients between the report of pain at the time of its occurrence, for example, a score on a rating scale, and the estimate of that score at a later time (the remembered pain). Accuracy refers to the extent of agreement between records of the original event and the corresponding memory for an individual. According to this distinction, memories may be reliable but not accurate.

Cross-References

► Pain Memory

ACE Inhibitors, Beta(β)-Blockers

Definition

Drugs used to lower blood pressure and relieve heart failure.

Cross-References

► Postoperative Pain, Acute Pain Management Principles

Acetaminophen

► Paracetamol
► Pharmacology of Cyclooxygenase Inhibitors
► Postoperative Pain, Paracetamol
► Simple Analgesics

Acetylation

Definition

The acetyl group of acetylsalicylic acid (aspirin) binds to serine 530 in the active site of COX-1 or serine 516 in the active site of COX-2. This prevents the access of arachidonic acid to the catalytic site of the cyclooxygenase.

Cross-References

► Cyclooxygenases in Biology and Disease
► Pharmacology of Cyclooxygenase Inhibitors

Acetylcholine

Synonyms

ACh

Definition

Acetylcholine is a neurotransmitter synthesized from choline and acetyl coenzyme A. It is localized in large reticular formation neurons and is the chemical mediator in the synapse of a motor endplate. The electrical signal of the motor nerve terminal causes release of many packets of acetylcholine. The packets are released into the synaptic cleft, where receptors in the postjunctional membrane of the striated muscle fiber membrane convert the chemical signal to an electrical signal (a propagated action potential), which can produce muscle contractile activity. Normally, an occasional acetylcholine packet is released spontaneously by the nerve terminal without a nerve signal. Each packet produces a miniature endplate potential in the muscle fiber, but its amplitude is too small to be propagated. Myofascial trigger points are associated with excessive spontaneous release of acetylcholine packets in affected endplates.

Cross-References

▶ Myofascial Trigger Points
▶ Thalamic Neurotransmitters and
  Neuromodulators

Acetylcholine Receptors

Definition

Receptors for the neurotransmitter acetylcholine, which can be distinguished into muscarinergic (G protein-coupled) and nicotinergic (ion channel) receptors.

ACh

▶ Acetylcholine

Ache in First Trimester or in Early Gestation

▶ Pain in Early Pregnancy

Acidosis

Definition

Acidosis is the disturbance of the acid–base balance, characterized by acidity (decreased pH) by accumulation of protons, caused by injury, inflammation, or ischemia. Acidosis is an important source of pain. In humans, it produces nonadapting nociceptor excitation and contributes to hyperalgesia and allodynia in inflammation.

Cross-References

▶ Acid-Sensing Ion Channels
▶ TRPV1, Regulation by Protons

Acid-Sensing Ion Channels

Eric Lingueglia and Michel Lazdunski
Institut de Pharmacologie Moléculaire et
Cellulaire (IPMC) CNRS/UNS UMR7275,
Valbonne-Sophia Antipolis, France

Synonyms

ASIC; ASIC1a; ASIC1b: ASICβ, BNaC2β; ASIC2a: mammalian degenerin 1 (MDEG1), brain sodium channel 1 (BNC1a, BNaC1a); ASIC2b: mammalian degenerin 2 (MDEG2), brain sodium channel 1 (BNaC1b); ASIC3: dorsal-root acid-sensing ion channel (DRASIC); ASICa, brain sodium channel 2 (BNC2, BNaC2a)

Definition

Acid-Sensing Ion Channels (ASICs) are membrane proteins that form depolarizing cation (principally sodium) channels. They are largely present on peripheral and central neurons. ASIC channels are opened by extracellular protons to produce a depolarization that can trigger action potential and/or modify neuronal excitability, depending on the amplitude and kinetics of pH change. Extracellular tissue acidosis is commonly associated with pain at the periphery in conditions like inflammation, ischemia, tumors, infections, lesions, or traumatic injuries. Peripheral ASICs have also been involved in cutaneous mechanical sensitivity and visceral mechanosensory function. In the central nervous system, these channels have been proposed to participate in the transmission and in the modulation of nociceptive signals in the spinal cord, besides their contribution to fear, anxiety, depression, and neuronal degeneration.

Characteristics

The three-dimensional structure of ASIC1 shows that three subunits, among the six characterized in rodents, need to assemble to form a channel (Fig. 1b). Four different genes encode these six subunits (ASIC1 to ASIC4, Fig. 1a), with two splice variants for ASIC1 (1a and 1b) and ASIC2 (2a and 2b) (Chen et al. 1998; Lingueglia et al. 1997; Waldmann et al. 1997a, b). Each subunit is 510–560 amino acids long, with two hydrophobic transmembrane regions flanking a large extracellular loop representing more than 50 % of the protein (Fig. 1b). The properties of ASIC channels (i.e., activation, inactivation and reactivation kinetics, pH sensitivity, ion selectivity, pharmacology) vary according to their subunit composition, which accounts for the large diversity of native ASIC-like currents observed in neurons (Krishtal and Pidoplichko 1980). ASIC channels are not voltage dependent and their $pH_{0.5}$ range from 4.0 to 6.8 with activation thresholds close to pH 7.0 for ASIC1 and ASIC3, i.e., well in the pathophysiological pH range (Table 1). They show steep pH dependence, making them very sensitive sensors of extracellular protons. Most ASIC channels are transiently activated and desensitize rapidly (within seconds). Moreover, ASIC3-containing channels have an additional slowly activating, sustained component that does not inactivate as long as the pH remains acidic (Fig. 1a). ASIC2b is not activated by acidic pH when expressed alone but can associate with other ASICs to modulate their properties and/or their regulation (Lingueglia et al. 1997).

Almost all ASIC isoforms are present in primary sensory neurons of the trigeminal, vagal, and dorsal root ganglia. ASIC1, ASIC2, and ASIC3 are expressed in nociceptors, including high-threshold mechanoreceptors. ASIC2 and

Acid-Sensing Ion Channels, Fig. 1 (a) Phylogenic tree of ASIC channels. There are four different genes encoding six isoforms in rodents, with two variants (a and b) for ASIC1 and ASIC2. Examples of current associated with each isoform and elicited by a change in external pH from 7.4 to 5.0 (indicated by the bar above the current traces) are represented. ASIC channels possess different properties (kinetics of activation, inactivation and reactivation, ionic selectivity, pH sensitivity) depending on the type of isoform and of their combination (see below). ASIC2b and ASIC4 are not activated by extracellular protons on their own, but can associate with other ASICs to modulate their properties and/or their expression. (b) Structure and expression in the nervous system. Functional Acid-Sensing Ion Channels (ASICs) are formed by the assembling of three identical subunits (homomers) or three different subunits (heteromers) (Jasti et al. 2007). The structure of one subunit is shown in inset. ASICs predominantly conduct Na^+ but homomeric ASIC1a also displays some calcium permeability. ASICs are largely expressed in the nervous system. Almost all subunits are present in the peripheral nervous system (PNS) with a strong expression of ASIC3. ASIC1a and ASIC2 (both variant a and b) are predominant in the central nervous system (CNS), including the spinal cord. ASIC4 is also present in the CNS but its exact function remains unknown

Acid-Sensing Ion Channels, Table 1 Properties of ASIC channels

Isoform	Tissue distribution	$pH_{0.5}$ activation	Inhibitors/ activators	Endogenous modulators	Physiology	Potential therapeutic implications
ASIC1a	Strong expression in PNS, brain, spinal cord, retina	6.2–6.8	Amiloride and derivatives benzamil and EIPA, NSAIDs (ibuprofen/flurbiprofen), A-317567, diarylamidines, Psalmotoxin 1 (PcTx1), mambalgins	Arachidonic acid, lactate, NO, RFamide peptides, dynorphins	Central modulation of pain (alone or in association with ASIC2a or ASIC2b), central sensitization to pain in the spinal cord	*Inhibition:* pain, psychiatric disorders (anxiety, panic, depression), stroke, neurodegenerative diseases
	Also present in taste cells, peripheral and central auditory system, immune cells, lung epithelial cells, bone, vascular smooth muscle cells, gliomas		Potently activated by MitTx		Also involved in primary hyperalgesia in muscle, visceral but not cutaneous mechanosensation, possible implication in chemotransduction of low pH by carotid body	*Activation:* seizure treatment
					Important role in the CNS: synaptic transmission and plasticity, learning and memory, innate and conditioned fear, chemosensory function of the amygdala, depression, visual transduction	
ASIC1b	PNS, taste cells, cochlear hair cells, carotid body, immune cells	5.1–6.2	Amiloride, diarylamidines, mambalgins	NO	Nociception	Pain
			Potently activated by MitTx			
ASIC2a	PNS (including specialized cutaneous mechanosensory structures), brain, spinal cord, retina, peripheral and central auditory system, taste cells, bone, vascular smooth muscle cells, some expression in carotid body, gliomas	4.1–5.0	Amiloride, diarylamidines, A-317567, mambalgins (when associated with ASIC1a)	NO, Zn^{2+}	Modulation of ASIC1a in the CNS, central modulation of pain (in association with ASIC1a), visual transduction, cutaneous and visceral mechanosensation, noise susceptibility of hearing, baroreceptor reflex, may be involved in sour taste perception, vascular smooth muscle cell migration	Pain, control of blood pressure
			Response to H^+ enhanced by MitTx			

(continued)

Acid-Sensing Ion Channels, Table 1 (continued)

Isoform	Tissue distribution	$pH_{0.5}$ activation	Inhibitors/ activators	Endogenous modulators	Physiology	Potential therapeutic implications
ASIC2b	PNS, brain, spinal cord, retina, taste cells	n/a	PcTx1 and mambalgins (when associated with ASIC1a), APETx2, salicylic acid and diclofenac (when associated with ASIC3)		Modulate the properties of other ASICs (ASIC1a, ASIC2a, and ASIC3)	Presumably pain in association with ASIC3 or ASIC1a
ASIC3	PNS (including specialized cutaneous mechanosensory structures), retina, taste cells, carotid body, inner ear, testis, bone, lung epithelial cells, vascular smooth muscle cells, dendritic cells	6.2–6.7	Amiloride, NSAIDs (aspirin/ diclofenac), A-317567, diarylamidines, APETx2 Activated by MitTx and GMQ	Arachidonic acid, lactate, NO, agmatine, hypertonicity, RFamide peptides	Cutaneous acidic and primary inflammatory pain, secondary mechanical hyperalgesia in muscle and joints, cutaneous and visceral mechanosensation and mechanonociception, neuronal mechanosensor for pressure-induced vasodilation, acid sensing in gastroesophageal afferents, cardiac pain, maintenance of retinal integrity, hearing, testosterone homeostasis, possible implication in chemotransduction of low pH by carotid body and local vascular control in muscle	*Inhibition:* inflammation related pain, postoperative pain, chronic muscular pain, angina, arthritis, gastritis, inflammatory and noninflammatory bowel disorders *Activation:* protection against pressure ulcers

n/a not applicable, *PNS* peripheral nervous system

ASIC3 have also been found in large non-nociceptive neurons that mostly correspond to low threshold mechanoreceptors. The proteins have been detected in the soma and in the peripheral nerve endings of DRG neurons, but their presence on central projections in the dorsal horn of the spinal cord is less clear. ASICs support most of the native proton-activated cation currents in sensory neurons, although a significant part of the sustained response to low pH (below pH 6.0) can be attributed to the capsaicin receptor TRPV1. However, the native ASIC-like responses are approximately tenfold more sensitive to changes in H^+ than the TRPV1 responses.

Peripheral ASIC channels are important for somatic and visceral pain. Their role in inflammatory pain is particularly noteworthy (Table 1). They display both an increase in expression and an upregulation of their activity by several components of the "inflammatory soup." ASIC3 is particularly important in this regard as a pain sensor (Deval et al. 2008, 2010) that integrates responses to several molecular signals produced during inflammation or ischemia (protons, hypertonicity, lactic and arachidonic acid, ATP, NO, agmatine, bradykinin, serotonin) to contribute to peripheral sensitization of nociceptors. ASIC3 is also important for joint and muscle pain (Sluka et al. 2003). ASICs transduce acid sensation in gastric sensory nerve endings (Wultsch et al. 2008) and play a role in gastrointestinal mechanosensory function (Page et al. 2005). ASIC1a and ASIC2 (both variant a and b) are largely expressed in spinal second-order sensory neurons that receive primary afferent inputs. Their expression is upregulated during peripheral inflammation and they are probably playing a role in the processing of noxious stimuli and in the generation of hyperalgesia and allodynia in persistent pain (Baron et al. 2008; Duan et al. 2007). ASIC1a + ASIC2a heteromeric channels seem to be particularly important (Diochot et al. 2012). Some ASICs are also important for pain modulation since blocking of the ASIC1a homomeric channels, and possibly the ASIC1a + ASIC2b heteromeric channels, at the spinal and/or supraspinal level results in an activation of the endogenous enkephalin pathway and in increased levels of Met-enkephalin in the cerebrospinal fluid (Mazzuca et al. 2007), producing strong analgesic effects.

ASICs have been involved in mechanosensory function (Table 1). Interestingly, the general ASIC contribution to mechanoreceptor function in visceral fibers (Page et al. 2005) is much larger and significantly different than the one observed in skin, where studies of simple knockout mice showed only subtle contribution of ASIC2 and ASIC3 to normal touch sensation (Chen et al. 2002; Price et al. 2000, 2001). However, ASIC3 is an essential neuronal sensor for the skin vasodilation response to direct pressure in both humans and rodents, and for protecting against pressure ulcers in mice (Fromy et al. 2012).

ASIC currents are inhibited by amiloride, a potassium-sparing diuretic, which acts as a poorly selective, nondiscriminative pore blocker of ASICs at the micromolar range of concentrations (Waldmann et al. 1997b). Amiloride also has an additional paradoxical enhancing effect on the sustained current component of ASIC3. The amidine A-317567 has been described as a more specific although still nondiscriminative inhibitor of ASICs, with IC_{50} between 2 and 30 mM on native ASIC currents in DRG neurons (Dube et al. 2005). ASIC1a and ASIC3 are also directly inhibited by therapeutic concentrations of nonsteroidal anti-inflammatory drugs (NSAIDs) (flurbiprofen and ibuprofen for ASIC1a, salicylic acid, aspirin, and diclofenac for ASIC3, $IC_{50}\sim$92–350 mM) (Voilley et al. 2001). Antiprotozoal diarylamidines (DAPI, diminazene, HSB, and pentamidine) have been reported to potently inhibit ASIC currents in primary cultures of hippocampal neurons with apparent affinities ranging from 300 nM to 38 μM (Chen et al. 2010).

Two small peptides that selectively and efficiently block ASICs in vitro and in vivo in a subtype-specific manner have been isolated from animal venoms (Diochot et al. 2007). The spider peptide Psalmotoxin 1 (PcTx1) blocks ASIC1a homomeric channel with an IC50 of ~0.9 nM (Escoubas et al. 2000). The sea anemone peptide APETx2 blocks homomeric ASIC3 with an IC50 of ~63 nM as well as ASIC3-containing heteromeric channels (IC50~0.1– 2 μM) (Diochot et al. 2004). Both PcTx1 and APETx2 evoke potent analgesic effects in rodents upon central and peripheral injection, respectively (Deval et al. 2008; Mazzuca et al. 2007). Mambalgins are peptides from black mamba venom that specifically inhibit all the channel subtypes expressed in the central nervous system (i.e., ASIC1a, ASIC1a + ASIC2a and ASIC1a + ASIC2b), as well as ASIC1b expressed in nociceptors (Diochot et al. 2012). Central injections of mambalgins evoke a naloxone-insensitive analgesia, which is different from the one produced by PcTx1 and operates through blockade of ASIC1a + ASIC2a channels, while the peripheral analgesic effect observed after intraplantar injection of mambalgins involves ASIC1b-containing channels. A Texas coral snake toxin that potently activates peripheral

ASICs in nociceptive neurons (and especially ASIC1a) has been shown to evoke pain in animals (Bohlen et al. 2011). Similarly, small molecules like the synthetic compound 2-guanidine-4-methylquinazoline (GMQ) or the endogenous arginine-metabolite agmatine (Yu et al. 2010) that act as nonproton activators and/or modulators of ASIC3 evoke pain-related behaviors in an ASIC3-dependent manner.

Neuropeptides can also modulate ASIC channels through direct interaction. FMRFamide, an invertebrate neuropeptide, and structurally related peptides present in the central nervous system of mammals, such as neuropeptide FF (NPFF, FLFQPQRFamide), potentiate H$^+$ gated currents of heterologously expressed ASIC1 and ASIC3, but not ASIC2a (EC50 ~ 10–50 μM; threshold ~1 μM) (Askwith et al. 2000; Catarsi et al. 2001; Deval et al. 2003). Dynorphin A and big dynorphin, two endogeneous opioid neuropeptides, have also been shown to enhance the ASIC1a channel activity (Sherwood and Askwith 2009).

Cross-References

▶ ASICs

References

Askwith, C. C., Cheng, C., Ikuma, M., Benson, C., Price, M. P., & Welsh, M. J. (2000). Neuropeptide FF and FMRFamide potentiate acid-evoked currents from sensory neurons and proton-gated DEG/ENaC channels. Neuron, 26(1), 133–141.

Baron, A., Voilley, N., Lazdunski, M., & Lingueglia, E. (2008). Acid sensing ion channels in dorsal spinal cord neurons. The Journal of Neuroscience, 28(6), 1498–1508.

Bohlen, C. J., Chesler, A. T., Sharif-Naeini, R., Medzihradszky, K. F., Zhou, S., King, D., et al. (2011). A heteromeric Texas coral snake toxin targets acid-sensing ion channels to produce pain. Nature, 479(7373), 410–414.

Catarsi, S., Babinski, K., & Seguela, P. (2001). Selective modulation of heteromeric ASIC proton-gated channels by neuropeptide FF. Neuropharmacology, 41(5), 592–600.

Chen, C. C., England, S., Akopian, A. N., & Wood, J. N. (1998). A sensory neuron-specific, proton-gated ion channel. Proceedings of the National Academy of Sciences of the United States of America, 95(17), 10240–10245.

Chen, C. C., Zimmer, A., Sun, W. H., Hall, J., Brownstein, M. J., & Zimmer, A. (2002). A role for ASIC3 in the modulation of high-intensity pain stimuli. Proceedings of the National Academy of Sciences of the United States of America, 99(13), 8992–8997.

Chen, X., Qiu, L., Li, M., Durrnagel, S., Orser, B. A., Xiong, Z. G., et al. (2010). Diarylamidines: high potency inhibitors of acid-sensing ion channels. Neuropharmacology, 58(7), 1045–1053.

Deval, E., Baron, A., Lingueglia, E., Mazarguil, H., Zajac, J. M., & Lazdunski, M. (2003). Effects of neuropeptide SF and related peptides on acid sensing ion channel 3 and sensory neuron excitability. Neuropharmacology, 44(5), 662–671.

Deval, E., Noel, J., Lay, N., Alloui, A., Diochot, S., Friend, V., et al. (2008). ASIC3, a sensor of acidic and primary inflammatory pain. The EMBO Journal, 27(22), 3047–3055.

Deval, E., Gasull, X., Noel, J., Salinas, M., Baron, A., Diochot, S., et al. (2010). Acid-sensing ion channels (ASICs): pharmacology and implication in pain. Pharmacology and Therapeutics, 128(3), 549–558.

Diochot, S., Baron, A., Rash, L. D., Deval, E., Escoubas, P., Scarzello, S., et al. (2004). A new sea anemone peptide, APETx2, inhibits ASIC3, a major acid-sensitive channel in sensory neurons. The EMBO Journal, 23(7), 1516–1525.

Diochot, S., Salinas, M., Baron, A., Escoubas, P., & Lazdunski, M. (2007). Peptides inhibitors of acid-sensing ion channels. Toxicon, 49(2), 271–284.

Diochot, S., Baron, A., Salinas, M., Douguet, D., Scarzello, S., Dabert-Gay, A. S., et al. (2012). Black mamba venom peptides target acid-sensing ion channels to abolish pain. Nature, 490, 552–555.

Duan, B., Wu, L. J., Yu, Y. Q., Ding, Y., Jing, L., Xu, L., et al. (2007). Upregulation of acid-sensing ion channel ASIC1a in spinal dorsal horn neurons contributes to inflammatory pain hypersensitivity. The Journal of Neuroscience, 27(41), 11139–11148.

Dube, G. R., Lehto, S. G., Breese, N. M., Baker, S. J., Wang, X., Matulenko, M. A., et al. (2005). Electrophysiological and in vivo characterization of A-317567, a novel blocker of acid sensing ion channels. Pain, 117(1–2), 88–96.

Escoubas, P., De Weille, J. R., Lecoq, A., Diochot, S., Waldmann, R., Champigny, G., et al. (2000). Isolation of a tarantula toxin specific for a class of proton-gated Na$^+$ channels. The Journal of Biological Chemistry, 275(33), 25116–25121.

Fromy, B., Lingueglia, E., Sigaudo-Roussel, D., Saumet, J. L., & Lazdunski, M. (2012). Asic3 is a neuronal mechanosensor for pressure-induced vasodilation that protects against pressure ulcers. Nature Medicine. 18, 1205–1207.

Jasti, J., Furukawa, H., Gonzales, E. B., & Gouaux, E. (2007). Structure of acid-sensing ion channel 1 at 1.9 A resolution and low pH. *Nature, 449*(7160), 316–323.

Krishtal, O. A., & Pidoplichko, V. I. (1980). A receptor for protons in the nerve cell membrane. *Neuroscience, 5*(12), 2325–2327.

Lingueglia, E., de Weille, J. R., Bassilana, F., Heurteaux, C., Sakai, H., Waldmann, R., et al. (1997). A modulatory subunit of acid sensing ion channels in brain and dorsal root ganglion cells. *The Journal of Biological Chemistry, 272*(47), 29778–29783.

Mazzuca, M., Heurteaux, C., Alloui, A., Diochot, S., Baron, A., Voilley, N., et al. (2007). A tarantula peptide against pain via ASIC1a channels and opioid mechanisms. *Nature Neuroscience, 10*(8), 943–945.

Page, A. J., Brierley, S. M., Martin, C. M., Price, M. P., Symonds, E., Butler, R., et al. (2005). Different contributions of ASIC channels 1a, 2, and 3 in gastrointestinal mechanosensory function. *Gut, 54*(10), 1408–1415.

Price, M. P., Lewin, G. R., McIlwrath, S. L., Cheng, C., Xie, J., Heppenstall, P. A., et al. (2000). The mammalian sodium channel BNC1 is required for normal touch sensation. *Nature, 407*(6807), 1007–1011.

Price, M. P., McIlwrath, S. L., Xie, J., Cheng, C., Qiao, J., Tarr, D. E., et al. (2001). The DRASIC cation channel contributes to the detection of cutaneous touch and acid stimuli in mice. *Neuron, 32*(6), 1071–1083.

Sherwood, T. W., & Askwith, C. C. (2009). Dynorphin opioid peptides enhance acid-sensing ion channel 1a activity and acidosis-induced neuronal death. *The Journal of Neuroscience, 29*(45), 14371–14380.

Sluka, K. A., Price, M. P., Breese, N. M., Stucky, C. L., Wemmie, J. A., & Welsh, M. J. (2003). Chronic hyperalgesia induced by repeated acid injections in muscle is abolished by the loss of ASIC3, but not ASIC1. *Pain, 106*(3), 229–239.

Voilley, N., de Weille, J., Mamet, J., & Lazdunski, M. (2001). Nonsteroid anti-inflammatory drugs inhibit both the activity and the inflammation-induced expression of acid-sensing ion channels in nociceptors. *The Journal of Neuroscience, 21*(20), 8026–8033.

Waldmann, R., Bassilana, F., De Weille, J. R., Champigny, G., Heurteaux, C., & Lazdunski, M. (1997a). Molecular cloning of a non-inactivating proton-gated Na^+ channel specific for sensory neurons. *The Journal of Biological Chemistry, 272,* 20975–20978.

Waldmann, R., Champigny, G., Bassilana, F., Heurteaux, C., & Lazdunski, M. (1997b). A proton-gated cation channel involved in acid-sensing. *Nature, 386*(6621), 173–177.

Wultsch, T., Painsipp, E., Shahbazian, A., Mitrovic, M., Edelsbrunner, M., Lazdunski, M., et al. (2008). Deletion of the acid-sensing ion channel ASIC3 prevents gastritis-induced acid hyperresponsiveness of the stomach-brainstem axis. *Pain, 134*(3), 245–253.

Yu, Y., Chen, Z., Li, W. G., Cao, H., Feng, E. G., Yu, F., et al. (2010). A nonproton ligand sensor in the acid-sensing ion channel. *Neuron, 68*(1), 61–72.

Acinar Cell Injury

▶ Visceral Pain Model, Pancreatic Pain

Acting-out

▶ Anger and Pain

Action Potential

Definition

Electrical potential actively generated by excitable cells. In nerve cells, the action potential is generated by a transient (less than 1 ms) increase in Na^+ and K^+ conductances, which brings the membrane potential to the equilibrium potential of Na^+. Immediately afterwards, the membrane repolarizes and becomes more negative than before, generating an afterhyperpolarization. In unmyelinated axons, the action potential propagates along the length of the axon through local depolarization of each neighboring patch of membrane. In myelinated axons, action potential is generated only in the Ranvier nodes and jumps rapidly between nodes increasing markedly the propagation speed.

Cross-References

▶ Demyelination
▶ Molecular Contributions to the Mechanism of Central Pain
▶ Nociceptor Generator Potential

Action Potential Conduction of C-Fibers

▶ Mechano-Insensitive C-Fibers, Biophysics

Actiq®

Definition

Actiq® is a transmucosal fentanyl system that provides significant pain relief for breakthrough cancer pain. It is also used for pain control in opioid-tolerant cancer patients.

Cross-References

▶ Postoperative Pain, Fentanyl

Actiq® Oral Transmucosal Fentanyl Citrate

▶ Postoperative Pain, Fentanyl

Activa®

Definition

It is the brand name (Medtronic, Minneapolis, USA) of a system of electrodes, connectors, and implantable pulse generators for the treatment of movement disorders, pain, and epilepsy, by stimulation of the basal ganglia, midbrain, and thalamus.

Cross-References

▶ Pain Treatment: Spinal Cord Stimulation

Activation/Reassurance

Geoffrey Harding
Sandgate Spinal Medicine Clinic, Sandgate, QLD, Australia

Synonyms

Reassurance and activation

Definition

Activation and reassurance are interventions that have been used for the treatment of acute low back pain. They involve having the practitioner gain the patient's confidence that they do not have a serious cause of pain, and that remaining active, or restoring activity, is beneficial for their recovery.

Characteristics

Systematic reviews have shown that bed rest is neither appropriate nor effective for acute low back pain (Koes and van den Hoogen 1994; Waddell et al. 1997). Bed rest offers no therapeutic advantages, and is less effective than alternative treatments in terms of rate of recovery, relief of pain, return to daily activities, and time lost from work. By inference, these results support keeping patients active.

Nevertheless, patients may harbor fears or misconceptions about their pain, which may inhibit their resumption of activities. Explanation and reassurance are required to overcome these fears.

Evidence

The study of Indahl et al. (1995) constitutes a landmark in the management of nonspecific musculoskeletal conditions. It was the first rigorously controlled trial to demonstrate long-term efficacy for an intervention based on reassurance and activation, with no passive interventions.

Patients were provided with a biological model of their painful condition. They were assured that light activity would not further injure the structures that were responsible for their pain, and was more likely to enhance the repair process. The link between emotions and musculoskeletal pain was explained as a muscular response. Patients were told that increased tension in the muscles for any reason would increase the pain and add to the problem. It was explained how long-standing pain and associated fear could create vicious cycles of muscular activity that caused pain to persist. It was strongly emphasized that the worst thing they could do would be to act in a guarded, overcautious way.

Regardless of clinical and radiographic findings, all patients were told to mobilize the affected parts by light, nonspecific exercise, within the limits of intense pain exacerbation. No fixed exercise goals were set, but patients were given guidelines and encouraged to set their own goals. Great emphasis was placed on the need to overcome fear about the condition and associated sickness behavior. Misunderstandings about musculoskeletal pain were dealt with.

The principal recommendation was to undertake light, normal activities, and move as flexibly as possible. Activities involving static work for the regional muscles were discouraged. No restrictions were placed on lifting, but twisting when bending was to be avoided. Acute episodes of pain in the affected region were to be treated as acute muscles spasm, with stretching and further light activity. Instruction was reinforced at 3 months and at 1 year.

The actively treated patients exhibited a clinically and statistically significant difference from the control group with respect to decrease in sickness leave. At 200 days, 60 % in the control group, but only 30 % in the intervention group, were still on sick leave. A 5-year follow-up demonstrated that these differences were maintained (Indahl et al. 1998). Only 19 % of the intervention group were still on sick leave at 5 years, compared with 34 % in the control group.

The results of Indahl et al. (1995) were corroborated by another study (McGuirk et al. 2001). The intervention was based on the principles set by Indahl, and focused on identifying the

patient's fears, providing explanation, motivating patients to resume activities, and helping them maintain those activities. This approach achieved greater reductions in pain than did usual care, with fewer patients progressing to chronic pain, less use of other health care, and greater patient satisfaction.

Principles

Providing reassurance and motivating patients into activity are skills that have to be learnt. It is not enough to simply give information in the form of test results, diagnoses, prognoses, or proposed treatments. The manner of the consultation and the doctor's ability to empathize with the anxious patient is a prerequisite to any "motivational interview" (McDonald and Daly 2001). In order to develop empathy, a long consultation may be required. However, reassurance can, nevertheless, be achieved through a systematic series of shorter consultations (Roberts et al. 2002).

Interviewing techniques can be adapted to achieve an "educational outcome" (Arborelius and Bremberg 1994). The process of consulting or interviewing in a motivational way has been detailed (Kurtz et al. 2005), and is quite different from a normal medical interview that is geared toward collecting and collating information in as short a time as possible. Naturally, the educational (or motivational) interview demands more time from the practitioner. However, it is more effective in terms of changing behavior toward self-motivation (Miller and Rollnick 2002).

The doctor must establish an initial rapport with the patient. In general, one should greet each patient as if they were a friend of a friend, not a complete stranger. The doctor should not give the impression of rushing.

The concerns with which patients present can be encapsulated by Watson's quartet (Watson 1999): "I hurt," "I can't move," "I can't work," and "I'm scared." The latter can be expanded to encompass: What has happened? Why has it happened? Why me? Why now? What would happen if nothing were done about it? What should I do about it, and who should I consult for further help?

It is useful to ask patients what they think has caused their problems – the answers given to these questions are often surprising, and can sometimes hold the key to guiding patients through a complex biopsychosocial landscape.

There are no routine responses to these issues and questions. The practitioner must be prepared to respond in an informed, convincing, and caring manner. One example of an explanation might be:

> Well, we don't actually know why you have developed this but there are many reasons, and some of them come down to just bad luck. It might be related to an event or an injury, but these are often hard to track down. At the end of the day I can say that there doesn't seem to be anything that you could have avoided, and the problem is one that is not serious – it is painful, but not harmful. It might happen again and it might not.
>
> There are lots of people who will tell you that it's 'this' or 'that' which has caused it, but frankly this is speculation in most cases. Some people will tell you that it's because you have weak muscles, but you know that the fittest athletes in the world get injured from time to time, and there are many people out of condition who never get injuries. Others might say that it is your posture. But you have presumably not altered your posture in many years and you have never had the problem before. So trying to fix your posture in a major way might be pointless at this stage. I can say that there is no disease process going on and there are no broken bones or things that the surgeons have to fix. It's not something that you will pass onto your children and it will not shorten your lifespan. It might be that you will have to look at the type of work you do, but we will get more of an idea about that as time goes on.

This sort of explanation takes an enormous amount of time, but shortchanging the patient will result in a less-than-effective consultation. The paradox of appearing to have shortage of time will result in no change accomplished, whereas appearing to have "all day" often results in a change occurring in a matter of minutes (Miller and Rollnick 2002).

As the patient raises issues, their narrative should be expanded, with the use of phrases such as "tell me more about that." Terms and expressions used by the patient should be checked for meaning, so that the doctor understands what the patient is communicating.

Developing rapport relies on the appropriate use of eye contact, expressing concern and understanding, and dealing sensitively with the patient during the physical examination.

A thorough examination is a necessary prerequisite for gaining the satisfaction (and thus the confidence) of the patient (McCracken et al. 2002). The reasons for examination procedures should be explained.

The practitioner can reassure patients by developing an "educational enterprise" (Daltroy 1993). Printed material is an effective reinforcer of tuition (see ▶ Patient Education). Models and pictures serve to explain concepts about normal structure and pathology. The language used should be appropriate to the patient and understood by them. Alarming and distressing terms should be avoided.

When recommending exercises, those exercises should be demonstrated, and the patient's ability to reproduce them should be observed and confirmed. The same confirmation should be obtained when advice is given about how the patient will undertake their desired activities. Checking their understanding is what converts the consultation from one in which instructions are simply issued, to one in which the patient is confident about that instruction.

References

Arborelius, E., & Bremberg, S. (1994). Prevention in practice. How do general practitioners discuss lifestyle issues with their patients? *Patient Education and Counseling, 23*, 23–31.

Daltroy, L. H. (1993). Consultations as educational experiences doctor-patient communication in rheumatological disorders. *Baillière's Clinical Rheumatology, 7*, 221–239.

Indahl, A., Velund, L., & Reikeraas, O. (1995). Good prognosis for low back pain when left untampered: A randomized clinical trial. *Spine, 20*, 473–477.

Indahl, A., Haldorsen, E. H., Holm, S., et al. (1998). Five-year follow-up study of a controlled clinical trial using light mobilization and an informative approach to low back pain. *Spine, 23*, 2625–2630.

Koes, B. W., & van den Hoogen, H. M. M. (1994). Efficacy of bed rest and orthoses of low back pain. A review of randomized clinical trials. *European Journal of Physical and Rehabilitation Medicine, 4*, 96–99.

Kurtz, S. M., Silverman, J. D., Benson, J., et al. (2005). Marrying content and process in clinical method teaching: Enhancing the Calgary-Cambridge guides. *Academic Medicine, 78*(8), 802–809.

McCracken, L. M., Evon, D., & Karapas, E. T. (2002). Satisfaction with treatment for chronic pain in a specialty service: Preliminary prospective results. *European Journal of Pain, 6*, 387–393.

McDonald, I. G., & Daly, J. (2001). On patient judgement. *Internal Medicine Journal, 31*, 184–187.

McGuirk, B., King, W., Govind, J., et al. (2001). The safety, efficacy, and cost-effectiveness of evidence-based guidelines for the management of acute low back pain in primary care. *Spine, 26*, 2615–2622.

Miller, W. R., & Rollnick, S. (2002). *Motivational interviewing: Preparing people for change* (2nd ed.). New York: Guilford Press.

Roberts, L., Little, P., Chapman, J., et al. (2002). Practitioner-supported leaflets may change back pain behaviour. *Spine, 27*, 1821–1828.

Waddell, G., Feder, G., & Lewis, M. (1997). Systematic reviews of bed rest and advice to stay active for acute low back pain. *The British Journal of General Practice, 47*, 647–652.

Watson, P. (1999). The MSM quartet. *Australasian Musculoskeletal Medicine, 4*, 8–9.

Active Inhibition

Definition

Active inhibition implies that nociceptive processing is suppressed by specific inhibitory mechanisms, as opposed to simply reflecting the absence of excitatory input.

Cross-References

▶ Formalin Test

Active Myofascial Trigger Point

Definition

An active trigger point is a myofascial trigger point that is causing, or contributing to, a clinical pain complaint. When it is compressed, the individual recognizes the induced referred pain as familiar and recently experienced.

Cross-References

▶ Dry Needling
▶ Myofascial Trigger Points

ActiveLocus

Synonyms

EPN locus

Definition

The motor component of a Myofascial Trigger Point is the active locus, or endplate-noisy locus (EPN locus). From this locus, spontaneous electrical activity, known as endplate noise (EPN), can be recorded. It is related to taut band formation in skeletal muscle fibers.

Cross-References

▶ Dry Needling

Activities of Daily Living

Definition

Activity: The execution of a task or action by an individual. Activities of daily living refer to normal physical activities, such as getting out of bed, walking (initially with support), sitting, and personal toileting.

Cross-References

▶ Physical Medicine and Rehabilitation, Team-Oriented Approach
▶ Postoperative Pain, Importance of Mobilization

Activity

Definition

Activity is described as the execution of a task or action by an individual. It represents the individual perspective of functioning. Difficulties an individual may have in executing activities are activity limitations.

Cross-References

▶ Functioning and Disability Definitions

Activity Limitation

▶ Impairment, Pain-Related

Activity Limitations

Definition

Difficulties an individual may have in executing activities.

Cross-References

▶ Impairment, Pain-Related
▶ Physical Medicine and Rehabilitation,
   Team-Oriented Approach

Activity Measurement

Definition

A measure of personal activities of daily living (e.g., showering, dressing, toileting, and feeding), independent activities of daily living (e.g., cleaning, cooking, shopping, and banking), and discretionary activities of daily living (e.g., driving, visiting, and leisure activities). This also includes the automated assessment of motor activity.

Cross-References

▶ Pain Assessment in the Elderly

Activity Mobilization

Definition

Strategies aimed at maximizing a chronic pain patient's participation in activities of daily living.

Cross-References

▶ Catastrophizing

Activity-dependent Slowing

▶ Nociceptors and Activity-Dependent Changes in Axonal Conduction Velocity

Acupuncture

Definition

A system of healing that is part of traditional Chinese medicine, also containing herbal therapy and tuina (a kind of massage). Acupuncture consists of the insertion and twirling of thin solid needles into specific points on the body that lie along the so-called channels or meridians, in order to treat different, usually chronic, symptoms. The needles are usually inserted into muscles but may also be inserted superficially into the skin. Neither the acupuncture points nor the meridians have been unequivocally verified in a biological sense but are considered of importance by most practitioners of acupuncture.

Cross-References

▶ Acupuncture Efficacy
▶ Acupuncture Mechanisms
▶ Alternative Medicine in Neuropathic Pain

Acupuncture Efficacy

Edzard Ernst
Complementary Medicine, Peninsula Medical
School Universities of Exeter and Plymouth,
Exeter, UK

Definition

▶ Acupuncture can be defined as the insertion of needles into the skin and underlying tissues at specific sites (acupuncture points) for therapeutic or preventative purposes (Ernst et al. 2001). Sometimes other forms of point stimulation are used: electrical current (electroacupuncture), pressure (acupressure), heat (moxibustion), or laser light (laser acupuncture). Acupuncture is part of the ancient Chinese medical tradition. In recent years, a new style (Western acupuncture) has emerged, which no longer adheres to the Taoist philosophies underpinning Chinese acupuncture but seeks explanations for its mode of action from modern concepts of neurophysiology and other branches of medical science.

Characteristics

The evidence for or against the efficacy (or effectiveness) of acupuncture is highly heterogeneous and often contradictory. Thus, single trials, even of good quality, may not provide a representative picture of the current evidence. The following section is therefore exclusively based on systematic reviews of controlled clinical trials, i.e., on the totality of the available trial data rather than on a possibly biased selection of it. Whenever more than one such publication is available, the most up-to-date one was chosen.

Any Chronic Pain

One landmark paper summarized the results of 51 randomized clinical trials testing the efficacy of acupuncture as a treatment of all forms of chronic pain (Ezzo et al. 2000). Any type of acupuncture was considered. The studies were rated for methodological rigor using the Jadad score (Jadad et al. 1996). The results revealed a significant association between lower-quality studies and positive outcomes. There was no clear evidence to demonstrate that acupuncture is superior to sham acupuncture or to standard treatment. Good evidence emerged that it is better than waiting list (i.e., no acupuncture). The quality of the review was rated "good" by independent assessors (Tait et al. 2002). Depending on one's viewpoint, one can interpret these findings differently. Acupuncture "fans" would claim that they demonstrate acupuncture to be as good as standard treatments, while skeptics would point out that the data suggest that acupuncture has no more than a placebo effect. Pooling the data for all types of chronic pain is perhaps an approach too insensitive to tease out effects on more defined types of pain. Other systematic reviews have therefore focussed on more specific targets.

Dental Pain

Sixteen controlled trials were available, 11 of which were randomized (Ernst and Pittler 1998). All studies of manual or electroacupuncture were included. Their methodological quality was assessed using the Jadad score (Jadad et al. 1996). The collective evidence suggested that acupuncture can alleviate dental pain, even when compared against sham acupuncture. The strength of the conclusion was, however, limited through the often low quality of the primary data. The quality of the review was rated by independent assessors as "satisfactory" (Tait et al. 2002). Since effective and safe methods for relieving dental pain exist, the clinical relevance of acupuncture for dental pain may be limited.

Headache

A Cochrane Review summarized the evidence from 26 randomized or quasi-randomized trials of any type of acupuncture (Linde et al. 2001). Their methodological quality was assessed using the Jadad score (Jadad et al. 1996). The overall results support the role of acupuncture for recurrent headaches but not for migraine or other types of headache. The conclusions were limited through the often low methodological quality of the primary studies. The review was independently rated to be of good quality (Tait et al. 2002).

Neck Pain

Fourteen randomized clinical trials of all types of acupuncture were included in a systematic review (White and Ernst 1999). Their rigor was evaluated using the Jadad score (Jadad et al. 1996) and found to be mixed. About half of the trials generated a positive result, while the other half could not confirm such a finding. Thus, the efficacy of acupuncture was not deemed to be established. The quality of the review was rated "good" (Tait et al. 2002).

Back Pain

A Cochrane Review assessed the effectiveness of manual acupuncture or electroacupuncture for nonspecific back pain (van Tulder et al. 2001). Eleven randomized trials were included and evaluated according to the Cochrane Back Review Group criteria. The results were mixed, but overall acupuncture was not found to be of proven effectiveness, not least because the quality of the primary studies was found to be wanting. This review was rated as of good quality (Tait et al. 2002). Other systematic reviews of these data have drawn different conclusions, e.g., Ernst and White (1998). An updated review on the subject including many new studies is now being conducted.

Fibromyalgia

A systematic review included four cohort studies and three randomized clinical trials of any type of acupuncture (Berman et al. 1999). Their methodological quality as assessed using the Jadad score (Jadad et al. 1996) was mixed, but in some cases good. The notion that acupuncture alleviates the pain of fibromyalgia patients was mainly based on one high-quality study and thus not fully convincing. The quality of the review was rated as "satisfactory" (Tait et al. 2002).

Osteoarthritis

A systematic review of controlled acupuncture trials for osteoarthritis of any joint included 13 studies (Ernst 1997). Their methodological quality was evaluated using the Jadad score (Jadad et al. 1996) and found to be highly variable. The methodologically sound studies tended to yield negative results. Sham acupuncture turned out to be as effective as real acupuncture in reducing pain. Thus, it was concluded that acupuncture has a powerful placebo effect. Whether or not it generates specific therapeutic effects was deemed uncertain.

Conclusion

These systematic reviews collectively provide tantalizing but not convincing evidence for acupuncture's pain reducing effects. The evidence is limited primarily by the paucity of studies and their often low methodological quality. The scarcity of research funds in this area is likely to perpetuate these problems. Since acupuncture is a relatively safe therapy (Ernst and White 2001), it deserves to be investigated in more detail and with more scientific rigor, e.g., using the novel sham needle devices (Park et al. 2002; Streitberger and Kleinhenz 1998) that have recently become available.

Cross-References

▶ Acupuncture Mechanisms

References

Berman, B., Ezzo, J., Hadhazy, V., et al. (1999). Is acupuncture effective in the treatment of fibromyalgia. *Journal of Family Practice, 48*, 213–218.

Ernst, E. (1997). Acupuncture as a symptomatic treatment of osteoarthritis. *Scandinavian Journal of Rheumatology, 26*, 444–447.

Ernst, E., & Pittler, M. H. (1998). The effectiveness of acupuncture in treating acute dental pain: A systematic review. *British Dental Journal, 184*, 443–447.

Ernst, E., & White, A. R. (1998). Acupuncture for back pain. A meta-analysis of randomized controlled trials. *Archives of Internal Medicine, 158*, 2235–2241.

Ernst, E., & White, A. R. (2001). Prospective studies of the safety of acupuncture: A systematic review. *American Journal of Medicine, 110*, 481–485.

Ernst, E., Pittler, M. H., Stevinson, C., et al. (2001). *The desktop guide to complementary and alternative medicine*. Edinburgh: Mosby.

Ezzo, J., et al. (2000). Is acupuncture effective for the treatment of chronic pain? A systematic review. *Pain, 86*, 217–225.

Jadad, A. R., et al. (1996). Assessing the quality of reports of randomized clinical trials -is blinding necessary? *Controlled Clinical Trials, 1996*(17), 1–12.

Linde, K., et al. (2001). Acupuncture for idiopathic headache (Cochrane review). In: *The Cochrane Library*, Issue 2. Update Software, (pp. 1–46). Oxford.

Park, J., White, A., Stevinson, C., et al. (2002). Validating a new non-penetrating sham acupuncture device: Two randomised controlled trials. *Acupuncture in Medicine, 20*, 168–174.

Streitberger, K., & Kleinhenz, J. (1998). Introducing a placebo needle into acupuncture research. *Lancet, 352*, 364–365.

Tait, P. L., Brooks, L., & Harstall, C. (2002). *Acupuncture: Evidence from systematic reviews and meta-analyses. Alberta heritage foundation for medical research*. Canada: Edmonton.

van Tulder, M. W., Cherkin, D. C., Berman, B., et al. (2001). Acupuncture for low back pain. Available: http://www.cochranelibrary.com

White, A. R., & Ernst, E. (1999). A systematic review of randomized controlled trials of acupuncture for neck pain. *Rheumatology, 38*, 143–147.

Acupuncture Mechanisms

Christer P. O. Carlsson
Rehabilitation Clinic, Lunds University Hospital, Lund, Sweden

Definition

▶ Acupuncture is a traditional Chinese therapeutic method for the treatment of different symptoms including pain. Thin, solid needles are inserted into proposed specific points on the body, called acupuncture points. The needles are inserted through the skin to varying depths, often into the underlying musculature. The needles are often twirled slowly for a short time, 30–60 s and may be left in place for a varying time, 2–30 min. Many modifications of the method have been described, and the concept of acupuncture is not well defined. The method of applying electrical stimulation via acupuncture needles, ▶ Electro-acupuncture (EA), was introduced in 1958.

The treatments are usually applied in series of 8–12 sessions, each treatment lasting 20–30 min and separated by 1/2–2 weeks. Needling is often performed with some needles near the source of pain (called local points) and some other needles on the forearms and lower legs (called distal points).

Common Clinical Observations Concerning Therapeutic Acupuncture for Chronic Pain

After the first few acupuncture treatments there may be some hours of pain relief or nothing at all happens. Often pain relief starts 1–2 days after treatment. Some patients even get worse and have a temporary aggravation of their symptoms for some days before they start to improve. This aggravation can be seen for 2–3 days or even for a week. For those responding to acupuncture, usually both the degree and duration of the pain relief increase after each treatment, a clinical observation that has gained some experimental support (Price et al. 1984).

Acupuncture is a Form of Sensory Afferent Stimulation

As acupuncture needles are inserted into the tissue and mostly down to the muscular layer, they excite receptors and nerve fibers, that is, the needles mechanically activate somatic afferents. Other forms of afferent sensory stimulation are trigger point needling or dry needling and transcutaneous electrical nerve stimulation (▶ TENS) as well as vibration. These methods may share some common features concerning mechanisms of action. A special method is painful sensory stimulation, which has been used through the centuries, an idea that a short but very painful stimulus would reduce pain. These methods have been called "▶ counter irritation" or "▶ hyperstimulation analgesia," and acupuncture is sometimes regarded as such. However, it is important to know that most patients who are treated with acupuncture describe the procedure as relaxing and pleasant but not painful.

The term acupuncture analgesia (AA) was used for electro-acupuncture (EA) used to get powerful and immediate pain relief during surgery, first used in China in 1958 but not described until 1973 (Foreign Languages Press 1973). A success rate of 90 % was claimed among those selected for the method. However, it soon became clear that only a minority of patients could develop so strong an analgesia as to tolerate surgery. Less than 10 % of the patients showed a satisfactory response in acupuncture trials (Bonica 1974). Among these 10 %, only one-third had acceptable analgesia according to Western standards. Even so, patient selection and psychological preparations were crucial, and often combinations with local anesthetics or other drugs were used.

Felix Mann (1974) reported 100 observations on patients receiving AA. In only 10 % of the experiments was the resulting analgesia considered adequate for surgery. He emphasized that in ▶ therapeutic acupuncture (TA) to treat different symptoms, a mild stimulus was all that was usually required. This was in contrast to that needed to obtain AA where the stimulation had to be continued for at least 20 min and had to be painful to the maximum level the patient could tolerate. He concluded that usually, the stimulus required to achieve AA was so intense that the resulting pain would be unacceptable to most Western patients. For the main differences between AA and TA, see Table 1.

Characteristics

The proposed AA effect on surgical pain initiated physiological research where the goal was to find an explanation for immediate and very strong analgesia. Consequently, physiological research during the last 25–35 years has concentrated on

Acupuncture Mechanisms, Table 1 Differences between acupuncture analgesia and therapeutic acupuncture

Acupuncture analgesia	Therapeutic acupuncture
Immediate and strong hypoalgesia is the goal	Immediate hypoalgesia is not the goal
Fast onset (minutes)	Slowly induced symptom relief after a number of treatments
	The effects gradually increase after additional treatments
Short term = minutes	Long term = days-weeks-months
The stimulation is felt very strongly. It is often painful and uncomfortable	The stimulation is felt rather weakly. It is rarely painful and often relaxing.
Used most often in different physiological experiments and for surgical hypoalgesia	Used for clinical pain relief and other symptom relief
Often electroacupuncture and pain threshold experiments on humans or animals	Most often manual acupuncture but can also be electroacupuncture

explaining a phenomenon that may only exist in about 3–10 % of the population and that may have little in common with therapeutic acupuncture.

The experimental acupuncture research has concentrated on very short-term effects (after a single treatment of EA) where pain thresholds and/or central neurochemicals (mostly endorphins) have been measured. The research groups have mostly used conscious animals where no special care has been taken to rule out stress-induced analgesia (▶ SIA) (Akil et al. 1984). In some studies it is explicitly noted that the animals showing obvious signs of discomfort during EA also had pain threshold elevations, but that this was not the case for those who were not distressed (e.g., Bossut and Mayer 1991; Galeano et al. 1979; Wang et al. 1992).

Conclusions from the Existing Acupuncture Experimental Data

Most acupuncture research on animals has been performed using (strong) EA, even though human therapeutic acupuncture is most often performed with gentle manual acupuncture. Much of the animal research on acupuncture probably only shows the consequences of nociceptive stimulation and the activation of ▶ SIA and ▶ Diffuse Noxious Inhibitory Control (DNIC). When manual acupuncture has been used in animal research, no pain threshold elevation has been described.

Pain threshold elevation in humans only seems to occur if the stimulation is painful and does not correspond at all with the clinical outcome after therapeutic acupuncture. Endorphins are partially involved in acupuncture analgesia in humans. Thus, AA in humans is believed to rely both on opioid and non-opioid mechanisms. However, whether endorphins are involved both locally (in the tissues) and within the central nervous system is not known (Price and Mayer 1995). Thus, the hitherto performed experimental acupuncture mechanism research is really only valid for acupuncture analgesia and not for therapeutic acupuncture.

Acupuncture Mechanisms: The Standard Neurophysiological Model

Several physiological mechanisms have been suggested to account for the pain-relieving effect of acupuncture. Spinal and supraspinal endorphin release has been proposed, as has the activation of diffuse noxious inhibitory control (DNIC) through bulbospinal paths. The involvement of neurochemicals like serotonin, noradrenalin, and different endorphins as well as hormones like Adrenocorticotrophic hormone (ACTH) and cortisone has been studied in detail.

Acupuncture physiology is often summarized in the following manner (Han 1987; Pomeranz 2000): For acupuncture needles inserted within the segment of pain:

- Spinal gate-control mechanism

For extrasegmental acupuncture:

- Activation of midbrain structures Periaqueductal Grey (PAG) and the descending pain-relieving system (involving endorphins, serotonin, and noradrenaline).
- Diffuse noxious inhibitory control (DNIC) is sometimes claimed to be involved.
- Activation of the hypothalamic pituitary adrenal axis (HPA) with increased levels (in the blood) of β-endorphin and ACTH/cortisone.

Problems with the Standard Neurophysiological Model to Explain Clinical Observations

The model can only explain very short-term pain relief after each stimulation period. The gate-control mechanism is only active during stimulation and the descending inhibitory system for up to perhaps 8 h.

The model cannot explain why, in some patients, pain relief starts some days after the treatment whether the patient is first worse or not. The gate control does not start some days after the stimulation and that does not hold for the descending pain-inhibitory systems either. The model cannot explain why there seems to be more prolonged pain relief after additional treatments and why there seems to be long-term pain relief after a course of 8–12 treatments in at least some patients. From clinical research, in which the author has been involved, the conclusion has been drawn that clinically relevant long-term (>6 months) pain relief from acupuncture can be seen in a proportion of patients with chronic nociceptive pain (Carlsson and Sjölund 1994; Carlsson and Sjölund 2001). Probably, the standard neurophysiological model can explain AA, but even so it should be realized that AA is mostly a painful stimulation – and if the gate-control mechanisms are implicated, then the stimulation should be non-painful.

For a summary of probable acupuncture mechanisms for both TA and AA, see below. For a full reference list to all of this section, see (Carlsson 2002).

Hypothesis on Different Mechanisms for Acupuncture

Acupuncture Induces Peripheral Events

These events might improve tissue function and induce local pain relief. These events are induced from all local needles in the tissue and not only from traditional acupuncture points.

Axon Reflexes, Neuropeptides, and Dichotomizing Nerve Fibers Just the insertion of a needle in the tissue induces changes close to the needle (in all different tissues penetrated) and through axon reflexes. The flare reaction (reddening, vasodilatation) is often seen locally around the acupuncture needles. This vasodilatation in the skin due to axon reflexes has been recognized for a very long time, and the mechanisms have been clarified in detail. The stimulation of Ad or C fibers releases vasoactive and pro-inflammatory neuropeptides (e.g., Calcitonin Gen Related Polypeptide (CGRP), Substance P (SP), Neurokinin A (NKA), opioids, galanin, somatostatin, and Vasoactive Intestinal Polypeptide (VIP)). The profound and prolonged vasodilatation is probably mediated mostly by CGRP. EA (and TENS) produces peripheral vasodilatation, in skin and muscle, both experimentally and clinically, that probably is caused by this neuropeptide release.

Remarkably, CGRP is pro-inflammatory, but it has also been shown that CGRP in low doses has a potent anti-inflammatory action. Thus, it seems like a form of balance exists between anti-inflammatory and pro-inflammatory effects (low dose or high dose of CGRP) in the tissue. This may have a practical consequence – strong stimulation leads to more damage in the tissue and thus more pain through higher levels of CGRP, while careful stimulation leads to smaller amounts of CGRP release and thus an anti-inflammatory, pain-relieving effect. This is a practical observed phenomenon when working a lot with acupuncture.

The identification of dichotomizing spinal nerves having branches to two different types of tissues might be relevant here for influences (through axon reflexes) on deeper tissues than only skin and muscles. For example, stimulation to the saphenous nerve gives rise to vasodilatation around the sciatic nerve, and nerve fibers to the intervertebral disc do also supply the groin skin. So, these dichotomizing nerve fibers may explain why sometimes effects in deeper tissues can be observed.

Neuropeptides and Trophic Effects Acupuncture can improve salivary flow, long term, for patients with xerostomia of different causes. Manual acupuncture has been shown to significantly increase the blood flow in the area overlying the parotid glands. Moreover, the concentration of the neuropeptides VIP and CGRP in saliva from xerostomic patients has been shown to increase after acupuncture. It is also known that neuropeptides like SP, VIP, and CGRP have a trophic effect on glandular tissues (leading to regeneration). A possible explanation for the long-term effects of acupuncture might be that the release of these neuropeptides induces regeneration of traumatized glandular tissue.

Local Endorphins Local endorphins and their different receptors have been found on nociceptive afferents in inflammatory conditions. The different endorphins are secreted from inflammatory cells in the tissue after an injury. A synthesis of endorphin receptors from the dorsal root ganglion starts as a response to the nociceptive input to the dorsal horn. The endorphins and receptors accumulate at the injury site after a few days. This accumulation may lead to a peripheral opioid analgesia some days after an injury. The penetration of acupuncture needles induces small tissue injuries; thus, there may, in some instances, be an increase in local endorphins after some days. This could be one explanation for pain relief coming 2–3 days after a treatment session and a possible reason why it appears to be so useful to use many local acupuncture needles.

Changes in Nociceptor Density in the Tissue Some pain conditions, tendinosis, seem to have an increased density of nociceptors peripherally in the tender painful area (e.g., Achilles tendinosis). It has been shown that there is an ingrowth of new nerves, nociceptors, in the area. Thus, the density of nociceptors is increased. That may be one explanation of these chronic pain conditions. The author has been involved in one study where we investigated the peripheral innervation before and after a series of very local subcutaneous acupuncture needlings. It was then shown, very surprisingly, that the number of peripheral nerve endings was reduced. Especially there was a reduction in the number of CGRP-containing nerve fibers. Thus, perhaps there are in the periphery a plastic downregulation of nociceptor density as a partial explanation for pain relief (Carlsson et al. 2006).

Spinal Mechanisms

These mechanisms can be stimulated from regional, segmental and extrasegmental needles, not only from traditional acupuncture points.

Gate Control The inhibition produced through this mechanism works fast and short term, with effects occurring mainly during stimulation. Thus, this mechanism only explains pain relief during the non-painful stimulation period and perhaps some hours afterwards.

Long-Term Potentiation (LTP) and Long-Term Depression (LTD) in the Dorsal Horn Long-term potentiation (LTP) is a form of synaptic plasticity where the synaptic strength increases. Long-term depression (LTD) is the opposite – a synaptic plasticity where the synaptic strength decreases. The LTP phenomenon is probably one background to central sensitization and memories in the pain transmission system. Central sensitization can occur within the Central Nervous System (CNS) from the dorsal horn level up to the thalamus and perhaps even to the cortex and is probably involved in some forms of chronic pain.

A long-term depression of C-fibre evoked mono- or polysynaptic excitatory postsynaptic potentials in the superficial spinal dorsal horn has been found upon low-frequency stimulation of afferent Aδ-fibers (mildly painful) in the same segment. At least in awake animals, this LTD may last for days and weeks and might thus explain pain relief with a little longer duration, than the gate-control mechanism does.

It is usual to note, clinically, that there is a great variability within and between patients even if the same stimulation parameters are used. This may depend on the balance that seems to exist between LTP and LTD. The result of an Aδ stimulation period appears to be dependent on the initial resting membrane potential of the WDR cells. If the WDR cell was hyperpolarized (e.g., activity in the segmental inhibitory system and/or descending inhibitory systems), the result of the stimulation was a LTD. If instead the cell was a little depolarized (e.g., ongoing pain, descending excitatory influences), then the result was a LTP. Thus, the same stimulation parameter sometimes induces LTP (with more pain) and sometimes LTD (with less pain), just dependent on the initial condition of the WDR cells.

Propriospinal Pain Inhibition (Extra-segmental Needles) Sandkuhler has pointed out that numerous propriospinal intersegmental systems exist to be involved in pain inhibition. These propriospinal neurons may be activated by afferent stimulation or by supraspinal pathways. With superperfusion techniques, it has been shown that axons from thoracic, cervical, or sacral levels can induce a lumbar antinociceptive effect. The efficacy of this propriospinal inhibition has been shown to be similar to the inhibition produced by stimulation of PAG. Thus, the propriospinal antinociceptive neurons may constitute a third component of endogenous antinociception in addition to segmental and supraspinal descending inhibition.

Supraspinal Mechanisms

These mechanisms can be stimulated from needles distributed all over the body, not only from traditional acupuncture points.

The Descending Pain-Inhibitory Systems This well-known descending pain-inhibitory system is best activated by noxious and/or stressful events.

As direct electrical stimulation to PAG only gives rise to short-term pain relief, it does not seems very realistic that acupuncture could give rise to long-term effects through this system. However, probably this system is involved in AA.

The DNIC System The DNIC gives only an extremely short-lasting pain-inhibitory effect activated by painful stimulation applied outside the segment of pain. Thus, this system, also, might be involved in AA.

The Sympathetic Nervous System and the HPA axis Low-intensity stimulation gives rise to reduced sympathetic outflow and reduced amount of adrenaline/noradrenaline from the

Acupuncture Mechanisms, Table 2 Probable acupuncture mechanisms

Summary of probable mechanisms for acupuncture	Therapeutic acupuncture: mostly gentle manual. Usual clinical use	Acupuncture analgesia: high-intensity electroacupuncture Physiological experiments and surgical analgesia
Local events in the tissue (**Local needles**)	Axon reflexes in the tissue around needles and deeper through dichotomizing fibers giving increased circulation and neuropeptide release	Tissue trauma around the needles giving rise to more local pain (CGRP in higher doses has pro-inflammatory actions)
	These can act as trophic factors (e.g., regeneration of glands). They can also have anti-inflammatory effects (like low dose of CGRP)	Increased local pain for some days
	Release of local endorphins to local receptors	
	Perhaps a downregulation of nociceptive afferents in the tissue surrounding the needles	
Segmental mechanisms and somato-autonomous reflexes (Regional needles)	Gate mechanism and perhaps long-term depression (LTD)	Gate mechanism and perhaps LTD
	Sympathetic inhibition with increased segmental circulation	Sympathetic stimulation with decreased segmental circulation
Central mechanisms (Distal, regional, and some local needles)	Sympathetic inhibition. Decreased levels of stress hormones, adrenaline, and cortisone in plasma.	Sympathetic stimulation. Increased levels of the stress hormones, ACTH, adrenaline, and cortisone in plasma
	Perhaps oxytocin is involved and induces long-term antistress-like effects	DNIC is activated. Descending pain inhibition from PAG with endorphins, serotonin, and noradrenaline

adrenals, while strong (noxious) stimulation gives rise to the opposite. ACTH/cortisone increases after painful or stressful acupuncture but not after non-painful or non-stressful. Thus, the intensity of stimulation seems to be very important for what system is activated.

Oxytocin Research has indicated that oxytocin probably is secreted in response to non-noxious sensory stimulation. This hormone seems to give rise to long-term effects, of an antistress nature, that resemble those after acupuncture. Indeed, in humans, intrathecal oxytocin has been shown to induce pain relief in lumbar pain. It has also been shown that different kinds of sensory stimulation (2Hz EA, thermal stimulation, massage, or vibration) increased oxytocin in plasma and CSF. Further, oxytocin has given rise to anxiolysis and sedation. Thus, oxytocin might be a candidate for inducing long-term effects after therapeutic acupuncture.

Cortical, Psychological, and Placebo Mechanisms (From the Treatment Sessions)

Since the mid-1990s different brain-imaging techniques (e.g., PET and fMRI) have been performed. These have shown the involvement of many brain areas after acupuncture. However, the picture is very complex. It seems like some areas are activated and some areas are deactivated. It seems like a broad network of areas are affected of acupuncture stimulation. These areas belong to the somatosensory cortex as well as areas involved in affective and cognitive processing. How the different results shall be interpreted is not clear at this time. Many areas involved after acupuncture are also involved in the pain processing as well as in the processing of placebo mechanisms. For details, see, for example, Huang et al. 2012.

Of course, acupuncture, like all other treatments, might have a placebo effect. This has also been proposed from the brain-imaging studies. Therapeutic acupuncture involves regular visits to a therapist for about 6–12 sessions. These sessions do not only consist of needle insertions but also discussions about possible changes, alternative diagnosis, and perhaps therapeutic alternatives. If there is a good patient-therapist relation, these regular visits will probably lead to advice (counseling) and reassurances, which probably reduce the anxiety. In support of this, it has been shown that among compared physical therapies (including placebos), those involving more time with the therapist or more treatments had a significant tendency for the best outcome. Thus, acupuncture treatment involves counseling and reassurance. This probably gives rise to anxiety reduction and thereby improves some different symptoms.

Summary

The mechanisms of therapeutic acupuncture are probably a mixture of physiological events from the most peripheral needling area to the spinal cord interactions and further to the brainstem pain processing areas as well as placebo and other mechanisms at the cortical level. For a summary, see also Table 2 below.

References

Akil, H., Watson, S. J., Young, E., et al. (1984). Endogenous opioids: Biology and function. *Annual Review of Neuroscience, 7,* 223–255.

Bonica, J. J. (1974). Acupuncture anesthesia in the People's Republic of China. Implications for American medicine. *Journal of the American Medical Association, 229,* 1317–1325.

Bossut, D. F., & Mayer, D. J. (1991). Electroacupuncture analgesia in rats: Naltrexone antagonism is dependent on previous exposure. *Brain Research, 549,* 47–51.

Carlsson, C. (2002). Acupuncture mechanisms for clinically relevant long-term effects -reconsideration and a hypothesis. *Acupuncture in Medicine, 20,* 82–99.

Carlsson, C. P. O., & Sjölund, B. (1994). Acupuncture and subtypes of chronic pain: Assessment of long-term results. *The Clinical Journal of Pain, 10,* 290–295.

Carlsson, C., & Sjölund, B. (2001). Acupuncture for chronic low back pain: A randomized placebo-controlled study with long-term follow-Up. *The Clinical Journal of Pain, 17,* 296–305.

Carlsson, C. P., Sundler, F., & Wallengren, J. (2006). Cutaneous innervation before and after one treatment period of acupuncture. *British Journal of Dermatology, 155*(5), 970–976.

Foreign Languages Press. (1973). *Acupuncture anaesthesia.* Beijing: Foreign Languages Press.

Galeano, C., Leung, C. Y., Robitaille, R., et al. (1979). Acupuncture analgesia in rabbits. *Pain, 6,* 71–81.

Han, J. S. (1987). *The neurochemical basis of pain relief by acupuncture. A collection of papers 1973–1987.* Beijing: Beijing Medical University.

Huang, W., Pach, D., Napadow, V., Park, K., Long, X., Neumann, J., Maeda, Y., Nierhaus, T., Liang, F., & Witt, C. M. (2012). Characterizing acupuncture stimuli using brain imaging with FMRI–a systematic review and meta-analysis of the literature. *PLoS One, 7*(4), 32960. doi:10.1371/journal.pone.0032960. Epub 2012 Apr 9.

Mann, F. (1974). Acupuncture analgesia. Report of 100 experiments. *British Journal of Anaesthesia, 46,* 361–364.

Pomeranz, B. (2000). Acupuncture analgesia -basic research. In G. Stux & R. Hammerschlag (Eds.), *Clinical acupuncture, scientific basis* (pp. 1–28). Berlin: Springer.

Price, D. D., & Mayer, D. J. (1995). Evidence for endogenous opiate analgesic mechanisms triggered by somatosensory stimulation (including acupuncture) in humans. *Pain Forum, 4,* 40–43.

Price, D. D., Rafii, A., Watkins, L. R., et al. (1984). A psychophysical analysis of acupuncture analgesia. *Pain, 19,* 27–42.

Wang, J. Q., Mao, L., & Han, J. S. (1992). Comparison of the antinociceptive effects induced by electroacupuncture and transcutaneous electrical nerve stimulation in the rat. *International Journal of Neuroscience, 65,* 117–129.

Acupuncture-Like TENS

Definition

The delivery of acupuncture-like TENS is to generate activity in small-diameter group III muscle afferents, leading to the release of opioid peptides in a similar manner to that suggested for acupuncture. TENS is administered using low-frequency train (1–4 Hz) bursts (5–8 pulses at 100 Hz) at a high, but non-painful, intensity to stimulate selectively large-diameter muscle afferents. This results in a "strong but comfortable" muscle twitch that elicits group III muscle afferent activity. The stimulation should go on for 30 min to provide temporary analgesia for 3–6 h. At variance with classical acupuncture, there is no long-term effect.

Cross-References

▶ Transcutaneous Electrical Nerve Stimulation (TENS) in Treatment of Muscle Pain
▶ Transcutaneous Electrical Nerve Stimulation Outcomes

Acute Backache

▶ Lower Back Pain, Acute

Acute Experimental Monoarthritis

▶ Arthritis Model, Kaolin-Carrageenan-Induced Arthritis (Knee)

Acute Experimental Synovitis

▶ Arthritis Model, Kaolin-Carrageenan-Induced Arthritis (Knee)

Acute Idiopathic Demyelinating Polyneuropathy

▶ Guillain-Barré Syndrome

Acute Idiopathic Demyelinating Polyradiculoneuropathy

▶ Guillain-Barré Syndrome

Acute Ischemia Test

▶ Tourniquet Test

Acute Knee Joint Inflammation

▶ Arthritis Model, Kaolin-Carrageenan-Induced Arthritis (Knee)

Acute Lumbago

▶ Lower Back Pain, Acute

Acute Pain

Ross MacPherson and Michael J. Cousins
Department of Anesthesia and Pain Management, Royal North Shore Hospital University of Sydney, Sydney, NSW, Australia

Characteristics

Why Should We Aim to Optimize the Management of Acute Pain?

Postoperative pain is a major marker of perioperative morbidity and mortality and its effective treatment should be a goal in every hospital and institution. We should all aim to control pain, not only for humanitarian reasons, but also to attenuate the psychological and physiological stress with which it is associated following trauma or surgery. While it is now recognized that adequate pain control alone is not sufficient to reduce surgical morbidity, it remains an important variable and one that is perhaps more readily controlled (Kehlet and Holte 2001).

Adequate management of postoperative pain is vital to attenuate the stress response to surgery and the accompanying pathophysiological changes in metabolism, respiratory, cardiac, sympathetic nervous system, and neuroendocrine functions. These effects (summarized in Neuroendocrine and metabolic responses to surgery after NH&MRC 1999) are wide ranging and have significant impact on homeostasis. Effects on the respiratory system are most prominent, as persistent pain will result in a reduction in respiratory effort that then leads to hypoxemia from significant ventilation/perfusion mismatching. Continuing hypoventilation predisposes to collapse of lung segments and the supervening infection that follows carries significant morbidity. Psychological and behavioral changes (e.g., yellow flags) also accompany pain states and may need to be recognized and managed. Not only will proper management of postoperative pain result in greater patient comfort and earlier discharge home, but the improved earlier mobilization and return to function will also reduce serious postoperative complications such as venous thromboembolism.

Neuroendocrine and Metabolic Responses to Surgery

Endocrine

- Catabolic – due to increase in ACTH, cortisol, ADH, GH, catecholamines, renin, angiotensin II, aldosterone, glucagon, interleukin-1 (After Macintyre et al. 2010)
- Anabolic – due to decrease in insulin, testosterone

Metabolic

- Carbohydrate – hyperglycemia, glucose intolerance, insulin resistance
- Due to increase in hepatic glycogenolysis (epinephrine, glucagon) and gluconeogenesis (cortisol, glucagon, growth hormone, epinephrine, free fatty acids)
- Due to decrease in insulin secretion/action
- Protein – muscle protein catabolism, increased synthesis of acute-phase proteins
- Due to increase in cortisol, epinephrine, glucagon, interleukin-1
- Fat – increased lipolysis and oxidation
- Due to increase in catecholamines, cortisol, glucagon, growth hormone

- Water and electrolyte flux – retention of H_2O and Na^+, increased excretion of K^+, decreased functional extracellular fluid with shifts to intracellular compartments
- Due to increase in catecholamines, aldosterone, ADH, cortisol, angiotensin II, prostaglandins, and other factors

However, despite the emergence of pain management as a specialty and the availability of a wide range of guidelines and templates for effective analgesia, pain continues to be poorly managed. Why this should be the case is a difficult question to answer, although there is clearly a wide range of possibilities (Cousins and Phillips 1986; Macintyre and Ready 1996).

As can be seen from "Reasons for ineffective analgesia (after NH&MRC 1999)," in some cases it may be simply the result of inadequate knowledge or equipment, but sometimes there can be more disturbing reasons. Macintyre (2001) has pointed out that some health service personnel are still concerned that pain relief can be "too efficacious" and thereby mask postoperative complications such as urinary retention, compartment syndrome, or even myocardial infarction. Another barrier to providing effective analgesia is a view held in some quarters that maintaining the patient in pain is somehow a useful way to aid diagnosis – a concept with no valid scientific basis (Attard et al. 1992; Zolte and Cust 1986).

Reasons for Ineffective Analgesia

- The common idea that pain is merely a symptom and not harmful in itself (After Macintyre et al. 2010)
- The mistaken impression that analgesia makes accurate diagnosis difficult or impossible
- Fear of the potential for addiction to opioids
- Concerns about respiratory depression and other opioid-related side effects such as nausea and vomiting
- The lack of understanding of the pharmacokinetics of various agents
- The lack of appreciation of variability in analgesic response to opioids
- Prescriptions for opioids, which include the use of inappropriate doses and/or dose intervals

- Misinterpretation of doctor's orders by nursing staff, including the use of lower ranges of opioid doses and delaying opioid administration
- The mistaken belief that patient weight is the best predictor of opioid requirement
- The mistaken belief that opioids must not be given more often than 4 hourly
- Patients' difficulties in communicating their need for analgesia

Mechanisms in Acute Pain

The manner in which pain signals are processed and modulated is a complex topic that is covered in detail elsewhere. However, the following brief overview is provided as a background to the sections that follow.

The traditional view of the processing of pain inputs is that they are first detected through nonspecific polymodal nociceptors that respond to a range of stimuli, including thermal, chemical, and mechanical alterations. It is a process designed to alert us to tissue damage. These inputs are then transmitted by A delta and C-type fibers to the spinal cord at speeds of between 2 m/s in the case of the C-type fibers and 10 m/s in the myelinated A delta fibers.

These peripheral nerves terminate in the dorsal horn of the spinal cord where they undergo considerable modulation both via neurotransmitters present at that site and through the action of descending tracts from higher centers, which usually have an inhibitory role. Following modulation, the nociceptive impulse is finally transmitted through tracts to supraspinal sites. Although a number of links are involved, the spinothalamic tract is perhaps the most prominent.

Having given this outline, it is now accepted that our nervous system is a "plastic" environment where stimuli or trauma in any one part of the body can invoke change within other body systems, especially that of the nervous system (Cousins and Power 1999). Changes in nerve function are particularly important and this plasticity can lead nerve fibers whose physiological role is not normally to transmit pain signals to act as nociceptors. For example, while A delta

and C fibers are traditionally seen as primary nociceptive fibers, A beta fibers can become nociceptive under certain circumstances.

Coincident with this is the development of peripheral sensitization. Trauma or other noxious stimuli to tissue result in a neurogenic inflammatory response that in turn leads to vasodilation, increased nerve excitability, and the eventual release of a range of inflammatory mediators such as serotonin, substance P, histamine, and cytokines – the so-called sensitizing soup. This altered environment leads to a modification in the way that input signals are processed with innocuous stimuli being sensed as noxious or painful stimuli, leading to the phenomena of ► hyperalgesia.

The Scope of Acute Pain Management

Acute pain management has developed into a subspecialty in its own right during the last decade with an ever-increasing range of activities. In the hospital setting, the major role of the acute pain team is in the area of postoperative pain management in the surgical patient, although their involvement must not be limited to these patients. In patients with burns, appropriate pain management will help in optimizing pain control both in the early stages where skin grafting and debridement are being carried out and later when the patient requires assistance to undergo physiotherapy. In the patient with spinal cord injury, the initial phase following the injury is often complicated by acute neuropathic pain where early intervention is critical, while in the oncology patient, acute pain can complicate therapy, as in the patient who develops mucositis as a complication of treatment.

Providing Comprehensive Acute Pain Management

Acute and postoperative pain is best managed by an acute pain team and there are a number of structural models of how these are best set up and operated (Rawall and Allvin 1996). While many are headed by consultant anesthetists, this is not always the case and often the day-to-day running of the team is managed by a specialist pain nurse, with medical staff used only for backup when necessary. Acute pain teams need to have clearly defined guidelines and major goals, which will be dictated in part by their institution and circumstances (see Clinical practice guidelines for Acute Pain teams, Cousins and Power 1999). Irrespective of how the team is organized, there must be an efficient method of referral of patients either from the operating theater or from the various surgical teams.

Clinical Practice Guidelines for Acute Pain Teams
Guidelines
- A collaborative, interdisciplinary approach to pain control, including all members of the healthcare team and input from the patient and the patient's family, when appropriate. An individualized proactive pain control plan developed preoperatively by patients and practitioners (since pain is easier to prevent than to treat) (Cousins and Power 1999)
- Assessment and frequent reassessment of the patients' pain
- The use of both drug and nondrug therapies to control and/or prevent pain
- A formal, institutional approach, with clear lines of responsibility

Major Goals
- Reduce the incidence and severity of patients' postoperative or post-traumatic pain
- Educate patients about the need to communicate regarding unrelieved pain, so they can receive prompt evaluation and effective treatment
- Enhance patient comfort and satisfaction
- Contribute to fewer postoperative complications and, in some cases, shorter stays after surgical procedures

Where possible, the pain team should also be involved in ► preoperative education of the elective surgical patient. At such a meeting, the patients' fears and anxieties about pain should be addressed, as there is considerable evidence to suggest that patients who have the opportunity to speak about their concerns about postoperative pain prior to surgery do better and use less medication than control groups. A number of studies have consistently pointed out that pain is usually

the major fear of patients undergoing surgery. During preoperative assessment, at least in the elective patient, it is important to obtain a full medication history especially in relation to use of analgesic agents and the duration of such therapy. Tolerance to opioids can develop quickly and identifying patients who attend for surgery with a history of oral opioid use is important, as they will most likely have different analgesic requirements when compared to the opioid-naïve individual.

The acute pain team also needs to be responsible for the overall postoperative management of the patient. This includes ensuring that regular monitoring and recording of physiological parameters occurs. Details such as oxygen saturation, respiratory rate, and pain status need to be recorded regularly and reviewed. Pain scores can be recorded either numerically or by descriptors. It is important to record pain levels both at rest and on movement, since treatment strategies for these problems will differ. Movement pain in particular is better treated with adjuvant agents rather than opioids.

Accurate recording of physiological data in patients being treated for acute pain is mandatory. Sedation scores and respiratory rate are important in reducing the incidence of opioid-induced toxicity. Pain management records or electronic data apparatus should also allow for the recording of any associated ► adverse events (such as nausea and vomiting) and record data in a form allowing regular or ongoing ► audit. Such audits of acute pain patients should, where possible, allow not only for examination of the parameters already described but also for ► outcome measures. The acute pain team should supervise the transition from a parenteral to an oral analgesic regime. Likewise, members of the acute pain service must recognize when a patient might be suffering a ► persistent acute pain state or undergoing transition from an acute to a chronic pain state and need referral to chronic pain specialists.

Postoperative care also involves being alert for warning signs, so-called ► red flags that might indicate developing complications of the surgery or trauma. In patients previously well controlled using a particular analgesic regime, continuing episodes of unexpected pain requiring increasing doses of medication should alert the practitioner. Under these circumstances, an investigation should be made to elicit the cause of these events, which might be a result of complications of surgery or trauma. This should be diagnosed and treated directly, rather than merely increasing doses of analgesic drugs (Cousins and Phillips 1986).

Preemptive Analgesia Much has been made of the usefulness of ▶ preemptive or preventive analgesia. The concept of providing analgesia prior to a surgical stimulus and thus reducing ▶ central sensitization seems to be a logical and useful proposition and generated a great deal of initial enthusiasm (Dahl and Kehlet 1993; Woolf and Chong 1993). Unfortunately, subsequent controlled trials have failed to consistently demonstrate that any of the commonly used strategies are effective in reducing postoperative pain or analgesic use. These include the preoperative administration of opioids, nonsteroidal anti-inflammatory drugs, and the provision of local analgesic neural blockade (Gill et al. 2001; Podder et al. 2000; Uzunkoy et al. 2001). Much research has been conducted in an effort to ascertain the reasons for this (Charlton 2002; Kehlet 1998; Kissin 1996). Some hypotheses that have been advanced include the suggestion that when local anesthesia is employed in a preemptive setting, any failure to provide complete blockade will still allow sensitization to occur (Lund et al. 1987). Another possibility is that the timing between placement of the blockade and the commencement of surgery is critical, with a time interval of at least 30 min being required between drug administration and surgery (Senturk et al. 2002). One question that has not been fully answered is whether the use of preemptive analgesia might lead to a reduction in the number of patients progressing from acute to chronic pain states. Early studies such as that of Bach et al. (1998) suggested that this may well be the case, and this has been supported by more recent reports (Obata et al. 1999).

Treatment Strategies: General The principles of management of acute nociceptive pain are generally ▶ multimodal. This implies using a number of agents, sometimes given by different routes, to maximize pain control. While pain control after some minor procedures can be controlled by non-opioids alone, opioids remain the main stay of moderate to severe pain management. The use of combinations of ▶ adjuvant analgesics, also known as ▶ balanced analgesia, allows for a reduction in opioid dosage and thus side effects, which can be useful in managing some aspects of pain that can be less responsive to opioids alone.

With regard to the selection of a route of drug administration, while the use of the oral route might initially seem easiest, it is rarely used in the first instance. The variable bioavailability of oral products coupled with postoperative attenuation of gastrointestinal function and the possibility of superimposed vomiting makes this route a poor choice initially. Parenteral administration is usually called for and the intravenous route is the preferred route of administration, often using ▶ patient-controlled analgesia (PCA) devices.

Patient-Controlled Analgesia PCA as a means of drug administration has to a degree revolutionized modern pain management. Although purchase of the devices represents a significant financial outlay, there are savings to be made in terms of medical and nursing staff time, as well as less tangible benefits, such as reducing the number of needle stick injuries. Importantly, patients generally feel positive about using PCAs (Chumbley et al. 1999), with most studies suggesting that the feeling of "being in control" was the most common reason for the high level of satisfaction (Albert and Talbott 1988). However, despite a number of inbuilt safety mechanisms, overdosage can still occur with these devices, and strict postoperative monitoring is imperative (Macintyre 2001). While the intramuscular route can be used for intermittent analgesia, the pharmacokinetics is often unattractive, requiring repeated injections. Furthermore, intramuscular analgesia is most often prescribed on a p.r.n. or "as required" basis, which perforce implies that the patient must be in a pain state before they request the medication – a situation that should be avoided. Finally, every intramuscular (or indeed subcutaneous) injection given presents a possibility for a needlestick injury to occur – another situation best avoided.

Epidural Analgesia Much has been written about the risks and benefits associated with the use of epidural analgesia in the postoperative period, and interpreting the results of these myriad studies conducted under varying circumstances is extremely difficult. There is no doubt that epidural analgesia provides a number of real advantages. It allows the use of drug combinations, which can be delivered close to appropriate receptor sites in the spinal cord (Schmid and Sandler 2000), it reduces the requirements of opioid analgesics (Niemi and Breivik 1998), and it generally allows for a faster return of physiological function, especially gastrointestinal and respiratory status in the postoperative period. The degree to which this occurs appears to be dependent, at least in part, on the nature of surgery performed (Young Park et al. 2001).

However, more recently, despite the fact that there are considerable benefits associated with the use of epidural infusions, attention has focussed on the nature and incidence of complications associated with epidural infusions (Horlocker and Wedel 2000; Rigg et al. 2002; Wheatley et al. 2001). These complications can range from local or systemic infection through to hematoma formation and local or permanent neurological sequelae. The rates of the most serious complications of permanent nerve defects or paraplegia are quoted as between 0.005 % and 0.03 % (Aromaa et al. 1997; Dahlgren and Tornebrandt 1995). Again analysis of these data is difficult because of the number of variables involved. For example, there is growing evidence that those people who develop epidural neurological complications frequently have significant preexisting pathologies, which may predispose them to such complications. Lastly, there has been considerable debate about the guidelines for epidural placement and removal in patients undergoing perioperative anticoagulation. This is

especially so when fractionated or low molecular weight heparin products are employed, because of the possibility of increased risk of development of epidural hematoma under these circumstances. Again, the evidence is conflicting (Bergqvist et al. 1992; Horlocker and Wedel 1998). Patient-controlled epidural analgesia is a means of pain management that combines the efficacy of epidurally administered drugs with the convenience of patient control. In many countries the use of epidural analgesia, especially for the control of postoperative pain, has declined, on account of the potentially serious adverse effects associated with its use, coupled with the increased use of targeted peripheral nerve blocks that provide specific areas of regional anesthesia (Grossi and Urmey 2003).

Intrathecal Analgesia The intrathecal route of drug administration can be useful both as a means of providing anesthesia and for postoperative analgesia. Both opioids and local anesthetic agents have been administered by this route. While the use of low doses of less lipophilic agents such as morphine is popular and gives prolonged postoperative care, the use of this route is not without risk, as there has been a rise in the number of cases of transient neurological symptoms following lignocaine use (Johnson 2000). Recent reviews have suggested that the use of intrathecal morphine for postoperative pain is efficacious, but that certain types of surgery are more responsive than others to the type of management (Eandi et al. 2003; Meylan et al. 2009).

Pharmacotherapies
Opioids With regard to the ▶ opioids, there has been an increase both in the range of drugs available and in their routes of administration. The traditional range of opioids such as morphine, pethidine, and fentanyl has been augmented by drugs such as ▶ oxycodone, dihydrocodeine, and ▶ hydromorphone. None of these drugs are actually "new," having been synthesized in some cases almost 100 years ago, but rather they have been rediscovered by a new generation of prescribers. Oxycodone in particular is

available in a sustained-release form that exhibits a useful biphasic pharmacokinetic profile. The role of pethidine (meperidine) in modern pain management continues to be problematic. While it still has a place under certain circumstances, it should be avoided as an agent for longer-term use, owing to its apparently increased abuse potential and the risk of accumulation of the excitatory metabolite norpethidine (Pattullo and MacPherson 2011). Codeine (when used alone) is also a drug of limited utility. Although it has good oral bioavailability, it is a poor analgesic, has a high degree of constipation, and is influenced by genetic polymorphism. The increased opioid armamentarium has also given scope for ▶ opioid rotation. Although this is a strategy primarily associated with chronic pain management, patients can develop a degree of tolerance to opioids even after a few days. Where continued opioid treatment is needed for whatever reason, switching opioids often results in enhanced pain control, often together with a reduction in dosage. Methadone is an interesting drug, which has generated some recent interest. Its unusual pharmacokinetic profile, with a long and unpredictable half-life of up to 72 h, makes it impracticable for use in the very early stages of acute pain. However, it can be used in later stages where a long-acting oral product is preferable. That the drug has activity at the NMDA receptor as well as the mu opioid receptor is well known. However, it has always been difficult to assess to what, if any, extent this contributes to its analgesic effect and the fact that it has been shown to be of benefit in the treatment of other pain states such as phantom limb pain (Bergmans et al. 2002).

Non-Opioids The non-opioids are a diverse group of drugs with differing modes of action and means of administration. Most show clear synergism with the opioids. Members of this group include tramadol, the nonsteroidal anti-inflammatory drugs (NSAIDS), COX-2 inhibitors, and ketamine.

Paracetamol ▶ Paracetamol should be almost the universal basis of acute and postoperative

pain control. A number of well-controlled trials have clearly demonstrated that regular paracetamol, when given in a dose of 1 g q.i.d., clearly reduces opioid requirements by up to 30 %. Side effects are minimal and the drug is very well tolerated. In most countries it is available in both oral and rectal forms, and in a small number, a parenteral prodrug propacetamol is also available.

The only real contraindication to the prescribing of paracetamol is impaired hepatic function, where the drug is probably best avoided. Much work has also been done on the efficacy of other drugs given in combination with paracetamol. In general, the analysis of trial data suggests that while the combination of codeine phosphate (60 mg) has benefits over paracetamol alone, the use of paracetamol with lower quantities seems to confer little benefit. Parenteral paracetamol has now been available for some years and, although more expensive than oral or rectal products, now provides a convenient parenteral form of administration. It has no specific advantages over the oral form and should be ceased as soon as the patient is able to commence enteral feeding. Although paracetamol generally offers a very favorable adverse effect profile, the dose must be reduced in elderly or cachectic patients, and a close monitor kept on liver function tests. Adverse effects to paracetamol use have still been reported when the drug has been used at "therapeutic" dosage.

Tramadol Tramadol is unique among analgesic agents in having a dual action. Its main activity probably lies in enhancing the action of noradrenaline and 5-hydroxytryptamine at the spinal cord level, while it also has a very weak agonist activity at the mu receptor at supraspinal sites. Tramadol is a very useful drug for the management of mild to moderate pain, and the fact that it can be given orally or by the intravenous or intramuscular routes further adds to its versatility.

A recent review of tramadol use in acute pain has confirmed its effectiveness in both nociceptive and neuropathic pain, making it a useful agent in cases of mixed pain or where the diagnosis is unclear (Duehmke et al. 2006; Moore and McQuay 1997).

Its low addiction potential makes it a good choice for long-term use. Because of the risk of precipitating serotonin syndrome, tramadol is probably best avoided in combination with many of the different antidepressant medications, especially the SSRIs, although in clinical practice the real risk seems quite low. Recent studies have confirmed that it possesses significant synergy when combined with paracetamol, and indeed a combination product is now available in some countries (Fricke et al. 2002). There are few studies available on the usefulness of combination of tramadol with opioids, although initial results appear encouraging (Webb et al. 2002).

Tramadol is also attractive because of its low abuse potential. Certainly in comparison to strong opioids, the incidence of abuse, dependence, and withdrawal is considerably lower (Cicero et al. 1999). However, a number of such cases have been reported, almost all of which were in patients with a preexisting history of drug or substance abuse (Brinker et al. 2002; Lange-Asschenfeldt et al. 2002).

In the management of postoperative pain, all efforts should be made to reduce the incidence of postoperative nausea and vomiting, which is not only uncomfortable for the patient but can also lead to fluid imbalance, impaired respiratory function, and electrolyte disturbances. In this regard the use of tramadol is somewhat problematic, as the incidence of nausea and vomiting is at least as high as with opioids (Silvasti et al. 2000; Stamer et al. 1997). However, some strategies have been suggested to attenuate this response including administration of an intraoperative loading dose (Pang et al. 2000) and slow IV administration (Petrone et al. 1999). Should management of tramadol-induced nausea and vomiting require pharmacological intervention, recent studies suggest that members of the butyrophenone class such as droperidol might be a better choice than 5HT3 antagonists such as ondansetron, which might not only be less effective but also antagonize tramadol's analgesic effects.

Nonsteroidal Anti-Inflammatory Drugs

NSAIDs (▶ NSAIDs, Survey) are widely used in acute pain management (Merry and Power 1995).

While they may be used as the sole agent in mild pain, they are primarily employed as adjunctive medications in combination with opioids in moderate to severe pain states. Here their action both at central and peripheral sites complements opioid activity, and they are especially useful in the management of pain associated with movement. There have always been concerns associated with the use of NSAIDs in the surgical patient because of the risk of the development of serious complications, especially renal impairment. However, careful patient selection and monitoring, the use of a product with a short half-life, and restricting the duration of treatment to about 3 days greatly reduce the danger.

The discovery of the two isoforms of the cyclooxygenase (COX) enzyme has more recently led to the development of COX-2-specific inhibitors such as celecoxib and rofecoxib, with the aim of developing potent NSAIDs without significant associated gastrointestinal side effects. The majority of studies on these drugs have been conducted in outpatient populations, and whether they offer any advantage over traditional NSAIDs in the management of postoperative pain is unclear. Even more recently, a parenteral COX-2 inhibitor (parecoxib) has been developed specifically for the management of postoperative pain, and initial results of studies are encouraging.

Unfortunately, the cardiovascular safety of these products has recently come under scrutiny resulting in the removal of some products from the marketplace, owing to an increase in thromboembolic events associated with its use (Solomon et al. 2004). There were initial concerns that this might have been a class effect of COX-2 inhibitors, but it now appears to be agent specific. Although the analgesia offered by these agents is not superior to that seen with conventional NSAIDS, nevertheless in long-term use they almost certainly offer improved safety over conventional NSAIDs (Tramèr et al. 2000).

Gabapentin and Pregabalin These antiepileptic agents have been used for some time in the management of neuropathic pain with excellent results. However, in recent times, there have been a number of studies involving their use in acute postoperative pain. What is interesting is that it appears that gabapentin administration, even for a few days in the perioperative period, can improve quality of analgesia and reduce opioid use. Clearly more work is needed in this area, but a recent systematic review certainly suggests that the perioperative use of either gabapentin or pregabalin improves the quality of postoperative analgesia and reduces opioid use (Elina et al. 2007).

▶ Ketamine is an important second-line drug in the pain physician's armamentarium. Well known as an anesthetic agent, it has in the last decade or so found use as an analgesic product when used in sub-anesthetic doses. The drug has some useful N-methyl-D-aspartate (NMDA) receptor antagonist activity and can also augment the action of opioids in the treatment of nociceptive pain. The usual psychomimetic effects of the drug are not usually a problem in the dosages employed, although the development and release of the S(+) might signal a resurgence in the interest of this drug. The clinical applications of ketamine have recently been briefly reviewed by Persson (2010), while an earlier paper by Elia (2005) has undertaken a systematic review of ketamine use in postoperative pain.

Neuropathic Pain Comprehensive acute pain management also entails the recognition and management of ▶ acute neuropathic pain. Neuropathic pain is most frequently seen as a sequela of long-term pathological states such as diabetes or herpes zoster infection (Bowsher 1991). However, this is not always the case and acute neuropathic pain can be seen immediately following surgical procedures where peripheral nerves have been disrupted, such as in the ▶ post-thoracotomy syndrome following specific events such as acute spinal cord injury or as evidenced by ▶ phantom limb pain following amputation. It is important to be alert for the signs or symptoms of neuropathic pain in the acute or postoperative phase (see Features suggestive of neuropathic pain after NH&MRC 1999). Failure to diagnose such a condition will result not only in prolonged

pain but also most probably in the patient being given increasing doses of opioid medication in a futile effort to control the condition (Hayes and Molloy 1997).

Features Suggestive of Neuropathic Pain
- Pain can be related to an event causing nerve damage (After Macintyre et al. 2010).
- Pain unrelated to ongoing tissue damage.
- Sometimes a delay between event and pain onset:
  - The pain is described as burning, stabbing, pulsing, or electric-shock-like.
  - Hyperalgesia.
  - Allodynia (indicative of central sensitization).
  - Dysesthesia.
- Poor response to opioids.
- The pain is usually paroxysmal and often worse at night.
- Pain persists in spite of the absence of ongoing tissue damage.

The management of neuropathic pain can be complex and much has been written on the usefulness of various pain strategies. A wide range of drugs with differing pharmacological targets such as ▶ anticonvulsant medications, notably ▶ gabapentin and carbamazepine; ▶ antidepressants; and ▶ membrane stabilizing agents such as ▶ mexiletine/Mexitil have all been employed with varying success. Local anesthetics such as lignocaine have all been found to be useful, especially in the acute case, where they can be administered as a subcutaneous infusion.

Specific Acute Pain States
There are some acute pain states that have been subject to more extensive research and whose symptomatology and pathogenesis follow recognized patterns. These include acute lower back pain, pain following chest trauma or thoracic surgery, compartment syndrome, and the acute presentation of ▶ complex regional pain syndrome. There have also been significant advances in our understanding of ▶ acute pain mechanisms and the differentiation between visceral and somatic (deep or superficial) pain.

Summary

There have been a number of significant improvements in the management of acute and postoperative pain management during the past decade. To some degree, this has been helped by the emergence of new drugs or, in some cases, whole new drug groups. However, in the main, advances in acute and postoperative pain management have come about by recognizing how to manage pain better with existing drugs, focussing on the use of drug combinations to maximize outcomes. There has also been a greater appreciation of the importance of diagnosing acute neuropathic pain, requiring a different approach. Those involved in pain management have embarked on a virtual crusade in an effort to convince health professionals that acute and postoperative pain can be and must be appropriately and successfully managed. Perhaps the most important lesson of all is an appreciation that all chronic pain must start as acute pain. Appropriate management of acute pain will therefore have the additional bonus of eventually reducing the worldwide burden of patients having to suffer debilitating chronic pain states.

References

Albert, J. M., & Talbott, T. M. (1988). Patient-controlled analgesia vs. conventional intramuscular analgesia following colon surgery. *Diseases of the Colon and Rectum, 31,* 83–86.

Aromaa, U., Lahdensuu, M., & Cozanitis, D. A. (1997). Severe complications associated with epidural and spinal anaesthesia in Finland 1987–1993. A study based on patient insurance claims. *Acta Anaesthesiologica Scandinavica, 41,* 445–452.

Attard, A. R., Corlett, M. J., Kinder, N. J., et al. (1992). Safety of early pain relief for acute abdominal pain. *BMJ, 305,* 554–556.

Bach, S., Noreng, M. F., & Tjellden, N. U. (1998). Phantom limb pain in amputees during the first 12 months following limb amputation, after preoperative lumbar epidural blockade. *Pain, 33,* 297–301.

Bergmans, L., Snijdelaar, D. G., Katz, J., et al. (2002). Methadone for phantom limb pain. *The Clinical Journal of Pain, 18,* 203–205.

Bergqvist, D., Lindblad, B., & Matzsch, T. (1992). Low molecular weight heparin for thromboprophylaxis and epidural / spinal anaesthesia: Is there a risk? *Acta Anaesthesiologica Scandinavica, 36,* 605–609.

Bowsher, D. (1991). Neurogenic pain syndromes and their management. *British Medical Bulletin, 47,* 644–666.

Brinker, A., Bonnel, R., & Beitz, J. (2002). Abuse, dependence or withdrawal associated with Tramadol. *The American Journal of Psychiatry, 159,* 881.

Charlton, J. E. (2002). Treatment of postoperative pain. In M. A. Giamberardino (Ed.), *Pain 2002 – An updated review: Refresher course syllabus.* Seattle: IASP Press.

Chumbley, G. M., Hall, G. M., & Salmon, P. (1999). Why do patients feel positive about patient-controlled analgesia? *Anaesthesia, 54,* 38638–38639.

Cicero, T., Adams, E., & Galler, A. (1999). Postmarketing surveillance program to monitor Ultram (tramadol hydrochloride) abuse in the United States. *Drug and Alcohol Dependence, 57,* 7–22.

Cousins, M. J., & Phillips, G. D. (1986). Acute pain management. In: Cousins MJ & Phillips GD (eds) *Clinics in Critical Care Medicine.* New York: Churchill Livingstone.

Cousins, M. J., & Power, I. (1999). Acute and postoperative pain. In P. D. Wall & R. Melzack (Eds.), *Textbook of pain* (4th ed., pp. 447–492). Edinburgh: Churchill Livingstone.

Cousins, M. J., Power, I., & Smith, G. (2000). 1996 Labat lecture: Pain – A persistent problem. *Regional Anesthesia and Pain Medicine, 25,* 6–21.

Dahl, J. B., & Kehlet, H. (1993). The value of pre-emptive analgesia in the treatment of postoperative pain. *British Journal of Anaesthesia, 70,* 434.

Dahlgren, N., & Tornebrandt, K. (1995). Neurological complications after anaesthesia. A follow up of 18,000 spinal and epidural anaesthetics performed over three years. *Acta Anaesthesiologica Scandinavica, 39,* 872–880.

Duehmke, R. M., Hollingshead, J., & Cornblath, D. R. (2006). Tramadol for neuropathic pain. *Cochrane Database of Systematic Reviews* (3) Art. No.: CD003726. doi: 10.1002/14651858.CD003726.pub3.

Eandi, J. A., de Vere White, R. W., Tunuguntla, H. S. G. R., Bohringer, C. H., & Evans, C. P. (2003). Can single dose preoperative intrathecal morphine sulfate provide cost-effective postoperative analgesia and patient satisfaction during radical prostatectomy in the current era of cost containment? *Prostate Cancer and Prostatic Disease, 5*(3), 226–230.

Elia, A. (2005). Ketamine and post-operative pain – A quantitative systematic review of randomised trials. *Pain, 113,* 61–70.

Elina, M., Tiippana, E. M., Hamunen, K., Kontinen, V. K., & Kalso, E. (2007). Do surgical patients benefit from perioperative gabapentin/pregabalin? A systemic review of efficacy and safety. *Anesthesia and Analgesia, 104,* 1545–1556.

Fricke, J. J., Karim, R., Jordan, D., et al. (2002). A double-blind, single-dose comparison of the analgesic efficacy of tramadol/acetaminophen combination tablets, hydrocodone/acetaminophen combination tablets, and placebo after oral surgery. *Clinical Therapeutics, 24*, 953–968.

Gill, P., Kiani, S., Victoria, B., & Atcheson, R. (2001). Pre-emptive analgesia with local anaesthetic for herniorrhaphy. *Anaesthesia, 56*, 414–417.

Grossi, P., & Urmey, W. F. (2003). Peripheral nerve blocks for anaesthesia and postoperative analgesia. *Curr Opin Anaesthesiol, 16*(5), 493–501.

Hayes, C., & Molloy, A. R. (1997). Neuropathic pain in the perioperative period. In A. R. Malloy & I. Power (Eds.), *International anesthesiology clinics. Acute and chronic pain* (pp. 67–81). Philadelphia: Lippincott-Raven.

Horlocker, T. T., & Wedel, D. J. (1998). Neuraxial blockade and low molecular weight heparins: Balancing perioperative analgesia and thromboprophylaxis. *Regional Anesthesia and Pain Medicine, 23*(Suppl 2), 164–177.

Horlocker, T. T., & Wedel, D. J. (2000). Neurologic complications of spinal and epidural anesthesia. *Regional Anesthesia and Pain Medicine, 25*, 83–98.

Johnson, M. (2000). Potential neurotoxicity of spinal anesthesia with lidocaine. *Mayo Clinic Proceedings, 75*, 921–932.

Kehlet, H. (1998). Modification of responses to surgery by neural blockade: Clinical implications. In M. J. Cousins & P. O. Bridenbaugh (Eds.), *Neural blockade in clinical anesthesia and management of pain* (pp. 129–178). Philadelphia: Lippincott-Raven.

Kehlet, H., & Holte, K. (2001). Effect of postoperative analgesia on surgical outcome. *British Journal of Anaesthesia, 87*, 62–72.

Kissin, I. (1996). Preemptive analgesia: Why its effect is not always obvious. *Anesthesiology, 84*, 1015–1019.

Lange-Asschenfeldt, C., Weigmann, H., Hiemke, C., et al. (2002). Serotonin syndrome as a result of fluoxetine in a patient with tramadol abuse: Plasma level-correlated symptomatology. *Journal of Clinical Psychopharmacology, 22*, 440–441.

Lund, C., Selmar, P., Hensen, O. B., et al. (1987). Effect of epidural bupivacaine on somatosensory evoked potentials after dermatomal stimulation. *Anesthesia and Analgesia, 66*, 343–348.

Macintyre, P. E. (2001). Safety and efficacy of patient-controlled analgesia. *British Journal of Anaesthesia, 87*, 36–46.

Macintyre, P. E., & Ready, L. B. (1996). *Acute pain management: A practical guide*. London: WB Saunders.

Macintyre, P. E., Schug, S. A., Scott, D. A., Visser, E. J., Walker, S. M., & APM:SE Working Group of the Australian and New Zealand College of Anaesthetists and Faculty of Pain Medicine. (2010a). *Acute pain management: Scientific evidence* (3rd ed.). Melbourne: ANZCA & FPM.

Macintyre, P. E., Schug, S. A., Scott, D. A., Visser, E. J., Walker, S. M., & APM:SE Working Group of the

Australian and New Zealand College of Anaesthetists and Faculty of Pain Medicine. (2010b).

Merry, A., & Power, I. (1995). Perioperative NSAIDs: Towards greater safety. *Pain Review, 2*, 268–291.

Meylan, N., Elia, N., Lysakowski, C., & Tramèr, M. R. (2009). Benefit and risk of intrathecal morphine without local anaesthetic in patients undergoing major surgery: Meta-analysis of randomized trials. *British Journal of Anaesthesia, 102*(2), 156–167.

Moore, R. A., & McQuay, H. J. (1997). Single-patient data meta-analysis of 3453 postoperative patients: Oral tramadol versus placebo, codeine and combination analgesics. *Pain, 69*, 287–294.

NH&MRC (1999). National Health and Medical Research Council. Acute pain management: Scientific evidence. Australian Government Printer.

Niemi, G., & Breivik, H. (1998). Adrenaline markedly improves thoracic epidural analgesia produced by a low-dose infusion of bupivacaine, fentanyl and adrenaline after major surgery. *Acta Anaesthesiologica Scandinavica, 42*, 897–909.

Obata, H., Saito, S., & Fujita, N. (1999). Epidural block with mepivacaine before surgery reduces long-term post-thoracotomy pain. *Canadian Journal of Anaesthesia, 46*, 1127–1132.

Ong, C. K. S., Lirk, P., Tan, C. H., & Seymour, R. A. (2007). An evidence-based update on nonsteroidal anti-inflammatory drugs. *Clinical Medicine & Research, 5*(1), 19–34.

Pang, W.-W., Mok, M., Huang, S., et al. (2000). Intraoperative loading attenuates nausea and vomiting of tramadol patient-controlled analgesia. *Canadian Journal of Anaesthesia, 47*, 968–973.

Pattullo, G., & MacPherson, R.D. (2011). *Pethidine: The case for its withdrawal* (pp. 17–27). Australian Anaesthesia Publ Australian and New Zealand College of Anaesthetists.

Persson, J. (2010). Wherefore ketamine? *Curr Opin Anaesthesiol. 23*(4), 455–460.

Petrone, D., Kamin, M., & Olson, W. (1999). Slowing the titration rate of tramadol HCI reduces the incidence of discontinuation due to nausea and/or vomiting: A double-blind randomized trial. *Journal of Clinical Pharmacy and Therapeutics, 24*, 115–123.

Podder, S., Wig, J., Malhotra, S., et al. (2000). Effect of pre-emptive analgesia on self-reported and biological measures of pain after tonsillectomy. *European Journal of Anaesthesiology, 17*, 319–324.

Rawall, N., & Allvin, R. (1996). Epidural and intrathecal opioids for postoperative pain management in Europe – A 17 nation questionnaire study of selected hospitals. *Acta Anaesthesiologica Scandinavica, 63*, 583–592.

Rigg, J. R. A., Jarmrozki, K., Myles, P. S., et al. (2002). Epidural anaesthesia and analgesia and outcome of major surgery: A randomised trial. *Lancet, 359*, 1276–1282.

Schmid, R. L., & Sandler, A. N. (2000). Use and efficacy of low-dose ketamine in the management of acute postoperative pain: Review of current techniques and outcomes. *Pain, 82*, 111–125.

Senturk, M., Ozcan, P.-E., Talu, G.-K., et al. (2002). The effects of three different analgesia techniques on long-term postthoracotomy pain. *Anesthesia and Analgesia, 94*, 11–15.

Silvasti, M., Svartling, N., Pitkanen, M., et al. (2000). Comparison of intravenous patient-controlled analgesia with tramadol versus morphine after microvascular breast reconstruction. *European Journal of Anaesthesiology, 17*, 448–455.

Solomon, D. H., Glynn, R. J., Levin, R., et al. (2004). Relationship between selective COX-2 inhibitors and acute myocardial infarction in older adults. *Circulation, 109*, 2068–2073.

Stamer, U., Maier, C., Grondt, S., et al. (1997). Tramadol in the management of post-operative pain: A double-blind, placebo- and active drug-controlled study. *European Journal of Anaesthesiology, 14*, 646–654.

Tramèr, M. R., Moore, R. A., Reynolds, D. J., & McQuay, H. J. (2000). Quantitative estimation of rare adverse events which follow a biological progression: A new model applied to chronic NSAID use. *Pain, 85*(1–2), 169–182.

Uzunkoy, A., Coskun, A., & Akinci, O. (2001). The value of pre-emptive analgesia in the treatment of postoperative pain after laparoscopic cholecystectomy. *European Surgical Research, 33*, 39–41.

Webb, A., Leong, S., Myles, P., et al. (2002). The addition of a tramadol infusion to morphine patient-controlled analgesia after abdominal surgery: A double-blinded, placebo-controlled randomized trial. *Anesthesia and Analgesia, 95*, 1713–1718.

Wheatley, R. G., Schug, S. A., & Watson, D. (2001). Safety and efficacy of postoperative epidural analgesia. *British Journal of Anaesthesia, 87*, 47–61.

Woolf, C. J., & Chong, M. S. (1993). Preemptive analgesia – Treating postoperative pain by preventing the establishment of central sensitization. *Anesthesia and Analgesia, 77*, 362–379.

Young Park, W., Thompson, J. S., & Lee, K. K. (2001). Effect of epidural anesthesia and analgesia on perioperative outcome: A randomized, controlled veterans affairs cooperative study. *Annals of Surgery, 234*, 560–571.

Zolte, N., & Cust, M. D. (1986). Analgesia in the acute abdomen. *Annals of the Royal College of Surgeons of England, 68*, 209–210.

Acute Pain Database

▶ Postoperative Pain, Data Gathering and Auditing

Acute Pain in Children, Postoperative

Jennifer Rabbitts and Cornelius Groenwald
Department of Anesthesiology and Pain Medicine, Seattle Children's Hospital, University of Washington School of Medicine, Seattle, WA, USA

Synonyms

Acute postoperative pain in children; Pediatric postsurgical pain

Definition

Surgical trauma from tissue destruction and musculoskeletal strain causes inflammatory and neurohumoral responses with release of vaso- and immunoreactive substrates (cytokines, complement, arachidonic acid, metabolites, nitric oxide, free oxygen radicals) that promote inflammation, vascular hyperpermeability, and pain.

The duration of postoperative pain that is considered acute is not clearly defined and is dependent on the amount of tissue trauma and the time taken for healing. In the perioperative setting, pain that persists after healing has occurred could be considered persistent pain.

Introduction

Despite advances in the safety of perioperative care, pain remains a problem for many children after surgery. Moderate-to-severe pain occurs in almost half of children hospitalized after an operation and is not limited to any specific surgeries with fairly similar rates occurring after orthopedic; general; ear, nose, and throat; and neuro- and plastic surgeries (Groenewald et al. 2012). Teenagers are more at risk than children, with males and females being equally affected. Moderate-to-severe pain in pediatric patients, once present, is difficult to treat and tends to persist in subsequent days, contributing to delayed recovery. Poorly controlled postoperative pain may also predispose patients to developing chronic pain syndromes (Macrae 2008).

Characteristics

Causes and Risk Factors

Pain occurs after surgery for a number of reasons. Direct trauma to tissue from incisions and handling, stretch from retraction, and insufflation for laparoscopy causes somatic and referred pain. Musculoskeletal pain may also occur from prolonged positioning. In addition, children may have pain from tissue ischemia, inflammation, and gastrointestinal disturbance (distension or spasms). Pain from complications, for example, compartment syndrome, must always be considered and ruled out. Children may also have pain from the underlying disease process or comorbid conditions and from painful procedures in the postoperative period (venipuncture, dressing changes, and continuous passive motion machine).

Several biological, psychological, and social processes have been identified as determinants of acute and chronic pain in children. In the context of surgery, psychological processes, particularly preoperative levels of anxiety, have been identified as an important risk factor for acute postoperative pain in children (Kain et al. 2006; Kotzer 2000). Postoperative sleep quality in children has also been shown to be linked to acute postoperative pain (Kain et al. 2002), but the relationships between sleep and pain in children after surgery are not yet clear. Moderate-to-severe pain present several days into the hospitalization tends to persist through the hospitalization (Groenewald et al. 2012). Risk factors for more severe and persistent postoperative pain in children are active areas of investigation.

Importance of Effective Management of Postoperative Pain

Postoperative pain has significant immediate and long-lasting effects and delays overall recovery. Pain has deleterious effects on pulmonary,

cardiovascular, endocrine, immunologic, coagulation, gastrointestinal, and musculoskeletal systems, delays mobilization, and is also associated with impaired wound healing. Pain is central in the surgical stress response which contributes to postoperative morbidity (Kehlet 1997). Fatigue and sleep disturbance are also contributory to the stress response and are also exacerbated by pain.

Pain is one of the main outcomes affecting quality of life in children after major surgery (Jacobsen et al. 2010). Moreover, inadequately treated pain can lead to central sensitization and further pain problems in the future. Pain is also one of the main factors affecting both the child's and their parents' satisfaction with their perioperative care (Bull and Grogan 2010; Cucchiaro et al. 2006b; Wagner et al. 2007).

On the other hand, many pharmacologic pain management strategies also have deleterious short- and long-term effects. All analgesic medications have potential side effects and complications, for example, opioid-induced nausea and vomiting, sedation, and ileus which delay postoperative recovery. In addition, opioid escalation may lead to tolerance and opioid-induced hyperalgesia (Lee et al. 2011). It is, therefore, important that a balanced multimodal approach to postoperative analgesia is used to enhance postoperative recovery (Kehlet and Dahl 1993).

Assessment Tools

Effective pain assessment requires measurement tools that are reliable (reproducible over time), valid (accurately measure pain), responsive (track changes in pain over time), and clearly interpretable and feasible (easy to use). In addition, pain measurements should be frequent, at minimum every 4 h, before and after pain relief interventions and as medically indicated. Pain measurement during activity is a better indicator of effective pain control and postoperative outcome than measurement at rest only (Kehlet and Dahl 1993). Various pediatric pain scales have been developed based on the child's age, developmental status, and verbal ability, and these can be broadly classified as being either observational (nonverbal children) or self-report (children able to report pain in terms of the scale used).

To standardize pain measurement tools for research trials the Ped-IMMPACT group (Pediatric Initiative on Methods, Measurement, and Pain Assessment in Clinical Trials) commissioned two systematic reviews in 2005 aimed at evaluating the use of existing pain measurement tools in children (McGrath et al. 2008; Stinson et al. 2006; von Baeyer and Spagrud 2007). Stinson et al. identified three age-specific self-report scales validated for acute postoperative pain measurement: The Poker Chip Tool for ages 3–4 years, the Faces Pain Scale-Revised for ages 4–12 years, and the Visual Analog Scale for ages 8 years and beyond (Stinson et al. 2006). All three are well established in both research and clinical practice and have been extensively validated and found reliable. Interestingly at the time of Stinson's review, the numerical rating scale (NRS), a widely used pain scale in clinical practice had not yet been adequately studied; however, its use for acute postoperative pain assessment was recently validated (Page et al. 2012). Observational pain scores are needed for children unable to self-report due to being too young, cognitively impaired, or critically ill. Von Baeyer and Spagrud reviewed all published reports of observational measures in patients 1–18 years of age (von Baeyer and Spagrud 2007). They recommend the Face, Legs, Arms, Cry, Consolability (FLACC) and Children's Hospital of Eastern Ontario Pain Scale (CHEOPS) pain scales for acute postoperative pain assessment in nonverbal children older than 1 year of age. Both of these have been extensively used in clinical research and have excellent reliability and responsiveness. For pain assessment at home by parents the Parents' Postoperative Pain Measure (PPPM) was recommended. The only pain measurement recommended for critically ill children is the COMFORT scale. The Ped-IMMPACT Group did not publish guidelines for the assessment of pain intensity in children younger than age 1 or for those cognitive disabilities; however, scales validated postoperatively for children younger than 1 year include the Neonatal Pain, Agitation and Sedation Scale (NPASS) for term neonates (Hummel et al. 2008) and the Premature Infant Pain Profile (PIPP) for preterm and term infants (Stevens et al. 2010).

Obstacles to adequate assessment of pain in children include communication and language barriers, unwillingness to participate, and disruptive behaviors (frustration, annoyance). In the postoperative setting, additional barriers include sedation, intubation, and difficulty differentiating pain from agitation or from discomfort from other sources (e.g., orthopedic casts, urinary catheter, intravenous lines, etc.). Differentiating behaviors, distress and pain is particularly difficult as pain itself has both emotional and sensory components (Blount and Loiselle 2009).

Strategies for Postoperative Pain Management

Multimodal opioid sparing analgesia is an important part of "fast-track" approaches to enhance postoperative function and recovery (Kehlet and Dahl 1993). Strategies target multiple levels of the nociceptive pathways to achieve effective analgesia through additive or synergistic effects of different analgesics, with fewer side effects.

Painful nociceptive stimuli from the site of tissue injury are conducted via peripheral nerves to the spinal cord where first-order neurons synapse in the dorsal horn. Ascending spinothalamic tract fibers travel to the cerebral cortex where the perception of pain is then further influenced by emotional factors, the memory of previous pain, and behavioral experiences. Descending inhibitory pathways then modulate pain conduction pathways at the spinal level. The practice of providing pain control by targeting several of these levels is termed multimodal or balanced analgesia. For example, pain can be treated at the local site of surgical trauma with local anesthetic agents and anti-inflammatory drugs, while at the spinal cord level it may be treated by opioids and local anesthetic agents and at the cortical level with opioids, acetaminophen, $\alpha2$-agonists, and non-pharmacological behavioral therapies. The goal of effective multimodal analgesia is the achievement of additive or synergistic effects among different analgesics thereby reducing the total dosing of individual medications, particularly opioids, with concomitant decreasing of side effects. The objective of multimodal postoperative analgesia is to hasten convalescence, improve surgical outcomes, and prevent postsurgical hypersensitization and persistent pain (Kehlet 1997; Kehlet and Dahl 1993).

The introduction of preemptive analgesia, that is, the amelioration of postoperative pain by preemptive administration of opioids or local anesthetics prior to surgical trauma, has remained popular in anesthetic practice. However, this technique remains controversial as there is no scientific evidence to support it (Dahl and Kehlet 2011).

Preventive analgesia, that is, the treatment of pain to prevent central sensitization and persistent postoperative pain, is separate from preemptive analgesia and is currently a focus of research (Dahl and Kehlet 2011).

Pharmacological Treatment of Postoperative Pain

Acetaminophen and nonsteroidal anti-inflammatory drugs (NSAIDs) are effective in managing mild-to-moderate postoperative pain. Acetaminophen is available in oral, rectal, and intravenous forms. The oral and intravenous doses are 10–15 mg/kg every 4–6 h, while the rectal dosing is 40 mg/kg, with repeated dosing 20 mg/kg every 6 h in children and every 12 h in neonates. Acetaminophen is remarkably safe when used within these dosing limitations; however, accidental overdose may lead to hepatic necrosis. Postoperatively due to restricted oral intake or nausea and vomiting, many children are unable to tolerate oral acetaminophen, and therefore, the recently introduced intravenous formulation of the drug is gaining popularity. Acetaminophen is efficacious in children of all ages, including infants; however, the dose should be reduced in neonates with unconjugated hyperbilirubinemia (Palmer et al. 2008). Acetaminophen is often combined with NSAIDs for superior postoperative analgesia. NSAIDs act via inhibition of the COX-system to reduce prostaglandin synthesis, thus reducing both pain and inflammation at the surgical site; unfortunately, the same mechanism may lead to significant side effects, including gastric ulceration, bleeding, and impaired renal function (Lynn et al. 2007). Contraindications to NSAIDs used in the postoperative setting include hypovolemia, acute renal impairment, thrombocytopenia, coagulopathy, and active hemorrhage. These conditions are fairly prevalent in hospitalized children and may be one of the reasons why NSAIDS are not routinely prescribed to all postoperative patients (Groenewald et al. 2012).

Opioids remain the mainstay of management of moderate-to-severe acute postoperative pain. Opioid agonists commonly used postoperatively include morphine, hydromorphone, fentanyl, methadone, codeine, hydrocodone, and oxycodone. These exert their analgesic action through $\mu 1$ receptors located in the central nervous system, and can be administered via various routes including enteral, intravenous, intrathecal, epidural, transmucosal, and transdermal routes. The agonist-antagonist opiates, such as nalbuphine and pentazocine, which are also used to treat moderate-to-severe pain, provide analgesia mainly by κ receptor-mediated interactions. Morphine has well-established pharmacokinetics in children (Duedahl and Hansen 2007); however, it has a 50 % incidence of side effects, including respiratory depression, sedation, nausea, vomiting, pruritus, urinary retention, ileus, and constipation. The presence of these side effects makes morphine and other opioid agonists undesirable for unimodal analgesia. Scheduled or as needed intravenous bolus doses of opioids may be sufficient to treat intermittent pain; however, for patients experiencing continuous and/or moderate-to-severe pain, opioids should be given as a continuous intravenous infusion or as patient controlled analgesia (PCA). Opioids should be tapered as recovery occurs and pain improves.

Local anesthetic agents can be administered as single agents or in combination with additives at the level of the neuraxis, the peripheral nerve or the surgical site as continuous infusions or single shot injections. Administration for postoperative pain in children is usually done while under anesthesia for the surgical procedure. Regional anesthesia is an integral part of postoperative pain management in adults and is gaining popularity in children. Many blocks have been extrapolated from adult practice and currently data is being collected from large numbers of children in both

Europe (French Language Societies of Pediatric Anesthesiologists or ADARPEF) and the USA (Pediatric Regional Anesthesia Network or PRAN). Information from these databases will hopefully resolve the many controversial issues in this area and provide evidence-based support for routine regional anesthesia in children undergoing surgery.

The most commonly performed neuraxial block is the caudal block. It is inserted with relative ease in children up to about age 5 and landmarks are readily palpable. Epidural infusions are commonly used for extensive lower extremity orthopedic surgery, abdominal surgery, and thoracic surgery. At the level of the neuraxis, the addition of opioids acts synergistically to improve analgesia; however, these can also be administered intravenously. Using clonidine as an additive in epidural anesthesia enhances the duration of the block (Cucchiaro et al. 2006a). Neonates have pharmacokinetic differences in local anesthetics and, therefore, should receive lower infusion rates and shorter acting drugs. In a national pediatric continuous epidural audit in Great Britain and Ireland, there were approximately 1 in 2,000 serious incidents, 1 in 1,100 intermediate severity incidents, and 1 in 189 low severity incidents. Residual effects from a grade 1 incident 12 months after surgery occurred in one out of the total of 10,633 cases (Llewellyn and Moriarty 2007), a similar rate to that reported in opioid infusions (Morton and Errera 2010).

Data from ADARPEF and PRAN show an increase in use of peripheral nerve blocks and declining neuraxial blocks since the advent of ultrasound guidance (Bosenberg 2012). Continuous peripheral nerve blockade has been shown to be equally effective for orthopedic surgeries as continuous epidural anesthesia, with some additional advantages (Ganesh et al. 2007). Peripheral techniques have the benefits of avoiding potential neuraxial complications, having fewer technical failures, reducing incidence of postoperative nausea and vomiting, avoiding bladder catheterization, interfering less with ambulation, and offering prolonged analgesia for outpatients. Truncal blocks (transversus abdominis plane block, rectus sheath block) are commonly used for analgesia after smaller abdominal procedures.

Ketamine is a potent anesthetic and analgesic that is particularly useful in treating acute postoperative pain in opioid-tolerant patients. Ketamine acts via inhibition of the N-methyl-D-aspartate receptor complex (NMDA) which plays a central role in pain sensitization (Himmelseher and Durieux 2005). Primary concerns for the use of ketamine include adverse neurocognitive effects (e.g., hallucinations) which are commonly seen when used at anesthetic doses (1–2 mg/kg). However, when used in the subanesthetic doses for perioperative pain management (0.1–0.15 mg/kg), there is a lower incidence of unpleasant side effects associated with ketamine. To effectively decrease postoperative pain and opioid requirements ketamine should be administered continuously during the perioperative period. Single doses prior to nociceptive stimulation have not been shown to improve pain management; however, anecdotally a single postoperative dose of ketamine (0.1 mg/kg) may significantly reduce analgesic requirements in opioid-resistant patients.

Gabapentin may be a useful adjunct to postoperative pain management. The perioperative administration of oral gabapentin has been found to decrease postoperative morphine requirements and reduce early pain scores in the immediate postoperative period in children and adolescents undergoing scoliosis repair surgery (Rusy et al. 2010). Despite these promising results, data on the perioperative benefits and use of gabapentin and pregabalin are still lacking and they are consequently not recommended as a standard component of multimodal analgesia.

Clonidine, an α_2-receptor agonist, is associated with many beneficial analgesic properties. Given orally it decreases postoperative pain and analgesia requirements, while it may also prolong and improve neuraxial and peripheral nerve blocks when co-administered with local anesthetics block (Cucchiaro et al. 2006a). Dexmedetomidine, a centrally acting α_2-receptor antagonist that is given intravenously, shares some of the same analgesic characteristics as clonidine and in addition provides excellent sedation (Kraemer and Rose 2009).

Summary

Postsurgical pain remains challenging to manage in children. This population has a high prevalence of moderate-to-severe pain. Accurate pain assessment requires the use of age and developmentally appropriate pain scales. The adequate management of acute pain is particularly important in the postsurgical population as uncontrolled pain may, in addition to unnecessary suffering, result in a variety of adverse consequences. Multimodal pain management techniques, the pharmacological targeting of multiple components of the pain conduction pathways, are superior to unimodal treatment strategies and enhance postoperative function and recovery.

References

Blount, R. L., & Loiselle, K. A. (2009). Behavioural assessment of pediatric pain. *Pain Research & Management, 14*(1), 47–52.

Bosenberg, A. (2012). Regional anesthesia in children: The future. *Paediatric Anaesthesia.* doi:10.1111/j.1460-9592.2012.03836.x.

Bull, J., & Grogan, S. (2010). Children having spinal surgery to correct scoliosis: A qualitative study of parents' experiences. *Journal of Health Psychology, 15*(2), 299–309.

Cucchiaro, G., Adzick, S. N., Rose, J. B., Maxwell, L., & Watcha, M. (2006a). A comparison of epidural bupivacaine-fentanyl and bupivacaine-clonidine in children undergoing the Nuss procedure. *Anesthesia and Analgesia, 103*(2), 322–327. doi:103/2/322 [pii]10.1213/01.ane.0000221047.68114.ad, table of contents.

Cucchiaro, G., Farrar, J. T., Guite, J. W., & Li, Y. (2006b). What postoperative outcomes matter to pediatric patients? *Anesthesia and Analgesia, 102*(5), 1376–1382.

Dahl, J. B., & Kehlet, H. (2011). Preventive analgesia. *Current Opinion in Anaesthesiology, 24*(3), 331–338. doi:10.1097/ACO.0b013e328345afd9.

Duedahl, T. H., & Hansen, E. H. (2007). A qualitative systematic review of morphine treatment in children with postoperative pain. *Paediatric Anaesthesia, 17*(8), 756–774. doi:PAN2213 [pii]10.1111/j.1460-9592.2007.02213.x.

Ganesh, A., Rose, J. B., Wells, L., Ganley, T., Gurnaney, H., Maxwell, L. G., DiMaggio, T., Milovcich, K., Scollon, M., Feldman, J. M., & Cucchiaro, G. (2007). Continuous peripheral nerve blockade for inpatient and outpatient postoperative analgesia in children. *Anesthesia and Analgesia, 105*(5), 1234–1242. doi:105/5/1234 [pii]10.1213/01.ane.0000284670.17412.b6, table of contents.

Groenewald, C. B., Rabbitts, J. A., Schroeder, D. R., & Harrison, T. E. (2012). Prevalence of moderate-severe pain in hospitalized children. *Paediatric Anaesthesia.* doi:10.1111/j.1460-9592.2012.03807.x.

Himmelseher, S., & Durieux, M. E. (2005). Ketamine for perioperative pain management. *Anesthesiology, 102*(1), 211–220. doi:00000542-200501000-00030 [pii].

Hummel, P., Puchalski, M., Creech, S. D., & Weiss, M. G. (2008). Clinical reliability and validity of the N-PASS: Neonatal pain, agitation and sedation scale with prolonged pain. *Journal of Perinatology, 28*(1), 55–60. doi:7211861 [pii]10.1038/sj.jp. 7211861.

Jacobsen, E. B., Thastum, M., Jeppesen, J. H., & Pilegaard, H. K. (2010). Health-related quality of life in children and adolescents undergoing surgery for pectus excavatum. *European Journal of Pediatric Surgery, 20*(2), 85–91.

Kain, Z. N., Mayes, L. C., Caldwell-Andrews, A. A., Alexander, G. M., Krivutza, D., Teague, B. A., & Wang, S. M. (2002). Sleeping characteristics of children undergoing outpatient elective surgery. *Anesthesiology, 97*(5), 1093–1101. doi:00000542-200211000-00010 [pii].

Kain, Z. N., Mayes, L. C., Caldwell-Andrews, A. A., Karas, D. E., & McClain, B. C. (2006). Preoperative anxiety, postoperative pain, and behavioral recovery in young children undergoing surgery. *Pediatrics, 118*(2), 651–658.

Kehlet, H. (1997). Multimodal approach to control postoperative pathophysiology and rehabilitation. *British Journal of Anaesthesia, 78*(5), 606–617.

Kehlet, H., & Dahl, J. B. (1993). The value of "multimodal" or "balanced analgesia" in postoperative pain treatment. *Anesthesia and Analgesia, 77*(5), 1048–1056.

Kotzer, A. M. (2000). Factors predicting postoperative pain in children and adolescents following spine fusion. *Issues in Comprehensive Pediatric Nursing, 23*(2), 83–102.

Kraemer, F. W., & Rose, J. B. (2009). Pharmacologic management of acute pediatric pain. *Anesthesiology Clinics, 27*(2), 241–268. doi:S1932-2275(09)00042-1 [pii]10.1016/j.anclin.2009.07.002.

Lee, M., Silverman, S. M., Hansen, H., Patel, V. B., & Manchikanti, L. (2011). A comprehensive review of opioid-induced hyperalgesia. *Pain Physician, 14*(2), 145–161.

Llewellyn, N., & Moriarty, A. (2007). The national pediatric epidural audit. *Paediatric Anaesthesia, 17*(6), 520–533. doi:PAN2230 [pii]10.1111/j.1460-9592.2007.02230.x.

Lynn, A. M., Bradford, H., Kantor, E. D., Seng, K. Y., Salinger, D. H., Chen, J., Ellenbogen, R. G., Vicini, P., & Anderson, G. D. (2007). Postoperative ketorolac tromethamine use in infants aged 6–18 months: The effect on morphine usage, safety assessment, and stereo-specific pharmacokinetics. *Anesthesia and Analgesia, 104*(5), 1040–1051. doi:104/5/1040 [pii] 10.1213/01.ane.0000260320.60867.6c, tables of contents.

Macrae, W. A. (2008). Chronic post-surgical pain: 10 years on. *British Journal of Anaesthesia, 101*(1), 77–86. doi:aen099 [pii]10.1093/bja/aen099.

McGrath, P. J., Walco, G. A., Turk, D. C., Dworkin, R. H., Brown, M. T., Davidson, K., & Zeltzer, L. (2008). Core outcome domains and measures for pediatric acute and chronic/recurrent pain clinical trials: PedIMMPACT recommendations. *The Journal of Pain, 9*(9), 771–783. doi:S1526-5900(08)00550-6 [pii]10.1016/j.jpain.2008.04.007.

Morton, N. S., & Errera, A. (2010). APA national audit of pediatric opioid infusions. *Paediatric Anaesthesia, 20*(2), 119–125. doi:PAN3187 [pii]10.1111/j.1460-9592.2009.03187.x.

Page, M. G., Katz, J., Stinson, J., Isaac, L., Martin-Pichora, A. L., & Campbell, F. (2012). Validation of the numerical rating scale for pain intensity and unpleasantness in pediatric acute postoperative pain: Sensitivity to change over time. *The Journal of Pain, 13*(4), 359–369. doi:S1526-5900(11)00957-6 [pii] 10.1016/j.jpain.2011.12.010.

Palmer, G. M., Atkins, M., Anderson, B. J., Smith, K. R., Culnane, T. J., McNally, C. M., Perkins, E. J., Chalkiadis, G. A., & Hunt, R. W. (2008). I. V. acetaminophen pharmacokinetics in neonates after multiple doses. *British Journal of Anaesthesia, 101*(4), 523–530. doi:aen208 [pii]10.1093/bja/aen208.

Rusy, L. M., Hainsworth, K. R., Nelson, T. J., Czarnecki, M. L., Tassone, J. C., Thometz, J. G., & Weisman, S. J. (2010). Gabapentin use in pediatric spinal fusion patients: A randomized, double-blind, controlled trial. *Anesthesia and Analgesia, 110*(5), 1393–1398. doi:110/5/1393 [pii]10.1213/ANE.0b013e3181d41dc2.

Stevens, B., Johnston, C., Taddio, A., Gibbins, S., & Yamada, J. (2010). The premature infant pain profile: Evaluation 13 years after development. *The Clinical Journal of Pain, 26*(9), 813–830. doi:10.1097/AJP.0b013e3181ed1070.

Stinson, J. N., Kavanagh, T., Yamada, J., Gill, N., & Stevens, B. (2006). Systematic review of the psychometric properties, interpretability and feasibility of self-report pain intensity measures for use in clinical trials in children and adolescents. *Pain, 125*(1–2), 143–157. doi:S0304-3959(06)00257-0 [pii]10.1016/j.pain.2006.05.006.

von Baeyer, C. L., & Spagrud, L. J. (2007). Systematic review of observational (behavioral) measures of pain for children and adolescents aged 3 to 18 years. *Pain, 127*(1–2), 140–150. doi:S0304-3959(06)00419-2 [pii] 10.1016/j.pain.2006.08.014.

Wagner, D. S., Yap, J. M., Bradley, K. M., & Voepel-Lewis, T. (2007). Assessing parents preferences for the avoidance of undesirable anesthesia side effects in their children undergoing surgical procedures. *Paediatric Anaesthesia, 17*(11), 1035–1042.

Acute Pain in Children, Procedural

Lindsey Cohen, Laura A. Cousins and Sarah R. Martin
Psychology, Georgia State University, Atlanta, GA, USA

Synonyms

Acute procedural pain in children; Pediatric pharmacological interventions; Pediatric psychological interventions

Definition

Acute procedural pain refers to the relatively brief pain that infants and children experience as a result of necessary ▶ invasive preventative, diagnostic, and therapeutic medical procedures. Procedural pain management is the pharmacological, psychological, and physical interventions used to prevent, reduce, or eliminate the affective and sensory components of pain that accompany many acute medical procedures.

Characteristics

Acute procedural pain can be defined as the affective and sensory distress that accompanies a brief but invasive medical event. This includes routine (e.g., venous access, immunization injections) and specialized (e.g., bone marrow aspiration, urethral catheterization, burn debridement) procedures. Pain is not a necessary part of the medical intervention, and this aversive affective and sensory side effect is often untreated, especially in children (Finley et al. 2005). A growing body of literature suggests that acute procedural pain in children can have significant negative effects in the short- and long-term both for the pediatric patient as well as the family (e.g., Taddio 1999). These and other concerns (e.g., social and economic) likely contributed to the Joint Commission on Accreditation of Healthcare

Organizations directing that pain be considered the "fifth vital sign" (Phillips 2000) and the International Association of the Study of Pain and the World Health Organization stating that "the relief of pain should be a human right" (see Brennan and Cousins 2004).

The aims of acute procedural pain management are to minimize the affective and sensory distress that occurs prior to, during, and immediately following the procedure. In addition to these immediate aspects of the procedure, acute procedural pain management efforts might target long-term effects (e.g., attitudes, memories, adherence) and self-efficacy (e.g., enhance coping with stressors) (e.g., Cohen et al. 2008b).

Invasive Procedures in Children

Starting soon after birth, children face a number of routine and specialized acute medical procedures. The most common routine procedures are vaccinations with children undergoing several dozen immunizations in their first 2 years of life, with most being administered via intramuscular injection. Some developing countries, such as those in northern Africa, require 15 immunizations within the first 24 months of life; Europe requires 19 within that period; and the United States requires 25 (World Health Organization 2011). Children who are born prematurely, who have been diagnosed with a disease, or who experience accidents or injuries will undergo additional acute painful procedures. Stevens et al. (2004) estimate that neonates born between 27 and 31 weeks of age undergo an average of 134 painful procedures within the first 2 weeks of their lives. For the 10 % of newborns at the lowest birth weights, Stevens et al. found that they receive an average of more than 300 painful procedures within the same period of time. Preterm infants are also at risk for increased pain sensitivity (for a review, see Beggs and Fitzgerald 2007) and this continued exposure to painful stimuli may lead to functional changes in pain and stress processing systems in the brain (Grunau et al. 2005; Slater et al. 2010).

Factors that Affect Procedural Pain in Children

There is significant variability in children's reaction to acute procedures with some children sleeping or showing little reaction, with others vomiting, panicking, or fainting. Clearly, some of the variance can be explained by the nature of the procedures – venipuncture is arguably qualitatively and quantitatively less painful than burn debridement. Individual differences also account for some of the differences in procedural pain; for example, age, gender, prior experience, culture, and temperament have been linked to pain response (e.g., Bustos et al. 2008; Chen et al. 2000; Fillingim et al. 2009; Kim et al. 2004; Lamberg 1998; Miller and Newton 2006; Piira et al. 2007; Ranger and Campbell-Yeo 2008; Sweet et al. 1999). Environmental factors (e.g., child-friendly atmosphere; e.g., Zempsky et al. 2004), physical positioning (e.g., Lacey et al. 2008; Sparks et al. 2007; Stephens et al. 1999), and other aspects of the procedure (e.g., needle gauge) have been shown to influence children's procedural response. Adult behavior has received significant attention given that it explains approximately 25–50 % of the variance in child acute procedural distress (Cohen et al. 2002; Frank et al. 1995). Cohen et al. (2008b) outlined the Stimulus-Response Model of Pain as well as a modified version of the Social Ecological Model (Bronfenbrenner 1979) to detail some of the important contextual variables influencing children's acute procedural pain. Given the complexity in children's response to acute procedures, an evidence-based practice perspective – integrating the best available research evidence; patient characteristics, preferences, and values; and clinical expertise of the health-care staff – should provide an optimal framework for approaching acute procedural pain in children.

Assessment of Procedural Pain in Children

The assessment of acute pain is challenging given that pain is a personal and thus subjective experience. This is complicated in children with unsophisticated vocabularies for describing their internal experiences. The accurate assessment of children's pain is vital for diagnosis and guiding interventions. Fortunately, there are a number of well-validated and reliable self-report, adult proxy-report, and behavior observational tools to assess pediatric acute pain (for reviews, see Cohen et al. 2006, 2008a; von Baeyer and Spagrud 2007).

Pharmacological Interventions for Procedural Pain in Children
Routine Medical Procedures
Data indicate that pharmacological interventions can provide some relief for children's routine needle pain (for reviews, see Baxter and Cohen 2009; Schechter et al. 2007; Zempsky 2008). The most common pharmacological agents are topical anesthetics (e.g., prilocaine, lidocaine, tetracaine), which provide brief dermal anesthesia. Although the literature generally support the pain relieving qualities of these agents, predominately for venous access (e.g., Zempsky 2008), there are downsides including cost and time of onset (e.g., 20–60 min for analgesia). There is mixed support for the use of quick-onset of vapocoolant spray (Cohen et al. 2009; Schechter et al. 2007; Shah et al. 2009). Recent advances in healthcare have focused on decreasing the time of onset of activation via technological methods (e.g., carbon dioxide to transmit buffered lidocaine; Luhmann et al. 2004) and needle-free options (e.g., Zempsky et al. 2008).

Specialized Medical Procedures
Procedures that are more severe (e.g., bone marrow aspirations, burn debridement) will require a wider array of pharmacological approaches to manage children's pain. Some pharmacological agents for the affective and sensory pain accompanying specialized medical procedures include topical anesthetics, nerve blocks, opioids, ketamine, nitrous oxide, and anxiolytics (Finley 2001). For example, pain management for pediatric cancer lumbar punctures might include lidocaine in addition to deep sedation or general anesthesia (Gorchynski and McLaughlin 2011; Iannalfi et al. 2005). However, there are risks that should be considered with sedation, anesthesia, and other pharmacologic approaches to pediatric pain management for specialized procedures (Finley 2001).

Psychological Interventions for Procedural Pain in Children
Routine Medical Procedures
Psychological interventions for routine procedures include relaxation, deep breathing, praise of appropriate behavior, and distraction (Blount et al. 2009; Pillai Riddell et al. 2011; Powers 1999). The most widely documented psychological intervention for routine pediatric procedural pain is distraction. Distraction functions via diverting the child's attention away from negative sensory stimuli (e.g., needles) toward more positive and engaging stimuli (e.g., cartoon movies; Cohen 2008; DeMore and Cohen 2005). Data support the efficacy of distraction for pediatric immunization pain relief in infants (Cohen 2002; Cohen et al. 2006), preschoolers (Berberich and Landman 2009), and preadolescents (Cohen et al. 1999). Distraction stimuli include movies, nonprocedural talk, imagery, toys, and music, with some data suggesting that the most effective approaches engage the child in multiple modalities (DeMore and Cohen 2005). A systematic review of randomized controlled trials for pediatric immunizations concluded that both child-directed (e.g., cartoons, music) and nurse-led (e.g., trained to provide distraction with movies or toys) distraction effectively reduce pain and distress, the benefits of parent-led distraction (e.g., trained to distract with movies or toys) remain inconclusive (Chambers et al. 2009; Uman et al. 2010). Despite the effectiveness of many studies using multifaceted distraction techniques, it is challenging to determine which components of the intervention are contributing the strongest effects (Cohen 2010). Combined cognitive-behavioral interventions and breathing exercises have also demonstrated reductions in children's self-reported pain during immunization (Chambers et al. 2009). Data equally support the benefit of distraction for children undergoing venous access (e.g., MacLaren and Cohen 2005; Bellieni et al. 2006; Wang et al. 2008; Yoo et al. 2011). In addition to directly targeting the child, data suggest that providing distraction education for parents prior to venous access procedures is beneficial (Kleiber et al. 2001).

Specialized Medical Procedures

Given that pharmacological approaches do not adequately address all of the pain associated with specialized procedures, such as burn care and psychological approaches are necessary (e.g., de Jong et al. 2007; Hoffman et al. 2004; for a review, see Morris et al. 2009). Due to the severity of pain, more engaging interventions have been deemed most effective in decreasing pain during dressing changes and burn treatment, including multimodal distraction in the form of a handheld device, "augmented reality" (using overlays of virtual images onto the actual world), and immersive virtual reality (Miller et al. 2010; Mott et al. 2008; Schmitt et al. 2011; Das et al. 2005). Unfortunately, cost and portability are current barriers to widespread dissemination of these approaches (Hoffman et al. 2011). Psychological approaches for pediatric oncology procedures include hypnosis and imagery techniques (Liossi et al. 2009; Richardson et al. 2006) as well as virtual and standard distraction (Gershon et al. 2004; Windich-Biermeier et al. 2007) and music distraction (Nguyen et al. 2010).

Summary

Pain is an unfortunate and common side effect of routine and specialized acute procedures that can result in negative short- and long-term consequences. Regardless of health status, it is likely that children will undergo numerous potentially painful procedures. Given that children's acute medical pain is often predictable, practitioners can be prepared to minimize, assess, and treat this negative emotional and physical event. Fortunately, there are a range of validated assessment tools available that can be used to quantify the experience and guide intervention approaches. The extant literature not only provides data to support a range of pharmacological and psychological methods to provide children with pain relief, but also provides suggestions on how to effectively combine and tailor these techniques to particular procedures and populations.

References

Baxter, A., & Cohen, L. L. (2009). Pain management. In G. R. Strange, W. R. Ahrens, R. W. Schaftermeyer, & R. A. Wiebe (Eds.), *Pediatric emergency medicine* (3rd ed., pp. 155–163). New York: McGraw Hill.

Beggs, S., & Fitzgerald, M. (2007). Development of peripheral and spinal nociceptive systems. In K. J. S. Anand, B. J. Stevens, & P. J. McGrath (Eds.), *Pain in neonates and infants: Pain research and clinical management* (3rd ed., pp. 11–24). Toronto: Elsevier.

Bellieni, C. V., Cordelli, D. M., Raffaelli, M., Ricci, B., Morgese, G., & Buonocore, G. (2006). Analgesic effect of watching TV during venipuncture. *Archives of Disease in Childhood, 91*, 1015–1017.

Berberich, F. R., & Landman, Z. (2009). Reducing immunization discomfort in 4- to 6-year-old children: A randomized clinical trial. *Pediatrics, 124*, e203–e209.

Blount, R. L., Zempsky, W. T., Jaaniste, T., Evans, S., Cohen, L. L., Devine, K. A., & Zeltzer, L. K. (2009). Management of pain and distress due to medical procedures. In M. C. Roberts & R. Steele (Eds.), *Handbook of pediatric psychology* (4th ed., pp. 171–188). New York: Guilford Press.

Brennan, F., & Cousins, M. J. (2004). Pain relief as a human right. *Pain: Clinical Updates, 12*(5), 1–4.

Bronfenbrenner, U. (1979). *The ecology of human development*. Cambridge, MA: Harvard University Press.

Bustos, T., Jaaniste, T., Salmon, K., & Champion, G. D. (2008). Evaluation of a brief parent intervention teaching coping-promoting behavior for the infant immunization context: A randomized controlled trail. *Behavior Modification, 32*, 450–467.

Chambers, C. T., Taddio, A., Uman, L. S., & McMurty, C. M. (2009). Psychological interventions for reducing pain and distress during routine childhood immunizations: A systematic review. *Clinical Therapeutics, 31*, S77–S107.

Chen, E., Craske, M. G., Katz, E. R., Schwartz, E., & Zeltzer, L. K. (2000). Pain-sensitive temperament: Does it predict procedural distress and response to psychological treatment among children with cancer? *Journal of Pediatric Psychology, 24*, 269–278.

Cohen, L. L. (2002). Reducing infant immunization distress through distraction. *Health Psychology, 21*, 207–211.

Cohen, L. L. (2008). Behavioral approaches to anxiety and pain management for pediatric venous access. *Pediatrics, 122*, S134–S139.

Cohen, L. L. (2010). A multifaceted distraction intervention may reduce pain and discomfort in children 4–6 years of age receiving immunisation. *Evidence-Based Nursing, 13*, 15–16.

Cohen, L. L., Blount, R. L., Cohen, R. J., Schaen, E. R., & Zaff, J. (1999). Comparative study of distraction versus topical anesthesia for pediatric pain management during immunizations. *Health Psychology, 18*, 591–598.

Cohen, L. L., Bernard, R. S., Greco, L. R., & McClellan, C. B. (2002). A child-focused intervention for coping with procedural pain: Are parent and nurse coaches necessary? *Journal of Pediatric Psychology, 27,* 749–757.

Cohen, L. L., MacLaren, J. E., Fortson, B. L., Friedman, A., DeMore, M., Lim, C. S., Shelton, E., & Gangaram, B. (2006). Randomized clinical trial of distraction for infant immunization pain. *Pain, 125,* 165–171.

Cohen, L. L., Lemanek, K., Blount, R. L., Dahlquist, L. M., Lim, C. S., Palermo, T. M., McKenna, K. D., & Weiss, K. E. (2008a). Evidence-Based assessment of pediatric pain. *Journal of Pediatric Psychology, 33,* 939–955.

Cohen, L. L., MacLaren, J. E., & Lim, C. S. (2008b). Pain and pain management. In R. G. Steele, T. D. Elkin, & M. C. Roberts (Eds.), *Handbook of evidence based therapies for children and adolescents: Bridging science and practice* (pp. 283–296). New York: Springer.

Cohen, L. L., MacLaren, J. E., DeMore, M., Fortson, B., Friedman, A., Lim, C. L., & Gangaram, B. (2009). A randomized controlled trial of vapocoolant for pediatric immunization distress relief. *The Clinical Journal of Pain, 25,* 490–494.

Das, D. A., Grimmer, K. A., Sparnon, A. L., McRae, S. E., & Thomas, B. H. (2005). The efficacy of playing a virtual reality game in modulating pain for children with acute burn injuries: A randomized controlled trial. *BMC Pediatrics, 5,* 1–10.

de Jong, A. E. E., Middelkoop, E., Faber, A. W., & Van Loey, N. E. E. (2007). Non-pharmacological nursing interventions for procedural pain relief in adults with burns: A systematic literature review. *Burns, 33,* 811–827.

DeMore, M., & Cohen, L. L. (2005). Distraction for pediatric immunization pain: A critical review. *Journal of Clinical Psychology in Medical Settings, 12,* 281–291.

Fillingim, R. B., King, C. D., Ribeiro-Dasilva, M. C., Rahim-Williams, B., & Riley, J. L. (2009). Sex, gender, and pain: A review of recent clinical and experimental findings. *The Journal of Pain, 10,* 447–485.

Finley, G. A. (2001). Pharmacological management of procedure pain. In G. A. Finley & P. J. McGrath (Eds.), *Acute and procedure pain in infants and children: Progress in pain research and management* (Vol. 20, pp. 57–76). Seattle: IASP Press.

Finley, G. A., Franck, L. S., Grunau, R. E., & von Baeyer, C. L. (2005). Why children's pain matters. *Pain: Clinical Updates, 13*(5), 1–6.

Frank, N. C., Blount, R. L., Smith, A. J., Manimala, M. R., & Martin, J. K. (1995). Parent and staff behavior, previous child medical experience, and maternal anxiety as they relate to child distress and coping. *Journal of Pediatric Psychology, 20,* 277–289.

Gershon, J., Zimand, E., Pickering, M., Rothbaum, B. O., & Hodges, L. (2004). A pilot and feasibility study of virtual reality as a distraction for children with cancer. *Journal of the Academy of Child and Adolescent Psychiatry, 43,* 1243–1249.

Gorchynski, J., & McLaughlin, T. (2011). The routine utilization of procedural pain management for pediatric lumbar punctures: Are we there yet? *Journal of Clinical Medicine Research, 26,* 164–167.

Grunau, R. E., Holsti, L., Haley, D. W., et al. (2005). Pain exposure predicts dampened cortisol, behavior and cardiac stress reactivity in preterm infants in the NICU. *Pain, 113,* 293–300.

Hoffman, H. G., Patterson, D. R., Magula, J., Carrougher, G. J., Zeltzer, K., Dagadakis, S., et al. (2004). Water-friendly virtual reality pain control during wound care. *Journal of Clinical Psychology, 60,* 189–195.

Hoffman, H. G., Chambers, G. T., Meyer, W. J., Arceneaux, L. L., Russell, W. J., Seibel, E. J., et al. (2011). Virtual reality as an adjunctive non-pharmacologic analgesic for acute burn pain during medical procedures. *Annals of Behavioral Medicine, 41,* 183–191.

Iannalfi, A., Bernini, G., Caprilli, S., Lippi, A., Tucci, F., & Messeri, A. (2005). Painful procedures in children with cancer: Comparison of moderate sedation and general anesthesia for lumbar puncture and bone marrow aspiration. *Pediatric Blood & Cancer, 45,* 933–938.

Kim, H., Neubert, J. K., San Miguel, A., Xu, K., Krishnaranju, R. K., Iadarola, M. J., Goldman, D., & Dionne, R. A. (2004). Genetic influence on variability in human acute experimental pain sensitivity associated with gender, ethnicity and psychological temperament. *Pain, 109,* 488 496.

Kleiber, C., Craft-Rosenberg, M., & Harper, D. C. (2001). Parents as distraction coaches during IV insertion: A randomized study. *Journal of Pain and Symptom Management, 22,* 851–861.

Lacey, C. M., Finkelstein, M., & THygeson, M. V. (2008). The impact of positioning on fear during immunizations: Supine versus sitting up. *Journal of Pediatric Nursing, 23,* 195–200.

Lamberg, L. (1998). Venus orbits closer to pain than Mars. Rx for one sex may not benefit the other. *Journal of the American Medical Association, 280,* 120–128.

Liossi, C., White, P., & Hatira, P. (2009). A randomized clinical trial of a brief hypnosis intervention to control venepuncture-related pain of paediatric cancer patients. *Pain, 142,* 255–263.

Luhmann, J. D., Hurt, S., Schootman, M., & Kennedy, R. M. (2004). A comparison of buffered lidocaine versus Ela-Max® before peripheral intravenous catheter IV insertions in children. *Pediatrics, 113,* e217–e220.

MacLaren, J. E., & Cohen, L. L. (2005). A comparison of distraction strategies for venipuncture distress in children. *Journal of Pediatric Psychology, 30,* 387–396.

Miller, C., & Newton, S. E. (2006). Pain perception and expression: The influence of gender, personal self-efficacy, and lifespan socialization. *Pain Management Nursing, 7,* 148–152.

Miller, K., Rodger, S., Bucolo, S., Greer, R., & Kimble, R. M. (2010). Multi-modal distraction. Using technology to combat pain in young children with burn injuries. *Burns, 36,* 647–658.

Morris, L. D., Louw, Q. A., & Grimmer-Somers, K. (2009). The effectiveness of virtual reality on reducing pain and anxiety in burn injury patients: A systematic review. *The Clinical Journal of Pain, 25,* 815–826.

Mott, J., Bucolo, S., Cuttle, L., Mill, J., Hilder, M., Miller, K., et al. (2008). The efficacy of an augmented virtual reality system to alleviate pain in children undergoing burns dressing changes: A randomised controlled trial. *Burns, 34,* 803–808.

Nguyen, T. N., Nilsson, S., Hellström, A. L., & Bengtson, A. (2010). Music therapy to reduce pain and anxiety in children with cancer undergoing lumbar puncture: A randomized clinical trial. *Journal of Pediatric Oncology Nursing, 27,* 146–155.

Phillips, D. M. (2000). JCAHO pain management standards are unveiled. *JAMA: The Journal of the American Medical Association, 284,* 428–429.

Piira, T., Champion, G. D., Bustos, T., Donelly, N., & Lui, K. (2007). Factors associated with infant pain response following an immunization injection. *Early Human Development, 83,* 319–326.

Pillai Riddell, R., Racine, N., Turcotte, K., Uman, L. S., Horton, R., Din Osmun, L., et al. (2011). Nonpharmacological management of procedural pain in infants and young children: An abridged cochrane review. *Pain Research & Management, 16,* 321–330.

Powers, S. W. (1999). Empirically supported treatments in pediatric psychology: Procedure-related pain. *Journal of Pediatric Psychology, 24,* 131–145.

Ranger, M., & Campbell-Yeo, M. (2008). Temperament and pain response: A review of the literature. *Pain Management Nursing, 9,* 2–9.

Richardson, J., Smith, J. E., McCall, G., & Pilkington, K. (2006). Hypnosis for procedure-related pain and distress in pediatric cancer patients: A systematic review of effectiveness and methodology related to hypnosis interventions. *Journal of Pain and Symptom Management, 31,* 70–84.

Schechter, N. L., Zempsky, W. T., Cohen, L. L., McGrath, P. J., Meghan McMurtry, C., & Bright, N. S. (2007). Pain reduction during pediatric immunizations: Evidence-based review and recommendations. *Pediatrics, 119,* e1184–e1198.

Schmitt, Y. S., Hoffman, H. G., Blough, D. K., Patterson, D. R., Jensen, M. P., Soltani, M., et al. (2011). A randomized, controlled trial of immersive virtual reality analgesia, during physical therapy for pediatric burns. *Burns, 37,* 61–68.

Shah, V., Taddio, A., & Rieder, M. J. (2009). Effectiveness and tolerability of pharmacological and combined interventions for reducing injection pain during routine childhood immunizations; Systematic review and meta-analyses. *Clinical Therapeutics, 31,* s104–s151.

Slater, R., Fabrizi, L., Worley, A., Meek, J., Boyd, S., & Fitzgerald, M. (2010). Premature infants display increased noxious-evoked neuronal activity in the brain compared to health age-matched term-born infants. *NeuroImage, 52*(2), 583–589.

Sparks, L. A., Setlik, J., & Luhman, J. (2007). Parental holding and positioning to decrease IV distress in young children: A randomized controlled trial. *Journal of Pediatric Nursing, 22,* 440–447.

Stephens, B. K., Barkey, M. E., & Hall, H. R. (1999). Techniques to comfort children during stressful procedures. *Accident and Emergency Nursing, 7*(4), 226–236.

Stevens, B., Yamada, J., & Ohlsson, A. (2004). Sucrose for analgesia in newborn infants undergoing painful procedures. *Cochrane Database System Revision,* (3), CD001069.

Sweet, S. D., McGrath, P. J., & Symons, D. (1999). The roles of child reactivity and parenting context in infant pain response. *Pain, 80,* 655–661.

Taddio, A. (1999). Effects of early pain experience: The human literature. In P. J. McGrath & G. A. Finley (Eds.), *Chronic and recurrent pain in children and adolescents: Progress in pain research and management* (Vol. 13, pp. 57–74). Seattle: IASP Press.

Uman, L. S., Chambers, C. T., McGrath, P. J., Kisely, S., Matthews, D., & Hayton, K. (2010). Assessing the quality of randomized controlled trials examining psychological interventions for pediatric procedural pain: Recommendations for quality improvement. *Journal of Pediatric Psychology, 35,* 693–703.

von Baeyer, C. L., & Spagrud, L. J. (2007). Systematic review of observational (behavioral) measures of pain for children and adolescents aged 3 to 18 years. *Pain, 127,* 140–150.

Wang, Z., Sun, L., & Chen, A. (2008). The efficacy of non-pharmacological methods of pain management in school age children receiving venepuncture in a paediatric department: A randomized controlled trial of audiovisual distraction and routine psychological intervention. *Swiss Medical Weekly, 138,* 579–584.

Windich-Biermeier, A., Sjoberg, I., Conkin Dale, J., Eshelman, D., & Guzzetta, C. E. (2007). Effects of distraction on pain, fear, and distress during venous port access and venipuncture in children and adolescents with cancer. *Journal of Pediatric Oncology Nursing, 24,* 8–19.

World Health Organization. (2011). Data, statistics and graphics by subject. Geneva, Switzerland. Available at: http://www.who.int/immunization_monitoring/data/data_subject/en/index.html. Accessed 25 Apr 2011.

Yoo, H., Kim, S., Hur, H., & Kim, H. (2011). The effects of an animation distraction intervention on pain response of preschool children during venipuncture. *Applied Nursing Research, 24,* 94–100.

Zempsky, W. T. (2008). Pharmacological approaches for reducing venous access pain in children. *Pediatrics, 122,* 140–153.

Zempsky, W. T., Cravero, J. P., & The Committee on Pediatric Emergency Medicine and Section on

Anesthesiology and Pain Medicine. (2004). Relief of pain and anxiety in pediatric patients in emergency medical systems. *Pediatrics, 114*, 1348–1356.

Zempsky, W. T., Bean-Lijewski, J., Kauffman, R. E., Koh, J. L., Malviva, S. V., Rose, J. B., Richards, P. T., & Gennevois, D. J. (2008). Needle-free powder lidocaine delivery system provides rapid effective analgesia for venipuncture or cannulation pain in children: Randomized, double-blind comparison of venipuncture and venous cannulation pain after fast-onset needle free powder lidocain or placebo treatment trial. *Pediatrics, 121*, 979–987.

Acute Pain Management in Infants

Rebecca Pillai Riddell[1], Nicole Racine[2] and Bonnie J. Stevens[3]
[1]Department of Psychology, York University, Toronto, ON, Canada
[2]Psychology, York University, Toronto, ON, Canada
[3]University of Toronto and The Hospital for Sick Children, Toronto, ON, Canada

Synonyms

Infant pain reduction/therapy/treatment; Infant pain therapy; Infant pain treatment; Pain management in infants

Definition

Infant pain management is defined as any strategy or technique administered to an infant experiencing pain with the intention to lessen pain sensation and/or perception. Pain management strategies include drugs and various nonpharmacological strategies (contextual, cognitive, and behavioral) interventions described in this entry. Pain management during infancy has been almost exclusively focused on acute procedural (including postoperative) pain, thus the emphasis throughout this entry will be on pain reduction strategies for procedural pain. A recent in-depth meta-analysis was conducted by examining 13 different nonpharmacological strategies for

infant pain management, which will serve as a key reference for this entry (Pillai Riddell et al. 2011).

Characteristics

Developmental Considerations

A sensitive appreciation of the infant in pain and their complete reliance on their caregiver is a fundamental starting point for approaching infant pain management (Als et al. 1994; Pillai Riddell and Racine 2009). Aside from the caregiver context, infants are also different from other stages due to: (a) greater sensitivity to noxious stimuli due to immature nervous system pathways, (b) immature cognitive ability to comprehend the purpose or predict the end of a painful procedure, (c) limited developmental motor competency to manage their pain, and (d) minimal communication abilities to alert a caregiver who can alleviate their pain. Complicating this issue is that even within infancy, there is a steep development trajectory that impacts the effectiveness and/or appropriateness of certain nonpharmacological interventions.

A recent Cochrane Review of nonpharmacological interventions for infant procedural pain purported that different intervention strategies may be differentially efficacious for infants of different ages (Pillai Riddell et al. 2011). From an age vantage point, the data suggested potential developmental trends of different strategies across the infant/early child stage. For example, while sufficient evidence showed that kangaroo care was one of the most efficacious strategies for preterm infants, limited evidence suggested this was not the case for neonates, and the impact on infants older than 1 month remains unknown. The age of the infant should be considered when selecting the most appropriate intervention for pain management.

An Integrated Approach to Acute Pain Management

Pain management during infancy should be multifaceted and integrated within every step of the decision-making process, from deciding whether a particular procedure is warranted to determining the least painful methods of undertaking the procedure to selecting the best pain management strategy. An informed understanding of drug therapies (e.g., topical anesthetics, sucrose) and nonpharmacological strategies are a crucial facet of integrated pain management.

Limit Exposure to Pain-Inducing Procedures
Often the routine care of an ill infant necessarily includes the infliction of pain for diagnostic or therapeutic purposes. However, recent guidelines recommend that health care providers attempt to limit the number of painful procedures performed on infants (Joint Fetus and Newborn Committee of the Canadian Paediatric Society and American Academy of Pediatrics 2000). The number and frequency of painful procedures, particularly those often repeated during an infant's hospitalization (e.g., heel lance), should be carefully considered within the developmental stage and health status of the infant. Before subjecting an infant to a painful procedure, caregivers should determine whether the procedure is warranted in relation to the potential benefit to the child's health status. Unnecessary procedures should be avoided and alternative nonpainful or less painful options should always be explored.

Select the Least Painful Diagnostic or Therapeutic Method
If a painful procedure is unavoidable, conducting the procedure in the least painful manner should be considered.

For example, venipuncture is recommended as less painful than heel lance for blood sampling in newborns (Shah and Ohlsson 2011). Another example may be found in the circumcision context. In addition to dorsal penile nerve blocks, the specific clamp used to hold the foreskin (i.e., Mogen) can moderate pain and distress by shortening procedure time (Brady-Fryer et al. 2009). For immunization, using the anterolateral thigh as the injection site, performing intramuscular injections rapidly without prior aspiration, using the least painful injectate formulation, and injecting the most painful vaccine last when

multiple vaccines are administered have all been shown to result in a less painful experience for infants (Taddio et al. 2010).

Nonpharmacological Strategies for Management of Infant Acute Pain
In addition to selecting the least distressing procedure possible, it is crucial to try to moderate the pain experience through direct pain management strategies – both pharmacological and nonpharmacological. The following is a review of key nonpharmacological strategies.

Contextual Strategies to Manage Infant Pain The context in which a painful procedure is conducted modifies behavioral and physiological aspects of infant pain. Context can refer to (a) the personal context of the infant, specifically that pain responses of infants are significantly increased with a history of numerous painful procedures and (b) the environmental context, most often the presence of stressful elements such as significant handling, unpredictable noises, multiple caregivers, and bright lights. Limiting and moderating painful experiences for the individual infant in the present may help prevent maladaptive pain experiences in the future (Taddio et al. 2010).

In terms of environmental context, evidence from a recent review (Pillai Riddell et al. 2011) suggests environmental modification techniques (i.e., low noise and lighting, clustering procedures to avoid overhandling, exposure to maternal voice) for preterms have equivocal efficacy, while exposing a preterm or neonate to calming and familiar smells may be efficacious but more methodologically rigorous trials must be conducted. With older infants, simply allowing the parent to be present and very basic parental coaching strategies have not been found to be efficacious in reducing infant pain. More developmentally informed parent management research is needed. In a separate review, the efficacy of music for pain relief in infants was presented as equivocal (Cepeda et al. 2010).

Cognitive Strategies Despite extensive evidence of the value of inhibitory mechanisms in pain control with older children and adults, researchers have only begun to consider the inhibitory cognitive capabilities of the infant in relation to pain (e.g., distraction). Any intervention that is suspected to have a mechanism of action that impacts an infant's abilities to perceive the pain experience is considered a cognitive strategy. The main intervention falling under this category is distraction. Two types of distractions were reviewed (Pillai Riddell et al. 2011) for older infants with toy-mediated distraction showing no efficacy and video-mediated distraction suggesting promise as a pain management technique for older infants.

Behavioral Strategies Most of the interventional pain research on infants has been conducted within this domain. These strategies involve either direct (e.g., rocking) or indirect (e.g., nonnutritive sucking) manipulation of the infant's body by a caregiver. Common strategies involve nonnutritive sucking (NNS, e.g., pacifiers), skin-to-skin contact (e.g., kangaroo care), swaddling or facilitative tucking, rocking/holding, touch/massage, or some combination of the above.

While the synergistic effect of sucrose and NNS has been demonstrated in a review on sucrose (Stevens et al. 2010), research has also simply investigated the nonpharmacological impact of nonnutritive sucking. Pillai Riddell et al. (2011) demonstrated that there is sufficient evidence to support the use of nonnutritive sucking for pain behaviors in preterm infants and neonates. Another behavioral strategy sharing the sucking mechanism would be breastfeeding. While the research on breast milk alone is equivocal, a recent review on the topic (Shah et al. 2009) concluded that, for term and pre-term neonates, breastfeeding is efficacious for reducing procedural pain.

In the Pillai Riddell et al. (2011) review, skin-to-skin contact (also known as kangaroo care) was noted to have received substantial research attention with preterm infants and was deemed efficacious in reducing pain reactivity and improving pain regulation. Another group of pain management strategies that were reviewed

also related to the positioning or containing of the infant during painful procedures. Swaddling and facilitative tucking appear to have sufficient evidence to support efficacy on their own for preterm infants, and emerging evidence suggests preliminary efficacy in neonates.

The effects of rocking, holding, or both were also reviewed for their effect on infant pain-related distress. Rocking is efficacious in improving the regulation of pain-related distress in neonates, but not the initial pain reactivity immediately after the painful stimuli for neonates or older infants. There is also evidence to suggest that simulated rocking and water are inefficacious in reducing pain for preterm infants. Other physical techniques, such as touch and massage, which provide counter-stimulation to the nociceptive input, have shown preliminary evidence suggesting inefficacy across infant stages (i.e. preterm born infants, neonates, and older infants).

Summary

Understanding that unrelieved pain during infancy can alter an individual's pain sensation and perception underscores the importance of infant caregivers' responsibility for being cognizant of the vast array of empirically supported nonpharmacological strategies available to appropriately manage infant pain. Infant age should be taken into consideration when selecting strategies to reduce pain.

References

Als, H., Lawhon, G., Duffy, F. H., et al. (1994). Individualized developmental care for the very low-birth-weight preterm infant. Medical and neurofunctional effects. *JAMA: The Journal of the American Medical Association, 272*, 853–858.

Brady-Fryer, B., Wiebe, N., & Lander, J. A. (2009). Pain relief for neonatal circumcision. *Cochrane Database of Systematic Reviews*, (1). Art. No.: CD004217. doi:10.1002/14651858.CD004217.pub2.

Cepeda, M. S., Carr, D. B., Lau, J., & Alvarez, H. (2010). Music for pain relief. *Cochrane Database of Systematic Reviews*, (8). Art. No.: CD004843. doi:10.1002/14651858.CD004843.pub2.

Joint Fetus and Newborn Committee of the Canadian Paediatric Society and American Academy of Pediatrics. (2000). Prevention and management of pain and stress in the neonate. *Pediatrics, 105*, 454–461.

Pillai Riddell, R., & Racine, N. (2009). Assessing pain in infancy: The caregiver context. *Pain Research & Management, 14*, 27–32.

Pillai Riddell, R. R., Racine, N. M., Turcotte, K., Uman, L. S., Horton, R. E., Din Osmun, L., Ahola Kohut, S., Hillgrove Stuart, J., Stevens, B., & Gerwitz-Stern, A. (2011). Non-pharmacological management of infant and young child procedural pain. *Cochrane Database of Systematic Reviews*, (10). Art. No.: CD006275. doi:10.1002/14651858.CD006275.pub2.

Shah, V. S., & Ohlsson, A. (2011). Venepuncture versus heel lance for blood sampling in term neonates. *Cochrane Database of Systematic Reviews*, (10). Art. No.: CD001452. doi:10.1002/14651858.CD001452.pub4.

Shah, P. S., Aliwalas, L. L., & Shah, V. S. (2009). Breastfeeding or breast milk for procedural pain in neonates. *Cochrane Database of Systematic Reviews*, (1). Art. No.: CD004950. doi:10.1002/14651858.CD004950.pub2.

Stevens, B., Yamada, J., & Ohlsson, A. (2010). Sucrose for analgesia in newborn infants undergoing painful procedures. *Cochrane Database of Systematic Reviews*, (1). Art. No.: CD001069. doi:10.1002/14651858.CD001069.pub3.

Taddio, A., Appleton, M., Bortolussi, R., Chambers, C., Dubey, V., Halperin, S., et al. (2010). Reducing the pain of childhood vaccination: an evidence-based clinical practice guidelines (summary). *Canadian Medical Association Journal, 182*(18), 1–7.

Acute Pain Mechanisms

Paul M. Murphy
Department of Anaesthesia and Pain Management, Royal North Shore Hospital, St. Leonards, NSW, Australia

Definition

Acute pain is defined as "pain of recent onset and probable limited duration. It usually has an identifiable temporal and causal relationship to injury or disease" (Ready and Edwards 1992). The perception of acute pain requires transduction of noxious mechanical, thermal, or chemical stimuli by nociceptive neurons, integration, and modulation at the level of the spinal cord and ultimately transmission to cortical centers.

Characteristics

Peripheral Nociception

▶ Nociceptors in the skin and other deeper somatic tissues such as periosteum are morphologically free nerve endings or simple receptor structures. A ▶ noxious stimulus activates the nociceptor depolarizing the membrane *via* a variety of stimulus–specific transduction mechanisms. C polymodal nociceptors are the most numerous of somatic nociceptors and respond to a full range of mechanical, chemical, and thermal noxious stimuli. Polymodal nociceptors are coupled to unmyelinated C fibers. Electrophysiological activity in these slow conduction C fibers is characteristically perceived as dull, burning pain. Faster conducting Aδ fibers are coupled to more selective thermal and mechanothermal receptors considered responsible for the perception of sharp or "stabbing" pain (Julius and Basbaum 2001).

Inflammatory–Induced Peripheral Sensitization

A complex interaction of molecules produced during the inflammation acting on nociceptors results in functional, morphological, and electrophysiological changes causing "primary hyperalgesia." Nociceptors are sensitized due to changes in the absolute numbers of Na^+ and K^+ channels and their relative "open-closed" kinetics. This results in neuronal activation in response to innocuous stimuli and spontaneous ectopic discharges. Inflammatory mediators also act to increase the activity of "silent" nociceptors normally unresponsive to even noxious stimuli. There is an increase in many ion channel subtypes, (particularly the ▶ tetrodotoxin (TTX)-resistant Na^+ channel) both on the axon and also in the dorsal root ganglion (DRG) (Kidd and Urban 2001). There is upregulation of receptor expression, including substance P and brain–derived growth factor (BDGF). Morphological changes including sprouting of unmyelinated nerve fibers have also been identified.

Spinal Cord Integration

The majority of somatic nociceptive neurons enter the dorsal horn spinal cord at their segmental level. A proportion of fibers pass either rostrally or caudally in ▶ Lissauer's tract. Somatic primary afferent fibers terminate predominantly in laminas I (marginal zone) and II (substantia gelatinosa) of the dorsal horn where they synapse with projection neurons and excitatory/inhibitory interneurons. Some Aδ fibres penetrate more deeply into lamina V. Projection neurons are of three types classified as nociceptive specific (NS), low threshold (LT), and wide dynamic range (WDR) neurons. The NS neurons are located predominantly in lamina I and respond exclusively to noxious stimuli. They are characterized by a small receptive field. LT neurons, which are located in laminae III and IV, respond to innocuous stimuli only. WDR neurons predominate in lamina V (also in I), display a large receptive field, and receive input from wide range of sensory afferents (C, Aβ) (Parent 1996).

Spinal Modulation and Central Sensitization

Glutamate and aspartate are the primary neurotransmitters involved in spinal excitatory transmission. Fast post-synaptic potentials generated via the action of glutamate on AMPA receptors are primarily involved in nociceptive transmission (Smullen et al. 1990). Prolonged C-fiber activation facilitates glutamate-mediated activation of ▶ NMDA receptors and subsequent prolonged depolarization of the WDR neuron (termed "▶ wind-up"). This is associated with removal of a Mg^+ plug from the NMDA-gated ion channel. The activation of this voltage-gated Ca^+ channel is associated with an increase in intracellular Ca^+ and upregulated neurotransmission (McBain and Mayer 1994). The peptidergic neurotransmitters substance P and calcitonin G-related peptide (CGRP) are co-produced in glutaminergic neurons and released with afferent stimulation. These transmitters appear to play a neuromodulatory role, facilitating the action of excitatory amino acids. A number of other molecules including glycine, GABA, somatostatin, endogenous opioids, and ▶ endocannabinoids play modulatory roles in spinal nociceptive transmission (Fürst 1999).

Projection Pathways

Nociceptive somatic input is relayed to higher cerebral centers via three main ascending pathways, the spinothalamic, spinoreticular, and spinomesencephalic tracts (Basbaum and Jessel 2000). The spinothalamic path originates in laminae I and V–VII and is composed of NS and WDR neuron axons. It projects to thalamus via lateral (▶ neospinothalamic tract), and medial or ▶ paleospinothalamic tracts. The lateral tract passes to the ventro-postero-medial nucleus and subserves discriminative components of pain, while the medial tract is responsible for the autonomic and emotional components of pain. Additional fibers pass to reticular activating system, where they are associated with the arousal response to pain and the periaqueductal gray matter (PAG) where ascending inputs interact with descending modulatory fibers. The spinoreticular pathway originates in laminae VII and VIII and terminates on the medial medullary reticular formation. The spinomesencephalic tract originates in laminae I and V and terminates in the superior colliculus. Additional projections pass to the mesencephalic PAG. It appears that this pathway is not essential for pain perception but plays an important role in the modulation of afferent inputs.

Cortical Representation

Multiple cortical areas are activated by nociceptive afferent input including the primary and secondary somatosensory cortex, the insula, the anterior cingulate cortex and the prefrontal cortex. Pain is a multidimensional experience with sensory-discriminative and affective-motivational components. Advances in functional brain imaging have allowed further understanding of the putative role of cortical structures in the pain experience (Treede et al. 1999).

Localization
1. Primary somatosensory cortex
2. Secondary somatosensory cortex
3. Insula
   Intensity
1. Prefrontal cortex
2. Right posterior cingulate cortex
3. Brain stem
4. Periventricular gray matter
   Affective Component
1. Left anterior cingulate cortex
   Threshold
1. Cingulate cortex
2. Left thalamus
3. Frontal inferior cortex

References

Basbaum, A. I., & Jessel, T. M. (2000). The perception of
   pain. In E. R. Kandel, J. H. Schwartz, & T. M. Jessell
   (Eds.), *Principles of neural science* (pp. 472–91). New
   York: McGraw-Hill.
Fürst, S. (1999). Transmitters involved in antinociception
   in the spinal cord. *Brain Res Bull, 48,* 129–41.
Julius, D., & Basbaum, A. I. (2001). Molecular mecha-
   nisms of nociception. *Nature, 413,* 203–7.
Kidd, B. L., & Urban, L. (2001). Mechanisms of inflam-
   matory pain. *Br J Anaesth, 87,* 3–11.
McBain, C. J., & Mayer, M. L. (1994). N-methyl-D-
   aspartic acid receptor structure and function. *Physiol
   Rev, 74,* 723–60.
Parent, A. (1996). *Carpenter's human neuroanatomy.*
   Baltimore: Williams & Wilkins.
Ready, L. B., & Edwards, W. T. (1992). *Management of
   acute pain: A practical guide. Taskforce on acute pain.*
   Seattle: IASP Publications.
Smullen, D. H., Skilling, S. R., & Larson, A. A. (1990).
   Interactions between substance P, calcitonin G related
   peptide taurine and excitatory amino acids in the spinal
   cord. *Pain, 42,* 93–101.
Treede, R.-D., Kenshalo, D. R., Gracely, R. H., et al. (1999).
   The cortical representation of pain. *Pain, 79,* 105–11.

Acute Pain Service

Synonyms

APS

Definition

Poor perioperative pain management is remedied not so much in the development of new techniques but by the development of Acute Pain Services (APS) to exploit existing expertise. APSs have been established in many countries. Most are headed up by anesthesiologists. An APS consists of anesthesiologist-supervised pain nurses and an ongoing educational program for patients and all health personnel involved in the care of surgical patients. The benefits of an APS include increased patient satisfaction and improved outcome after surgery. It raises the standards of pain management throughout the hospital. Optimal use of basic pharmacological analgesia improves the relief of postoperative pain for most surgical patients. More advanced approaches, such as well-tailored epidural analgesia, are used to relieve severe dynamic pain (e.g., when coughing). This may markedly reduce risks of complications in patients at high risk of developing postoperative respiratory infections and cardiac ischemic events. Chronic pain is common after surgery. Better acute pain relief offered by an APS may reduce this distressing long-term complication of surgery.

Cross-References

▶ Multimodal Analgesia in Postoperative Pain

Acute Pain Team

Synonyms

APT

Definition

A team of nurse(s) and doctors (usually anesthesiologist(s)) that specializes in preventing and treating acute pain after surgery, trauma, due to medical conditions, and in some hospitals also labor pain.

Cross-References

▶ Postoperative Pain, Acute Pain Management Principles
▶ Postoperative Pain, Acute Pain Team

Acute Pain Therapy

Definition

Acute pain therapy is the therapy of acute pain and best defined as "pain of recent onset and probable limited duration. It usually has an identifiable temporal and causal relationship to injury or disease" (Ready and Edwards 1992).

The most common manifestation of acute pain in the medical setting is postoperative pain.

Acute pain therapy is aimed at rapid pain relief and increasingly performed by acute pain services (Macintyre and Scott 2009). The management is guided by guidelines published in a number of countries (Macintyre et al. 2010; Practice guidelines for acute pain management in the perioperative setting: An updated report by the American Society of Anesthesiologists Task Force on Acute Pain Management 2012).

References

Macintyre, P. E., & Scott, D. A. (2009). Acute pain management and acute pain services. In M. J. Cousins, D. B. Carr, T. T. Horlocker, & P. O. Bridenbaugh (Eds.), *Neural blockade in clinical anesthesia and pain medicine*. Philadelphia: Wolters Kluwer, Lippincott, Williams and Wilkins.
Macintyre, P. E., Schug, S. A., Scott, D. A., Visser, E. J., & Walker, S. M. (2010). *Acute pain management: Scientific evidence*. Melbourne: ANZCA&FPM.
Practice guidelines for acute pain management in the perioperative setting: An updated report by the American Society of Anesthesiologists Task Force on Acute Pain Management. (2012). [Practice Guideline].

Anesthesiology, 116(2), 248–273. doi: 10.1097/ALN.0b013e31823c1030
Ready, L. B., & Edwards, W. T. (1992). *Management of acute pain: A practical guide. Taskforce on acute pain*. Seattle: IASP Publications.

Acute Pain, Subacute Pain, and Chronic Pain

Wade King
Department of Clinical Research, Royal Newcastle Centre University of Newcastle, Newcastle, NSW, Australia

Synonyms

Pain of recent origin; Persisting pain

Definition

Acute pain is pain that has been present for less than 3 months (Merskey 1979; Merskey and Bogduk 1994). Chronic pain is pain that has been present for more than 3 months (Merskey 1979; Merskey and Bogduk 1994). Subacute pain is a subset of acute pain: It is pain that has been present for at least 6 weeks but less than 3 months (van Tulder et al. 1997).

Characteristics

Acute pain, **subacute** pain, and **chronic** pain are defined by units of time, but the concepts on which they are based are more fundamentally related to causation and prognosis. This entry discusses what the definitions imply and the clinical significance of classifying pain into these categories.

Acute pain was first defined by Bonica, in his textbook published in 1953, as "a complex constellation of unpleasant sensory, perceptual and emotional experiences and certain associated autonomic, physiologic, emotional and behavioural responses" (Bonica 1953). Bonica went on to say "invariably, acute pain and these associated responses are provoked by ... injury and/or disease ... or abnormal function." Thus, acute pain was originally defined as a combination of biological and psychological phenomena resulting from bodily impairment. Pain was recognized as playing the important physiological role of making an individual aware of impairment so that they can respond appropriately. Responses include withdrawal from the stimulus causing the pain, to avoid further impairment, and behaviors that minimize the impact of the impairment and facilitate recovery. For example, if a person suffers from a fracture, the resultant pain warns them so that they can limit activities that further deform the injured part. In this way, acute pain is fundamentally associated with the early stage of a condition and with the healing process. It can be expected to last for as long as the healing process takes to restore the impaired tissue.

Chronic pain was originally defined by Bonica as "pain that persists a month beyond the usual course of an acute disease or ... time for an injury to heal or that is associated with a chronic pathologic process." The implication was that if pain persists beyond the time in which an impaired tissue usually heals, the condition involves more than a simple insult to the tissue. Possible biological explanations for persistent pain would be (1) that the original insult caused damage beyond the capacity of the natural healing process to repair, (2) that the insult was repetitive, causing a series of impairments with sequential healing times, (3) that the condition involved a chronic pathological process which continued to impair tissue over a long period, or (4) that the biological impairment was prolonged by the application of inappropriate interventions. Possible psychological explanations for persistent pain invoke endogenous factors such as cognitions and behaviors that inhibit recovery. Recognition of these endogenous factors led Engel to develop the biopsychosocial model of chronic pain (Engel 1977), recognizing the interplay of prognostic risk factors that influence the duration of pain.

The time factor ascribed by Bonica, 1 month longer than the usual time of recovery, would vary from condition to condition. In order to standardize the definitions of acute and chronic pain, attempts were made to ascribe finite durations to them. In 1974, Sternbach (1974) suggested 6 months as an arbitrary limit, so pain present for up to 6 months would be classed as acute and that present for more than 6 months would be deemed chronic. Others felt 6 months was too long, and discussion ensued. The International Association for the Study of Pain (I.A.S.P.) formed a committee chaired by Merskey to consider such issues and it determined, in 1979 in a publication defining pain terms, that "three months is the most convenient point of division..." (Merskey 1979).

Thus, we have the current definitions of acute and chronic pain as pain present for less than, and more than, 3 months. The 3-month period is arbitrary, but it operationalizes the definitions so that pains can be classified readily and systematically as acute or chronic.

The definition of subacute pain has not been addressed so deliberately. The term "subacute" evolved to describe longer-lasting acute pain and has been quoted in the literature (van Tulder et al. 1997; Hippocrates. Of the epidemics) as pain present for between 6 weeks and 3 months. As such, it forms a subset of acute pain. The main division between acute and chronic pain remains at 3 months.

The pragmatism of the time-based definitions should not be allowed to obscure the concept, from which they were derived, that different types of condition give rise to acute and chronic pain. Acute pain should be considered primarily as pain due to a condition that is likely to resolve spontaneously by natural healing. Chronic pain should be considered as signifying a condition unlikely to resolve spontaneously by natural healing. The clinical significance of the categories of pain flows from the implicit likelihood of spontaneous recovery which is crucial to management and prognosis.

The rational management of acute pain is clear when the condition is understood as likely to resolve within a short time by natural healing.

No therapeutic intervention is necessary for recovery, by definition, so rational management involves helping the patient understand the situation, reassuring them, and simply allowing natural healing to proceed. The only active intervention that might be needed is something to ease the pain while healing occurs, and the least invasive measure for that purpose is to be preferred. Such an approach carries the least risk of iatrogenic disturbance of the healing process. It fits nicely with Hippocrates's aphorism of "first, do no harm," (Cochrane 1977) to which doctors have subscribed for centuries. Cochrane promoted this approach in his farsighted work that led to the formal development of evidence-based medicine; he wrote of "the relative unimportance of therapy in comparison with the recuperative power of the human body" (Indahl et al. 1995) and wondered "how many things are done in modern medicine because they can be, rather than because they should be" (Indahl et al. 1995). The effectiveness of the approach has been shown by Indahl et al. (Hippocrates. Of the epidemics) in the management of subacute low back pain and by McGuirk et al. (2001) in the management of acute low back pain.

The rational management of chronic pain is quite different. As the conditions giving rise to chronic pain will not resolve spontaneously, intervention is indicated in virtually every case. The key to the problem is valid diagnosis. Psychosocial factors are important in chronic pain, but their roles are usually secondary to what began and often persists as a biological impairment. If the treating clinician can identify an underlying biological mechanism (a causative injury or disease), many chronic conditions have specific treatments that will control the pain effectively (Lord et al. 1996; Govind et al. 2003). Psychosocial factors must always be considered too and addressed if necessary in the management of the condition, but not to the exclusion of the fundamental (biological) cause. Pursuing the diagnosis of a disorder so as to address its cause seems obvious and is standard practice in other fields of medicine but for some

reason is controversial in pain medicine. Chronic low back pain and chronic neck pain in particular are often managed as if precise diagnosis is impossible, which in these days is untrue in the majority of cases (Bogduk 2004). If specific treatment is applied and the pain controlled, associated psychosocial problems can also be expected to remit; there is sound evidence (Wallis et al. 1997) to show this happens and no sound evidence to show that when pain is controlled effectively, pain-related psychosocial problems persist.

References

Bogduk, N. (Ed.). (2004). *Practice guidelines for spinal diagnostic and treatment procedures*. San Francisco: International Spine Intervention Society.

Bonica, J. J. (1953). *The management of pain*. Philadelphia: Lea & Febiger.

Cochrane, A. L. (1977). *Effectiveness and efficiency. Random reflections on health services* (p. 5). Cambridge: Cambridge University Press.

Engel, G. (1977). The need for a new medical model: A challenge for biomedicine. *Science, 196*, 129–136.

Govind, J., King, W., Bailey, B., & Bogduk, N. (2003). Radiofrequency neurotomy for the treatment of third occipital headache. *Journal of Neurology, Neurosurgery, and Psychiatry, 74*, 88–93.

Hippocrates. Of the epidemics, written 400 B.C.E translated by Francis Adams. http://classics.mit.edu/Hippocrates/epidemics.html.

Indahl, A., Indahl, A., Velund, L., & Reikerås, O. (1995). Good prognosis for low back pain when left untampered. *Spine, 20*, 473–477.

Lord, S. M., Barnsley, L., Wallis, B. J., McDonald, G. J., & Bogduk, N. (1996). Percutaneous radiofrequency neurotomy for the treatment of chronic cervical zygapophysial joint pain: A randomized, double-blind controlled trial. *New England Journal of Medicine, 335*, 1721–1726.

McGuirk, B., King, W., Govind, J., Lowry, J., & Bogduk, N. (2001). Safety, efficacy and cost-effectiveness of evidence-based guidelines for the management of acute low back pain in primary care. *Spine, 26*, 2615–2622.

Merskey, H. (1979). Pain terms: A list with definitions and notes on usage recommended by the IASP Subcommittee on Taxonomy. *Pain, 6*, 249–252.

Merskey, H., & Bogduk, N. (Eds.). (1994). *Classification of chronic pain. Descriptions of chronic pain syndromes and definitions of pain terms* (2nd ed., p. xi). Seattle: IASP Press.

Sternbach, R. A. (1974). *Pain patients: Traits and treatment*. New York: Academic.

van Tulder, M. W., Koes, B. W., & Bouter, L. M. (1997). Conservative treatment of acute and chronic nonspecific low back pain. A systematic review of randomized controlled trials of the most common interventions. *Spine, 22*, 2128–2156.

Wallis, B. J., Lord, S. M., & Bogduk, N. (1997). Resolution of psychological distress of whiplash patients following treatment by radiofrequency neurotomy: A randomised, double-blind, placebo-controlled trial. *Pain, 73*, 15–22.

Acute Painful Diabetic Neuropathy

▶ Diabetic Neuropathies

Acute Pelvic Pain

▶ Gynecological Pain and Sexual Functioning

Acute Phase Protein

Definition

Liver proteins whose synthesis increases in inflammation and trauma.

Cross-References

▶ Pain Control in Children with Burns

Acute Postinfectious Polyradiculoneuropathy

▶ Guillain-Barré Syndrome

Acute Postoperative Pain in Children

▶ Acute Pain in Children, Postoperative

Acute Postoperative Pain Therapy

▶ Postoperative Pain, Thoracic and Cardiac Surgery

Acute Procedural Pain in Children

▶ Acute Pain in Children, Procedural

Acute Salpingitis

▶ Chronic Pelvic Pain, Pelvic Inflammatory Disease and Adhesions

Acute Sciatica

▶ Lower Back Pain, Acute

Acute Stress Disorder

Definition

A psychiatric disorder whose onset is within 1 month of exposure to trauma, and whose symptoms are similar to posttraumatic distress disorder. They include reexperiencing the event as with flashbacks and nightmares, dissociative symptoms like numbing, avoidance of any reminder of the trauma, and hyperarousal or increased generalized anxiety.

Cross-References

▶ Pain Control in Children with Burns

Acute-Recurrent Pain

▶ Chronic Postsurgical Pain (CPSP)

Adaptation

Definition

Adaptation refers to a decrease in the firing rate of action potentials in the face of continuing excitation.

Cross-References

▶ Coping and Pain
▶ Mechanonociceptors

Adaptation Phase

Definition

A phase of the psychophysiological assessment designed to permit patients to become acclimated.

Cross-References

▶ Psychophysiological Assessment of Pain

ADD Protocol

▶ Assessment of Discomfort in Dementia Protocol

Addiction

Definition

Addiction is a biopsychosocial disease, with a strong genetic component, characterized by the maladaptive use of a substance in a manner that reflects: (1) loss of control over use; (2) compulsive use; (3) continued use despite physical, psychological, or social harm; and (4) craving. Although the word "dependence" often is used as a synonym for addiction, it is imprecise; physical dependence, a phenomenon defined by the occurrence of withdrawal following abrupt discontinuation or dose reduction, or administration of an antagonist, is not addiction, but a physiological phenomenon. Tolerance also is a physiological response, defined by the need for a higher dose to obtain the same effect.

Cross-References

▶ Cancer Pain, Evaluation of Relevant Comorbidities and Impact
▶ Cancer Pain Management
▶ Opioids, Clinical Opioid Tolerance
▶ Opioid Receptors
▶ Opioid Therapy in Cancer Patients with Substance Abuse Disorders, Management
▶ Postoperative Pain, Opioids
▶ Psychiatric Aspects of the Management of Cancer Pain

Adduction

Definition

Movement of a body part toward the midline of the body.

Cross-References

▶ Cancer Pain Management, Orthopedic Surgery

Adenoassociated Virus Vectors

Synonyms

AAV

Definition

Adenoassociated virus (AAV)-based vectors are derived from a nonpathogenic parvovirus. AAVs are thought to be naturally defective, because of their requirement for coinfection with a helper virus, such as Ad or HSV, for a productive infection. The single-stranded 4.7-kB DNA genome is packaged in a 20-nm particle. AAV is not associated with any known disease and induces very little immune reaction when used as a vector. For applications requiring a relatively small transgene, AAV vectors are very attractive, but the small insert capacity limits their utility for applications requiring a large transgene.

Cross-References

▶ Opioids and Gene Therapy

Adenoma

Definition

Adenoma is a benign growth starting in the glandular tissue. Adenomas can originate from many organs including the colon, adrenal, and thyroid. In the majority of cases, these neoplasms stay benign, but some transform to malignancy over time.

Cross-References

▶ NSAIDs and Cancer

Adenomyosis

Definition

The growth of endometrial glands and stroma into the uterine myometrium, to a depth of at least 2.5 mm from the basalis layer of the endometrium.

Cross-References

▶ Dyspareunia and Vaginismus
▶ Gynecological Pain, Neural Mechanisms

Adenomyosis Model

▶ Visceral Pain Models, Female Reproductive Organ Pain

Adenosine 5′ Triphosphate

Synonyms

ATP

Definition

ATP is one of the five nucleotides that serve as building blocks of nucleic acids. Structurally, adenine and guanine nucleotides are purines, whereas cytosine, thymine, and uracil are pyrimidines. ATP is also the main energy source for cells. More recently it has been recognized that ATP, some of its metabolites, as well as some other nucleotides play a role as extracellular signaling molecules by activating specific cell surface receptors.

Cross-References

▶ Purine Receptor Targets in the Treatment of Neuropathic Pain

Adenoviral Vectors

Definition

Adenoviral (Ad) vectors are based on a relatively nonpathogenic virus that causes respiratory infections. The 36 kb linear, double-stranded Ad DNA is packaged in a 100-nm-diameter capsid. In first-generation Ad vectors, the early region 1 (E1) gene was deleted to generate a replication-defective vector and to create space for an inserted gene coding for a marker or therapeutic protein. A cell line that complements the E1 gene deletion allows propagation of the viral vector in cultured cells. These first-generation Ad vectors can accommodate up to approximately 8 kb of insert DNA. In high-capacity Ad vectors, the entire Ad vector genome is "gutted" (hence the alternative name "gutted Ad vector"), removing all viral genes and providing 30 kb of insert cloning capacity.

Cross-References

▶ Opioids and Gene Therapy

Adequate Stimulus

Definition

A term coined by Sherrington in the 1890s to define the optimal stimulus for the activation of a particular nervous system structure. For nociceptive systems in humans, it is simply defined as "a pain-producing stimulus"; for animal studies it has been defined as a stimulus that produces, or threatens to produce, tissue damage. This is valid for studies of skin sensation, but may not be valid for deep tissues such as viscera.

Cross-References

▶ Nocifensive Behaviors of the Urinary Bladder
▶ Visceral Pain Model, Urinary Bladder Pain (Irritants or Distension)

Adherence

Definition

The active, voluntary, collaborative involvement of a patient in a mutually acceptable course of behavior to produce a desired therapeutic result.

Cross-References

▶ Multidisciplinary Pain Centers, Rehabilitation

Adhesion Molecules

Definition

Circulating leukocytes migrate to injured tissue and the classical steps of rolling, firm adhesion, and chemotaxis depend on a class of cell surface molecules that are known as adhesion molecules. The major families of cell adhesion molecules include the selectins, the immunoglobulin superfamily, and the integrins. The selectins are important for leukocyte homing to particular tissues. Selectins can be expressed on leukocytes (L-selectin (CD62L)), vascular endothelium (P-selectin (CD62P) and E-selectin (CD62E)), and platelets (P-selectin). The initial step of leukocyte migration, rolling, is mediated by selectins on leukocytes (L-selectin) and endothelium (P- and E-selectin). The immunoglobulin superfamily includes intercellular cell adhesion molecule-1 (ICAM-1), ICAM-2, and vascular cell adhesion molecule-1 (VCAM-1). Integrins are a large family of proteins that mediate cell adhesion, migration, activation, embryogenesis, growth, and differentiation by interacting with immunoglobulin superfamily members such as intercellular adhesion molecule-1. Leukocytes migrate through the vessel wall, directed by platelet-endothelial cell adhesion molecule-1 and other immunoglobulin ligands. Many cell–cell interactions are dependent on adhesion and signal transduction via adhesion molecules, in particular the integrins. Interruption of this cascade can block immunocyte extravasation.

Cross-References

▶ Opioids in the Periphery and Analgesia

Adjunctive Drugs

Definition

Adjunctive drugs are medications that may be used to manage the side effects of analgesics or potentiate analgesic effects.

Cross-References

▶ Analgesic Guidelines for Infants and Children

Adjusted Odds Ratio

Definition

"Adjusted Odds Ratio" is the expression of probability after taking into account possible confounding variables.

Cross-References

▶ Psychiatric Aspects of the Epidemiology of Pain

Adjustment Disorder

Definition

Adjustment disorder, defined by DSM-IV, includes significant depressive symptoms or

anxiety symptoms after an identifiable stress, for example, a painful illness, injury, or hospitalization that do not meet severity or duration criteria for a mood or anxiety disorder.

Cross-References

▶ Somatization

Adjuvant

Definition

An add-on intervention that is used to enhance the benefit of an existing therapy.

Cross-References

▶ Cancer Pain Management, Adjuvant Analgesics in Management of Pain due to Bowel Obstruction

Adjuvant Analgesic

Definition

Drugs that are added to a traditional analgesic, such as an opioid, to enhance the analgesic effects, or alternatively, drugs that have a primary indication other than pain, but are analgesic in some painful conditions. Examples include some antidepressants and anticonvulsants.

Cross-References

▶ Analgesic Guidelines for Infants and Children
▶ Cancer Pain Management, Adjuvant Analgesics in Management of Pain Due to Bowel Obstruction

▶ Cancer Pain Management, Nonopioid Analgesics
▶ Cancer Pain Management, Principles of Opioid Therapy, Drug Selection
▶ Opioid Rotation in Cancer Pain Management

Adjuvant Analgesics in Management of Cancer-Related Bone Pain

Keri L. Fakata[1] and Arthur G. Lipman[2]
[1]College of Pharmacy and Pain Management Center, University of Utah, Salt Lake City, USA
[2]Departments of Pharmscotherpay and Anesthesiology and Pain Management Center, University of Utah, Salt Lake City, UT, USA

Synonyms

Boney pain; Cancer-related bone pain; Malignant bone pain

Definition

▶ Adjuvant ▶ analgesics in the management of cancer-related bone pain are supplemental treatments that are added to the primary analgesics, usually NSAIDs and opioids. These additional analgesic interventions include radiation, using either palliative ▶ radiotherapy or ▶ radiopharmaceuticals, and two classes of medications, ▶ bisphosphonates and steroids.

Characteristics

Normal bone undergoes constant remodeling in which resorption or formation of bone occurs. The cells involved in these processes are ▶ osteoblasts and ▶ osteoclasts, respectively. These cells respond to signals from several types of mediators, including hormones, prostaglandins, and ▶ cytokines. Tumor cells invade bone and interrupt the balance between osteoblastic and

osteoclastic activity, alter bone integrity, and produce pain (Mercadante 1997).

Boney cancers can be exquisitely painful. The severity of pain does not always correlate with radiographic findings. Primary and metastatic bone tumors produce severe pain in about 90 % of patients who develop such tumors. Therefore, aggressive and effective treatment of boney cancer pain is important to maintain patients' quality of life.

Boney metastases occur in approximately 60–85 % of patients who develop metastatic disease from some of the more common cancers, e.g., bredast, prostate, and lung. Bone is one of the most common metastatic sites. There are also primary bone cancers, e.g., myeloma, osteosarcoma, and Ewing's sarcoma (Mercadante 1997).

When tumors metastasize to bone, they can either be osteolytic, causing boney destruction, or osteoblastic, producing sclerotic boney changes (1). Figure 1 illustrates bone changes in cancer. Examples of these processes are prostatic cancer stimulating osteoblasts to lay down boney material and breast cancer causing osteolysis from stimulation of osteoclasts. Mixed osteoblastic-osteoclastic states also can occur.

Chemical mediators, most notably prostaglandins and cytokines, are released in areas of tumor infiltration. These mediators stimulate osteoclasts or osteoblasts and nociceptors (Payne 1997). When tumor invasion occurs, the highly innervated periosteum that surrounds bone is disturbed, and microfractures may occur within the trabeculae (Payne 1997). Nerve entrapment can also occur as disease progresses, due either to direct tumor effects or to collapse of the skeletal structure (Mercadante 1997; Payne 1997; Benjamin 2002).

Radiopharmaceuticals and bisphosphonates are very effective at treating boney pain; some clinicians consider these first-line therapies. The combination of the two may be additive or synergistic in the treatment of bone pain and dose sparing to lessen dose-related complications of opioid therapy (Hoskin 2003).

Radiotherapy and radiopharmaceuticals are often underutilized therapies for treating bone pain. These two methods of delivering

a   b   c

Adjuvant Analgesics in Management of Cancer-Related Bone Pain, Fig. 1 Cancer effects on bone. (a) Normal bone (balance between formation and remodeling). (b) Osteolytic bone (unbalanced – increase in osteoclastic activity). (c) Osteoblastic bone (unbalanced – increase in bone formation)

radionuclides have comparable efficacy as analgesics. A systematic review of 20 trials (12 using external field radiation and 8 using radioisotopes) showed that 1 in 4 patients received complete pain relief in 1 month and 1 in 3 patients achieved at least 50 % pain relief. For radiotherapy, no differences in efficacy or adverse events were reported with single or multiple fractional dosing in the external field trials. Radiotherapy has been reported to be up to 80 % effective for the treatment of boney pain (McQuay et al. 2000). Radiation can be delivered by localized or widespread external beam radiation and also by systemic bone-seeking radioisotopes. For widespread painful boney metastases, external ▶ hemibody radiation may be administered. With radiation administered above the diaphragm, pneumonitis is a risk (Mercadante 1997). Below the diaphragm administration commonly causes nausea, vomiting, and diarrhea. If whole body radiation is the goal, a period of 4–6 weeks between treatments must occur to allow bone marrow recovery.

An alternative to systemic delivery is the use of radioisotopes that target bone. There are four such agents available: ^{89}strontium (^{89}Sr), ^{32}phosphorous (^{32}P), ^{186}rhenium (^{186}Re), and ^{153}samarium (^{153}Sm). ^{89}Sr is the most commonly used due to its greater specificity for bone. All of these agents target osteoblastic activity. They emit beta particles and are associated with less systemic toxicity than hemibody

radiation. However, bone marrow suppression is still a risk. Use of these radiopharmaceuticals is limited due to the expense of the drugs and by storage and disposal requirements (Hoskin 2003). Current radioisotope research is focusing on low-energy electron emitters over the current energetic β emitters to produce therapeutic benefit without bone marrow suppression (Bouchet et al. 2000).

Local irradiation is the treatment of choice for localized bone pain because this method is associated with a low incidence of local toxicity and virtually no systemic toxicity. Radiotherapy often provides relatively prompt pain relief, which is probably due to reduced effects of local inflammatory cells responsible for the release of inflammatory mediators, not tumor regression alone.

Bisphosphonates are another form of systemic treatment for bone pain. A recent meta-analysis of 30 randomized controlled trials, to evaluate relief of pain from bone metastases, supports the use of bisphosphonates as adjunct therapy when primary analgesics and/or radiotherapy is inadequate to treat the pain (Wong and Wiffen 2002). Evidence is lacking for the use of bisphosphonates as first-line therapy for immediate relief of bone pain.

Two bisphosphonates are currently approved for the treatment of painful boney metastasis in the United States: pamidronate and zoledronic acid. Both are intravenous preparations. Doses of 90 mg pamidronate administered over 2–4 h and 4 mg zoledronic acid administered over 15 min every 3–4 weeks have comparable effectiveness in reducing the need for radiotherapy, decreasing the occurrence of fractures, and reducing pain scores (Lucas and Lipman 2002). The most common adverse effects of both agents include bone pain, anorexia, nausea, myalgia, fever, and injection site reaction. Bisphosphonates have been associated with renal toxicity. Bisphosphonates bind strongly to the bone surface and are taken up by osteoclasts during bone resorption. The osteoclasts are then inhibited and apoptosis is induced. The reduction in the number of osteoclasts inhibits boney metastasis. The bisphosphonates also have an antitumor effect, possibly due to drug uptake in tumor cells (Green and Clezardin 2002).

Although NSAIDs are generally considered first-line drugs for mild cancer pain, their specific role in boney pain is currently being investigated. A recent study in mice evaluated a cycloxygenase-2 (COX-2) selective NSAID on movement-evoked cancer bone pain and tumor burden. A decrease of ongoing and movement-evoked pain was seen in acutely treated mice (day 14 post-tumor implantation), and the same decrease in pain was expressed as well as decreased tumor burden, osteoclastogenesis, and bone destruction, by 50 % of chronically treated mice (day 6 post-tumor implantation) (Sabino et al. 2002). Tumors that invade bone express COX-2, possibly as a mechanism for implantation. This work supports the inhibition of prostaglandin synthesis as being the mechanism of action of the drugs in cancer-related bone pain.

Systemic steroids can also be useful adjuvants in cancer-related bone pain due to their broad-spectrum anti-inflammatory properties. They are most commonly used for spinal cord compression due to collapse of vertebrae or pressure by the tumor itself. Approximately 90 % of prostatic metastases involve the spine, with the lumbar region most commonly affected. Early diagnosis of spinal cord compression is critical. It presents as localized back pain in 90–95 % of patients; muscle weakness, autonomic dysfunction, and sensory loss will follow if untreated (Benjamin 2002). Intravenous dexamethasone is a steroid of choice due to its high potency, low mineralocorticoid activity, and low cost.

When primary analgesics, i.e., nonsteroidal anti-inflammatory drugs (NSAIDs) and opioids, no longer control boney pain adequately, adjuvants should be considered. Local radiation should be used when pain is localized, and fractures are ruled out. Pain due to solid tumors tends to respond greater to radiotherapy than bisphosphonates. Generally, as the disease progresses, patients will have received both of these modalities. The role of their use together has yet to be evaluated. To forestall neurological complications of spinal cord compression, steroids are indicated and should be started promptly upon suspicion.

References

Benjamin, R. (2002). Neurologic complications of prostate cancer. *American Family Physician, 65,* 1834–1840.

Bouchet, L. G., Bolch, W. E., Goddu, S. M., et al. (2000). Considerations in the selection of radiopharmaceuticals for palliation of bone pain from metastatic osseous lesions. *Journal of Nuclear Medicine, 41,* 682–687.

Green, J. R., & Clezardin, P. (2002). Mechanisms of bisphosphonate effects on osteoclasts, tumor cell growth, and metastasis. *American Journal of Clinical Oncology, 25,* 3–9.

Hoskin, P. J. (2003). Bisphosphonates and radiation therapy for palliation of metastatic bone disease. *Cancer Treatment Reviews, 29,* 321–327.

Lucas, L. K., & Lipman, A. G. (2002). Recent advances in pharmacotherapy for cancer pain management. *Cancer Practice, 10,* 14–20.

McQuay, H. J., Collins, S. L., Carroll, D., et al. (2000). Radiotherapy for the palliation of painful bone metastases. *Cochrane Database of Systematic Reviews,* CD001793.

Mercadante, S. (1997). Malignant bone pain: Pathophysiology and treatment. *Pain, 69,* 1–18.

Payne, R. (1997). Mechanisms and management of bone pain. *Cancer, 80,* 1608–1613.

Sabino, M. A., Ghilardi, J. R., Jongen, J. L., et al. (2002). Simultaneous reduction in cancer pain, bone destruction, and tumor growth by selective inhibition of cyclooxygenase-2. *Cancer Research, 62,* 7343–7349.

Wong, R., & Wiffen, P. J. (2002). Bisphosphonates for the relief of pain secondary to bone metastases. *Cochrane Database of Systematic Reviews,* CD002068.

Adjuvant Analgesics in Management of Cancer-Related Neuropathic Pain

Russell K. Portenoy[1] and Ebtesam Ahmed[2]
[1]Department of Pain Medicine and Palliative Care, Beth Israel Medical Center, New York, NY, USA
[2]College of Pharmacy and Health Sciences, St. John's University, Department of Pain Medicine and Palliative Care, Beth Israel Medical Center, New York, NY, USA

Definition

An adjuvant analgesic (see "▶ Adjuvant Analgesic") is any drug that has a primary indication other than pain but is analgesic in some painful conditions.

Characteristics

Although numerous barriers may lead to undertreatment (Van den Beuken-van Everdingen et al. 2007), opioid-based therapy is widely accepted as the first-line strategy for moderate or severe chronic pain due to active cancer and usually is considered to be a second-line approach for those with limited or remote disease. Other categories of analgesic drugs, which include the nonopioids (e.g., the NSAIDs) and the so-called adjuvant analgesics, can be combined with opioid therapy or used independently in selected patients.

Historically, the term "adjuvant analgesic" has referred to a drug that has a primary indication other than pain but may be analgesic in specific painful conditions (Deandrea et al. 2008). This definition is likely to evolve, however, as more of these drugs are used as primary analgesics for various types of pain. When combined with opioid therapy, these drugs collectively may be termed "coanalgesics."

In the treatment of pain related to active cancer, adjuvant analgesics usually are added to opioid therapy after the opioid dose has been titrated to optimize the balance between analgesics and side effects. The occurrence of troublesome side effects before satisfactory analgesia occurs characterizes the pain as "poorly responsive" to the specific opioid and route of administration, and a trial of one or more of these drugs is a common strategy to address this scenario (Lussier and Portenoy 2010).

The decision to offer a trial of an adjuvant analgesic to address poor opioid responsiveness must be based on the findings of a comprehensive assessment of the pain and the patient (Saarto and Wiffen 2007). Rational decisions require an understanding of the nature of the pain, status of the underlying disease and its treatment, salient medical and psychiatric comorbidities, psychosocial factors, and the values and preferences of the patient and family. All this information informs the clinical recommendations and the goals that are pursued.

Types of Adjuvant Analgesics

There have been few studies of the adjuvant analgesics in populations with cancer pain and their use has been based largely on data obtained from studies in other populations and clinical experience (Lussier and Portenoy 2010). Based on this information, some drug classes appear to have potential utility in heterogeneous painful conditions and have been described as "multipurpose" adjuvant analgesics. The drugs in this category are usually combined with opioid therapy to treat patients with opioid-refractory neuropathic pain. The most important groups are the corticosteroids and the analgesic antidepressants. Other adjuvant analgesics only have evidence of efficacy in neuropathic pain and are used in cancer populations for this purpose. The most prominent are anticonvulsants. Other groups of adjuvant analgesics are used to treat bone pain, musculoskeletal pain, or pain and other symptoms in bowel obstruction (Lussier and Portenoy 2010).

First-Line Adjuvant Analgesics in Cancer Pain

Corticosteroids. In populations with advanced cancer, corticosteroids have been widely used for symptoms such as pain, nausea, fatigue, poor appetite, malaise (Loblaw et al. 2005). Although evidence from adequate clinical trials is lacking, there is extensive clinical experience that suggests benefit for varied pain syndromes, including neuropathic pain resulting from nerve compression, bone pain, pain associated with capsular expansion or duct obstruction, pain from bowel obstruction, pain caused by lymphedema, and headache caused by increased intracranial pressure. The mechanism of action is unknown, but may relateto the reduction of peritumoral edema, anti-inflammatory effects, and direct effects on nociceptive neural systems.

Although many clinicians favor dexamethasone, presumably because of its relatively low mineralocorticoid effects, there are no comparative trials of the drugs in this class. Prednisone and methylprednisolone are acceptable alternatives. Based on clinical experience in the treatment of emergent spinal cord compression (Loblaw et al. 2005), a high-dose dexamethasone regimen has been used to treat very severe and escalating pain (sometimes called "pain emergencies"). The more common low-dose regimen comprises 1–2 mg per day, with dose escalation as needed to obtain symptom control.

Antidepressants. Analgesic efficacy is best established for some of the tricyclic compounds and the serotonin-norepinephrine reuptake inhibitors (SNRIs). The analgesic mode of action presumably relates to the enhanced availability of monoamines at synapses in pathways that are part of the descending pain modulating system (Saarto and Wiffen 2007).

The tricyclic antidepressants include tertiary amine compounds, such as amitriptyline, imipramine, and doxepin, and secondary amine compounds, such as nortriptyline and desipramine. The latter drugs are less anticholinergic and sedating and generally are preferred for use in patients with cancer. All the tricyclic compounds are relatively contraindicated in patients with serious heart disease, severe prostatic hypertrophy, and narrow-angle glaucoma.

The SNRIs include duloxetine, minalcipran, venlafaxine, and desvenlafaxine. Evidence of analgesic efficacy is best for duloxetine (Arnold et al. 2005; Wernicke et al. 2006), but like other drugs, trials in cancer pain are lacking. The risk profile of the SNRIs is more favorable than that of the tricyclic antidepressants, and clinicians commonly initiate a trial of an analgesic antidepressant with one of these drugs. The SNRIs side effects include nausea, sexual dysfunction, and somnolence or mental clouding.With the exception of bupropion and two SSRIs (paroxetine and citalopram), analgesic efficacy has not been demonstrated for other antidepressants (Semenchuk et al. 2001; Sindrup et al. 1990, 1992). Bupropion, which is a dopamine and norepinephrine reuptake blocker, is distinguished by its tendency to be activating. Patients with pain complicated by fatigue or somnolence may be considered for a trial of this drug.

Anticonvulsants. The gabapentinoids, gabapentin and pregabalin, often are considered first-line therapy for neuropathic pain of diverse types in patients without major depression (Dworkin et al. 2007). They are usually considered after an antidepressant trial if pain is complicated by depressed mood. In the cancer population, these drugs are first-line co-analgesics, along with the corticosteroids and antidepressants.

Both gabapentin and pregabalin act by binding to the alpha-2 delta protein modulator of the N-type, voltage-gated calcium channel. Binding to this protein reduces calcium influx into the neuron and lessens the likelihood of depolarization. The main difference between the drugs is pharmacokinetic. Absorption of gabapentin is facilitated by a saturable transporter in the small bowel and central nervous system. At relatively higher doses (approximately 1,800 mg per day), kinetics becomes nonlinear and less absorption occurs with each dose increase. This means that gabapentin has a pharmacokinetic ceiling, which coexists with a possible pharmacodynamic ceiling of the type observed with all adjuvant analgesics. Pregabalin has linear kinetics, which simplify dose titration.

Gabapentin and pregabalin have a relatively good safety profile. They do not require hepatic metabolism and there are no known drug-drug interactions. They are renally excreted and the dose must be adjusted in the setting of renal failure. Their common side effects are mental clouding, dizziness, somnolence, and peripheral edema.

In the medically frail cancer patient, gabapentin often is initiated at a dose of approximately of 100–300 mg per day. The dose is gradually escalated while monitoring analgesia and side effects. If pain relief does not occur, dose escalation in the absence of an analgesic ceiling or adverse effects typically extends to 3,600 mg per day and sometimes higher. The starting dose of pregabalin is usually 50–75 mg per day (or lower, 25 mg, in those with advanced age or frailty) and escalation to the usual effective dose of 150–300 mg twice daily typically is accomplished in 2–3 steps over a week or two. Although head-to-head prospective studies comparing the relative efficacy of gabapentin and pregabalin are lacking, it is appropriate to consider a follow-on trial of the alternate drug if an attempt with the first does not yield a benefit.

Other anticonvulsants also have been studied as analgesics. Divalproex and phenytoin are older drugs and have a long history of use for neuropathic pain. Newer anticonvulsants generally have more favorable side effect and safety profiles, however, and are preferred. Although evidence of analgesic effects is very limited for all of these drugs, trials of oxcarbazepine, topiramate, and lamotrigine are usually considered before others.

Other Multipurpose Drugs Used for Neuropathic Pain

Alpha-2 Adrenergic Agonists. Clonidine and tizanidine are alpha-2 adrenergic agonists. Clonidine has been used in diverse types of chronic pain; intraspinal clonidine has been shown to reduce neuropathic pain in patients with severe cancer pain partly responding to opioids (Eisenach et al. 1995). Tizanidine has demonstrated analgesic efficacy in myofascial pain syndrome and chronic headache. The use of these drugs is limited by their side effects, which include dry mouth, somnolence, and orthostatic hypotension. A trial usually is considered only after other adjuvant analgesics have proved ineffective. Tizanidine has less hypotensive effects and may be preferred over clonidine for a trial in medically frail patients with opioid-refractory pain. It may be initiated at 1–2 mg at night, and

the dose is then gradually escalated while monitoring analgesia and adverse effects.

Cannabinoids. Cannabinoids interact with an endogenous system that includes cannabinoid-like ligands, the endocannabinoids, and multiple receptors in both the periphery and central nervous system. There are several drugs commercially available and others under study. An oromucosal spray containing tetrahydrocannabinol (THC) plus cannabidiol (and smaller concentrations of other compounds), known as nabiximols, is undergoing worldwide development and has already been approved in several countries for spasticity due to multiple sclerosis and opioid-refractory pain due to cancer (Russo et al. 2007).

Topical Analgesics. Although topical analgesics have been used for neuropathic pain, they have the potential for broader application. Creams and patches containing local anesthetics, capsaicin preparations, nonsteroidal anti-inflammatory drugs, tricyclic compounds, or other drugs are available commercially or may be compounded, singly or in combination. The most widely used topical therapies for pain contain local anesthetics. A lidocaine 5 % patch now is approved for the treatment of postherpetic neuralgia and now is widely used in focal and regional pains of all types. Topical analgesics containing local anesthetics are also available in creams and gels.

The mechanism of capsaicin is to release and then deplete substance P from the terminals of afferent C-fibers. Treatment with low-concentration (e.g., 0.1 %), commercially available creams has been demonstrated to yield weak to moderate analgesic effects in controlled trials of various types of neuropathic and arthropathic pain (Knotkova et al. 2008). A high-concentration (8 %) capsaicin patch has become available in some countries for the treatment of postherpetic neuralgia; this patch is applied for 1–2 h and if effective, can yield analgesia that lasts months.

Other Drugs

Several other drug classes also are considered for refractory cancer-related neuropathic pain.

Sodium channel blockade has been recognized as an analgesic mechanism for decades. Oral agents, such as mexiletine, and brief infusion of parenteral lidocaine, may have utility for pain. These drugs are generally safe (Wymer et al. 2009), but require cautious administration in medically ill populations. A new drug, lacosamide, has a unique mechanism involving sodium channel modulation and also may be considered for a trial in refractory neuropathic pain(Ben-Ari et al. 2007).

The N-methyl-D-aspartate (NMDA) receptor is involved in both the sensitization of central neurons and the functioning of the opioid receptor. Evidence that NMDA receptor antagonists are analgesic in clinical pain is very limited. In populations with cancer-related neuropathic pain, particularly those with advanced illness, ketamine often is considered (Chizh and Headley 2005). This drug is a dissociative anesthetic and has a side effect profile that includes psychotomimetic effects and a hypersympathetic state. For refractory pain, it may be tried in subanesthetic doses as a brief infusion (repeated as necessary), a more prolonged infusion, or oral therapy. Treatment must be carefully monitored and coadministration of a sedative-hypnotic drug is commonly used to reduce the risk of psychotomimetic side effects.

Other NMDA receptor antagonists, such as memantine, amantadine, and dextromethorphan, also have been studied in diverse types of neuropathic pain, but results have been mixed (Hugel et al. 2003). They are rarely considered for trials in cancer-related neuropathic pain that has not responded to other agents.

GABA receptors may be involved in pain processing. Among the GABA receptor inhibitors are the benzodiazepines, which affect the $GABA_A$ receptor subtype, and baclofen, which affects the $GABA_B$ subtype. Evidence supporting the use of benzodiazepines as analgesics is very limited. Based on clinical experience, a trial of clonazepam often is considered for refractory neuropathic pain, especially if pain is accompanied by anxiety (Fromm et al. 1984). Baclofen is

an antispasticity drug with established efficacy in trigeminal neuralgia (Fromm et al. 1984). It has also been used anecdotally for neuropathic pain of other types.

Summary

The development of adjuvant analgesics during recent decades has been rapid, and there now are numerous drugs in many classes. The clinical approach to the selection and administration of these drugs in populations with cancer pain remains largely empirical, based on data obtained in other populations and experience. Studies that compare the safety and effectiveness of these drugs in various indications are badly needed.

References

Arnold, L. M., Rosen, A., Pritchett, Y. L., et al. (2005). A randomized, double-blind, placebo-controlled trial of duloxetine in the treatment of women with fibromyalgia with or without major depressive disorder. *Pain, 119*, 5–15.

Ben-Ari, A., Lewis, M. C., & Davidson, E. (2007). Chronic administration of ketamine for analgesia. *Journal of Pain & Palliative Care Pharmacotherapy, 21*(1), 7–14.

Chizh, B. A., & Headley, P. M. (2005). NMDA antagonists and neuropathic pain–Multiple drug targets and multiple uses. *Current Pharmaceutical Design, 11*(23), 2977–2994.

Deandrea, S., Montanari, M., Moja, L., & Apolone, G. (2008). Prevalence of undertreatment in cancer pain. A review of published literature. *Annals of Oncology, 19*(12), 1985–1991.

Dworkin, R. H., O'Connor, A. B., Backonja, M., et al. (2007). Pharmacologic management of neuropathic pain: Evidence-based recommendations. *Pain, 132*, 237–251.

Eisenach, J. C., DuPen, S., Dubois, M., Miguel, R., & Allin, D. (1995). Epidural clonidine analgesia for intractable cancer pain: The Epidural Clonidine Study Group. *Pain, 61*(3), 391–399.

Fromm, G. H., Terrence, C. F., & Chattha, A. S. (1984). Baclofen in the treatment of trigeminal neuralgia: Double-blind study and long-term follow-up. *Annals of Neurology, 15*, 240–244.

Hugel, H., Ellershaw, J. E., & Dickman, A. (2003). Clonazepam as an adjuvant analgesic in patients with cancer-related neuropathic pain. *Journal of Pain and Symptom Management, 26,* 1073–1074.

Knotkova, H., Pappagallo, M., & Szallasi, A. (2008). Capsaicin (TRPV1 Agonist) therapy for pain relief: Farewell or revival? *The Clinical Journal of Pain, 24*(2), 142–154.

Loblaw, D. A., Perry, J., & Chambers, A. (2005). Systematic review of the diagnosis and management of malignant extradural spinal cord compression: The Cancer Care Ontario Practice Guidelines Initiative's Neuro-Oncology Disease Site Group. *Journal of Clinical Oncology, 23,* 2028–2037.

Lussier, D., & Portenoy, R. K. (2010). Adjuvant analgesics in pain management. In G. Hanks, N. I. Cherny, N. Christakis, M. Fallon, S. Kaasa, & R. K. Portenoy (Eds.), *Oxford textbook of palliative medicine* (4th ed., pp. 706–734). Oxford, England: Oxford University Press.

Russo, E. B., Guy, G. W., & Robson, P. J. (2007). Cannabis, pain, and sleep: Lessons from therapeutic clinical trials of Sativex, a cannabis-based medicine. *Chemistry & Biodiversity, 4,* 1729–1743.

Saarto, T., & Wiffen, P. J. (2007). Antidepressants for neuropathic pain. *Cochrane Database System Review,* CD005454.

Semenchuk, M. R., Sherman, S., & Davis, B. (2001). Double blind, randomized trial of Bupropion SR for the treatment of neuropathic pain. *Neurology, 57,* 1583–1588.

Sindrup, S. H., Bjerre, U., & Dejgaard, A. (1992). The SSRI citalopram relieves the symptoms of diabetic neuropathy. *Clinical Pharmacology & Therapeutics, 52,* 547–552.

Sindrup, S. H., Gram, L. F., & Brosen, K. (1990). The SSRI paroxetine is effective in the treatment of diabetic peripheral neuropathy symptoms. *Pain, 42,* 135–144.

Van den Beuken-van Everdingen, M. H., de Rijke, J. M., Kessels, A. G., et al. (2007). Prevalence of pain in patients with cancer: A systematic review of the past 40 years. *Annals of Oncology, 18*(9), 1437–1449.

Wernicke, J. F., Pritchett, Y. L., & D'Souza, D. N. (2006). A randomized controlled trial of duloxetine in diabetic peripheral neuropathic pain. *Neurology, 67*(2), 1411–1420.

Wymer, J. P., Simpson, J., Sen, D., et al. (2009). Efficacy and safety of lacosamide in diabetic neuropathic pain: An 18-week double-blind placebo-controlled trial of fixed-dose regimens. *The Clinical Journal of Pain, 25*(5), 376–385.

Adjuvant Arthritis

▶ Arthritis Model, Adjuvant-Induced Arthritis

Adjuvant-induced Arthritis

▶ Arthritis Model, Adjuvant-Induced Arthritis

Adolescent

▶ Pain in Children

Adrenergic Agonist

Definition

An adrenergic agonist is a ligand that binds to adrenergic receptors and elicits a response in the cell or tissue in which the receptors are present.

Cross-References

▶ Adrenergic Antagonist
▶ Sympathetically Maintained Pain, Clinical Pharmacological Tests

Adrenergic Antagonist

Definition

An adrenergic antagonist is either a drug that prevents other ligands from binding to adrenergic receptors in the case of a competitive antagonist or, in the case of a noncompetitive antagonist, a ligand that when bound to the receptor prevents agonists from eliciting a response.

Cross-References

▶ Adrenergic Agonist
▶ Sympathetically Maintained Pain, Clinical Pharmacological Tests

Adrenergic Receptors

Definition

Adrenoceptors (adrenergic receptors) are G protein-coupled receptors activated by release of norepinephrine (noradrenaline), epinephrine (adrenaline), and various adrenergic agonists. They are expressed by neurons in the central, peripheral sensory, and sympathetic nervous systems as well as in end organs innervated by sympathetic fibers. Activation of α_1- or $\beta_{1\ or\ 4}$-adrenoceptors produces neuronal excitation or smooth muscle contraction, whereas activation of α_2-adrenoceptors inhibits neuronal firing or neurotransmitter release, resulting in sedation, analgesia, or hypotension. Noradrenergic sympathetic postganglionic neurons express α_2-adrenoceptors, the activation of which presynaptically inhibits norepinephrine release; nociceptive primary sensory neurons also express α_2-adrenoceptors, the activation of which inhibits release of nociceptive neurotransmitters in spinal cord dorsal horn. Activation of $\beta_{2\ or\ 3}$-adrenoceptors produces smooth muscle relaxation or cardioinhibition; β-adrenoceptors appear less involved in pain than α-adrenoceptors.

Cross-References

▶ Sympathetically Maintained Pain in CRPS I, Human Experimentation

Adult Respiratory Distress Syndrome

▶ ARDS

Adverse Effects

Definition

Unwanted side effects of drug treatment.

Cross-References

▶ NSAIDs, Adverse Effects

Adverse Neural Tension

Definition

Adverse neural tension is defined as abnormal physiological and mechanical responses created by the nervous system components, when their normal range of motion and stretch capabilities are tested.

Cross-References

▶ Chronic Pelvic Pain, Physical Therapy Approaches, and Myofascial Abnormalities

Adverse Selection

Definition

Adverse selection can defined as a situation when people who have worse than average risks are most likely to acquire insurance.

Aerobic Exercise

▶ Exercise

Affective

Definition

Category of experiences associated with emotions that range from pleasant to unpleasant.

Cross-References

▶ McGill Pain Questionnaire

Affective Analgesia

Definition

Affective analgesia is the preferential suppression of the emotional reaction of humans and animals to noxious stimulation.

Cross-References

▶ Thalamo-Amygdala Interactions and Pain

Affective Component (Aspect, Dimension) of Pain

Definition

Affective component of pain refers to that quality of the pain experience that causes pain to be unpleasant or aversive. It may be involved in the "suffering" component of persistent pain, and could also involve separate neural pathways in the brain than those involved in the sensory-discriminative component of pain (discrimination and localization of a nociceptive stimulus).

Cross-References

▶ Amygdala, Pain Processing and Behavior in Animals
▶ Hypnotic Analgesia
▶ Primary Somatosensory Cortex (S1), Effect on Pain-Related Behavior in Humans
▶ Primary Somatosensory Cortex (SI)
▶ Thalamo-Amygdala Interactions and Pain

Affective Pain Processing

▶ Thalamo-Amygdala Interactions and Pain

Affective Responses

Definition

Changes in mood or emotion-related behaviors elicited by noxious stimuli. Examples of these responses include aggressive behavior and freezing.

Cross-References

▶ Spinohypothalamic Tract, Anatomical Organization, and Response Properties
▶ Spinothalamic Neuron

Affective-Motivational

Definition

Relating to affect and forces that drive behavior.

Cross-References

▶ Secondary Somatosensory Cortex (S2) and Insula, Effect on Pain-Related Behavior in Animals and Humans

Affective-Motivational Dimension of Pain

Definition

A component of the pain experience that signals the unpleasant hedonic qualities and emotional reactions to noxious stimulation and generates the motivational drive to escape from or terminate such stimulation. This corresponds to the subjective experience of the immediate unpleasantness of pain and the urge to respond behaviorally.

Cross-References

▶ Pain Processing in the Cingulate Cortex, Behavioral Studies in Humans
▶ Thalamo-Amygdala Interactions and Pain

Afferent Fiber/Afferent Neuron

Definition

Afferent fibers are any of the nerve fibers that bring information in form of action potentials (spikes) to a target neuron. The term is most often used in relation to the peripheral nervous system. In this case afferent fibers carry information toward the central nervous system. The cell bodies of afferent fibers in the peripheral nerves reside in dorsal root, trigeminal or nodose ganglia.

Afferent Projections

Definition

See ▶ afferent fibers.

Cross-References

▶ Afferent Fiber/Afferent Neuron

Afferent Signal

Definition

An afferent signal is a neuronal signal in the form of action potentials (spikes) that are carried

toward target neurons. In peripheral nerves the afferent signal is carried toward the central nervous system.

Cross-References

▶ Afferent Fiber/Afferent Neuron

Afterdischarge(s)

Definition

Afterdischarge is the continued neuronal response after an inciting stimulus has terminated, indicating a prolonged reaction.

Cross-References

▶ Molecular Contributions to the Mechanism of Central Pain
▶ Spinal Cord Injury Pain Model, Contusion Injury Model
▶ Trigeminal Neuralgia: Diagnosis and Treatment

Afterhyperpolarization

Synonyms

AHP

Definition

The term refers to the membrane potential of an axon or whole neuron during recovery following an action potential. AHP indicates a more negative potential compared to the resting or prestimulus membrane potential. AHPs in different neurons may have different time constants and different molecular mechanisms. AHPs influence the conduction velocity and excitability of axons.

Cross-References

▶ Mechano-insensitive C-Fibers, Biophysics
▶ Molecular Contributions to the Mechanism of Central Pain

Afterpains

▶ Postpartum Pain

Age and Chronicity

▶ Pain in the Workplace, Risk Factors for Chronicity, and Demographics

Age Regression

Definition

This refers to the use of hypnotic suggestion to return to an earlier time of life in imagination. This technique is used in the context of psychotherapy utilizing hypnosis and may be an exploratory or therapeutic technique. Studies suggest that age regression is extremely unreliable in retrieving accurate information about the past, but that it can be considered part of the individual's life narrative.

Cross-References

▶ Therapy of Pain, Hypnosis

Age-Related Pain Diagnoses

Definition

Pain diagnoses that are more frequent in the elderly, like osteoarthritis, zoster, arteritis,

polymyalgia rheumatica, or artherosclerotic peripheral vascular disease.

Cross-References

▶ Psychological Treatment of Pain in Older Populations

Aggression

▶ Anger and Pain

Agonist

Definition

An agonist is an endogenous or exogenous substance that can interact with and activate a receptor, initiating a physiological or a pharmacological response characteristic of that receptor.

Cross-References

▶ Postoperative Pain, Appropriate Management

Agreed Medical Examination

▶ Independent Medical Examinations

AHP

▶ Afterhyperpolarization

AIDS and Pain

▶ Pain in Human Immunodeficiency Virus Infection and Acquired Immune Deficiency Syndrome

Alcock's Canal

Definition

This is the space within the obturator internus fascia lining the lateral wall of the ischiorectal fossa that transmits the pudendal vessels and nerves.

Cross-References

▶ Clitoral Pain

Alcock's Canal Syndrome

▶ Pudendal Neuralgia in Women

Alcohol-induced Pancreatitis

▶ Visceral Pain Model, Pancreatic Pain

Alcoholism

▶ Metabolic and Nutritional Neuropathies

Alfentanil

Definition

This is a very potent short-acting opioid.

Cross-References

▶ CRPS-1 in Children

Algesia

▶ Hyperalgesia

Algesic Agent/Algesic Chemical

Definition

A chemical substance that elicits pain when administered (or released from pathologically altered tissue) in a concentration that excites nociceptors. Examples are serotonin (5-hydroxytryptamine) and bradykinin (a nonapeptide).

Cross-References

▶ Sensitization of Muscular and Articular Nociceptors
▶ Visceral Pain Model, Angina Pain

Algodystrophy

▶ Complex Regional Pain Syndromes, Clinical Aspects
▶ CRPS, Evidence-Based Treatment
▶ Neuropathic Pain Models, CRPS-I Neuropathy Model
▶ Postoperative Pain, Acute Presentation of Complex Regional Pain Syndrome
▶ Sympathetically Maintained Pain in CRPS I, Human Experimentation

Algogen

Definition

Chemical substance with the ability to induce pain and hyperalgesia.

Cross-References

▶ Polymodal Nociceptors, Heat Transduction
▶ UV-Induced Erythema

Algogenic Actions of Protons

Definition

Lowering muscle pH causes acute ischemia pain since protons produce non-adapting excitation of muscle nociceptors.

Cross-References

▶ Tourniquet Test

Algometer

Definition

An algometer is a calibrated device that can apply painful stimuli of graded intensities. A commonly used device is the pressure algometer, which is used to evaluate deep tissue pain threshold (i.e., muscle, tendon, periosteum).

Cross-References

▶ Threshold Determination Protocols

Algoneuron

Definition

Algoneurons are peripheral afferents that, when activated, evoke a sensation of pain.
  Note: In contrast to "nociceptor," "algoneuron" focuses on the *sensory effect* of the

afferent's signal and *not its receptive field properties*. Under normal circumstances, Aδ- and C-nociceptors are algoneurons of high threshold. Peripheral sensitization might cause these to become low-threshold algoneurons. In the presence of central sensitization, Aβ low-threshold mechanoreceptors (LTMs) can function as algoneurons.

Alice-in-Wonderland Syndrome

Definition

A disorder of perception where visual disturbances occur. It was given its name due to the fact that the syndrome's symptoms are remarkably similar to the distortions in body image and shape as experienced by the main character in Lewis Carrol's 1865 novel "Alice in Wonderland." Objects either appear to be much larger (macropsia) or smaller (micropsia) than normal, and there is also usually an impaired perception of time and place.

Cross-References

▶ Migraine, Childhood Syndromes

ALIF

Synonyms

Anterior lumbar interbody fusion

Definition

Anterior lumbar interbody fusions are grafts/cages placed between the vertebral bodies by anterior approach.

Cross-References

▶ Spinal Fusion for Chronic Back Pain

Allele Dosage Study

▶ Association Study

Alleles

Definition

Alternate forms of a gene or genetic locus; the basic unit of genetic variability. Organisms inherit two alleles (maternal and paternal) of every gene, which may or may not be identical. Different alleles may produce protein isozymes (i.e., proteins with different amino acid sequences), alter expression levels of proteins, or have no effect whatsoever.

Cross-References

▶ Cell Therapy in the Treatment of Central Pain
▶ Opioid Analgesia, Strain Differences

Allocortex

Definition

The allocortex is a three-layered cortex. In the hippocampus, the three layers are the stratum oriens, the stratum pyramidale, and the molecular zone consisting of the stratum radiatum and stratum lacunosum-moleculare.

Cross-References

▶ Nociceptive Processing in the Hippocampus and Entorhinal Cortex, Neurophysiology and Pharmacology

Allodynia

▶ Calcium Channels in the Spinal Processing of Nociceptive Input

Allodynia (Clinical, Experimental)

Rolf-Detlef Treede
Institute of Physiology and Pathophysiology,
Johannes Gutenberg University, Mainz,
Germany

Synonyms

Dynamic mechanical hyperalgesia; Obsolete: hyperesthesia; Touch-evoked pain;

Definition

The term "allodynia" was introduced to describe a puzzling clinical phenomenon; in some patients, gentle touch may induce a pronounced pain sensation ("touch-evoked pain"). In the current taxonomy of the International Association for the Study of Pain (IASP), allodynia is defined as pain induced by stimuli that are not normally painful.

If taken literally, this definition means that any reduction in pain threshold would be called "allodynia." According to the IASP taxonomy, increases in pain to suprathreshold stimuli are called "▶ hyperalgesia." Because the neural mechanisms of ▶ sensitization typically cause a leftward shift in the stimulus-response function that encompasses both reduced thresholds and increased suprathreshold responses, these definitions have been controversial ever since their introduction. Moreover, behavioral studies in animals often use withdrawal threshold measures without any suprathreshold tests, leading to an inflationary use of the term "allodynia" in studies that often bear no resemblance to the initial clinical phenomenon. An alternative definition that captures the spirit of the original clinical observations (Merskey 1982; Treede et al. 2004) defines allodynia as pain due to a non-nociceptive stimulus.

This definition implies that allodynia is pain in the absence of the adequate stimulus for nociceptive afferents (touch is not a "nociceptive stimulus"). Operationally, the presence of mechanical allodynia can be tested with stimulators that do not activate nociceptive afferents (e.g., a soft brush). The situation is less clear for other stimulus modalities such as cooling stimuli. For those cases, where it is not clinically possible to determine whether or not the test stimuli activate nociceptive afferents, "hyperalgesia" is useful as an umbrella term for all types of increased pain sensitivity.

Characteristics

Some patients – particularly after peripheral nerve lesions – experience pain from gentle touch to their skin, a faint current of air, or mild cooling from evaporation of a drop of alcohol. Touch-evoked pain may adapt during constant skin contact but is readily apparent for all stimuli applied in a stroking movement across the skin (Fig. 1). Touch-evoked pain is also called dynamic mechanical allodynia (Ochoa and Yarnitsky 1993). Reaction times of touch-evoked pain are too short for C-fiber latencies, and it can be abolished by an A-fiber conduction block (Campbell et al. 1988). Moreover, both mechanical and electrical pain thresholds in those patients are often identical to the normal tactile detection thresholds (Gracely et al. 1992). These lines of evidence suggest that this strange pain sensation is mediated by Aβ-fiber low-threshold mechanoreceptors (touch receptors).

It was difficult to find the correct term to describe this clinical phenomenon. Because of the altered perceived quality of tactile stimuli, it was called "painful tactile dysesthesia." Due to the increased perception in response to a tactile stimulus, it was also called "hyperesthesia" defined as "a state in which a stimulus, which does not cause pain in normally innervated

tissues, does cause pain in the affected region" (Noordenbos 1959; quoted from Loh and Nathan (1978), who added that this was typically a very slight stimulus). This definition, however, ignored the change in perceived quality (from tactile to painful). According to the perceived quality, this phenomenon should have been called "mechanical hyperalgesia."

At the time when most of the clinical characteristics of allodynia had been established, the only known neurobiological mechanism of hyperalgesia was peripheral sensitization of nociceptive afferents (Raja et al. 1999), leading to heat hyperalgesia at an injury site (primary hyperalgesia). Peripheral sensitization differs from the clinical phenomenon described above in many characteristics; it is spatially restricted to injured skin, and the enhanced sensitivity is for heat stimuli, not for mechanical stimuli. The concept of central sensitization was introduced much later than the concept of peripheral sensitization (Woolf 1983). Thus, hyperalgesia also appeared to be an inadequate term at that time. As a consequence, a new word was introduced, "allodynia," indicating "a different type of pain" (Merskey 1982).

Dynamic mechanical allodynia occurs in a variety of clinical situations: secondary hyperalgesia surrounding an injury site, postoperative pain, joint and bone pain, visceral pain and delayed onset muscle soreness, as well as many ▶ neuropathic pain states.

Mechanisms of Allodynia

The fact that both nociceptive and tactile primary afferents converge on one class of central nociceptive neurons (WDR, wide dynamic range) led to the proposal that central sensitization of WDR neurons to their normal synaptic input may be the mechanism behind dynamic mechanical allodynia. These mechanisms were elucidated in an experimental surrogate model (▶ secondary hyperalgesia surrounding a site of capsaicin injection). Parallel experiments in humans and monkeys showed that capsaicin injection induced dynamic mechanical allodynia (LaMotte et al. 1991) without any changes in the mechanical response properties of nociceptive afferents

Allodynia (Clinical, Experimental), Fig. 1 Assessment of dynamic mechanical allodynia. A 57-year-old male patient with a plexus lesion following abdominal surgery on the left side. (a) Gentle tactile stimuli that do not activate nociceptive afferents were moderately painful on the affected left leg (*filled circles*), whereas they elicited normal non-painful touch sensation on the unaffected right leg (*open circles*). Note that the intensity of allodynia was independent of the pressure exerted by the three stimulators that were stroked across the skin at the same speed. *CW* cotton wisp, *QT* cotton-tipped applicator, *BR* brush. Mean ± SEM across five measurements. (b) Photograph of the three stimulators used for the assessment of dynamic mechanical allodynia in the quantitative sensory testing (QST) protocol of the German Research Network on Neuropathic Pain (Rolke et al. 2006) and video of their mode of application

(Baumann et al. 1991). The responses of spinal cord WDR neurons to brushing, however, were increased following capsaicin injection; in addition, nociceptive-specific HT neurons became responsive to brushing stimuli (Simone et al. 1991). Thus, ▶ central sensitization consisted of enhanced responses of central nociceptive neurons to a normal peripheral input. This was confirmed in humans by electrical microstimulation of tactile Aβ-fibers that evoked a sensation of touch in normal skin but touch plus pain in hyperalgesic skin (Torebjörk et al. 1992). Central sensitization resembles long-term potentiation of excitatory synaptic transmission in other neural systems (Sandkühler 2000). High-frequency electrical stimulation patterns that induce long-term potentiation of synaptic transmission in the dorsal horn also induce mechanical allodynia in human subjects that may outlast the conditioning stimulus for several hours (Klein et al. 2004). Chronic maintenance of the central sensitization leading to allodynia, however, appears to depend on a continuous peripheral nociceptive input that can be dynamically modulated, e.g., by heating and cooling the skin (Gracely et al. 1992; Koltzenburg et al. 1994).

Conflicting Terminology and the Inflationary Use of "Allodynia"

After the introduction of the word "allodynia," there were two terms that could describe a state of increased pain sensitivity, hyperalgesia and allodynia. Researchers and clinicians alike started to wonder when to use which term. The 1994 edition of the IASP pain taxonomy addressed this issue by reserving the word "hyperalgesia" for an enhanced response to a stimulus that is normally painful. Pain induced by stimuli that are not normally painful was to be called "allodynia." Technically, this means that any reduction in pain threshold shall be called "allodynia" (Cervero and Laird 1996).

Allodynia (Clinical, Experimental), Table 1 Peripheral and central sensitization, allodynia, and hyperalgesia

Clinical phenomenon	Input	Peripheral sensitization	Central sensitization	IASP taxonomy 1994		Proposed taxonomy	
				Allodynia	Hyperalgesia	Allodynia	Hyperalgesia
Touch-evoked pain	Tactile Aβ-fibers		X	X		X	(X[a])
Reduced threshold to pinprick pain	Aδ-nociceptors		X	X			X
Increased response to pinprick pain	Aδ-nociceptors		X		X		X
Reduced threshold to heat pain	Aδ- and C-nociceptors	X		X			X
Increased response to heat pain	Aδ- and C-nociceptors	X			X		X

[a]Hyperalgesia is proposed to be used as an umbrella term for all types of enhanced pain sensitivity

Table 1 illustrates why this definition was controversial ever since its introduction. ▶ Peripheral sensitization leads to a leftward shift of the stimulus-response function for heat stimuli, consisting of both a reduction in threshold and an increase in response to suprathreshold stimuli (Raja et al. 1999). The psychophysical correlate, ▶ primary hyperalgesia to heat, now needs to be described with two different terms, simply depending on how it is being tested; if a researcher decides to determine heat pain threshold, its reduction is called "heat allodynia"; if the researcher decides to use suprathreshold stimuli, the increase in perceived pain is called "heat hyperalgesia." Thus, the 1994 IASP taxonomy led to the paradoxical situation that two different names are used to describe a unitary phenomenon, the psychophysical correlate of peripheral sensitization. Likewise, secondary hyperalgesia to pinprick stimuli as a psychophysical correlate of central sensitization to A-fiber nociceptor input is also characterized by reduced pain threshold plus increased suprathreshold pain (Treede et al. 2004).

The 1994 IASP taxonomy was only reluctantly accepted in the scientific community, since time-honored terms such as primary and secondary hyperalgesia (for review see Treede et al. 1992) were artificially fractionated. In the recent past, allodynia was used for an increasing number of phenomena, particularly in animal studies, simply because it is often less difficult to obtain a threshold measure than a suprathreshold measure. This excessive use of the term allodynia, however, has distracted from its original clinical implications. The mechanisms of reduced heat pain threshold have nothing in common with touch-evoked pain, yet both are being called allodynia. In fact, most of the animal studies that use the term "allodynia" are irrelevant for clinical allodynia, because they study reduced withdrawal thresholds for nociceptive stimuli (heat or pinprick).

Instead of artificially dividing two sub-phenomena that by mechanisms of sensitization are intimately linked (threshold and suprathreshold changes), the terms allodynia and hyperalgesia should provide guidance toward a mechanism-based classification of pain. Contrary to the intentions of the authors of the IASP taxonomy, the inflationary use of "allodynia" was also counterproductive for furthering the understanding of the clinical phenomenon that it was originally conceived for, touch-evoked pain.

Clinical Implications and a Unifying Proposal

Semantically, the term "allodynia" implies pain by a stimulus that is alien to the nociceptive system (αλλοσ, Greek for "other"). Thus, allodynia should only be used when the mode of testing allows inference to a pain mechanism that relies on activation of a non-nociceptive input (e.g., low-threshold mechanoreceptors). If pain is reported to stroking the skin with gentle tactile stimuli, this mechanism is strongly implied, and such tests are easily employed in clinical trials as well as in daily practice. The distinction whether enhanced pain sensitivity is due to facilitation of nociceptive or non-nociceptive input is less clear for other stimuli. For example, pain due to gentle cooling, which is a frequent finding in some neuropathic pain states, is still enigmatic and so is the distinction of whether it should be called hyperalgesia or allodynia to cold. Peripheral sensitization of nociceptive afferents, central sensitization to non-nociceptive cold fiber input, or central disinhibition by selective loss of a sensory channel specific for non-noxious cold that exerts a tonic inhibition of nociceptive channels are valid alternatives (Wasner et al. 2004).

Thus, in many cases, the mechanism of enhanced pain sensitivity may be unknown, and it will not be evident whether or not a test stimulus activates nociceptive afferents. For these situations, it is useful to have an umbrella term that does not imply any specific mechanism. Hyperalgesia traditionally was such an umbrella term, corresponding to the leftward shift in the stimulus-response function relating magnitude of pain to stimulus intensity. Parallel to the definition of sensitization, hyperalgesia was characterized by a decrease in pain threshold, increased pain to suprathreshold stimuli, and spontaneous pain. We have therefore suggested the reinstitution of hyperalgesia as the umbrella term for increased pain sensitivity in general (as the antonym to ▶ hypoalgesia) and returning the term allodynia to its old definition, i.e., describing a state of altered somatosensory signal processing wherein activation of non-nociceptive afferents causes pain (Treede et al. 2004).

References

Baumann, T. K., Simone, D. A., Shain, C. N., et al. (1991). Neurogenic hyperalgesia: The search for the primary cutaneous afferent fibers that contribute to capsaicin-induced pain and hyperalgesia. *Journal of Neurophysiology, 66*, 212–227.

Campbell, J. N., Raja, S. N., Meyer, R. A., et al. (1988). Myelinated afferents signal the hyperalgesia associated with nerve injury. *Pain, 32*, 89–94.

Cervero, F., & Laird, J. M. A. (1996). Mechanisms of touch-evoked pain (allodynia): A new model. *Pain, 68*, 13–23.

Gracely, R. H., Lynch, S. A., & Bennett, G. J. (1992). Painful neuropathy: Altered central processing, maintained dynamically by peripheral input. *Pain, 51*, 175–194.

Klein, T., Magerl, W., Hopf, H. C., et al. (2004). Perceptual correlates of nociceptive long-term potentiation and long-term depression in humans. *Journal of Neuroscience, 24*, 964–971.

Koltzenburg, M., Torebjörk, H. E., & Wahren, L. K. (1994). Nociceptor modulated central sensitization causes mechanical hyperalgesia in acute chemogenic and chronic neuropathic pain. *Brain, 117*, 579–591.

LaMotte, R. H., Shain, C. N., Simone, D. A., et al. (1991). Neurogenic hyperalgesia: Psychophysical studies of underlying mechanisms. *Journal of Neurophysiology, 66*, 190–211.

Loh, L., & Nathan, P. W. (1978). Painful peripheral states and sympathetic blocks. *Journal of Neurology, Neurosurgery & Psychiatry, 41*, 664–671.

Merskey, H. (1982). Pain terms: A supplementary note. *Pain, 14*, 205–206.

Noordenbos, W. (1959). *PAIN: Problems pertaining to the transmission of nerve impulses which give rise to pain*. London: Elsevier.

Ochoa, J. L., & Yarnitsky, D. (1993). Mechanical hyperalgesias in neuropathic pain patients: Dynamic and static subtypes. *Annals of Neurology, 33*, 465–472.

Raja, S. N., Meyer, R. A., Ringkamp, M., et al. (1999). Peripheral neural mechanisms of nociception. In P. D. Wall & R. Melzack (Eds.), *Textbook of pain* (4th ed., pp. 11–57). Edinburgh: Churchill Livingstone.

Rolke, R., Magerl, W., Campbell, K. A., et al. (2006). Quantitative sensory testing: A comprehensive protocol for clinical trials. *European Journal of Pain, 10*, 77–88.

Sandkühler, J. (2000). Learning and memory in pain pathways. *Pain, 88*, 113–118.

Simone, D. A., Sorkin, L. S., Oh, U., et al. (1991). Neurogenic hyperalgesia: Central neural correlates in responses of spinothalamic tract neurons. *Journal of Neurophysiology, 66*, 228–246.

Torebjörk, H. E., Lundberg, L. E. R., & LaMotte, R. H. (1992). Central changes in processing of mechanoreceptive input in capsaicin-induced secondary hyperalgesia. *The Journal of Physiology, 448*, 765–780.

Treede, R. D., Handwerker, H. O., Baumgärtner, U., et al. (2004). Hyperalgesia and allodynia: Taxonomy, assessment, and mechanisms. In K. Brune & H. O. Handwerker (Eds.), *Hyperalgesia: Molecular mechanisms and clinical implications* (pp. 1–15). Seattle: IASP Press.

Treede, R. D., Meyer, R. A., Raja, S. N., et al. (1992). Peripheral and central mechanisms of cutaneous hyperalgesia. *Progress in Neurobiology, 38*, 397–421.

Wasner, G., Schattschneider, J., Binder, A., et al. (2004). Topical menthol – A human model for cold pain by activation and sensitization of C nociceptors. *Brain*, *127*, 1159–1171.

Woolf, C. J. (1983). Evidence for a central component of post-injury pain hypersensitivity. *Nature*, *306*, 686–688.

Allodynia and Alloknesis

Robert H. LaMotte
Department of Anesthesiology, Yale School of Medicine, New Haven, CT, USA

Synonyms

Alloknesis and allodynia

Definition

Allodynia and alloknesis are abnormal sensory states wherein normally innocuous stimuli elicit unpleasant sensations or aversive responses.

▶ Allodynia is the nociceptive sensation or aversive response evoked by a stimulus that is normally non-nociceptive ("allo" – "other"; "dynia" – pain). For example, a light stroking of the skin produced by the lateral motion of clothing, or the heat produced by the body are stimuli that do not elicit nociceptive sensations or responses under normal circumstances. However, these stimuli may become nociceptive after a cutaneous injury produced, for example, by sunburn. In contrast, ▶ hyperalgesia is defined as the abnormal nociceptive state in which a normally painful stimulus such as the prick of a needle elicits a greater than normal duration and/or magnitude of pain.

Alloknesis is the itch or ▶ pruriceptive sensation (from the Latin word *prurire*, to itch) or scratching behavior evoked by a stimulus that is normally non-pruriceptive ("allo", and "knesis", an ancient Greek word for itching). For example, a light stroking of the skin normally evokes the sensation of touch and perhaps tickle but not itch. However, when cutaneous alloknesis develops within the vicinity of a mosquito bite, or is present in an area of dermatitis, a light stroking of the skin can evoke an itch or exacerbate an ongoing itch. In contrast, ▶ hyperknesis is defined as the abnormal pruriceptive state in which a normally pruritic stimulus (such as a fine diameter hair which can elicit a prickle sensation followed by an itch) elicits a greater than normal duration and/or magnitude of itch. The cutaneous areas of enhanced itch (alloknesis and hyperknesis) are also referred to as "▶ itchy skin."

The abnormal sensory states of allodynia, alloknesis, hyperalgesia, and hyperknesis that are initiated by an inflammatory or irritating stimulus can exist both within the area directly exposed to the stimulus (in which case they are termed "primary") and can sometimes extend well beyond the area (in which case the sensory states outside the area are termed "secondary"). For example, when the skin receives a local, first-degree burn, primary allodynia and hyperalgesia may exist within the burned skin and secondary allodynia and hyperalgesia in the skin immediately surrounding the burn.

Characteristics

Allodynia is exhibited in a variety of forms such as the tenderness of the skin to combing the hair during a migraine headache, the discomfort of normal movements of the gut with irritable bowel syndrome, the soreness of muscles accompanying musculoskeletal inflammation or trauma, and the chronic tenderness to touch or to gentle warming of the skin associated with trauma or inflammatory diseases of the peripheral or central nervous system.

Allodynia can also be experimentally produced by the application of a noxious or irritant thermal, mechanical or chemical stimulus to the skin. For example, an intradermal injection into the forearm of capsaicin, the irritant agent in hot peppers, elicits not only a burning pain in the immediate vicinity of the injection site but allodynia and hyperalgesia to mechanical stimulation in the surrounding skin not in contact with the irritant (LaMotte et al. 1991) (Fig. 1a).

Allodynia and Alloknesis, Fig. 1 Abnormal sensory states produced by algesic or pruritic chemicals applied to the volar forearm in human. (**a**) The borders of punctate hyperalgesia and allodynia to stroking after an intradermal injection of capsaicin (100 µg). (**b**) The wheal and the borders of hyperalgesia and hyperknesis to punctate stimulation and alloknesis to stroking after an intradermal injection of histamine (20 µg). (**c**) The borders of hyperknesis, hyperalgesia, and alloknesis after the insertion a few cowhage spicules into the skin. Capsaicin and histamine evoked a flare (not shown), but cowhage did not. A different subject was used in each experiment

Alloknesis

Itchy skin and/or itch are characteristic of many cutaneous disorders such as atopic, allergic, and irritant contact dermatitis and can accompany such systemic diseases as renal insufficiency, cholestasis, Hodgkin's disease, polycythemia vera, tumors, and HIV infection.

Alloknesis can be experimentally produced in human volunteers by the iontophoresis (Magerl et al. 1990) or intradermal injection (Simone et al. 1991b) of histamine into the skin. The histamine evokes a sensation of itch accompanied by local cutaneous reactions consisting of a flare (redness of the skin mediated by a local axon reflex wherein vasodilatory neuropeptides are released by collaterals of activated nerve endings) and a wheal (local edema) (Simone et al. 1991b) (Fig. 1b). Within the wheal and within the surrounding skin that is not exposed to histamine, there develops alloknesis to lightly stroking the skin and hyperknesis and hyperalgesia to

mechanical indentation of the skin with a fine prickly filament (Simone et al. 1991b; Atanassoff et al. 1999; LaMotte et al. 2009).

Itch and alloknesis can also be produced in the absence of a flare or wheal by single spicules of cowhage (*Mucuna pruriens*), a tropical legume (LaMotte et al. 2009) (Fig. 1c). Because the wheal and flare are elicited in response to histamine, the absence of these reactions in response to cowhage suggests that itch and itchy skin can be elicited by histamine-independent mechanisms, as is the case in most kinds of clinical pruritus.

Interactions Between Pain and Itch

Pain and hyperalgesia have an inhibitory effect on itch and itchy skin. The enhanced itch and itchy skin resulting from injecting histamine into an anesthetic bleb of skin (as opposed to a bleb of saline) have been explained on the basis of a reduced activation of histamine responsive nociceptive neurons (Atanassoff et al. 1999). In contrast, histamine-induced itch and itchy skin are absent or attenuated in the hyperalgesic skin surrounding a capsaicin injection (Brull et al. 1999). Thus, even though alloknesis and hyperknesis co-exist with the area of mild hyperalgesia induced by histamine (Fig. 1b), they are suppressed or prevented from developing when the hyperalgesia becomes sufficiently intense, as is the case after the injection of capsaicin. Similarly, cowhage spicules produced neither itch nor alloknesis within an area of hyperalgesia produced by a heat injury of the skin (Graham et al. 1951). Observations such as these confirm the existence of functional interactions between pruriceptive and nociceptive neural systems and lend support to the hypothesis that the mechanisms of itch and itchy skin are inhibited centrally by mechanisms that underlie pain and hyperalgesia (Brull et al. 1999; Nilsson et al. 1997; Ward et al. 1996).

Neural Mechanisms of Allodynia and Alloknesis

Allodynia and hyperalgesia from an intradermal injection of capsaicin are believed to be initiated as a result of activity in a subpopulation of mechanically insensitive nociceptive afferent peripheral neurons (MIAs) (LaMotte 1992; Schmelz et al. 2003). A working model of the neural mechanisms of capsaicin-induced allodynia and hyperalgesia posits that capsaicin-responsive MIAs release neurochemicals that sensitize nociceptive neurons in the dorsal horn of the spinal cord. These neurons, in turn, receive convergent input from (a) low-threshold primary afferents with thickly myelinated axons mediating the sense of touch and (b) nociceptive afferents with thinly myelinated axons mediating the sense of mechanically evoked pricking pain. The sensitized neurons exhibit a de novo or greater than normal response to innocuous tactile stimuli, as well as an enhanced response to noxious punctate stimulation, thereby accounting for allodynia and hyperalgesia, respectively. In support of this is the reported sensitization of nociceptive spinothalamic tract (STT) neurons, recorded electrophysiologically in animals, to innocuous touch and to noxious punctate stimulation after an intradermal injection of capsaicin (Simone et al. 1991a) via a mechanism called ▶ central

sensitization (see also Fig. 2 in ▶ Ectopia, Spontaneous regarding possible chronic central sensitization leading to allodynia and hyperalgesia after injury of peripheral sensory neurons).

Alloknesis and hyperknesis might be explained using a similar mechanistic model (LaMotte 1992). That is, there may exist pruriceptive STT neurons that can become sensitized to light mechanical touch and to punctate stimulation with a fine filament, after an application of histamine or cowhage to the skin, thereby accounting for alloknesis and hyperknesis respectively (Davidson et al. 2012).

A subpopulation of nociceptive peripheral MIAs, in humans, responds to histamine but not to cowhage (Schmelz et al. 2003; Namer et al. 2009). Also, subpopulations of mechanosensitive nociceptive neurons, some with unmyelinated axons and others with thinly myelinated axons, respond to histamine and/or to cowhage in humans, monkeys, and mice (Han et al. 2012; Johanek et al. 2008; Namer et al. 2009; Ringkamp et al. 2011). Other nociceptive peripheral neurons respond to thermal, mechanical, or chemical noxious stimuli but not to pruritic chemicals such as histamine or cowhage.

Some of these neurons in human and monkeys exhibited responses that were comparable in time course to the sensation of itch reported by humans in response to the same stimuli (LaMotte et al. 2009; Ringkamp et al. 2011).

In monkeys, a subpopulation of nociceptive STT neurons responded either to histamine or to cowhage but rarely to both (Davidson et al. 2012). But the majority of nociceptive STT neurons did not respond to either pruritic agent.

Scratching can reduce the ongoing histamine evoked activity in pruriceptive STT neurons (Davidson et al. 2009). Thus, chemically evoked itch may occur when there is a selective activation of pruriceptive neurons in the absence of sufficient activity in non-pruriceptive nociceptive neurons such as those normally activated by noxious mechanical stimuli. Similarly, alloknesis may be chemically evoked via activity in pruriceptive neurons but presumably in the absence of the sensitization of the non-pruriceptive central neurons that mediate allodynia.

References

Atanassoff, P. G., Brull, S. J., Zhang, J. M., et al. (1999). Enhancement of experimental pruritus and mechanically evoked dysesthesias with local anesthesia. *Somatosensory and Motor Research, 16,* 299–303.

Brull, S. J., Atanassoff, P. G., Silverman, D. G., et al. (1999). Attenuation of experimental pruritus and mechanically evoked dysesthesias in an area of cutaneous allodynia. *Somatosensory and Motor Research, 16,* 291–298.

Davidson, S., Zhang, X., Khasabov, S. G., Simone, D. A., Giesler, G. J. Jr. (2009). Relief of itch by scratching: State-dependent inhibition of primate spinothalamic tract neurons. Nature Neuroscience, *12*(5), 544–546.

Davidson, S., Zhang, X., Khasabov, S. G., Moser, H. R., Honda, C. N., Simone, D. A., & Giesler, G. J. Jr. (2012). Pruriceptive spinothalamic tract neurons: physiological properties and projection targets in the primate. *Journal of Neurophysiology, 108*(6), 1711–1723.

Graham, D. T., Goodell, H., & Wolff, H. G. (1951). Neural mechanisms involved in itch, "itchy skin," and tickle sensations. *The Journal of Clinical Investigation, 30,* 37–49.

Han, L., Ma, C., Liu, Q., Weng, H. J., Cui, Y., Tang, Z., Kim, Y., Nie, H., Qu, L., Patel, K. N., Li, Z., McNeil, B., He, S., Guan, Y., Xiao, B., LaMotte, R. H., & Dong, X. (2012). A subpopulation of nociceptors specifically linked to itch. *Nature Neuroscience, 16*(2), 174–182. PMCID: PMC3557753.

Johanek, L. M., Meyer, R. A., Friedman, R., Greenquist, K. W., Shim, G., Borzan, J., Hartke, T.V., LaMotte, R. H., & Ringkamp, M. (2008). A role for polymodal C-fiber afferents in non-histaminergic itch. *Journal of Neuroscience, 28*(30), 7659–7669. PMC2564794.

LaMotte, R. H. (1992). Subpopulations of "nocifensor neurons" contributing to pain and allodynia, itch and alloknesis. *American Pain Society Journal, 1,* 115–126.

LaMotte, R. H., Shain, C. N., Simone, D. A., et al. (1991). Neurogenic hyperalgesia: Psychophysical studies of underlying mechanisms. *Journal of Neurophysiology, 66,* 190–211.

LaMotte, R. H., Shimada, S. G., Green, B. G., & Zelterman, D. (2009). Pruritic and nociceptive sensations and dysesthesias from a spicule of cowhage. *Journal of Neurophysiology, 101*(3), 1430–1443. PMCID: PMC2666414.

Magerl, W., Westerman, R. A., Mohner, B., et al. (1990). Properties of transdermal histamine iontophoresis: Differential effects of season, gender and body region. *The Journal of Investigative Dermatology, 94,* 347–352.

Namer, B., Carr, R., Johanek, L. M., Schmelz, M., Handwerker, H. O., & Ringkamp, M. (2008). Separate peripheral pathways for pruritus in man. *Journal of Neurophysiology, 100,* 2062–2069.

Nilsson, H.-J. A., Levinsson, A., & Schouenborg, J. (1997). Cutaneous field stimulation (CFS) – a new powerful method to combat itch. *Pain, 71,* 49–55.

Ringkamp, M., Schepers, R. J., Shimada, S. G., Johanek, L. M., Hartke, T. V., Borzan, J., Shim, B., LaMotte, R. H., & Meyer, R. A. (2011). A role for nociceptive, myelinated nerve fibers in itch sensation. *Journal of Neuroscience, 31*(42), 14841–14849. PMC3218799.

Schmelz, M., Schmidt, R., Weidner, C., et al. (2003). Chemical response pattern of different classes of C-nociceptors to pruritogens and algogens. *Journal of Neurophysiology, 89,* 2441–2448.

Simone, D. A., Oh, U., Sorkin, L. S., et al. (1991a). Neurogenic hyperalgesia: Central neural correlates in responses of spinothalamic tract neurons. *Journal of Neurophysiology, 66,* 228–246.

Simone, D. A., Alreja, M., & LaMotte, R. H. (1991b). Psychophysical studies of the itch sensation and itchy skin ("alloknesis") produced by intracutaneous injection of histamine. *Somatosensory and Motor Research, 8,* 271–279.

Ward, L., Wright, E., & McMahon, S. B. (1996). A comparison of the effects of noxious and innocuous counterstimuli on experimentally induced itch and pain. *Pain, 64,* 129–138.

Allodynia in Fibromyalgia

Definition

A lowered pain threshold that characterizes clinical examination findings in fibromyalgia.

Cross-References

▶ Muscle Pain, Fibromyalgia Syndrome (Primary, Secondary)

Allodynia Test, Mechanical and Cold Allodynia

Kyungsoon Chung
Department of Neuroscience and Cell Biology,
University of Texas Medical Branch,
Galveston, TX, USA

Synonyms

Cold allodynia test; Mechanical allodynia test

Definition

Allodynia is defined as "pain due to a stimulus which does not normally provoke pain" by the International Association for the Study of Pain (Lindblom et al. 1986). It is important to recognize that allodynia involves a change in the quality of a sensation, since the original modality is normally non-painful but the response is painful. There is, thus, a loss of specificity of a sensory modality.

Characteristics

Because allodynia is an evoked pain, testing requires an external stimulation of non-painful quality. Mainly three different types of stimulation have been used to test allodynia in animal models of neuropathic pain: mechanical touch, cold, and air blow. All testing methods rely on withdrawal response of the affected body parts, such as the foot or face, to stimulus, based on the premise that the animal's avoidance of touching, cooling, or air blowing is an allodynic reaction.

Mechanical (Tactile) Allodynia Test: Foot Withdrawal Response to Von Frey Filament Stimulus

Since cutaneous tactile allodynia is a major complaint of neuropathic pain patients (Bonica 1990), testing for signs of mechanical (tactile) allodynia is an important aspect of behavioral tests for neuropathic pain. Mechanical allodynia is often tested by quantifying tactile sensitivity, using a set of von Frey filaments (also referred to as Semmes-Weinstein [S-W] monofilaments, a series of nylon monofilaments of increasing stiffness that exert defined levels of force as they are pressed to the point where they bend; Stoelting Co., Wood Dale, IL). Mechanical sensitivity is quantified either by determining mechanical threshold (Baik et al. 2003; Chaplan et al. 1994; Tal and Bennett 1994), or by measuring response frequency (Hashizume et al. 2000; Kim and Chung 1992).

Measurement of Mechanical Thresholds in Rodents

Although there are several ways of measuring mechanical thresholds, measurement of foot withdrawal thresholds to mechanical stimuli by using the up-down method has been utilized most extensively (Baik et al. 2003; Chaplan et al. 1994). The rats are placed under a transparent plastic dome (85 × 80 × 280 mm) on a metal wire mesh floor. A series of 8 von Frey (VF) filaments with approximately equal logarithmic incremental (0.22) VF values (3.65, 3.87, 4.10, 4.31, 4.52, 4.74, 4.92, and 5.16) are used to determine the threshold stiffness required for 50 % paw withdrawal. Because VF values are logarithmically related to gram (g) values [$VF=\log(1000\times g)$], the chosen VF numbers are equivalent to 0.45, 0.74, 1.26, 2.04, 3.31, 5.50, 8.32, and 14.45 in gram value, respectively. Starting with filament 4.31, VF filaments are applied perpendicular to the plantar surface of the hind paw and depressed until they bend for 2–3 s. Whenever a positive response to a stimulus occurs, the next smaller VF filament is applied. Whenever a negative response occurs, the next higher one is applied. The test is continued until the response of six stimuli, after the first change in response, has been obtained or until the test reaches either end of the spectrum of the VF set. The 50 % threshold value is calculated by using the formula of Dixon (Dixon 1980) : 50 % threshold=X+kd, where X is the value of the final VF filament used (in log units), k is the tabular value for the pattern of positive/negative responses, and d is the mean difference between stimuli in log units (0.22). In the case where continuous positive or negative responses are observed all the way out to the end of the stimulus spectrum, values of 3.54 or 5.27 are assigned, respectively, by assuming a value of ±0.5 for k. The outcome of behavioral data is expressed as VF values (maximum range, 3.54·5.27) and plotted in a linear scale. Because VF values are logarithmically related to gram values, plotting in gram values requires logarithmic plots. The mechanical threshold for foot withdrawal in a normal rat is usually a VF value of 5.27 (18.62 g) (Baik et al. 2003). After L5 spinal nerve ligation, mechanical thresholds decline to around 3.54 (0.35 g) by the third day, and this level is maintained for weeks (Park et al. 2000). Since thresholds of most nociceptors are higher than 1.5 g (Leem et al. 1993), foot withdrawals elicited lower than this value can be assumed to be mechanical allodynia when the foot does not show a sign of inflammation and thus peripheral sensitization.

The same procedures can be applied to mice with a different set of VF filaments (VF values of 2.44, 2.63, 3.22, 3.61, 4.0, 4.35, and 4.74 that are equivalent to 0.03, 0.07, 0.2, 0.4, 1.0, 2.7, and 5.5 g force, respectively). In mice, the baseline mechanical sensitivity as well as the levels of developed hypersensitivity after nerve injury varied among different strains of mice (Mogil et al. 1999).

Another method has also been used to determine mechanical thresholds based on foot withdrawal reflex responses to VF filament stimulation. In this experimental paradigm, a series of VF filaments whose stiffness are within a non-painful stimulus range are selected, based on the testing locations. The VF filaments are applied perpendicular to the skin and depressed until they bend, flexor withdrawal reflexes are then observed. Starting from the weakest filament, the von Frey filaments are tested in order of increasing stiffness. The minimum force (determined as the stiffness of VF filament) required to elicit a flexor withdrawal reflex is recorded as the mechanical threshold. Depending on each specific experiment, the number of applications with each VF filament, times of intervals between stimuli, and the criteria of threshold determination were somewhat variable. For example, the first filament in the series that evoked at least one response from five applications was designated as the threshold by Tal and Bennett (1994), while Ma and Woolf (1996) determined that the minimum force required to elicit a reproducible flexor withdrawal reflex on each of three applications of the VF filaments would be recorded as the threshold.

Measurements of Paw Withdrawal Frequencies
The general method of stimulus application with VF filaments, and recording positive or negative

withdrawal reflex responses, are the same as the method used for the threshold measurement. The differences are:
1. Sensitivity testing is done by repeated stimuli with each defined VF filament.
2. Frequency of positive response is measured and used as an indicator of mechanical sensitivity.

In one experiment, mechanical stimuli are applied to the plantar surface of the hind paw with six different von Frey filaments ranging from 0.86 to 19.0 g (0.86, 1.4, 2.5, 5.6, 10.2, 19.0 g). The 0.86 and 19.0 g filaments produce a faint sense of touch and a sense of pressure, respectively, when tested on our own palm. A single trial of stimuli consisted of 6–8 applications of a von Frey filament within a 2–3 s period; each trial is repeated five times at approximately 3-min. intervals on each hind paw. The occurrence of foot withdrawal in each of five trials was expressed as a percent response frequency [number of foot withdrawals/5 (number of trials) $\times$ 100 = % response frequency], and this percentage is used as an indication of mechanical sensitivity. For a given test day, the same procedure is repeated for the remaining five different von Frey filaments, in ascending order starting from the weakest. In the sham operated control rat, the strongest VF filament (19.0 g) produces a 10 % response, but none of the other filaments produced any response (0 %). Seven days after L5/6 spinal nerve ligation, response frequency increases to 40 % and 80 % by stimuli with 0.86 and 19.0 g filaments, respectively (Kim and Chung 1992).

Cold Allodynia Test: Foot Withdrawal Response to Acetone or Cold Plate
Two different methods have been used for cold allodynia testing in animal models of neuropathic pain: the acetone test and the cold plate test.

Acetone Test
The rat is placed under a transparent plastic dome on a metal mesh floor and acetone is applied to the plantar surface of the foot. Application of acetone is done by an acetone bubble formed at the tip of a piece of polyethylene

A

tubing (1/16 ID), which is connected to a syringe. The bubble is then gently touched to the heel. The acetone quickly spreads over the proximal region of the plantar surface of the foot and evaporates. On our own palm surface of the hand, this stimulus produces a strong but non-painful cooling sensation as the acetone evaporates. Normal rats either ignore the stimulus, or it produces a very brief withdrawal response. After L5/6 spinal nerve ligation, rats briskly withdraw the hind foot after some delay (about 0.2–0.3 s) and subsequently shake, tap, or lick the hind paw in response to acetone application to the affected paw. For quantification of cold allodynic behavior, acetone is applied five times (once every 5 min) to each paw. The frequency of foot withdrawal is expressed as a percent: (number of trials accompanied by brisk foot withdrawal) × 100/(number of total trials). As a control, warm water (30 °C) is applied in the same manner as acetone. A significant increase in the frequency of foot withdrawals in response to acetone application was interpreted as cold allodynia (Choi et al. 1994).

In another experiment, 0.15 ml of acetone was sprayed onto the plantar surface of the hind paw for assaying cold allodynia. As in the acetone bubble test, normal rats either ignore the stimulus or it produces a very brief and small withdrawal reflex. Rats with sciatic neuritis reacted with a large and prolonged withdrawal response. Approximately one-half of the neuritic rats displayed cold allodynia while almost all rats with chronic constriction injury to the sciatic nerve showed cold allodynia (Bennett 1999).

Cold Plate Test

In the cold plate test, rats are confined beneath an inverted, clear plastic cage (18 × 28 × 13 cm) placed upon a metal floor (e.g., aluminum plate), which is chilled to 4 °C by an underlying water bath. While exposed to the cold floor for 20 min, the animals' behavior is monitored, and the frequency of hind paw withdrawals and the duration the hind paw held above the floor (i.e., hind paw withdrawals related to stepping are not counted) are measured. The 4 °C floor does not produce any pain when our palm of the hand is immobilized on it for 20 min, and it does not evoke any pain-related responses from unoperated control rats. In neuropathic rats with sciatic chronic constriction injury, the average frequency and cumulative duration of hind paw withdrawals on the nerve-damaged side increases about fivefold and twofold, respectively, compared to that of normal rats. In addition, some rats also demonstrate vague, scratching-like movements and also lick the affected hind paw (Bennett and Xie 1988). This method is based on the premise that the animal's avoidance of touching the cold plate is an allodynic reaction. One notable fact, however, is that a complete denervation of the foot does not change this behavior (Choi et al. 1994), making it questionable that the foot lifting behavior is related to allodynia, since allodynia would require the presence of functioning sensory receptors.

Air-Puff Allodynia Test: Withdrawal Response to Air-Puff Stimuli

The air-puff testing has been used to quantify the mechanical allodynia of the orofacial region in the animal model of trigeminal neuralgia (Ahn et al. 2004, 2005). The testing is based on face withdrawal behavior in response to constant air puffs of graded pressures applied to the affected orofacial area. A custom-designed, pressure-regulated air-puff stimulator is used to provide air stimulation (Yeomans and Klukinov 2012). The withdrawal thresholds were estimated from 10 consecutive trials of constant-pressure air puffs, 4-s duration, a 10-s interstimulus interval (Ahn et al. 2005), are performed with a ramp of air-puff pressures. In normal naïve rats, the air-puff pressure that produces withdrawal behaviors is approximately 25 psi, but this pressure decreased to 5 psi 30 min after a subcutaneous injection of IL-1β into the vibrissa. This allodynia persisted at least 180 min (Ahn et al. 2004).

References

Ahn, D. K., Jung, C. Y., Lee, H. J., Choi, H. S., Ju, J. S., & Bae, Y. C. (2004). Peripheral glutamate receptors participate in interleukin-1β-induced mechanical allodynia in the ororfacial area of rats. *Neuroscience Letters, 357*, 203–206.

Ahn, D. K., Kim, K. H., Jung, C. Y., Choi, H. S., Lim, E. J., Youn, D. H., & Bae, Y. C. (2005). Role of peripheral group I and II metabotropic glutamate receptors in IL-1β-induced mechanical allodynia in the ororfacial area of concious rats. *Pain, 118*, 53–60.

Baik, E. J., Chung, J. M., & Chung, K. (2003). Peripheral norepinephrine exacerbates neuritis-induced hyperalgesia. *The Journal of Pain, 4*, 212–221.

Bennett, G. J. (1999). Does a neuroimmune interaction contribute to the genesis of painful peripheral neuropathies? *Proceedings of the National Academy of Sciences of the United States of America, 96*, 7737–7738.

Bennett, G. J., & Xie, Y.-K. (1988). A peripheral mononeuropathy in Rat that produces disorders of pain sensation like those seen in man. *Pain, 33*, 87–107.

Bonica, J. J. (1990). Causalgia and other reflex sympathetic dystrophies. In J. J. Bonica (Ed.), *The management of pain* (pp. 220–243). Philadelphia: Lea & Febiger.

Chaplan, S. R., Bach, F. W., Pogrel, J. W., Chung, J. M., & Yaksh, T. L. (1994). Quantitative assessment of tactile allodynia in the rat paw. *Journal of Neuroscience Methods, 53*, 55–63.

Choi, Y., Yoon, Y. W., Na, H. S., Kim, S. H., & Chung, J. M. (1994). Behavioral signs of ongoing pain and cold allodynia in a rat model of neuropathic pain. *Pain, 59*, 369–376.

Dixon, W. J. (1980). Efficient analysis of experimental observations. *Annual Review of Pharmacology and Toxicology, 20*, 441–462.

Hashizume, H., Rutkowski, M. D., Weinstein, J. N., & DeLeo, J. A. (2000). Central administration of methotrexate reduces mechanical allodynia in an animal model of radiculopathy/sciatica. *Pain, 87*, 159–169.

Kim, S. H., & Chung, J. M. (1992). An experimental model for peripheral neuropathy produced by segmental spinal nerve ligation in the rat. *Pain, 50*, 355–363.

Leem, J. W., Willis, W. D., & Chung, J. M. (1993). Cutaneous sensory receptors in the rat foot. *Journal of Neurophysiology, 69*, 1684–1699.

Lindblom, U., Merskey, H., Mumford, J. M., Nathan, P. W., Noordenbos, W., & Sunderland, S. (1986). Pain terms, a current list with definitions and notes on usage. *Pain, 3*, 215–221.

Ma, Q.-P., & Woolf, C. J. (1996). Progressive tactile hypersensitivity: An inflammation-induced incremental increase in the excitability of the spinal cord. *Pain, 67*, 97–106.

Mogil, J. S., Wilson, S. G., Bon, K., Lee, S. E., Chung, K., Raber, P., Pieper, J. O., Hain, H. S., Belknap, J. K., Hubert, L., Elmer, G. I., Chung, J. M., & Devor, M. (1999). Heritability of nociception I: Responses of 11 inbred mouse strains on 12 measures of nociception. *Pain, 80*, 67–82.

Park, S. K., Chung, K., & Chung, J. M. (2000). Effects of purinergic and adrenergic antagonists in a rat model of painful peripheral neuropathy. *Pain, 87*, 171–179.

Tal, M., & Bennett, G. J. (1994). Extra-territorial pain in rats with a peripheral mononeuropathy: Mechano-hyperalgesia and mechano-allodynia in the territory of an uninjured nerve. *Pain, 57*, 375–382.

Yeomans D. C, & Klukinov, M. (2012). A rodent model of trigeminal neuralgia. Chapter 8, In Z. D. Luo (Ed.), *Pain research: Methods and protocols, methods in molecular biology* Vl. 851 (pp. 121–131), Springer Science LLC.

Allodynia: General

Definition

The International Association for the Study of Pain (IASP) defines allodynia as: "Pain due to a stimulus that dose not normally provoke pain." This may occur in any somatosensory modality: mechanical, thermal, or chemical, and so tactile allodynia, heat allodynia, etc. Any reduction in pain threshold is "allodynia," whereas increased pain to suprathreshold stimuli is referred to as "hyperalgesia." Allodynia is a neologism derived from the Greek "allo" = "other" and "dynia" = pain.

Cross-References

▶ Allodynia and Alloknesis
▶ Anesthesia Dolorosa Model, Autotomy
▶ Calcium Channels in the Spinal Processing of Nociceptive Input
▶ Chronic Pelvic Pain, Musculoskeletal Syndromes
▶ Clitoral Pain
▶ Cognitive-Behavior Treatment of Pain
▶ Complex Regional Pain Syndromes, Clinical Aspects
▶ CRPS, Evidence-Based Treatment
▶ CRPS-1 in Children
▶ Deafferentation Pain
▶ Descending Circuits in the Forebrain, Imaging
▶ Diagnosis and Assessment of Clinical Characteristics of Central Pain
▶ Dietary Variables in Neuropathic Pain
▶ Drugs Targeting Voltage-Gated Sodium and Calcium Channels
▶ Drugs with Mixed Action and Combinations: Emphasis on Tramadol

Allodynia: Skin

Definition

See also "▶ Primary and Secondary Hyperalgesia."

Allodynia means pain due to a stimulus that does not normally provoke pain and is mostly used as a synonym for "Touch evoked hyperalgesia" (e.g., pain evoked by stroking the skin with a wisp of cotton wool). This kind of stimulation does not excite nociceptors – even in their sensitized state. It excites sensitive mechanoreceptor units belonging to "another sensory modality." It typically occurs in an inflamed area, but also in a surrounding secondary zone, and therefore has been attributed to altered central synaptic processes. However, it often depends on the ongoing barrage to projection neurons in the spinal cord by impulses of sensitized and spontaneously active nociceptors. For example, pricking capsaicin into the skin induces allodynia at the injection site and in a surrounding secondary zone. If the spontaneous activity of nociceptors is brought to subside by cooling the injection site, the allodynia vanishes.

Cross-References

Alloknesis and Allodynia

Alpha(α) 2-Adrenoceptor Agonists

Definition

Drugs that activate α2-adrenoceptors.

Cross-References

Alpha(2)-Adrenergic Receptors

Definition

The sympathetic nervous system is an involuntary system that plays an important role in normal physiological functions, such as control of body temperature and regulation of blood flow to various tissues in the body. These nerves release a chemical called norepinephrine that activates specific receptors, called adrenergic receptors or adrenoceptors. There are two main subtypes of adrenoceptors – one of which is the alpha adrenoceptors.

Cross-References

▶ Sympathetically Maintained Pain in CRPS II, Human Experimentation

Alpha(α) 2-Adrenergic Agonists in Pain Treatment

Benjamin Thomas[1], Carsten Bantel[2], Laura Stone[3] and George L. Wilcox[4]
[1]Chelsea & Westminster Hospital Foundation Trust, London, UK
[2]Department of Surgery & Cancer? Anaesthetics Section, Imperial College of Science and Medicine Chelsea and Westminster Campus, London, UK
[3]Faculty of Dentistry, Alan Edwards Centre for Research on Pain, Montreal, QC, Canada
[4]Department of Neuroscience, Pharmacology and Dermatology, University of Minnesota Medical School, Minneapolis, MN, USA

Synonyms

Alpha(α) 2-adrenergic receptor agonists; Alpha (α) 2-agonists; α_{2A} or α_{2C}-adrenoceptor agonists; α_2-adrenergic agonists; α_2-receptor agonists; α-agonists

Definition

Alpha$_2$-adrenergic agonists are drugs that mediate their analgesic (antinociceptive) effects by acting on α_2-adrenoceptors (α_{2A}, α_{2B}, α_{2C}) in the peripheral and central nervous system.

Characteristics

Indications and Patients

Alpha$_2$-adrenoceptor (α_2AR) agonists are used for treatment of acute (intra- and postoperative) as well as chronic (neuropathic) pain states. They are effective in patients of all age groups. α_2AR agonists have also been safely used in pregnancy, labor, and during cesarean sections. Furthermore, there is evidence that they provide hemodynamic stability in patients with coexisting cardiovascular diseases during phases of noxious stimulation (e.g., orotracheal intubation) by attenuating the sympathetic response.

Dose and Route of Administration

With ▶ clonidine being the prototypical α_2AR agonist, these drugs have been administered in different doses and by a wide variety of routes: systemic, peripheral, regional, neuraxial, and central (Table 1). They have been used as premedication, in combination with other drugs, or as sole analgesic during and after surgery, and in the treatment of chronic pain either by bolus or continuous infusions or as part of a ▶ patient-controlled analgesia (PCA) regimen.

Drug Interactions

Preclinical and clinical studies investigating the antinociceptive effect of α_2AR agonists and their interactions with other drug classes have demonstrated synergistic interaction with opioids as well as opioid-sparing effects. The rare observation of synergy between drugs of the same class has also been made with coadministration of clonidine and dexmedetomidine intrathecally in mice. Possible mechanisms are discussed below. It has also been demonstrated that α_2AR agonists have been demonstrated to reduce the ▶ minimal alveolar concentration (MAC) of volatile

Alpha(α) 2-Adrenergic Agonists in Pain Treatment, Table 1 Alpha 2-adrenergic drug dosing: clonidine

Route		Dose	Duration
Premedication			
Children		2–4 µg/kg	
Elderly patients		1–2 µg/kg	
Perioperative analgesia			
Intrathecal		Max 1 µg/kg	
		Up to 75 µg	
Epidural	With opioid	1–4 µg/kg	Long; dose dependent
Caudal	With mepivacaine	0.75–3 µg/kg	Up to 24 h
Peripheral nerve block		0.1–0.5 µg/kg	
Bier block		1–2 µg/kg	
Postoperative analgesia			
Epidural	Clonidine alone	1–4 µg/kg	
		100–150 µg/h	
Analgesia for labor pain			
Intrathecal	With bupivacaine	50–200 µg	
		max 1 µg/kg	
Epidural	With bupivacaine + fentanyl	30–150 µg	
		75 µg	
Chronic pain			
Epidural	Infusion	100–900 µg	8 h
		30 µg/h	Up to 2 weeks

anesthetics and attenuate the pain from propofol injection. Numerous studies have shown that combining α_2AR agonists with local anesthetics both prolongs the sensory blockade and also improves the quality of the block. Subsequently, α_2AR agonists may be considered as an adjuvant therapy for both general and local anesthesia.

Other Effects

Compared to opioids, far less respiratory depression is seen with α_2AR agonists. Drugs of this class produce sedation by an action that originates in the brainstem and converges on the endogenous pathways responsible for non-REM sleep. Dose-dependent effects of α_2AR agonists are also noted in the cardiovascular system. At low doses, these drugs induce hypotension through actions on locus coeruleus and nucleus tractus solitarius, which results in a decrease in sympathetic outflow. At higher doses, α_2AR agonists induce vasoconstriction in the periphery and can result in a rise in systemic blood pressure. A combination of

sympatholytic and vagomimetic actions of α_2AR agonists cause a decrease in heart rate. Additional features that are useful in the perioperative period include the ability of α_2AR agonistic drugs to produce xerostomia (dry mouth) and anxiolysis.

Analgesic (Antinociceptive) Sites of Action

The α_2ARs are G-protein-coupled receptors associated with $G_{i/o}$ heterotrimeric G proteins. They exist in three subtypes: A, B, and C (α_{2A}AR, α_{2B}AR, α_{2C}AR). They are present on peripheral nerves, in the spinal cord, and at supraspinal pain-modulating centers. They have therefore been applied to all parts of the nervous system in an effort to generate analgesia in patients or antinociception in animals (for review, see Fairbanks et al. 2009b).

Periphery

Although in preclinical models, peripheral injections or topical application of α_2AR agonists appeared promising for pain control (Li et al. 2007), the utility of local peripheral administration has proven to be inconsistent in clinical studies. These inconsistencies may be due to the patient population examined, as topical clonidine has been shown to be antihyperalgesic in the subset of neuropathic pain patients with sympathetically maintained pain. Consistent with that inference, a very recent clinical trial demonstrated that patients suffering from diabetic neuropathy and retaining some nociceptive fibers in their feet received excellent therapeutic benefits from topical treatment with the α_2AR agonist clonidine (Campbell et al. 2012).

Peripheral α_2ARs are found on sympathetic and sensory nerves, where they have been proposed to act as autoreceptors to inhibit neuronal excitability and transmitter release. There is a growing body of evidence that an inflammatory response might be prerequisite for the peripheral site of action of α_2AR receptor agonists. This has been hypothesized because of the demonstration of α_2ARs on inflammatory cells, especially macrophages. Perineural application of the α_2AR agonist clonidine reduced nerve injury-induced release of the proinflammatory cytokine TNFα, and the time course of this action was paralleled by a clear antinociceptive effect in an animal model of ▶ neuropathic pain. Hence, it is now suggested that macrophages invade the site of traumatic nerve damage and contribute to an inflammation-maintained pathogenic mechanism through the release of proinflammatory cytokines and that α_2AR agonists attenuate this process by reducing the inflammatory response rather than by direct action on peripheral nerves (Lavand'homme and Eisenach 2003).

Spinal Cord

From recent data, the spinal cord dorsal horn has clearly emerged as a pivotal site of α_2AR analgesic action. Administration of α_2AR agonists result in antinociception and analgesia in animal models and human subjects by both pre- and postsynaptic actions. These spinal analgesic actions of α_2AR agonists are largely mediated by the α_{2A}AR subtype, and presynaptic α_{2A}ARs on primary afferent nociceptive Aδ- and C-fibers are positioned to directly modulate pain processing through attenuation of excitatory synaptic transmission

(Stone et al. 1997, 1998; Riedl et al. 2009). This has been supported by results showing an inhibitory effect of α_2AR on spinal glutamate release in synaptosomal and electrophysiological experiments (Kawasaki et al. 2003; Li and Eisenach 2001). The mechanisms underlying the actions of the α_{2A}AR subtype have been investigated recognizing the known synergy of opioids and a_2AR agonists. Coexpression of the δ-opioid receptor (DOP) and the α_{2A}AR occurs on spinal primary afferent fibers and inhibition of calcitonin gene-related peptide (CRGP) release occurs on agonism of both. Using both the DOP agonist deltorphin II (DELT) and clonidine, possible mechanisms of signal transmission were evaluated in spinal slices (Overland et al. 2009). Coadministering inhibitors of phospholipase C (PLC), protein kinase A, and protein kinase C (PKC), the effects of individual agonist administration were found to be dependent only on PLC while synergy resulted from the activation of both PLC, and PKC together. PKA was found not to be involved in the action of either DELT or clonidine. Additionally, this CGRP release inhibition was maintained in the presence of tetrodotoxin, suggesting synergy occurs within a single subcellular compartment. This in vitro PKC-dependent system was further found to translate to synergistic antinoception in vivo when using a thermal antinociceptive test in mice. Worth noting simultaneously are the findings that antinociceptive synergy is seen preclinically when coadministering the two α_2AR agonists clonidine and dexmedetomidine intrathecally in mice (Fairbanks et al. 2009a). Both agonists have been thought to act via the α_{2A}AR subtype but observations made in knockout mice have shown that the α_{2C}AR subtype, likely on secondary dorsal horn neurons (Olave and Maxwell 2003), is simultaneously required for synergy to occur. Direct hyperpolarization of postsynaptic spinal neurons by α_2AR agonists may also play an important role in the spinal analgesic action of α_2AR agonists (Sonohata et al. 2004). Direct antinociceptive actions at the spinal cord level are mediated by α_{2A} and α_{2C} receptors. However, these actions are supported via descending noradrenergic pathways that may involve α_{2B} receptors as well.

There is also a growing body of evidence showing plasticity in the analgesic effects of α_2AR agonists, especially in hypersensitivity-maintained pain states. For example, α_2AR agonists have a greater efficacy under circumstances of neuropathic pain. This may be due to the upregulation of the α_{2C}AR subtype following nerve injury, resulting in an alteration of the α_2AR-agonist site of action, and the involvement of different pathways in the generation of α_2AR-induced antinociception (Duflo et al. 2002; Paqueron et al. 2003; Stone et al. 1999).

It has been suggested that the antihyperalgesic effect of α_2AR-agonists in hypersensitivity-maintained pain states (e.g., neuropathic pain) is mediated, at least in part, through non-α_{2A}ARs. Furthermore, under those conditions, antihyperalgesia against mechanical but not thermal stimuli seems to be dependent on cholinergic mechanisms. This is supported by most recent data indicating that α_2AR-agonists exert their action via cholinergic neurons, which have been modulated by the interaction of ▶ nerve growth factor (NGF) with its low-affinity p75 receptor. It has further been hypothesized that α_2-adrenergic agonists facilitate the release of acetylcholine (Ach). The released Ach has been shown to act mainly on muscarinergic and to a lesser extent on nicotinic ▶ acetylcholine receptors to induce the release of nitric oxide (NO) and thereby antinociception (Pan et al. 1999). The recent discovery that α_2AR agonists can suppress itch in mouse models implies alpha adrenoceptor involvement in the dysesthesia of itch and a subsequent effect of the descending noradrenergic system inhibition on signaling in the spinal cord (Gotoh et al. 2011).

Supraspinal Sites
The catecholaminergic cell groups A5, A6 (locus coeruleus, LC), and A7 in the dorsolateral pons of the brainstem have been identified as the most important supraspinal sites for α_2AR-mediated antinociception. These areas express α_2ARs and send and receive projections to and from other pain-modulating parts of the brain, for instance, the periaqueductal gray (PAG) and the rostral ventromedial medulla (RVM). Therefore, they act as important relay stations for pain-modulating pathways. They are also centers from which ▶ descending inhibitory noradrenergic (NA) pathways originate. These pathways terminate in parts of the spinal cord dorsal horn that modulate spinal pain processing.

Normally, tonic firing in LC neurons suppresses activity in A5/7 cell groups; consequently, the noradrenergic outflow through the descending NA pathways is inhibited (Bie et al. 2003; Nuseir and Proudfit 2000). Activation of α_2ARs in the LC can inhibit activity in certain cells resulting in behavioral changes, which are in accordance with antinociceptive actions of the injected drugs. These effects could be completely reversed by ▶ intrathecal application of an α_2AR antagonist, suggesting a mechanism of action involving increased spinal NA release in response to the supraspinal agonist injection (Dawson et al. 2004).

From these results, it has been suggested that α_2AR agonists, in decreasing the activity of LC neurons, disinhibit the A5/A7 cell groups and, therefore, indirectly activate the descending inhibitory noradrenergic pathways with the resultant increased spinal NA release. Evidence from recent studies suggests that the released NA acts on α_{2B}ARs in the spinal cord, which are not located on primary afferents; instead, these may be located on interneurons or ascending excitatory pathways to mediate antinociception (Dawson et al. 2004; Kingery et al. 2002). In addition to antinociception, the LC also mediates the sedative actions of α_2AR agonists by inhibition of cell firing in some LC neurons.

The possible importance of these noradrenergic pathways under circumstances of chronic pain has also recently been suggested. Data obtained from an animal model of neuropathic pain, for example, showed an increased expression of key enzymes of catecholamine synthesis, tyrosine hydroxylase and dopamine β-hydroxylase, in the LC and spinal cord. This increased expression has been interpreted as a reflection of an enhanced activity in the descending NA system, with an increased noradrenaline turnover in response to the ongoing activity in nociceptive pathways (Ma and Eisenach 2003).

Clinical Applications

The most commonly used α_2AR agonists clinically are clonidine and dexmedetomidine, and they have been deployed in all age groups in oral, intravenous, intramuscular, epidural, and intrathecal modalities. Due to increased cost and decreased availability, dexmedetomidine is the less frequently used. While the most common administrative uses are likely to be for analgesia and sedation, there is evidence that they are opioid sparing and can reduce perioperative myocardial ischemia, intraoperative blood loss, and postoperative nausea and vomiting (Wu et al. 2004). The clinical administration of α_2AR agonists will require balance of the required effect with their sedative properties. Where the desired effect is sedation, α_2AR agonists have found mixed response largely relating to their concurrent effects of hypotension and bradycardia. Usage varies in the intensive care setting; one study, including 240 German ICUs, documented two-thirds of local sedation regimens as including clonidine. The agonists are useful where sympathetic mediation results, for instance, in alcohol withdrawal, although the interactions of chronic alcohol use with α_{2A}ARs have resulted in prolonged sedation profiles (Berggren et al. 2002). Perioperative administration is another area of use. Oral clonidine has been associated with reduced perioperative blood loss in ENT and enhanced recovery programs, and these seem likely findings secondary to controlled hypotension and localized vasculature response (Mohseni and Ebneshahidi 2011). Postoperative analgesic regimes can include α_2AR agonists in bolus form or infusions under machine, nurse, or patient control. Evidence suggests that postoperative administration has no effect in enhancing the speed of onset or effectiveness of opioid analgesia, although the synergy of the two groups and opioid-sparing effects are well described. The recently described α_2AR agonist synergy between clonidine and dexmedetomidine is yet to be investigated in humans. The other major area of interest relates to the suggested findings of reduced myocardial ischemia following clonidine usage. This potentially significant finding is currently being investigated further by a randomized, international, double-blinded trial which could have global consequences (POISE-2 trial).

References

Berggren, U., Eriksson, M., Fahlke, C., Sundkler, A., & Balldin, J. (2002). Extremely long recovery time for the sedative effect of clonidine in male type 1 alcohol-dependent subjects in full sustained remission. *Alcohol, 28*(3), 181–187.

Bie, B., Fields, H. L., Williams, J. T., et al. (2003). Roles of alpha1- and alpha2-adrenoceptors in the nucleus raphe Magnus in opioid analgesia and opioid abstinence-induced hyperalgesia. *Journal of Neuroscience, 23*, 7950–7957.

Campbell, C. M., Kipnes, M. S., Stouch, B. C., Brady, K. L., Kelly, M., Schmidt, W. K., Petersen, K. L., Rowbotham, M. C., & Campbell, J. N., (2012). Randomized control trial of topical clonidine for treatment of painful diabetic neuropathy. *PAIN, 153*(9), 1815–1823.

Dawson, C., Ma, D., Chow, A., et al. (2004). Dexmedetomidine enhances analgesic action of nitrous oxide: Mechanisms of action. *Anesthesiology, 100*, 894–9045.

Duflo, F., Li, X., Bantel, C., et al. (2002). Peripheral nerve injury alters the alpha2 adrenoceptor subtype activated by clonidine for analgesia. *Anesthesiology, 97*, 636–641.

Fairbanks, C. A., Kitto, K. F., Nguyen, O., et al. (2009a). Clonidine and dexmedetomidine produce antinociceptive synergy in mouse spinal cord. *Anesthesiology, 110*, 638–647.

Fairbanks, C. A., Stone, L. S., & Wilcox, G. L. (2009b). Pharmacological profiles of alpha 2 adrenergic receptor agonists identified using genetically altered mice and isobolographic analysis. *Pharmacology and Therapeutics, 123*(2), 224–238.

Gotoh, Y., Andoh, T., & Kuraishi, Y. (2011). Noradrenergic regulation of itch transmission in the spinal cord mediated by α-adrenoceptors. *Neuropharmacology, 61*(4), 825–831.

Kawasaki, Y., Kumamoto, E., Furue, H., et al. (2003). Alpha 2 adrenoceptor-mediated presynaptic inhibition of primary afferent glutamatergic transmission in rat substantia gelatinosa neurons. *Anesthesiology, 98*, 682–689.

Kingery, W. S., Agashe, G. S., Guo, T. Z., et al. (2002). Isoflurane and nociception: Spinal alpha2a adrenoceptors mediate antinociception while supraspinal alpha1 adrenoceptors mediate pronociception. *Anesthesiology, 96*, 367–374.

Lavand'homme, P. M., & Eisenach, J. C. (2003). Perioperative administration of the alpha2-adrenoceptor agonist clonidine at the site of nerve injury reduces the development of mechanical hypersensitivity and modulates local cytokine expression. *Pain, 105*, 247–254.

Li, X. H., & Eisenach, J. C. (2001). a2A-adrenoceptor stimulation reduces capsaicin-induced glutamate release from spinal cord synaptosomes. *Journal of Pharmacology and Experimental Therapeutics, 299*, 939–944.

Li, C., Sekiyama, H., Hayashida, M., et al. (2007). Effects of topical application of clonidine cream on pain behaviors and spinal Fos protein expression in rat models of neuropathic pain, postoperative pain, and inflammatory pain. *Anesthesiology, 107*, 486–494.

Ma, W., & Eisenach, J. C. (2003). Chronic constriction injury of sciatic nerve induces the up-regulation of descending inhibitory noradrenergic innervation to the lumbar dorsal horn of mice. *Brain Research, 970*, 110–118.

Mohseni, M., & Ebneshahidi, A. (2011). The effect of oral clonidine premedication on blood loss and the quality of the surgical field during endoscopic sinus surgery: A placebo-controlled clinical trial. *Journal of Anesthesia, 25*(4), 614–617.

Nuseir, K., & Proudfit, H. K. (2000). Bidirectional modulation of nociception by GABA neurons in the dorsolateral pontine tegmentum that tonically inhibit spinally projecting noradrenergic A7 neurons. *Neuroscience, 96*, 773–783.

Olave, M. J., & Maxwell, D. J. (2003). Neurokinin-1 projection cells in the rat dorsal horn receive synaptic contacts from axons that possess alpha2C-adrenergic receptors. *Journal of Neuroscience, 23*, 6837–6846.

Overland, A. C., Kitto, K. F., Chabot-Doré, A., et al. (2009). Protein kinase c mediates the synergistic interaction between agonists acting at alpha2-adrenergic and delta-opioid receptors in spinal cord. *Journal of Neuroscience, 29*(42), 13264–13273.

Pan, H. L., Chen, S. R., & Eisenach, J. C. (1999). Intrathecal clonidine alleviates allodynia in neuropathic rats: Interaction with spinal muscarinic and nicotinic receptors. *Anesthesiology, 90*, 509–514.

Paqueron, X., Conklin, D., & Eisenach, J. C. (2003). Plasticity in action of intrathecal clonidine to mechanical but not thermal nociception after peripheral nerve injury. *Anesthesiology, 99*, 199–204.

Riedl, M. S., Schnell, S. A., Overland, A. C., Chabot-Dore, A. J., Taylor, A. M., Ribeiro-Da-Silva, A., Elde, R. P., Wilcox, G. L., & Stone, L. S. (2009). Coexpression of alpha(2A)-adrenergic and delta-opioid receptors in substance P-containing terminals in rat dorsal horn. *The Journal of Comparative Neurology, 513*, 385–398.

Sonohata, M., Furue, H., Katafuchi, T., et al. (2004). Actions of noradrenaline on substantia gelatinosa neurones in the rat spinal cord revealed by in vivo patch recording. *The Journal of Physiology, 555*, 515–526.

Stone, L. S., Macmillan, L., Kitto, K. F., et al. (1997). The α2a-adrenergic receptor subtype mediates spinal analgesia evoked by î ± 2 agonists and is necessary for spinal adrenergic/opioid synergy. *Journal of Neuroscience, 17*, 7157–7165.

Stone, L. S., Broberger, C., Vulchanova, L., et al. (1998). Differential distribution of alpha2a and alpha2c adrenergic receptor immunoreactivity in the rat spinal cord. *Journal of Neuroscience, 18*, 5928–5937.

Stone, L. S., Vulchanova, L., Riedl, M. S., Wang, J., Williams, F. G., Wilcox, G. L., & Elde, R. (1999).

Effects of peripheral nerve injury on alpha-2a and alpha-2c adrenergic receptor immunoreactivity in the rat spinal cord. *Neuroscience, 93*, 1399–1407.

Wu, C. T., Jao, S. W., Borel, C. O., et al. (2004). The effect of epidural clonidine on perioperative cytokine response, postoperative pain and bowel function in patients undergoing colorectal surgery. *Anesth Analg, 99*, 502–509.

Alpha(α) 2-Adrenergic Receptor Agonists

▶ Alpha(α) 2-Adrenergic Agonists in Pain Treatment

Alpha(α) 2-Agonists

▶ Alpha(α) 2-Adrenergic Agonists in Pain Treatment

Alpha(α) EEG Wave Intrusion

Definition

The intrusion of fast-frequency EEG alpha (7.5–11 Hz) activity into slow wave sleep (SWS). The SWS is dominated by large and slow EEG waves of delta type (0.5–4.0 Hz); it also characterizes sleep stages three and four.

Cross-References

▶ Orofacial Pain, Sleep Disturbance

Alpha(α)-D Galactose

Definition

Lectins are proteins that bind to the carbohydrate portion of glycoproteins and glycolipids.

The isolectin Griffonia simplicifolia I-B4 (IB4) binds specifically to terminal α-galactose, the terminal sugar on galactose-α1,3-galactose carbohydrates on glycoproteins and glycolipids. The IB4 lectin labels about one half of the small- and medium-diameter DRG neurons in rat and mouse. It is not yet clear which proteins or lipids in DRG neurons account for the majority of labeling by IB4 binding.

Cross-References

► Immunocytochemistry of Nociceptors

Alpha(α)-Delta(δ) Sleep

Definition

Simultaneous recordings of delta and alpha brainwaves during sleep.

Cross-References

► Fibromyalgia

Alpha(α)-I-Acid Glycoprotein

Definition

The most important serum binding protein for opioids and local anesthetics.

Cross-References

► Acute Pain in Children, Postoperative

AL-TENS (Acupuncture-like TENS)

► Transcutaneous Electrical Nerve Stimulation

Alternative Medicine

► Alternative Medicine in Neuropathic Pain

Alternative Medicine in Neuropathic Pain

Miles Belgrade[1], Robyn Drach[2] and Sherri Haas[2]
[1]Fairview Pain Management Center, University of Minnesota Medical Center, Minneapolis, MN, USA
[2]Fairview Pain Management Center, Minneapolis, MN, USA

Synonyms

Alternative medicine; Alternative therapies; Complementary medicine; Complementary therapies; Holistic medicine; Nontraditional medicine; Unconventional medicine

Definition

In 1993, Eisenberg utilized a working definition of alternative medicine as interventions that are not taught widely in medical schools and that are not generally available in US hospitals (Eisenberg et al. 1993). However, there has been a rise in availability of complementary medical practices in Western-based medical institutions and more medical schools are incorporating unconventional therapies into their curricula. A broader definition of alternative and complementary medicine would be *those medical systems, practices, interventions, applications, theories or claims that are not part of the dominant or conventional medical system of that society* (National Institutes of Health on Alternative Medical systems and Practices in the United States). This definition is flexible in that it recognizes alternative and complementary medicine as culturally based. This definition also allows for changes in what constitutes alternative or complementary practices as a society evolves or changes.

The concept of *alternative medicine* implies practices used instead of conventional medical practice, whereas *complementary medicine* refers to practices that are integrated with conventional care. Neither of these terms accurately reflects the most common way in which unconventional practices are incorporated into treatment. Most of the time, physicians are unaware of their patients' use of alternative health practices that are applied simultaneously with conventional treatment. Thus, these practices are neither instead of, nor integrated with, conventional treatment. They are simply a separate, dual track of care.

Introduction

This chapter will provide a framework for understanding and organizing the wide array of alternative and complementary therapies, and will provide a perspective on those therapies that have been used to treat various neuropathic pain states. Wherever possible, evidence-based therapies will be highlighted and reviewed.

Characteristics

Medical conditions that have effective and well-tolerated treatments generally do not motivate a search for alternatives – especially when such alternatives may be based on theoretical constructs that are foreign to the patient and their physician. Complex pain problems, like chronic neuropathic pain, that have multiple mechanisms are hard to treat even with the availability of newer pharmacological modulators. Many of the conventional therapies for neuropathic pain have adverse effects that interfere substantially with quality of life. It is not surprising that patients suffering from neuropathic pain would look outside conventional medicine for more effective and better-tolerated treatments.

▶ Acupuncture, ▶ chiropractic, ▶ homeopathy, herbal medicine, traditional Chinese medicine, massage, ▶ biofeedback, the list of complementary and alternative therapies is seemingly limitless. Just as we categorize conventional medical practice into pharmacological, surgical, physical rehabilitative, and behavioral techniques, it is helpful to organize the broad array of alternative medicine practices into categories that allow practitioners to better understand the options available and how they differ from each other. It is convenient to separate all of CAM into three broad groups (Fig. 1):

1. World medicine systems
2. Other comprehensive systems of medicine that are not culturally based
3. Individual therapies

A system of medicine such as homeopathy or chiropractic consists of both a diagnostic and a therapeutic approach to a wide array of symptoms, illnesses, and diseases. It is based on a philosophy of health and disease that gives rise to the types of treatments that are utilized. A world medicine system like traditional Chinese medicine or Ayurvedic medicine is a system of medicine that is based on the traditions and philosophy of a world culture. Individual therapies are not linked to a culture or a complete medical system and are generally used to treat a certain subset of symptoms or problems. Examples include biofeedback, massage, and vitamin therapy. All therapies can be further subdivided into one or more of seven functional groups:

1. Meditative/mindful
2. Spiritual
3. Energy based
4. Stimulation based
5. Movement based
6. Mechanical or manipulative
7. ▶ Nutriceutical

Mindful or meditative therapies utilize the mind to produce changes in physical and emotional status. Meditation, hypnosis, and yoga can fall into this category. Spiritual therapies on the other hand, utilize a letting go of the mind and giving up control to a higher power as in prayer. Energy-based therapies rely on a construct of vital energy or an energy field that must be in proper balance to maintain health. Traditional acupuncture, healing touch, and yoga all use the concept of vital energy. Acupuncture can also be considered a stimulation-based therapy.

Alternative Medicine in Neuropathic Pain, Fig. 1   Organizational chart of alternative and complementary therapies (From Belgrade 2003)

Alternative Medicine in Neuropathic Pain, Table 1   Examples of complementary and alternative therapies organized into functional groups (From Belgrade 2003)

Mindful	Spiritual	Energy based	Stimulation based	Movement based	Mechanical/ manipulative	Nutriceutical
Hypnosis		Massage	TENS	Exercise		
Imagery	Prayer	Therapeutic touch	Acupuncture	Dance therapy	Chiropractic	Vitamins
Meditation	Spiritual healing	Homeopathy	Massage	Alexander technique	Osteopathy	Diet
Relaxation	Psychic healing	Acupuncture	Aromatherapy	Tai Chi	Massage	Herbal medicine
Biofeedback	Yoga	Qi Gong	Therapeutic touch	Qi Gong	Craniosacral therapy	Homeopathy
Yoga		Yoga	Music	Yoga	Rolfing	Aromatherapy

Thus, many practices fall into more than one functional category (Table 1).

Prevalence and Cost

Several large surveys in the United States, Europe, and Australia demonstrate extensive use of alternative and complementary therapies by the public. Prevalence estimates are confounded by what practices are included as unconventional. For example, are ice, heat, and prayer to be included when they are so commonly utilized? Aside from such universal practices, 42 % of the US population made use of alternative treatments as of 1997 (Eisenberg et al. 1998). Fifteen percent of Canadians visited an alternative health practitioner in the previous 12 months (Millar 1997). In Europe, prevalence of alternative health care use varies from 23 % in Denmark to 49 % in France (Fisher and Ward 1994). Alternative medicine use in Australia has also been estimated to be 49 % (MacLennon et al. 1996).

Brunelli and Gorson surveyed 180 consecutive patients with peripheral neuropathy about their use of complementary and alternative medicine (CAM) (Brunelli and Gorson 2004). Forty-three percent of patients reported using at least one type of CAM. Patients with burning neuropathic pain used CAM at a significantly higher rate than those without such pain. Diabetic neuropathy patients were also significantly more likely to use CAM. Other predictors of CAM use were younger age and college educated. Types of treatments employed by patients were megavitamins (35 %), magnets (30 %), acupuncture (30 %), herbal remedies (22 %), and chiropractic (21 %). Lack of pain control was the most common reason for CAM use and nearly half of the patients did not discuss it with their physician.

The United States spends $27 billion each year on alternative medicine. That figure reflects out-of-pocket expenses alone and is nearly equal to the cost of physician services and triple the cost of hospitalizations (Eisenberg et al. 1998). Health benefit payers are facing the quandary of determining which alternative services are worthy of coverage and to what extent. The question of standards of care for the various alternative forms of therapy represents a quagmire that confronts everyone, patients, physicians, health benefit administrators, and the alternative practitioners themselves.

Acupuncture and Other Stimulation-Based Therapies

Acupuncture is one component of traditional Chinese medicine. As such, it has its theoretical roots in Taoist ideas about the universe, living systems, health, and disease. Modern scientific scrutiny has already yielded more information about acupuncture mechanisms than for any other alternative therapy. The discovery of opioid receptors and ▶ endorphins has led to a large number of investigations into the role these receptors and ▶ ligands play in producing acupuncture analgesia. Nearly all such studies support the conclusion that acupuncture analgesia is mediated in part by the opioid system. Acupuncture analgesia can be reversed with administration of ▶ naloxone (Meyer et al. 1977; Pomeranz and Cheng 1979; Tsunoda et al. 1980). Increased levels of endogenous opioid following acupuncture have been directly measured in humans (Clement-Jones et al. 1980; Pert et al. 1984). Antiserum to opioid receptors applied to the periaqueductal gray matter has been shown to block experimental acupuncture analgesia in primates.

Han and Terenius reviewed a number of studies that demonstrate the importance of biogenic amines in acupuncture analgesia (Han and Terenius 1982). Ablating the ▶ descending inhibitory pathway for pain at the dorsal and medial raphe nuclei blunted acupuncture analgesia. Blocking serotonin receptors in rabbits and rats also diminished acupuncture analgesia. Administering a serotonin precursor potentiates acupuncture analgesia. Serotonin and its by-products are increased in the lower brainstem during acupuncture analgesia. Other neurochemical mediators of experimental acupuncture analgesia have been implicated in preliminary investigations including ▶ substance P, ▶ CGRP, ▶ CCK, and ▶ C-fos (Belgrade 1994).

That stimulation of tissue, including neural tissue, produces analgesia has only recently gained acceptance in conventional medicine. Neurosurgeon Norman Shealy pioneered the use of transcutaneous electrical nerve stimulation (TENS) in the 1970s – less than a decade after Melzack and Wall published their gate theory of pain modulation that postulated a competitive inhibition of pain by non-noxious stimuli. Wallin and colleagues showed that spinal cord stimulation inhibits ▶ long-term potentiation of spinal ▶ wide dynamic range neurons (Wallin et al. 2003). Hanai (2000) demonstrated a similar response to peripheral nerve stimulation.

Clinical Studies

In one extensive multicenter randomized controlled trial of acupuncture, amitriptyline or placebo for HIV-related neuropathic pain, no differences were found between groups; but all groups showed significant reductions in pain (Shlay et al. 1998). Using an electroacupuncture-like treatment, Hamza and colleagues showed a substantial reduction in pain scores and analgesic use and improvement in quality of life measures among patients with Type II diabetes and painful neuropathy in a sham-controlled crossover trial of 50 patients (Hamza et al. 2000). More recently, Schroeder et al. reported a pilot study of acupuncture compared with conventional treatment for chemotherapy-induced peripheral neuropathy. Those patients receiving acupuncture tended to show more improvement in nerve conduction studies than the control group (Schroeder et al. 2011). Earlier, Schoreder had shown that acupuncture was effective for controlling pain and improving nerve conduction in idiopathic peripheral neuropathy (Schroder et al. 2007). In a randomized controlled trial of acupuncture for severe acute pain in Herpes Zoster, acupuncture was as effective as pharmacological treatment at controlling pain (Ursini et al. 2011).

In a multicenter randomized placebo-controlled study using static magnetic fields in the form of magnetized insoles for diabetic peripheral neuropathy, Weintraub et al. showed statistically significant reductions in burning, numbness, and tingling after 3–4 months (Weintraub et al. 2003). Cortical stimulation for neuropathic pain has also been reported. In a small case series, Rainov and Heidecke report a sustained >50 % reduction in trigeminal and glossopharyngeal neuralgia for 72 months with motor cortex stimulation using a quadripolar electrode contralateral to the side of pain (Rainov and Heidecke 2003).

Although clinical studies are lacking for specific neuropathic pain conditions, meditative and mindful therapies such as hypnosis have been utilized for pain management for more than a century. Rainville and colleagues used PET scanning in normal subjects to show that pain unpleasantness is mediated in the anterior cingulate, anterior insula, and posterior cerebellum (Rainville et al. 1997). He used hypnosis to reduce the unpleasantness of an experimental pain stimulus and to distinguish it from pain intensity, localizing the two components functionally in the brain. The growing understanding of unpleasantness as distinct from pain intensity leads one to conclude that many nonspecific therapies that "quiet" the nervous system's emotional, anticipatory component of pain can play just as important a role as analgesics. In this way, many alternative and complementary therapies can be beneficial. Obviously, much clinical research is needed to define the scope and value of these therapies as well as their mechanisms of action. In the meantime, the prevalence and

popularity of CAM among patients with neuropathic pain requires that the physician be acquainted with these therapies and guide patients toward the better studied, safest, and most appropriate techniques for the neurological condition.

Manual Therapies

According to the 2007 National Health Interview Survey, manipulation ranked in the top ten CAM therapies among both adults and children. The survey found that 8.6 % of adults and 2.8 % of children had used manipulation (http://nccam.nih.gov/health/backgrounds/manipulative.htm 2007). Osteopathic manipulative treatment is a modality used by osteopathic physicians to complement conventional treatment. Osteopathic manipulative treatment is defined in the Glossary of Osteopathic Terminology as "the therapeutic application of manually guided forces by an osteopathic physician to improve physiologic function and/or support homeostasis that has been altered by somatic dysfunction" (Clinical Guideline Subcommittee on Low Back Pain: American Osteopathic Association 2010). Somatic dysfunction may be associated with pain generators that interfere with the body's innate ability to heal itself. OMT may be delivered to reduce or remove the identified somatic dysfunction and to modulate central and peripheral mechanisms involved in pain generation (Kuchera 2007).

Osteopathy is a comprehensive system of medicine that incorporates diagnostic and therapeutic strategies to address body unity issues, enhance homeostatic mechanisms, and maximize structure-function interrelationships (Seffinger et al. 2007). Although osteopathy is recognized by the National Institutes of Health (NIH) in the United States as mainstream medicine, OMT is classified by the NIH's National Center of Complementary and Alternative Medicine as one of several promising complementary manipulative/body-based practices in addition to chiropractic and massage therapy (http://nccam.nih.gov/health/backgrounds/manipulative.htm 2007).

Research regarding manual treatment for patients with neuropathic pain is lacking. The little evidence that does exist consists primarily of small studies or case reports. In 1999, Raison-Peyron et al. demonstrated sustained improvements with paraspinal physiotherapy and manipulation in four of six patients with varying degrees of relief from 1 to 9 years. There has been one report of spinal manipulation being effective for patients with brachioradial pruritis (Richardson et al. 2009).

The Active Release Technique Soft Tissue Management System has proven to be clinically promising in treating conditions related to overuse. In a case study of posterior interosseous nerve syndrome using soft tissue manipulation therapy, overall improvement in symptoms was observed with restoration of motor deficits and decrease in pain. Positive results persisted 1 and 6 months following treatment (Saratsiotis and Myriokefalitakis 2010).

The technique of neural gliding attempts to take the nerve throughout the available range of motion to improve the actual excursion of the nerve, decrease adhesions, and reduce symptoms by allowing the nerve to move freely. The most current and best available evidence in regard to neural gliding exercises for treatment of carpal tunnel syndrome comes from a few uncontrolled, randomized clinical trials, incomplete systematic reviews and anecdotal clinical evidence. A systematic review of this evidence revealed a possible trend toward pain and symptom reduction, improved sensation, and improved function and strength (Medina McKeon and Yancosek 2008).

In a case report of a patient with notalgia paresthetica, a chronic sensory neuropathy affecting one of the cutaneous dorsal rami of the upper thoracic region, OMT techniques were applied to relieve pressure on the exiting dorsal rami thereby providing symptomatic relief (Richardson et al. 2009).

Although it has been suggested that manipulation is a safe and effective treatment of radicular symptoms, there is little evidence for its efficacy and it is considered to be contraindicated by some. In a prospective single-blind randomized controlled trial comparing OMT with chemonucleolysis for treatment of symptomatic lumbar disc herniation, outcomes of leg pain, back pain, and self-reported disability improved in both groups. There were no instances of major complications from either treatment (Burton et al. 2000). In a study of 40 patients randomized to either surgical microdiskectomy or standardized chiropractic spinal manipulation, significant improvement in both treatment groups compared to baseline was observed in all outcome measures. Sixty percent of patients with sciatica who had failed other medical management benefited from spinal manipulation to the same degree as if they underwent surgical intervention (McMorland et al. 2010). In a randomized double-blind trial comparing active and simulated manipulations to treat acute back pain and sciatica, manipulations appear more effective on the basis of percentage of pain-free cases and number of days with moderate or severe pain (Santilli et al. 2006).

Nutraceuticals Included in this section are those nutraceuticals that have shown promising results in clinical trials for treatment of neuropathic pain. While there are several others that have been studied, the evidence is not as convincing.

Alpha-Lipoic Acid (ALA) has been used as a treatment for peripheral neuropathy in Europe for decades. Three large-scale, double-blind, placebo-controlled trials – the Alpha-Lipoic Acid in Diabetic Neuropathy (ALADIN) studies – have examined various routes, dosages, and neurological effects of ALA. The first ALADIN study found ALA 600 or 1,200 mg IV for 3 weeks was superior to placebo for reduction of neuropathic symptoms. The 600 mg dose yielded slightly better results with fewer side effects. A meta-analysis of four placebo-controlled trials ($n = 1,258$), all with the same protocol of ALA 600 mg IV for 3 weeks, found a continuous daily improvement in symptoms beginning on the eighth day of treatment. In an uncontrolled study of 26 patients with diabetic neuropathy, an oral dose of 600 mg ALA for 3 months resulted in reversal from symptomatic to asymptomatic neuropathy (Head 2006).

Acetyl-L-Carnitine (ALC) has been tested in clinical studies for treatment of peripheral neuropathy. In two multicenter randomized controlled trials, ($n = 1,679$) patients with diabetic neuropathy were treated with ALC 1,500–3,000 mg orally for 1 year. Reductions in neuropathic pain were reported in both studies and nerve regeneration was documented in one study (Head 2006; Evans et al. 2008; Pitter and Ernst 2008). In a study of patients with chemotherapy-induced neuropathy, 26 patients with cisplatin- or paclitaxel-induced peripheral neuropathy were given ALC 1 g IV for 10–20 days. All cisplatin-treated patients experienced improvement in neuropathy symptoms, while 8 of 12 patients in the paclitaxel group and 8 of 10 in the combination group experienced improvement in symptoms (Head 2006). In a study of patients with HIV-associated antiretroviral toxic neuropathy, ALC 1,500 mg twice daily resulted in improvement of neuropathic pain in 15 of 21 patients (Hart et al. 2004).

Benfotiamine is the most extensively studied form of thiamine for treatment of peripheral neuropathy. In a double-blind, randomized placebo-controlled study, 40 patients with diabetic neuropathy were treated with benfotiamine 100 mg four times daily or placebo for 3 weeks. A statistically significant improvement in neuropathy was reported in the treatment group compared to placebo. A small study found benefit of benfotiamine for alcoholic neuropathy treated with 450 mg benfotiamine daily for 2 weeks, followed by 300 mg daily for an additional 4 weeks (Head 2006).

Vitamin B12 deficiency has been associated with peripheral neuropathy. In a double-blind, placebo-controlled trial, 21 patients with diabetic neuropathy were given oral methylcobalamin 500 mcg three times daily for 4 months, while 22 patients received placebo. Significant improvements in symptoms were reported in the treatment group compared to placebo (Head 2006).

The amino acid N-acetylcysteine (NAC) is a potent antioxidant. Fourteen colon cancer patients were randomly assigned to 1,200 mg oral NAC or placebo daily. After four cycles of chemotherapy, two of five in the NAC group and seven of nine in the placebo group experienced neuropathy; after eight cycles no one in the NAC group experienced neuropathy compared to five of nine in the placebo group. Only 1 in the NAC group, but 8 of 9 in the placebo group experienced neuropathy after 12 cycles (Head 2006).

Zinc is abundant in the central nervous system and regulates pain. In an animal study, zinc was found to alleviate pain through high-affinity binding to the *NMDA* receptor NR2A subunit (Nozaki et al. 2011). A clinical study examined the effect of zinc for diabetic neuropathy. Thirty patients with diabetic neuropathy were supplemented with 660 mg zinc sulfate daily or placebo for 6 weeks. Nerve conduction velocity significantly increased in the group supplemented with zinc (Head 2006).

Another cationic compound, agmatine (decarboxylated arginine), has recently been introduced as a nutraceutical and tested in about 60 patients with disc-associated radiculopathy using an open label dose escalation design followed by a placebo-controlled double-blind design (Keynan et al. 2010). In the first study, mild gastrointestinal side effects were noted in three patients at the highest doses, but these were reversed upon cessation of treatment. Continuous improvement of symptoms was evident in both groups of the latter study with no adverse events, but the agmatine group's symptoms improved more (70 % vs. 27 % relative to baseline).

Fatty acids are essential for nerve membrane structure. Patients with diabetes appear to have impaired conversion of linoleic acid to gamma-linolenic acid (GLA). In a double-blind study, 22 patients with diabetic neuropathy were assigned to 360 mg GLA or placebo for 6 months. Subjects treated with GLA exhibited significantly better neuropathy scores, nerve conduction velocity, and action potentials compared to placebo (Head 2006).

Geranium (*Pelargonium*) oil was tested for postherpetic neuropathy in a double-blind randomized controlled trial. It reported a dose-dependent reduction in pain compared with placebo.

Magnets It has been hypothesized that electromagnetic fields may benefit peripheral neuropathy by polarization of neurons that may be firing ectopically. Pulsed magnetic field therapy was used to treat 24 subjects with peripheral neuropathy. Average pain scores decreased by 21 % at the end of nine treatments and by 49 % at a follow-up assessment. In a randomized controlled trial on 375 patients with diabetic neuropathy, subjects wore multipolar, static, 450 G magnetic insoles or placebo insoles in their shoes for 4 months. There was a significant reduction in neuropathic symptoms in the magnetic-insole group compared to placebo (Pitter and Ernst 2008).

Mindful, Meditative, and Psychological Therapies for Neuropathic Pain

Within the mindful and meditative therapies category of the complementary and alternative medicine (CAM) literature, various psychological techniques are often listed such as hypnosis, imagery, meditation, relaxation, and biofeedback. In actuality, there are many forms of psychological theories and techniques utilized in multidisciplinary chronic pain treatment centers by health psychologists and other therapists. While these theories and techniques have been labeled as CAM treatments by health professionals or the general public, pain specialists often consider them mainstream or under the label Integrative Medicine, depending on the definition. Regardless, there is strong empirical evidence to support the use of these techniques for chronic pain management. However, less research has been conducted demonstrating their efficacy for neuropathic pain.

As reviewed in Turk et al. (2010), since the initial publication of behavioral treatment for chronic pain by Fordyce et al. in 1968 (Fordyce 1976), a large number of clinical trials have been conducted demonstrating efficacy for psychological treatments for chronic pain. Cognitive-behavioral therapy (CBT) and operant conditioning have been the psychological approaches most commonly researched. Specific psychological techniques are also evaluated in the literature with cognitive therapy, supportive counseling, hypnosis, relaxation, biofeedback, and meditation being the most cited. It is also important to note that various psychological techniques can also be incorporated into CBT therapy. CBT also incorporates the behavioral

Alternative Medicine in Neuropathic Pain, Table 2 Examples of psychological approaches and techniques used with chronic pain patients

Psychological approaches and techniques	
Cognitive behavioral therapy	Using behavioral and cognitive techniques to achieve changes in behavior, thoughts, and emotions
Operant behavioral therapy	Using principles of behavioral psychology to reduce pain behaviors and increase appropriate ones
Mindfulness meditation	Intentional self-regulation of awareness, a systematic focus on inner and outer experiences
Acceptance and commitment therapy	Observing thoughts and feelings as they are, and acting in ways consistent with valued goals and life directions.
Motivational interviewing	General therapeutic approach and techniques to increase belief in and possibility to engage in adaptive behaviors
Eye movement desensitization and reprocessing (EMDR)	An information processing psychotherapy that uses an eight phase approach to address the experiential contributors to chronic pain
Supportive counseling	Empathetic listening, encouragement, and acknowledgement of distress
Cognitive therapy	Changing cognitive activity to change behavior, thoughts, and emotions
Hypnosis	An altered state of awareness designed to focus attention to change participants experience of pain
Relaxation	For example, progressive muscle relaxation (PMR), autogenic training
Biofeedback	Measuring psychophysiological data by noninvasive electrical devices, and providing real-time feedback to raise awareness and conscious control of the autonomic nervous system
Imagery	Using imagination to reduce stress, relieve pain, and stimulate healing responses in your body

principles of operant conditioning. Common psychological theories and techniques are listed in Table 2. While the article by Turk et al. focused on CBT, since theoretically based approaches have better quality research, our focus will be on all psychotherapeutic research with neuropathic pain patients.

Psychological interventions as a whole have been found to be effective for many pain diagnoses in improvements for measures of pain, mood/affect, cognitive coping and appraisal, health-related quality of life, and depression (Eccleston et al. 2009; Morley et al. 1999; Hoffman et al. 2007). However, further evidence is needed to clarify long-term efficacy and vocationally relevant outcomes (Eccleston et al. 2009; Hoffman et al. 2007). In a meta-analysis of psychological treatment for chronic low back pain, CBT and self-regulatory treatments (e.g., biofeedback, relaxation, or hypnosis) were found to be specifically effective for pain intensity (Hoffman et al. 2007). Hypnosis has been found to reduce pain relative to no treatment or standard medical interventions, and results have been similar to interventions such as progressive muscle relaxation (PMR) and autogenic training (Elkins et al. 2007; Jensen and Patteson 2006). Further research is needed to compare the effects of hypnosis for chronic pain to similar therapies.

Recent literature has also focused on newer psychological approaches to chronic pain including mindfulness-based stress reduction (MBSR) or mindfulness meditation and acceptance and commitment therapy (ACT) (McCracken and Thompson 2011). Research on mindfulness meditation for chronic pain has been promising and found to improve mental and physical aspect of chronic pain in several studies (Gardner-Nix 2009). Initial studies of ACT was found to be related to lower disability and distress, and overall functioning in patients (McCracken and Thompson 2011; McCracken and Zhao-O'Brien 2010; McCracken and Vellerman 2010).

Clinical Studies in Neuropathic Pain *Cognitive-Behavioral Therapy*: In a recent systematic review (Mehta et al. 2011), two randomized controlled trials, six prospective controlled trials, and one cohort study of CBT for patients with spinal cord injuries (SCI) were analyzed. CBT was found to be moderately effective for depression, coping, and adjustment in adults following SCI. Pain was not specifically addressed in this review. Two studies, a prospective interventional study and nonrandomized parallel cohort design, were used to assess a comprehensive CBT pain management program for SCI (Perry et al. 2010; Budh et al. 2006). These studies found that a CBT group demonstrated improvements in mood, life interference, anxiety, pain catastrophizing, and sleep.

Behavioral Therapy: Minimal research was found specifically for behavioral therapy or operant techniques and neuropathic pain, other than case studies on the behavioral treatment of reflex sympathetic dystrophy (RSD) (Alioto 1981). These studies suggested that behavioral therapy may decrease pain, medication use, and improve mood after increased engagement in the treatment program.

Hypnosis: Multiple case studies were conducted showing promising effects of hypnosis for reflex sympathetic dystrophy (Gainer 1992) and phantom limb pain (Oakley et al. 2002). In a review (Oakley et al. 2002), research for hypnotic imagery for phantom limb pain has fallen into two categories, ipsative/imagery-based and movement/imagery-based. The ipsative/imagery approach takes into account the way an individual represents their pain and attempts to modify it, while the movement/imagery approach focuses on "moving" or taking control over the amputation. Both approaches appear to have promise from limited case study data.

A case series design was conducted (Jensen et al. 2005) using hypnosis in 33 patients with chronic pain from the following conditions: SCI, multiple sclerosis, amputation, cerebral palsy, post-polio syndrome, and Charcot Marie Tooth Syndrome. There was a decrease in pain intensity that lasted for at least 3 months, but no effect on mood or pain interference variables was found.

EMDR: Several case studies and a case series design were found for the effect of EMDR for phantom limb pain (Schneider et al. 2008). Schneider reported that for five cases, phantom pain was decreased using EMDR. This effect lasted from 14 months to 2 years.

Relaxation Training: A few case studies have indicated that autogenic training may be beneficial for RSD (Kawano et al. 1989; Fialka et al. 1996). One small randomized pilot study (Fialka 1996) compared an autogenic training group to controls. Both groups participated in home therapy. A significant decrease in pain was found for both groups. The only intragroup difference found was for skin temperature. Hough and Kleinginna (2002) evaluated a variety of relaxation techniques (e.g., meditation, visual imagery, PMR) that were used for six patients with SCI of varying injury levels. Relaxation techniques were found to be beneficial for pain and mood. The authors concluded that clinicians should consider level of spinal cord injury and individual differences when selecting relaxation techniques.

Biofeedback: Recent attention has been given to the use of thermal biofeedback in patients with diabetes. Thermal biofeedback has been suggested to be helpful in diabetic patients to improve circulation, pain, neuropathy, ulcer healing, ambulatory activity, and quality of life (Galper et al. 2003). Patients who underwent thermal biofeedback and relaxation training or self-hypnosis training were found to have increased peripheral temperature (Needham et al. 1993; Rice and Schindler 1992) and decreased levels of depression (Rice and Schindler 1992). In a randomized controlled study, Rice et al. found that thermal biofeedback combined with autogenic training, PMR, and breathing exercises increased foot ulcer healing, sensory nerve function, patient activity, and blood flow to the feet (Rice et al. 2001).

There was some preliminary evidence that thermal biofeedback combined with other relaxation techniques may decrease pain in patients with phantom limb pain (Sherman et al. 1979; Belleggia and Birbaumer 2001; Harden et al. 2005) and reflex sympathetic dystrophy (Blanchard 1979; Barowsky et al. 1987).

Comparative Studies: Three preliminary studies, two randomized controlled trials and one quasi-experimental study, compared psychological treatments in patients with neuropathic pain.

Self-hypnosis and EMG biofeedback were found to decrease pain intensity in 37 patients with SCI (Jensen et al. 2009a). The self-hypnosis group maintained gains at 3 months follow-up. In the second study, self-hypnosis was found to decrease pain levels and increase physical functioning in patients with multiple sclerosis for up to 3 months compared to a PMR relaxation group (Jensen et al. 2009b). CBT and supportive psychotherapy demonstrated decreased pain, but the CBT group also showed decreased pain interference and distress in a group of patients with peripheral neuropathy due to HIV (Evans et al. 2003). There was one study that suggested that a combination of psychological treatments such as hypnosis, imagery, and cognitive therapy helped to decrease pain in 25 amputees for up to 6 months (Bamford 2006).

Psychological therapies do appear to have a beneficial impact on chronic pain and its related psychosocial issues. Most of the studies are limited to case series, uncontrolled trials, or quasi-experimental designs. Neuropathic pain is represented by a smaller number of these trials but shows similar benefit and the larger studies are often in heterogeneous populations of pain patients.

References

Alioto, J. T. (1981). Behavioral treatment of reflex sympathetic dystrophy. *Psychosomatics, 22*(6), 539–540.
Bamford, C. (2006). A multifaceted approach to the treatment of phantom limb pain using hypnosis. *Comtemporary Hypnosis, 23*(3), 115–126.
Barowsky, E. I., Zweig, J. B., & Moskowitz, J. (1987). Thermal biofeedback in the treatment of symptoms associated with reflex sympathetic dystrophy. *Journal of Child Neurology, 2*, 229–232.
Belgrade, M. (1994). Two decades after ping-pong diplomacy: Is there a role for acupuncture in American Pain Medicine? *APS Journal, 3*(2), 73–83.
Belgrade, M. (2003). Alternative and complementary therapies. In A. Rice, C. A. Warfield, D. Justins, & C. Eccleston (Eds.), *Clinical pain management* (Chronic pain part 1st ed., Vol. 2). London: Arnold Press.
Belleggia, G., & Birbaumer, N. (2001). Treatment of phantom limb pain with combined EMG and thermal biofeedback: A case report. *Applied Psychophysiology and Biofeedback, 26*(2), 141–146.
Blanchard, E. B. (1979). The use of temperature biofeedback in the treatment of chronic pain due to causalgia. *Biofeedback and Self-Regulation, 4*(2), 183–188.

Brunelli, B., & Gorson, K. C. (2004). The use of complementary and alternative medicines by patients with peripheral neuropathy. *Journal of Neurological Sciences, 218*(1–2), 59–66.

Budh, C. N., Kowalski, J., & Lundeberg, T. (2006). A comprehensive pain management programme comprising educational, cognitive, and behavioural interventions for neuropathic pain following spinal cord injury. *Journal of Rehabilitation Medicine, 38,* 172–180.

Burton, K. A., Tillotson, M. K., & Cleary, J. (2000). Single-blind randomized controlled trial of chemonucleolysis and manipulation in the treatment of symptomatic lumbar disc herniation. *European Spine Journal, 9,* 202–207.

Clement-Jones, V., et al. (1980). Increased beta-endorphin but not met-enkephalin levels in human cerebrospinal fluid after acupuncture for recurrent pain. *The Lancet, 2,* 946–949.

Clinical Guideline Subcommittee on Low Back Pain: American Osteopathic Association (2010). American osteopathic association guidelines for osteopathic manipulative treatment (OMT) for patients with low back pain. *Journal of the American Osteopathic Association, 110,* 653–666.

Eccleston, C., Williams, A. C. D. C., Morley, S. (2009). Psychological therapies for the management of chronic pain (excluding headache) in adults (Review). Cochrane Database Systems Review; Apr 15;(2):1–81.

Eisenberg, D. M., et al. (1993). Unconventional medicine in the United States. *New England Journal of Medicine, 328,* 246–252.

Eisenberg, D. M., et al. (1998). Trends in alternative medicine use in the United States, 1990–1997: Results of a follow-up national survey. *Journal of the American Medical Association, 280,* 1569–1575.

Elkins, G., Jensen, M. P., & Patterson, D. R. (2007). Hypnotherapy for the management of chronic pain. *International Journal of Experimental Hypnosis, 55*(3), 275–287.

Evans, S., Fishman, B., Spielman, L., & Haley, A. (2003). Randomized trial of cognitive behavioral therapy versus supportiv e psychotherapy for HIV-related pheripheral neuropathic pain. *Psychosomatics, 44*(1), 44–50.

Evans, J. D., Jacobs, T. F., & Evans, E. W. (2008). Role of acetyl-*L*-carnitine in the treatment of diabetic peripheral neuropathy. *The Annals of Pharmacotherapy, 42,* 1686–1691.

Fialka, V., Korpan, M., Saradeth, T., Paternostro-Slugo, T., Hexel, O., Frischenschlager, O., & Ernst, E. (1996). Autogenic training for reflex sympathetic dystrophy: a pilot study. *Complementary Therapies in Medicine, 4,* 103–105.

Fisher, P., & Ward, A. (1994). Complementary medicine in Europe. *BMJ, 309,* 107–111.

Fordyce, W. E. (1976). *Behavioral methods for chronic pain and illness.* St. Louis: Mosby.

Gainer, M. J. (1992). Hypnotherapy for reflex sympathetic dystrophy. *American Journal of Clinical Hypnosis, 34*(4), 227–232.

Galper, D. I., Taylor, A. G., & Cox, D. J. (2003). Current status of mind-body interventions for vascular complications of diabetes. *Family & Community Health, 26*(1), 34–40.

Gardner-Nix, J. (2009). Mindfulness-based stress reduction for chronic pain management. *Clinical handbook of mindfulness* (Vol. 43, pp. 369–381). New York: Springer Science and Buisness Media LLC NY.

Hamza, M. A., et al. (2000). Percutaneous electrical nerve stimulation: a novel analgesic therapy for diabetic neuropathic pain. *Diabetes Care, 23*(3), 365–370.

Han, J. S., & Terenius, L. (1982). Neurochemical basis of acupuncture analgesia. *Annual Review of Pharmacology and Toxicology, 22,* 193–220.

Hanai, F. (2000). Effect of electrical stimulation of peripheral nerves on neuropathic pain. *Spine, 25*(15), 1886–1892.

Harden, R. N., Houle, T. T., Green, S., Remble, T. A., Weinland, S. R., Colio, S., Lauzon, J., & Kuiken, T. (2005). Biofeedback in the treatment of phantom limb pain: A time-series analysis. *Applied Psychophysiology and Biofeedback, 30*(1), 83–93.

Hart, A. M., Wilson, A. D. H., Montovani, C., Smith, C., Johnson, M., Terenghi, G., & Youle, M. (2004). Acetyl-l-carnitine: A pathogenesis based treatment for HIV-associated antiretroviral toxic neuropathy. *AIDS, 18,* 1549–1560.

Head, K. A. (2006). Peripheral neuropathy: Pathogenic mechanisms and alternative therapies. *Alternative Medicine Review, 11,* 294–329.

Hoffman, B. M., Papas, R. K., Chatoff, D. K., & Kerns, R. D. (2007). Meta-analysis of psychological interventions for chronic low back pain. *Health Psychology, 26*(1), 1–9.

Hough, S., & Kleingianna, C. (2002). Individualizing relaxation training in spinal cord injury: Importance of injury level and person factors. *Rehabilitation Psychology, 47*(4), 415–425.

Jensen, M., & Patteson, D. R. (2006). Hypnotic treatment of chronic pain. *Journal of Behavioral Medicine, 29*(1), 95–124.

Jensen, M. A., Hanley, M. A., Engel, J. M., Romano, J. M., Barber, J., Cardenas, D. D., Kraft, G. H., Hoffman, A. H., & Patterson, D. R. (2005). Hypnotic analgesia for chronic pain in persons with disabilities: A case series. *Journal of Clinical and Experimental Hypnosis, 53*(2), 198–228.

Jensen, M. P., Barber, J., Romano, J. M., Hanley, M. A., Raichle, K. A., Molton, I. R., Engel, J. M., Osborne, T. L., Stoelb, B. L., Cardenas, D. D., & Patterson, D. (2009a). Effects of self-hypnosis training and EMG biofeedback relaxation training on chronic pain in persons with spinal cord injury. *International Journal of Clinical and Experimental Hypnosis, 57*(3), 239–268.

Jensen, M. P., Barber, J., Romano, J. M., Molton, I. R., Raichle, K. A., Osborne, T. L., Engel, J. M., Stoelb, B. L., Kraft, G. H., & Patterson, D. R. (2009b). A comparison of self-hypnosis vs progressive muscle relaxation in patients with multiple scherosis and chronic pain. *International Journal of Clinical and Experimental Hypnosis, 57*(2), 198–221.

Kawano, M., Matsuoka, M., Kurokawa, T., Tomita, S., Mizuno, Y., & Ueda, K. (1989). *Acta Paediatrica Japonica, 31,* 500–503.

Keynan, O., Mirovsky, Y., Dekel, S., Gilad, V. H., & Gilad, G. M. (2010). Safety and efficacy of dietary agmatine sulfate in lumbar disc-associated radiculopathy. An open-label, dose-escalating study followed by a randomized, double-blind, placebo-controlled trial. *Pain Medicine, 11,* 356–368.

Kuchera, M. L. (2007). Applying osteopathic principles to formulate treatment for patients with chronic pain. *Journal of the American Osteopathic Association, 107*(Suppl 6), ES28–ES38.

MacLennon, A. H., Wilson, D. H., & Taylor, A. W. (1996). Prevalence and cost of alternative medicine in Australia. *The Lancet, 347,* 569–573.

McCracken, L. M., & Thompson, M. (2011). Psychological advances in chronic pain: A concise selective review of research from 2010. *Current Opinion in Supportive and Palliative Care, 5,* 122–126.

McCracken, L. M., & Vellerman, S. C. (2010). Psychological flexibility in adults with chronic pain: a study of acceptance, mindfulness, and values-based action in primary care. *Pain, 148,* 141–147.

McCracken, L. M., & Zhao-O'Brien, J. (2010). General psychological acceptance and chronic pain: there is more to accept that the pain itself. *European Journal of Pain, 14,* 170–175.

McMorland, G., Suter, E., Casha, S., du Plessis, S. J., & Hurlbert, R. J. (2010). Manipulation or microdiskectomy for sciatica? A prospective randomized clinical study. *Journal of Manipulative and Physiological Therapeutics, 33,* 576–584.

Medina McKeon, J. M., & Yancosek, K. E. (2008). Neural gliding techniques for the treatment of carpal tunnel syndrome: A systematic review. *Journal of Sport Rehabilitation, 17,* 324–341.

Mehta, S. M., Orenczuk, S., Hansen, K., Aubut, J. L., Hitzig, S. L., Legassic, M., & Teasell, R. W. (2011). An evidence-based review of the effectiveness of cognitive behavioral therapy for psychosocial issues post spinal cord injury. *Rehabilitation Psychology, 56*(1), 15–25.

Meyer, D. J., et al. (1977). Antagonism of acupuncture analgesia in man by the narcotic antagonist naloxone. *Brain Research, 121,* 368–372.

Millar, W. J. (1997). Use of alternative health care practitioners by Canadians. *Canadian Journal of Public Health, 88,* 154–158.

Morley, S., Eccleston, C., & Williams, A. (1999). Systematic review and meta-analysis of randomized controlled trials of cognitive behavioral therapy and behaviour therapy for chronic pain in adults, excluding headache. *Pain, 80,* 1–13.

National Institutes of Health. (1994). Alternative medicine: Expanding horizons: A report to the National Institutes of Health on Alternative Medical systems and Practices in the United States. Washington, DC: US Government Printing Office (017-040-00537-7).

Needham, W. E., Eldridge, L. S., Harabedian, B., & Crawford, D. G. (1993). Blindness, diabetes, and amputation: Alleviatino of depression and pain through thermal biofeedback therapy. *Journal of Visual Impairment and Blindness, 87*(9), 368–370.

Nozaki, C., Vergnano, A. M., Filliol, D., Ouagazzal, A. M., Le Goff, A., Carvalho, S., Reiss, D., Gaveriaux-Fuff, C., Neyton, J., Paoletti, P., & Kieffer, B. L. (2011). Zinc alleviates pain through high-affinity binding to the NMDA receptor NR2A subunit. *Nature Neuroscience, 14,* 1017–1022.

Oakley, D. A., Whitman, L. G., & Halligan, P. W. (2002). Hypnotic imagery as a treatment for phantom limb pain: Two case reports and a review. *Clinical Rehabilitation, 16,* 368–377.

Perry, K. N., Nicholas, M. K., & Middleton, J. W. (2010). Comparison of a pain management program with usual care in a pain management center for people with spinal cord injury-related chronic pain. *The Clinical Journal of Pain, 26*(3), 206–216.

Pert, A., et al. (1984). Alteration in rat central nervous system endorphins following transauricular electroacupuncture analgesia in the periaqueductal gray of the rabbit. *Brain Research, 322,* 289–296.

Pitter, M. H., & Ernst, E. (2008). Complementary therapies for neuropathic and neuralgic pain systematic review. *The Clinical Journal of Pain, 24,* 731–733.

Pomeranz, B., & Cheng, R. (1979). Suppression of noxious responses in single neurons of cat spinal cord by electroacupuncture and reversal by the opiate antagonist naloxone. *Experimental Neurology, 64,* 327–349.

Rainov, N. G., & Heidecke, V. (2003). Motor cortex stimulation for neuropathic facial pain. *Neurological Research, 25*(2), 157–161.

Rainville, P., Duncan, G. H., Price, D. D., et al. (1997). Pain affect encoded in human cingulate but not somatosensory cortex. *Science, 277,* 968–971.

Rice, B. I., & Schindler, J. V. (1992). Effect of thermal biofeedback-assisted relaxation training on blood circulation in the lower extremities of a population with diabetes. *Diabetes Care, 15,* 853–858.

Rice, B. I., Kalker, A. J., Schindler, J. V., & Dixon, R. M. (2001). Increased foot ulcers healing with biofeedback-assisted relaxation training. *Journal of American Podiatry Medical Associates, 91,* 132–141.

Richardson, B. S., Way, B. V., & Speece, A. J. (2009). Osteopathic manipulative treatment in the management of Notalgia Paresthetica. *Journal of the American Osteopathic Association, 109,* 605–608.

Santilli, V., Beghi, E., & Finucci, S. (2006). Chiropractic manipulation in the treatment of acute back pain and sciatica with disc protrusion: a randomized double-blind clinical trial of active and simulated spinal manipulations. *The Spine Journal, 6,* 131–137.

Saratsiotis, J., & Myriokefalitakis, E. (2010). Diagnosis and treatment of posterior interosseous nerve syndrome using soft tissue manipulation therapy: A case study. *Journal of Bodywork and Movement Therapies, 14*, 397–402.

Schneider, J., Hofmann, A., Rost, C., & Schapiro, F. (2008). EMDR in the treatment of chronic phantom limb pain. *Pain Medicine, 9*(1), 76–82.

Schroder, S., Liepert, J., Remppis, A., & Greten, J. H. (2007). Acupuncture treatment improves nerve conduction in peripheral neuropathy. *European Journal of Neurology, 14*(3), 276–281.

Schroeder, S., Meyer-Hamme, G., & Epplee, S. (2011). Acupuncture for chemotherapy-induced peripheral neuropathy (CIPN): A pilot study using neurography. *Acupuncture in Medicine.* (Epub ahead of print).

Sherman, R. A., Gall, N., & Gormly, J. (1979). Treatment of phantom limb pain with muscular relaxation training to disrupt the pain-anxiety-tension cycle. *Pain, 6*, 47–55.

Shlay, J. C., et al. (1998). Acupuncture and amitriptyline for pain due to HIV-related peripheral neuropathy: A randomized controlled trial. *Journal of the American Medical Association, 280*(18), 1590–1595.

Tsunoda, Y., et al. (1980). Antagonism of acupuncture analgesia by naloxone in unconscious man. *The Bulletin of Tokyo Medical and Dental University, 27*, 89–94.

Turk, D. C., Audette, J., Levy, R. M., Mackey, S. C., & Stanos, S. (2010). Assessment and treatment of psychosocial coorbidities in patients with neuropathic pain. *Mayo Clinic Proceedings, 83*(Suppl. 3), S42–S50.

Ursini, T., Tontodonati, M., Lamberto, M., et al. (2011). Acupuncture for the treatment of severe acute pain in herpes zoster: Results of a nested, open label, randomized trial in the VZV Pain Study. *BMC complementary and Alternative Medicine, 11*(46). http://www.biomedcentral.com/1472-6882/11/46.

Wallin, J., et al. (2003). Spinal cord stimulation inhibits long-term potentiation of spinal wide dynamic range neurons. *Brain Research, 973*(1), 39–43.

Weintraub, M. I., et al. (2003). Static magnetic field therapy for symptomatic diabetic neuropathy: A randomized, double-blind, placebo-controlled trial. *Archieves of Physical Medicine and Rehabilitation, 84*(5), 736–746.

Alternative Rat Models of Ureteric Nociceptive Stimulation In Vivo

Definition

Noxious stimulation of the ureter in previous studies has been achieved using stimulus modalities other than experimentally induced "stones." For example:

- Electrical stimulation of the ureter has been employed in the unanesthetized rat (Giamberardino et al. 1988). Although the stimulus was not natural, this model offered the advantage of a stimulus that could be readily controlled and modulated in frequency, intensity, and duration. Unfortunately, nocifensive behaviors and referred muscle hyperalgesia associated with stimulation were inconstant.

- In another study, distension of the renal pelvis after cannulation of the ureteric-pelvic junction was evaluated. This stimulus produced rather variable pseud-affective responses that were unrelated to stimulus intensity (Brasch and Zetler 1982).

- More recently, the ureter in an anesthetized rat was cannulated close to the bladder and responses (cardiovascular changes) to graded ureter distension were characterized (Roza and Laird 1995). Responses to stimuli less than 25 mmHg were never observed whereas suprathreshold pressures evoked responses proportional to stimulus intensity. The stimulus-response curve was dose-dependently attenuated by morphine in a naloxone reversible manner. The authors concluded that the characteristics of the responses observed correlated well with pain sensations in man, and with the properties of ureteric primary afferent neurons in animals. This model fulfils most of the criteria proposed as ideal for a noxious visceral stimulus: the experiments are reproducible, the results consistent, and the responses proportional to stimulus intensity. However, the procedure is invasive and can only be applied in an anesthetized rat; it is therefore not suitable for behavioral studies. On the other hand, it is ideal for electrophysiological studies, not only in normal animals but also in animals with calculosis, allowing the comparison of neural processing of acute visceral noxious stimulation in normal animals with that of animals with chronic visceral pain and referred hyperalgesia using the same stimulation technique.

Cross-References

▶ Visceral Pain Model, Kidney Stone Pain

References

Brasch, H., & Zetler, G. (1982). Caerulein and morphine in a model of visceral pain: effects on the hypotensive response to renal pelvis distension in the rat. *Naunyn-Schmiedeberg's Archives of Pharmacology, 319*, 161–167.

Giamberardino, M. A., Rampin, O., Laplace, J. P., Vecchiet, L., & Albe-Fessard, D. (1988). Muscular hyperalgesia and hypoalgesia after stimulation of the ureter in rats. *Neuroscience Letters, 87*, 29–34.

Roza, C., & Laird, J. (1995). Pressor responses to distension of the ureter in anesthetised rats: characterisation of a model of acute visceral pain. *Neuroscience Letters, 198*, 9–12.

Alternative Therapies

▶ Alternative Medicine in Neuropathic Pain

Ambiguity

▶ Impairment Rating, Ambiguity
▶ Impairment Rating, Ambiguity, IAIABC System

Amelioration

Definition

The improvement or bettering of the meaning of a word through semantic change. The opposite of pejoration.

Cross-References

▶ Lower Back Pain, Physical Examination

Amenorrhea

Definition

Amenorrhea is the absence of menstruation, which is normal before puberty, during pregnancy, or after menopause. Congenital abnormalities of the reproductive tract, metabolic disorders (such as diabetes or obesity), and endocrine disorders (including altered pituitary, thyroid, or ovarian function) are the most common causes of amenorrhea. Medications that alter hormonal status, including opioids, can also lead to amenorrhea. In some cases, emotional disorders can lead to a cessation of menses.

Cross-References

▶ Cancer Pain Management, Opioid Side Effects, Endocrine Changes, and Sexual Dysfunction

American Society of Anesthesiologists' Status Category

Definition

Each status category/class gives an overall impression of the complexity of the patient's medical condition. If the procedure is performed as an emergency, an "E" is added to the category/class:

Class 1 – a healthy patient
Class 2 – a patient with mild systemic disease
Class 3 – a patient with severe systemic disease that limits activity but is not incapacitating
Class 4 – a patient with incapacitating systemic disease that is a constant threat to life
Class 5 – a moribund patient not expected to survive 24 h with or without surgery

Cross-References

▶ Postoperative Pain, Preoperative Education

Amide Anesthetic

Definition

A member of one of the two major chemical classes of local anesthetics, differentiated by the intermediate chain linking a lipophilic group and an ionizable group (usually a tertiary amine). The pharmacologic class of agents comprised of lidocaine, bupivacaine, ropivacaine, mepivacaine, prilocaine, and etidocaine.

The other major class is ester anesthetic.

Cross-References

► Acute Pain in Children, Postoperative
► Drugs with Mixed Action and Combinations: Emphasis on Tramadol
► Postoperative Pain, Methadone

Amidine

► Postoperative Pain, Methadone

Amidine Hydrochloride

► Postoperative Pain, Methadone

Aminobisphosphonate

Definition

A class of drugs that blocks bone-resorbing cells (osteoclasts) and prevents bone loss.

Cross-References

► Cancer Pain Management, Orthopedic Surgery

Aminomethyl-Cyclohexane-Acetic Acid

► Postoperative Pain, Gabapentin

Amitriptyline

Definition

A tricyclic antidepressant drug utilized for the treatment of chronic pain, particularly effective in the craniofacial region. Its antinociceptive effect is independent of its antidepressive activity in that they occur at doses tenfold lower (10–25 mg/day) than antidepressant doses, possibly involving a blockade of voltage-gated sodium channels. Amitriptyline's prophylactic efficacy in chronic daily headache and migraine is limited.

Cross-References

► Atypical Facial Pain: Etiology, Pathogenesis, and Management
► Fibromyalgia, Mechanisms, and Treatment
► Preventive Migraine Therapy

AMPA Glutamate Receptor (AMPA Receptor)

Definition

A type of ionotropic glutamate receptor that is activated by the specific agonist *alpha*-Amino-3-hydroxy-5-methyl-4-isoxazolepropionate (AMPA). AMPA receptors comprise several subunits (GluR1, GluR2, GluR3, GluR4) that form a heteromeric receptor-ion channel complex, the composition of which affects the kinetic properties of the receptor-ion channel. AMPA receptors mediate the majority of fast synaptic transmission in the central nervous system.

Cross-References

▶ Metabotropic Glutamate Receptors in the Thalamus
▶ Opiates During Development

Amphibian Peptides

Definition

Amphibian skin contains a wide variety of peptides that are often homologous or even identical to the gastrointestinal hormones and neurotransmitters of the Mammalia.

Striking examples are cerulein, the amphibian counterpart of mammalian cholecystokinin and gastrin; physalemin and kassinin, counterparts of the mammalian neuropeptides substance P and neurokinins; the amphibian bombesins and litorins, which heralded the discovery of the gastrin-releasing peptides (mammalian bombesin) and neuromedin B; and finally, sauvagine, whose structure elucidation preceded that of the analogous, hypothalamic corticotropin-releasing hormone. Other peptide families common to amphibian skin and mammalian tissues are bradykinins, angiotensins, somatostatins, and the thyrotropin-releasing hormone. Opioid peptides have so far only been in the skin of the hylid frog of the Phyllomedusine stock. During his long scientific life, the pharmacologist Vittorio Erspamer sought biologically active molecules in more than 500 amphibian species from all over the world and showed that the amphibian skin and its secretions offer an inexhaustible supply of biologically active peptides for pharmacological research.

Cross-References

▶ Opioid Peptides from the Amphibian Skin

Amphipathic

Definition

An amphipathic segment is a segment with opposing hydrophobic and hydrophilic faces, oriented spatially along the axis of the segment

Cross-References

▶ Capsaicin Receptor
▶ Thalamus, Clinical Pain, Human Imaging

Amygdala

Definition

A prominent group of neurons forming an almond-shaped structure at the level of the temporal cortex in primates, and form part of the limbic system. In the rat, the amygdala is ventrolateral, close to both the temporal and perirhinal cortices. It is divided schematically into four groups: cortical and basal (main olfactory), medial (accessory olfactory), central (autonomic), basolateral and lateral (frontotemporal and temporal cortices). The precise role of this region remains incompletely understood. It seems that one of its roles is to mark perceptions with an affective label that provides an appropriate significance in the environment of the species. In the framework of pain, it triggers an aversive reaction and fear that causes the organism to avoid dangerous stimuli. It also plays a role in the development of memories with an emotional component.

Cross-References

▶ Amygdala, Pain Processing and Behavior in Animals

▶ Arthritis Model, Kaolin-Carrageenan-Induced Arthritis (Knee)
▶ Parabrachial Hypothalamic and Amygdaloid Projections

Amygdala, Functional Imaging

Ute Habel and Frank Schneider
Department of Psychiatry and Psychotherapy,
RWTH Aachen University, Aachen, Germany

Synonyms

Functional magnetic resonance imaging (fMRI); Positron emission tomography (PET)

Definition

The amygdala is an essential key structure in the cerebral limbic network underlying emotion processing. As such, it is suggested to be part of the brain circuit involved in the processing of pain, which is known to include strong affective components. Neuroimaging studies pointing to amygdala involvement during pain processing are currently increasing. The amygdala is a small almond shape structure in the anterior temporal lobe with a variety of functions for emotion processing together with learning and memory. It is supposed to execute an evaluative associative function, combining external cues with internal responses, thereby assessing and defining the valence, relevance, and significance of stimuli. It is its extensive connectivity with various cortical and subcortical areas that enables fast automatic, but also more conscious deliberate, responses. Its role in pain processing is however less clear.

Characteristics

Negative affect is typically evoked by acute pain. Key structures of the ▶ limbic system have been

identified that play an important role in regulating affective behavior; among the most important are the subcortical and cortical areas, the anterior cingulate, the insula, and the prefrontal cortex. Most notably, assessment of emotional valence of stimuli and the provocation of distinct emotional reactions are mediated by the amygdala. This central role in emotion processing can be executed due to a broad cortical and subcortical network in which the amygdala is located and which is able to provide it with raw information via the short thalamus route but also with highly processed polymodal input from sensory cortices. Finally, the amygdala is not a unitary structure, but consists of several nuclei exerting different functions. It is believed to have a major role in pain because of the strong association and interaction between pain and emotion but also because of the specific nociceptive inputs to the latero-capsular part of the central nucleus, the major output system within the amygdala, indicating that, within this accumulation of nuclei, this part may represent the "nociceptive amygdala" (Neugebauer et al. 2004). For ▶ fMRI, mapping of activation within this region is, however, critical posing technical and methodological problems, which often call into question the validity and reliability of imaging results reporting amygdala activation. This may possibly be one of the reasons, why early neuroimaging findings mostly failed to demonstrate clear amygdala activation during pain perception. FMRI of this deep subcortical region is confronted with a set of difficulties, such as movement, respiratory, inflow, and susceptibility artifacts (see ▶ Inflow Artifacts) and nonetheless the rapid habituation of amygdala responses to repeated stimulus presentations. This is of special relevance for experimental pain studies, which mostly rely on the application of ▶ block designs, which are especially prone to habituation. Recent methodological advances in neuroimaging may have partly overcome these inherent mapping difficulties, accounting for the increase in pain studies successfully demonstrating amygdala participation (Bingel et al. 2002; Bornhövd et al. 2002).

Amygdala, Functional Imaging, Table 1 Overview of pain studies reporting amygdala activation

Author	Imaging method	Painful stimulation	Number of subjects	Amygdala activation/deactivation
Beccera et al. 1999	fMRI (1,5 T)	Thermal stimulation (Peltier-based thermode) 46 °C	2 groups of 6 healthy subjects	Deactivation of the amygdala
Becerra et al. 2001	fMRI (1,5 T)	Thermal stimulation (Peltier-based thermode) 46 °C compared to 41 °C	8 healthy subjects	Activation in the sublenticular extended amygdala in the early phase
Bingel et al. 2002	fMRI (1,5 T)	YAG infrared laser stimulation	14 healthy subjects	Bilateral activation to unilateral stimulation
Bornhövd et al. 2002	fMRI (1,5 T)	YAG infrared laser stimulation	9 healthy subjects	Activation increasing with stimulus intensity
Derbyshire et al. 1997	PET (H_2 ^{15}O)	CO_2 laser (mild/moderate pain vs. warm)	12 healthy subjects	Decreased rCBF
Evans et al. 2002	fMRI (1,5 T)	Mechanical ventilation at 12–14 breaths/min	6 healthy subjects	Activation
		Air hunger vs. baseline		
Hsieh et al. 1995	PET (^{15}O Butanol)	Intracutaneous injection of a minute amount of ethanol vs. saline	4 healthy subjects	Nonsignificant activation
Lu et al. 2004	fMRI (3 T)	Fundus balloon distension (17.0 0.8 mmHg) vs. baseline	10 healthy subjects	Activation
Petrovic et al. 2004	PET (H_2 ^{15}O)	Cold pressure test (0–1 °C water with ice or glycol) vs. cold water (19 °C)	10 healthy subjects	Deactivation in response to context manipulations increasing anticipated pain duration
Schneider et al. 2001	fMRI (1,5 T)	Balloon dilatation of a dorsal foot vein	6 healthy subjects	Amygdala activation correlated with subjective online pain ratings
Wilder-Smith et al. 2004	fMRI (1,5 T)	Rectal balloon distention alone or with painful heterotopic stimulation of the foot with ice water	10 patients with irritable bowel syndrome, 10 healthy subjects	Amygdala activation in patients with irritable bowel syndrome (constipation) during heterotopic stimulation

Alternatively, it is also conceivable that the majority of pain stimulation techniques failed to evoke pain that provoked strong emotional responses, hence falling short of observing amygdala involvement. The frequent failure of these early studies to report changes in autonomic arousal during painful stimulation corroborates this assumption. In an attempt to model acute traumatic nociceptive pain, a ▶ PET study used intracutaneous injection of ethanol (Hsieh et al. 1995). Affective and heart rate changes were described in subjects, and cerebral activation was found in subcortical structures, specifically the hypothalamus and the periaqueductal gray. These regions are taken to constitute the brain defense system which functions as a modulator for aversive states. Although signal increases in the amygdala were detected by the authors, they failed to be significant.

Despite more recent neuroimaging findings reporting amygdala involvement in pain processing, a full characterization of its function during pain perception is still lacking and at first sight results seem to be equivocal, pointing to activations as well as deactivations of the amygdala in this context (Table 1).

One fMRI investigation applied painful stimulation with a strong affective component to measure pain related changes in cerebral activity (Schneider et al. 2001). By inflating an indwelling balloon catheter, a dorsal foot vein of healthy volunteers was stretched to a noxious distress physical level, which induced vascular pain associated with a particularly strong negative effect. Since the sensory innervation of veins exclusively subserves nociception, non-painful cosensations were excluded. Additionally, brief stimulations of only a few minutes produce vascular pain that escapes adaptation and is generally reported as particularly aching in character. During noxious stimulation, the subjects continuously rated perceived pain intensity on a pneumatically coupled visual analogue scale, which was used as permanent feedback to adjust balloon expansion so that the pain intensity could be kept at intended values at all times. The analysis strategy that focused primarily on correlations of signal changes with these subjective ratings, rather than the generally applied signal variations to a stimulation-based reference function (▶ Boxcar Design), facilitated producing evidence for amygdala activation (Fig. 1). Hence, these results indicated a relevant role of the amygdala in the subjective component of painful experiences and suggested that in the widespread cerebral network of pain perception, the limbic system and especially the amygdala may be instrumental in the affective aspects of pain. Supporting evidence for these conclusions come from neuroimaging findings during air hunger (Evans et al. 2002) or fundus balloon distension (Lu et al. 2004). Dyspnea was induced in healthy subjects by mechanical ventilation until a sensation of "urge to breathe" and "starved for air" was reached and compared to mild hypocapnia. This pain is also very afflicted with strong negative effect. Correspondingly, a network of limbic and paralimbic nodes was activated, including anterior insula, anterior cingulate, operculum, thalamus, cerebellum, basal ganglia, and also amygdala, that is, the majority of regions forming part of the limbic network also involved in emotion processing. Similarly, moderate gastric pain was induced in ten healthy subjects using fundus balloon distension (Lu et al. 2004) and resulted in a widespread activation pattern of subcortical as well as cortical regions, among them insula and amygdala. This may once again point especially to the strong affective component of visceral pain. Since visceral pain may be indicative of an urgent and marked system imbalance possibly endangering survival, strong affective responses with the objective of initiating adequate adaptations and reactions seem to have an evolutionary purpose and be necessary.

Amygdala, Functional Imaging, Fig. 1 Individual signal intensities in the amygdala following correlation with subjective ratings of the six individual participants (From Schneider et al. 2001)

Amygdala activation is however not restricted to visceral pain but also visible during other kinds of painful stimulation in animals as well as humans (Bingel et al. 2002). Unilateral laser that evoked painful stimuli of either side, which also avoided concomitant tactile stimulation and anticipation as well as habituation, successfully demonstrated bilateral amygdala activation, most probably representing the affective pain component (Fig. 2). In contrast, basal ganglia and cerebellum displayed corresponding unilateral activation and may probably be related to defensive and withdrawal behavior. RCBF (regional cerebral blood flow) changes were also found in limbic structures of rats during noxious formalin nociception (Morrow et al. 1998).

Hence, the role of the amygdala as a "sensory gateway to the emotions" (Aggleton and Mishkin 1986) with an evaluative function seems to extend to pain perception as well. An increasing number of studies supported the notion of a common evaluative system with a central role of the amygdala in the processing of painful but also non-painful or novel stimuli. The amygdala demonstrated not only coding of the pain amount by showing a linearly increasing response to augmenting painfulness (Fig. 3) but also significant responses during uncertain trials in which the stimulus was not perceived, and hence,

Amygdala, Functional Imaging, Fig. 2 Amygdala activation emerged bilaterally in response to painful unilateral laser stimulation. *Left*: Fitted responses applied to the *left* (*blue line*) or *right* (*red line*) hand for the left (*left graph*) and right (*right graph*) hemispheres. The *dotted lines* show the standard error of the mean (SEM) (From Bingel et al. 2002)

Amygdala, Functional Imaging, Fig. 3 Picture: Bilateral amygdala activation ($p = 0.001$) on a coronal slide. Graphs: *Left side* entails regression coefficients indicating amount of response for each trial (P0–P4). *Right side* depicts amount of signal change in the amygdala as a function of peristimulus time separately for all stimuli (P0–P4; from Bornhövd et al. 2002)

Amygdala, Functional Imaging, Fig. 4 Coronal slices showing sublenticular extended amygdala (SLEA) activation in the early (*left*) and late phases (*middle*) in response to a noxious thermal stimulation (46 °C). Overlap (*white*) of early (*yellow/red*) and late (*blue*) phases (*right*)

a judgment on the nature and valence is required (Bornhövd et al. 2002). Furthermore, the amygdala, here more specifically the sublenticular extended amygdala, seems to be characterized by early responses (to noxious thermal stimuli) in contrast to regions activated later and associated specifically to somatosensory processing, such as the thalamus, somatosensory cortex, and insula (Becerra et al. 2001) (Fig. 4). This is in accordance with the activation characteristic of the amygdala during ▶ classical conditioning (Büchel et al. 1998), in which a rapid adaptation to the conditioned stimulus has been observed, pointing to a major role of the amygdala during the early phase of learning, during the establishment of an association between the neutral stimulus and the (un)conditioned response. Hence, the early response during pain seems to reflect the association between the painful stimulus and an adequate internal response determining the negative valence of the stimulus.

However, sometimes deactivation as opposed to activation has been observed in the amygdala during painful stimulation, for example, with fMRI in response to thermal stimuli (45 °C)

(Beccera et al. 1999). In this study only six subjects were investigated and changes were low level. Similar deactivations have also been reported using PET during mild or moderate pain due to CO_2 laser stimulation compared to non-painful warm sensations (Derbyshire et al. 1997). Hence, a possible moderating variable for activations and deactivations may be the specific thermal pain sensation, which was similar during both experiments. Alternatively, the deactivation may reflect another functional activation characteristic of the amygdala under certain circumstances. Hence, the deactivation may simply be the consequence of the nature of the experimental pain stimulus. An early activation in the amygdala for purposes of evaluation and affective judgment may be followed by a deactivation, possibly representing the attempt to regulate and cope with the affective aspects of the painful experience as well as the painful sensation itself that cannot be escaped in this special experimental setup. This interpretation is supported by recent PET findings. Petrovic et al. (2004) investigated the influence of context manipulations before the painful stimulation on the activation pattern during noxious (cold pressure) stimulation. Subjects were informed prior to stimulation if it was going to be painful or not and if it would last for 1 or 2 min. Anticipating that the pain was going to last longer was accompanied by a decrease in amygdala activation and changes in autonomic parameters but also cognitive processes in the majority of subjects that consisted of strategies to cope with the stressful but unavoidable pain. This amygdala deactivation was paralleled by activation in the anterior cingulate, pointing to interactions within this limbic network constituting the brain's pain matrix responsible for the development and modulation as well as coverage and termination of the affective noxious events.

This study also highlights some methodological problems of pain imaging studies in general and those with a special focus on the amygdala. Anticipation may alter amygdala response characteristics and may lead to deactivations instead of activations. Furthermore, the individual variability in pain responses and several methodological factors, such as imaging method, data analysis, and control condition used for comparison with pain condition, influence results as well as their interpretation.

However, further indications that the amygdala serves coping functions during pain perception come from clinical trials. Here, visceral pain hypersensitivity is discussed as a possible relevant pathogenic factor in various chronic pain syndromes, such as ▶ irritable bowel syndrome (IBS). Reduced signals in the amygdala (as well as in further limbic network nodes such as insula and striatum) have also been observed in patients with irritable bowel syndrome during rectal pain stimulation (Bonaz et al. 2002) and are in accordance with the interpretation of deactivations found in healthy controls. It may be suggested that deactivations in patients may correspond to the effort to modulate and control the strong affective components of the painful experiences. Unfortunately this study failed to include healthy controls, and hence, a conclusion on the dysfunctional or compensatory aspects of these activations in patients remains elusive. Interestingly, a recent fMRI study (Wilder-Smith et al. 2004) investigating rectal pain alone or accompanied by painful foot stimulation (ice water, activating endogenous pain inhibitory mechanisms) in patients with irritable bowel syndrome as well as healthy controls found differential activations between groups in the amygdala (activation in constipated patients) as well as further affective-limbic regions (hippocampus, insula, anterior cingulate, prefrontal cortex, etc.) during heterotopic stimulation.

Hence, the amygdala is not only implicated in the affective aspects of pain processing, including both the appraisal of a painful stimulation and the initiation of adequate responses, and the experiential affective aspects, such as stress, fear, or anxiety, but also the modification, attenuation, and coping of these affective experiential aspects. This multiple functionality is supported by behavioral findings demonstrating amygdala activation during enhancement as well as inhibition of pain (Neugebauer et al. 2004). First, it may be a protective mechanism to detect a possible harmful stimulus, hence amplifying the painful experience; however, in case of unavoidable harm or pain, it may be the most suitable response to reduce the painfulness by inhibition (e.g., via the periaqueductal gray). Finally, the central role in pain and emotion makes it highly likely that it may also be involved in the dysfunctional aspects of chronic (visceral) pain. For example, the involvement of the amygdala during memory and learning may be relevant facets for the development of chronic pain.

However, the diversity of functions exerted by the amygdala as indicated by the different imaging studies on experimental and chronic pain, such as affective painful experience, but also modulation of this experience as an evolutionary sensible warning and evaluative survival system, including an effective adaptation mechanism in case of inescapable painful stimulation, suggests the involvement of other brain regions as well. Hence, the function of the amygdala cannot be determined alone but only within a greater cortical and subcortical network. Despite its relevance, it is only the continuous and intensive interconnections, interactions, and feedback mechanisms with other brain regions that account for the complex and intact function of this structure in pain and emotion.

References

Aggleton, J. P., & Mishkin, M. (1986). The amygdala: Sensory gateway to the emotions. In R. Plutchik & H. Kellermann (Eds.), *Emotion: Theory, research, and experience* (Vol. 3, pp. 281–299). New York: Academic.

Beccera, L. R., Breiter, H. C., Stojanovic, M., et al. (1999). Human brain activation under controlled thermal stimulation and habituation to the noxious heat: An fMRI study. *Magnetic Resonance in Medicine, 41*, 1044–1057.

Becerra, L., Breiter, H. C., Wise, R., et al. (2001). Reward circuitry activation by noxious thermal stimuli. *Neuron, 32*, 927–946.

Bingel, U., Quante, M., Knab, R., et al. (2002). Subcortical structures involved in pain processing: Evidence from single-trial fMRI. *Pain, 99*, 313–321.

Bonaz, B., Bacin, M., Papillon, E., et al. (2002). Central processing of rectal pain in patients with irritable bowel syndrome: An fMRI study. *The American Journal of Gastroenterology, 97*, 654–661.

Bornhövd, K., Quante, M., Glauche, V., et al. (2002). Painful stimuli evoke different stimulus-response functions in the amygdala, prefrontal, insula and somatosensory cortex: A single-trial fMRI study. *Brain, 125*, 1326–1336.

Büchel, C., Morris, J., Dolan, R. J., et al. (1998). Brain systems mediating aversive conditioning: An event-related fMRI study. *Neuron, 20*, 947–957.

Derbyshire, S. W., Jones, A. K., Gyulai, F., et al. (1997). Pain processing during three levels of noxious stimulation produces differential patterns of central activity. *Pain, 73*, 431–445.

Evans, K. C., Banzett, R. B., Adams, L., et al. (2002). BOLD fMRI identifies limbic, paralimbic, and cerebellar activation during air hunger. *Journal of Neurophysiology, 88*, 1500–1511.

Hsieh, J. C., Ståhle-Bäckdahl, M., Hägermark, Ö., et al. (1995). Traumatic nociceptive pain activates the hypothalamus and the periaqueductal gray: A positron emission tomography study. *Pain, 64*, 303–314.

Lu, C. L., Wu, Y. T., Yeh, T. C., et al. (2004). Neuronal correlates of gastric pain induced by fundus distension: A 3 T-fMRI study. *Neurogastroenterology and Motility, 16*, 575–587.

Morrow, T. J., Paulson, P. E., Danneman, P. J., et al. (1998). Regional changes in forebrain activation during the early and late phase of formalin nociception: Analysis using cerebral blood flow in the rat. *Pain, 75*, 355–365.

Neugebauer, V., Li, W., Bird, G. C., et al. (2004). The amygdala and persistent pain. *The Neuroscientist, 10*, 221–234.

Petrovic, P., Carlsson, K., Petersson, K. M., et al. (2004). Context-dependent deactivation of the amygdala during pain. *Journal of Cognitive Neuroscience, 16*, 1289–1301.

Schneider, F., Habel, U., Holthusen, H., et al. (2001). Subjective ratings of pain correlate with subcortical-limbic blood flow: An fMRI study. *Neuropsychobiology, 43*, 175–185.

Wilder-Smith, C. H., Schindler, D., Lovblad, K., et al. (2004). Brain functional magnetic resonance imaging of rectal pain and activation of endogenous inhibitory mechanisms in irritable bowel syndrome patient subgroups and healthy controls. *Gut, 53*, 1595–1601.

Amygdala, Nociceptive Processing

▶ Nociceptive Processing in the Amygdala: Neurophysiology and Neuropharmacology

Amygdala, Pain Processing and Behavior in Animals

Barton H. Manning
Amgen Inc., Thousand Oaks, CA, USA

Synonyms

Amygdaloid complex; Nociceptive processing in the amygdala, behavioral and pharmacological studies

Definition

The ▶ amygdala is an almond-shaped structure in the ventromedial temporal lobe that constitutes part of the brain's limbic system. It comprises several neuroanatomically and functionally distinct nuclei with widespread connections to and from a variety of cortical and subcortical brain regions.

Characteristics

In a general sense, the amygdala plays a prominent role in the coordination of defense reactions to environmental threats (LeDoux 2003).

The hypothesized role of the amygdala in emotional information processing represents one component in this overall role. Clearly, environmental threats are diverse and include the animate (e.g., extraspecies predators, intraspecies rivals) and inanimate (e.g., thorns or spines on plants). Stimuli signaling the presence of threats can be "natural" elicitors of the psychological state of fear such as a sudden, novel sound or the presence of a larger animal. Or previously "neutral" stimuli (discrete sensory cues or distinct environmental contexts) can come to elicit defense reactions following occasions in which they coincided in time with an occurrence of injury or the presence of a natural threat (i.e., through ▶ classical conditioning processes). Such "conditioned" stimuli can elicit either acute fear or the qualitatively different state of ▶ anxiety, which is a more future-oriented psychological state that readies the animal for a potential environmental threat.

The amygdala is well connected to coordinate reactions to stimuli that signal potential danger. By way of incoming neuroanatomical connections to its central and basolateral subdivisions, the amygdala receives information from the organism's internal environment (▶ viscerosensation) and information from the external environment consisting of simple sensory inputs and complex ▶ multisensory perceptions. This information has already been highly processed by various subcortical and cortical brain structures (e.g., cortical sensory association areas), but the amygdala serves the purpose of attaching emotional significance to the input. By way of its outgoing neuroanatomical connections, the amygdala communicates with brain areas involved in motor preparation/ action and autonomic responses. When sensory information arrives relating to environmental danger, the amygdala probably is involved both in the generation of emotional states (e.g., fear, anxiety) and the coordination of appropriate ▶ autonomic and behavioral changes that enhance the chance of survival (e.g., defensive fight or flight, subsequent avoidance behaviors, submissive postures, tonic immobilization, autonomic arousal, and hypoalgesia or ▶ hyperalgesia).

Since pain can signal injury or the potential for injury, it should not be surprising that the processing of nociceptive information by the amygdala can be one of the triggers of these events. Electrophysiological studies show that individual amygdala neurons, particularly in the central nucleus of the amygdala (CeA), respond to brief nociceptive thermal and mechanical stimulation of the skin and or nociceptive mechanical stimulation of deeper (knee joint) tissue (Bernard et al. 1996; Neugebauer et al. 2004). Many CeA neurons have large receptive fields, with some neurons being excited by and others inhibited by nociceptive stimulation. The lateral capsular and, to a lesser extent, the lateral division of the CeA have been termed the "nociceptive amygdala" and receive nociceptive input from lamina I of the spinal and trigeminal ▶ dorsal horns. This lamina I input arrives at the CeA via several different routes (Gauriau and Bernard 2002): (1) indirectly, from relays in the lateral and external medial areas of the brainstem parabrachial complex (lamina I → PB → CeA); (2) indirectly, from the posterior triangular nucleus of the thalamus (PoT) to the amygdalostriatal transition area (AStr), which overlaps partly with the CeA (lamina I → PoT →AStr/CeA); (3) indirectly, from the ▶ insular cortex by way of the PoT (lamina I → PoT → IC → CeA); and (4) to a much lesser extent, from direct, monosynaptic projections (lamina I → CeA). The basolateral complex of the amygdala also probably receives highly processed nociceptive information from unimodal and polymodal sensory areas of the cerebral cortex (Shi and Cassell 1998).

Human functional ▶ neuroimaging studies have supported a role for the amygdala in nociceptive processing by correlating changes in neural activity in the amygdala with the perception of brief painful stimuli. In a manner analogous to the different responses of individual CeA neurons described above, presentation of a painful thermal stimulus to skin of healthy human subjects can result in increases or decreases in neural activity in the amygdala as measured by ▶ positron emission tomography (PET) or functional ▶ magnetic resonance imaging (fMRI), depending on the stimulation parameters employed. These changes appear to be linearly related to stimulus intensity (Bornhovd et al. 2002; Derbyshire et al. 1997).

In addition to brief pain, neuroplastic changes in amygdala neurons may contribute to the induction and maintenance of ▶ chronic pain states. Rodent studies utilizing indirect measures of neuronal activation in the forebrain (e.g., ▶ immediate early gene expression or changes in regional cerebral blood flow) have suggested increases in neural activity in the amygdala that correlate with behavioral indices of persistent pain. Several groups have analyzed patterns of Fos protein-like immunoreactivity (Fos-LI) in the rat forebrain after hind paw injection of formalin (i.e., the formalin test). The formalin test involves injecting a small volume of dilute formalin into a hind paw, resulting in an array of pain-related behaviors (paw lifting, licking, and flinching) that persists for 11/2–2 h. Behavioral indices of formalin-induced ▶ nociception correlate with appearance of Fos-LI in the basolateral amygdala (Nakagawa et al. 2003). Fos-LI also appears in the basolateral amygdala and CeA following stimulation of the trigeminal ▶ receptive field in conscious rats with ▶ capsaicin (Ter Horst et al. 2001) or after prolonged, nociceptive colonic distension (Monnikes et al. 2003). In a rat model of ▶ neuropathic pain (the chronic constriction injury, or CCI, model), a significant increase in regional cerebral blood flow (rCBF) is seen in the basolateral amygdala after 8 or 12 weeks, but not 2 weeks following CCI surgery (Paulson et al. 2002).

The response characteristics of individual CeA neurons have been studied in vivo in rats with or without experimental arthritis in a knee joint (Neugebauer et al. 2004). Prolonged nociception produced by injection of ▶ carrageenan and ▶ kaolin into the knee joint results in enhancement of both receptive field size and responsiveness to mechanical stimulation of a subset of CeA neurons. Infusion, by ▶ microdialysis, of a selective ▶ NMDA receptor antagonist (AP5) or an mGluR1 receptor antagonist (CPCCOEt) into the CeA inhibits the increased responses to nociceptive and normally innocuous mechanical stimuli more potently in the arthritic versus the control condition. By contrast, infusion of a non-NMDA (AMPA/kainate) receptor antagonist (NBQX) or an mGluR5 receptor antagonist (MPEP) inhibits background activity and evoked responses under both normal control and arthritic conditions. These data suggest a change in mGluR1 and NMDA receptor function and activation in the amygdala during pain-related sensitization, whereas mGluR5 and non-NMDA receptors probably are involved in brief as well as prolonged nociception.

In vitro brain slice ▶ electrophysiology has provided additional insights (Neugebauer et al. 2004). It is possible to study properties of synaptic transmission (using ▶ whole cell patch clamp recordings) in brain slices taken from control rats versus rats with persistent pain. In the nociceptive CeA of such rats, it is possible to study ▶ monosynaptic excitatory postsynaptic currents (EPSCs) evoked by electrical stimulation of afferents from the parabrachial complex or from the basolateral amygdala. In rats with experimental arthritis, enhanced synaptic transmission (larger amplitude of evoked monosynaptic EPSCs) is observed at both the nociceptive PB-CeA ▶ synapse and the polymodal (including nociceptive) BLA-CeA synapse as compared with control rats. CeA neurons from arthritic rats also develop an increase in excitability. Induction of experimental ▶ colitis (by intracolonic injection of ▶ zymosan) produces similar effects, except for the fact that enhanced synaptic transmission is observed only at the nociceptive PB-CeA synapse. In the arthritis model, synaptic plasticity in the amygdala is accompanied by an increase in ▶ presynaptic mGluR1 function. Both the selective mGluR1 antagonist CPCCOEt and the group III mGluR agonist LAP4 decrease the amplitude of EPSCs more potently in CeA neurons from arthritic rats than in control animals. The selective group III mGluR antagonist UBP1112 reverses the inhibitory effect of LAP4. During the application of LAP4, paired-pulse facilitation was increased, while no significant changes in slope conductance and action potential firing rate of CeA neurons were

Amygdala, Pain Processing and Behavior in Animals, Fig. 1 A simplified illustration of major nociceptive pathways to the amygdala and possible consequences of stimulation of these pathways. Abbreviations: *IC* insular cortex, *PB* parabrachial complex, *PoT* posterior triangular nucleus of the thalamus

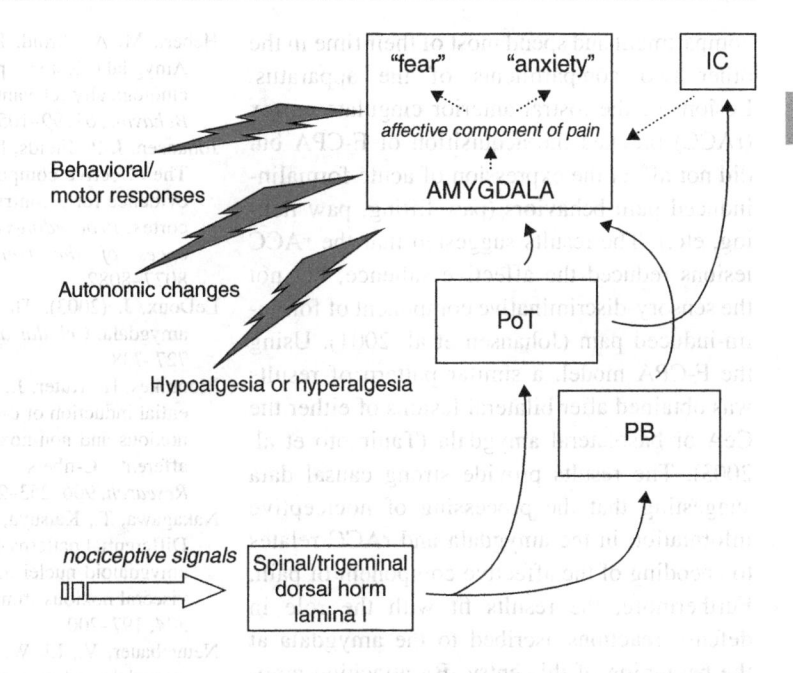

observed. These data suggest that presynaptic mGluR1 receptors and group III mGluRs regulate synaptic plasticity in the amygdala in a rat model of arthritis.

Human neuroimaging studies have provided additional supporting evidence by correlating changes in neural activity in the amygdala with the perception of persistent pain. In patients suffering from ▸ irritable bowel syndrome (IBS), Wilder-Smith et al. (2005) demonstrated a bilateral decrease in neural activity in the amygdala during episodes of experimentally induced rectal pain.

Neuroimaging techniques, measurement of immediate early gene responses, and in vivo electrophysiological studies are useful for identifying brain regions with activity that covaries with the presence or absence of pain or nociception, but such studies are limited with respect to mechanistic insights and determining cause versus effect. On the contrary, rodent behavioral studies have been highly informative in this regard. Such studies provide evidence that the amygdala is involved in encoding the affective or aversive component of pain. Hebert et al. (1999) used an alley-shaped apparatus with an array of protruding, sharp pins situated in the middle of the alley to investigate this issue. During 10 min test

sessions, the behavioral patterns of normal rats were characterized by voluntary contact with the pins followed by periods of avoidance and risk assessment (referred to by the investigators as "stretch attend" and "stretch approach" behaviors). Of the group of normal rats tested, few actually crossed the array of pins. In contrast, rats with bilateral lesions of the amygdala showed a significant increase in both the number of crossings of the pin array and the amount of time spent on the pins as compared with normal rats. The results suggest that the aversive quality of the painful mechanical stimulation imparted by the pin array is encoded at least partly by the amygdala.

The affective/aversive quality of pain in rodents also has been studied using a variation of the place-conditioning paradigm. In 2001, Johansen et al. introduced the formalin-induced condition place avoidance model (F-CPA). By pairing the experience of formalin-induced pain with a distinct environmental context/compartment within a place-conditioning apparatus, the investigators hoped to establish a behavioral endpoint that is directly related to the negative ▸ affective component of pain. After two pairings of formalin-induced pain (1 h) with the compartment, rats learned to avoid the compartment and spend most of their time in the other two compartments of the apparatus. Lesions of the rostral anterior cingulate cortex (rACC) blocked the acquisition of F-CPA but did not affect the expression of acute formalin-induced pain behaviors (paw lifting, paw licking, etc.). The results suggested that the rACC lesions reduced the affective salience, but not the sensory-discriminative component of formalin-induced pain (Johansen et al. 2001). Using the F-CPA model, a similar pattern of results was obtained after bilateral lesions of either the CeA or basolateral amygdala (Tanimoto et al. 2003). The results provide strong causal data suggesting that the processing of nociceptive information in the amygdala and rACC relates to encoding of the affective component of pain. Furthermore, the results fit with the role in defense reactions ascribed to the amygdala at the beginning of this entry. By attaching emotional significance to a stimulus signaling danger (in this case the pain associated with formalin), the amygdala sets the stage for coordination of appropriate acute and delayed responses to the stimulus by way of its multitude of connections with other brain regions and neural circuitry (Fig. 1). These responses include acute protective behaviors and autonomic responses followed by avoidance of the environment in which the pain was experienced.

References

Bernard, J. F., Bester, H., & Besson, J. M. (1996). Involvement of the spino-parabrachio-amygdaloid and hypothalamic pathways in the autonomic and affective emotional aspects of pain. *Progress in Brain Research, 107*, 243–255.

Bornhovd, K., Quante, M., Glauche, V., et al. (2002). Painful stimuli evoke different stimulus-response functions in the amygdala, prefrontal, insula and somatosensory cortex: A single-trial fMRI study. *Brain, 125*, 1326–1336.

Derbyshire, S. W., Jones, A. K., Gyulai, F., et al. (1997). Pain processing during three levels of noxious stimulation produces differential patterns of central activity. *Pain, 73*, 431–445.

Gauriau, C., & Bernard, J. F. (2002). Pain pathways and parabrachial circuits in the rat. *Experimental Physiology, 87*, 251–258.

Hebert, M. A., Ardid, D., Henrie, J. A., et al. (1999). Amygdala lesions produce analgesia in a novel, ethologically relevant acute pain test. *Physiology and Behavior, 67*, 99–105.

Johansen, J. P., Fields, H. L., & Manning, B. H. (2001). The affective component of pain in rodents: Direct evidence for a contribution of the anterior cingulate cortex. *Proceedings of the National Academy of Sciences of the United States of America, 98*, 8077–8082.

LeDoux, J. (2003). The emotional brain, fear, and the amygdala. *Cellular and Molecular Neurobiology, 23*, 727–738.

Monnikes, H., Ruter, J., Konig, M., et al. (2003). Differential induction of c-fos expression in brain nuclei by noxious and non-noxious colonic distension: Role of afferent C-fibers and 5-HT3 receptors. *Brain Research, 966*, 253–264.

Nakagawa, T., Katsuya, A., Tanimoto, S., et al. (2003). Differential patterns of c-fos mRNA expression in the amygdaloid nuclei induced by chemical somatic and visceral noxious stimuli in rats. *Neuroscience Letters, 344*, 197–200.

Neugebauer, V., Li, W., Bird, G. C., et al. (2004). The amygdala and persistent pain. *The Neuroscientist, 10*, 221–234.

Paulson, P. E., Casey, K. L., & Morrow, T. J. (2002). Long-term changes in behavior and regional cerebral blood flow associated with painful peripheral mononeuropathy in the rat. *Pain, 95*, 31–40.

Shi, C.-J., & Cassell, M. D. (1998). Cascade projections from somatosensory cortex to the rat basolateral amygdala via the parietal insular cortex. *The Journal of Comparative Neurology, 399*, 469–491.

Tanimoto, S., Nakagawa, T., Yamauchi, Y., et al. (2003). Differential contributions of the basolateral and central nuclei of the amygdala in the negative affective component of chemical somatic and visceral pains in rats. *European Journal of Neuroscience, 18*, 2343–2350.

Ter Horst, G. J., Meijler, W. J., Korf, J., et al. (2001). Trigeminal nociception-induced cerebral Fos expression in the conscious rat. *Cephalalgia, 21*, 963–975.

Wilder-Smith, C. H., Schindler, D., Lovblad, K., et al. (2005). Brain functional magnetic resonance imaging of rectal pain and activation of endogenous inhibitory mechanisms in irritable bowel syndrome patient subgroups and healthy controls. *Gut, 53*, 1595–1601.

Amygdaloid Complex

▶ Amygdala, Pain Processing and Behavior in Animals

Anaerobic Glycolysis

Definition

Glycolysis is a metabolic process that yields energy by converting glucose into lactic acid. It occurs in skeletal muscle when the blood supply is not sufficient for aerobic metabolism. The process is less effective than the aerobic metabolism (yields less ATP per mol. of glucose).

Cross-References

▶ Muscle Pain Model, Ischemia-Induced and Hypertonic Saline-Induced

Analgesia

Definition

A reduction or absence of pain. Also, a reduced response to a stimulation that would normally be painful. In humans, a decrease in the experience of pain (ongoing or spontaneous). It is sometimes described by increased thresholds to elicit sensation of pain following an evoked response. In the experimental (i.e., preclinical) setting, it can also be described as a situation in which the intensity of the stimulus required to evoke an escape or avoidance response is increased above normal, or the time required to respond to a noxious stimulus is increased above normal.

Cross-References

▶ Cancer Pain, Assessment in the Cognitively Impaired
▶ Cathepsin and Microglia
▶ Cytokine Modulation of Opioid Action
▶ Descending Circuitry, Opioids
▶ Lateral Thalamic Lesions, Pain Behavior in Animals
▶ Postsynaptic Dorsal Column Projection, Anatomical Organization

Analgesia During Labor and Delivery

Yaakov Beilin[1] and Bryan Mahony[2]
[1]Department of Anesthesiology, and Obstetrics, Mount Sinai School of Medicine, New York, NY, USA
[2]Department of Anesthesiology, Ohio State University Medical Center, Mount Sinai Hospital, New York, NY, USA

Characteristics

Analgesia for labor and delivery is now safer than ever. Anesthesia-related maternal mortality has decreased from 4.3 per million live births during the years 1979–1981 to 1.1 per million live births during the years 1995–2005. The increased use of regional anesthesia for the parturient is in part responsible for this decrease in mortality (Li et al. 2009). Safety is the primary goal of obstetrical anesthesia. For labor analgesia, a secondary goal is the minimization or elimination of maternal lower extremity muscle weakness associated with epidural and subarachnoid local anesthetics. While it has been widely accepted that no increase in the rate of cesarean sections can be attributed to epidural anesthesia (American College of Obstetricians and Gynecologists Committee on Obstetric Practice 2006), patients with less motor block are more satisfied with their anesthetic experience, and although controversial, motor blockade related to labor epidural analgesia has been implicated as a factor in an increase in forceps delivery. However, avoiding or limiting the motor block may attenuate or eliminate these effects (Chestnut 1997). In addition to epidural analgesia, anesthesiologists also provide spinal anesthesia and the combined spinal-epidural technique for labor analgesia. The purpose of this entry is to review analgesic techniques that are currently used to provide labor analgesia.

Epidural analgesia has been the most popular technique for the relief of labor pain. Its popularity is related to its efficacy and safety. Women can obtain almost complete relief from the pain of labor. From the anesthesiologist's perspective, because a catheter is threaded into the epidural space allowing variable duration and intensity of analgesia, it is also a versatile technique. During the earlier stages of labor, dilute solutions of local anesthetic can be used to achieve analgesia. As labor progresses, a more concentrated solution of local anesthetic can be used, or an adjunct, such as an opioid, can be added. Additionally, the epidural catheter can be utilized to maintain a low dermatomal level of analgesia for labor (thoracic 10–lumbar 1), and, if needed, the dermatomal level can be raised to thoracic 4 for cesarean delivery.

The agent most commonly utilized for labor epidural analgesia is a local anesthetic. Opioids are commonly added to the local anesthetic to decrease the motor block. However, unless large doses of opioids are used, they do not on their own confer adequate analgesia for labor pain. Continuous infusions of epidural local anesthetic combined with an opioid are frequently employed during labor. Continuous infusions provide a more stable level of analgesia than that provided by intermittent bolus techniques. This effect translates into decreased work load for the anesthesiologist and better analgesia for the mother. Furthermore, without the frequent bolus injections, there may be less risk of maternal hypotension. Currently used continuous infusion solutions contain 0.04–0.125 % of a local anesthetic (bupivacaine or ropivacaine or levobupivacaine) with the addition of an opioid (fentanyl or sufentanil).

Some anesthesiologists use patient-controlled epidural analgesia (PCEA). This technique allows the patient to regulate the amount of anesthetic agent they receive, controlling their analgesic level. Despite the fact that there are many well-controlled studies regarding PCEA, the optimal dosing regimens have not been determined. Many current strategies have explored the use of mandatory scheduled boluses in conjunction with patient-controlled intermittent boluses; however, inadequate research has been conducted to confer an obvious advantage. Compared with continuous infusion or intermittent bolus techniques, PCEA is associated with fewer anesthetist

interventions, lower total dose of local anesthetic used, and less motor block. Less anesthetic also decreases the frequency of maternal hypotension (Halpern and Carvalho 2009). A commonly used PCEA regimen is bupivacaine 0.0625 % with fentanyl 2 ug/cc at the following PCEA settings: 10 mL/h basal rate, 5 mL bolus dose, 10 min lockout, and a 30 mL/h maximum limit. This author is not aware of any reported complications to the parturient with PCEA use. But theoretical risks include those that have been seen in the general surgical patient including high dermatomal level or overdose from excessive self-administration, from a helpful family member, or secondary to a catheter that has migrated into the subarachnoid space.

The safety of epidural opioids has been well documented. Despite decreased neonatal neurobehavioral scores shortly after delivery, epidural fentanyl has not been linked to any long-term (4 years) developmental effects (Ounsted et al. 1978). The clinical relevance of lower neurobehavioral scores around the time of delivery is unknown. There is some data that epidural fentanyl at doses >150 ug may influence breastfeeding (Beilin et al. 2005), but these findings need to be confirmed. Respiratory depression in the neonate is also of little concern with epidural fentanyl. Respiratory parameters of neonates whose mothers received epidural fentanyl (up to 400 ug) are similar to neonates whose mothers did not receive any fentanyl.

There are a number of problems with labor epidural analgesia that have prompted some to seek alternative techniques. First, the time from epidural catheter placement until the patient is comfortable is variable but, depending on the local anesthetic used, can take up to 30 min. Other disadvantages of labor epidural analgesia include maternal hypotension, inadequate analgesia (15–20 % of cases), and, even with the very dilute local anesthetic solutions, motor blockade.

Subarachnoid opioids offer rapid, intense analgesia with minimal changes in blood pressure or motor function. The opioid is usually administered as part of a combined spinal-epidural (CSE) technique. After locating the epidural space in the usual manner, a long small-gauge spinal needle with a pencil-point design is inserted through the epidural needle into the subarachnoid space. A subarachnoid opioid alone or in combination with local anesthetic is injected. The spinal needle is removed and an epidural catheter threaded for future use. Analgesia begins within 3–5 min and lasts 1–1.5 h. A continuous epidural infusion of dilute local anesthetic/opioid solution is immediately started after securing the epidural catheter. Starting the epidural infusion immediately, versus waiting for pain to recur, prolongs the spinal medication by approximately 60 min with minimal side effects (Beilin et al. 2002). It would be tempting to thread a catheter into the subarachnoid space to enable administration of repeated doses of opioid into this space; however, an increased risk of cauda equina syndrome has been associated with the placement of subarachnoid catheters, especially microcatheters. Therefore, this technique is infrequently used now with epidural catheters in rare instances such as morbid obesity, significant cardiac disease, the presence of a difficult airway, or previous extensive spine surgery (Palmer 2010).

Fentanyl or sufentanil are the most commonly utilized subarachnoid opioids with the CSE technique. Differences between the two drugs are subtle, and choice of one over the other is based on personal preference. However, the cost of sufentanil is greater than that of fentanyl. Most anesthesiologists use between 10 and 25 ug of fentanyl and between 2 and 5 ug of sufentanil. Adding 1 mL of bupivacaine 0.25 % to either fentanyl or sufentanil prolongs the duration by about 20 min for fentanyl and 30 min for sufentanil. Side effects of adding bupivacaine are minimal and may protect the patient from developing pruritus (Asokumar et al. 1998). Whether this added duration is worthwhile is based on personal preference. At Mount Sinai, we commonly use fentanyl 25 ug with 1 mL of bupivacaine 0.25 % for the subarachnoid dose.

There are several advantages to the CSE technique. The primary advantage is the rapid (3–5 min) onset of analgesia. Additionally,

patients have less motor block and greater patient satisfaction with the CSE technique versus the "standard" epidural technique of bupivacaine 0.25 %. The greater satisfaction is related to the faster onset of action and less motor block (Simmons et al. 2007).

There are some concerns about the CSE technique, most of which are only theoretical. There is no increased risk of subarachnoid catheter migration of the epidural catheter, and metallic particles are not produced as a result of passing one needle through another. The incidence of post-dural puncture headache is not increased with the CSE technique. An increase in end-tidal carbon dioxide has been reported in women who received subarachnoid sufentanil, but the risk of clinically significant respiratory depression is extremely rare (Carvalho 2008). The risk of hypotension is also not greater with the CSE technique than with standard epidural regimens. The most common side effect of subarachnoid opioids is pruritus, with a reported incidence as great as 95 % that is easily treated with either an antihistamine or naloxone.

It is possible that the epidural catheter may not actually have been placed in the epidural space after the CSE technique is performed, and this may not be detected until the analgesia from the subarachnoid opioid has dissipated (1–2 h). If, during this time period, the woman requires an emergent cesarean delivery, the catheter may fail and the patient may require a general anesthetic. Lee et al. (2009) found that the risk of failed epidural catheters was lower in women who received CSE analgesia than those who received epidural analgesia, likely due to the confirmation of having reached the epidural space that is achieved with the recognition of free-flowing CSF in the spinal portion of the combined technique. However, it may be prudent not to use the CSE technique in a woman who is a poor risk for general analgesia, e.g., one with a bad airway or obesity, so that the epidural catheter can be immediately tested and effectiveness confirmed.

Many studies have reported fetal bradycardia associated with uterine hypertonus after subarachnoid opioid injection. One proposed theory for increased uterine tone is related to the rapid decrease in maternal catecholamines associated with the onset of pain relief. With the decrease in circulating beta adrenergic agonists, there is a predominance of alpha activity that leads to uterine contractions. Most studies, prospective and retrospective, find neonatal outcome based on umbilical artery pH and base excess similar, if not improved, with neuraxial analgesia (CSE or standard epidural) versus other analgesic strategies. Further, a review of literature comparing cesarean rate between standard epidural technique and CSE has shown no difference in the rate if cesarean delivery (Van de Velde 2005). If hypertonus occurs, treatment should include subcutaneous terbutaline or intravenous nitroglycerin.

The term walking epidural has become popular in the lay community. The term walking epidural refers to any epidural or spinal technique that allows ambulation with analgesia. In most centers, few patients choose to ambulate. Most choose to rest or sleep once they have adequate labor analgesia. However, if patients do not want to ambulate, using a technique that produces minimal motor blockade will still improve maternal satisfaction. Both epidural analgesia using dilute local anesthetic/opioid solutions and a CSE technique can achieve this goal. Several precautions should be taken before allowing a parturient to walk during epidural or CSE analgesia. These women should be candidates for intermittent fetal heart rate monitoring. Maternal blood pressure and fetal heart rate should be monitored for 30–60 min after induction. Even small doses of subarachnoid and epidural local anesthetics can produce some motor deficits. Assess motor function by having the parturient perform a modified deep knee bend or stepping up and down on a stool. The patient must have an escort at all times. Fetal heart rate and maternal blood pressure should be reassessed at least every 30 min.

In summary, techniques and drugs available to the modern-day obstetric anesthesiologist approach the objectives of an ideal labor anesthetic. The future of obstetric anesthesia lies in further refining the pharmacologic agents utilized and our neuraxial techniques to make obstetric anesthesia safer and more efficacious as we work to better care for our patients.

References

American College of Obstetricians and Gynecologists Committee on Obstetric Practice. (2006). ACOG committee opinion. No. 339: Analgesia and cesarean delivery rates. *Obstetrics and Gynecology, 107*(6), 1487–1488.

Asokumar, B., Newman, L. M., McCarthy, R. J., Ivankovich, A. D., & Tuman, K. J. (1998). Intrathecal bupivacaine reduces pruritus and prolongs duration of fentanyl analgesia during labor: A prospective, randomized controlled trial. *Anesthesia and Analgesia, 87*, 1309–1315.

Beilin, Y., Nair, A., Arnold, I., Bernstein, H. H., Zahn, J., Hossain, S., & Bodian, C. A. (2002). A comparison of epidural infusions in the combined spinal/epidural technique for labor analgesia. *Anesthesia and Analgesia, 94*, 927–932.

Beilin, Y., Bodian, C. A., Weiser, J., Hossani, S., Arnold, I., Feierman, D. E., Martin, G., & Holzman, I. (2005). Effect of labor epidural analgesia with and without fentanyl on infant breast-feeding. *Anesthesiology, 103*, 1211–1217.

Carvalho, B. (2008). Respiratory depression after neuraxial opioids in the obstetric setting. *Anesthesia and Analgesia, 107*(3), 956–961.

Chestnut, D. H. (1997). Does epidural analgesia during labor affect the incidence of cesarean delivery? *Regional Anesthesia, 22*, 495–499.

Halpern, S. H., & Carvalho, B. (2009). Patient-controlled epidural analgesia for labor. *Anesthesia and Analgesia, 108*, 921–928.

Lee, S., Lew, E., Lim, Y., & Sia, A. T. (2009). Failure of augmentation of labor epidural analgesia for intrapartum cesarean delivery: A retrospective review. *Anesthesia and Analgesia, 108*, 252–254.

Li, G., Warner, M., Lang, B. H., Huang, L., & Sun, L. S. (2009). Epidemiology of anesthesia-related mortality in the United States, 1999–2005. *Anesthesiology, 110*(4), 759–765.

Ounsted, M. K., Boyd, P. A., Hendrick, A. M., et al. (1978). Induction of labour by different methods in primiparous women. II. Neuro-behavioural status of the infants. *Early Human Development, 2*, 241–253.

Palmer, C. M. (2010). Continuous spinal anesthesia and analgesia in obstetrics. *Anesthesia and Analgesia, 111*(6), 1476–1479.

Simmons, S., Cyna, A., Dennis, A., & Hughes, D. (2007). Combined spinal epidural versus epidural analgesia in labour. *Cochrane Database of Systematic Reviews, 3*, CD003401.

Van de Velde, M. (2005). Neuraxial analgesia and fetal bradycardia. *Current Opinion in Anesthesiology, 18*, 253–256.

Analgesia for Labor and Delivery

Definition

Pain relief during labor and delivery can be administered intravenously or via the neuraxis as an epidural or spinal block. Intravenous medication is usually not adequate, as it only attenuates the pain but does not eliminate the pain. Epidural or spinal analgesia, generally administered by an anesthesiologist, virtually eliminates labor pain without a loss of consciousness.

Cross-References

▶ Analgesia During Labor and Delivery

Analgesic Effect of Oxycodone

Definition

The analgesic effect of oxycodone is mainly mediated by the parent compound.

Cross-References

▶ Postoperative Pain, Oxycodone

Analgesic Gap

Definition

The increase in pain sometimes associated with reduction in analgesic dosage or analgesic strategies.

Cross-References

▶ Postoperative Pain, Transition from Parenteral to Oral Drugs

Analgesic Guidelines for Infants and Children

Stephen C. Brown
Department of Anesthesia, The Hospital for Sick Children, Toronto, ON, Canada

Synonyms

Drug guidelines; Pediatric dosing guidelines

Definition

The goal of administering analgesia is to relieve pain without intentionally producing a sedated state.

Characteristics

Oral Analgesics

Analgesics include acetaminophen, nonsteroidal anti-inflammatory drugs, and opioids. While acetaminophen and opioids remain the cornerstone for providing analgesia for our youngest patients, the scope and diversity of drugs expand as those patients grow older. ▶ Adjuvant analgesics include a variety of drugs with analgesic properties that were initially developed to treat other health problems. These adjuvant analgesics (such as anticonvulsants and antidepressants) have become a cornerstone of pain control for children with chronic pain, especially when pain has a neuropathic component.

Pain control should include regular pain assessments, appropriate analgesics and adjuvant analgesics administered at regular dosing intervals, adjunctive drug therapy for symptom and side effects control, and nondrug therapies to modify the situational factors that can exacerbate pain and suffering. The guiding principles of analgesic administration are "by the ladder," "by the clock," "by the child," and "by the mouth." "By the ladder" refers to a three-step approach for selecting drugs according to their analgesic potency based on the child's pain level – acetaminophen to control mild pain, codeine to control moderate pain, and morphine for strong pain (World Health Organization 1990; Zernikow et al. 2006). The ladder approach was based on our scientific understanding of how analgesics affect pain of nociceptive origins (▶ nociceptive pain). If pain persists despite starting with the appropriate drug, recommended doses, and dosing schedule, move up the ladder and administer the next more potent analgesic. Even when children require opioid analgesics, they should continue to receive acetaminophen (and nonsteroidal anti-inflammatory drugs, if appropriate) as supplemental analgesics. The analgesic ladder approach is based on the premise that acetaminophen, codeine, and morphine should be available in all countries and that doctors and healthcare providers can relieve pain in the majority of children with a few drugs. Although there is a recommendation to delete codeine from the WHO list of essential medicines for children, at present the use of codeine remains an important element in the WHO pain ladder, and thus it is included in these opioid analgesic guidelines. It is well known that codeine is metabolized by CYP2D6 to morphine. If this enzyme is inhibited though, its analgesic efficacy will be diminished. A more recent and greater concern are the reported fatalities in patients who are ultrarapid metabolizers of codeine to morphine resulting in a greater production of morphine which could thus lead to respiratory depression (Kelly et al. 2012).

However, increasing attention is focusing on "thinking beyond the ladder" in accordance with our improved understanding of pain of neuropathic origins (Krane et al. 2003). Children should receive adjuvant analgesics to more specifically target neuropathic mechanisms. Regrettably, two of the main classes of adjuvant analgesics, antidepressants and anticonvulsants, have unfortunate names. Proper education of healthcare providers, parents, and children should lead to a wider acceptance and use of these medications for pain management. For example, amitriptyline may require 4–6 weeks to affect depression, but often requires only

1–2 weeks to affect pain. The newer classes of antidepressants, the selective serotonin reuptake inhibitors (SSRIs), may be beneficial to treat depression in a child with pain, but have not been shown to be beneficial for pain management. Selective serotonin and norepinephrine reuptake inhibitors (SSNRIs) may be of greater benefit, but convincing clinical trials are not available. The other main class of adjuvant analgesics is the anticonvulsants. The two principal medications used for this purpose in pediatrics are carbamazepine and gabapentin. With gabapentin, the main dose limiting side effect is sedation, so that a slow titration to maximal dose is required. Because of its greater number of significant side effects, the use of carbamazepine has decreased and the use of gabapentin has increased. We still await published studies to support the wide use of gabapentin in children.

Nonsteroidal anti-inflammatory drugs (NSAIDs) are similar in potency to aspirin. NSAIDs are used primarily to treat inflammatory disorders and to lessen mild to moderate acute pain. They should be used with caution in children with hepatic or renal impairment, compromised cardiac function, hypertension (since they may cause fluid retention and edema) and a history of GI bleeding or ulcers. NSAIDs may also inhibit platelet aggregation and thus must be monitored closely in patients with prolonged bleeding times. NSAIDs have been used for many years in pediatrics and with their minimal side effects and many advantages (no effect on ventilation, no physical dependence, morphine sparing effect, etc.), their use should be encouraged.

The specific drugs and doses are determined by the needs of each child. The drugs listed in this entry are based on guidelines from our institution (The Hospital for Sick Children 2011–2012), and do not necessarily reflect results of evidence-based reviews. Recommended starting doses for analgesic medications to control children's disease-related pain are listed in Tables 1 and 2; starting doses for adjuvant analgesic medications to control pain, drug-related side effects, and other symptoms are listed in Table 3. (For further

Analgesic Guidelines for Infants and Children, Table 1 Non-opioid drugs to control pain in children

Drug	Dosage	Comments
Acetaminophen	10–15 mg kg^{-1} PO, every 4–6 h	Lacks gastrointestinal and hematological side effects; lacks anti-inflammatory effects (may mask infection-associated fever)
		Dose limit of 65 mg kg^{-1} day^{-1} or 4 g day^{-1}, whichever is less
Ibuprofen	5–10 mg kg^{-1} PO, every 6–8 h	Anti-inflammatory activity
		Use with caution in patients with hepatic or renal impairment, compromised cardiac function or hypertension (may cause fluid retention, edema), history of GI bleeding or ulcers, may inhibit platelet aggregation
		Dose limit of 40 mg kg^{-1} day^{-1}; max dose of 2,400 mg day^{-1}
Naproxen	10–20 mg kg^{-1} day^{-1} PO, divided every 12 h	Anti-inflammatory activity. Use with caution and monitor closely in patients with impaired renal function. Avoid in patients with severe renal impairment
		Dose limit of 1 g day^{-1}
Diclofenac	1 mg kg^{-1} PO, every 8–12 h	Anti-inflammatory activity. Similar GI, renal, and hepatic precautions as noted above for ibuprofen and naproxen
		Dose limit of 50 mg/dose
Ketorolac	0.5 mg/kg/dose IV q6h PRN	Maximum duration of IV Ketorolac is 2 days. Do not use in tonsillectomy patients due to increase risk of bleeding, Do not use in patients with Impaired renal function.
		Dose limit <16yrs: 15 mg/dose >16yrs: 30mg/dose

Note: Increasing the dose of non-opioids beyond the recommended therapeutic level produces a "ceiling effect," in that there is no additional analgesia but there are major increases in toxicity and side effects. Abbreviations: *PO* by mouth, *GI* gastrointestinal (Modified from Brown and McGrath 2010)

review of analgesics and adjuvant analgesics in children, see Brown and McGrath (2010); Brown et al. (2010); Schechter et al. (2003)).

Oral Analgesic Dosing Schedules

Children should receive analgesics at regular times, "by the clock," to provide consistent pain relief and prevent breakthrough pain. The specific drug schedule (e.g., every 4 or 6 h) is based on the drug's duration of action and the child's pain severity. Although breakthrough pain episodes have been recognized as a problem in adult pain control, they may represent an even more serious problem for children. Unlike adults, who generally realize that they can demand more potent analgesic medications or demand more frequent dosing intervals, children have little control, little awareness of alternatives, and fear that their pain cannot be controlled. They may become progressively frightened, upset, and preoccupied with their symptoms. Thus, it is essential to establish and maintain a therapeutic window of pain relief for children.

Analgesic doses should be adjusted "by the child." There is no one dose that will be appropriate for all children with pain. The goal is to select a dose that prevents children from experiencing pain before they receive the next dose. It is essential to monitor the child's pain regularly and adjust analgesic doses as necessary to control the pain. The effective opioid dose to relieve pain varies widely among different children or in the same child at different times. Some children require large opioid doses at frequent intervals to control their pain. If such doses are necessary for effective pain control and the side effects can be managed by adjunctive medication (▶ adjunctive drugs) so that children are comfortable, then the doses are appropriate. Children receiving opioids may develop altered sleep patterns so that they are awake at night fearful and complaining about pain and sleep intermittently throughout the day. They should receive adequate analgesics at night with antidepressants or hypnotics as necessary to enable them to sleep throughout the night. To relieve ongoing pain, opioid doses should

Analgesic Guidelines for Infants and Children, Table 2 Opioid analgesics: usual starting doses for children

Drug	Equianalgesic dose (parenteral)	Starting dose IV	IV:PO ratio	Starting dose PO/ Transdermal	Duration of action
Morphine	10 mg	Bolus dose – 0.05–0.1 mg kg^{-1} every 2–4 h. Continuous infusion – 0.01–0.04 mg kg^{-1} h^{-1}	1:3	0.15–0.3 mg kg^{-1}/dose every 4 h	3–4 h
Hydromorphone	1.5 mg	0.015–0.02 mg kg^{-1} every 4 h	1:5	0.06 mg kg^{-1} every 3–4 h	2–4 h
Codeine	120 mg	Not recommended		1.0 mg kg^{-1} every 4 h (dose limit 1.5 mg kg^{-1}/dose)	3–4 h
Oxycodone	5–10 mg	Not recommended		0.1–0.2 mg kg^{-1} every 3–4 h	3–4 h
Fentanyl[a]	100 µg	1–2 µg kg^{-1} h^{-1} as continuous infusion		12 µg patch for children, dependent on weight and previous opioid use	72 h (patch)
Methadone	10 mg	0.1 mg kg^{-1} every 4–8 h	1:2	0.2 mg kg^{-1} every 4–8 h	12–50 h
Tramadol	100 mg	2.0 mg kg^{-1} every 4–6 h		1.0 mg kg^{-1} every 4–6 h (dose limit 400 mg day^{-1})	4–6 h

Note: Doses are for opioid naïve patients. For infants under 6 months, start at one-quarter to one-third the suggested dose and titrate to effect
Principles of opioid administration:
1. If inadequate pain relief and no toxicity at peak onset of opioid action, increase dose in 50 % increments
2. Avoid IM administration
3. Whenever using continuous infusion, plan for hourly rescue doses with short onset opioids if needed. Rescue dose is usually 50–200 % of continuous hourly dose. If greater than six rescues are necessary in 24 h period, increase daily infusion total by the total amount of rescues for previous 24 h ÷ 24. An alternative is to increase infusion by 50 %
4. To change opioids – because of incomplete cross-tolerance, if changing between opioids with short duration of action, start new opioid at 50 % of equianalgesic dose. Titrate to effect. If changing between opioids from short to long duration of action (i.e., morphine to methadone), start at 25 % of equianalgesic dose and titrate to effect
5. To taper opioids – anyone on opioids over 1 week must be tapered to avoid withdrawal – taper by 50 % for 2 days and then decrease by 25 % every 2 days. When the dose is equianalgesic to an oral morphine dose of 0.6 mg kg^{-1} day^{-1}, it may be stopped. Some patients on opioids for prolonged periods may require much slower weaning
Abbreviations: *PO* by mouth, *IV* intravenous
[a]Potentially highly toxic. Not for use in acute pain control
Modified from Brown and McGrath (2010)

be increased steadily until comfort is achieved, unless the child experiences unacceptable side effects, such as somnolence or respiratory depression (Table 4).

"By the mouth" refers to the oral route of drug administration. Medication should be administered to children by the simplest and most effective route, usually by mouth. Since children are afraid of painful injections they may deny that they have pain or they may not request medication. When possible, children should receive medications through routes that do not cause additional pain. Although optimal analgesic administration for children requires flexibility in selecting routes

according to children's needs, parenteral administration is often the most efficient route for providing direct and rapid pain relief. Since intravenous, intramuscular, and subcutaneous routes cause additional pain for children, serious efforts have been expended on developing more pain-free modes of administration that still provide relatively direct and rapid analgesia. Attention has focused on improving the effectiveness of oral routes.

Intravenous Analgesia
Many hospitals have restricted the use of intramuscular injections because they are painful and

Analgesic Guidelines for Infants and Children, Table 3   Adjuvant analgesics: doses for children

Drug category	Drug, dosage	Indications	Comments
Antidepressants	Amitriptyline, 0.2–0.5 mg kg^{-1} PO	Neuropathic pain (i.e., vincristine-induced, radiation plexopathy, tumor invasion, CRPS-1)	Usually improved sleep and pain relief within 3–5 days
	Titrate upward by 0.25 mg kg^{-1} every 2–3 days. Maintenance: 0.2–3.0 mg k^{-1}	Insomnia	
	Alternatives: nortriptyline, doxepin, imipramine		Anticholinergic side effects are dose limiting. Use with caution for children with increased risk for cardiac dysfunction
Anticonvulsants	Gabapentin, 5 mg kg^{-1} day^{-1} PO	Neuropathic pain, especially shooting, stabbing pain	Side effects: gastrointestinal upset, ataxia, dizziness, disorientation, and somnolence
	Titrate upward over 3–7 days Maintenance: up to 15–50 mg kg^{-1} day^{-1} PO divided TID		
	Carbamazepine, initial dosing: 10 mg kg^{-1} day^{-1} PO divided OD or BID. Maintenance: up to 20–30 mg kg^{-1} day^{-1} PO divided every 8 h. Increase dose gradually over 2–4 weeks		Monitor for hematological, hepatic, and allergic reactions with carbamazepine
	Alternatives: pregabalin, oxcarbazepine		
Sedatives, hypnotics, anxiolytics	Diazepam, 0.025–0.2 mg kg^{-1} PO every 6 h	Acute anxiety, muscle spasm	Sedative effect may limit opioid use. Other side effects include depression and dependence with prolonged use
	Lorazepam, 0.05 mg kg^{-1}/dose SL	Premedication for painful procedures	
	Midazolam, 0.5 mg kg^{-1}/dose PO administered 15–30 min prior to procedure; 0.05 mg kg^{-1}/dose IV for sedation		
Antihistamines	Hydroxyzine, 0.5 mg kg^{-1} PO every 6 h	Opioid-induced pruritus, anxiety, nausea	Sedative side effects may be helpful
	Diphenhydramine, 0.5–1.0 mg kg^{-1} PO/IV every 6 h		
Psychostimulants	Dextroamphetamine, Methylphenidate, 0.1–0.2 mg kg^{-1} BID	Opioid-induced somnolence	Side effects include agitation, sleep disturbance, and anorexia
	Escalate to 0.3–0.5 mg kg^{-1} as needed	Potentiation of opioid analgesia	Administer second dose in afternoon to avoid sleep disturbances
Corticosteroids	Prednisone, prednisolone, and dexamethasone dosage depends on clinical situation (i.e., dexamethasone initial dosing: 0.2 mg kg^{-1} I.V. Dose limit 10 mg. Subsequent dose 0.3 mg kg^{-1} day^{-1} I.V. divided every 6 h.)	Headache from increased intracranial pressure, spinal or nerve compression; widespread metastases	Side effects include edema, dyspeptic symptoms, and occasional gastrointestinal bleeding

Abbreviations: *CRPS-1* complex regional pain syndrome, Type 1, *PO* by mouth, *I.V.*, intravenous, *SL* sublingual
Modified from Brown and McGrath (2010)

Analgesic Guidelines for Infants and Children, Table 4 Opioid side effects

Side effect	Management
Respiratory depression	Reduction in opioid dose by 50 %, titrate to maintain pain relief without respiratory depression
Respiratory arrest	Naloxone, titrate to effect with 0.01 mg kg^{-1}/dose I.V./ETT increments or 0.1 mg kg^{-1}/dose I.V./ETT, repeat PRN. Small frequent doses of diluted naloxone or naloxone drip are preferable for patients on chronic opioid therapy to avoid severe, painful withdrawal syndrome. Repeated doses are often required until opioid side effect subsides
Drowsiness/sedation	Frequently subsides after a few days without dosage reduction; methylphenidate or dextroamphetamine (0.1 mg kg^{-1} administered twice daily, in the morning and midday so as not to interfere with nighttime sleep). The dose can be escalated in increments of 0.05–0.1 mg kg^{-1} to a maximum of 10 mg/dose for dextroamphetamine and 20 mg/dose for methylphenidate
Constipation	Increased fluids and bulk, prophylactic laxatives as indicated
Nausea/vomiting	Administer an antiemetic (e.g., ondansetron, 0.1 mg kg^{-1} I.V./PO every 8 h)
	Antihistamines (e.g., dimenhydrinate 0.5 mg kg^{-1}/dose every 4–6 h I.V./PO) may be used. Pre-chemotherapy, Nabilone 0.5–1.0 mg PO and then every 12 h may also be used
Confusion, nightmares, hallucinations	Reassurance only, if symptoms mild. A reduced dosage of opioid or a change to a different opioid or add neuroleptic (e.g., haloperidol 0.1 mg kg^{-1} PO/I.V. every 8 h to a maximum of 30 mg day^{-1})
Multifocal myoclonus; seizures	Generally occur only during extremely high dose therapy; reduction in opioid dose indicated if possible. Add a benzodiazepine (e.g., clonazepam 0.05 mg kg^{-1} day^{-1} divided BID or TID increasing by 0.05 mg kg^{-1} day^{-1} every 3 days PRN up to 0.2 mg kg^{-1} day^{-1}. Dose limit of 20 mg day^{-1})
Urinary retention	Rule out bladder outlet obstruction, neurogenic bladder, and other precipitating drug (e.g., tricyclic antidepressant). Particularly common with epidural opioids. Change of opioid, route of administration, and dose may relieve symptom. Bethanechol or catheter may be required

I.V. intravenous, *PO* by mouth, *ETT* endotracheal tube, *PRN* as needed
Reprinted from Brown and McGrath (2010)

drug absorption is not reliable; they advocate the use of intravenous lines into which drugs can be administered directly without causing further pain. Topical anesthetic creams should also be applied prior to the insertion of intravenous lines in children. The use of ▶ portacatheters has become the gold standard in pediatrics, particularly for children with cancer under the care of the physician, who require administration of multiple drugs at weekly intervals.

Continuous infusion has several advantages over intermittent subcutaneous, intramuscular, or intravenous routes. This method circumvents repetitive injections, prevents delays in analgesic drug administration, and provides continuous levels of pain control without children experiencing increased side effects at peak level and pain breakthroughs at trough level. Continuous infusion should be considered when children have pain for which oral and intermittent parenteral opioids do not provide satisfactory pain control, when intractable vomiting prevents the use of oral medications and when intravenous lines are not desirable. Children receiving a continuous infusion should continue to receive "rescue doses" to control breakthrough pain, as necessary. As outlined in Table 2, the rescue doses should be 50–200 % of the continuous infusion hourly dose. If children experience repeated breakthrough pain, the basal rate can be increased by 50 % or by the total amount of morphine administered through the rescue doses over a 24 h period (divided by 24 h).

Patient-Controlled Analgesia

▶ Patient-controlled analgesia (PCA) enables children to administer analgesic doses according to their pain level. PCA provides children with a continuum of analgesia that is prompt, economical, not nurse dependent and results in a lower overall narcotic use (Rodgers et al. 1988; Schechter et al. 2003). It has a high degree of safety, allows for wide variability between patients and removes delay in analgesic administration (for a review, see Berde and Solodiuk 2003). It can now be regarded as a standard for the delivery of analgesia in children aged >5 years (McDonald and Cooper 2001). However, there are opposing views about the use of ▶ background infusions with PCA. Although they may improve efficacy, they may increase the occurrence of adverse effects such as nausea and respiratory depression. In a comparison of PCA with and without a background infusion for children having lower extremity surgery, the total morphine requirements were reduced in the PCA only group and the background infusion offered no advantage (McNeely and Trentadue 1997). In another study comparing background infusion and PCA, children between 9 and 15 achieved better pain relief with PCA, while children between 5 and 8 showed no difference (Bray et al. 1996). Our current standard is to add a background infusion to the PCA if the pain is not controlled adequately with PCA alone. The selection of opioid used in PCA is perhaps less critical than the appropriate selection of parameters such as bolus dose, lockout, and background infusion rate. The opioid choice may be based on adverse effect profile rather than efficacy. Clearly, patient-controlled analgesia offers special advantages to children who have little control and who are extremely frightened about uncontrolled pain. PCA is, as it states, patient-controlled analgesia. When special circumstances require that alternate people administer the medication, we do allow both nurse and parent-controlled analgesia. Under these circumstances, parents require our nurse educators to fully educate them on the use of PCA. In an alert by the Joint Commission on Accreditation of Healthcare Organizations (JCAHO), they advise that serious adverse events can result when family members, caregivers, or clinicians who are not authorized become involved in administering the analgesia for the patient "by proxy" (Sentinel Event Alert 2004).

Transdermal Fentanyl

Fentanyl is a potent synthetic opioid, which like morphine binds to mu receptors. However, fentanyl is 75–100 × more potent than morphine. The intravenous preparation of fentanyl has been used extensively in children. A transdermal preparation of fentanyl was introduced in 1991 for use with chronic pain. This route provides a noninvasive but continuously controlled delivery system. Although limited data is available on transdermal fentanyl (TF) in children, its use is increasing for children with pain. In a 2001 study, TF was well tolerated with effective pain relief in 11 of 13 children and provided an ideal approach for children where compliance with oral analgesics was problematic (Noyes and Irving 2001). In another study, when children were converted from oral morphine doses to TF, the investigators noted diminished side effects and improved convenience with TF (Hunt et al. 2001). The majority of parents and investigators considered TF to be better than previous treatment. No serious adverse events were attributed to fentanyl, suggesting that TF was both effective and acceptable for children and their families. Similarly, no adverse effects were noted in a study of TF for children with pain due to sickle cell crisis (Christensen et al. 1996). This study showed a significant relationship between TF dose and fentanyl concentration; pain control with the use of TF was improved in 7 of 10 patients in comparison to PCA alone. In a multicenter crossover study in adults, TF caused significantly less constipation and less daytime drowsiness in comparison to morphine, but greater sleep disturbance and shorter sleep duration (Ahmedzai and Brooks 1997). Of those patients able to express a preference, significantly more preferred fentanyl patches. As with all opioids, fatal adult complications have been noted with the use of multiple transdermal patches. In a review of TF use in the pediatric population (Zernikow et al. 2007), though there are no pediatric randomized or controlled cohort studies, pharmacokinetic studies in children show that the time to reach steady state serum drug concentrations seems to be longer, clearance higher, and elimination half-life shorter when compared to adults. TF may be associated with less constipation compared to morphine use. Parents and medical professionals were satisfied with TF to a higher degree than with individual analgesic pretreatment regimens. An approximate conversion factor of 45 mg day/day oral morphine to 12 µg/h TF is used for initial therapy dose estimation for children receiving long-term morphine therapy. The 12 µg/h patch is the lowest dose patch available. I note that supplemental taping is often required to secure the patches in children and that I often recommend the changing of patches at 48 h instead of the manufacturers recommendation of 72 h. Evidence of the superiority of TF treatment over conventional opioid administration in children remains scarce and of low quality.

Summary

I have guided you through the basics of the administration of analgesics for the pediatric patient from the oral route through to intravenous and the PCA route and finally discussed an analgesic administered by the transdermal route. By the use of these drugs as examples and with the simple principles discussed to apply them, we can hopefully attain the goal of decreasing the intensity of pain in children – no matter what the setting.

References

Ahmedzai, S., & Brooks, D. (1997). Transdermal fentanyl versus sustained-release oral morphine in cancer pain: Preference, efficacy, and quality of life. The TTS-Fentanyl Comparative Trial Group. *Journal of Pain and Symptom Management, 13*, 254–261.

Berde, C. B., & Solodiuk, J. (2003). Multidisciplinary programs for management of acute and chronic pain in children. In N. L. Schechter, C. B. Berde, & M. Yaster (Eds.), *Pain in infants, children, and adolescents* (2nd ed., pp. 471–486). Philadelphia: Lippincott Williams and Wilkins.

Bray, R. J., Woodhams, A. M., Vallis, C. J., et al. (1996). A double-blind comparison of morphine infusion and patient controlled analgesia in children. *Paediatric Anaesthesia, 6*, 121–127.

Brown, S. C., & McGrath, P. A. (2010). Paediatric palliative medicine – Pain control. In G. Hanks, N.

Cherny, et al. (Eds.), *Oxford textbook of palliative medicine* (4th ed., pp. 1318–1336). Oxford: Oxford University Press.

Brown, S. C., Taddio, A., & McGrath, P. A. (2010). Pharmacological considerations in infants and children. In P. Beaulieu, D. Lussier, et al. (Eds.), *Pharmacology of pain* (pp. 529–545). Seattle: IASP Press.

Christensen, M. L., Wang, W. C., Harris, S., et al. (1996). Transdermal fentanyl administration in children and adolescents with sickle cell pain crisis. *Journal of Pediatric Hematology/Oncology, 18*, 372–376.

Hunt, A., Goldman, A., Devine, T., et al. (2001). Transdermal fentanyl for pain relief in a paediatric palliative care population. *Palliative Medicine, 15*, 405–412.

Kelly, L. E., Reider, M., van den Anker, J., et al. (2012). More codeine fatalities after tonsillectomy in North American children. *Pediatrics, 129*(5), e1343–e1347.

Krane, E. J., Leong, M. S., Golianu, B., et al. (2003). Treatment of pediatric pain with nonconventional analgesics. In N. L. Schechter, C. B. Berde, & M. Yaster (Eds.), *Pain in infants, children, and adolescents* (2nd ed., pp. 225–241). Philadelphia: Lippincott Williams and Wilkins.

McDonald, A. J., & Cooper, M. G. (2001). Patient-controlled analgesia: An appropriate method of pain control in children. *Paediatric Drugs, 3*, 273–284.

McNeely, J. K., & Trentadue, N. C. (1997). Comparison of patient-controlled analgesia with and without nighttime morphine infusion following lower extremity surgery in children. *Journal of Pain and Symptom Management, 13*, 268–273.

Noyes, M., & Irving, H. (2001). The use of transdermal fentanyl in pediatric oncology palliative care. *The American Journal of Hospice & Palliative Care, 18*, 411–416.

Rodgers, B. M., Webb, C. J., Stergios, D., et al. (1988). Patient-controlled analgesia in pediatric surgery. *Journal of Pediatric Surgery, 23*, 259–262.

Schechter, N. L., Berde, C. B., & Yaster, M. (2003). *Pain in infants, children, and adolescents* (2nd ed.). Philadelphia: Lippincott Williams& Wilkins.

Sentinel Event Alert. (2004). Patient controlled analgesia by proxy: Joint Commission of Healthcare Organizations 20;33:1–2

The Hospital for Sick Children. (2011). *The 2011–2012 formulary* (30th ed.). Toronto: The Hospital for Sick Children.

World Health Organization. (1990). *Cancer pain relief and palliative care*. Geneva: World Health Organization.

Zernikow, B., Smale, H., Michel, E., et al. (2006). Paediatric cancer pain management using the WHO analgesic ladder – results of a prospective analysis from 2265 treatment days during a quality improvement study. *European Journal of Pain, 10*, 587–595.

Zernikow, B., Michel, E., & Anderson, B. (2007). Transdermal fentanyl in childhood and adolescence: A comprehensive literature review. *The Journal of Pain, 8*, 187–207.

Analgesic Ladder

Definition

In 1986, WHO proposed a three step analgesic ladder. Although the specific elements are no longer viewed as a guideline, it remains important as an indication of an international consensus in favor of the use of opioid drugs as the mainstay in the treatment of moderate to severe cancer pain. The ladder suggests that non-opioids (step 1) be administered in case of mild pain. If this is not enough, or pain is moderate, so-called weak opioids (step 2) should be used, often in combination with a NSAID and an adjuvant drug if needed. If pain is severe, or step 2 drugs are not effective, treatment with a single-entity opioid should be used, the prototype of which is morphine. Again, co-treatment with a NSAID or adjuvant drug is used as appropriate.

Cross-References

▶ Cancer Pain

Analgesic Tolerance

▶ Opioids, Clinical Opioid Tolerance

Analgesic Treatment

Definition

A treatment used to reduce pain without causing loss of consciousness.

Cross-References

▶ Adjuvant Analgesics in Management of Cancer-Related Bone Pain
▶ Cancer Pain Management, Cancer-Related Breakthrough Pain, Therapy

Analgesics

Definition

Analgesics are drugs (pharmacological agents) that provide pain relief.

Cross-References

▶ History of Analgesics
▶ NSAIDs, COX-Independent Actions
▶ Opioids in Geriatric Application
▶ Opioids, Clinical Opioid Tolerance

Analysis of Pain Behavior

▶ Assessment of Pain Behaviors

Anaphylactic Reaction

Synonyms

Anaphylaxis

Definition

A severe allergic reaction that starts when the immune system mistakenly responds to a relatively harmless substance as if it were a serious threat.

Cross-References

▶ Diencephalic Mast Cells

Anaphylaxis

▶ Anaphylactic Reaction

Anesthesia

Definition

Loss of sensation and usually of consciousness without loss of vital functions, artificially produced by the administration of one or more agents that block the passage of pain impulses along nerve pathways to the brain

Cross-References

▶ Thalamic Nuclei Involved in Pain: Cat and Rat

Anesthesia Dolorosa

Definition

Literally "painful anesthesia," anesthesia dolorosa is spontaneous pain felt in a body part that has been denervated or deafferented. The limb itself is numb and unresponsive to applied stimuli but is nonetheless painful. This can occur as a result of a surgical lesion of peripheral nerves, sensory ganglia, or sensory roots, usually intended to relieve pain. Because of overlap of innervation territories, more than one adjacent nerve, ganglion, or root needs to be damaged. It also occurs following traumatic brachial plexus avulsion. Anesthesia dolorosa is similar to phantom limb pain (PLP) except that in PLP the painful limb is no longer present, due to amputation.

Cross-References

▶ Central Nervous System Stimulation for Pain
▶ Dorsal Root Ganglionectomy and Dorsal Rhizotomy
▶ Neuropathic Pain Model, Spared Nerve Injury
▶ Peripheral Neuropathic Pain

Anesthesia Dolorosa Due to Plexus Avulsion

▶ Plexus Injuries and Deafferentation Pain

Anesthesia Dolorosa Model, Autotomy

Marshall Devor
Department of Cell & Developmental Biology, Institute of Life Sciences and Center for Research on Pain, The Hebrew University of Jerusalem, Jerusalem, Givat Ram, Israel

Synonyms

Animal model for phantom limb pain; Animal model of spontaneous neuropathic pain; Autotomy model of neuropathic pain; Deafferentation model of neuropathic pain; Denervation model of neuropathic pain; Neuroma model of neuropathic pain

Definition

Anesthesia dolorosa (Latin for "painful numbness") is a seemingly paradoxical chronic pain state in which, despite the presence of ongoing pain, the painful body part is completely numb and insensate. Applied stimuli are not felt. To create this state in animals a limb is made insensate either (1) by cutting all peripheral nerves that serve it (▶ Denervation), or (2) by cutting the corresponding dorsal roots (▶ Deafferentation). Hence the animal model of anesthesia dolorosa is actually a family of models. The presence of ongoing pain is inferred from the observation of "autotomy" behavior or its consequences.
▶ Autotomy is a behavior pattern in which the animal licks, bites, and chews its denervated limb (self-mutilation). Quantification is usually based on the amount of tissue lost from the extremity as a function of time after the surgical denervation/deafferentation, or on the number of days required

to reach a criterion amount of tissue loss (Wall et al. 1979). This behavioral model is thought to emulate anesthesia dolorosa and phantom limb pain in humans. Anesthesia dolorosa (e.g., after brachial plexus avulsion) differs from phantom limb pain (e.g., after leg amputation) in that the body part in which the pain is felt is still present; it has not been amputated. Despite the continued presence of the limb the autotomy model is useful for studying both conditions, and spontaneous neuropathy in general. The key component is the nerve injury. It is unlikely that the presence of the insensate limb in the model, and in clinical anesthesia dolorosa, contributes materially to the spontaneous pain mechanism.

Characteristics

The presence of spontaneous ▶ dysesthesia and pain is probably the most common and troublesome symptom of painful neuropathies. It occurs in nearly all neuropathic pain patients either as an isolated symptom or in combination with exaggerated response to applied stimuli (▶ Allodynia and ▶ Hyperalgesia). In addition to being of great clinical significance, the presence of pain in an insensate limb is paradoxical and represents a challenge for theoretical understanding. Since its development, the denervation/deafferentation model has proved to be an important tool in identifying the biological mechanism(s) underlying neuropathic pain, and in evaluating the mode of action of therapeutic agents (Devor and Seltzer 1999; Devor 2006). It has also been an important platform for the discovery of genetic polymorphisms that affect susceptibility to developing neuropathic pain (Devor and Raber 1990; Nissenbaum et al. 2010). Although in recent years it has been largely superseded by partial denervation models which use as a behavioral endpoint the evaluation of allodynia and hyperalgesia (rather than autotomy), the model remains a useful behavioral probe of ongoing (spontaneous) painful dysesthesia in experimental animals.

Background and Ethical Considerations

Although it had been recognized previously that animals, rodents to primates, tend to lick, scratch, and bite an insensate body part, this autotomy behavior was not recognized as a potential indicator of ongoing pain until the mid-1970s (Basbaum 1974; Wall et al. 1979). Investigators rarely witness actual autotomy behavior. Rather, the accumulated amount of tissue loss is scored. Long-term observations and video monitoring indicate that autotomy may occur either in brief "attacks" or in quiet bouts of deliberate chewing, separated by hours or days in which no further tissue loss occurs. This suggests that autotomy may reflect transient, sometimes paroxysmal peaks of pain perhaps overlaid on a continuous ongoing pain. Across-strains genetic analysis in mice indicates that autotomy behavior is part of a pain "type" that includes thermal nociception (Mogil et al. 1999). Perhaps, in mice at least, the ongoing pain has a burning quality.

It is essential to understand that, since the limb is entirely numb, autotomy behavior per se is not painful, even when the self-inflicted tissue loss includes entire digits. Pain arises spontaneously in association with the underlying neural injury. The ongoing pain remains even when steps are taken to prevent autotomy behavior itself, such as with the use of a protective ruff placed around the animal's neck, or when a foul-tasting substance is painted on the limb (Devor and Seltzer 1999). If reports from human patients can serve as guidance, it is safe to assume that most animals that suffer allodynia and hyperalgesia in partially denervated limbs also have spontaneous pain. The only reason that autotomy does not occur along with allodynia and hyperalgesia in the partial nerve injury models is that the very act of licking and biting the limb provokes pain (Koplovitch et al. 2012). Autotomy is prevented by "nociceptive sensory cover." The absence of autotomy in a nerve-injured animal with residual sensation in the limb should, therefore, *not* be taken as evidence for the absence of ongoing pain. In the chronic constriction injury (CCI) model of neuropathic pain, for example, there may be patches of complete skin denervation and these are targets for autotomy behavior (Bennett and Xie 1988). Esthetic considerations aside, ethical constraints on the use of lesions which trigger autotomy are no different in kind from those associated with the use of other neuropathic pain models.

Does Autotomy Behavior Reflect Pain?

Pain is a private experience (1st person) that cannot be felt by another, only inferred through context and the observation of nocifensive behavior (e.g., escape, distress vocalization, spoken language). Drawing inferences about ongoing pain from spontaneously emitted behavior, such as autotomy, is intrinsically more uncertain than concluding that pain is felt when an animal shows distress in response to an applied stimulus. Skeptics have questioned the proposition that autotomy reflects pain with two main arguments. First, anesthesia dolorosa does not typically trigger self-injurious behavior in human patients, and second, autotomy may reflect an animal's attempt to rid itself of a useless, insensate, but pain-free limb. The first critique is weak as socialization, and the anticipation of consequences are expected to prevent self-mutilation in humans with normal mental capacity, but not in animals. Moreover, compulsive autotomy-like behavior does occur in some people with ongoing dysesthesias (including itch) and pain (Mailis 1996; Symons 2011). As for the second critique, rendering a limb numb by sustained local anesthetic nerve block does not trigger autotomy (Blumenkopf and Lipman 1991). Likewise, when denervation is carried out in two stages, the latent time before autotomy behavior begins is linked to the moment of the initial nerve injury, even though it produces only partial denervation, and not the moment that the limb was finally rendered numb. The ability of a prior nerve lesion to accelerate the onset of autotomy when nociceptive sensory cover is ultimately removed is called the "priming effect" (Koplovitch et al. 2012).

There are many positive indicators that autotomy reflects spontaneous pain (Coderre et al. 1986; Levitt 1985; Seltzer 1995). These include the following:

- Limb denervation and deafferentation frequently cause ongoing neuropathic pain in humans. As in humans, palpating neuromas in rats evoke distress vocalization and struggling (Dorsi et al. 2008).
- Neural injuries that are followed by autotomy behavior trigger massive barrages of spontaneous discharge in injured afferents. There is a suggestive temporal correspondence between this discharge and autotomy, particularly for ectopia in nociceptive C-fibers (Devor and Seltzer 1999).
- Depletion of C-fibers with neonatal capsaicin treatment suppresses autotomy (Devor et al. 1982), and resecting neuromas in adults delays autotomy until a new spontaneously active neuroma forms (Seltzer 1995).
- Different forms of nerve section (cut, freeze, cautery, crush, etc.) produce identical anesthesia, but yield different degrees of autotomy, presumably because of differences in the resulting ectopia (Zeltser et al. 2000).
- Autotomy is suppressed in a dose-dependent manner by drugs that reduce ectopic firing and/or relieve neuropathic pain in humans (e.g., anticonvulsants, local anesthetics, tricyclics, NMDA receptor antagonists). Likewise, analgesics minimally effective against neuropathic pain, such as NSAIDs, do not suppress autotomy (Coderre et al. 1986; Seltzer 1995; Devor and Seltzer 1999).
- Spinal injection of excitants, such as strychnine, tetanus toxin, alumina cream, penicillin, and substance P, which almost certainly causes pain, induces scratching and biting of the corresponding limb, and sometimes frank autotomy (Coderre et al. 1986; Devor and Seltzer 1999).
- Blockade of descending antinociceptive control by appropriate brain stem or spinal tract lesions augments autotomy (Coderre et al. 1986; Saade et al. 1993), while midbrain or dorsal column stimulation and dorsal root entry zone (DREZ) lesions suppress it (Kauppila and Pertovaara 1991; Rossitch et al. 1993; Devor and Seltzer 1999; Guenot et al. 2002).
- Autotomy is accompanied by paw guarding, protective gait, sleep disturbances, sometimes weight loss, and stress-related increase in plasma corticosterone levels. It is augmented

by stressful conditions, such as isolation and cold stress, and reduced by taming and social contact (Coderre et al. 1986; Kauppila and Pertovaara 1991; Seltzer 1995; Devor and Seltzer 1999; Raber and Devor 2002).

- There is clear evidence that genes, as well as environmental factors, determine the level of autotomy behavior. For example, there are consistent differences in autotomy among strains of mice and rats, despite identical denervation and sensory loss. Likewise, genetic selection and interbreeding experiments indicate a high degree of heritability with at least one gene of major effect (Devor and Raber 1990; Mogil et al. 1999; Seltzer et al. 2001).

- An autotomy pain susceptibility gene located on mouse chromosome 15 has recently been identified as *Cacng2*. Polymorphisms in the human homolog of this gene, CACNG2, affect the likelihood that neuropathic pain will develop in women who have undergone partial or complete amputation of a breast (Nissenbaum et al. 2010).

Mechanisms of Ongoing Pain in the Denervation/Deafferentation Model

A limb may be rendered insensate by denervation or deafferentation and both situations may produce anesthesia dolorosa in humans and autotomy behavior in animals. The terms "denervation" and "deafferentation" are frequently confused and misused; they do not mean the same thing. ▶ Denervation in the present context refers to severing sensory axons that innervate the limb. Sensory endings rapidly degenerate in the process of anterograde (Wallerian) degeneration. However, sensory cell bodies in the DRG, their connectivity with the spinal cord, and the ability of (ectopically generated) afferent impulses to reach the CNS are largely preserved. ▶ Deafferentation refers to blocking the arrival of afferent impulses into the CNS by severing dorsal roots (▶ Dorsal Rhizotomy). The sensory neurons in the dorsal root ganglion (DRG) mostly survive, as do sensory endings in the skin. The limb is *not*

denervated. It is very likely that pain due to denervation and deafferentation results from different mechanisms (Devor 2006).

Pain and autotomy after nerve injury is probably due to abnormal spontaneous afferent discharge generated ectopically at the nerve injury site, and in axotomized DRG neurons. There might also be a contribution by residual intact neurons that continue to innervate adjacent skin. The ectopic firing plays two roles. First, it constitutes a primary nociceptive afferent signal. Second, it probably triggers central sensitization in the spinal cord dorsal horn, and perhaps also in the brain. The sensitized CNS amplifies and augments pain sensation due to the spontaneous afferent discharge. It also renders light tactile input from residual neighboring afferents painful, yielding tactile allodynia in the skin bordering on the denervated zone (Defrin et al. 1996; Devor 2006).

Pain and autotomy after deafferentation must be due to another mechanism as dorsal rhizotomy does not trigger massive ectopia in axotomized afferents, and even if it did, the impulses would have no access to the CNS. Pain following rhizotomy is, therefore, presumed to be due to impulses that originate within the deafferented CNS itself. Deafferentation triggers many structural and neurochemical changes in the CNS, and abnormal bursting discharges have been recorded in deafferented spinal dorsal horn in animals and in humans (see ▶ Central Changes After Peripheral Nerve Injury). The possibility that deafferentation pain is indeed due to this activity is supported by the observation that surgical destruction of the abnormal dorsal horn tissue by ▶ DREZotomy often relieves this type of pain (Rossitch et al. 1993; Guenot et al. 2002).

References

Basbaum, A. I. (1974). Effects of central lesions on disorders produced by dorsal rhizotomy in rats. *Experimental Neurology, 42*, 490–501.

Bennett, G., & Xie, Y.-K. (1988). A peripheral mononeuropathy in rat that produces disorders of pain sensation like those seen in man. *Pain, 33*, 87–107.

Blumenkopf, B., & Lipman, J. J. (1991). Studies in autotomy: Its pathophysiology and usefulness as a model of chronic pain. *Pain, 45*, 203–210.

Coderre, T. J., Grimes, R. W., & Melzack, R. (1986). Deafferentation and chronic pain in animals: An evaluation of evidence suggesting autotomy is related to pain. *Pain, 26*, 61–84.

Defrin, R., Zeitoun, I., & Urca, G. (1996). Strain differences in autotomy level in mice: Relation to spinal excitability. *Brain Research, 711*, 241–244.

Devor, M. (2006). Response of nerves to injury in relation to neuropathic pain. Chapter 58. In S. L. McMahon & M. Koltzenburg (Eds.), *Wall and Melzack's textbook of pain* (5th ed., pp. 905–927). London: Churchill Livingstone.

Devor, M., & Raber, P. (1990). Heritability of symptoms in an experimental model of neuropathic pain. *Pain, 42*, 51–67.

Devor, M., & Seltzer, Z. (1999). Pathophysiology of damaged nerves in relation to chronic pain. In P. D. Wall & R. Melzack (Eds.), *Textbook of pain* (4th ed., pp. 129–164). London: Churchill Livingstone.

Devor, M., Inbal, R., & Govrin-Lippmann, R. (1982). Genetic factors in the development of chronic pain. In I. Lieblich (Ed.), *The genetics of the brain* (pp. 273–296). Amsterdam: Elsevier-North Holland.

Dorsi, M. J., Chen, L., Murinson, B. B., Pogatzki-Zahn, E. M., Meyer, R. A., & Belzberg, A. J. (2008). The tibial neuroma transposition (TNT) model of neuroma pain and hyperalgesia. *Pain, 134*, 320–334.

Guenot, M., Bullier, J., & Sindou, M. (2002). Clinical and electrophysiological expression of deafferentation pain alleviated by dorsal root entry zone lesions in rats. *Journal of Neurosurgery, 97*, 1402–1409.

Kauppila, T., & Pertovaara, A. (1991). Effects of different sensory and behavioral manipulations on autotomy caused by a sciatic lesion in rats. *Experimental Neurology, 111*, 128–130.

Koplovitch, P., Minert, A., & Devor, M. (2012). Spontaneous pain in partial nerve injury models of neuropathy and the role of nociceptive sensory cover. *Experimental Neurology, 236*, 103–111.

Levitt, M. (1985). Dysesthesias and self-mutilation in humans and subhumans: A review of clinical and experimental studies. *Brain Research Reviews, 10*, 247–290.

Mailis, A. (1996). Compulsive targeted self-injurious behaviour in humans with neuropathic pain: A counterpart of animal autotomy? Four case reports and literature review. *Pain, 64*, 569–578.

Mogil, J. S., Wilson, S. G., Bon, K., Pieper, J. O., Lee, S. E., Chung, K., Raber, P., Hain, H. S., Belknap, J. K., Hubert, L., Elmer, G. I., Chung, J. M., & Devor, M. (1999). Heritability of nociception. II. "Types" of nociception revealed by genetic correlation analysis. *Pain, 80*, 83–93.

Nissenbaum, J., Devor, M., Seltzer, Z., Gebauer, M., Michaelis, M., Tal, M., Dorfman, R., Abitbul-Yarkoni, M., Lu, Y., Elahipanah, T., delCanho, S., Minert, A., Fried, K., Persson, A. K., Shpigler, H., Shabo, E., Yakir, B., Pisante, A., & Darvasi, A. (2010). Susceptibility to chronic pain following nerve injury is genetically affected by CACNG2. *Genome Research, 20*, 1180–1190.

Raber, P., & Devor, M. (2002). Social variables affect phenotype in the neuroma model of neuropathic pain. *Pain, 97*, 139–150.

Rossitch, E. J., Abdulhak, M., Olvelmen-Levitt, J., Levitt, M., & Nashold, B. J. (1993). The expression of deafferentation dysesthesias reduced by dorsal root entry zone lesions in the rat. *Journal of Neurosurgery, 78*, 598–602.

Saade, N. E., Ibrahim, M. Z. M., Atweh, S. F., & Jabbur, S. J. (1993). Explosive autotomy imduced by simultaneous dorsal column lesion and limb denervation: a possible model for acute deafferentation pain. *Experimental Neurology, 119*, 272–279.

Seltzer, Z. (1995). The relevance of animal neuropathy models for chronic pain in humans. *Seminars in Neuroscience, 7*, 31–39.

Seltzer, Z., Max, M. B., & Diehl, S. R. (2001). Mapping a gene for neuropathic pain-related behavior following peripheral neurectomy in the mouse. *Pain, 93*, 101–106.

Symons, F. J. (2011). Self-injurious behavior in neurodevelopmental disorders: Relevance of nociceptive and immune mechanisms. *Neuroscience and Biobehavioral Reviews, 35*, 1266–1274.

Wall, P. D., Devor, M., Inbal, R., Scadding, J. W., Schonfeld, D., Seltzer, Z., & Tomkiewicz, M. M. (1979). Autotomy following peripheral nerve lesions: Experimental anaesthesia dolorosa. *Pain, 7*, 103–111.

Zeltser, R., Zaslansky, R., Beilin, B., & Seltzer, Z. (2000). Comparison of neuropathic pain induced in rats by various clinically-used neurectomy methods. *Pain, 89*, 19–24.

Anesthesiologic Interventions

▶ Cancer Pain Management, Anesthesiologic Interventions, Spinal Cord Stimulation, and Neuraxial Infusion

Anesthesiological Interventions

▶ Cancer Pain Management, Anesthesiologic Interventions, and Neural Blockade

Anesthesiologist

Definition

A medical doctor specializing in preventing and treating pain during surgery (general anesthesia (sleeping patient) or local or regional anesthesia (with part of the body made numb and feeling no pain)). Anesthesiologists also take care of critically ill patients in the intensive care units, in emergency, and pre-hospital settings.

Cross-References

▶ Postoperative Pain, Acute Pain Management Principles
▶ Postoperative Pain, Acute Pain Team

Anesthetic Block

▶ Cancer Pain Management, Anesthesiologic Interventions

Anesthetic Blockade

Definition

Injection of local anesthetics in a nerve branch or plexus

Cross-References

▶ Deafferentation Pain

Aneurysm

Definition

An aneurysm is a localized dilatation of a blood vessel, commonly an artery, which may cause symptoms by enlarging or bleeding.

Cross-References

▶ Primary Cough Headache

Anger and Pain

Akiko Okifuji
Department of Anesthesiology, University of Utah, Salt Lake City, UT, USA

Synonyms

Acting-out; Aggression; Anger-in; Frustration; Hostility

Definition

Anger is an emotional experience involving cognitive appraisal and action tendency (Smedslund 1992). There have been numerous anecdotal reports since the early days of pain medicine, suggesting that anger may be an associated or resultant emotional experience of pain. There are several terms that are used interchangeably. For the purpose of clarification, in this entry, the following definitions will be applied:

- Frustration: affective state that arises when one's effort has been blocked, thwarted
- Anger: strong feeling of displeasure associated with cognitive appraisal that injustice has occurred and action tendency to remedy the perceived injustice

- Hostility: unfriendly attitudinal disposition with tendency to become angry
- Aggression: behavioral actualization of the action tendency associated with anger

Characteristics

Anger is a common emotional experience associated with pain, particularly chronic pain. Pain, by virtue of its aversive phenomenological nature, frequently brings on the perception of injustice and frustration. Additionally, the sense of injustice may come with having to undergo multiple diagnostic tests without finding fruitful findings. This often raises a question of legitimacy of pain, leading to interpersonal hardship. Functional limitations associated with chronic pain may severely impair the patient's ability to be a productive member of a workforce, enjoy the recreations they used to engage in, and nurture their personal relationships with friends and families.

Parameters of Anger
Anger is a multidimensional construct. It involves the temporary parameters of the experience such as frequency, recurrence, duration, intensity of the experience, expression styles, and target of the anger. The earlier studies mostly focused on anger levels (state) or one's tendency to become angry (trait). Those earlier studies generally demonstrated the relationship between anger and pain severity in chronic pain patients (Wade et al. 1990). The directionality of the relationship has been a topic of much debate. Literature supports both, suggesting that the relationship is likely dynamic, reciprocal, and complex. For example, anger exacerbates pain severity in chronic pain patients (Gaskin et al. 1992). Pain also triggers angry reaction; the experimental studies show higher frustration and hostility as a result of noxious stimulation (Berkowitz and Thome 1987).

Another parameter of anger is how anger is experienced and expressed ("anger management style"). One of the two styles that have been most studied is "anger-in," in which people are aware

of the presence of anger but the expression is suppressed. The concept of "trait" anger-in is thought to be confounded with a more general concept of negative affectivity (Burns et al. 2008), and the delineation of the true relationship between a tendency to inhibit anger and pain is difficult. Typically, trait anger-in and pain intensity in chronic pain patients are modest and positively related (Kerns et al. 1994; van Middendorp et al. 2010). Experimental manipulation of anger inhibition, on the other hand, consistently seems to heighten pain sensitivity in healthy people (Quartana and Burns 2007) as well as in chronic pain patients (Burns and Holly 2008). The other style is "anger-out," in which angry feelings are overtly expressed. The high degree of anger-out suggests undercontrolled, excessive demonstrations of anger. Anger-out seems to be implicated with higher pain sensitivity in various situations. It is correlated with greater sensitivity to acute pain in healthy people and pain patients (Bruehl et al. 2002) as well as greater pain in postsurgical patients (Bruehl et al. 2006). Clinical pain among chronic pain patients may also be related to their tendency to express anger (Bruehl et al. 2003).

Finally, a target parameter of anger may be important in understanding pain patients. Anger is generally a provoked feeling and requires a specific target, object, or person with whom one feels angry. The degree to which anger is related to pain may greatly differ, depending upon targets with which people experience anger. Self and healthcare providers appear to be common targets of anger among chronic pain patients. Interestingly, anger at self is related to depression, whereas anger at healthcare providers is related to the perceived level of functional disability (Okifuji et al. 1999). The results suggest that the assessment of specific targets with which patients experience anger may be important in understanding the overall clinical picture of their pain condition.

Mechanisms
Psychodynamic Model
The role of anger in medically unexplained pain complaints plays a central role in

a psychodynamic conceptualization. When a person experiences anger, the person's psyche classifies the emotion as unacceptable, and it channels the feeling into somatic symptoms. In the early days of research evaluating the etiology of chronic pain, the high prevalence of depressed mood in chronic pain patients led the psychodynamic paradigm to propose the notion of "masked depression," in which chronic pain was considered as a somatically expressed form of depression. Depression, in turn, was considered as "anger turned inward," with a person holding a self-depreciating view of self. However, empirical support for the psychodynamic model is limited to the correlational association between pain and negative moods.

Psychosocial-Behavioral Model
Behaviorally, anger, when it is poorly managed, may contribute to the suffering of a person living with pain. Functional limitations that often accompany their condition significantly compromise the quality of life, leading to frustration, and persistent irritability further compromises interpersonal relationships. Inhibition of anger expression has been found to be related to depression especially for chronic pain patients reporting severe pain (Estlander et al. 2008). Moreover, when expressed inappropriately, anger may interfere with how the person interacts with healthcare providers. Intense anger may jeopardize the cooperative relationship between the providers and the patient or decrease the patient's willingness to comply with the regimen; as a result, the patient may not receive the optimal benefit from the treatment.

Psychophysiological-Neurological Model
Anger may also contribute to pain via autonomic activation. Anger is associated with the general elevation in the sympathetic responses. The orchestrated arousal of the sympathetic tone is analogous to what we experience in response to a stressor. Such stress responses, particularly muscle tension, are known to be potentially problematic for pain patients. Pain patients exhibit a greater level of muscle tension in the pain-afflicted region than in other non-

affected areas in response to a stressor (Flor and Turk 1989), suggesting that the elevation of muscle tension associated with anger may play a role in perpetuating the stress-tension-pain cycle. On the other hand, when patients who tend to suppress their anger reexperience/recall anger, they seem to show reduced paraspinal muscle reactivity (Burns 1997). More recently, it has been suggested that muscle tension may show marked elevation where there is a mismatch between a personal style of anger regulation and situational demand (e.g., high anger-out people trying to suppress anger) (Burns and Holly 2008).

Cardiovascular response to anger may also play a role in pain experience. Janssen et al. (2001) has shown that acutely induced anger might lead to elevation of blood pressure and increased pain thresholds, possibly indicating the presence of stress-induced analgesia. Similarly, a recent study (Quartana et al. 2010) showed that effort to suppress provoked anger attenuated blood pressure response to pain and was positively related to greater pain report, again suggestive of a possible role of autonomic stress response in how anger may modulate pain experience.

More recently, it has been suggested that the dysregulation in the endogenous opioid function may mediate the relationship between anger and pain. Expressed anger seems to compromise the endogenous opioid reactivity to experimentally induced pain (Bruehl et al. 2002; Bruehl et al. 2003). Reduced release of beta endorphin in response to pain has also been observed in those with high degree of anger-out (see Bruehl et al. (2009) for an excellent review on this topic).

Treatment Implications
Treatment of pain, particularly chronic pain, requires cooperation and active participation from patients. Anger, if not properly managed, is likely to interfere with treatment efficacy. It is reasonable to assume that angry patients may be reluctant to follow the regimen. Angry patients with suboptimal coping skills may also find it difficult to adaptively change their lifestyles to

accommodate rehabilitation. At this time, very little is known about how anger interacts with rehabilitative efforts for pain patients. Burns et al. (1998) reported that male patients showed the inverse relationship between the pretreatment level of anger suppression and improvement in mood and self-reported level of activity. The result from their subsequent study suggests (Burns et al. 1999) that patients with a high degree of anger-out may not develop a sense of rapport with their healthcare providers.

Anger is not necessarily maladaptive. Anger can be an adaptive emotional response to the injustice that patients perceive. However, the accumulation of research suggests that poorly managed anger exacerbates pain and disability and interferes with the treatment efforts. Effective self-management of anger may be essential for the successful rehabilitative effort of pain patients. Psychoeducational approaches help patients to better understand the concept and how poorly managed anger may contribute to their pain. Fernandez (2002) suggests several approaches to help patients acquire better anger coping skills via cognitive reappraisal, behavioral modification, and appropriate affective disclosure. There has been a proliferation of research demonstrating the salient effects of poorly managed anger on pain experience in the past decade. By contrast, however, very little has been done to test the efficacy of anger management in improving pain therapy outcomes. Future research is warranted to evaluate the enhancement effects of such approaches for pain rehabilitation.

References

Berkowitz, L., & Thome, P. (1987). Pain, expectation, negative affect, and angry aggression. *Motivation and Emotion, 11*, 183–193.

Bruehl, S., Burns, J. W., et al. (2002). Anger and pain sensitivity in chronic low back pain patients and pain-free controls: The role of endogenous opioids. *Pain, 99*, 223–233.

Bruehl, S., Chung, O. Y., et al. (2003). The Association between anger expression and chronic pain intensity: Evidence for partial mediation by endogenous opioid dysfunction. *Pain, 106*, 317–324.

Bruehl, S., Chung, O. Y., et al. (2006). Anger regulation style, postoperative pain, and relationship to the A118G mu opioid receptor gene polymorphism: A preliminary study. *Journal of Behavioral Medicine, 29*, 161–169.

Bruehl, S., Burns, J. W., et al. (2009). Pain-related effects of trait anger expression: Neural substrates and the role of endogenous opioid mechanisms. *Neuroscience and Biobehavioral Reviews, 33*, 475–491.

Burns, J. W. (1997). Anger management style and hostility: Predicting symptom-specific physiological reactivity among chronic low back pain patients. *Journal of Behavioral Medicine, 20*, 505–522.

Burns, J. W., & Holly, A. (2008). Trait anger management style moderates effects of actual ("State") anger regulation on symptom-specific reactivity and recovery among chronic low back pain patients. *Psychosomatic Medicine, 70*, 898–905.

Burns, J. W., Johnson, B. J., et al. (1998). Anger management style and the prediction of treatment outcome among male and female chronic pain patients. *Behaviour Research and Therapy, 36*, 1051–1062.

Burns, J., Higdon, L., et al. (1999). Relationships among patient hostility, anger expression, depression and the working alliance in a work hardening program. *Annals of Behavioral Medicine, 21*, 77–82.

Burns, J. W., Quartana, P., et al. (2008). Effects of anger suppression on pain severity and pain behaviors among chronic pain patients: Evaluation of an ironic process model. *Health Psychology, 27*, 645–652.

Estlander, A. M., Knaster, P., et al. (2008). Pain intensity influences the relationship between anger management style and depression. *Pain, 140*, 387–392.

Fernandez, E. (2002). *Anxiety, depression, and anger in pain*. Toronto: University of Toronto Press.

Flor, H., & Turk, D. C. (1989). Psychophysiology of chronic pain: Do chronic pain patients exhibit symptom-specific psychophysiological responses? *Psychological Bulletin, 105*, 215–259.

Gaskin, M. E., Greene, A. F., et al. (1992). Negative affect and the experience of chronic pain. *Journal of Psychosomatic Research, 36*(8), 707–713.

Janssen, S. A., Spinhoven, P., et al. (2001). Experimentally induced anger, cardiovascular reactivity, and pain sensitivity. *Journal of Psychosomatic Research, 51*, 479–485.

Kerns, R. D., Rosenberg, R., et al. (1994). Anger expression and chronic pain. *Journal of Behavioral Medicine, 17*, 57–67.

Okifuji, A. D., Turk, D. C., et al. (1999). Anger in chronic pain: Investigations of anger targets and intensity. *Journal of Psychosomatic Research, 47*, 1–12.

Quartana, P. J., Bounds, S., et al. (2010). Anger suppression predicts pain, emotional, and cardiovascular responses to the cold pressor. *Annals of Behavioral Medicine, 39*(3), 211–221.

Quartana, P. J., & Burns, J. W. (2007). Painful consequences of anger suppression. *Emotion, 7*, 400–414.

Smedslund, J. (1992). How shall the concept of anger be defined? *Theory Psychology, 3*, 5–34.

van Middendorp, H., Lumley, M. A., et al. (2010). Effects of anger and anger regulation styles on pain in daily life of women with fibromyalgia: A diary study. *European Journal of Pain, 14*, 176–182.

Wade, J. B., Price, D. D., et al. (1990). An emotional component analysis of chronic pain. *Pain, 40*, 303–310.

Anger-in

▶ Anger and Pain

Angiitis of the CNS

▶ Headache Due to Arteritis

Angina Pectoris

Definition

Severe chest discomfort usually caused by inadequate blood flow through the blood vessels of the heart as a result of cardiac disease resulting in myocardial ischemia (inadequate oxygen supply to the heart). It is often treated by medical means or by surgical or angioplastic revascularization. It is rarely treated by spinal cord stimulation. Angina is often accompanied by shortness of breath, sweating, nausea, and dizziness.

Cross-References

▶ Pain Treatment: Spinal Cord Stimulation
▶ Spinothalamic Tract Neurons, Visceral Input
▶ Thalamus
▶ Thalamus and Visceral Pain Processing (Human Imaging)
▶ Thalamus, Clinical Visceral Pain, Human Imaging
▶ Visceral Pain Model, Angina Pain

Angina Pectoris, Neurophysiology and Psychophysics

Nirit Weiss[1], S. Ohara[2], Stanley W. Anderson[2] and Frederick A. Lenz[3]
[1]Department of Neurosurgery, Mount Sinai Medical Center, New York, NY, USA
[2]Department of Neurosurgery, Johns Hopkins Hospital, Johns Hopkins Medical Institutions, Johns Hopkins University, Baltimore, MD, USA
[3]Department of Neurosurgery, Johns Hopkins Hospital, Baltimore, MD, USA

Synonyms

Interoceptive sensation; Sympathetic afferents; Visceral sensation

Definition

The role of the somatic sensory ▶ thalamus in angina related to cardiac disease is demonstrated by stimulation of thalamus in patients with a history of angina and by the presence of cells projecting to monkey thalamus that respond to cardiac stimulation.

Characteristics

The sensory mechanisms of angina are poorly understood, although it is a common, clinically significant symptom. Recent evidence suggests that the perception of angina is correlated with central nervous system activity encoding cardiac injury. Noxious cardiac stimuli evoke activity in sympathetic afferent nerves (Foreman et al. 1986), in ascending spinal pathways (spinothalamic (STT) and ▶ dorsal column (DC) pathways) and in cells of the principal sensory nucleus of the ▶ thalamus (Horie and Yokota 1990).

STT cells in the upper thoracic spinal cord projecting to the region of VP respond to coronary artery occlusion (Blair et al. 1984) and intracardiac injection of bradykinin (Blair et al. 1982). Cells at the posteroinferior aspect of VP in the cat respond to intracardiac injections of bradykinin (Horie and Yokota 1990) and to stimulation of cardiac sympathetic nerves (Taguchi et al. 1987). Neurons in the thalamic principal sensory nucleus also encode visceral inputs from gastrointestinal and genitourinary systems in monkeys (Bruggemann et al. 1998). Therefore, experimental studies suggest that cells in the region of VP encode noxious visceral and cardiac stimuli.

The spinothalamic tract sends a dense projection particularly to the posterior inferior lateral aspect of monkey VP (Apkarian and Hodge 1989). Projections from the spinothalamic tract are also found posterior and inferior to VP in the posterior nucleus and in the ventral posterior inferior nucleus. VP projects to primary somatosensory cortex while the region posterior and inferior to VP projects to secondary somatosensory cortex, insular cortex, and retroinsular cortex (Jones 1985).

Involvement of sympathetics in the perception of angina is based upon evidence that stimulation of the superior cervical ganglion produces pain and that lesions of the sympathetic ganglia and dorsal roots relieve angina (reviewed by Meller and Gebhart 1992). Involvement of thalamus in the sensation of angina is suggested by the case of a patient with angina successfully treated by balloon angioplasty (Lenz et al. 1994). During thalamic exploration for implantation of a stimulating electrode, microstimulation evoked a pain "almost identical" to her angina, except that it began and stopped instantaneously with stimulation. This time course of this sensation was typical of sensations evoked by thalamic microstimulation but not those evoked by cardiac disease (Lenz et al. 1993). Stimulation-associated angina was not accompanied by the cardiac indices of angina in the setting of myocardial infarction. A biochemical link between the experience of angina pectoris and activation of thalamic neurons may have been elucidated during experimentation with dipyridamole stress testing. Vagal release of adenosine to the A(1) receptors in thalamus and prefrontal cortex appears to play a role (Ito et al. 2002).

The description of her typical angina and stimulation-evoked angina included words with a strong affective dimension from a questionnaire. In a similar setting the atypical chest pain of panic disorder was "almost identical" to that produced by microstimulation in the same thalamic area as the present case (Lenz et al. 1995). Stimulation-evoked pain without an affective dimension was observed in a retrospective analysis of patients without experience of spontaneous chest pain with a strong affective dimension (Lenz et al. 1994, 1995). Therefore, stimulation-evoked chest pain included an affective dimension as a result of conditioning by the prior experience of angina of cardiac origin.

The affective dimension of stimulation-associated pain might be analogous to emotional phenomena evoked by stimulation of amygdala in patients with epilepsy who have prior experience of these phenomena during the aura of their seizures (Halgren et al. 1978). The region posterior to Vc is linked to nociceptive cortical areas that project to the amygdala. Vcpc projects to anterior insular cortex (Mehler 1962) whereas Vcpor projects to the inferior parietal lobule, including the parietal operculum and secondary somatosensory cortex (SII) (Locke et al. 1961) which project, directly or indirectly, to the amygdala. Noxious sensory input to these cortical areas is demonstrated by evoked potentials in response to tooth pulp stimulation (Chatrian et al. 1975). Lesions of SII interfere with discrimination of noxious stimuli (Greenspan and Winfield 1992) while lesions of insula impair emotional responses to painful stimuli (Berthier et al. 1988). Thus, there is good evidence that cortical areas receiving input from Vcpc and Vcpor are involved in pain processing.

SII and insular cortical areas involved in pain processing also satisfy criteria for areas involved in corticolimbic connections (see ▶ corticolimbic circuits). In monkeys, a nociceptive sub-modality selective area has been found within SII (Dong et al. 1989). SII cortex projects to insular areas that project to the amygdala (Friedman et al. 1986). SII and insular cortex have bilateral primary noxious sensory input (Chatrian et al. 1975), and cells in these areas responding to noxious stimuli have bilateral representation (Dong et al. 1989; Chatrian et al. 1975). Therefore, cortical areas receiving input from Vcpc and Vcpor may be involved in memory for pain through corticolimbic connections (Mishkin 1979).

References

Apkarian, A. V., & Hodge, C. J. (1989). Primate spinothalamic pathways: III. Thalamic terminations of the dorsolateral and ventral spinothalamic pathways. *The Journal of Comparative Neurology, 288*, 493–511.

Berthier, M., Starkstein, S., & Leiguarda, R. (1988). Asymbolia for pain: A sensory-limbic disconnection syndrome. *Annals of Neurology, 24*, 41–49.

Blair, R. W., Weber, N., & Foreman, R. D. (1982). Responses of thoracic spinothalamic neurons to intracardiac injection of bradykinin in the monkey. *Circulation Research, 51*, 83–94.

Blair, R. W., Ammons, W. S., & Foreman, R. D. (1984). Responses of thoracic spinothalamic and spinoreticular cells to coronary artery occlusion. *Journal of Neurophysiology, 51*, 636–648.

Bruggemann, J., Shi, T., & Apkarian, A. V. (1998). Viscerosomatic interactions in the thalamic ventral posterolateral nucleus (VPL) of the squirrel monkey. *Brain Research, 787*, 269–276.

Chatrian, G. E., Canfield, R. C., Knauss, T. A., et al. (1975). Cerebral responses to electrical tooth pulp stimulation in man. An objective correlate of acute experimental pain. *Neurology, 25*, 745–757.

Dong, W. K., Salonen, L. D., Kawakami, Y., et al. (1989). Nociceptive responses of trigeminal neurons in SII-7b cortex of awake monkeys. *Brain Research, 484*, 314–324.

Foreman, R. D., Blair, R. W., & Ammons, W. S. (1986). Neural mechanisms of cardiac pain. In F. Cervero & J. F. B. Morrison (Eds.), *Progress in brain research* (pp. 227–243). New York: Elsevier Science Publishers BV.

Friedman, D. P., Murray, E. A., O'Neill, J. B., et al. (1986). Cortical connections of the somatosensory fields of the lateral sulcus of macaques: Evidence for a corticolimbic pathway for touch. *The Journal of Comparative Neurology, 252*, 323–347.

Greenspan, J. D., & Winfield, J. A. (1992). Reversible pain and tactile deficits associated with a cerebral tumor compressing the posterior insula and parietal operculum. *Pain, 50*, 29–39.

Halgren, E., Walter, R. D., Cherlow, D. G., et al. (1978). Mental phenomena evoked by electrical stimulation of the human hippocampal formation and amygdala. *Brain, 101*, 83–117.

Horie, H., & Yokota, T. (1990). Responses of nociceptive VPL neurons to intracardiac injection of bradykinin in the cat. *Brain Research, 516*, 161–164.

Ito, H., Yokoyama, I., Tamura, Y., Kinoshita, T., Hatazawa, J., Kawashima, R., & Iida, H. (2002). Regional changes in human cerebral blood flow during dipyridamole stress: Neural activation in the thalamus and prefrontal cortex. *NeuroImage, 16*(3 pt. 1), 788–793.

Jones, E. G. (1985). *The thalamus*. New York: Plenum.

Lenz, F. A., Seike, M., Richardson, R. T., et al. (1993). Thermal and pain sensations evoked by microstimulation in the area of human ventrocaudal nucleus. *Journal of Neurophysiology, 70*, 200–212.

Lenz, F. A., Gracely, R. H., Hope, E. J., et al. (1994). The sensation of angina can be evoked by stimulation of the human thalamus. *Pain, 59*, 119–125.

Lenz, F. A., Gracely, R. H., Romanoski, A. J., et al. (1995). Stimulation in the human somatosensory thalamus can reproduce both the affective and sensory dimensions of previously experienced pain. *Nature Medicine, 1*, 910–913.

Locke, S., Angevine, J. B., & Marin, O. S. M. (1961). Projection of magnocellular medial geniculate nucleus in man. *The Anatomical Record, 139*, 249–250.

Mehler, W. R. (1962). The anatomy of the so-called "pain tract" in man: An analysis of the course and distribution of the ascending fibers of the fasciculus anterolateralis. In J. D. French & R. W. Porter (Eds.), *Basic research in paraplegia* (pp. 26–55). Springfield: Thomas.

Meller, S. T., & Gebhart, G. F. (1992). A critical review of the afferent pathways and the potential chemical mediators involved in cardiac pain. *Journal of Neuroscience, 48*, 501–524.

Mishkin, M. (1979). Analogous neural models for tactual and visual learning. *Journal of Neuropsychopharmacology, 17*, 139–151.

Taguchi, H., Masuda, T., & Yokota, T. (1987). Cardiac sympathetic afferent input onto neurons in nucleus ventralis posterolateralis in cat thalamus. *Brain Research, 436*, 240.

Angiogenesis

Definition

Angiogenesis refers to the growth of new blood vessels, which is an important naturally occurring process in the organism, both in normal and tumor tissue. In the case of cancer, the new vessels provide oxygen and nutrition for the tumor cells and allow tumor cells to escape into the circulation and lodge into other organs (tumor metastases).

Cross-References

▶ NSAIDs and Cancer

Angiography

Definition

An angiography is necessary for the diagnosis of CNS vasculitides.

Cross-References

▶ Headache Due to Arteritis

Animal Model for Phantom Limb Pain

▶ Anesthesia Dolorosa Model, Autotomy

Animal Model of Spontaneous Neuropathic Pain

▶ Anesthesia Dolorosa Model, Autotomy

Animal Models

▶ Animal Models of Inflammatory Bowel Disease

Animal Models and Experimental Tests to Study Nociception and Pain

Jin Mo Chung
Department of Neuroscience and Cell Biology, University Chair in Neuroscience, University of Texas Medical Branch, Galveston, TX, USA

Introduction

Animal models and experimental tests to study nociception and pain are important because they are the major tools that make studying nociception and pain possible. It would not be an exaggeration to say that progress on pain research has been made only to the degree that these essential research tools are available.

Among the oldest and the most commonly used nociceptive tests would be the ▶ tail flick test that was developed by D'Amour and Smith in 1941 (D'Amour and Smith 1941). This is a test for acute pain in normal rodents, and since then many other tests and models, focused primarily on chronic or persistent pain, have been developed. Availability of these tests and models enabled research on persistent pain to flourish during recent decades. The present section documents the majority of commonly used animal models and experimental tests.

Characteristics

Overview of Topics
Tests for Nociception and Pain
Nociceptive tests utilize observations of animal behavior after delivering noxious mechanical, heat, or chemical stimuli to a defined body part. In this section, we will address a variety of tests used to study nociception and pain. Two of these tests, the ▶ allodynia test, mechanical and cold allodynia, and the ▶ Randall-Selitto paw pressure test, use a mechanical stimulus to elicit responses. The tail flick test, the ▶ thermal nociception test, and the ▶ hot plate test (assay) use noxious heat as the stimulus. There are a number of ways to apply chemical stimuli to elicit pain behaviors. However, one of the most common methods is an injection of formalin into the paw of a rodent – the ▶ formalin test.

The thermal hyperalgesia test and the allodynia test, in particular, have been widely used in recent years. The thermal hyperalgesia test, which was developed by Hargreaves et al. (1988), uses the latency of escape behaviors of a rodent after application of a noxious heat stimulus to estimate changes in the heat pain threshold. This test has hence been frequently used to quantify the development of heat hyperalgesia in various painful conditions.

The allodynia test in a neuropathic pain model using von Frey filaments was first conducted by Seltzer et al. (1990) and quantifies changes in mechanical threshold for pain behavior. Kim and Chung (1991, 1992) subsequently used von Frey filaments extensively to quantify mechanical allodynia in their model of neuropathic pain. All these tests are for quantification of pain behavior in various pain models.

Animal Models

Numerous good animal models representing various pain syndromes have been developed in the past, particularly during the last couple of decades. These include various musculoskeletal pain models (► arthritis model, kaolin-carrageenan-induced arthritis (knee); ► arthritis model, adjuvant-induced arthritis; ► cancer pain model, bone cancer pain model; ► muscle pain model, ischemia-induced and hypertonic saline-induced; ► animal models of inflammatory muscle pain; ► sprained ankle pain model); and visceral pain models (pain originating from various parts of the gastrointestinal tract, heart, kidney, pancreas, urinary bladder, and female reproductive organs). In addition, there are a number of neuropathic pain models – produced by injuries to either the peripheral or the central nervous system. Many of these models are described in this section.

There has been an explosion in the development of peripheral neuropathic pain models in recent years, as well as in studies conducted using them. The field of neuropathic pain was revolutionized by the initial development of a model by Bennett and Xie (1988), which was followed by other models (Seltzer et al. 1990; Kim and Chung 1992; Na et al. 1994; Decosterd and Woolf 2000). All these models have in common: They produce a partial nerve injury so that an area of the skin is partially denervated but a part of the innervation is left intact. Direct comparison of multiple models in a single study is rare, but Kim et al. (1997) compared three neuropathic pain models, the chronic constriction injury (CCI) (► neuropathic pain model, chronic constriction injury), partial sciatic nerve ligation (PSL) (► neuropathic pain model, partial sciatic

nerve ligation model), and spinal nerve ligation (SNL) (► neuropathic pain model, spinal nerve ligation model) models. They found that these three models displayed similar behavioral patterns with minor differences in specific features, presumably due to the difference in populations and numbers of afferent fibers that are denervated versus those left intact in each model. For example, the CCI model showed a relatively larger magnitude of behavioral signs representing ongoing pain whereas the spinal nerve ligation (SNL) model displayed more robust mechanical allodynia. As far as the nature of injury is concerned, the SNL model is highly artificial in that it produces an injury to one or two spinal nerves selectively, whereas the PSL model closely resembles the nerve injury produced by gunshot wounds, on which the description of classical causalgia was based (Mitchell 1872). On the other hand, if one wants to reduce the variability between animals, a stereotyped injury such as SNL would be beneficial. Therefore, having such a variety of models provides a good opportunity to select and use a model depending on the questions posed and the given circumstances.

Another area of animal modeling that has flourished in recent years is the area of central neuropathic pain, particularly spinal cord injury pain models. In the central nervous system, post-stroke pain models of the cerebral cortex as well as of the thalamus are available. In the spinal cord, we now have models for contusion, ischemic and focal injuries produced by mechanical as well as by chemical means. These are described in this section.

Visceral pain is a clinically important topic. There are animal models representing pain arising from various visceral organs, ranging from the heart to the kidney, pancreas, urinary bladder, and various parts of the gastrointestinal tract. All of these models attempt to imitate a clinical situation that causes pain, such as ischemia (e.g., angina), overdistension of the gastrointestinal tract, or chemical/mechanical irritation of ductal structures.

Discussion and Future Direction

Although this section describes a number of animal models and experimental tests used to study nociception and pain, there are a large number of already developed tests and models that are not included here. We hope to be able to include them as their usage becomes more widespread. At the same time, there are a number of models and tests that need to be developed and these will be included in this section as they become available. Therefore, this section is expected to grow rapidly as we progress in pain research.

Ethical considerations in animal welfare are very important issues for all animal research, but this is especially important in pain studies because these require using painful stimuli; yet, the pain and stress of the animals must be minimized. Therefore, although the nature of the studies calls for inducing some levels of painful stimuli, pain and stress must be kept at the minimal level. Fortunately, animals in most models do not display signs of severe chronic pain and discomfort as evidenced by normal weight gain and grooming. However, should the animal show signs of unbearable discomfort, the experiment must be terminated by humanely euthanizing the animals. Maintaining pain and discomfort at the minimum level is important not only for the humane treatment of experimental animals but also for obtaining the most reliable scientific data without contamination by undesirable stress-induced factors.

How to define what a good animal model is can be debatable. However, a good animal model should at least: (1) replicate a human disease condition faithfully, (2) show little variability between investigators and between laboratories, and (3) be easy to produce. Most of the models presented in this section satisfy these criteria; however, some are better in one aspect and worse in others and some are the other way around. A good model should also replicate the most important aspect of a human pain condition and employ animal tests most relevant to these aspects. A model may employ a testing method that is designed to be convenient for experimenters but which does not necessarily test the most relevant aspect of pain in patients. This is a shortcoming which should be corrected.

It is sometimes difficult to relate the results of tests in animal models to human diseases. For example, in the case of a disease with motor deficits, a question may arise as to whether motor deficits seen in animals would be the same as those seen in human patients. This is particularly a problem in pain research because animals cannot verbally express sensory experience and we have to rely on their behaviors and our interpretation of them. Such an indirect approach leaves much room for a subjective interpretation. Therefore, we must pay particular attention to this problem when we deal with animal models and testing in animals.

As mentioned above, animals in all the models described in this section display pain behavior, but the intensity of the pain seems to be much less than the pain that is intended to be modeled. For example, although it is common for humans to lose their appetite and to lose weight while suffering from chronic pain, in most animal models of pain, the animals seldom show signs of severe suffering for an extended period. Another example would be that many neuropathic pain sufferers have excruciating sensitivity to tactile stimuli so that even gentle movement of hair will cause pain. However, rats in all neuropathic pain models can be handled and the affected areas touched without too much of a response. Furthermore, these rats usually bear some weight during locomotion although they invariably have some motor deficits. Why is there such a difference in the intensity of pain? It is possible that none of the developed models truly represent a severe human pain condition. It is also possible that animals react differently from humans to the same intensity of pain and that the models may still be valid. We can argue for one or the other with no definite answer, but this is something we need to consider when we deal with animal models.

Frequently, most of the animals used in a given animal model may consistently show signs of pain. Such consistency is a good thing in a pain model since there will be less variability between animals. On the other hand, this can be viewed as a bad feature in a model that represents a pain syndrome, since it is rare for all patients with a particular disease state to develop pain. For example, only 10–15 % of patients with a peripheral nerve injury develop neuropathic pain, yet virtually all rats in neuropathic pain models show pain behaviors. Why is this true? Are these still good models? These are difficult questions to answer.

One explanation commonly used is that a genetic factor may play a role so that some patients may have a genetic makeup prone to develop pain after peripheral nerve injury. In support of this contention, there are vast differences in pain behavioral responses to a peripheral nerve injury among different strains of rats or mice. However, there is no direct proof indicating that this is the true explanation. This is a factor we need to keep in mind as well when conducting studies of animal models.

Recently operant behavioral assays, which test "pain" in animals beyond simple motor reflex changes, have been developed. These tests are thought to give a different and possibly superior assay of clinical pain, especially tonic pain (Mogil 2009; King et al. 2009). This is an important emerging approach, and further development of operant assay methods is thus desirable.

Although animal models for many painful conditions are described in this section, more good animal models are needed for common painful conditions, such as lower back pain, headaches, and myofascial pain. The main reason for the lack of such models is that it is technically difficult to develop them. However, it is imperative to develop animal models for these clinically common painful conditions so that we can make scientific progress in understanding these important pain conditions.

Summary

Many good clinically relevant animal models for various painful conditions are available now and their availability provides powerful tools for scientific studies, as well as for the development of new analgesic drugs. Undoubtedly, we will need to refine existing models and to develop new models as well as test methods representing painful conditions faithfully.

References

Bennett, G. J., & Xie, Y.-K. (1988). A peripheral mononeuropathy in rat that produces disorders of pain sensation like those seen in man. *Pain, 33*, 87–107.

D'Amour, F. E., & Smith, D. L. (1941). A method for determining loss of pain sensation. *Journal of Pharmacology and Experimental Therapeutics, 72*, 74–79.

Decosterd, I., & Woolf, C. J. (2000). Spared nerve injury: An animal model of persistent peripheral neuropathic pain. *Pain, 87*, 149–158.

Hargreaves, K., Dubner, R., Brown, F., et al. (1988). A new and sensitive method for measuring thermal nociception in cutaneous hyperalgesia. *Pain, 32*, 77–88.

Kim, S. H., & Chung, J. M. (1991). Sympathectomy alleviates mechanical allodynia in an experimental animal model for neuropathy in the rat. *Neuroscience Letters, 134*, 131–134.

Kim, S. H., & Chung, J. M. (1992). An experimental model for peripheral neuropathy produced by segmental spinal nerve ligation in the rat. *Pain, 50*, 355–363.

Kim, K. J., Yoon, Y. W., & Chung, J. M. (1997). Comparison of three rodent neuropathic pain models. *Experimental Brain Research, 113*, 200–206.

King, T., Vera-Portocarrero, L., Gutierrez, T., Vanderah, T. W., Dussor, G., Lai, J., Fields, H. L., & Porreca, F. (2009). *Unmasking the tonic-aversive state in neuropathic pain, nature neuroscience 12, 1364-1366Mitchell SW (1872) injuries of nerves and their consequences* (pp. 252–281). Philadelphia: JB Lippincott.

Mitchell, S. W. (1872). *Injuries of nerves and their consequences* (p. 252) Philadelphia, PA: J.B. Lippincott.

Mogil, J. S. (2009). Animal models of pain: Progress and challenges. *Nature Reviews Neuroscience, 10*, 283–294.

Na, H. S., Han, J. S., Ko, K. H., et al. (1994). A behavioral model for peripheral neuropathy produced in rat's tail by inferior caudal trunk injury. *Neuroscience Letters, 177*, 50–52.

Seltzer, Z., Dubner, R., & Shir, Y. (1990). A novel behavioral model of neuropathic pain disorders produced in rats by partial sciatic nerve injury. *Pain, 43*, 205–218.

Animal Models of Inflammatory Bowel Disease

Julie A. Christianson[1] and Brian Davis[2]
[1]Department of Anatomy and Cell Biology, University of Kansas Medical Center, Kansas City, KS, USA
[2]Department of Neurobiology, University of Pittsburgh School of Medicine, Pittsburgh, PA, USA

Synonyms

Animal models; Colitis; Crohn's disease; Inflammatory bowel disease; Ulcerative colitis

Definition

Inflammatory bowel disease (IBD) manifests as a complex chronic inflammatory disorder thought to be caused by a combination of environmental and genetic factors. Clinically, IBD presents as either ulcerative colitis (UC) or Crohn's disease (CD), which predominantly affects the colon and/or the distal small intestine, respectively (Hendrickson et al. 2002).

Characteristics

Approximately 600,000 Americans suffer from IBD, with the majority of patients diagnosed with the disease during their third decade of life. The most common symptoms include diarrhea, abdominal pain, fever, weight loss, ▶ arthralgias, and arthritis (Hendrickson et al. 2002). While the exact causes of IBD remain unknown, certain environmental and genetic factors have been shown to play a role in the development of IBD. Environmental factors may include smoking, diet, physical activity, childhood infections, microbial agents, and stress (Fiocchi 1998). The familial incidence of both CD and UC is remarkably high. The frequency of IBD in first-degree relations has been reported to be as high as 40 %. Among populations, IBD is most common among whites of European descent, although it is present in all races and ethnic groups (Fiocchi 1998).

The onset of IBD is generally thought to arise from T lymphocytes infiltrating a weakened epithelial lining and thereby initiating a pathological immune response within the bowel (Bhan et al. 1999; Blumberg et al. 1999). For IBD, the focus of research has been on CD4 T cells, also known as T-helper cells. These cells are capable of secreting large amounts of ▶ cytokine or ▶ growth factors that affect other immune cells and interacting tissues. Mature CD4 cells can be divided into Th1 and Th2 cells based on the complement of cytokines they produce. Th1 cells secrete IL-2, IFNγ, and TNF, whereas Th2 cells secrete IL-4, IL-5, IL-13, and TGFβ cytokines (Fig. 1). A balance between

these two cell types appears to be required for immunological homeostasis, as disruption of this balance can lead to pathological inflammation. Interestingly, Th1 and Th2 cell types predominate in CD and UC, respectively, and because of this, many animal models of IBD employ genetic deficiencies or antibodies against the cytokines associated with Th1 or Th2 CD4 cells (Bhan et al. 1999; Blumberg et al. 1999).

The important role played by these immune cells does not mean that environmental factors do not contribute significantly to the onset of IBD. Multiple studies have shown that animals housed in a pathogen-free environment do not develop IBD (Kim and Berstad 1992; Wirtz and Neurath 2000). This indicates that while alterations in the immune system are important for development of IBD, in most cases the development of pathology requires an environmental trigger that may include pathogens, stress, and nutrition.

Animal Models

Animal models of IBD can be separated into four main categories: spontaneous colitis, inducible colitis, adoptive transfer, and genetically engineered. Each model offers a novel approach to studying IBD; however none presents an exact model of the human condition.

Spontaneous Colitis

Symptoms of IBD occur naturally and are studied in C3H/HeJBir mice, SAMP1/Yit mice, cotton-top tamarins, and juvenile rhesus macaques. C3H/HeJBir mice can develop inflammation in the colon that peaks between 3 and 6 weeks of age and generally resolves by 10–12 weeks of age (Wirtz and Neurath 2000; Hendrickson et al. 2002). In contrast, SAMP1/Yit mice develop inflammation in the distal small intestine and cecum by 20 weeks of age with increasing lesion severity and incidence (Blumberg et al. 1999; Wirtz and Neurath 2000; Hendrickson et al. 2002). Cotton-top tamarins and juvenile rhesus monkeys are both primate models of UC, hallmarked by mucosal inflammation occurring only in the colon (Kim and Berstad 1992; Ribbons et al. 1997; Wirtz and Neurath 2000). Considering the high frequency of familial IBD,

Animal Models of Inflammatory Bowel Disease, Fig. 1 Imbalance between T-helper cells may contribute to CD and UC. T-helper cell type 1 (Th1) and T-helper cell type 2 (Th2) participate in cell-mediated immunity and antibody-mediated immunity, respectively. Antigen-presenting cells (APC) produce IL-12 in response to specific antigens, and this induces undifferentiated T-helper cells (Th0) to become Th1 cells. The cytokine(s) responsible for inducing Th0 cells to become Th2 cells have not been identified. Th1 and Th2 cells both secrete specific cytokines that act through positive and negative feedback loops. Th1 cytokines (IL-2, IFNγ) enhance Th1 cell proliferation while also inhibiting Th2 cell proliferation. IFNγ also functions to increase Th1 cell differentiation by upregulating IL-12 production. Th2 cytokines both increase (IL-4) and decrease (TGFβ) the proliferation of Th2 cells. TGFβ and IL-10 both suppress cytokine production by Th1 cells, and IL-10 also decreases the differentiation of Th1 cells by downregulating IL-12 production

spontaneous models of IBD are highly useful in the study of genetic susceptibility to IBD.

Inducible Colitis

Interruption of the mucosal barrier of the bowel can lead to transient or chronic inflammation. Various agents can induce IBD in this manner including formalin, acetic acid, carrageenan, dextran sulfate sodium (DSS), 2,4,6-trinitrobenzene sulfonic acid (TNBS), dinitrobenzene sulfonic acid (DNBS), indomethacin, oxazolone, or ethanol.

Intracolonic administration of dilute formalin or acetic acid induces a transient inflammation in the colon of rats or mice. Their effects occur very quickly and have been used extensively in the study of visceral pain (Kim and Berstad 1992). In contrast, chronic inflammation can be induced by oral ingestion of carrageenan or DSS, subcutaneous injection of indomethacin, or intracolonic administration of TNBS, DNBS, oxazolone, or ethanol. Carrageenan induces an early mucosal inflammation of the cecum with subsequent mucosal inflammation of the colon of rodents (Kim and Berstad 1992). Ingestion of DSS initially results in lesions and crypt formation within the mucosal lining of the colon of both rats and mice. This is followed by a secondary inflammation and infiltration of cytokines (Kim and Berstad 1992; Mahler et al. 1998). In mice, rats, and rabbits, intracolonic administration of TNBS or DNBS results in epithelial necrosis that leads to increased mucosal permeability and transmural inflammation (Kim and Berstad 1992; Elson et al. 1996). It is interesting to note that mouse strain differences exist regarding the susceptibility to either DDS or TNBS-induced IBD (Mahler et al. 1998). C3H/HeJ mice are highly susceptible to both DSS and TNBS, whereas C57Bl/6 and DBA/2 mice are less vulnerable to DSS and resistant to TNBS, again emphasizing the importance of genetic factors in the occurrence of IBD (Elson et al. 1996; Mahler et al. 1998). In rats, subcutaneous injection of indomethacin produces both an acute and a chronic inflammation within the small bowel, as well as epithelial injury measurable by mucosal permeability (Yamada et al. 1993). Mice or rats given intrarectal oxazolone develop a severe mucosal inflammation of the distal colon (Wirtz and Neurath 2000). Similarly, intrarectal ethanol results in destruction of the surface epithelium and necrosis extending throughout the mucosal layer of both mice and rats (Kamp et al. 2003). Inducible models are important in the study of IBD in that they establish a mechanical or chemical disruption of the mucosal barrier within the bowel, thereby providing an adequate model to study the chain of events that occur during the initial activation of the mucosal immune system.

Adoptive Transfer

IBD can be generated by transferring activated immune cells from normal animals into immunocompromised host animals. The most common model involves transferring CD4-positive T cells with a high expression of CD45RB (CD4$^+$CD45RB^{high}) from wild-type animals into severe combined ▶ immunodeficient (SCID) or recombination activating gene (RAG) knockout mice (Wirtz and Neurath 2000; Hendrickson et al. 2002). CD4$^+$CD45RB^{high} T cells produce high levels of Th1 cytokines, which have been shown to play a role in the induction and maintenance of IBD, in particular CD (Bhan et al. 1999; Blumberg et al. 1999). IBD can also be induced by introducing activated hsp60-specific CD8$^+$ T cells into immunodeficient or T cell receptor (TCR) β−/− mice (Wirtz and Neurath 2000). This results in degeneration of the mucosal epithelium in the small bowel with massive leukocytic infiltration within the ▶ lamina propria and epithelial layers. The adoptive transfer models have provided an excellent paradigm for gaining a better understanding for the role of T cells in the development and maintenance of IBD.

Genetically Engineered

The use of genetically altered mice has provided an excellent approach to studying the roles of specific immune cells and cytokines in IBD. As mentioned previously, an imbalance of Th1- and Th2-type cytokines has been shown to play a role in IBD (Bhan et al. 1999; Blumberg et al. 1999). Several transgenic and knockout mouse models have been generated to study the roles of Th1 and Th2 cytokines in IBD. Overexpression of HLA-B27 or STAT-4 both increases the production of Th1-type cytokines, including TNFα and IFNγ, most likely through the activation of IL-12 (Wirtz and Neurath 2000; Hendrickson et al. 2002). Similarly, mice with a deletion of IL-2, IL-2Rα, IL-10, CRF2-4, G$_{i \alpha}$2, STAT-3, or the AU-rich region of TNF overproduce Th1 cytokines and develop symptoms of IBD (Wirtz and Neurath 2000; Hendrickson et al. 2002). On the other hand, overexpression of IL-7 or a deletion of TCR-α results in a Th2-mediated IBD, mostly due to an increased production of Th1 cells (Wirtz and Neurath 2000).

Genetic models have also been generated to investigate aspects of IBD other than cytokine production. Disruption in the integrity of the intestinal epithelium has been implicated in IBD. This has been demonstrated in a mouse model that overexpresses a dominant negative form of N-cadherin using a small intestine-specific promoter (Wirtz and Neurath 2000). Similarly, deletion of the multiple drug-resistant gene (*mdr1a*) resulted in IBD, solely due to the lack of *mdr1 a* expression on intestinal epithelial cells (Wirtz and Neurath 2000). Intestinal trefoil factor (ITF) is luminally secreted after inflammation and is thought to aid in maintaining the barrier function of mucosal surfaces and facilitating healing processes after injury. Mice with a genetic deletion of ITF are significantly more susceptible to induction of IBD by DSS, indicated by increased colonic ulceration and morbidity (Mashimo et al. 1996; Wirtz and Neurath 2000). To investigate the role of enteric ganglia cells in IBD, a mouse model was developed that expresses herpes simplex virus (HSV) thymidine kinase (TK), driven by the glia-specific glial fibrillary acidic protein (GFAP) promoter.

When the antiviral agent ganciclovir (GCV) is injected subcutaneously, the HSV-TK metabolizes the GCV into toxic nucleotide analogs that induce cell death within their host cells, in this case enteric glial cells. Disruption of ileal and jejunal glial cells resulted in overt inflammation of the small bowel; however the colon remained unaffected (Bush et al. 1998). Genetic models have provided an excellent tool for investigating the possible roles of specific cytokines and structural proteins in IBD.

Implications for Pain Studies

As previously mentioned, patients with IBD often suffer from abdominal pain. Animal models, primarily models of inducible colitis, are often used to investigate changes in nociceptive processing that arise from IBD. The two most common methods for assessing visceral pain in animals are colorectal distension (CRD) and the acetic acid writhing test. CRD uses balloon distension of the distal colon to induce activation of both first- and second-order sensory afferents and contraction of abdominal muscles (visceromotor response), both of which can be quantifiably measured to determine visceral sensitivity (Kamp et al. 2003). In the writing test, ▶ intraperitoneal injections of acetic acid induce abdominal contractions along the length of the torso with corresponding arching of the back (Martinez et al. 1999). The mechanisms underlying the acetic acid writhing test are relatively unknown; therefore CRD is a much more reliable and consequently more widely used test for visceral hypersensitivity.

Several studies have used CRD as a means to study the effects of acute and chronic colon inflammation in rodents. Intracolonic application of acetic acid or ethanol was shown to significantly increase the number of abdominal contractions as well as the visceromotor response, during CRD (Martinez et al. 1999; Kamp et al. 2003). Similar results were observed in a TNBS-induced model of IBD (Sengupta et al. 1999). Visceral hyperalgesia has largely gone unstudied in genetic models of IBD. This is unfortunate as these models present an excellent opportunity for investigating the possible roles that cytokines and other molecules may play in the genesis of visceral hyperalgesia associated with IBD.

Animal models of IBD provide researchers with the tools to investigate specific aspects of the disease in an in vivo setting. While none of the models wholly represents the disease as it appears in humans, they each provide a useful tool with which to study specific aspects of the disease, including the manifestation of visceral hyperalgesia.

References

Bhan, A. K., Mizoguchi, E., Smith, R. N., et al. (1999). Colitis in transgenic and knockout animals as models of human inflammatory bowel disease. *Immunology Reviews, 169*, 195–207.

Blumberg, R. S., Saubermann, L. J., & Strober, W. (1999). Animal models of mucosal inflammation and their relation to human inflammatory bowel disease. *Current Opinion in Immunology, 11*, 648–656.

Bush, T. G., Savidge, T. C., Freeman, T. C., et al. (1998). Fulminant jejuno-ileitis following ablation of enteric glia in adult transgenic mice. *Cell, 93*, 189–201.

Elson, C. O., Beagley, K. W., Sharmanov, A. T., et al. (1996). Hapten-induced model of murine inflammatory bowel disease: Mucosa immune responses and protection by tolerance. *Journal of Immunology, 157*, 2174–2185.

Fiocchi, C. (1998). Inflammatory bowel disease: Etiology and pathogenesis. *Gastroenterology, 115*, 182–205.

Hendrickson, B. A., Gokhale, R., & Cho, J. H. (2002). Clinical aspects and pathophysiology of inflammatory bowel disease. *Clinical Microbiology Reviews, 15*, 79–94.

Kamp, E. H., Jones, R. C., 3rd, Tillman, S. R., et al. (2003). Quantitative assessment and characterization of visceral nociception and hyperalgesia in mice. *American Journal of Physiology. Gastrointestinal and Liver Physiology, 284*, 434–444.

Kim, H. S., & Berstad, A. (1992). Experimental colitis in animal models. *Scandinavian Journal of Gastroenterology, 27*, 529–537.

Mahler, M., Bristol, I. J., Leiter, E. H., et al. (1998). Differential susceptibility of inbred mouse strains to dextran sulfate sodium-induced colitis. *American Journal of Physiology, 274*, G544–G551.

Martinez, V., Thakur, S., Mogil, J. S., et al. (1999). Differential effects of chemical and mechanical colonic irritation on behavioral pain response to intraperitoneal acetic acid in mice. *Pain, 81*, 179–186.

Mashimo, H., Wu, D. C., Podolsky, D. K., et al. (1996). Impaired defense of intestinal mucosa in mice lacking intestinal trefoil factor. *Science, 274*, 262–265.

Ribbons, K. A., Currie, M. G., Connor, J. R., et al. (1997). The effect of inhibitors of inducible nitric oxide synthase on chronic colitis in the rhesus monkey. *Journal of Pharmacology and Experimental Therapeutics, 280*, 1008–1015.

Sengupta, J. N., Snider, A., Su, X., et al. (1999). Effects of kappa opioids in the inflamed rat colon. *Pain, 79*, 175–185.

Wirtz, S., & Neurath, M. F. (2000). Animal models of intestinal inflammation: New insights into the molecular pathogenesis and immunotherapy of inflammatory bowel disease. *International Journal of Colorectal Disease, 15*, 144–160.

Yamada, T., Deitch, E., Specian, R. D., et al. (1993). Mechanisms of acute and chronic intestinal inflammation induced by indomethacin. *Inflammation, 17*, 641–662.

Animal Models of Inflammatory Muscle Pain

▶ Muscle Pain Model, Inflammatory Agents-Induced

Animal Models of Inflammatory Myalgia

▶ Muscle Pain Model, Inflammatory Agents-Induced

ANKTM1

▶ TRPA1 Channel

Ankylosing Spondylitis

Definition

Ankylosing spondylitis is an inflammatory joint disease that is characterized by enthesitis, an inflammation at points of attachment of tendons to bone. The vertebrae may become linked by bony bridging (bamboo spine).

Cross-References

▶ Chronic Low Back Pain: Definitions and Diagnosis
▶ NSAIDs and Their Indications
▶ Sacroiliac Joint Pain

Ankylosis

Definition

Bony ankylosis occurs when bone remodeling as a result of inflammation or damage occurs, resulting in a fusion of the joint. This causes joint immobility. Fibrous ankylosis occurs when inflammation of fibrous or connective tissues of the joint results in proliferation of tissue and in reduced mobility or stiffness of the joint.

Cross-References

▶ Arthritis Model, Adjuvant-Induced Arthritis

Annulus Fibrosus

Definition

Annulus fibrosus is an outer anatomical structure of the intervertebral disc composed of fibrocartilage and fibrous tissue, delimiting the nucleus pulposus. The annulus fibrosus has a nociceptive innervation.

Cross-References

▶ Lumbar Traction
▶ Whiplash

Anorexia

Definition

Loss of appetite.

Cross-References

► Clinical Migraine with Aura

Antecedents and Consequences of Behavior

Definition

The set of factors that occurred temporally before and after a behavioral event or experience. The antecedents may contribute to an individual's expectations for the future, and the behavioral responses that are received in close proximity to an event can serve to influence subsequent responses and experiences. Thus, antecedents and consequences play a role in determining the onset, maintenance, and exacerbation of inappropriate behaviors, or contribute to appropriate and adaptive responses to similar situations and sensations in the future.

Cross-References

► Psychological [Behavioral Health] Assessment of Pain

Anterior Cingulate Cortex

Synonyms

ACC

Definition

The anterior cingulate cortex (ACC), a component of the limbic system, is an area of the brain located just above the corpus callosum. The ACC is involved in many functions, including attention, emotion, and response selection, among others. Its descending connections to the medial thalamic nuclei and the periaqueductal gray, along with evidence from brain imaging studies, also support a role for the ACC in the descending modulation of pain.

Cross-References

► Cingulate Cortex, Nociceptive Processing, Behavioral Studies in Animals
► Descending Circuits in the Forebrain, Imaging

Anterior Lumbar Interbody Fusion

► ALIF

Anterior Primary Ramus

Definition

The anterior branch of a spinal nerve that provides the nerve supply to the extremities (e.g., brachial plexus) and the chest wall.

Cross-References

► Pain Treatment: Spinal Nerve Blocks

Anterior Pulvinar Nucleus

Definition

The anterior pulvinar nucleus extends from the medial pulvinar and posterior nuclei, situated between the center median and ventral posterior nuclei.

Cross-References

▶ Thalamic Nuclei Involved in Pain, Human and Monkey

Anterior Spinothalamic Tract

▶ Paleospinothalamic Tract

Anterograde Axonal Tracer (Anterograde Labeling)

Definition

A substance (protein, enzyme) that is injected at the level of the neuronal soma. It is incorporated within the soma and then conveyed in an anterograde (orthodromic) direction in the axon up to the endings. The tracer is generally colored with a histochemical reaction, with or without an earlier immune amplification reaction.

Cross-References

▶ Parabrachial Hypothalamic and Amygdaloid Projections
▶ Spinal Dorsal Horn Pathways, Dorsal Column (Visceral)
▶ Spinohypothalamic Tract, Anatomical Organization, and Response Properties

---

Anterograde Transport

Definition

Anterograde transport is the movement of proteins away from the cell body.

Cross-References

▶ Opioid Receptor Trafficking in Pain States

Anterolateral Cordotomy

Definition

Ablation of the spinothalamic tract by open surgical section or through the application of a thermal coagulation probe.

Cross-References

▶ Cancer Pain Management, Neurosurgical Interventions
▶ Spinothalamic Neuron

Antiarrhythmics

▶ Drugs Targeting Voltage-Gated Sodium and Calcium Channels

Anticholinergics

Definition

A class of drugs also referred to as antimuscarinics that are used as smooth muscle antispasmodics and antisecretory drugs. Anticholinergic medications include the natural belladonna alkaloids (atropine and hyoscine) and synthetic and semisynthetic derivatives. The synthetic and semisynthetic derivatives are separated into tertiary amines (i.e., dicyclomine) and quaternary ammonium compounds (i.e., hyoscine butylbromide and glycopyrrolate). The quaternary ammonium compounds are less lipid soluble than the natural alkaloids and are, therefore, less likely to cross the blood-brain barrier and cause side effects such as agitation and hallucinations.

Cross-References

▶ Cancer Pain Management, Adjuvant Analgesics in Management of Pain Due to Bowel Obstruction

Anticipatory Anxiety

Definition

Anticipatory anxiety refers to fear experienced before encountering a threatening situation or experience. In relation to experimental pain, anticipatory anxiety relates to a child's perception of the extent to which the upcoming pain stimulus may lead to harm or damage to one's physical integrity.

Cross-References

▶ Experimental Pain in Children
▶ Respondent Conditioning of Chronic Pain

Anticoagulants

▶ NSAID-Induced Lesions of the Gastrointestinal Tract

Anticonvulsant (Agent)

Definition

Antiepileptics are agents that prevent or arrest seizures, and are primarily used in the management of epilepsy.

Cross-References

▶ Drugs Targeting Voltage-Gated Sodium and Calcium Channels
▶ Postoperative Pain, Anticonvulsant Medications

▶ Post-Seizure Headache
▶ Preventive Migraine Therapy

Anticonvulsant Medication

▶ Postoperative Pain, Anticonvulsant Medications

Anticonvulsants

▶ Drugs Targeting Voltage-Gated Sodium and Calcium Channels

Antidepressant Analgesics in Pain Management

J. H. Atkinson[1] and Mark D. Sullivan[2]
[1]Psychiatry Service, VA San Diego Healthcare System and Department of Psychiatry, University of California, San Diego, CA, USA
[2]Psychiatry and Behavioral Sciences, University of Washington, Seattle, WA, USA

Synonyms

First-generation antidepressants; Heterocyclic antidepressants; Monoamine oxidase inhibitors; Norepinephrine-dopamine reuptake inhibitors; Norepinephrine-dopamine transporter inhibitors; Second-generation antidepressants; Serotonin antagonist reuptake inhibitors (SARIs); Serotonin selective reuptake inhibitors (SSRIs); Serotonin-norepinephrine reuptake inhibitors (SNRIs); Tricyclic antidepressants

Definition

Prevailing dogma holds that standard antidepressant drugs exert their primary therapeutic effects either through inhibition of one or more of the high-affinity neuronal plasma membrane monoamine postsynaptic transporters, such as the norepinephrine transporter and the serotonin transporter, by blocking presynaptic receptors, or by inhibiting enzymes involved in monoamine metabolism. Whether by blocking reuptake into the presynaptic neuron, by increasing release, or inhibiting degradation, the end result is increased availability of key monoamines at the synapse.

Antidepressants can be characterized by structure (e.g., tricyclic or heterocyclic rings), function (e.g., transporter inhibition), or chronology. The advantage of definition by function is that it can assist with anticipation of adverse effects, and therefore with matching patients to a safer and more tolerable agent. Using this approach the tricyclic antidepressants (e.g., amitriptyline, imipramine) are nonselective serotonin-norepinephrine reuptake inhibitors because they block both presynaptic transport of these monoamines and they also have antimuscarinic-anticholinergic, a_1-adrenergic antagonist, and histamine (H_1) antihistaminic properties. The selective serotonin reuptake inhibitors (e.g., citalopram, fluoxetine, sertraline) primarily bind to the serotonin transporter; selective dual serotonin-norepinephrine reuptake inhibitors block both serotonin and norepinephrine reuptake transporters (e.g., venlafaxine, duloxetine) but have limited effects on other receptors; and the norepinephrine agonist/dopamine reuptake inhibitors (e.g., bupropion) selectively block these respective receptors. Another major class is the alpha 2 adrenergic receptor antagonist/serotonin receptor blocker (e.g., mirtazapine) which increase serotonin and norepinephrine neurotransmission by novel blocking of adrenergic presynaptic receptors. The final major class is the monoamine oxidase (MAO) inhibitors (e.g., selegiline) which increase serotonin, norepinephrine, and dopamine neurotransmission by preventing enzymatic degradation of these key monoamines. In terms of classification by chronology, the tricyclics and monoamine oxidase inhibitors were the first antidepressants to be routinely prescribed in clinical care, and therefore are customarily referred to as "first-generation" agents. The newer more selective drugs are described as "second generation."

Characteristics

Historical

The historical context of antidepressant analgesia was thoroughly reviewed in the First Edition (2004) and is only briefly updated here. The MAO inhibitors, the first clinically effective antidepressants, were synthesized in the early 1950s after improvement in mood was observed in depressed patients treated for tuberculosis by iproniazid, an agent with some monoamine oxidase inhibitory effects. Throughout the 1960s, they were employed in open-label trials for migraine headache based on a rationale that diminished serotonin neurotransmission was associated with cerebral vasodilation and subsequent migraine attacks. The open label nature of the trials made it difficult to evaluate efficacy, but it was argued that substantial and sustained improvement was achieved in patients who had a long record of prior treatment failures (e.g., Anthony and Lance 1969). Early agents (e.g., phenelzine) were nonselective inhibitors of MAO A and B, and the need for dietary restriction to avoid hypertensive crises, along with adverse drug-drug interactions with opioid analgesics, mitigated against their widespread use, and they were replaced in pain management by tricyclic antidepressants. Dietary restriction is not necessary with newer, selective MAO inhibitors (e.g., selegiline) but even these more tolerable agents have not been widely used or trialed for chronic pain.

The tricyclic antidepressants, so-named because their chemical structure contains three rings, were synthesized in the early 1950s and tested first as antipsychotics, based on the efficacy of other three-ringed molecules (i.e., chlorpromazine) in schizophrenia. Although they had no antipsychotic properties, an antidepressant effect was serendipitously noted, and these drugs came into psychiatric practice (e.g., imipramine, amitriptyline) by the early 1960s. Almost simultaneously case series in the French literature reported efficacy in chronic pain syndromes. Reports of efficacy in open-label and randomized trials began appearing in the mid-to late 1960s for various syndromes, including chronic "tension headache" (e.g., Lance and Curran 1964), postherpetic neuralgia (Woodforde et al. 1965), atypical facial pain (Lascelles 1966), migraine headache (e.g., Dalessio 1967), and "mixed arthritic disorders" (e.g., McDonald-Scott 1969). In the 1970s and early 1980s, initial trials reporting efficacy emerged for chronic low back pain (Jenkins 1976) and painful diabetic polyneuropathy (Turkington 1980). Several important developments characterize work from the late 1980s and throughout the 1990s to the present. First, scientific rigor of trials steadily improved. Second, it became clear that the antidepressant and analgesic effects of these agents were independent, and that treatment was efficacious for nondepressed individuals with certain chronic pain states. Third, studies were conducted to assess comparative efficacy of classical tricyclics with SSRIs (e.g., Max et al. 1992). Relatedly, clinical trials examined the effects of newly available SNRIs and NDRIs. There also was renewed interest in concentration-response effects of antidepressants, aimed at determining if higher doses or concentrations enhanced efficacy, or if instead there were therapeutic "windows" of efficacy. Fourth, the list of other potentially responsive conditions grew, to include functional bowel syndromes (e.g., see Drossman et al. 2009) and fibromyalgia (e.g., see Clauw 2008). Finally, in the last decade it has been possible to conduct systematic reviews and meta-analyses aimed at determining which chronic pain syndromes are responsive, to what drugs, and to what degree.

Pharmacodynamics

Antidepressants are agents with complex effects on neurotransmitter receptors or enzymes. A complete and adequate explanation of their mechanism of action is lacking. One theory is that antidepressants in part exert therapeutic effects by downregulating some postsynaptic receptors (e.g., β-adrenergic receptors, 5HT2a receptors), enhancing 5HT1a receptor transmission, and decreasing firing of auto-inhibitory monoamine neurons. Other evidence suggests antidepressants may have neuroprotective effects, and promote neurogenesis and synaptic

connectivity. Some antidepressants have effects on cholinergic transmission, on acid-sensing sodium and calcium channels, on glutamate and N-methyl-D-aspartate (NMDA) receptors, on sigma 1 and 2 receptors, and on histone acetylation. With regard to both pain and mood, the clinical relevance of these many effects is uncertain. Site(s) of mechanistic action are also unclear and may involve inhibitory pain pathways descending from the dorsal raphe (serotonergic) or locus coeruleus (noradrenergic) to the dorsal horn of the spinal cord.

Because of their widespread effects on multiple neurotransmitter systems, the tricyclic antidepressants, compared to more selective agents, are much more likely to be associated with adverse effects. Within the tricyclic class, the parent drugs with a tertiary amine structure (e.g., amitriptyline, imipramine) are more likely than their metabolite, the secondary amines (e.g., nortriptyline, desipramine), to have adverse effects.

There is disagreement regarding the comparative efficacy among antidepressants with selective noradrenergic effects, serotonergic effects, or a balanced combination of both actions. Historically, most research up to the early 1990s employed tricyclic agents. The parent compounds (e.g., amitriptyline, imipramine) can be said to have relatively "balanced" effects, although their primary metabolites (e.g., nortriptyline, desipramine) are predominantly noradrenergic. An influential early meta-analyses (Onghena and Van Houdenhove 1992) concluded overall that agents with "balanced" serotonin and noradrenaline action are more efficacious than selective drugs. This conclusion still echoes in the literature, but in retrospect may be an oversimplification, since the analysis included few studies of serotonin selective agents.

There are two schools of thought with reference to differential efficacy across diagnostic categories antidepressant mechanisms. One argument is that chronic pain is a homogenous condition because all pain must have a common final pathway. It follows that efficacy demonstrated in one condition can be generalized to others. An alternative argument is that pain syndromes are diagnostically distinct and that discussions of efficacy and of comparative efficacy must reference diagnostically homogeneous pain disorders. Differential responsiveness across pain disorders as determined by estimates of efficacy (e.g., number-needed-to-treat) provides evidence supporting this position. In this regard, studies of diabetic neuropathy often suggest superiority of both selective noradrenergic (desipramine) and balanced (amitriptyline, imipramine, clomipramine) agents over selective serotonergic drugs (e.g., fluoxetine, citalopram). Nevertheless, other research finds at least some analgesic action of serotonin selective agents for these same neuropathic pain diagnosis. Studies in chronic back pain suggest efficacy for more selective noradrenergic agents (nortriptyline, maprotiline) whereas an SSRI (paroxetine) is indistinguishable from placebo. Yet work in migraine headache argues for at least some efficacy of SSRIs, albeit inferior to agents with both noradrenergic and serotonergic actions. These data can make a case for the proposition that noradrenergic action is important if not essential to analgesic efficacy across various pain syndromes. But because of limitations in the field, as discussed below, it is difficult to appraise this hypothesis, or that dual action is fundamentally superior to action of either noradrenergic or serotonergic systems alone.

A related issue is whether there are dose-or concentration-response effects of antidepressant analgesia. Possibilities include (1) monotonic increases in analgesia as dose or concentration increases; (2) "thresholds" (i.e., a concentration which must be exceeded for analgesia); or (3) therapeutic "windows" of efficacy (i.e., boundaries of concentrations for efficacy, with diminished efficacy either above or below these boundaries). The question of dose-response is complicated by at least two considerations. One is that the pharmacodynamic properties of some antidepressants vary by dose: For example, venlafaxine usually is a very selective serotonin transport blocker at lower doses, but reliably also has noradrenergic effects commencing at higher doses (e.g., 225–375 mg); similarly, paroxetine is an SSRI at usual dosages, but also may have

noradrenergic effects at doses of exceeding 40 mg daily. The other consideration involves pharmacokinetics: tricyclics and SSRIs, for example, are metabolized by the Cytochrome (CYP) P450 enzyme system. Because of genetic variability in rates of metabolism (slow, medium, fast), a given dose may result in tenfold differences in concentration across individuals.

Research does not consistently reveal dose- or concentration-related effects, a failure that has been attributed to study design, since most pain trials were not designed for clinical psychopharmacology (Max 1994). Fixed dose designs are limited by CYP 450 polymorphism. Individualized dosing to response is limited by the problem that patients achieving relief at low doses or concentrations would stay at that level, while patients with severe pain would be escalated to maximum tolerable dose, thus possibly masking a concentration-response correlation. Forced titration to maximal dosage or concentration is thought to be limited by the possibility that response may occur at lower concentrations, but may be attributed to the higher doses or concentrations. The most rigorous approach is concentration-controlled trials in which individuals are randomly allocated to preassigned low, medium, or high concentrations of the study drug, and therapeutic drug monitoring is used to confirm assignments. This design is complex and rarely implemented. It is not surprising then that the main question has not been answered. It is possible that concentration-response relationships may vary by pain disorder or pathophysiology. Considering the most rigorous studies, there is some evidence for a positive concentration-response relationship in diabetic neuropathy for imipramine and paroxetine, as well as evidence for a low concentration window for both postherpetic neuralgia with amitriptyline, and for desipramine in chronic back pain. Research on the newer "second-generation" antidepressants is scant. There is some evidence for a positive concentration-response relationship for venlafaxine in diabetic neuropathy, but studies of duloxetine show no correlations for diabetic neuropathy, back pain, or fibromyalgia.

The evidence base for efficacy of antidepressants in various chronic pain states is reviewed below. There are important gaps, however, in this literature (Kroenke et al. 2009). There are very few head-to-head trials – either within antidepressant classes or between antidepressants and other analgesics; thus, little is known of comparative efficacy. There is scant data to guide initial treatment, either in terms of which antidepressants or when and at what step in the analgesic "ladder" should antidepressants be employed. For example, we know far less than we should about the efficacy of the SSRIs in particular: There has been almost no follow-up after initial studies suggested potential efficacy for certain syndromes (e.g., diabetic neuropathy). Little research examines efficacy of combination treatment with antidepressants and other analgesics (e.g., opioids, nonsteroidals, anticonvulsants), or if combination treatment allows for dose reductions compared to monotherapy. Finally, almost all studies are very short term, rarely lasting over 3 months.

Evidence-Based Studies
The evidence base described here draws on systematic reviews and meta-analyses. When summarizing comparative efficacy between antidepressants, or between antidepressants and other drug classes (e.g., anticonvulsants), both indirect and direct meta-analyses are presented, where possible. Indirect evidence consists of trials that evaluate different drugs versus placebo. Direct evidence consists of studies of head-to-head trials of the competing agents. It is argued that direct and indirect comparisons usually agree, but may diverge if there are differences between trials in terms of methodological rigor, or target populations, measurement, or other factors (Chou and Huffman 2007).

Acute Pain Studies
Early studies of first-generation antidepressants for acute experimentally induced pain and postoperative dental extraction pain yielded negative or inconsistent results. One intriguing study in nondepressed patients with acute/subacute back pain (mean duration 60 days) suggested efficacy

for high-dose amitriptyline (150 mg daily) compared to acetaminophen, but unfortunately this line of research has not been extended (Stein et al. 1996). The availability of effective analgesics for acute pain has perhaps limited interest in exploring possible acutely analgesic effects of newer antidepressants.

Cancer Pain
Cancer pain is of diverse and often complex etiology, often not diagnosable as a purely neuropathic, somatic, or visceral pain syndrome. Any one patient may experience pain both due to direct effects of tumor (e.g., invasion of viscera or nerve, metastasis to bone) or from consequences of surgery (e.g., deafferentation), radiotherapy, or chemotherapy. Most research in antidepressant analgesia in cancers focuses on neuropathic pain, but the body of research is exceedingly limited. Although there are a few small sample size studies suggesting efficacy of amitriptyline or venlafaxine for postmastectomy pain, these results are inconsistent, and there is other data suggesting no benefit for chemotherapy-induced (i.e., cisplatin) neuropathy and other neuropathic cancer pain. It is unfortunately the case that use of antidepressants for neuropathic cancer pain must be largely based on the unproven assumption that efficacy in noncancer pain (e.g., painful diabetic neuropathy or postherpetic neuralgia) is generalizable.

Chronic Nonmalignant, Nonneuropathic Pain
Chronic Low Back Pain
Chronic low back pain is one of the most prevalent and disabling conditions in clinics and the workplace. Its etiopathogenesis is uncertain, but it is thought to be a heterogeneous condition, with persistent pain attributed to both peripheral phenomenon like muscle inflammation and injury, or degenerative disease of the spine, as well as to central sensitization. Less than 5 % of chronic low back pain patients experience "sciatica" or lumbosacral radiculopathy, but this condition can markedly impair function and is treatment resistant.

Evidence for efficacy of first- and second-generation antidepressants in chronic low back pain is mixed: Expert systematic reviews and meta-analyses, often even after examining virtually the same literature, may come to differing conclusions. Some suggest there is negligible benefit with these agents (White et al. 2011), some identify only minimal effects, and others find moderate efficacy with tricyclic antidepressants (e.g., Salerno et al. 2002; Chou and Huffman 2007; Kroenke et al. 2009) but little efficacy for SSRIs. One of the newer SNRIs, duloxetine, may be effective (Kroenke et al. 2009). Only a few studies specifically assess response rates in patients with sciatica or lumbar radiculopathy, and the results are not encouraging. This uncertainty in the literature regarding the overall efficacy in chronic back pain may in part be due to difficulties in sample specification – the unclear etiology of chronic back pain is compounded by the failure of many studies to attempt to rigorously diagnose or define the study sample. This likely diagnostic heterogeneity across samples may mean some relatively more – or less – responsive syndromes (e.g., sciatica) are combined and compared, with a weak treatment effect noted. A second problem is that some reviews and meta-analyses consider all antidepressants as a single class, thus obscuring possible differential response between agents. Another obvious problem is the very small sample size in most studies, meaning that results are unstable.

Fibromyalgia

Fibromyalgia is a complex disorder characterized by persistent widespread and decreased pain as well symptoms of fatigue, stiffness, insomnia, and mood and anxiety disorders. Two meta-analyses (e.g., see Arnold 2007) suggest efficacy of antidepressants in fibromyalgia pain and overall symptom burden, but careful drug selection may be important to outcome. One meta-analysis of antidepressants in the treatment of fibromyalgia (O'Malley et al. 2000) included 13 trials of antidepressants, including amitriptyline ($N = 8$), clomipramine, and maprotiline ($N = 1$); SSRIs (fluoxetine [$N = 2$]); citalopram [$N = 1$]; and a reversible inhibitor of the MAO-A enzyme, moclobemide [$N = 1$]. Outcomes included the number of tender points, and patient ratings of

pain, sleep, fatigue, and overall well-being. The pooled results showed significant symptomatic benefit of antidepressants that was moderate for sleep, overall well-being, and pain severity, and mild for fatigue and number of tender points. Patients treated with antidepressants were more than four times as likely to improve as those on placebo. The overall positive impact of a broad spectrum of antidepressants as determined by these meta-analyses must be tempered by the problem that most studies did not exclude individuals with mood or anxiety disorders, which could confound outcomes. Nevertheless, the more careful studies which focus on patients with pain without psychiatric comorbidity still show positive effects.

The tricyclics (e.g., amitriptyline) are often described as more efficacious than second-generation agents in terms of NNT. But recent reviews suggest this apparent advantage may be an artifact of methodological deficits and a smaller number of participants enrolled in the earlier studies (Hauser et al. 2011). These reviews (Hauser et al. 2011) suggest SNRIs like duloxetine (60 or 120 mg) and milnacipran (100–200 mg) are comparable in efficacy to first-generation agents and are more tolerable. A trial of venlafaxine was negative, but the dose (<75 mg/day) was in the range where only serotonin transporter blockade would be expected. The role of SSRIs in general is unclear. One very methodologically sound trial of fluoxetine reported efficacy for nondepressed patients treated with a flexible dosing strategy (20–80 mg), but two trials of citalopram and one trial of paroxetine were negative. There is the suggestion that higher doses of SSRIs might be required. The role of MAOIs is still unclear. One randomized trial of moclobemide was negative, but a pilot trial of pirlindole suggested efficacy.

Irritable Bowel Syndrome

Irritable bowel syndrome (IBS) is a chronic relapsing and remitting disorder characterized by abdominal pain and change in bowel habit (diarrhea or constipation). The etiology is uncertain, and may involve visceral hypersensitivity to distention, gut inflammatory or immune dysfunction, stress-induced hyperalgesia, or autonomic dysfunction (Ford et al. 2009).

The overall quality of clinical trials of antidepressants for IBS has been judged to be moderately good. One recent meta-analysis reporting on both first- and second-generation agents noted that in 9 studies comparing TCAs (amitriptyline, desipramine) to placebo, TCAs were beneficial with an overall NNT of 4. Doses were generally in the lower range used in other pain syndromes (e.g., amitriptyline or nortriptyline 10–50 mg, imipramine or desipramine 10–50 mg daily).

There were five studies of SSRIs (citalopram, fluoxetine, paroxetine) compared to placebo. SSRIs were efficacious, with an NNT of 3.5. Daily doses were generally in the range used to treat major depression (e.g., citalopram 20–40 mg, fluoxetine 20–40 mg, paroxetine 20–40 mg).

Depression and anxiety symptoms often accompany IBS, but improvement in pain symptoms was not correlated with scores on depression measures, suggesting that treatment exerted an analgesic effect independent of an antidepressant effect.

There are too few studies in children and adolescents to support the use of antidepressants for irritable bowel syndrome in this population.

Rheumatoid Arthritis, Ankylosing Spondylitis, and Osteoarthritis

Pain is a prominent symptom in inflammatory and degenerative arthritides like rheumatoid arthritis and hip and knee osteoarthritis. Despite the prevalence and impact of these illnesses, only a handful of studies have assessed the analgesic effects of antidepressants: This lack is thought to be attributable to the efficacy of standard, first-line anti-inflammatory regimens. Unfortunately, most research has combined patients with any of these syndromes into heterogeneous groups, and there has been work exclusively assessing effects on hip or knee osteoarthritis, despite their growing importance in an aging population. Results suggest that first-generation drugs (amitriptyline, dothiepin, trimipramine) in

low doses may be efficacious. But most research includes patients with prominent depressive symptoms (e.g., fatigue, insomnia), making it unclear whether improvement is due to analgesic effects or antidepressive effects.

Neuropathic Pain (NP)

Neuropathic pain is an important, disabling condition affecting 5 % or more of the general population in industrialized countries. It arises from injury or disease of the peripheral or central nervous system, or a mixture of both. The most prevalent neuropathic pain syndrome is probably lumbosacral radiculopathy (its therapy has been addressed above, in Chronic Back Pain). Apart from this disorder, the most frequent conditions associated with neuropathic pain are diabetes mellitus, herpes zoster infection, HIV disease, physical trauma (e.g., spinal cord injury, amputation), multiple sclerosis, and direct or indirect (treatment-related) effects of neoplasm. Most research, and the strongest data on efficacy of selected antidepressants, addresses painful diabetic peripheral neuropathy and postherpetic neuralgia. There is much less research – and unfortunately only limited evidence of a benefit of antidepressants – for other neuropathic pain syndromes.

Painful Diabetic Peripheral Neuropathy and Postherpetic Neuralgia

As reviewed in the first edition of this encyclopedia, controlled studies conducted in the 1990s firmly established the efficacy of tricyclics as one of the "first-line" analgesics for postherpetic neuralgia and painful diabetic neuropathy (Dworkin et al. 2007). The number-needed-to-treat (NNT = 3) also established these syndromes as among the most treatment responsive of all chronic pain states. Subsequent work demonstrated that second-generation antidepressants with noradrenergic effects (e.g., SNRIs like duloxetine, higher dose venlafaxine) probably have almost comparable efficacy, and that bupropion may also be effective (Dworkin et al. 2007).

The efficacy of SSRIs is less certain. Some work in diabetic neuropathy suggests fluoxetine is equivalent to placebo; other studies in this disorder having similarly small sample sizes report efficacy with paroxetine and citalopram. It is unfortunate that the role SSRIs have not been more definitively assessed, given their safety and tolerability.

Other Neuropathic Pain: HIV Neuropathy, Spinal Cord Injury, Phantom Limb Pain

TCAs have not differed significantly from placebo in RCTs of patients with HIV neuropathy, spinal cord injury-related neuropathic pain, and phantom limb pain. Despite the importance of each of these conditions, there are very few studies using first-generation drugs and none using second-generation antidepressants.

In summary, the most responsive neuropathic pain syndromes are all small fiber neuropathies (i.e., diabetic peripheral neuropathy, postherpetic neuralgia). Large fiber neuropathies (e.g., lumbosacral radiculopathy, cisplatin-induced neuropathy, and some HIV myeloneuropathy) appear to be treatment resistant.

Practical Guidelines
Patient-Specific Factors

Enhancing the likelihood of successful treatment requires consideration of patient-specific and drug-specific factors. Key patient factors include pain diagnosis, comorbid medical conditions, concurrent medications, and expectancy. Consideration of diagnosis and co-occurring medical conditions and treatment will assist in understanding the risk-benefit ratio of a treatment trial, such that antidepressants are prescribed only for pain syndromes most likely to be responsive, and potential adverse effects are understood. Evaluation of patient expectancy and therapeutic alliance is essential so that there is agreement on a course of action and a definition of "success." Some patients are biased against antidepressant pharmacotherapy because of stigma about "psychiatric" or "mind-altering" drugs, or concerns that such prescriptions means the physician believes pain is "imaginary." Other individuals have very high expectations for analgesia (e.g., complete pain relief). Education about possible analgesic mechanism of action of antidepressants, and their role as adjuncts to a comprehensive plan to reduce pain impact on life quality and function, may assist in promoting adherence and an alliance.

Drug-Specific Factors

The initial treatment decision is whether to select a first- or second-generation agent, taking into account efficacy, cost, and adverse effects. In terms of efficacy, the first-generation agents have a longer track record of success and lower cost, but a greater likelihood of adverse effects compared to second-generation agents. Among first-generation agents, tertiary amine tricyclics (e.g., amitriptyline, imipramine) are often recommended as standard therapy for responsive syndromes (e.g., painful diabetic neuropathy), but secondary amine tricyclics (e.g., nortriptyline, desipramine) have a lower risk of adverse effects and equivalent therapeutic effects. Individuals over age 50 or with history of cardiovascular disease should have a pretreatment electrocardiogram. TCAs are associated with repolarization abnormalities (QT interval prolongation), bundle branch block, hypotension, atrial and ventricular arrhythmias, and risk of myocardial infarction. Anticholinergic effects of these agents may cause urinary retention, especially in individuals who also use opioid analgesics.

Dosing for secondary tricyclic amines is usually low (e.g., nortriptyline or desipramine 25–75 mg daily), but may need to reach full antidepressant doses (e.g., nortriptyline or desipramine 100–150 mg daily).

In the medically ill or elderly, second-generation drugs like duloxetine or higher dose venlafaxine may be preferred alternatives. Duloxetine is contraindicated in hepatic insufficiency; doses of venlafaxine should be reduced by 25–50 % in renal or hepatic insufficiency. Both duloxetine and venlafaxine may increase blood pressure. There is much less data supporting the use of SSRIs in most chronic pain syndromes, but their favorable adverse effect profile make them worthy of consideration as second- or third-line treatment. An exception to this statement may be in IBS, where efficacy of SSRIs seems on par with TCAs. The SSRIs are generally safe in renal or hepatic insufficiency; a main risk of SSRIs is rare induction of mania in individuals with bipolar disorder. Citalopram in dosages above 40 mg daily (20 mg in patients over age 60) is associated with cardiac conduction (OT_c interval) delay, suggesting other SSRIs may have this liability. When used in the third trimester of pregnancy, both SNRIs and SSRIs are associated with neonatal respiratory distress and other symptoms suggesting direct toxic or withdrawal syndromes.

Dosing for duloxetine is usually 60–120 mg daily, milnacipran 100–200 mg daily, and venlafaxine 150–325 mg daily.

Because of their ability to inhibit CYP 450 2D6, SNRIs and SSRIs may interfere with analgesic effects of codeine and may increase TCA concentrations. Migraine headache is prevalent and may be a co-occurring pain syndrome. Triptan antimigraine medications are serotonin 1 D receptor agonists and may interact with SSRIs or SNRIs to cause a serotonin syndrome of muscular rigidity, autonomic instability, and altered consciousness.

Elderly patients are at elevated risk for adverse effects due to antimuscarinic effects (e.g., delirium) or alpha-1 antagonism (e.g., hypotension) and cardiac conduction delays, as well as SSRI-induced hyponatremia.

There is insufficient evidence to support use of MAO inhibitors in chronic pain.

Management of Adverse Effects

Adverse effects may be time dependent (start immediately but often diminish over 1–2 weeks) or dose dependent (increase as dose increases). Most authorities agree there is no reason to preemptively treat anticipated adverse effects. The preferred strategy is to start antidepressants with low doses, increase slowly after 2–4 weeks, and wait for resolution of side effects. If adverse effects persist, the choice is whether to switch to a potentially more tolerable drug, or lastly, to treat the adverse effect. Effects due to alpha-1 antagonism (orthostatic hypotension) may be addressed with 9-alpha-fluorohydrocortison 0.025–0.050 mg qd; some due to alpha-2 presynaptic agonism (erectile dysfunction) may be

managed with sildenafil 50–100 mg or similar agents, while others (delayed ejaculation, anorgasmia, diminished libido) may require bupropion 50–100 mg qd; histamine (H_1) antagonism (sedation) requires bedtime dosing, switching, or dose reduction; muscarinic antagonism (dry mouth) may be reduced by sugar free candies or gum; others (constipation) may respond to stool softeners (docusate sodium 100–200 mg qd), and some (urinary retention) may require bethanechol 10–50 mg qd to qid. Although most adverse effects diminish with time, hypotension and sexual adverse effects usually do not.

Drug Switching, Combination Treatments

Chronic pain by definition is treatment resistant. Depending upon the pain diagnosis, perhaps only 50 % of patients prescribed an antidepressant will achieve satisfactory analgesia (usually defined as a 30–50 % reduction in pain intensity) on the first treatment trial. Discontinuation of an ineffective regimen of a TCA, SNRI, or SSRI is best done by tapering by 25–50 % every 5 days to reduce likelihood of anticholinergic (TCA) rebound or serotonin withdrawal syndrome. One exception is that fluoxetine can be abruptly discontinued because of its very long half-life.

There is almost no data-based guidance on whether failure of one category of antidepressants (e.g., TCA or SNRI) implies likely failure of antidepressants as a class. Most authorities would advocate switching to another class of drugs altogether if there is a clearly suitable alternative (e.g., from an antidepressant to an anticonvulsant in neuropathic pain).

Likewise there is almost no data on efficacy of combination therapy (e.g., antidepressants with anticonvulsants or with opioids). In practice polypharmacy is the norm, and the usual rationale is to build upon incremental gains achieved using one class by adding another class of drugs concurrently. There is no rationale for combining within the antidepressant class (e.g., TCA + SNRI or SSRI), and such combinations may be associated with drug interactions and marked elevations of TCA concentrations due to CYP 450 2 D6 inhibition.

Treatment resistance (i.e., failure on two or more classes of agents) may indicate nonadherence, failure to diagnose comorbid psychiatric or substance use disorder, or misdiagnosis of the target pain syndrome.

Conclusions

Selected antidepressants with noradrenergic transporter blockade are effective in selected chronic pain syndromes, and most of these antidepressants also exhibit at least some serotonin transporter blockade. Interestingly the SSRIs seem to be less often efficacious, but this category of antidepressant also has a role in certain disorders, and the field would greatly benefit by additional investigations focusing on their role. More research also is needed assess the possible role of therapeutic drug monitoring, and to identify the most effective sequences of antidepressant treatment and the efficacy of combination therapies.

References

Arnold, L. M. (2007). Duloxetine and other antidepressants in the treatment of patients with fibromyalgia. *Pain Medicine, 8*(Suppl 2), S63–S74.

Anthony, M., & Lance. (1969). Monoamine oxidase inhibition in the treatment of migraine. *Archives of Neurology, 21,* 263–268 http://onlinelibrary.wiley.com/doi/10.1002/ana.410100102/pdf.

Chou, R., & Huffman, L. H. (2007). Medications for acute and chronic low back pain: A review of the evidence for an American pain society/American college of physicians clinical practice guideline. *Annals of Internal Medicine, 147*(7), 505–514.

Clauw, D. J. (2008). Pharmacotherapy for patients with fibromyalgia. *The Journal of Clinical Psychiatry, 69* (Suppl 2), 25–29.

Dalessio, D. J. (1967). Chronic pain syndromes and disordered cortical inhibition: effects of tricyclic compounds. Dis Nerv Syst, 28, 325.

Drossman, D. A., Chey, W. D., Johanson, J. F., Fass, R., Scott, C., Panas, R., & Ueno, R. (2009). Clinical trial: lubiprostone in patients with constipation-associated irritable bowel syndrome–results of two randomized, placebo-controlled studies. *Alimentary Pharmacology & Therapeutics, 29*(3), 329–341. http://www.ncbi.nlm.nih.gov/pubmed/19006537.

Dworkin, R. H., O'Connor, A. B., Backonja, M., Farrar, J. T., Finnerup, N. B., Jensen, T. S., Kalso, E. A.,

Loeser, J. D., Miaskowski, C., Nurmikko, T. J., Portenoy, R. K., Rice, A. S., Stacey, B. R., Treede, R. D., Turk, D. C., & Wallace, M. S. (2007). Pharmacoligical management of neuropathic pain: Evidence-based recommendations. *Pain, 132*(3), 237–251.

Ford, A. C., Talley, N. J., Schoenfeld, P. S., Quigley, E. M., & Moayyedi, P. (2009). Efficacy of antidepressants and psychological therapies in irritable bowel syndrome: Systematic review and meta-analysis. *Gut, 58*(3), 367–378.

Hauser, W., Petzke, F., Uceyler, N., & Sommer, C. (2011). Comparative efficacy and acceptability of amitriptyline, duloxetine and milnacipran in fibromyalgia syndrome: A systematic review with meta-analysis. *Rheumatology, 50,* 532–543.

Jenkins, D. G., Ebbutt, A. F., & Evans, C. D. (1976). Tofranil in the treatment of low back pain. *Journal of International Medical Research, 4*(2 Suppl), 28–40. http://www.ncbi.nlm.nih.gov/pubmed/140827.

Kroenke, K., Krebs, E. E., & Bair, M. J. (2009). Pharmacotherapy of chronic pain: A synthesis of recommendations from systematic reviews. *General Hospital Psychiatry, 31*(3), 206–219.

Lance, J. W., & Curran, D. A. (1964). Treatment of chronic tension headache. *Lancet, 1*(7345), 1236–9. http://www.ncbi.nlm.nih.gov/pubmed/14152626.

Lascelles, R. G. (1966). Atypical facial pain and depression. Br J Psychiatry, 112(488), 651–659. http://www.ncbi.nlm.nih.gov/pmc/articles/PMC1840754/.

Max, M. B. (1994). Antidepressants as analgesics. In H. L. Fields & L. C. Liebeskind (Eds.), *Pharmacological approaches to the treatment of chronic pain: New concepts and critical issues* (pp. 229–246). Seattle: IASP Press.

Max, M. B., Lynch, S. A., Muir, J., Shoaf, S. E., Smoller, B., & Dubner, R. (1992). Effects of desipramine, amitriptyline, and fluoxetine on pain in diabetic neuropathy. N Engl J Med, 326(19), 1250–6.

McDonald-Scott, W. A. (1969). The relief of pain with an antidepressant in arthritis. *Practitioner, 202,* 802–7.

O'Malley, P. G., Balden, E., Tomkins, G., Santoro, J., Kroenke, K., Jackson, J. L. (2000). Treatment of fibromyalgia with antidepressants: a meta-analysis. *Journal of General Internal Medicine, 15,* 659–666.

Onghena, P., & Van Houdenhove, B. (1992). Antidepressant-induced analgesia in chronic pain: A meta-analysis of 39 placebo-controlled studies. *Pain, 49,* 205–219.

Salerno, S. M., Browning, R., & Jackson, J. L. (2002). The effect of antidepressant treatment on chronic back pain: A meta-analysis. *Archives of Internal Medicine, 162*(1), 19–24.

Stein, C., Pflüger, M., Yassouridis, A., et al. (1996). No Tolerance to Peripheral Morphine Analgesia in Presence of Opioid Expression in Inflamed Synovia. *Journal of Clinical Investigation 98,*793–799. http://www.jci.org/articles/view/118852/files/pdf.

Turkington, R. W. (1980). Depression masquerading as diabetic neuropathy. *Journal of the American Medical Association, 243*, 1147–1140.

White, A. P., Arnold, P. M., Norvell, D. C., Ecker, T., & Fehlings, M. G. (2011). Pharmacological management of chronic low back pain. *Spine, 36*, S131–S143.

Woodforde, J. M., Dwyer, B., McEwen, B. W., et al. (1965). The treatment of postherpetic neuralgia. *Medical Journal of Australia, 2*, 866–872.

Antidepressant Drugs

Definition

Antidepressant drugs are primarily used in the management of depressive disorders.

Cross-References

▶ Diabetic Neuropathy, Treatment
▶ Postoperative Pain, Antidepressants
▶ Preventive Migraine Therapy

Antidepressants in Neuropathic Pain

Soeren H. Sindrup[1] and Troels Staehelin Jensen[2]
[1]Department of Neurology, Odense University Hospital, Odense, Denmark
[2]Danish Pain Research Center, Department of Neurology, Aarhus University Hospital, Aarhus, Denmark

Definition

Neuropathic pain is pain arising as a direct consequence of a lesion or disease affecting the somatosensory nervous system. In peripheral neuropathic pain, the lesion is located in the peripheral nervous system, and painful polyneuropathies (diabetic and non-diabetic), post-herpetic neuralgia and chronic pain after surgery (e.g., post-mastectomy pain syndrome) are prominent examples of this category of neuropathic pain. Post-stroke pain, pain after spinal cord injury, and pain in multiple sclerosis represent examples of central neuropathic pain conditions.

Antidepressants are drugs primarily developed to treat depression. The antidepressants that have been found to relieve neuropathic pain are ▶ tricyclic antidepressants (TCAs), serotonin noradrenaline reuptake inhibitors (SNRIs), selective serotonin reuptake inhibitors (SSRIs) and a dopamine noradrenaline reuptake inhibitor (DNRI). Within the pain field, the important drugs in these categories are TCAs: amitriptyline, imipramine, clomipramine, nortriptyline, desipramine and maprotyline; SNRIs: venlafaxine and duloxetine; SSRIs: paroxetine, fluoxetine, citalopram and escitalopram (see Table 1).

Characteristics

TCAs were among the first ▶ evidence-based treatments for neuropathic pain and this drug class is still, together with anticonvulsants, the mainstay of treatment for this type of pain. TCAs have been tested in various peripheral and central neuropathic pain conditions and there are also data on SNRIs, SSRIs and a DNRI for some of the conditions (Sindrup et al. 2005; Finnerup et al. 2010).

Pharmacology of Antidepressants

TCAs have a genuine analgesic effect, since (1) they have analgesic efficacy in experimental pain in humans and animals; (2) relieve neuropathic pain in patients both with and without concomitant depression; and (3) have a more prompt effect at lower doses in pain than in depression (Sindrup et al. 2005). The pharmacological actions of TCAs are numerous (Table 1): inhibition of ▶ presynaptic reuptake of serotonin and noradrenaline, postsynaptic blockade of α-adrenergic and NMDA receptors, and blockade of sodium and possibly also calcium channels (Sindrup et al. 2005). All of these actions have a potential for relief of neuropathic pain, due to the specific mechanisms of this type of pain (Woolf and Mannion 1999) (Fig. 1). However, it is thought that the pain-relieving effect is mainly

Antidepressants in Neuropathic Pain, Table 1 Pharmacological profile of antidepressant drugs tried in neuropathic pain

		TCA		SNRI	DNRI	SSRI
						Fluoxetine
		Amitriptyline	Nortriptyline			
				Venlafaxine		Paroxetine
		Imipramine	Desipramine		Bupropion	
				Duloxetine		Citalopram
		Clomipramine	Maprotiline			
						escitalopram
Reuptake inhibition	Serotonin	+	−/(+)	+	−	+
	Noradrenaline	+	+	+	+	−
	Dopamine	−	−	−	+	−
Receptor Blockade	α-adrenergic	+	+	−	−	−
	H¹-histaminergic	+	+	−	−	−
	Musc. cholinergic	+	+	−	−	−
	NMDA	+	+	−	?	−
Ion channel blockade	Sodium	+	+	−/(+)	?	−/(+)/?
	Calcium	+	+	?	?	?

attributed to the TCA action on monoamines and sodium channels. The more selective antidepressants, SNRIs, SSRIs and one DNRI (bupropion), have an effect on the reuptake of amines, apparently without other actions. Therefore, the latter drug classes may only interfere with parts of the neuropathic pain mechanisms (Table 1).

Evidence

Numerous ▶ randomised, ▶ double-blind, placebo-controlled clinical trials have shown that TCAs relieve painful polyneuropathies and post-herpetic neuralgia, and a few trials have indicated that TCAs also have the potential to relieve central post-stroke pain and post-mastectomy pain syndrome (Finnerup et al. 2010). Lack of effect of the TCA amitriptyline in spinal cord injury pain in a single trial may have been caused by insufficient dosing, and a negative outcome in a study on amitriptyline in post-amputation pain could be related to inclusion of a number of patients with minimal pain. Thus, TCAs appear to be effective in central and peripheral neuropathic pain. The SNRIs venlafaxine and duloxetine relieve painful diabetic polyneuropathy, and SSRIs also apparently have a weak effect in this condition (Finnerup et al. 2010). In a study including a mixture of different types of peripheral neuropathic pain, bupropion provided astonishing pain relief (Semenchuk et al. 2001).

Efficacy of Antidepressants in Neuropathic Pain

▶ Numbers needed to treat (NNT) to obtain one patient with more than 50 % pain relief, calculated from pooled data from randomised placebo-controlled trials, is used to give a rough estimate of the efficacy of different antidepressants in peripheral and central neuropathic pain and some of their subcategories (Table 2) (Sindrup et al. 2005; Finnerup et al. 2010). For TCAs, the NNT is 4.0 (CI 2.6–8.5) in central pain and 2.3 (2.1–2.7) in peripheral neuropathic pain, and there are only minor differences between the efficacy of TCAs in different peripheral neuropathic pain conditions. The SNRIs venlafaxine and duloxetine seem to have slightly lower efficacy than TCA in painful polyneuropathy. The SSRIs have been tested in painful diabetic polyneuropathy and appear to have rather low efficacy, with an NNT value of 6.8.

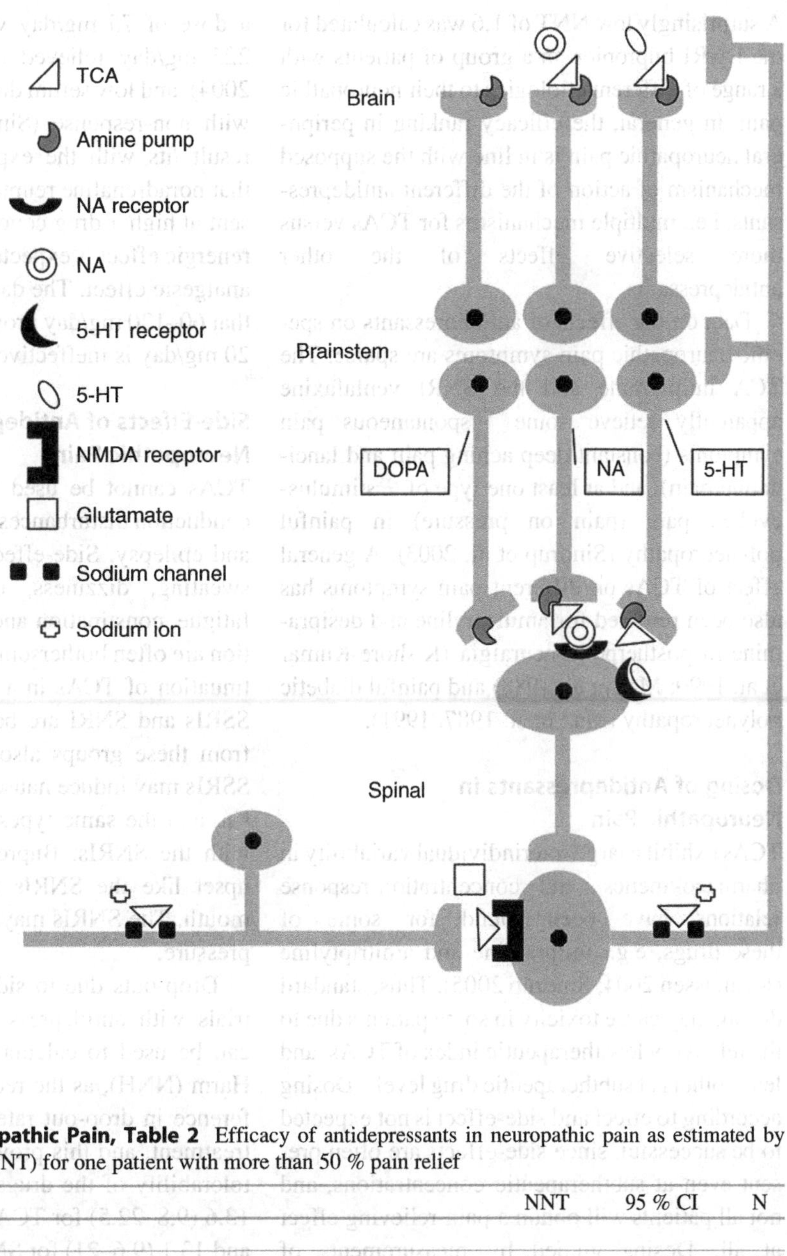

**Antidepressants in Neuropathic Pain,
Fig. 1** Mechanisms and
sites of action of tricyclic
antidepressants (*TCA*) in
neuropathic pain on
peripheral nerves, in the
dorsal horn of the spinal
cord and at supraspinal
levels. *NA* noradrenaline;
5-HT serotonin; *DOPA*
dopamine; *NMDA*
N-methyl-D-aspartate

◿ TCA

🌛 Amine pump

◡ NA receptor

◎ NA

🌙 5-HT receptor

◖ 5-HT

▐ NMDA receptor

▢ Glutamate

▪▪ Sodium channel

◌ Sodium ion

Brain

Brainstem

DOPA / NA 5-HT

Spinal

Antidepressants in Neuropathic Pain, Table 2 Efficacy of antidepressants in neuropathic pain as estimated by
Numbers Needed to Treat (NNT) for one patient with more than 50 % pain relief

	NNT	95 % CI	N
Peripheral neuropathic pain			
TCAs	2.3	2.1–2.7	397
Serotonergic and noradrenergic TCAs (Amitriptyline, imipramine, clomipramine)	2.2	1.9–2.6	232
Noradrenergic TCAs (desipramine, nortriptyline, maprotiline)	2.5	2.1–3.3	165
DNRI (bupropion)	1.6	1.3–2.1	41
SNRI (venlafaxine, duloxetine)	5.0	3.9–6.8	435
SSRI (fluoxetine, paroxetine, citalopram, escitalopram)	6.8	3.9–27	122
Central neuropathic pain			
TCAs	4.0	2.6–8.5	59

TCA tricyclic antidepressants, *DNRI* dopamine and noradrenaline reuptake inhibitor, *SNRI* serotonin noradrenaline
reuptake inhibitor, *SSRI* selective serotonin reuptake inhibitor, *N* number of patients exposed to active treatment in the
underlying trials

A surprisingly low NNT of 1.6 was calculated for the DNRI bupropion in a group of patients with a range of different etiologies to their neuropathic pain. In general, the efficacy ranking in peripheral neuropathic pain is in line with the supposed mechanism of action of the different antidepressants, i.e., multiple mechanisms for TCAs versus more selective effects of the other antidepressants.

Data on the effects of antidepressants on specific neuropathic pain symptoms are sparse. The TCA imipramine and the SNRI venlafaxine apparently relieve some ► spontaneous pain symptoms (constant deep aching pain and lancinating pain), and at least one type of ► stimulus-evoked pain (pain on pressure) in painful polyneuropathy (Sindrup et al. 2003). A general effect of TCAs on different pain symptoms has also been reported for amitriptyline and desipramine in postherpetic neuralgia (Kishore-Kumar et al. 1990; Max et al. 1988) and painful diabetic polyneuropathy (Max et al. 1987, 1991).

Dosing of Antidepressants in Neuropathic Pain

TCAs exhibit a large interindividual variability in pharmacokinetics, and concentration-response relations have been found for some of these drugs, e.g., imipramine and amitriptyline (Rasmussen 2004; Sindrup 2005). Thus, standard dosing may cause toxicity in some patients due to the relatively low therapeutic index of TCAs, and leave others at subtherapeutic drug levels. Dosing according to effect and side-effect is not expected to be successful, since side-effects are often present even at subtherapeutic concentrations, and not all patients will obtain a pain-relieving effect at all. Dosing guided by measurements of serum drug concentrations (► therapeutic drug monitoring) is suggested to improve therapeutic outcome, i.e., a start dose of 25–50 mg/day and dose adjustment according to a drug level measured after 2–3 weeks on the start dose.

The pharmacokinetics of SNRIs, DNRI and SSRIs show less interindividual variability and the therapeutic index is probably higher. Dosing according to effect and side-effects is therefore feasible. The studies on venlafaxine showed that a dose of 75 mg/day was ineffective, whereas 225 mg/day relieved pain (Rowbotham et al. 2004), and low serum drug levels were associated with non-response (Sindrup et al. 2003). This result fits with the experimental data showing that noradrenaline reuptake inhibition is first present at higher drug concentration, and the noradrenergic effect is expected to be important for the analgesic effect. The data on duloxetine indicate that 60–120 mg/day provides pain relief, whereas 20 mg/day is ineffective (Goldstein et al. 2005).

Side-Effects of Antidepressants in Neuropathic Pain

TCAs cannot be used in patients with cardiac conduction disturbances, cardiac incompensation and epilepsy. Side-effects including dry mouth, sweating, dizziness, orthostatic hypotension, fatigue, constipation and problems with micturition are often bothersome and will lead to discontinuation of TCAs in a number of patients. The SSRIs and SNRI are better tolerated, but drugs from these groups also cause side-effects. The SSRIs may induce nausea, vomiting and dyspepsia, and the same types of side-effects are seen with the SNRIs. Bupropion may cause gastric upset like the SNRIs and like the TCAs dry mouth. The SNRIs may also lead to rising blood pressure.

Drop-outs due to side-effects during clinical trials with antidepressants in neuropathic pain can be used to calculate ► Number Needed to Harm (NNH), as the reciprocal value of the difference in drop-out rates on active and placebo treatment, and this provides a rough estimate of tolerability of the drugs. The overall NNHs are 13.6 (9.8–22.5) for TCAs, 19 (8.1–∞) for SSRIs and 13.1 (9.6–21) for SNRIs. The somewhat better tolerability of SSRIs and SNRIs than of TCAs is reflected in these figures. Treatment discontinuation may be more frequent in daily clinical practice than in the setting of a clinical trial.

Discussion and Conclusion

To summarize, TCAs and SNRIs are evidence-based treatments of peripheral neuropathic pain and TCAs appear to be more efficacious than SNRIs. SSRIs relieve peripheral neuropathic pain with low efficacy, whereas a limited amount of data indicates that the DNRI bupropion could be very effective for this type of pain. TCAs may work for central pain, whereas none of the other antidepressants have been tried for this category of neuropathic pain. Thus, antidepressants are, together with anticonvulsants, first line treatments for peripheral (TCAs and SNRIs) and central (TCAs) neuropathic pain. Our present knowledge does not allow us to predict which patients with neuropathic pain will respond to treatment with antidepressants.

References

Finnerup, N. B., Sindrup, S. H., & Jensen, T. S. (2010). The evidence for pharmacological treatment of neuropathic pain. *Pain, 150*, 573–581.

Goldstein, D. J., Lu, Y., Detke, M. J., et al. (2005). Duloxetine vs. placebo in patients with painful diabetic neuropathy. *Pain, 116*, 109–118.

Kishore-Kumar, R., Max, M. B., Schafer, S. C., et al. (1990). Desipramine relieves postherpetic neuralgia. *Neurology, 47*, 305–312.

Max, M. B., Culnane, M., Schafer, S. C., et al. (1987). Amitriptyline relieves diabetic neuropathy in patients with normal and depressed mood. *Neurology, 37*, 589–596.

Max, M. B., Schafer, S. C., Culnane, M., et al. (1988). Amitriptyline, but not lorazepam, relieves postherpetic neuralgia. *Neurology, 38*, 1427–1432.

Max, M. B., Kishore-Kumar, R., Schafer, S. C., et al. (1991). Efficacy of desipramine in painful diabetic neuropathy: A placebo-controlled trial. *Pain, 45*, 3–9.

Rasmussen, P. V., Jensen, T. S., Sindrup, S. H., et al. (2004). TDM-based imipramine treatment in neuropathic pain. *Therapeutic Drug Monitoring, 26*, 352–360.

Rowbotham, M. C., Goli, V., Kunz, N. R., et al. (2004). Venlafaxine extended release in the treatment of painful diabetic polyneuropathy: A double-blind, placebo-controlled study. *Pain, 110*, 697–706.

Semenchuk, M. R., Sherman, S., & Davis, B. (2001). Double-blind, randomized trial of bupropion SR for the treatment of neuropathic pain. *Neurology, 57*, 1583–1588.

Sindrup, S. H., Bach, F. W., Madsen, C., et al. (2003). Venlafaxine versus imipramine in painful polyneuropathy. A randomized, controlled trial. *Neurology, 60*, 1284–1289.

Sindrup, S. H., Otto, M., Finnerup, N. B., et al. (2005). Antidepressants in neuropathic pain. *Basic and Clinical Pharmacology and Toxicology, 96*, 399–409.

Woolf, C. J., & Mannion, R. J. (1999). Neuropathic pain. Aetiology, symptoms, mechanisms and management. *Lancet, 353*, 1959–1964.

Antidromic Activation/Invasion

Definition

Eliciting action potentials in the axon of a neuron, which propagate toward the cell body to invade the soma and the dendrites, in an opposite direction to that observed when the neurons are naturally excited (orthodromic direction). The stimulation of an axonal ending triggers a potential that is conveyed in the antidromic direction. The recognition of an antidromic potential on three criteria (latency stability, ability to follow high-frequency stimulation, and observation of collision between orthodromic and antidromic potential) permitted the identification of one projection of a recorded neuron.

Cross-References

- ▶ Corticothalamic and Thalamocortical Interactions
- ▶ Nociceptor, Fatigue
- ▶ Nociceptors in the Orofacial Region (Temporomandibular Joint and Masseter Muscle)
- ▶ Parabrachial Hypothalamic and Amygdaloid Projections
- ▶ Spinohypothalamic Tract, Anatomical Organization, and Response Properties
- ▶ Spinothalamic Neuron

Antidromic Microstimulation Mapping

Definition

Antidromic microstimulation is a technique that can be used to map the locations of the cell bodies of origin of a nervous system pathway. An electrical stimulus is applied through a microelectrode that is inserted into a nervous system region of interest. The stimulus intensity is kept minimal to prevent stimulus spread. A series of microelectrode tracks are made transversely across a region suspected to contain the cells of origin of the pathway terminating near the stimulating electrode. Recordings are made through this electrode so that antidromically activated neurons can be identified. The stimulating and recording sites are reconstructed after the experiment, often with the assistance of electrolytic lesions or other types of marks made by passing current through the electrodes.

Cross-References

▶ Spinothalamic Input: Cells of Origin (Monkey)

Antiepileptic

▶ Postoperative Pain, Gabapentin

Antiepileptic Agents

▶ Postoperative Pain, Anticonvulsant Medications

Antiepileptic Drugs (Agents)

Definition

Antiepileptic drugs are primarily used in the management of epilepsy.

Cross-References

▶ Diabetic Neuropathy, Treatment
▶ Postoperative Pain, Anticonvulsant Medications
▶ Postoperative Pain, Gabapentin

Antihyperalgesic Effect

Definition

An effect leading to the attenuation of hyperalgesia, usually produced by surgical or pharmacological methods.

Cross-References

▶ Muscle Pain Model, Inflammatory Agents-Induced
▶ NSAIDs, Mode of Action
▶ Opioid Modulation of Nociceptive Afferents in Vivo

Anti-Inflammatories

▶ NSAIDs, Survey

Anti-inflammatory Cytokines

Definition

Cytokines involved in negatively regulating the inflammatory response.

Cross-References

▶ Cytokines, Regulation in Inflammation

Antinociception

Definition

Attenuation of nociceptive processing in the nervous system, due either to an attenuation of synaptic nociceptive transmission or an increased inhibition of nociceptive transmission.

Cross-References

▶ Cell Therapy in the Treatment of Central Pain
▶ Cytokines, Effects on Nociceptors
▶ Dietary Variables in Neuropathic Pain
▶ GABA Mechanisms and Descending Inhibitory Mechanisms
▶ Nitrous Oxide Antinociception and Opioid Receptors
▶ Opioids in the Spinal Cord and Modulation of Ascending Pathways (N. gracilis)
▶ Secondary Somatosensory Cortex (S2) and Insula, Effect on Pain-Related Behavior in Animals and Humans
▶ Stimulation-Produced Analgesia
▶ Vagal Input and Descending Modulation

Antinociceptive Effects of General Anesthetics

Definition

Nociceptors are inhibited to varying degrees when under anesthesia.

Cross-References

▶ Thalamic Nuclei Involved in Pain: Cat and Rat

Antinociceptive Models

Definition

More properly named "experimental pain models." A model might consist of tests performed in animals or in healthy human volunteers. It is useful for testing antinociception when the test-results are indicative of antinociceptive actions, e.g., by drugs. Animal models of experimental pain include the tail flick test, ▶ Hot Plate Test (Assay), hot water tail withdrawal, paw pressure, and others. In all cases, a measured nociceptive stimulus of a thermal, chemical, or

pressure nature is applied and the response of the animal or human subject is monitored. The comparison of behavioral reactions under medication and placebo serves as measure of antinociception.

Cross-References

▶ Nitrous Oxide Antinociception and Opioid Receptors

Antiphospholipid Syndrome

Definition

Diagnosis with the detection of lupus anticoagulant and IgG-anticardiolipin antibodies; primary or secondary in collagen vascular disease (SLE).

Cross-References

▶ Headache Due to Arteritis

Antipyretic Analgesics

▶ NSAIDs and Their Indications

Antisense Oligonucleotide

Synonyms

ASO

Definition

A DNA sequence, typically 15–25 nucleotides in length, designed to bind to a complementary sequence on a target RNA molecule. As a result, the protein product coded by that particular RNA is not synthesized. ASO can be delivered in vitro or in vivo to reversibly inhibit the synthesis of a protein of interest.

Cross-References

▶ Purine Receptor Targets in the Treatment of Neuropathic Pain

Anxiety

Definition

Anxiety is the subjective feeling of apprehension, dread, or foreboding accompanied by a variety of autonomic signs and symptoms, with or without a stressful situation. The focus of anticipated danger may be internal or external. The state of anxiety places defensive physiological mechanisms in a heightened state, facilitating the fight-flight response in case the threatening event occurs. Anxiety is often distinguished from fear of a concrete danger. Anxiety is more reflective of a threat that is not immediately apparent.

Cross-References

▶ Amygdala, Pain Processing and Behavior in Animals
▶ Fear and Pain
▶ Pain in the Workplace, Risk Factors for Chronicity, and Psychosocial Factors

Anxiety Sensitivity

Definition

Anxiety sensitivity refers to the fear of anxiety symptoms arising from the belief that anxiety has harmful somatic, psychological, and social consequences.

Cross-References

▶ Fear and Pain

Anxiolysis

▶ Minimal Sedation

Apamin

Definition

Bee venom blocking the SK type of Ca^{2+}-activated K^+ channels.

Cross-References

▶ Mechano-insensitive C-Fibers, Biophysics

Apophyseal Joint

▶ Zygapophyseal Joint

Apoptosis

Synonyms

Programmed cell death

Definition

Apoptosis is a type of cell death in which the cell uses a specialized cellular machinery to kill itself; it is also called programmed cell death. It is a physiological process of the organism to eliminate 'damaged or overaged cells.

Cross-References

▶ NSAIDs and Cancer
▶ NSAIDs, COX-Independent Actions

Apoptotic Degeneration

Definition

Programmed cell death, which involves a tightly controlled death pathway. It helps avoid tissue inflammation, which usually accompanies cell death through cell damage.

Cross-References

▶ GABA and Glycine in Spinal Nociceptive Processing

Application Route

▶ Opioid Therapy in Cancer Pain Management, Route of Administration

Appraisal

Definition

The mental act of evaluating the significance of a particular symptom, situation, or outcome, or the assessment of the threat value of a particular symptom or stimulus.

Cross-References

▶ Catastrophizing
▶ Psychology of Pain, Assessment of Cognitive Variables
▶ Stress and Pain

APS

▶ Acute Pain Service

APT

▶ Acute Pain Team

Arachidonic Acid

Definition

Arachidonic acid is a C_{20} carboxylic acid with four isolated double bonds at positions 5, 8, 11, and 14. This esterified fatty acid is released from phospholipids in cell membranes by the action of phospholipase A_2, activated by pro-inflammatory cytokines. Further enzymatic processing of arachidonic acid results in the production of a range of prostanoids (prostaglandins and thromboxanes). This includes PGE_2 (this has a role in limiting inflammation by inhibiting production of some cytokines such as interleukin-1) and TXA_2 (involved in platelet aggregation and hemostasis). Metabolites are named "eicosanoids" referring to the common structural feature of 20 carbon atoms.

Cross-References

▶ Coxibs and Novel Compounds, Chemistry
▶ Cyclooxygenases in Biology and Disease
▶ NSAIDs, Chemical Structure and Molecular Mode of Action
▶ Postoperative Pain, COX-2 Inhibitors

Arachnoid Membrane

Definition

The arachnoid membrane is a delicate, nonvascular membrane that is closely attached to the outermost layer, the dura mater. The epidural space surrounds the dura mater sac.

Cross-References

▶ Postoperative Pain, Intrathecal Drug
  Administration

Archispinothalamic Tract

Definition

Part of the paleospinothalamic tract, it is an intersegmental nerve fiber tract that travels for 2–4 segments.

Cross-References

▶ Parafascicular Nucleus, Pain Modulation

ARDS

Synonyms

Adult respiratory distress syndrome

Definition

ARDS is a severe form of acute lung failure requiring mechanical ventilation.

Cross-References

▶ Pain Control in Children with Burns

Area Postrema

Definition

One of the circumventricular organs interfacing between the brain and cerebral spinal fluid. Receives nerve fibers from the solitary nucleus, spinal cord, and adjacent areas of the medulla.

Cross-References

▶ Brainstem Subnucleus Reticularis Dorsalis
  Neuron

Arousal

Definition

Arousal is both a behavioral and an electroencephalographic response to a variety of strong stimuli, including painful ones. During arousal, there is a heightened level of conscious awareness.

Cross-References

▶ Spinothalamic Tract Neurons, Descending
  Control by Brain Stem Neurons

Arterial Spasm

Definition

Arterial constriction, vasospasm.

Cross-References

▶ Primary Exertional Headache

Arthralgias

Definition

Neuralgic pain in a joint or joints.

Cross-References

Arthritis

Definition

Arthritis is defined as inflammation of a joint, usually a synovial joint. Arthritis can be acute or chronic. Examples are gout and rheumatoid arthritis. Symptoms of arthritis are swelling, warmth, pain, and loss of function, in particular in arthritis with destructive components. Gout and rheumatoid arthritis are primarily inflammatory diseases which are characterized by synovitis. Gout is usually observed in a single joint, whereas rheumatoid arthritis is a polyarthritis that involves peripheral and proximal joints with a symmetric distribution. An inflammatory component is also typical for many cases of osteoarthritis. The hallmark of osteoarthritis is the progressive destruction and loss of cartilage, but more recently strong inflammatory components are described in osteoarthritis. However, while rheumatoid arthritis is a systematic inflammation, osteoarthritis is usually localized to single joints. Histologically, arthritic joints show an inflammatory cell infiltrate, and the types of infiltrating cells depend on the stage and the type of inflammation. At acute stages the infiltrate consists usually of granulocytes but at the chronic stage monocytes may prevail. Depending on the type of arthritis the bone shows considerable changes. While osteoarthritis is characterized by new bone formation, rheumatoid arthritis is characterized by bone loss near the joint.

Cross-References

Arthritis Model, Adjuvant-Induced Arthritis

N. L. Chillingworth[1] and L. F. Donaldson[2]
[1]School of Physiology and Pharmacology, University of Bristol, Bristol, UK
[2]School of Life Sciences, University of Nottingham, Nottingham, UK

Synonyms

Adjuvant arthritis; Adjuvant-induced arthritis

Definition

Adjuvant-induced ▶ arthritis is a model of chronic immune-mediated joint inflammation that is induced by injection, usually sub- or intradermally, of a suspension of heat killed *Mycobacterium tuberculosis* (▶ Mycobacterium Species) in oil (▶ Freund's Complete Adjuvant or FCA).

Characteristics

The classical model of adjuvant-induced polyarthritis is induced in rats, using an intradermal injection of mycobacterium tuberculosis suspension in paraffin oil at the tail base. The reaction to adjuvant injection is generally one of systemic illness, with inflammation affecting tarsal, carpal, phalangeal, and spinal joints after 11–16 days (Pearson and Wood 1959). Arthritis is accompanied by lesions of the eyes, ears, nose, skin, and genitals, in addition to anorexia and profound weight loss. The disease follows a relapsing-remitting course after the initial 2 weeks and may persist for several months (Pearson and Wood 1959).

The appearance of the arthritis is very similar to that of rheumatoid arthritis in humans, and for this reason this model has been used as an animal model of rheumatoid arthritis, in studies of both disease mechanisms and in the development of potential analgesic drugs (Rainsford 1982). Gross lesions in animals with adjuvant arthritis are seen as edematous swellings of multiple joints, particularly the tibiotarsal joints of the hindpaws. As the disease progresses, periarticular swellings develop in the hind limbs and tail. Persistent disease over several months may ultimately result in chronic joint deformation. Microscopic features of adjuvant arthritis are apparent before the gross lesions. As the disease progresses there are signs of joint destruction, with joints showing new bone formation, synovitis, inflammation of the bone marrow, and fibrous and bony ▶ ankylosis. Joint destruction is thought to be a result of the production of autoantibodies, possibly as a result of cross-reactivity of antibodies against mycobacterial proteins with host proteoglycans (van Eden et al. 1985) in response to the FCA injection.

Behaviorally rats show weight loss, reduced mobility, increased vocalization, and irritability (Pearson and Wood 1959; De Castro Costa et al. 1981). Animals also exhibit signs of chronic pain, such as altered ▶ nociceptive thresholds and increased self-administration of analgesic drugs (Colpaert et al. 1982). Adjuvant arthritis has also been used as a model of chronic stress as animals show increased corticosterone secretion, loss of diurnal rhythm of secretion, and other parameters of increased physiological stress, such as increased adrenal and splenic weight, and decreased thymic weight (Sarlis et al. 1992).

Although classical adjuvant polyarthritis has been considered to be a good model of rheumatoid arthritis, the original model has been modified by several groups to reduce the severity of the disease and hence the potential suffering of the animals, in line with ethical recommendations on the reduction in the severity of animal models of human disease.

Adjuvant arthritis has been modified by (a) reduction of the amount of mycobacterium injected and (b) the route of injection of the adjuvant. Injection of adjuvant into one footpad has been used to induce a localized arthritis, but this model can result in more widespread inflammation if not carefully controlled. Refinement of classical adjuvant arthritis has led to

definition of models of unilateral arthritis that affects only one joint, rather than the polyarthritis seen in the original model. This type of model has several advantages, in that principally it enables study of a limited arthritis without the complications of systemic disease seen in polyarthritis. The advantage of an internal uninflamed control joint contralateral to the arthritic joint was thought to be an added advantage of this model, until the limitations of this approach were identified (see below).

Modified Adjuvant-Induced Arthritis Models

One of the most commonly used refined models of adjuvant arthritis is that in which FCA is injected locally around a joint in which arthritis is to be induced (Donaldson et al. 1993). Intra-articular injection of FCA is also possible and also results in a stable and reliable monoarthritis (Butler et al. 1992), however, when the tibiotarsal joint is used, such intra-articular injection is complicated, as the joint space is small. Intra-articular injection of FCA in larger joints, such as the knee joint, also gives a reliable arthritis.

Injection of FCA into the skin around the tibiotarsal joint results in a reproducible arthritis after 14 days, which is maintained as a unilateral arthritis for at least 60 days postinjection (Donaldson et al. 1993). Gross features of this monoarthritis include tibiotarsal joint swelling, often resulting in a near doubling in the circumference of the joint (Fig. 1a), with cutaneous erythema and occasional breakdown of the skin over the joint. Mobility of the animals and use of the inflamed paw is only slightly altered, and most animals continue to show normal exploratory behaviors, although there is significant mechanical allodynia in the inflamed paw (Fig. 1b). Weight gain of the animals is also normal.

Histologically, the affected joint shows most of the features seen in classical adjuvant arthritis, except for the more severe aspects such as ankylosis. Inflammatory infiltrate into bone marrow, joint space, and synovium is seen, as is synovial hyperplasia and pannus (Fig. 2). There are no obvious changes in the contralateral tibiotarsal

A

Arthritis Model, Adjuvant-Induced Arthritis, Fig. 1 (a) Joint circumferences of rats injected with FCA around one tibiotarsal joint (■v) or control animals injected with vehicle (○μ). FCA results in a significant increase in joint circumference ($^\dagger p < 0.001$) over the 10 days of study. (b) Change in 50 % mechanical withdrawal threshold in FCA-induced monoarthritis. Graphs show the withdrawal thresholds in the FCA-injected joint (v■), the contralateral uninflamed paw (θ□), and in control, uninjected rats (μ○). There is a significant decrease in withdrawal threshold seen in FCA-induced arthritis ($^\dagger p < 0.001$). (c) Contralateral joint circumferences in rats with FCA-induced monoarthritis. The contralateral joint remains unaffected by arthritis until day 10, when there is a significant increase in joint swelling in FCA-injected (■v) but not vehicle-injected rats (○μ). Note that the degree of swelling is not as great as in the FCA-injected limb ($^\dagger p < 0.001$)

joint, either at the gross or histological level (but see also below).

This modified adjuvant monoarthritis also results in decreased mechanical nociceptive thresholds in the inflamed limb (Fig. 1b), but not in the contralateral limb, and thus this model has been used in studies of chronic pain. Not surprisingly, in an animal in chronic discomfort, rats do exhibit some signs of stress, but these are extremely mild and only include a loss of diurnal variation in corticosterone secretion with no effect on other parameters associated with chronic hypothalamic-pituitary-adrenal axis activation (Donaldson et al. 1994).

Thus, a modified adjuvant monoarthritis is a commonly used alternative to the classical adjuvant polyarthritis. Monoarthritis has similar features to polyarthritis and models rheumatoid arthritis well, but with fewer confounding features.

Arthritis Model, Adjuvant-Induced Arthritis, Fig. 2 Histological appearance of FCA-induced monoarthritis and control joint. (**a**) Photomicrograph of the right (uninjected) tibiotarsal joint showing normal cartilage (*C*), subchondral bone (*B*), and synovium (*S*). Bone marrow spaces (*BM*) can also be seen.

(**b**) Photomicrograph of the left injected tibiotarsal joint showing hyperplastic synovium (*HS*) and inflammatory infiltrate in the joint (*I*). There is early cartilage and subchondral bone destruction evident (*arrow*) with more advanced pannus seen invading subchondral bone (*P*). Scale bars = 100 μm

Neurogenic Inflammation in Adjuvant Arthritis

The injection of FCA into the tail base of rats results in arthritis that affects multiple joints. The spread of arthritis is not due to a spread of mycobacterium from the site of injection to the joints, but rather to an activation of the immune system resulting in a systemic delayed hypersensitivity reaction. Whole body irradiation and ablation of active T lymphocytes delays the onset of adjuvant polyarthritis (Wakesman et al. 1960), but does not abolish it completely.

However, it has been hypothesized that immune activation alone cannot explain the precise symmetry often seen in both clinical and experimental arthritis. Damage to the peripheral nervous system in adjuvant polyarthritis can result in the sparing of specific joints, implying that the development of arthritis in this model is dependent on an interaction between an intact nervous system and the immune system (Donaldson et al. 1995). This suggests that the involvement of multiple joints in the tail base

model is not purely an immune-mediated effect, but that ▶ neurogenic inflammation is involved in arthritis. The precise nervous pathway through which signals are transmitted, which results in the spread of arthritis from one joint to another, is not yet known. It is, however, known that in addition to the peripheral nerves being integral to this effect (Donaldson et al. 1995), spinal mechanisms are also important as damage to the appropriate spinal cord segment will also stop contralateral joint damage (Decaris et al. 1999).

In modified models of adjuvant arthritis, local injection of FCA around the joint can also result in neurogenic spread of arthritis to the contralateral tibiotarsal joint after 10–14 days (Fig. 1c). This effect is dependent on the amount of adjuvant injected, that is, greater amounts of adjuvant result in a more distant spread of disease (Donaldson et al. 1993). For this reason, the use of the contralateral limb/joint as an internal control is often inappropriate in monoarthritic models, as there may be covert arthritis in the contralateral joint that may affect behavioral (pain behaviors) or physiological parameters (neuronal activity).

Adjuvant Arthritis in Other Experimental Animals

FCA is used as an immunological adjuvant in other species to enhance autoimmune reactions to co-injected antigens, such as in ovalbumin-induced arthritis in rabbits (Pettipher and Henderson 1988), where cell-mediated immunity is required for full development of the disease. Adjuvant polyarthritis or monoarthritis has been very difficult to induce in species other than the rat using FCA alone, rather than as an adjuvant for immunization. Guinea pigs form granulomas at the site of adjuvant injection and do not develop polyarthritis, but do develop a monoarthritis when FCA is injected into the hindpaw (Hood et al. 2001).

The mouse is a species in which it has been notoriously difficult to induce adjuvant arthritis. There are very few reports of adjuvant arthritis in mice, and those that have attempted to induce arthritis in this species have had limited success. Tail base injection of FCA does not induce widespread arthritis in mice (Larson et al. 1986), and local FCA injection in mice does not reliably induce arthritis in all animals (Ratkay et al. 1994). In addition, altered nociception in adjuvant inflammation in mice is also inconsistent (Larson et al. 1986). Recent work has, however, defined an adjuvant arthritis model in mice that is reliable both in terms of the consistent induction of arthritis in all animals and in which all animals exhibit thermal hyperalgesia and mechanical allodynia similar to that seen in rats (Chillingworth and Donaldson 2003; Gauldie et al. 2004). In adjuvant arthritis in mice, thermal hyperalgesia and mechanical allodynia develop very rapidly (within 24 h) and are maintained for at least 15 days (Chillingworth and Donaldson 2003; Gauldie et al. 2004). This model requires the use of very much higher concentrations of FCA than those usually used to induce monoarthritis in rats (25 mgkg^{-1} in mice vs. 0.6 mgkg^{-1} in rats). Probably as a result of the relative

resistance of mice to immune stimulation by FCA, arthritis remains unilateral in mice for at least 20 days, despite the much higher concentration of FCA used, with no apparent signs of contralateral inflammation. The reasons for this apparent resistance in mice and other species to the arthritic effects of FCA are unknown, but it is probably attributable to differences in the immune reactions to the mycobacterium antigens (Audibert and Chedid 1976).

Thus, a reliable model of adjuvant arthritis now also exists in mice that can be used for similar purposes as that in rats, but can also be used to extend studies on disease progression and modification to include the use of genetically modified mice.

Adjuvant polyarthritis is still used as a model of rheumatoid arthritis, but has confounding features such as poor animal health. Modifications of adjuvant polyarthritis to limit the disease to a single joint have improved this model, from both the animal welfare perspective and in the ease of data interpretation. This model is now established in both rats and mice, allowing study of arthritis and inflammatory nociception in the two most commonly used experimental species.

References

Audibert, F., & Chedid, L. (1976). Adjuvant disease induced by mycobacteria, determinants of arthritogenicity. *Agents and Actions, 6*, 75–85.

Butler, S. H., Godefroy, F., Besson, J. M., & Weil-Fugazza, J. (1992). A limited arthritic model for chronic pain studies in the rat. *Pain, 48*, 73–81.

Chillingworth, N. L., & Donaldson, L. F. (2003). Characterisation of a Freund's complete adjuvant-induced model of chronic arthritis in mice. *Journal of Neuroscience Methods, 128*, 45–52.

Colpaert, F. C., Meert, T., De Witte, P., & Schmitt, P. (1982). Further evidence validating adjuvant arthritis as an experimental model of chronic pain in the rat. *Life Sciences, 31*, 67–75.

De Castro Costa, M., De Sutter, P., Gybels, J., & Van Hees, J. (1981). Adjuvant-induced arthritis in rats: A possible animal model of chronic pain. *Pain, 10*, 73–185.

Decaris, E., Guingamp, C., Chat, M., Philippe, L., Grillasca, J. P., Abid, A., et al. (1999). Evidence for neurogenic transmission inducing degenerative cartilage damage distant from local inflammation. *Arthritis and Rheumatism, 42*, 1951–1960.

Donaldson, L. F., Seckl, J. R., & McQueen, D. S. (1993). A discrete adjuvant-induced monoarthritis in the rat: Effects of adjuvant dose. *Journal of Neuroscience Methods, 49*, 5–10.

Donaldson, L. F., McQueen, D. S., & Seckl, J. R. (1994). Endogenous glucocorticoids and the induction and spread of monoarthritis in the rat. *Journal of Neuroendocrinology, 6*, 649–654.

Donaldson, L. F., McQueen, D. S., & Seckl, J. R. (1995). Neuropeptide gene expression and capsaicin-sensitive primary afferents: Maintenance and spread of adjuvant arthritis in the rat. *The Journal of Physiology, 486*, 473–482.

Gauldie, S. D., McQueen, D. S., Clarke, C. J., & Chessell, I. P. (2004). A robust model of adjuvant-induced chronic unilateral arthritis in two mouse strains. *Journal of Neuroscience Methods, 139*(2), 281–291.

Hood, V. C., Cruwys, S. C., Urban, L., & Kidd, B. L. (2001). The neurogenic contribution to synovial leucocyte infiltration and other outcome measures in a guinea pig model of arthritis. *Neuroscience Letters, 299*, 201–204.

Larson, A. A., Brown, D. R., el-Atrash, S., & Walser, M. M. (1986). Pain threshold changes in adjuvant-induced inflammation: A possible model of chronic pain in the mouse. *Pharmacology, Biochemistry, and Behavior, 24*, 49–53.

Pearson, C. M., & Wood, F. D. (1959). Studies of polyarthritis and other lesions in rats by injection of mycobacterial adjuvant. I. General clinical and pathological characteristics and some modifying factors. *Arthritis and Rheumatism, 2*, 440–445.

Pettipher, E. R., & Henderson, B. (1988). The relationship between cell-mediated immunity and cartilage degradation in antigen-induced arthritis in the rabbit. *British Journal of Experimental Pathology, 69*, 113–122.

Rainsford, K. D. (1982). Adjuvant polyarthritis in rats: Is this a satisfactory model for screening anti-arthritic drugs? *Agents and Actions, 12*, 452–458.

Ratkay, L. G., Tait, B., Tonzetich, J., & Waterfield, J. D. (1994). Lpr and MRL background gene involvement in the control of adjuvant enhanced arthritis in MRL-lpr mice. *Journal of Autoimmunity, 7*, 561–573.

Sarlis, N. J., Chowdrey, H. S., Stephanou, A., & Lightman, S. L. (1992). Chronic activation of the hypothalamo-pituitary-adrenal axis and loss of circadian rhythm during adjuvant-induced arthritis in the rat. *Endocrinology, 130*, 1775–1779.

van Eden, W., Holoshitz, J., Nevo, Z., Frenkel, A., Klajman, A., & Cohen, I. R. (1985). Arthritis induced by a T-lymphocyte clone that responds to mycobacterium tuberculosis and to cartilage proteoglycans. *Proceedings of the National Academy of Sciences of the United States of America, 82*, 511701–515120.

Wakesman, B. H., Pearson, C. M., & Sharp, J. T. (1960). Studies of arthritis and other lesions induced in rats by injection of mycobacterial adjuvant. II. Evidence that the disease is a disseminated immunologic response to exogenous antigen. *Proceedings of the National Academy of Sciences of the United States of America, 85*, 403–417.

Arthritis Model, Kaolin-Carrageenan-Induced Arthritis (Knee)

Volker Neugebauer
Department of Neuroscience & Cell Biology, University of Texas Medical Branch, Galveston, TX, USA

Synonyms

Acute experimental monoarthritis; Acute experimental synovitis; Acute knee joint inflammation; Kaolin-Carrageenan-induced arthritis; K/C arthritis

Definition

Aseptic inflammatory monoarthritis induced by injections of kaolin and carrageenan into the synovial cavity of one knee joint results in damage to the cartilage, inflammation of the synovia and synovial fluid exudates, as well as pain behavior and neuroplastic changes in the peripheral and central nervous system.

Characteristics

The K/C arthritis is a well-established model of an acute onset monoarthritis resembling osteoarthritis, which is characterized by degeneration of hyaline articular cartilage and subsequent inflammation and pain. The K/C arthritis model has been used in cats (Coggeshall et al. 1983; Schaible and Grubb 1993), monkeys (Dougherty et al. 1992), rats (Neugebauer et al. 1993;

Sluka and Westlund 1993), mice (Zhang et al. 2001; Gatta et al. 2012), and guinea pigs (Gomis et al. 2007) to study pain mechanisms in the peripheral and central nervous systems. The K/C arthritis model produces inflammation, behavioral changes, and ▶ neuroplasticity with a distinct time-course of acute onset (1–3 h) and plateau phase (after 5–6 h) that persists for at least 1 week (Neugebauer et al. 2007).

Induction of Arthritis

This experimental arthritis is induced in one knee by intra-articular injections of kaolin and carrageenan into the knee joint. The inorganic kaolin (bolus alba, "China clay"), which is hydrated aluminum silicate ($H_2Al_2Si_2O_8.H_2O$) (Merck 13, 5300), is used to inflict mechanical damage to the cartilage as in osteoarthritis and, as an adjuvant, to increase the effectiveness of the active inflammatory compound carrageenan. Lambda carrageenan type IV (name derived from the Irish coastal town of Carragheen) is a mixture of sulfated polysaccharides extracted from the red seaweed *Gigartina* (Merck 13, 1878). Although subcutaneous, intramuscular, and intra-articular injections of each compound alone can produce inflammation, the combination of kaolin and carrageenan results in a more robust and longer lasting inflammation with a more constant and highly reproducible time-course. Several experimental protocols exist for the induction of the K/C arthritis in different species.

Monkey

0.5 ml of a solution containing 5 % carrageenan plus 5 % kaolin is injected into the knee joint cavity through the lateral aspect of the leg. The knee joint is then repeatedly flexed and extended for 15 min (Dougherty et al. 1992).

Cat

0.4–0.5 ml of a 4 % kaolin suspension is injected into the synovial cavity through the lateral aspect of the knee joint. After alternating flexions and extensions of the knee for 15 min, 0.3 ml of a 2 % carrageenan solution is injected intra-articularly, and the knee is flexed and extended for 5 min. The movements facilitate the damage to the cartilage

and the development of inflammation (Coggeshall et al. 1983; Schaible and Grubb 1993).

Rat

Kaolin and carrageenan are injected either sequentially or together according to the following protocols: (1) 80–100 µl of a 4 % kaolin suspension is injected into the joint cavity through the patellar ligament. After repetitive flexions and extensions of the knee for 15 min, a carrageenan solution (2 %, 80–100 µl) is injected into the knee joint cavity and the leg is flexed and extended for another 5 min (Neugebauer et al. 1993). (2) 100 µl of a solution of 3 % kaolin and 3 % carrageenan is injected into the knee joint cavity, and the knee joint is flexed and extended for 1 min (14) or 5–10 min (Sluka and Westlund 1993), 50 µl of a mixture of 3 % kaolin and 3 % carrageenan is injected into one knee joint.

Mouse

As in the rat, two protocols are used: (1) 40 µl of a 4 % kaolin suspension is injected into the articular cavity through the patellar ligament. After flexing and extending the joint for 15 min, 40 µl of a carrageenan solution (2 %) is injected and the leg is moved for another 5 min (Gatta et al. 2012). (2) A mixture of kaolin (3 %) and carrageenan (3 %) in 50 µl of saline is injected into the knee through the patellar ligament (Zhang et al. 2001).

Histopathology

Intra-articular injections of kaolin and carrageenan cause a unilateral aseptic inflammation with the following characteristics: swelling of the knee joint (measured as increased circumference of the knee), increased intra-articular pressure, hyperthermia of the knee, and edema with marked cellular infiltration (polymorphonuclear leucocytes) (Schaible and Grubb 1993; Schaible et al. 2002; Sluka and Westlund 1993).

Pain Behavior

The K/C arthritis model is accompanied by spontaneous pain behavior in awake freely moving animals, including limping, guarding of the leg with the arthritic knee, avoidance of joint movements, decreased weight bearing on the leg with the arthritic knee, and reduced exploratory behavior

(Neugebauer et al. 2003; Schaible and Grubb 1993; Sluka and Westlund 1993). Awake arthritic animals also show increased evoked pain behavior (Neugebauer et al. 2003; Schaible et al. 2002; Sluka 1996; Sluka and Westlund 1993; Urban et al. 1999; Yang et al. 1996; Yu et al. 2002; Zhang et al. 2001): primary mechanical ▶ allodynia (reduced vocalization threshold to mechanical stimulation of the arthritic knee); secondary allodynia and ▶ hyperalgesia (reduced paw withdrawal threshold and latency, respectively) for mechanical and thermal stimuli applied to the hindpaw. Frequency and total duration of vocalizations in the audible and ultrasonic ranges are increased (Neugebauer et al. 2007). Arthritic animals also show increased anxiety-like behavior in the elevated plus maze test (Neugebauer et al. 2007).

Neurochemical Changes
Inflammatory mediators, neuropeptides, and excitatory amino acids accumulate in the inflamed tissue of the knee and the synovial fluid (Lawand et al. 2000; Schaible and Grubb 1993; Schaible et al. 2002). Sources include immune cells, inflammatory cells, serum (plasma extravasation), and articular nerve fibers (neurogenic component). These substances play an important role in the "▶ peripheral sensitization" of articular afferent nerve fibers (see below and Fig. 1), which results in the enhanced production and release of various neurotransmitters (amino acids) and neuromodulators (peptides) into the spinal cord. Changes and mechanisms in the K/C arthritis pain model are listed below:

Inflammation
- Edema (increased knee joint circumference ipsilateral but not contralateral)
- Increased intra-articular pressure
- Increased temperature of arthritic (but not contralateral) knee
- Cellular infiltration (neutrophils)

Pain Behavior
- Spontaneous pain behavior
- Primary allodynia (mechanical)
- Secondary hyperalgesia (mechanical and thermal)
- Secondary allodynia (mechanical and thermal)

Neurochemistry
- Periphery
  Prostaglandins (PGE2, PGI2)
  Bradykinin
  Leukotrienes
  Histamine
  Serotonin
  Excitatory amino acids (EAA; glutamate but not aspartate)
  Nitric oxide (NO) metabolites (arginine, citrulline)
  Substance P
  Calcitonin gene–related peptide (CGRP)
  Galanin
  NPY
  Somatostatin (SST)
- Spinal Cord
  EAA (Glutamate, glutamine, aspartate)
  NO metabolites (citrulline)
  Substance P
  NKA
  Calcitonin gene–related peptide (CGRP)
  Prostaglandins (PGE2)
  Endocannabinoid COX-2 metabolites (prostamide $F_{2\alpha}$ $PMF_{2\alpha}$)

Electrophysiology
- Periphery
  Sensitization of groups II, III, and IV (Aβ, Aδ, and C) articular afferent nerve fibers, including silent nociceptors
  Dorsal root reflexes in groups II, III, and IV (Aβ, Aδ, and C) articular afferent fibers
- Spinal Cord
  Sensitization of spinal neurons in the superficial and deep dorsal horn and in the ventral horn (nociceptive-specific, wide-dynamic-range, inhibited, nonresponsive types)
- Brainstem/Brain
  Increased descending inhibition and facilitation
  Sensitization of neurons in the amygdala (multireceptive neurons in the basolateral and central nuclei but not nociceptive-specific neurons in the central nucleus)
  Decreased responsiveness of neurons in the infra- and pre-limbic regions of the medial prefrontal cortex (mPFC)

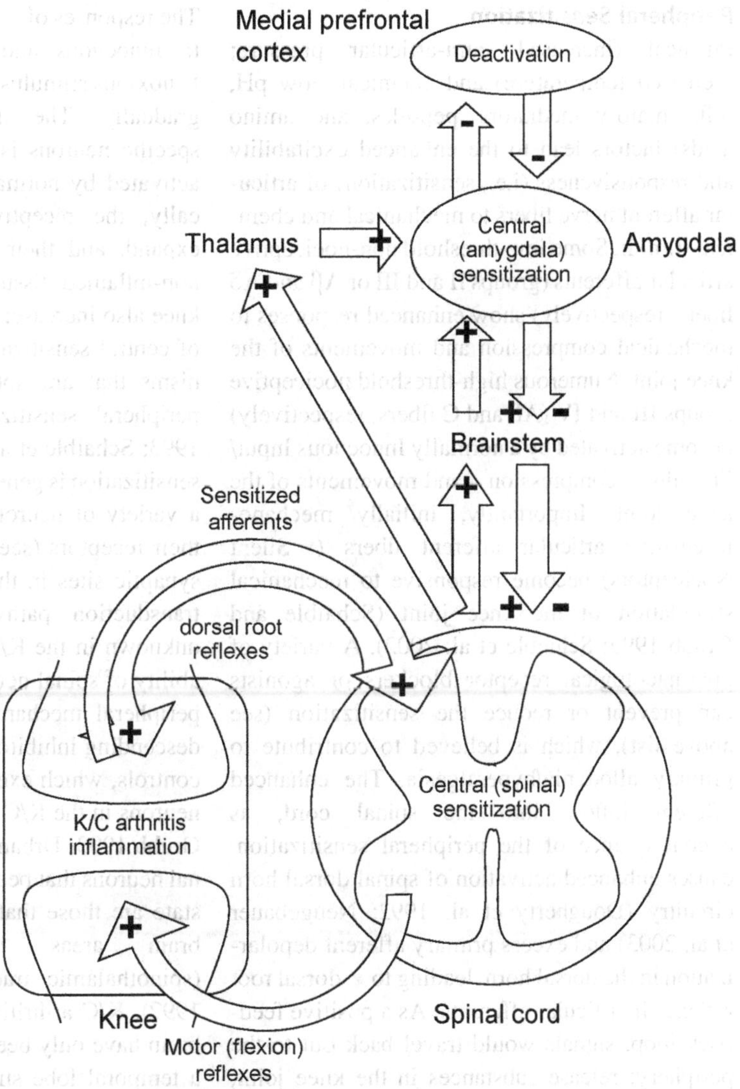

Arthritis Model, Kaolin-Carrageenan-Induced Arthritis (Knee),

Fig. 1 Peripheral and central pain mechanisms in the K/C arthritis model. Knee joint inflammation causes sensitization of articular afferent nerve fibers, which results in enhanced input to the spinal cord, dorsal root reflexes back to the arthritic knee, increased flexion reflexes, and central sensitization of spinal neurons. Sensitized spinal neurons send their enhanced signals through pain pathways to the brainstem and brain to cause central sensitization of amygdala neurons and deactivation of medial prefrontal cortical pyramidal cells (plasticity in other brain areas remains to be determined in the K/C model). The amygdala can modulate pain behavior by activating or inhibiting descending facilitation or descending inhibition

Pharmacology
- Periphery

  Excitatory: NMDA receptor, Non-NMDA receptor, Neurokinin 1 (NK1) receptor, Neurokinin 2 (NK2) receptor, NO/NOS

  Inhibitory: Galanin receptor, Opioid receptors (mu, kappa, ORL1), Somatostatin receptor
- Spinal Cord

  Excitatory: NMDA receptor, Non-NMDA receptor, Metabotropic glutamate receptor (mGluR) group I, Neurokinin 1 (NK1) receptor, Neurokinin 2 (NK2) receptor, Calcitonin gene–related peptide (CGRP1)

  receptor, Calcium channels (L-, N-, P-type), Prostaglandins (PGE$_2$), Prostamide F$_{2\alpha}$ (PMF$_{2\alpha}$), Nicotinic cholinergic receptor

  Inhibitory: GABA$_A$ (but not GABA$_B$) receptor, Cannabinoid CB-1 receptor
- Amygdala

  Excitatory: NMDA receptor, non-NMDA receptor, Metabotropic glutamate receptor (mGluR) group I, Corticotropin-releasing factor (CRF), Calcitonin gene–related peptide (CGRP)

  Inhibitory: Metabotropic glutamate receptor (mGluR) groups II and III

Peripheral Sensitization

Physical (increased intra-articular pressure; increased temperature) and chemical (low pH, inflammatory mediators, peptides, and amino acids) factors lead to the enhanced excitability and responsiveness (i.e., sensitization) of articular afferent nerve fibers to mechanical and chemical stimuli. Some low-threshold non-nociceptive articular afferents (groups II and III or Aβ and Aδ fibers, respectively) show enhanced responses to mechanical compression and movements of the knee joint. Numerous high-threshold nociceptive groups III and IV (Aδ and C fibers, respectively) become activated by a normally Innocuous Input/Stimulus – compression – and movements of the knee joint. Importantly, initially mechano-insensitive articular afferent fibers (▶ Silent Nociceptors) become responsive to mechanical stimulation of the knee joint (Schaible and Grubb 1993; Schaible et al. 2002). A variety of pharmacological receptor blockers or agonists can prevent or reduce the sensitization (see above list), which is believed to contribute to primary allodynia/hyperalgesia. The enhanced afferent inflow into the spinal cord, as a consequence of the peripheral sensitization, causes enhanced activation of spinal dorsal horn circuitry (Dougherty et al. 1992; Neugebauer et al. 2003) and excess primary afferent depolarization in the dorsal horn, leading to ▶ dorsal root reflexes in articular afferents. As a positive feedback loop, signals would travel back out to the periphery; release substances in the knee joint, and contribute to the inflammation (Sluka et al. 1995). Importantly, the sympathetic nervous system does not seem to contribute to the inflammatory, behavioral, and peripheral electrophysiological changes in the K/C arthritis pain model (Schaible and Grubb 1993; Schaible et al. 2002; Sluka 1996).

Central Sensitization

Enhanced incoming signals in articular afferents from the arthritic knee result in the intraspinal release of various substances (transmitters, modulators), and trigger the development of neuroplastic changes of spinal neurons (Dougherty et al. 1992; Neugebauer et al. 1993).

The responses of ▶ wide-dynamic-range neurons to innocuous and noxious compression (see ▶ noxious stimulus) of the arthritic joint increase gradually. The threshold of ▶ nociceptive-specific neurons is lowered, such that they are activated by normally innocuous stimuli. Typically, the receptive fields of these neurons expand, and their responses to stimulation of non-inflamed tissue remote from the arthritic knee also increase; both are considered evidence of central sensitization (i.e., spinal pain mechanisms that are not simply a reflection of the peripheral sensitization; Schaible and Grubb 1993; Schaible et al. 2002; Sluka 1996). Central sensitization is generated and maintained through a variety of neurotransmitters, modulators, and their receptors (see above list) at pre- and post-synaptic sites in the spinal cord, but the signal transduction pathways involved are largely unknown in the K/C arthritis model. The excitability of spinal neurons is not only regulated by peripheral mechanisms but also through tonic descending inhibitory and excitatory supraspinal controls, which exert enhanced effects on spinal neurons in the K/C arthritis model (Schaible and Grubb 1993; Urban et al. 1999). Among the spinal neurons that become sensitized in the arthritis state are those that send their axons to various brain areas including the thalamus (spinothalamic tract cells) (Dougherty et al. 1992). K/C arthritis pain-related changes in the brain have only been studied in the ▶ amygdala, a temporal lobe structure, which as part of the ▶ limbic system plays a key role in emotionality and negative affective states, and is believed to be a neural substrate of the reciprocal relationship between emotion and pain. Two major subpopulations of neurons in the latero-capsular part of the central nucleus of the amygdala ("nociceptive amygdala") develop nociceptive plasticity in the K/C arthritis pain model: multireceptive neurons (comparable to spinal wide-dynamic-range neurons) and nonresponsive neurons without a receptive field, but not nociceptive-specific neurons (Neugebauer and Li 2003). Synaptic transmission and neuronal excitability are enhanced in amygdala neurons in brain slices from rats with K/C arthritis, suggesting that plasticity in the amygdala can be maintained independently of afferent input from the arthritic knee (Neugebauer et al. 2003). Both purely nociceptive inputs from the spino-parabrachio-amygdaloid pain pathway, and highly integrated polymodal inputs from the fear/anxiety-circuitry in the lateral and basolateral amygdala are required to produce these plastic changes, which is consistent with a role of the amygdala as the interface between pain and affect (Neugebauer and Li 2003; Neugebauer et al. 2003). Multireceptive neurons in the basolateral amygdala also show central sensitization and synaptic plasticity (Ji et al. 2010). Enhanced nociceptive processing and increased neuronal excitability in the amygdala in the K/C arthritis model critically depend on the upregulation of presynaptic G-protein-coupled metabotropic glutamate receptors of the mGluR1 subtype, activation of postsynaptic corticotropin-releasing factor CRF1 receptors, and enhanced function of postsynaptic N-methyl-D-aspartate (NMDA) receptors through protein kinase A (PKA)-dependent phosphorylation. The amygdala is closely interconnected with other forebrain structures and brainstem centers known to be part of the endogenous pain control system. In contrast to amygdala hyperactivity, pyramidal neurons in the infra- and pre-limbic regions of the medial prefrontal cortex (mPFC) show decreased background and evoked activity in the K/C model, which may contribute to cognitive decision-making deficits (Ji et al. 2010). Pain-related plasticity in other brain areas remains to be determined in the K/C arthritis model.

References

Coggeshall, R. E., Hong, K. A., Langford, L. A., et al. (1983). Discharge characteristics of fine medial articular afferents at rest and during passive movements of inflamed knee joints. *Brain Research, 272*, 185–188.

Dougherty, P. M., Sluka, K. A., Sorkin, L. S., et al. (1992). Neural changes in acute arthritis in monkeys. I. Parallel enhancement of responses of spinothalamic tract neurons to mechanical stimulation and excitatory amino acids. *Brain Research Reviews, 55*, 1–13.

Gatta, L., Piscitelli, F., Giordano, C., Boccella, S., Lichtman, A., Maione, S., & Di Marzo, V. (2012). Discovery of prostamide F2alpha and its role in inflammatory pain and dorsal horn nociceptive neuron hyperexcitability. *PLoS One, 7*, e31111.

Gomis, A., Miralles, A., Schmidt, R. F., & Belmonte, C. (2007). Nociceptive nerve activity in an experimental model of knee joint osteoarthritis of the guinea pig: Effect of intra-articular hyaluronan application. *Pain, 130*, 126–136.

Ji, G., Sun, H., Fu, Y., Li, Z., Pais-Vieira, M., Galhardo, V., & Neugebauer, V. (2010). Cognitive impairment in pain through amygdala-driven prefrontal cortical deactivation. *The Journal of Neuroscience, 30*, 5451–5464.

Lawand, N. B., McNearney, T., & Westlund, K. N. (2000). Amino acid release into the knee joint: Key role in nociception and inflammation. *Pain, 86*, 69–74.

Neugebauer, V., & Li, W. (2003). Differential sensitization of amygdala neurons to afferent inputs in a model of arthritic pain. *Journal of Neurophysiology, 89*, 716–727.

Neugebauer, V., Lucke, T., & Schaible, H.-G. (1993). N-methyl-D-aspartate (NMDA) and Non-NMDA receptor antagonists block the hyperexcitability of dorsal horn neurons during development of acute arthritis in rat's knee joint. *Journal of Neurophysiology, 70*, 1365–1377.

Neugebauer, V., Li, W., Bird, G. C., et al. (2003). Synaptic plasticity in the amygdala in a model of arthritic pain: Differential roles of metabotropic glutamate receptors 1 and 5. *The Journal of Neuroscience, 23*, 52–63.

Neugebauer, V., Han, J. S., Adwanikar, H., Fu, Y., & Ji, G. (2007). Techniques for assessing knee joint pain in arthritis. *Molecular Pain, 3*, 8–20.

Schaible, H.-G., & Grubb, B. (1993). Afferent and spinal mechanisms of joint pain. *Pain, 55*, 5–54.

Schaible, H.-G., Ebersberger, A., & von Banchet, G. S. (2002). Mechanisms of pain in arthritis. *Annals of the New York Academy of Sciences, 966*, 343–354.

Sluka, K. A. (1996). Pain mechanisms involved in musculoskeletal disorders. J Orthop Sports Phys Ther 24: 240–254.

Sluka, K. A., & Westlund, K. N. (1993). Behavioral and immunohistochemical changes in an experimental arthritis model in rats. *Pain, 55*, 367–377.

Sluka, K. A., Rees, H., Westlund, K. N., et al. (1995). Fiber types contributing to dorsal root reflexes induced by joint inflammation in cats and monkeys. *Journal of Neurophysiology, 74*, 981–989.

Urban, M. O., Zahn, P. K., & Gebhart, G. F. (1999). Descending facilitatory influences from the rostral medial medulla mediate secondary, but not primary hyperalgesia in the rat. *The Journal of Neuroscience, 90*, 349–352.

Yang, L. C., Marsala, M., & Yaksh, T. L. (1996). Characterization of time course of spinal amino acids, citrulline and PGE$_2$ release after carrageenan/kaolin-induced knee joint inflammation: A chronic microdialysis study. *Pain, 67*, 345–354.

Yu, Y. C., Koo, S. T., Kim, C. H., et al. (2002). Two variables that can be used as pain indices in experimental animal models of arthritis. *Journal of Neuroscience Methods, 115*, 107–113.

Zhang, L., Hoff, A. O., Wimalawansa, S. J., et al. (2001). Arthritic calcitonin/[alpha] calcitonin gene-related peptide knockout mice have reduced nociceptive hypersensitivity. *Pain, 89*, 265–273.

Arthritis Model, Osteoarthritis

James D. Pomonis
Algos Preclinical Services, Roseville, MN, USA

Synonyms

Arthritis; Degenerative joint disease; Osteoarthritis; Osteoarthritis model

Definition

Osteoarthritis is a condition in which physical or biological damage to the cartilage of ▶ synovial joints leads to destruction of the ▶ cartilage and remodeling of the bone underneath the affected cartilage.

Introduction

Osteoarthritis (OA) is a progressive disease of the joints characterized by the loss of articular cartilage, damage to subchondral bone, and joint space narrowing. According to the Centers for Disease Control, the prevalence of OA in the United States is estimated to be almost 14 % in adults age 25 and older and greater than 33 % in adults over the age of 65. However, it has also been estimated that 80–90 % of adults 65 and older have evidence of OA. Age is the greatest risk factor for the development of OA, and as the population over the age of 65 is rising dramatically, the prevalence of the disease is likewise expected to increase (Elders 2000; Hinton et al. 2002).

OA is frequently associated with severe pain, but the source of this pain is not fully understood. A clue to understanding the source of OA-related pain may well be found in the pathology associated with the condition. This pathology is characterized by damage to the cartilage in synovial joints, alterations in the physiology of chondrocytes, and profound changes in the subchondral bone including sclerosis, cyst formation, and the development of osteophytes or bony spurs beneath the affected cartilage (Bendele and Hulman 1988). Over time, these changes lead to radiologic evidence for the presence of OA, and in fact, this is the primary method for diagnosis of OA in humans. OA tends to be most common in specific joints, namely, the knees, hips, small joints of the hands, and the cervical and lumbar spine (Cushnaghan et al. 1990).

Current treatment for OA-related pain can be divided into three categories: physical/occupational therapy, pharmacological treatment, and surgical intervention. Physical therapy can maintain muscle strength around the joint and provide a benefit by increasing joint stability (Altma et al. 2000). Therapeutic intervention is most commonly achieved by using nonsteroidal anti-inflammatory drugs (NSAIDs). Although these drugs have been shown clinically to provide pain relief in OA patients, this pain relief is often incomplete, and their use is often accompanied by unwanted side effects including the induction of ulcers (Altma et al. 2000; Scheiman 2003; Marnett 2009). The use of cyclooxygenase-2 selective inhibitors such as celecoxib has mitigated the frequency of GI-related adverse events, but these therapies have been under scrutiny for concerns related to cardiovascular events. Other pharmacological therapies include injections of steroids or high molecular weight hyaluronate into the affected joints. Finally, the incidence of surgical intervention has grown rapidly over the past 30 years but is typically considered when the pain associated with the OA has become intractable. As such, the need for safer and more effective analgesics for the treatment of OA-related pain represents an area of significant research need.

The necessity to discover new therapeutics for the treatment of OA-related pain logically requires the development of animal models of OA that are more refined in terms of the pathology and/or the etiology of the disease as seen in the clinical population. To this end, three main approaches have been pursued: genetic models, surgical models, and chemical models (Moskowitz et al. 1979; Sandy et al. 1984; Ameye and Young 2006).

Spontaneous models of OA include the Dunkin Hartley guinea pig model of OA, in which animals show mild degeneration of articular cartilage beginning at approximately 3 months of age that continues to progress as the animals age (McDougall et al. 2009). Strains of mice have also been reported to show evidence of OA with age (Rostand et al. 1986). Surgical models of OA typically involve damage to the meniscus, often via a partial meniscectomy, and may also involve transection of the anterior cruciate ligament (van der Kraan et al. 1989; Fernihough et al. 2004). Chemical models of OA involve intra-articular injection of substances such as papain or monosodium iodoacetate (MIA) that are known to disrupt chondrocyte metabolism and hence promote joint degeneration (Guingamp et al. 1997; Guzman et al. 2003).

The vast majority of the initial studies using any of these models primarily investigated the pathological characteristics of the models, but little or no attention was given to investigating the pain behavior associated with the models. That changed when Guingamp and colleagues (1997) used biotelemetry to examine the effects of intra-articular injection of MIA on spontaneous mobility. In that study, they found a strong, positive correlation of the amount of MIA injected, the resultant pathological changes in the affected knee joints, and decreased mobility. Although these data indicated that MIA-induced OA produces pain-related behaviors in rodents, it had not been demonstrated whether these behavioral changes could be reversed with administration of analgesics. Furthermore, the use of biotelemetry limited the number of laboratories that were capable of performing this assay. The next significant advance in the

establishment of the MIA model came from studies involving the use of hind limb weight bearing both as a measure of the pain due to MIA injection as well as the effect of various analgesic compounds on these altered.

These advances in the ability to assess pain-related behaviors in the various animal models of OA led to significant effort related to discerning which model would be the model of choice for the study of novel analgesics. Early comparative studies between chemical and surgical models of OA as well as the various chemical models of OA indicated that the MIA model of OA was preferable as it showed appropriate, chronic pathology and was both highly reproducible and feasible across laboratories (Fernihough et al. 2004; Pomonis et al. 2005). This remainder of this chapter will focus on the MIA model of OA in rats. It is important to note that it is best to consider this model as a model for OA-related pain, as it is well suited for the study of potential therapies for the treatment of this pain. However, its suitability for assessing potential disease modifying therapies may be lessened due to the nature and onset of the pathology.

Characteristics

Induction of OA using MIA is a relatively simple process involving a single injection of MIA solution into the articular space of one knee joint (Marker and Pomonis 2012). The amount of MIA injected may vary, but most frequently ranges from 1 mg to 3 mg in a 25–50 μL volume. The injection is most often unilateral, but bilateral injections have also been used. MIA is a metabolic inhibitor that blocks the activity of glyceraldehyde-3-phosphate dehydrogenase, preventing glycolysis. As such, this inhibition ultimately leads to the death of the cells that are exposed to the MIA. When injected into a joint space, MIA preferentially acts on chondrocytes, causing damage to the cartilage of the joint.

Following the MIA injection, there is a time-dependent, biphasic response that is observed both in terms of the cellular effects

Arthritis Model, Osteoarthritis, Fig. 1 Hematoxylin and eosin stained sections of normal and osteoarthritic knee joint of the rat. Shown are the articular surface and subchondral bone regions of femoral condyles after saline injection (*left*) or MIA injection (*right*). Necrotic chondrocytes are shrunken in their lacunae and a dense fibrous matrix fills the area where bone spicules have degenerated and lost their continuity with the articular surface (*double headed arrow*). The bone marrow is severely damaged in this region

and the resultant behaviors displayed by the animals. Initially, there is an acute inflammatory response that peaks approximately 3 days after MIA injection and resolves within 7 days of the injection. While pain-related behaviors are present during this early phase, they are unlikely to represent pain associated with OA per se, as there is little or no pathology observed indicating that the behaviors observed during the first 7 days are a result of any OA-like condition. While there is noticeable edema and infiltration of various inflammatory cells, the articular cartilage is still relatively intact and the subchondral bone does not begin to show effects of the MIA injection until day 7 post-injection or later (Guzman et al. 2003).

However, by 14 days after MIA injection, certain key pathologies are observed that are consistent with the presence of OA in these animals including changes to the subchondral bone, often manifest as replacement of bone marrow with spindle cells, particularly beneath damaged cartilage (Fig. 1). It is at this time that radiographic evidence of OA becomes visible before becoming extremely pronounced in the following weeks (Fig. 2). The changes to the subchondral bone progress with time and extensive bone remodeling can be observed. These changes in the subchondral bone may play a very significant role in the development and maintenance of OA-related pain, as subchondral bone and bone marrow represent a likely source for the mechanistic drivers of this pain. Cartilage, the tissue that is most profoundly affected in OA, is not innervated. However, cortical and subchondral bone are innervated by sensory as

Arthritis Model, Osteoarthritis, Fig. 2 Radiograph showing the effect of iodoacetate on the bones of the rat knee joint. Injection of 3 mg iodoacetate (*right panel*) leads to profound deformation of the bones relative to saline-injected control of the knee joint (*left panel*). These changes are present as rough edges of the bone, apparent loss of bone density, and displacement of the patella (kneecap), indicating joint swelling

well as sympathetic fibers (Mach et al. 2002). Although this innervation does not establish the link between subchondral bone and the mechanisms of OA-related pain, MRI imaging showed a significant, positive correlation between the presence of bone marrow lesions and pain in patients with radiographically confirmed OA of the knee (Felson et al. 2001). Further evidence for the involvement of subchondral bone is supported by the observation that administration of bisphosphonates – a class of compounds that inhibit osteoclast-mediated bone resorption – can prevent and even reverse pain-related behaviors in the MIA model (Strassle et al. 2010).

The efficacy of NSAIDs for the management of OA-related pain indicates that inflammation plays a role in the generation and maintenance of this pain. However, there is also evidence that neuropathy may also play a prominent role. Expression of ATF-3 (a marker of neuronal injury) in DRG neurons that innervate the knee increases after injection of MIA (Ivanavicius et al. 2007; Orita et al. 2011; Ferreira Gomes et al. 2012), astrocytes and microglia in the spinal cord become activated as the disease progresses (Sagar et al., 2011), and deep dorsal horn neurons

show increased sensitivity to mechanical stimulation in their receptive field, an effect which is reversed by administration of pregabalin (Harvey and Dickenson 2009; Thakur et al. 2012). While it is unclear whether a similar neuropathic pathology is present in human OA, it is worth noting that duloxetine is approved for the treatment of chronic musculoskeletal pain due to osteoarthritis, suggesting a neuropathic component to human OA pain.

Behavioral assessment of pain associated with models of OA significantly lagged behind the pathological and mechanistic studies that were being performed in the various models. To further complicate the picture, the OA models were different from many of the other models of inflammatory and neuropathic pain that are characterized by primary or secondary hyperalgesia or allodynia in a hind paw. The OA models involved pain at the tibiofemoral joint, an anatomical site for which no standardized testing approaches had yet been established. However, as the assessment of pain behaviors in rodents became more refined, researchers were able to apply new techniques to the models of OA to determine if the pathology seen in these

Arthritis Model, Osteoarthritis, Fig. 3 Time course of the development of OA-related pain behaviors in the MIA model of OA in the rat. Intra-articular injection of 2 mg MIA produces significant and prolonged behavioral changes indicative of OA-related pain including alterations in hind limb weight bearing (**a**), decreased hind limb grip strength (**b**), and primary mechanical hyperalgesia at the affected knee (**c**)

models correlated in any way with alterations in behavior. In general, these approaches led to the assessment of movement-evoked or use-related pain as opposed to mechanical or thermal sensitivity. This in and of itself was desirable to researchers as it was believed that these assessments would more closely mimic spontaneous pain and therefore would have more predictive validity to human clinical trials.

The assessment of hind limb weight bearing was one of the first behavioral assays to be widely used to study OA-related pain and is very likely the most widely used assay for this purpose to date (Bove et al. 2003; Kobayashi et al. 2003). The assay is performed by using an incapacitance meter to measure the amount of weight borne by each hind limb individually when the animal is in a semi-rearing position. Following injection of MIA, there is a significant reduction in the weight borne on the affected hind limb that peaks approximately 3 days after MIA injection

and at least partially resolves by 7–10 days post-injection. Beginning around day 10 post-injection, the deficits in weight bearing begin to become apparent again and progress over the subsequent weeks (Fernihough et al. 2004; Pomonis et al. 2005). These MIA-induced alterations in weight bearing are not only time-dependent but also vary with the amount of MIA injected. Injection of 0.3 mg MIA does not produce significant changes in hind limb weight bearing, but injections of 1, 2, or 3 mg MIA produce graded changes in behavior (Fig. 3a).

At present, there is no consensus on what aspect of pain-related behavior the weight bearing assay is measuring – it may assess use-related pain, ongoing pain, mechanical hyperalgesia, or a learned behavior to avoid use of the affected limb. However, it is clear that the alterations in weight bearing induced by MIA injection can be normalized by administration of a range of analgesic compounds including opiates, NSAIDs, TRPV1 antagonists, and monoamine reuptake inhibitors.

MIA-induced OA is also characterized by deficits in hind limb grip strength, another potential assay for use-related pain (Chandran et al. 2009). As seen with measurements of hind limb weight bearing following MIA injection, assessment of hind limb grip strength shows progressive nature over time (Fig. 3b), and the deficits in grip strength can be restored with a range of analgesics including NSAIDs, antiepileptic drugs such as lamotrigine and gabapentin, and the monoamine reuptake inhibitor duloxetine. A unique feature of using the hind limb grip strength assay to assess the efficacy of therapeutics is the impact of motor impairing side effects. Unlike most stimulus-response assays in which sedative or motor impairing side effects lead to false-positives, any such effects in the grip strength assay will lead to an apparent lack of efficacy.

Along with the use- and movement-related pain behaviors seen in rats after MIA injection, mechanical sensitivity is also present, manifest as primary mechanical hyperalgesia at the arthritic knee and secondary mechanical allodynia at the hind paw ipsilateral to the injected knee (Marker and Pomonis 2012). As with the other behaviors seen following MIA injection, both of these traits are dose- and time-dependent (Fig. 3c). Although secondary mechanical allodynia is not a commonly reported symptom of OA patients, primary mechanical sensitivity has been reported in patients with painful OA. These behaviors also are sensitive to pharmacological reversal with clinically relevant compounds such NSAIDs and duloxetine, indicating that this behavioral readout may be relevant for the assessment of novel therapeutics.

As the study of the MIA model has continued, new insights have been gained regarding additional behavioral measures of the pain associated with joint damage. One important consequence of OA in humans is the effect the condition has on functional status. Indeed, OA is the most common cause for disability claims in the United States. As such, if the MIA model is to be considered to have significant face validity, it should result in the suppression of certain spontaneous behaviors. This effect has been confirmed using two separate approaches. MIA injection suppresses spontaneous home cage activity and also inhibits wheel running (Stevenson et al. 2011) in the subgroup of rats that showed a relatively high level of baseline wheel running. Further evidence for spontaneous pain in the MIA model comes from studies using the conditioned place preference test (Liu et al. 2011; Okun et al. 2012).

Summary

The MIA model of OA-related pain has several features and characteristics that have made it the model of choice for many groups investigating the efficacy of novel therapeutics for the treatment of OA pain. The model is technically simple in terms of induction, is highly reproducible in terms of the onset and magnitude of pain-related behaviors, and offers several behavioral assessments that can be used to test potential therapeutics. The pathology present in the model shows many similarities with the pathology of advanced OA in humans, and while the model may not be appropriate for the assessment of disease modifying therapies, it does represent a logical, practical tool especially for drug discovery.

References

Altma, R., Hochberg, M., Moskowitcz, R. W., & Schnitzer, T. J. (2000). Recommendations for the medical management of osteoarthritis of the hip and knee: 2000 update. *Arthritis and Rheumatism, 43*, 1905–1915.

Ameye, L. G., & Young, M. F. (2006). Animal models of osteoarthritis: Lessons learned while seeking the "holy grail". *Current Opinion in Rheumatology, 18*(5), 537–547.

Bendele, A. M., & Hulman, J. F. (1988). Spontaneous cartilage degeneration in guinea pigs. *Arthritis and Rheumatism, 31*, 561–565.

Bove, S., Calcaterra, S., et al. (2003). Weight bearing as a measure of disease progression and efficacy of anti-inflammatory compounds in a model of monosodium iodoacetate-induced osteoarthritis. *Osteoarthritis and Cartilage, 11*, 821–830.

Chandran, P., Pai, M., Blomme, E. A., Hsieh, G. C., Decker, M. W., & Honore, P. (2009). Pharmacological modulation of movement-evoked pain in a rat model of osteoarthritis. *European Journal of Pharmacology, 613*(1–3), 39–45.

Cushnaghan, J., Cooper, C., et al. (1990). Clinical assessment of osteoarthritis of the knee. *Annals of the Rheumatic Diseases, 49*, 768–770.

Elders, M. J. (2000). The increasing impact of arthritis on public health. *Journal of Rheumatology, 60*, 6–8.

Felson, D. T., Chaisson, C. E., Hill, C. L., Totterman, S. M., Gale, M. E., Skinner, K. M., et al. (2001). The association of bone marrow lesions with pain in knee osteoarthritis. *Annals of Internal Medicine, 134*(7), 541–549.

Fernihough, J., Gentry, C., et al. (2004). Pain related behaviour in two models of osteoarthritis in the rat knee. *Pain, 112*, 83–93.

Ferreira-Gomes, J., Adaes, S., Meireles Sousa, R., Mendonca, M., & Castro-Lopes, J. M. (2012). Dose-dependent expression of neuronal injury markers during experimental osteoarthritis induced by monoiodoacetate in the rat. *Molecular Pain, 8*, 50. www.molecularpain.com/content/8/1/50.

Guingamp, C., Gefout-Pottie, P., et al. (1997). Mono-iodoacetate-induced experimental osteoarthritis: A dose-response study of loss of mobility, morphology, and biochemistry. *Arthritis and Rheumatism, 40*, 1670–1679.

Guzman, R., Evans, M., et al. (2003). Mono-iodoacetate-induced histologic changes in subchondral bone and articular cartilage of rat femorotibial joints: An animal model of osteoarthritis. *Toxicologic Pathology, 31*, 619–624.

Harvey, V. L., & Dickenson, A. H. (2009) Behavioural and electrophysiological characterization of experimentally induced osteoarthritis and neuropathy in C57Bl/6 mice. *Molecular Pain, 5*, 18. www.molecularpain.com/content/5/1/18

Hinton, R., Moody, R. L., Davis, A. W., & Thomas, S. F. (2002). Osteoarthritis: Diagnosis and therapeutic considerations. *American Family Physician, 65*(5), 841–848.

Ivanavicius, S. P., Ball, A. D., Heapy, C. G., Westwood, F. R., Murray, F., & Read, S. J. (2007). Structural pathology in a rodent model of osteoarthritis is associated with neuropathic pain: Increased expression of ATF-3 and pharmacological characterisation. *Pain, 128*, 272–282.

Kelly, S., Dunham, J. P., Murray, F., Read, S., Donaldson, L. F., & Lawson, S. N. (2012). Spontaneous firing in C-fibers and increased mechanical sensitivity in A-fibers of knee joint-associated mechanoreceptive primary afferent neurons during MIA-induced osteoarthritis in the rat. *Osteoarthritis and Cartilage, 20*, 305–313.

Kobayashi, K., Imaizumi, R., et al. (2003). Sodium iodoacetate-induced experimental osteoarthritis and associated pain model in rats. *Journal of Veterinary Medical Science, 65*(11), 1195–1199.

Liu, P., Okun, A., Ren, J., Guo, R., Ossipov, M. H., Xie, J., et al. (2011). Ongoing pain in the MIA model of osteoarthritis. *Neuroscience Letters, 493*(3), 72–75.

Mach, D. B., Rosgers, S. D., Sabino, M. C., Luger, N. M., Schwei, M. J., Pomonis, J. D., et al. (2002). Origins of skeletal pain: Sensory and sympathetic innervation of the mouse femur. *Neuroscience, 113*(1), 155–166.

Marker, C. L., & Pomonis, J. D. (2012). The monosodium iodoacetate model of osteoarthritis pain in the rat. *Methods in Molecular Biology, 851*, 239–248.

Marnett, L. J. (2009). The COXIB experience: A look in the rearview mirror. *Annual Review of Pharmacology and Toxicology, 49*, 265–290.

McDougall, J. J., Andruski, B., Schuelert, N., Hallgrimsson, B., & Matyas, J. R. (2009). Unraveling the relationship between age, nociception and joint destruction in naturally occurring osteoarthritis of Dunkin Hartley guinea pigs. *Pain, 141*, 222–232.

Moskowitz, R. W., Howell, D. S., Goldberg, V. M., Muniz, O., & Pita, J. C. (1979). Cartilage proteoglycan alterations in an experimentally induced model of rabbit arthritis. *Arthritis and Rheumatism, 22*, 155–163.

Okun, A., Liu, P., Davis, P., Ren, J., Remeniuk, B., Brion, T., Ossipov, M. H., Xie, J., Dussor, G. O., King, T., & Porreca, F. (2012). Afferent drive elicits ongoing pain in a model of advanced osteoarthritis. *Pain, 153*, 924–933.

Orita, S. Ishikawa, T., Miyagi, M., Ochiai, N., Inoue, G., Eguchi, Y., Kamoda, H., Arai, G., Toyone, T., Aoki, Y., Kubo, T., Takahashi, K., & Ohtori, S. (2011). Pain-related sensory innervation in monoiodoacetate-induced osteoarthritis in rat knees that gradually develops neuronal injury in addition to inflammatory pain. *BMC Musculoskel Disorders, 12*, 134. www.biomedcentral.com/147-2474/12/134

Pomonis, J. D., Boulet, J., et al. (2005). Development and pharmacological characterization of a rat model of osteoarthritis pain. *Pain, 114*, 339–346.

Rostand, K. S., Baker, J. R., Caterson, B., & Christner, J. E. (1986). Articular cartilage proteoglycans from normal and osteoarthritic mice. *Arthritis and Rheumatism, 29*, 95–105.

Sagar, D. R., Burston, J. J., Hathway, G. J., Woodhams, S. G., Pearson, R. G., Bennett, A. J., Kendall, D. a., Scammell, B. E., & Chapman, V. E. (2011). The contribution of spinal glial cells to chronic pain behavior in the monosodium iodoacetate model of osteoarthritic pain. *Molecular Pain, 7*(88). www.molecularpain.com/content/7/1/88

Sandy, J. D., Adams, M. E., Billinghan, M. E. J., Plaas, A., & Muir, H. (1984). In vivo and in vitro stimulation of chondrocyte biosynthetic activity in early experimental osteoarthritis. *Arthritis and Rheumatism, 27*, 388–397.

Scheiman, J. M. (2003). Gastroduodenal safety of cyclooxygenase-2 inhibitors. *Current Pharmaceutical Design, 9*, 2197–2206.

Stevenson, G. W., Mercer, H., Cormier, J., Dunbar, C., Benoit, L., Adams, C., Jezierski, J., Luginbuhl, A., & Bilsky, E. J. (2011). Monosodium iodoacetate induced osteoarthritis produces pain depressed wheel running in rats: implications for preclinical behavioral assessment of chronic pain. *Pharmacology, Biochemistry & Behavior, 98*(1), 35–42.

Strassle, B. W., Mark, L., Leventhal, L., Piesla, M. J., Jian Li, X., Kennedy, J. D., et al. (2010). Inhibition of osteoclasts prevents cartilage loss and pain in a rat model of degenerative joint disease. *Osteoarthritis and Cartilage, 18*, 1319–1328.

Thakur, M., Rahman, W., Hobbs, C., Dickenson, A. H., & Bennett, D. L. H. (2012). Characterisation of a peripheral neuropathic component of the rat monoiodoacetate model of osteoarthritis. *PLoS ONE, 7*(3), e33730. www.plosone.org

van der Kraan, P., Vitters, E., et al. (1989). Development of osteoarthritic lesions in mice by "metabolic" and "mechanical" alterations in the knee joints. *American Journal of Pathology, 135*, 1001–1014.

Arthritis Urethritica

▶ Reiter's Syndrome

Arthritogenic Pain

▶ Muscle Pain in Systemic Inflammation (Polymyalgia Rheumatica, Giant Cell Arteritis, Rheumatoid Arthritis)

Articular

Definition

Pertaining to the joints.

Cross-References

▶ Sacroiliac Joint Pain

Articular Afferents, Morphology

Karl Messlinger
University of Erlangen-Nürnberg, Erlangen, Germany

Synonyms

Articular sensory receptors; Sensory endings in joint tissues

Definition

Afferent nerve fibers innervating articular tissues, in a narrow sense the sensory endings of afferent fibers in joint tissues, particularly in joint capsules and articular ligaments. Thick myelinated afferents form corpuscular nerve endings, and thin myelinated and unmyelinated afferents are without a corpuscular end structure (noncorpuscular endings, free nerve endings).

Characteristics

Remarks on the Classification of Sensory Endings

Sensory receptors are classified either according to their morphological appearance, which is thought to be correlated with functional properties, or according to the electrophysiological properties of their sensory axons within peripheral nerves. The velocity of action potentials running along the afferent fiber is used for a basic classification, dividing the afferents into slowly conducting ($A\delta$ or group III, and C or group IV) and fast conducting groups ($A\beta$ or group II). Whereas the $A\beta$ fibers possess corpuscular nerve endings, the thinly myelinated $A\delta$ and the unmyelinated C fibers terminate as so-called free nerve endings in peripheral tissues. This term has been used since the late nineteenth century, when new staining techniques enabled anatomists to visualize fine nerve endings that were obviously not enclosed or accompanied by specific cellular end structures (Hinsey 1927). In the middle of the twentieth century, researchers combined psycho-physiological and histological methods, and found that spot-like areas in human skin and deep tissues, the noxious stimulation of which was painful, contained nothing more than "free nerve endings" (Weddell and Harpman 1940). This led to the general conclusion that "free nerve endings" are the sensory end structures of nociceptors. Subsequent studies have shown, however, that sensory receptors signaling (innoc-uous) warm and cold, as well as many low-threshold mechanoreceptors, are also slowly conducting. New trials to correlate electrophysi-ological and histological data on slowly conducting afferents have not been much more conclusive with respect to morpho-functional characteristics (see below), whereas many addi-tional structural details have been acquired by selective marking and high-resolution techniques, such as confocal and electron micros-copy. We now know that those (previously described) unmyelinated peripheral nerve fibers and "free nerve endings," which are visible with the light microscope, are usually Remak bundles composed of several unmyelinated sensory axons and Schwann cells.

Composition of Articular Nerves

The distribution of different types of nerve fibers has been studied in detail in the medial and posterior articular nerves (MAN and PAN) innervating the knee joint in cat, rat, and mouse, using electron microscopy (Langford and Schmidt 1983; Hildebrandt et al. 1991; Salo and Theriault 1997; Ebinger et al. 2001). The composition of these nerves is similar, with a proportion of about 80 % being unmyelinated fibers, although the total number of nerve fibers differs between the three species (e.g., in the PAN: cat about 1,200, rat 400–600, mouse about 200). Comparisons between normal and sympathectomized animals have revealed that up to 50 % of unmyelinated fibers are afferent (Langford and Schmidt 1983; Salo and Theriault 1997). The diameters of myelinated sensory axons range from 1 to 8 µm in the rat (Hildebrandt et al. 1991), and from 1 to 18 µm in the cat. The majority of myelinated fibers in the MAN have diameters below 6 µm (belonging to group III), whereas in the PAN, thicker nerve fibers (group II) predominate (Heppelmann et al. 1988).

Neuropeptide Content of Articular Afferents

The proportion of neuropeptide-containing ▶ dorsal root ganglion (DRG) cells innervating the cat knee joint has been determined with ret-rograde labeling techniques and immunohisto-chemistry. ▶ Calcitonin gene-related peptide (CGRP) immunoreactivity was found in 35 % and ▶ substance P (SP) immunoreactivity in 17 % of labeled afferents from the MAN (Hanesch et al. 1991). In the PAN, the propor-tions are similar. Whereas all SP immunoreactive DRG neurons were found to be small or medium-sized cells, CGRP immunoreactivity was also detected in some large neurons with a diameter of more than 50 µm (Hanesch et al. 1991). ▶ Somatostatin (Som) seems to be colocalized with SP in nearly all MAN afferents (Hanesch et al. 1995). An acute experimental knee joint arthritis caused significant upregulation of the number of CGRP – but not SP-positive neurons (Hanesch et al. 1997), whereas in the human, arthritic painful hip joint upregulation of both neuropeptides has been observed (Saxler et al. 2007). In the rat knee joint, the proportion of CGRP and SP immunoreactive afferents was 33 % and 10 %, respectively (Salo and Theriault 1997). In the dog, the respective proportions were 29 % for CGRP and 17 % for SP; in 10 % of afferents these two neuropeptides seemed to be colocalized (Tamura et al. 1998). A major pro-portion of peptidergic mouse (knee and ankle) joint afferents have been found to express the vanilloid TRPV1 receptor, an important nocicep-tive transduction channel (Cho and Valtschanoff 2008).

Ultrastructure of Fine Sensory Endings in Articular Tissues

The fine sensory innervation of the cat knee joint has been studied by electron microscopy followed by quantitative analysis of structures and 3D-reconstruction (Heppelmann et al. 1990, 1995) and reviewed with respect to their presumptive functional role (Messlinger 1996). In this paragraph, general ultrastructural features are discussed that probably apply to all non-corpuscular sensory endings in deep tissues. As fine sensory endings lack a corpuscular structure, the only morphological landmark that may indicate the end of the conductive part of the nerve fibers, and the beginning of the sensory endings, is the termination of the perineurial sheath surrounding the peripheral nerve (Fig. 1). In group III fibers, an additional transition zone exists in which the nerve fibers have lost their myelin sheath but are still enclosed by the perineurium. Distal to the perineurial sheath, the sensory axons maintain their accompanying ▶ Schwann cells. Individual group III fibers are usually encased by their own Schwann cell, whereas several group IV axons are frequently bound together sharing a common Schwann cell, as is the case within the peripheral nerve (Remak bundles). These bundles of group IV, as well as the individual group III fibers, ramify several times to form tree-like sensory endings (Fig. 1). Apart from the larger mean diameter of group III compared to group IV fibers, the sensory endings of group III fibers can be identified by a characteristic "neurofilament core," a bundle of centrally arranged neurofilaments that run along the whole length of the sensory axon up to their terminal branches. There is morphological evidence that the afferents are receptive along their entire tree-like sensory endings, which can measure up to several hundred μm in length. Each sensory axon forms periodically arranged varicose segments that are characterized by bare areas, where there are gaps in the cover formed by the Schwann cell so that the axon membrane is partly exposed to the surrounding tissue. Membrane channels may be concentrated here and exposed to the extracellular space. Accumulated mitochondria, glycogen particles, and some vesicles are regularly found in the varicosities. The axoplasm beneath the bare areas has an electron dense substructure, which has been described as a "receptor matrix" in various

Articular Afferents, Morphology, Fig. 1 Three-dimensional reconstruction, based on serial electron microscopic images, of the sensory endings of a group III fiber in the cat knee joint (*left*) with a terminal branch at higher magnification (*right*). The branched endings arise from a peripheral nerve ensheathed by perineurium (*P*) and extend between venous vessels (*V*) and fat cells (*F*). They are wrapped up to their terminals by Schwann cells (*SC*) that form gaps in which the sensory axon (*A*) is exposed to the surrounding (*right*)

types of sensory nerve endings (Andres and von Düring 1973). These specialized areas, presumably receptive in nature, stretch along the whole sensory branches of non-corpuscular endings.

Topography of Fine Sensory Endings in the Knee Joint

Sensory endings of group III and group IV fibers are found in nearly all tissues of the knee joint, in particular the articular capsule, the superficial layers of the ligaments, and the tendons and muscles that insert at the joint. Most of the fine sensory endings are located within vascularized layers of the articular capsule running along venous vessels, while others extend into dense connective tissue or between fat cells. In search of a functional consequence of this differential topography, trials have been made to combine electrophysiological and morphological techniques (Messlinger et al. 1995). In these experiments, the sensory endings of functionally characterized group III units were marked with fine needles within their receptive fields, the positioning of which was guided by impulse responses to the needles. The results of this study were fairly conclusive with respect to the mechanical and chemical sensitivity of units. The sensory endings of high-threshold afferents that can be regarded as mechano-nociceptors were most frequently located in structures of dense connective tissue (ligaments, tendons, collagenous layers of the articular capsule), whereas the endings of low-threshold afferents were usually found innervating vascularized and soft connective tissues. Secondly, nociceptors that terminated in dense connective tissues were clearly less chemosensitive to the close arterial application of bradykinin or prostaglandins compared to the low mechanical threshold afferents that innervated vascularized tissues. It is not yet clear if these functional differences are determined by intracellular modifications, differences in the receptor equipment of the sensory endings, or whether they are simply dependent on the surrounding tissues.

Corpuscular Nerve Endings in Articular Tissues

Corpuscular sensory endings in the joint are Ruffini- and Pacini-like corpuscles that have been classified as type I and type II endings in early morphological studies (Freeman and Wyke 1967). Ruffini-like corpuscles are found in the fibrous joint capsule and within ligaments in different joints, while Golgi tendon organs (type III) can be found in muscles inserting at the joint (review by Zimny 1988). Morphologically these two types are very similar. Ruffini-like corpuscles have a globular or ovoid shape, are enclosed by a capsule of several cell layers, and are supplied by a myelinated nerve fiber of 5–8 μm in diameter. One nerve fiber can innervate up to six corpuscles. Within the capsule, the sensory axon ramifies forming several unmyelinated branches that wind around bundles of collagen fibers. The intracorpuscular collagen fibers may be connected with the extracorpuscular network of collagen to conduct mechanical distension to the corpuscle (Andres and von Düring 1973).

According to ultrastructural data from different species including man, there is a broad variety of corpuscular form and size, and there are also transient corpuscle types, for which the corpuscle may be incomplete or absent (Halata et al. 1985). Functionally, Ruffini-like corpuscular receptors are low-threshold, slowly adapting mechanoreceptors that respond to distension of articular structures. Pacini-like corpuscles (type II endings) are localized in the joint capsule and in periarticular fatty tissue (Freeman and Wyke 1967). They have an oval or longish form, a capsule composed of many cell layers (derived from fibroblasts and perineurial cells), and they are supplied by a thick myelinated nerve fiber, which is not ramified but runs through the long axis of the corpuscle as a central cylinder. Detailed electron microscopic studies were made on Pacini-like corpuscles in the knee joint of different species including man (Halata et al. 1985). Functionally, Pacini-like corpuscles are rapidly adapting mechanoreceptors with a very low threshold for movement and vibratory stimuli. Corpuscular sensory endings play only a minor, if any, role in articular nociception and pain. It is likely, however, that they regulate reflexes and posture programs that are modulated by nociceptive inputs.

References

Andres, K. H., & von Düring, M. (1973). Morphology of cutaneous receptors. In A. Iggo (Ed.), *Handbook of sensory physiology* (Somatosensory system, Vol. 2, pp. 3–28). Berlin: Springer.

Cho, W. G., & Valtschanoff, J. G. (2008). Vanilloid receptor TRPV1-positive sensory afferents in the mouse ankle and knee joints. *Brain Research, 1219*, 59–65.

Ebinger, M., Schmidt, R. F., & Heppelmann, B. (2001). Composition of the medial and posterior articular nerves of the mouse knee joint. *Somatosensory and Motor Research, 18*, 62–65.

Freeman, M. A. R., & Wyke, B. (1967). The innervation of the knee joint. An anatomical and histological study in the cat. *Journal of Anatomy, 101*, 505–532.

Halata, Z., Rettig, T., & Schulze, W. (1985). The ultrastructure of sensory nerve endings in the human knee joint capsule. *Anatomy and Embryology, 172*, 265–275.

Hanesch, U., Heppelmann, B., & Schmidt, R. F. (1991). Substance P- and calcitonin gene-related peptide immunoreactivity in primary afferent neurons of the cat's knee joint. *Neuroscience, 45*, 185–193.

Hanesch, U., Heppelmann, B., & Schmidt, R. F. (1995). Somatostatin-like immunoreactivity in primary afferents of the medial articular nerve and colocalization with substance P in the cat. *The Journal of Comparative Neurology, 354*, 345–352.

Hanesch, U., Heppelmann, B., & Schmidt, R. F. (1997). Quantification of cat's articular afferents containing calcitonin gene-related peptide or substance P innervating normal and acutely inflamed knee joint. *Neuroscience Letters, 233*, 105–108.

Heppelmann, B., Heuss, C., & Schmidt, R. F. (1988). Fiber size distribution of myelinated and unmyelinated axons in the medial and posterior articular nerves of the cat's knee joint. *Somatosensory Research, 5*, 273–281.

Heppelmann, B., Messlinger, K., Neiss, W. F., et al. (1990). Ultrastructural three-dimensional reconstruction of group III and group IV sensory nerve endings ("Free Nerve Endings") in the knee joint capsule of the cat: Evidence for multiple receptive sites. *The Journal of Comparative Neurology, 292*, 103–116.

Heppelmann, B., Messlinger, K., Neiss, W. F., et al. (1995). Fine sensory innervation of the knee joint capsule by group III and group IV nerve fibers in the cat. *The Journal of Comparative Neurology, 351*, 415–428.

Hildebrandt, C., Oqvist, G., Brax, L., et al. (1991). Anatomy of the rat knee joint and fibre composition of a major articular nerve. *Anatomical Record, 229*, 545–555.

Hinsey, J. C. (1927). Some observations on the innervation of skeletal muscle of the cat. *The Journal of Comparative Neurology, 44*, 87–195.

Langford, L. A., & Schmidt, R. F. (1983). Afferent and efferent axons in the medial and posterior articular nerves of the cat. *Anatomical Record, 206*, 71–78.

Messlinger, K. (1996). Functional morphology of nociceptive and other fine sensory endings ("Free Nerve Endings") in different tissues. In T. Kumazawa, L. Kruger, & K. Mizumura (Eds.), *The polymodal receptor – A gateway to pathological pain* (Progress in brain research, Vol. 113, pp. 273–298). Amsterdam: Elsevier.

Messlinger, K., Pawlak, M., Steinbach, H., et al. (1995). A new combination of methods for localization, identification, and three-dimensional reconstruction of the sensory endings of articular afferents characterized by electrophysiology. *Cell and Tissue Research, 281*, 283–294.

Salo, P. T., & Theriault, E. (1997). Number, distribution and neuropeptide content of rat knee joint afferents. *Journal of Anatomy, 190*, 515–522.

Saxler, G., Löer, F., Skumavc, M., Pförtner, J., & Hanesch, U. (2007). Localization of SP- and CGRP-immunopositive nerve fibers in the hip joint of patients with painful osteoarthritis and of patients with painless failed total hip arthroplasties. *European Journal of Pain, 11*, 67–74.

Tamura, R., Hanesch, U., Schmidt, R. F., et al. (1998). Examination of colocalization of calcitonin gene-related peptide- and substance P-like immunoreactivity in the knee joint of the dog. *Neuroscience Letters, 254*, 53–56.

Weddell, G., & Harpman, J. A. (1940). The neurohistological basis for the sensation of pain provoked from deep fascia, tendon, and periosteum. *The Journal of Neurology, Neurosurgery, and Psychiatry, 3*, 319–328.

Zimny, M. L. (1988). Mechanoreceptors in articular tissues. *The American Journal of Anatomy, 182*, 16–32.

Articular Nociceptors

Hans-Georg Schaible
Department of Physiology - Neurophysiology, Friedrich-Schiller-University of Jena, Jena, Germany

Synonyms

Joint nociceptors

Definition

▶ Articular nociceptors are primary afferent neurons in joint nerves (or primary afferent neurons supplying joints) that signal and encode the impact of noxious stimuli to joints. In the normal joint, articular nociceptors are mainly or exclusively activated by noxious mechanical stimuli applied to the joint. Articular nociceptors are sensitized for mechanical stimulation in the process of joint inflammation.

Such processes are involved in inflammatory joint diseases such as rheumatid arthritis as well as in osteoarthritis.

Characteristics

Pain in Joints

Pain in a normal joint is commonly elicited by twisting or hitting the joint. In conscious humans pain in the normal joint can be elicited when noxious mechanical or chemical stimuli are applied to the fibrous structures, such as ligaments and fibrous capsule. No pain is elicited by stimulation of cartilage. Stimulation of normal synovial tissue rarely evokes pain. Stimulation of fibrous structures with innocuous mechanical stimulation can evoke pressure sensations.

Joint inflammation such as rheumatoid arthritis or gout is characterized by hyperalgesia and persistent pain at rest. Noxious stimuli cause stronger pain than normal, and pain is even evoked by mechanical stimuli whose intensity does not normally elicit pain, i.e., movements in the working range and gentle pressure, e.g., during palpation. Discharge properties of joint nociceptors correspond to these characteristic phenomena of joint pain (Schaible and Grubb 1993; Schaible 2006).

The most frequent cause of joint pain is osteoarthritis (OA) (Felson 2005). In the initial stages, pain in the osteoarthritic joints is elicited by movements and loading of the joint but at later stages resting pain may occur (Scott 2006). It is conceivable that pain during OA has a strong inflammatory component because severe inflammatory processes such as infiltration of osteoarthritic joints with inflammatory cells may occur, and the profile of elevated cytokine production may be similar to that of inflammatory diseases (Schaible 2012).

Anatomy of Joint Innervation

Joints are innervated by branches descending from main nerve trunks or their muscular, cutaneous, and periosteal branches. A typical joint nerve contains thick myelinated Aβ (group II), thinly myelinated Aδ (group III), and a high proportion (~80 %) of unmyelinated C (group IV) fibers. The latter are either sensory afferents or sympathetic efferents (each ~50 %). While Aβ fibers with corpuscular endings of the Ruffini, Golgi, and Pacini type in fibrous capsule, articular ligaments, menisci, and adjacent periosteum are not nociceptive, numerous articular Aδ and C fibers are nociceptive. Aδ and C fibers terminate as unencapsulated free nerve endings in the fibrous capsule, adipose tissue, ligaments, menisci, and the periosteum. Using staining for

nerve fibers and neuropeptides, endings were also identified in the synovial layer. The major neuropeptides in joint nerves are ▶ substance P, ▶ CGRP, and ▶ somatostatin. Neurokinin A, ▶ galanin, enkephalins, and neuropeptide Y have also been localized in joint afferents. The cartilage is not innervated (Schaible 2006, 2013).

Figure 1 shows the reconstruction of peripheral nerve endings of joint afferents of cat's knee joint. Typical free nerve endings of joint afferents are ensheathed by ▶ Schwann cells, and only some sites are not covered. These exposed areas appear as a string of beads. It is assumed that these exposed areas are receptive sites along the fibers (Heppelmann et al. 1990).

Mechanosensitivity of Joint Afferents

Joint afferents have been mainly recorded in articular nerves supplying cat knee and rat knee and ankle joints. They were characterized after their responses to innocuous and noxious mechanical stimuli. Light to moderate pressure applied to the joint and movements within the working range of the joint are innocuous stimuli which are not normally painful. Noxious stimuli are strong pressure at intensities that are felt as pain and movements exceeding the working range of the joint, such as twisting against the resistance of the tissue.

Figure 2 shows four typical joint afferents of cat's knee joint with different sensitivities to movements. Figure 2a displays a low-threshold Aδ fiber. This fiber had two receptive fields in the fibrous capsule (dots). It responded phasically to extension (ext) of the knee, and it was strongly activated by inward rotation (IR) within the working range of the knee joint. This fiber was thus activated by innocuous movements, i.e., the threshold was in the innocuous range. However, strongest responses were elicited by noxious movements such as noxious inward rotation (n.IR). Typically, these neurons are also activated by light pressure applied to the receptive field. Such a response pattern is also seen in many low-threshold Aβ fibers in the fibrous capsule and in ligaments including the anterior cruciate ligament. Although these units have their

Articular Nociceptors, Fig. 1 Schematic drawings of a group III (Aδ) and group IV (C) fiber sensory ending in the knee joint capsule of the cat. A terminal tree is formed by several long and short branches. The sensory axons consist of periodically arranged thick and thin segments forming spindle-shaped beads. The axolemma is not completely unsheathed by its accompanying Schwann cells; the bare areas are presumably receptive sites (From Heppelmann et al. (1990))

SENSORY ENDING

terminal tree

III   IV

III: TERMINAL BRANCH

end bulb

axonal bead

IV: TERMINAL BRANCH

end bulb

axonal bead

AXONAL BEAD

Schwann cell

sensority axon

bare area (recept.site)

axonal bulge

strongest response in the noxious range they are not considered nociceptive neurons. The discharge rate seems to encode the strength of a stimulus from the innocuous to the noxious range, but it does not encode the presence of a noxious stimulus *per se*. In fact, the most adequate innocuous mechanical stimulus can evoke a stronger response than a noxious mechanical stimulus, e.g., a noxious movement in another direction.

The other fiber types shown in Fig. 2 respond mainly or exclusively to noxious mechanical stimuli. Figure 2b shows an Aδ fiber with a receptive field in the patellar ligament (dot). This unit responded weakly, with a few spikes only, during outward rotation in the working range (OR), but it had a strong response to noxious outward rotation (n.OR). The C fiber in Fig. 2c, with a receptive field in the fibrous capsule, was exclusively activated by noxious movements. It did not respond to any innocuous movement, but showed pronounced responses when the joint was twisted (noxious outward rotation, n.OR). These neurons require also high pressure intensity to elicit a response by probing the receptive field. Figure 2d displays an Aδ fiber with a receptive field in the anterior capsule that did not respond to any innocuous or noxious movement, but responded to noxious pressure onto the receptive field.

Not shown in Fig. 2 are sensory neurons which are mechanoinsensitive under normal conditions. These neurons can be identified by electrical stimulation of the joint nerve, but under normal

Articular Nociceptors, Fig. 2 Four different articular afferents of cats knee joint exemplifying classes of afferents according to their responses to passive movements. *Dots* in the insets: receptive fields identified by probing the joint. *Ext* extension, *IR* inward rotation (pronation), *OR* outward rotation (supination), *n.IR* and *n.OR* noxious IR and OR; mid pos, mid (resting) position (From Schaible and Grubb (1993))

conditions, no receptive field is found, and no response is elicited by any innocuous and noxious movement. They respond to injection of KCl into the joint artery, and some seem to respond to inflammatory mediators. However, a proportion of these neurons is rendered mechanosensitive during inflammation in the joint (see below), and therefore these units were called silent nociceptors. An estimation is that about one-third of the sensory C fibers and a small proportion of Aδ fibers in the joint nerve are mechanoinsensitive. The proportion of silent nociceptors can be different in different joint nerves. For example, the posterior articular nerve of cat's knee seems to contain many more silent nociceptors than the medial articular nerve.

Figure 3 shows for the medial articular nerve of cat's knee joint the proportions of Aβ, Aδ, and C fibers in the categories defined in Fig. 2. Only those neurons that had a detectable receptive field and that were activated by innocuous and/or noxious mechanical stimuli applied to the normal joint are included (initially mechanoinsensitive sensory neurons are not included). It is shown that most Aβ fibers were either strongly or weakly activated by innocuous stimuli. More than 50 % of the Aδ fibers and about 70 % of the sensory C fibers were classified as high-threshold units (Schaible and Grubb 1993; Schaible 2006).

Changes of Mechanosensitivity of Joint Afferents During Inflammation

It has been pointed out above that an inflamed joint hurts during movements in the working range and during palpation, and that pain may occur under resting conditions. It is a characteristic feature of joint nociceptors that their mechanosensitivity is increased during inflammation. Many low-threshold Aδ and C fibers show increased responses to movements in the working range. Most strikingly, a large proportion of high-threshold afferents (see Fig. 2c, d)

Group II
21–60 m/s

Group III
5.5–20 m/s

Group IV
<2.5 m/s

14%

54%

32%

24%

33.5%

30.5%

12%

10%

36.5%

19.5%

34%

⫽⫽ Activated by non noxious movements

■ Weakly activated by non noxious movements

〰 Activated only by noxious movements

▦ Not activated by movements

Articular Nociceptors, Fig. 3 Mechanosensitivity of primary afferent neurons supplying normal cat's knee joint. The graph shows the proportions of $A\beta$, $A\delta$, and C fibers in the different sensitivity classes (From Schaible and Grubb (1993))

are sensitized such that they respond to movements in the working range of the joint.

Furthermore, initially mechanoinsensitive afferents (silent nociceptors) are sensitized and become mechanosensitive (Schaible and Grubb 1993). Thus, silent nociceptors are recruited for the encoding of noxious events during an inflammatory process. Increased mechanosensitivity has also been noted during chronic forms of arthritis, suggesting that mechanical sensitization is an important neuronal basis for chronic persistent hyperalgesia of the inflamed joint (Guilbaud et al. 1985).

Chemosensitivity of Joint Afferents

The vast majority of sensory $A\delta$ and C fibers in the joint nerve are chemosensitive for endogenous compounds that are produced and released during pathophysiological conditions and that contribute to the inflammatory process itself. Such mediators are able to excite and/or sensitize primary afferent neurons for mechanical stimuli and for chemical stimuli. The following general aspects should be noted. First, these mediators only affect $A\delta$ and C fibers of the joint nerve, not $A\beta$ fibers. Second, an effect is typically elicited only in subpopulations of the articular units (i.e., not all units express the full range of chemosensitivity). Third, high-threshold

(nociceptive) as well as low-threshold $A\delta$ and C articular afferents (not nociceptive specific) are affected or not affected by a certain mediator. Thus, the chemosensitivity of a unit is not strictly correlated to its mechanosensitivity (Schaible and Grubb 1993; Schaible 2006).

Mediators such as bradykinin, prostaglandins E_2 and I_2, and serotonin can induce firing of neurons and/or increase their responses to movements. After bolus injections of the compounds into the joint artery, differences in the pattern of effects were noted. Bradykinin excited joint afferents less than 1 min, but sensitized them for mechanical stimuli for several minutes even if bradykinin did not excite the neuron. Both PGE_2 and PGI_2 cause ongoing discharges and/or sensitization to mechanical stimulation of the joint. The effect of PGE_2 has a slow onset and duration of minutes, the action of PGI_2 begins quickly and has a short duration. In the rat ankle joint PGI_2 excites and sensitizes a much larger proportion of units than PGE_2. In addition, these PGs sensitize joint afferents to the effects of bradykinin, whether or not they have an excitatory effect by themselves. PGE_2 and bradykinin together can cause a stronger sensitization to mechanical stimulation than bradykinin or PGE_2 alone.

Bradykinin receptors are also involved in mechanical sensitization evoked by the proteinase-activated receptor-4 (Russell et al. 2010). Aδ- and C-fibers of the joint are also sensitized to mechanical stimuli by serotonin (Schaible and Grubb 1993; Schaible 2006), substance P, and VIP (Herbert and Schmidt 2001). Acid-sensing ion channel 3 may be involved in the increase of mechanosensitivity because ASIC3 gene knockout mice showed less arthritis-induced mechanical hyperalgesia (Yuan et al. 2010). Mainly excitatory effects were elicited by capsaicin, ATP, and adenosin (Dowd et al. 1998a, b). The peptides galanin (Heppelmann et al. 2000), neuropeptide Y (Just and Heppelmann 2001), and nociceptin (McDougall et al. 2000) sensitized some neurons and reduced responses in other neurons. However, some mediators are just inhibitory (see below).

Recordings from joint nociceptors of rat knee joint showed that a single intraarticular injection of proinflammatory cytokines causes a slow increase in mechanosensitivity (Schaible et al. 2010). The cytokines tumor necrosis factor-α (TNF-α), interleukin-1ß (IL-1ß), and interleukin-6 (IL-6) play a major role in the pathogenesis of immune-mediated arthritis, and they are involved in human RA and OA models (Schaible 2013). Administration of TNF- or IL-6 into the normal knee joint sensitized Aδ- (TNF-α) and C-fibers (TNF-α and IL-6) to innocuous and noxious rotation of the knee joint. The mechanical sensitization by IL-6 was enhanced by coadministration of the soluble receptor sIL-6R. Increases in the mechanosensitivity of C-fibers were also generated by interleukin-1ß (IL-1ß) and interleukin-17A (IL-17A) (Ebbinghaus et al. 2012; Richter et al. 2012). The so-induced mechanical sensitization persisted for hours which contrasts to the short-lasting mechanosensitivity increases after intraartertial injection of "classical mediators" (see above). However, IL-1ß significantly reduced the mechanosensitivity of Aδ-fibers (Ebbinghaus et al. 2012).

A number of compounds reduce the (enhanced) mechanosensitivity of Aδ- and C-fibers of the joint. Intraarticular opioids reduce the discharges of joint afferents (Russell et al. 1987) and peripheral opioids produce profound analgesia (Stein et al. 2009). A reduction of mechanosensitivity of joint afferents is also produced by somatostatin (Heppelmann and Pawlak 1997; Pawlak and Schmidt 2004). The activation of somatostatin receptors may be an interesting option for future pain therapy (Helyes et al. 2004; Imhof et al. 2011; Yao et al. 2008). Cannabinoid receptor agonists reduce mechanosensitivity which may be of therapeutical relevance (Schuelert and McDougall 2008). However, the CB1 receptor agonist anandamide may also activate TRPV1 receptors (Schuelert and McDougall 2008), and a cannabinoid 2 receptor agonist may even cause excitation of joint afferents of the osteoarthritic joint (Schuelert et al. 2010).

References

Dowd, E., McQueen, D. S., Chessell, I. P., & Humphrey, P. P. A. (1998a). P2X receptor-mediated excitation of nociceptive afferents in the normal and arthritic rat knee joint. *British Journal of Pharmacology, 125,* 341–346.

Dowd, E., McQueen, D. S., Chessell, I. P., & Humphrey, P. P. A. (1998b). Adenosine A1 receptor-mediated excitation of nociceptive afferents innervating the normal and arthritic rat knee joint. *British Journal of Pharmacology, 125,* 1267–1271.

Ebbinghaus, M., Uhlig, B., Richter, F., Segond von Banchet, G., Gajda, M., Bräuer, R., & Schaible, H.-G. (2012). The role of interleukin-1ß in arthritic pain: Main involvement in thermal but not in mechanical hyperalgesia in rat antigen-induced arthritis. *Arthritis Rheumatism, 64,* 3897–3907.

Felson, D. T. (2005). The sources of pain in knee osteoarthritis. *Current Opinion Rheumatology, 17,* 624–628.

Guilbaud, G., Iggo, A., & Tegner, R. (1985). Sensory receptors in ankle joint capsules of normal and arthritic rats. *Experimental Brain Research, 58,* 29–40.

Helyes, Z., Szabó, A., Németh, J., Jakab, B., Pintér, E., Bánvölgyi, A., Kereskai, L., Kéri, G., & Szolcsányi, J. (2004). Antiinflammatory and analgesic effects of somatostatin released from capsaicin-sensitive sensory nerve terminals in a Freund´s adjuvant-induced chronic arthritis model in the rat. *Arthritis Rheumatism, 50,* 1677–1685.

Heppelmann, B., & Pawlak, M. (1997). Inhibitory effect of somatostatin on the mechanosensitivity of articular afferents in normal and inflamed knee joints of the rat. *Pain, 73,* 377–382.

Heppelmann, B., Messlinger, K., Neiss, W., & Schmidt, R. F. (1990). Ultrastructural three-dimensional reconstruction of group III and group IV sensory nerve endings (free nerve endings) in the knee joint capsule of the rat: Evidence for multiple receptive sites. *The Journal of Comparative Neurology, 292,* 103–116.

Heppelmann, B., Just, S., & Pawlak, M. (2000). Galanin influences the mechanosensitivity of sensory endings in the rat knee joint. *European Journal of Neuroscience, 12*, 1567–1572.

Herbert, M. K., & Schmidt, R. F. (2001). Sensitisation of group III articular afferents to mechanical stimuli by substance P. *Inflammation Research, 50*, 275–282.

Imhof, A.-K., Glück, L., Gajda, M., Luck, A., Bräuer, R., Schaible, H.-G., & Schulz, S. (2011). Differential antiinflammatory and antinociceptive effects of octreotide and pasireotide in a model of immune-mediated arthritis. *Arthritis Rheumatism, 63*, 2352–2362.

Just, S., & Heppelmann, B. (2001). Neuropeptide y changes the excitability of fine afferent units in the rat knee joint. *British Journal of Pharmacology, 132*, 703–708.

McDougall, J. J., Pawlak, M., Hanesch, U., & Schmidt, R. F. (2000). Peripheral modulation of rat knee joint afferent mechanosensitivity by Nociceptin/Orphanin FQ. *Neuroscience Letters, 288*, 123–126.

Pawlak, M., & Schmidt, R. F. (2004). Octreotide, a somatostatin analogue, attenuates movement evoked discharges of fine afferent units from inflamed knee joints of rats. *Neuroscience Letters, 361*, 80–183.

Richter, F., Natura, G., Ebbinghaus, M., Segond von Banchet, G., Hensellek, S., König, C., Bräuer, R., & Schaible, H.-G. (2012). Interleukin-17 sensitizes joint nociceptors for mechanical stimuli and contributes to arthritic pain through neuronal IL-17 receptors in rodents. *Arthritis Rheumatism, 64*, 4125–4134.

Russell, N. J. W., Schaible, H.-G., & Schmidt, R. F. (1987). Opiates inhibit the discharges of fine afferent units from inflamed knee joint of the cat. *Neuroscience Letters, 76*, 107–112.

Russell, F. A., Veldhoen, V. E., Tchitchkan, D., & McDougall, J. J. (2010). Proteinase-activated receptor-4 (PAR4) activation leads to sensitization of rat joint primary afferents via a bradykiin B2 receptor-dependent mechanism. *Journal of Neurophysiology, 103*, 155–163.

Schaible, H. G. (2006). Basic mechanisms of deep somatic pain. In S. B. McMahon & M. Koltzenburg (Eds.), Textbook of Pain (5th ed.). London: Elsevier Churchill Livingstone.

Schaible, H.-G. (2012). Mechanisms of chronic pain in osteoarthritis. *Current Rheumatology Reports, 14*, 549–556.

Schaible, H.-G. (2013). Joint pain: Basic mechanisms. In S. B. McMahon & M. Koltzenburg (Eds.), *Textbook of pain* (6th ed.). London: Churchill Livingstone.

Schaible, H.-G., & Grubb, B. D. (1993). Afferent and spinal mechanisms of joint pain. *Pain, 55*, 5–54.

Schaible, H.-G., & Schmidt, R. F. (1983a). Activation of groups III and IV sensory units in medial articular nerve by local mechanical stimulation of knee joint. *Journal of Neurophysiology, 49*, 35–44.

Schaible, H.-G., & Schmidt, R. F. (1983b). Responses of fine medial articular nerve afferents to passive movements of knee joint. *Journal of Neurophysiology, 49*, 1118–1126.

Schaible, H.-G., & Schmidt, R. F. (1985). Effects of an experimental arthritis on the sensory properties of fine articular afferent units. *Journal of Neurophysiology, 54*, 1109–1122.

Schaible, H.-G., & Schmidt, R. F. (1988). Time course of mechanosensitivity changes in articular afferents during a developing experimental arthritis. *Journal of Neurophysiology, 60*, 2180–2195.

Schaible, H.-G., Segond von Banchet, G., Boettger, M. K., Bräuer, R., Gajda, M., Richter, F., Hensellek, S., Brenn, D., & Natura, G. (2010). The role of proinflammatory cytokines in the generation and maintenance of joint pain. *Annals New York Academy of Sciences, 1193*, 60–69.

Schuelert, N., & McDougall, J. J. (2008). Cannabinoid-mediated antinociception is enhanced in rat osteoarthritic knees. *Arthritis Rheumatism, 58*, 145–153.

Schuelert, N., Zhang, C., Mogg, A. J., Broad, L. M., Hepburn, D. L., Nisenbaum, E. S., Johnson, M. P., & McDougall, J. J. (2010). Paradoxical effects of the cannabinoid CB(2) receptor agonist GW405833 on rat osteoarthritic knee joint pain. *Osteoarthritis and Cartilage, 18*, 1536–1543.

Scott, D. L. (2006). Osteoarthritis and rheumatoid arthritis. In S. B. McMahon & M. Koltzenburg (Eds.), *Textbook of pain* (5th ed.). London: Churchill Livingstone.

Stein, C., Clark, J. D., Oh, U., Vasko, M. R., Wilcox, G. L., Overland, A. C., Vanderah, T. W., & Spencer, R. H. (2009). Peripheral mechanisms of pain and analgesia. *Brain Research Reviews, 60*, 90–113.

Yao, F., Guo, Y., Lu, S., Sun, C., Zhang, Q., Wang, H., & Zhao, Y. (2008). Mechanical hyperalgesia is attenuated by local administration of octreotide in pristine-induced arthritis in Dark-Agouti rats. *Life Sciences, 83*, 732–738.

Yuan, F.-L., Chen, F.-H., Lu, W.-G., & Li, X. (2010). Acid sensing ion channels 3: A potential therapeutic target for pain treatment in arthritis. *Molecular Biology Reports, 37*, 3233–3238.

Articular Sensory Receptors

▶ Articular Afferents, Morphology

As Needed Dosage Regimen

Definition

A method of dose titration where the dose of drug is fixed and the interval between administered doses is determined by the response of the patient.

In analgesic therapy using "as needed" regimen, the dose of an opioid is not repeated until the patient reports the return of some intensity of pain.

Cross-References

▶ Opioid Rotation

Ascending Nociceptive Pathways

Glenn J. Giesler
Department of Neuroscience, University of Minnesota, Minneapolis, MN, USA

Introduction

Nociceptive neurons in the spinal cord and trigeminal nuclei send their axons to terminate within a large number of regions in the upper cervical spinal cord, brain stem, and diencephalon. The precise roles of each of these pathways in nociception have not yet been established with certainty, and it is likely that their roles vary among species. This entry presents a summary of prominent findings on several of the most thoroughly examined ascending nociceptive projections. Many specific topics are dealt with in more detail by individual contributors to the Encyclopedic Reference of Pain. These are referred to throughout this entry.

Characteristics

Spinothalamic Tract

The most widely studied ascending nociceptive pathway originating in the spinal cord is the spinothalamic tract (STT). Nearly 100 years ago, anatomical studies indicated that lesions of the spinal cord caused the degeneration of axons within the thalamus. These early studies were performed in a variety of species including

Ascending Nociceptive Pathways, Fig. 1 Schematic representation of spinothalamic tract and thalamic projection to primary somatic sensory cortex from the ventral posterior lateral nucleus of thalamus

primates. The first description of spinothalamic tract axons has been attributed to Edinger (Willis and Coggeshall 2004). In the early 1970s, antidromic activation techniques were first used to identify and functionally characterize spinothalamic tract neurons. These methods have been used in a large number of studies in the intervening years to examine many facets of the function and organization of this pathway. Both anterograde and retrograde tracing techniques have also been used extensively to determine the locations and numbers of the cells of origin of the spinothalamic tract in several species as well as the areas of termination of spinothalamic tract axons. Several of the prominent features of the spinothalamic tract are schematically illustrated in Fig. 1. The cells of origin of this pathway are found within the spinal gray matter at all levels of the cord. STT neurons comprise a small percentage of spinal cord neurons. It has been estimated, based on retrograde tracing studies, that there are between 10,000 and 20,000 STT neurons in the spinal cord of primates (Apkarian and Hodge 1989; Willis and Coggeshall 2004). Particularly high concentrations have been described in upper cervical segments. Within the gray matter, STT neurons are concentrated in the marginal zone (see ▶ Spinothalamic Tract Neurons, Morphology) and within the deep dorsal horn (lamina V; see ▶ Spinothalamic Tract Neurons, in Deep Dorsal Horn). A sizeable number of STT neurons are also located within the intermediate gray zone and the ventral horn (Willis et al. 1979; Giesler et al. 1981). STT cell bodies and dendrites receive glutamatergic inputs (see ▶ Spinothalamic Tract Neurons, Glutamatergic input) and several types of peptidergic inputs including substance P and CGRP (see ▶ Spinothalamic Tract Neurons, Peptidergic input). Nitric oxide appears to play an important role in modulating the activity of STT neurons (see ▶ Spinothalamic Tract Neurons, Descending Control by Brain Stem Neurons). Axons of STT neurons decussate at a level near the cell body, and the majority turns and ascends within the ventral half of the lateral funiculus. Several groups of investigators have shown that STT axons originating in marginal zone neurons ascend in a position that is dorsal to STT axons originating in neurons within the deep dorsal horn. Within thoracic levels, STT axons of marginal zone neurons are generally located dorsal to the denticulate ligament in the dorsal lateral funiculus, whereas the axons of lamina V neurons are found within the ventral part of the lateral funiculus (Apkarian and Hodge 1989; Zhang et al. 2000). There is also a somatotopic organization of STT axons. Axons from lumbosacral levels ascend on the periphery of the lateral funiculus, whereas STT axons from rostral levels are located closer to the gray matter (Applebaum et al. 1975).

STT axons ascend through the lateral and ventral brain stem. Collateral branches are frequently given off by these axons supplying nociceptive sensory information to a number of nuclei, particularly within the reticular formation, throughout the length of the brain stem.

Several early clinical cases in which injury to the spinal cord had blocked the sense of pain in patients indicated that the axons carrying nociceptive information crossed within the spinal cord and ascended within the ventral or anterior half (see ▶ Cordotomy Effects on Humans and Animal Models; Willis and Coggeshall 2004). These observations led to the first surgical attempts to relieve chronic pain by cutting the anterolateral quadrant of the spinal cord, the area that carries the overwhelming majority of spinothalamic tract axons. This procedure, cordotomy, can very effectively produce pain relief for patients, but the positive effects are short lived and pain frequently returns within a few months or a year. It is not known which tracts begin to carry the nociceptive information following a cordotomy. More studies are needed on this important phenomenon.

Although cordotomies are infrequently used now in the United States to relieve pain (they have generally been replaced with the use of opiates), they continue to be used by neurosurgeons in many countries. In the early years, laminectomies were performed to allow the lesions to be made. Cordotomies are now frequently done percutaneously under local anesthesia.

STT axons terminate in three principal regions of the thalamus including the ventral posterior lateral (VPL), central lateral and adjacent parts of the medial dorsal nucleus, and posterior thalamic nuclei (Mehler 1969; Cliffer et al. 1989; Craig 2004; Craig et al. 1994; Graziano and Jones 2004). STT terminations within VPL are somatotopically organized. Axons ascending from lumbosacral levels terminate within the lateral part of VPL; those from cervical levels end within the medial part of the nucleus. Within VPL of primates, STT terminals are concentrated within small areas that are surrounded by large regions that are dominated by the endings of medial lemniscal axons. In carnivores, STT axons are concentrated within the periphery of VPL. It has been shown that a high percentage of nociceptive neurons within the primate VPL can be antidromically activated from SI parietal cortex, indicating that nociceptive input to VPL neurons via STT axons is transmitted to the cortex.

A second area of termination of the STT is the central lateral nucleus and the adjacent lateral region of the medial dorsal nucleus. There does not appear to be a somatotopic organization to this termination. Many of the nociceptive neurons within this area of the thalamus have large, bilateral, even whole-body receptive fields. Thus, it is unlikely that this region is involved in localization of nociceptive stimuli. It appears more likely that STT inputs and thalamic neurons within this region are involved in the production of affective/emotional and motor responses to nociceptive stimulation via their widespread cortical projections. It has been shown that STT neurons that project to this region are frequently located within the intermediate zone and ventral horn of the spinal cord. Many of these STT neurons have whole-body receptive fields including the face.

Responses of STT neurons to a variety of somatic and visceral stimuli have been examined (Willis and Coggeshall 2004). In primates, the vast majority of STT neurons have been classified as nociceptive, responding either preferentially (wide dynamic range, WDR) or specifically (high threshold, HT) to noxious stimuli. In most (but not all) studies, higher percentages of HT-STT neurons have been found in the marginal zone and more WDR neurons within the deep dorsal horn. Receptive fields of neurons in the marginal zone tend to be smaller sometimes being restricted to a single toe. The receptive fields of deeper neurons often cover much of the ipsilateral leg. Many STT cells are powerfully activated by noxious thermal stimulation of their receptive fields. Response thresholds to noxious heat stimuli are often between 45 °C and 55 °C (Kenshalo et al. 1979). Repeated applications of noxious heat stimuli leads to sensitization including reduced responses thresholds, increased responses to identical noxious heat stimuli, and the production of ongoing activity (see ▶ Spinothalamic Tract Neurons, Central Sensitization). STT neurons also receive nociceptive input from muscles and joints (Foreman et al. 1979), and they are activated by stimulation

using noxious chemicals (▶ Spinothalamic Tract Neurons, Responses to Chemical Stimulation). Nociceptive information originating from receptors on the face in the oral and nasal cavities is carried to the ventral posterior medial nucleus of thalamus by trigeminothalamic tract projections (see ▶ Trigeminothalamic Tract Projections).

In a large number of studies, STT neurons at a number of levels of the spinal cord have been examined for possible input from visceral structures. It has been shown that STT neurons can be activated by noxious stimulation of the heart, esophagus, urinary bladder, testicles, vagina, colon, rectum, gall bladder, and bile duct (see ▶ Spinothalamic Tract Neurons, Visceral Input). In almost all cases, STT neurons that respond to stimulation of a visceral organ have somatic receptive fields as well. Frequently, the somatic receptive fields were found to be located in areas to which noxious stimulation of the examined organ produced referred pain in human studies. These findings indicate that STT axons are capable of carrying nociceptive visceral information and that the convergence of somatic and visceral nociceptive input probably contributes to the production of referred pain.

Spinohypothalamic Tract (SHT)
In the late 1980s, Burstein et al. (1987) noted that spinal cord neurons could be antidromically activated using small amplitude current pulses delivered through electrodes located within the hypothalamus of rats. In addition, injections of anterograde tracers into the spinal cord labeled axons within several areas of the hypothalamus including the lateral, posterior, and ventromedial hypothalamus. Injections of retrograde tracers that were restricted to the hypothalamus labeled thousands of neurons within the spinal cords of rats. SHT cell bodies are located in the marginal zone and the deep dorsal horn. SHT axons have been shown to ascend to the posterior thalamus, then turn ventrally and laterally entering the supraoptic decussation. These axons continue to ascend in a position just dorsal to the optic tract and enter the hypothalamus. Many SHT axons ascend to the level of the optic chiasm where they decussate, turn posteriorly, and descend

within the supraoptic decussation on the side ipsilateral to the cell body. SHT axons have been shown to end in the ipsilateral hypothalamus, posterior thalamus, and brain stem. Some have even been shown to descend as far as the level of the medulla (Zhang et al. 1995). SHT neurons are frequently nociceptive. Some also receive an apparent input from innocuous thermoreceptors. It has been suggested that through their complex, bilateral projections and frequent branches, SHT axons could provide nociceptive input to a variety of areas of the brain stem and forebrain that are involved in nociceptive processing (see ▶ Spinohypothalamic Tract, Anatomical Organization, and Response Properties). SHT neurons have also been identified and characterized in monkeys. Large numbers of neurons within all divisions of the trigeminal complex and upper cervical segments also send axonal projections to the hypothalamus (see ▶ Trigeminohypothalamic Tract).

Spinoreticular Tract (SRT)

The SRT is a direct projection from spinal cord neurons to the reticular formation of medulla, pons, and midbrain (Fig. 2; see ▶ Spinomesencephalic Tract). Regions that receive these direct spinal afferent fibers include the nucleus gigantocellularis and nucleus dorsalis both within the medulla and the cuneiform nucleus of the midbrain. Since several of these regions in the reticular formation in turn send ascending nociceptive projections to the forebrain, it is believed that the SRT is part of a multisynaptic projection system to the thalamus and probably is involved in providing nociceptive information that is used in producing cortical arousal.

It is difficult to identify the cells of origin of this projection with certainty since it is known that at least some axons ascending to higher levels of the brain stem or diencephalon pass through or near the reticular formation without giving off collaterals within it. Such axons would not be considered as part of the SRT since they do not provide information to neurons within the reticular formation. Injections of retrograde tracers into the reticular formation could be taken up by such axons as well as by SRT axons. In addition, such axons may

Ascending Nociceptive Pathways, Fig. 2 Schematic illustration of spinoreticular tract

be activated in studies in which antidromic activation techniques are used to examine SRT neurons. Measures have been taken such as stimulating with high amplitude current pulses at higher levels of the neuraxis to insure that examined axons do not ascend beyond the area of interest. Studies in which antidromic methods have been used have shown that many SRT neurons are nociceptive (Fields et al. 1975; Haber et al. 1982; Yezierski and Schwartz 1986). These neurons have been frequently recorded deep within the spinal gray matter and have large complex receptive fields frequently including the face. Retrograde tracing studies indicate that SRT neurons are found within the marginal zone and deep dorsal horn, but a large percentage are located within the intermediate zone and ventral horn (Menetrey et al. 1983).

Spinoparabrachial Tract

Somatic sensory and nociceptive information ascends directly from the spinal cord to several

Ascending Nociceptive Pathways, Fig. 3 Schematic representation of spinoparabrachial tract projection

sub-nuclei of the parabrachial nucleus, which is located lateral to the superior cerebellar peduncle within the rostral pons and caudal midbrain (Fig. 3; see ▶ Spinoparabrachial Tract and ▶ Parabrachial Hypothalamic and Amygdaloid Projections). The locations of the cells of origin of the spinoparabrachial tract have been established using electrophysiological and anatomical techniques. Injections of retrograde tracers that are restricted to the parabrachial nucleus label a large number of spinal neurons at all levels of the spinal cord of rats and cats. Although spinoparabrachial tract neurons are found throughout much of the gray matter, the fact that they are highly concentrated within the marginal zone has attracted a great deal of interest in this projection. Anterograde tracing studies indicate that neurons in the marginal zone send a large projection via the dorsal part of the lateral funiculus to the parabrachial nuclei on both sides (Bernard et al. 1989, 1995). Studies in which

antidromic activation has been used to identify spinoparabrachial tract neurons in cats indicate that the overwhelming majority is activated by noxious stimuli (Hylden et al. 1986; Light et al. 1993). The parabrachial nuclei are known to have large projections to several areas of the forebrain that are involved in nociception including the hypothalamus and the amygdala (Bernard et al. 1989; see ▶ Parabrachial Hypothalamic and Amygdaloid Projections). Therefore, this projection appears well suited for providing nociceptive information that is used for producing cognitive, emotional, or affective responses to pain.

Spinocervicothalamic Tract (SCT)

The spinocervicothalamic projection is schematically depicted in Fig. 4. Spinocervical tract neurons are located throughout the length of the spinal cord (Craig 1978). These neurons send their ascending axons into the dorsal part of the ipsilateral lateral funiculus. SCT axons ascend to upper cervical segments where they terminate within the lateral cervical nucleus, an island of neurons located with the dorsal lateral funiculus. The lateral cervical nucleus (LCN) extends from segment C3 through C1. The number of neurons that form the LCN varies greatly among species. The LCN is a large prominent nucleus in carnivores (Truex et al. 1970) and can contain as many as 10,000 neurons. In cats, lesions of the dorsal lateral funiculus have been reported to reduce nociceptive responses. A. G. Brown and colleagues (1981) performed an elegant and thorough series of studies of SCT neurons in cats. SCT neurons are located in the deep dorsal horn (laminae III-V). Many receive a powerful afferent input from innocuous mechanoreceptors. Evidence from a number of studies indicates that as many as half of SCT neurons also receive a nociceptive input (Cervero et al. 1977; Brown 1981; Kniffki et al. 1977). These SCT neurons respond to noxious mechanical and thermal stimuli.

In rodents, the LCN has been shown to be at least an order of magnitude smaller. The LCN is also comparatively small in monkeys although precise cell counts are not available. The LCN

Thalamus

Pons

Medulla

Lateral Cervical
Nucleus

Spinal Cord

axons in
ipsi DLF

Ascending Nociceptive Pathways, Fig. 4 Schematic
representation of the spinocervicothalamic tract

has been examined in humans, and it has been
reported to be highly variable. Truex et al. (1970)
reported that some individuals appear to have
a prominent LCN on one side and few if any
LCN neurons on the other. No evidence in
Nissl-stained material could be found for an
LCN in several other individuals. Other individuals appeared to have a clear LCN on both sides.
These findings suggest a lesser, variable role for
the spinocervicothalamic tract in nociception in
individual humans.

Physiological studies have indicated that
roughly half of LCN neurons in carnivores are
nociceptive (Kajander and Giesler 1987). LCN
neurons have been shown to respond specifically
or preferentially to noxious mechanical stimuli.
Many of these neurons can also be activated by
noxious heat stimuli. LCN neurons that receive
mechanoreceptive or nociceptive input are
somatotopically organized; neurons in the lateral
LCN receive input from lumbosacral segments,

whereas neurons in the medial LCN receive input
from cervical levels. Craig and Tapper (1978)
reported that a small number of neurons in the
medial LCN have nociceptive whole-body receptive fields. Axons of LCN decussate in upper
cervical spinal cord and ascend to terminate in
the contralateral VPL (Boivie 1970). As many as
half of ascending axons of LCN neurons give off
branches that terminate within the midbrain.

**Postsynaptic Dorsal Column
Projection (PSDC)**
Early evidence for the existence of the PSDC
projection was discovered in electrophysiological experiments by Uddenberg (1968). He noted
that some axons that were recorded in the dorsal
columns responded to stimulation of ipsilateral
peripheral nerves with multiple spike discharges,
an indication that at least one synapse intervened
between the stimulated and recorded axons.
A schematic drawing illustrating the basic organization of this projection is illustrated in Fig. 5.
Injections of retrograde tracers into the dorsal
column nuclei of cats, rats, and monkeys label
large numbers of neurons throughout the length
of the spinal cord (Bennett et al. 1983; Giesler
et al. 1984). Many of these are located in nucleus
proprius or laminae III and IV. A smaller number
are found near the central canal (see ▶ Postsynaptic Dorsal Column Projection, Functional
Characteristics). Anterograde tracing studies
indicate that most axons of this type ascend
within the ipsilateral dorsal columns but some
appear to ascend within the dorsal lateral funiculus (Cliffer and Giesler 1989). Such studies also
show that the terminals of this projection were
somewhat separated from the endings of primary
afferent axons within the dorsal column nuclei. In
cats, PSDC axons frequently terminate in the
periphery of the nuclei and primary afferent fibers
often terminate in the cores of the two nuclei. In
rats, the terminations of these projections appear
to overlap more substantially. It is difficult to
determine the frequencies with which PSDC
axons terminate on neurons within the dorsal
column nuclei that project to the contralateral
VPL. In cats, several electrophysiological
studies have shown that roughly half of PSDC

Ascending Nociceptive Pathways, Fig. 5 The post-synaptic dorsal column pathway

neurons can be driven exclusively by innocuous mechanical stimulation and the remainder can be classified as WDR neurons (Uddenberg 1968; Angaut-Petit 1975), indicating that this projection in cats is capable of conveying nociceptive information. These cells have been shown to be powerfully activated by noxious mechanical and heat stimuli. PSDC neurons have not been systematically examined in monkeys, but the presence of nociceptive neurons within the dorsal column nuclei in monkeys is consistent with the idea that PSDC neurons are nociceptive primates. Accurate functional classification in rats is less clear. In one early study, it was concluded that few, if any, PSDC neurons were conclusively nociceptive in rats. On the other hand, several lines of evidence indicate that noci-ceptive visceral information is carried by this projection in rats, monkeys, and possibly humans. Willis, Al-Chaer, and colleagues

(Al-Chaer et al. 1996, 1998; see ▶ Postsynaptic Dorsal Column Neurons, Responses to Visceral Input) have performed an elegant series of studies showing that PSDC neurons convey nociceptive visceral information that reaches the thalamus. They have also pointed out that surgical section of the medial area of the dorsal columns can relieve chronic visceral pain in patients. This result would appear to indicate that axons carrying nociceptive visceral information within the lateral funiculus (e.g., spinothalamic, spinoreticular, spinohypothalamic tract axons) are not sufficient to maintain visceral nociception since these axons are spared when the dorsal columns are sectioned. This seems surprising since many spinothalamic tract axons carry nociceptive visceral information and anterolateral cordotomies have been used for nearly 100 years to relieve chronic visceral pain. More studies are needed to resolve the precise roles of these path-ways in carrying nociceptive information from the viscera, particularly in primates including humans.

Spinosolitary Tract
Several types of information indicate that a number of spinal cord neurons send a direct projection to the solitary nucleus in the medulla. Injections of anterograde tracers into the spinal cord gray matter label small numbers of axons within the solitary nucleus (Cliffer et al. 1989). Injections of retrograde tracers restricted to the solitary nucleus label neurons at all segmental levels in rats (Menetrey and Basbaum 1987; Esteves et al. 1993). Spinosolitary neurons were found in the marginal zone, lamina V, and the area around the central canal, the primary areas of the spinal gray matter in which nociceptive processing occurs. At this time, the neurons in the spinal cord that project to the solitary nuclei have not been physiologically identified and characterized. Therefore, it has not been established beyond doubt that they carry nocicep-tive information. Thus, the role of this projection is not certain. Many neurons within the solitary nuclei have ascending projections, suggesting that this polysynaptic projection could contribute to nociceptive processing.

References

Al-Chaer, E. D., Lawand, N. B., Westlund, K. N., & Willis, W. D. (1996). Pelvic visceral input into the nucleus gracilis is largely mediated by the postsynaptic dorsal column pathway. *Journal of Neurophysiology, 76*, 2675–2690.

Al-Chaer, E. D., Feng, Y., & Willis, W. D. (1998). A role for the dorsal column in nociceptive visceral input into the thalamus of primates. *Journal of Neurophysiology, 79*, 3143–3150.

Angaut-Petit, D. (1975). The dorsal column system: II. Functional properties and bulbar relay of the postsynaptic fibres of the cat's fasciculus gracilis. *Experimental Brain Research, 22*, 471–493.

Apkarian, A. V., & Hodge, C. J. (1989). Primate spinothalamic pathways: I. A quantitative study of the cells of origin of the spinothalamic pathway. *The Journal of Comparative Neurology, 288*, 447–473.

Applebaum, A. E., Beall, J. E., Foreman, R. D., & Willis, W. D. (1975). Organization and receptive fields of primate spinothalamic tract neurons. *Journal of Neurophysiology, 38*, 572–586.

Bennett, G. J., Seltzer, Z., Lu, G.-W., Nishikawa, N., & Dubner, R. (1983). The cells of origin of the dorsal column postsynaptic projection in the lumbosacral enlargement of cats and monkeys. *Somatosensory Research, 1*, 131–149.

Bernard, J. F., Peschanski, M., & Besson, J. M. (1989). A possible spino(trigemino)-ponto-amygdaloid pathway for pain. *Neuroscience Letters, 100*, 83–88.

Bernard, J. F., Dallel, R., Raboisson, P., Villanueva, L., & Le Bars, D. (1995). Organization of the efferent projections from the spinal cervical enlargement to the parabrachial area and periaqueductal gray: A PHA-L study in the rat. *The Journal of Comparative Neurology, 353*, 480–505.

Boivie, J. (1970). The termination of the cervicothalamic tract in the cat. An experimental study with silver impregnation methods. *Brain Research, 19*, 333–360.

Brown, A. G. (1981). *Organization in the spinal cord: The anatomy and physiology of identified neurones* (pp. 17–72). Berlin: Springer.

Burstein, R., Cliffer, K. D., & Giesler, G. J., Jr. (1987). Direct somatosensory projections from the spinal cord to the hypothalamus and telecephalon. *Journal of Neuroscience, 7*, 4159–4161.

Cervero, F., Iggo, A., & Molony, V. (1977). Responses of spinocervical tract neurones to noxious stimulation of the skin. *Journal of Physiology (London), 267*, 537–558.

Cliffer, K. D., & Giesler, G. J., Jr. (1989). Postsynaptic dorsal column pathway of the rat. III. Distribution of ascending afferent fibers. *Journal of Neuroscience, 9*, 3146–3168.

Craig, A. D., Jr. (1978). Spinal and medullary input to the lateral cervical nucleus. *The Journal of Comparative Neurology, 181*, 729–744.

Craig, A. D. (2004). Distribution of trigeminothalamic and spinothalamic lamina I terminations in the macaque monkey. *The Journal of Comparative Neurology, 477*, 119–148.

Craig, A. D., Jr., & Tapper, D. N. (1978). Lateral cervical nucleus in the cat: Functional organization and characteristics. *Journal of Neurophysiology, 41*, 1511–1534.

Craig, A. D., Bushnell, M. C., Zhang, E. T., & Blomqvist, A. (1994). A thalamic nucleus specific for pain and temperature sensation. *Nature, 372*, 770–773.

Esteves, F., Lima, D., & Coimbra, A. (1993). Structural types of spinal cord marginal (lamina I) neurons projecting to the nucleus of the tractus solitarus in the rat. *Somatosensory and Motor Res., 10*, 203–216.

Fields, H. L., Wagner, G. M., & Anderson, S. D. (1975). Some properties of spinal neurons projecting to the medial brainstem reticular formation. *Experimental Neurology, 47*, 118–134.

Foreman, R. D., Schmidt, R. F., & Willis, W. D. (1979). Effects of mechanical and chemical stimulation of fine muscle afferents upon primate spinothalamic tract cells. *Journal of Physiology (London), 286*, 215–231.

Giesler, G. J., Jr., Yezierski, R. P., Gerhart, K. D., & Willis, W. D. (1981). Spinothalamic tract neurons that project to medial and/or lateral thalamic nuclei: Evidence for a physiologically novel population of spinal cord neurons. *Journal of Neurophysiology, 46*, 1285–1308.

Giesler, G. J., Jr., Nahin, R. L., & Madsen, A. M. (1984). Postsynaptic dorsal column pathway of the rat. I. Anatomical studies. *Journal of Neurophysiology, 51*, 260–275.

Graziano, A., & Jones, E. G. (2004). Widespread thalamic terminations of fibers arising in the superficial medullary dorsal horn of monkeys and their relation to calbindin immunoreactivity. *Journal of Neuroscience, 24*, 248–256.

Haber, L. H., Moore, B. D., & Willis, W. D. (1982). Electrophysiological response properties of spinoreticular neurons in the monkey. *The Journal of Comparative Neurology, 207*, 75–84.

Hylden, J. H. K., Hayashi, H., Dubner, R., & Bennett, G. J. (1986). Physiology and morphology of the lamina I spinomesencephalic projection. *The Journal of Comparative Neurology, 247*, 505–515.

Kajander, K. C., & Giesler, G. J., Jr. (1987). Responses of neurons in the lateral cervical nucleus of the cat to noxious cutaneous stimulation. *Journal of Neurophysiology, 57*, 1686–1702.

Kenshalo, D. R., Leonard, R. B., Chung, J. M., & Willis, W. D. (1979). Responses of primate spinothalamic neurons to graded and to repeated noxious heat stimuli. *Journal of Neurophysiology, 42*, 1370–1389.

Kniffki, K.-D., Mense, S., & Schmidt, R.-F. (1977). The spinocervical tract as a possible pathway for muscular nociception. *Journal of Physiology, Paris, 73*, 359–366.

Light, A. R., Sedivec, M. J., Casale, E. J., & Jones, S. L. (1993). Physiological and morphological characteristics of spinal neurons projecting to the parabrachial region of the cat. *Somatosensory and Motor Research*, *10*, 309–325.

Mehler, W. R. (1969). Some neurological species differences – A posteriori. *Annals of the New York Academy of Sciences, 167*, 424–468.

Menetrey, D., & Basbaum, A. I. (1987). Spinal and trigeminal projection to the nucleus of the solitary tract: A possible substrate for somatovisceral and viscerovisceral reflex activation. *The Journal of Comparative Neurology, 255*, 439–450.

Menetrey, D., Roudier, F., & Besson, J. M. (1983). Spinal neurons reaching the lateral reticular nucleus as studied in the rat by retrograde transport of horseradish peroxidase. *The Journal of Comparative Neurology, 220*, 439–452.

Truex, R. C., Taylor, M. J., Smythe, M. Q., & Gildenberg, P. L. (1970). The lateral cervical nucleus of the cat, dog, and man. *The Journal of Comparative Neurology, 139*, 93–104.

Uddenberg, N. (1968). Functional organization of long, second-order afferents in the dorsal funiculus. *Experimental Brain Research, 4*, 377–382.

Willis, W. D., Jr., & Coggeshall, R. E. (2004). *Sensory mechanisms of the spinal cord* (3rd ed.). New York: Kluwer Academic/Plenum.

Willis, W. D., Kenshhalo, D. R., Jr., & Leonard, R. B. (1979). The cells of origin of the primate spinothalamic tract. *The Journal of Comparative Neurology, 188*, 543–573.

Yezierski, R. P., & Schwartz, R. H. (1986). Response and receptive-field properties of spinomesencephalic tract cells in the cat. *Journal of Neurophysiology, 55*, 76–95.

Zhang, X., Kostarczyk, E., & Giesler, G. J., Jr. (1995). Spinohypothalamic tract neurons in the cervical enlargement of rats: Descending axons in the ipsilateral brain. *Journal of Neuroscience, 15*, 8393–8407.

Zhang, X., Wenk, H. N., Honda, C. N., & Giesler, G. J., Jr. (2000). Locations of spinothalamic tract axons in cervical and thoracic spinal cord white matter in monkeys. *Journal of Neurophysiology, 83*, 2869–2880.

Aseptic Meningitis

▶ Headache in Aseptic Meningitis

ASIC

▶ Acid-Sensing Ion Channels

ASIC1a

▶ Acid-Sensing Ion Channels

ASIC1b: ASICβ, BNaC2β

▶ Acid-Sensing Ion Channels

ASIC2a: Mammalian Degenerin 1 (MDEG1), Brain Sodium Channel 1 (BNC1a, BNaC1a)

▶ Acid-Sensing Ion Channels

ASIC2b: Mammalian Degenerin 2 (MDEG2), Brain Sodium Channel 1 (BNaC1b)

▶ Acid-Sensing Ion Channels

ASIC3: Dorsal-Root Acid-Sensing Ion Channel (DRASIC)

▶ Acid-Sensing Ion Channels

ASICa, Brain Sodium Channel 2 (BNC2, BNaC2a)

▶ Acid-Sensing Ion Channels

ASICs

Synonyms

Acid-sensing ion channels

Definition

A family of proteins combined to form *acid-sensing ion channels* (ASIC) of the degenerin family. The channels are gated by acidity (threshold pH 6.8) and are often found in nociceptive afferents. There are several subtypes of ASICs including ASIC1, ASIC2, ASIC3, and ASIC4. ASICs are expressed throughout the central and peripheral nervous system. ASIC channels are related to epithelial sodium channels in the kidney (ENaC) and to degenerins in the model organism *Caenorhabditis elegans*. ASIC channels may play a role in mediating cardiac ischemic pain by sensing extracellular acidification. Mice and *C. elegans* worms, deficient in ASIC subunits, show deficits in mechanosensation.

Cross-References

▶ Acid-Sensing Ion Channels
▶ Nociceptors in the Orofacial Region (Skin/Mucosa)
▶ TRPV1, Regulation by Protons
▶ Visceral Pain Model; Esophageal Pain

ASO

▶ Antisense Oligonucleotide

Aspartate

Definition

Aspartate is an excitatory amino-acid neurotransmitter.

Cross-References

▶ Somatic Pain

Aspirin-like Drugs

▶ Nonsteroidal Anti-Inflammatory Drugs (NSAIDs)
▶ NSAIDs and Their Indications
▶ NSAIDs, Mode of Action

Assessment

Definition

An assessment is a comprehensive description of a patient's condition designed to constitute a basis for treating or otherwise managing that condition. One systematic approach to assessment requires identifying the patient's physical, psychological, social, and vocational complaints, problems, or disabilities. Having been identified, these may be targeted individually and separately, or collectively, for treatment.

An assessment may be formulated in the absence of a diagnosis, and is thereby a substitute for a diagnosis; but it can also complement a diagnosis. In some instances, although a diagnosis may be available, it may not be possible to cure or to rectify the condition responsible for a patient's pain. In that event, formulating an assessment allows treatment to target the pain and its consequences instead of the actual cause.

Some practitioners might prefer to restrict the term "assessment" to apply to the act or process of obtaining information about a patient, and use the term "formulation" to apply to the actual description that results from this process.

Cross-References

▶ Psychological [Behavioral Health] Assessment of Pain

Assessment of Discomfort in Dementia Protocol

Synonyms

ADD protocol

Definition

Assessment of Discomfort in Dementia Protocol is an algorithm approach involving exclusion of common physical causes for discomfort in adults with dementia.

Cross-References

▶ Cancer Pain, Assessment in the Cognitively Impaired

Assessment of Hypoalgesia

▶ Hypoalgesia, Assessment

Assessment of Pain Behaviors

Francis J. Keefe and Anava A. Wren
Pain Prevention and Treatment Research, Department of Psychiatry and Behavioral Sciences, Duke University Medical Center, Durham, NC, USA

Synonyms

Analysis of Pain Behavior; Observation of Pain Behavior; Recording of Pain Behavior

Definition

Patients who have pain exhibit a variety of behaviors that serve to communicate the fact

that pain is being experienced. These behaviors have been termed "pain behaviors" (Fordyce 1976). Pain behaviors can be verbal (e.g., verbal descriptions of the intensity, location, and quality of pain; vocalizations of distress; moaning, or complaining) or nonverbal (e.g., withdrawing from activities, taking pain medication, pain-related body postures or facial expressions). Fordyce (1976) was one of the first to address the importance of pain behaviors. According to Fordyce's operant behavioral model, pain behaviors that initially occur in response to acute injury are sometimes maintained over much longer periods of time because they lead to reinforcing consequences. For example, a brief period of bed rest can be adaptive in response to ▶ acute pain, but when pain persists, excessive bed rest can promote deconditioning and decrease a person's tolerance for pain. In addition, attention from a concerned spouse may initially be helpful for someone coping with pain, but if that spouse becomes overly ▶ solicitous, such behavior may actually increase physical and psychological disability in the person experiencing pain (Fordyce 1976; Keefe and Lefebvre 1994). Pain behavior assessment allows one to identify problem pain behaviors and analyze the variables controlling those behaviors.

In the operant behavioral model (depicted in Fig. 1, adapted from Fordyce 1979), pain behaviors reflect the influence of three important factors:

1. Nociception: nervous system responses that produce aversive input
2. Pain: the conscious perception of nociception
3. Suffering: the negative emotional responses to nociception and pain

This model has several important implications. First, the model maintains that pain and pain behavior may be related and influence each other but are not necessarily synonymous. Thus, careful assessments of persons with pain should focus not only on underlying biological factors (e.g., nociception) but also on overt behaviors (e.g., verbalizations of pain, time spent in bed, pain-related body postures). Second, the model serves to guide treatment efforts designed to improve adjustment to persistent pain by

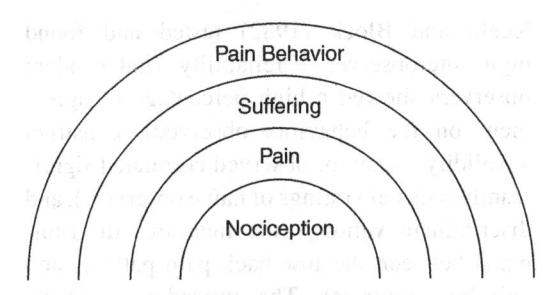

Assessment of Pain Behaviors, Fig. 1 Fordyce's (1979) behavioral model of pain

Assessment of Pain Behaviors, Fig. 2 Keefe and Lefebvre's (1994) systems model of pain behavior

modifying pain behavior. Behavioral treatments based on this model include graded activation programs in which:

- Patients learn to gradually increase their activity level
- Pain medications are switched from a ▶ pro re nata (prn) basis to time-contingent basis

Behavioral treatment protocols based on these methods have been found to be effective in randomized clinical trials (see Turner 1996 for a review of this literature).

Recently, behavioral theorists (e.g., Keefe and Lefebvre 1999) have developed more comprehensive models of pain behavior based on ▶ systems theory. As illustrated by Fig. 2, these models maintain that pain behavior can influence and be influenced by an array of environmental, psychological, and behavioral factors. For example, social reinforcement for engaging in exercise (an environmental factor) could increase ▶ self-efficacy (a psychological factor), which, in turn, could decrease emotional arousal (a biological factor) related to engaging in painful activities. Consistent with this model, researchers have identified a number of psychological and social variables related to pain behaviors.

Patients who are depressed, for example, have been shown to exhibit higher levels of pain behavior (Keefe et al. 1986), while those who report a high degree of self-efficacy, or confidence in their ability to control pain, exhibit lower levels of pain behavior (Buckelew et al. 1994).

Characteristics

There are three basic methods of pain behavior assessment: self-monitoring, automated recording, and direct observation.

Self-monitoring

In self-monitoring, patients directly record their own behavior, including key, pain-related behaviors. This is often done using a daily diary, similar to that initially used by Fordyce (1976), which asked the individuals to record on an hourly basis the amount of time they spent sitting, standing or walking, and reclining, along with their pain medication intake. Diary data can often be examined and analyzed in treatment sessions using simple graphs. Patients, for example, who show a very low level of uptime (time spent up and out of the reclining position) may benefit from behavioral and physical therapy interventions designed to increase their level and range of activity. One concern about using self-monitoring is the degree to which these records are reliable and valid. However, recent research indicates that high-quality data can be obtained from daily diary recording methods if patients receive systematic training (Keefe et al. 1997).

The major strengths of self-monitoring are that it is simple and inexpensive, can be used over long time periods, provides a better real-time measure of behavior than retrospective reports or questionnaires, and can increase a patient's awareness of his or her own behavior.

Automated Recording

Several electromechanical devices have been developed to automatically record important behaviors, such as time spent up and out of bed or activity level. Recently, actigraphy has been used to monitor activity level (Sugimoto et al. 1997). A commercially available device, the Actigraph, monitors activity level by using an advanced ▶ accelerometer to detect motion and a microprocessor to control how data on such motion is collected and stored. The device can be worn by patients in their natural environment to provide continuous, objective information concerning their overall activity level (Sugimoto et al. 1997).

Direct Observation

In direct observation, observers who are trained in the coding of pain behavior carefully watch patients as they engage in daily activities and record the pain behaviors that are observed. Two approaches to direct observation have been used: standard behavior sampling and naturalistic observation.

Standard Behavior Sampling

Clinical observations suggest that the level of pain behavior varies depending on what activities a patient is engaged in. To standardize the conditions under which pain behavior is sampled, researchers have asked patients to engage in a series of standard activities, and then observed the pain behaviors that occur. A good example of this strategy is the observation method developed by Keefe and Block (1982) for recording pain behavior in chronic low back pain patients. Patients participated in a 10-min session in which they were asked to sit, stand, walk, and recline for 1–2 min, each in randomized order. The sessions were videotaped and then scored by trained observers using an interval recording strategy, in which the observer watches for 20 s and records for 10 s. The observers coded five pain behaviors:

1. Guarding: abnormally slow, stiff, interrupted, or rigid movement
2. Bracing: stiff pain avoidant static position
3. Rubbing: touching or holding of pain area
4. Grimacing: obvious pain-related facial expression
5. Sighing

Standard behavioral sampling can yield data that is both highly reliable and highly valid.

Keefe and Block (1982) tested and found high interobserver ▶ reliability (independent observers showed a high percentage of agreement on the behaviors observed), construct ▶ validity (behavior observed correlated significantly with pain ratings of naïve observers), and discriminate validity (the measures discriminated between the low back pain patients and pain-free controls). The procedure used by Keefe and Block has been modified to record the pain behaviors of arthritis patients and has been shown to be similarly reliable and valid (McDaniel et al. 1986).

Standard behavior sampling is a useful method of pain behavior assessment. Pain behaviors have been shown to be more frequently observed when a patient is moving, than when in a static position (Keefe and Block 1982). One can thus structure a standardized situation to elicit more pain behaviors than might otherwise be observed. Clinicians can use standard behavior sampling to evaluate treatment effects, by comparing the pain behaviors observed before and after treatment is received. By standardizing the situation under which pain behavior is observed, it is possible to analyze the social and psychological variables that contribute to those behaviors. For example, patients who report higher levels of pain when in the room with their spouse, as opposed to a neutral observer, may have an overly solicitous spouse who contributes to their display of pain behaviors. Romano et al. (1992) videotaped 50 chronic pain patients and their spouses as they jointly preformed specified tasks. It was found that spouse solicitous behaviors were significantly more likely than chance to both precede and follow nonverbal pain behaviors.

Naturalistic Observation

It is often desirable to observe and record pain behavior in naturalistic clinical settings, such as an inpatient unit or physical examination. Keefe et al. (1987) developed an observation method for recording the pain behaviors and activity level of patients in inpatient pain management units. Their method was designed to be performed by the nursing staff as part of their normal duties.

Daily graphs of activity level and pain behavior generated from these observations were used to identify problem behaviors, make treatment decisions, and evaluate patient progress. Pain behavior assessment can also be conducted during a physical examination. For example, Keefe et al. (1984) recorded the pain behavior exhibited by low back pain patients during a physical examination. A higher level of pain behavior was significantly correlated with a greater number of mechanical and neurological findings.

It is possible to combine elements of naturalistic and standard behavior sampling. For example, Richards et al. (1982) designed a standardized observation method, the University of Alabama at Birmingham (UAB) Pain Behavior Scale, which is intended to be used in a naturalistic setting. During morning rounds, the patient is briefly observed walking, standing, and moving from sitting to standing and from standing to sitting. Behaviors are recorded and rated as to their frequency and severity on a three-point scale. Reliability between observers is generally quite high, and the method requires minimal training.

In sum, over the past 25 years, clinical researchers have developed and refined a number of methods for pain behavior assessment. These methods have been shown to be reliable and valid, and are now being widely used in the assessment of patients suffering from ▶ chronic pain and persistent, disease-related pain. In clinical settings, pain behavior assessment is an important component of any comprehensive assessment of patients suffering from chronic pain.

References

Buckelew, S. P., Parker, J. C., Keefe, F. J., et al. (1994). Self-efficacy and pain behavior among subjects with fibromyalgia. *Pain, 56*, 191–201.

Fordyce, W. E. (1976). *Behavioral methods for chronic pain and illness*. Saint Louis: CV Mosby.

Fordyce, W. E. (1979). Environmental factors in the genesis of low back pain. In: Bonica JJ, Liebeskind JE, Albe-Fessard DG (eds) Proceedings of the second world congress on pain (pp 659–666). Adv. Pain Res. Therapy, vol 3. New York: Raven Press.

Keefe, F. J., Affleck, G., Lefebvre, J. C., et al. (1997). Coping strategies and coping efficacy in rheumatoid arthritis: A daily process analysis. *Pain, 69*, 43–48.

Keefe, F. J., & Block, A. R. (1982). Development of an observation method for assessing pain behavior in chronic low back pain patients. *Behavior Therapy, 13*, 363–375.

Keefe, F. J., Crisson, J. E., & Trainor, M. (1987). Observational methods for assessing pain: A practical guide. In J. A. Blumenthal & D. C. McKee (Eds.), *Applications for behavioral medicine and health psychology* (pp. 67–94). Sarasota: Professional Resources Exchange.

Keefe, F. J., & Lefebvre, J. (1994). Pain behavior concepts: Controversies, current status, and future directions. In G. Gebhart, D. L. Hammond, & T. S. Jensen (Eds.), *Proceedings of the VIIth world congress on pain* (pp. 127–148). New York: Elsevier.

Keefe, F. J., & Lefebvre, J. (1999). Behavioral therapy. In R. Melzack & P. Wall (Eds.), *Textbook of pain* (4th ed., pp. 1445–1461). London: Churchill Livingstone.

Keefe, F. J., Wilkins, R. H., & Cook, W. A. (1984). Direct observations of pain behavior in low back pain patients during physical examination. *Pain, 20*, 59–68.

Keefe, F. J., Wilkins, R. H., Cook, W. A., et al. (1986). Depression, pain, and pain behavior. *Journal of Consulting and Clinical Psychology, 54*, 665–669.

McDaniel, L. K., Anderson, K. O., Bradley, L. A., et al. (1986). Development of an observation method for assessing pain behavior in rheumatoid arthritis patients. *Pain, 24*, 185–184.

Richards, R., Nepomuceno, C., Riles, M., et al. (1982). Assessing pain behavior: The UAB pain behavior scale. *Pain, 14*, 393–398.

Romano, J. M., Turner, J. A., Friedman, L. S., et al. (1992). Sequential analysis of chronic pain behaviors and spouse responses. *Journal of Consulting and Clinical Psychology, 60*, 777–782.

Sugimoto, A., Hara, Y., Findley, T. W., et al. (1997). A useful method for measuring daily physical activity by a three-direction monitor. *Scandinavian Journal of Rehabilitation Medicine, 29*, 37–42.

Turner, J. A. (1996). Educational and behavioral interventions for back pain in primary care. *Spine, 21*, 2851–2857.

Assimilation

Definition

Giving up the values, beliefs, material culture, and practices of their native group and adopting those of the host culture

Cross-References

► Cancer Pain, Assessment of Cultural Aspects

Association Study

Synonyms

Allele dosage study

Definition

Also known as an allele dosage study, this involves comparing the frequencies of alleles of genes or DNA markers between different phenotypic groups (e.g., those with a disease versus those without). If allele frequencies differ between the groups, the gene examined (or one very nearby) is implicated in the trait in question.

Cross-References

► Alleles
► Opioid Analgesia, Strain Differences

Astrocytes

Synonyms

Astroglia

Definition

Astrocytes are star-shaped glial cells integrally involved in synaptic communication by providing nutrients, support, and insulation for neurons of the central nervous system. As immunocompetent cells, astrocytes can be activated by bacteria and viruses to release classical immune products. Current studies suggest that astrocytes maintain exaggerated pain in pathological pain models.

Cross-References

► Cord Glial Activation
► Diencephalic Mast Cells
► Glial Cells in Orofacial Pathological Pain Mechanisms

Astroglia

► Astrocytes

Asymmetric Junctions

Definition

Gray (1962) divided the central synapse into two types, based on the differences in synaptic density between the pre- and postsynaptic membranes, asymmetric (type I) and symmetric (type II).

Cross-References

► Trigeminal Brainstem Nuclear Complex, Anatomy

Ataxia

Definition

Imbalance and poor control of various parts of the body; may reflect damage to the large sensory neurons that subserve joint position sense and coordination.

Cross-References

▶ Diabetic Neuropathies

Atenolol

Definition

Beta-blocker

Cross-References

▶ Preventive Migraine Therapy

At-Level Neuropathic Pain

Definition

Neuropathic pain located in the segments adjacent to the level of the spinal cord lesion. Also referred to as border-reaction, end-zone, segmental, or radicular pain.

Cross-References

▶ Spinal Cord Injury Pain Model, Contusion Injury Model

At-Level Phenomena

Definition

Alterations in sensations or spontaneous sensations that are referred to the body region, which is represented by the region of the spinal cord that is damaged by spinal injury

Cross-References

▶ Spinal Cord Injury Pain Model, Cordotomy Model

ATP

▶ Adenosine 5′ Triphosphate

ATP-Dependent Na+/K+ Pump

Definition

The principal primary active transport system in neurons, Na/K-ATPase, utilizes energy to maintain cation cellular concentrations by extruding Na and accumulating K ions, thus creating an electrical potential across the neuronal cell membrane. It is estimated that 25–40 % of brain energy utilization may be related to Na/K-ATPase activity. Abnormalities in the pump may lead to neuronal dysfunction, although the exact relationship to familial hemiplegic migraine is not known.

Cross-References

▶ Migraine, Childhood Syndromes

ATrPw

▶ Attachment Trigger Point

Attachment Trigger Point

Synonyms

ATrPw

Definition

An attachment trigger point is pathogenetically distinct from, and secondary to, a central myofascial trigger point. It is a region of inflammatory-type reaction (an enthesopathy) at the musculotendinous junction, or at the bony attachment of the muscle where the taut band fibers attach and produce increased sustained tension.

Cross-References

▶ Myofascial Trigger Points

Attentional Bias

Definition

The tendency to selectively attend to threatening information in comparison to neutral information

Cross-References

▶ Hypervigilance and Attention to Pain

Attentional Mechanisms

Definition

Attentional mechanisms are cognitive processes that focus sensory processing on particular inputs and de-emphasize other inputs.

Cross-References

▶ Spinothalamic Tract Neurons, Descending Control by Brain Stem Neurons

Attitude

▶ Chronic Gynecological Pain, Doctor-Patient Interaction

Attributable Effect

▶ Effect Size

Attributable Effect and Number Needed to Treat

Nikolai Bogduk
University of Newcastle, Newcastle Bone & Joint Institute, Royal Newcastle Centre, Newcastle, NSW, Australia

Synonyms

Effectiveness; Efficacy; NTT; Number needed to treat

Definition

The ▶ attributable effect of a treatment is the extent to which it achieves its outcomes, beyond that achieved by nonspecific effects of the intervention. It is the extent to which outcomes can be attributed to the specific components of a treatment by which it is purported to work.

The ▶ number needed to treat (NNT) is a measure of how effective a treatment is. Specifically, it is the number of patients who must achieve a particular outcome before one of those patients, on average, can be claimed to have responded because of the specific effects of the treatment (as opposed to having responded to the nonspecific effects of treatment). As a measure of the power of a treatment, NNT effectively discounts the apparent power by the extent to which outcomes are achieved by nonspecific effects. The larger the number, the more the treatment works by nonspecific effects. The smaller the number, the more the treatment has a specific effect.

Characteristics

The attributable effect is derived from categorical data. It requires data on whether the treatment has worked or not, in comparison with a control treatment, in the form shown in Table 1. Ideally, the control treatment should be one with no specific effects, i.e., a ▶ placebo.

For the purposes of the initial explanation, it does not matter what the definitions are of success or failure; that comes later. All that is important is that (somehow) a decision is made as to whether the treatment has been successful or not.

The proportion of patients who succeeded with the index treatment is $a/(a + b)$. Let this proportion be P_{index}, which is expressed as a decimal.

The proportion of patients who succeeded with the control treatment is $c/(c + d)$. Let this proportion be $P_{control}$, which is expressed as a decimal.

The attributable effect (AE) of the index treatment is the extent to which its success rate exceeds that achieved by the control treatment. The argument is that the control treatment provides nonspecific effects, but these are also a component of the index treatment. The attributable effect of the index treatment is what remains when the success rate of the index treatment is discounted for these nonspecific effects.

Mathematically,

$$AE = P_{index} - P_{control}$$

Since P_{index} and $P_{control}$ are both proportions, AE is also a proportion. It stipulates the proportion of patients treated, whose successful

Attributable Effect and Number Needed to Treat, Table 1 The categorical results of a clinical trial of an index treatment

Treatment	Result	
	Success	Failure
Index	a	b
Control	c	d

outcome can be legitimately attributed to the effects of the index treatment, above and beyond any nonspecific effects.

Thus, if N patients are subjected to the index treatment, one would expect that $(N \times P_{index})$ patients would have a successful outcome. However, $(N \times P_{control})$ of these patients would, on average, have responded because of nonspecific effects of the treatment.

Therefore, only $N(P_{index} - P_{control})$ would have responded because of the specific effects of the treatment, i.e., $(N \times AE)$. In other words, when N patients are treated, only $(N \times AE)$ patients respond to the attributable effect of the treatment.

Any group of patients who achieve a successful outcome will consist of those who responded to the attributable effect of the treatment and those who responded to nonspecific effects of treatment. There is no way of determining which particular patient or patients responded to the attributable effect or to nonspecific effects, but outcome data from large samples of patients can be used to show how many, on average, would have responded to the attributable effect. The number needed to treat (NNT) is used to indicate this proportion.

Let N_S be the number of patients who achieve a successful outcome. This number consists of two types of patients: those whose outcome was due to the attributable effect (N_{AE}) and those who had nonspecific responses (N_{NS}), i.e.,

$$N_S = N_{AE} + N_{NS}$$

But

$$N_{AE} = N_S \times AE$$

And

$$N_{NS} = N_S \times (1 - AE)$$

Wherefore,

$$N_S = [N_S \times AE] + [N_S \times (1 - AE)]$$

In large studies, N_S will be large, and both $[N_S \times AE]$ and $[N_S \times (1 - AE)]$ will be large. In small studies, N_S will be small, and both $[N_S \times AE]$ and $[N_S \times (1 - AE)]$ will be correspondingly small. Nevertheless, the proportion between $[N_S \times AE]$ and N_S will be the same.

Clearly, that proportion is mathematically simply the attributable effect, i.e.,

$$[N_S \times AE]/N_S = AE$$

However, this is an abstract number, with no immediate, apparent relationship to clinical practice.

A way of expressing the proportion more meaningfully is to express it in terms of whole patients.

Since $[N_S \times AE]$ will always be smaller than N_S, the smallest sample size in which the proportion can be expressed in terms of whole patients is one in which $[N_S \times AE]$ equals 1:

$$[N_S \times AE] = 1.$$

In which case,

$$N_S = 1/AE$$

Under these conditions, N_S becomes the number needed to treat (Cook and Sackett 1995, Laupacis et al. 1988), i.e.,

$$NNT = 1/AE$$

Under these conditions, for every NNT patient with a successful outcome, 1 will have responded to the attributable effect, and the remainder will have responded to nonspecific effects. All that is required to determine the ratio between attributable and nonspecific effects is a knowledge of the attributable effect.

For example, if an index treatment has a success rate of 78 %, and a control treatment has a success rate of 45 %,

$$P_{index} = 0.78$$

$$P_{control} = 0.45$$

$$P_{index} - P_{control} = 0.33$$

$$AE = 0.33$$

$$1/AE = 3$$

$$NNT = 3$$

Thus, for every three patients with a successful outcome, 1 can be attributed to the specific effects of the treatment, while the other 2 outcomes were due to nonspecific effects. If 30 patients were to have a successful outcome, 10 would be due to the attributable effect and 20 due to nonspecific effects.

Consider another example, in which the success rate of a treatment is 56 % and that of the control treatment is 36 %:

$$P_{index} = 0.56$$

$$P_{control} = 0.36$$

$$P_{index} - P_{control} = 0.20$$

$$AE = 0.20$$

$$1/AE = 5$$

$$NNT = 5$$

For every five patients with a successful outcome, 1 is due to the specific effects of treatment, and 4 are due to nonspecific effects.

Clearly, the larger the NNT, the weaker the treatment is, for a greater proportion of patients who appear to respond do so because of nonspecific effects. Conversely, the smaller the NNT, the more powerful the treatment is. As a benchmark, an NNT of 3 or less is considered to be good.

However, NNT is not a measure of how effective a treatment is overall. It is a measure of how much a particular outcome is due to specific effects of the treatment, compared with nonspecific effects. In these terms, a powerful treatment is one in which much, or most, of the outcome is due to specific effects, i.e., the attributable effect, as opposed to nonspecific effects. Less powerful treatments may nevertheless be effective, but their outcomes are due less to the attributable effect and more to nonspecific effects. NNT reveals the proportion between these two types of effect.

The implication of a large NNT is that most of the outcome observed is due to nonspecific effects and could be achieved without using the index treatment at all. Accordingly, NNT is an index of the utility of a treatment.

If a treatment is costly, or carries a high risk of complications, and has a large NNT, its use can be called into question. The large NNT indicates that because most of the effect is nonspecific, the cost and risk of the index treatment may not be justified, because the same, or similar, outcome might be achieved by other means.

If the NNT is high, it means that doctors will need to treat large numbers of patients before they get an attributable effect. This consumes time and effort. The doctors might consider if this large effort is worthwhile and whether their efforts might not be better spent using another treatment.

A large NNT also means that funds are being expended on large numbers of patients in order to get gains in a minority. Doctors might reflect as to whether these funds might be better spent differently or if as good a result might be achieved, on average, by using less expensive treatments.

For example, the NNT for epidural steroids is about 11 (McQuay and Moore 1996). Effectively, this means that for every 11 patients who get a successful outcome, only one can be claimed to have responded to the specific effects of the injections. The cost of that one success is not just the time and expense required for that one case but also the costs incurred for the other 10 patients.

Subscripts

The NNT is not a single measure of all of the effects of a treatment. It measures the power of a treatment only with respect to the outcomes specified in the original table of data from which the NNT was derived. Therefore, the pedantic but accurate use of NNT requires that the outcome be specified. This might be done as a subscript but is usually omitted in practice because of the typographic impositions incurred. Nevertheless, the concept is conveyed by this notation.

If the success in question is "ability to walk 1 km in 10 min," the NNT for that outcome would be recorded as

$$NNT_{\text{ability to walk 1 km in 10 min}}$$

If the success in question is "achieving a reduction of at least 50 % in VAS score," the NNT for that outcome would be

$$NNT_{\text{reduction in VAS by 50 \%}}$$

No one uses this notation, but it is taken as understood. Readers should understand that authors leave this implicit. They expect readers to have noticed what outcome they are addressing. Therefore, readers should consult the methods and results sections of any study to find out what the subscript would have been had the authors used this complete notation.

This is not an example of academic pedantry or an idiosyncrasy. It is an important realization lest NNT be abused. As a number, an NNT might look good and might be used to extol a treatment as successful and useful. However, the treatment might not be as good as it sounds if the reader realizes that the NNT pertains to an unconvincing or uncompelling outcome.

For example, the NNT for many drug therapies in pain medicine is about 3, which is considered a good score. Readers might, however, care to ask: Exactly what was the outcome measure? The risk is of readers being lulled into believing that with an NNT of 3, they could expect that for every three patients that they treat, one will be totally cured. This is not the case, for the NNT in question actually refers to "patients lowering their VAS by 50 %." It says nothing about patients being completely relieved. In actual fact, in this instance, an NNT of 3 means that for every three patients who obtained greater than 50 % relief of their pain, only one achieved this because of the effects of the drug used.

For NNT to be meaningful, the subscript must be specified. For a complete picture of how powerful a treatment is, authors should indicate the NNT for each outcome, e.g., $NNT_{complete\ relief}$, $NNT_{50\ \%\ relief}$, and $NNT_{return\ to\ work}$.

References

Cook, R. J., & Sackett, D. L. (1995). The number needed to treat: A clinically useful measure of treatment effect. *BMJ, 310*, 452–454.

Laupacis, A., Sackett, D. L., & Roberts, R. S. (1988). An assessment of clinically useful measures of the consequences of treatment. *The New England Journal of Medicine, 318*, 1728–1733.

McQuay, H. S., & Moore, A. (1996). Epidural steroids for sciatica. *Anaesthesia and Intensive Care, 24*, 284–286.

Atypical Facial Neuralgia

▶ Atypical Facial Pain: Etiology, Pathogenesis, and Management

Atypical Odontalgia

Yair Sharav
Department of Oral Medicine, School of Dental Medicine, Hebrew University-Hadassah, Jerusalem, Israel

Definition

The IASP (International Association for the Study of Pain) has defined atypical odontalgia as a severe throbbing pain without major pathology, or it may be defined as pain of dental origin without a definitive organic cause (Woda and Pionchon 1999).

Characteristics

Pain is localized to a tooth, or sometimes more than one tooth, which shows no dental pathology. Pain may be spontaneous or evoked by hot or cold foods and is usually strong and may throb (Czerninsky et al. 1999).

Etiology

Marbach (1978) postulated that pain is the result of previous trauma, such as tooth extraction or tooth pulp extirpation, which interferes with the central nervous system pain modulatory mechanisms and coined the name "phantom tooth pain." This idea is supported by the observation that experimental tooth extraction produces brainstem lesions in the trigeminal nucleus caudalis and that more extensive tooth pulp injury is associated with heightened excitability changes of trigeminal brainstem neurons (Hu et al. 1990). Although far from proven, a ▶ deafferentation associated with peripheral nerve injury may be responsible for some types of atypical facial pain. Vascular changes are other possible underlying mechanisms for atypical facial pain. ▶ Neurovascular orofacial pain (NVOP) may be especially relevant to the diagnosis of atypical intraoral pain (atypical odontalgia). NVOP, possibly a new diagnostic entity (Sharav and Benoliel 2008), shares many of the signs and symptoms common to other ▶ neurovascular-type craniofacial pain. It was found to be associated with atypical toothache (Benoliel et al. 1997) and to mimic ▶ pulpitis (Czerninsky et al. 1999). The onset of VOP is around 40–50 years of age, and it affects females at a rate of 2.5 times more than males.

It is possible that cases with neurovascular or other undiagnosed orofacial pain metamorphose into and coexist with trigeminal neuropathy as a result of nerve injury from repeated dental interventions aimed at pain relief. This theory is supported by findings that chronic orofacial pain patients undergo extensive but often misguided surgical interventions (Sharav and Benoliel 2008).

While pulp extirpation may eliminate the pain for a short time, pain tends to recur in another tooth (Czerninsky et al. 1999). The prophylactic use of beta-blockers or tricyclic antidepressants is usually beneficial (Benoliel et al. 1997; Czerninsky et al. 1999; Sharav and Benoliel 2008, and see Chap. 9).

Cross-References

▶ Atypical Facial Pain: Etiology, Pathogenesis, and Management

References

Benoliel, R., Elishoov, H., & Sharav, Y. (1997). Orofacial pain with vascular-type features. *Oral Surgery, Oral Medicine, Oral Pathology, Oral Radiology, and Endodontics, 84*, 506–512.

Czerninsky, R., Benoliel, R., & Sharav, Y. (1999). Odontalgia in vascular orofacial pain. *Journal of Orofacial Pain, 13*(3):196–200.

Hu, J. W., Sharav, Y., & Sessle, B. J. (1990). Effect of one – or two-stage deafferentation of mandibular and maxillary tooth pulps on the functional properties of trigeminal brainstem neurons. *Brain Research, 516*, 271–279.

Marbach, J. J. (1978). Phantom tooth pain. *Journal of Endodontics, 4*, 362–372.

Sharav, Y., & Benoliel, R. (2008). *Orofacial pain & headache*. Edinburgh: Mosby Elsevier.

Woda, A., & Pionchon, P. A. (1999). Unified concept of idiopathic orofacial pain: Clinical features. *Journal of Orofacial Pain, 13*, 172–184.

Atypical Facial Pain: Etiology, Pathogenesis, and Management

Yair Sharav
Department of Oral Medicine, School of Dental Medicine, Hebrew University-Hadassah, Jerusalem, Israel

Synonyms

Atypical facial neuralgia; Atypical odontalgia; Burning mouth syndrome; Complex regional pain syndrome; Idiopathic orofacial pain; Phantom tooth pain; Stomatodynia

Definition

The original term, atypical facial pain, is a historical counterpart to the "typical" presentation of trigeminal neuralgia. This ill-defined chronic facial pain condition is employed as a "wastebasket" definition, applied by elimination, of facial pain "not fulfilling other criteria." Recently, attempts have been made to define atypical facial pain in a more positive way and not merely by elimination (Woda and Pionchon 1999; Sharav and Benoliel 2008). Atypical facial pain can be described as chronic facial pain of constant intensity, which usually has a burning quality and occasionally intensifies to produce a throbbing sensation. Pain does not wake the patient from sleep and is not triggered by remote stimuli, but may be intensified by stimulation of the painful area itself. No local signs are present that can be related to the pain. No etiological factors are identifiable in the orofacial region.

Characteristics

This chronic intraoral or facial pain may start in one quadrant of the mouth and often spreads across the midline. Changes in pain location are frequent and may result in extensive dental work, alcohol nerve blocks, and surgery that do not usually alleviate the pain. Pain location is often ill-defined. Pain is usually constant and of moderate intensity and has a burning quality that occasionally intensifies to produce a throbbing sensation. However, pain does not wake the patient from sleep. The pain is not triggered by remote stimuli, but may be intensified by stimulation of the painful area itself. Accompanying ▶ autonomic phenomena are not observed. Typically, there is a lack of objective signs in most of these patients. The age range at examination is wide (20–82 years); the mean age of patients with ▶ atypical odontalgia is around 45–50 years (Marbach 1978; Vickers et al. 1998) and of patients with ▶ burning mouth syndrome is around 55 years (Grushka et al. 1987). All reports of atypical oral and facial pain indicate that an overwhelming majority are females (82–100 %).

Atypical facial pain should be differentiated from pains associated with a causative lesion. The symptoms of chronic atypical facial pain may be observed secondary to a slow-growing cerebellopontine angle tumor. The most common intraoral presentations of atypical facial pain are ► atypical odontalgia and ► burning mouth syndrome, and these are therefore discussed separately below.

Etiology

There is no identifiable uniform etiology of atypical oral and facial pain (Loeser 1989). Several underlying mechanisms have been proposed. A number of reports have suggested that atypical facial pain is a psychiatric disorder (Feinmann et al. 1984). Depression is considered the most likely diagnosis and is explained based on the catecholamine hypothesis of affective disorders. However, Sharav et al. (1987) showed that only 2 of their 28 patients with chronic facial pain were cortisol non-suppressors on the dexamethasone suppression test and that half the patients were not depressed at all. Grushka et al. (1987) conclude that the personality characteristics of patients with burning mouth syndrome are similar to those seen in other chronic pain patients and that these personality disturbances tend to increase with increased pain. The major logical fallacy for the purely psychological model of orofacial pain is that it assumes that failure to find a biological marker for a particular orofacial pain condition implies that psychological factors must be primary. However, pain researchers increasingly recognize that the association between observable tissue damage or other identifiable pathological process and the presence or extent of pain is weak at best (Sharav and Benoliel 2008). Vickers et al. (1998) suggested a possible neuropathic pain mechanism, but pointed out that it cannot explain all cases, and suggested that some may fit the diagnosis of ► complex regional pain syndrome.

Treatment

While various treatment modalities are used for atypical oral and facial pain, the predominant trends are clear. All authors firmly recommend avoiding surgical or dental interventions for the relief of pain (Loeser 1989) since such interventions usually exacerbate the condition. Reassurance, psychological counseling, and the use of antidepressants, particularly from the tricyclic group, have been found to be a promising mode of therapy. Two double-blind controlled studies demonstrated that tricyclic antidepressant drugs were superior to placebo in reducing chronic facial pain (Feinmann et al. 1984; Sharav et al. 1987). Furthermore, Sharav et al. (1987) showed that ► amitriptyline was effective, for most chronic facial pain states, in a daily dose of 30 mg or less, and that the relief of pain was independent of the antidepressant activity. As with other chronic pain conditions, particularly those with unclear etiologies, it has been associated with underlying psychological distress, and these patients may particularly benefit from cognitive behavioral therapy (CBT) (Sharav and Benoliel 2008).

References

Feinmann, C., Harris, M., & Cawley, R. (1984). Psychogenic facial pain: Presentation and treatment. *BMJ, 288*, 436–438.

Grushka, M., Sessle, B. J., & Miller, R. (1987). Pain and personality profiles in burning mouth syndrome. *Pain, 28*, 155–167.

Loeser, J. D. (1989). Tic douloureux and atypical facial pain. In P. D. Wall & R. Melzack (Eds.), *Textbook of pain* (2nd ed., pp. 535–543). London: Churchill Livingstone.

Marbach, J. J. (1978). Phantom tooth pain. *Journal of Endodontics, 4*, 362–372.

Sharav, Y., & Benoliel, R. (2008). *Orofacial pain & headache*. Edinburgh: Mosby Elsevier.

Sharav, Y., Singer, E., Schmidt, E., Dionne, R. A., & Dubner, R. (1987). The analgesic effect of amitriptyline on chronic facial pain. *Pain, 31*, 199–209.

Vickers, E. R., Cousins, M. J., Walker, S., & Chisholm, K. (1998). Analysis of 50 patients with atypical odontalgia. *Oral Surgery, Oral Medicine, Oral Pathology, Oral Radiology, and Endodontics, 85*, 24–32.

Woda, A., & Pionchon, P. A. (1999). Unified concept of idiopathic orofacial pain: Clinical features. *Journal of Orofacial Pain, 13*, 172–184.

Audit

► Postoperative Pain, Data Gathering and Auditing

Audit Report

Definition

An audit report refers to conclusions made by grouping data according to criteria, so that inferences can be made from them.

Cross-References

▶ Postoperative Pain, Data Gathering and Auditing

Aura

Definition

In the context of migraine, a recurrent disorder manifesting in attacks of reversible focal neurological symptoms that usually develop gradually over 5–20 min and last for less than 60 min. The most common aura is visual consisting of a bright zig-zag pattern of visual disturbance moving slowly across the filed and leaving after it an area of visual loss.

Cross-References

▶ Hemicrania Continua
▶ Migraine, Pathophysiology
▶ New Daily Persistent Headache

Autacoids

▶ Prostaglandins, Spinal Effects

Autobiographical Memory

Definition

Autobiographical memory is memory of one's life. It is central to the establishment and maintenance of the self-concept and one's personal and social identity, i.e., the sense of who you are. Autobiographical memory is considered to comprise of two types of memories: memories about specific experiences at specific times and knowledge of facts relevant to the self, e.g., one's date of birth, information about family relationships, schooling, and historical events that have occurred in one's lifetime.

Cross-References

▶ Pain Memory

Autogenic Feedback

Definition

A combination of autogenic training and thermal biofeedback, used for the purpose of promoting hand warming and generalized relaxation.

Cross-References

▶ Biofeedback in the Treatment of Pain

Autogenic Training

Definition

A form of relaxation training where verbal cues (e.g., "my hands are heavy and warm") are paired with physiological aspects of the relaxation process.

Cross-References

▶ Relaxation in the Treatment of Pain
▶ Relaxation Training
▶ Therapy of Pain, Hypnosis

Autologous Graft

Definition

Transplant tissue or cell source that is taken from the same or genetically identical individual

Cross-References

► Cell Therapy in the Treatment of Central Pain

Autonomic

Definition

Pertaining to the part of the vertebrate nervous system that regulates involuntary action, as of the intestines, heart, and glands. This is divided into the sympathetic nervous system, the parasympathetic and the enteric (Furness 2006; Jänig 2006) nervous system (Furness 2006; Jänig 2006).

Cross-References

► Psychiatric Aspects of Visceral Pain

References

Furness, J. B. (2006). The enteric nervous system. Blackwell Science Ltd, Oxford.
Jänig, W. (2006). The integrative action of the autonomic nervous system. Cambridge University Press, Cambridge, New York.

Autonomic Dysreflexia

Definition

Autonomic dysreflexia is a syndrome in persons with spinal cord injury (mostly complete) above the midthoracic segmental level. This syndrome is characterized by hypertension (increase of diastolic and systolic arterial blood pressure), increased sweating, piloerection, decrease of blood flow through skin, nasal congestion, bradycardia, and headache. It is generated by reflex activation of sympathetic systems to stimulation of spinal afferent neurons innervating viscera (e.g., during distension or contraction of urinary bladder or rectum) or deep somatic structures (joints, skeletal muscle). Bradycardia is secondary to stimulation of arterial baroreceptors via reflex activation of parasympathetic cardiomotor neurons (Jänig 2006; Mathias and Bannister 2012).

Cross-References

► Spinal Cord Injury Pain

References

Jänig, W. (2006). *The integrative action of the autonomic nervous system: Neurobiology of homeostasis*. Cambridge/New York: Cambridge University Press.
Mathias, C. J., & Bannister, R. (Eds.). (2012). *Autonomic failure* (5th ed.). Oxford: Oxford University Press.

Autonomic Features

Definition

Symptoms referable to activation or inhibition of the peripheral pathways of the autonomic nervous system. These symptoms may include increase or decrease of blood flow, blood pressure, and sweating, changes in functioning of visceral organs, ptosis, miosis, lacrimation, conjunctival injection, nasal congestion, and rhinorrhea. For example, the latter autonomic symptoms classically accompany attacks of cluster headache but may also occur during painful exacerbations of hemicrania continua.

Cross-References

▶ Hemicrania Continua

Autonomic Functions

Definition

Nervous regulation of the homeostasis of blood pressure, cardiac rhythm, blood circulation, blood fluid balance, respiration, pupil diameter, visceral motility, exocrine and neuroendocrine secretions, energy metabolism, and internal temperature. This is achieved by a constant interaction between the central and peripheral nervous systems.

Cross-References

▶ Hypothalamus and Nociceptive Pathways
▶ Pain Processing in the Cingulate Cortex, Behavioral Studies in Humans

Autonomic Nervous System

Definition

The autonomic nervous system is that part of the nervous system that controls body functions not under our direct voluntary control, such as the cardiovascular system, gastrointestinal tract, pelvic organs, body temperature, and some other functions. It includes in the periphery the sympathetic, parasympathetic, and enteric nervous system (Jänig 2006).

Cross-References

▶ Amygdala, Pain Processing and Behavior in Animals
▶ Diabetic Neuropathies
▶ Hereditary Neuropathies

References

Jänig, W. (2006). *The integrative action of the autonomic nervous system: Neurobiology of homeostasis*. Cambridge/New York: Cambridge University Press.

Autonomic Phenomena

Definition

Signs and symptoms pertaining to the functioning of the autonomic nervous system. For example, local autonomic signs in vascular-type craniofacial pain include tearing, redness of eye, nasal congestion, rhinorrhea, and cheek swelling.

Autonomic Reactions/Symptoms

Definition

Changes of blood pressure, heart rate, blood flow, sweating, pupil diameter, visceral motility, visceral secretion, etc., mediated by autonomic systems (mostly sympathetic when triggered by a noxious stimulus).

Cross-References

▶ Autonomic Functions
▶ Hypnic Headache
▶ Parabrachial Hypothalamic and Amygdaloid Projections

Autoregulation of the Cerebral Vessels

Definition

Autoregulation of the cerebral vessels refers to the capability of the cerebral vascular system to hold the cerebral perfusion stable during a wide range of systemic blood pressure.

Cross-References

▶ Primary Exertional Headache

Autotomy

Definition

Self-injurious behavior in which a body part that has been denervated or deafferented is compulsively licked, bitten, and chewed (self-mutilation). This is commonly observed in animals following neurectomy or dorsal rhizotomy, and is widely (although not universally) presumed to reflect the animal's response to dysesthesias or pain in the affected limb (anesthesia dolorosa). In the animal model of peripheral neurectomy, animals typically begin biting off nails and digits on the denervated paw within a few days following nerve injury.

Cross-References

▶ Anesthesia Dolorosa Model, Autotomy
▶ Dietary Variables in Neuropathic Pain

Autotomy Model of Neuropathic Pain

▶ Anesthesia Dolorosa Model, Autotomy

Autotraction

▶ Traction

Avocado-Soybean Unsaponifiables

▶ Nutraceuticals

Avoidance Behavior

Definition

Behavior aimed at avoiding or postponing undesirable situations or experiences. In chronic low back pain patients, avoidance behavior often consists of avoiding those activities that are believed to promote pain and/or (re)injury.

Cross-References

▶ Disability, Fear of Movement
▶ Hypervigilance and Attention to Pain
▶ Muscle Pain, Fear-Avoidance Model

Avulsion

Definition

Avulsion refers to the traumatic disconnection of nerve root from the spinal cord.

Cross-References

▶ Plexus Injuries and Deafferentation Pain

Avulsion Fracture

Definition

A fracture caused by a muscle pulling off a piece of bone from the area of the muscle's attachment.

Cross-References

▶ Cancer Pain Management, Orthopedic Surgery

Awakening

Definition

An increase in EEG and heart rate frequency with a rise in muscle tone that lasts more than 10 s. Subjects remain in sleep and are not usually aware of external influences. They could complain of non-refreshing sleep on the following day.

Cross-References

▶ Orofacial Pain, Sleep Disturbance

Awareness

▶ Consciousness and Pain

Axillary Block

Definition

Injection of local anesthetic into the axillary brachial plexus sheath resulting in sensory blockade of the hand. The forearm and inner aspect of the arm may be incompletely blocked due to inadequate blockade of the musculocutaneous and median nerves.

Cross-References

▶ Acute Pain in Children, Postoperative

Axolemma

Definition

Membrane around the nerve cell; the membranous sheath that encloses the long thin extension of a nerve cell (axon)

Cross-References

▶ Perireceptor Elements

Axon

Definition

An appendix of a neuron which carries action potentials either from the cell body away or, in case of afferent neurons in the peripheral nervous system, from the peripheral axon terminals toward the central nervous system. Axons can be very long, in case of peripheral nerves more than a meter, and might be branched. Branches are called axon collaterals. In contrast, dendrites, the other type of appendices of neurons, carry information toward the cell body of the neuron (in some rare cases also as action potentials).

By some authors axons have been simply defined as neuronal appendices carrying information from the cell body away. This excludes, however, the peripheral afferent nerves from this definition. Otherwise, they have much in common with other axons.

Cross-References

▶ Nociceptors, Action Potentials, and Post-Firing Excitability Changes
▶ Spinothalamic Tract Neurons, Morphology
▶ Toxic Neuropathies
▶ Wallerian Degeneration

Axon Reflex

Definition

The activation of a nociceptive C-fiber results not only in the transmission of action potentials toward the CNS but also along its branching fibers back toward the skin. This leads to the release of the vasoactive substances CGRP (calcitonin-gene-related peptide) and substance P. CGRP induces vasodilatation by binding to receptors at precapillary arterioles (flare reaction), Substance P induces plasma extravasation from venoles. These reactions are confined to the extension of the arborization of the excited C-unit(s) and have been called "neurogenic inflammation."

Cross-References

▶ Nociceptor, Axonal Branching
▶ Nociceptors in the Dental Pulp

Axonal Action Potentials

▶ Nociceptors, Action Potentials, and Post-Firing Excitability Changes

Axonal Arborization

Definition

All branches of an axon can be seen as an axonal arborization. Branches may depart from a stem axon often, but not always, in the periphery, close to the terminals.

Cross-References

▶ Spinothalamic Tract Neurons, Morphology

Axonal Degeneration

Definition

The pathologic term to describe disintegration of nerve fibers (axons). If an axon is cut, the distal part separated from the neuronal cell body will always degenerate. Degeneration may also occur in some forms of neuropathies, usually starting

from the peripheral terminals which are the first to lose their receptive and conductile properties (dying back neuropathy).

Cross-References

▶ Toxic Neuropathies

Axonal Sprouting

▶ Sympathetic and Sensory Neurons After Nerve Lesions, Structural Basis for Interactions

Axonal/Axoplasmic Transport

Definition

Anterograde axonal (or axoplasmic) transport is the energy-dependent mechanism by which materials synthesized in the cell body are moved to distal regions of neuronal processes. It is broadly divided into fast axonal transport, which involves the movement of materials within vesicles such as neurotransmitters, and slow axonal transport, incorporating movement of cytoskeletal proteins and cytoplasmic constituents. Retrograde axonal transport is the movement of materials such as proteins destined for degradation or molecules acquired from the external environment back to the cell body. Impaired axonal transport has been implicated in many neuropathies, including diabetic neuropathy, as it would be likely to starve peripheral parts of the axons of critical materials and also disrupt the delivery of factors from the environment back to the cell body.

Cross-References

▶ Dietary Variables in Neuropathic Pain
▶ Neuropathic Pain Model, Diabetic Neuropathy Model
▶ Opioids and Inflammatory Pain
▶ Toxic Neuropathies

Axotomy

Definition

Separation of an axon from its cell body (perikaryon). Axotomy of a nerve or a dorsal root is frequently used as an animal model for nerve injury and neuropathic pain, particularly in rats or mice. When the axon of a neurone is separated from the cell body, the neurone is said to be axotomized. While the distal part of the lesion degenerates, the proximal stump often becomes swollen with accumulated organelles originally destined for the nerve terminals. The soma responds within ~6 h, by changing the synthesis of proteins from transmitter synthesizing enzymes, to those associated with regeneration of axonal membrane and other structural components. If regeneration is prevented, e.g., by scar formation, a neuroma forms.

Cross-References

▶ Immunocytochemistry of Nociceptors
▶ Peptides in Neuropathic Pain States

▶ Retrograde Cellular Changes After Nerve Injury
▶ Sympathetic and Sensory Neurons after Nerve Lesions, Structural Basis for Interactions

Azathioprine

Definition

Azathioprine is an immunosuppressive agent, purine derivate, steroid-sparing agent in cranial arteritis, and treatment of choice in Behçet's disease.

Cross-References

▶ Headache Due to Arteritis
▶ Vascular Neuropathies

B

Babies or Infants

▶ Pain Assessment in Neonates

Back Disorders

▶ Back Pain, Associated Disability, and Work Loss

Back Injuries

▶ Back Pain, Associated Disability, and Work Loss

Back Pain

▶ Chronic Back Pain and Spinal Instability
▶ Chronic Musculoskeletal Pain
▶ Radiculopathies

Back Pain in the Workplace

John D. Loeser
Department of Neurological Surgery,
University of Washington, Seattle, WA, USA

Synonyms

Nonspecific low back pain; Work intolerance

Definition

Report of the International Association for the Study of Pain Taskforce on Back Pain in the Workplace chaired by Professor Wilbert E. Fordyce.

Characteristics

The International Association of the Study of Pain (IASP), under the leadership of its President, Michael J. Cousins, established a Taskforce on Back Pain in the Workplace in 1989. Wilbert E. Fordyce was the Chairperson; Professor Cousins appointed 25 Taskforce

G.F. Gebhart, R.F. Schmidt (eds.), *Encyclopedia of Pain*, DOI 10.1007/978-3-642-28753-4,
© Springer-Verlag Berlin Heidelberg 2013

members from a wide variety of professions: medicine, physical therapy, economics, labor, business, insurance, and administration. All were from developed countries including the United States, Australia, Canada, the United Kingdom, and Sweden. The Taskforce report was approved by the IASP Council and published in 1995. The 77-page Report consisted of an introduction and eight chapters, along with an extensive bibliography. Professor Fordyce composed the final draft of the report, after multiple reviews and contributions by all of the Taskforce members. The final paragraph of the report was:

> We conclude that existing disability programs are a major contributor to the explosion in disability ascribed to nonspecific low back pain. We call for a new paradigm for the assessment and management of disability ascribed to low back pain. The new program should have a primary goal of reducing worker pain, suffering, and loss of economic and personal rewards of gainful employment. We believe that the costs of such a program should accrue to both worker and employer in a fashion that provides appropriate incentives for each to reduce injuries and disability. Major changes are also needed in health care delivery for low back pain. Disability has complex social, personal, and physical antecedents. It is time to change our strategies for containment. (Fordyce 1995)

The first chapter set the stage for the succeeding sections, by identifying nonspecific low back pain (NSLBP) as the focus of the report. It defined this condition and other terms such as disability and impairment, and emphasized that disability is always based upon the interaction between the person and his or her environment. The second chapter reviewed the literature on the epidemiology of NSLBP and its costs. The third chapter built upon Loeser's model of pain (Loeser 1980) and elucidated the biomedical and biopsychosocial models of pain. The fourth chapter discussed the ambiguities associated with the diagnosis and treatment of chronic low back pain and the medicalization of suffering. The fifth chapter looked at disability programs and their intent and distinguished physical from psychological factors that can cause disability. The sixth chapter discussed the prevention of

disability in the workplace. The seventh chapter excerpted the treatment guidelines from the United States Agency for Health Care Policy and Research (AHCPR) Guidelines (1994), and a similar report from the UK (1994). It emphasized the need for rehabilitation rather than cure. The eighth chapter proposed changes in disability systems, and the ninth chapter contained a proposal for the management of long-term disability.

The Taskforce report called for new approaches to the treatment of nonspecific low back pain and the disability that was ascribed to it. It called for the recognition of NSLBP as a problem of activity intolerance, not a biomedical issue. It made heavy use of the diagnostic and treatment guidelines that had recently been published in the United States and the United Kingdom. Medical management was to be on a time-contingent, not pain-contingent basis. The focus of interventions for NSLBP should be shifted from the clinic to the worksite. The Report called for comprehensive reevaluation of any worker whose function was not restored within 6 weeks. Those who failed to return to work with restoration of function after 6 weeks should be classified as unemployed, and not medically disabled, due to NSLBP. Other recommendations included the limitation of permanent disability status to those who had clear-cut impairment that was irremediable; temporary disability status should not require permanent impairment. The authors also stated that psychological and psychiatric conditions ascribed to chronic pain should not lead to disability ascribed to pain. A strong emphasis on vocational rehabilitation permeated this report. Finally, a complete overhaul of all of the existing administrative policies, relating to disability and impairment from NSLBP, was recommended.

No other publication of the IASP Press has stirred up so much controversy, and certainly, none has led to actions such as the changes in disability policies that have been implemented in many countries and states in the United States. The report succeeded in threatening those with vested interests in many aspects of the disability process, especially those who earned their living providing interminable health care to people with NSLBP who rarely seemed to return to work. Labor interests claimed that the report was an attempt to take away benefits that had been earned through difficult negotiations. Business interests thought that costs to them could be increased by such a plan. Insurance companies worried about their costs and administrators saw their empires threatened. Vituperative and erroneous letters to the editors of pain journals persisted for several years. The members of the Taskforce in general, and its chairperson in particular, were called to task for even thinking such thoughts, let alone publishing them. IASP Council members, every one of whom had been sent a draft copy of the report prior to its publication, were chastised for approving its publication.

The primary argument leveled against the Report was: The idea that its plan would deprive injured working people of health care, and that labeling them as unemployed after 6 weeks would damage both their self-esteem and financial well-being. Some felt that "work intolerance" was not a valid concept, and that all of the workers who complained of back pain harbored an occult disease, which was amenable to more therapy of the sort that they provided. The authors of this report were accused of having no sympathy for injured working people who were entitled to whatever benefits they now received, as if such benefits were decreed by some higher being. The Report did suggest that changing from a biomedical model to a biopsychosocial model would reduce health-care costs and hasten return to gainful employment, which would be an advantage to both the worker and the employer, as well as those who fund disability systems.

Seventeen years later, what has been the effect of this Taskforce report? Some governmental agencies have attempted to implement its recommendations with mixed results. Some have returned to the old systems. Not many health-care providers have changed their concepts of NSLBP, but the entity is at least more visible in debates about how to solve the disability crisis. Western societies have not come to grips with the huge impact of NSLBP on their economies, in part because of the political impact of shifting many of those who are said to be disabled by low back pain to the group of those that are considered unemployed. To do so would more than double the unemployment rate in most countries! Those who earn a living providing health care to patients with NSLBP have not accepted the rather conclusive evidence that none of their interventions have been shown to improve the outcome from an episode of acute low back pain. It is clear at this time that the Taskforce report failed in its attempt to simplify a very complex problem by dramatically altering how people think about disability. Many of its proposals were not practical and many were far too radical for the political processes that surround disability.

Nonetheless, this report has had a perceptible effect upon thinking about the management of low back pain and disability. First, it served as yet another resource for those who have long been concerned about the excessive amounts of useless health care that have been provided to sufferers with nonspecific low back pain. By concurring with US and UK expert reports (and other countries later), a much more cogent argument in favor of a new approach to NSLBP has been generated. The need for evidence that a treatment was effective was also strongly supported. Many compensation systems have moved away from acute symptomatic interventions, to comprehensive, multidisciplinary approaches to patients with chronic nonspecific low back pain.

Second, it has served as a focus for a debate on what disability programs are designed to accomplish, and how they should be constructed. The principle behind the Taskforce recommendations was that a properly constructed disability program should facilitate return to gainful employment, not impede it. The incorporation of behavioral principles into the fabric of health-care and disability programs was seen as a step toward better outcomes at lower costs.

Third, it demonstrated that multidisciplinary, multinational scientific organizations, such as the IASP, could generate meaningful reports that could serve as a focal point for discussion and lead to social change.

Change comes slowly in social and political issues, such as the award of disability status to those who claim to be unable to work because of NSLBP. It is unlikely that a cataclysmic event will occur in spite of mounting costs and additional evidence that the way Western societies handle this problem is counterproductive for both the worker and society. The specifics of the proposal set forth in this report are not too important; the conceptual changes that it called for are the primary reason for its value. For those who look back on health-care and disability programs in 50 years, I predict that it will be seen as prescient.

References

Clinical Standards Advisory Group. (1994). *Back pain: Report of a CSAG Committee on back pain*. London: HMSO.

Bigos, S. J. (1994). *Acute low back problems in adults, clinical practice guideline*. AHCPR publication no. 95–0642. Rockville: U.S. Department of Health and Human Services, Public Health Service, Agency for Health Care Policy and Research.

Fordyce, W. E. (1995). *Back pain in the workplace* (p. 67). Seattle: IASP Press.

Loeser, J. D. (1980). Perspectives on pain. In P. Turner (Ed.), *Proceedings of the first World Congress on clinical pharmacology and therapeutics* (pp. 316–326). London: Macmillan.

Back Pain, Associated Disability, and Work Loss

Ernest Volinn
Pain Research Center, University of Utah, Salt Lake City, UT, USA

Synonyms

Back disorders; Back injuries

Definition

Disability associated with back pain essentially comprises two different clinical entities,
"back pain" and "disability," each of which initially seems to be readily definable but upon scrutiny remains elusive. Even so, definitions will be examined.

Back Pain

Other terms are sometimes used in place of "back pain," but they do little to clarify the meaning of this term, and, in fact, may further obscure it. "Back injury" is a term commonly found in the literature. Unless a sudden, traumatic event, an identifiable accident, is substantiated and directly linked to the onset or aggravation of pain, "back injury" is a misattributed term. That criterion would exclude a large proportion of workers as well as others who seek health care for what here is referred to as back pain. "Back disorders" is another term commonly found in the literature, although its constituents often are left unstated. Apparently, this is a generic term that subsumes different types of back problems, such as self-reports of back pain, self-reports of specifically work-related back pain, physicians' records of back examinations, and tabulations of worker's compensation claims or other administrative data. Each of these back problems may have different predictors and the aggregation of them into the single entity subsumed by "back disorders" may weaken or altogether undermine an analysis – e.g., predictors of self-reported back pain may differ from predictors of worker's compensation claims for back pain (Volinn et al. 2005). As problematic as the term "back pain" may be, no other term seems to be an improvement on it.

To turn to back pain itself, a team of experts used a modified Delphi approach to arrive at its "minimal definition" (Dionne et al. 2008), which soon became widely cited in the literature. This definition is embedded in two questions well suited for prevalence surveys. Figure 1 displays these questions.

Note that the second question excludes minor, fleeting pain and, in so doing, introduces physical function into the definition of back pain (i.e., "pain bad enough to limit your usual activities or change your daily routine"). As a number of studies have found, physical function and,

a For face-to-face interviews and paper or online question naires:

In the past 4 weeks, have you had pain in your low back (in the area shown on the diagram)?

Yes ☐   No ☐

If yes, was this pain bad enough to limit your usual activities or change your daily routine for more than one day?

Yes ☐   No ☐

Low Back

b For telephone surveys:

In the past 4 weeks, have you had pain in your low back?

Yes ☐   No ☐

If yes, was this pain bad enough to limit your usual activities or change your daily routine for more than one day?

Yes ☐   No ☐

Back Pain, Associated Disability, and Work Loss, Fig. 1 Minimal definition of low back pain (From Dionne et al. 2008. Figure used with permission of Wolters Kluwer Health)

on the other hand, "pain intensity" (i.e., magnitude of felt pain) are distinct phenomena, with correlations between them typically in the 0.3 to 0.4 range (Volinn et al. 2010). Additionally, the definition in Fig. 1 pertains to the low back region. The extension of physical function to pain intensity and relevance only to the low back do not invalidate the definition in Fig. 1, but, rather, suggest that the definition of "back pain" will likely continue to be a matter of discussion.

A reason why defining back pain has remained so problematic is that its origin is still largely obscure. More than a generation ago, Nachemson stated:

> Having been engaged in research in this field for nearly 25 years and having been clinically engaged in back problems for nearly the same period of time and as a member of and scientific adviser to several international back associations, I can only state that for the majority of our patients, the true cause of low back pain is unknown.

Even though the pace of research on back pain has accelerated, few back pain specialists would dispute Nachemson's statement were it made today. According to panels of experts, a definite cause of back pain in more than 85 % of all cases is still unknown (Airaksinen et al. 2005; Hoy et al. 2010).

"*Chronic* back pain" is a term well entrenched in the literature, but the time dimension adds still more ambiguity. Chronic pain obviously lasts a long time, but the question is just when does acute (shorter-lasting) pain become chronic. In the literature, the most frequent threshold for when pain becomes chronic is 3 months, although 6 months is also common, with the underlying supposition that the thresholds demarcate when pain "persists past the normal time of healing" (Merskey and Bogduk 1994). Arbitrarily established thresholds of this nature lead to still more questions. Does empirical evidence substantiate a threshold before which normal healing in fact takes place? Systematic reviews show that most of those who undergo back pain are not pain free 3 or 6 months after its onset; rather, they continue to experience recurrences or long-lasting pain (the terminology varies), although the pain may be less intense than at its onset (Airaksinen et al. 2005; Hoy et al. 2010). How is such pain to be classified?

In contrast to a strict timeframe imposed by researchers and clinicians, there is another option, which, although commonsensical, is overlooked in the literature. It is to delineate chronic back pain according to the subjective experience of those undergoing it themselves. Seen from the perspective of a succession of responses on numerical rating scales, the assumption may be that the natural history of pain intensity is to subside roughly in proportion to the passage of time. The relationship at the outset may not be linear, but, especially as time passes (weeks, months, years), progressive reduction of pain intensity may be expected. Viewed as such, back pain may be characterized as chronic if those undergoing it indicate that it is present at multiple points in time following an index assessment, and that it either worsens or fails to improve over time.

Back Pain *Disability*

Of several prominent definitions of disability, none predominates in the literature. Such terminological indecisiveness permits the latitude to choose the particular definition of disability with the most potential to elucidate the lives of those with back pain. If those undergoing back pain present their symptoms to a clinician at all, they are within the clinic only a minuscule amount of time in relation to the time they are outside it. One frequently cited definition of disability stands out because it is conducive to a more encompassing view of the effects of back pain. According to this definition, disability is a "form of inability or limitation in performing roles and tasks expected of an individual within a social environment" (Nagi 1979). Defined as such, disability directs attention outward from the body and toward the larger social environment. Patients present themselves in a certain role within the clinic, whereas included in the above definition of disability are their other roles in the social environment beyond the clinic.

The two commonest measures pertaining specifically to back pain disability are the Oswestry Disability Index and the Roland-Morris Disability Questionnaire (Roland and Fairbank 2000). Both of these measures elicit self-reports on capability to perform everyday activities, such as physical activity, sleeping, and personal care.

Seemingly, according to the above definition, loss of the work role would be integral to "disability." Nevertheless, the two common measures of back pain disability omit the work role. The omission is puzzling, because of the centrality of work to people's health and well-being (Volinn et al. 2009). Loss of work due to back pain, furthermore, has a demonstrable impact on the economies of entire societies (Hoy et al. 2010; van Tulder et al. 2002). Common measures of disability are moderately correlated with loss of the work role (e.g., 0.44 (Waddell et al. 2002)) but, even so, for the most part one of these phenomena varies independently of the other. In other words, assessment of loss of the work role requires a separate measure, and other aspects of disability may not serve as a proxy for it. The measurement of loss of the work role, however, is multidimensional and, rather than a question or two, it requires a series of questions and careful forethought (Table 1).

Back Pain, Associated Disability, and Work Loss, Table 1 Questions to consider in measuring work disability associated with back pain

Did workers remain at work while still undergoing back pain of some severity? If remaining at work, were they at less than full capacity due to back pain (i.e., presenteeism)?

Was time off work reported in the first place? Workers may have missed work due to back pain but not reported doing so. Even if their back pain was work related and they were eligible for worker's compensation, they may have received medical care but not reported to their health-care provider that the back pain was work related.

Was the start of time off work accurately reported? Was the onset of back pain sudden, resulting from of injury or an identifiable event? If its onset was gradual and happened over the course of time, what instigated time off work?

Was the end of time off work accurately reported? Did workers resume work because their back pain subsided, or for some other reason (e.g., cessation of worker's compensation or the threat of its cessation)?

If workers missed work and then returned, did they return at less than full capacity? Did they return but at reduced hours?

After missing work due to back pain, did workers return to the same job they held before they went off work, or did they return to one with different job tasks?

Did workers return to work only to miss work again shortly thereafter due to back pain?

Did workers not return to work for other reasons besides back pain? Examples of other reasons include no job was available for them, or some other condition or symptom besides back pain prevented return to work.

Characteristics

Epidemiology of Work Disability Associated with Back Pain

Loss of the work role, of course, may not be relevant to those out of the labor force, such as retirees and full-time students. Otherwise, loss of the work role is central to disability. For this reason, the section that follows pertains to work loss. A diagram of a funnel (or wineglass) placed on its side may be used to illustrate critical epidemiological aspects of work loss associated with back pain (Fig. 2). Viewed as such, all workers would enter the wide end of the funnel and most workers would proceed to its first stage, with a progressively smaller proportion of them reaching each successive stage.

The first stage after entering the model would be self-reported back pain, either accompanied or unaccompanied by disability. Back pain approaches death and taxes in its inevitability, with a lifetime prevalence (back pain ever) of upward of 84 % (Airaksinen et al. 2005). In a systematic review of studies conducted in several countries, the mean point prevalence (back pain now) was 18 % and the mean 1-year prevalence was 38 % (Hoy et al. 2010).

Far more is known about the epidemiology of back pain in developed countries than in developing countries, although, of course, the great majority of the world's population resides in developing countries. According to a review of studies, however, the 1-year prevalence of back pain in several African countries averages 50 % (Louw et al. 2007). The 1-year prevalence of back pain exceeds 50 % in studies of workers with physically demanding jobs in China (Volinn et al. 2010). In urban areas of developing countries, increasingly the sites of sweatshop labor and other types of physically demanding jobs, the need for research on the epidemiology of back pain is critical.

The next stage of the funnel-shaped model is missed work due to back pain. That is a far less frequently occurring event than the experience of back pain itself. For example, of 65 % of the construction workers in New York City who reported the symptom of back pain during a 1-year period, only 12 % reported they missed work because of it (Goldsheyder et al. 2002).

Still less frequent than missing work due to back pain, and at a stage closer to the narrow end of the funnel, is administratively reported back pain. Several studies have shown that administrative reports compiled from worksites grossly underestimate the prevalence of workers' back pain (Volinn et al. 2001). Accordingly, even though the U.S. Bureau of Labor Statistics is mandated to collect data on work-related injuries and illnesses, back problems are routinely underreported (Volinn et al. 2001). Worker's compensation claims for back pain constitute another type of administrative report. These too are relatively rare events in comparison with the proportion of workers who report back pain in

**Back Pain,
Associated Disability,
and Work Loss,
Fig. 2** Epidemiology,
work disability associated
with back pain

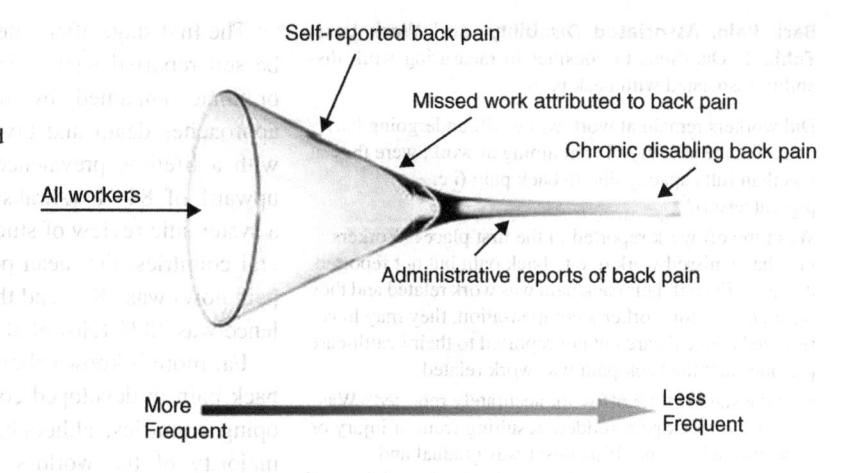

a questionnaire or interview. The annual rate in the USA is less than 2 back pain claims per 100 workers eligible for worker's compensation (Murphy and Volinn 1999). That rate is to be contrasted with 9 % of all workers in the USA who report they had back pain during a 1-year period that was both attributable to work and severe enough to last a week or more (Park et al. 1993). In other words, the rate of worker's compensation claims for back pain is less than one-fourth the rate with which workers report work-related back pain of some severity.

The last stage of the model, at the narrow end to represent how infrequently workers reach it, is chronic disabling back pain, which is here represented by total loss of the work role for 90 days or more. Worker's compensation data from the USA illustrate the disproportionate costs accrued by those who reach this stage (Volinn et al. 2001). Of about 2 % of all workers who file a compensation claim each year, about 10 % are off work for 90 days or more as indicated by time loss payments. In a given year, then, about 0.2 % of all workers eligible for worker's compensation – or 2 out of 1,000 workers – file a back pain claim that persists into chronicity. These relatively few workers, however, account for about 90 % of all worker's compensation costs for back pain.

Costs of chronic claims for back pain are mainly "indirect costs" that cover loss of the work role, as distinct from "direct costs" that cover medical care. Indirect costs include wage

compensation payments and payments for early retirement. In European countries such as the UK, Sweden, and the Netherlands, about 90 % of all back pain costs (direct + indirect) are indirect (van Tulder et al. 2002). Among worker's compensation claimants of back pain in the USA, the proportion of indirect costs in relation to all back pain costs is less but still exceeds 50 % (Murphy and Volinn 1999). The minuscule proportion of all workers whose back pain and loss of the work role persist into chronicity largely accounts for what is known as the "back pain problem."

References

Airaksinen, O., Hildebrandt, J., Mannion, A. F., et al. (2005). European guidelines for the management of chronic non-specific low back pain (pp. 30–33). Available at www.kovacs.org/Imagenes/EuropeanGuidelinesCHRONIC.LBP.pdf. Accessed 10 Mar 2012.

Dionne, C. E., Dunn, K. M., Croft, P. R., et al. (2008). A consensus approach toward the standardization of back pain definitions for use in prevalence studies. *Spine, 33*, 95–103.

Goldsheyder, D., Nordin, M., Schecter Weiner, S., et al. (2002). Musculoskeletal symptom survey among mason tenders. *American Journal of Industrial Medicine, 42*, 348–396.

Hoy, D., Brooks, P., Blyth, F., et al. (2010). The epidemiology of low back pain. *Best Practice & Research. Clinical Rheumatology, 24*, 769–781.

Louw, Q., Morris, L. D., & Grimmer-Somers, K. (2007). The prevalence of low back pain in Africa: A systematic review. *BMC Musculoskeletal Disorders, 8*, 105.

Merskey, H., & Bogduk, N. (1994). *Classification of chronic pain: Descriptions of chronic pain syndromes and definitions of pain terms* (2nd ed., pp. XI–XIII). Seattle: IASP Press.

Murphy, P. L., & Volinn, E. (1999). Is occupational low back pain on the rise? *Spine, 24*, 691–697.

Nagi, S. Z. (1979). The concept and measurement of disability. In E. D. Berkowitz (Eds.), *Disability policies and government programs* (pp. 1–15). New York: Praeger.

Park, C. H., Wagener, D. K., Winn, D. M., et al. (1993). *Health conditions among the currently employed: United States, 1988* (National Center for Health Statistics, Vital and Health Statistics, Series 10, No 186, pp. 7–22). Washington, DC: U.S. Government Printing Office.

Roland, M., & Fairbank, J. (2000). The Roland-Morris disability questionnaire and the Oswestry disability questionnaire. *Spine, 25*, 3115–3124.

van Tulder, M. W., Koes, B. W., & Bombardier, C. (2002). Low back pain. Best practice and research. *Clinical Rheumatology, 16*, 761–775.

Volinn, E., Spratt, K. F., Magnusson, M. L., et al. (2001). The Boeing prospective study and beyond. *Spine, 26*, 1613–1622.

Volinn, E., Nishikitani, M., Volinn, W., et al. (2005). Back pain claim rates in Japan and the U.S.: Framing the puzzle. *Spine, 30*, 697–704.

Volinn, E., Fargo, J. D., & Fine, P. G. (2009). Opioid therapy for nonspecific low back pain and the outcome of chronic work loss. *Pain, 142*, 194–201.

Volinn, E., Yang, B. X., He, J., et al. (2010). West China hospital set of measures in Chinese to evaluate back pain treatment. *Pain Medicine, 11*, 637–647.

Waddell, G., Aylward, M., & Sawney, P. (2002). *Back pain, incapacity for work, and social security benefits: An international literature review and analysis* (pp 9–16). London: Royal Society of Medicine Press.

Back Schools

▶ Information and Psychoeducation in the Early Management of Persistent Pain

Background Activity/Firing

Definition

Background activity is the spontaneous discharge of neurons that is independent of afferent input or any other direct stimulation.

Cross-References

▶ Molecular Contributions to the Mechanism of Central Pain

▶ Postsynaptic Dorsal Column Projection, Functional Characteristics

Background Infusion

Definition

Background infusion is a constant intravenous infusion of an analgesic medication that can be used in conjunction with a PCA.

Cross-References

▶ Analgesic Guidelines for Infants and Children

Baclofen

Definition

Baclofen is an agonist at G-protein-coupled $GABA_B$ receptors.

Cross-References

▶ GABA and Glycine in Spinal Nociceptive Processing

Bad Back

▶ Lower Back Pain, Acute

Balanced Analgesia

▶ Multimodal Analgesia

▶ Multimodal Analgesia in Postoperative Pain

Balanced Analgesic Regime

Synonyms

Multimodal analgesia

Definition

This term refers to the use of multiple analgesics with different mechanisms or sites of action. It can also be referred to as "▶ multimodal analgesia." There is evidence that this can improve pain relief and reduce opioid consumption and thereby their adverse effects. There has been particular interest in using this as part of an overall approach to hasten recovery rates after surgery and to reduce in-patient stay.

Cross-References

▶ Postoperative Pain, COX-2 Inhibitors

References

Buvanendran, A., & Kroin, J. S. (2009). Multimodal analgesia for controlling acute postoperative pain. *Current Opinion in Anaesthesiology, 22*, 588–593.
Kehlet, H. (1997). Multimodal approach to control postoperative pathophysiology and rehabilitation. *British Journal of Anaesthesia, 78*, 606–617.
Kehlet, H., & Dahl, J. B. (1993). The value of "multimodal" or "balanced analgesia" in postoperative pain treatment. *Anesthesia and Analgesia, 77*, 1048–1056.

Balneotherapy

Definition

The term comes from the Latin balneum (bath).

Cross-References

▶ Spa Treatment

Bariatric Surgery

▶ Metabolic and Nutritional Neuropathies

Baricity

Definition

Baricity is the ratio comparing the density of one solution to another. In order to make a drug hypobaric to cerebrospinal fluid (CSF), it must be less dense than the CSF having a baricity less than 1.0000 or a specific gravity appreciably less than 1.0069 (the mean value of CSF specific gravity). Intrathecal local anesthetic solutions can be hypobaric, isobaric, or hyperbaric. The predictability of the spread of intrathecal local anesthesia can be improved by altering both the baricity of the solution and the position of the patient during the intrathecal local anesthetic injection.

Cross-References

▶ Postoperative Pain, Intrathecal Drug Administration

References

Pitkanen, M., & Rosenberg, P. H. (2003). Local anaesthetics and additives for spinal anaesthesia–characteristics and factors influencing the spread and duration of the block. *Best Practice & Research. Clinical Anaesthesiology, 17*, 305–322.

Barotraumas

Definition

Barotraumas are lung injuries due to high airway pressure.

Cross-References

▶ Pain Control in Children with Burns

Barreloids

Definition

Groups of neurons in the thalamus that are responsive to whisker stimulation. Individual barreloids are driven by specific whiskers.

Cross-References

▶ Thalamic Plasticity and Chronic Pain

Basal Ganglia

Definition

The basal ganglia belong to subcortical structures of the brain and comprise of a group of nuclei associated with motor control, motivation, and learning functions.

Cross-References

▶ Functional Imaging of Cutaneous Pain

Basal Lamina

Definition

Extracellular matrix characteristically found under epithelial cells. The basal lamina, immediately adjacent to the cells, is a product of the epithelial cells themselves and contains collagen type IV.

Cross-References

▶ Perireceptor Elements

Basal Ventral Medial Nuclei

Synonyms

VM, VMb

Definition

Part of the ventral posterior complex.

Cross-References

▶ Spinothalamic Terminations, Core and Matrix
▶ Thalamic Nuclei Involved in Pain, Human and Monkey

Baseline Phase

Definition

A phase of the psychophysiological assessment involving collection of initial data to help design treatment targets and assess progress over time.

Cross-References

▶ Psychophysiological Assessment of Pain

Basic Body Awareness Therapy

Definition

A method that focuses on stability in postural function combined with freedom of movement, breathing, and body experiences considered crucial for total self-identity. The aim of the therapy is to increase sensory motor awareness and locomotor control, through exercise that focuses on a stable relation to the ground, posture, gait and truncal positioning, and the maintenance of balance related to the central axis of the body.

Cross-References

▶ Body Awareness Therapies

Basic Work Activities

Definition

The abilities and aptitudes necessary to do most jobs. Basic work activities include physical functions, such as walking, standing, sitting, lifting, pushing, pulling, reaching, carrying, or handling; the capacities for seeing, hearing, and speaking; and the mental abilities for understanding, carrying out, and remembering simple instructions; use of judgment; responding appropriately to supervision, co-workers, and usual work situations; and dealing with changes in a routine work setting.

Cross-References

▶ Disability Evaluation in the Social Security Administration

Basilar-Type Migraine

Definition

Previously a migraine aura in which the clinical symptoms can be attributed to brainstem dysfunction. The symptoms can last from 5 to 60 min and may include ataxia, vertigo, diplopia, bitemporal or binasal visual field loss, dysarthria, tinnitus, hyperacusia, bilateral sensory changes, and alterations in consciousness, but no motor weakness in peripheral limbs. Originally, this was felt to be due to constriction of the basilar artery, hence the name basilar migraine. The International Classification of Headache Disorders II (ICHD-II) has classified this condition as basilar-type migraine preserving the historical link, but acknowledging the fact that the basilar artery may not be involved at all in the pathogenesis of this condition. ICHD-III will replace this term with brainstem aura.

Cross-References

▶ Migraine Without Aura
▶ Migraine, Childhood Syndromes

Bath Headache

▶ Primary Exertional Headache

Bcl-2

Definition

Bcl-2 is a proto-oncogene activated by chromosome translocation in human B-cell lymphomas. It encodes for a plasma membrane protein. The gene product inhibits programmed cell death (apoptosis).

Cross-References

▶ NSAIDs and Cancer

BDNF

▶ Brain-Derived Neurotrophic Factor

Bed Nucleus of the Stria Terminalis

Definition

Group of neurons located in the forebrain, chiefly at the end of the stria terminalis, a fasciculus that links this nucleus with the amygdala. The lateral portion of this nucleus has a functional relationship with the central nucleus of the amygdala (central extended amygdala). It may play a critical role in autonomic reactions linked to diffuse stress.

Cross-References

▶ Nociceptive Processing in the Amygdala: Neurophysiology and Neuropharmacology
▶ Parabrachial Hypothalamic and Amygdaloid Projections

Behavior Modification

Definition

Involves helping parents learn strategies to promote well behaviors and extinguish overt pain behaviors such as crying, guarding, limping, and restricting activities due to the pain.

Cross-References

▶ Complex Chronic Pain in Children, Interdisciplinary Treatment

Behavior Therapies

Definition

Systematic application of learning theory principles to treat behavioral problems or disorders.

Treatment focus is on observable and measurable physiological or environmental events and on the patient's resultant behaviors.

Cross-References

▶ Psychiatric Aspects of the Management of Cancer Pain

Behavioral

Definition

Pertaining to the outwardly observed behavior of the individual.

Cross-References

▶ Pain as a Cause of Psychiatric Illness
▶ Psychological Treatment in Acute Pain

Behavioral Assessment

Definition

Evaluation of the response of an individual, group, or species to its environment.

Cross-References

▶ Disability, Functional Capacity Evaluations

Behavioral Descriptions of Symptoms

▶ Nonorganic Symptoms and Signs

Behavioral Experiments

Definition

A behavioral experiment is a therapeutic tool used to test the dysfunctional beliefs or misconceptions about the relation between certain stimuli and aversive consequences. The goal is to decrease the subjective credibility of the dysfunctional interpretations and to increase the credibility of the alternative interpretations.

Cross-References

▶ Disability, Fear of Movement
▶ Fear Reduction Through Exposure in Vivo

Behavioral Measure

▶ Central Pain: Outcome Measures in Preclinical Research

Behavioral Pain Scale

Definition

Indirect measure of pain in which presence or intensity of pain is inferred by noting changes from baseline levels for certain observable behaviors such as crying, facial expression, and limb movements rate.

Cross-References

▶ Pain Assessment in Children

Behavioral Responses to Examination

▶ Nonorganic Symptoms and Signs

Behavioral Self-management

Definition

Skills that are employed by the patient to increase positive adaptation to the experience of pain (e.g., exercise and appropriate use of analgesic medication).

Cross-References

▶ Psychological Treatment of Headache

Behavioral Studies in Animals

▶ Cingulate Cortex, Nociceptive Processing, Behavioral Studies in Animals

Behavioral Therapies to Reduce Disability

Michael J. L. Sullivan and Whitney Scott
Department of Psychology, McGill University, Montreal, QC, Canada

Synonyms

Exposure techniques for reducing activity avoidance; Graded activity approaches to chronic pain; Operant conditioning approaches to chronic pain; Pain management programs

Definition

Behavior therapy refers to a host of techniques that are based on principles of learning theory. Behavior therapy approaches to pain and disability proceed from the assumption that disability is a form of behavior, which is governed by the same environmental forces that influence all forms of human behavior. Current approaches vary with respect to their relative emphasis on different components of learning theory such as ▶ classical conditioning instrumental conditioning and ▶ social learning. Disability is typically construed as a reduction in physical or psychological capacity, from a pre-injury or pre-illness level. The primary goal of behavior therapy is to modify environmental contingencies such that disability is minimized.

Characteristics

According to the principles of operant conditioning, behaviors are influenced by their consequences. A particular behavior (e.g., moaning) that is followed by a positive consequence (e.g., empathic attention) will have a higher probability of being emitted in the future, regardless of the level of pain. In this case, "moaning" becomes instrumental in achieving empathic attention.

The association between physical activity and pain symptoms has also been discussed as a significant factor in the development and maintenance of pain behavior (Philips 1987; Vlaeyen and Linton 2000). Individuals with musculoskeletal pain conditions typically experience increases in pain following activity, and decreases in pain upon activity cessation. This relation between activity and pain sets a stage ideal for the learning of escape or avoidance behavior (Vlaeyen et al. 1995). Activity cessation or activity avoidance is reinforced by the reduction in experienced pain. Since activity cessation and activity avoidance essentially define "disability," learning principles have been discussed as fundamental to the development of disability associated with pain (Vlaeyen and Linton 2000). It has been suggested that disability can persist solely on the basis of an individual's expectation that a certain behavior will result in pain (Philips 1987). Since avoidance of activity does not allow for disconfirmation of the individual's expectation or belief that pain will result from activity, disability can be maintained long after a pain condition has resolved (Turk et al. 1983; Vlaeyen and Linton 2000).

Behavior therapy initially emerged from empirical research on the determinants of learning (Skinner 1953). In the 1960s and 1970s, *Wilbert* Fordyce and his colleagues applied the concepts of learning theory to the problem of chronic pain (Fordyce 1976; Fordyce et al. 1968). The focus of Fordyce's approach to treatment was not on reducing the experience of pain, but on reducing the overt display of pain. The targets selected for treatment were pain behaviors such as distress vocalizations, facial grimacing, limping, guarding, medication intake, activity withdrawal, and activity avoidance.

The first behavioral approaches to the management of pain and disability were conducted within inpatient settings that permitted systematic observation of pain behaviors, as well control over environmental contingencies influencing pain behavior (Fordyce 1976). Staff were trained to monitor pain behavior, and to selectively reinforce "well behaviors" and selectively ignore "pain behaviors." Results of several studies revealed that the manipulation of reinforcement contingencies could exert powerful influence on the frequency of display of pain behaviors (Fordyce et al. 1985). The manipulation of reinforcement contingencies was also applied to other domains of pain-related behavior and shown to be effective in reducing medication intake, reducing downtime, and maximizing participation in goal-directed activity.

A number of clinical trials on the efficacy of behavioral treatments for the reduction of pain and disability yielded positive findings (Sanders 1996). However, given the significant resources required to implement contingency management interventions, issues concerning the cost-efficacy of behavioral therapy for pain and disability were raised. Concern was also raised over the maintenance of treatment gains since reinforcement contingencies outside the clinic setting could not be readily controlled. In order to increase access and reduce costs, behavioral treatments were modified to permit their administration on an outpatient basis. This change in delivery format compromised to some degree the control over environmental contingencies, and required greater reliance on self-monitoring and self-report measures (Sanders 1996).

The early work of Fordyce and his colleagues was at once novel and contentious with its focus on reducing pain behavior as opposed to ameliorating the pain condition. Critics voiced their concerns that behavioral treatments might only be effective in training stoicism, but were not dealing with the underlying problem. In response, Fordyce pointed to literature indicating that the magnitude of the relation between pain and disability was modest at best, and that treatments aimed at reducing pain often had no appreciable effect on level of disability (Fordyce et al. 1985). According to Fordyce, to effectively treat disability, disability had to be targeted directly.

There has been a recent resurgence of interest in the application of behavioral interventions for the management of pain-related disability. Emerging programs differ from traditional approaches with greater emphasis on increasing activity involvement and reducing avoidance and escape behaviors. Recent programs have also tended to adopt return to work as the primary goal of the intervention (e.g., Sullivan and Adams 2010).

Graded activity is a form of behavioral intervention that reduces disability through principles of operant conditioning (Lindström et al. 1992). Treatment begins by establishing patients' tolerances for various activities where they are instructed to continue until pain or fatigue interferes. Their performance on an activity in response to these instructions is called a baseline. During subsequent training sessions, patients are given a performance quota for each session that is established by the treatment team. They are instructed to make a sincere effort to meet the quota even if they are feeling uncomfortable, and are also instructed not to exceed the quota if they are feeling especially good. To optimize chances of success, the quota set during the initial training session is below a patient's baseline. Quotas are gradually increased during subsequent training sessions, so that by the end of the treatment program, they are substantially higher than the patient's baseline. Graded activity has been associated with improvements on self-reported disability and pain-related work absences. (Leeuw et al. 2008; Staal et al. 2004).

Based on principles of classical conditioning, Vlaeyen and colleagues (2002) have developed a behavioral treatment to reduce fears of pain, movement, and reinjury. The treatment program is based on the premise that pain-related disability arises from an individual's fears that movement or activity may precipitate increases in pain or injury exacerbation. Clients enrolled in this treatment are first asked to rate the intensity of their fears of engaging in a variety of different physical activities (Kugler et al. 1999). A fear hierarchy is then established and in vivo exposure techniques are used to desensitize the individual to the feared activities. A number of randomized controlled trials have highlighted the efficacy of this approach for reducing pain-related fear and minimizing the degree of disability associated with pain (Leeuw et al. 2008; Linton et al. 2008; Woods and Asmundson 2008).

The Progressive Goal Attainment Program (PGAP) is another recently developed treatment that utilizes a variety of behavioral techniques with the principal goal of reducing occupational disability (Sullivan et al. 2006). PGAP is a community-based intervention that aims to increase life-role activity involvement and minimize psychological risk factors for prolonged disability, including fear of movement/reinjury, catastrophizing, perceived disability, and depression. Behavioral strategies, such as goal setting, activity monitoring, activity scheduling, and activity exposure, are used to maximize individuals' participation in daily activities. The goals of these techniques include reducing fears of movement, and challenging the clients' beliefs about their level of disability. PGAP outcome studies reveal significant reductions on risk factors for prolonged disability and increases in the likelihood of returning to work (Sullivan et al. 2006, 2010). Recent findings also suggest the utility of a telephonic version of PGAP for reducing pain-related disability (Sullivan and Simon, 2012).

Several additional treatment programs contain behavioral elements which might be useful for targeting pain-related disability. Acceptance-based interventions aim to shift behavioral efforts from pain control to participation in valued activities. Techniques include experiential exposure, values clarification, and committed action exercises. Acceptance-based interventions have been associated with significant improvements on a number of indices of disability (McCracken et al. 2005; Vowles and McCracken 2008). Behaviorally based motivational techniques, including shaping, modeling, and altering reinforcement contingencies and outcome expectancies of functional behavior, might also hold promise (Jensen et al. 2003). A motivational program using behavioral techniques showed significant long-term improvements on pain-related disability (Friedrich et al. 2005).

In summary, the literature reveals a wide range of interventions that have been developed to modify pain behavior. Behavior therapies for reducing disability have enjoyed a resurgence of interest in recent years. This interest is partly borne out of current clinical perspectives emphasizing the importance of targeting disability as the primary goal of intervention. The interest in behavioral therapies for persistent pain conditions is likely to be maintained as clinical studies continue to yield positive results in terms of observable outcomes such as increased activity and return to work. Such outcomes are being increasingly recognized as potentially more meaningful, whether assessed in terms of personal, social, or societal benefit, than subjective outcomes such as pain reduction or subjective well-being.

References

Fordyce, W. E. (1976). *Behavioral methods in chronic pain and illness*. St. Louis: CV Mosby.

Fordyce, W. E., Fowler, R. S., Lehmann, J. F., & De Lateur, B. J. (1968). Some implications of learning in problems of chronic pain. *Journal of Chronic Diseases, 21*, 179–190.

Fordyce, W. E., Roberts, A. H., & Sternbach, R. A. (1985). The behavioral management of chronic pain: A response to critics. *Pain, 22*, 113–125.

Friedrich, M., Gittler, G., Arendasy, M., & Friedrich, K. M. (2005). Long-term effect of a combined exercise and motivational program on the level of disability of patients with chronic low back pain. *Spine (Phila Pa 1976), 30*, 995–1000.

Jensen, M. P., Nielson, W. R., & Kerns, R. D. (2003). Towards the development of a motivational model of pain self-management. *The Journal of Pain, 4*, 477–492.

Kugler, K., Wijn, J., Geilen, M., et al. (1999). *The Photograph Series of Daily Activities (PHODA). CD ROM version 1.0*. Heerlen, The Netherlands: Institute for Rehabilitation Research and School for Physiotherapy.

Leeuw, M., Goossens, M. E. J. B., van Breukelen, G. J. P., et al. (2008). Exposure in vivo versus operant graded activity in chronic low back pain patients: Results of a randomized controlled trial. *Pain, 138*, 192–207.

Lindström, I., Öhlund, C., Eek, C., Wallin, L., Peterson, L. E., Fordyce, W. E., & Nachemson, A. L. (1992). The effect of graded activity on patients with subacute low back pain: A randomized prospective clinical study with an operant-conditioning behavioral approach. *Physical Therapy, 72*, 279–290.

Linton, S. J., Boersma, K., Jansson, M., et al. (2008). A randomized controlled trial of exposure in vivo for patients with spinal pain reporting fear of work-related activities. *The Journal of Pain, 12*, 722–730.

McCracken, L. M., Vowles, K. E., & Eccleston, C. (2005). Acceptance-based treatment for persons with complex, long-standing chronic pain: A preliminary analysis of treatment outcome in comparison to a waiting phase. *Behaviour Research and Therapy, 43*, 1335–1346.

Philips, H. C. (1987). Avoidance behavior and it's role in sustaining chronic pain. *Behaviour Research and Therapy, 25*, 273–279.

Sanders, S. H. (1996). Operant conditioning with chronic pain: Back to basics. In R. J. Gatchel & D. C. Turk (Eds.), *Psychological approaches to pain management: A practitioner's handbook* (pp. 112–130). New York: Guilford Press.

Skinner, B. F. (1953). *Science and human behavior*. New York: MacMillen.

Staal, J. B., Hlobil, H., Twisk, J. W. R., et al. (2004). Graded activity for low back pain in occupational health care: A randomized, controlled trial. *Annals of Internal Medicine, 140*, 77–84.

Sullivan, M. J. L., & Adams, H. (2010). Psychosocial treatment techniques to augment the impact of physiotherapy interventions for low back pain. *Physiotherapy Canada, 62*, 180–189.

Sullivan, M. J. L., & Simon, G. (2012). A telephonic intervention for promoting occupational re-integration in work-disabled individuals with musculoskeletal pain. *Translational Behavioral Medicine, 2*, 149–158.

Sullivan, M. J. L., Adams, H., Rhodenizer, T., & Stanish, W. D. (2006). A psychosocial risk factor-targeted intervention for the prevention of chronic pain and disability following whiplash injury. *Physical Therapy, 86*, 8–18.

Turk, D. C., Meichenbaum, D., & Genest, M. (1983). *Pain and behavioral medicine. A cognitive-behavioral perspective*. New York: Guilford Press.

Vlaeyen, J. W. S., & Linton, S. J. (2000). Fear-avoidance
and its consequences in chronic musculoskeletal pain:
A state of the art. *Pain, 85*, 317–332.

Vlaeyen, J. W. S., de Jong, J., Geilen, M., Heuts,
P. H. T. G., & van Breukelen, G. (2002). The treatment
of fear of movement/re-injury in chronic low back
pain: Further evidence on the effectiveness of exposure in vivo. *The Clinical Journal of Pain, 18*,
251–261.

Vlaeyen, J. W., Kole-Snijders, A. M., Boeren, R. G., van
Eek, H. (1995). Fear of movement/(re)injury in
chronic low back pain and its relation to behavioral
performance. *Pain, 62*(3), 363–372.

Vowles, K. E., & McCracken, L. M. (2008).
Acceptance and values-based action in chronic
pain: A study of treatment effectiveness and process.
Journal of Consulting and Clinical Psychology, 76,
397–407.

Woods, M. P., & Asmundson, G. J. G. (2008). Evaluating
the efficacy of graded in vivo exposure for the treatment of fear in patients with chronic back pain:
A randomized controlled clinical trial. *Pain, 136*,
271–280.

Behavioral Treatment

▶ Operant Treatment of Chronic Pain
▶ Psychological Treatment of Headache

Behavioral/Cognitive Perspective

▶ Cognitive-Behavioral Perspective of Pain

Behçet's Disease

Definition

Systemic vasculitis of the veins presenting with
iridocyclitis and oral and genital ulcers.

Cross-References

▶ Headache Due to Arteritis

Below-Level Neuropathic Pain

Definition

Neuropathic pain occurring in a more diffuse
distribution below the level of the spinal cord
lesion. Also referred to as central dysesthesia
syndrome, remote, or phantom pain.

Cross-References

▶ Cognitive-Behavior Treatment of Pain
▶ Spinal Cord Injury Pain Model, Contusion
Injury Model

Benign Cough Headache

▶ Primary Cough Headache

Bennett Model

▶ Neuropathic Pain Model, Chronic Constriction
Injury
▶ Retrograde Cellular Changes After Nerve
Injury

Beriberi Disease

▶ Metabolic and Nutritional Neuropathies

Beta(β) Blockers

Definition

Beta-adrenergic receptor antagonists.

Cross-References

▶ Preventive Migraine Therapy

Beta(β)-Adrenergic Sympathetic System

Definition

It belongs to the autonomic nervous system and is typically activated during stress, including pain.

Cross-References

▶ Placebo Analgesia and Descending Opioid Modulation

Beta(β)-Endorphin

Definition

β-Endorphin is a 31 amino acid peptide produced from proopiomelanocortin, a pituitary precursor also found in the arcuate nucleus of the hypothalamus.

Cross-References

▶ Opiates During Development

Beta(β)-Thromboglobulin

Definition

Constituent of platelet alpha-granules that is secreted from the platelet upon stimulation. This protein has been considered platelet specific. It is involved in the negative regulation of normal and pathologic megakaryocytopoiesis and seems to have chemotactic activity for human fibroblasts.

BFB

▶ Biofeedback

BGT

Definition

BGT is a plasma membrane GABA transporter; it also transports betaine.

Cross-References

▶ GABA and Glycine in Spinal Nociceptive Processing

Bias

Definition

Systematic deviations of the results from the truth.

Cross-References

▶ Lumbar Traction

Bicuculline

Definition

Antagonist at GABA$_A$ receptors.

Cross-References

▶ GABA and Glycine in Spinal Nociceptive Processing

Binding Medical Examination

▶ Independent Medical Examinations

Binding Studies

Definition

Receptor binding studies, in the context of positron emission tomography (PET), may serve as an indirect indicator of neurotransmitter release. Specific radioligand binding is of a competitive nature, and if opioids induce release of endogenous CCK, for example, a decreased binding of a CCK-radioligand could be expected.

Cross-References

► Opioid-Induced Release of CCK

Bioavailability

Definition

Bioavailability is defined as a measure of the extent and rate of active drug that reaches the systemic circulation and its subsequent availability at the site of action after administration of a dosage. It is the fraction of the administered dose that has reached the system and is then available to exert effects. Factors that may influence the bioavailability of a drug are the release from the pharmaceutical product, the absorption from the gastrointestinal tract, the transformation due to the hepatic first-pass metabolism, and, last but not least, the degradation processes prior to reaching system circulation.

Cross-References

► NSAIDs, Pharmacokinetics

Biobehavioral Factors

► Stress and Pain

Biobehavioral Model

► Chronic Pain, Diathesis-Stress Model of

Biofeedback

Synonyms

BFB

Definition

Biofeedback (BFB) refers to a group of nonpharmacological interventions that, by providing external (computer-aided) feedback about physiological responses (such as blood pressure, hand temperature, or muscle tension a person would not normally be aware of), aims at improving self-regulation of physiological responses that typically are not voluntarily perceived and controlled. The success of BFB in the treatment of chronic pain has been attributed to physiological and/or psychological factors. Physiological accounts are based on the assumption that a dysregulation of specific physiological processes (e.g., elevated muscle tension) underlies the pain problem. Hence, learned control of the relevant physiological process should lead to corresponding pain relief. Yet, empirical studies have not provided consistent and convincing support for this hypothesis. The literature on biofeedback for headache and back pain suggests that cognitive factors, such as an increase in perceived self-efficacy and a decrease in negative pain-related cognitions, may mediate the effectiveness of biofeedback in the treatment of chronic pain.

Cross-References

► Alternative Medicine in Neuropathic Pain
► Biofeedback in the Treatment of Pain
► Chronic Back Pain and Spinal Instability

Biofeedback in the Treatment of Pain

Frank Andrasik[1] and Beverly E. Thorn[2]
[1]Department of Psychology, Applied Psychophysiology and Biofeedback, The University of Memphis, Memphis, TN, USA
[2]Department of Psychology, University of Alabama, Tuscaloosa, AL, USA

Synonyms

Pain therapy; Self-control; Self-regulation; Voluntary control

Definition

▶ Biofeedback is a ▶ nonpharmacological treatment that uses electrical instruments to help people become aware of, and then exert control over, physiological responses (Andrasik and Lords 2004). It involves (1) detecting and amplifying a biological response by using certain measurement devices (or transducers) and electronic amplifiers, (2) converting these bioelectrical signals to a form that can be easily understood and processed by the patient, and (3) providing relatively immediate feedback of the signal to enable the patient to influence the response. The feedback signal is most often presented in visual and/or auditory modalities,

in either binary (signal on or signal off at a specified threshold) or continuous proportional fashion (as muscle tension decreases, the tone or click rate decreases, signaling progress at the goal of lowering); on occasion, combinations of both are used. Several theories have been used to account for biofeedback, ranging from operant learning to cognitive and expectancy models (Schwartz and Schwartz 2003). In all cases, the goal of biofeedback is to acquire the ability to regulate the targeted physiology in the desired direction (e.g., decrease excessive muscle tension, increase blood flow).

Characteristics

Approaches to Biofeedback

Three different rationales have been advanced for the use of biofeedback in the treatment of pain (Andrasik 2005; Andrasik and Lords 2004; Belar and Kibrick 1986; Flor 2001). For simplicity, we term them *general*, *specific*, and *indirect*.

General Approach to Biofeedback

The "general" approach employs biofeedback (and related procedures, such as relaxation; see Thorn and Andrasik in this volume) to lower overall arousal and promote a generalized state of relaxation. Several assumptions underlie this use: (1) Reducing general arousal results in a concurrent reduction in central processing of peripheral sensory inputs; (2) the second assumption derives from the observed relationship between negative affect and pain, wherein negative affect is associated with decreased pain tolerance and increased reports of pain (Fernandez 2002). Achieving a more relaxed state leads to reductions in negative affect, which in turn enhances pain tolerance and decreases pain reports. (3) A final consideration is based on the association of stress and pain, which Melzack (1999) has discussed in detail. Prolonged activation of the stress system increases cortisol levels, which is related to a number of physiological changes that can give rise to pain. Stress is also directly linked to the negative affective triad. Based on these considerations, a case can be made that most pain patients could benefit from some form of relaxation and arousal reduction, especially when the pain is associated with increased muscle tension. This approach is probably the most common (and it also requires the least technical proficiency).

Examples of Biofeedback as a General Aid to Relaxation

Any response modality indicative of heightened arousal can theoretically serve as a target for promoting relaxation, but three are used most often – muscle tension (electromyographic or EMG), skin conductance (or sweat gland activity), and peripheral temperature. These modalities, termed as the "workhorses" of the biofeedback general practitioner (Andrasik and Lords 2004), are easily collected, quantified, and interpreted and are discussed below. Other responses, such as heart rate, respiration, blood volume, and electroencephalogram, can be useful, but these are not commonly employed and will not be addressed further (see Flor 2001, for discussion).

EMG-Assisted Relaxation

The rationale for employing muscle tension feedback to facilitate relaxation is straightforward. The basis of the EMG signal is the small electrochemical changes that occur when a muscle contracts, and these changes can be monitored by placing a series of electrodes along the muscle fibers (to assess the muscle action potentials associated with the ion exchange across the membrane of the muscles). When EMG is used for generalized relaxation, sensors are usually placed on the forehead, neck, or shoulders. When other sites remain tense (such as arms), clinicians may also provide training for them.

Skin Conductance-Assisted Relaxation

Electrical activity of the skin or sweating has long been thought to be associated with arousal. It was used in the late 1800s to facilitate understanding when working with cases of hysterical anesthesias, and Carl Jung used it in the early 1900s as a way to "read the mind" in word-association experiments. Sensors are placed on body surface areas that are most densely populated with "eccrine" sweat glands (such as the palm of the hand or the fingers), as these respond primarily to psychological stimulation and are innervated by the sympathetic branch of the autonomic nervous system (Boucsein 1992; Stern et al. 2001). Conductance measures are preferred in clinical application because they have a linear relationship to the number of sweat glands that are activated. This permits a straightforward explanation to patients (as arousal increases, so does skin conductance; focusing on decreasing skin conductance helps to lower arousal and to achieve a state of relaxation).

Skin Temperature-Assisted Relaxation

Peripheral skin temperature is of value because it provides an indirect measure of blood flow and of sympathetic nervous activity (constriction of peripheral blood flow is under control of the sympathetic branch of the nervous system, decreases in sympathetic outflow lead to increased vasodilation and a resultant rise in peripheral temperature due to the warmth of the blood). Thus, temperature feedback may best be thought of at the moment as yet another way to facilitate general relaxation. When used for purposes of general relaxation, it is typically augmented by components of autogenic training, leading to a procedure termed "▶ autogenic feedback."

Recent research with functional MRI has shown that distraction, a common component of relaxation, leads to significant activation within the periaqueductal gray region, a site recognized for higher cortical control of pain (Tracey et al. 2002). Thus, these treatments may be impacting central mechanisms as well.

These approaches have been used extensively and successfully with pain patients (see Andrasik and Lords 2004).

Specific Approach

When utilizing the specific approach, a detailed ▶ psychophysiological assessment is conducted first to identify the physiological dysfunction or response modalities assumed to be relevant to the pain condition and to do so under varied stimulus conditions (psychological and physical) that mimic work and rest (reclining, bending, stooping, lifting, working a keyboard, simulated stressors, etc.) in order to guide treatment efforts and gauge progress. It typically incorporates the following phases: adaptation, resting baseline, self-control baseline, stress induction, stress recovery, etc. This approach is described more fully in Andrasik, Thorn, and Flor (▶ Psychophysiological Assessment of Pain), as well as in Arena and Schwartz (2003) and Flor (2001). When biofeedback-assisted relaxation is the main focus, the psychophysiological assessment is much briefer, if done at all.

Examples of the Specific Application

Most research and practice has focused on biofeedback as a general approach to decrease stress, tension, and pain. With patients with certain characteristics, more specific approaches are emerging as either alternative or preferred treatments. A brief example is provided here for purposes of illustration.

Arena (cited within Arena and Blanchard 2002) describes a straightforward approach to a more individualized biofeedback treatment for chronic low back pain. Treatment begins with EMG biofeedback-assisted relaxation, initially from the frontal or forehead area, which is then followed by feedback of the trapezius muscles, all performed with the patient seated comfortably. As the patient becomes proficient, positions are changed in order to facilitate generalization of training effects. Now, practice is performed in a comfortable office chair (with arms supported), then moves to an office chair without arm support, and then to a standing position. This phase of training continues for 12–16 sessions.

If improvement is insufficient and the patient has not had a prior course of general relaxation training, then this may be pursued (▶ Relaxation in the Treatment of Pain). If this is unwarranted or has been unsuccessful, then an abbreviated psychophysiological assessment, based on the logic of biomechanical theory, is conducted to analyze the problem further. EMG sensors are

placed bilaterally on the paraspinals (L4–L5) and the biceps femoris (back of the thigh). Recordings are made in at least two positions: sitting with back supported in a recliner and standing with arms by the side. These sites are used because they have been found to provide greater information than other sites (such as quadriceps femoris or gastrocnemius) in prior examinations.

The data will reveal one of three patterns of abnormality: (1) unusually low muscle tension levels (which Arena states most typically occurs when nerve damage and muscle atrophy are present), (2) unusually high muscle tension levels (which Arena states is the most common finding), or (3) left-right asymmetry, wherein one side has normal muscle values and the other side is either abnormally high or low. Treatment focuses on returning EMG values to normal levels. Arena notes that much can be learned by examining gait and posture and also by correcting faulty positions. Sella (2003) has also commented on these postural aspects.

The approach that Arena describes is appealing because of its simplicity. The difficulty is in determining normal versus abnormal values. Experience with a considerable number of patients is necessary for this. Some researchers have been approaching this problem in a more systematic manner and have developed detailed normative data banks for multiple muscle sites to help identify aberrant values, as these may be suggestive of bracing or favoring of a position or posture, and, thus, targets for EMG treatment.

In this more expanded approach, multiple bilateral recordings need to be made in fairly rapid succession. To make this feasible, Cram (1990) uses two handheld "post" electrodes to obtain brief (around 2 s per site) sequential bilateral recordings while the patient is sitting and standing (rather than having to apply multiple leads or apply and reapply the same set of leads). Although seemingly straightforward, this approach is actually complex because a number of factors can influence the results (angle and force of sensor application, amount of adipose tissue present, as fat acts as an insulator and dampens the signal, and the exactness of sensor placement). Further, results will vary as a function of band-pass settings, making it difficult to compare across data sets with different settings.

Indirect Approach

Belar and Kibrick (1986) describe what they term the "indirect" approach to biofeedback. This approach is based more upon clinical than empirical foundations, and it involves using biofeedback as a means for facilitating psychosomatic therapy. Consider a patient who steadfastly holds to a purely somatic view, refusing to accept the notion that other factors (emotional, behavioral, or environmental) may be precipitating, perpetuating, or exacerbating pain and somatic symptoms. In this instance, a referral for biofeedback is likely to be less threatening (it is construed as a "physical" treatment for a "physical" problem) and to at least open the door for help. As "physiological insight" is acquired, the patient may begin to grasp the broader picture, i.e., the interplay of physical and psychological factors. It is not uncommon for this type of patient, who denies psychological factors upon entry to therapy, to make a request like the following after just a few sessions of biofeedback: "Doc, how about turning off the biofeedback equipment today. I want to talk about a few things." From this point on, session time may be divided between biofeedback and psychotherapy.

Some Clinical Consideration

Few difficulties have been reported when using biofeedback as a general relaxation procedure. A small portion of clients may experience what has been termed "▶ relaxation-induced anxiety," noted to be a sudden increase in anxiety during deep relaxation, which can range from mild to moderate intensity and approach the level of a minor panic attack (▶ Relaxation in the Treatment of Pain). It is important for the therapist to remain calm, reassure the patient that the episode will pass and, when possible, have the patient sit up for a few minutes or even walk about the office when this happens. With patients who are believed to be at risk for relaxation-induced

anxiety, it may be helpful to instruct them to focus more on the somatic aspects as opposed to the cognitive aspects of training (Arena and Blanchard 2002). See Schwartz et al. (2003) for discussion of other problems and solutions.

References

Andrasik, F. (2005). Relaxation and biofeedback self-management for pain. In M. V. Boswell, B. E. Cole, & R. S. Weiner (Eds.), *Weiner's pain management: A practical guide for clinicians* (7th ed.). Boca Raton: CRC Press.

Andrasik, F., & Lords, A. O. (2004). Biofeedback. In L. Freeman (Ed.), *Mosby's complementary & alternative medicine: A research-based approach* (2nd ed., pp. 207–235). Philadelphia: Elsevier Science.

Arena, J. G., & Blanchard, E. B. (2002). Biofeedback training for chronic pain disorders: A primer. In R. J. Gatchel & D. C. Turk (Eds.), *Psychological approaches to pain management: A practitioner's handbook* (2nd ed., pp. 159–187). New York: Guilford Press.

Arena, J. G., & Schwartz, M. S. (2003). Psychophysiological assessment and biofeedback baselines: A primer. In M. S. Schwartz & F. Andrasik (Eds.), *Biofeedback: A practitioner's guide* (3rd ed., pp. 128–158). New York: Guilford Press.

Belar, C. D., & Kibrick, S. A. (1986). Biofeedback in the treatment of chronic back pain. In A. D. Holzman & D. C. Turk (Eds.), *Pain management: A handbook of psychological treatment approaches* (pp. 131–150). New York: Pergamon Press.

Boucsein, W. (1992). Plenum University Press. New York: Electrodermal activity.

Cram, J. R. (1990). EMG muscle scanning and diagnostic manual for surface recordings. In J. R. Cram et al. (Eds.), *Clinical EMG for surface recordings* (Vol. 2, pp. 1–141). Nevada City: Clinical Resources.

Fernandez, E. (2002). *Anxiety, depression, and anger in pain: Research findings and clinical options*. Dallas: Advanced Psychological Resources.

Flor, H. (2001). Psychophysiological assessment of the patient with chronic pain. In D. C. Turk & R. Melzack (Eds.), *Handbook of pain assessment* (2nd ed., pp. 76–96). New York: Guilford Press.

Melzack, R. (1999). Pain and stress: A new perspective. In R. J. Gatchel & D. C. Turk (Eds.), *Psychosocial factors and pain: Critical perspectives* (pp. 89–106). New York: Guilford Press.

Schwartz, N. M., & Schwartz, M. S. (2003). Definitions of biofeedback and applied psychophysiology. In M. S. Schwartz & F. Andrasik (Eds.), *Biofeedback: A practitioner's guide* (3rd ed., pp. 27–39). New York: Guilford Press.

Schwartz, M. S., Schwartz, N. M., & Monastra, V. J. (2003). Problems with relaxation and biofeedback-assisted relaxation, and guidelines for management. In M. S. Schwartz & F. Andrasik (Eds.), *Biofeedback: A practitioner's guide* (3rd ed., pp. 251–264). New York: Guilford Press.

Sella, G. E. (2003). Neuropathology considerations: Clinical and SEMG/biofeedback applications. *Applied Psychophysiology and Biofeedback, 28*, 93–105.

Stern, J. E. (2001). Electrophysiological and morphological properties of pre-autonomic neurones in the rat hypothalamic paraventricular nucleus. *J Physiol, 537*, 161–177.

Tracey, I., Proghaus, A., Gati, J. S., Clare, S., Smith, S., Menon, R. S., & Matthews, P. M. (2002). Imaging attentional modulation of pain in the periaqueductal gray in humans. *Journal of Neuroscience, 22*, 2748–2752.

Biogenic Amine Theory

Definition

Biogenic amine theory of the relationship between chronic pain and depression hypothesizes that both mood and pain regulation are mediated by bioamine systems including dopaminergic, noradrenergic, and serotonergic neurotransmitter systems.

Cross-References

▶ Depression and Pain

Biological Clock

Definition

An area of the brain thought to be the hypothalamus that dictates biological rhythms such as the sleep-wake cycle and menstruation.

Cross-References

▶ Hypnic Headache

Biological Rhythms

Definition

Biological rhythms are periodic and predictable-in-time variations of biological phenomena.

Cross-References

▶ Diurnal Variations of Pain in Humans

Biomagnetometer

▶ Magnetoencephalography in Assessment of Pain in Humans

Biopsy

Definition

Necessary for most vasculitides; leptomeningeal and parenchymal biopsy are necessary for the diagnosis of IAN, and temporal artery biopsy is mandatory for the diagnosis of temporal arteritis.

Cross-References

▶ Headache Due to Arteritis

Biopsychosocial

▶ Ethics of Pain, Culture, and Ethnicity

Biopsychosocial Model

Definition

In contrast to the biomedical model that emphasizes physical pathology and disease, the biopsychosocial model acknowledges the complex interaction of psychological and social factors as well as biological (physical) ones in the understanding and expression of disease and illness. From this perspective on medical problems, the diversity in illness expression is accounted for by the interrelationships among biological changes, psychological states, and social and cultural contexts. All of these shape the individual's perception of and response to physiological perturbations.

Cross-References

▶ Chronic Pain, Diathesis-Stress Model of
▶ Diathesis-Stress Model of Chronic Pain
▶ Multiaxial Assessment of Pain
▶ Multidisciplinary Pain Centers, Rehabilitation
▶ Psychological Treatment of Chronic Pain, Prediction of Outcome

Biopsychosocial Perspective

Definition

In contrast to the traditional but outdated biomedical model and its emphasis on disease, the more heuristic biopsychosocial perspective focuses on illness, which is the result of the complex interaction of physiological, psychological, and social variables.

Cross-References

▶ Disability Assessment, Psychological/ Psychiatric Evaluation
▶ Psychological Treatment of Chronic Pain, Prediction of Outcome

Bischof Myelotomy

▶ Midline Myelotomy

Bisphosphonates

Nikolai Bogduk
University of Newcastle, Newcastle Bone & Joint Institute, Royal Newcastle Centre, Newcastle, NSW, Australia

Definition

Bisphosphonates are chemicals with a structure similar to that of the mineral pyrophosphate, which occurs in bone. As drugs, they are used to treat a variety of bone diseases, some of which are associated with pain.

Characteristics

Bisphosphonates were discovered in the nineteenth century and were used as corrosion inhibitors and complexing agents in the textile, oil, and fertilizer industries. Their potential for use in medicine arose when it was recognized that pyrophosphate inhibited precipitation of calcium. Pyrophosphate was, therefore, considered as a drug that might be used to control calcium metabolism but was found not to be active orally and was rapidly hydrolyzed if administered parenterally. The similar chemical structure of bisphosphonates to pyrophosphate (Fig. 1) made them a suitable substitute for pyrophosphate as a drug.

A substantial number of bisphosphonates have been developed. These include pamidronate, clodronate, etidronate, tiludronate, alendronate, and ibondronate. They each share the same basic chemical structure but differ in their substitution radicals (Fig. 2). Pharmacologically, they differ in their relative potencies for different conditions.

Mechanism

Upon being absorbed into bone, bisphosphonates inhibit the release of calcium from the bone matrix and are toxic to osteoclasts. By both mechanisms, bisphosphonates prevent bone resorption.

HO R_1 OH 　 HO OH
$O=P-C-P=O$ 　 $O=P-O-P=O$
HO R_2 OH 　 HO OH

Bisphosphonates　　**Pyrophosphate**

Bisphosphonates, Fig. 1 The structural formulae of bisphosphonates and pyrophosphate. Instead of oxygen, carbon is the central atom of the bisphosphonate molecule; and it subtends two radicles (R_1 and R_2)

　　　　　NH_2
　　　　　CH_2
HO CH_3 OH 　 HO CH_2 OH
$O=P-C-P=O$ 　 $O=P-C-P=O$
HO OH OH 　 HO OH OH

Etidronate　　**Pamidronate**

Cl　　　　　　NH_2
　　　　　　　CH_2
　　　　　　　CH_2
HO S OH 　 HO CH_2 OH
$O=P-C-P=O$ 　 $O=P-C-P=O$
HO H OH 　 HO OH OH

Tiludronate　　**Alendronate**

Bisphosphonates, Fig. 2 The structural formulae of a selection of bisphosphonates

Applications

Due to their ability to inhibit bone resorption, bisphosphonates have been used in the treatment of disease in which bone resorption is a major feature. Their principal application has been as a treatment for osteoporosis. They have also been used to treat Paget's disease of bone.

Cancer Pain

In pain medicine, bisphosphonates were first used to treat painful tumors of bone, in which osteolysis was a feature. They have been particularly effective for reducing the pathological effects of myeloma and metastases, such as fractures (Ross et al. 2004; Djulbegovic et al. 2002). In these conditions, bisphosphonates both inhibit the progression of the cancer and relieve its pain. For cancer bone pain, they have proved to be a valuable adjunctive therapy (Wong and Wiffen 2002).

Bisphosphonates may be used orally or as an intravenous infusion. Intravenous infusions appear to be more effective and are administered once every 1 or 2 weeks. Relief of pain usually occurs within 1 week of starting treatment. During that period, other analgesics, such as ▶ opioids, need to be used to provide relief of pain; but as and once the effect of bisphosphonates occurs, other analgesics can be tapered.

CRPS

Focal, juxta-articular osteoporosis is a feature of ▶ complex regional pain syndromes, general aspects (CRPS). This feature prompted the exploration of bisphosphonates for the treatment of CRPS. However, although they were used to treat the osteoporosis, they had a fortuitous effect on pain.

Open studies announced good relief of pain following treatment with pamidronate (Cortet et al. 1997; Kubalek et al. 2001). Indeed, one study reported that 86 % of patients treated with oral pamidronate for 3 days were completely relieved of pain within 1–5 weeks (Kubalek et al. 2001). Controlled trials have verified this ▶ effectiveness. Intravenous alendronate (Adami et al. 1997) and intravenous clodronate (Varenna et al. 2000) are both significantly more effective than placebo. Following treatment with clodronate, patients achieved an average of 93 % reduction in pain (Varenna et al. 2000). Intravenous pamidronate is more effective than placebo, but its efficacy seems to be less than that of other bisphosphonates (Robinson et al. 2004).

Side Effects

Due to their actions, bisphosphonates can lower serum calcium levels. Oral preparations can cause gastric ulceration. Intravenous preparations can cause fever, bone pain, and transient lymphopenia. Intravenous pamidronate can cause anterior uveitis, which resolves after cessation of treatment.

References

Adami, S., Fossaluzza, V., Gatti, D., Fracassi, E., & Braga, V. (1997). Bisphosphonate therapy of reflex sympathetic dystrophy syndrome. *Annals of the Rheumatic Diseases, 56*, 201–204.

Cortet, B., Flipa, R. M., Coquerelle, P., Duquesnoy, B., & Delcanbre, B. (1997). Treatment of severe, recalcitrant reflex sympathetic dystrophy: Assessment of efficacy and safety of the second generation bisphosphonate pamidronate. *Clinical Rheumatology, 16*, 51–56.

Djulbegovic, B., Wheatley, K., Ross, J., Clark, O., Bos, G., Goldschmidt, H., Cremer, F., Alsina, M., & Glasmacher, A. (2002). Bisphosphonates in multiple myeloma. *Cochrane Database of Systematic Reviews*, (4), CDC003188.

Kubalek, I., Fain, O., Paries, J., Kettaneh, A., & Thomas, M. (2001). Treatment of reflex sympathetic dystrophy with pamidronate: 29 cases. *Rheumatology, 40*, 1394–1397.

Robinson, J. N., Sandom, J., & Chapman, P. T. (2004). Efficacy of pamidronate in complex regional pain syndrome type 1. *Pain Medicine, 5*, 276–280.

Ross, J. R., Saunders, Y., Edmonds, P. M., Wonderling, D., Normand, C., & Broadley, K. (2004). A systematic review of the role of bisphosphonates in metastatic disease. *Health Technology Assessment, 8*, 1–176.

Varenna, M., Zucchi, F., Ghiringhelli, D., Binelli, L., Bevilacqua, M., Bettica, P., & Sinigaglia, L. (2000). Intravenous clodronate in the treatment of reflex sympathetic dystrophy syndrome. A randomized, double blind, placebo controlled study. *Journal of Rheumatology, 27*, 1477–1483.

Wong, R., & Wiffen, P. J. (2002). Bisphosphonates for the relief of pain secondary to bone metastases. *Cochrane Database of Systematic Reviews*, (2), CDC002068.

Black Disc Disease

▶ Discogenic Back Pain

Black Flags

Definition

Black Flags refer to organizational obstacles to recovery including objective work characteristics and conditions of employment. Black Flags are not a matter of perception, and affect all workers equally. They include both nationally established policy concerning conditions of employment and sickness policy, and working conditions specific to a particular organization.

Cross-References

▶ Blue Flags
▶ Yellow Flags

Bladder Pain Syndrome

▶ Interstitial Cystitis and Chronic Pelvic Pain

Blink Reflex

Definition

Excitatory reflex elicited in the orbicularis oculi muscle by innocuous stimuli in the supraorbital region and consisting of an early (R1) and late (R2) component.

Cross-References

▶ Jaw-Muscle Silent Periods (Exteroceptive Suppression)

Block Design

Synonyms

Boxcar design

Definition

Experimental design, in which stimuli are presented in a fixed sequence independently of subjective responses of the subject, resulting in a fixed sequence of baseline and activation periods repeated frequently. For data analysis, each block is regarded as a unit; hence, all stimuli or tasks of this block should pertain to only one condition.

Cross-References

▶ Amygdala, Functional Imaging

Blood Brain Barrier

Definition

A cellular barrier that separates blood from brain and protects the latter from the chemical messenger systems flowing around the body; see also Blood Nerve Barrier.

Cross-References

▶ Blood Nerve Barrier
▶ Diencephalic Mast Cells

Blood Clot

▶ Deep Venous Thrombosis

Blood Clot in Brain

▶ Headache Due to Intracranial Bleeding

Blood Nerve Barrier

Definition

A naturally occurring barrier created by the modification of capillaries (as by reduction in fenestration and formation of tight cell-to-cell

contacts) that prevents many substances from leaving the blood and crossing the capillary walls into the nervous system; see also Blood Brain Barrier.

Cross-References

▶ Blood Brain Barrier
▶ Cytokines, Regulation in Inflammation

Blood Oxygenation Level Dependent

▶ BOLD

Blood Patch

Definition

Epidural injection of 10–15 ml of the patient's own blood. It improves low CSF pressure headaches immediately by simple volume replacement, and subsequently by sealing spinal leaks.

Cross-References

▶ Headache Due to Low Cerebrospinal Fluid Pressure

Blue Flags

Definition

Blue Flags are perceived features of work, generally associated with higher rates of symptoms, ill-health, and work loss. They are characterized by features such as high demand/low control, unhelpful management style, poor social support from colleagues, perceived time pressure, and lack of job satisfaction.

Cross-References

▶ Black Flags
▶ Yellow Flags

BMD (Burning Mouth Disorder)

▶ Burning Mouth Syndrome

BMS

▶ Burning Mouth Syndrome

Body Awareness

Definition

An overall concept for the awareness of the body. It includes body consciousness, body management, and deepened body experience.

Cross-References

▶ Body Awareness Therapies

Body Awareness Therapies

Alice Kvåle
Department of Public Health and Primary Health Care, Faculty of Medicine and Dentistry
University of Bergen, Bergen, Norway

Synonyms

Basic Body Awareness Therapy; Body-Mind Approaches; Psychomotor Physiotherapy; Psychosomatic and Psychiatric Physiotherapy; Relaxation Therapy

Definition

The term ▶ "body awareness" is applied as an overall concept to the awareness between body and mind. It includes body consciousness, body management, and deepened body experience. Body awareness is dependent on the capacity to perceive and integrate information coming from all sense organs, including proprioceptors of muscles and joints. The concept also concerns the relation between one's own body and feelings, and the perception of the body as "me." Body awareness therapies focus on the whole person, rather than on separate parts. The body is seen as a functional and interacting entity and psyche and soma are considered indivisible. Through body awareness techniques, the aim is to reduce tension and guarded movements and to increase sensory motor awareness and locomotor control. The techniques range through relaxation, massage, and active exercises, in order to improve and normalize muscular control and help the patient to become aware of how the mind and body interact.

Characteristics

Body awareness therapies are based upon an assumption that the body reacts to both physical and psychological strain, over time affecting flexibility and ability to relax, muscle tension, respiration, and posture (Bunkan and Thornquist 1990; Sundsvold and Vaglum 1985). A healthy person generally has a good ability to perceive the whole of his body as well as parts of it and has intact and good body awareness. The healthy person has a realistic self-image, adequate muscle tension, good stability and balance, flexibility in movements, unrestricted respiration, and the ability to listen to signals from the body. Many patients are unaware of how they hold and use their body, ways that may perpetuate and worsen their problems, over time resulting in a number of secondary physical aberrations. Fear of movement in the early stages of pain may lead to avoidance behavior, resulting in reduced activities and guarded movement patterns (Vlaeyen and Linton 2000).

Patients with long-lasting musculoskeletal pain conditions have been found to have postural and respiratory changes, as well as reduced flexibility and ability to relax and palpable changes in muscles and skin (Kvåle et al. 2003). Furthermore, a close relationship between physical and psychological aberrations has been documented most prominently in patients with widespread pain (Kvåle et al. 2001). Posture, respiration, movements, and muscles function interactively and a common pitfall is to overlook this in treatment, only concentrating upon the body part that the patient has reported as most painful. In patients with long-lasting pain conditions, the whole body should therefore be examined and treated, because changes in one region may have widespread effects in the rest of the body. Furthermore, promotion of the body-mind mutuality may enable a person to gain a new perspective over the health problem, leading to less helplessness, better coping, and pain relief.

Treatments that focus on pathoanatomical aspects and on single physiotherapy interventions have been shown to have few or no long-term effects in patients with long-lasting musculoskeletal pain conditions (van Tulder et al. 1997). Multidisciplinary rehabilitation interventions including cognitive and behavioral approaches have been found to be superior to many other treatment interventions in persistent pain problems (van Tulder et al. 2001). Most of these programs include elements common to body awareness therapies, such as ▶ body awareness training and relaxation. However, it is unclear which type and part of such programs are most effective. Outcome studies and randomized controlled trials have become the golden standard for examining the efficiency of therapies. However, within treatment of long-lasting pain problems, a dilemma still exists between experience-based knowledge and the so-called evidence-based knowledge. Likewise, knowledge of the effect of body awareness used alone or as part of other therapies is still limited. Patients with pain disorders represent a heterogeneous group; in such a group, the effect of the treatment on one subgroup may not be visible when patients with several different characteristics are pooled (Haldorsen et al. 2002).

Body and Mind Therapies

Psychomotor Physiotherapy

Since the late 1940s, a branch of physiotherapy called ▶ psychomotor physiotherapy or ▶ psychiatric and psychosomatic physiotherapy has been developed in the Scandinavian countries (Bunkan and Thornquist 1990). This direction recognizes the body as a functional and interacting entity and considers psyche and soma as indivisible. Posture and respiration may express repressed emotions and give an indication of the general tension of the person, aspects that are taken into account during treatment. Flexibility of movement and elasticity of muscles bear the key to proper functioning, physically as well as psychologically. Excessive muscular work when carrying out movements, poor coordination, and restrained respiration may result from both pain and strain, with negative consequences for body flow and rhythm. Through psychomotor physiotherapy body awareness techniques, the aim is to improve and normalize muscular control and help the patient to become aware of how the mind and body interact. During therapy, patient contact is mainly obtained through the body, listening and talking being secondary. Different techniques ranging from stimulating massage to active exercise are part of the therapy sessions. Common to all approaches is teaching the patient to register tension and to notice the difference between contracted and relaxed muscles. Furthermore, the therapy focuses on how respiration can influence muscle tension and guarded movements. Body sensitivity training can be done individually or in small groups, depending on the individual's degree of problem and needs. Exercises can focus directly or indirectly on respiration, they can include rhythmic swinging and exercises with rotation and exercises in sitting, standing, and walking. Included in the therapy is teaching the patients relaxation techniques (see below). Recent smaller studies have shown that psychomotor physiotherapy may reduce pain and tender points, symptoms of subjective health complaints, depression, anxiety, insomnia, fatigue, and improve quality of life, although the process takes time (Anderson et al. 2007; Breitve et al. 2010).

Some therapists integrate special exercises such as Feldenkreis, Pilates, Laban, or Alexander techniques (Braverman et al. 2003) into psychomotor physiotherapy and some might occasionally add music. If music is used, the therapist must ensure that it does not distract attention from body consciousness. Common to the above-mentioned techniques is the reeducation of the body to interrupt poor posture and movement patterns, to develop smooth coordinated actions and to perform functional energy-efficient tasks without unnecessary muscular contractions and restricted breathing.

In Norway, physiotherapists (PTs) must take 2 years of additional education at postgraduate level in order to qualify as licensed psychiatric and psychosomatic physiotherapists to work with psychiatric patients. Parts of these body and mind techniques are, however, included in basic physiotherapy education in the Scandinavian countries. Further experience in the techniques is often offered as shorter postgraduate courses for qualified PTs. For PTs working with pain patients, this is considered adequate. As body-mind therapies influence the way patients feel and behave, appropriate training and ethical practice are necessary and essential.

Basic Body Awareness Therapy (Basic BAT)

Inspired by the French movement therapist and psychotherapist Dropsy (Dropsy 1993), the Swedish physiotherapist, Roxendal, developed a method called ► basic body awareness therapy (Roxendal 1985), which is widely used today in Scandinavia in patients with pain problems (Gyllensten 2001; Malmgren-Olsson 2002). Stability in postural function combined with freedom of movement and breathing are also essential elements in this therapy approach and the body experiences are considered crucial for total self-identity. In therapy, the aim is to increase sensory motor awareness and locomotor control, through exercise that focuses on a stable relation to the ground, posture, gait, and truncal positioning and the maintenance of balance related to the central axis of the body. The emphasis is in particular on the development of the mental presence and awareness that underlies

all that is practiced in the movement exercises. The exercises consist of simple basic movements of daily living such as lying, sitting, standing, and walking, used to normalize postural control and coordination, breathing, and muscular tension. Patients are usually instructed in small groups. Sometimes different massage techniques can be used by the therapist or in pair exercises in the group. During the sessions, the patients have the opportunity to talk about their experience of movements and to reflect on the interaction between pain and environmental factors. Elements from T'ai-chi and Zen meditation as well as those from Feldenkreis, Laban, and other body-oriented programms are often integrated when practicing basic BAT and in the promotion of awareness of muscular tension and muscular relaxation (Gyllensten 2001).

The Mensendieck System (MS)

In the Netherlands, Denmark, Norway, Sweden, and the United States, another direction of body awareness program has been in use for many decades. This method, developed by the Dutch American physician Bess Mensendieck in the early 1900s, is a pedagogically designed system of functional exercises based on an analysis of human anatomy, physiology, and biomechanics (Mensendieck 1937). The MS is taught primarily to individuals with musculoskeletal problems attributable to imbalances in muscle function with primary or secondary muscular tension and pain. The aims of MS are to improve the individual's functional capacity in daily life by increasing the understanding of anatomically and physiologically correct movements and improving the ability to interpret body signals.

Only small studies of the mentioned therapies are available and unfortunately often without control groups. In a recent study, BAT and Feldenkreis approaches were found to be more beneficial for patients with nonspecific musculoskeletal disorders than individual physiotherapy (Malmgren-Olsson 2002). This is in line with what has been found concerning cognitive behavioral treatment. In another study concerning patients with fibromyalgia, a MS intervention was found to be more effective than a BAT intervention and had a significant effect on pain, function, and coping at an 18 months follow-up too (Kendall et al. 2000).

Relaxation and Meditation Programs

Standardized instructions or meditative techniques are used to elicit a relaxation response, characterized by generalized decrease in the sympathetic nervous system and metabolic activity and an altered state of consciousness, described as a subjective experience of well-being (Kabat-Zinn et al. 1985). The techniques include autogenic training (patients are taught to achieve feelings of heaviness and warmth in their limbs), transcendental meditation or yoga (deep relaxation accompanied by subjective experiences such as peace of mind and a sense of well-being), and Jacobson's method of progressive relaxation (people are taught to relax individual muscle groups in progression, after first contracting them). Yoga may also incorporate flexibility and strengthening exercises, as well as awareness of how nutrition, physical health practices, and ethical and social conduct affect the body and mind.

One review of several stress management studies showed that relaxation and behavioral skills are helpful and that such group methods are both more cost-effective and more beneficial than individual counseling (Sims 1997). In a few studies, relaxation techniques are compared with other forms of conservative treatments. In one small study, patients with chronic low back pain were randomly assigned to three different kinds of treatment. Eight sessions with relaxation training, consisting of progressive relaxation, breathing techniques, autogenic training, and visual imagery, gave better results in reducing EMG and pain and in increasing relaxation and activity, than either EMG biofeedback alone or placebo (Stuckey et al. 1986). Relaxation training should, however, be used in addition to exercise and be combined with coping skills training and stress management (Sandstrom and Keefe 1998).

Final Remarks

Since the gate control theory became known, cognitive-behavioral and body awareness therapies seem to have gained increasing acceptance

as part of conventional pain management programs (Braverman et al. 2003). Although there is a need for more evidence-based support for the effect of body awareness therapies, combination of therapies may provide patients with long-lasting pain conditions, with the skills and knowledge needed to increase the sense of control over pain and possible perpetuating factors. The integration of physiotherapy, psychological techniques and, if necessary, appropriate pharmacotherapeutic regimens maximizes effectiveness in managing long-lasting pain.

References

Anderson, B., Strand, L. I., & Råheim, M. (2007). The effect of long-term body awareness training succeeding a multimodal cognitive behavior program for patients with widespread pain. *J. Musculoskel. Pain, 15*(3), 19–29.

Braverman, D. L., Ericksen, J. J., Shah, R. V., et al. (2003). Interventions in chronic pain management. 3. New frontiers in pain management: Complementary techniques. *Arch. Phys. Med. Rehab., 84*, 45–49.

Breitve, M., Hynninen, M.J., & Kvåle, A. (2010). The effect of psychomotor physical therapy on subjective health complaints and psychological symptoms. *Physiother Res Int., 15*(4), 212–221.

Bunkan, B. H., & Thornquist, E. (1990). Psychomotor therapy: an approach to the evaluation and treatment of psychosomatic disorders. In T. Hegna & M. Sveram (Eds.), *International perspectives in physical therapy 5: Psychological and psychosomatic problems* (pp. 45–74). London: Churchill Livingstone.

Dropsy, J. (1993). Leva i sin kropp. Kroppsuttryck och mänsklig kontakt. (Vivre dans son corps). Natur och Kultur, Stockholm

Gyllensten, A. L. (2001). *Basic body awareness therapy - Assessment, treatment and interaction.* Sweden, Lund: Lund University. Doctoral thesis.

Haldorsen, E. M. H., Grasdal, A. L., Skouen, J. S., et al. (2002). Is there a right treatment for a particular patient group? Comparison of ordinary treatment, light multidisciplinary treatment, and extensive multidisciplinary treatment for long-term sick-listed employees with musculoskeletal pain. *Pain, 95*, 49–63.

Kabat-Zinn, J., Lipworth, L., & Burney, R. (1985). The clinical use of mindfulness meditation for the self-regulation of chronic pain. *J. Behav. Med., 8*, 163–190.

Kendall, S. A., Brolin-Magnusson, K., Sören, B., et al. (2000). A pilot study of body awareness programs in the treatment of fibromyalgia syndrome. *Arthritis Care Res., 13*, 304–311.

Kvåle, A., Ellertsen, B., & Skouen, J. S. (2001). Relationships between physical findings (GPE-78) and psychological profiles (MMPI-2) in patients with long-lasting musculoskeletal pain. *Nord. J. Psychiatry, 55,* 177–184.

Kvåle, A., Skouen, J. S., & Ljunggren, A. E. (2003). Discriminative validity of the Global Physiotherapy Examination (GPE-52) in patients with long-lasting musculoskeletal pain versus healthy persons. *J. Musculoske. Pain, 11,* 23–35.

Malmgren-Olsson, E.-B. (2002). A comparison between three physiotherapy approaches with regard to health-related factors in patients with non-specific musculoskeletal disorders. *Disabil. Rehabil., 24,* 181–189.

Mensendieck, B. M. (1937). *The Mensendieck system of functional exercises.* Portland: Southworth-Anthoensen Press.

Roxendal, G. (1985). *Body awareness therapy and the body awareness scale, treatment and evaluation in psychiatric physiotherapy.* Gothenburgh, Sweden: Gothenburgh University. Doctoral thesis.

Sandstrom, M. J., & Keefe, F. J. (1998). Self-management of fibromyalgia: the role of formal coping skills training and physical exercise training programs. *Arthritis Care Res., 11,* 432–447.

Sims, J. (1997). The evaluation of stress management strategies in general practice: An evidence-led approach. *Br. J. Gen. Pract., 47,* 577–582.

Stuckey, S. J., Jacobs, A., & Goldfarb, J. (1986). EMG biofeedback training, relaxation training, and placebo for the relief of chronic back pain. *Percept. Mot. Skills, 63,* 1023–1036.

Sundsvold, M. Ø., & Vaglum, P. (1985). Muscular pains and psychopathology: Evaluation by the GPM method. In T. H. Michel (Ed.), *International perspectives in physical therapy. 1. Pain* (pp. 18–47). London: Churchill Livingstone.

van Tulder, M. W., Koes, B. W., & Bouter, L. M. (1997). Conservative treatment of acute and chronic nonspecific low back pain. A systematic review of randomized controlled trials of the most common interventions. *Spine, 22,* 2128–2156.

van Tulder, M. W., Ostelo, R., Vlaeyen, J. W. S., et al. (2001). Behavioral treatment for chronic low back pain - A systematic review within the framework of the Cochrane Back Review Group. *Spine, 26,* 270–281.

Vlaeyen, J. W., & Linton, S. J. (2000). Fear-avoidance and its consequences in chronic musculoskeletal pain: A state of the art. *Pain, 85,* 317–332.

Body Awareness Training

Definition

Includes different body awareness techniques; also called body-mind approaches, which focus on the whole person, rather than on separate body parts. The body is seen as a functional and interacting entity, and psyche and soma are considered indivisible. The aims of the awareness training are to reduce tension and guarded movements, and to increase sensory motor awareness and locomotor control. Techniques range from relaxation, massage, and active exercises in order to improve and normalize muscular control and help the patient to become aware of how the mind and body interact.

Cross-References

▶ Body Awareness Therapies

Body Functions

Definition

The physiological functions of body systems (including psychological functions).

Cross-References

▶ Functioning and Disability Definitions
▶ Physical Medicine and Rehabilitation, Team-Oriented Approach

Body of the Clitoris

Definition

The body is the portion of the clitoris, 2–3 cm in length, consisting of erectile tissue, and hangs downward from the pubic bone, connecting the glans and the crura.

Cross-References

▶ Clitoral Pain

Body Structures

Definition

Body structures are anatomical parts of the body such as organs, limbs, and their components.

Cross-References

▶ Functioning and Disability Definitions

Body-Mind Approaches

Definition

Includes a variety of methods focusing on body consciousness, body management, and body experience, and is defined synonymously with body awareness therapies.

Cross-References

▶ Body Awareness Therapies

Body-Self Neuromatrix

Definition

A model proposed by R. Melzack to explain how acute and/or long-term stress responses modulate the multidimensional pain experience. Three parallel neural networks, which contribute to the sensory-discriminative, affective-motivational, and evaluative-cognitive dimensions of the pain experience, receive input from a number of sources, including somatic, visual, cognitive, emotional, as well as the stress-regulation systems (including endocrine, immune, opioid, HPA, and LC-NE).

Cross-References

▶ Stress and Pain

BOLD

Synonyms

Blood oxygenation level dependent

Definition

The BOLD signal arises due to the higher ratio of oxy- to deoxyhemoglobin in local draining venules and veins in response to neuronal activity.

Cross-References

▶ Thalamus, Clinical Pain, Human Imaging

BOLD Activity

▶ Thalamus, Clinical Pain, Human Imaging

Bone Cancer Pain

▶ Cancer Pain, Animal Models

Bone Marrow

Definition

The soft sponge-like material that is found inside the bones and that produces blood cells from immature cells called stem cells.

Cross-References

▶ Diencephalic Mast Cells

Bone Metastases

Definition

Spread of cancer to bone.

Cross-References

▶ Cancer Pain Management Radiotherapy

Bone Pain

Definition

Bone pain is frequently caused by growth of blood-borne tumor cells in the bone. Primary bone tumors can also cause pain. Tumors most often producing painful bone metastases are cancer of breast, prostate, bronchus, and kidney. Tumor cells release a number of osteoclast-activating factors (OAFs) which include prostaglandins, kinins, substance P, and parathyroid-hormone-related peptides. OAFs are responsible not only for bone destruction but also for sensitization of the nerve endings and pain. Bone pain consists of inflammatory, instability, and neuropathic components. Bone pain causes in the spinal cord similar activation of the glial astrocytes as seen in neuropathic pain. The bone pain is usually sensitive to prostaglandin inhibitors and only partially sensitive to opioids. Bisphosphonates may be helpful in restoring bone integrity and reducing pain.

Cross-References

▶ Adjuvant Analgesics in Management of Cancer-Related Bone Pain
▶ Cancer Pain
▶ Cancer Pain Management, Orthopedic Surgery
▶ Cancer Pain Model, Bone Cancer Pain Model

Bone Scan

Peter Lau
Department of Clinical Research, Royal Newcastle Hospital, University of Newcastle, Newcastle, NSW, Australia

Synonyms

Bone scintigraphy

Definition

Bone scan is a diagnostic test to detect focal changes in bone metabolism. It involves administering a radioactive substance that accumulates in bone in proportion to its metabolism. The emitted radiation constitutes the basis for the scan.

Characteristics

If bone metabolism is increased, it will draw a greater than normal supply of blood. If a substance that is used by bone is rendered radioactive and is administered into the bloodstream, a greater than normal concentration of the substance will accumulate in the region of bone that has the increased blood supply and increased metabolic activity. The focal concentration of the substance can be detected from the increased radiation that it emits from where it is concentrated. Any disease process that increases blood flow will result in a positive bone scan. Conversely, if a disease decreases bone metabolism, it will be manifest as less than normal radioactive activity.

Because of its similarity to pyrophosphate, technetium-99 m methylene diphosphonate (MDP) is a radioactive tracer that is absorbed by bone. If injected intravenously it accumulates in all bones. The radiation that it emits can be detected by a scanner and an image can be produced of the distribution and concentration

Bone Scan, Fig. 1 The arterial phase of a bone scan. Radioactive tracer is seen flowing through the arteries of the lower limbs and into the peripheral tissues (Images provided by courtesy of Dr John Booker of Hunter Imaging Group and Sonic Health, Newcastle, Australia)

of the tracer. In a normal bone, the tracer is distributed reasonably uniformly throughout the bone, and the radiation emitted will cast a photographic image of the shape of the bone on a digital detector. Increased uptake of radio-isotope is seen as a "hot" spot and decreased uptake as a "cold" spot.

There are three phases of the bone scan study. The initial phase occurs immediately after the intravenous injection of the radioactive tracer. During this phase, the tracer is distributed into the bloodstream. During the second phase, the tracer is delivered to bone through the arterial system (Fig. 1). Scans obtained during this phase reflect the perfusion of bones by blood. The third phase reflects the accumulation of the tracer in the bone and its binding to it. Uptake is maximal at 2 h. At this time, a significant propor-tion of the unbound tracer will have been excreted by the kidneys.

The extent to which a lesion accumulates the tracer is a function of both the local blood flow and the extraction efficiency, i.e., the rate at which the bone removes tracer from the blood. The extent to which these two factors contribute to accumulation differs in different conditions.

Application

Increased uptake of radioactive tracer can be observed in benign conditions, such as fracture, osteomyelitis, and myositis ossificans, in which local inflammation stimulates an increased blood flow. Increased uptake occurs in malignant, primary bone lesions such as osteogenic sarco-mas. Metastases can manifest as either increased or decreased uptake, depending on the nature of the tumor (Fig. 2).

Decreased uptake occurs in disorders such as Perthe's disease and avascular necrosis (e.g., of the femur, lunate, or scaphoid). In Paget's disease

Bone Scan, Fig. 3 A bone scan of Paget's disease. There is increased uptake of tracer in the bones of the skull (Image provided by courtesy of Dr John Booker of Hunter Imaging Group and Sonic Health, Newcastle, Australia)

Bone Scan, Fig. 2 A whole-body bone scan. The white areas are sites of increased uptake of tracer throughout the skeletal system, indicative of disseminated metastases (Images provided by courtesy of Dr John Booker of Hunter Imaging Group and Sonic Health, Newcastle, Australia)

there is characteristically an increased radioisotope concentration in the extracellular fluid by passive diffusion as well as increased vascularity. This results in an intense uptake of technetium (Fig. 3).

Stress Fractures

The application of bone scanning in pain medicine lies in the detection of stress fractures in patients with leg pain and foot pain precipitated by prolonged activity, and in athletes with back pain. The particular advantage of bone scanning is that it can detect stressed bone before it actually fractures. The appearance on the scan is referred to as a stress reaction (Fig. 4). Early detection allows

rest for the offending activity to be implemented, with good chances of averting fracture.

The utility of bone scanning once a fracture has occurred is more contentious. Classical teaching maintains that a positive scan would indicate a recent fracture, which would implicate the fracture as the source of pain. However, in the case of fractures of the pars interarticularis of lumbar vertebrae, the relationships between bone scan, pars defect, and symptoms are imperfect, although a positive scan is likely to be associated with pain. The scan is negative in the majority of patients with pain (Lowe et al. 1984). In patients with a radiologically evident pars fracture, the bone scan is just as likely to be negative as positive (Elliot et al. 1988).

Necrosis

Bone infarction and aseptic necrosis can affect the femoral head, humeral head, and tibia. These are benign lesions but may cause pain. Although uncommon, they are an important differential diagnosis of pain in the hip, shoulder, and knee. Early detection is paramount so that surgical treatment can be implemented before the joint is irreparably damaged. Plain radiography in the early stage may be normal or show only minor periosteal reaction, but a bone scan will typically

Bone Scan, Fig. 4 A bone scan of the lower limbs showing a stress reaction in the left tibia, which appears as a white spot of increased uptake of tracer (*arrow*) (Images provided by courtesy of Dr John Booker of Hunter Imaging Group and Sonic Health, Newcastle, Australia)

show decreased uptake. The differential diagnosis includes osteomyelitis or bone tumor.

To distinguish avascular necrosis from infection of bone, a gallium scan may be used. Not to be confused with a bone scan, a gallium scan uses radioactive gallium citrate as the tracer. It usually takes a few days to accumulate gallium citrate in the areas of the body tissues where white blood cells accumulate. If the problem is caused by infection, certain inflammatory diseases, or neoplasms such as lymphoma, the accumulated gallium will show up on the scan as a "hot" spot.

Metastases

Metastatic bone disease is a common cause of bone pain in cancer patients. Breast, lung, and prostate cancer account for more than 50 % of the primary cancer that commonly has extensive bony metastasis. More than 50 % of patients with this primary cancer will eventually develop bone metastasis (Malawer and Delaney 1993). Lymphoma can also involve bone. Pain may arise from the bone because of the invasion by tumor, or it may arise because of pathological fracture.

Malignancy is more common in elderly patients, but age-related degenerative changes may confound the diagnosis. In the spine, symmetry of findings is the basis for distinguishing tumor from asymptomatic degenerative changes. Symmetrical uptake around the zygapophysial joins is more likely to be due to osteoarthrosis. Focal, increased uptake in the vertebral bodies or pedicles is highly specific for metastatic spine disease. Nevertheless, plain radiography and CT or MRI is required to confirm the diagnosis.

Bone scan is definitely more sensitive than skeletal X-ray in detecting metastatic bone disease, except in myeloma, renal cell carcinoma, and thyroid cancer. The metastases of these latter tumors are purely lytic. Therefore, they are clearly apparent on plain radiographs. Meanwhile, they are not associated with significant new bone formation, which renders them difficult to detect on bone scan.

With advances in chemotherapy and hormonal treatment, patients with breast or prostate cancer can have a prolonged life expectancy. As a result, their chances of developing bony metastasis are high compared with those of patients with lung cancer, from which the survival rate is less than 6 months.

Postoperative Fusion

Spinal fusion is used to treat spinal pain. A pronounced focal increase in uptake within the fusion mass, at more than 1 year postoperatively, is a highly suspicious sign for pseudoarthrosis. Bone scan has a sensitivity of 78 % and a specificity of 83 % in the detection of pseudoarthrosis, which is superior to plain radiography (43 % and 50 %, respectively)

(Even-Sapir et al. 1994; Lusins et al. 1989). Such abnormalities, however, are not a sign of pain, for they have also been identified in asymptomatic patients.

Asymmetric uptake in the sacroiliac region has also been frequent noted in patients who have undergone spinal fusion (up to 75 % of cases). It occurs on the side from which the bone graft has been harvested. Increased uptake can occur in the sacroiliac joints themselves and has been inferred to reflect alterations in spinal mechanics (Even-Sapir et al. 1994; Lusins et al. 1989).

Joint Prostheses

Increased uptake can occur around joint prostheses (Fig. 5). This can be due to aseptic loosening or to infection. *Staphylococcus epidermidis* and *S. aureus* are the most common organisms (Love et al. 2001). Septic prostheses require resection arthroplasty, a long course of antibiotics and a lengthy hospitalization. In aseptic loosening, the patient typically undergoes a single stage revision arthroplasty. Differentiation between infection and aseptic loosening is difficult, but extremely important.

With cemented hip prostheses, if the uptake about the acetabular component of the prosthesis exceeds that of the activity of the iliac crest, loosening is strongly suspected.

With porous-coated prostheses the distinction between infection and loosening can be very difficult. Nevertheless bone scan is useful as a screening test because a negative bone scan excludes a sinister lesion. A sensitivity of 65 % and a specificity of 70 % have been quoted for infected prosthesis (Love et al. 2001).

Indium scanning with In-11^{1-} WBC (white blood cells) is used to differentiate aseptic loosening and hip prosthesis loosening. Abnormalities with activity greater than that in the iliac crest should be considered suspicious for infection (Rand and Brown 1990).

Knee Prostheses

Bone scan is not sensitive for infected prostheses of the knee, because 60 % of the femoral components and 90 % of the tibial components

Bone Scan, Fig. 5 A bone scan of the lower limbs showing increased uptake of tracer in the upper right tibia, indicative of loosening of z knee prosthesis (Images provided by courtesy of Dr John Booker of Hunter Imaging Group and Sonic Health, Newcastle, Australia)

demonstrate increased periprosthetic activity more than 1 year after surgery in asymptomatic patients (Palestro 1995). However, white blood cell imaging has a sensitivity, specificity and accuracy of approximately 85 % in the diagnosis of infected knee prosthesis (Love et al. 2001). False positive examinations have been reported in association with active rheumatoid arthritis and with massive osteolysis of the adjacent femur and tibia (Love et al. 2001).

References

Elliot, S., Huitson, A., & Wastie, M. L. (1988). Bone scintigraphy in the assessment of spondylolysis in patients attending a sports injury clinic. *Clinical Radiology, 39,* 269–272.

Even-Sapir, E., Martin, R. H., Mitchell, M. J., et al. (1994). Assessment of painful late effects of lumbar spinal fusion with SPECT. *Journal of Nuclear Medicine, 35,* 416–422.

Love, C., Thomas, M. B., Marwin, S. E., et al. (2001). Role of nuclear medicine in diagnosis of the infected joint replacement. *Radiographics, 21,* 12290–1238.

Lowe, J., Schachner, E., Hirschberg, E., et al. (1984). Significance of bone scintigraphy in symptomatic spondylolysis. *Spine, 9,* 654–655.

Lusins, J. O., Danielski, E. F., & Goldsmith, S. J. (1989). Bone SPECT in patients with persistent back pain after lumbar spine surgery. *Journal of Nuclear Medicine, 30*, 490–496.

Malawer, M. M., & Delaney, T. F. (1993). Treatment of metastatic cancer to bone. In V. T. DeVita, S. Hellman, & S. A. Rosenberg (Eds.), *Cancer: Principles and practice of oncology* (4th ed., pp. 2225–2245). Philadelphia: JB Lippincott.

Palestro, C. J. (1995). Radionuclide imaging after skeletal interventional procedures. *Seminars in Nuclear Medicine, 25*, 3–14.

Rand, J. A., & Brown, M. L. (1990). The value of indium 111 leukocyte scanning in the evaluation of painful or infected total knee arthroplasties. *Clinical Orthopaedics, 259*, 179–182.

Bone Scintigraphy

▶ Bone Scan

Boney Pain

▶ Adjuvant Analgesics in Management of Cancer-Related Bone Pain

Borderline Leprosy

Definition

Borderline leprosy is an immunologically unstable form of leprosy, which is transitional between the tuberculoid and lepromatous forms and has clinical and histological features of both types.

Cross-References

▶ Hansen's Disease

Botox®™

▶ Botulinum Toxin

Bottom of Form

Definition

Validity – the extent to which a measurement, test, or study measures what it claims to measure.

Cross-References

▶ Assessment of Pain Behaviors

Botulinum Toxin

Jayantilal Govind
Pain Clinic, Department of Anaesthesia, University of New South Wales, Sydney, NSW, Australia

Synonyms

Botox®™

Definition

Botulinum neurotoxin (Botox®™, BoNT) is a purified neurotoxin derived from the anaerobic spore-forming bacterium, *Clostridium botulinum*. The toxin has been advocated for the treatment of acute and chronic pain occurring in a heterogeneous group of disorders.

Eight immunologically distinct serotypes have been identified, each produced by a distinct strain of *Clostridium botulinum*. With the exception of C2, A–G inclusive are all neurotoxins. To date, only type A and type B have been introduced into clinical practice.

Characteristics

Mechanism

Botulinum toxin is well known for its effects on the motor system, where it inhibits the release of acetylcholine (ACh) from cholinergic terminals of motor neurons (Simpson 2000). It also inhibits release of ACh from pre-ganglion sympathetic and parasympathetic neurons and postganglionic parasympathetic nerves (Simpson 2000). It produces a flaccid paralysis in the motor system (Humeau et al. 2000). Elsewhere, it produces autonomic dysfunction and various disturbances of central nervous system function (Hatheway 1995; Tyler 1963; Santini et al. 1999), which can occur during botulism or following peripheral administration of the toxin.

The effect is temporary. While synaptic activity is eradicated, nerve endings are preserved. Recovery of neurotransmission does occur eventually. Blockade by botulinum toxin at the neuromuscular junction induces the formation of an extensive network of nerve terminal sprouts and a number of other proteins essential for neurotransmission, including voltage-sensitive sodium and calcium channels for action potential conduction and nerve-evoked transmitter release. Sprouting is rapid and the sprouts are only transitory, because a second distinct phase of recovery occurs with the return of synaptic activity to the original nerve terminals, accompanied by elimination of the then superfluous sprouts (Paiva et al. 1999). The original terminals remain unable to undergo measurable exoendocytosis until almost 2 months after injection of toxin A (Paiva et al. 1999).

The mechanism of botulinum toxin in the management of pain is not known. Proponents of its use in pain are actively pursuing a mechanism distinct from its effects on motor nerves. Although cholinergic neurons are believed to mediate nociceptive pathways in the central nervous system, a central action of botulinum toxin seems unlikely, because its effectiveness appears to be related to local administration at the site of pain.

Technique
The agent is administered either by subcutaneous, intradermal, or intramuscular injection. (For dosage schedule, the reader is advised to consult with the manufacturer's recommendations. With respect to potency, the neurotoxins are not interchangeable given the lack of international units).

Applications
In certain jurisdictions, the toxin has been approved for the treatment of blepharospasm, spasticity due to juvenile cerebral palsy, cervical dystonia (spasmodic torticollis), hemifacial spasm, glabellar lines, and primary hyperhidrosis of the axilla. This is consistent with its known mechanisms of action.

For the treatment of pain, the use of BoNT is still considered off-label in certain jurisdictions and, therefore, cannot be publicly recommended.

Indications
The toxin has been actively promoted for the management of a number of diverse conditions including achalasia, anismus, benign prostatic hypertrophy, dysphonia, essential tremor, migraine, tension-type headaches, myofascial pain, pancreatitis, pelvic floor disorders, rectal fissures, temporomandibular joint syndrome, and urinary sphincter dysfunction (Munchau and Bhatia 2000).

Efficacy
Advocating BoNT for a diverse and heterogeneous group of disorders presupposes a common pain pathophysiology. Published observational studies suffer severe methodological flaws including small sample size, lack of randomization, poor outcome measures, and construct validity.

In the treatment of ▶ chronic low back pain, BoNT ostensibly relieves pain due to muscle spasm. A randomized double-blind study, in which 15 patients received the toxin, 60 % reported at least 50 % pain relief at 8 weeks, compared to 12 % who received saline (Foster et al. 2001). The study is limited by its small sample size and large confidence intervals. None of the recipients reported complete relief of pain. The results attenuate over time, and its efficacy beyond 8 weeks is not known. While it may offer temporary palliation, its short duration of action limits its utility.

Two randomized, placebo-controlled trials showed that BoNT was not clinically beneficial for the treatment of neck pain and whiplash-associated disorders. Statistically significant effects were not achieved in a study of 15 participants (Freund and Schwartz 2000).

In a subsequent study of 25 participants, no specific benefits could be identified. Participants treated with normal saline showed comparable outcomes to those treated with BoNT over 4 months (Wheeler et al. 2001).

No studies have been reported that establish either the ▶ effectiveness or the ▶ efficacy of the toxin for the treatment of myofascial pain syndrome or ▶ fibromyalgia (Smith et al. 2002).

Contradictory results have been reported for the treatment of headache. Silberstein et al. (2000) enrolled 123 patients with episodic migraine to randomly receive either 25 units or 75 units of BoNT, or placebo, injected at multiple fixed symmetrical pericranial sites. At 3 months, postinjection patients receiving 25 units showed significant reduction in headache frequency, headache severity, acute medication use, and migraine-associated vomiting compared to placebo. However, the outcomes of participants receiving 75 units did differ from those receiving placebo. Thirteen cases of ptosis and two cases of diplopia were reported. It is possible the study may not have been truly double blinded, and whether the toxin exerted a true anti-migraine effect remains speculative.

Evaluating the studies published thus far according to evidence-based medicine criteria, Evers et al. (2002) reported no sufficient positive evidence for the general treatment of idiopathic and cervicogenic headaches with BoNT.

Contraindications

Known contraindications include hypersensitivity to any ingredient in the formulation, myasthenia gravis, amyotrophic lateral sclerosis, infection, or the Eaton-Lambert syndrome. The toxin may be potentiated by aminoglycoside antibiotics, spectinomycin, or any other drug that might interfere with the neuromuscular transmission, e.g., tubocurarine, tetracyclines, and lincomycin, or drugs that interfere with the intraneuronal concentrations of calcium (Dysport; Botox 2002).

Side Effects and Complications

The product information provided by manufacturers lists a variety of possible side effects and complications (Dysport; Botox 2002). These include skin reactions, gastrointestinal disturbances, respiratory symptoms, and neurological, cardiovascular, and ocular disturbances. The product also contains a small amount of human albumin, and hence, the risk of transmission of viral infection cannot be excluded with absolute certainty.

Acknowledgement Dr. Jayantilal Govind passed away on June 20, 2009 at age 68. We thank Prof. Nik Bogduk for the update of his entry.

References

Botox (2002). Product information. Allergan Australia Pty Ltd. Gordon, New South Wales, Australia.

Dysport. Product information. IPSEN Pty Ltd. Glen Waverley, Victoria, Australia.

Evers, S., Rahmann, A., Vollmer-Haase, J., & Husstedt, I. W. (2002). Treatment of headache with botulinum toxin A – A review according to evidence-based medicine criteria. *Cephalalgia, 22*, 699 710.

Foster, L., Clapp, L., Erickson, M., & Jabbari, B. (2001). Botulinum toxin A and chronic low back pain. A randomized, double-blind study. *Neurology, 56*, 1290–1293.

Freund, B., & Schwartz, M. (2000). Treatment of whiplash associated with neck pain with botulinum toxin-A: A pilot study. *Journal of Rheumatology, 27*, 481–484.

Hatheway, C. L. (1995). Botulism: The present status of the disease. *Current Topics in Microbiology and Immunology, 195*, 53–75.

Humeau, Y., Doussau, F., Grant, N. J., & Poulain, B. (2000). How botulinum and tetanus neurotoxins block neurotransmitter release. *Biochimie, 82*, 427–446.

Munchau, A., & Bhatia, K. P. (2000). Uses of botulinum toxin injection in medicine today. *British Medical Journal, 320*, 161–165.

Paiva, A., Meunier, F. A., Molgo, J., Aoki, K. R., & Dolly, J. (1999). Functional repair of motor endplates after botulinum neurotoxin type-A poisoning: Biphasic switch of synaptic activity between nerve sprouts and their parent terminals. *Proceedings of the National Academy of Sciences, 96*, 3200–3205.

Santini, M., Fabri, S., Sagnelli, P., Manfredi, M., & Francia, A. (1999). Botulism: A case associated with pyramidal signs. *European Journal of Neurology, 6*, 91–93.

Silberstein, S., Mathew, N., Saper, J., & Jenkins, S. (2000). Botulinum toxin type-A as a migraine preventive treatment. *Headache, 40*, 445–450.

Simpson, L. L. (2000). Identification of the characteristics that underlie botulinum toxin potency: Implication for designing novel drugs. *Biochimie, 82*, 943–953.

Smith, H. S., Audette, J., & Royal, M. A. (2002). Botulinum toxin in pain management of soft tissue syndromes. *The Clinical Journal of Pain, 18*, S147–S154.

Tyler, H. R. (1963). Botulinum toxin: Effect on the central nervous system of man. *Science, 139*, 847–848.

Wheeler, A. H., Goolkasian, P., & Gretz, S. S. (2001). Botulinum toxin A for the treatment of chronic neck pain. *Pain, 94*, 255–260.

Boxcar Design

▶ Block Design

Brachial and/or Lumbar Plexopathy

Definition

This is a disorder with compression, entrapment, infiltration or other lesion to the nerves of the plexus. Symptoms include motor loss, sensory deficits, and pain. In oncological patients, brachial plexopathy may be caused by a Pancoast tumor growing in the lung top. Suprascapular nerve entrapment and accompanying shoulder pain may be seen in cancer patients. Lumbar plexopathy in cancer patients is more often caused by metastases to the lumbar and pelvic regions.

Cross-References

▶ Cancer Pain

Brachial Plexus

Definition

The lowest four cervical and first thoracic spinal cord nerve roots (C5–T1) unite and branch to form the brachial plexus in the lower part of the neck and behind the clavicle. The brachial plexus is responsible for the innervation of the upper extremity.

Cross-References

▶ Acute Pain in Children, Postoperative
▶ Plexus Injuries and Deafferentation Pain
▶ Postoperative Pain, Regional Blocks

Brachial Plexus Avulsion

Definition

This is a traumatic lesion consisting of detachment of cervical root(s) corresponding to the brachial plexus (i.e., C5 to Th1) from the spinal cord and its main mechanism is stretching. It frequently results in a syndrome of constant burning or aching pain punctuated by paroxysms of crushing pain.

Cross-References

▶ Brachial Plexus Avulsion and Surgery in the Dorsal Root Entry Zone (DREZ)
▶ DREZ Procedures

Brachial Plexus Avulsion and Surgery in the Dorsal Root Entry Zone (DREZ)

Marc Sindou
Department of Neurosurgery, Hopital Neurologique Pierre Wertheimer, University of Lyon 1, Lyon, France

Synonyms

Brachial plexus injury; Cervical root avulsion; Dorsal root entry zone and brachial plexus avulsion; Dorsal root-spinal cord junction; DREZ lesions; Microsurgical DREZotomy; Neurosurgery for pain in the DREZ

Definition

The vast majority of Brachial Plexus Injuries (BPI) occurs as a consequence of motorcycle accidents. In most BPI injuries, structures of the plexus are stretched down and laterally (Sunderland 1968), resulting in multiple lesions along the neural fibers, within their extra- and/or intraspinal portions, at the entry zone into the spinal cord, up to the Lissauer's tract and the superficial layers of the dorsal horn. This leads to the appearance of intramedullary hemorrhages and necrosis, then to gliosis and microcysts, intraoperatively observed. BPI are followed by secondary chronic pain in the deafferented area of the upper limb in 30–90 % of the patients, according to publications (Wynn-Parry 1980). In the particular group of patients with severe supraganglionic lesions, that is with true avulsion (BPA), the rate of patients with pain has been reported at as much as 90 % (Narakas 1988). Microelectrode recordings of Dorsal Horn neurons in animal models (Guenot et al. 2002), and also in humans, during surgery (Guenot et al. 1999; Jeanmonod et al. 1989) have evidenced spontaneous electrophysiological hyperactivities, which are hypothesized to be due to deafferentation phenomena, together with ectopic generation of spikes in the gliotic scar tissue. The resulting neuropathic pain represents a clinical stereotype: a continuous pain background described as burning, throbbing and/or aching sensations, associated with electrical shooting-like paroxysms.

Characteristics

Topography of pain most frequently covers several dermatomes of the upper limb. In most patients, pain distribution is predominant in the lowest radicular territories, which is in the hand and forearm. This is well in concordance with the observed anatomical lesions. According to a recent study in our series, dorsal root damages extended to all five roots from C5 to Th1 in 43 % of the patients, from C6 to Th1 in 14 %, C7 to Th1 in 14 %, and C8 to Th1 in 4 % and were diversely

grouped in the other 25 %. C8 was the most frequently (in 90 %) and severely (avulsion almost always) damaged DR. C5 was the less frequently (in 50 %) and less severely (avulsion in only 25 %) damaged root. Interesting, lesions in the ventral roots were quite similarly distributed, with the exception of the C5 root level, where it appeared that the ventral root was more frequently and more severely impaired than the corresponding C5 dorsal root.

After BPA, level of pain intensity is generally between 7/10 and 10/10 on the VAS scoring system, in spite of analgesic medications including opioids at maximum doses.

Direct repair or neurotization with donor nerves (accessory, intercostal descending cervical plexus) have been reported to may have a more or less preventing effect on appearance of pain. Direct reimplantation of avulsed roots into the spinal cord itself is still in the preliminary phase of clinical trials. Medical treatments, whatever the therapeutic class used – including anticonvulsants and antidepressants – are usually irrelevant with time. Transcutaneous and Spinal Cord Stimulation techniques are ineffective; as a matter of fact, after avulsion the fibers to stimulate degenerate up to the brain stem nuclei (Sindou et al. 2003). Deep Brain Stimulation or Motor Cortex Stimulation generally does not give lasting pain relief. Classical ablative procedures – namely, cervical anterolateral cordotomy, spinothalamic mesencephalic tractotomy – reveal very little effective, and not exempt from side-effects. The recently introduced Motor Cortex Stimulation do not seem to have the same promising results in pain after BPI as for pain after stroke.

▶ Surgery in the DREZ introduced in the seventies (Sindou 1972; Sindou et al. 1974; Nashold and Ostdahl 1979) demonstrates efficacy long-term for relieving pain in the patients with BP as well as in patients affected with segmentally generated pain after ▶ Spinal cord Injury (Sindou et al. 2001). Therefore DREZ-surgery is currently used for those indications in dedicated centers. Detailed descriptions of the ▶ Microsurgical DREZotomy procedure have recently been given in several books on

Neurosurgical Techniques (Sindou 2002). Briefly, after ipsilateral hemilaminectomy and dural opening over the avulsed segments of the cervical spinal cord, the cord is freed from arachnoid adhesions. Incision of the dorsolateral sulcus is made with a micro-knife, 2 mm in depth and oriented at 35 ° medially and ventrally, that is the axis of the dorsal horn. Then, dotted microcoagulations are performed inside the dorsal horn down to approximately the Vth layer of the Rexed classification, which is at 3–4 mm deep. The microcoagulations are made under magnified vision with a sharp graduated bipolar forceps. Special care must be taken to locate the microcoagulations strictly inside the deafferented dorsal horn, in between the cuneate fasciculus of the dorsal column, on the medial side, and the corticospinal tract, on the lateral side. Important is avoidance of impairment of the sensory and motor pathways, respectively (Jeanmonod and Sindou 1993).

A recent survey of our 95 patients, with a follow-up ranging from 1 to 18 years (6 years on average), shows the following results: 85 % had their pain relieved without the need of any medical treatment two-third, one-third with additional moderate doses of non-opioid analgesics; 71 % benefited from an improvement in their activity level. Mortality was nil. Neurological complications consisted of long-tract deficits: permanent ataxia in 3.6 % and permanent weakness in 3.6 % in the ipsilateral lower limb, and genitourinary disturbances in 1.3 %.

Our results (Sindou et al. 2005) are approximately the same as those reported in the literature with other DREZ-lesion makers, namely, RF-thermocoagulation (Nashold and Ostdahl 1979), laser-beam (Levy et al. 1983), and ultrasonic probe (Dreval 1993).

Conclusions

1. Gross anatomical lesions, observed in patients referred for pain after severe ▶ brachial plexus injury, are mostly supraganglionic. In our series, 78 % of the dorsal roots corresponding to the brachial plexus (that is C5 to Th1) were found impaired. Of those, 79 % were totally avulsed and 7.5 % avulsed partially; 13.5 % had their rootlets present but with an important atrophic aspect. These findings correspond to a clear-cut deafferentation state of the spinal cord and constitute a consistent substrate for considering neuropathic pain after BPA as pain from deafferentation origin.
2. Study of long-term outcomes after therapeutic DREZ-lesioning shows that surgery is able to obtain lasting pain relief in two-thirds of the patients.
3. The positive effects of destructive surgery on the DREZ target favors the hypothesis that the main mechanism of pain after BPA is the appearance of pain generators in the neurons and/or imbalance in fiber circuitry of the Dorsal Horn at the level of the avulsed cervical segments.

References

Dreval, O. N. (1993). Ultrasonic DREZ-operations for treatment of pain due to brachial plexus avulsion. *Acta Neurochirurgica, 122*, 76–81.

Guenot, M., Hupe, J. M., Mertens, P., et al. (1999). A new type of microelectrode for obtaining unitary recordings in the human spinal cord. *Journal of Neurosurgery, 91*, 25–32.

Guenot, M., Bullier, J., & Sindou, M. (2002). Clinical and electrophysiological expression of deafferentation pain alleviated by dorsal root entry zone lesions in rats. *Journal of Neurosurgery, 97*, 1402–1409.

Jeanmonod, D., & Sindou, M. (1993). Somatosensory function following dorsal root entry zone lesions in patients with neurogenic pain or spasticity. *Journal of Neurosurgery, 74*, 916–932.

Jeanmonod, D., Sindou, M., Magnin, M., et al. (1989). Intra-operative unit recording in the human dorsal horn with a simplified floating microelectrode. *Electroencephalography and Clinical Neurophysiology, 74*, 450–454.

Levy, W. J., Nutkiewicz, A., Ditmore, M., et al. (1983). Laser-induced dorsal root entry zone lesions for pain control. Report of three cases. *Journal of Neurosurgery, 59*, 884–886.

Narakas, A. (1988). Pain syndromes in brachial plexus injuries. In G. Brunelli (Ed.), *Textbook of microsurgery* (pp. 809–816). Paris: Masson.

Nashold, B. S., Jr., & Ostdahl, R. H. (1979). Dorsal root entry zone lesions for pain relief. *Journal of Neurosurgery, 51*, 59–69.

Sindou. (2002). Dorsal root entry zone lesions. In K. J. Burchiel (Ed.), *Surgical management of pain* (pp. 701–713). New York: Thieme.

Sindou, M. (1972). *Study of the dorsal root entry zone. The selective MicroDREZtomy for pain surgery.* Lyon: University Med Thesis Press.

Sindou, M., Blondet, E., Emery, E., & Mertens, P. (2005). Microsurgical lesioning in the dorsal root entry zone for pain due to brachial plexus avulsion: A prospective series of 55 patients. *Journal of Neurosurgery, 102,* 1018–1028.

Sindou, M., Quoex, C., & Baleydier, C. (1974). Fiber organization at the posterior spinal cord-rootlet junction in man. *The Journal of Comparative Neurology, 153,* 15–26.

Sindou, M., Mertens, P., Bendavid, U., et al. (2003). Predictive value of somato-sensory evoked potentials (central conduction time) for patients' selection to spinal cord stimulation for neuropathic pain. *Neurosurgery, 52,* 1374–1384.

Sindou, M., Mertens, P., & Wael, M. (2001). Microsurgical DREZotomy for pain due to spinal cord and/or cauda equina injuries: Long-term results in a series of 44 patients. *Pain, 92,* 159–171.

Sunderland, S. (1968). *Nerves and nerve injuries. The painful sequelae of injuries to peripheral nerves. Causalgia* (pp. 411–430). Edinburgh: Churchill-Livingstone.

Wynn-Parry, C. B. (1980). Pain in avulsion lesions of the brachial plexus. *Pain, 9,* 41–53.

Brachial Plexus Avulsion Injury

▶ Plexus Injuries and Deafferentation Pain

Brachial Plexus Compression

▶ Thoracic Outlet Syndrome

Brachial Plexus Injury

Definition

Trauma occurring at any level of the brachial plexus, from spinal cord to peripheral branches; may be due to direct trauma or indirectly to stretching.

Cross-References

▶ Brachial Plexus Avulsion and Surgery in the Dorsal Root Entry Zone (DREZ)

Brachialgia

▶ Radicular Pain, Diagnosis

Bradykinin

Definition

A potent inflammatory peptide messenger, consisting of nine amino acids, which is generated from a protein precursor (kallidin) through the action of specific enzyme kallikrein. Bradykinin is released from damaged tissue during injury or inflammation. It is also released from mast cell and also produced in the blood, where it serves as a vasodilator and increases vessel permeability. It also stimulates prostacyclin formation. BK not only excites a high percentage of nociceptors but also sensitizes them to other noxious stimuli through activation of B1 and B2 receptors. It is one of the most potent pain-producing substances.

Cross-References

▶ ERK Regulation in Sensory Neurons During Inflammation
▶ Mechanonociceptors
▶ Nociceptors in the Orofacial Region (Skin/Mucosa)
▶ Postsynaptic Dorsal Column Projection, Anatomical Organization
▶ Satellite Glial Cells and Chronic Pain
▶ Spinal Dorsal Horn Pathways, Dorsal Column (Visceral)
▶ TRPV1 Modulation by p2Y Receptors
▶ TRPV1, Regulation by Protons
▶ Visceral Pain Model, Angina Pain

Bradykinin Receptors

Definition

Two types of molecular receptors for bradykinin are known presently, namely the B2 and B1 receptor. Both receptors are coupled to a G protein, i.e., they do not control an ion channel, but are involved in the synthesis of second messengers in the axoplasm of the nociceptive ending. In intact tissue, BKN binds to the B2 receptor. If the tissue is inflamed or otherwise pathologically altered, the B1 receptor molecule is synthesized in the soma of the nociceptor and transported to the ending where the receptor is built in the membrane. The de-novo synthesis of the B1 receptor is an important aspect of the sensitization of the ending.

Cross-References

▶ Muscle Pain Model, Ischemia-Induced and Hypertonic Saline-Induced

---

Brain Based Interventions

G. Lorimer Moseley
Sansom Institute for Health Research, University of South Australia Neuroscience Research Australia, Adelaide, Australia

Synonyms

Cortical retraining; Sensory motor retuning

Definition

Brain based interventions for chronic pain are those that aim to reverse maladaptive changes in cortical organization that are observed in people with chronic pain.

Background to the Definition

Systematic changes in the response profile of neurons in the primary sensory and motor areas is called cortical reorganization. Cortical reorganization occurs in people with chronic pain. Rehabilitation strategies that target the reversal of these changes appear to offer chronic pain patients functional and symptomatic improvement. These strategies can be grouped together according to what type of cortical representation they are targeting: cognitive and conceptual representations are targeted with cognitive and behavioural strategies, sensory representations are targeted with sensory discrimination strategies, motor representations are targeted with cross-modal movement strategies.

Introduction

That the brain plays an ubiquitous role in pain has been accepted for centuries (Descartes 1664) – no brain, no pain. That the brain changes when pain persists and that these changes in the brain may actually contribute to chronic pain, are relatively new findings. Evidence of these changes has sparked interest in the brain as a potential target for treatment of people in pain. This chapter will outline the current state of the art in this growing area.

Characteristics

The Brain Responds to the Perceived Reality not the Actual Reality

Nociception is neither sufficient nor necessary for pain, although it is undoubtedly the most common trigger. Tactile perception, pain and other bodily feelings can be thought of as outputs of the brain that are based on an informed interpretation of the information coming from one's body. Pain emerges according to the apparent danger to body tissues and the need for a concerted response from the individual, not according to activity in nociceptive fibres or the actual state of the tissues. This differentiation between nociception and pain is often

ignored such that the terms are used synonymously in the lay and scientific literature.

The Brain Changes when Pain Persists

There seem to be three main changes that occur in
the brain when pain persists. The first seems
consistent with the fundamental property of
biological systems to adapt according to use.
That such adaptation occurs in neural systems is
well understood – learning is one example –
'practice makes perfect' – central sensitisation
is another (Woolf and Salter 2006). The clinical
manifestation of central sensitisation, allodynia
and hyperalgesia, offer a clear biological advantage by increasing the sensitivity to peripheral
inputs of both the non-nociceptive and the nociceptive systems, thereby optimising the likelihood of tissue healing.

Sensitisation also occurs at the supraspinal
level. In people with chronic pain, seeing the
painful limb being touched can be enough to
evoke pain and swelling even when the limb
was not in fact touched (Acerra and Moseley
2005). Imagined movements of the painful limb
can increase pain and swelling in the absence of
detectable muscle activity or limb movement
(Moseley et al. 2008a). It is important to note
that this sensitization will not necessarily be confined to one sensory modality but can extend to
others, most likely by associative learning
processes, and may thus lead to easy activation
of pain through non-nociceptive channels
(Schneider et al. 2004).

The second type of change that occurs in the
brain when pain persists involves changes in the
response profile of brain cells. This type of change,
coined 'cortical reorganisation', denotes plastic
changes in the organisation of cortical representations. Cortical representations are networks of
brain cells, the activation of which is associated
with a particular output. For our purposes, we can
categorise cortical representations as:

(i) Cognitive or conceptual – 'How is the mean-
    ing of pain represented in the brain?'
(ii) Sensory – 'How is the size, shape and loca-
    tion of the body represented in the brain?
(iii) Motor – 'How is movement represented in
    the brain?'

Cognitive or Conceptual Reorganisation

As we have noted, the meaning of pain is critical
in determining its impact and severity. The meaning of pain is represented in the human brain and
as pain persists, these representations can shift in
a way that emphasises the threat value of nociceptive input. Non-noxious cues become more
potent triggers of pain and generalise so that
more and varied cues can evoke pain.

Sensory Reorganisation

That nervous system injury could induce
profound reorganisation of the primary sensory
cortex in humans was illustrated when amputees
were stimulated on the lip – the cortical representation of the lip (i.e., in sensory cortex contralateral to the amputation) had invaded the area that
should represent the amputated hand (Elbert et al.
1994; Yang et al. 1994). Cortical reorganisation
was then linked to pain by comparing amputees
with phantom limb pain to those without (Flor
et al. 1995). In that study, representation of the lip
was only misplaced (into the hand area) for the
amputated side (i.e. contralateral to the amputation). Importantly, no such shift was observed in
those amputees without phantom limb pain
(i.e. the shift was not simply a result of amputation) and in those with phantom limb pain, the
intensity of pain strongly correlated with the
magnitude of the shift of the lip representation.
A similar pattern of primary sensory cortex
reorganisation, and its relationship with pain,
has since been observed in people with a range
of chronic pain disorders (for review see (Flor
et al. 2006)).

Primary motor cortex also undergoes
reorganisation when pain persists. The functional
significance of the reorganisation seems similar to
that observed in primary sensory cortex – lip
movements invade hand movement area in motor
cortex in amputees with phantom limb pain, but
not in those without phantom limb pain and the
magnitude of the shift is again related to pain (Karl
et al. 2001; Lotze et al. 2001; Lotze et al. 1999).

Higher order body maps have also been examined in people with chronic pain. For example,
the maps the brain uses for movement – the
so-called working body schema, have been

investigated using timed motor imagery tasks such as the left/right hand judgement and left/right trunk rotation judgement tasks (Schwoebel et al. 2002). Amputees with phantom limb pain make more errors when asked to judge whether a pictured hand is a left or a right hand (Nico et al. 2004).

Brain Plasticity: From Barriers to Opportunities

The changes that occur in the brain when pain persists clearly present barriers to successful recovery. However, the plasticity that underpins them is also a potential target for treatment.

Cognitive and Behavioural Strategies

We previously noted that pain emerges according to the apparent danger to body tissues and the need for a concerted response from the individual, not according to the actual danger to body tissues. This means that *anything* that increases the apparent danger should facilitate pain. There is a very large amount of literature on the influence of beliefs and attitudes, knowledge and expectations on the way people cope with pain (Williams and Keefe 1991; Tota-Faucette et al. 1993; Turk 1996) and a growing literature on how these factors influence pain itself (Moseley and Arntz 2007). Many of these influences may be implicit and occur entirely outside the awareness of the person (Kleinbohl et al. 1999). According to pain's underlying biology, modulation of these factors should also modulate pain.

Operant behavioural treatment aims to decrease pain behaviours in an effort to extinguish pain, to increase activity levels and healthy behaviours related to work, leisure time and the family, to reduce and manage medication and to modify the behaviour of significant others. The overall goal is to reduce disability by reducing pain and increasing healthy behaviours. Medication is switched from a *prn* basis to a fixed time schedule, where medication is given at certain times of the day to avoid negative reinforcement learning from occurring. Similar principles are applied to the enhancement of activity and the reduction of inactivity and

disability. This approach is effective in patients with chronic back pain as well as other pain syndromes (Fordyce 1973; Thieme et al. 2003, 2006; van Tulder et al. 2001).

Cognitive treatment includes reconceptualising the meaning of pain by targeted explanation and training – "*Explaining pain*", coping skills training, attention diversion, use of imagery or relaxation, increasing self efficacy. These approaches improve pain and function across chronic pain disorders (Moseley et al. 2004; Thorn and Kuhajda 2006; Turk et al. 1998).

Strategies to Normalise Sensory Representations

The clarity and size of sensory representations is use-dependent. Therefore, to induce changes in sensory cortex, one must stimulate the sensory system accordingly. Stimulation is most likely to induce changes in sensory cortical representation if the characteristics of a stimuli are important, for example playing a musical instrument or reading Braille, or the objective of the task is important, for example unwrapping food (Braun et al. 2002; Jenkins et al. 1990; Wang et al. 1995). This principle was applied to a training program in which amputees with phantom limb pain were required to discriminate between electrocutaneous stimuli of different frequencies and at different locations on their stump (Flor et al. 2001). Eight electrodes were attached to the residual limb and provided high intensity non-painful electric stimulation of varying intensity and location that led to the experience of intense phantoms. The patients were trained to discriminate the location or the frequency (alternating trials) of the stimulation and received feedback on the correct responses. The training was conducted for 90 min per day and was spread over a time of 2 weeks (10 days of training). The trained patients showed significantly better discrimination ability on the stump than an untrained control group. They also experienced a more than 60 % reduction of phantom limb pain and a significant reversal of cortical reorganization. The alterations in discrimination ability, pain and cortical reorganization were highly significantly correlated, which lends support to the notion that cortical reorganisation might actually contribute to phantom limb pain. The protocol successfully applied to phantom limb pain has also been adapted for use with people with other chronic pain disorders and shows similar effects e.g. (Moseley et al. 2008b).

In amputees, wearing a prosthesis affects phantom limb pain and cortical reorganization (Lotze et al. 1999). Patients who systematically used a myoelectric prosthesis that provides sensory and visual as well as motor feedback to the brain reported much less phantom limb pain and showed less cortical reorganization than patients who used either a cosmetic prosthesis or no prosthesis at all. In fact, the relationship between phantom limb pain and cortical reorganization was strongly mediated by prosthesis use. This suggests that sensory input to the brain region that formerly represented the now absent limb may be beneficial in reducing phantom limb pain.

Strategies to Normalise Motor Representations

The most 'famous' strategy that aims to normalise motor representations is mirror therapy – some studies are positive, others are not. Graded motor imagery, which incorporates mirror therapy, seems effective for CRPS and phantom limb pain after amputation or brachial plexus avulsion injury (Daly and Bialocerkowski 2009). Graded motor imagery is a three stage program commencing with hourly sessions of left/right limb judgement tasks, followed by imagined limb movements and then mirror therapy. The order of components appears to be important for the effect (Moseley 2005). One theory that might explain any effect of mirror therapy or graded motor imagery is that these treatments correct cortical body maps so as to remove the incongruence between motor commands and sensory feedback (McCabe et al. 2008). In line with this idea, McCabe et al. (2005) used a mirror to create a discrepancy between actual and seen movement (movement seen in the mirror is discrepant from the movement behind the mirror if both arms perform asynchronous movements). These authors reported the presence of painful and non-painful paraesthesias as a consequence of

the incongruent movement condition and suggested that sensorimotor incongruence might cause the abnormal sensations seen in many neuropathic pain syndromes.

Summary

There is little doubt that neuroplastic changes occur in the central nervous system when pain persists. There is a growing body of literature that suggests that these changes play an important role in the development and maintenance of chronic pain. Longitudinal studies and controlled outcome studies are still needed to elucidate in greater detail the efficacy and mechanisms of feedback-based interventions designed to alter cortical pain memories.

References

Acerra, N. E., & Moseley, G. L. (2005). Dysynchiria: Watching the mirror image of the unaffected limb elicits pain on the affected side. *Neurology, 65*(5), 751–753.

Braun, C., Haug, M., et al. (2002). Functional organization of primary somatosensory cortex depends on the focus of attention. *NeuroImage, 17*(3), 1451–1458.

Daly, A. E., & Bialocerkowski, A. E. (2009). Does evidence support physiotherapy management of adult complex regional pain syndrome type one? A systematic review. *European Journal of Pain, 13*(4), 339–353.

Descartes, R. (1664). *L'Homme*, Paris. (In French.) Digitized photographic reproduction available online (PDF and TIFF). Reprinted in Adam and Tannery vol. XI. Partial English translation in CSM, vol. I. Complete English translation in Hall 1972.

Elbert, T., Flor, H., et al. (1994). Extensive reorganisation of the somatosensory cortex in adult humans after nervous system injury. *Neuroreport, 5*, 2597–2597.

Flor, H., Denke, C., et al. (2001). Effect of sensory discrimination training on cortical reorganisation and phantom limb pain. *Lancet, 357*(9270), 1763–1764.

Flor, H., Elbert, T., et al. (1995). Phantom-limb pain as a perceptual correlate of cortical reorganization following arm amputation. *Nature, 375*(6531), 482–484.

Flor, H., Nikolajsen, L., et al. (2006). Phantom limb pain: A case of maladaptive CNS plasticity? *Nature Reviews Neuroscience, 7*(11), 873–881.

Fordyce, W. E. (1973). An operant conditioning method for managing chronic pain. *Postgraduate Medicine, 53*(6), 123–128.

Jenkins, W. M., Merzenich, M. M., et al. (1990). Functional reorganization of primary somatosensory cortex in adult owl monkeys after behaviorally controlled tactile stimulation. *Journal of Neurophysiology, 63*(1), 82–104.

Karl, A., Birbaumer, N., et al. (2001). Reorganization of motor and somatosensory cortex in upper extremity amputees with phantom limb pain. *Journal of Neuroscience, 21*(10), 3609–3618.

Kleinbohl, D., Holzl, R., et al. (1999). Psychophysical measures of sensitization to tonic heat discriminate chronic pain patients. *Pain, 81*(1–2), 35–43.

Lotze, M., Flor, H., et al. (2001). Phantom movements and pain – An MRI study in upper limb amputees. *Brain, 124,* 2268–2277.

Lotze, M., Grodd, W., et al. (1999). Does use of a myoelectric prosthesis prevent cortical reorganization and phantom limb pain? *Nature Neuroscience, 2*(6), 501–502.

McCabe, C. S., Haigh, R. C., et al. (2005). Simulating sensory-motor incongruence in healthy volunteers: Implications for a cortical model of pain. *Rheumatology, 44*(4), 509–516.

McCabe, C. S., Haigh, R. C., et al. (2008). Mirror visual feedback for the treatment of complex regional pain syndrome (Type 1). *Current Pain and Headache Reports, 12*(2), 103–107.

Moseley, G. L. (2005). Is successful rehabilitation of complex regional pain syndrome due to sustained attention to the affected limb? A randomised clinical trial. *Pain, 114*(1–2), 54–61.

Moseley, G. L., & Arntz, A. (2007). The context of a noxious stimulus affects the pain it evokes. *Pain, 133,* 64–71.

Moseley, G. L., Nicholas, M. K., et al. (2004). A randomized controlled trial of intensive neurophysiology education in chronic low back pain. *The Clinical Journal of Pain, 20*(5), 324–330.

Moseley, G. L., Zalucki, N., et al. (2008a). Thinking about movement hurts: The effect of motor imagery on pain and swelling in people with chronic arm pain. *Arthritis Care and Research, 59*(5), 623–631.

Moseley, G. L., Zalucki, N. M., et al. (2008b). Tactile discrimination, but not tactile stimulation alone, reduces chronic limb pain. *Pain, 137*(3), 600–608.

Nico, D., Daprati, E., et al. (2004). Left and right hand recognition in upper limb amputees. *Brain, 127*(1), 120–132.

Schneider, C., Palomba, D., et al. (2004). Pavlovian conditioning of muscular responses in chronic pain patients: Central and peripheral correlates. *Pain, 112*(3), 239–247.

Schwoebel, J., Coslett, H. B., et al. (2002). Pain and the body schema: Effects of pain severity on mental representations of movement. *Neurology, 59*(5), 775–777.

Thieme, K., Flor, H., et al. (2006). Psychological pain treatment in fibromyalgia syndrome: Efficacy of operant behavioural and cognitive behavioural treatments. *Arthritis Research & Therapy, 8*(4), R121.

Thieme, K., Gromnica-Ihle, E., et al. (2003). Operant behavioral treatment of fibromyalgia: A controlled study. *Arthritis and Rheumatism, 49*(3), 314–320.

Thorn, B. E., & Kuhajda, M. C. (2006). Group cognitive therapy for chronic pain. *Journal of Clinical Psychology, 62*(11), 1355–1366.

Tota-Faucette, M. E., Gil, K. M., et al. (1993). Predictors of response to pain management treatment. The role of family environment and changes in cognitive processes. *The Clinical Journal of Pain, 9*(2), 115–123.

Turk, D. C. (1996). Biopsychosocial perspective on chronic pain. Psychological approaches to pain management. In R. Gatchel & D. C. Turk (Eds.), *A practitioners handbook* (pp. 3–32). New York: The Guildford Press.

Turk, D. C., Okifuji, A., et al. (1998). Interdisciplinary treatment for fibromyalgia syndrome: Clinical and statistical significance. *Arthritis Care and Research, 11*(3), 186–195.

van Tulder, M. W., Ostelo, R., et al. (2001). Behavioral treatment for chronic low back pain: A systematic review within the framework of the Cochrane back review group. *Spine (Phila Pa 1976), 26*(3), 270–281.

Wang, X., Merzenich, M. M., et al. (1995). Remodelling of hand representation in adult cortex determined by timing of tactile stimulation. *Nature, 378*(6552), 71–75.

Williams, D. A., & Keefe, F. J. (1991). Pain beliefs and the use of cognitive-behavioral coping strategies. *Pain, 46*(2), 185–190.

Woolf, C. J., & Salter, M. (2006). Plasticity and pain: The role of the dorsal horn. In S. B. McMahon & M. Koltzenburg (Eds.), *Textbook of pain* (pp. 91–107). London: Elsevier.

Yang, T. T., Gallen, C., et al. (1994). Sensory maps in the human brain. *Nature, 368*(6472), 592–593.

Brain Electrode

▶ Deep Brain Stimulation

Brain Stem

Definition

The brain stem is the posterior region of the brain and is structurally continuous with the spinal cord and consists of the medulla, pons, and midbrain. It contains the major motor and sensory innervation to the face and neck via the cranial nerves and their nuclei. The ▶ trigeminal brain stem nuclear complex (TBNC) comprises an important component of the brain stem and the major relay and processing center for orofacial pain. It also contains the PAG and RVM, important in descending modulation of pain, and controls many basic and unconscious functions, including regulation of the cardiovascular and respiratory systems.

Cross-References

- ▶ Descending Circuitry, Transmitters and Receptors
- ▶ Forebrain Modulation of the Periaqueductal Gray and Its Role in Pain
- ▶ Migraine, Pathophysiology
- ▶ Nociceptive Processing in the Brain Stem
- ▶ Trigeminal Brainstem Nuclear Complex, Anatomy

Brain Stem Respiratory Centers

Definition

Brain Stem Respiratory Centers are neuron assemblies located in the Pons and Medulla that control involuntary respiration. Nociceptive as well as other somatosensory inputs, and non-somatosensory (e.g., visceral) inputs, can access the centers and thereby modify breathing.

Cross-References

- ▶ Pain Treatment, Implantable Pumps for Drug Delivery

Brain Stimulation Analgesia

- ▶ Stimulation-Produced Analgesia

Brain-Derived Neurotrophic Factor

Synonyms

BDNF

Definition

A member of the neurotrophin family that signals via the trkB receptor. **Brain-derived neurotrophic factor** (BDNF) is upregulated in trkA-expressing DRG neurons that are exposed to inflammatory mediators such as NGF. BDNF has roles in neuronal survival and development, and has been suggested to function as a neuromodulator of synaptic transmission and spinal nociception, especially during inflammatory states. BDNF, synthesized in the DRG, is transported to the central terminals of the primary afferents, released into the spinal dorsal horn, and binds to trkB receptors on second-order sensory neurons.

Cross-References

- ▶ ERK Regulation in Sensory Neurons During Inflammation
- ▶ Nerve Growth Factor, Sensitizing Action on Nociceptors
- ▶ Spinal Cord Nociception, Neurotrophins

Brain-Gut Axis

Definition

Brain-gut axis refers to the bidirectional flow of information between the gut (gastrointestinal tract) and brain and consequent feedback and modulation associated with that flow of information, most commonly effected by the gut's enteric nervous system.

Cross-References

▶ Recurrent Abdominal Pain

Brainstem Subnucleus Reticularis Dorsalis Neuron

Luis Villanueva[1], Claude Robert[2] and
Charles-Daniel Arreto[2]
[1]INSERM UMR 894, Centre de Psychiatrie et
Neurosciences, Paris, France
[2]Université Paris Descartes, INSERM UMR 894,
France

**Brainstem Subnucleus Reticularis Dorsalis Neuron,
Fig. 1** Bright-field image of a section of the medulla
caudal to the obex stained with the Kluver and Barrera
technique. The *dotted line* represents the delimitation of
the SRD area with regard to the surrounding trigeminal
(Sp5), cuneate (Cu), nucleus of the solitary tract (Sol), and
subnucleus reticularis ventralis (SRV) regions. *pyx*
pyramidal decussation

Introduction

Neurons whose cell body and dendrites are
located in the medullary Subnucleus Reticularis
Dorsalis (SRD) and whose axons issue collaterals
to both the thalamus and the dorsal horn of the
spinal cord. SRD neurons respond selectively to,
and encode, noxious stimuli from any part of the
body. Because the SRD innervates several areas
involved in motor processing and receives strong,
direct influences from several cortical regions, it
could provide a structural basis for the processing
of nociceptive and motor activities.

Characteristics

Several groups have shown that widespread areas
throughout the brainstem reticular formation contain neurons responsive to noxious stimuli
(Bowsher 1976; Villanueva et al. 1988; Dugast
et al. 2003) and that focal stimulation of some
bulbar reticular areas can elicit escape behavior
(Casey 1969). However, the way in which the
reticular structures participate in the processing
of nociceptive information was not clear. This
was because reticular units activated by noxious
stimulation showed irregular responses and
changes in excitability, had receptive fields
which were difficult to define and presented
some degree of heterosensory convergence.

As a result, it was stated that the reticular formation did not play a specific role in the processing
of pain.

This proposal has been challenged by data
obtained in the rat showing that a well-delimited
area within the caudal-most aspect of the
medulla, the *Subnucleus Reticularis Dorsalis*
(SRD), could play a selective role in processing
cutaneous and visceral nociceptive inputs
(Villanueva et al. 1988). The SRD region extends
caudo-rostrally from the spino-medullary junction to the level of the area postrema and lies
ventral to the cuneate nucleus, medial to the
magnocellular layer of trigeminal nucleus
caudalis, and is separated from the subnucleus
reticularis ventralis by an acellular boundary
extending from the solitary tract to the dorsal
border of the lateral reticular nucleus (Fig. 1).
The SRD contains neurons that respond and
encode selectively noxious stimuli from any
part of the body (Fig. 2; Villanueva et al. 1996)
and their C-fiber components exhibit the "wind-
up" phenomenon during repetitive stimulation.
Additional data obtained in the monkey has
demonstrated that there are neurons with similar
features to those described in the rat SRD.
Medullary units recorded in monkeys exhibit
convergence of nociceptive inputs from

Brainstem Subnucleus Reticularis Dorsalis Neuron, Fig. 2 (a) Camera lucida drawing from successive frontal sections of a SRD neuron labeled with a juxtacellular injection of biotin-dextran. Note that its axonal processes give both an ascending and descending collateral (Adapted from Monconduit et al. (2002)). (b) Single sweep recordings showing Aδ- and C-fiber evoked responses of a SRD neuron following supramaximal percutaneous electrical stimulation (square-wave pulses, 2 ms duration) of different areas of the body (*arrows*). Note that massive Aδ- and C-fiber responses were evoked from all body areas (Adapted from Villanueva et al. (1988))

widespread areas of the body and encode the intensities of peripheral noxious stimuli (Villanueva et al. 1990). Thus, in different species, dorsal medullary reticular neurones constitute a morpho-functional entity which processes nociceptive inputs.

SRD neurons regulate the flow of sensory information in spinal and thalamic nociceptive pathways and project to several areas of the central nervous system that are involved in motor processing. These areas include the deep dorsal horn and the more dorsal part of the ventral horn at all levels of the spinal cord (Villanueva et al. 1995), brainstem motor nuclei, the giganto- and parvocellular reticular formation (Bernard et al. 1990), and motor-related areas of the forebrain such as the ventromedial, parafascicular thalamic nuclei and the ventral zona incerta

(Villanueva et al. 1998). Conversely, various regions of the deep dorsal horn (laminae V-VII) that contain neurons that respond to noxious cutaneous and/or visceral stimuli provide the main input to the dorsal-most aspect of the SRD (Villanueva et al. 1991; Raboisson et al. 1996). The SRD, via its reciprocal connections, exerts both inhibitory (Bouhassira et al. 1992) and excitatory (Dugast et al. 2003) influences on spinal cord outflow. Behavioral studies in rats have shown modifications of motor responses such as decreases in the latency of tail-flick withdrawal reactions, following glutamate injections in the SRD. Moreover, the same studies showed that lesions which included the dorsal aspect of the SRD induced an increase in the latencies of the hot-plate, tail-flick, and formalin-elicited withdrawal reactions (Almeida et al. 1999).

Interestingly, SRD neurons receive strong, direct influences from numerous cortical regions (Desbois et al. 1999) and provide monosynaptic connections to both the thalamus and spinal cord (Monconduit et al. 2002). From a general point of view, it has been suggested that the pyramidal system might implement and refine basic motor patterns generated by brainstem reticulospinal circuits and this could apply to those relayed at the level of the SRD (Canedo 1997). Various corticofugal influences could, via SRD collaterals, reach rostral and caudal sites in the CNS almost simultaneously. Perhaps, one of the advantages conferred by single SRD cells projecting to two different CNS regions is that it allows for more precise temporal synchrony between the two target areas.

The SRD is a medullary substrate for the processing of nociceptive and motor activities. The cortex might, via pyramidal influences onto SRD neurons, coordinate the nociceptive information that ascends to it (Zhang et al. 2005). Corticofugal mechanisms could allow the cortex to select its own input by suppressing or augmenting transmission of signals through SRD-hindbrain/forebrain pathways or by coordinating activities in spino-SRD-spinal circuits, and thus selecting the relevant information caused by the noxious stimulus itself.

References

Almeida, A., Storkson, R., Lima, D., et al. (1999). The medullary dorsal reticular nucleus facilitates pain behaviour induced by formalin in the rat. *European Journal of Neuroscience, 11*, 110–122.

Bernard, J. F., Villanueva, L., Carroue, J., et al. (1990). Efferent projections from the subnucleus reticularis dorsalis (SRD): A phaseoleus vulgaris leucoagglutinin study in the rat. *Neuroscience Letters, 122*, 257–262.

Bouhassira, D., Villanueva, L., Bing, Z., et al. (1992). Involvement of the subnucleus reticularis dorsalis in diffuse noxious inhibitory controls in the rat. *Brain Research, 595*, 353–357.

Bowsher, D. (1976). Role of the reticular formation in responses to noxious stimulation. *Pain, 2*, 361–378.

Canedo, A. (1997). Primary motor cortex influences on the descending and ascending systems. *Progress in Neurobiology, 51*, 287–335.

Casey, K. L. (1969). Somatosensory responses of bulboreticular units in the awake cat: Relation to escape producing stimuli. *Science, 173*, 77–80.

Desbois, C., Le Bars, D., & Villanueva, L. (1999). Organization of cortical projections to the medullary subnucleus reticularis dorsalis: A retrograde and anterograde tracing study in the rat. *The Journal of Comparative Neurology, 410*, 178–196.

Dugast, C., Almeida, A., & Lima, D. (2003). The medullary dorsal reticular nucleus enhances the responsiveness of spinal nociceptive neurons to peripheral stimulation in the rat. *European Journal of Neuroscience, 18*, 580–588.

Monconduit, L., Desbois, C., & Villanueva, L. (2002). The integrative role of the rat medullary subnucleus reticularis dorsalis in nociception. *European Journal of Neuroscience, 16*, 937–944.

Raboisson, P., Dallel, R., Bernard, J. F., et al. (1996). Organization of efferent projections from the spinal cervical enlargement to the medullary subnucleus reticularis dorsalis and the adjacent cuneate nucleus: A PHA-L study in the rat. *The Journal of Comparative Neurology, 367*, 503–517.

Villanueva, L., Bernard, J. F., & Le Bars, D. (1995). Distribution of spinal cord projections from the medullary subnucleus reticularis dorsalis and the adjacent cuneate nucleus: A phaseolus vulgaris leucoagglutinin (PHA-L) study in the rat. *The Journal of Comparative Neurology, 352*, 11–32.

Villanueva, L., Bouhassira, D., Bing, Z., et al. (1988). Convergence of heterotopic nociceptive information onto subnucleus reticularis dorsalis neurons in the rat medulla. *Journal of Neurophysiology, 60*, 980–1009.

Villanueva, L., Cliffer, K. D., Sorkin, L., et al. (1990). Convergence of heterotopic nociceptive information onto neurons of the caudal medullary reticular formation in the monkey (Macaca fascicularis). *Journal of Neurophysiology, 63*, 1118–1127.

Villanueva, L., De Pommery, J., Menétrey, D., & Le Bars, D. (1991). Spinal afferent projections to subnucleus reticularis dorsalis in the rat. *Neuroscience Letters, 134*, 98–102.

Villanueva, L., Desbois, C., Le Bars, D., & Bernard, J. F. (1998). Organization of diencephalic projections from the medullary subnucleus reticularis dorsalis and the adjacent cuneate nucleus: A retrograde and anterograde tracer study in the rat. *The Journal of Comparative Neurology, 390*, 133–160.

Zhang, L., Zhang, Y., & Zhao, Z. Q. (2005). Anterior cingulate cortex contributes to the descending facilitatory modulation of pain via dorsal reticular nucleus. *European Journal of Neuroscience, 22*, 1141–1148.

Breakthrough Pain

▶ Evoked and Movement-Related Neuropathic Pain

Breakthrough Pain (BP)

Synonyms

Pain flares

Definition

Pain flares that are experienced in the setting of controlled background pain are defined as breakthrough pain (BP). It typically has a rapid onset and short duration (median 30 min). Most frequently, BP is encountered during voluntary movements. Some BPs are spontaneous; however, they are not related to any other precipitating factor. Yet another category of BP is the end-of-dose pain, or pain appearing at the end of the dose interval of analgesics. The cause of BP should be vigorously investigated and specifically treated. Episodic pain or temporary pain is used as synonym.

Cross-References

▶ Cancer Pain
▶ Cancer Pain Management, Anesthesiologic Interventions, and Neural Blockade
▶ Cancer Pain Management, Principles of Opioid Therapy, Dosing Guidelines
▶ Evoked and Movement-Related Neuropathic Pain
▶ Incident Pain
▶ Opioids in Geriatric Application
▶ Opioid Responsiveness in Cancer Pain Management
▶ Opioid Therapy in Cancer Pain Management, Route of Administration
▶ Rest and Movement Pain

Brennan Pain Model

▶ Nick Model of Cutaneous Pain and Hyperalgesia

Brodmann Areas

Definition

A system developed by Brodmann for the classification of brain cortical areas based on the specific organization of neurons. When examined perpendicular to its surface, the cerebral cortex can be subdivided into layers, each containing subpopulations of neurons that can be distinguished based on their morphology and connectivity with other cortical and subcortical areas. Different Brodmann areas are distinguished by the relative importance of each cortical layer.

Cross-References

▶ Cingulate Cortex, Functional Imaging
▶ Pain Processing in the Cingulate Cortex, Behavioral Studies in Humans

Bruxism

Definition

Grinding movements of the teeth leading to lesions of the teeth and pain in the masticatory muscles or the temporomandibular joint.

Cross-References

▶ Orofacial Pain, Movement Disorders
▶ Sensitization of Muscular and Articular Nociceptors

Bulbospinal Modulation of Nociceptive Inputs

▶ Descending Facilitation and Inhibition in Neuropathic Pain

Bulbospinal Pathways

▶ Descending Circuitry, Opioids

Buprenorphine

Definition

A semisynthetic opiate possessing both narcotic agonist (μ opioid receptor) and antagonist (κ opioid receptor) activity. It has limited activity and low intrinsic efficacy at the μ opioid receptor in comparison to full opioid receptor agonists. Buprenorphine displays a ceiling effect to analgesia.

Cross-References

▶ Opioids and Inflammatory Pain

Burn Pain Control In Children

▶ Pain Control in Children with Burns

Burning Feet Syndrome

▶ Diabetic Neuropathies

Burning Mouth Syndrome

Yair Sharav
Department of Oral Medicine, School of Dental Medicine, Hebrew University-Hadassah, Jerusalem, Israel

Synonyms

BMD (burning mouth disorder); BMS; Oral dysesthesia; Stomatodynia; Stomatopyrosis

Definition

The ▶ burning mouth syndrome (BMS) is an intraoral pain disorder not accompanied by distinct clinical signs. Its vague definition and unknown etiology warrant its inclusion under atypical oral and facial pain. Inclusion criteria of patients with BMS may differ in different studies, e.g., the presence of systemic disorders may or may not be an exclusion criterion (Grushka 1987). Although clear diagnostic criteria have been proposed, these should be field tested and validated (see Table 11.11 in Sharav and Benoliel 2008). Burning mouth syndrome (BMS) is a poorly understood pain condition that is most probably ▶ neuropathic with a ▶ central component. The condition is characterized by a burning mucosal pain with no major visible signs, often accompanied by dysgeusia and xerostomia. The term "symptomatic" burning mouth refers to symptoms associated with local or systemic causes and is distinguished from the "idiopathic" BMS presently discussed.

Characteristics

Burning pain often occurs at more than one oral site, with the anterior two-thirds of the tongue, the anterior hard palate, and the mucosal aspect of the lower lip most frequently affected (e.g., Grushka et al. 1987a). The pain is as intense as toothache but differs from toothache in pain quality. Burning pain is constant throughout the day or begins by mid-morning and reaches maximum intensity by early evening but is usually not present at night and does not disturb sleep (Grushka 1987). More than two-thirds of the patients complain of altered taste sensation (dysgeusia) accompanying the burning sensation; in many cases the alteration is described as a spontaneous metallic taste. Grushka (1987) reported no significant difference between BMS and controls in any clinical oral features including number of teeth, oral mucosal conditions, presence of Candida infection, and parafunctional habits. Several subsequent studies have reported similar findings (for review see Sharav and Benoliel 2008; Jääskeläinen 2012).

B

Psychophysical Assessment

There is evidence for taste dysfunction in BMS, especially in those individuals with self-reported dysgeusia. Sweet thresholds were significantly higher for BMS than control subjects. In one extensive study, no differences were found between BMS and control subjects in somatosensory modalities such as two-point discrimination, temperature perception, and stereognostic ability (Grushka et al. 1987b); Grushka et al. (1987b) did, however, find that heat pain tolerance was significantly reduced at the tip of the tongue of BMS subjects and suggested that hyperalgesia in these patients may depend on prolonged temporal or spatial central summation. There have been other more recent reports of somatosensory as well as gustatory changes in BMS subjects (for review see Jääskeläinen 2012).

Etiology

The etiology of BMS is not clear. Many possible etiologies have been suggested and include local, intraoral factors as well as general, systemic ones. Accumulating data suggest that BMS may be induced by damage to the taste system at the level of the chorda tympani nerve. This damage could result in reduced inhibition of trigeminal nerve inputs into the central nervous system (CNS) that in turn leads to an intensified response to oral irritants. Recent evidence suggests neuropathic changes may occur in intraoral nerve fibers, and there are also reports of CNS changes (for review see Jääskeläinen 2012). There is still need for more research into the underlying mechanisms (for review see Sharav and Benoliel 2008; Jääskeläinen 2012).

However, once a clear causative factor can be identified, and its elimination stops the burning symptoms, the condition is no more that of idiopathic BMS. Looking for these factors is therefore essential before the condition can be defined as BMS.

Local Factors

These include galvanic currents, denture allergy and mechanical irritation, and decreased salivary secretion or change in saliva composition. Most studies have not supported an allergic or mechanical irritation cause for BMS. Most salivary flow rate studies have not demonstrated a significant decrease in salivary output. On the other hand, some studies found significant alterations in salivary components such as proteins, immunoglobulins, and phosphates, as well as differences in saliva pH buffering capacity. Whether these alterations in salivary composition are a causal or a coincidental event in BMS is unknown.

Systemic Factors

Among these are menopause and hormonal imbalance, nutritional deficiencies, and psychogenic factors. Significantly more oral discomfort was found in menopausal women; however, oral discomfort was not associated with any of the vasomotor symptoms of menopause nor did it show any relationship to mucosal health. The presence of oral discomfort bore no relationship to follicle-stimulating hormone (FSH) or estradiol levels measured in menopausal women. In spite of the conflicting data on the effect of menopause and estrogen replacement therapy on oral discomfort, the high frequency of oral complaints in menopausal women clearly indicates a significant, although poorly understood, association between menopause and BMS.

Numerous studies have used psychological questionnaires and psychiatric interviews to demonstrate psychological disturbances such as depression, anxiety, and irritability in patients with BMS (e.g., Grushka et al. 1987a). BMS patients showed elevation in certain personality characteristics similar to those seen in other chronic pain patients (Grushka et al. 1987a). However, it is unclear whether depression and other personality characteristics are causative or the result of the pain.

In conclusion, there is no clear etiology of BMS available today, and it is possible that a further subclassification of this "syndrome" is needed. A more rigorous definition of inclusion and exclusion criteria may help in future studies of etiology and possible therapy.

Treatment

Before treatment is instigated, local and systemic underlying factors should be ruled out and treated. Unfortunately, in many instances these corrections may not improve the burning sensation and oral discomfort. Symptomatic treatment with psychotropic drugs, such as amitriptyline or clonazepam, may be of some benefit, but no good controlled studies are available to demonstrate a real effect of these drugs. The local application of various medications was recently examined. Interestingly, burning sensation increased after local application of topical anesthesia, while dysgeusia symptoms were more likely to decrease (Ship et al. 1995). Ship et al. (1995) suggested a centrally based neuropathic condition for the burning sensation and that the topical anesthesia may be releasing peripheral inhibition of central sensory pathways. Application of ► capsaicin to oral mucosa was beneficial in about 50 % of patients in an open study, but double-blind controlled studies are warranted before this treatment can be recommended for BMS. Other topically applied medications have shown some success (e.g., clonazepam). Therapy-resistant BMS has been associated with underlying psychological distress, and these patients may particularly benefit from cognitive behavioral therapy (CBT) (for review see Sharav and Benoliel 2008; Jääskeläinen 2012).

Cross-References

► Atypical Facial Pain: Etiology, Pathogenesis, and Management
► Nociception in Nose and Oral Mucosa
► Orofacial Pain

References

Grushka, M. (1987). Clinical features of burning mouth syndrome. *Oral Surgery, 63*, 30–36.
Grushka, M., Sessle, B. J., & Miller, R. (1987a). Pain and personality profiles in burning mouth syndrome. *Pain, 28*, 155–167.
Grushka, M., Sessle, B. J., & Howley, T. P. (1987b). Psychophysical assessment of tactile pain and thermal sensory functions in burning mouth syndrome. *Pain, 28*, 169–184.
Jääskeläinen, S. K. (2012). Pathophysiology of primary burning mouth syndrome. *Clinical Neurophysiology, 123*, 71–77.
Sharav, Y., & Benoliel, R. (2008). *Orofacial pain & headache*. Edinburgh: Mosby Elsevier.
Ship, J. A., Grushka, M., Lipton, J. A., et al. (1995). Burning mouth syndrome: An update. *Journal of the American Dental Association, 126*(7), 842–853.

Burning Pain

Definition

Burning pain is the typical sensation when intense heat affects the skin. It is also felt upon stimulation with chemical irritants, e.g., capsaicin, binding to TRPV1 receptors. This receptor molecule is activated by heat, protons and by chemicals. However, burning pain is also characteristic of different forms of neuropathic pain in which other mechanisms induce the pain sensation.

Cross-References

► First and Second Pain Assessment (First Pain, Pricking Pain, Pin-Prick Pain, Second Pain, Burning Pain)

Bursitis

Definition

A painful bursae, usually resulting from unaccustomed physical activity, injury to the area, or constant pressure for extended periods of time such as the weight of sleeping on a specific structure. Classified as a localized STP.

Cross-References

► Muscle Pain, Fibromyalgia Syndrome (Primary, Secondary)

Burst Activity in Thalamus and Pain

Jonathan O. Dostrovsky
Department of Physiology, University of
Toronto, Toronto, ON, Canada
Faculty of Dentistry, University of Toronto,
Toronto, ON, Canada

Synonyms

Bursting activity; Calcium spike bursts; Low-threshold calcium spike (LTS) burst; Thalamic bursting activity

Definition

Thalamic neurons normally fire in two distinct modes – the tonic mode and the bursting mode. The latter firing mode is usually only observed during sleep. This "▶ bursting activity" is due to the intrinsic membrane properties of thalamic neurons. When they are in a hyperpolarized state, T-type calcium channels become de-inactivated and can then lead to calcium entry (a ▶ low-threshold calcium spike LTS) when they are activated. This calcium spike gives rise to a short high-frequency burst of action potentials. Thalamic neurons can also fire in bursts which are due to other mechanisms, and this type of bursting activity differs in terms of its detailed characteristics from the LTS bursts. Several studies have reported the existence of LTS bursting activity in awake chronic pain patients and have suggested that this activity may be related to their pain. Animal studies have also reported pain-related alterations in bursting activity.

Introduction

Deafferentation and damage to CNS regions involved in the somatosensory system have been reported to give rise to abnormal bursting firing patterns in some neurons at various levels of the somatosensory system and in particular in

spinal cord dorsal horn and thalamus (Albe-Fessard and Longden 1983; Black 1974; Loeser et al. 1968; Lombard and Larabi 1983; Lenz et al. 1994). These early observations have led investigators to propose that such spontaneous bursting activity is due to pathology and may give rise to chronic central pain (e.g., see Boivie 1994; Lenz and Dougherty 1997 and ▶ Thalamic Bursting Activity, Chronic Pain).

Characteristics

Thalamocortical neurons are characterized by several intrinsic voltage- and time-dependent ionic conductances that give rise to two major firing states. During sleep, thalamic neurons tend to fire in a characteristic bursting firing pattern (the burst mode) which ceases during wakefulness to be replaced by a more regular firing pattern (the tonic mode) (see Steriade et al. 1997), although bursting activity is also observed during wakefulness (Ramcharan et al. 2005). This bursting activity occurs when the cell is in a hyperpolarized state which results in de-inactivation of low-threshold T-type calcium channels, and which subsequently open when the cell depolarizes slightly, thus generating a low-threshold calcium spike (LTS) associated with a burst of sodium-generated action potentials. The burst of action potentials induced by the LTS has a very characteristic pattern, which consists of progressively increasing interspike intervals in each burst, and a first interspike interval that decreases with increasing burst size and thus can be identified in extracellular recordings (Domich et al. 1986). This type of bursting, which has been extensively studied in experimental animals, has also been shown to occur in the lateral thalamus of patients when they fall asleep (Tsoukatos et al. 1997; Zirh et al. 1998).

It is frequently assumed that LTS-induced bursting normally occurs only during sleep and drowsiness and thus that its presence in an awake patient signifies a pathological condition. Several groups have reported the existence of thalamic neurons in awake chronic pain patients, which fire in a bursting pattern similar to the low-threshold calcium spike-mediated bursting activity that has been reported during sleep in animal and human studies (Jeanmonod et al. 1996; Lenz and Dougherty 1997; Lenz et al. 1994; Rinaldi et al. 1991). Although such activity is common in medial thalamus, it is also observed in lateral thalamus, including Vc (VPL and VPM). It has been proposed that such firing may give rise to, or at least contribute to, the patients' chronic pain (Lenz and Dougherty 1997; Llinas et al. 1999). In support of this, it has been reported that in pain patients with central pain secondary to spinal cord trans-action, there is a higher likelihood of finding bursting activity in the border zone/anesthetic area of the deafferented somatosensory representation in Vc (Lenz et al. 1994). This is an attractive hypothesis but difficult to prove since the bursting neurons do not have receptive fields, and so it is unclear whether they are part of the nociceptive system. Furthermore, bursting activity in the awake patient is by no means limited to pain patients, and has been reported in movement disorder and epilepsy patients, and can be found in regions that are unlikely to be related to the somatosensory system (Ohara et al. 2007; Radhakrishnan et al. 1999). Rhythmic LTS bursting activity in thalamus has been proposed by Llinas et al. (1999) to be the basis of many neurological disorders including pain and has been termed thalamocortical dysrhythmia. In a study on thalamic neurons in awake patients who had undergone an amputation and had phantom or stump pain, bursting activity was noted, but it was not of the LTS bursting type (Lenz et al. 1998). Recent animal studies have reported pain and/or injury-related alterations in thalamic neuronal firing and in particular increased bursting activity (e.g., Gerke et al. 2003; Hains et al. 2006; Weng et al. 2000) some of which is LTS-type bursting.

Cross-References

▶ Thalamic Bursting Activity, Chronic Pain

References

Albe-Fessard, D., & Longden, A. (1983). Use of an animal model to evaluate the origin of and protection against deafferentation pain. In J. J. Bonica, U. Lindblom, & A. Iggo (Eds.), *Advances in pain research and therapy* (pp. 691–700). New York: Raven.

Black, R. G. (1974). A laboratory model for trigeminal neuralgia. *Advances in Neurology, 4*, 651–659.

Boivie, J. (1994). Central pain. In P. D. Wall (Ed.), *Textbook of pain* (pp. 871–902). Philadelphia: Churchill Livingstone.

Domich, L., Oakson, G., & Steriade, M. (1986). Thalamic burst patterns in the naturally sleeping cat: A comparison between cortically projecting and reticularis neurones. *The Journal of Physiology, 379*, 429–449.

Gerke, M. B., Duggan, A. W., Xu, L., & Siddall, P. J. (2003). Thalamic neuronal activity in rats with mechanical allodynia following contusive spinal cord injury. *Neuroscience, 117*, 715–722.

Hains, B. C., Saab, C. Y., & Waxman, S. G. (2006). Alterations in burst firing of thalamic VPL neurons and reversal by Na(v)1.3 antisense after spinal cord injury. *Journal of Neurophysiology, 95*, 3343–3352.

Jeanmonod, D., Magnin, M., & Morel, A. (1996). Low-threshold calcium spike bursts in the human thalamus. Common physiopathology for sensory, motor and limbic positive symptoms. *Brain, 119*, 363–375.

Lenz, F. A., & Dougherty, P. M. (1997). Pain processing in the human thalamus. In M. Steriade, E. G. Jones, & D. A. McCormick (Eds.), *Thalamus* (Experimental and clinical aspects, Vol. II, pp. 617–652). Amsterdam: Elsevier.

Lenz, F. A., Kwan, H. C., Martin, R., et al. (1994). Characteristics of somatotopic organization and spontaneous neuronal activity in the region of the thalamic principal sensory nucleus in patients with spinal cord transection. *Journal of Neurophysiology, 72*, 1570–1587.

Lenz, F. A., Zirh, A. T., Garonzik, I. M., et al. (1998). Neuronal activity in the region of the principal sensory nucleus of human thalamus (ventralis caudalis) in patients with pain following amputations. *Neuroscience, 86*, 1065–1081.

Llinas, R. R., Ribary, U., Jeanmonod, D., et al. (1999). Thalamocortical dysrhythmia: A neurological and neuropsychiatric syndrome characterized by magneto-encephalography. *Proceedings of the National Academy of Sciences of the United States of America, 96*, 15222–15227.

Loeser, J. D., Ward, A. A. J., & White, L. E., Jr. (1968). Chronic deafferentation of human spinal cord neurons. *Journal of Neurosurgery, 29*, 48–50.

Lombard, M. C., & Larabi, Y. (1983). Electrophysiological study of cervical dorsal horn cells in partially deafferented rats. In J. J. Bonica, U. Lindblom, & A. Iggo (Eds.), *Advances in pain research and therapy* (pp. 147–154). New York: Raven.

Ohara, S., Taghva, A., Kim, J. H., & Lenz, F. A. (2007). Spontaneous low threshold spike bursting in awake humans is different in different lateral thalamic nuclei. *Experimental Brain Research, 180*, 281–288.

Radhakrishnan, V., Tsoukatos, J., Davis, K. D., et al. (1999). A comparison of the burst activity of lateral thalamic neurons in chronic pain and non-pain patients. *Pain, 80*, 567–575.

Ramcharan, E. J., Gnadt, J. W., & Sherman, S. M. (2005). Higher-order thalamic relays burst more than first-order relays. *Proceedings of the National Academy of Sciences of the United States of America, 102*, 12236–12241.

Rinaldi, P. C., Young, R. F., Albe-Fessard, D., et al. (1991). Spontaneous neuronal hyperactivity in the medial and intralaminar thalamic nuclei of patients with deafferentation pain. *Journal of Neurosurgery, 74*, 415–421.

Steriade, M., Jones, E. G., & McCormick, D. A. (1997). *Thalamus*. Oxford: Elsevier Science.

Tsoukatos, J., Kiss, Z. H., Davis, K. D., et al. (1997). Patterns of neuronal firing in the human lateral thalamus during sleep and wakefulness. *Experimental Brain Research, 113*, 273–282.

Weng, H. R., Lee, J. I., Lenz, F. A., Schwartz, A., Vierck, C., Rowland, L., & Dougherty, P. M. (2000). Functional plasticity in primate somatosensory thalamus following chronic lesion of the ventral lateral spinal cord. *Neuroscience, 101*, 393–401.

Zirh, T. A., Lenz, F. A., Reich, S. G., et al. (1998). Patterns of bursting occurring in thalamic cells during parkinsonian tremor. *Neuroscience, 83*, 107–121.

Burst Firing Mode

Synonyms

Calcium spike bursting; Low-threshold calcium spike bursting (LTS)

Definition

A pattern of spontaneous action potential firing, demonstrated by thalamic neurons in which a clear pattern of interspike intervals exists. This type of bursting is caused by de-inactivation of T-type calcium channels when neurons become hyperpolarized as frequently occurs during sleep. This mode contrasts with the so-called tonic firing mode generally present during wakefulness. It is important to note, however, that bursting activity in the thalamus can also be caused by other mechanisms and those bursts have different features.

Cross-References

▶ Burst Activity in Thalamus and Pain
▶ Spinothalamocortical Projections to Ventromedial and Parafascicular Nuclei

Bursting Activity

Definition

A neuronal firing pattern characterized by short clusters of high-frequency firing (short interspike intervals), in comparison to the mean firing rate of the neuron interspersed by time periods with minimal activity. In thalamus, bursting activity is frequently caused by a low-threshold calcium spike, with the interspike intervals typically ranging from 2 to 5 ms and pre-burst silent intervals of at least 100 ms. However, not all thalamic bursts are of this type.

Cross-References

▶ Burst Activity in Thalamus and Pain
▶ Central Pain, Human Studies of Physiology
▶ Chronic Pain
▶ Corticothalamic and Thalamocortical Interactions
▶ Thalamic Bursting Activity
▶ Thalamic Plasticity and Chronic Pain

Butterfly Erythema

Definition

Of the face, typical symptom of systemic lupus erythematosus (SLE).

Cross-References

▶ Headache Due to Arteritis

Cross-References

► Burst Activity in Thalamus and Pain
► Spinothalamocortical Projections to Ventromedial and Parafascicular Nuclei

Bursting Activity

Definition

A neuronal firing pattern characterized by bursts of high-frequency firing (short interspike intervals, in comparison to the mean firing rate of the neuron interspersed by fast periods with quiescent activity. In thalamus, bursting activity is primarily caused by low-threshold calcium spikes with the interspike intervals typically ranging from 2–5 ms and put burst silent intervals of about 100 ms. However, not all thalamic bursts are of this type.

Cross-References

► Burst Activity in Thalamus and Pain
► Central Pain, Human Studies of Physiology
► Thalamic Bursting Activity, Chronic Pain
► Thalamocortical and Thalamocortical Interactions
► Thalamic Bursting Activity
► Thalamic Plasticity and Chronic Pain

Butterfly Erythema

Definition

Of the face, typical symptom of systemic lupus erythematosus (SLE).

Cross-References

► Headache Due to Arteritis

neurons is different in different lateral thalamic nuclei. Experimental Brain Research, 120, 261–288.

Radhakrishnan, V., Tsoukatos, J., Davis, K. D., et al. (1999). A comparison of the burst activity of lateral thalamic neurons in chronic pain and non-pain patients. Pain, 80, 567–575.

Ramcharan, E. J., Gnadt, J. W., & Sherman, S. M. (2000). Higher-order thalamic relays burst more than first-order relays. Proceedings of the National Academy of Sciences of the USA, 102, 12236–12241.

Rinaldi, P. C., Young, R. F., Albe-Fessard, D., et al. (1991). Spontaneous neuronal hyperactivity in the medial and intralaminar thalamic nuclei of patients with deafferentation pain. Journal of Neurosurgery, 74, 415–421.

Sherman, S. M., & Guillery, R. W. (1996). Functional organization of thalamocortical relays. Journal of Neurophysiology, 76, 1367–1395.

Steriade, M., Jones, E. G., & McCormick, D. A. (1997). Thalamus. Oxford: Elsevier.

Swadlow, H. A., & Gusev, A. G. (2001). The impact of "bursting" thalamic impulses at a neocortical synapse. Nature Neuroscience, 4, 402–408.

Weng, H. R., Lee, J. I., Lenz, F. A., Schwartz, A., Vierck, C., Rowland, L., & Dougherty, P. M. (2000). Functional plasticity in primate somatosensory thalamus following chronic lesion of the spinal cord. Neuroscience, 101, 393–401.

Zirh, T. A., Lenz, F. A., Reich, S. G., et al. (1998). Patterns of bursting occurring in thalamic cells during parkinsonian tremor. Neuroscience, 83, 107–121.

Burst Firing Mode

Synonyms

Calcium spike bursting; Low threshold calcium spike bursting (LTS)

Definition

A pattern of spontaneous neuron potential firing during which neurons fire bursts of action potentials in which a clear pattern of interspike intervals exist. This type of bursting is caused by the inactivation of T-type calcium channels when neurons are hyperpolarized. Bursts frequently occur during sleep. Burst mode contrasts with the more typical tonic firing mode, although present when these bursts are important to note, however that bursting neurons in the thalamus can also be caused by other mechanisms and those bursts have different features.

C

3CH134

▶ Phosphatases and Glia

C Afferent Axons/Fibers

Synonyms

C Fibers

Definition

See "C and A fibers" and "Axon Reflex."

Cross-References

▶ A Fibers (A-Fibers)
▶ Alpha(α) 2-Adrenergic Agonists in Pain Treatment
▶ Infant Pain Mechanisms
▶ Insular Cortex, Neurophysiology, and Functional Imaging of Nociceptive Processing
▶ Magnetoencephalography in Assessment of Pain in Humans
▶ Morphology, Intraspinal Organization of Visceral Afferents
▶ Nociceptor, Categorization
▶ Nociceptors in the Orofacial Region (Temporomandibular Joint and Masseter Muscle)
▶ Nociceptor(s)
▶ Opiates During Development
▶ Spinothalamic Tract Neurons, in Deep Dorsal Horn
▶ Thalamus, Dynamics of Nociception
▶ Vagal Input and Descending Modulation

C and A Fibers

Definition

The nomenclature A, B, and C-fibers was introduced by Erlanger and Gasser (1924) for classifying nerve fibers in the peripheral nervous system according to their conduction velocities. The term "B-fibers" for thin myelinated preganglionic sympathetic efferent fibers is rarely used, whereas the denomination "A-fiber" is widely used to refer to myelinated afferent or efferent peripheral axons with conduction velocities between 2 to greater than 100 m/s. The term "C-fiber" refers to unmyelinated peripheral axons with conduction velocities less than 2 m/s. Anatomically A and B fiber types are different. Each A-fiber has dedicated Schwann cells forming the myelin sheath around it. This sheath has a high content of lipids which increase membrane resistance and contributes to the fast conduction velocity of action potentials which are carried from one to the next intersection between two myelin covered segments (Ranvier Node). This form of nerve conduction is called "saltatory." In contrast, C-fibers have no myelin

G.F. Gebhart, R.F. Schmidt (eds.), *Encyclopedia of Pain*, DOI 10.1007/978-3-642-28753-4,
© Springer-Verlag Berlin Heidelberg 2013

sheaths and conduct action potentials continuously. They are covered in groups by a Schwann cell; such groups contain a variable number of C-fibers are called Remak Bundles. In peripheral nerves, C-fibers and are much more abundant than A-fibers.

In the PNS, nociceptive input is conveyed from the peripheral terminals to the spinal cord, by C-fibers and by thinly myelinated A-fibers.

Unmyelinated and myelinated axons are also present in the central nervous system. However, in the CNS the myelin sheath is formed by central glia and not by Schwann cells.

Cross-References

▶ Opioids in the Periphery and Analgesia

References

Erlanger, J., & Gasser, H. S. (1924). The compound nature of the action current of nerve as disclosed by the cathode ray oscilloscope. *The American Journal of Physiology, 70*, 624–666.

C Fibers

▶ C Afferent Axons/Fibers

Cachexia

Definition

General ill health and malnutrition that presents as muscle wasting, dehydration, and reduced behavioral activities in untreated diabetic animals.

Cross-References

▶ Neuropathic Pain Model, Diabetic Neuropathy Model

CACNA1A

Definition

CACNA1A is a gene encoding the α1A subunit of a voltage-sensitive calcium channel abundantly expressed in neuronal tissue. It is activated by high voltage and gives rise to P/Q-type calcium currents. Mutations in the human gene are related to diseases like spinocerebellar ataxia type 6, familial hemiplegic migraine, and episodic ataxia type 2. In mice, mutations in the orthologous gene are responsible for the "leaner" and "tottering" phenotypes.

Cross-References

▶ Calcitonin Gene-Related Peptide and Migraine Headaches

Calbindin

Definition

Vitamin D–dependent calcium-binding protein that is present in specific sensory neuronal cell types.

Cross-References

▶ Spinothalamic Terminations, Core and Matrix
▶ Thalamus, Receptive Fields, Projected Fields, Human

Calbindin D-28k

Definition

A member of the EF-hand family of calcium-binding proteins, which buffer intracellular calcium concentration and mediate a variety of cellular functions. Calbindin D-28k has six EF-hand domains, but only four of them bind to calcium. Calbindin D-28k has been used as a marker of nerve cells in neuroanatomical studies, since it selectively distributes in subpopulations of central nervous system neurons in specific regions. In the monkey thalamus, calbindin D-28k antibodies selectively label the matrix domain of the medial ventroposterior nucleus of the thalamus, which is related to trigeminothalamic projection from the caudal spinal trigeminal nucleus subnucleus caudalis.

Cross-References

▶ Trigeminothalamic Tract Projections

Calbindin-Immunoreceptive Matrix Cells

Definition

Sensory neurons that stain positively for the presence of vitamin D-dependent calcium-binding protein.

Cross-References

▶ Thalamic Nuclei Involved in Pain, Human and Monkey

Calcitonin Gene-Related Peptide and Migraine Headaches

Kirsten Arndt, Henri Doods and Stefan Just
CNS Pharmacology, Boehringer Ingelheim
Pharma GmbH & Co. KG, Biberach, Germany

Synonyms

CGRP

Definition

▶ Migraine is a complex, multi-symptom disease affecting 10–16 % of the western population. It has a higher prevalence in women than in men. Migraine characteristics are its episodic appearance and symptoms such as unilateral headache, phono- and/or photophobia, facial mechanical allodynia, and nausea and vomiting (Headache Classification Subcommittee of the International Headache Society 2004). Preceding the headache, 20–30 % of the patients experience focal neurological symptoms termed aura. Based on the presence or absence of an aura, migraines are classified as either "classical" migraine (migraine with aura) or "common" migraine (migraine without aura). Generally, a migraine attack can be subdivided into different phases that include the premonitory phase, headache, and the postdrome. The prevalence of the different symptoms varies over the phases with, e.g., being tired and weary as approximately equally prominent in all phases and "stiff neck" being most prevalent in the headache and postdrome phases. Typically, a migraine attack lasts from 4 h to 72 h.

▶ Calcitonin gene-related peptide (CGRP) is a neuropeptide and important vasodilator. It is released during migraine attacks. Blockade of CGRP receptors in humans alleviates migraine pain.

Characteristics

Underlying Causes of a Migraine Attack
Despite many approaches to understand the initiation of migraine attacks, "the migraine trigger" or the physiological starting point has not yet been identified. To date it is also not possible to come up with a primary underlying cause for migraines, and probably there is more than one. Several genetic factors have been discussed as rendering a person more susceptible to developing migraines, as well as disturbances of central neuronal function or cranial changes in vasodynamics. Many of these hypotheses evolved from studies of one subtype of migraine

patients, the ones experiencing classical migraines. Potential genetic defects which might cause classical migraines include missense mutations in the ▶ CACNA1A gene (a1 subunit of the voltage-dependent P-/Q-type calcium channel) and the ATP1A2 gene (a2 isoform of the Na,K-ATPase), both accounting for cases of familial hemiplegic migraine.

Besides the genetic factors, local disturbances of central neuronal function, i.e., the cortical spreading depression (CSD) of Leão or activity changes in brainstem/midbrain neuronal systems have been hypothesized to be a main cause underlying migraine attacks. In the forebrain, CSD is a wave of excitation that propagates across the cortex, followed by a wave of suppression. Evidence suggests that CSD is the physiological phenomenon causing the aura phenomenon of classical migraines (Hadjikhani et al. 2001). It is not clear if CSD also occurs in common migraines. Neuronal dysfunction of brainstem/midbrain neuronal systems involved in pain inhibition/facilitation has also been suggested (Knight and Goadsby 2001; May 2003). Human fMRI studies during migraine attacks indicate changed neuronal activity in certain brainstem and midbrain areas (Weiller et al. 1995). The anatomical location suggests that the function of these regions is associated with adjusting nociceptive information entering the brain by either inhibiting or facilitating the responses of neurons and sensory terminals in the brainstem trigeminal nucleus caudalis (TNC). Furthermore, extensive changes in vasodynamics have long been implicated in the pathogenesis of migraines, e.g., alterations in intracranial vessel diameters followed by reactive changes in extracranial vessel diameters were already investigated in 1938 by Graham and Wolff. Also, the first specific treatment employed, the ergotamines, works as vasoconstrictors. More recent support for the importance of vasodilatation is coming from observations employing transcranial laser Doppler measurements, which suggest that migraine attacks are associated with intracranial large arterial dilatation on the headache side.

Together the possible causes described above provide links to the observed migraine symptomatology and potential underlying mechanisms. They deliver the biological "hardware" which contributes to migraine pathogenesis. The focus is on the cranial vasculature and the trigeminal sensory system controlling the vasculature and transmitting information to brainstem and midbrain nuclei, which in turn monitor and adjust the incoming information. During a migraine attack, adjustments made by these systems might be reflected by certain symptoms. As an example, the symptom unilateral throbbing pain at the temples seems to be caused by mechanisms like sensitization of sensory afferents in the periphery. In addition, facial mechanical allodynia might be initiated by neuronal sensitization processes advancing from sensory to central neurons (Burstein et al. 2000). Understanding migraine symptoms and the potential mechanisms driving an attack is an important basis for understanding the disease and getting a handle on the essential molecular drivers that direct the biological systems towards a migraine attack.

The Neuropeptide CGRP: An Important Molecular Player in Migraine Attacks

Several molecules have been identified that trigger headache responses fulfilling the criteria of migraine attacks. Among them are NO released from glyceryl trinitrate (GTN), histamine, PGE2, and CGRP. These molecules have in common that they induce vasodilatation and affect sensory neuron function. The 37-amino acid neuropeptide CGRP has been shown to be one of the most potent vasodilators in human and animal tissues and is widely distributed throughout the body. Two isoforms, α and β CGRP, are known, which originate from different genes. The isoforms differ by only one and three amino acids in rats and humans, respectively. The peptides are expressed by primary afferent neurons, which mostly fire in the C- and Aδ fiber range, as well as by motor neurons, the autonomous nervous system, and central neurons. The CGRP receptor mediating CGRP function consists of three components, a G protein-coupled seven-transmembrane receptor element known as calcitonin-like receptor (CLR), a receptor-associated membrane protein 1 (RAMP1), and a protein termed receptor component protein (RCP) (Poyner et al. 2002). While CLR and RAMP1 are membrane constituents, RCP is a cytoplasmic protein shown to be crucial for the efficient intracellular coupling of the G protein and adenylate cyclase and thus for the production of cAMP. Constituents of the CGRP receptor have been shown to be expressed by peripheral and central neurons as well as by smooth muscle cells in the vascular system.

Ample evidence that CGRP could play an important role in migraine pathophysiology comes from various studies investigating the transmitter content of external jugular vein blood during migraine attacks (Goadsby and Edvinsson 1994; Sarchielli et al. 2000). These studies demonstrate that levels of CGRP normally found to be in the lower pM range and below are approximately twofold increased during acute migraine attacks. Also, increasing the plasma levels by a short infusion of CGRP triggers immediate headaches followed by a delayed severe headache in migraineurs. The symptoms of the delayed headache are indistinguishable from that of a migraine. Furthermore, current gold standard migraine treatment with sumatriptan, a 5HT1B/1D agonist, reduces blood levels of CGRP in humans and in animal experiments. A relationship between pain intensity and plasma CGRP levels has been suggested. Finally, more direct evidence for the contribution of CGRP to migraine pain was introduced by a phase II clinical trial. Olesen et al. (2004) investigated the effectiveness of the small molecule CGRP antagonist BIBN4096 in reducing migraine pain. In this multicenter, double blind, randomized clinical trial, a dose-dependent relief from migraine pain was observed after BIBN4096 intravenous administration. The 2.5 mg dose represented the dose group with the highest patient number. It displayed a response rate of approximately 66 % over 27 % for placebo (Fig. 1). A general pain relieving effect was already observed 30 min after application of BIBN4096. Significant efficacy over placebo was also observed in other migraine specific secondary endpoints, including the pain-free rate, the 24-h recurrence rate and typical migraine associated symptoms like nausea and phono- and/or photophobia. The drug was

Calcitonin Gene-Related Peptide and Migraine Headaches, Fig. 1 Two-hour headache response after intravenous administration of several doses BIBN4096 in the phase II clinical trial

well tolerated and no serious adverse events were observed. Efficacy of several orally available CGRP antagonists for migraine treatment has been confirmed in more recent clinical trials (BI 44370 TA, MK-3207, MK-0974) (Ho et al. 2008; Hewitt et al. 2011; Diener et al. 2011).

In summary, evidence like increased CGRP levels during an attack, the induction of migraine attacks by infusion of CGRP, CGRP levels showing a relationship to pain intensity, and high CGRP levels detected early in an attack, as well as current and future treatment affecting CGRP function, point to a significant contribution of the neuropeptide to the pathogenesis of migraines. CGRP might be not only a marker but also an important driver of processes underlying migraine symptomatology. Nevertheless, to date its mechanism of action in the migraine setting is still not fully understood.

Potential Function of CGRP in Migraine

Some insights into the potential function of CGRP come from animal studies investigating the activity of the highly selective and competitive CGRP antagonist BIBN4096 in the vascular and neuronal system. BIBN4096 has a high affinity for the human CGRP receptor (14.4 pM) and potently reverses CGRP-induced vasodilatation in various rat and guinea pig vascular tissues as well as human cerebral arteries (Doods et al. 2000). Furthermore, in an in vivo model where an increase in facial blood flow is induced by unilateral electrical stimulation of the trigeminal ganglion, BIBN4096 dose

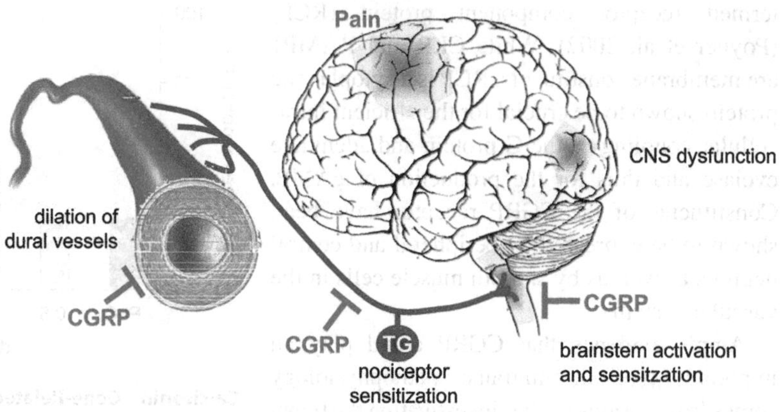

**Calcitonin Gene-Related
Peptide and Migraine
Headaches,
Fig. 2** Possible points of
action of CGRP in migraine
pathogenesis

dependently reduces the evoked blood flow with an ID_{50} of 0.003 mg/kg in marmoset monkeys. More detailed investigations into the mechanism of action of the antagonist in humans showed that in healthy volunteers BIBN4096 prevented CGRP-induced extracranial dilatation and concomitantly reduced CGRP-induced headaches (Petersen et al. 2005). Furthermore, MK-3702, an orally available CGRP antagonist, has been studied in a rhesus monkey capsaicin-induced dermal vasodilation model and displayed a potent inhibition of induced blood flow with an ED50 of about 0.8 nM (Salvatore et al. 2010). Together these data support the significance of CGRP-induced vasodynamic changes in migraine headaches.

CGRP and its receptor system are not only expressed by the vasculature and mononuclear and glial cells, but also by second order neurons in brainstem TNC as well as to small extend at the central terminals of trigeminal sensory neurons (Lennerz et al. 2008). Besides affecting vasodynamics, it might therefore well be that during migraine attacks increased CGRP levels influence trigeminal neuronal information processing via direct neuronal action. In addition trigeminal neuronal activity could be modulated by a more complex indirect mechanism via modulation of mononuclear and glial cells (Ho et al. 2010; Raddant and Russo 2011). Although the role of CGRP on peripheral trigeminal neurons is not fully understood, it has been shown that direct application of CGRP into the TNC increases the firing rate of central trigeminal neurons. In this setting, intravenous administration of the CGRP

antagonist BIBN4096 dose dependently normalizes the activity of TNC neurons. This finding indicates that CGRP participates in sensitizing second order neurons in the TNC, a mechanism that might contribute to the complex neurovascular mechanisms of generation of migraine pain (Fig. 2).

Summary

In summary, it has been shown that CGRP can affect nociceptive processing in the trigeminal system. Increased CGRP levels being present during migraine attacks could imply a role of CGRP in sensitization of primary and/or central neurons besides inducing vasodynamic changes in the cranial vasculature (Fig. 2).

Although migraine pathogenesis still offers vagueness with respect to the trigger(s) and numerous hypotheses on causes and the biological processes behind the symptoms, the testing of new molecular principles like that of the CGRP antagonists opens new paths for the understanding of underlying mechanisms and the identification of important contributors to this complex disease.

References

Burstein, R., Yarnitsky, D., Goor-Aryeh, I., Ransil, B. J., & Bajwa, Z. H. (2000). An association between migraine and cutaneous allodynia. *Annals of Neurology, 47*(5), 614–624.
Diener, H. C., Barbanti, P., Dahlöf, C., Reuter, U., Habeck, J., & Podhorna, J. (2011). BI 44370 TA, an

oral CGRP antagonist for the treatment of acute migraine attacks: Results from a phase II study. *Cephalalgia, 31*(5), 573–584.

Doods, H., Hallermayer, G., Wu, D., Entzeroth, M., Rudolf, K., Engel, W., & Eberlein, W. (2000). Pharmacological profile of BIBN4096BS, the first selective small molecule CGRP antagonist. *British Journal of Pharmacology, 129*(3), 420–423.

Goadsby, P. J., & Edvinsson, L. (1994). Neuropeptides in migraine and cluster headache. *Cephalalgia, 14*(5), 320–327.

Hadjikhani, N., Sanchez Del Rio, M., Wu, O., Schwartz, D., Bakker, D., Fischl, B., Kwong, K. K., Cutrer, F. M., Rosen, B. R., Tootell, R. B., Sorensen, A. G., & Moskowitz, M. A. (2001). Mechanisms of migraine aura revealed by functional MRI in human visual cortex. *Proceedings of the National Academy of Sciences of the United States of America, 98*(8), 4687–4692.

Headache Classification Subcommittee of the International Headache Society. (2004). The international classification of headache disorders. *Cephalalgia, 24* (Suppl. 1), 9–160.

Hewitt, D. J., Aurora, S. K., Dodick, D. W., Goadsby, P. J., Ge, Y. J., Bachman, R., Taraborelli, D., Fan, X., Assaid, C., Lines, C., & Ho, T. W. (2011). Randomized controlled trial of the CGRP receptor antagonist MK-3207 in the acute treatment of migraine. *Cephalalgia, 31*(6), 712–722.

Ho, T. W., Mannix, L. K., Fan, X., Assaid, C., Furtek, C., Jones, C. J., Lines, C. R., Rapoport, A. M., & MK-0974 Protocol 004 Study Group. (2008). Randomized controlled trial of an oral CGRP receptor antagonist, MK-0974, in acute treatment of migraine. *Neurology, 70*(16), 1304–1312.

Ho, T. W., Edvinsson, L., & Goadsby, P. J. (2010). CGRP and its receptors provide new insights into migraine pathophysiology. *Nature Reviews. Neurology, 6*(10), 573–582.

Knight, Y. E., & Goadsby, P. J. (2001). The periaqueductal grey matter modulates trigeminovascular input: A role in migraine? *Neuroscience, 106*(4), 793–800.

Lennerz, J. K., Rühle, V., Ceppa, E. P., Neuhuber, W. L., Bunnett, N. W., Grady, E. F., & Messlinger, K. (2008). Calcitonin receptor-like receptor (CLR), receptor activity-modifying protein 1 (RAMP1), and calcitonin gene-related peptide (CGRP) immunoreactivity in the rat trigeminovascular system: Differences between peripheral and central CGRP receptor distribution. *The Journal of Comparative Neurology, 507*(3), 1277–1299.

May, A. (2003). Headache: Lessons learned from functional imaging. *British Medical Bulletin, 65*, 223–234.

Olesen, J., Diener, H. C., Husstedt, I. W., Goadsby, P. J., Hall, D., Meier, U., Pollentier, S., Lesko, L. M., & BIBN 4096 BS Clinical Proof of Concept Study Group. (2004). Calcitonin gene-related peptide receptor antagonist BIBN 4096 BS for the acute treatment of migraine. *The New England Journal of Medicine, 350*(11), 1104–1110.

Petersen, K. A., Lassen, L. H., Birk, S., Lesko, L., & Olesen. (2005). BIBN4096BS antagonizes human alpha-calcitonin gene related peptide-induced headache and extracerebral artery dilatation. *Journal of Clinical Pharmacology and Therapeutics, 77*(3), 202–213.

Poyner, D. R., Sexton, P. M., Marshall, I., Smith, D. M., Quirion, R., Born, W., Muff, R., Fischer, J. A., Foord, S. M., & International Union of Pharmacology. XXXII. (2002). The mammalian calcitonin gene-related peptides, adrenomedullin, amylin, and calcitonin receptors. *Pharmacological Reviews, 54*(2), 233–246.

Raddant, A. C., & Russo, A. F. (2011). Calcitonin gene-related peptide in migraine: Intersection of peripheral inflammation and central modulation. *Expert Reviews in Molecular Medicine, 29*, 13.

Salvatore, D. P., Liliana, M., Michele, C., et al. (2010). Sorafenib plus octreotide is an effective and safe treatment in advanced hepatocellular carcinoma: multicenter phase II So. LAR. study. *Cancer Chemotherapy and Pharmacology, 66*, 837–44.

Sarchielli, P., Alberti, A., Codini, M., Floridi, A., & Gallai, V. (2000). Nitric oxide metabolites, prostaglandins and trigeminal vasoactive peptides in internal jugular vein blood during spontaneous migraine attacks. *Cephalalgia, 20*(10), 907–918.

Weiller, C., May, A., Limmroth, V., Juptner, M., Kaube, H., Schayck, R. V., et al. (1995). Brain stem activation in spontaneous human migraine attacks. *Nature Medicine, 1*(7), 658–660.

Calcitonin Gene-related Peptide and Spinal Cord Nociception

▶ CGRP and Spinal Cord Nociception

Calcitonin-Gene-Related Peptide

Synonyms

CGRP

Definition

CGRP is a 37-amino acid peptide that is produced by tissue-specific processing of the calcitonin gene. It belongs to the Calcitonin/CGRP family which includes other peptides like calcitonin, amylin, adrenomedullin, and intermedin. CGRP is comprised of at least two isoforms: αCGRP and βCGRP. The human βCGRP differs by three amino acids from αCGRP and has similar effects. Whereas αCGRP is present in a subset of polymodal nociceptive afferents in nearly all tissues, βCGRP is predominantly found in the enteric nervous system. CGRP is also expressed in motor neurons and seems to stimulate neuromuscular transmission. CGRP receptors are heteromers of the calcitonin receptor-like receptor (CLR), a receptor activity-modifying protein (RAMP), and the receptor component protein (RCP). Coupling of CLR with RAMP1 gives rise to high affinity for CGRP, while coupling with RAMP2 and RAMP3 favor binding of adrenomedullin. CGRP receptors couple to a Gs-protein and increase adenylate cyclase activity. In the central nervous system CGRP functions primarily as a neuromodulator. Release of CGRP from the peripheral terminals of sensory neurons contributes to neurogenic inflammation. During this process, CGRP is the most potent vasodilator in the microcirculation identified so far and acts by relaxing small arteries and arterioles. Furthermore, CGRP acts together with substance P to potentiate plasma extravasation where proteins from the blood stream pass into the surrounding tissue. Release of CGRP from the central terminals of DRG neurons modulates spinal cord neurons, in part by enhancing the actions of substance P. The role of CGRP in pain processing is subject of ongoing research.

Cross-References

▶ Alternative Medicine in Neuropathic Pain
▶ Calcitonin Gene-Related Peptide and Migraine Headaches
▶ Clinical Migraine Without Aura
▶ Immunocytochemistry of Nociceptors
▶ Migraine, Pathophysiology
▶ Nociceptor, Categorization
▶ Nociceptors in the Orofacial Region (Meningeal/Cerebrovascular)
▶ Opioid Modulation of Nociceptive Afferents In Vivo

▶ Opioids in the Periphery and Analgesia
▶ Spinal Cord Nociception, Neurotrophins
▶ Thalamus, Visceral Representation

Calcium Channel

▶ NMDA Receptors in Spinal Nociceptive Processing

Calcium Channel Blockers

Definition

A class of drugs with the capacity to prevent calcium ions from passing through biologic membranes. These agents are used to treat hypertension, angina pectoris, and cardiac arrhythmias; examples include nifedipine, diltiazem, verapamil, and amlodipene.

Cross-References

▶ Headache Attributed to a Substance or Its Withdrawal

Calcium Channels in the Spinal Processing of Nociceptive Input

Horacio Vanegas
Instituto Venezolano de Investigaciones
Científicas (IVIC), Caracas, Venezuela

Synonyms

Allodynia; High-threshold calcium channels; High-threshold VDCCs; High-voltage calcium channels; HVCCs; Hyperalgesia; Low-threshold calcium channels; Low-threshold VDCCs; Low-voltage calcium channels; LVCCs; Spinal dorsal horn; Spinal nociceptive transmission;

VDCCs; VGCCs; Voltage-dependent calcium channels; Voltage-sensitive calcium channels; VSCCs

Definition

Calcium channels that open upon depolarization (voltage-dependent calcium channels, VDCCs) enable calcium ions to enter neurons. VDCCs are thereby involved in presynaptic release of neurotransmitters, postsynaptic neuron excitability, intracellular regulation of second and third messengers, and expression of genes. Spinal VDCCs that are opened by relatively large depolarizations (high-threshold VDCCs) or by small depolarizations (low-threshold VDCCs) are involved in normal nociception as well as in the hyperalgesia and allodynia that result from inflammatory and mechanical lesions of peripheral tissues or from lesions of primary afferents fibers (Cao 2006; Vanegas and Schaible 2000).

Characteristics

Voltage-Dependent Calcium Channels

VDCCs are classified according to their electrophysiological properties and their sensitivity to specific antagonists (Miljanich and Ramachandran 1995). Traditionally labeled L-, N-, P/Q-, and R-type are high-threshold, and T-type are low-threshold, VDCCs (Catterall et al. 2003). A rational nomenclature (Ca$_v$1-3), based on the characteristics of the α_1 subunit of the VDCC molecule, has been introduced (Catterall et al. 2003); however, the traditional nomenclature is still widely used (Table 1). The α_1 subunit is the main component of the VDCC because it contains the conduction pore, the voltage sensor, and the gating apparatus. The other subunits are called $\alpha_2\delta$, β, and γ. VDCCs may be found in the membrane of presynaptic terminals, where their activation triggers the release of neurotransmitters, as well as in the neuronal soma and dendrites. T-type VDCCs are found mostly on the soma and dendrites, and

Calcium Channels in the Spinal Processing of Nociceptive Input, Table 1 Voltage-dependent calcium channels and some antagonists (Catterall et al. 2003; Vanegas and Schaible 2000)

Channel type	Antagonist
L Ca$_v$1.1-3	*Benzothiazepines*: diltiazem *Dihydropyridines*: nicardipine, nifedipine, nimodipine, nitrendipine *Phenylalkylamines*: verapamil
P/Q Ca$_v$2.1	*w-Agapeptides*: w-agatoxin-IVA, w-agatoxin-IVB
N Ca$_v$2.2	*w-Conopeptides*: *Natural*: w-conotoxin-GVIA, w-conotoxin-CIVD (AM336). *Synthetic*: SNX-111 (ziconotide, equivalent to w-conotoxin-MVIIA), SNX-124 (equivalent to w-conotoxin-GVIA), SNX-159, SNX-239
R Ca$_v$2.3	SNX482
T Ca$_v$3.1-3	Mibefradil, kurtoxin

participate in the modulation of neuronal excitability. Most of what is known regarding the role of spinal VDCCs in pain mechanisms (Cao 2006; Vanegas and Schaible 2000) derives from studies based on the use of specific channel antagonists or blockers. Some of these are shown in Table 1, and new compounds are constantly being developed (Cao 2006; Carstens et al. 2011). Selective blockers for R-type VDCCs were found rather recently; their actions upon pain mechanisms are therefore less well known. One and the same neuron may express several types of VDCC. An antagonist to only one channel type therefore blocks only a fraction of the VDCCs present in a neuron or a neuronal ensemble such as the spinal cord, and the effect of one antagonist may be additive to the effect of antagonists to other channel types.

It must be emphasized that VDCCs in neurons are under constant change due to regulation by, e.g., GABAergic, glutamatergic, and opioidergic synaptic receptors, intracellular calcium, calmodulin and kinases, as well as to endocytosis (Adams et al. 2012; Cao 2006; Evans and Zamponi 2006).

Animal Models for the Study of VDCCs in Nociception

On the one hand, there are the models for "acute" nociception, in which brief and intense stimuli are applied to normal tissues (Vanegas and Schaible 2000). These stimuli include noxious heat or noxious pressure as applied to skin or joints; application or injection of algogenic substances such as capsaicin, mustard oil, formalin, or acetic acid; and noxious distension of hollow viscera. Upon stimulus application, withdrawal reflexes and other protective behaviors can be measured, or the response of dorsal spinal nociceptive neurons can be recorded, prior to and during the action of specific VDCC antagonists. Also spinal neuronal responses to electrical stimulation of nociceptive primary afferent fibers may serve as a measure of nociception.

On the other hand, animal models of persistent damage include inflammation of skin or joints, surgical wounds, long-lasting hyperexcitability induced by peripheral application or injection of capsaicin, mustard oil, or formalin, ligation of peripheral nerves, administration of anticancer drugs, diabetic neuropathy, etc. (Vanegas and Schaible 2000). These manipulations induce peripheral and central sensitization, and the experimental animals thus respond in an exaggerated manner to the acute stimuli mentioned above. This is akin to hyperalgesia and allodynia, and the effect of specific VDCC antagonists on these exaggerated responses can be investigated.

It is now possible to generate mice that lack one of the molecular subunits of specific calcium channels, and to study their nociceptive responses both under normal and under sensitized conditions. The VDCC defect in these animals, however, is not restricted to the spinal cord.

Role of Spinal VDCCs in Nociception

Acute Nociception

In awake and in anesthetized rats, L- and N-type antagonists may or may not depress responses to mechanical, thermal, and visceral noxious stimuli, or responses to electrical stimulation of nociceptive afferents (Malmberg and Yaksh 1994; Neugebauer et al. 1996). As expected, the action of these antagonists may be exerted

pre- and/or postsynaptically (Heinke et al. 2004) and may involve excitatory as well as inhibitory synaptic activity. Blockade of spinal P/Q-type channels causes a slight increase in neuronal responses, thus suggesting that they normally participate in predominantly inhibitory mechanisms (Cao 2006; Ebersberger et al. 2004; Matthews and Dickenson 2001a; Nebe et al. 1999). T-type VDCCs contribute to the long-term potentiation (LTP) of spinal nociceptive neurons induced by the arrival of low frequency action potentials in nociceptive primary afferents (Drdla and Sandkühler 2008). Low frequency firing is the typical response of nociceptive primary afferents, and LTP is one of the mechanisms of central sensitization (see below). On the other hand, blockade of spinal T-type channels causes an inhibition of spinal neuronal responses to electrical stimulation of nociceptive afferents as well as to low- and high-intensity mechanical and thermal stimuli (Matthews and Dickenson 2001b).

Mice with a genetically induced lack of N-type VDCCs may or may not show decreased responses to acute thermal and/or mechanical noxious stimuli (Hatakeyama et al. 2001; Kim et al. 2001; Murakami et al. 2002; Saegusa et al. 2001). Mice that lack the α_{1E} subunit (which may be part of the R- or the T-type channel) have normal responses to thermal and mechanical noxious stimuli (Saegusa et al. 2001).

Sensitized Nociception

As mentioned above, sustained arrival of nociceptive impulses induces a spinal state of central sensitization that underlies the hyperalgesia and allodynia typical of chronic pain. In animal models which utilize nociceptive stimulation by means of capsaicin or mustard oil (Vanegas and Schaible 2000), blockade of spinal N-type channels always prevents the subsequent exaggeration of responses ("primary and secondary hyperalgesia and allodynia") to acute test stimuli. Also, blockade of spinal L- or P/Q-type channels generally prevents secondary hyperalgesia and allodynia.

As regards responses to test stimuli during inflammation or inflammation-like processes such as surgical wounds, blockade of N-type channels in animals has never failed to prevent or attenuate primary hyperalgesia, secondary hyperalgesia and allodynia, or the late, "inflammatory" response to formalin injection (Vanegas and Schaible 2000). The most effective doses of N-type antagonists, however, cause motor disturbances after 30–60 min. In human patients, one intrathecally administered N-type channel blocker alleviated postsurgical pain but produced severe adverse effects (Atanassoff et al. 2000). On the other hand, blockade of spinal L-type channels has generally prevented the late phase of the formalin response and has attenuated primary and secondary mechanical hyperalgesia and allodynia (but not thermal hyperalgesia) due to knee inflammation (Vanegas and Schaible 2000). Finally, blockade of spinal P/Q-type channels before and during induction of inflammation prevents the exaggeration of responses to stimulation of the inflamed knee or the uninflamed ankle. However, blockade of P/Q-type channels when central sensitization is already established attenuates only responses to stimulation of the sensitized nociceptors in the knee (primary hyperalgesia and allodynia) but has no influence upon responses to stimulation of normal nociceptors in the uninflamed ankle (secondary hyperalgesia and allodynia) (Nebe et al. 1997). The effect of a given compound on the prevention of a painful condition may therefore be different from its effect on the alleviation of the already established condition.

Mice with a genetically induced lack of either N-type VDCCs or the α_{1E} subunit of VDCCs show an attenuation of the late phase of the formalin response (Hatakeyama et al. 2001; Kim et al. 2001; Saegusa et al. 2000, 2001). The writhing response to intraperitoneal acetic acid, a model of visceral nociception, may (Kim et al. 2001) or may not (Saegusa et al. 2000, 2001) be attenuated.

In animals with peripheral nerve damage, spinal application of antagonists to either L- or P/Q-type VDCCs does not alter the hyperalgesia and allodynia, nor the increased spinal neuronal responses to noxious heat, pressure, or electrical stimulation of nociceptive afferents (Matthews

and Dickenson 2001a; Vanegas and Schaible 2000). In contrast, spinal application of N-type channel antagonists has proven effective against neuropathic nociceptive behavior in animals (Vanegas and Schaible 2000) and against the increased spinal neuronal responses to noxious heat, pressure, or electrical stimulation of nociceptive afferents induced by nerve damage (Matthews and Dickenson 2001a). Also, mice with a genetically induced lack of N-type VDCCs fail to develop neuropathic behavioral responses after peripheral nerve damage (Saegusa et al. 2001). Intrathecal administration of one N-type antagonist alleviated neuropathic pain and allodynia in human patients, although with considerable adverse effects (Brose et al. 1997; Penn and Paice 2000). Finally, a T-type VDCC antagonist inhibited spinal neuronal responses to noxious heat, pressure, or electrical stimulation of nociceptive afferents, but this remained unaltered after development of neuropathy (Matthews and Dickenson 2001b).

In recent years, considerable attention has been given to the $\alpha_2\delta$-1 subunit of VDCCs (Cao 2006; Taylor 2009) especially because it is upregulated mainly presynaptically in the spinal cord after peripheral nerve injury (Li et al. 2004; Luo et al. 2001) and because it is the only high-affinity molecular target for gabapentin and pregabalin. These two drugs consequently reduce calcium influx via N-type and P/Q-type VDCCs, and diminish the release of glutamate, GABA, substance P, CGRP, acetylcholine, and glycine (Taylor 2009). Genetically induced overexpression of the $\alpha_2\delta$-1 subunit leads to increased presynaptic release of glutamate and increased AMPA and NMDA receptor activation in the spinal cord as well as to behavioral allodynia (Nguyen et al. 2009). In these animals, gabapentin blocked the enhanced VDCC calcium inflow (Li et al. 2006). Upregulation of the $\alpha_2\delta$-1 subunit after nerve injury correlates with enhanced excitability of nociceptive spinal neurons and to hyperalgesia and allodynia in animal models (Li et al. 2004; Li et al. 2006). Also experimental spinal cord injury induces allodynia that correlates with an upregulation of the $\alpha_2\delta$-1 subunit and can be reversed by gabapentin (Boroujerdi et al. 2011). Gabapentin and pregabalin have proven clinically useful against neuropathic pain (Finnerup et al. 2010).

In Summary

Pharmacological or genetic reduction of VDCC function may or may not have an effect upon normal nociceptive mechanisms. In situations where inflammation or inflammation-like processes have induced an enhancement of spinal nociceptive phenomena (central sensitization), the participation of VDCCs in the spinal processing of nociceptive input becomes more obvious, and reduction of L-, P/Q-, T- and, particularly, N-type VDCC function attenuates the exaggeration of responses to noxious and innocuous stimulation. N-type VDCC antagonists also attenuate the spontaneous pain, the hyperalgesia, and the allodynia which result from damage to primary afferents, yet with considerable adverse effects. A reduction of VDCC function by gabapentin and pregabalin via the $\alpha_2\delta$-1 subunit has gathered much interest particularly in experiments where the $\alpha_2\delta$-1 subunit is upregulated, such as peripheral nerve damage or spinal cord injury, and in cases of human neuropathic pain.

Potential Therapeutic Use of VDCC Antagonists

Any hope of using VDCC antagonists for alleviating hyperalgesia and allodynia in a clinical setting must reckon with several drawbacks. All studies with spinal VDCC antagonists have used the intrathecal route of administration, which is problematic, even more so for compounds with low water solubility. On the other hand, systemically administered VDCC antagonists may have undesirable effects on a variety of organs. Sufficiently beneficial effects with VDCC antagonists are attained mostly with doses which already cause unwanted effects. At the same time, VDCC antagonists, or modulators like the gabapentanoids, have some positive attributes. They may block the synaptic release of several mediators involved in nociceptive transmission, hyperalgesia, and allodynia. This would be an advantage over the use of individual antagonists to, e.g., glutamate,

substance P, neurokinin A, and CGRP. In contrast with opioids, which may also broadly decrease synaptic release, VDCC antagonists or modulators do not seem to give rise to tolerance. In some cases, normal somesthesia and motricity have been spared by doses which are effective against hyperalgesia and allodynia. Specific antagonists to various VDCC types can be combined in submaximal doses to achieve a summated effect. The gabapentanoids can be administered orally. Finally, VDCC antagonists or modulators may synergize with other analgesics, antidepressants, and local anesthetics.

References

Adams, D. J., Callaghan, B., & Berecki, G. (2012). Analgesic conotoxins: Block and G protein-coupled receptor modulation of N-type (Ca$_v$2.2) calcium channels. *British Journal of Pharmacology, 166,* 486–500.

Atanassoff, P. G., Hartmannsgruber, M. W., Thrasher, J., et al. (2000). Ziconotide, a new N-type calcium channel blocker, administered intrathecally for acute postoperative pain. *Regional Anesthesia and Pain Medicine, 25,* 274–278.

Boroujerdi, A., Zeng, J., Sharp, K., Kim, D., Steward, O., & Luo, Z. D. (2011). Calcium channel alpha-2-delta-1 protein upregulation in dorsal spinal cord mediates spinal cord injury-induced neuropathic pain states. *Pain, 152,* 649–655.

Brose, W. G., Gutlove, D. P., Luther, R. R., et al. (1997). Use of intrathecal SNX-111, a novel, N-type, voltage-sensitive, calcium channel blocker, in the management of intractable brachial plexus avulsion pain. *The Clinical Journal of Pain, 13,* 256–259.

Cao, Y. -Q. (2006). Voltage-gated calcium channels and pain. *Pain, 126,* 5–9.

Carstens, B. B., Clark, R. J., Daly, N. L., Harvey, P. J., Kaas, Q., & Craik, D. J. (2011). Engineering of conotoxins for the treatment of pain. *Current Pharmaceutical Design, 17,* 4242–4253.

Catterall, W. A., Striessnig, J., Snutch, T. P., & Perez-Reyes, E. (2003). International Union of pharmacology. XL. Compendium of voltage-gated ion channels: Calcium channels. *Pharmacological Reviews, 55,* 579–581.

Drdla, R., & Sandkühler, J. (2008). Long-term potentiation at C-fibre synapses by low-level presynaptic activity *in vivo*. *Molecular Pain, 4,* 18.

Ebersberger, A., Portz, S., Meissner, W., Schaible, H.-G., & Richter, F. (2004). Effects of N-, P/Q- and L-type calcium channel blockers on nociceptive neurones of the trigeminal nucleus with input from the dura. *Cephalalgia, 24,* 250–261.

Evans, R. M., & Zamponi, G. W. (2006). Presynaptic Ca^{2+} channels – integration centers for neuronal signaling pathways. *Trends in Neuroscience, 29,* 617–624.

Finnerup, N. B., Sindrup, S. H., & Jensen, T. S. (2010). The evidence for pharmacological treatment of neuropathic pain. *Pain, 150,* 573–581.

Hatakeyama, S., Wakamori, M., Ino, M., et al. (2001). Differential nociceptive responses in mice lacking the α_{1B} subunit of N-type Ca^{2+} channels. *NeuroReport, 12,* 2423–2427.

Heinke, B., Balzer, E., & Sandkühler, J. (2004). Pre- and postsynaptic contributions of voltage-dependent Ca^{2+} channels to nociceptive transmission in rat spinal lamina I neurons. *European Journal of Neuroscience, 19,* 103–111.

Kim, C., Jun, K., Lee, T., et al. (2001). Altered nociceptive responses in mice deficient in the α_{1B} subunit of the voltage-dependent calcium channel. *Molecular and Cellular Neurosciences, 18,* 235–245.

Li, C.-Y., Song, Y.-H., Higuera, E. S., & Luo, Z. D. (2004). Spinal dorsal horn calcium channel a_2d-1 subunit upregulation contributes to peripheral nerve injury-induced tactile allodynia. *Journal of Neuroscience, 24,* 8494–8499.

Li, C. Y., Zhang, X. L., Matthews, E. A., Li, K. W., Kurwa, A., Boroujerdi, A., Gross, J., Gold, M. S., Dickenson, A. H., Feng, G., & Luo, Z. D. (2006). Calcium channel alpha2delta1 subunit mediates spinal hyperexcitability in pain modulation. *Pain, 125,* 20–34.

Luo, Z. D., Chaplan, S. R., Higuera, E. S., Sorkin, L. S., Stauderman, K. A., Williams, M. E., & Yaksh, T. L. (2001). Upregulation of dorsal root ganglion a2d calcium channel subunit and its correlation with allodynia in spinal nerve-injured rats. *Journal of Neuroscience, 21,* 1868–1875.

Malmberg, A. B., & Yaksh, T. L. (1994). Voltage-sensitive calcium channels in spinal nociceptive processing: Blockade of N- and P-type channels inhibits formalin-induced nociception. *Journal of Neuroscience, 14,* 4882–4890.

Matthews, E. A., & Dickenson, A. H. (2001a). Effects of spinally delivered N- and P-type voltage-dependent calcium channel antagonists on dorsal horn neuronal responses in a rat model of neuropathy. *Pain, 92,* 235–246.

Matthews, E. A., & Dickenson, A. H. (2001b). Effects of ethosuximide, a T-type Ca^{2+} channel blocker, on dorsal horn neuronal responses in rats. *European Journal of Pharmacology, 415,* 141–149.

Miljanich, G. P., & Ramachandran, J. (1995). Antagonists of neuronal calcium channels: Structure, function, and therapeutic implications. *Annual Review of Pharmacology and Toxicology, 35,* 707–734.

Murakami, M., Fleischmann, B., De Felipe, C., et al. (2002). Pain perception in mice lacking the β_3 subunit of voltage-activated calcium channels. *Journal of Biological Chemistry, 277,* 40342–40351.

Nebe, J., Vanegas, H., Neugebauer, V., et al. (1997). ω-agatoxin IVA, a P-type calcium channel antagonist, reduces nociceptive processing in spinal cord neurons with input from the inflamed but not from the normal knee joint -an electrophysiological study in the rat in vivo. *European Journal of Neuroscience, 9,* 2193–2201.

Nebe, J., Ebersberger, A., Vanegas, H., et al. (1999). Effect of ω-agatoxin IVA, a P-type calcium channel antagonist, on the development of spinal neuronal hyperexcitability caused by knee inflammation in rats. *Journal of Neurophysiology, 81,* 2620–2626.

Neugebauer, V., Vanegas, H., Nebe, J., et al. (1996). Effects of N- and L-type calcium channel antagonists on the responses of nociceptive spinal cord neurons to mechanical stimulation of the normal and inflamed knee joint. *Journal of Neurophysiology, 76,* 3740–3749.

Nguyen, D., Deng, P., Matthews, E. A., Kim, D. S., Feng, G., Dickenson, A. H., Xu, Z. C., & Luo, Z. D. (2009). Enhanced pre-synaptic glutamate release in deep-dorsal horn contributes to calcium channel alpha-2-delta-1 protein-mediated spinal sensitization and behavioral hypersensitivity. *Molecular Pain, 5,* 6.

Penn, R. D., & Paice, J. A. (2000). Adverse effects associated with the intrathecal administration of ziconotide. *Pain, 85,* 291–296.

Saegusa, H., Kurihara, T., Zong, S., et al. (2000). Altered pain responses in mice lacking the α_{1E} subunit of the voltage-dependent Ca^{2+} channel. *Proceedings of the National Academy of Sciences USA, 97,* 6132–6137.

Saegusa, H., Kurihara, T., Zong, S., et al. (2001). Suppression of inflammatory and neuropathic pain symptoms in mice lacking the N-type Ca^{2+} channel. *EMBO Journal, 20,* 2349–2356.

Taylor, C. P. (2009). Mechanisms of analgesia by gabapentin and pregabalin -Calcium channel a_2-d [$Ca_V a_2$-d] ligands. *Pain, 142,* 13–16.

Vanegas, H., & Schaible, H.-G. (2000). Effects of antagonists to high-threshold calcium channels upon spinal mechanisms of pain, hyperalgesia and allodynia. *Pain, 85,* 9–18.

Calcium Spike Bursting

▶ Burst Firing Mode

Calcium Spike Bursts

▶ Burst Activity in Thalamus and Pain

CAMs

▶ Cellular Adhesion Molecules

Cancer Pain

Malgorzata Krajnik[1] and Zbigniew Zylicz[2]
[1]Chair of Palliative Care, Nicolaus Copernicus University in Torun, Collegium Medicum in Bydgoszcz, Bydgoszcz, Poland
[2]Hildegard Hospiz, Basel Switzerland, Basel, Switzerland

Synonyms

Malignant pain; Oncological pain; Pain due to cancer

Definition

"Cancer pain" is a conglomerate name for all kinds of pain symptoms experienced in the course of a malignant disease. The common denominator of pain associated with cancer is that the suffering experienced by the patient is a combination of many different kinds of pain, for example, other symptoms of the disease as well as psychological, spiritual, and existential factors.

Characteristics

Pain is experienced by 20–50 % of cancer patients at the time of diagnosis. This prevalence increases up to 65 % in the case of patients with advanced stages of the disease (Van den Beuken-van Everdingen et al. 2007). Within these statistics, half of the patients normally experience moderate or severe pain, whereas 20–30 % experience very severe or excruciating pain. Usually, cancer patients experience pain in more than one location and of more than one kind. According to Twycross

and Fairfield (1982), only 20 % of patients have one single localization of pain. Almost 25 % of the patients experienced two or more kinds of pain. A large majority of the patients (>92 %) had one or more pains caused directly by the cancer while more than 20 % of patients had one or more pains caused by nonmalignant causes including cancer therapies (Caraceni and Portenoy 1999).

According to Grond et al. (1996) cancer pain may be related to the following:
- Tumor growth: 85 %
- Tumor treatment: 17 %
- Progressive debility: 9 %
- Concurrent disorder: 9 %

With improvement of oncological treatments and patients' longer survival, it can be expected that the number of pains due directly to cancer may decrease but number of pains due to the treatment, debility, and concurrent disorders may increase. These non-cancer pains may be less sensitive to opioids.

Cancer pain may originate in many locations and have multiple, intertwined mechanisms. The top 10 pains among 211 patients (Twycross and Fairfield 1982) with advanced cancer were:
- Bone
- Visceral
- Neuropathic
- Soft tissue
- Immobility
- Constipation
- Myofascial
- Cramp
- Esophagitis

It therefore follows that careful evaluation and examination is necessary to make a diagnosis of cancer pain. Evaluation of pain is based primarily on probability and pattern recognition. Factors that may be helpful in the diagnosis of cancer pain are:

The natural history of the disease (e.g., breast cancer often causes bone metastases and bone pain, ovarian cancer may rarely cause brain metastases and headache).

A time course of the pain symptoms may be important. Brachial or lumbar plexopathy that resulted in pain even before radiation therapy is probably caused by the growth of the tumor, but may be exacerbated by the radiation therapy. Lack of pain two or three weeks after irradiation may be not related to the opioids administered, but to the radiation therapy.

Plain radiographic imaging is helpful in discovering the origin of bone pain. Pain originating in soft tissues and neural tissues needs more sophisticated techniques (MRI). Some types of pain (e.g., neuropathic pain) may be "invisible" to radiographic imaging. Neuropathic pain may frequently be a component of another pain (Portenoy and Lesage 1999), it may be a complication of therapy, but may also be a paraneoplastic symptom. An evaluation of neurological deficit may frequently lead to better understanding and diagnosis of pain. Spinal cord compression (SCC) occurs in 3 % of all cancer patients. In some tumors, like prostate cancer, it may be much more frequent.

Matching results of radiological imaging with patients' complaints is essential: When the patient suffers of severe pain in the pelvic crest, and the scan shows bone abnormalities in a different location, a syndrome of nerve compression should be looked for, instead of bone metastasis which needs radiotherapy.

Nerve compression or entrapment usually gives a specific syndrome (e.g., vertebral collapse may cause nerve root compression). However, there are many peripheral nerve entrapment syndromes (e.g., meralgia paresthetica) which are usually overlooked as they are invisible with modern imaging techniques.

Response to analgesics and pattern of adverse effects may give a clue to which analgesics will be tolerated and whether there is hope that increase of their dose will successfully control pain. When the patient does not tolerate even 1 mg of morphine and this dose does not control his pain, it is usually wise to change the opioid but not to increase the dose of the first. Full history of the pharmacological and non-pharmacological treatment of the pain should be taken.

Emotional, social, and spiritual factors may make the precise diagnosis of pain difficult, and should be recognized and addressed specifically.

A distinction should be made between pain while at rest and pain during movement. Pain may have also a circadian character. Many types of cancer pains are constant, but some may appear to have an exploding character (breakthrough pain), even when the optimal analgesia is provided (Caraceni and Portenoy 1999).

Patients who describe their pain as throbbing, lancinating, and burning, and exacerbated by light touch during examination, may appear to have mechanical allodynia, which is a sign of neuropathic pain.

Knowledge of the so-called referred pain syndromes (Giamberardino and Vecchiet 1995) can be helpful in establishing diagnosis. (e.g., unilateral facial or ear pain may be associated with mediastinal involvement by bronchial carcinoma or reflux esophagitis. Right shoulder pain may be due to an enlarged liver.) Metastases in the lower lumbar region may result in pain localized in the sacroiliacal joint.

Principles of cancer pain treatments were elaborated by the WHO (Ventafridda et al. 1985). Early clinical trials showed high efficacy of this approach in 95–98 % of pain symptoms. It is unclear whether this is still valid now as the patients with cancer needing pain therapy may be different from those four decades ago.

Cancer pain is not usually the only symptom of the disease. It is frequently accompanied by other symptoms like delirium, nausea and vomiting, weakness, fatigue, weight loss, dyspnea, dry mouth, constipation or diarrhea, pruritus, and probably many others. Treatment of pain alone may decrease its intensity but increase intensity of other symptoms. So cancer pain should always be seen in the context of other symptoms and in the context of the whole person.

Cognitive failure progressing to delirium, but also some other adverse effects that emerge in the course of treatment, may compromise the patient's quality of life more seriously than pain. So the aim of the treatment is the balanced control of all symptoms, not only pain control. Sometimes making pain and other symptoms bearable, but not alleviating them fully, is the only viable option. The principles of WHO are as follows:

Cancer Pain, Table 1 The WHO analgesic ladder

Step 1	Step 2	Step 3
Non-opioids	Weak opioids	Strong opioids
Paracetamol, various NSAIDs +/− adjuvants	Codeine, tramadol (low-dose morphine, low-dose oxycodone, low-dose hydromorphone) +/− non-opioid analgesic +/− adjuvants	Morphine, oxycodone, hydromorphone, fentanyl, buprenorphine, methadone +/− non-opioid analgesic +/− adjuvants

Cancer Pain, Table 2 Adjuvants

Indication	Drugs used
Neuropathic pain	Amitryptyline, gabapentine, pregabaline, venlafaxine
Insufficient analgesic effects	Ketamine
Nerve compression	Dexamethasone
Reflux esophagitis	Proton pump inhibitors
Abdominal cramp, bladder cramp	Butyl-scopolamine
Muscle cramp	Benzodiazepine, baclofen
Respiratory depression, breathlessness, sedation, tiredness	Methylphenidate
Constipation	Lactulose and sennosides, or macrogol
Nausea and vomiting	Metoclopramide, haloperidol, levomepromazine, cyclizine
Dry mouth	Pilocarpine
Pruritus	Paroxetine

- Preferably give the medication by mouth.
- Give the next dose of analgesics before the effect of the previous dose ceases.
- Give drugs according to the analgesic ladder (Tables 1 and 2).

Cancer pain is usually a dynamic and complex phenomenon that should be assessed regularly. Prolonged use of analgesics may induce plastic changes in the central nervous system which are similar to those induced by pain itself. This phenomenon is called opioid-induced hyperalgesia. (Zylicz and Twycross 2008) Also, prolonged use of opioid drugs for pain may induce

tolerance, while the adverse effects may increase. To prevent this, more and more combinations of the various drugs are used. With proper choice of drugs, continuous assessment, and adjustment of medication, it is possible to control pain with optimal preservation of alertness (although not till the end) in most cancer patients. Controlling cancer pain is of importance for the dying patient, as well as for the family of the dying who need to carry on normal living insofar as that is possible, and to adapt to the loss of a loved one.

Cross-References

▶ Cancer Pain Management, Treatment of Neuropathic Components
▶ Pain Treatment, Intracranial Ablative Procedures
▶ Psychiatric Aspects of the Management of Cancer Pain

References

Caraceni, A., & Portenoy, R. K. (1999). An international survey of cancer pain characteristics and syndromes. IASP Task Force on Cancer Pain. International Association for the Study of Pain. *Pain, 82,* 263–274.

Giamberardino, M. A., & Vecchiet, L. (1995). Visceral pain, referred hyperalgesia and outcome: New concepts. *European Journal of Anaesthesiology. Supplement, 10,* 61–66.

Grond, S., Zech, D., Diefenbach, C., Radbruch, L., & Lehmann, K. A. (1996). Assessment of cancer pain: a prospective evaluation in 2266 cancer patients referred to a pain service. *Pain, 64,* 107–14.

Portenoy, R. K., & Lesage, P. (1999). Management of cancer pain. *Lancet, 353,* 1695–1700.

Twycross, R. G., & Fairfield, S. (1982). Pain in far-advanced cancer. *Pain, 14,* 303–310.

Van den Beuken-van Everdingen, M. H., de Rijke, J. M., Kessels, A. G., Schouten, H. C., van Kleef, M., & Patijn, J. (2007). Prevalence of pain in patients with cancer: A systematic review of the past 40 years. *Annals of Oncology, 18,* 1437–1449.

Ventafridda, V., Saita, L., Ripamonti, C., & De Conno, F. (1985). WHO guidelines for the use of analgesics in cancer pain. *International Journal of Tissue Reactions, 7,* 93–96.

Zylicz, Z., & Twycross, R. (2008). Opioid-induced hyperalgesia may be more frequent than previously thought. *J Clin Oncol, 26,* 1564.

Cancer Pain and Pain in HIV/AIDS

Russell K. Portenoy[1] and Ebtesam Ahmed[2]
[1]Department of Pain Medicine and Palliative Care, Beth Israel Medical Center, New York, NY, USA
[2]College of Pharmacy and Health Sciences, St. John's University, Department of Pain Medicine and Palliative Care, Beth Israel Medical Center, New York, NY, USA

Synonyms

Cancer-related pain; HIV/AIDS-related pain; Pain in HIV/AIDS

Definition

The terms "cancer pain" and "pain in HIV/AIDS" refer to the assessment and management of acute or chronic pain syndromes that are either directly related to a malignant neoplasm or to HIV infection, respectively, or to the treatments that are used to manage these diseases.

Introduction

Chronic pain, a highly prevalent symptom in populations with cancer or HIV/AIDS, may be associated with loss of function, compromised quality of life, and profound suffering. Although there are important differences between cancer and HIV/AIDS and each disorder itself is extraordinarily diverse, there are broad commonalities in the approach to pain assessment and management. The specific issues encountered in the management of HIV/AIDS are addressed in the entry on ▶ Pain in HIV/AIDS. The remaining entries in this section, which focus on cancer pain, may be understood to apply to both the cancer and HIV/AIDS populations.

Characteristics

Epidemiology

Studies of cancer pain epidemiology reveal that approximately 30–50 % of patients undergoing antineoplastic therapy and 75–90 % of patients with advanced disease have chronic pain severe enough to warrant opioid therapy (Vainio and Auvinen 1996). Although the prevalence of pain in HIV/AIDS has probably declined with the advent of highly active antiretroviral therapy, recent surveys suggest that chronic pain affects about half of this population (Dobalian et al. 2004).

It is widely accepted that a large majority of patients with cancer pain can attain satisfactory relief with available therapies. Unfortunately, surveys indicate a high rate of undertreatment (Anderson et al. 2004). Undertreatment is a complex phenomenon that may result from a variety of patient-related barriers and clinician-related barriers or from distortions in the health care system that limit access to treatment. Recent studies underscore the influence on these barriers of cultural factors and race and ethnicity (Anderson et al. 2004; Green et al. 2003). Education of patients, families and professional staff, and system level strategies such as quality improvement activities are needed to address the problem of undertreatment.

Models of Care

In populations with life-threatening illnesses, the management of pain should be incorporated into a broader effort to ameliorate the physical, psychosocial, and spiritual issues that undermine quality of life or worsen suffering for the patient or family. The therapeutic approach that addresses these issues is known as palliative care. The relationship between cancer pain and palliative care (see ▶ Cancer Pain Management, Interface between Cancer Pain Management and Palliative Care) must be understood to optimize the care of these patients, particularly those with advanced disease.

Palliative care is a therapeutic approach to the care of patients with life-threatening illnesses and their families. It is focused on maintaining quality of life throughout the course of the illness and addressing the challenging needs of patients who are approaching the end of life. The goals of this model include control of pain and other symptoms; management of psychological distress, comorbid psychiatric disorders, and spiritual distress; support for effective communication and decision-making; provision of practical help in the home; and ongoing support for the family and management of the dying process in a manner that allows a comfortable and dignified death and effective grieving on the part of the family.

Palliative care is now considered an approach that should be implemented at a generalist level by every physician who cares for those with serious medical illness. It should also be available at a specialist level for those patients and families who warrant this level of care. The need for specialist level palliative care, which typically occurs in the setting of advanced illness, is being met in the United States by a growing number of institution-based palliative care programs (www.nationalconsensusproject.org) and by more than 3,300 hospice programs providing palliative care at the end of life. All of these programs are underutilized and poorly understood by patients, families, and professional staff.

Evaluation of Pain

The goals of pain assessment include characterization of the pain complaint, evaluation of the etiology, syndrome, and putative mechanisms sustaining the pain, and the assessment of the impact of the pain and relevant comorbidities.

Pain Characteristics

A comprehensive evaluation should begin with an assessment of the pain characteristics. The temporal features comprise onset, duration, course, and fluctuation. Most patients with chronic pain related to a progressive illness experience pain that begins insidiously and fluctuates broadly but gradually progresses. Periodic short-lived flares of pain, or "▶ breakthrough pain" (Caraceni et al. 2004), are very common and should be separately assessed.

Topographic features include primary location and radiation. Pain may be referred from any structure, including nerve, bone, muscle, soft tissue, and viscera, and knowledge of pain referral patterns is needed to guide the evaluation. For example, pain in the inguinal crease may require evaluation of numerous structures to identify the underlying lesion, including the pelvic bones and hip joint, pelvic sidewall, paraspinal gutter at an upper lumbar spinal level, and intraspinal region at the upper lumbar level.

Measurement of pain intensity may be accomplished using a numeric scale (0–10), verbal rating scale (none, mild, moderate, severe), or a visual analogue scale. Pictorial scales are particularly useful when assessing pain in the pediatric cancer (see ▶ Cancer Pain, Assessment in Children) population and pain in the cognitively impaired (see ▶ Cancer Pain, Assessment in the Cognitively Impaired). The particular scale is less important than its consistent use to monitor and document the status of a specific pain (e.g., "worst pain during the past week") over time. In assessing pain intensity, it also is important to obtain information about factors that increase or decrease the pain. The quality of the pain is assessed using verbal descriptors, such as aching, sharp, throbbing, burning, or stabbing. Combined with other information, these descriptors allow broad inferences about the type of pathophysiology that sustains the pain.

Inferred Pathophysiology, Etiology, and Syndromes

Characterization of the pain complaint must be complemented by a physical examination and appropriate laboratory tests and imaging to provide the information necessary to elucidate the likely etiology of the pain, the pain syndrome, and the inferred type of pathophysiology sustaining the pain. In populations with cancer or HIV/AIDS, the etiology often relates to an identifiable structural lesion, such as neoplastic invasion of bone. Identification of the etiology often informs the overall treatment of the patient by defining the extent of disease.

Numerous pain syndromes have been characterized in the cancer (Caraceni et al. 1999) and HIV/AIDS (Hewitt et al. 1997) populations. Syndrome identification helps to define the underlying etiology and prognosis and also may suggest the need for additional testing or specific therapies.

Although inferences about the type of pathophysiology sustaining the pain represent a simplification of very dynamic processes occurring in both the periphery and in the central nervous system, these inferences have been incorporated into clinical practice because they have utility when deciding on a treatment strategies. The labels that are applied to the pathophysiological categories include nociceptive, neuropathic, psychogenic, and mixed.

Nociceptive pain refers to pain that is believed to be sustained by ongoing activation of pain-sensitive primary afferent neurons by injury to tissue. In the setting of cancer, nociceptive pain usually is due to direct invasion by the neoplasm (Caraceni et al. 1999). When somatic structures are involved, the pain is termed "somatic pain" and the pain is typically aching, throbbing, stabbing, and familiar. Bone pain is the most common type. "Visceral pain" occurs when the nociceptive lesion involves visceral structures; it is usually gnawing or crampy when arising from obstruction of a hollow viscus and aching or stabbing when arising from damage to organ capsules.

Pain is labeled neuropathic if it is believed to be sustained by abnormal somatosensory processing in the peripheral or central nervous systems (CNS). Neuropathic mechanisms are involved in approximately 40 % of cancer pain syndromes and may be caused by disease or by treatment (Caraceni et al. 1999). Dysesthesia or abnormal uncomfortable sensations that may be described using words such as "burning," "shock-like" or "electrical" are suggestive of neuropathic mechanisms, as is the presence of abnormal findings on the sensory examination, such as allodynia (pain induced by nonpainful stimuli) or hyperalgesia (increased perception of painful stimuli).

The term "psychogenic pain" is a generic label referring to pain that is believed to be sustained predominantly by psychological factors. Pain of this type may be more precisely characterized through the widely accepted taxonomy of somatoform disorders proposed by the American Psychiatric Association (American Psychiatric Association 1994). These pains share an assessment that reveals positive evidence for the psychopathology that is believed to be causally related to the pain. Although psychological influences are profoundly important in the presentation of the pain and the patient's ability to adapt and function, psychogenic pain itself appears to be distinctly uncommon in the cancer and HIV/AIDS populations.

Occasionally, pain occurs that defies clinical characterization according to a clear etiology or syndrome. In the absence of positive evidence for any distinctive type of pain, it is usually best to label the symptom as "idiopathic." In the setting of progressive medical illnesses, idiopathic pains require regular reassessment in the hope that an explanatory process will become clear over time.

Impact and Comorbidities
The pain evaluation also must include an assessment of impact and relevant comorbidities (see ▶ Cancer Pain, Evaluation of Relevant Comorbidities and Impact). The impact of the pain may be considered from the perspective of varied physical, psychosocial, and spiritual domains. As appropriate, these domains and specific medical and psychiatric comorbidities must be specifically assessed to fully understand the targets for treatment.

Pain Management
The overall treatment strategy should proceed from the broader therapeutic perspective of palliative care. The multidimensional assessment guides treatment that is often multimodal and best implemented through the efforts of professionals in varied disciplines. The immediate goal of pain relief should be pursued in tandem with interventions that address other sources of distress.

Analgesic approaches can be broadly divided into (1) primary therapies directed against the etiology of the pain and (2) symptomatic therapies. Primary treatments for the pain include antineoplastic therapies – ▶ radiotherapy, ▶ chemotherapy, immunotherapy, and surgery (see ▶ Palliative Surgery in Cancer Pain Management) – and other interventions directed at structural pathology. Orthopedic surgery interventions (see ▶ Cancer Pain Management, Orthopedic Surgery) to address bony metastases exemplify the potential for primary therapy directed at specific structural pathology. In the HIV/AIDS population, primary treatment may include antiretroviral therapy and other interventions. If it is feasible and clinically appropriate to provide a primary therapy that can effectively treat the source of the pain, the analgesic consequences can be profound. Most patients have pain that cannot be addressed solely by a primary disease modifying approach. The most important symptomatic therapy is an opioid-based analgesic drug regimen, which may be complemented by a large number of other treatments.

Pharmacotherapy

Prospective trials indicate that more than 70 % of patients can achieve adequate relief of cancer pain using a pharmacologic approach (Schug et al. 1990). Effective pain management requires expertise in the use of the nonsteroidal anti-inflammatory drugs (NSAIDs), opioid analgesics, and adjuvant analgesics. The term "adjuvant analgesic" is applied to a diverse group of drugs, most of which have primary indications other than pain but can be effective analgesics in specific disorders such as neuropathic pain.

There is a broad consensus in favor of an approach to cancer pain management that was developed by an expert panel of the World Health Organization almost two decades ago and was termed the "analgesic ladder" (World Health Organization 1996). This approach has been highly influential, reinforcing the consensus view that persistent moderate to severe cancer pain should be treated with an opioid-based drug regimen. The details of the model have evolved over time, but it remains useful as a tool for educating clinicians and policymakers.

According to the analgesic ladder approach, mild to moderate cancer pain is first treated with a nonopioid analgesic (see ▶ Cancer Pain Management, Nonopioid Analgesics), such as acetaminophen or a nonsteroidal anti-inflammatory drug (NSAID). This drug is combined with an adjuvant drug that can be selected either to provide additional analgesia (i.e., an adjuvant analgesic) or to treat a side effect of the analgesic or a coexisting symptom. Patients who present with moderate to severe pain or who do not achieve adequate relief after a trial of a NSAID should be treated with an opioid, often combined with a NSAID or adjuvant drugs.

Nonsteroidal Anti-inflammatory Drugs

NSAIDs appear to be especially useful in patients with bone pain or pain related to grossly inflammatory lesions and relatively less useful in patients with neuropathic pain (Wallenstein and Portenoy 2002). These drugs may have an opioid sparing effect that can limit the potential for dose-related opioid side effects. The use of the NSAIDs may be limited by their toxicities and a maximal efficacy that is insufficient to address most cancer pain. All NSAIDs have the potential for nephrotoxicity, with effects that range from peripheral edema to acute or chronic renal failure. All these drugs increase the risk of gastrointestinal ulcers and bleeding. The selective cyclooxygenase (COX)-2 inhibitors have a relatively reduced risk of these outcomes, and many clinicians recommend these drugs as first-line therapy for all patients at relatively high risk of ulcer, including the elderly, those concurrently receiving a corticosteroid, and those with a prior history of peptic ulcer disease or NSAID-induced gastroduodenopathy (Wallenstein and Portenoy 2002). The risk is also reduced by coadministration of a proton pump inhibitor, the prostaglandin analogue misoprostol, and possibly high-dose H2 blockers. Recent data have raised concerns about an increased risk of cardiovascular events among those treated with the selective COX-2 inhibitors (Solomon et al. 2004), but the relative risks and benefits compared to nonselective drugs in medically ill populations have yet to be defined. There are also no data in the medical ill from which to judge the relative outcomes associated with administration of selective COX-2 inhibitors alone *versus* nonselective NSAIDs plus a gastroprotective agent. At the present time, the approach is a matter of clinical judgment.

Adjuvant Analgesics

The adjuvant analgesics comprise numerous drugs in diverse classes (Lussier and Portenoy 2004). Treatment with one of these drugs is generally considered if an optimally administered opioid regimen fails to provide a satisfactory balance between pain relief and side effects. These drugs are particularly useful in the treatment of neuropathic pain, bone pain, and pain related to bowel obstruction.

Corticosteroids are multipurpose adjuvant analgesics and also are used to improve anorexia, nausea, and fatigue. In populations with cancer pain or pain due to HIV/AIDS, the first-line adjuvant analgesics for neuropathic pain (see ▶ Adjuvant Analgesics in Management of Cancer-Related Neuropathic Pain) are the corticosteroids, anticonvulsants, and antidepressants. Use of the anticonvulsants and antidepressants is supported by numerous controlled trials in varied populations (Lussier and Portenoy 2004). Other drugs considered for neuropathic pain comprise the GABA agonist baclofen, alpha-2 adrenergic agonists, and various N-methyl-D-aspartate inhibitors. Topical agents, such as the lidocaine patch, represent another strategy for these pain syndromes.

The most commonly used adjuvant analgesics for bone pain (see ▶ Adjuvant Analgesics in Management of Cancer-Related Bone Pain) are the bisphosphonates. These drugs also have been demonstrated to reduce skeletal morbidity, such as fractures. Other drugs used for bone pain include radiopharmaceuticals and calcitonin. Adjuvant analgesics for bowel obstruction (see ▶ Cancer Pain Management, Adjuvant Analgesics in Management of Pain Due to Bowel Obstruction) can often control pain and other symptoms and obviate the need for drainage procedures. Treatment usually involves the combination of anticholinergic drugs, octreotide, and corticosteroids (Lussier and Portenoy 2004).

Opioid Analgesics

Opioid pharmacotherapy is the mainstay approach for the management of moderate to severe pain in populations with cancer or HIV/AIDS. Guidelines for opioid selection (see ▶ Cancer Pain Management, Principles of Opioid Therapy, Drug Selection) have evolved from the WHO analgesic ladder approach, which originally labeled some opioids as "weak" (for moderate pain) and some as "strong" (for severe pain) (World Health Organization 1996). This distinction is based on conventional practice and not pharmacology. In the United States, drugs typically administered to address moderate pain in the opioid naïve patient include codeine, hydrocodone (combined with acetaminophen or ibuprofen), dihydrocodeine (combined with aspirin), oxycodone (combined with aspirin, acetaminophen or ibuprofen), propoxyphene, and occasionally, meperidine. Tramadol, a unique centrally acting analgesic with a mechanism that is partly opioid, is also generally included in this group. These drugs are conventionally used for moderate pain either because of dose dependent toxicity or because the combination products contain a nonopioid analgesic with a maximum safe dose.

Opioids conventionally selected for severe pain, particularly when patients have already been exposed to the short acting drugs on the first rung of the analgesic ladder, include morphine, fentanyl, oxycodone (without acetaminophen or aspirin), hydromorphone, oxymorphone, levorphanol and methadone. Historically, morphine was described as the preferred first-line drug, but there is large variation in the response to different opioids and it is best to select an initial trial based on the available formulations, cost, and prior experience. Although the role of methadone has expanded in recent years because of its low cost, long half-life, and unexpectedly high potency (presumably related to the D-isomer in the racemate, which is a N-methyl-D-aspartate blocker), it poses challenges in dose selection, dose adjustment, and monitoring that are not shared by the other pure mu agonist opioids. Experience is needed to use this drug safely.

Opioids may be delivered by any of numerous routes of administration (see ▶ Opioid Therapy in Cancer Pain Management, Route of Administration). Long-term dosing is best accomplished by an oral or transdermal route. Numerous oral formulations are available, including modified release forms of morphine, oxycodone, or hydromorphone. Other drugs, such as oxymorphone, may become available. The latter drugs provide effective analgesia with a prolonged dosing interval, increasing convenience, and potentially adherence to therapy. The transdermal route is available for fentanyl and offers a 48–72- h dosing interval. Other transdermal formulations are in development. The oral transmucosal form of fentanyl has been shown to be safe and efficacious when used to treat breakthrough pain in cancer patients (Christie et al. 1998). Rectal formulations of many opioids, such as oxymorphone, hydromorphone, and morphine, are available but are seldom used for long-term administration.

Long-term parenteral dosing is possible for patients who are poor candidates for oral or transdermal formulations. Continuous subcutaneous infusion or continuous intravenous infusion (if the patient has an indwelling central venous port) can be implemented with any opioid available in an injectable formulation. Opioids and other drugs may also be delivered into the epidural or intrathecal spaces. The strongest indication for neuraxial infusion is the presence of intolerable somnolence or confusion during systemic opioid therapy.

The most important principle of opioid administration is individualization of the dose (Jacox et al. 1994). In all cases, the dose of an opioid should be gradually increased until acceptable analgesia is produced or unmanageable side effects supervene. The absolute dose of the opioid is immaterial as long as the balance between analgesia and side effects remains acceptable to the patient. Most patients achieve a favorable outcome and remain on a stable dose until pain recurs as a result of disease progression. Recurrent pain following a period of dose stability usually requires reevaluation of the patient and another period of dose titration.

For persistent or frequently recurring pain, the best outcome is achieved by a fixed, around-the-clock dosing schedule. Long acting opioids are often used because of the convenience and the likelihood that treatment adherence will be better than with frequent daily doses. Breakthrough pain commonly is managed by coadministration of short-acting, as needed, "rescue doses."

Dose titration usually yields a favorable balance between analgesia and side effects. In some cases, however, treatment-limiting side effects occur and render the treatment ineffective. This scenario is known as poor ▶ opioid responsiveness, a phenomenon that should be viewed as individual to the patient, the drug and route, and the moment in time.

Patients with pain that is poorly responsive to the opioid therapy must undergo a change in treatment. There are no comparative trials to guide clinical practice. Four strategies should be considered and a specific approach selected based on the assessment and clinical judgment. Given the individual variation in the response to different drugs, one strategy is to switch to an alternative opioid, an approach called ▶ opioid rotation. A second strategy is to coadminister one or more treatments for the side effect that limits dose escalation. Sophisticated approaches are now available to address cognitive dysfunction and gastrointestinal effects, the two most common types of opioid-related toxicity. A third strategy involves the use of a pharmacological approach that would potentially allow reduction in the requirement for the systemic opioid. Coadministration of a NSAID or adjuvant analgesic or a trial of neuraxial infusion might be considered. Finally, a strategy using a nonpharmacological approach that would potentially reduce the opioid requirement might be considered. This might involve an invasive therapy, such as neural blockade or any of a variety of other approaches.

Patients who are treated for a prolonged period with opioids must be continually reassessed for side effects. In addition to common toxicities, opioids may produce a variety of uncommon effects or outcomes that are yet poorly recognized. For example, endocrine changes (see ▶ Cancer Pain Management, Opioid Side Effects, Endocrine Changes, and Sexual Dysfunction) associated with opioid use, such as hypogonadism, may be associated with fatigue, sexual dysfunction, or osteoporosis and should be assessed in some populations.

Opioids are potentially abusable drugs and the clinicians who prescribe them for chronic pain should be familiar with the principles of addiction medicine. Although abuse and addiction during opioid therapy for pain are very uncommon in the population of cancer patients with no prior history of substance abuse, these outcomes should always be assessed and are significant concerns in subpopulations with substance use disorders. Understanding the special considerations posed by opioid use in patients with substance use disorders (see ► Opioid Therapy in Cancer Patients with Substance Abuse Disorders, Management) provides essential information that must be applied in the treatment of all patients.

Other Approaches in the Management of Chronic Pain

There are numerous alternative strategies that may be considered in the treatment of pain related to cancer or HIV/AIDS. In almost every situation, the use of these approaches has been extrapolated from experience in the populations with chronic noncancer pain syndromes. Noninvasive analgesic strategies can be broadly categorized into psychological interventions, rehabilitative treatments, and complementary or alternative medicine approaches. Specific psychological approaches have been applied successfully in the management of pain and related symptoms (Breitbart et al. 2004). Treatments include relaxation training, distraction, hypnosis, and biofeedback. Although many of these techniques require experienced personnel to implement, several forms of relaxation training can be taught by the nonspecialist. Behavioral interventions like the use of an activities diary to improve physical functioning have achieved wide acceptance in the management of noncancer pain and are occasionally considered for medically ill patients. A variety of psychoeducational and psychotherapeutic approaches may be implemented in an effort to improve coping, adaptation, family integrity, functioning, and quality of life.

Rehabilitative therapies (see ► Cancer Pain Management, Rehabilitative Therapies), such as therapeutic exercise, use of orthotics, and modalities such as heat, cold, and electrical stimulation, may be useful in selected patients. Refractory, movement-induced pain, such as that related to bone metastases, may be partially relieved by bracing the painful part, and a well-fitting prosthesis may reduce stump pain. Therapeutic exercise may lessen pain associated with immobility, trigger points in muscle, and ankylosis. Although all the rehabilitative approaches remain inadequately studied, clinical experience is favorable.

Complementary and alternative medicine (CAM) approaches are commonly pursued by medically ill patients. Some of these interventions, such as meditation and other mind-body approaches, nutritional support, acupuncture, and massage, are widely used for pain and are generally considered mainstream strategies (Pan et al. 2000). Others, such as homeopathy and naturopathy, have little scientific support. Clinicians should provide whatever data are available about these approaches and support informed decision-making.

Invasive strategies for pain include neurosurgical interventions (see ► Cancer Pain Management, Neurosurgical Interventions), such as cordotomy; a variety of injection therapies, including neural blockade; and the implantable therapies of neuraxial infusion and spinal cord stimulation. Although the injection therapies and implants are sometimes described as anesthesiological therapies (see ► Cancer Pain Management, Anesthesiologic Interventions, and Neural Blockade), they are now performed by pain specialists in a variety of disciplines.

► Neural blockade includes a diverse group of procedures that transiently or permanently block sympathetic nerves, somatic nerves, or both (Swarm et al. 2004). Injections may be diagnostic, prognostic, or therapeutic. The solutions injected may be local anesthetics, in which case the effects typically are short-lived only, or neurolytic substances. Neurolytic blockade generally is seldom employed and considered in the context of advanced illness. One exception is the celiac plexus blockade for pancreatic cancer, experience with which is sufficiently favorable to apply early in the treatment of pain.

In medically ill populations, neuraxial infusion and stimulation (see ▶ Cancer Pain Management, Anesthesiologic Interventions, Spinal Cord Stimulation, and Neuraxial Infusion) are seldom considered. The role of neuraxial infusion may evolve as a result of a controlled trial that demonstrated benefits of an implanted pump for intrathecal drug delivery over conventional systemic opioid therapy in a population with cancer (Smith et al. 2002). Other techniques for neuraxial infusion include a tunneled percutaneous epidural catheter and an implanted epidural catheter connected to a subcutaneous portal. The latter techniques are usually preferred for patients with life expectancies shorter than 3 months.

Conclusion

Pain is a common complication of cancer and HIV/AIDS. Treatment of pain from a broader perspective of palliative care is likely to yield the best outcomes. Relatively simple therapeutic approaches can provide effective pain relief in a large majority of patients. Pain relief must be considered an aspect of best clinical practice for all clinicians who treat patients with these illnesses.

References

American Psychiatric Association. (1994). Somatoform disorders. In *Diagnostic and statistical manual of mental disorders (DSM-IV)* (4th ed., pp. 445–471). Washington, DC: American Psychiatric Association.

Anderson, K. O., Mendoza, T. R., Payne, R., et al. (2004). Pain education for underserved minority cancer patients: A randomized controlled trial. *Journal of Clinical Oncology, 22,* 4918–4925.

Breitbart, W., Payne, D., & Passik, S. (2004). Psychological and psychiatric interventions in pain control. In D. Doyle, G. Hanks, N. I. Cherny, et al. (Eds.), *Oxford textbook of palliative medicine* (3rd ed., pp. 424–438). Oxford: Oxford University Press.

Caraceni, A., Portenoy, R. K., & Working Group of the IASP Task Force on Cancer Pain. (1999). An international survey of cancer pain characteristics and syndromes. *Pain, 82,* 263–274.

Caraceni, A., Martini, C., Zecca, E., et al. (2004). Breakthrough pain characteristics and syndromes in patients with cancer pain. An international survey. *Palliative Medicine, 18,* 177–183.

Christie, J. M., Simmonds, M., Patt, P., et al. (1998). A dose-titration, multicenter study of oral transmucosal fentanyl citrate (OTFC) for the treatment of breakthrough pain in cancer patients using transdermal fentanyl for persistent pain. *Journal of Clinical Oncology, 16,* 3238–3245.

Dobalian, A., Tsao, J. C., & Duncan, R. P. (2004). Pain and the use of outpatient services among persons with HIV: Results from a nationally representative survey. *Medical Care, 42,* 129–138.

Green, C. R., Anderson, K. O., Baker, T. A., et al. (2003). The unequal burden of pain: Confronting racial and ethnic disparities in pain. *Pain Medicine, 4,* 277–294.

Hewitt, D. J., McDonald, M., Portenoy, R. K., et al. (1997). Pain syndromes and etiologies in ambulatory AIDS patients. *Pain, 70,* 117–124.

Jacox, A., Carr, D. B., Payne, R., et al. (1994). *Management of cancer pain. AHCPR publication No. 94-0592: Clinical practice guideline No. 9.* Rockville: U.S. Department of Health and Human Services, Public Health Service.

Lussier, D., & Portenoy, R. K. (2004). Adjuvant analgesics. In D. Doyle, G. Hanks, N. I. Cherny, et al. (Eds.), *Oxford textbook of palliative medicine* (3rd ed., pp. 349–377). Oxford: Oxford University Press.

Pan, C. X., Morrison, R. S., Ness, J., et al. (2000). Complementary and alternative medicine in the management of pain, dyspnea, and nausea and vomiting near the end of life. A systematic review. *Journal of Pain and Symptom Management, 20,* 374–387.

Schug, S. A., Zech, O., & Dorr, U. (1990). Cancer pain management according to WHO analgesic guidelines. *Journal of Pain and Symptom Management, 5,* 27–32.

Smith, T. J., Staats, P. S., Stearns, L. J., et al. (2002). Randomized clinical trial of an implantable drug delivery system compared with comprehensive medical management for refractory cancer pain: Impact on pain, drug-related toxicity, and survival. *Journal of Clinical Oncology, 19,* 4040–4049.

Solomon, D. H., Schneeweiss, S., Glynn, R. J., et al. (2004). Relationship between selective cyclooxygenase-2 inhibitors and acute myocardial infarction in older adults. *Circulation, 109,* 2068–2073.

Swarm, R. A., Karanikolas, M., & Cousins, M. J. (2004). Anaesthetic techniques for pain control. In D. Doyle, G. Hanks, N. I. Cherny, et al. (Eds.), *Oxford textbook of palliative medicine* (3rd ed., pp. 378–395). Oxford: Oxford University Press.

Vainio, A., & Auvinen, A. (1996). Prevalence of symptoms among patients with advanced cancer: An international collaborative study. *Journal of Pain and Symptom Management, 12,* 3–10.

Wallenstein, D. J., & Portenoy, R. K. (2002). Nonopioid and adjuvant analgesics. In A. M. Berger, R. K. Portenoy, & D. E. Weissman (Eds.), *Principles and practice of palliative care and supportive oncology* (pp. 84–97). Philadelphia: Lippincott-Raven.

World Health Organization. (1996). *Cancer pain relief* (2nd ed.). Geneva: World Health Organization.

www.nationalconsensusproject.org

Cancer Pain Management

Definition

The assessment and treatment – both pharmacological, and nonpharmacological – of pain in patients who have a diagnosis of cancer.

Cross-References

▶ Cancer Pain Management, Adjuvant Analgesics in Management of Pain Due To Bowel Obstruction
▶ Cancer Pain Management, Anesthesiologic Interventions, and Neural Blockade
▶ Cancer Pain Management, Anesthesiologic Interventions, Spinal Cord Stimulation, and Neuraxial Infusion
▶ Cancer Pain Management, Cancer-Related Breakthrough Pain, Therapy
▶ Cancer Pain Management, Chemotherapy
▶ Cancer Pain Management, Interface between Cancer Pain Management and Palliative Care
▶ Cancer Pain Management, Neurosurgical Interventions
▶ Cancer Pain Management, Nonopioid Analgesics
▶ Cancer Pain Management, Opioid Side Effects, Cognitive Dysfunction
▶ Cancer Pain Management, Opioid Side Effects, Endocrine Changes, and Sexual Dysfunction
▶ Cancer Pain Management, Orthopedic Surgery
▶ Cancer Pain Management, Patient-Related Barriers
▶ Cancer Pain Management, Principles of Opioid Therapy, Dosing Guidelines
▶ Cancer Pain Management, Principles of Opioid Therapy, Drug Selection
▶ Cancer Pain Management Radiotherapy
▶ Cancer Pain Management, Rehabilitative Therapies
▶ Cancer Pain Management, Treatment of Neuropathic Components
▶ Cancer Pain Management, Undertreatment and Clinician-Related Barriers
▶ Palliative Surgery in Cancer Pain Management

Cancer Pain Management Radiotherapy

Edward Chow[1], Nora Janjan[2] and Stephen Lutz[3]
[1]Department of Radiation Oncology, Sunnybrook Health Sciences Centre Odette Cancer Centre (T2-182), Toronto, ON, Canada
[2]University of Texas M. D. Anderson Cancer Center, Houston, TX, USA
[3]Blanchard Valley Regional Cancer Center, Findlay, OH, USA

Definition

Radiation is used to treat cancer and other diseases. The radiation dose is described in terms of centigray (cGy) or gray (Gy) (100 cGy = 1 Gy). For curative radiation, 5,000–8,000 cGy is generally prescribed over five treatments each week for 5–8 weeks of radiation. This means that 180–200 cGy of radiation is given each day for a total of 25–40 fractions or treatments. When radiation is given with palliative intent, the dose usually ranges from a single 8 Gy fraction to 15 fractions of 2.5 Gy over 3 weeks.

There are many ways that radiation can be administered, which are broadly classified as external radiation and brachytherapy. External radiation and brachytherapy each have many forms of treatment. The goal of any type of radiation treatment is to limit the radiation dose only to the involved areas.

External Beam Radiation

External radiation treats the site of disease by using different types of x-ray beams derived from a machine called a linear accelerator, which is built on the principles used in an x-ray machine. This is the most common way that radiation is administered. Through the use of specialized blocks, a linear accelerator focuses the radiation only on the area to be treated with millimeter precision and blocks radiation from affecting adjacent tissues.

The choice of the type of radiation used in a radiation treatment plan depends on the location of the tumor and critical structures near the tumor. The concept of integral dose is used to describe the selection of the type of radiation in order to maximize the radiation dose in the tumor yet minimize treatment of uninvolved adjacent tumors.

There are many types of external beam radiation that are derived from a linear accelerator. These types of radiation include gamma rays and electron beam radiation. Gamma rays are used to treat tumors located deep in the body, while electron beam radiation treats superficial cancers.

Radiation loses energy as it passes through tissue, and the characteristic of a radiation beam is described by the amount of energy that is lost as it passes through the tissue. The term *Dmax* refers to the depth of the tissue where 100 % of the radiation dose is deposited. Only 50 % of the prescribed radiation dose from a single beam from a Cobalt unit may penetrate to a tumor located 10 cm from the skin surface. To overcome this, many beams of radiation are routinely used to treat a tumor. The sum of the radiation doses at the tumor gives a high total radiation dose. Additionally, higher-energy linear accelerators were developed that deliver very little radiation dose near the skin surface and while much higher percentages of the prescribed radiation reach tumors located 10 cm or more in depth from the skin surface.

Three-dimensional conformal radiation therapy and intensity-modulated radiation therapy apply the principles of integral dose by targeting the tumor through multiple low-dose radiation beams. All the radiation energy from these beams adds to give the tumor a very high radiation dose, but the radiation dose from the individual beams is so low that there are minimal effects from the radiation in the adjacent normal tissues.

Proton radiation applies the Bragg peak principle to minimize integral dose. With the Bragg peak principle, virtually all the radiation dose is deposited over a specified area, in a peak-like distribution, below the skin surface. Because of adjacent critical structures, proton radiation is commonly applied in CNS and pediatric tumors.

Gamma Rays

Gamma rays are penetrating forms of x-rays, and many energy levels of gamma rays are available for treatment (Tables 1 and 2). In general:

- High-energy gamma rays deeply penetrate the tissues but spare the skin from radiation reactions like erythema and moist desquamation. This type of radiation is used to treat deep-seated tumors in the abdominal and pelvic areas. Due to the characteristics of high-energy gamma rays, the skin receives little to no dose.
- An example of a high-energy gamma radiation beam is an 18 MV photon beam (a photon is a package of energy). If 100 cGy is prescribed from one radiation beam, 100 cGy is deposited 3.5 cm from the skin surface (Dmax), and 85 % of the radiation dose (85 cGy) is deposited at 10 cm of tissue below the skin surface.
- Low-energy gamma rays do not penetrate the tissues well and cause more skin reactions. This type of radiation is used to treat head and neck cancers and breast cancers where there may be tumor infiltration of the skin.
- An example of a low-energy gamma radiation beam is a cobalt-60 beam or a 6 MV photon beam. If 100 cGy is prescribed from one 6 MV radiation beam, 100 cGy is deposited 1.5 cm from the skin surface (Dmax), and 60 % of the radiation dose (60 cGy) is deposited at 10 cm of tissue below the skin surface.

Dmax is the depth in cm below the skin surface where 100 % of the prescribed radiation dose is given. The % depth dose at 10 cm is the % of the prescribed radiation that penetrates to a tumor located 10 cm below the skin surface = 100 cGy/% depth dose. Two thousand centigray will be given to the tumor (located

Cancer Pain Management Radiotherapy, Table 1 Gamma radiation – prescribing 100 cGy for a tumor located 10 cm below the skin

Type of gamma beam	Dmax (cm below skin surface)	% depth dose at 10 cm below the skin surface	Dose of radiation (cGy) needed to give 100 cGy @ 10 cm below the skin	Dose of radiation at Dmax with each radiation fraction(cGy)	Total dose of radiation to the skin at Dmax (20 radiation fractions)
Cobalt-60	0.5	57	175	175	3,500 cGy ≅ 0.5 cm
6 MV photons	1.5	65	154	154	3,080 cGy ≅ 1.5 cm
18 MV photons	3.5	80	125	125	2,500 cGy ≅ 3.5 cm

Cancer Pain Management Radiotherapy, Table 2 Electron beam radiation – radiation is deposited over a short distance from the skin's surface

Type of electron beam (MeV)	Dmax (cm below skin surface)	80 % of radiation dose deposited within(cm)	95 % of radiation dose deposited within (cm)
6	1.5	2	3
9	2.3	3	4.5
12	3	4	6

As an example, if a 100 cGy of radiation is administered with 6 MeV electron beam, 100 cGy would be given at 1.5 cm below the skin, only 20 cGy would be delivered 2 cm below the skin surface (80 % of the radiation dose deposited within 2 cm of the skin surface) and only 5 cGy would be delivered 3 cm below the skin surface (95 % of the radiation dose delivered within 3 cm of the skin surface). Therefore, 80 % of the radiation dose is given between 1.5 and 2 cm from the skin surface. Less than 5 % of the radiation dose reaches structures more than 3 cm from the skin surface. The use of electron beam radiation to treat posterior cervical lymph nodes in head and neck cancer is particularly critical to spare the underlying spinal cord

10 cm from the skin surface) from an anterior radiation field.

Another 2,000 cGy will be given to the tumor from a posterior radiation field. This will deliver 4,000 cGy to the tumor. Each radiation fraction will equal 200 cGy (100 cGy from the anterior and 100 cGy from the posterior field), and 20 fractions will be required to give a total radiation dose of 4,000 cGy. Eighteen MV photons are more effective in treating abdominal and pelvic tumors, and they cause little skin reaction. Cobalt-60 and 6 MV photons are effective in treating more superficial tumors like head and neck and breast cancers that

may have tumor infiltration of the skin; in this case, a higher radiation dose to the skin is essential to control the tumor.

Electron Beam Radiation

Electron beam radiation is used to treat cancers located in the superficial tissues, like lymph nodes in the neck, the chest wall after a mastectomy, and skin cancers. It provides an important alternative to gamma rays, because underlying critical structures like the heart and spinal cord may not receive any radiation at all from electron beam therapy.

- The depth of penetration in centimeters of tissue from an electron beam energy can be generalized by the following rule:
- Eighty percent of the radiation dose will be deposited if the electron beam energy is divided by three, and 95 % of the radiation dose will be deposited if the electron beam energy is divided by two.
  - For example, a 9 MeV electron beam deposits 80 % of the dose 3 cm from the skin surface and 95 % of the dose 4.5 cm from the skin surface.

As an example, if a 100 cGy of radiation is administered with 6 MeV electron beam, 100 cGy would be given at 1.5 cm below the skin, only 20 cGy would be delivered 2 cm below the skin surface (80 % of the radiation dose deposited within 2 cm of the skin surface), and only 5 cGy would be delivered 3 cm below the skin surface (95 % of the radiation dose delivered within 3 cm of the skin surface). Therefore, 80 % of the radiation dose is given between 1.5 and 2 cm from the skin surface. Less than 5 % of the radiation dose reaches structures more than 3 cm from the skin surface. The use of electron beam radiation to treat posterior cervical lymph nodes in head and neck cancer is particularly critical to spare the underlying spinal cord.

Summary: External Beam Radiation

External beam treatment planning involves the specific selection of radiation beams to target the tumor and minimize treatment of adjacent normal tissues. Limiting the dose of radiation to adjacent tissues is critical to avoid side effects from radiation and spare critical structures like the spinal cord. These treatment-related side effects, especially those to mucosal surfaces that result in mucositis, nausea, and diarrhea, can significantly impact on the functional status of the patient. This is of significant concern in patients who are already debilitated from their disease.

Brachytherapy

Brachytherapy is the placement of radioactive sources in or adjacent to a tumor bed. These radioactive sources can be permanently implanted in a tumor or placed temporarily near the tumor location. Permanently implanted radiation sources are used to treat prostate cancer. The radiation dose gradually decays until no further radiation is emitted. For lung cancer, radioactive sources that are placed in a catheter are temporarily adjacent to a tumor blocking the bronchus for a prescribed amount of time to deliver a specific radiation dose. After the prescribed amount of time, the radiation sources are removed. Brachytherapy also uses high-dose rate (a very strong radioactive source that gives radiation over a few minutes) and low-dose rate (a radioactive source that gives a low-dose of radiation over several hours).

The choice of brachytherapy used primarily depends on the tumor location. The radiation dose is extremely localized because of the inverse-square law (radiation dose $= 1/d^2$). Applying this principle, the radiation dose drops off rapidly as the distance from the radiation source increases. For example, the radioactive dose 2 cm away from the radioactive source is only one-fourth the radiation dose next to the source. Brachytherapy is an important option in palliative care because it can deliver radiation over a few minutes to days instead of several weeks. As the radiation is well localized, brachytherapy can be used to re-irradiate tissues that have previously been irradiated. Treatment-related side effects with brachytherapy are limited.

Radioactive isotopes, which have affinity to specific areas like the bone or thyroid, can also be injected into the blood stream and are a form of brachytherapy. The effects of the radiation are localized to the site of deposition, and little radiation is administered to adjacent tissues. Like more conventional brachytherapy, ▶ radioisotopes allow re-irradiation of previously irradiated areas because the radiation dose is concentrated in the site of disease.

Mechanisms of Action

Radiotherapy results in damage to cellular DNA. Radiotherapy causes both direct and indirect damage to the reproductive DNA material of the cell. Direct damage takes place in the form of base deletions, single and double strand breaks in the DNA chain. Indirect damage occurs by the interaction of radiotherapy with water molecules in the cell, which releases toxic-free radicals.

▶ Palliative radiotherapy is well established for the treatment of symptomatic ▶ bone metastases. The exact mechanism of its action is still uncertain; however, the absence of a dose–response relationship, rapid responses, and poor correlation of symptomatic relief with radiosensitivity suggest that an effect on host mechanisms of ▶ pain could also be important. Markers of bone remodeling have been suppressed by antineoplastic therapy, and the response of these bone markers has been applied to monitoring therapy for bone metastases.

In the UK Bone Pain Radiotherapy Trial (Bone Pain Trial Working Party 1999), 22 patients were entered into a supplementary study to establish the effects of local radiotherapy for metastatic bone pain on markers of osteoclast activity, particularly the pyridinium cross-links pyridinoline and deoxypyridinoline, the latter being specific for bone turnover. Urine samples were collected before and 1 month after radiotherapy. Patients were treated with either a single 8 Gy or 20 Gy in five daily fractions. Pain response was scored with validated pain charts completed by patients. Urinary pyridinium concentrations were correlated with pain response. In patients who did not respond to palliative radiation (nonresponders), baseline concentrations of both pyridinoline and deoxypyridinoline were higher than those who responded (responders) and rose further after treatment, whereas in responders, the mean values remained unchanged (Fig. 1). This resulted in significant differences between responders and nonresponders for both indices after treatment ($P = 0.027$). The authors concluded that radiotherapy-mediated inhibition of bone resorption, and thus osteoclastic activity, could be a predictor for pain response. They also proposed that tumor cell killing reduces the production of osteoclast-activating factors or there is a direct effect upon osteoclasts within the radiation volume, distinct from tumor shrinkage. Their study supports the results from randomized trials that high-dose radiotherapy is not necessary for pain relief and that single low doses of treatment are more than adequate for most patients (Hoskin et al. 2000).

Measurement of urinary N-telopeptide [Ntx] and bone alkaline phosphatase [BAP] was performed in three large randomized trials with 1,824 patients treated with zoledronic acid. Patients with high and moderate Ntx (>50 nmol/mmol creatinine) and BAP ($\geq$146 U/L) had a twofold increase in their risk of skeletal complications and disease progression. High Ntx levels in solid tumor patients were associated with a four- to sixfold increased

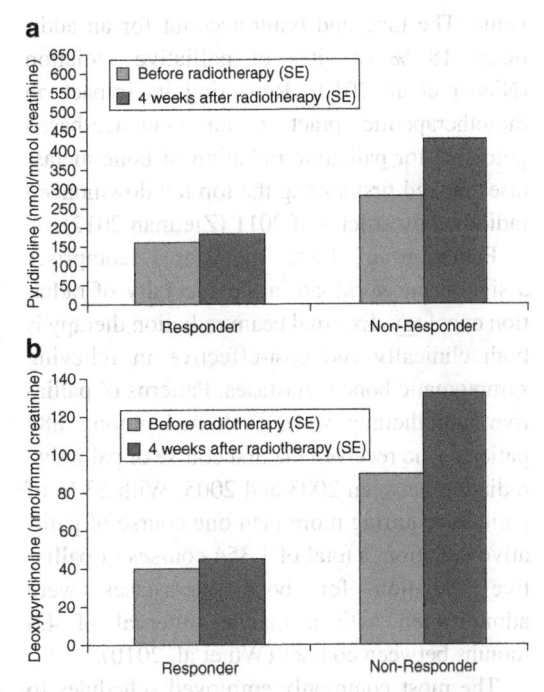

Cancer Pain Management Radiotherapy, Fig. 1 Effect of radiotherapy on urinary markers of osteoclast activity related to pain response. Total: 22 patients, 8 with breast cancer and 14 with prostate cancer; 5 patients showed no response, 9 a partial response, and 8 a complete response

risk of death on study. BAP also showed some correlation with risk of negative clinical outcomes (Coleman et al. 2005). Higher pretreatment serum beta C-terminal telopeptide was a significant predictor of shorter recurrence-free survival for bone-only metastasis (Lipton et al. 2011).

Characteristics

External Beam Local Radiotherapy

Radiation therapy has long been employed in the management of bone metastases, including relief of bone pain, and prevention of impending fractures. Stabilization of bony destruction occurs in 80 %, and re-ossification takes place in 50 % of patients after radiotherapy. Approximately half of all administered radiotherapy is with palliative intent. Bone metastases account for 63 % of all palliative treatments; spinal metastases constitute 26 % and pelvic/hip metastases 20 % of this value. The lung and brain account for an additional 18 % of sites of palliative radiation (Nieder et al. 2010). Reflecting its impact on radiotherapeutic practice, an evidence-based guideline for palliative radiation of bone metastases ranked first among the top ten downloaded radiotherapy articles of 2011 (Zietman 2012).

Palliation of bone metastases comprises a significant workload in the specialty of radiation oncology. External beam radiation therapy is both clinically and cost-effective in relieving symptomatic bone metastases. Patterns of palliative radiotherapy were evaluated among 887 patients who received the first course of palliative radiation between 2003 and 2005. With 33 % of patients requiring more than one course of palliative radiation, a total of 1,354 courses of palliative radiation for bone metastases were administered with a median interval of 4.0 months between courses (Wu et al. 2010).

The most commonly employed schedules to treat bone metastases include a single 8 Gy fraction, 20 Gy in 5 daily fractions, 30 Gy in 10 daily fractions, and 40 Gy in 20 daily fractions. Numerous randomized trials have been conducted on dose-fractionation schedules of palliative radiotherapy (Hartsell et al. 2005). While equally effective in relieving pain, the re-treatment rate with multiple fractions is lower (8 % vs. 20 %), but a single fraction optimizes patient and caregiver convenience (Lutz et al. 2011). Pain relief, however, is achieved in almost 60 % of re-irradiated bone metastases (Huisman et al. 2012).

Retrospective series have documented prompt improvement in pain in 80–90 % of patients with various dose fractionations, without causing significant hematological or gastrointestinal side effects. One of the first randomized studies on bone metastases was conducted by the Radiation Therapy Oncology Group (RTOG). The initial analysis of this trial (RTOG 74-02) concluded that a low-dose short-course schedule was as effective in pain relief as more aggressive high-dose protracted schedules (Tong et al. 1982). However, this study was criticized for using physician-based pain assessment. A reanalysis of the same set of data, grouping solitary and multiple bone metastases, using the endpoint of

pain relief, and taking into account analgesic intake and re-treatment, concluded that the number of radiation fractions was significantly associated with complete pain relief. This conclusion was directly contrary to the initial report (Blitzer 1985), making the choice of endpoints very important in defining the outcome of clinical trials (Chow et al. 2002). International Consensus has been reached on palliative radiotherapy endpoints in response definitions, eligibility criteria for future trials, criteria constituting re-irradiation, radiation techniques, and parameters of follow-up (Chow et al. 2012).

Several prospective randomized trials have subsequently been performed that compared the efficacy of different dose-fractionation schedules. They included the large-scale multicenter trials comparing the efficacy of a single 8 Gy treatment against multiple treatments. The UK Bone Pain Working Party found no difference in the degree and duration of pain relief in a study of 765 patients randomized to receive either a single treatment or five fractionated treatments (Bone Pain Trial Working Party 1999). The Dutch Bone Metastasis Study included 1,171 patients and found no difference in pain relief or the quality of life following a single 8 or 24 Gy in six daily radiation treatments (Steenland et al. 1999). The definitive prospective randomized trial in the United States was the Radiation Therapy Oncology Group 97-14 study of single fraction versus 30 Gy in 10 fractions. The ASTRO evidence-based guideline has concluded that no additional trials are needed to confirm the effectiveness of single-fraction radiation for bone metastases (Lutz et al. 2011). Assessed at baseline, 4, 8, and 12 weeks after randomization, there were two functional interference clusters at baseline. These two clusters were physical interference (58 % of total; normal work and walking ability) and psychosocial interference (13 % of relating to others and sleep). The functional interference persisted among nonresponders but significantly decreased at week 12 in responders (Chow et al. 2010).

One critical review on the subject of radiation dose fractionation suggested that protracted fractionated radiotherapy, given over 2–4 weeks, results in more complete and durable pain relief (Ratanatharathorn et al. 1999). This review was performed because of concerns regarding the influence of length of survival on the durability of pain relief. It was unclear whether higher radiation doses were necessary for durable pain relief in patients who survived longer. A meta-analysis showed no significant difference in complete and overall pain relief between single and multi-fraction palliative radiotherapy for bone metastases. No dose–response relationship could be detected. The meta-analysis reported that the complete response rates (absence of pain after radiotherapy) were 33.4 % and 32.3 % after single and multi-fraction radiation treatment, respectively, while the overall response rates were 62.1 % and 58.7 %, respectively. The latter became 72.7 % and 72.5 %, respectively, when the analysis was restricted to evaluated patients alone (Wu et al. 2002). Most patients will experience pain relief in the first 2–4 weeks after radiotherapy, be it single or multiple fractionations. Another meta-analysis from 1986 onward of 16 randomized trials not only found no difference in the extent of pain relief, overall rates of response, but no significant difference was also identified in pathological fracture or spinal cord compression (Chow et al. 2007).

How, then, are radiation oncologists to prescribe treatment? Despite the equivalence of single and multiple fractionations, multiple surveys on the patterns of practice of radiation oncologists show the underuse of single fractionation in daily practice (Ben-Josef et al. 1998; Chow et al. 2000; Roos 2000). Independently predictive factors of the use of single-fraction radiation were country of training, location of practice, practice type (academic vs. private practice), and clinical presentation. The primary clinical factors influencing the selection of a single fraction were prognosis, risk of spinal cord compression, and performance status (Fairchild et al. 2009).

Implementation of clinical pathways within a comprehensive cancer center did, however, influence practice patterns. With just a pathway between 2003 and 2008 specifying 1–5 fractions, with an option of 10–14 fractions for symptomatic bone metastases, there was a significantly greater selection of 1–5 fractions. In 2009, when pathway management decisions were peer-reviewed, there was greater compliance with the guidelines and the mean number of radiation fractions declined further. Academic practices were more like likely than community practices (63 % vs. 23 %; $p < 0.0001$) to deliver one to five radiation fractions, presumably due to economic pressures (Beriwal et al. 2012).

There is no doubt that in patients with short life expectancy, protracted schedules are a burden. However, in patients with a longer expected survival, such as breast and prostate cancer patients with bone metastases only, other parameters need to be taken into account. The anatomic location to be treated may also influence the radiation schedule that is used. Smaller daily doses of radiation (250–300 cGy) are generally better tolerated when larger treatment fields are needed especially over mucosal surfaces like those in the head and neck region, esophagus, abdomen, and pelvis.

Since re-treatment rates are known to be higher following a single versus multiple fractions, patients with good performance status may also wish to share the decision-making process. A survey of patients with bone metastases suggested that patients are not prepared to trade off long-term outcomes in favor of a shorter treatment course, with durability of pain relief and the avoidance of re-treatment more important than short-term convenience factors (Barton et al. 2001). Patients who have pain relief from palliative radiation also have improved quality of life with the parameters of physical functioning and daily activities (Zeng et al. 2012). However, they need to be aware of the potential physician bias of more readiness to retreat after single fraction, accounting for the difference in re-treatment rates in the trials.

Re-irradiation

As effective systemic treatment and better supportive care result in improved survival, certain subsets of patients with bone metastases have longer life expectancies than before. There are at least three scenarios where re-irradiation may be considered. Response to re-irradiation may be different for each of these scenarios: (1) no pain relief or pain progression after initial radiotherapy, (2) partial response with initial radiotherapy and the hope to achieve further pain reduction with more radiotherapy, and (3) partial or complete response with initial radiotherapy but subsequent recurrence of pain.

Available data support the re-irradiation of sites of metastatic bone pain following initial irradiation, particularly where this follows an initial period of response. There is also limited evidence that a proportion of nonresponders would respond to re-irradiation. However, there remains a small group of patients who appear to be nonresponsive to any amount of palliative radiotherapy. Although the data do support the clinical practice of re-irradiation, the preferred dose fractionation at the time of re-irradiation is unknown.

Wide-Field/Half-Body External Beam Radiation (HBI)

Patients with bone metastases can have multiple sites of disease and present with diffuse symptoms affecting several sites. ▶ Wide-field radiation/half-body external beam radiation (HBI) has been used to treat patients with multiple painful bone metastases. Single-fraction HBI has been shown to provide pain relief in 70–80 % of patients. Pain relief is apparent within 24–48 h. Toxicities include minor bone marrow suppression and gastrointestinal side effects, such as nausea and vomiting in upper-abdominal radiation. Pulmonary toxicity is minimal provided the lung dose is limited to 6 Gy (corrected dose). Fractionated HBI was investigated in a phase II study that compared a single fraction ($n = 14$) with fractionated HBI (25–30 Gy in 9–10 fractions) ($n = 15$). Pain relief was achieved in over 94 % of patients. At 1 year, 70 % in the fractionated and 15 % in the single-fraction group had pain control; re-treatment was required in 71 % and 13 % for the single and fractionated group, respectively.

Wide-field radiation has been compared to radiation localized to the bone metastases. A randomized trial of 499 patients compared local radiation alone versus local radiation plus

a single fraction of HBI. The study documented a lower incidence of new bone metastases (50 % vs. 68 %) and fewer patients requiring further local radiotherapy at 1 year after HBI (60 % vs. 76 %) (Poulter et al. 1992). Between 1999 and 2002, 653 patients were randomized to HBI or localized radiation. Pain scores and analgesic consumption were recorded at baseline and weeks 1,2,4,8, and 12. No significant difference was identified in pain scores based on radiation field area except for partial responders at week 4; this solitary difference resolved with the correlation of analgesic intake. The International Consensus definition for radiation response only showed a significant correlation between radiation field size and partial pain relief week 2 after radiation. Failing to demonstrate clinical advantage, wide-field radiation is not utilized given the potential toxicities (Chow et al. 2009).

Systemic Radiation: Radioisotopes

An alternative approach to the palliation of multiple bone metastases is the administration of a bone-seeking radioactive isotope that is taken up at sites of bone metastases with osteoblastic activity. The isotopes deliver the radiation dose through the release of beta particles; their range of radiation is only a few millimeters, thereby concentrating their dose within the bone metastases and delivering little dose to the adjacent normal bone marrow. The radiation half-life is about 50 days in the metastasis and only 14 days in the normal bone marrow. Some radioisotopes also have a gamma component, so that the uptake can be measured by conventional bone scan techniques.

Systemic radiation using radioisotopes such as strontium, samarium, and rhenium has been increasingly used to palliate bone pain and improve the quality of life in these patients. These agents are particularly useful among patients with multiple symptomatic metastases. As the radiation is localized to the metastases, radioisotopes can be used for re-irradiation when there are concerns about normal tissue tolerance to further external beam radiation. Also, improved outcomes and cost-effectiveness have been demonstrated when radioisotopes have been combined with external beam radiation. The TransCanada study, comparing external beam radiation alone or in combination with strontium-89 for bone metastases, demonstrated better pain relief and a reduced need for external beam radiation to other sites in the latter group (Porter et al. 1993). Convenience to the patient is the other key advantage because radioisotope therapy is administered as a single injection. With normal bone marrow parameters, re-treatment with radioisotope therapy is also possible because the adjacent normal tissues do not receive any radiation dose.

Recently, an alpha particle emitter, 223Ra-chloride (Alpharadin), has been approved for the treatment of bone metastases. The advantage of alpha particles involves tissue range and number of DNA hits needed to kill a cell. Alpharadin complexes with hydroxyapatite, gets incorporated into the bony matrix, and has a half-life of 11.4 days (Cheetham and Petrylak 2012).

A flare of pain, like that seen with hormonal therapy in breast cancer, may precede the relief of symptoms. It may take up to 1–2 months before there is relief of symptoms. Due to this, patients should have a life expectancy that allows the benefit of the relief of symptoms. Contraindications to bone-seeking radioisotopes include compromised bone marrow tolerance defined as a platelet count <60,000/mL, white blood cell count <2.5 × 10^3/mL, disseminated intravascular coagulation, or myelosuppressive chemotherapy within previous month. As bone-seeking radioisotopes do not penetrate to soft tissues, they are also contraindicated for treatment of soft tissue metastases or epidural extension with ▶ spinal cord compression. As with external beam radiation, an impending patho-logical fracture should be evaluated for surgical stabilization. Impending or established patho-logical fractures and cord compressions are acute emergencies to be managed by surgical or radiation oncology teams. Once the fracture or compression is stabilized, radioisotopes may be appropriate therapy for ongoing palliation of pain.

Controlling Side Effects of Treatment

Radiation treatment planning is the most critical aspect of reducing radiation side effects. Management of the acute effects of radiotherapy requires attentive medical management that prevents expected side effect. Radiation side effects are specific to the area treated. No side effects are observed when a femoral bone metastasis is treated by radiation. Careful radiation treatment planning that avoids critical structures like mucosal surfaces can prevent most side effects.

Patients should be reassured that the unavoidable side effects that they experience will resolve following the completion of radiotherapy. Skin reactions are usually minimal during radiotherapy for bone metastases and are limited only to the radiation portal. Nausea and vomiting, resulting from a radiation portal that includes the abdomen, will usually respond to antiemetic therapy. Diarrhea, resulting from abdominopelvic radiation, will also usually respond to treatment. Local irritation from mucositis of the oropharyngeal region may be relieved by soluble aspirin, analgesics, or benzydamine mouthwashes. Secondary infections, like Candida, should be treated.

The side effects of electron beam radiation are more limited because they only treat superficial structures like the ribs, skin lesions, and superficial lymph nodes. Underlying structures are spared with the selection of the proper electron beam energy. This characteristic is especially important with re-irradiation to avoid injury to critical structures like the spinal cord. The most prominent side effect of electron beam radiation is an erythematous skin reaction and possible moist desquamation. Other side effects listed above do not occur with electron beam radiation because the radiation beam does not penetrate to these structures.

No side effects, other than a possible flare of pain in the first 2 weeks after administration, are observed with systemic radioisotope therapy because all the radiation is localized to the bone. This is a significant consideration for patients who have symptoms of the disease or other treatments. Side effects from external beam radiation are also more severe when the radiation fields are large because more normal tissues are treated. Systemic radioisotopes can have significant advantage over large external beam radiation fields by reducing the risk for side effects like nausea and diarrhea.

Radiotherapy for Complications of Bone Metastases

Pathological and Impending Fractures
Pathological fractures are handled with orthopedic stabilization whenever possible. Surgery rapidly controls pain and returns the patient to mobility. As elective orthopedic stabilization has reportedly resulted in good pain relief and sustained mobility in up to 90 % of patients, early identification of patients with a high risk of fracture is especially important. Prophylactic orthopedic fixation is often advised to avoid the trauma of a pathological fracture.

The criteria often used to determine fracture risk in long bones include:
- Persistent or increasing local pain despite radiotherapy, particularly when aggravated by functional loading
- A solitary, well-defined lytic lesion greater than 2.5 cm
- A solitary, well-defined lesion circumferentially involving more than 50 % of the cortical bone
- Metastatic involvement of the proximal femur associated with a fracture of the lesser trochanter

Although radiotherapy provides pain relief and tumor control, it does not restore bone stability. Postoperative radiotherapy is usually recommended after surgical stabilization of a pathologic fracture. Patients who are without visceral metastases and who have a relatively long expected survival (e.g., >3 months) are more likely to benefit from post-op radiotherapy. As the entire bone is at risk for microscopic involvement and the procedure involved in rod placement may seed the bone at other sites, the length of the entire rod used for bone stabilization should be included in the radiation field.

When the radiation fields are more limited, instability of the rod, resulting in pain and need for reoperation, can result from recurrent osteolytic metastases outside the radiation portal.

Spinal Cord Compression and Cauda Equina Compression
Malignant spinal cord and cauda equina compression is a devastating compression of advanced malignancy. Early diagnosis is essential. Presenting symptoms include radicular pain, paresis, paralysis, paresthesia, and bowel/bladder dysfunction. Surgery, radiation, and steroids are the standard treatment options in this condition (Loblaw et al. 2003). Radiotherapy results in pain relief in over 75 % of patients.

Radiotherapy is indicated in patients without spinal instability or bone compression, when surgery is medically hazardous or technically difficult, and in patients who refuse surgery. When compared to short-course radiotherapy (1×8 Gy or 5×4 Gy), long-course radiotherapy courses resulted in better 1-year local control rates (61 % vs. 81 %), but there were similar functional outcomes and survival (Rades et al. 2011). Follow-up after radiotherapy is critical, as salvage therapies should be introduced before significant neurologic deficits occur (Loblaw et al. 2012). When surgery is not an option, salvage stereotactic body radiotherapy [SBRT] is effective with 65 % having pain relief, 93 % stable or improved disease, and a median progression-free survival of 9 months (Mahadevan et al. 2011).

The outcome of treatment depends mostly on the speed of diagnosis and neurological status at initiation of treatment. Over 70 % of patients are still ambulatory following radiation if they are ambulatory on presentation. However, for those who are paralyzed when they present for treatment, less than 30 % will regain neurologic function. Preexisting comorbidity, pretreatment ambulatory status, the presence of bone compression and spinal instability, and patient preferences should be considered in clinical decision-making. Improved motor function is associated with better performance status, favorable tumor type, and slower development of motor deficits.

Survival is associated with better performance status, no visceral metastases, involvement of one to three vertebrae, ambulatory status, and bisphosphonate administration after radiotherapy (Rades et al. 2011).

Conclusion

Many forms of radiation therapy are possible in the treatment of bone metastases. Meeting the goal of palliative care, suffering can be effectively and efficiently relieved with the use of radiation. Ongoing trials continue to refine therapeutic approaches to determine the optimal radiation schedule and modalities used. Treatment-related side effects can be minimized through the use of radiation treatment planning and anticipating and preventing known side effects.

Compared to other approaches, like systemic chemotherapy, hormonal therapy, and bisphosphonates, that administer treatment on an ongoing basis, all forms of radiation therapy are completed within a day to several weeks with durable control of symptoms. Cost-benefit analyses demonstrate the benefit of radiation over other forms of therapy for bone metastases. Furthermore, there is no evidence to support a survival benefit for the administration of systemic chemotherapy, hormonal therapy, or bisphosphonates in metastatic disease. Considering quality-of-life issues of time spent under therapy, toxicities of therapy, and socioeconomic cost, radiotherapy continues to be underutilized in the treatment of bone metastases.

References

Barton, M. B., Dawson, R., Jacob, S., et al. (2001). Palliative radiotherapy of bone metastases: An evaluation of outcome measures. *Journal of Evaluation in Clinical Practice, 7*, 47–64.

Ben-Josef, E., Shamsa, F., Williams, A., et al. (1998). Radiotherapeutic management of osseous metastases: A survey of current patterns of care. *International Journal of Radiation Oncology, Biology, Physics, 40*, 915–921.

Beriwal, S., Rajagopalan, M. S., Flickinger, J. C., Rakfal, S. M., Rodgers, E., & Heron, D. E. (2012). How effective are clinical pathways with and without online peer-review? An analysis of bone metastases pathway in a large, integrated National Cancer Institute-Designated Comprehensive Cancer Center Network. *International Journal of Radiation Oncology, Biology, Physics, 83*, 1246–1251.

Blitzer, P. (1985). Reanalysis of the RTOG study of the palliation of symptomatic osseous metastases. *Cancer, 55*, 1468–1472.

Bone Pain Trial Working Party. (1999). 8 Gy single fraction radiotherapy for the treatment of metastatic skeletal pain: Randomized comparison with multi-fraction schedule over 12 months of patient follow-up. *Radiotherapy and Oncology, 52*, 111–121.

Cheetham, P. J., & Petrylak, D. P. (2012). Alpha particles as radiopharmaceuticals in the treatment of bone metastases: Mechanisms of action of radium-223 chloride [Alpharadin] and radiation protection. *Oncology, 26*, 330–337.

Chow, E., Danjoux, C., Wong, R., et al. (2000). Palliation of bone metastases: A survey of patterns of practice among Canadian radiation oncologists. *Radiotherapy and Oncology, 56*, 305–314.

Chow, E., Wu, J. S., Hoskin, P., et al. (2002). International consensus on palliative radiotherapy endpoints for future clinical trials in bone metastases. *Radiotherapy and Oncology, 64*, 275–280.

Chow, E., Harris, K., Fan, G., Tsao, M., & Sze, W. M. (2007). Palliative radiotherapy trials for bone metastases: A systemic review. *Journal of Clinical Oncology, 25*, 1423–1436.

Chow, E., Makhani, L., Culleton, S., Makhani, N., Davis, L., Campos, S., & Sinclair, E. (2009). Would larger radiation fields lead to a faster onset of pain relief in the palliation of bone metastases? *International Journal of Radiation Oncology, Biology, Physics, 74*, 1563–1566.

Chow, E., James, J., Barsevick, A., Hartsell, W., Ratcliffe, S., Scarantino, C., Ivker, R., Roach, M., Suh, J., Petersen, I., Konski, A., Demas, W., & Bruner, D. (2010). Functional interference clusters in cancer patients with bone metastases: A secondary analysis of RTOG 9714. *International Journal of Radiation Oncology, Biology, Physics, 76*, 1507–1511.

Chow, E., Hoskin, P., Mitera, G., Zeng, L., Lutz, S., Roos, D., Hahn, C., van der Linden, Y., Hartsell, W., & Kumar, E. (2012). Update of the international consensus on palliative radiotherapy endpoints for future clinical trials in bone metastases. *International Journal of Radiation Oncology, Biology, Physics, 82*, 1730–1737.

Coleman, R. E., Major, P., Lipton, A., Brown, J. E., Lee, K.-A., Smith, M., Saad, F., Zheng, M., Jiang Hei, Y., Seaman, J., & Cook, R. (2005). Predictive value of bone resorption and formation markers in cancer patients with bone metastases receiving the bisphosphonate zoledronic acid. *Journal of Clinical Oncology, 23*, 4925–4935.

Fairchild, A., Barnes, E., Ghosh, S., Ben-Josef, E., Roos, D., Hartsell, W., Holt, T., Wu, J., Janjan, N., & Chow, E. (2009). International patterns of practice in palliative radiotherapy for painful bone metastases: Evidence-based practice? *International Journal of Radiation Oncology, Biology, Physics, 75*, 1501–1510.

Hartsell, W. F., Scott, C. B., Watkins-Bruner, D., Watkins-Bruner, D., Scarantino, C. W., Scarantino, C. W., Ivker, R. A., Roach, M., Suh, J. H., Demas, W. F., Movsas, B., Peterson, I. A., Konski, A. A., Cleeland, C. S., & Janjan, N. A. (2005). Phase III randomized trial of 8Gy in 1 fraction vs. 30 Gy in 10 fractions for palliation of painful bone metastases: Analysis of RTOG 97-14. *Journal of the National Cancer Institute, 97*, 798–804.

Hoskin, P. J., Stratford, M. R. L., Folkes, L. K., et al. (2000). Effect of local radiotherapy for bone pain on urinary markers of osteoclast activity. *The Lancet, 355*, 1428–1429.

Huisman, M., van den Bosch, M. A. A. J., Wijlemans, J. W., van Vulpen, V., van der Linden, Y. M., & Verkooijen, H. M. (2012). Effectiveness of reirradiation for painful bone metastases: A systematic review and meta-analysis. *International Journal of Radiation Oncology, Biology, Physics, 84*, 8–14.

Lipton, A., Chapman, J.-A. W., Demers, L., Shepherd, L. E., Han, L., Wilson, C. F., Pritchard, K. I., Leitzel, K. E., Ali, S. M., & Pollak, M. (2011). Elevated bone turnover predicts for bone metastasis in postmenopausal breast cancer: Results of NCIC CTG MA.14. *Journal of Clinical Oncology, 29*, 3605–3611.

Loblaw, D. A., Laperriere, N. J., Chambers, A. et al., & Group at CCOPGIN-ODS. (2003). Diagnosis and management of malignant epidural spinal cord compression (Evidence Summary Report No. 9-9). Cancer Care Ontario. http://www.ccopebc.ca./neucpg.html. Accessed 4 Apr 2003.

Loblaw, D. A., Mitera, G., Ford, M., & Laperriere, N. J. (2012). A 2011 updated systematic review and clinical practice guideline for the management of malignant extradural spinal cord compression. *International Journal of Radiation Oncology, Biology, Physics, 84*(2), 312–317.

Lutz, S., Berk, L., Chang, E., Chow, E., Hahn, C., Hoskin, P., Howell, D., Konski, A., Kachnic, L., Lo, S., Sahgal, A., Silverman, L., von Gunten, C., Mendel, E., Vassil, A., Bruner, D. W., & Hartsell, W. (2011). Palliative radiotherapy for bone metastases: An ASTRO evidence-based guideline. *International Journal of Radiation Oncology, Biology, Physics, 79*, 965–976.

Mahadevan, A., Floyd, S., Wong, E., Jeyapalan, S., Groff, M., & Kasper, E. (2011). Stereotactic body radiotherapy reirradiation for recurrent epidural spinal metastases. *International Journal of Radiation Oncology, Biology, Physics, 81*, 1500–1505.

Nieder, C., Pawinski, A., Haukland, E., Dokmo, R., Phillipi, I., & Dalhaug, A. (2010). Estimating need for palliative external beam radiotherapy in adult cancer patients. *International Journal of Radiation Oncology, Biology, Physics, 76*, 207–211.

Porter, A., McEwan, A., & Powe, J. (1993). Results of a randomized phase III trial to evaluate the efficacy of strontium-89 adjuvant to local field external beam irradiation in the management of endocrine resistant metastatic prostate cancer. *International Journal of Radiation Oncology, Biology, Physics, 25*, 805–813.

Poulter, C., Cosmatos, D., Rubin, P., et al. (1992). A report of RTOG 8206: A phase III study of whether the addition of single dose hemibody irradiation to standard fractionated local field irradiation is more effective than local field irradiation alone in the treatment of symptomatic osseous metastases. *International Journal of Radiation Oncology, Biology, Physics, 23*, 207–214.

Rades, D., Lange, M., Veninga, T., Stalpers, L. J. A., Bajrovic, A., Adamietz, I. A., Rudat, V., & Schild, S. E. (2011). Final results of a prospective study comparing the local control of short-course and long-course radiotherapy for metastatic spinal cord compression. *International Journal of Radiation Oncology, Biology, Physics, 79*, 524–530.

Ratanatharathorn, V., Powers, W., Moss, W., et al. (1999). Bone metastasis: Review and critical analysis of random allocation trials of local field treatment. *International Journal of Radiation Oncology, Biology, Physics, 44*, 1–18.

Roos, D. (2000). Continuing reluctance to use single fractions of radiotherapy for metastatic bone pain: An Australian and New Zealand practice survey and literature review. *Radiotherapy and Oncology, 56*, 315–322.

Steenland, E., Leer, J., van Houwelingen, H., et al. (1999). The effect of a single fraction compared to multiple fractions on painful bone metastases: A global analysis of the Dutch bone metastasis study. *Radiotherapy and Oncology, 52*, 101–109.

Tong, D., Gillick, L., & Hendrickson, F. (1982). The palliation of symptomatic osseous metastases: Final results of the Study By The Radiation Therapy Oncology Group. *Cancer, 50*, 893–899.

Wu, J., Wong, R., Johnston, M., Bezjak, A., & Whelan, T. (2002). Meta-analysis of dose-fractionation radiotherapy trials for the palliation of painful bone metastases. *International Journal of Radiation Oncology, Biology, Physics, 55*, 594–605.

Wu, J. S. Y., Kerba, M., Wong, R. K. S., Mckimmon, E., Eigl, B., & Hagen, N. A. (2010). Patterns of practice in palliative radiotherapy for painful bone metastases: Impact of a regional rapid access clinic on access to care. *International Journal of Radiation Oncology, Biology, Physics, 78*, 533–538.

Zeng, L., Chow, E., Bedard, G., Zhang, L., Fairchild, A., Vassiliou, V., Alm El-Din, M. A., Jesus-Garcia, R., Kumar, A., Forges, F., Tseng, L.-M., Hou, M.-F., Chie, W.-C., & Bottomley, A. (2012). Quality of life after palliative radiation therapy for patients with painful bone metastases: Results of an international study validating the EORTC QLC-BM22. *International Journal of Radiation Oncology, Biology, Physics*, 84(3), e337–e342.

Zietman, A. (2012). The Red Journal's top 10 most downloaded articles of 2011. *International Journal of Radiation Oncology, Biology, Physics*, 83, 1073–1074.

Cancer Pain Management, Adjuvant Analgesics in Management of Pain due to Bowel Obstruction

Paul C. Rousseau and Leigh M. Vaughan
Medical University of South Carolina, Charleston, SC, USA

Definition

Non-opioid analgesics utilized in the management of pain and colic in malignant bowel obstruction. Three types of medication are commonly used: (1) anticholinergics; (2) corticosteroids; and (3) the synthetic ▶ somatostatin analogue, octreotide.

Characteristics

The pharmacological management of pain associated with ▶ malignant bowel obstruction addresses three critical pathophysiological mechanisms: (1) intestinal distention and resultant bowel ischemia; (2) paroxysmal intestinal hypermotility; and (3) increased intestinal secretions that perpetuate a vicious cycle of distention-intestinal secretion-peristalsis (Basson et al. 1989; Rousseau 1998).

Opioids are the mainstay of pain management in malignant bowel obstruction, and frequently assuage the pain associated with intestinal distention and bowel ischemia. The pain that occurs secondary to intestinal hypermotility is often referred to as colic, and is precipitated by paroxysms of increased peristalsis that thrust against the mechanical resistance of a malignant obstruction (Baines 1998). Colicky pain often

requires the use of anticholinergic medications in addition to opioids to mitigate the excruciating discomfort of intestinal hypermotility (Rousseau 1998; Baines 1998; Krouse et al. 2002; Lublin and Schwartzentruber 2002; Ripamonti et al. 2000; Ripamonti et al. 2001).

Anticholinergic medications decrease peristalsis of the gastrointestinal tract through a competitive inhibition of the muscarinic receptors of intestinal smooth muscle, and an impairment of ganglionic neural transmission in the bowel wall (Mercadante 1997; Baines 1994). Although there are few well-controlled studies evaluating their efficacy in reducing abdominal colic in malignant bowel obstruction, anecdotal experience supports their ability to relieve malignant colic, and they are commonly used in the management of bowel obstruction. Recommended medications include glycopyrrolate, hyoscine butylbromide, scopolamine, hyoscyamine sulfate, and atropine. Scopolamine is a tertiary amine that crosses the blood-brain-barrier and may be associated with systemic side effects such as sedation or confusion. The transdermal patch is a commonly used route of administration for scopolamine and is ideal for patients who cannot tolerate oral intake, but it has a prolonged time for peak effect and adjuvant anticholinergics may be needed in the initial 24 hours. Glycopyrrolate, a quaternary ammonium anticholinergic, and hyoscine butylbromide (not available in the United States) both exhibit limited lipid solubility (Rousseau 1998; Krouse et al. 2002; Davis and Furste 1999). They penetrate the blood–brain barrier less than other anticholinergics, and as a result, reduce central nervous system side effects such as agitation and hallucinations (Rousseau 1998; Baines 1998; Mercadante 1997). Aside from agitation and hallucinations, anticholinergic drugs can cause dry mouth and eyes, drowsiness, tachycardia, hypotension, and urinary retention; dry mouth can be reduced by sucking ice cubes and drinking small sips of water, while dry eyes can be alleviated with physiologic eye drops (Table 1).

Corticosteroids have been used empirically in malignant bowel obstruction to reduce peritumor and perineural inflammatory edema (Rousseau 1998; Baines 1998; Krouse et al. 2002;

Cancer Pain Management, Adjuvant Analgesics in Management of Pain due to Bowel Obstruction, Table 1 ▸ Adjuvant Analgesics in malignant bowel obstruction[a]

Drug	Dose
Glycopyrrolate[b]	1–2 mg tid-qid OR
	0.1–0.2 mg tid-qid SC/IV
Hyoscine butylbromide	40–120 mg daily SC/IV
Scopolamine	400–800 mcg OR every 4–6 h
	1–2 TDP every 48–72 h
	0.8–2 mg via CSI every 24 h
Atropine	1–2 drops ophthalmic solution SL every 4 h
	0.4 mg OR every 4 h
	0.4 mg SC every 4 h
Dexamethasone	4–8 mg daily OR
	2–8 mg every 6 h SC/IV
	2–12 mg via CSI every 24 h
Prednisone	5–20 mg every 4–6 h
Octreotide	150 mcg tid SC
	0.2–0.9 mg via CSI every 24 h

[a]List is not inclusive, dosages are suggested initial doses that may be prescribed as scheduled and/or prn doses, may need to be titrated to effect dependent upon patient's clinical condition and expected life expectancy
[b]Reduce dose in renal insufficiency
OR oral; *SC* subcutaneous; *IV* intravenous; *TDP* transdermal patch; *SL* sublingual; *CSI* continuous subcutaneous infusion; *prn* as needed

Mercadante 1997), and thereby improve gastrointestinal transport and colicky pain (Ripamonti 1994). They act both as antiemetics and analgesics, the latter ostensibly due to prostaglandin inhibition, suppression of migration of polymorphonuclear leukocytes, and reversal of increased capillary permeability. In an inclusive meta-analysis of published articles and studies on the use of corticosteroids in malignant bowel obstruction, secondary to advanced gynecological and gastrointestinal cancers, Feur and Broadley noted a trend for improvement in bowel obstruction for patients treated with corticosteroids, but this trend did not reach statistical significance. Although corticosteroids did appear to improve the symptoms of bowel obstruction, Feur and Broadley did not explicitly mention their ability to relieve pain (Feur and Broadley 1999). In another study on the benefits of

corticosteroids in malignant bowel obstruction, 68 % of patients without a nasogastric tube prescribed corticosteroids, versus 33 % without a nasogastric tube prescribed a placebo, achieved symptom relief ($p = 0.047$); however, in patients who already had a nasogastric tube in place, the results were less significant (60 % vs. 33 %, $p = 0.080$). Unfortunately, the sample size of this study was small, and while the authors mentioned colic as a symptom in 62 % of study participants, they did not specify if colic was relieved with corticosteroids (Laval et al. 2000). In a systematic review in 2007 included the review of Laval et al., and an addition study with 39 patients, randomized to corticosteroids or placebo (Hardy 1998), but because of the crossover design and flawed methodology, the authors concluded that corticosteroids still lack sufficient efficacy data (Mercadante 2007). Side effects of corticosteroids are minimal when used in malignant bowel obstruction, with oral candidiasis the most commonly reported adverse event (Mercadante 1997).

Octreotide, a synthetic analogue of somatostatin, reduces the gastrointestinal secretion of gastrin, vasoactive intestinal peptide (VIP), insulin, pepsin, glucagon, and pancreatic bicarbonate and enzymes; reduces mesenteric blood flow (Mercadante 1997; Pandha and Waxman 1996); exhibits a proabsorptive effect on water and ions (Rousseau 1998); and inhibits intestinal motility (Rousseau 1998; Baines 1998; Krouse et al. 2002; Pandha and Waxman 1996). By reducing intestinal secretions and inhibiting gastrointestinal motility, octreotide positively modulates gastrointestinal function and controls emesis in 70–100 % of patients (Baines 1998; Krouse et al. 2002). Abdominal distention and colic are similarly controlled in a significant number of patients (Baines 1998). These statistics are supported by a prospective randomized controlled trial comparing the antisecretory effects of octreotide and hyoscine butylbromide in patients with inoperable malignant bowel obstruction who have nasogastric tubes. In this study, octreotide was very effective in reducing the amount of gastrointestinal secretions in patients with nasogastric tubes, and as an implicit consequence, though not specifically stated, reducing colicky pain (Ripamonti et al. 2000). Octreotide has been shown to be superior to anticholinergics alone for the pharmacologic management of inoperable malignant bowel obstruction (Mercadante 2007). The toxicity and side-effect profile of octreotide is excellent (aside from discomfort at the site of injection), with no major side effects reported, even at a dose of 1.2 mg daily (Riley and Fallon 1994).

Mercadante et al. in a prospective study in 2004 treated 15 patients who were admitted to a palliative care unit for symptomatic management of an inoperable malignant bowel obstruction using a combination of propulsive and antisecretive agents. The regimen included metoclopramide 60 mg/day, octreotide 0.3 mg/day, and dexamethasone 12 mg daily, followed by an osmotic agent. Intestinal transit was restored in 14 of 15 patients within 1–5 days and symptom control was prolonged (Mercadante 2004). In a 20 year review of Octreotide, with 15 trials of various designs (randomized controlled, observational and prospective) and 281 patients, the authors concluded that octreotide has a success rate of 60–90 %, and based on octreotide's proven efficacy and low side effect profile, it is recommended as first line therapy for symptomatic management of malignant bowel obstruction (Mercadante 2012).

References

Baines, M. (1994). Intestinal obstruction. In G. W. Hanks (Ed.), *Palliative medicine: Problem areas in pain and symptom management* (pp. 147–156). Plainview, NY: Cold Spring Harbor Laboratory Press.

Baines, M. J. (1998). The pathophysiology and management of malignant intestinal obstruction. In D. Doyle, G. W. C. Hanks, & N. MacDonald (Eds.), *Oxford textbook of palliative medicine* (2nd ed., pp. 526–534). Oxford: Oxford University Press.

Basson, M. D., et al. (1989). Does vasoactive intestinal polypeptide mediate the pathophysiology of bowel obstruction? *American Journal of Surgery, 157*, 109–115.

Davis, M. P., & Furste, A. (1999). Glycopyrrolate: A useful drug in the palliation of mechanical bowel obstruction. *Journal of Pain and Symptom Management, 18*, 153–154.

Feur, D. J., & Broadley, K. E. (1999). Systematic review and met-analysis of corticosteroids for the resolution of malignant bowel obstruction in advanced gynaecological and gastrointestinal cancers. *Annals of Oncology, 10*, 1035–1041.

Hardy, J. (1998). Pitfalls in placebo-controlled trials in palliative care: dexamethasone for the palliation of malignant bowel obstruction. *Palliative Medicine, 12*, 437–442.

Krouse, R. S., et al. (2002). When the sun can set on an unoperated bowel obstruction: Management of malignant bowel obstruction. *Journal of the American College of Surgery, 195*, 117–128.

Laval, G., et al. (2000). The use of steroids in the management of inoperable intestinal obstruction in terminal cancer patients: Do they remove the obstruction? *Palliative Medicine, 14*, 3–10.

Lublin, M., & Schwartzentruber, D. J. (2002). Bowel obstruction. In A. Berger, R. K. Portenoy, & D. E. Weissman (Eds.), *Principles and practice of palliative care and supportive oncology* (pp. 250–263). Philadelphia: Lippincott Williams & Wilkins.

Mercadante, S. (1997). Assessment and management of mechanical bowel obstruction. In R. K. Portenoy & E. Bruera (Eds.), *Topics in palliative care* (Vol. 1, pp. 113–130). New York: Oxford University Press.

Mercadante, S. (2004). Aggressive pharmacological treatment for reversing malignant bowel obstruction. *Journal of Pain and Symptom Management, 28*, 412–416.

Mercadante, S. (2007). Medical treatment for inoperable malignant bowel obstruction: A qualitative systematic review. *Journal of Pain and Symptom Management, 33*, 217–223.

Mercadante, S. (2012). Octreotide for malignant bowel obstruction: Twenty years after. *Critical Reviews in Oncology/Hematology, 83*(3), 388–392.

Pandha, H. S., & Waxman, J. (1996). Octreotide in malignant intestinal obstruction. *Anti-Cancer drugs, 7*, 5–10.

Riley, J., & Fallon, M. T. (1994). Octreotide in terminal malignant obstruction of the gastrointestinal tract. *European Journal of Palliative Care, 1*, 23–25.

Ripamonti, C. (1994). Management of bowel obstruction in advanced cancer patients. *Journal of Pain and Symptom Management, 9*, 193–200.

Ripamonti, C., et al. (2000). Role of octreotide, scopolamine butylbromide, and hydration in symptom control of patients with inoperable bowel obstruction and nasogastric tubes: A prospective randomized trial. *Journal of Pain and Symptom Management, 19*, 23–34.

Ripamonti, C., et al. (2001). Clinical-practice recommendations for the management of bowel obstruction in patients with end-stage cancer. *Supportive Care in Cancer, 9*, 223–233.

Rousseau, P. C. (1998). Management of malignant bowel obstruction in advanced cancer: A brief review. *Journal of Palliative Medicine, 1*, 65–72.

Cancer Pain Management, Anesthesiologic Interventions

Paul A. Sloan
Department of Anesthesiology, University of
Kentucky, Lexington, KY, USA

Synonyms

Anesthetic block; Nerve block

Definition

Information about noxious events is transmitted to the brain via neurons of the peripheral and central nervous system. Blockade of this nerve conduction using applied local anesthetics to temporarily block pain transmission, spinal medications (e.g., opioids) to modulate pain transmission, or neurolytic substances to destroy the pain-transmitting nerves is the domain of anesthesiologic interventions for cancer pain management.

Characteristics

Most patients with cancer pain can have their pain well managed using standard opioids and other analgesics via the oral or parenteral routes. However, approximately 5–15 % of cancer patients with pain will experience severe pain resistant to traditional analgesic therapies (Sloan and Melzack 1999). For these unfortunate patients, anesthesiologic techniques often provide welcome pain relief. The use of alcohol to destroy nerve tissue in the treatment of painful conditions is almost 80 years old, and, along with phenol, became popular in the 1950s and 1960s for the treatment of cancer pain (Maher 1955). With more recent acceptance and use of chronic opioid therapy for cancer pain, the current role of neuroablative techniques has chiefly been for patients with terminal disease and pain unresponsive to traditional analgesics (Sloan 2007). The use of local anesthetics and spinal opioids to modulate pain transmission has gained popularity in the past decade, because of the low risk involved and the ability to reverse the treatment course at any time.

The use of local anesthetics to provide long-term pain relief for cancer patients has principally involved infiltration around peripheral nerves and ▶ epidural or ▶ intrathecal administration. Local anesthetics injected close to a peripheral nerve or the spinal cord will block nerve impulses and provide degrees of nociceptive, sensory, or motor nerve blockade. Anesthesiologists have searched for the correct combination of local anesthetic concentration, infusion site, and rate of infusion to achieve analgesia without the significant side effects of motor or sensory blockade. Two common sites of peripheral nerve block with local anesthetic infusion are the brachial plexus (Fig. 1) and lumbar plexus. Tumors originating in lung, head and neck, or other sites may invade the brachial plexus, resulting in severe neuropathic pain, and pelvic tumors may similarly invade the lumbar plexus. The percutaneous insertion of catheters for the infusion of local anesthetics has provided excellent and long-term pain relief among such cancer patients (Buchanan et al. 2009; Douglas and Bush 1999). A basal infusion may be physician prescribed and augmented by the use of patient-controlled boluses for breakthrough pain. Portable infusion pumps (small enough to be worn on the body) allow the patient to return home for as long as possible. Typical local anesthetics used are bupivacaine (1–5 mg/ml), lidocaine (5–10 mg/ml), or ropivacaine (1–5 mg/ml), with infusion rates of 3–8 ml/h.

The spinal cord is bathed by cerebrospinal fluid (CSF) that is contained in a strong protective membrane, the dura mater. Local anesthetics applied outside the dura mater (epidural) or within the CSF (intrathecal) have been used to control cancer pain (Fig. 2). A catheter is usually placed percutaneously into the epidural or intrathecal space for the chronic infusion of local anesthetics. The catheter may be tunneled under the skin and exit to an external portable pump, or remain buried and attached to an implanted pump. These procedures are usually performed using local anesthesia and intravenous sedatives. The pumps allow for the constant infusion of local anesthetics such as bupivacaine (Dahm et al. 2000) and are often combined with spinal opioids to provide a synergistic analgesic effect (Sloan 2004). Cancer patients with lower

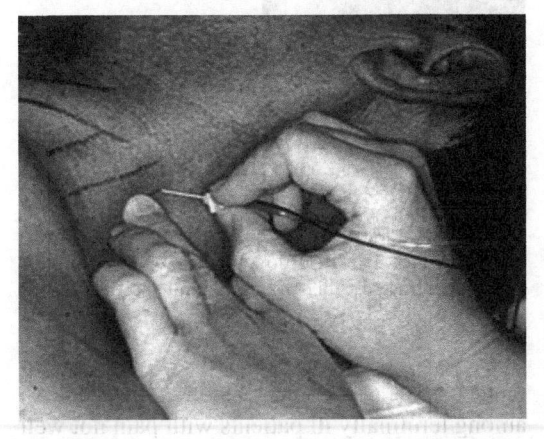

Cancer Pain Management, Anesthesiologic Interventions, Fig. 1 Interscalene approach to brachial plexus block

extremity or truncal pain to the waist are best suited to these approaches; however, upper thoracic and cervical placement of spinal catheters is sometimes used. Possible complications include local anesthetic toxicity (uncommon), hypotension, tachyphylaxis, infection, and motor weakness. Despite the side effect potential, most patients tolerate chronic infusion of local anesthetics without difficulty and can be monitored at home. In addition to spinal administration, local anesthetics have been infused chronically into the pleural space via a percutaneous catheter for the relief of refractory lung cancer pain (Aguilar et al. 1992).

Spinal administration of opioids for pain relief was first applied to patients in 1979 (Behar et al. 1979) and has been advocated for the relief of intractable cancer pain. Spinal opioids can be given by both the epidural and intrathecal routes of administration, with the epidural dose being approximately 10 times the intrathecal dose. It is believed that spinally applied opioids modulate pain transmission by direct action on specific opioid receptors in the dorsal horn of the spinal cord (Bedder 1997). In this

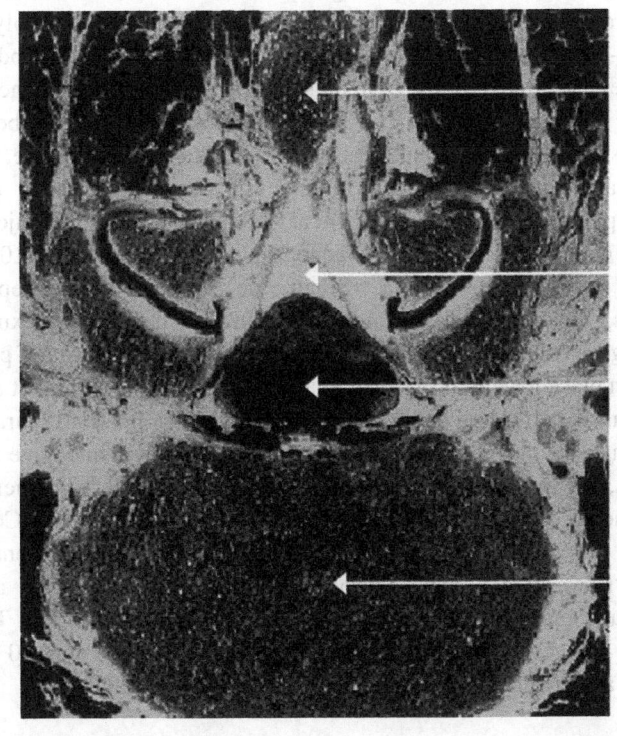

Spinous process

Epidural space

Intrathecal space

Vertebral body

Cancer Pain Management, Anesthesiologic Interventions, Fig. 2 Cross section through a lumbar vertebra

way, there may be fewer and/or less intense systemic opioid side effects, and a better quality of analgesia may be obtained. Many opioids have been used to achieve successful analgesia, with morphine remaining the most common. A single injection of spinal morphine will provide 12–24 h of pain relief. However, a continuous infusion of opioid is often used and can be managed safely at home (Deer et al. 2011). The effective dose is titrated to analgesia or intolerable side effects. A useful starting dose for epidural morphine is 10 % of the current daily oral morphine dose, and the final daily epidural doses reported in the literature have ranged from 2 to over 1,000 mg.

The combination of opioids and local anesthetics applied to the spinal cord in animals has resulted in a synergistic effect (Akerman et al. 1988) and led to the clinical practice of combining spinal analgesics to achieve powerful analgesia with the fewest possible side effects. The coadministration of these drugs has gained widespread acceptance and has been shown to limit morphine dose progression during long-term spinal infusion (van Dongen et al. 1999). Other spinal drugs that have provided some analgesia include the neuronal calcium channel blocker ziconotide, alpha-adrenergic agonists (clonidine, dexmedetomidine), and the GABA A agonist, midazolam.

Complications (10 % of all patients) associated with spinal analgesics include local infection, CSF leak, pain on injection, mechanical problems with catheter function, headache, meningitis, and epidural hematoma (Sloan 2007) or abscess (Fig. 3). Local infection can usually be managed with antibiotics, and catheters can remain in place for months (Aprili et al. 2009). The typical side effects of opioids may be seen, including nausea and vomiting, sedation, urinary retention, and pruritus. Patient accommodation to these side effects of opioids typically occurs.

The historic "nerve block" using neurolytic agents to physically modify or "deaden" nerve tissue to produce analgesia has now become the least prevalent anesthetic technique for the

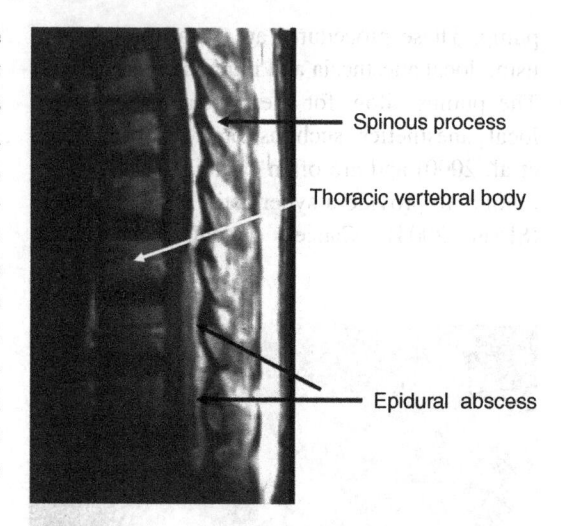

Spinous process

Thoracic vertebral body

Epidural abscess

Cancer Pain Management, Anesthesiologic Interventions, Fig. 3 MRI of low thoracic epidural abscess in a patient treated with chronic epidural analgesics

management of cancer pain. Nonetheless, among terminally ill patients with pain not well controlled by other measures, certain neurolytic injections can provide excellent pain relief with a minimum of complications. The destruction of the celiac plexus in the upper abdomen is the most common neurolytic block performed (Sloan 2006). Upper abdominal pain associated with cancer of the pancreas, distal esophagus, stomach, or liver may be treated using a simple percutaneous technique. Imaging studies (x-ray, CT scan, ultrasound) guide accurate needle placement, and the injection of 25–50 ml of 50 % alcohol or 6–10 % phenol results in destruction of nociceptive neurons passing through the celiac plexus, thus producing pain relief in the majority of patients that may last for many months (Wong et al. 2004). While significant complications (paraplegia, pneumothorax) are infrequent and have been reduced by radiographic needle placement, they have not been entirely eliminated. Common complications include postural hypotension and diarrhea; however, patients usually accommodate to these symptoms within 48 h. The celiac plexus block can be repeated at 6–10 weeks to provide additional pain relief.

Additional neurolytic blocks occasionally used to treat cancer pain include lumbar sympathetic block for urologic pain, superior hypogastric block for pelvic pain, subarachnoid block for chest wall pain, subarachnoid saddle block for perineal pain, and cranial nerve block for head or neck pain. Noninvasive ► Gamma Knife destruction of the pituitary has also been tried for intractable bilateral cancer pain (Sloan et al. 1996).

Other innovations may prove valuable in the future. Ziconotide, a neuronal-specific calcium channel blocker, has become available and is recently recommended as a first line intrathecal drug for the management of cancer pain (Deer et al. 2011). Frequent side effects require a slow and careful dose titration to analgesia. Some evidence exists for the analgesic effect of intrathecal ketorolac, as well as the benzodiazepine, midazolam. Ultra-low doses of opioid antagonists have been proposed to enhance opioid analgesia in one case report (Hamann et al. 2008) Additional prospective studies for all these agents are required to determine efficacy and side effect profiles.

Anesthetic interventions for pain management are well established, and new therapies are evolving. The appropriate use of anesthesiologic interventions to help manage acute and chronic pain among cancer patients will ensure pain relief and improved quality of life for our patients.

References

Aguilar, J. L., Montes, A., Samper, D., et al. (1992). Interpleural analgesia through a dupen catheter for lung cancer pain. *Cancer, 70*, 2621–2623.

Akerman, B., Arwestrom, E., & Post, C. (1988). Local anesthetics potentiate spinal morphine antinociception. *Anesthesia and Analgesia, 67*, 943–948.

Aprili, D., Bandschapp, O., Rochlitz, C., et al. (2009). Serious complications associated with external intrathecal catheters used in cancer pain patients: A systematic review and meta-analysis. *Anesthesiology, 111*, 1346–1355.

Bedder, M. D. (1997). Neuraxial analgesic blockade: Physiology, pharmacology, and complications. In W. C. V. Parris, H. W. Foster, & R. Melzack (Eds.),

Cancer pain management: Principles and practice (pp. 171–180). Newton: Butterworth-Heinemann.

Behar, M., Magora, F., Olshwang, D., et al. (1979). Epidural morphine in treatment of pain. *Lancet, 1*, 527–529.

Buchanan, D., Brown, E., Millar, F., et al. (2009). Outpatient continuous interscalene brachial plexus block in cancer-related pain. *Journal of Pain and Symptom Management, 38*, 629–634.

Dahm, P., Lundborg, C., Janson, M., et al. (2000). Comparison of 0.5 % intrathecal bupivacaine with 0.5 % intrathecal ropivacaine in the treatment of refractory cancer and noncancer pain conditions: Results from a prospective, crossover, double-blind, randomized study. *Regional Anesthesia and Pain Medicine, 25*, 480–487.

Deer, T. R., Smith, H. S., Burton, A. W., et al. (2011). Comprehensive consensus based guidelines on intrathecal drug delivery systems in the treatment of pain caused by cancer pain. *Pain Physician, 14*, E283–E312.

Douglas, I., & Bush, D. (1999). The use of patient-controlled boluses of local anaesthetic via a psoas sheath catheter in the management of malignant pain. *Pain, 82*, 105–107.

Hamann, S. R., Sloan, P. A., & Witt, W. O. (2008). Low-dose intrathecal naloxone to enhance intrathecal morphine analgesia: A case report. *Journal of opioid management, 4*, 251–254.

Maher, R. M. (1955). Relief of pain in incurable cancer. *Lancet, 1*, 18.

Sloan, P. A. (2004). The evolving role of interventional pain management in oncology. *Journal of Supportive Oncology, 2*, 491–506.

Sloan, P. A. (2006). The role of gastrointestinal endoscopic ultrasound-guided celiac plexus neurolytic block for pancreatic cancer pain. *Journal of Supportive Oncology, 4*, 463–464.

Sloan, P. A. (2007). Neuraxial pain relief for intractable cancer pain. *Current Pain and Headache Reports, 11*, 283–289.

Sloan, P. A., & Melzack, R. (1999). Long-term patterns of morphine dosage and pain intensity among cancer patients. *The Hospice Journal, 14*, 35–47.

Sloan, P. A., Hodes, J., & John, W. (1996). Radiosurgical pituitary ablation for cancer pain. *Journal of Palliative Care, 12*, 51–53.

van Dongen, R. T., Crul, B. J., & van Egmond, J. (1999). Intrathecal coadministration of bupivacaine diminishes morphine dose progression during long-term intrathecal infusion in cancer patients. *The Clinical Journal of Pain, 15*, 166–172.

Wong, G. Y., Schroeder, D. R., Carns, P. E., et al. (2004). Effect of neurolytic celiac plexus block on pain relief, quality of life, and survival in patients with unresectable pancreatic cancer: A randomized controlled trial. *Journal of the American Medical Association, 291*, 1092–1099.

Cancer Pain Management, Anesthesiologic Interventions, and Neural Blockade

Harald Breivik[1] and Gunnvald Kvarstein[2]
[1]Department of Pain Management and Research, Oslo University Hospital, Rikshospitalet and University of Oslo, Oslo, Norway
[2]Department of Pain Management and Pain Research, Oslo University Hospital, Rikshospitalet, Oslo, Norway

Synonyms

Anesthesiological interventions; Neural blockade

Definition

Anesthesiologic interventions are transient or long-lasting interruptions of pain impulse conduction in peripheral nerves, nerve roots, or the spinal cord. This will hinder the conscious perception of pain and lessen the suffering associated with cancer and its destruction of normal tissues.

Introduction

When opioid analgesics and co-analgesic drugs became better known and more widely used from the late 1980s onward, invasive anesthesiological methods for "intractable" cancer pain (Swerdlow 1983) were not needed as much as before this paradigm shift. However, analgesic drugs do not always relieve pain well enough and often have adverse effects such as severe nausea, dizziness, and cognitive dysfunctions. Increased burden of suffering results for patients with advanced cancer disease. Therefore, invasive anesthesiological methods are of interest again in palliative medicine (Breivik 2000; Burton et al. 2009). The present entry is an overview of neurological blockades for cancer pain. Detailed descriptions

of "how to do it" are available in comprehensive textbooks for practitioners of neural blocks (Breivik 2000; Campbell 2008; Burton et al. 2009; Cousins et al. 2009). Radiofrequency lesioning may be alternative methods and we refer the readers to chapters covering this method and to Crul et al. (2008). Radiographic documentation of needle placement, and the use of ultra sound equipment for localizing nerve structures are explained in other chapters. These methods may add precision and extra safety to invasive procedures for cancer pain treatment. They do require more resources and training to be helpful.

Characteristics

Anesthesiologic interventions may be performed with ▶ local anesthetic drugs, which stop pain impulse conduction along nerves for up to several hours, or with ▶ neurolytic drugs, which block transmission for up to several months. These interventions will remove pain from the locally anesthetized part of the body, which may become numb and partly paralyzed. Interruptions of pain impulses conducted through visceral nerves, e.g., the celiac plexus of the upper abdomen, do not cause numbness or skeletal muscle weakness. Severe forms of cancer pain can be relieved effectively by skillfully applied anesthesiologic interventions (Breivik 2000; Burton et al. 2009). Risks of adverse effects exist and must be understood, prevented, and minimized when considering these approaches (Campbell 2008; Rathmell 2009).

Reversible Blockade of Pain Impulses by Local Anesthetic Drugs

Infiltration of a painful primary or metastatic tumor, or the peripheral nerves innervating a painful cancerous growth, with a local anesthetic drug results in local or regional anesthesia and pain relief lasting from 1 to 12 h. The duration can be prolonged with repeated injections or infusion into ▶ perineural catheters. Local anesthetic infiltrations and ▶ simple nerve blocks should be used liberally (Twycross 1994). Relief of acute, overwhelmingly intense

▶ breakthrough pain from a pathological fracture or an acute tumor infarct can be relieved effectively with a regional anesthetic block.

Local anesthetic nerve blocks are useful for localizing nerves to be exposed to neurolysis, but they should not be used to predict the effect of neurolytic blocks or other neurodestructive treatment, such as ▶ thermocoagulation, radiofrequency lesioning, or ▶ surgical denervation (Hogan and Abrams 2009). Neurolytic interventions are less complete and generally cause less intense pain relief than local anaesthetic blocks.

Spinal Cord Analgesia by Epidural or Subarachnoid Catheter Infusion of Drugs

Epidural Analgesia Potent pain relief can occur when low concentrations of a local anesthetic drug, an opioid, and an adrenergic drug are administered epidurally (Breivik 2008a). With an epidural catheter at the appropriate ▶ spinal cord segmental level (cervical to the upper lumbar), a reliable infusion pump, and trained nurses, analgesia can be maintained for prolonged periods at home, in a hospice, or in a palliative care institution. Stable and sterile drug mixtures are needed (Breivik 2012). Details of epidural catheter placement and maintenance, monitoring, prevention and management of complications are described in textbooks (Breivik 2008a; 2012).

Prolonged Infusion into the Spinal Subarachnoid Space

Insertion of a catheter into the subarachnoid space allows infusion of a local anesthetic solution containing an opioid and clonidine (Whyte and Lauder 2011) or adrenaline (Breivik 2012) directly into the ▶ cerebrospinal fluid. This may give satisfactory analgesia when most other methods have failed (Nitescu et al. 2002). Prolonged, effective, and safe pain relief is possible, even when the patient is living at home (Nitescu et al. 2002).

Inflammatory Pain Treated with Injections of Local Anesthetic and Corticosteroid Drugs

Corticosteroids have analgesic effects when administered epidurally, perineurally, or intralesionally. These are in part systemic effects, in part local effects on nociceptive nerve endings, and in part effects on inflammatory or neuroplastic pain mechanisms (Twycross 1994; Breivik 2000; Kidd 2007).

Segmental epidural applications of local anesthetic drugs combined with methylprednisolone, dexamethasone, or triamcinolone may relieve ▶ radicular nerve root pain (Breivik 2012; Ramos-Galvez and Goodall 2008), including that caused by a tumor impinging on spinal nerve roots (Twycross 1994). Perineural injections of a mixture of a local anesthetic and a corticosteroid may cause immediate and prolonged pain relief in nerves made hyperexcitable from local trauma, infiltration of tumor, or inflammatory reactions (Twycross 1994).

Intralesional injection of local anesthetics and depot corticosteroids similarly may cause immediate and, in some patients, prolonged pain relief, e.g., in multiple myeloma lesions or metastases from breast cancer in ribs (Twycross 1994).

Neurolytic Nerve Blocks with Ethanol or Phenol

The advantage of neurolytic blocks is apparent when a single intervention can replace or reduce the need for daily administration of multiple drugs in high doses, causing severe side effects. However, a misplaced injection or injection of an inappropriate dose may cause severe complications. Duration of analgesia after a neurolytic block is limited although some blocks can be repeated with success. Due to the possibility of severe post-denervation neuropathic pain, neurolytic blocks of peripheral nerves should be reserved for patients with life expectancy of less than about 1 year. ▶ Visceral sympathetic blocks may be considered in patients with longer life expectancy (Breivik 2008b).

Celiac Plexus Block

Celiac plexus block relieves pain from cancer in the upper part of the abdomen (Breivik 2000). Randomized studies of celiac plexus neurolysis demonstrated improved pain control, reduction in opioid analgesic consumption, improved quality of life, and in some studies also prolonged survival (Lillemo et al. 1993; Eisenberg et al. 1995; Gunaratnam et al. 2001; Puli et al. 2009; Wong et al. 2004). The block can be repeated with success if pain recurs after weeks to months. Pneumothorax and shoulder pain from ethanol irritation of the diaphragmatic peritoneum are short-lived side effects. The sympatholysis that results may cause ▶ orthostatic hypotension, urinary retention and diarrhea for a few days. However, paraplegia and intestinal infarction may result if the neurolytic solution is injected into arteries or veins. Severely misplaced epidural, paravertebral, and subarachnoid injections have been reported, with tragic consequences (Breivik 2000; Rathmell 2009).

Indications for Celiac Plexus Block

Cancer of the pancreas is the classic indication. However, cancer of the stomach, duodenum, liver, gallbladder, and choledocal duct, causing visceral type pain in the upper abdomen, are indications as well. When abdominal wall involvement occurs, pain impulses will also travel through somatic afferent nociceptor fibers of the intercostal nerves. Neurolytic block of the visceral afferent nociceptor fibers passing through the celiac plexus will not relieve this pain.

Techniques of Celiac Plexus and Splanchnic Nerve Blockade

The celiac plexus is the prevertebral sympathetic plexus that supplies the upper abdominal viscera with autonomic sympathetic innervations. It is located in the retroperitoneal tissue space, anterior to the body of the first lumbar and twelfth thoracic vertebrae. The aorta lies behind, while the inferior vena cava and right renal vessels lie in front of the celiac plexus. The greater, the lesser, and the least splanchnic nerves connect the celiac plexus via the spinal nerve roots to the thoracic spinal cord. These nerves can be interrupted by radio-frequency denervation in the thoracic region, an approach favored by some (Burton et al. 2009).

The celiac plexus can be injected intraoperatively when the surgeon has determined that the tumor is unresectable (Lillemo et al. 1993). However, the block has most often been performed with needles through the skin from the back or with an anterior approach with needle guidance from computerized tomography scanning (CT) (Burton et al. 2009). In the traditional technique (Breivik 2000, 2008b; Campbell 2008; Burton et al. 2009) the patient is prone, and a needle is inserted on each side under the midpoint of the 12th rib. With local anesthetic infiltration, the needle is advanced toward the anterolateral corner of the body of the first lumbar vertebra. When this bone is contacted, the depth marker is moved 1.5–2 cm from the skin level toward the hub of the needle. The needle is now withdrawn to the subcutaneous tissue, redirected, and advanced 1.5–2 cm beyond the anterolateral corner of the vertebral body, preferably under radiological guidance. If the aorta is punctured on the left side, or the vena cava on the right side, the needle is withdrawn until blood stops coming on aspiration. At this point, 5 ml of lidocaine 20 mg/ml is injected in each needle. The patient is soon able to tell whether his pain has disappeared. He will also be able to inform the anesthesiologist of any signs of intraspinal or intravascular injection. This is sufficient information to be able to go ahead with an injection of 50 ml of 50 % ethanol (96 % ethanol diluted with bupivacaine 2.5 mg/ml) in each of the two needles.

Although radiological guidance may increase safety, it should be noted that radiographic imaging of the needle position is no guarantee that the injected neurolytic agent will not cause somatic neurological deficit. Two reported cases of paraplegia following celiac plexus block in spite of radiographic needle control underscore this important point (Breivik 2000). If this block is performed with the patient under general anesthesia, the indications of correct needle positioning from a test dose of a small, concentrated dose of local anesthetic are lost. Some of the reports of poor or short-lasting pain relief from celiac plexus block may be due to an insufficient injected volume or concentration of ethanol, which should not be less than 50 ml of 50 %. Phenol 5–8 % is used by some, but the risk of vascular complications may be greater (Burton et al. 2009). Endoscopic ultrasonography guided neurolysis of the celiac plexus may increase safety even more (Gunaratnam et al. 2001; Puli et al. 2009).

Side Effects of Celiac Plexus Block with Ethanol

An immediate side effect of celiac plexus block is a drop in blood pressure and also orthostatic hypotension for some time following the block. If this occurs, it may be treated with a head-low position, ephedrine and intravenous colloid-containing fluids. Due to the sympathetic denervation of the major part of the gastrointestinal tract, diarrhea may also result, often a desirable effect in patients who have been on oral opioids for some time. The patient should be warned that transient shoulder pain may be a result of the ethanol reaching the diaphragm. Too rapid a discontinuation of morphine may cause withdrawal symptoms. Some patients may need a reduced dose of oral morphine for residual pain or for pain stemming from tumor invasion of the abdominal wall (Breivik 2000).

Blocks of Superior Hypogastric Plexus, Bilateral Lumbar Sympathetic Chains, and Ganglion Impar

Phenol blocks of the superior hypogastric plexus or bilateral lumbar sympathetic block may relieve patients with visceral pain from pelvic organs and sigmoid colon (Campbell 2008; Burton et al. 2009). Local pelvic infiltration of cervical or prostatic carcinoma can cause somatic nociceptive pain that limit the effectiveness of sympathetic blocks. The same is true for perineal pain, when mainly of somatic origin and pain impulses travel via sacral and pudendal nerves. Perineal pain with sympathetic components can be relieved by blocking the ganglion impar (Burton et al. 2009).

Subarachnoid Spinal Nerve Root Neurolytic Block

Segmental, one-sided denervation of sacral nerve roots can be obtained by spinal puncture close to the affected nerve roots, followed by subarachnoid injection of 0.3–0.6 ml of hypobaric ethanol (96 %) with the patient's painful side up, or hyperbaric phenol (5 % in glycerol) with the painful side down (Campbell 2008). This technique is used less nowadays due to the dissatisfaction with the degree and duration of relief, and because of risks of severe side effects (Twycross 1994; Campbell 2008; Burton et al. 2009). Continuous catheter infusion of local anesthetics with opioids is a more effective alternative and has less risk of severe complications (Nitescu et al. 2002). However, under circumstances with very limited resources, where other pain relief is unavailable, the simple methods of subarachnoid ethanol or phenol block may still have a place. It can be performed at the bedside with a spinal needle and a syringe (Stovner and Endresen 1972). It is superior to no pain relief at all in a terminally ill cancer patient who is suffering from excruciating pain (Kuzucu et al. 1966; Swerdlow 1983; Drechsel 1984).

Subdural and Epidural Neurolytic Blocks with Phenol

These techniques have been used for cancer pain in the cervical and upper thoracic region where subarachnoid block is less effective (Burton et al. 2009).

Neurolytic Blocks of Selected Peripheral and Paravertebral Nerves

Trigeminal nerve block may be indicated for pain from head and neck cancer with trigeminal nerve involvement. Less than 0.5 ml ethanol 96 % is needed to anesthetize an appropriate branch (Swerdlow 1983; Campbell 2008).

Glossopharyngeal nerve block may give good pain relief when pharmacological therapy fails in painful conditions of the mouth and throat. Ethanol is injected at the jugular foramen and may easily also block the vagus, accessory, and hypoglossal nerves, causing dysphagia if bilateral block is required (Montgomery and Cousins 1972). Radiofrequency lesioning of ganglion Gasseri, ganglion sphenopalatinum may be effective and appropriate alternative techniques (Crul et al. 2008).

Intercostal nerve block with 1.0 ml of ethanol 96 % or phenol 5–6 % often gives excellent pain relief, lasting up to a few weeks (Campbell 2008).

Paravertebral Spinal Nerve Blocks

For well-localized somatic pain, a paravertebral spinal nerve block can be performed, with needle insertion 2–3 cm lateral to the spinous process (of the same vertebra in the lumbar region, but of the vertebra above the segmental nerve to be blocked in the thoracic region). After contact with the transverse process, the needle is redirected below the transverse process, under radiographic or ultrasound guidance when available, and 3–5 ml of the solution is injected. Pain relief may be good, but the duration is unpredictable. The block can be considered in debilitated and bedridden patients with severe pain not responding well to pharmacological treatment. Unintended subarachnoid injection, or injection into a radicular artery to the spinal cord, and the common epidural spread of a paravertebral injection all may cause severe neurological complications, e.g., paraplegia (Breivik 2000; Norum and Breivik 2010).

Sacral Nerve Blocks (Posterior Transsacral Blocks)

When pelvic metastases cause nerve compression pain which is not controlled by pharmacological means, sacral nerve blocks may be useful. Radiographic guidance is helpful, but reliable anatomical landmarks exist: The second sacral foramen lies about 1 cm below and medial to the posterior superior iliac spine. The fifth sacral foramen lies about 1 cm inferior and lateral to the sacral cornua. The third and fourth foramina lie on the line between the first and the fifth (Hill 2008). The needle is advanced 0.5–1.5 cm beyond the opening of the foramina, and about 3 ml of the solution is injected at each site.

Summary

Invasive anesthesiological techniques are still useful for severe cancer pain when pharmacological regimens do not control pain or cause intolerable adverse effects. Pain specialists with anesthesiological background and training must be able to perform a variety of such interventions. Imaging techniques and ultrasound devices are helpful, but in depth knowledge of the anatomy is still important for success. This entry is a review of the most useful interventions. Excellent textbooks and manuals now exist for stepwise, in detail prescriptions for how to do such blocks and interventions (Hill 2008; Campbell 2008; Cousins et al. 2009).

References

Breivik, H. (2000). Nerve blocks – Simple injections, epidurals, spinals and more complex blocks. In K. H. Simpson & K. Budd (Eds.), *Cancer pain management. A comprehensive approach* (pp. 84–98). Oxford: Oxford University Press.

Breivik, H. (2012). Local anaesthetic blocks and epidurals. In S. B. McMahon & M. Kolzenburg (Eds.), *Wall and Melzack's textbook of pain* (6th ed., Chapter 38). London: Elsevier Churchill Livingstone.

Breivik, H. (2008a). Epidural analgesia. In H. Breivik, W. Campbell, & M. Nicholas (Eds.), *Clinical pain management. Practice and procedures* (pp. 311–321). London: Hodder Arnold.

Breivik, H. (2008b). Sympathetic blocks. In H. Breivik, W. Campbell, & M. Nicholas (Eds.), *Clinical pain management. Practice and procedures* (pp. 322–326). London: Hodder Arnold.

Burton, A. W., Phan, P. C., & Cousins, M. J. (2009). Treatment of cancer pain: Role of neural blockade and neuromodulation. In M. J. Cousins, D. B. Carr, T. T. Horlocker, & P. O. Bridenbaugh (Eds.), *Cousins' and Bridenbaugh's neural blockade in clinical anaesthesia and pain medicine* (4th ed., pp. 1111–1153). Philadelphia: Wolters Kluwer/Lippincott Williams & Wilkins.

Campbell, W. I. (2008). Neurolytic blocks. In H. Breivik, W. Campbell, & M. Nicholas (Eds.), *Clinical pain management. Practice and procedures* (pp. 337–348). London: Hodder Arnold.

Cousins, M. J., Carr, D. B., Horlocker, T. T., & Bridenbaugh, P. O. (2009). *Cousins' and Bridenbaugh's Neural blockade in clinical anaesthesia and pain medicine* (4th ed.). Philadelphia: Wolters Kluwer/Lippincott Williams & Wilkins.

Crul, B. J. P., van Zundert, J. H. M., & van Kleef, M. (2008). Radiofrequency lesioning and treatment of chronic pain. In H. Breivik, W. Campbell, M. Nicholas (Eds.), *Clinical pain management. Practice and procedures* (pp. 389–403). London: Hodder Arnold.

Drechsel, U. (1984). Treatment of cancer pain with neurolytic agents. *Recent Results of Cancer Research, 89*, 137–147.

Eisenberg, E., Carr, D. B., & Chalmers, T. C. (1995). Neurolytic celiac plexus block for treatment of cancer pain: A meta-analysis. *Anesthesia and Analgesia, 80*, 290–295.

Gunaratnam, N. T., Sarma, A. V., Norton, I. D., & Wiersema, M. J. (2001). A prospective study of EUS-guided celiac plexus neurolysis for pancreatic cancer pain. *Gastrointestinal Endoscopy, 54,* 316–324.

Hill, D. (2008). Peripheral nerve blocks. In H. Breivik, W. Campbell, M. Nicholas (Eds.), *Clinical pain management. Practice and procedures* (pp. 255–292). London: Hodder Arnold.

Hogan, Q. H., & Abrams, S. E. (2009). Diagnostic and prognostic neural blockade. In M. J. Cousins, D. B. Carr, T. T. Horlocker, & P. O. Bridenbaugh (Eds.), *Cousins' and Bridenbaugh's neural blockade in clinical anaesthesia and pain medicine* (4th ed., pp. 811–847). Philadelphia: Wolters Kluwer/Lippincott Williams & Wilkins.

Kidd, B. L. (2007). The mechanisms of chronic pain. In H. Breivik & M. Shipley (Eds.), *Pain best practice and research compendium* (pp. 17–24). London: Elsevier.

Kuzucu, E. Y., Derrick, W. S., & Wilber, S. A. (1966). Control of intractable pain with subarachnoid alcohol block. *The Journal of the American Medical Association, 195,* 541–544.

Lillemo, K. D., Cameron, J. L., & Kaufman, H. S. (1993). Chemical splanchnicectomy in patients with unresectable pancreatic cancer. *Annals of Surgery, 217,* 447–457.

Montgomery, W., & Cousins, M. J. (1972). Aspects of management of chronic pain illustrated by ninth nerve block. *British Journal of Anaesthesia, 44,* 383–385.

Nitescu, P. V., Apelgren, L., & Curelaru, I. A. (2002). Long-term intrathecal and intracisternal treatment of malignant and nonmalignant pain using external pumps. In H. Breivik, W. Campbell, & C. Eccleston (Eds.), *Clinical pain management. Practical applications and procedures* (pp. 285–306). London: Arnold.

Norum, H. M., & Breivik, H. (2010). A systematic review of comparative studies indicates that paravertebral block is neither superior nor safer than epidural analgesia for pain after thoracotomy. *Scandinavian Journal of Pain, 1,* 12–23.

Puli, S. R., Reddy, J. B., Bechtold, M. L., Antillon, M. R., & Brugge, W. R. (2009). EUS-guided celiac plexus neurolysis for pain due to chronic pancreatitis or pancreatic cancer pain: A meta-analysis and systematic review. *Digestive Diseases Science, 54,* 2330–2337.

Ramos-Galvez, I. N., & Goodall, I. D. (2008). Epidural (interlaminar, intraforaminal, and caudal) steroid injections for back pain and sciatica. In H. Breivik, W. Campbell, & M. Nicholas (Eds.), *Clinical pain management. Practice and procedures* (pp. 416–426). London: HodderArnold.

Rathmell, J. P. (2009). Complications of pain medicine. In M. J. Cousins, D. B. Carr, T. T. Horlocker, & P. O. Bridenbaugh (Eds.), *Cousins' and Bridenbaugh's neural blockade in clinical anaesthesia and pain medicine* (4th ed., pp. 1111–1153). Philadelphia: Wolters Kluwer/Lippincott Williams & Wilkins.

Stovner, J., & Endresen, R. (1972). Intrathecal phenol for cancer pain. *Acta Anaesthesiologica Scandinavica, 16,* 17–21.

Swerdlow, M. (1983). *Relief of intractable pain* (3rd ed.). Amsterdam: Excerpta Medica.

Twycross, R. (1994). *Pain relief in advanced cancer.* Edinburgh: Churchill Livingstone.

Whyte, E., & Lauder, G. (2011). Intrathecal infusion of bupivacaine and clonidine provides effective analgesia in a terminally ill child. *Paediatric Anaesthesia, 33,* 549–551.

Wong, G. Y., Schroeder, D. R., Carns, P. E., Wilson, J. L., Martin, D. P., Kinney, M. O., Mantilla, C. B., & Warner, D. O. (2004). Effect of neurolytic celiac plexus block on pain relief, quality of life, and survival in patients with unresectable pancreatic cancer: A randomized controlled trial. *Journal of the American Medical Association, 291,* 1092–1099.

Cancer Pain Management, Anesthesiologic Interventions, Spinal Cord Stimulation, and Neuraxial Infusion

Muthukumar Vaidyaraman and Peter S. Staats
Division of Pain Medicine, The Johns Hopkins University, Baltimore, MD, USA

Synonyms

Anesthesiologic interventions; Neuraxial infusion; Spinal cord stimulation

Definition

Anesthesiologic interventions, ▶ spinal cord stimulation, and ▶ neuraxial infusion are types of interventional therapies that may be used in the treatment of cancer pain.

Characteristics

Cancer is the second most frequent cause of death in the United States. Cancer patients fear pain more than other symptoms, and at least a third suffer pain at the time of diagnosis (Jacox et al. 1994). Pain is heterogeneous (Bonica 1990), and the prevalence and complexity of pain increases in populations with advanced disease.

Humane treatment demands aggressive treatment of cancer pain. To this end, the World Health Organization (World Health Organization 1990) and the Agency for Health Care Policy and Research (Jacox et al. 1994) have published guidelines on the management of cancer-related pain. These guidelines, however, either fail to mention or minimize the importance of the role of interventional therapy. Although opioids continue to be a mainstay of cancer pain treatment, patients frequently experience side effects, such as sedation and constipation. Many experience only imperfect cancer pain relief.

The Role of Interventional Therapies
Interventional therapies include neural blockade, spinal drug delivery, and spinal cord stimulation (Staats 1998). Clinicians should consider using these therapies in cancer patients who suffer intractable opioid side effects or in whom opioids provide inadequate relief. Altering pain conduction with neural blockade techniques or delivering lower doses of agents directly to the spinal cord can often improve the balance between analgesia and side effects. Interventional therapies may also be cost-effective (Erdine and Talu 1998); for example, long-term (more than 3 months) administration of high-dose opioids delivered at home via patient-controlled analgesia is much more expensive on a strictly medical cost basis than implantable drug delivery. In appropriate patients, interventional therapies can make it possible to achieve the goal of decreasing pain and the side effects associated with higher doses of systemic therapy while also reducing health care and associated costs.

Neural Blockade
The effect of a nerve block can be short-term (using local anesthetics) or long-term (with chemical, thermal, or surgical applications).

Nerve blocks can be peripheral, visceral, or intraspinal. Intraspinal blocks are now rarely performed because advances in spinal analgesia offer alternatives, and increased life expectancy in cancer patients makes the risk of its associated complications unacceptable.

Short-acting local anesthetic blocks carry a very low risk and are usually diagnostic, guiding clinical decisions about the use of longer-acting methods. For example, the physician may anesthetize the painful areas prior to a neurodestructive technique. This offers the patient the opportunity to experience the feeling of a permanent procedure. If anesthesia is not associated with pain relief, it also helps the physician prognosticate the permanent neurolytic procedure. In some cases, a series of blocks with local anesthetic (e.g., sympathetic blockade) can achieve lasting relief of neuropathic pain.

Although clinicians commonly refer to neurodestructive blocks as permanent, the nerve may regenerate, leading to a return of pain. Neurolysis typically provides 3–6 months of relief.

Anatomic Location of Nerve Blocks
Among the many possible peripheral nerve blocks, clinicians most frequently block intercostal (Bolotin et al. 2000) and trigeminal nerves to treat somatic pain. A visceral nerve block can be performed to address visceral pain. The most common is the celiac plexus block (Polati et al. 1998). The superior hypogastric plexus block is indicated for the visceral pain that may be prominent in patients with various pelvic organ cancers, even in advanced stages (de Leon-Casasola et al. 1993; Plancarte et al. 1997), and ganglion impar (Walther's) blockade mitigates the burning visceral pain in the perineal area and can have a beneficial impact on urgency (Plancarte et al. 1990).

Neuromodulation
Spinal Cord Stimulation (SCS)
SCS may be efficacious in patients whose pain has a neuropathic component (Grabow et al. 2003) and may benefit certain cancer patients, for example, those with post-thoracotomy pain,

postherpetic neuralgia, and/or radicular lower extremity pain. With this approach, an electrical current is applied to the dorsal aspect of the spinal cord through an implanted wire. The current is created either by an implanted generator or an external generator that transmits to an implanted antenna. There have been no controlled trials of SCS in cancer patients, but anecdotal experience has been favorable in carefully selected patients. It is important to recognize that with the current technology it is a relative contraindication to implant a device if repeated MRIs to monitor the disease are necessary. However, with newer technology and alternative imaging techniques, it may be possible in certain situations.

Neuraxial Infusion

Epidural delivery of morphine with an external pump is widely used to treat cancer pain, especially in the postoperative setting. The timing of doses of epidural analgesia can be patient-controlled, and patients may continue to use this method of pain control after they are discharged from the hospital.

Compared with epidural delivery, intrathecal delivery of opioids has the advantages of providing analgesia with a much smaller dose of the drug and a reduced chance of infection with implanted systems. Intrathecal drug delivery should be considered in those patients with chronic pain inadequately relieved by maximum medical management, or relieved at the expense of intolerable side effects (Paice et al. 1996). Only preservative-free morphine is approved to treat pain intrathecally. Additional widely used agents include hydromorphone, fentanyl, clonidine, and bupivacaine. Ziconotide, a novel n-type calcium channel blocker, is under investigation and has been demonstrated to be efficacious in a controlled trial (Staats et al. 2004). Screening for intrathecal treatment includes assessment and treatment of comorbid psychological conditions (Olson 1992) and a trial with either single bolus or infusion of intrathecal agents. Multiple agents are now recommended in the treatment of cancer pain (Deer et al. 2012).

Pain Treatment and Survival in Cancer Patients

Investigators are beginning to report findings of increased survival rates in cancer patients whose pain is controlled in a manner that reduces side effects and improves quality of life. In one study, patients with unresectable pancreatic cancer who received an alcohol celiac block survived longer than those who received a placebo block (Staats et al. 2001). Another study that compared optimal medical management alone with optimal medical management plus intrathecal drug delivery, in a randomized trial involving 200 patients with refractory cancer pain, intrathecal therapy reduced the incidence of drug-related toxicity and reliance on systemic analgesics while improving pain scores, quality of life for patients and caregivers, and survival rates (Smith et al. 2002). The possibility that controlling pain will lengthen the life of terminally ill patients is intriguing and should add to our urgency in promoting the consideration of pain as a disease that must be treated as aggressively as possible.

References

Bolotin, G., Lazarovici, H., Uretzky, G., et al. (2000). The efficacy of intraoperative internal intercostal nerve block during video-assisted thoracic surgery on postoperative pain. *The Annals of Thoracic Surgery, 70*, 1872–1875.

Bonica, J. J. (1990). Cancer pain. In J. J. Bonica (Ed.), *The management of pain* (2nd ed., pp. 200–460). Philadelphia: Lea and Febiger. Vol. I.

de Leon-Casasola, O. A., Kent, E., & Lema, M. J. (1993). Neurolytic superior hypogastric plexus block for chronic pelvic pain associated with cancer. *Pain, 54*, 145–151.

Deer, T. R., Prager, J., Levy, R., et al. (2012). Polyanalgesic consensus conference 2012: Recommendations for the management of pain by intrathecal (intraspinal) drug delivery: Report of an interdisciplinary expert panel. *Neuromodulation: Technology at the Neural Interface, 15*, 436–466.

Erdine, S., & Talu, G. K. (1998). Cost effectiveness of implantable devices versus tunneled catheters. *Current Review of Pain, 2*, 157–162.

Grabow, T. S., Tella, P. K., & Raja, S. N. (2003). Spinal cord stimulation for complex regional pain syndrome: An evidence-based medicine review of the literature. *The Clinical Journal of Pain, 19*, 371–383.

Jacox, A., Carr, D. B., & Payne, R. (1994). New clinical practice guidelines for the management of cancer pain. *The New England Journal of Medicine, 330*, 169–173.

Olson, K. (1992). *An approach to psychological assessment of chronic pain patients*. Minneapolis: NCS Assessments.

Paice, J. A., Penn, R. D., & Shott, S. (1996). Intraspinal morphine for chronic pain: A retrospective multicenter study. *Journal of Pain and Symptom Management, 11*, 71.

Plancarte, R., Amescua, C., & Patt, R. B. (1990). Presacral blockade of the ganglion of Walther (ganglion impar). *Anesthesiology, 73*, A751.

Plancarte, R., de Leon-Casasola, O. A., El-Helaly, M., et al. (1997). Neurolytic superior hypogastric plexus block for chronic pelvic pain associated with cancer. *Regional Anesthesia, 22*, 562–568.

Polati, E., Finco, G., Gottin, L., et al. (1998). Prospective randomized double-blind trial of neurolytic celiac plexus block in patients with pancreatic cancer. *British Journal of Surgery, 85*, 199–201.

Smith, T. J., Staats, P. S., Pool, G., et al. (2002). Randomized clinical trial of an implantable drug delivery system compared to comprehensive medical management for refractory cancer pain: Impact on pain, drug-related toxicity, and survival. *Journal of Clinical Oncology, 20*, 4040–4049.

Staats, P. S. (1998). Cancer pain: Beyond the ladder. *Journal of Back and Musculoskeletal Rehabilitation, 10*, 69–80.

Staats, P. S., Hekmat, H., Sauter, P., & Lillemoe, K. (2001). The effects of alcohol celiac plexus block, pain, and mood on longevity in patients with unresectable pancreatic cancer: A double-blind, randomized, placebo-controlled study. *Pain Medicine, 2*, 28–34.

Staats, P. S., Yearwood, T., Charapata, S., et al. (2004). Intrathecal ziconotide in the treatment of refractory malignant pain: A controlled clinical trial. *JAMA: The Journal of the American Medical Association, 291*, 63–70.

World Health Organization. (1990). *Cancer pain relief*. Geneva: Author.

Cancer Pain Management, Cancer-Related Breakthrough Pain, Therapy

Giovambattista Zeppetella
St. Clare Hospice, Hastingwood, UK

Synonyms

Pain flares

Definition

Breakthrough pain is a transient increase in pain intensity over background pain. It occurs commonly in cancer patients and is a heterogeneous phenomenon that may be incapacitating or debilitating or significantly impact quality of life. Breakthrough pain is a distinct component of cancer pain and requires specific management.

Characteristics

People with cancer-related pain often report that their pain varies during the course of the day. Two patterns of pain can be identified: continuous background pain, which may respond well to around-the-clock (ATC) ▶ analgesics (Fig. 1), and transitory exacerbations of pain, which break through the ATC analgesics (Fig. 2). Transient increases in pain in a cancer patient who has stable persistent pain treated with ▶ opioids may be defined as breakthrough pain (Portenoy and Hagen 1990). The definition of breakthrough pain has been the subject of much discussion, with some investigators classifying episodic pains as breakthrough pain irrespective of the analgesic regimen, and others classifying episodic pains in this way irrespective of whether background pain is controlled. Even the term "breakthrough" is not one that is universally accepted (Mercadante et al. 2002).

A number of studies have evaluated the characteristics of breakthrough pain in patients attending cancer centers or pain clinics, and in hospice inpatients or outpatients. These studies have varied in their sampling procedures and in inclusion and exclusion criteria. Breakthrough pain is usually characterized according to its location, severity, temporal characteristics, relationship to regular ▶ analgesia, precipitating factors, predictability, pathophysiology, etiology, and palliative factors (Portenoy 1997). The reported prevalence of breakthrough pain has varied from 20 % to 95 % (Zeppetella and Ribeiro 2003). Pain is typically of fast onset and short duration and feels similar to background pain, except that it may be more severe. Like background pain, the pathophysiology of breakthrough pain may be visceral, somatic, or neuropathic, and the etiology may be related directly to cancer or cancer treatment or unrelated to the cancer.

Cancer Pain Management, Cancer-Related Breakthrough Pain, Therapy, Fig. 1 Background pain. Successful management of background pain with regular analgesia and without serious adverse effects

Cancer Pain Management, Cancer-Related Breakthrough Pain, Therapy, Fig. 2 Breakthrough pain. A transient increase over background pain that breaks through the regular analgesia; pain may vary in frequency, intensity, and duration

Breakthrough pain is a heterogeneous phenomenon that may be different for each patient; patients may have more than one pain type. Two types of breakthrough pain have been described: incident pain, which is precipitated by volitional factors such as movement or non-volitional such as bladder spasm, and spontaneous pain, which occurs in the absence of a specific trigger. End-of-dose failure is sometimes included as a breakthrough pain subtype. It occurs because either the analgesic is described at

an inadequate dose or the interval between administrations is too long. As background pain is not controlled, end-of dose pain is not breakthrough pain.

Despite the self-limiting nature of breakthrough pain, it can have a profound impact on both patients' and carers' quality of life. Patients with breakthrough pain are often less satisfied with their analgesic therapy, have decreased functioning because of their pain, and have an increased level of anxiety and depression

(Portenoy et al. 1999). Unrelieved breakthrough pain also increases economic burden placed on the healthcare system (Fortner et al. 2002). Effective management is, therefore, essential and can only be achieved through meticulous assessment, good communication, and patient and caregiver participation. Failure to take these factors into account can lead to ineffective analgesia, unwanted adverse effects, and poor adherence to therapy.

Management of breakthrough pain should be part of a holistic framework and appropriate to the stage of the patent's disease. A combination of pharmacological and non-pharmacological treatment strategies may be required. Three principles of management have been proposed (Portenoy 1997). First, implementation of primary therapies can lead to improvement in both background and breakthrough pain. These interventions include those that modify the pathological process (e.g., radiotherapy, chemotherapy, and surgery), and those that manage reversible problems (e.g., antitussives for cough, laxatives for constipation, and antispasmodics for bladder spasm). Second is optimizing of the scheduled analgesic regimen using the World Health Organization analgesic ladder as a basis for therapy (World Health Organization 1996). Patients may require a combination of opioid and non-opioid analgesics and, in some cases, ▶ adjuvant analgesics such as antidepressants, anticonvulsants, bisphosphonates, and corticosteroids. Third is the use of specific non-pharmacological or pharmacological interventions for breakthrough pain. Non-pharmacological interventions include physiotherapy, cognitive techniques, and orthopedic procedures. Pharmacological management is usually in the form of supplemental analgesia also known as rescue medication. End-of-dose failure responds best to increasing ATC medication, whereas other breakthrough pain subtypes are usually managed with rescue medication, as increasing the ATC medication to cover all incident and spontaneous breakthrough pains can lead to adverse effects.

Successful use of rescue medication requires adequate analgesia to be obtained without excessive adverse effects. When using rescue medication, a number of issues should be considered (What do we give? How much, when in relation to the pain? By what route and how frequently?). Rescue medication is best administered before or soon after the onset of breakthrough pain. As breakthrough pain is typically of fast onset and short duration, rescue medication should ideally be potent, absorbed and excreted rapidly, easy to administer and produce minimal adverse effects (Fig. 3). Opioids are most commonly used, and the current approach involves giving an additional dose based on the patient's ATC analgesia. Opioids may be administered by several routes, including oral, oral transmucosal, rectal, intravenous, subcutaneous, and intraspinal (intrathecal and epidural). The route of administration is important, as this can influence the onset of analgesia and duration of effect. Oral medication is generally used as it is convenient and often the most acceptable (Walker et al. 2003). When using morphine as rescue medication, the "normal-release" formulation (which has a 4-hourly duration of action) is used; the current recommendation by the European Association for Palliative Care is to use the same as the 4-hourly dose of normal-release morphine (approximately 16 % of the daily dose) when necessary (Hanks et al. 2001). However, as breakthrough pain can vary in etiology, intensity, and duration, the dose of rescue medication may also vary (Fig. 3).

Patients with advanced disease may require more than one route of drug administration due to difficulty in swallowing, nausea, or other gastrointestinal problems (Coyle et al. 1990). Furthermore, the onset of effect with most currently available opioids does not match the rapid onset of a typical breakthrough pain episode. Parenteral opioids (subcutaneous, intravenous, or intramuscular) achieve a peak effect more rapidly than oral opioids but are invasive, inconvenient, costly, and can be uncomfortable; the same is true of intraspinal opioids. Sublingual administration allows rapid absorption and is convenient, accessible, and generally well accepted, but absorption can be poor or irregular. Rectal administration, although

Cancer Pain Management, Cancer-Related Breakthrough Pain, Therapy, Fig. 3 Ideal rescue medication. The plasma level of medication follows the profile of breakthrough pain without excessive adverse effects

generally safe, effective, and inexpensive, may be inconvenient and unacceptable to some patients, and absorption can be unreliable. Oral transmucosal administration provides a simple and convenient route, which offers the potential for more rapid absorption and onset of action than the oral route and avoids hepatic first pass metabolism. Oral transmucosal fentanyl citrate (OTFC) is a fentanyl-impregnated lozenge, specifically developed for the management of breakthrough pain. OTFC is rapidly absorbed through the oral mucosa and in controlled studies has been shown to be safe and effective (Coluzzi et al. 2001; Farrar et al. 1998). OTFC should be individually titrated to a successful dose that is not predicted from the ATC opioid dose.

References

Coluzzi, P. H., Schwartzberg, L., Conroy, J. D., et al. (2001). Breakthrough cancer pain: A randomized trial comparing Oral Transmucosal Fentanyl Citrate (OTFC) and Morphine Sulphate Immediate Release (MSIR). *Pain, 91*, 123–130.

Coyle, N., Adelhardt, J., Foley, K. M., et al. (1990). Character of terminal illness in the advanced cancer patient: Pain and other symptoms during the last four weeks of life. *Journal of Pain and Symptom Management, 5*, 83–93.

Farrar, J. T., Cleary, J., Rauch, R., et al. (1998). Oral Transmucosal fentanyl citrate: Randomized, double-blind, placebo-controlled trial for the treatment of breakthrough pain in cancer patients. *Journal of the National Cancer Institute, 90*, 611–616.

Fortner, B. V., Okon, T. A., & Portenoy, R. K. (2002). A survey of pain-related hospitalizations, emergency department visits, and physician office visits reported by cancer patients with and without history of breakthrough pain. *Journal of Pain, 3*, 38–44.

Hanks, G. W., De Conno, F., Cherny, N., et al. (2001). Morphine and alternative opioids in cancer pain: The EAPC recommendations. *British Journal of Cancer, 84*, 587–93.

Mercadante, S., Radbruch, L., Caraceni, A., et al. (2002). Episodic (breakthrough) pain: Consensus Conference of an Expert Working Group of the European Association for palliative care. *Cancer, 94*, 832–839.

Portenoy, R. K. (1997). Treatment of temporal variations in chronic cancer pain. *Seminars in Oncology, 24*, S16-7–S16-12.

Portenoy, R. K., & Hagen, N. A. (1990). Breakthrough pain: Definition, prevalence and characteristics. *Pain, 41*, 273–281.

Portenoy, R. K., Payne, D., & Jacobsen, P. (1999). Breakthrough pain: Characteristics and impact in patients with cancer pain. *Pain, 81*, 129–134.

Walker, G., Wilcock, A., Manderson, C., et al. (2003). The acceptability of different routes of administration of analgesia for breakthrough pain. *Palliative Medicine, 17*, 219–221.

World Health Organization. (1996). *Cancer pain relief.* Geneva: Author.

Zeppetella, G., & Ribeiro, M. D. C. (2003). The pharmacotherapy of episodic pain. *Expert Opinion on Pharmacotherapy, 4*, 493–502.

Cancer Pain Management, Chemotherapy

Seth Resnick
Silver Hill Hospital, New York University
School of Medicine, USA

Synonyms

Chemotherapy

Definition

Chemotherapy, the use of chemicals as therapy, is, along with surgery and radiation, one of the most common treatments for cancer. Chemotherapy is generally administered intravenously, but oral agents are becoming increasingly available.

Characteristics

Over the last generation, chemotherapy has evolved based on the principles of cytotoxicity in cellular systems, empiric observation in preclinical models, and incremental improvements achieved with clinical trials. Chemotherapeutic agents can be classified by their presumed mechanism of action, e.g., alkylator of DNA, antimetabolite, mitotic inhibitor, and repair enzyme inhibitor. Chemotherapy indications are based on clinical trials data, with the optimum therapeutic ratio determined by establishing the dose based on maximum therapeutic benefit with acceptable toxicity. Many chemotherapeutic agents are schedule-dependent, both for efficacy and toxicity. Combination therapy usually has a therapeutic advantage over single agent treatments.

Optimum benefit for cancer treatment is achieved with a basic understanding of the tumor site-specific clinical characteristics, including prognostic factors and patterns of spread, coupled with knowledge of the role and sequencing of the treatment modalities. Depending on the tumor type and the presentation of the disease in an individual patient, multidisciplinary planning results in treatment that might involve a single modality, such as surgery; the concurrent use of radiation and/or chemotherapy; or sequential treatments, such as surgery followed by the addition of radiation and/or chemotherapy.

Prospective clinical trials, often randomized, have established standards of practice. Therapeutic endpoints vary. In clinical situations where cure is the expected outcome, survival is conventionally used as the therapeutic endpoint. In clinical situations where cure is a less likely outcome, then clinical endpoints such as progression-free survival, symptom-free interval, or overall survival are commonly used. The ▶ RECIST criteria have been established as an objective reproducible measurement of reduction in tumor; these criteria generally correlate with improvements in quality of life and survival.

When the therapeutic intent is predominantly palliative, then other therapeutic endpoints are used. Commonly used palliative clinical trials endpoints include a >50 % decrease in pain score (measured using validated methodologies), >50 % decrease in opioid usage, and improvement in performance scale, all lasting more than 4 weeks.

Cancer chemotherapy with established benefits is unfortunately associated with what can be significant toxicities. With the increasing availability of interventions to manage symptoms, chemotherapy can be better tolerated. Fatigue is one of the most commonly encountered chemotherapy-associated toxicities. Although often multifactorial, fatigue associated with chemotherapy is often caused by anemia. There is a strong body of evidence demonstrating the value of the ▶ erythropoietins to increase the hemoglobin in patients receiving chemotherapy. This erythropoietin-induced increase in hemoglobin is correlated with improvement in quality of life and reversal of fatigue associated with the cancer and cancer treatment.

▶ Granulocytopenia is also a common chemotherapy toxicity. With the availability of recombinant hematopoietic growth factors, the associated morbidities of fever, sepsis, and antibiotic use, which may require hospitalization, are significantly reduced.

Chemotherapy-induced nausea and vomiting, formerly a dose-limiting toxicity, is almost always preventable or at least greatly reduced by treatment. An array of effective antiemetic agents is now available, including ▶ serotonin blockers, benzodiazepines, steroids, and the recently available ▶ NKδ1 blockers.

In addition to these chemotherapy-associated toxicities, consideration needs to be given to the effects of chemotherapy-associated organ damage. Examples are the cumulative myocardial damage associated with anthracycline chemotherapy, the pulmonary toxicity associated with bleomycin, the nephrotoxicity associated with the platinols, and the peripheral neuropathy associated with the vinca alkaloids. Other considerations for patients receiving chemotherapy revolve around the establishment of safe and easy venous access for the administration of chemotherapy and the maintenance of nutrition.

Clinical Decision-Making

Given the reality that chemotherapy is often of limited benefit and associated with significant toxicities, clinical decision-making is critically important. The decision whether or not to use chemotherapy for palliative intent must consider a number of important perspectives. From the provider prospective, medical decisions are usually based on both evidence and the "expert" opinion of the provider. The provider preferences are influenced by many factors, including what could be globally defined as self-serving interests, practical issues, and issues surrounding reimbursement. From the patient prospective, preferences are influenced by the understanding and expectation of outcomes, by psychosocial issues including developmental stage and beliefs, and by practical and financial issues. Financial issues include not only reimbursement for medical expenses but what can be burdensome, out-of-pocket expenses for such things as transportation and parking, and the loss of income because of the patient or care partner not being able to work. Clinical decision-making should be a shared medical decision-making process, which

assumes that both provider and patient develop an understanding of the relative balance between the autonomy of the physician and the preferences of the patient.

A recent study (Koedoot et al. 2002) used clinical vignettes to identify variables that influence provider preference for "watchful waiting," that is, deferring the introduction of palliative chemotherapy. Based on information gathered from questionnaires administered to more than 1,000 oncologists, age was the strongest predictor, followed by the patient's wish to be treated, and the expected survival gain. One of the conclusions of this study is that oncologists' recommendations are consistent and based on objective criteria (Koedoot et al. 2002).

Probably the most important determinant of provider preference is the available evidence of palliative benefit of chemotherapy for a specific cancer diagnosis. Although high-level evidence from clinical trials is limited, these trials form the basis for the general assumption that chemotherapy can be a standard of care for palliation of symptoms associated with most malignancies. The risk/benefit determination of treatment versus no treatment must be highly individualized and based on the tumor type, the symptoms associated with the tumor, overall clinical status, and, of course, coupled with patient preference.

Pancreatic Cancer

Perhaps the strongest argument for an evidence basis for the palliative benefit of chemotherapy is for pancreatic cancer. In a landmark prospective study that led to the licensing of gemcitabine, 126 symptomatic pancreas cancer patients were enrolled in a study that began with a lead-in period during which patients were stabilized with analgesics. Patients were then randomized to receive what was at that point the experimental agent gemcitabine, or what was at that time a standard of care, 5-fluorouracil. The primary endpoints of the study were pain intensity and related analgesic consumption, Karnofsky performance status, and weight. In order to meet the criteria of objective benefit, the required improvement in these symptoms had to be sustained for at least 4 weeks, without worsening of any symptom during the observation period. The symptom improvement was correlated to objective tumor response, time to tumor progression, and survival. There was evident clinical benefit in 24 % of patients who received gemcitabine and 5 % treated with 5-fluorouracil ($P = 0.0022$); median survival durations were 6 versus 4 months ($P = 0.0025$), and the 12-month survival rate was 18 % versus 2 % (Burris et al. 1997).

Colon Cancer

Patients with progressive metastatic colon cancer were randomized to test supportive care with or without the chemotherapeutic agent irinotecan. The primary endpoint of survival demonstrated an improvement from 14 % at 1 year to 36 % ($P = 0.001$). Palliative benefit was evident with demonstration in the secondary endpoints, with a longer duration of pain-free survival, longer duration to any significant decrease in performance status, and time to more than 5 % weight loss. Corroborating data was evident in the results of quality of life assessments (Cunningham et al. 1998).

Breast Cancer

The hypothesis that there is a relationship between tumor shrinkage and improvement in disease-related symptoms was evaluated in a prospective randomized trial of 300 women with metastatic breast cancer, during which symptoms were assessed for change over time associated with the cancer chemotherapy (Geels et al. 2000). Utilizing established quality of life questionnaires and what are now known as the ► Common Toxicity Criteria, the authors were able to demonstrate a clear correlation between patients' symptoms and objective tumor response.

Lung Cancer Chemotherapy

For non-small cell lung cancer, the benefits of palliative chemotherapy are evident from a meta-analysis of 52 randomized clinical trials of chemotherapy. Comparing best supportive care to chemotherapy in a group of patients who had not had prior chemotherapy for metastatic

disease, there was a 10 % absolute improvement and survival at 1 year with chemotherapy; expressed as a hazard ratio, this was equal to 0.73 (Thongprasert et al. 1999). In a specific chemotherapy trial of docetaxel versus best supportive care in a group of 103 patients who had had previous treatment with platinum-based chemotherapy, there was a significant difference between the two groups in their requirement for opioid analgesics and other medications for symptom management (Shepherd et al. 2000).

Prostate

For prostate cancer, the best evidence is from a prospective randomized trial of prednisone with or without the chemotherapeutic agent, mitoxantrone, in 161 men with metastatic prostate cancer. The primary endpoints were improvement in health-related quality of life assessed by questionnaire. Both groups demonstrated improvement in quality of life, but the important parameters of physical functioning and pain were significantly better in the mitoxantrone group, with longer duration of response compared to the prednisone alone group (Tannock et al. 1996).

Ovary Cancer

For ovarian cancer, the data are not as well established but implied from objective response data. Among 27 women, only seven achieved an objective response, and there were improvements in symptom endpoints (Doyle et al. 2001).

References

Burris, H. A., III, Moore, M. J., Anderson, J., et al. (1997). Improvements in survival and clinical benefit with gemcitabine as first line therapy for patients with advanced pancreas cancer: A randomized trial. *Journal of Clinical Oncology, 15*, 2403–2413.

Cunningham, D., Pyrohonen, S., James, R. D., et al. (1998). Randomised trial of irinotecan plus supportive care versus supportive care alone after fluorouracil failure for patients with metastatic colorectal cancer. *Lancet, 352*, 1413–1418.

Doyle, C., Crump, M., Pintilie, M., et al. (2001). Does palliative chemotherapy palliate? Evaluation of expectations, outcomes, and costs in women receiving chemotherapy for advanced ovarian cancer. *Journal of Clinical Oncology, 19,* 1266–1274.

Geels, P., Eisenhauer, E., Bezjak, A., et al. (2000). Palliative effect of chemotherapy: Objective tumor response is associated with symptom improvement in patients with metastatic breast cancer. *Journal of Clinical Oncology, 18,* 2395–2405.

Koedoot, C. G., de Haes, J. C., Heisterkamp, S. H., et al. (2002). Palliative chemotherapy or watchful waiting? A vignettes study among oncologists. *Journal of Clinical Oncology, 20,* 3658–3664.

Shepherd, F., Dancey, J., Ramlau, R., et al. (2000). Prospective randomized trial of docetaxel versus best supportive care in patients with non-small cell lung cancer previously treated with platinum-based chemotherapy. *Journal of Clinical Oncology, 18,* 2095–2103.

Tannock, I. F., Osoba, D., Stockler, M. R., et al. (1996). Chemotherapy with mitoxantrone plus prednisone or prednisone alone for symptomatic hormone-resistant prostate cancer: A Canadian randomized trial with palliative end points. *Journal of Clinical Oncology, 14,* 1756–1764.

Thongprasert, S., Sanguanmitra, P., Juthapan, W., et al. (1999). Relationship between quality of life and clinical outcomes in advanced non-small cell lung cancer: Best supportive care (BSC) versus BSC plus chemotherapy. *Lung Cancer, 24,* 17–24.

Cancer Pain Management, Interface Between Cancer Pain Management and Palliative Care

Peter G. Lawlor
Our Lady's Hospice, Medical Department, Dublin, Ireland

Synonyms

Hospice care; Palliative care and cancer pain management; Supportive care

Definition

The recently revised World Health Organization definition of palliative care reads: "Palliative care is an approach that improves the ▶ quality of life of patients and their families facing the problems associated with life-threatening illness, through the prevention and relief of ▶ suffering by means of early identification and impeccable assessment and treatment of pain and other problems, physical, psychosocial and spiritual" (http://www.who.int/cancer/palliative/en/).

Characteristics

Cancer Pain and Palliative Care: A Global Perspective and Introductory Outline

Cancer is one of those life-threatening illnesses referred to in the WHO definition of palliative care. Approximately one-third of the population in developed countries will be diagnosed with cancer. The estimated worldwide number of new cases each year is expected to rise from 10 million in the year 2000 to 15 million in 2020. The number of annual worldwide cancer-related deaths is expected to rise from 6 million to 10 million over the same period (World Health Organization and International Union against Cancer 2003). Much of the projected increase in cancer mortality relates to an increase in the elderly population, whose age is associated with an increased risk of developing cancer, and who are likely to have multiple comorbidities and increasing general care needs. Despite a survival rate of approximately 50 % in developed countries for all cancers combined, about 70 % of all cancer patients are estimated to need palliative care. In developing countries, this proportion rises to around 80 % (World Health Organization 2002). Hence, for the majority of cancer patients, treatment with a curative outcome proves to be either ultimately elusive or an unrealistic goal from the time of diagnosis.

Pain is present in 20–50 % of cancer patients at diagnosis and in at least 75 % of those patients with advanced disease, often in association with other distressing symptoms such as fatigue and anorexia (Donnelly and Walsh 1995; World Health Organization 1997). A survey of cancer patients suggested that pain was directly related to the cancer in 85 %; anticancer treatments such as surgery, chemotherapy, and radiotherapy in 17 %; and other comorbidities in 9 % (Grond et al. 1996).

A consensus exists among pain management specialists regarding the multidimensional nature of cancer pain, and scientific evidence supports the concept (World Health Organization 1997; Lawlor 2003; Ahles et al. 1983; Portenoy and Lesage 1999). The ultimate expression of pain intensity represents not only the perception of the basic physiological input from peripheral nociceptor activation but varying levels of multiple other inputs, which may relate to psychological state, cognitive status (the presence of delirium or dementia), the meaning of pain (for example, fears of disease progression), cultural norms, and distress of an existential and spiritual nature. The multidisciplinary palliative care model with its broad holistic principles recognizes these dimensions, and embodies a multidimensional assessment approach as an integral part of cancer pain management (Lawlor 2003).

To enable the reader to appreciate the special role of palliative care, and to assist in understanding interactive roles in the interfaces of palliative care and cancer pain management, four broad aspects are described: firstly, the historical background of the palliative care model and its interfaces; secondly, the fundamental practice principles and service delivery levels of modern palliative care; thirdly, issues in palliative care delivery in the interface with other specialties and health care personnel at the various locations and stages of cancer care; and fourthly, bridging strategies at the aforementioned interfaces.

Palliative Care Interfaces: Historical Background

Palliative care originated from the hospice model of care (MacDonald 1993). Although the term "hospice" has medieval origins referring to a place of shelter or rest, its nineteenth century use referred mainly to a site of care, but in addition, many of these sites also included "hospice for the dying" as part of their name. Historically, "hospice" therefore has a strong association with the terminal phase of illness, and the terms "▶ hospice care" and "terminal care" have been used interchangeably. Up to the twentieth century, most medical care interventions were not curative but provided symptom relief, and as such

were essentially palliative. Medical advances in the first half of the twentieth century resulted in a shift to obtaining a cure and waging a "war" against cancer. The biomedical aspects of care largely became the focus of care, often at the cost of ignoring the total illness experience from the patient and family perspective, an experience that invariably generates substantial psychosocial and spiritual needs. The more modern hospice movement, incorporating the bulk of today's palliative care principles, and typified by St Christopher's Hospice in London, was born in the late 1960s mainly to address these needs (MacDonald 1993). In the USA, hospice care eligibility is restricted by requirements for an estimated 6 month prognosis and willingness to forego life-prolonging treatment. Generally, palliative care is broader in its scope than either hospice or ▶ supportive care, and is advocated earlier in the disease trajectory. ▶ Palliative medicine now has medical specialty status in the UK, Australia, and Ireland.

Practice Principles and Service Delivery Levels of Modern Palliative Care

The complex web of palliative care with its broad holistic purview is summarized through a schematic matrix in Fig. 1. For many patients with cancer pain, progression of the disease process affects functioning in the physical, psychological, spiritual, and social domains, thereby reducing overall quality of life (QOL). Examples of problems in the QOL domains include: reduced mobility; loss of independence; depression; anger; fear; anxiety; guilt; anticipatory grieving; financial hardship; family stress; and exhaustion. The palliative care approach recognizes the distress generated in the main QOL domains, and aims to support patients and families in coping with the burden of advancing disease, in striving to achieve optimal QOL, and in adjusting over time to their inevitable demise. This recognition and intervention is especially important in the case of suffering (Cassell 1982) or total pain (Kearney 1996), where the expression of pain is attributable only in part to nociceptor activation, but perhaps in greater proportion to psychosocial and spiritual distress.

Cancer Pain Management, Interface Between Cancer Pain Management and Palliative Care,
Fig. 1 Cancer pain and the multidisciplinary palliative care approach (Adapted from http://www.who.int/cancer/palliative/en/)

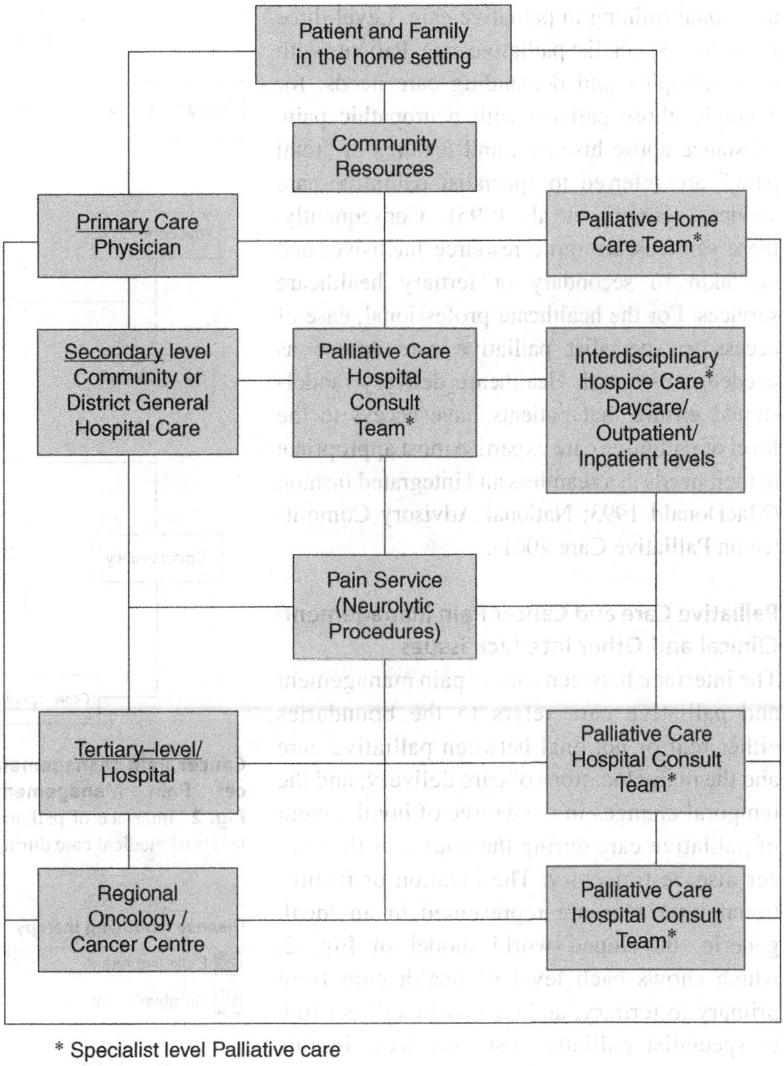

* Specialist level Palliative care

The WHO definition of palliative care states that it is "applicable early in the course of illness, in conjunction with other therapies that are intended to prolong life, such as chemotherapy or radiation therapy, and includes those investigations needed to better understand and manage distressing clinical complications" (World Health Organization 2002). This distinguishes the modern palliative care approach as being active and not necessarily passive. Nonetheless, the stigma of passivity often persists and probably reflects the more traditional origins of palliative care, especially hospice care.

In developed countries, especially those where palliative medicine is recognized as a specialty, palliative care services are often tiered from level one to three on the basis of the specialization level and expertise of the professionals delivering palliative care (National Advisory Committee on Palliative Care 2001). Level one refers to the practice of the basic "palliative care approach." This embodies a set of principles with which all health care professionals should be familiar and be capable of adopting in their practice. Level two or generalist palliative care refers to that delivered by professionals who are not practicing full-time in palliative care, but who have some additional training in palliative care. Level three refers to ▶ specialist palliative care. Patients with more complex and demanding care needs, for example, those patients with neuropathic pain, substance abuse histories, and features of "total pain," are referred to specialist palliative care services (Bruera et al. 1995). Consequently, these services are more resource intensive, and are akin to secondary or tertiary healthcare services. For the healthcare professional, ease of access to specialist palliative care advice as needed is essential. Healthcare delivery models should ensure that patients have access to the level of palliative care expertise most appropriate to their needs in a seamless and integrated fashion (MacDonald 1993; National Advisory Committee on Palliative Care 2001).

Palliative Care and Cancer Pain Management: Clinical and Other Interface Issues

The interface between cancer pain management and palliative care refers to the boundaries either real or notional between palliative care and the many locations of care delivery, and the temporal changes in the degree of involvement of palliative care during the course of the cancer disease trajectory. The location or institutional interfaces are represented in an ideal, generic, developed world model in Fig. 2, which shows each level of health care from primary to tertiary, and each with a direct link to specialist palliative care services. In this model, some international or geographic differences may occur regarding the level of interaction between and within various levels, and also regarding the degree of provision of specialist palliative care.

The temporal interface of palliative care and disease-modifying therapies, and their respective levels of use in the progressive disease trajectory (from diagnosis to death) of a hypothetical cancer patient is represented in schematic form in Fig. 3. Disease-modifying therapies such as surgery, chemotherapy, and radiation can be offered, depending on the relative burden/benefit associated with the treatment. This is done with either curative or palliative intent depending on the stage of disease. The curative intent phase for this

Cancer Pain Management, Interface Between Cancer Pain Management and Palliative Care, Fig. 2 Interface of palliative care and specific sites and levels of medical care during the cancer disease trajectory

Cancer Pain Management, Interface Between Cancer Pain Management and Palliative Care, Fig. 3 Palliative care and its temporal involvement compared to other therapies in the cancer disease trajectory

patient is relatively brief, is followed by a phase where there is modest use of palliative disease-modifying interventions such as radiation therapy. This phase ends rather abruptly prior to a short terminal or hospice phase of care. The role of palliative or supportive care progressively increases in association with disease progression, and finally envelopes the hospice contribution to the terminal phase of care.

Although the high level of need for palliative care is generally well recognized in advanced cancer, its delivery to patients and their families is often inadequate. This shortfall may be associated with various health service delivery factors of a political and/or financial nature, especially in the case of developing countries, where palliative care services are often only rudimentary or sometimes nonexistent (World Health Organization 2002). In developed countries, the shortfall in palliative care delivery may relate to professional factors such as lack of knowledge of pain management; multiple boundary issues such as the fear of losing or separating from the patient, or fear that exposure to palliative care in itself will hasten a patient's demise, or concern that the patient could feel abandoned if palliative care is consulted; and a state of relative denial of disease progression. This denial is often reflected by the relentless pursuit of burdensome treatments, whose outcomes often have a deleterious effect on quality of life, and are perceived as being unnecessarily aggressive by the palliative care professional. In addition, patient and caregiver denial may result in varying degrees of reticence to accept the palliative care approach. Cultural norms may result in a "conspiracy of silence" (denial reflected by a nondiscussion of disease status), or a "conspiracy of words" (denial reflected by limited discussion and use of euphemistic terms, for example, "spot" or "shadow" to evasively describe the presence of cancer). This must be sensitively recognized as a potential challenge to both the delivery and acceptance of palliative care in its most idealistic format (World Health Organization 1997).

Palliative Care and Cancer Pain Management: Interface Bridging Strategies

A number of strategies can allow a functionally smooth and readily traversable interface between palliative care and other specialties in the management of cancer pain and associated symptoms (MacDonald 1993).

Firstly, the need for mutual appreciation of each other's roles is of pivotal importance, for example, the palliative care physician needs to appreciate the potential palliative benefit associated with chemotherapy, radiation, or surgery, and the oncologist or surgeon needs to appreciate the benefit of early palliative care involvement for symptom control advice, support for adjustment to disease progression, and assistance with the planning of care, such as instituting palliative home care support or discussing hospice placement. Such a premise will often allow a shared care model, and thereby, the shared goal of achieving optimal QOL for patients and their families rightly takes precedence over "patient ownership" or territorial concerns.

Secondly, consistently good communication is essential. The shared, systematic use of validated, low burden assessment tools for pain and other symptoms throughout all levels of health service delivery is of great assistance in communication regarding cancer pain management (MacDonald 1993; Chang et al. 2000). Other ways to facilitate communication include person-to-person corridor conversations, joint rounds or joint clinics, and user-friendly technology. For the purpose of maintaining continuity of care, both the mutual appreciation and enhanced communication paradigms should ensure that the patient's general practitioner or family physician, and palliative home nurse are not disenfranchised, especially in the case of patients who are discharged to the community.

Other valuable bridging strategies include mutually shared research agendas; mutually shared educational rounds, ideally with some multidisciplinary input; integrated treatment sites; integrated administrative input, for example, shared participation in regional council bodies that might advise on care delivery and its development; and shared use of resources. The benefits for the patient with cancer pain as a result of smooth negotiation across the palliative care interfaces include ease of access to services, for example, fast-tracking of radiation oncology referrals; integration and continuity of care, which diminishes the risk of a sense of abandonment, a perception held by some patients when curative therapy is no longer possible.

Summary and Conclusions

In conclusion, a multidimensional assessment of cancer pain is paramount and constitutes an integral component of the palliative care approach. Optimal cancer pain management must recognize unique individual patient and family care needs, whatever the location and wherever this may occur in the cancer disease trajectory. Although the need for specialist palliative care services varies in relation to the temporal trajectory and location of cancer care, the basic palliative care approach is a fundamental requisite that essentially spans all stages and sites of care. Although various political, cultural, geographical, administrative, and financial factors will clearly influence the degree of development and support of the palliative care model in different areas, healthcare planning must aim to achieve a seamless integration of the different aspects of palliative care service delivery and other areas of cancer care, and to offer flexibility for patients to access the different levels of care (each level with a link to specialist palliative care services) as determined by their individual and specific needs.

References

Ahles, T. A., Blanchard, E. B., & Ruckdeschel, J. C. (1983). The Multidimensional Nature of Cancer-Related Pain. *Pain, 17*, 277–288.

Bruera, E., Schoeller, T., Wenk, R., et al. (1995). A prospective multicenter assessment of the Edmonton staging system for cancer pain. *Journal of Pain and Symptom Management, 10*, 348–355.

Cassell, E. J. (1982). The nature of suffering and the goals of medicine. *The New England Journal of Medicine, 306*, 639–645.

Chang, V. T., Hwang, S. S., & Feuerman, M. (2000). Validation of the Edmonton symptom assessment scale. *Cancer, 88*, 2164–2171.

Donnelly, S., & Walsh, D. (1995). The symptoms of advanced cancer. *Seminars in Oncology, 22*, 67–72.

Grond, S., Zech, D., Diefenbach, C., et al. (1996). Assessment of cancer pain: A prospective evaluation in 2266 cancer patients referred to a pain service. *Pain, 64*, 107–114.

http://www.who.int/cancer/palliative/en/. Accessed 4 November 2003.

Kearney, M. (1996). *Mortally wounded stories of soul pain, death, and healing.* New York: Scribner.

Lawlor, P. G. (2003). Multidimensional assessment: Pain and palliative care. In E. Bruera & R. K. Portenoy (Eds.), *Cancer Pain* (pp. 67–88). New York: Cambridge University Press.

MacDonald, N. (1993). The interface between oncology and palliative medicine. In D. Doyle, G. W. C. Hanks, & N. MacDonald (Eds.), *Oxford textbook of palliative medicine* (pp. 11–17). Oxford: Oxford University Press.

National Advisory Committee on Palliative Care. (2001). Palliative Care – An overview. In Report of the National Advisory Committee on Palliative Care. Dublin: Department of Health and Children.

Portenoy, R. K., & Lesage, P. (1999). Management of cancer pain. *Lancet, 353*, 1695–1700.

World Health Organization. (1997). *Looking forward to cancer pain relief for all. International consensus on the management of cancer pain* (pp. 1–70). Oxford: WHO.

World Health Organization. (2002). *Pain relief and palliative care* (pp. 83–91). Geneva: WHO.

World Health Organization and International Union against Cancer. (2003). *Global action against cancer* (pp. 1–24). Geneva: WHO and UICC.

Cancer Pain Management, Neurosurgical Interventions

Richard B. North[1] and Peter S. Staats[2]
[1]The Johns Hopkins University, Baltimore, MD, USA
[2]Division of Pain Medicine, The Johns Hopkins University, Baltimore, MD, USA

Definition

Cancer pain arises from the presence, progression, or treatment of cancer or from an unidentifiable source in patients with cancer. Cancer pain may be nociceptive, neuropathic, or a mixture of both.

A neurosurgical intervention uses an ablative, augmentative, or anatomic surgical technique, or a combination of these techniques, to relieve pain.

Characteristics

In humans, cancer assumes many guises, can affect every physiologic system and tissue, and strikes with various degrees of virulence. Cancer and its treatment cause pain in most patients, with the severity and impact of this pain dictated by a host of influences (Patt 2002).

Medical management, which is often sufficient to treat cancer pain, sometimes fails to provide adequate relief or results in side effects that substantially reduce the quality of the patient's life (nausea, constipation, vomiting, incontinence, mood changes, sedation, diarrhea, confusion, etc.) (McNicol et al. 2003). It might seem reasonable to wait until medical management fails before pursuing neurosurgical intervention, but in certain cases, for example, when a patient foreseeably will need an implanted device or ablative procedure, neurosurgeons and interventional pain physicians should intervene before the patient's condition deteriorates too far to support the intervention.

When neurosurgical techniques result in pain relief, they can enhance a patient's quality of life by increasing functional capacity and offering freedom from troublesome side effects. Compared with medical management, neurosurgical techniques can improve continuity of relief, reduce the patient's and physician's time spent adjusting dosages, and minimize the development of tolerance or adverse effects from opioids. It is possible that, in some patients, neurosurgical intervention for pain relief will prolong survival (Smith et al. 2002; Staats et al. 2001).

Patient Selection
Patient selection for neurosurgical intervention is based on a consideration of the nature of the disease, the impact of the disease on the patient's life, and the nature of the pain. One major factor in assessing disease impact and therapeutic options is the patient's prognosis, which involves quantitative and qualitative factors, such as age, life expectancy, type of cancer, the intervention's cost-effectiveness, risk/benefit ratio, duration of effect and the patient's functional capacity, personal values/wishes, and living/family

conditions. A multimodal assessment will reveal if a patient has a good (expected to survival at least a year and to be active for most of that time) or a poor prognosis (expected to survive less than 6 months and to be inactive). Pain type and severity are also important indicators of the appropriate intervention. Relative contraindications for interventional pain therapy include the presence of an untreated comorbid psychological disorder or current drug abuse (North 2002; Levy 2002).

Treatment Options
Interventional treatment options range from simple injections of short-acting local anesthetics (and, occasionally, of an adjuvant steroid or lytic agent) to complex anatomic, ablative, and augmentative neurosurgical techniques.

Anatomic procedures address structural problems causing pain – for example, tumor removal or debulking. Spinal reconstructive surgery (decompression, stabilization) for metastatic disease is a common example. To the extent that this addresses the cause of pain directly, it has obvious appeal, but the cause of pain may not be completely clear, and there may be alternatives (such as radiation therapy alone). Patients with advanced disease and limited life expectancy may not be candidates for reconstructive surgery.

Ablative procedures destroy portions of the peripheral or central nervous system to block pain transmission. This may be achieved through chemical (the direct application of alcohol or phenol), thermal (► Cryoablation or radiofrequency, see ► Radiofrequency Ablation), surgical (cutting), or radiosurgical means.

Peripheral ablative procedures are applicable to pain generators in the distribution of specific peripheral nerves, if they may be sacrificed without incurring an unacceptable deficit. For example, a peripheral ► neurectomy might relieve the pain in the distribution of the supra- or infraorbital nerve. Percutaneous radiofrequency or open ► dorsal rhizotomy has a role in cases where a tumor involves the chest wall. Sympathetic ganglionectomy involving several adjacent levels will denervate somatic and/or visceral tissue in the trunk or abdomen or in a limb. All of these may be emulated by reversible local anesthetic injections, to predict the results of a permanent procedure and thereby aid in patient selection.

Intraspinal ablative techniques interrupt input or rostral transmission of nociceptive signals in patients with severe pain and a poor prognosis. These techniques include open or percutaneous ▶ anterolateral cordotomy for intermittent and/or evoked neuropathic pain affecting one, or sometimes, two limbs (Tasker 1995), and ▶ midline commissural myelotomy for diffuse lower body pain. Cordotomy is associated with significant risks, especially with bilateral application for midline or axial pain. Cordotomy is often ruled out when pain is above C5 or the patient suffers from pulmonary dysfunction. Myelotomy addresses bilateral and midline pain in a single procedure and is best reserved for cancer patients with bladder, bowel, and sexual dysfunction. Most patients achieve nearly complete pain relief after cordotomy and myelotomy, but some experience recurrent pain months later.

Intracranial ablative therapies interrupt or change the perception of pain transmission. ▶ Stereotactic cingulotomy is appropriate for severe pain in diffuse areas of the body and provides at least 3 months relief for most cancer patients (Hassenbusch 1996). The less-frequently used ▶ stereotactic medial thalamotomy relieves pain for as many as 50 % of patients with a good or poor prognosis and severe, otherwise intractable, nociceptive pain that is widespread, midline, bilateral, or located in the head or neck; it also may help approximately 30 % with neuropathic pain (intermittent and evoked neuropathic pain responds more readily than continuous pain). ▶ Hypophysectomy is an appropriate treatment for diffuse pain associated with widespread cancer and provides relief through an unknown mechanism of action in 45–95 % of patients. These procedures may be performed not only by stereotactic probe placement but also by radiosurgery (see ▶ Stereotactic Radiosurgery).

Augmentative procedures modulate activity in the intact nervous system by electrical stimulation or continuous drug infusion to change pain perception. They have the advantages of reversibility, titration of dose, and of a trial or test phase

which emulates the definitive procedure exactly. This is not the case for anatomic or ablative procedures. Due to the high initial expense, physicians reserve implanted devices for patients expected to survive at least 3 months.

The usual indication for somatosensory stimulation is neuropathic pain restricted to a specific area, and this stimulation can target the spinal cord, a peripheral nerve, or the thalamus (to treat continuous neuropathic pain, such as Post-Radiation Plexopathy). The goal of such electrical stimulation is to induce a ▶ paresthesia that covers the painful area, effectively replacing the pain with a tolerable, non-noxious sensation. Nociceptive pain may also be treated by periaqueductal or periventricular gray stimulation. Currently there is a relative contraindication for implantation of electrical stimulation devices in those requiring repeated MRIs, so consideration of future care is important.

The best-established indication for continuous drug infusion, in particular morphine, is diffuse midline or bilateral nociceptive cancer pain. The infusion catheter enters the epidural, intrathecal, or intraventricular space, and the drug delivery system includes an implanted or an external pump (Staats and Luthardt 2004). Multiple drugs are now used intrathecally to modulate pain transmission that may be more effective than systemic medications alone (Staats and Luthardt 2004; Deer et al. 2012).

References

Deer, T. R., et al. (2012). Polyanalgesic consensus conference: Recommendations for the management of pain by intrathecal (intraspinal) drug delivery: Report of an interdisciplinary expert panel. Neuromodulation: Technology at the Neural Interface.

Hassenbusch, S. I. (1996). Intracranial ablative procedures for pain. In J. R. Youmans (Ed.), *Neurological surgery* (4th ed., pp. 3541–3551). Philadelphia: WB Saunders.

Levy, R. M. (2002). Intrathecal opioids: Patient selection. In K. Burchiel (Ed.), *Surgical management of pain* (pp. 469–484). New York: Thieme.

McNicol, E., Horowicz-Mehler, N., Fisk, R. A., American Pain Society, et al. (2003). Management of opioid side effects in cancer-related and chronic noncancer pain: A systematic review. *The Journal of Pain, 4*, 231–256.

North, R. B. (2002). Spinal cord stimulation: Patient selection. In K. Burchiel (Ed.), *Surgical management of pain* (pp. 469–484). New York: Thieme.

Patt, R. B. (2002). Cancer pain. In K. Burchiel (Ed.), *Surgical management of pain* (pp. 469–484). New York: Thieme.

Smith, T. J., Staats, P. S., Pool, G., et al. (2002). Randomized clinical trial of an implantable drug delivery system compared to comprehensive medical management for refractory cancer pain: Impact on pain, drug-related toxicity, and survival. *Journal of Clinical Oncology, 20,* 4040–4049.

Staats, P. S., Hekmat, H., Sauter, P., & Lillemoe, K. (2001). The effects of alcohol celiac plexus block, pain, and mood on longevity in patients with unresectable pancreatic cancer: A double-blind, randomized placebo-controlled study. *Pain Medicine, 2,* 28–34.

Staats, P. S., & Luthardt, F. (2004). *Intrathecal therapy for cancer pain. Just the facts pain medicine.* New York: McGraw Hill.

Tasker, R. R. (1995). Percutaneous cordotomy. In H. H. Schmidek & W. H. Sweet (Eds.), *Operative neurosurgical techniques: Indications methods and results* (pp. 1595–1611). Philadelphia: WB Saunders.

Cancer Pain Management, Nonopioid Analgesics

Alison Murray and Jose Pereira
Foothills Medical Center, Calgary, AB, Canada

Definition

Nonopioid analgesics comprise of all those medications prescribed for pain relief that do not fall into the opioid class. These include paracetamol (acetaminophen), aspirin (acetylsalicylic acid), and the nonsteroidal anti-inflammatory drugs (NSAIDS). Paracetamol is an analgesic-antipyretic with weak anti-inflammatory activity. Aspirin is a derivative of salicylic acid and has analgesic, antipyretic, and anti-inflammatory activities. Nonsteroidal anti-inflammatory drugs are a heterogeneous group of compounds. They share the same therapeutic action and side-effect profile as aspirin. Nonopioid analgesics are recommended for use in the case of mild cancer pain, as described in Step I of the World Health Organization Analgesic Ladder for the treatment

of cancer pain (World Health Organization 1986). Nonopioids were used on 11 % of treatment days in a large prospective study involving 2,118 cancer patients, while the so-called weak opioids (WHO Step II) were used on 31 % of days and the so-called strong opioids (Step III) on 49 % of treatment days (Zech et al. 1995). In another study, 52 % of cancer patients started on a nonopioid analgesic needed to be switched to a weak or strong opioid because of escalating pain (Ventafridda et al. 1987). Nonopioid analgesics are also used as adjuvant medications to opioids for the management of moderate-to-severe pain (Step II and III of the ▶ WHO Analgesic Ladder).

Characteristics

Paracetamol (Acetaminophen)
Mode of Action
Paracetamol has analgesic and antipyretic effects but only weak anti-inflammatory effects. Its mechanism of action is unknown, although it is thought to act centrally rather than peripherally (Flower et al. 1980). Paracetamol may be given orally or rectally.

Pharmacokinetics
It is metabolized by the liver.

Efficacy
Paracetamol is considered to be a weak analgesic without anti-inflammatory effects. There is little evidence supporting the use of paracetamol alone in cancer pain, although there are some studies which support its use when combined with an opioid (codeine or oxycodone) for moderate pain (Carlson et al. 1990). Paracetamol 600 mg plus codeine 60 mg has been found to be equivalent to ketorolac 10 mg, and superior to placebo in reducing cancer pain in a double-blind randomized controlled study.

There is insufficient evidence to support the addition of paracetamol to strong opioids for moderate-to-severe cancer pain (McNicol et al. 2005; Naba et al. 2012).

Adverse Effects

Side effects include rash, urticaria, and nausea. Paracetamol may induce hepatotoxicity after acute ingestion of large doses (> 10 g) or chronic ingestion of daily doses exceeding 4 g. Other serious but rare side effects include nephrotoxicity, blood dyscrasias, pancreatitis, and angioedema. Caution should be used in patients with impaired liver function/chronic alcohol use, or impaired renal function (Flower et al. 1980). Most clinicians will consider paracetamol safe to use in patients with liver metastases on the condition that overt hepatic failure is not present. There are no studies to support or refute this practice.

Nonsteroidal Anti-Inflammatory Drugs (NSAIDs)

Mode of Action

NSAIDs inhibit peripheral prostaglandin synthesis through the inhibition of cyclooxygenase enzymes, COX-1 (found in normal cells) and COX-2 (induced during the inflammatory process). Inhibition of COX-2 produces anti-inflammatory effects: decreased release of inflammatory mediators (such as substance P and cytokines), which leads to decreased stimulation of peripheral ▶ nociceptors (Jenkins and Bruera 1999). COX-1 inhibition can lead to many of the known side effects of NSAIDs, in particular gastrointestinal ulceration and inhibition of platelet aggregation. It has also been suggested that NSAIDs may have a central effect on pain perception by reducing NMDA-mediated hyperalgesia (Jenkins and Bruera 1999).

The older generation of NSAIDs (including aspirin) has variable effects on both COX-1 and COX-2. A new generation of NSAIDs (etodolac, meloxicam, nabumetone, celecoxib, valdecoxib, and rofecoxib) shows COX-2-selective inhibition (Jenkins and Bruera 1999), and those created most recently (the coxibs: celecoxib, valdecoxib, and rofecoxib) are highly COX-2 selective. Rofecoxib was removed from the world market in 2004 due to increased cardiovascular risk profile. There are currently ongoing trials investigating a new role for COX-2 inhibitors in cancer prevention and treatment.

Pharmacokinetics

NSAIDS are given by the oral or the rectal route, with the exception of ketorolac, which is also available in a parenteral preparation. NSAIDs are metabolized by the liver and excreted in the urine and feces (Flower et al. 1980). A topical diclofenac gel is available in North America and Europe.

Efficacy

In multiple dose trials, NSAIDs have been found to be as effective as weak opioids or weak opioid/paracetamol preparations (Goudas et al. 2001). No individual NSAID or class of NSAID has been shown to be more effective in the control of pain (Goudas et al. 2001; Mercadante 2001). NSAIDs appear to have a dose–response relationship with respect to analgesic efficacy (Eisenberg et al. 1994), and they also demonstrate a ceiling effect (supramaximal doses do not demonstrate superiority over recommended doses) (Eisenberg et al. 1994). The efficacy of NSAIDs does not vary with the route of administration (Tramer et al. 1998). It has been common practice to prescribe NSAIDs for certain cancer pain syndromes (in particular, metastatic bone pain), but the evidence to support this is largely lacking (Goudas et al. 2001; Mercadante 2001; Eisenberg et al. 1994).

There are no current trials available involving the use of COX-2-selective inhibitors in cancer pain. Data from the nonmalignant pain literature (largely osteoarthritis and rheumatoid arthritis populations) demonstrate that coxibs are as effective as nonselective NSAIDs in terms of pain relief (Kuritzky and Weaver 2003).

There are no available studies on the use of topical NSAID gel and cancer pain.

When NSAIDs are coadministered with opioids as ▶ adjuvant analgesics, they produce an opioid-sparing effect, but there is no consistent reduction in side effects (Jenkins and Bruera 1999; Goudas et al. 2001; Mercadante 2001). A systematic review concluded that there is weak evidence to support the addition of NSAIDs to strong opioids for improved analgesia or opioid-sparing effect (Naba et al. 2012).

Side Effects

The incidence of side effects with NSAIDs demonstrates a dose–response relationship and rises significantly after multiple dosing over 7–10 days (Eisenberg et al. 1994). One of the major side effects of nonselective NSAIDs is gastrointestinal ulceration (resulting in clinically significant gastrointestinal bleeding) (Mercadante 2001). The majority of studies have been performed in noncancer populations. NSAID users have three times the risk of gastrointestinal ulceration compared to nonusers (Jenkins and Bruera 1999).Advanced age, history of peptic ulcer disease, and concurrent corticosteroid or anticoagulant therapy are known risk factors for NSAID-induced gastrointestinal bleeding and perforation (Hernandez-Diaz and Rodriguez 2000). Omeprazole has been shown to decrease the incidence of NSAID-induced gastrointestinal ulceration. Eradication of *Helicobacter pylori* prior to NSAID administration may reduce the risk in the general population (Jenkins and Bruera 1999). In uncontrolled studies of cancer populations, dyspepsia has been reported by 7–13 % of users, and 5–20 % of users discontinued NSAIDs due to an adverse event (Goudas et al. 2001). In cancer patients, advanced disease and the presence of liver disease (primary cancer or metastases) have been associated with a higher rate of gastrointestinal bleeding events (Mercadante et al. 2000).

A second side effect is impaired renal function. Acute renal dysfunction is usually reversible and improves after discontinuation of the NSAID. It is caused by a decrease in prostaglandin-dependent renal plasma flow. Older age, hypertension, concomitant use of diuretics, preexisting renal failure, diabetes, and dehydration are risk factors for acute NSAID-induced renal failure (De Broe and Elseviers 1998). Chronic and permanent renal failure secondary to NSAIDs is much rarer, and is probably due to acute tubular necrosis (Mercadante 2001).

Other known side effects of NSAIDs include platelet inhibition, blood dyscrasias, hepatic damage, and CNS toxicity (tinnitus, visual disturbances, dizziness, etc.) (Mercadante 2001).

There are no trials comparing the side effects of COX-2 inhibitors compared to nonselective NSAIDs in cancer populations (Goudas et al. 2001). The use of COX-2 inhibitors is associated with fewer endoscopic ulcers and fewer gastrointestinal events compared to nonselective NSAIDs in the general population (Laine 2003), but they have a similar rate of nephrotoxic events. Coxibs do not inhibit platelet function as nonselective NSAIDs do. Overall, there is not sufficient evidence to support the use of selective COX-2 inhibitors over the use of nonselective NSAIDS in the treatment of cancer pain (Davis et al. 2005).

COX-2 inhibitors have been associated with an increased risk of cardiovascular events, primarily myocardial infarction (Solomon et al. 2004). This led to the withdrawal of rofecoxib from the world market in 2004. A systematic review of the cardiovascular risk of the nonselective NSAIDS shows that ibuprophen and naproxen have the lowest risk of cardiovascular events (McGettigan and Henry 2011).

Dipyrone

Dipyrone has anti-inflammatory, antipyretic, and analgesic effects. Its mechanism of action is unknown. It is used in Europe, but it is not available in North America due to concerns about agranulocytosis (Flower et al. 1980).

References

Carlson, R. W., Borrison, R. A., Sher, H. B., et al. (1990). A multi-institutional evaluation of the analgesic efficacy and safety of ketorolac tromethamine, acetaminophen plus codeine, and placebo in cancer pain. *Pharmacotherapy, 10*, 211–216.

Davis, M. P., Walsh, D., Lagman, R., & LeGrand, S. B. (2005). Controversies in pharmacotherapy of pain management. *The Lancet Oncology, 6*, 696–704.

De Broe, M. E., & Elseviers, M. M. (1998). Analgesic nephropathy. *The New England Journal of Medicine, 338*, 446–452.

Eisenberg, E., Berkey, C. S., Carr, D., et al. (1994). Efficacy and safety of nonsteroidal anti-inflammatory drugs for cancer pain: A meta-analysis. *Journal of Clinical Oncology, 12*, 2756–2765.

Flower, R. J. M. S., Vane, J. R., & Moncada, S. (1980). Analgesic-antipyretics and anti-inflammatory agents; Drugs employed in the treatment of gout. In L. S. Goodman & A. Gilman (Eds.), *Goodman and Gilman's the pharmacological basis of therapeutics* (6th ed.). New York: Macmillan.

Goudas, L., Carr, D. B., Bloch, R., et al. (2001). Management of cancer pain. *Evidence Report/Technology Assessment, 35,* 1–5.

Hernandez-Diaz, S., & Rodriguez, L. A. (2000). Association between nonsteroidal anti-inflammatory drugs and upper gastrointestinal tract bleeding/perforation: An overview of epidemiologic studies published in the 1990's. *Archives of Internal Medicine, 160,* 2093–2099.

Jenkins, C. A., & Bruera, E. (1999). Nonsteroidal anti-inflammatory drugs as adjuvant analgesics in cancer patients. *Palliative Medicine, 13,* 183–196.

Kuritzky, L., & Weaver, A. (2003). Advances in rheumatology: Coxibs and beyond. *Journal of Pain and Symptom Management, 25,* 6–20.

Laine, L. (2003). Gastrointestinal effects of NSAIDs and coxibs. *Journal of Pain and Symptom Management, 25,* 32–40.

McGettigan, P., Henry, D. (2011). Cardiovascular risk with nonsteroidal anti-inflammatory drugs: systematic review of population-based controlled observational studies. *PLoS Med.* doi:10.1371/journal.pmed.1001098. Available at: http://www.plosmedicine.org.

McNicol, E., Strassels, SA., Goudas, L., Lau, J., & Carr, DB. (2005). NSAIDs or paracetamol, alone or combined with opioids for cancer pain. *Cochrane Database Syst Rev, 25*(1), CD005180.

Mercadante, S. (2001). The use of anti-inflammatory drugs in cancer pain. *Cancer Treatment Reviews, 27,* 51–61.

Mercadante, S., Barresi, L., Cassucio, A., et al. (2000). Gastrointestinal bleeding in advanced cancer patients. *Journal of Pain and Symptom Management, 19,* 160–162.

Naba, M., Librada, S., Redondo, M. J., Pigni, A., Brunelli, C., & Caraceni, A. (2012). The role of paracetamol and nonsteroidal anti-inflammatory drugs in addition to WHO Step III opioids in the control of pain in advanced cancer. A systematic review of the literature. *Palliative Medicine, 26*(4), 305–312.

Solomon, D. H., Schneeweiss, D. S., Glynn, R. J., Kiyota, Y., Levin, R., Mogun, H., & Avorn, J. (2004). Relationship between selective cyclooxygenase-2 inhibitors and acute myocardial infarction in older adults. *Circulation, 109*(17), 2068–2073.

Tramer, M. R., Williams, J. E., Carroll, D., et al. (1998). Comparing analgesic efficacy of non-steroidal anti-inflammatory drugs given by different routes in acute and chronic pain: A qualitative systematic review. *Acta Anaesthesiologica Scandinavica, 42,* 71–79.

Ventafridda, V., Tamburini, M., Caraceni, A., et al. (1987). A validation study of the WHO method for cancer pain relief. *Cancer, 59,* 850–856.

World Health Organization. (1986). *Cancer pain relief.* Geneva: World Health Organization.

Zech, D. F., Grond, S., Lynch, J., et al. (1995). Validation of world health organization guidelines for cancer pain relief: A 10-year prospective trial. *Pain, 63,* 65–76.

Cancer Pain Management, Opioid Side Effects, Cognitive Dysfunction

Kimberson Cochien Tanco, Eduardo Bruera and Sriram Yennurajalingam
Department of Palliative Care and Rehabilitation Medicine, The University of Texas M. D. Anderson Cancer Center, Houston, TX, USA

Introduction

Opioid prescriptions have increased markedly in recent years driven by a more concentrated focus on pain management. Cancer pain occurs in 33 % of patients after curative treatment, 59 % of patients on anticancer therapies, and 64 % of patients with metastatic, advanced, or terminal cancer (Ripamonti et al. 2012). It has become equally as important to recognize the neurological effects of these medications.

Definition

► Cognitive dysfunction refers to changes in consciousness, higher cortical functions, mood, or perception that may be induced by any of numerous neurological or systemic diseases or by ingestion of substances, including drugs, that have the potential for central nervous system toxicity.

Characteristics

Impaired cognition may be seen in the earlier stages of cancer. This may manifest as a slight decrease in comprehension, loss of abstract thinking, difficulty in word finding, forgetfulness, mental fatigue, difficulty in concentrating, and irritability. Cancer-related impairment can be difficult to distinguish from opioid-induced impairment. Symptom clusters like uncontrolled pain, anxiety, depression, and fatigue also affect the expression and severity of cognitive dysfunction. Furthermore, other drugs and modalities of cancer treatment such as chemotherapy also have cognitive effects (Kurita and Lundorff 2009). Finally, other comorbidities like metabolic disorders, infection, dehydration, and dementia need to be considered (Lawlor 2002). Lung cancer, daily morphine equivalent doses of greater than or equal to 400 mg, older age, low performance status, and time since diagnosis fewer than 15 months were associated with increased risk of cognitive dysfunction. On the other hand, the presence of breakthrough pain had a negative correlation (Kurita et al. 2011).

Opioid-induced sedation usually occurs when opioid dosing is initiated or when the dose is increased. Approximately 7–10 % of patients receiving strong opioids for cancer pain have persistent sedation related to opioids medication (Bruera et al. 1995). In some cases, this sedation is multifactorial; the opioid may contribute but may not be the primary cause. Some patients, however, appear to have a very narrow or nonexistent therapeutic window, and the persistent sedation can be ascribed directly to the drug.

Opioid-induced neurotoxicity may take the form of a syndrome that may include, in addition to sedation, cognitive impairment, hallucinations, delirium, myoclonus, and even seizures. Generalized hyperalgesia and allodynia may also occur. This syndrome, which can occur in milder and partial forms, appears to be dose related and also potentially associated with preexisting cognitive impairment or delirium or metabolic disturbances such as dehydration or renal failure. When these changes occur in morphine-treated patients, accumulation of morphine metabolites, specifically morphine-3-glucuronide (M3G), may be causative.

Chronic opioid exposure has been documented to be associated with deficits in verbal working memory, cognitive impulsivity or risk taking, and cognitive flexibility or verbal fluency (Baldacchino et al. 2012). Other neuropsychological domains shown to be affected include strategic planning, perseveration, attention, visuospatial ability, long-term memory, and learning (Ersche and Sahakian 2007; Verdejo-Garcia and Perez-Garcia 2007).

Aside from symptomatic and behavioral-related cognitive changes, neuroplastic changes in specific areas of the brain have been observed. A volumetric decrease in gray matter was seen in the right amygdala, which is responsible for drug-induced associative learning, drug craving, reinforcement, dependence, and acute withdrawal. Moreover, volumetric increase in gray matter was observed in areas outside the reward-processing networks including the right hypothalamus, left inferior frontal gyrus, right ventral posterior cingulate, and right caudal pons. These areas are rich in mu-opioid receptors. Likewise, the hypothalamic hypocretin system, which is integral in cognitive arousal and attention, is suppressed. Follow-up scans after an average of 4.7 months after opioid cessation showed that these changes were permanent (Younger et al. 2011).

Minor cognitive deficits are often precursors of more profound cognitive dysfunction. Regular objective cognitive monitoring is essential to circumvent the possibility of overlooking or misdiagnosing delirium or cognitive deficits as other psychiatric or systemic disorder like depression.

The management of severe neurotoxicity related to opioids incorporates the same strategies considered when the side effects are less severe. This begins with a detailed assessment to identify treatable causes. Management of opioid-related cognitive dysfunction includes use of opioid-sparing regimens, dose reduction, opioid rotation, discontinuation of contributory drugs like hypnotics, and addition of a drug targeting the opioid-induced adverse effect (Strassels 2008). Opioid-sparing medications include adjuvant analgesics like NSAIDs, corticosteroids, and drugs focused on neuropathic pain like gabapentin, amitriptyline, or pregabalin. However, caution must also be used in utilizing these adjuvants as they may also cause sedation and other complications. Psychostimulants, such as dextroamphetamine, methylphenidate, or modafinil, may help to counteract fatigue and sedation from opioids. Patients who are not candidates for psychostimulant therapy may be tried on an anticholinesterase inhibitor, such as donepezil (Slatkin et al. 2001). In patients with persistent sedation, a change of opioid also may be helpful. Symptomatic treatment with haloperidol or another neuroleptic can be given in cases of delirium. Benzodiazepines may be needed in cases of persistent myoclonus.

References

Baldacchino, A., Balfour, D. J. K., Passetti, F., Humphris, G., & Matthews, K. (2012). Neuropsychological consequences of chronic opioid use: A quantitative review and metaanalysis. *Neuroscience and Biobehavioral Reviews, 26*, 2056–2068.

Bruera, E., Watanabe, S., Faisinger, R. L., et al. (1995). Custom-made capsules and suppositories of methadone for patients on high dose opioids for cancer pain. *Pain, 62*, 141–146.

Ersche, K. D., & Sahakian, B. J. (2007). The neuropsychology of amphetamine and opiate dependence: Implications for treatment. *Neuropsychology Review, 17*, 317–336.

Kurita, G. P., & Lundorff, L. (2009). The cognitive effects of opioids in cancer: A systematic review. *Supportive Care in Cancer, 17*, 11–21.

Kurita, G. P., Sjogren, P., Ekholm, O., Kaasa, S., Loge, J. H., Poviloniene, I., & Klepstad, P. (2011). Prevalence and predictors of cognitive dysfunction in opioid-treated patients with cancer: A multinational study. *Journal of Clinical Oncology, 29*(10), 1297–1303.

Lawlor, P. (2002). The panorama of opioid-related cognitive dysfunction in patients with cancer. *Cancer, 94*(6), 1836–1853.

Ripamonti, C. I., Santini, D., Maranzano, E., Berti, M., & Roila, F. (2012). Management of cancer pain: ESMO clinical practice guidelines. *Annals of Oncology, 23* (Suppl. 7), vii139–vii154.

Slatkin, N. E., Rhiner, M., & Maluso Bolton, T. (2001). Donepezil in the treatment of opioid-induced sedation: Report of six cases. *Journal of Pain and Symptom Management, 21*, 425–438.

Strassels, S. A. (2008). Cognitive effects of opioids. *Current Pain and Headache Reports, 12*, 32–36.

Verdejo-Garcia, A. J., & Perez-Garcia, M. (2007). Profile of executive deficits in cocaine and heroin polysubstance users: Common and differential effects on separate executive components. *Psychopharmacology, 190*, 517–530.

Younger, J. W., Chu, L. F., D'Arcy, N. T., Trott, K. E., Jastrzab, L. E., & Mackey, S. C. (2011). Prescription opioid analgesics rapidly change the human brain. *Pain, 152*, 1803–1810.

Cancer Pain Management, Opioid Side Effects, Endocrine Changes, and Sexual Dysfunction

Russell K. Portenoy[1] and Ebtesam Ahmed[2]
[1]Department of Pain Medicine and Palliative Care, Beth Israel Medical Center, New York, NY, USA
[2]College of Pharmacy and Health Sciences, St. John's University, Department of Pain Medicine and Palliative Care, Beth Israel Medical Center, New York, NY, USA

Introduction

Opioids change endocrine function in a large proportion of patients. They reduce the production of sex hormones in both sexes and increase production of prolactin. These changes may cause clinical hypogonadism in both men and women, with varied clinical consequences.

Characteristics

The effect of opioid compounds on hormonal function has been termed opioid endocrinopathy, or in case of androgen hormones, opioid-induced androgen deficiency. Hormonal effects may be produced by any mu agonist drug administered by any route of administration (Rajagopal 2003; Bliesener 2005; Oltmanns 2005). For example, opioids suppress plasma testosterone levels in persons using heroin, methadone, or other opioids (Finch et al. 2000; Paice et al. 1994; Mendelson et al. 1975), and testosterone levels return to normal after stopping opioid therapy (Finch et al. 2000). Both systemic and intraspinal opioid delivery lower hormone levels (Abs et al. 2000).

In addition to direct effects on the secretion of sex hormones, opioids increased the level of prolactin, which secondarily affects testosterone and other hormones. Prolactin is increased during heroin use (Ellingboe et al. 1980), in normal subjects given intravenous injections of morphine

(Delitala et al. 1983), and in cancer patients given intraventricular injections of morphine (Su et al. 1987). Presumably, opioids block the normal hypothalamic mechanisms for tonic suppression of prolactin release from the anterior pituitary, a process-mediated dopamine release. Prolactin reduces levels of luteinizing hormone (LH) and follicle stimulating hormone (FSH) in men and women, which subsequently depress testosterone and other sex hormones; in females, prolactin also stimulates lactation.

In patients undergoing short-term treatment of pain with an opioid, endocrine effects are not clinically important. During long-term therapy, however, hypogonadism has many potential consequences. In both men and women, hypogonadism may be associated with sexual dysfunction or fertility problems; disturbances in the menstrual cycle are common among premenopausal women treated with opioids. Hypogonadism may cause fatigue and depressed mood, symptoms that may be multiply determined in chronic pain patients. Long-term changes in bone (increased osteoporosis) and muscle (diminished muscle mass) also may result.

Given the potential for adverse clinical consequences of opioid-induced hypogonadism during long-term therapy, clinicians have begun to address this side effect. There is yet no evidence of the benefits of hormonal treatment in opioid-induced hypogonadism, and the decision to treat is made case-by-case after weighing risks and benefits. In men, many clinicians now will measure testosterone and free testosterone during opioid treatment; a check of hormone levels may be particularly justified if patients report sexual dysfunction, change in secondary sex characteristics, or symptoms of fatigue or depressed mood. The primary treatment for male opioid-induced testosterone deficiency is testosterone supplementation (Daniell 2006), preferably by topical cream, gel, or patch. Topical medications are preferred over injections because they provide relatively stable testosterone concentrations not easily attainable with intramuscular injections (Abs et al. 2000). Erectile dysfunction also may be addressed by the use of sildenafil, tadalafil, or vardenafil.

There is less agreement about the appropriate clinical response to potential hypogonadism in women. In premenopausal women, changes in the menstrual cycle accompanied by symptoms such as fatigue may warrant a check of hormone levels, followed by a trial of estrogen therapy if appropriate. In postmenopausal women, the risks of long-term hormone replacement therapy are well known, but measurement of hormone levels may nonetheless be appropriate if pain is accompanied by sexual dysfunction, fatigue or depression, or the finding of osteoporosis during screening. Estrogen replacement, or for some types of sexual dysfunction, trials of low-dose testosterone or a phosphodiesterase type 5 inhibitor like sildenafil, are considered in some cases. Expert advice usually is needed to help rationalize a plan of care.

Occasionally, opioid-induced hypogonadism is associated with hyperprolactinemia, severe enough to warrant treatment. Like repletion therapy with testosterone or estrogen, there are no data from adequate clinical trials to help inform clinical decisions about the risks and benefits in the setting of opioid-induced endocrinopathy. Recommendations based on clinical judgment usually require the input of an endocrine specialist. Treatment for hyperprolactinemia with bromocriptine, pergolide, or cabergoline can normalize serum prolactin levels and restore a more normal response to testosterone (Fleming and Paice 2001).

Additional research is needed to determine the prevalence and clinical impact of opioid-induced endocrinopathy. At present, hypogonadism should be recognized as among the most common potential adverse effects of long-term opioid therapy. In the absence of data, decisions about treatment must be made on a case-by-case basis following a careful assessment of the pain and comorbidities, the opioid regimen, and other conditions that may be associated with endocrine dysfunction. Measurement of testosterone and free testosterone in the setting of long-term opioid therapy for pain is often performed in men, and trials of testosterone replacement are becoming more common. Treatment of postmenopausal women is more fraught due to the recognized risks of long-term hormone replacement therapy, but the potential benefits may be such that a treatment trial is indicated. The input of endocrine specialists is valuable in making these judgments.

References

Abs, R., Verhelst, J., Maeyaert, J., et al. (2000). Endocrine consequences of long-term intrathecal administration of opioids. *Journal of Clinical Endocrinology and Metabolism, 85*, 2215–2222.

Bliesener, N., Albrecht, S., Schwager, A., Weckbecker, K., Lichtermann, D., & Klingmuller, D. (2005). Plasma testosterone and sexual function in men receiving buprenorphine maintenance for opioid dependence. *Journal of Clinical Endocrinology and Metabolism, 90*, 203–206.

Daniell, H. W., Lentz, R., & Mazer, N. A. (2006). Open label pilot study of testosterone patch therapy in men with opioid-induced androgen deficiency. *Journal of Pain, 7*, 200–210.

Delitala, G., Grossman, A., & Besser, G. M. (1983). The participation of hypothalamic dopamine in morphine-induced prolactin release in man. *Clinical Endocrinology, 19*, 437–444.

Ellingboe, J., Mendelson, J. H., & Kuehnle, J. C. (1980). Effects of heroin and naltrexone on plasma prolactin levels in man. *Pharmacology Biochemistry and Behavior, 12*, 163–165.

Finch, P. M., Roberts, L. J., Price, L., et al. (2000). Hypogonadism in patients treated with intrathecal morphine. *The Clinical Journal of Pain, 16*, 251–254.

Fleming, M. P., & Paice, J. A. (2001). Sexuality and chronic pain. *Journal of Sex Education and Therapy, 26*, 204–214.

Mendelson, J. H., Meyer, R. E., Ellingboe, J., et al. (1975). Effects of heroin and methadone on plasma cortisol and testosterone. *Journal of Pharmacology and Experimental Therapeutics, 195*, 296–302.

Oltmanns, K. M., Fehm, H. L., & Peters, A. (2005). Chronic fentanyl application induces adrenocortical insufficiency. *Journal of Internal Medicine, 257*, 478–480.

Paice, J. A., Penn, R. D., & Ryan, W. (1994). Altered sexual function and decreased testosterone in patients receiving intraspinal opioids. *Journal of Pain and Symptom Management, 9*, 126–131.

Rajagopal, A., Vassilopoulou-Sellin, R., Palmer, J. L., Kaur, G., & Bruera, E. (2003). Hypogonadism and sexual dysfunction in male cancer survivors receiving chronic opioid therapy. *Journal of Pain and Symptom Management, 26*, 1055–1061.

Su, C. F., Liu, M. Y., & Lin, M. T. (1987). Intraventricular morphine produces pain relief, hypothermia, hyperglycaemia and increased prolactin and growth hormone levels in patients with cancer pain. *Journal of Neurology, 235*, 105–108.

Cancer Pain Management, Orthopedic Surgery

John H. Healey[1] and Scott R. Miller[2]
[1]Orthopaedic Service, Department of Surgery, Memorial Sloan-Kettering Cancer Center Weill Medical College of Cornell University, New York, NY, USA
[2]Memorial Sloan-Kettering Cancer Center, New York, USA

Definition

Orthopedic cancer pain includes biologic or mechanical pain that affects the musculoskeletal system and compromises the function and independence of cancer patients. The neurochemical basis of bone pain differs from that of inflammatory or neuropathic pain. It is often multifactorial in origin and requires treatment of all contributing factors. Options for management include medication, radiation, chemotherapy, and surgical stabilization.

Characteristics

Differentiating Pain in the Cancer Patient
The management of cancer pain requires an accurate diagnosis and emphasizes an appropriate clinical work-up to eliminate alternative causes of pain. Factors to consider that are of particular importance in the cancer patient include cancer diagnosis, timing of symptoms, and oncologic treatments. This information assists in differentiating cancer-associated pain from concomitant disease processes such as spinal cord compression or arthritis. Comprehensive management of bone pain requires specific treatment to address each cause.

Radicular pain occurs in patients without cancer but may also be the result of tumors of the spine. Radicular pain (radiculopathy) is experienced in the distribution of specific spinal nerves, causing upper- or lower-extremity pain. Pericapsular or arm pain can result from cervical tumors, while lumbar tumors may cause buttock or lower-limb pain. This pain may or may not be associated with neurologic symptoms of a sensory or motor nature.

Referred pain arises from irritation to a dermatome and typically is caused by a lesion affecting the same bone proximal to the symptomatic site. The most common example is distal femur and knee pain due to a proximal femur or hip lesion. Appropriate physical examination is of paramount importance. The limp that is due to a knee problem differs significantly from that of a hip problem and should alert the clinician to problems proximal to the area of perceived pain. In the case of the proximal femoral lesion, certain provocative tests have compelling diagnostic value. For example, passive abduction or active adduction of the hip isolates the adductor musculature and its pubic origin as the source of pain. Hence, by ruling out intrinsic hip pathology, an unnecessary hip replacement operation can be avoided. Another common source of pain is an avulsion fracture of the anterior inferior iliac spine. Since this is the origin of the rectus femoris, activities that stress the muscle (such as raising a straightened leg from a supine position) can precipitate pain, whereas hip flexion from the seated position does not.

Degenerative arthritis of the hip, knee, spine, or shoulder may be present in older cancer patients. Avascular necrosis of the hip or shoulder can develop from steroids, chemotherapy, or radiation therapy utilized to treat the underlying oncologic process. Simple radiographs are the best way to look for these traditional, non-oncologic causes of bone pain in the cancer patient. It is important not to attribute all pain to the patient's underlying tumor. Occasionally, paraneoplastic syndromes may cause bone or joint pain and elude detection by the usual diagnostic tests. This problem can occur with any cancer but seems to be most common among patients with small cell lung cancer and HIV-related malignancies. Infection must also be excluded, particularly in patients with hematologic malignancies or treatment-related neutropenia.

Cancer itself within the bone may cause pain. Breast cancer cells trigger an increase in calcitonin gene-related product (CGRP) expression in nerves within the bone and induce the growth of nerve fibers within the bone. As a result of increased nerve growth and neurotransmitter levels, metastatic cancer can produce pain (Bloom et al. 2011).

During the course of chemotherapy treatment, patients often receive granulocyte colony-stimulating factor (GCSF), which produces localized and diffuse bone pain. The cause is poorly defined, but it seems to be the result of acute pressure that develops in the bone marrow from marrow expansion and possibly from the evolution of nociceptors and cytokines in response to GCSF.

Classically, orthopedic cancer pain is a *mechanical* pain. Excess force or weight-bearing bends the mechanically insufficient bone. This stretches the periosteum, stimulates pain receptors, and produces pain. Prior to fracture, an otherwise normal stimulus (i.e., walking or lifting) can trigger significant pain. This suggests an impending pathologic fracture and requires immediate attention. Once the fracture occurs, there are associated periosteal tears, hemorrhage, and soft tissue trauma that lead to further pain.

Diagnostics

A full work-up for an oncologic lesion is prudent. In patients with known primary disease, this is best accomplished with a full-body bone scan and subsequent biopsy of an accessible lesion. In those patients without an existing cancer diagnosis, radiographic study selection is tailored to the most likely primary source (computed tomography [CT] of the chest, abdomen, pelvis) as suggested by history, physical examination, and laboratory data. Plain radiographs remain the most efficient way to assess the global integrity of the bone in the appendicular skeleton and to determine whether surgery is a consideration to treat the mechanical pain. CT imaging should be considered in lesions involving the pelvic girdle and the spine. Magnetic resonance imaging (MRI) can be effective in detecting otherwise occult lesions. It is also absolutely necessary when a primary bone malignancy is in the differential diagnosis (Weber 2010).

Cancer Pain at the Molecular Level

Pain in the periphery is processed as such at the site of insult and signaled to the appropriate centers in the brain by primary afferent sensory nerves. There are two subtypes: A-fibers (myelinated) and C-fibers (unmyelinated). Most C-fibers are nociceptors and convey harmful peripheral stimuli to the central nervous system. These nerves detect chemical (endothelins, prostaglandins, bradykinin, nerve growth factor) and mechanical changes in the vicinity. They can gradually undergo peripheral sensitization, resulting in a lower stimulus-response threshold. For example, in an animal model of bone cancer, normal, "sub-painful" levels of mechanical stress to the femur result in the release of substance P from nociceptors (a "pain" signal). In the local environment of the tumor, various cells secrete numerous molecules, including interleukin [IL]-1, IL-6, transforming growth factor (TGF)-alpha and -beta, platelet-derived growth factor (PDGF), epidermal growth factor (EGF), prostaglandins, and endothelins, all of which can contribute to the peripheral sensitization of afferent nerves and an increased pain burden (Mantyh et al. 2002).

Biologically, rapidly growing tumors increase intraosseous pressure (by blocking venous blood return) and release pain-mediating substances. Many neurotransmitters and nerve fibers that are involved in normal development of pathologic conditions have been identified within the bone (Bergstrom et al. 2003). Substance P, CGRP, vasointestinal peptide, and other compounds that stimulate nociceptors, augment local perfusion, and/or have direct metabolic effects on bone further contribute to the pain generation. Nerve fibers that are histologically positive for CGRP are widely distributed in the bone, periosteum, and marrow. They directly regulate local bone remodeling during growth and repair by elevating the CGRP concentration in the cell microenvironment (Irie et al. 2002; Mantyh et al. 2002).

Mechanical and structural stability also play a substantial role in bone pain. The same microenvironment that contributes to nerve sensitization has a major impact on structural stability. The pH of the tumor microenvironment is known to be acidic. In normal bone physiology, osteoclasts and osteoblasts function together in maintaining bone, creating new tissue and remodeling as necessary. Osteoclasts resorb bone in a locally acidic environment. The acidity of the tumor microenvironment ultimately promotes significant bone resorption, though there is variability in different primary cancers. Pain will occur when the structure has been affected to such a degree that microfractures propagate or the nerve supply within the periosteum is activated (Yoneda et al. 2011; Weilbaecher et al. 2011).

The neurotransmitter profile of oncologic bone pain is different from that of both inflammatory and neuropathic pain (Clohisy and Mantyh 2003). Substance P and CGRP are both upregulated in inflammatory pain, downregulated in neuropathic pain, and unchanged in states of bone metastasis. Glial fibrillary acidic protein (GFAP) is significantly upregulated in the spinal cord in bone cancer pain but not in inflammatory or neuropathic pain. Several other changes that lead to central sensitization occur in the spinal cord. Experiments in an animal model of bone cancer have shown that mechanical stress leads to the upregulation of c-Fos (a marker of afferent nerve activation), reflecting increased neuronal activity. Astrocytosis was induced, which increased the extracellular glutamate level and contributed to CNS excitotoxicity. In addition, increased levels of the peptide dynorphin, which lowers the stimulus threshold centrally for noxious signals, were detected (Weilbaecher et al. 2011).

Biomechanical Considerations

Bone is the weakest and most likely to fracture when strained in torque (twisting on a firmly planted foot, for example). It is relatively stronger when stress is axial and the forces are in compression or distraction (tension). Metastases and erosions weaken the bone to well-defined degrees. A 50 % cortical defect in the femur causes 60–90 % reduction in bone strength

(Michaeli et al. 1999). The fracture risk posed by a bone lesion was, historically, a contested probability. Mirel has put forth a clinical/radiograph scoring system (pain, lesion type/location/size) that predicted fracture risk (Mirels 1989). While not perfect, the system is widely used when deciding whether to prophylactically fix an impending fracture. Bone mineral density can also be used as a major predictor of fracture risk (Michaeli et al. 1999). Recent work with quantitative computed tomography and engineering-beam structural analysis has shown promise in predicting bone failure (Tanck et al. 2009; Lee 2007). In spite of these numbers and predictions, not all lesions are associated with pain. However, the presence of bone pain is usually associated with a bony defect and requires treatment to improve the biomechanical stability of the bone.

Treatment Options

Prevention of bone resorption (i.e., osteoclast function) prevents bone pain and improves patient quality of life. Clinical studies have proven that aminobisphosphonates reduce bone resorption, pain, and the need for radiation therapy. A large meta-analysis of randomized studies demonstrated a reduced odds ratio for fractures (vertebral 0.69, 95 % confidence interval 0.57–0.84, $P < 0.0001$; non-vertebral 0.65, 0.57–0.79, $P < 0.0001$), need for radiotherapy (0.67, 0.57–0.79, $P < 0.0001$), and hypercalcemia (0.54, 0.36–0.81, $P = 0.003$). While aminobisphosphonate use for less than 6 months was not associated with a significant decrease in the incidence of orthopedic surgery, administration longer than 1 year resulted in a significant reduction (0.59, 0.39–0.88, $P = 0.009$) (Ross et al. 2003). This class of drugs causes apoptosis of osteoclasts. Along the same lines, attention has turned to developing drugs to target various receptors/channels that govern osteoclast function, including osteoprotegerin (OPG), vanilloid receptor-1 (VR1) antagonist, acid-sensing ion channel (ASIC) blockers, calcitonin, and denosumab (RANKL neutralizing antibody). Alternative targets in cancer pain management include TGF-β, bone matrix proteins, and osteoblasts. Animal models have demonstrated that an elevated TGF-β level in bone promotes bone resorption, while suppression of TGF-β decreases metastases. Osteoblasts have also been targeted, particularly in those tumors with a sclerotic appearance (bone-forming tumors such as prostate cancer), via endothelin-1 (Weilbaecher et al. 2011).

Despite the clear benefits of optimal medical management, many patients will still have bone pain and sustain fractures. Mechanical pain is alleviated by reducing the load on the bone, externally or internally (Healey and Brown 2000). External support and protection comes from protected weight-bearing with a mobility aid (i.e., cane, crutches, or walker) or bracing (i.e., lumbar support or functional cast brace for the humerus). Internal fixation reinforces the bone with metal, bypassing the bony defect and sharing the stress with the intact bone. Fixation of an orthopedic implant in normal bone is needed proximal and distal to the deficiency. When the defect is close to the end of the bone and it is impossible to meet the goal of rigid fixation through the metaphysis or epiphysis, the joint must be sacrificed, and a hemiarthroplasty replacement is needed. This surgery is very successful in alleviating mechanical cancer pain acutely (Bickels et al. 2009). It works well long-term if antitumor therapies prevent local bone destruction and destabilization of the implant. This is frequently done with adjuvant radiation therapy to the affected area. In about 80 % of patients, the use of radiation is associated with a significant decrease in pain burden and a relatively low complication rate (Falkmer et al. 2003). Studies suggest that this pain relief is from a direct action on cancer cells, leading to a decrease in cancer size and ultimately decreased osteolysis (Goblirsch et al. 2005). It may also affect intraosseous nerves and neurotransmitters and give more pain relief than would be expected based on the antineoplastic effect alone (Bourgeois et al. 2011). Single-dose fractionation works very well at palliating the bone cancer pain. Combining antineoplastic drug treatment with antiresorptive therapies may give better pain and disease control (Michailidou and Holen 2012).

The pharmacologic and surgical management of orthopedic cancer pain continues to evolve.

While there is little argument over whether to prophylactically fix a painful lesion at risk for pathologic fracture, there is no prospective study comparing fixation options. Treatment consists of reinforcing the bone, stopping the cancer progression, and preventing new sites of symptomatic disease involvement.

References

Bergstrom, J., Ahmed, M., Kreicbergs, A., & Nylander, I. (2003). Purification and quantification of opioid peptides in bone and joint tissues-a methodological study in the rat. *Journal of Orthopaedic Research, 21*(3), 465–469.

Bickels, J., Dadia, S., & Lidar, Z. (2009). Surgical management of metastatic bone disease. *Journal of Bone and Joint Surgery America, 91*(6), 1503–1516.

Bloom, A. P., Jimenez-Andrade, J. M., Taylor, R. N., Castañeda-Corral, G., Kaczmarska, M. J., Freeman, K. T., Coughlin, K. A., Ghilardi, J. R., Kuskowski, M. A., & Mantyh, P. W. (2011). Breast cancer-induced bone remodeling, skeletal pain, and sprouting of sensory nerve fibers. *The Journal of Pain, 12*(6), 698–711.

Bourgeois, D. J., Kraus, S., Maaloof, B. N., & Sartor, O. (2011). Radiation for bone metastases. *Current Opinion in Supportive and Palliative Care, 5*(3), 227–232.

Clohisy, D., & Mantyh, P. W. (2003). Bone cancer pain. *Cancer, 97*(Suppl. 415), 866–873.

Falkmer, U., Jarhult, J., Wersall, P., & Cavallin-Stahl, E. (2003). A systematic overview of radiation therapy effects in skeletal metastases. *Acta Oncologica, 42* (5–6), 620–633.

Goblirsch, M., Lynch, C., Mathews, W., Manivel, J. C., Mantyh, P. W., & Clohisy, D. R. (2005). Radiation treatment decreases bone cancer pain through direct effect on tumor cells. *Radiation Research, 164*(4 Pt. 1), 400–408.

Healey, J. H., & Brown, H. K. (2000). Complications of bone metastases: Surgical management. *Cancer, 88* (Suppl. 12), 2940–2951.

Irie, K., Hara-Irie, F., Ozawa, H., & Yajima, T. (2002). Calcitonin gene-related peptide (CGRP)-containing nerve fibers in bone tissue and their involvement in bone remodeling. *Microscopy Research and Technique, 58*(2), 85–90.

Lee, T. (2007). Predicting failure load of the femur with simulated osteolytic defects using noninvasive imaging technique in a simplified load case. *Annals of Biomedical Engineering, 35*(4), 642–650.

Mantyh, P. W., Clohisy, D. R., Koltzenburg, M., & Hunt, S. P. (2002). Molecular mechanisms of cancer pain. *Nature Reviews. Cancer, 2*(3), 201–209.

Michaeli, D. A., Inoue, K., Hayes, W. C., & Hipp, J. A. (1999). Density predicts the activity-dependent failure load of proximal femora with defects. *Skeletal Radiology, 28*(2), 90–95.

Michailidou, M., & Holen, I. (2012). Combinations of bisphosphonates and classical anticancer drugs: A preclinical perspective. *Recent Results in Cancer Research, 192*, 145–169.

Mirels, H. (1989). Metastatic disease in long bones: A proposed scoring system for diagnosing impending pathologic fractures. *Clinical Orthopaedics and Related Research, 249*, 256–264.

Ross, J. R., Saunders, Y., Edmonds, P. M., Patel, S., Broadley, K. E., & Johnston, S. R. (2003). Systematic review of role of bisphosphonates on skeletal morbidity in metastatic cancer. *British Medical Journal, 327*(7413), 469–472.

Tanck, E., van Aken, J. B., van der Linden, Y. M., Schreuder, H. W., Binkowski, M., Huizenga, H., & Verdonschot, N. (2009). Pathological fracture prediction in patients with metastatic lesions can be improved with quantitative computed tomography based computer models. *Bone, 45*(4), 777–783.

Weber, K. L. (2010). Evaluation of the adult patient (aged >40 years) with a destructive bone lesion. *The Journal of the American Academy of Orthopaedic Surgeons, 18*, 169–179.

Weilbaecher, K. N., Guise, T. A., & McCauley, L. K. (2011). Cancer to bone: A fatal attraction. *Nature Reviews. Cancer, 11*(6), 411–425.

Yoneda, T., Hata, K., Nakanishi, M., Nagae, M., Nagayama, T., Wakabayashi, H., Nishisho, T., Sakurai, T., & Hiraga, T. (2011). Molecular events of acid-induced bone pain. *International Bone & Mineral Society BoneKEy, 8*(4), 195–204.

Cancer Pain Management, Patient-Related Barriers

Cielito Reyes-Gibby and Karen O. Anderson
The University of Texas MD Anderson Cancer Center, Houston, TX, USA

Definition

Patients' limited knowledge of, beliefs about, and attitudes toward pain and pain treatment that detract from optimal pain management.

Characteristics

Patients, as consumers of health care, play an important role in the successful management of pain. Critical components for the successful clinical management of pain include the patients' ability to communicate their need for pain control, provide feedback on the effectiveness of treatment, and adherence to the requirements of therapy. Patients' limited knowledge of, beliefs about, and attitudes toward pain and pain treatment may therefore, detract from optimal pain management. Several barriers (Cleeland et al. 1997; Ward et al. 1993) to the practice of good pain management have been identified. They include those associated with patients (reluctance to report pain and to take pain medications), the health-care system (low priority given to pain control, limited health insurance coverage for pain treatments), and health-care professionals (inadequate knowledge, reluctance to treat, fear of controlled-substance regulations).

Among cancer patients, studies have shown that patients' reluctance to report pain is a primary reason for inadequate pain control. Von Roenn et al. (1993), Larson et al. (1993) and Breuer et al. (2011) have demonstrated that both oncologists and oncology nurses identified patient reluctance to report pain as one of the major barriers to the adequate control of cancer pain.

Studies have shown a number of reasons why patients are reluctant to report pain to their physicians or nurses. Often rated as a number one concern is *fear of addiction and dependence*. Opioids are the cornerstone of cancer pain therapy. First described in the World Health Organization's guidelines for cancer pain relief (World Health Organization (WHO) 1986), and more recently in the National Comprehensive Cancer Network (NCCN) guidelines (Benedetti et al. 2000), the protocol for pain treatment provides simple recommendations for the use of oral analgesics and adjuvants: Non-opioids and adjuvants for mild pain; weak opioids, often combined with a non-opioid and adjuvants, for moderate pain; and strong opioids, combined with weak opioids and adjuvants, for moderate to severe pain. There is, however, a common misconception among patients that opioids cause addiction. Often, patients and their family caregivers mistake the signs of withdrawal for addiction rather than physical dependence, perpetuating the belief

that addiction is a sequela of taking opioids. Family members also fear that the patient will be thought of as an addict, adding substantial pressure on the patient to refrain from taking opioids.

The mistaken notion that pain medications taken early will not be effective when pain gets worse further compounds the problem. Some patients believe that if they take pain medication they will become tolerant to the effects of analgesics when their disease progresses (Cleeland et al. 1997). Because of this *fear of drug tolerance*, many withhold or "save" their medication until they can no longer tolerate their pain.

Concern about side effects from pain medicines is another commonly reported barrier to adequate pain management. Many believe that side effects from analgesics are even more bothersome than the pain itself. In a study of underserved cancer patients, Anderson and colleagues (2002) found that a majority (75 %) reported problems with side effects from pain medicines. Commonly reported side effects include constipation, sedation, and nausea.

Belief that increasing pain signifies disease progression causes patients who are unwilling to face this possibility to deny their cancer pain. Patients are also afraid to bother their health-care providers with symptoms they consider to be an expected part of their disease. Because of the *belief that pain is an inevitable part of cancer* and that nothing can be done about it, patients are willing to accept and endure pain. In a population-based survey (Levin et al. 1985), approximately 50 % of the respondents considered cancer to be an extremely painful disease. Approximately 40 % believed that cancer treatment is extremely painful and 37 % rated cancer treatment as moderately painful. Furthermore, over 70 % thought that cancer pain could be so severe that one would consider suicide and more than 60 % believed that cancer patients usually die a painful death. These findings document the misconceptions surrounding cancer pain.

Patients also think that if they complain of symptoms, their *health-care providers will be distracted from their efforts to cure the disease*. A study specifically investigated whether patients' self-reports of pain vary by treatment setting (Reyes-Gibby et al. 2003) in this case an outpatient chemotherapy clinic and an outpatient breast clinic. Medical charts of patients seen during the same day in both the outpatient chemotherapy clinic and the outpatient breast clinic were reviewed and pain ratings were abstracted. Statistically significant differences in patients' self-reports of pain were observed in the two treatment settings. Fifty-one percent of patients' self-reports of pain differed between the two treatment settings, with 38 % reporting a pain score ≥ 4 in the outpatient breast clinic and 0 in the outpatient chemotherapy clinic. The authors suggested that the results may indicate that patients were reluctant to report pain in the chemotherapy clinic to avoid delaying their treatment. Another possible explanation for this finding is that patients may have believed that the chemotherapy clinic was not set up for pain intervention (to contact a physician for pain medication) and were thus reluctant to distract the clinic health-care professional from administering chemotherapy.

Wanting to be labeled as "good patients" and not as "complainers" also influences patients' reporting of pain. Many patients think that their behavior will influence the quality of their care. Families may compound the problem by trying to work out with the patient what is important to tell their physicians (Cleeland 1987).

Not surprisingly, other frequently reported barriers are *forgetting to take pain medications, reluctance to take pain medications as often as prescribed* (Ersek et al. 1999), and the *belief that one should be able to tolerate pain without medication* (Thomason et al. 1998). In addition, the patient's willingness to take analgesic medications may be influenced by past experiences with pain relief and analgesic side effects, and previous interactions with their providers and the health-care system (Meghani and Keane 2007).

While most of the studies on patient-related barriers to pain were conducted with non-Hispanic whites, studies in recent years explored the influence of sociocultural factors on patient-related barriers to adequate pain management. There are many definitions of culture, but one

that is helpful for understanding the effects of culture on pain was offered by Cecil Helman (1990). He defined culture as "a set of guidelines (both explicit and implicit) that individuals inherit as members of a particular society, that tells them how to view the world, how to experience it emotionally, and how to behave in it in relation to other people, and to natural and supernatural forces." Helman offered specific propositions as to the pervasive effects of culture on pain: (1) Not all social or cultural groups respond to pain in the same way. (2) How people perceive and respond to pain, both in themselves and in others, can be largely influenced by their cultural background. (3) How, and whether, people communicate their pain to health professionals and to others can also be influenced by cultural factors. These propositions underlie the commonly held assertion that there may be significant cultural and linguistic differences in the way people feel, view, and report pain.

Over 15 years of studies on cancer-related pain indicate that racial and ethnic minority patients are at risk for undertreatment of pain, as compared to non-Hispanic whites (Reyes-Gibby et al. 2007; Meghani et al. 2012). Patient-related barriers, as well as provider and system barriers, contribute to cancer pain disparities. For example, research has indicated that culture impacts people's behavioral responses to pain and their interpretations of the meaning of pain. A study by Cleeland and colleagues (Cleeland et al. 1997) of Hispanics and African-Americans who attended minority outpatient clinics showed that Hispanics reported more concerns about taking too much pain medication and more concerns about possible side effects from analgesics, as compared to African-American patients.

A study by Anderson and colleagues (2002) noted that, although the reasons for not taking pain medications do not significantly vary among African-Americans and Hispanic patients, differences exist between the two groups in terms of the meaning of cancer pain. African-Americans talked about the sensory component of pain, describing pain as "hurt" and as limited activity and impaired function. In contrast, Hispanics tended to focus more on the emotional component of pain, describing pain as "emotional suffering." In terms of patient-related barriers to cancer pain management, commonly stated barriers for both groups were the following: the need to be strong and not lean on pain medications; concern about addiction and the possible development of tolerance to pain medications; concern about side effects; and family reactions to their use of pain medications. Another important finding of this study was that a majority of patients in both ethnic groups would wait until their pain severity was a 10 on a 10-point scale before calling their health-care provider.

Other studies have identified religious beliefs, folk healers, nondrug interventions, and assistance from family members as important for pain management. A study of patients receiving analgesics from home health or hospice agencies found that Hispanic patients were more likely than Caucasian patients to report beliefs (e.g., take pain medicines only when pain is severe) that could hinder effective pain management (Juarez et al. 1998a). A qualitative study of Hispanic cancer patients receiving home health or hospice care also found that many patients believed that pain should be approached with stoicism (Juarez et al. 1998b). Similarly, semistructured interviews of African-American patients with solid tumors and pain indicated that stoicism, faith, and finding meaning in the cancer pain experience were common approaches for coping with pain (Meghani and Keane 2007).

While not directly related to patients' attitudes and beliefs, other notable barriers patients face include the cost and limited availability of analgesics. Difficulty with co-payments or incidental costs (e.g., transportation) associated with obtaining their prescriptions is especially relevant for those who lack insurance coverage or who have a limited income. In terms of availability, studies of pharmacies in New York and Michigan found that pharmacies in minority communities stocked insufficient opioids for pain management as compared to pharmacies in nonminority communities (Morrison et al. 2000; Green et al. 2005).

When patients' attitudes and beliefs interfere with effective pain management, patients can

benefit from a number of interventions. Educational interventions that provide accurate information about pain and pain treatments have shown some positive results in studies of cancer-related pain (Oliver et al. 2001; Rimer et al. 1987; Syrjala et al. 2008). For example, Syrjala and colleagues evaluated the efficacy of an educational intervention that consisted of video plus print materials in a randomized clinical trial of patients with cancer-related pain. Each patient in the intervention group also received one session with a nurse who reviewed the educational materials and discussed the patient's individual concerns. As compared to a control group that received nutrition education, the pain education group reported decreased pain intensity, reduced barriers to pain relief, and greater opioid use. The results of this study provide support for the importance of educational interventions, augmented with brief health-care professional contact, in changing pain-related beliefs and improving pain management. Other recent studies have confirmed the importance of providing individualized education on cancer pain that addresses the patient's specific concerns and beliefs regarding pain treatment (Lin et al. 2006; Kalauokalani et al. 2007).

References

Anderson, K. O., Richman, S. P., Hurley, J., Palos, G., Valero, V., Mendoza, T. R., Gning, I., & Cleeland, C. S. (2002). Cancer pain management among underserved minority outpatients: Perceived needs and barriers to optimal control. *Cancer, 94*(8), 2295–2304.

Benedetti, C., Brock, C., Cleeland, C., Coyle, N., Dube, J. E., Ferrell, B. et al. (2000). NCCN practice guidelines for cancer pain. *Oncology (Huntington), 14*(11A), 135–150.

Breuer, B., Fleishman, S. B., Cruciani, R. A., & Portenoy, R. K. (2011). Medical oncologists' attitudes and practice in cancer pain management: A national survey. *Journal of Clinical Oncology, 29*(36), 4769–4775. Epub 2011 Nov 14. PubMed PMID: 22084372.

Cleeland, C. S. (1987). Barriers to the management of cancer pain. *Oncology, 1*(Suppl. 2), 19–26.

Cleeland, C. S., Gonin, R., Baez, L., Loehrer, P., & Pandya, K. J. (1997). Pain and treatment of pain in minority patients with cancer. The Eastern Cooperative Oncology Group Minority Outpatient Pain Study. *Annals of Internal Medicine, 127*(9), 813–816.

Ersek, M., Kraybill, B. M., & Pen, A. D. (1999). Factors hindering patients' use of medications for cancer pain. *Cancer Practice, 7*(5), 226–232. PubMed PMID: 10687591.

Green, C. R., Ndao-Brumblay, S. K., West, B., & Washington, T. (2005). Differences in prescription opioid analgesic availability: Comparing minority and white pharmacies across Michigan. *The Journal of Pain, 6*(10), 689–699.

Hellman, C. (1990). *Culture, health and illness: An introduction for health professionals.* London: Wright.

Juarez, G., Ferrell, B., & Borneman, T. (1998a). Influence of culture on cancer pain management in Hispanic patients. *Cancer Practice, 6*(5), 262–269.

Juarez, G., Ferrell, B., & Borneman, T. (1998b). Perceptions of quality of life in Hispanic patients with cancer. *Cancer Practice, 6*(6), 318–324.

Kalauokalani, D., Franks, P., Oliver, J. W., Meyers, F. J., & Kravitz, R. L. (2007). Can patient coaching reduce racial/ethnic disparities in cancer pain control? Secondary analysis of a randomized controlled trial. *Pain Medicine, 8*(1), 17–24. PubMed PMID: 17244100.

Larson, P. J., Viele, C. S., Coleman, S., Dibble, S. L., & Cebulski, C. (1993). Comparison of perceived symptoms of patients undergoing bone marrow transplant and the nurses caring for them. *Oncology Nursing Forum, 20*(1), 81–87.

Levin, D. N., Cleeland, C. S., & Dar, R. (1985). Public attitudes toward cancer pain. *Cancer, 56*(9), 2337–2339.

Lin, C. C., Chou, P. L., Wu, S. L., Chang, Y. C., & Lai, Y. L. (2006). Long-term effectiveness of a patient and family pain education program on overcoming barriers to management of cancer pain. *Pain, 122*(3), 271–281. Epub 2006 Mar 20. PubMed PMID: 16545909.

Meghani, S. H., & Keane, A. (2007). Preference for analgesic treatment for cancer pain among African Americans. *Journal of Pain and Symptom Management, 34*(2), 136–147. Epub 2007 May 25. PubMed PMID: 17531436.

Meghani, S. H., Byun, E., & Gallagher, R. M. (2012). Time to take stock: A meta-analysis and systematic review of analgesic treatment disparities for pain in the United States. *Pain Medicine, 13*(2), 150–174. doi:10.1111/j.1526-4637.2011.01310.x. Epub 2012 Jan 13.

Morrison, R. S., Wallenstein, S., Natale, D. K., Senzel, R. S., & Huang, L. L. (2000). "We don't carry that"–failure of pharmacies in predominantly nonwhite neighborhoods to stock opioid analgesics. *The New England Journal of Medicine, 342*(14), 1023–1026.

Oliver, J. W., Kravitz, R. L., Kaplan, S. H., & Meyers, F. J. (2001). Individualized patient education and coaching to improve pain control among cancer outpatients. *Journal of Clinical Oncology, 19*(8), 2206–2212.

Reyes-Gibby, C. C., McCrory, L. L., & Cleeland, C. S. (2003). Variations in patients' self-report of pain by treatment setting. *Journal of Pain and Symptom Management, 25*(5), 444–448.

Reyes-Gibby, C. C., Aday, L. A., Todd, K. H., Cleeland, C. S., & Anderson, K. O. (2007). Pain in aging community-dwelling adults in the United States: Non-Hispanic whites, non-Hispanic blacks, and Hispanics. *The Journal of Pain, 8*(1), 75–84. Epub 2006 Sep 1. PubMed PMID: 16949874; PubMed Central PMCID: PMC1974880.

Rimer, B., Levy, M. H., Keintz, M. K., Fox, L., Engstrom, P. F., & MacElwee, N. (1987). Enhancing cancer pain control regimens through patient education. *Patient Education and Counseling, 10*(3), 267–277.

Syrjala, K. L., Abrams, J. R., Polissar, N. L., Hansberry, J., Robison, J., DuPen, S., Stillman, M., Fredrickson, M., Rivkin, S., Feldman, E., Gralow, J., Rieke, J. W., Raish, R. J., Lee, D. J., Cleeland, C. S., & DuPen, A. (2008). Patient training in cancer pain management using integrated print and video materials: A multisite randomized controlled trial. *Pain, 135*(1–2), 175–186. Epub 2008 Jan 8. PubMed PMID: 18093738; PubMed Central PMCID: PMC2396560.

Thomason, T. E., McCune, J. S., Bernard, S. A., Winer, E. P., Tremont, S., & Lindley, C. M. (1998). Cancer pain survey: Patient-centered issues in control. *Journal of Pain and Symptom Management, 15*(5), 275–284.

Von Roenn, J. H., Cleeland, C. S., Gonin, R., Hatfield, A. K., & Pandya, K. J. (1993). Physician attitudes and practice in cancer pain management. A survey from the Eastern Cooperative Oncology Group. *Annals of Internal Medicine, 119*(2), 121–126.

Ward, S. E., Goldberg, N., Miller-McCauley, V., Mueller, C., Nolan, A., Pawlik-Plank, D., Robbins, A., Stormoen, D., & Weissman, D. E. (1993). Patient-related barriers to management of cancer pain. *Pain, 52*(3), 319–324.

World Health Organization (WHO). (1986). *Cancer pain relief.* Geneva: WHO.

Cancer Pain Management, Principles of Opioid Therapy, Dosing Guidelines

James F. Cleary
Pain & Policy Studies Group, UW Carbone Cancer Center, Madison, WI, USA

Characteristics

▶ Opioids are the mainstay of cancer pain management. All patients with moderate to severe cancer pain should have access to opioids (Cleary 2000). The primary goal of opioid dosing is to establish pain relief with regular dosing within a therapeutic window, the dose that maximizes pain control with minimal side effects. Opioids can be administered by various routes: oral (immediate-release or modified-release products); transdermal; transmucosal; intravenous; subcutaneous or intraspinal.

Initial Dosing

In most cases, the initial treatment with opioids begins with short-acting formulations. This may be in the form of opioid-only products (codeine, oxycodone, or morphine) or oral combination products (e.g., hydrocodone/acetaminophen or oxycodone/acetaminophen). The dose-limiting factor in these combination products is acetaminophen; the total daily dose should be limited to 4 g per day, placing a limitation of 8–12 tablets per day, depending on the milligrams of acetaminophen per tablet. Short-acting opioids are normally administered every 3–4 h, but products such as immediate-release oxycodone or morphine may be given as often as hourly in the situation of uncontrolled pain.

Modified- or slow-release opioid products should be used cautiously as an initial opioid treatment, given the potential for side effects and the longer elimination with these compounds. This is particularly true with the elderly and those with renal impairment, in whom opioid clearance may be decreased, causing increased and prolonged side effects with low doses of these products (Osborne et al. 1986).

Around-the-Clock Dosing

The administration of opioids at regular intervals has been an important part of cancer pain management (Saunders 1963). In registration studies, investigators have found analgesic equivalence between immediate-release products and modified-release products at identical daily doses (Walsh et al. 1992). Improved patient compliance, and, therefore, improved pain control, is anticipated with modified-release products. However, in the situation of limited resources, patients can be prescribed regularly administered, immediate-release oral products with the anticipation of satisfactory pain control.

Patients can be converted from immediate-release products to modified-release products. Some modified-release products have a biphasic release mechanism that results in both immediate and more sustained pain relief. The dose of the modified-release formulation should be calculated by converting the daily dose of immediate-release opioids (including those administered parenterally) to the modified-release formulation, which is then administered over the 24-h period according to the medication's dosing schedule.

There are now oral opioids approved for daily and twice daily dosing. The transdermal fentanyl patch has a duration of 2–3 days and transdermal bruprenorphine, a duration of 7 days (Davis 2005). With modified-release compounds, dose escalation is usually best accomplished by increasing the dose rather than shortening the dosing interval. However, a careful history should be taken to explore the concept of "end of dose failure." A patient may report effective analgesia for 8–10 h following the administration of a modified-release opioid on a twice daily schedule. Rather than increasing the dose and therefore increasing the risk of side effects, it may be preferable to decrease the dosing interval to every 8 h. This also applies to once daily oral products, with which twice daily dosing may be necessary. With the transdermal fentanyl patch, patients may report only one good day out of three despite increases in the dose delivered per hour. Changing the patch every 48 h may result in more effective pain control (Payne et al. 1995).

▶ Breakthrough pain can usually be managed by providing a short-acting drug at a dose equal to 10–15 % of 24-h daily dosing (Portenoy and Hagen 1990). Experience with oral transmucosal fentanyl citrate, however, suggests that there is not a clear correlation between the around-the-clock opioid dose and the breakthrough pain dose (Farrar et al. 1998).

Dose Escalation

If pain is not well controlled and side effects are not severe, the dose of opioid should be increased. The percentage escalation should be based on the severity of the pain, the side effects, and other medical factors. If pain is moderate, the increase should be 25–50 % of the daily dose (Jacox et al. 1994). If pain is severe, a dose escalation of 50–100 % of the daily dose should be made. With most oral products, this escalation can safely take place every 24 h. In the case of the transdermal fentanyl patch, dose escalation should not take place more often than every 48 h, although in inpatient settings with close clinical supervision, dose increases have been safely made every 24 h when needed (Payne et al. 1995). When opioid infusions are used in the treatment of pain, dose escalations should take place no more often than the time required for four half-lives, unless there is careful monitoring. In the case of morphine, this is every 10–16 h. Escalating the infusion rate more often than this may result in increased drug toxicity. Patients may use patient-controlled analgesia together with clinician-administered boluses to effect better pain control, but will still require 4–5 half-lives to reach steady state.

Pain Emergency

In the case of a pain emergency, a loading dose may be the most effective way to treat the pain (Davis 2004). The loading dose should be established with frequent dosing administered either intravenously or subcutaneously. Many clinicians commence an opioid infusion alone, but it will require 4–5 half-lives before maximal pain relief can be achieved. With the half-life of morphine being 2.5–4 h, it would take 10–16 h to reach steady state, and therefore potentially leave the patient with uncontrolled pain. If a loading dose is added to an opioid infusion, the ongoing opioid dose should be the sum of the previous opioid regimen together with that calculated from the loading dose.

Dosing in Special Situations

In the presence of renal impairment (Osborne et al. 1986) and in the elderly (Cleary and Carbone 1997), the clearance of either primary drug or metabolites may be decreased. Therefore, both more cautious titration and longer dosing intervals may be necessary. Near the end of life, it may also be appropriate to decrease the dose of morphine administered to reduce the risk of toxicity.

References

Cleary, J. F. (2000). Cancer pain management. *Cancer Control, 7*, 120–131.
Cleary, J. F., & Carbone, P. P. (1997). Palliative medicine in the elderly. *Cancer, 80*, 1335–1347.
Davis, M. P. (2004). Acute pain in advanced cancer: An opioid dosing strategy and illustration. *The American Journal of Hospice & Palliative Care, 21*, 47–50.
Davis, M. D. (2005). Bruprenophine in cancer pain. *Supportive Care in Cancer, 13*(11), 878–887.
Farrar, J. T., Cleary, J., Rauck, R., et al. (1998). Oral transmucosal fentanyl citrate: Randomized, double-blinded, placebo-controlled trial for treatment of breakthrough pain in cancer patients. *Journal of the National Cancer Institute, 90*, 611–616.
Jacox, A., Carr, D. B., Payne, R., et al. (1994). Management of cancer pain: Clinical practice guideline No 9. Agency for Health Care Policy and Research. US Dept of Health and Human Services, Public Health Service AHCPr Publication 94–052.
Osborne, R. J., Joel, S. P., & Slevin, M. L. (1986). Morphine intoxication in renal failure: The role of Morphine-6-Glucuronide. *British Medical Journal, 292*, 1548–1549.
Payne, R., Chandler, S., & Einhaus, M. (1995). Guidelines for the clinical use of transdermal fentanyl. *Anti-Cancer Drugs, 6*, 50–53.
Portenoy, R. K., & Hagen, N. A. (1990). Breakthrough pain: Definition, prevalence and characteristics. *Pain, 41*, 273–281.
Saunders, C. (1963). The treatment of intractable pain in terminal cancer. *Proceedings of the Royal Society of Medicine, 56*, 195–197.
Walsh, T. D., MacDonald, N., Bruera, E., et al. (1992). A controlled study of sustained-release morphine sulfate tablets in chronic pain from advanced cancer. *American Journal of Clinical Oncology, 15*, 268–272.

Cancer Pain Management, Principles of Opioid Therapy, Drug Selection

Marie Fallon
Department of Oncology, Edinburgh Cancer Center University of Edinburgh, Edinburgh, UK

Synonyms

Equianalgesic dose; Opioid dose titration; Opioid pseudo-pharmacological ceiling dose; Opioid responsiveness; Simple analgesic; WHO analgesic ladder

Definition

▶ Opioid is a general term that includes naturally occurring, semisynthetic, and synthetic drugs that produce their effects by combining with opioid receptors and are stereospecifically antagonized by naloxone. Opioid drugs are the mainstay in the treatment of moderate to severe cancer pain.

Characteristics

Although concurrent use of other approaches and interventions may be appropriate in many patients and necessary in some, analgesic drugs are needed in almost every patient with cancer pain. The keystone of cancer pain management is a good assessment. Insight into the complex multidimensional features of cancer pain and an appreciation of the patient's clinical and psychosocial characteristics are fundamental for the success of any pain management strategy, pharmacological, or otherwise.

Drugs whose primary clinical action is the relief of pain are conventionally classified on the basis of their activity at opioid receptors, as either opioid or non-opioid analgesics. A third class, ▶ adjuvant analgesics, are drugs with other primary indications that can be effective analgesics in specific circumstances. The major group of drugs used in cancer pain management are the opioid analgesics.

Analgesic therapy with opioids, non-opioids, and adjuvant analgesics is developed for the individual patient through a process of continuous evaluation, so that a favorable balance between pain relief and adverse pharmacological effects is maintained. The World Health Organization (WHO) structured approach to drug selection for cancer pain, known as the ▶ WHO analgesic ladder, when combined with appropriate dosing guidelines is capable of providing adequate relief to 70–90 % of patients (World Health Organization 1986; Ventafridda et al. 1987; Takeda 1990; Walker et al. 1983; Schug et al. 1990; Zech et al. 1995). This approach emphasizes that the intensity of pain, rather than its specific etiology, should be the prime consideration in analgesic selection.

Cancer Pain Management, Principles of Opioid Therapy, Drug Selection, Fig. 1 WHO analgesic ladder

The WHO ladder was never intended to be used in isolation. Rather, it should be integrated with other approaches to cancer pain management such as radiotherapy, chemotherapy, anesthetic interventions, physiotherapy, and relaxation techniques.

The WHO ladder approach advocates three basic steps (Fig. 1): Patients with mild cancer-related pain should be treated with a non-opioid analgesic (also known as a ▶ simple analgesic), which should be combined with adjuvant drugs, if a specific indication for these exists. For example, a patient with mild to moderate arm pain caused by radiation-induced brachial plexopathy may benefit when a tricyclic antidepressant is added to paracetamol (acetaminophen) (McQuay and Moore 1997; Kalso et al. 1998).

Patients who are relatively non-tolerant and present with moderate pain, or who fail to achieve adequate relief after a trial of a non-opioid analgesic, should be treated with an opioid conventionally used for mild to moderate pain (formerly known as a "weak" opioid). This treatment is typically accomplished using a combination product containing a non-opioid (e.g., aspirin or paracetamol (acetaminophen)) and an opioid (such as codeine, oxycodone, or propoxyphene). This combination can also be coadministered with an adjuvant analgesic. The doses of these combination products can be increased until the maximum dose of the non-opioid analgesic is attained (e.g., 4,000–6,000 mg paracetamol (acetaminophen)); beyond this dose, the opioid contained in the combination product could be increased as a single agent, or the patient could be switched to an opioid conventionally used in Step 3.

Patients who present with severe pain, or who fail to achieve adequate relief following appropriate administration of drugs on the second step of the analgesic ladder, should receive an opioid conventionally used for moderate to severe pain (formerly known as a "strong" opioid). This group includes morphine, diamorphine, fentanyl, oxycodone, phenazocine, hydromorphone, methadone, levorphanol, and oxymorphone. These drugs may also be combined with a non-opioid analgesic or an adjuvant drug. Clearly, the boundary between opioids used in the second and third steps of the analgesic ladder is somewhat artificial, since low doses of morphine or other opioids for severe pain can be less effective than high doses of codeine or propoxyphene. Accordingly, some debate surrounds the usefulness of Step 2 of the WHO ladder. This is currently being examined in clinical studies.

According to the WHO guidelines, a trial of opioid therapy should be given to all patients with chronic pain of moderate or greater severity. This approach has been subject to criticism concerning its evidence base and its use over time. Much of the evidence is based on components of the ladder examined in nonmalignant pain, e.g., an opioid combined with a non-opioid is more effective than either alone, or systematic reviews of adjuvant analgesics (which appear to suggest that the number needed to treat [NNT] for most of the commonly used drugs is about three (McQuay and Moore 1998)).

Some of the criticism may be related to the common misconception that the WHO ladder is a "recipe" for cancer pain management. In fact, it is a set of principles that need to be applied appropriately with other non-pharmacological treatments to each individual situation.

The morphine-like agonist drugs are widely used to manage cancer pain. Although they may differ from morphine in quantitative characteristics, they qualitatively mimic the pharmacological profile of morphine, including both desirable and undesirable effects.

Comparative trials of opioids in cancer pain are extremely difficult to perform. While good quality randomized trials to provide evidence for pharmacotherapy of cancer pain are preferred, clinical decisions are currently based on limited trial evidence; basic pharmacology of the opioid and particular properties relating to renal, hepatic, and cognitive impairment; and progress in basic science which gives foundation to our belief of genetic variability in response to opioid analgesia (Rossi et al. 1997). No strong evidence speaks for the superiority of one opioid over another. For this reason, the aforementioned factors, along with practical clinical issues such as possible routes of drug administration, determine drug selection.

Although most cancer pain is responsive to opioid analgesia to some extent, the side effects experienced by an individual are the practical limiting factors in what is termed "▶ opioid responsiveness." While pure opioid agonists have no pharmacological ceiling dose, an "▶ opioid pseudo-pharmacological ceiling dose" exists in some situations because of dose-limiting side effects. The common side effects of all opioids are constipation, dry mouth, sedation, nausea, and vomiting. The first two often persist, but tolerance usually develops to sedation, nausea, and vomiting. Sedation, however, tends to be the common limiting factor during ▶ opioid dose titration, particularly in genetically susceptible individuals or in pain syndromes that usually require larger doses of opioid, e.g., neuropathic pain (Portenoy et al. 1990).

In individuals troubled with side effects, especially sedation, with one opioid, a switch to an alternative opioid may result in an improved balance between analgesia and unwanted side effects (Fallon 1997). It is impossible to predict such an improvement, but clinical experience with opioids makes this an acceptable strategy. In the case of neuropathic pain, use and titration

Cancer Pain Management, Principles of Opioid Therapy, Drug Selection, Table 1 Opioid analgesics (pure mu agonists) used for the treatment of chronic pain

Morphine-like agonists	Equianalgesic doses[a]	Half-life (h)	Peak effect (h)	Duration (h)	Toxicity	Comments	Oral bioavailability (%)	Active metabolites
Morphine	10 s.c.	2–3	0.5–1	3–6	Constipation, nausea, sedation most common; respiratory depression rare in cancer patients	Standard comparison for opioids; multiple routes available	20–30	M6G
	20–60 p.o.[b]	2–3	1.5–2	4–7				
Controlled-release morphine	20–60 p.o.[b]	2–3	3–4	8–12			20–30	M6G
Sustained-release morphine	20–60 p.o.[b]	2–3	4–6	24		Once-a-day morphine approved in some countries	20–30	M6G
Hydromorphone	1.5 s.c.	2–3	0.5–1	3–4	Same as morphine	Used for multiple routes	35–80	No
	7.5 p.o.	2–3	1–2	3–4				
Oxycodone	20–30	2–3	1	3–6	Same as morphine	Combined with aspirin or acetaminophen for moderate pain in USA; available orally without coanalgesic for severe pain	60–90	Oxymorphone
Controlled-release oxycodone	20–30	2–3	3–4	8–12				Oxymorphone
Oxymorphone	1 s.c.	–	0.5–1	3–6	Same as morphine	No oral formulation		Glucuronides
	10 p.r.	–	1.5–3	4–6				
Meperidine (pethidine)	75 s.c.	2–3	0.5–1	3–4	Same as morphine + CNS excitation; contraindicated in those on MAO inhibitors	Not used for cancer pain due to toxicity in higher doses and short half-life	30–60	Norpethidine
Diamorphine	5 s.c.	0.5	0.5–1	4–5	Same as morphine	Analgesic action due to metabolites, predominantly morphine; only available in some countries		Morphine
Levorphanol	2 s.c.	12–16	0.5–1	4–6	Same as morphine	With long half-life, accumulation occurs after beginning or increasing dose		No
	4 p.o.							

(continued)

Cancer Pain Management, Principles of Opioid Therapy, Drug Selection, Table 1 (continued)

Morphine-like agonists	Equianalgesic doses[a]	Half-life (h)	Peak effect (h)	Duration (h)	Toxicity	Comments	Oral bioavailability (%)	Active metabolites
Methadone[c]	10 s.c. / 20 p.o. (see text)	12 ≥150	0.5–1.5	4–8	Same as morphine	Risk of delayed toxicity due to accumulation; useful to start dosing on p.r.n.	60–90	No
Codeine	130 s.c. / 200 p.o.	2–3	1.5–2	3–6	Same as morphine	Usually combined with non-opioid	60–90	Morphine
Propoxyphene HCl (dextropropoxyphene)	–	12	1.5–2	3–6	Same as morphine plus seizures with overdose	Toxic metabolite accumulates but not significant at doses used clinically; usually combined with non-opioid	40	Norpropoxyphene
Propoxyphene napsylate (dextropropoxyphene)	–	12	1.5–2	3–6	Same as hydrochloride	Same as hydrochloride	40	Norpropoxyphene
Hydrocodone	–	2–4	0.5–1	3–4	Same as morphine	Only available combined with acetaminophen; only available in some countries		Hydromorphone
Dihydrocodone	–	2–4	0.5–1	3–4	Same as morphine	Only available combined with aspirin or acetaminophen in some countries	20	Morphine
Fentanyl	–	3–12	–	–	Same as morphine	Can be administered as a continuous I.V. or s.c. infusion; based on clinical experience, 100 mcg/h is roughly equianalgesic to morphine 4 mg/h I.V.	25/buccal <2/oral	No
Fentanyl transdermal system	–	13–22	–	48–72	Same as morphine	Based on clinical experience, 100 mcg/h is roughly equianalgesic to morphine 4 mg/h; recent study indicates a ratio of oral morphine: transdermal fentanyl of 100:1	90/transdermal	No

[a] Dose that provides analgesia equivalent to 10 mg i.m. morphine. These ratios are useful guidelines when switching drugs or routes of administration

[b] Extensive survey data suggest that the relative potency of i.m.:p.o. or s.c.:p.o. morphine of 1:6 changes to 1:2–3 with chronic dosing

[c] When switching from another opioid to methadone, the potency of methadone is much greater than indicated on this table

of an appropriate adjuvant analgesic would seem another critical step in improving the balance between analgesia and side effects.

There are no good data that examine a route switch rather than an ▶ opioid switch as a beneficial strategy in the management of ▶ opioid adverse effects. Clinically, the main benefit in switch to the spinal route of delivery is the possibility of adding drugs to manage neuropathic pain and incident pain, such as local anesthetic.

The transdermal route can be useful in patients having swallowing difficulties and in some cases of resistant constipation with oral opioids. Studies with transdermal fentanyl show a trend towards less constipation (Wong et al. 1997).

There is one randomized, controlled trial comparing oxycodone favorably with morphine with regard to hallucinations; however, the numbers are small and larger trials would be needed to say there is good evidence for the superiority of oxycodone (Kalso and Vainio 1990). Hydromorphone is another alternative to morphine if toxicity or drowsiness is problematic, but again, there is no high-quality evidence suggesting that one drug is superior to another.

Methadone is a synthetic opioid with unusual qualities and different pharmacological properties from the other opioids used in cancer pain. Steady-state plasma concentration is usually reached at 1 week but may take up to 28 days. Serious adverse effects are avoided if the initial period of dosing is accomplished with "as needed" administration, which is not the usual recommendation in cancer pain dosing regimens. When steady state has been achieved, scheduled dose frequency should be determined by the duration of analgesia following each dose.

The ▶ equianalgesic dose ratio of morphine to methadone has been the subject of controversy. Recent data suggest that the ratio correlates with the total opioid dose administered before switching to methadone. Among patients receiving low doses of morphine, the ratio is 4:1. In contrast, for patients receiving more than 300 mg of oral morphine, the ratio is approximately 10:1 or 12:1 (Ripamonti et al. 1998).

The therapeutic armamentarium for opioids has expanded over time, and familiarity with a range of opioid agonists and with the use of equianalgesic tables to convert doses if switching opioids is necessary (Table 1). It is clear that patients are at risk of under- or overdosing by virtue of individual sensitivities; hence, a logical, systematic approach to reassessment of the entire clinical situation is the key to managing opioid adverse effects and/or uncontrolled cancer pain, rather than an automatic switch to another opioid.

References

Fallon, M. T. (1997). Opioid rotation – Does it have a role? *Palliative Medicine, 11*, 176–178.

Kalso, E., Tramer, M. R., et al. (1998). Systemic local anaesthetic-type drugs in chronic pain: A systematic review. *European Journal of Pain, 2*, 3–14.

Kalso, E., & Vainio, A. (1990). Morphine and oxycodone hydrochloride in the management of cancer pain. *Clinical Pharmacology and Therapeutics, 47*, 639–646.

McQuay, H. J., & Moore, R. A. (1997). Antidepressants and chronic pain. *British Medical Journal, 314*, 763–764.

McQuay, H., & Moore, A. (1998). *An evidence-based resource for pain relief.* Oxford: Oxford Medical Publications.

Portenoy, R. K., Foley, K. M., & Inturrisi, C. E. (1990). The nature of opioid responsiveness and its implications for neuropathic pain: New hypotheses derived from studies of opioid infusions. *Pain, 43*, 273–286.

Ripamonti, C., De Conno, F., et al. (1998). Equianalgesic dose/ratio between methadone and other opioid agonists in cancer pain: Comparison of two clinical experiences. *Annals of Oncology, 9*, 79–83.

Rossi, G. C., Leventhal, L., Pan, Y. X., et al. (1997). Antisense mapping of MOR-1 in rats. Distinguishing between morphine and morphine-6 beta-glucuronide antinociception. *Journal of Pharmacology and Experimental Therapeutics, 281*, 109–114.

Schug, S. A., Zech, D., et al. (1990). Cancer pain management according to WHO analgesic guidelines. *Journal of Pain and Symptom Management, 5*, 27–32.

Takeda, F. (1990). Japan's WHO cancer pain relief program. In K. M. Foley, J. J. Bonica, V. Ventafridda, et al. (Eds.), *Advances in pain research and therapy* (Vol. 16, pp. 475–483). New York: Raven Press.

Ventafridda, V., Tamburini, M., Caraceni, A., et al. (1987). A validation study of the WHO method for cancer pain relief. *Cancer, 59*, 851–856.

Walker, V. A., Hoskin, P. J., et al. (1983). Evaluation of WHO analgesic guidelines for cancer pain in a hospital-based palliative care unit. *Journal of Pain and Symptom Management, 3*, 145–149.

Wong, J. O., Chui, G. L., Tsao, C. J., et al. (1997). Comparison of oral controlled-release morphine with transdermal fentanyl in terminal cancer pain. *Acta Anaesthesiologica Sinica, 35*, 25–32.

World Health Organization. (1986). *Cancer pain relief.* Geneva: Author.

Zech, D. F. J., Grond, S., Lynch, J., et al. (1995). Validation of world health organization guidelines for cancer pain relief. A 10 year prospective study. *Pain, 63*, 65–76.

Cancer Pain Management, Rehabilitative Therapies

Andrea Cheville
Department of Physical Medicine and Rehabilitation, Mayo Clinic, Rochester, MN, USA

Synonyms

Palliative rehabilitation; Preventative rehabilitation; Restorative rehabilitation; Supportive rehabilitation

Definition

The potential role of rehabilitation in pain management is often overlooked. Rehabilitative techniques include modalities that can directly influence pain (e.g., topical cold and desensitization techniques) and interventions that preserve and restore function. The latter are the focus of this essay.

Characteristics

Rehabilitative Goal Setting in Pain Management

As with any clinical intervention, the therapeutic goals of rehabilitation must be established prior to the initiation of therapy to facilitate reassessment at future time points. Dietz developed a structured approach to goal setting in cancer rehabilitation that is extremely useful and applicable in pain management

(Dietz 1985). He identified four broad categories of rehabilitation goals that can be used to define the purpose of interventions and to guide their strategic integration for optimal results. As outlined in Table 1, these include (1) preventative rehabilitation that attempts to preclude or mitigate functional morbidity resulting from pain, the pathophysiological process driving it, or its treatment; (2) restorative rehabilitation that describes the effort to restore the patient to a premorbid level of function when little or no long-term impairment is anticipated; (3) supportive rehabilitation that attempts to maximize function when long-term impairment, disability, and handicap result from the pain, its source, or its treatment; and (4) palliative rehabilitation, which decreases dependency in mobility and self-care in association with the provision of comfort and emotional support. Many interventions (e.g., resistive and aerobic exercise) may fall into more than one category, depending on the motivation for their use. For example, resistive exercise can be used preventatively to avoid deconditioning, restoratively to reverse it, and supportively to minimize it. The specifics of the therapeutic prescription and anticipated duration of therapy may vary widely, contingent on goal definition.

The effort to precisely define how these four general types of goals may apply to each patient is critical for a number of reasons. It ensures that potentially beneficial therapies will not be overlooked. It clarifies for both the clinician and the patient the purpose for which therapies are prescribed. This allows objective future assessment of whether a therapy has been successful. If therapeutic goals have not been met after an adequate trial, the rationale for discontinuation can be made evident to the patient. Definition of goals is also useful in justifying therapy to third-party payers and in anticipating the duration of therapy. The profound heterogeneity of presentation encountered in pain management confounds a rigidly algorithmic approach to the prescription of rehabilitative therapies. The four broad categories of clinical goals offer practitioners a flexible structure in which to develop a comprehensive and integrated therapeutic plan for functional preservation.

Cancer Pain Management, Rehabilitative Therapies, Table 1 Examples of goals and interventions that can be classified within the general therapeutic categories of preventive, restorative, supportive, and palliative rehabilitation

Preventative rehabilitation	The effort to restore the patient to a premorbid level of function when little or no long-term impairment is anticipated	
	Goal	Possible intervention
	Avoid development of secondary pain generators	Normalize motor recruitment patterns
	Minimize adverse effects of immobility:	
	(a) Deconditioning	Aerobic and resistive exercise
	(b) Contractures	Stretching, positioning
	(c) Osteopenia	Strategic loading of axial and appendicular structures
	Correct maladaptive biomechanical patterns	Posture and gait modification
Restorative rehabilitation	The effort to restore the patient to a premorbid level of function when little or no long-term impairment is anticipated	
	Goal	Possible intervention
	Eliminate musculoskeletal pain generators	Myofascial release techniques
	Restore power postoperatively to compromised muscle groups	Progressive resistive exercises
Supportive rehabilitation	The attempt to maximize function when long-term impairment, disability, and handicap are anticipated	
	Goal	Possible intervention
	Optimize mobility status	Provide patient with can, walker, etc.
	Optimize ADL independence	Instruction in compensatory strategies
Palliative rehabilitation	The effort to decrease dependency in mobility and self-care in association with the provision of comfort and support	
	Goal	Possible intervention
	Reduce dependency in toileting and grooming	Provide appropriate ADL assistive devices
	Preserve community integration	Prescription of wheelchair or scooter

Rehabilitation, Musculoskeletal System, and Pain

The musculoskeletal system is the primary focus of virtually all pain-oriented rehabilitation approaches. Dysfunction in the musculoskeletal system can (1) produce a primary pain generator, (2) function as a primary pain generator, (3) produce secondary pain generators, (4) function as secondary pain generators, or (5) have no role in the pain syndrome but undermine patients' functional status. Examples of the first four situations are given in Table 2. It is important to note that many musculoskeletal structures (e.g., the rotator cuff) can function in each of these categories. Rehabilitation is based on the fact that muscles, fascia, and bones respond to external forces that can be manipulated with therapeutic intent. For example, muscles can be stretched, strengthened, aerobically

conditioned, or rendered more proprioceptively responsive, depending on the therapeutic demands to which they are subjected. The correct choice of type, location, and intensity of pressure (s) requires accurate identification, not only of the involved anatomy and pathophysiological processes but determination of precisely how these may be contributing to a patient's global pain experience and functional decline. It is critical that permissive factors (e.g., laxity, contractures, biomechanical malalignment) be identified and definitively addressed in order to prevent recurrence of the primary and development of secondary pain generators.

Pain, particularly if associated with movement, engenders inactivity. The many adverse consequences of inactivity have been well documented. They include reduced cardiovascular endurance, diminished muscle strength and

Cancer Pain Management, Rehabilitative Therapies, Table 2 Examples of musculoskeletal structures functioning as (1) primary pain generators, (2) contributors to primary pain generator, (3) secondary pain generators, and (4) contributors to secondary pain generators effects of inactivity

Primary pain generator	Contributor to primary pain generator	Secondary pain generator	Contributor to secondary pain generator
Structures that become painful due to trauma, overuse, or inflammation	Permissive factors (e.g., muscle weakness or tightness or dysfunctional biomechanics) that impose stress on the primary pain generator	Structures that become painful due to spasm, overuse, or inflammation related to the primary pain generator	Permissive biomechanical factors that arise consequent to the primary pain generator
Rotator cuff tendonitis	Weakness of scapular stabilizers	Trapezius myofascial pain	Premature upper trapezius recruitment
Osteoarthritis of the hip joint	Flexibility deficits of hip muscles	Greater trochanteric bursitis	Iliotibial band tightness
Myofascial pain of scapular retractors	Pectoralis muscle tightness	Rotator cuff tendonitis	Altered scapular biomechanics
Discogenic lumbar nerve root compression	Weakness of the abdominal muscles	Lumbar paraspinal muscle spasm	Lumbar hyperextension to reduce pain

stamina, osteopenia with reduced fracture threshold, reduced periarticular distensibility, articular cartilage degeneration, compromise of neural patterns required for coordinated activity, reduced plasma volume, diminished proprioceptive acuity, and chemical alterations in connective tissue (e.g., ligaments, tendons) causing failure with reduced loading. Studies characterizing these physiological parameters have found that the rate of loss far exceeds the rate of recovery once therapeutic activity has been initiated (Saltin et al. 1968; Noyes et al. 1974; Beckman and Buchanan 1995). Some changes, particularly those in articular cartilage, may be irreversible. As inactivity has such profound and widespread adverse effects and because its consequences may be slow and difficult to reverse, it is essential that patients preserve their activity levels within the limits imposed by their pain. The prescription of preventative rehabilitation strategies can greatly facilitate this goal.

Scope of Rehabilitation

Rehabilitative interventions encompass a broad and highly varied collection of therapies. Most can be used with supportive, restorative, preventive, or palliative intent. The overarching goal of rehabilitation is functional restoration and preservation. However, many approaches can be used to definitively address pain generators, as with manual techniques for myofascial pain. Interventions can be grouped into the following categories: modalities, manual approaches, therapeutic exercise, provision of assistive devices, education in compensatory strategies, and orthotics.

Modalities

Although few blinded, prospective, randomized clinical trials have been conducted to establish the efficacy of most modalities, their routine integration into rehabilitative programs is the current standard of care. A review of the few randomized controlled trials and the many observational studies (Philadelphia Panel 2001) found evidence in support of a small, transient treatment effect from the application of heat. The delivery of therapeutic heat is often characterized by depth of penetration (superficial and deep) or mechanism of transmission (conduction, convection, radiation, evaporation, and conversion). Superficial heat increases the temperature in skin and subcutaneous fat to a depth of approximately 1 cm. Superficial modalities include hot packs, heating pads, fluidotherapy, and paraffin baths. Deep heating modalities increase temperature at a depth of approximately 3.5–7.0 cm and can, therefore, influence muscles, tendons, ligaments, and bones while sparing the skin and subcutaneous fat. Although ultrasound is the most commonly used deep heating modality, reviews offer little support of its efficacy, and its clinical utility remains a subject of debate (Baker et al. 2001; Robertson and Baker 2001).

Cryotherapy is a second modality used in the treatment of cancer pain (Lehmann 1990). The majority of cryotherapeutic modalities use superficial conduction. Cold packs come in the form of ice packs, hydrocollator packs, and endothermic packs that rely on chemical reactions to lower temperature. Ice massage is the application of ice directly to the skin's surface. The phases of reaction are coolness, aching, hypesthesia, and analgesia. Cold (5–13 °C) water immersion is a convection modality, and vapocoolant sprays rely on evaporative cooling. Medical contraindications to cold therapy include, but are not limited to, arterial insufficiency, Raynaud's phenomenon, cryoglobulinemia, cold hypersensitivity, and paroxysmal cold hemoglobinuria.

Hydrotherapy involves the external application of water of any temperature to achieve therapeutic goals. Hydrotherapeutic modalities are often used for wound or skin debridement and cleansing, as well as in the supportive and palliative care of pain-related to arthritis, chronic inflammatory states, and myofascial syndromes.

Traction, the use of strategic displacing force to stretch soft tissue and separate articular surfaces, has been used for nonmalignant conditions, particularly those associated with nerve root impingement. Despite decades of research regarding the clinical benefits of spinal traction, there is no consensus regarding its utility. Primary malignancies of bone or spinal cord, osteomyelitis or discitis, unstable spinal fractures, end-stage osteoporosis, central disc herniations, carotid or vertebral artery disease, rheumatoid arthritis (cervical), and pregnancy (lumbar) are absolute contraindications to the use of traction.

Functional electrical stimulation (FES) applies electrical stimulation via pad electrodes to depolarize motor nerves, at either the axon or neuromuscular junction. It has been used in patients with potentially painful conditions such as spinal cord injury (SCI), peripheral nerve injury, spasticity, and cardiopulmonary deconditioning. It has not been used in the management of cancer pain.

Transcutaneous electrical nerve stimulation (TENS) influences pain by stimulating large-diameter, myelinated Aβ nerve fibers. There is a large clinical experience suggesting that a subgroup does benefit. Multiple electrode and stimulation configurations must be tested to ensure an adequate trial. Studies of the technique have yielded inconsistent results (Fargas-Babjak 2001). For theoretical reasons, patients are cautioned against stimulating directly over tumors.

Iontophoresis allows charged molecules to penetrate cell membranes and enter tissue through the application of an electric field. Negative, positive, and ground electrodes are secured to the patient, and a 10–30 m current is used to transfer medication from the electrode into the surrounding tissues. This modality creates the potential for medication delivery in the treatment of spasticity, chronic inflammatory states, and myofascial pain syndromes. There is no experience in the use of this approach in medically ill patients.

Phonophoresis also facilitates the transdermal delivery of topical medications but uses ultrasound to facilitate medication delivery. The technique has been used to treat soft tissue inflammation. Again, there is no experience in the cancer pain population.

Manual Therapies

Manual therapies refer to a vast array of hands-on techniques designed to normalize soft tissue and joint mobility. Types of manual therapies include everything from massage to myofascial release, acupressure, and joint range of motion. Common manual techniques include:

- Massage
- Myofascial release
- Soft tissue mobilization
- Manual lymphatic drainage
- Acupressure
- Shiatsu
- Rolfing

- Reflexology
- Craniosacral therapy
- Osteopathic manipulation
- Joint mobilization
- Muscle energy techniques
- Passive range of motion

The benefits of manual approaches are generally short-lived if patients do not comply with a concurrent stretching, strengthening, and conditioning regimen. Physical therapists, osteopathic physicians, massage therapists, acupuncturists, as well as a host of complementary and alternative practitioners use manual techniques for pain control.

When chronic pain affects the musculoskeletal system, a pain-spasm cycle begins. Nociception causes reflexive muscle contraction, which in turn increases nociception, and the cycle is set in motion leading ultimately to painful, chronic muscle hypertonicity. This results in muscle weakness, joint contractures, aberrant biomechanics, and dysfunctional kinesthetic patterns. As patients attempt to perform normal daily activities, the compromised system becomes overused, exacerbating the pain cycle and producing secondary pain generators. By normalizing joint and soft tissue physiology, manual techniques can be used to break this pain cycle and reestablish function biomechanics. Ideally, patient referral for manual techniques should not be delayed until the pain-muscle-spasm cycle is well established.

Massage focuses on decreasing pain through increased circulation and mechanical movement of tissues. Techniques vary in the variety of hand strokes, applied pressure, and direction of force. Many beneficial physiological effects have been associated with massage (Field 2002). In addition to its use in pain management, it has been adapted to the treatment of lymphedema and contractures.

Myofascial release is a technique that purportedly facilitates normal movement within the fascial system. Fascia is the connective tissue that provides support throughout the body, and fascia that is injured or contracted could contribute to pain. Practitioners use vigorous "hands-on" compression and stretching techniques to alter the mobility of affected tissues.

Several manual techniques are used to affect joint physiology and normalize accessory movement. The most basic technique is passive range of motion (PROM). A practitioner using PROM will move a joint through its physiologic range of motion in an effort to maintain and/or restore motion. In addition, PROM also stretches the soft tissues surrounding a joint and normalizes accessory movements in a joint. By restoring normal joint mechanics on the accessory level, normal physiologic motion can follow. There are five grades of joint mobilization based on the amplitude of movement through the full range of accessory movement. Grades I and II provide the smallest amount of movement. These techniques are used to prevent contractures and pain relief. Grades III and IV provide increased pressure within the joint's range of normal motion. These techniques can help to promote normal physiologic motion as well as providing pain relief. Grade V joint mobilization is known as a "manipulation." Grade V manipulations are performed at high velocity and intensity to the end of the accessory range of motion. Physical therapists, osteopathic physicians, and chiropractors use joint mobilization. In most cases, only chiropractors use manipulations.

Muscle energy techniques (MET) are used to restore normal joint and soft tissue movement through patients' own force. METs are based on the theory of joint mobilization to facilitate movement but recruit patients' own muscular force. The practitioner places the patient in a specific position, and the patient is asked to push against the practitioner's counterforce in order to facilitate movement in joints and soft tissues.

Acupressure involves the strategic application of pressure to trigger points and derives from the theories of acupuncture. The use of pressure to relieve trigger points has been incorporated into many soft tissue techniques.

Therapeutic Exercise

The strategic use of exercise to enhance strength, coordination, stamina, and flexibility is perhaps the most powerful intervention in rehabilitation medicine. The fact that muscles and fascia respond predictably to imposed external demands underlies this therapeutic approach. Muscle changes that can be achieved through therapeutic exercise include increased capillary density, enhanced neuromuscular responsiveness, normalization of muscle length-tension relationships, altered resting tone, and increased elaboration of mitochondrial and sarcoplasmic proteins (de Lateur 1996).

For optimal benefit, the type, intensity, and frequency of exercise must be rationally chosen following a comprehensive clinical examination. The exercise prescription will be determined by the presence, severity, and distribution of flexibility and strength deficits; the degree of deconditioning; and the presence abnormal segmental biomechanics. An exercise program combining stretching, aerobic conditioning, and strengthening should be tailored to each patient's unique requirements.

Patients may protest that the intensity of their pain precludes participation in an exercise program. Even minimal resistive exercises, however, can lead to significant improvements in strength and functional status. Brief isometric muscle contractions can be performed in bed or in a chair against gravity or with gravity eliminated. One brief isometric contraction per day was demonstrated to prevent loss of strength in bedridden rheumatoid arthritis patients (Atha 1981).

Stretching, or flexibility activities, can influence muscles by altering their length-tension relationships and resting tone. Stretching is an integral part of the treatment of myofascial pain (Travell and Simons 1983) and pain associated with muscle spasms. It is commonly used to achieve adequate range of motion for performance of daily activities. Flexibility assessment must consider patients' overall level of muscle tightness, their required range of motion given their activity profile, and the impact of contracture-related asymmetry on movement patterns. Two stretching techniques predominate in rehabilitation medicine: ballistic and static. Ballistic stretching involves repetitious bouncing movements at the end range of joint range. Static stretching, in contrast, involves slow, steady soft tissue distension, which is maintained for several

seconds. Comparisons suggest that static is superior to ballistic stretching and that meaningful benefit can be achieved through three to five sessions per week (Sady et al. 1982). Pain patients require a slowly progressive approach to static stretching. Ballistic stretching is indicated only in unusual cases and should be performed under the care of a physical medicine and rehabilitation specialist.

Resistive exercise is used to strengthen selected muscles or muscle groups by forcing them to contract repeatedly against resistance. Resistive training can be utilized with restorative, supportive, and/or preventative intent. Goals must consider patients' current level of conditioning, the prognosis for their pain and related medical comorbidities, their past fitness histories, and the rigor of their anticipated activity profile (e.g., vocational, avocational). To induce strength gains, an intensity of at least 60 % of the one-repetition maximum must be used. Table 3 outlines common parameters utilized in prescribing resistive exercise for generalized or focal weakness.

Aerobic conditioning differs from resistive exercise in its emphasis on continuous rhythmic contraction of large muscle groups. Jogging and cycling are common examples. Aerobic training allows patients to perform daily activities with less effort, maintain greater independence, and enjoy an enhanced sense of well-being (NIH Consensus Developments Panel on Physical Activity and Cardiovascular Health 1996). If the medical condition permits, regular physical activity may improve the ability to perform physical activities and attenuate the psychological morbidity associated with chronic pain. A prescription for exercise should consider baseline fitness levels, associated motor impairments and comorbidities, and patient tolerance. For deconditioned patients with chronic pain, the initial aerobic training may entail walking slowly for 5–10 min. Exercise can also be used to enhance coordination, posture, proprioception, balance, and performance of integrated movement patterns. Much of the literature establishing the value of exercise for these applications derives from sports and performing arts medicine. Techniques to improve proprioception

Cancer Pain Management, Rehabilitative Therapies, Table 3 Considerations in prescription of therapeutic resistive exercise

Exercise parameter	Considerations and recommendations
Choice of exercise	Exercises strength deficient muscle groups in multiple planes and across a range of length-tension relationships
Order of exercise	Begin with large muscle group exercises. For circuit training, start with legs
Number of sets	Begin with one set and progress to three or more sets of each exercise
Rest between sets	3 min for heavy resistance, 2–3 for moderate resistance, 1–2 for light resistance
Intensity	60 % one-repetition maximum, 6–15 repetitions
Rate of progression	Increase 2.5–5 % when level of resistance is perceived as "moderate"
Program variation	Variations in intensity, positioning, order of exercises, and choice of exercise should be adopted weekly to avoid overtraining
Speed specificity	Intermediate velocity unless training goals involve rapid or slow velocity activities
Contraction specificity	Isotonic training unless joint pain is prohibitive, and then initiate training with isometric activities. Isometrics can be used to prevent loss of strength
Joint angle specificity	Loading should be maintained throughout entire functional range

include use of a tilt or wobble board, sideways walking or running, and agility drills. Balance-enhancing activities include tossing and catching a ball while standing on heels, toes, or one leg. Therapy balls can be utilized to address deficits in truncal stability. Generally, patients sit on the ball while shifting their weight in different directions or lifting their legs. There are many variations on balance activities. Their selection should consider the patients' deficits and desired activity profile. The adage that the best training for an activity is the activity itself also holds true for coordination training.

Orthotics

Orthotics are braces designed to alter articular mechanics when their integrity has been compromised by pain, weak muscles, impaired sensation, or other anatomical disruption. Orthotics may be used therapeutically to provide support, restore normal alignment, protect vulnerable structures, address soft tissue contractures, substitute for weak muscles, or maintain joints in positions of least pain. Many orthotics are available prefabricated, "off the shelf." While such braces often suffice, patients may require more expensive custom orthoses for optimal benefit. Orthotics can be used to address pathology in virtually any articular structure. The use of these devices for pain patients must be tempered by concern for engendering long-term dependency and validating patients' impairments. Orthotics are most often used in pain management on a transient basis to keep joints in a fixed position, to allow resolution of local inflammation, or to rest painful muscles and/or tendons that act on the joint. They also can be used preventatively, when joints or osseous structures are at risk of injury or contracture, and patients are incapable, despite resistive exercise and proprioceptive enhancement, of protecting them. Such circumstances often arise in the context of systemic illness. For example, spinal extension orthoses such as the Jewett or Cash brace limit spinal flexion and thereby prevent excessive loading and compression fracture of the anterior vertebral bodies. Palliative splinting is used to optimize comfort when function is no longer a primary concern. An excellent example is the use of slings to keep flaccid upper extremities tethered near the body and out of harm's way. Referral to an orthotist or physical medicine and rehabilitation specialist will ensure that patients receive appropriate orthoses. However, many of these professionals lack experience with pain patients. It is important to communicate the goals of treatment and the precise reason(s) why the orthotic is being prescribed to the rehabilitation professional. In this way, patients have the best chance of receiving an orthotist suited to their unique requirements.

Many patients with chronic pain, particularly pain related to chronic medical illness, require adaptive devices to enhance their safety, comfort, and autonomy while moving about the home or community. Ready access to such devices is essential if pain patients are to remain socially integrated within their communities. Adaptive equipment designed to augment mobility ranges from prefabricated single-point canes to complex motorized wheelchair systems. Handheld assistive devices are generally variations on canes, crutches, and walkers. Devices can be customized to distribute weight bearing onto intact structures to minimize the pain. Patients with severe deconditioning, paresis, osseous instability, or other sources of impaired mobility may require a wheelchair or scooter. Even when deficits are presumed transient, a wheelchair can sustain community integration and fragile social connections. Wheelchair tolerance and utilization depend on the prescription of an appropriate model.

Assistive devices have also been developed to maximize patients' independence and reduce pain during performance of activities of daily living (ADLs). Dependence for self-care has been shown to erode quality of life across many medical diagnoses. Devices are available to assist patients with independent dressing, grooming, toileting, as well as performance of more complex activities such as cooking and housekeeping.

Conclusion

Clinicians must become discriminating consumers of rehabilitation services if they are to optimally benefit their patients. This need arises from the fact that the delivery of rehabilitation services occurs within a socioeconomic context that imposes fiscal pressures upon its providers. It is essential that clinicians question patients regarding the particulars of their treatment and relay any concerns to the treating therapist. Including specific requirements on the therapy prescription can also help to ensure that patients receive appropriate care. Clinicians must recognize that rehabilitation professionals vary widely in their levels of experience, biomechanical acumen, interpersonal skill, and mastery of manual techniques. Ideally, clinicians should refer patients to therapists with whom they are familiar and can readily enter into clinical dialogue. Most therapists welcome

guidance from physicians and nurses and are highly motivated to cultivate the skills required to optimally serve pain patients.

References

Atha, J. (1981). Strengthening muscle. *Exercise and Sport Sciences Reviews, 9*, 1–73.

Baker, K., Robertson, V., & Duck, F. (2001). A review of therapeutic ultrasound: Biophysical effects. *Physical Therapy, 81*, 1351–1358.

Beckman, S., & Buchanan, T. (1995). Ankle inversion injury and hypermobility: Effect on hip and ankle muscle electromyography onset latency. *Archives of Physical Medicine and Rehabilitation, 76*, 1138–1143.

de Lateur, B. J. (1996). Therapeutic exercise (pp. 401–419).

Dietz, J. J. (1985). Rehabilitation of the patient with cancer. In P. Calabresi, P. S. Schein, & S. A. Rosenberg (Eds.), *Medical oncology* (pp. 1501–1522). New York: Macmillan.

Fargas-Babjak, A. (2001). Acupuncture, transcutaneous electrical nerve stimulation, and laser therapy in chronic pain. *The Clinical Journal of Pain, 17*, 105–113.

Field, T. (2002). Massage therapy. *The Medical Clinics of North America, 86*, 163–171.

Lehmann, J. F. (1990). *Therapeutic heat and cold.* Baltimore: Williams & Wilkins.

NIH Consensus Developments Panel on Physical Activity and Cardiovascular Health. (1996). Physical activity and cardiovascular health. *Journal of the American Medical Association, 276*, 241–246.

Noyes, F. R., Torvik, P. J., Hyde, W. B., et al. (1974). Biomechanics of ligament failure II. An analysis of immobilization, exercise, and reconditioning effects in primates. *The Journal of Bone & Joint Surgery, 56*, 1406–1418.

Philadelphia Panel. (2001). Philadelphia panel evidence-based clinical practice guidelines on selected rehabilitation for knee pain. *Physical Therapy, 81*, 1675–1700.

Robertson, V., & Baker, K. (2001). A review of therapeutic ultrasound: Effectiveness studies. *Physical Therapy, 81*, 1339–1350.

Sady, S., Wortman, M., & Blanke, D. (1982). Flexibility training: Ballistic, static or proprioceptive neuromuscular facilitation? *Archives of Physical Medicine and Rehabilitation, 63*, 261–263.

Saltin, B., Blomqvist, G., Mitchell, J. H., et al. (1968). Response to exercise after bed rest and after training: A longitudinal study of adaptive changes in oxygen transport and body composition. *Circulation, 38*, VII1–VII78.

Travell, J., & Simons, D. (1983). *Myofascial pain and dysfunction. The trigger point manual.* Baltimore: Williams and Wilkins.

Cancer Pain Management, Treatment of Neuropathic Components

Paul W. Wacnik[1] and Richard L. Boortz-Marx[2]
[1]Neuromodulation Research, Medtronic Inc., Minneapolis, MN, USA
[2]Division of Pain Medicine, Department of Anesthesiology, University of North Carolina, Chapel Hill, North Carolina, USA

Definition

Pain due to pathological functioning of either the peripheral nervous system (PNS) or the central nervous system (CNS) is classified as neuropathic. These processes may directly stimulate the pain system or damage nociceptive pathways to shift the balance between painful and nonpainful inputs to the CNS (Merskey and Bogduk 1994). Neurological symptoms of ▶ neuropathic pain in cancer include continuous, burning, itching, aching, and cramping pain evoked by mechanical, chemical, or thermal stimuli. Such symptoms can be accounted for by an intact, normally functioning nervous system sensing a noxious stimulus (in this case, a ▶ tumor) manifesting as ▶ somatic pain or by a component of the nervous system damaged by the impact of previous antineoplastic therapies, such as surgery, chemotherapeutic agents, radiation oncology, or by progression of the disease (Portenoy 1991). Distinguishing between the two etiologies is often problematic.

Characteristics

Pain is but one symptom of many experienced with cancer. Uncontrolled pain can profoundly compromise quality of life and may interfere with antineoplastic treatment (Portenoy and Lesage 1999). Pain in cancer can be either constant or variable in character, owing to the inevitability that cancer will involve tumor growth and progression, changes in the tissue surrounding the tumor, and therapeutic interventions intended to control tumor growth. Although no overall "cure" exists for most cancers, advances in anti-neoplastic therapies have allowed patients to live longer with their disease, making long-term ▶ cancer pain therapy a consideration of increasing importance. The incidence of cancer pain is high in patients with advanced disease as well as in patients undergoing active treatment for solid tumors (30–50 %; Portenoy and Lesage 1999; Foley 2000). The intensity of this pain is often overwhelming. In a multisite cancer pain study, two-thirds of patients rated their worst pain, using an 11-point numerical rating scale during a given day, at 7 out of a maximum 10, with an average pain level of 4.7 throughout the day, even though 91 % of these patients were receiving opioid analgesics (Caraceni and Portenoy 1999). This observation demonstrates that most patients, even those receiving analgesic therapy, live with moderate to severe daily pain. One complication in understanding and treating cancer pain is its variable and complex nature (see Table 1). More than 25 % of patients have nerve injury that occurs in tandem with damage to other structures and the pain has a "mixed" pathophysiology with more than one type of pain, most frequently somatic-nociceptive pain. ▶ Nociceptive pain involves direct ongoing activation of intact nociceptors (pain sensitive neurons) in either somatic or

Cancer Pain Management, Treatment of Neuropathic Components, Table 1 A compilation of pain types due to cancer from a multisite study (Adapted from Caraceni and Portenoy 1999)

Cancer pain pathophysiology	% Occurrence
Somatic nociceptive only	32.2
Somatic and neuropathic	23.3
Visceral nociceptive only	15.2
Somatic and visceral	10.8
Neuropathic only	7.7
Somatic, visceral, and neuropathic	5.2
Visceral and neuropathic	3.6
Unknown only	1.7
Other with psychogenic	1.5
Psychogenic only	0.3

visceral tissue. In essence, this is the intact, normally functioning nervous system sensing a noxious stimulus (in this case a tumor). On the other hand, pain due to pathological function of either the peripheral or central nervous system is classified as neuropathic. In this case, the presence of cancer induces phenotypic changes in the pain sensing system, increasing the sensitivity of nerves to normally innocuous stimuli, or exaggerating the response to noxious stimuli.

The findings of Boortz-Marx's group in 2004 were similar with 15 % of cancer patients having neuropathic pain, 25 % showing somatic-nociceptive pain and 60 % showing "mixed" pain characteristics. This outcome underscores the complexity of the alterations in pain sensing systems in cancer. Although pain due to the neoplasm is the focus of much investigation and treatment, there are considerable instances of pain observed in cancer patients related to noxious interventions, including pain associated with diagnostic and therapeutic interventions (e.g., surgical procedures and lumbar punctures), analgesic techniques, ▶ chemotherapy toxicity, hormonal therapy, and radiotherapy. The fact that cancer pain can evolve from one form to another with progression of the disease implies that analgesic treatment regimens must also evolve to match a changing etiology.

Diagnoses of Neuropathic Components of Cancer Pain

The diagnosis of neuropathic cancer pain is a clinical diagnosis. Based on the clinical presentation of the character and quality of pain, one is led to a provisional diagnosis of neuropathic cancer pain, which appropriately initiates therapeutic options, which may possibly include anti-neuropathic pharmacotherapy. Some types of neuropathic injury produce aching, stabbing, or throbbing pain, but these syndromes often present with an unfamiliar quality or sensory distortion. Burning, shooting, and tingling are suggestive of nerve involvement, but not sufficient to make the diagnosis. Accompanying abnormal sensations are often found on examination, including hypesthesia (increased threshold), hyperesthesia (decreased threshold), paresthesias (spontaneous

non-painful threshold), dysesthesia (spontaneous pain threshold), hyperpathia (prolonged stimulus response), and allodynia (pain from a normally innocuous stimulus). The presence of changes in small fibers on biopsy of skin or peripheral nerves may be a useful confirmation, but has proven largely unnecessary in several clinical studies. Nonetheless, the diagnosis of neuropathic pain is based on clinical presentation.

Pharmacological Treatment of Neuropathic Components of Cancer Pain

Traditionally opioids have been utilized in the treatment of cancer pain. However, their efficacy for neuropathic pain states is controversial (McQuay 1999). The "mixed pain" state lends itself to effective treatment with opioids (several preparations of long- and short-acting opioids to be administered by different routes). Routes of administration include oral, submucosal, parenteral, subcutaneous, transdermal, rectal, epidural, and intrathecal. Undesirable side effects of opioids (cognitive impairment, somnolence, constipation, fatigue, hallucinations, endocrine dysfunction, disruptive sleep architecture and myoclonus) may be present without adequate analgesia being achieved.

The adjuvant group of analgesics (pharmacotherapy designed for other purposes, but producing analgesia in certain circumstances) has proved to be effective in the treatment of neuropathic cancer pain. Traditionally, the tricyclic antidepressants (amitriptyline, nortriptyline, imipramine, desipramine, and doxepin) have demonstrated efficacy in the treatment of neuropathic pain states. The side effect profile of this group has led individuals not to consider these medications as first-line therapy. The predominance of anticholinergic side effects (dry mouth, somnolence, cognitive impairment, cardiac arrhythmias, urinary retention, increased intra-ocular pressure, and constipation) is often dose related, but can have advantages in patients afflicted with disrupted sleep architecture. Often doses needed for treatment of neuropathic pain states are much lower than those needed for the treatment of depression. Other antidepressants such as the selective serotonin reuptake inhibitors, atypical antidepressants, and mono-amine oxidase inhibitors (MAOIs) have not proven efficacious unless predominant depression accompanies the neuropathic pain state. Most recently, two selective norepinephrine reuptake inhibitors (SNRI) received approval for treatment of painful diabetic peripheral neuropathy.

Several anticonvulsant medications constitute the mainstay in the treatment of most neuropathic pain states, including neuropathic pain accompanying cancer. The traditional anticonvulsants (phenytoin, carbamazepine and valproate) have fallen from favor because of their toxic side effects related to bone marrow suppression; this side effect is particularly unwelcome in a patient population that has already received a toxic impact from chemotherapeutic agents used in the treatment of their disease state. Newer classes of anticonvulsants have recently evolved that employ varying modes of activity relating to voltage-gated channels (alpha-2/delta subunit of voltage-gated calcium channels and voltage-gated sodium channels) and glutamate-gated channels of the AMPA subtype. These gabapentinoids agents include pregabalin (Lyrica) and gabapentin (Neurontin and Gralise) and have an indication for neuropathic pain. Others, such as topiramate (Topamax), tiagabine (Gabitril), levetiracetam (Keppra), lamotrigine (Lamictal), oxcarbazepine (Trileptal), zonisamide (Zonegran), and lacosamide (Vimpat) which target the sodium channels and CRMP-2 (collapsing response mediator protein-2) glutamate receptors, have off-label uses; however, their mechanism of action suggests they have a place in the treatment of neuropathic pain. Overall, drugs such as the gabapentinoids have few drug-drug interactions, are well tolerated (starting at a low dose and escalating slowly), and have produced favorable outcomes.

Other adjuvant medications used in the treatment of neuropathic cancer pain have included the alpha-2 adrenergic agonists (tizanidine – only available as an oral compound) local anesthetics including mexiletine (oral), lidocaine, bupivacaine, ropivacaine, tetracaine which may be administered perineurally, parenterally, epidurally, or intrathecally; NMDA receptor antagonists (dextromethorphan and ketamine) administered orally, parenterally, or intrathecally (ketamine only); GABA agonists (e.g., baclofen) or facilitators (e.g., diazepam) administered orally or intrathecally; corticosteroids (prednisone and dexamethasone) administered parenterally or orally; topical agents (EMLA, gabapentin, morphine, lidocaine, and capsaicin); and phenol or alcohol (chemical neurolysis in specific cases) administered around nerves or ganglia (Chong et al. 1997; Eisenach et al. 1995; Stubhaug and Breivik 1997; Rowbotham 1994; Galer et al. 1999). Boortz-Marx and colleagues published results of a multicenter study looking at the impact of intrathecal drug delivery on cancer pain (Smith et al. 2002). Mixtures of drugs including opioid agonists, local anesthetics, and alpha-2 adrenergic agonists were applied as intrathecal preparations. Patients enrolled in this study reported reduced pain, reduced toxic side effects, improved quality of life, and a trend toward extended life expectancy.

Non-pharmacological Treatment of Neuropathic Components of Cancer Pain

Non-pharmacological treatment options serve as successful adjuncts in the treatment of neuropathic cancer pain. The International Association for the Study of Pain (IASP) has characterized chronic pain as a "bio-psycho-social-spiritual" process. Chronic pain affects all patients at all stages of their cancer disease. Non-pharmacological strategies including patient psychological education, supportive psychotherapy, and cognitive-behavioral interventions have been demonstrated to be effective in the cancer pain patient. These therapies lead to patient empowerment, improved stress management, improved patient recognition, and modification of factors contributing to physical and emotional distress (Thomas and Weiss 2000). Other major non-pharmacological modalities that have demonstrated efficacy include therapeutic touch, massage, aromatherapy, reflexology, relaxation, guided imagery, visualization, meditation, biofeedback, acupuncture, and music therapy (O'Collaghan 1996; Penson and Fisher 1995).

Modeling Aspects of Persistent Cancer Pain

In order to understand the basic mechanisms involved in cancer pain and ultimately to provide insight into mechanism-based therapies, several groups have developed rodent models of tumor-induced hyperalgesia. The classification of pain in cancer can be either neuropathic or nociceptive (somatic), but is often mixed (Caraceni and Portenoy 1999). This distinction is a familiar theme, not only clinically, but also in pain models. Neuropathic pain models in rodents have mostly focused on mechanical trauma to the sciatic or spinal nerves; similarly, in animal models of cancer pain, the tumor mass itself may impose structural damage on nearby nerve bundles. See Pacharinsak and Beitz (2008) for a review of animal models of cancer pain.

Neural Plasticity in Cancer Pain Models

Plasticity in pain processing within the CNS can lead to pathological pain states that manifest not only as increased nociceptive sensitivity at the injury site, but also as secondary hyperalgesia and referred pain. In models of cancer pain, the potential for both peripheral and central plasticity is likely, particularly when tumor cells are implanted into the distal femur, where hyperalgesia both at the tumor site (Schwei et al. 1999) and secondarily in the paw (Wacnik et al. 2001) are measurable. Schwei et al. (1999) and Shimoyama et al. (2005) found increased immunoreactivity for dynorphin and c-fos protein in the dorsal horn correlated with tumor growth, possibly signaling neuroplastic changes. In a rat model of cancer pain where tumors are implanted in the tibia, Urch et al. (2003) demonstrated increased responses to mechanical, thermal, and electrical (A beta, C-fiber, and post-discharge evoked response) stimuli in wide dynamic range dorsal horn neurons in tumor-bearing animals, but no changes in nociceptive specific neurons. Tumor-induced peripheral neuropathies include aberrant firing of peripheral nociceptors adjacent to the tumor, identified by spontaneous activity in 34 % of cutaneous C fibers and an increase in epidermal nerve branching concomitant with a decrease in the actual number of fibers in skin overlying the tumor (Cain et al. 2001). Furthermore, immunohistochemical analysis of tumors revealed innervation of tumors with CGRP-immunoreactive nerve fibers. The density of these tumor-nerve appositions was positively correlated with the level of hyperalgesia, whereas that of blood vessels was inversely correlated (Wacnik et al. 2005). Also see review by Schmidt et al. (2010).

Models of Chemotherapy-Induced Neuropathic Pain

In cancer patients, neuropathic pain is frequently associated with direct tumor invasion of the peripheral nerve or spinal cord or secondarily caused by cancer chemotherapy. Several rodent models have been developed to model painful chemotherapy-induced neuropathy, using, for example, vincristine or paclitaxel. In the periphery, vincristine has been shown to enhance C-fiber responsiveness and to induce structural changes to large-diameter sensory neurons and myelinated axons (Topp et al. 2000). In the CNS, vincristine promotes central sensitization in the dorsal horn of the spinal cord, as indicated by increased spontaneous activity, increased responsiveness to C- and Aδ-fiber activity, and abnormal "windup" in response to afferent C-fiber activity (Weng et al. 2003). Also see review by Bennett (2010).

Conclusions

Pain categories say little about the actual underlying mechanisms. Whether the category is neuropathic or nociceptive, the tumor or antecedent chemotherapy is ultimately the genesis of the pain. Damage to several different tissue types is part of tumorigenesis, and this damage in turn may make different contributions to the resulting pain. There is a high likelihood that a tumor could damage the nerve through compression, stretching, or infiltration (i.e., neuropathic), as well as by invading somatic or visceral structures and releasing mediators that activate nociceptive fibers (i.e., nociceptive somatic/visceral). Although the malignant mass begins as a single entity, it may manifest pain via several means and via complex interactions. Accordingly, work toward a mechanistic classification of cancer pain must take these various means and interactions into account.

References

Bennett, G. J. (2010). Pathophysiology and animal models of cancer-related painful peripheral neuropathy. *The Oncologist, 15*(Suppl. 2), 9–12.

Cain, D. M., Wacnik, P. W., Turner, M., et al. (2001). Functional interactions between tumor and peripheral nerve: Changes in excitability and morphology of primary afferent fibers in a murine model of cancer pain. *Journal of Neuroscience, 21*, 9367–9376.

Caraceni, A., & Portenoy, R. (1999). An international survey of cancer pain characteristics and syndromes. *Pain, 82*, 263–274.

Chong, S. F., Bretsher, M. E., Mailliard, J. A., et al. (1997). Pilot study evaluating local anesthetics administered systemically for treatment of pain in patient with advanced cancer. *Journal of Pain and Symptom Management, 13*, 112–117.

Eisenach, J. C., DuPen, S., Dubois, M., et al. (1995). Epidural clonidine analgesia for intractable cancer pain. The Epidural Clonidine Study Group. *Pain, 61*, 391–399.

Foley, K. M. (2000). Controlling cancer pain. *Hospital Practice (Office Ed.), 35*, 101–108. 111–102.

Galer, B. S., Rowbotham, M. C., Pcrander, J., et al. (1999). Topical lidocaine patch release post-herpetic neuralgia more effectively than a vehicle topical patch: Results of an enriched enrollment study. *Pain, 80*, 533–538.

McQuay, H. (1999). Opioids in pain management. *The Lancet, 353*, 2229–2232.

Merskey, H., & Bogduk, N. (1994). *Classification of chronic pain* (2nd ed.). Seattle: IASP Press.

O'Collaghan, C. C. (1996). Complimentary therapies in terminal care: Pain, music, creativity, and music therapy in palliative care. *American Journal of Hospice and Palliative Medicine, 13*, 43–49.

Pacharinsak, C., & Beitz, A. (2008). Animal models of cancer pain. *Comparative Medicine, 58*(3), 220–233.

Penson, J., & Fisher, R. (1995). *Complimentary therapies in palliative care for people with cancer* (pp. 233–245). London: Arnold.

Portenoy, R. K. (1991). Cancer pain general design issues. In M. Max, R. Portenoy, & E. Laska (Eds.), *Advances in pain research and therapy* (pp. 233–266). New York: Raven.

Portenoy, R. K., & Lesage, P. (1999). Management of cancer pain. *Lancet, 353*(9165), 1695–1700.

Rowbotham, M. C. (1994). *Pharmacological approaches to the treatment of chronic pain: New concepts in critical issues*. Seattle: IASP Press.

Schmidt, B. L., Hamamoto, D. T., Simone, D. A., et al. (2010). Mechanism of cancer pain. *Molecular Interventions, 10*(3), 164–178.

Schwei, M. J., Honore, P., Rogers, S. D., et al. (1999). Neurochemical and cellular reorganization of the spinal cord in a murine model of bone cancer pain. *Journal of Neuroscience, 19*, 10886–10897.

Shimoyama, M., Tatsuoka, H., Ohtori, S., et al. (2005). Change of dorsal horn neurochemistry in a mouse model of neuropathic cancer pain. *Pain, 114*, 221–230.

Smith, T. J., Staats, P. S., Pool, G., et al. (2002). An implantable drug delivery system for refractory cancer pain improves pain control, drug-related toxicity, and survival compared to comprehensive medical management. American Society of Clinical Oncology, May 2002. *Proceedings of the American Society of Clinical Oncology, 21*, 360a.

Stubhaug, A., & Breivik, H. (1997). Long-term treatment of chronic neuropathic pain with NMDA (N-Methyl-D-Aspartate) receptor antagonists ketamine. *Acta Anaesthesiologica Scandinavica, 41*, 329–331.

Thomas, E. M., & Weiss, S. M. (2000). Non-pharmacological interventions with chronic cancer pain in adults. *Cancer Control, 7*, 157–164.

Topp, K. S., Tanner, K. D., & Levine, J. D. (2000). Damage to the cytoskeleton of large diameter sensory neurons and myelinated axons in vincristine-induced painful peripheral neuropathy in the rat. *Journal of Comparative Neurology, 424*(4), 563–76.

Urch, C. E., Donovan-Rodriguez, T., & Dickenson, A. H. (2003). Alterations in dorsal horn neurones in a rat model of cancer-induced bone pain. *Pain, 106*, 347–356.

Wacnik, P. W., Eikmeier, L. J., Ruggles, T. R., et al. (2001). Functional interactions between tumor and peripheral nerve: Morphology, algogen identification, and behavioral characterization of a new murine model of cancer pain. *Journal of Neuroscience, 21*, 9355–9366.

Wacnik, P. W., Baker, C., Blazar, B. R., et al. (2005). Tumor-induced mechanical hyperalgesia involves CGRP receptors and altered innervation and vascularization of DsRed2 fluorescent hindpaw tumors. *Pain, 115*, 95–106.

Weng, H. R., Cordella, J. V., & Dougherty, P. M. (2003). Changes in sensory processing in the spinal dorsal horn accompany vincristine-induced hyperalgesia and allodynia. *Pain, 103*, 131–138.

Cancer Pain Management, Undertreatment and Clinician-Related Barriers

Sean O'Mahony
Palliative medicine, Rush University Medical Center, Chicago, IL, USA

Definition

Cancer pain is pain that can be attributed to a malignancy, or complication of a malignancy, or its treatment.

Characteristics

Cancer pain affects 17 million people worldwide (Coyle et al. 1990; Daut and Cleeland 1982). Currently, the WHO estimates that 6.2 million cancer and HIV/AIDS patients experience moderate or severe pain annually, because they do not receive opioid analgesia. Prevalence rates of 30–40 % are reported for patients receiving active treatment and 70–90 % for patients with advanced cancer (Kelsen et al. 1995). Cancer pain occurs at multiple sites; in one study of 2,266 cancer patients, 70 % of patients had pain at two or more sites (Zeppetella et al. 2000). The duration of cancer pain varies, but it can extend to several months or years (Petzke et al. 1999). In the USA, an Eastern Cooperative Oncology Group survey of 1,308 ambulatory cancer patients found that 67 % reported recent pain, and 36 % reported their pain severity as sufficient to interfere with their function (Von Roenn et al. 1994). Cancer pain is an independent predictor for survival time in patients undergoing treatment for cancer (Quinten 2009).

Undertreatment of Cancer Pain

Although reviews of the literature confirm that cancer pain may be relieved in 70–90 % of patients, an increasing body of evidence suggests that cancer pain remains undertreated internationally. Deficits in the treatment of cancer pain extend across specialty and level of experience. A French national questionnaire study of general practitioners and specialists indicated that only 10 % of patients treated by general practitioners, and 21 % of patients treated by specialists, were receiving treatment regimens appropriate to their pain severity (Vainio 1995; Liang 2011). An analysis of the computerized medical records of more than one million German patients revealed that only 1.9 % of patients with cancer were receiving prescriptions for strong opioid medications, and many patients with cancer were receiving medications at inappropriate intervals and often on an "as needed" basis (Zenz et al. 1995).

Fear of addiction and respiratory depression appear to limit physician's use of strong opioids. In a study of 13,625 US nursing home residents

with cancer and daily pain, factors that were predictive of undertreatment included poor cognitive status, polypharmacy, and advanced age (age > 85) (Bernabei et al. 1998). Factors that were predictive of undertreatment of pain in the ECOG study of ambulatory oncology patients included minority status, discrepancy between patient and physician rating of pain severity, age, female sex, and poor performance status and type of treatment center (Apolone 2009). Introduction of daily pain measurement into the care of patients receiving cancer treatment reduces undertreatment of cancer pain (Giovanis 2010). Much of this undertreatment relates to a failure to prescribe breakthrough medication (Greco 2011).

Undertreatment of Cancer Pain in Special Populations

Elderly and Minority Patients

Undertreatment of cancer pain has been reported in minorities and the elderly in ambulatory settings, hospitals and nursing homes (Von Roenn et al. 1994; Bernabei et al. 1998). Elderly patients with cancer pain often have comorbidities. In conjunction with polypharmacy, this can account for a greater suceptibility to adverse drug events. Cognitive impairment and impaired communication also render some elderly patients with pain susceptible to undertreatment. Despite many published guidelines on pain assessment in nonverbal cognitively impaired persons, assessments are frequently not completed or documented (Curtiss 2010).

Despite the high prevalence of cancer pain in the elderly, they tend to be excluded from analgesic trials. Between 1987 and 1990, 83 randomized clinical trials of anti-inflammatory drugs in 10,000 patients included only 203 patients over the age of 65. The occurrence of complications of cancer pain in the elderly has not been extensively studied, including gait disturbance, falls, delayed rehabilitation, malnutrition, polypharmacy, and cognitive impairment.

Ambulatory cancer patients treated at centers that predominantly treat minority patients in the USA have been reported to be three times less likely than patients treated elsewhere to receive adequate cancer pain management (Von Roenn et al. 1994). Access to analgesic medication is impaired by lower availability of opioid medications in pharmacies located in predominantly minority neighborhoods. Clinician adherence to racial stereotypes is implicated in reports of disparities in morphine equivalent doses for different ethnic groups. Preference for analgesics for cancer pain among African Americans has been linked to a number of covert factors such as meaning of cancer pain treatment, past experience with pain relief and analgesic side effects, fears of dependency and tolerance, and past experience with providers and the health system (Meghani 2007).

Cancer Pain in the Developing World
It has been predicted that 10 of the 15 million new cases of cancer worldwide in the year 2015 will occur in the Third World; as many as 90 % of these will continue to present with advanced disease. The World Health Organization assesses progress in cancer pain management by national per capita morphine consumption. Fifty percent of countries use little or no morphine (UN ▶ International Narcotics Control Board (United Nations International Narcotics Control Board 1992)). The ten countries consuming 57 % of all morphine in 1991 have ranked highest in morphine consumption for many years. However, 100 of the poorest countries with the majority of the world's population used just 14 % of all morphine. In 2007, North American and European populations alone consumed 92 % of the world's morphine supply, while the remaining 8 % of morphine was shared among 80 % of humanity. INCB quotas are extremely inequitable: the quota for morphine is 10 kg in Nigeria (population 130 million) versus 5,200 kg in Belgium (population 10.5 million) – a 6,000-fold difference per capita. Meeting these patients' analgesic needs is rendered impossible by the INCB's quotas which make opioids scarce. The west's futile and socially destructive war on drugs not only fuels the war in Afghanistan through illegal distribution of opium but denies a potentially economic source of opioids for the many persons living with undertreated pain in developing countries (Attaran 2011).

Many of these countries lack economic resources and medical infrastructure to produce and distribute oral opioid medications. Restrictions exist on the duration that a patient can receive oral morphine or the locations where opioid medications can be received. When available orally, the available formulations of morphine (e.g., 10 mg controlled-release formulations in the Philippines) may predispose to underdosing.

Barriers to Effective Pain Management
Clinician Educational Needs
Barriers to effective pain management are often conceptualized in terms of healthcare provider, patient, family, institutional, and societal factors. About 50 % of physicians are reported as having erroneous assumptions about the use of opioids for cancer pain (Fife et al. 1993; Pflughaupt et al. 2010). These misconceptions include concerns about tolerance, addiction, role of various routes of administration, and prevalence and management of side effects. As many as 20 % of physicians are reported to regard cancer pain as inevitable and something that cannot be effectively managed (Fife et al. 1993). These misconceptions extend across disciplines and medical specialties. More than one third of doctors and nurses in a US survey of 971 clinicians believed that the use of opioid medications should be restricted based on the stage of a patient's illness (Elliott and Elliott 1992). Knowledge deficits do not appear to correlate with the level of exposure to cancer pain and training in palliative care. Despite their widespread availability, physicians continue to be reluctant to use validated assessment instruments.

Patient and Family Barriers
Many patients and families have unrealistic concerns about the use of pain medications and inadequate knowledge of cancer pain management (Fortier 2012). Patients often regard opioids as a last resort to be reserved for intolerable pain, which they believe will be an inevitable sequela of their illness. Patient barriers are stronger in older patients and in patients with lower educational and income levels.

Economic Considerations

Internationally, reimbursement of patients for analgesic medications is limited. Reimbursement by prevailing medical payers also does not cover operational costs for the provision of comprehensive cancer pain management services. Palliative care is funded, in most countries, through a combination of public and private funding sources. However, there is heavy dependence on philanthropy and community fund-raising. In an attempt to limit the impact on healthcare systems of soaring costs for medications, many healthcare insurance benefits limit coverage of medications or place restrictions on the number of refills on prescriptions, the number of dosage units of a medication, or the number of medications that a patient can receive. An unintentional effect of such policies may be the premature admission of chronically ill patients to skilled long-term nursing care facilities (Soumerai et al. 1987; Abernathy 2008). In many countries, oral opioids and long-acting formulations of opioid medications, in particular, are prohibitively expensive when consideration is given to average incomes.

Even in wealthy countries, availability of inpatient palliative care beds is limited. In Germany, it was estimated in 1996 that the national need for palliative care beds was about 4,000 beds. Only 230 were available, and at the same time, patients described spending an average of 2 years with cancer pain prior to having access to a pain clinic (Strumpf et al. 1996).

Historically, in the USA, healthcare insurance companies have paid for procedures at the expense of comprehensive medication coverage. Medicare, the prevailing carrier for patients with chronic illness and for the elderly, will pay for the cost of home infusions of opioid medications at a cost of US $250–300 per day. However, the same medications administered orally are often not reimbursed (Witteveen et al. 1999).

Inconsistent access to effective pain management resources commonly results in unnecessary hospital admissions for pain and symptom management. At one hospital alone, it was reported that 4 % of the hospital's admissions were for uncontrolled pain, at an annual cost of US $5.1 million. Little work exists on the impact of

▶ cost-shifting to patients and their families, in terms of family income or lost work days for cancer pain management in the community.

The rising cost of opioid medications, for long-acting formulations in particular, may place severe strains on the budgets of hospice organizations. Hospices may also be negatively impacted by requirements to destroy opioid medications after a patient dies. Recommended changes include alterations in federal guidelines to enable pharmacies to partially dispense if a patient is resident in a long-term care facility or has a documented terminal illness.

References

Abernathy, A. P., Wheeler, J. L., & Fortner, B. V. (2008). A health economic model of breakthrough pain. *The American Journal of Managed Care, 14*(5 Suppl. 1), s129–s140.

Apolone, G., Corli, O., & Caraceni, A. (2009). Cancer pain outcome research study group (CPOR SG) investigators. Patterns and quality of cancer pain management. Results from the cancer pain outcome research study group. *British Journal of Cancer, 100*(10), 1566–1574.

Attaran, A., & Boozary, A. (2011). For peace and pain: The medical legitimization of Afghanistan's poppy crop. *Journal of Epidemiology and Community Health, 65*(5), 396–398.

Bernabei, R., Gambassi, I., & Lapane, K. F. (1998). Pain management in elderly patients with cancer. *Journal of the American Medical Association, 279*(23), 1877–1882.

Coyle, N., Adelhardt, J., Foley, K. M., et al. (1990). Character of terminal illness in the advanced cancer patient: Pain and other symptoms in the last 4 weeks of life. *Journal of Pain and Symptom Management, 5*, 83–93.

Curtiss, C. P. (2010). Challenges in pain assessment in cognitively intact and cognitively impaired older adults with cancer. *Oncology Nursing Forum, 37* (Suppl.), 7–16.

Daut, R. L., & Cleeland, C. S. (1982). The prevalence and severity of pain in cancer. *Cancer, 50*, 13–18.

Elliott, T. E., & Elliott, B. A. (1992). Physician attitudes and beliefs about use of morphine for cancer pain. *Journal of Pain and Symptom Management, 7*, 141–148.

Fife, B. L., Irick, N., & Painter, J. D. (1993). A comparative study of the attitudes of physicians and nurses towards the management of cancer pain. *Journal of Pain and Symptom Management, 8*, 132–139.

Fortier, M. A., Wahi, A., Maurer, E. L., et al. (2012). Attitudes regarding analgesic use and pain expression in parents of children with cancer. *Journal of Pediatric Hematology/Oncology, 34*, 257–262. [Epub ahead of print].

Giovanis, P., De Leonardis, G., Garna, A., et al. (2010). Clinical governance benchmarking issues in oncology: Aggressiveness of cancer care and consumption of strong opioids. A single-center experience on measurement of quality of care. *Tumori, 96*(3), 443–447.

Greco, M. T., Corli, O., Montanari, M., & Cancer Pain Outcome Research Study Group (CPOR SG) Investigators. (2011). Epidemiology and pattern of care of breakthrough cancer pain in a longitudinal sample of cancer patients: Results from the cancer pain outcome research study group. *The Clinical Journal of Pain, 27*(1), 9–18.

Kelsen, D. P., Portenoy, R. K., Thaler, H. T., et al. (1995). Pain and depression in patients with newly diagnosed pancreas cancer. *Journal of Clinical Oncology, 13*, 748–755.

Liang, S. Y., Li, C. C., Wu, S. F., et al. (2011). The prevalence and impact of pain among Taiwanese oncology outpatients. *Pain Management Nursing, 12*(4), 197–205.

Meghani, S. H., & Keane, A. (2007). Preference for analgesic treatment for cancer pain among African Americans. *Journal of Pain and Symptom Management, 34*(2), 136–147.

Petzke, F., Radbruch, L., Zech, D., Loick, G., & Grond, S. (1999). Temporal presentation of chronic cancer pain: Transitory pains on admission to a multidisciplinary pain clinic. *Journal of Pain and Symptom Management, 17*, 391–401.

Pflughaupt, M., Scharnagel, R., Gossrau, G., et al. (2010). Physicians' knowledge and attitudes concerning the use of opioids in the treatment of chronic cancer and non-cancer pain. *Schmerz, 24*(3), 267–275.

Soumerai, S. B., Avorn, J., Ross-Degnan, D., et al. (1987). Payment restrictions for prescription medications under medicaid. *The New England Journal of Medicine, 317*, 550–556.

Strumpf, M., Zenz, M., & Donner, B. (1996). Germany: Status of cancer pain and palliative care. *Journal of Pain and Symptom Management, 12*, 109–111.

United Nations International Narcotics Control Board. (1992). *Narcotic drugs: Estimated world requirements for 1993, statistics for 1991*. Vienna: United Nations.

Vainio, A. (1995). Treatment of terminal cancer pain in France: A questionnaire study. *Pain, 62*, 155–162.

Von Roenn, J. H., Cleeland, C. S., Gonin, R., et al. (1994). Pain and its treatment in outpatients with metastatic cancer. *The New England Journal of Medicine, 330*, 592–596.

Witteveen, P. O., Van Groenestijn, M. A. C., Blijham, G. H., et al. (1999). Use of resources and costs of palliative care with parenteral fluids and analgesics in the home setting for patients with end-stage cancer. *Annals of Oncology, 10*, 161–165.

Zenz, M., Zenz, T., Tryba, M., et al. (1995). Severe under-treatment of cancer pain: A 3 year survey of the German situation. *Journal of Pain and Symptom Management, 10*, 187–190.

Zeppetella, G., O'Doherty, C. A., & Collins, S. (2000). Prevalence and characteristics of breakthrough pain in cancer patients admitted to a hospice. *Journal of Pain and Symptom Management, 20*, 87–95.

Cancer Pain Model, Bone Cancer Pain Model

Darryl T. Hamamoto and Donald A. Simone
Diagnostic and Biological Sciences, University of Minnesota, Minneapolis, MN, USA

Definition

Bone is the third most common site for tumor metastases (Fitzgibbon and Loeser 2010), and pain is the most frequent symptom for patients with metastatic bone cancer (Pecherstorfer and Vesely 2000). Bone cancer pain is often difficult to treat (Jimenez-Andrade et al. 2010). Animal models have recently been developed to elucidate the underlying mechanisms of bone cancer pain. A better understanding of these mechanisms may lead to the development of novel and more effective approaches to treating bone cancer pain.

Characteristics

Animal models of bone cancer pain have only been developed in rodents (i.e., mice and rats). One advantage of murine models is that mice are commonly used to study cancer biology, so that many cancer cell lines are available for use in mice. Many strains of mice have been genetically modified, and these strains may be used to determine the role of specific biochemicals in bone cancer pain. The advantage of using rats in a model of bone cancer pain is that rats are commonly used in pain research, so their nociceptive systems have been well studied. Many ▶ nocifensive behavioral assays used to test rodents with bone cancer pain were initially developed in rats and have been used in models

Cancer Pain Model, Bone Cancer Pain Model, Fig. 1 Responses of nociceptive dorsal horn neurons to mechanical stimuli of the hind paw of mice implanted with sarcoma cells into and around the calcaneus bone. (a) Examples of the responses of WDR neurons from a control and a tumor-bearing (cancer) mouse to mechanical stimuli. (b) Mechanical stimuli evoked greater mean (±SEM) number of impulses from WDR neurons in tumor-bearing compared to control mice. (c) Examples of the responses of HT neurons from a control and a tumor-bearing mouse to mechanical stimuli. (d) There were no differences in the mean (±SEM) number of impulses from HT neurons evoked by the mechanical stimuli between tumor-bearing and control mice. $*P < 0.05$, $****P < 0.0001$ for number of impulses between tumor-bearing and control mice for each mechanical stimulus (Modified from Khasabov et al. 2007)

of inflammatory and neuropathic pain (Pacharinsak and Beitz 2008). Thus, comparisons between the mechanisms underlying inflammatory, neuropathic, and bone cancer pain can be made more easily. Finally, because of its larger size, it is easier to perform surgical procedures on a rat than a mouse.

In these rodent models of bone cancer pain, tumor cells are implanted into different bones. In three of the four models, tumor cells are implanted into bones of the hind limb, specifically the femur, tibia, and calcaneus bones. Use of the hind limb is advantageous because many behavioral tests of ▶ nociception apply the stimulus to the hind paw of rats, and thus the responses to these stimuli have been well characterized. It is easier to apply stimuli to the hind paw than the forepaw in rodents because the hind paw is larger and away from the animal's line of sight. The lumbar enlargement of the spinal cord is easier to access than the cervical enlargement, and so it is easier to apply drugs to or record

▶ electrophysiological activity from ▶ dorsal horn neurons. The lumbar and cervical enlargements are where ▶ primary afferents/neurons that innervate the hind limb and forelimb terminate, respectively. As a result, the electrophysiological responses and neurochemical characteristics of the dorsal horn neurons that have receptive fields on the hind limb have been better characterized than those innervating the forelimb. For example, Fig. 1 shows the responses of wide dynamic range dorsal horn neurons to mechanical stimulation of the receptive field on the hind paws of naive (control) and tumor-bearing (cancer) mice (Khasabov et al. 2007). As the hind limb is longer than the forelimb in rodents, surgical preparation for electrophysiological recording from primary afferent fibers is easier. Thus, the electrophysiological responses of the primary afferent fibers that innervate the hind paw, and the neurochemical characteristics of their associated ▶ dorsal root ganglion neurons, have been better studied.

Cancer Pain Model, Bone Cancer Pain Model, Fig. 2 Bone destruction scoring system for tumor-induced osteolysis after intrafemoral injection of tumor cells. (**a**) Normal bone marrow and intact bone in control femora. (**b**) Tumor-bearing femora show replacement of bone marrow with tumor cells and osteolysis (arrowhead). (**c**) Radiographs illustrate the bone destruction scoring system by showing control bone (*0*, no loss) and progressive bone loss in tumor-bearing bones (*1–3*). Progressive bone loss was evaluated as follows: (1) intramedullary osteolysis without full-thickness cortical osteolysis, (2) intramedullary osteolysis and full-thickness cortical erosion through one cortex, and (3) intramedullary osteolysis and full-thickness cortical erosion through both cortices. Arrowheads indicate regions of osteolysis (From Jimenez-Andrade and Mantyh 2010)

Femur Model

The first reported animal model of bone cancer pain used NCTC 2472 cells, derived from a spontaneous connective tissue tumor, which were implanted through an arthrotomy in the knee joint into the medullary cavity of one femur of C3H/he mice (Jimenez-Andrade and Mantyh 2010). The hole in the articular surface of the distal femur, through which the sarcoma cells are implanted, is sealed to keep the cells within the medullary cavity. Implantation of sarcoma cells produces an increase in both the number of osteoclasts and in osteoclast activation, which results in osteolytic lesions in the implanted femur and invasion of the sarcoma cells into the adjacent soft tissues (Fig. 2).

Mice with sarcoma cells implanted into the femur exhibit spontaneous guarding and flinches, movement-evoked nocifensive behaviors during spontaneous or forced ambulation, and mechanical ► allodynia, as evidenced by flinching, guarding, fighting, and vocalization produced by normally non-noxious palpation of the affected limb (Jimenez-Andrade and Mantyh 2010; Pacharinsak and Beitz 2008). The frequency of palpation-induced nocifensive behaviors was correlated with the extent of bone destruction.

One feature of this model is that the sarcoma cells are sealed in the bone. This is similar to the clinical situation in which tumor cells metastasize to the medullary cavity of a bone. Also, nociceptive stimuli are applied to the distal femur, the location where the sarcoma cells erode through the bone, suggesting that the nociceptive behaviors are due to excitation of nociceptors in the area. However, it is not clear whether excitation of nociceptors located in bone, muscle, or skin is responsible for evoking the nocifensive behaviors. Interestingly, implantation of sarcoma cells into the femur also produced mechanical allodynia at the plantar surface of the hind paw, as shown by a decrease in the threshold force required to evoke a hind paw withdrawal (Pacharinsak and Beitz 2008). This finding suggests that the sarcoma cells may be injuring nerve trunks passing through the area, resulting in a neuropathic pain condition as well as releasing potential algogens (Hamamoto and Simone 2009). Alternatively, excitation of nociceptors located near the sarcoma cells might produce sensitization of dorsal horn neurons (i.e., ► central sensitization) that also have cutaneous receptive fields on the hind paw.

Tibia Model

A model of bone cancer pain in rats was developed in which mammary gland carcinoma cells (MRMT-1) were implanted through an incision in the skin into the tibia, 5 mm distal to the knee joint of one hind limb in female (Medhurst et al. 2002) or male (Urch et al. 2003) Sprague-Dawley rats. The hole through which the breast carcinoma cells were implanted is sealed. Implantation of these breast carcinoma cells produces an increase in the number of osteoclasts-like cells and time-dependent bone destruction. Tumor-bearing rats exhibit mechanical ▶ hyperalgesia (i.e., paw pressure) and allodynia (i.e., von Frey stimulation to the plantar surface of the hind paw; Fig. 3), cold allodynia (i.e., acetone applied to the plantar surface of the hind paw), a decrease in weight bearing on the implanted limb, and ambulatory-evoked pain (i.e., limping and guarding on a Rotarod treadmill). Thus, this model of bone cancer pain in rats produces histological and behavioral changes that were similar to those found in the femur model in mice.

A murine model of bone cancer pain, produced by implantation of NCTC 2472 sarcoma cells through a skin incision into the tibia of C3H/he mice, has also been reported (Menendez et al. 2003). The hole in the tibia through which the sarcoma cells are implanted is not sealed. Implantation of sarcoma cells results in an increase in the number of osteoclasts. At the time when the sarcoma cells erode through the bone and grow into the surrounding soft tissue, mice exhibit thermal hypoalgesia (i.e., increased paw withdrawal latencies when on a hot plate). As the tumor mass grows and the extent of bone destruction increases, mice exhibit thermal hyperalgesia.

In both of these models, test stimuli are applied to the hind paw, a site distant from the location of the tumor cells. Increased responses to stimuli applied distant to the site of implantation of the tumor cells suggesting a neuropathic injury or central sensitization may underlie these tumor-evoked behaviors.

Calcaneus Model

Implantation of NCTC 2472 sarcoma cells percutaneously into and around the calcaneus bone

Cancer Pain Model, Bone Cancer Pain Model, Fig. 3 (a) Changes in response frequency to von Frey filament stimulation following intra-tibial injections of MRMT-1 cells. Rats received intra-tibial injections of 3×10^3 syngeneic MRMT-1 breast cancer cells ($n = 6$/group). Data are means $\pm$ SEM. (*$P \leq 0{:}05$ compared to day 0). (b) Changes in response frequency to von Frey filament stimulation 19 days following intra-tibial injections of MRMT-1 cells (3×10^3 cells/bone), heat-killed cells, or vehicle. Rats received intra-tibial injections of syngeneic MRMT-1 breast cancer cells, heat-killed cells, or vehicle ($n = 6$/group). Data are means $\pm$ SEM. *$P \leq 0{:}05$ compared to vehicle (From Medhurst et al. 2002)

of C3H/he mice produces osteolysis and evokes mechanical hyperalgesia (Pacharinsak and Beitz 2008). Mechanical hyperalgesia was observed upon stimulation of the plantar surface of the ipsilateral hind paw with ▶ von Frey monofilaments. These mice also exhibited cold hyperalgesia (i.e., enhanced paw withdrawal responses when placed on a 5 °C plate). Histological examination shows that sarcoma cells are found under the skin of the plantar surface of the hind paw and that the number of nerve fibers in the epidermis (i.e., epidermal nerve fibers) above the tumor decreases (Cain et al. 2001; Hamamoto and Simone 2009). Interestingly, the proportion of neuropeptide containing primary afferent fibers increases suggesting that the sarcoma cells affect subclasses of primary afferent fibers differently.

One advantage of this model of bone cancer pain is that the effect of the sarcoma cells on the electrophysiological responses of primary afferent fibers innervating the tissue near the tumor can be determined. In fact, a proportion of nociceptive C-fibers innervating the plantar surface of the hind paw exhibits spontaneous activity (Fig. 4), which may produce central sensitization and contribute to the mechanical hyperalgesia observed (Cain et al. 2001; Hamamoto and Simone 2009). Thus, interactions between the sarcoma cells and primary afferent fibers can be examined using both histological and electrophysiological methods.

This model differs from the models using the femur and tibia in that the sarcoma cells are not sealed in the bone but are implanted both in and around the calcaneus bone. As a practical issue, the calcaneus bone in mice is small and the volume of its medullary cavity is low, making it difficult to keep the cell suspension within the medullary cavity. Additionally, sarcoma cells are implanted without making an incision through the skin. To seal the hole would require an incision to expose the calcaneus bone, which might produce inflammation and mechanical hyperalgesia. Moreover, implantation of sarcoma cells into the tissues around the calcaneus bone produced less mechanical and cold hyperalgesia than when the sarcoma cells were implanted into the bone (Pacharinsak and Beitz 2008). These findings suggest that interaction of the sarcoma cells with the calcaneus bone is important for full development of the mechanical and cold hyperalgesia observed in this model.

Humerus Model

Movement-related cancer pain is thought to be a predictor of poor response to routine pharmacotherapy in cancer patients (Jimenez-Andrade

Cancer Pain Model, Bone Cancer Pain Model, Fig. 4 Representative examples of spontaneous activity and conduction latency of three C-fibers recorded in skin overlying a fibrosarcoma tumor. (**a**, *top*) Spontaneous activity of this nociceptor occurred generally at a steady rate of _1 Hz. (**a**, *bottom*) Three superimposed oscilloscope traces showing constant conduction latency evoked by electrical stimulation at the RF. (**b**) A second C-fiber with a slower (0.15 Hz) rate of spontaneous activity. (**c**) Spontaneous activity of this nociceptor occurred in bursts of three to five action potentials. Arrows indicate stimulus artifact. The 10 s calibration below the spontaneous activity in C applies to all three traces of ongoing activity. Calibrations of individual conduction latencies are shown under the overlaid traces of constant conduction velocity (From Cain et al. 2001)

et al. 2010). Thus, a model of deep tissue cancer pain may be ideal for studying this specific characteristic of cancer pain. Implantation of NCTC 2472 sarcoma cells through the proximal end of

Cancer Pain Model, Bone Cancer Pain Model, Fig. 5 Morphine dose-dependently inhibits deep tissue hyperalgesia measured by grip force. Systemically administered morphine produces dose-dependent attenuation of deep tissue hyperalgesia (30 m postdrug) in both tumor-bearing and carrageenan-injected C3H/He mice ðn ($n = 4$–20/group). Dose-response curves of the inhibition of deep tissue hyperalgesia by morphine in 4 % carrageenan-injected mice were evaluated on day 2 post-carrageenan injection. Conversely two separate groups of tumor-implanted mice were tested for morphine attenuation of grip force hyperalgesia on day 7 or day 10 postimplantation. Inset: % maximum possible effect shows a significant 2.7–3.5-fold decrease in morphine potency for mice implanted both 7 and 10 days earlier (From Wacnik et al. 2003)

each humerus bone in C3H/he mice produces movement-related hyperalgesia, observed as a reduction in forelimb ▶ grip force (Wacnik et al. 2003) (see ▶ Muscle Pain Model, Inflammatory Agents-Induced). This reduction in grip force can be attenuated with morphine (Wacnik et al. 2003) or cannabinoids (Kehl et al. 2003) and is possibly due to sensitization of nociceptors in the triceps muscles.

An interesting advantage of the humerus model is that the effectiveness of an analgesic treatment is indicated by an increase in response. That is,

as the tumor grows, mice exhibit a decrease in grip force that is reversed by effective analgesics (Fig. 5). In other nocifensive behavioral tests, the animal shows an increase in responsiveness with tissue injury that is reversed by effective analgesic treatment. However, a decrease in paw withdrawal response or vocalization could also be due to sedation or motor effects of the test treatment. In the humerus model, sedation and motor impairment would be likely to produce a decrease in grip force, resulting in the analgesic effects of a treatment being more easily separated from any sedative effects or motor impairment.

References

Cain, D. M., Wacnik, P. W., Turner, M., et al. (2001). Functional interactions between tumor and peripheral nerve: Changes in excitability and morphology of primary afferent fibers in a murine model of cancer pain. *The Journal of Neuroscience, 21*, 9367–9376.

Fitzgibbon, D. R., & Loeser, J. D. (2010). Bone metastases. In *Cancer pain: Assessment, diagnosis, and management* (pp. 76–85). Philadelphia: Wolters Kluwer/Lippincott Williams & Wilkins.

Hamamoto, D. T., & Simone, D. A. (2009). Neuropathic pain produced by cancer and chemotherapy. In M. Dobretsov & J. M. Zhang (Eds.), *Mechanisms of pain in peripheral neuropathy* (pp. 219–254). Kerala: Research Signpost.

Jimenez-Andrade, J. M., & Mantyh, P. W. (2010). Cancer pain: From the development of mouse models to human clinical trials. In L. Kruger & A. R. Light (Eds.), *Translational pain research: From mouse to man* (pp. 77–98). Boca Raton: CRC Press.

Jimenez-Andrade, J. M., Mantyn, W. G., Bloom, A. P., Ferng, A. S., Geffre, C. P., & Mantyh, P. W. (2010). Bone cancer pain. *Annals of the New York Academy of Sciences, 1198*, 173–181.

Kehl, L. J., Hamamoto, D. T., Wacnik, P. W., Croft, D. L., Norsted, B. D., Wilcox, G. L., & Simone, D. A. (2003). A cannabinoid agonist differentially attenuates deep tissue hyperalgesia in animal models of cancer and inflammatory muscle pain. *Pain, 103*, 175–186.

Khasabov, S. G., Hamamoto, D. T., Harding-Rose, C., & Simone, D. A. (2007). Tumor-evoked hyperalgesia and sensitization of nociceptive dorsal horn neurons in a murine model of cancer pain. *Brain Research, 1180*, 7–19.

Medhurst, S. J., Walker, K., Bowes, M., et al. (2002). A rat model of bone cancer pain. *Pain, 96*, 129–140.

Menendez, L., Lastra, A., Fresno, M. F., et al. (2003). Initial thermal heat hypoalgesia and delayed hyperalgesia in a murine model of bone cancer pain. *Brain Research, 969*, 102–109.

Pacharinsak, C., & Beitz, A. (2008). Animal models of cancer pain. *Comparative Medicine, 58*, 220–233.

Pecherstorfer, M., & Vesely, M. (2000). Diagnosis and monitoring of bone metastases: clinical means. In Body J-J (Eds.), *Tumor Bone Diseases and Osteoporosis in Cancer Patients* (97–129). New York: Marcel Dekker, Inc.

Urch, C. E., Donovan-Rodriguez, T., & Dickenson, A. H. (2003). Alterations in dorsal horn neurones in a rat model of cancer-induced bone pain. *Pain, 106*, 347–356.

Wacnik, P. W., Kehl, L. J., Trempe, T. M., et al. (2003). Tumor implantation in mouse humerus evokes movement-related hyperalgesia exceeding that evoked by intramuscular Carrageenan. *Pain, 101*, 175–186.

Cancer Pain Models

Definition

A clinically relevant model typically developed in animals and used to study the mechanisms and related neurobiology of cancer-induced pain. Often, but not exclusively, these models are developed in rodents.

Cross-References

▶ Cancer Pain, Animal Models
▶ Cancer Pain Management, Treatment of Neuropathic Components
▶ Cancer Pain Model, Bone Cancer Pain Model

Cancer Pain, Animal Models

Patrick W. Mantyh[1] and Mary Ann C. Sabino[2]
[1]Department of Pharmacology, The University of Arizona, AZ, USA
[2]Oral and Maxillofacial Surgery, Hennepin County Medical Center University of Minnesota, Minneapolis, MN, USA

Synonyms

Bone cancer pain; Malignant pain

Definition

Cancer pain, distinguished from nonmalignant pain, arises from products derived from tumor and/or supporting cells which infiltrate soft tissue and/or bony structures. Cancer pain may also arise as a result of therapies used to treat cancer such as radiation and/or chemotherapy.

Introduction

Pain is the first sign of cancer and 30–50 % of all cancer patients will experience moderate to severe pain. Cancer-associated pain can be present at any time during the course of the disease, but the frequency and intensity of cancer pain tends to increase with advancing stages of cancer. Seventy-five to ninety-five percent of patients with metastatic or advanced stage cancer will experience significant amounts of cancer-induced pain (Portenoy et al. 1999).

Although there is significant variability in the type, severity, and evolution of cancer pain, two major components are generally recognized. The first component, known as ▶ ongoing pain, is most often the first to present, described as a dull ache or throbbing in character and usually increases in severity with disease progression. A second component of bone cancer pain frequently emerges over time and is more acute in nature. This second pain is known as ▶ incident or ▶ breakthrough pain, as it frequently occurs either spontaneously, with intermittent exacerbations of pain, or by movement of the cancerous bone (Mercadante and Arcuri 1998).

In recent years, animal models of cancer pain have been developed. These models have led to the discovery of products derived from cancer cells, inflammatory cells, immune cells, and bone cells involved in the generation of cancer pain. Identification of key proteins/neurotransmitters and their cellular targets has led to the development and testing of novel therapies which hold significant promise in managing a difficult condition.

Characteristics

Models of Cancer Pain
Osteolytic Bone Cancer Model

The first animal model of bone cancer pain involved the injection of murine osteolytic sarcoma cells into the intramedullary space of the murine femur (Fig. 1). A critical component of this model is that the tumor cells are confined within the marrow space of the injected femur, and the tumor cells do not invade adjacent soft tissues (Schwei et al. 1999). The elimination of soft tissue infiltration of tumor cells distinguishes this animal model from other models of bone metastases such as intracardiac delivery of tumor cells or intrafemoral/intratibial injection *without* placement of a barrier to prevent extraosseous tumor invasion. The confinement of tumor cells within the bony cavity allows researchers to isolate behavioral and neurochemical alterations without the confound of pain arising from extrabony sites.

Following tumor injection, the fluorescent cancer cells proliferate, and both ongoing and movement-evoked pain-related behaviors increase in severity as the tumor develops. These pain behaviors correlate with the progressive tumor-induced bone destruction that ensues,

which appears to mimic the condition seen in patients with primary or metastatic bone cancer. In this model, peripheral nerve destruction as well as alterations in neurochemical markers implicated in pain transmission has been observed (Fig. 2). Injection of different tumor cell lines such as breast cancer, colon cancer, and melanoma into the femur of immunocompromised mice has also been described (Sabino et al. 2003). Distinct differences in pain-related behaviors, osteolysis, tumor progression, and CNS changes were observed thereby suggesting that not all bone cancers are alike.

Osteoblastic Prostate Model

An osteoblastic (bone-forming) model was subsequently developed to define the behavioral, histopathologic, and neurochemical basis of blastic bone cancers such as metastatic prostate cancer (Halvorson et al. 2006). Canine prostate tumor cells were injected into nude mice, and both ongoing and movement evoked pain behaviors were evident as early as 9 days after inoculation. The prostate osteoblastic model differs from the murine osteolytic model both histologically and radiologically (Fig. 3). An increase in osteoblasts, the principle bone-forming cell, was seen in this model. Additionally, while the

Cancer Pain, Animal Models, Fig. 1 Bone cancer pain animal model. (**a**) Radiograph of lower half of adult mouse demonstrating the femur through which osteolytic fluorescent sarcoma cells were injected. Tumor cells are confined within the intramedullary space by placement of an amalgam plug (*arrow*). Two weeks later, femurs can be assessed at the whole bone level for extraosseous invasion (none noted in **b**), tumor burden (using excitation filters to visualize green fluorescent protein expressed by tumor cells) (**c**), and bone destruction (high-power radiographic imaging, arrow denotes extensive bone destruction) (**d**)

Cancer Pain, Animal Models, Fig. 2 Neurochemical changes in the spinal cord and dorsal root ganglia (DRG) in bone cancer pain. (**a**) Confocal imaging of glial fibrillary acidic protein (GFAP) expressed by astrocytes in a spinal cord of a tumor-bearing mouse. Note increased expression only on side ipsilateral to tumorous limb. (**b**) High-power magnification of spinal cord showing hypertrophy of astrocytes (*green*) without changes in neuronal numbers (*red*, stained with neuronal marker, NeuN). (**c**) Confocal image of dorsal root ganglia labeling injured neurons labeled with activating transcription factor 3 (*magenta*, ATF-3). (**d**) Non-myelinating schwann cells (*green*, GFAP) and macrophage infiltration (*yellow*, CD68) can also be seen

osteolytic sarcoma cell line has been shown to destroy primary afferent fibers as they infiltrate the intramedullary canal and periosteum of bone, there is simultaneous injury and sprouting of sensory fibers in bones injected with the prostate cell line. Prostate tumor cells are found in small colonies or nests that are separated histologically by bridges of woven bone. New woven bone formation, accompanied by regions of bone destruction, distinguishes this osteoblastic model and highlights its correlation with the human condition.

Soft Tissue Models of Cancer Pain
Following the publication of the first animal model of cancer pain, several rodent models of soft tissue and bone cancer pain have been described. A rat model of bone cancer pain was developed (Medhurst et al. 2002) whereby MRMT-1 mammary carcinoma cells were injected into the tibiae of syngeneic rats. Morphine-reversible pain behaviors and ▶ osteolysis were evident 2–3 weeks following inoculation. Mouse models of soft tissue sarcoma (Wacnik et al. 2001) and soft tissue oral carcinoma (Pickering et al. 2008) and rat models of neuropathic cancer (Eliav et al. 2004) were also developed and involved growth of tumor cells in soft tissues surrounding bones and alongside sciatic nerves, respectively. Mechanical and thermal ▶ hyperalgesia, demyelination, progressive tumor infiltration, and nerve destruction were seen in these models.

A spontaneous model of cancer pain has also been described and involves the use of transgenic mice that spontaneously develop

Cancer Pain, Animal Models, Fig. 3 Bone remodeling and tumor growth in the sarcoma and prostate carcinoma-injected femurs have different characteristics depending on the osteolytic or osteoblastic component of the tumor cells as assessed by μCT imaging and hematoxylin and eosin (H&E) staining. Non-tumor-bearing femurs present relative absence of bone formation or bone destruction (**a, d**). The sarcoma-injected femurs display a primarily osteolytic appearance visible as regions absent of trabecular bone at the proximal and distal heads (**b**) as well as replacement of normal hematopoietic cells by tumor cells (**e**). The prostate carcinoma-injected femurs mainly present an osteoblastic appearance which is characterized by pathologic bone formation in the intramedullary space (**c**) surrounding pockets of tumor cells which generate diaphyseal bridging structures (**f**). (**a**)–(**f**) Scale bar: 0.5 mm. *T* tumor, *H* normal hematopoietic cells, *WB* prostate cancer-induced woven bone formation

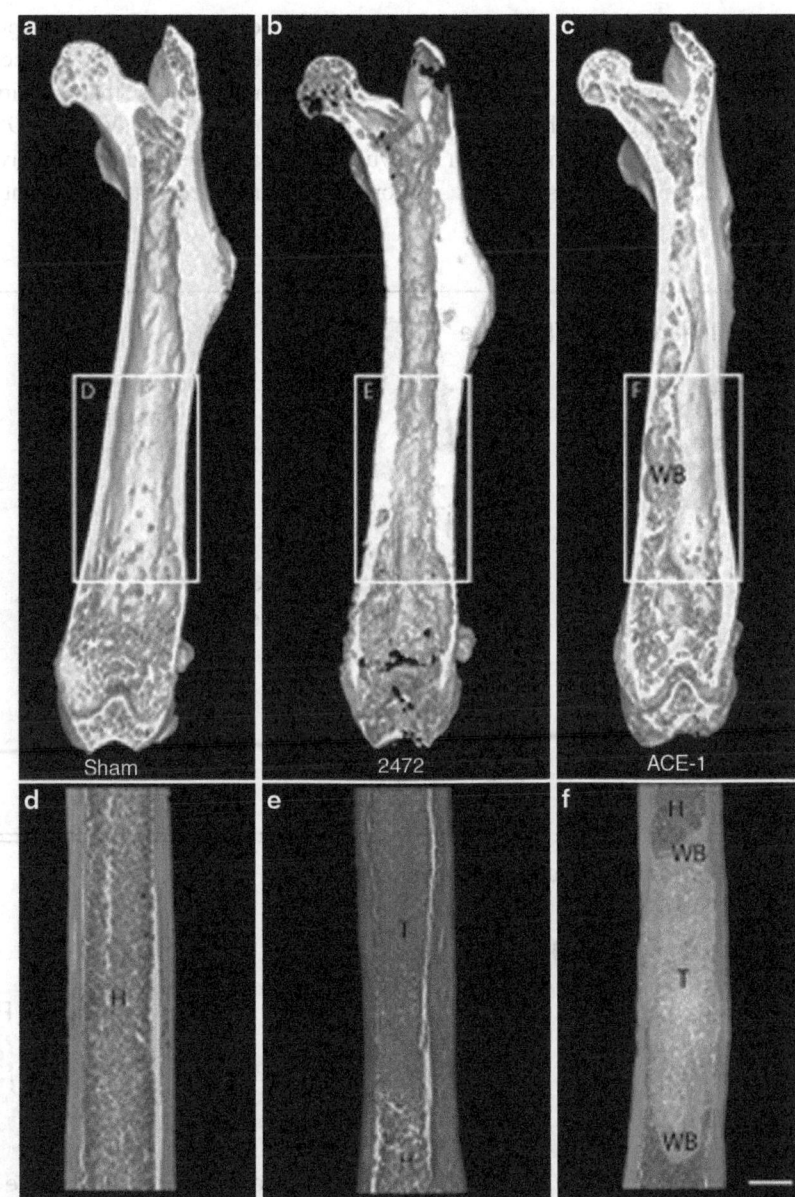

pancreatic cancer (Lindsay et al. 2005; Sevcik et al. 2006). Lesions were detectable as early as 9–12 weeks with progression to late stage pancreatic cancer by 14–24 weeks. Increases in vascularity, inflammatory infiltrate, and sensory and sympathetic innervation were seen as early as 5–7 weeks; however, pain-related behaviors were not detectable until late stages of the disease process. These pain-related behaviors were attenuated by morphine. These clinical, histologic, and behavioral findings are consistent with human pancreatic cancers which typically are not detected until pain is evident during later stages of the disease.

Using Models to Define Mechanisms of Bone Cancer Pain and Test Current and Novel Therapies

Given the development of animal models of cancer pain, significant advances have been made in understanding the molecular and cellular mechanisms which drive cancer pain (Fig. 4).

Prostaglandins
Cancer cells and tumor-associated macrophages have both been shown to express high levels of COX-2, leading to high levels of ▸ prostaglandins. Prostaglandins have been shown to be involved in the sensitization and/or direct excitation of ▸ nociceptors, by binding to several prostanoid receptors expressed by nociceptors that sensitize or directly excite nociceptors (Vasko 1995). Treatment of cancer animals with selective COX-2 inhibitors resulted in significant attenuation of pain

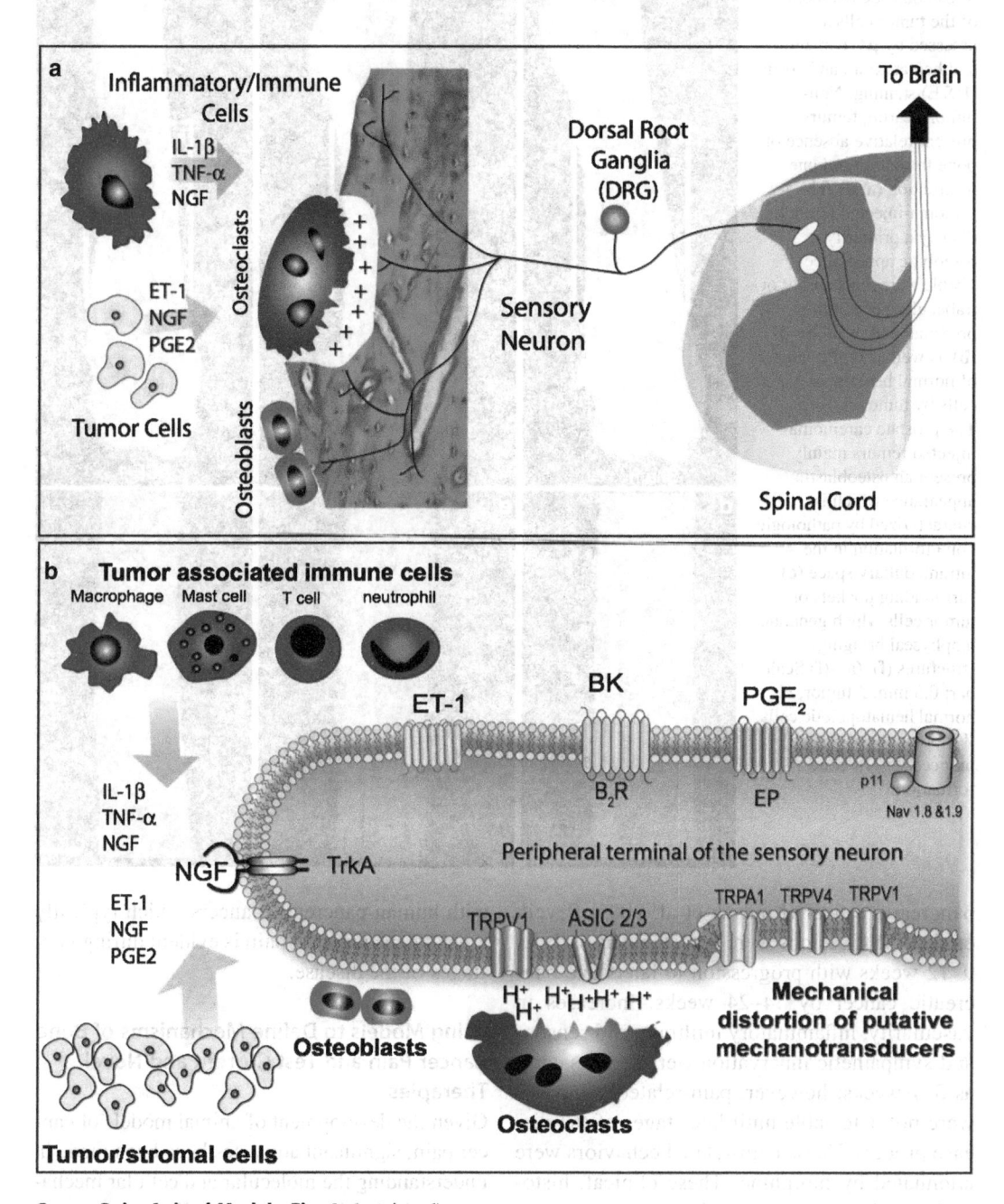

Cancer Pain, Animal Models, Fig. 4 (continued)

behaviors, as well as many of the neurochemical changes suggestive of both ▶ peripheral sensitization and central sensitization (Medhurst et al. 2002; Sabino et al. 2003).

Endothelins

▶ Endothelins have been found in cancer patients (Nelson et al. 1995) and in cancer-bearing mice (Wacnik et al. 2001; Pickering et al. 2008), and their levels have been correlated with the severity of pain (Nelson et al. 1995). In fact, endothelin concentration was found to correlate more with pain-related behaviors than tumor volume (Pickering et al. 2008). Endothelins could contribute to cancer pain by directly sensitizing or exciting nociceptors, as a subset of small unmyelinated primary afferent neurons express endothelin A receptors (Pomonis et al. 2001). Both ongoing and movement-evoked pain behaviors are attenuated with systemic administration of antagonists to the endothelin A but not B receptors (Peters et al. 2004). Furthermore, direct application of endothelin to peripheral nerves induces activation of primary afferent fibers and an induction of pain behaviors (Davar et al. 1998). These findings indicate endothelin antagonists may be useful in inhibiting cancer pain.

Acidosis

Studies have demonstrated that subsets of sensory neurons express different pH sensitive ion channels (Julius and Basbaum 2001). Two channels important in the transmission of nociceptive or painful stimuli are transient receptor potential vanilloid 1 (TRPV1, formerly known as vanilloid receptor 1 or VR-1) and acid-sensing ion channel-3 (ASIC3). Both of these channels are activated and sensitized in acidic environments (pH 4.0–5.0) created by osteoclasts and within the ischemic tumor stroma. ▶ Osteoclasts, the body's principal bone resorbing cell, play an essential role in cancer-induced bone loss and contribute to the etiology of bone cancer pain (Honore et al. 2000; Luger et al. 2001; Sabino et al. 2002; Sevcik et al. 2004). Potent antiresorptive compounds such as bisphosphonates and denosumab (formerly known as osteoprotegerin, OPG) are currently being used clinically to prevent the progression of metastatic bone cancers. Both antiresorptive compounds were highly potent in causing osteoclast apoptosis thereby reducing both cancer-induced bone destruction and the development of pain-related behaviors, suggesting an important role for osteoclasts in the development and maintenance of bone cancer pain.

TRPV1 is expressed in primary afferent fibers, but not sarcoma cells, located within bone. Direct antagonism of the TRPV1 receptor or disruption of the TRPV1 gene results in a significant reduction in nocifensive bone cancer pain-related behaviors, however, had no effect on tumor progression or tumor growth (Ghilardi et al. 2005). Thus, TRPV1 may serve as a potential target for the palliative management of bone cancer patients.

Cancer Pain, Animal Models, Fig. 4 Schematic showing factors in bone (**a**) and receptors/channels expressed by nociceptors that innervate the skeleton (**b**) that drive bone cancer pain. A variety of cells (tumor cells and stromal cells including inflammatory/immune cells, osteoclasts, and osteoblasts) drive bone cancer pain (**a**). Nociceptors that innervate the bone use several different types of receptors to detect and transmit noxious stimuli that are produced by cancer cells (*yellow*), tumor-associated immune cells (*blue*), or other aspects of the tumor microenvironment. There are multiple factors that may contribute to the pain associated with cancer (**b**). The transient receptor potential vanilloid receptor-1 (TRPV1) and acid-sensing ion channels (ASICs) detect extracellular protons produced by tumor-induced tissue damage or abnormal osteoclast-mediated bone resorption. Tumor cells and associated inflammatory (immune) cells produce a variety of chemical mediators including prostaglandins (PGE2), nerve growth factor (NGF), endothelins (ET-1), and bradykinin (BK). Several of these proinflammatory mediators have receptors on peripheral terminals and can directly activate or sensitize nociceptors. It is suggested that movement-evoked breakthrough pain in cancer patients is partially due to the tumor-induced loss of the mechanical strength and stability of the tumor-bearing bone so that normally innocuousmechanical stress can now produce distortion of the putative mechanotransducers (TRPV1, TRPV4, and TRPA1) that innervate the bone (Adapted from Jimenez-Andrade et al. 2010b)

Growth Factors

Recent studies suggest that primary afferent fibers that innervate bone are highly plastic and undergo continuous morphologic change. This appears to be mediated, in part, by nerve growth factor (NGF) which has been shown to induce marked sprouting in sensory nerve fibers expressing its cognate receptor, tropomyosin receptor kinase A (TrkA) (Jimenez-Andrade et al. 2010a; Mantyh et al. 2010). NGF is differentially expressed in tumor cell lines. The osteosarcoma cell line expresses high levels of NGF (Sevcik et al. 2005) whereas the prostate osteoblastic cell line does not (Halvorson et al. 2005). Despite this difference, early and sustained administration of an anti-NGF antibody significantly attenuated both pain-related behaviors and prevented both sympathetic and sensory nerve sprouting in both osteolytic and osteoblastic animal models. Interestingly, both pain behaviors and sprouting are unchanged if anti-NGF was administered in the acute, late setting. In the context of tumor cells such as the prostate cell line that do not express NGF, other potential sources of NGF that may induce sprouting include inflammatory and/or stromal cells within the cancerous bone. Currently, a monoclonal antibody against NGF (Tanezumab, Pfizer) is undergoing Phase II clinical studies for the relief of osteoarthritis, and Phase II studies using Tanezumab as an adjunct to opioids in patients with painful metastatic bone diseases have recently been completed (www.clinicaltrials.gov). Publication of results is pending.

Summary

Insights into the mechanisms that induce cancer pain are now coming from animal models. These models have begun to provide a glimpse into the mechanisms by which tumors cause pain and how this sensory information is processed. Chemicals derived from tumor cells and inflammatory cells and cells derived from bone appear to be simultaneously involved in driving this frequently difficult to control pain state. Understanding the mechanisms involved in the pathophysiology of cancer pain will improve our ability to provide mechanism-based treatments and therapies and improve the quality of life of cancer patients. Insights such as this promise to fundamentally change the way cancer pain is controlled.

References

Davar, G., Hans, G., Fareed, M. U., et al. (1998). Behavioral signs of acute pain produced by application of endothelin-1 to rat sciatic nerve. *Neuroreport, 9*, 2279–2283.

Eliav, E., Tal, M., & Benoliel, R. (2004). Experimental malignancy in the rat induces early hypersensitivity indicative of neuritis. *Pain, 110*, 727–737.

Ghilardi, J. R., Rohrich, H., Lindsay, T. H., et al. (2005). Selective blockade of the capsaicin receptor TRPV1 attenuates bone cancer pain. *The Journal of Neuroscience, 25*, 3126–3131.

Halvorson, K. G., Kubota, K., Sevcik, M. A., et al. (2005). A blocking antibody to NGF attenuates skeletal pain induced by prostate tumor cells growing in bone. *Cancer Research, 65*, 9426–9435.

Halvorson, K. G., Sevcik, M. A., Ghilardi, J. R., et al. (2006). Similarities and differences in tumor growth, skeletal remodeling and pain in an osteolytic and osteoblastic model of bone cancer. *The Clinical Journal of Pain, 22*, 587–600.

Honore, P., Luger, N. M., Sabino, M. A., et al. (2000). Osteoprotegerin blocks bone cancer-induced skeletal destruction, skeletal pain and pain-related neurochemical reorganization of the spinal cord. *Nature Medicine, 6*, 521–528.

Jimenez-Andrade, J. M., Bloom, A. P., Stake, J. I., et al. (2010a). Pathological sprouting of adult nociceptors in chronic prostate cancer-induced bone pain. *The Journal of Neuroscience, 30*, 14649–14656.

Jimenez-Andrade, J. M., Mantyh, W. G., Bloom, A. P., et al. (2010b). Bone cancer pain. *Annals of the New York Academy of Sciences, 1198*, 173–181.

Julius, D., & Basbaum, A. I. (2001). Molecular mechanisms of nociception. *Nature, 413*, 203–210.

Lindsay, T. H., Jonas, B. M., Sevcik, M. A., et al. (2005). Pancreatic cancer pain and its correlation with changes in tumor vasculature, macrophage infiltration, neuronal innervation, body weight and disease progression. *Pain, 119*, 233–246.

Luger, N. M., Honore, P., Sabino, M. A., et al. (2001). Osteoprotegerin diminishes advanced bone cancer pain. *Cancer Research, 61*, 4038–4047.

Mantyh, W. G., Jimenez-Andrade, J. M., Stake, J. I., et al. (2010). Blockade of nerve sprouting and neuroma formation markedly attenuates the development of late stage cancer pain. *Neuroscience, 171*, 588–598.

Medhurst, S. J., Walker, K., Bowes, M., et al. (2002). A rat model of bone cancer pain. *Pain, 96*, 129–140.

Mercadante, S., & Arcuri, E. (1998). Breakthrough pain in cancer patients: Pathophysiology and treatment. *Cancer Treatment Reviews, 24*, 425–432.

Nelson, J. B., Hedican, S. P., George, D. J., et al. (1995). Identification of endothelin-1 in the pathophysiology of metastatic adenocarcinoma of the prostate. *Nature Medicine, 1*, 944–949.

Peters, C. M., Lindsay, T. H., Pomonis, J. D., et al. (2004). Endothelin and the tumorigenic component of bone cancer pain. *Neuroscience, 126*, 1043–1052.

Pickering, V., Jay Gupta, R., Quang, P., et al. (2008). Effect of peripheral endothelin-1 concentration on carcinoma-induced pain in mice. *European Journal of Pain, 12*, 293–300.

Pomonis, J. D., Rogers, S. D., Peters, C. M., et al. (2001). Expression and localization of endothelin receptors: Implication for the involvement of peripheral glia in nociception. *The Journal of Neuroscience, 21*, 999–1006.

Portenoy, R. K., Payne, D., & Jacobsen, P. (1999). Breakthrough pain: Characteristics and impact in patients with cancer pain. *Pain, 81*, 129–134.

Sabino, M. A., Ghilardi, J. R., Jongen, J. L., et al. (2002). Simultaneous reduction in cancer pain, bone destruction, and tumor growth by selective inhibition of cyclooxygenase-2. *Cancer Research, 62*, 7343–7349.

Sabino, M. A., Luger, N. M., Mach, D. B., et al. (2003). Different tumors in bone each give rise to a distinct pattern of skeletal destruction, bone cancer-related pain behaviors and neurochemical changes in the central nervous system. *International Journal of Cancer, 104*, 550–558.

Schwei, M. J., Honore, P., Rogers, S. D., et al. (1999). Neurochemical and cellular reorganization of the spinal cord in a murine model of bone cancer pain. *The Journal of Neuroscience, 19*, 10886–10897.

Sevcik, M. A., Luger, N. M., Mach, D. B., et al. (2004). Bone cancer pain: The effects of the bisphosphonate alendronate on pain, skeletal remodeling, tumor growth and tumor necrosis. *Pain, 111*, 169–180.

Sevcik, M. A., Ghilardi, J. R., Peters, C. M., et al. (2005). Anti-NGF therapy profoundly reduces bone cancer pain and the accompanying increase in markers of peripheral and central sensitization. *Pain, 115*, 128–141.

Sevcik, M. A., Jonas, B. M., Lindsay, T. H., et al. (2006). Endogenous opioids inhibit early-stage pancreatic pain in a mouse model of pancreatic cancer. *Gastroenterology, 131*, 900–910.

Vasko, M. R. (1995). Prostaglandin-induced neuropeptide release from spinal cord. *Progress in Brain Research, 104*, 367–380.

Wacnik, P. W., Eikmeier, L. J., Ruggles, T. R., et al. (2001). Functional interactions between tumor and peripheral nerve: Morphology, algogen identification, and behavioral characterization of a new murine model of cancer pain. *The Journal of Neuroscience, 21*, 9355–9366.

Cancer Pain, Assessment in Children

Kathleen McGinn
Anesthesiology & Pain Medicine, Seattle
Children's Hospital, Seattle, WA, USA

Definition

Cancer pain in children is associated with three major etiologies: (a) ▸ procedural distress, (b) ▸ iatrogenic effects of treatment, and (c) disease-related pain. Chronic pain among long-term cancer survivors (see ▸ Cancer Survivorship) is increasingly recognized as surveillance of these patients continues. Each has unique components that dictate the nature of the assessments to be made and treatments that may ensue.

Characteristics

Procedural Distress

For many of the cancers most common in the pediatric population (e.g., leukemia, lymphoma), invasive procedures are integral to the diagnostic and treatment process. Included here are bone marrow aspirations and biopsies, for diagnostic purposes, and lumbar punctures, both for diagnostic purposes and to deliver intrathecal chemotherapy as prophylaxis against or treatment for central nervous system disease (Balis et al. 2002). The latter occurs on a recurrent basis over the course of several months of treatment. Among survivors of pediatric cancer, the trauma of undertreated procedural distress is often the most disturbing element of their entire cancer experience, sometimes severe enough to leave children and parents with post-traumatic stress symptoms (Stuber et al. 1998). Interventions combining sedating and analgesic agents along with preparatory and cognitive-behavioral strategies (see ▸ Cognitive-Behavior Treatment of Pain) have greatly reduced the pain and anxiety associated with invasive procedures (Berde et al. 2002; Conte et al. 1999). Assessment strategies involve three modalities: self-report of the child, behavioral observations, and physiological indicators of stress (Walco et al. 2005). As these do not necessarily correlate with one another, it is imperative to specify the target outcome(s) of the intervention (Walco et al. 2005). Ultimately the goal would be to implement an algorithm optimizing the match between patient needs and available interventions, so that comfort may be assured while risk and cost are minimized.

Iatrogenic Effects

Treatments for cancer involve chemotherapy, radiation, surgery, and stem cell transplantation. Each potentially leads to iatrogenic pain problems (Collins and Weisman 2003). Symptoms resulting from chemotherapeutic and radiation treatment most commonly include nausea and vomiting, fatigue, neutropenia and related vulnerability to infection, and pain. Common iatrogenic pain syndromes in pediatric cancer are related to mucositis, infectious process, and ▶ neuropathic pain (from ▶ peripheral neuropathy) associated with agents such as vincristine. Pain secondary to radiation is also seen in some instances. Painful sequelae of surgical interventions occur, including phantom limb pain related to amputations. Acute abdominal pain, severe enough to require administration of opioids, has been noted among patients who have undergone allogeneic bone marrow transplantation. Finally, as follow-up programs for long-term survivors of childhood cancer provide data, it appears that chronic pain syndromes may be of concern. For example, aseptic necrosis of the bone is a known long-term effect of high-dose steroid usage and does not manifest until well after the child has completed standard treatment protocols.

Disease-Related Pain

Disease-related pain in pediatric cancer is not as common as in neoplasms common in the adult population. Pain may be present at the time of diagnosis, as pain in the long bones is associated with leukemia, severe headache is a symptom of brain tumors, solid tumors may elicit somatic pain in the focal area of tumor growth or infiltration, and abdominal tumors may generate visceral pain. In contrast to adult patients with such tumors, children with cancer usually experience a significant remission of pain within 2 weeks of beginning treatment (Miser et al. 1987). Unfortunately, the other circumstance where pain is problematic is in advanced illness, often in the context of end-of-life care. Thus, many of the same types of syndromes seen at presentation, where pain is of high intensity and often unrelenting, recur in advanced illness and should be addressed in service of maximizing quality of life (Wolfe and Grier 2002).

Assessment Strategies

Specific pain assessment strategies depend on the age or ▶ developmental level of the child and the nature of the pain in question (McGrath and Gillespie 2001). Whenever possible, it is optimal to use patients' verbal reports to assess their pain experience, skills that emerge in children between the ages of 3 and 7 years (American Academy of Pediatrics Committee on Psychosocial Aspects of Child and Family Health, American Pain Society Task Force on Pain in Infants, Children, and Adolescents 2001). In preverbal children, pain is assessed through physiological indicators of distress (Sweet and McGrath 1998), through specific behavioral indicators (McGrath 1998; Lilley et al. 1997) or through some combination thereof (Franck et al. 2000). During the toddler and preschool years, as children begin to use language meaningfully, increased emphasis may be placed on verbal report, but one must take care to use language and terminology that is familiar to the child. Parental ratings of pain may also be helpful in preverbal children.

For procedural distress, assessment strategies should include two major components, anxiety and pain. The term "distress" is used as a way to represent the combined effect of these two factors, as they are separable only in concept, not in practice. Behavior rating instruments typically define specific behaviors, and frequency of occurrence, duration, or intensity is scored during a specified period. Typically, this includes a segment prior to, during, and after the procedure. Likewise, physiological indicators, including heart rate, blood pressure, vagal tone, and cortisol responses, have been used as indicators of acute distress. Self-report measures focus on the subjective experience of the child and typically encompass concerns about pain, anxiety, and perceived self-efficacy in coping (Walco et al. 2005).

For pain related to both disease and the iatrogenic effects of treatment, self-report measures help ascertain the location of pain, intensity, sensory, affective, and evaluative components, as well as the impact of pain on functioning. Especially when confronting pain that is more chronic or recurrent in nature, assessing the contextual factors associated with the pain experience (developmental level, temperament, characteristic pain responsiveness, family issues, impact of pain, etc.) becomes very important.

References

American Academy of Pediatrics Committee on Psychosocial Aspects of Child and Family Health, American Pain Society Task Force on Pain in Infants, Children, and Adolescents. (2001). Policy statement: The assessment and management of acute pain in infants, children and adolescents. *Pediatrics, 108*, 793–797.

Balis, F. M., Holcenberg, J. S., & Blaney, S. M. (2002). General principles of chemotherapy. In P. A. Pizzo & D. G. Poplack (Eds.), *Principles and practice of pediatric oncology* (4th ed., pp. 237–308). Philadelphia: Lippincott, Williams & Wilkins.

Berde, C. B., Billett, A. L., & Collins, J. J. (2002). Symptom management in supportive care. In P. A. Pizzo & D. G. Poplack (Eds.), *Principles and practice of pediatric oncology* (4th ed., pp. 1301–1332). Philadelphia: Lippincott, Williams & Wilkins.

Collins, J. J., & Weisman, S. J. (2003). Management of pain in childhood cancer. In N. L. Schechter, C. B. Berde, & M. Yaster (Eds.), *Pain in infants, children, and adolescents* (2nd ed., pp. 517–538). Philadelphia: Lippincott, Williams & Wilkins.

Conte, P. M., Walco, G. A., Sterling, C. M., et al. (1999). Procedural pain management in pediatric oncology: A review of the literature. *Cancer Investigation, 17*, 448–459.

Franck, L. S., Greenberg, C. S., & Stevens, B. (2000). Pain assessment in infants and children. *Pediatric Clinics of North America, 47*, 487–512.

Lilley, C. M., Craig, K. D., & Grunau, R. E. (1997). The expression of pain in infants and toddlers: Developmental changes in facial action. *Pain, 72*, 161–70.

McGrath, P. J. (1998). Behavioral measures of pain. In G. A. Finley & P. J. McGrath (Eds.), *Measurement of pain in infants and children. Progress in pain research and management* (Vol. 10, pp. 83–102). Seattle: IASP Press.

McGrath, P. A., & Gillespie, J. (2001). Pain assessment in children and adolescents. In D. C. Turk & R. Melzack (Eds.), *Handbook of pain assessment* (2nd ed., pp. 97–118). New York: Guilford.

Miser, A., Dothage, J., Wesley, R., et al. (1987). The prevalence of pain in a pediatric and young adult cancer population. *Pain, 29*, 73–83.

Stuber, M. L., Kazak, A. E., Meeske, K., et al. (1998). Is posttraumatic stress a viable model for understanding responses to childhood cancer? *Child and Adolescent Psychiatric Clinics of North America, 7*, 169–182.

Sweet, S. D., & McGrath, P. J. (1998). Physiological measures of pain. In G. A. Finley & P. J. McGrath (Eds.), *Measurement of pain in infants and children. Progress in pain research and management* (Vol. 10, pp. 59–81). Seattle: IASP Press.

Walco, G. A., Conte, P. M., Labay, L. E., et al. (2005). Procedural distress in children with cancer: Self-report, behavioral observations, and physiological parameters. *The Clinical Journal of Pain, 21*, 484–90.

Wolfe, J., & Grier, H. E. (2002). Care of the dying child. In P. A. Pizzo & D. G. Poplack (Eds.), *Principles and practice of pediatric oncology* (4th ed., pp. 1477–1493). Philadelphia: Lippincott, Williams & Wilkins.

Cancer Pain, Assessment in the Cognitively Impaired

Wendy M. Stein
San Diego Hospice and Palliative Care, UCSD, San Diego, CA, USA

Synonyms

Cognitive impairment; Delirium; Dementia

Definition

Cognitive impairment or altered mental state is a change in the patient's usual premorbid state of mind, which can include delirium and dementia as well as altered emotions and behaviors (Folstein and Folstein 1994).

Characteristics

This section will focus on the challenging assessment of cancer pain in the communicative and non-communicative cognitively impaired patient. Acute and subacute changes in mental status have been well documented in as many as 20–30 % of medical inpatients and 50–90 % of nursing home residents (Folstein and Folstein 1994) and can substantially impair a patient's ability to participate actively in both the initial pain assessment and subsequent evaluation of treatment efficacy. The initial step is to establish the patient's premorbid mental status, and the nature and association of any clinical changes that may have occurred. Vision and hearing impairments can further complicate both the assessment and treatment of pain in the cognitively impaired and need to be adequately compensated.

Population-based studies of community dwelling elderly have shown a twofold increase in pain problems significant enough to impair daily function among those over age 60 (Crook et al. 1984). Older patients have a higher prevalence of a number of specific pain syndromes including most types of cancer, peripheral vascular disease, ▶ temporal arteritis and ▶ polymyalgia rheumatica, ▶ peripheral neuropathies, ▶ herpes zoster, and subsequent ▶ postherpetic neuralgia. One might then reasonably expect that individuals suffering from dementia and delirium would also be disproportionately affected by these syndromes, resulting in a large number of cognitively impaired elderly in desperate need of aggressive and appropriate pain assessment and treatment.

In addition, older patients as a group are at high risk for undertreatment of cancer pain. Cleeland et al. showed that those over the age of 70 were at higher risk of receiving inadequate pain management (Cleeland 1998). Bernabei et al. documented significant numbers of nursing home patients with cancer suffering from daily pain, and those over age 85 were at particular risk of having no ▶ analgesia at all (Bernabei et al. 1998).

Dementia Versus Delirium

Folstein and Folstein define dementia as "a syndrome characterized by a decline in multiple cognitive functions occurring in clear consciousness" (Folstein and Folstein 1994). Dementia has been estimated in 10–15 % of elderly surgical patients, one-third of elderly medical inpatients, and more than 50 % of nursing home residents (Folstein and Folstein 1994). Delirium, on the other hand, is noted for its acute or subacute presentation with waxing and waning levels of consciousness, global cognitive impairment, and disorganized wake-sleep cycles. Delirium has been well described in the medical literature in patients with cancer (Derogatis et al. 1983); however, it also occurs commonly in the elderly, especially in those with underlying dementias. Rates of delirium have ranged as high as 50 % in hospitalized patients with previously diagnosed dementia (Ouslander et al. 1997); in reality the figures may be even larger, as in the community dementia is often well hidden by well meaning family and friends who compensate for gradually increasing mental deficits over time.

Studies of Pain in the Communicative Cognitively Impaired

The reader should keep in mind that most pain studies of assessment in the cognitively impaired have not focused specifically on cancer pain, but on all types of pain in general. Parmalee et al. (1996) studied self-reported pain in 758 elders with mild to moderate levels of cognitive impairment. This study showed that pain complaints decreased with increasing levels of cognitive impairment over time, i.e., that patients became less good at reporting their pain. However, when cognitively impaired individuals reported pain, their pain complaints were no less valid. Ferrell et al. (1995) studied patients with moderate to severe degrees of cognitive impairment in ten community nursing homes. The researchers presented five commonly used pain scales in enlarged format and with adequate amounts of light and with hearing augmentation if required. Of those presented, the ▶ Present Pain Intensity (PPI) subscale had the highest rate of completion (65 %). Although only 32 % could complete all five scales presented, 83 % of these patients with significant cognitive impairment could complete at least one of the scales presented. These findings validate the utility of offering a variety of scales when using standardized measures with the cognitively impaired, and then continuing with what works best for that patient.

Pain in Non-communicative Cognitively Impaired Patients

Unfortunately, many early studies of pain assessment in this population focused on the use of pediatric instruments, and specifically those developed for neonates, without first establishing data in this population. These tools were developed to measure acute, and often externally induced procedure related pain. In the cognitively impaired elderly, clinicians are most often evaluating chronic or acute on pain, which is likely to present differently. There have now been multiple studies to validate the use of standardized tools specifically adapted and/or developed for use in this population such as the ▶ Faces Pain Scale (Herr et al. 1998). However, global evaluations of a combination of nonverbal behavior, vocalizations, changes in function, and caregiver reports are most often used as indicators of pain in the non-communicative cognitively impaired elderly (AGS Panel on Chronic Pain in Older Persons 2002). Marzinski (1991) studied 60 patients living in a dementia unit, of which 43 % had potentially painful conditions based on chart review. The most clinically interesting finding of this study was that the nursing staff could clearly identify what amounted to normative behavior for individual patients, and once aberrations were identified, to act upon them.

Several studies have sought to link assessment with initiating pain interventions empirically in non-communicative cognitively impaired elders. The most comprehensive project of this nature is that of Kovach et al. (1999). The research team worked with 104 non-communicative demented elderly who had signs or symptoms of pain or discomfort at 32 nursing homes. The nursing staffs were instructed in the use of the ▶ Assessment of Discomfort in Dementia (ADD) Protocol if a patient displayed signs or symptoms of distress. After excluding common physical causes for discomfort such as occult infection, or impaction by thorough physical assessment, and review of history including consultations with family and/or physicians, nonpharmacologic comfort interventions were undertaken. If unsuccessful, the nurses were educated to administer "as needed" non-opioid analgesics and continue up the WHO ladder depending on response and discussion with physicians. The sample population decreased behavioral symptoms associated discomfort from an average of 32.85–23.47 symptoms on protocol, which was a statistically significant change. Of note was the fact that 88 % of the staff felt that the protocol was beneficial.

General Recommendations

The clinician needs to create an optimal environment in order to adequately assess pain in the communicative cognitively impaired. Limited attention span may demand dividing up the initial assessment into several shorter sessions. Additional time should be allowed for the patient to assimilate questions to allow for receptive or expressive aphasia. Ensuring that large print cards, adequate lighting, and hearing aids or pocket amplification devices and refractive lenses are available is vital. Qualitative descriptions of pain should bear equal weight and be encouraged. If using standardized tools, multiple tools should be presented upon initial assessment so as to discern which is easiest for that particular patient, document this in the clinical record, and communicate this with the care team (Stein 1996). Copies of this tool can be made available (without identifying data on it) on the chart and at the bedside, ensuring that each subsequent assessment is performed exactly the same way regardless of the team member asking the questions. As the cognitively impaired elderly are often unable to report on previous pain, assessment should be conducted more frequently than would be necessary with a cognitively intact patient (Stein 2001). Family members and caregivers can assist in this process by helping to journal the patient's pain when they visit, thus providing a comprehensive picture.

Deviation from what is considered baseline or "normal" behavior for the non-communicative cognitively impaired patient remains the clinical imperative (Stein 2001). This should trigger a comprehensive bedside physical examination coupled with laboratory and imaging studies, consistent with the benefits and burdens likely to be incurred and the patient and/or family's wishes. Physical examination should specifically focus on occult sources or atypical presentations of infection such as pneumonia or urinary tract infection, as well as fecal impaction. The clinical involvement of the entire interdisciplinary team should be sought to elicit alterations in sleeping, eating, or elimination as sources of change in behavior. Chart review should exclude changes in medications, doses, or dose intervals as potential causes. After all of the above have been carefully completed and reviewed with family members and staff and informed consent obtained from involved family members, there may be a role for a well-defined empiric trial of short-acting pain medication with consistent and continual reassessment (Stein 2001).

References

AGS Panel on Chronic Pain in Older Persons. (2002). The management of chronic pain in older persons. *Journal of the American Geriatrics Society, 50*, 1–20.

Bernabei, R., Gambassi, G., Lapane, K., et al. (1998). Management of pain in elderly patients with cancer. SAGE study group. Systematic assessment of geriatric drug Use via epidemiology. *JAMA: The Journal of the American Medical Association, 279*, 1877–1882.

Cleeland, C. S. (1998). Undertreatment of cancer pain in the elderly. *JAMA: The Journal of the American Medical Association, 279*, 1914–1915.

Crook, J., Rideout, E., & Browne, G. (1984). The prevalence of pain complaints among a general population. *Pain, 18*, 299–314.

Derogatis, L. R., Morrow, G. R., Fetting, J., et al. (1983). The prevalence of psychiatric disorders among cancer patients. *JAMA: The Journal of the American Medical Association, 249*, 751–757.

Ferrell, B. A., Ferrell, B. R., & Rivera, L. (1995). Pain in cognitively impaired nursing home patients. *Journal of Pain and Symptom Management, 10*, 591–598.

Folstein, M. F., & Folstein, S. E. (1994). Syndromes of altered mental state. In W. B. Hazzard, E. L. Bierman, & J. P. Blass (Eds.), *Principles of geriatric medicine and gerontology* (pp. 1197–1204). New York: McGraw-Hill.

Herr, K. A., Mobily, P. R., Kohout, F. J., et al. (1998). Evaluation of the faces pain scale for use with the elderly. *The Clinical Journal of Pain, 14*, 29–38.

Kovach, C. R., Weissman, D. E., Griffi, J., et al. (1999). Assessment and treatment of discomfort for people with late-stage dementia. *Journal of Pain and Symptom Management, 18*, 412–419.

Marzinski, L. R. (1991). The tragedy of dementia: Clinically assessing pain in the confused, nonverbal elderly. *Journal of Gerontological Nursing, 17*, 25–28.

Ouslander, J. G., Osterweil, D., & Morley, J. (1997). Delirium and dementia. In W. B. Hazzard et al. (Eds.), *Medical care in the nursing home* (pp. 147–162). New York: McGraw-Hill.

Parmalee, P. A. (1996). Pain in cognitively impaired older persons. In B. A. Ferrell (Ed.), *Clinics in geriatric medicine* (pp. 473–485). Philadelphia: WB Saunders.

Stein, W. M. (1996). Cancer pain in the elderly. In B. R. Ferrell & B. A. Ferrell (Eds.), *Pain in the elderly* (pp. 69–80). Seattle: IASP Press.

Stein, W. M. (2001). Assessment of symptoms in the cognitively impaired. In E. Bruera & R. K. Portenoy (Eds.), *Topics in palliative care* (Vol. 5, pp. 123–133). New York: Oxford University Press.

Cancer Pain, Assessment of Cultural Aspects

Terry Altilio
Department of Pain Medicine and Palliative Care, Beth Israel Medical Center, New York, USA

Definition

Culture is defined as learned values, beliefs, behaviors, and systems of meaning which guide a worldview and inform decisions. Culture is constantly redefined, evolving, and negotiated, and respectful, comprehensive care requires combining theoretical knowledge with a meaningful exploration of the "patients unique history, family constellation, and socioeconomic status" (Koenig and Gates-Williams 1995).

Introduction

Developing a consciousness and respectful curiosity about the culture of patients and families is essential to understanding their responses to cancer, to pain, and to cancer pain as a combined entity. This same curiosity is applicable to the culture and consequent attitudes and beliefs of clinicians and institutions which inform the context in which patients and families receive care. Beyond the culture of patients and families, we are mandated to examine and, when necessary, challenge the cultural nuances which may influence treatment decisions as well as the attitudes, spoken and unspoken, which pervade the relationships within which pain care is provided.

Immigration and the changing ethnic composition of countries around the world drive the mandate to provide culturally sensitive care and to develop tools and educational materials which are linguistically and culturally tailored. As clinicians, researchers and policy advocates recognize the need to respond to diverse languages, histories, and psychosocial-spiritual perspectives, concurrently, there are increased time constraints and resource limitations which challenge the best of intentions of health-care professionals. Yet, if cultural and linguistic variables are treated in an insouciant manner, care may be ineffective, conflictual, and potentially harmful (Lasch 2000). For example, the use of family members for medical interpretation while expedient and often necessary in an emergency may create an untenable emotional conflict for family members and a risk for health-care providers who can never be assured that the medical information provided was interpreted with accuracy. While the challenge of cultural diversity increases demands on providers, the richness implicit in cultural difference provides both an opportunity for personal and professional growth and allows for the crafting of interventions that diminish pain and suffering in a manner that is imminently respectful of the uniqueness of the person and their pain experience.

Characteristics

Although knowledge of specific cultures can be instructive, cultural identification alone is not predictive of specific behaviors, values, or beliefs. Intracultural variation emanating from such factors as gender, age, geographic location,

ties to country of origin, ▶ assimilation, and acculturation contribute to the dynamic nature of culture. Culture infuses the experience of the patient and family, and practitioners are more or less influenced by the culture of the health-care system, their profession, the disease, and the larger society. It is at this interface that understanding, collaboration, and negotiation need to occur (Koenig and Gates-Williams 1995; Kagawa-Singer and Blackhall 2001; Kagawa-Singer 1996).

Studies of the relationship of culture and pain date back to 1950 and include both clinical and laboratory research. Studies have focused on such varied aspects as pain threshold, pain perception and description, beliefs, attitudes and behaviors, undertreatment in minority patients, coping strategies, use of opioid medications, ascribed meaning and outcomes, pharmacodynamics, and pharmacokinetics (Edwards et al. 2001; Zatzick and Dimsdale 1990). As far back as 1952, Zborowski studied the influence of ▶ ethnicity on pain response patterns, attitudes, beliefs, and meaning (Zborowski 1952). In 1966, Zola studied interethnic differences in the response and attitudes toward symptoms such as pain (Zola 1966). In 1980, Bates proposed a biocultural model which integrates the gate control theory of Melzack and Wall with social learning theory and sociocultural variations to create a framework that considers family, community and environmental factors that might influence physical, cognitive, and psychological processes, which potentially might modulate pain and impact perception (Al-Atiyyat 2009; Bates 1987; Melzach and Wall 1982).

Cultural aspects of pain such as meaning, attribution, description, and health-seeking behaviors become increasingly relevant as research points to the influence of these factors on outcomes. For example, a study of chest pain in a population from the Southern United States revealed that cultural differences resulted in atypical descriptions, which have the potential to mislead clinicians who depend on the description of the character of chest pain as they triage for ischemic heart disease (Summers et al. 1999). A study of indigenous and nonindigenous people with persistent chest pain in the Northern Territory of Australia revealed that delays in presentation affected both rural and urban indigenous peoples, as well as rural nonindigenous people; these delays influence management options for acute myocardial infarction (Ong and Weeramanthri 2000). Racial and ethnic minorities with cancer pain have been found to be at greater risk for ineffective pain management (Cleeland et al. 1994, 1997). Such information assists clinicians and health educators to know that signs and symptoms, and pain-related behaviors, may not be the same for all cultural groups. Pain related to a diagnosis of cancer is further complicated by the values, attributions, and beliefs surrounding the disease itself, which may vary and include such responses as optimism, hopelessness, shame, or acceptance. These variations can provoke clinician curiosity, vigilance, and thoughtfulness and provide direction to culturally informed assessment, education, and outreach efforts.

Research on culture and pain continues to expand and requires both creativity and resources to study biology as well as the interactions with the environment, culture, and ethnicity which inform perspectives and interactions (Al-Atiyyat 2009). Often, samples are small and convenience based and exclude non-English speakers. There is considerable variation in study populations, theoretical formulations, study design, and findings. Some fail to distinguish race, culture, and ethnicity, and others have mixed populations of immigrants and later generations. It is unclear whether instruments are interpreted the same way across cultures and whether the differences noted are due to culture, learning, or psychosocial factors (Rollman 1998; Im et al. 2007). Assessment and study challenges include cross-cultural validity of behavioral cues to indicate pain level, cross-cultural validity of assessment tools, and expanding study designs to include variables such as intracultural differences and influence of clinician culture on outcome. Assessment tools such as the Brief Pain Inventory and McGill Pain Questionnaire have been validated and translated into languages other than English. In addition, acceptance of unique cultures requires

understanding that linear assessment tools may have little relevance in cultures where numbers have no relationship to pain rating, or where narrative or pictures are essential to communication (Kagawa-Singer and Blackhall 2001; Hallenbeck and Goldstein 1999). Qualitative studies have the potential to provide a lens into the patient's environment and their interpretation of their pain experience, creating a context for further assessment and intervention beyond the physical. For example, a study employed qualitative interviews of black Caribbean and white British patients and identified attributed meanings to cancer pain which included pain as challenge, enemy, test of faith, and punishment (Koffman et al. 2008). The interacting facets of cancer, pain, and culture are like a kaleidoscope wherein various aspects of the pain experience can be explored and studied.

Consideration of cultural issues in the healthcare setting means that practitioners have to reconcile the possibility that their personal beliefs and values, and the values of the medical system, may not be shared by the patient and family (Hallenbeck 2002). Consequently, clinicians need to develop an awareness of the culture and traditions of the population they treat. This awareness becomes a hypothesis and serves only as background to guide inquiry and exploration. The subjective nature of pain requires that each patient's pain description, report and behaviors be explored and understood in the context of cultural variation related to pain. Clinicians bring their own personal and professional cultural perspectives into the medical setting, and it is at this interface that the assessment of pain, clinical judgment, and related treatment interventions takes place. As early as 1977, Davitz et al. reported on the emergency room care of a 45-year-old male whose pain behaviors were observed by three nurses of different ▸ ethnic backgrounds (Davitz et al. 1977). Each nurse had different reactions to and judgments about his expression of pain. A nurse from northern Europe described the patient as very emotional and overreactive. A nurse from Puerto Rico felt the patient's outbursts were understandable, and an American nurse expected that an adult male of his age and background would be stronger and more stoic. A follow-up study found that nurses from various countries differed in their judgments about the degree of patient suffering. These findings are an alert to all disciplines of the importance of self-awareness, as clinician beliefs and values can significantly impact attitudes and practice (Bonham 2001).

Characteristics of pain assessment and management may be influenced by culture. While categorized as discrete aspects, these variables are integrated and overlapping and related to the larger historical, political, class, and socioeconomic issues (Koenig and Gates-Williams 1995). For example, in some countries, the experience of marginalization and discrimination of minority patients creates an environment of mistrust and/or hopelessness, which requires clinicians to spend additional time and ongoing effort to establish beginning rapport and credibility. The undertreatment of pain, including cancer pain, in minority populations has been identified in a variety of studies, and further work is needed to isolate the multidimensional factors that influence these outcomes (Cleeland et al. 1994, 1997; Zborowski 1969).

Expression and Language of Pain

The presence of pain can be conveyed using verbal, facial, or body language. For many, how one expresses and responds to pain is a reflection of the responses, attitudes, and anxieties learned within family beginning in childhood. In addition to the sensory component of pain, emotional, cognitive, psychological, social, and spiritual factors converge to create a pain experience. For example, in cultures where expressions of emotion are unacceptable, physical symptoms may become the route for expression of these feelings. These variables and relationships may be influenced by culture as well as family dynamics and over time may be reinforced in adulthood or modified by alternate experience or by learning.

Culture may be looked at as a blueprint for behaviors, influencing how, if, and to whom the subjective pain experience is expressed and whether it is avoided, ignored or accepted, and described in direct, substitute, or metaphoric terms

(Al-Atiyyat 2009). In cultures where control is valued, endurance and stoicism would be expected and outward expression of pain viewed as a sign of weakness, possibly resulting in guilt or shame. Similar behaviors are not always based on the same value system. For example, stoicism may reflect a reluctance to impose or disrupt others rather than a sign of endurance. This is especially important with a diagnosis of cancer, where historically emphasis has been on curing the disease with pain and quality-of-life issues perceived as a distraction from that primary goal. Expressions of pain, such as crying, cursing, or moaning, can be the learned and expected way to express distress and seek care and are not always correlated with the degree of pain. These behaviors may relate to feared consequences rather than the pain itself (Mims 1989). In patients with cancer, emerging or increasing pain may be interpreted as a sign of progressing disease, and this symbolic significance becomes a source of suffering and distress. Dramatic expressions of pain can serve an important purpose in systems and situations where people have to compete for care, or where health-care professionals have minimized or challenged reports of pain. Cultures that value patience and choose not to disrupt others need outreach from health-care staff to create an environment where patients feel their comfort to be an important focus of care. Another variable that influences the expression of pain and related content is time orientation. In societies that are future-oriented, patients may project ahead, and rather than focus on the sensation of pain and relief, will anticipate and worry about the future impacts of pain and disease (Mims 1989).

Meaning of Pain

The meaning of pain may have an important relationship to response, behaviors, and coping. Cultural traditions often prepare individuals to anticipate pain in certain situations, which can increase feelings of control and lessen anxiety and distress related to the unknown. Pain that has been experienced before has different meaning than pain that is related to a life-threatening illness, or is unexpected, and all will be integrated in accordance with personal and cultural beliefs and values. The narratives and descriptions not only reflect pain but also attempt to make sense of it within a cultural context and within the context of a known diagnosis (Hallenbeck 2002; Davitz et al. 1977; Lipson et al. 2000). Health may be considered a gift from God or a reflection of internal equilibrium. Cancer and pain may be interpreted as a sign of internal imbalance, divine displeasure, a curse, witchcraft, or, perhaps, the result of transgressions in prior life. If pain is considered part of God's plan, a test of faith or an opportunity for character building and redemption, then acceptance and the manner of coping may be a demonstration of courage and faith. Where pain is an extension of God's will, or when suffering is felt to be an expected part of life, there may be a sense of fatalism that results in a feeling of powerlessness with no anticipation of relief or acceptance of treatment (Hallenbeck 2002; Davitz et al. 1977; Lipson et al. 2000).

Expectations of Health-Care Professionals

The expectations of health-care professionals are highly influenced by cultural variables. Respect for clinicians may be expressed by behaviors such as not making direct eye contact nor challenging or questioning practitioners. Many traditions support a hierarchical system, and patients do not expect a consultative model where they are asked for feedback and given options. Expecting to be guided by experts, patients may interpret the consultative approach as a reflection of incompetence. Health-care relationships are a reflection of the larger culture, which may be high or low context. Low-context culture, similar to Western biomedicine, places emphasis on individuality and a more linear verbal communication, whereas a high-context environment emphasizes the relational aspects that are important to care, decision making, and inclusion of family and community (Hallenbeck 2002; Davitz et al. 1977). Consultation with a health-care professional may follow the primary healing efforts of family or healers. Acknowledging and integrating these traditional interventions can be signs of respect for patients and their support systems and may enhance outcome. In situations where the language of patient and clinician differ, perceptions of patient and family about the use of medical translators need

to be explored. Family members are untrained in the use of medical terminology and when acting as interpreters may experience role conflict and confusion (Zhou et al. 1993). Clinicians are in the unique position of being unaware of the actual content and accuracy of the communication when family and friends translate through their own cognitive and emotional filters.

Response to and Acceptance of Treatments

Treatment response is based on a range of values and beliefs. In addition to the ongoing investigation of interethnic differences in drug response, culture impacts acceptance of medications, interventions, and side effects. Sedation and cognitive impairment may be very troubling to patients whose clarity of mind and contact with others are essential to spiritual, emotional, and philosophical well-being. Treatments that are internal, such as pills or injections, may be preferred by some, while external interventions, such as massage and salves may be preferable to others. In cultures where acupuncture equates with positive outcome, the sensation of insertion of needles may have positive association. Certain medications, such as morphine or methadone, may have symbolic significance and be associated with addiction or death. When medications are feared, patients may discontinue them quickly. In high-context cultures, treatment plans and options may need to be discussed with family and community, which may include healers, elders, or clergy (Hallenbeck and Goldstein 1999).

Caregiver Responses

The responses of caregivers are multifaceted and often relate to how care is provided to a sick family or community member. The expected roles and responses of patients, parents, siblings, family, and community may evolve from an acute illness model. For some, a cancer diagnosis may yield helplessness and passivity as the accepted and expected response to illness and pain (Hallenbeck and Goldstein 1999). Transition to a chronic disease state can challenge traditional beliefs and behaviors, as the support must become long term, and there is a need to accommodate to a person who receives ongoing treatments and may or may not be disabled (Lipson et al. 2000). Traditional beliefs, which encourage rest and dependence, can be very responsive to situations of pain in life threatening illness and less helpful in the setting of chronic cancer pain, where goals of care might include recovering function and encouraging participation beyond the role of patient.

Summary

The convergence of culture, pain, and the diagnosis of cancer creates a tapestry woven with unique colors and textures. These variables alone and together have the potential to yield endless studies and hypotheses. As the research develops, the tasks at hand include providing respectful care, individualizing assessment and intervention, and maximizing the support, comfort, and continuity that comes from culture which can sustain patients and families through medical and life transitions, crises, and beyond.

References

Al-Atiyyat, N. M. H. (2009). Cultural diversity and cancer pain. *Journal of Hospice and Palliative Nursing, 11*(3), 154–164.

Bates, M. S. (1987). Ethnicity and pain: A biocultural model. *Social Science and Medicine, 24*(1), 47–50.

Bonham, V. L. (2001). Race, ethnicity and pain treatment: Striving to understand the causes and solutions to the disparities in pain treatment. *Journal of Law, Medicine and Ethics, 29*, 52–68.

Cleeland, C. S., Gonin, R., Harfield, A. K., Edmonson, J. H., Blum, R., Stewart, J. A., & Pandya, K. J. (1994). Pain and its treatment in outpatients with metastatic cancer. *New England Journal of Medicine, 330*, 592–596.

Cleeland, C. S., Gonin, R., Baez, L., et al. (1997). Pain and pain treatment in minority outpatients with metastatic cancer: The eastern cooperative oncology group minority outpatient pain study. *Annals of Internal Medicine, 127*, 813–816.

Davitz, L. L., Davitz, J. R., & Higuchi, Y. (1977). Cross-cultural inferences of physical pain and psychological distress. *Nursing Times, 73*, 521–558.

Edwards, C. L., Fillingim, R. B., & Keefe, F. (2001). Race, ethnicity and pain - topical review. *Pain, 94*, 133–137.

Hallenbeck, J. L. (2002). Cross-cultural issues. In A. M. Berger, R. K. Portenoy, & D. E. Weissman (Eds.), *Principles and practice of palliative care* (pp. 661–672). Philadelphia: Lippincott, Williams & Wilkins.

Hallenbeck, J. L., & Goldstein, M. K. (1999). Decisions at the end of life – Cultural considerations beyond medical ethics. *Generations, 23*(1), 24–29.

Im, E., Chee, W., Guevara, E., et al. (2007). Gender and ethnic differences in cancer pain experience: A multiethnic study in the United States. *Nursing Research, 56*(5), 296–306.

Kagawa-Singer, M. (1996). Cultural systems. In R. McCorkle, M. Grant, & M. Frank-Stromborg (Eds.), *Cancer nursing* (2nd ed., pp. 38–52). Philadelphia: WB Saunders.

Kagawa-Singer, M., & Blackhall, L. J. (2001). Negotiating cross-cultural issues at the end of life "You got to go where he lives". *JAMA, 286*, 2993–3001.

Koenig, B. A., & Gates-Williams, J. (1995). Understanding cultural differences in caring for dying patients. *Western Journal of Medicine, 163*, 244–249.

Koffman, J., Morgan, M., Edmonds, F., Speck, P., & Higginson, I. (2008). Cultural meanings of pain: A qualitative study of black Caribbean and white British patients with advanced cancer. *Journal of Palliative Medicine, 22*, 350–359.

Lasch, K. E. (2000). Culture, pain and culturally sensitive pain care. *Pain Management Nursing, 1*(3), 16–32.

Lipson, J. G., Dibble, S. L., & Minarik, P. A. (Eds.). (2000). *Culture and nursing care: A pocket guide* (pp. 3–4). San Francisco: UCSF Nursing Press.

Melzach, R., & Wall, P. D. (1982). *The challenge of pain* (p. 173). England: Penguin Books.

Mims, B. C. (1989). Sociologic and cultural aspects of pain. In C. D. Tollison (Ed.), *Handbook of chronic pain management* (pp. 17–25). Baltimore: Williams and Wilkins.

Ong, M. A., & Weeramanthri, T. S. (2000). Delay times and management of acute myocardial infarction in indigenous and non-indigenous people in the northern territory. *Medical Journal of Australia, 173*, 173–174.

Rollman, G. B. (1998). Culture and pain. In S. S. Kazarian & D. R. Evans (Eds.), *Cultural clinical psychology, theory, research and practice* (pp. 267–286). New York: Oxford University Press.

Summers, R. L., Cooper, G. J., Carlton, F. B., et al. (1999). Prevalence of chest pain descriptions in a population from the southern United States. *American Journal of Medical Science, 318*, 142–145.

Zatzick, D. F., & Dimsdale, J. E. (1990). Cultural variation in response to painful stimuli. *Psychosomatic Medicine, 52*, 544–557.

Zborowski, M. (1952). Cultural components in responses to pain. *Journal of Social Issues, 8*, 16–30.

Zborowski, M. (1969). *People in pain* (pp. 30–41). San Francisco: Jossey-Bass.

Zhou, H. H., Sheller, J. R., Nu, H., et al. (1993). Ethnic differences in response to morphine. *Clinical Pharmacology and Therapeutics, 54*, 507–513.

Zola, J. (1966). Culture and symptoms, an analysis of patients presenting complaints. *American Sociological Review, 31*, 615.

Cancer Pain, Epidemiology

Cielito Reyes-Gibby and Karen O. Anderson
The University of Texas MD Anderson Cancer
Center, Houston, TX, USA

Definition

Epidemiology provides a framework for measuring the prevalence, incidence, and risks associated with pain in cancer patients. Epidemiological studies use the paradigm of exposure-disease assessment. Classical epidemiological research focuses on the distribution of diseases/disease conditions and their determinants within populations and relies on field-tested measures of exposure (usually self-reported) in defining disease-exposure associations. More recently, advances in molecular technology have made possible the measurement of genetic markers of disease, prognosis, and therapeutic response. This newer molecular epidemiology paradigm represents the confluence of sophisticated advances in molecular biology and field-tested epidemiological methodologies. Thus, in the past, epidemiology provided research methods that could be used to answer the following questions: What is the occurrence of pain in people with cancer? At what stage of the disease does pain become a problem? Are cancer patients at a higher risk of pain compared to patients with chronic conditions such as diabetes or cardiovascular disease? To what extent does the prevalence of cancer pain vary by age, gender, or race? What factors are associated with the prevalence or incidence of pain? Today, epidemiological methods can also help answer the following: Do genetic variants modulate the expression of pain and the response to pain treatment?

Characteristics

Estimates of the number of cancer patients experiencing pain vary widely, mainly because of a lack of uniformity (or standardization) in the definitions and assessment measures used

(i.e., there is no gold standard) and the heterogeneity of pain conditions (nociceptive vs. neuropathic). Other factors contributing to the wide variability of results include, but are not limited to, the heterogeneity of cancer types (breast, lung, etc.), the stage of disease, and the type of treatment settings (outpatient vs. inpatient or palliative) in which studies were conducted. In many instances, the issue of potential confounding factors has not been adequately addressed. There is also no distinction in the various measures of disease occurrence and risk used in measuring the distribution, determinants, and natural history of pain. For example, *prevalence measures* which quantify the proportion of a population with a condition at a given point in time (point prevalence rate), the proportion of persons affected by the condition during a defined period of time (period prevalence rate), or the proportion of a population affected over their lifetime (lifetime prevalence rate), are commonly confused with *incidence measures* (which quantify the probability or rate of onset of a condition among persons with no prior history). A recent report of the National Cancer Institute (Patrick et al. 2004), while emphasizing the need for better symptom epidemiologic studies, elaborates the methodological issues and challenges facing researchers who study the epidemiology of pain. In this entry, we report on studies on the prevalence of pain among cancer patients and the risk factors associated with pain and its undertreatment. We also offer recommendations for future studies on the epidemiology of pain.

Prevalence of Cancer Pain

Pain is prevalent for large numbers of patients with cancer (Reyes-Gibby et al. 2002; Cleeland et al. 1997; Portenoy et al. 1994; Von Roenn et al. 1993). Approximately 55 % of outpatients with metastatic cancer have disease-related pain, and 36 % have pain of sufficient severity to impair their functioning and quality of life. Significant pain is rarely a problem at diagnosis, but as disease becomes metastatic, a majority of patients will have pain. Pain prevalence and severity obviously vary with the adequacy of pain management. However, despite the existence of national and international pain management guidelines, many patients with pain are not prescribed an analgesic appropriate to the severity of their pain. Multicenter studies indicate that approximately 40 % of patients with cancer pain are not prescribed analgesics potent enough to manage their pain, with additional patients receiving insufficient doses of the analgesic prescribed.

Studies of pain following treatments such as surgery, chemotherapy, or radiotherapy show pain as prevalent treatment sequelae. Despite the thousands of cancer patients who undergo treatment such as curative surgery, very little is known about the prevalence, duration, and functional impact of pain in patient populations who receive specific cancer treatments. Surgical patients not only suffer from the side effects of the chemotherapy or radiotherapy they may receive before surgery for their cancer, but may also develop serious postsurgical symptoms. For example, acute pain due to therapy for head and neck cancer is prevalent secondary to ablative surgery, chemo- and/or radiotherapy. Pain due to cancer treatment usually involves an acute inflammatory response that may be associated with concomitant nerve injury. In one study, head and neck cancer patients ($n = 124$) receiving radiation therapy reported pain was present at study entry, highest at a 2-week follow-up, declining toward the end of treatment, and persisting at 3-month follow-up (Epstein et al. 2009). The most common (neuropathic) pain descriptors chosen were aching (20 %) and burning (27 %); nociceptive words chosen were dull (22 %), sore (32 %), tender (35 %), and throbbing (23 %). However, to date, limited data exist on the extent to which acute pain develops to chronic pain in patients with head and neck cancer.

Among breast cancer patients, 41 % of those who had conservative breast surgery with radiotherapy reported pain as a consequence of treatment. Pain generally started within 3 months after the completion of therapy, was localized in the axillary region, and was intermittent (Amichetti and Caffo 2003). The pain was mainly described as aching (59 %), tender (51 %), and cramping (43 %). In comparison to the patients who did not experience pain, those who suffered from pain had significantly worse scores in physical, psychological, and other quality-of-life measures. Other studies have found that 5–20 % of women experience postmastectomy pain, a syndrome that is more common among patients who undergo axillary node dissection or experience postoperative complications (Carpenter et al. 1998; Stevens et al. 1995; Vecht 1990).

It is less recognized that many cancer patients continue to experience pain well after treatment has ended. Chaplin and Morton (1999) assessed 93 head and neck cancer patients and found that 48 % had head and neck pain when first seen, and 25 % and 26 % had pain that persisted at 12 and 24 months, respectively. The prevalence of shoulder and arm pain was greater after treatment, increasing from 14 % at diagnosis to 37 % at 1 year and 26 % at 24 months. Any pain (pain in either the head and neck or shoulder and arm or both) at 2 years was strongly predicted by earlier posttreatment pain (at 3 months or at 12 months). Shoulder and arm pain at 2 years was strongly correlated with surgical treatment of the neck, although no difference in pain experience was noted between those who had radical neck dissections and those who had more conservative procedures.

Among breast cancer patients treated with paclitaxel therapy, as many as 64 % experienced chemotherapy-induced peripheral neuropathy (CIPN) during paclitaxel treatment. Follow-up survey data, 9.5 years after paclitaxel therapy, revealed that those with CIPN during treatment were three times more likely to be diagnosed with neuropathic pain (Reyes-Gibby et al. 2009a).

Studies have examined the prevalence of pain among cancer patients in advanced stage. As part of the SUPPORT studies, McCarthy et al. (2000) evaluated over 1,000 cancer patients 3–6 months before death, at 1–3 months before death, and during the 3 days before death.. As expected, as patients progressed toward death, their 6 month prognosis deteriorated significantly and the severity of their disease worsened. Cancer patients experienced significantly more pain and confusion as death approached. Severe pain was common; more than one quarter of patients with cancer experienced significant pain 3–6 months before death and more than 40 % were in significant pain during their last 3 days of life.

While it is recognized that pain is prevalent for the majority of cancer patients, evidence suggests that the elderly and patients from minority groups may have an even greater risk for pain and poor pain management (Reyes-Gibby et al. 2006; Anderson et al. 2000; Cleeland et al. 1997). Undertreatment of pain was shown to be higher among patients seen at clinics serving minority clients, with minority patients three times more likely to be undermedicated with analgesics than patients treated in nonminority treatment settings (Cleeland et al. 1994). In a subsequent study (Cleeland et al. 1997), patients treated at centers seeing primarily African Americans, Hispanics, or both, as well as patients treated at university centers were more likely to receive inadequate analgesia (77 %) than patients treated in nonminority community treatment settings (52 %). Minority patients had the severity of their pain underestimated by their physicians, reported that they needed stronger pain medication, and felt that they needed to take more analgesics than their doctors had prescribed. Assessing differences between minority groups, more Hispanic patients reported lower levels of pain relief than African Americans. Furthermore, a study of a nationally representative sample of adults in the United States showed that Latino (Hispanic) and Anglos (non-Hispanic Whites) reported lower rates of activity impairment due to pain than did African Americans (non-Hispanic blacks). Also, more African Americans than Anglos and Latinos reported functional impairment as a result of pain (Reyes-Gibby et al. 2007a). A survey of a multiethnic sample of 480 cancer patients found ethnic and racial differences in self-reported pain (Im et al. 2007). The Asian Americans reported the lowest pain scores on multiple types of scales. The Hispanics reported the highest scores on visual analog, pictorial, and numerical pain scales. Surprisingly, the visual analog and verbal descriptor pain scores of the

African Americans were significantly lower than those of the white and Hispanic participants. Given that all participants were recruited from community sites and internet support groups, the samples may not have been representative of larger population groups. A recent descriptive study of 281 outpatients with solid tumors and cancer-related pain found African-American patients reported higher pain intensity ratings as compared to white patients (Vallerand et al. 2005).

When minority patients are medically underserved due to a lack of health insurance, underinsurance, and/or limited financial resources, they may be particularly at risk for pain and undertreatment of pain. For example, a study focused on the pain treatment of low income Hispanic and African-American patients found that 65 % of the patients reported severe pain (Anderson et al. 2002). The high levels of pain may have been due to inadequate dosages or patient's nonadherence to analgesic regimens. Most patients in both minority groups reported taking their analgesics less often than prescribed by their oncologists. A study of 116 low-income women who had undergone breast cancer treatment found African-American and Hispanic women were more likely than white women to report pain (Eversley et al. 2005). In multivariate analyses, an increased number of reported symptoms were associated with lower income, being Hispanic, having a mastectomy, and receiving chemotherapy.

Only a few longitudinal studies of cancer-related pain have examined whether patients' experiences of pain differed by race or ethnicity. A sample of over 1,000 women with metastatic breast cancer and bone metastases who were enrolled in a clinical trial completed regular pain assessments for 1 year (Castel et al. 2008). The results indicated pain severity worsened more rapidly among nonwhite than white women. A longitudinal study of men aged 65 or older with newly diagnosed prostate cancer found that African-American men reported more severe pain and urinary symptoms as compared to white men at a 1-year follow-up (Jayadevappa et al. 2006). A recent longitudinal study examining breakthrough pain in white and nonwhite patients with advanced cancer found nonwhites experienced increased pain severity and more breakthrough pain episodes (Green et al. 2009; Montague et al. 2009).

Although few disagree that painful conditions should be treated regardless of age, studies document undertreatment of cancer pain in older populations. In a study of 4,003 elderly cancer patients (24 %, 29 %, and 38 % of those aged 85 years or older, 75–84 years, and 65–74 years, respectively), Bernabei et al. (1998) found that more than a quarter (26 %) of patients experiencing daily pain received no analgesic agent. Patients aged 85 years or older were found less likely to receive morphine or other strong opiates than those aged 65–74 years (13 % vs. 38 %, respectively), and were more likely to receive no analgesia at all. African-American elderly patients were also at a higher risk for undertreatment, suggesting the disparity in pain treatment is compounded for elderly minority patients. A study that looked at undertreatment of pain for ambulatory patients with metastatic cancer (Cleeland et al. 1994) also showed that older age (70 years or older) was a significant predictor of inadequate analgesia, according to the World Health Organization guidelines (World Health Organization 1986) for treatment of patients with cancer pain. Increased pain complaints with advancing age are associated with increased pathological load, as older adults are at greater risk for diseases that cause pain, e.g., cancer, arthritis, etc. Furthermore, although they are more likely to experience pain than younger individuals, older adults tend to be less likely to complain of pain. Factors such as other medical problems, cognitive and sensory impairments, and depression are possible contributors to the underreporting of pain among older adults (Reyes-Gibby et al. 2002).

Factors Associated with Inadequate Treatment and Control of Pain

Many cancer specialists recognize that pain control is often suboptimal. Medical oncologists were surveyed about their treatment of cancer pain in a study conducted by the Eastern Cooperative Oncology Group (ECOG) (Von Roenn et al. 1993). Only half of the physicians surveyed indicated that cancer pain control was "good" or "very good" in their practice setting. Seventy-five percent of the physicians indicated that the most important barrier to cancer pain management was inadequate pain assessment. Over 60 % reported that physicians' reluctance to prescribe analgesics and patient's unwillingness to report pain or take opioids were barriers to adequate pain treatment. Inadequate knowledge about cancer pain management was reported by over 50 % of the ECOG physicians surveyed. The survey documents that a substandard level of education about cancer pain management and a reluctance to address it in practice existed at all levels of professional health care. Cleeland et al. (2000) repeated the ECOG study format with physician members of the Radiation Therapy Oncology Group. Although there has been some improvement in the use of stronger analgesics, many barriers to good pain control remain, and poor pain assessment is still seen as the major barrier to good pain management.

Many cancer specialists recognize that pain control is often suboptimal. Medical oncologists were surveyed about their treatment of cancer pain in a study conducted by the Eastern Cooperative Oncology Group (ECOG) (Von Roenn et al. 1993). Only half of the physicians surveyed indicated that cancer pain control was "good" or "very good" in their practice setting. Seventy-five percent of the physicians indicated that the most important barrier to cancer pain management was inadequate pain assessment. Over 60 % reported that physicians' reluctance to prescribe analgesics and patient unwillingness to report pain or take opioids were barriers to adequate pain treatment. Inadequate knowledge about cancer pain management was reported by over 50 % of the ECOG physicians surveyed. The survey documents that a substandard level of education about cancer pain management and a reluctance to address it in practice existed at all levels of professional health care The ECOG study format was repeated with physician members of the Radiation Therapy Oncology Group

(Cleeland et al. 2000) and with physicians and nurses treating low income minority patients (Anderson et al. 2000). The results of both studies indicated that many barriers to good pain control remain, and poor pain assessment is still seen as the major barrier to good pain management.

Improving Control of Cancer Pain

In spite of the recent concerns over symptom management, there is substantial evidence that symptoms that could, in principle, be well managed are nonetheless undertreated. Pain could be more adequately controlled if we systematically applied the knowledge that we now have about its management.

There is at least preliminary evidence that improving assessment can improve pain management, and may improve the management of other symptoms as well. Trowbridge et al. (1997) conducted a randomized controlled trial of 320 patients and 13 oncologists. Patients were asked while in the clinic to complete assessments of their pain, their pain regimens, and the degrees of relief received. Follow-up was conducted 4 weeks later by mail-in survey. The intervention group's clinical charts contained a summary of the completed pain scales and the oncologists who treated these patients were instructed to review the summary sheet prior to an evaluation. This summary was not available for the oncologists treating the patients in the control group. Results showed a significant difference ($p = .016$) in the physicians' prescription patterns. In the control group, prescriptions for 86 % of the patients did not change, with no increase in analgesic prescriptions. In the intervention group, analgesic prescriptions changed for 25 % of the patients, decreasing for 5 %, and increasing for 20 %. A decrease in the incidence of pain described as "more than life's usual aches and pains" was found at follow-up for the intervention group ($p = .05$). The findings from this study suggest that standardized pain assessment leads to improved cancer pain management.

Beginning with the publication of the World Health Organization's Cancer Pain Management guidelines in 1986, several guidelines for the practice of cancer pain management have been issued. There is, however, only one published study that evaluates the effectiveness of adherence to a pain management guideline for cancer pain (Du Pen et al. 1999). In this study, 81 cancer patients were enrolled in a prospective, longitudinal, randomized controlled study from 26 medical oncologist outpatient clinics in western Washington. A multilevel treatment algorithm based on the AHRQ Guidelines for Cancer Pain Management was compared with standard? practice (control) therapies for pain and symptom management used by community oncologists. Patients randomized to the pain algorithm group achieved a statistically significant reduction in pain intensity when compared with standard community practice.

In a recently completed Eastern Cooperative Oncology Group study (Cleeland et al. 2005), we found that the institutional use of a protocol for pain management improved pain control in lung and prostate cancer patients, but failed to improve the pain management of other patients (breast cancer and myeloma patients) in the same institutions who were *not* specifically treated by the pain protocol. Forty-eight percent of patients with lung and prostate cancer in the institutions that used a protocol for pain management reported statistically significant reduction in moderate and severe pain ($p < .01$), compared with 15 % of lung and prostate patients treated as usual in the control institutions.

A system improvement approach was used to improve pain management in a national healthcare system (Cleeland et al. 2003), with some reported success. A joint collaborative effort of the Veterans Health Administration and the Institute for Healthcare Improvement (VHA-IHI) was initiated, the focus of the project was to achieve rapid improvements in pain management in the clinical and operational areas in the national VHA system. An underlying premise was the spread of existing knowledge to multiple sites. A total of 70 teams were formed and were asked to identify their goals and to implement changes in their practice setting. Importantly, faculty involved in the initiative developed the attributes of an ideal pain management system. Changes that were implemented included the development of new assessment forms to increase the number of patients with documented assessment and follow-up for treatment plan; development of guidelines and/or protocols; formation of one-on-one, group-provider, or computer-based educational methods to improve both assessment and follow-up, and/or ways to effectively link provider education with other system changes such as drug ordering or assessment processes; and development of innovative ways to link primary care physicians with pain specialists.

The following outcomes were reported: Moderate or severe pain on study units dropped from 24 % to 17 %; pain assessment increased from 75 % to 85 %; pain care plans for patients with at least mild pain increased from 58 % to 78 %; and number of patients provided with pain educational materials increased from 35 % to 62 %. The results of these studies, taken together, suggest that improved assessment should improve symptom management for cancer patients.

Genetic Association Studies of Pain

Whereas classical epidemiology approaches have resulted in the identification of subgroups of individuals that are at higher risk for pain (e.g., by age and racial/ethnic groups), molecular methods in epidemiology integrate the use of biological markers that specify events at the physiological, cellular, and molecular levels. Thus, while studies have generally established that individual and group risk assessments are possible using classical or traditional epidemiology methods, integrating the use of molecular epidemiology methods augments individual and group risk assessments by providing more person-specific information (genetic profile). Furthermore, the use of molecular epidemiological methods can also reduce misclassification of exposure (avoiding recall bias in exposure identification) and provide information about when interventions can be most effective (pharmacogenetics). Importantly, because *host genetic polymorphisms are stable markers, genetic association research*

will enable early identification of cancer patients at high risk for severe pain and will help facilitate early pain interventions.

The classic central dogma regarding the genome is that DNA in the nucleus directs the production of proteins. Proteins carry out multiple life functions through the RNA. By analyzing the DNA, we are able to understand the variability in the genome and the extent to which this variability contributes to physiological traits, including disease and symptom susceptibility. Among the several types of variations that exist in the human genome, single nucleotide polymorphisms (SNPs) are the most common and easiest to measure. Given that pain is a complex trait, a review of published genetic association studies of pain in cancer patients suggests that several genes (*OPRM1; COMT, ABCB1/MDR1, IL6, IL8, TNF, PTGS2*) participate in the epidemiology of cancer-related pain.

Located on human chromosome 6q24-q25(23), the mu opioid receptor is considered the most important target for morphine. Thus, polymorphisms of the mu opioid receptor (*OPRM1*) gene have been prime candidates for genetic influences on the efficacy of opioids. In particular, polymorphism in the A118G OPRM1 has been significantly associated with pressure pain sensitivity and variability in opioid efficacy. Klepstad and colleagues (2004) assessed the influence of the A118G OPRM1 polymorphism in a heterogeneous sample of Norwegian cancer inpatients who were on stable morphine doses for 3 days. When they evaluated the genetic profile of these patients, they found that patients who were homozygous for the variant 118 G allele required about twice the morphine doses relative to wild-type patients in order to achieve adequate pain relief (AA = 97 mg/24 h; 95 % CI = 77,119 vs. GG = 225 mg/24 h; 95 % CI = 136,313; $p < 0.006$). Further research is needed to validate this finding.

Another extensively studied pain candidate gene is the catechol-*O*-methyltransferase (COMT) gene located on chromosome 22q11.21. The COMT enzyme metabolizes the catecholamines dopamine, epinephrine, and norepinephrine; is known to be a key modulator of dopaminergic and adrenergic neurotransmissions; and is hypothesized to influence opioid receptor density. The Val158Met polymorphism was found to influence the activity of the *COMT* enzyme. Using the same Norwegian patient sample, Klepstad and colleagues (Rakvåg et al. 2005) demonstrated that patients with the Val/Val genotype ($n = 44$) needed more morphine (155 mg/24 h; 95 % CI = 106,203) than those with the Val/Met (117 mg/24; 95 % CI = 97,137; $n = 96$) and the Met/Met genotype (95 mg/24 h; 95 % CI = 71,119 n = 67; $p < 0.025$).

With the assumption that pain is a complex trait and that multiple genes impact the efficacy of opioids, Reyes-Gibby and colleagues assessed the potential joint effects of the COMT and OPRM1 gene variants in the clinical efficacy of morphine (Reyes-Gibby et al. 2007b) in the same Norwegian patient sample. They found that carriers of the OPRM1 AA and COMT Met/Met genotypes needed the lowest morphine dose to achieve pain relief (87 mg/24 h; 95 % CI = 57,116) and those with neither Met/Met nor AA genotype required the highest morphine dose (147 mg/24 h; 95 % CI — 100, 180). Even after controlling for demographic and clinical variables in the multivariable analyses, the significant joint effects for the Met/Met and AA genotypes (OR = 0.278; 95 % CI = 0.102. 0.756; $p < 0.012$) persisted.

The role of multiple genes in the epidemiology of cancer pain was also studied by Campa and colleagues (2008). In particular, they assessed the importance of the OPRM1 A118G polymorphism and of the C3435T SNP of ABCB1/ MDR1 gene, a major determinant of morphine bioavailability, in pain relief from morphine in 137 Italian cancer patients. They found that AA carriers of OPRM1 A118G reported a significantly greater reduction in pain than GG carriers. They also found that TT carriers of C3435T SNP of ABCB1/ MDR1 gene experienced significantly greater pain relief than homozygous wild-type CC carriers. A significant joint effect of these two genes is as follows: Carriers of OPRM1AA and ABCB1TT genotypes report greater pain reduction (4.8; 95 % CI = 4.21, 5.39; $p < 0.0001$) as compared to carriers of OPRM1AA and ABCB1 CC or CT

genotypes (3.1; 95 % CI = 2.72, 3.48) and of other allele combinations (1.3; 95 % CI = 0.65, 1.95).

Ample evidence exists that polymorphisms in cytokine genes also play a role in cancer and cancer pain. In a study of lung cancer patients, Reyes-Gibby and colleagues (Reyes-Gibby et al. 2007) genotyped functional single nucleotide polymorphisms in tumor necrosis factor-A (TNF-a — 308 G/A), interleukin-6 (IL-6) – 174 G/C, and IL-8 – 251 T/A and determined their associations with pain severity. After controlling for epidemiologic (age and sex), clinical (stage of disease, comorbidities), and symptom (depressed mood and fatigue) variables known to influence pain severity, variant alleles in IL-8 – 251 T/A were significant predictors of severe pain for White patients. A follow-up study(Reyes-Gibby et al. 2008) assessed if the same genetic variants would influence analgesic dose in lung cancer patients who were referred for pain management. They found that variant alleles in TNFa — 308 G/A remained significantly associated with pain severity ($b = 0.226; p = 0.036$), and carriers of the IL-6 – 174C/C genotypes required 4.7 times higher dose of opioids for pain relief relative to GG and GC genotypes. A study of pancreatic cancer patients (Reyes-Gibby et al. 2009b) further confirmed that select polymorphisms in IL8 were related to pain severity scores. Other studies have demonstrated the influence of genetic polymorphisms in inflammation genes on the severity of cancer-related symptoms (Rausch et al. 2010; Collado-Hidalgo et al. 2008). These studies indicate that genetic factors play a key role in patients' pain and their response to opioids. In the future, we expect that genetic profiling could become an essential component of an individualized pain treatment program for patients with cancer.

Future Directions

Overall, studies of the incidence, prevalence, severity, and treatment of cancer-related pain have a number of shortcomings. Often, pain is evaluated and reported without stratifying for heterogeneity with respect to pain type, disease type and stage, disease treatment, and response to disease treatment. Durations of pain may be days, months, or years, but assessments often are performed cross-sectionally rather than longitudinally, thus failing to assess patterns and trajectories.

We need to explore whether pain and other symptoms experienced by cancer patients differ quantitatively and qualitatively with those of noncancer populations. Cohort studies that will provide clinicians with information regarding the incidence, severity, and duration of pain and related symptoms after a diagnosis of cancer should also be useful in determining the epidemiology of cancer pain. Finally, we need to integrate the use of molecular information in studies of individuals at risk for severe pain, given disease and treatment information.

References

Amichetti, M., & Caffo, O. (2003). Pain after quadrantectomy and radiotherapy for early-stage breast cancer: Incidence, characteristics and influence on quality of life. Results from a retrospective study. *Oncology, 65*(1), 23–28.

Anderson, K. O., Mendoza, T. R., Valero, V., Richman, S. P., Russell, C., Hurley, J., DeLeon, C., Washington, P., Palos, G., Payne, R., & Cleeland, C. S. (2000). Minority cancer patients and their providers: Pain management attitudes and practice. *Cancer, 88*(8), 1929–1938.

Anderson, K. O., Richman, S. P., Hurley, J., Palos, G., Valero, V., Mendoza, T. R., Gning, I., Cleeland, C. S. (2002). Cancer pain management among underserved minority outpatients: Perceived needs and barriers to optimal control. *Cancer, 94*(8), 2295–2304.

Bernabei, R., Gambassi, G., Lapane, K., Landi, F., Gatsonis, C., Dunlop, R., Lipsitz, L., Steel, K., & Mor, V. (1998). Management of pain in older adults patients with cancer. SAGE study group. Systematic assessment of geriatric drug Use via epidemiology. *Journal of the American Medical Association, 279*(23), 1877–1882.

Campa, D., Gioia, A., Tomei, A., Poli, P., & Barale, R. (2008). Association of ABCB1/MDR1 and OPRM1 gene polymorphisms with morphine pain relief. *Clinical Pharmacology & Therapeutics, 83*(4), 559–566.

Carpenter, J. S., Andrykowski, M. A., Sloan, P., Cunningham, L., Cordova, M. J., Studts, J. L., McGrath, P. C., Sloan, D., & Kenady, D. E. (1998). Postmastectomy/postlumpectomy pain in breast cancer survivors. *Journal of Clinical Epidemiology, 51*(12), 1285–1292.

Castel, L. D., Saville, B. R., Depuy, V., Godley, P. A., Hartmann, K. E., & Abernethy, A. P. (2008). Racial differences in pain during 1 year among women with metastatic breast cancer: A hazards analysis of interval-censored data. *Cancer, 112*(1), 162–170.

Chaplin, J. M., & Morton, R. P. (1999). A prospective, longitudinal study of pain in head and neck cancer patients. *Head & Neck, 21*(6), 531–537.

Cleeland, C. S., Gonin, R., Hatfield, A. K., Edmonson, J. H., Blum, R. H., Stewart, J. A., & Pandya, K. J. (1994). Pain and its treatment in outpatients with metastatic cancer. *The New England Journal of Medicine, 330*(9), 592–596.

Cleeland, C. S., Gonin, R., Baez, L., Loehrer, P., & Pandya, K. J. (1997). Pain and treatment of pain in minority patients with cancer. The eastern cooperative oncology group minority outpatient pain study. *Annals of Internal Medicine, 127*(9), 813–816.

Cleeland, C. S., Janjan, N. A., Scott, C. B., Seiferheld, W. F., & Curran, W. J. (2000). Cancer pain management by radiotherapists: A survey of radiation therapy oncology group physicians. *International Journal of Radiation Oncology, Biology, and Physics, 47*(1), 203–208.

Cleeland, C. S., Reyes-Gibby, C. C., Schall, M., Nolan, K., Paice, J., Rosenberg, J. M., Tollett, J. H., & Kerns, R. D. (2003). Rapid improvement in pain management: The veterans health administration and the institute for healthcare improvement collaborative. *The Clinical Journal of Pain, 19*(5), 298–305.

Collado-Hidalgo, A., Bower, J. E., Ganz, P. A., Irwin, M. R., & Cole, S. W. (2008). Cytokine gene polymorphisms and fatigue in breast cancer survivors: Early findings. *Brain, Behavior, and Immunity, 22*, 1197–1200.

Du Pen, S. L., Du Pen, A. R., Polissar, N., Hansberry, J., Kraybill, B. M., Stillman, M., Panke, J., Everly, R., & Syrjala, K. (1999). Implementing guidelines for cancer pain management: Results of a randomized controlled clinical trial. *Journal of Clinical Oncology, 17*(1), 361–370.

Epstein, J. B., Wilkie, D. J., Fischer, D. J., Kim, Y. O., & Villines, D. (2009). Neuropathic and nociceptive pain in head and neck cancer patients receiving radiation therapy. *Head & Neck Oncology, 1*(1), 26.

Eversley, R., Estrin, D., Dibble, S., Wardlaw, L., Pedrosa, M., & Favila-Penney, W. (2005). Post-treatment symptoms among ethnic minority breast cancer survivors. *Oncology Nursing Forum, 32*(2), 250–256.

Green, C. R., Montague, L., & Hart-Johnson, T. A. (2009). Consistent and breakthrough pain in diverse advanced cancer patients: A longitudinal examination. *Journal of Pain and Symptom Management, 37*(5), 831–847.

Im, E. O., Chee, W., Guevara, E., Liu, Y., Lim, H. J., Tsai, H. M., Clark, M., Bender, M., Suk Kim, K., Hee Kim, Y., & Shin, H. (2007). Gender and ethnic differences in cancer pain experience: A multiethnic survey in the United States. *Nursing Research, 56*(5), 296–306.

Jayadevappa, R., Chhatre, S., Whittington, R., Bloom, B. S., Wein, A. J., & Malkowicz, S. B. (2006). Health-related quality of life and satisfaction with care among older men treated for prostate cancer with either radical prostatectomy or external beam radiation therapy. *British Journal of Urology International, 97*(5), 955–962.

Klepstad, P., Rakvåg, T. T., Kaasa, S., Holthe, M., Dale, O., Borchgrevink, P. C., Baar, C., Vikan, T., Krokan, H. E., & Skorpen, F. (2004). The 118 A > G polymorphism in the human mu-opioid receptor gene may increase morphine requirements in patients with pain caused by malignant disease. *Acta Anaesthesiologica Scandinavica, 48*(10), 1232–1239.

McCarthy, E. P., Phillips, R. S., Zhong, Z., Drews, R. E., & Lynn, J. (2000). Dying with cancer: patients' Function, symptoms, and care preferences as death approaches. *Journal of the American Geriatrics Society, 48*(5 Suppl.), S110–S121.

Patrick, D. L., Ferketich, S. L., Frame, P. S., Harris, J. J., Hendricks, C. B., Levin, B., Link, M. P., Lustig, C., McLaughlin, J., Reid, L. D., Turrisi, A. T. 3rd, Unutzer, J., & Vernon, S. W. (2004). National Institutes of Health State-of-the-Science Panel. National Institutes of Health State-of-the-Science Conference Statement: Symptom management in cancer: pain, depression, and fatigue, July 15–17, 2002. *Journal of the National Cancer Institute Monographs*, (32), 9–16.

Portenoy, R. K., Thaler, H. T., Kornblith, A. B., Lepore, J. M., Friedlander-Klar, H., Coyle, N., Smart-Curley, T., Kemeny, N., Norton, L., Hoskins, W., et al. (1994). Symptom prevalence, characteristics and distress in a cancer population. *Quality of Life Research, 3*(3), 183–189.

Rakvåg, T. T., Klepstad, P., Baar, C., Kvam, T. M., Dale, O., Kaasa, S., Krokan, H. E., & Skorpen, F. (2005). The Val158Met polymorphism of the human catechol-O-methyltransferase (COMT) gene may influence morphine requirements in cancer pain patients. *Pain, 116*(1–2), 73–78.

Rausch, S. M., Clark, M. M., Patten, C., Liu, H., Felten, S., Li, Y., et al. (2010). Relationship between cytokine gene single nucleotide polymorphisms and symptom burden and quality of life in lung cancer survivors. *Cancer, 116*, 4103–4113.

Reyes-Gibby, C. C, Aday, L., & Cleeland, C. (2002). Impact of pain on self-rated health in the community-dwelling older adults. *Pain, 95*(1–2), 75–82.

Reyes-Gibby, C. C., Aday, L. A., Anderson, K. O., Mendoza, T. R., & Cleeland, C. S. (2006). Pain, depression, and fatigue in community-dwelling adults with and without a history of cancer. *Journal of Pain and Symptom Management, 32*(2), 118–128.

Reyes-Gibby, C. C., Aday, L. A., Todd, K. H., Cleeland, C. S., & Anderson, K. O. (2007a). Pain in aging community-dwelling adults in the United States: non-Hispanic whites, non-Hispanic blacks, and Hispanics. *Journal of Pain, 8* (1), 75–84. Epub 2006 Sep 1. PubMed PMID: 16949874; PubMed Central PMCID: PMC1974880.

Reyes-Gibby, C. C., Shete, S., Rakvåg, T., Bhat, S. V., Skorpen, F., Bruera, E., Kaasa, S., & Klepstad, P. (2007b). Exploring joint effects of genes and the clinical efficacy of morphine for cancer pain: OPRM1 and COMT gene. *Pain, 130*(1–2), 25–30.

Reyes-Gibby, C. C., El, O. B., Spitz, M. R., Parsons, H., Kurzrock, R., Wu, X., et al. (2008). The influence of tumor necrosis factor-alpha −308 G/a and IL-6–174 G/C on pain and analgesia response in lung cancer patients receiving supportive care. *Cancer Epidemiology, Biomarkers & Prevention, 17*, 3262–3267.

Reyes-Gibby, C. C., Morrow, P. K., Buzdar, A., & Shete, S. (2009a). Chemotherapy-induced peripheral neuropathy as a predictor of neuropathic pain in breast cancer patients previously treated with paclitaxel. *Journal of Pain, 10*(11), 1146–1150.

Reyes-Gibby, C. C., Shete, S., Yennurajalingam, S., Frazier, M., Bruera, E., Kurzrock, R., et al. (2009b). Genetic and nongenetic covariates of pain severity in patients with adenocarcinoma of the pancreas: Assessing the influence of cytokine genes. *Journal of Pain and Symptom Management, 38*, 894–902.

Stevens, P. E., Dibble, S. L., & Miaskowski, C. (1995). Prevalence, characteristics, and impact of postmastectomy pain syndrome: An investigation of women's experiences. *Pain, 61*(1), 61–68.

Trowbridge, R., Dugan, W., Jay, S. J., Littrell, D., Casebeer, L. L., Edgerton, S., Anderson, J., & O'Toole, J. B. (1997). Determining the effectiveness of a clinical-practice intervention in improving the control of pain in outpatients with cancer. *Academic Medicine, 72*(9), 798–800.

Vallerand, A. H., Hasenau, S., Templin, T., & Collins-Bohler, D. (2005). Disparities between black and white patients with cancer pain: The effect of perception of control over pain. *Pain Medicine, 6*(3), 242–250.

Vecht, C. J. (1990). Arm pain in the patient with breast cancer. *Journal of Pain Symptom and Management, 5*(2), 109–117.

Von Roenn, J. H., Cleeland, C. S., Gonin, R., Hatfield, A. K., & Pandya, K. J. (1993). Physician attitudes and practice in cancer pain management. A survey from the eastern cooperative oncology group. *Annals of Internal Medicine, 119*(2), 121–126.

World Health Organization. (1986). *Cancer pain relief.* Geneva: WHO.

Cancer Pain, Evaluation of Relevant Comorbidities and Impact

Myra Glajchen
Department of Pain Medicine and Palliative Care, Beth Israel Medical Center Albert Einstein College of Medicine, New York, NY, USA

Definition

The assessment of patients with cancer pain often reveals the existence of important physical or psychiatric comorbidities. These comorbidities may require treatment and also may influence the selection of pain therapy. Relevant comorbidities in cancer pain include fatigue, arthritis, sleep disorders, gastrointestinal complications, psychiatric conditions such as anxiety, adjustment disorder, depression, cognitive impairment, and substance abuse.

Characteristics

In some cases, a comorbid condition is both an independent medical issue and a potential etiology of the pain. For example, major depression may be expressed, in part, by persistent pain. If the patient presents in this way, treatment for the depression must be given priority and included among the analgesic strategies pursued. In other circumstances, a comorbid condition may be unrelated to the pain itself, but may impair function and undermine treatment strategies intended to restore functioning. This is commonly encountered in the cancer patient with persistent gastrointestinal distress following chemotherapy.

A useful taxonomy for evaluating the impact of relevant comorbidities on cancer pain is listed below. The coexisting symptom or condition should be evaluated for its:

- Direct contribution to the pain
- Predisposition to cause adverse treatment effects
- Compromise of rehabilitative efforts
- Ability to undermine quality of life
- Ability to increase risk of serious adverse medical outcomes

Fatigue is a common comorbidity affecting 50–96 % of people with cancer (Cella et al. 2002; Campos et al. 2011). Generally, fatigue is characterized by decrease in energy, distress, and decreased functional status (Montazeri 2009). Fatigue in cancer patients can result from the course of the disease, preexisting physical or psychological conditions, effects of medication, or lack of exercise. Fatigue may also result from surgery, chemotherapy, and radiation therapy. Because fatigue can severely undermine sense of well-being, activities of daily living,

relationships with family and friends, and treatment compliance, prompt diagnosis and treatment are recommended (Weis 2011).

Gastrointestinal complications (constipation, impaction, bowel obstruction, nausea, vomiting, and diarrhea) are common for oncology patients. The growth and spread of cancer, as well as related treatment, contribute to these conditions. Despite advances in management, nausea and vomiting remain among the most distressing side effects for patients and families (Glare et al. 2004). In addition, gastrointestinal comorbidities undermine quality of life and interfere with treatment (Roscoe et al. 2009).

Alterations in nutritional status begin at diagnosis and proceed through treatment. ▶ Protein-calorie malnutrition (PCM) is common and can lead to progressive wasting, weakness, and debilitation. Anorexia, the loss of appetite, is present in 15 % to 25 % of patients at diagnosis and may also occur as a treatment side effect (Hopkinson et al. 2008). Depression, loss of hope, and anxiety are all associated with anorexia (Hopkinson et al. 2008). Anorexia can hasten the course of cachexia, a progressive wasting syndrome (Donohoe et al. 2011) marked by weakness and progressive loss of body weight, fat, and muscle. Pain treatment in the context of anorexia and cachexia is complex.

About 44 % of cancer patients report anxiety during treatment, while 23 % report high anxiety (Stark et al. 2002; Waller et al. 2012). Anxiety that is unusually intense or persistent may suggest an adjustment disorder. Factors that increase the likelihood of developing adjustment disorders include pain, a history of anxiety disorders, functional limitations, advancing disease, and absence of social support (Miovic and Block 2007; Ganz et al. 2003; Breitbart 1995). Patients can become immobilized by anxiety and unable to comply with treatment. Therefore, prompt diagnosis and early intervention are optimal.

Depression is a disabling condition affecting 15–25 % of cancer patients (Schag and Heinrich 1989; Breitbart et al. 2000). While some sadness is normal in reaction to cancer, it is important to distinguish normal sadness from a depressive disorder (Block 2000). Clinicians should assess mood carefully and be proactive in initiating therapy. Early intervention is indicated if symptoms persist or recur, if patients present with a history of depression, poor social support, a poor prognosis, and a high level of disability. A combination of psychotherapy and antidepressant drugs represents the most effective strategy for pain-related depression. Antidepressants with established analgesic effectiveness may be preferred (Breitbart et al. 2000; Block 2000).

Delirium, agitation, and cognitive impairment occur in 28–48 % of patients with advanced cancer (Waller et al. 2012; Lawlor et al. 2000). The causes of delirium are multifactorial, and may include direct effects of the disease on the CNS, metabolic or electrolyte disturbances, treatment, and medications (Lawlor et al. 2000; Morita et al. 2002). The presence of delirium poses a major barrier to doctor-patient communication, family distress, and treatment compliance (Mori et al. 2011). Prompt diagnosis and treatment are therefore optimal.

Nearly one third of the population of the United States has used illicit drugs, and an estimated 6 % to 27 % have a ▶ substance abuse problem of some type (Colliver and Kopstein 1991). Because of the prevalence of substance abuse, (Weissman and Haddox 1989) problems related to addiction may be encountered in the oncology setting. A history of substance abuse can increase the risk of morbidity or mortality among patients with advanced disease, complicate the selection of opioid therapies, and undermine treatment compliance. This can be mitigated by a therapeutic strategy that addresses drug-taking behavior and associated risks while implementing other therapies. Knowledge about the basic concepts of ▶ addiction medicine and collaboration with addiction specialists may also be helpful (Substance Abuse and Mental Health Services Administration 2010).

The goal of a comprehensive approach to the patient with cancer pain is to organize a therapeutic strategy that can enhance comfort and function and ameliorate comorbidities – with the therapeutic endpoint of reducing suffering and improving quality of life.

References

Block, S. D. (2000). Assessing and managing depression in the terminally ill patient. ACP-ASIM End-of-Life Care Consensus Panel. American College of Physicians - American Society of Internal Medicine. Ann Intern Med, 132(3), 209–218.

Breitbart, W. (1995). Identifying patients at risk for, and treatment of major psychiatric complications of cancer. Support Care Cancer, 3(1), 45–60.

Breitbart, W., Rosenfeld, B., Pessin, H., Kaim, M., Funesti-Esch, J., Galietta, M., Nelson, C. J., & Brescia, R. (2000). Depression, hopelessness, and desire for hastened death in terminally ill patients with cancer. JAMA, 284(22), 2907–2911.

Campos, M. P., Hassan, B. J., Riechelmann, R., & Del Giglio, A. (2011). Cancer-related fatigue: a practical review. Ann Oncol., 22(6), 1273–1279. Jun.

Cella, D., Lai, J. S., Chang, C. H., et al. (2002). Fatigue in cancer patients compared with fatigue in the general United States population. Cancer, 94(2), 528–538.

Colliver, J. D., & Kopstein, A. N. (1991). Trends in cocaine abuse reflected in emergency room episodes reported to DAWN. Public Health Rep, 106(1), 59–68.

Donohoe CL, Ryan AM, Reynolds JV. (2011). Cancer cachexia: mechanisms and clinical implications. Gastroenterol Res Pract. Vol. 2011, Article ID 601434, 13 pages, doi:10.1155/2011/601434

Ganz, P. A., Guadagnoli, E., Landrum, M. B., Lash, T. L., Rakowski, W., & Silliman, R. A. (2003). Breast cancer in older women: quality of life and psychosocial adjustment in the 15 months after diagnosis. J Clin Oncol, 21(21), 4027–4033.

Glare, P., Pereira, G., Kristjanson, L. J., Stockler, M., & Tattersall, M. (2004). Systematic review of the efficacy of antiemetics in the treatment of nausea in patients with far-advanced cancer. Support Care Cancer., 12(6), 432–440. Jun.

Hopkinson, J. B., Wright, D. N., & Foster, C. (2008). Management of weight loss and anorexia. Ann Oncol., 19(Suppl 7), vii289–vii293. 2008 Sep.

Lawlor, P. G., Gagnon, B., Mancini, I. L., et al. (2000). Occurrence, causes, and outcome of delirium in patients with advanced cancer: a prospective study. Arch Intern Med, 160(6), 786–794.

Miovic, M., & Block, S. (2007). Psychiatric disorders in advanced cancer. Cancer., 110(8), 1665–1676. Oct 15.

Montazeri, A. (2009). Quality of life data as prognostic indicators of survival in cancer patients: an overview of the literature from 1982 to 2008. Health Qual Life Outcomes., 23(7), 102. Dec 23.

Mori, M., Parsons, H. A., De la Cruz, M., Elsayem, A., Palla, S. L., Liu, J., Li, Z., Palmer, L., Bruera, E., & Fadul, N. A. (2011). Changes in symptoms and inpatient mortality: a study in advanced cancer patients admitted to an acute palliative care unit in a comprehensive cancer center. J Palliat Med., 14(9), 1034–1041. Sept.

Morita, T., Tei, Y., Tsunoda, J., et al. (2002). Underlying pathologies and their associations with clinical features in terminal delirium of cancer patients. *J Pain Symptom Manage, 23*(2), 107–113.

Roscoe, J. A., Morrow, G. R., Colagiuri, B., Heckler, C. E., Pudlo, B. D., Colman, L., Hoelzer, K., & Jacobs, A. (2009). Insight in the prediction of chemotherapy-induced nausea. *Support Care Cancer., 18*(7), 869–876. Jul 2010.

Schag, C. A., & Heinrich, R. L. (1989). Anxiety in medical situations: adult cancer patients. *J Clin Psychol, 45*(1), 20–27.

Stark, D., Kiely, M., Smith, A., et al. (2002). Anxiety disorders in cancer patients: their nature, associations, and relation to quality of life. *J Clin Oncol, 20*(14), 3137–3148.

Substance Abuse and Mental Health Services Administration. *Results from the 2010 National Survey on Drug Use and Health: Summary of National Findings.* NSDUH Series H-41 HHS Publication No. (SMA) 11–4658. Rockville, MD: Substance Abuse and Mental Health Services Administration; 2011.

Waller, A., Williams, A., Groff, S. L., Bultz, B. D., Carlson, L. E. (2013). Screening for distress, the sixth vital sign: examining self-referral in people with cancer over a one-year period. *Psychooncology, 22*(2), 388–395.

Weis, J. (2011). Cancer-related fatigue: prevalence, assessment and treatment strategies. *Expert Rev Pharmacoecon Outcomes Res., 11*(4), 441–446. Aug.

Weissman, D. E., & Haddox, J. D. (1989). Opioid pseudoaddiction–an iatrogenic syndrome. *Pain, 36*(3), 363–366.

Cancer Pain, Goals of a Comprehensive Assessment

Pauline Lesage
Department of Pain Medicine and Palliative Care, Beth Israel Medical Center, New York, USA

Definition

The assessment of cancer pain requires an understanding of the disease and its extent, in addition to its diverse comorbidities, to clarify the nature of the pain and its physical, psychological, and social disturbances. The list below describes a stepwise assessment of patients with cancer pain. This understanding will guide the treatment plan, focused on providing comfort and improving quality of life.

Step 1: Data collection
- Pain-related history
  - Other relevant history
  - Disease related
  - Other symptoms
  - Psychiatric history
  - Social resources
- Concurrent quality-of-life concerns
  - Other symptoms
  - Concerns in physical, psychological, social, and spiritual domains
  - Other concerns (e.g., financial)
  - Assessment of the family
- Available laboratory and imaging data
  - Radiographs and scans
  - Tumor markers
  - Hematologic parameters
  - Biochemical parameters

Step 2: Interpret the findings
- Pain-related constructs
  - Etiology
  - Inferred pathophysiology
  - Syndrome identification
- Extent of disease
- Goals of care
  - Prolonging survival
  - Optimizing function
  - Optimizing comfort

Step 3: Formulating a treatment strategy
- Further evaluation, if needed
- Multimodality approach to the pain
- Treatment of the pain etiology, if possible
- Treatments for concurrent concerns
- Patient/family education

Characteristics

The comprehensive assessment of pain characteristics should include information about (1) temporal characteristics (onset and duration), (2) course (stable, improving, worsening, or widely fluctuating), (3) severity (both average and worst), (4) location, (5) quality, and (6) provocative and palliative factors.

Acute pain usually has a well-defined onset and a readily identifiable cause (e.g., surgical incision). It may be associated with anxiety, overt pain behaviors (moaning or grimacing), and signs of sympathetic hyperactivity (including tachycardia, hypertension, and diaphoresis). In contrast, chronic pain is usually characterized by an ill-defined onset and association with persistent focus of pathology. The course of the pain may be linked to the disease and may change with the response to primary antineoplastic therapies. With chronic pain, overt pain behaviors and sympathetic hyperactivity are typically absent, and vegetative signs, including lassitude, sleep disturbance, and anorexia, may be present. A clinical depression evolves in some patients.

Most patients with chronic cancer pain also experience periodic flares of pain, or breakthrough pain (Haugen 2010). An important subtype of breakthrough pain is known as ▶ incident pain. Diagnosis of breakthrough pain may suggest specific interventions, such as the use of supplemental medication that specifically targets the cause.

Pain may be focal, multifocal, referred, or generalized. Pain may be referred from a lesion involving any of a large group of structures, including nerve, bone, muscle or other soft tissue, and viscera. Various subtypes of ▶ referred pain can be distinguished. Pain may be referred anywhere along the course of an injured peripheral nerve (such as pain in the thigh or knee from a lumbar plexus lesion) or nerve root (known as radicular pain). Pain may also be referred to a site remote from a nociceptive lesion and outside the dermatome affected by the lesion (e.g., shoulder pain from diaphragmatic irritation).

Measurement of pain intensity is an essential element in the pain assessment. Intensity may be monitored using a numeric scale (0–10), verbal rating scale (none, mild, moderate, severe), or a visual analogue scale. The particular scale used is less important than its regular use to record the status of the pain over time.

Interpreting the Assessment

The pain history, combined with the information gleaned from a physical examination and appropriate imaging and laboratory studies, usually provides sufficient data to generate meaningful conclusions concerning the etiology of the pain, its pathophysiology, and its defining syndrome. The latter constructs are very useful in clarifying the need for additional evaluation, prognosticating outcomes, and developing an efficient therapeutic strategy.

Etiology

The etiology of acute pain is usually clear-cut. Further evaluation to determine the underlying lesion is not indicated unless the course varies from the expected. In contrast, the etiology of chronic cancer-related pain may be more difficult to characterize. In most cases, pain is due to direct invasion of pain-sensitive structures by the neoplasm (Cherny 2007). The structures most often involved are bone and neural tissue, but pain can also occur when there is an obstruction of hollow viscus, distention of organ capsules, distortion or occlusion of blood vessels, and infiltration of soft tissues. In about one-quarter of patients, the etiology relates to an antineoplastic treatment, and fewer than 10 % have pain unrelated to the neoplasm or its treatment (Foley 1996). Many patients, particularly those with advanced illness, have multiple etiologies and several sources of pain.

Given the association between pain and the underlying neoplasm, clarification of the specific relationship between the pain and the disease is an essential element of the assessment. A survey of patients referred to a pain service in a major cancer hospital noted that previously unsuspected lesions were identified in 63 % of patients (Gonzalez et al. 1991). This outcome altered the known extent of disease in virtually all patients, changed the prognosis for some, and provided an opportunity for a primary antineoplastic therapy in approximately 15 %.

Pathophysiology

Although inferences about the predominating mechanisms underlying pain are certainly gross simplifications of very complex and dynamic processes, and can never be proved in the clinical setting, this classification has utility in treatment planning. Pain syndromes may be labeled nociceptive, neuropathic, psychogenic, and mixed (Merskey and Bogduk 1994).

The quality of somatic ▶ nociceptive pain is typically described as aching, stabbing, throbbing, or pressure-like. The quality of visceral nociceptive pain is usually gnawing or crampy when related to obstruction of hollow viscus, and aching or stabbing when associated with injury to other visceral structures.

Neuropathic mechanisms are involved in approximately 40 % of cancer pain syndromes and can be disease-related (e.g., tumor invasion of nerve plexus) or treatment-related (e.g., postmastectomy syndrome or chemotherapy-induced painful polyneuropathy) (Caraceni et al. 1999). Among those with metastatic disease, ▶ neuropathic pain usually results from neoplastic injury to peripheral nerves (peripheral neuropathic pain). Other, less common, subtypes include (1) those sustained by CNS processes (sometimes generically termed deafferentation pain) and (2) those in which the pain is believed to be maintained by efferent activity in the sympathetic nervous system (so-called sympathetically maintained pain) (Attal et al. 2008).

Neuropathic pain is diagnosed on the basis of the patient's verbal description of the pain, the findings on examination, and evidence of nerve injury. If patients describe pain in dysesthetic terms, such as "burning," "shock-like," or "electrical," this is suggestive of neuropathic mechanisms (Grond et al. 1999). Likewise, the existence of neuropathic pain may be inferred by the findings of abnormal sensations on physical examination, most notably ▶ allodynia and ▶ hyperalgesia.

Pain may be labeled psychogenic if there is positive evidence from the assessment that psychological factors predominate in sustaining the symptom. ▶ Psychogenic pain may be more formally characterized according to the widely accepted taxonomy of the American Psychiatric Association (one of the Somatoform Disorders described in the *Diagnostic and Statistical Manual*, 4th Revision). Although psychological factors commonly influence the expression and severity of pain in the cancer population, it is rare to identify a pain as psychogenic. If a credible pathophysiologic diagnosis is not apparent, and there is no strong evidence of a predominating psychological cause, it is best to label the pain idiopathic.

Assessment of Related Constructs

Chronic cancer-related pain is best addressed with consideration of a set of broader constructs, including suffering, quality of life, symptom distress, and the goals of care. These understandings continually inform decision-making about interventions to manage pain.

Pain and Suffering

Suffering, a global construct intricately related to the experience of pain (Fig. 1), has been described as a perceived threat to the integrity of the person, as a type of "total pain," or as overall impairment in quality of life (Cassell 1982). Suffering may be primarily related to symptoms such as pain or physical losses; to psychiatric disorders such as depression or psychological processes such as disturbed body image; to social issues such as loss of role functioning, familial disruption, or tension in the relationship with a significant other; or to spiritual concerns such as loss of faith. Other factors, such as financial concerns, may be prominent. Psychosocial and spiritual strengths and weaknesses that long predated the cancer may be important.

Comprehensive pain assessment must address issues related to suffering. An understanding of the physical, psychological, social, and spiritual issues important to the patient is fundamental to this process. A therapeutic approach focused only on pain may not meaningfully benefit a patient whose suffering is caused by other disturbances.

Pain and Quality of Life

Quality of life is related to the construct of suffering but has been more formally characterized for a research context. Numerous instruments have been created to measure quality of life, and these have been developed from the perspective of two major characteristics: subjectivity and multidimensionality (Owen et al. 2007).

The inherent subjectivity of quality of life has implications for assessment. Although clinicians commonly make inferences about patients'

Cancer Pain, Goals of a Comprehensive Assessment,
Fig. 1 Distinctions and interactions between nociception, pain, and suffering (Reproduced with permission from Portenoy RK (1992) Cancer pain: pathophysiology and syndromes. Lancet 339:1026)

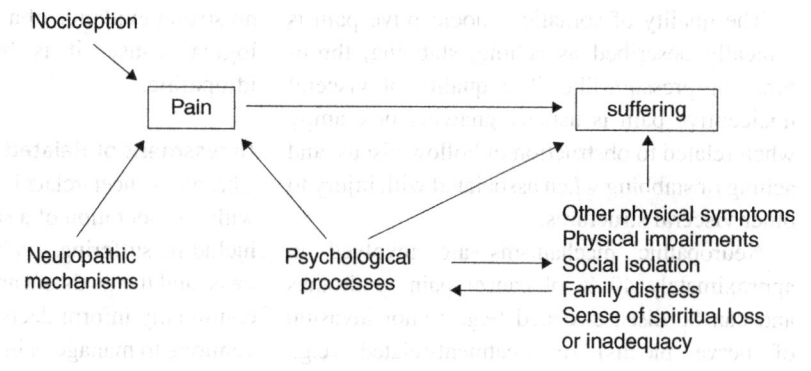

well-being, or rely on proxy evaluation by family or others, the likelihood of inaccuracy in such appraisals must be appreciated (Grossman et al. 2001). For the evaluation of quality of life, and each of its contributing concerns, including pain, the "gold standard" of assessment is self-report.

Like pain itself, quality of life is multidimensional. Although a single question can screen for overall well-being, a fuller understanding of the factors that must be addressed to improve quality of life requires probing of issues related to physical, psychological, social, and spiritual functioning.

Pain and Symptom Distress

Most patients with chronic cancer pain experience other symptoms concurrently. Studies have demonstrated that pain, fatigue, and psychological distress are most prevalent across populations with advanced cancer. Global symptom distress is a useful construct that characterizes overall symptom burden (Portenoy et al. 1994).

Goals of Care

In developing a therapeutic approach to address pain and other quality-of-life concerns in patients with cancer, it is essential to clarify the goals of care. At any point in time, treatment may be guided by one or more overriding goals: (1) to cure or prolong life, (2) to maintain function, and (3) to provide comfort. These goals are strongly influenced by many factors, including the status of the disease, the availability and burdens of therapy, psychosocial comorbidity, and the degree of spiritual or existential distress. Goals are very dynamic and often evolve over time with

the vagaries of the disease, the availability of treatment, and optimal palliative care. Current goals continually influence therapeutic decision-making.

References

Attal, N., Fermanian, C., Fermanian, J., Lanteri-Minet, M., Alkhaar, H., & Bouhassira, D. (2008). Neuropathic pain: Are these distinct subtypes depending on the aetiology or anatomical lesion? *Pain, 138*(2), 343–353.

Caraceni, A., Portenoy, R. K., & A Working Group of the IASP Task Force on Cancer Pain. (1999). An international survey of cancer pain characteristics and syndromes. *Pain, 82,* 263–274.

Cassell, E. J. (1982). The nature of suffering and the goals of medicine. *The New England Journal of Medicine, 306,* 639–645.

Cherny, N. I. (2007). Cancer pain: Principles of assessment and syndromes. In A. M. Berger, J. L. Shuster, & J. H. Von Roenn (Eds.), *Principles and practice of palliative care and supportive oncology* (3rd ed.). Philadelphia: Lippincott Williams &Wilkins.

Foley, K. M. (1996). Pain syndromes in patients with cancer. In R. K. Portenoy & R. M. Kranner (Eds.), *Pain management: Theory and practice* (p. 191). Philadelphia: FA Davis.

Gonzalez, G. R., Elliott, K. J., Portenoy, R. K., & Foley, K. M. (1991). The impact of a comprehensive evaluation in the management of cancer pain. *Pain, 47,* 141–144.

Grond, S., Radbruch, L., Meuser, T., et al. (1999). Assessment and treatment of neuropathic cancer pain following WHO guidelines. *Pain, 79,* 15–20.

Grossman, S. A., Sheidler, V. R., Sweden, K., et al. (2001). Correlation of patient and caregiver ratings of cancer pain. *Journal of Pain and Symptom Management, 6,* 53–57.

Haugen, D. F. (2010). Assessment and classification of cancer breakthrough pain: A systemic review. *Pain, 149*(3), 476–482.

Merskey, H., & Bogduk, N. (1994). *Classification of chronic pain: Descriptions of chronic pain syndromes and definitions of pain terms* (2nd ed.). Seattle: IASP Press.

Owen, J. E., Boxley, L., & Klapow, J. C. (2007). Measurement of quality of life outcomes. In Am Berger, J. L. Shuster, & J. H. Vpon Roenn (Eds.), *Principles and practice of palliative care and supportive oncology*. Philadelphia: Lippincott Williams & Wilkins.

Portenoy, R. K., Thaler, H. T., Kornblith, A. B., et al. (1994). Patients with cancer: The memorial symptom assessment scale: An instrument for the evaluation of symptom prevalence, characteristics and distress. *European Journal of Cancer, 30A*, 1326–1336.

Cancer Pain, Palliative Care in Children

Stephen C. Brown
Department of Anesthesia, The Hospital for Sick Children, Toronto, ON, Canada

Synonyms

Dying child; End-of-life care; Pain control in children with cancer; Palliative care in children

Definition

Pain control is an integral component of caring for all children with cancer. Children with cancer may experience many different types of pain from invasive procedures, the cumulative effects of toxic therapies, progressive disease, or from psychological factors. While the goal in cancer care is to minimize morbidity on the road to cure, at times even with the best of care, we fail to achieve this goal, and thus we also address the issue of palliative pain management. ▶ Palliative care is the management of patients with active, progressive, far-advanced disease, for whom prognosis is limited and the focus of care is the quality of life.

Characteristics

Pediatric cancer pain encompasses four broad etiologic categories – procedural pain, the cumulative effects of toxic therapies, progressive disease, and psychological factors. Children's cancer pain is often complex with multiple sources, comprised of nociceptive and neuropathic components (Collin and Weisman 2003; Brown and McGrath 2010; Miaskowski et al. 2005; World Health Organization 1998; American Academy of Pediatrics 2000). In addition, several ▶ situational factors usually contribute to children's pain, distress, and disability. Thus, to treat pain in these children adequately, we must evaluate the primary pain sources and also ascertain which situational factors are relevant for which children and families. Treatment emphasis should shift accordingly from an exclusive disease-centered framework to a more child-centered focus.

In 1993, the World Health Organization (WHO) and the International Association for the Study of Pain (IASP) invited experts in the fields of oncology, palliative care, anesthesiology, neurology, pediatrics, nursing, palliative care, psychiatry, psychology, and pastoral care to write guidelines for the management of pediatric cancer pain and palliative care, resulting in the 1998 publication "Cancer pain relief and palliative care in children."

A child-centered framework is required for understanding and controlling pain for these children. Pain control should include regular pain assessments, appropriate analgesics administered at regular dosing intervals, adjunctive drug therapy for symptom and side effects control, and nondrug interventions to modify the situational factors that can exacerbate pain and suffering. Parents concerned about medication centered on their children's comfort are calling for improved communication, standardization of nursing procedures and techniques, and a guide for a clear understanding of what to expect and from whom (Sobo et al. 2002). Basic information on pathophysiology, pharmacology, analgesic guidelines for children, and management of procedural pain is presented in other entries in this Encyclopedic Reference of Pain. This entry provides a complementary focus by describing the unique nature of children's cancer pain including the primary factors that affect their pain and quality of life, providing references to guidelines for selecting and administering drug therapy in accordance with the nociceptive and neuropathic components and recommending practical nondrug therapies for integration within a hospital, home, or hospice setting.

The Nature of a Child's Cancer Pain

Throughout the last decade, we have gained an increasing appreciation of the ▶ plasticity and complexity of a child's pain. As with adults, a child's pain is often initiated by tissue damage caused by noxious stimulation, but the consequent pain is neither simply nor directly related to the amount of tissue damage. Perhaps even more than in adults, differing pain responses to the same tissue damage are noted. The eventual pain evoked by a relatively constant noxious stimulus can be different depending on children's expectations, perceived control, or the significance that they attach to the pain. Children do not sustain tissue damage in an isolated manner, devoid of a particular context, but actively interpret the strength and quality of any pain sensations, determine the relevance of any hurting, and learn how to interpret the pain by observing the general environment, especially the behavior of other people. Children's perceptions of pain are defined by their age and cognitive level, their previous pain experiences against which they evaluate each new pain, the relevance of the pain or disease-causing pain, their expectations for obtaining eventual recovery and pain relief, and their ability to control the pain themselves. While plasticity and complexity are critical features for all pain perception, plasticity seems an even more important feature for controlling children's pain.

Much research has been conducted to identify the critical factors responsible for the plasticity of pain perception. PET and functional MRI studies have demonstrated that painful stimulation activates different cortical regions – depending on an individual's expectations and attention (Price and Bushnell 2004; Tracey and Bushnell 2009; Apkarian et al. 2005). Human studies evaluating the impact of environmental and psychological factors on the perception of experimentally induced pain have been conducted primarily in adults. However, results from the few laboratory studies conducted with children are consistent with those from adult studies (see ▶ Experimental Pain in Children). In addition, much compelling evidence about the powerful mediating role of psychological factors in children's pain derives from clinical studies of acute, recurrent, and chronic pain. These studies highlight the need to recognize and evaluate the mediating impact of these factors in order to optimally control children's pain.

Certain child factors are stable such as gender, temperament, and cultural background, while other factors change progressively such as age, cognitive level, previous pain experience, and family learning. These child characteristics shape how children generally interpret and experience the various sensations caused by tissue damage. In contrast, various cognitive, behavioral, and emotional factors are not stable. They represent a unique interaction between the child and the situation in which the pain is experienced. These situational factors can vary dynamically throughout the course of a child's illness, depending on the specific circumstances in which children experience pain. For example, a child receiving treatment for cancer will have repeated injections, central port access, and lumbar punctures – all of which may cause some pain (depending on the analgesics, anesthetics, or sedatives used). Even though the tissue damage from these procedures is the same each time, the particular set of situational factors for each treatment is unique for a child – depending on a child's (and parent's) expectations, a child's (and parent's and staff's) behaviors, and on a child's (and family's) emotional state. Although the causal relationship between an injury and a consequent pain seems direct and obvious, what children understand, what they do, and how they feel all affect their pain. Certain factors can intensify pain, exacerbate suffering, or affect adversely a child's quality of life. While parents and health care providers may be unable to change the more stable child characteristics, they can modify situational factors and dramatically improve children's pain and lives.

The Impact of Situational Factors on a Child's Pain

Cognitive factors include children's understanding about the pain source, their ability to control what will happen, their expectations regarding the quality and strength of pain sensations that they will experience, their primary focus of attention (that is distracted away from or focused primarily on the pain), and their knowledge of pain control strategies. In general, children's pain can be lessened by providing accurate age-appropriate information about pain, for example, emphasizing the specific sensations that children will experience (such as the stinging quality of an injection, rather than the general hurting aspects), by increasing their control and choices, by explaining the rationale for what can be done to reduce pain, and by teaching them some independent pain-reducing strategies (Brown and McGrath 2005). Behavioral factors refer to the specific behaviors of children, parents, and staff when children experience pain and also encompass parents' and children's wider behaviors in response to a course of repeated painful treatments, chronic pain, or progressive illness. Common behavioral factors include children's distress or coping reactions (e.g., crying, using a pain control strategy, withdrawing from life) and parents' and health staff's subsequent reactions to them (e.g., displaying frustration, calmly providing encouragement for children to use pain control strategies, engaging them in conversation and activities). Emotional factors include parents' and children's feelings in response to pain and to the daily effects of the underlying illness or condition. Children's emotions affect their ability to understand what is happening, their ability to cope positively, their behaviors, and ultimately their pain.

There are dynamic interactions among cognitive, behavioral, and emotional factors. Health staff can significantly lessen children's pain, not only by administering potent analgesics but also by increasing children's understanding and control, by decreasing their emotional distress and by teaching children some simple physical, behavioral, and cognitive methods to complement analgesic medications and further reduce their pain.

If the disease progresses and treatment emphasis shifts from curative to palliative, cognitive and emotional factors become the most salient situational factors that affect pain for children. Children probably have endured a prolonged period of intermittent pain, physical disability, and multiple aversive treatments. Children receiving curative therapies are more focused on the future consequences of their disease. Their thoughts, behaviors, and feelings change if their care becomes palliative. If this should happen, the type of support, information, and guidance children require changes. While the impact of palliation is profound for all children and families, each child and family is unique with respect to their specific psychological, medical, social, and spiritual needs. All families experience anguish and grief, but they may also experience denial, anxiety, anger, guilt, frustration, and depression. It is essential that health professionals listen attentively and observe carefully, not only to ensure that all the needs of both the child and family are met, but also to resolve the myriad factors that can exacerbate children's pain and suffering. The primary situational factors in pediatric palliative care are listed (see below).

Situational Factors in Pediatric Palliative Care

Cognitive Factors
- Meaning of death
- Inaccurate understanding: impact of situational factors on pain and quality of life
- Course of disease
- Palliative *versus* curative therapy
- Little independent control over pain
- Limited choices
- Expectation for continuing pain and suffering
- Misunderstanding of drug therapy: opioids
  - Dosing and administration
  - Criteria for evaluating effectiveness

Behavioral Factors
- Social withdrawal
- Physical inactivity
- Passive approach to pain control
- Secondary gains: stress reduction
  - Emotional denial
  - Parent or staff attention

- Inappropriate drug management: choice or mode of drug administration
- Failure to treat opioid-related side effects aggressively
- Failure to evaluate pain sources and document pain level
- Failure to use effective nondrug therapies

Emotional
- Anxiety about dying and death
  - Suffering
  - Meaning of life
- Fear of separation
  - Inadequate pain control
  - Increasing adverse symptoms
  - Impact on family
- Anger
- Sadness or depression
- Distancing by staff and friends

A shift in care from curative to palliative therapies may signify to some children and families that health professionals are giving up on the child. Children and families must understand that stopping ineffective therapies is not giving up but represents a rational decision based on children's best interests. Pain control is an essential component of cancer care and of palliative care (Chaffee 2001; Galloway and Yaster 2000; Goldman 1998; Goldman et al. 2003; Gregoire and Frager 2006). Children and parents should not fear that health professionals have given up on controlling pain and aversive symptoms. Pain and all symptoms must be treated aggressively from the dual perspective of targeting the primary source of tissue damage and modifying the secondary contributing factors. Although most children receive accurate information about their disease and required treatments, few children or their parents receive concrete information about their pain, the factors that can attenuate or exacerbate it, a rationale for the interventions they receive, and training in effective nonpharmacological pain control techniques. The latter may be particularly important for children in palliative care, who have diminishing control in their lives. Children and their parents do not know that prescribed pain control treatments may vary in efficacy at different times throughout their treatment due to variations in disease activity or situational factors. Without this knowledge, their confidence in certain therapies can decrease, even though these therapies would effectively alleviate pain at another time. The fear of inadequate pain control places an enormous emotional burden on an already distressed child and family and can create a situation in which their pain and disability intensifies.

Children seem to know intuitively, even when dying has not been discussed directly with them. They fear separation and abandonment; some children may fear that their illness is a punishment. Dying children may feel frightened, isolated, and guilty unless they are able to openly express and resolve their concerns. Many observers have noted that children who are dying have a level of maturity far beyond their years. It is essential to acknowledge and resolve their fears. Children should receive accurate information, consistent with their religious beliefs, presented in a calm reassuring manner. They may need concrete reassurance that they will not suffer when they die, that they will not be alone, and that their families will remember them. Unresolved emotions add anguish and may intensify their pain.

Treating a Child's Pain

A treatment algorithm illustrating a practical approach for treating a child's pain is illustrated in Fig. 1. Pain assessment is an integral component of diagnosis and treatment. The differential diagnosis of a child's pain is a dynamic process that guides clinical management. We should select specific therapies to target the responsible central and peripheral mechanisms and to mitigate the pain-exacerbating impact of situational factors, recognizing that the multiple causes and contributing factors will vary over time. Drug therapies – analgesics, adjuvant analgesics, and anesthetics – are essential for pain control, but nondrug therapies – cognitive, physical, and behavioral – are also essential. As we monitor the child's improvement in response to the therapies initiated, we refine our pain diagnosis and treatment plan accordingly. Drug therapies are covered in the entry "▶ Analgesic Guidelines

Cancer Pain, Palliative Care in Children, Fig. 1 Treatment algorithm for pain management

1. Evaluate the Child with Pain

- Assess sensory characteristics of pain
- Conduct medical examination and appropriate diagnostic tests
- Evaluate probable involvement of nociceptive and neuropathic mechanisms
- Appraise situational factors contributing to child's pain

2. Diagnose the Primary and Secondary Causes

- Current nociceptive and neuropathic components
- Attenuating physical symptoms
- Relevance of key cognitive, behavioural, and emotional factors

3. Select Appropriate Therapies

Drugs
- Analgesics
- Adjunct analgesics
- Anesthetics

Non-Drugs
- Psychological
- Physical
- Behavioural

4. Implement Pain Management Plan

- Provide feedback on causes and contributing factors to parents (and child)
- Provide rationale for integrated treatment plan
- Measure child's pain regularly
- Evaluate effectiveness of treatment plan
- Revise plan as needed

for Infants and Children"; anesthetic blocks are covered in the entry "▶ Acute Pain in Children, Postoperative"; and acute procedural pain is covered in the entry by Cohen.

The optimal relief of pain in pediatric palliative care begins with the recognition that you are assessing and treating an individual child with pain, not managing pain as a symptom apart from the child. A thorough medical history, physical examination, and assessment of pain characteristics and contributing factors are necessary to establish a correct clinical diagnosis. Specific interventions should be selected and administered to children as part of a comprehensive pain program, in the same manner as the most appropriate analgesics are selected and administered in adequate doses, at regular dosing intervals, through the most efficient routes.

Children seem to possess an enhanced ability to absorb themselves completely in a task, game, or imagined event and thus might be more able than adults to trigger endogenous pain-inhibitory mechanisms. Even very young children can easily learn to use a variety of practical pain control methods to lessen their anxiety, distress, and pain. The specific methods depend on the age of the child, the type of pain experienced, and the resources available. Deep breathing, alternately tightening and relaxing their fists, squeezing their mother's hand, listening to stories or music, and imagining that they are in a pleasant setting can be very effective for reducing procedural-related pain, when used with appropriate analgesics. Families should understand that what they think, how they behave, and how they feel affect their children's pain. Then they can begin to work independently and with staff to create additional nondrug pain control methods based on the child's interest, the cultural setting, and the availability of resources.

Summary

Optimal pain control for children with cancer and for children receiving palliative care requires an integrated treatment plan with both drug and nondrug therapies (Frager 1997; Hooke et al. 2002; Liben et al. 2008; Brown and McGrath 2010). Comprehensive care includes curative therapies when available, pain and symptom management, and compassionate support for children and their families. It is essential to focus not only on the medical management of children's disease but also on the psychosocial and spiritual factors that affect children's pain and suffering.

Specific interventions must be selected after determination of the primary and secondary sources of noxious stimulation and after a thorough assessment of the unique situational, behavioral, emotional, and familial factors that affect a child's pain. All analgesics should be selected "by the ladder" and administered "by the clock," "by the child," and by an effective and painless route. Dosing intervals should be frequent enough to control pain adequately so that children do not experience an alternating cycle of pain, drowsy analgesia, pain, etc. Special problems in palliative pain control may arise when children die at home, unless parents and medical and nursing teams communicate openly about the availability of potent analgesics and the flexibility of dosing routes and regimens. The choice for pain control is not merely "drug versus nondrug therapy" but rather a therapy that mitigates both the causative and contributing factors for pain. We have the knowledge and thus the obligation to ensure that children receive adequate pain control from the time that they are diagnosed with cancer throughout their treatment protocol, including those cases that proceed to palliative care.

References

American Academy of Pediatrics. (2000). American Academy of Pediatric's Committee on Bioethics and Committee on Hospital Care. Palliative care for children. *Pediatrics, 106*, 351–357 (Reaffirmed February 2012).

Apkarian, A. V., Bushnell, M. C., Treede, R. D., et al. (2005). Human brain mechanisms of pain perception and regulation in health and disease. *European Journal of Pain, 9*, 463–484.

Brown, S. C., & McGrath, P. A. (2005). Pain in children. In M. Pappagallo (Ed.), *The neurological basis of pain*. New York: McGraw-Hill.

Brown, S. C., & McGrath, P. A. (2010). Paediatric palliative medicine – Pain control. In G. Hanks, N. Cherny, et al. (Eds.), *Oxford textbook of palliative medicine* (4th ed., pp. 1318–1336). Oxford: Oxford University Press.

Chaffee, S. (2001). Pediatric palliative care. *Primary Care, 28*, 365–390.

Collin, J. J., & Weisman, S. J. (2003). Management of pain in childhood cancer. In N. L. Schechter, C. B. Berde, & M. Yaster (Eds.), *Pain in infants, children, and adolescents* (pp. 517–538). Philadelphia: Lippincott Williams & Wilkins.

Frager, G. (1997). Palliative care and terminal care of children. *Child and Adolescent Psychiatric Clinics of North America, 6*, 889–909.

Galloway, K. S., & Yaster, M. (2000). Pain and symptom control in terminally ill children. *Pediatric Clinics of North America, 47*, 711–747.

Goldman, A. (1998). ABC of palliative care. Special problems of children. *BMJ, 316*, 49–52.

Goldman, A., Frager, G., & Pomietto, M. (2003). Pain and palliative care. In N. L. Schechter, C. B. Berde, & M. Yaster (Eds.), *Pain in infants, children, and adolescents* (pp. 539–562). Philadelphia: Lippincott Williams & Wilkins.

Gregoire, M.-C., & Frager, G. (2006). Ensuring pain relief for children at the end of life. *Pain Research & Management, 11*, 163–171.

Hooke, C., Hellsten, M. B., Stutzer, C., et al. (2002). Pain management for the child with cancer in end-of-life care: APON position paper. *Journal of Pediatric Oncology Nursing, 19*, 43–47.

Liben, S., Papadatou, D., & Wolfe, J. (2008). Paediatric palliative care: Challenges and emerging ideas. *Lancet, 371*, 852–864.

Miaskowski, C., Cleary, J., Burney, J., et al. (2005). *Guideline for the management of cancer pain in adults and children. APS Clinical Practice Guidelines Series, No. 3*. Glenview: American Pain Society.

Price, D. D., & Bushnell, M. C. (2004). *Psychological modulation of pain: Integrating basic science and clinical perspectives*. Seattle: IASP Press.

Sobo, E. J., Billman, G., Lim, L., et al. (2002). A rapid interview protocol supporting patient-centered quality improvement: Hearing the parent's voice in a pediatric cancer unit. *The Joint Commission Journal on Quality Improvement, 28*, 498–509.

Tracey, I., & Bushnell, M. C. (2009). How neuroimaging studies have changed us to rethink: Is chronic pain a disease? *The Journal of Pain, 10*, 1113–1120.

World Health Organization. (1998). *Cancer pain relief and palliative care in children*. Geneva: World Health Organization.

Cancer Survivorship

Definition

Sometimes defined as the time extending from the moment of diagnosis forward; alternatively defined as the period after cancer is in remission or cured, or converted into a long-term chronic illness.

Cross-References

▶ Cancer Pain, Assessment in Children

Cancer Therapy

Definition

Refers to anticancer chemotherapy, radiation treatment, surgery, or endocrine/hormonal treatment to cure or control cancer and its progression.

Cross-References

▶ Cancer Pain Management
▶ Psychiatric Aspects of the Management of Cancer Pain

Cancer-related Bone Pain

▶ Adjuvant Analgesics in Management of Cancer-Related Bone Pain

Cancer-related Pain

▶ Cancer Pain and Pain in HIV/AIDS

Cannabinoid

Definition

Cannabinoids are derivatives of Δ9-tetrahydrocannabinol (Δ9-THC), the constituent of marijuana that is responsible for its psychoactive effects. Cannabinoids can be synthetic molecules that act through cannabinoid receptors.

Cross-References

▶ Evoked and Movement-Related Neuropathic Pain

Cannabinoid Receptors

Definition

Receptors that are activated by the active constituent of *Cannabis sativa*, Δ9-tetrahydrocannabinol (Δ9-THC). These receptors can also be activated by endogenous ligands such as anandamide (so-called endocannabinoids). Two types of receptor have been identified, CB1 and CB2, which are G-protein coupled. CB1 receptors are found in several brain areas.

Capacity

Definition

An individual's ability to execute a task or an action, the highest probable level of functioning that a person may reach in a given domain, at a given moment.

Cross-References

▶ Disability, Functional Capacity Evaluations

Capitated Care

Definition

A provider receives a set fee for each patient assigned to the practice through an insurance carrier. The fee is often called a per-member-per-month, and is independent of the care the patient requires.

Cross-References

▶ Disability Management in Managed Care System

Capsaicin

Definition

The pungent ingredient of chili peppers (8-methyl N-vanillyl 6-nonenamide) from the *capsicum* family, which can be used to selectively activate nociceptive sensory neurons via its activation of a cation channel TRPV1 (from the family of "transient receptor potential," TRP channels). This channel can also be activated by heat and the low pH of inflamed tissues. Not all nociceptive afferents express the TRPV1, but this subgroup of nociceptors often also produces neuropeptides and is responsible for the neurogenic inflammation.

Whereas most of the mammalian species are sensitive to capsaicin, birds have a variant of the TRPV1 receptor which cannot be activated by capsaicin.

Capsaicin has been applied to the skin in order to treat neuropathic pain, such as postherpetic neuralgia. The therapeutic mechanism is supposed to be a desensitization of irritated nociceptors.

Cross-References

▶ Amygdala, Pain Processing and Behavior in Animals
▶ Atypical Facial Pain: Etiology, Pathogenesis, and Management

▶ Exogenous Muscle Pain
▶ Freezing Model of Cutaneous Hyperalgesia
▶ Human Thalamic Response to Experimental Pain (Neuroimaging)
▶ Inflammation, Modulation by Peripheral Cannabinoid Receptors
▶ Mechano-Insensitive C-Fibers, Biophysics
▶ Muscle Pain Model, Ischemia-Induced and Hypertonic Saline-Induced
▶ Nociceptors in the Orofacial Region (Skin/Mucosa)
▶ Opioid Modulation of Nociceptive Afferents In Vivo
▶ PET and fMRI Imaging in Parietal Cortex (SI, SII, Inferior Parietal Cortex BA40)
▶ Polymodal Nociceptors, Heat Transduction
▶ Sensitization of Muscular and Articular Nociceptors
▶ Species Differences in Skin Nociception
▶ Spinothalamic Tract Neurons, Glutamatergic Input
▶ Spinothalamic Tract Neurons, Role of Nitric Oxide
▶ Sympathetically maintained Pain and Inflammation, Human Experimentation
▶ Toxic Neuropathies
▶ TRPV1 Modulation by p2Y Receptors
▶ TRPV1, Regulation by Nerve Growth Factor
▶ TRPV1, Regulation by Protons

Capsaicin Receptor

Antonio Vicente Ferrer-Montiel
Institute of Molecular and Cell Biology, University Miguel Hernández, Elche, Alicante, Spain

Synonyms

Heat sensor; TRPV1; Vanilloid receptor subunit 1

Definition

The capsaicin receptor is a calcium permeable nonselective cation channel that is gated by noxious heat ($\geq$ 42 °C), voltage, vanilloid compounds such as capsaicin and protons. In primary sensory neurons, the capsaicin receptor transduces noxious chemical and thermal stimuli into action potentials at the peripheral endings. Thus, the capsaicin receptor is a key molecular component of the pain pathway.

Characteristics

The capsaicin receptor is a member of the transient receptor potential (TRP) mammalian gene superfamily (Caterina and Julius 2001). These channels are considered molecular gateways in sensory systems, since several of these proteins transduce chemical and physical stimuli into neuronal activity, i.e., membrane potential changes. The capsaicin receptor gave its name to the vanilloid subfamily (TRPV) of TRP channels. This sensory receptor is an integrator of noxious thermal stimuli as well as of irritant chemicals such as vanilloids, protons, and pro-algesic substances (Caterina and Julius 2001; Szallasi and Appendino 2004). Temperature gates the channel by shifting the voltage-dependent activation toward the neuronal resting potential (Voets et al. 2004). In contrast, chemical activators of the receptor such as capsaicinoid and vanilloid-like molecules and acidosis act as gating modifiers, reducing the temperature threshold of channel gating from 42 °C to below body temperature (36 °C). This is the underlying mechanism for the characteristic pungency of capsaicin and related molecules.

Molecularly, the functional channel is a tetrameric membrane protein with four identical subunits assembled around a central aqueous pore. Each receptor subunit displays a membrane domain composed of six transmembrane segments (S1-S6) with an ► amphipathic region between the fifth and sixth segment that forms the channel conductive pore (Caterina and Julius 2001; Ferrer-Montiel et al. 2004). The protein also has cytoplasmic N- and C-termini (Fig. 1). In the N-terminus, TRPV1 channels exhibit three ankyrin domains that mediate protein-protein interactions with cytosolic proteins and consensus sequences for protein kinases. The protein displays a cytosolic C-terminus domain containing phosphoinositide, and calmodulin-binding (CAM) domains, as well as phosphorylation sites (Caterina and Julius 2001; Ferrer-Montiel et al. 2004; Nagy et al. 2004). In addition, the C-end has a TRP-like motif that functions as an association domain for receptor subunits (Garcia-Sanz et al. 2004). The receptor has two abundant single nucleotide ► polymorphisms that produce amino acid substitutions, one in codon 315 (Met^{315} Ile) at the N-terminus domain and the other at amino acid 585 (Ile^{585} Val) located in the fifth transmembrane segment. Interestingly, gender, ethnicity, and temperament seem to contribute to individual variation in thermal and cold pain sensitivity by interactions in part with these TRPV1 single nucleotide polymorphisms (Kim et al. 2004).

The capsaicin receptor is widely expressed in neuronal and non-neuronal cells of both endodermal and mesodermal origin, implying that the receptor is involved in diverse physiological functions. These include thermosensory transduction, as well as chemical signaling presumably mediated by ► endovanilloid compounds. In situ hybridization, immunocytochemical analysis, and drug-binding assays have shown TRPV1 expression in $\approx$ 50 % of dorsal root and trigeminal ganglion neurons, in the dorsal horn of the spinal cord and in the caudal nucleus of the spinal trigeminal complex. The majority of TRPV1 positive neurons also colocalize with the nerve growth factor (NGF) receptor trkA, the lectin IB4, and the neuropeptides involved in nociceptive transmission such as substance P (SP) and calcitonin gene-related peptide (CGRP) (Caterina and Julius 2001; Nagy et al. 2004; Szallasi and Appendino 2004). Vanilloid sensitive nociceptors are peptidergic, small diameter neurons that give rise to unmyelinated C fibers, although some Aδ fibers are responsive to vanilloid derivatives (Nagy et al. 2004; Szallasi and Appendino 2004). Somatic and visceral primary afferents express TRPV1 at both the spinal and peripheral terminals. In addition to a subset of nociceptors, the capsaicin receptor is present in neurons of the central nervous system and in non-neuronal cells.

Capsaicin Receptor, Fig. 1 Molecular model of the capsaicin receptor subunit. The figure displays a membrane domain composed of six transmembrane segments (S1–S6) with an *amphipathic* region between the fifth and sixth segment that forms the channel conductive pore, which also contains the glutamic acids (E600 and E648) responsible for the pH-induced channel gating. The protein also has cytoplasmic N- and C-termini. In the N-terminus, TRPV1 channels exhibit three ankyrin domains and consensus sequences for PKA and Src.

The capsaicin-binding site is located at the N-end of the S3 segment, near S502, a serine residue phosphorylated by PKA, PKC, and CaMKII. The protein displays a cytosolic C-terminus domain containing phosphoinositide (PIP₂) and calmodulin-binding (CAM) domains and phosphorylation sites for PKC (T704, S800) and CaMKII (T704). In addition, the C-end has a TRP domain that may contribute to the association of receptor subunits. The higher Ca²⁺ permeability with respect to Na⁺ and K⁺ is also depicted.

For instance, TRPV1 mRNA or protein is widely expressed in brain regions such as the olfactory nuclei, cerebral cortex, dentate gyrus, central amygdala, striatum, centromedian and paraventricular thalamic nuclei, hypothalamus, substantia nigra, reticular formation, locus coeruleus, inferior olive, and cerebellar cortex (Nagy et al. 2004). Furthermore, the receptor is expressed in a variety of epithelial tissues such as the skin, human hair follicles, lungs, uroepithelium of the urinary bladder, the vascular system, and the inner ear (Nagy et al. 2004). Taken together, all these observations underscore the notion that TRPV1 is a widely expressed protein whose function is critical for diverse physiological conditions.

The pivotal role of TRPV1 in nociceptive transduction has suggested a contribution of the channel to diverse pathophysiological processes. In particular, cumulative evidence is substantiating the tenet that ▶ nociceptor sensitisation by pro-inflammatory agents is primarily achieved by the capsaicin receptor. This protein is the endpoint target of intracellular signaling cascades triggered by inflammatory mediators that lead to remarkable potentiation of its channel activity which, in turn, promotes the hyperexcitability of nociceptors (Planells-Cases et al. 2005; Julius and Basbaum 2001; Davis et al. 2000). Enhancement of TRPV1 function by pro-algesic agents may be accomplished either by direct activation of the channel or by its posttranslational modification by intracellular metabolic cascades (Planells-Cases et al. 2005; Julius and Basbaum 2001). Direct activation of TRPV1 responses has been reported for lipid mediators such as arachidonic acid metabolites including anandamide, N-arachidonyl-dopamine (NADA), and N-oleyldopamine. Similarly, several eicosanoids, particularly those derived from the enzymatic action of 5-lipoxygenase or 12-lipoxygenase, are capable of activating TRPV1. In particular, 12-(hyperoxy)eicosatetraenoic acid (12-HPETE) and leukotriene B_4 (LTB_4) have exhibited the most potent agonistic activity. In addition, the acidosis that develops in inflamed tissues is also a direct activator of the TRPV1 channel activity (van der Stelt and Di Marzo 2004). The potency and efficacy of each singular mediator is quite low but in inflammatory conditions several of these modulators are simultaneously released and act synergistically. Therefore, direct gating of TRPV1 responses by inflammatory agents acting as channel agonists notably increases the excitability of nociceptors, resulting in a hyperalgesic condition. Prolonged activation of the receptor, leads to an intracellular $[Ca^{2+}]$ rise that, in turn, activates intracellular signaling that triggers the release of pro-inflammatory agents at peripheral terminals. This dual action further increases the excitability of the nociceptors. In addition, inflammation-evoked activation of intracellular protein networks results in TRPV1 phosphorylation, release of tonically inhibited receptors and an increment in the surface expression of functional channels, all being major events underlying the nociceptor activation and sensitization that leads to ▶ hyperalgesia (Planells-Cases et al. 2005). Indeed, TRPV1 expression is upregulated in tissue samples from patients with inflammatory bowel disease and Crohn's disease and also in patients with rectal hypersensitivity, as well as in those affected by vulvodynia (see ref. in Nagy et al. 2004; Szallasi and Appendino 2004; Planells-Cases et al. 2005). Thus, TRPV1 receptors are critical determinants of the sensitization of primary afferents after injury or inflammation. The involvement of TRPV1 in heat hypersensitivity is underscored by the reduced thermal hyperalgesia of TRPV1 null mice (Caterina et al. 2000), and by the attenuation of this inflammatory associated phenomenon by noncompetitive antagonists of the TRPV1 channel (García-Martínez et al. 2002). Accordingly, the important contribution of TRPV1 receptor to the onset and maintenance of neurogenic inflammation has validated it as a therapeutic target for inflammatory pain management and a tremendous effort is being carried out to develop clinically useful modulators of the receptor dysfunction characteristic of human diseases (Szallasi and Blumberg 1999).

Cross-References

▶ Polymodal Nociceptors, Heat Transduction
▶ TRPV1 Receptor, Species Variability
▶ TRPV1, Regulation by Nerve Growth Factor
▶ TRPV1, Regulation by Protons

References

Caterina, M. J., & Julius, D. (2001). The vanilloid receptor: A molecular gateway to the pain pathway. *Annual Review of Neuroscience, 24*, 487–517.

Caterina, M. J., Leffler, A., Malmberg, A. B., et al. (2000). Impaired nociception and pain sensation in mice lacking the capsaicin receptor. *Science, 288*, 306–313.

Davis, J. B., Gray, J., Gunthorpe, M. J., et al. (2000). Vanilloid receptor-1 is essential for inflammatory thermal hyperalgesia. *Nature, 405*, 183–187.

Ferrer-Montiel, A., Garcia-Martinez, C., Morenilla-Palao, C., et al. (2004). Molecular architecture of the vanilloid receptor. Insights for drug design. *European Journal of Biochemistry, 271*, 1820–1826.

García-Martínez, C., Humet, M., Planells-Cases, R., et al. (2002). Attenuation of thermal nociception and hyperalgesia by VR1 blockers. *Proceedings of the National Academy of Sciences of the United States of America, 99*, 2374–2379.

Garcia-Sanz, N., Ferández-Carvajal, A., Morenilla-Palao, C., et al. (2004). Identification of tetramerization domain in the C-terminus of the vanilloid receptor. *Journal of Neuroscience, 24*, 5306–5314.

Julius, D., & Basbaum, A. I. (2001). Molecular mechanisms of nociception. *Nature, 413*, 203–210.

Kim, H., Neubert, J. K., San Miguel, A., et al. (2004). Genetic influence on variability in human acute experimental pain sensitivity associated with gender, ethnicity and psychological temperament. *Pain, 109*, 488–496.

Nagy, I., Sántha, P., Jancsó, G., et al. (2004). The role of the vanilloid (capsaicin) receptor (TRPV1) in physiology and pathology. *European Journal of Pharmacology, 500*, 351–369.

Planells-Cases, R., García-Sanz, N., Morenilla-Palao, C., et al. (2005). Functional aspects and mechanisms of TRPV1 involvement in neurogenic inflammation that leads to thermal hyperalgesia. *Pflugers Archive European Journal of Physiology, 451*(1), 151–159.

Szallasi, A., & Appendino, G. (2004). Vanilloid receptor TRPV1 antagonists as the next generation of painkillers. Are we putting the cart before the horse? *Journal of Medicinal Chemistry, 47*, 1–7.

Szallasi, A., & Blumberg, P. M. (1999). Vanilloid (capsaicin) receptors and mechanisms. *Pharmacological Reviews, 51*, 159–211.

van der Stelt, M., & Di Marzo, V. (2004). Endovanilloids. Putative endogenous ligands of transient receptor potential vanilloid 1 channels. *European Journal of Biochemistry, 271*, 1827–1834.

Voets, T., Droogmans, G., Wissenbach, U., et al. (2004). The principle of temperature-dependent gating in cold- and heat-sensitive TRP channels. *Nature, 430*, 748–754.

Capsaicin Receptor, Regulation by NGF

▶ TRPV1, Regulation by Nerve Growth Factor

Capsaicin Receptor, Regulation by Protons

▶ TRPV1, Regulation by Protons

Capsaicin Receptors

▶ Polymodal Nociceptors, Heat Transduction

Capsazepine

Definition

Capsazepine is a synthetic analogue of capsaicin. It binds to the TRPV1 receptor and inhibits the activation of the neuron by capsaicin.

Cross-References

▶ Capsaicin Receptors
▶ TRPV1, Regulation by Protons

Carbenoxolone

Definition

Carbenoxolone is a gap junction decoupler. When administered onto the spinal cord, it disrupts gap junctions between. In addition to decoupling gap junctions, carbenoxolone can also exert nonspecific effects, including inhibition of one beta-hydroxysteroid dehydrogenase, at higher doses. Peri-spinal administration of carbenoxolone blocks mirror-image pain.

Cross-References

▶ Cord Glial Activation

Cardiac Stress Response

▶ Postoperative Pain, Pathophysiological Changes in Cardiovascular Function in Response to Acute Pain

Cardiac Surgery

Definition

All cardiac surgical procedures performed with or without cardiopulmonary bypass.

Cross-References

▶ Postoperative Pain, Thoracic and Cardiac Surgery

Career Assessment

▶ Vocational Assessment in Chronic Pain

Caregiver

Definition

Any person who assesses and provides care to the individual experiencing pain (health care professionals, family member, or friend). As it pertains to the newborn infants, the maternal caregiving relationship during infancy is considered a key mediator and moderator of risk factors on infant development.

Cross-References

▶ Pain Assessment in Neonates

Carotid Arteries

Definition

The common carotid artery divides into the internal and external carotid arteries in the neck. The former supplies the forebrain with blood, and the latter supplies the face and scalp.

Cross-References

▶ Primary Cough Headache

Carotidynia

Definition

A syndrome of pain in the distribution of the carotid artery in the neck due to any cause.

Cross-References

▶ Headache Due to Dissection

Carpal Tunnel Syndrome

A. Lee Dellon
Department of Plastic Surgery and Neurosurgery,
Johns Hopkins University, Baltimore, USA
Dellon Institutes for Peripheral Nerve Surgery,
Baltimore, MD, USA

Synonyms

Entrapment neuropathies, carpal tunnel syndrome; Median nerve compression at the wrist

Definition

▶ Carpal tunnel syndrome is a combination of patient complaints related to chronic compression of the median nerve within the carpal tunnel. Since the carpal tunnel is at the wrist, the painful symptoms of which the patient complains are related to the small and large myelinated nerve fibers that supply the palmar, but not thenar, skin along an axis that is radial to the longitudinal axis of the ring finger and the distal dorsal tips of the index and middle finger. The thenar skin is innervated by the palmar cutaneous branch of the median nerve, which arises 5 cm proximal to the wrist, and therefore abnormal sensibility of the thenar skin is not included in the definition of carpal tunnel syndrome. Motor symptoms include complaints of clumsiness in the use of the thumb, as the median nerve's motor branch innervates the abductor pollicis brevis, the opponens pollicis, and the short head of the flexor pollicis brevis. When the wrist is flexed, pressure increases upon the median nerve, causing decreased blood flow within the median nerve, and the resulting ischemia causes transmission of neural impulses interpreted as numbness or tingling (paresthesia) and, in some patients, actually as pain, causing them to awaken at night. Acute onset of pain in the distribution of the median nerve is not included in the definition of carpal tunnel syndrome and indicates acute compression with axonal loss rather than chronic compression and demyelination.

Characteristics

Carpal tunnel syndrome is the most common example of chronic nerve compression, with a prevalence of at least 2 % in the general population, 14 % in diabetics without peripheral neuropathy, and 30 % in diabetics with peripheral neuropathy (Perkins et al. 2002). The physical examination findings required to confirm a diagnosis in the first patient reported to have a carpal decompression were blisters on the tips of the thumb, index, and middle finger, due to anesthesia, and wasting of the thenar muscles (Woltman 1941). The patient was an acromegalic with hypertrophic neuropathy as the cause of the compression of the median nerve, and James Learmonth, a neurosurgeon at the Mayo Clinic, divided the transverse carpal ligament. In 1950, George Phalen, a surgeon in Cleveland, described a series of patients with carpal tunnel syndrome, which has become the classic description. Today a history of numbness or paresthesias in the thumb and index finger, associated with nighttime awakening, has become the classic history. The classic physical examination findings are a positive Phalen sign (symptoms provoked with wrist flexion for 1 min) or a positive Tinel sign (distally radiating sensory phenomenon when the median nerve is tapped at the wrist) (Mackinnon and Dellon 1988). It is unusual today for a patient to have sufficient chronicity of symptoms or severity of compression to have thenar muscle wasting at the time or presentation.

Documentation of carpal tunnel syndrome requires either ▶ electrodiagnostic testing (EDT) or ▶ neurosensory testing (NST). EDT is objective, but remains with significant number of false negative findings, such that a meta-analysis done by the American Neurologic Association (AAEM 1993) found that just 66 % of patients using clinical symptoms and findings as the gold

standard had positive EDT. In contrast, quantitative sensory testing has been demonstrated to be valid, reliable, to correlate well with patient symptoms, and to be painless (Arezzo et al. 1993). Thermal threshold testing documents the presence of small nerve fiber pathology and does not become abnormal until late in the pathology of carpal tunnel syndrome. Vibratory threshold testing documents the presence of large fiber pathology, but, because the stimulus is a wave, it can give ambiguous information when used to stimulate the index finger or the thumb: these fingers are innervated by both the radial sensory and the median nerve, both of which will be stimulated by the waveform. NST is a form of quantitative sensory testing that measures the cutaneous pressure threshold for one- and two-point moving and static-touch. This testing found a clear distinction between the 99 % upper confidence limit in the normal age-matched population and those patients with even a mild degree of carpal tunnel syndrome (Dellon and Keller 1997), suggesting that this methodology would have a high sensitivity in evaluating patients with chronic nerve compression. This was confirmed in a blinded, prospective study evaluating the ability of EDT versus NST with the Pressure-Specified Sensory Device to identify patients with carpal tunnel syndrome from an asymptomatic population (Weber et al. 2000). That study found that the sensitivity of NST versus EDT was 92 % versus 81 %, and the specificity of NST versus EDT was 82 % versus 77 %. Since NST is painless, the patient is willing to have repeated studies in those clinical situations in which they may be necessary, e.g., when nonoperative treatment is first prescribed or there is workplace environment modification (ergonomics), treatment failure, or evaluation of impairment for a disability rating.

Another indication for NST is the differential diagnosis of failure to improve following carpal tunnel decompression, in which a proximal source of median nerve compression, such as the pronator syndrome, is considered. EDT has a high false negative percentage with this nerve compression, with up to 75 % false negative in many studies, whereas NST, by measuring the cutaneous pressure threshold of the thenar eminence, can document this proximal site of compression (Rosenberg et al. 2001). Evaluation of the cutaneous pressure threshold over the dorsoradial aspect of the hand can identify the presence of radial sensory nerve entrapment, another nerve compression to be considered in the differential diagnosis of "failed carpal tunnel decompression." In those patients in whom a C6 radiculopathy is being considered in the differential diagnosis, the electromyographic component of EDT is still the critical method required for documentation.

The treatment of carpal tunnel syndrome is clearly defined based upon staging the degree of compression (Dellon 2001). Measurement of peripheral nerve function permits a numerical grading system which incorporates both the motor and sensory systems and can be applied to other upper extremity nerve compressions like the ulnar nerve at the elbow, cubital tunnel syndrome, or the lower extremity tibial nerve compression, the tarsal tunnel syndrome (Aszmann et al. 2002). This numerical grading system permits statistical comparison of treatment options for the wide range of clinical symptoms and findings usually seen in carpal tunnel syndrome or other nerve entrapment syndromes. For the mild degrees of compression, nonoperative treatment is appropriate with splinting of the wrist in a neutral position, nonsteroidal anti-inflammatory medication, change of activities of daily living, and cortisone injection. Failure of this approach, manifested by persistence or progression of symptoms, is an indication for surgical decompression of the median nerve at the wrist. Another indication for surgery is if more severe degrees of compression are present initially, such as the axonal loss associated with abnormal two-point discrimination or muscle atrophy.

Randomized prospective studies (Gerritsen et al. 2001; Trumble et al. 2002), in general, agree that 80–90 % of patients can achieve good to excellent symptomatic relief through either a traditional open or the newer endoscopic approach. In experienced hands, the complications for each procedure are similar. These complications include failure of the procedure to achieve the desired result, injury to the median nerve or its branches, and a painful incision. For certain occupations, the endoscopic approach appears to permit a slightly earlier return to work.

Based upon preoperative staging, a routine intraoperative internal neurolysis does not improve results (Mackinnon et al. 1991); however, adding an internal microsurgical neurolysis based upon intraoperative pathology has not been evaluated in a prospective study. An ▶ internal neurolysis may be indicated, therefore, and is utilized by this author, for intraneural fibrosis typically identified in the setting of recurrent median nerve entrapment (Chang and Dellon 1993) or that observed in diabetics with superimposed nerve compression (Aszmann et al. 2002; Dellon 1992). Results of median nerve decompression in diabetics give excellent results in the majority of patients. In the presence of neuropathy, EDT cannot reliably identify the presence of carpal tunnel syndrome, and clinical decision making, based upon the presence of the Phalen and Tinel sign, is still considered valid (Perkins et al. 2002).

References

AAEM Quality Assurance Committee. (1993). Literature review of the usefulness of nerve conduction studies and electromyography for the evaluation of patients with carpal tunnel syndrome. *Muscle & Nerve, 16*, 1392–1414.

Arezzo, J. C., Bolton, C. F., Boulton, A., et al. (1993). Quantitative sensory testing: A consensus report from the peripheral neuropathy association. *Neurology, 43*, 1050–1052.

Aszmann, O. C., Kress, K., & Dellon, A. L. (2002). Results of decompression of peripheral nerves in diabetics: A prospective, blinded study utilizing computer-assisted sensorimotor testing. *Plastic and Reconstructive Surgery, 106*, 816–822.

Chang, B., & Dellon, A. L. (1993). Surgical management of recurrent carpal tunnel syndrome. *The Journal of Hand Surgery, 18*, 467–470.

Dellon, A. L. (1992). Treatment of symptoms of diabetic neuropathy by peripheral nerve decompression. *Plastic and Reconstructive Surgery, 89*, 689–697.

Dellon, A. L. (2001). Clinical grading of peripheral nerve problems. *Neurosurgery Clinics of North America, 12*, 229–240.

Dellon, A. L., & Keller, K. M. (1997). Computer-assisted quantitative sensory testing in carpal and cubital tunnel syndromes. *Annals of Plastic Surgery, 38*, 493–502.

Gerritsen, A. A. M., Uitdehaag, B. M. J., van Geldere, D., et al. (2001). Systematic review of randomized clinical trials of surgical treatment for carpal tunnel syndrome. *The British Journal of Surgery, 88*, 1285–1295.

Mackinnon, S. E., & Dellon, A. L. (1988). Carpal tunnel syndrome. In S. E. Mackinnon & A. L. Dellon (Eds.), *Surgery of the peripheral nerve* (pp. 149–170). New York: Thieme.

Mackinnon, S. E., McCabe, S., Murray, J. F., et al. (1991). Internal neurolysis fails to improve results of primary carpal tunnel decompression. *The Journal of Hand Surgery, 16*, 211–216.

Perkins, B. A., Olaaleye, D., & Bril, V. (2002). Carpal tunnel syndrome in patients with diabetic polyneuropathy. *Diabetes Care, 25*, 565–569.

Rosenberg, D., Conolley, J., & Dellon, A. L. (2001). Thenar eminence quantitative sensory testing in diagnosis of proximal median nerve compression. *Journal of Hand Therapy, 14*, 258–265.

Trumble, T. E., Diao, E., Abrams, R. A., et al. (2002). Single-portal endoscopic carpal tunnel release compared with open release. *The Journal of Bone and Joint Surgery, 84*, 1107–1115.

Weber, R., Weber, R. A., Schuchmann, J. A., et al. (2000). A prospective blinded evaluation of nerve conduction velocity versus pressure-specified sensory testing in carpal tunnel syndrome. *Annals of Plastic Surgery, 45*, 252–257.

Woltman, H. W. (1941). Neuritis associated with acromegaly. *Archives of Neurology & Psychiatry, 45*, 680–682.

Carpal Tunnel Syndrome (CTS)

Definition

Chronic compression of the median nerve as it passes through the carpal tunnel in the wrist. The pressure placed on the median nerve could be caused by excessive pressure due to tendon or tissue inflammation and excessive fluid in the wrist. The condition normally results in reduced nerve conduction velocity, and pain and numbness in the thumb, index, and middle fingers, sometimes the ring finger, but not the little finger. With use of the hand, these symptoms can become disturbing during the daytime. Carpal tunnel syndrome has been associated with repetitive work and frequent forceful exertions, as well as several systemic health conditions such as diabetes and pregnancy. The symptoms are over the palmar side of the hand and not the dorsum. These include nighttime awakening with numbness or tingling.

Cross-References

▶ Carpal Tunnel Syndrome
▶ Ergonomics Essay
▶ Neuropathic Pain Model, Chronic Constriction Injury

Carrageenan

Definition

A colloidal extract from seaweed and other red algae that causes inflammation with associated pain. See also "▶ Carrageenan Inflammation."

Cross-References

▶ Amygdala, Pain Processing and Behavior in Animals

Carrageenan Inflammation

Definition

Carrageenan is an inflammatory irritant utilized to cause inflammatory pain. Carrageenan can be injected into a paw, muscle, or joint. Carrageenan inflammation is associated with heat and mechanical hyperalgesia.

Cross-References

▶ Arthritis Model, Kaolin-Carrageenan-Induced Arthritis (Knee)
▶ TENS, Mechanisms of Action

Cartilage

Definition

Connective tissue that surrounds the ends of bones as they meet to form a joint. The cartilage serves to form a smooth surface around the bones, providing for cushioning and allowing for smooth and easy movement of the joint. Degradation of cartilage is a key characteristic of osteoarthritis.

Cross-References

▶ Arthritis Model, Osteoarthritis

Case Rate

Definition

Flat fee paid for a patient's treatment, based on the diagnosis and/or presenting problem.

Cross-References

▶ Disability Management in Managed Care System

Case-Control Study

Definition

A study that starts with the identification of persons with the disease (or outcome variable) of interest, and a suitable control (comparison) group of persons without the disease (Last 2001).

Cross-References

▶ Prevalence of Chronic Pain Disorders in Children

References

Last, JM (Ed). A Dictionary of Epidemiology, Fourth Edition. Oxford University Press, 2001.

Caspases

Definition

Caspases are a family of cysteine proteases that cleave proteins after aspartic acid residues. They are the main executors of apoptosis or programmed cell death (PCD) and cause the characteristic morphological changes of the cell during apoptosis such as shrinkage, chromatin condensation, and DNA fragmentation.

Cross-References

▶ NSAIDs and Cancer

CAT Scan

▶ CT Scanning

Catabolism

▶ Postoperative Pain, Pathophysiological Changes in Metabolism in Response to Acute Pain

Catalogue

▶ Taxonomy

Catastrophic Cognitions

▶ Catastrophizing

Catastrophic Thinking

▶ Catastrophizing

Catastrophizing

Michael J. L. Sullivan[1] and Heather Adams[2]
[1]Department of Psychology, McGill University, Montreal, QC, Canada
[2]Department of Surgery, McGill University Health Centre, Montreal, QC, Canada

Synonyms

Catastrophic cognitions; Catastrophic thinking; Maladaptive coping

Definition

Pain catastrophizing has been defined as an exaggerated negative "mental set" that is brought to bear during an actual or anticipated pain experience (Sullivan et al. 2001). Pain catastrophizing is considered to be a multidimensional construct that includes elements of ▶ rumination (i.e., excessive focus on pain sensations), ▶ magnification (i.e., exaggerating the threat value of pain sensations), and ▶ helplessness (i.e., perceiving oneself as unable to cope with pain symptoms).

Characteristics

Catastrophizing, Pain, and Disability

To date, numerous studies have been published documenting a relation between catastrophizing and pain. A relation between catastrophizing and pain has been observed in diverse clinical and experimental populations (Sullivan et al. 2001; Quartana et al. 2009). Catastrophizing has been associated with increased pain and ▶ pain behavior, increased use of health-care services, longer hospital stays, and increased use of analgesic medication. In chronic pain patients, catastrophizing has been associated with heightened disability, predicting the risk of chronicity and the severity of disability better than illness-related variables or pain itself (Keefe et al. 1989; Sullivan et al. 2001, 2011). A relation between catastrophizing and adverse pain outcomes has been observed in children as young as 7 years (Crombez et al. 2003). For reviews of current research and theory on pain catastrophizing, the reader is referred to Sullivan et al. (2001), Sullivan (2012), and Quartana et al. (2009).

Several experimental and clinical prospective studies suggest that catastrophizing might play a causal role in the amplification and persistence of pain and disability (Keefe et al. 1989; Sullivan 2012; Sullivan et al. 2004). These findings do not rule out the possibility that catastrophizing may be reactive to variations in pain experience, but they do nevertheless highlight that the assessment of catastrophizing can permit prediction of future pain and pain-related disability.

Functions of Catastrophizing

Theoretical conceptualizations of pain catastrophizing have appealed to cognitive constructs such as ▶ appraisals, ▶ cognitive errors, and ▶ schema activation to account for the relation between catastrophizing and pain (Sullivan et al. 1995). It has been suggested that, as a function of a learning history characterized by heightened pain experience, catastrophizers may develop expectancies about the high threat value of painful stimuli (i.e., primary appraisal) and about their inability to effectively manage the stress associated with painful experiences (i.e., secondary appraisal). Once activated, catastrophizers' "pain schema" might influence emotional or cognitive functioning in a manner that contributes to heightened pain experience (Leeuw et al. 2007).

Catastrophizing has also been discussed as a key factor in the development of pain-related fear. It has been suggested that following injury, individuals who engage in catastrophic thinking are likely to develop heightened fears of pain, movement and reinjury (Vlaeyen and Linton 2000). By contributing to the development of pain-related fears, catastrophizing might heighten the risk for different aspects of disability, such as activity discontinuation and activity avoidance (Leeuw et al. 2007).

It has also been suggested that catastrophizers might engage in exaggerated displays of pain behavior as a means of coping with pain (Sullivan 2012). This "communal coping" model of pain catastrophizing draws on theoretical frameworks addressing the interpersonal dimensions of coping. Specifically, the Communal Coping Model of catastrophizing suggests that individuals differ in the degree to which they adopt social or relational goals in their efforts to cope with stress. Catastrophizers' expressive displays of distress might be used, consciously or unconsciously, as a strategy to maximize proximity, or to solicit assistance or empathic responses from others in their social environment (Sullivan 2012).

Mechanisms of Action

Research points to the potential roles of expectancies, attention, and coping as possible avenues through which catastrophizing might heighten the probability of the persistence of pain and disability following injury (Sullivan et al. 2011; Crombez et al. 1997). Specifically, research suggests that catastrophizers have negative expectancies for pain outcomes, and they show deficits in their ability to disengage their attention from pain stimuli (Van Damme et al. 2004).

Catastrophizing has been shown to compromise the effectiveness of rehabilitative and pharmacological interventions for pain. It has been suggested that catastrophizing might be associated with endogenous opioid dysregulation (Quartana et al. 2009). It has also been shown that catastrophizing is associated with greater inflammatory response to acute injury, more pronounced muscle fatigue in response to exertion, and greater activation of brain areas responsible for modulation of affective, attentional, and motor aspects of pain (Quartana et al. 2009; Sullivan et al. 2011).

Assessment

Several assessment instruments have been developed to assess pain catastrophizing. Considerable research on catastrophizing has used the Coping Strategies Questionnaire (CSQ) (Rosenstiel and Keefe 1983). The CSQ consists of seven coping subscales, including a six-item catastrophizing subscale. The Pain Catastrophizing Scale (PCS) (Sullivan et al. 1995) is another widely used measure of catastrophizing that adopts a multidimensional view of the construct. The PCS is a 13-item self-report questionnaire that assesses three dimensions of catastrophizing: rumination ("I can't stop thinking about how much it hurts"), magnification ("I worry that something serious may happen"), and helplessness ("It's awful and I feel that it overwhelms me"). Electronic copies of the PCS in various languages can be downloaded from the following site: http://sullivan-painresearch.mcgill.ca.

Self-report measures have also been developed for assessing catastrophizing in children and adolescents (Gil et al. 1993; Crombez et al. 2003). Interview methods have also been used; however, their application to clinical settings has been limited.

Treatment

The robust relation between catastrophizing and pain has prompted a growing number of clinicians and researchers to identify catastrophizing as a central factor in the clinical management of disabling pain conditions. Following multidisciplinary treatment for pain, reductions in catastrophizing are often noted (Burns et al. 2003; Smeets et al. 2006). A few intervention programs have been developed that specifically target catastrophic thinking as a primary goal of treatment. Thorn et al. (2002) have described a 10-week, cognitive behavioral intervention designed to reduce catastrophic thinking in headache sufferers. In this treatment program, thought recording and cognitive restructuring techniques are used as a means of monitoring and modifying catastrophic thoughts. Riddle et al. (2011) described an 8-week coping skills intervention designed to reduce catastrophizing in order to promote recovery following joint replacement surgery. Sullivan et al. (2006) described a 10-week life-role reintegration program designed to facilitate return-to-work following occupational injury. In the latter program, disclosure techniques, ► activity mobilization strategies, and cognitive restructuring are used to minimize catastrophic thinking and facilitate progress toward occupational reintegration. Across a number of studies, reduction in catastrophizing has been shown to be a significant determinant of treatment success (Smeets et al. 2006; Sullivan et al. 2006; Sullivan and Stanish 2003).

Cross-References

▶ Cognitive-Behavioral Perspective of Pain
▶ Coping and Pain
▶ Ethics of Pain, Culture, and Ethnicity
▶ Psychological Aspects of Pain in Women
▶ Psychological Treatment in Acute Pain
▶ Psychological Treatment of Headache

References

Burns, J. W., Kubilus, A., Bruehl, S., Harden, R. N., & Lofland, K. (2003). Do changes in cognitive factors influence outcome following multidisciplinary treatment for chronic pain? A cross-lagged panel analysis. *Journal of Consulting and Clinical Psychology, 71,* 81–91.
Crombez, G., Eccleston, C., Baeyens, F., & Eelen, P. (1997). When somatic information threatens, catastrophic thinking enhances attentional interference. *Pain, 74,* 230–237.
Crombez, G., Bijttebier, P., Eccleston, E., Mascagni, G. M., Goubert, L., & Verstraeten, K. (2003). The child version of the pain catastrophizing scale (PCS-C): A preliminary validation. *Pain, 104,* 639–646.
Gil, K. M., Thompson, R. J., Keith, B. R., Tota-Faucette, M., Noll, S., & Kinney, T. R. (1993). Sickle cell disease pain in children and adolescents: Change in pain frequency and coping strategies over time. *Journal of Pediatric Psychology, 18,* 621–637.
Keefe, F. J., Brown, G. K., Wallston, K. A., & Caldwell, D. S. (1989). Coping with rheumatoid arthritis: Catastrophizing as a maladaptive strategy. *Pain, 37,* 51–56.
Leeuw, M., Goossens, M. E., et al. (2007). The fear-avoidance model of musculoskeletal pain: Current state of scientific evidence. *Journal of Behavioral Medicine, 30,* 77–94.
Quartana, P. J., Campbell, C. M., et al. (2009). Pain catastrophizing: A critical review. *Expert Review of Neurotherapeutics, 9,* 745–758.
Riddle, D. L., Keefe, F. J., Nay, W. T., McKee, D., Attarian, D. E., & Jensen, M. P. (2011). Pain coping skills training for patients with elevated pain catastrophizing who are scheduled for knee arthroplasty: A quasi-experimental study. *Archives of Physical Medicine and Rehabilitation, 92,* 859–865.
Rosenstiel, A. K., & Keefe, F. J. (1983). The use of coping strategies in chronic low back pain patients: Relationship to patient characteristics and current adjustment. *Pain, 17,* 33–44.
Smeets, R. J., Vlaeyen, J. W. S., et al. (2006). Reduction of pain catastrophizing mediates the outcome of both physical and cognitive-behavioral treatment in chronic low back pain. *The Journal of Pain, 7,* 261–271.
Sullivan, M. J. L. (2012). The communal coping model of pain catastrophizing: Clinical and research implications. *Canadian Psychology, 53,* 31–42.
Sullivan, M. J. L., & Stanish, W. D. (2003). Psychologically based occupational rehabilitation: The pain-disability prevention program. *The Clinical Journal of Pain, 19,* 97–104.
Sullivan, M. J. L., Bishop, S. R., & Pivik, J. (1995). The pain catastrophizing scale: Development and validation. *Psychological Assessment, 7,* 524–532.
Sullivan, M. J. L., Thorn, B., Haythornthwaite, J. A., Keefe, F. J., Martin, M., Bradley, L. A., & Lefebvre, J. C. (2001). Theoretical perspectives on the relation between catastrophizing and pain. *The Clinical Journal of Pain, 17,* 52–64.
Sullivan, M. J. L., Thorn, B., Rodgers, W., & Ward, C. (2004). Path model of psychological antecedents of pain experience: Experimental and clinical findings. *The Clinical Journal of Pain, 20,* 164–173.
Sullivan, M. J. L., Adams, H., et al. (2006). A psychosocial risk factor–targeted intervention for the prevention of chronic pain and disability following whiplash injury. *Physical Therapy, 86,* 8–18.
Sullivan, M. J. L., Adams, H., Martel, M.-O., Scott, W., & Wideman, T. H. (2011). Catastrophizing and perceived injustice: Risk factors for the transition to chronicity following whiplash injury. *Spine, 36,* S244–S249.
Thorn, B. E., Boothy, J., & Sullivan, M. J. L. (2002). Targeted treatment of catastrophizing in the management of chronic pain. *Cognitive and Behavioral Practice, 9,* 127–138.
Van Damme, S., Crombez, G., & Eccleston, C. (2004). Disengagement from pain: The role of catastrophic thinking about pain. *Pain, 107,* 70–76.
Vlaeyen, J. W. S., & Linton, S. J. (2000). Fear-avoidance and its consequences in chronic musculoskeletal pain: A state of the Art. *Pain, 85,* 317–332.

Cathepsin and Microglia

Anna K. Clark, Andrea C. Ogbonna and Marzia Malcangio
Wolfson Centre for Age Related Diseases, King's College London, London, UK

Synonyms

Analgesia; Chemokine; Fractalkine (CX3CL1); Microglia; Neuropathic pain; P2X7 receptor; Peripheral nerve injury; Spinal cord microglia

Definition

The cathepsins are lysosomal proteases that are ubiquitously expressed. Cathepsin S is a cysteine protease that displays tissue-specific distribution in antigen-presenting cells (APCs). Cathepsin S (CatS) is critically involved in the process of antigen presentation and is therefore essential for host immune function. Secretion of CatS by spinal microglia contributes to enhanced nociceptive transmission in rodent models of peripheral nerve injury. Microglia are cells abundant in CNS parenchyma which belong to the myelomonocytic lineage also including monocytes and macrophages. Quiescent microglia perform immune surveillance of the nervous system by elaborate branch structures which rarely overlap with nearby microglia cells. Microglia are rapidly activated by direct insults to the central nervous system (CNS), but they also react acutely to peripheral nervous system (PNS) insults. Activated microglia show amoeboid morphology, increased phagocytic activity, enhanced migratory capacity within the CNS, and increased expression of cell surface glycoproteins, including CD45 and class II major histocompatibility complex (MHC-II).

Characteristics

Cathepsin S (CatS) is a lysosomal enzyme belonging to the papain family of cysteine proteases. CatS was initially purified from bovine lymph nodes in 1975 (Turnsek et al. 1975). Since then, CatS has been isolated from the spleen and lymph nodes of a number of species, including bovine, rabbit, rat, and human.

The first member of the cathepsin family to be identified was cathepsin C (CatC) in 1948 (Gutmann and Fruton 1948). Other cathepsins were identified more than 20 years later, during the 1970s, when cathepsin B (CatB), cathepsin H (CatH), and cathepsin L (CatL) were described. The first amino acid sequence of a cathepsin was reported in 1983 when rat CatB and CatH were shown to share high sequence homology with the plant protease papain (Takio et al. 1983),

commonly used as a meat tenderizer. The sequence of chicken CatL followed several years later (Wada et al. 1987). During the 1990s, molecular biology techniques led to increased productivity throughout the cathepsin research field and the amino acid sequence of rat CatC was reported in 1991 (Ishidoh et al. 1991). In the same year, the primary sequence of bovine CatS was reported by two independent groups (Wiederanders et al. 1991; Ritonja et al. 1991), followed by human (Shi et al. 1992) and rat (Petanceska and Devi 1992) CatS shortly afterward. Cathepsins are classified according to their structure and type of catalytic activity. To date, there are 11 cysteine cathepsins, as well as two aspartic cathepsins (D and E) and two serine cathepsins (A and G).

Almost all cells express some level of papain-like lysosomal cysteine proteases. Of these, CatB is the most abundant member of the cathepsin family. However, several of the cathepsins have been found to display tissue-specific distribution. Specifically, CatS is expressed in a variety of professional, bone marrow-derived APCs, including macrophages (Liuzzo et al. 1999b; Shi et al. 1992; Barclay et al. 2007), B-lymphocytes (Riese et al. 1996), dendritic cells (DCs) (Shi et al. 1999), and microglia (Petanceska et al. 1994; Liuzzo et al. 1999a, b; Gresser et al. 2001; Clark et al. 2010a), as well as in nonprofessional APCs, such as epithelial cells (Beers et al. 2005) and keratinocytes (Schwarz et al. 2002).

The cathepsins are synthesized as preproenzymes (inactive zymogens). The prepeptide is then cleaved from the proenzyme during transport through the membrane of the ER. The N-terminal propeptide is responsible for the stability and correct folding of the cathepsin (Tao et al. 1994) and also for targeting to the lysosomal compartments (Cuozzo et al. 1995). In addition, the propeptide blocks the active site of the enzyme (Podobnik et al. 1997; Coulombe et al. 1996), preventing substrate binding from occurring, therefore preventing premature activation of the cathepsin during transport through the ER and Golgi apparatus (Tao et al. 1994). Once inside the acidic environment of the lysosome, the propeptide is removed, resulting in the production of mature active enzyme (Kominami et al. 1988; Turk et al. 1999). The proteolytic removal of the propeptide may be accomplished by a number of proteases, including cathepsin D (CatD) (Rowan et al. 1992) as well as autocatalytic mechanisms, which can be triggered by the acidic pH of the lysosome (Mason et al. 1987; Rozman et al. 1999; Vasiljeva et al. 2005).

The regulation of cysteine protease activity, such as CatS activity, is achieved via a number of factors. One major regulator of protease over activity is the narrow pH range at which the lysosomal proteases are able to function. Cysteine proteases function optimally at the acidic pH found in intracellular compartments, such as in lysosomes, and are unstable at neutral pH. Such instability acts as a mechanism for limiting proteolytic damage if an enzyme escapes from its intracellular compartment. One exception to this rule is CatS. Once secreted into the neutral pH of extracellular environment, CatS is extremely stable, and remains functionally active (Vasiljeva et al. 2005; Liuzzo et al. 1999a; Petanceska and Devi 1992; Shi et al. 1992). This unique characteristic may account for specific roles of CatS outside of lysosomes, in particular the role of CatS in nociceptive transmission.

The major means of regulation of cysteine proteases cathepsins is by their endogenous protein inhibitors, the cystatins. The cystatins are a large superfamily of naturally occurring molecules that inhibit cysteine proteases by forming a tight, reversible complex with the active site of the enzyme. The cystatins are subdivided into three main groups based on their sequence homology: stefins, cystatins, and kininogens. Of these subfamilies, the kininogens have long been recognized as precursor proteins of the vasoactive kinins (e.g., bradykinin) as well as having functional roles as participants in inflammation (Levine and Reichling 1999). Of particular interest with regard to CatS is cystatin C. Cystatin C was originally isolated from the serum of patients with autoimmune diseases (Brzin et al. 1984) and has since been isolated from a number of species including the rat (Cole et al. 1989). Cystatin C expression

modulates the enzymatic activity of CatS. CatS plays a vital role in antigen presentation and the intracellular expression of cystatin C appears to tightly regulate CatS activity during the maturation of DCs (Pierre and Mellman 1998). As the cells mature, the expression of cystatin C in lysosomes steadily decreases. However, the expression of CatS remains constant throughout the maturation process, suggesting that cystatin C is regulating the enzymatic activity of CatS in immature DCs. Interest in cystatin C in relation to nociception was sparked in 2003 when a study suggested that it may represent a potential CSF biomarker for pain in humans (Mannes et al. 2003). However, this suggestion was refuted the following year when a second larger study found no changes in cystatin C levels in the CSF of patients with neuropathic pain compared to controls (Eisenach et al. 2004).

Intracellular Roles of Cathepsin S

Cathepsin S plays a critical role in antigen presentation by which short peptide fragments of pathogen-derived proteins are displayed on the cell surface of specialized APCs for T-lymphocyte recognition. Class II major histocompatibility complex (MHC) molecules sample antigenic peptides from lysosomal compartments and display them on the cell surface of APCs for T-lymphocyte recognition. The initial stages of antigen presentation, during which MHC class II molecules are transported from their site of synthesis to lysosomal compartments, is heavily regulated by a chaperone molecule, the invariant chain (Ii). During transport, a specific region of the Ii binds to the peptide groove of the $\alpha\beta$ heterodimer (Ghosh et al. 1995), preventing premature peptide loading into the MHC class II peptide groove before transport to the lysosome. Once inside the lysosomal compartment, degradation of the Ii is vital for antigenic peptide fragments to access the MHC class II binding groove and this process is critically dependent on CatS.

A large number of studies have implicated CatS in proteolysis of Ii. Incubation of various APCs with the specific CatS inhibitor LHVS (Morpholinurea-leucine-homophenylalanine-vinyl sulfone-phenyl) results in the

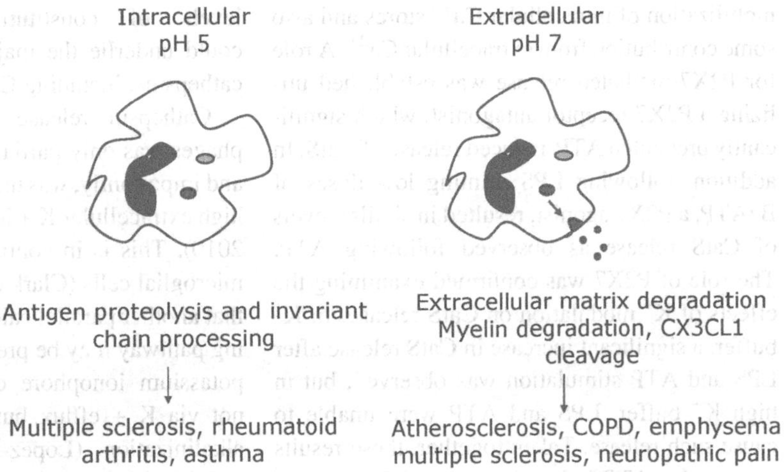

Cathepsin and Microglia,
Fig. 1 Intracellular and
extracellular roles of
cathepsin S

Intracellular
pH 5

Extracellular
pH 7

Antigen proteolysis and invariant
chain processing

Extracellular matrix degradation
Myelin degradation, CX3CL1
cleavage

Multiple sclerosis, rheumatoid
arthritis, asthma

Atherosclerosis, COPD, emphysema
multiple sclerosis, neuropathic pain

accumulation of Ii fragments, and addition of recombinant CatS protein to these cells restored digestion of Ii (Riese et al. 1996; Villadangos et al. 1997). In addition, in vivo administration of LHVS to mice resulted in deficits in Ii degradation, antigen presentation, and immune responses following immunization with exogenous antigen (Riese et al. 1998). More recently, the use of knockout mice has determined that CatS is vital for degrading Ii fragments to CLIP in vivo. B lymphocytes and DCs from mice deficient in CatS are unable to degrade Ii, resulting in impaired antigen presentation (Shi et al. 1999; Nakagawa et al. 1999; Driessen et al. 1999). As the inhibition of intracellular CatS should result in immunosuppression, CatS inhibitors are being evaluated for the treatment of multiple sclerosis, rheumatoid arthritis, and asthma (Fig. 1).

Extracellular Roles of Cathepsin S

The activity of CatS is not restricted to intracellular compartments, as the secretion of CatS has been observed from a number of cell types, including macrophages and microglia (Liuzzo et al. 1999a, b; Petanceska et al. 1996; Clark et al. 2010a). CatS has been observed to cleave some extracellular matrix proteins and may therefore play a role in tissue remodeling. In vitro CatS is able to degrade extracellular matrix proteins, including laminin, fibronectin, and several collagens (Petanceska et al. 1996), chondroitin sulfate

(Petanceska et al. 1996), heparan sulfate, and basement membrane proteoglycans (Liuzzo et al. 1999b; Petanceska et al. 1996), myelin basic protein, and amyloid beta peptide (Liuzzo et al. 1999a). This role in extracellular matrix, as well as myelin degradation, provides scope for investigating therapeutic potential of CatS inhibitors in atherosclerosis, chronic obstructive pulmonary disease (COPD), emphysema, and multiple sclerosis (Fig. 1). We have recently identified a new role for extracellular CatS in facilitating pain mechanisms in the spinal cord (see below).

Mechanisms of Cathepsin S release

The secretion of CatS from both macrophages and microglia is enhanced by inflammatory mediators, including lipopolysaccharide (LPS), ATP, pro-inflammatory cytokines, and neurotrophins (Liuzzo et al. 1999a, b; Petanceska et al. 1996; Riese et al. 1996; Clark et al. 2010a).

Recently, we have demonstrated that primary microglia in culture release CatS in a P2X7 receptor-dependent mechanism (Clark et al. 2010a). Stimulation of microglia with high concentrations of ATP results in significant release of active CatS. No such release occurred with a low concentration of ATP, suggesting a lack of a role for high-affinity receptors for ATP such as P2X4. Additionally, for significant ATP-induced CatS release to occur, microglial cells first required priming by incubation with LPS, and requires an increase in cytosolic calcium following mobilization of intracellular Ca^{2+} stores and also some contribution from extracellular Ca^{2+}. A role for P2X7-mediated release was established utilizing a P2X7 receptor antagonist, which significantly prevented ATP-induced release of CatS. In addition, following LPS priming low doses of BzATP, a P2X7 agonist, resulted in similar levels of CatS release as observed following ATP. The role of P2X7 was confirmed examining the effects of K^+ modulation on CatS release. In K^+ buffer, a significant increase in CatS release after LPS and ATP stimulation was observed, but in high K^+ buffer, LPS and ATP were unable to cause such release. Taken together, these results suggest that ATP-induced release of CatS occurs through activation of P2X7 receptor and resulting in K^+ efflux. Determination of the intracellular signaling pathways downstream of P2X7 activation revealed that both phospholipase C and phospholipase A2, as well as p38 mitogen-activated protein kinase (MAPK) are critical for ATP-induced release of CatS.

This mechanism of CatS release is consistent with the delineated pathway for interleukin-1β (IL-1β) release from monocytes (Andrei et al. 2004), and LPS is known to regulate ATP-induced IL-1β release from microglial cells through P2X7 receptor activation (Di irgilio 2007; Ferrari et al. 2006). Indeed, we observed concomitant release of both IL-1β and CatS proteins in microglial cell media after ATP stimulation of LPS primed cells, suggestive of a shared mechanism of release via P2X7 receptor activation (Clark et al. 2010a).

P2X7 receptor-mediated release of CatS has also been reported from both human lung macrophages and mouse bone marrow-derived macrophages (BMDM) (Lopez-Castejon et al. 2010). In contrast to primary microglial cells, ATP stimulation induced release of active CatS from unprimed macrophages and was independent of IL-1β release. Activation of P2X7 receptor is still critical in the mechanism of CatS release as ATP stimulation induced release of cathepsins from BMDM from wild-type mice not P2X7R knockout mice. Interestingly, there was a significant decrease in basal levels of cathepsin release from P2X7R knockout BMDMs suggesting low

levels of constitutive P2X7R activation could underlie the majority of basal release of cathepsins, including CatS from macrophages.

Cathepsin release from unprimed macrophages was only partially dependent on calcium and importantly, was unaltered by the presence of high extracellular K + ions (Lopez-Castejon et al. 2010). This is in contrast to CatS release from microglial cells (Clark et al. 2010a) and suggests that another parallel calcium-independent signaling pathway may be present. Nigericin, a proton/potassium ionophore caused cathepsin release not via K + efflux but by inducing lysosomal alkalinization (Lopez-Castejon et al. 2010), which is known to precede lysosome destabilization (Boya and Kroemer 2008). As such, it is proposed that activation of P2X7 receptor in macrophages induces release of cathepsins, including CatS through both calcium-dependent direct exocytosis and calcium-independent pH-dependent lysosome destabilization.

Microglia in Nociception

Microglia have emerged as important contributors to chronic pain mechanisms in numerous preclinical models of chronic pain (McMahon and Malcangio 2009). Following peripheral nerve or tissue injury, the central terminals of primary afferent fibers release ATP and chemokines, which activate receptors on dorsal horn microglia inducing enhanced response states. Upon entering such states, microglia release a number of neuromodulatory factors, including CatS that contribute to nociceptive processing. Pharmacological, molecular, and genetic manipulations of the function or expression of microglial mediators significantly influence pain responses in a number of rodent models of neuropathic pain and to a lesser extent models of inflammatory pain (Inoue and Tsuda 2009; McMahon and Malcangio 2009).

Cathepsin S in Nociception

The critical contribution of microglial cells to chronic pain, in conjunction with an established role of CatS in APCs and high stability of CatS at neutral pH, led us to hypothesis that one possible extracellular function of CatS may be an involvement in nociceptive transmission. CatS was initially identified as a possible mediator of chronic pain and a potential target for therapeutic intervention following the observation of increased CatS mRNA levels in the lumbar DRG in two models of peripheral nerve injury (Barclay et al. 2007). Upregulation of the CatS gene had also been reported in the sciatic nerve (Kubo et al. 2002) and DRG following axotomy (Costigan et al. 2002), and in the DRG following spinal nerve ligation (SNL) (Valder et al. 2003). Immunohistochemical studies revealed that following PNL, in the periphery, CatS expression is observed in infiltrating macrophages both in the DRG and at the injury site (Barclay et al. 2007). In the spinal cord, CatS expression is observed in microglia under naïve conditions, and is extensively upregulated following partial sciatic nerve ligation (PNL) in the ipsilateral dorsal horn (Clark et al. 2007, 2009). Following PNL, the temporal profile of CatS expression in both macrophages and spinal microglia correlates with the maintenance of neuropathy, rather than its development (Barclay et al. 2007; Clark et al. 2007).

Pharmacological inhibition of CatS has elucidated a role for the protease in chronic pain. Spinal administration of LHVS, an irreversible CatS inhibitor (Riese et al. 1998), is able to reverse established pain behaviors following PNL. A single intrathecal administration of LHVS transiently reverses allodynia in 7-day and 14-day neuropathic rats, when CatS expression in the dorsal horn is high, but is ineffective at 3 days post-injury when CatS expression is submaximal (Clark et al. 2007). Interestingly, in 14-day neuropathic rats, LHVS dose-dependently reverses established mechanical hyperalgesia suggesting that CatS contained in spinal microglial cells is a major contributor to the maintenance of neuropathic pain rather than to the initial development of mechanical hypersensitivity. LHVS preferentially reverses hyperalgesia in comparison to allodynia when administered acutely. However, chronic inhibition of CatS appears to be more efficacious in the reversal of mechanical allodynia. Chronic intrathecal administration of LHVS from day 0–7 post-PNL fails to prevent the development of mechanical allodynia, whereas treatment from day 7–14 post-PNL attenuates already established allodynia (Clark et al. 2007). Systemic administration of LHVS also reverses mechanical hyperalgesia but is unable to attenuate established allodynia (Barclay et al. 2007). The central microglial source of the enzyme is therefore essential for the persistence of neuropathic allodynia. As well as attenuation of behavioral hypersensitivity, the extent of ipsilateral microglial activation following PNL is significantly reduced following chronic intrathecal delivery of LHVS, suggesting that this enzyme is essential for microglial cells to remain in their hyperactive phenotype associated with an enhanced pain-related response. Similarly, chronic intrathecal delivery of LHVS is also able to reverse established pain behaviors associated with spinal cord injury and rheumatoid arthritis, via a reduction in microglial response (Clark et al. 2010c; 2012).

Importantly, several highly selective CatS inhibitors have been shown to be efficacious in attenuating neuropathic pain behaviors following a single administration as well as after chronic dosing (Irie et al. 2008; Dalrymple et al. 2009; Holsinger et al. 2010), suggesting that these newly developed compounds may be beneficial in the treatment of neuropathic pain.

A number of observations substantiate the pronociceptive role of CatS. Intrathecal delivery of activated recombinant CatS evokes dose-dependent mechanical hypersensitivity that develops within 30 min of administration (Clark et al. 2007). The development of hyperalgesia is specific to CatS since the related enzymes Cathepsin B and Cathepsin L do not alter paw withdrawal thresholds. The effects of spinally administered CatS are completely prevented by prior LHVS administration, indicating a local, spinal site of action for the enzyme (Clark et al. 2007). The finding that exogenous CatS is pro-nociceptive clearly points to an extracellular site of action and is consistent with the observations that the behavioral effects are rapid and maximal at the first time point studied, 30 min after enzyme injection. Indeed, unlike other cathepsins, CatS is extremely stable and functionally active at neutral extracellular pH

(Vasiljeva et al. 2005; Liuzzo et al. 1999a; Petanceska and Devi 1992; Shi et al. 1992). Similarly, intraplantar CatS is also able to induce pain like behaviors when administered to rodents (Barclay et al. 2007).

Cathepsin S Pro-nociceptive Effects Are Mediated via Fractalkine Cleavage

Chemokines encompass a large family of cytokines that at present number over 50 individual chemokines, signaling via 19 receptors (reviewed in Bajetto et al. 2002; Murphy et al. 2000), which are subdivided into four groups based on the position of their first N-terminal cysteine residues. Fractalkine (FKN; CX3CL1) is at present the only member of the CX3C class of chemokines (Bazan et al. 1997; Pan et al. 1997). The structure of FKN means that, in contrast to most other chemokines FKN can exist as both transmembrane and soluble forms (Bazan et al. 1997). The binding of chemokines to their receptors is highly promiscuous. However, FKN only signals via one chemokine receptor, the CX3CR1 receptor. A large number of chemokines and their receptors have been identified in the CNS under both normal and pathological conditions, and evidence suggests that chemokines play a role in the regulation of neuronal activity and synaptic plasticity.

In particular, a role for FKN in nociceptive transmission has been proposed. FKN is expressed by neurons in the spinal cord (Verge et al. 2004; Lindia et al. 2005; Clark et al. 2009) and brain (Harrison et al. 1998; Nishiyori et al. 1998; Schwaeble et al. 1998; Hughes et al. 2002). In contrast, its receptor, CX3CR1, is expressed by microglia (Verge et al. 2004; Lindia et al. 2005; Harrison et al. 1998; Nishiyori et al. 1998; Hughes et al. 2002). Following a peripheral nerve injury, upregulation of CX3CR1 is observed in microglia (Verge et al. 2004; Lindia et al. 2005; Zhuang et al. 2007; Holmes et al. 2008; Staniland et al. 2010). Interestingly, *de novo* expression of FKN is observed in astrocytes following SNL (Lindia et al. 2005), but not in other models of neuropathy (Verge et al. 2004; Clark et al. 2009).

Intrathecal administration of FKN is pro-nociceptive, inducing thermal (Milligan et al. 2004, 2005) and mechanical (Milligan et al. 2004, 2005; Clark et al. 2007) hypersensitivity, which is prevented by pretreatment with a neutralizing agent against CX3CR1 (Milligan et al. 2004, 2005) or FKN itself (Clark et al. 2007). Importantly, in wild-type mice, intrathecal FKN induces pain behaviors which are absent in CX3CR1 knockout mice (Clark et al. 2007). Indeed, FKN induces nociceptive behaviors via the CX3CR1 receptor on microglia, which induces activation of p38 MAPK-mediated pathways (Clark et al. 2007; Zhuang et al. 2007), and are mediated in part by microglial-derived cytokines (Milligan et al. 2005). Impairment of FKN/CX3CR1 signaling following several models of nerve injury using neutralizing antibodies against either CX3CR1 (Milligan et al. 2004; Zhuang et al. 2007) or FKN itself (Clark et al. 2007) attenuates neuropathic pain behaviors. In addition, we have recently shown that CX3CR1 knockout mice have deficits in both neuropathic and inflammatory pain as a result of reduced microglial response (Staniland et al. 2010).

We utilized a bioinformatics approach which identified the FKN extracellular domain as a possible cleavage target for CatS. The predicted cleavage sites for FKN would liberate the soluble, biologically active extracellular domain of this chemokine. A number of observations provide evidence that the molecular substrate for CatS is neuronal FKN and that the mechanism responsible for the pro-nociceptive effects of CatS is FKN shedding from neuronal membranes. Firstly, intrathecal administration of a FKN neutralizing antibody is able to prevent the development of mechanical hyperalgesia following either FKN chemokine domain (FKN-CD) or CatS (Clark et al. 2007). Extracellular CatS indirectly enhances the microglial response via p38 MAPK following proteolytic liberation of the soluble form of neuronal FKN. Intrathecal administration of both FKN-CD (Clark et al. 2007; Zhuang et al. 2007) and CatS (Clark et al. 2007) results in the phosphorylation of p38 MAPK, a marker of rapid microglial cell activation, and this phosphorylation is attenuated following pre-administration of FKN neutralizing antibody (Clark et al. 2007). Secondly, CatS-induced mechanical allodynia is

Cathepsin and Microglia, Fig. 2 Schematic of proposed pro-nociceptive mechanism of CatS. In the dorsal horn area innervated by damaged fibers P2X7 receptor activation on microglia (1) release CatS (2) which then liberates soluble FKN from neurons (3). FKN feeds back onto the microglial cells via the CX3CR1 receptor (4) to activate the p38 MAPK pathway (5) and release inflammatory mediators that activate neurons and result in chronic pain

absent in mice lacking the CX3CR1 receptor, indicating that CatS pro-nociceptive effects are mediated via cleavage of FKN and activation of the CX3CR1 receptor (Clark et al. 2007). Thirdly, incubation of cultured dorsal root ganglia neurons with CatS significantly reduces the levels of cell-associated FKN (Clark et al. 2007). Finally, levels of soluble FKN in CSF are significantly elevated alongside increased CatS activity in neuropathic conditions (Clark et al. 2009). Using an ex vivo dorsal horn with dorsal roots attached slice preparation, we confirmed that neuronal FKN cleavage occurs in the dorsal horn under conditions in which microglia are in an activated state, such as after nerve injury or following chemical stimulation with LPS, and is entirely dependent on CatS activity (Clark et al. 2009).

Concluding Remarks

Following peripheral nerve injury, the following schematic is proposed (Fig. 2). CatS expressing spinal cord microglia in the area of the dorsal horn innervated by damaged fibers release CatS following activation of the P2X7 receptor by high extracellular concentration of ATP (Clark et al. 2010a). Extracellular CatS then liberates sFKN from the membranes of spinal neurons. The released sFKN then feeds back onto the microglial cells via the CX3CR1 receptor to activate the p38 MAPK pathway. Activation of this intracellular pathway contributes to neuropathic pain (Jin et al. 2003; Tsuda et al. 2004) by modulating the synthesis and release of pro-nociceptive mediators such as cytokines, including IL-1β (Clark et al. 2006, 2010b). In summary, a role for the CatS/FKN pair in the maintenance of neuropathic hypersensitivity suggests that CatS inhibition constitutes a novel therapeutic approach for the treatment of chronic pain.

References

Andrei, C., Margiocco, P., Poggi, A., Lotti, L. V., Torrisi, M. R., & Rubartelli, A. (2004). From the cover: Phospholipases C and A2 control lysosome-mediated IL-1{beta} secretion: Implications for inflammatory processes. *PNAS, 101,* 9745–9750.
Bajetto, A., Bonavia, R., Barbero, S., & Schettini, G. (2002). Characterization of chemokines and their receptors in the central nervous system:

Physiopathological implications. *Journal of Neurochemistry, 82*, 1311–1329.

Barclay, J., Clark, A. K., Ganju, P., Gentry, C., Patel, S., Wotherspoon, G., Buxton, F., Song, C., Ullah, J., Winter, J., Fox, A., Bevan, S., & Malcangio, M. (2007). Role of the cysteine protease cathepsin S in neuropathic hyperalgesia. *Pain, 130*, 225–234.

Bazan, J. F., Bacon, K. B., Hardiman, G., Wang, W., Soo, K., Rossi, D., Greaves, D. R., Zlotnik, A., & Schall, T. J. (1997). A new class of membrane-bound chemokine with a CX3C motif. *Nature, 385*, 640–644.

Beers, C., Burich, A., Kleijmeer, M. J., Griffith, J. M., Wong, P., & Rudensky, A. Y. (2005). Cathepsin S controls MHC class II-mediated antigen presentation by epithelial cells in vivo. *Journal of Immunology, 174*, 1205–1212.

Boya, P., & Kroemer, G. (2008). Lysosomal membrane permeabilization in cell death. *Oncogene, 27*, 6434–6451.

Brzin, J., Popovic, T., Turk, V., Borchart, U., & Machleidt, W. (1984). Human cystatin, a new protein inhibitor of cysteine proteinases. *Biochemical and Biophysical Research Communications, 118*, 103–109.

Clark, A. K., D'Aquisto, F., Gentry, C., Marchand, F., McMahon, S. B., & Malcangio, M. (2006). Rapid co-release of interleukin 1beta and caspase 1 in spinal cord inflammation. *Journal of Neurochemistry, 99*, 868–880.

Clark, A. K., Yip, P. K., Grist, J., Gentry, C., Staniland, A. A., Marchand, F., Dehvari, M., Wotherspoon, G., Winter, J., Ullah, J., Bevan, S., & Malcangio, M. (2007). Inhibition of spinal microglial cathepsin S for the reversal of neuropathic pain. *Proceedings of the National Academy of Sciences of the United States of America, 104*, 10655–10660.

Clark, A. K., Yip, P. K., & Malcangio, M. (2009). The liberation of fractalkine in the dorsal horn requires microglial cathepsin S. *Journal of Neuroscience, 29*, 6945–6954.

Clark, A. K., Wodarski, R., Guida, F., Sasso, O., & Malcangio, M. (2010a). Cathepsin S release from primary cultured microglia is regulated by the P2X7 receptor. *Glia, 58*, 1710–1726.

Clark, A. K., Staniland, A. A., Marchand, F., Kaan, T. K. Y., McMahon, S. B., & Malcangio, M. (2010b). P2X7-Dependent release of interleukin-1beta and nociception in the spinal cord following lipopolysaccharide. *Journal of Neuroscience, 30*, 573–582.

Clark, A. K., Marchand, F., D'Auria, M., Davies, M., Grist, J., Malcangio, M., & McMahon, S. B. (2010c). Cathepsin S inhibitor attenuates neuropathic pain and microglial response associated with spinal cord injury. *The Open Pain Journal, 3*, 117–122.

Clark, A. K., Grist, J., Al-Kashi, A., Perretti, M., & Malcangio, M. (2012). Microglial Cathepsin S and CX3CR1 contribute to chronic pain in collagen induced arthritis. *Arthritis and Rheumatism, 64*, 2038–2047.

Cole, T., Dickson, P. W., Esnard, F., Averill, S., Risbridger, G. P., Gauthier, F., & Schreiber, G. (1989). The cDNA structure and expression analysis of the genes for the cysteine proteinase inhibitor cystatin C and for beta2-microglobulin in rat brain. *European Journal of Biochemistry, 186*, 35–42.

Costigan, M., Befort, K., Karchewski, L., Griffin, R. S., D'Urso, D., Allchorne, A., Sitarski, J., Mannion, J. W., Pratt, R. E., & Woolf, C. J. (2002). Replicate high-density rat genome oligonucleotide microarrays reveal hundreds of regulated genes in the dorsal root ganglion after peripheral nerve injury. *BMC Neuroscience, 16*, 16.

Coulombe, R., Grochulski, P., Sivaraman, J., Menard, R., Mort, J. S., & Cygler, M. (1996). Structure of human procathepsin L reveals the molecular basis of inhibition by the prosegment. *EMBO Journal, 15*, 5492–5503.

Cuozzo, J. W., Tao, K., Ql, W., Young, W., & Sahagian, G. G. (1995). Lysine-based structure in the proregion of procathepsin L is the recognition site for mannose phosphorylation. *Journal of Biological Chemistry, 270*, 15611–15619.

Dalrymple, S. A, Grist, J., Clark, A. K., & Malcangio, M. (2009). Systemic inhibition of cathepsin S attenuates vincristine-induced neuropathic hypersensitivity.

Di Virgilio, F. (2007). Liaisons dangereuses: P2X7 and the inflammasome. *Trends in Pharmacological Sciences, 28*, 465–472.

Driessen, C., Bryant, R. A. R., Lennon-Dumenil, A. M., Villadangos, J. A., Bryant, P. W., Shi, G. P., Chapman, H. A., & Ploegh, H. L. (1999). Cathepsin S controls the trafficking and maturation of MHC class II molecules in dendritic cells. *The Journal of Cell Biology, 147*, 775–790.

Eisenach, J. C., Thomas, J. A., Rauck, R. L., Curry, R., & Li, X. (2004). Cystatin C in cerebrospinal fluid is not a diagnostic test for pain in humans. *Pain, 107*, 207–212.

Ferrari, D., Pizzirani, C., Adinolfi, E., Lemoli, R. M., Curti, A., Idzko, M., Panther, E., & Di Virgilio, F. (2006). The P2X7 receptor: A key player in IL-1 processing and release. *Journal of Immunology, 176*, 3877–3883.

Ghosh, P., Amaya, M., Mellins, E., & Wiley, D. C. (1995). The structure of an intermediate in class II MHC maturation: CLIP bound to HLA-DR3. *Nature, 378*, 457–462.

Gresser, O., Weber, E., Hellwig, A., Riese, S., & Regnier-Vigouroux, A. (2001). Immunocompetent astrocytes and microglia display major differences in the processing of the invariant chain and in the expression of active cathepsin L and cathepsin S. *European Journal of Immunology, 31*, 1813–1824.

Gutmann, H. R., & Fruton, J. S. (1948). On the proteolytic enzymes of animal tissues VIII. An intracellular enzyme related to chymotrypsin. *Journal of Biological Chemistry, 174*, 851–858.

Harrison, J. K., Jiang, Y., Chen, S., Xia, Y., Maciejewski, D., McNamara, R. K., Streit, W. J., Salafranca, M. N.,

Adhikari, S., Thompson, D. A., Botti, P., Bacon, K. B., & Feng, L. (1998). Role for neuronally derived fractalkine in mediating interactions between neurons and CX3CR1-expressing microglia. *PNAS, 95*, 10896–10901.

Holmes, F. E., Arnott, N., Vanderplank, P., Kerr, N. C., Longbrake, E. E., Popovich, P. G., Imai, T., Combadiere, C., Murphy, P. M., & Wynick, D. (2008). Intra-neural administration of fractalkine attenuates neuropathic pain-related behaviour. *Journal of Neurochemistry, 106*, 640–649.

Holsinger, L. I., Grist, J., Clark, A. K., Booth, R., Malcangio, M., & Dalrymple, S. A. (2010). Systemic inhibition of cathepsin S attenuates hypersensitivity in animal models of neuropathic and inflammatory pain.

Hughes, P. M., Botham, M. S., Frentzel, S., Mir, A., & Perry, V. H. (2002). Expression of fractalkine (CX3CL1) and its receptor, CX3CR1, during acute and chronic inflammation in the rodent CNS. *Glia, 37*, 314–327.

Inoue, K., & Tsuda, M. (2009). Microglia and neuropathic pain. *Glia, 57*, 1469–1479.

Irie, O., et al. (2008). Discovery of orally bioavailable cathepsin S inhibitors for the reversal of neuropathic pain. *Journal of Medicinal Chemistry, 51*, 5502–5505.

Ishidoh, K., Muno, D., Sato, N., & Kominami, E. (1991). Molecular cloning of cDNA for rat cathepsin C. Cathepsin C, a cysteine proteinase with an extremely long propeptide. *Journal of Biological Chemistry, 266*, 16312–16317.

Jin, S. X., Zhuang, Z. Y., Woolf, C. J., & Ji, R. R. (2003). p38 mitogen-activated protein kinase is activated after a spinal nerve ligation in spinal cord microglia and dorsal root ganglion neurons and contributes to the generation of neuropathic pain. *Journal of Neuroscience, 23*, 4017–4022.

Kominami, E., Tsukahara, T., Hara, K., & Katunuma, N. (1988). Biosyntheses and processing of lysosomal cysteine proteinases in rat macrophages. *FEBS Letters, 231*, 225–228.

Kubo, T., Yamashita, T., Yamaguchi, A., Hosokawa, K., & Tohyama, M. (2002). Analysis of genes induced in peripheral nerve after axotomy using cDNA microarrays. *Journal of Neurochemistry, 82*, 1129–1136.

Levine, J. D., & Reichling, D. B. (1999). Peripheral mechanisms of inflammatory pain. In P. D. Wall & R. Melzack (Eds.), *Texbook of pain* (pp. 59–84). Livingstone: Churchill.

Lindia, J. A., McGowan, E., Jochnowitz, N., & Abbadie, C. (2005). Induction of CX3CL1 expression in astrocytes and CX3CR1 in microglia in the spinal cord of a rat model of neuropathic pain. *The Journal of Pain, 6*, 434–438.

Liuzzo, J. P., Petanceska, S. S., & Devi, L. A. (1999a). Neurotrophic factors regulate cathepsin S in macrophages and microglia: A role in the degradation of myelin basic protein and amyloid beta peptide. *Molecular Medicine, 5*, 334–343.

Liuzzo, J. P., Petanceska, S. S., Moscatelli, D., & Devi, L. A. (1999b). Inflammatory mediators regulate cathepsin S in macrophages and microglia: A role in attenuating heparan sulfate interactions. *Molecular Medicine, 5*, 320–333.

Lopez-Castejon, G., Theaker, J., Pelegrin, P., Clifton, A. D., Braddock, M., & Surprenant, A. (2010). P2X(7) receptor-mediated release of cathepsins from macrophages is a cytokine-independent mechanism potentially involved in joint diseases. *Journal of Immunology, 185*, 2611–2619.

Mannes, A. J., Martin, B. M., Yang, H. Y., Keller, J. M., Lewin, S., Gaiser, R. R., & Iadarola, M. J. (2003). Cystatin C as a cerebrospinal fluid biomarker for pain in humans. *Pain, 102*, 251–256.

Mason, R. W., Gal, S., & Gottesman, M. M. (1987). The identification of the major excreted protein (MEP) from a transformed mouse fibroblast cell line as a catalytically active precursor form of cathepsin L. *Biochemical Journal, 248*, 449–454.

McMahon, S. B., & Malcangio, M. (2009). Current challenges in glia-pain biology. *Neuron, 64*, 46–54.

Milligan, E. D., Zapata, V., Chacur, M., Schoeniger, D., Biedenkapp, J., O'Connor, K. A., Verge, G. M., Chapman, G., Green, P., Foster, A. C., Naeve, G. S., Maier, S. F., & Watkins, L. R. (2004). Evidence that exogenous and endogenous fractalkine can induce spinal nociceptive facilitation in rats. *European Journal of Neuroscience, 20*, 2294–2302.

Milligan, E., Zapata, V., Schoeniger, D., Chacur, M., Green, P., Poole, S., Martin, D., Maier, S. F., & Watkins, L. R. (2005). An initial investigation of spinal mechanisms underlying pain enhancement induced by fractalkine, a neuronally released chemokine. *European Journal of Neuroscience, 22*, 2775–2782.

Murphy, P. M., Baggiolini, M., Charo, I. F., Hebert, C. A., Horuk, R., Matsushima, K., Miller, L. H., Oppenheim, J. J., & Power, C. A. (2000). International union of pharmacology XXII. Nomenclature for chemokine receptors. *Pharmacology Review, 52*, 145–176.

Nakagawa, T. Y., Brissette, W. H., Lira, P. D., Griffiths, R. J., Petrushova, N., Stock, J., McNeish, J. D., Eastman, S. E., Howard, E. D., & Clarke, S. R. M. (1999). Impaired invariant chain degradation and antigen presentation and diminished collagen-induced arthritis in cathepsin S null mice. *Immunity, 10*, 207–217.

Nishiyori, A., Minami, M., Ohtani, Y., Takami, S., Yamamoto, J., Kawaguchi, N., Kume, T., Akaike, A., & Satoh, M. (1998). Localization of fractalkine and CX3CR1 mRNAs in rat brain: Does fractalkine play a role in signaling from neuron to microglia? *FEBS Letters, 429*, 167–172.

Pan, Y., Lloyd, C., Zhou, H., Dolich, S., Deeds, J., Gonzalo, J. A., Vath, J., Gosselin, M., Ma, J., Dussault, B., Woolf, E., Alperin, G., Culpepper, J., Gutierrez-Ramos, J. C., & Gearing, D. (1997). Neurotactin, a membrane-anchored chemokine upregulated in brain inflammation. *Nature, 387,* 611–617.

Petanceska, S., & Devi, L. (1992). Sequence analysis, tissue distribution, and expression of rat cathepsin S. *Journal of Biological Chemistry, 267,* 26038–26043.

Petanceska, S., Burke, S., Watson, S. J., & Devi, L. (1994). Differential distribution of messenger RNAs for cathepsins B, L and S in adult rat brain: An in situ hybridization study. *Neuroscience, 59,* 729–738.

Petanceska, S., Canoll, P., & Devi, L. A. (1996). Expression of rat cathepsin S in phagocytic cells. *Journal of Biological Chemistry, 271,* 4403–4409.

Pierre, P., & Mellman, I. (1998). Developmental regulation of invariant chain proteolysis controls MHC class II trafficking in mouse dendritic cells. *Cell, 93,* 1135–1145.

Podobnik, M., Kuhelj, R., Turk, V., & Turk, D. (1997). Crystal structure of the wild-type human procathepsin B at 2.5 A resolution reveals the native active site of a papain-like cysteine protease zymogen. *Journal of Molecular Biology, 271,* 774–788.

Riese, R. J., Wolf, P. R., Bromme, D., Natkin, L. R., Villadangos, J. A., Ploegh, H. L., & Chapman, H. A. (1996). Essential role for cathepsin S in MHC class II-associated invariant chain processing and peptide loading. *Immunity, 4,* 357–366.

Riese, R. J., Mitchell, R. N., Villadangos, J. A., Shi, G. P., Palmer, J. T., Karp, E. R., De Sanctis, G. T., Ploegh, H. L., & Chapman, H. A. (1998). Cathepsin S activity regulates antigen presentation and immunity. *The Journal of Clinical Investigation, 101,* 2351–2363.

Ritonja, A., Colic, A., Dolenc, I., Ogrinc, T., Podobnik, M., & Turk, V. (1991). The complete amino acid sequence of bovine cathepsin S and a partial sequence of bovine cathepsin L. *FEBS Letters, 283,* 329–331.

Rowan, A. D., Mason, P., Mach, L., & Mort, J. S. (1992). Rat procathepsin B. Proteolytic processing to the mature form in vitro. *Journal of Biological Chemistry, 267,* 15993–15999.

Rozman, J., Stojan, J., Kuhelj, R., Turk, V., & Turk, B. (1999). Autocatalytic processing of recombinant human procathepsin B is a bimolecular process. *FEBS Letters, 459,* 358–362.

Schwaeble, W. J., Stover, C. M., Schall, T. J., Dairaghi, D. J., Trinder, P. K. E., Linington, C., Iglesias, A., Schubart, A., Lynch, N. J., Weihe, E., & Schafer, M. K. (1998). Neuronal expression of fractalkine in the presence and absence of inflammation. *FEBS Letters, 439,* 203–207.

Schwarz, G., Boehncke, W. H., Braun, M., Schroter, C. J., Burster, T., Flad, T., Dressel, D., Weber, E., Schmid, H., & Kalbacher, H. (2002). Cathepsin S activity is detectable in human keratinocytes and is selectively upregulated upon stimulation with interferon-gamma. *The Journal of Investigative Dermatology, 119,* 44–49.

Shi, G. P., Munger, J. S., Meara, J. P., Rich, D. H., & Chapman, H. A. (1992). Molecular cloning and expression of human alveolar macrophage cathepsin S, an elastinolytic cysteine protease. *Journal of Biological Chemistry, 267,* 7258–7262.

Shi, G. P., Villadangos, J. A., Dranoff, G., Small, C., Gu, L., Haley, K. J., Riese, R., Ploegh, H. L., & Chapman, H. A. (1999). Cathepsin S required for normal MHC class II peptide loading and germinal center development. *Immunity, 10,* 197–206.

Staniland, A. A., Clark, A. K., Wodarski, R., Sasso, O., Maione, F., D'Acquisto, F., & Malcangio, M. (2010). Reduced inflammatory and neuropathic pain and decreased spinal microglial response in fractalkine receptor (CX3CR1) knockout mice. *Journal of Neurochemistry, 114,* 1143–1157.

Takio, K., Towatari, T., Katunuma, N., Teller, D. C., & Titani, K. (1983). Homology of amino acid sequences of rat liver cathepsins B and H with that of papain. *PNAS, 80,* 3666–3670.

Tao, K., Stearns, N. A., Dong, J. M., Wu, Q. L., & Sahagian, G. G. (1994). The proregion of cathepsin L is required for proper folding, stability, and ER exit. *Archives of Biochemistry and Biophysics, 311,* 19–27.

Tsuda, M., Mizokoshi, A., Shigemoto-Mogami, Y., Koizumi, S., & Inoue, K. (2004). Activation of p38 mitogen-activated protein kinase in spinal hyperactive microglia contributes to pain hypersensitivity following peripheral nerve injury. *Glia, 45,* 89–95.

Turk, B., Dolenc, I., Lenarcic, B., Krizaj, I., Turk, V., Bieth, J. G., & Bjork, I. (1999). Acidic pH as a physiological regulator of human cathepsin L activity. *European Journal of Biochemistry, 259,* 926–932.

Turnsek, T., Kregar, I., & Lebez, D. (1975). Acid sulphydryl protease from calf lymph nodes. *Biochimica et Biophysica Acta (BBA) Enzymology, 403,* 514–520.

Valder, C. R., Liu, J. J., Song, Y. H., & Luo, Z. D. (2003). Coupling gene chip analyses and rat genetic variances in identifying potential target genes that may contribute to neuropathic allodynia development. *Journal of Neurochemistry, 87,* 560–573.

Vasiljeva, O., Dolinar, M., Pungercar, J. R., Turk, V., & Turk, B. (2005). Recombinant human procathepsin S is capable of autocatalytic processing at neutral pH in the presence of glycosaminoglycans. *FEBS Letters, 579,* 1285–1290.

Verge, G. M., Milligan, E. D., Maier, S. F., Watkins, L. R., Naeve, G. S., & Foster, A. C. (2004). Fractalkine (CX3CL1) and fractalkine receptor (CX3CR1) distribution in spinal cord and dorsal root ganglia under basal and neuropathic pain conditions. *European Journal of Neuroscience, 20,* 1150–1160.

Villadangos, J. A., Riese, R. J., Peters, C., Chapman, H. A., & Ploegh, H. L. (1997). Degradation of mouse invariant chain: Roles of cathepsins S and D and the influence of major histocompatibility complex polymorphism. *The Journal of Experimental Medicine, 186,* 549–560.

Wada, K., Takai, T., & Tanabe, T. (1987). Amino acid sequence of chicken liver cathepsin L. *European Journal of Biochemistry, 167,* 13–18.

Wiederanders, B., Broemme, D., Kirschke, H., Kalkkinen, N., Rinne, A., Paquette, T., & Toothman, P. (1991). Primary structure of bovine cathepsin S comparison to cathepsins L, H, B and papain. *FEBS Letters, 286*, 189–192.

Zhuang, Z. Y., Kawasaki, Y., Tan, P. H., Wen, Y. R., Huang, J., & Ji, R. R. (2007). Role of the CX3CR1/p38 MAPK pathway in spinal microglia for the development of neuropathic pain following nerve injury-induced cleavage of fractalkine. *Brain, Behavior, and Immunity, 21*, 642–651.

Cauda Equina

Definition

A syndrome of fluctuating weakness and sensory loss caused by ischemia of the lumbosacral roots in a narrow spinal canal. It also refers to a collection of spinal roots descending from the lower spinal cord and occupying the vertebral canal below the cord.

Cross-References

▶ Chronic Back Pain and Spinal Instability
▶ Radiculopathies

Caudal Analgesia or Anesthesia

Definition

Regional anesthesia by injection of local anesthetic solution or other drugs into the epidural space via sacral hiatus. It is mainly used to provide anesthesia and analgesia in pediatric cases.

Cross-References

▶ Caudal Epidural Blocks
▶ Postoperative Pain, Epidural Infusions

References

Johr, M., & Berger, T. M. (2012). Caudal blocks. *Paediatric Anaesthesia, 22*, 44–50.

Caudal Epidural Blocks

Definition

Sensory and motor block of the thoracic, lumbar, or sacral nerves achieved by injecting local anesthetic via needle or catheter inserted through the sacral hiatus into the epidural space.

Cross-References

▶ Caudal Analgesia or Anesthesia

Caudal Epidural Steroids

▶ Epidural Steroid Injections

Caudal Injection

▶ Epidural Steroid Injections for Chronic Back Pain

Caudalis, Vc

▶ Subnucleus Caudalis

Caudate Nucleus

Definition

One of the main nuclei of the basal ganglia; part of the striatum connected principally with prefrontal and other association areas of cortex.

Cross-References

► Nociceptive Processing in the Nucleus Accumbens, Neurophysiology and Behavioral Studies
► Parafascicular Nucleus, Pain Modulation

Causalgia

Synonyms

CRPS2; Traumatic peripheral neuropathy

Definition

A syndrome of sustained burning pain, allodynia, and hyperpathia after a traumatic nerve lesion, often combined with vasomotor and sudomotor dysfunction and later trophic changes.

Cross-References

► Causalgia: Assessment
► Complex Regional Pain Syndrome
► Complex Regional Pain Syndrome and the Sympathetic Nervous System
► Complex Regional Pain Syndromes, Clinical Aspects
► CRPS, Evidence-Based Treatment
► Postoperative Pain, Acute Presentation of Complex Regional Pain Syndrome
► Sympathetically Maintained Pain in CRPS II, Human Experimentation

Causalgia and Reflex Sympathetic Dystrophy

► Opioids in the Management of Complex Regional Pain Syndrome

Causalgia: Assessment

Ellen Jørum
The Laboratory of Clinical Neurophysiology, Rikshospitalet University, Oslo, Norway

Definition

► Causalgia is defined as "A syndrome of sustained burning pain, ► allodynia and ► hyperpathia after a traumatic nerve lesion, often combined with vasomotor and ► sudomotor dysfunction and later trophic changes" (Classification of chronic pain 1994). This is an old term, and causalgia will now be called ► complex regional pain syndrome type 2 – CRPS-II (with nerve lesion).

Causalgia, or complex regional pain syndrome, is as the name implies a complex pain syndrome, being characterized by the presence of ► spontaneous pain and alterations in sensibility including allodynia and hyperpathia, as well as symptoms and signs of autonomic dysfunction.

Since this entry will be dealing with "assessments," the complicated neurophysiological mechanisms will not be mentioned. The etiology is given since we are talking about a traumatic nerve lesion. This may occur in relation to all sorts of injuries.

Characteristics

Assessment of Nerve Lesion

This is a pain syndrome occurring after a traumatic nerve lesion, and much emphasis will have to be put on the identification of a specific nerve lesion.

The best way to detect and to assess a peripheral nerve lesion is to perform a detailed clinical neurological examination of the patient, including investigation of motor and sensory function and myotatic reflexes. Since pain may arise following either a sensory or a mixed motor and sensory nerve, a detailed mapping of motor and sensory deficits is crucial. In most cases, there will be a lesion of sensory nerve fibers, and a careful study of the innervation territory of sensory deficit is warranted. However, because of the involvement of central sensitization mechanisms, the sensory deficit may sometimes exceed the innervation territory of the nerve; one may proceed by admitting the patient to a clinical neurophysiologist for EMG (electromyography) and neurography. By neurography, measurements of conduction velocities and other variables such as distal delay, motor and sensory amplitudes, and latency of late volleys such as the H and F waves are measured. Nerve conduction studies play an important role in precisely delineating the extent and distribution of a peripheral nerve lesion and give some indication of nerve-root pathology (by evaluation of late reflexes) (Jørum and Arendt-Nielsen 2003). ▶ Electrodiagnostic studies are capable of demonstrating the peripheral nerve injury of causalgia or CRPS-II (Devor 1983). Neurography does not evaluate the function of thin nerve fibers such as A δ mediating cold/sharp pain and C-fibers mediating the sensation of heat, heat pain, and some forms of tactile pain. These nerve fibers may be evaluated by quantitative sensory testing (QST) (Gracely et al. 1996). It is important to note that the number of nerves available for neurography is restricted. Neurography will routinely be performed on the major nerves of the upper extremity (median, ulnar, and radial nerves) and in the lower extremity (peroneal, tibial, and sural nerves); it is also possible, to some extent, to perform neurography of a few proximal sensory nerves in both upper and lower extremity. EMG is also a helpful method in evaluating neurogenic affection of muscles, and since it may be performed in a large extent of muscles, more nerves may hereby be examined.

Assessment of Clinical Pain

The pain in causalgia will have the same possible characteristics as pain following neuropathic pain in general, including the presence of spontaneous and evoked pain. Although pain following nerve lesions has often been described as burning, there are no pain descriptors which are specific to neuropathic pain (and thereby to causalgia).

Pain in causalgia may be described by a large number of adjectives, of which burning, throbbing, and aching are only a few examples. Spontaneous pain may be ongoing and/or paroxysmal, and the intensity may be evaluated by a visual analogue score (0–10 cm) or (0–100 mm), where 0 represents no pain and 10 or 100 the worst thinkable pain. Pain intensity may spontaneously vary in intensity but is often aggravated by physical activity and exposure to cold. Lightly touching the painful area will, in most cases, provoke a severe pain, but pain may also be evoked by thermal stimuli (frequently cold, not so frequently heat).

Assessment of Allodynia/Hyperpathia

Since causalgia may be characterized by the presence of allodynia and hyperpathia, an assessment of these phenomena is important. Both are described elsewhere, but it is briefly repeated here that allodynia is defined as "pain due to a stimulus which does not normally provoke pain" and hyperpathia as "a painful syndrome characterized by an abnormally painful reaction to a stimulus, especially a repetitive stimulus, as well as an increased threshold" (Classification of chronic pain 1994). Allodynia and hyperpathia to both ▶ tactile stimuli and ▶ thermal stimuli may be tested. In clinical practice, allodynia to light touch is the easiest to perform. One may apply a cotton swab or a brush, gently moving it over the painful area, and record whether this normally nonpainful stimulus evokes pain or not. Hyperpathia to light touch may also be present, but in general, the two phenomena may coexist. For the determination of allodynia and hyperpathia to thermal stimuli, special equipment is generally needed. One may obtain some indication of the presence or not of allodynia to cold or heat by applying thermorollers over the skin, rollers which are set at fixed temperatures, i.e., 25 °C and 40 °C. The rollers are first and foremost manufactured for testing reduced sensibility, and since heat pain is normally perceived at temperatures around 42–43 °C, it may be difficult to conclude with heat allodynia if pain is perceived at 40 °C. Cold allodynia, on the other hand, may be easier to demonstrate. In the upper extremity, cold pain will normally be seen at a threshold of 10–15 °C, while in the lower extremity, it will be

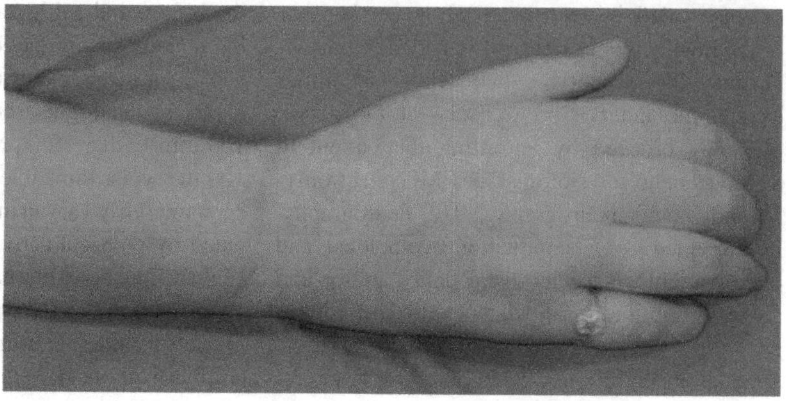

Causalgia: Assessment,
Fig. 1 Edema in the right
hand in a patient with CRPS

Causalgia: Assessment, Fig. 1 Edema in the right hand in a patient with CRPS

below 10 °C. For more accurate determinations of thresholds for cold allodynia or heat allodynia, quantitative sensory testing (QST) may be performed with the use of a thermotest. Cold allodynia will be the most prominent finding and will not be different from cold allodynia in other cases of neuropathic pain (Jørum et al. 2003).

Heat allodynia may exist but is seen less frequently. Hyperalgesia/hyperpathia to pinprick may also be demonstrated, and as for allodynia to light touch, it may be of value to map the area by moving the von Frey hair or a needle from normal skin centripetally towards the area of hyperalgesia.

Assessment of Vasomotor and Sudomotor Dysfunction

The inclusion of possible vasomotor and sudomotor dysfunction is essential for the diagnosis of CRPS-II or causalgia. The clinician should look for signs like edema (Fig. 1), sweat dysregulation (usually increased sweating but also possibly reduced sweating), alterations in skin temperature (cooler or warmer) reflecting vasomotor changes, and trophic changes of the skin, hair, and nails. The diagnosis of a full-developed CRPS is not difficult. However, the diagnosis of milder cases may prove difficult, especially since the patients' clinical picture may change over time (vasodilatation at first, then vasoconstriction, and finally dystrophic changes), and the dynamic alterations may also include diurnal fluctuations. The evaluation of autonomic dysfunction may in most cases be performed by a clinical examination, but laboratory tests may prove helpful. These can measure changes in autonomic function with higher sensitivity and more objectivity. Most laboratory studies of autonomic dysfunction in patients with CRPS have been conducted on patients with CRPS type 1 (formerly reflex sympathetic dystrophy). One must add that a general assumption is that CRPS-I or CRPS-II does not differ in the changes believed to be dependent on the sympathetic nervous system (Stanton-Hicks et al. 1995). Various tests may be employed to assess the function of the autonomic nervous system, both well-validated routine tests and more experimental procedures. For the assessment of sudomotor function, sympathetic skin response and quantitative sudomotor axon reflex test (QSART) may be employed. Recordings of skin potentials from the foot or hand may be made following a stimulus such as electric shock, a noise, a cough, or an inspiratory gasp. The advantage of this method is that it is easy to perform in a routine clinical neurophysiological laboratory. The disadvantage is its large variability, the tendency of the responses to habituate, and that it has little sensitivity, especially compared with the more sophisticated QSART. In the latter test, the sweat output in response to iontophoretic application of 10 % acetylcholine is recorded by a sudorometer. The major advantage is that the test is sensitive and reproducible in controls and in patients with neuropathy (Low 2003). The disadvantage is that the equipment has only recently become commercially available and is not yet represented in many laboratories. In a study of 102 patients with CRPS-I by Sandroni and coworkers (1998), they found that some of the indices that correlated most reliably with clinical data, and with each other, were QSART and skin temperature reductions. The authors computed the sudomotor index from the change in sweat volume, latency, and persistent sweat activity. QSART was positive at a single site if the affected side showed changes 50 % greater than the nonaffected side.

The assessment of vasomotor (vasoconstriction/vasodilatation) dysfunction may be performed by indirect methods, such as measuring of skin temperature or thermography, and by more sophisticated laser Doppler examinations. The measurement of skin temperature is an easy but indirect way to detect changes in vasomotor function, where a difference between the affected and nonaffected extremity exceeding 1 °C will be regarded as significant. Thermography is also easy to perform, but has the disadvantage that it will be regarded as an indirect way of testing vasomotor function. Laser Doppler investigations are also easy to perform, by measuring flow at a limited area or by scanning a larger area, but interpretations of data related to pathophysiology may be difficult. Elegant examinations are performed with laser Doppler examination, as described in a paper by Weber et al. (2001), in combination with transcutaneous electrical stimulation. They found that the vasodilatation as a response to electrical stimulation of the skin increased significantly (more pronounced vasodilatation) in a group of mainly CRPS-type-1 patients compared to normal controls. There have been many critical comments to the IASP diagnostic criteria for CRPS in general, and many authors have questioned the specificity of the criteria.

In a study by Bruehl and coworkers (1999) of 117 patients meeting the IASP criteria for CRPS and 43 patients with neuropathic pain from diabetic neuropathy, they found that signs or symptoms of edema versus color changes versus sweat dysregulation satisfied three criteria and discriminated between the groups. Although diagnostic sensitivity was high at 98 %, specificity was poor at 36 %. The diagnosis of CRPS was likely to be correct in only 40 % of the cases, and the results of this study suggested that the criteria used by IASP had inadequate specificity and were likely to lead to overdiagnosis. However, it was again emphasized that this study was performed on patients with CRPS of both type 1 (approximately two-thirds) and type 2.

References

Bruehl, S., Harden, R. N., Galer, B. S., Saltz, S., Betram, M., Backonja, M., Gayles, R., Rudin, N., Bhugra, M. K., & Stanton-Hicks, M. (1999). External validation of iasp diagnostic criteria for complex regional pain syndrome and proposed research diagnostic criteria. Pain, 81, 147–154.

Devor, M. (1983). Nerve pathophysiology and mechanisms of pain in causalgia. Journal of the Autonomic Nervous System, 7, 371–384.

Gracely, R. H., Price, D. D., Rogers, W. J., & Bennett, G. J. (1996). Quantitative sensory testing in patients with complex regional pain syndrome (CRPS) l and ll. In W. Jänig & M. Stanton-Hicks (Eds.), Reflex sympathetic dystrophy: A reappraisal, progress in pain research and management (Vol. 6, p. 151). Seattle: IASP Press.

Jørum, E., Warncke, T., & Stubhaug, A. (2003). Cold allodynia and hyperalgesia in neuropathic pain: The effect of N-methyl-D-aspartate (NMDA) receptor antagonist ketamine: A double-blind, cross-over comparison with alfentanil and placebo. Pain, 101, 229–235.

Low, P. (2003). Testing the autonomic nervous system. Seminars in Neurology, 23, 407–421.

Sandroni, P., Low, P. A., Ferrer, T., Opfer-Gehrking, T., Willner, C., & Wilson, P. R. (1998). Complex regional pain syndrome (CRPS 1): Prospective study and laboratory evaluation. The Clinical Journal of Pain, 14, 282–289.

Stanton-Hicks, M., Jänig, W., Hassenbusch, S., et al. (1995). Reflex sympathetic dystrophy: Changing concepts and taxonomy. Pain, 63, 127–133.

Weber, M., Birklein, F., Neundörfer, B., & Schmelz, M. (2001). Facilitated neurogenic inflammation in complex regional pain syndrome. Pain, 91, 251–257.

Cawthorne and Cookseys' Eye-Head Exercises

▶ Coordination Exercises in the Treatment of Cervical Dizziness

CBT

▶ Cognitive-Behavioral Therapy

CCI

▶ Neuropathic Pain Model, Chronic Constriction Injury

CCI Model

▶ Chronic Constriction Injury Model

CCK

▶ Cholecystokinin

CDH (Chronic Daily Headache)

▶ Chronic Daily Headache in Children

Celiac Plexus Block

Definition

The celiac plexus is involved in nociceptive pain transmission from intra-abdominal organs from

the distal esophagus to the transverse, and sometimes to the descending colon. Interruption of nociceptive signals by injection of phenol or alcohol onto the plexus may provide excellent and prolonged pain relief for patients suffering particularly from gastric and pancreatic carcinoma.

Cross-References

▶ Cancer Pain Management, Anesthesiologic Interventions, and Neural Blockade

Cell Adhesion

Definition

An intercellular connection by which cells stick to other cells or noncellular components of their environment. Cell adhesion generally requires special protein complexes at the surface of cells.

Cross-References

▶ NSAIDs and Cancer

Cell Grafts

▶ Cell Therapy in the Treatment of Central Pain

Cell Minipumps

▶ Cell Therapy in the Treatment of Central Pain

Cell Therapy in the Treatment of Central Pain

Mary J. Eaton
Research Service (151), Miami VA Health Care System, Miami VAMC, Miami, FL, USA

Synonyms

Cell grafts; Cell minipumps; Cell transplantation

Definition

Cell therapy is the use of transplant of cells from primary or immortalized sources to reverse or reduce the symptoms or causes of pain that arise from injury to the central nervous system. These cell grafts can be placed near the original injury, such as in or near the spinal cord, or further away depending on how they are intended to affect that injury or the symptoms of the resultant pain. Primary cell sources are derived from a single cohort donor animal and cannot be expanded or kept in vitro for extended periods. Immortalized cell sources are derived from any animal source, but have been engineered to be, or are naturally, expandable in vitro, and provide multiple transplant sources.

Introduction

Even with continuous improvements in surgical management, physical therapy, and the availability of newer pharmacological agents, many patients following peripheral and central neural injuries continue to suffer from difficult-to-treat chronic pain. Newer treatments to modulate and reduce central pain include the use of cell grafts that release antinociceptive molecules synthesized by those cell transplants. Cell therapy to treat neuropathic pain after spinal cord injury (SCI) is in its infancy, but the development of cellular strategies that would replace or be used as an adjunct to current pharmacologic treatments for neuropathic pain have

progressed tremendously over the past 20 years. A variety of molecules are released by such grafts for ▶ antinociception to the spinal cord. Presumably, such cell grafts act as "cellular minipumps" which are able to release neuroactive antinociceptive molecules to the spinal cord pain processing pathways. Cell lines, rather than primary cell grafts, can also offer a renewable and possibly safe-to-use source of cells. Grafts of either primary or immortalized sources could be expected to reduce or eliminate side effects associated with the large doses of pharmacological agents required for centrally acting pain-reducing agents, such as opioids or antidepressants.

Characteristics

The earliest studies using cell transplants for pain were developed from the concept of descending or local spinal inhibitory neurotransmitter modulation of sensory information, and that these same agents released by cell grafts after injury could provide antinociception. A variety of ▶ neurotransmitters, peptides, opioids, and more recently ▶ neurotrophins have been implicated in spinal inhibition. These include the endogenous neurotransmitters serotonin (5-HT), noradrenaline, and γ-aminobutyric acid (GABA); the endogenous opioids β-endorphin and enkephalins; endogenous peptides such galanin; and neurotrophins such as brain-derived neurotrophic factor (BDNF). Many of the commonly used pharmacologic therapies target these agents' receptors and reuptake mechanisms to increase or imitate their presence in acute and chronic pain. But it was recognized as early as 1980s that these agents could be supplied by grafts of ▶ autologous primary adrenal medullary chromaffin cells after nerve injury. Chromaffin cells contain a cocktail of antinociceptive agents, peptides, and neurotrophins (Wilson et al. 1981). To be able to use chromaffin cell therapy in humans, adrenal chromaffin cell grafts were prepared from xenogeneic bovine sources and tested for antinociception after nerve injury (Sagen et al. 1993). There have been a variety of animal studies using primary medullary tissue or disso-ciated chromaffin cultures as a ▶ subarachnoid graft source to reduce behavioral hypersensitivity in models of SCI-induced pain (Brewer and Yezierski 1998). But nonhuman, xenogeneic tissue sources are not likely to be used clinically, even if they are more abundant, given their increased risk of antigenicity and rapid rejection by the human host. Similarly, adult human chromaffin tissue has been transplanted and tested in humans for terminal cancer pain (Pappas et al. 1997), which is often neuropathic pain. But when the immune response in the human host was examined after the grafts, it was concluded that further purification or ▶ immunoisolation of grafts would be needed to use such tissues in multiple transplants, given the antigenicity of the diverse cell sources. There is a recent report (Bhattacharya et al. 2002) of successful human fetal adrenal transplant to treat pain associated with rheumatoid arthritis, certainly suggesting that fetal or precursor chromaffin tissue could be used as an antinociceptive source (Bes et al. 2001). But such primary tissue sources for the purification and use of chromaffin cells are not likely to be homogeneous since they are often obtained from multiple donors. The ability to use and manipulate stable antinociceptive cell lines as a defined source has provided a rich literature for their experimental use in cell therapy.

To that end, a number of immortalized cell lines were derived from the rat brainstem to model how such cell line grafts might function in models of pain after grafting near the spinal cord. These cells express an oncogene, such as the temperature-sensitive ▶ allele of large T antigen (tsTag), that confers immortalization and is able to cause expansion of the cells at low temperatures in vitro. But this oncogene is downregulated at the higher transplant temperatures in the animal. One of these rat neural precursor lines was isolated from embryonic day 12.5 (E12.5) rat brainstem and immortalized with the tsTag sequence. This cell line has been made to synthesize and secrete the neurotrophin BDNF, by the addition of the sequence for rat BDNF to its ▶ genome, causing these cells to have improved survival in vitro and in vivo, and

develop a permanent serotonergic phenotype. Since additional 5-HT was postulated to have a beneficial antinociceptive effect on neuropathic pain, cells were placed in a lumbar subarachnoid location after sciatic nerve injury and pain. Grafts of these serotonergic cells placed 1 week after nerve injury and the development of severe hypersensitivity to thermal and tactile stimuli were able to potently and permanently reverse the symptoms of neuropathic pain (Eaton et al. 1997). Transplants of similar murine cell lines genetically engineered to synthesize and secrete potentially antinociceptive molecules such the inhibitory peptide galanin, the neurotrophin BDNF, and the inhibitory neurotransmitter GABA have been tested in the same partial nerve injury pain model (Eaton 2000).

This same bioengineered rat serotonergic cell line has been successfully used to reduce neuropathic pain and improve locomotor function after grafting (Hains et al. 2001b) in the SCI dorsal hemisection pain model. Pain reduction requires that the cells be placed in the subarachnoid space, where they can affect the dorsal horn pathways and reduce spinal neuronal hyperexcitability (Hains et al. 2003a), probably modulated through specific 5-HT receptors.

A similar bioengineering approach was used to immortalize primary embryonic rat and bovine chromaffin cells using tsTag for immortalization. Grafts of these cells placed in the subarachnoid space in a model of neuropathic pain after nerve injury were able to reduce that pain without forming tumors in the host animal (Eaton et al. 2000). But without complete removal of the oncogene before grafting, such an approach would not ever be completely safe as a clinical method for cell therapy for pain. The next advance in creation of cell lines for therapeutic use has been the development of reversibly immortalized cell lines, as modeled in rat chromaffin cell lines with an excisable oncogene.

▶ Studies exploiting site-specific ▶ DNA recombination and Cre/lox excision have suggested that cells can be targeted in vitro and in vivo for removal of deleterious genes, including the large T antigen. Reversible immortalization with Tag and Cre/lox technology was first reported with human fibroblasts by Westerman and Leblouch (Westerman and Leboulch 1996) and more recently with human myogenic cells and hepatocytes (Kobayashi et al. 2000). Introduction of the gene for Cre recombinase into a cell's genome allows for Cre to excise any recombinant sequences present that are flanked by small loxP sequences. Rat chromaffin cells have been immortalized with an oncogenic tsTag construct, utilizing retroviral infection of these early chromaffin precursors, where the tsTag construct was flanked by loxP sequences (floxed). Following isolation of immortalized cells, the cells were further infected with a retrovirus expressing the CrePR1 gene, which encodes for a fusion protein which combines Cre activity plus the mutant human steroid receptor, hPRB891. After incubation of the cells with the synthetic steroid RU486, the fusion protein is translocated to the cell nucleus which allows for Cre to excise the tsTag oncogene and the chromaffin cells are effectively dis-immortalized in vitro. Such reversibly immortalized chromaffin cells, which express many features of the primary chromaffin cell, are able to reverse neuropathic pain after transplant in a nerve injury model (Eaton et al. 2002). Such studies model the use of reversibly immortalized cell lines for cell therapy for human antinociception.

An entirely different approach with cell lines is the use of human neural lines that contain an oncogene that can be downregulated with agents such as retinoic acid (RA). An example of such a human neural cell line is the NT2 line (Andrews 1984), which after ▶ differentiation with RA for greater than 2 weeks in vitro, differentiates into stable human neural cells that never form tumors after grafting and have been used to ameliorate a variety of traumatic and neurodegenerative problems (Trojanowski et al. 1997). Such cells represent immortalized stem cells or neural progenitors that have been spontaneously immortalized and only differentiate after RA into the nontumor, neural phenotype. These cells offer great potential for the treatment of central pain, since they appear safe and can be easily used in the clinical setting, being placed via spinal tap into the subarachnoid space for pain.

Neuronal Precursors: NT2-Derived Cell Lines for Pain – Since the NT2 cell line contains a mixed phenotype population of cells, many of which would likely be antinociceptive based on multiple studies with rat cell lines by this author and others using central and peripheral models of neuropathic pain (Cejas et al. 2000; Duplan et al. 2004; Eaton et al. 1997, 1999a, b; Hains et al. 2001a, b, 2002b, 2003a; Stubley et al. 2001), it was considered likely that individual cell lines could be subcloned from the NT2 parental cell lines using ordinary subcloning techniques involving isolation of individual cells plated sparsely, allowing them to grow into colonies, surrounding these with cloning rings, and removing these colonies to establish individual cell lines. This rather laborious process resulted in a number of well-growing, morphologically and immunohistochemically distinct NT2 subclonal cell lines, numbered consecutively as they were isolated. Two of the cell lines were chosen for their potential to function as sources of neurotransmitters which might prove useful in further testing in animal models of pain, the hNT2.17 (Eaton et al. 2007) and hNT2.19 (Eaton et al. 2008) cell lines. Since they are derived from the neuroprogenitor parent cell line NT2, these are considered to be human neuroprogenitor cell lines as well, resulting in a neuronal-limited phenotype. These human progenitor cell lines are being developed for clinical use. Their characterization and use in animal models reflect what will be required of any similar regenerative cell therapy for FDA approval (Chen et al. 2007; Sagen 2003).

The Human Neuronal GABA hNT2.17 Cell Line – Centrally induced excitotoxic SCI has been developed as one of the models of central neuropathic pain (Yezierski et al. 1998; Yezierski and Park 1998). Intraspinal injection of quisqualic acid (QUIS), a mixed AMPA/metabotropic receptor agonist, produces injury with pathological characteristics similar to those associated with ischemic and traumatic SCI (Liu et al. 1997). In addition to the pathological changes that this SCI induces, significant mechanical allodynia and thermal hyperalgesia have been shown to be important behavioral components, without the additional motor dysfunction seen in other SCI models (Dietz and Curt 2006). Each of these sensory behaviors is indicative of altered sensory function and/or pain, similar to that reported after SCI. After spinal transplantation of primary adrenal tissue grafts following QUIS injections, pain related behaviors, including the hypersensitivity to mechanical stimuli and "excessive grooming," were significantly reduced (Brewer and Yezierski 1998), demonstrating that cell therapy can be easily tested in this SCI model. Given the related loss of GABA inhibition that seems to accompany SCI and the induction of neuropathic pain (Meisner et al. 2010; Qin et al. 2009; Stiller et al. 1995), the excitotoxic SCI model was used to examine the hNT2-derived GABAergic hNT2.17 cell transplant into the lumbar subarachnoid space following injury and the ability of those grafts to reverse behavioral hypersensitivity (Eaton et al. 2007). These cells cease to express tumor genes; express an exclusively neuronal, GABAergic, and glycinergic phenotype; and synthesize, secrete, and release GABA and glycine into the extracellular environment with differentiation (Eaton et al. 2007). Their morphology is similar to the GABA/glycine spinal interneurons found in the dorsal horn sensory laminae (Sutherland et al. 2002), and such characteristics are stable in more than 10 years of use in transplant studies. These inhibitory human neurons additionally colocalize GABA and glycine and the vesicular inhibitory amino acid transporter (VIATT, VGAT), especially along the neurite outgrowths in vitro, suggesting that this molecular machinery allows corelease in hNT2.17 cells (Wojcik et al. 2006), without the need for a separate glycine transporter. When differentiated hNT2.17 cells are placed 2 weeks after the QUIS SCI, mechanical allodynia and thermal hyperalgesia are potently and permanently attenuated, with no greater effect when twice the normal transplant dose (one million cells/i.t. injection) is used (Eaton and Wolfe 2009; Wolfe et al. 2007), or grafts are placed in the cervical subdural space (Eaton and Wolfe 2009). Besides transplant dose and graft placement, the immunosuppression regimen and

transplant time after SCI were also optimized (Eaton and Wolfe 2009). The same optimal transplant dose was only moderately effective when placed in chronic SCI, 6 weeks after SCI, compared to 100 % effectiveness when placed in an acute SCI, 2 weeks after injury. Additionally, maximal graft effectiveness required 2 weeks of immunosuppression with cyclosporine A (CsA; 10 mg/Kg), immediately following transplant. No immunosuppression or less lengthy exposure to CsA provided minimal or no attenuation (Eaton and Wolfe 2009). The "excessive grooming" behaviors associated with this model were also examined. When excessively grooming rats that had been transplanted with either viable or nonviable hNT2.17 cells and exposed to different immunosuppression regimens were examined for development, resolution, worsening, or no change of excessive grooming, a trend toward improvement was associated with viable grafts and at least 1 week of accompanying CsA immunosuppression. When transplant was delayed to 6 weeks, no improvement in excessive grooming was seen. This last finding duplicates what was seen in the QUIS SCI model and graft of antinociceptive adrenal medullary tissue (Brewer and Yezierski 1998), suggesting that potent reversal of behavioral hypersensitivity may have a neuroprotective effect on the progression of spinal-excitotoxicity-associated spinal lesions. Much as we (Eaton et al. 1998) and others (Ibuki et al. 1997) have seen in the unilateral chronic constriction injury (CCI) peripheral sciatic nerve injury model and antinociceptive cell grafts, Vaysse and colleagues (Vaysse et al. 2011) reported that the decrease in GABA expression in the spinal dorsal horn of CCI-injured animals is concomitant with a decline of its synthetic enzyme GAD67 immunoreactivity (ir) and mRNA but not GAD65. In hNT2.17-transplanted animals, a strong induction of GAD67 mRNA 1 week after graft was seen, which was followed by a recovery of GAD67 and GABA ir. This effect paralleled a reduction of hindpaw hypersensitivity and thermal hyperalgesia induced by CCI. These results suggest not only that hNT2.17 GABA cells can modulate neuropathic pain after CCI by minimizing the imbalance and restoring the cellular GABAergic pathway but that such a mechanism may be associated with any potent antinociceptive cell graft, at least in the CCI model. Disinhibition and loss of spinal GABA modulation are also well reported in SCI pain (Meisner et al. 2010). But evidence for a GABAergic mechanism associated with hNT2.17 transplant and antinociception in these other models of pain awaits further studies.

The Human Neuronal 5-HT hNT2.19 cell line- Current understanding of central and supraspinal (Shapiro 1997) mechanisms for the induction and maintenance of chronic pain after SCI also suggest a major role for the hypofunction of serotonergic (5-HT) inhibitory systems (Hains et al. 2002a, 2003b; Zhang et al. 1993). This same SCI leads to the loss of descending serotonergic excitatory inputs caudal to the lesion site and altered neurotransmitter status within the ventral horn a-motoneurons, which also contributes to motor dysfunction after SCI (Hains et al. 2001b; Saruhashi et al. 1996). A variety of animal studies have used a 5-HT rat cell line (Eaton et al. 1997; Hains et al. 2001a, b, 2003a) or 5-HT raphe transplants (Feraboli-Lohnherr et al. 1997; Ribotta et al. 2000) as a means to ameliorate some of these problems. Supplemental cell therapy can also work to create a spinal environment to ameliorate local damage and simultaneously promote a regenerative response in multiple axonal populations, including descending spinal serotonin fibers (Ramer et al. 2004), or reverse chronic pain after SCI by reversing the hyperexcitability in the dorsal horn pain processing centers (Hains et al. 2003a). The use of 5-HT cell therapy with a rat 5-HT cell line that is able to permanently reverse neuropathic pain that develops after partial nerve injury (Eaton et al. 1997) and hemisection SCI (Hains et al. 2002b) has been described. Similarly, the human neuronal 5-HT hNT2.19 cells used for cell therapy after severe contusive SCI reverses behavioral hypersensitivity (Eaton et al. 2011), without affecting motor dysfunction when grafts are placed intrathecally. These same cell grafts modestly recover motor function when placed intraspinally (Eaton et al. 2008) in the same severe contusion SCI model of

chronic pain and motor dysfunction. Additionally, grafts of hNT2.19 cells attenuate tactile allodynia and thermal hyperalgesia in the excitotoxic SCI QUIS model, much like grafts of hNT2.17 cells. In fact, lumbar intrathecal 5-HT hNT2.19 and GABA hNT2.19 grafts are equally nociceptive no matter which SCI pain model is used, excitotoxic or contusive, suggesting that these cells may affect the same or similar mechanism of action that is common to both models that initiate behavioral hypersensitivity. We have already shown a GABAergic mechanism of action for grafts of hNT2.17 cells (Vaysse et al. 2011) and suggested it may be common in SCI pain. Grafts of a 5-HT rat neuronal cell line is antinociceptive after hemisection SCI (Hains et al. 2001b) and depends on the graft location (Hains et al. 2002b), much like grafts of hNT2.19 cells. It does so by attenuating bilateral hyperexcitability of dorsal horn neurons (Hains et al. 2003a), restoring spinal serotonin, downregulating the serotonin transporter, and increasing BDNF tissue content in the spinal cord (Hains et al. 2001a). These same 5-HT rat cell line grafts also induce a GABAergic mechanism of action in the CCI model of nerve injury and neuropathic pain (Eaton et al. 1998). Obviously, it will be important in future studies to understand how each separate human neuronal cell line provides antinociception in each PNS or CNS pain model, but the same or similar mechanisms are not out of the question. Since the 5-HT hNT2.19 cells, like the hNT2.17 cell line, is exclusively neuronal, although with a different neurotransmitter phenotype, and equally nontumorigenic before and after transplant (Eaton et al. 2011), this human progenitor cell line is equally appropriate to develop as a clinical tool not only to treat neuropathic pain but also motor dysfunction, especially after SCI (Hernándeza et al. 2011; Ruff et al. 2012; Walker et al. 2010).

Stem Cells for Pain – There are few published successful methods to treat human SCI pain with stem cells. Their promise lies in the future and will likely require a degree of genetic or laboratory manipulation. Regenerative medicine and the use of human embryonic stem cells also currently engender ethical considerations. But, with rapid advances in the knowledge about the biology and ability to manipulate embryonic stem and precursor cells derived from CNS and other tissues, the future will likely have a place for this approach for cell therapy uses.

Recent examples include a recent report (Lin et al. 2002) utilizing the partial nerve injury with CCI to induce neuropathic pain; rat spinal embryonic progenitor cells (SPC) that used basic fibroblast growth factor-2 (FGF-2) for proliferation of the SPC in vitro were able to reduce thermal hyperalgesia after intrathecal transplant. Presumably, grafted cells had been induced to a GABAergic phenotype by FGF-2 in vitro and survived in its absence after transplant, maintaining their phenotype to modulate the neuropathic pain. The authors suggest that the grafts also increased the glycine content in the CSF of grafted animals, suggesting that if precursors could be induced to a phenotype that provides nociceptive inhibition, they would function much like cell minipumps, surviving in the intrathecal space. Also in the CCI model of peripheral pain (Klass et al. 2007), freshly isolated syngeneic marrow mononuclear cells were injected i.v. following the unilateral nerve injury and tactile allodynia and thermal hyperalgesia evaluated weekly. Marrow transplantation did not prevent pain, and 5 days after CCI, all animals were equivalently lesioned. However, 10 days after CCI, rats that received marrow transplants demonstrated paw withdrawal response and paw withdrawal latency patterns indicating recovery from pain, whereas untreated rats continued to have significant pain behavior patterns. The mechanisms underlying this improvement following bone marrow injection are unknown. The authors speculate that the marrow cells functioned as anti-inflammatory, neuroprotective, and proangiogenic, modulating ischemic, inflammatory, and cytotoxic events in the pain that follows nerve constriction in this model. However, marrow transplants are also known to exacerbate diabetic neuropathy in a different model of pain (Terashima et al. 2005). In this case, marrow cells fused with peripheral neurons, stimulating apoptosis.

One cause of severe neuropathic pain in traumatic injury that involves SCI is spinal root avulsion, and replacement of DRG neurons could reduce that pain. A recent study investigated whether human neural stem/progenitor cells (hNSPCs) transplanted to the DRG cavity can serve as a source for repairing lost peripheral sensory connections (Akesson et al. 2008). The hNSPCs robustly differentiate to neurons, which survive long-term transplantation. The neuronal population in the transplants was tightly surrounded by astrocytes, suggesting their active role in neuron survival. Furthermore, 3 months after grafting, hNSPCs were found in the dorsal root transitional zone (DRTZ) and within the spinal cord. The level of differentiation of transplanted cells was high in the core of the transplants, whereas cells that migrated to the DRTZ and spinal cord were undifferentiated, nestin-expressing precursors. However, hNSPCs are not sufficient to restore normal sensory function; additional factors are required to guide their differentiation to the desired type of neurons.

Another approach to pain reduction with regenerative methods is in the low back pain model or intervertebral disc replacement (Hiyama et al. 2008). Because disc disease progenitor cells existing in degenerate human discs are accompanied by a loss of cellularity, there is considerable interest in regeneration of cells of both the anulus fibrosus (AF) and nucleus pulposus (NP). Apparently, progenitor cells exist in degenerate human discs (Risbud et al. 2007), a result confirmed in degenerate discs extracted from rats. Cells isolated from degenerate human tissues expressed CD105, CD166, CD63, CD49a, CD90, CD73, p75 low affinity nerve growth factor receptor, and CD133/1, proteins that are characteristic of marrow mesenchymal stem cells. This suggests that these endogenous progenitors might be used to repair the intervertebral disc and ultimately reduce low back pain. In another study, human-bone-marrow-derived CD34(+) (hematopoietic progenitor cells) and CD34(+) (nonhematopoietic progenitor cells, including mesenchymal stem cells) cells were isolated, fluorescent labeled, and injected into rat coccygeal discs. Only the CD34(+) cells survived and differentiated within the relative immune privileged nucleus pulposus of intervertebral disc degeneration in this rat disc degeneration model (Wei et al. 2009). Olfactory cells, adult tissue-derived stem cell derived from the olfactory mucosa, may be another useful source for disc repair (Murrell et al. 2009) if they can differentiate into a nucleus pulposus chondrocyte phenotype in vivo after transplantation into the injured intervertebral disc. The donor-derived cells expressing CT2 and CSPG markers, specific for the nucleus pulposus chondrocyte phenotype, were achieved in after delivery to the nucleus pulposus region of rats whose discs had previously been lesioned before transplant, suggesting their potential for relief of human disc regeneration and back pain relief. But all these limited applications suggest that pain relief needs a more well-defined approach, modeled in the development of neural progenitors that have been recently developed.

Acknowledgments This material was based on work supported by the Department of Veterans Affairs, Veterans Health Administration, Rehabilitation Research and Development Service, grant B4438I and grant B4862R, both awarded to Dr. Mary J. Eaton; the Miami VA Healthcare System; the American Syringomyelia & Chiari Alliance Project; the Column of Hope, Chiari and Syringomyelia Research Foundation; and the continuing support of The Miami Project to Cure Paralysis and Department of Neurosurgery, Miller School of Medicine, Miami, Florida. The intellectual/technical assistance of Drs. Eva Widerström-Noga, Stacey Quintero Wolfe, and Alice Holohean and from assistants Miguel Martinez, Massiel Perez, and Yadira Salgueiro is gratefully appreciated.

References

Akesson, E., Sandelin, M., Kanaykina, N., Aldskogius, H., & Kozlova, E. N. (2008). Long-term survival, robust neuronal differentiation, and extensive migration of human forebrain stem/progenitor cells transplanted to the adult rat dorsal root ganglion cavity. *Cell Transplantation, 17*, 1115–1123.

Andrews, P. W. (1984). Retinoic acid induces neuronal differentiation of a cloned human embryonal carcinoma cell line in vitro. *Developmental Biology, 103*, 285–293.

Bes, J. C., Frydel, B. R., Potter, E. D., Walters, W. M., Castellanos, D. A., Brunschwig, J. P., & Sagen, J.

(2001). Human embryonic and fetal adrenal glands as sources of neural precursors for possible transplantation strategies. *American Society for Neural Transplantation and Repair, 8*, 61.

Bhattacharya, N., Chhetri, M. K., Mukherjee, K. L., Das, S. P., Mukherjee, M. A., Bhattacharya, M., & Bhattacharya, S. (2002). Human fetal adrenal transplant: A possible role in relieving intractable pain in advanced rheumatoid arthritis. *Clinical and Experimental Obstetrics and Gynecology, 29*, 197–206.

Brewer, K. L., & Yezierski, R. P. (1998). Effects of adrenal medullary transplants on pain-related behaviors following excitotoxic spinal cord injury. *Brain Research, 798*, 83–92.

Cejas, P. J., Martinez, M., Karmally, S., McKillop, M., McKillop, J., Plunkett, J. A., Oudega, M., & Eaton, M. J. (2000). Lumbar transplant of neurons genetically modified to secrete brain-derived neurotrophic factor attenuate allodynia and hyperalgesia after sciatic nerve constriction. *Pain, 86*, 195–210.

Chen, H. I., Bakshi, A., Royo, N. C., Magge, S. N., & Watson, D. J. (2007). Neural stem cells as biological minipumps: A faster route to cell therapy for the CNS? *Current Stem Cell Research & Therapy, 2*, 13–22.

Dietz, V., & Curt, A. (2006). Neurological aspects of spinal-cord repair: Promises and challenges. *Lancet Neurology, 5*, 688–694.

Duplan, H., Li, R. Y., Vue, C., Zhou, H., Emorine, L., Herman, J. P., Tafani, M., Lazorthes, Y., & Eaton, M. J. (2004). Grafts of immortalized chromaffin cells bio-engineered to improve met-enkephalin release also reduce formalin-evoked c-fos expression in rat spinal cord. *Neuroscience Letters, 370*, 1–6.

Eaton, M. J. (2000). Emerging cell and molecular strategies for the study and treatment of painful peripheral neuropathies. *Journal of the Peripheral Nervous System, 5*, 59–74.

Eaton, M. J., & Wolfe, S. Q. (2009). Clinical feasibility for cell therapy utilizing a human neuronal cell line to treat neuropathic pain-like behaviors following SCI in the rat. *Journal of Rehabilitation Research and Development, 46*, 145–166.

Eaton, M. J., Dancausse, H. R., Santiago, D. I., & Whittemore, S. R. (1997). Lumbar transplants of immortalized serotonergic neurons alleviates chronic neuropathic pain. *Pain, 72*(1–2), 59–69.

Eaton, M. J., Plunkett, J. A., Martinez, M. A., Karmally, S., Montanez, K., & Rabade, J. (1998). Changes in GAD and GABA immunoreactivity in the spinal dorsal horn after peripheral nerve injury and promotion of recovery by lumbar transplant of immortalized serotonergic neurons. *Journal of Chemical Neuroanatomy, 16*, 57–72.

Eaton, M. J., Karmally, S., Martinez, M. A., Plunkett, J. A., Lopez, T., & Cejas, P. (1999a). Lumbar transplants of neurons genetically modified to secrete galanin reverse pain-like behaviors after partial sciatic nerve injury. *Journal of the Peripheral Nervous System, 4*, 245–257.

Eaton, M. J., Plunkett, J. A., Martinez, M. A., Lopez, T., Karmally, S., Cejas, P., & Whittemore, S. R. (1999b). Transplants of neuronal cells bio-engineered to synthesize GABA alleviate chronic neuropathic pain. *Cell Transplantation, 8*, 87–101.

Eaton, M. J., Martinez, M., Frydel, B., Karmally, S., Lopez, T., & Sagen, J. (2000). Initial characterization of the transplant of immortalized chromaffin cells for the attenuation of chronic neuropathic pain. *Cell Transplantation, 9*, 637–656.

Eaton, M. J., Herman, J. P., Jullien, N., Lopez, T., Martinez, M., & Huang, J. (2002). Immortalized chromaffin cells disimmortalized with cre/lox site-directed recombination for use in cell therapy for pain. *Experimental Neurology, 175*, 49–60.

Eaton, M. J., Wolfe, S. Q., Martinez, M. A., Hernandez, M., Furst, C., Huang, J., Frydel, B. R., & Gomez-Marin, O. (2007). Subarachnoid transplant of a human neuronal cell line attenuates chronic allodynia and hyperalgesia after excitotoxic SCI in the rat. *The Journal of Pain, 8*, 33–50.

Eaton, M. J., Pearse, D. D., McBroom, J. S., & Berrocal, Y. A. (2008). The combination of human neuronal serotonergic cell implants and environmental enrichment after contusive SCI improves motor recovery over each individual strategy. *Behavioural Brain Research, 194*, 236–241.

Eaton, M. J., Widerström-Noga, E., & Wolfe, S. Q. (2011). Subarachnoid transplant of a human neuronal serotonergic cell line attenuates chronic allodynia and hyperalgesia without affecting motor dysfunction after severe contusive spinal cord injury. *Neurology Research International, 24*, 891605. Epub.

Feraboli-Lohnherr, D., Orsal, D., Yakovleff, A., Gimenez, Y., Ribotta, M., & Privat, A. (1997). Recovery of locomotor activity in the adult chronic spinal rat after sublesional transplantation of embryonic nervous cells: Specific role of serotonergic neurons. *Experimental Brain Research, 113*, 443–454.

Hains, B. C., Fullwood, S. D., Eaton, M. J., & Hulsebosch, C. E. (2001a). Subdural engraftment of serotonergic neurons following spinal hemisection restores spinal serotonin, downregulates the serotonin transporter, and increases BDNF tissue content in the rat. *Brain Research, 913*, 35–46.

Hains, B. C., Johnson, K. M., McAdoo, D. J., Eaton, M. J., & Hulsebosch, C. E. (2001b). Engraftment of immortalized serotonergic neurons enhances locomotor function and attenuates pain-like behavior following spinal hemisection injury in the rat. *Experimental Neurology, 171*, 361–378.

Hains, B. C., Everhart, A. W., Fullwood, S. D., & Hulsebosch, C. E. (2002a). Changes in serotonin, serotonin transporter expression and serotonin denervation supersensitivity: Involvement in chronic central pain after spinal hemisection in the rat. *Experimental Neurology, 175*, 347–362.

Hains, B. C., Yucra, J. A., Eaton, M. J., & Hulsebosch, C. E. (2002b). Intralesion transplantation of serotonergic precursors enhances locomotor recovery but has no effect on development of chronic central pain following hemisection injury in rats. *Neuroscience Letters, 324*, 222–226.

Hains, B. C., Johnson, K. M., Eaton, M. J., Willis, W. D., & Hulsebosch, C. E. (2003a). Serotonergic neural precursor cell grafts attenuate bilateral hyperexcitability of dorsal horn neurons after spinal hemisection in rat. *Neuroscience, 116*, 1097–1110.

Hains, B. C., Willis, W. D., & Hulsebosch, C. E. (2003b). Serotonin receptors 5-HT1A and 5-HT3 reduce hyperexcitability of dorsal horn neurons after chronic spinal cord hemisection injury in rat. *Experimental Brain Research, 149*, 174–186.

Hernándeza, Y., Torres-Espína, A., & Navarro, X. (2011). Adult stem cell transplants for spinal cord injury repair current state in preclinical research. *Current Stem Cell Research & Therapy, 6*(3), 273–287.

Hiyama, A., Mochida, J., & Sakai, D. (2008). Stem cell applications in intervertebral disc repair. *Cellular and Molecular Biology, 54*, 24–32 (Noisy-le-grand).

Ibuki, T., Hama, A. T., Wang, X. T., Pappas, G. D., & Sagen, J. (1997). Loss of GABA immunoreactivity in the spinal dorsal horn of rats with peripheral nerve injury and promotion of recovery by adrenal medullary grafts. *Neuroscience, 76*, 845–858.

Klass, M., Gavrikov, V., Drury, D., Stewart, B., Hunter, S., Denson, D. D., Hord, A., & Csete, M. (2007). Intravenous mononuclear marrow cells reverse neuropathic pain from experimental mononeuropathy. *Anesthesia and Analgesia, 104*, 944–948.

Kobayashi, N., Miyazaki, M., Fukaya, K., Inoue, Y., Sakaguchi, M., Noguchi, H., Matsumura, T., Watanabe, T., Totsugawa, T., Tanaka, N., & Namba, M. (2000). Treatment of surgically induced acute liver failure with transplantation of highly differentiated immortalized human hepatocytes. *Cell Transplantation, 9*, 733–735.

Lin, C. R., Wu, P. G., Shih, H. C., Cheng, J. T., Lu, C. Y., Chou, A. C., & Yang, L. C. (2002). Intrathecal spinal progenitor cell transplantation for the treatment of neuropathic pain. *Cell Transplantation, 11*, 17–24.

Liu, S., Ruenes, G. L., & Yezierski, R. P. (1997). NMDA and non-NMDA receptor antagonists protect against excitotoxic injury in the rat spinal cord. *Brain Research, 756*, 160–167.

Meisner, J. G., Marsh, A. D., & Marsh, D. R. (2010). Loss of GABAergic interneurons in laminae I–III of the spinal cord dorsal horn contributes to reduced GABAergic tone and neuropathic pain after spinal cord injury. *Journal of Neurotrauma, 27*, 729–737.

Murrell, W., Sanford, E., Anderberg, L., Cavanagh, B., & Mackay-Sim, A. (2009). Olfactory stem cells can be induced to express chondrogenic phenotype in a rat intervertebral disc injury model. *The Spine Journal, 9*(7), 585–594.

Pappas, G. D., Lazorthes, Y., Bes, J. C., Tafani, M., & Winnie, A. P. (1997). Relief of intractable cancer pain by human chromaffin cell transplants:

Experience at two medical centers. *Neurological Research, 19*, 71–77.

Qin, C., Du, J. Q., Tang, J. S., & Foreman, R. D. (2009). Bradykinin is involved in the mediation of cardiac nociception during ischemia through upper thoracic spinal neurons. *Current Neurovascular Research, 6*, 89–94.

Ramer, L. M., Au, E., Richter, M. W., Liu, J., Tetzlaff, W., & Roskams, A. J. (2004). Peripheral olfactory ensheathing cells reduce scar and cavity formation and promote regeneration after spinal cord injury. *The Journal of Comparative Neurology, 473*, 1–15.

Ribotta, M. G., Provencher, J., Feraboli-Lohnherr, D., Rossignol, S., Privat, A., & Orsal, D. (2000). Activation of locomotion in adult chronic spinal rats is achieved by transplantation of embryonic raphe cells reinnervating a precise lumbar level. *Journal of Neuroscience, 20*, 5144–5152.

Risbud, M. V., Guttapalli, A., Tsai, T. T., Lee, J. Y., Danielson, K. G., Vaccaro, A. R., Albert, T. J., Gazit, Z., Gazit, D., & Shapiro, I. M. (2007). Evidence for skeletal progenitor cells in the degenerate human intervertebral disc. *Spine, 32*, 2537–2544.

Ruff, C. A., Wilcox, J. T., & Fehlings, M. G. (2012). Cell-based transplantation strategies to promote plasticity following spinal cord injury. *Experimental Neurology, 235*(1), 78–90.

Sagen, J. (2003). Cellular therapies for spinal cord injury: What will the FDA need to approve moving from the laboratory to the human? *Journal of Rehabilitation Research and Development, 40*(Suppl. 1), 71–79.

Sagen, J., Wang, H., Tresco, P. A., & Aebischer, P. (1993). Transplants of immunologically isolated xenogeneic chromaffin cells provide a long-term source of pain-reducing neuroactive substances. *Journal of Neuroscience, 13*, 2415–2423.

Saruhashi, Y., Young, W., & Perkins, R. (1996). The recovery of 5-HT immunoreactivity in lumbosacral spinal cord and locomotor function after thoracic hemisection. *Experimental Neurology, 139*, 203–213.

Shapiro, S. (1997). Neurotransmission by neurons that use serotonin, noradrenaline, glutamate, glycine, and gamma-aminobutyric acid in the normal and injured spinal cord. *Neurosurgery, 40*, 168–176.

Stiller, C. O., Cui, J. G., O'Connor, W. T., Brodin, E., Meyerson, B. A., & Linderoth, B. (1995). GABA-ergic mechanisms may be involved in the spinal cord stimulation induced reversion of allodynia: Animal studies with microdialysis. *American Pain Society Abstracts, 95*, 801.

Stubley, L. A., Martinez, M., Karmally, S., Lopez, T., & Eaton, M. J. (2001). Only early GABA cell therapy is able to reverse neuropathic pain after partial nerve injury. *Journal of Neurotrauma, 18*, 471–477.

Sutherland, F. I., Bannatyne, B. A., Kerr, R., Riddell, J. S., & Maxwell, D. J. (2002). Inhibitory amino acid transmitters associated with axons in presynaptic apposition to cutaneous primary afferent axons in the cat spinal cord. *The Journal of Comparative Neurology, 452*, 154–162.

Terashima, T., Kojima, H., Fujimiya, M., Matsumura, K., Oi, J., Hara, M., Kashiwagi, A., Kimura, H., Yasuda, H., & Chan, L. (2005). The fusion of bone-marrow-derived proinsulin-expressing cells with nerve cells underlies diabetic neuropathy. *Proceedings of the National Academy of Sciences of the United States of America, 102*, 12525–12530.

Trojanowski, J. Q., Kleppner, S. R., Hartley, R. S., Miyazono, M., Fraser, N. W., Kesari, S., & Lee, V. M. (1997). Transfectable and transplantable postmitotic human neurons: Potential "platform" for gene therapy of nervous system diseases. *Experimental Neurology, 144*, 92–97.

Vaysse, L., Sol, J. C., Lazorthes, Y., Courtade-Saidi, M., Eaton, M. J., & Jozan, S. (2011). GABAergic pathway in a rat model of chronic neuropathic pain: Modulation after intrathecal cell transplantation of an inhibitory human neuronal cell line. *Neuroscience Research, 69*, 111–120.

Walker, P. A., Harting, M. T., Shah, S. K., Day, M. C., El Khoury, R., Savitz, S. I., Baumgartner, J., & Cox, C. S. (2010). Progenitor cell therapy for the treatment of central nervous system injury: A review of the state of current clinical trials. *Stem Cells International, 2010*, 369578.

Wei, A., Tao, H., Chung, S. A., Brisby, H., Ma, D. D., & Diwan, A. D. (2009). The fate of transplanted xenogeneic bone marrow-derived stem cells in rat intervertebral discs. *Journal of Orthopaedic Research, 27*, 374–379.

Westerman, K. A., & Leboulch, P. (1996). Reversible immortalization of mammalian cells mediated by retroviral transfer and site-specific recombination. *Proceedings of the National Academy of Sciences of the United States of America, 93*, 8971–8976.

Wilson, S., Chang, K., & Viveros, O. (1981). Opioid peptide synthesis in bovine and human adrenal chromaffin cells. *Peptides, 2*(Suppl 1), 83–88.

Wojcik, S. M., Katsurabayashi, S., Guillemain, I., Friauf, E., Rosenmund, C., Brose, N., & Rhee, J. S. (2006). A shared vesicular carrier allows synaptic corelease of GABA and glycine. *Neuron, 50*, 575–587.

Wolfe, S. Q., Salzberg, M., Cumberbatch, N. M. A., Garg, M., Furst, C., Martinez, M., Hernandez, M., Reimers, R., Berrocal, Y., & Eaton, M. J. (2007). Optimizing the transplant dose of a human neuronal cell line to treat SCI pain in the rat. *Neuroscience Letters, 414*, 121–125.

Yezierski, R. P., & Park, S. H. (1998). The mechanosensitivity of spinal sensory neurons following intraspinal injections of quisqualic acid in the rat. *Neuroscience Letters, 157*, 115–119.

Yezierski, R. P., Liu, S., Ruenes, G. L., Kajander, K. J., & Brewer, K. L. (1998). Excitotoxic spinal cord injury: Behavioral and morphological characteristics of a central pain model. *Pain, 75*, 141–155.

Zhang, B., Goldberger, M. E., & Murray, M. (1993). Proliferation of SP- and 5-HT-containing terminals in lamina II of rat spinal cord following dorsal rhizotomy: Quantitative EM-immunocytochemical studies. *Experimental Neurology, 123*, 51–62.

Cell Transplantation

▶ Cell Therapy in the Treatment of Central Pain

Cell-Mediated Immunity

Definition

An arm of the immune system that recognizes cell-associated antigens and consists of T-cells, phagocytes, and NK cells as cellular effectors.

Cross-References

▶ Viral Neuropathies

Cellular Adhesion Molecules

Synonyms

CAMs

Definition

Cellular adhesion molecules (CAMs) are cell surface proteins involved in the binding of cells, usually leukocytes, to each other, to endothelial cells, or to extracellular matrix. Most of the CAMs characterized so far fall into three general families of proteins: the immunoglobulin (Ig) superfamily, the integrin family, or the selectin family. The Ig superfamily of adhesion molecules, including ICAM-1, ICAM-2, ICAM-3, VCAM-1, and MadCAM-1, binds to integrins on leukocytes and mediates their flattening onto the blood vessel wall, with their subsequent extravasation into the surrounding tissue.

Cross-References

▶ Cytokine Modulation of Opioid Action

Cellular and Receptor Mechanisms Underlying the Development of Persistent Pain

▶ Descending Circuitry, Molecular Mechanisms of Activity-Dependent Plasticity

Cellular Targets of Substance P

Definition

Possible sources of NK1 receptor stimulated NGF biosynthesis including mast cells, which have been reported to be prominent sources of skin NGF, although the expression of NK1 receptors on these cells is unsure. There are only a few reports suggesting the presence of NK1 receptors on mast cells. However, it has to be taken into account that, depending on anatomic site, mast cells show variations in cell size, cytoplasmic granule ultrastructure, mediator content, sensitivity to stimulation by secretagogues, and in their susceptibility to various pharmacological agents. Thus, it has been shown recently that functional NK1 receptors are induced by IL-4 and stem cell factor, suggesting that under certain conditions, like those accompanying inflammation, mast cells could gain increased responsiveness to NK1 agonists. Keratinocytes have been shown to express beta adrenoceptors and to produce NGF in response to substance P. However, there are diverging reports as to the type of tachykinin receptor primarily expressed by murine keratinocytes. Other possible sources of NGF include macrophages/monocytes that express NK1 receptors and can be stimulated by substance P to produce cytokines.

Cross-References

► NGF, Regulation During Inflammation

Cementum

Definition

Cementum is the mineralized tissue that covers the root of a tooth. At the level where it abuts the enamel of the crown, it is very thin and often abraded, exposing the underlying sensitive dentin.

Cross-References

► Dental Pain, Etiology, Pathogenesis, and Management

Central Changes After Peripheral Nerve Injury

Marshall Devor
Department of Cell & Developmental Biology, Institute of Life Sciences and Center for Research on Pain, The Hebrew University of Jerusalem, Jerusalem, Givat Ram, Israel

Synonyms

Central hyperexcitability state; Central sensitization; Centralization; Transganglionic changes after peripheral nerve injury; Transition to chronicity

Definition

Following peripheral nerve injury numerous structural, neurochemical, and electrophysiological parameters change in the central nervous system (CNS), especially in the spinal cord and brain stem areas that receive direct primary afferent input. At least some of these central changes contribute to chronic neuropathic pain either directly, by generating pain-signaling impulses of CNS origin, or indirectly by amplifying or otherwise modulating afferent signals entering from the periphery. Changes have also long been noted supraspinally, including in the cerebral cortex (Campbell 1905; Kim and Nabekura 2011). In most instances, we are not certain whether or not a particular central change plays an important role in neuropathic pain. Spinal deafferentation due to sensory ganglion damage or dorsal rhizotomy also causes central changes. These tend to be more severe. Effects of deafferentation will not be reviewed here.

Characteristics

"Central Sensitization," "Pain Centralization," and "Transition to Chronicity"

"Central sensitization" refers to an altered state of central neural processing in which afferent signals that enter the CNS from the periphery are amplified or otherwise modified. A unique type of modification, that goes beyond simple amplification, is that light brush and touch signals carried centrally by low-threshold mechanoreceptive Aβ afferents can evoke a sensation of pain (Campbell et al. 1988; Devor et al. 1991; Woolf 1983). The concept that pain hypersensibility in inflamed tissue and in neuropathy may be due to abnormal signal processing in the CNS and not only to changes in the responsiveness of afferents in the periphery (PNS) has a long history in neurology. The obvious link between indicators of inflammation (reddening, warming, and edema) and hypersensitivity gave rise to the idea that nociceptive afferent endings may become sensitized in the periphery, causing "primary hyperalgesia" (Lewis 1942). The fact that a region of "secondary hyperalgesia" surrounding the primary region does not show obvious inflammation, but nonetheless becomes hypersensitive, promoted the idea of amplification within the CNS (Hardy et al. 1952).

The striking symptom in the secondary area is "dynamic tactile allodynia," i.e., tenderness to gentle brushing. Although marginalized at the time, the conviction that the CNS makes an important contribution to pain amplification was revived in the 1980s. The revival was partly under the influence of Melzack and Wall's "Gate control theory" which legitimized the idea of CNS modulation in general and partly due to new evidence from animal models and experimental observations in humans. Particularly notable is the previously heretical concept that non-nociceptive Aβ touch afferents can signal pain. The broad acceptance of this concept represents a revolution in thinking on chronic pain mechanisms.

When first introduced, central sensitization referred to a spinal pain hypersensitivity state triggered by afferent input entering the CNS on nociceptive C-afferents and perhaps also nociceptive Aδ-afferents (Woolf 1983). This hypersensitivity state comes on rapidly following onset of the nociceptive stimulus (seconds or minutes), and rapidly dissipates on cessation of the stimulus (minutes or hours). It can be maintained indefinitely, however, by a continuous barrage of nociceptive input such as may occur in chronic inflammatory conditions and neuropathy. Thus, as originally conceived central sensitization is a labile, dynamic state dependent on ongoing nociceptive afferent drive (Gracely et al. 1992; Ji et al. 2003; Koltzenburg et al. 1994; Torebjork et al. 1992). A number of neural mechanisms were put forward to explain the sensitization including NMDA-R activation and long-term potentiation (LTP), all correspondingly labile.

Further research into underlying neural mechanisms, however, revealed many additional central changes triggered by nerve injury including some which were both irreversible, such as loss of inhibitory interneurons, and independent of ongoing afferent input. The discovery of durable central changes coincided with the revival of another classical concept, "pain centralization." Believers in pain centralization claim that persistent severe pain of peripheral origin can "burn itself into the CNS" in the same way that a torrential stream can carve a canyon through solid rock. That is, in time persistent pain creates a central hyperexcitability state that becomes independent of afferent input entering from the periphery. Acute pain is said to have undergone a "transition to chronicity" (Kalso 1997). This concept goes far beyond what was originally meant by central sensitization. If true it is important, as it implies that pain relief has a deadline. If pain is not relieved soon enough it may centralize and become intractable, inaccessible to existing treatments, and perhaps permanent.

Implications for Therapeutic Management

The idea that pain, if allowed to persist, can transition to chronicity with a gradual centralization of the pain mechanism is widely accepted in the medical community despite the tentative nature of the supporting evidence. There are more parsimonious explanations of chronic pain. For example, the generator of pain at the acute stage may simply persist despite medical efforts to remove it. In principle, peripheral generators can persist indefinitely. Indeed, there is good reason to doubt that prolonged pain, per se, ever becomes independent of peripheral drivers. For example, if it did, then severe pain arising from a well-defined persistent source would remain indefinitely even after removal of the source. This clearly does not occur following removal of pain-provoking sources by childbirth, passage of a kidney stone, or replacement of an osteoarthritic hip. The logic, of course, presumes that the peripheral source has indeed been removed. For hip replacement this is usual, but it is probably not the case in all patients. However, even when the actual peripheral pain driver cannot be precisely identified or removed it is usually possible to transiently prevent the peripheral nociceptive impulses from reaching the CNS by blocking the corresponding nerve, plexus, or sensory ganglia. In fact, such diagnostic blocks reliably stop the pain for the duration of the block. This is true for both spontaneous pain and pain evoked by gently brushing the numb skin or the skin adjacent to it (allodynia). Although not impossible, it remains to be proven that persistent pain can actually leave a permanent trace by creating independent pain generators within the CNS proper (Devor 2006). This caution applies even to frank peripheral nerve injury. Where persistent pain *can* cause permanent mischief even after it has been definitively extinguished is when it induces a downward spiral of psychosocial deterioration, dependence, and financial distress in the chronic pain sufferer.

In the minds of many investigators the idea of durable centralization has merged with the original concept of dynamic, labile central sensitization. For these individuals "central sensitization" has become an umbrella term that covers all peripherally evoked central changes believed to contribute to chronic pain, labile and durable. This is unfortunate as the labile processes that bring about central hyperexcitability can be treated by methods directed at the peripheral driver as well as by methods directed at the CNS amplification process. Pain due to strictly CNS mechanisms is inherently more difficult to address.

Spatial Aspects of Hypersensibility due to Central Changes

Central sensitization due to a focal peripheral driver is regionally circumscribed. Thus, following a localized burn tactile allodynia spreads circumferentially beyond the area of primary injury in an annulus of secondary hyperalgesia (Raja et al. 1984). Likewise, if precipitated by injury to a spinal nerve or sensory ganglion, central sensitization generates hypersensitivity in the distribution of neighboring nerves, or nearby dermatomes (Kirk and Denny-Brown 1970). Pain extending beyond the primary driving source is sometimes given the odd name "extraterritorial pain." The concept can be extended still further. Pathology in one visceral organ, for example, can cause hyperesthesia in neighboring organs (Giamberardino 2008), and meningeal inflammation in migraine can cause tenderness on the scalp (Burstein et al. 2000). Some authors have gone even further, positing that broad expanses of the central somatosensory representation can become persistently hyperexcitable. This is a popular explanation for widely distributed global pain symptoms such as in fibromyalgia (Banic et al. 2004; McDermid et al. 1996).

Varieties of Central Change Induced by Neuropathy

A bewildering number of central changes have been reported in animal models of neuropathic (and inflammatory) pain. In fact, there are so many that it challenges credulity that all contribute to central sensitization (or centralization, if it indeed occurs). Reasons for the uncertainty are discussed below. What follows is a list of central changes induced by peripheral nerve injury that might reasonably be predicted to affect pain processing. They are grouped by type. Most have been documented in one, or only a few neuropathic pain models, or are inferred from models of inflammation, and may not be universal. Some appear paradoxical. For example, a priori, depletion of an excitatory transmitter or increased expression of an inhibitory one should suppress pain, not cause pain. However, since these neurotransmitters might be acting on inhibitory interneurons, a contribution to painful hypersensitivity cannot be ruled out. The list is presented with a minimum of annotation and with insufficient references (due to editorial limitations). Some changes may overlap, be redundant, or describe the same process under different headings or using different markers. The list is certainly incomplete; newly identified changes continue to be reported at frequent intervals.

Changes in the Neurochemistry of Primary Afferent Neurons and Their Terminals in the Spinal Cord After Peripheral Nerve Injury

- Expression of substance P (SP) and CGRP, two key excitatory peptide neurotransmitters/ neuromodulators, is decreased in small diameter nociceptive DRG neurons and their central terminals following nerve injury. But there is a concomitant increase in medium and large diameter DRG neurons. In the case of CGRP, the increase may be of particular importance for pain (Nitzan-Luques et al. 2011).

- Expression of the inhibitory peptide neurotransmitter/neuromodulator galanin is increased.
- Cellular content of the excitatory peptide neurotransmitters/neuromodulators neuropeptide Y (NPY) and vasoactive intestinal peptide (VIP) is increased in injured afferents, as is that of a variety of proinflammatory cytokines.
- Expression of the mu opioid morphine receptor gene (MOR) is reduced in DRG neurons, and mu receptor content is depleted in afferent terminals.
- The Ca^{2+} channel subunit $\alpha 2\delta 1$, the target of the analgesics gabapentin and pregabalin, is upregulated in axotomized primary afferent neurons and their central terminals, perhaps enhancing synaptic release.
- Expression of the transducer/ion channels TRPV1 and P2X3 is depressed in afferent terminals.
- Tissue plasminogen activator is induced in primary afferent neurons by axotomy and released from their terminal endings. This increases the excitability of dorsal horn neurons.
- Expression of the TTX-S Na^+ channel Nav1.3 is upregulated in axotomized DRG neurons while that of TTX-R Nav1.8 and Nav1.9 is downregulated. All three channels contribute to the membrane excitability of primary afferent neurons and their central terminals.
- The preceding are only high-profile examples of changes in gene expression in DRG neurons, and in gene product density in afferent terminals, following axotomy (Hokfelt et al. 1997). These are the tip of the iceberg. Recent studies using oligonucleotide arrays and RNA sequencing indicate that literally thousands of transcripts are up- or downregulated in axotomized DRG. "Intact" neighboring neurons that have not been axotomized may also undergo changes in gene expression (Hammer et al. 2010; Persson et al. 2009). Each of these represents a candidate pain mechanism. Only a few, however, have been championed by an investigator or subjected to even a minimum of functional analysis.

Changes in the Neurochemistry and Gene Expression of CNS Neurons and Glia After Peripheral Nerve Injury

- As with primary sensory neurons, studies using oligonucleotide arrays suggest that very large numbers of transcripts are up- or downregulated in the spinal cord as a consequence of nerve injury. Again, few of these have been subjected to even minimal functional analysis.
- A number of activity-regulated immediate early genes are upregulated in postsynaptic neurons in the dorsal horn, including *c-fos* and *jun-B*. These tend to be transcription factors and hence probably affect the expression of numerous other downstream genes.
- The Ca^{+2}-sensing protein DREAM has been identified as a spinal cord transcriptional repressor of endogenous opioids that modulate pain (Cheng et al. 2002).
- Many transmembrane and intracellular signaling cascades in postsynaptic neurons are activated by the phosphorylation of protein kinases, such as ERK, CREB, and MAP protein kinases (notably p38 MAPK). More than 500 protein kinase genes are present in the mammalian genome (Manning et al. 2002).
- Nerve injury is associated with reduced GABA release in the dorsal horn following electrical nerve stimulation and hence disinhibition.
- Levels of cyclooxygenase (COX-1 and COX-2) in the dorsal horn are altered with consequent changes in arachidonic acid metabolites including (excitatory) prostanoids and leukotrienes.
- A decrease in mu opiate receptors on postsynaptic dorsal horn neurons following nerve injury may occasion decreased intrinsic spinal inhibition.
- Several purinergic receptors are upregulated in dorsal horn microglia, potentially enhancing response to the excitatory neurotransmitter ATP (McMahon and Malcangio 2009).
- Concentrations of reactive oxygen species (ROSs) increase in the dorsal horn following

nerve injury, reducing spinal inhibition and exciting spinal neurons (Yowtak et al. 2011).

- There is an increase in the spinal content of pro-inflammatory cytokines including IL1β, IL6, TNFα, but also in the anti-inflammatory cytokine IL10. These compounds are synthesized in activated astrocytes and microglia and are released into the extracellular space (Watkins and Maier 2002). Some are also produced in neurons. Proinflammatory cytokines can sensitize and directly excite postsynaptic dorsal horn neurons.

- Sphingolipid metabolism in the spinal cord is altered following nerve injury. Some species of sphingolipid, at least, induce neuronal hyperexcitability (Patti et al. 2012).

- Increased nocistatin expression decreases spinal GABA inhibition.

- There is an alteration in the content of many neurotransmitters in postsynaptic dorsal horn neurons, some inhibitory (e.g., 5-HT, NA, GABA, glycine) and some prohyperalgesic (e.g., dynorphin). There is also an increase in the spinal content of many bioactive molecules of uncertain function in pain processing, such as certain lectins and GAP43.

Structural Changes in the CNS After Peripheral Nerve Injury

- Central terminals of axotomized primary afferent neurons show a morphologically distinct "degeneration atrophy." The functional significance is uncertain, but the change might be associated with altered synaptic release or even degeneration.

- Time-dependent loss of many functionally unidentified neurons in the dorsal horn has been reported (neurodegeneration), although the magnitude and significance of this effect has been disputed (Azkue et al. 1998; Whiteside and Munglani 2001).

- Transsynaptic atrophy and cell loss have also been reported in somatotopically appropriate supraspinal projection systems. A striking example is the primary somatosensory cortex in long-term amputees (Campbell 1905; Draganski et al. 2006).

- There are reports that intraspinal terminals of low-threshold mechanoreceptive Aβ afferents enter a growth mode, extending sprouts dorsally into spinal laminae 1 and 2 where they may form ectopic synaptic contacts with pain signaling spinal neurons. Endings of neighboring non-injured afferents may also sprout. Both findings are controversial, however, as they might simply reflect axotomy-induced enhancement of the visualization of afferent connections present normally.

- There is a loss of neurons that immunolabel for the inhibitory neurotransmitters glycine and GABA (Castro-Lopes et al. 1993). This may reflect permanent loss of inhibitory interneurons.

- Large numbers of astrocytes and microglia in the dorsal horn are "activated," shortly after nerve injury, showing hypertrophy, increased numbers (hyperplasia) and altered expression of neuroactive molecules including proinflammatory cytokines, chemokines, neurotrophins, and their receptors (McMahon and Malcangio 2009; Watkins and Maier 2002).

- Nerve injury, although not inflammation, increases the permeability of the blood–brain barrier (BBB) in the spinal cord permitting the entry of blood-borne immune cells and other moieties and activation of spinal glia and neurons (Costigan et al. 2009; Gordh et al. 2006).

- ATP, an excitatory neurotransmitter, is released from astroglia activated following nerve injury. ATP is also hydrolyzed into the inhibitory neuroactive transmitter adenosine.

Electrophysiological and Other Functional Changes in the CNS After Peripheral Nerve Injury

- The pitter-patter of background spontaneous afferent activity (ectopia) depolarizes postsynaptic dorsal horn neurons, bringing them closer to their spike threshold. This increases the level of spontaneous activity generated within the dorsal horn and enhances the response of dorsal horn neurons to weak residual inputs from the periphery.

- The Gate Control Theory predicts that selective loss of input from large diameter low-threshold afferent neurons will bias the spinal gate toward augmented nociception by reducing presynaptic inhibition.
- The dorsal root potential (DRP) and primary afferent depolarization (PAD), two measures of spinal presynaptic inhibition, are suppressed following nerve injury. Reduced inhibition increases spinal response (Wall and Devor 1981).
- Receptive fields (RF) of dorsal horn neurons expand, increasing the RF overlap between neighboring neurons. A given peripheral stimulus now activates more spinal cord neurons.
- Somatotopic maps in the spinal cord, forebrain and cortex become reorganized along with numerous changes in CNS activation patters detected by fMRI.
- The overall impulse volley that ascends in the spinal cord toward the brain upon electrical stimulation of a peripheral nerve is reduced by about 50 % beginning 1–2 weeks following nerve injury. However, it is possible that some specific components of the ascending volley are enhanced (Wall and Devor 1981).
- NMDA type glutamate receptors on postsynaptic dorsal horn pain signaling neurons are normally blocked and nonfunctional at resting membrane potential. Primary afferent nociceptive input produces a prolonged, shallow depolarization, probably due to the release of SP (and other peptides). This relieves the block, enables the NMDA-Rs, and transiently enhances postsynaptic response to Aβ touch input.
- Nociceptive afferent input induces PKC-dependent phosphorylation of NMDA-R subunits, perhaps contributing to prolonged spinal pain hypersensitivity.
- There is also activation of non-NMDA glutamate receptors (AMPA/kainate receptors) in the spinal cord following nociceptive afferent input.
- Long-term potentiation (LTP) is facilitated by nerve injury (Sandkuhler and Gruber-Schoffnegger 2012).

- Repetitive stimulation at low frequency reveals homosynaptic facilitation ("windup"). This is augmented after nerve injury.
- As noted, nociceptive mediators SP, CGRP, NPY, and BDNF come to be expressed in low-threshold mechanoreceptive Aβ afferents. Activity in these afferents can directly activate pain signaling dorsal horn neurons. These afferents may also acquire the ability to trigger and maintain central sensitization (Devor 2009).
- There is increased release of the excitatory neurotransmitter glutamate in the dorsal horn. "Spillover" facilitates the response of NMDA receptors to glutamate, including glutamate released from Aβ touch afferents.
- Suppression of the Cl⁻ pump and the consequent depolarizing shift of the Cl⁻ reversal potential can cause the normally inhibitory neurotransmitter GABA to yield excitation (Coull et al. 2003).
- Engagement of protease-activated receptors, notably PAR-1, has been implicated in central sensitization (Narita et al. 2005).
- Many synapses present on dorsal horn neurons are relatively ineffective at driving the postsynaptic neuron ("silent synapses"). These can be strengthened by a variety of mechanisms, opening new functional pathways including Aβ access to ascending nociceptive circuitry.
- BDNF released from afferent terminals, and perhaps synthesized locally, sensitizes postsynaptic neurons in the superficial dorsal horn.
- The duration of spinal postsynaptic potentiation is augmented by nerve injury.
- Injury discharge triggered by acute transection of primary afferent neurons may selectively damage inhibitory spinal interneurons, perhaps by the sudden release of high levels of glutamate (excitotoxic cell death).
- Spontaneous discharge, particularly in a burst mode, is augmented in dorsal horn postsynaptic neurons and also in supraspinal relays. Although uncertain, some of this activity may be generated within the CNS rather than reflecting elevated peripheral drive.

Changes in Spinal Pain Modulation Descending from the Brain

- Background brain stem descending inhibition may be reduced following nerve injury (Yarnitsky et al. 2012; De Felice et al. 2011).
- Deafferentation by dorsal rhizotomy, and perhaps nerve injury as well, triggers sprouting and the proliferation of synaptic terminals of descending inhibitory reticulospinal 5HT-containing neurons within the dorsal horn (Polistina et al. 1990).
- Brain stem descending facilitation may be enhanced following nerve injury (Vera-Portocarrero et al. 2006).

Reservations Concerning the Multiplicity of Proposed Mechanisms

In principle, it ought to be possible to determine the relative contribution of each central change by spinal delivery of agents that counter the changes, one at a time. Each such agent should block a fraction of the neuropathic pain symptoms, and the appropriate combination of agents should block the pain entirely. In practice, however, this approach has produced confusing results. In some cases, appropriate blocking drugs are not available. In many others, the application of drugs intended to reverse individual central changes has been claimed to eliminate neuropathic pain entirely. This suggests that experimental results might have been exaggerated, or tested under highly specific, idiosyncratic circumstances. Perhaps many processed are aligned in series so that blocking one blocks them all. Alternatively, pain symptoms may have a threshold such that partial suppression of many independent processes indeed yields complete pain suppression.

Another problem is with the agents themselves. Pharmacological agents that show a high degree of specificity when tested under specific in vitro conditions often prove to have unanticipated effects when tested in vivo in complex behavioral paradigms.

For example ketamine, famous as an NMDA-R antagonist, also blocks Na^+ channels as does amitriptyline—a tricyclic antidepressant, famous as a catecholamine reuptake inhibitor

(Song et al. 2000; Zhou and Zhao 2000). This problem sometimes extends also to newer transgenic technologies. Finally, few authors check whether the agents they deliver to the spinal cord actually act there. Particularly in animal models of neuropathy, ectopic impulse discharge originating in the DRG is thought to play an important role in the initiation and maintenance of central sensitization (Devor 2006). Since the DRG shares the epidural and the intrathecal space with the spinal cord, spinally delivered drugs access primary sensory neurons as well as CNS neurons. It is essential to document that the spinally administered drug being tested does not silence peripheral ectopia, because this alone would be expected to relieve pain symptoms, irrespective of the central process being tested. Such confirmation is rarely done.

How Does Nerve Injury Trigger Central Change?

Little is known with confidence about the mechanisms whereby peripheral nerve injury triggers and maintains central changes. There are three fundamental possibilities:

1. *Depolarization due to impulse traffic per se*: The resting potential of postsynaptic neurons is determined, in part, by the constant barrage of excitatory and inhibitory postsynaptic potentials impinging on their dendritic arbor (spatial and temporal summation). Ectopic afferent activity in neuropathy enhances the barrage and depolarizes the neurons, bringing their resting potential closer to firing threshold. A result is increased spontaneous firing and response to inputs.

2. *Other actions of transmitters released by afferent impulse traffic*: Neurotransmitter and neuromodulator molecules released from afferent terminals during spike activity may have postsynaptic effects beyond the moment-to-moment modulation of the membrane potential. Coupling to transmembrane signaling pathways can have multiple short, intermediate, and long-term effects on the behavior of postsynaptic neurons. Block of

the ectopic afferent input using local anesthetics or other means, for example, prevents the development of many central changes in neurons and glia (Pitcher and Henry 2008; Wen et al. 2007).

3. *Trophic interactions*: More speculatively, nerve injury might bring about central changes completely independent of impulse traffic and synaptic release. During embryonic development, the very survival of primary sensory and second-order CNS neurons is dependent on mutual neurotrophic interactions. Beyond a critical period the neurons lose their acute dependence on neurotrophic support. However, neuronal phenotype continues to be affected by trophic molecules, and excitability can be altered by changes in the provision of neurotrophins (Boucher and McMahon 2001). Membrane-bound recognition molecules could also play a trophic role (e.g., NCAMs and ephrin). The release of these signaling molecules or their incorporation into the presynaptic membrane is regulated by spike-evoked exocytosis. But it might also be regulated by constitutive processes unrelated to afferent impulse traffic (Battaglia et al. 2003; Fields et al. 2001).

Perspective

Only a generation ago there was little concept of the central processes that might underlie neuropathic pain. The situation has since reversed so that today we are awash with candidate theories. It is a high priority to develop strategies for prioritizing central changes in terms of their relative contribution to pain symptomatology. Likewise, it is essential to establish the mechanism(s) by which nerve injury triggers CNS changes. If all or most central changes are due to abnormal primary afferent input it may be possible to prevent or reverse the central changes by controlling afferent input. Alternatively, key central changes might offer opportunities for direct therapeutic intervention aimed at the CNS.

Acknowledgment Thanks to Linda Watkins and Zsuzsanna Wiesenfeld-Hallin for helpful comments on the manuscript (first edition). I wish to apologize to the numerous colleagues whose work was noted in this entry but not specifically cited because of editorial restrictions on the number of references permitted.

References

Azkue, J. J., Zimmermann, M., Hsieh, T. F., & Herdegen, T. (1998). Peripheral nerve insult induces NMDA receptor-mediated, delayed degeneration in spinal neurons. *European Journal of Neuroscience, 10*, 2204–2206.

Banic, B., Petersen-Felix, S., Andersen, O. K., Radanov, B. P., Villiger, P. M., Arendt-Nielsen, L., & Curatolo, M. (2004). Evidence for spinal cord hypersensitivity in chronic pain after whiplash injury and in fibromyalgia. *Pain, 107*, 7–15.

Battaglia, A. A., Sehayek, K., Grist, J., McMahon, S. B., & Gavazzi, I. (2003). EphB receptors and ephrin-B ligands regulate spinal sensory connectivity and modulate pain processing. *Nature Neuroscience, 6*, 339–340.

Boucher, T. J., & McMahon, S. B. (2001). Neurotrophic factors and neuropathic pain. *Current Opinion in Pharmacology, 1*, 66–72.

Burstein, R., Yarnitsky, D., Goor-Aryeh, I., Ransil, B. J., & Bajwa, Z. H. (2000). An association between migraine and cutaneous allodynia. *Annals of Neurology, 47*, 614–624.

Campbell, A. W. (1905). *Histological studies on the localization of cerebral function*. Cambridge: Cambridge University Press.

Campbell, J. N., Raja, S. N., Meyer, R. A., & MacKinnon, S. E. (1988). Myelinated afferents signal the hyperalgesia associated with nerve injury. *Pain, 32*, 89–94.

Castro-Lopes, J. M., Tavares, I., & Coimbra, A. (1993). GABA decreases in the spinal cord dorsal horn after peripheral neurectomy. *Brain Research, 620*, 287–291.

Cheng, H. Y., Pitcher, G. M., Laviolette, S. R., Whishaw, I. Q., Tong, K. I., Kockeritz, L. K., Wada, T., Joza, N. A., Crackower, M., Goncalves, J., Sarosi, I., Woodgett, J. R., Oliveira-dos-Santos, A. J., Ikura, M., van der Kooy, D., Salter, M. W., & Penninger, J. M. (2002). DREAM is a critical transcriptional repressor for pain modulation. *Cell, 108*, 31–43.

Costigan, M., Moss, A., Latremoliere, A., Johnston, C., Verma-Gandhu, M., Herbert, T. A., Barrett, L., Brenner, G. J., Vardeh, D., Woolf, C. J., & Fitzgerald, M. (2009). T-cell infiltration and signaling in the adult dorsal spinal cord is a major contributor to neuropathic pain-like hypersensitivity. *Journal of Neuroscience, 29*, 14415–14422.

Coull, J. A., Boudreau, D., Bachand, K., Prescott, S. A., Nault, F., Sik, A., De Koninck, P., & De Koninck, Y. (2003). Trans-synaptic shift in anion gradient in spinal lamina I neurons as a mechanism of neuropathic pain. *Nature, 424*, 938–942.

De Felice, M., Sanoja, R., Wang, R., Vera-Portocarrero, L., Oyarzo, J., King, T., Ossipov, M. H., Vanderah, T. W., Lai, J., Dussor, G. O., Fields, H. L., Price, T. J., & Porreca, F. (2011). Engagement of descending inhibition from the rostral ventromedial medulla protects against chronic neuropathic pain. *Pain, 152*, 2701–2709.

Devor, M. (2006). Response of nerves to injury in relation to neuropathic pain. Chapter 58. In S. L. McMahon & M. Koltzenburg (Eds.), *Wall and Melzack's textbook of pain* (5th ed., pp. 905–927). London: Churchill Livingstone.

Devor, M. (2009). Ectopic discharge in A-beta afferents as a source of neuropathic pain. *Experimental Brain Research, 196*, 115–128.

Devor, M., Basbaum, A., Bennett, G., Blumberg, H., Campbell, J., Dembowsky, K., Guilbaud, G., Janig, W., Koltzenberg, M., Levine, J., Otten, U., & Portenoy, R. (1991). Mechanisms of neuropathic pain following peripheral injury. In A. Basbaum & J.-M. Besson (Eds.), *Towards a new pharmacology of pain* (pp. 417–440). Chichester: Dahlem Konferenzen/Wiley.

Draganski, B., Moser, T., Lummel, N., Gansshauer, S., Bogdahn, U., Haas, F., & May, A. (2006). Decrease of thalamic gray matter following limb amputation. *NeuroImage, 31*, 951–957.

Fields, R. D., Eshete, F., Dudek, S., Ozsarac, N., & Stevens, B. (2001). Regulation of gene expression by action potentials: Dependence on complexity in cellular information processing. *Novartis Foundation Symposium, 239*, 160–172.

Giamberardino, M. A. (2008). Women and visceral pain: Are the reproductive organs the main protagonists? Mini-review at the occasion of the "European Week Against Pain in Women 2007." *European Journal of Pain, 12*, 257–260.

Gordh, T., Chu, H., & Sharma, H. S. (2006). Spinal nerve lesion alters blood-spinal cord barrier function and activates astrocytes in the rat. *Pain, 124*, 211–221.

Gracely, R., Lynch, S., & Bennett, G. (1992). Painful neuropathy: Altered central processing, maintained dynamically by peripheral input. *Pain, 51*, 175–194.

Hammer, P., Banck, M. S., Amberg, R., Wang, C., Petznick, G., Luo, S., Khrebtukova, I., Schroth, G. P., Beyerlein, P., & Beutler, A. S. (2010). mRNA-seq with agnostic splice site discovery for nervous system transcriptomics tested in chronic pain. *Genome Research, 20*, 847–860.

Hardy, J. D., Wolf, H. G., & Goodell, H. (1952). *Pain sensations and reactions.* New York: William and Wilkins.

Hokfelt, T., Zhang, X., Xu, Z.-Q., Ji, R.-R., Shi, T., Corness, J., Kerekes, N., Landry, R.C., Holmberg, K., & Broberger, C. (1997). Phenotype regulation in dorsal root ganglion neurons after nerve injury: Focus on peptides and their receptors. In D. Borsook (Ed.), *Molecular neurobiology of pain* (pp. 115–143). Seattle, WA: IASP Press.

Ji, R. R., Kohno, T., Moore, K. A., & Woolf, C. J. (2003). Central sensitization and LTP: Do pain and memory share similar mechanisms? *Trends in Neurosciences, 26*, 696–705.

Kalso, E., (1997). Prevention of chronicity. In T. S. Jensen, J. A. Turner, & Z. H. Wiesenfeld-Hallin (Eds.), *Proceedings of the 8th World Congress on pain. Progress in pain research and management* (pp. 215–230, Vol. 8). Seattle: IASP Press.

Kim, S. K., & Nabekura, J. (2011). Rapid synaptic remodeling in the adult somatosensory cortex following peripheral nerve injury and its association with neuropathic pain. *Journal of Neuroscience, 31*, 5477–5482.

Kirk, E. J., & Denny-Brown, D. (1970). Functional variation in dermatomes in the macaque monkey following dorsal root lesions. *Journal of Comparative Neurology, 139*, 307–320.

Koltzenburg, M., Torebjork, H., & Wahren, L. (1994). Nociceptor modulated central sensitization causes mechanical hyperalgesia in acute chemogenic and chronic neuropathic pain. *Brain, 117*, 579–591.

Lewis, T. (1942). *Pain.* London: Macmillan. 192 pp.

Manning, G., Whyte, D. B., Martinez, R., Hunter, T., & Sudarsanam, S. (2002). The protein kinase complement of the human genome. *Science, 298*, 1912–1934.

McDermid, A. J., Rollman, G. B., & McCain, G. A. (1996). Generalized hypervigilance in fibromyalgia: Evidence of perceptual amplification. *Pain, 66*, 133–144.

McMahon, S. B., & Malcangio, M. (2009). Current challenges in glia-pain biology. *Neuron, 64*, 46–54.

Narita, M., Usui, A., Niikura, K., Nozaki, H., Khotib, J., Nagumo, Y., Yajima, Y., & Suzuki, T. (2005). Protease-activated receptor-1 and platelet-derived growth factor in spinal cord neurons are implicated in neuropathic pain after nerve injury. *Journal of Neuroscience, 25*, 10000–10009.

Nitzan-Luques, A., Devor, M., & Tal, M. (2011). Genotype-selective phenotypic switch in primary afferent neurons contributes to neuropathic pain. *Pain, 152*, 2413–2426.

Patti, G. J., Yanes, O., Shriver, L. P., Courade, J. P., Tautenhahn, R., Manchester, M., & Siuzdak, G. (2012). Metabolomics implicates altered sphingolipids in chronic pain of neuropathic origin. *Nature Chemical Biology, 8*, 232–234.

Persson, A. K., Gebauer, M., Jordan, S., Metz-Weidmann, C., Schulte, A.M., Schneider, H.C., Ding-Pfennigdorff, D., Thun, J., Xu, X. J., Wiesenfeld-Hallin, Z., Darvasi, A., Fried, K., & Devor, M. (2009). Correlational analysis for identifying genes whose regulation contributes to chronic neuropathic pain. *Molecular Pain, 5*, 7.

Pitcher, G. M., & Henry, J. L. (2008). Governing role of primary afferent drive in increased excitation of spinal nociceptive neurons in a model of sciatic neuropathy. *Experimental Neurology, 214,* 219–228.

Polistina, D. C., Murray, M., & Goldberger, M. E. (1990). Plasticity of dorsal root and descending serotoninergic projections after partial deafferentation of the adult rat spinal cord. *The Journal of Comparative Neurology, 299,* 349–363.

Raja, S. N., Campbell, J. N., & Meyer, R. A. (1984). Evidence for the different mechanisms of primary and secondary hyperalgesia following heat injury to the glabrous skin. *Brain, 107,* 1179–1188.

Sandkuhler, J., & Gruber-Schoffnegger, D. (2012). Hyperalgesia by synaptic long-term potentiation (LTP): An update. *Current Opinion in Pharmacology, 12,* 18–27.

Song, J., Ham, S., Shin, Y., & Lee, C. (2000). Amitriptyline modulation of Na(+) channels in rat dorsal root ganglion neurons. *European Journal of Pharmacology, 401,* 297–305.

Torebjork, H., Lundberg, L., & LaMotte, R. (1992). Central changes in processing of mechanoreceptive input in capsaicin-induced secondary hyperalgesia in humans. *Journal of Physiology (London), 448,* 765–780.

Vera-Portocarrero, L. P., Zhang, E. T., Ossipov, M. H., Xie, J. Y., King, T., Lai, J., & Porreca, F. (2006). Descending facilitation from the rostral ventromedial medulla maintains nerve injury-induced central sensitization. *Neuroscience, 140,* 1311–1320.

Wall, P., & Devor, M. (1981). The effect of peripheral nerve injury on dorsal root potentials and on the transmission of afferent signals into the spinal cord. *Brain Research, 209,* 95–111.

Watkins, L. R., & Maier, S. F. (2002). Beyond neurons: Evidence that immune and glial cells contribute to pathological pain states. *Physiological Reviews, 82,* 981–1011.

Wen, Y. R., Suter, M. R., Kawasaki, Y., Huang, J., Pertin, M., Kohno, T., Berde, C. B., Decosterd, I., & Ji, R. R. (2007). Nerve conduction blockade in the sciatic nerve prevents but does not reverse the activation of p38 mitogen-activated protein kinase in spinal microglia in the rat spared nerve injury model. *Anesthesiology, 107,* 312–321.

Whiteside, G. T., & Munglani, R. (2001). Cell death in the superficial dorsal horn in a model of neuropathic pain. *Journal of Neuroscience Research, 64,* 168–173.

Woolf, C. J. (1983). Evidence for a central component of postinjury pain hypersensitivity. *Nature, 306,* 686–688.

Yarnitsky, D., Granot, M., Nahman-Averbuch, H., Khamaisi, M., & Granovsky, Y. (2012). Conditioned pain modulation predicts duloxetine efficacy in painful diabetic neuropathy. *Pain, 153,* 1193–1198.

Yowtak, J., Lee, K. Y., Kim, H. Y., Wang, J., Kim, H. K., Chung, K., & Chung, J. M. (2011). Reactive oxygen species contribute to neuropathic pain by reducing spinal GABA release. *Pain, 152,* 844–852.

Zhou, Z. S., & Zhao, Z. Q. (2000). Ketamine blockage of both tetrodotoxin (TTX)-sensitive and TTX-resistant sodium channels of rat dorsal root ganglion neurons. *Brain Research Bulletin, 52,* 427–433.

Central Gray

▶ Opioid Electrophysiology in PAG

Central Gray/Central Grey

▶ Opioid Electrophysiology in PAG

Central Hyperexcitability

▶ Visceral Pain Model; Esophageal Pain

Central Hyperexcitability State

▶ Central Changes After Peripheral Nerve Injury

Central Hyperexcitabiltiy

▶ Neuronal Hyperexcitability

Central Lateral Nucleus (CL)

Definition

The intralaminar complex is a group of thalamic nuclei composed of neurons that are located within the internal medullary lamina, a nerve fiber sheet that can be used to subdivide different parts of the thalamus. The central lateral nucleus is one of the rostral group of intralaminar nuclei of the primate thalamus. It receives input from the spinothalamic tract and projects broadly to the sensorimotor cortex, as well as to the striatum.

Cross-References

► Spinothalamic Input: Cells of Origin (Monkey)
► Spinothalamic Terminations, Core and Matrix
► Thalamotomy for Human Pain Relief
► Thalamus, Visceral Representation

Central Lobe

► Insular Cortex, Neurophysiology, and Functional Imaging of Nociceptive Processing

Central Medial Nucleus (CM)

Definition

A nucleus within the internal medullary lamina, located ventrally to the central lateral nucleus and lateral to the parafascicular nucleus.

Cross-References

► Thalamus, Visceral Representation

Central Nervous System Map

Definition

The organization of locations on or in the brain or spinal cord that represent the characteristics of a stimulus, such as the receptive field, or of motor output, such as stimulation-evoked movement.

Cross-References

► Thalamus, Receptive Fields, Projected Fields, Human

Central Nervous System Pain Disorder

► Neuropathic Pain in Children

Central Nervous System Portion of a Cranial Nerve

Definition

The proximal portion of a cranial nerve over which myelin is associated with glial (oligodendroglial) cells rather than with the Schwann cells, which are associated with myelin in the periphery.

Cross-References

► Trigeminal, Glossopharyngeal, and Geniculate Neuralgias

Central Nervous System Stimulation for Pain

Mary O. Adedoyin, Peter T. Ohara and Luc Jasmin
Department of Anatomy, University of California San Francisco, San Francisco, CA, USA

Synonyms

CNS stimulation in treatment of neuropathic pain

Definition

Electrical stimulation of the central nervous system is a nondestructive and reversible therapy for certain forms of difficult-to-treat chronic pain. Candidates have to fail a trial of the more conventional therapies. An electrode is placed over the spinal cord, cerebral cortex, or in the thalamus, hypothalamus, or in the central gray matter. Analgesia is produced when a small current is delivered through this electrode, and usually persists for some time after the current is turned off.

Characteristics

Both clinical and experimental studies have determined that electrical stimulation to specific areas of the spinal cord or brain is analgesic. Considering the complexity of chronic pain it is often perplexing that such a simple technique results in relieving pain syndromes often deemed "intractable," while complex pharmaceutical approaches, often because their side effects, end in failure. Since the 1960s, the technology associated with electrical stimulation for pain has remained essentially the same, although the quality of the components has improved. The two main components are the electrode and the current generator, both of which are internalized (i.e., all is inside the patient). The electrode (4–20 contacts) is usually placed over the dorsal aspect (dorsal columns) of the spinal cord, in the brain parenchyma (thalamus, hypothalamus, or periaqueductal gray matter [PAG]), or over the surface of the motor cortex. The current generator or implantable pulse generator (IPG) typically has a thin profile of 10 mm and is approximately 1 ounce. The current generator (includes a rechargeable battery) is concealed under the skin and connected to the electrode through subcutaneous wires. Telemetry programming using a handheld computer either wirelessly or with an antenna, laid over the appropriate skin area, allows adjusting the stimulation for each patient. Voltage, current, frequency, pulse width, and times of the day that the system is on can be adjusted after the system is implanted. The current generator can be used to deliver tonic or burst pulses stimulation to the spinal cord.

Spinal Cord Stimulation (SCS)

Spinal cord stimulation (SCS) was first done in 1967 and since then well over a hundred thousand patients have received this treatment worldwide. About 10,000 individuals will be implanted in the United States alone this year. While the mechanism remains to be fully elucidated, a widely regarded explanation is the gate theory. This theory suggests that nociceptive information entering the CNS is reduced by the activity of innocuous (low threshold large fibers) sensory afferents or brain activity. The effect of electrical stimulating neural pathways capable of inhibiting nociceptive information is to "close the gate" on the noxious input and result in analgesia. Accordingly, the electrode is positioned over the dorsal columns of the spinal cord to stimulate the lemniscal fibers. There, only the superficial fibers are stimulated; this is a technological limitation. The spinal level at which the electrodes are placed correspond to the site where the fibers from the affected dermatome/viscerotome are closest to the surface. Interestingly, the analgesia produced by stimulation often lasts long after the cessation of the electric current (minutes to hours). This persistent analgesia is thought to depend not only on an effect on neurons adjacent to the stimulating electrode, but also on long neural loops linking the spinal cord to the brain that result in the inhibition of spinothalamic projection neurons (Gerhart et al. 1984). Obviously, when the underlying disease involves a loss of large fibers activity (as in multiple sclerosis or post-cordotomy dysesthesia), stimulation is less likely to work. Another often cited mechanism is that spinal cord stimulation blocks sympathetic nervous system fibers. This would explain the favorable results obtained in complex regional pain syndrome (CRPS) and in angina pectoris. Other indications are neuropathic leg pain (not back pain) associated with failed back syndrome, diabetic neuropathy, ischemic leg pain not treatable by vascular surgery, phantom limb pain, postherpetic neuralgia, urinary voiding dysfunction symptoms, spinal cord injury pain, tabes dorsalis, atypical stiff limb syndrome, and spinal cord injury pain. Long-term, moderate relief can be obtained in many patients, predominantly adults; however, complete relief has been reported in children with CRPS (Olsson et al. 2008).

In spite of the use of electrical stimulation to control pain in large number of patients, there have been only a few placebo-controlled evaluation of this treatment (Mailis-Gagnon et al. 2004). A significant placebo effect might exist in patients utilizing SCS, and the relief might be related more to the unpleasantness than to the intensity of the pain sensation. These findings are somewhat undermined by the difficulty in controlling for placebo response in a therapy that depends on the production of a sensory phenomenon to work (i.e., patients feel a tingling or paresthesia in the stimulated area when the current generator is on). De Ridder and colleagues (De Ridder et al. 2010) report that in their patient sample population paresthesia-free burst stimulation system was more effective at suppressing neuropathic pain compared to the standard tonic stimulation parameter. As with any new system, more rigorously controlled studies are required to fully understand the therapeutic benefits and any significant mechanistic difference if any between the modes of stimulation.

Currently, there is a dearth of clinical studies reporting functional assessment as a secondary measure to support patient feedback, which creates an unavoidable inherent bias. Kemler and colleagues (Kemler et al. 2004), however, have showed that patients report pain relief with SCS but with no observable significant change in clinical functional status. Although it is important to understand the exact mechanism of pain relief, it will likely not decrease the value of this therapy to the patient whether it is via placebo or some other mechanism. While all agree that better clinical studies are needed to confirm the effectiveness of SCS, especially since the benefits appear moderate, the continued use appears justified since it decreases the cost of treating pain (Taylor et al. 2005, 2010). In addition, the timing of SCS intervention may be an important factor determining the degree of efficacy experienced by the patient; Truin and colleagues (Truin et al. 2011) show in their neuropathic rat model that early use of SCS resulted in a higher responders/nonresponders ratio and prolongation of the system's anti-allodynic therapeutic effect compared to animals who received stimulation at a later time post injury. Based on this finding, they proposed that early as opposed to late therapeutic interventions that include SCS could be critical to modulating underlying mechanisms that play a role in the onset and severity of chronic neuropathic pain.

Deep Brain Stimulation (DBS)

Deep brain stimulation (DBS) has been out of favor since 1999 when Medtronic (Medtronic Inc. Minneapolis, MN) decided not to pursue FDA approval for this therapy due to the lack of conclusive clinical data (Coffey 2001). In spite of this, DBS is still used "off-label" here and in other countries for selected patients (Nandi and Aziz 2004). Electrodes are implanted in the ventropostero-lateral thalamus, posterior limb of the internal capsule, periventricular gray including the posterior hypothalamus and PAG. Stimulation in these areas is reported as fairly successful in patients with failed back syndrome, trigeminal neuropathy other than tic douloureux, certain forms of central pain (post-stroke pain), peripheral neuropathy, anesthesia dolorosa, and post-cordotomy dysesthesia. Other indications outside of pain include movement disorders, obsessive-compulsive disorder, binge eating-related obesity, and cerebral palsy.

Similar to dorsal column stimulation, the neural mechanism underlying DBS-induced analgesia is a matter of speculation. Marchand and colleagues (Marchand et al. 2003) studied the effect of placebo stimulation in patients with thalamic electrodes installed for pain control. Their key finding is that placebo analgesia is a significant element of DBS and that this is reinforced by the degree of paresthesia felt during the stimulation. (Fontaine et al. 2010) performed a randomized placebo-controlled double-blind prospective crossover study, and reported DBS efficacy in the open phase but not the randomized phase. Unlike the study of Marchand and colleagues (2003), Fontaine and colleagues (2010) stimulated the hypothalamus, which has been demonstrated not to induce perceptible sensations (paresthesia); however, the caveat with this study is the crossover feature of the study does not eliminate possible sustained placebo effect. Both studies support the contention that further controlled investigations are needed to understand the mechanisms by which stimulation of the central nervous system produces analgesia.

Analgesia produced by stimulation of the PAG may be via the endogenous opioid system; however, this needs to be further investigated. An interaction between the autonomic and the pain modulatory system is another possible mechanism. Pereira and colleagues (Pereira et al. 2010) report that in patients with chronic neuropathic pain, stimulation of the ventral PAG reduced the high-frequency power of the heart variability ratio, which correlated positively with analgesia, an effect that was absent after dorsal PAG stimulation. Based on this finding, they proposed that ventral PAG stimulation increases parasympathetic output to reduce pain via anatomical connections distinct from dorsal PAG, which has been shown to contribute to pain reduction via reduction of the sympathetic drive (Green et al. 2006). Thus, descending vagal control appears to be involved in modulating analgesic efficacy.

Recently, DBS has experienced a rebirth based on the finding that stimulation of the hypothalamus can reduce the occurrence of cluster headaches (Horton's headaches). Cluster headaches are a type of vascular pain, occurring more commonly in males, and characterized by unilateral headaches lasting 30–90 min, and recurring one to six times a day for several weeks. Between attacks, patients can be asymptomatic for months. The pain is reported as excruciating and characteristically located around the eye, which is visibly congested and tearing. While some measures such as avoiding alcohol and tobacco reduce the frequency of the attacks, in many patients the prevention and treatment of cluster headaches presents a challenge. In 1998,

the positron emission tomography (PET) observation that the symptomatic phase of the cluster headaches was accompanied by the activation of a region of the ipsilateral medio-caudal hypothalamus led to the idea that stimulation of this area might be of therapeutic value (Ekbom and Waldenlind 2004; Leone et al. 2004). Unilateral or bilateral electrodes and high-frequency stimulation have been effective in reducing the occurrence and duration of the attacks. PET imaging in cluster headache has revealed that during hypothalamic stimulation, some pain activated areas such as the ipsilateral anterior cingulate, primary somatosensory cortices, and the insular cortex bilaterally are switched off (May et al. 2003). DBS does not bring a cure to individuals with cluster headaches, and more recent studies suggest that occipital nerve stimulation might achieve equivalent if not better results while being much less invasive.

Motor Cortex Stimulation (MCS)

Motor cortex stimulation (MCS) has also yielded positive results in treating patients with central pain following ischemic brain or spinal injury, and with deafferentation pain in the trigeminal or spinal territories. Here an electrode is laid over the part of motor cortex (usually over the dura rather than directly on the pia matter) that corresponds to the area were the pain is felt on the opposite side of the body (according to the homunculus).

The rationale that guided surgeons to attempt cortical stimulation was based on the results of many decades of experimentation which demonstrated that electrical stimulation of the cerebral cortex affects spinal and trigeminal afferent sensory transmission and that this effect is, in part, mediated by presynaptic inputs onto somatic afferents including nociceptive afferents (Abdelmoumene et al. 1970). The effect of stimulation of the cerebral cortex, however, can be either inhibitory, excitatory, or both, on spinothalamic neurons (Yezierski et al. 1983). Because the latency of excitation is significantly shorter than the latency of inhibition, the latter could be polysynaptic via inhibitory interneurons and/or presynaptic processes on primary afferent terminals. Experiments in cats (Tsubokawa et al. 1991) have shown that motor cortex stimulation blocks spontaneous burst activity induced by spinothalamic deafferentation. In rats, motor cortex stimulation causes activation of limbic system structure (PAG, amgydala, anterior cingulated cortex), and endogenous opiate centers, modulating the descending pain inhibitory system (Pagano et al. 2011).

The most common indication for cortical stimulation is central pain syndrome. Dejerine recognized the syndrome a century ago, while an intern at the Salpêtrière Hospital in Paris. Because all of his patients with this previously unrecognized pain syndrome were found to have strokes in the posterolateral thalamic area at autopsy, Dejerine and Roussy, coined the term "Thalamic Syndrome." A complete thalamic syndrome is uncommon and subsequently, the term "thalamic pain" was used for all pain conditions that arose from both thalamic and non-thalamic CNS lesions. Since the majority of central pain syndromes occur after an ischemic stroke, the designation "central post-stroke pain" (CPSP) is often employed. The current definition of central pain has become quite broad and accommodates etiologies of central pain such as Parkinson's disease or epilepsy in which there is no thalamic lesion or interruption of thalamic afferents. It should be noted that "central pain" differs from "centralized pain," which is considered to result from remodeling of the CNS as a consequence of a peripheral nerve injury. One example of centralized pain for which cortical stimulation has recently been shown to be successful is trigeminal deafferentation pain (Brown and Pilitsis 2005).

Although it is important to optimize patient selection for motor cortex stimulation, there is no unequivocal way to preoperatively separate the responders from the nonresponders. Only patients with intact motor cortex and corticospinal projections should be chosen. In patients with large lesions of motor cortex or pyramidal tract on the side opposite to the symptoms, stimulation of the ipsilateral motor cortex has been shown to produce analgesia. Stimulation is usually ineffective, however, in subjects with profound sensory loss. Imaging studies have also been recommended prior to surgery in order to confirm the presence of cortical hypoperfusion and ascertain the side of implantation, since cortical stimulation is associated with a cortical reperfusion. Appropriate electrode placement is critical to obtaining pain relief in patients; therefore, different methods have been attempted for placement over appropriate neural targets. Recently, Esfahani and colleagues (Esfahani et al. 2011) reported three cases of intractable neuropathic facial pain in which preoperative fMRI was used to localize facial areas before electrode implantation surgery. In their study, all three patients reported either absolute or near-absolute pain relief post electrode implantation at their last follow-up session. When implemented, the use of brain mapping such as fMRI or neuronavigation allows for accurate localization and control for individual cortical plasticity. Another strategy to select suitable patients is by using a trial of noninvasive stimulation, using a transcranial magnetic coil over the motor cortex could also be a helpful way of selecting the potential responders.

Noninvasive cortical brain stimulation is showing promise as treatment for drug-resistant neuropathic pain, although the mechanism is not clear. In patients with medically retractable neuropathic pain, sham-controlled double-blind studies have demonstrated clinical efficacy of repetitive transcranial magnetic stimulation (rTMS) and trancranial direct current stimulation (tDCS). Defrin and colleagues (Defrin et al. 2007) reported a 30 % reduction in pain intensity and a significant 4 °C increase in heat-pain threshold in their six subjects that received rTMS compared to the sham group. Similarly, Soler and colleagues (Soler et al. 2010) report that of the 10 patients treated with real tDCS; 30 % reported "much" and 50 % reported "minimal" improvement of pain intensity perception.

Vagal Nerve Stimulation (VNS)

Vagal nerve stimulation (VNS) acts indirectly to stimulate the CNS through sensory afferents to the caudal brainstem. VNS is currently used to treat certain forms of epilepsy, cardiac effect, depression, and is under investigation as possible therapy for improving cognition, and multiple sclerosis. It has become an accepted method for treating cluster headaches and migraine (Mauskop 2005), and possibly other pain disorders. VNS was reported to be clinically therapeutic in a patient with an active DBS who suffered head trauma post first implantation when modifications of the DBS parameters did not produce adequate relief (Franzini et al. 2009); of note there was no crosstalk or signal interference between the two systems, and the patient did not experience any significant side effects. Although this is a single case study, it exemplifies the multifaceted complex nature of pain and the importance of further investigations to better understand the mechanisms underlying neurostimulation analgesic effect. Because VNS can decrease the nociceptive threshold in depressed individuals (Borckardt et al. 2005) and depression often accompanies pain, patient selection for this technique will be critical.

Given the role played by the autonomic system in pain modulation, one can understand the role played by vagal nerve stimulation. However, pain relief has also been demonstrated to occur via occipital nerve stimulation (ONS), although its mechanism is less clear. Nevertheless, occipital nerve stimulation has gained favor with surgeons as a first pass to treat pain associated with drug-resistant cluster headaches. Its minimal invasiveness and therapeutic efficacy in patients makes it an attractive alternative to VNS in patients that meet the criteria for selection.

Conclusion

Further clinical studies are needed to validate CNS stimulation as an effective treatment of pain, mainly because most of the evidence is anecdotal or retrospective. Minimal side effects usually related to the stimulation itself subside with time, and are generally well tolerated by patients. Basic research is critically needed to establish the mechanism of action, and it is possible that stimulation-induced analgesia might open the door to a new

understanding of the means by which pain operates. DBS has also gained popularity for the treatment of movement disorders, a field where much greater understanding of the neural circuitry has fostered unique cooperation between clinicians and basic scientists. The same type of teamwork could benefit the development of new ideas and advances in the field of pain.

References

Abdelmoumene, M., Besson, J. M., & Aleonard, P. (1970). Cortical areas exerting presynaptic inhibitory action on the spinal cord in cat and monkey. *Brain Research, 20,* 327–329.

Borckardt, J. J., Kozel, F. A., Anderson, B., Walker, A., & George, M. S. (2005). Vagus nerve stimulation affects pain perception in depressed adults. *Pain Research & Management, 10,* 9–14.

Brown, J. A., & Pilitsis, J. G. (2005). Motor cortex stimulation for central and neuropathic facial pain: A prospective study of 10 patients and observations of enhanced sensory and motor function during stimulation. *Neurosurgery, 56,* 290–297; discussion 290–297.

Coffey, R. J. (2001). Deep brain stimulation for chronic pain: Results of two multicenter trials and a structured review. *Pain Medicine, 2,* 183–192.

De Ridder, D., Vanneste, S., Plazier, M., van der Loo, E., & Menovsky, T. (2010). Burst spinal cord stimulation: Toward paresthesia-free pain suppression. *Neurosurgery, 66,* 986–990.

Defrin, R., Grunhaus, L., Zamir, D., & Zeilig, G. (2007). The effect of a series of repetitive transcranial magnetic stimulations of the motor cortex on central pain after spinal cord injury. *Archives of Physical Medicine and Rehabilitation, 88,* 1574–1580.

Ekbom, K., & Waldenlind, E. (2004). Cluster headache: The history of the Cluster Club and a review of recent clinical research. *Functional Neurology, 19,* 73–81.

Esfahani, D. R., Pisansky, M. T., Dafer, R. M., & Anderson, D. E. (2011). Motor cortex stimulation: Functional magnetic resonance imaging-localized treatment for three sources of intractable facial pain. *Journal of Neurosurgery, 114,* 189–195.

Fontaine, D., Lazorthes, Y., Mertens, P., Blond, S., Geraud, G., Fabre, N., Navez, M., Lucas, C., Dubois, F., Gonfrier, S., Paquis, P., & Lanteri-Minet, M. (2010). Safety and efficacy of deep brain stimulation in refractory cluster headache: A randomized placebo-controlled double-blind trial followed by a 1-year open extension. *The Journal of Headache and Pain, 11,* 23–31.

Franzini, A., Messina, G., Leone, M., Cecchini, A. P., Broggi, G., & Bussone, G. (2009). Feasibility of simultaneous vagal nerve and deep brain stimulation in chronic cluster headache: Case report and considerations. *Neurological Sciences, 30*(Suppl 1), S137–S139.

Gerhart, K. D., Yezierski, R. P., Wilcox, T. K., & Willis, W. D. (1984). Inhibition of primate spinothalamic tract neurons by stimulation in periaqueductal gray or adjacent midbrain reticular formation. *Journal of Neurophysiology, 51*, 450–466.

Green, A. L., Wang, S., Owen, S. L., Xie, K., Bittar, R. G., Stein, J. F., Paterson, D. J., & Aziz, T. Z. (2006). Stimulating the human midbrain to reveal the link between pain and blood pressure. *Pain, 124*, 349–359.

Kemler, M. A., De Vet, H. C., Barendse, G. A., Van Den Wildenberg, F. A., & Van Kleef, M. (2004). The effect of spinal cord stimulation in patients with chronic reflex sympathetic dystrophy: Two years' follow-up of the randomized controlled trial. *Annals of Neurology, 55*, 13–18.

Leone, M., Franzini, A., Broggi, G., May, A., & Bussone, G. (2004). Long-term follow-up of bilateral hypothalamic stimulation for intractable cluster headache. *Brain, 127*, 2259–2264.

Mailis-Gagnon, A., Furlan, A. D., Sandoval, J. A., & Taylor, R. (2004). Spinal cord stimulation for chronic pain. *Cochrane Database of Systematic Reviews*, (3), CD003783.

Marchand, S., Kupers, R. C., Bushnell, M. C., & Duncan, G. H. (2003). Analgesic and placebo effects of thalamic stimulation. *Pain, 105*, 481–488.

Mauskop, A. (2005). Vagus nerve stimulation relieves chronic refractory migraine and cluster headaches. *Cephalalgia, 25*, 82–86.

May, A., Leone, M., Boeker, H., Bussone, G., Sprenger, T., & Draganski, B. (2003). Deep brain stimulation in cluster headache: Preventing intractable pain by activating the pain network. *Cephalalgia, 23*, 656.

Nandi, D., & Aziz, T. Z. (2004). Deep brain stimulation in the management of neuropathic pain and multiple sclerosis tremor. *Journal of Clinical Neurophysiology, 21*, 31–39.

Olsson, G. L., Meyerson, B. A., & Linderoth, B. (2008). Spinal cord stimulation in adolescents with complex regional pain syndrome type I (CRPS-I). *European Journal of Pain, 12*, 53–59.

Pagano, R. L., Assis, D. V., Clara, J. A., Alves, A. S., Dale, C. S., Teixeira, M. J., Fonoff, E. T., & Britto, L. R. (2011). Transdural motor cortex stimulation reverses neuropathic pain in rats: A profile of neuronal activation. *European Journal of Pain, 15*, 268.e1–268.e14.

Pereira, E. A., Lu, G., Wang, S., Schweder, P. M., Hyam, J. A., Stein, J. F., Paterson, D. J., Aziz, T. Z., & Green, A. L. (2010). Ventral periaqueductal grey stimulation alters heart rate variability in humans with chronic pain. *Experimental Neurology, 223*, 574–581.

Soler, M. D., Kumru, H., Pelayo, R., Vidal, J., Tormos, J. M., Fregni, F., Navarro, X., & Pascual-Leone, A. (2010). Effectiveness of transcranial direct current stimulation and visual illusion on neuropathic pain in spinal cord injury. *Brain, 133*, 2565–2577.

Taylor, R. S., Ryan, J., O'Donnell, R., Eldabe, S., Kumar, K., & North, R. B. (2010). The cost-effectiveness of spinal cord stimulation in the treatment of failed back surgery syndrome. *The Clinical Journal of Pain, 26*, 463–469.

Taylor, R. S., Van Buyten, J. P., & Buchser, E. (2005). Spinal cord stimulation for chronic back and leg pain and failed back surgery syndrome: A systematic review and analysis of prognostic factors. *Spine, 30*, 152–160.

Truin, M., van Kleef, M., Linderoth, B., Smits, H., Janssen, S. P., & Joosten, E. A. (2011). Increased efficacy of early spinal cord stimulation in an animal model of neuropathic pain. *European Journal of Pain, 15*, 111–117.

Tsubokawa, T., Katayama, Y., Yamamoto, T., & Koyama, S. (1991). Chronic motor cortex stimulation for the treatment of central pain. *Acta Neurochir (Wien), 52*(Suppl), 137–9.

Yezierski, R. P., Gerhart, K. D., Schrock, B. J., & Willis, W. D. (1983). A further examination of effects of cortical stimulation on primate spinothalamic tract cells. *Journal of Neurophysiology, 49*, 424–441.

Central Neuropathic Pain

David Bowsher
Pain Research Institute, University Hospital Aintree, Liverpool, UK

Introduction

Dejerine and Roussy (1906) described three cases of pain following strokes involving the thalamus and named the condition "thalamic syndrome" – a name that has, unfortunately, remained in the literature – because it was later shown that similar pain follows infarction of other, nonthalamic, telencephalic areas such as the cortex (Foix et al. 1927). Infarction in brainstem or the anterolateral medulla oblongata can also cause pain in the condition known as Wallenberg's syndrome (Ajuriaguerra 1937).

The onset of pain following central lesions is not restricted to supraspinal pathology. Unfortunately, recognition of the fact that central pains following spinal cord damage, such as anterolateral cordotomy (White and Sweet 1969), and of course spinal cord injury (SCI) are similar to those of supraspinal origin was slow to develop.

Definition

Leijon et al. (1989) proposed the name "central post-stroke pain" (CPSP) to cover all these contingencies – however, this name does not encompass cases caused by conditions other than stroke. The postmortem anatomical pathology of 11 cases in which lesions of the central nervous system, including cortical lesions, two tumors, and one multiple aneurysm, resulted in the onset of pain was described in detail by Davison and Schick (1935). Subarachnoid hemorrhage, multiple sclerosis, tumor, cerebral abscess, Behçet's disease, Parkinson's disease, and arteriovenous aneurysm have all been implicated as causes of central pain. It is well known that syringomyelia is associated with central pain (Madsen et al. 1994), and central pain caused by spinal cord injury (SCI) has been described in a number of reports (Yezierski 1996). Painful postcordotomy dysesthesia was recognized by White and Sweet (1955), who stated that 20 % of patients undergoing anterolateral cordotomy and surviving for more than a year developed painful dysesthesia. Siddal et al. (1999) concluded that 50 % of patients with spinal cord injury (SCI) suffer central neuropathic pain, as do most patients with syringomyelia.

It can convincingly be argued that the seat of pathophysiology in neuropathic pains due to insult of peripheral nerves (e.g., painful diabetic neuropathy, complex regional pain syndromes, and postherpetic neuralgia) is in the central nervous system; the symptomatology certainly fulfills the criterion enunciated in the next paragraph. However, they will not be dealt with in detail in this chapter. Thus, it would perhaps be better to recognize a category of central neuropathic pain (CNP) due to (unspecified) damage to any part of the neuraxis. That such pains be recognized as neuropathic requires one more criterion: the pain occurs in an area of sensory change (total or, much more frequently, partial deficit).

Incidence

While it has long been known that not all cases of spinal cord injury or Wallenberg's syndrome necessarily suffer spontaneous neuropathic pain, it is important to emphasize the findings of Bogousslavsky and his colleagues (1988): only 25 % of patients with thalamic infarcts involving the thalamic somatosensory relay nucleus (VPL, Vc) actually develop pain. In a prospective study, Andersen (1995) reported that about 8 % of all surviving stroke patients develop CNP. In a population of 250 million (e.g., USA), there are 250,000 strokes p.a., of whom about 170,000 survive. Thus, the annual incidence of CPSP is approximately 11,000 new cases. Since most patients with CPSP do not have very severe or life-threatening motor impairment (only 8 % of 111 of our CPSP patients were plegic, and only 29 % paretic), the prevalence must be very much higher. At least 30 % of MS patients have central pain, and in spinal cord injury (SCI), CNP is present in about two thirds of patients. SCI pain has been subdivided into pain at, above, and below the level of injury in an attempt to develop a more meaningful taxonomy (Siddall and Loeser 2001). Thus, despite the fact that far from all patients with injury to the CNS develops CNP, the overall prevalence is very high – yet it is regarded as a rarity by most members of the medical and allied professions.

Characteristics

Central pain is greatly influenced by autonomic factors. Seventy-nine CPSP patients were asked to identify factors that exacerbated or alleviated their pain: 59 % found their pain was exacerbated by cold and 57 % by emotional stress, while 38 % were alleviated by relaxation (which is why most patients with CNP can fall asleep without difficulty, even though they may wake in pain) and 15 % by warmth (Bowsher 1996a). The most important feature of CNP, as noted above, is that it occurs in an area of sensory change (partial or total deficit of one or more somatosensory modalities); this is indeed an essential criterion. The pain is experienced within the area of sensory change, i.e., it is smaller than the area exhibiting sensory change. Of course, patients do not present with sensory loss, but with pain; so this will be dealt with first. About 40 % of

CPSP patients begin to experience pain immediately after their stroke; the other 60 % have a later onset, which may be up to 2 years after stroke, but the median is 3 months (Bowsher 1996a). In the case of spinal cord injury, a similar pattern is observed: pain is immediate in about a third of cases, with later onsets up to 2 years but a median of 3 months (Widerström-Noga 2003).

Early descriptions emphasized burning pain as a prominent characteristic of central pain. Close interrogation reveals that this is most frequently paradoxical burning (ice burn). While this type of pain, when it occurs, is indeed characteristic of central pain and is much emphasized because patients say "It's like nothing I ever experienced before," burning pain is not a sine qua non of neuropathic pain. Only 47 % of 111 personal CPSP patients complained of burning pain (Bowsher 1996a). Others experienced aching or throbbing pain (35 %) or shooting/stabbing pain (7 %); no type of pain was perceived to be more intense than any other (▶ visual analogue scale, VAS). Background pain was exacerbated by emotional stress or environmental cold in about half of patients (28 % by both). Following spinal cord injury, burning and aching pain were found in almost equal proportions (Widerström-Noga 2002).

Sensory Change
Some somatosensory modalities may be entirely lost, while in others, the sensation may still be present but with a raised threshold compared to the mirror-image area on the unaffected side. There is frequently dissociation between various somatosensory submodalities. From the point of view of spontaneous pain in CPSP, the modalities most concerned appear to be those subserving innocuous thermal sensations (Bowsher 1996b) and sharpness discrimination (tested with weighted needles); pain intensity correlated with thermal (particularly warmth) and sharpness discrimination threshold elevation. Patients with aching pain had a significantly higher perception threshold for tactile (von Frey) stimuli than those with burning pain, while the latter had much higher thresholds for innocuous warmth and cold (but not for painful heat). Both had higher

thresholds for sharpness and innocuous thermal modalities than patients suffering from strokes with sensory loss but no pain, and additionally, patients with burning pain, but not those with aching pain, had a very much higher threshold for warmth than did pain-free stroke patients (Bowsher 1996a).

Eide et al. (1996) found that CNP in SCI patients was correlated with intensity of sensory deficit. Central pain does not accompany loss of only Aβ modalities (touch, innocuous pressure, vibration), as those subserved by smaller peripheral fibers (Aδ, C) are also compromised. It was noted that in cordotomized patients with malignant disease, where nonneuropathic pain due to the neoplasm returned, pinprick (sharpness) threshold also returned toward that on the unaffected side, while in those patients who developed neuropathic postcordotomy pain, the deficit in sharpness sensation remained (Lahuerta et al. 1994). Indeed, so far as central pain is concerned, it is irrelevant whether or not sensations subserved by Aβ fibers are affected. Contrarily, stroke patients with sensory deficits but no pain frequently have a very marked tactile deficit but much less extensive thermal and sharpness deficits.

▶ Allodynia is defined as pain produced by innocuous stimulation. As it only occurs in patients with peripheral or central neuropathic pain, it is pathognomonic when found. However, unfortunately for the diagnostician, it is found in only about half of patients with supraspinal CNP. By far, the commonest form of allodynia is dynamic mechanical or tactile: allodynia caused by a moving tactile stimulus subserved by Aβ fibers. The provoking stimulus may be the movement of clothes across the skin or a breeze on the face. Other forms of stimulus that may produce allodynia are thermal (particularly cold, which unlike other forms of allodynia is twice as common in males as in females; warmth-provoked allodynia is rare). The threshold for innocuous warmth is significantly higher in CPSP patients with allodynia (all forms) than in those without. Movement-related allodynia also occurs in CPSP, provoked by active or passive movement, so presumably initiated from stretch receptors.

While the pain of allodynia usually occurs in the area stimulated, it may occur in a remote area, even one which itself is neither spontaneously painful nor allodynic. When somatosensory thresholds in patients with mechanical allodynia are compared with thresholds in pain-free stroke patients, it is found that the significance of the difference in thresholds between affected and unaffected mirror-image areas between the two groups is 0.02 for sharpness, 0.007 for warmth, 0.047 for cold, and 0.004 for heat pain, but it is nonsignificant for mechanical modalities (Bowsher 1996a).

Treatment

The mainstay of treatment until recently has been adrenergically active ▶ tricyclic antidepressants, i.e., principally ami- or nortriptyline, in relatively low doses (Leijon and Boivie 1989). Spontaneous recovery undoubtedly occurs (usually unreported), sometimes as a result of a further stroke (which may also suppress other neurogenic pains such as postherpetic neuralgia). Successful treatment with tricyclic antidepressants (TCAs) is time dependent, as with several other neuropathic pains – i.e., if treatment is initiated within 6 months of pain onset (NOT of stroke occurrence), 89 % of our patients gained relief; within 12 months, 67 %; but thereafter, less than 50 %. Ami- and nortriptyline are poorly tolerated and produce disagreeable side effects. Success has been reported with a more recent and less noxious antidepressant, venlafaxine. This is also adrenergically active; it should be noted that selective serotonin reuptake inhibitors (SSRIs), which have no SNRI activity, are ineffective in neuropathic pains (e.g., Max et al. 1992). Unlike peripheral neuropathic pain (PHN), the presence or absence of allodynia does not influence the therapeutic response to tricyclics.

Older ▶ anticonvulsant agents, notably carbamazepine, have proven themselves ineffective (Leijon and Boivie 1989) in the treatment of central neuropathic pain. Membrane-stabilizing drugs such as lignocaine (infusions) and mexiletine (Awerbuch and Sandyk (1990)) have been effective; mexiletine added to a TCA has been shown to be beneficial in CPSP (Bowsher

1995b) in some cases unresponsive to TCAs alone. Lignocaine (lidocaine) is a local anesthetic that blocks sodium channels. Boas et al. (1982) reported its analgesic effect when given systemically (I.V.) in conditions with neuropathic pain. Among newer anticonvulsants, which do not block sodium channels, gabapentin monotherapy is now widely used, though no statistics are yet available. Lamotrigine has also shown promise, perhaps especially when added to a TCA. Of the opioids, dextromethorphan, in combination with gabapentin, is the most widely used; favorable claims have also been made for methadone. NMDA receptor antagonists (a minor additional property of dextromethorphan and methadone) are said to be effective, as shown by use of the short-acting ketamine; relief of CPSP by oral ketamine has been reported (Vick and Lamer 2001), and there is a report of relief of syringomyelic pain by intravenous ketamine (Cohen and DeJesus 2004).

Surgical treatment has varied from thalamic lesioning to the implantation of stimulating electrodes – thalamic, periaqueductal, or spinal – and more recently, on the motor cortex (Tsubokawa et al. 1993), which is the most promising of the stimulation methods, with a reported success rate of 60–70 %. Additional treatments of pain associated with spinal cord injury are dealt with in the volume edited by Yezierski and Burchiel (2002). Relaxation therapy, which has a known beneficial effect, should not be overlooked in all forms of CNP.

Possible Mechanisms of Neuropathic Pain

We have to explain a condition that differentially and variably affects somatosensory submodalities and autonomic function; which follows insult to the central or peripheral nervous system, but not ineluctably – indeed, in only a minority of cases; and in which the sensory change may or may not be accompanied by neuropathic pain (although, as discussed above, the presence or absence of pain does appear to depend on the intensity of innocuous thermal loss, particularly for warmth).

It strikes the present author that the cardinal fact about neuropathic pain (central or peripheral) is that only a minority of individuals who suffer apparently appropriate insult to the central or peripheral nervous system actually develop neuropathic pain. This is equally true, it would seem, of nerve ligation experiments in animals, by no means all of which show signs or symptoms of neuropathic pain.

It was suggested earlier (Bowsher 1995b) that the best-fit theoretical model for central pain would appear to be one in which a widely distributed transmitter/ligand in the central nervous system and/or its specific receptors may become depleted; possible changes in receptor function were also mentioned by Siddall and Loeser (2001), specifically in relation to spinal cord injury. It has been known for a long time that some transmitters, such as serotonin, have both a global function, disturbance of which may be reflected in some psychiatric disorders, and a specific one, disturbance of which is seen in particular "focal" conditions such as migraine; enhanced pain sensitivity following injury may be regulated by spinal NK1 receptor expressing neurons (Suzuki et al. 2002), while spinal 5HT3 receptors mediate descending excitatory controls on spinal neurons activated in some neuropathic pain states (McCleane et al. 2003). Although we still have little idea what the transmitter(s) or receptor(s) concerned with central pain may be, recent developments in the field lend some support to this type of argument. For example, it has been shown in the NMDA system that presynaptic transmitters and postsynaptic receptors may be present in varying quantities (concentrations, densities) in the central nervous system (Yu and Salter 1999). Another relevant observation is that ubiquitin C-terminal hydrolase is upregulated in rats with sciatic nerve constriction injuries (Moss et al. 2002). Wang et al. (2002) have described as many as 148 genes which are up- or downregulated in the dorsal root ganglia of neuropathic rats.

It may, therefore, be suggested that changes in transmitter concentration and/or receptor density, as either up- or downregulation, may occur following nervous system injury. Following appropriate insult, transmitters and/or receptors may

undergo sudden and massive depletion, leading to immediate onset of CNP, or one or other or both may deplete slowly, giving rise to later onset of pain; they may recover their original levels/concentrations/density so that the pain "spontaneously" disappears; fluctuant recovery may account for fluctuating degrees of CNP.

However, if this hypothesis is even partly valid, there are a number of unexplained phenomena, among which are the following:

1. Why, in what is apparently the same condition, does one form of therapy succeed in some cases but fail in others, in which another form of therapy (a different drug) is effective (e.g., TCAs, which act on serotonergic and adrenergic systems, versus gabapentin, which acts on subunits of voltage-dependent Ca^{++} channels)?

2. It is fairly widely reported that patients with CPSP may have their pain alleviated by a second stroke; we have also seen the pains of both postherpetic and trigeminal neuralgias relieved by a subsequent stroke. Such events are hardly going to increase levels of transmitter X or densities of receptor Y!

Although progress has been made in understanding the mechanisms responsible for central pain, there are clearly additional questions to address, the answers to which will hopefully provide new insights into more effective treatment strategies.

Cross-References

▶ Spinal Cord Injury Pain Model, Hemisection Model

References

Andersen, G. (1995). Incidence of central post-stroke pain. *Pain, 61*, 187–194.

Awerbuch, G. I., & Sandyk, R. (1990). Mexiletine for thalamic pain syndrome. *International Journal of Neuroscience, 55*, 129–133.

Boas, R. A., Covino, B. G., & Shanharian, A. (1982). Analgesic response to i. v. lignocaine. *British Journal of Anaesthesia, 54*, 501–505.

Bogousslavsky, J., Regli, F., & Uske, A. (1988). Thalamic infarcts: Clinical syndromes, etiology, and prognosis. *Neurology, 38*, 837–848.

Bowsher, D. (1995a). The management of central post-stroke pain. *Postgraduate Medical Journal, 71*, 598–604.

Bowsher, D. (1995b). Central pain. *Pain Reviews, 2*, 175–186.

Bowsher, D. (1996a). Central pain: Clinical and physiological characteristics. *Journal of Neurology, Neurosurgery and Psychiatry, 61*, 62–69.

Bowsher, D. (1996b). Central pain of spinal origin. *Spinal Cord, 34*, 707–710.

Bowsher, D., Leijon, G., & Thomas, K. A. (1998). Central post-stroke pain: Correlation of magnetic resonance imaging with clinical pain characteristics and sensory abnormalities. *Neurology, 51*, 1352–1358.

Cohen, S. P., & DeJesus, M. (2004). Ketamine patient-controlled analgesia for dysesthetic central pain. *Spinal Cord, 42*, 425–428.

Davison, C., & Schick, W. (1935). Spontaneous pain and other sensory disturbances. *Archives of Neurology and Psychiatry (Chicago), 34*, 1204–1237.

de Ajuriaguerra, J. (1937). *La Douleur dans les Affections du Système Nerveux Central*. Paris: Doin.

Dejerine, J., & Roussy, J. (1906). Le syndrome thalamique. *Revue Neurologique, 14*, 521–532.

Eide, P. K., Jörum, E., & Stenejhem, A. E. (1996). Somatosensory findings in spinal cord injury patients with central dysaesthesia pain. *Journal of Neurology, Neurosurgery and Psychiatry, 60*, 411–415.

Foix, C., Chavany, J. A., & Lévy, M. (1927). Syndrome pseudo-thalamique d'origine pariétale. Lésion de l'artère du sillon interpariétal (Pa P1P2 antérieures, petit territoire insulo-capsulaire). *Revue Neurologique (Paris), 35*, 68–76.

Lahuerta, J., Bowsher, D., Buxton, P. H., & Lipton, S. (1994). Percutaneous cervical cordotomy: A review of 181 operations in 146 patients, including a study on the location of "pain fibers" in the second cervical spinal cord segment of 29 cases. *Journal of Neurosurgery, 80*, 975–985.

Leijon, G., Boivie, J., & Johansson, I. (1989). Central post-stroke pain – A study of the mechanisms through analyses of the sensory abnormalities. *Pain, 37*, 173–185.

Madsen, P. W., Yezierski, R. P., & Holets, V. R. (1994). Syringomyelia: Clinical observations and experimental studies. *Journal of Neurotrauma, 11*, 241–254.

Max, M., Lynch, S. A., Muir, J., et al. (1992). Effects of desipramine, amitriptyline, and fluoxetine on pain in diabetic neuropathy. *New England Journal of Medicine, 326*, 1250–1256.

McCleane, G. J., Suzuki, R., & Dickenson, A. H. (2003). Does a single intravenous injection of 5HT3 receptor antagonist ondansetron have an analgesic effect in neuropathic pain? a double-blinded, placebo-controlled cross-over study. *Anesthesia and Analgesia, 97*, 1474–1478.

Moss, A., Blackburn-Munro, G., Garry, E. M., et al. (2002). A role of the ubiquitin-proteosome system in neuropathic pain. *Journal of Neuroscience, 22*, 1363–1372.

Siddall, P. J., & Loeser, J. D. (2001). Pain following spinal cord injury. *Spinal Cord, 39*, 63–73.

Siddall, P. J., Taylor, D. A., McClelland, J. M., et al. (1999). Pain report and the relationship of pain to physical factors in the first 6 months following spinal cord injury. *Pain, 81*, 187–197.

Suzuki, R., Morcuende, S., Webber, M., et al. (2002). Superficial NK^1-expressing neurons control spinal excitability through activation of descending pathways. *Nature Neuroscience, 5*, 1319–1326.

Taylor, C. P., Gee, N. S., Su, T. Z., et al. (1998). A summary of mechanistic hypotheses of gabapentin pharmacology. *Epilepsy Research, 29*, 233–249.

Tsubokawa, T., Katayama, Y., Yamamoto, T., Hirayama, T., & Koyama, S. (1993). Chronic motor cortex stimulation in patients with thalamic pain. *Journal of Neurosurgery, 78*, 393–401.

Vick, P. G., & Lamer, T. J. (2001). Treatment of central post-stroke pain with oral ketamine. *Pain, 92*, 311–313.

Wang, H., Sun, H., Della Penna, K., et al. (2002). Chronic neuropathic pain is accompanied by global changes in gene expression and shares pathobiology with neurodegenerative diseases. *Neuroscience, 114*, 529–546.

White, J. C., & Sweet, W. H. (1955). *Pain: Its mechanisms and neurosurgical control*. Springfield Ill: CC Thomas.

White, J. C., & Sweet, W. H. (1969). *Pain and the neurosurgeon. A 40 year experience*. Springfield Ill: CC Thomas.

Widerström-Noga, E. G. (2002). Evaluation of clinical characteristics of pain and psychosocial factors after spinal cord Injury. In R. P. Yezierski & K. J. Rurchiel (Eds.), *Spinal cord injury pain: Assessment, mechanisms, management* (pp. 53–82). Seattle: IASP Press.

Widerström-Noga, E. G. (2003). Chronic pain and non-painful sensations following SCI: Is there a relationship? *The Clinical Journal of Pain, 19*, 39–47.

Yezierski, R. P. (1996). Pain following spinal cord injury: The clinical problem and experimental studies. *Pain, 73*, 115–119.

Yezierski, R. P., & Burchiel, K. J. (2002). *Spinal cord injury pain: Assessment, mechanisms, management*. Seattle: IASP.

Yu, X. M., & Salter, M. W. (1999). Src, a molecular switch governing gain control of synaptic transmission mediated by N-methyl-D-aspartate receptors. *Proceedings of the National Academy of Science USA, 96*, 7697–7704.

Central Neuropathic Pain from Spinal Cord Injury

Philip J. Siddall
Department of Pain Management, Greenwich
Hospital, University of Sydney, Sydney, NSW,
Australia

Definition

Central pain from spinal cord injury (SCI) refers to ▶ neuropathic pain that occurs following traumatic or atraumatic injury to the spinal cord. It may be due to dysfunction occurring at spinal and/or supraspinal levels. Two main types of central neuropathic pain are described following SCI. These are ▶ at-level neuropathic pain and ▶ below-level neuropathic pain (Siddall et al. 2002).

Introduction

Central neuropathic pain is a major problem following spinal cord injury. It is a common consequence of damage to the spinal cord. However, the mechanisms contributing to the development of pain are incompletely understood, and successful treatment of the pain is often difficult. Despite this, there have been major advances in our understanding of some of the mechanisms that may underlie the problem, and this is leading to the development of new treatment strategies that offer the promise of improved management.

Characteristics

Prevalence

The prevalence of central neuropathic pain following SCI is relatively high. In the first 6 months following SCI, it has been reported that 35 % of patients have at-level neuropathic pain and 25 % of patients have below-level neuropathic pain. At 5 years following injury, there is little change in these numbers, with 42 % having at-level neuropathic pain and 34 % having below-level neuropathic pain (Siddall et al. 2003). This increase in numbers of people reporting pain reflects the lack of success in alleviating the pain. It also reflects the late onset of neuropathic pain, even years following injury, in many people. The prospective study by Siddall et al. (2003) also indicates a strong correlation between the presence of both types of neuropathic pain within 6 months and at 5 years following injury. This unfortunately suggests that if either of these pain types is present at 6 months, then there is a strong likelihood that it will be present at 5 years.

Diagnosis

A taxonomy of pain following SCI was proposed by the International Association for the Study of Pain (IASP) Task Force on SCI pain and is in widespread use (Siddall et al. 2002). Since then, an international working group has sought to achieve further consensus on SCI pain taxonomy (Widerström-Noga et al. 2008). The final report of this group is now published in the journal Spinal Cord (Bryce et al. 2012). It identifies four main types of pain following SCI. These are musculoskeletal, visceral, at-level neuropathic, and below-level neuropathic pains. The presence of neuropathic pain is suggested by descriptors such as electric-like, shooting, and burning with pain located in or adjacent to a region of sensory loss.

At-level and below-level neuropathic pains have distinctive features.

At-level neuropathic pain occurs as a band of pain in the dermatomes adjacent to the level of injury and is therefore sometimes referred to as segmental, end-zone, or border-zone pain. It may be due to damage to nerve roots and therefore be a form of peripheral neuropathic pain. However, animal models have clearly demonstrated that at-level neuropathic pain may occur in the absence of nerve root damage and therefore it may be central in origin.

Below-level neuropathic pain occurs more diffusely, in a bilateral distribution below the level of the spinal cord lesion in the region of sensory disturbance, and is sometimes referred to as central dysesthesia syndrome. It is most likely due to changes in the spinal cord and brain following SCI. Therefore, below-level neuropathic pain is generally regarded as a central neuropathic pain.

Mechanisms

At-level neuropathic pain may be due to nerve root compression. The mechanisms of pain associated with nerve root compression are similar to other forms of peripheral neuropathic pain and are described elsewhere. However, at-level neuropathic pain may also be due to changes within the spinal cord itself as a consequence of injury. The specific mechanisms underlying both at-level and below-level neuropathic pain are incompletely understood. However, there are a number of secondary changes that occur as a consequence of spinal cord damage, which may result in the generation or amplification of nociceptive signals (Yezierski 2009). These include:

1. Damage to the spinal cord may result in increased levels of glutamate which activates N-methyl D-aspartate (NMDA), non-NMDA, and metabotropic glutamate receptors. This activation of glutamate receptors results in activation of a cascade of secondary processes within neurons, which ultimately results in increased neuronal excitability. There is also upregulation of other channels such as the Na_v 1.3 sodium channel which may contribute to neuronal hyperexcitability.

2. Alternatively, increased neuronal excitability may be a consequence of reduced inhibition within the spinal cord. Several mechanisms have been proposed, including reduced function of inhibitory neurotransmitters and receptors such as γ aminobutyric acid (GABA) and glycine. This may occur through reduction in inputs from surrounding regions that normally exert an inhibitory action on the region, which has lost inputs (inhibitory surround). Loss of inhibition may also occur through disruption to inhibitory neurotransmitter production, release, or uptake as a consequence of spinal cord damage. A reduction in inhibition may also be due to a decrease in the levels of inhibitory controls exerted by descending antinociceptive pathways.

3. Injury to the spinal cord will also initiate inflammatory and immune responses, which will have both direct and indirect effects on the long-term integrity of spinal cord structures, as well as functional changes in sensory processing.

4. As well as the changes above, damage to the spinal cord may also result in microglial activation which may further contribute to sensitization. Structural changes within the dorsal horn, with remodeling of primary afferent terminals, have also been demonstrated.

Both increased excitation and loss of inhibition may give rise to a population of neurons close to the site of injury that have an increased responsiveness to peripheral stimulation and may even fire spontaneously. These alterations in the properties of spinal neurons may give rise to the phenomenon of hyperesthesia (► allodynia and ► hyperalgesia) and spontaneous pain, respectively.

As well as alterations at a spinal level, alterations in the chemistry and firing properties of supraspinal neurons have also been demonstrated. The main site that has been investigated is the thalamus (Ralston et al. 2000; Ohara et al. 2002). As well as thalamic changes, it has been demonstrated that central neuropathic pain following SCI is associated with reorganization of the somatosensory cortex.

Although these changes at a spinal and supraspinal level appear to underlie the development of pain, there may also be a peripheral contribution. There is evidence that, even in those with clinically complete injuries, there is some preservation of ascending tracts which may transmit information from the periphery and contribute to the presence of pain.

Treatment

There are relatively few studies that have specifically examined the effectiveness of treatments for pain following SCI, and in these situations, subject numbers are generally low. Therefore, there is little definitive evidence to guide management. In the few randomized controlled trials that have been done, many of the treatments were no more effective than the placebo, and therefore adequate control of central neuropathic SCI pain is generally difficult (Finnerup et al. 2001). Principles of treatment are derived largely from the treatment of other neuropathic pain conditions (Siddall and Middleton 2006).

It is traditionally stated that opioids are relatively ineffective for the treatment of neuropathic pain. However, there is increasing evidence that they may be effective if an appropriate dose is used. A randomized controlled trial of intravenous morphine (9–30 mg) found a significant reduction in brush-evoked allodynia, but no significant effect on spontaneous neuropathic SCI pain (Attal et al. 2002). Long-term use of opioids may also be a problem because of side effects such as constipation, which may be more of an issue in individuals with SCI. Tramadol may also be an option because of its serotonergic and noradrenergic effects, and there is evidence from a randomized controlled trial of its efficacy in central neuropathic SCI pain although it has a relatively high incidence of adverse events.

Intravenous or subcutaneous infusion of local anesthetics, such as lidocaine (lignocaine), is also widely used for the treatment of neuropathic pain and may be effective for neuropathic SCI pain. One of the actions of local anesthetics is to produce sodium channel blockade. This reduces the amount of ectopic impulses generated by activity at these receptors. Relatively low concentrations of local anesthetic are required to reduce ectopic neural activity in damaged nerves. There is evidence from randomized controlled trials supporting the efficacy of intravenous lidocaine in treating neuropathic SCI pain. However, the effect is short term. Although its mode of action is different from local anesthetics, propofol is another agent that may be administered parenterally and has been shown to provide effective relief of neuropathic SCI pain (Canavero et al. 1995).

As mentioned above, SCI may result in an increased release of glutamate and activation of NMDA receptors, resulting in central neuronal hyperexcitability. NMDA receptor antagonists such as ketamine have been used as a treatment for neuropathic pain following SCI, and several randomized controlled trials support its efficacy

although again the effect is short term. Administration is generally by infusion via the intravenous or subcutaneous route. One of the main problems with the use of ketamine is the occurrence of disturbing side effects such as hallucinations, although benzodiazepines may help reduce these symptoms. Although careful monitoring can help to minimize the rate of occurrence of these side effects, they can be distressing to the person when they do occur.

Oral rather than parenteral administration of a number of drugs is also possible and may be preferred. Unfortunately, most studies suggest that these oral agents are less effective than drugs used parenterally. Most of the evidence for the effectiveness of opioids for the treatment of neuropathic pain comes from studies using acute intravenous administration. Local anesthetics are not available in oral form, and the local anesthetic congener mexiletine does not appear to be as effective as lidocaine in reducing neuropathic SCI pain (Chiou Tan et al. 1996). Similarly, ketamine is difficult to administer orally, and other NMDA antagonists available for administration via the oral route, such as dextromethorphan, are also not as effective.

Several agents that are used widely for the treatment of persistent neuropathic pain are not available for parenteral administration but are available in the oral form. These include the tricyclic antidepressants such as amitriptyline and nortriptyline and anticonvulsants such as carbamazepine, valproate, lamotrigine, gabapentin, and pregabalin. However, specific evidence of efficacy in the treatment of neuropathic SCI pain is limited.

Randomized controlled trials of trazodone and amitriptyline (Cardenas et al. 2002) both failed to find an effect greater than placebo, although numbers in the trazodone study were low and the amitriptyline study contained subjects who had both musculoskeletal and neuropathic pains. A more recent randomized controlled trial of amitriptyline found that it was effective in those with depressive symptoms. Newer selective agents such as duloxetine may have a place despite a negative finding in the only randomized controlled trial to date.

Gabapentin is used widely for the treatment of neuropathic pain, and there are several small randomized controlled trials that indicate efficacy. Two randomized controlled trials also demonstrate the effectiveness of pregabalin. A randomized controlled trial of sodium valproate also failed to demonstrate a significant analgesic effect. Lamotrigine has been demonstrated to be effective, but only for the evoked component of neuropathic SCI pain (Finnerup et al. 2002).

Consideration may be given to spinal administration of drugs if oral approaches are unsuccessful. Intrathecal administration of morphine and clonidine has been helpful in some people with neuropathic SCI pain (Siddall et al. 2000). There has also been interest in noninvasive brain stimulation techniques with a randomized controlled trial demonstrating superiority over placebo.

Spinal cord injury requires a major psychological adjustment. Awareness of these issues is important in the evaluation of the person with pain. As with any pain condition, psychological factors may contribute to the perception and expression of pain. Pain report may be an expression of difficulty in adjustment, and therefore psychological approaches that attempt to deal with these issues may be helpful. Utilization of pain management strategies and cognitive behavioral therapy (CBT) may be helpful in achieving optimal pain management (Umlauf 1992).

Summary

Central neuropathic pain has a number of contributing mechanisms with a number of demonstrated changes at both spinal and supraspinal levels. The relative contribution of these contributors almost certainly varies from person to person. This complexity and variability of contributors is reflected in response to treatment which in general is only partially successful. Treatment relies on appropriate recognition of the type of pain, systematic use of available treatments, and a holistic approach.

Acknowledgment Philip Siddall is supported by grants from the Australian and New Zealand College of Anaesthetists (Grant ID 12/010) and the Australian National Health and Medical Research Council (Grant ID 1002354).

References

Attal, N., Guirimand, F., Brasseur, L., Gaude, V., Chauvin, M., & Bouhassira, D. (2002). Effects of IV morphine in central pain – A randomized placebo-controlled study. *Neurology, 58*, 554–563.

Bryce, T. N., Biering-Sorensen, F., Finnerup, N. B., Cardenas, D. D., Defrin, R., Lundeberg, T., Norrbrink, C., Richards, J. S., Siddall, P., Stripling, T., Treede, R. D., Waxman, S. G., Widerstrom-Noga, E., Yezierski, R. P., & Dijkers, M. (2012). International Spinal Cord Injury Pain Classification: part I. Background and description. *Spinal Cord, 50*, 413–417.

Canavero, S., Bonicalzi, V., Pagni, C. A., et al. (1995). Propofol analgesia in central pain – Preliminary clinical observations. *Journal of Neurology, 242*, 561–567.

Cardenas, D. D., Warms, C. A., Turner, J. A., Marshall, H., Brooke, M. M., & Loeser, J. D. (2002). Efficacy of amitriptyline for relief of pain in spinal cord injury: Results of a randomized controlled trial. *Pain, 96*, 365–373.

Chiou Tan, F. Y., Tuel, S. M., Johnson, J. C., Priebe, M. M., Hirsh, D. D., & Strayer, J. R. (1996). Effect of mexiletine on spinal cord injury dysesthetic pain. *American Journal of Physical Medicine & Rehabilitation, 75*, 84–87.

Finnerup, N. B., Yezierski, R. P., Sang, C. N., Burchiel, K. J., & Jensen, T. S. (2001). Treatment of spinal cord injury pain. *Pain Clinical Updates, 9*, 1–6.

Finnerup, N. B., Sindrup, S. H., Flemming, W. B., Johannesen, I. L., & Jensen, T. S. (2002). Lamotrigine in spinal cord injury pain: A randomized controlled trial. *Pain, 96*, 375–383.

Ohara, S., Garonzik, I., Hua, S., & Lenz, F. A. (2002). Microelectrode studies of the thalamus in patients with central pain and in control patients with movement disorders. In R. P. Yezierski & K. J. Rurchiel (Eds.), *Spinal cord injury pain: Assessment, mechanisms, management* (pp. 219–236). Seattle: IASP Press.

Ralston, D. D., Dougherty, P. M., Lenz, F. A., Weng, H. R., Vierck, C. J., & Ralston, H. J. (2000). Plasticity of the inhibitory circuitry of the primate ventrobasal thalamus following lesions of somatosensory pathways. In M. Devor, M.C. Rowbotham, & Z. Wiesenfeld-Hallin (Eds.), *Proceedings of the 9th world congress on pain* (pp. 427–434). Seattle: IASP Press.

Siddall, P. J., & Middleton, J. W. (2006). A proposed algorithm for the management of pain following spinal cord injury. *Spinal Cord, 44*, 67–77.

Siddall, P. J., Molloy, A. R., Walker, S., Mather, L. E., Rutkowski, S. B., & Cousins, M. J. (2000). Efficacy of intrathecal morphine and clonidine in the treatment of neuropathic pain following spinal cord injury. *Anesthesia and Analgesia, 91*, 1493–1498.

Siddall, P. J., Yezierski, R. P., & Loeser, J. D. (2002). Taxonomy and epidemiology of spinal cord injury pain. In R. P. Yezierski & K. J. Rurchiel (Eds.), *Spinal cord injury pain: Assessment, mechanisms, management* (pp. 9–24). Seattle: IASP Press.

Siddall, P. J., McClelland, J. M., Rutkowski, S. B., & Cousins, M. J. (2003). A longitudinal study of the prevalence and characteristics of pain in the first 5 years following spinal cord injury. *Pain, 103*, 249–257.

Umlauf, R. L. (1992). Psychological interventions for chronic pain following spinal cord injury. *The Clinical Journal of Pain, 8*, 111–118.

Widerström-Noga, E., Biering-Sorensen, F., Bryce, T., Cardenas, D. D., Finnerup, N. B., Jensen, M. P., et al. (2008). The international spinal cord injury pain basic data set. *Spinal Cord, 46*, 818–823.

Yezierski, R. P. (2009). Spinal cord injury pain: Spinal and supraspinal mechanisms. *Journal of Rehabilitation Research and Development, 46*, 95–108.

Central Pain

Definition

Central pain is defined by the International Association for the Study of Pain (IASP) as "Pain caused by a lesion or disease of the central somatosensory nervous system."

Cross-References

- ▶ Cell Therapy in the Treatment of Central Pain
- ▶ Central Nervous System Stimulation for Pain
- ▶ Central Pain in Multiple Sclerosis
- ▶ Central Pain Syndrome
- ▶ Central Pain, Human Studies of Physiology
- ▶ Central Pain, Outcome Measures in Clinical Trials
- ▶ Central Pain, Pharmacological Treatments
- ▶ Diagnosis and Assessment of Clinical Characteristics of Central Pain
- ▶ DREZ Procedures
- ▶ Functional Changes in Sensory Neurons Following Spinal Cord Injury in Central Pain
- ▶ Pain Treatment, Motor Cortex Stimulation
- ▶ Poststroke Pain Model, Thalamic Pain (Lesion)
- ▶ Secondary Somatosensory Cortex (S2) and Insula, Effect on Pain-Related Behavior in Animals and Humans
- ▶ Stimulation Treatments of Central Pain

Central Pain and Cancer

Paolo L. Manfredi and Gilbert R. Gonzales
Department of Neurology, Memorial Sloan-Kettering Cancer Center, New York, USA

Definition

In order to be diagnosed with central pain (CP) associated with cancer, patients must have a lesion within the central nervous system (CNS) caused by cancer or its treatment, pain in a distribution compatible with the CNS lesion, and no other lesions that could potentially cause pain in the same area (Casey 1991; Gonzales and Casey 2003; Gonzales et al. 2003).

This definition includes pathology in the spinal cord, brain stem, or cerebral hemispheres. Lesions in the peripheral nervous system can produce CNS changes, but these are secondary CNS changes and are not categorized as CP syndromes (Casey 1991; Gonzales and Casey 2003). The most common causes of CNS injuries that result in CP are stroke, spinal cord trauma, and multiple sclerosis (Casey 1991). However, cancer and its treatment can also cause CP (Gonzales et al. 2003).

Characteristics

Although CP occurs infrequently, over 15 % of patients with systemic cancer have metastases to the brain or spinal cord (Clouston et al. 1992), making it possible for some of these patients to go on and develop CP. Central pain caused by cancer is mostly described through case reports, such as the study of thalamic tumors by Cheek and Taveras (Cheek and Taveras 1966). Pagni and Canavero (1993) and Gan and Choksey (2001) have reported CP from extra-axial tumors such as meningiomas. Delattre and colleagues have reported CP caused by leptomeningeal disease (Delattre et al. 1989). A more recent study by Gonzales et al. described the prevalence and causes of central pain in patients with cancer (Gonzales et al. 2003). In this study conducted in a cancer center, a relatively high number of general oncology patients admitted to the neurology service (2 %) or seen in consultation by the pain service (4 %) were found to have CP. It is important to underscore that the prevalence of central pain seen in these patients is not representative of the true prevalence of central pain in hospitalized patients with cancer, as these patients were selected by their referral to the pain service or admission to the neurology ward. In this study, spinal cord lesions were by far more likely to cause CP compared to brain and brainstem lesions, as is the case in patients with noncancer causes of central pain (Gonzales et al. 2003).

When a patient with cancer has radiological documentation of a lesion within the CNS, pain in a distribution compatible with the CNS lesion, and no other lesions that could potentially cause pain in the same area, a diagnosis of CP can be made. In order to be classified as CP related to cancer, the inciting CNS lesion must be caused by cancer or its treatment.

Pain descriptors suggestive of CP, such as burning, numb, cold, pins and needles, electric shock, can also help with the diagnosis.

On physical examination it may be possible to elicit different neurological abnormalities, depending on the location and size of the CNS lesion. The finding of altered temperature sensation in the painful area is consistently seen in all CP patients with cancer, as expected from the experience with nonmalignant causes of CP (Gonzales and Casey 2003; Gonzales et al. 2003). A detailed sensory examination is therefore essential.

Central pain may be delayed by days to years after CNS injury (Gonzales 1995). In one cancer patient, CP was found to be delayed by up to 6 years after the diagnosis and treatment of the spinal cord tumor (Gonzales et al. 2003). This is much longer than the delay usually seen in nonmalignant causes of spinal cord CP.

The treatment strategies in patients with cancer and CP may include antitumor therapies such as radiation, chemotherapy, surgical resection, and steroids to decrease edema. Aside from addressing the treatment of the tumor, the treatment of CP in patients with cancer can be approached as with nonmalignant CP (Gonzales and Casey 2003; Gonzales et al. 2003) and include antidepressants, anticonvulsants, opioids, clonidine, baclofen, acetaminophen, and NSAIDs.

References

Casey, K. L. (1991). Pain and central nervous system disease: A summary and overview. In K. L. Casey (Ed.), *Pain and central nervous system disease: The central pain syndromes*. New York: Raven.

Cheek, W. R., & Taveras, J. M. (1966). Thalamic tumors. *Journal of Neurosurgery, 24*, 505–513.

Clouston, P. D., DeAngelis, L. M., & Posner, J. (1992). The spectrum of neurological disease in patients with systemic cancer. *Annals of Neurology, 31*, 268–273.

Delattre, J. Y., Walker, R. W., & Rosenblum, M. K. (1989). Leptomeningeal gliomatosis with spinal cord or cauda equina compression: A completion of supratentorial gliomas in adults. *Acta Neurologica Scandinavica, 79*, 133–139.

Gan, Y. C., & Choksey, M. S. (2001). Parafalcine meningioma presenting with facial pain: Evidence for cortical theory or pain? *British Journal of Neurosurgery, 15*, 350–352.

Gonzales, G. R. (1995). Central pain: Diagnosis and treatment strategies. *Neurology, 45*, 11–16.

Gonzales, G. R., & Casey, K. L. (2003). Central pain syndromes. In A. D. S. Rice, C. A. Washeld, D. Justins, et al. (Eds.), *Clinical pain management chronic pain*. London: Arnold Press.

Gonzales, G. R., Tuttle, S., Thaler, H. T., et al. (2003). Central pain in patients with cancer. *The Journal of Pain, 4*, 351–354.

Pagni, C. A., & Canavero, S. (1993). Paroxysmal perineal pain resembling tic doubureux, only symptom of a Dorsal meningioma. *Italian Journal of Neurological Sciences, 14*, 323–324.

Central Pain Following Traumatic Brain Injury

Ruth Defrin
Physical Therapy, Faculty of Medicine, Tel-Aviv
University, Tel-Aviv, Israel

Definition

The definition of ▶ central pain according to the International Association for the study of Pain (Merskey and Bogduk 1994) is: "pain initiated or caused by a primary lesion or dysfunction in the central nervous system." Recently, a slightly different definition was proposed: "pain arising as a direct consequence of a lesion or disease affecting the somatosensory system" (Treede et al. 2008). Such conditions may include, but are not restricted to spinal cord injury, brain infarcts, brain hemorrhages, syringomyelia, tumors, and traumatic brain injury (TBI).

   Traumatic brain injury (TBI) is a nondegenerative, noncongenital insult to the brain from an external mechanical force, leading to permanent or temporary impairment of cognitive, physical, and/or psychosocial functions, with an associated diminished or altered state of consciousness. TBI can be classified based on severity (mild, moderate, severe), mechanism (closed/penetrating head injury), or other features. Its outcome can range from complete recovery to permanent disability or death.

Introduction

Scope of the Pain Problem After TBI

Chronic pain is one of the known consequences of a TBI with prevalence rates varying between 22 and 95 % (Uomoto and Esselman 1993; Beetar et al. 1996; Lahz and Bryant 1996; Nampiaparampil 2008). Chronic pain in patients with TBI can originate from a number of sources. In some instances, the pain is a direct result of tissue injuries inflicted during the event

causing the TBI, such as bone fractures and dislocations, cervical injuries, soft tissue trauma, and peripheral nerve or plexus damage. Chronic pain can also originate from secondary consequences of the trauma including pressure palsies, periarticular new bone formation, spasticity, deep venous thrombosis, and abnormal posture (e.g., Cooper 1993; Ivanhoe and Hartman 2004). In most of these instances, the pain is located in body regions that are associated with known tissue damage or abnormality and is relatively easy to diagnose. However, there are instances in which chronic pain is located in body regions that have not been injured during the trauma or were not associated with any observed pathology. The chronic pain in these body regions may be of central origin unless psychiatric underlying reasons have been identified.

Characteristics

Characteristics of Central Pain After TBI

Central Pain (CP) After TBI: An Underestimated Problem

Unfortunately, the reports on central pain after TBI are so scarce that the prevalence rate of the problem has not been established yet (Boivie 2006). From our experience and that of others in the field, it appears that the scope of the problem is underestimated possibly due to the plethora of comorbidities after moderate-severe TBI and due to a decreased awareness to the possibility of central pain after TBI. To date, there are only two studies in the literature in which individuals with CP after TBI are characterized. One is a case study of a 43-year-old man with severe TBI who underwent motor cortex stimulation for the treatment of CP (Son et al. 2006). In the other, a systematic psychophysical evaluation was conducted in individuals with TBI, with and without CP and healthy controls (Ofek and Defrin 2007). The individuals with TBI had sustained moderate to severe injuries. The characteristics below are based on these studies and on clinical experience.

Onset and Duration of CP

CP usually appears several weeks to months after the TBI with an average onset of 6 months (range 0.5–30 m) and remains for years. About 40 % of individuals reported that the pain had developed within the first month after injury, 50 % reported that pain had developed between 2 and 12 months after injury, and in 10 % of individuals pain had emerged after 12 months. Regardless of the onset of pain, it remains for years after the TBI.

Location of Pain

Figure 1 shows the location of CP in 15 individuals with TBI. In 14 individuals (93 %), pain was restricted to one body side. One patient reported having pain more centrally in the lower back region in addition to the pain in one body side (patient no. 6). The pain is usually present in the body side exhibiting a more severe motor and sensory dysfunction (see below) and it may encompass a single or several body regions with no known dermatomal arrangement. The body regions that were most frequently reported as painful were: knee (93 %), shoulder (80 %), foot (73 %), hand (53 %), thigh (46 %), lower back (46 %), upper back (40 %), head and face (40 %), and arm (33 %). The entire body side can also be painful.

Intensity and Quality of Pain

Pain intensity may fluctuate and it may range between 1 and 5 (on a 0–5 scale), i.e., it may be very intense. Many individuals with TBI report that the pain prevents them from concentrating on work ("it doesn't let the brain work"), that the pain is exhausting and excruciating ("like real torture") and irritating ("like non-stop exertion"). According to the McGill pain questionnaire, individuals with TBI who suffer from CP often select the following descriptors for pain: "pricking," "pounding/throbbing," "pressing," "burning," "cutting," "cold," "freezing," "troublesome," "numb," and "wretched."

Aggravating and Alleviating and Factors

All (100 %) individuals with TBI who suffer from CP reported that physical effort (active movement) exacerbates the pain. Additional aggravating factors were: cold weather (60 %), fatigue (40 %), touch (26 %), tension (26 %), immobilization (20 %), and electrical nerve stimulation (20 %). Pain was reported to be alleviated by relaxation or rest (60 %), heating or warm weather (40 %), and massage (20 %). Four patients (26 %) reported that pain medications were helpful.

Sensory Profile of Individuals with Central Pain After TBI

The sensory profile of an individual is evaluated with psychophysical measurements. The results of such measurements can reveal the nature of the sensory loss as a result of the injury and infer on the mechanism of pain. Psychophysical measurements are based on the administration of various stimuli and recording the response to these stimuli. There are two major sensory systems within the nervous system: (1) the pain and temperature pathways (the anterolateral spinothalamic and thalamocortical pathways) and (2) the touch and vibration pathways (the dorsal column-medial lemniscal system and thalamocortical pathways). Evaluating these two systems require the administering of thermal and tactile stimuli, respectively.

Thermal and Tactile Sensibility

Although individuals with severe-moderate TBI both with and without CP exhibit disrupted sensibility compared with healthy controls due to the brain injury, the former TBI group has some distinct features. First, in individuals with CP, the detection thresholds for thermal sensations (warm, cold, and heat-pain) measured in the affected, painful body side (which is the side contralateral to the TBI) were significantly higher than those of individuals without CP. Second, whereas all (100 %) individuals with CP exhibited abnormally high thermal thresholds in their affected body regions, only 7 (44 %) of those without CP exhibited these features. Third, in individuals with TBI who suffer from CP, thermal thresholds of the affected side were significantly higher than those of their nonaffected side, whereas in individuals with TBI who do not suffer from CP there were no

**Central Pain Following
Traumatic Brain Injury,
Fig. 1** The location of
chronic pain among
individuals with TBI

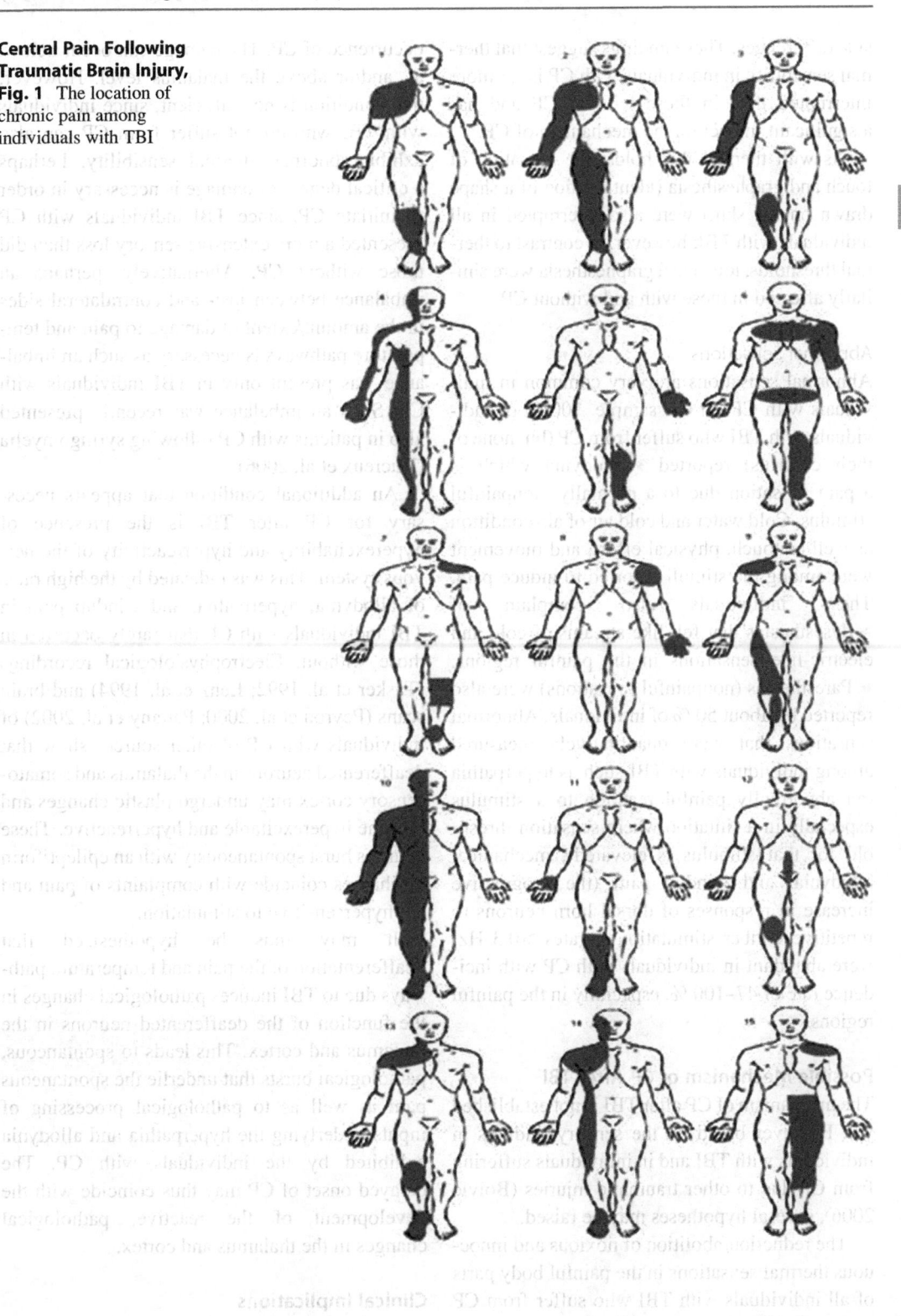

side differences. These findings suggest that thermal sensibility in individuals with CP is far more interrupted than in those without CP and has a significant impact on the mechanism of CP.

As with thermal thresholds, the sensation of touch and graphesthesia (identification of a shape drawn on the skin) were also interrupted in all individuals with TBI; however, in contrast to thermal thresholds, touch and graphesthesia were similarly affected in those with and without CP.

Abnormal Sensations

Abnormal sensations are very common in individuals with CP. In our sample, 100 % of individuals with TBI who suffer from CP (but none of their controls) reported ▶ allodynia which is a pain sensation due to a normally, nonpainful stimulus. Cold water and cold air of air-condition as well as touch, physical effort, and movement were among the stimuli reported to induce pain. These individuals also complain of ▶ dysesthesias that felt like streams of cold and electric-like sensations in the painful regions. ▶ Paresthesias (nonpainful sensations) were also reported by about 50 % of individuals. Abnormal sensations that were quantitatively measured among individuals with TBI such as hyperpathia (an abnormally painful reaction to a stimulus especially in a situation where sensation threshold for that stimulus is elevated), mechanical allodynia, and windup pain (the progressive increase in responses of dorsal horn neurons to repetitive C-fiber stimulation at rates >0.3 Hz) were abundant in individuals with CP with incidence rate of 47–100 %, especially in the painful regions.

Possible Mechanism of CP After TBI

The mechanism of CP after TBI is not established yet. However based on the sensory findings in individuals with TBI and in individuals suffering from CP due to other traumatic injuries (Boivie 2006), several hypotheses may be raised.

The reduction/abolition of noxious and innocuous thermal sensations in the painful body parts of all individuals with TBI who suffer from CP suggest that damage to the pain and temperature system is a necessary condition for the

occurrence of CP. The damage can occur below, at, and/or above the thalamus level. However, this condition is not sufficient, since individuals with TBI who do not suffer from CP can also exhibit abnormal thermal sensibility. Perhaps a critical degree of damage is necessary in order to initiate CP, since TBI individuals with CP presented a more extensive sensory loss than did those without CP. Alternatively, perhaps an imbalance between ipsi- and contralateral sides in the amount/extent of damage to pain and temperature pathways is necessary as such an imbalance was present only in TBI individuals with CP. Such an imbalance was recently presented also in patients with CP following syringomyelia (Ducreux et al. 2006).

An additional condition that appears necessary for CP after TBI is the presence of hyperexcitability and hyperreactivity of the nervous system. This was indicated by the high rates of allodynia, hyperpathia, and windup pain in TBI individuals with CP that rarely occurred in those without. Electrophysiological recordings (Tasker et al. 1992; Lenz et al. 1994) and brain scans (Peyron et al. 2000; Pattany et al. 2002) of individuals with CP of other sources show that deafferented neurons in the thalamus and somatosensory cortex may undergo plastic changes and become hyperexcitable and hyperreactive. These neurons burst spontaneously with an epileptiform discharges coincide with complaints of pain and are hypersensitive to stimulation.

It may thus be hypothesized that deafferentation of the pain and temperature pathways due to TBI induces pathological changes in the function of the deafferented neurons in the thalamus and cortex. This leads to spontaneous, pathological bursts that underlie the spontaneous pain as well as to pathological processing of inputs underlying the hyperpathia and allodynia exhibited by the individuals with CP. The delayed onset of CP may thus coincide with the development of the reactive, pathological changes in the thalamus and cortex.

Clinical Implications

Individuals with TBI, especially those with moderate and severe injuries, suffer from a variety of problems including motor dysfunction, cognitive deficits, behavioral changes, and emotional distress (Vogelbaum et al. 1998; Sherman et al. 2006). Due to the complex rehabilitation of these individuals, the attention to and management of chronic pain may be overlooked, especially considering that CP occurs in seemingly intact body regions. Often, chronic pain is deemed as a symptom of a cognitive/affective disorder and rarely does the possibility of CP arise. This may lead to unnecessary suffering and to interference in the rehabilitation process.

We thus propose that a diagnosis of CP should always be considered when individuals with TBI present with clinical features similar to those presented here. Several lines of treatment can be offered including antidepressants, anticonvulsants, and lidocaine (for reviews, see Finnerup et al. 2005; Hulsebosch 2005). It should be emphasized, however, that the treatment of CP is difficult and patients should be carefully and systematically monitored for efficacy and side effects. It should also be pointed out that chronic pain may also have a psychogenic origin and that such a possibility should also be considered.

Chronic Posttraumatic Headache After TBI: Possibly a Form of CP

It is impossible to ignore chronic posttraumatic headache (CPTHA) when discussing CP after TBI. CPTHA is perhaps the most prevalent type of pain after mild TBI with a prevalence rate of 47–95 % compared to 20 % in moderate-severe TBI (Uomoto and Esselman 1993; Couch and Bears 2001; Walker et al. 2005). As in CP after TBI, the mechanism of CPTHA is unknown. A recent study suggests that CPTHA bear features characteristic of CP (Defrin et al. 2010).

Sensory testing done on individuals with TBI with and without CPTHA revealed that although all individuals had altered sensibility compared with healthy controls, those with CPTHA had a more considerable sensory loss. Thus, individuals with CPTHA had significantly higher heat-pain threshold in both the head and hands but significantly lower pressure-pain threshold in the painful regions of the head compared with pain-free TBI individuals. The generalized thermal hypoalgesia seen in the head and hands of all (100 %) individuals with CPTHA indicated the existence of a central damage to the pain and temperature system, similar to that occurring in individuals suffering from CP.

Other features exhibited by individuals with CPTHA like mechanical allodynia in the painful regions, aggravation of pain by stress/tension, and sharp and pressing pain are also characteristics of CP. The authors thus concluded that CP may be one of the contributing factors to CPTHA. Nevertheless, the cranial hyperalgesia and the high rates of posttraumatic symptoms found in the individuals with CPTHA (Defrin et al. 2010) suggest that CPTHA is a complex syndrome possibly resulting from a combination of injury to both peripheral and central structures as well of psychological factors.

References

Beetar, J. T., Guilmette, T. J., & Sparadeo, F. R. (1996). Sleep and pain complaints in symptomatic traumatic brain injury and neurological populations. *Archives of Physical Medicine and Rehabilitation, 77*, 1298–1302.

Boivie, J. (2006). Central pain. In S. B. McMahon & M. Koltzenburg (Eds.), *Textbook of pain* (pp. 1057–1074). New York: Churchill Livingston Elsevier.

Cooper, P. R. (Ed.). (1993). *Head injury* (3rd ed., pp. 1–27). Baltimore: Williams & Wilkins.

Couch, J. R., & Bears, C. (2001). Chronic daily headache in the posttrauma syndrome: Relation to extent of head injury. *Headache, 41*, 559–564.

Defrin, R., Gruener, H., Schreiber, S., & Pick, C. G. (2010). Quantitative somatosensory testing of subjects with chronic post-traumatic headache: Implications on its mechanisms. *European Journal of Pain, 14*, 924–931.

Ducreux, D., Attal, N., Parker, F., & Bouhassira, D. (2006). Mechanisms of central neuropathic pain: A combined psychophysical and fMRI study in syringomyelia. *Brain, 129*, 963–976.

Finnerup, N. B., Otto, M., McQuay, H. J., Jensen, T. S., & Sindrup, S. H. (2005). Algorithm for neuropathic pain treatment: An evidence based proposal. *Pain, 118*, 289–305.

Hulsebosch, C. E. (2005). From discovery to clinical trials: Treatment strategies for central neuropathic pain after spinal cord injury. *Current Pharmaceutical Design, 11*, 1411–1420.

Ivanhoe, C. B., & Hartman, E. T. (2004). Clinical caveats on medical assessment and treatment of pain after TBI. *The Journal of Head Trauma Rehabilitation, 19*, 29–39.

Lahz, S., & Bryant, R. A. (1996). Incidence of chronic pain following traumatic brain injury. *Archives of Physical Medicine and Rehabilitation, 77,* 889–891.

Lenz, F., Martin, R., Tasker, R., & Dostrovsky, J. D. (1994). Characterization of somatotopic organization and spontaneous neuronal activity in the region of the thalamic principle sensory nucleus in patients with spinal cord transection. *Journal of Neurophysiology, 72,* 1570–1587.

Merskey, H., & Bogduk, N. (1994). *Classification of chronic pain, description of chronic pain syndromes and definitions of pain terms.* Seattle: IASP Press.

Nampiaparampil, D. E. (2008). Prevalence of chronic pain after traumatic brain injury: Systematic review. *Journal of the American Medical Association, 300,* 711–719.

Ofek, H., & Defrin, R. (2007). The characteristics of chronic central pain after traumatic brain injury. *Pain, 131,* 330–340.

Pattany, P. M., Yezierski, R. P., Widerstrom-Noga, E. G., Bowen, B. C., Martinez-Arizala, A., Garcia, B. R., & Quencer, R. M. (2002). Proton magnetic resonance spectroscopy of the thalamus in patients with chronic neuropathic pain after spinal cord injury. *American Journal of Neuroradiology, 23,* 901–905.

Peyron, R., Garcia-Larrea, L., Gregoire, M. C., Convers, P., Richard, A., Lavenne, F., Barral, F. G., Mauguiere, F., Michel, D., & Laurent, B. (2000). Parietal and cingulate processes in central pain. A combined positron emission tomography (PET) and functional magnetic resonance imaging (fMRI) study of an unusual case. *Pain, 84,* 77–87.

Sherman, K. B., Goldberg, M., & Bell, K. R. (2006). Traumatic brain injury and pain. *Physical Medicine and Rehabilitation Clinics of North America, 17,* 473–490.

Son, B. C., Lee, S. W., Choi, E. S., Sung, J. H., & Hong, J. T. (2006). Motor cortex stimulation for central pain following a traumatic brain injury. *Pain, 123,* 210–216.

Tasker, R. R., DeCarvalho, G. T., & Dolan, E. J. (1992). Intractable pain of spinal cord origin: Clinical features and implication for surgery. *Journal of Neuroscience, 77,* 373–378.

Treede, R. D., Jensen, T. S., Campbell, J. N., Cruccu, G., Dostrovsky, J. O., Griffin, J. W., Hansson, P., Hughes, R., Nurmikko, T., & Serra, J. (2008). Neuropathic pain: Redefinition and a grading system for clinical and research purposes. *Neurology, 70,* 1630–1635.

Uomoto, J. M., & Esselman, P. C. (1993). Traumatic brain injury and chronic pain: Differential types and rates by head injury severity. *Archives of Physical Medicine and Rehabilitation, 74,* 61–64.

Vogelbaum, M. A., Vollmer, D. G., & Dacey, R. G. (1998). Craniocerebral trauma. *Clinical Neurology, 3,* 1–85.

Walker, W. C., Seel, R. T., Curtiss, G., & Warden, D. L. (2005). Headache after moderate and severe traumatic brain injury: A longitudinal analysis. *Archives of Physical Medicine and Rehabilitation, 86,* 1793–1800.

Central Pain in Multiple Sclerosis

Jorgen Boivie
Department of Neurology, University Hospital, Linköping, Sweden

Introduction

Central neuropathic pain in multiple sclerosis. Previously equated with dysesthetic pain, since there was a belief that all central pain in multiple sclerosis (MS) was of dysesthetic quality, this has been shown to be incorrect.

Definition

▶ Central pain (CP) is neuropathic pain caused by primary lesions (e.g., MS lesions) in the central nervous system (CNS), either in the brain or spinal cord.

Characteristics

Epidemiology

Contrary to what was previously claimed in the literature, several studies from the last two decades have shown that many patients with MS have pain and that pain is a major problem for many MS patients. In the early literature, it was recognized that a small minority of MS patients have ▶ trigeminal neuralgia (TN) and even fewer ▶ painful tonic seizures. Furthermore, some MS patients have pain caused by spasticity, while others report pain is not a common problem. According to recent research, however, 44–79 % of all MS patients have problems with pain (Ehde et al. 2003; Österberg 2005; Svendsen 2003), and about half of these patients have central pain, according to the only study in which central pain has been specifically investigated (Österberg et al. 2005).

In two postal surveys about pain, responses were received from 442 and 508 MS patients (Ehde et al. 2003; Svendsen et al. 2003). In the

Central Pain in Multiple Sclerosis, Table 1 Results from prevalence studies on pain in MS. Ages and durations in years. Prevalence in percent of the total population. The prevalence figures for central pain are estimates made from descriptions of pain and were not calculated by the authors themselves

Study	Vermote 1986	Kassirer 1987	Moulin 1988	Stenager 1991	Indaco 1994	Stenager 1995	Österberg 2005
Nr of pats	83	28	159	117	122	49	364
Mean age	–	49	47	43	38	–	54
MS duration	–	29	13	8	13	–	23
All pain	54	75	55	65	57	86	65
CP incl. TN	31	64*	29	–	–	–	28
TN	3,6	18**	5,9	3	9***	14	4,9
Dysesthetic pain	–	–	29	–	22	20	–
Pain in extremities	15	64	–	22	–	55	21
Spasm-induced pain	–	53	13	21	19	4	1
Nontrigeminal paroxysmal pain	4	–	5	7	5	41	2
Pain qualities	Burning Pricking Stabbing Dull	Burning Tingling	Burning Tingling Aching	–	–	–	Aching Burning Pricking Stabbing Smarting Squeezing

TN trigeminal neuralgia. *Neurogenic origin, **including atypical facial pain and ***neuralgic pain (face and head)

American study, 44 % reported persistent, bothersome pain as a result of MS. Pain was of moderate to severe intensity in 63 % and interfered with life moderately or severely in 49 % (Ehde et al. 2003). In the Danish study, no significant difference in the total proportion of individuals with pain was found between MS patients and controls (79 % and 75 %), but when responses were analyzed, it became clear that pain was more severe in MS patients (Svendsen et al. 2003). In this study, the impact of pain on daily life ranged from moderate to severe in 45 % and 7 % of the patients, respectively. Results from studies over the last 20 years, in which neurologists have interviewed and examined MS with regard to pain, point in the same direction. Seven studies from the last 20 years have shown the prevalence of pain in MS patients has been found to be 54–86 % (Table 1).

In a systematic study of central pain in MS, the prevalence of CP in a population of 364 MS patients was found to be 27.5 %, including 4.9 % with TN (Österberg et al. 2005). An additional 15 patients had possible CP. If the patients with probable CP had been included, the outcome would have been a prevalence of 31.6 %. These figures compare well with results of previous studies (Moulin 1989; Vermonte et al. 1986).

Features of Central Pain

In the study by Österberg et al. (2005), many aspects of CP in MS were investigated, including the following:

- The prevalence of CP increased with age and disease duration with peaks between 40 and 60 years of age and 10–20 years of disease duration, but not with a higher degree of disability. The prevalence of CP was as high as 31 % 10 years after onset of MS and almost the same after 11–20 years of disease; thereafter, the prevalence of CP decreased to 14–18 %. Thus, it appears that neither age nor duration of MS increases the risk of developing CP. This partly contradicts the results of previous studies, where it was found that the pain prevalence increased with age (Clifford and Trotter 1984; Stenager et al. 1991), disease duration (Kassirer and Osterberg 1987), and disability (Stenager et al. 1995). Prospective studies are needed to give more reliable information on these matters.

- There was a large span in the time interval between clinical onset of MS and the onset of CP, ranging from 7 years before other symptoms to 25 years after other symptoms. In 57 % of patients with CP, pain started within 5 years of onset of the disease, and after 10 years the figure was 73 %.
- In some patients, CP was the first symptom of MS before any other symptom, while in others CP appeared together with other symptoms. CP preceded other symptoms in 6 % of patients, and it was part of the onset symptoms in 20 % of these patients and in 5.5 % of all MS patients.
- A large majority of patients experienced daily pain (88 %); only 30 % had pain-free moments, lasting minutes to hours.
- The intensity of pain varied somewhat, and 44 % of patients experienced a constant intensity. This and the irritating quality of the pain contribute to the fact that patients rate their pain as a heavy burden (Ehde et al. 2003).
- More than one-third of patients experienced CP in two to four separate pain loci, often with differing modality, time of onset, and intensity (TN excluded).
- The most common location of CP was in the lower extremities (87 %) and in the upper extremities (3 %).
- Half the patients experienced pain both superficially in skin and in deeper parts of the body, which is similar to that found in central poststroke pain (Leijon et al. 1989).
- More than 80 % of MS patients with CP experienced two or more pain qualities, which is similar to that found by Leijon et al. (1989) for central poststroke pain. The most common qualities were burning (40 %) and aching (40 %). Thus, no pain qualities or combination of pain qualities are pathognomonic for CP in MS.
- Out of 364 patients, only 2 % had pain caused by spasticity. Instead, it was found that many patients with CP also have spasticity, but it was not the cause of pain. This conclusion is shared by many clinicians.

As for other MS symptoms, CP can be one of several symptoms in a relapse, or the only symptom. Clifford and Trotter (1984) reported this phenomenon, in two patients with temporary burning pain during a relapse of MS. The distribution of the three forms of MS in patients with CP does not generally differ among MS patients (relapsing-remitting, secondary progressive, primary progressive). From the literature, it is known that emotional stress, light touch, cold, and physical activity can increase CP (Boivie 1999). This aspect has not been systematically studied in MS, but it has been noted that many MS patients experience worse pain after physical activity.

Some patients with MS who have pareses, spasticity, and dyscoordination of movement will develop nociceptive musculoskeletal pain. In a recent study, 21 % of MS patients were found to have nociceptive pain (Österberg et al. 2005), which is in the same range found in previous studies (14–39 %; Kassirer and Osterberg 1987; Moulin 1989; Vermote et al. 1986).

Sensory Abnormalities

The most common neurological sign in patients with nontrigeminal CP is sensory disturbance. Almost all patients have at least one abnormal finding in the sensory examination, with a decrease in sensibility to cold occurring more often than any other submodalities (Österberg and Boivie, in preparation). In this study of 62 patients with nontrigeminal CP, both clinical and quantitative methods were used to test sensibility. There was a large variation between patients with regard to degree and submodality of abnormalities. Some patients had severe defects in all submodalities, whereas only one or two were affected in others. In the quantitative tests, all patients except two (97 %) had abnormal sensibility to temperature and/or pain. Significant differences in abnormalities were found between regions with CP and regions without CP for the following perception thresholds: difference limen (i.e., innoxious temperature), warmth, cold, and cold pain. No significant differences were found in the thresholds for heat pain.

Among patients who did not perceive nonnoxious warmth at all, but could feel heat pain, the burning sensation of heat struck patients suddenly and with high intensity as the temperature reached threshold. This was observed in 19 % of patients. A corresponding sensation did not appear with cold. Eight patients had noxious cold-evoked paradoxical heat pain. Nonpainful dysesthesias were commonly evoked by noxious heat or cold.

The results from the sensory tests indicate that most MS patients with CP have lesions affecting the spinothalamocortical pathways (temperature and pain), but to a lesser degree affect the medial lemniscal pathways (tactile, position sense, and vibration).

Mechanisms
The cellular mechanisms underlying central pain in general, and definitely for MS, are largely unknown. Several investigators have reported that CP develops as a result of lesions affecting the spino- and quintothalamic pathways, that is, pathways most important for the sensibility of pain and temperature. Furthermore, lesions of these pathways can be located at any level of the neuraxis (for references see Boivie 1999). The results from examination of the sensibility in MS patients with CP support this hypothesis. It has been proposed that the crucial lesion is one that affects neospinothalamic projections, that is, projections to the ventroposterior thalamic region (Bowsher 1996). The effects of such a lesion involve neurones of the spinothalamic pathway which become hyperexcitable due to reduced tonic inhibition.

Based on results from experimental studies, Craig (1998) proposed a similar hypothesis about the mechanisms of central pain stating that "central pain is due to the disruption of thermosensory integration and the loss of cold inhibition of burning pain," which in turn is caused by a lesion of the spinothalamic projection activated by cold receptors in the periphery. The disrupted fibers are thought to tonically inhibit nociceptive thalamocortical neurones, increasing discharge and producing pain. Like several other hypotheses, this might be applicable in some patients, but not others, because of the location of the lesions and character of the pain.

One can only speculate about the location of lesions responsible for the development of CP in MS, because, as shown with MRI, practically all patients have disseminated lesions in both the brain and spinal cord. However, based on clinical grounds, it is proposed that much of the CP located in the lower extremities is due to lesions in the spinal cord. The bilateral nature of pain supports this idea.

Treatment
The treatment of MS with interferons and similar substances does not appear to have any symptomatic effect on central pain or on any other symptom.

Several treatment modalities are used in the management of CP in general, but almost no controlled clinical trials have been performed in MS and only a few in other forms of CP. The only exception is the study of oral cannabinoid dronabinol in 28 MS patients. A statistically significant, but weak, effect was found (Svendsen et al. 2004).

Treatments that are used for CP are listed above, but most of them are based on clinical experience and tradition, rather than on results from controlled clinical trials. This means that no evidence-based recommendations can be made for the management of CP in MS. Tricyclic antidepressants have been found to be effective for many patients with Central post stroke pain (see essay on this "Central post stroke pain"), but the experience is that many MS patients get severe side effects from these drugs.

Treatment Modalities Used for Central Pain
Among antidepressants and antiepileptics, the most frequently used are listed.
- Antidepressant drugs (AD)
  - Amitriptyline
  - Desipramine
  - Doxepin
  - Imipramine
  - Nortriptyline
- Antiepileptic drugs (AED)
  - Carbamazepine
  - Gabapentin
  - Lamotrigine

- Oxcarbazepine
- Pregabalin
• Analgesics
  - Morphine
  - OxyContin
  - Codeine
• Sensory stimulation
  - Transcutaneous electrical stimulation (TENS)
  - Spinal cord stimulation (SCS)
  - Deep brain stimulation (DBS)
  - Motor cortex stimulation (MCS)

Among the antiepileptic drugs, carbamazepine is effective for trigeminal neuralgia associated with MS, but does not appear to relieve nontrigeminal central pain (Österberg and Boivie, in preparation). Lamotrigine was shown to relieve CP in stroke, but it has not been tested in MS patients. Many neurologists have tried gabapentin with some success in nontrigeminal CP, but in the literature, only case reports support its use.

In one study with IV morphine, ten patients with CP from MS and five patients from stroke were tested. Only a trend to pain relief was found, and during the following 12-week open treatment period, only three patients experienced a positive effect (Attal et al. 2002). From clinical experience, it appears that some MS patients obtain long-term relief from weak opioids, but no systematic observations support this. The experience with TENS for CP is meager, and positive results have not been reported. The same is true for spinal cord stimulation and deep brain stimulation.

References

Attal, N., Guirimand, F., Brasseur, L., Gaude, V., Chauvin, M., & Bouhassira, D. (2002). Effects of IV morphine in central pain: A randomized placebo-controlled study. *Neurology, 58*, 554–563.
Boivie, J. (1999). Central pain. In P. D. Wall & R. Melzack (Eds.), *Textbook of pain* (4th ed., pp. 879–914). Edinburgh: Churchill Livingstone.
Bowsher, D. (1996). Central pain: clinical and physiological characteristics. *Journal of Neurology, Neurosurgery and Psychiatry, 61*, 62–69.
Clifford, D. B., & Trotter, J. L. (1984). Pain in multiple sclerosis. *Archives of Neurology, 41*, 1270–1272.
Craig, A. D. (1998). A new version of the thalamic disinhibition hypothesis of central pain. *Pain Forum, 7*, 1–14.
Ehde, D. M., Gibbons, L. E., & Chwastiak, L. (2003). Chronic pain in a large community sample of persons with multiple sclerosis. *Multiple Sclerosis, 9*, 605–611.
Kassirer, M., & Osterberg, D. (1987). Pain in chronic multiple sclerosis. *Journal of Pain and Symptom Management, 2*, 95–97.
Leijon, G., Boivie, J., & Johansson, I. (1989). Central post-stroke pain–neurological symptoms and pain characteristics. *Pain, 36*, 13–25.
Moulin, D. E. (1989). Pain in multiple sclerosis. *Neurologic Clinics, 7*, 321–331.
Österberg, A., Boivie, J., & Thuomas, K. Å. (2005). Central pain in multiple sclerosis - prevalence, clinical characteristics and mechanisms. *European Journal of Pain, 9*, 531–542.
Stenager, E., Knudsen, L., & Jensen, K. (1991). Acute and chronic pain syndromes in multiple sclerosis. *Acta Neurologica Scandinavica, 84*, 197–200.
Stenager, K., Knudsen, L., & Jensen, K. (1995). Acute and chronic pain syndromes in multiple sclerosis. A 5-year follow-up study. *Italian Journal of Neurological Sciences, 16*, 629–632.
Svendsen, K., Jensen, T., & Overvald, K. (2003). Pain in patients with multiple sclerosis: A population-based study. *Archives of Neurology, 60*, 1089–1094.
Svendsen, K., Jensen, T., & Bach, F. (2004). Does the cannabinoid dronabinol reduce central pain in multiple sclerosis? Randomised double blind controlled crossover trial. *British Medical Journal, 329*, 253.
Vermote, R., Ketelaer, P., & Carton, H. (1986). Pain in multiple sclerosis patients. *Clinical Neurology and Neurosurgery, 88*, 87–93.

Central Pain Mechanisms

▶ Molecular Contributions to the Mechanism of Central Pain

Central Pain Model

▶ Spinal Cord Injury Pain Model, Ischemia Model

Central Pain Pathways

Definition

The pathways that carry information about noxious stimuli to the brain, includes the spinothalamic tract and trigeminal system.

Cross-References

▶ Thalamic Nuclei Involved in Pain, Human and Monkey

Central Pain Syndrome

A neurological condition caused by a lesion or disease of the central somatosensory nervous system. See also ▶ Central Pain.

Cross-References

▶ Lateral Thalamic Lesions, Pain Behavior in Animals

Central Pain, Diagnosis

David Vivian
Metro Spinal Clinic, South Caulfield, VIC, Australia

Synonyms

Deafferentation pain; Dejerine-Roussy syndrome; Thalamic pain

Definition

Central pain is pain whose source lies in the central nervous system, i.e., the brain, brainstem,

or spinal cord. The cardinal defining feature is that the pain is not evoked by neural activity in peripheral nerves.

Characteristics

In terms of clinical features, central pain has particular characteristics. The pain is typically spontaneous and burning in quality, associated with abnormal sensations. The latter may include:

- Hyperaesthesia (increased sensitivity to touch)
- Hyperalgesia (increased sensitivity to noxious stimuli)
- Allodynia (touch and brush are perceived as painful)
- Paraesthesiae (sensation of pins and needles)
- Formication (sensation of ants crawling on skin) and
- Diminished topoesthesia (ability to locate a sensation somatotopically)

In these respects, central pain resembles neuropathic pain, and the taxonomic distinction between the two conditions is not always clear. Some forms of neuropathic pain are likely to be central in origin rather than arising from the damaged peripheral nerve. The distinction is most clear when the precipitating injury of disease is obviously in the central nervous system. In such cases, the term "central pain" applies unambiguously.

Examples

Classical examples of central pain are the pain of brachial plexus avulsion, spinal cord injury pain, pain after stroke, and pain due to infarction of ▶ thalamic nuclei. The latter is known as thalamic pain syndrome or the Dejerine-Roussy syndrome. Complex regional pain syndrome is the most florid example of central pain.

Others examples include postherpetic neuralgia, peripheral nerve injury, and painful peripheral neuropathy. The latter, however, are contentious, for it is not always evident the extent to which the pain is central in origin or due to peripheral mechanisms such as neuroma, or ectopic impulse generation in peripheral nerves or their ▶ dorsal root ganglia.

Mechanism

Central pain is believed to result from deafferentation: when neurons in the central nervous system lose their accustomed afferent input, either from a peripheral nerve or from an ascending sensory tract. In particular, partial deafferentation is considered most likely to be associated with the development of central pain.

Deafferentation seems to induce plastic changes in the central nervous system. Numerous theories describe these plastic changes in terms of: removal of local inhibition, changes in neuronal membrane excitability, and synaptic reorganization. These changes occur rapidly after central nervous system damage. However, it is typical for central pain to occur sometime after such an injury (days to weeks).

Experiments in animals have been conducted in which recordings are established from dorsal horn neurons that respond to input from particular peripheral nerves. If those nerves are then severed, the dorsal horn neurons no longer respond to peripheral stimuli but exhibit a variety of changes (Anderson et al. 1971; Macon 1979; 3). They become spontaneously active, and eventually no longer become responsive to typical neurotransmitters. These features suggest that their membranes become unstable, and the neurons no longer maintain receptors when they are denied of their accustomed input. The latter feature probably underlies the notorious resistance of central pain to pharmacological therapy.

Spontaneous activity has been confirmed in humans with central pain. When subjects with spinal cord injury pain have been explored with electrodes, spontaneously active neurons have been found in the spinal cord immediately above the level of injury (Loeser et al. 1968). Similar activity has been recorded after experimental spinal cord injury in cats (Loeser and Ward 1967).

The abnormal sensations associated with central pain are most likely due to disinhibition. Sensory perception is normally subject to a variety of central excitatory and inhibitory controls. These are mediated by the dorsolateral tract and by tracts in the anterior funiculus of the spinal cord. When peripheral nerves are severed, the balance between these central modulating influences is disturbed, sometimes with contrasting effects.

Disinhibition results in sensitization of ▶ Second-Order Neurons. Sensations mediated by intact nerves become exaggerated. This is manifest in the form of hyperaesthesia, hyperalgesia, and allodynia.

In experimental animals, the effects of sectioning the trigeminal nerve (Denny-Brown et al. 1973), spinal nerves (Denny-Brown and Yanagisawa 1973), or the spinal cord (Denny-Brown 1979) can be modulated pharmacologically and surgically. Discrete sectioning of the lateral portion of the dorsolateral tract results in shrinkage of the area of sensory nerve loss (Denny-Brown and Yanagisawa 1973; Denny-Brown et al. 1973). Conversely, sectioning the medial portion of the dorsolateral tract increases the area (Denny-Brown and Yanagisawa 1973; Denny-Brown et al. 1973). Sectioning both anterior funiculi reverses the sensory loss caused by spinothalamic tractomy, but the restored sensation is hyperaesthetic (Denny-Brown 1979). Administering L-dopa reduces sensory loss (Denny-Brown et al. 1973), as does a subconvulsive dose of strychnine: an antagonist of the inhibitory transmitter – glycine (Denny-Brown 1979; Denny-Brown and Yanagisawa 1973; Denny-Brown et al. 1973).

In humans who have undergone dorsal root section for pain but whose pain recurs, administration of L-dopa increases their pain but decreases the area of cutaneous anesthesia (Hodge and King 1976). Reciprocally, administration of methyldopa decreases pain but increases numbness. Similarly, tryptophan – a serotonin precursor – reduces pain but increases anesthesia (King 1980).

These phenomena indicate that the effects of deafferentation are not fixed, but are subject to a complex variety of controls. Release of these controls, following peripheral nerve injury or central nervous injury, underlies the varied appearance of pain and altered sensations associated with central pain.

For the central pain of thalamic syndrome, a variety of explanations have been advanced; but they, too, revolve around deafferentation and disinhibition. Cells in the ventroposterior thalamic nuclei become spontaneously active, and produce pain in the area of the body that they subtend (Lenz et al. 1987, 1989). Experimental stimulation of these cells evokes pain in the deafferented region (Lenz et al. 1988). The abnormal activity is believed to arise because of loss of inhibition of medial thalamic nuclei by the reticular nucleus (Cesaro et al. 1991; Mauguiere and Desmedt 1988). Damage to the spinothalamic tract seems to be the precipitating factor for these changes.

Diagnosis

The diagnosis of central pain rests largely on the recognition of the clinical features and history of the pain. Spontaneous burning pain associated with abnormal, exaggerated sensations implies a central mechanism. If the history indicates disease or injury to the central nervous system, the diagnosis is confirmed. The diagnosis is less certain if injury to the central nervous system is not evident, or when the injury or disease affects peripheral nerves.

A presumptive test for central pain is the administration of lignocaine by systemic intravenous infusion. Such an infusion is believed to suppress the spontaneous activity in the central nervous system believed to be responsible for central pain.

Treatment

Central pain is notoriously difficult to treat. Standard analgesics seem to have little or no effect, which correlates with the lack of receptors found in animals with spontaneously active neurons after deafferentation.

Agents that suppress ectopic activity, or which stabilize nerve-cell membranes, are more likely to relieve central pain, provided that side effects can be tolerated. Such agents include local anesthetic agents, administered either by systemic infusion or orally, and membrane-stabilizing agents such as gabapentin and lamotrigine.

Otherwise, central pain can be treated, with reasonable success, by neuroaugmentive surgical procedures such as dorsal column stimulation, dorsal root entry zone lesioning, and deep brain stimulation.

References

Anderson, L. S., Black, R. G., Abraham, J., et al. (1971). Neuronal hyperactivity in experimental trigeminal deafferentation. *Journal of Neurosurgery, 35*, 444–451.

Cesaro, P., Mann, M. W., Moretti, J. L., et al. (1991). Central pain and thalamic hyperactivity: A single photon emission computerized tomographic study. *Pain, 47*, 329–336.

Denny-Brown, D. (1979). The enigma of crossed sensory loss with cord hemisection. In J. J. Bonica et al. (Eds.), *Advances in pain research and therapy* (Vol. 3, pp. 889–895). New York: Raven.

Denny-Brown, D., & Yanagisawa, N. (1973). The function of the descending root of the fifth nerve. *Brain, 96*, 783–814.

Denny-Brown, D., Kirk, E. J., & Yanagisawa, N. (1973). The tract of lissauer in relation to sensory transmission in the dorsal horn of spinal cord in the macaque monkey. *The Journal of Comparative Neurology, 151*, 175–199.

Hodge, C. J., & King, R. B. (1976). Medical modification of sensation. *Journal of Neurosurgery, 44*, 21–28.

King, R. B. (1980). Pain and tryptophan. *Journal of Neurosurgery, 53*, 44–52.

Lenz, F. A., Tasker, R. R., & Dostrovsky, J. O. (1987). Abnormal single unit activity recorded in the somatosensory thalamus of a quadriplegic patient with central pain. *Pain, 31*, 225–236.

Lenz, F. A., Dostrovsky, J. O., Tasker, R. R., et al. (1988). Single-unit analysis of the human ventral thalamic nuclear group: Somatosensory responses. *Journal of Neurophysiology, 59*, 299–316.

Lenz, F. A., Kwan, H. C., Dostrovsky, J. O., et al. (1989). Characteristics of the bursting pattern of action potential that occurs in the thalamus of patients with central pain. *Brain Research, 496*, 357–360.

Loeser, J. D., & Ward, A. A. (1967). Some effects of deafferentation on neurons of the cat spinal cord. *Archives of Neurology, 17*, 629–636.

Loeser, J. D., Ward, A. A., & White, L. E. (1968). Chronic deafferentation of human spinal cord neurons. *Journal of Neurosurgery, 29*, 48–50.

Macon, J. B. (1979). Deafferentation hyperactivity in the monkey spinal trigeminal nucleus: Neuronal responses to amino acid iontophoresis. *Brain Research, 161*, 549–554.

Mauguiere, F., & Desmedt, J. E. (1988). Thalamic pain syndrome of Dejerine-Roussy: Differentiation of four subtypes assisted by somatosensory evoked patients data. *Archives of Neurology, 45*, 1312–1320.

Central Pain, Human Studies of Physiology

Frederick A. Lenz[1], Andres Fernandez[1],
S. H. Patel[1], A. Taghva[1] and Nirit Weiss[2]
[1]Department of Neurosurgery, Johns Hopkins
Hospital, Baltimore, MD, USA
[2]Department of Neurosurgery, Mount Sinai
Medical Center, New York, NY, USA

Definition

Spontaneous thalamic cellular activity is often categorized as either ▶ bursting activity (▶ spike-bursts, bursting mode) or as ▶ tonic firing mode (tonic mode) (Steriade et al. 1990). Many studies have suggested that increased spike-bursting occurs in the thalamus of patients with chronic neuropathic pain (Jeanmonod et al. 1994; Lenz et al. 1994, 1998; Rinaldi et al. 1991). Thalamic bursting has also been reported in monkeys with interruption of the spinothalamic tract (STT), which sometimes develop sensory abnormalities similar to those seen in patients with similar lesions (Weng et al. 2000). This bursting is certainly associated with lesions of the somatic sensory pathways to thalamus and is perhaps associated with the pain that develops following such lesions.

Characteristics

The Thalamic Region of Vc and Its Importance in Pain Processing

Several lines of evidence demonstrate that the ventral caudal nucleus of the human sensory thalamus (Vc), the human analog of monkey ventral posterior (VP) nucleus (Hirai and Jones 1989), is an important component in human pain-signaling pathways. Studies of patients at autopsy following lesions of the STT show the densest STT termination in the Vc region including posterior and inferior subnuclei of Vc, suprageniculate,

and posterior subnuclei (Bowsher 1957; Mehler 1962; Walker 1943). In monkeys, the STT originating in dorsal horn lamina I, in part, terminates in Vmpo (Craig et al. 1994; Graziano and Jones 2004). In humans, cells responding to noxious and temperature stimuli can be located in all of these areas (Davis et al. 1999; Lee et al. 1999; Lenz et al. 1993). Thus, both anatomic and physiologic data demonstrate the presence of a distributed group of thalamic nuclei with pain-related activity.

Is Thalamic Functional Mode Altered in Chronic Pain States?

Spike-bursting activity refers to a particular pattern of ▶ interspike intervals (ISI) between action potentials, such that a spike-burst begins after a relatively long ISI, and is comprised of a series of action potentials with a short ISI (typically <6 ms) (Lenz et al. 1994; Steriade et al. 1990). Thereafter, the ISIs progressively increase in length so that the cell's firing decelerates throughout the spike-burst.

In patients with spinal transection, the highest rate of bursting occurs in cells that do not have peripheral receptive fields and that are located in the thalamic representation of the anesthetic part of the body. Since the pain is also in the anesthetic part of the body, this bursting may be due to loss of sensory input or be the cause of pain or both. These cells also have the lowest firing rates in the interval between bursts (principal event rate) (Lenz et al. 1994). The low firing rates suggest that these cells have decreased tonic excitatory drive and are hyperpolarized, perhaps due to loss of excitatory input from the STT (Blomqvist et al. 1996; Dougherty et al. 1996; Eaton and Salt 1990). Therefore, the available evidence suggests that thalamic cells deafferentated by spinal transection (lesions) are dominated by spike-bursting and low firing rates between bursts, consistent with membrane hyperpolarization (Lenz et al. 1998; Steriade et al. 1990).

Spike-bursting activity is maximal in the region posterior and inferior to the core nucleus of Vc (Table 4 in Lenz et al. 1994). Stimulation in this area may evoke the sensation of pain more frequently than does stimulation in the core of Vc (Dostrovsky et al. 1991; Hassler 1970; Ohara and Lenz 2003). Thus, increased spike-bursting activity may be correlated with some aspects of abnormal sensations (e.g., ▶ dysesthesia or pain) that these patients experience. However, in patients with spinal transection, the painful area and the area of sensory loss overlap (Lenz et al. 1994). Thus, the bursting activity might be related to deafferentation of the thalamus from the input from the STT, rather than to pain.

These findings about spike-bursting activity in spinal patients have been called into question by a recent study in patients with chronic pain (Radhakrishnan et al. 1999). It has been reported that the number of bursting cells per trajectory in patients with movement disorders (controls) is not different from that in patients with chronic pain. However, there are significant differences between the two studies (Lenz et al. 1994; Radhakrishnan et al. 1999) in terms of (1) patient population (spinal cord injury vs. mixed chronic pain), (2) location of cells studied (Vc vs. anterior and posterior to Vc), and (3) analysis methods (incidence of bursting cells vs. bursting parameters). Clearly, the increase in bursting activity demonstrated in the earlier study is more applicable to the region of the principal somatic sensory nucleus in patients with central pain from spinal transection (Lenz et al. 1994).

Further support for increased spike-bursts occurring in spinal cord injured patients is found in thalamic recordings from monkeys with thoracic anterolateral cordotomies (Weng et al. 2000). Some of these animals showed increased responsiveness to electrocutaneous stimuli and thus may represent a model of central pain (Vierck 1991). The most pronounced changes in firing pattern were found in thalamic multireceptive cells, which respond to both cutaneous brushing and compressive stimuli, with activity that is not graded into the noxious range. In comparison with normal controls, multireceptive cells in monkeys with cordotomies showed significant increases in the number of bursts occurring spontaneously or in response to brushing or compressive stimuli. The changes in bursting behavior were widespread,

occurring in the thalamic representation of upper and lower extremities, both ipsilateral and contralateral to the cordotomy.

Although there is an increase in spike-burst activity in central pain secondary to spinal injury, there does not appear to be a direct relationship between spike-burst firing and pain. Spike-bursts are also found in the thalamic representation of the monkey's upper extremity and of the representation of the arm and leg ipsilateral to the cordotomy. Pain is not typically experienced in these parts of the body in patients with thoracic spinal cord transection or cordotomy (Beric et al. 1988). Spike-bursts are increased in frequency during slow-wave sleep and drowsiness in all mammals studied (Steriade et al. 1990) including man in the absence of pain (Zirh et al. 1997). However, such bursting could cause pain if stimulation in the vicinity of the bursting cells produced the sensation of pain. This finding has been reported in a study of sensations evoked by stimulation of the region of Vc in patients with central pain, including those with spinal cord injuries (Lenz et al. 1998). Thus, there is evidence from both human and animal studies for a correlation between central pain following spinal cord injury and an altered thalamic neuronal action potential firing pattern. It appears that there is an increase in spike-burst firing in patients with pain following spinal injury. The exact physiologic relationships which link the pattern of thalamic firing to the human perception of pain in this condition are still unclear.

Acknowledgment Supported by grants to FAL from the NIH: NS39498 and NS40059.

References

Beric, A., Dimitrijevic, M. R., & Lindblom, U. (1988). Central dysesthesia syndrome in spinal cord injury patients. *Pain, 34,* 109–116.

Blomqvist, A., Ericson, A. C., Craig, A. D., et al. (1996). Evidence for glutamate as a neurotransmitter in spinothalamic tract terminals in the posterior region of owl monkeys. *Experimental Brain Research, 108,* 33–44.

Bowsher, D. (1957). Termination of the central pain pathway in man: The conscious appreciation of pain. *Brain, 80,* 606–620.

Craig, A. D., Bushnell, M. C., Zhang, E. T., et al. (1994). A thalamic nucleus specific for pain and temperature sensation. *Nature, 372*, 770–773.

Davis, K. D., Lozano, A. M., Manduch, M., et al. (1999). Thalamic relay site for cold perception in humans. *Journal of Neurophysiology, 81*, 1970–1973.

Dostrovsky, J. O., Wells, F. E. B., & Tasker, R. R. (1991). Pain evoked by stimulation in human thalamus. In Y. Sjigenaga (Ed.), *International symposium on processing nociceptive information* (pp. 115–120). Amsterdam: Elsevier.

Dougherty, P. M., Li, Y. J., Lenz, F. A., et al. (1996). Evidence that excitatory amino acids mediate afferent input to the primate somatosensory thalamus. *Brain Research, 278*, 267–273.

Eaton, S. A., & Salt, T. E. (1990). Thalamic NMDA receptors and nociceptive sensory synaptic transmission. *Neuroscience Letters, 110*, 297–302.

Graziano, A., & Jones, E. G. (2004). Widespread thalamic terminations of fibers arising in the superficial medullary dorsal horn of monkeys and their relation to calbindin immunoreactivity. *Journal of Neuroscience, 24*, 248–256.

Hassler, R. (1970). Dichotomy of facial pain conduction in the diencephalon. In A. E. Walker (Ed.), *Trigeminal neuralgia* (pp. 123–138). Philadelphia: Saunders.

Hirai, T., & Jones, E. G. (1989). A new parcellation of the human thalamus on the basis of histochemical staining. *Brain Research Reviews, 14*, 1–34.

Jeanmonod, D., Magnin, M., & Morel, A. (1994). A thalamic concept of neurogenic pain. In G. F. Gebhart, D. L. Hammond, & T. S. Jensen (Eds.), *Proceedings of the 7th world congress on pain. Progress in pain research and management*, Vol. 2 (pp. 767–787). Seattle: IASP Press.

Lee, J.-I., Antezanna, D., Dougherty, P. M., et al. (1999). Responses of neurons in the region of the thalamic somatosensory nucleus to mechanical and thermal stimuli graded into the painful range. *The Journal of Comparative Neurology, 410*, 541–555.

Lenz, F. A., Seike, M., Lin, Y. C., et al. (1993). Neurons in the area of human thalamic nucleus ventralis caudalis respond to painful heat stimuli. *Brain Research, 623*, 235–240.

Lenz, F. A., Kwan, H. C., Martin, R., et al. (1994). Characteristics of somatotopic organization and spontaneous neuronal activity in the region of the thalamic principal sensory nucleus in patients with spinal cord transection. *Journal of Neurophysiology, 72*, 1570–1587.

Lenz, F. A., Gracely, R. H., Baker, F. H., et al. (1998). Reorganization of sensory modalities evoked by stimulation in the region of the principal sensory nucleus (ventral caudal-Vc) in patients with pain secondary to neural injury. *The Journal of Comparative Neurology, 399*, 125–138.

Mehler, W. R. (1962). The anatomy of the so-called "pain tract" in man: An analysis of the course and distribution of the ascending fibers of the fasciculus

anterolateralis. In J. D. French & R. W. Porter (Eds.), *Basic research in paraplegia* (pp. 26–55). Springfield: Thomas.

Ohara, S., & Lenz, F. A. (2003). Medial lateral extent of thermal and pain sensations evoked by microstimulation in somatic sensory nuclei of human thalamus. *Journal of Neurophysiology, 90*, 2367–2377.

Radhakrishnan, V., Tsoukatos, J., Davis, K. D., et al. (1999). A comparison of the burst activity of lateral thalamic neurons in chronic pain and non-pain patients. *Pain, 80*, 567–575.

Rinaldi, P. C., Young, R. F., Albe-Fessard, D. G., et al. (1991). Spontaneous neuronal hyperactivity in the medial and intralaminar thalamic nuclei in patients with deafferentation pain. *Journal of Neurosurgery, 74*, 415–421.

Steriade, M., Jones, E. G., & Llinas, R. R. (1990). *Thalamic oscillations and signaling*. New York: Wiley.

Vierck, C. J. (1991). Can mechanisms of central pain syndromes be investigated in animal models? In K. L. Casey (Ed.), *Pain and central nervous system disease: The central pain syndromes* (pp. 129–141). New York: Raven.

Walker, A. E. (1943). Central representation of pain. *Research Publications – Association for Research in Nervous and Mental Disease, 23*, 63–85.

Weng, H. R., Lee, J. I., Lenz, F. A., et al. (2000). Functional plasticity in primate somatosensory thalamus following chronic lesion of the ventral lateral spinal cord. *Neuroscience, 101*, 393–401.

Zirh, A. T., Lenz, F. A., Reich, S. G., et al. (1997). Patterns of bursting occurring in thalamic cells during Parkinsonian tremor. *Neuroscience, 83*, 107–121.

Central Pain, Outcome Measures in Clinical Trials

Eva Widerström-Noga
The Miami Project to Cure Paralysis, Department of Neurological Surgery, University of Miami Miller, School of Medicine, Miami, FL, USA

Synonyms

Effectiveness measure

Definition

An *outcome assessment* is a performance indicator that assesses a person's symptom severity, and other health status subsequent to, and resulting from, a treatment, procedure, or other therapeutic interventions.

Introduction

Central neuropathic pain (CNP) is a neurological condition and the result of trauma or neurological disease involving the central nervous system (CNS). This type of pain can be a prominent feature in the complex clinical picture associated with disease or trauma involving the CNS, e.g., stroke, multiple sclerosis, epilepsy, tumors, syringomyelia, brain or spinal cord injury, or Parkinson's disease (Boivie 2003). CNP is commonly associated with both spontaneous and evoked non-painful and painful sensations (Boivie 2003; Treede et al. 2008), which contribute to its unpleasant character. Due to the persistent nature of CNP (Bowsher 1999), there is an increased risk of significantly decreased health-related quality of life (HRQoL). Therefore, effective pain management strategies are urgently needed for these pain patient populations. Although CNP is common after certain CNS injuries such as spinal cord injury (Siddall et al. 2003), the overall prevalence of CNP in the general population is relatively low (Boivie 2003), and therefore, the progress of large-scale clinical pain trials examining the effects of therapeutic strategies is hampered. Due to the small CNP patient populations, there is a need for standardized diagnostic and outcome measures that can facilitate multicenter trials and the interpretation and application of research results to improve the management of CNP. Measures that evaluate multiple aspects of the CNP phenotype that are generalizable across multiple trauma and disease etiologies might lead to more unified approaches to evaluating CNP and its relationship to underlying pain mechanisms and hence mechanisms-based treatments (Woolf and Max 2001).

Characteristics

Several groups have provided recommendations regarding pain outcome measures in heterogeneous

pain populations and for neuropathic pain. For instance, the Initiative on Methods, Measurement, and Pain Assessment in Clinical Trials group (IMMPACT; www.immpact.org) has provided guidelines regarding useful outcome domains in pain trials across heterogeneous pain populations (Dworkin et al. 2005). The underlying idea is that a broad assessment of several important outcome domains would generate more complete reports of clinical trial results and, therefore, improve the overall risk-benefit evaluation. The outcome domains suggested by the IMMPACT group include pain severity, physical functioning, emotional functioning, participants' ratings of improvement and satisfaction with treatment, symptoms and adverse effects, and participant disposition. These domains and their application to CNP will be briefly discussed in this entry.

Other international expert groups (e.g., the German Research Network on Neuropathic Pain (Geber et al. 2011)) and the Neuropathic Pain Special Interest group of the International Association for the Study of Pain (Haanpää et al. 2011) have focused on developing recommendations regarding the use of sensory measures. One important purpose of these efforts is to develop mechanism-based diagnoses of neuropathic pain as an attempt to advance the development of treatments targeting specific underlying mechanisms. Thus, sensory measures are very important in multimodal assessment strategies and can be used in combination with the outcome domains proposed by the IMMPACT group. This entry will discuss the role of sensory measures as part of a multimodal outcome measurement strategy rather than on their role as diagnostic measures.

Multimodal Pain Assessment

Since persistent pain, regardless of type, depends on multiple pathophysiological mechanisms and a wide variety of psychosocial factors, a comprehensive pain evaluation should include the pain characteristics, sensory status, and relevant psychosocial factors in addition to a clinical examination.

Clinical Pain Characteristics

Ratings of pain intensity or pain severity are the most widely used primary outcome measures in clinical pain trials (Farrar et al. 2001). Several different rating scales have proven to be valid for assessing pain intensity, including the Numerical Rating Scale (NRS), the Verbal Rating Scale (VRS), and the Visual Analogue Scale (VAS). The 0–10 NRS has been recommended by the IMMPACT consensus group for use in pain clinical trials (Dworkin et al. 2005), and by a NIDDR-sponsored consensus group regarding SCI-related pain (Bryce et al. 2007).

Other clinical features of pain (i.e., location, quality, temporal pattern) are commonly evaluated in the pain history and this information can be used for both the evaluation of treatment outcome and for diagnostic purposes. For example, information regarding the temporal characteristics of pain such as constancy may provide important information regarding the burden of pain. A recent study in SCI showed that constancy of pain was one of the factors that significantly contributed to particularly disturbing pain (Felix et al 2007).

Sensory Status

In the classification of neuropathic pain, a combination of clinical pain symptoms and signs of neurological function are utilized (Haanpää et al. 2011). The criteria include positive sensory signs such as allodynia or hyperalgesia, and/or negative signs (sensory deficits) and clinical pain symptoms, such as "burning" or "electric" pain.

Self-report Instruments

There are several self-report instruments available that can be used in clinical or research settings as part of the neuropathic pain assessment. For example, the Neuropathic Pain Symptom Inventory (NPSI) (Bouhassira et al. 2004) includes questions about the severity of "burning," "squeezing," "pressure," "electric," or "stabbing" pain, pain evoked by "brushing," "pressure," or "cold," and other non-painful abnormal sensations (i.e., "pins and needles" or "tingling"). The validity of the NPSI and thus its usefulness in neuropathic pain conditions is supported by both clinical experience and research showing that "burning" pain, "electric"

pain, sensations of "numbness," and "pins and needles" are commonly experienced by people with neuropathic pain. Other instruments also assess neuropathic pain severity but are primarily designed to identify the severity of neuropathic pain characteristics. One example is the Neuropathic Pain Scale (NPS) (Galer and Jensen 1997) which includes ratings of pain intensity and unpleasantness, questions regarding the temporal aspects of pain; severity ratings of "sharp," "dull," "sensitive," "hot," "cold," and "itchy" sensations; and severity ratings of "deep" and "surface" pain. Each question can be used to determine treatment effects on a specific neuropathic symptom, or the sum score can be used to determine the overall pain relieving effect. Other instruments, such as the Leeds Assessment of Neuropathic Symptoms and Signs (LANSS) (Bennett 2001), which includes five questions and a sensory examination, are primarily designed to determine the presence of specific neuropathic symptoms, rather than their severity.

Quantitative Sensory Testing (QST)

The evaluation of neurological function, including the quantification and determination of sensory, motor, and autonomic functions, is essential in neuropathic pain conditions (Geber et al. 2011). Moreover, QST is particularly important in translational research, since it facilitates the comparison with common basic pain research outcomes such as evoked nociceptive behavioral responses. Psychometric evaluations of QST methodology suggest that QST is a sensitive measure of outcome in clinical trials involving persons with neuropathic pain (Vranken et al. 2011), stable over time (test-retest reliability) and valid for SCI-related chronic neuropathic pain (Finnerup et al. 2003; Felix and Widerström-Noga 2009), and for neuropathic pain in general (Geber et al. 2011). The measurement of thermal and thermal pain thresholds quantifies small-fiber and spinothalamic tract–mediated function, whereas dorsal column–mediated function can be quantified by measuring the sensitivity to tactile and vibratory stimuli. Sensory profiles are essential components of clinical neuropathic pain phenotypes and potentially useful in facilitating mechanism-based treatments. However, more research is needed to clarify the utility of QST as a clinical outcome measure and for the development of mechanism-based treatment strategies for CNP populations.

Psychosocial Factors
Health-Related Quality of Life (HRQoL)
In CNP, complete remission of pain is unlikely to occur either spontaneously or due to treatment (Bowsher 1999). Therefore, measures that assess factors that may influence HRQoL are of importance. HRQoL is subjective and based on personal preferences and values concerning multiple dimensions of life including well-being and enjoyment of life. The IMMPACT group suggested that in addition to pain severity, assessing some of these core HRQoL domains, (e.g., physical and emotional functioning) would provide a basis for a multidimensional evaluation of pain. However, in CNP populations that may also experience physical impairments, these domains may not only be influenced by pain but by physical impairments and general health issues.

Pain Impact
Although a clinical trial may be designed primarily to reduce the severity of the CNP, an improvement in physical and emotional function may be caused by a decrease in the severity of a less refractory pain type rather than a direct effect on CNP. Different types of pain may also influence pain-related disability and function to various degrees (Marshall et al. 2002; Widerström-Noga et al. 2009). In addition, people who experience CNP have different ways to cope and deal with their pain condition and coping styles and adaptation may significantly modulate the impact of CNP (Widerström-Noga et al. 2006).

Physical Functioning
Because CNP is usually associated with some form of neurological disease or trauma, physical functioning is often impaired. Although a general measure of physical function applicable to heterogeneous pain populations would allow for

optimal comparisons across various pain populations, physical functioning in chronic pain populations afflicted with neurological disease or trauma may be more influenced by the neurological impairment per se than by chronic pain. Commonly used measures for the evaluation of physical function in disabled populations may not capture a decrease in physical function specifically due to pain.

In order to evaluate the impact of pain on physical activity in populations with physical impairments, pain interference measures can be used. These instruments assess the interference specifically caused by pain and, may therefore, provide a more useful alternative or complement to instruments that assess general functional ability in populations afflicted with physical impairment (Cruz-Almeida et al. 2009). Common pain interference measures include the Brief Pain Inventory (BPI) (Cleeland and Ryan 1994) and the Life Interference subscale of Multidimensional Pain Inventory (Kerns et al. 1985). Both these measures have demonstrated adequate psychometric properties in samples of persons with SCI and CNP, and therefore, they have been recommended as useful measures of pain interference after SCI (Bryce et al. 2007). The psychometric properties of the MPI-SCI Life Interference (LI) subscale have been established for the SCI chronic pain population (Widerström-Noga et al. 2006) and include adequate convergent construct validity (r = 0.61), and excellent internal consistency (r = 0.90) and test-retest reliability (r = 0.81).

Emotional Functioning
Research and clinical experience suggest that psychosocial factors are critical for both the severity of pain and the response to treatment. Affective distress, such as depressed mood, anxiety, and anger, is closely related to the experience of chronic pain in a variety of heterogeneous populations (Banks and Kerns 1996). Although the causality of these relationships is unclear, individuals who experience persistent pain, report more psychological distress and excessive fatigue than those who do not experience pain. Anxiety and depression levels have been found to

exhibit strong relationships with both pain severity and pain interference (Nicholson et al. 2009). After SCI, research has identified three different psychosocial subgroups associated with SCI-related chronic pain (Widerström-Noga et al. 2009). Two of these subgroups, the *Dysfunctional* with higher levels of pain severity and life interference, and the *Adaptive Copers* with lower levels of pain severity and life interference, closely resembled subgroups observed in other chronic pain populations. Similar to physical functioning, it is difficult to discern to what extent pain contributes to the emotional distress in pain syndromes associated with a neurological injury or disease. It is also difficult to determine to what extent a particular pain (in a person with several pain types) influences emotional function. However, since affective distress is a very important factor, integral to the pain experience and one that may have a profound impact on HRQoL, it is an important domain to assess despite these limitations.

Differentiated Pain Evaluation of Pain After SCI: The International Spinal Cord Injury Basic Pain Data Set (ISCIBPDS)

People who experience CNP may have several simultaneous pain types (Siddall et al. 2003). In these cases, separate evaluations need to be performed for each different pain problem. This is important since different pain types presumably have different underlying mechanisms and thus pain-specific treatment responses should be expected. A differentiated assessment strategy for longitudinal data collection was recently suggested for persons with SCI in the ISCIBPDS (Widerström-Noga et al. 2008).

The ISCIBPDS is a result of a collaborative effort by members of the International Spinal Cord Society, the American Spinal Injury Association, the International Association for the Study of Pain, and the American Pain Society. The ISCIBPDS was developed to provide a simple, differentiated, multidimensional, and standardized pain assessment for pain after SCI. The underlying idea was that comparable sets of outcome measures in clinical practice and in clinical trials would facilitate collaboration,

translation of basic research, interpretation, and application of results to improve the clinical management of SCI-related pain including CNP.

The recommendations made by the IMMPACT group (Dworkin et al. 2005) was used as a framework for the ISCIBPDS, which includes assessments of pain severity, and physical and emotional functioning in order to capture the multidimensional nature of pain. The items of the ISCIBPDS are adapted to important issues related to SCI (i.e., several simultaneous different pain problems, physical impairments due to CNS injury, etc.). The validity of a self-administered version of the ISCIBPDS components measuring pain interference, pain intensity, pain location, frequency, and temporal aspects of pain was recently demonstrated (Jensen et al. 2010).

The ISCIBPDS is designed to collect information regarding as many as three separate and current pain problems since most people with SCI experience three or fewer pain problems (Siddall et al. 2003; Felix et al. 2007). The information collected concerns location, intensity, temporal pattern of pain, and a pain classification including several types of neuropathic pain that may be experienced after a SCI. The pain classification scheme incorporates elements of previous SCI pain taxonomies, but it is primarily based upon the IASP taxonomy (Siddall et al. 2000). The pain categories include all persistent pain types that a person with SCI may experience: (1) Musculoskeletal (Nociceptive); (2) Visceral (Nociceptive); (3) Other (Nociceptive); (4) At-level (Neuropathic); (5) Below-level (Neuropathic); and (6) Other (Neuropathic) or "Unknown" and should be classified in combination with the ASIA Impairment Scale (Marino et al. 2003) to define the pain location relative to the neurological level of injury.

To address the psychosocial impact of pain, the ISCIPDS includes questions regarding pain interference, which is the extent that pain negatively affects or hinders an individual's daily activities. Pain interference is an important and sensitive (Cruz-Almeida et al. 2009) outcome in clinical pain trials. In a recent study, a high level of pain interference was significantly predictive of a particular pain being viewed as "most disturbing" (Felix et al. 2007). The pain interference questions of the ISCIBPDS contains three items from the Life Interference (LI) subscale of the Multidimensional Pain Inventory SCI (MPI-SCI) version (Widerström-Noga et al. 2006) evaluating impact of pain on activities in general and on recreational, social and family-related activities, and three items specifically asking about pain interference with general activities, such as mood and sleep.

The ISCIBPDS is designed for use in both clinical and research settings. It is likely that similar approaches could be developed for use in other CNP populations. The ISCIBPDS is part of the NIH/NINDS Common Data Elements (http://www.commondataelements.ninds.nih.gov/SCI.aspx) "to provide a foundation of variables that are common across studies in order to increase the efficiency and consistency of data collection and to facilitate data sharing."

Participants' Ratings of Global Improvement
The perception of how much a particular treatment has improved a person's pain is an interesting and important complement to other outcomes. The Patient Global Impression of Change (PGIC) (Farrar et al. 2001) measures the individual's global perspective on the pain-relieving effect of a treatment intervention. The PGIC is evaluated by one question that is answered on a 7-point categorical scale ranging from 1 = "very much improved" to 7 = "very much worse." This rating reflects dimensions that are unique for each individual and, therefore, provides important information incorporating the person's own perspective about the overall benefit of a treatment.

Adverse Effects
The assessment of adverse effects in a clinical trial is made in order to assess the risk-benefit of a treatment. Adverse effects are either directly caused by the treatment or indirectly caused by the treatment on a preexisting illness or a physical impairment. In people with neurological trauma or disease, the latter situation may need special consideration since adverse effects that have a relatively minor impact in populations without physical impairments can cause significant

problems in a physically impaired pain population. For example, an additional negative impact on cognitive ability caused by a pharmacological intervention in a person who has a preexisting cognitive impairment due to a traumatic brain injury, or further decreased bowel function in a person who has impaired bowel function due to SCI, may cause difficulties that interfere with adequate dosing, adherence to treatment, and the risk-benefit ratio. Therefore, it is important to monitor not only severity of adverse effects but also impact on the preexisting conditions associated with the neurological disease or trauma.

Participant Disposition
To adequately interpret the results of a clinical trial and to determine whether the results are applicable to a larger population, details concerning participants screened for enrollment (e.g., reasons for dropout, non-compliance), need to be provided (for details, see the ▶ CONSORT Statement (Moher et al. 2001)). In CNP populations, this is particularly important since the reasons for withdrawal and non-adherence may be disease specific and not related to CNP but to the other consequences of neurological disease or trauma.

Conclusion
The assessment domains recommended by the IMMPACT group in combination with a differentiated and comprehensive evaluation of pain symptoms and signs appear to be appropriate also for CNP populations. Due to the fact that people who have CNP frequently have varying degrees of physical impairment, specific assessment of pain-related interference with physical and emotional functioning may be more useful than general measures of physical and emotional function. In the selection of specific instruments to be used as core outcome measures, not only must validity and reliability be considered, but also whether the instrument can be used with diverse CNP populations associated with a variety of diseases and traumas.

References

Banks, S. M., & Kerns, R. D. (1996). Explaining high rates of depression in chronic pain: A diathesis-stress framework. *Psychological Bulletin, 119*, 95–110.

Bennett, M. (2001). The LANSS pain scale: The leeds assessment of neuropathic symptoms and signs. *Pain, 92*(1–2), 147–157.

Boivie, J. (2003). Central pain and the role of quantitative sensory testing (QST) in research and diagnosis. *European Journal of Pain, 7*, 339–343.

Bouhassira, D., Attal, N., Fermanian, J., Alchaar, H., Gautron, M., Masquelier, E., Rostaing, S., Lanteri-Minet, M., Collin, E., Grisart, J., et al. (2004). Development and validation of the neuropathic pain symptom inventory. *Pain, 108*(3), 248–257.

Bowsher, D. (1999). Central pain following spinal and supraspinal lesions. *Spinal Cord, 37*, 235–238.

Bryce, T. N., Budh, C. N., Cardenas, D. D., Dijkers, M., Felix, E. R., Finnerup, N. B., et al. (2007). Outcome measures for pain after spinal cord injury: An evaluation of reliability and validity. *The Journal of Spinal Cord Medicine, 30*, 421–440.

Cleeland, C. S., & Ryan, K. M. (1994). Pain assessment: Global use of the brief pain inventory. *Annals of the Academy of Medicine, Singapore, 23*(2), 129–138.

Cruz-Almeida, Y., Alameda, G., & Widerstrom-Noga, E. G. (2009). Differentiation between pain-related interference and interference caused by the functional impairments of spinal cord injury. *Spinal Cord, 47*(5), 390–395.

Dworkin, R. H., Turk, D. C., Farrar, J. T., Haythornthwaite, J. A., Jensen, M. P., Katz, N. P., Kerns, R. D., Stucki, G., Allen, R. R., Bellamy, N., et al. (2005). Core outcome measures for chronic pain clinical trials: IMMPACT recommendations. *Pain, 113*(1–2), 9–19.

Farrar, J. T., Young, J. P., Jr., LaMoreaux, L., Werth, J. L., & Poole, R. M. (2001). Clinical importance of changes in chronic pain intensity measured on an 11-point numerical pain rating scale. *Pain, 94*, 149–158.

Felix, E. R., & Widerström-Noga, E. G. (2009). Reliability and validity of quantitative sensory testing in persons with spinal cord injury and neuropathic pain. *Journal of Rehabilitation Research and Development, 46*(1), 69–83.

Felix, E. R., Cruz-Almeida, Y., & Widerstrom-Noga, E. G. (2007). Chronic pain after spinal cord injury: What characteristics make some pains more disturbing than others? *Journal of Rehabilitation Research and Development, 44*(5), 703–716.

Finnerup, N. B., Johannesen, I. L., Fuglsang-Frederiksen, A., Bach, F. W., & Jensen, T. S. (2003). Sensory function in spinal cord injury patients with and without central pain. *Brain, 126*(Pt 1), 57–70.

Galer, B. S., & Jensen, M. P. (1997). Development and preliminary validation of a pain measure specific to neuropathic pain: The neuropathic pain scale. *Neurology, 48*(2), 332–338.

Geber, C., Klein, T., Azad, S., Birklein, F., Gierthmühlen, J., Huge, V., Lauchart, M., Nitzsche, D., Stengel, M., Valet, M., et al. (2011). Test-retest and interobserver reliability of quantitative sensory testing according to the protocol of the German Research Network on Neuropathic Pain (DFNS): A multi-centre study. *Pain, 152*(3), 548–556.

Haanpää, M., Attal, N., Backonja, M., Baron, R., Bennett, M., Bouhassira, D., Cruccu, G., Hansson, P., Haythornthwaite, J. A., Iannetti, G. D., et al. (2011). NeuPSIG guidelines on neuropathic pain assessment. *Pain, 152*(1), 14–27.

Jensen, M. P., Widerstrom-Noga, E., Richards, J. S., Finnerup, N. B., Biering-Sorensen, F., & Cardenas, D. D. (2010). Reliability and validity of the international spinal cord injury basic pain data set items as self-report measures. *Spinal Cord, 48*(3), 230–238.

Kerns, R. D., Turk, D. C., & Rudy, T. E. (1985). The west Haven-Yale multidimensional pain inventory (WHYMPI). *Pain, 23*, 345–356.

Marino, R. J., Barros, T., Biering-Sorensen, F., Burns, S. P., Donovan, W. H., Graves, D. E., Haak, M., Hudson, L. M., & Priebe, M. M. (2003). International standards for neurological classification of spinal cord injury. *The Journal of Spinal Cord Medicine, 26*(Suppl 1), S50–S56.

Marshall, H. M., Jensen, M. P., Ehde, D. M., & Campbell, K. M. (2002). Pain site and impairment in individuals with amputation pain. *Archives of Physical Medicine and Rehabilitation, 83*, 1116–1119.

Moher, D., Schulz, K. F., & Altman, D. G. (2001). The CONSORT statement: Revised recommendations for improving the quality of reports of parallel-group randomised trials. *Lancet, 357*, 1191–1194.

Nicholson, P. K., Nicholas, M. K., & Middleton, J. (2009). Spinal cord injury-related pain in rehabilitation: A cross-sectional study of relationships with cognitions, mood and physical function. *European Journal of Pain, 13*(5), 511–517.

Siddall, P. J., McClelland, J. M., Rutkowski, S. B., & Cousins, M. J. (2003). A longitudinal study of the prevalence and characteristics of pain in the first 5 years following spinal cord injury. *Pain, 103*, 249–257.

Siddall, P. J., Yezierski, R. P., & Loeser, J. D. (2000). Pain following spinal cord injury: Clinical features, prevalence, and taxonomy. *International Association for the Study of Pain Newsletter, 2000–3*, 3–7.

Treede, R. D., Jensen, T. S., Campbell, J. N., Cruccu, G., Dostrovsky, J. O., Griffin, J. W., Hansson, P., Hughes, R., Nurmikko, T., & Serra, J. (2008). Neuropathic pain: Redefinition and a grading system for clinical and research purposes. *Neurology, 70*(18), 1630–1635.

Vranken, J. H., Hollmann, M. W., van der Vegt, M. H., Kruis, M. R., Heesen, M., Vos, K., Pijl, A. J., & Dijkgraaf, M. G. (2011). Duloxetine in patients with central neuropathic pain caused by spinal cord injury or stroke: A randomized, double-blind, placebo-controlled trial. *Pain, 152*(2), 267–273.

Widerström-Noga, E., Biering-Sorensen, F., Bryce, T., Cardenas, D. D., Finnerup, N. B., Jensen, M. P., Richards, J. S., & Siddall, P. J. (2008). The international spinal cord injury pain basic data set. *Spinal Cord, 46*(12), 818–823.

Widerström-Noga, E. G., Cruz-Almeida, Y., Martinez-Arizala, A., & Turk, D. C. (2006). Internal consistency, stability, and validity of the spinal cord injury version of the multidimensional pain inventory. *Archives of Physical Medicine and Rehabilitation, 87*, 516–523.

Widerström-Noga, E. G., Cruz-Almeida, Y., Felix, E. R., & Adcock, J. P. (2009). Relationship between pain characteristics and pain adaptation type in persons with SCI. *Journal of Rehabilitation Research and Development, 46*(1), 43–56.

Woolf, C. J., & Max, M. B. (2001). Mechanism-based pain diagnosis: Issues for analgesic drug development. *Anesthesiology, 95*, 241–249.

Central Pain, Pharmacological Treatments

Nanna Brix Finnerup[1] and Soeren H. Sindrup[2]
[1]Aarhus University Hospital, Danish Pain Research Center, Aarhus, Denmark
[2]Department of Neurology, Odense University Hospital, Odense, Denmark

Definition

▶ Central pain can be a consequence of different diseases and includes central poststroke pain (CPSP), central pain in spinal cord injury (SCI), multiple sclerosis, perhaps Parkinson's and Huntington's disease, AIDS, and brain trauma. In spinal cord injury, central neuropathic pain is experienced at and/or below the level of a lesion and may be difficult to separate from peripheral neuropathic pain components caused by root lesions. Pain associated with this pathology is generally felt at the level of injury. Since different types of SCI pain are usually not separated, this review includes all trials on the treatment of pain associated with spinal cord injury.

Any lesion along the spinothalamocortical pathway may lead to central pain, and ▶ neuronal hyperexcitability caused by increased excitation and/or decreased inhibition is an additional mechanism, which provides the mechanistic basis for the use of drugs for neuropathic pain. Most pharmacological agents developed for treatment of this condition act by depressing neuronal activity; modulating sodium or calcium channels; increasing inhibition with γ-aminobutyric acid (GABA) or serotonergic, noradrenergic, or enkephalinergic agonists; or decreasing activation via glutamate receptors, especially the N-methyl-D-aspartate (NMDA) receptor.

Characteristics

To suggest an evidence-based treatment algorithm based on ▶ randomized double-blind placebo-controlled trials on central pain is difficult, considering the fact that although new trials are emerging, there are still only a few, small-sized studies on central pain (Tables 1 and 2). ▶ Number needed to treat (NNT) and number needed to harm (NNH) (in this text defined as treatment-related withdrawals) are used to compare efficacy and harm of individual drugs.

Antidepressants

Tricyclic antidepressants block the reuptake of norepinephrine or serotonin, but activation on NMDA receptors and sodium channels may also play a role in their analgesic actions. The tricyclic antidepressant amitriptyline has been studied in two controlled trials. In a three-way cross-over study, amitriptyline 75 mg daily was effective in relieving pain (Leijon and Boivie 1989). The pain-relieving effect correlated well with total plasma concentration (a) with high number of responders and (b) having plasma concentrations exceeding 300 nmol/L. In patients with SCI, amitriptyline 10–125 mg daily had no effect on different types of pain, including nociceptive pain (Cardenas et al. 2002), but the average amitriptyline dose was low, as were serum concentrations (mean 92 ng/ml), i.e., below the level associated with response in CPSP. Amitriptyline relieved SCIP compared to an active placebo in a trial comparing the effect to both placebo and

Central Pain, Pharmacological Treatments, Table 1 Randomized, double-blind, placebo-controlled trials on oral drugs in central pain

Active drug, daily dose	Study	Condition	Design, no. of patients	Outcome	NNT (95 % CI)	NNH (95 % CI)
Amitriptyline, 75 mg	Leijon and Boivie (1989)	CPSP	Cross-over 15	Ami > pla	1.7 (1.2–3.1)	∞
Amitriptyline, 10–125 mg	Cardenas et al. (2002)	SCI pain	Parallel 84	Ami = pla	–	9.2 (4.2–∞)
Amitriptyline, 150 mg	Rintala et al. (2007)	SCI pain	Cross-over 38	Ami > pla	∞	∞
Trazodone, 150 mg	Davidoff et al. (1987)	SCI pain	Parallel 18	Tra = pla	–	NA
Duloxetine, 120 mg	Vranken et al. (2011)	CPSP or SCI pain	Parallel 48	Dul = pla	–	
Carbamazepine, 800 mg	Leijon and Boivie (1989)	CPSP	Cross-over 15	Carb = pla	–	15.0 (5.2–∞)
Lamotrigine, 200 mg	Vestergaard et al. (2001)	CPSP	Cross-over 30	Ltg > pla	NA	10.0 (4.8–∞)
Lamotrigine, 200–400 mg	Finnerup et al. (2002)	SCI pain	Cross-over 22	Ltg = pla	–	∞
Lamotrigine, 400 mg	Breuer et al. (2007)	Multiple sclerosis	Cross-over 12	Lam = pla	–	∞
Valproate, 600–2,400 mg	Drewes et al. (1994)	SCI pain	Cross-over 20	Val = pla	–	∞
Gabapentin, up to 1,800 mg	Tai et al. (2002)	SCI pain	Cross-over 7	Gab = pla	–	14.0 (4.8–∞)
Gabapentin, 3,600 mg	Rintala et al. (2007)	SCI pain	Cross-over 38	Gab = pla	–	∞
Gabapentin, up to 3,600 mg	Levendoglu et al. (2004)	SCI pain	Cross-over 20	Gab > pla	NA	∞
Pregabalin, 600 mg	Siddall et al. (2006)	SCI pain	Parallel 146	Pre > pla	7.0 (3.9–37)	∞
Pregabalin, 600 mg	Vranken et al. (2007)	CPSP or SCI pain	Parallel 40	Pre > pla	3.3 (1.9–14)	∞
Pregabalin, 600 mg	Kim et al. (2011)	CPSP	Parallel 219	Pre = pla	–	∞
Levetiracetam, 3,000 mg	Finnerup et al. (2009)	SCI pain	Cross-over 34	Lev = pla	–	∞
Mexiletine, 450 mg	Chiou-Tan et al. (1996)	SCI pain	Cross-over 11	Mex = pla	–	∞
Tramadol, 40 mg	Norrbrink and Lundeberg (2009)	SCI pain	Parallel 35	Tra > pla	∞	∞
Dronabinol, 5–10 mg	Svendsen et al. (2004)	Multiple sclerosis	Cross-over 24	Can > pla	3.4 (1.8–23.4)	∞
THC:CBD, 25.9 mg:24.0 mg	Rog et al. (2005)	Multiple sclerosis	Parallel 66	Can > pla	3.7 (2.2–13)	∞

CPSP central post-stroke pain, *SCI* spinal cord injury, *CI* confidence interval

gabapentin (Rintala et al. 2007). The serotonin-noradrenaline reuptake inhibitor (SNRI) duloxetine was studied in patients with SCI or CPSP. There was no significant effect on the primary efficacy measure, but an effect of several secondary outcome measures, and it is possible that the trial was too small to show an effect (Vranken et al. 2011). The heterocyclic antidepressant trazodone 150 mg daily had no effect on neuropathic pain in spinal injury (Davidoff et al. 1987). The possibility of preventing CPSP was studied using amitriptyline (10 mg the first day after the onset of stroke was diagnosed, titrated to 75 mg within 3 weeks) or placebo administered to 39 stroke patients for 1 year (Lampl et al. 2002). Within this year, CPSP developed in three patients receiving amitriptyline (VAS 5.0) and in four receiving placebo (VAS 5.4). Two patients in the amitriptyline group and three patients in the placebo group developed allodynia. With an expected 8 % incidence of CPSP,

Central Pain, Pharmacological Treatments, Table 2 Randomized, double-blind, placebo-controlled trials on nonoral drugs in central pain

Active drug, dose administration	Study	Condition	Design, no. of patients	Outcome
Lidocaine IV 5 mg/kg	Attal et al. (2000)	CPSP/ SCI pain	Cross-over 16	Lid > pla
Lidocaine IV 2.5 mg/kg	Kvarnstrøm et al. (2004)	SCI pain	Cross-over 10	Lid = pla
Lidocaine SA 50–100 mg	Loubser and Donovan (1991)	SCI pain	Cross-over 21	Lid > pla
Lidocaine IV 5 mg/kg	Finnerup et al. (2005)	SCI pain	Cross-over 24	Lid > pla
Ketamine IV 60 μg/kg + 6 μg/kg/min	Eide et al. (1995)	SCI pain	Cross-over 9	Ket > pla
Ketamine IV 0.4 mg/kg	Kvarnstrøm et al. (2004)	SCI pain	Cross-over 10	Ket > pla
Alfentanil IV 7 μg/kg + 0.6 μg/kg/min	Eide et al. (1995)	SCI pain	Cross-over 9	Alf > pla
Propofol IV 0.2 mg/kg	Canavero and Bonicalzi (2004)	CPSP/SCI pain	Cross-over 44	Pro > pla
Morphine IV 9–30 mg	Attal et al. (2002)	CPSP/SCI pain	Cross-over 15	Mor = pla
Morphine IT 0.2–1.5 mg	Siddall et al. (2000)	SCI pain	Cross-over 15	Mor = pla
Clonidine IT 50–100 μg or 300–500 μg				Clo = pla Mor + clo > pla
Naloxone IV up to 8 mg	Bainton et al. (1992)	CPSP	Cross-over 20	Nal = pla
Baclofen 50 μg	Herman et al. (1992)	SCI pain	Cross-over 6	Bac > pla

CPSP central poststroke pain, *SCI* spinal cord injury, *IV* intravenous, *SA* subarachoidal, *IT* intrathecal

this sample size is probably too small to detect an effect, but this was the first study of its kind and more are encouraged.

Antiepileptic Drugs

Antiepileptic drugs include a broad spectrum of drugs used in the management of epilepsy, and exert their analgesic actions through multiple mechanisms, by either reducing excitation and/ or enhancing inhibition. Carbamazepine blocks voltage-dependent sodium channels and may have minor effects on calcium channels and serotonergic systems. In a three-way cross-over study of amitriptyline, carbamazepine 800 mg, and placebo, carbamazepine did not reduce CPSP compared to placebo (Leijon and Boivie 1989). However, both amitriptyline and carbamazepine treatments gave a 20 % lower mean pain intensity score during the last week of treatment. Based on the relatively small number of patients in this study, a significant effect of carbamazepine in CPSP cannot be excluded. Valproate has several pharmacological effects, including GABAergic

and antiglutamatergic actions. The effect of valproate 600–2,400 mg daily was examined in a cross-over study in patients with spinal injury (Drewes et al. 1994). Although a trend toward improvement was observed, valproate was not significantly better than placebo in relieving pain. Lamotrigine inhibits voltage-dependent sodium channels to stabilize neuronal membranes and inhibits release of excitatory amino acids, principally glutamate. In CPSP, lamotrigine 200 mg daily reduced pain with a mean reduction of 30 % (Vestergaard et al. 2001). Lamotrigine also reduced cold evoked ▶ allodynia assessed by an acetone droplet. In spinal cord injury pain (SCIP), lamotrigine 200–400 mg daily was not more effective than placebo in reducing pain, although a subgroup of patients with incomplete injury and evoked pain reported an effect on spontaneous pain (Finnerup et al. 2002). Lamotrigine also failed to relieve neuropathic pain following multiple sclerosis (Breuer et al. 2007). Gabapentin, which is thought to exert its analgesic actions by binding to an $\alpha_2\delta$ subunit of voltage-gated calcium channels, has been studied in three trials in SCIP. In a small study with seven patients, gabapentin up to 1,800 mg had no significant effect on pain intensity (Tai et al. 2002). There was, however, a trend toward improvement, and a significant effect on unpleasant feeling. In another study, gabapentin up to 3,600 mg reduced the intensity and frequency of pain and several pain descriptors in 20 paraplegics with complete SCI (Levendoglu et al. 2004) but failed to relieve SCIP in another trial (Rintala et al. 2007). Gabapentin in combination with the NMDA antagonist dextromethorphan was found to be superior to placebo, and to either component alone, in patients with neuropathic pain following spinal injury (Sang et al. 2001). Pregabalin (similar mechanism as gabapentin) have been shown to reduce SCIP and CPSP (Siddall et al. 2006, Vranken et al. 2007). Levetiracetam failed to relieve SCIP (Finnerup et al. 2009). Oxcarbazepine (similar mechanism as carbamazepine) and other newer anticonvulsants such as tiagabine and zonisamide have not been tested in controlled trials in central pain.

Other Oral Drugs
Opioids also have an effect on neuropathic pain. Tramadol was examined in SCIP, and despite a high withdrawal rate, tramadol significantly relieved SCIP (Norrbrink and Lundeberg 2009). Titration should be slow and individual. The cannabinoid dronabinol (a synthetic δ-9-tetrahydrocannabinol) has been studied in 24 patients with central pain caused by multiple sclerosis. It significantly relieved central pain (Svendsen et al. 2004). Also whole-plant cannabis-based medicine relieved central pain caused by multiple sclerosis. Mexiletine, a sodium channel blocker, did not relieve pain in 11 patients with SCIP in doses of 450 mg daily (Chiou-Tan et al. 1996).

Nonoral Drugs
Sodium channel blockers may play a role in the treatment of central pain. Lidocaine in doses of 2.5 mg/kg, administered intravenously over 40 min, had no pain-relieving effect on pain in spinal cord injury patients (Kvarnström et al. 2004),

while 5 mg/kg administered intravenously over 30 min significantly decreased spontaneous ongoing pain, brush-evoked allodynia, and static mechanical ▶ hyperalgesia but was no better than placebo against thermal allodynia and hyperalgesia in patients with CPSP or SCIP (Attal et al. 2000). It was also found that lidocaine 5 mg/kg over 30 min relieved spontaneous pain in spinal cord injury patients with ($n = 12$) and without ($n = 12$) evoked pain and that lidocaine relieved pain felt at and below the level of injury (Finnerup et al. 2005). Subarachnoid infusion of lidocaine was significantly better than placebo in relieving SCIP (Loubser and Donovan 1991). Adequate spinal anesthesia, proximal to the level of spinal injury, seems important for a positive response to lidocaine, suggesting the existence of a "pain generator" in the spinal cord of some patients.

NMDA receptor antagonists given intravenously were reported to relieve SCIP in two studies (Eide et al. 1995; Kvarnström et al. 2004). Studies on opioids in central pain trials have yielded diverging results. Intravenous morphine was reported to have an effect on brush-evoked allodynia, but not on spontaneous pain in CPSP and SCIP patients (Attal et al. 2002); intravenous alfentanil was effective in relieving SCIP (Eide et al. 1995); and finally, morphine given intrathecally was effective in SCIP patients, but only in combination with clonidine (an α_2-adrenergic agonist) (Siddall et al. 2000). No effect of naloxone in CPSP was found (Bainton et al. 1992). Propofol, a GABA$_A$-receptor agonist, injected as a single intravenous bolus of 0.2 mg/kg, relieved spontaneous pain and allodynia in 44 patients with spinal cord injury and poststroke pain (Canavero and Bonicalzi 2004). Intrathecal baclofen, a GABA$_B$-receptor agonist, was also reported to relieve dysesthesia in six patients with SCI or multiple sclerosis (Herman et al. 1992).

Conclusions

Tricyclic antidepressants, sodium channel blockers, NMDA antagonists, GABA agonists,

Central Pain, Pharmacological Treatments, Table 3 First-line treatment options for central pain

Drug class and name	Dosage	Common side effects and cautions
Tricyclic antidepressants, e.g., imipramine or amitriptyline	25 mg daily initially, increasing by 25 mg every 2 weeks, usually up to 150 mg daily in one to two divided doses. Plasma drug levels should be monitored (optimal plasma levels of imipramine plus desipramine is 300–750 nM)	Dry mouth, constipation, and urinary retention, orthostatic hypotension, sedation, and increased spasticity are reported. Contraindicated in patients with heart failure, cardiac conduction blocks (ECG before start) and epilepsy
Gabapentin	300 mg daily initially, increasing by 300 mg every third day, to 1,800–4,800 mg daily	Dizziness, sedation, ataxia, and occasional peripheral edema. Renal impairment requires dosage adjustment
Pregabalin	75 mg daily initially, increasing up to 600 mg daily	Dizziness, sedation, ataxia. Renal impairment requires dosage adjustment

calcium channel blockers, tramadol, and cannabinoids are shown to relieve central pain. In randomized controlled trials on oral treatment, amitriptyline, pregabalin, tramadol, and cannabinoids have been effective in relieving pain, but large-scale randomized controlled studies are lacking, and the treatment algorithm for central pain is still based on effective treatments for peripheral neuropathic pain. Antidepressants and anticonvulsants (Table 3) are first-line drugs for central pain. In many cases, treatment provides only partial or no relief, and other types of drugs, combination therapy, intrathecal therapy, and different nonpharmacological approaches may be considered in these cases.

References

Attal, N., Gaude, V., Brasseur, L., Dupuy, M., Guirimand, F., Parker, F., & Bouhassira, D. (2000). Intravenous lidocaine in central pain: A double-blind, placebo-controlled, psychophysical study. *Neurology, 54*, 564–574.

Attal, N., Guirimand, F., Brasseur, L., Gaude, V., Chauvin, M., & Bouhassira, D. (2002). Effects of IV morphine in central pain: A randomized placebo-controlled study. *Neurology, 58*, 554–563.

Bainton, T., Fox, M., Bowsher, D., & Wells, C. (1992). A double-blind trial of naloxone in central post-stroke pain. *Pain, 48*, 159–162.

Breuer, B., Pappagallo, M., Knotkova, H., Guleyupoglu, N., Wallenstein, S., & Portenoy, R. K. (2007). A randomized, double-blind, placebo-controlled, two-period, cross-over, pilot trial of lamotrigine in patients with central pain due to multiple sclerosis. *Clinical Therapeutics, 29*, 2022–2030.

Canavero, S., & Bonicalzi, V. (2004). Intravenous subhypnotic propofol in central pain: A double-blind, placebo-controlled, cross-over study. *Clinical Neuropharmacology, 27*, 182–186.

Cardenas, D. D., Warms, C. A., Turner, J. A., Marshall, H., Brooke, M. M., & Loeser, J. D. (2002). Efficacy of amitriptyline for relief of pain in spinal cord injury: Results of a randomized controlled trial. *Pain, 96*, 365–373.

Chiou-Tan, F. Y., Tuel, S. M., Johnson, J. C., Priebe, M. M., et al. (1996). Effect of mexiletine on spinal cord injury dysesthetic pain. *American Journal of Physical Medicine & Rehabilitation, 75*, 84–87.

Davidoff, G., Guarracini, M., Roth, E., Sliwa, J., & Yarkony, G. (1987). Trazodone hydrochloride in the treatment of dysesthetic pain in traumatic myelopathy: A randomized, double-blind, placebo-controlled study. *Pain, 29*, 151–161.

Drewes, A. M., Andreasen, A., & Poulsen, L. H. (1994). Valproate for treatment of chronic central pain after spinal cord injury. A double-blind cross-over study. *Paraplegia, 32*, 565–569.

Eide, P. K., Stubhaug, A., & Stenehjem, A. E. (1995). Central dysesthesia pain after traumatic spinal cord injury is dependent on N-methyl-D-aspartate receptor activation. *Neurosurgery, 37*, 1080–1087.

Finnerup, N. B., Sindrup, S. H., Bach, F. W., Johannesen, I. L., & Jensen, T. S. (2002). Lamotrigine in spinal cord injury pain: A randomized controlled trial. *Pain, 96*, 375–383.

Finnerup, N. B., Biering-Sorensen, F., Johannesen, I. L., et al. (2005). Intravenous lidocaine relieves spinal cord injury pain: A randomized controlled trial. *Anesthesiology, 102*, 1023–1030.

Finnerup, N. B., Grydehoj, J., Bing, J., Johannesen, I. L., Biering-Sorensen, F., Sindrup, S. H., & Jensen, T. S. (2009). Levetiracetam in spinal cord injury pain: A randomized controlled trial. *Spinal Cord, 47*, 861–867.

Herman, R. M., D'Luzansky, S. C., & Ippolito, R. (1992). Intrathecal baclofen suppresses central pain in patients with spinal lesions. A pilot study. *The Clinical Journal of Pain, 8,* 338–345.

Kim, J. S., Basford, G., Murphy, T. K., Martin, A., Dror, V., & Cheung, R. (2011). Safety and efficacy of pregabalin in central post-stroke pain. *Pain, 152,* 1018–1023.

Kvarnström, A., Karlsten, R., Quiding, H., & Gordh, T. (2004). The analgesic effect of intravenous ketamine and lidocaine on pain after spinal cord injury. *Acta Anaesthesiologica Scandinavica, 48,* 498–506.

Lampl, C., Yazdi, K., & Röper, C. (2002). Amitriptyline in the prophylaxis of central poststroke pain. *Stroke, 33,* 3030–3032.

Leijon, G., & Boivie, J. (1989). Central post-stroke pain – a controlled trial of amitriptyline and carbamazepine. *Pain, 36,* 27–36.

Levendoglu, F., Ögün, C. Ö., Özerbil, Ö., Ögun, T. C., & Ugurlu, H. (2004). Gabapentin is a first line drug for the treatment of neuropathic pain in spinal cord injury. *Spine, 29,* 743–751.

Loubser, P. G., & Donovan, W. H. (1991). Diagnostic spinal anaesthesia in chronic spinal cord injury pain. *Paraplegia, 29,* 25–36.

Norrbrink, C., & Lundeberg, T. (2009). Tramadol in neuropathic pain after spinal cord injury: A randomized, double-blind, placebo-controlled trial. *The Clinical Journal of Pain, 25,* 177–184.

Rintala, D. H., Holmes, S. A., Courtade, D., Fiess, R. N., Tastard, L. V., & Loubser, P. G. (2007). Comparison of the effectiveness of amitriptyline and gabapentin on chronic neuropathic pain in persons with spinal cord injury. *Archives of Physical Medicine and Rehabilitation, 88,* 1547–1560.

Rog, D. J., Nurmikko, T. J., Friede, T., & Young, C. A. (2005). Randomized, controlled trial of cannabis-based medicine in central pain in multiple sclerosis. *Neurology, 65,* 812–819.

Sang, C. N., Dobosh, L., Miller, V., & Brown, R. (2001). Combination therapy for refractory pain following spinal cord injury using the low affinity N-methyl-D-aspartate (NMDA) receptor antagonist dextromethorphan and gabapentin. *Abstract 20th Annual Scientific Meeting APS. Journal of Pain* 2:10.

Siddall, P. J., Molloy, A. R., Walker, S., & Rutkowski, S. B. (2000). The efficacy of intrathecal morphine and clonidine in the treatment of pain after spinal cord injury. *Anesthesia and Analgesia, 91,* 1–6.

Siddall, P. J., Cousins, M. J., Otte, A., Griesing, T., Chambers, R., & Murphy, T. K. (2006). Pregabalin in central neuropathic pain associated with spinal cord injury: A placebo-controlled trial. *Neurology, 67,* 1792–1800.

Svendsen, K. B., Jensen, T. S., & Bach, F. (2004). Does the cannabinoid dronabinol reduce central pain in multiple sclerosis? Randomised double blind placebo controlled crossover trial. *BMJ, 329,* 253–258.

Tai, Q., Kirshblum, S., Chen, B., Millis, S., et al. (2002). Gabapentin in the treatment of neuropathic pain after spinal cord injury: A prospective, randomized, double-blind, cross-over trial. *The Journal of Spinal Cord Medicine, 25,* 100–105.

Vestergaard, K., Andersen, G., Gottrup, H., Kristensen, B. T., & Jensen, T. S. (2001). Lamotrigine for central poststroke pain: A randomized controlled trial. *Neurology, 56,* 184–190.

Vranken, J. H., Dijkgraaf, M. G., Kruis, R. M., van der Vegt, M. H., Hollmann, M. W., & Heesen, M. (2007). Pregabalin in patients with central neuropathic pain: A randomized, double-blind, placebo-controlled trial of a flexible-dose regimen. *Pain, 136,* 150–157.

Vranken, J. H., Hollmann, M. W., van der Vegt, M. H., Kruis, R. M., Heesen, M., Pijl, A. J., & Dijkgraaf, M. G. (2011). Duloxetine in patients with central neuropathic pain caused by spinal cord injury or stroke: A randomized, double-blind, placebo-controlled trial. *Pain, 152,* 267–273.

Central Pain: Outcome Measures in Preclinical Research

Cathrine Baastrup
Danish Pain Research Center, Aarhus University
Aarhus University Hospital, Aarhus, Denmark

Synonyms

Behavioral measure; Effectiveness measure; Pain assay; Pain signal; Quantification of pain

Definition

An outcome measure is a performance indicator that assesses the status of specific parameters subsequent to or resulting from a procedure, intervention, treatment, or overtime in a group of animals.

Introduction

▶ Central pain, like all other types of pain, is not readily measured in animal models. Ideally, such measures should comprise assays to evaluate the level of spontaneous and ongoing pain and assays to measure the areas and intensities of allodynia and hyperalgesia. When working with animal models of pain, it is important to remember that all measurements have an inherent restriction in face validity and will at best be surrogate measures of pain (including ▸ allodynia and ▸ hyperalgesia) and should thus be translated with care. Terms like hypersensitivity or increased responses with a clear definition of such responses are thus recommended over terms like allodynia and hyperalgesia.

Another incongruence between clinical and preclinical central pain measurements is that while central pain patients report ongoing and spontaneous pain to be the most difficult to live with, measurements of evoked responses are predominant in preclinical research.

In the following section, outcome measures that have been used to evaluate pain-like behavior in preclinical models of central pain (e.g., ▸ spinal cord injury due to ▸ contusion, ▸ cordotomy, or ▸ hemisection, ▸ ischemia, electrochemical, ▸ excitotoxicity and the ▸ thalamic or ▸ cortical post-stroke models) will be mentioned.

Characteristics

Measuring Evoked Responses
Reflex Withdrawal Responses
Assays used to evaluate pain-like behavior in conscious animals have historically relied on ▸ spinal reflex-based responses to peripheral stimuli. The most frequently applied method consists of stimulating an animal on the plantar surface of the hindpaw with increasing strengths of von Frey monofilaments until the animal ▸ withdraws the paw from the stimulus. This method is also sometimes called mechanical allodynia, based on the premise that the animal's withdrawal ideally represents an allodynic reaction. The majority of methods are derived from this paradigm, alternating the type of stimulation (thermal or mechanical), site of stimulation (tail or hindpaw), and response (threshold, frequency, or latency). The methods are described in more detail in the section "▸ Allodynia test, mechanical and cold allodynia" and "▸ Animal models and experimental test to study nociception and pain."

These methods are also frequently used for evaluation of central pain models. The following types of stimulations are used:
1. Cold (▸ acetone, cold plate, ice probe)
2. Heat (▸ light diode, infrared beam, ▸ hot plate, ▸ hot water) – probably less relevant than cold based on the clinical picture of central pain
3. Mechanical (▸ von Frey filaments, pinprick, soft brush, cotton swab, ▸ pressure)

The sites of stimulation are commonly:
1. Hindpaw
2. Tail
3. Trunk – particularly in thoracic spinal cord injury models (▸ at-level stimulation)

The compelling advantages of these methods are that they are quick to perform, can be done after limited training, require little equipment and facilities, and are reproducible and that extensive data for comparison has been published.

The three major disadvantages for the use of evoked spinal reflexes as a preclinical outcome measure in central pain models will be presented in the following:
1. *Paralysis/motor impairment and decreased sensory function may interfere with the ability to perform a withdrawal.*
2. *Increased spinal reflexes due to spastic syndrome.*

An injury to the central nervous system, particularly to the spinal cord, often results in symptoms of the spastic syndrome. The spastic syndrome includes a large range of positive symptoms and signs of the upper motor neuron lesion, such as ▸ spasticity, increased nociceptive flexor/withdrawal reflexes, flexor and extensor ▸ spasms, clonus, and dyssynergic movement patterns (Vierck et al. 1990).

In animals, spinal lesions and complete ▸ spinal cord transsection similarly lead to the development of the spastic syndrome (Nielsen et al. 2007), with facilitated spinal reflexes, expanded nociceptor receptive fields, and increased muscle tone, thus making a distinction between the spastic syndrome and the nociceptive reflex problematic (see also "▸ Opioids and Reflexes"). Unfortunately, only very few studies actually perform an assessment to verify the presence of spasms

or spasticity in central pain models. Furthermore, the assessment tests for the spastic syndrome are very similar to the ones used for the
assessment of hypersensitivity described
above (withdrawal, tail flick), which complicates the picture. It should also be noted that
symptoms of the spastic syndrome like
increased spinal reflexes may be subtle and
thus may not present itself as an obvious bias.

3. *Reflex response ensures no cerebral
   processing.*

Withdrawal reflexes like tail flick and
hindpaw withdrawal are regulated by segmentally organized spinal mechanisms and are
present in spinalized animals (Ghorpade and
Advokat 1994). The study of excitability
changes of spinal sensory and motor neurons
in the spinal cord can therefore be evaluated
with reflex-based assessment strategies, but
they do not ensure supraspinal processing
and are particularly problematic in models of
SCI due to the development of a hyperactive
reflex circuitry as part of the spastic syndrome
(Baastrup et al. 2010).

In summary, assays relying on spinal
reflexes like paw withdrawal and tail flick are
not well suited as outcome measures in preclinical central pain research due to risk of
spastic syndrome.

Supraspinal Behavior

One way of reducing the risk of bias from
the spastic syndrome and ensuring some level
of supraspinal processing is to substitute the
simple reflex responses (paw withdrawal and
▶ tail flick) with responses that require supraspinal
involvement, e.g., licking, guarding, ▶ struggling,
▶ vocalizing, jumping, and biting (Vierck 2006).
These responses to nociceptive input depend on
spinal-brainstem-spinal circuits and are present in
decerebrate animals (Woolf 1984).

In order to study nociceptive responses in
animals, appropriate behavioral measures that
engage supraspinal processes are required.
Measuring the supraspinal behaviors mentioned
above still does not necessarily ensure cerebral
processing. Behavioral outcomes that require
cerebral involvement rely on more complex

behavior, i.e., processing of sensory information,
decisions based on environmental cues, and
initiation of behavioral responses to nociceptive
stimuli (Vierck et al. 2008), and are thus more
likely to also measure and integrate the ▶ affect
perceived by the animal.

These assays include the operant escape and the
place escape/avoidance paradigm (PEAP), which
have both been used in models of central pain.

In the ▶ operant escape paradigm, the latencies and durations of an animal escape from
a compartment where the floor is heated or cooled
to a platform at neutral temperature in an
adjacent compartment are measured (Mauderli
et al. 2000). A similar paradigm is available
for testing thermal and mechanical sensitivity
of an animal's face (Nolan et al. 2011).

The place escape/avoidance paradigm (PEAP)
(LaBuda and Fuchs 2000; Baastrup et al. 2011)
relies on escape/avoidance learning to a novel
aversive environment. The animals are stimulated continuously with a von Frey filament in
the test side (black – natural preference) of
a biased box, and the time spent in the escape
side (white – natural avoidance) is measured.

Predictability to Clinical Efficacy

Even though a correlation between spontaneous
▶ below-level pain and evoked ▶ at-level pain in
▶ human spinal cord injury has been suggested
(Finnerup et al. 2003), the predictable validity of
preclinical evoked responses to human spontaneous pain remains unclear.

Measuring Spontaneous Pain-Like Behavior

Indicative Signs

Different strategies have been attempted in the
search for a preclinical spontaneous pain-like
signal. So far, general signs of welfare such as
weight, fur condition and grooming, open field
activity, and porphyrin production (red tears) are
used as indications of spontaneous pain.

The Next Step

The mice pain face, published by Langford et al.
in 2010, unfortunately does not apply to rodent
neuropathic pain models, i.e., central pain –
either due to lack of actual spontaneous pain in

these models or because chronic pain does not manifest itself in this way in animals. Furthermore, this is comparable to the fact that chronic pain in humans often is not visible either.

The first publication using a novel approach to capture perceived pain-like affect in an animal model of central pain (spinal cord injury), a modified ▶ conditioned place preference, was published by Davoody et al. in 2011. Administration of an analgesic in one compartment of a biased box compared to a reference compound in the reference compartment resulted in a preference for the analgesic compartment, and thus pain relief was rewarding for the animal (Davoody et al. 2011). Further investigation in these types of outcome measures will be of significant value for preclinical central pain evaluation in the future.

Pain-Related Behavior

Indirect measurements of ▶ pain-like distress in animals may possibly comprise measurements of common clinical pain ▶ comorbidities like ▶ anxiety- and ▶ depression-like behavior, which may be useful as supplementary information, but the role for such outcome measures in models of central pain is still unclear.

Possible Biases to Consider in Central Pain Models

- Motor and balance deficits, paralysis, musculoskeletal pain
- Bowl and bladder problems, visceral pain
- Face and construct validity of the available models

See also "▶ Animal Models and Experimental Tests to Study Nociception and Pain" (section "Discussion and Future Direction").

References

Baastrup, C., Maersk-Moller, C. C., Nyengaard, J. R., Jensen, T. S., & Finnerup, N. B. (2010). Spinal-, brainstem- and cerebrally mediated responses at- and below-level of a spinal cord contusion in rats: Evaluation of pain-like behavior. *Pain, 151,* 670–679.

Baastrup, C., Jensen, T. S., & Finnerup, N. B. (2011). Pregabalin attenuates place escape/avoidance behavior in a rat model of spinal cord injury. *Brain Research, 1370,* 129–135. Epub 2010 Nov 9, 129–135.

Davoody, L., Quiton, R. L., Lucas, J. M., Ji, Y., Keller, A., & Masri, R. (2011). Conditioned place preference reveals tonic pain in an animal model of central pain. *The Journal of Pain, 12,* 868–874.

Finnerup, N. B., Johannesen, I. L., Fuglsang-Frederiksen, A., Bach, F. W., & Jensen, T. S. (2003). Sensory function in spinal cord injury patients with and without central pain. *Brain, 126,* 57–70.

Ghorpade, A., & Advokat, C. (1994). Evidence of a role for N-methyl-D-aspartate (NMDA) receptors in the facilitation of tail withdrawal after spinal transection. *Pharmacology Biochemistry and Behavior, 48,* 175–181.

LaBuda, C. J., & Fuchs, P. N. (2000). A behavioral test paradigm to measure the aversive quality of inflammatory and neuropathic pain in rats. *Experimental Neurology, 163,* 490–494.

Langford, D. J., Bailey, A. L., Chanda, M. L., Clarke, S. E., Drummond, T. E., Echols, S., et al. (2010). Coding of facial expressions of pain in the laboratory mouse. *Nature Methods, 7,* 447–449.

Mauderli, A. P., costa-Rua, A., & Vierck, C. J. (2000). An operant assay of thermal pain in conscious, unrestrained rats. *Journal of Neuroscience Methods, 97,* 19–29.

Nielsen, J. B., Crone, C., & Hultborn, H. (2007). The spinal pathophysiology of spasticity–from a basic science point of view. *Acta Physiologica (Oxford, England), 189,* 171–180.

Nolan, T. A., Hester, J., Bokrand-Donatelli, Y., Caudle, R. M., & Neubert, J. K. (2011). Adaptation of a novel operant orofacial testing system to characterize both mechanical and thermal pain. *Behavioural Brain Research, 217,* 477–480.

Vierck, C. J. (2006). Animal models of pain. In S. McMahon & M. Koltzenburg (Eds.), *Wall and Melzack's textbook of pain* (5th ed., pp. 175–186). Philadelphia: Elsevier Churchill Livingstone.

Vierck, C. J., Greenspan, J. D., & Ritz, L. A. (1990). Long-term changes in purposive and reflexive responses to nociceptive stimulation following anterolateral chordotomy. *Journal of Neuroscience, 10,* 2077–2095.

Vierck, C. J., Hansson, P. T., & Yezierski, R. P. (2008). Clinical and pre-clinical pain assessment: are we measuring the same thing? *Pain, 135,* 7–10.

Woolf, C. J. (1984). Long term alterations in the excitability of the flexion reflex produced by peripheral tissue injury in the chronic decerebrate rat. *Pain, 18,* 325–343.

Central Pattern Generator

Definition

Cellular networks in the brainstem that are organized to initiate and maintain motor activity through pattern generation and rhythm generation.

Cross-References

▶ Orofacial Pain, Movement Disorders

Central Poststroke Pain

Henriette Klit
Danish Pain Research Center, Aarhus University
Hospital, Aarhus C, Denmark

Synonyms

CPSP; Dejerine-Roussy syndrome; Thalamic pain

Definition

Central poststroke pain is a ▶ central neuropathic pain condition following a stroke (cerebrovascular accident). Central poststroke pain (CPSP) is characterized by pain within the area of sensory abnormalities in the part of the body that is affected by the lesion in the central nervous system (CNS).

Introduction

Stroke, previously known as a cerebrovascular accident, is defined by WHO as a "…neurological deficit of cerebrovascular cause that persists beyond 24 hours." Stroke can be caused by both ischemic and hemorrhagic lesions of the central nervous system (CNS). The incidence of stroke rises with age. Stroke is the third leading cause of death in the Western world and a major cause of long-term disabilities, including paralysis, aphasia, cognitive dysfunction, and sensory deficits. Although less well known, pain can also be a consequence of stroke with reported prevalence rates varying between 11 % and 55 %, clearly demonstrating that poststroke pain is common in stroke survivors (Lundstrom et al. 2009; Jonsson et al. 2006;

Klit et al. 2011). The most common type of poststroke pain is musculoskeletal pain, which includes shoulder pain and pain due to spasticity, but also central neuropathic pain can be seen following stroke (Widar et al. 2002). Patients may present with several types of poststroke pain concomitantly.

Central poststroke pain is a central neuropathic pain condition caused by a stroke. Other common causes of central pain include multiple sclerosis and traumatic spinal cord injury. Although central pain is not as common following stroke as in these other conditions, stroke is still the leading cause of central pain due to the fact that stroke is a very prevalent disease (Boivie 2006b), with an estimated 1.1 million new stroke events per year and 6 million stroke survivors in Europe (Truelsen et al. 2006).

Central poststroke pain (CPSP) is characterized by pain within the area of sensory abnormalities in the part of the body that is affected by the lesion in the central nervous system (CNS).

Characteristics

History of CPSP

The concept of central pain was introduced by the German neurologist Ludwig Edinger in 1891 (Edinger 1891). One of the first descriptions of CPSP is from the French neurologist Dejerine and the Swiss-French neuropathologist Roussy in their famous paper from 1906: "Le Syndrome Thalamic" (Dejerine and Roussy 1906). The syndrome was previously known as the Dejerine-Roussy or thalamic pain syndrome, but since it has been shown that extra-thalamic lesions can also cause this pain condition, the term "CPSP" is preferred (Henry et al. 2007).

There are no pathognomonic symptoms or signs of CPSP, i.e., no symptoms or signs that can delineate CPSP from other poststroke pain conditions with certainty. There is no official definition of CPSP, but it can be defined as a central neuropathic pain condition, i.e., pain caused by a lesion or disease of the central somatosensory nervous system (IASP pain terminology), resulting from a stroke. We have

Central Poststroke Pain,
Fig. 1 Area of pain shown
in red in three patients with
central poststroke pain

suggested the following criteria for defining
CPSP (Klit et al. 2009):

Mandatory criteria:

- Pain within an area of the body corresponding
  to the lesion of the CNS.
- A history suggestive of a stroke and onset of
  pain at or after stroke onset.
- Confirmation of a CNS lesion by imaging and/
  or negative or positive sensory signs confined
  to the area of the body corresponding to the
  lesion of the CNS.
- Other causes of pain such as nociceptive or
  peripheral neuropathic have been excluded or
  considered highly unlikely.

  Supportive criteria:

- No primary relation to movement, inflammation, or other local tissue damage.
- Descriptors such as burning, painful cold,
  electric shocks, aching, pressing, stinging,
  and pins and needles, although all pain
  descriptors can apply.
- Allodynia or dysesthesia to touch.

Incidence/Prevalence

The incidence of CPSP is reported to be between
1 % and 11 % (Jonsson et al. 2006; Bowsher
2001). In two different prospective studies of
consecutively admitted stroke patients, 7–8 % of
the stroke survivors developed CPSP (Andersen
et al. 1995; Klit et al. 2011). CPSP can develop
after ischemic or hemorrhagic lesions anywhere
in the somatosensory part of the central nervous

system, i.e., after lesions in the brain stem, thalamus, thalamocortical projections, subcortical,
and possibly cortical areas (Kumar et al. 2009).
There is some evidence to suggest that the incidence of CPSP may be higher following lesions
in certain parts of the thalamus (ventral-posterior
parts) and the brain stem (lateral medulla,
Wallenberg's syndrome) (Lampl et al. 2002;
MacGowan et al. 1997). There are no uniform
signs indicating the origin of pain with respect to
onset, character, and intensity.

The onset of pain varies, but only a minority of
the patients has an acute onset, while most
patients develop pain within the first 6 months
after the stroke. In a few cases, very late onset has
been reported (Canavero and Bonicalzi 2007).

Location of Pain

The area of pain can vary in size. The most
common presentation is a hemibody distribution
with involvement of both the arm and leg on one
side and with or without involvement of the same
side of the trunk and/or the face (Fig. 1) (Klit et al.
2011). But the pain can also be located to smaller
areas such as the hand and the foot or just around
the eye. After lesions of the lateral medulla, some
patients may present with pain located to one side
of the face and the opposite side of the body.

Clinical Characteristics

CPSP shares the same characteristics as other
neuropathic pain conditions. There is often

a combination of so-called positive and negative symptoms and signs, i.e., gain of function, such as hypersensitivity, ▶ allodynia, or ▶ hyperalgesia; and loss of function, such as ▶ anesthesia or hyposensitivity. The description of the pain varies greatly between patients and often includes words such as burning, aching, pricking, lacerating, shooting, squeezing, throbbing, and painful pins and needles. The affective descriptors included troublesome, annoying, and tiring (Leijon et al. 1989; Boivie 2006a; Vestergaard et al. 1995).

The pain can be described as either superficial, deep, or both. The intensity of pain varies from slight to intolerable. The pain is often constant, but it can be intermittent. The pain can be both spontaneous and/or evoked. Evoked pain can be triggered by external or internal stimuli, such as touch, cold, movement, or stress (Bowsher 1996). The pain seems to be long standing, often lifelong, but at present, there are no good prospective studies to illustrate the natural history of the disease.

All patients present with some kind of sensory abnormality, and their pain is always located within an area of sensory abnormality. There are (almost) always findings of abnormal sensation to pain (pinprick) and/or temperature (Leijon and Boivie 1989; Vestergaard et al. 1995). The sensory processing of temperature and pain occurs via the spinothalamic tract (STT) and the spinotrigeminothalamic tracks projecting to the thalamus. Thus, there is an indication that lesions in these pathways are implicated in the pathophysiological mechanisms behind the development of CPSP. But many patients with lesions in the CNS but without central pain also have deficits in pain and/or temperature function. It is therefore often stated that a lesion of the STT is necessary but not sufficient to cause CPSP.

As in other neuropathic pain conditions, hypersensitivity including allodynia to, e.g., touch or cold is a common finding (Vestergaard et al. 1995), as are punctuate hyperalgesia and temporal summation (Kalita et al. 2011).

Imaging and Neurophysiology

PET studies have shown flow changes in the thalamus, with hypoactivity at rest and hyperactivity during evoked allodynia (Garcia-Larrea et al. 2006; Peyron et al. 2000). MRI and SPECT studies have also shown flow abnormalities with both hyper- and hypoperfusion in the thalamus and the parietal cortex. In a recent study using these techniques, the authors could not differentiate between patients with and without allodynia (Kalita et al. 2011). Neurophysiological studies using microelectrodes have shown increased abnormal spontaneous and evoked activity in the ventral-caudal nucleus of the thalamus of patients with central pain (Lenz et al. 1987; Hirayama et al. 1989).

Pathophysiology

The pathophysiology behind CPSP is not known, and we still do not understand why some patients develop CPSP and others do not. From the clinical findings, we may find some clues as to the underlying mechanisms. As already mentioned, the common findings of abnormal temperature and pain sensation indicate a lesion of the SST. The fact that the pain is always located within an area of sensory loss seems to indicate that deafferentation is necessary. Hypersensitivity phenomena, such as ▶ dysesthesia and ▶ allodynia, are common and could reflect disinhibition or central sensitization, while thalamic hyperactivity points to functional or plastic changes in the thalamus or other cortical regions.

Several theories about the possible mechanisms behind CPSP have been put toward. In 1911, Head and Holmes hypothesized that loss of STT input to the posterior lateral part of the thalamus causes disinhibition of the medial thalamus leading to pain. An elaboration of this theory is that loss of normal inhibition from the rapidly conducting "neospinothalamic" or lateral STT projections causes a disinhibition of the slowly conducting polysynaptic or paleospinoreticulothalamic or medial STT projections, thus resulting in pain (Casey et al. 1996; Tasker et al. 1991; Cesaro et al. 1991; Pagni 1989). Another theory states that central pain may be caused by hyperactive bursting in the thalamus caused by low-threshold calcium spikes due to deafferentation of the ascending pathways to the thalamus (Wang and Thompson 2008;

Central Poststroke Pain, Table 1 Class 1 randomised, double-blind placebo-controlled trials for treatment of central poststroke pain

Drug	Dosage/day	Outcome	Number of patients	Publication
Amitriptyline	75 mg	Positive	15	Leijon and Boivie (1989)
Carbamazepine	800 mg	Negative	14	Leijon and Boivie (1989)
Lamotrigine	200 mg	Positive	30	Vestergaard et al. (2001)
Pregabalin	300–600 mg	Positive	40 (mixed)	Vranken et al. (2008)
Pregabalin	Up to 600 mg	Negative	219	Kim et al. (2011)
Duloxetine	60–120 mg	Negative	48 (mixed CP)	Vranken et al. (2011)
Transdermal ketamine	50–75 mg	Negative	33 (mixed CP)	Vranken et al. (2005)
IV Morphine	9–30 mg	Negative	15 (mixed CP)	Attal et al. (2002)
IV Lidocaine	5 mg/kg	Positive	16 (mixed CP)	Attal et al. (2000)
IV Propofol	0.2 mg/kg	Positive	44 (mixed CP)	Canavero and Bonicalzi (2004)
IV Naloxone	8 mg	Negative	20	Bainton et al. (1992)

Weng et al. 2000). Craig (1998) has suggested that central pain is due to thermosensory disinhibition, where, in short, a lesion of the lateral cool-signaling spinothalamocortical projections to the thermosensory area of the insula through the posterior part of the ventral-medial nucleus (VMpo) causes disinhibition of a medial limbic network involving the parabrachial nucleus and the periaqueductal gray of the brain stem, the medial thalamus, and the ACC. In the dynamic reverberation theory by Canavero (1994), pain is hypothesized to be caused by a lesion of the STT by creating an imbalance in the normal oscillatory "dialogue" between the cortex and the thalamus.

Diagnosing CPSP

The diagnosis of CPSP is based on the pain and stroke history and a thorough clinical neurological examination with emphasis on the sensory findings. A pain drawing may be useful. In some cases, a quantitative sensory testing may be useful to confirm the sensory abnormalities. Neurophysiological investigations or imaging can be used in order to rule out other causes of pain.

Treatment

The treatment of CPSP is often difficult due to limited efficacy and dose-limiting side effects. Patients are often elderly and have other concomitant medical conditions that must be taken into account when planning the treatment, especially

with tricyclic antidepressants. The class 1 randomized, double-blinded, placebo-controlled trials in CPSP are summarized in Table 1. In summary, the only positive studies of oral treatment in CPSP are studies of amitriptyline, lamotrigine, and pregabalin, but recently a large negative trial of pregabalin has been published (Kim et al. 2011). According to the EFNS treatment guidelines for neuropathic pain (Attal et al. 2010), treatment of central neuropathic pain includes amitriptyline and gabapentin as first-line drugs, tramadol and stronger opioids as second- and third-line drugs, while lamotrigine may be considered for CPSP.

Nonpharmacological treatment strategies for treatment-resistant cases include stimulation therapy, including transcutaneous electrical nerve stimulation, repetitive transcranial magnetic stimulation (rTMS), motor cortex stimulation, spinal cord stimulation, and deep brain stimulation (Kim 2009), although there is limited evidence for these treatments. As in other chronic pain conditions, patients may benefit from other treatments including cognitive behavioral therapy, physiotherapy, and educational programs.

Summary

In conclusion, central poststroke pain is a central neuropathic pain condition caused by a stroke and characterized by pain within an area of sensory deficits corresponding to the lesion in the CNS. CPSP is seen in approximately 8 % of stroke survivors. Treatment of CPSP can be challenging.

References

Andersen, G., Vestergaard, K., Ingeman-Nielsen, M., & Jensen, T. S. (1995). Incidence of central post-stroke pain. *Pain, 61*, 187–193.

Attal, N., Gaude, V., Brasseur, L., Dupuy, M., Guirimand, F., Parker, F., et al. (2000). Intravenous lidocaine in central pain: A double-blind, placebo-controlled, psychophysical study. *Neurology, 54*, 564–574.

Attal, N., Guirimand, F., Brasseur, L., Gaude, V., Chauvin, M., & Bouhassira, D. (2002). Effects of IV morphine in central pain: A randomized placebo-controlled study. *Neurology, 58*, 554–563.

Attal, N., Cruccu, G., Baron, R., Haanpää, M., Hansson, P., Jensen, T. S., et al. (2010). EFNS guidelines on the pharmacological treatment of neuropathic pain: 2010 revision. *European Journal of Neurology, 17*, 1113–1e88.

Bainton, T., Fox, M., Bowsher, D., & Wells, C. (1992). A double-blind trial of naloxone in central post-stroke pain. *Pain, 48*, 159–162.

Boivie, J. (2006a). Central pain. In S. B. McMahon & M. Koltzenburg (Eds.), *Wall and Melzack's textbook of pain* (5th ed., pp. 1057–1074). Philadelphia: Elsevier/ Churchill Livingstone.

Boivie, J. (2006b). Central post-stroke pain. In F. Cervero & T. S. Jensen (Eds.), *Handbook of clinical neurology, 81*, (pp. 715–730). Edinboruogh: Elsevier B.V.

Bowsher, D. (1996). Central pain: Clinical and physiological characteristics. *Journal of Neurology, Neurosurgery, and Psychiatry, 61*, 62–69.

Bowsher, D. (2001). Stroke and central poststroke pain in an elderly population. *The Journal of Pain, 2*, 258–261.

Canavero, S. (1994). Dynamic reverberation. A unified mechanism for central and phantom pain. *Medical Hypotheses, 42*, 203–207.

Canavero, S., & Bonicalzi, V. (2004). Intravenous subhypnotic propofol in central pain: A double-blind, placebo-controlled, crossover study. *Clinical Neuropharmacology, 27*, 182–186.

Canavero, S., & Bonicalzi, V. (2007). Central pain syndrome: Elucidation of genesis and treatment. *Expert Review of Neurotherapeutics, 7*, 1485–1497.

Casey, K. L., Beydoun, A., Boivie, J., Sjolund, B., Holmgren, H., Leijon, G., et al. (1996). Laser-evoked cerebral potentials and sensory function in patients with central pain. *Pain, 64*, 485–491.

Cesaro, P., Mann, M. W., Moretti, J. L., Defer, G., Roualdes, B., Nguyen, J. P., et al. (1991). Central pain and thalamic hyperactivity: A single photon emission computerized tomographic study. *Pain, 47*, 329–336.

Craig, A. D. (1998). A new version of the thalamic disinhibition hypothesis of central pain. *Pain Forum, 7*, 1–14.

Dejerine, J., & Roussy, G. (1906). La syndrome thalamique. *Revue Neurologique (Paris), 14*, 521–532.

Edinger, L. (1891). Giebt es central antstehender Schmerzen? *Dtsch Z Nervenheilk, 1*, 262–282.

Garcia-Larrea, L., Maarrawi, J., Peyron, R., Costes, N., Mertens, P., Magnin, M., et al. (2006). On the relation between sensory deafferentation, pain and thalamic activity in Wallenberg's syndrome: A PET-scan study before and after motor cortex stimulation. *European Journal of Pain, 10*, 677–688.

Henry, J. L., Yashpal, K., & Lalloo, C. (2007). Central poststroke pain: A persective. In J. L. Henry, A. Panju, & K. Yashpal (Eds.), *Central neuropathic pain: Focus on poststroke pain* (pp. 3–5). Seattle: IASP Press.

Hirayama, T., Dostrovsky, J. O., Gorecki, J., Tasker, R. R., & Lenz, F. A. (1989). Recordings of abnormal activity in patients with deafferentation and central pain. *Stereotactic and Functional Neurosurgery, 52*, 120–126.

Jonsson, A. C., Lindgren, I., Hallstrom, B., Norrving, B., & Lindgren, A. (2006). Prevalence and intensity of pain after stroke: A population based study focusing on patients' perspectives. *Journal of Neurology, Neurosurgery, and Psychiatry, 77*, 590–595.

Kalita, J., Kumar, B., Misra, U. K., & Pradhan, P. K. (2011). Central post stroke pain: Clinical, MRI, and SPECT correlation. *Pain Medicine, 12*(2), 282–288.

Kim, J. S. (2009). Post-stroke pain'. *Expert Review of Neurotherapeutics, 9*, 711–721.

Kim, J. S., Bashford, G., Murphy, T. K., Martin, A., Dror, V., & Cheung, R. (2011). Safety and efficacy of pregabalin in patients with central post-stroke pain. *Pain, 152*, 1018–1023.

Klit, H., Finnerup, N. B., & Jensen, T. S. (2009). Central post-stroke pain: Clinical characteristics, pathophysiology, and management. *Lancet Neurology, 8*, 857–868.

Klit, H., Finnerup, N. B., Andersen, G., & Jensen, T. S. (2011). Central poststroke pain: A population-based study. *Pain, 152*, 818–824.

Kumar, B., Kalita, J., Kumar, G., & Misra, U. K. (2009). Central poststroke pain: A review of pathophysiology and treatment. *Anesthesia and Analgesia, 108*, 1645–1657.

Lampl, C., Yazdi, K., & Roper, C. (2002). Amitriptyline in the prophylaxis of central poststroke pain. Preliminary results of 39 patients in a placebo-controlled long-term study. *Stroke, 33*, 3030–3032.

Leijon, G., & Boivie, J. (1989). Central post-stroke pain–a controlled trial of amitriptyline and carbamazepine. *Pain, 36*, 27–36.

Leijon, G., Boivie, J., & Johansson, I. (1989). Central post-stroke pain–neurological symptoms and pain characteristics. *Pain, 36*, 13–25.

Lenz, F. A., Tasker, R. R., Dostrovsky, J. O., Kwan, H. C., Gorecki, J., Hirayama, T., et al. (1987). Abnormal single-unit activity recorded in the somatosensory thalamus of a quadriplegic patient with central pain. *Pain, 31,* 225–236.

Lundstrom, E., Smits, A., Terent, A., & Borg, J. (2009). Risk factors for stroke-related pain 1 year after first-ever stroke. *European Journal of Neurology, 16,* 188–193.

MacGowan, D. J., Janal, M. N., Clark, W. C., Wharton, R. N., Lazar, R. M., Sacco, R. L., et al. (1997). Central poststroke pain and Wallenberg's lateral medullary infarction: Frequency, character, and determinants in 63 patients. *Neurology, 49,* 120–125.

Pagni, C. A. (1989). Central pain due to spinal cord and brain stem damage. In P. D. Wall & R. Melzack (Eds.), *Textbook of pain* (2nd ed., pp. 634–655). Edinburgh: Churchill Livingstone.

Peyron, R., Garcia-Larrea, L., Gregoire, M. C., Convers, P., Richard, A., Lavenne, F., et al. (2000). Parietal and cingulate processes in central pain. A combined positron emission tomography (PET) and functional magnetic resonance imaging (fMRI) study of an unusual case. *Pain, 84,* 77–87.

Tasker, R. R., de Carvalho, G., & Dostrovsky, J. O. (1991). The history of central pain syndromes, with observations concerning pathophysiology and treatment. In K. L. Casey (Ed.), *Pain and central nervous system disease: The central pain syndromes* (1st ed., pp. 31–58). New York: Raven.

Truelsen, T., Piechowski-Jóźwiak, B., Bonita, R., Mathers, C., Bogousslavsky, J., & Boysen, G. (2006). Stroke incidence and prevalence in Europe: A review of available data. *European Journal of Neurology, 13,* 581–598.

Vestergaard, K., Nielsen, J., Andersen, G., Ingeman-Nielsen, M., Arendt-Nielsen, L., & Jensen, T. S. (1995). Sensory abnormalities in consecutive, unselected patients with central post-stroke pain. *Pain, 61,* 177–186.

Vestergaard, K., Andersen, G., Gottrup, H., Kristensen, B. T., & Jensen, T. S. (2001). Lamotrigine for central poststroke pain: A randomized controlled trial. *Neurology, 56,* 184–190.

Vranken, J. H., Dijkgraaf, M. G., Kruis, M. R., van Dasselaar, N. T., & van der Vegt, V. (2005). Iontophoretic administration of S(+)-ketamine in patients with intractable central pain: A placebo-controlled trial. *Pain, 118,* 224–231.

Vranken, J. H., Dijkgraaf, M. G., Kruis, M. R., van der Vegt, V., Hollmann, M. W., & Heesen, M. (2008). Pregabalin in patients with central neuropathic pain: A randomized, double-blind, placebo-controlled trial of a flexible-dose regimen. *Pain, 136,* 150–157.

Vranken, J. H., Hollmann, M. W., van der Vegt, M. H., Kruis, M. R., Heesen, M., Vos, K., et al. (2011). Duloxetine in patients with central neuropathic pain caused by spinal cord injury or stroke: A randomized, double-blind, placebo-controlled trial. *Pain, 152,* 267–273.

Wang, G., & Thompson, S. M. (2008). Maladaptive homeostatic plasticity in a rodent model of central pain syndrome: Thalamic hyperexcitability after spinothalamic tract lesions. *The Journal of Neuroscience, 28,* 11959–11969.

Weng, H. R., Lee, J. I., Lenz, F. A., Schwartz, A., Vierck, C., Rowland, L., et al. (2000). Functional plasticity in primate somatosensory thalamus following chronic lesion of the ventral lateral spinal cord. *Neuroscience, 101,* 393–401.

Widar, M., Samuelsson, L., Karlsson-Tivenius, S., & Ahlstrom, G. (2002). Long-term pain conditions after a stroke. *Journal of Rehabilitation Medicine, 34,* 165–170.

Central Sensitization

Definition

Central sensitization is an umbrella term for a number of phenomena, all of which are characterized by an increase in the responsiveness of nociceptive neurons in the central nervous system, best characterized in the spinal dorsal horn, to sensory stimulation. Central sensitization may be induced by conditioning noxious stimulation such as trauma, inflammation, nerve injury, or electrical stimulation of sensory nerves at C-fiber strength. It is considered to contribute to afferent-induced forms of hyperalgesia and allodynia. Proposed spinal mechanisms include enhanced strength at excitatory synapses in pain pathways, synaptic long-term potentiation, excessive primary afferent depolarization (PAD), and reduced inhibition.

Cross-References

▶ Arthritis Model, Kaolin-Carrageenan-Induced Arthritis (Knee)
▶ Cancer Pain Model, Bone Cancer Pain Model
▶ Central Changes After Peripheral Nerve Injury
▶ Chronic Pelvic Pain, Musculoskeletal Syndromes
▶ Drugs Targeting Voltage-Gated Sodium and Calcium Channels

▶ Exogenous Muscle Pain
▶ Formalin Test
▶ GABA Mechanisms and Descending
   Inhibitory Mechanisms
▶ Glial Cells in Orofacial Pathological Pain
   Mechanisms
▶ Gynecological Pain, Neural Mechanisms
▶ Long-Term Potentiation and Long-Term
   Depression in the Spinal Cord
▶ Metabotropic Glutamate Receptors in Spinal
   Nociceptive Processing
▶ Muscle Pain Model, Inflammatory Agents-
   Induced
▶ Nociceptive Processing in the Brain Stem
▶ Pain Mechanisms in Endometriosis
▶ Pain-Modulatory Systems, History of
   Discovery
▶ Postherpetic Neuralgia, Pharmacological and
   Nonpharmacological Treatment Options
▶ Postoperative Pain, Acute Neuropathic Pain
▶ Psychiatric Aspects of Pain and Dentistry
▶ Psychological Treatment of Headache
▶ Quantitative Sensory Testing
▶ Referred Muscle Pain, Assessment
▶ Restless Legs Syndrome
▶ Spinothalamic Neuron
▶ Spinothalamic Tract Neurons, Role of Nitric
   Oxide
▶ Transition from Acute to Chronic Pain

Central Sulcus

Synonyms

Rolandic sulcus

Definition

The convolutions of the cerebral cortex have
a general organization that is similar for all
humans. One constant and readily recognizable
sulcus is the central sulcus (of Rolando), and
marks the division between the frontal and parietal lobes.

Cross-References

▶ Motor Cortex, Effect on Pain-Related
   Behavior

Central Trigger Point

Synonyms

CTrP

Definition

Clinically, central trigger point is characteristically a very tender, circumscribed nodule-like
spot in the mid-portion of a palpable taut band
of skeletal muscle fibers, and it usually refers pain
when compressed. This trigger point may be
active or latent and can induce attachment trigger
points.

Cross-References

▶ Myofascial Trigger Points

Central Venous Thrombosis

▶ Headache Due to Venous Sinus Thrombosis

Centralization

▶ Central Changes After Peripheral Nerve Injury

Central-Peripheral Distal Axonopathy

▶ Distal Axonopathy

Centrifugal Control of Nociceptive Processing

▶ GABA Mechanisms and Descending Inhibitory Mechanisms
▶ Spinothalamic Tract Neurons, Descending Control by Brain Stem Neurons

Centrifugal Control of Sensory Inputs

Definition

The central nervous system regulates access of sensory information by neural pathways that inhibit or facilitate sensory processing. Such regulatory pathways can be intrinsic to the spinal cord (and trigeminal nuclei) or can originate from a variety of structures in the brain.

Cross-References

▶ Spinothalamic Tract Neurons, Descending Control by Brain Stem Neurons

Cephalalgia

▶ Headache

Ceramide

Definition

An intracellular signaling molecule liberated by activation of the sphingomyelin pathway. This pathway is activated by NGF via its action on the p75 receptor.

Cross-References

▶ Nerve Growth Factor, Sensitizing Action on Nociceptors

Cerebellum

Definition

The cerebellum is located dorsal to the brainstem and pons, and inferior to the occipital lobe. It mainly serves sensory-motor integration, by providing constant feedback signals to adapt fine-tune movements according to momentary muscle length and tone and body posture.

Cross-References

▶ Functional Imaging of Cutaneous Pain

Cerebral Cortex

Definition

The cerebral cortex is the thin, convoluted surface layer of nerve cell bodies (also called gray matter) of the cerebral hemispheres responsible for receiving and analyzing sensory information, for the execution of voluntary muscle movement, thought, reasoning, and memory.

Cross-References

▶ Cingulate Cortex, Functional Imaging
▶ Clinical Migraine with Aura
▶ Descending Circuitry, Transmitters and Receptors
▶ PET and fMRI Imaging in Parietal Cortex (SI, SII, Inferior Parietal Cortex BA40)

Cerebral Venous Thrombosis

▶ Headache Due to Venous Sinus Thrombosis

Cerebrospinal Fluid

Synonyms

CSF

Definition

Fluid within the four brain ventricles, mainly produced by the choroid plexus. The average pressure (in lateral recumbent position) is 150–250 mmH$_2$O, depending on CSF secretion and absorption, intracranial arterial and venous pressure, hydrostatic pressure, brain bulk, and status of surrounding coverings.

Cross-References

▶ Cancer Pain Management, Anesthesiologic Interventions, and Neural Blockade
▶ Headache due to Low Cerebrospinal Fluid Pressure

Cervical Discogram

▶ Cervical Discography

Cervical Discography

David Diamant
Neurological and Spinal Surgery, Lincoln, NE, USA

Synonyms

Cervical discogram; Provocation discogram; Provocative discography

Definition

Cervical discography is a diagnostic procedure designed to determine if a cervical intervertebral disc is the source of a patient's neck pain. It involves injecting contrast medium into the disc in an attempt to reproduce the patient's pain.

Characteristics

Principles

The cervical intervertebral discs are innervated by nociceptive fibers from the cervical sinuvertebral nerves, the vertebral nerves, and the cervical sympathetic trunks (Bogduk et al. 1989; Groen et al. 1990; Mendel et al. 1992). Being endowed with a nerve supply, the cervical discs are potentially a source of neck pain.

There are no conventional means by which to determine if a patient's neck pain arises from a cervical intervertebral disc. There are no signs on ▶ musculoskeletal examination by which this can be established and no signs on medical imaging. Abnormalities evident on ▶ magnetic resonance imaging (MRI) correlate poorly with whether the disc is painful or not (Parfenchuck and Janssen 1994; Schellhas et al. 1996). Furthermore, fissures that may be present across the posterior aspect of cervical discs are a normal age change (Oda et al. 1988) and do not constitute a painful lesion (Parfenchuck and Janssen 1994; Schellhas et al. 1996).

Provocation discography is the only means by which to test if a cervical disc is painful or not. The procedure involves injecting contrast medium into the nucleus pulposus of the disc, in an effort to reproduce the patient's pain. Although the contrast medium outlines the internal structure of the disc, this is not seminal to the diagnosis. The critical component of the procedure is reproduction of the patient's pain.

Studies in normal volunteers and in patients have demonstrated that cervical discs can produce neck pain, under experimental conditions (Schellhas et al. 1996; Cloward 1959; Grubb and Kelly 2000). Referred pain patterns encompass areas that are topographically separated from the site of pathology. Furthermore, discs at particular segmental levels produce pain in fairly consistent regions (Schellhas et al. 1996; Grubb and Kelly 2000) (Fig. 1). The C2–3 disc typically refers to the occiput. The C3–4 disc typically refers to the area of the C7 spinous process, with spread toward the side of the neck. The C4–5 disc typically refers toward the superior angle of the scapula but may spread from the base of the neck to the top of the shoulder. The C5–6 disc refers to the center of the scapular border, and the C6–7 disc refers to the

inferior angle of the scapula, but both may spread over the entire scapula, across the shoulder, and into the proximal upper limb. These pain patterns can be used to plan which segmental levels should be targeted for investigation.

If stimulation of the disc reproduces the patient's pain pattern (concordant pain), it may be presumed that such is their pain generator. If stimulating the disc is not painful, or produces an atypical (non-concordant) pain pattern, this disc is presumed not to be the pain generator.

Technique

The patient lies supine and the neck is prepared for an aseptic procedure. The operator inserts a needle through the skin of the neck and into the anterolateral aspect of the target disc, until it reaches the center of the nucleus (Fig. 2). Thereupon, contrast medium is injected, both to verify correct placement (Fig. 3) and to test for reproduction of pain. The nucleus pulposus of a typical cervical intervertebral disc will admit 0.2–0.4 cc of injectate (Kambin et al. 1980). Whether the patient develops concordant pain or not is the critical component of the procedure.

Validity

In order to be valid, the International Association for the Study of Pain (Merskey and Bogduk 1994) recommends that cervical discography be subjected to anatomical controls. Not only does provocation

Cervical Discography, Fig. 1 Patterns of distribution of pain after stimulation of cervical intervertebral discs at the segments indicated (Reproduced courtesy of the International Spinal Intervention Society)

Cervical Discography, Fig. 2 Radiographs showing needles placed into cervical intervertebral discs in preparation for discography. (**a**) Anterior view. (**b**) Lateral view (Reproduced courtesy of the International Spinal Intervention Society)

Cervical Discography, Fig. 3 Radiographs of cervical discography after injecting of contrast medium. (**a**) Anterior view. (**b**) Lateral view (Reproduced courtesy of the International Spinal Intervention Society)

of the intervertebral disc need to reproduce the individual's pain concordantly, but also provocation at adjacent levels must not reproduce such pain. Additional criteria have been recommended by the International Spinal Intervention Society (2004) pertaining to the intensity of the pain produced during disc stimulation and the potential role of other structures in the generation of pain.

Cervical discs in normal, asymptomatic volunteers can be made to hurt by discography (Schellhas et al. 1996). However, in such individuals the evoked pain is not severe. They rate the pain as typically less than 6 on a 10-point scale. In contrast, patients with disc pain report reproduction of moderate or severe pain, which they typically rate as greater than seven (Schellhas et al. 1996). Accordingly, it is recommended that for cervical disc stimulation to be considered positive, the evoked pain must have an intensity of seven or greater (International Spinal Intervention Society 2004). This criterion serves to prevent minor pain from an asymptomatic disc, being considered positive.

Provocation of the intervertebral disc may elicit concordant pain, yet other sources of such pain can be the actual pain generator. As such, the issue of diagnostic specificity of this procedure is questionable. As they share a similar segmental innervation, the cervical zygapophysial joints can refer pain to similar regions as the intervertebral discs (Bogduk and Aprill 1993). Consequently, cervical discography can be false-positive in patients whose pain originates from the zygapophysial joints at the same segment as the disc stimulated. Some 40 %

of patients with positive responses to discography have their pain relieved by blocks of the cervical zygapophysial joints (see ▶ Cervical Medial Branch Blocks), which is not compatible with the disc being the primary source of their pain. Accordingly, it has been recommended that cervical discography be undertaken only when the cervical zygapophysial joints, at the areas of concern, have been excluded as the source of the patient's pain (International Spinal Intervention Society 2004).

Additionally, there are other pitfalls that may compromise the validity of cervical discography. Technical errors will compromise the diagnostic validity of cervical discography. The needle tip must be in the nucleus pulposus; otherwise, stimulation of the anulus fibrosus will be likely to yield a painful response, regardless of whether that disc is the pain generator or not.

Applications

The primary purpose of cervical discography is to determine if cervical discs are the source of patient's neck pain. It is indicated, therefore, in patients whose cause of pain cannot be established by other means, who have not benefited from conservative therapy, and for whom a diagnosis is desired or required.

A secondary purpose of cervical discography is to help physicians plan interventional management. One treatment option is anterior cervical discectomy and fusion. For this procedure, surgeons are interested not only if a disc hurts but also if other discs hurt. The greater the numbers of discs that appear painful, the less inclined surgeons are to undertake surgery. Not only is multi-level fusion more technically demanding for the surgeons, and more hazardous to the patient, its outcomes are less favorable than fusion at a single level.

Utility

Cervical discography was originally developed with the prospect of finding one or perhaps only two discs that were painful so that fusion might be undertaken to relieve the patient's pain. Accordingly, cervical discography was expected to have positive predictive value. Subsequent studies, however, have thwarted this aspiration.

It has become evident that cervical discs are infrequently symptomatic at single levels (Grubb and Kelly 2000). More commonly, discs at three levels, and even four or more levels, are symptomatic. This pattern essentially precludes surgical therapy. Consequently, in practice, cervical discography has more of a negative predictive value. It serves far more often to prevent surgery than to encourage it. Indeed, in one series, only 10 % of patients proceeded to surgery in the light of their responses to cervical discography (Grubb and Kelly 2000).

References

Bogduk, N., & April, C. (1993). On the nature of neck pain, discography and cervical zygapophysial joint blocks. *Pain, 54*, 213–217.

Bogduk, N., Windsor, M., & Inglis, A. (1989). The innervation of the cervical intervertebral discs. *Spine, 13*, 2–8.

Cloward, R. B. (1959). Cervical diskography. A contribution to the aetiology and mechanism of neck, shoulder and arm pain. *Annals of Surgery, 130*, 1052–1064.

Groen, G. J., Baljet, B., & Drukker, J. (1990). Nerves and nerve plexuses of the human vertebral column. *American Journal of Anatomy, 188*, 282–296.

Grubb, S. A., & Kelly, C. K. (2000). Cervical discography: Clinical implications from 12 years of experience. *Spine, 25*, 1382–1389.

International Spinal Intervention Society. (2004). Cervical discography. In N. Bogduk (Ed.), *Practice guidelines for spinal diagnostic and treatment procedures*. San Francisco: Author.

Kambin, P., Abda, S., & Kurpicki, F. (1980). Intradiskal pressure and volume recording: Evaluation of normal and abnormal cervical disks. *Clinical Orthopaedics, 146*, 144–147.

Mendel, T., Wink, C. S., & Zimny, M. L. (1992). Neural elements in human cervical intervertebral discs. *Spine, 17*, 132–135.

Merskey, H., & Bogduk, N. (1994). *Classification of chronic pain. Descriptions of chronic pain syndromes and definition of pain terms* (2nd ed., p. 108). Seattle: IASP Press.

Oda, J., Tanaka, H., & Tsuzuki, N. (1988). Intervertebral disc changes with aging of human cervical vertebra from the neonate to the eighties. *Spine, 13*, 1205–1211.

Parfenchuck, T. A., & Janssen, M. E. (1994). A correlation of cervical magnetic resonance imaging and discography/computed tomographic discograms. *Spine, 19*, 2819–2825.

Schellhas, K. P., Smith, M. D., Gundry, C. R., & Pollei, S. R. (1996). Cervical discogenic pain. prospective correlation of magnetic resonance imaging and discography in asymptomatic subjects and pain sufferers. *Spine, 21*, 300–312.

Cervical Facet Blocks

▶ Cervical Medial Branch Blocks

Cervical Facet Denervation

▶ Cervical Medial Branch Neurotomy

Cervical MBBs

▶ Cervical Medial Branch Blocks

Cervical Medial Branch Blocks

Jayantilal Govind[1] and Nikolai Bogduk[2]
[1]Pain Clinic, Department of Anaesthesia, University of New South Wales, Sydney, NSW, Australia
[2]University of Newcastle, Newcastle Bone & Joint Institute, Royal Newcastle Centre, Newcastle, NSW, Australia

Synonyms

Cervical facet blocks; Cervical MBBs; Cervical zygapophyseal joint blocks; Z joint blocks

Jayantilal Govind: deceased.

Definition

Cervical medial branch blocks (MBBs) are a diagnostic test to determine if a patient's neck pain is mediated by one or more of the medial branches of the cervical dorsal rami. This is achieved by anesthetizing the target nerve with a minute volume of local anesthetic. In the absence of evidence to the contrary, a positive response to MBBs implies that the patient's pain stems from the zygapophyseal joint innervated by the nerves anesthetized.

Characteristics

Rationale

At typical cervical levels, the medial branches of the cervical dorsal rami pass across the waist of the ipsisegmental articular pillar (Fig. 1). They innervate the zygapophyseal joints above and below, before supplying the posterior muscles of the neck (Bogduk 1982). The third occipital nerve and the C7 medial branch cross the joint that they supply.

The zygapophyseal ("Z") joints are the only structures supplied by these nerves that are affected by disorders that can be a source of chronic pain (Barnsley and Bogduk 1993). These disorders cannot be diagnosed by musculoskeletal examination or by medical imaging. Diagnostic blocks are the only validated means by which the Z joints can be implicated or excluded as the source of pain. In order to anesthetize a given joint, both nerves that innervate it must be blocked.

Technique

The blocks are performed with the patient lying in a comfortable position, prone, supine, or on their side. Under fluoroscopic guidance, a fine needle is inserted through the skin and muscles of the neck and onto the articular pillar where the target nerve lies (Fig. 2). The nerve can be anesthetized with as little as 0.3 ml of local anesthetic.

Principles

The primary objective of cervical MBBs is to establish if anesthetizing the target nerves

Cervical Medial Branch Blocks, Fig. 1 A lateral radiograph of the cervical spine. The course of the third occipital nerve and the medial branches of the *C3* to *C7* dorsal rami across the articular pillars is indicated with *dotted lines*

Cervical Medial Branch Blocks, Fig. 2 A lateral radiograph of a cervical spine, showing a needle in position for a *C5* medial branch block

relieves the patient's pain. If the pain is not relieved, the targeted joint can be excluded as the source of pain and a new source considered, such as a joint at another segmental level or an intervertebral disk. If pain is relieved, the response constitutes prima facie evidence that the targeted nerves are mediating the patient's pain and that it arises from the joint that they supply.

In order to be positive, the blocks must produce complete relief of pain. Partial reduction of pain does not constitute a positive response.

In some patients, however, their pain may arise from more than one joint. They may experience pain from both joints at the same segment, from consecutive joints on the same side, or from joints at separate and displaced segments. Typical patterns are C5–6 on both sides, C5–6 and C6–7 ipsilaterally, and C5–6 and C2–3 on the same side.

In such patients, anesthetizing one joint will not relieve all of their pain. However, blocking that joint will completely relieve pain in the particular region to which that joint refers its pain. Similarly, blocking the other joint will relieve pain in the remaining area. The subtlety of the diagnostic criterion is that the patient obtains complete relief of pain in a particular topographical distribution. This is not the same as the patient obtaining partial relief of their pain overall (International Spine Intervention Society 2004).

Validity

Cervical MBBs are target specific, and the use of minute volumes of local anesthetic precludes other structures being anesthetized (Barnsley and Bogduk 1993). To avoid false-positive responses, the blocks must be controlled. Single diagnostic blocks are associated with an unacceptably high rate of false-positive responses (Barnsley et al. 1993a). Although placebo controls can be used, these may not be practical under conventional circumstances, but comparative local anesthetic blocks can be used (International Spine Intervention Society 2004; Barnsley et al. 1993b; Lord et al. 1995).

On separate occasions, the same block is repeated using local anesthetics with different durations of action. The test is negative if the repeat block fails to relieve pain. If both blocks

relieve the pain, the relief may be concordant or discordant and can be either prolonged or not (Barnsley et al. 1993b).

Concordant relief is a short-lasting relief when a short-acting agent is used and a long-lasting relief when a long-acting agent is used. When relief is longer after the short-acting agent is used, the response is classed as discordant. If the relief substantially outlasts the expected duration of action of either agent, the response is classed as prolonged. Discordant and prolonged responses are probably due to local anesthetics, particularly lignocaine, acting on "open" sodium channels (Butterworth and Strichartz 1990).

Concordant responses are the ideal. They have a sensitivity of 54 % and specificity of 88 % (Lord et al. 1995). The high specificity means that concordant responses are very unlikely to be false. The low sensitivity, however, means that not all patients with zygapophyseal joint pain will be detected.

If discordant responses are accepted as positive, the sensitivity rises to 100 %, but the specificity drops to 65 % (Lord et al. 1995). Thus, all patients with zygapophyseal joint pain will be detected, but some will be false-positive.

Epidemiology

The zygapophyseal joints are the single most common source of chronic neck pain. They are the source of pain in at least 50 % (Barnsley et al. 1995; Lord et al. 1996a) and up to 88 % (Gibson et al. 2000) of patients with neck pain after whiplash. In 53 % of patients with headache after whiplash, the pain can be traced to the C2–3 joint (Lord et al. 1994).

Indications

Neck pain for which a diagnosis is required is the primary indication for cervical MBBs. To date, they have been used only for the investigation of patients with chronic neck pain, but their judicious application in patients with subacute pain would be worthy of exploration. Isolating the source of pain and instituting appropriate treatment expeditiously could serve to prevent chronicity.

Patient Selection

Studies in normal volunteers (Dwyer et al. 1990) and in patients (Fukui et al. 1996) have shown that the cervical zygapophyseal joints generate somatic referred pain in characteristic regions, specific to the segmental location of the joint stimulated (Fig. 3). These patterns can be used to select the joints and nerves most likely to respond to blocks (International Spine Intervention Society 2004; Aprill et al. 1990).

To optimize their efficiency, MBBs should be performed in patients with discrete areas of neck pain, which correspond to one or more of these areas of referred pain. Patients with more diffuse patterns of pain are less likely to have identifiable joints as the source of pain.

Contraindications

Absolute contraindications include localized or systemic infection, and possible pregnancy. Relative contraindications may include an allergy to contrast media or local anesthetics, the concurrent treatment with nonsteroidal anti-inflammatory medication including aspirin, and neurological signs. ▶ Radicular pain and chronic neck pain may coexist. While cervical medial branch blocks may alleviate neck pain and any ▶ somatic referred pain, they will not relieve radicular pain.

Evaluation

The value of diagnostic blocks lies in the information that they provide (International Spine Intervention Society 2004). For the blocks to be valid, the patient must be able to cooperate fully. They must understand that the procedure is a diagnostic one, and it is neither designed nor intended to be therapeutic. When multiple sources of pain are suspected, patients should understand that relief may occur in only a particular topographic area; and they must be able to determine if they obtain relief in a discrete area. They should also understand the use of a visual analogue scale or numerical pain rating scale.

Blocks should not be performed when the patient's pain is minimal, lest they be unable to distinguish the effects of a block from natural

Cervical Medial Branch Blocks, Fig. 3 A map of the referred pain patterns from cervical zygapophyseal joints at the segments indicated

fluctuations in pain. Cervical MBBs are not recommended in patients whose typical pain is less than 40 on a 100 mm scale.

The response to diagnostic blocks should be evaluated immediately after the procedure and for some time afterwards, at the location at which the block was performed, and by an independent observer using validated and objective instruments and tools. Doing so avoids potential errors such as observer bias, patient or operator's expectations, and recall bias. For a response to be judged positive, relief of pain should be accompanied by restoration of activities that are normally limited by pain.

Utility

Cervical MBBs have diagnostic utility, in that they can pinpoint the source of the patient's pain. Establishing a firm diagnosis prevents the futile pursuit of a diagnosis by other means. MBBs also have therapeutic utility. A positive response to blocks predicts a favorable outcome from ▶ radiofrequency neurotomy (Lord et al. 1996b).

Acknowledgement Dr. Jayantilal Govind passed away on June 20, 2009 at age 68. We thank Prof. Nik Bogduk for the update of his entry.

References

Aprill, C., Dwyer, A., & Bogduk, N. (1990). Cervical zygapophyseal joint pain patterns – II: A clinical evaluation. *Spine, 15*, 458–461.

Barnsley, L., & Bogduk, N. (1993). Medial branch blocks are specific for the diagnosis of cervical zygapophyseal joint pain. *Regional-Anaesthesie, 18*, 343–350.

Barnsley, L., Lord, S., Wallis, B., & Bogduk, N. (1993a). False-positive rates of cervical zygapophysial joint blocks. *The Clinical Journal of Pain, 9*, 124–130.

Barnsley, L., Lord, S., & Bogduk, N. (1993b). Comparative local anaesthetic blocks in the diagnosis of cervical zygapophyseal joint. *Pain, 55*, 99–106.

Barnsley, L., Lord, S. M., Wallis, B. J., & Bogduk, N. (1995). The prevalence of chronic cervical zygapophyseal joint pain after whiplash. *Spine, 20*, 20–26.

Bogduk, N. (1982). The clinical anatomy of the cervical dorsal rami. *Spine, 7*, 319–330.

Butterworth, J. F., & Strichartz, G. R. (1990). Molecular mechanisms of local anesthesia: A review. *Anesthesiology, 72*, 711–734.

Dwyer, A., Aprill, C., & Bogduk, N. (1990). Cervical zygapophyseal joint pain patterns – I: A study in normal volunteers. *Spine, 15*, 453–457.

Fukui, S., Ohseto, K., Shiotani, M., et al. (1996). Referred pain distribution of the cervical zygapophyseal joints and the cervical dorsal rami. *Pain, 68*, 79–83.

Gibson, T., Bogduk, N., MacPherson, J., & McIntosh, A. (2000). Crash characteristics of whiplash associated chronic neck pain. *Journal of Musculoskeletal Pain, 8*, 87–95.

International Spine Intervention Society. (2004). Cervical medial branch blocks. In N. Bogduk (Ed.), *Practice guidelines for spinal diagnostic and treatment procedures*. San Francisco: International Spine Intervention Society.

Lord, S., Barnsley, L., Wallis, B., & Bogduk, N. (1994). Third occipital nerve headache: A prevalent study. *Journal of Neurology, Neurosurgery & Psychiatry, 57*, 1187–1190.

Lord, S. M., Barnsley, L., & Bogduk, N. (1995). The utility of comparative local anaesthetic blocks versus placebo controlled blocks for the diagnosis of cervical zygapophyseal joint pain. *The Clinical Journal of Pain, 11*, 208–213.

Lord, S., Barnsley, L., Wallis, B. J., & Bogduk, N. (1996a). Chronic cervical zygapophyseal joint pain after whiplash: Placebo controlled prevalence study. *Spine, 21*, 1737–1745.

Lord, S. M., Barnsley, L., Wallis, B. J., McDonald, G. J., & Bogduk, N. (1996b). Percutaneous radiofrequency neurotomy for chronic cervical zygapophyseal joint pain. *The New England Journal of Medicine, 335*, 1721–1726.

Cervical Medial Branch Neurotomy

Jayantilal Govind
Pain Clinic, Department of Anaesthesia,
University of New South Wales, Sydney,
NSW, Australia

Synonyms

Cervical facet denervation; Cervical radiofrequency neurotomy

Definition

Cervical medial branch neurotomy is a treatment for neck pain or headache, stemming from one or more of the zygapophysial joints of the cervical spine. It involves coagulating the nerves that innervate the painful joint, or joints, with an electrode inserted onto the nerves through the skin and muscles of the back of the neck.

Characteristics

Mechanism

Medial branch neurotomy achieves relief of pain by interrupting the transmission of nociceptive information from a painful zygapophysial joint, by generating a heat lesion in the nerves that mediate the pain (see ▶ Radiofrequency Neurotomy, Electrophysiological Principles).

Indications

Cervical medial branch neurotomy is not a treatment for any form of neck pain. It is explicitly and solely designed to relieve pain from the zygapophysial joints. Therefore, the singular indication for the procedure is complete, or near complete, relief of pain following controlled, ▶ diagnostic blocks of the nerves from the painful joint or joints, i.e., ▶ cervical medial branch blocks. These blocks must be controlled, because the false-positive rate of uncontrolled blocks is such that responses to single blocks will be false in up to 30 % of patients, and those patients will not benefit from the denervation procedure (Barnsley et al. 1993; Bogduk and Holmes 2000).

Technique

A detailed protocol has been produced by the International Spinal Injection Society (International Spinal Intervention Society 2004). In essence, the procedure is performed under local anesthesia, in a room equipped with a fluoroscope and the necessary equipment to generate the lesions. Sedation should be avoided so that the patient can be alert to any problems that might occur during the procedure and which threaten their safety. The objective is to make a lesion as long as possible along the course of the target nerve. In order to capture the curved course of each cervical medial branch, the electrode must be introduced in both of two ways (Fig. 1). An oblique pass, about 30° lateral from the sagittal plane, is used to reach the proximal part of the nerve, where it lays anterolaterally to the articular pillar (or anterolaterally to the C2/3 zygapophysial joint, in the case of the third occipital nerve). The second pass is along the sagittal plane to reach the nerve where it lies laterally to the articular pillar.

Along both the oblique pass and the sagittal pass, the electrode is introduced through the skin and muscles of the posterior neck, so that its tip lies parallel to the target nerve and against the articular pillar (or the C2–3 joint). At typical cervical levels, this requires inserting the electrode upwards from below the target level, for the nerves course downwards as well as backwards (Fig. 2). The third occipital nerve runs transversely and so can be approached along a transverse plane instead of an inclined one (Fig. 3).

At each target point, two or three lesions need to be made, in order to accommodate possible variations in the course of the nerve (International Spinal Intervention Society 2004; Lord et al. 1998). Each lesion is made by increasing the heating current gradually by about 1 °C per second. Raising the temperature slowly provides time for both the patient and

Cervical Medial Branch Neurotomy, Fig. 1 A sketch of a top view of the course of a cervical medial branch. As the nerve follows a curved path around the articular pillar, electrodes must be introduced along a sagittal plane and along a 30° oblique plane

Cervical Medial Branch Neurotomy, Fig. 2 Lateral radiograph of an electrode, inserted along a sagittal path, in position to coagulate a *C5* medial branch. The course of the nerve is depicted by a *dotted line*

the physician to react if any untoward sensations arise, either because the electrode has dislodged or because the target site has not been adequately anesthetized, and for the physician to respond before any injury occurs to the patient. Once a temperature of 80–85 °C has been achieved, it is maintained for about 90 s to ensure adequate coagulation of the nerve.

Cervical Medial Branch Neurotomy, Fig. 3 Lateral radiograph of an electrode, inserted along a sagittal path, in position to coagulate the third occipital nerve. The course of the nerve is depicted by a *dotted line*

The procedure is repeated for all nerves that were anesthetized, in order to produce relief of pain during the prior conduct of diagnostic cervical medial branch blocks.

Variants

The optimal technique requires that the electrode be placed parallel to the target nerves and that multiple lesions be made to accommodate variations in the course of the nerve (International Spinal Intervention Society 2004; Lord et al. 1998; Lord et al. 1996a; McDonald et al. 1999). Only this technique has been tested and shown to produce complete and lasting relief of pain.

Variants are used by some operators, ostensibly in the belief that the procedure is faster or easier. These variants, however, have not been tested and their efficacy is not known (Bogduk 2002).

Efficacy

A controlled trial has shown that the effects of cervical medial branch neurotomy cannot be attributed to a placebo response (Lord et al. 1996a). When correctly performed, the efficacy of cervical medial branch neurotomy is genuine.

Provided that patients are correctly selected using controlled cervical medial branch blocks, and provided that the optimal technique is used,

good outcome can be expected from cervical medial branch neurotomy. All patients should obtain relief of their pain, which should be evident as soon as the anesthesia for the procedure wears off and any postoperative pain abates. If such immediate relief is not evident, an error will have occurred either during the diagnostic blocks or in the conduct of the procedure, both of which should then be reviewed.

When neck pain is the target complaint, and joints below C2/3 are treated, some 70 % of patients obtain complete relief of their pain (Lord et al. 1998; Lord et al. 1996a; McDonald et al. 1999). When headache is the target complaint, and the third occipital nerve is targeted, some 85 % of patients obtain complete relief (Govind et al. 2003). The cardinal reasons for initial failures are suboptimal placement of electrodes or failure, during the conduct of diagnostic blocks, to recognize a source of pain from an adjacent joint.

If complete relief of pain is achieved, it is attended by restoration of activities of normal living and no need for other health care for the pain (Lord et al. 1996a, 1998; McDonald et al. 1999; Govind et al. 2003). Furthermore, it is associated with complete resolution of psychological distress, without any need for psychological treatment (Wallis et al. 1997). The procedure is equally effective in patients who have compensation claims as those who do not (McDonald et al. 1999; Bogduk 2002; Govind et al. 2003; Sapir and Gorup 2001).

Relief of pain is not permanent. In time, the coagulated nerves regenerate and may again transmit nociceptive information from the painful joint or joints. The time that it takes for this regeneration to occur depends on how accurately and how thoroughly the nerves were coagulated. After medial branch neurotomy at typical cervical levels, patients can expect complete relief of pain for at least 9 months and up to 12 months or longer (Lord et al. 1996a, 1998 ; McDonald et al. 1999). The duration of relief after third occipital neurotomy for headache is, on the average, slightly shorter; some patients may maintain relief for only 6 months, although durations longer than 12 months have been reported (Govind et al. 2003).

Pain usually returns gradually, although it may not return to its former intensity. If pain recurs and becomes sufficiently intense again as to warrant treatment, cervical medial branch neurotomy can be repeated in order to reinstate relief. Patients can have multiple repetitions without prejudicing their response (Lord et al. 1998; McDonald et al. 1999; Govind et al. 2003).

Complications

Provided that the correct technique is used, no complications are associated with this procedure. Side effects are uncommon when the procedure is performed at typical cervical levels (Lord et al. 1998) but are more common when the third occipital nerve is coagulated. They include dysesthesiae when medial branches with cutaneous distributions are coagulated and ataxia when the third occipital nerve is coagulated. The dysesthesiae are self-limiting and do not require treatment, as a rule. The ataxia is accommodated by the patient relying on visual cues for balance and is readily tolerated in exchange for the relief from headache.

Whereas it may be believed by some that denervating a joint will create a neuropathic joint (Charcot's arthropathy), there is no evidence that this occurs, and no grounds for believing that it would occur (Lord and Bogduk 1997). Charcot's arthropathy occurs in limbs that have been completely denervated, in which potentially unstable joints are not protected by muscle activity. In contrast, the zygapophysial joints are intrinsically stable; they are stabilized further by the intervertebral disc, and most of the muscles that act on the affected segment remain functional.

Utility

Cervical medial branch neurotomy is the singular means by which pain from cervical zygapophysial joints can be eliminated. No other forms of treatment have been shown to be as effective for the treatment of proven cervical zygapophysial joint pain. No other form of treatment has been shown to consistently provide complete relief of neck pain or cervicogenic headache.

Given that the prevalence of cervical zygapophysial joint pain is in excess of 50 % in patients with chronic neck pain after whiplash (Lord et al. 1994, 1996b; Barnsley et al. 1995; Gibson et al. 2000), and given that no other form of treatment has been shown to be effective for these patients, cervical medial branch neurotomy has a potentially enormous application in practice.

Acknowledgement   Dr. Jayantilal Govind passed away on June 20, 2009, at age 68. We thank Prof. Nik Bogduk for the update of his entry.

References

Barnsley, L., Lord, S., Wallis, B., & Bogduk, N. (1993). False-positive rates of cervical zygapophysial joint blocks. *The Clinical Journal of Pain, 9*, 124–130.

Barnsley, L., Lord, S. M., Wallis, B. J., & Bogduk, N. (1995). The prevalence of chronic cervical zygapophysial joint pain after whiplash. *Spine, 20*, 20–26.

Bogduk, N. (2002). Radiofrequency treatment in Australia. *Pain Practice, 2*, 180–182.

Bogduk, N., & Holmes, S. (2000). Controlled zygapophysial joint blocks: The travesty of cost-effectiveness. *Pain Medicine, 1*, 25–34.

Gibson, T., Bogduk, N., Macpherson, J., & McIntosh, A. (2000). Crash characteristics of whiplash associated chronic neck pain. *Journal of Musculoskeletal Pain, 8*, 87–95.

Govind, J., King, W., Bailey, B., & Bogduk, N. (2003). Radiofrequency neurotomy for the treatment of third occipital headache. *Journal of Neurology, Neurosurgery & Psychiatry, 74*, 88–93.

International Spinal Intervention Society. (2004). Cervical medial branch neurotomy. In N. Bogduk (Ed.), *Practice guidelines for spinal diagnostic and treatment procedures*. San Francisco: International Spinal Intervention Society.

Lord, S. M., & Bogduk, N. (1997). Treatment of chronic cervical zygapophysial joint pain. *The New England Journal of Medicine, 336*, 1531.

Lord, S., Barnsley, L., Wallis, B., & Bogduk, N. (1994). Third occipital nerve headache: A prevalence study. *Journal of Neurology, Neurosurgery & Psychiatry, 57*, 1187–1190.

Lord, S., Barnsley, L., Wallis, B. J., & Bogduk, N. (1996a). Chronic cervical zygapophysial joint pain after whiplash: A placebo-controlled prevalence study. *Spine, 21*, 1737–1745.

Lord, S. M., Barnsley, L., Wallis, B., McDonald, G. M., & Bogduk, N. (1996b). Percutaneous radio-frequency neurotomy for chronic cervical zygapophyseal joint pain. *The New England Journal of Medicine, 335*, 1721–1726.

Lord, S. M., McDonald, G. J., & Bogduk, N. (1998). Percutaneous radiofrequency neurotomy of the cervical medial branches: A validated treatment for cervical zygapophyseal joint pain. *Neurosurgery Quarterly, 8*, 288–308.

McDonald, G. J., Lord, S. M., & Bogduk, N. (1999). Long term follow-up of patients treated with cervical radiofrequency neurotomy for chronic neck pain. *Neurosurgery, 45*, 61–68.

Sapir, D. A., & Gorup, J. M. (2001). Radiofrequency medial branch neurotomy in litigant and non-litigant patients with cervical whiplash. *Spine, 26*, E268–E273.

Wallis, B. J., Lord, S. M., & Bogduk, N. (1997). Resolution of psychological distress of whiplash patients following treatment by radiofrequency neurotomy: A randomised, double-blind, placebo-controlled trial. *Pain, 73*, 15–22.

Cervical Periradicular Epidural Steroid Injection

▶ Cervical Transforaminal Injection of Steroids

Cervical Radiofrequency Neurotomy

▶ Cervical Medial Branch Neurotomy

Cervical Root Avulsion

Definition

Traumatic lesion of ventral and/or dorsal root, at the cervical level, consisting of detachment of the constituting rootlets of the root from the spinal cord; main mechanism is stretching.

Cross-References

▶ Brachial Plexus Avulsion and Surgery in the Dorsal Root Entry Zone (DREZ)

Cervical Selective Nerve Root Injection

▶ Cervical Transforaminal Injection of Steroids

Cervical Transforaminal Injection of Corticosteroids

Definition

The directed deposition of corticosteroid into the cervical intervertebral neuroforamen.

Cross-References

▶ Cervical Transforaminal Injection of Steroids

Cervical Transforaminal Injection of Steroids

James P. Rathmell
Department of Anesthesiology, University of Vermont College of Medicine, Burlington, USA

Synonyms

Cervical periradicular epidural steroid injection; Cervical selective nerve root injection

Definition

▶ Cervical transforaminal injection of corticosteroids is a treatment for cervical ▶ radicular pain in which corticosteroids are delivered into a cervical intervertebral neuroforamen.

Characteristics

Cervical radicular pain affects about one person per 1,000 of population, per year (Radhakrishnan et al. 1994), and is most often caused by a disc herniation or foraminal stenosis. Its natural history can be favorable (Bogduk et al. 1999), but not all patients recover naturally. For relieving cervical radicular pain, conservative therapy, typically including graduated exercise and oral analgesics, is supported only by observational studies, which have not controlled for natural history or nonspecific effects of treatment. The controlled studies that have been conducted have shown no significant benefit for traction or exercises (British Association of Physical Medicine 1966; Goldie and Landquist 1970; Klaber et al. 1990). Surgery is the mainstay of treatment if conservative therapy fails (Chestnut et al. 1992; Ahlgren and Garfin 1996). Surgery, however, is not without risks and constitutes a major undertaking for patients.

CTFIS constitutes an option for treatment, instead of surgery, when conservative therapy does not result in resolution of symptoms, and pain is the sole indication for treatment.

Rationale

The rationale for injecting steroids is that they suppress inflammation of the nerve, which is believed to be the basis for radicular pain. The rationale for using a transforaminal route of injection, rather than an interlaminar route, is that the injectate is delivered directly onto the target nerve. This ensures that the medication reaches the target area in maximum concentration at the site of the suspected pathology.

Indications

Cervical radicular pain is the only indication for cervical transforaminal injection of steroids. Radicular pain is recognized by its dynatomal distribution, which is distinctly different from the ▶ dermatomes of the same nerve (Slipman et al. 1998). Confidence in the diagnosis is enhanced if the patient also has ▶ radiculopathy, although this may not always be evident. Paraesthesiae, segmental numbness, weakness, and loss of reflexes are reliable and valid signs of radiculopathy that allow the diagnosis to be made clinically, without recourse to investigations. Disc protrusion and foraminal stenosis are the most common causes, but diagnostic imaging is required to exclude tumors and other infrequent causes such as infection, trauma, or inflammatory arthritides (Boyce and Wang 2003).

Cervical Transforaminal Injection of Steroids, Fig. 1 A drawing of an axial view of the cervical intervertebral foramen and adjacent structures at the level of C6, with a needle inserted parallel to the axis of the foramen along its posterior wall. Note the proximity of adjacent structures. *IJV*, internal jugular vein; *CA*, common carotid artery; *VA*, vertebral artery; *C6*, vertebral body of C6; *ScA*, anterior scalene muscle, *ScM*, middle scalene muscle; *sap*, superior articular process of C5–6 zygapophysial joint (Reproduced with permission from Rathmell et al. 2004)

Anatomy

The C3–C8 spinal nerves lie in the lower half of their respective intervertebral foramina. These foramina face anterolaterally. The vertebral artery lies just anterior to the exiting nerve (Fig. 1). Radicular branches from the vertebral artery lie adjacent to the spinal nerve and its roots to the spinal cord.

Technique

The patient lies supine, and a correct oblique view of the target foramen is obtained with a fluoroscope (Fig. 2). The correct oblique view is critical because, in less oblique views, the vertebral artery may lie along the course of the needle (Rathmell et al. 2004).

Through a puncture point overlying the posterior half of the target foramen, a 25 gauge, 21/2–31/2 in. needle is passed into the neck and then carefully readjusted to enter the foramen immediately in front of the anterior aspect of the superior articular process at the midpoint of the foramen (Fig. 3a). Above this level, the needle may encounter veins; below it, the needle may

Cervical Transforaminal Injection of Steroids, Fig. 2 A sketch showing patient position and fluoroscope orientation to obtain an oblique view of the cervical intervertebral foramina

Cervical Transforaminal Injection of Steroids, Fig. 3 Right anterior oblique radiograph demonstrating a needle in position along the posterior aspect of the right C6–7 intervertebral foramen. Inset of mid-portion of image with bony structures labeled: *SAP*, superior articular process; *La*, lamina; *Ped*, pedicle; *IAP*, inferior articular process; *SpP*, spinous process; *C6*, C6 vertebral body; *C7*, C7 vertebral body. (**b**) Anteroposterior radiograph demonstrating needle in final position within the right C6–7 intervertebral foramen. The needle lies halfway between the medial and lateral borders of the articular pillars. Inset of mid-portion of image with bony structures labeled: *SpP*, spinous processes of C5, C6, and C7; *Facets*, medial and lateral aspect of the facet column; *TrP (T1)*, transverse process of T1 (Reproduced with permission from Rathmell et al. 2004)

encounter the spinal nerve and its arteries. The needle must stay in contact with the posterior wall to avoid the vertebral artery. On anteroposterior view, the tip of the needle should not be advanced past the midpoint of the articular pillar (Fig. 3b). Insertion beyond this depth risks puncturing the dural sleeve or thecal sac.

Under direct, real-time fluoroscopy, a small volume of nonionic contrast medium (1.0 ml or less) is injected. The solution should outline the proximal end of the spinal nerve and spread centrally toward the epidural space (Fig. 4). Real-time fluoroscopy is essential to check for inadvertent intra-arterial injection, which may

Cervical Transforaminal Injection of Steroids, Fig. 4 Anteroposterior radiograph demonstrating needle in final position within the right C6–7 intervertebral foramen after injection of 1 ml of radiographic contrast medium (iohexol 180 mg/ml). Contrast outlines the spinal nerve and extends along the lateral aspect of the epidural space above the foramen (*arrows*) (Reproduced with permission from Rathmell et al. 2004)

occur even if the needle is correctly placed. Intra-arterial injection is manifested by very rapid clearance of the injected contrast material. In a vertebral artery, the contrast material will streak in a cephalized direction. In a radicular artery, it will blush briefly in a transverse direction medially toward the spinal cord. In either instance, the needle should be withdrawn, and the procedure should be postponed until after a period long enough for the puncture to have healed.

Sometimes the contrast medium may fill epiradicular veins. These are recognized by the slow clearance of the contrast medium, characteristic of venous flow. In that event, the needle should be adjusted by either slightly withdrawing the needle or redirecting it to a position slightly lower on the posterior wall of the foramen.

Only a small volume of contrast medium (1.0 ml or less) is required to outline the dural sleeve of the spinal nerve. As it spreads onto the thecal sac, the contrast medium will assume a linear configuration (Fig. 4). Rapid dilution of the contrast medium implies subarachnoid spread, which may occur if the needle has punctured the

thecal sac when there is lateral dilatation of the dural root sleeve into the ▶ intervertebral foramen. In that event, the procedure should be abandoned and rescheduled to avoid potential subarachnoid deposition of local anesthetic or steroid. Once the target nerve has been correctly outlined, a small volume of a short-acting local anesthetic (lidocaine 1 %, 0.5–1.5 ml) is injected in order to anesthetize the target nerve. While ensuring that the needle has not displaced, the procedure is completed by injecting a small dose of corticosteroid (betamethasone, 3–6 mg or triamcinolone 20–40 mg).

Complications

A case report has detailed fatal spinal cord infarction following CTFIS (Brouwers et al. 2001). Another report referred to several unpublished cases in Australia, Europe, and the USA in which patients suffered severe neurologic sequelae (Rathmell et al. 2004). Injection of corticosteroid into a radicular artery is one plausible mechanism for neurologic injury to occur (Baker et al. 2002). Meticulous attention to real-time fluoroscopic imaging is required to avoid such complications (Rathmell et al. 2004; Baker et al. 2002).

Outcomes

There are no randomized controlled trials comparing CTFIS to placebo or other treatments. The literature is limited to three small observational studies.

Bush and Hillier (1996) treated patients with three different forms of injection therapy: cervical or brachial plexus block, CTFIS, and x-ray-guided, interlaminar epidural steroid injection. They reported that 76 % of patients achieved complete relief of arm pain, but it is not possible from their report to derive what proportion responded to transforaminal injections.

Slipman et al. (2000) reported good or excellent results, at 12–45-month follow-up, in 60 % of 20 patients treated with an average of 2.2 injections. They did not, however, provide separate results for each category of outcome.

Vallee et al. (2001) studied 34 patients with cervical radiculopathy with refractory symptoms after 2 months of medical management.

Good or excellent results were reported in 53 % of 32 patients at 6 months, after an average of 1.3 injections. At 3 months, 29 % of patients had complete relief of pain. This proportion persisted at 6 months, but diminished to 20 % at 12 months.

These studies appear to paint an encouraging picture of the role for CTFIS. However, this sentiment must be tempered by the relatively low level of evidence these studies provide.

Caveats

CTFIS is an emerging therapy whose efficacy has not been corroborated by controlled studies. Yet it is associated with serious complications whose incidence is not properly known. Although a possible option for cervical radicular pain, it should only be undertaken by operators familiar with the relevant anatomy, with sufficient experience and skill to maximize the safety of the procedure.

References

Ahlgren, B. R., & Garfin, S. R. (1996). Cervical radiculopathy. *The Orthopedic Clinics of North America, 27*, 253–263.

Baker, R., Dreyfuss, P., Mercer, S., & Bogduk, N. (2002). Cervical transforaminal injection of corticosteroids into a radicular artery: A possible mechanism for spinal cord injury. *Pain, 103*, 211–215.

Bogduk, N. (1999). *Medical management of acute cervical radicular pain. An evidence-based approach*. Newcastle Australia: Newcastle Bone and Joint Institute.

Boyce, B. H., & Wang, J. C. (2003). Evaluation of neck pain, radiculopathy, and myelopathy: Imaging, conservative treatment, and surgical indications. *Instructional Course Lectures, 52*, 489–495.

British Association of Physical Medicine. (1966). Pain in the neck and arm: A multicentre trial of the effects of physiotherapy. *British Medical Journal, 1*, 253–258.

Brouwers, P. J. A. M., Kottnik, E. J. B. L., Simon, M. A. M., & Prevo, R. L. (2001). A cervical anterior spinal artery syndrome after diagnostic blockade of the right C6-nerve root. *Pain, 91*, 397–399.

Bush, K., & Hillier, S. (1996). Outcome of cervical radiculopathy treated with periradicular/epidural corticosteroid injections: A prospective study with independent clinical review. *European Spine Journal, 5*, 319–325.

Chestnut, R. M., Abithol, J. J., & Garfin, S. R. (1992). Surgical management of cervical radiculopathy. *The Orthopedic Clinics of North America, 23*, 461–474.

Goldie, I., & Landquist, A. (1970). Evaluation of the effects of different forms of physiotherapy in cervical pain. *Scandinavian Journal of Rehabilitation Medicine, 2–3*, 117–121.

Klaber Moffett, J. A., Hughes, G. I., & Griffiths, P. (1990). An investigation of the effects of cervical traction. Part 1: Clinical effectiveness. *Clinical Rehabilitation, 4*, 205–211.

Radhakrishnan, K., Litchy, W. J., O'Fallon, W. M., & Kurland, L. T. (1994). Epidemiology of cervical radiculopathy. A population-based study of Rochester, Minnesota, 1976–1990. *Brain, 117*, 325–335.

Rathmell, J. P., Aprill, C., & Bogduk, N. (2004). Cervical transforaminal injection of steroids. *Anesthesiology, 100*, 1595–1600.

Slipman, C. W., Lipetz, J. S., Jackson, H. B., Rogers, D. P., & Vresilovic, E. J. (2000). Therapeutic selective nerve root block in the non-surgical treatment of atraumatic cervical spondylotic radicular pain: A retrospective analysis with independent clinical review. *Archives of Physical Medicine and Rehabilitation, 81*, 741–746.

Slipman, C. W., Plastaras, C. T., Palmitier, R. A., Huston, C. W., & Sterenfeld, E. B. (1998). Symptom provocation of fluoroscopically guided cervical nerve root stimulation. Are dynatomal maps identical to dermatomal maps? *Spine, 23*, 2235–2242.

Vallee, J. N., Feydy, A., Carlier, R. Y., Mutschler, C., Mompoint, D., & Vallee, C. A. (2001). Chronic cervical radiculopathy: Lateral approach periradicular corticosteroid injection. *Radiology, 218*, 886–892.

Cervical Zygapophyseal Joint Blocks

▶ Cervical Medial Branch Blocks

CFA

Synonyms

Complete Freund's Adjuvant

Definition

Complete Freund's Adjuvant is used in animal experiments to induce inflammation.

Cross-References

▶ Complete Freund's Adjuvant
▶ Substance P Regulation in Inflammation

CFA or FCA

▶ Complete Freund's Adjuvant

c-Fos

Definition

A gene that codes for a transcription factor (Fos). *c-Fos* can be switched on rapidly as a result of various stimuli, and its product regulates the expression of various other genes in the cell. Fos protein, which can be detected by immunocytochemistry, can be used to demonstrate that a neuron has been activated, and is therefore used as a marker to map neuronal recruitment to stimuli, including noxious stimuli. In addition, c-Fos may play a role in activating "late-response" genes.

Cross-References

▶ Alternative Medicine in Neuropathic Pain
▶ c-Fos Immediate-Early Gene Expression
▶ Freezing Model of Cutaneous Hyperalgesia
▶ Nociceptive Processing in the Hippocampus and Entorhinal Cortex, Neurophysiology and Pharmacology
▶ Opioid Receptors at Postsynaptic Sites
▶ Spinothalamic Tract Neurons, Role of Nitric Oxide

c-Fos Immediate-Early Gene Expression

Definition

c-Fos is one of a family of genes, called "immediate-early genes," which are expressed very soon after a salient environmental event (e.g., pain). The proteins encoded by immediate-early genes act as transcription factors to affect the expression of other genes. By using immunohistochemistry for Fos, the protein product of *c-Fos*, one can identify with single-cell resolution those neurons that "responded" to the noxious stimulus, see also c-Fos.

Cross-References

▶ c-Fos

CFS

▶ Cutaneous Field Stimulation

CGRP

▶ Calcitonin-Gene-Related Peptide
▶ Calcitonin Gene-Related Peptide and Migraine Headaches

CGRP and Spinal Cord Nociception

Andrea Ebersberger
Department of Physiology, Friedrich Schiller University of Jena, Jena, Germany

Synonyms

Calcitonin gene-related peptide and spinal cord nociception; Spinal cord nociception and CGRP

Definition

▶ Calcitonin gene-related peptide (CGRP) is a 37 amino acid peptide of the calcitonin family with two isoforms, α and β CGRP, with similar biological functions (Poyner 1992). The neuropeptide is synthesized in up to 50 % of the small- and medium-sized dorsal root ganglion neurons (DRGs) and is transported along the axon to the peripheral endings (Donnerer et al. 1992) and to the central endings of the neuron in the dorsal horn of the spinal cord. In addition to its role in nociception, CGRP is involved in a number of other functions including vasodilation.

Characteristics

Localization of Spinal CGRP

Fibers showing CGRP-immunoreactivity are located in the dorsal horn of the spinal cord in laminae I and II at high density and in lamina V at lower density (Wiesenfeld-Hallin et al. 1984). CGRP-like immunoreactivity and CGRP mRNA are also localized in motoneurons of the ventral horn.

Localization of CGRP Receptors

The term "CGRP receptor" is used only for the combination of the calcitonin-like receptor (CLR) and the receptor activity-modifying protein (RAMP1). For functionality, an additional protein called receptor component protein (RCP) is required (Barwell et al. 2012).

The density of CGRP receptors is highest in the superficial and deep dorsal horn, but there are also receptors in the ventral horn. CLR- and RAMP1-ir were expressed in neuronal cell bodies but lacking in central glia of the dorsal horn (Marvizón et al. 2007; Lennerz et al. 2008). The distribution of CGRP binding sites is similar to the distribution of CGRP-positive fibers; however, there was no weak colocalization of CGRP and its receptor (Eftekhari and Edvinsson 2011). CGRP receptors may even be located at sites where no CGRP-containing fibers are terminating. These receptors might be reached by the neuropeptide by diffusion within the tissue.

At a closer look CLR is located in glutamatergic presynaptic terminals in the dorsal horn that contain alpha (2C) adrenergic receptors and opioids (Marvizón 2007). Only part of these terminals contains RAMP1, which may form CGRP receptors with CLR. CLR may form other receptors with other receptor activity-modifying proteins.

Under pathophysiological conditions such as peripheral inflammation, increases and decreases in binding sites for CGRP in the spinal cord have been observed (Galeazza et al. 1992). Not only the receptor but also its accessory protein (RCP) can be regulated in inflammatory or neuropathic pain states (Ma et al. 2003).

Release of CGRP

There is a high basal release of CGRP from central endings of afferents in the spinal cord, which was shown with antibody-coated microprobes (Schaible et al. 1994; Morton and Hutchison 1989). This release is not influenced by activating peripheral afferents with innocuous stimuli or motor activity in the physiological range. Noxious mechanical stimuli, however, increase this basal release (Morton and Hutchison 1989). The situation changes during peripheral inflammation. During acute knee inflammation, mechanical stimuli of innocuous intensity to the knee induce spinal CGRP release (Schaible et al. 1994). This is probably caused by a sensitization of the afferents to mechanical stimuli. Additionally basal and evoked release is much higher in inflamed than in normal animals (Collin et al. 1993; Ogbonna et al. 2013). This may result from an upregulation of the synthesis of CGRP that has been observed in the acute and chronic stage of inflammation.

CGRP Receptor Agonists and Antagonists

Both isoforms of CGRP bind to the CRL unit of the CGRP receptor, while RAMP is necessary for CGRP recognition. CGRP also binds to adrenomedullin receptors and amylin receptors.

The classical antagonist CGRP(8-37) binds to the C terminal of the CRL unit of the CGRP receptor (Barwell et al. 2012). BIBN4096BS, the first potent non-peptide antagonist, interacts with the recognition site for CGRP on RAMP. The substance proved to be beneficial in the treatment of migraine. Therefore, a huge number of various CGRP antagonists with various binding sites were created (Durham and Vause 2010).

Up to now, there is no inhibitor of RCP known.

Effect of Spinal CGRP

Released CGRP can exert its action directly by binding to CGRP receptors. However, it also

interacts with the release and metabolism of substance P. CGRP can facilitate release of substance P, and, in addition, CGRP controls the
amount of substance P by inhibiting the
enzymatic degradation of substance P (Duggan
et al. 1992).

Behavioral Experiments. In behavioral
experiments, intrathecally applied CGRP
facilitated the responses to noxious stimulation
(e.g., Wiesenfeld-Hallin et al. 1984), and
antagonization of endogenous CGRP with
$CGRP_{8-37}$, a $CGRP_1$ receptor antagonist, was
antinociceptive. CGRP was also shown to support
the generation and maintenance of ▶ mechanical
allodynia and ▶ hyperalgesia in rats. $CGRP_{8-37}$
alleviated mechanical and ▶ thermal allodynia in
chronic central pain. Thus, CGRP has a role in
normal nociception, but it is also involved in pain
states (reviewed by Schaible et al. 2004).

Effect on Neuronal Activity. Application of
CGRP in the vicinity of spinal cord neurons
caused no, or only weak, excitation of the neurons
(Ryu et al. 1988). In some cases background
activity was enhanced. But CGRP has
a facilitatory effect on evoked activity in
spinal cord neurons, e.g., activities evoked by
innocuous or noxious mechanical stimulation or
substance P (Biella et al. 1991; for review see
Neugebauer 2009). CRGP is also involved in the
development and maintenance of spinal
hyperexcitability. This effect was shown in
a model of knee joint inflammation where
CGRP or the antagonist CGRP8-37 had been
ionophoretically applied in the vicinity of spinal
cord neurons that responded to mechanical
stimulation of the leg (Fig. 1). CGRP-induced
sensitization of nociceptive spinal dorsal horn
neurons includes PKA- and PKA-dependent
mechanisms (Sun et al. 2004).

Since ▶ glutamate receptors are of major
importance in the excitation of spinal nociceptive
neurons, a possible interaction between the
responses of spinal cord neurons to CGRP and
▶ NMDA or ▶ AMPA was investigated.
Coadministration of CGRP and AMPA or
NMDA enhanced the responses to the excitatory
amino acids (Ebersberger et al. 2000). In conclusion, one explanation for the spinal effect of

CGRP and Spinal Cord Nociception, Fig. 1 The development of knee joint inflammation is paralleled by the
development of spinal hyperexcitability measured as an
increase in neuronal responses to repetitive noxious
pressure applied to the knee joint (*upper trace*). When
$CGRP_{8-37}$ was ionophoretically applied during and
90 min after induction of inflammation in the vicinity of
the recorded neurons, the development of spinal
hyperexcitability was prevented (Neugebauer et al. 1996)

CGRP is its influence on ▶ glutamatergic
neuronal transmission. This is corroborated by
the finding that CGRP increases EPSPs and
synaptic plasticity in the dorsal horn (Bird
et al. 2006). In addition, CGRP inhibits voltage
gated K^+ channels thus regulating neuronal excitability (Hou and Yu 2010). For further information see Neugebauer (2009). The role of CGRP in
chronic pain models remains to be determined.

References

Barwell, J., Wootten, D., Simms, J., Hay, D. L., & Poyner,
    D. R. (2012). RAMPs and CGRP receptors. *Advances
    in Experimental Medicine and Biology, 744*, 13–24.
Biella, G., Panara, C., Pecile, A., & Sotgiu, M. L. (1991).
    Facilitatory role of calcitonin gene-related peptide
    (CGRP) on excitation induced by substance P (SP)
    and noxious stimuli in rat spinal dorsal horn neurons.
    Brain Research, 559, 352–356.
Bird, G. C., Han, J. S., Fu, Y., Adwanikar, H., Willis,
    W. D., & Neugebauer, V. (2006). Pain-related synaptic
    plasticity in spinal dorsal horn neurons: Role of CGRP.
    Molecular Pain, 2, 31.

Collin, E., Frechilla, D., Pohl, M., et al. (1993). Opioid control of the release of calcitonin gene-related peptide-like material from the rat spinal cord in vivo. *Brain Res, 609*, 211–222.

Donnerer, J., Schuligoi, R., & Stein, C. (1992). Increased content and transport of substance P and calcitonin gene-related peptide in sensory nerves innervation inflamed tissue: Evidence for a regulatory function of nerve growth factor in vivo. *Neuroscience, 49*, 693–698.

Duggan, A. W., Schaible, H. G., Hope, P. J., & Lang, C. W. (1992). Effect of peptidase inhibition on the pattern of intraspinally released immunoreactive substance P detected with antibody microprobes. *Brain Research, 579*, 261–269.

Durham, P. L., & Vause, C. V. (2010). Calcitonin gene-related peptide (CGRP) receptor antagonists in the treatment of migraine. *CNS Drugs, 24*, 539–548.

Ebersberger, A., Charbel Issa, P., Venegas, H., & Schaible, H. G. (2000). Differential effects of calcitonin gene-related peptide and calcitonin gene-related peptide 8–37 responses to *N*-methyl-D-aspartate or (R, S)-α-amino-3-hydroxy-5-methylisoxazole-4-propionate in spinal nociceptive neurons with knee joint input in the rat. *Neuroscience, 99*, 171–178.

Eftekhari, S., & Edvinsson, L. (2011). Calcitonin gene-related peptide (CGRP) and its receptor components in human and rat spinal trigeminal nucleus and spinal cord at C1-level. *BMC Neuroscience, 12*, 112.

Galeazza, M. T., Stucky, C. L., & Seybold, V. S. (1992). Changes in 125IhCGRP binding sites in rat spinal cord in an experimental model of acute, peripheral inflammation. *Brain Research, 591*, 198–208.

Hou, J. F., & Yu, L. C. (2010). Blockade effect of BIBN409BS on CGRP-induced inhibition on whole cell K+currents in spinal dorsal horn neurons of rats. *Neuroscience Letters, 469*, 15–28.

Lennerz, J. K., Rühle, V., Ceppa, E. P., Neuhuber, W. L., Bunnett, N. W., Grady, E. F., et al. (2008). Calcitonin receptor-like receptor (CLR), receptor activity-modifying protein 1 (RAMP1), and calcitonin gene-related peptide (CGRP) immunoreactivity in the rat trigeminovascular system: Differences between peripheral and central CGRP receptor distribution. *The Journal of Comparative Neurology, 507*, 1277–1299.

Ma, W., Chabot, J.-G., Powell, K. J., et al. (2003). Localization and modulation of calcitonin gene-related peptide-receptor component protein-immunoreactive cells in the rat central and peripheral nervous system. *Neuroscience, 120*, 677–694.

Marvizón, J. C., Pérez, O. A., Song, B., Chen, W., Bunnett, N. W., Grady, E. F., et al. (2007). Calcitonin receptor-like receptor and receptor activity modifying protein 1 in the rat dorsal horn: Localization in glutamatergic presynaptic terminals containing opioids and adrenergic alpha2C receptors. *Neuroscience, 148*, 250–265.

Morton, C. R., & Hutchison, W. D. (1989). Release of sensory neuropeptides in the spinal cord: Studies with calcitonin-gene related peptide and galanin. *Neuroscience, 31*, 807–815.

Neugebauer, V. (2009). CGRP in spinal cord pain mechanisms. In M. Malcangio (Ed.), *Synaptic plasticity in pain* (pp. 175–197). New York: Springer.

Neugebauer, V., Rühmenapp, P., & Schaible, H.-G. (1996). Calcitonin gene related peptide is involved in the generation and maintenance of hyperexcitability of dorsal horn neurons observed during development of acute inflammation in rat's knee joint. *Neuroscience, 71*, 1095–1109.

Ogbonna, A. C., Clark, A. K., Gentry, C., Hobbs, C., & Malcangio, M. (2013). Pain-like behaviour and spinal changes in the monosodium iodoacetate model of osteoarthritis in C57Bl/6 mice. *European Journal of Pain, 17*(4), 514–526.

Poyner, D. R. (1992). Calcitonin gene related peptide: Multiple actions, multiple receptors. *Pharmacy Therapy, 56*, 23–51.

Ryu, P. D., Gerber, G., Murase, K., & Randic, M. (1988). Actions of calcitonin gene-related peptide on rat spinal dorsal horn neurons. *Brain Research, 441*, 357–361.

Schaible, H.-G., Freudenberger, U., Neugebauer, V., et al. (1994). Intraspinal release of immunoreactive calcitonin gene-related peptide during development of inflammation in the joint in vivo – A study with antibody microprobes in cat and rat. *Neuroscience, 62*, 1293–1304.

Schaible, H.-G., et al. (2004). Involvement of CGRP in nociceptive processing and hyperalgesia: Effects of CGRP on spinal and dorsal root ganglion neurons. In K. Brune & H. O. Handwerker (Eds.), *Hyperalgesia: Molecular mechanisms and clinical implications. Progress in brain research and management* (Vol. 30, pp. 201–227). Seattle: IASP Press.

Sun, R. Q., Tu, Y. J., Lawand, N. B., Yan, J. Y., Lin, Q., & Willis, W. D. (2004). Calcitonin gene-related peptide receptor activation produces PKA- and PKC-dependent mechanical hyperalgesia and central sensitization. *Journal of Neurophysiology, 92*, 2859–2866.

Wiesenfeld-Hallin, S., HÄkfelt, T., Lundberg, J. M., et al. (1984). Immunoreactive calcitonin-gene related peptide and substance P coexist in sensory neurons in the spinal cord and interact in spinal behavioral responses of the rat. *Neuroscience Letters, 52*, 199–204.

Changes in Thalamic Physiology Occurring in Patients with Chronic Pain

▶ Thalamus, Receptive Fields, Projected Fields, Human

Channel Inactivation

Definition

A process whereby a channel enters a nonconducting state. This can occur from an open state and is reflected by current decay during a prolonged voltage step in the case of voltage-gated channels. Inactivation from an open state is also referred to as p-type inactivation. Inactivation can also occur from a closed state a process also referred to as steady-state, slow, or C-type inactivation. A period of membrane hyperpolarization is required to remove voltage-gated channels from an inactivated state.

Cross-References

▶ Painful Channelopathies

Channelopathies

Definition

A channelopathy is the result of a mutation(s) in the gene encoding the channel, producing changes in channel properties sufficient to produce clinical symptoms. These changes are generally classified according to whether they result in an increase in channel activity (gain of function) or a decrease in channel activity (loss of function).

Cross-References

▶ Migraine, Pathophysiology
▶ Painful Channelopathies

Charcot-Marie-Tooth Disease

Synonyms

CMT

Definition

CMT is any inherited neuropathy that is not part of a syndrome.

Cross-References

▶ Hereditary Neuropathies

Charcot-Marie-Tooth Disease (CMT)

▶ Hereditary Neuropathies

C-Heat Receptor

▶ Polymodal Nociceptors, Heat Transduction

Chemesthesis

Definition

Sensitivity to all chemicals that produce sensations other than (or in addition to) taste or smell.

Cross-References

▶ Nociception in Nose and Oral Mucosa

Chemical Lesion

Definition

Selective lesion of neuronal cell bodies by chemical means. The method is to inject a concentrated solution of an excitatory amino acid that excites certain neurons. At high concentrations, amino acids (e.g., glutamic, kainic, quisqualic, ibotenic acids) can produce selective neuronal death by excitotoxicity.

Cross-References

▶ Thalamotomy, Pain Behavior in Animals

Chemical Name: N-Phenyl-N-(1-(2-phenylethyl)-4-Piperidinyl-Procainamide

▶ Postoperative Pain, Fentanyl

Chemical Sympathectomy

▶ Sympathetic Blocks

Chemical Transmitter

▶ Thalamic Neurotransmitters and Neuromodulators

Chemoattractants

Definition

Chemoattractants have been divided into two categories. One category is represented by the classical chemoattractants like platelet-activating factor (PAF), leukotriene B4 (LTB4), formyl-methionyl-leucyl-phenylalanine (FMLP), and complement protein C5a. The second category consists of compounds belonging to the chemokine group. Both classical chemoattractants and chemokines act on target cells through seven-transmembrane domain receptors that are coupled to heteromeric G-proteins and elicit chemotactic responses.

Cross-References

▶ Neutrophils in Inflammatory Pain

Chemokine

▶ Cathepsin and Microglia

Chemokines

Synonyms

Chemotactic cytokines

Definition

Chemokines are cytokines with chemoattractant properties, inducing cells with the appropriate receptors to migrate toward the source of the chemokines, which includes the family of proinflammatory activation-inducible cytokines. These proteins are mainly chemotactic for different cell types. Based on chromosomal locations of individual genes, two different subfamilies of chemokines are distinguished as CXC-chemokines (CXCL; also known as alpha-chemokines or 4q chemokine family) and CC-chemokines (CCL; also known as beta-chemokines or 17q chemokine family). In the earliest phases of inflammation, chemokines are released and induce direct chemotaxis in nearby responsive cells. Together with intercellular adhesion molecules, chemokines and their receptors serve to localize and enhance the inflammatory reaction at the site of tissue damage.

Cross-References

▶ Cytokine Modulation of Opioid Action
▶ Cytokines as Targets in the Treatment of Neuropathic Pain
▶ Cytokines, Effects on Nociceptors
▶ Neutrophils in Inflammatory Pain

Chemo-Nociceptors

▶ Histology of Nociceptors

Chemosensation

Definition

Sensations initiated through chemicals, e.g., gustatory- or olfactory-mediated sensations or sensations mediated through the intranasal trigeminal nerves.

Cross-References

▶ Nociception in Nose and Oral Mucosa

Chemosensitive Sympathetic Afferent Fibers

Definition

Afferent fibers that transmit information resulting from a variety of chemicals that are released during myocardial ischemia.

Chemotactic

Definition

A chemical (sodium morrhuate) used in prolotherapy solutions, which acts by attracting inflammatory cells.

Cross-References

▶ Prolotherapy

Chemotactic Cytokines

▶ Chemokines

Chemotherapy

Definition

Treatment with drugs that kill cancer cells that may be given by one or more of the following methods: orally, by venous or arterial injection (through a catheter or port), or topically.

Cross-References

▶ Cancer Pain Management, Chemotherapy
▶ Cancer Pain Management, Treatment of Neuropathic Components

Chemotherapy-Induced Neuropathy

Definition

Many of the drugs used in chemotherapeutic treatment of cancer induce neuropathic changes in peripheral nerve fibers. These changes may induce sensory loss and numbness, but they are also frequently associated with spontaneous pain and/or hypersensibility to applied stimuli. Examples of such drugs are colchicine, vinblastine, vincristine, cisplatin, carboplatin, taxol, and lenalidomide. These drugs typically disrupt cytoskeletal elements, such as microtubules, and interfere with axoplasmic transport. The resulting interruption of molecular trafficking within the neuron is the likely pathophysiological cause of the neuropathy, at least for some of the agents.

Cross-References

▶ Ulceration, Prevention by Nerve
Decompression

Chest Pain

Definition

Pain perceived at the thoracic level, caused by
disease of a thoracic viscus or by disorders of
muscles, bones, and other pain-sensitive structures of the chest wall. It can also be referred
from structures outside the chest.

When derived from a thoracic viscus, it is often
caused by coronary artery disease, but can also
originate from non-cardiac structures such as the
esophagus. Visceral chest pain is usually described
as a sensation of pressure, squeezing, and/or
crushing, accompanied by neurovegetative signs
such as pallor, sweating, nausea, and/or vomiting.
Chest pain of non-visceral origin is sharper in
quality and better defined spatially.

Cross-References

▶ Visceral Pain Model, Angina Pain

Chiari Type I Malformation

Definition

This is a protrusion of the cerebellar tonsils,
below the foramen magnum, which can cause
a valve-like obstruction to the flow of cerebrospinal fluid.

Cross-References

▶ Primary Cough Headache

Child

▶ Pain in Children

Childhood Chronic Pain

▶ Prevalence of Chronic Pain Disorders in
Children

Childhood Migraine

▶ Migraine, Childhood Syndromes

Childhood Sexual Abuse

Definition

Any child below the age of consent may be
deemed to have been sexually abused when
another sexually mature person has, by design
or by neglect of their usual societal or specific
responsibilities in relation to that child,
engaged or permitted the engagement of that
person in any activity of a sexual nature that is
intended to lead to the sexual gratification of
the sexually mature person. This definition
pertains whether or not it involves genital
contact or physical contact, and whether or
not there is discernible harmful outcome in
the short term.

Cross-References

▶ Chronic Pelvic Pain, Physical and Sexual
Abuse

Children

▶ Experimental Pain in Children

Chiropractic

Definition

Therapeutic manipulation of the spine to treat a wide variety of conditions by correcting dysfunction in spinal alignment.

Cross-References

▶ Alternative Medicine in Neuropathic Pain
▶ Spinal Manipulation, Characteristics
▶ Spinal Manipulation, Pain Management

Chiropractics

▶ Spinal Manipulation, Characteristics

Chloride Transporter

Synonyms

Cl⁻ transporter

Definition

A chloride transporter is a membrane protein that assists in the movement of ions across a cell membrane. Different Cl⁻ transporters can move Cl⁻ against its concentration gradient (i.e., from a low concentration to a high concentration) or its electrical gradient (i.e., from the electrically neutral extracellular space to the more negatively charged intracellular space).

Cross-References

▶ GABA Mechanisms and Descending Inhibitory Mechanisms

Chloro-Phenyl-2-Methylaminocyclohexanone-Hydrochloride

▶ Postoperative Pain, Ketamine

Cholecystokinin

Synonyms

CCK

Definition

An eight amino acid peptide present in the gastrointestinal tract and in the nervous system that modulates pain sensation as well as other neuronal processes. It was named for its effects on the gall bladder, but also thought to have pro-nociceptive effects in the spinal cord.

Cross-References

▶ Alternative Medicine in Neuropathic Pain
▶ Pain-Modulatory Systems, History of Discovery

▶ Peptides in Neuropathic Pain States
▶ Placebo Analgesia and Descending Opioid Modulation

Chondrocytes

Definition

Chondrocytes are cells that produce cartilage by secretion of the cartilaginous matrix. The secretion of the matrix by chondrocytes leads to their encapsulation in this matrix where they eventually undergo programmed cell death or apoptosis. As a consequence, new chondrocytes constantly arise from their precursor cells. Chondrocytes arise from chondroblasts, which arise from mesenchymal cells.

Cross-References

▶ Arthritis Model, Osteoarthritis

Chondroitin

▶ Nutraceuticals

Chromosomes

Definition

Chromosomes contain the cell's genetic information, and are structured as compact intertwined molecules of DNA located in the nucleus of cells.

Cross-References

▶ NSAIDs, Pharmacogenetics

Chronic Abdominal Pain

▶ Functional Abdominal Pain in Children

Chronic Back Pain and Spinal Instability

Daniel Lubelski[1], Kene Ugokwe[2], Ajit Krishnaney[3], Michael P. Steinmetz[4] and Edward C. Benzel[5]
[1]Cleveland Clinic Lerner College of Medicine, Cleveland, OH, USA
[2]Department of Neurosurgery, St Elizabeth Health Center, Youngstown, OH, USA
[3]Department of Neurosurgery, Cleveland Clinic, Cleveland, OH, USA
[4]Department of Neurosurgery, MetroHealth Medical Center, Case Western Reserve University School of Medicine, Cleveland, OH, USA
[5]Department of Neurosurgery, Center for Spine Health Neurological Institute Cleveland Clinic, Cleveland, OH, USA

Synonyms

Back pain; Discogenic pain; Dysfunctional segmental motion; Lumbago; Mechanical low back pain; Muscle spasm; Myofascial pain syndrome; Painful disk syndrome

Definition

Chronic back pain is defined as pain in the dorsal aspect of the trunk (from the neck to the pelvis) that persists for more than 12 weeks (Gatchel et al. 1986). It may be related to degenerative, neoplastic, traumatic, or infectious conditions. Chronic back pain may also be related to spinal ▶ instability. Spinal instability is defined as the inability of the spine to limit patterns of movement or displacement that may lead to deformity or pain. Other entities that are particularly worthy of definition are mechanical back pain and myofascial pain syndrome, discogenic pain, pain of soft tissue injury origin, and pain of bony tissue injury origin:

- Mechanical back pain – a deep and agonizing pain that increases with activity, such as the assumption of the upright posture (loading) and decreases with inactivity, such as assuming the supine position (unloading).
- Myofascial pain syndrome – synonymous with muscle spasm or strain. It is usually self-limiting. There often exists an underlying cause.
- Discogenic low back pain – pain that originates from the intervertebral disk and the disk space.
- Pain of soft tissue injury origin – such pain originates from the damage or destruction of richly innervated soft tissue. It may occur after surgery and is also seen with muscle tear.

Characteristics

Chronic back pain is common. Eighty five percent of cases are idiopathic. Potential anatomical sources of back pain include muscle, ligaments, tendons, bones, ► facet joints and disks. In many cases, it is difficult to determine the exact cause of back pain because of significant overlap in the nerve supply to the aforementioned structures. Approximately 80 % of Americans experience clinically significant back pain. Eighty to ninety percent of the attacks resolve within 6 weeks (Bigos et al. 1994). It is the second most common reason for which people seek medical attention (Cypress 1983). Back pain accounts for 15 % of all sick leaves and it is the most common cause of disability for people less than 45 years of age (Cunningham & Kelsey, 1984). Along with headache, back pain is the most common cause of missed days of work and lost productive time at work due to health-related reduced performance. This has accounted for a mean of 5.28 h per worker per week of lost productive time.(Stewart et al. 2003) An estimated $61.2 billion is lost to pain-related decreased productivity among the US workforce (Stewart et al. 2003).

The incidence of chronic back pain has been increasing over the past number of years. Freburger et al. (2009) surveyed a representative sample of North Carolina households and found that between 1992 and 2006, the incidence of chronic, impairing lower back pain rose from 3.9 % to 10.2 %. Additionally, while the mean number of visits to all health care providers remained the same between 1992 and 2006, the percentage of individuals who had seen a health care provider for lower back pain in the past year rose from 73.1 % to 84 %. Using data from more than 10,000 respondents to the 1999–2002 National Health and Nutrition Examination Survey, Hardt et al. (2008) found that chronic pain in the back was more frequently cited as the region of non-minor pain relative to pain in any other region.

Spinal instability is a common cause of back pain. Some of the causes of spinal instability include age-related degenerative changes, prior spinal surgery, physically dependent occupations, sedentary lifestyles, obesity, poor posture, and certain sporting activities. In a patient with chronic back pain, a diagnosis to consider is spinal tumor. Pain at night that is relieved by aspirin may be suggestive of an osteoid osteoma or a benign ► osteoblastoma. Infections such as discitis (an infection of the disk space between the vertebral bodies) must also be ruled out. Spinal compression fractures may also be the source of severe back pain, particularly in the elderly. Other potential etiologies of chronic back pain include:

Degenerative Conditions

- Degenerative spondylolisthesis: slippage of one vertebral body over another (Pearcy and Shepherd, 1983).
- ► Spinal stenosis: a narrowing of the spinal canal, nerve root canals, or intervertebral foramina; the acquired form is caused by degenerative conditions (Arnoldi et al. 1976).
- Lateral recess syndrome: The lateral recess is the channel alongside the pedicle where the exiting nerve root resides.

Spondyloarthropathies

- Paget's disease: This is a condition characterized by areas of abnormal bone growth and an increased rate of bone resorption.
- Ankylosing spondylitis: This is a connective tissue disease characterized by inflammation of the spine and joints resulting in pain and stiffness.

It is important to note that all of the aforementioned conditions contribute to or cause spinal instability. Certain patients, however, have no organic disease and their back pain is psychogenic in nature. This may be as a result of secondary gain, i.e., financial or emotional gain (Waddell et al. 1980).

Spinal Instability

There are two categories of spinal instability: (1) acute spinal instability and (2) chronic spinal instability. Acute spinal instability may be further divided into overt and limited instability, while chronic instability is subcategorized into glacial instability and dysfunctional segmental motion (Benzel 2001). Denis described the three-column concept of identifying criteria for instability of the spine (Denis 1983).

The Three-Column Concept of Spinal Integrity and Stability

The spine is divided into three columns.

- Anterior column – composed of the ventral half of the disk and the vertebral body, including the anterior longitudinal ligament.
- Middle column – composed of the dorsal half of the disk and vertebral body and the posterior longitudinal ligament.
- Posterior column – the dorsal bony complex (posterior arch) and the dorsal ligamentous complex, including the supraspinous and interspinous ligaments and the ligamentum flavum.

Many authors have used a point system approach to quantify the extent of acute instability. White & Panjabi (1990); Panjabi & Lydon (1994) described the accumulation of 5 or more points as being indicative of spinal instability (Panjabi 1994; White 1990) (Table 1). They also described a stretch test in which the progressive addition of cervical traction weight was accompanied by clinical assessments and radiographs. The test was positive for instability when a disk interspace separation of greater than 1.7 mm, or a change in angle greater than 7.5° between pre- and post-stretch measurements, was observed. Most clinicians do not employ this method due to the risks involved

Chronic Back Pain and Spinal Instability, Table 1 Quantitation of Acute Instability for Subaxial Cervical, Thoracic, and Lumbar Injuries (The Point System)

Condition	Points Assigned
Loss of integrity of anterior and middle column	2
Loss of integrity of posterior columns	2
Acute resting translational deformity	2
Acute resting angulation deformity	2
Acute dynamic translation deformity exaggeration	2
Acute dynamic angulation deformity exaggeration	2
Neural element injury	3
Acute disk narrowing at the level of suspected pathology	1
Dangerous loading anticipated	1

(Panjabi 1994; White 1990)

and its cumbersome nature. Flexion and extension radiographs or MRI Dvorak and Panjabi (1991) may also be helpful in determining the degree of instability.

Acute Instability

Overt Instability

Overt instability is defined as the inability of the spine to support the torso during normal activity. This is usually acute in nature, e.g., after a trauma. It also has a chronic component, and may also occur in the setting of tumor, infection, or ▶ degenerative disease. It is characterized by circumferential loss of spinal integrity. Treatment may involve surgical stabilization, with or without decompression. The back pain experienced with overt instability is usually associated with soft tissue injury and muscle spasms (Fig. 1a).

Limited Instability

Limited instability is characterized by loss of either ventral or dorsal spinal integrity. Posterior column disruption is not always associated with instability, unless the posterior longitudinal ligament and the middle column are disrupted. Failure of the middle column represents an unstable injury (Denis 1983). Chronic forms of both overt

Chronic Back Pain and Spinal Instability, Fig. 1 (a) Fracture dislocation in the thoracolumbar spine representative of overt instability. (b) Wedge compression fracture elucidating limited instability

Chronic Back Pain and Spinal Instability, Fig. 2 (a) Spondylolisthesis in the lumbar spine which is an example of glacial instability. (b) Lumbar canal stenosis and degenerative disk disease showing dysfunctional segmental motion

and limited instability exist, especially when overt and limited forms do not heal adequately (Fig. 1b).

Chronic Instability

Glacial Instability

Glacial instability is defined as spinal instability that is neither overt nor limited. It does not pose a significant risk for the rapid development of a spinal deformity. The deformity progresses gradually, like a glacier moving down a mountain. Glacial instability is associated with

pain that is mechanical in nature. In managing this type of instability, one must factor in the degree of progression of deformity and the subjective complaint of pain. Causes include trauma, tumors, and congenital defects (Fig. 2a).

Dysfunctional Segmental Motion

Dysfunctional segmental motion is a type of instability that is related to disk interspace or vertebral body degenerative changes, tumor, or infection. A deep and agonizing pain that is usually worsened by activity and improved by inactivity usually characterizes dysfunctional segmental motion. This type of pain is similar to that observed with glacial instability. This pain results from the exaggeration of reflex muscle activity, which picks up the slack from the inadequate intrinsic stability from the spine. The associated pain syndrome, as described here, is commonly known as mechanical low back pain (Fig. 2b). The criteria for this diagnosis are unclear, and while MRI and radiographs have been suggested as effective methods to determine the presence of dysfunctional segmental motion, there is no objective data to conclusively make this determination.

Treatment

The best mode of treatment for chronic back pain has remained unclear. While there are zealous proponents for specific treatments, there has been little evidence that unequivocally suggested one method's superiority. The various conservative treatments include medications, physical therapy, and injections; the gold standard for the surgical approach has remained the spine fusion. A prospective 2-year multi-country study of more than 2,000 patients concluded that the common medical procedures used to treat chronic low back pain were ineffective. (Hansson and Hansson 2000) The studies on the efficacy of surgical treatment have been equivocal as well (Szpalski et al. 2007).

When patients present with chronic low back pain, the initial management usually consists of nonsurgical therapy; except in the presence of ▶ cauda equina syndrome, progressive neurologic deficit, or profound motor weakness. Also, one may proceed directly to surgery in the presence of severe pain that is not sufficiently controlled with pain medications. For the majority of patients suffering from chronic back pain, however, more studies are needed to determine who would benefit from each treatment.

Nonsurgical Treatment

- *Bed rest* helps reduce pressure on nerve roots and intradiscal pressure, which is lowest in the supine semi-Fowler position (Nachemson 1992). Severe pain may be decreased by lying in a fetal position or by lying supine with knees and hips flexed on a firm mattress with a soft top layer (Slipman et al. 2007). Bed rest, however, has been shown in subsequent studies to be a relatively ineffective form of treatment (Malmivaara et al. 1995). In fact, a study by Rozenberg et al. (2002) of 281 patients with lower back pain who were treated with either 4 days of bed rest or continuation of normal activities showed that while pain intensity and spinal mobility outcomes were similar in both groups, the bed rest group had an increased number of subsequent days of sick leave.
- *Exercise*, including physical therapy with low stress aerobic exercise, may help relieve symptoms and strengthen back muscles. Low stress aerobic exercise can minimize debility due to inactivity. Conditioning exercises for trunk muscles are also helpful if symptoms persist. There are proponents of individual-targeted therapeutic exercises, such as stretching and strengthening, using the McKenzie Method. This approach uses "centralization" techniques and identifies patients' directional preferences to develop appropriate exercises. A recent meta-analysis demonstrated that this method may be more effective than passive physical therapy for acute back pain, but there is limited evidence for chronic back pain. (Machado et al. 2006).
- *Analgesics* (e.g., NSAIDs) in the initial short-term period may be helpful. Opioids may be required for more intense pain.
- *Muscle relaxants*, such as cyclobenzaprine and methocarbamol, may help with pain of muscle spasm origin (myofascial component of pain).
- *Epidural steroid injections* may be of assistance in some cases of chronic pain.
- *Modality (physical) treatments*, including transcutaneous electrical nerve stimulation (TENS) and traction, qualify as physical treatments but may only provide minor relief for some patients. Cryotherapy and thermotherapy, by applying to the skin either cold packs or heating pads, respectively, have been shown to provide temporary analgesia. These methods are often combined with

stretching and other exercises to provide pain relief. ▶ Biofeedback has been advocated for chronic low back pain (Bush et al. 1985).

• *Patient education*, which is also described as back schools of cognitive therapy have been used as a way to prevent recurrent episodes of back pain. Successful programs in Sweden, Canada, and California were described since 1977, touting their ability to reduce back pain incidence. They involved classes in which the patients were educated on spinal anatomy and biomechanics, along with receiving specific exercise programs and what they can and cannot do in their daily lives (Slipman et al. 2007).

• *Injection therapy*, including trigger point and ligament injections, are controversial and are of equivocal efficacy. Acupuncture has been studied in randomized clinical trials for chronic low back pain. The studies have been mostly contradictory. This, however, should not discount the fact that they may be effective for a subset of people.

Surgical Treatment

In patients with compressive lesions, surgery (decompression; e.g., ▶ laminectomy) is the next step when conservative therapy fails (Holdsworth 1963). Fusion with or without instrumentation is appropriate in the refractory patient with mechanical low back pain. The goal of surgical intervention for mechanical low back pain is stabilization of unstable spinal segments. Spinal fusion is an accepted therapy for fracture/dislocation or acute instability that may result from tumor or infection. It also may be used in selected patients with glacial instability or dysfunctional segmental motion. It is important to note that the use of spinal instrumentation increases the fusion rate (Lorenz & Zindrick, 1991), but not necessarily the clinical outcome.

References

Arnoldi, C. C., Brodsky, A. E., Cauchoix, J., et al. (1976). Lumbar spinal stenosis and nerve root entrapment syndromes. Definition and classification. *Clinical Orthopaedics and Related Research, 115*, 4–5.

Benzel, E. C. (2001). *Biomechanics of Spine Stabilization: Stability and Instability of the Spine.* (pp. 29–43). AANS Press. Rolling Meadows, Illinois

Bigos, S., Bowyer, O., Braen, G., et al. (1994). *Acute Low back problems in adults clinical practice guideline 14.* AHCPR publication No 95–0642. Rockville, MD: Agency for Healthcare Policy and Research, Public Health Service, US Department of Health and Human Services.

Bush, C., Ditto, B., & Feuerstein, M. (1985). A controlled evaluation of paraspinal EMG biofeedback in the treatment of chronic Low back pain. *Health Psychology, 4*, 307–321.

Cunningham, L. S., & Kelsey, J. L. (1984). Epidemiology of musculoskeletal impairments and associated disability. *American Journal of Public Health, 74*, 574–579.

Cypress, B. K. (1983). Characteristics of physician visits for back symptoms: A national perspective. *American Journal of Public Health, 73*, 389–395.

Denis, F. (1983). The three column spine and its significance in the classification of acute thoracolumbar spine injuries. *Spine, 8*, 817–831.

Dvorak, J., & Panjabi, M. M. (1991). Functional radiographic diagnosis of the lumbar spine: Flexion-extension and lateral bending. *Spine, 16*, 562–571.

Freburger, J. K., Holmes, G. M., Agans, R. P., et al. (2009). The rising prevalence of chronic low back pain. *Archives of Internal Medicine, 169*(3), 251–258.

Gatchel, R. J., Mayer, T. G., Capra, P., et al. (1986). Quantification of lumbar function, VI: The Use of psychological measures in guiding physical functional restoration. *Spine, 11*, 36–42.

Hansson, T. H., & Hansson, E. K. (2000). The effects of common medical interventions on pain, back function, and work resumption in patients with chronic low back pain: A prospective 2 year cohort study in six countries. *Spine, 25*(23), 3055–3064.

Hardt, J., Jacobsen, C., Goldberg, J., Nickel, R., & Buchwald, D. (2008). Prevalence of chronic pain in a representative sample in the united states. *Pain Medicine, 9*(7), 803–812.

Holdsworth, F. W. (1963). Fractures, dislocations, and fracture-dislocations of the spine. *Journal of Bone & Joint Surgery, 45B*, 6–20.

Lorenz, M., & Zindrick, M. (1991). A comparison of single level fusion with and without hardware. *Spine, 16*, 455–458.

Machado, L. A. C., de Souza, M. V. S., Ferreira, P. H., & Ferreira, M. L. (2006). The McKenzie method for low back pain: A systematic review of the literature with a meta-analysis approach. *Spine, 31*(9), E254–E262.

Malmivaara, A., Hakkinen, U., & Aro, T. (1995). The treatment of acute Low back pain - Bed rest, exercises, or ordinary activity? *The New England Journal of Medicine, 322*, 351–355.

Nachemson, A. L. (1992). Newest knowledge of Low back pain. A critical look. *Clinical Orthopaedics, 279*, 8–20.

Panjabi, M. M., & Lydon, C. (1994). On the understanding of clinical instability. *Spine, 19*, 2642–2650.

Pearcy, M., & Shepherd, J. (1983). Is there instability in spondylolisthesis? *Spine, 10*, 461–473.

Rozenberg, S., Delval, C., Rezvani, Y., et al. (2002). Bed rest or normal activity for patients with acute low back pain: A randomized controlled trial. *Spine, 27*(14), 1487–1493.

Slipman, C. W., Derby, R., Simeone, F. A., & Mayer, T. G. (2007). *Interventional spine: An algorithmic approach* (1st ed.). Philadelphia: Saunders.

Stewart, W. F., Ricci, J. A., Chee, E., Morganstein, D., & Lipton, R. (2003). Lost productive time and cost due to common pain conditions in the US workforce. *Journal of the American Medical Association, 290*(18), 2443–2454.

Szpalski, M., Robert, G., Huec, L., & Charles, J. (2007). *Nonfusion technologies in spine surgery*. Philadelphia: Lippincott Williams & Wilkins.

Waddell, G., McCulloch, J. A., Kummel, E., et al. (1980). Non-organic physical signs in Low back pain. *Spine, 5*, 117–125.

White, A. A., & Panjabi, M. M. (1990). *Clinical biomechanics of the spine* (2nd ed., pp. 30–342). Philadelphia: Lippincott.

Chronic Central Pain Models

▶ Contusion Model of Chronic Central Pain After Spinal Cord Injury

▶ Spinal Cord Injury Pain Model, Hemisection Model

Chronic Constriction Injury Model

Synonyms

CCI model

Definition

CCI is one of the first widely used animal models of persistent neuropathic pain, introduced by Bennett and Xie in 1998. It consists of tying several ligatures of chromic gut suture material loosely around the sciatic nerve. The chromic gut causes inflammation and gradual swelling of the nerve within the ligatures and hence induces gradual constriction injury of the nerve. Most myelinated axons are severed by the constriction, although many unmyelinated axons survive. The nerve injury causes spontaneous ectopic firing, ectopic mechanosensitivity at the injury site, and behavioral signs of spontaneous pain, allodynia and hyperalgesia. After several months the nerve swelling subsides, the severed myelinated axons regenerate, and the pain resolves. CCI-type injury has been applied in rodents to other nerves with similar effects.

Cross-References

▶ Neuropathic Pain Model, Chronic Constriction Injury

▶ Neuropathic Pain Model, Partial Sciatic Nerve Ligation Model

▶ Nociceptive Processing in the Hippocampus and Entorhinal Cortex, Neurophysiology and Pharmacology

▶ Peptides in Neuropathic Pain States

▶ Purine Receptor Targets in the Treatment of Neuropathic Pain

References

Bennett, G. J., & Xie, Y. K. (1988). A peripheral mononeuropathy in rat that produces disorders of pain sensation like those seen in man. *Pain, 33*, 87–107.

Chronic Constrictive Injury

▶ Neuropathic Pain Model, Chronic Constriction Injury

Chronic Daily Headache

▶ New Daily Persistent Headache

Chronic Daily Headache in Children

Mark Connelly and Jennifer Bickel
Integrative Pain Management, Children's Mercy
Hospitals and Clinics, University of Missouri-
Kansas City School of Medicine, Kansas City,
MO, USA

Synonyms

CDH (chronic daily headache); Chronic
migraine; Chronic tension-type headache; Headache; Transformed migraine

Definition

The concept of "chronic daily headache" (CDH)
was first introduced in the adult headache literature about 30 years ago to characterize headache occurring almost every day often after
a history of episodic migraine (Mathew et al.
1982). The first publication on CDH in the pediatric literature appeared in 1994 (Holden et al.
1994). Since that time, definition criteria remain
actively debated. Chronic daily headache most
often is defined as headaches lasting at least 4 h
per episode occurring at a frequency of 15 days
or more per month for at least 3 months (Seshia
et al. 2010; Gladstein and Rothner 2010).
Included in this definition are headaches that
are perceived to be continuous (occurring all
day every day). Typically, the term is applied
to headaches for which other etiologies have
been ruled out (e.g., head/neck trauma, vascular
disorder, intracranial mass, or increased intracranial pressure), and primary CDH is the focus
of the remainder of this entry. However, the
term CDH may still be applied if the underlying
headache type is thought to be secondary to
another condition (Seshia 2012). In recent
years, the Headache Classification Committee
has advised replacing the term "chronic daily
headache" with "chronic headache," but this
recommendation has not been globally accepted
(Seshia 2012).

Characteristics

Characteristics and Classification

Characteristics of CDH may vary by subtype but
in general are comparable between pediatric and
adult presentations (Seshia 2012). However, children may use different language to describe their
headaches or may express their pain differently
from adults depending on their cognitive development. Children also may have greater difficulty
articulating when headaches start and stop.

Figure 1 shows prototypical symptom patterns
plotted over time for three typical CDH subtypes in
children. Frequently in children with CDH, headaches that initially could have been classified as
episodic migraine have transformed over time into
a syndrome in which intermittent headache episodes with strong migraine features occur along
with (usually) less intense continuous, daily, or
near-daily headaches. This subtype of CDH has
been referred to as "transformed migraine" but is
more formally classified as "chronic migraine"
(Headache Classification Committee et al. 2006),
with some experts arguing that the presence or
absence of tension-type headache features should
be explicitly designated when classifying the headache type (Seshia et al. 2010). In this presentation

Chronic Daily Headache in Children, Fig. 1 Chronic
daily headache subtype symptom patterns

of CDH, children may describe the frequent but less severe headaches as a "band-like" or "pressure" type of pain, whereas often children describe the more severe intermittent episodes as a "throbbing," "pounding," "stabbing," or "knife-like" sensation. Approximately 85 % of the time pain location is bilateral, with pain being perceived to be most severe bifrontally (Seshia 2012). The frequency of the more severe episodes may range from once per month to a few times per week. Typically, these episodes are associated with migrainous features such as nausea, vomiting, and sensory sensitivities (i.e., sensitivity to light, noise, and/or smells). At times children also will report a particular type of sensitivity to normal touch sensations, wherein light touch or pressure applied to the head is perceived as painful ("allodynia"). Further, a small percentage of children with this type of headache disorder may additionally report brief and intermittent severe stabbing ("ice-pick") pain sensations at multiple locations around the head occurring many times per day (Mack 2011).

In other cases of CDH, there is no report of current or past migrainous features (e.g., sensitivity to lights or sounds, nausea/vomiting) associated with headaches. Instead, children report a history consistent with episodic tension-type headache and over time the headache has progressed in frequency to occur daily or near daily. This subtype of CDH debatably is classified as "chronic tension-type headache" (Gladstein and Rothner 2010; Seshia et al. 2009).

In a third variation of chronic daily headache in children, a child with no or minimal history of headaches suddenly develops a continuous headache within a span of 24–72 h. This subtype of headache typically is classified as "new daily persistent headache" (Seshia et al. 2010; Baron and Rothner 2010). Often this type of CDH is particularly alarming to children and their families given the relative absence of prior experiences with headache. Frequently the child or parent can recall the exact date of onset, and in some cases the date of onset is proximal to a viral illness or trauma (Baron and Rothner 2010). The pain typically is moderate, "pressing," and poorly localized. Usually the headache is not associated with migrainous features.

A fourth variation of chronic daily headache presents as a unilateral steady frontal headache with frequent daily severe pain exacerbations that are accompanied with cranial autonomic features (e.g., conjunctival injection and/or tears, nasal congestion, eyelid edema and/or drooping, and forehead or facial sweating). This presentation is classified as "hemicrania continua" and is unique in its significant responsiveness to a particular medicine (indomethacin) (Gladstein and Rothner 2010). This type of CDH is quite uncommon in the pediatric population and thus often is omitted in the literature on pediatric CDH.

Epidemiology

In the past decade, there has been a considerable increase in knowledge of the epidemiology of CDH in children, albeit limited by variations in classification and diagnostic rigor. Extant studies suggest that CDH is a common chronic pain syndrome in children and adolescents regardless of geographic location, affecting about 1–8 % of the pediatric population depending on the study (Lipton et al. 2011). Girls are affected as much as three times as frequently as boys across the pediatric age range (Wang et al. 2006; Arruda et al. 2010). Although CDH has been reported to occur in children below 6 years of age (Raieli et al. 2005), typically its prevalence increases from early childhood to preadolescence and adolescence. In headache clinic samples, children with CDH comprise approximately 30 % of patients being evaluated (Seshia et al. 2010; Gladstein and Rothner 2010).

Genetics

A strong family history of headache often has been documented with pediatric primary headaches, albeit not specifically with CDH. Although family history of headache does not exclude the role of shared environment, there has been some speculation regarding the extent of genetic contribution to chronic headache in children. Research into the genetics of pediatric CDH is rendered difficult by wide phenotypic variability within each age group and between ages, significant comorbidities, the absence of known specific clinical and biological markers, and the potential role of modifying genes (Lisi et al. 2005). Overall, the research is mixed regarding the influence of genes associated with such processes and structures as serotonin transport, catecholamine metabolism, folate metabolism, and potassium, sodium, and calcium channels (Hershey 2008). Currently, there are no specific identified genetic risk factors for CDH in children.

Comorbid Conditions

Other pain syndromes. Children with CDH may report co-occurrence of persistent pain at other body sites, such as abdominal pain, back pain, and/or widespread musculoskeletal pain. Although headache often has been shown to be comorbid with other pain syndromes in children and adults and may become part of a central sensitization disorder (Scher et al. 2006), data are unavailable on the specific association of pediatric CDH with other pain syndromes.

Sleep disturbances. Studies on the association between sleep problems and CDH in children typically have suggested a robust connection that may be even stronger in children with migraine features relative to tension-type (Seshia et al. 2010). In particular, CDH often is associated with increased sleep onset latency and poor sleep maintenance. However, it remains unclear if children with CDH have identifiable sleep disorders versus general disturbed sleep. It also is unclear if chronic headaches precede or follow from the development of sleep disturbance. Results of polysomnographic studies have suggested that children with tension-type headache may be likely to have bruxism and that those with chronic migraine are likely to have disrupted sleep architecture (reduced rapid eye movement and slow-wave sleep) (Seshia et al. 2010; Vendrame et al. 2008). Additional studies are warranted prior to making conclusions about an association between sleep disorders per se and CDH in children.

Psychiatric comorbidity. Community-based studies typically have documented increased likelihood of symptoms of anxiety and depressed mood in children with CDH relative to healthy controls, with up to about one-third of pediatric patients with CDH having significant anxiety or depressive symptoms (Seshia 2012). Whether these symptoms are more likely to predate headache development or develop as a consequence of living with frequent headaches remains debated,

but the association likely is bidirectional. The mood disturbance of a child with CDH may resolve during headache treatment without headache improving or headaches may improve while mood does not, suggesting some independence of the conditions. Biological parents of children with CDH also are more likely to have an anxiety or depressive disorder (Seshia et al. 2008).

With regard to other psychiatric comorbidities, a recent review of psychopathology in children with migraine in clinical settings suggested that children with migraine have no greater thought problems, social problems, or externalizing behavior problems than healthy children (Bruijn et al. 2010). Other studies of children with CDH, however, have found greater rates of psychiatric disorders (e.g., adjustment disorder, conversion and other somatoform disorders, histrionic traits) relative to normative data or matched controls (Seshia et al. 2010). Some studies also have found a link between CDH in adolescents and increased risk of suicidal thoughts (not necessarily attempts) even when controlling for depression (Wang et al. 2007; van Tilburg et al. 2011).

Obesity. Adult research has shown obesity to be a risk factor for chronification of migraine, but it is unclear if the incidence of obesity is greater in children with chronic daily headache relative to the general population (Seshia et al. 2010; Bigal and Rapoport 2012). Some studies have identified that obesity may be more likely to be comorbid in females with chronic headaches compared to males (Seshia et al. 2010).

Orthostatic symptoms. Children with CDH often identify symptoms of dizziness, weakness, or near syncope that occurs both during and between episodes of severe headache (Mack 2011). Children with CDH may demonstrate tachycardia and/or decreased blood pressure upon standing from a sitting position. These signs and symptoms may be related to broader autonomic nervous system dysfunction that may precede or result from chronic headache. Typically, orthostatic symptoms will resolve with increased fluid and salt intake.

Pathophysiology

Although several speculations exist regarding the pathophysiology of primary CDH in children, the mechanisms remain unclear (Gladstein and Rothner 2010). There are no definitive genetic, biochemical, or environmental markers of the disorder, which complicates identification of pathophysiological mechanisms. Sensitization of both central and peripheral nervous systems are implicated in the development of CDH but their relative contribution is debated. Pathological changes in the trigeminal system, neuroimmunity, and pain transmission and modulation systems also are thought to be central to the development of CDH (Seshia et al. 2010). For CDH with migraine features, the pathophysiology may include components of neurogenic inflammation, vascular reactivity, and serotonin dysregulation that have been thought to contribute to episodic migraine, whereas musculoskeletal abnormalities and pericranial muscle tenderness may be more relevant to the development of chronic tension-type headache.

In an effort to elucidate possible pathophysiological mechanisms in childhood CDH, investigators often have focused on establishing risk factors. Although in many cases of CDH there is no identifiable antecedent problem, some studies support the potential etiological significance of minor head trauma, viral infection, other illnesses, surgery, suboptimal health habits (poor nutrition, irregular sleep schedule, limited physical activity, and frequent caffeine consumption), anxiety, and depression (Gladstein and Rothner 2010). Stressors in particular appear to contribute to chronification of headaches in up to half of children with CDH (Seshia et al. 2008, 2010). Common stressors include family discord, social rejection, bullying (including "cyber bullying"), and school difficulties. Stress associated with early childhood maltreatment also has been identified as having potential etiological significance in the development of CDH.

Frequent use of analgesic medicine has been implicated as a risk factor for headache chronification in adults and children (Seshia et al. 2010). Although "medication overuse headache" is considered a separate diagnostic entity from CDH and rarely do children meet the criteria (taking a given medicine 10 or more days per month for at least 3 months), using certain medicines such as nonsteroidal anti-inflammatories, opioids, or triptans more than a few times per week may increase likelihood of developing and maintaining CDH in children.

Impact

Chronic headaches can have a substantial emotional and financial impact on the life of a child and family. Headaches often adversely affect school attendance (and therefore parent's work attendance) and/or participation in other typical activities of childhood and the cost of evaluation and treatment of headaches can be excessive. However, few studies have formally assessed the burden of headaches specifically within children with CDH. In a recent population-based study of adolescents identified as having chronic migraine (Lipton et al. 2011), the majority of respondents scored in the severe range of disability measures. From our clinical experience, it is a minority of children with CDH that have substantial disability often most likely manifested as missing on average at least 1 day of school per week for headaches. Children with severe disability associated with headaches often require greater time and personnel resources for optimal evaluation and treatment (e.g., an interdisciplinary team comprising a physician, psychologist, and other allied health professionals).

Evaluation of CDH in Children

Evaluation of CDH should be rooted in a biopsychosocial framework to determine not only the nature of the physical symptoms but the broader context in which the symptoms have developed and been maintained. The goal of the evaluation is to facilitate classification of headache, determine what factors may contribute to occurrence and maintenance, set the stage for the treatment plan, and determine potential barriers to a positive prognosis (e.g., comorbid disorders). The evaluation should include a past medical history and detailed headache history (length of time with headaches, severity, quality, location), family history (headache or other pain syndromes in the family, family mental health history), inquiry into academic, social, and family functioning (e.g., intensity of extracurricular activities, school attendance, grades, experiences of bullying/social exclusion or rejection, quality of relationships with family members, maltreatment experience), perceived stressors, and evaluation of mood. When evaluating mood, questioning the child about suicidality is important; suicidal ideation often is increased in the population of children with CDH and may prompt the selection of a different medication that carries less risk of intensifying suicidal thoughts. Questions about history of self-harm (e.g., self-cutting or prior suicide attempts), frequency of thoughts about suicide, extent of planning, and supports (e.g., strong connection to family, future goals) all are important for a basic suicide screening. Often it can be helpful to interview children separately from the parent/guardian for at least a portion of the initial clinic visit in order to discuss these more sensitive topics in private.

Evaluation of CDH also should include a systematic and detailed physical examination, with a focus on the neurological exam. Components of the evaluation should include at minimum a fundoscopic examination, an examination of motor coordination and symmetry of motor movements, and observation of eye movements, gait, and reflexes (including plantar responses).

The majority of children with CDH will not require further specific investigation. In only about 10 % of cases of children with CDH or less is there an identifiable secondary cause and often this is apparent from history and physical alone (Seshia et al. 2010). Additional evaluation through bloodwork or imaging typically only is indicated if the neurological examination reveals concerns, elements of the history suggest other causes of headache that have yet to be worked up (e.g., trauma or infection), or the caregiver or child are highly distressed and would benefit from reassurance through testing. Flags for potential secondary headache include persistent unilateral headache, intractable headache, focal neurological symptoms, headache worsening with posture change or coughing, personality change, progressive deterioration of school work, and headache awakening the child from sleep (Seshia et al. 2009). If other causes of CDH are suspected, then MRI of the head (including MRA and MRV) and spine should be considered rather than computed tomography due to less risk to the child and greater yield of information

(Seshia et al. 2010). However, sometimes findings such as white matter abnormalities, arachnoid cysts, or pineal cysts will be seen on MRI and become a source of concern to the family despite being of no likely clinical significance; preempting this by letting families know of this potential prior to the MRI is recommended. If idiopathic increased intracranial pressure (pseudotumor cerebri) is suspected based on signs (papilledema and sixth nerve palsy) or symptoms (diplopia, tinnitus, or eye pain), then a lumbar puncture can be considered but should be used judiciously given the risk of a post lumbar puncture headache. Additional investigations are determined by the differential diagnosis being considered (e.g., thyroid study, sedimentation rate, antinuclear antibody concentration, or viral tests).

Management of CDH

Despite CDH being a relatively prevalent chronic medical condition in children, only a small minority of those affected present to a healthcare provider or have taken medicines to prevent headaches (Lipton et al. 2011). Although CDH may self-resolve without treatment (Gladstein and Rothner 2010; Wang et al. 2009), it is likely that a significant number of children experience chronic headache and associated disability for much of their lives without ever being treated. This may be partly facilitated by the myth that CDH in children is a permanent condition for which there are no effective or safe treatment options.

Management options for CDH include both pharmacological treatment (e.g., medications and supplements) and nonpharmacological treatment (e.g., cognitive-behavioral therapy, relaxation training, manual therapies, massage therapy, chiropractic or osteopathic manipulation, and acupuncture). In general, the empirical evidence base for any particular treatment modality for CDH in children is lacking, but accumulated experience at different treatment centers suggests that a combination of medication, education, and adjuvant therapies often is needed to optimize outcomes (Seshia 2012).

Education. Children with primary CDH and their parents/caregivers often find themselves both hoping for and dreading an underlying medical diagnosis that will explain all symptoms.

A practitioner therefore must strike a balance between providing reassurance about the absence of a serious cause, while explaining that the headache itself *is* the disorder and is influenced by multiple contributing factors. It can be quite difficult for families to understand how a headache can persist without being a symptom of some underlying problem. Unless sufficient time is taken early on to address this concern, families may "doctor shop" in search of an answer or instant cure.

Laying a foundation for the treatment plan for CDH also is a critical aspect of education, including emphasizing with the child and family the importance of attending to stressors and health behaviors (diet, sleep, physical activity) to facilitate a greater likelihood of other therapies being effective. Some families may need education on the rationale and steps for discontinuing the overuse of certain medications that may be contributing to headache maintenance. Education around the importance of a gradual return to (or continuation of) participating in normal activities, particularly school, also is critical.

Pharmacological treatment. Despite the high prevalence of CDH in children and adolescents, there is a significant lack of well-designed trials to guide medication management. For several years, individuals with CDH were excluded from headache prevention trials. Currently, there are no FDA-approved treatments for CDH in the pediatric population. The following recommendations are mainly derived from adult studies and clinical experience at our institution.

The pharmacological treatment of headache generally utilizes both abortive and preventative medications. NSAIDS, triptans, and antiemetics all have efficacy in aborting headaches. However, if an individual has a high frequency of headaches, abortive medicines are unlikely to be effective. Focus instead primarily should be placed on decreasing the overall frequency and severity of the headaches, not on stopping each individual headache. If abortive mediations are used, it is recommended to keep the use at 8 days or less per month.

Opioid analgesics and butalbital-containing medications have essentially no role in the management of CDH. These medications often are ineffective and increase the risk of developing medication overuse headaches (i.e., rebound headaches). It is a common misperception that "rebound headaches" only occur with the daily use of pain relievers. However, recent studies have suggested that it is not a "rebound" from abortive medications that leads to high frequency headaches. Instead, the use of certain acute medications increases the likelihood of migraine chronification. Opioids increase the risk of migraine chronification beginning with exposure of approximately 8 days per month, and butalbital increases the risk with exposure of as little as 5 days per month. NSAIDS and triptans may be somewhat less likely to lead to migraine progression (Bigal and Lipton 2008).

The goal of preventative medicines when treating CDH is not headache freedom. Instead, the purpose is a gradual decrease in the severity and frequency of headaches until headaches are no longer affecting daily functioning and quality of life. It is typical for a patient to respond first with a decrease in severity of daily headaches, later followed by an increase in headache-free days. Medications may take up to 3 months to show full effectiveness. It is often best to start with a low dose of the medication with gradual increases based on efficacy and tolerability. If realistic goals are not first discussed with the family prior to starting the medication, the family likely will voice frustration and impatience and/or become non-adherent.

There is no first line pharmaceutical treatment in the management of chronic headaches. The medication is chosen based on headache classification, side effect profile, and patient preference. For example, a competitive high school athlete may wish to avoid beta-blockers because of risk of exercise intolerance. Topiramate may be ideal for the overweight child with migraines given its effect on weight reduction. Also, "natural" supplements or vitamins may be recommended for families that prefer to avoid daily medications. For children for whom risk of suicide is elevated, certain medicines should be avoided or at least used with caution and frequent monitoring (e.g., anticonvulsants and serotonin reuptake inhibitors).

The table below provides a listing of medicines used in the preventative management of CDH in children, along with dosing information and comments. Medicines and their doses are based on extrapolation from published adult CDH trials and clinical experience at our institution. In addition to the medicines listed in the table, onabotulinumtoxinA might be considered in certain cases of pediatric CDH. In 2010, onabotulinumtoxinA became the first treatment to receive FDA approval in adults for management of CDH. Although pediatric studies are limited, most pediatric headache centers offer onabotulinumtoxinA injections to patients who have failed to respond to two or more adequate trials of traditional headache prevention.

Medicine	Target dose[a]	Comments
Beta-blockers		
Propranolol	2–3 mg/kg/day	Avoid use in competitive high school athletes and children with history of asthma
Calcium channel blockers		
Flunarizine	5 mg (above age 5)	Not available in the United States. Data are unclear if other calcium channel blockers are equally effective
Tricyclic antidepressants		
Amitriptyline	1 mg/kg/day	If too sedating, nortriptyline may be considered
SSRIs/SNRIs		
Fluoxetine	20 mg daily	Consider in adolescents with comorbid depressive mood if suicide risk is considered low and regular follow-up is possible
Venlafaxine XR	150 mg daily	
Anticonvulsants		
Topiramate	2–3 mg/kg/day	Consider in overweight/obese children
Divalproex sodium	15–45 mg/kg/day	Use with caution in females of childbearing age
Gabapentin	300–400 mg TID	
Antihistamines		
Cyproheptadine	2–4 mg q 8–12 h	Weight gain and sedation may be an issue in older children
Supplements/vitamins		

(*continued*)

Medicine	Target dose[a]	Comments
Magnesium oxide or gluconate	9 mg/kg/daily	Consider in children who want to avoid prescription medicines
Riboflavin	200–400 mg daily	Consider in young children
Petasites	50–75 mg BID	Purified extract from butterbur plant. Avoid non-purified formulations which have been linked to liver toxicity

[a]Note: Medicines are best tolerated if started at approximately 20 % of target dose with gradual titration as tolerated.

Nonpharmacological treatment. Most well studied among the nonpharmacological treatment options for headache in children are relaxation training and cognitive-behavioral therapy. Relaxation training involves learning specific strategies for inducing a relaxation response, which helps reduce the adverse impact of sympathetic activation on the initiation and maintenance of headaches and also can help improve associated symptoms such as sleep disruption. Common techniques include diaphragmatic breathing, progressive muscle relaxation, visualization/imagery, and meditation (Sieberg et al. 2012). Acquiring relaxation skills may be facilitated through the use of biofeedback equipment that uses visual and/or auditory media to indicate the extent to which the relaxation skills are having the desired autonomic effect; however, biofeedback is not necessary to the efficacy of relaxation strategies. Cognitive-behavioral therapy often incorporates some element of relaxation training but is a broader term that incorporates cognitive coping techniques (i.e., identifying, challenging, and modifying beliefs and thought habits that may contribute to pain, disability, and/or distress) and other behavioral strategies (e.g., identifying and modifying responses to headaches from parents/caregivers that contribute to less successful child coping and greater disability). A recent meta-analysis that included 16 trials of psychological therapies for headaches in children found that these therapies led to clinically significant improvements in pain but did not significantly reduce disability or improve emotional functioning beyond chance (Palermo et al. 2010). All studies to date have not explicitly focused on CDH, however, such that there is no definitive evidence about the efficacy of psychological therapies specifically for CDH. We recommend considering suggesting self- or therapist-guided relaxation training to most children with CDH and referring children with high disability and/or comorbid psychiatric disorders for broader psychological assessment and therapy.

The efficacy of other nonpharmacological treatment options for children with CDH remains largely unknown. One study found support for the efficacy of laser acupuncture in reducing primary headaches in children but the sample was not specific to CDH (Seshia 2012). Although manual therapies such as massage and chiropractic have been used to treat chronic headache in children, there have been no trials published on their safety or efficacy in this population.

Inpatient pain rehabilitation. Under some circumstances, admitting children with CDH to an inpatient pain rehabilitation program may be considered and has some empirical evidence of efficacy (Hechler et al. 2009; Lanzi et al. 2007). Typically, this is reserved for children with severe disability (e.g., not attending school), psychiatric comorbidity, and/or significant psychosocial factors presumed to be maintaining headaches who are not improving with outpatient therapies alone (Seshia 2012). The inpatient program often continues for a few weeks and involves medical management, more intensive psychological assessment and management, and adjuvant therapies (e.g., physical and occupational therapy, schoolwork assistance by specially trained teachers) directed at improving functioning with pain during the hospitalization and beyond. Observations during the inpatient stay also can be particularly helpful with conceptualizing the child's pain by changing the context in which it occurs (e.g., observing whether functioning changes in the presence or absence of parents/caregivers).

Outcomes

Outcome data are sparse for children treated for CDH, and length of follow-up for published studies varies greatly (1–10 years). Most studies report that headaches have decreased in frequency and/or associated disability by the point of follow-up

(Seshia et al. 2010). In one study on the long-term (10-year) follow-up of children with CDH, 86 % were improved (Guidetti et al. 2009). However, CDH may endure as a disabling problem for a clinically significant number of children, with reported risk factors for poorer outcomes being female sex, migraine headache, medication overuse, younger age at onset of CDH, greater duration of CDH, and comorbid psychiatric conditions (Seshia et al. 2010).

Summary

Chronic daily headache is one of the most common pain disorders of childhood, affecting at least 1 % of school-aged children and potentially persisting into adulthood. Although the definition and classification of CDH continues to be debated, the condition often presents as a daily or near-daily baseline headache and severe intermittent headaches that may be accompanied with sleep disruption, dizziness or weakness, and anxious and/or depressed mood. Frequently CDH disrupts normal child and family activities but only a minority of patients have substantial associated disability or seek treatment. The evaluation of children with CDH should proceed from a biopsychosocial framework to determine differential diagnoses and identify psychological and social factors that may be relevant to the presentation and prognosis. Successful approaches to treatment typically are multimodal, including education, use of preventive medications, judicious use of abortive medicines if indicated, nonpharmacological management strategies such as lifestyle modification and relaxation training, and in some cases inpatient pain rehabilitation. The majority of children with CDH improve with time and treatment, but some will have enduring headaches and disability into adulthood. More research is needed on which treatments work best for which children with CDH in order to optimize short- and long-term outcomes.

References

Arruda, M. A., Guidetti, V., Galli, F., Albuquerque, R. C., & Bigal, M. E. (2010). Frequent headaches in the preadolescent pediatric population: A population-based study. *Neurology, 74*(11), 903–908.

Baron, E. P., & Rothner, A. D. (2010). New daily persistent headache in children and adolescents. *Current Neurology and Neuroscience Reports, 10*(2), 127–132.

Bigal, M. E., & Lipton, R. B. (2008). Excessive acute migraine medication use and migraine progression. *Neurology, 71*(22), 1821–1828.

Bigal, M. E., & Rapoport, A. M. (2012). Obesity and chronic daily headache. *Current Pain and Headache Reports, 16*(1), 101–109.

Bruijn, J., Locher, H., Passchier, J., Dijkstra, N., & Arts, W. F. (2010). Psychopathology in children and adolescents with migraine in clinical studies: A systematic review. *Pediatrics, 126*(2), 323–332.

Gladstein, J., & Rothner, A. D. (2010). Chronic daily headaches in children and adolescents. *Seminars in Pediatric Neurology, 17*, 88–92.

Guidetti, V., Mittica, P., & Galli, F. (2009). Evolution of chronic daily headache in children and adolescents: A 10-year follow-up study. *Cephalalgia, 29*(Suppl. 1), 105.

Headache Classification Committee, Olesen, J., Bousser, M. G., Diener, H. C., Dodick, D., First, M., & Steiner, T. J. (2006). New appendix criteria open for a broader concept of chronic migraine. *Cephalalgia, 26*(6), 742–746.

Hechler, T., Dobe, M., Kosfelder, J., Damschen, U., Hübner, B., Blankenburg, M., & Zernikow, B. (2009). Effectiveness of a 3-week multimodal inpatient pain treatment for adolescents suffering from chronic pain: Statistical and clinical significance. *The Clinical Journal of Pain, 25*(2), 156–166.

Hershey, A. (2008). Genetics of headache in children. Where are we headed? *Current Pain and Headache Reports, 12*, 367–372.

Holden, E. W., Gladstein, J., Trulsen, M., & Wall, B. (1994). Chronic daily headache in children and adolescents. *Headache, 34*(9), 508–514.

Lanzi, G., D'Arrigo, S., Termine, C., Rossi, M., Ferrari-Ginevra, O., Mongelli, A., & Beghi, E. (2007). The effectiveness of hospitalization in the treatment of paediatric idiopathic headache patients. *Psychopathology, 40*(1), 1–7.

Lipton, R. B., Manack, A., Ricci, J. A., Chee, E., Turkel, C. C., & Winner, P. (2011). Prevalence and burden of chronic migraine in adolescents: Results of the chronic daily headache in adolescents study (C-dAS). *Headache, 51*(5), 693–706.

Lisi, V., Garbo, G., Miccichè, F., Stecca, A., Terrazzino, S., Franzoi, M., & Battistella, P. A. (2005). Genetic risk factors in primary paediatric versus adult headache: Complexities and problematics. *The Journal of Headache and Pain, 6*(4), 179–181.

Mack, K. J. (2011). Chronic daily headache in children. In J. S. Hyams & L. G. Feld (Eds.), *Consensus in pediatrics, frequent headaches in children and adolescents: Diagnosis and management* (pp. 1–6). Evansville, IL: Mead Johnson & Company.

Mathew, N. T., Stubits, E., & Nigam, M. P. (1982). Transformation of episodic migraine into daily headache: Analysis of factors. *Headache, 22*(2), 66–68.

Palermo, T. M., Eccleston, C., Lewandowski, A. S., Williams, A. C., & Morley, S. (2010). Randomized controlled trials of psychological therapies for management of chronic pain in children and adolescents: An updated meta-analytic review. *Pain, 148*(3), 387–397.

Raieli, V., Eliseo, M., Pandolfi, E., La Vecchia, M., La Franca, G., Puma, D., & Ragusa, D. (2005). Recurrent and chronic headaches in children below 6 years of age. *The Journal of Headache and Pain, 6*(3), 135–142.

Scher, A. I., Stewart, W. F., & Lipton, R. B. (2006). The comorbidity of headache with other pain syndromes. *Headache, 46*(9), 1416–1423.

Seshia, S. S. (2012). Chronic daily headache in children and adolescents. *Current Pain and Headache Reports, 16*, 60–72.

Seshia, S. S., Phillips, D. F., & von Baeyer, C. L. (2008). Childhood chronic daily headache: A biopsychosocial perspective. *Developmental Medicine and Child Neurology, 50*(7), 541–545.

Seshia, S. S., Abu-Arafeh, I., & Hershey, A. D. (2009). Tension-type headache in children: The Cinderella of headache disorders! *The Canadian Journal of Neurological Sciences, 36*(6), 687–695.

Seshia, S. S., Wang, S. J., Abu-Arafeh, I., Hershey, A. D., Guidetti, V., Winner, P., & Wöber-Bingöl, C. (2010). Chronic daily headache in children and adolescents: A multi-faceted syndrome. *The Canadian Journal of Neurological Sciences, 37*(6), 769–778.

Sieberg, C. B., Huguet, A., von Baeyer, C. L., & Seshia, S. (2012). Psychological interventions for headache in children and adolescents. *The Canadian Journal of Neurological Sciences, 39*(1), 26–34.

van Tilburg, M. A., Spence, N. J., Whitehead, W. E., Bangdiwala, S., & Goldston, D. B. (2011). Chronic pain in adolescents is associated with suicidal thoughts and behaviors. *The Journal of Pain, 12*(10), 1032–1039.

Vendrame, M., Kaleyias, J., Valencia, I., Legido, A., & Kothare, S. V. (2008). Polysomnographic findings in children with headaches. *Pediatric Neurology, 39*(1), 6–11.

Wang, S. J., Fuh, J. L., Lu, S. R., & Juang, K. D. (2006). Chronic daily headache in adolescents: Prevalence, impact, and medication overuse. *Neurology, 66*(2), 193–197.

Wang, S. J., Juang, K. D., Fuh, J. L., & Lu, S. R. (2007). Psychiatric comorbidity and suicide risk in adolescents with chronic daily headache. *Neurology, 68*(18), 1468–1473.

Wang, S. J., Fuh, J. L., & Lu, S. R. (2009). Chronic daily headache in adolescents: An 8-year follow-up study. *Neurology, 73*(6), 416–422.

Chronic Fibromyalgia

▶ Psychiatric Aspects of the Epidemiology of Pain

Chronic Gynecological Pain, Doctor-Patient Interaction

William Stones and Francis Githae Muriithi
Department of Obstetrics and Gynaecology,
Aga Khan University, Nairobi, Kenya

Synonyms

Attitude; Doctor-patient communication; Therapeutic relationships

Definition

Chronic pelvic pain (CPP) can be defined as a constant or intermittent pain in the lower abdomen or pelvis which is not exclusively related to menstrual period (dysmenorrhea) or sexual intercourse (dyspareunia) and which lasts for at least 3 months (Williams et al. 2004).

The proposition that patient outcomes in ▶ chronic pelvic pain are influenced by the quality of interaction between patient and doctor.

Characteristics

While it is a truism that across the spectrum of illness and disease, patients link their outcomes to experiences in consultations with particular doctors, there are specific factors to consider in the case of chronic gynecological pain. The nature of the condition and associated symptoms mean that women are often distressed and anxious about associated issues such as fertility and sexuality. Women frequently report unsatisfactory and dismissive attitudes during consultations for pain (Grace 1995), and this has been ascribed to inappropriately persistent mechanistic views of pain causation (Grace 1998). Gynecologists trained into a technical diagnostic and therapeutic role experience a dissonance between patient expectations for explanation and support and their own approach of detecting disease and

disengaging once its apparent absence is established, despite patients' ongoing symptoms of pain.

What Do Doctors Think About Women with Pelvic Pain?

To obtain primary data on medical attitudes, focus group discussions were conducted with groups of general practitioners, gynecologists, and patients (Selfe et al. 1998a). Themes common to all groups were expression of the need to find a pathological cause for the pain, something that would provide a diagnosis. Issues relating to time were discussed by all groups, and this was particularly related by hospital gynecologists and GPs to aspects of communication. Both groups of doctors were aware of the possible stress-related and psychological influences of pain, although patients seemed to avoid direct discussion of psychological factors which may have been related to their pain. Themes particular to general practitioners were diagnosis by exclusion, especially with respect to ▶ irritable bowel syndrome (IBS) and ▶ pelvic inflammatory disease (PID). They stated that the lack of a firm diagnosis meant that dealing with patients with CPP could be difficult, particularly those patients who have experienced a "negative" laparoscopy. If no visible pathology was seen at laparoscopy to account for the pain, GPs found it difficult to know how to proceed with persistent patients. A provisional diagnosis, even if it was considered imprecise, could provide a label by which to justify the patient's symptoms. This applied especially to IBS and "pelvic congestion."

GPs recognized the value of effective communication but acknowledged that a good rapport with a patient was unlikely to develop after one short visit. GPs also commented on the complexity of a complaint in which it was necessary to sort out "the emotional components of the problem."

Hospital gynecologists implied that identifying pathology would somehow validate the pain as "real." When visible pathology had been excluded by previous referrals to other specialities such as gastroenterology, comments included "some patients are going to be afraid that you're going to think it's a psychological problem" and "they

Chronic Gynecological Pain, Doctor-Patient Interaction, Fig. 1 Fitted probabilities of pain resolution and women's rating of initial consultation, with and without exercise impairment (Reproduced with permission from Selfe et al. (1998b))

[the patients] come in and they say, I'm not mad you know." There was a clear awareness that an anxious patient may make diagnosis difficult and that patients are alarmed by the suggestion of a psychological diagnosis because they will be "labelled." Hospital gynecologists were acutely aware of the time sometimes necessary to deal effectively with patients suffering from CPP. Development of rapport was found to be "difficult to establish in one outpatient session" If needed, consultation time was extended, but doctors were concerned that "you're filling up the waiting room" and there was anxiety that if, for example, a patient starts to divulge adverse psychosexual experiences, it may well become a lengthy consultation; the doctor thinks "Oh hell! I've got three patients in the waiting room." After a negative laparoscopy, hospital gynecologists felt the need to look for other causes, asking the question "Have I missed something?" They recognized this as problematic, as the patient is disheartened by not having a diagnosis, a "label" by which to define her problem.

It was hypothesized that a propensity to inwardly directed ► hostility might underlie some of the difficulty experienced by some patients in establishing rapport (Fry and Stones 1996). However, further work highlighted the importance of the individual clinician and consultation in influencing outcomes. Disease factors are of course also important; in statistical models, the interaction between disease states and the consultation setting was identified. In women referred to a hospital gynecology department for symptoms present for at least 6 months, continuing pain 6 months after an initial consultation was predicted by the presence of endometriosis but also by the patient's initial report of pain interfering with exercise, her rating of the initial medical consultation as less satisfactory, and the individual clinician undertaking the consultation. Interestingly, the outcomes were not affected by the doctor's grade or gender (Selfe et al. 1998b), although there has been much attention in women's health literature to the gender imbalance among gynecologists and the mismatch with women's preference to see a female doctor in many countries. In this study there was an interaction between the impact of pain on exercise and the extent to which the patient's rating of the quality of the consultation was associated with probability of pain resolution. For the group where exercise was not impaired, there was a significant association, whereas this was not the case among those in whom exercise was impaired by pain (Fig. 1). It may be that those with less impairment are more amenable to the therapeutic effect of a good consultation.

What Are the Expectations Regarding the Doctor-Patient Relationship?

A review "Qualitative research as the basis for a biopsychosocial approach to women with chronic pelvic pain" (Souza et al. 2011) revealed that patients expressed dissatisfaction and a strain of the doctor-patient relationship from a lack of clear diagnosis. Frustration is further aggravated by miscommunication arising from an ineffectiveness of the current methods of investigation and treatment. Failure to allay anxiety at initial appointment, rejection or downplaying of their complaints by health professionals, and defining their symptoms based on test results alone all have negative effects. The dissatisfaction of women apparently arises from the lack of a holistic approach to the problem, because health professionals tend to focus on the physical symptoms. Chronic pelvic pain patients reported difficulty in describing their pain objectively as well as in understanding the explanations and instructions received from health professionals during the appointments, which contributes to their feeling of not being believed. The instructions given may be by inappropriate terminology, leading to the miscommunication. One of the major expectations that pelvic pain patients have regarding their contact with health professionals is that their pain will be legitimized or validated, not through a specific diagnosis but through the weight given to the symptoms reported and to the impact that the pain has on various aspects of their lives. It is therefore essential that the pain that these women feel be acknowledged.

A failure to acknowledge these facts leads to a loss of incentive to cope with the disease or lead to patients withdrawing from seeking professional help.

A qualitative study assessing how comfortable general practitioners are in diagnosing chronic pelvic pain in the primary care setting showed an aspect of therapeutic nihilism (McGowan et al. 2010). This finding has a direct impact on the doctor-patient relationship and subsequent consultation.

Lasting Influences of Consultations

The dimensions within which women recalled consultations 6 months later were examined. Lasting impressions of the doctor-patient interaction that had taken place could be characterized in terms of the constructs "affect," "expectation," and "cognition" (Stones et al. 2006). The first describes social interaction and the emotional response felt by the patient. It relates to whether the doctor is recalled as approachable and expressing a friendly interest in and concern for the patient. Opposite recollections would not necessarily mean that the doctor had been unpleasant or rude to the patient but may simply reflect a businesslike manner, a hurried consultation, or the use of technical jargon. The construct "cognition" relates to memory and understanding, a prerequisite for adherence to treatment. Often, patients do not understand what they have been told, are afraid to ask questions, and forget much of what has been said during a consultation. Under such circumstances, it is not unreasonable that patients who were given little information or could not remember much of the detail of the encounter would find it unsatisfactory.

The items loading to the subscale denoted "expectation" relate to the patient's wish to understand her problem and what the implications may be for the future. Women expected to be given a diagnosis and as a result a cure. If this expectation was not fulfilled, patients described the situation as "disappointing" or felt that they had been "fobbed off." In the present study, the item "I have received enough information about my condition" loaded to "expectation" rather than to "cognition." This suggests that, rather than the quantity, the appropriateness of information given in relation to expectations is important in determining satisfaction.

Lower current levels of pain at follow-up were associated with highly significantly more favorable recall of the original consultation in terms of the "expectation" subscale. However, the impact of an initial consultation rated positively by the patient on the subscales "affect" and "expectation," but not "cognition," was still evident after controlling for current pain intensity. Thus, the experience of the initial consultation appeared to have an influence on the patient 6 months later irrespective of pain outcome. It is concluded that the doctor's affect and the extent to which expectations were met rather than simple information provision are the elements of communication through which this influence is mediated.

To avoid frustration to the patient and doctor in clinical practice, a structured assessment of women suffering from chronic pelvic pain using a cognitive behavioral model has been suggested. This type of assessment provides information about the impact of chronic pelvic pain on a particular patient's daily life. It also facilitates referral for pain management (Weijenborg et al. 2009).

Caregivers need to be well prepared to handle psychosomatic issues and if necessary make prompt referrals so that the women seeking advice could be led to effective diagnostics and therapy faster than before (Neis and Neis 2009).

Reassurance

Relief of symptoms by provision of reassurance is likely to be one of the doctor's aims following negative investigations, such as laparoscopy, but can also arise from explanation of normal findings at an initial consultation, for example, where a careful history and physical examination indicates a minimal likelihood of infection or endometriosis. However, there are mixed findings from research on the impact of assessment as therapy. In a randomized trial design, an ultrasound scan used as a basis for counselling about normal findings was found to be helpful in relieving symptoms (Ghaly 1994). The reassuring effect of laparoscopy has been reviewed (Price and Blake 1999). However, this approach can be counterproductive, as shown in a study where women were shown photographs of laparoscopic findings in order to reinforce counselling, where no benefit was seen (Onwude et al. 2004).

Summary

In conclusion, doctors need to be aware of the importance of establishing a therapeutic relationship through effective consulting behavior that meets patients' expectations and exploits opportunities for provision of reassurance and support. Gaining women's trust and developing a strong patient-physician relationship are of utmost importance for the long-term outcome of care (Vercellini et al. 2009).

References

Fry, R. P. W., & Stones, R. W. (1996). Hostility and doctor-patient interaction in chronic pelvic pain. *Psychotherapy and Psychosomatics, 65*, 253–257.

Ghaly, A. F. F. (1994). The psychological and physical benefits of pelvic ultrasonography in patients with chronic pelvic pain and negative laparoscopy. A random allocation trial. *Journal of Obstetrics and Gynaecology, 14*, 269–271.

Grace, V. M. (1995). Problems of communication, diagnosis, and treatment experienced by women using the New Zealand health services for chronic pelvic pain: A quantitative analysis. *Health Care for Women International, 16*, 521–535.

Grace, V. M. (1998). Mind/body dualism in medicine: The case of chronic pelvic pain without organic pathology – A critical review of the literature. *International Journal of Health Services, 28*, 127–151.

McGowan, L., Escott, D., Luker, K., et al. (2010). Is chronic pelvic pain a comfortable diagnosis for primary care practitioners: A qualitative study. *BMC Family Practice, 11*, 7.

Neis, K. J., & Neis, F. (2009). Chronic pelvic pain: Cause, diagnosis and therapy from a gynaecologist's and an endoscopist's point of view. *Gynecological Endocrinology, 25*(11), 757–761.

Onwude, J. L., Thornton, J. G., Morley, S., et al. (2004). A randomised trial of photographic reinforcement during postoperative counselling after diagnostic laparoscopy for pelvic pain. *European Journal of Obstetrics, Gynecology, and Reproductive Biology, 112*, 89–94.

Price, J. R., & Blake, F. (1999). Chronic pelvic pain: The assessment as therapy. *Journal of Psychosomatic Research, 46*, 7–14.

Selfe, S. A., Matthews, Z., & Stones, R. W. (1998a). Factors influencing outcome in consultations for chronic pelvic pain. *Journal of Women's Health, 7*, 1041–1048.

Selfe, S. A., van Vugt, M., & Stones, R. W. (1998b). Chronic gynaecological pain: An exploration of medical attitudes. *Pain, 77*, 215–225.

Souza, P. P., Romão, A. S., Rosa-e-Silva, J. C., et al. (2011). Qualitative research as the basis for a biopsychosocial approach to women with chronic pelvic pain. *Journal of Psychosomatic Obstetrics and Gynecology, 32*(4), 165–172.

Stones, R. W., Lawrence, W. T., & Selfe, S. A. (2006). Lasting impressions: Influence of the initial hospital consultation for chronic pelvic pain on dimensions of patient satisfaction at follow-up. *Journal of Psychosomatic Research, 60*, 163–167.

Vercellini, P., Somigliana, E., Vigano, P., et al. (2009). Chronic pelvic pain in women: Etiology, pathogenesis and diagnostic approach. *Gynecological Endocrinology, 25*(3), 149–158.

Weijenborg, P. T. M., Ter Kuile, M. M., & Stones, W. (2009). A cognitive behavioural based assessment of women with chronic pelvic pain. *Journal of Psychosomatic Obstetrics and Gynecology, 30*(4), 262–268.

Williams, R. E., Hartmann, K. E., & Steege, J. F. (2004). Documenting the current definitions of chronic pelvic pain: Implications for research. *Obstetrics and Gynecology, 103*, 686–691.

Chronic Illness Problem Inventory

Definition

The Chronic Illness Problem Inventory is a 65-item instrument developed to measure patient functioning in the areas of physical limitations, psychosocial functioning, health care behaviors, and marital adjustment.

Cross-References

▶ Pain Inventories

Chronic Inflammatory Demyelinating Polyneuropathy

Synonyms

Chronic relapsing polyneuropathy; CIDP

Definition

Chronic inflammatory demyelinating polyneuropathy is caused by demyelination due to an immune reaction. It is characterized by progressive weakness and impaired sensory function in the arms and legs. It is related to

Guillain–Barré syndrome, and is sometimes considered to be the chronic version of that disease.

Cross-References

▶ Inflammatory Neuritis

Chronic Joint Pain

▶ Chronic Musculoskeletal Pain

Chronic Low Back Pain: Definitions and Diagnosis

Jenna E. Zauk[1] and Christopher J. Winfree[2]
[1]St. George's University, School of Medicine, Grenada, West Indies
[2]Department of Neurological Surgery, Neurological Institute Columbia University, New York, NY, USA

Synonyms

Ankylosing spondylitis; Chronic pain; CT myelography; Diagnostic block; Discogenic pain; Facet joint; Fibromyalgia; Internal disc disruption; Intervertebral disc; Low back pain; Paget's disease; Pathological fracture; Provocative discography; Reiter's syndrome; Sacroiliac joint; Spondyloarthropathy; Spondylolisthesis; Zygapophyseal joint

Definition

Chronic low back pain is defined as pain localized to the region bordered by L1 above, S1 below, and the erector spinae muscles laterally (Merskey and Bogduk 1994), that lasts for more than 3 months (Merskey 1986). Chronic low back pain is often associated with pain radiating to the buttock or legs; however, it is useful to distinguish between these distinct pain syndromes, as they are typically treated differently.

Characteristics

General
Men and women are affected equally, most commonly during the four to six decades (Deyo and Weinstein 2001). Chronic low back pain is the most costly of the work-related disabilities, and about 3–4 % of individuals in the United States are affected annually (Wheeler 2010). The lumbar spine is a complex anatomical region, with a number of structures that may serve as pain generators (discussed below).

Neurophysiological and psychological factors, such as damaged tissues and nerves, perception of pain, emotional stress, and behavioral patterns, contribute to this complex disability (Wheeler and Murrey 2005; Wheeler 2010). Because of these cognitive/behavioral and physiological elements, it is highly recommended to treat this medical condition with interdisciplinary rehabilitation (Chou et al. 2009). However, patients must be willing to commit a great deal of time and effort to this therapy, which can amount to over 100 h. Based upon recent studies, this form of rehabilitation is successful in long-term pain relief for non-radicular low back pain (van Geen et al. 2007). However, common inconveniences of these programs include expensive fees, restricted insurance coverage, and limited locations (Wheeler 2010).

History
The initial assessment of the chronic low back pain patient includes a careful medical history. The duration and severity of symptoms, predisposing factors, exacerbating factors, alleviating factors, and treatments may all be explored in the patient interview. Information regarding occupational history, disability, litigation, psychosocial influences, and secondary gain may also be gleaned during the initial discussion. It is important to inquire about neurological function, such as weakness, bowel and bladder function, and sexual function, to establish whether neurological compromise is present, as treatment of these patients will likely proceed in a more urgent manner. Patients with "red flags" such as a history of malignancy, advanced age, immunosuppression, or recent trauma also require a more urgent workup. It is also important to document a baseline level of pain upon which to judge efficacy of interventions, if any.

Physical Exam
The physical exam is generally helpful but not typically sensitive or specific in the setting of chronic low back pain. The astute observer may discern peculiar movements or postures characteristic of the back pain patient. For example, these patients sometimes sit bent forward at the waist, as hyperextension often aggravates low back symptoms. They may also experience pain while raising a straight leg in the supine position, or when light pressure is applied to the patient's head (Wheeler 2010). Some patients fail to get comfortable, and spend the time during the interview switching positions, or alternating standing and sitting.

Imaging
Conventional roentgenography is rapid, inexpensive, and best suited for inspection of the osseous structures. Vertebral fracture, osteoporosis, loss of disc height, end plate changes, facet hypertrophy, and alignment deformity are all readily seen on conventional X-rays. Characteristic bony changes of selected spondyloarthropathies are also readily identifiable with this modality. The typical static images include anteroposterior and lateral lumbar and sacral views, left and right lateral lumbar oblique (for inspection of the neural foraminae), and a cone-down lateral sacral view. Dynamic images to assist in the evaluation of spinal stability include lateral flexion and extension views.

Magnetic resonance imaging (MRI) is the best modality available for evaluation of the soft tissue structures of the spine. The intervertebral discs, ligaments, and neural structures are all well seen. Typically, non-contrast imaging is all that is required; however, the use of contrast may be useful in the setting of tumor, malignancy, or previous surgery. Caution must be exercised in the interpretation of positive findings on MRI, as over half of asymptomatic subjects will have at least one abnormality on MRI imaging (Jensen et al. 1994).

Computerized tomography (CT) scanning is useful to delineate the bony anatomy of the spine, and to clarify positioning of metallic instrumentation. It is also a vital part of provocative discography, providing detailed anatomical data on the morphology of the intervertebral discs in question. CT myelography provides the optimum contrast between neural and non-neural structures, and is especially useful when MRI scanning is compromised by artifact, as in the setting of previous spinal instrumentation, or if the patient is unable to undergo MRI scanning. Subtle nerve root compression or spinal canal encroachment invisible on MRI may be obvious on CT myelography.

Bone scanning has limited but specific uses in the imaging of chronic back pain. It may be useful in the evaluation of suspected pseudoarthrosis at a previous fusion site, or also of malignancy or infection. Nerve root blocks, electromyography (EMG) and somatosensory-evoked potential (SSEP) testing are useful diagnostic studies when imaging tests have proven to be unreliable. These studies will detect nerve damage by assessing the electrical transmission and the speed at which nerve signals travel (Wheeler 2010).

Discogenic Pain

Discogenic pain is defined as pain from the disc itself and not surrounding structures compressed by the disc. It is probably the most common cause of chronic low back pain, with internal disc disruption present in approximately 40 % of cases (Schwarzer et al. 1995a). These patients often describe pain localized to the low back, often accompanied by pain referred to the hip, buttock,

or groin. Pain that radiates to the lower extremities may be patchy or "sclerotomal" in nature rather that dermatomal, as seen in herniated discs. These patients typically report that their pain is aggravated by standing or sitting and relieved by lying down, presumably since this position takes the pressure off the intervertebral discs. The intervertebral disc receives innervation along at least the outer third of the annulus (Coppes et al. 1997). The disc may become painful when it undergoes disruption; however, nonspecific degenerative changes are not thought to be specifically painful (Moneta et al. 1994). These changes may be obvious on conventional imaging studies, and include loss of height, rupture, end plate changes, and dehydration. Other, subtler changes, such as internal disc disruption (IDD), may only be revealed as small radial fissures on CT imaging following discography (Vanharanta et al. 1988).

Provocative discography consists of both physiologic and radiographic assessment, and is used to establish whether a degenerated disc is a pain generator in the chronic low back pain patient. It involves the fluoroscopic placement of a needle into the intervertebral disc of the awake patient, and subsequent injection of contrast dye. Intradiscal insertion and injection pressures are measured. A positive discogram occurs when the patient's back pain is reproduced when dye is injected into a particular disc at modest pressures (concordant pain). A negative discogram occurs when the pain on injection only occurs at extremely high pressures, or does not match the patient's normal pain. A complete discogram always includes the injection of a normal control disc, to exclude the possibility of nonspecific patient pain responses to injection. Once the procedure is complete, high-resolution CT scanning is performed to illustrate the morphology of the injected discs. Typically, positive discs show some evidence of internal disruption, with extravasation of contrast beyond the inner confines of the annulus. The key diagnostic criteria are the presence of concordant pain in a radiographically demonstrated degenerated disc.

When careful technique is performed using these rigid criteria, the false-positive rate for provocative discography approaches 0 % (Walsh et al. 1990).

A recent study (Willems et al. 2004) determined that discitis is a rare consequence of discography procedures with or without the use of prophylactic antibiotics. Their findings determined an occurrence rate of 0.25 % based on the number of patients (n = 200), and 0.094 % based on the number of disc injections (Willems et al. 2004). However, in a recent review of randomized controlled trials, the American Pain Society was hesitant to recommend provocative discography as an accurate procedure for detecting discogenic pain in patients with chronic non-radicular low back pain (Chou et al. 2009). Their reasoning was based upon the insufficient amount of data to prove that discography procedures enhance patients' outcomes (Chou et al. 2009). For instance, degenerative disc disease can be asymptomatic; thus, discogram images will not reveal evidence of the disease in those particular cases (Chou et al. 2009). Also, patients without severe back pain but who have somatization or prior back surgery can result in negative discograms (Carragee et al. 2000).

A randomized, controlled study in 2004 ($n = 49$) involving interventional therapies for discogenic back pain demonstrated that patients unresponsive to intradiscal electrothermal therapy (IDET) showed positive results with radiofrequency denervation of the ramus communicans nerves (Oh and Shim 2004). Mean visual analog scale (VAS) pain scores (3.8 vs. 6.3 on a 0–10 scale, $P < 0.05$), and physical function scores (58.9 vs. 46.5, $P < 0.05$) were significantly improved with radiofrequency neurotomy when compared to lidocaine injection treatments. Several studies revealed that intradiscal corticosteroid injections and percutaneous intradiscal radiofrequency thermocoagulation (PIRFT) are ineffective therapies for chronic back pain. Few placebo-controlled studies have been conducted; thus, evidence is limited (Oh and Shim 2004; Chou et al. 2009).

Zygapophyseal Joint Pain

The lumbar facet joints, or zygapophyseal joints, are synovial joints that receive their segmental innervation from the medial facet branch of the posterior division of the respective spinal nerve. The role of facet joint arthropathy in the pathogenesis of low back pain was recognized early in the twentieth century (Ghormley 1933). These patients typically have low back pain that may radiate across the buttock or thigh. Many patients with cervical, thoracic and lumbar pain typically exhibit facet pain as well (Wheeler 2010). The facet joints stiffen and degenerate and may eventually suffer from osteoarthritis (Wheeler 2010). One estimate of its incidence in a low back pain population was 15 %, based upon diagnostic blockade (Schwarzer et al. 1994). As in other forms of low back pain, there are no reliable, reproducible diagnostic criteria (Schwarzer et al. 1994). Injection of the joint or its innervating fibers using a double block technique to help reduce the influence of the placebo effect, and to increase sensitivity and specificity, may represent the most reliable diagnostic method (Schwarzer et al. 1994). This technique involves the injection of facet joint (or the medial facet branch) on one visit, followed by a second confirmatory injection at a separate visit. A positive response occurs when the patient reports >50 % relief with both blocks. Patients may demonstrate significant pain relief shortly after intra-articular lidocaine block treatment; however, the presence of long-term pain relief varies from trial to trial and the evidence is limited (Boswell et al. 2007; Wheeler 2010).

Sacroiliac Joint Pain

The sacroiliac joint is a diarthrodial joint with limited mobility, receiving its innervation from the lumbosacral roots. Thought to be an important contributor to chronic low back pain, one estimate of its incidence in a selected population of low back pain was 19 % (Maigne et al. 1996). In chronic low back pain patients without disc herniation, the incidence of sacroiliac joint pain is <10 % (Manchikanti et al. 2009). Pain may be located across the low back with radiation to the buttock or groin, as in other forms of low back pain. Some patients have tenderness over the sacroiliac joint. Despite the myriad of specific physical maneuvers designed to provoke a positive response in patients with sacroiliac joint pain, no test alone or in combination with others is reliable when compared with diagnostic blockade of the joint itself (Dreyfuss et al. 1996). Radiographic imaging is likewise unhelpful (Schwarzer et al. 1995b).

Spinal Instability

Spondylolisthesis consists of facet joint arthropathy combined with disc degeneration. This results in one vertebral body subluxing over another with the potential for spinal nerve damage (Mardjetko et al. 1994). Common at L4–5 and L5–S1, such movement may be due to trauma, degenerative changes such as osteoporosis, congenital defects, or a pathological fracture. X-rays readily demonstrate spondylolisthesis, and may further show movement of the vertebral bodies in relation to each other on dynamic views.

Other Causes

The spondyloarthropathies are a group of rheumatologic diseases that typically generate chronic low back pain that is worst upon awakening, and improves with activity (McCowin et al. 1991). Ankylosing spondylitis commonly affects young adult males, causes low back pain, decreased spinal range of motion, and has characteristic bony changes on spine imaging. Reiter's syndrome, also known as reactive arthritis, is another common cause of arthritis and low back pain in men. It consists of the triad of arthritis, conjunctivitis, and urethritis. HIV-positive males have a high incidence rate of approximately 75 %. The etiology of this syndrome is unclear; yet, it has been identified as a sexually transmitted and dysenteric disease (Schwartz et al. 2010). Paget's disease is a common cause of back pain in the elderly, and alters bone metabolism

to produce fragile and deformed bones (Mazanec 1999; AAOS 2007). Characteristic findings on history are deep, arthritic pain, and on imaging are localized bony overgrowths. Fibromyalgia is a common disorder causing back pain, associated with other conditions such as headache and irritable bowel syndrome. Radiographic investigations are typically normal. Diagnosis is based upon the symptoms of widespread pain and discreet trigger points (Mazanec 1999). According to the SDI healthcare analytics organization, females compromise 91 % of all fibromyalgia patients (Dussias et al. 2010).

References

American Academy of Orthopaedic Surgeons. (Oct 2007). http://orthoinfo.aaos.org/topic.cfm?topic=A00076
Boswell, M. V., Trescot, A. M., Datta, S., Schultz, D. M., Hansen, H. C., & Abdi, S. (2007). Interventional techniques: Evidence-based practice guidelines in the management of chronic spinal pain. *Pain Physician, 10*, 7–111.
Carragee, E. J., Tanner, C. M., Khurana, S., Hayward, C., Welsh, J., Date, E., Truong, T., Rossi, M., & Hagle, C. (2000). The rates of false-positive lumbar discography in select patients without low back symptoms. *Spine, 25*, 1373–1380.
Chou, R., Loeser, J., Owens, D., Rosenquist, R., Atlas, S., Baisden, J., Carragee, E., Grabois, M., Murphy, D., Resnick, D., Stanos, S., Shaffer, W., & Wall, E. (2009). Interventional therapies, surgery, and interdisciplinary rehabilitation for low back pain, an evidence-based clinical practice guideline from the American pain society. *Spine, 34*, 1066–1077.
Coppes, M. H., Marani, E., Thomeer, R. T., & Groen, G. J. (1997). Innervation of "painful" lumbar discs. *Spine, 22*, 2342–2350.
Deyo, R. A., & Weinstein, J. N. (2001). Low back pain. *The New England Journal of Medicine, 344*, 363–370.
Dreyfuss, P., Michaelsen, M., Pauza, K., McLarty, J., & Bogduk, N. (1996). The value of medical history and physical examination in diagnosing sacroiliac joint pain. *Spine, 21*, 2594–2602.
Dussias, P., Kalali, A., & Staud, R. (2010). Treatment of fibromyalgia. *Psychiatry (Edgmont), 7*, 15–18.
Ghormley, R. K. (1933). Low back pain: With special reference to the articular facets, with presentation of an operative procedure. *Journal of the American Medical Association, 101*, 1773–1777.

Jensen, M. C., Brant-Zawadzkie, M. N., Obuchowski, N., Modic, M. T., Malkasian, D., & Ross, J. S. (1994). Magnetic resonance imaging of the lumbar spine in people without back pain. *The New England Journal of Medicine, 331*, 69–73.

Maigne, J.-Y., Aivaliklis, A., & Pfefer, F. (1996). Results of sacroiliac joint double block and value of sacroiliac pain provocation tests in 54 patients with low back pain. *Spine, 21*, 1889–1892.

Manchikanti, L., Helm, S., Singh, V., Benyamin, R., Datta, S., Hayek, S., Fellows, B., & Boswell, M. (2009). An algorithmic approach for clinical management of chronic spinal pain. *Pain Physician, 12*, E225–E264.

Mardjetko, S. M., Connolly, P. J., & Shott, S. (1994). Degenerative lumbar spondylolisthesis: A meta-analysis of literature 1970-1993. *Spine, 19*, 2256S–2265S.

Mazanec, D. J. (1999). Evaluating back pain in the elderly. *Cleveland Clinic Journal of Medicine, 66*, 89–99.

McCowin, P. R., Borenstein, D., & Wiesel, S. W. (1991). The current approach to the medical diagnosis of low back pain. *Orthopedic Clinics of North America, 22*, 315–325.

Merskey, H. (1986). Classification of chronic pain: Descriptions of chronic pain syndromes and definitions of pain terms. *Pain Supplement, 3*, S1–S225.

Merskey, H., & Bogduk, N. (1994). *Classification of chronic pain: Descriptions of chronic pain syndromes and definitions of pain terms* (2nd ed., pp. 11–36). Seattle: IASP Press.

Moneta, G. B., Videman, T., Kaivanto, K., Aprill, C., Spivey, M., Vanharanta, H., Sachs, B., Guyer, R., Hochschuler, S., Raschbaum, R., & Mooney, V. (1994). Reported pain during lumbar discography as a function of annular ruptures and disc degeneration: A re-analysis of 833 discograms. *Spine, 19*, 1968–1974.

Oh, W. S., & Shim, J. C. (2004). A randomized controlled trial of radio frequency denervation of the rammus communicans nerve for chronic discogenic low back pain. *The Clinical Journal of Pain, 20*, 55–60.

Schwartz, R., Romani, J., & Puig, L. (June 2010). http://emedicine.medscape.com/article/1107206-print

Schwarzer, A. C., Aprill, C. N., Derby, R., Fortin, J., Kine, G., & Bogduk, N. (1994). Clinical features of patients with pain stemming from the lumbar zygapophysial joints: Is the lumbar facet syndrome a clinical entity? *Spine, 19*, 1132–1137.

Schwarzer, A. C., Aprill, C. N., Derby, R., Fortin, J., Kine, F., & Bogduk, N. (1995a). The prevalence and clinical features of internal disc disruption in patients with chronic low back pain. *Spine, 20*, 1878–1883.

Schwarzer, A. C., Aprill, C. N., Charles, N., & Bogduk, N. (1995b). The sacroiliac joint in chronic low back pain. *Spine, 20*, 31–37.

van Geen, J., Edelaar, M. J., Janssen, M. E., & van Eijk, J. (2007). The long-term effect of multidisciplinary back training. *Spine, 32*, 249–255.

Vanharanta, H., Sachs, B. L., Spivey, M., Hochschuler, S. H., Guyer, R. D., Rashbaum, R. F., Ohnmeiss, D. D., & Mooney, V. (1988). A comparison of CT/discography, pain response and radiographic disc height. *Spine, 13*, 321–324.

Walsh, T. R., Weinstein, J. N., Spratt, K. F., Lehmann, T. R., Aprill, C., & Sayre, H. (1990). Lumbar discography in normal subjects: A controlled, prospective study. *The Journal of Bone and Joint Surgery American Volume, 72A*, 1081–1088.

Wheeler, A. H. (May 2010). Low back pain and sciatica. http://emedicine.medscape.com/article/1144130-overview

Wheeler, A. H., & Murrey, D. B. (2005). Spinal pain: Pathogenesis, evolutionary mechanisms, and management. In M. Pappagallo (Ed.), *The neurological basis of pain* (pp. 421–452). New York: McGraw-Hill.

Willems, P., Jacobs, W., Duinkerke, E., & De Kleuver, M. (2004). Lumbar discography: Should we use prophylactic antibiotics? A study of 435 consecutive discograms and a systematic review of the literature. *Journal of Spinal Disorders & Techniques, 17*, 243–247.

Chronic Migraine

Definition

Headache occurring on 15 days or more per month on a basis on a biological basis of migraine. Typically, a patient has a remote history of episodic migraine which slowly increases in frequency over time to a near-daily headache with migraine features. Patients with frequent headache are often overusing analgesics.

Cross-References

▶ Chronic Daily Headache in Children
▶ New Daily Persistent Headache

Chronic Musculoskeletal Pain

Jacqui Clinch
Royal National Hospital for Rheumatic
Diseases NHS Foundation Trust, Bath, UK

Synonyms

Back pain; Chronic joint pain; Chronic widespread pain; Limb pain; Persistent musculoskeletal pain

Introduction

Primary and secondary care physicians review a large number of children who have a wide variety of musculoskeletal pains (Sherry 2000; Malleson and Clinch 2003). A minority of these will have identifiable pathology where there is direct impact on the musculoskeletal system, including Juvenile Idiopathic Arthritis (JIA), or conditions such as cerebral palsy where musculoskeletal pain is a consequence of other underlying pathology. The majority of presenting children, however, do not have easily identifiable inflammatory or other obvious disease processes that can help us understand and explain the cause of pain. However, for most young people presenting to clinic, regardless of whether the cause of their pain is well-known, the chronic experience of pain has often had a large and wholly negative impact on the physical and psychological well-being of the young person and their family (Malleson and Clinch 2003). We know that, irrespective of aetiology, pain is a multidimensional phenomenon with sensory, physiological, cognitive, affective, spiritual, and behavioral components. Emotions related to pain, behavioral responses to pain, beliefs, attitudes, spiritual and cultural attitudes about pain and pain control all alter how pain is experienced by modifying the transmission of noxious stimuli to the brain.

It is now widely understood that children who suffer persistent musculoskeletal pain and other symptoms have a significant chance of developing chronic widespread pain continuing to have pain and pain-associated disability in adult life (Jones et al. 2007).

This review will address the different childhood musculoskeletal conditions that may present and then move onto discussing key areas of management. As always there will be much that is a reflection of best practice rather than robustly evidence based. This, one hopes, will expose areas that others will want to investigate further.

Characteristics

Childhood Musculoskeletal Presentations

A frequent observation among pediatricians (particularly pediatric rheumatologists) is that increased numbers of young people are presenting with ongoing musculoskeletal pain (Clinch and Eccleston 2009). We know that pain is a common feature of childhood. One study showed that 83 % of school-aged children had experienced an episode of pain during the preceding 3 months. In this same study sample, 30.8 % of the children and adolescents stated that the pains had been present for over 6 months. Musculoskeletal pains accounted for 64 % of all the pains that were reported (Roth-Isigkeit 2005). A separate study followed 1,756 schoolchildren (mean age 10.8) and identified 564 (32.1 %) children with musculoskeletal pain. The children were re-evaluated after 1 year, and 4 years (at adolescence) using the same pain questionnaire. At 1-year follow-up, 53.8 % (95 % CI 48.8–58.8) of the children reported pain persistence (persistent preadolescent musculoskeletal pain). At 4-year follow-up, 63.5 % (95 % CI 58.7–68.1) of them had musculoskeletal pain (El–Metwally et al. 2004). Those with persistent preadolescent musculoskeletal pain had approximately three times higher risk of pain recurrence.

There are a number of conditions that are reviewed in the clinical setting:

Chronic Widespread Pain

Chronic widespread pain refers to generalized pain which often involves large areas of the musculoskeletal system. There are many other labels given to this condition including juvenile fibromyalgia (Buskila 2009) and Diffuse Idiopathic Pain (Malleson 2001). Onset is often gradual. There may have been an initial insult (infections including streptococcus and Epstein Barr) or a co-diagnosis of benign musculoskeletal condition (such as hypermobility) (Gedalia et al. 1993); but often there is no obvious trigger and only vague recollections of the time of onset. There may be areas of extreme bodily hypersensitivity (allodynia), but there is often an absence of the autonomic changes that we see in more localized pain conditions (Malleson and Clinch 2003). What is striking in the young people with diffuse pain is co-occurring fatigue, poor sleep patterns, and negative mood (Buskila 1996). It is widely believed, however, that associated negative mood in children is in reaction to the pain-associated disability rather than a primary depression. This is in contrast to adults with fibromyalgia where primary depression is frequently seen (Buskila et al. 1995).

Juvenile Hypermobility

Joint hypermobility, as currently diagnosed using adult criteria, is very common in children. A recent UK study showed that 19.8 % of a cohort of 6,038 13-year-old schoolchildren would receive a diagnosis of generalized hypermobility using the Beighton scoring system (Clinch et al. 2009). There is conflict in the literature regarding the association of hypermobility with musculoskeletal pain (Gedalia et al. 1993, Leone et al. 2009); it is certainly true that clinicians do find evidence of hypermobility or ligamentous laxity of joints in many children presenting with joint pain (Murray 2006) (conditions that include chronic widespread pain, recurrent lower-limb arthralgia, anterior knee pain syndromes, and back pain), but we remain unsure of the relevance of this. The challenge remains to interpret symptoms correctly as being related to the hypermobility and to predict why such children become symptomatic.

Complex Regional Pain Syndrome (CRPS)

CRPS is not uncommonly seen in pediatric rheumatology clinics. The differential diagnostic list is long but includes juvenile idiopathic arthritis, bone malignancy, and trauma. The diagnosis of CRPS remains a clinical one. There may be precipitating trauma. In children, the lower limb is more commonly involved than the upper limb (Murray et al. 2000). The pain is usually out of proportion to the inciting event and accompanied by allodynia. Autonomic changes are present; these include swelling, reduced cutaneous perfusion, and thermodynamic instability. There is also a marked reduction in range of movement and, in severe cases, trophic changes. The International Association for the Study of Pain (IASP) has diagnostic criteria for adults with CRPS. Although these diagnostic criteria may hold true for children, it is widely believed that the dystrophic changes and long-term disability are less common when compared with adults (Stanton-Hicks 2010).

Back Pain

Low back pain specifically is commonly reported in the adolescent population. In a recent cross-sectional study, the prevalence of low back pain that lasted for at least 1 month was 24 % (Watson et al. 2002). Often this is thought to be related to lifestyle influences on a developing spine, such as postural habit (slouching), load bearing on the back (e.g., school bags), or engagement in sedentary activity (e.g., computer use). There are insufficient data to bring clarity to these arguments although it is fair to say that there is no conclusive evidence for any one of these factors being significantly related to the increase in back pain.

Neck pain is also frequently reported in the schoolchild population with 11.5 % of a Dutch cohort (comprising over 3,000 children) complaining of persistent pain in the neck and shoulders. This study showed no relation between persistent pain and physical activity or mechanical loads (including bags) (Ståhl et al. 2008).

Childhood Disease and Chronic Pain

Musculoskeletal pain can complicate almost any other chronic childhood illness (Varni and Walco 1988) including juvenile arthritis, inflammatory bowel disease (Buskila et al. 1999), cerebral palsy, cancer (Jay et al. 1986), sickle cell disease (Stinson and Naser 2003) and muscular dystrophies (Engel et al. 2005). This can be a direct effect of disease (particularly in rheumatic and neurological conditions) or indirectly through other disease sequelae (such as infection in cystic fibrosis and prolonged immobility in any of the chronic diseases). One must not forget that the musculoskeletal system is constantly developing in children, thus any condition that leads to immobility or asymmetry will cause an imbalance or shift of mechanical load.

The relationship between juvenile arthritis and chronic pain is well recognized. As with other diseases the degree of disabling pain does not always mirror inflammatory joint activity. A growing body of research in rheumatic diseases, such as JIA, highlights the importance of environmental and cognitive behavioural influences in the pain experience of children in addition to the contribution of disease activity (Anthony and Schanberg 2007).

Childhood cancers can directly cause localised musculoskeletal pain (osteosarcomas, Ewings Sarcoma, etc.) or indirectly (phantom limb pain can develop following amputation/ tumour removal). Certain chemotherapy agents may lead to problems with neuropathic pain (Moore and Pinkerton 2009).

Assessing Chronic Childhood Musculoskeletal Pain

Ostensibly the features of the history taking are similar to other conditions: previous medical history, specific history leading to pain, family circumstances, and medication review. An outline of key areas of the history (with rationale) is given in Table 1.

Unfortunately pain can have a direct affect on other systems, leading to complaints that can be as disabling as the pain itself. An outline of the more common signs and symptoms is given in Table 2.

For the physician the goal of the history taking is often to exclude serious possible causes, identify key problems, build a trusting relationship with the patient (and family), and, if possible, identify a treatment plan. However, the average child with persistent pain not only has a chronic experience of pain and distress, but a chronic experience of medical settings, feelings of being doubted, having the validity of their pain contested, and, often a history of failed and pain exacerbating interventions(Clinch and Eccleston 2009).

Physical Examination

The physical examination should be thorough. Time spent at this stage may prevent repetition and unnecessary, distressing investigations at a later date. Recently a musculoskeletal screening examination (called the pGALS) has been validated; this is a quick, informative examination technique (Foster et al. 2006). If there is concern regarding the diagnosis then this is the time to order all investigations and ensure these are followed up. Undue delay can lead to increased patient and parental fear and worry about missed diagnoses.

Psychometric Instruments

A number of questionnaire or interview techniques have been used in researching the impact of pain on the lives of children. There has been recent systematic review of all of the instruments that have been used to assess aspects of adolescent chronic pain (Eccleston et al. 2006) and all of the instruments used to assess the impact of adolescent chronic pain on parents (Jordan et al. 2006). There are some well validated and commonly used tools, most notably the Varni/ Thompson Paediatric Pain Questionnaire

Chronic Musculoskeletal Pain, Table 1 Outline of key areas to be addressed while taking a history

Areas and Domains to Assess	Rationale for Assessment
Onset of Pain When did the pain start? Where did it start? Was there a preceding infection, trauma or operation? How was it treated initially What was happening around the time the pain started	Identify any nerve injury, muscular trauma or intercurrent infection that may have triggered pain. Therapeutic interventions that include long periods of immobilisation, accelerated physical rehabilitation and repeat operative investigations may all be factors in pain amplification. Environmental events (school, home, other medical episodes) may be factors in the onset of pain.
Characteristics of the pain Has the pain spread from original place? How would you describe the pain? How severe is the pain on a good day and on a bad day? Do you suffer pins and needles? Has the pain got any better or worse? Is there variation in the pain during the day? Does the pain alter at night? Does it wake you up? What makes it worse? What makes it better? Is it painful to lightly touch the area that is painful? Does that area look unusual?	It is important to establish the nature of the pain episodes as this may direct management Neuropathic pains are often described as burning, shooting, deep, electric and occasionally episodic. There may be pins and needles and areas of numbness. Complex regional pain syndrome has neuropathic features and also extreme hypersensitivity (allodynia) over affected regions - "I cannot have a duvet on my leg at night". The affected area may appear swollen and shiny. Young people often feel the limb is more swollen than it actually is. Erythromelalgia is usually worse at night, with painful burning sensations in hands and/or feet. Movement usually leads to an escalation of pain. Many young people avoid weightbearing or transferring for fear of a pain flare.
Other Symptoms Is there a fever, rash or weight loss? Has menstruation altered? Is there altered bowel habit? Do you have nausea? Do you suffer abdominal pain? Is there any muscle weakness? Do you have any areas that are numb? Do you suffer dizziness? Have you passed out or suddenly fallen to the floor? Is fatigue a problem? Do you suffer headaches/migraines? Do you feel colder/warmer than previously? Do you suffer from blurred vision? Has your mood been affected?	It is important to ensure there are no red flags indicating missed pathology. Nocturnal pain, extreme weight loss, hard neurological features are examples of symptoms and signs that need further paediatric work-up. Common pain associated symptoms are included in this list on the left. On occasion these become as disabling as the pain and need to be included in the rehabilitation plan.
Effect of pain on daily living Do you find it hard to get to sleep/stay asleep? Do you 'catnap' in the day? What can you do on a 'bad'/good day? How much school have you managed over the past 6 months? How is your concentration/memory? On 'bad/good days' how is your mood? How has the pain affected your fitness/hobbies? Do you need help in areas where you were previously independent? How has this affected your family? What do you think is causing this? Do you have fears about a particular illness? Do the parents have fears concerning the pain that they feel has not been addressed?	Pain can have a significant effect on sleep, mood, independence and general functioning. It is important to know how well a young person is sleeping (and where - it is not unusual for a previously independent teenager to go back into the parents' bed). Establishing how catastrophic the effect on lifestyle has been is also key; children may have a history of lying on the sofa reading all day or they may have been competing for their country in hockey - by understanding expectations of quality of life you can tailor the rehabilitation process.

(continued)

Chronic Musculoskeletal Pain, Table 1 (continued)

Past and family history of illness Have you suffered painful conditions previously? Have you suffered fatigue, sleeplessness or anxiety previously? Any operations or illnesses as a younger child? Any neonatal problems? Have any family members suffered illness? Is there a family history of painful conditions?	There is much recent research into past painful events on the developing nociceptive system. This is particularly relevant to children who have been premature and received intensive care. Other family illness experience is relevant; there may be a direct genetic link with pain experienced and/or environmental aspects that are factors in the child developing a pain condition.
Family, emotional and social circumstances Who currently lives at home? What are the occupations of the main carers? Have one or both carers changed/stopped their job since the pain condition started? Can you identify any stressors in school, family or peer groups?	Having an idea of the family environment is important. Parents often adapt their home life to care for the child and employment patterns change. Siblings' relationships and friendships may alter. School often becomes challenging both physically and emotionally (previously able children may not know how to communicate their conditions to peers and have differing degrees of help from school staff).

(Varni et al. 1987) focusing on pain, and the Functional disability Inventory (Walker and Greene 1991), focusing on disability.

Pain intensity in children has long been measured using simple severity measurement tools such as the Visual Analogue Scale (Carlsson 1983). Whilst these are useful measures prospectively they give no indication of the impact of pain on the young person's life. Two multidisciplinary tools have recently been developed, one specifically for use in measuring the impact of pain on adolescents with chronic pain (Bath Adolescent Pain Questionnaire) (Eccleston et al. 2005) and one specifically for use in measuring the impact of adolescent chronic pain on parents (Bath Adolescent Pain Questionnaire for Parents) (Jordan et al. 2008).

Physiological Measures of Pain Related Indexes

There are little data on the role or utility of measuring physiological processes in the context of childhood musculoskeletal chronic pain. The role of thermography and bone scintigraphy in complex regional pain syndrome has been studied (Lightman et al. 1987; Goldsmith et al. 1989), evaluating use in the early and late stages of this pain condition. Opinion is divided on their usefulness. There is a large, currently unmet need for paediatric research in this area.

Rehabilitation for Children with Musculoskeletal Pain

Clearly, if assessment shows an ongoing undertreated medical condition, this must be addressed first. For example, children with JIA frequently suffer persistent joint pain and some of this will be ameliorated with improved control of the underlying inflammatory process.

Assuming that any underlying disease process has been addressed the health professional then has to move to symptom management and psychosocial rehabilitation. Although it may be tempting to focus solely on the amelioration of the suffering of the child, one of the most important aspects of rehabilitation is "inclusion". A dedicated team that works consistently with the adolescent and family will facilitate communication, ensure effective delivery of therapy, reduce iatrogenic influences, and enable goals to be reached earlier (Clinch and Eccleston 2009; Anthony and Schanberg 2007). Most approaches to rehabilitation share common features including physical therapy, education, symptom control and behavioural intervention.

Physical Therapy

Despite physical therapy interventions being seen as the 'front line' in treating musculoskeletal pain, there is very little robust evidence to show their effect for most musculoskeletal pain

Chronic Musculoskeletal Pain, Table 2 Common musculoskeletal pain-related symptoms/signs

Hypervigilance and hypersensitivity
Patients often report a heightened awareness of pain and pain associated cues. It is unclear whether this is caused by fear of pain, or hypersensitivity of pain receptors leading to allodynia and hyperalgesic ranges. Clinically this presents as young people describing unbearable pain with minimal skin contact, and heightened fear of being touched, for example on examination.
Perceived thermodysregulation
This is more common in adolescent girls. Limbs are particularly cool and mottled. Occasionally there will be areas that are very red and hot to touch on a background of the mottled skin.
There may also be an abnormal perception of temperature (Leone et al. 2009) with an increase in thermal pain sensitivity (Murray 2006).
Autonomic dysfunction
Pain is a powerful stressor. Continuous pain signals, immobility and fatigue act directly on the autonomic system (Murray et al. 2000). In an environment of physical and emotional anxiety the sympathetic system is more active. This leads to tachycardia, hyperventilation (compounded with panic attacks), cold sweats, blurred vision, abdominal pain and extreme pallor. Girls particularly complain of nausea, dizziness and episodes of feeling faint. Children look unwell during these episodes of increased pain. It is not unusual for attending paediatricians to investigate cardiovascular, neurological and gastrointestinal systems in an attempt to elicit pathology.
Musculoskeletal disequilibrium
These young people are still growing, often in their peripubertal growth spurt. By not moving a painful limb the brain has an altered representation of that area. Proprioceptive signals from the joints are reduced and the limb held in a rigid, fixed position. Legs may 'give way'. Knees and hips are held flexed, feet are inverted and hands are clenched with flexed wrists. These positions are often described as the most comfortable. Muscles and tendons quickly tighten and this complicates the pain and disability. The adaptive positioning of a young person with leg or abdominal pains particularly affects the gait and resting positions and thus alters the loads on the spine and pelvis (Stanton-Hicks 2010).

conditions. For example, in a small study evaluating physical therapy interventions in treating hypermobility and pain, an effect was found when comparing general physiotherapy to no therapy, but no differences were found between generalised and targeted therapy in this group (Kemp et al. 2010).

In some conditions, such as CRPS, several studies report the beneficial effect of early intensive physiotherapy (Sherry 2000). The aim of intensive physiotherapy is accelerated mobilisation. Many cases of diffuse pain will require a gentle, paced approach. In all cases the increase of activity should be consistent despite the pain (and in the absence of pain relief). With musculoskeletal pains the more active the musculoskeletal system becomes the more likely the muscle spasms and tightening are to reduce. In turn, proprioception improves and autonomic changes subside (Lee et al. 2002).

Where possible, the young person should work to devise his or her own 'fitness plan'. Using a local gym rather than a hospital physiotherapy gym allows youth to start to return to a more normal environment (Eccleston et al. 2003a). Working in this consistent, paced manner can be extremely hard for the young person and their parents. Parental anxiety is high (Eccleston et al. 2004) and there is a fear that bodily damage or injury will be done. Psychological support during this time is key to successful outcomes. The young person will need help setting goals, learning how to communicate pain to peers and family, keeping up motivation on 'bad days', managing low mood, dealing with anger and frustration and overcoming fears. Often they have not been at school for a long period of time and need help in preparing again for this difficult environment. In some cases there may be other mental health needs that can be identified and appropriately treated.

Other Therapies

There are ongoing studies, specifically with CRPS patients, on visual counterstimulation using mirrors that show promise. Mirror therapy focusing on the hypothesis that incongruence between motor output and sensory input produces CRPS is underway with adults (McCabe et al. 2003). There is no evidence as yet supporting the use of mirror therapy in children. In addition, other therapies such as Spinal Cord Stimulation and Transcutaneous electrical nerve stimulation do not have an evidence base for management of chronic musculoskeletal pain in children.

Complementary therapies are commonly utilised by patients with chronic pain (Tsao and Zeltzer 2005). There is little evidence supporting many of these therapies in children and adolescents with musculoskeletal pain.

Education

One should never underestimate that the average patient's understanding of anatomy and physiology is at best partial, and more typically fantastical. A critical first step in all rehabilitation is to offer, re-offer, and reinforce an understanding of how one's body may be working to maintain pain. With musculoskeletal pain this is best done visually, showing the effects of immobility and avoidant physical behaviour on the developing system.

Education about pain may be difficult to grasp because there is a dominant cultural view of pain as a warning sign that a disease process or abnormality is present. There is a need for the young person and family to find a cause for pain, and often scepticism from health professionals that the pains are present at all.

Children and their parents often have inflexible or rigid ideas not only about the cause of pain, but also of pain as an important signal of damage or disease. Rhetorical devices (metaphors, stories, examples, pictures) that counter this rigid thinking can be helpful. For example, we often give the common example of headaches (rarely due to malignancy, infection or stroke - but still disabling); back pain is rarely related to malignancy or arthritis. Explaining the fascinating case of phantom limb experiences can also be helpful, to introduce flexibility to the idea that brain signals must signify ongoing peripheral damage.

Pharmacotherapy

The number of analgesics and interventions used is a sign that there are no well-controlled

therapeutic trials in the arena of childhood chronic pain. It is becoming widely accepted, however, that any analgesic intervention should be used alongside multidisciplinary therapy (Stanton-Hicks 1998; Kashikar-Zuck 2006). Oral treatments described in adult pain literature include tricyclic antidepressants, NSAIDs, opioids, anticonvulsants and glucocorticoids (Eland 1988; Mailis and Furlan 2003). Sympathetic blockade and botulinum toxin injections have been used in localised muscular pain. For CRPS, anticonvulsants such as gabapentin and pregabalin have been postulated as helpful for the neuropathic pain. It must be stated that the evidence supporting the efficacy of any of these pharmacotherapies in children is lacking (Eccleston and Malleson 2002). Further research is needed including a comprehensive and systematic review on the existing evidence of pharmacotherapies for childhood chronic musculoskeletal pain, and a programme of trials to establish the evidence base.

Medication review is an essential task of the clinician in this population. Children with co-existing disease are often on other medications. Polypharmacy, risk of dosing and combination errors, and confusion over the utility of previously and currently prescribed medication is common. Drug withdrawal, detoxification, and replacement should be carefully managed.

Psychological Therapies

One could postulate that for any child who has persistent musculoskeletal pain, cognitive and behavioral therapies targeting processes that allow the young person to re-engage with age appropriate activities would prove invaluable. There is evidence that an interdisciplinary programme for adolescents complexly disabled with chronic pain has a positive significant impact (Eccleston et al. 2003b). The targets of interdisciplinary therapies are multiple, in fitting with the widespread impact of chronic pain. Disengagement from the pursuit of pain relief and engagement with normal life activities is the overall goal. Its effectiveness is measured by success in return to normal schooling, reduced depression and anxiety, reduction in parental stress, and increased activity. The role of the physician is often as an educator, to oversee analgesic withdrawal, and to support the overall message that it is safe to increase activity despite pain and reduce reliance on medical support.

Family involvement in any child who has persistent musculoskeletal pain is widely believed to be essential. When engaged in rehabilitation programmes with their child it has been shown that parents/carers become less anxious and develop improved coping strategies. This further facilitates the rehabilitation and reduction of functional disability of their child (Eccleston et al. 2003b).

Summary

Within rheumatology, increasing numbers of children are presenting to clinics with chronic widespread pain, hypermobility and chronic joint pain, CRPS, pain as part of a disease process (e.g., JIA) and other localised limb pains. This increase may be in part due to the growing societal expectation for permanent and total wellbeing, or by an increasingly sedentary lifestyle, but the data cannot yet give us a clear direction for any target of public health intervention. Whatever the cause, we have a population of approximately 1–2 % of children with severe disabling chronic pain beginning to find their way to clinic.

Early assessment and intervention may lead to reassurance and tailoring of subsequent multidisciplinary rehabilitation. There are specific areas of progress in musculoskeletal pain assessment and management. Well-developed assessment techniques and tools exist, and we have a good understanding of the context of assessment and the importance of a broad multidimensional perspective. Promising treatments aimed at the self-management of disability and distress and return of children and families to normal functioning are emerging. We are now gaining insight into the role that pain may play in affecting neurodevelopmental/nocidevelopmental

pathways of the evolving system in a child(Fitzgerald and Walker 2009).

There are still hurdles in optimising our treatment for children with musculoskeletal pain. Early appropriate pharmacotherapy may be helpful in altering the trajectory of the child developing persistent and intractable musculoskeletal pain; but, as the research is scant the guidance is still lacking. We know that untreated complex pain is personally, socially and financially burdensome for individuals, families and societies, but there are few psychologists and physiotherapists in paediatric settings with training in the rehabilitation of the child with chronic musculoskeletal pain to offer treatment.

Untreated, a significant number of these children with musculoskeletal chronic pain will become adults with chronic pain who are dependent on the healthcare and social care services. By identifying problems early, instigating multidisciplinary therapy and supporting collaborative research the outcome for these children will continue to improve.

References

Anthony, K. K., & Schanberg, L. E. (2007). Assessment and management of pain syndromes and arthritis pain in children and adolescents. *Rheumatic Diseases Clinics of North America, 33*(3), 625–660.
Buskila, D. (1996). Fibromyalgia in children–lessons from assessing nonarticular tenderness. *Journal of Rheumatology, 23*(12), 2017–2019.
Buskila, D. (2009). Pediatric fibromyalgia. *Rheumatic Diseases Clinics of North America, 35*(2), 253–261.
Buskila, D., et al. (1995). Fibromyalgia syndrome in children-an outcome study. *Journal of Rheumatology, 22*(3), 525–528.
Buskila, D., Odes, L. R., Neuman, L., & Odes, H. S. (1999). Fibromyalgia in inflammatory bowel disease. *Journal of Rheumatology, 26*(5), 1167–1171.
Carlsson, A. M. (1983). Assessment of chronic pain. I. Aspects of the reliability and validity of the visual analogue scale. *Pain, 16*, 87–101.
Clinch, J., & Eccleston, C. (2009). Chronic musculoskeletal pain in children: Assessment and management. *Rheumatology (Oxford, England), 48*(5), 466–474.
Clinch, J., Doerner, R., Clark, E., Palmer, S. & Tobias, J. (2009). The prevalence of hypermobility in UK schoolchildren; Where should the line be drawn? *Rheumatology, 48*(suppl 1).

Eccleston, C., & Malleson, P. (2002). Managing chronic pain in children and adolescents We need to address the embarrassing lack of data for this common problem. *BMJ, 326*(7404), 1408–1409.

Eccleston, C., Stourbout, C., Connell, H., McCracken, L., & Clinch, J. (2003a). Chronic pain in adolescents: Evaluation of a programme of interdisciplinary cognitive behaviour therapy. *Archives of Disease in Childhood, 88*(10), 881–885.

Eccleston, C., Malleson, P., Clinch, J., Connell, H., & Stourbout, C. (2003b). Chronic pain in adolescents: Evaluation of a programme of interdisciplinary cognitive behaviour therapy. *Archives of Disease in Childhood, 88*(10), 881–885.

Eccleston, C., Crombez, G., Scotford, A., Connell, H., & Clinch, J. (2004). Adolescent chronic pain: Patterns and predictors of emotional distress in adolescents with chronic pain and their parents. *Pain, 108*(3), 221–229.

Eccleston, C., Jordan, A., McCracken, L. M., Connell, H., & Clinch, J. (2005). The bath adolescent pain questionnaire (BAPQ): Development and preliminary psychometric evaluation of an instrument to assess the impact of chronic pain on adolescents. *Pain, 118*, 263–270.

Eccleston, C., Jordan, A. L., & Crombez, G. (2006). The impact of chronic pain on adolescents: A review of previously used measures. *Journal of Pediatric Psychology, 31*, 684–697.

Eland, J. M. (1988). Pharmacologic management of acute and chronic pediatric pain. *Issues in Comprehensive Pediatric Nursing, 11*(2–3), 93–111.

El-Metwally, A., Salminen, J. J., Auvinen, A., Kautiainen, H., & Mikkelsson, M. (2004). Prognosis of non-specific musculoskeletal pain in preadolescents: A prospective 4-year follow-up study till adolescence. Prognosis of non-specific musculoskeletal pain in preadolescents: A prospective 4-year follow-up study till adolescence. *Pain, 110*(3), 550–559.

Engel, J. M., Kartin, D., & Jaffe, K. M. (2005). Exploring chronic pain in youths with duchenne muscular dystrophy: A model for pediatric neuromuscular disease. *Physical Medicine and Rehabilitation Clinics of North America, 16*(4), 1113–1124.

Fitzgerald, M., & Walker, S. M. (2009). Infant pain management: A developmental neurobiological approach. *Nature Clinical Practice Neurology, 5*(1), 35–50.

Foster, H. E., Kay, L. J., Friswell, M., Coady, D., & Myers, A. (2006). Musculoskeletal screening examination (pGALS) for school-age children based on the adult GALS screen. *Arthritis and Rheumatism, 55*(6), 981.

Gedalia, A., et al. (1993). Joint hypermobility and fibromyalgia in schoolchildren. *Annals of the Rheumatic Diseases, 52*(7), 494–496.

Goldsmith, D. P., Vivino, F. B., Heyman, S., et al. (1989). Nuclear imaging and clinical features of childhood reflex neurovascular dystrophy:

Comparison with adults. *Arthritis and Rheumatism, 32*, 480–485.

Jay, S. M., Elliott, C., & Varni, J. W. (1986). Acute and chronic pain in adults and children with cancer. *Journal of Consulting and Clinical Psychology, 54*(5), 601–607.

Jones, G. T., Silman, A. J., Power, C., & Macfarlane, G. J. (2007). Are common symptoms in childhood associated with chronic widespread body pain in adulthood? Results from the 1958 British birth cohort study. *Arthritis and Rheumatism, 56*(5), 1669–1675.

Jordan, A. L., Eccleston, C., & Crombez, G. (2006). The impact of adolescent chronic pain on parents: A review of previously used measures. *Journal of Pediatric Psychology, 31*(7), 684–697.

Jordan, A., Eccleston, C., McCracken, L. M., Connell, H., & Clinch, J. (2008). The bath adolescent pain - parental impact questionnaire (BAP-PIQ): Development and preliminary psychometric evaluation of an instrument to assess the impact of parenting an adolescent with chronic pain. *Pain, 137*, 478–487.

Kashikar-Zuck, S. (2006). Treatment of children with unexplained chronic pain. *Lancet, 367*(9508), 380–382.

Kemp, S., Roberts, I., Gamble, C., et al. (2010). A randomized, comparative trial of generalized vs targeted physiotherapy in the management of childhood hypermobility. *Rheumatology (Oxford, England), 49*(2), 315–325.

Lee, B. H., Scharff, L., Sethna, N. F., et al. (2002). Physical therapy and cognitive-behavioral treatment for complex regional pain syndromes. *Journal of Pediatrics, 141*, 135–140.

Leone, V., Tornese, G., Zerial, M., Locatelli, C., Ciambra, R., Bensa, M., & Pocecco, M. (2009). Joint hypermobility and its relationship to musculoskeletal pain in schoolchildren: A cross-sectional study. *Archives of Disease in Childhood, 94*(8), 627–632.

Lightman, H. I., Pochaczevsky, R., Aprin, H., & Ilowite, N. (1987). Thermography in childhood reflex sympathetic dystrophy. *Journal of Pediatrics, 111*, 551–555.

Mailis, A. & Furlan, A. (2003). Sympathectomy for neuropathic pain. *Cochrane Database Syst Rev*, (2): CD002918.

Malleson, P. N. (2001). Chronic musculoskeletal and other idiopathic pain syndromes. *Archives of Disease in Childhood, 84*(3), 189–192.

Malleson, P., & Clinch, J. (2003). Pain syndromes in children. *Current Opinion in Rheumatology, 15*(5), 572–580.

McCabe, C. S., Haigh, R. C., Ring, E. F., Halligan, P. W., Wall, P. D., & Blake, D. R. (2003). A controlled pilot study of the utility of mirror visual feedback in the treatment of complex regional pain syndrome (type 1). *Rheumatology (Oxford, England), 42*(1), 97–101.

Moore, A., & Pinkerton, R. (2009). Vincristine: Can its therapeutic index be enhanced? *Pediatric Blood & Cancer, 53*(7), 1180–1187.

Murray, K. J. (2006). Hypermobility disorders in children and adolescents. *Best Practice & Research. Clinical Rheumatology, 20*(2), 329–351.

Murray, C. S., et al. (2000). Morbidity in reflex sympathetic dystrophy. *Archives of Disease in Childhood, 82*(3), 231–233.

Roth-Isigkeit, A. (2005). Pain among children and adolescents: Restrictions in daily living and triggering factors. *Pediatrics, 115*, 152–162.

Sherry, D. D. (2000). Pain syndromes in children. *Current Rheumatology Reports, 2*(4), 337–342.

Ståhl, M., Kautiainen, H., El-Metwally, A., et al. (2008). Non-specific neck pain in schoolchildren: Prognosis and risk factors for occurrence and persistence. A 4-year follow-up study. *Pain, 137*(2), 316–322.

Stanton-Hicks, M. (1998). Complex regional pain syndromes: Guidelines for therapy. *The Clinical Journal of Pain, 14*(2), 155–166.

Stanton-Hicks, M. (2010). Plasticity of complex regional pain syndrome (CRPS) in children. *Pain Medicine, 11*(8), 1216–1220.

Stinson, J., & Naser, B. (2003). Pain management in children with sickle cell disease. *Paediatric Drugs, 5*(4), 229–241.

Tsao, J. C., & Zeltzer, L. K. (2005). Complementary and alternative medicine approaches for pediatric pain: A review of the state-of-the-science. *Evidence-Based Complementary and Alternative Medicine, 2*(2), 149–159.

Varni, J. W., & Walco, G. A. (1988). Chronic and recurrent pain associated with pediatric chronic diseases. *Issues in Comprehensive Pediatric Nursing, 11*(2–3), 145–158.

Varni, J. W., Thompson, K. L., & Hanson, V. (1987). The Varni/Thompson pediatric pain questionnaire I chronic musculoskeletal pain in juvenile rheumatoid arthritis. *Pain, 28*, 27–38.

Walker, L. S., & Greene, J. W. (1991). The functional disability inventory: Measuring a neglected dimension of child health status. *Journal of Pediatric Psychology, 16*, 39–58.

Watson, K. D., Papageorgiou, A. C., Jones, G. T., et al. (2002). Low back pain in schoolchildren: Occurrence and characteristics. *Pain, 97*(1–2), 87–92.

Chronic Nerve Pain

▶ Chronic Neuropathic Pain

Chronic Neural Sensitization

▶ Visceral Pain Model, Irritable Bowel Syndrome Model

Chronic Neurogenic Pain

▶ Chronic Neuropathic Pain

Chronic Neuropathic Pain

Synonyms

Chronic nerve pain; Chronic neurogenic pain; Neuropathic pain

Definition

Neuropathic pain is pain caused by a lesion or disease of the somatosensory nervous system. Trauma may provoke a sudden, acute pain that resolves quickly. More prolonged but still transient pain, lasting weeks or months, may develop following nerve inflammation, or nerve lesions such as crush or freeze that do not disrupt the continuity of the endoneurium and perineurium. The pain resolves after the injured nerve fibers have regenerated, or the inflammation has cleared. Frequently, however, neuropathic pain persists far beyond the time of wound healing and then it is considered to be a chronic condition. The lesion or disease may affect the peripheral nervous system (peripheral neuropathic pain) or the central nervous system (central neuropathic pain, or simply "central pain").

Cross-References

▶ Pain Treatment, Implantable Pumps for Drug Delivery

Chronic Pain

Definition

Pain that persists on a constant basis for 3 months or more. Some researchers and clinicians use a timeline of 6 months or more. Chronic pain rarely has a known cause, and may be hard to localize in the body. Treatment is usually limited to general measures of pain control. Emotive symptoms are commonly an important component of the suffering associated with chronic pain. The character of a chronic pain signal is often misinterpreted or grossly distorted. There is often a measurable imbalance between pronociceptive and antinociceptive forces consistent with the perception of pain. No definition of chronic pain exists for newborns, but prolonged pain has been defined as that lasting several hours or days.

Cross-References

▶ Acute Pain, Subacute Pain, and Chronic Pain
▶ Amygdala, Pain Processing and Behavior in Animals
▶ Assessment of Pain Behaviors
▶ Chronic Low Back Pain: Definitions and Diagnosis
▶ Depression and Pain
▶ Impact of Familial Factors on Children's Chronic Pain
▶ Muscle Pain, Fibromyalgia Syndrome (Primary, Secondary)
▶ Pain Assessment in Neonates
▶ Pain in the Workplace, Risk factors for Chronicity, Job Demands
▶ Pain Inventories
▶ Pain Treatment, Intracranial Ablative Procedures
▶ Pain Treatment, Motor Cortex Stimulation
▶ Physical Exercise
▶ Postoperative Pain, Persistent Acute Pain
▶ Prevalence of Chronic Pain Disorders in Children
▶ Psychiatric Aspects of Pain and Dentistry
▶ Psychological Treatment of Chronic Pain, Prediction of Outcome
▶ Sleep and Pain
▶ Thalamus, Clinical Pain, Human Imaging

Chronic Pain Disorder (DSM-IV)

▶ Psychiatric Aspects of the Epidemiology of Pain

Chronic Pain in Children, Physical Medicine and Rehabilitation

Susan Tupper
Pediatrics, University of Saskatchewan, Saskatoon, SK, Canada

Synonyms

Occupational therapy in children; Physical therapy; Rehabilitation

Definition

Physical medicine and rehabilitation for children with chronic pain is a multidisciplinary approach that incorporates goal-oriented assessment and treatment aimed at maximizing independent function in necessary or valued activities. Many health-care providers may be involved in the rehabilitation process including physical therapists (PTs), occupational therapists (OTs), physicians, exercise therapists, and recreation therapists. This entry will focus on PT and OT roles and interventions.

Characteristics

Chronic pain in childhood may develop as a consequence of an injury or identifiable disease processes (e.g., spinal cord injury, burn, or arthritis), or for an unidentifiable reason (e.g., chronic daily headache, complex regional pain syndrome, fibromyalgia). Chronic pain, regardless of etiology, can have an extensive impact on many aspects of a child's life with potentially lifelong and devastating consequences.

Rehabilitation Settings

Rehabilitation of the child with chronic pain takes place in a variety of settings, including inpatient or outpatient hospital departments, community outpatient clinics, day care, patients' homes, and schools.

Rehabilitation Framework

Treatment of chronic pain has evolved from a disease-focused framework to a family-centered model in which the child and his or her family are vital to the decision-making process. The World Health Organization's International Classification of Functioning Disability and Health (ICF) is a useful conceptual framework for evaluation and treatment planning that is aimed at maximizing participation in valued activities (World Health Organization 2007). The ICF conceptualization of disability is the interaction between the individual's impairments in body structures and functioning (e.g., pain, loss of range or motion, loss of strength) and dynamic contextual factors (e.g., built environments, social environments) experienced throughout the lifespan.

Assessment

There are three components to a family-centered initial assessment of a child with a chronic pain condition: (1) medical and social history and interview with the patient and her family, (2) physical examination, and (3) environmental examination. During the history and interview, if the child with pain has sufficient cognitive and communication abilities, she should be provided an opportunity to tell the rehabilitation team about her pain and how it has impacted her life. This will assist the team in identifying important activities and roles that have been affected by pain, and will form the basis of development of treatment goals and intervention planning. The history will include questions about primary domains of pain (i.e., pain intensity, location, quality, triggering, and relieving factors), and secondary domains that are influenced by pain (i.e., physical functioning, emotional

functioning, role functioning, and sleep). Pain assessment in children who lack the cognitive or communication capacity to reliably self-report requires the input of primary caregivers to identify pain and comfort behaviors that can be monitored during rehabilitation.

The physical examination will be specific to the musculoskeletal, cardiovascular, or neurological structures and functions affected. Assessment of physical function may include both observation of performance of physical tasks and use of standardized self-report or proxy report assessment tools that measure difficulty with specific activities. Depending on the limitations reported during the history and interview, the rehabilitation team may choose to observe the patient perform activities of daily living, standardized functional tests (e.g., 6 min walk, timed stair climbing, Gross Motor Function Measure), or specific activities related to the treatment goals (e.g., sport-specific skills).

The initial assessment may include an environmental examination to determine if adaptations to the activities or environments are necessary to enable the patient to fully participate in activities. The environmental examination may include discussions with parents, day-care staff, teachers, school administrators, or coaches as well as observation of performance of functional activities within the natural environments. Adaptations to the environment may include prescription of assistive devices (e.g., braces or ambulation aides), equipment (e.g., specialized chair, desk or bed, lap-top computer), or accommodations to the usual routine (e.g., permission to move during class to avoid prolonged sitting, extra time for examinations to accommodate for impaired attention due to pain).

Interventions

Physical therapists (PTs) typically provide noninvasive physical and behavioral interventions to address impairments in body functions and structures identified during the physical assessment. Occupational therapists (OTs) facilitate participation in everyday activities through recovery and maintenance of function and compensation for permanent loss of function with the use of adaptive equipment, modifications to activities and environments.

Physical interventions are a critical component of management of chronic pain in children, along with pharmaceutical and psychological interventions (Ayling-Campos et al. 2011). Children with chronic pain conditions tend to have lower physical activity participation, muscle strength, cardiovascular conditioning, and health-related quality of life than their healthy peers (Maggio et al. 2010).

Interventions are selected to meet the functional goals based on the impairments and limitations identified in the assessment. Treatment goals are developed through a collaborative, educational process whereby the therapists provide information about the nature of the condition and treatment options available, their risks and potential benefits. The child and family are central to the process in that they provide information on the child's functional abilities prior to the onset of the pain condition, and select the course of treatment based on the recommended options to meet treatment goals. Successful rehabilitation depends on agreement between the therapists and the patient on treatment goals and approach (Daniels and Wearden 2011).

Physical interventions for rehabilitation of the child with chronic pain include both active and passive therapies. Active therapies include therapeutic exercises and recommendations for participation in social, school, and leisure activities. Specific therapeutic exercises are selected to address the impairments in body functions and structures identified in the physical examination. For example, loss of range of motion, loss of strength, or lower cardiovascular fitness would be addressed, respectively, with stretching, strengthening, and conditioning exercises. All children, even those who do not present with specific physical impairments, are encouraged to engage in regular physical activity in order to

achieve the associated general health benefits (Tremblay et al. 2011).

The World Health Organization recommends that children between the age of 5 and 18 participate in at least 1 h of moderate (e.g., brisk walk) to vigorous (e.g., running) physical activity daily for general health benefits (World Health Organization 2012). The majority of children with chronic pain can work toward this goal; however, children with chronic pain often present with mobility impairments and fear that movement will cause tissue damage. An essential component of rehabilitation for children with chronic pain is education about the nature of their condition, and an interactive educational process whereby the patient's and family's beliefs about pain and activity are elicited and erroneous beliefs addressed with accurate information. Due to mobility impairments, fears, and pain with physical activity, children with chronic pain often benefit from a graded physical activity program (Ayling-Campos et al. 2011). There are many barriers to physical activity participation; however, pain and fear of movement are added barriers for children with chronic pain. Rehabilitation therapists work with children with chronic pain to reduce barriers to activity participation and encourage and support the child as she engages in increased activity.

Passive therapies for rehabilitation of the child with chronic pain include manual therapy (e.g., massage, joint mobilizations, or manipulations), therapeutic modalities (e.g., cryotherapy, thermotherapy, transcutaneous electrical nerve stimulation [TENS], low-level laser), braces, orthotics, and splints.

There is limited evidence to support the safety or effectiveness of joint mobilizations or manipulations for the treatment of chronic pain in children. Therapeutic effectiveness has been reported for adults in several randomized controlled trials (RCTs) for nonspecific back pain without peripheral radiation (Assendelft et al. 2004) and chronic neck pain (Bronfort et al. 2004). Spinal manipulations were found to be no better than sham manipulations for reducing pain in children with cervicogenic headache (Borusiak et al. 2010). There is some evidence to support the use of massage therapy for children with chronic pain (Suresh et al. 2008).

TENS is an electrical current delivered by a portable device through electrodes attached to the skin. Pain relief from TENS is very rapid (i.e., less than 1 h after onset of the stimulation), but also very short-lived after removal of the stimulation. Adults with chronic nociceptive and neuropathic pain report short-term reduction of pain intensity with TENS; however, there is little evidence for the use of TENS for management of chronic pain in children (Prentice 2011).

Cryotherapy (e.g., ice packs, ice massage, ice baths) and thermotherapy (e.g., hot packs, fluidotherapy, whirlpool, or contrast baths) may be useful for short-term relief of pain, reduction of muscle spasticity, and to be used in combination with active exercises. For example, a therapist may recommend use of a warm whirlpool bath prior to or during range of motion exercises to facilitate muscle relaxation and pain relief during stretching exercises.

Other therapeutic modalities frequently used for reduction of pain in adults include therapeutic ultrasound and low-level laser. Therapeutic ultrasound is used to deliver high-frequency sound waves which produce deep tissue thermal and nonthermal physiologic effects (Prentice 2011). Precaution is warranted for use of ultrasound over active epiphyses (Houghton et al. 2010). No evidence was found to support the use of therapeutic ultrasound for treatment of chronic pain in children. Low-level laser delivers an electromagnetic (light) energy that produces physiologic effects with little or no thermal energy production. Low-level laser has been used in the management of chronic pain conditions in adults (Chow et al. 2009); however, no research was found for management of chronic pain in children with low-level laser.

Resting and functional splints are used to maintain good joint position, protect inflamed joints, stretch or prevent contractures, relieve pain, and improve range of motion. Custom made orthotics have been shown to have a small advantage in relieving pain and reducing disability compared to ready-made shoe inserts and supportive athletic shoes in children with arthritis (Powell et al. 2005).

Passive therapies may be appropriate for use in some children with chronic pain if there are no contraindications for use (e.g., vascular or sensory impairments, active infection, communication impairments). However, there is limited evidence to support the effectiveness and safety of many passive therapies in children, and administration of these should be done only with appropriate assessment and close monitoring for adverse effects. Use of passive therapies is not a substitute for active therapies that aim to preserve or recover function.

Cognitive-Behavioral Interventions and Education

Cognitive-behavioral interventions delivered by the rehabilitation team include relaxation training, stress-management, activity pacing information, and pain coping strategies such as distraction. Relaxation training strategies include progressive muscle relaxation, guided imagery, and diaphragmatic breathing (von Baeyer and Tupper 2010). Cognitive-behavioral interventions, in combination with active therapies are effective for management of chronic pain in children (Clinch and Eccleston 2009).

Example

An adolescent with an idiopathic chronic pain condition reports that since the initiation of the pain complaint, she attends school sporadically, has withdrawn from dance class and no longer socializes with friends. The student complains of pain with prolonged sitting in the classroom and is unable to carry her textbooks to and from school. She finds dance class fatiguing and has pain during specific dance moves. She has withdrawn from social activities for fear of being incapacitated by pain in front of her peers. From the physical examination, the rehabilitation team will identify specific impairments that are impacting the patient's ability to participate in school and leisure activities. Therapeutic exercises will be provided that address the patient's movement and strength impairments and build tolerance to prolonged sitting. The patient may be instructed on modifications to posture and performance of functional activities. The rehabilitation team will inform the patient about appropriate passive interventions to address these impairments and monitor exercise progressions over time. With approval of the patient and her family, the rehabilitation team may also speak with the patient's teachers and dance instructor to determine if modifications to the school furniture, equipment, and dance classes can be made to enable a safe return to school and dance class. The rehabilitation team will work with the patient to identify opportunities for social engagement with peers that she is able to manage. A gradual progression back to full participation may be possible with appropriate psychological and physical treatment.

Summary

Physical medicine and rehabilitation for chronic pain in children involves a multidisciplinary team of health-care practitioners that provide comprehensive assessment and management of pain conditions through a goal-oriented approach that aims to maximize independent functioning in necessary and valued activities. The patient and family are integral to the rehabilitation process. Assessment of pain and function are ongoing throughout rehabilitation and include self-report or observational assessment of pain and function, physical examination, and environmental assessments. The focus of rehabilitation is noninvasive including active and passive physical interventions such as therapeutic exercise, general activity participation, manual therapy, therapeutic modalities, braces, splints and orthotics, environmental adaptations, and cognitive-behavioral interventions to promote relaxation and coping with pain.

References

Assendelft, W. J., Morton, S. C., Yu, E. I., Suttorp, M. J., & Shekelle, P. G. (2004). Spinal manipulative therapy for low back pain. *Cochrane Database of Systematic Reviews*, (1):CD000447.

Ayling-Campos, A., Amaria, K., Campbell, F., & McGrath, P. A. (2011). Clinical impact and evidence base for physiotherapy in treating childhood chronic pain. *Physiotherapy Canada*, 63(1), 21–33.

Borusiak, P., Biedermann, H., Bosserhoff, S., & Opp, J. (2010). Lack of efficacy of manual therapy in children and adolescents with suspected cervicogenic headache: Results of a prospective, randomized, placebo-controlled, and blinded trial. *Headache*, 50(2), 224–230.

Bronfort, G., Haas, M., Evans, R. L., & Bouter, L. M. (2004). Efficacy of spinal manipulation and mobilization for low back pain and neck pain: A systematic review and best evidence synthesis. *Spine Journal: Official Journal of the North American Spine Society*, 4(3), 335–356.

Chow, R. T., Johnson, M. I., Lopes-Martins, R. A., & Bjordal, J. M. (2009). Efficacy of low-level laser therapy in the management of neck pain: A systematic review and meta-analysis of randomised placebo or active-treatment controlled trials. *Lancet*, 374(9705), 1897–1908.

Clinch, J., & Eccleston, C. (2009). Chronic musculoskeletal pain in children: Assessment and management. *Rheumatology*, 48(5), 466–474.

Daniels, J., & Wearden, A. J. (2011). Socialization to the model: The active component in the therapeutic alliance? A preliminary study. *Behavioural and Cognitive Psychotherapy*, 39(2), 221–227.

Houghton, P. E., Nussbaum, E. L., & Hoens, A. M. (2010). Electrophysical agents – Contraindications and precautions: An evidence-based approach to clinical decision making in physical therapy. *Physiotherapy Canada*, 62(5), 1–80.

Maggio, A. B., Hofer, M. F., Martin, X. E., Marchand, L. M., Beghetti, M., & Farpour-Lambert, N. J. (2010). Reduced physical activity level and cardio-respiratory fitness in children with chronic diseases. *European Journal of Pediatrics*, 169(10), 1187–1193.

Powell, M., Seid, M., & Szer, I. S. (2005). Efficacy of custom foot orthotics in improving pain and functional status in children with juvenile idiopathic arthritis: A randomized trial. *Journal of Rheumatology*, 32(5), 943–950.

Prentice, W. E. (2011). *Therapeutic modalities in rehabilitation* (4th ed.). New York: McGraw Hill Medical.

Suresh, S., Wang, S., Porfyris, S., Kamasinski-Sol, R., & Steinhorn, D. M. (2008). Massage therapy in outpatient pediatric chronic pain patients: Do they facilitate significant reductions in levels of distress, pain, tension, discomfort, and mood alterations? *Paediatric Anaesthesia, 18*(9), 884–887.

Tremblay, M. S., Warburton, D. E., Janssen, I., Paterson, D. H., Latimer, A. E., Rhodes, R. E., & Duggan, M. (2011). New Canadian physical activity guidelines. *Applied Physiology, Nutrition, and Metabolism, 36*(1), 36–58.

von Baeyer, C. L., & Tupper, S. M. (2010). Procedural pain management for children receiving physiotherapy. *Physiotherapy Canada, 62*(4), 327–337.

World Health Organization. (2007). *ICF-CY: International classification of functioning, disability and health children and youth version*. Geneva: World Health Organization.

World Health Organization. (2012). *Global strategy on diet, physical activity and health: Physical activity and young people*. Retrieved from http://www.who.int/dietphysicalactivity/factsheet_young_people/en/index.html

Chronic Pain Prevalence

▶ Prevalence of Chronic Pain Disorders in Children

Chronic Pain, Diathesis-Stress Model of

Herta Flor
Department of Cognitive and Clinical Neuroscience, Central Institute of Mental Health, Medical Faculty Mannheim, Heidelberg University, Mannheim, Germany

Synonyms

Biobehavioral model; Biopsychosocial model; Multidimensional model; Psychobiological model

Definition

The diathesis-stress model views pain as a result of the interaction of predisposing and stress-related factors and assumes that social, psychological, and physiological factors are equally important in the understanding of chronic pain states.

Characteristics

Several authors (e.g., Apkarian et al. 2009; Apkarian et al. 2011; Flor 2012; Flor and Turk 2011) have formulated biobehavioral conceptualizations of chronic pain states that incorporate a multifactorial dynamic view of chronic pain with a mutual interrelationship of physiological and psychological factors and the change in these interrelationships over time. The diathesis-stress model views pain as a response with physiological, behavioral, and subjective components, that may or may not have an underlying organic pathological basis in the sense of a structural change but that will always have physiological antecedents and consequences. That is, behavior is also physiological and physiological processes have behavioral expressions. Thus, these three interrelated levels are incorporated in one model. The physiological level is comprised of ascending and descending neuronal connections, supraspinal and cortical mechanisms, as well as biochemical processes on all levels. The verbal-subjective modality consists of thoughts, feelings, and images. The behavioral-motor level takes into consideration pain behaviors ranging from medication intake to grimacing, limping, and use of the health care system. Interactions between the levels are continuous – learning occurs in physiological mechanisms, and physiological mechanisms are modified by behavioral and subjective changes. Thus, pain behaviors may be motivated by physiological, cognitive, and behavioral antecedents and consequences, all of which need to be considered in the analysis of pain. This view is consistent with the IASP definition that, to reiterate, views pain as an "unpleasant sensory and emotional experience associated with actual or potential tissue damage, or described in terms of such damage" (Merskey and Bogduk 1994, pp. 209–214) but in addition, emphasizes cognitive and learning parameters as integral to the experience of pain as a dynamic process. Although everyone experiences acute pain, only a small percentage of people develop chronic pain syndromes. These preconditions include predisposing factors, precipitating stimuli, precipitating responses, and maintaining factors (Fig. 1).

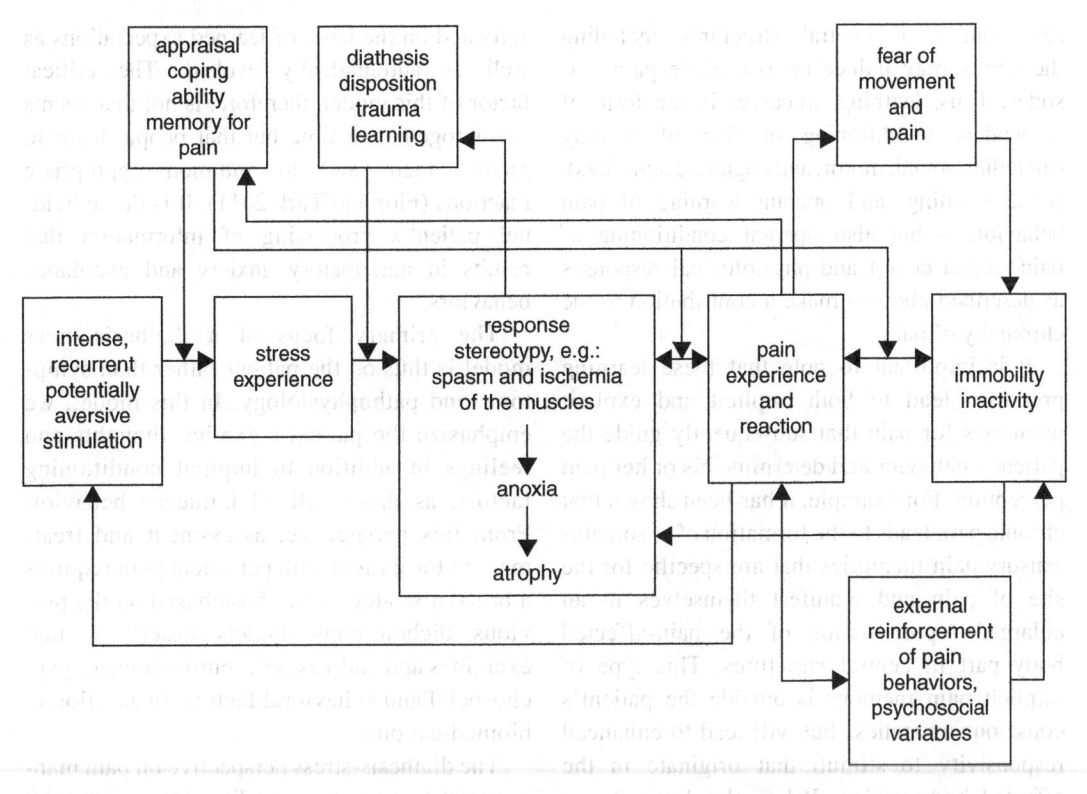

Chronic Pain, Diathesis-Stress Model of, Fig. 1 A diathesis-stress model delineating the main factors contributing to the development and maintenance of chronic pain

The existence of a physiological predisposition or diathesis involving a specific body system is the first component of the ▶ diathesis-stress model. This predisposition consists of a reduced threshold for nociceptive activation that may be related to genetic variables, previous trauma, or social learning experiences and results in a physiological response stereotypy of the specific body system. The existence of persistent aversive external or internal stimuli (pain-related or other stressors) with negative meaning activate the sympathetic nervous system and muscular processes (e.g., various aversive emotional stimuli such as familial conflicts or pressures related to employment) as unconditioned and conditioned stimuli and motivate avoidance responses. Aversive stimuli may be characterized by "excessive" intensity, duration, or frequency of an external or internal stimulus. "Inadequate" or "maladaptive" behavioral, cognitive, or physiological

repertoires of the individual to reduce the impact of these aversive environmental or internal stimuli are among the precipitating responses. Operant and respondent learning of behavioral, verbal-subjective, and physiological pain responses may maintain the pain experiences.

An important role is played by the cognitive processing of external or internal stimuli related to the experience of stress and pain, for example, increased perception, preoccupation, and overinterpretation of physical symptoms or inadequate perception of internal stimuli such as muscle tension levels. Moreover, the nature of the coping response – active avoidance, passive tolerance, or depressive withdrawal – may determine the type of problem that develops as well as the course of the illness. Subsequent maladaptive physiological responding such as increased and persistent sympathetic arousal and increased and persistent muscular reactivity as well as sensitization of central structures including the cortex may induce or exacerbate pain episodes. Thus, learning processes in the form of respondent conditioning of fear of activity (including social, motor, and cognitive activities), social learning, and operant learning of pain behaviors – but also operant conditioning of pain-related covert and physiological responses as described above – make a contribution to the chronicity of pain.

It is important to note that these learning processes lead to both implicit and explicit memories for pain that subsequently guide the patient's behavior and determine his or her pain perception. For example, it has been shown that chronic pain leads to the formation of ▶ somatosensory pain memories that are specific for the site of pain and manifest themselves in an enlarged representation of the pain-affected body part in central structures. This type of implicit pain memory is outside the patient's conscious awareness but will lead to enhanced responsivity to stimuli that originate in the affected body region. It has also been shown that the explicit recall of a pain-related episode leads to the activation of a large cortical network subserving pain as shown by the enhanced dimensional complexity of the EEG in chronic pain patients. Thus, learnt pain memories (i.e., psychological processes) directly influence the physiological processing of pain (cf. Apkarian 2008; Flor and Turk 2011; Sandkühler 2000).

In short, a diathesis-stress model places greatest emphasis on the role of learning factors in the onset, exacerbation, and maintenance of pain for those patients with persistent pain problems. It was suggested that a range of factors predispose individuals to develop chronic or recurrent acute pain; however, the predisposition is necessary but not sufficient. In addition to anticipation, avoidance, and contingencies of reinforcement, cognitive factors, in particular expectations, are also of central importance in a diathesis-stress model of chronic pain. Conditioned reactions are viewed as self-activated on the basis of learned expectations as well as automatically evoked. The critical factor of this model, therefore, is not that events occur together in time but that people learn to predict them and to summon appropriate reactions (Flor and Turk 2011). It is the individual patient's processing of information that results in anticipatory anxiety and avoidance behaviors.

The primary focus of a diathesis-stress model is thus on the patient rather than symptoms and pathophysiology. In this model, we emphasize the patient's explicit thoughts and feelings in addition to implicit conditioning factors, as these will all influence behavior. From this perspective, assessment and treatment of the patient with persistent pain requires a broader strategy than those based on the previous dichotomous models described, that examines and addresses the entire range of psychosocial and behavioral factors, in addition to biomedical ones.

The diathesis-stress perspective on pain management focuses on providing the patient with techniques to gain a sense of control over the effects of pain on his or her life as well as actually modifying the affective, behavioral, cognitive, and sensory facets of the experience (Flor and Turk 2011). Behavioral experiences help to show patients that they are capable of more than they assumed, increasing their sense of personal competence. ▶ Cognitive techniques help to place affective, behavioral, cognitive, and sensory responses under the patient's control. Our assumption is that long-term maintenance of behavioral changes will occur only if the patient has learned to attribute success to his or her own efforts. There are suggestions that these treatments can result in changes of beliefs about pain, coping style, and reported pain severity, as well as direct behavior changes. Further, treatment that results in increases in perceived control over pain and decreased catastrophizing are also associated with decreases in pain severity ratings and functional disability.

References

Apkarian, A. V. (2008). Pain perception in relation to emotional learning. *Current Opinion in Neurobiology, 18*(4), 464–468.

Apkarian, A. V., Baliki, M. N., & Geha, P. Y. (2009). Towards a theory of chronic pain. *Progress in Neurobiology, 87*(2), 81–97.

Apkarian, A. V., Hashmi, J. A., & Baliki, M. N. (2011). Pain and the brain: Specificity and plasticity of the brain in clinical chronic pain. *Pain, 152*(3 Suppl.), S49–S64.

Flor, H. (2012). New developments in the understanding and management of persistent pain. *Current Opinion in Psychiatry, 25*(2), 109–113.

Flor, H., & Turk, D. C. (2011). *Chronic pain: An integrated biobehavioral approach.* Seattle: IASP Press.

Merskey, H., & Bogduk, N. (Eds.). (1994). Classification of chronic pain. Descriptions of chronic pain syndromes and definitions of pain terms (2nd ed.). Seattle, WA: IASP Press.

Sandkühler, J. (2000). Learning and memory in pain pathways. *Pain, 88*(2), 113–118.

Chronic Pain, Patient-Therapist Interaction

Mark P. Jensen
Department of Rehabilitation Medicine,
University of Washington, Seattle, WA, USA

Synonyms

Relationship; Therapeutic alliance

Definition

▶ Patient-therapist interaction refers to the verbal and nonverbal interactions between health-care providers and their patients or clients. These interactions may include, but are not necessarily limited to, communication and negotiations concerning the patient's history, diagnosis, and clinical care.

Characteristics

Patient-Therapist Interaction and Pain Treatment Satisfaction

Research shows that pain intensity has only a weak to moderate association with pain treatment satisfaction, due primarily to the fact that patients tend to report relatively high levels of satisfaction with pain management care, regardless of pain severity levels (Dawson et al. 2002; McCracken et al. 2002; Miaskowski et al. 1994; Ward and Gordon 1996). On the other hand, factors related to the interactions between patients and clinicians show consistent associations with treatment satisfaction across measures and samples. For example, Sherwood and colleagues (Sherwood et al. 2000) found that treatment satisfaction with pain treatment was higher when patients felt that their pain was addressed with the patient as an informed partner and was lower when the providers appeared uncaring, were slow to respond to pain complaints, or were perceived to lack knowledge and skills. Similarly, Dawson and colleagues (Dawson et al. 2002) found that many patients who reported severe pain, and who still reported that they were satisfied with pain care, attributed their satisfaction to their belief that their health-care provider was making efforts to provide pain relief. Riley and colleagues (Riley et al. 2001) found that patient-perceived quality of caregiver communication predicted satisfaction with individualized pain treatment plans in a sample of 107 patients seen in an orofacial pain clinic. McCracken and colleagues (McCracken et al. 1997) identified confidence and trust in the treatment provider, pain reduction, and time spent waiting in the clinic as predictors of satisfaction with chronic pain treatment. In short, patients who are satisfied with their pain care seem to be those who see themselves as having a collaborative relationship with a treatment provider who works closely with the patient to address the pain problem. While pain reduction may be associated with treatment satisfaction in some instances, pain reduction does not appear to be necessary for patients to be satisfied with their care.

Patient-Therapist Interaction and Treatment Adherence and Outcomes

Research also shows that ▶ treatment adherence is related to patient-therapist interactions, in particular, the existence of a collaborative relationship between the patient and clinician. While very little of this research has been performed in samples of persons with chronic pain (for an exception, see discussion of Alamo et al. (2002) and Glombiewski et al. (2010), below), this finding is consistent across a wide variety of patient populations and treatments (Cruz and Pincus 2002). For example, Gavin et al. (1999) showed that a measure of ▶ treatment alliance (measuring the extent to which patients and clinicians agree on treatment goals and hold each other in positive regard) was associated with medication adherence and treatment outcome in a sample of adolescents with severe asthma. Weiss et al. (2002) similarly found that a measure of working alliance (which assessed patient-therapist agreement of treatment goals and the tasks to be used to work toward these goals and the degree to which a bond exists between the therapist and patient) was associated with future medication adherence and treatment outcome in a sample of patients with psychotic disorders. In a sample of patients with schizophrenia, Frank and Gunderson (1990) found that patients who were able to form a good treatment alliance with their therapists within the first 6 months of treatment adhered more to their medication regimen and achieved better outcomes after 2 years (with less medication) than patients who did not form a good treatment alliance.

Alamo et al. (2002) found evidence for the efficacy of a ▶ patient-centered approach in a sample of patients with chronic pain. In this study, a group of family physicians were trained in the use of a patient-centered approach to pain treatment, which included the following components, among others: listening to the patient without interrupting in the first moments of the consultation, being supportive and ▶ empathic, allowing and encouraging patients to ask questions, trying to reach an agreement about the nature of the problem, and trying to find common ground about the management plan. Patients of the physicians who were trained in and used the patient-centered approach reported significantly greater improvements after 1 year in psychological distress and number of tender points than the patients of family physicians not trained in the patient-centered approach. Although not statistically significant, there was also a trend for patients in the patient-centered condition to report greater decreases in pain intensity than patients who received usual care. In a separate study, Glombiewski and colleagues (Glombiewski et al. 2010) identified low treatment satisfaction as one factor that was associated with dropout from CBT treatment of chronic pain.

Implications for Improving Pain Management Care

The findings briefly reviewed above support the conclusion that patient-therapist interaction factors play a role in patient treatment satisfaction and adherence, as well as in long-term treatment outcome. Clinicians would therefore be wise to communicate with patients in ways that will maximize therapeutic alliance. Attention to three specific areas may be particularly useful: (1) communicate efforts to manage pain, (2) make an effort to build and maintain rapport, and (3) encourage a collaborative relationship.

Communicate Efforts to Manage Pain

Most patients come to realize, by the time their pain becomes chronic, that there is no quick fix for their pain problem. However, many patients still hope that a new treatment will be discovered that will provide pain relief. Also, there exist many treatments for pain that subgroups of patients may find effective, even if these treatments are not necessarily effective for the majority of patients (Engel et al. 2002). Clinicians could improve their relationships with patients and increase patient treatment satisfaction, if they communicated that (1) they are vigilant in their efforts to identify effective pain treatments as these treatments are reported in the research literature and (2) they are willing to work with the patient to find the specific combination of treatments and approaches that are most effective for each individual patient.

Make an Effort to Build and Maintain Rapport

One of the most effective ways to build rapport is to incorporate ► reflective listening in interactions with patients. Reflective listening involves making statements that indicate that you understand what the patient is saying. Such statements differ substantially from usual clinician responses, such as questions that call for or require a response or statements that merely provide information (Miller and Rollnick 2002; Rollnick et al. 1999; Rollnick et al. 2007). Examples of reflective responses to the statement "I am sick and tired of this pain" would include "You are frustrated that it is taking so long for us to find an approach that effectively reduces your pain" or "You are ready to try something new to get this pain under better control." The immediate concern of some clinicians, when they consider using reflective listening during a patient encounter, is that such statements might open a Pandora's box of patient talking, which may then use up the (limited) time devoted to the encounter. However, clinicians need not make every statement a reflective one. Moreover, encounters that include reflective listening can actually be more efficient than encounters that do not, because once patients feel understood, they may feel less compelled to interrupt and ask questions. In support of this idea, one study found that being responsive to patient concerns (as opposed to ignoring patient concerns) was associated with a decrease in the time of the encounter (Levinson et al. 2000).

Encourage a Collaborative Relationship

For the most part, adequate management of chronic pain involves active patient participation and involvement, and good long-term adjustment depends much more on what patients do than on what is done to them (Jensen et al. 2003). As reviewed above, patients are more likely to actively participate in treatment and adhere to treatment recommendations when there is a good therapeutic alliance and when there is a collaborative patient-therapist relationship. Developing and maintaining a collaborative relationship with patients does not have to involve any special skills, other than a willingness to avoid lecturing patients or telling them what to do and a willingness to provide the patient with options and choices.

Lecturing patients can create resentment and ► resistance (Miller and Rollnick 2002). It is more effective to provide guidance, options, and advice while at the same time communicating the expectation that the patient can decide which parts of the recommended treatment plan he or she will follow. As choice increases adherence, the collaborative clinician offers patients choices whenever possible (Miller and Rollnick 2002; Rollnick et al. 2007). For example, if both the clinician and patient decide that a graduated exercise program is worth trying, the patient should then be offered different methods for starting and maintaining such a program. The patient could be offered sessions with a physical therapist who could help them to develop a home program. Another option would be to allow the patient to develop a simple reactivation program at home, under the guidance of his or her physician, without the need to visit a physical therapist. The home program could be highly structured, with a series of specific exercises that the patient decides to engage in on a daily basis; or it could be less structured, with the patient setting goals to increase general activity levels by a certain percent every week. As there are so many paths that may be taken to reach any one treatment goal, a treatment plan can, and should, be developed that is tailored to the patient's own situation, goals, and interests and that is developed in close collaboration with the patient. Such collaboration will increase the chances that the patient will adhere to the treatment plan and ultimately increase the chances of a successful treatment outcome.

References

Alamo, M. M., Moral, R. R., & Pérula de Torres, L. A. (2002). Evaluation of a patient-centred approach in generalized musculoskeletal chronic pain/fibromyalgia patients in primary care. *Patient Education and Counseling, 48*, 23–31.

Cruz, M., & Pincus, H. A. (2002). Research on the influence that communication in psychiatric encounters has on treatment. *Psychiatric Services, 53*, 1253–1265.

Dawson, R., Spross, J. A., Jablonski, E. S., Hoyer, D. R., Sellers, D. E., & Solomon, M. Z. (2002). Probing the paradox of patients' satisfaction with inadequate pain management. *Journal of Pain and Symptom Management, 23*, 211–220.

Engel, J. M., Kartin, D., & Jensen, M. P. (2002). Pain treatment and health care in persons with cerebral palsy: Frequency and helpfulness. *American Journal of Physical Medicine & Rehabilitation, 81*, 291–296.

Frank, A. F., & Gunderson, J. G. (1990). The Role of the therapeutic alliance in the treatment of schizophrenia. *Archives of General Psychiatry, 47*, 228–236.

Gavin, L. A., Wamboldt, M. Z., Sorokin, N., Levy, S. Y., & Wamboldt, F. S. (1999). Treatment alliance and its association with family functioning, adherence, and medical outcome in adolescents with severe, chronic asthma. *Journal of Pediatric Psychology, 24*, 355–365.

Glombiewski, J. A., Hartwich-Tersek, J., & Rief, W. (2010). Attrition in cognitive-behavioral treatment of chronic back pain. *The Clinical Journal of Pain, 26*, 593–601.

Jensen, M. P., Nielson, W. R., & Kerns, R. D. (2003). Toward the development of a motivational model of pain self-management. *The Journal of Pain, 4*, 477–492.

Levinson, W., Gorawara-Bhat, R., & Lamb, J. (2000). A study of patient clues and physician responses in primary care and surgical settings. *Journal of the American Medical Association, 284*, 1021–1027.

McCracken, L. M., Klock, P. A., Mingay, D. J., Asbury, J. K., & Sinclair, D. M. (1997). Assessment of satisfaction with treatment for chronic pain. *Journal of Pain and Symptom Management, 14*, 292–299.

McCracken, L. M., Evon, D., & Karapas, E. T. (2002). Satisfaction with treatment for chronic pain in a specialty service: Preliminary prospective results. *European Journal of Pain, 6*, 387–393.

Miaskowski, C., Nichols, R., Brody, R., & Synold, T. (1994). Assessment of patient satisfaction utilizing the American Pain Society's quality assurance standards on acute and cancer-related pain. *Journal of Pain and Symptom Management, 9*, 5–11.

Miller, W., & Rollnick, S. (2002). *Motivational interviewing: Preparing people to change* (2nd ed.). New York: Guilford Press.

Riley, J. L., Meyers, C. D., Robinson, M. E., Bulcourf, B., & Gremillion, H. A. (2001). Factors predicting orofacial pain patient satisfaction with improvement. *Journal of Orofacial Pain, 15*, 29–35.

Rollnick, S., Mason, P., & Butler, C. (1999). *Health behavior change: A guide for practitioners*. Edinburgh: Churchill Livingstone.

Rollnick, S. P., Miller, E. R., & Butler, C. C. (2007). *Motivational interviewing in health care: Helping patients change behavior*. New York: Guilford Press.

Sherwood, G., Adams-McNeill, J., Starck, P. L., Nieto, B., & Thompson, C. J. (2000). Qualitative assessment of hospitalized patients' satisfaction with pain management. *Research in Nursing & Health, 23*, 486–495.

Ward, S. E., & Gordon, D. B. (1996). Patient satisfaction and pain severity as outcomes in pain management: A longitudinal view of one settings' experience. *Journal of Pain and Symptom Management, 11*, 242–251.

Weiss, K. A., Smith, T. E., Hull, J. W., Piper, A. C., & Huppert, J. D. (2002). Predictors of risk of nonadherence in outpatients with schizophrenia and other psychotic disorders. *Schizophrenia Bulletin, 28*, 341–349.

Chronic Pain, Persistent Pain

▶ Psychiatric Aspects of the Epidemiology of Pain

Chronic Pain, Thalamic Plasticity

▶ Thalamic Plasticity and Chronic Pain

Chronic Paroxysmal Hemicrania

Synonyms

CPH

Definition

Chronic paroxysmal hemicrania is a trigemino-autonomic syndrome with frequent repeated facial pains, each lasting for 3 min or more. Attacks can recur over >1 year without remission periods or with remission periods lasting <1 month.

Cross-References

▶ Paroxysmal Hemicrania
▶ Primary Stabbing Headache

Chronic Pelvic Inflammatory Disease

▶ Chronic Pelvic Pain, Pelvic Inflammatory Disease and Adhesions

Chronic Pelvic Pain

▶ Dyspareunia and Vaginismus
▶ Interstitial Cystitis and Chronic Pelvic Pain
▶ Gynecological Pain and Sexual Functioning
▶ Gynecological Pain, Neural Mechanisms
▶ Pain Mechanisms in Endometriosis

Chronic Pelvic Pain Model

▶ Visceral Pain Models, Female Reproductive Organ Pain

Chronic Pelvic Pain Syndrome

Definition

Chronic pelvic pain syndrome (CPPS) is the occurrence of chronic pelvic pain where there is no proven infection or other obvious local pathology that may account for the pain. It is often associated with negative cognitive, behavioral, sexual, or emotional consequences as well as with symptoms suggestive of lower urinary tract, sexual, bowel or gynecological dysfunction. CPPS is a subdivision of chronic pelvic pain.

Cross-References

▶ Chronic Gynecological Pain, Doctor-Patient Interaction
▶ Chronic Pelvic Pain, Physical and Sexual Abuse
▶ Chronic Pelvic Pain, Physical Therapy Approaches, and Myofascial Abnormalities
▶ Gynecological Pain and Sexual Functioning
▶ Gynecological Pain, Neural Mechanisms
▶ Taxonomy of Pelvic Pain

Chronic Pelvic Pain, Endometriosis

Fred M. Howard
University of Rochester, Rochester, NY, USA

Synonyms

Endometriosis externa.

Definition

Endometriosis is the presence of ectopic endometrial glands and stroma, that is, endometrium located outside the endometrial cavity (Fig. 1). Occasionally only endometrial glands are present in cases of endometriosis. Hemosiderin-laden macrophages may be present also, but their solitary presence without endometrial glands or stroma does not confirm endometriosis. Sampson first applied the name "endometriosis" to ectopic endometrium in 1921 (Sampson 1921).

Characteristics

Endometriosis is most often found within the peritoneal cavity in the pelvis but may also occur in numerous other, even remote, locations (Table 1). There is still a great deal about it that is unclear and controversial, and it remains an enigmatic disorder in that the etiology, the natural history, and the

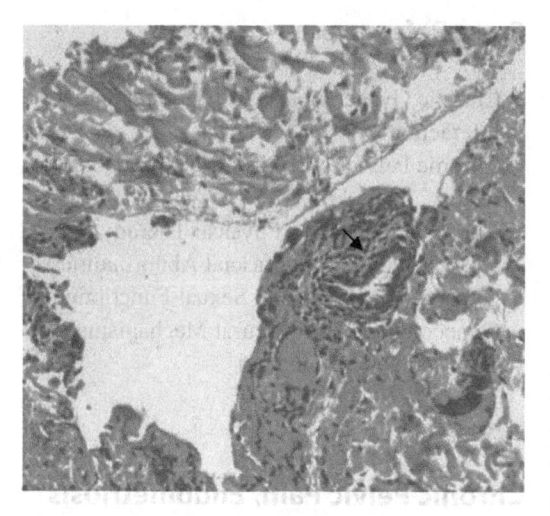

**Chronic   Pelvic   Pain,   Endometriosis,
Fig. 1** Endometriosis of the pelvic peritoneum

Chronic Pelvic Pain, Endometriosis, Table 1 Possible sites of endometriosis

Common sites	Less common or rare sites
Ovaries	Umbilicus
Round ligaments	Laparotomy scars
Broad ligaments	Hernial sacs
Uterosacral ligaments	Small intestine
Rectovaginal septum	Rectum
Appendix	Sigmoid
Pelvic peritoneum	Ureters
Bladder	Vulva
Pelvic lymph nodes	Extremities
Cervix	Pleural cavity
Vagina	Lung
Fallopian tubes	Nasal mucosa

precise mechanisms by which it causes pain are not completely understood. In some patients, it behaves almost like a malignancy, spreading rapidly and widely, yet in others, it is a seemingly irrelevant, insignificant, incidental histological diagnosis. Grossly, endometriosis has a variety of appearances; it may be red, purple, blue, white, yellow, brown, clear or black lesions, peritoneal pockets or windows, and adhesions or chocolate cysts. Endometriosis is primarily a disease of women of reproductive age.

Prevalence

As endometriosis is accurately diagnosed only by surgical biopsy with histological confirmation, accurate prevalence is difficult to determine but is estimated to be about 7 % (Barbieri 1990). In women who undergo a laparoscopy to evaluate chronic pelvic pain (CPP), the prevalence of endometriosis is about 33 % (Howard 1993). In patients undergoing laparoscopy for infertility, the prevalence is about 40 %. Additionally it appears that about 70 % of women with endometriosis have some type of pelvic pain symptoms. It has been observed that the severity of pain frequently does not correlate with the severity of endometriosis (Vercellini 1991).

Etiology

None of several theories of etiology alone explain the protean manifestations of endometriosis or the predilection of some women but not others to develop symptomatic endometriosis. Sampson's theory is that endometriosis is due to retrograde flow of menstrual effluent through the fallopian tubes into the peritoneal cavity. However, most women experience some degree of retrograde menstruation, so there is more to the development of endometriosis than just retrograde flow. Metaplasia of coelomic epithelium, the epithelium from which the Mullerian duct is derived, can result in endometriosis. Metaplasia needs an induction phenomenon or factor, which might be menstrual debris, estrogen, or progesterone. The theory of lymphatic and vascular metastases is invoked to explain the occurrence of endometriosis in remote locations such as the pleura, nose, and spinal column. It has been reported that endometriosis is present in pelvic lymph nodes in 30 % of women with pelvic endometriosis, which supports this theory. A defect of the immunological system is supported by a good deal of research and helps to explain why not all women develop endometriosis secondary to retrograde menstruation. It also ties into the theory of genetic predisposition, as an immunologic disorder may be inherited. Finally, recent research has focused on the endometrial cells that are present in endometriosis and found a number of abnormalities that may contribute either to initiation or development of endometriosis. For example, endometriosis cells have been found to have abnormal production of aromatase cytochrome P_{450}, an enzyme that is not present in normal endometrium and is integral to the conversion of androstenedione and testosterone to estrogen. This ability to produce estrogen locally may directly stimulate the growth of endometrial cells of endometriotic lesions. Endometriosis cells have also been observed to have increased amounts of vascular endothelial growth factor and interleukin-6.

It is worthwhile to especially note the role of estrogen, as there are several clinical observations that suggest that the development and persistence of endometriosis are estrogen dependent. First, endometriosis is rare before puberty or after menopause unless the woman is on estrogen replacement therapy. Second, bilateral oophorectomy typically results in regression of endometriotic lesions. Third, decreased levels of estradiol via GnRH agonist treatment result in regression of endometriosis. Fourth, endometriosis can develop in the prostatic utricle of a male with DES treatment for prostatic cancer. Finally, immunohistochemical studies show virtually all endometriosis lesions contain estrogen receptors.

The etiology of symptoms in some women but not in others is also not understood. Endometriosis might produce symptoms due to swelling of the tissue with hormonal stimulation, plus extravasation of blood and menstrual debris into surrounding tissues, with production of prostaglandins as possible chemical mediators of pain and inflammation. It is also hypothesized that lesions produce prostaglandins and cause functional pain symptoms, like dysmenorrhea, via direct production of prostaglandins (Vernon 1985). Finally, it has been suggested that lesions cause pain via nociceptor stimulation by mechanical pressure or by stretching tissue. Such a mechanism would predict that larger lesions and more deeply infiltrating lesions would cause more frequent or more severe pain (Cornillie 1990).

Symptoms and Signs

Classically the woman with endometriosis presents with one or more of the following: an adnexal mass (endometrioma), infertility, pelvic pain, or dysmenorrhea. Estimates are that up to 40 % of women with endometriosis have chronic pelvic pain. Pelvic pain most often starts as dysmenorrhea, and at least 75 % of women with endometriosis-associated pelvic pain have dysmenorrhea as an initial component of their pain. Dyspareunia with deep penetration is also a frequent component of endometriosis-associated pain, occurring in about 33 % of cases. Although CPP, dyspareunia, and dysmenorrhea are significantly more common in women with endometriosis than in women with a normal pelvis, these pain symptoms are not as specific nor diagnostic for endometriosis as is commonly thought and by themselves do not justify a diagnosis of endometriosis.

Intestinal involvement, usually of the appendix, rectosigmoid, or anterior rectum, may cause abdominal pain, dyspareunia, tenesmus, dyschezia, constipation, diarrhea, low back pain, and, rarely, hematochezia or symptoms of bowel obstruction. Urinary tract involvement, most often at the bladder peritoneum and anterior cul-de-sac, may cause frequency, pressure, dysuria, or hematuria. Involvement of the distal one-third of the ureter may lead in rare cases to symptoms of ureteral obstruction. With lung involvement, endometriosis may rarely cause dyspnea on exertion, pleural effusion, and lung collapse. Catamenial hemothorax has been reported. A cyclically bleeding or a cyclically tender mass in an incisional scar may also occur with endometriosis.

The physical examination is often normal, but there is tenderness, especially during the menses, in many women with endometriosis-associated pelvic pain. A fixed retroverted uterus with posterior tenderness is particularly suggestive of endometriosis. Tender nodularity of the uterosacral ligaments and cul-de-sac is a classically described finding with endometriosis. Narrowing of the posterior vaginal fornix or lateral deviation of the position of the cervix may rarely be present. In patients with endometriomas a tender adnexal mass may be noted.

Diagnostic Studies

At this time, an accurate diagnosis can only be made by surgical and histological confirmation. Before the introduction of diagnostic laparoscopy, this required an exploratory laparotomy. Laparoscopy is a much less invasive procedure and is currently the recommended diagnostic procedure for any patient suspected of having endometriosis.

A preoperative ultrasound is worthwhile as 15–20 % of women with endometriosis have endometriomas. These generally range from 3 to 8 cm in size and are not always palpable by physical examination. Measurement of serum Ca-125 levels has a low sensitivity and specificity, although it is more reliable for advanced stage disease (Lanzone 1991). Ca-125 levels are also elevated with cancers of the ovary, endometrium, gastrointestinal tract, fallopian tube, and breast as well as pelvic inflammatory disease, pregnancy, menses, and leiomyomata. Adolescents with endometriosis should be evaluated for obstructive anomalies of the reproductive system using magnetic resonance imaging, ultrasound, and hysterography. About 10 % will have such an anomaly.

In patients in whom intestinal endometriosis is suspected, evaluation of the intestinal tract is usually normal, as most patients have involvement of the serosa or muscularis, not the mucosa. Sigmoidoscopy may show a bluish submucosal mass in some of these cases. Mucosal lesions may be diagnosed by sigmoidoscopy, colonoscopy, or barium enema studies, but these evaluations are certainly not necessary on a routine basis.

In patients with suspected urinary tract involvement, cystoscopy and intravenous pyelography or computerized tomography of the kidney, ureters, and bladder may be indicated.

Medical Treatment

There are many options for medical and surgical treatment of endometriosis-associated pelvic pain (see list below).

General treatment options for women with endometriosis-associated pelvic pain:

- Observation with palliative treatment
- Conservative surgery
- Hormonal suppression
- Combined medical and surgical treatments
- Definitive extirpative surgery (radical surgery)

Danazol, a 17-ethinyl-testosterone derivative, was approved by the FDA for treatment of endometriosis in 1976 (Greenblatt et al. 1971). It is mildly androgenic and anabolic, properties that account for many of its side effects. Side effects include acne, edema, weight gain, hirsutism, voice changes, hot flushes, abnormal uterine bleeding, decreased breast size, decreased libido, vaginal dryness, nausea, weakness, and muscle cramps. Danazol does not significantly affect LH and FSH levels in premenopausal women, but lowers estrogen levels by directly inhibiting steroidogenesis at the ovarian and adrenal levels. Sixty percent to 83 % of patients obtain significant relief of pelvic pain with danazol therapy, and the number needed to treat is 1.7 (Barbieri et al. 1982). Danazol is contraindicated in patients with abnormal uterine bleeding, pregnancy, and breastfeeding or impaired renal, cardiac, or hepatic function.

Gonadotropin-releasing hormone (GnRH) agonists are analogues of naturally occurring gonadotropin-releasing hormone and are the most commonly prescribed medical treatment for endometriosis in the USA. Examples of GnRH agonists are nafarelin, leuprolide, and goserelin. They all work at the hypothalamic-pituitary level to shut down LH and FSH production and release, resulting in a dramatic decline in estradiol levels. Pain relief with GnRH agonists is about the same as with danazol. Side effects of the GnRH agonists are loss of bone density, hot flashes, vaginal dryness, decreased libido, headaches, emotional lability, acne, and reduced breast size (Bergqvist et al. 1998; Kennedy et al. 1990; Wheeler et al. 1992). Because of concerns about loss of bone density, add-back therapy with 5 mg day-1 of norethindrone acetate, with or without 0.625 conjugated equine estrogen, is recommended for up to 1 year of treatment

(Hornstein et al. 1998). After discontinuation of treatment, symptoms tend to recur in most patients, with mean time to recurrence between 8 and 11 months.

A number of progestagens, particularly medroxyprogesterone acetate and norethindrone acetate, are used to treat endometriosis. Breakthrough bleeding, prolonged amenorrhea, mood changes, depression, weight gain, and irritability are common side effects. Gestrinone, a 19-nortestosterone derivative with mostly progestagenic and low androgenic activity, has also been used to treat endometriosis (not available in the USA) (The Gestrinone Italian Study Group 1996).

Oral contraceptives are commonly used to treat endometriosis. They appear to be less effective than GnRH agonists in relieving dysmenorrhea and dyspareunia but comparable in relief of non-menstrual pelvic pain (Vercellini et al. 1993). Side effects are weight gain, breast tenderness, nausea, chloasma, abnormal uterine bleeding, enlargement of myomas, thrombophlebitis and thromboembolism, increased appetite, irritability, depression, edema, hypertension, and increased vaginal discharge. They are contraindicated in women with a history of or high risk of thrombosis or a history of breast cancer and relatively contraindicated with diabetes, collagen vascular disease, or hypertension.

Surgical Treatment

Surgical treatment of endometriosis may be conservative (i.e., without extirpation of the uterus, tubes, or ovaries) or radical (i.e., with extirpation of the uterus or one or both ovaries). Radical surgery appears to be more effective in achieving pain relief, but conservative surgery is more appropriate in younger women, especially if preservation of reproductive potential is important. Also, conservative surgery can be done at the time of laparoscopic diagnosis (Howard 1994). The core of either conservative or radical surgical treatment is the removal or destruction of all endometriotic lesions. A randomized clinical trial of conservative surgery for endometriosis shows that on average there is a 50 % decrease of pain 6 months postoperatively and the number needed to treat is 2.5 (Sutton et al. 1994).

Both presacral neurectomy (resection of the superior hypogastric plexus) and uterosacral neurectomy (transection of the uterosacral ligament) have been recommended for relief of CPP associated with endometriosis, but results of randomized clinical trials show efficacy only for presacral neurectomy, (Candiani et al. 1992; Zullo et al. 2003) not uterosacral neurectomy (Sutton et al. 2001; Vercellini et al. 2003). Presacral neurectomy is most effective for the treatment specifically of midline dysmenorrhea. There appears to be a small effect, if any, on non-menstrual pelvic pain or dyspareunia.

If fertility is not desired, then hysterectomy, with or without bilateral salpingo-oophorectomy, is often recommended for endometriosis-associated pelvic pain. There is no consensus as to the advisability of removal of both ovaries if one or both are not directly involved by endometriosis, but recurrence of pain when one or both ovaries are preserved has been reported to be increased (relative risk for pain recurrence of 6.1, confidence interval 2.5-14.6) (Namnoum et al. 1995). Although uncommon, endometriosis has been reported to recur after hysterectomy and bilateral salpingo-oophorectomy, with and without estrogen replacement therapy. Finally, currently available data do not allow a conclusion as to whether medical or conservative surgical treatment is more effective in the treatment of endometriosis-associated pelvic pain.

Combined Medical and Surgical Treatment

There are no clinical trials of preoperative medical treatment. There are, however, at least three randomized, placebo-controlled clinical trials of postoperative medical treatment (Hornstein et al. 1997; Parazzini et al. 1994; Telimaa et al. 1987). These trials suggest that pain is decreased with postoperative medical treatment while patients are on the medications, but within 6–12 months after discontinuation of medications, pain levels are similar in postoperative patients whether or not they received medical treatment. A reasonable way to apply these data is to initiate medical treatment after conservative surgical debulking therapy if patients have persistent or recurrent pain, rather than treat all patients postoperatively.

References

Barbieri, R. L. (1990). Etiology and epidemiology of endometriosis. *American Journal of Obstetrics and Gynecology, 162*, 565–567.

Barbieri, R. L., Evans, S., & Kistner, R. W. (1982). Danazol in the treatment of endometriosis: Analysis of 100 cases with 4-year follow-up. *Fertility and Sterility, 37*, 737–746.

Bergqvist, A., Bergh, T., Hogstrom, L., et al. (1998). Effects of triptorelin versus placebo on the symptoms of endometriosis. *Fertility and Sterility, 69*, 702–708.

Candiani, G. B., Fedele, L., Vercellini, P., et al. (1992). Presacral neurectomy for the treatment of pelvic pain associated with endometriosis: A controlled study. *American Journal of Obstetrics and Gynecology, 167*, 100–103.

Cornillie, F. J., Oosterlynck, D., Lauweryns, J. M., et al. (1990). Deeply infiltrating pelvic endometriosis: Histology and clinical significance. *Fertility and Sterility, 53*, 978–983.

Greenblatt, R. B., Dmowski, W. P., Mahesh, V. B., et al. (1971). Clinical studies with an antigonadotropin – Danazol. *Fertility and Sterility, 22*, 102–112.

Hornstein, M. D., Hemmings, R., Yuzpe, A. A., et al. (1997). Use of nafarelin versus placebo after reductive laparoscopic surgery for endometriosis. *Fertility and Sterility, 68*, 860–864.

Hornstein, M. D., Surrey, E. S., Weisberg, G. W., et al. (1998). Leuprolide acetate depot and hormonal add-back in endometriosis: A 12-month study. *Obstetrics and Gynecology, 91*, 16–24.

Howard, F. M. (1993). The role of laparoscopy in chronic pelvic pain: Promise and pitfalls. *Obstetrics Gynecology Survey, 48*, 357–387.

Howard, F. M. (1994). Laparoscopic evaluation and treatment of women with chronic pelvic pain. *Journal of America Association Gynecology Laparoscopic, 1*, 325–331.

Kennedy, S. H., Williams, I. A., Brodribb, J., et al. (1990). A comparison of nafarelin acetate and danazol in the treatment of endometriosis. *Fertility and Sterility, 53*, 998–1003.

Lanzone, A., Marane, R., Muscatello, R., et al. (1991). Serum Ca-125 levels in the diagnosis and management of endometriosis. *The Journal of Reproductive Medicine, 36*, 603.

Namnoum, A. B., Hickman, T. N., Goodman, S. B., et al. (1995). Incidence of symptom recurrence after hysterectomy for endometriosis. *Fertility and Sterility, 64*, 898–902.

Parazzini, F., Fedele, L., Busacca, M., et al. (1994). Postsurgical medical treatment of advanced endometriosis: Results of a randomized clinical trial. *American Journal of Obstetrics and Gynecology, 171*, 1205–1207.

Sampson, J. A. (1921). Perforating hemorrhagic (chocolate) cysts of the ovary. *Archives of Surgery, 3*, 245.

Sutton, C. J. G., Ewen, S. P., Whitelaw, N., et al. (1994). Prospective, randomized, double-blind trial of laser laparoscopy in the treatment of pelvic pain associated with minimal, mild, and moderate endometriosis. *Fertility and Sterility, 62*, 696–700.

Sutton, C., Pooley, A. S., Jones, K. D., et al. (2001). A prospective, randomized, double-blind controlled trial of laparoscopic uterine nerve ablation in the treatment of pelvic pain associated with endometriosis. *Gynaecologiae Endoscopy, 10*, 217–222.

Telimaa, S., Ronnberg, L., & Kauppila, A. (1987). Placebo-controlled comparison of danazol and high-dose medroxyprogesterone acetate in the treatment of endometriosis after conservative surgery. *Gynecological Endocrinology, 1*, 363–371.

The Gestrinone Italian Study Group. (1996). Gestrinone versus a gonadotropin-releasing hormone agonist for the treatment of pelvic pain associated with endometriosis: A multicenter, randomized, double-blind study. *Fertility and Sterility, 66*, 911–919.

Vercellini, P., Aimi, G., Busacca, M., et al. (2003). Laparoscopic uterosacral ligament resection for dysmenorrhea associated with endometriosis: Results of a randomized, controlled trial. *Fertility and Sterility, 80*, 310–319.

Vercellini, P., Bocciolone, L., Vendola, N., et al. (1991). Peritoneal endometriosis: Morphologic appearance in women with chronic pelvic pain. *The Journal of Reproductive Medicine, 36*, 533.

Vercellini, P., Trespidi, L., Colombo, A., et al. (1993). A gonadotropin-releasing hormone agonist versus a low-dose oral contraceptive for pelvic pain associated with endometriosis. *Fertility and Sterility, 60*, 75–79.

Vernon, M. W., Beard, J. S., Graves, K., et al. (1985). Classification of endometriotic implants by morphological appearance and capacity to synthesize prostaglandin F. *Fertility and Sterility, 46*, 801–806.

Wheeler, J. M., Knittle, J. D., & Miller, J. D. (1992). Depot leuprolide versus danazol in treatment of women with symptomatic endometriosis. I. Efficacy results. *American Journal of Obstetrics and Gynecology, 167*, 1367–1371.

Zullo, F., Palomba, S., Zupi, E., et al. (2003). Effectiveness of presacral neurectomy in women with severe dysmenorrhea caused by endometriosis who were treated with laparoscopic conservative surgery: A 1-year prospective randomized double-blind trial. *American Journal of Obstetrics and Gynecology, 189*, 5–10.

Chronic Pelvic Pain, Epidemiology

▶ Epidemiology of Chronic Pelvic Pain

Chronic Pelvic Pain, Laparoscopic Pain Mapping

Fred M. Howard
University of Rochester, Rochester, NY, USA

Synonyms

Conscious laparoscopic pain mapping; Conscious pain mapping; Laparoscopic pain mapping; Laparoscopy (for pain) under local anesthesia; Pain mapping; Patient-assisted laparoscopy

Definition

Laparoscopic pain mapping is a diagnostic laparoscopy under local anesthesia, with or without conscious sedation, performed with the goal of identifying sources or generators of pain in women with chronic pelvic pain. It has been suggested that laparoscopic pain mapping can lead to the treatment of subtle or atypical areas of disease that might have been overlooked if the procedure had been done under general anesthesia (Almeida and Val-Gallas 1997). Also, it may help to avoid surgical interventions when there are no surgically treatable pain generators present. However, there are limited data confirming these potential benefits (Demco 1997).

Laparoscopy under local anesthesia is not new, but its use as a diagnostic modality to localize areas of tenderness potentially responsible for chronic pelvic pain is a relatively new technique. It was first described in 1996 in a series of 11 patients with pelvic pain, ten of whom were found to have diffuse visceroperitoneal ▶ tenderness (Palter and Olive 1996). It has also been used to evaluate women with endometriosis-associated pelvic pain (Demco 1998).

The technique used to map the pelvis is a gentle probing or tractioning of tissues and organs with a blunt probe or forceps passed through a secondary cannula site. During a systematic evaluation of the entire pelvis, the patient is asked to note the presence or absence of tenderness and pain and to rate its severity when present. In particular, the patient is to note if there is replication of her usual or presenting pain. Diagnosis of an etiologic lesion or organ should be based on the severity and quality of pain elicited, especially reproduction of the patient's presenting pain. It has been suggested that applying or injecting local anesthetic to sites of focal tenderness may block the pain response and possibly improve the predictability that surgical excision will be therapeutic. Also, superior hypogastric plexus block can be done at the time of laparoscopic pain mapping and may help to predict the efficacy of presacral neurectomy (Steege 1998).

It has been shown that when laparoscopy is done under local anesthesia in women without pelvic pain, there is no significant pain or tenderness to probing of the uterus, ovaries, omentum, or bowel (Zupi et al. 1999). Table 1 summarizes the pain levels of viscera in women with chronic pelvic pain in one unpublished series of laparoscopies under local anesthesia.

Characteristics

Not all women with chronic pelvic pain are candidates for laparoscopic pain mapping. Patients who are morbidly obese (BMI greater than 30), have significant anesthesia risk, have anxiety disorders or psychiatric disease, or are known or suspected to have severe adhesive disease are not ideal candidates for the procedure. Counseling and preparation of the patient is also crucial, because chronic pelvic pain patients

Chronic Pelvic Pain, Laparoscopic Pain Mapping, Table 1 Visual analogue scores (VAS) of various anatomic structures at the time of laparoscopic pain mapping

Anatomic structure	Median VAS when structure is not site of pain	Median VAS when structure is site of pain
Uterus	2	9
Ovaries	3	8
Fallopian tubes	2	–
Round ligaments	1	–
Uterosacral ligaments and Cul-de-sac	1	9
Intestines	1	–

Unpublished data courtesy of P. Reginald

experience greater pain with laparoscopy under local anesthesia than non-pain patients. Intolerable pain is the most common reason for failures with laparoscopic pain mapping. Other reasons for failures may include adhesions that obliterate the visual field, inability to access the peritoneal cavity, and inability to visualize the entire pelvis. Published success rates range from 70 % to 100 %.

As with traditional diagnostic laparoscopy, endometriosis and adhesions are the most common diagnoses made with laparoscopic pain mapping; i.e., they are the lesions that most frequently elicit pain and tenderness at the time of pain mapping. However, even in patients with tender endometriosis or adhesions, other endometriotic lesions or adhesions may not be tender to probing at the time of laparoscopic pain mapping.

Other less common lesions found to map pain at the time of laparoscopic pain mapping include sciatic hernia, leiomyoma, hernia repair site, postoperative peritoneal cyst, colon carcinoma, chronic ileal disease, staple at the ureter, pseudo stone secondary to gallbladder spillage, peritoneal puckering, herniorrhaphy site, and peritoneal scarring. Individual pelvic viscera have also been mapped as painful, including the uterus, ovary, fallopian tube, round ligament, appendix, bladder, and vaginal apex.

Endometriosis

In one published series, there were 15 cases with successful conscious pain mapping and a visual diagnosis of endometriosis, and in these cases, endometriotic lesions were mapped as painful sites in seven. In all seven of these cases, there was histological confirmation of the diagnosis. In the remaining eight successfully mapped cases in which endometriotic lesions did not map positively, there was histological confirmation of the diagnosis in only two cases. Thus, seven of nine cases with histologically confirmed endometriosis mapped pain to endometriotic lesions versus none of six cases in which the visual diagnosis of endometriosis was not histologically confirmed ($P = 0.007$, Fischer's exact test). These findings emphasize the importance of the histologically confirmed presence of endometrial glands and stroma to generation of pain by suspected endometriotic lesions. Another published series correlating the appearance of endometriotic lesions with tenderness showed that 84 % of red lesions, 76 % of clear lesions, 44 % of white scar lesions, and 22 % of black lesions are painful (Demco 1998). Findings at laparoscopic pain mapping have also suggested that up to as much as a 3 cm area of normal-appearing peritoneum surrounding endometriotic lesions may be tender.

Adhesions

Whether adhesions can produce pain is controversial. All of the published series of conscious pain mapping have shown that adhesions can be tender and that their stimulation can reproduce a patient's pain. This strongly supports the hypothesis that some adhesions cause abdominopelvic pain. As appears to be the case with endometriosis, these data also suggest that adhesions are not always a source of pain in women with chronic pelvic pain and adhesions.

Chronic Visceral Pain

Laparoscopic pain mapping not infrequently shows generalized pelvic visceral and peritoneal hypersensitivity, suggesting such patients may have chronic visceral pain syndrome of visceral neuropathic etiology. We believe that our finding of patients with diffuse pelvic visceral and peritoneal tenderness may represent a way to confirm the diagnosis of chronic visceral pain syndrome, a diagnosis that has been suggested for chronic pain syndromes believed to be of visceral origin (Wesselman 1999). Unpublished data suggest that reproductive viscera tend to be tenderer than peritoneum or intestines at the time of laparoscopic pain mapping. It may be that many of the patients with no apparent diagnosis at the time of diagnostic laparoscopy under general anesthesia may have chronic visceral pain syndrome. Clearly further evaluations are needed to confirm these proposals.

Summary

Chronic pelvic pain is a multifactorial and complicated disorder. It is premature to assume that the findings with laparoscopic pain mapping translate directly into cause and cure. For example, comparing series of patients from our center who were evaluated with traditional diagnostic laparoscopy to those evaluated with laparoscopic pain mapping showed similar prevalences of endometriosis and adhesions in both groups (endometriosis in 38 % vs. 40 % and adhesions in 34 % vs. 54 %, respectively) (Howard 1994). Treatment based on the findings at the time of laparoscopic pain mapping did not change the outcomes, however, compared to the outcomes based on traditional diagnostic laparoscopy. With laparoscopic evaluation and treatment without pain mapping, 78 % of patients had decreased pain and 45 % were pain-free. With laparoscopic pain mapping, 44 % had decreased pain and 16 % were pain-free. The patients in the two groups were not identical, as only one-half of the patients in the traditional diagnostic laparoscopy series had undergone prior evaluations and treatments for chronic pelvic pain, compared to all of the patients evaluated by laparoscopic pain mapping.

The clinical value of conscious pain mapping both diagnostically and therapeutically cannot be stated yet, as only observational studies are available. Current data suggest it does help to avoid unnecessary operative laparoscopies in some cases.

Whether it improves outcomes in women with chronic pelvic pain by leading to more specific medical and surgical treatments requires more study and probably a prospective, randomized trial.

References

Almeida, O. D., Jr., & Val-Gallas, J. M. (1997). Conscious pain mapping. *The Journal of the American Association of Gynecologic Laparoscopists, 4*, 587–590.

Demco, L. A. (1997). Effect on negative laparoscopy rate in chronic pelvic pain patients using patient assisted laparoscopy. *The Journal Society Laparoendoscopic Surgery, 1*, 319–321.

Demco, L. A. (1998). Mapping the source and character of pain due to endometriosis by patient-assisted laparoscopy. *The Journal of the American Association of Gynecologic Laparoscopists, 5*, 241–245.

Howard, F. M. (1994). Laparoscopic evaluation and treatment of women with chronic pelvic pain. *The Journal of the American Association of Gynecologic Laparoscopists, 1*, 325–331.

Palter, S. F., & Olive, D. L. (1996). Office microlaparoscopy under local anesthesia for chronic pelvic pain. *The Journal of the American Association of Gynecologic Laparoscopists, 3*, 359–364.

Steege, J. F. (1998). Superior hypogastric block during microlaparoscopic pain mapping. *The Journal of the American Association of Gynecologic Laparoscopists, 5*, 265–267.

Wesselman, U. (1999). A call for recognizing, legitimizing, and treating chronic visceral pain syndromes. *Pain Forum, 8*, 146–150.

Zupi, E., Sbracia, M., Marconi, D., et al. (1999). Pain mapping during minilaparoscopy in infertile patients without pathology. *The Journal of the American Association of Gynecologic Laparoscopists, 6*, 51–54.

Chronic Pelvic Pain, Musculoskeletal Syndromes

Robert D. Gerwin
Department of Neurology, Johns Hopkins University, Bethesda, MD, USA

Synonyms

Myofascial pain syndromes; Myofascial trigger points; Viscerosomatic pain syndromes

Definition

Chronic pelvic pain syndromes (CPPS) and even acute pelvic pain syndromes (APPS) can have both visceral and body wall (somatic) representations, referred to as viscerosomatic and musculoskeletal or myofascial pain syndromes, respectively.

Characteristics

Pain can arise from a disorder of a visceral organ, as in inflammatory bowel disease or endometriosis, ureteral calculus, or interstitial cystitis. The presenting complaint, however, can be ▶ somatic pain, such as lower abdominal wall pain in irritable bowel syndrome, celiac disease, or colitis; flank pain in renal colic and pelvic floor or pelvic region; and ▶ myofascial pain syndromes in interstitial cystitis or endometriosis. The visceral component can be difficult to identify or can be overlooked, because visceral pain is diffuse and poorly localized. One must have a high degree of suspicion in order to identify the visceral component of pain when the complaint is somatic pain like a myofascial pelvic floor muscle pain syndrome. The visceral syndromes that are common in women include endometriosis and interstitial cystitis, irritable bowel syndrome, and irritable bladder syndrome, whereas prostatitis or ▶ prostadynia, interstitial cystitis, and irritable bowel syndrome are common in men (Vecchiet et al. 1992; Wesselmann 1999; Lukban et al. 2001; Verne et al. 2001; Nadler 2002).

Myofascial pain syndromes can also refer pain to a region where visceral pain is common, either as local or as referred pain. The attendant trigger point formation and the referred pain can cause or aggravate visceral organ dysfunction. This is readily seen in bladder or bowel dysfunction in the presence of trigger points in the levator ani and other pelvic floor muscles. Abdominal wall and pelvic floor trigger points can persist long after an initiating visceral insult has passed. This is seen very strikingly when abdominal wall ▶ trigger points form (in the abdominal oblique muscles) in response to ureteral colic

and persist for weeks after the stone has passed, or recur weeks or months later like renal colic, when in fact there is only somatic trigger point pain. Treatment of the trigger points eliminates the local and the referred pain.

Visceral Pain

▶ Visceral nociception and pain can result from such nontissue injuring mechanisms as distention or increased capsular pressure or from inflammation, ischemia, or obstruction. Thus, visceral pain can arise from conditions that produce or threaten tissue destruction or from conditions that are benign. Interstitial cystitis, irritable bowel syndrome, chronic prostatitis, and endometriosis are common causes of pelvic pain with pelvic region muscular pain representation. As many as 20–30 % of cases of chronic prostatic-related pain conditions are not due to inflammatory or infectious causes (Krieger et al. 2002; Schaeffer et al. 2002) and could be called prostadynia, a condition largely related to pelvic floor myofascial pain syndromes. Vulvodynia and vulvar vestibulitis are not visceral pain syndromes, in the strict sense that they do not arise from hollow viscera, but they are nonmuscular sources of chronic pelvic pain that are associated with musculoskeletal pain syndromes.

Pain from pelvic region visceral and genital structures is transmitted to the central nervous system through the visceral afferent sensory fibers of both sympathetic nerves (the splanchnic nerves) and parasympathetic nerves (the pelvic nerve), involving both the parasympathetic and sympathetic nerves in the superior and inferior hypogastric plexuses. Transmission of pelvic region pain is by both the dorsal spinal column to the thalamus and via the lateral columns to the medullary lateral reticular nucleus (Ness 2000). Pseudoaffective responses associated with heart rate and blood pressure changes are mediated via the lateral column.

In addition to direct nociceptive pathways from visceral organs as well as from muscle to the spinal cord and brain, there are psychophysiological responses as well. These responses are associated with central and perhaps peripheral sensitization. They are not uniformly found in all hollow viscera. Rectal distention does not result in an accelerated response with repeated stimulation as is found in the colon or in the urinary bladder. Repeated stimulation of the urinary bladder by a distending stimulus results in lowering of the pressure threshold needed to evoke sensory responses and an acceleration of autonomic responses (Ness et al. 1998). The same effect has been noted following repeated distention of the colon. Moreover, the area of sensitization or the region of the body where discomfort is experienced widens with successive stimulation.

Referred Pain from Viscera

Studies of subjects exposed to repeated distention of hollow organs (colon, urinary bladder) have shown that pain from visceral organs is experienced in the body wall. Referred pain from the gall bladder to the right shoulder area, esophageal pain to the chest, and cardiac pain to the neck and arm are well known, as are prostatic and labor pains referred to the perineum. Visceral organs in general have both a somatic representation and a cutaneous representation. These representations are related to the local spread of incoming nociceptive stimulation to the spinal cord. Thus, referred pain from genitourinary structures is felt in pelvic region musculoskeletal structures rather than thoracic or lower extremity structures.

The mechanisms underlying central sensitization have been well studied over the past quarter century. Cutaneous pain and somatic referred pain are both clinically experienced. Pelvic floor pain, lower abdominal and flank pain, and proximal thigh pain are all examples of myofascial referred pain syndromes. There is an interaction between primary visceral pain, primary myofascial pain, and referred pain to and from these regions. Pain that originates as a myofascial pain syndrome from nonvisceral causes can be experienced as visceral pain with symptoms characteristic of interstitial cystitis, irritable bowel syndrome, or genital dysfunction. As discussed above, visceral pain can result in muscular pain. In other words, the symptoms of referred pain go in both directions, visceral to somatic and somatic to visceral (Gerwin 2002). Both the

visceral and somatic (muscular) pain syndromes need to be treated in order to produce a positive outcome.

The mechanisms of visceral-induced hypersensitization include both peripheral nerve and central nervous system activation. Sensitization lowers the pain threshold, magnifies pain, and results in a larger area of the body being perceived to be painful. An example of a mechanism resulting in peripheral sensitization is the increase in rectal nerve fibers that are immunoreactive to the heat and capsaicin receptor vanilloid receptor 1 (Chan et al. 2003). ► Central sensitization results in part through activation of the dorsal horn neuron by such means as N-methyl D-aspartate receptor activation or by the convergence of two afferent nociceptive nerve fibers on one dorsal horn neuron. A recently discovered mechanism of central hypersensitization is the increase in the number of spinal lamina 1 neurons that express substance P receptor, seen in both noninflamed distended rat colon and in the inflamed rat colon (Honoré et al. 2002). Visceral inflammation markedly increased the number and the rostrocaudal extent of lamina 1 substance P receptor neurons that were activated by normally nonnoxious and noxious distention of the colon. Neuroplastic changes in the central nervous system take place very rapidly in response to nociceptive afferent input from viscera and from muscle, resulting in ► hypersensitivity, ► allodynia, and an increase in the receptive fields of individual dorsal horn neurons. Tenderness develops in response to nonnociceptive stimulation and increased perceived pain in response to nociceptive stimulation, and this is perhaps part of the basis for referred pain.

Another phenomenon that is important for pain referred from viscera to somatic tissues and from abdominal and pelvic myofascial trigger points to viscera is the extent of spread of afferent input in the spinal cord. Spinal cord spread of nonnociceptive sensory afferent fibers in the classical sensory system is over about four segments, two segments upward and two segments downward. Spinal cord spread of nociceptive sensory afferent fibers is 7–10 segments in extent. The more extensive spread of nociceptive sensory afferent fibers increases the possibility of referred pain at a greater distance from the source of the nociceptive stimulus. Nociceptive neuronal activation in the dorsal horn spreads from neuron to neuron through the excitation of cell surface receptors and the unmasking and activation of inactive or nonfunctional afferent connections from normally nonfunctional receptive fields. Activating these quiescent afferent connections results in a given dorsal horn neuron responding to stimulation of receptive fields that normally belong to other dorsal horn neurons. Thus, chronic painful stimulation from either pelvic region viscera or from abdominal and pelvic region myofascial trigger points results in pain referral either to the somatic muscles or to visceral structures or regions, respectively. Viscerosomatic pain referral is segmental and is most common locally. Thus, pain from interstitial cystitis is most often felt in the pubic region and in the ventral portion of the pelvic floor muscles, like the levator ani muscle. Pain from endometriosis is often felt in the suprapubic and pubic region and in the low back. Pain in irritable bowel syndrome is often felt in the dorsal or posterior portion of the levator ani and other pelvic floor muscles like the piriformis muscles. Trophic changes can also be seen in the region of referred pain from visceral trigger points. Neurogenic plasma extravasation causing cutaneous edema was seen in the region of referred pain in the rat in which uterine inflammation was produced (Wesselmann and Lai 1997).

Abdominal and Pelvic Floor Myofascial Pain Syndromes

Myofascial trigger points are painful areas within muscle that are composed of tender regions or zones on tight or taut bands of muscle that can be palpated in accessible muscles (Simons et al. 1999). The trigger zones within the muscle can be quiescent and not painful until the muscle is activated by use, or they can be spontaneously painful even at rest. A property of myofascial trigger points that arises from central sensitization of dorsal horn neurons in the spinal cord is the development of referred pain zones in

a predominantly localized, segmental manner, similar to the segmental spread of referred pain from viscera. Trigger points in the muscles discussed below are relevant to chronic pelvic pain syndromes when stimulation of the trigger points reproduces symptoms commonly experienced by the patient. Visceral dysfunction, such as bladder or bowel irritability or dyspareunia, can occur as a result of myofascial trigger points. The classical chronic pelvic pain triad of endometriosis or prostadynia together with interstitial cystitis and irritable bowel syndrome is usually accompanied by abdominal and pelvic floor muscle trigger points that serve to perpetuate and aggravate the pelvic visceral syndromes.

The following muscle trigger point referred patterns are of importance in pelvic visceral pain syndromes:

1. Lateral abdominal wall myofascial trigger points refer pain to the lower abdominal quadrant and to the groin. These muscles are affected by chronic colitis, Crohn's disease, and renal and ureteral chronic pain conditions and can be associated with abdominal cramping pain, diarrhea, or constipation.

2. Thoracolumbar paraspinal muscle trigger points, especially the lower thoracic, lumbar, and multifidi muscle trigger points, refer pain to the buttock, sacral, and coccygeal region and are associated with both rectal-colonic dysfunction and genitourinary dysfunction.

3. Pelvic floor muscles, including the pubococcygeus and iliococcygeus muscles of the levator ani, are associated with pelvic organ dysfunction, the anterior or ventral portion with genitourinary dysfunction, and the posterior or dorsal portion with colon and rectal/anal dysfunction. The bulbospongiosus, ischiocavernosus, and transverse perinei muscles develop trigger points in association with genital dysfunction and give rise to ► dyspareunia in women and penile, scrotal, and prostatic pain in men.

4. Obturator internus trigger points often accompany chronic pelvic pain conditions involving bowel, bladder, and genital organs, but are particularly associated with rectal pain.

5. The gluteal muscles, including the gluteus maximus, gluteus medius, gluteus minimus, and piriformis muscle, give rise to trigger points in association with pain from any of the pelvic organs.

6. Adductor magnus medial thigh muscle trigger points refer pain deep within the pelvis and can simulate genital, bladder, or rectal pain.

In addition to the referred pain that results from visceral pain referral to the body wall, there are more widespread pain representations associated with CPP syndromes. Pain pressure thresholds in males with CPP show lower pain thresholds (increased tenderness) at remote sites, including the deltoid (Davis et al. 2010).

Diagnosis of these myofascial pain syndromes is made by palpation of the appropriate muscles for hardened or taut bands within the muscle. These bands generally run from one tendinous insertion to the other, will be tender to palpation, and may refer pain to areas commonly experienced as painful by the patient. Referred pain phenomena usually take about 5–10 s of firm pressure to develop.

Treatment

Treatment of CPP is difficult. Studies to provide adequate guidelines are lacking. However, it seems clear that multimodal approaches using pharmacotherapy when appropriate, dietary therapy (there is some evidence for antioxidants in diet in urological pain conditions), physical and other manual therapies, acupuncture, cognitive behavioral therapy, surgical approaches, and direct treatment of muscle wall trigger points are required. Radiofrequency thermocoagulation of ganglion impar showed favorable results in a preliminary small study (Demircay et al. 2010), but large-scale studies of this procedure are lacking.

Treatment is directed toward inactivating the myofascial trigger points and preventing their return. Inactivation of trigger points requires reducing or eliminating the pain from the trigger point and restoring normal length to the shortened, contracted, taut, or hard band of muscle that harbors the trigger point. This can be done manually through the use of trigger point compression, local stretching, and then stretching the entire muscle. In chronic pelvic pain conditions, this may involve working within the pelvis through the vagina or the rectum, even though the majority of manual work can be done externally.

In chronic, recurrent abdominal pain and CPP associated with endometriosis and recurrent ureterolithiasis, treatment of abdominal wall pain by trigger point needling or injection can be quite effective in relieving pain. Trigger points can be reduced and pain relieved by physical therapy modalities such as electrical stimulation, including percutaneous electrical stimulation. Ultrasound may release superficial trigger points, but does not penetrate far enough to be effective in most deep trigger points. Trigger point injection with local anesthetic or needling the trigger zone with an acupuncture needle without instilling local anesthetic is a highly effective way of inactivating trigger points. A local twitch response elicited from the trigger zone (which is followed by partial or complete relaxation of the taut or hardened band and a reduction in trigger point tenderness) is the confirmatory sign that the trigger zone has actually been entered and injected. There is no evidence to support the injection of substances other than local anesthetics, such as steroids, cyanocobalamin, or ketorolac. Injection or dry needling can be done diagnostically to see if a trigger point is indeed causing a problem, to treat acute pain and to facilitate physical therapy. Trigger point injections in the pelvic region can almost always be done externally, including injections of the levator ani, piriformis, and obturator internus muscles. Only rarely is it actually necessary to inject vaginally. Conditions that create or perpetuate pelvic pain and pelvic region myofascial trigger points must be addressed and corrected. Thus, vulvar vestibulitis, interstitial cystitis, and structural factors such as pelvic torsion, pubic symphysis shear, and sacroiliac joint dysfunction must be treated and corrected. Muscle energy techniques are often adequate for the correction of the structural perpetuating factors. Treatment of trigger points in the muscles that refer pain to the pelvis will often result in dramatic improvement of chronic pelvic pain states and of pelvic organ dysfunction.

References

Chan, C. H. L., Facer, P., Davis, J. B., et al. (2003). Sensory fibres expressing capsaicin receptor TRPV1 in patients with rectal hypersensitivity and faecal urgency. *Lancet, 361*, 385–391.

Davis, S. N., Maykut, C. A., Binik, Y. M., Amsel, R., & Carrier, S. (2010). Tenderness as measured by pressure pain thresholds extend beyond the pelvis in chronic pelvic pain syndrome in men. *Journal of Sexual Medicine, 8*(1), 232–239.

Demircay, E., Kabatas, C. T., Yilmaz, C., Tuncay, C., & Altinors, N. (2010). Radiofrequency thermocoagulation of ganglion impar in the management of coccydynia: Preliminary results. *Turkish Neurosurgery, 20*, 328–333.

Gerwin, R. D. (2002). Myofascial and visceral pain syndromes: Visceral-somatic pain representations. *Journal of Musculoskeletal Pain, 10*, 165–175.

Honoré, P., Kamp, E. H., Rogers, S. D., Gebhart, G. F., & Mantyh, P. W. (2002). Activation of lamina 1 spinal cord neurons that express the substance p receptor in visceral nociception and hyperalgesia. *The Journal of Pain, 3*, 3–11.

Krieger, J. N., Ross, S. O., Deutsch, L., et al. (2002). The NIH consensus concept of chronic prostatitis/chronic pelvic pain syndrome compared with traditional concepts of nonbacterial prostatitis and prostatodynia. *Current Urology Reports, 3*, 301–306.

Lukban, J. C., Parkin, J. V., Holzberg, A. S., et al. (2001). Interstitial cystitis and pelvic floor dysfunction: A comprehensive review. *Pain Medicine, 2*, 60–71.

Nadler, R. B. (2002). Bladder training biofeedback and pelvic floor myalgia. *Urology, 60*(S6), 42–43.

Ness, T. J. (2000). Evidence for ascending visceral nociceptive information in the dorsal midline and lateral spinal cord. *Pain, 87*, 83–88.

Ness, T. J., Richter, H. E., Varner, R. E., et al. (1998). A psychophysiological study of discomfort produced by repeated filling of the urinary bladder. *Pain, 76*, 61–69.

Schaeffer, A. J., Knauss, J. S., Landis, J. R., et al. (2002). Chronic prostatitis collaborative research network study group. Leukocyte and bacterial counts do not correlate with severity of symptoms in men with chronic prostatitis: The National Institutes of Health Chronic Prostatitis Cohort Study. *Journal of Urology, 168*, 1048–1053.

Simons, D. G., Travell, J. G., & Simons, L. S. (1999). *Myofascial pain and dysfunction: The trigger point manual* (2nd ed., pp. 11–93). Baltimore: Williams and Wilkins.

Vecchiet, L., Giamberardino, M. A., & de Bigontina, P. (1992). Referred pain from viscera. In F. Sicuteri, L. Tenenius, L. Vecchiet, et al. (Eds.), *Pain versus man, advances in pain research and therapy* (Vol. 20, pp. 101–110). New York: Raven.

Verne, G. N., Robinson, M. E., & Price, D. D. (2001). Hypersensitivity to visceral and cutaneous pain in the irritable bowel syndrome. *Pain, 93*, 7–14.

Wesselmann, U. (1999). Pain – The neglected aspect of visceral disease. *European Journal of Pain, 3*, 189–191.

Wesselmann, U., & Lai, J. (1997). Mechanisms of referred visceral pain: Uterine inflammation in the adult virgin rat results in neurogenic plasma extravasation in the skin. *Pain, 73*, 309–317.

Chronic Pelvic Pain, Pelvic Inflammatory Disease and Adhesions

Fred M. Howard
University of Rochester, Rochester, NY, USA

Synonyms

Acute salpingitis; Chronic pelvic inflammatory disease; Pelvic inflammatory disease; Salpingitis-oophoritis-peritonitis; Tubo-ovarian complex

Definition

Pelvic inflammatory disease (PID) comprises a spectrum of inflammatory disorders of the upper female genital tract, including any combination of endometritis, salpingitis, tubo-ovarian abscess, and pelvic peritonitis (Centers for Disease Control 1997). Chronic abdominopelvic pain (CPP) and adhesive disease are significant sequelae of pelvic inflammatory disease (Safrin et al. 1992).

PID is most often due to ascent of organisms from the vagina and cervix, but it may also be from contiguous spread of organisms (e.g., from appendicitis) or from lymphatic or hematogenous spread (e.g., tuberculosis). PID may be iatrogenic after invasive diagnostic or therapeutic procedures, but more often it is spontaneously associated with sexual activity. Infection with gonococcus or chlamydia increases the risk of PID.

In the United States, there are an estimated 1.2 million visits per year to physicians' offices (Curran 1980), and about 280,000 women per year are hospitalized for PID (Jones 1980; Washington 1984). In addition, an estimated 150,000 surgical procedures are performed annually for complications of salpingitis. The estimated incidence of PID is 14.2 per 100 women.

Characteristics

Etiology

PID is a polymicrobial infection. Microorganisms that have been recovered from the upper genital tracts of women with PID include *N. gonorrhoeae*, *C. trachomatis*, mycoplasmas, and anaerobic and aerobic bacteria from the endogenous vaginal floor such as *Bacteroides*, *Peptostreptococcus*, *Gardnerella vaginalis*, *Escherichia coli*, *Haemophilus influenza*, and aerobic *Streptococci* (Eschenbach et al. 1975; Monif et al. 1976). PID is often precipitated by gonococcal or chlamydial infections of the cervix.

Chronic pelvic pain develops in 18–36 % of women subsequent to PID and may occur in up to 67 % after three or more episodes of PID (Haggerty 2003; Westrom 1980). (The etiology of CPP after PID is not known but is generally thought to be due to adhesive disease and to injury of the fallopian tubes and ovaries by the infection (Westrom 1988)). One possible mechanism for CPP may be related to chronic inflammation due to host immunological responses to the acute infection. Another may be recurrent infections due to repeated infectious exposures and weakened host defenses secondary to previous damage (Moller et al. 1995; Patton et al. 1983). At second-look laparoscopy, 88 % of post-PID women with CPP had morphological changes in the fallopian tubes or ovaries or both, and the severity of chronic pelvic pain was highly correlated with the extensiveness of pelvic adhesions (Table 1) (Westrom 1992).

Chronic Pelvic Pain, Pelvic Inflammatory Disease and Adhesions, Table 1 Peritubal adhesions and distal tubal pathology among PID patient with chronic pain

Women with chronic pain			
	Total women	Number	Rate (%)
Adhesions			
None	57	5	9
Slight	23	3	13
Moderate	40	25	63
Extensive	32	29	91
Fimbriated ends			
Normal	74	23	31
Phimotic	18	8	44
Clubbed	41	23	56

Signs and Symptoms

PID is difficult to diagnosis because of wide variation in symptoms and signs. Overall, the clinical diagnosis of symptomatic PID has a positive predictive value of 65–90 %, yet many patients with mild symptoms may go unrecognized. Even though PID is an infectious disease, 30–50 % of patients are afebrile. Lower abdominal and pelvic pain are usually present and of less than 2 weeks duration. Physical examination usually shows bilateral lower abdominal tenderness, bilateral adnexal tenderness, and cervical motion tenderness. Abnormal cervical or vaginal discharge is characteristic. Laboratory testing may show elevated erythrocyte sedimentation rate, elevated C-reactive protein, leukocytosis, and positive cultures for *N. gonorrhoeae* or *C. trachomatis*. There is a strong correlation between exposure to sexually transmitted organisms and PID.

Laparoscopy is the current gold standard for the diagnosis of acute PID. However, routine laparoscopy is logistically and economically impractical for all patients suspected of having acute PID. Other useful studies for diagnosing acute PID include histopathological evidence of endometritis on endometrial biopsy (Paavonen et al. 1985) and transvaginal sonography or other imaging techniques showing thickened fluid-filled tubes with or without pelvic fluid or tubo-ovarian complex (Cacciatore et al. 1992).

CPP associated with prior PID is generally thought to be due to adhesions or tubal disease, but usually there are no specific findings on examination that allow a diagnosis of adhesions or tubal damage. Occasionally with dense uterine adhesions, the uterus is found in a fixed, immobile retroverted position. With hydrosalpinges or tubo-ovarian complexes, a tender adnexal mass may be palpable. Laparoscopy for CPP may show fallopian tubes that are tortuous, clubbed, or phimotic and may show tubo-ovarian adhesions in women with prior PID. In many patients laparoscopic findings may be normal.

Treatment

Acute PID should be treated empirically with broad-spectrum antibiotics to cover the variety of likely aerobic and anaerobic pathogens. Oral versus parenteral treatment does not appear to alter outcomes nor change the likelihood of developing CPP. Hospitalization should occur if the diagnosis is uncertain and the patient seems ill, the patient is pregnant, the patient does not respond clinically to oral antimicrobial therapy, the patient is unable to follow or tolerate an outpatient oral regimen, the patient has severe illness, nausea and vomiting, or high fever, the patient has a tubo-ovarian abscess, or the patient is immunodeficient.

The ideal treatment of women with CPP secondary to PID is not known. Empirical antibiotic therapy is often tried in women with CPP in whom prior, chronic, or recurrent PID is suspected. Unfortunately there are no studies supporting efficacy for this approach. Surgical treatment by adhesiolysis or even removal of severely damaged organs, such as hydrosalpinges, is utilized for conservative management, but only observational and anecdotal data are available to support this approach. Hysterectomy with bilateral salpingo-oophorectomy is commonly performed for CPP with pelvic findings suggestive of prior PID, although there is not much published information documenting efficacy for relief of CPP.

References

Cacciatore, B., Leminen, A., Ingman-Friberg, S., et al. (1992). Transvaginal sonographic findings in ambulatory patients with suspected PID. *Obstetrics and Gynecology, 80,* 912–916.

Centers for Disease Control. (1997). 1998 guidelines for treatment of sexually transmitted diseases. *MMWR Recommendations and Reports, 47,* 79–86.

Curran, J. W. (1980). Economic consequences of pelvic inflammatory disease in the United States. *American Journal of Obstetrics and Gynecology, 1080,* 138–905.

Eschenbach, D. A., Buchanan, T., Pollock, H. M., et al. (1975). Polymicrobial etiology of acute pelvic inflammatory disease. *The New England Journal of Medicine, 293,* 166.

Haggerty, C. L., Schulz, R., & Ness, R. B. (2003). Lower quality of life among women with chronic pelvic pain after pelvic inflammatory disease. *Obstetrics and Gynecology, 102,* 934–939.

Jones, O. G., Saida, A. A., & St John, R. K. (1980). Frequency and distribution of salpingitis and pelvic inflammatory disease in short stay in hospitals in the United States. *American Journal of Obstetrics and Gynecology, 138,* 905.

Moller, B. R., Krostiansen, F. V., Thorsen, P., et al. (1995). Sterility of the uterine cavity. *Acta Obstetricia et Gynecologica Scandinavica, 74,* 216–219.

Monif, G. R. G., Welkos, S. L., Baer, H., et al. (1976). Cul de-sac isolates from patients with endometritis, salpingitis, peritonitis and gonococcal endocervicitis. *American Journal of Obstetrics and Gynecology, 126,* 158.

Paavonen, J., Aine, R., Teisala, K., et al. (1985). Comparison of endometrial biopsy and peritoneal fluid cytology with laparoscopy in the diagnosis of acute PID. *American Journal of Obstetrics and Gynecology, 151,* 645–650.

Patton, D. L., Halbert, S. A., Kuo, C. C., et al. (1983). Host response to Chlamydia trachomatis infection of the fallopian tube in pig-tailed monkeys. *Fertility and Sterility, 40,* 829–840.

Safrin, S., Schachter, J., Dahrouge, D., et al. (1992). Long-term sequelae of acute PID. *American Journal of Obstetrics and Gynecology, 166,* 1300–1305.

Washington, A. E., Cates, W., & Sadi, A. A. (1984). Hospitalizations for pelvic inflammatory disease. Epidemiology and trends in the United States 1975 and 1981. *The Journal of the American Medical Association, 251,* 2529–2533.

Westrom, L. (1980). Incidence, prevalence and trends of acute pelvic inflammatory disease and its consequences in industrialized countries. *American Journal of Obstetrics and Gynecology, 138,* 880–892.

Westrom, L. (1988). Chronic pain after acute PID. In P. Belfort, J. A. Piatti, & T. K. A. B. Eskes (Eds.),

Advances in gynecology and obstetrics. Proceedings of the XIIth World Congress Gynecol Obstet. Rio de Janerio.

Westrom, L. V., & Berger, G. S. (1992). In G. S. Berger & L. V. Westrom (Eds.), *Consequences of pelvic inflammatory disease* (pp. 101–114). New York: Raven Press.

Chronic Pelvic Pain, Physical and Sexual Abuse

Beverly J. Collett[1], Christine Cordle[1] and Hayley Poole[2]
[1]University Hospitals of Leicester, Leicester, UK
[2]Hadley House, Leicester General Hospital, Leicester, UK

Definition

Physical Abuse

Hitting, shaking, throwing, poisoning, burning or scalding, drowning, suffocating, or otherwise causing physical harm to another person (Department of Health 2000).

Sexual Abuse

Forcing or enticing another person to take part in sexual activities, whether or not they are aware of what is happening. The activities may involve physical contact, including penetrative (e.g., rape or buggery) and non-penetrative acts. They may include noncontact activities, such as involving another person in looking at, or in the production of, pornographic material or watching sexual activities or encouraging another person to behave in sexually inappropriate ways (Department of Health 2000).

The issue of defining sexual abuse in practice is both complex and problematical.

Characteristics

The role of abusive experiences in ▶ chronic pelvic pain began to be the focus of investigation in the early 1980s. Since then, there has been growing documentation of prevalence rates for sexual and ▶ physical abuse during childhood and adulthood in patients presenting with a variety of chronic medical conditions including chronic pelvic pain.

More attention has been given to ▶ sexual abuse than physical abuse in the literature. Reported prevalence rates for sexual abuse in female patients with chronic pelvic pain range from 26–64 % (Collett et al. 1998; Reiter and Gambone 1990; Toomey et al. 1993; Walker et al. 1988; Walling et al. 1994) and from 11 % to 50 % for physical abuse (Rapkin et al. 1990; Toomey et al. 1993). This wide range is a reflection of the significant methodological problems characterizing this area of research. One of the main problems is with definitions. Sexual abuse is defined in various ways by different investigators. Some studies have only included contact sexual abuse, whereas others have used a wider definition of sexual abuse and have included episodes of threatened or attempted abuse without any direct genital contact. Other studies have excluded isolated incidents of abuse. Some have distinguished between childhood abuse (usually below the age of 14 years) and abuse in adulthood, whereas others have not made this distinction.

Similar problems exist with definitions of physical abuse. Some studies only include life-threatening abuse, where there is intent to seriously harm or kill, whereas others have defined it as any pattern of physical discipline or punishment, performed by a more powerful other individual to a person aged under 18 years, resulting in physical injury such as marks, bruises, and welts.

It is clearly important for investigators to define sexual and physical abuse in a more uniform way, taking into account the age at the time of abuse, the nature, the range and duration of the abuse, and the relationship of the perpetrator to the abused person. There are also problems with defining chronic pelvic pain. Different definitions exist, based on duration, location and type of pain, and relation to menstruation and sexual activity.

The other factor known to influence reported prevalence rates is the method of data collection. Most data is collected using self-administered questionnaires. It has been suggested that in-depth interviews carried out in a certain style are more likely to result in abuse disclosure (Wyatt and Peters 1986).

Many of the early studies did not include control groups, and although this has been addressed in more recent research, there is still a lack of agreement over what constitutes an appropriate control group. Groups are rarely matched for such factors as chronicity of pain and age. Age may have a role in abuse reporting. Briere found that older women reported less sexual abuse than younger women, and this was attributed to the older woman's discomfort with speaking about sexuality (Briere 1992). Many studies are limited by small sample sizes and are based on a population of patients referred to speciality pain clinics. The findings from such clinics may not be generalizable to patients seen in routine gynecology clinics or in the primary care setting. There is a need for prospective studies with larger sample groups, although these are very difficult to carry out. The methodological problems in this area of research make it difficult to draw conclusions from the data presented.

A number of controlled studies, comparing women with chronic pelvic pain with women with pain in a different site and women with no pain, have demonstrated an increased prevalence of lifetime sexual abuse in the chronic pelvic pain group (Collett et al. 1998). The association between sexual abuse and pelvic pain complaints has been shown to be stronger if the abuse occurred before the age of 15 years and was repeated in adulthood (Jamieson and Steege 1997; Lampe et al. 2000). In a comprehensive review of relevant articles, multivariable meta-regression analysis showed that sexual abuse was not associated with any particular type of chronic pelvic pain. In this analysis, poor-quality studies showed more prominent associations between abuse and pelvic pain than good-quality studies (Latthe et al. 2006).

Less information is available for physical abuse and chronic pelvic pain. One often-quoted study by Rapkin et al. (1990) found that women with chronic pelvic pain were significantly more likely to report a history of childhood physical abuse than women receiving routine gynecology care or women with pain in another site. Other workers have suggested that a history of physical abuse may be more relevant than a history of sexual abuse in women with chronic pelvic pain (Walling et al. 1994).

Both physical and sexual abuse have been shown to be associated with significantly more psychological distress in women with chronic pelvic pain when compared with women with no history of abuse. The association is stronger for women reporting physical abuse. Sexual abuse has also been shown to be one of a number of risk factors for the development of sexual dysfunction in women with chronic pelvic pain.

There has been considerable speculation in the literature as to the possible mechanisms linking abuse with chronic pelvic pain. It has been suggested by several workers that any type of childhood trauma, including violence, separations, illnesses, and neglect, would be a predictor for pain proneness in adulthood. It is argued that traumatized individuals are more likely to feel vulnerable about their body and their health. Some psychologists conceptualize the pain among victims of sexual abuse as a defense against the overwhelming emotions connected to the traumatic experience. Linton et al. (1996) suggested an indirect relationship, in that abuse may affect pain by altering perception and one's ability to cope with pain. Support for this model comes from the knowledge that abused patients have higher levels of depression and affective distress, and lower levels of perceived control, compared to non-abused patients. It may be that women with an abuse history have a tendency to report more symptoms of any kind as a function of psychological disturbance. Although the data are consistent with the idea that abuse may be instrumental in the development of chronic pain problems, we do not understand the mechanisms involved with this. It has been pointed out that abuse may be a correlate rather than a causal variable, since abuse may be related to many lifestyle risk factors.

A prospective investigation of childhood sexual and physical abuse and neglect found no association with adult pain symptoms. However, unexplained pain symptoms were found to be associated with retrospective self-reports of all types of childhood victimization. Perhaps, patients who recall childhood abuse and neglect perceive themselves to be trapped by a history that they cannot undo, and this in turn leads to helplessness and passivity (Raphael et al. 2001).

In view of the high prevalence of lifetime abuse in women with chronic pelvic pain, it has been argued that such women should be routinely asked about any abuse history as part of their assessment. A number of studies have shown that women with chronic pelvic pain are rarely asked about assault, and only a small minority of women volunteer this information to a gynecologist or physician. When women were asked whether they thought they should be asked, 90 percent answered in the affirmative (Robohm and Buttenheim 1996). In addition to asking about any past abuse history, it is important to ensure that the women are not involved in any ongoing abuse. Interpersonal sensitivity in treating these women is particularly important. Concerns about trust in the doctor-patient relationship are common, and women with a history of ▶ childhood sexual abuse report more anxiety, embarrassment, and vulnerability associated with a gynecological examination than other women (Robohm and Buttenheim 1996).

Linton found that more than 85 percent of the women in his sample did not believe that their history of abuse affected their pain or their sex lives. Interventions, therefore, oriented directly toward abuse may be difficult to incorporate into general pain treatment. Specific therapy for the abuse may be the best way to deal with the problem if the patient has unresolved issues around the abuse, which are affecting her current functioning.

Little is known about the association of abuse and pelvic pain in men. A history of physical and emotional abuse, particularly in adulthood, has been shown to significantly increase the likelihood of experiencing symptoms suggestive of chronic prostatitis/chronic pelvic pain syndrome (Hu et al. 2007).

In summary, there are complex interactions between child and adult abuse, depression, stressful life events, and the occurrence of chronic pelvic pain, and clinicians need to take these psychosocial factors into consideration when assessing and treating women with chronic pelvic pain.

Little research into physical and sexual abuse among patients with chronic vulvar pain syndromes or ▶ vulvodynia has been undertaken. However, studies that have been undertaken show that there is a significantly higher incidence of sexual abuse and psychological distress in chronic pelvic pain patients, compared to women with chronic vulvar pain and to gynecological controls (Bodden-Heidrich et al. 1999; Reed et al. 2000).

References

Bodden-Heidrich, R., Kuppers, V., Beckmann, M. W., Rechenberger, I., & Bender, H. G. (1999). Chronic pelvic pain syndrome (CPPS) and chronic vulvar pain syndrome (CVPS): Evaluation of psychosomatic aspects. *Journal of Psychosomatic Obstetrics and Gynecology, 20,* 145–151.

Briere, J. (1992). *Child abuse trauma. Theory and treatment of the lasting effects.* London: Sage.

Collett, B. J., Cordle, C. J., Stewart, C. R., & Jagger, C. (1998). A comparative study of women with chronic pelvic pain, chronic non-pelvic pain and those with no history of pain attending general practitioners. *British Journal of Obstetrics and Gynaecology, 105,* 87–92.

Department of Health. (2000). *Working together to safeguard children in need and their families* (p. 5). London: The Stationary Office.

Hu, J. C., Link, C. L., McNaughton-Collins, M., Barry, M. J., & McKinlay, B. (2007). The association of abuse and smptoms suggestive of chronic prostatitis/chronic pelvic pain syndrome: Results from the boston area community health survey. *Journal of General International Medicine, 22*(11), 1532–1537.

Jamieson, D. J., & Steege, J. F. (1997). The association of sexual abuse with pelvic pain complaints in a primary care population. *American Journal of Obstetrics and Gynecology, 177,* 1408–1412.

Lampe, A., Solder, E., Ennemoser, A., Schubert, C., Rumpold, G., & Sollner, W. (2000). Chronic pelvic pain and previous sexual abuse. *Obstetrics and Gynecology, 96,* 929–933.

Latthe, P., Mignini, L., Gray, R., Hills, R., & Khan, K. (2006). Factors predisposing women to chronic pelvic pain: Systematic review. *BMJ.* doi:10.1136/bmj.38748.697465.55.

Linton, S. J., Larden, M., & Gillow, A. M. (1996). Sexual abuse and chronic musculoskeletal pain: Prevalence and psychological factors. *The Clinical Journal of Pain, 12*, 215–221.

Raphael, K. G., Widom, C. S., & Lange, G. (2001). Childhood victimization and pain in adulthood: A prospective investigation. *Pain, 92*, 283–293.

Rapkin, A. J., Kames, L. D., Darke, L. L., Stampler, F. M., & Nabiloff, B. D. (1990). History of physical and sexual abuse in women with chronic pelvic pain. *Obstetrics and Gynecology, 76*, 92–95.

Reed, B. D., Haefner, H. K., Punch, M. R., Roth, R. S., Gorenflo, D. W., & Gillespie, B. W. (2000). Psychosocial and sexual functioning in women with vulvodynia and chronic pelvic pain. *Journal of Reproductive Medicine, 45*, 624–632.

Reiter, R. C., & Gambone, J. C. (1990). Demographic and historic variables in women with idiopathic chronic pelvic pain. *Obstetrics and Gynecology, 75*, 428–432.

Robohm, J. S., & Buttenheim, M. (1996). The gynaecological care experiences of adult survivors of childhood sexual abuse: A preliminary investigation. *Women & Health, 24*, 59–75.

Toomey, T. C., Hernandez, J. T., Gittelman, D. F., & Hulka, J. F. (1993). Relationship of sexual and physical abuse to pain and psychological assessment variables in chronic pelvic pain. *Pain, 53*, 105–109.

Walker, E., Katon, W., Harrop-Griffiths, J., Holm, L., Russo, J., & Hickok, L. R. (1988). Relationship of chronic pelvic pain to psychiatric diagnoses and childhood sexual abuse. *The American Journal of Psychiatry, 145*, 75–80.

Walling, M. K., Reiter, R. C., O'Hara, M. W., Milburn, A. K., Lilly, G., & Vincent, S. D. (1994). Abuse history and chronic pain in women: Prevalence of sexual abuse and physical abuse. *The Obstetrician and Gynaecologist, 84*, 193–199.

Wyatt, G. E., & Peters, S. D. (1986). Methodological considerations in research on the prevalence of child sexual abuse. *Child Abuse & Neglect, 10*, 241–251.

Chronic Pelvic Pain, Physical Therapy Approaches, and Myofascial Abnormalities

Rhonda K. Kotarinos
Oakbrook Terrace, IL, USA

Synonyms

Chronic pelvic pain syndrome; Myofascial pain syndrome; Myofascial trigger point pain syndrome; Myofascial trigger point

Definition

Myofascial trigger point: A hyperirritable spot in skeletal muscle that is associated with a hypersensitive palpable nodule in a taut band (Simons et al. 1999).

Myofascial pain syndrome/myofascial trigger point pain syndrome: The sensory, motor, and autonomic symptoms caused by myofascial trigger points (Simons et al. 1999).

Chronic pelvic pain: The occurrence of persistent or recurrent episodic pelvic pain associated with symptoms suggestive of lower urinary tract, sexual, bowel, or gynecological dysfunction with the absence of proven infection or obvious pathology (Abram et al. 2003).

Introduction

In today's environment of evidence-based medicine, the highest quality of level I evidence is desired when evaluating patient care strategies. This is difficult to achieve even when conditions are perfect. Quality research is less complicated and easier to execute when there are clear and decisive definitions related to the subject being studied. Nebulous and spurious definitions limit the ability to perform epidemiologic research, randomized trials, and evaluation of therapeutic management. Standardized definitions for conditions that are primarily defined by a core group of symptoms are difficult to develop and get an adequate consensus agreement among the representative health providers.

Chronic pelvic pain and myofascial pain syndrome are two conditions that are difficult to study secondary to the lack of consensus in their definitions. A major problem with defining chronic pelvic pain is that it is not gender specific; both sexes have a pelvis. There is much overlap in their symptoms. Females usually will consult their obstetrician/gynecologist who will then refer them to a urologist or urogynecologist, whereas, the male will initiate care with the urologist. In either case, the patient presents to the physician with a combination of symptoms, frequently with an overlap of lower urinary tract symptoms. Problems with definition also develop because the pelvis is the home of more than one organ system: reproductive, urinary, and colorectal.

Defining myofascial pain syndrome, at times also referred to as myofascial trigger point syndrome, is equally fraught with confusion among involved health-care providers. Centric to either term in establishing the definition is the presence of a myofascial trigger point, which at this time also has its own set of problems. Currently, there are no objective imaging studies or laboratory tests to facilitate a trigger point diagnosis and subsequently the myofascial pain syndrome diagnosis. Within the research, there have been two reports that may prove promising for future objective diagnostic options (Shah et al. 2005; Sikdar et al. 2009).

A recent review found that the major obstacle to defining myofascial pain syndrome is the lack of agreement of the diagnostic criteria for a myofascial trigger point (Tough et al. 2007). The four most frequently chosen criteria to diagnose a myofascial trigger point were "tender spot in a taut band of skeletal muscle," "patient pain recognition," "predicted referral pattern," and "a local twitch response." An additional criterion of "limited range of motion," when added to the above four, was the most common combination of diagnostic findings found in the review (Tough et al.). Travell and Simons along with Andrew Fischer are considered the leading authoritative experts on trigger point and myofascial pain syndrome diagnosis.

Simons et al. refined the diagnostic criteria in the hope of facilitating diagnostic sensitivity by adding the term nodule to describe the finding in the taut skeletal muscle band and the qualifier that the limited range of motion was painful. (Simons et al. 1999) The most recent diagnostic criteria by Mense and Simons state that "active trigger points are circumscribed spots of tenderness in a nodule of a palpable taut band and patient recognition of the pain, evoked by pressure on the tender spot as being familiar" (Mense and Simons 2001). According to Simons et al., myofascial pain syndrome is "the sensory, motor and autonomic symptoms caused by myofascial trigger points" (Simons et al. 1999). Only trigger points within a muscle belly, rather than any other tissue, are responsible for a myofascial pain syndrome, and the specific muscle or muscle group should by identified when the diagnosis is made (Simons et al. 1999; Mense and Simons 2001).

Location and duration of the pain appear to be the most basic criteria in the definition of chronic pelvic pain in women but are seldom defined (FitzGerald et al. 2006). According to Williams et al., the typical description of the location of pain is below the umbilicus or in the female organs. The duration of pain utilized in definitions of female pelvic pain ranges from 3 to 6 months with the American College of Obstetricians and Gynecologists stating in their technical bulletin 6 or more months as the significant criteria of duration (Kavadias and Baessler 2011; ACOG Practice Bulletin 2004).

Various diagnoses can fall under the umbrella term of chronic pelvic pain. These include but are not limited to dyspareunia, vulvar pain syndromes, urethral syndrome, interstitial cystitis/painful bladder syndrome, irritable bowel syndrome, chronic prostatitis, pelvic floor tension myalgia, levator ani syndrome, proctalgia fugax, urgency, coccygodynia, testicular pain, and ejaculatory or post-ejaculatory pain. Chronic prostatitis/chronic pelvic pain syndrome describes urological pain complaints and voiding symptoms in men with exclusion criteria such as active infection, urogenital cancer, urinary tract disease, functionally limiting urethral stricture, and neurological disease (Kreiger et al. 1999). Other pelvic organ systems are recognized as potential sources of pain in this syndrome.

A definition that incorporates both sexes and the diverse symptomatology of chronic pelvic pain is preferred. The International Continence Society's most recent publication defines pelvic pain syndrome as the occurrence of persistent or recurrent episodic pelvic pain associated with symptoms suggestive of lower urinary tract, sexual, bowel, and gynecological dysfunction with absence of proven infection or obvious pathology (Abram et al. 2003).

Characteristics

In 1992, Baskin and Tanagho published a case series entitled "Pelvic Pain Without Pelvic Organs." The most profound conclusion of that article was that "the pelvic muscular element, which could well be the source of pain must be evaluated" (Baskin and Tanagho 1992). Slowly but surely, the potential involvement of muscles in the pelvis and elsewhere is considered and evaluated in patients presenting with chronic pelvic pain and its assorted symptoms. The primary goal for the health-care provider in managing the chronic pelvic pain patient is to decrease their symptoms which will translate into an improved quality of life. Pharmacologic therapies for chronic pelvic pain are not disease specific and are instituted in a trial and error manner resulting in few randomized, placebo-controlled studies that support their efficacy. Treatment choices usually start with the safest and least invasive options and then progress to other treatments that may have more potential for morbidity if the initial treatments are ineffective. Consequently, this is the clinical challenge for health-care providers treating pelvic pain disorders.

The variations of myofascial abnormalities found in patients with chronic pelvic pain have not been well described. Yet, the recognition of the role of myofascial abnormalities of the pelvis has led to several reports of chronic pelvic pain symptom relief by therapeutic efforts directed at the muscular and connective tissue abnormalities found in these patients. A small case series of patients with a diagnosis of interstitial cystitis reported a significant decrease in VAS pain scores after physical therapy treatment (Kotarinos 1995). The treatment consisted of myofascial manipulation, manual trigger point release, connective tissue manipulation to address the connective tissue changes associated with trigger points, and a viscerosomatic reflex and a pelvic floor reeducation program focused on normalizing the length-strength relationship of the pelvic floor muscles. Weiss

reported on the effectiveness of manual physical therapy to treat the symptoms of interstitial cystitis, urethral syndrome, and urgency-frequency with or without pelvic pain in both men and women. Seventy percent of the interstitial cystitis patients and 83 % of the urgency/frequency patients had moderate to marked improvement in their symptoms. Clemens treated men with chronic pelvic pain with biofeedback-directed pelvic floor reeducation and bladder training that resulted in a significant decrease in pain, urgency, and general bother (Clemens et al. 2000). More recently, Cornel et al. reported significant decreases in symptom scores and pelvic floor hypertonus after treatment with biofeedback-assisted pelvic floor muscle reeducation. Anderson et al. also report that moderate or marked improvement in symptoms was noted in 72 % of their study population of men who underwent manual myofascial trigger point release and paradoxical relaxation training (Anderson et al. 2005).

Myofascial trigger points are treated usually with manual techniques and can be combined with dry needling or trigger point injections. Langford et al. studied the effect of trigger point injections on 18 female patients with chronic pelvic pain of at least 6 months duration (Langford 2007). Trigger points were identified manually by transvaginal palpation of the levator ani bilaterally. Pelvic floor muscle exercises were included in the treatment after the injection. Thirteen of the 18 women (72 %) had a decrease in pain measured by a VAS of 50 % or more as compared to pre-injection VAS. Complete resolution of the pain occurred in 33 % of the 18 women. Trigger point injections can be an effective option in treating chronic pelvic pain. FitzGerald et al. reported in a case series of 37 women with painful bladder syndrome that treatment by myofascial release and connective tissue manipulation decreased pain and urinary symptom scores (Fitzgerald 2006). Two reports utilizing nonspecific soft tissue transvaginal techniques and manual treatment of the SI joints and hip girdle also had significant decreases in symptom scores rating pain, urgency, and disability (Lukban et al. 2001; Oyama et al. 2004).

Although widely practiced and apparently very effective, there had been no reports of level I or II evidence to support the use of specialized external and internal manual pelvic physical therapy for the myofascial component of chronic pelvic pain syndromes. Recognizing that there was no strong evidence to support the use of manual physical therapy with this patient population, a randomized single-blind clinical trial was performed to establish the feasibility of comparing two manual therapies, specialized manual physical therapy and nonspecific global therapeutic massage for treatment of interstitial cystitis/painful bladder syndrome and chronic prostatitis/chronic pelvic pain syndrome (Fitzgerald et al. 2009). This is the first published randomized clinical trial comparing manual physical therapy and global therapeutic massage for the treatment of urological chronic pelvic pain syndromes. The main objective was achieved by demonstrating that a clinical trial of this nature was feasible. There were several criteria to assess the feasibility of the study including the ability of the therapists to adhere to the protocol as determined by the records of treatment, adverse events, and the rate of response to therapy as assessed by the patient global response assessment. The 57 % positive responder rate was the highest ever for a treatment trial for urological pelvic pain syndromes in a level 1 study. As a result of the impressive results from the feasibility study, a second randomized controlled treatment trial was conducted that resulted in a 59 % positive responder rate (Fitzgerald et al. 2012).

Summary

In spite of the fact that definitions for either chronic pelvic pain or myofascial pain syndrome cannot be agreed upon in the medical community,

there appears to be a very close relationship between the two. Anecdotal evidence is plentiful to support treating the myofascial abnormalities found in patients with chronic pelvic pain complaints. Two level I studies are robust indicators that the manual physical therapy to address the myofascial dysfunctions of chronic pelvic pain should be considered as a treatment option (Fitzgerald et al. 2009; Fitzgerald et al. 2012). The somatic abnormalities associated with the myofascial pain syndrome component of chronic pelvic pain may be the primary abnormality in some patients and secondary in others. In either situation, they should be sought out and actively treated.

References

Abram, P., Cardozo, L., Fall, M., et al. (2003). The standardisation of terminology in lower urinary tract dysfunction: Report from the standardisation sub-committee of the International Continence Society. *Urology, 61*, 37–49.

ACOG Practice Bulletin. (2004). Chronic pelvic pain. *Obstetrics and Gynecology, 103*, 589–605.

Anderson, R., Wise, D., & Sawyer, T. (2005). Integration of myofascial trigger point release and paradoxical relaxation training treatment of chronic pelvic pain in men. *Journal of Urology, 174*, 155–160.

Baskin, L., & Tanagho, E. (1992). Pelvic pain without pelvic organs. *Obstetrics and Gynecology, 147*, 683–686.

Clemens, J., Nadler, R., Schaeffer, A., et al. (2000). Biofeedback, pelvic floor re-education, and bladder training for male chronic pelvic pain syndrome. *Urology, 56*, 951–955.

FitzGerald, M., Ervan, K., Arcilla, G., et al. (2006). Physical therapy can reduce symptoms of painful bladder syndrome. Accepted abstract AUGS.

Fitzgerald, M., Anderson, R., & Potts, J. (2009). Randomized multicenter feasibility trial of myofascial physical therapy for the treatment of urological chronic pelvic pain syndromes. *Journal of Urology, 182*, 570–580.

FitzGerald, M. P., Payne, P .K., Lukacz, E. S., et al. (2012) Randomized multicenter clinical trial of myofascial physical therapy in women with interstitial cystitis/painful bladder syndrome and pelvic floor tenderness. *Journal of Urology, 187*, 2113–2118.

Kavadias, T., & Baessler, K. (2011). Pelvic pain in urogynecology. Part I: Evaluation, definitions and diagnoses. *International Urogynecology Journal, 22*, 385–393.

Kotarinos, R., (1995). Until the answer is found. Abstract presentation interstitial cystitis symposium National Institiutes of Health, Bethsda, Maryland.

Kreiger, J. N., Nyberg, L., & Nickel, J. (1999). NIH consensus definition and classification of prostatitis. *Journal of the American Medical Association, 282*, 236–237.

Langford, C., Nagy, S., & Ghoniem, G. (2007). Levator ani trigger point injections: An underutilized treatment for chronic pelvic pain. *Neurology and Urodynamics, 26*, 59–62.

Lukban, J., Whitmore, K., Kellogg-Spadt, S., et al. (2001). The effect of manual therapy in patients diagnosed with interstitial cystitis, high tone pelvic floor dysfunction and sacroiliac dysfunction. *Urology, 57*, 121.

Mense, S., & Simons, D. (2001). *Muscle pain understanding its nature, diagnosis, and treatment.* Baltimore: Lippincott Williams & Wilkins.

Oyama, I., Rejb, A., Lukban, J., et al. (2004). Modified Thiele massage as a therapeutic intervention for female patients with interstitial cystitis and high-tone pelvic floor dysfunction. *Urology, 64*, 862.

Shah, J., Phillips, T., Danoff, J., et al. (2005). An in vivo microanalytical technique for measuring the local biochemical milieu of human skeletal muscle. *Journal of Applied Physiology, 99*, 1977–1984.

Sikdar, S., Shah, J., Gebreah, B., et al. (2009). Novel applications of ultrasound technology to visualize and characterize myofascial trigger points and surrounding soft tissue. *Archives of Physical Medicine and Rehabilitation, 90*, 1829–1838.

Simons, D. G., Travell, J. G., Simons, L. S., et al. (1999). *Myofascial pain and dysfunction: The trigger point manual* (2nd ed., Vol. 1). Baltimore: Williams & Wilkins.

Tough, E., White, A., Richard, S., et al. (2007). Variability of criteria used to diagnose myofascial trigger point syndrome – evidence from a review of the literature. *The Clinical Journal of Pain, 23*, 2780286.

Chronic Postoperative Pain

▶ Iatrogenic Causes of Neuropathy

Chronic Postsurgical Neuralgia

▶ Iatrogenic Causes of Neuropathy

Chronic Postsurgical Pain (CPSP)

Stephan A. Schug[1] and W. A. Williams[2]
[1]School of Medicine and Pharmacology, Royal Perth Hospital, Faculty of Medicine, Dentistry & Health Sciences, The University of Western Australia, UWA Anaesthesiology, Perth, WA, Australia
[2]Royal Perth Hospital and University of Western Australia, Perth, WA, Australia

Synonyms

Acute-recurrent pain; Long-lasting acute pain; Persistent acute pain; Persistent postsurgical pain; Prolonged acute pain

Definition

A publication by the International Association for the Study of Pain (IASP) defines persistent postsurgical pain as pain that develops after surgical intervention and lasts at least 2 months; other causes for the pain have to be excluded, in particular pain from a condition preceding the surgery (Macrae and Davies 1999). This definition has been criticized as overly simplistic.

Introduction

Chronic postsurgical pain is an understimated complication of surgery. The overall incidence after major surgery is in the range of 20–50 % depending on type of surgery and in particular risk of nerve injury. Severe disabling pain occurs in 2–10 % (Kehlet et al. 2006; Macrae 2008). This has been confirmed in a recent population epidemiological study (Johansen et al. 2012).

Characteristics

Chronic postsurgical pain is commonly of neuropathic nature, although ongoing nociception and inflammatory processes may play a role in some patients (Kehlet et al. 2006).

Summary

Chronic postsurgical pain is an underestimated but relatively common and significant complication of surgery. It has often features of neuropathic pain and is linked to intraoperative nerve injury. Approaches to prevent chronic postsurgical pain need further study; preliminary data suggest that techniques of regional anesthesia, perioperative use of α-2-δ modulators (gabapentin, pregabalin), and NMDA antagonists might play a role here. But avoiding unnecessary surgery, performing surgery minimally invasive, and providing good postoperative analgesia may also reduce the incidence of this complication.

References

Johansen, A., Romundstad, L., Nielsen, C. S., Schirmer, H., & Stubhaug, A. (2012). Persistent postsurgical pain in a general population: Prevalence and predictors in the Tromso study. *Pain, 153*(7), 1390–1396.

Kehlet, H., Jensen, T. S., & Woolf, C. J. (2006). Persistent postsurgical pain: Risk factors and prevention. *Lancet, 367*(9522), 1618–1625.

Macrae, W. A. (2008). Chronic post-surgical pain: 10 years on. *British Journal of Anaesthesia, 101*(1), 77–86.

Macrae, W. A., & Davies, H. T. O. (1999). Chronic post-surgical pain. In I. K. Crombie, S. Linton, P. Croft, M. Von Korff, & L. LeResche (Eds.), *Epidemiology of pain* (pp. 125–142). Seattle: International Association for the Study of Pain.

Chronic Postsurgical Pain Syndromes

Synonyms

CPSP; Persistent postsurgical pain

Definition

A pain lasting more than 3 months, persisting after healing from a surgical procedure. Chronic

postsurgical pain syndromes are most often the consequence of a peripheral nerve injury directly or indirectly caused by the surgical intervention, subsequently leading to central sensitization. However, not all CPSP is of neuropathic nature and ongoing inflammation and nociception have to be considered as causes, too.

Cross-References

▶ Iatrogenic Causes of Neuropathy

References

Kehlet H, Jensen TS, Woolf CJ. Persistent postsurgical pain: risk factors and prevention. Lancet. 2006. 367:1618–25.

Chronic Relapsing Polyneuropathy

▶ Chronic Inflammatory Demyelinating Polyneuropathy

Chronic Tension-Type Headache

Definition

Classification of International Headache Society describing a subtype of chronic daily headache as very frequent headaches ($\geq$15 days per month for at least 3 months, $\geq$180 days per year). Pain has two of the following characteristics: pressing or tightening quality, mild or moderate intensity, bilateral location, and not aggravated by routine physical activity. Headache should be featureless. Headache is not attributed to another disorder, an individual's use of analgesics, or other acute medication, and is $\leq$10 days per month.

Cross-References

▶ Chronic Daily Headache in Children
▶ New Daily Persistent Headache

Chronic Vulvar Pain

▶ Gynecological Pain and Sexual Functioning

Chronic Widespread Allodynia

▶ Muscle Pain, Fibromyalgia Syndrome (Primary, Secondary)

Chronic Widespread Pain

Synonyms

CWP

Definition

Chronic widespread pain refers to pain in the left side of the body, pain in the right side of the body, pain above the waist, and pain below the waist. In addition, axial pain (cervical, thoracic, or low back) must be present and pain must have persisted for at least 3 months (Wolfe et al. 1990).

Cross-References

▶ Chronic Musculoskeletal Pain
▶ Fibromyalgia
▶ Physical Exercise
▶ Psychiatric Aspects of the Epidemiology of Pain

References

Wolfe, F., Smythe, H., Yunus, M. B., Bennett, R. M., Bombardier, C., Goldenberg, D., Tugwell, A., Campbell, S. M., Abeles, M., Clark, P., Fan, A. G., Farber, J. J., Franklin, C. M., Gatter, R. A., Mamaty, D., Lessard, J., Lichtbroun, A. S., Masi, A. T., McCain, G. F. A., Reynolds, W. J., Romano, T. J., Russell, I. J., & Sheon, R. P. (1990). The American College of Rheumatolgy Criteria for the Classification of Fibromyalgia. *Arthritis and Rheumatism, 33*(2), 160–172.

Chronicity, Prevention

Michael K. Nicholas
Royal North Shore Hospital, University of Sydney, St. Leonards, NSW, Australia

Synonyms

Secondary prevention

Definition

Interventions aimed at reducing the risk of a recently injured person developing chronic pain and disability.

Characteristics

Natural History

The natural history of pain following injury, especially nonspecific injuries like low back pain, is that such pain typically eases to minimal levels within 1–12 weeks. If pain persists beyond 12 weeks, it has been reclassified as "chronic" (Merskey and Bogduk 1994) and may last for prolonged periods. However, recurrences of low back pain, once it has settled, are common. A systematic review of published evidence by Pengel et al. (2003) found that recurrences of low back pain in the first year after initial onset were in the region of 70 %. Epidemiological findings of chronic pain, using the 3-month definition, indicate that around 20 % of adults in western industrialized countries report experiencing such pain (e.g., Blyth et al. 2001). Importantly, however, at least half of this group report that persisting pain was having little impact on their lives, indicating that having persisting pain does not, by itself, mean that the person will be disabled by it. In contrast, the remaining half of those with persisting pain reported that it was interfering in their lives to some degree, a proportion reporting substantial impact on their lifestyles due to pain.

Psychosocial Interventions to Prevent Chronic, Disabling Pain

There is accumulating evidence that the transition from acute to chronic pain, and the associated development of disability, can be modulated by psychological and environmental (or social) factors (Chou and Shekelle 2010; Mallen et al. 2007). This has a number of implications, especially as many of these modulating factors are potentially changeable through interventions. A number of studies have described attempts to modify these psychosocial factors to prevent the development of secondary disability. Some interventions have addressed psychological risk factors, such as beliefs, fears, coping strategies, and avoidance and escape behaviors. Other interventions have addressed environmental or social risk factors, such as workplace arrangements, or compensation factors. At least 18 controlled treatment studies, addressing these risk factors in people with acute and subacute musculoskeletal pain conditions, have been reported (Nicholas et al. 2011).

Waddell and Burton (2000), in their systematic review of the treatment literature for acute and subacute low back pain, concluded that there is strong evidence that advice to continue ordinary activities of daily living as normally as possible despite the pain can give equivalent or faster symptomatic recovery from acute

symptoms and leads to shorter periods of work loss, fewer recurrences, and less work loss the following year than "traditional" medical treatment. Specifically, they reported that interventions containing education (aimed at managing pain and disability, as opposed to simply anatomy/physiology), reassurance and advice (to stay active), progressive fitness exercises, and pain management advice (using behavioral principles) were likely to reduce the risks of developing chronic, disabling pain. Waddell and Burton also found that there was moderate evidence that such approaches were more effective, if they were combined with organizational (workplace) arrangements aimed at facilitating return to work.

Importance of Selecting Those at Risk

Recently, a review of studies of controlled psychosocial interventions (Nicholas et al. 2011) concluded that not everyone with an acute musculoskeletal injury needed specific psychosocial interventions and that usual care at the primary health level would suffice in those identified as being at low risk of developing chronic and disabling pain. However, those identified as being at high risk of chronic, disabling pain do benefit from interventions targeted at these psychosocial risk factors. This conclusion is consistent with that of Jellema et al. (2005) after their review of a psychosocial intervention by general medical practitioners in primary care who found no difference with usual care. In that study, no assessment was made of the patients' risk of becoming chronic cases.

A similar example is provided by Haldorsen et al. (2002) who compared three different treatment modalities with a large sample of injured workers with a range of musculoskeletal conditions who still had jobs, but had been sick-listed for at least 8-weeks in the previous 2-years. The patients were first classified according to three prognostic categories for likely return to work (based on a combination of questionnaire evaluation of psychosocial factors and physiotherapy assessment). Patients in each category were then randomly assigned to one of the different treatments. The treatments consisted of "ordinary treatment" (referral to a general practitioner combined with some physiotherapy); "light multidisciplinary treatment" (like a light mobilization program or encouragement to resume normal activities, similar to that reported by Indahl et al. (1995)); and "extensive multidisciplinary treatment" (4-weeks of an intensive program similar to that described by Bendix et al. (1998)).

The results reported by Haldorsen et al. indicated that injured workers assessed to have a good prognosis of return to work achieved good outcomes, regardless of whichever level of intervention they received. While those with only medium prognostic profiles benefited equally from the light and intensive multidisciplinary programs, but not from "standard or ordinary" care. However, those assessed to have a poor prognosis responded best to the intensive multidisciplinary treatment, with significantly better return to work rates up to 14-months post-treatment than light mobilization (55 % vs. 37 %).

Overall, there is strong evidence that in musculoskeletal (soft-tissue) injuries, where there are no indications of major pathology that might require surgical intervention, encouragement to resume normal activities or exercises, despite the presence of pain, is effective in minimizing subsequent chronic and disabling pain. However, it is likely to be most cost-effective if such interventions are reserved for those assessed as being at high risk of poor outcomes. This means that primary health care providers should include assessment of the risk in their initial assessment and treat accordingly.

References

Bendix, B. T., Bendix, A., Labriola, M., & Boekgaard, P. (1998). Functional restoration for chronic low back pain two year follow-up of two randomized clinical trials. *Spine, 23*, 717–725.

Blyth, F. M., March, L. M., Brnabic, A. J. M., Jorm, L. R., Williamson, M., & Cousins, M. J. (2001). Chronic pain in Australia: A prevalence study. *Pain, 89*, 127–134.

Chou, R., & Shekelle, P. (2010). Will this patient develop persistent disabling low back pain? *JAMA, 303*(13), 1295–1302.

Haldorsen, E. M., Grasdal, A. L., Skouen, J. S., Risa, A. E., Kronholm, K., & Ursin, H. (2002). Is there a right treatment for a particular patient group? Comparison of ordinary treatment, light multidisciplinary treatment, and extensive multidisciplinary treatment for long-term sick-listed employees with musculoskeletal pain. *Pain, 95*, 49–63.

Indahl, A., Velund, L., & Reikeraas, O. (1995). Good prognosis for low back pain when left untampered: A randomized clinical trial. *Spine, 20*, 473–477.

Jellema, P., van der Windt, D. A. M. W., van der Horst, H. E., Blankenstein, A. H., Bouter, L. M., & Stalman, W. A. B. (2005). Why is a treatment aimed at psychosocial factors not effective in patients with (sub)acute low back pain? *Pain, 118*, 350–359.

Mallen, C. D., Peat, G., Thomas, E., Dunn, K., & Croft, P. R. (2007). Prognostic factors for musculoskeletal pain in primary care: A systematic review. *British Journal of General Practice, 57*, 655–661.

Merskey, H., & Bogduk, N. (1994). *Classification of chronic pain. Descriptions of chronic pain syndromes and definitions of pain terms* (2nd ed.). Seattle: IASP Press.

Nicholas, M. K., Linton, S. J., Watson, P. J., & Main, C. J. (2011). The early identification and management of psychological risk factors (Yellow Flags) in patients with low back pain: A reappraisal. *Physical Therapy, 11*(91), 737–753.

Pengel, L. H., Herbert, R., Maher, C. G., & Refshauge, K. M. (2003). Acute low back pain: Systematic review of its prognosis. *British Medical Journal, 327*, 323–327.

Waddell, G., & Burton, K. (2000). Evidence review. In J. T. Carter & L. N. Birrell (Eds.), *Occupational health guidelines for the management of low back pain at work - principal recommendations*. London: Faculty of Occupational Medicine.

Chung Model

▶ Retrograde Cellular Changes After Nerve Injury
▶ Spinal Nerve Ligation Model

Encyclopedia of Pain

Gerald F. Gebhart • Robert F. Schmidt
Editors

Encyclopedia of Pain

Second Edition

Volume 2

Ci–F

With 795 Figures and 217 Tables

🐎 **Springer** Reference

Editors
Gerald F. Gebhart
Department of Anesthesiology
Center for Pain Research
University of Pittsburgh
Pittsburgh, PA, USA

Robert F. Schmidt
Physiologisches Institut der
Universität Würzburg
Würzburg, Germany

ISBN 978-3-642-28752-7   ISBN 978-3-642-28753-4 (eBook)
ISBN 978-3-642-28754-1 (print and electronic bundle)
DOI 10.1007/978-3-642-28753-4
Springer Heidelberg New York Dordrecht London

Library of Congress Control Number: 2013951551

Printed on acid-free paper

Springer is part of Springer Science+Business Media (www.springer.com)

Preface to the Second Edition

As was pointed out in the Preface to the first edition, pain is the principal reason individuals seek medical and dental attention. Fortunately, acute tissue insults associated with pain – and typically inflammation – readily resolve with appropriate treatment and care. Chronic pain states, in contrast, are notoriously difficult to manage and are commonly associated with comorbidities (e.g., depression, cross-organ sensitization) and reduced quality of life. Indeed, chronic pain has come to be considered by some as a disease in its own right. This edition of the *Encyclopedia of Pain* updates the content of the 1st edition and introduces new topics consistent with advances in our knowledge of underlying mechanisms of pain. Thus, new content on channelopathies, expanded and updated material on the role of glia and immune-competent cell interactions with nociceptive neurons, and advances in imaging and changes in brain structure in chronic pain states are included in this edition.

The cardinal objective of research on pain is the translation of knowledge between the laboratory and the clinic. The present prevailing refrain is "bench to bedside," but in fact many ideas for research in pain derive from clinical observations and the absence of knowledge to explain clinical pain states. In the recent past, basic and clinical science research has seen the application of imaging techniques to visualize areas of the brain that are involved in both the sensory and affective aspects of pain, cloning of pain-related channels and receptors, and the application of elegant genetic manipulations that selectively "label" functionally distinct dorsal root ganglion sensory and spinal neurons, two-photon imaging, and optogenetic approaches.

Accordingly, much is new in this second edition of the *Encyclopedia of Pain*. The number of authors who have contributed to this edition is close to 800. In addition to a print edition, the publisher, Springer-Verlag, will produce an electronic version that will make comprehensive searches easy to manage. The electronic version, to be made available on the online publishing platform SpringerReference, can be updated regularly to ensure that the content remains current.

October 2013
Gerald F. Gebhart   Robert F. Schmidt
Pittsburgh, PA, USA   Würzburg, Germany

Preface to the First Edition

As all medical students know, pain is the most common reason for a person to consult a physician. Under ordinary circumstances, acute pain has a useful, protective function. It discourages the individual from activities that aggravate the pain, allowing faster recovery from tissue damage. The physician can often tell from the nature of the pain what its source is. In most cases, treatment of the underlying condition resolves the pain. By contrast, children born with congenital insensitivity to pain suffer repeated physical damage and die young (see Sweet WH (1981) Pain 10:275).

Pain resulting from difficult to treat or untreatable conditions can become persistent. Chronic pain "never has a biologic function but is a malefic force that often imposes severe emotional, physical, economic, and social stresses on the patient and on the family..." (Bonica JJ (1990) The Management of Pain, vol 1, 2nd edn. Lea & Febiger, Philadelphia, p 19). Chronic pain can be considered a disease in its own right.

Pain is a complex phenomenon. It has been defined by the Taxonomy Committee of the International Association for the Study of Pain as "An unpleasant emotional and sensory experience associated with actual or potential tissue damage, or described in terms of such damage" (Merskey H and Bogduk N (1994) Classification of Chronic Pain, 2nd edn. IASP Press, Seattle). It is often ongoing, but in some cases it may be evoked by stimuli. Hyperalgesia occurs when there is an increase in pain intensity in response to stimuli that are normally painful. Allodynia is pain that is evoked by stimuli that are normally non-painful.

Acute pain is generally attributable to the activation of primary efferent neurons called nociceptors (Sherrington CS (1906) The Integrative Action of the Nervous System. Yale University Press, New Haven; 2nd edn, 1947). These sensory nerve fibers have high thresholds and respond to strong stimuli that threaten or cause injury to tissues of the body. Chronic pain may result from continuous or repeated activation of nociceptors, as in some forms of cancer or in chronic inflammatory states, such as arthritis.

However, chronic pain can also be produced by damage to nervous tissue. If peripheral nerves are injured, peripheral neuropathic pain may develop. Damage to certain parts of the central nervous system may result in central neuropathic pain. Examples of conditions that can cause central neuropathic pain include spinal cord injury, cerebrovascular accidents, and multiple sclerosis.

Research on pain in humans has been an important clinical topic for many years. Basic science studies were relatively few in number until experimental work on pain accelerated following detailed descriptions of peripheral nociceptors and central nociceptive neurons that were made in the 1960s and 1970s, by the discovery of the endogenous opioid compounds and the descending pain control systems in the 1970s and the application of modern imaging techniques to visualize areas of the brain that are affected by pain in the 1990s. Accompanying these advances has been the development of a number of animal models of human pain states, with the goal of using these to examine pain mechanisms and also to test analgesic drugs or non-pharmacologic interventions that might prove useful for the treatment of pain in humans. Basic research on pain now emphasizes multidisciplinary approaches, including behavioral testing, electrophysiology, and the application of many of the techniques of modern cell and molecular biology, including the use of transgenic animals.

The *Encyclopedia of Pain* is meant to provide a source of information that spans contemporary basic and clinical research on pain and pain therapy. It should be useful not only to researchers in these fields but also to practicing physicians and other health care professionals and to health care educators and administrators. The work is subdivided into 35 fields, and the Field Editor of each of these describes the areas covered in each in a brief review entry. The topics included in a field are the subject of a series of short essays, accompanied by key words, definitions, illustrations, and a list of significant references. The number of authors who have contributed to the encyclopedia exceeds 550. The plan of the publisher, Springer-Verlag, is to produce both print and electronic versions of this encyclopedia. Numerous links within the electronic version should make comprehensive searches easy to manage. The electronic version will be updated at sufficiently short intervals to ensure that the content remains current.

The editors thank the staff at Springer-Verlag who have provided oversight for this project, including Rolf Lange, Thomas Mager, Claudia Lange, Natasja Sheriff, and Michaela Bilic. Working with these outstanding individuals has been a pleasure.

July 2006
ROBERT F. SCHMIDT   WILLIAM D. WILLIS
Würzburg, Germany   Galveston, Texas, USA

About the Editors

Gerald F. Gebhart received his PhD from The University of Iowa at Iowa City, Iowa, USA, after which he completed a 2-year fellowship at the Université de Montréal, Montréal, Canada, under the direction of Herbert H. Jasper. He began his academic career in the Department of Pharmacology at The University of Iowa in 1973, where he developed a productive research program and led a Pain Interest Group. His career includes a sabbatical year in Heidelberg, Germany (1981–1982), and leadership as head of the Department of Pharmacology from 1996 to 2006. In 2006, Dr. Gebhart was named director of the Center for Pain Research, Department of Anesthesiology, the University of Pittsburgh at Pittsburgh, Pennsylvania, USA. Dr. Gebhart has served the pain community throughout his career. He was elected to the Board of Directors of the American Pain Society (1991–1994) and subsequently elected president of the APS (1997). In 2000, he became the founding editor of *The Journal of Pain*, the official journal of the APS, serving as editor-in-chief until 2010. Dr. Gebhart also served on the Council and later as president-elect, president, and past president of the International Association for the Study of Pain (2005–2012).

Dr. Gebhart is best known for his research contributions in descending modulation of pain and mechanisms of visceral pain, and was designated in 2004 by Thomson ISI as a Highly Cited Researcher (Neuroscience). His scientific contributions have been recognized with several awards, including a 5-year Bristol Myers-Squibb Award for Excellence in Pain Research (1989–1994), a 10-year MERIT Award from the National Institutes of Health (1993–2003),

the Frederick W.L. Kerr Award from the American Pain Society (1994), the Purdue Pharma Prize for Pain Research (2004), the Janssen Award in Gastroenterology (2005), the Founders Award from the American Academy of Pain Medicine (2006), and the Patrick D. Wall Award from the British Pain Society (2012).

Robert F. Schmidt studied medicine at Heidelberg University (Germany). He graduated in 1959 and was promoted to Dr.med. in the same year. After residencies in clinical disciplines, Dr. Schmidt went to the John Curtin School of Medical Research of the Australian National University in Canberra, ACT, where he worked under the head of the Department of Neurophysiology, Sir John C. Eccles. He received his Ph.D. in 1963, the same year Sir John was awarded the Nobel Prize for Physiology and Medicine.

Returning to Heidelberg Medical School, Dr. Schmidt continued his academic career. He completed his habilitation in 1964, becoming a tenured lecturer (Wissenschaftlicher Rat) in 1966 and an associate professor in 1970. After an 8-month sabbatical with Sir John at the State University of New York, Dr. Schmidt became professor and director of the Department of Physiology at the University of Kiel, Germany. In 1982, he moved to the University of Würzburg to take the same position in the Department of Physiology. In 2000, Dr. Schmidt became professor emeritus at this department.

The research work of Dr. Schmidt and his associates in both Kiel and Würzburg focused on the neurophysiology of fine sensory afferents and their connections and functions in the spinal cord. In Würzburg, he concentrated on the processes leading to the changes in response characteristics of nociceptive afferents (i.e., peripheral sensitization) during inflammation of joints, discovering in the process silent nociceptors, which have subsequently been described in all vertebrates studied, including humans. In addition, Dr. Schmidt has edited, coedited, authored, and coauthored numerous textbooks and monographs, most of them having several editions and translations.

In recognition of his research contributions, Dr. Schmidt received many prizes, awards, and honors, including a D.Sc. honoris causa from the University of New South Wales, Sydney, Australia, and an Honorary Professorship from the University of Tübingen, Germany. He was elected to membership in the Akademie der Wissenschaften und der Literatur, Mainz, Germany. Dr. Schmidt is an Honorary Member of the Colombian Association for the Study of Pain, the Japan Physiological Society, the German Society for the Study of Pain, the German Pain Association, the International Association for the Study of Pain (IASP), and the German Physiological Society (in which he served as president in 1979). In 2000, he was awarded the Bundesverdienstkreuz 1st Class (Order of Merit of the German Federal Republic).

Section Editors

Ebtesam Ahmed College of Pharmacy and Health Sciences, St. John's University, Department of Pain Medicine and Palliative Care, Beth Israel Medical Center, New York, NY, USA

A. Vania Apkarian Department of Physiology, Northwestern University Feinberg School of Medicine, Chicago, IL, USA

Lars Arendt-Nielsen Center for Sensory-Motor Interaction, Department of Health Science and Technology, School of Medicine, Aalborg University, Aalborg, Denmark

Carlos Belmonte Instituto de Neurociencias, Universidad Miguel Hernandez-CSIC, San Juan de Alicante, Spain

Niels Birbaumer Institute of Medical Psychology and Behavioral Neurobiology, University of Tübingen, Tübingen, Germany

Nikolai Bogduk University of Newcastle, Newcastle Bone & Joint Institute, Royal Newcastle Centre, Newcastle, NSW, Australia

Jin Mo Chung Department of Neuroscience and Cell Biology, University
Chair in Neuroscience, University of Texas Medical Branch, Galveston,
TX, USA

Joyce DeLeo Dartmouth Hitchcock Medical Center, Dartmouth Medical
School, Hanover, NH, USA

Marshall Devor Department of Cell & Developmental Biology, Institute of Life Sciences and Center for Research on Pain, The Hebrew University of Jerusalem, Jerusalem, Givat Ram, Israel

Hans-Christoph Diener Department of Neurology, University Hospital Essen, University Duisburg-Essen, Essen, Germany

Jonathan O. Dostrovsky Department of Physiology, University of Toronto, Toronto, ON, Canada

Ronald Dubner Department of Neural and Pain Sciences, University of Maryland Dental School, Baltimore, MD, USA

Michel Y. Dubois NYU Medical School, NYU Langone Medical Centers, Center for Research & Treatment of Pain, NYU Ambulatory Care Center, New York, NY, USA

Nanna Brix Finnerup Aarhus University Hospital, Danish Pain Research Center, Aarhus, Denmark

Herta Flor Department of Cognitive and Clinical Neuroscience, Central Institute of Mental Health, Medical Faculty Mannheim, Heidelberg University, Mannheim, Germany

Gerald F. Gebhart Department of Anesthesiology, Center for Pain Research, University of Pittsburgh, Pittsburgh, PA, USA

Maria Adele Giamberardino Pathophysiology of Pain Laboratory and Centre for Fibromyalgia and Musculoskeletal Pain, Department of Medicine and Science of Aging, University of Chieti, Chieti, Italy

Glenn J. Giesler Department of Neuroscience, University of Minnesota, Minneapolis, MN, USA

Peter Goadsby Headache Center, Department of Neurology, University of California, San Francisco, San Francisco, CA, USA

Michael S. Gold Department of Anesthesiology, Center for Pain Research, University of Pittsburgh, Pittsburgh, PA, USA

Hermann O. Handwerker Department of Physiology and Pathophysiology, University of Erlangen/Nürnberg, Erlangen, Germany

Burkhard Hinz Institute of Toxicology and Pharmacology, University of Rostock, Rostock, Germany

Wilfrid Jaenig Physiologisches Institut, Christian-Albrechts-Universität zu
Kiel, Kiel, Germany

Michaela Kress Department für Physiologie und Medizinische Physik,
Medizinische Universität Innsbruck, Innsbruck, Austria

Frederick A. Lenz Department of Neurosurgery, Johns Hopkins Hospital, Baltimore, MD, USA

Tonya M. Palermo Seattle Children's Hospital Research Institute, CW8-6 - Child Health, Behavior and Development, University of Washington, Seattle, WA, USA

Frank Porreca Department of Pharmacology, University of Arizona Health Sciences Center, Tucson, AZ, USA

Russell K. Portenoy Department of Pain Medicine and Palliative Care, Beth Israel Medical Center, New York, NY, USA

James P. Robinson Department of Anesthesiology & Pain Medicine, University of Washington, Seattle, WA, USA

Hans-Georg Schaible Department of Physiology - Neurophysiology, Friedrich-Schiller-University of Jena, Jena, Germany

Robert F. Schmidt Physiologisches Institut der Universität Würzburg, Würzburg, Germany

Stephan A. Schug School of Medicine and Pharmacology, Royal Perth Hospital, Faculty of Medicine, Dentistry & Health Sciences, The University of Western Australia, UWA Anaesthesiology, Perth, Australia

Barry J. Sessle Faculty of Dentistry, University of Toronto, Toronto, ON, Canada

Bengt H. Sjölund Department of Community Medicine and Rehabilitation Medicine, Umea University, Umea, Sweden

Mark D. Sullivan Psychiatry and Behavioral Sciences, University of Washington, Seattle, WA, USA

Irene Tracey Pain Imaging Neuroscience (PaIN) Group, Department of Physiology, Anatomy and Genetics, Oxford University, Oxford, UK

Dennis C. Turk Department of Anesthesiology & Pain Research, University of Washington, Seattle, WA, USA

Félix Viana Instituto de Neurociencias, Universidad Miguel Hernandez-CSIC, San Juan de Alicante, Alicante, Spain

Katy Vincent Nuffield Department of Obstetrics and Gynaecology, The Women's Centre, John Radcliffe Hospital, University of Oxford, Oxford, UK

Gary A. Walco Department of Anesthesiology and Pain Medicine, University of Washington School of Medicine, Seattle Children's Hospital, Seattle, WA, USA

George L. Wilcox Department of Neuroscience, Pharmacology and Dermatology, University of Minnesota Medical School, Minneapolis, MN, USA

Katja Wiech Nuffield Department of Clinical Neurosciences, John Radcliffe Hospital, Oxford, UK

George L. Wilcox, Department of Neuroscience, Pharmacology, and Dermatology, University of Minnesota Medical School, Minneapolis, MN, USA

Katja Wiech, Nuffield Department of ... Clinical Neurosciences ... John Radcliffe Hospital, Oxford UK

Contributors

Tipu Aamir The Auckland Regional Pain Service, Auckland Hospital, Auckland, New Zealand

Catherine Abbadie Department of Pharmacology, Merck Research Laboratories, Rahway, NJ, USA

Frances V. Abbott Departments of Psychiatry and Psychology, McGill University, Montreal, QC, Canada

Heather Adams Department of Surgery, McGill University Health Centre, Montreal, QC, Canada

Mary O. Adedoyin Department of Anatomy, University of California San Francisco, San Francisco, CA, USA

Rolf H. Adler University of Berne Medical School, Kehrsatz, Switzerland

Claire D. Advokat Department of Psychology, Louisiana State University, Baton Rouge, LA, USA

Reto M. Agosti Headache Center Hirlsanden, Zurich, Switzerland

Ebtesam Ahmed College of Pharmacy and Health Sciences, St. John's University, Department of Pain Medicine and Palliative Care, Beth Israel Medical Center, New York, NY, USA

Marie-Claire Albanese Department of Psychology, McGill University, Montreal, QC, Canada

Kathryn M. Albers Department of Neurobiology, University of Pittsburgh School of Medicine, Pittsburgh, PA, USA

Phillip J. Albrecht Center for Neuropharmacology and Neuroscience, Albany Medical College, Albany, NY, USA

Elie D. Al-Chaer Center for Pain Research, University of Arkansas for Medical Sciences, Little Rock, AR, USA

Terry Altilio Department of Pain Medicine and Palliative Care, Beth Israel Medical Center, New York, USA

Rainer Amann Medical University Graz, Graz, Austria

Ron Amir Institute of Life Sciences, The Hebrew University of Jerusalem, Jerusalem, Israel

Praveen Anand Peripheral Neuropathy Unit, Imperial College London, London, UK

Karen O. Anderson The University of Texas MD Anderson Cancer Center, Houston, TX, USA

Ole K. Andersen Department of Health Science & Technology, Aalborg University, Aalborg, Denmark

Stanley W. Anderson Department of Neurosurgery, Johns Hopkins Hospital, Johns Hopkins Medical Institutions, Johns Hopkins University, Baltimore, MD, USA

Trevor Anderson Fairview Pain & Palliative Care, University of Minnesota Medical Center, Minneapolis, MN, USA

Karl-Erik Andersson Wake Forest Institute for Regenerative Medicine, Wake Forest University School of Medicine, Winston Salem, USA

Frank Andrasik Department of Psychology, Applied Psychophysiology and Biofeedback, The University of Memphis, Memphis, TN, USA

A. Vania Apkarian Department of Physiology, Northwestern University Feinberg School of Medicine, Chicago, IL, USA

Lars Arendt-Nielsen Center for Sensory-Motor Interaction, Department of Health Science and Technology, School of Medicine, Aalborg University, Aalborg, Denmark

Kirsten Arndt CNS Pharmacology, Boehringer Ingelheim Pharma GmbH & Co. KG, Biberach, Germany

Charles-Daniel Arreto Université Paris Descartes, INSERM UMR 894, France

Erik Askenasy Neurobiology and Anatomy, Health Science Center, University of Texas, Houston, TX, USA

J. H. Atkinson Psychiatry Service, VA San Diego Healthcare System and Department of Psychiatry, University of California, San Diego, CA, USA

Nadine Attal Centre d'Evaluation et de Traitement de la Douleur, Hôpital Ambroise Paré, INSERM U 987 APHP, Boulogne-Billancourt, France

Christoph Aufenberg Functional Neurosurgery, Neurosurgical Clinic University Hospital, Zürich, Switzerland

Beate Averbeck Sanofi-Aventis Deutschland GmbH, Frankfurt, Germany

Qasim Aziz Section of GI Sciences, Hope Hospital University of Manchester, Salford, UK

Cathrine Baastrup Danish Pain Research Center, Aarhus University Aarhus University Hospital, Aarhus, Denmark

F. W. Bach Danish Pain Research Center, Department of Neurology, Aarhus University Hospital, Aarhus, Denmark

S. K. Back Neuroscience Research Institute and Department of Physiology, Korea University, College of Medicine, Seoul, Republic of Korea

Misha-Miroslav Backonja Department of Neurology, University of Wisconsin-Madison, Madison, WI, USA

Banani Banerjee Division of Gastroenterology and Hepatology and Pediatric Gastroenterology, Medical College of Wisconsin, Milwaukee, WI, USA

John D. Banja Department of Rehabilitation Medicine, Center for Ethics Emory University, Atlanta, GA, USA

Carsten Bantel Department of Surgery & Cancer? Anaesthetics Section, Imperial College of Science and Medicine Chelsea and Westminster Campus, London, UK

Andrew Paul Baranowski The Pain Management Centre, National Hospital for Neurology and Neurosurgery, UCL Hospitals NHS Foundation Trust, London, UK

Alex Barling Department of Neurology, Birmingham Muscle and Nerve Centre Queen Elizabeth Hospital, Birmingham, UK

Ralf Baron Department of Neurological Pain Research and Therapy, Christian Albrechts University Kiel, Kiel, Germany

Gordon A. Barr Department of Anesthesiology and Critical Care Medicine, Children's Hospital of Philadelphia University of Pennsylvania, Philadelphia, PA, USA

Emily J. Bartley Department of Psychology, The University of Tulsa, Tulsa, OK, USA

Jeffrey R. Basford Department of Physical Medicine and Rehabilitation, Mayo Clinic and Mayo Foundation, Rochester, NY, USA

Michele C. Battie Department of Physical Therapy, University of Alberta, Edmonton, AB, Canada

Ulf Baumgartner Chair of Neurophysiology, Center for Biomedicine and Medical Technology Mannheim, Heidelberg University, Mannheim, Germany

Nicola Beale Nuffield Department of Anaesthetics, Oxford University Hospitals NHS Trust, UK

Alain Beaudet Montreal Neurological Institute, McGill University, Montreal, QC, Canada

Lino Becerra P.A.I.N. Group, Department of Anesthesiology, Perioperative and Pain Medicine, Boston Children's Hospital Department of Radiology Massachusetts General Hospital Center for Pain and the Brain, Harvard Medicals, Boston, MA, USA

Yaakov Beilin Department of Anesthesiology, and Obstetrics, Mount Sinai School of Medicine, New York, NY, USA

Alvin J. Beitz Department of Veterinary and Biomedical Sciences, University of Minnesota, St. Paul, MN, USA

Miles Belgrade Fairview Pain Management Center, University of Minnesota Medical Center, Minneapolis, MN, USA

Carlo Valerio Bellieni Neonatal Intensive Care Unit, University Hospital, Siena, Italy

Carlos Belmonte Instituto de Neurociencias, Universidad Miguel Hernandez-CSIC, San Juan de Alicante, Spain

Allan J. Belzberg Department of Neurosurgery, Johns Hopkins School of Medicine, Baltimore, MD, USA

Matt Bender Department of Neurosurgery, Johns Hopkins Hospital, Baltimore, MA, USA

Fabrizio Benedetti Department of Neuroscience, Clinical and Applied Physiology Program, University of Turin Medical School, Turin, Italy

David L. H. Bennett Wolfson CARD, Kings College London, London, UK

Nuffield Department of Clinical Neurosciences, John Radcliffe Hospital, The University of Oxford, Oxford, UK

Robert Bennett Department of Medicine, Oregon Health and Science University, Portland, USA

Edward C. Benzel Department of Neurosurgery, Center for Spine Health Neurological Institute Cleveland Clinic, Cleveland, OH, USA

David A. Bereiter Department of Diagnostic and Biological Sciences, University of Minnesota School of Dentistry, Minneapolis, MN, USA

Elmar Berendes Klinik für Anästhesiologie, Operative Intensivmedizin und Schmerztherapie, Klinikum Krefeld, Krefeld, Germany

Karen J. Berkley Program in Neuroscience, Florida State University, Tallahassee, FL, USA

Peter Berlit Department of Neurology, Alfried Krupp Hospital, Essen, Germany

Jean-François Bernard INSERM UMR-894, UPMC Site Pitié-Salpêtrière, Paris, France

Anne K. Bertelsen Neurology, Aleris Hospital, Bergen, Norway

Jennifer Bickel Integrative Pain Management, Children's Mercy Hospitals and Clinics, University of Missouri-Kansas City School of Medicine, Kansas City, MO, USA

Marcelo E. Bigal Department of Neurology and Montefiore Headache Unit, Albert Einstein College of Medicine, Bronx, NY, USA

Yitzchak M. Binik McGill University, Montreal, QC, Canada

Niels Birbaumer Institute of Medical Psychology and Behavioral Neurobiology, University of Tübingen, Tübingen, Germany

Frank Birklein Department of Neurology, University of Mainz, Mainz, Germany

Kathryn A. Birnie Department of Psychology, Dalhousie University, Halifax, NS, Canada

Thomas Bishop Centre for Neuroscience, King's College London, London, UK

Henning Bliddal The Parker Institute, Frederiksberg Hospital, Copenhagen, Denmark

Nikolai Bogduk University of Newcastle, Newcastle Bone & Joint Institute, Royal Newcastle Centre, Newcastle, NSW, Australia

Jorgen Boivie Department of Neurology, University Hospital, Linköping, Sweden

Michael R. Bond University of Glasgow, Glasgow, UK

Richard L. Boortz-Marx Division of Pain Medicine, Department of Anesthesiology, University of North Carolina, Chapel Hill, North Carolina, USA

James M. Borowczyk Department of Orthopaedics and Musculoskeletal Medicine, University of Otago, Christchurch, Christchurch, New Zealand

David Borsook P.A.I.N. Group, Department of Anesthesiology, Perioperative and Pain Medicine, Boston Children's Hospital Department of Radiology Massachusetts General Hospital Center for Pain and the Brain, Harvard Medicals, Boston, MA, USA

George S. Borszcz Department of Psychology, Wayne State University, Detroit, USA

Jasenka Borzan Department of Anesthesiology and Critical Care Medicine, Johns Hopkins University, Baltimore, USA

Regina M. Botting London, UK

Didier Bouhassira Centre d'Evaluation et de Traitement de la Douleur, Hôpital Ambroise Paré, INSERM U 987 APHP, Boulogne-Billancourt, France

Laurence Bourgeais INSERM UMR 894, Centre de Psychiatrie et Neurosciences, Paris, France

David Bowsher Pain Research Institute, University Hospital Aintree, Liverpool, UK

Emma L. Brandon Anaesthesiology and Pain Medicine, University of Western Australia Royal Perth Hospital, Perth, WA, Australia

Jennifer Brawn P.A.I.N. Group, Center for Pain and the Brain, Boston Children's Hospital, Harvard Medical School, Waltham, MA, USA

Harald Breivik Department of Pain Management and Research, Oslo University Hospital, Rikshospitalet and University of Oslo, Oslo, Norway

Kori L. Brewer Department of Emergency Medicine, Brody School of Medicine East Carolina University, Greenville, NC, USA

Christopher R. Brigham Brigham and Associates, Inc., Portland, USA

Abigail Broderick Nuffield Department of Obstetrics and Gynaecology, University of Oxford, Oxford, UK

Jeffrey A. Brown Department of Neurological Surgery, Wayne State University School of Medicine, Detroit, MI, USA

Stephen C. Brown Department of Anesthesia, The Hospital for Sick Children, Toronto, ON, Canada

Eduardo Bruera Department of Palliative Care and Rehabilitation Medicine, The University of Texas M. D. Anderson Cancer Center, Houston, TX, USA

Pablo Brumovsky Faculty of Biomedical Sciences, Austral University, Buenos Aires, Argentina
Consejo Nacional de Investigaciones Científicas y Técnicas (CONICET), Buenos Aires, Argentina

Kay Brune Department of Experimental and Clinical Pharmacology and Toxicology, Friedrich Alexander University Erlangen-Nürnberg, Erlangen, Germany

Maria Burian Department of General, Visceral and Transplantation Surgery, University of Heidelberg, Heidelberg, Germany

Dan Buskila Rheumatic Disease Unit, Soroka Medical Center, Beer Sheva, Israel

Catherine M. Cahill Anesthesiology & Perioperative Care, University of California, Irvine, Irvine, CA, USA

Brian E. Cairns Faculty of Pharmaceutical Sciences, University of British Columbia, Vancouver, BC, Canada

Nigel A. Calcutt Department of Pathology, University of California San Diego, La Jolla, CA, USA

Margarita Calvo Wolfson CARD, Kings College London, London, UK

James N. Campbell Johns Hopkins University School of Medicine, Baltimore, MD, USA

Lauren Campbell Department of Psychology, York University, Toronto, ON, Canada

Ling Cao Department of Biomedical Sciences, College of Osteopathic Medicine, University of New England, ME, USA

Adam Carinci Department of Anesthesia, Critical Care and Pain Medicine, Massachusetts General Hospital Harvard Medical School, Boston, MA, USA

Giancarlo Carli Department of Physiology, University of Siena, Siena, Italy

Christer P. O. Carlsson Rehabilitation Clinic, Lunds University Hospital, Lund, Sweden

Susan M. Carlton Neuroscience and Cell Biology, University of Texas Medical Branch, Galveston, TX, USA

Richard Carr Clinic for Anaesthesiology, Heidelberg University, Mannheim, Baden-Wuerttemberg, Germany

Benjamin S. Carson Department of Neurosurgery, Johns Hopkins Hospital, Baltimore, MA, USA

Brian Carter University of Missouri at Kansas City Children's Mercy Hospital and Bioethics Center, Kansas City, MO, USA

Kenneth L. Casey Department of Neurology, University of Michigan Medical Center, Ann Arbor, MI, USA

Christine T. Chambers Departments of Pediatrics and Psychology, Dalhousie University, Halifax, NS, Canada

Sandra Chaplan Janssen Research and Development, LLC, San Diego, CA, USA

C. Richard Chapman Department of Anesthesiology, University of Utah School of Medicine, Salt Lake City, UT, USA

Santosh K. Chaturvedi Psychiatry, National Institute of Mental Health & Neurosciences, Bangalore, India

Andrew C. N. Chen Human Brain Mapping and Cortical Imaging Laboratory, Aalborg University, Aalborg, Denmark

Hsinlin Thomas Cheng Department of Neurology, University of Michigan, Ann Arbor, MI, USA

Pavel Cherkas Faculty of Dentistry, University of Toronto, Toronto, ON, Canada

Andrea Cheville Department of Physical Medicine and Rehabilitation, Mayo Clinic, Rochester, MN, USA

Chen Yu Chiang Faculty of Dentistry, University of Toronto, Toronto, ON, Canada

N. L. Chillingworth School of Physiology and Pharmacology, University of Bristol, Bristol, UK

Edward Chow Department of Radiation Oncology, Sunnybrook Health Sciences Centre Odette Cancer Centre (T2-182), Toronto, ON, Canada

Julie A. Christianson Department of Anatomy and Cell Biology, University of Kansas Medical Center, Kansas City, KS, USA

MacDonald J. Christie Discipline of Pharmacology, University of Sydney, Sydney, Australia

Jin Mo Chung Department of Neuroscience and Cell Biology, University Chair in Neuroscience, University of Texas Medical Branch, Galveston, TX, USA

Kyungsoon Chung Department of Neuroscience and Cell Biology, University of Texas Medical Branch, Galveston, TX, USA

Anna K. Clark Wolfson Centre for Age Related Diseases, King's College London, London, UK

W. Crawford Clark College of Physicians and Surgeons, Columbia University, New York, USA

James F. Cleary Pain & Policy Studies Group, UW Carbone Cancer Center, Madison, WI, USA

John De Clercq Department of Rehabilitation Sciences and Physiotherapy, Ghent University Hospital, Ghent, Belgium

Jacqui Clinch Royal National Hospital for Rheumatic Diseases NHS Foundation Trust, Bath, UK

Terence J. Coderre Departments of Anesthesia, Neurology & Neurosurgery and Psychology, McGill University, Montreal, QC, Canada

Robert Coghill Department of Neurobiology and Anatomy, Wake Forest University School of Medicine, Winston-Salem, NC, USA

Lindsey Cohen Psychology, Georgia State University, Atlanta, GA, USA

Alan L. Colledge Utah Labor Commission, International Association of Industrial Accident Boards and Commissions Central Utah Medical Clinic, South Salt Lake City, UT, USA

Beverly J. Collett University Hospitals of Leicester, Leicester, UK

Lesley A. Colvin Department of Anaesthesia, Critical Care and Pain Medicine Western General Hospital, Edinburgh, UK

Kevin C. Conlon Trinity College Dublin, University of Dublin, Dublin, Ireland

Mark Connelly Integrative Pain Management, Children's Mercy Hospitals and Clinics, University of Missouri-Kansas City School of Medicine, Kansas City, MO, USA

Brian Y. Cooper Department of Oral and Maxillofacial Surgery, University of Florida, Gainesville, FL, USA

Christine Cordle University Hospitals of Leicester, Leicester, UK

Laura A. Cousins Psychology, Georgia State University, Atlanta, GA, USA

Michael J. Cousins Department of Anesthesia and Pain Management, Royal North Shore Hospital University of Sydney, Sydney, NSW, Australia

Verne C. Cox Department of Psychology, University of Texas at Arlington, Arlington, TX, USA

George A. Crabill Department of Neurosurgery, Johns Hopkins Hospital, Baltimore, MA, USA

Rebecca M. Craft Psychology, Washington State University, Pullman, WA, USA

Kenneth D. Craig Department of Psychology, University of British Columbia, Vancouver, BC, Canada

Richard Crevenna Medizinische Universität Wien, Wien, Austria

Geert Crombez Department of Experimental Clinical and Health Psychology, Ghent University, Ghent, Belgium

Joan Crook School of Nursing, Faculty of Health Sciences, McMaster University, Hamilton, ON, Canada

Giorgio Cruccu Department of Neurological Sciences, La Sapienza University, Rome, Italy

Michele Curatolo Department of Anesthesiology, Division of Pain Therapy, University Hospital of Bern Inselspital, Bern, Switzerland

Nachum Dafny Neurobiology and Anatomy, Health Science Center, University of Texas, Houston, TX, USA

Jorgen B. Dahl Department of Anaesthesiology, Glostrup University Hospital, Herlev, Denmark

Yi Dai Department of Anatomy and Neuroscience, Hyogo College of Medicine, Hyogo, Japan

Stefaan Van Damme Department of Experimental Clinical and Health Psychology, Ghent University, Ghent, Belgium

Alexandre F. M. DaSilva Biologic & Materials Sciences, University of Michigan Dental School, MI, USA

Steve Davidson Pain Center & Department of Anesthesiology, Washington University School of Medicine, MO, USA

Brian Davis Department of Neurobiology, University of Pittsburgh School of Medicine, Pittsburgh, PA, USA

Karen D. Davis Brain, Imaging and Behaviour – Systems Neuroscience, Toronto Western Research Institute, Toronto, ON, Canada

Department of Surgery, University of Toronto, Toronto, ON, Canada

Richard O. Day Department of Clinical Pharmacology and Toxicology, St Vincent's Hospital, Sydney, NSW, Australia

Isabelle Decosterd Pain Center, University Hospital Center and University of Lausanne, Lausanne, Switzerland

Ruth Defrin Physical Therapy, Faculty of Medicine, Tel-Aviv University, Tel-Aviv, Israel

Antoon F. C. De Laat Department of Oral/Maxillofacial Surgery, Catholic University of Leuven, Leuven, Belgium

Joyce DeLeo Dartmouth Hitchcock Medical Center, Dartmouth Medical School, Hanover, NH, USA

A. Lee Dellon Department of Plastic Surgery and Neurosurgery, Johns Hopkins University, Baltimore, USA

Dellon Institutes for Peripheral Nerve Surgery, Baltimore, MD, USA

Marshall Devor Department of Cell & Developmental Biology, Institute of Life Sciences and Center for Research on Pain, The Hebrew University of Jerusalem, Jerusalem, Givat Ram, Israel

Perry Dhaliwal Department of Neurosurgery, Cleveland Clinic Spine Institute, Cleveland, OH, USA

David Diamant Neurological and Spinal Surgery, Lincoln, NE, USA

Anthony H. Dickenson Department of Pharmacology, University College London, London, UK

Hans-Christoph Diener Department of Neurology, University Hospital Essen, University Duisburg-Essen, Essen, Germany

Natalia Dmitrieva Program in Neuroscience, Florida State University, Tallahassee, FL, USA

L. F. Donaldson School of Life Sciences, University of Nottingham, Nottingham, UK

Henri Doods CNS Pharmacology, Boehringer Ingelheim Pharma GmbH & Co. KG, Biberach, Germany

Victoria Dorf Social Security Administration, Baltimore, MD, USA

Michael J. Dorsi Department of Neurosurgery, Johns Hopkins School of Medicine, Baltimore, MD, USA

Jonathan O. Dostrovsky Department of Physiology, University of Toronto, Toronto, ON, Canada
Faculty of Dentistry, University of Toronto, Toronto, ON, Canada

Robyn Drach Fairview Pain Management Center, Minneapolis, MN, USA

Ron Dressler Utah Labor Commission, International Association of Industrial Accident Boards and Commissions Central Utah Medical Clinic, South Salt Lake City, UT, USA

Paul Dreyfuss Department of Rehabilitation Medicine, University of Washington, Seattle, USA

Peter D. Drummond School of Psychology, Murdoch University, Perth, Australia

Ronald Dubner Department of Neural and Pain Sciences, University of Maryland Dental School, Baltimore, MD, USA

Anne Ducros Headache Emergency Department, INSERM, Paris, France

Gary H. Duncan Département de stomatologie, Faculté de médecine dentaire, Université de Montréal, Montreal, QC, Canada
Department of Neurology and Neurosurgery, McGill University, Montreal, QC, Canada

Mary J. Eaton Research Service (151), Miami VA Health Care System, Miami VAMC, Miami, FL, USA

Andrea Ebersberger Department of Physiology, Friedrich Schiller University of Jena, Jena, Germany

Christopher Eccleston Pain Management Unit, University of Bath, Bath, UK

John Edmeads Sunnybrook and Woman's College Health Sciences Centre, University of Toronto, Toronto, Canada

Edzard Ernst Complementary Medicine, Peninsula Medical School Universities of Exeter and Plymouth, Exeter, UK

Reuben Escorpizo Swiss Paraplegic Research, Nottwil, Switzerland
Department of Health Sciences and Health Policy, University of Lucerne, Lucerne, Switzerland
ICF Research Branch in cooperation with the World Health Organization (WHO) Collaborating Centre for the Family of International Classifications in Germany, Nottwil, Switzerland

Schonstein Eva RehabCo, Fyshwick, ACT, Australia

Joshua C. Eyer Department of Psychology, University of Alabama, Tuscaloosa, AL, USA

Marie Faaborg-Andersen McGill University, Montreal, QC, Canada

Keri L. Fakata College of Pharmacy and Pain Management Center, University of Utah, Salt Lake City, USA

Marie Fallon Department of Oncology, Edinburgh Cancer Center University of Edinburgh, Edinburgh, UK

Melissa Farmer Department of Physiology / Feinberg School of Medicine, Northwestern University, Chicago, IL, USA

Samantha R. Fashler Department of Psychology, University of British Columbia, Vancouver, BC, Canada

Jean-Baptiste Fassier Department of Occupational Health and Medicine, UMRESTTE-Université Claude Bernard Lyon 1, Lyon, France

Charlotte Feinmann Eastman Dental Institute London, University College London, London, UK

Elizabeth Roy Felix The Miami Project to Cure Paralysis, University of Miami Miller School of Medicine, Miami, FL, USA

Bin Feng Center for Pain Research, Department of Anesthesiology, School of Medicine, University of Pittsburgh, Pittsburgh, Pennsylvania, USA

Yi Feng Department of Anesthesiology, Peking University People's Hospital, Peking, China

Andres Fernandez Department of Neurosurgery, Johns Hopkins Hospital, Baltimore, MD, USA

Antonio Vicente Ferrer-Montiel Institute of Molecular and Cell Biology, University Miguel Hernández, Elche, Alicante, Spain

Perry G. Fine Pain Research Center, School of Medicine, University of Utah, Salt Lake City, UT, USA

David J. Fink Department of Neurology, University of Michigan and VA Ann Arbor Healthcare System, Ann Arbor, MI, USA

Nanna Brix Finnerup Aarhus University Hospital, Danish Pain Research Center, Aarhus, Denmark

David A. Fishbain Department of Psychiatry, University of Miami, Miami, FL, USA

Maria Fitzgerald Neuroscience, Physiology & Pharmacology, University College London, London, UK

Per Flisberg Department of Anaesthesia and Intensive Care, Lund University Hospital, Lund, Sweden

Herta Flor Department of Cognitive and Clinical Neuroscience, Central Institute of Mental Health, Medical Faculty Mannheim, Heidelberg University, Mannheim, Germany

Christopher M. Flores Drug Discovery, Johnson and Johnson Pharmaceutical Research and Development, Spring House, PA, USA

Wilbert E. Fordyce Department of Rehabilitation, University of Washington School of Medicine, Seattle, WA, USA

Robert D. Foreman Department of Physiology, College of Medicine, University of Oklahoma Health Sciences Center, Oklahoma City, OK, USA

Gary M. Franklin Department of Environmental and Occupational Health Sciences, University of Washington, Seattle, WA, USA

Department of Health Services, University of Washington, Seattle, WA, USA

Jan Fridén Department of Hand Surgery, Sahlgrenska University Hospital, Gothenborg, Sweden

Justin Friedlander The Arthur Smith Institute for Urology, North Shore-LIJ Healthcare System, Hofstra North Shore LIJ School of Medicine, New Hyde Park, NY, USA

Allan H. Friedman Division of Neurosurgery, Duke University Medical Center, Durham, NC, USA

Perry N. Fuchs Department of Psychology, University of Texas at Arlington, Arlington, TX, USA

Deborah Fulton-Kehoe Department of Environmental and Occupational Health Sciences, University of Washington, Seattle, WA, USA

Andrea D. Furlan Institute for Work & Health, Toronto, ON, Canada

Vasco Galhardo Departamento de Biologia Experimental, Faculdade de Medicina, Universidade do Porto, Portugal

Andreas R. Gantenbein RehaClinic Bad Zurzach, Bad Zurzach, Switzerland

I. Garonzik Department of Neurosurgery, Johns Hopkins Medical Institutions, Baltimore, MD, USA

Robert Gassin Musculoskeletal Medicine Clinic, Frankston, VIC, Australia

Robert J. Gatchel Department of Psychology, The University of Texas, Arlington, TX, USA

Gerald F. Gebhart Department of Anesthesiology, Center for Pain Research, University of Pittsburgh, Pittsburgh, PA, USA

Gerd Geisslinger Department of Clinical Pharmacology, University Frankfurt am Main, Frankfurt, Germany

Robert D. Gerwin Department of Neurology, Johns Hopkins University, Bethesda, MD, USA

Maria Adele Giamberardino Pathophysiology of Pain Laboratory and Centre for Fibromyalgia and Musculoskeletal Pain, Department of Medicine and Science of Aging, University of Chieti, Chieti, Italy

Stephen Gibson National Ageing Research Institute Department of Medicine, University of Melbourne, Parkville, VIC, Australia

Glenn J. Giesler Department of Neuroscience, University of Minnesota, Minneapolis, MN, USA

Philip L. Gildenberg Baylor Medical College, Houston, TX, USA

Myra Glajchen Department of Pain Medicine and Palliative Care, Beth Israel Medical Center Albert Einstein College of Medicine, New York, NY, USA

Hartmut Göbel Kiel Migraine and Headache Center, Kiel Pain Center, Kiel, Germany

Peter Goadsby Headache Center, Department of Neurology, University of California, San Francisco, San Francisco, CA, USA

Philippe Goffaux Faculty of Medicine, University of Sherbrooke, Sherbrooke, QC, Canada

Ana Gomis Instituto de Neurociencias de Alicante, Universidad Miguel Hernández-CSIC, San Juan de Alicante, Spain

Gilbert R. Gonzales Department of Neurology, Memorial Sloan-Kettering Cancer Center, New York, USA

Allan Gordon Wasser Pain Management Centre, Mount Sinai Hospital University of Toronto, Toronto, ON, Canada

Liesbet Goubert Department of Experimental Clinical and Health Psychology, Ghent University, Ghent, Belgium

Jayantilal Govind Pain Clinic, Department of Anaesthesia, University of New South Wales, Sydney, NSW, Australia

Richard H. Gracely Departments of Medicine-Rheumatology and Neurology, University of Michigan Health System, Ann Arbor, MI, USA

Garry G. Graham Department of Pharmacology, School of Medical Sciences, University of New South Wales, Sydney, NSW, Australia

David K. Grandy Department of Physiology and Pharmacology, Oregon Health and Science University, Portland, OR, USA

Thomas Graven-Nielsen Laboratory for Musculoskeletal Pain and Motor Control, Center for Sensory-Motor Interaction, Aalborg University, Aalborg, Denmark

Barry Green Pierce Laboratory, Yale School of Medicine, New Haven, CT, USA

Paul G. Green Oral & Maxillofacial Surgery, University of California San Francisco, San Francisco, CA, USA

Joel D. Greenspan Department of Neural and Pain Sciences, University of Maryland School of Dentistry, Baltimore, MD, USA

John W. Griffin Department of Neurology, Johns Hopkins University School of Medicine, Baltimore, MD, USA

Cornelius Groenwald Department of Anesthesiology and Pain Medicine, Seattle Children's Hospital, University of Washington School of Medicine, Seattle, WA, USA

Douglas P. Gross Department of Physical Therapy, University of Alberta WCB-Alberta Millard Health, Edmonton, AB, Canada

Sabine Grösch Institute of Clinical Pharmacology, Clinical Center of the Goethe University, Frankfurt, Germany

Blair D. Grubb Department of Cell Physiology and Pharmacology, University of Leicester, Leicester, UK

Ruth Eckstein Grunau Pediatrics Dept, University of British Columbia Developmental Neurosciences & Child Health, Child & Family Research Institute, Vancouver, BC, Canada

Christoph Gutenbrunner Department of Physical Medicine and Rehabilitation, Hanover Medical School, Hanover, Germany

Maija Haanpää Department of Neurosurgery, Helsinki University Central Hospital, Helsinki, Finland

Sherri Haas Fairview Pain Management Center, Minneapolis, MN, USA

Ute Habel Department of Psychiatry and Psychotherapy, RWTH Aachen University, Aachen, Germany

Thomas Hachenberg Clinic for Anaesthesiology and Intensive Therapy, Otto-von-Guericke University, Magdeburg, Germany

Nouchine Hadjikhani Martinos Center for Biomedical Imaging, Massachusetts General Hospital Harvard Medical School, Charlestown, MA, USA

Ole Haegg Department of Orthopedic Surgery, Sahlgren University Hospital, Goteborg, Sweden

Bryan C. Hains Department of Neurology, Yale University School of Medicine, West Haven, CT, USA

Else K. B. Hals University of Oslo, Faculty of Dentistry, Oslo, Norway

Darryl T. Hamamoto Diagnostic and Biological Sciences, University of Minnesota, Minneapolis, MN, USA

Donna L. Hammond Department of Anesthesia, University of Iowa, Iowa City, IA, USA

Menachem Hanani Hadassah-Hebrew University Medical Center Hebrew University Medical School, Jerusalem, Israel

Hermann O. Handwerker Department of Physiology and Pathophysiology, University of Erlangen/Nürnberg, Erlangen, Germany

George M. Hanna Department of Anesthesia, Critical Care and Pain Medicine, Massachusetts General Hospital Harvard Medical School, Boston, MA, USA

Jing-Xia Hao Department of Physiology and Pharmacology, Karolinska Institutet, Stockholm, Sweden

Eleni G. Hapidou Department of Psychiatry and Behavioural Neurosciences, McMaster University and Hamilton Health Sciences Chronic Pain Management Unit, Chedoke Hospital, Hamilton, ON, Canada

Geoffrey Harding Sandgate Spinal Medicine Clinic, Sandgate, QLD, Australia

Louise Harding Hospitals NHS Foundation Trust, University College London, London, UK

Kenneth M. Hargreaves Department of Endodontics, University of Texas Health Science Center at San Antonio, San Antonio, USA

D. Paul Harries Pain Treatment Center, Samaritan Hospital Lexington, Lexington, KY, USA

Denise Harrison Nursing, Children's Hospital of Eastern Ontario (CHEO), University of Ottawa, Ottawa, ON, Canada

Steven Harte Anesthesiology and Internal Medicine/Rheumatology, University of Michigan VA Ann Arbor Healthcare System, Ann Arbor, MI, USA

Barbara A. Hastie Community Dentistry & Behavioral Science, University of Florida College of Dentistry, Gainesville, FL, USA

Jennifer A. Haythornthwaite Psychiatry & Behavioral Sciences, Johns Hopkins University, Baltimore, MD, USA

Heinz-Joachim Häbler FH Bonn-Rhein-Sieg, Rheinbach, Germany

John H. Healey Orthopaedic Service, Department of Surgery, Memorial Sloan-Kettering Cancer Center Weill Medical College of Cornell University, New York, NY, USA

Alicia A. Heapy VA Connecticut Healthcare System, Westhaven, CT, USA

Anne M. Heffernan The Adelaide and Meath Hospital Incorporating The National Children's Hospital, Dublin, Ireland

Geert Van der Heijden Julius Center, Department of Epidemiology, University Medical Center Utrecht, Utrecht, The Netherlands

Mary M. Heinricher Departments of Neurological Surgery and Behavioral Neuroscience, Oregon Health & Science University, Portland, USA

Robert D. Helme Department of Medicine, Royal Melbourne Hospital, University of Melbourne, Melbourne, Australia

Kim Helsen Research Group on Health Psychology, KU Leuven Katholieke Universiteit Leuven, Leuven, Belgium

Department of Experimental–Clinical and Health Psychology, Ghent University, Ghent, Belgium

Amin S. Herati The Arthur Smith Institute for Urology, North Shore-LIJ Healthcare System, Hofstra North Shore LIJ School of Medicine, New Hyde Park, NY, USA

Robert D. Herbert Neuroscience Research Australia, Sydney, Australia

Christiane Hermann Department of Clinical Psychology, Justus-Liebig University Giessen, Giessen, Germany

Juan F. Herrero Department of Physiology, Faculty of Medicine University of Alcala, Alcalá de Henares Madrid, Spain

Aki J. Hietaharju Department of Neurology and Rehabilitation, Pain Clinic Tampere University Hospital, Tampere, Finland

Karin Westlund High Department of Physiology, University of Kentucky, Lexington, KY, USA

Marita Hilliges Dalarna University, Halmstad, Sweden

Max J. Hilz University of Erlangen-Nuremberg, Erlangen, Germany

Burkhard Hinz Institute of Toxicology and Pharmacology, University of Rostock, Rostock, Germany

Anthony R. Hobson Section of GI Sciences, Hope Hospital University of Manchester, Salford, UK

Edward Högestätt Department of Laboratory Medicine, Lund University, Division of Clinical Chemistry and Pharmacology, Lund, Sweden

Tomas Hökfelt Department of Neuroscience, Karolinska Institute, Stockholm, Sweden

Sarah V. Holdridge Department of Pharmacology and Toxicology, Queen's University, Kinston, Canada

Kjell Hole University of Bergen, Bergen, Norway

Graham R. Holland Department of Cariology, University of Michigan, Ann Arbor, MI, USA

Chang-Zern Hong Department of Physical Medicine and Rehabilitation, University of California, Irvine, CA, USA

Minoru Hoshiyama Department of Integrative Physiology, National Institute for Physiological Sciences, Okazaki, Japan

Fred M. Howard University of Rochester, Rochester, NY, USA

Sung-Tsang Hsieh Department of Anatomy and Cell Biology, Department of Neurology, National Taiwan University Hospital, National Taiwan University, Taipei, Taiwan

James W. Hu Faculty of Dentistry, University of Toronto, Toronto, Canada

Claire E. Hulsebosch Spinal Cord Injury Research, University of Texas Medical Branch, Galveston, TX, USA

Thomas Hummel Department of Otorhinolaryngology, Technical University of Dresden Medical School, Dresden, Germany

Mark R. Hutchinson School of Medical Sciences, University of Adelaide, Adelaide, SA, Australia

Charles E. Inturrisi Department of Pharmacology, Weill Cornell Medical College, New York, USA

Koji Inui Department of Integrative Physiology, National Institute for Physiological Sciences, Okazaki, Japan

Koichi Iwata Department of Physiology, Nihon University School of Dentistry, Chiyoda-ku, Japan

Suhayl J. Jabbur Anatomy, Cell Biology and Physiology, Faculty of Medicine, American University of Beirut, Beirut, Lebanon

Neuroscience Program, Faculty of Medicine, American University of Beirut, Beirut, Lebanon

Nora Janjan University of Texas M. D. Anderson Cancer Center, Houston, TX, USA

Luc Jasmin Department of Anatomy, University of California San Francisco, San Francisco, CA, USA

Anne Jaumees Royal North Shore Hospital, St Leonards, NSW, Australia

Wilfrid Jaenig Physiologisches Institut, Christian-Albrechts-Universität zu Kiel, Kiel, Germany

Daniel Jeanmonod Functional Neurosurgery, The Center of Ultrasound Functional Neurosurgery, Solothurn, Switzerland

Andrew Jeffreys Director Department of Anaesthesia & Pain Medicine, Western Health, Melbourne, VIC, Australia

Mark P. Jensen Department of Rehabilitation Medicine, University of Washington, Seattle, WA, USA

Troels Staehelin Jensen Danish Pain Research Center, Department of Neurology, Aarhus University Hospital, Aarhus, Denmark

Kurt L. Johnson Department of Rehabilitation Medicine, University of Washington, Seattle, WA, USA

Mark I. Johnson Faculty of Health and Social Sciences, Leeds Pallium Research Group, Leeds Metropolitan University, Leeds, UK

Mark Johnston Hibiscus Coast Highway, Orewa, New Zealand

Edward Jones Center for Neuroscience, University of California, Davies, CA, USA

Jeroen R. De Jong Department of Rehabilitation, University Hospital Maastricht, Maastricht, The Netherlands

Sven-Eric Jordt Department of Pharmacology, Yale University School of Medicine, New Haven, CT, USA

Ellen Jørum The Laboratory of Clinical Neurophysiology, Rikshospitalet University, Oslo, Norway

Stefan Just CNS Pharmacology, Boehringer Ingelheim Pharma GmbH & Co. KG, Biberach, Germany

Timothy K. Y. Kaan Department of Anatomy, University of California San Francisco, San Francisco, CA, USA

Noufissa Kabli Department of Pharmacology and Toxicology, Queen's University, Kinston, Canada

Ryusuke Kakigi Department of Integrative Physiology, National Institute for Physiological Sciences, Okazaki, Japan

Peter C. A. Kam Anaesthesia, Central Clinical School, The University of Sydney, Sydney, Australia

Giresh Kanji The Sports and Pain Clinic, Wellington, New Zealand

Joel Katz Department of Psychology, McGill University, Montreal, QC, Canada

Valerie Kayser NSERM U677 91Bd de l'Hôpital, Paris, France

Francis J. Keefe Pain Prevention and Treatment Research, Department of Psychiatry and Behavioral Sciences, Duke University Medical Center, Durham, NC, USA

Lois J. Kehl Department of Anesthesiology, University of Minnesota, Minneapolis, MN, USA

Maureen Kelley Department of Pediatrics, Bioethics Division, University of Washington School of Medicine, WA, USA

Seattle Children's Hospital, Seattle, WA, USA

Stephen Kennedy Nuffield Department of Obstetrics and Gynaecology, University of Oxford, Oxford, UK

Edmund Keogh Department of Psychology, University of Bath, Bath, UK

Robert D. Kerns VA Connecticut Healthcare System, Westhaven, CT, USA

Sanjay Khanna Department of Physiology, National University of Singapore, Singapore

Kok Eng Khor Department of Pain Management, Prince of Wales Hospital University of New South Wales, Randwick, NSW, Australia

Mina Khoshnejad Department of Neuroscience, University of Montreal, Montreal, QC, Canada

H. J. Kim Department of Life Science, Yonsei University, Wonju, Republic of Korea

Sara King Faculty of Education (School Psychology), Mount Saint Vincent University, Halifax, NS, Canada

Wade King Department of Clinical Research, Royal Newcastle Centre University of Newcastle, Newcastle, NSW, Australia

K. L. Kirsh Behavioral Medicine, The Pain Treatment Center of the Blue grass, Lexington, KY, USA

Henriette Klit Danish Pain Research Center, Aarhus University Hospital, Aarhus C, Denmark

Ricardo J. Komotar Department of Neurological Surgery, Neurological Institute Columbia University, New York, NY, USA

Sungtae Koo Division of Meridian and Structural Medicine, School of Korean Medicine, Pusan National University, Yangsan, Korea

Rhonda K. Kotarinos Oakbrook Terrace, IL, USA

Katalin J. Kovacs Department of Veterinary and Biomedical Sciences, University of Minnesota, St. Paul, MN, USA

Malgorzata Krajnik Chair of Palliative Care, Nicolaus Copernicus University in Torun, Collegium Medicum in Bydgoszcz, Bydgoszcz, Poland

Elliot Krane Anesthesia and Pediatrics, Stanford University School of Medicine Lucile Packard Children's Hospital at Stanford, Stanford, CA, USA

Ajit Krishnaney Department of Neurosurgery, Cleveland Clinic, Cleveland, OH, USA

Greg Krohm International Association of Industrial Accident Boards and Commissions (IAIABC), Utah Valley University, USA

Lawrence Kruger UCLA Geffen School of Medicine, Los Angeles, CA, USA

M. M. ter Kuile Department of Gynecology, Leiden University Medical Center, Leiden, The Netherlands

Promil Kukreja Department of Anesthesiology, University of Alabama at Birmingham, Birmingham, AL, USA

Alice Kvåle Department of Public Health and Primary Health Care, Faculty of Medicine and Dentistry University of Bergen, Bergen, Norway

Gunnvald Kvarstein Department of Pain Management and Pain Research, Oslo University Hospital, Rikshospitalet, Oslo, Norway

Jean-Jacques Labat Neurology and Rehabilitation, Clinique Urologique, Nantes, France

Gaston Labrecque Faculty of Pharmacy, University of Laval, Quebec City, QC, Canada

Marco Lacerenza Pain Medicine Center, Casa di Cura S. Pio X, Opera San Camillo Foundation, Milan, Italy

Uri Ladabaum Gastroenterology & Hepatology, Stanford Cancer Institute, Redwood City, CA, USA

Marie Andree Lahaie McGill University, Montreal, QC, Canada

Miguel J. A. Lainez-Andres Department of Neurology, University of Valencia, Valencia, Spain

Robert H. LaMotte Department of Anesthesiology, Yale School of Medicine, New Haven, CT, USA

James W. Lance Institute of Neurological Sciences, University of New South Wales, Sydney, Australia

Robert G. Large The Auckland Regional Pain Service, Auckland Hospital, Auckland, New Zealand

Alice A. Larson Department of Veterinary and Biomedical Sciences, University of Minnesota, St. Paul, MN, USA

Roberta Lattanzi Department of Physiology and Pharmacology, Sapienza University of Rome, Rome, Italy

Peter Lau Department of Clinical Research, Royal Newcastle Hospital, University of Newcastle, Newcastle, NSW, Australia

Stefan A. Laufer Institute of Pharmacy, Eberhard-Karls University of Tuebingen, Tuebingen, Germany

Stefan Lautenbacher Physiologische Psychologie, Otto-Friedrich-Universität, Bamberg, Germany

Gilles J. Lavigne Facultés de Médecine Dentaire et de Médecine, Université de Montréal, Hopital Sacre Coeur de Montreal. Surgery-trauma dept, Montréal, QC, Canada

Emily Law Center for Child Health, Behavior, and Development, Seattle Children's Research Institute, Seattle, WA, USA

Peter G. Lawlor Our Lady's Hospice, Medical Department, Dublin, Ireland

Michel Lazdunski Institut de Pharmacologie Moléculaire et Cellulaire (IPMC) CNRS/UNS UMR7275, Valbonne-Sophia Antipolis, France

Vincenzo Di Lazzaro Institute of Neurology, Campus Biomedico University, Rome, Italy

Alyssa Lebel Boston Children's Hospital, Boston, MA, USA

Frank Lee Johns Hopkins University, Baltimore, MD, USA

Maaike Leeuw Department of Medical, Clinical and Experimental Psychology, Maastricht University, Maastricht, The Netherlands

Siri Graff Leknes Research Fellow, Psykologisk institutt, Oslo, Norway

Frederick A. Lenz Department of Neurosurgery, Johns Hopkins Hospital, Baltimore, MD, USA

Guillaume Léonard Faculty of Medicine, University of Sherbrooke, Sherbrooke, QC, Canada

Pauline Lesage Department of Pain Medicine and Palliative Care, Beth Israel Medical Center, New York, USA

Isobel Lever Neural Plasticity Unit, Institute of Child Health University College London, London, UK

Khan W. Li Department of Neurosurgery, Johns Hopkins Hospital, Baltimore, MD, USA

Alan R. Light Department of Anesthesiology, University of Utah, Salt Lake City, UT, USA

Michael Lim Department of Neurosurgery and Department of Neurology, Johns Hopkins Hospital, Baltimore, MA, USA

Deolinda Lima Department of Experimental Biology, Faculdade de Medicina da Universidade do Porto, Porto, Portugal

Volker Limmroth Department of Neurology, University of Essen, Essen, Germany

Eric Lingueglia Institut de Pharmacologie Moléculaire et Cellulaire (IPMC) CNRS/UNS UMR7275, Valbonne-Sophia Antipolis, France

Steven J. Linton Department of Occupational and Environmental Medicine, Orebro Medical Center, Orebro, Sweden

Arthur G. Lipman Departments of Pharmscotherpay and Anesthesiology and Pain Management Center, University of Utah, Salt Lake City, UT, USA

Richard B. Lipton Departments of Neurology, Epidemiology and Population Health, Albert Einstein College of Medicine, Bronx, NY, USA

Diana Lisi Department of Psychology, York University, Toronto, ON, Canada

Kenneth M. Little Division of Neurosurgery, Duke University Medical Center, Durham, NC, USA

Edward Littleton Queen Elizabeth Hospital, Birmingham, UK

Spencer S. Liu Virginia Mason Medical Center, Seattle, WA, USA

Anne Elisabeth Ljunggren Section for Physiotherapy Science, Department of Public Health and Primary Health, Faculty of Medicine, University of Bergen, Bergen, Norway

John D. Loeser Department of Neurological Surgery, University of Washington, Seattle, WA, USA

Joern Loetsch Institute for Clinical Pharmacology, Pharmaceutical Center Frankfurt, Johann Wolfgang Goethe University, Frankfurt, Germany

Patrick Loisel Disability Prevention Research and Training Centre, Université de Sherbrooke, Longueuil, Canada

Elisa Lopez-Dolado Servicio de Rehabilitacion, National Hospital of Paraplèjicos SESCAM, Toledo, Spain

Jürgen Lorenz Faculty of Life Science, Competence Center Gesundheit, University of Applied Science Hamburg, Hamburg, Germany

Dayna R. Loyd US Army Institute of Surgical Research, Fort Sam Houston, TX, USA

Daniel Lubelski Cleveland Clinic Lerner College of Medicine, Cleveland, OH, USA

Stephen Lutz Blanchard Valley Regional Cancer Center, Findlay, OH, USA

Bruce Lynn Department of Physiology, University College London, London, UK

Ross MacPherson Department of Anesthesia and Pain Management, Royal North Shore Hospital University of Sydney, Sydney, NSW, Australia

Christophe Maes Manager Innovatie AZ Alma, Belgium

Michel Magnin INSERM, Neurological Hospital, Lyon, France

Bryan Mahony Department of Anesthesiology, Ohio State University Medical Center, Mount Sinai Hospital, New York, NY, USA

Steven F. Maier Department of Psychology and Neuroscience, and The Center for Neuroscience, University of Colorado at Boulder, Boulder, CO, USA

Thorsten J. Maier Institute of Pharmaceutical Chemistry, Goethe University, Frankfurt, Germany

Chris J. Main Keele University, Keele, Staffordshire, UK

Marzia Malcangio Wolfson Centre for Age Related Diseases, King's College London, London, UK

Paolo L. Manfredi Department of Neurology, Memorial Sloan-Kettering Cancer Center, New York, USA

Kaisa Mannerkorpi Department of Rheumatology and Inflammation Research, Institute of Medicine, University of Gothenburg, Göteborg, Sweden

Barton H. Manning Amgen Inc., Thousand Oaks, CA, USA

Patrick W. Mantyh Department of Pharmacology, The University of Arizona, AZ, USA

Jianren Mao Department of Anesthesia and Critical Care, Harvard Medical School, Boston, MA, USA

Serge Marchand Faculty of Medicine, University of Sherbrooke, Sherbrooke, QC, Canada

Paolo Marchettini Pain Medicine Center, Scientific Institute San Raffaele, Milan, Italy

Sarah R. Martin Psychology, Georgia State University, Atlanta, GA, USA

Peggy Mason Department of Neurobiology, Pharmacology and Physiology, University of Chicago, Chicago, IL, USA

Scott Masters Caloundra Spinal and Sports Medicine Centre, Caloundra, QLD, Australia

Laurence E. Mather Department of Anaesthesia and Pain Management, University of Sydney at Royal North Shore Hospital, St. Leonards, Sydney, NSW, Australia

Bill McCarberg Pain Services, Kaiser Permanente, San Diego, USA

Lance McCracken Health Psychology/Psychology Department, King's College London, London, UK

Kathleen McGinn Anesthesiology & Pain Medicine, Seattle Children's Hospital, Seattle, WA, USA

Patricia A. McGrath The Hospital for Sick Children Pain Research Center, Toronto, ON, Canada

Patrick McGrath Departments of Psychology, Pediatrics and Psychiatry, Dalhousie University, Canada

Greg McIntosh Canadian Back Institute, Toronto, ON, Canada

Peter McIntyre Department of Pharmacology, University of Melbourne, Melbourne, Australia

Elspeth M. McLachlan Neuroscience Research Australia, University of New South Wales, Randwick, NSW, Australia

University of Glasgow, Glasgow, UK

Peter A. McNaughton Department of Pharmacology, University of Cambridge, Cambridge, UK

J. Mark Melhorn Section of Orthopaedics, Department of Surgery, University of Kansas School of Medicine, Wichita, KS, USA

Ronald Melzack Department of Psychology, McGill University, Montreal, QC, Canada

Lorne M. Mendell Department of Neurobiology and Behavior, Stony Brook University, Stony Brook, NY, USA

George Mendelson Department of Psychological Medicine, Monash University, Caulfield, VIC, Australia

Siegfried Mense Neuroanatomy, Heidelberg University, Medical Faculty Mannheim, CBTM, Mannheim, Germany

Daniel Menétrey Biomédicale, Paris, France

Sebastiano Mercadante Pain Relief & Palliative Care, La Maddalena Cancer Center University of Palermo, Palermo, Italy

Susan Mercer School of Biomedical Sciences, The University of Queensland, Brisbane, Australia

Karl Messlinger University of Erlangen-Nürnberg, Erlangen, Germany

Richard A. Meyer Department of Neurosurgergy, School of Medicine Johns Hopkins University, Baltimore, MD, USA

Walter J. Meyer III Shriners Hospitals for Children, Shriners Burns Hospital, Galveston, TX, USA

Department of Psychiatry, University of Texas Medical Branch, Galveston, TX, USA

Björn A. Meyerson Department of Clinical Neuroscience, Section of Neurosurgery, Karolinska Institute/University Hospital, Stockholm, Sweden

Martin Michaelis Sanofi-Aventis Deutschland GmbH, Frankurt am Main, Germany

Joseph Mignogna Michael E. DeBakey VA Medical Center, Houston VA HSR&D Center of Excellence (152), Houston, TX, USA

Judith A. Mikacich Department of Obstetrics and Gynecology, University of California Los Angeles Medical Center, Los Angeles, CA, USA

Kensaku Miki Department of Integrative Physiology, National Institute for Physiological Sciences, Okazaki, Japan

Vjekoslav Miletic School of Medicine and Public Health, University of Wisconsin, Madison, WI, USA

Scott R. Miller Memorial Sloan-Kettering Cancer Center, New York, USA

Erin D. Milligan Department of Neurosciences, University of New Mexico, Albuquerque, NM, USA

Michael S. Minett Molecular Nociception Group, Wolfson Institute for Biomedical Research, University College London, London, WC, UK

J. Mocco Department of Neurological Surgery, Neurological Institute Columbia University, New York, NY, USA

Jeffrey S. Mogil Department of Psychology, McGill University, Montreal, QC, Canada

Harvey Moldofsky Sleep Disorders Clinic of the Centre for Sleep and Chronobiology, Toronto, ON, Canada

Robert M. Moldwin The Arthur Smith Institute for Urology, North Shore-LIJ Healthcare System, Hofstra North Shore LIJ School of Medicine, New Hyde Park, NY, USA

Derek C. Molliver Department of Neurobiology, University of Pittsburgh School of Medicine, Pittsburgh, PA, USA

Robert D. Mootz Washington State Department of Labor and Industries, Olympia, WA, USA

Laura Mordillo-Mateos FENNSI Group, Hospital Nacional de Paraplégicos, Toledo, Spain

Anne Morel Laboratory for Functional Neurosurgery, Neurosurgical Clinic University Hospital Zürich, Zürich, Switzerland

Anne Morinville Montreal Neurological Institute, McGill University, Montreal, QC, Canada

Stephen Morley Academic Unit of Psychiatry and Behavioral Sciences, University of Leeds, Leeds, UK

David B. Morris University of Virginia, Charlottesville, VA, USA

G. Lorimer Moseley Sansom Institute for Health Research, University of South Australia Neuroscience Research Australia, Adelaide, Australia

Joachim Mossner Medical Clinic and Policlinic II, University Clinical Center Leipzig AöR, Leipzig, Germany

Eric A. Moulton P.A.I.N. Group, Anesthesiology, Boston Children's Hospital, Harvard Medical School, Waltham, MA, USA

Francis Githae Muriithi Department of Obstetrics and Gynaecology, Aga Khan University, Nairobi, Kenya

Anne Z. Murphy Neuroscience Institute, Georgia State University, Atlanta, GA, USA

Paul M. Murphy Department of Anaesthesia and Pain Management, Royal North Shore Hospital, St. Leonards, NSW, Australia

Alison Murray Foothills Medical Center, Calgary, AB, Canada

H. S. Na Neuroscience Research Institute and Department of Physiology, Korea University, College of Medicine, Seoul, Republic of Korea

Daisuke Naka Department of Integrative Physiology, National Institute for Physiological Sciences, Okazaki, Japan

Yoshio Nakamura Pain Research Center, Department of Anesthesiology University of Utah School of Medicine, Salt Lake City, UT, USA

Barbara Namer Department of Physiology & Pathophysiology, FAU, Erlangen/Nuremberg, Bavaria, Germany

Matti V. O. Närhi Department of Biomedicine/Physiology, Department of Dentistry, University of Eastern Finland, Institute of Medicine, Kuopio, Finland

H. J. W. Nauta University of Texas Medical Branch, Galveston, TX, USA

Lucia Negri Department of Physiology and Pharmacology, Sapienza University of Rome, Rome, Italy

Gary Nesbit Department of Radiology, Oregon Health and Science University, Portland, OR, USA

Timothy Ness Department of Anesthesiology, University of Alabama at Birmingham, Birmingham, AL, USA

Volker Neugebauer Department of Neuroscience & Cell Biology, University of Texas Medical Branch, Galveston, TX, USA

Lawrence C. Newman Albert Einstein College of Medicine, New York, NY, USA

Michael K. Nicholas Royal North Shore Hospital, University of Sydney, St. Leonards, NSW, Australia

Ellen Niederberger Pharmazentrum frankfurt, Institute of Clinical Pharmacology, Goethe-University Hospital, Frankfurt, Germany

Lone Nikolajsen Department of Anaesthesiology, Danish Pain Research Center Aarhus University Hospital, Aarhus, Denmark

Katie Nixdorf Comprehensive Pain Center, Oregon Health and Science University, Portland, OR, USA

Donald R. Nixdorf Department of Diagnostic and Biological Sciences, School of Dentistry University of Minnesota, Minneapolis, MN, USA

Carl E. Noe University of Texas Southwestern Medical Center, Lubbock, TX, USA

Koichi Noguchi Department of Anatomy and Neuroscience, Hyogo College of Medicine, Hyogo, Japan

Richard B. North The Johns Hopkins University, Baltimore, MD, USA

Turo J. Nurmikko Department of Neurological Science, University of Liverpool, and Pain Research Institute, The Walton Centre NHS Foundation Trust, Liverpool, UK

Anne Louise Oaklander Nerve Injury Unit, Departments of Neurology and Pathology, Massachusetts General Hospital Harvard Medical School, Boston, MA, USA

Koichi Obata Department of Anatomy and Neuroscience, Hyogo College of Medicine, Hyogo, Japan

Tim F. Oberlander Division of Developmental Pediatrics, University of British Columbia, Vancouver, BC, Canada

Bruno Oertel Institute for Clinical Pharmacology, Pharmaceutical Center Frankfurt, Johann Wolfgang Goethe University, Frankfurt, Germany

Andrea C. Ogbonna Wolfson Centre for Age Related Diseases, King's College London, London, UK

Alfred T. Ogden Department of Neurological Surgery, Neurological Institute Columbia University, New York, NY, USA

S. Ohara Department of Neurosurgery, Johns Hopkins Hospital, Johns Hopkins Medical Institutions, Johns Hopkins University, Baltimore, MD, USA

Peter T. Ohara Department of Anatomy, University of California San Francisco, San Francisco, CA, USA

Akiko Okifuji Department of Anesthesiology, University of Utah, Salt Lake City, UT, USA

Jean-Louis Oliveras Inserm, Paris, France

Antonio Oliviero FENNSI Group, National Hospital of Parapl´ejicos, Toledo, Spain

Sean O'Mahony Palliative medicine, Rush University Medical Center, Chicago, IL, USA

Peregrine B. Osborne Dept Anatomy & Neuroscience, University of Melbourne, Melbourne, VIC, Australia

Michael H. Ossipov Department of Pharmacology, University of Arizona College of Medicine, Tucson, AZ, USA

Tonya M. Palermo Seattle Children's Hospital Research Institute, CW8-6 - Child Health, Behavior and Development, University of Washington, Seattle, WA, USA

Shreela Palit Department of Psychology, The University of Tulsa, Tulsa, OK, USA

Juan A. Pareja Department of Neurology, Fundación Hospital Alcorcón, Madrid, Spain

Steven D. Passik Psychiatry and Anesthesiology, Vanderbilt University, Nashville, TN, USA

Gavril W. Pasternak Laboratory of Molecular Neuropharmacology, Memorial Sloan-Kettering Cancer Center, New York, USA

S. H. Patel Department of Neurosurgery, Johns Hopkins Hospital, Baltimore, MD, USA

Gavin Pattullo Department of Anesthesia and Pain Management, Royal North Shore Hospital, University of Sydney, St. Leonards, NSW, Australia

Kevin Pauza Tyler Spine and Joint Hospital, Tyler, TX, USA

Eric M. Pearlman Pediatrics, Mercer University School of Medicine, The Children's Hospital at Memorial University Medical Center, Savannah, USA

Elvira De la Pena Instituto de Neurociencias, Universidad Miguel Hernández-CSIC, Alicante, Spain

Yuan Bo Peng Department of Psychology, University of Texas at Arlington, Arlington, TX, USA

Jose Pereira Foothills Medical Center, Calgary, AB, Canada

Edward Perl Department of Cell and Molecular Physiology, School of Medicine University of North Carolina, Chapel Hill, NC, USA

Tiffany Grace Perry Center for Spine Health, The Cleveland Clinic Foundation, Cleveland, OH, USA

Marie Pertin Pain Center, University Hospital Center and University of Lausanne, Lausanne, Switzerland

Antti Pertovaara Department of Physiology, Institute of Biomedicine University of Helsinki, Helsinki, Finland

Peter Peter Department of Neurobiology and Anatomy, University of Rochester Medical Center, Rochester, NY, USA

Lisa M. Peters Department of Anesthesiology and Pain Medicine, Seattle Children's Hospital, Seattle, WA, USA

Pramit Phal Department of Radiology, Oregon Health and Science University, Portland, OR, USA

Rebecca Pillai Riddell Department of Psychology, York University, Toronto, ON, Canada

Department of Psychiatry, Hospital for Sick Children, Toronto, ON, Canada

Issy Pilowsky University of Adelaide, Adelaide, Australia

Leon Plaghki Institute of Neuroscience (IoNS), Université Catholique de Louvain, Brussels, Belgium

Markus Ploner Department of Neurology, Technische Universität München, Munich, Germany

Esther M. Pogatzki-Zahn Department of Anesthesiology and Intensive Care Medicine, University Muenster, Muenster, Germany

Michael Polydefkis Department of Neurology, Johns Hopkins University School of Medicine, Baltimore, MD, USA

James D. Pomonis Algos Preclinical Services, Roseville, MN, USA

Dieter Pongratz Friedrich Baur Institute, Ludwig Maximilians University, Munich, Germany

Hayley Poole Hadley House, Leicester General Hospital, Leicester, UK

Frank Porreca Department of Pharmacology, University of Arizona Health Sciences Center, Tucson, AZ, USA

Russell K. Portenoy Department of Pain Medicine and Palliative Care, Beth Israel Medical Center, New York, NY, USA

Ian Power Critical Care and Pain Medicine, The University of Edinburgh, Edinburgh, UK

Donald D. Price Oral Surgery and Neuroscience, McKnight Brain Institute University of Florida, Gainesville, FL, USA

Daryl Pullman Medical Ethics, Memorial University of Newfoundland, St. John's, NL, Canada

Yunhai Qiu Department of Integrative Physiology, National Institute for Physiological Sciences, Okazaki, Japan

Michael Quittan Department of Physical Medicine and Rehabilitation, Kaiser-Franz-Joseph Hospital, Vienna, Austria

Raymond M. Quock Department of Pharmaceutical Sciences, Washington State University College of Pharmacy, Pullman, WA, USA

Jennifer Rabbitts Department of Anesthesiology and Pain Medicine, Seattle Children's Hospital, University of Washington School of Medicine, Seattle, WA, USA

Nicole Racine Psychology, York University, Toronto, ON, Canada

Gabor B. Racz Texas Tech University Health Sciences Center, Lubbock, TX, USA

Lukas Radbruch Department of Palliative Medicine, University of Bonn, Bonn, Germany

Robert B. Raffa Department of Pharmaceutical Sciences, Temple University School of Pharmacy, Philadelphia, PA, USA

Vasudeva Raghavendra Algos Therapeutics Inc., Saint Paul, MN, USA

Pierre Rainville Department of Stomatology, Faculty of Dental Medicine, University of Montreal, Montreal, QC, Canada

Srinivasa N. Raja Department of Anesthesiology and Critical Care Medicine, Johns Hopkins University, Baltimore, USA

Alan Randich University of Alabama, Birmingham, AL, USA

Andrea J. Rapkin Department of Obstetrics and Gynecology, University of California Los Angeles Medical Center, Los Angeles, CA, USA

Z. Harry Rappaport Department of Neurosurgery, Rabin Medical Center Tel-Aviv University, Petah Tiqva, Israel

Matthew N. Rasband Department of Molecular and Cellular Biology, Baylor College of Medicine, Farmington, Houston TX, USA

University of Connecticut Health Center, Farmington, CT, USA

Douglas Rasmusson Department of Physiology and Biophysics, Dalhousie University, Halifax, Canada

James P. Rathmell Department of Anesthesiology, University of Vermont College of Medicine, Burlington, USA

K. Ravishankar The Headache and Migraine Clinic, Jaslok Hospital and Research Centre Lilavati Hospital and Research Centre, Mumbai, India

Ke Ren Department of Neural and Pain Sciences, University of Maryland Dental School, Baltimore, MD, USA

Seth Resnick Silver Hill Hospital, New York University School of Medicine, USA

Cielito Reyes-Gibby The University of Texas MD Anderson Cancer Center, Houston, TX, USA

Jamie L. Rhudy Department of Psychology, The University of Tulsa, Tulsa, OK, USA

Alfredo Ribeiro-da-Silva Department of Pharmacology and Therapeutics, McGill University, Montreal, QC, Canada

Frank L. Rice Center for Neuropharmacology and Neuroscience, Albany Medical College, Albany, NY, USA

Matthias Ringkamp Department of Neurosurgergy, School of Medicine Johns Hopkins University, Baltimore, MD, USA

Pierre J. M. Rivière Ferring Research Institute, San Diego, CA, USA

Claude Robert Université Paris Descartes, INSERM UMR 894, France

Roger Robert Neurosurgery, Hotel Dieu, Nantes, CHU, France

James P. Robinson Department of Anesthesiology & Pain Medicine, University of Washington, Seattle, WA, USA

John Robinson Pain Management Centre, Burwood Hospital, Christchurch, New Zealand

Alfonso Romero-Sandoval Department of Pharmaceuticals and Administrative Sciences, Presbyterian College School of Pharmacy, Clinton, SC, USA

David Roselt Aberdovy Clinic, Bundaberg, QLD, Australia

Jackie Ross Early Pregnancy Unit, King's College Hospital Early Pregnancy Unit, London, UK

Shlomo Rotshenker Department of Medical Neurobiology, IMRIC, Faculty of Medicine Hebrew University of Jerusalem, Jerusalem, Israel

Paul C. Rousseau Medical University of South Carolina, Charleston, SC, USA

Ranjan Roy Department of Clinical Health Psychology, Faculty of Medicine, University of Manitoba, Winnipeg, MB, Canada

Carolina Roza Biología de Sistemas (Unidad Fisiología), Universidad de Alcalá, Alcalá de Henares, Madrid, Spain

Todd D. Rozen Department of Neurology, Geisinger Health System, Wilkes-Barre, PA, USA

Roman Rukwied Department of Anaesthesiology and Intensive Care Medicine, Faculty of Clinical Medicine Mannheim University of Heidelberg, Heidelberg, Germany

Irwin Jon Russell Division of Clinical Immunology and Rheumatology, Department of Medicine, The University of Texas Health Science Center, San Antonio, TX, USA

Nayef E. Saadé Anatomy, Cell Biology and Physiology, Faculty of Medicine, American University of Beirut, Beirut, Lebanon

Mary Ann C. Sabino Oral and Maxillofacial Surgery, Hennepin County Medical Center University of Minnesota, Minneapolis, MN, USA

Thomas E. Salt Institute of Ophthalmology, University College London, London, UK

A. Samdani Department of Neurosurgery, Johns Hopkins Medical Institutions, Baltimore, MD, USA

Yasmin Sana King's College Hospital, London, UK

Margarita Sanchez-del-Rio Neurology Department, Hospital Ruber Internacional, Madrid, Spain

Jürgen Sandkühler Department of Neurophysiology, Center for Brain Research Medical University of Vienna, Vienna, Austria

Peter S. Sándor RehaClinic, Cantonal Hospital Baden, Baden, Switzerland

Johannes Sarnthein Functional Neurosurgery, Neurosurgical Clinic University Hospital, Zürich, Switzerland

Susanne K. Sauer Department of Physiology and Pathophysiology, University of Erlangen, Erlangen, Germany

Steven Savvas Biomedical Division, National Ageing Research Institute, Melbourne, VIC, Australia

Jana Sawynok Department of Pharmacology, Dalhousie University, Halifax, Canada

John W. Scadding The National Hospital for Neurology and Neurosurgery, London, UK

Frederieke Geraldine Schaafsma Department of Public and Occupational Health, EMGO+ Institute/VU Medical Center, Amsterdam, MB, The Netherlands

Hans-Georg Schaible Department of Physiology - Neurophysiology, Friedrich-Schiller-University of Jena, Jena, Germany

Steven S. Scherer Department of Neurology, The Perelman School of Medicine at the University of Pennsylvania, Philadelphia, PA, USA

Michael Schäfer Department of Anesthesiology and Intensive Care Medicine, Charité University, Berlin, Germany

Martin Schmelz Department of Anesthesiology in Mannheim, University of Heidelberg, Mannheim, Germany

Robert F. Schmidt Physiologisches Institut der Universität Würzburg, Würzburg, Germany

Suzan Schneeweiss The Hospital for Sick Children, Toronto, ON, Canada

Frank Schneider Department of Psychiatry and Psychotherapy, RWTH Aachen University, Aachen, Germany

Alfons Schnitzler Department of Neurology, Heinrich Heine University, Düsseldorf, Germany

Jean Schoenen Department of Neurology, University of Liège, CHR Citadelle, Belgium

Department of Neuroanatomy, University of Liège, CHR Citadelle, Belgium

Jens Schouenborg Neuronano Research Center, Lund University, Lund, Sweden

Stephan A. Schug School of Medicine and Pharmacology, Royal Perth Hospital, Faculty of Medicine, Dentistry & Health Sciences, The University of Western Australia, UWA Anaesthesiology, Perth, Australia

Patrick Schuss Department of Neurosurgery, University-Hospital Bonn, Bonn, Germany

Anthony C. Schwarzer Faculty of Medicine and Health Sciences, The University of Newcastle, Newcastle, Australia

Whitney Scott Department of Psychology, McGill University, Montreal, QC, Canada

Laura C. Seidman Pediatric Pain Program, Department of Pediatrics David Geffen School of Medicine at UCLA, Los Angeles, CA, USA

Ze'ev Seltzer Faculty of Dentistry, University of Toronto Centre for the Study of Pain, Toronto, ON, Canada

Peter Selwyn Albert Einstein College of Medicine, Bronx, NY, USA

Patrick B. Senatus Department of Neurological Surgery, Neurological Institute Columbia University, New York, NY, USA

Jyoti N. Sengupta Division of Gastroenterology and Hepatology and Pediatric Gastroenterology, Medical College of Wisconsin, Milwaukee, WI, USA

Mariano Serrao Department of Neurology and Otorinolaringoiatry, University of Rome La Sapienza, Rome, Italy

Barry J. Sessle Faculty of Dentistry, University of Toronto, Toronto, ON, Canada

Faculty of Dentistry, University of Toronto, Toronto, ON, Canada

Virginia S. Seybold Department of Neuroscience, University of Minnesota, Minneapolis, MN, USA

T. Shah Royal Perth Hospital and University of Western Australia, Crawley, WA, Australia

Yair Sharav Department of Oral Medicine, School of Dental Medicine, Hebrew University-Hadassah, Jerusalem, Israel

S. Murray Sherman Department of Neurobiology, University of Chicago, Chicago, IL, USA

Yoshio Shigenaga Department of Oral Anatomy and Neurobiology, Osaka University, Osaka, Japan

Susan Shin Neurology, Mount Sinai School of Medicine, New York, NY, USA

Edward A. Shipton Department of Anaesthesia, Christchurch School of Medicine and Health Sciences, University of Otago, Christchurch, New Zealand

Yoram Shir The Alan Edwards Pain Management Unit, McGill University Health Centre, Montreal, QC, Canada

Peter Shrager Department of Neurobiology and Anatomy, University of Rochester Medical Center, Rochester, NY, USA

Marc J. Shulman VA Connecticut Healthcare System, Yale University, New Haven, CT, USA

David M. Sibell Comprehensive Pain Center/Spine Center, Oregon Health and Science University, Portland, OR, USA

Philip J. Siddall Department of Pain Management, Greenwich Hospital, University of Sydney, Sydney, NSW, Australia

Stephen D. Silberstein Jefferson Medical College, Thomas Jefferson University and Jefferson Headache Center, Thomas Jefferson University Hospital, Philadelphia, PA, USA

Donald A. Simone Diagnostic and Biological Sciences, University of Minnesota, Minneapolis, MN, USA

David G. Simons Department of Rehabilitation Medicine, Emory University, Atlanta, GA, USA

David Simpson Clinical Neurophysiology Laboratories, Mount Sinai Medical Center, New York, NY, USA

Marc Sindou Department of Neurosurgery, Hopital Neurologique Pierre Wertheimer, University of Lyon 1, Lyon, France

Soeren H. Sindrup Department of Neurology, Odense University Hospital, Odense, Denmark

Bengt H. Sjölund Department of Community Medicine and Rehabilitation Medicine, Umea University, Umea, Sweden

Carsten Skarke Institute for Translational Medicine and Therapeutics (ITMAT), University of Pennsylvania Perelman School of Medicine, Philadelphia, PA, USA

Paul A. Sloan Department of Anesthesiology, University of Kentucky, Lexington, KY, USA

Kathleen A. Sluka Physical Therapy and Rehabilitation Science Graduate Program, University of Iowa, Iowa City, IA, USA

Seymour Solomon Montefiore Medical Center, Albert Einstein College of Medicine, Bronx, NY, USA

Claudia Sommer Department of Neurology, University of Würzburg, Würzburg, Germany

Linda S. Sorkin Occupational Therapy and Rehabilitation, Anesthesia Research Laboratories, University of California, San Diego, CA, USA

Ingrid Söderback Uppsala University, Uppsala, Sweden

Kari Sørensen Department of Pain Management and Research, Oslo University Hospital, Rikshospitalet, Oslo, Norway

Michael Spaeth Friedrich-Baur-Institute, University of Munich, Munich, Germany

Peregrin O. Spielholz Washington State Department of Labor and Industries, SHARP Program, Olympia, WA, USA

Peter S. Staats Division of Pain Medicine, The Johns Hopkins University, Baltimore, MD, USA

Brett R. Stacey Comprehensive Pain Center, Oregon Health and Science University, Portland, OR, USA

Wendy M. Stein San Diego Hospice and Palliative Care, UCSD, San Diego, CA, USA

Christoph Stein Department of Anesthesiology and Intensive Care Medicine, Freie Universitaet Berlin Charité Campus Benjamin Franklin, Berlin, Germany

Michael P. Steinmetz Department of Neurosurgery, MetroHealth Medical Center, Case Western Reserve University School of Medicine, Cleveland, OH, USA

Jair Stern Functional Neurosurgery, Neurosurgical Clinic University Hospital, Zürich, Switzerland

Bonnie J. Stevens University of Toronto and The Hospital for Sick Children, Toronto, ON, Canada

Carl-Olav Stiller Division of Clinical Pharmacology and Department of Medicine, Karolinska University Hospital, Stockholm, Sweden

Jennifer N. Stinson The Hospital for Sick Children, Toronto, ON, Canada

Henry Stockbridge University of Washington School of Public Health and Community Medicine, Seattle, WA, USA

Christian S. Stohler Baltimore College of Dental Surgery, University of Maryland, Baltimore, MD, USA

Laura Stone Faculty of Dentistry, Alan Edwards Centre for Research on Pain, Montreal, QC, Canada

William Stones Department of Obstetrics and Gynaecology, Aga Khan University, Nairobi, Kenya

Andreas Straube Department of Neurology, Ludwig Maximilians University, Munich, Germany

Jenny Strong University of Queensland, Brisbane, QLD, Australia

Audun Stubhaug Department of Pain Management and Research, Oslo University Hospital, Rikshospitalet and University of Oslo, Oslo, Norway

Gerold Stucki Swiss Paraplegic Research, Nottwil, Switzerland

Department of Health Sciences and Health Policy, University of Lucerne, Lucerne, Switzerland

ICF Research Branch in cooperation with the World Health Organization (WHO) Collaborating Centre for the Family of International Classifications in Germany, Nottwil, Switzerland

Cheryl L. Stucky Department of Cell Biology Neurobiology and Anatomy, Medical College of Wisconsin, Milwaukee, WI, USA

Mathias Sturzenegger Department of Neurology, University Hospital, Inselspital, University of Bern, Bern, Switzerland

Yasuo Sugiura School of Human Care Study, Nagoya University of Arts and Sciences, Nissin City, Aichi, Japan

Mark D. Sullivan Psychiatry and Behavioral Sciences, University of Washington, Seattle, WA, USA

Michael J. L. Sullivan Department of Psychology, McGill University, Montreal, QC, Canada

Rie Suzuki Department of Pharmacology, University College London, London, UK

Kristina B. Svendsen Danish Pain Research Center, Aarhus University Hospital, Aarhus, Denmark

Peter Svensson Clinical Oral Physiology, Aarhus University, Aarhus, Denmark

Peter Szucs Spinal Neuronal Networks, Instituto de Biologia Molecular e Celular - IBMC, Porto, Portugal

A. Taghva Department of Neurosurgery, Johns Hopkins Hospital, Baltimore, MD, USA

Kimberson Cochien Tanco Department of Palliative Care and Rehabilitation Medicine, The University of Texas M. D. Anderson Cancer Center, Houston, TX, USA

Kersi Taraporewalla Royal Brisbane and Womens' Hospital, University of Queensland, Brisbane, QLD, Australia

Irmgard Tegeder Pharmazentrum frankfurt, Institute of Clinical Pharmacology, Goethe-University Hospital, Frankfurt, Germany

Astrid Juhl Terkelsen Danish Pain Research Center, Department of Neurology, Aarhus University Hospital, Aarhus, Denmark

Gregory W. Terman Department of Anesthesiology and the Graduate Program in Neurobiology and Behavior, University of Washington, Seattle, WA, USA

Alison Thomas Lismore, NSW, Australia

Benjamin Thomas Chelsea & Westminster Hospital Foundation Trust, London, UK

Jacob Thomas School of Medical Sciences, University of Adelaide, Adelaide, SA, Australia

Miles Thompson Goldsmiths, University of London, London, UK

Beverly E. Thorn Department of Psychology, University of Alabama, Tuscaloosa, AL, USA

Cindy L. Thurston-Stanfield Department of Biomedical Sciences, University of South Alabama, Mobile, AL, USA

Arne Tjølsen Department of Bomedicine, University of Bergen, Haukeland University Hospital, Bergen, Norway

Andrew J. Todd Institute of Neuroscience and Psychology, College of Medical, Veterinary and Life Sciences, University of Glasgow, Glasgow, UK

Makoto Tominaga Department of Cellular and Molecular Physiology, Mie University School of Medicine, Tsu Mie, Japan

Victor Tortorici Instituto Venezolano de Investigaciones Cientificas (IVIC), Caracas, Venezuela

Irene Tracey Pain Imaging Neuroscience (PaIN) Group, Department of Physiology, Anatomy and Genetics, Oxford University, Oxford, UK

Diep Tuan Tran Department of Integrative Physiology, National Institute for Physiological Sciences, Okazaki, Japan

Rolf-Detlef Treede Institute of Physiology and Pathophysiology, Johannes Gutenberg University, Mainz, Germany

Jennie C. I. Tsao Pediatric Pain Program, Department of Pediatrics David Geffen School of Medicine at UCLA, Los Angeles, CA, USA

Lawrence Tsen Department of Anesthesiology, Perioperative and Pain Medicine, Brigham and Women's Hospital Harvard Medical School, Boston, MA, USA

Jeanna Tsenter Department of Physical Medicine and Rehabilitation, Hadassah University Hospital, Jerusalem, Israel

Maurits Van Tulder Institute for Research in Extramural Medicine, Free University Amsterdam, Amsterdam, The Netherlands

Eldon Tunks Hamilton Health Sciences Corporation, Hamilton, ON, Canada

Susan Tupper Pediatrics, University of Saskatchewan, Saskatoon, SK, Canada

Dennis C. Turk Department of Anesthesiology & Pain Research, University of Washington, Seattle, WA, USA

Judith A. Turner Department of Psychiatry and Behavioral Sciences, University of Washington, Seattle, WA, USA

Department of Rehabilitation Medicine, University of Washington, Seattle, WA, USA

Stephen P. Tyrer Royal Victoria Infirmary, University of Newcastle upon Tyne, Newcastle-on-Tyne, UK

Alexander Tzabazis Department of Anesthesia, Stanford University, Stanford, CA, USA

Kene Ugokwe Department of Neurosurgery, St Elizabeth Health Center, Youngstown, OH, USA

Lindsay S. Uman Department of Psychology, IWK Health Centre, Halifax, NS, Canada

Christiane Vahle-Hinz Department of Neurophysiology and Pathophysiology, University Medical Center Hamburg-Eppendorf, Hamburg, Germany

Muthukumar Vaidyaraman Division of Pain Medicine, The Johns Hopkins University, Baltimore, MD, USA

Todd W. Vanderah Department of Pharmacology, College of Medicine University of Arizona, Tucson, AZ, USA

Guy Vanderstraeten Department of Rehabilitation Sciences and Physiotherapy, Ghent University Hospital, Ghent, Belgium

Horacio Vanegas Instituto Venezolano de Investigaciones Cientificas (IVIC), Caracas, Venezuela

Jean-Jacques Vatine Outpatient and Research Division, Reuth Medical Center, Sackler Faculty of Medicine, Tel Aviv University, Tel Aviv, Israel

Leigh M. Vaughan Medical University of South Carolina, Charleston, SC, USA

Linda K. Vaughn Department of Biomedical Sciences, Marquette University, Milwaukee, WI, USA

Enrique Vazquez Instituto Venezolano de Investigaciones Cientificas (IVIC), Caracas, Venezuela

Louis Vera-Portocarrero Department of Pharmacology, College of Medicine University of Arizona, Tucson, AZ, USA

Paul Verrills Metropolitan Spinal Clinic, Prahran, VIC, Australia

Tine Vervoort Department of Experimental Clinical and Health Psychology, Ghent University, Ghent, Belgium

Félix Viana Instituto de Neurociencias, Universidad Miguel Hernández-CSIC, San Juan de Alicante, Alicante, Spain

Charles J. Vierck Jr. Department of Neuroscience and McKnight Brain Institute, University of Florida College of Medicine, Gainesville, FL, USA

Luis Villanueva INSERM UMR 894, Centre de Psychiatrie et Neurosciences, Paris, France

Marcelo Villar Faculty of Biomedical Sciences, Austral University, Buenos Aires, Argentina

Katy Vincent Nuffield Department of Obstetrics and Gynaecology, The Women's Centre, John Radcliffe Hospital, University of Oxford, Oxford, UK

David Vivian Metro Spinal Clinic, South Caulfield, VIC, Australia

Johan W. S. Vlaeyen Research Group Health Psychology, Catholic University of Leuven, Leuven, Belgium

Department of Clinical Psychological Science, Maastricht University, Maastricht, The Netherlands

Stéphanie Volders Research Group on Health Psychology, Leuven, Belgium

Ernest Volinn Pain Research Center, University of Utah, Salt Lake City, UT, USA

Lucy Vulchanova Department of Veterinary and Biomedical Sciences, University of Minnesota, St. Paul, MN, USA

Paul W. Wacnik Neuromodulation Research, Medtronic Inc., Minneapolis, MN, USA

Gordon Waddell University of Glasgow, Glasgow, UK

Gary A. Walco Department of Anesthesiology and Pain Medicine, University of Washington School of Medicine, Seattle Children's Hospital, Seattle, WA, USA

Suellen M. Walker University of Sydney Pain Management and Research Centre, Royal North Shore Hospital, NSW, Australia

Victoria C. J. Wallace Department of Anaesthetics, Pain Medicine and Intensive Care Chelsea and Westminster Hospital Campus, London, UK

Xiaohong Wang Department of Integrative Physiology, National Institute for Physiological Sciences, Okazaki, Japan

Gunnar Wasner Department of Neurological Pain Research and Therapy, Neurological Clinic, Kiel, Germany

Shoko Watanabe Department of Integrative Physiology, National Institute for Physiological Sciences, Okazaki, Japan

Linda R. Watkins Department of Psychology and Neuroscience, and The Center for Neuroscience, University of Colorado at Boulder, Boulder, CO, USA

Peter N. Watson Department of Medicine, University of Toronto, Toronto, ON, Canada

James Watt North Shore Hospital, Auckland, New Zealand

Stephen G. Waxman Department of Neurology, Yale School of Medicine, New Haven, CT, USA

Inge Wegner Julius Center, Department of Epidemiology, University Medical Center Utrecht, Utrecht, The Netherlands

Feng Wei Department of Neural and Pain Sciences, University of Maryland Dental School, Baltimore, MD, USA

Christian Weidner Department of Physiology and Experimental Pathophysiology, University Erlangen, Erlangen, Germany

P. T. M. Weijenborg Department of Gynecology, Leiden University Medical Center, Leiden, The Netherlands

Robin Weir School of Nursing, Faculty of Health Sciences, McMaster University, Hamilton, ON, Canada

Nirit Weiss Department of Neurosurgery, Mount Sinai Medical Center, New York, NY, USA

Ursula Wesselmann Departments of Neurology and Anesthesiology / Division of Pain Management, University of Alabama at Birmingham, Birmingham, AL, USA

Dagmar Westerling Department of Anesthesiology, Central Hospital of Kristianstad, Kristianstad, Sweden

Karin N. Westlund Department of Physiology, University of Kentucky, Lexington, KY, USA

Thomas M. Wickizer Health Services Management and Policy, Ohio State University, Colombus, OH, USA

Eva Widerström-Noga The Miami Project to Cure Paralysis, Department of Neurological Surgery, University of Miami Miller, School of Medicine, Miami, FL, USA

Julie Wieseler Department of Psychology and Neuroscience, and The Center for Neuroscience, University of Colorado at Boulder, Boulder, CO, USA

Zsuzsanna Wiesenfeld-Hallin Karolinska Institute, Stockholm, Sweden

Ann Wigham McCoull Clinic, Prudhoe Hospital, Northumberland, UK

George L. Wilcox Department of Neuroscience, Pharmacology and Dermatology, University of Minnesota Medical School, Minneapolis, MN, USA

Oliver H. G. Wilder-Smith Pain Centre, Department of Anaesthesiology, Radboud University Nijmegen Medical Centre, Nijmegen, The Netherlands

Kjersti Wilhelmsen Section for Physiotherapy Science, Department of Public Health and Primary Health, Faculty of Medicine, University of Bergen, Bergen, Norway

Victor Wilk Brighton Spinal Group, Brighton, VIC, Australia

W. A. Williams Royal Perth Hospital and University of Western Australia, Perth, WA, Australia

Sara E. Williams Division of Pediatric Gastroenterology, Medical College of Wisconsin, Milwaukee, WI, USA

William D. Willis Department of Neuroscience and Cell Biology, University of Texas Medical Branch, Galveston, TX, USA

John B. Winer Department of Neurology, Birmingham Muscle and Nerve Centre Queen Elizabeth Hospital, Birmingham, UK

Christopher J. Winfree Department of Neurological Surgery, Neurological Institute Columbia University, New York, NY, USA

Janet Winter Novartis Institute for Medical Science, London, UK

Alain Woda Faculté de Chirurgie Dentaire, University Clermont-Ferrand, Clermont-Ferrand, France

John N. Wood Molecular Nociception Group, Wolfson Institute for Biomedical Research, University College London, London, WC, UK

Lee Woodson Shriners Hospitals for Children, Shriners Burns Hospital, Galveston, TX, USA

Department of Anesthesia, University of Texas Medical Branch, Galveston, TX, USA

Anava A. Wren Pain Prevention and Treatment Research, Department of Psychiatry and Behavioral Sciences, Duke University Medical Center, Durham, NC, USA

Xiao-Jun Xu Department of Physiology and Pharmacology, Section of Integrative Pain Rsearch Karolinska Institutet, Stockholm, Sweden

Hiroshi Yamasaki Department of Integrative Physiology, National Institute for Physiological Sciences, Okazaki, Japan

Michael Yelland School of Medicine, Logan campus, Griffith University, Logan, Australia

Sriram Yennurajalingam Department of Palliative Care and Rehabilitation Medicine, The University of Texas M. D. Anderson Cancer Center, Houston, TX, USA

David C. Yeomans Department of Anesthesia, Stanford University, Stanford, CA, USA

Robert P. Yezierski Comprehensive Center for Pain Research and Department of Neuroscience, The McKnight Brain Institute University of Florida, Gainesville, FL, USA

Way Yin Department of Anesthesiology, University of Washington, Seattle, WA, USA

Atsushi Yoshida Department of Oral Anatomy and Neurobiology, Osaka University, Osaka, Japan

Joanna M. Zakrzewska Eastman Dental Hospital UCLH NHS Foundation Trust, London, UK

Jenna E. Zauk St. George's University, School of Medicine, Grenada, West Indies

Hanns Ulrich Zeilhofer Institute for Pharmacology and Toxicology, University of Zurich, and Swiss Federal Institute of Technology (ETH) Zurich, Zürich, Switzerland

Lonnie K. Zeltzer Pediatric Pain Program, Department of Pediatrics David Geffen School of Medicine at UCLA, Los Angeles, CA, USA

Giovambattista Zeppetella St. Clare Hospice, Hastingwood, UK

Xijing Zhang Department of Neuroscience, University of Minnesota, Minneapolis, MN, USA

Min Zhuo Department of Physiology, University of Toronto, Toronto, ON, Canada

C. Zöllner Department of Anesthesiology and Intensive Care Medicine, Charité University, Berlin, Germany

Krina Zondervan Genetic and Genomic Epidemiology Unit, Nuffield Dept of Obstetrics and Gynaecology, Wellcome Trust Centre for Human Genetics, Oxford, UK

Peter Zygmunt Department of Laboratory Medicine, Lund University, Division of Clinical Chemistry and Pharmacology, Lund, Sweden

Zbigniew Zylicz Hildegard Hospiz, Basel Switzerland, Basel, Switzerland

David C. Yeomans Department of Anesthesia, Stanford University, Stanford, CA, USA

Robert P. Yezierski Comprehensive Center for Pain Research and Department of Neuroscience, The McKnight Brain Institute, University of Florida, Gainesville, FL, USA

Wei Yin Department of Anesthesiology, University of Washington, Seattle, WA, USA

Atsushi Yoshida Department of Oral Anatomy and Neurobiology, Osaka University, Osaka, Japan

Joanna M. Zakrzewska Eastman Dental Hospital UCLH NHS Foundation Trust, London, UK

Jania K. Zaak St. George's University School of Medicine, Grenada, West Indies

Hanns Ulrich Zeilhofer Institute for Pharmacology and Toxicology, University of Zurich, and Swiss Federal Institute of Technology (ETH), Zurich, Zurich, Switzerland

Lonnie K. Zeltzer Pediatric Pain Program, Department of Pediatrics, David Geffen School of Medicine at UCLA, Los Angeles, CA, USA

Giovambattista Zeppetella St. Clare Hospice, Hastingwood, UK

Xijing Zhang Department of Neuroscience, University of Minnesota, Minneapolis, MN, USA

Min Zhuo Department of Physiology, University of Toronto, Toronto, ON, Canada

C. Zöllner Department of Anaesthesiology and Intensive Care Medicine, Charité University, Berlin, Germany

Krina Zondervan Genetic and Genomic Epidemiology Unit, Nuffield Department of Obstetrics and Gynaecology, Wellcome Trust Centre for Human Genetics, Oxford, UK

Peter Zygmunt Department of Laboratory Medicine, Lund University, Division of Clinical Chemistry and Pharmacology, Lund, Sweden

Zbigniew Zylicz Hildegard Hospice, Basel Switzerland, Basel, Switzerland

CIDP

▶ Chronic Inflammatory Demyelinating Polyneuropathy

Cingulate Cortex

Definition

Cortical tissue also referred to as Limbic Cortex or Cingulate Gyrus that encircles the hippocampus and other structures of the limbic region. It is sometimes interrupted as treatment for obsessive compulsive disorder or pain.

Cross-References

▶ Cingulate Cortex, Nociceptive Processing, Behavioral Studies in Animals
▶ Pain Treatment, Intracranial Ablative Procedures

Cingulate Cortex, Functional Imaging

Pierre Rainville
Department of Stomatology, Faculty of Dental Medicine, University of Montreal, Montreal, QC, Canada

Definition

The cingulate cortex is a part of the medial frontal and parietal ▶ cerebral cortices situated immediately above the ▶ corpus callosum and curving around its most anterior part. The human cingulate cortex can be subdivided into several subregions as shown in Fig. 1. Histologically, the anterior cingulate cortex (ACC) comprises ▶ Brodmann area 25 (BA25) below the rostrum of the corpus callosum and BA24, BA32, and BA33 above the body and curving around the

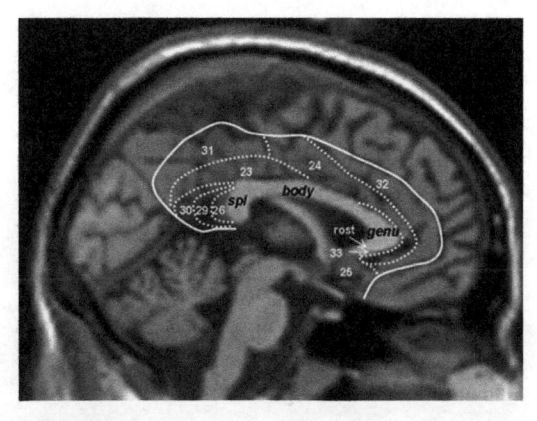

Cingulate Cortex, Functional Imaging, Fig. 1 Anatomy of the human cingulate cortex. The cingulate cortex is delimited ventrally by the corpus callosum. The *full line* corresponds to its approximate anterior, posterior, and dorsal borders. The subregions of the cingulate cortex are delimited by *dotted lines* according to the corresponding Brodmann areas identified by numbers. Regions of the cingulate cortex can also be divided based on their location relative to the body of the corpus callosum (dorsal = supracallosal; ventral = subcallosal) and to the anterior (rost = rostrum and genu) or posterior (spl = splenium) parts of the corpus callosum. The subgenual area corresponds to BA25, the perigenual area corresponds to BA33 and to the anterior part of BA24 and BA32, and the retrosplenial area corresponds to BA26, BA29, and BA30

genu of the corpus callosum. The posterior region of the cingulate gyrus corresponds to BA23, BA31, BA30, BA29, and BA26. Traditionally, the cingulate cortex is considered part of the ▶ limbic system, which is associated with emotions. Furthermore, functional brain imaging studies have shown that subregions of the cingulate cortex, in particular within the anterior cingulate cortex (ACC), are activated in response to painful stimuli and therefore may be involved in pain-related functions. Seeking to explain these findings, several theoretical models have placed the cingulate cortex at the interface of the sensory, affective, cognitive, skeletomotor, and visceromotor systems (e.g., Vogt 2005).

Characteristics

Nociceptive Activation of the Cingulate Cortex

Almost all functional brain-imaging studies of pain using ▶ positron emission tomography (PET) or

► functional magnetic resonance imaging (fMRI) have found pain-related activation of the ACC (Apkarian et al. 2005). In these studies, experimental pain is produced using noxious thermal, mechanical, electrical, or chemical stimuli applied to the skin or viscera of normal human volunteers. Peaks of activation are found most reliably in the caudal part of the supracallosal ACC and occasionally in the perigenual area. This activation is consistent with the projection of nociceptive neurons from the medial and intralaminar nuclei of the thalamus to BA24 (Vogt 2005).

Functional Role of the ACC in Pain and Pain Modulation

Pain is a complex experience described along sensory-discriminative and ► affective-motivational dimensions influenced by cognitive processes and associated with somatomotor and visceromotor responses. The ACC may contribute to each of these aspects. Levels of activation in the supracallosal ACC have been correlated with the subjective reports of pain (Coghill et al. 2003). More specifically, activity in certain regions of the ACC may reflect the ► affective-motivational dimension of pain, as the modulation of pain unpleasantness, independently of pain intensity, changes ACC activity (Rainville 2002). Furthermore, the affective-motivational dimension of pain is strongly influenced by cognitive-evaluative processes shown to be highly dependent upon the function of the ACC (Price 2000).

The supracallosal ACC is part of higher-order systems involved in the voluntary regulation of behavioral and motor responses. Therefore, pain-related activation in this area probably reflects motivational aspects of pain behaviors and conscious processes involved in pain-related responses. Similarly, the perigenual region of the ACC is involved in physiological arousal, a fundamental aspect of emotion and a critical component of pain-related affective responses. Changes in activity within the perigenual area have been associated with physiological arousal during emotional experiences, the performance of motor and cognitive tasks, and the anticipation of pain (see Rainville 2002). However, the

precise nature of the relationship between pain-related activity in the ACC and physiological arousal remains to be demonstrated. Finally, posterior regions of the cingulate cortex may be involved in the representation of viscerosomatic changes, but the functional role of this area in pain is unclear.

The ACC is also critically involved in cognitive processes, as demonstrated by innumerable brain imaging studies showing activation in this area during the performance of cognitive tasks (see Rainville 2002). However, this activation is typically observed in a more dorsal region, within BA32, whereas pain-related activation is more ventral, in BA24. Nonetheless, regions of the ACC involved in cognitive processes may be involved in the modulation of pain. Studies investigating the modulation of pain by hypnosis (Rainville et al. 1999), placebo analgesia (Petrovic et al. 2002; Wager et al. 2004), and distraction (Valet et al. 2004) have in fact shown some activation in the dorsal anterior region of the ACC (typically in BA32). This activation is associated with a decrease in the response to painful stimuli in other pain-related areas such as the somatosensory cortices and is consistent with a reduction in the pain reported by subjects. In some of these studies, activation of the ACC has also been found to correlate with activity in the brain stem, consistent with the activation of the periaqueductal grey area (PAG), a key structure of descending pain inhibitory mechanisms.

A role for the ACC in pain modulation is also supported by studies investigating the opioid system. The μ-opioid receptor is found in many brain areas including the ACC. In a PET study, the systemic administration of the μ-opioid agonist fentanyl increased activity in the ACC in a control condition without pain, and reduced pain evoked activity within the thalamus and cortical areas, including the ACC (Casey et al. 2000). This decrease in pain-related ACC activity was associated with a decrease in reported pain intensity and pain unpleasantness, as well as with pain-related heart rate responses. Endogenous activation of the μ-opioid system in the ACC and other areas has also been associated with decreased affective ratings of pain

(Zubieta et al. 2001) and suggested as an underlying mechanism of placebo analgesia (Petrovic et al. 2002).

Beyond Nociception: The ACC and the Mental Representation of Pain

In addition to its roles in the experience of pain and in the modulation of pain, the ACC may also be activated in the context of pain, in the absence of a noxious stimulus. This has been suggested by studies showing that ACC activation may occur during the anticipation of pain, prior to the presentation of a painful stimulus (Rainville 2002). However, pain-related activation of the ACC may also be observed when the subject is merely thinking about, but not expecting or anticipating, an actual painful stimulus to the self. For example, ACC activity has been reported in response to visual stimuli depicting a hand or foot receiving a painful stimulus (Morrison et al. 2004; Jackson et al. 2005), videos showing a patient displaying pain behaviors (Botvinick et al. 2005), or cues indicating that a loved one is receiving a painful stimulus (Singer et al. 2004). ACC activation during both pain experiences and in response to cues signaling impending pain in the self or pain in others may reflect a general involvement of this structure in the mental representation of pain.

A Multifunctional and Integrative Role for the ACC in Pain

From this brief overview of brain imaging studies, it is clear that the role of the ACC in pain is multifaceted. The ACC receives ascending nociceptive input and is in an excellent position to integrate this information with signals related to the motivational, emotional, and underlying viscerosomatic state of the organism. The ACC is also well positioned to modify cognitive and behavioral priorities based on the biological significance of pain signals. In addition, the ACC may contribute to higher-order processes involved in pain modulation. Finally, the ACC may play a critical role in the mental representation of pain in self and in others. This may provide a fundamental neurobiological basis for social interactions and pain empathy.

References

Apkarian, A. V., Bushnell, M. C., Treede, R. D., & Zubieta, J. K. (2005). Human brain mechanisms of pain perception and regulation in health and disease. *European Journal of Pain, 9*(4), 463–484.

Botvinick, M., Jha, A. P., Bylsma, L. M., et al. (2005). Viewing facial expressions of pain engages cortical areas involved in the direct experience of pain. *NeuroImage, 25*, 312–319.

Casey, K. L., Svensson, P., Morrow, T. J., et al. (2000). Selective opiate modulation of nociceptive processing in the human brain. *Journal of Neurophysiology, 84*, 525–533.

Coghill, R. C., McHaffie, J. G., & Yen, Y. F. (2003). Neural correlates of interindividual differences in the subjective experience of pain. *Proceedings of the National Academy of Sciences of the United States of America, 100*, 8538–8542.

Jackson, P. L., Meltzoff, A. N., & Decety, J. (2005). How do we perceive the pain of others? A window into the neural processes involved in empathy. *NeuroImage, 24*, 771–779.

Morrison, I., Lloyd, D., di Pellegrino, G., et al. (2004). Vicarious responses to pain in anterior cingulate cortex: Is empathy a multisensory issue? *Cognitive, Affective, & Behavioral Neuroscience, 4*, 270–278.

Petrovic, P., Kalso, E., Petersson, K. M., et al. (2002). Placebo and opioid analgesia – Imaging a shared neuronal network. *Science, 295*, 1737–1740.

Price, D. D. (2000). Psychological and neural mechanisms of the affective dimension of pain. *Science, 288*, 1769–1772.

Rainville, P. (2002). Brain mechanisms of pain affect and pain modulation. *Current Opinion in Neurobiology, 12*, 195–204.

Rainville, P., Hofbauer, R. K., Paus, T., et al. (1999). Cerebral mechanisms of hypnotic induction and suggestion. *Journal of Cognitive Neuroscience, 11*, 110–125.

Singer, T., Seymour, B., O'Doherty, J., et al. (2004). Empathy for pain involves the affective but not sensory components of pain. *Science, 303*, 1157–1162.

Valet, M., Sprenger, T., Boecker, H., et al. (2004). Distraction modulates connectivity of the cingulo-frontal cortex and the midbrain during pain – An fMRI analysis. *Pain, 109*, 399–408.

Vogt, B. A. (2005). Pain and emotion interactions in subregions of the cingulate gyrus. *Nature Reviews Neuroscience, 6*, 533–544.

Wager, T. D., Rilling, J. K., Smith, E. E., et al. (2004). Placebo-induced changes in FMRI in the anticipation and experience of pain. *Science, 303*, 1162–1167.

Zubieta, J. K., Smith, Y. R., Bueller, J. A., et al. (2001). Regional mu opioid receptor regulation of sensory and affective dimensions of pain. *Science, 293*, 311–315.

Cingulate Cortex, Nociceptive Processing, Behavioral Studies in Animals

Perry N. Fuchs and Verne C. Cox
Department of Psychology, University of Texas
at Arlington, Arlington, TX, USA

Synonyms

Anterior cingulate cortex; Behavioral studies in
animals; Nociceptive processing in the cingulate
cortex; Pain affect

Definition

Nociceptive processing in the ▶ cingulate cortex is
believed to reflect the affective component of pain
and is probably a critical structure in the neural
network that contributes to the suffering that
accompanies persistent pain states. Behavioral
studies in animals have utilized paradigms that
measure escape and/or avoidance of a noxious
stimulus. It is assumed that escape/avoidance
behavior is evidence that animals find the noxious
stimulus aversive. With the use of these paradigms,
the precise nature of the neuroanatomical and neurochemical mechanisms underlying the processing
of higher-order pain processing in rodents is beginning to be understood.

Characteristics

The experience of pain consists of affective and
sensory components. It is the affective component that underlies the suffering that accompanies many persistent pain states. An
understanding of the neural substrates mediating
the affective processing of nociceptive stimulation is essential for the development of successful pain therapies. The ▶ anterior cingulate
cortex (ACC) is one brain region that has been
implicated in the affective component of pain
with relatively little involvement in sensory

processing (Vogt 1985). For instance, ACC neuronal activity increases in direct response to and/
or anticipation of noxious, but not nonnoxious
chemical, mechanical, and thermal stimuli
(Hutchison et al. 1999; Koyama et al. 2001).
Surgical cingulotomy and cingulectomy, or
transection of the cingulum bundle and cingulate cortex, respectively, decrease the affective
response to noxious stimuli, but do not alter
ability to localize the unpleasant stimulus.
Brain imaging studies consistently report
increased ACC neuronal activity preceding and
during the presentation of an acute noxious stimulus or during persistent pain conditions. Hypnotic suggestions to selectively decrease pain
affect prior to and during noxious stimulation
resulted in decreased ratings of pain unpleasantness, but not pain intensity (Rainville et al.
1997). Manipulating pain unpleasantness by
hypnotic suggestion also changed the regional
cerebral blood flow (rCBF) in the ACC, but not
in the somatosensory cortex, providing evidence
that the ACC is involved in the processing of
pain-related affect, but not in the sensory
processing of noxious stimulation (Rainville
et al. 1997).

Most of the studies that have examined
supraspinal focal brain stimulation or
▶ microinjection of drugs into discrete nuclei
have measured reflexive behavioral response to
acute noxious stimulation. Indeed, activation of
various subcortical, brainstem, and spinal cord
systems produces antinociception, as revealed
by an increase in the threshold or latency to
respond to noxious stimulation. In most studies,
it is assumed that manipulations of limbic system
structures alter pain processing through selective
modulation of pain affect (Donahue et al. 2001).
However, the majority of past and current
behavioral paradigms cannot provide definitive
information about the aversive and unpleasant
qualities of a persistent pain condition. Thus,
it has been difficult to examine the affective/
motivational and sensory/discriminative components of pain processing in animal models.
In addition, any study that attempts to examine
higher-order processing of noxious input in
animals must address the exact nature of

pain affect. For instance, the affective (i.e., worrisome, cruel, fearful, terrifying, etc.) nature of chronic pain in humans is dissociable from the sensory (i.e., shooting, stabbing, pinching, cramping, etc.) nature of the condition by the descriptors that patients select on the ▶ McGill Pain Questionnaire. In animal studies, the precise nature of pain affect is more difficult to define. Affect, as it relates to noxious input, is most certainly a negative hedonic state that can be identified in one way as aversion to a noxious stimulus.

With this in mind, a number of investigators have developed new behavioral methodologies to study the complexity of nociceptive processing in the ACC of animals. The underlying feature of these methodologies is that escape and avoidance of a noxious stimulus are clear indications that animals find the stimulus aversive (Fuchs 2000). For instance, Johansen et al. (2001) utilized the ▶ formalin test with the place-conditioning paradigm. In this paradigm, formalin injection elicits an acute nociceptive response and induces conditioned place avoidance (F-CPA) to the compartment of the apparatus that is associated with the formalin injection. Another recent paradigm utilizes place escape/avoidance in which animals must associate the application of the mechanical stimulus to the hyperalgesic paw with the preferred dark area of the test chamber (LaBuda and Fuchs 2000). Escape/avoidance behavior is measured as a shift from the preferred dark area of the chamber to increased time spent within the non-preferred light area of the chamber. In both instances, damage to the ACC, by either neurotoxic or electrolytic procedures, decreases behavioral evidence that the noxious stimulus is aversive (Johansen et al. 2001; LaGraize et al. 2004). At this time, the most parsimonious explanation of the behavioral studies in rodents is that the ACC is involved in the modulation and processing of pain affect and that manipulations of the ACC decrease the affective component of pain processing.

Behavioral paradigms that permit the separate assessment of affective and sensory components of the pain experience have led to examination of the underlying neurochemical processes that might be involved in modulating pain affect. The avoidance behavior as measured by the place escape/avoidance paradigm is positively related to c-fos activation in the region of the ACC (Uhelski et al. 2012). One possible mechanism of action by which the ACC selectively modulates pain affect is via the mu-opioid receptor system. The ACC has a high density of opioid receptors, and activation of the ACC mu-opioid receptor system during sustained pain is negatively correlated with McGill Pain Questionnaire affective scores (Zubieta et al. 2001). In animals, morphine microinjection into the ACC produces a selective decrease in escape/avoidance to mechanical stimulation of the hyperalgesic paw (LaGraize et al. 2006). Additional contributions of the NMDA (King et al. 2012; Lei et al. 2004; Quintero et al. 2011), cholecystokinin (Erel et al. 2004), glutamatergic (Johansen and Fields 2004; Yan et al. 2012), estrogen (Xiao et al. 2012), and astrocyte (Chen et al. 2012; Lu et al. 2011) systems have been reported.

It is unlikely that the above findings can be attributed to the role of the ACC in learning and memory processes, that is, the possibility that ACC manipulation interferes with acquisition and retention of an escape/avoidance response rather than a change in the negative hedonic value of the mechanical stimulus. Morphine has been found to impair performance in various tests of memory such as the Morris water maze and the ▶ radial arm maze. However, the effect of morphine on the radial arm maze requires chronic high-dose administration (up to 40 mg/kg) that almost certainly is associated with sedation and impairment in task performance rather than with interference in memory. Other investigators report biphasic results in rats such that lower doses of morphine enhance, while higher doses impair, memory. Anatomically, impairment of learning and memory function is typically associated with manipulations to the more posterior regions of the cingulate cortex.

It is inappropriate to examine the role of the ACC in nociceptive processing in animals with the use of reflexive behaviors to examine the aversive nature of persistent pain conditions. Experiments using only quantified mechanical thresholds or acute formalin injection behaviors can lead to the erroneous conclusion that the ACC does not effect supraspinal pain processing. Clearly, functional alterations of the ACC by lesion and neurochemical methods reduce the avoidance of noxious mechanical hind paw stimulation as measured using the place escape/avoidance paradigm and the formalin conditioned place avoidance paradigm. Sensory mechanisms of pain processing are clearly important but fail to highlight the mechanisms underlying the affect that accompanies many persistent pain conditions. Clinically, the sensation of pain can be treated by reducing the sensory input as well as by manipulating affective-motivational and cognitive factors (Melzack and Casey 1968). Therefore, an understanding of the neural substrates mediating the affective processing of nociceptive stimulation should advance our knowledge of pain processing and contribute to advances in therapeutic interventions to reduce the affective component of pain that accounts for the suffering so frequently seen in clinical conditions.

References

Chen, F.-L., Dong, Y.-L., Zhang, Z.-J., et al. (2012). Activation of astrocytes in the anterior cingulate cortex contributes to the affective component of pain in an inflammatory pain model. *Brain Research Bulletin, 87,* 60–66.

Donahue, R. R., LaGraize, S. C., & Fuchs, P. N. (2001). Electrolytic lesion of the anterior cingulate cortex decreases inflammatory, but not neuropathic pain in rats. *Brain Research, 897,* 131–138.

Erel, U., Arborelius, L., & Brodin, E. (2004). Increased cholecystokinin release in the rat anterior cingulate cortex during carrageenan-induced arthritis. *Brain Research, 1022,* 39–46.

Fuchs, P. N. (2000). Beyond reflexive measures to examine higher order pain processing in rats. *Pain Research & Management, 5,* 215–219.

Hutchison, W. D., Davis, K. D., Lozano, A. M., et al. (1999). Pain-related neurons in human cingulate cortex. *Nature Neuroscience, 2,* 403–405.

Johansen, J. P., & Fields, H. L. (2004). Glutamatergic activation of anterior cingulate cortex produces an aversive teaching signal. *Nature Neuroscience, 7,* 398–403.

Johansen, J. P., Fields, H. L., & Manning, B. H. (2001). The affective component of pain in rodents: Direct

evidence for a contribution of the anterior cingulate cortex. *Proceedings of the National Academy of Sciences of the United States of America, 98,* 8077–8808.

King, T., Qu, C., Okun, A., et al. (2012). Contribution of PKMζ-dependent and independent amplification to components of experimental neuropathic pain. *Pain, 153,* 1263–1273.

Koyama, T., Kato, K., Tanaka, Y. Z., et al. (2001). Anterior cingulate activity during pain-avoidance and reward tasks in monkeys. *Neuroscience Research, 39,* 421–430.

LaBuda, C. J., & Fuchs, P. N. (2000). A behavioral test paradigm to measure the aversive quality of inflammatory and neuropathic pain in rats. *Experimental Neurology, 163,* 490–494.

LaGraize, S. C., LaBuda, C. J., Rutledge, M. A., et al. (2004). Differential effect of anterior cingulate cortex lesion on mechanical hypersensitivity and escape/avoidance behavior in an animal model of neuropathic pain. *Experimental Neurology, 188,* 139–148.

LaGraize, S. C., Borzan, J., Peng, Y. B., et al. (2006). Selective regulation of pain affect following activation of the opioid anterior cingulate cortex system. *Experimental Neurology, 197,* 22–30.

Lei, L.-G., Sun, S., Gao, Y.-J., et al. (2004). NMDA receptors in the anterior cingulate cortex mediate pain-related aversion. *Experimental Neurology, 189,* 413–421.

Lu, Y., Zhu, L., & Gao, Y.-J. (2011). Pain-related aversion induces astrocytic reaction and proinflammatory cytokine expression in the anterior cingulate cortex in rats. *Brain Research Bulletin, 84,* 178–182.

Melzack, R., & Casey, K. L. (1968). Sensory, motivational, and central control determinants of pain. In D. R. Kenshalo (Ed.), *The skin senses* (pp. 423–439). Springfield: CC Thomas.

Quintero, G. C., Herrera, J., & Bethancourt, J. (2011). Cortical NR2B NMDA subunit antagonism reduces inflammatory pain in male and female rats. *Journal of Pain Research, 4,* 301–308.

Rainville, P., Duncan, G. H., Price, D. D., et al. (1997). Pain affect encoded in human anterior cingulate but not somatosensory cortex. *Science, 277,* 968–971.

Uhelski, M. L., Morris-Bobzean, S. A., Dennis, T. S., et al. (2012). Evaluating underlying neuronal activity associated with escape/avoidance behavior in response to noxious stimulation in adult rats. *Brain Research, 1433,* 56–61.

Vogt, B. A. (1985). Cingulate cortex. In A. Peters & E. G. Johns (Eds.), *Cerebral cortex* (pp. 80–149). New York: Plenum Publishing Corporation.

Xiao, X., Yang, Y., Zhang, Y., et al. (2012). Estrogen in the anterior cingulate cortex contributes to pain-related aversion. *Cerebral Cortex.* doi: 10.1093/cercor/bhs201.

Yan, N., Cao, B., Xu, J., et al. (2012). Glutamatergic activation of anterior cingulate cortex mediates the affective component of visceral pain memory in rats. *Neurobiology of Learning and Memory, 97,* 156–164.

Zubieta, J.-K., Smith, Y. R., Bueller, J. A., et al. (2001). Regional mu opioid receptor regulation of sensory and affective dimensions of pain. *Science, 293,* 311–315.

Cingulotomy

Definition

The creation of lesions in the cingulate gyrus, usually for the relief of intractable pain and in the treatment of psychiatric disorders and addiction.

Cross-References

▶ Pain Processing in the Cingulate Cortex, Behavioral Studies in Humans
▶ Pain Treatment, Intracranial Ablative Procedures

CIPA

▶ Congenital Insensitivity to Pain with Anhidrosis

Circadian Rhythms/Variations

Definition

Rhythms of parameters with a frequency close to, but not exactly, 24-h.

Cross-References

▶ Diurnal Variations of Pain in Humans

Circadian Variations in Pain Level

▶ Diurnal Variations of Pain in Humans

Circumventricular Brain Structures

Definition

Specific regions within the brain where the blood-brain barrier is weak or absent.

Cross-References

▶ Proinflammatory Cytokines

Cl⁻ Transporter

▶ Chloride Transporter

CL100

▶ Phosphatases and Glia

Classic or Classical Migraine

▶ Clinical Migraine with Aura

Classical Conditioning

▶ Pavlovian Conditioning
▶ Respondent Conditioning of Chronic Pain

Classical Conditioning and Chronic Pain Syndromes

Synonyms

Pavlovian Conditioning; Respondent Conditioning

Definition

Also called "Pavlovian conditioning" or "respondent conditioning," it is a type of learning that results from the association of stimuli with reflex responses, whereby a neutral stimulus (usually called the conditioned stimulus, CS) acquires the response eliciting properties of a previously potent stimulus (the unconditioned stimulus, US). Through repeated pairings of the CS (e.g., tone) with the US (e.g., shock), the CS can be used to elicit the same or a similar response, the conditioned response (CR, e.g., withdrawal) as the US.

Cross-References

▶ Amygdala, Functional Imaging
▶ Amygdala, Pain Processing and Behavior in Animals
▶ Behavioral Therapies to Reduce Disability
▶ Muscle Pain, Fear-Avoidance Model
▶ Pavlovian Conditioning
▶ Psychology of Pain and Psychological Treatment
▶ Respondent Conditioning of Chronic Pain

Classical Massage

▶ Massage, Basic Considerations

Classification

▶ Functioning and Disability Definitions
▶ Taxonomy

Claudication

Definition

Claudication is characterized by leg pain and weakness brought on by walking, with the disappearance of the symptoms following a brief rest.

Cross-References

▶ Vascular Neuropathies

Clearance

Definition

The clearance describes the hypothetical volume of fluid that is cleared from the unchanged drug per unit of time by any elimination pathway. The clearance can be further categorized as the total clearance (CL) (all elimination pathways) or the organic clearance (e.g., renal clearance (CL_R), hepatic clearance (CL_H), etc.).

Cross-References

▶ NSAIDs, Pharmacokinetics

Clinical Decision Making

▶ Empathy and Pain

Clinical Migraine with Aura

Peter Goadsby
Headache Center, Department of Neurology
University of California, San Francisco,
San Francisco, CA, USA

Synonyms

Previously used terms: Classic or classical migraine; Migraine accompagnée; Ophthalmic migraine

Definition

The International Headache Society in the second edition of the International Classification of Headache Disorders (ICHD-II; www.ihs-classification.org; Headache Classification Subcommittee of the International Headache, S. 2004) defines migraine with aura as "Recurrent disorder manifesting in attacks of reversible focal neurological symptoms that usually develop gradually over 5–20 minutes and last for less than 60 minutes. Headache with the features of migraine without aura usually follows the aura symptoms. Less commonly, headache lacks migrainous features or is completely absent."

Introduction

Migraine is a primary headache disorder affecting about 18 % of women and 6 % of men in USA and Western Europe (Lipton et al. 2007; Stovner et al. 2007). Migraine is essentially a central nervous system disturbance of sensory perception with disabling headache, characterized by moderate to severe head pain, usually accompanied by nausea, photophobia, phonophobia which may be preceded by focal neurologic symptoms which are called aura. About 30 % of migraine patients experience aura in relation with their headaches. The aura consists of fully reversible visual, sensory or language, more rarely motor or brainstem symptoms, which usually last from 5 to 60 min and precede the appearance of a migraine-like headache. Sometimes a typical aura can be followed by a headache lacking of migrainous features or may be even experienced without any ensuing headache. Both the visual and sensory symptoms (the most common ones) often have a bimodal progression with positive features (shimmering lights, zig-zag lines, prickling paresthesias) followed by negative ones (blind spot or numbness). Usually the duration of each aura symptoms range between 5 and 60 min, yet some patients can easily experience an aura lasting for more than 1 h, especially if non visual symptoms are present, when it is called prolonged aura if up to 7 days in duration, and persistent aura when lasting longer than 7 days.

One of the main features of the aura, which differentiates it from a transient ischemic attack or an epileptic seizure, is the slow progression of symptoms. This character was such important to understand the basis of the migraine aura pathophysiology, suggesting to Living in 1873 that those symptoms were due to a "nerve storm" (Liveing 1873). Lashley in 1941 charted his own visual auras and concluded that the symptomatology reflected a cortical process progressing with a speed of 3 mm/min across the primary visual cortex (Lashley 1941). In 1944 the Brazilian neurophysiologist Leão described in rabbit brain a wave of inhibition of electrical activity, which spread across the cortical mantle at a speed of 2–3 mm per minute calling it cortical spreading depression (CSD) and noticed its similarity with the migraine aura (Leao 1944). For some time a hypothesis of ischemia followed by hyperemia was considered the basis for migraine, the so-called *Vascular Theory* of Wolff. In 1974 the advent of a technique of measurement of regional cerebral blood flow (rCBF) directly in humans made it possible to detect spreading oligemia, a mild reduction of the rCBF not to an ischemic threshold, during migraine aura. Lauritzen et al. detected during a migraine aura a wave of oligemia starting from the occipital area, progressing onward at 2–3 mm per minute not respecting arterial territories (Lauritzen et al. 1983). The data argued contradicted the ischemic hypothesis and suggested that aura is primarily a neuronal event accompanied by vascular changes. At present all the pathophysiological mechanisms underlying migraine aura are not still understood but it seems clear that CSD, a primarily neuronal event, is in some important parts responsible for aura.

Characteristics

A typical migraine aura can be characterized by visual, sensory and language disturbances. Visual symptoms are very common in migraine aura. In about 30–40 % of cases they can be followed by sensory symptoms and in the 10–20 % by language symptoms. In less than 1 % of patients aura is characterized also by motor symptoms, loss of strength in at least one limb, and/or brainstem (previously called basilar-type) symptoms. These latter conditions are respectively called hemiplegic migraine and migraine with brainstem aura (previously called basilar-type migraine).

Visual Symptoms
Visual symptoms can include positive features (e.g., flickering lights, spots or lines) and/or negative features (i.e., loss of vision). They are typically homonymous, which means involving the

same half of the visual field. They often present as a fortification spectrum: a zigzag figure near the point of fixation that may gradually spread right or left and assume a laterally convex shape with an angulated scintillating edge, leaving variable degrees of absolute or relative scotoma in its wake. These figures are also called *teichopsia*, term deriving from Greek *teichos*, meaning "wall", as they appear as a fortification. In other cases a blind spot without positive phenomena (scotoma) or unformed flashes of lights (photopsia) may occur. Less frequently more complex visual phenomena can be experienced during a visual aura. They consist of a distortion of the vision or illusions and are described as "kaleidoscopic vision", "illusory splitting" (object or persons appearing to be split, along fracture lines of different form and orientations, into two or more parts that may be displaced and separated from each other), "cinematic visions" (loss of smooth movements of observed scenes), metamorphopsia (altered perception of shape), micropsia/macropsia (objects are perceived to be smaller/bigger than they actually are) or "Alice in wonderland syndrome" (perceptual distortion of one's body size and shape).

Sensory Symptoms

Typically they consist of paresthesias (pins and needles) moving at a slow march from the point of origin and affecting a greater or smaller part of one side of the body and face. Numbness may occur in its wake, but it may also be the only symptom. Sensory symptoms classically involve one side of the body. The hand (96 % of cases) and face (67 %) are the body parts most often affected, whereas the leg (24 %) and trunk (18 %) are less commonly involved (Russell and Olesen 1996). The typical hand-mouth distribution is also called cheiro-oral. Frequently the disturbances have atypical march, starting in the thumb and gradually spreading to the whole hand and the perioral region.

Language Symptoms

Paraphasic errors (substitution of one word or sound for the intended word or sound) and other types of impaired language productions are the most common speech disturbances during a migraine aura (occurring relatively in 76 % and 72 % of cases), less frequently impaired comprehension also occur (38 %) (Russell and Olesen 1996). It can sometimes be difficult to dissect language problems due to concentration impairment and attentional issues, which are common in migraine, from primary language dysfunction.

Headache Phase

As described above a typical migraine aura can be followed by a migraine headache, a non-migraine headache, or even occur alone. All these subtypes are coded in the ICHD-II and are respectively named: typical aura with migraine headache (code 1.2.1), typical aura with non-migraine headache (code 1.2.2), typical aura without headache (code 1.2.3) (Headache Classification Subcommittee of the International Headache, S. 2004). The ICHD-II diagnostic criteria of these three different subtypes are reported in the boxes below. Patients experiencing a migraine aura can frequently experience these three conditions during their life. Yet, when aura symptoms occur in the absence of a migraine headache and in a patient who has never suffered from migraine, it is important to exclude other conditions that may mimic a migraine aura, especially when the symptoms are very short, or very long duration, they are primarily negative, or they begin after the age of 40.

> **Box 1. ICHD-II Diagnostic Criteria for Typical Aura with Migraine Headache: Code 1.2.1**
> A. At least 2 attacks fulfilling criteria B–D
> B. Aura consisting of at least one of the following, but no motor weakness:
>   1. fully reversible visual symptoms including positive features (e.g., flickering lights, spots or lines) and/or negative features (i.e., loss of vision)
>   2. fully reversible sensory symptoms including positive features (i.e., pins and needles) and/or negative features (i.e., numbness)
>   3. fully reversible dysphasic speech disturbance

C. At least two of the following:
1. homonymous visual symptoms (additional loss or blurring of central vision may occur) and/or unilateral sensory symptoms
2. at least one aura symptom develops gradually over ≥5 min and/or different aura symptoms occur in succession over ≥5 min
3. each symptom lasts ≥5 and ≤60 min
D. Headache fulfilling criteria B-D for 1.1 Migraine without aura begins during the aura or follows aura within 60 min

---

Box 2. ICHD-II Diagnostic Criteria for Typical Aura with Non-migraine Headache: Code 1.2.2

A. At least 2 attacks fulfilling criteria B-D
B. Aura consisting of at least one of the following, but no motor weakness:
1. fully reversible visual symptoms including positive features (e.g., flickering lights, spots or lines) and/or negative features (i.e., loss of vision)
2. fully reversible sensory symptoms including positive features (i.e., pins and needles) and/or negative features (i.e., numbness)
3. fully reversible dysphasic speech disturbance
C. At least two of the following:
1. homonymous visual symptoms (additional loss or blurring of central vision may occur) and/or unilateral sensory symptoms
2. at least one aura symptom develops gradually over ≥5 min and/or different aura symptoms occur in succession over ≥5 min
3. each symptom lasts ≥5 and ≤60 min
D. Headache that does not fulfil criteria B-D for 1.1 Migraine without aura begins during the aura or follows aura within 60 min

---

Box 3. ICHD-II Diagnostic Criteria for Typical Aura Without Headache: Code 1.2.3

A. At least 2 attacks fulfilling criteria B-D
B. Aura consisting of at least one of the following, but no motor weakness:
1. fully reversible visual symptoms including positive features (e.g., flickering lights, spots or lines) and/or negative features (i.e., loss of vision)
2. fully reversible sensory symptoms including positive features (i.e., pins and needles) and/or negative features (i.e., numbness)
3. fully reversible dysphasic speech disturbance
C. At least two of the following:
1. homonymous visual symptoms (additional loss or blurring of central vision may occur) and/or unilateral sensory symptoms
2. at least one aura symptom develops gradually over ≥5 min and/or different aura symptoms occur in succession over ≥5 min
3. each symptom lasts ≥5 and ≤60 min
D. Headache does not occur during aura nor follow aura within 60 min

Other Subtype of Migraine Aura

Hemiplegic Migraine (HM)

Migraine with aura including motor weakness. If at least one first- or second-degree relative has migraine aura including motor weakness than the patient can be diagnosed with a familiar hemiplegic migraine (FHM), otherwise the diagnosis is of sporadic hemiplegic migraine (SHM). Boxes 4 and 5 have ICHD-II criteria for FHM and SHM. The prevalence of the sporadic form is at least 0.002 % (Thomsen et al. 2003) and that the prevalence of the familial form is at least 0.003 % (Haan et al. 1994). Specific genetic mutations for three forms of FHM have been identified: in FHM1 mutations are found in the CACNA1A gene on chromosome 9, in FHM2, mutations occur in the ATP1A1 gene on chromosome 1; and in FHM3 mutations are present in the SCN1A gene on chromosome 2. CACNA1A gene encodes for pore-forming α1 subunit of neuronal CaV2.1 (P/Q type) voltage-gated calcium channels, ATP1A1 for catalytic α2 subunit of a glial and neuronal sodium-potassium pump, and SCN1A for pore-forming α1 subunit of neuronal NaV1.1 voltage-gated sodium channels (Russell and Ducros 2011). Attacks are similar in SHM and FHM, although these episodes have a notable variability among patients, which is partly explained by the genetic heterogeneity alongside probable modifying environmental and genetic factors (Ducros et al. 2001). Hemiplegic migraine usually involved several different aura symptoms, whereas typical migraine aura frequently involves only visual symptoms. The most common other aura symptoms in HM are sensory disturbances (98 %, that usually occur on the same areas affected by the motor deficit), visual symptoms (89 %) and speech disturbances (79 %) (Thomsen et al. 2002). Brainstem, or Basilar-type, aura symptoms (see below) are also present in 69 % of patients with FHM and in 72 % of patients with SHM (Kirchmann et al. 2006). The aura usually starts with progressive sensory or visual symptoms. The degree of motor deficit is highly variable, ranging from mild clumsiness to total hemiplegia. Weakness may affect either side of the body, sometimes alternately in successive episodes. When one-sided it involves at least the upper limb (Ducros and Thomsen 2006). HM usually lasts much longer than a typical aura. Even the same visual and sensory aura symptoms were considerably prolonged in HM compared with migraine with typical aura: mean duration of about two hours for visual symptoms and four hours for sensory symptoms. Duration range of motor symptoms is very wide, with mean values of 5 h 36 min for SHM and 7h05min for FHM. Motor weakness usually clears up over a period of 24 h, however, weakness may lasts 2–3 days in up to 20 % of patients with FHM1 and FHM2 (Russell and Ducros 2011). Headache onset generally follows the aura. The attack characteristics are usually those of a typical attack of migraine without aura, i.e., a severe pulsating unilateral headache with nausea, vomiting, phonophobia, and photophobia. As for typical migraine aura, some patients with HM have mild headaches associated with the aura. Less than 1 % of patients with hemiplegic migraine never had accompanying headache. Aura-like symptoms with motor weakness not followed by a headache require diagnostic caution. Taking the history is critical because a sensory loss of a typical non-motor aura can be easily mistaken as a slight weakness by the patient and the symptoms have most often already subsided at the time of the clinical examination.

Box 4. ICHD-II Diagnostic Criteria for Familiar Hemiplegic Migraine (FHM): Code 1.2.4

A. At least 2 attacks fulfilling criteria B and C

B. Aura consisting of fully reversible motor weakness and at least one of the following:
1. fully reversible visual symptoms including positive features (e.g., flickering lights, spots or lines) and/or negative features (i.e., loss of vision)
2. fully reversible sensory symptoms including positive features (i.e., pins and needles) and/or negative features (i.e., numbness)
3. fully reversible dysphasic speech disturbance

C. At least two of the following:
1. at least one aura symptom develops gradually over ≥5 min and/or different aura symptoms occur in succession over ≥5 min
2. each aura symptom lasts ≥5 min and <24 h
3. headache fulfilling criteria B-D for 1.1 Migraine without aura begins during the aura or follows onset of aura within 60 min

D. At least one first- or second-degree relative has had attacks fulfilling these criteria A-E

Box 5. ICHD-II Diagnostic Criteria for Sporadic Hemiplegic Migraine (SHM): Code 1.2.5

A. At least 2 attacks fulfilling criteria B and C

B. Aura consisting of fully reversible motor weakness and at least one of the following:
   1. fully reversible visual symptoms including positive features (e.g., flickering lights, spots or lines) and/or negative features (i.e., loss of vision)
   2. fully reversible sensory symptoms including positive features (i.e., pins and needles) and/or negative features (i.e., numbness)
   3. fully reversible dysphasic speech disturbance

C. At least two of the following:
   1. at least one aura symptom develops gradually over ≥5 min and/or different aura symptoms occur in succession over ≥5 min
   2. each aura symptom lasts ≥5 min and <24 h
   3. headache fulfilling criteria B-D for 1.1 Migraine without aura begins during the aura or follows onset of aura within 60 min

D. No first- or second-degree relative has attacks fulfilling these criteria A-E

Brainstem (Basilar-Type) Migraine (BM)

Migraine with aura symptoms clearly originating from the brainstem and/or from both hemispheres simultaneously affected, but no motor weakness.

The ICHD-II criteria define that migraine aura should include at least two of the following symptoms: dysarthria, vertigo, tinnitus, hypacusia, diplopia, visual symptoms simultaneously in both temporal and nasal fields of both eyes, ataxia, decreased level of consciousness, simultaneously bilateral paraesthesias (see Box 6 for diagnostic criteria). In addition, most patients have typical visual, sensory, or aphasic aura during attacks of BM like in migraine with typical aura.

Originally Bickerstaff proposed the term basilar artery migraine or basilar migraine for patients in whom the aura symptoms seemed indicate a dysfunction of the basilar artery territory (Bickerstaff 1961). Yet, since involvement of the basilar artery territory is uncertain, the term basilar-type migraine (BTM) was subsequently preferred. Moreover, some authors are skeptical about the fact that BTM is an independent disease entity different from migraine with typical aura (MTA). A recent study conducted by Kirchmann and colleagues on patients with migraine with aura found that BTM occurred in 10 % (38/362) of patients with MTA. The authors concluded that BTM seemingly may occur at times in any patient with MTA and that there is no firm clinical, epidemiologic, or genetic evidence that basilar-type migraine is an independent disease entity different from migraine with typical aura (Kirchmann et al. 2006).

Box 6. ICHD-II Diagnostic Criteria for Basilar-Type Migraine: Code 1.2.6

A. At least 2 attacks fulfilling criteria B-D

B. Aura consisting of at least two of the following fully reversible symptoms, but no motor weakness:
   1. dysarthria
   2. vertigo
   3. tinnitus
   4. hypacusia
   5. diplopia
   6. visual symptoms simultaneously in both temporal and nasal fields of both eyes
   7. ataxia
   8. decreased level of consciousness
   9. simultaneously bilateral paraesthesias

C. At least one of the following:
   1. at least one aura symptom develops gradually over ≥5 min and/or different aura symptoms occur in succession over ≥5 min
   2. each aura symptom lasts ≥5 and ≤60 min

D. Headache fulfilling criteria B-D for 1.1 Migraine without aura begins during the aura or follows aura within 60 min

Retinal Migraine

Repeated attacks of monocular visual disturbance, including scintillations, scotoma or blindness, associated with migraine headache (ICHD-II diagnostic criteria in Box 7). The prevalence of retinal migraine has been estimated as 1 in 200 persons with migraine (Troost and Zagami 2000). Epidemiologic studies of retinal migraine are unavailable, and extreme caution is required when exclusively diagnosing retinal migraine. Compression of the optic nerve and transient ischemic attacks arising from the ipsilateral carotid artery must be excluded, particularly when monocular visual loss is not followed by headache (Lance and Goadsby 2005).

Box 7. ICHD-II Diagnostic Criteria for Retinal Migraine: Code 1.4

A. At least 2 attacks fulfilling criteria B and C
B. Fully reversible monocular positive and/or negative visual phenomena (e.g., scintillations, scotomata or blindness) confirmed by examination during an attack or (after proper instruction) by the patient's drawing of a monocular field defect during an attack
C. Headache fulfilling criteria B-D for 1.1 Migraine without aura begins during the visual symptoms or follows them within 60 min
D. Normal ophthalmological examination between attacks

Therapy

With respect to the symptomatic treatment, while for the headache phase common acute therapy for migraine (or non-migraine headache) can be used, at present there is no reliable therapeutic agent to abort the aura. There have been isolated case reports where drugs appear to have aborted migraine aura (Goldner and Levitt 1987; Rothrock 1997; Kowacs et al. 1999; Rozen 2000; Rozen 2003); none of these studies were controlled. A small open label study of ketamine suggested that ketamine may reduce the severity and duration of aura in humans

(Kaube et al. 2000). Ketamine indeed has been shown to block CSD in animals (Rashidy-Pour et al. 1995). A randomized parallel group active comparator controlled study seems to demonstrate for the first time a reduction in clinical severity of prolonged migraine aura with ketamine (Afridi et al. 2012). This study offers a pharmacological parallel between animal experimental work on cortical spreading depression and the clinical problem (Andreou and Goadsby 2009).

No studies have addressed the problem of when to start a preventive treatment in patients who experience migraine with aura. Considering that patients who experience migraine with aura frequently have also migraine without aura, a general evaluation of migraine impact on the patient quality of life should be made by the physician before prescribing a preventive treatment (see ▶ Migraine Without Aura). Currently, no evidence exists about a specific preventive medication for migraine with aura. Indeed international and national guidelines make no distinction between the preventive treatment of migraine with and without aura (Evers et al. 2009). For pharmacological and non-pharmacological preventive treatment please see migraine without aura.

Summary

About 30 % of patient with migraine suffer by episodes of migraine with aura. Migraine with aura is characterized by transient focal neurological symptoms, usually visual, sensory or dysphasic, more rarely motor or brainstem, called "aura" that are usually followed by a migraine-like headache. The symptoms of the aura phase develop usually progressively over minutes and generally last for 5–60 min, although in some circumstances can be longer. The presumed substrate of migraine with aura is cortical spreading depression (CSD), a primarily neuronal phenomenon that causes a progressive inhibition of cortical function, starting from the occipital area and spreading onward at a speed of 2–3 mm per minute. The susceptibility to generate a CSD seems to be an intrinsic property of the migrainous brain. Patients can be prescribed with effective preventive treatments used for migraine without aura as well as common symptomatic treatments of the headache phase. On the other hand at present there is no reliable therapeutic agent to abort the aura.

References

Afridi, S., Giffin, N.J., Kaube, H., & Goadsby, P.J. (2012) A randomized controlled trial of intranasal ketamine in migraine with prolonged aura. *Neurology, 78* (in press).

Andreou, A. P., & Goadsby, P. J. (2009). Therapeutic potential of novel glutamate receptor antagonists in migraine. *Expert Review on Investigational Drugs, 18*, 789–804.

Bickerstaff, E. R. (1961). Basilar artery migraine. *Lancet, 1*, 15–17.

Ducros, A., & Thomsen, L. L. (2006). Sporadic and familial hemiplegic migraine. In J. Olesen, P. J. Goadsby, N. Ramadan, P. C. Tfelt-Hansen, & K. M. A. Welch (Eds.), *The headaches* (pp. 577–587). Philadelphia: Lippincott Williams & Wilkins.

Ducros, A., Denier, C., et al. (2001). The clinical spectrum of familial hemiplegic migraine associated with mutations in a neuronal calcium channel. *The New England Journal of Medicine, 345*(1), 17–24.

Evers, S., Afra, J., et al. (2009). EFNS guideline on the drug treatment of migraine–revised report of an EFNS task force. *European Journal of Neurology, 16*(9), 968–981.

Goldner, J. A., & Levitt, L. P. (1987). Treatment of complicated migraine with sublingual nifedipine. *Headache, 27*(9), 484–486.

Haan, J., Terwindt, G. M., et al. (1994). Familial hemiplegic migraine in The Netherlands. Dutch migraine genetics research group. *Clinical Neurology and Neurosurgery, 96*(3), 244–249.

Headache Classification Subcommittee of the International Headache, S. (2004). The international classification of headache disorders: 2nd edition. *Cephalalgia, 24*(Suppl 1), 9–160.

Kaube, H., Herzog, J., et al. (2000). Aura in some patients with familial hemiplegic migraine can be stopped by intranasal ketamine. *Neurology, 55*(1), 139–141.

Kirchmann, M., Thomsen, L. L., et al. (2006). Basilar-type migraine: Clinical, epidemiologic, and genetic features. *Neurology, 66*(6), 880–886.

Kowacs, P. A., Piovesan, E. J., et al. (1999). Prolonged migraine aura without headache arrested by sumatriptan. A case report with further considerations. *Cephalalgia, 19*(4), 241–242.

Lance, J. W., & Goadsby, P. J. (2005). *Mechanism and management of headache*. Philadelphia: Elsevier/Butterworth/Heinemann.

Lashley, K. S. (1941). Patterns of cerebral integration indicated by the scotomas of migraine. *Archives of Neurology and Psychiatry, 46*, 331–339.

Lauritzen, M., Skyhoj Olsen, T., et al. (1983). Changes in regional cerebral blood flow during the course of classic migraine attacks. *Annals of Neurology, 13*(6), 633–641.

Leao, A. A. P. (1944). Spreading depression of activity in cerebral cortex. *Journal of Neurophysiology, 7*, 359–390.

Lipton, R. B., Bigal, M. E., et al. (2007). Migraine prevalence, disease burden, and the need for preventive therapy. *Neurology, 68*(5), 343–349.

Liveing, E. (1873). *On megrin, sick-headache, and some allied dis- orders*. London: J & A Churchill.

Rashidy-Pour, A., Motaghed-Larijani, Z., et al. (1995). Tolerance to ketamine-induced blockade of cortical spreading depression transfers to MK-801 but not to AP5 in rats. *Brain Research, 693*(1–2), 64–69.

Rothrock, J. F. (1997). Successful treatment of persistent migraine aura with divalproex sodium. *Neurology, 48*(1), 261–262.

Rozen, T. D. (2000). Treatment of a prolonged migrainous aura with intravenous furosemide. *Neurology, 55*(5), 732–733.

Rozen, T. D. (2003). Aborting a prolonged migrainous aura with intravenous prochlorperazine and magnesium sulfate. *Headache, 43*(8), 901–903.

Russell, M. B., & Ducros, A. (2011). Sporadic and familial hemiplegic migraine: Pathophysiological mechanisms, clinical characteristics, diagnosis, and management. *Lancet Neurology, 10*(5), 457–470.

Russell, M. B., & Olesen, J. (1996). A nosographic analysis of the migraine aura in a general population. *Brain, 119*(Pt 2), 355–361.

Stovner, L., Hagen, K., et al. (2007). The global burden of headache: A documentation of headache prevalence and disability worldwide. *Cephalalgia, 27*(3), 193–210.

Thomsen, L. L., Eriksen, M. K., et al. (2002). A population-based study of familial hemiplegic migraine suggests revised diagnostic criteria. *Brain, 125*(Pt 6), 1379–1391.

Thomsen, L. L., Ostergaard, E., et al. (2003). Evidence for a separate type of migraine with aura: Sporadic hemiplegic migraine. *Neurology, 60*(4), 595–601.

Troost, B. T., & Zagami, A. S. (2000). Ophthalmoplegic migraine and retinal migraine. In J. Olesen, P. C. Tfelt-Hansen, & K. M. Welch (Eds.), *The headaches* (pp. 513–515). Philadelphia: Lippincott, Williams and Wilkins.

Clinical Migraine Without Aura

▶ Migraine Without Aura

Clinical Pain

Definition

"Physiological" pain is pain felt in the intact, non-injured organism. "Clinical" pain is pain felt after a tissue injury has induced sensitization of the peripheral and central nervous system.

Cross-References

▶ Postoperative Pain, Preemptive or Preventive Analgesia

Clinical Signs

Definition

Clinical signs are abnormal responses elicited by clinicians during the physical examination.

Cross-References

▶ Hypoalgesia, Assessment

Clinical Trial

Definition

A clinical trial is a research study designed to test the safety and/or effectiveness of drugs, devices, treatments, or preventive measures in humans. Clinical trials can usually be divided into four categories or "phases."

Cross-References

▶ Central Pain, Outcome Measures in Clinical Trials

Clitoral Pain

Allan Gordon
Wasser Pain Management Centre, Mount Sinai Hospital University of Toronto, Toronto, ON, Canada

Introduction

"Why is it that the very places in my body that should give me pleasure give me pain"?

Quote from a woman with a painful clitoris and painful nipples...The PNPC Syndrome (Gordon 2000).

Characteristics

Description of the Clitoris

A good place to find out about the clitoris is the website www.the-clitoris.com. Here the clitoris is described in clear terms with excellent graphics.

The major parts are the prepuce or hood, the glans, the body, and the crura (O'Connell et al. 1998; O'Connell et al. 2005; www.the-clitoris. com). Figure 1 (courtesy Ainsley Savage) shows the anatomy of the vulva.

The major function of the clitoris is sexual pleasure and orgasm. The clitoris has up to 8,000 nerve endings making it extremely sensitive

Clitoral Pain, Fig. 1 Anatomy of the vulva (Courtesy Ainsley Savage)

Clitoral Pain, Fig. 2 The clitoris (Courtesy The Clitoris.com)

(www.the-clitoris.com). Figure 2 shows the glans, crura and vestibular bulbs (courtesy The Clitoris.com)

The clitoris is innervated by the dorsal nerve of the clitoris off the pudendal nerve. This nerve transmits somatic sensation including pain, temperature, vibration, and proprioception. The clitoris also receives sympathetic and parasympathetic fibers for sexual function.

A clear description of the newer concepts of anatomy is contained in O'Connell et al. (2005). This discusses the rather extensive anatomical reach of the clitoris, the erectile tissue, and the nervous function.

What is Clitoral Pain?
Clitoral pain is pain either present at rest and/or induced by touch or pressure on the glans, body of the clitoris, and/or crura of the clitoris. Occasionally pain is felt in, next to, or above the clitoral hood or just posterior and/or inferior to the glans and body. It may be there on its own or more commonly associated with vulvar pain or urethral pain and also deeper muscular pain. The pain may be referred into the urethra from other sites, e.g., the urethra or vice versa.

Perry's definition (Perry 2000, 2004) was a "specific subset of vulvodynia produced by neuralgia of the anterior division of the pudendal nerve." However, that may be too exclusive a definition since there is such a variety of causes possible.

Many women with genital pain including vulvodynia and vulvar vestibulitis may not mention their clitoral pain unless they have been so involved with the healthcare system that they lose their reserve. More importantly, many practitioners do not ask or examine for it. For instance, many women with fibromyalgia or irritable bowel may also have vulvar pain and/or clitoral pain, but do not volunteer it or are not asked about it. This is certainly the case in our pain clinic where such pain is described mainly upon direct questioning.

Some women have a "sensitive" clitoris (see classification below) that is uncomfortable at rest or to touch but it does not interfere with sexual activity. Others may have a true pain in the clitoris, enough so that they do not like it touched. These terms need to be better defined in further population health studies.

In an internet survey of 423 women with vulvar pain, only 3.3 % had clitoral pain, and none had it alone (Gordon et al. 2003).

Questions To Ask During the History

In evaluating individuals for clitoral pain, one needs to ask the following:

- Does the woman spontaneously complain about CP, with questioning or after examination?
- Is it pain or sensitivity? This may be difficult to sort out.
- Is the pain present at rest, with being touched (provoked) or both?
- Is arousal and orgasm heightened, spontaneous, or reduced?
- Is there nipple sensitivity?
- Does intercourse or other sexual activity occur?
- Is sexual activity painful?
- Does and/or can orgasm occur?
- Is orgasm painful?
- Was there a specific inciting event such as trauma or psychological stress?
- Is/was there genital trauma?
- Was there excessive bicycle riding or horseback riding?
- Is there urinary urgency or pain?
- Is there urethral pain?
- Is there anorectal pain?
- Was there unwanted sexual abuse especially at an early age?
- Was there a preceding surgical event?
- Is there infection or another vulvar disease (lichen sclerosus)?
- Is there evidence of vulvodynia, either dysesthetic or provoked vestibulodynia?
- Is there a suggestion of pudendal nerve entrapment (Labat et al. 2008)?

Examination for Clitoral Pain

To examine for clitoral pain, one needs to expose the glans of the clitoris and lift the hood. We carefully use a Q-tip and ask the patient if there is pain on touching the glans, the hood, between the hoods and the glans, the crura, between the glans and the urethra, and above the hood on the labia. Sometimes, looking for hyperalgesia a sharper object is used.

One should look for swelling, atrophy, scarring, redness, or white lesion.

It is not until they are examined that the patient may realize or volunteer that pain felt during intercourse or other sexual activity could be related to a painful clitoris.

Proposed Classification of Types of Clitoral Pain

1. Clitoral sensitivity ... uncomfortable sensitivity but not pain at rest or while being touched.
2. Provoked clitoral pain.
   - Clitoral allodynia ... pain in or around the clitoris induced by light touch
   - Clitoral hyperalgesia ... increased pain in or around the clitoris when touched by a pointed object.
3. Clitoral dysesthesia ... burning pain at rest

A large population study is necessary in the "normal" female population. These kinds of symptoms may be more common than is thought.

Literature and Internet Search

Wesselman and Burnett (Wesselman and Burnett 1999) refer to very few cases and talk mainly about pain post-ritualistic clitoral excision. It is not mentioned in "Chronic Pelvic Pain: An Integrated Approach" (Steege et al. 1998).

The International Pelvic Pain Society website http://www.pelvicpain.org (Perry 2004) contains an article on clitoral pain that was also published in a textbook (Perry 2000) and in the Fall 2000 edition of the National Vulvodynia Association Newsletter. Perry mentions metabolic (e.g., diabetes), violent stimulation, tight clothing, or laser vaporization as known causes of clitoral pain.

A PubMed search for keywords "clitoris pain" yields articles on FGM (female genital mutilation) (Okonofy et al. 2002), painful priapism

(Medina 2002), clitoral neuroma (Fernandes-Aguilar and Noel 2003), hair tourniquet (Kuo et al. 2002), and phimosis (Munarriz et al. 2002) as well as various tumors.

The author's paper in 2002 represents the only significant collection of cases (Gordon 2002), which reports the clinical features of 21 women with clitoral pain. It was expanded upon in the first edition of this encyclopedia.

Hibner et al. (2012) report subpubic bone entrapment of the dorsal nerve to the clitoris. Sedner (2008) also reports this phenomenon.

Glazer et al. (2010, 2011) report that a small number of Somali Canadian women subjected to female genital cutting have chronic pain of multiple body regions particularly the vulva. Glazer (2012) has studied this population.

Foldes et al. (2012) document clitoral pain after female genital mutilation.

Chi et al. (2012) document pain including clitoral pain seen with lichen sclerosus.

Abdulcadir et al. (2012) document a painful clitoral neuroma after FGM, an unusual complication.

Lee et al. (2011) reported that low estrogen oral contraceptives do not seem to be associated with clitoral pain.

In a recent article by Facelle et al. (2012), it is reported that 71 % of women with persistent genital arousal syndrome report pain in the clitoris.

An Internet search, with a search engine such as Google however, will pick up information on chat line discussions, sex information sites, and/or sites that appear pornographic.

My Patient Population
In our paper (Gordon 2002), we reported on 21 individuals, 7 from my own practice and 14 who had contacted the author online through various sources. We have added to the numbers and are in the process of reporting a larger group of 55 patients. Some patients were followed for up to 4 years. The National Vulvodynia Association also referred female patients.

Recently we have done a review of patients looking for symptoms of persistent genital arousal disorder. The PGAD data has been reported at the Canadian Pain Society Meeting in Whistler in 2012 and is being prepared for publication awaiting analysis of further cases. At least 9 patients with clitoral pain had symptoms of PGAD.

The following represents information from the initial paper and personal observations and communication looking at 50+ more individuals the author has seen.

Clinical Points
- The rest pain is often described as mild to severe and to be burning, stinging, or aching. Many of our examined clinic patients have rest pain.
- There is pain on touching or contacting the clitoris, and the pain is moderate to severe in intensity. The women examined may have allodynia, hyperalgesia, or pain on firmer touch or pressure. The allodynia and hyperalgesia suggests a neuropathic pain process.
- Allodynia of the clitoris and introitus often coexist as well as periurethral touch-induced pain.
- Intercourse is almost always painful even with the use of techniques to avoid touching the clitoris. This interferes greatly with intimate relations. What may not be recognized is that intercourse pain may be caused by the clitoral pain.
- Primary anorgasmia is relatively uncommon (Gordon 2002), but secondary anorgasmia may occur because of the pain and unwillingness to try because of pain. Most continue to be orgasmic, some also achieving orgasm through a vaginal approach.
- Many women avoid intercourse or sexual activity because of the pain. Sexual desire may not be decreased despite the pain but more commonly desire is suppressed. Non-penetrative sex may also be painful.
- Clitoral pain is usually associated with vulvar pain both provoked and at rest.
- The pain may be part of established diseases including lichen sclerosus (Chi et al. 2012). Our article describes five such patients (Gordon 2002). We have recently seen others who also have associated PGAD.

- Some patients have experienced pain for 10 or more years, up to 18 years in practice and 25 years online.
- In others it was much shorter in duration, sometimes just months.
- Relapse and remissions may occur and spontaneous recovery may also occur.
- Trauma is not uncommon and we have at least six cases that began after surgery such as laparoscopy, bladder lifts (Milou et al. 2012), or delivery.
- It has been seen associated with vulvar and pelvic varices (we have 2 such patients).
- In at least 11 of our recent cases, it is linked with pudendal neuralgia.
- We know of four women who had remission with pregnancy, with the remission persisting in four postpartum. Another one worsened after pregnancy.
- We had at least 10 women who complained of painful or very highly sensitive nipples associated with the clitoral pain (Gordon 2000, 2002). The reason for this association is unclear. In our recent patients this was seen much less commonly seen but still present.
- Clitoral pain may be associated with vulvar pain, vaginismus, urethral pain, and occasionally anal pain as well as interstitial cystitis, painful urethral syndromes, and urethral dyssynergia.
- There is often a fruitless search for infections.
- Other comorbidities include endometriosis, fibromyalgia, anorexia, and back pain including spondylolisthesis. Depression was common.
- Painful clitoral priapism was seen in two of our patients.
- Clitoral pain may rarely be associated with a constant state of arousal.
- Clitoral pain may be the presenting feature in some and the most disconcerting, but more commonly there are other genital pain issues.
- In a small number, sexual trauma may have preceded or precipitated the clitoral pain.
- Women with clitoral pain can be desperate for explanations and relief, are concerned about being labeled as psychogenic, and look back in regret to events, particularly traumatic or surgical events that may have precipitated or aggravated their condition.

Causes of Clitoral Pain in Our Groups

1. Direct "sexual" trauma (e.g., vigorous masturbation, vigorous cunnilingus, a slipped vibrator, and vigorous sexual intercourse were all reported to us).
2. Direct "nonsexual" trauma (e.g., doing the splits on a beam while doing gymnastics, injury during car accident from direct trauma).
3. Injury to lumbar spine and/or coccyx and possible lower sacral root irritation.
4. Pelvic fracture and indirect trauma.
5. Multiple sclerosis with other urogenital complaints of pain, altered libido, and voiding dysfunction due to central plaques.
6. Post-hysterectomy especially with vascular misadventure (two with recurrent engorgement).
7. Postspinal fusion procedure with insertion of a cage.
8. Peripheral polyneuropathy.
9. Part of a unilateral or bilateral pudendal nerve entrapment syndrome.
10. Associated with lichen sclerosus with deforming tissue destruction.
11. Following explained or unexplained infection, such as yeast, vaginosis, or herpes.
12. Primary clitoral pain of unknown cause
13. Part of an unexplained genital pain syndrome involving clitoris, urethra, vulva, and anus.
14. Part of a severe chronic pelvic pain disorder with diffuse muscle and genital pain.
15. Related to sexual abuse.
16. A feature of vulvodynia.
17. A feature of a vulvar/pelvic vascular malformation.

Common Mechanism

While yeast or other infections might suggest an acute nociceptive mechanism, most patients have evidence of a chronic neuropathic pain. There is clitoral allodynia and at times hyperalgesia. While documented or undocumented pudendal nerve entrapment may present in some in Alcock's canal, this would not explain pain seen in direct trauma. Clitoral pain seems to be an example of neuropathic pain and seen in central pain (MS), polyneuropathy or lichen sclerosus. Ultimately, whatever the insult, there must be peripheral and central sensitization. Clitoral pain deserves more study as an example of neuropathic pain derived from many basic causes, be it post-infectious, posttraumatic, iatrogenic, direct injury, or chronic nerve entrapment.

Clinical Evaluation
This must include the following:
- Clinical history
- Sexual history
- Psychosocial history and evaluation
- Alcohol, addiction, and abuse history
- Treatments tried
- Neurological, musculoskeletal, and gynecological exam
- Detailed vulvar exam including Q-tip and pain and temperature and retraction of prepuce
- Pelvic and rectal exam checking for deeper pain and anal pain
- Pelvic and spinal imaging when clinically appropriate with ultrasound, CT scan, or magnetic resonance scan
- Pudendal nerve latencies if available
- Quantitative sensory testing of genital area
- Cultures of vulva, vagina, and urine
- Urogynecological evaluation if urinary symptoms

Treatment
The treatment depends upon clinical diagnosis and comorbidity. There are no published clinical trials and even very few anecdotal reports on clitoral pain. Management within a broad paradigm such as the five pillars of pain management (Gordon 2012) is necessary. Awareness of the condition, asking the right questions in the right way, and validating the patients pain are key, and a detailed multidisciplinary or transdisciplinary examination is necessary: Management is best in a multidisciplinary pain clinic where chronic pelvic region pain is regularly treated. Use of guidelines for neuropathic pain may be helpful (e.g., Moulin et al. 2007).

Treatments to Try and Comments
1. Oral tricyclic antidepressants such as amitriptyline. Occasionally effective.
2. Oral anticonvulsants such as gabapentin. Variable results.
3. Topical medications such as amitriptyline, gabapentin, ketamine, and even capsaicin (which burns). More studies needed with these agents. Clinical trials are needed.
4. Opiates. They are often tried because of other comorbid pain conditions, and experience suggests that they are not particularly effective in clitoral pain.
5. Topical cocaine. Usually associated with addictive behavior.
6. Anesthetic block of dorsal nerve of clitoris. No literature studies (Perry 2004).
7. Pudendal nerve block. Of diagnostic value in pudendal neuralgia and no specific studies for clitoral pain.
8. Pudendal nerve decompression. Must be well documented and in experienced hands only.
9. Physical therapy. This may allow therapy of the spine and coccyx, and internal and external muscles. Some therapists advocate pudendal nerve release through a transvaginal approach.
10. Specific treatment of underlying, associated, or comorbid conditions such as vulvodynia, lichen sclerosus, and depression.
11. Psychological therapies including cognitive behavioral therapy should be considered.
12. Pelvic floor surface EMG may be tried only if there is evidence of pelvic floor disturbance and if the probe does not hurt.
13. Sexual therapies. This assists in developing alternate techniques and exploring attitudes, preferences, and experiences.
14. Cannabis: no trials but anecdotal reports from patients does suggest some benefit.
15. Sacral nerve stimulation (Marcelissen et al. 2010) has been reported as effective. There is a need for a clinical trial however.

The Relationship with Persistent Genital Arousal Disorder
This syndrome was first reported by Leiblum and Nathan in 2001. Facelle et al. (2012) published a recent review of the condition in which there is sexual arousal unwanted and without sexual stimulation that persists for an extended period of time and do not subside completely on

their own. We are currently studying a group of women with this condition and find that it may be associated with genital pain. Fecelle et al. (2013) reports that 71 % have clitoral pain.

Summary

Clinical trials will clearly be necessary in order to understand what really works. In the meantime, conservative multimodal treatments should be considered. The relative rarity of clitoral pain, the many associated conditions, and the fact that most practitioners are not prepared to evaluate these patients has meant that trials are not carried out. However, it is also clear that we need to know more about the clinical and natural history of clitoral pain, as well as understand more about the pathophysiology and the various associated and/or comorbid features. That can be achieved by detailed study including the use of neurophysiologic techniques and the creation of an observational database. But it all starts with a detailed questioning of women with pelvic pain and careful, appropriate examination.

Clitoral pain is best reframed as an example of neuropathic pain and studied in the same way as other forms of neuropathic pain. A search is necessary for pudendal neuralgia and other genital pain syndromes.

References

Abdulcadir, J., Pusztaszeri, M., Vilarino, R., Dubuisson, J. B., & Vlastos, A. T. (2012). Clitoral neuroma after female genital mutilation/cutting: A rare but possible event. *The Journal of Sexual Medicine, 9*(4), 1220–1225.

Bekker, M. D. MD, PhD, Hogewoning, C. R. C. MSc, Wallner, C. MD, PhD, Elzevier, H. W. MD, PhD, & DeRuiter, M. C. PhD. (2012). The somatic and autonomic innervation of the clitoris; Preliminary evidence of sexual dysfunction after minimally invasive slings. *The Journal of Sexual Medicine, 9*, 1566–1578.

Chi, C. C., Kirtschig, G., Baldo, M., Lewis, F., Wang, S. H., & Wojnarowska, F. (2012). Systematic review and meta-analysis of randomized controlled trials on topical interventions for genital lichen sclerosus. *Journal of the American Academy of Dermatology, 67*(2), 305–312.

Facelle, T. M., Sadeghi-Nejad, H., & Goldmeier, D. (2013). Persistent genital arousal disorder: Characterization, etiology, and management. *The Journal of Sexual Medicine, 10*(2), 439–450.

Fernandes-Aguilar, S., & Noel, J. C. (2003). Neuroma of the clitoris after female genital cutting. *Obstetrics and Gynecology, 101*, 1053–1054.

Foldès, P., Cuzin, B., & Andro, A. (2012). Reconstructive surgery after female genital mutilation: A prospective cohort study. *Lancet, 380*(9837), 134–141.

Glazer, E. (2012). Embodiment, pain, and circumcision in Somali-Canadian women. A thesis submitted in conformity with the requirements for the degree of Master of Science, Graduate Department of the Institute of Medical Science University of Toronto Copyright 2012.

Glazer, E., Blom, K., Duplessis, D., Mason, R., Du Mont, J., Angus, J., Gordon, A., Dunn, S., Pukall, C., Einstein, G. (2010). Socio-cultural embodiment, pain, and circumcision in Somali women in Toronto. *Gender Medicine, 7*(5), (original article in vol 7(5), 516).

Glazer, E., Karim, N., Gordon, A., Pukall, C., & Einstein, G. (2011). Chronic pain in Somali-Canadian women with infibulation. *Society for Neuroscience* (Abstr).

Gordon, A. S. (2000). The PNPC syndrome. Platform presentation at PUGO Special Interest Group meeting (Sept 23, 2000) Hamburg, Germany.

Gordon, A. S. (2002). Clitoral pain: The great unexplored pain in women. *Journal of Sex & Marital Therapy, 28*, 181–185.

Gordon, A. (2012). The five pillars of pain management. *Pain Management, 2*, 335–344.

Gordon, A. S., Panahian-Jand, M., McComb, F., et al. (2003). Characteristics of women with vulvar pain disorders: Response to a web based survey. *Journal of Sex & Marital Therapy, 29*, 25.58.

Hibner, M., Castellanos, M. E., Drachman, D., & Balducci, J. (2012). Repeat operation for treatment of persistent pudendal nerve entrapment after pudendal neurolysis. *Journal of Minimally Invasive Gynecology, 19*(3), 325–330.

Kuo, J. H., Smith, L. M., & Berkowitz, C. D. (2002). A hair tourniquet resulting in strangulation and amputation of the clitoris. *Obstetrics and Gynecology, 99*, 939–941.

Labat, J. J., Riant, T., Robert, R., Amarenco, G., Lefaucheur, J. P., & Rigaud, J. (2008). Diagnostic criteria for pudendal neuralgia by pudendal nerve entrapment (Nantes criteria). *Neurourology and Urodynamics, 27*(4), 306–310.

Marcelissen, T., Van Kerrebroeck, P., & de Wachter, S. (2010). Sacral neuromodulation as a treatment for neuropathic pain after abdominal hysterectomy. *International Urogynecology Journal, 21*(10), 1305–1307. doi:10.1007/s00192-010-1145-x.

Medina, C. A. (2002). Clitoral priapism: A rare condition presenting as a cause of vulvar pain. *Obstetrics and Gynecology, 100*, 1089–1091.

Moulin, D. E., Clark, A. J., Gilron, I., Ware, M. A., Watson, C. P., Sessle, B. J., Coderre, T., Morley-Forster, P. K., Stinson, J., Boulanger, A., Peng, P., Finley, G. A., Taenzer, P., Squire, P., Dion, D., Cholkan, A., Gilani, A., Gordon, A., Henry, J., Jovey, R., Lynch, M., Mailis-Gagnon, A., Panju, A., Rollman, G. B., Velly, A., & Canadian Pain Society. (2007). Pharmacological management of chronic neuropathic pain. *Pain Research & Management, 12*(1), 13–21.

Munarriz, R., Talakoub, L., Kuohung, W., et al. (2002). The prevalence of phimosis of the clitoris in women presenting to the sexual dysfunction clinic: Lack of correlation to disorders of desire, arousal and orgasm. *Journal of Sex & Marital Therapy, 28,* 181–185.

O'Connell, H. M., Hutson, J. M., Anderson, C. R., et al. (1998). Anatomical relationship between Urethra.

O'Connell, H. M., Sanjeevan, K. V., & Hutson, J. M. (2005). Anatomy of the clitoris. *The Journal of Urology, 174*(4 Part 1), 1189–1195.

Okonofy, F. E., Larsen, U., Oronsaye, F., et al. (2002). The association between female genital cutting and correlates of sexual and gynecological morbidity in Edo State, Nigeria. *BJOG: An International Journal of Obstetrics and Gynaecology, 109,* 1089–1096.

Perry, C. P. (2000). Vulvodynia. In F. M. Howard, C. P. Perry, J. E. Carter, et al. (Eds.), *Pelvic pain: Diagnosis and management* (pp. 204–210). Philadelphia: Lippincott Williams & Wilkins.

Perry, C. P. (2004). Clitoral pain. http://www.pelvicpain.org/

Sedy, J. (2008). The entrapment of the dorsal nerve of penis, clitoris under the pubis: An alternative source of pudendal neuralgia. *Pain Physician, 11*(3):381 (author reply 382)

Steege, J. F., Metzger, D. A., & Levy, B. S. (Eds.). (1998). *Chronic pelvic pain: An integrated approach.* Philadelphia: WB Saunders.

Wesselman, U., & Burnett, A. L. (1999). Genitourinary pain. In P. Wall & R. Melzack (Eds.), *Textbook of pain* (4th ed., pp. 689–709). Edinburg: Churchill Livingstone.

Clofazimine

Definition

Clofazimine is a substituted iminophenazine dye with antibacterial properties, which is used in the treatment of leprosy; it is administered orally.

Cross-References

▶ Hansen's Disease

Clonidine

Definition

Clonidine is a prototypical α_2-adrenoceptor agonist with potent sympatholytic effects. It activates neuronal cell membrane potassium channels causing hyperpolarization, reducing the rate of neuronal firing. It also inhibits neurotransmitter release by inhibiting calcium conductance through N-type calcium channels. Clinically, it has antihypertensive, sedative, anxiolytic, and analgesic actions, which may also be useful in the management of complex pain states.

Cross-References

▶ Acute Pain in Children, Postoperative
▶ Alpha(α) 2-Adrenergic Agonists in Pain Treatment
▶ Postoperative Pain, Appropriate Management

Cluster Analysis

▶ Pain Measurement by Questionnaires, Psychophysical Procedures, and Multivariate Analysis

Cluster Headache

Definition

A relatively uncommon headache disorder with a striking male predilection. They are called cluster headaches because attacks occur in a group or "cluster" during which time patients suffer from 1 to 8 headaches daily. Individual headaches last between 15 and 180 min each, are excruciatingly severe, and are associated with one or more autonomic features such as conjunctival injection, lacrimation, nasal congestion, rhinorrhea, forehead and facial sweating, miosis, ptosis, and eyelid edema. Cluster cycles last 2–8 weeks each and are followed by remission periods of 6–12 months, during which time patients are pain-free. Attacks are strictly unilateral orbitally, supraorbitally, and/or temporally.

Cross-References

▶ Hemicrania Continua
▶ Human Thalamic Response to Experimental Pain (Neuroimaging)
▶ Paroxysmal Hemicrania

Clustering

▶ Trafficking and Localization of Ion Channels

C-Mechanoheat Receptor

▶ Polymodal Nociceptors, Heat Transduction

C-mechanoreceptor

▶ Mechanonociceptors

CM-PF Complex

▶ Parafascicular Nucleus, Pain Modulation

CMT

▶ Charcot-Marie-Tooth Disease

CNS Stimulation in Treatment of Neuropathic Pain

▶ Central Nervous System Stimulation for Pain

CNV

▶ Contingent Negative Variation

CO Staining

▶ Cytochrome Oxidase Staining

CO2 Laser

Definition

A type of laser device which allows to stimulate selected skin spots with coherent high-energy laser light. This type of heat stimulation has been frequently used in experimental pain research, since time course and quantity of the energy emission can be precisely controlled.

Cross-References

▶ Magnetoencephalography in Assessment of Pain in Humans

Cochlea

Definition

The cochlea is the coiled structure within the inner ear where vibrations caused by sound are transduced into neural impulses by the Organ of Corti.

Cross-References

▶ Perireceptor Elements

Cognition

Definition

Cognition is the mental act or process by which knowledge may be acquired or analyzed. A distinction is typically made between the content of cognition such as beliefs, attitudes, or memories and the process of cognition involving acts such as perception, interpretation, inference, retrieval, or problem solving.

Cross-References

▶ Pain Assessment in the Elderly
▶ Psychology of Pain, Assessment of Cognitive Variables

Cognitive

Definition

Pertaining to internal thoughts, images, and constructs of the individual.

Cross-References

▶ Pain as a Cause of Psychiatric Illness
▶ Psychological Treatment in Acute Pain

Cognitive Appraisal

Definition

Cognitive appraisal is an evaluation of environmental stimuli. Resources available to cope with a potential stressor, and its ability to offset the threatening stimuli, determine the degree and intensity of the stress response.

Cross-References

▶ Stress and Pain

Cognitive Aspect of Pain

Definition

Perceptual component of pain perception including pain-related learning and memory and the recognition of the painful nature of the stimulus.

Cross-References

▶ Primary Somatosensory Cortex (S1), Effect on Pain-Related Behavior in Humans

Cognitive Assessment

▶ Psychology of Pain, Assessment of Cognitive Variables

Cognitive Coping Training

Definition

Cognitive coping training includes, but is not limited to, attention diversion strategies, reinterpreting pain sensations, calming self-statements, and imagery techniques.

Cross-References

▶ Psychological Treatment of Headache

Cognitive Dysfunction

Definition

Changes in consciousness, higher cortical functions, mood, or perception that may be induced by any of numerous neurological or systemic diseases or by ingestion of substances, including drugs that have the potential for central nervous system toxicity.

Cross-References

► Cancer Pain Management, Opioid Side Effects, Cognitive Dysfunction

Cognitive Error

Definition

A discrepancy between actual and perceived aspects of a particular situation.

Cross-References

► Catastrophizing

Cognitive Factors

Definition

Cognitive factors are the beliefs, appraisals, expectations, and meanings through which individuals filter information. These cognitive factors will have an influence on emotional and physiological arousal and behavior.

Cross-References

► Cognitive-Behavioral Perspective of Pain
► Psychological [Behavioral Health] Assessment of Pain

Cognitive Impairment

► Cancer Pain, Assessment in the Cognitively Impaired

Cognitive Model of Pain

► Cognitive-Behavioral Perspective of Pain

Cognitive Modulation of Pain

► Descending Circuits in the Forebrain, Imaging

Cognitive Restructuring

Definition

Cognitive restructuring is a process whereby patients are taught to monitor their thoughts and feelings, and to evaluate whether these are appropriate in light of the circumstances and evidence available. Patients are encouraged to become aware of their automatic responses and to consider the accuracy of the assumptions on which these responses are based. They are instructed to evaluate the assumptions that result in specific behaviors and to consider alternatives that may be more adaptive.

Cross-References

► Catastrophizing
► Multidisciplinary Pain Centers, Rehabilitation
► Psychological Treatment in Acute Pain
► Psychological Treatment of Headache

Cognitive Schemata

Definition

Cognitive schemata are models that people create to assist them to structure and make sense of life circumstances, constructs, and categories. These are efficient templates that help people navigate in their world and filter new information. Newly acquired information must be assimilated into these existing cognitive structures or will contribute to a modification (accommodation) of the schemata to incorporate the new data of experience.

Cross-References

▶ Cognitive-Behavioral Perspective of Pain

Cognitive Tasks

Definition

Procedures used to engage participants in a particular cognitive activity. These include (but are not restricted to) tasks such as the retrieval of words or memories, reaction times to identify specific stimulus features or to complete different mental operations, or the interpretation of ambiguous stimuli. The operation of specific cognitive processes is typically inferred on the basis of different patterns of cognitive task performance.

Cross-References

▶ Psychology of Pain, Assessment of Cognitive Variables

Cognitive Therapy

▶ Cognitive-Behavior Treatment of Pain

Cognitive-behavior Modification

▶ Cognitive-Behavior Treatment of Pain

Cognitive-Behavior Therapy

▶ Psychological Treatment of Pain in Older Populations

Cognitive-behavior Therapy (CBT)

▶ Cognitive-Behavior Treatment of Pain

Cognitive-Behavior Treatment of Pain

Dennis C. Turk
Department of Anesthesiology & Pain Research,
University of Washington, Seattle, WA, USA

Synonyms

Cognitive Therapy; Cognitive-Behavior Modification; Cognitive-Behavior Therapy (CBT)

Definition

Cognitive-Behavior Therapy (CBT) or cognitive-behavior modification (CBM) are generic terms that incorporate a wide range of treatment modalities (e.g., stress management, distraction, relaxation, problem-solving, cognitive restructuring), all of which are designed to enhance coping, facilitate self-management, and improve emotional and physical functioning. The primary goals are to help patients understand the effects that thoughts, feelings, and behaviors have on pain, the potential of patients to exert some control over their symptoms, and education and practice in the use of cognitive and behavioral coping skills.

Characteristics

The cognitive-behavioral (CB) approach to pain management evolved from research on a number of mental health problems (e.g., anxiety, depression, phobias). Following the initial empirical research on CB techniques in the early 1970s, there have been a large number of research and clinical applications. The common denominators across different CB approaches include:

1. Interest in the nature and modification of a patient's thoughts, feelings, and beliefs, as well as behaviors
2. Some commitment to behavior therapy procedures in promoting change (such as graded practice, homework assignments, relaxation, relapse prevention training, social skills training) (Turk et al. 1983)

In general, the CB therapist is concerned with using environmental manipulations, as are behavior (operant conditioning) therapists, but for the CB therapist such manipulations represent informational feedback trials that provide an opportunity for the patient to question, reappraise, and acquire self-control over maladaptive thoughts, feelings, behaviors, and physiological responses. Although the CB approach was originally developed for the treatment of mental health disorders, the perspective has much in common with the multidimensional conceptualizations of pain, which emphasize the contributions of cognitive and affective, as well as sensory phenomena (Melzack and Wall 1965). The CB perspective emphasizes the important contribution of psychological variables such as the perception of control, the meaning of pain to the patient, and dysphoric affect (Flor and Turk 2011).

It is important to differentiate the CB perspective from CB treatments. The CB perspective is based on five central assumptions, and can be superimposed upon any treatment approach used with chronic pain patients:

- People are active processors of information and not passive reactors.
- Thoughts (e.g., appraisals, expectancies, beliefs) can elicit and influence mood, affect physiological processes, have social

consequences and can also serve as an impetus for behavior; conversely, mood, physiology, environmental factors, and behavior can influence the nature and content of thought processes.
- Behavior is reciprocally determined by both the individual and environmental factors.
- People can learn more adaptive ways of thinking, feeling, and behaving.
- People should be active collaborative agents in changing their maladaptive thoughts, feelings, and behaviors.

In many cases, the perspective is as important as the content of the therapeutic modalities employed, somatic as well as psychological (Turk 1997).

The application of the CB perspective to the treatment of chronic pain involves a complex clinical interaction and makes use of a wide range of tactics and techniques. Despite the specific techniques used, all CB treatment approaches are characterized by being present, focused, active, time limited, and structured. Therapists are not simply conveyers of information acting on passive patients, but serve as educators, coaches, and trainers. They work in concert with the patient (and sometimes family members) to achieve mutually agreed upon goals.

A growing body of research has demonstrated the important roles that cognitive factors (appraisals, beliefs, expectancies) play in exacerbating pain and suffering, contributing to disability and influencing response to treatment (Flor and Turk 2011). Thus, CB interventions are designed to help patients identify maladaptive patterns and acquire, develop, and practice more adaptive ways of responding. Patients are encouraged to become aware of, and monitor, the impact that negative pain-engendering thoughts and feelings play in the maintenance of maladaptive overt and covert behaviors. Additionally, patients are taught to recognize the connections linking cognitions, affective, behavioral, and physiological responses together with their joint consequences. Finally, patients are encouraged to undertake "personal experiments," and to test the effects of their appraisals, expectations, and beliefs by means of selected homework assignments.

The CB therapist is concerned not only with the role that patients' thoughts play in contributing to their disorders but, equally important, with the nature and adequacy of the patients' behavioral repertoire, since this affects resultant interpersonal situations. A CB treatment program for pain patients is multifaceted. Treatment may be conducted for individuals or groups on an inpatient or outpatient basis.

The CB perspective should be considered not merely as a set of methods designed to address the psychological components of pain and disability, but as an organizing strategy for more comprehensive rehabilitation (Turk 1997). For example, patients' difficulties arising during physical therapy may be associated not only with physical limitations, but also with the fear engendered by anticipation of increased pain or concern about injury. Therefore, from a CB perspective, physical therapists need to address not only the patient's performance of physical therapy exercises and the accompanying attention to body mechanics, but also the patient's expectancies and fears as they will affect the amount of effort, perseverance in the face of difficulties and adherence with the treatment plan (Meichenbaum and Turk 1987). These cognitive and affective processes, including self-management concerns, may be impediments to rehabilitation and thus need to be considered and addressed, along with traditional instructions regarding the proper performance of exercise. The attention paid to the individual's thoughts and expectancies by a behavioral health practitioner should be adopted by all members of the interdisciplinary treatment team. In short, CB treatment should not be viewed as totally scripted. Therapists must realize that flexibility and clinical skills have to be brought to bear throughout the treatment program.

The CB treatment consists of five overlapping phases:
1. Initial assessment
2. Collaborative reconceptualization of the patient's views of pain

3. Skills acquisition and skills consolidation, including cognitive and behavioral rehearsal
4. Generalization, maintenance, and relapse prevention
5. Booster sessions and follow-up

Although the five treatment phases are listed separately, it is important to appreciate that they overlap. The distinction between phases is designed to highlight the different components of the multidimensional treatment. Moreover, although the treatment, as presented, follows a logical sequence, it should be implemented in a flexible, individually tailored fashion. Patients proceed at varying paces and the therapist must be sensitive to these individual differences.

The CB treatment is not designed to eliminate patients' pain per se, although the intensity and frequency of their pain may be reduced as a result of increased activity, physical reconditioning achieved during physical therapy, and the acquisition of various cognitive and behavioral coping skills. Rather, the treatment is designed to help patients learn to live more effective and satisfying lives, despite the presence of varying levels of discomfort that may persist. Other goals include the reduction of excessive reliance on the healthcare system, reduced dependence on analgesic medications, increased functional capacity and, whenever feasible, return to employment or usual household activities. The primary objectives of CB treatment are:

- To combat demoralization by assisting patients to change their view of their pain and suffering from overwhelming to manageable
- To assist patients to reconceptualize their view of themselves from being passive, reactive and helpless to being active, resourceful, and competent
- To teach patients that there are coping techniques and skills that can be used to help them to adapt and respond to pain and the resultant problems
- To help patients learn the associations between thoughts, feelings, and their behavior, and subsequently to identify and alter automatic, maladaptive patterns

- To teach patients specific coping skills and, moreover, when and how to utilize these more adaptive responses
- To bolster self-confidence and to encourage patients to attribute successful outcomes to their own efforts
- To help patients anticipate problems proactively and generate solutions, thereby facilitating maintenance and generalization

The overriding message of the CB approach is that people are not helpless in dealing with their pain, nor need they view pain as an all-encompassing determinant of their lives. Rather, a variety of resources are available for confronting pain, and pain will come to be viewed by patients in a more differentiated manner. The treatment encourages patients to maintain a problem-solving orientation and to develop a sense of resourcefulness, instead of the feelings of helplessness and withdrawal that revolve around bed, physicians, and pharmacists.

The CB approach offers promise for use with a variety of chronic pain syndromes across all developmental levels (Turk and Okifuji 1999). CB approaches have been evaluated in a number of clinical pain studies. The results tend to support the effectiveness of CB therapy in reducing pain and improving functional activities (Morley et al. 1999). The American Psychological Association Task Force on Treatment Efficacy designated CBT for chronic pain as one of 20 applications of psychological treatments for which there was significant empirical support. CBT is frequently used as a complement to other treatment modalities including information, exercise, and medication in the treatment of chronic pain patients, and is incorporated within interdisciplinary and multidisciplinary programs (Flor et al. 1992).

Taken as an aggregate, the available evidence suggests that the CB approach has a good deal of potential as a treatment approach by itself and in conjunction with other treatments. The CB perspective is a reasonable way for healthcare providers to think about and deal with their patients, regardless of the therapeutic modalities utilized.

References

Flor, H., Fydrich, T., & Turk, D. C. (1992). Efficacy of multidisciplinary pain treatment centers: A meta-analytic review. *Pain, 49,* 221–230.

Flor, H., & Turk, D. C. (2011). Chronic pan: An integrated biobehavioral approach. New York: IASP Press

Meichenbaum, D., & Turk, D. C. (1987). *Facilitating treatment adherence: A practitioner's guidebook* New York: Plenum Press.

Melzack, R., & Wall, P. D. (1965). Pain mechanisms: A new theory. *Science, 150,* 971–979.

Morley, S., Eccleston, C., & de Williams, A. C. (1999). Systematic review and meta-analysis of randomized controlled trials of cognitive behavior therapy and behavior therapy for chronic pain in adults, excluding headache. *Pain, 80,* 1–13.

Turk, D. C. (1997). Psychological aspects of pain. In P. Bakule (Ed.), *Expert pain management* (pp. 124–178). Springhouse, PA: Springhouse.

Turk, D. C., & Okifuji, A. (1999). A cognitive-behavioral approach to pain management. In P. D. Wall & R. Melzack (Eds.), *Textbook of pain* (4th ed., pp. 1431–1444). London: Churchill Livingstone.

Turk, D. C., Meichenbaum, D., & Genest, M. (1983). *Pain and behavioral medicine: A cognitive-behavioral perspective.* New York: Guilford Press.

Cognitive-Behavioral Model

Definition

In the context of bodily sensations, this model states that (mis)interpretations (cognitions) of bodily sensations have a profound impact on how patients perceive their body, cope, and behave. Often a vicious circle is assumed between bodily sensations, cognitions, and behavior.

Cross-References

▶ Muscle Pain, Fear-Avoidance Model

Cognitive-Behavioral Model of Chronic Low Back Pain

Definition

Cognitive behavioral models of chronic low back pain state that catastrophic misinterpretations in response to acute pain lead to fear of movement/ (re)injury. As a consequence, a vicious cycle develops in which subsequent avoidance of activities and hypervigilance lead to an increase in functional disability, pain, depression, and a decrease in physical fitness, thereby further advancing chronicity.

Cross-References

▶ Disability, Fear of Movement

▶ Fear and Pain

Cognitive-Behavioral Perspective of Pain

Dennis C. Turk
Department of Anesthesiology & Pain Research, University of Washington, Seattle, WA, USA

Synonyms

Behavioral/Cognitive Perspective; Cognitive Model of Pain

Definition

The cognitive-behavioral (CB) model of pain suggests that ▶ cognitive factors (beliefs, appraisals), in particular, and expectations rather than conditioning factors are of central importance. Thus, the model is a hyphenated one. It does not ignore the important role of contextual factors and principles of learning theory, but rather incorporates them within an integrated perspective on the individual experiencing pain and pain management. The CB model proposes that so-called conditioned reactions are largely self-activated on the basis of learned expectations rather than automatically evoked. The critical factor for the CB model, therefore, is not that events occur together in time or are operantly reinforced, but that people

learn to predict them based on experiences and information processing. They filter information through their preexisting knowledge and organized representations of knowledge (Turk and Salovey 1985; Flor and Turk 2011), and react accordingly. Their responses, consequently, are based not on objective reality but their idiosyncratic interpretations of reality. As interaction with the environment is not a static process, attention is given to the ongoing reciprocal relationships among physical, cognitive, affective, social, and behavioral factors.

Characteristics

According to the CB model then, it is the person with pain's perspective, based on their idiosyncratic attitudes, beliefs, appraisals, and expectations that filter and interact reciprocally with emotional factors, social influences, behavioral responses, as well as sensory phenomena. Moreover, individuals' behaviors elicit responses from significant others (including health-care professionals) that can reinforce both adaptive and maladaptive modes of thinking, feeling, and behaving.

There are five central assumptions that characterize the CB perspective on pain (Flor and Turk 2011). The first assumption is that all people are active processors of information rather than passive reactors to environmental contingencies. People attempt to make sense of stimuli from the external environment by filtering information through organized schema derived from their prior learning histories, and by general strategies that guide the processing of information. People's responses (overt as well as covert) are based on appraisals and subsequent expectations and are not totally dependent on the actual consequences of their behaviors (i.e., positive and negative reinforcements and punishments). Thus, from this perspective, anticipated consequences are as important in guiding behavior as actual consequences.

A second assumption is that one's thoughts (e.g., appraisals, attributions, expectancies) can elicit or modulate affect and physiological arousal, both of which may serve as impetuses for behavior. Conversely, affect, physiology, and behavior can instigate or influence one's thinking processes. Thus, the causal priority is dependent upon where in the cycle one chooses to begin. Causal priority may be less of a concern than the view of an interactive process that extends over time.

Unlike orthodox operant model that emphasizes the influence of the environment on behavior, the CB perspective focuses on the reciprocal effects of the person on the environment, as well as the influence of environment on behavior. The third assumption of the CB perspective, therefore, is that both the environment and the individual reciprocally determine behavior. People not only passively respond to their environment but elicit environmental responses by their behavior. In a very real sense, people create their environments. The person who becomes aware of physical events (symptoms) and decides the symptom requires attention from a health-care provider, initiates a set of circumstances different from the person with the same symptom who chooses to self-manage (▶ self-management).

A fourth assumption is that if people have learned maladaptive ways of thinking, feeling, and responding, then successful interventions designed to alter behavior should focus on maladaptive thoughts, feelings, physiology, and behaviors, and not one to the exclusion of the others. There is no expectancy that changing thoughts, feelings, or behaviors will necessarily result in the other two following.

The final assumption of the CB perspective is that in the same way as people are instrumental in developing and maintaining maladaptive thoughts, feelings, and behaviors, they can, are, and should be considered active agents of change of their maladaptive modes of responding. People with chronic pain, no matter how severe, despite their common beliefs to the contrary, are not helpless pawns of fate. They can and should become instrumental in learning and carrying out more effective modes of responding to their environment and their plight.

From the CB perspective, people with pain are viewed as having negative expectations about their own ability to control certain motor skills without pain. Moreover, people with pain tend to believe they have limited ability to exert any control over their pain. Such negative, maladaptive appraisals about the situation and personal efficacy may reinforce the experience of demoralization, inactivity, and overreaction to nociceptive stimulation. These cognitive appraisals and expectations are postulated as having an effect on behavior, leading to reduced efforts and activity that may contribute to increased psychological distress (▶ helplessness) and subsequently physical limitations. If we accept that pain is a complex, subjective phenomenon that is uniquely experienced by each person, then knowledge about idiosyncratic beliefs, appraisals, and ▶ coping repertoires become critical for optimal treatment planning, and for accurately evaluating treatment outcome (Turk et al. 1983; Flor and Turk 2011).

Biomedical factors that may have initiated the original report of pain play less and less of a role in disability over time, although secondary problems associated with deconditioning may exacerbate and serve to maintain the problem. Inactivity leads to increased focus on and preoccupation with the body and pain, and these cognitive-attentional changes increase the likelihood of misinterpreting symptoms, overemphasis on symptoms, and the perception of oneself as being disabled. Reduction of activity, fear of reinjury, pain, loss of compensation, and an environment that, perhaps, unwittingly supports the pain-patient role can each impede alleviation of pain, successful rehabilitation, reduction of disability, and improvement in adjustment. As has been noted, cognitive factors may not only affect the patient's behavior and indirectly their pain, but may actually have a direct effect of physiological factors believed to be associated with the experience of pain (e.g., Flor and Turk 2011).

Internal Representation
People respond to medical conditions in part based on their subjective representations of illness and symptoms (▶ cognitive schemata). When confronted with new stimuli, the person engages in a meaning analysis that is guided by the schemata that best match the attributes of the stimulus. It is on the basis of the person's unique schema that incoming stimuli are interpreted, labeled, and acted on.

People build fairly elaborate representations of their physical state, and these representations provide the basis for action plans and coping. Beliefs about the meaning of pain and one's ability to function despite discomfort are important aspects of the cognitive schemata of pain. These representations are used to construct causal, covariational, and consequential information from their symptoms. For example, a cognitive schema that one has a very serious, debilitating condition, that disability is a necessary aspect of pain, that activity is dangerous, and that pain is an acceptable excuse for neglecting responsibilities will likely result in maladaptive responses. Similarly, if patients believe they have a serious condition that is quite fragile and a high risk for reinjury, they may fear engaging in physical activities. Through a process of stimulus generalization, patients may avoid more and more activities, become more physically deconditioned, and more disabled.

People's beliefs, appraisals, and expectations about pain, their ability to cope, social supports, their disorder, the medicolegal system, the health-care system, and their employers are all important as they may facilitate or disrupt the patient's ► sense of control. These factors also influence patients' investment in treatment, acceptance of responsibility, perceptions of disability, support from significant others, expectancies for treatment, acceptance of treatment rationale, and adherence to treatment.

Cognitive interpretations will also affect how individuals present symptoms to significant others, including health-care providers and employers. Overt communication of pain, suffering, and distress will enlist responses that may reinforce pain behaviors and impressions about the seriousness, severity, and uncontrollability of the pain. That is, reports of pain may lead physicians to prescribe more potent medications, order additional diagnostic tests, and, in some cases perform surgery (e.g., Turk and Okifuji 1997; Martell et al. 2007. Family members may express sympathy, excuse the patient from usual responsibilities, and encourage passivity thereby fostering further ► physical deconditioning. It should be obvious that the CB perspective integrates the operant conditioning emphasis on external reinforcement and the respondent's view of learned avoidance within the framework of information processing.

Expectations

People with persistent pain often have negative expectations about their own ability and responsibility to exert any control over their pain, and they avoid activities that they believe will exacerbate their pain or contribute to additional injury (Vlaeyen and Linton 2000; Leeuw et al. 2007). Moreover, they often view themselves as helpless. Such negative, maladaptive appraisals about their condition, situation, and their personal efficacy in controlling their pain and problems associated with pain serve to reinforce their overreaction to nociceptive stimulation, inactivity, and experience of demoralization. These cognitive appraisals are posed as having an effect on behavior leading to reduced effort, reduced perseverance in the face of difficulty and activities, and increased psychological distress.

The specific thoughts and feelings that patients experience prior to exacerbations of pain, during an exacerbation or intense episode of pain, as well as following a pain episode can greatly influence the experience of pain, subsequent pain episodes, and response to treatment (e.g., Burns et al. 2003; Merrick and Sjolund 2009). Moreover, the methods patients use to control their emotional arousal and symptoms are important predictors of both cognitive and behavioral responses.

Self-control

The CB perspective on pain management focuses on providing patients with a repertoire of techniques to help them gain a sense of control over the effects of pain on their lives, as well as actually modifying the affective, behavioral, cognitive, and sensory facets of the experience (Flor and Turk 2011). Behavioral experiences help to show patients that they are capable of more than they assumed, increasing their sense of personal competence. Cognitive techniques (e.g., problem solving and coping skills training) help to place affective, behavioral, cognitive, and sensory responses under the patient's control. The assumption is that long-term maintenance of behavioral changes will only occur if the patient has learned to attribute success to his or her own efforts. There are suggestions that these treatments can result in changes of beliefs about pain, coping style, and reported pain severity, as well as direct behavior changes. Further, treatment that results in increases in perceived control over pain and decreased ▶ catastrophizing also are associated with decreases in pain severity ratings and functional disability (Sullivan et al. 2001; Turner and Aaron 2001). The most important factor in poor coping may be the presence of catastrophizing, rather than differences in the nature of specific adaptive coping strategies (Jensen et al. 1999; Keefe et al. 2004).

In summary, people have prior learning histories that precede the development of their current pain. Based on these experiences, they have developed cognitive schema that consist of all information acquired over their lifetime. These cognitive schemata serve as the filters through which all subsequent experiences are perceived and to which they are responded. Thus, it is essential that people with chronic pain be viewed as integrated wholes, not body parts that are damaged and requiring repair. Failure to attend to cognitive, affective, and behavioral influences either by a narrow focus on physiology and anatomy, as is the case in the traditional medical model, or exclusively on environmental influences, as is emphasized in operant models, will prove to be inadequate. People are more than physical parameters or pawns of reinforcement contingences. Rather, they observe, anticipate, and interpret internal and external stimuli. Moreover, people exist in a social environment and contextual factors also play an important role in the pain experience. Since chronic pain is by virtue of its sole defining characteristic chronic, and since there is no cure, it is essential that treatment focuses on their adaptation to the symptoms and accompanying problems and not just the symptoms and physical pathology.

References

Burns, J. W., Kubilus, A., Bruehl, S., Harden, R. N., & Lofland, K. (2003). Do changes in cognitive factors influence outcome following multidisciplinary treatment for chronic pain? A cross-lagged panel analysis. *J. Consult. Clin. Psychol., 71*, 81–91.

Flor, H., & Turk, D. C. (2011). *Chronic pain: An integrated biobehavioral approach.* Seattle: IASP Press.

Jensen, M. P., Romano, J. M., Turner, J. A., et al. (1999). Patient beliefs predict patient functioning: Further support for a cognitive-behavioral model of chronic pain. *Pain, 81*, 95–104.

Keefe, F. J., Affleck, G., France, C. R., et al. (2004). Gender differences in pain, coping, and mood in individuals having osteoarthritic knee pain: A within day analysis. *Pain, 110*, 571–577.

Leeuw, M., Goossens, M. E., Linton, S. J., Crombez, G., Boersma, K., & Vlaeyen, J. W. (2007). The fear-avoidance model of musculoskeletal pain: Current state of scientific evidence. *J. Behav. Med., 20*, 77–94.

Martell, B. A., O'Connor, P. G., Kerns, R. D., Becker, W. C., Morales, L., Knashawn, H., Kosten, T. R., & Fiellin, D. A. (2007). Systematic review: opioid treatment for chronic back pain: prevalene, efficacy, and association with addiction. *Ann. Intern. Med., 146*, 116–127.

Merrick, D., & Sjolund, B. H. (2009). Patients' pretreatment beliefs about recovery influence outcomes of a pain rehabilitation program. *Eur. J. Phys. Rehabil. Med., 45*, 391–401.

Sullivan, M. J. L., Thorn, B., Haythornthwaite, J. A., et al. (2001). Theoretical perspectives on the relation between catastrophizing and pain. *Clin. J. Pain, 17*, 52–64.

Turk, D. C., Meichenbaum, D., & Genest, M. (1983). *Pain and behavioral medicine: A cognitive-behavioral perspective.* Guilford: New York.

Turk, D. C., & Okifuji, A. (1997). What factors affect physicians' decisions to prescribe opioids for chronic non-cancer pain patients? *Clin. J. Pain, 13*, 330–336.

Turk, D. C., & Salovey, P. (1985). Cognitive structures, cognitive processes, and cognitive-behavioral modification: I. Client issues. *Cogn. Ther. Res., 9*, 1–17.

Turner, J. A., & Aaron, L. A. (2001). Pain-related catastrophizing: What is it? *Clin. J. Pain, 17*, 65–71.

Vlaeyen, J. W. S., & Linton, S. J. (2000). Fear-avoidance and its consequences in chronic musculoskeletal pain: A state of the art. *Pain, 85*, 317–332.

Cognitive-Behavioral Programs

▶ Interdisciplinary Pain Rehabilitation

Cognitive-Behavioral Theory and Therapy

Definition

Cognitive-behavioral theories of the relationship between pain and depression suggest that a person's negative cognitive appraisals of pain may contribute to the development of depression.

Cross-References

▶ Depression and Pain

Cognitive-Behavioral Therapy

Synonyms

CBT

Definition

Cognitive-behavioral therapy refers to a class of psychotherapeutic techniques capable of addressing a range of psychological problems including chronic pain, anxiety, depression, or specific symptoms of psychosis. A central assumption of CBT is that a person's emotional and behavioral reactions to adverse circumstances are determined by their cognitive representation of the circumstances, and their appraisal of them, thus being able to influence a person's mood or behavior. The main aim of therapy is therefore to change the way in which the individual represents the circumstance and its appraisal. This is achieved through active elicitation of a person's thoughts and beliefs, and active behavioral and cognitive tasks that are designed to enable the person to reevaluate the circumstances in a manner that is likely to be beneficial for him or her.

Cross-References

▶ Cognitive-Behavioral Perspective of Pain
▶ Cognitive-Behavior Treatment of Pain
▶ Coping and Pain
▶ Dyspareunia and Vaginismus
▶ Impact of Familial Factors on Children's Chronic Pain
▶ Premenstrual Syndrome
▶ Psychiatric Aspects of Visceral Pain
▶ Psychological Treatment in Acute Pain
▶ Psychological Treatment of Headache

▶ Psychological Treatment of Pain in Children
▶ Psychology of Pain, Efficacy of Cognitive-Behavioral Treatment
▶ Recurrent Abdominal Pain
▶ Relaxation in the Treatment of Pain

Cognitive-Behavioral Transactional Model of Family Functioning

Definition

Cognitive-behavioral transactional model of family functioning emphasizes the role of cognitions and beliefs in the appraisal of pain and pain-related behavioral interactions.

Cross-References

▶ Spouse, Role in Chronic Pain

Cognitive-Behavioral Treatment

▶ Psychological Treatment of Headache

Cohort

Definition

Cohort refers to a group, band, or body of people who are followed over time.

Cross-References

▶ Prevalence of Chronic Pain Disorders in Children

Cohort Study

▶ Longitudinal Study

Coital Pain

▶ Dyspareunia and Vaginismus

Cold Allodynia

Definition

Pain produced by a normally non-painful cold stimulus.

Cross-References

▶ Neuropathic Pain Model, Partial Sciatic Nerve Ligation Model
▶ Neuropathic Pain Model, Tail Nerve Transection Model

Cold Allodynia Test

Definition

One or two drops of acetone are applied to the plantar aspect of the skin. Rapid evaporation of the acetone causes modest cooling. In rats, this procedure does not normally produce a significant withdrawal or avoidance reaction. However, in the presence of neuropathic hypersensibility to cooling (e.g., in the presence of mononeuropathy), a vigorous nocifensive reaction may be induced, lasting for up to 20 s.

Cross-References

▶ Allodynia Test, Mechanical and Cold Allodynia
▶ Thalamotomy, Pain Behavior in Animals

Cold Hyperalgesia

Definition

Intense pain induced by a cooling stimulus that normally evokes only modest pain. This is a common symptom in neuropathic pain patients.

Cross-References

▶ Nociceptors, Cold Thermotransduction

Cold Nociception

▶ Nociceptors, Cold Thermotransduction

Cold Nociceptors

Donald A. Simone
Diagnostic and Biological Sciences, University of Minnesota, Minneapolis, MN, USA

Definition

Nociceptors are sensory receptors located on peripheral nerve endings that are excited by noxious or painful stimuli. Subgroups of nociceptors are excited by cold and encode the intensity of cold stimuli.

Characteristics

In humans, reduction of skin temperature to approximately 15 °C or less evokes a sensation of pain. Interestingly, pain from noxious cold can have different qualities, including cold, burning, pricking, and aching, depending on the intensity, duration, rate of stimulation, and location of the cold stimulus (Chéry-Croze 1983; Davis and Pope 2002; Kunkle 1949). This suggests that different classes of nociceptors are excited by noxious cold stimuli.

Early studies documented that a portion of cutaneous Aδ and C nociceptors that are responsive to mechanical stimuli are excited by noxious cold (Bessou and Perl 1969; Burgess and Perl 1967; Georgopoulos 1977; LaMotte and Thalhammer 1982; Saumet et al. 1985). Many of the nociceptors excited by cold are also excited by heat and are considered to be polymodal nociceptors. However, the exact proportion of Aδ and C nociceptors excited by cold varied among the studies because stimulus temperatures and species used varied. For example, nociceptors in monkeys were not excited by stimuli to 15 °C (Perl 1968), whereas 78 % of Aδ and C nociceptors were excited by the application of ice to the skin (LaMotte and Thalhammer 1982). In an extensive study of the response properties of the various subtypes of nociceptors in rats, Leem et al. (1993) found that approximately 10 % or less of Aδ and C nociceptors were excited by cold. Using stimulus temperatures between 27 °C and 12 °C, they found that response thresholds of nociceptors were between 22 °C and 12 °C. However, when a wider range of stimulus temperatures was used, it was found that nearly all mechanosensitive nociceptors in rat hairy skin were excited by noxious cold (Simone and Kajander 1996; Simone and Kajander 1997). Response thresholds for Aδ nociceptors ranged from 14 °C to −18 °C; 79 % of Aδ nociceptors had cold response thresholds at or below 0 °C. Most C nociceptors exhibited response thresholds above 0 °C and thresholds ranged from 12 °C to –6 °C. A wide range of cold response thresholds was also observed for nociceptors in mouse skin (Cain et al. 2001).

Sensitivity of nociceptors to cold has also been described in humans (Campero et al. 1996; Torebjörk 1974). Using stimulus intensities between 19 °C and 0 °C, it was found that 40 % of C nociceptors that were sensitive to mechanical and heat stimuli were also excited by noxious cold stimuli.

A few studies have examined the intensity encoding properties of nociceptors for noxious cold stimuli (Georgopoulos 1977; Simone and Kajander 1996; Simone and Kajander 1997). Using a range of stimulus temperatures from 20 °C to −12 °C (or sometimes to −12 °C), each for a duration of 10 s, it was found that responses of Aδ and C nociceptors in rats increased as stimulus temperature decreased (Simone and Kajander 1996; Simone and Kajander 1997). The number of evoked impulses, the average discharge rate, and the peak discharge rate each increased as stimulus temperature decreased. Discharge rates evoked by stimulus temperatures greater than 0 °C were typically low (less than 1 impulse per second) and increased with colder temperatures. Power functions were generated to determine the stimulus-response relationship for Aδ nociceptors, and the slopes of the power functions ranged from 0.12 to 2.28.

In summary, the majority of cutaneous nociceptors that are excited by mechanical stimuli, including polymodal nociceptors, are excited by noxious cold stimuli. Cutaneous nociceptors exhibit a wide range of cold response thresholds and encode stimulus intensity.

References

Bessou, P., & Perl, E. R. (1969). Response of cutaneous sensory units with unmyelinated fibers to noxious stimuli. *Journal of Neurophysiology, 32*, 1025–1043.

Burgess, P. R., & Perl, E. R. (1967). Myelinated afferent fibres responding specifically to noxious stimulation of the skin. *The Journal of Physiology, 190*, 541–562.

Cain, D. M., Khasabov, S. G., & Simone, D. A. (2001). Response properties of mechanoreceptors and nociceptors in mouse glabrous skin: An in vivo study. *Journal of Neurophysiology, 85*, 1561–1574.

Campero, M., Serra, J., & Ochoa, J. L. (1996). C-polymodal nociceptors activated by noxious low temperature in human skin. *The Journal of Physiology, 497*, 565–572.

Chéry-Croze, S. (1983). Painful sensation induced by a thermal cutaneous stimulus. *Pain, 17*, 109–137.

Davis, K. D., & Pope, G. E. (2002). Noxious cold evokes multiple sensations with distinct time courses. *Pain, 98*, 179–185.

Georgopoulos, A. (1977). Stimulus-response relations in high-threshold mechanothermal fibers innervating primate glabrous skin. *Brain Research, 128,* 547–552.

Kunkle, E. C. (1949). Phasic pains induced by cold. *Journal of Applied Physiology, 1,* 811–824.

LaMotte, R. H., & Thalhammer, J. G. (1982). Response properties of high-threshold cutaneous cold receptors in the primate. *Brain Research, 244,* 279–287.

Leem, J. W., Willis, W. D., & Chung, J. M. (1993). Cutaneous sensory receptors in the rat foot. *Journal of Neurophysiology, 69,* 1684–1699.

Perl, E. R. (1968). Myelinated afferent fibers innervating the primate skin and their response to noxious stimuli. *The Journal of Physiology (London), 197,* 593–615.

Saumet, J.-L., Chéry-Croze, S., & Duclaux, R. (1985). Response of cat skin mechanothermal nociceptors to cold stimulation. *Brain Research Bulletin, 15,* 529–532.

Simone, D. A., & Kajander, K. C. (1996). Excitation of rat cutaneous nociceptors by noxious cold. *Neuroscience Letters, 213,* 53–56.

Simone, D. A., & Kajander, K. C. (1997). Responses of cutaneous A-fiber nociceptors in noxious cold. *Journal of Neurophysiology, 77,* 2049–2060.

Torebjörk, H. E. (1974). Afferent C units responding to mechanical, thermal and chemical stimuli in human non-glabrous skin. *Acta Physiologica Scandinavica, 92,* 374–390.

Cold Pressure

▶ Pain in Humans, Thermal Stimulation (Skin, Muscle, Viscera), Laser, Peltier, Cold (Cold Pressure), Radiant, Contact

Cold Stimulation

▶ Pain in Humans, Thermal Stimulation (Skin, Muscle, Viscera), Laser, Peltier, Cold (Cold Pressure), Radiant, Contact

Cold Therapy

▶ Therapeutic Heat, Microwaves, and Cold

Cold Thermoreceptor

Definition

An afferent nerve fiber characterized by nerve terminals which are excited in a graded way by cooling the skin. High- and low-threshold cold receptors have been distinguished. Most of the former belong to the C-fiber class, among the sensitive cold receptors in human skin are many A delta fibers. A characteristic transduction molecule of cold fibers is the TRPM8 receptor which can also be activated by the chemical menthol.

Cross-References

▶ Cold Nociceptors
▶ Nociceptors, Cold Thermotransduction

Cold Thermotransduction

▶ Nociceptors, Cold Thermotransduction

Cold-Induced Hyperalgesia

▶ Freezing Model of Cutaneous Hyperalgesia

Cold-Pressor Task

▶ Experimental Pain in Children

Colitis

Definition

Inflammation of the large intestine, or colon, often caused by a primary disease, irritation of the bowel, antibiotic use, or ulceration. Symptoms may include abdominal pain, bloating, diarrhea, dehydration, and increased intestinal gas. Acute colitis can be produced experimentally in animals by the intracolonic instillation of an irritant, such as acetic acid, mustard oil, trinitrobenzene sulfonic acid, etc. Colitis in animals is usually quantified histologically or by myeloperoxidase activity and is behaviorally assessed as colorectal hypersensitivity.

Cross-References

► Amygdala, Pain Processing and Behavior in Animals
► Animal Models of Inflammatory Bowel Disease
► Visceral Pain Model, Lower Gastrointestinal Tract Pain

Collateral Sprouting

Definition

Induction of new axonal sprouts from an intact nerve by nerve growth factor. The sprouts grow into an area of an adjacent, injured, degenerating, peripheral nerve. Cortical sensations related to this phenomenon can be related to either the injury itself, or the response to dividing the injured peripheral nerve to treat a painful neuroma. Collateral sprouting pain can be misinterpreted as failure of the original treatment for the painful scar or painful neuroma.

Cross-References

► Painful Scars

Colorectal Distension

Synonyms

CRD

Definition

Controlled innocuous or noxious visceral stimulus applied by distension of a balloon placed in the lumen of the descending colon and rectum. In humans, under normal conditions, distension pressures below 40 mmHg are felt as non-painful, whereas distension pressures greater than 40 mmHg are felt as painful.

Cross-References

► Descending Modulation of Visceral Pain
► Postsynaptic Dorsal Column Neurons, Responses to Visceral Input
► Spinal Dorsal Horn Pathways, Dorsal Column (Visceral)
► Visceral Pain Model, Lower Gastrointestinal Tract Pain

Colposcopy

Definition

Colposcopy is a diagnostic tool aimed at verifying the cause of abnormalities found in Pap smears. It involves a visual examination of the cervix, genitals, and vagina as well as the application of acetic acid to identify abnormal cells.

Cross-References

► Dyspareunia and Vaginismus

Combined Spinal Epidural (CSE) Technique

Definition

An anesthetic technique where both spinal and epidural anesthesia is administered to the same patient as one technique. It offers the advantages of a spinal anesthetic (intense analgesia) with the advantages of an epidural anesthetic (flexibility and long duration of action).

Cross-References

► Analgesia During Labor and Delivery

Commissural Myelotomy

▶ Midline Myelotomy

Common Fears

Definition

More specific fears directed at an apparent object (such as spider phobia, agoraphobia, fear of pain). They arise as the result of an interaction between the fundamental fears and learning experiences, and can thus be logically derived from these three fundamental fears.

Cross-References

▶ Fear and Pain

Common Migraine

▶ Migraine Without Aura

Common Toxicity Criteria

Definition

WHO and National Cancer Institute method of assessing the morbidity associated with cancer and cancer treatment by organ system.

Cross-References

▶ Cancer Pain Management, Chemotherapy

Communication Limitations or Neurological Impairment

▶ Pain in Children with Disabilities

Comorbid Pain Conditions

▶ Pain Mechanisms in Endometriosis

Comorbidities

▶ NSAID-Induced Lesions of the Gastrointestinal Tract

Comorbidity

Definition

Comorbidity defines an association between two or more disorders or diagnoses at one time that are more than coincidental.

Cross-References

▶ Preventive Migraine Therapy
▶ Psychiatric Aspects of the Epidemiology of Pain

Compartmental Syndrome

▶ Postoperative Pain, Compartment Syndrome

Compensation, Disability, and Pain in the Workplace

James P. Robinson[1] and Dennis C. Turk[2]
[1]Department of Anesthesiology & Pain Medicine, University of Washington, Seattle, WA, USA
[2]Department of Anesthesiology & Pain Research, University of Washington, Seattle, WA, USA

Introduction

Pain is the epitome of a private experience. Only the person with pain has direct access to the experience. At the same time however, pain has a public face, because people in pain communicate their suffering to others either verbally or by nonverbal behavior and often request assistance. Observers of someone in pain are challenged because they are not able to experience the other's pain directly. Thus, they are forced to make inferences and may well have difficulty interpreting the communications of the pain sufferer.

Social interactions related to pain occur in both formal and informal settings. As an example of the latter, a person with pain will communicate his or her suffering to family members. Research has demonstrated that the family members, in turn, respond to the pain sufferer in fairly predictable ways (Gauthier et al. 2011; Issner JB et al. 2012). But not only family members respond to the behavior of those who experience pain. Coworkers and supervisors and health-care providers are also asked to respond by making workplace accommodations in the case of the former and prescribing treatment in the latter. Health-care providers are also frequently asked to make recommendations regarding the impairments of individuals who report incapacitation because of pain. These may serve as the basis for financial compensation and accommodations on the job. The present field of this volume addresses social interactions related to pain that occur in a formal context – in particular, it addresses interactions between pain sufferers and (1) health-care providers who offer treatment and make decisions about disability and (2) societal agencies that provide benefits for work disability. Examples of such societal agencies include workers' compensation programs (which compensate individuals only if their medical problems were caused by work) and programs such as the United States Social Security Administration (which compensate disabled people regardless of the cause of their disability).

The problems of impairment and disability (these two interrelated concepts are defined and differentiated below) secondary to pain are elusive and yield conflicts and contradictions at multiple levels of analysis. The central problem relates to the fact that there is currently no objective measure of pain. As a consequence, decisions regarding the presence and impact of pain are based on inferences derived from prior experience and assumptions related to the amount of pain that might be expected given objective indications of pathology or disease. At a clinical level, conflicts and distrust frequently develop between disability applicants and the health-care provider and disability adjudicators who are involved in their claims. Independent medical examiners and attorneys often become participants in these conflicts. People experiencing pain often feel that physicians and others discount their reports; physicians and adjudicators may experience doubt or outright skepticism about reports of pain in the absence of corroborating physical findings. The entry ▶ Ethics of Pain-Related Disability Evaluations explores ethical and conceptual issues related to the task of making inferences about the pain of another person.

Characteristics

Disability Systems
Communities frequently provide assistance to people who are incapacitated. This type of helping behavior can be seen not only in modern societies, but also in primitive ones and even in communities of infrahuman primates (Fabrega 1997). Tolerance toward and care of the sick, infirm, injured, and disabled appears to be a fundamental feature of society dating back to prehistoric times (Ranavaya and Rondinelli 2000). Formal disability compensation systems were recorded as many as 4,000 years ago by the Babylonians, who provided compensation for loss of life or a body part incurred in the service of the state. During the time of the ancient Egyptians and Greeks, the state provided compensation for injuries caused by a wrongful act or occurring in the context of military service, respectively. Contemporary approaches to the issue of assistance for incapacitated people can be traced back to social insurance programs instituted in Germany in the late nineteenth century (Ritter 1983).

The social insurance programs that are most relevant to the present field of this volume are those that provide income maintenance benefits (also referred to as "wage replacement" or "time loss" benefits) for persons who are incapacitated from work. As noted in a comprehensive survey of social insurance programs in 138 countries and territories, programs to assist citizens who are incapacitated exist in most countries (Social Security Administration, Office of Research, Evaluation and Statistics 1999). Several of these are discussed as illustrations in the entries ▶ Impairment Rating, Ambiguity; ▶ Impairment Rating, Ambiguity, IAIABC System; ▶ Disability Evaluation in the Social Security Administration, and ▶ Rating Impairment due to Pain in a Workers' Compensation System.

Among income maintenance programs, a basic distinction can be made between work injury (or workers' compensation) programs and disability programs. Workers' compensation programs provide benefits for individuals who are disabled because they have sustained injuries or developed diseases "out of and in the course of employment" (p. 10) (Williams 1991). Disability programs in contrast provide benefits for people who are incapable of working for any reason. Thus, both workers' compensation programs and disability programs provide income maintenance only if an individual is judged to be unable to work because of a medical condition, but workers' compensation programs impose an additional criterion – that the medical condition was caused by the person's work activities.

A central thesis of this field in the Encyclopedic Reference of Pain is that important problems and significant ambiguities emerge as individuals with painful disorders interact with agencies that administer income maintenance programs. A corollary of this thesis is that the problems that emerge in these interactions will depend on specific features of disability programs. It is beyond the scope of this entry to discuss the complexities of disability programs, but even a cursory review of them reveals enormous variation from country to country in the manner in which programs are financed and integrated into a network of social security programs, and in the eligibility criteria and procedures they use to evaluate applicants (Social Security Administration, Office of Research, Evaluation and Statistics 1999). Some of the entries (e.g., the ones on patient credibility and pain behavior; see ▶ Credibility, Assessment and ▶ Nonorganic Symptoms and Signs) are relevant to any system of income maintenance for persons who allege work incapacity because of pain. However, many of the other entries discuss attempts by specific disability agencies to address problems presented by pain-related incapacitation. These entries should be viewed as providing examples of ways in which specific disability agencies have addressed problems associated with pain, rather than as comprehensive discussions of the ways in which these problems might be addressed.

At an administrative level, bureaucracies that have specific imperatives operate disability programs. For example, the Social Security Administration in the United States, which operates the two largest disability programs, strives for uniformity, objectivity, and cost containment in its disability evaluation procedures (Derthick 1990). One of the entries, ▶ Impairment, Pain-Related, points out that however sensible these goals are from a bureaucratic standpoint, they are often at odds with the clinical realities of patients with chronic pain (Osterweis et al. 1987; Robinson et al. 2004).

At a societal level, additional concerns and perspectives emerge. As noted previously, disability programs reflect an ethical commitment to support citizens who are incapable of working. If a program is too restrictive, it may consign needy and worthy applicants to poverty; if it is too lax, it may encourage disability applications from people who are actually capable of working. Even if a program is meeting its ethical mandate quite well, it may strain the financial resources of the payer to the breaking point. This is precisely the concern that is now being voiced about many of the programs run by the Social Security Administration in the United States.

At a scientific level, issues related to disability are usually unanswerable, because the research underpinnings for "disability evaluation science" are so meager. Moreover, the search for scientifically valid conclusions about disability may be compromised by the adversarial settings in which disability evaluations are conducted, and the enormous financial stakes involved (see ▶ Independent Medical Examinations).

Finally, at a philosophical level, the problem of disability evaluation in chronic pain rests on an epistemological dilemma – the information available to an individual suffering pain is fundamentally different from the information available to any external observer who attempts to assess the pain. Scarry succinctly captured this dilemma when she asserted: "To have great pain is to have certainty; to hear that another person has pain is to have doubt"(p. 7) (Scarry 1985).

In clinical situations, patients frequently communicate the sense that they are severely incapacitated by pain and ask health-care providers to support their claims. Typically, the health-care providers who evaluate these patients cannot identify tissue damage or organ pathology that makes the limitations communicated by the patient seem inevitable, proportional, or even plausible. The health-care provider then has the dilemma of integrating the patient's subjective report with the objective evidence of tissue damage and organ pathology to establish some final judgment about the extent to which the patient really is incapacitated. At one extreme, a health-care provider might simply ignore a patient's self-assessments and make a disability determination based strictly on objective findings of tissue damage or organ pathology. At the opposite extreme, the health-care provider might accept the patient's description of the situation, and provide a disability rating that is congruent with the patient's self-assessment. Adjudicators at disability agencies face the same dilemma when they decide whether or not to award benefits to disability applicants. It has proved extremely difficult to find some intermediate position in which both objective evidence and self-assessments by patients can be incorporated into disability evaluations (see ▶ Impairment Rating, Ambiguity;

▶ Impairment Rating, Ambiguity, IAIABC System; ▶ Impairment, Pain-Related).

The issues discussed in this field are particularly relevant to five groups of professionals: (1) clinicians who treat patients with chronic pain, (2) physicians who conduct independent medical examinations (see ▶ Independent Medical Examinations), (3) allied professionals who participate in disability mitigation programs (e.g., vocational ▶ Rehabilitation counselors), (4) individuals involved in disability policy, and (5) psychologists interested in the study of interpersonal perception and the ways in which social interactions influence pain.

Definitions: Disability and Impairment

"Disability" and "impairment" are fundamental concepts that provide the conceptual cornerstones for disability programs. Disability can be informally defined as the inability to carry out certain activities because of a medical problem (an abnormality in the structure or function of an organ or body part). There are different kinds of disability, since there are different types of activities that might be affected by a physical disorder. For example, a person might be disabled in the sense of being unable to work, in the sense of being unable to manage his or her personal finances or in the sense of being unable to perform activities of daily living independently. The current field focuses specifically on work disability – the inability to engage in substantial, gainful employment.

Unfortunately, there is no unique formal definition of work disability, since various agencies that administer disability programs define the term slightly differently. These differences reflect a fundamental reality about disability agencies and the criteria used for making decisions. For a disability agency, definitions serve the practical function of identifying the criteria that applicants must meet in order to be eligible for benefits. Thus, agencies have different definitions, because they have different mandates and different eligibility criteria.

The United States Social Security Administration defines disability as: "the inability to engage in any substantial gainful activity...by reason of any medically determinable physical or mental impairment that can be expected to result in death or that has lasted or can be expected to last for a continuous period of not less than 12 months" (p. 2) (SSA Publication 1994). This definition reflects three facts about eligibility criteria for the Social Security Disability Insurance and Supplemental Security Income programs: (1) applicants must be totally disabled from work, (2) the work disability must be "permanent" (or at least long-term), and (3) causation is irrelevant – that is, an individual is eligible for benefits regardless of how or why he or she became disabled. In contrast, the American Medical Association's Guides to the Evaluation of Permanent Impairment, 5th edition (AMA Guides) defines disability as "an alteration of an individual's capacity to meet personal, social, or occupational demands or statutory or regulatory requirements because of an impairment" (p. 8) (Cocchiarella and Andersson 2001). This very broad definition reflects the fact that the AMA Guides describe an evaluation system that is relevant to many kinds of disability (rather than just work disability) and that permits gradations in disability to be identified (rather than just the two categories of totally disabled vs. nondisabled).

An impairment can best be understood as a deficiency in the functioning of an organ or body part that leads to incapacitation or disadvantage in various arenas (such as inability to work or inability to do routine activities at home). "Impairment" is defined in a similar way by different disability agencies and expert panels. As an example, the formal definition of impairment given in the AMA Guides is: "a loss, loss of use or derangement of any body part, organ system or organ function" (p. 2) (Cocchiarella and Andersson 2001). Thus, one might say: "Mr. Jones' heart has been impaired since he suffered a myocardial infarction" or "Mrs. Brown's right hand is impaired because of her carpal tunnel syndrome." The entry ▶ Impairment, Pain-Related addresses problems that arise when the AMA Guides construct of impairment is applied to individuals who report incapacitation because of pain. Additional

discussions of the definitions and the relationships among body function and structures, activities, and participation as conceptualized by the World Health Organization are included in the entries ▶ Functioning and Disability Definitions and ▶ WHO System on Impairment and Disability.

Disability agencies assume significant linkages between impairment and disability. First, they construe impairment as a necessary condition for disability. The logic underlying this requirement is simple. Disability programs are designed to assist people who are incapable of competing in the workplace because of a medical condition. In essence, disability programs attempt to partition persons who fail in the workplace into two large groups, ones who fail because of a medical condition and ones who fail for other reasons (e.g., lack of demand for their skills). Therefore, disability programs require evidence that an applicant actually has a medical problem underlying his or her workplace failure. Impairment provides the necessary evidence, since it can be viewed as a marker that an individual has a medical problem that diminishes his or her capability. Conversely, if a person has no apparent impairment, this means that he or she does not have limitations due to a medical condition.

Disability agencies typically assume that the severity of a person's impairment is highly correlated with the severity (or probability) of his or her being disabled from work. Thus, even when an agency grants awards only on the basis of work disability, the agency will often seek information about a person's impairment in order to rationalize its decision about whether or not to award disability benefits. Studies have shown, however, that the association between physical impairments and work disability is far from perfect (Waddell 1987). As the above discussion suggests, the practical question facing a disability agency is to determine whether or not an applicant is sufficiently incapacitated to be eligible for disability benefits. Impairment evaluation should be viewed as an intermediate step in making this determination. Many of the entries in the field of compensation, disability, and pain in the workplace focus on disability and disability determination.

When Should a Disorder Be Construed as a Pain Problem?

Pain occurs in a wide range of medical disorders. It is usually conceptualized as a "component" of a disorder that is typically inextricably interwoven with other manifestations of disease or injury. From this perspective, it is somewhat arbitrary to abstract pain from other manifestations of disorders, just as it would be arbitrary to isolate fevers from malaria or shortness of breath from myocardial infarction.

A related point is that many painful disorders are treated by specialists of an affected organ system rather than by pain specialists. For example, chest pain is a cardinal symptom of cardiac ischemia, but a patient with chest pain of cardiac origin is much more likely to be treated by a cardiologist or a thoracic surgeon rather than by a pain specialist. Specialists in organ pathology typically view pain as a symptom of a biological abnormality that is important as a guide to appropriate diagnosis and treatment. In this model, symptoms and signs are used to diagnose disorders and treatment is directed toward the abnormal pathophysiology that comprises a disorder, rather than toward the symptoms of the disorder per se. The assumption is that once the underlying biological abnormality has been corrected, the symptoms – including pain – will abate.

The discussion in the previous paragraph raises a practical question. When should disorders that occur in the workplace be construed as pain problems? Two related criteria are suggested. A disorder is appropriately construed as a pain problem when either (1) patients report pain, but do not demonstrate structural or functional abnormalities that unequivocally explain their pain or (2) patients demonstrate limitations in activities that are not inevitable consequences of biological abnormalities, but rather appear to be consequences of pain that the patients experience as they engage in activities. One or both of these circumstances occur sufficiently frequently to warrant a thorough discussion of pain in the workplace.

Basic Problems Involving the Interface Between Disability Systems and Injured Workers with Chronic Pain

Objective Factors

Disability agencies strive to use disability evaluation procedures that are based on objective findings of incapacitation. This administrative objective seems innocuous enough; however, it has profound implications. In essence, disability agencies must make significant simplifying assumptions about disability in order to achieve their administrative goal. The most fundamental simplifying assumption is that impairment and disability should be "transparent" to an experienced physician, namely, that activity limitations as described or demonstrated by patients should be highly correlated with evidence of tissue damage or organ dysfunction that can be objectively assessed by a physician. This assumption underlies the routine demand that disability adjudicators make for physicians to rely on "objective findings" when they proffer conclusions about activity limitations of claimants.

The assumption that impairment and disability can be objectively assessed is so pervasive that most physicians – and essentially all disability adjudicators – accept it without question. The assumption is valid for certain medical conditions. For example, physicians have tools to quantify impairment stemming from amputations or complete spinal cord injuries. However, in many medical conditions – including most chronic pain syndromes – physicians cannot objectively identify impairments that rationalize the activity limitations that patients report. In a monograph on the disability programs administered by the Social Security Administration in the United States, Osterweis et al. (1987) summarize the problem as follows.

The notion that all impairments should be verifiable by objective evidence is administratively necessary for an entitlement program. Yet this notion is fundamentally at odds with a realistic understanding of how disease and injury operate to incapacitate people. Except for a very few conditions, such as the loss of a limb, blindness, deafness, paralysis or coma, most diseases and injuries do not prevent people from working by mechanical failure. Rather, people are incapacitated by a variety of unbearable sensations when they try to work (p. 28).

In essence, this statement indicates that in many medical disorders disability is related to subjective factors – especially pain – rather than to objectively measurable "mechanical failure" of body parts or organs. As discussed in several entries (▶ Impairment Rating, Ambiguity; ▶ Impairment Rating, Ambiguity, IAIABC System; ▶ Disability Evaluation in the Social Security Administration; ▶ Disability, Effect of Physician Communication; ▶ Rating Impairment due to Pain in a Workers' Compensation System), disability agencies have developed different solutions to the problem of reconciling the administrative imperative of basing disability determinations on objective findings with the reality that subjective factors are the proximate cause of work incapacitation for many patients.

Credibility

Many disability agencies have at least implicitly accepted the premise that determinations about eligibility for disability benefits should rely, at least in part, on claimants' reports about pain and other subjective experiences. To the extent that disability determination relies on subjective reports by claimants about the difficulties they have when working, agencies must be concerned about how to assess the integrity of these reports. In the extreme case, it is possible for malingerers to simulate incapacitation and receive disability benefits when they have essentially no work limitations. The problem of malingering is addressed in ▶ Malingering, Primary and Secondary Gain.

A more common scenario is that claimants report dramatic activity restrictions because of pain, and demonstrate severe functional limitations during physical examination. Examining physicians and disability agencies face a quandary when evaluating these patients, since, as noted above, it is often not possible to identify evidence of organ damage that makes the alleged activity restrictions seem inevitable, proportional, or even plausible. The entry ▶ Credibility, Assessment addresses general

problems in the assessment of credibility and veracity among patients with chronic pain. In ▶ Nonorganic Symptoms and Signs, the assessment of "nonorganic signs" on physical examination is specifically addressed, namely, the assessment of examination findings that indicate something other than well-described organic pathology.

Severity of Impairment and Disability

As discussed, disability agencies attempt to base decisions on objective evidence of organ or body part dysfunction. The assessment of disability secondary to pain is in conflict with this administrative imperative. People with chronic pain typically attribute their pain and activity limitations to dysfunction of an organ or body part. But these subjective reports are difficult to assess precisely because examination of the involved organ or body part often does not identify abnormalities that make the pain reports inevitable (Robinson et al. 2004). It often appears to an examiner that the affected organ or body part is capable of functioning, but that the claimant does not use it normally because of pain. Thus, the examiner has the challenge of determining how much weight to place on the claimant's subjective reports, as opposed to objective findings of organ or body part dysfunction (see ▶ Impairment, Pain-Related and ▶ Disability Evaluation in the Social Security Administration for discussions of attempts to integrated subjective reports with objective evidence in impairment/disability determination).

A comparison between rheumatoid arthritis (RA) and fibromyalgia syndrome (FMS) provides a dramatic example of the difficulty of assessing severity of impairment disability in a person with chronic pain. People with RA and FMS typically report comparably severe activity limitations. However, the former typically have observable evidence of joint inflammation or destruction, whereas physical and laboratory examinations for the latter are often completely normal (except for tenderness to palpation). The question for a disability agency is should the activity limitations reported by people diagnosed with FMS be given as much weight as those reported by people with RA for purposes of awarding disability benefits? This type of dilemma occurs routinely when claimants allege incapacitation secondary to pain. Three entries in this field cover specific examples of syndromes in which patients complain of severe pain, but have little or no evidence of tissue pathology that rationalizes the complaints – low back pain (Epidemiology of Work Disability, Back Pain), FMS (▶ Disability in Fibromyalgia Patients), and upper extremity injuries (▶ Disability, Upper Extremity).

Causation

Causation is important when a person is seeking disability benefits from an agency that is responsible only for medical conditions that arise in certain circumstances. In particular, workers' compensation carriers are responsible only for conditions that arise out of employment. The assessment of causation is thorny at both the conceptual (Kramer and Lane 1992; Lakoff and Johnson 1999) and practical levels. The difficulty in determining whether a medical condition has been caused by work is by no means limited to painful conditions. For example, controversy rages about whether and when hearing loss should be construed as work related (Dembe 1996). However, issues of causation are often particularly vexing for painful conditions that are commonly attributed to work. The difficulty stems in part from the fact that many disorders that are commonly attributed to work (e.g., non-radicular low back pain) often reflect a combination of degenerative changes and inciting physical loads. As a result, it is difficult to weigh the influence of work exposures with nonwork exposures. But another reason for the difficulty is that an examiner typically has no objective method for determining the severity of incapacitation in a patient with LBP. The claimant may allege that his LBP started or became worse during the course of activities at work, but the examiner will not be able to confirm or deny such a statement on the basis of objective medical information. Thus, the problem of determining causation in people with chronic pain overlaps with the problem of determining severity of disability – in both instances, the examiner must decide how much weight to assign to claimants' reports.

Disability agencies differ significantly in the standard they set for establishing causation. Some agencies follow the principle that in order for an index injury to be accepted as the cause of a claimant's impairment, the injury must be the significant factor contributing to the impairment. Others adopt a much lower standard of causation that has been described as "lighting up." When this standard applies, an index injury may be viewed as the cause of an impairment even when the injury is minor and impairment is severe. For example, consider a person with a multiply operated knee who falls at work, develops an effusion in the knee and is told by an orthopedist that he needs a total knee replacement. If the individual's workers' compensation carrier operated under the "lighting up" standard of causation, this person's knee symptoms and need for a total knee replacement would be viewed as caused by his slip and fall. The differences between disability agencies in the criteria required for establishing causation highlight the general point that the rules and regulations of different disability agencies vary greatly.

Iatrogenesis

Most clinicians who treat injured workers believe that some of them report symptoms that are not explained biologically, but rather reflect influences of the disability system as filtered by psychological tendencies. The term "disability syndrome" is often used to describe these people. A disability syndrome is conceptualized as a set of dysfunctional attitudes and beliefs that evolves over time following an injury (Robinson et al. 1997). One probable contributor to a disability syndrome is the injured worker's adaptation to nonwork roles; for example, an injured worker might take over childcare duties, while his wife enters the work force to make up for family income losses. Another likely contributor is the dysfunctional interactions that may occur between an injured worker and a disability agency.

As Hadler (1996) has pointed out, injured workers who claim ongoing work disability must run a gauntlet of challenges by their claims manager, threats to their benefits, and independent medical examinations. In these situations, they must convincingly portray themselves as incapacitated in order to maintain their benefits. Hadler argues that after trying so hard to convince others of their incapacity, injured workers are likely to have difficulty conceptualizing themselves as fit for work or to approach return to work with confidence. In addition, workers may develop resentment toward the disability agency with which they interact and thereby become more resistant to rehabilitation.

The disability syndrome construct is difficult to prove empirically. For example, it is possible that injured workers who fail to recover in a timely way had medical or psychosocial risk factors that existed at the time of their injury, but were recognized only after the workers showed a delayed recovery (Mustard and Hertzman 2001). However, the construct provides a plausible explanation for the common observation that an injured worker, who seems straightforward and highly motivated just after an injury, often becomes more resistant to rehabilitation at a later point.

The possibility that disability policies and agencies inadvertently promote disability is a cause for great concern. The concern is similar to that expressed by critics of welfare programs in the United States during the 1990s. The criticism of those programs was that by providing long-term financial support for unemployed, indigent members of the society, welfare programs actually promoted continued indigence and unemployment.

Epidemiology of Pain in the Workplace

As discussed above, there is ambiguity about the circumstances in which it is appropriate to label a medical disorder as a pain disorder. This ambiguity clouds the interpretation of data on the epidemiology of pain in the workplace. The ambiguity is increased further by the fact that the databases of workers' compensation systems and disability agencies generally do not provide data in a manner that highlights the role that pain plays in various injuries.

However, a few simplifying assumptions make it possible to obtain at least approximate

data regarding the frequency of work related disorders in which pain is a major problem, but objective evidence of biological injury is either minimal or bears an equivocal relation to the pain that injured workers report. First, many musculoskeletal disorders meet these criteria. The discussion below will focus on these disorders, and in particular on two types of musculoskeletal problems – LBP and upper extremity disorders associated with repetitive motion (see ▶ Disability, Upper Extremity). Second, although sprains can be associated with unequivocal evidence of a ▶ Structural Lesion (e.g., a complete tear of the anterior cruciate ligament would be coded as a sprain), workers' compensation carriers routinely use the designation sprain/strain to describe musculoskeletal complaints in the absence of definite evidence of a significant structural lesion. Thus, the designation "sprain or strain" in a workers' compensation database can be taken as a proxy for a disorder characterized primarily by musculoskeletal pain in the absence of a definite structural lesion. Third, many of the disorders coded as upper extremity repetitive motion disorders meet the above criteria, since they are not associated with definite evidence of structural lesions (Miller and Topliss 1988).

Common Work Injuries and Illness

Workers' compensation carriers generally distinguish between work injuries and occupational illnesses (Blessman 1991). Conceptually, an injury involves a single event, whereas an illness arises gradually as a consequence of repeated work exposures. The distinction between a work injury and an occupational illness is obvious if a logger who is crushed by a tree is compared to a carpenter who develops a mesothelioma after years of exposure to asbestos. However, the distinction becomes opaque when it is applied to musculoskeletal disorders. For example, episodes of LBP are coded as injuries, whereas carpal tunnel syndrome and other upper extremity disorders associated with repetitive motion are coded as occupational illnesses.

Compensation, Disability, and Pain in the Workplace, Table 1 1988 Washington State injury profile by nature of injury

Nature of injury	I All claims (% of total claims)	II % of claims in column 1 that led to time loss >120 days	III # of claims with time loss >120 days (% of total claims with time loss >120 days)
Contusions, bruises, cuts, lacerations, punctures, scratches, abrasions	75,713 (47.6 %)	0.8 %	644 (9.6 %)
Sprains, Strains	59,729 (37.6 %)	7.5 %	4,456 (66.1 %)
Fractures	7,424 (4.7 %)	11.9 %	882 (13.1 %)
Burns	5,777 (3.6 %)	0.6 %	38 (0.6 %)
Multiple injuries	2,689 (1.7 %)	10.2 %	273 (4.0 %)
Dislocations	1,483 (0.9 %)	19.0 %	282 (4.2 %)
Amputations	237 (0.1 %)	18.1 %	43 (0.6 %)
Other	5,918 (3.7 %)	2.1 %	123 (1.8 %)
Total	158,970 (100 %)	–	6,741 (100 %)

Washington State Department of Labor and Industries (unpublished)

Tables 1 and 2 provide unpublished data about injuries that occurred among workers covered by the Department of Labor and Industries in Washington State of the United States in 1988. Table 1 provides a breakdown of injuries according to the mechanism of injury. Column 1 indicates that the largest proportion of injuries were classified as contusions, bruises, cuts, lacerations, punctures, scratches or abrasions. The second largest group consisted of sprains and strains. Table 2 provides a breakdown of conditions (injuries and illnesses combined) according to the body part affected. The most commonly affected body part is the hand or fingers. However, over 27,000 injuries involved the back, with an additional 8,427 involving the back and neck. Approximately 90 % of the back injuries were diagnosed as sprains or strains.

Data from the Bureau of Labor Statistics in the United States (US Department of Labor 1995) indicate that approximately 7 % of all claims filed during 1992 were for occupational illnesses rather than injuries. Approximately 60 % of the claims for illnesses involved repetitive trauma, usually of the upper extremity. (Carpal tunnel syndrome is a subset of upper extremity illnesses caused by repetitive trauma). The second and third largest groups are skin disorders (14 %) and respiratory conditions due to toxic agents (5 %). The most frequent types of repetitive activities associated with occupational illnesses are repetitive grasping or moving of objects (other than tools) and repetitive use of tools. As one would expect from the types of repetitive activities that produce occupational illnesses, the overwhelming majority of them occur in manufacturing jobs. In fact, the 25 industries with the highest rates of repetitive trauma occupational illnesses are all in the manufacturing sector, with meat packing plants and manufacturers of motor vehicles leading the way. The Bureau of Labor Statistics also identifies keyboarding as a significant cause of repetitive trauma induced occupational illnesses, but it is much less important than repetitive grasping or repetitive use of tools.

The Effect of Workplace Factors
An enormous amount of research has been conducted to determine work place factors that affect the probability that workers will become injured. The entry ► Pain in the Workplace, Risk Factors for Chronicity, and Workplace Factors reviews many of the work place factors that have been investigated. The entry ► Ergonomics provides a historical overview of the field of ergonomics and summarizes research on risk factors for upper extremity work related musculoskeletal disorders and LBP. This research has supported the conclusion that the risk of upper extremity musculoskeletal disorders is increases when jobs require forceful, repetitive motions, especially

Compensation, Disability, and Pain in the Workplace, Table 2   1988 Washington State injury and illness profile by body part

Body part	I All claims (% of total claims)	II % of claims in column 1 that led to time loss >120 days	III # of claims with time loss >120 days (% of total claims with time loss >120 days)
Finger	29,721 (17.7 %)	0.7 %	207 (2.8 %)
Back	27,196 (16.2 %)	9.6 %	2,598 (34.6 %)
Eyes	17,981 (10.7 %)	0.1 %	19 (0.2 %)
Hand/wrist	17,444 (10.4 %)	3.7 %	643 (8.5 %)
Back and Neck	8,427 (5.0 %)	8.7 %	734 (9.8 %)
Knee	7,478 (4.4 %)	8.1 %	609 (8.1 %)
Foot	6,611 (3.9 %)	2.6 %	173 (2.3 %)
Multiple	5,583 (3.3 %)	10.2 %	571 (7.6 %)
Other	47,616 (28.3 %)	4.1 %	1,945 (25.9 %)
Total	168,057 (100 %)	–	7,499 (100 %)

Washington State Department of Labor and Industries (unpublished)

when workers are also subjected to vibration and have to maintain awkward positions. Research has also found four factors associated with an increased risk of LBP, (1) lifting and forceful movements, (2) awkward postures, (3) heavy physical work, and (4) whole body vibration.

The entry ▶ Pain in the Workplace, Risk Factors for Chronicity, Job Demands also addresses workplace factors that influence the frequency of work injuries. It considers ergonomic factors briefly and also reviews several psychosocial work place factors that influence injury rates, including monotonous work, lack of personal control, and job dissatisfaction.

Conditions Associated with Prolonged Time Off Work

Bureau of Labor Statistics data (US Department of Labor 1995) in the United States indicate that 66 % of all workers who filed injury or occupational illness claims in 1992 did not miss time from work. Among claims that are associated with work disability, it is important to distinguish ones associated with short periods off work versus ones associated with prolonged time loss. Although any injured worker who reports that he or she is unable to work must be viewed with concern, most injuries involving time off work end uneventfully with the worker returning to his or her job after a short time period. For example, among workers in 1992 who went off work because of a back injury, the median time off work was 7 days (US Department of Labor 1995).

As far as long-term work disability is concerned, Column 1 of Table 1 shows that contusions, bruises, cuts, lacerations, punctures, scratches, and abrasions represented the largest category of industrial injuries in Washington State during 1988, but Column 2 indicates that only 0.8 % of patients with these conditions lost more than 120 days from work. As a result, these patients represented only 9.6 % of the total of patients with time loss of more than 120 days (column 3). Further examination of Columns 2 and 3 reveals that the probability of protracted time loss is high for dislocations, amputations, fractures, and multiple injuries. It is moderately high for sprains and strains. However, since sprains and strains are relatively common to begin with, patients with these conditions represent 66 % of the total number of patients with time loss greater than 120 days.

Columns 2 and 3 of Table 2 provide similar information for industrial conditions in Washington State as a function of affected body part (injuries and illnesses are combined in this table). Inspection reveals that 9.6 % of patients with back disorders remain off work more than 120 days. Since back problems occur frequently, fully 34.6 % of the group who miss more than 120 days from work do so because of back problems.

Summary

Many vexing and interrelated issues are associated with the assessment and compensation of individuals who develop painful conditions in the workplace. The entries included in this field were designed to be evocative and to offer insightful perspectives on the important nuances that come into play when judgments are made about such individuals. Some of the entries describe efforts that selected disability agencies have made to address problems associated with the assessment of impairment and disability in injured workers with painful conditions. But rather than providing clear and definitive answers to the difficult issues surrounding pain in the workplace, the entries should sensitize readers to the complexities and ambiguities in this area.

References

Blessman, J. E. (1991). Differential treatment of occupational disease v occupational injury by workers' compensation in Washington State. *Journal of Occupational Medicine, 33*, 121–126.

Cocchiarella, L., & Andersson, G. B. J. (2001). *Guides to the evaluation of permanent impairment* (5th ed.). Chicago: AMA Press.

Dembe, A. E. (1996). *Occupation and disease*. New Haven: Yale University Press.

Derthick, M. (1990). *Agency under stress*. Washington, DC: The Brookings Institution.

Fabrega, H. J. (1997). *Evolution of sickness and healing*. Berkeley: University of California Press.

Gauthier, N., Thibault, P., Sullivan, M. J. (2011). Catastrophizers with chronic pain display more pain behaviour when in a relationship with a low catastrophizing spouse. *Pain Res Manag, 16*(5), 293–299.

Hadler, N. M. (1996). If you have to prove you are ill, you can't get well. The object lesson of fibromyalgia. *Spine, 21*, 2397–2400.

Issner, J. B., Cano, A., Leonard, M. T., Williams, A. M. (2012). How do I empathize with you? Let me count the ways: relations between facets of pain-related empathy. *Journal of Pain, 13*(2):167–175.

Kramer, M. S., & Lane, D. A. (1992). Causal propositions in clinical research and practice. *Journal of Clinical Epidemiology, 45*, 639–649.

Lakoff, G., & Johnson, M. (1999). *Philosophy in the flesh*. New York: Basic Books.

Miller, M. H., & Topliss, D. J. (1988). Chronic upper limb pain syndrome (repetitive strain injury) in the Australian workforce: A systematic cross sectional rheumatological study of 229 patients. *Journal of Rheumatology, 11*, 1705–1712.

Mustard, C., & Hertzman, C. (2001). Relationship between health services outcomes and social and economic outcomes in workplace injury and disease: Data sources and methods. *American Journal of Industrial Medicine, 40*, 335–343.

Osterweis, M., Kleinman, A., & Mechanic, D. (Eds.). (1987). *Pain and disability*. Washington, DC: National Academy Press.

Ranavaya, M. I., & Rondinelli, R. D. (2000). The major U.S. disability and compensation systems: Origins and historical overview. In R. D. Rondinelli & R. T. Katz (Eds.), *Impairment rating and disability evaluation* (pp. 3–16). Philadelphia: WB Saunders.

Ritter, G. A. (1983). *Social welfare in Germany and Britain*. New York: Berg.

Robinson, J. P., Rondinelli, R. D., & Scheer, S. J. (1997). Industrial rehabilitation medicine 1: Why is industrial rehabilitation medicine unique? *Archives of Physical Medicine and Rehabilitation, 78*, 3–9.

Robinson, J. P., Turk, D. C., & Loeser, J. D. (2004). Pain, impairment, and disability in the AMA guides. *The Journal of Law, Medicine & Ethics, 32*, 315–326.

Scarry, E. (1985). *The body in pain*. New York: Oxford University Press.

Social Security Administration, Office of Research, Evaluation and Statistics. (1994). *Disability evaluation under social security* (SSA Publication No. 64–039). Washington, DC: Government Printing Office.

Social Security Administration, Office of Research, Evaluation and Statistics. (1999). *Social security programs throughout the world – 1999* (SSA Publication no. 13–11805). Washington, DC: Government Printing Office.

U.S. Department of Labor, Bureau of Labor Statistics. (1995). *Occupational injuries and illnesses: Counts, rates, and characteristics, 1992*. Washington, DC: GPO.

Waddell, G. (1987). Volvo award in clinical sciences. A new clinical model for the treatment of low-back pain. *Spine, 12*, 632–644.

Williams, C. A. (1991). *An International comparison of workers' compensation*. Boston: Kluwer.

Competitive Agonist or Antagonist

Definition

Compound that interacts with the endogenous ligand-binding site of the receptor.

Cross-References

▶ Nociceptive Processing in the Amygdala: Neurophysiology and Neuropharmacology

Complement

Definition

Nonspecific mediator of humoral immunity. Many agents (antigens) trigger the complement cascade, which may ultimately result in the generation of membrane attack complex and lysis of cells presenting the triggering agent.

Cross-References

▶ Inflammatory Neuritis

Complementary Medicine

▶ Alternative Medicine in Neuropathic Pain

Complementary Therapies

▶ Alternative Medicine in Neuropathic Pain

Complete Freund's Adjuvant

Synonyms

CFA or FCA

Definition

An oil emulsion that contains a dead mycobacterium that elicits an inflammatory and immune response in vivo. Injection of CFA into the skin of the hind paw is frequently used as an experimental model of peripheral inflammation in mice or rats.

Cross-References

▶ CFA
▶ IB4-Positive Neurons, Role in Inflammatory Pain
▶ Nerve Growth Factor Overexpressing Mice as Models of Inflammatory Pain

Complex Adaptive System

Definition

A complex adaptive system is a large set of units that interact with each other, and with an external environment, to produce overall patterns that are significantly more complex than the behaviors of the individual entities comprising the system.

Cross-References

▶ Consciousness and Pain

Complex Chronic Pain in Children, Interdisciplinary Treatment

Jennifer N. Stinson
The Hospital for Sick Children, Toronto, ON, Canada

Synonyms

Multidisciplinary treatment; Pediatric chronic pain management; Pain-related disability

Definition

Chronic pain in children is defined as any recurring (e.g., headaches, abdominal, and limb pain) or persistent pain (e.g., back pain, cancer pain, ▶ complex regional pain syndrome). Chronic pain may begin as acute pain, but it continues beyond the normal time expected for resolution of the problem or persists or recurs for other reasons (APS 2012). While most children with chronic pain function quite well, approximately 5–8 % will develop more complex chronic pain conditions with associated distress and disability (Huguet and Miro 2008). These children typically present with subjective ratings of pain out of proportion to the objective physical findings and are more disabled than one would expect.

COGNITIVE FACTORS	BEHAVIOURAL FACTORS	EMOTIONAL FACTORS
Information about cause and prognosis Knowledge about practical drug and nondrug therapy Treatment expectations Identification of pain triggers Recognition of or knowledge about how to resolve stress	Distress responses Use drug and nondrug therapy Child/family responses Resolution of stressful situations Participation in routine activities	Anticipatory Anxiety Heightened distress Fear Situation-specific stress Frustration Underlying anxiety and or depression

Tissue damage or Stressful situation

CHILD FACTORS
Age
Cognitive level
Gender
Temperament
Previous pain experience
Family learning
Culture

Pain

Complex Chronic Pain in Children, Interdisciplinary Treatment, Fig. 1 Model of situational and child factors that modify pain and disability

Introduction

Complex chronic pain may adversely affect all aspects of children's lives. They stop going to school, refrain from exercising or playing sports for fear of increasing the pain and withdraw from their peers. These children often have impaired sleep, suffer from anxiety and or depression, and have disrupted family relationships (Gold et al. 2009; Tsao et al. 2007). The pain becomes the focus of the children's lives and a vicious cycle of pain and disability develops. Furthermore, many factors such as how the child thinks, feels, and behaves in response to the pain can intensify the pain and distress and prolong the disability. Some of the common situational and child factors that intensify pain and prolong disability are outlined in Fig. 1 (McGrath and Hillier 2003). These children also tend to have components of ▸ nociceptive (normal pain) and ▸ neuropathic

pain (nerve pain), which make them more difficult to treat. Therefore, ▸ interdisciplinary chronic pain teams are necessary to treat complex chronic pain conditions adequately in order to ensure that assessments and interventions are child-centered (e.g., tailored to the needs of the individual child) rather than just focused on the underlying disease or condition.

Characteristics

Given that these complex chronic pain conditions have multiple causes; children must be treated from an interdisciplinary, ▸ multimodal, rehabilitation perspective. Drug, physical, and psychological therapies should be incorporated into a flexible, child-friendly program with particular attention to the cognitive, behavioral, and emotional factors contributing to the underlying problem. Because of the complexity of chronic pain, no single discipline has the expertise to assess and manage it independently. Therefore, specialized interdisciplinary chronic pain teams are now considered the standard of care for children with complex chronic pain conditions.

Composition of Interdisciplinary Pain Teams

Chronic pain teams for children generally include specialized physicians (e.g., anesthesiologists, neurologists, psychiatrists), nurses, psychologists, and physical therapists. More recently, teams may also include complementary and alternative medicine practitioners (e.g., acupuncturists, massage therapists) (Peng et al. 2007). The specific team members involved in any one case depend on the individual needs of the child and family. A child's initial consultation typically includes either a joint team interview and physical examination or separate interviews with each healthcare professional. Comprehensive physical and psychosocial assessment may typically last a few hours to a full day, depending on the child's previous diagnostic tests and the team's core assessment battery (i.e., standardized sensory testing, questionnaires). The team then meets to formulate the child's pain diagnosis and treatment plan. The treatment plan should include: the diagnosis (underlying causes and contributing factors), rationale for a rehabilitative approach with a clear description of the specific treatment options and an opportunity for the family to help fine-tune the plan. It is essential to educate children and their families about the nature of chronic pain (e.g., different from acute pain where there is a single cause and a single treatment) and the factors that intensify it, as well as the drug and nondrug strategies they can use to control the pain. Some children's clinics also offer inpatient, day, or residential treatment programs (Hechler et al. 2009).

Goals of Interdisciplinary Chronic Pain Treatment

Interdisciplinary chronic pain programs use a rehabilitative approach where children's pain is treated with the most appropriate drug and nondrug therapies and where the team assists children and their parents to improve children's function despite their pain. In some instances, teams work with families to help them understand that their child's pain might not be eradicated fully, so that efforts are directed toward improving function and quality of life. Interdisciplinary treatment goals include:

- Comprehensive physical and psychosocial assessment of the child with pain and his/her family to evaluate etiology and contributing factors.
- Design and implementation of a flexible child-centered treatment approach that addresses all causative factors.
- The treatment plan typically with pharmacological, psychological, and physical therapies and in some cases medical (e.g., nerve blocks) intervention.
- The specific goals of the treatment plan including: (a) increasing independent function in terms of activities of daily living, school, social, and physical activity; (b) facilitating adaptive problem solving, communication, and coping skills; (c) treating specific problems identified from the comprehensive assessment (e.g., depression, anxiety), and (d) helping children and their families to understand the nature of pain, the pain condition, and its treatment from a holistic perspective.
- Ongoing assessment and reevaluation of the treatment plan. One way to monitor children with chronic pain is through the use of electronic pain diaries. Stinson and colleagues (2008) have developed and established the validity of a new electronic (phone-based) multidimensional chronic pain measure for children called the e-Ouch.

Pharmacological Therapies

Pharmacologic methods are an important component of an integrated, flexible approach that combines psychological and physical strategies. However, the empirical evidence for the pharmacological treatment of chronic pain in children is drawn from adult studies. The choice of medication depends in part on the source of pain (e.g., nociceptive, neuropathic, or mixed). Pain medications are tailored to the individual needs of each child based on the results of their assessment. Drug therapies are divided broadly into analgesics and adjuvant medication. Analgesics are administered in a stepwise approach and are recommended for pain conditions with characteristics of nociceptive or mixed pain. Simple analgesics, such as acetaminophen and nonsteroidal anti-inflammatory drugs (e.g., ibuprofen) in adequate doses, are effective for some children. Opioids may be added to the analgesics regimen when these mild analgesics do not alleviate the pain. In contrast, pain conditions with characteristics of neuropathic pain are often resistant to drug therapies that typically relieve nociceptive pain (e.g., opioids). Therefore, adjuvant pain medications such as anticonvulsants and tricyclic antidepressants are used (Finnerup et al. 2010). Gabapentin is the most commonly used anticonvulsant, as it is safe and well tolerated. Amitriptyline is the most commonly used tricyclic antidepressant and is often recommended for children whose sleep is disturbed. Antidepressants are also helpful for children who have chronic pain and who are depressed (see ▶ Analgesic Guidelines for Infants and Children).

Psychological Therapies
Many psychological therapies are available to treat chronic pain in children. These treatments include counseling, ▶ relaxation therapy, ▶ biofeedback, ▶ behavioral modification, and cognitive strategies including ▶ hypnosis and ▶ psychotherapy. Often these therapies are integrated into a comprehensive cognitive-behavioral therapy (CBT) program that is directed at identifying and ameliorating the thinking, behaving, and feeling factors that affect a child's pain and disability. A recent systematic review documented the efficacy of CBT for chronic headache, abdominal pain, and fibromyalgia in children (Palermo et al. 2010). CBT is often organized into a program of therapy that is delivered by various members of the chronic pain team. The content of the programs varies across clinics but usually includes teaching children specific pain and life coping skills, encouraging positive family responses for resuming typical activities, reframing or changing the family's beliefs that may impede rehabilitation, exercise therapy, education, and self-management strategies. The goal of these psychological therapies is to help children regain their lives and take back control from the pain. Finally, there is strong evidence that these psychological treatments can be effective without a therapist being physically present, using alternative models of service delivery such as the Internet (Palermo et al. 2010). For example, we have shown that youth with chronic pain due to arthritis had significantly lower pain following a 12 week Internet-based treatment program compared to a control group (Stinson et al. 2010). These types of programs might help to overcome some of the traditional barriers to treatment (e.g., stigma associated with these therapies, acceptability, and accessibility).

Physical Therapies
Chronic pain often leads a child to avoid physical activity due to fear of reinjury or because it exacerbates the pain. Lack of muscle use leads to loss of muscle strength, flexibility, and endurance and overall deconditioning. Therefore, physical therapies are an integral component, and in certain instances the cornerstone, of treatment for children with complex chronic pain problems. Exercise therapy, physiotherapy, thermal (heat and cold) and sensory (▶ desensitization, ▶ transcutaneous electrical nerve stimulation or TENS) stimulation, and massage are the most commonly used physical modalities. They are frequently used in combination. Regular exercise (e.g., 20 min 3 × per week) should help improve sleep, mood, self-esteem, and energy levels. However, maintaining normal daily activities such as school, sports, and play are often as effective as a formal exercise program. Some children will benefit from intensive physiotherapy. Physiotherapy is usually administered on an outpatient basis with the ultimate goal of teaching the child to execute the program at home. A program the child enjoys (e.g., swimming) and one where the amount of time spent in the activity is gradually increased is one the child is more likely to continue with (McCarthy et al. 2003).

Summary

In summary, children with complex chronic pain conditions experience prolonged suffering and disability. The pain adversely impacts all aspects of children's lives in terms of physical, psychological, social, and role functioning. Many sensory, cognitive, behavioral, and emotional factors may intensify the pain and prolong pain-related disability. Moreover, these complex pain conditions tend to have components of nociceptive and neuropathic pain, which makes them more difficult to treat. Given this complexity, unidisciplinary and unimodal treatments are rarely successful. Therefore, children with complex chronic pain conditions must be treated from an interdisciplinary, multimodal, rehabilitation perspective. Pharmacological, physical, and psychological therapies should be incorporated into a flexible, child-centered program.

References

American Pain Society. (2012). Assessment and Management of Children with Chronic pain: A Position Statement from the American Pain Society. From: http://www.ampainsoc.org/advocacy/downloads/aps12-pcp.pdf

Finnerup, N. B., Sindrup, S. H., & Jensen, T. S. (2010). The evidence for pharmacological treatment of neuropathic pain. *Pain, 150*, 573–581.

Gold, J. I., Ak, Y., Mahrer, N. E., et al. (2009). Pediatric chronic pain and health-related quality of life. *Journal of Pediatric Nursing, 24*, 141–150.

Hechler, T., Dobe, M., Kosfedler, J., et al. (2009). Effectiveness of a 3 week multimodal inpatient pain treatment for adolescents suffering from chronic pain: Statistical and clinical significance. *The Clinical Journal of Pain, 25*, 156–166.

Huguet, A., & Miro, J. (2008). The severity of pediatric chronic pain: An epidemiological study. *The Journal of Pain, 9*, 226–236.

McCarthy, C. F., Shea, A. M., & Sullivan, P. (2003). Physical therapy management of pain in children. In N. L. Schechter, C. B. Berde, & M. Yaster (Eds.), *Pain in infants, children and adolescents* (pp. 434–448). Philadelphia: Lippincott Williams and Wilkens.

McGrath, P. A., & Hillier, L. M. (2003). Modifying the psychological factors that intensify children's pain and prolong disability. In N. L. Schechter, C. B. Berde, & M. Yaster (Eds.), *Pain in infants, children and adolescents* (pp. 85–104). Philadelphia: Lippincott Williams and Wilkens.

Palermo, T. M., Eccleston, C., Lewandowski, A. S., Williams A. C. & Morley, S. (2010). Randomized controlled trials of psychological therapies for management of chronic pain in children and adolescents: an updated meta-analytic review, *Pain, 148*(3), 205–213.

Peng, P., Stinson, J. N., Choiniere, M., et al. (2007). Dedicated multidisciplinary pain management centres for children in Canada: The current status. *Canadian Journal of Anaesthesia, 54*, 985–991.

Stinson, J., Stevens, B. J., Feldman, B. J., et al. (2008). Construct validity of a multidimensional electronic pain diary for adolescents with arthritis. *Pain, 136*, 281–292.

Stinson, J. N., McGrath, P. J., Hodnett, E., et al. (2010). An internet-based self-management program with telephone support for adolescents with juvenile idiopathic arthritis: A pilot randomized controlled trial. *Journal of Rheumatology, 37*(9), 1944–1952.

Tsao, C. I., Meldrum, M., Kim, S. C., & Zeltzer, L. K. (2007). Anxiety sensitivity and health-related quality of life in children with chronic pain. *The Journal of Pain, 8*, 814–823.

Complex Regional Pain Syndrome

Synonyms

Complex regional pain syndrome type I (CRPS I) or Reflex sympathetic dystrophy (RSD); Complex regional pain syndrome type 2 (CRPS2) or Causalgia; CRPS; Sudeck's atrophy

Definition

CRPS is a medical syndrome characterized by chronic regionalized pain, usually including extreme tactile allodynia, coupled with a variable combination of other symptoms including edema, blood flow, color and temperature changes, sudomotor changes, dysfunction, and bone turnover changes, all disproportionate in degree to what is expected from the noxious inciting event (minor tissue injury, limb splinting, etc.). CRPS, which is a relatively new designation for this kind of disorder, occurs in two types. In CRPS 1, which was formerly known as reflex sympathetic dystrophy (RSD), there is no identifiable precipitating nerve injury. In CRPS 2, formerly known as causalgia, a precipitating nerve injury is present.

Cross-References

Complex Regional Pain Syndrome (CRPS)

▶ Neurosurgical Treatment of Pain

Complex Regional Pain Syndrome and the Sympathetic Nervous System

James N. Campbell
Johns Hopkins University School of Medicine, Baltimore, MD, USA

Synonyms

Causalgia; Reflex sympathetic dystrophy; Sympathetically maintained pain

Definition

▶ Sympathetically maintained pain (SMP) refers to pain that is dependent on neural activity in the sympathetic nervous system and is present in some patients with ▶ complex regional pain syndrome (CRPS).

Characteristics

Activity in nociceptors induces an increase in sympathetic discharge. It is in this way, in part, that noxious (painful) stimuli induce a rise in blood pressure. Usually, the converse is not true: sympathetic activity does not impact the discharge of nociceptive neurons. In certain patients with pain, however, nociceptors acquire sensitivity to norepinephrine released by sympathetic efferents. Pain dependent on activity in the sympathetic nervous system is referred to as sympathetically maintained pain (SMP).

This linkage with the sympathetic nervous system may, in some patients, be a dominant mechanism for pain. SMP, in particular, is noted in many cases of complex regional pain syndrome (CRPS). CRPS typically occurs after trauma that may or may not result in nerve injury. The distal extremities, areas rich in sympathetic innervation, are usually affected. The patients present with edema (at least in the early stages) and striking hyperalgesia. Patients with CRPS often have motor disability with difficulty moving the affected painful body parts, regardless of an intact sensory/motor pathway. The skin may be very cool and sweaty, or sometimes very warm compared to the opposite normal side. The disorder appears to spread in some patients from distal to proximal parts of the extremity and may in fact spread to other extremities as well.

SMP Is a Receptor Disorder
The dramatic relief of pain that occurs with selective blockage of the sympathetic nervous system defines SMP. Three general ideas have been advanced to explain this phenomenon: (1) the anesthetic blocks "pain" fibers that course with the sympathetic efferent fibers, (2) the

sympathetic nervous system is overactive and thus induces pain, (3) the sympathetic nervous system acquires the capacity to activate nociceptors and hence induce pain. Evidence from both human and animal studies supports the latter hypothesis.

Stimulation of the sympathetic chain induces pain in patients with causalgia (Walker and Nulson 1948; White and Sweet 1969), even when the sympathetic chain is disconnected from the spinal cord. Thus, efferent actions of the sympathetic nervous system account for SMP. Injection of noradrenaline around stump neuromas or skin in patients with postherpetic neuralgia induces an increase in spontaneous pain (Chabal et al. 1992; Raja et al. 1998; Choi and Rowbotham 1997). In patients relieved of pain after a sympathetic block, intradermal injection of norepinephrine into the previously hyperalgesic area in physiological concentrations induces pain (Ali et al. 2000). Norepinephrine injected into normal subjects evokes little or no pain. This evidence strongly suggests that SMP does not arise from an excess of epinephrine, but rather from the presence of adrenergic receptors coupled to cutaneous nociceptors. Therefore, in SMP, norepinephrine that is normally released from the sympathetic terminals acquires the capacity to evoke pain by activating nociceptors.

Nerve Injury Induces Catechol Sensitization in Nociceptors

In a primate model, the L6 root was lesioned leaving the dorsum of the foot partly denervated (Ali et al. 1999). Intact nociceptors from adjacent uninjured roots developed spontaneous activity and a response to an α_1-adrenergic agonist, phenylephrine, applied to the receptive field. In monkeys where no spinal nerve lesion was applied, little or no catechol sensitivity and spontaneous activity was present. Using a somewhat different model, studies in rats have also demonstrated catechol sensitization of intact nociceptors after nerve injury of companion fibers (Sato and Perl 1991). C fibers ending in a neuroma also display adrenergic sensitivity (Devor and Jänig 1981; Häbler et al. 1988; Wall and Gutnick 1974; Scadding 1981). These data

indicate that SMP is likely to arise from expression of α_1-adrenergic receptors on the terminals of nociceptors.

Phentolamine Infusion as a Test for SMP

The gold standard for diagnosing SMP has been determination of the response to blockade of the appropriate level of the sympathetic chain with local anesthetic. Phentolamine, given systemically, has proven to be safe and is now considered (by some at least) to be a more specific way to diagnose SMP. After infusion of saline to hydrate the patient, and after administration of 1–2 mg of propranolol to block the development of reflex tachycardia, phentolamine in a dose of 1 mg/kg is given over 10 min. Skin temperature is monitored. If the skin temperature does not rise, then a higher dose of phentolamine may have to be given. The test can be blinded, such that the patient does not know when the drug is given. A positive result is the finding that systemic phentolamine and α-adrenergic antagonist relieve pain when given to patients with SMP (Raja et al. 1991; Arner 1991).

Alpha 1 or Alpha 2 Adrenergic Receptors

To sort out the adrenergic mechanisms in humans, further investigators have applied topical clonidine, an α_2-adrenergic agonist, to the painful skin in patients with SMP. Relief of hyperalgesia in the painful area resulted (Davis et al. 1991). Activation of α_2-adrenergic receptors, located on sympathetic terminals, blocks norepinephrine release. Thus, clonidine appears to relieve pain by blocking norepinephrine release. When phenylephrine, a selective α_1-adrenergic agonist, was applied to the clonidine-treated area, pain was rekindled in patients with SMP (Davis et al. 1991). Thus, clinical data, as with the primate physiological data, suggest that the α_1-adrenergic receptor plays a pivotal role in SMP. Whether a phenotypic change or other change explains this nociceptor chemical sensitization is unanswered. It is of interest that the density of α_1-adrenoceptors in the epidermis of hyperalgesic skin of patients with complex regional pain syndrome is increased (Drummond et al. 1996).

Treatment of Sympathetically Maintained Pain

By definition, SMP is relieved by performance of a sympathetic block. It has been frequently observed that in some patients the pain relief outlives the pharmacological action of the anesthetic block. A similar long-lasting pain relief has also been reported following systemic phentolamine infusion (Galer et al. 1992). A series of sympathetic blocks may lead to successful resolution of the pain problem.

In cases where sympathetic blockade fails to provide enduring pain relief, other strategies are needed. Surgical ▶ sympathectomy provides permanent sympathetic denervation and offers lasting pain relief (Singh et al. 2003). In the case of the upper extremity, the T2, 3, and 4 ganglia are removed. The stellate ganglion provides no innervation of the hand, and injury of this ganglion is avoided in order to prevent development of a Horner's syndrome. The thoraco-endoscopic approach affords excellent visualization of the sympathetic chain and provides a minimally invasive technique by which to achieve a thoracic sympathectomy. For the lumbar area, it is necessary to remove the sympathetic chain from L1 to L5. Typically, at least three ganglia are removed, and the length of the excised chain is 10 cm.

Signs of sympathetic innervation to the foot may return after weeks or months. This may be due to crossed innervation. In other words, sympathetic fibers from the contralateral side may reach the foot and provide sufficient innervation that the SMP returns. The test for this is to perform a contralateral sympathetic block. The block should lead to a striking increase in temperature in the contralateral foot (this does not occur customarily) and be associated with pain relief for at least the duration of the block. In this case, a contralateral lumbar sympathectomy may be performed, which will provide enduring pain relief.

Complications of sympathectomy include new pain attributable to the surgery. These patients appear to have a vulnerability to develop new pain with trauma. Compensatory hyperhidrosis may develop, but this is more of a problem in patients in whom the sympathectomy is done to treat hyperhidrosis.

Spinal cord electrical stimulation is an alternative treatment for SMP (Kemler et al. 2000). Of course ▶ spinal cord stimulation may be effective for treatment of sympathetically independent pain as well. The advantage of this technique is that the morbidity is low. The disadvantage is that continued stimulation and maintenance of the device is required to maintain the therapeutic effect. Notably, use of one modality does not preclude the later use of the other modality.

References

Ali, Z., Ringkamp, M., Hartke, T. V., et al. (1999). Uninjured C-fiber nociceptors develop spontaneous activity and alpha adrenergic sensitivity following L6 spinal nerve ligation in the monkey. *Journal of Neurophysiology, 81*, 455–466.

Ali, Z., Raja, S. N., Wesselmann, U., et al. (2000). Intradermal injection of norepinephrine evokes pain in patients with sympathetically maintained pain. *Pain, 88*, 161–168.

Arner, S. (1991). Intravenous phentolamine test: Diagnostic and prognostic use in reflex sympathetic dystrophy. *Pain, 46*, 17–22.

Chabal, C., Jacobson, L., Russell, L. C., et al. (1992). Pain response to perineuromal injection of normal saline, epinephrine, and lidocaine in humans. *Pain, 49*, 9–12.

Choi, B., & Rowbotham, M. C. (1997). Effect of adrenergic receptor activation on post-herpetic neuralgia pain and sensory disturbances. *Pain, 69*, 55–63.

Davis, K. D., Treede, R.-D., Raja, S. N., et al. (1991). Topical application of clonidine relieves hyperalgesia in patients with sympathetically maintained pain. *Pain, 47*, 309–317.

Devor, M., & Jänig, W. (1981). Activation of myelinated afferents ending in a neuroma by stimulation of the sympathetic supply in the rat. *Neuroscience Letters, 24*, 43–47.

Drummond, P. D., Skipworth, S., & Finch, P. M. (1996). Alpha [1]-adrenoceptors in normal and hyperalgesic human skin. *Clinical Science (Colch), 91*, 73–77.

Galer, B. S., Rowbotham, M. C., Von Miller, K., et al. (1992). Treatment of inflammatory, neuropathic and sympathetically maintained pain in a patient with Sjögren's syndrome. *Pain, 50*, 205–208.

Häbler, H.-J., Jänig, W., & Koltzenburg, M. (1988). A novel type of unmyelinated chemosensitive nociceptor in the acutely inflamed urinary bladder. *Agents and Actions, 25*, 219–221.

Kemler, M. A., Barendse, G. A., van Kleef, M., et al. (2000). Spinal cord stimulation in patients with chronic reflex sympathetic dystrophy. *The New England Journal of Medicine, 343*, 618–624.

Raja, S. N., Treede, R.-D., Davis, K. D., et al. (1991). Systemic alpha-adrenergic blockade with phentolamine: A diagnostic test for sympathetically maintained pain. *Anesthesiology, 74*, 691–698.

Raja, S. N., Abatzis, V., & Frank, S. (1998). Role of alpha-adrenoceptors in neuroma pain in amputees. *American Society of Anesthesiologists Abstracts*.

Sato, J., & Perl, E. R. (1991). Adrenergic excitation of cutaneous pain receptors induced by peripheral nerve injury. *Science, 251*, 1608–1610.

Scadding, J. W. (1981). Development of ongoing activity, mechanosensitivity and adrenaline sensitivity in severed peripheral nerve axons. *Experimental Neurology, 73*, 345–364.

Singh, B., Moodley, J., Shaik, A. S., et al. (2003). Sympathectomy for complex regional pain syndrome. *Journal of Vascular Surgery, 37*, 508–511.

Walker, A. E., & Nulson, F. (1948). Electrical stimulation of the upper thoracic portion of the sympathetic chain in man. *Archives of Neurology and Psychiatry, 59*, 559–560.

Wall, P. D., & Gutnick, M. (1974). Ongoing activity in peripheral nerves: The physiology and pharmacology of impulses originating from a neuroma. *Experimental Neurology, 43*, 580–593.

White, J. C., & Sweet, W. H. (1969). *Pain and the neurosurgeon: A forty year experience*. Springfield: Charles C. Thomas.

Complex Regional Pain Syndrome II

▶ Sympathetically Maintained Pain in CRPS II, Human Experimentation

Complex Regional Pain Syndrome Type 1

▶ CRPS-1 in Children

Complex Regional Pain Syndrome Type 2 (CRPS2)

▶ Complex Regional Pain Syndrome

Complex Regional Pain Syndrome Type I (CRPS I)

▶ Complex Regional Pain Syndrome

Complex Regional Pain Syndromes, Clinical Aspects

Ralf Baron
Department of Neurological Pain Research and Therapy, Christian Albrechts University Kiel, Kiel, Germany

Synonyms

Algodystrophy; CRPS; Causalgia; Morbus Sudeck; Reflex sympathetic dystrophy

Definition

The "IASP classification of chronic pain 2012" defines CRPS as a "syndrome characterized by a continuing (spontaneous and/or evoked) regional pain that is seemingly disproportionate in time or degree to the usual course of pain after trauma or other lesion. The pain is regional (not in a specific nerve territory or dermatome) and usually has a distal predominance of abnormal sensory, motor, sudomotor, vasomotor edema, and/or trophic findings. The syndrome shows variable progression over time." These chronic pain syndromes comprise of different additional clinical features including ▶ spontaneous pain, ▶ allodynia, ▶ hyperalgesia, edema, autonomic abnormalities, and trophic signs. In ▶ CRPS type I (reflex sympathetic dystrophy), minor injuries or fractures of a limb precede the onset of symptoms. ▶ CRPS type II (causalgia) develops after injury to a major peripheral nerve (Merskey and Bogduk 1994; Harden et al. 2001, 2010; Jänig and Baron 2003; Marinus et al. 2011).

Characteristics

CRPS Type I (Reflex Sympathetic Dystrophy)

The most common precipitating event is a trauma affecting the distal part of an extremity (65 %), especially fractures, postsurgical conditions, contusions, and strain or sprain. Less common occurrences are central nervous system lesions, like spinal cord injuries and cerebrovascular accidents as well as cardiac ischemia (Allen et al. 1999).

CRPS I patients develop asymmetrical distal extremity pain and swelling without presenting with a demonstrable nerve lesion. These patients often report a spontaneous burning pain felt in the distal part of the affected extremity. Characteristically, the pain is disproportionate in intensity to the inciting event. The pain usually increases when the extremity is in a dependent position. Stimulus-evoked pains are a striking clinical feature; they include ▶ mechanical allodynia and ▶ thermal allodynia and/or hyperalgesia (Gierthmühlen et al. 2012). These sensory abnormalities often appear early, are most pronounced distally, and have no consistent spatial relationship to individual nerve territories or to the site of the inciting lesion (Sieweke et al. 1999). Movement of, and pressure on, the joints (deep somatic allodynia) can elicit pain, even if the inciting lesion does not directly affect these. Autonomic abnormalities include swelling and changes in sweating and skin blood flow (Chelimsky et al. 1995; Birklein et al. 1998; Wasner et al. 2001, 2002). In the acute stages of CRPS I, the affected limb is often warmer than the contralateral limb. Sweating abnormalities, either hypohidrosis or, more frequently, hyperhidrosis are present in nearly all CRPS I patients. The acute distal swelling of the affected limb depends very critically on aggravating stimuli. Since it diminishes after sympathetic blocks, it is likely that it is maintained by sympathetic activity.

Trophic changes such as abnormal nail growth, increased or decreased hair growth, fibrosis, thin glossy skin, and osteoporosis may be present, particularly in chronic stages. Restrictions of passive movement are often present in long-standing cases, and may be related to both functional motor disturbances and trophic changes of joints and tendons (Baron et al. 1996) . Weakness of all muscles of the affected distal extremity is often present. Small accurate movements are characteristically impaired. Nerve conduction and electromyography studies are normal, except in patients in very chronic and advanced stages. About half of the patients have a postural or action tremor representing an increased physiological tremor (Deuschl et al. 1991). In about 10 % of cases, dystonia of the affected hand or foot develops (Bhatia et al. 1993).

CRPS Type II (Causalgia)

The symptoms of CRPS II are similar to those of CRPS I. The only exception is that a lesion of peripheral nerve structures and subsequent focal deficits are mandatory for the diagnosis. The symptoms and signs spread beyond the innervation territory of the injured peripheral nerve and often occur remote from the site of injury; however, a restriction to the territory is not in conflict with the current definition.

Sympathetically Maintained Pain (SMP)

Neuropathic pain patients presenting with similar clinical signs and symptoms can clearly be divided into two groups, by the negative or positive effect of selective sympathetic blockade, selective activation of sympathetic activity or antagonism of alpha adrenoceptor mechanisms (Raja et al. 1991; Baron et al. 2002).

The pain component that is relieved by specific sympatholytic procedures is considered "▶ sympathetically maintained pain" (SMP). Thus, SMP is now defined to be a symptom or the underlying mechanism in a subset of patients with neuropathic disorders and not a clinical entity. The positive effect of a sympathetic blockade is not essential for the diagnosis. On the other hand, the only possibility of differentiating between SMP and "sympathetically independent pain" (SIP) is the efficacy of a correctly applied sympatholytic intervention (Stanton-Hicks et al. 1995; Harden et al. 2007).

Diagnostic Procedure

The diagnosis of CRPS I and II follows the IASP clinical criteria (IASP taxonomy 2012; Harden et al. 2007):

1. Continuing pain, which is disproportionate to any inciting event.
2. Must report at least *one symptom in three of the four following categories:*
   Sensory: Reports of hyperalgesia and/or allodynia.
   Vasomotor: Reports of temperature asymmetry and/or skin color changes and/or skin color asymmetry.
   Sudomotor/Edema: Reports of edema and/or sweating changes and/or sweating asymmetry.
   Motor/Trophic: Reports of decreased range of motion and/or motor dysfunction (weakness, tremor, dystonia) and/or trophic changes (hair, nail, skin).
3. Must display at least *one sign* at time of evaluation in two or more of the following categories:*
   Sensory: Evidence of hyperalgesia (to pinprick) and/or allodynia (to light touch and/or deep somatic pressure and/or joint movement).
   Vasomotor: Evidence of temperature asymmetry and/or skin color changes and/or asymmetry.
   Sudomotor/Edema: Evidence of edema and/or sweating changes and/or sweating asymmetry.
   Motor/Trophic: Evidence of decreased range of motion and/or motor dysfunction (weakness, tremor, dystonia) and/or trophic changes (hair, nail, skin).
4. There is no other diagnosis that better explains the signs and symptoms.
   *A sign is counted only if it is observed at the time of diagnosis.
   A research criterion for CRPS is recommended that is more specific, but less sensitive than the clinical criteria; it requires that four of the symptom categories and at least two sign categories be present.

Subtypes of CRPS
CRPS I (old name: Reflex Sympathetic Dystrophy): As defined above, CRPS II (old name: Causalgia): Defined as above with electrodiagnostic or physical evidence of a major nerve lesion.

CRPS-NOS (Not Otherwise Specified): Partially meets CRPS criteria, not better explained by any other condition. This subtype was added to capture any patients previously diagnosed with CRPS who now do not meet criteria as elaborated above.

Possible inciting events of CRPS include the following:

CRPS I
- Peripheral tissues
  - Fractures and dislocations
  - Soft-tissue injury
  - Fasciitis
  - Tendonitis
  - Bursitis
  - Ligamentous strain
  - Arthritis
  - Mastectomy
  - Deep vein thrombosis
  - Immobilization
- Idiopathic
  - CRPS often follows minor trauma that could not be remembered by the patient. Also some patients negate any inciting event
- Viscera
  - Abdominal disease
  - Myocardial infarction
- Central nervous system
  - Spinal cord lesions
  - Head injury
  - Cerebral infarction
  - Cerebral tumor

CRPS II
- Peripheral nerve and dorsal root
  - Peripheral nerve trauma
  - Brachial plexus lesions
  - Root lesions

Treatment Algorithm
Treatment should be immediate and most importantly directed toward restoration of full function of the extremity. This objective is best attained in a comprehensive interdisciplinary setting, with particular emphasis on pain management and functional restoration

(treatment algorithm (see Stanton-Hicks et al. 1998)). The pain specialists should include neurologists, anesthesiologists, orthopedics, physiotherapists, psychologists, and the general practitioner.

Destructive surgery on the peripheral or central afferent nervous system in cases of CRPS always implicates further deafferentation, and thereby provides an increased risk for persistent deafferentation type of pain.

The severity of the disease determines the therapeutical regime. The reduction of pain is the precondition that all other interventions have to comply with. At the acute stage of CRPS, when the patients still suffer from severe pain, it is often impossible to carry out intensive active therapy. Painful interventions, and in particular vigorous physical therapy, at this stage leads to deterioration. Therefore, immobilization of the affected extremity and careful contralateral physical therapy should be the acute treatment of choice and pain treatment should be initiated immediately. There are only a few controlled treatment studies on CRPS. The first choice analgesic drugs, in which efficacy has been shown by clinical studies on other neuropathic pain syndromes, include antidepressants, gabapentin, pregabalin, opioids, and topical medication. Additionally, systemic corticosteroids treatment is frequently used. Calcium-regulating agents (calcitonin, bisphosphonates) are used in case of refractory pain. Sympatholytic procedures for SMP, preferably sympathetic ganglion blocks, are used if no or an insufficient pain relief is achieved and should be perpetuated in case of efficacy. If resting pain subsides, first passive physical therapy, later active isometric followed by active isotonic training should be performed in combination with sensory desensitization programs until restitution of complete motor function. Psychological treatment has to flank the regime to strengthen coping strategies and discover contributing factors. In refractory cases, spinal cord stimulation could be considered. If refractory dystonia develops, intrathecal baclofen application is worth considering.

References

Allen, G., Galer, B. S., & Schwartz, L. (1999). Epidemiology of complex regional pain syndrome: A retrospective chart review of 134 patients. *Pain, 80*, 539–544.

Baron, R., Blumberg, H., & Jänig, W. (1996). Clinical characteristics of patients with complex regional pain syndrome in Germany with special emphasis on vasomotor function. In W. Jänig, & M. Stanton-Hicks (Eds.), *Reflex sympathetic dystrophy: A reappraisal* (pp. 25–48). Seattle: IASP Press.

Baron, R., Schattschneider, J., Binder, A., Siebrecht, D., & Wasner, G. (2002). Relation between sympathetic vasoconstrictor activity and pain and hyperalgesia in complex regional pain syndromes: A case-control study. *Lancet, 359*, 1655–1660.

Bhatia, K. P., Bhatt, M. H., & Marsden, C. D. (1993). The causalgia-dystonia syndrome. *Brain, 116*(Pt 4), 843–851.

Birklein, F., Riedl, B., Neundörfer, B., & Handwerker, H. O. (1998). Sym-pathetic vasoconstrictor reflex pattern in patients with complex regional pain syndrome. *Pain, 75*, 93–100.

Chelimsky, T. C., Low, P. A., Naessens, J. M., Wilson, P. R., Amadio, P. C., & Ot'Brien, P. C. (1995). Value of autonomic testing in reflex Sym- pathetic dystrophy. *Mayo Clinic Proceedings, 70*, 1029–1040.

Deuschl, G., Blumberg, H., & Lücking, C. H. (1991). Tremor in reflex sympathetic dystrophy. *Archives of Neurology, 48*, 1247–1252.

Gierthmühlen, J., Maier, C., Baron, R., Tölle, T., RD, T., Birbaumer, N., Huge, V., Koroschetz, J., Krumova, E. K., Lauchart, M., Maihöfner, C., Richter, H., Westermann, A., & German Research Network on Neuropathic Pain (DFNS) study group. (2012). Sensory signs in complex regional pain syndrome and peripheral nerve injury. *Pain, 153*, 765–774.

Harden, R. N., Baron, R., & Jänig, W. (2001). *Complex regional pain syndrome*. Seattle: IASP Press.

Harden, R. N., Bruehl, S., Stanton-Hicks, M., & Wilson, P. R. (2007). Proposed new diagnostic criteria for complex regional pain syndrome. *Pain Medicine, 8*, 326–331.

Harden, R. N., Bruehl, S., Perez, R. S., Birklein, F., Marinus, J., Maihofner, C., Lubenow, T., Buvanendran, A., Mackey, S., Graciosa, J., Mogilevski, M., Ramsden, C., Chont, M., & Vatine, J. J. (2010). Validation of proposed diagnostic criteria (the "Budapest Criteria") for complex regional pain syndrome. *Pain, 150*, 268–274.

Jänig, W., & Baron, R. (2003). Complex regional pain syndrome: Mystery explained? *Lancet Neurology, 2*, 687–697.

Marinus, J., Moseley, G. L., Birklein, F., Baron, R., Maihöfner, C., Kingery, W. S., & van Hilten, J. J. (2011). Clinical features and pathophysiology of complex regional pain syndrome. *Lancet Neurology, 10*, 637–648.

Merskey, H., & Bogduk, N. (1994). *Classification of chronic pain: Descriptions of chronic pain syndromes and definition of terms* (2nd ed.). Seattle: IASP Press.

Raja, S. N., Treede, R. D., Davis, K. D., & Campbell, J. N. (1991). Systemic alpha-adrenergic blockade with phentolamine: A diagnostic test for sympathetically maintained pain. *Anesthesiology, 74*, 691–698.

Sieweke, N., Birklein, F., Riedl, B., Neundorfer, B., & Handwerker, H. O. (1999). Patterns of hyperalgesia in complex regional pain syndrome. *Pain, 80*, 171–177.

Stanton-Hicks, M., Jänig, W., Hassenbusch, S., Haddox, J. D., Boas, R., & Wilson, P. (1995). Reflex sympathetic dystrophy: Changing concepts and taxonomy. *Pain, 63*, 127–133.

Stanton-Hicks, M., Baron, R., Boas, R., Gordh, T., Harden, N., Hendler, N., Koltzenburg, M., Raj, P., & Wilder, R. (1998). Complex regional pain syndromes: Guidelines for therapy. *The Clinical Journal of Pain, 14*, 155–166.

Wasner, G., Schattschneider, J., Heckmann, K., Maier, C., & Baron, R. (2001). Vascular abnormalities in reflex sympathetic dystrophy (CRPS I): Mechanisms and diagnostic value. *Brain, 124*, 587–599.

Wasner, G., Schattschneider, J., & Baron, R. (2002). Skin temperature side differences – a diagnostic tool for CRPS? *Pain, 98*(19–26), 16.

Compliance

Definition

The extent to which patients are obedient and follow the instructions and prescriptions of a health care provider.

Cross-References

▶ Multidisciplinary Pain Centers, Rehabilitation

Complications

▶ Postoperative Pain, Adverse Events (Associated with Acute Pain Management)

Comprehensive Assessment

▶ Multiaxial Assessment of Pain

Computed Radiography

▶ Plain Radiography

Computer Applications

▶ Remote Management of Pediatric Pain

Computerized Axial Tomography

▶ CT Scanning

Conditioned Analgesia

Definition

Conditioned analgesia is a learned activation of pain-inhibitory systems.

Cross-References

▶ Pain-Modulatory Systems, History of Discovery

Conditioned Place, Behavioral Strategies in Pain Research

Definition

Conditioned place pairing is a classical paradigm that determines whether an animal prefers or avoids a context paired with a stimulus or condition. This assay is commonly used in addiction studies to study reward. In pain research, conditioned place aversion (CPA) can be elicited by pairing a context with a treatment that enhances pain and conditioned place preference (CPP) can result from pairing an environment with a drug that produces pain relief. The context is paired with the treatment one or more times (conditioning) and preference is determined at a later time, in the absence of the treatment.

Cross-References

▶ Nociceptive Processing in the Hippocampus and Entorhinal Cortex, Neurophysiology and Pharmacology

Conditioning

Definition

Mechanism through which repeated associations between two stimuli induce a new learned response. In particular, by pairing a neutral stimulus (conditioned stimulus) with an unconditioned stimulus (that induces a physiological response) many times, the neutral stimulus alone will be capable of producing a conditioned physiological response.

Cross-References

▶ Operant Perspective of Pain
▶ Placebo Analgesia and Descending Opioid Modulation

Conductance

Definition

A term taken from the physics of electricity referring to the movement of charge. It is the inverse of resistance and is defined by Ohm's Law as the current, "I" divided by the driving force or voltage, on the charge in question. In the nervous system, the "charge" that moves are the ions primarily Na^+, K^+, Cl^-, or $Ca++$, where "movement" is across a cell membrane

via specific channels. Conductance is generally used to describe the maximum number of channels open in response to a specific manipulation such as a voltage step for voltage-gated ion channels or a ligand for ligand-gated ion channels. The term is also applied to a single channel, where the single-channel conductance is a property of the channel that reflects the maximum current that can flow through the channel when it is open.

Cross-References

▶ Descending Circuitry, Opioids

Conduction Velocity

Definition

Speed of propagation of an action potential along a nerve fiber.

Cross-References

▶ Nociceptor, Categorization

Confidence in Coping Abilities

▶ Psychology of Pain, Self-efficacy

Confusional Migraine

Definition

A term not formally defined and whose use is discouraged. The syndrome is of sudden onset of agitation, confusion, and disorientation presumed to be due to migraine aura. Other aura symptoms may be present and headache may follow. The differential diagnosis includes transient global amnesia, complex partial seizure, intoxication/drug ingestion, and encephalitis.

Cross-References

▶ Migraine, Childhood Syndromes

Congener

Definition

A congener is one of two or more things of the same kind, as of animal or plant, with respect to classification.

Cross-References

▶ Headache Attributed to a Substance or its Withdrawal

Congenital Insensitivity to Pain with Anhidrosis

Max J. Hilz
University of Erlangen-Nuremberg, Erlangen, Germany

Synonyms

CIPA; Congenital sensory neuropathy with anhidrosis; Hereditary sensory and autonomic neuropathy type IV, HSAN IV, HSAN 4

Definition

Congenital insensitivity to pain with anhidrosis (CIPA) is a rare autosomal recessive disorder characterized by a lack of pain sensation; recurrent episodes of high, unexplained fever; anhidrosis, i.e., absence of sweating; self-mutilating behavior; and mental retardation (Swanson 1963; Pinsky and DiGeorge 1966; Dyck 1993; Axelrod 2002; Hilz et al. 1999).

Characteristics

CIPA is the second most frequent among the five HSANs classified by (Dyck 1993). The combination of distinctive anhidrosis and insensitivity to deep as well as superficial pain, with manifestation already in early childhood, is among the most prominent findings differentiating CIPA from the other HSANs, in particular from the most common of the five disorders, HSAN III, also called Riley-Day syndrome or familial dysautonomia. Familial dysautonomia is characterized by excessive sweating during autonomic crises and by significantly impaired perception of superficial pain but preserved deep pain perception (Axelrod 2002; Hilz 2002).

CIPA manifests in infancy or early childhood with frequent bouts of extreme, at first sight unexplained, fever and heat intolerance.

Anhidrosis is present on the trunk and upper extremities in all CIPA patients, while other body parts may be variably affected. Particularly with high environmental temperatures, anhidrosis induces hyperpyrexia, which in turn can lead to recurrent febrile convulsions and has been reported to account for death within the first 3 years in up to 20 % of CIPA children (Axelrod 2002; Indo 2002).

Anhidrosis also contributes to the development of a calloused, thickened skin with lichenification, i.e., a leathery induration and thickening of the epidermis with exaggeration of normal skin lines, giving it a bark-like appearance and hyperkeratosis particularly at palms and soles (Fig. 1) (Indo 2002; Hilz 2002). The skin is dry and warm and frequently shows deep and persistent ulcers at the heels (Fig. 2); nails are dystrophic (Axelrod 2002; Indo 2002; Pinsky and DiGeorge 1966).

Initially insensitivity to pain may not be apparent. Yet with increasing mobility, children sustain severe, frequently unrecognized injuries

Congenital Insensitivity to Pain with Anhidrosis, Fig. 1 Hyperkeratosis and leathery induration of the skin with exaggeration of normal skin lines in the palm of a CIPA patient. Fingernails are dystrophic. Fingertips show signs of chronic inflammation due to repeated self-mutilation

Congenital Insensitivity to Pain with Anhidrosis, Fig. 2 Deep persistent ulcer and chronic inflammation at the area of the heel in a CIPA patient

Congenital Insensitivity to Pain with Anhidrosis, Fig. 3 Neuropathic ankle joint (Charcot joint) due to repetitive fractures. Tissue swelling as a result of chronic inflammation and osteomyelitis. Chronic ulcer at the bottom of the heel. Mutilation of lower left lip due to repeated lip biting

without complaint. They incur repeated bruises or inadvertently self-inflicted burn injuries and cuts and multiple scars. Insensitivity to pain also accounts for often self-inflicted corneal scarring with opacities and even perforation of the cornea (Axelrod 2002; Indo 2002). With dentition, children start to bite their tongue, lips, and fingers, leading to a mutilated, bifid tongue with decubital ulcers or absent tip of the tongue, mutilated lips (Fig. 3), jaw malformation with edentate areas due to self-extraction of teeth (Bodner et al. 2002), or to amputated finger tips. Open wounds with poor healing; continuous self-mutilation; and frequent, often unnoticed fractures and joint injuries lead to osteomyelitis and – over the years – to grotesque joint deformities with neurogenic arthropathy (Fig. 3) or amputations (Axelrod 2002). Even with active or passive movement of fractured joints or bones, there is no tenderness or discomfort (Swanson 1963). Radiograms frequently show numerous fractures, particularly of the weight-bearing bones with neuropathic Charcot joints (Indo 2002).

Chronic inflammation may lead to secondary amyloidosis (Axelrod 2002; Indo 2002). Local and systemic infections often induce sepsis, which accounted for 20 % fatalities in one patient group (Shorer et al. 2001).

Most CIPA children are mentally retarded and have learning problems. Fifty percent show irritability, hyperactivity, and a tendency to rages and tantrums (Axelrod 2002).

Clinical examination confirms the absence of responses to superficial or deep, visceral pain, e.g., with electrical shock during neurophysiologic examination, intramuscular injection, or urinary catheterization (Swanson 1963), Moreover, temperature perception and discrimination between hot and cold stimuli is impaired (Indo 2002; Hilz 2002). Similarly, discrimination of sharp and dull stimuli is compromised, while light touch, vibration, and position senses are normal. Deep tendon reflexes are preserved, and there are no pathological reflexes. Corneal reflexes may be inconsistent. In contrast to HSAN III, emotional tear flow and fungiform papillae of the tongue are present (Axelrod 2002).

Motor and sensory nerve conduction, somatosensory, visual, or brainstem-evoked potentials are usually normal, while quantitative sensory testing of warm, cold, or heat perception is highly abnormal (Hilz 2002). The ▶ sympathetic skin response (SSR) is abnormal in all CIPA patients (Hilz et al. 1999; Shorer et al. 2001).

Intradermal injection of diluted histamine does not induce the typical ▶ triple response of a C-nerve fiber axon reflex-mediated vasodilatation with diffuse erythema and itching pain but only results in a circumscribed wheal at the site of injection (Axelrod 2002; Indo 2002).

Similarly, local sweating cannot be induced by injection of pilocarpine or methacholine (Pinsky and DiGeorge 1966).

Apart from anhidrosis, autonomic dysfunction is not a predominant characteristic of CIPA, though present. There may be orthostatic hypotension with preserved reflex tachycardia and supersensitivity to exogenous vasopressor drugs (Pinsky and DiGeorge 1966). Pupillary responses to cocaine are absent yet preserved with epinephrine instillation into the conjunctival sac, also suggesting sympathetic denervation (Indo 2002). Subcutaneous injection of cholinergic methacholine or neostigmine in a dosage that would normally induce tearing does not result in lacrimation in CIPA patients although emotional tearing is preserved (Pinsky and DiGeorge 1966).

The deficiency in pain and temperature perception and in sweating can be explained by the bioptic findings of absence of small myelinated and unmyelinated nerve fibers in the cutaneous branch of the radial nerve and a significant

reduction of these fibers in the sural nerve (for review see Indo 2002; Shorer et al. 2013).

Skeletal muscles biopsies obtained phenotypic variation in the muscles of CIPA patients, which may not related to their genetic mutation (Shorer et al. 2013). Skin biopsies show preserved sweat glands but absent innervation with loss of unmyelinated sudomotor fibers (for review see Indo 2002.)

Absence of intradermal C- and A-delta fibers appears to be the morphological basis of insensitivity to pain and anhidrosis (for review see Indo 2002).

Verzé et al. found no nerve branches or endings in the epidermis and only a few nerve fibers in deeper layers of the dermis (for review see Indo 2002).

Autopsy of one CIPA patient showed absence of small neurons in dorsal root ganglia, absence of small nerve fibers in the dorsal roots, absence of Lissauer's tract, and paucity of small fibers in the spinal tract of the trigeminal nerve (Swanson et al. 1965), i.e., lack of afferent pain and temperature mediating structures (Swanson 1963).

The genetic basis for CIPA are not mutations in the genes encoding ▶ neurotrophins but mutations in the gene encoding the ▶ tyrosine kinase A receptor (TRKA/NTRK1) located on chromosome 1 ($1q2^{1-}q22$) (Indo et al. 1996).

So far, 62 mutations have been identified in CIPA patients of various ethnicities, including 22 missense 11 nonsense, 19 frameshift, 9 splice-site mutations, and 1 gross deletion mutation (Indo 2002; Li et al. 2012).

The similarities of anatomical and clinical changes in TrkA knockout mice and in CIPA patients led Indo and coworkers to the discovery that mutations of the trkA (NTRK1) gene are the genetic basis of CIPA. Defect in NGF signal transduction at the trkA receptor causes failure to survive and probably apoptosis of developing small nerve fiber neurons (Indo et al. 1996; Indo 2002).

The only tool for prenatal diagnosis of CIPA is the identification of trkA (NTRK1) mutations (Indo 2002), but there are numerous mutations which prevent simple DNA diagnosis of the disease (Axelrod 2002).

There is no specific therapy, but treatment remains supportive and attempts to control hyperthermia, to prevent injuries, self-mutilation, and orthopedic or dental complications and to modify behavioral problems such as hyperactivity or rages (Axelrod 2002; Hilz 2002).

References

Axelrod, F. B. (2002). Hereditary sensory and autonomic neuropathies. Familial dysautonomia and other HSANs. *Clinical Autonomic Research, 12*(Suppl. 1), I/2–I/14.

Bodner, L., Woldenberg, Y., Pinsk, V., et al. (2002). Orofacial manifestations of congenital insensitivity to pain with anhidrosis: A report of 24 cases. *ASDC Journal of Dentistry for Children, 69*, 293–296.

Dyck, P. J. (1993). Neuronal atrophy and degeneration predominantly affecting peripheral sensory and autonomic neurons. In P. J. Dyck, P. K. Griffin, P. A. Low, et al. (Eds.), *Peripheral neuropathy* (pp. 1065–1093). Philadelphia: WB Saunders.

Hilz, M. J. (2002). Assessment and evaluation of hereditary sensory and autonomic neuropathies with autonomic and neurophysiological examinations. *Clinical Autonomic Research, 12*(Suppl. 1), I33–I43.

Hilz, M. J., Stemper, B., & Axelrod, F. B. (1999). Sympathetic skin response differentiates hereditary sensory autonomic neuropathies III and IV. *Neurology, 52*, 1652–1657.

Indo, Y. (2002). Genetics of congenital insensitivity to pain with anhidrosis (CIPA) or hereditary sensory and autonomic neuropathy type IV. Clinical, biological and molecular aspects of mutations in TRKA (NTRK1) gene encoding the receptor tyrosine kinase for nerve growth factor. *Clinical Autonomic Research, 12*(Suppl. 1), I20–I32.

Indo, Y., Tsuruta, M., Hayashida, Y., et al. (1996). Mutations in the *TRKA*/NGF receptor gene in patients with congenital insensitivity to pain with anhidrosis. *Nature Genetics, 13*, 485–488.

Li, M., Liang, J. Y., Sun, Z. H., et al. (2012). Novel nonsense and frameshift NTRK1 gene mutations in Chinese patients with congenital insensitivity to pain with anhidrosis. *Genetics and Molecular Research, 11*, 2156–2162.

Pinsky, L., & DiGeorge, A. M. (1966). Congenital familial sensory neuropathy with anhidrosis. *The Journal of Pediatrics, 68*, 1–13.

Shorer, Z., Moses, S. W., Hershkovitz, E., et al. (2001). Neurophysiologic studies in congenital insensitivity to pain with anhidrosis. *Pediatric Neurology, 25*, 397–400.

Shorer, Z., Shaco-Levy, R., Pinsk, V., et al. (2013). Variation of muscular structure in congenital insensitivity to pain and anhidrosis. *Pediatric Neurology, 48*, 311–313.

Swanson, A. G. (1963). Congenital insensitivity to pain with anhidrosis. *Archives of Neurology, 8*, 299–306.

Swanson, A. G., Buchan, G. C., & Alvord, E. C. (1965). Anatomic changes in congenital insensitivity to pain. *Archives of Neurology, 12*, 12–18.

Congenital Sensory Neuropathy with Anhidrosis

▶ Congenital Insensitivity to Pain with Anhidrosis

Conjunctival Injection

Definition

Redness of the conjunctiva (white) of the eye.

Cross-References

▶ Hemicrania Continua

Connective Tissue Disease

Definition

The autoimmune disease is characterized by an abnormal structure or function of one or more of the elements of connective tissue, i.e., collagen, elastin, or the mucopolysaccharides, including, for example, rheumatoid arthritis, systemic lupus erythematosus, and scleroderma.

Cross-References

▶ Vascular Neuropathies

Connective Tissue Manipulation

Definition

A superficial form of myofascial manipulation is the simple definition of connective manipulation. It involves manipulating the skin and superficial connective tissues above muscle to achieve reflex and local effects, which is believed to be based on a viscerocutaneous reflex. Reflex reactions include vasodilatation and diffuse or localized increase in sudomotor activity.

Cross-References

▶ Chronic Pelvic Pain, Physical Therapy Approaches, and Myofascial Abnormalities

Conscious Laparoscopic Pain Mapping

▶ Chronic Pelvic Pain, Laparoscopic Pain Mapping

Conscious Pain Mapping

▶ Chronic Pelvic Pain, Laparoscopic Pain Mapping

Consciousness and Pain

C. Richard Chapman[1] and Yoshio Nakamura[2]
[1]Department of Anesthesiology, University of Utah School of Medicine, Salt Lake City, UT, USA
[2]Pain Research Center, Department of Anesthesiology University of Utah School of Medicine, Salt Lake City, UT, USA

Synonyms

Awareness

Definition

Pain as Conscious Experience

Pain is the unique meaning of a tissue injury event that the brain, as a ▶ complex adaptive system, constructs. Pain emerges from large-scale patterns of brain activity that reflect sensory, emotional, and cognitive processing. As it is an aspect of consciousness, pain is distinct from ▶ nociception.

Consciousness

Consciousness is an emergent, self-organizing feature of brain activity that makes complex, adaptive interactions with the internal and external environments and self-reference in those environments possible (Chapman and Nakamura 2003).

Complex Adaptive System

A complex adaptive system is a large set of units that interact with each other and with an external environment, to produce overall patterns that are significantly more complex than the behaviors of the individual entities comprising the system. A complex adaptive system changes according to three key principles: order is emergent as opposed to predetermined, the system's history is irreversible, and the system's future is often unpredictable. Simpler complex systems nest within more elaborate complex systems (Morowitz and Singer 1995; Dooley 1996).

The essence of a complex adaptive system is that it organizes itself to optimize its adaptation to its environment. Its dynamic instability gives it the flexibility it needs to accommodate quickly to environmental change. Complex adaptive systems are dynamic in that they constantly change and adjust to disturbances. Such a system does not always undergo linear changes when it reorganizes to accommodate a disturbance; instead, it may demonstrate abrupt transitions in organizational patterns known as state transitions.

Self-Organization

▶ Self-organization is a process whereby a pattern at the global level of a system emerges solely from interactions among lower-level components of the system. The global pattern is an emergent property of the system itself. A self-organizing, adaptive system tends to take on lifelike qualities such as self-directedness, self-correction, self-preservation, and intelligence (Camazine et al. 2001).

Emergence

▶ Emergence is the process by which a system of interacting elements spontaneously acquires a qualitatively new pattern and structure, which is unpredictable from knowledge of the individual elements (Camazine et al. 2001). For example, combining hydrogen and oxygen gases at room temperature produces liquidity.

Intentionality

A complex adaptive system has ▶ intentionality when it exhibits directedness toward some future state or goal. Intent comprises the endogenous initiation, construction, and direction of perception, action, and goal-directed behavior (Freeman 1995, 2000).

Schemata

A ▶ schema is a perceptual hypothesis that serves as the fundamental unit for constructing awareness (Marcel 1983). ▶ Schemata are fuzzy, preconscious, and dynamical patterns, roughly related to dynamically stable patterns in neural networks (Rumelhart et al. 1986). The brain, as a complex adaptive system, adapts to the world and the body in which it dwells, by constantly forming, evaluating, and refining these global perceptual hypotheses. Ongoing awareness of the body and the world depends upon a continual process of reconstruction based on schemata. Simpler schemata nest within more complex schemata (for review see Martin 1994).

Characteristics

Pain and Nociception

Pain is an unpleasant, compelling aspect of somatic awareness. Like other aspects of consciousness, it is emergent, dynamic, intentional, and constructed. Also, like other aspects of consciousness, pain is the product of the brain operating as a pattern-forming, self-organized complex adaptive system.

Nociceptive traffic within the nervous system has many consequences (Willis and Westlund 1997; Craig 2003). Nociception is a reflexive motor stimulus and a trigger for autonomic arousal. In addition to activating sensory processes, nociceptive traffic activates the limbic brain via multiple pathways. Functional brain imaging studies reveal that complex patterns of processing occur in mesencephalic, limbic, and cortical structures during the emergence of pain

as a conscious experience. These processes appear to be complex, dynamic self-organization that makes the construction of subjective experience possible (Kelso 1995). Nociception is normally necessary, but not sufficient, for pain, which as an aspect of consciousness is an emergent feature of brain activity.

Properties of consciousness in general are also properties of pain. These properties include the following. Consciousness:

- Is personal, with a necessarily limited point of view
- Has mental contents that are stable for short periods and vary over longer intervals
- Has mental contents that are unified at any one moment and continuous over time
- Is selective, with a foreground and background, and has a limited capacity at a given moment
- Is intentional because it is directed at the world or the body

Pain and the Construction of Consciousness

The brain deals not with reality as the physical sciences study it but rather with an internal, autonomous representation of reality that it builds and revises from moment to moment, using sensory information and schemata that involve networks of association in memory. Subjective reality undergoes constant self-organized revision, which integrates sensory information, emotion, ratiocination, and other aspects of cognition to produce meaning. Therefore, brain representations of external objects and bodily events are dynamic constructions, not replications, of the external world or the body (Mountcastle 1998). Pain is not a static entity that an individual possesses, but rather a dynamic construction of bodily awareness, typically based on nociception (Legrain 2011).

The construction of consciousness, at any given moment, proceeds in response to the intentional and situational imperatives of each person. The intentionality of the person perceiving, feeling, and emoting constrains and drives the construction of consciousness (Freeman 2000). This process involves integrating sensory signals, memory and prior experience, expectations, and immediate and long-term goals and plans. It weaves all of these into a coherent, stable macro-emergent pattern that seems to underlie awareness of the world and one's self. Far from being a passive entity that merely registers information coming in from various sensory channels, the brain is an active, adaptive system that constantly models the world and the body in which it dwells. This modeling can change gradually from moment to moment or in abrupt state transitions.

Acute Versus Chronic Pain

In the consciousness studies framework, pain is a state of the brain. Nociceptive signaling from acute tissue injury can perturb the complex adaptive system and force non-stable reorganization. It tends to provoke a state transition from ordinary purposeful activity focused on immediate goals or needs to somatic preoccupation and protection. However, the system will tend to return to its normal, habitual patterns when the nociception decreases or terminates.

When pain becomes chronic, the complex adaptive system undergoes reorganization to a stable, global state characterized by low dynamic instability that makes the system resistant to change. The stable state is a complex adaptation to the physical health of the body, the social environment and role expectations, interpersonal relationships, and mood. Patients tend to demonstrate somatic preoccupation, conservation of energy, and illness behavior.

The consciousness studies perspective holds that chronic pain is difficult to treat, because an effective intervention must do more than eliminate the element of nociceptive signaling. It must disturb the complex adaptive system sufficiently to permit adaptive reorganization to a stable state of well-being, normal function, and optimal somatic awareness. The termination of nociceptive activity often fails to disturb the system sufficiently to enable reorganization in the desired direction. Psychological intervention, social adjustments, and increased physical activity are often necessary adjuncts for chronic pain therapy, because the synergistic effect of these combined has a better chance of perturbing the complex adaptive system than a single intervention alone.

Individual Differences

Individuals differ markedly from one another in the pain they experience in response to virtually identical tissue trauma. Although this stems in part from genetic factors that influence nociceptive transduction and/or transmission, most interindividual variance reflects each individual's unique construction of the pain experience. Personal past experience and personal interpretation of the immediate situation generate highly varied schemata, and these in turn construct highly individual personal experiences of pain. Put more simply, pain is the unique meaning of a tissue injury event that the brain, as a complex adaptive system, constructs. Different people construct different meanings in highly similar tissue injury situations.

Pain and Empathy

Humans are social by nature, and the boundaries of consciousness sometimes extend beyond the individual to others through empathy. Numerous brain imaging studies reveal similar patterns of brain activity when subjects feel their own emotions or observe the same emotions in others, and this holds for pain as well (Lamm et al. 2011; Decety 2010). In "synesthesia for pain," a person not only empathizes with another's pain but experiences the observed pain as if it were his or her own, leading some investigators to propose that this is an abnormal form of empathy for pain (Fitzgibbon et al. 2010). Patients with congenital sensitivity to pain have never experienced pain, and yet they yield functional brain imaging responses that are normal for experiencing pain when one observes another in pain (Danziger et al. 2009). The human brain's ability to construct pain, even when the injury belongs to another, is a powerful adaptive feature that might once have been instrumental for survival of the species.

Culture and Pain

Conscious experience is subject to cultural influences because the brain, as a complex adaptive system, is embedded within larger social and cultural complex adaptive systems. Consequently, cultural meaning systems may determine the possible range of states of consciousness. Some cultures may predispose those they influence to experience pain differently than others in a particular situation.

Pain Measurement
As pain is a personal experience, subjective report is the only way to access it. Conventional ▶ pain measurement practice assumes that patients and research subjects can exercise introspection to gauge the magnitude of some feature of pain, assign numbers accordingly, and report such numbers without bias. In principle, different people reporting pain should use a common scale. From the consciousness studies perspective, pain is the meaning of an injury event that the brain constructs. In contrived research contexts, pain reports behave more or less as conventional practice requires, albeit with substantial measurement error. It is less clear if clinical pain states, as personal meanings, possess the assumed feature of numerical scalability. Future progress in pain measurement must reexamine conventional assumptions about pain measurement, recognize the uniqueness of conscious individual experience as meaning, and evolve improved methods for pain assessment (Nakamura and Chapman 2002).

References

Camazine, S., Deneubourg, J., Franks, N. R., Sneyd, J., Theraulaz, G., & Bonabeau, E. (2001). *Self-organization in biological systems*. Princeton, NJ: Princeton Univ Press.
Chapman, C. R., & Nakamura, Y. (2003). Feeling pain: A constructivist perspective. In E. Kalso, A. Estlander, & M. Klockars (Eds.), *Psyche, soma and pain* (pp. 115–123). Helsinki, Finland: The Signe and Ane Gyllenberg Foundation.
Craig, A. D. (2003). Interoception: The sense of the physiological condition of the body. *Current Opinion in Neurobiology, 13*, 500–505.
Danziger, N., Faillenot, I., & Peyron, R. (2009). Can we share a pain we never felt? Neural correlates of empathy in patients with congenital insensitivity to pain. *Neuron, 61*, 203–212.
Decety, J. (2010). The neurodevelopment of empathy in humans. *Developmental Neuroscience, 32*, 257–267.
Dooley, K. (1996). A nominal definition of complex adaptive systems. *The Chaos Network, 8*, 2–3.
Fitzgibbon, B. M., Giummarra, M. J., Georgiou-Karistianis, N., Enticott, P. G., & Bradshaw, J. L. (2010). Shared pain: From empathy to synaesthesia.

Neuroscience and Biobehavioral Reviews, 34, 500–512.
Freeman, W. J. (1995). Societies of brains: A study in the neuroscience of love and hate. Hillsdale, NJ: Lawrence Erlbaum Associates.
Freeman, W. J. (2000). *How brains make up their minds*. New York: Columbia University Press.
Kelso, J. A. S. (1995). *Dynamic patterns: The self-organization of brain and behavior*. Cambridge, MA: MIT Press.
Lamm, C., Decety, J., & Singer, T. (2011). Meta-analytic evidence for common and distinct neural networks associated with directly experienced pain and empathy for pain. *NeuroImage, 54*, 2492.
Legrain, V. (2011). Where is my pain? *Pain, 152*, 467–468.
Marcel, A. (1983). Conscious and unconscious perception: Experiments on visual masking and word recognition. *Cognitive Psychology, 15*, 197–237.
Martin, B. (1994). The schema. In G. Cowan, D. Pines, & D. Meltzer (Eds.), *Complexity: Metaphors, Models and Reality* (pp. 263–285). Reading, MA: Addison-Wesley.
Morowitz, H. J., & Singer, J. L. (1995). The mind, the brain, and complex adaptive systems. In H. Morowitz, & J. L. Singer, (Eds.), (p. 237). Addison-Wesley Pub. Co: Reading, MA.
Mountcastle, V. B. (1998). Brain science at the century's ebb. *Daedalus, 127*, 1–36.
Nakamura, Y., & Chapman, C. R. (2002). Measuring pain: An introspective look at introspection. *Consciousness and Cognition, 11*, 582–592.
Rumelhart, D. D., Smolensky, P., McClelland, J. L., & Hinton, G. E. (1986). Schemata and sequential thought processes in PDP models. In J. L. McCleeland & D. E. Rumelhart (Eds.), *Parallel distributed processing: Explorations in the microstructure of cognition* (Psychological and biological models, Vol. 2, pp. 7–57). Cambridge, MA: MIT Press.
Willis, W. D., & Westlund, K. N. (1997). Neuroanatomy of the pain system and the pathways that modulate pain. *Journal of Clinical Neurophysiology, 14*, 2–31.

Consolidated Standards of Reporting Trials Statement

▶ CONSORT Statement

CONSORT Statement

Synonyms

Consolidated standards of reporting trials statement

Definition

A group of scientists and editors developed the CONSORT (Consolidated Standards of Reporting Trials) statement to improve the quality of reporting of clinical trials. The statement consists of a checklist and flow diagram that authors can use for reporting a trial. Many leading medical journals and major international editorial groups have adopted the CONSORT statement. The CONSORT statement facilitates critical appraisal and interpretation of trials by providing guidance to authors about how to improve the reporting of their trials.

Cross-References

▶ Central Pain, Outcome Measures in Clinical Trials

Constitutive Gene

Definition

A gene that is continuously expressed without regulation, i.e., the transcription can be neither suppressed nor encouraged. Constitutive genes often encode proteins with housekeeping functions.

Cross-References

▶ COX-1 and COX-2 in Pain
▶ NSAIDs, Adverse Effects

Construction of Consciousness

Definition

The construction of consciousness, at any given moment, proceeds in response to the intentional and situational imperatives of each person. The intentionality of the person perceiving, feeling, and acting drives the construction of consciousness. Consciousness is not a passive recipient of sensory experience, but an active, adaptive system that constantly models the world and the body.

Cross-References

▶ Consciousness and Pain

Contact

▶ Pain in Humans, Thermal Stimulation (Skin, Muscle, Viscera), Laser, Peltier, Cold (Cold Pressure), Radiant, Contact

Contact Stimulator

Definition

The most commonly used heat/cold contact stimulators rely on the Peltier principle (a thermoelectric effect in which passage of an electric current through a junction between two different solids causes heat to be produced or absorbed at that junction, according to the direction of current).

Cross-References

▶ Pain in Humans, Thermal Stimulation (Skin, Muscle, Viscera), Laser, Peltier, Cold (Cold Pressure), Radiant, Contact

Contemplation

Definition

A readiness to change stage, in which a person recognizes the potential importance of making a change in behavior, but has not yet made a commitment to changing behavior.

Cross-References

▶ Motivational Aspects of Pain

Contextual Factors

Definition

Contextual factors consist of two components: environmental factors and personal factors. Contextual factors are factors that together constitute the complete context of an individual's life, and in particular the background against which health states are classified in the International Classification of Functioning, Disability and Health (ICF).

Cross-References

▶ Functioning and Disability Definitions

Contingencies

Definition

Contingencies are events that occur in response to behavior. According to the operant model, contingencies can be reinforcing, punishing, or neutral with respect to their effects on behavior. Reinforcing contingencies strengthen (increase the probability of future occurrence of) behavior, punishing contingencies weaken behavior, and neutral contingencies have no impact on the future occurrence of behavior.

Cross-References

▶ Cognitive-Behavioral Perspective of Pain
▶ Motivational Aspects of Pain

Contingencies of Reinforcement

Definition

The nature of the responses that the behaviors of individuals elicit from their environment. Patterns of reinforcement will influence the occurrence or nonoccurrence of behaviors, with reinforcement leading to an increase in the behaviors and punishment contributing to a reduction in behaviors. The proximity, strength, desirability, and frequency of responses by others will influence the behaviors.

Cross-References

▶ Cognitive-Behavioral Perspective of Pain
▶ Motivational Aspects of Pain

Contingency Management

Definition

A system in which an individuals successful task completion is consistently rewarded.

Cross-References

▶ Chronic Pain in Children, Physical Medicine and Rehabilitation
▶ Training by Quotas

Contingent Negative Variation

Synonyms

CNV

Definition

The CNV is a negative slow cortical potential that is typically observed between a warning signal and a following imperative signal requiring a response. The CNV reflects attention, arousal, expectancy, and response preparation. For time intervals between the warning and the imperative signal that exceed 3 s, an early CNV component during the first 500 ms after the warning signal and a late CNV component starting at about 1 s prior to the imperative signal can be distinguished. The early CNV component presumably reflects an orienting response and contingency processing, while the late CNV is closely related to response preparation. In migraine, an enhanced CNV amplitude and an impaired habituation of the CNV have been observed, which has been interpreted as evidence for abnormal cortical information processing in migraine.

Cross-References

▶ Modeling, Social Learning in Pain
▶ Psychological Treatment of Pain in Children

Continuous Video Recordings

Definition

Non-stop video recording of animal behavior/activity. Data are commonly captured on magnetic tape or digitally for subsequent off-line scoring of behavior or activity. Infrared lighting permits filming during the dark phase.

Cross-References

▶ Visceral Pain Model, Kidney Stone Pain

Contract-Release

Definition

Contract-release is a manual technique used for the treatment of myofascial trigger points that can

serve as a patient self-treatment. It is a repeated series of brief (2 or 3 s), gentle (10 % of maximum effort), and voluntary contractions of the muscle harboring the trigger point. This is followed immediately by either passive or active stretching of the same muscle.

Cross-References

▶ Myofascial Trigger Points

Contracture

Definition

Involuntary muscle contractions are called contracture. There are three types of contractures: antalgic contractures, painful contractures, and myotatic contractures. Antalgic contractures mainly consist of a compensating phenomenon, related to a polysynaptic reflex, secondary to pain (e.g., the contracture of paraspinal muscles due to discal herniation), and usually last from several days to weeks. By contrast, painful contractures (also called contractures) include various types of cramps.

Cross-References

▶ Chronic Pain in Children, Physical Medicine and Rehabilitation
▶ Muscular Cramps

Contralateral

Definition

On or relating to the opposite side of the body.

The output of the primary motor cortex is principally to the opposite side of the body. Thus, to relieve pain restricted to one side of the motor cortex, the opposite side is stimulated.

Cross-References

► Motor Cortex, Effect on Pain-Related Behavior
► Opioid Receptor Trafficking in Pain States
► PET and fMRI Imaging in Parietal Cortex (SI, SII, Inferior Parietal Cortex BA40)
► Spinothalamocortical Projections from SM

Controlled Trial

Definition

A clinical trial that compares the effect of the drug to be tested with either a standard treatment or a placebo treatment.

Cross-References

► Antidepressants in Neuropathic Pain

Contusion Injury Model

► Spinal Cord Injury Pain Model, Contusion Injury Model

Contusion Model of Chronic Central Pain After Spinal Cord Injury

Claire E. Hulsebosch
Spinal Cord Injury Research, University of Texas Medical Branch, Galveston, TX, USA

Synonyms

Chronic central pain models; Spinal cord injury, central neuropathic pain

Definition

► Spinal Contusion Model of Chronic Central Pain is a spinal injury made with an impact device (2–2.5 mm diameter, 10 g. rod dropped from specific heights, the New York University Impactor – available through Rutgers University, the Ohio State Impactor and the Infinite Horizons Impactor – Precision Systems & Instrumentation, Lexington, Kentucky) after surgical procedures to remove the laminae at the selected site of injury in rats. The devices yield objective measures of the physics of the injury device such as compression depth, velocity of drop and, in the last device, a dwell time can be added in which the impactor remains at the specific depth of compression for a preset number of seconds. Injury sites commonly used include centrally placed injuries to one spinal segment (from T9 to T11) to model paraplegia or unilaterally placed lesions (C2 or C3) to model quadriplegia. The lesion that results is a progressive one in which the central gray matter undergoes necrosis immediately, followed by a progressive loss of gray and white matter until at 30–40 days postinjury, only a thin rim of white matter remains at the lesion site, the amount of spared tissue being inversely proportional to the severity of the lesion. Mechanical allodynia develops over 4–5 weeks bilaterally in the dermatome of the segment of lesion as well as bilaterally in both the forelimbs and hindlimbs, while thermal allodynia (allodynia – since lower temperature than that of presurgery evokes response after surgery) develops in these same regions but a week or so later than mechanical allodynia. Mice typically display a more rapid response; that is, in less than half the time course exhibited by rats. The mechanical and thermal allodynia persist for the life of the animal.

Characteristics

SCI is a very important clinical problem in that 10,000 cases a year occur. The population that most frequently incurs such injuries is between the ages of 16–25; that is, a young active population. Acute treatments have progressed as a result of better emergency medicine and improved hospital care so that these injuries are no longer viewed as immediately life threatening but at great costs. While the estimated yearly financial cost of long-term specialized care for paralyzed patients exceeds $10 billion, the personal costs to patients and their families are immeasurable. The majority of these patients, already physically debilitated, develop chronic central neuropathic pain (CNP) syndromes months to years after injury (Finnerup and Jensen 2004; reviewed by Christensen and Hulsebosch 1997). This pain so greatly affects the quality of life that depression and suicide frequently result (Cairns et al. 1996; Rintala et al. 1998; Segatore 1994; Widerström-Noga et al. 2001a, b). Research focused on improving recovery of function, including the reduction of CNP, is essential.

The study of chronic pain after SCI has been an area of neglect. A problematic issue in the development of CNP in the majority of SCI patients is that patients suffering from such symptoms are given psychiatric referrals, which do not address the pathophysiological mechanisms that underlie the condition. Since few models exist, little attention has been given to cellular mechanisms that will bring insight to reducing chronic pain in SCI. While advances in the management of acute SCI have been successful yet limited, the most recent being methylprednisolone (Bracken et al. 1992; Bracken and Holford 1993), there has been little fruitful progress in efforts to bring an understanding of the pathophysiology of CNP toward development of therapeutic approaches in the SCI patient. Consequently, acute pharmacological interventions that would inhibit the development of abnormal pain sensations are a desirable goal for improved outcome in the development of CNP in SCI patients.

There are several animal models of chronic pain after SCI: (1) the ischemic model developed by Hao, Xu, and Wiesenfeld-Hallin, an intravascular photochemical reaction occludes blood vessels in the thoracic cord, thereby producing spinal cord ischemia with a resultant band of mechanical allodynia on the trunk at the lesion site ("girdle" region, at level); (2) unilateral single quisqualate injections into the spinal dorsal horn of T10 to L4 spinal segments that

eventually results in overgrooming and mechanical allodynia in the peripheral dermatomes of segments that received the injection (at level); (3) the spinal contusion model that results in changes in spontaneous activity (Mills et al. 2001), as well as thermal and mechanical allodynia in both forelimbs and hindlimbs (above level, below level) and demonstrates the development of mechanical allodynia at the segment of injury that extends rostral for several dermatomes, "girdle" allodynia (at level, Hulsebosch et al. 2000), (4) selective surgical lesions done in C. Vierck's laboratory that included anterolateral lesions of white matter and both white and gray matter in monkeys and rats that produce overgrooming and mechanical allodynia in both rostral (above level) and caudal segments (below level), (5) a clip compression model developed in L. Weaver's laboratory in which the lower thoracic spinal cord is compressed by a 35 or 50 g clip and rats demonstrate the development of mechanical allodynia in the hindlimbs (below level), and (6) a spinal hemisection model in rats in which the spinal cord is unilaterally cut dorsoventrally with the development overtime of "girdle" pattern of mechanical allodynia (at level), thermal and mechanical allodynia in both forelimbs and hindlimbs (above and below level, Christensen and Hulsebosch 1997) and alterations in spontaneous activity (Hulsebosch 2002a, b).

The rodent contusion lesion parallels the injury profile described in human spinal cord injury (Bunge et al. 1993; Bunge 1994; Basso et al. 1996; Constantini and Young 1994). The use of rodents allows a tremendous advantage over other models since the number of experimental animals that can be generated due to a rapid time course of recovery allows a mechanistic approach to functional recovery in SCI as a result of experimental interventions. Considerable effort has been made by several laboratories to develop a standardized rodent SCI model – the contusion model was produced by the NYU impactor (after the developer, Dr. Wise Young who was then at New York University). Later, the Kentucky group, led by Stephen Scheff, developed a second-generation impactor, now called the Infinite Horizons Impactor. Both of these devices yield reproducible contusion lesions, which is critical for controlled modeling and testing therapeutic interventions. The contusion model has become a standard in many SCI studies. The rodent contusion model is further validated by the improved behavior and increased white matter sparing observed following administration of methylprednisolone, MP, (Behrmann et al. 1994) which corresponds to improvements observed following MP treatment in the patient SCI population. The BBB locomotor scoring method (Basso et al. 1996) and white matter sparing endpoints were selected as "gold standards" of recovery of function by the Multicenter Animal Spinal Cord Injury Study (MASCIS). For the purpose of somatosensory changes, the development and maintenance of mechanical and thermal allodynia are the "gold standard" of measure for "pain" studies (Hulsebosch et al. 2000).

Spinal cord injury (SCI) with the development of various pain states continues to present a significant challenge to physicians treating these disease entities. It is disturbing that embarrassingly little is known concerning the pathophysiology of pain following CNS trauma. Clearly, the attention devoted to the treatment of chronic central pain is underrepresented in terms of research, and thus treatment options. The definition of central pain, according to the International Association for the Society of Pain, is "pain initiated or caused by a primary lesion or dysfunction in the central nervous system," for example, as occurs following SCI. Chronic pain is pain that persists beyond the period of wound healing. Thus, chronic central pain (CCP) after SCI is pain that persists long after the SCI healing has occurred.

Central pain syndromes and dysesthesias (an unpleasant abnormal sensation that may or may not be painful) can be divided into two broad categories based on the dependency of the pain to peripheral stimuli: (1) persistent pain – occurs independent of peripheral stimuli, occurs spontaneously, and increases intermittently and is described as numbness, burning, cutting, piercing, or electric-like; (2) peripherally evoked pain – occurs in response to either a normally

nonnoxious or noxious stimuli. In the case that the peripherally evoked pain is in response to a normally nonnoxious stimuli, the pain state developed is considered to be allodynia. In the case that the peripherally evoked stimulus is in response to a normally noxious stimuli but the response is exaggerated, the pain state developed is considered to be hyperalgesia. A subtle but important definition is the state of increased sensitivity to stimulation which may or may not be painful which is considered to be hyperesthesia. Chronic central pain syndromes are characterized by the presence of persistent pain with concomitant changes in peripheral somatosensory responses. The International Association for the Study of Pain proposed clear classifications in the case of CCP (or central neuropathic pain, CNP) after spinal cord injury since there are clear differences with regard to kinds of pain presented in the clinic (Siddall et al. 2000).

References

Basso, D. M., Beattie, M. S., & Bresnahan, J. C. (1996). Graded histological and locomotor outcomes after spinal cord contusion using the NYU weight-drop device versus transection. *Experimental Neurology, 139*, 244–256.

Behrmann, D. L., Bresnahan, J. C., & Beattie, M. S. (1994). Modeling of acute spinal cord injury in the rat: Neuroprotection and enhanced recovery with methylprednisolone, U-74006 and YM-14673. *Experimental Neurology, 126*, 61–75.

Bracken, M. B., & Holford, T. R. (1993). Effects of timing of methylprednisolone or naloxone administration on recovery of segmental and long-tract neurological function in NASCIS 2. *Journal of Neurosurgery, 79*, 500–507.

Bracken, M. B., Shepard, M. J., Collins, W. F., Holford, T. R., Baskin, D. S., Eisenberg, H. M. et al. (1992). Methylprednisolone or maloxone treatment after acute spinal cord injury: 1-year follow-up data. *Journal of Neurosurgery, 76*, 23–31.

Bunge, R. P. (1994). Clinical implications of recent advances in neurotrauma research. In S. K. Salzman & A. I. Faden (Eds.), *The neurobiology of central nervous system trauma* (pp. 328–339). New York: Oxford University Press.

Bunge, R. P., Puckett, W. R., Becerra, J. L., Marcillo, A., & Quencer, R. M. (1993). Observations on the pathology of human spinal cord injury. A review and classification of 22 new cases with details from a case of chronic cord compression with extensive focal demyelination. In F. J. Seil (Ed.), *Advances in neurology* (Vol. 59, pp. 59–75). New York: Raven.

Cairns, D. M., Adkins, R. H., & Scott, M. D. (1996). Pain and depression in acute traumatic spinal cord injury: Origins of chronic problematic pain? *Archives of Physical Medicine and Rehabilitation, 77*, 329–335.

Christensen, M. D., & Hulsebosch, C. E. (1997). Chronic central pain after spinal cord injury. *Journal of Neurotrauma, 14*, 517–537.

Constantini, S., & Young, W. (1994). The effects of methylprednisolone and the ganglioside GM1 on acute spinal cord injury in rats. *Journal of Neurosurgery, 80*, 97–111.

Finnerup, N. B., & Jensen, T. S. (2004). Spinal cord injury pain – Mechanisms and treatment. *European Journal of Neurology, 11*, 73–82.

Hulsebosch, C. E. (2002a). Pharmacology of chronic pain after spinal cord injury: Novel acute and chronic intervention strategies. In R. P. Yezierski & K. J. Burchiel (Eds.), *Spinal cord injury pain: Assessment, mechanisms, management* (Progress in pain research and management, Vol. 23, pp. 189–204). Seattle, WA: IASP Press.

Hulsebosch, C. E. (2002b). Recent advances in pathophysiology and treatment of spinal cord injury. *Advances in Physiology Education, 26*, 238–255.

Hulsebosch, C. E., Hains, B. C., Waldrep, K., & Young, W. (2000). Bridging the gap: From discovery to clinical trials in spinal cord injury. *Journal of Neurotrauma, 17*, 1117–1128.

Mills, C. D., Hains, B. C., Johnson, K. M., & Hulsebosch, C. E. (2001). Strain and model differences in behavioral outcomes after spinal cord injury in rat. *Journal of Neurotrauma, 18*, 743–756.

Rintala, D. H., Loubser, P. G., Castro, J., Hart, K. A., & Fisher, M. J. (1998). Chronic pain in a community-based sample of men with spinal cord injury: Prevalence, severity, and relationship with impairment, disability, handicap, and subjective well-being. *Archives of Physical Medicine and Rehabilitation, 79*, 604–614.

Segatore, M. (1994). Understanding chronic pain after spinal cord injury. *Journal of Neuroscience Nursing, 26*, 230–236.

Siddall, P. J., Yezierski, R. P., & Loeser, J. D. (2000). *Pain following spinal cord injury: Clinical features, prevalence and taxonomy* (IASP Newsletter 3). Seattle, WA: IASP Press.

Widerström-Noga, E. G., Felipe-Cuervo, E., Yezierski, R. P. (2001a). Relationships among clinical characteristics of chronic pain after spinal cord injury. *Archives of Physical Medicine and Rehabilitation, 82*(9), 1191–1197.

Widerström-Noga, E. G., Felipe-Cuervo, E., Yezierski, R. P. (2001b). Chronic pain after spinal injury: interference with sleep and daily activities. *Archives of Physical Medicine and Rehabilitation, 82*(11), 1571–1577.

Conventional TENS

Definition

The delivery of TENS to stimulate selectively large diameter non-noxious cutaneous afferents, without concurrently activating small diameter nociceptive afferents or muscle efferents. This is achieved by high frequency (50–120 Hz), low (sensory) intensity pulsed currents to generate a "strong but comfortable" non-painful electrical paresthesia.

Cross-References

▶ Transcutaneous Electrical Nerve Stimulation (TENS) in Treatment of Muscle Pain
▶ Transcutaneous Electrical Nerve Stimulation Outcomes

Convergent

Definition

Related to the phenomenon of convergence, e.g., of visceral and somatic afferent fibers onto the same neurons in the CNS.

Cross-References

▶ Muscle Pain, Referred Pain
▶ Postsynaptic Dorsal Column Projection, Functional Characteristics

Conversion Disorder

Definition

One or more symptoms occur, involving voluntary motor or sensory functions that suggest a neurological or general medical condition (e.g., inability to move an arm or leg). The beginning or worsening of the symptoms must be linked to a psychological stressor. In children, pseudoseizures are the most common conversion disorder manifestations. Conversion Disorders occur more frequently in girls than boys, and more frequently post-puberty than pre-puberty. Unlike adults with conversion disorder who show La Belle Indifference, children show concern for their symptoms.

Coordination Exercises in the Treatment of Cervical Dizziness

Kjersti Wilhelmsen and Anne
Elisabeth Ljunggren
Section for Physiotherapy Science, Department of Public Health and Primary Health, Faculty of Medicine, University of Bergen, Bergen, Norway

Synonyms

Cawthorne and Cookseys' eye-head exercises; Vestibular adaptation exercises; Vestibular compensation exercises; Vestibular habituation exercises; Vestibular rehabilitation

Definition

Vestibular rehabilitation (VR) is an intervention aiming to promote central vestibular compensation through movements and exercises in patients with vestibular lesions. The intervention(s) is designed to decrease dizziness, improve balance function, and increase activity level (Herdman and Whitney 2007; Hillier and McDonnell 2011)

Vertigo is the sensation of self-motion when no self-motion is occurring or the sensation of distorted self-motion during an otherwise normal head movement (Bisdorff et al. 2009). Dizziness is the sensation of disturbed or impaired spatial orientation without a false or distorted sense of motion (Bisdorff et al. 2009). It describes an altered orientation in space, reflecting a discrepancy between internal sensation and external reality, creating sensory conflicts. The

conflicts can occur between any of the vestibular, visual, or somatosensory systems, or dizziness may be caused by central problems involving the integration and weighting of the different modalities and their relation to memory (Berthoz and Viaud-Delmon 1999). Dizziness may include sensations of light-headedness, faintness, giddiness, unsteadiness, imbalance, falling, waving, or floating (Wrisley et al. 2000).

Cervical vertigo (old terminology) can be defined as a syndrome of imbalance arising from disturbance of the cervical joint receptors and associated with a sensation of dizziness (Brown 1992). Cervicogenic dizziness is a nonspecific sensation of altered orientation in space and disequilibrium originating from abnormal afferent activities in the neck (Wrisley et al. 2000).

Characteristics

Cervicogenic Dizziness
A large number of patients seek physiotherapy for neck disorders, some of which are associated with dizziness (Clendaniel and Landel 2007). Different synonyms exist denoting these conditions; the most common term today being cervicogenic dizziness (Wrisley et al. 2000; Clendaniel and Landel 2007; Malmstrøm et al. 2007). The patients' complaints incorporate dizziness and problems with balance as well as limited range of motion and pain in the cervical region.

Cervicogenic dizziness is described as a condition of dizziness following neck trauma, usually caused by flexion-extension movement of the neck (cf. whiplash trauma). Neck pain is associated with the injury and precedes all other symptoms (Brown 1992; Wrisley et al. 2000). True vertigo, which is found in patients with acute vestibular conditions, is rare in patients with cervicogenic dizziness (Brown 1992). The sensation of dizziness is vague and nonspecific in character; the same type of dizziness is often described by patients with chronic vestibular deficits and by patients with neurological disorders (Clendaniel and Landel 2007). Patients may use words like light-headedness, faintness, giddiness, unsteadiness, imbalance, falling, waving, and floating. Other symptoms and/or findings are disequilibrium, ataxia, pain and limited range of motion in the cervical region, and visual disturbances that are aggravated by head movements and/or movements in the environment (Clendaniel and Landel 2007). Headache and tinnitus as well as hearing loss may also be present (Brown 1992; Clendaniel and Landel 2007). According to some authors, cervical vertigo is classified as a central vestibular disorder (Whitney and Rossi 2000). There is dispute as to whether the condition exists as an independent diagnosis; no definite diagnostic test is available (Wrisley et al. 2000; Clendaniel and Landel 2007; Malmstrøm et al. 2007).

The cause of cervicogenic dizziness is still debated (Clendaniel and Landel 2007). There are usually normal responses on vestibular function tests (Wrisley et al. 2000; Clendaniel and Landel 2007). Posturography has identified patients with cervicogenic dizziness (Karlberg et al. 1996a). There seems to be consensus regarding the neck's contribution to postural control and balance (Brown 1992; Karlberg et al. 1996a, Clendaniel and Landel 2007; Malmstrøm et al. 2007). Controversies exist whether dysfunctional cervical proprioceptors influence the function of the vestibular system (Clendaniel and Landel 2007). Sustained pain might alter proprioception, muscular function and structure (Clendaniel and Landel 2007; Falla and Farina 2007) and thereby contribute to alterations in joint position sense, which could contribute to dizziness. It may therefore be of interest for patients with cervicogenic dizziness to focus on dizziness and related balance problems in treatment.

Traditionally, treatment of cervicogenic dizziness has been directed toward relieving pain and consequences of pain (Furman and Whitney 2000). Neck pain and duration of complaints are strong predictors of dizziness in patients with vestibular and related disorders (Wilhelmsen et al. 2009). Karlberg et al. (1996b) have shown that traditional physiotherapy decreases frequency and intensity of dizziness in these patients, also in the long term (Malmstrøm et al.

2007). Based on the musculoskeletal hypothesis, vestibular rehabilitation may enhance potential effects of traditional physiotherapy in patients with cervicogenic dizziness.

Vestibular Rehabilitation (VR)

VR was developed in England in the 1940s by Cawthorne and Cooksey. Patients with postconcussion syndrome and vestibular deficiencies were introduced to exercises designed to train the eyes, muscle, and joint sense to compensate for vestibular dysfunction and to restore balance (Cooksey 1946). The program incorporated head and neck exercises, which the patient practiced in different starting positions (Dix 1974).

The basic principles developed by Cawthorne and Cooksey are still relevant today, although it may take on different forms. Increased knowledge of the brain's ability to compensate for deficient structures, as well as development of theories of motor control and learning, has given rise to an understanding of the vestibular compensatory mechanisms also as learning processes (Curthoys and Halmagyi 2007). Vestibular compensation is the process by which vestibular symptoms are reduced or brought to resolution. Complete recovery depends on the restoration of symmetry between the incoming signals to the different components constituting the balance system, that is, vision, somatosensory, and vestibular systems. The patient may take advantage of different strategies to enhance compensation, the overall goal of which is to achieve a stable vision and postural control, in static as well as more challenging dynamic situations.

The central compensatory process seems to involve both physical and chemical changes in the brain. It can be divided into acute and chronic stages (Herdman and Whitney 2007). Four neurophysiological mechanisms are described. ▶ Tonic rebalancing (spontaneous, physiological process by which the system tries to restore symmetry at the level of the vestibular nuclei) is the most important process in the acute stage. Vestibular adaptation (the long-term adjustments in the basic vestibulo-ocular and vestibulospinal reflexes) and vestibular habituation (the long-term reduction in the neurological responses to a noxious stimulus) (Shepard and Telian 1996) are compensatory processes in the chronic phase. Vestibular adaptation and habituation are facilitated by error signals, that is, dizziness, which can be brought on in different ways. The facilitating stimulus in adaptation is termed ▶ retinal slips, that is, movements of images across the retina. The error signal is brought on by a combination of visual fixation and head movements. The brain tries to minimize the error signal by increasing the gain of the vestibulo-ocular reflexes (Herdman and Whitney 2007). With habituation, dizziness is provoked by repeated exposure to movements provoking it, that is, performance of certain movements will produce dizziness (Shepard and Telian 1996). The last mechanism described is substitution, that is, alternative strategies are chosen to compensate for deficient vestibular functions. The latter mechanism comes into action when optimal compensation is lacking.

In the acute phase, compensation (the tonic rebalancing process) is a pathophysiological process occurring independently of the patient's activities. In the chronic phase, activities and exposure to visual input are necessary prerequisites in order to promote and enhance the recovery process. For some patients, a systematic and structured program of exercises referred to as VR may be required. The exercises can be divided into general (e.g., balance exercises) and specific exercises. Based on the underlying neurophysiological understanding of the compensatory processes, adaptation or habituation exercises can be identified, of which involve head movements. Basically, if there is a visually provoked vertigo, exercises that promote adaptation might be choice number one, while habituation may be used with patients whose dizziness primarily is produced by a change in head and/or bodily positions. However, preliminary conclusions in a recent study (Clendaniel 2010) indicate that head movements producing dizziness is the central issue, not the type of exercise, when it comes to reducing sensations of dizziness.

Examples of gaze adaptation exercises are:

- Exercise 1: Look at a stationary target placed on the wall. Move the head from side to side as fast as possible, keeping the target in focus. Repeat after a short rest, moving the head up and down.
- Exercise 2: Hold a card in the hand at arm length. Move the head and card in opposite directions, again keeping the target in focus.

Exercise 1 can also be performed holding a card in the hand. This reduces the distance between the eyes and target and stresses another part of the visual-vestibular system. Varying the tasks with regard to tempo, starting positions, and surroundings is a way to progress; the guiding lines should always be the patient's ability to keep the visual target in focus (Herdman and Whitney 2007).

A program of habituation exercises involves exposing the patient to positional changes that cause dizziness. Systematic assessment of various positional changes should be carried out, and movements provoking the dizziness make up the program. If, in lying, turning from supine and on to the side provokes dizziness, this movement should be included in the patient's program. The actual number of movements included should not exceed three to four at the time (see Shepard and Telian 1996).

There still seem to be only a few articles describing the use of VR as a supplement in the treatment of cervicogenic dizziness (Lystad et al. 2011). Wrisley et al. (2000) present the successful treatment of two patients with cervical dizziness having used a combination of traditional physiotherapy (such as ice treatment, massage, mobilization, and exercises) to decrease pain and increase the range of movement, VR in the form of adaptation (exercises 1 and 2 as presented above), and balance exercises. Tjell (2001) uses a habituation approach in the treatment of whiplash-associated disorders. The program comprises gaze and walk-gaze exercises, and progression is based on the patient's experience of pain following each treatment session. It is a slow process, but improvement in pain and dizziness is reported. A program with special focus on eye-neck coordination was administered to a group of patients with chronic neck pain (Revel et al. 1994). The aim of the program was to improve neck proprioception and to decrease pain and discomfort. The program included passive and active movements of eye and head, gaze stability, and cervical repositioning exercises. The exercises were performed in different starting positions over a period of 8 weeks. The authors suggest that the program could be useful for patients with cervicogenic dizziness.

It is our experience that a comprehensive approach is needed in the rehabilitation of patients with long-lasting dizziness, also when the condition is identified as cervicogenic dizziness. The musculoskeletal hypothesis is central in our intervention which is given as group treatment and based on a cognitive-behavioral approach promoting direct and indirect learning processes (Bandura 1986). A variety of adaptation, habituation, balance, and functional exercises are incorporated into the program aiming at provoking dizziness in line with suggestions by Clendaniel (2010). Body awareness, which is a core element in the intervention and described elsewhere in this volume (see "Body Awareness Therapies"), is stressed throughout to promote the functional interaction between body and mind (Kvåle et al. 2008). The strategy has been found useful in patients with sequel following vestibular lesions (Kvåle et al. 2008; Wilhelmsen et al. 2010) and has also been used for patients with cervicogenic types of dizziness.

References

Bandura, A. (1986). *Models of human nature and causality. Social foundation of thought and action, a social cognitive theory* (pp. 1–46). Englewood Cliffs, NJ: Prentice-Hall.

Berthoz, A., & Viaud-Delmon, I. (1999). Multisensory integration in spatial orientation. *Current Opinion in Neurobiology, 9*, 708–712.

Bisdorff, A., Von Brevern, M., Lempert, T., et al. (2009). Classification of vestibular symptoms: Towards an international classification of vestibular disorders. *Journal of Vestibular Research, 19*, 1–13.

Brown, J. J. (1992). Cervical contribution to balance: Cervical vertigo. In W. Berthoz, P. Graf, & P. Vidal (Eds.), *The head-neck sensory motor system* (pp. 644–647). New York, NY: Oxford University Press.

Clendaniel, R. A. (2010). The effects of habituation and gaze stability exercises in the treatment of unilateral vestibular hypofunction: A preliminary results. *Journal of Neurologic Physical Therapy, 34*, 111–116.

Clendaniel, R. A., & Landel, R. (2007). Non-vestibular diagnosis and imbalance: Cervical vertigo. In S. Herdman (Ed.), *Vestibular rehabilitation* (pp. 467–484). Philadelphia, PA: FA Davies.

Cooksey, F. S. (1946). Rehabilitation in vestibular injuries. *Proceedings of the Royal Society of Medicine, 39*, 273–277.

Curthoys, I. S., & Halmagyi, G. M. (2007). Vestibular compensation: Clinical changes in vestibular function with time after unilateral loss. In S. Herdman (Ed.), *Vestibular rehabilitation* (pp. 76–97). Philadelphia, PA: FA Davies.

Dix, M. R. (1974). Treatment of vertigo. *Physiotherapy, 60*, 380–384.

Furman, J. M., & Whitney, S. L. (2000). Central causes of dizziness. *Physical Therapy, 80*(1), 79–187.

Falla, D., & Farina, D. (2007). Neural and muscular factors associated with motor impairment in neck pain. *Current Rheumatology Reports, 9*, 497–502.

Herdman, S. J., & Whitney, S. L. (2007). Interventions for the patient with vestibular hypofunction. In S. Herdman (Ed.), *Vestibular rehabilitation* (pp. 309–359). Philadelphia, PA: FA Davies.

Hillier, S. L., & McDonnell, M. (2011). Vestibular rehabilitation for unilateral peripheral vestibular dysfunction. *Cochrane Database of Systematic Reviews, 2*, CD005397.

Karlberg, M., Johansson, R., Magnusson, M., et al. (1996a). Dizziness of suspected cervical origin distinguished by posturographic assessment of human postural dynamics. *Journal of Vestibular Research, 6*, 37–47.

Karlberg, M., Magnusson, M., Malmstrom, E. M., et al. (1996b). Postural and symptomatic improvement after physiotherapy in patients with dizziness of suspected cervical origin. *Archives of Physical Medicine and Rehabilitation, 77*, 874–882.

Kvåle, A., Wilhelmsen, K., & Fiske, H. A. (2008). Physical findings in patients with dizziness undergoing a group exercise programme. *Physiotherapy Research International, 13*, 162–175.

Lystad, R. P., Bell, G., Bonnevie Svendsen, M., et al. (2011). Manual therapy with and without vestibular reahibilitation for cervikogenic dizziness; a systematic review. *BMC Chiropractic & Manual Therapies, 19*, 21.

Malmström, E. M., Karlberg, M., Melander, A., et al. (2007). Cervicogenic dizziness – Musculoskeletal findings before and after treatment and long-term outcome. *Disability and Rehabilitation, 29*, 1193–1205.

Revel, M., Minguet, M., Gregoy, P., et al. (1994). Changes in cervicocephalic kinesthesia after a proprioceptive rehabilitation program in patients with neck pain: A randomized controlled study. *Archives of Physical Medicine and Rehabilitation, 75*, 895–899.

Shepard, N. T., & Telian, S. A. (1996). Vestibular reha-
    bilitation programs. In N. T. Shepard & S. A. Telian
    (Eds.), *Practical management of the balance disorder
    patient* (pp. 169–185). San Diego, CA: Singular.
Whitney, S. L., & Rossi, M. M. (2000). Efficacy of ves-
    tibular rehabilitation. *Otolaryngologic Clinics of
    North America, 33*, 659–672.
Wilhelmsen, K., Glad Nordahl, S. H., & Moe-Nilssen, R.
    (2010). Attenuation of trunk acceleration during walk-
    ing in patients with unilateral vestibular deficiencies.
    Journal of Vestibular Research, 20, 439–446.
Wilhelmsen, K., Ljunggren, A. E., Goplen, F., et al.
    (2009). Long-term symptoms in dizzy patients exam-
    ined in a university clinic. *BMC Ear Nose and Throat
    Disorders, 9*, 1.
Wrisley, D. M., Sparto, P. J., Whitney, S. L., et al. (2000).
    Cervicogenic dizziness: A review of diagnosis and
    treatment. *The Journal of Orthopaedic and Sports
    Physical Therapy, 30*, 755–766.

Coping

Definition

Psychological or behavioral strategies invoked in
order to minimize the impact of specific stressors on
an individual's physical or emotional well-being.

Cross-References

▶ Catastrophizing
▶ Coping and Pain
▶ Psychological [Behavioral Health] Assessment
  of Pain
▶ Psychology of Pain, Assessment of Cognitive
  Variables
▶ Stress and Pain

Coping and Pain

Jennifer A. Haythornthwaite
Psychiatry & Behavioral Sciences, Johns
Hopkins University, Baltimore, MD, USA

Synonyms

Adaptation; Coping strategies; Coping style

Definition

Cognitive and behavioral efforts to manage pain
or pain-related experiences, such as disability or
negative mood.

Characteristics

How individuals cope with pain is a consistent
predictor of various dimensions of the experience
of pain, including pain intensity, physical func-
tion/dysfunction, and psychological adaptation.
Catastrophizing, covered in its own section, is by
far the most consistent correlate of pain-related
outcomes. Much of the pain coping literature has
a strong theoretical foundation in the larger literature examining how individuals cope with stress.
Coping is usually defined as "constantly changing
cognitive and behavioral efforts to manage specific external and/or internal demands that are
appraised as taxing or exceeding the resources of
the person" (p. 141; Lazarus and Folkman 1984).
Note that this definition suggests that intention
underlies the efforts made by the individual and
emphasizes a process that may or may not be
successful. Complex issues arise in determining
the "adaptive" or "maladaptive" nature of any
coping strategy used to manage pain, and when
pain becomes chronic, coping attempts necessarily
broaden to encompass not only pain itself but also
negative emotions, social connectedness, and
pain-related disability across multiple domains of
function. A seemingly maladaptive coping strategy at one point in time may be adaptive at another
time; similarly, an adaptive strategy in one domain
(e.g., physical function) may be maladaptive in
another domain (e.g., social function; see Van
Damme et al. 2008). For example, the person
with low back pain who spends the weekend in
bed during a pain flare may experience less pain in
the short run but may also become deconditioned
if this use of rest persists across weeks or months.
Time spent in bed may reduce pain but may endan-
ger employment or social relationships. Recent
conceptualizations of coping extend the inherent
focus on motivation to consider the function of
coping behavior: assimilative coping attempts to

solve problems blocking the accomplishment of personal goals, and accommodative coping attempts to adapt personal goals to accommodate limitations or losses (Van Damme et al. 2008). The potential value of this conceptualization lies in its emphasis on the function of the coping behavior, the importance of developmental context, and the dynamic nature of problem appraisal (Van Damme et al. 2008).

Style versus Strategy

Much of the coping literature uses measures of pain coping strategies that encompass an indeterminate time frame and thereby measure general use of a strategy – style – as opposed to specific use of a coping strategy at a given point in time and in response to a specific targeted problem (e.g., pain vs. disability). While this approach has provided valuable data on stylistic use of pain coping strategies, the loss of information about the dynamic process of coping may explain many inconsistencies in the literature. Recent technological developments offer improved methods for assessing the short-term impact of specific pain coping strategies. Since coping is inherently a dynamic process, daily diaries now collected through electronic monitoring, often using a method called (real-time) momentary assessment (Sorbi et al. 2006), yield valuable information about the interplay of coping, pain, mood, and disability over time. This technology also offers the potential for identifying social interactions and environmental characteristics that influence these evolving processes.

Active versus Passive Coping

Early conceptualizations of pain coping in rheumatoid arthritis distinguished between active and passive coping. Active coping strategies are strategies used to directly control pain or pain-related dysfunction (e.g., distracting one's attention), whereas passive strategies relinquish control to others or limit activities (Brown and Nicassio 1987; e.g., the use of rest). Although originally conceptualized as adaptive (active) versus maladaptive (passive), more recent work has refined this characterization to include more detailed coping strategies that are associated with both positive and negative adjustment, although the

distinction between active and passive coping remains valuable (Jensen et al. 2001). A recent systematic review of prospective studies of musculoskeletal pain and disability found that greater use of passive coping strategies predicted poor functional outcomes and lower use of active coping strategies predicted increases in pain over time (Laisne et al. 2012).

Specific Pain Coping Strategies
Distraction, Ignoring Pain, or Distancing

Distraction – directing attention to something other than pain – may be more useful for managing acute pain than chronic pain (Boothby et al. 1999). Even with acute pain, however, distraction may not relieve the severe levels of pain often seen in clinical settings (Haythornthwaite et al. 2001), although recent findings using virtual reality distraction tools are quite exciting (Malloy and Milling 2010). Although some studies indicate distraction can improve function and well-being in people with chronic pain, others suggest distraction correlates with greater pain, pain-related interference, and distress. Ignoring pain is conceptually similar to distraction but focuses more on blocking out the pain; it has been associated with lower ratings of pain and improved pain-related functioning (Jensen 2011), as well as with perceptions that pain is controllable (Haythornthwaite et al. 1998). Ignoring pain is reported more frequently in individuals with a longer duration of chronic pain (Peters et al. 2000), suggesting that this strategy may be one that increases as individuals gain experience dealing with the challenges of chronic pain. Distancing includes thinking of the pain as outside one's body (Robinson et al. 1997) and overlaps with reinterpreting pain as another sensation (e.g., numbness). Although inconsistently related to positive outcomes, distancing may contribute to the perception that pain is controllable (Haythornthwaite et al. 1998). Increases in the use of these strategies seen following multidisciplinary treatment (Jensen et al. 2001) are consistent with the cognitive-behavioral formulation that these strategies are generally adaptive, contributing to the individual's perception that pain is controllable.

Positive Coping Statements

Coping self-statements refer to the positive, encouraging thoughts people use to manage pain. Certain studies have found positive self-statements to be associated with the perception that pain is controllable (Haythornthwaite et al. 1998). Reviews conclude that positive coping self-statements do not generally show a consistent relationship with reduced pain or improved functioning (Boothby et al. 1999). However, these coping self-statements are an integral component of most psychological interventions for pain management and show changes with treatment (Jensen et al. 2001).

Spirituality/Religiosity

Although spiritual beliefs and religious activities are often used to cope with stress and painful chronic illnesses, this domain has generally received surprisingly little attention in the pain literature. Historically, the primary method of measuring this domain has relied on a subscale of the Coping Strategies Questionnaire (Rosenstiel and Keefe 1983) that assesses passive prayer and hope. Higher scores on this scale are consistently associated with greater pain severity, disability, and distress across a number of studies. The measurement of praying and hoping has been empirically linked to catastrophizing, probably due to the shared passive quality of these scales. Since this strategy may be a response to difficult times, longitudinal or diary studies are needed to confirm its "maladaptive" function. In fact, more detailed assessments of day-to-day spirituality as a coping strategy find that daily spiritual experiences and positive spiritual coping are associated with greater positive mood and lower negative mood, but this coping strategy does not impact on pain reports (Keefe et al. 2001). Further research investigating the spiritual dimensions of coping may discover that actively seeking spiritual experiences forms a key dimension of accepting pain.

Social Support

When social interactions, including some expressions of pain, are intended to elicit help in managing pain or pain-related difficulties, these social interactions are conceptualized as coping strategies. The measurement of social support usually focuses on the individual's overall perceptions of support, typically incorporating two primary domains: *instrumental support*, the perception that the social environment provides resources or information, completes tasks to help the individual, or facilitates problem-solving, and *emotional support*, the perception that the social network provides sympathy, distraction, and sources of positive affect. As is the case with other coping strategies, the beneficial and/or harmful effects of social support may differ across domains of functioning and may interact with other coping strategies, such as catastrophizing. Recent prospective analyses indicate that perceived social support predicts reductions in pain-related disability and depression over 5 months following a limb amputation (Jensen et al. 2002). Diary analyses provide further details about the process by which these changes may transpire. Provision of social support is associated with lower same-day negative mood and lower next-day pain (Feldman et al. 1999). In the context of high pain, social support reduced depressed mood on the next day, suggesting a buffering role for social support. Very little research has examined what happens during these supportive exchanges, but qualitative information suggests that the sources of support are affirming their commitment to help, encouraging active coping – coping self-statements in particular – and discouraging helplessness and catastrophizing (Feldman et al. 1999). These responses contrast with more general social responses to pain that derive from the operant literature and are conceptualized as negative, or punishing, responses to pain expression (i.e., anger, irritation, or ignoring) from a significant other or attentive and solicitous responses (i.e., encouraging rest or taking over responsibilities), the latter being associated with poorer adaptation following a limb amputation (Jensen et al. 2002).

Rest

Resting in response to pain is a pain coping strategy that derives largely from the operant pain literature and has been consistently related to poorer outcomes (Jensen et al. 1991) when studied in heterogenous samples of patients typically seen in multidisciplinary specialty clinics. Most of the studies examining the deleterious effects of rest on pain-related interference and mood are cross-sectional (Jensen et al. 2002). Although use of rest following a limb amputation is not prospectively associated with poorer function (Jensen et al. 2002), reductions in rest are seen following multidisciplinary treatment and remain an important target for these interventions (Jensen et al. 2001).

Summary

To date, coping research has focused on pain coping strategies that influence pain and pain-related outcomes. Little attention, however, has been paid to the possibility that strategies for managing pain-related disability or negative mood likely differ from strategies for managing pain or that people suffering from chronic pain may not report use of these strategies when oriented to describe how they cope with pain. Recent conceptualizations of coping as motivated behavior that adapts to changing appraisals and personal goals over time (Van Damme et al. 2008), including data indicating that the use of pain coping strategies differs by pain duration (Peters et al. 2000) and site of pain (Heinberg et al. 2004), highlight the difficulties of summarizing this complex literature across these and other yet-to-be-identified dimensions.

References

Boothby, J. L., Thorn, B. E., Stroud, M. W., & Jensen, M. P. (1999). Coping with pain. In R. J. Gatchel & D. C. Turk (Eds.), Psychosocial factors in pain (pp. 343–359). New York: Guilford Press.

Brown, G. K., & Nicassio, P. M. (1987). Development of a questionnaire for the assessment of active and passive coping strategies in chronic pain patients. Pain, 31, 53–64.

Feldman, S. I., Downey, G., & Schaffer-Neitz, R. (1999). Pain, negative mood, and perceived support in chronic pain patients: A daily diary study of people with reflex sympathetic dystrophy syndrome. Journal of Consulting and Clinical Psychology, 67, 776–785.

Haythornthwaite, J. A., Menefee, L. A., Heinberg, L. J., & Clark, M. R. (1998). Pain coping strategies predict perceived control over pain. Pain, 77, 33–39.

Haythornthwaite, J. A., Lawrence, J. W., & Fauerbach, J. A. (2001). Brief cognitive interventions for burn pain. Annals of Behavioral Medicine, 23, 42–49.

Heinberg, L. J., Fisher, B. J., Wesselmann, U., Reed, J., & Haythornthwaite, J. A. (2004). Psychological factors in pelvic/urogenital pain: The influence of site of pain versus sex. Pain, 108, 88–94.

Jensen, M. P. (2011). Psychosocial approaches to pain management: An organizational framework. Pain, 152, 717–725.

Jensen, M. P., Turner, J. A., Romano, J. M., & Karoly, P. (1991). Coping with chronic pain: A critical review of the literature. Pain, 47, 249–283.

Jensen, M. P., Turner, J. A., & Romano, J. M. (2001). Changes in beliefs, catastrophizing, and coping are associated with improvement in multidisciplinary pain treatment. Journal of Consulting and Clinical Psychology, 69, 655–662.

Jensen, M. P., Ehde, D. M., Hoffman, A. J., Patterson, D. R., Czerniecki, J. M., & Robinson, L. R. (2002). Cognitions, coping and social environment predict adjustment to phantom limb pain. Pain, 95, 133–142.

Keefe, F. J., Affleck, G., Lefebvre, J., Underwood, L., Caldwell, D. S., Drew, J., Egert, J., Gibson, J., & Pargament, K. (2001). Living with rheumatoid arthritis: The role of daily spirituality and daily religious and spiritual coping. The Journal of Pain, 2, 101–110.

Laisne, F., Lecomte, C., & Corbiere, M. (2012). Biopsychosocial predictors of prognosis in musculoskeletal disorders: A systematic review of the literature. Disability and Rehabilitation, 34, 1912–1941.

Lazarus, R. S., & Folkman, S. (1984). Stress, appraisal, and coping. New York: Springer Publishing Company.

Malloy, K. M., & Milling, L. S. (2010). The effectiveness of virtual reality distraction for pain reduction: A systematic review. Clinical Psychology Review, 30, 1011–1018.

Peters, M. L., Sorbi, M. J., Kruise, D. A., Kerssens, J. J., Verhaak, P. F., & Bensing, J. M. (2000). Electronic diary assessment of pain, disability and psychological adaptation in patients differing in duration of pain. Pain, 84, 181–192.

Robinson, M. E., Riley, J. L., III, Myers, C. D., Sadler, I. J., Kvaal, S. A., Geisser, M. E., & Keefe, F. J. (1997). The coping strategies questionnaire: A large sample, item level factor analysis. The Clinical Journal of Pain, 13, 43–49.

Rosenstiel, A. K., & Keefe, F. J. (1983). The use of coping strategies in chronic low back pain patients: Relationship to patient characteristics and current adjustment. Pain, 3, 1–8.

Sorbi, M. J., Peters, M. L., Kruise, D. A., Maas, C. J., Kerssens, J. J., Verhaak, P. F., & Bendisn, J. M. (2006). Electronic momentary assessment in chronic pain II: Pain and psychological pain responses as predictors of pain disability. The Clinical Journal of Pain, 22, 67–81.

Van Damme, S., Crombez, G., & Eccleston, C. (2008). Coping with pain: A motivational perspective. Pain, 139, 1–4.

Coping Repertoire

Definition

Resources that individuals have for responding to problems and sources of stress and distress. These consist of methods to change the situation, adjust to emotional arousal and uncontrollable circumstances, and acceptance of limitations. The coping repertoire includes the internal resources, the external supports that are available, and the instrumental means that facilitate response such as financial means to assist in accommodation.

Cross-References

▶ Cognitive-Behavioral Perspective of Pain

Coping Strategies

▶ Coping and Pain

Coping Strategy/Style

Definition

A cognitive-behavioral response to a stressor (e.g., problem solving, recruiting social support) that mediates a person's response to a stressor and can impact the stress-pain relationship.

Cross-References

▶ Coping and Pain
▶ Stress and Pain

Coping Style

▶ Coping and Pain

Coprevalence

Definition

Coprevalence refers to the rates of co-occurrence of two specific disorders or clinical problems.

Cross-References

▶ Depression and Pain

Cord Glial Activation

Steven F. Maier[1], Linda R. Watkins[1],
Erin D. Milligan[2], Rolf H. Adler[3] and
Julie Wieseler[1]
[1]Department of Psychology and Neuroscience, and The Center for Neuroscience, University of Colorado at Boulder, Boulder, CO, USA
[2]Department of Neurosciences, University of New Mexico, Albuquerque, NM, USA
[3]University of Berne Medical School, Kehrsatz, Switzerland

Synonyms

Microglia activation; Spinal cord astrocyte

Definition

Cord glial activation refers to the activation of dorsal spinal cord ▶ microglia and ▶ astrocytes, which induces exaggerated pain via the release of substances that excite nocisponsive neurons (reviewed in Watkins et al. 2001). Activation is evidenced by upregulation of immunohistochemically detectable cell-type-specific activation markers, and increased release of neuroactive glial products including ▶ proinflammatory cytokines, ▶ excitatory amino acids (EAA), ▶ nitric oxide (NO), adenosine triphosphate (ATP), and prostaglandins. These released substances then act on

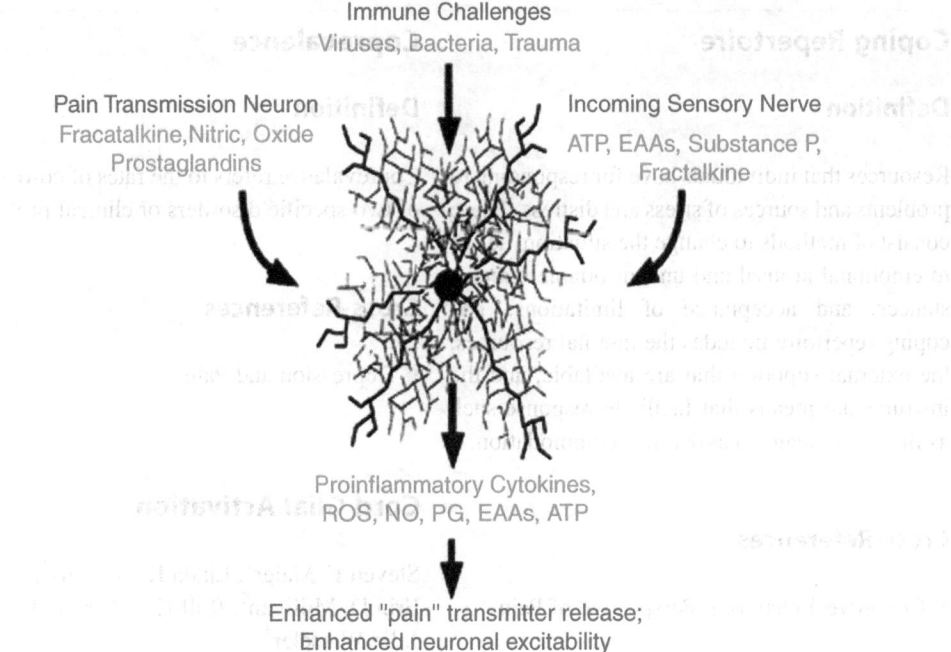

Immune Challenges
Viruses, Bacteria, Trauma

Pain Transmission Neuron
Fracatalkine, Nitric, Oxide
Prostaglandins

Incoming Sensory Nerve
ATP, EAAs, Substance P,
Fractalkine

Proinflammatory Cytokines,
ROS, NO, PG, EAAs, ATP

Enhanced "pain" transmitter release;
Enhanced neuronal excitability

Cord Glial Activation, Fig. 1 Glia are activated by three sources: (1) bacteria and viruses which bind specific activation receptors expressed by microglia and astrocytes, (2) substance P, excitatory amino acids (EAAs), and adenosine triphosphate (ATP) released by either A-delta or C fiber presynaptic terminals or by brain to spinal cord pain enhancement pathways, and (3) nitric oxide (NO), prostaglandins (PGs), and fractalkine released from pain transmission neurons. Following activation, microglia and astrocytes cause pain transmission neuron hyperexcitability and exaggerated release of substance P and EAAs from presynaptic terminals. These changes are created by the glial release of NO, EAAs, PGs, reactive oxygen species (ROS), and proinflammatory cytokines. Reprinted with permission from Watkins et al. 2001

neurons, thereby amplifying the pain response. Some evidence points to the activation of microglia, leading to the activation of astrocytes (Raghavendra et al. 2003).

Characteristics

Glia are non-neuronal cells traditionally thought of as "supportive cells" for neurons (Araque et al. 1999). Many of their functions involve the release of growth factors, protecting neurons from excitotoxicity, regulating extracellular ion concentrations, and removal of cellular debris (Fig. 1).

As discussed here, the term "glia" refers to astrocytes and microglia. While these are distinct cell types, most studies investigating exaggerated pain and glia are not able to isolate the individual contribution of the two cell types. The glia in the dorsal spinal cord differ from glia in other regions of the central nervous system (CNS). Several lines of evidence point to a striking heterogeneity among glial cells, which is influenced by the regional biochemical milieu. That is, spinal cord glia express receptors not found on glia derived from various brain regions and vice versa. For example, spinal cord is one of the few CNS regions in which glia express receptors for substance P (Beaujouan et al. 1990). Differences have also been reported between dorsal spinal cord and ventral spinal cord, and even between superficial dorsal horn and the remainder of the gray matter (Ochalski et al. 1997).

Glia, specifically astrocytes, are intimately entangled with neurons insofar as they encapsulate neuronal synapses, as well as wrap their processes around neuronal somas, axons, and dendrites (Araque et al. 1999). This proximity

Cord Glial Activation, Fig. 2 Example of microglial activation in response to a stimulus that creates exaggerated pain states. Peri-spinal administration of HIV-1 envelope glycoprotein gp120 creates exaggerated responses to both thermal and touch/pressure stimuli. Disruption of glial activation abolishes the pain changes. Furthermore, these pain changes are correlated with anatomical evidence of activation of both astrocytes and microglia. An example of such microglial activation is illustrated here. Panel A illustrates the dorsal horn of a rat injected peri-spinally with vehicle. Panel B is identical except that the rat received peri-spinal HIV-1 gp120 at a dose that creates exaggerated pain states. These photomicrographs are from the tissues collected for analysis reported in Milligan et al. 2001. Activation of microglia induces these cells to upregulate their expression of complement type 3 receptors. Thus, enzyme-labeled antibodies directed against complement receptor type 3 (OX-42 monoclonal antibodies) can be used to detect microglial activation by light microscopy. In the photomicrographs shown here, the bound enzyme-labeled antibodies catalyzed a reaction leading to precipitation of a colored reaction product. Thus, more antibodies bound (and hence more colored reaction product produced) reflects upregulation of the complement receptor type 3 microglial activation marker. By this method, activated microglia appear darker and more densely stained in accordance with their increased expression of OX-42 antibody-bound receptors. Scale bar, 25 μm in (**a**) and (**b**). (Reproduced from Milligan et al. 2001 with permission)

puts astrocytes in a unique position to modulate the activity of nocisponsive neurons. Garrison et al. were the first to investigate the relationship between exaggerated pain and ► glial activation (1994). Using the sciatic chronic constriction injury (CCI) model, they compared the expression of CCI-induced hyperalgesia and immunohistochemical markers of astrocyte activation (Garrison et al. 1991). In addition, they evaluated whether CCI-induced astrocyte activation would be affected by an ► N-methyl-D-aspartate (NMDA) receptor antagonist (MK801), known to inhibit CCI-induced hyperalgesia (Garrison et al. 1994). They found that (1) hyperalgesia induced by CCI is associated with astrocyte activation, and (2) this glial activation (like CCI-induced hyperalgesia) is inhibited by MK801. Thus, these studies revealed a predictive relationship between glial activation and exaggerated pain. Since then, positive correlations between exaggerated pain and activation of spinal cord microglia (see caption of Fig. 2 for discussion) and astrocytes have been revealed by diverse animal models of exaggerated pain, including those induced by inflammation and/or trauma to tissues, peripheral nerves, spinal roots, and/or spinal cord (DeLeo and Colburn 1999; Watkins et al. 2001). Such data naturally raise the question of whether glial activation is necessary and/or sufficient for induction and maintenance of exaggerated pain.

That glial activation plays a prominent role in the induction of exaggerated pain has been demonstrated using glial inhibitors. ► Fluorocitrate selectively blocks glial metabolic activity by inhibiting aconitase, a Krebs cycle enzyme (Paulsen et al. 1987; Hassel et al. 1992). Blocking the activation of spinal cord glial cells with fluorocitrate blocks exaggerated pain states, including thermal and mechanical hyperalgesia and mechanical allodynia (Meller et al. 1994; Milligan et al. 2000). Recently, a similar pattern has also been observed with the microglia-specific inhibitor, ► minocycline. Minocycline selectively inhibits the release of several neuroexcitatory substances from microglia, including inflammatory mediators. Administration of minocycline attenuates the development of mechanical allodynia and thermal hyperalgesia in response to spinal nerve transection, peripheral nerve inflammation, or spinal cord inflammation (Raghavendra et al. 2003). Thus, microglial activation plays a significant role in the development of exaggerated pain. Intriguingly, minocycline failed to reverse, or minimally reversed, these exaggerated pain states once they were fully expressed. These data, supported by parallel analyses of glial activation markers, revealing upregulation of GFAP (astrocyte activation marker), suggest that microglial activation may lead to astrocyte activation. Astrocytes may then play a more significant role in the maintenance of the pain facilitatory process. Taken together, these data suggest that glial activation is important for both the induction and maintenance of exaggerated pain.

It is important to note another key finding from studies of minocycline and fluorocitrate. That is, glia are not involved in normal pain processing (Milligan et al. 2000). Normally, glia are quiescent, releasing nothing that enhances pain. However, in situations leading to exaggerated pain states, glia become activated. Once activated, they begin releasing substances that contribute to the amplification of pain. Therefore, the involvement of glia in pain is determined by their state of activation.

The data strongly suggest the glial activation is important for the expression of exaggerated pain, but is it sufficient? To assess this question, glia alone need to be directly activated. While glia can be activated by classical pain neurotransmitters, including EAA, NO, prostaglandins, and extracellular ATP, these substances also activate neurons. To activate spinal cord glia without activating neurons, the portions of bacteria and viruses that bind to and activate the immune-like glial cells have been used. Administration of these substances induces mechanical allodynia and thermal hyperalgesia (Meller et al. 1994; Milligan et al. 2000). Thus, taken together, spinal cord glial activation is implicated in both the creation and maintenance of pain facilitation.

Bidirectional Neuron-Glia Communication

The discussion above describes glia as integral to the creation and maintenance of exaggerated pain. The question remains as to what signal from the periphery induces glial activation in the dorsal horn of the spinal cord, thereby inducing exaggerated pain. As described, one way is via the release of classical pain neurotransmitters/neuromodulators including EAA, substance P, prostaglandins, ATP, and NO, among others, by incoming sensory neurons and/or by intrinsic dorsal horn neurons in proximity to glia. While each of these substances excites spinal cord neurons in the pain pathway, they also activate astrocytes and microglia. Thus, these traditional neuronal mediators may exert their pain-modulating effects, at least in part, via glial activation

Once glia are activated, the question becomes what signal(s) maintains the persistent glial activation. One possible answer is that the signal is something released from dying or degenerating neurons associated with some models of chronic pain (Sugimoto et al. 1990; Coggeshall et al. 2001; Moore et al. 2002). It is well known that neuronal injury and degeneration activate glia via different pathways. Degenerating neurons have porous membranes, through which proteins such as heat-shock proteins can pass. These proteins may serve as signals of dying cells, thereby activating glia. Additionally, neurons inhibit glial activation via cell-to-cell contact (Chang et al. 2001; Neumann 2001). When a neuron dies, this contact is disrupted, leading to glial activation. Thus, it seems possible that persistent glial activation may be the end result.

There are a number of peripheral signals that may potentially trigger glial activation, such as altered neuronal activity, transported molecules from sensory neurons, or the removal of tonic inhibitory signals. However, little is known about the signal resulting in glial activation. One signal currently being explored is the putative neuron-to-glia signal ► fractalkine, a protein expressed on the extracellular surface of sensory neurons as well as dorsal horn neurons. Fractalkine is tethered to neuronal membranes by a mucin stalk (Chapman et al. 2000). When neurons are sufficiently activated, the mucin stalks break, releasing fractalkine into the extracellular fluid. Within the spinal cord, only

Cord Glial Activation, Fig. 3 Mixed astrocyte and microglia culture from neonatal rat pups showing selectivity of fractalkine for microglia. (**a**) Astrocytes labeled with glial fibrillary acidic protein (GFAP) visualized by a fluorescent antibody conjugated to Alexa 488. GFAP is a cytoskeletal protein that is highly specific to astrocytes. Note that despite these astrocytes being in a mixed culture containing microglia, no microglia are labeled for GFAP. (**b**) Microglia labeled with fractalkine, which is directly conjugated to the Cy3b fluorochrome. In the dorsal spinal cord, fractalkine receptors are found exclusively on microglia. Note that despite these microglia being in a culture containing astrocytes, no astrocytes are labeled by fractalkine. (**c**) Overlay of (**a**) and (**b**) to provide the true picture of the mixed culture

microglia express receptors for fractalkine, supporting the concept of fractalkine as a neuron-to-glia signal (Fig. 3).

Spinal administration of fractalkine induces mechanical allodynia and thermal hyperalgesia, while spinal administration of a fractalkine receptor antagonist blocks exaggerated pain, including neuropathic pain (Watkins and Maier 2003). Whether fractalkine-induced microglial activation leads, in turn, to astrocyte activation is a question that remains to be explored.

These neuron-to-glia signals point to a means by which sensory nocisponsive neurons and dorsal horn neurons communicate to glia, thereby inducing activation. In response to neuron-to-glia signals, glia release substances that then excite other glia and neurons. For example, microglial binding of fractalkine rapidly induces the release of the neuro- and glial-excitatory substances (including interleukin-1beta and nitric oxide) from dorsal spinal cord (Watkins and Maier 2003). The release of these excitatory substances induces exaggerated pain.

As ▶ immunocompetent cells, activated glia release classical immune products in response to neuronal signals. These include proinflammatory cytokines such as interleukin-1beta, interleukin-6, and ▶ tumor necrosis factor alpha (TNF-α) reduced. In their role as classical immune signals, these cytokines are "proinflammatory" because they orchestrate the early immune response to infection and damage by recruiting immune cells to the site and activating them. In spinal cord, these proinflammatory cytokines are signal molecules that activate neurons as well as glia, via binding to specific receptors on these cells. This is one means by which glia can signal neurons. Proinflammatory cytokines are not constitutively released in the spinal cord, and pharmacologically blocking their activity does not alter normal pain responsivity. However, in conditions leading to exaggerated pain, including CCI, sciatic inflammatory neuropathy (SIN), and spinal nerve transection, blocking proinflammatory cytokine activity in spinal cord blocks mechanical allodynia and thermal hyperalgesia (reviewed in DeLeo and Colburn 1999). Proinflammatory cytokines are, therefore, necessary for these exaggerated pain states. Mechanical allodynia and mechanical and thermal hyperalgesia are induced by spinal administration of proinflammatory cytokines (reviewed in Watkins et al. 2001). Thus,

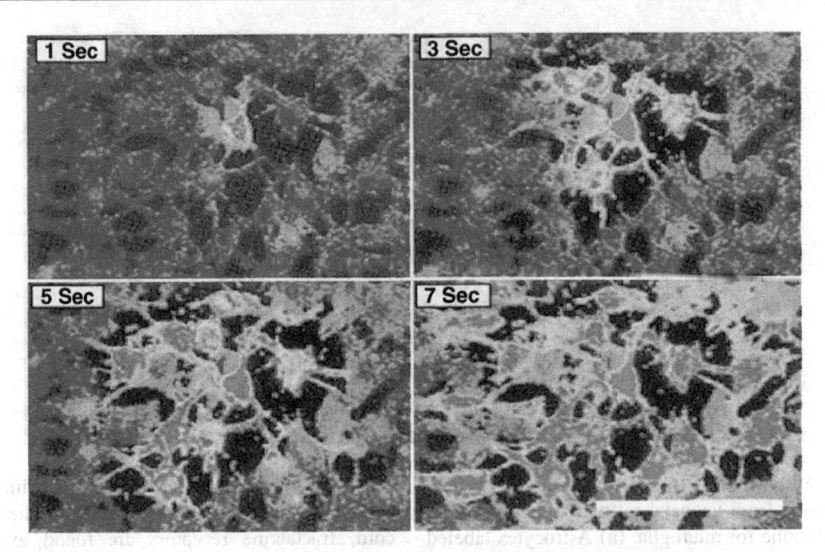

Cord Glial Activation, Fig. 4 Illustration of the spread of a calcium wave among astrocytes. Astrocytes are electrically coupled by gap junctions into vast networks. These electrical couplings allow astrocytes to rapidly communicate with one another over widespread areas. These images show a sequence of time-lapse photographic images, illustrating the rapid spread of excitation that occurs after the activation of a single glial cell at time 0. These newly activated, distant cells can then release pain-enhancing substances, thereby activating nearby pain neurons. Such gap junctional spread of excitation has recently been linked to pain facilitation, especially to mirror-image pain. (Modified from Araque et al. 1999. Reprinted with permission.)

proinflammatory cytokines are sufficient for exaggerated pain.

In addition to released signals acting on glia, astrocytes are also able to communicate with other astrocytes via ▶ gap junctions. This spread of excitation leads to the activation of distant glia (Fig. 4) and their release of neuroexcitatory substances (Araque et al. 1999).

The importance of this type of astrocyte-to-astrocyte communication, causing excitation of distant glia, raises the question of whether gap junctions may potentially provide an explanation for the spread of human pain beyond the original site of trauma, that is, mirror-image pain and extraterritorial pain. Such a communication network could, theoretically at least, lead to release of pain-enhancing substances by newly activated glia distant from the initial site of gap junctional activation. While no data yet exist in humans, the spread of pain in animal models is indeed blocked by spinal administration of the gap junction decoupler ▶ carbenoxolone. Carbenoxolone blocks the spread of mechanical allodynia in CCI and SIN (Spataro et al. 2004). Control studies, using glycyrrhetinic acid, verify that carbenoxolone-induced disruption of CCI and SIN-induced mirror-image pains are indeed due to its disruption of gap junctions. This is the first example of gap junctions, most likely of glial origin, being implicated in pain. Thus, gap junctional communication among astrocytes provides an intriguing and novel explanation for pain that arises in otherwise healthy body parts.

In summary, cord glial activation is integrally involved in diverse exaggerated pain phenomena. The powerful role that glia play in pain facilitation suggests that drugs targeting this non-neuronal cell type may hold promise for effective clinical pain control.

References

Araque, A., Parpura, V., Sanzgiri, R. P., & Haydon, P. G. (1999). Tripartite synapses: Glia, the unacknowledged partner. *Trends in Neurosciences, 22*, 208–215.
Beaujouan, J. C., DaguetdeMontety, M. C., Torrens, Y., Saffroy, M., Dietl, M., & Glowinski, J. (1990). Marked regional heterogeneity of 125I-bolton hunter substance

P binding and substance P-induced activation of phospholipase C in astrocyte cultures from the embryonic or newborn rat. *Journal of Neurochemistry, 54*, 669–675.

Chang, R. C., Chen, W., Hudson, P., Wilson, B., Han, D. S., & Hong, J. S. (2001). Neurons reduce glial responses to lipopolysaccharide (LPS) and prevent injury of microglial cells from over-activation by LPS. *Journal of Neurochemistry, 76*, 1042–1049.

Chapman, G. A., Moores, K., Harrison, D., Campbell, C. A., Steward, B. R., Strijbos, P. J. L. M. (2000). Fractalkine cleavage from neuronal membranes represents an acute event in the inflammatory response to excitotoxic brain damage. *Journal of Neuroscience, 20*: RC87, 81–85.

Coggeshall, R. E., Lekan, H. A., White, F. A., & Woolf, C. J. (2001). A-fiber sensory input induces neuronal cell death in the dorsal horn of the adult rat spinal cord. *The Journal of Comparative Neurology, 435*, 276–282.

DeLeo, J. A., & Colburn, R. W. (1999). Proinflammatory cytokines and glial cells: Their role in neuropathic pain. In L. Watkins (Ed.), *Cytokines and pain* (pp. 159–182). Basel: Birkhauser.

Garrison, C. J., Dougherty, P. M., Kajander, K. C., & Carlton, S. M. (1991). Staining of glial fibrillary acidic protein (GFAP) in lumbar spinal cord increases following a sciatic nerve constriction injury. *Brain Research, 565*, 1–7.

Garrison, C. J., Dougherty, P. M., & Carlton, S. M. (1994). GFAP expression in lumbar spinal cord of naive and neuropathic rats treated with MK-801. *Experimental Neurology, 129*, 237–243.

Hassel, B., Paulsen, R. E., Johnson, A., & Fonnum, F. (1992). Selective inhibition of glial cell metabolism by fluorocitrate. *Brain Research, 249*, 120–124.

Meller, S. T., Dykstra, C., Grzybycki, D., Murphy, S., & Gebhart, G. F. (1994). The possible role of glia in nociceptive processing and hyperalgesia in the spinal cord of the rat. *Neuropharmacology, 33*, 1471–1478.

Milligan, E. D., Mehmert, K. K., Hinde, J. L., Harvey, L. O., Martin, D., Tracey, K. J., Maier, S. F., & Watkins, L. R. (2000). Thermal hyperalgesia and mechanical allodynia produced by intrathecal administration of the human immunodeficiency virus-1 (HIV-1) envelope glycoprotein, gp120. *Brain Research, 861*, 105–116.

Moore, K. A., Kohno, T., Karchewski, L. A., Scholz, J., Baba, H., & Woolf, C. J. (2002). Partial peripheral nerve injury promotes a selective loss of GABAergic inhibition in the superficial dorsal horn of the spinal cord. *Journal of Neuroscience, 22*, 6724–6731.

Neumann, H. (2001). Control of glial immune function by neurons. *Glia, 36*, 191–199.

Ochalski, P. A., Frankenstein, U. N., Hertzberg, E. L., & Nagy, J. I. (1997). Connexin-43 in rat spinal cord: Localization in astrocytes and identification of heterotypic astro-oligodendrocytic gap junctions. *Neuroscience, 76*, 931–945.

Paulsen, R. E., Contestabile, A., Villani, L., & Fonnum, F. (1987). An in vivo model for studying function of brain tissue temporarily devoid of glial cell metabolism: The use of fluorocitrate. *Journal of Neurochemistry, 48*, 1377–1385.

Raghavendra, V., Tanga, F., & DeLeo, J. A. (2003). Inhibition of microglial activation attenuates the development but not existing hypersensitivity in a rat model of neuropathy. *Journal of Pharmacology and Experimental Therapeutics, 306*, 624–630.

Spataro, L. E., Sloane, E., Milligan, E. D., Maier, S. F., Wieseler-Frank, J., Schoeniger, D., Jekrich, B. M., Barrientos, R. M., & Watkins, L. R. (2004). Spinal gap junctions: Potential involvement in pain facilitation. *The Journal of Pain, 5*, 392–405.

Sugimoto, T., Bennett, G. J., & Kajander, K. C. (1990). Transsynaptic degeneration in the superficial dorsal horn after sciatic nerve injury: Effects of a chronic constriction injury, transection, and strychnine. *Pain, 42*, 205–213.

Watkins, L. R., & Maier, S. F. (2003). Glia: A novel drug discovery target for clinical pain. *Nature Reviews Drug Discovery, 2*, 973–985.

Watkins, L. R., Milligan, E. D., & Maier, S. F. (2001). Glial activation: A driving force for pathological pain. *Trends in Neurosciences, 24*, 450–455.

Cordotomy

Definition

Anterolateral cordotomy is a surgical procedure to sever the spinothalamic tract at a spinal level above dermatomes to which chronic pain is referred. Cordotomy is most useful in the management of intractable, unilateral lower extremity cancer pain.

Cross-References

▶ Central Nervous System Stimulation for Pain
▶ Neurosurgical Treatment of Pain
▶ Postherpetic Neuralgia, Pharmacological and Nonpharmacological Treatment Options
▶ Spinal Cord Injury Pain Model, Cordotomy Model
▶ Thalamus, Receptive Fields, Projected Fields, Human

Cordotomy Effects on Humans and Animal Models

Charles J. Vierck Jr.[1] and Alan R. Light[2]
[1]Department of Neuroscience and McKnight Brain Institute, University of Florida College of Medicine, Gainesville, FL, USA
[2]Department of Anesthesiology, University of Utah, Salt Lake City, UT, USA

Definition

Cordotomy is a surgical procedure introduced in the early 1900s for reduction of chronic intractable pain in humans. An anterolateral column of spinal white matter is cut or coagulated at a spinal level above and contralateral to a source of clinical pain, intentionally interrupting the ▶ spinothalamic tract. For weeks, months, or years after successful interruption of ascending pathways in the anterolateral column, pain and temperature sensations are diminished contralaterally over dermatomes beginning several segments below (caudal) to the lesion. The reduction of pain is apparent for clinical pain and for nociceptive stimulation of humans, other primates, and mammals (Vierck et al. 1986, 1995; White and Sweet 1969).

Characteristics

The intent of surgical section of anterolateral spinal white matter is to interrupt axons of the spinothalamic tract that originate predominantly in the contralateral gray matter in spinal segments below the lesion. The targeted projection cells receive input from nociceptive peripheral afferents selectively sensitive to stimuli that elicit pain or temperature sensations, and some are excited by non-nociceptive afferents. Other axons situated posteriorly in the spinal white matter, ipsilateral to their cells of origin, convey non-nociceptive information that subserves sensations of touch and ▶ proprioception. Therefore, cordotomy reduces pain and temperature sensitivity without producing clinically relevant deficits of touch or proprioception. Sensations reported to be lost or diminished by cordotomy are pain, itch, discrimination of sharpness with or without a painful quality, wetness, cold, warmth, and voluptuous sensations elicited by genital stimulation (Vierck et al. 1986). It is important to recognize that these effects of cordotomy probably result from combined interruption of ascending pathways to cerebral sites that include numerous thalamus nuclei, the hypothalamus, and the amygdaloid complex (Giesler et al. 1994).

Often chronic clinical pain is referred to deep tissues and appears to be particularly attenuated by cordotomy (Nathan and Smith 1979). The upper segmental levels for reduction of clinical pain and of pain from deep pressure stimulation have been reported to be well correlated and lower than the upper level of hypoalgesia for cutaneous stimulation. Similarly, visceral pain is reported to be well controlled by cordotomy (White and Sweet 1969), but the upper level of visceral pain reduction can differ from the upper level of diminished pain elicited by cutaneous pin prick testing. These differences in "sensory level" for deep and superficial pain are probably based on different entry levels for visceral and somatic afferents and on an extensive rostrocaudal dispersion in the spinal cord of afferent input from deep somatic structures. The effectiveness of anterolateral cordotomy for attenuation of visceral pain is surprising in view of evidence that interruption of a pathway located in the dorsal spinal columns can attenuate visceral pain (Nauta et al. 2000). It appears that different aspects (qualitative features?) of visceral sensation depend on rostral conduction by dorsal and ventral spinal pathways.

The dorsoventral and mediolateral extents of lateral and ventral column damage are important determinants of the sensory effects of cordotomy. A somatotopic organization exists dorsoventrally, with sacral dermatomes dorsal and rostral dermatomes ventral. Accordingly restriction of cordotomy to the anterolateral column can miss the most dorsal spinothalamic axons and result in sacral sparing. In addition, there are differences in dorsoventral distribution of rostrally projecting axons with distinguishable origins. For example, spinothalamic axons originating in the deep dorsal horn are more ventrally located in the anterolateral and ventral column, relative to cells from the superficial dorsal horn (Zhang et al. 2000). This appears to be the case, in general, for cells originating in the superficial and deep dorsal horn, which have overlapping, but differential, functional properties and projections to targets in the brain stem and thalamus (e.g., Craig 1998). These relationships could influence the long-term effects of a cordotomy.

For the mediolateral dimension of the anterolateral column, clinical investigators have postulated that pain from mechanical stimulation of the skin depends upon peripherally located axonal projections, while temperature sensations and pain from stimulation of deep tissues depend upon more medially located axons (Walker 1940). These suppositions encouraged a clinical strategy involving medially extensive anterolateral cordotomy, in order to maximally attenuate pain from deep tissues. Given this surgical approach, the expectation is that a successful unilateral cordotomy would produce contralateral analgesia, particularly for cutaneous stimulation. However, human and animal studies involving nociceptive cutaneous stimulation within dermatomes maximally affected by cordotomy indicate that pain sensitivity is not eliminated (King 1957; Vierck et al. 1990, 1995; Vierck and Luck 1979).

Not only is pain sensitivity partially preserved contralaterally, but chronic pain returns and pain sensitivity increases over time for some but not all cases after an initially successful cordotomy. Early on, the explanation for this was that the anterolateral lesion for cases of recovery was incomplete dorsally, ventrally, or medially. However, this possibility is contradicted by human and animal studies with histological confirmation of highly restricted lesions (superficially located in the anterolateral column) that produced very long-term hypoalgesia (Nathan and Smith 1979; Vierck and Luck 1979). Direct comparisons of cordotomy effects in monkeys indicate that recovery of pain sensitivity over time is more substantial for extensive

lesions (primarily medially) than for superficial lesions (Vierck et al. 1990). Explanations for these results may be (1) that all spinothalamic axons shift laterally with ascension of the spinal cord (Zhang et al. 2000) and (2) that the more extensive lesions partially damage spinal projection systems other than the spinothalamic tract (Fig. 1) (see ▸ Spinal Cord Injury Pain Model, Cordotomy Model).

When chronic pain returns after cordotomy, it appears to be qualitatively distinct from that experienced before surgery. Discrimination between qualitative features of chronic pain is difficult, but recovered pain can be more diffuse in character and bilaterally distributed (White and Sweet 1969). Several possible mechanisms for this are related to (1) ▸ deafferentation of neurons normally receiving input from rostrally projecting axons sectioned by cordotomy, and (2) enhanced transmission of nociception by diffusely projecting ascending pathways spared by cordotomy.

Return of clinical pain and of painful sensations elicited by stimulation after cordotomy is consistent with development of chronic pain that can occur following strokes that produce deficits of elicited pain and temperature sensations and, by implication, damage the spinothalamic tract at some level of the neuraxis (Boivie et al. 1989). Numerous human studies involving thalamic recording have demonstrated abnormal spontaneous activity in regions deafferented by these and other lesions (Lenz 1992). The presence of abnormally high levels of activity among populations of neurons that previously received nociceptive input can potentially be interpreted as pain. If this occurs, the distribution of pain would be expected to involve much of the deafferented region, rather than the more restricted site of pain experienced before cordotomy. Furthermore, recordings from the thalamus of monkeys after medially extensive cordotomy have shown that abnormal spontaneous activity is not restricted to the region of primary deafferentation within the nucleus ventralis posterolateralis (Weng et al. 2000). That is, the effects of deafferentation can extend beyond the somatotopically appropriate region of

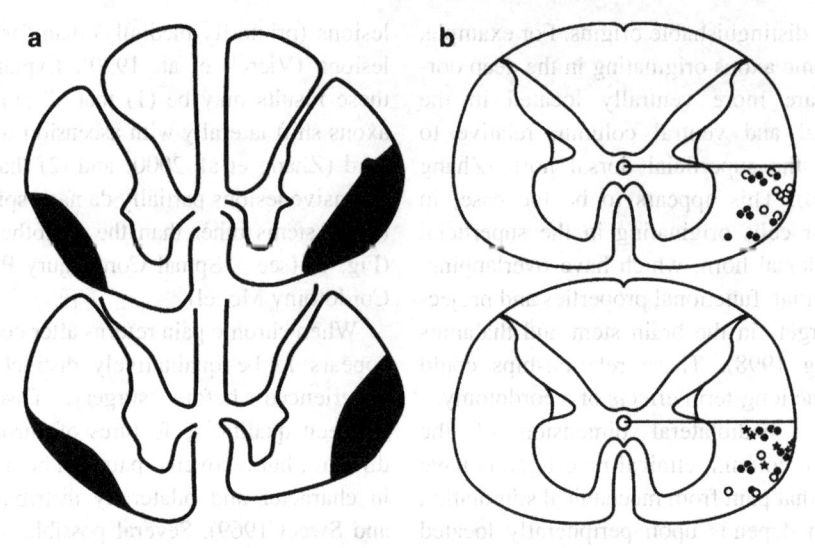

Cordotomy Effects on Humans and Animal Models, Fig. 1 (a) Diagrams of four unilateral cervical cordotomies in humans that produced enduring reductions of chronic clinical pain and sensitivity to nociceptive stimulation. Note that the lesions are restricted to the peripheral portions of the anterolateral white matter (After Nathan and Smith 1979). (b) Plots of the uniformly peripheral distribution of anterolateral axons (at the C2 segment of monkeys) that project to the ipsilateral thalamus from cells of origin in the contralateral gray matter. *Upper diagram*: *solid circles*: axons from cells in the superficial dorsal horn; *open circles*: axons from cells in the deep dorsal horn; overall, axons from superficial cells are distributed more dorsally, especially at thoracic levels (not shown). *Lower diagram*: cells with input from different populations of peripheral afferents: *solid circles*: wide dynamic range neurons; *open circles*: low threshold neurons; *stars*: high threshold neurons (From Zhang et al. 2000)

a principal target of spinothalamic projections to the thalamus.

Other clues relevant to an eventual return of clinical pain and pain sensitivity are offered by the stimuli capable of eliciting pain sensations after cordotomy. Thresholds for elicitation of pain are clearly elevated, but strong suprathreshold stimulation is perceived as painful (King 1957). This is apparent for electrical stimulation of the skin or a peripheral nerve, which progressively activates A-delta and C afferents as nociceptive stimulus intensity increases. A cordotomy that eliminates pain elicited by stimulation of A-delta afferents can preserve pain from electrical stimulation at intensities sufficient to excite C nociceptors (Collins et al. 1960). Similarly, pain elicited by the standard pinprick test involving single stimuli can be eliminated by a cordotomy, but repetitive pinprick stimulation at the same location produces pain (Nathan and Smith 1979; Walker 1940), presumably because of temporal summation. Temporal

summation by repetitive stimulation has been shown to depend upon stimulation of C nociceptors (e.g., Vierck et al. 1997). Also, when pain from application of focal pressure to deep tissues is eliminated by cordotomy, pain from muscular cramps can be perceived (White 1968). In this example, spatial summation from simultaneous stimulation of many C nociceptors may be responsible for breakthrough pain.

The varieties of pain retained after cordotomy suggest that pathways of rostral projection outside the contralateral lateral column are particularly sensitive to C nociceptor input. The following evidence indicates that this conduction system is diffusely distributed in the anterolateral and ventral spinal cord and is responsible for the return of pain sensitivity after cordotomy. Bilateral cordotomy does not prevent recovery of pain or eliminate pain that has developed after interruption of one anterolateral column (White and Sweet 1969). Therefore, it is unlikely that sparing of spinothalamic axons underlies the return of pain after cordotomy. Also, secondary lesions of dorsal spinal pathways in monkeys do not attenuate pain sensitivity that has returned after cordotomy (Vierck and Luck 1979). However, complete interruption of both ventral spinal quadrants (ventral hemisection) appears to eliminate pain sensitivity without recovery (Vierck and Luck 1979). Obvious candidates for diffuse rostral pain transmission are the spinoreticular pathways, with bilateral relays to cerebral systems involved in pain coding and modulation (e.g., spinomesencephalic and spinoreticulothalamic).

In summary, clinical and laboratory animal studies with histological verification of spinal lesions indicate that interruption of spinothalamic axons located superficially in the anterolateral spinal column effectively reduces chronic pain and pain sensitivity, long term. However, partial interruption of diffuse ventral spinal projection systems by extensive lesions can initiate a recovery process with adverse effects on pain sensitivity. Complete interruption of the bilateral ventral projection system and the spinothalamic tract produces analgesia, in contrast to the hypoalgesia produced by anterolateral cordotomy. Unfortunately, ventral hemisection also produces profound motor deficits by interrupting a number of descending spinal pathways. Therefore, when employment of spinal surgery is considered for alleviation of chronic, intractable pain, available evidence favors interruption of superficial axons in the anterolateral column.

References

Boivie, J., Leijon, G., & Johansson, I. (1989). Central post-stroke pain – a study of the mechanisms through analyses of the sensory abnormalities. *Pain, 37,* 173–185.

Collins, W., Nulsen, F., & Randt, C. (1960). Relation of peripheral nerve fibre size and sensation in man. *Archives of Neurology, 5,* 381–385.

Craig, A. D. (1998). A new version of the thalamic disinhibition hypothesis of central pain. *Pain Forum, 7,* 1–14.

Giesler, G. J., Jr., Katter, J. T., & Dado, R. J. (1994). Direct spinal pathways to the limbic system for nociceptive information. *Trends in Neurosciences, 17,* 244–250.

King, R. B. (1957). Postchordotomy studies of pain threshold. *Neurology, 7,* 610–614.

Lenz, F. A. (1992). The ventral posterior nucleus of thalamus is involved in the generation of central pain syndromes. *APS Journal, 1,* 42–51.

Nathan, P. W., & Smith, M. C. (1979). Clinico-anatomical correlation in anterolateral cordotomy. *Advances in Pain Research and Therapy, 3,* 921–926.

Nauta, H. J. W., Soukup, V. M., Fabian, R. H., et al. (2000). Punctate midline myelotomy for the relief of visceral cancer pain. *Journal of Neurosurgery, 92,* 125–130.

Vierck, C. J., Jr., & Luck, M. M. (1979). Loss and recovery of reactivity to noxious stimuli in monkeys with primary spinothalamic chordotomies, followed by secondary and tertiary lesions of other cord sectors. *Brain, 102,* 233–248.

Vierck, C. J., Jr., Greenspan, J. D., Ritz, L. A., et al. (1986). The spinal pathways contributing to the ascending conduction and the descending modulation of pain sensations and reactions. In T. L. Yaksh (Ed.), *Spinal systems of afferent processing* (pp. 275–329). New York: Plenum Press.

Vierck, C. J., Jr., Greenspan, J. D., & Ritz, L. A. (1990). Long term changes in purposive and reflexive responses to nociceptive stimulation in monkeys following anterolateral chordotomy. *Journal of Neuroscience, 10,* 2077–2095.

Vierck, C. J., Jr., Lee, C. L., Willcockson, H. H., et al. (1995). Effects of anterolateral spinal lesions on escape responses of rats to hindpaw stimulation. *Somatosensory & Motor Research, 12,* 163–174.

Vierck, C. J., Jr., Cannon, R. L., Fry, G., et al. (1997). Characteristics of temporal summation of second pain sensations elicited by brief contact of glabrous skin by a preheated thermode. *Journal of Neurophysiology, 78,* 992–1002.

Walker, A. E. (1940). The spinothalamic tract in man. *Archives of Neurology and Psychiatry, 43,* 284–298.

Weng, H.-R., Lee, J., Lenz, F., et al. (2000). Functional plasticity in primate somatosensory thalamus following chronic lesion of the ventral lateral spinal cord. *Neuroscience, 101,* 393–401.

White, J. (1968). Operations for the relief of pain in the torso and extremities: Evaluation of their effectiveness over long periods. In A. Soulairac, J. Cahn, & J. Charpentier (Eds.), *Pain* (pp. 503–519). New York: Academic.

White, J. C., & Sweet, W. H. (1969). *Pain and the neurosurgeon: A 40 year experience.* Springfield: Thomas CC.

Zhang, X., Honda, C., & Giesler, G. J., Jr. (2000). Position of spinothalamic tract axons in upper cervical spinal cord of monkeys. *Journal of Neurophysiology, 84,* 1180–1185.

Core and Matrix

▶ Spinothalamic Terminations, Core and Matrix

Coronary Artery Disease

▶ Visceral Pain Model, Angina Pain

Coronary Insufficiency

▶ Visceral Pain Model, Angina Pain

Corpus Callosum

Definition

The largest bundle of fibers connecting the two hemispheres. The corpus callosum is subdivided into four subregions with the splenium located in the most posterior region, the body extending anterior from the splenium, the genu consisting in the most anterior part, and the rostrum below the genu. Each region of the corpus callosum connects specific regions of one hemisphere of the cerebral cortex to the homologous area in the other hemisphere.

Cross-References

▶ Cingulate Cortex, Functional Imaging
▶ Pain Processing in the Cingulate Cortex, Behavioral Studies in Humans

Correct Rejection

Definition

Correct rejection is the probability of response "B" when event B has occurred.

Correlation

Definition

The degree to which variables change together.

Cross-References

▶ Pain in the Workplace, Risk Factors for Chronicity, Job Demands

Cortex and Pain

Definition

Cortical areas (e.g., anterior cingulate cortex, parietal operculum, insula, primary somatic sensory cortex) that show pain-related activity, usually based on imaging studies.

Cross-References

▶ Pain Treatment, Motor Cortex Stimulation

Cortical and Limbic Mechanisms Mediating Pain and Pain-Related Behavior

Kenneth L. Casey
Department of Neurology, University of
Michigan Medical Center, Ann Arbor, MI, USA

Characteristics

Scope of This Review

This entry will briefly review pain-related functions of the cerebral cortex and limbic system. For the purposes of this entry, the limbic system is defined as the medial and orbital portions of the frontal, parietal, and temporal lobes that form a ring around the upper brainstem and diencephalon. The cingulate gyrus is a defining structure of the limbic system, the major components of which have direct or oligosynaptic connections with cingulate cortex (see also ▶ Cingulate Cortex, Functional Imaging; ▶ Pain Processing in the Cingulate Cortex, Behavioral Studies in Humans; ▶ Cingulate Cortex, Nociceptive Processing, Behavioral Studies in Animals) and with the hypothalamus (MacLean 1955; Papez 1937). The subcortical limbic structures such as the amygdala (see also ▶ Amygdala, Functional Imaging; ▶ Nociceptive Processing in the Amygdala: Neurophysiology and Neuropharmacology), which is covered thoroughly in the essay in this Section by Volker Neugebaurer, will not be discussed. Modulatory mechanisms mediated by pathways descending from the cortex and brainstem are discussed elsewhere in this volume and will not be discussed here.

The neurobiological response to injury complicates the analysis of cortical function during pain. Peripheral or central nervous system injuries trigger complex biochemical mechanisms that result in varying degrees of anatomical and functional reorganization of the cerebral cortex (Kaas et al. 1997). In this entry, cortical functions will be considered only as mediated normally in the absence of clinically significant reorganization.

Conceptual Issues

"Function" refers to the mix of psychological, behavioral, and physiological manifestations of cerebral cortical activity, including somatic and autonomic responses that may be independent of the fully developed sensation of pain and the sensory, affective-motivational, and cognitive components of pain (Melzack and Casey 1968; Price 1988, 1999) as recognized by the International Association for the Study of Pain (IASP) (Merskey and Bogduk 1994) and in recent reviews of the cerebral cortical mechanisms mediating pain (Treede et al. 1999).

Spatial Distribution of Function

The clinical observation that pain can be relieved, at least partially, by spinothalamic tractotomy (Spiller and Martin 1912) supports the concept of an anatomically distinct pain pathway, but subsequent clinical and neurosurgical experience has emphasized the long-term unpredictability of the results obtained following most ablative procedures, especially those involving cerebral structures (Sweet 1982). Current methods of analyzing brain function show that all cortical and limbic functions are mediated by networks of interconnected cortical areas, each with a specific, although incompletely characterized, neuronal function. This concept is in accord with the results of current functional imaging studies of pain (Casey 1999, 2000; Davis 2000; Derbyshire 2000; Peyron et al. 2000) and the contemporary view that distributed parallel systems mediate cortical functions (Mesulam 1990).

Temporal Distribution of Function

Cortical and limbic functions are also distributed over time. This concept is important for considering cortical functions as they affect both acute and chronically painful conditions. How cortical functions influence the perception of pain and pain-related behavior may change with time from injury or, in the experimental setting, with time following a noxious stimulus. Some temporal variance may be due to the sequential activation of different cortical areas. Imaging studies of pain, for example, show that changes in the cortical response to noxious stimulation occur over a time span of approximately 45 s (Casey et al. 2001). Other temporal variation in cortical influence may occur simply because an injury or the context surrounding the injury has changed over time, so that the perceptual and behavioral impact of that function changes. All cortical areas are active to some degree throughout all time periods during which pain is experienced. Neurophysiological and functional brain imaging studies (cited below) show that each of the cortical structures listed here is active within milliseconds following a noxious stimulus. However, the clinically relevant impact of specific cortical structures on pain and pain behaviors deserves emphasis within particular time frames. Figure 1 is intended to summarize this hypothesized temporal distribution of the influence and impact of cortical and limbic

	EARLY (msec)	INT.MED. (msec-min)	LATE (min-years)
S1	X	x	x
S2	X	x	x
AI	X	x	x
ACC	X	X	X
PCC	X	X	?
IPL	x	X	x
PrMot	x	X	x
PI	x	X	x
MPC	x	X	X
OFC	x	X	X
HIPP	x	X	X
DLPFC	x	X	X

SENSORY	AFFECTIVE	COGNITIVE

Cortical and Limbic Mechanisms Mediating Pain and Pain-Related Behavior, Fig. 1 Relative influence of cortical function on pain and pain-related behavior. This figure summarizes the distribution of the influence of cerebral cortical and limbic functions on pain perception and pain-related behavior over time. All structures listed in the *left* column have some influence on pain and behavior throughout the experience of pain, but the degree of this influence and its clinical and behavioral impact varies over time as indicated by the size and boldness of the X marks. The sensory-discriminative function (*blue*) is largely limited to the primary (S1) somatosensory cortex. Affective function is indicated by shades of *red* and cognitive function by *green*. Hippocampal (and entorhinal cortex) function is a mixture of affective and cognitive elements. *S1* primary somatosensory cortex, *S2* secondary somatosensory cortex, *AI* anterior insula, *ACC* anterior cingulate cortex, *PCC* posterior cingulate cortex, *IPL* inferior parietal lobule, *Pr. Mot.* premotor cortex, *PI* posterior insula, *MPC* medial prefrontal cortex, *OFC* orbitofrontal cortex, *HIPP* hippocampus and entorhinal cortex, *DLPFC* dorsolateral prefrontal cortex

functions over time. Unfortunately, information about cortical and limbic function over periods ranging from hours to years is available only in the clinical setting and only when long-term follow-up is available.

Early Cortical and Limbic Processing (Milliseconds)

Primary Somatosensory Cortex (S1)
The S1 cortex (see also ▶ Primary Somatosensory Cortex (S1), Effect on Pain-Related Behavior in Humans) within the postcentral gyrus is composed of Brodmann areas 1, 2, 3a, and 3b, the latter two areas lying in the depths of the central sulcus and generally considered to be the major recipient of spatially refined cutaneous somatosensory input.

Lesions
There is evidence from clinical, neurophysiological, and functional brain imaging that the S1 cortex is a major site for identifying noxious stimuli along the temporospatial and intensity domains. Early clinical analysis based on wartime missile wounds suggested that small lesions within the S1 cortex could produce a somatotopically restricted hypoalgesia along with an impairment of somesthetic spatial and temporal discriminative functions (Marshall 1951; Russell 1945). However, direct anatomical corroboration was not available at the time, and subsequent studies have shown that even relatively modest impact trauma to the cortex can result in otherwise unrecognized and clinically significant subcortical lesions that could interrupt deep nociceptive pathways (Lighthall 1988; Lighthall et al. 1989).

Selective surgical lesions of S1 cortex for the relief of clinical pain have produced poor results over time (Sweet 1982). Furthermore, infarctions limited to the territory of the postcentral gyrus and superior-posterior parietal region produce a somatotopically limited impairment of tactile and kinesthetic discriminative functions with sparing of pain and temperature sensations (Bassetti et al. 1993). This suggests that the S1 cortex is not essential for mediating the perceptual aspects of clinically relevant chronic pain states.

Stimulation
Despite the strong evidence for nociceptive information reaching the S1 cortex (see below), electrical stimulation of the postcentral gyrus rarely, if ever, evokes pain in the conscious or partially sedated human (Nii et al. 1996; Sweet 1982). However, these stimuli do not typically activate neurons in the sulcal depths of S1 cortex, where nociresponsive neurons may be located.

Electrophysiology

The detailed somatotopic organization of neurons responding to tactile stimuli (Whitsel et al. 1971) has been confirmed in numerous investigations (Kaas et al. 1979; Kaas 1993). Single-cell recordings from the monkey reveal that there are nociceptive neurons within S1 cortex and that these have restricted receptive fields and responses that could mediate thermal nociceptive discrimination in the monkey (Kenshalo and Isensee 1983; Kenshalo et al. 1988, 2000). In humans, the earliest nociceptive information to reach the cerebral cortex, as determined by magnetoencephalographic (MEG) (see also ► Insular Cortex, Neurophysiology, and Functional Imaging of Nociceptive Processing) and evoked potential (EP) (see also ► Insular Cortex, Neurophysiology, and Functional Imaging of Nociceptive Processing, ► Nociceptive Processing in the Hippocampus and Entorhinal Cortex, Neurophysiology and Pharmacology) studies with noxious laser or brief mechanical stimulation, arrives nearly simultaneously in the primary somatosensory (S1), secondary somatosensory (S2) (see also ► Secondary Somatosensory Cortex (S2) and Insula, Effect on Pain-Related Behavior in Animals and Humans), anterior insular (AI) (see also ► Insular Cortex, Neurophysiology, and Functional Imaging of Nociceptive Processing), and the rostral anterior cingulate cortex (ACC) (see also ► Cingulate Cortex, Nociceptive Processing, Behavioral Studies in Animals, ► Pain Processing in the Cingulate Cortex, Behavioral Studies in Humans, ► Cingulate Cortex, Functional Imaging) (Arendt-Nielsen et al. 1999; Kakigi et al. 1995; Ploner et al. 1999; Schnitzler and Ploner 2000). Kanda and colleagues used recordings from the scalp and from implanted subdural electrodes to show that painful infrared laser stimulation evokes a potential with a peak latency of 220 ms in human S1 cortex (Kanda et al. 2000). A recent MEG study with laser-evoked selective stimulation of C fibers found nearly simultaneous responses within approximately 750 ms in both the S1 and S2 cortices (Tran et al. 2002).

Functional Imaging

Functional imaging studies support the concept that the S1 cortex participates in the sensory-discriminative aspect of pain, although cognitive variables modify the intensity of the response significantly (Bushnell et al. 1999; Hofbauer et al. 2001). High-resolution optical imaging studies show unique and possibly specific responses to noxious heating in subdivisions of area 3a (Tommerdahl et al. 1996). However, functional magnetic resonance imaging (fMRI) studies (see also ► PET and fMRI Imaging in Parietal Cortex (Si, Sii, Inferior Parietal Cortex BA40)) show that responses to noxious stimuli are also likely to be obtained from area 1 of the S1 cortex (Gelnar et al. 1999), a finding in general agreement with the results of cellular recording (Kenshalo and Isensee 1983; Kenshalo et al. 1988).

In summary, the weight of evidence from a variety of sources favors the view that S1 cortical neurons are specialized to engage in the earliest processes mediating the discriminative aspects of somatic sensation, including pain. Clinical observations suggest that these neurons are essential for nociceptive discriminative functions but are less likely to be essential for mediating or modulating the affective or cognitive aspects of chronically painful conditions.

Secondary Somatosensory Cortex (S2)

The S2 cortex occupies the posterior parietal operculum over the lateral (Sylvian) fissure and is adjacent to the posterior insula (see also ► Insular Cortex, Neurophysiology, and Functional Imaging of Nociceptive Processing, ► Secondary Somatosensory Cortex (S2) and Insula, Effect on Pain-Related Behavior in Animals and Humans, ► Nociceptive Processing in the Secondary Somatosensory Cortex). Because the S2 cortex receives input via the spinothalamic tract (Stevens et al. 1993) and has strong projections to the insula, it is in a position to transmit nociceptive information to limbic cortical structures, such as the cingulate, medial prefrontal (see also ► Prefrontal Cortex, Effects on Pain-Related Behavior), and orbital frontal cortices, via insular connections (Mesulam and Mufson 1982a, b).

Lesions

Lesions involving the human S2 cortex or the adjacent subcortical white matter have been associated with a central pain syndrome and a clinically detectable contralateral hypalgesia (Horiuchi et al. 1996; Schmahmann and Leifer 1992), but a contributing effect of injury to the insular cortex could not be completely eliminated. Greenspan and colleagues (1999) found that lesions involving the posterior opercular (S2) cortex were associated with a clinically detectable elevation of pain threshold in humans, while patients with lesions sparing S2 cortex had normal heat pain thresholds even when the lesion also involved the insula (Greenspan et al. 1999).

Stimulation

Recent studies using imaging methods to confirm the location of implanted depth electrodes have not thus far reported pain sensations or pain-related behavior during direct stimulation of the S2 cortex in humans (Ostrowsky et al. 2000, 2002). However, these investigations were focused on stimulation of the insular cortex, so a more definitive assessment awaits a systematic exploration of the human S2 cortex.

Electrophysiology

Cellular recordings from S2 cortex in anesthetized primates reveal very few neurons that respond to noxious stimuli. The somatotopic organization of primate S2 is quite coarse compared to that of S1 cortex, and the receptive fields are typically large – often bilateral or covering most of a single limb (Burton et al. 1995). In contrast with the observations in recordings from animals, noxious stimulation regularly evokes distinctive electrical and magnetic responses in human S2 cortex (Arendt-Nielsen et al. 1999; Bromm 2001; Kakigi et al. 2000; Ploner et al. 2002). Nociceptive information arrives nearly simultaneously at the human S2 and S1 cortex (approximately 150–200 ms to peak), but, following a brief stimulus, the duration of neuronal activity is significantly longer in S2 than in the S1 cortex (Inui et al. 2003; Kanda et al. 2000;

Ploner et al. 2002). Selective C fiber stimulation also evokes MEG and EP responses in S2 cortex (Opsommer et al. 2001; Tran et al. 2002). Frot and colleagues (2001) revealed differences in the latency and amplitude of S2 cortical responses to innocuous electrical and noxious (laser) stimulation during depth electrode recording in humans, suggesting that this cortex integrates both tactile and nociceptive inputs (Frot et al. 2001).

Functional Imaging

The S2 cortex is one of the most consistently activated structures in positron emission tomographic (PET) and fMRI studies (see also ▶ PET and fMRI Imaging in Parietal Cortex (SI, SII, Inferior Parietal Cortex BA40)) designed to distinguish among responses to noxious and innocuous stimuli (Burton et al. 1993; Casey 1999; Davis 2000; Derbyshire 2003; Peyron et al. 2000). An fMRI study revealed a coarse somatotopic organization within the S2 cortex (Ruben et al. 2001). However, Peyron and colleagues (2002), using a multimodality approach involving PET, fMRI, and scalp LEP recording with dipole localization in the same subjects, could not determine whether the pain-related activations they obtained were localized to the S2 cortex, the insula, or both structures. Their results led the authors to suggest the term "operculoinsular" cortex to refer to responses within the S2-insular region (Peyron et al. 2002). However, the recent results obtained from detailed depth recording favor the interpretation that both the S2 and insular cortices are independent generators of responses to noxious stimuli (Frot and Mauguiere 2003).

In summary, the evidence supports the view that human S2 cortex is a critical component of the cortical network mediating pain. Although there is a crude somatotopic organization within this cortex, it seems unlikely that it is critical for spatial discrimination. Rather, the S2 cortex appears to be involved in the early identification of noxious events that are combined with the localizing information provided through the S1 cortex and transmitted to other cortical areas for further analysis.

Anterior Insula

The anterior insula (see also ▶ Insular Cortex, Neurophysiology, and Functional Imaging of Nociceptive Processing, ▶ Secondary Somatosensory Cortex (S2) and Insula, Effect on Pain-Related Behavior in Animals and Humans, ▶ Nociceptive Processing in the Secondary Somatosensory Cortex) lies rostral to the most lateral point of the central sulcus, which also approximately defines the vertical plane extending through the anterior commissure perpendicular to the sagittal plane. The cortex of the anteroventral part of the insula is agranular, receives input from entorhinal cortex (see also ▶ Nociceptive Processing in the Hippocampus and Entorhinal Cortex, Neurophysiology and Pharmacology, ▶ Hippocampus and Entorhinal Complex: Functional Imaging), and sends projections to limbic cortical structures such as the entorhinal, periamygdaloid, and anterior cingulate cortices (Augustine 1996).

Lesions

There are no studies of the effects of localized anterior insular lesions specifically on pain perception. Greenspan and colleagues, however, identified two patients with lesions involving the anterior insula, both of whom had normal heat pain thresholds. Their three patients with anterior insular sparing and involvement of both S2 and posterior insula, however, had contralateral elevated thresholds for heat or mechanically induced pain (Greenspan et al. 1999). These observations suggest that the anterior insula is not an essential component of the cortical network mediating pain.

Stimulation

In a study confirming the site of stimulation in conscious humans, stimulation of the anterior insula produced visceral sensory experiences and visceral motor responses, but not reports of pain (Ostrowsky et al. 2000).

Electrophysiology

A neurophysiological study directed specifically at the anterior (granular) portion of the primate insula found that all responsive neurons had large receptive fields and were excited by innocuous somatic stimuli; however, the investigators searched for responses only with innocuous stimuli (Schneider et al. 1993).

Functional Imaging

PET imaging shows that the anterior insula is active during the early phase of a series of repetitive noxious heat stimuli, but not after the stimulation continues for 45 s (Casey et al. 2001). This is consistent with the findings of Ploghaus and colleagues who showed, using fMRI, that the anterior insula was active specifically during the anticipation of experimentally induced pain rather than during the experience of pain itself (Ploghaus et al. 1999). However, Porro and colleagues showed that activity in the anterior insula, together with the S1, anterior cingulate, and medial prefrontal cortices, increases during the anticipation of pain and is also correlated with perceived pain intensity (Porro et al. 2002). Nonetheless, a recent fMRI study reveals that the anterior insula is among the frontal and temporal brain structures responding specifically to stimulus novelty (Downar et al. 2002). These results confirm and elaborate on earlier imaging studies showing pain-related activity in the anterior insula (Casey et al. 1996; Coghill et al. 1994; Davis et al. 1998; Hsieh et al. 1994; Svensson et al. 1997).

In summary, the results suggest that the anterior insula is an essential component of the cortical network mediating some early aspects of pain perception including the anticipation of pain. Because anticipation implies the influence of past experience with pain, the anatomical connections of the anterior insula with limbic structures, such as the entorhinal cortex of the temporal lobe (see also ▶ Hippocampus and Entorhinal Complex: Functional Imaging ▶ Nociceptive Processing in the Hippocampus and Entorhinal Cortex, Neurophysiology and Pharmacology), are likely to be of critical importance in mediating this function.

Cingulate Cortex

Based on clinical observations and recent experimental studies, Vogt and colleagues have proposed that the anterior cingulate cortex (ACC; rostral to the plane of the anterior commissure) mediates primarily executive functions related to the emotional control of visceral, skeletal, and endocrine outflow; the posterior cingulate cortex (PCC) (see also ▶ Cingulate Cortex, Nociceptive Processing, Behavioral Studies in Animals, ▶ Pain Processing in the Cingulate Cortex, Behavioral Studies in Humans, ▶ Cingulate Cortex, Functional Imaging), however, is thought to subserve evaluative functions such as monitoring sensory events and personal behavior in relation to spatial orientation and memory (Vogt et al. 1992). The results of a more recent PET study suggested further that increased responses (activation) in subdivisions of the ACC signal participation in response selection and affective elaboration, while reduced responses (deactivation) in the PCC may reflect disengagement from visually guided processes (Vogt et al. 1996). However, other studies have shown strong neurovascular (Gelnar et al. 1999) and laser-evoked responses (▶ Insular Cortex, Neurophysiology, and Functional Imaging of Nociceptive Processing, ▶ Nociceptive Processing in the Secondary Somatosensory Cortex) in the PCC (Bentley et al. 2003).

Anatomical data shows that sectors of the ACC and a mid-cingulate area just below the supplementary motor area are strongly connected to the ventral horns of the spinal cord, thus providing direct access to motor mechanisms and the direct modulation of voluntary responses to noxious stimuli (Dum and Strick 1993; Hutchins et al. 1988).

Anterior Cingulate Cortex (ACC)

Lesions

Anterior cingulate lesions do not interfere with the ability of humans to recognize or respond to acute noxious stimuli. This is most clearly demonstrated in clinical cases. There are very few studies in which the location and extent of the lesion is known to be confined to the anterior cingulate cortex and the response to pain is tested specifically. However, Cohen and colleagues (Cohen et al. 1999) examined 12 patients, all with normal neurological examinations except

for peripheral findings related to their pain syndrome. Each underwent cingulotomy (see also ▶ Cingulate Cortex, Nociceptive Processing, Behavioral Studies in Animals, ▶ Pain Processing in the Cingulate Cortex, Behavioral Studies in Humans, ▶ Cingulate Cortex, Functional Imaging) for intractable pain and was examined pre- and postoperatively. The lesions were limited to the supracallosal mid-anterior cingulate cortex. The response to acute experimental pain was not tested. However, the patients reported only a modest relief of the intensity of their clinical pain but a significant reduction in the degree to which the pain interfered with their daily behavior and social function.

Animal studies of rostral ACC lesions reveal deficits in avoidance learning and nociceptively conditioned place avoidance, but the normal response to acute pain appears unimpaired (Gabriel et al. 1991; Johansen et al. 2001). Lesions involving the rodent mid-anterior, but not the pregenual cingulate cortex, however, impair the execution of escape responses to gradually increasing heat while sparing other nocifensive behaviors (Pastoriza et al. 1996).

Overall, the results show that different sectors of the ACC participate in the acquisition of learned responses to noxious stimuli, the association of negative hedonic attributes with acute noxious stimuli, and the execution of motor responses to acute noxious stimuli. The longer-term effects of ACC lesions will be considered in subsequent sections of this review.

Stimulation

The most consistent responses during electrical stimulation in humans are visceromotor changes associated with nausea, sensations of fullness, changes in blood pressure, and heart rate and cutaneous flushing (Talairach et al. 1973). Unpleasant emotional reactions are occasionally reported (Laitinen 1979; von Cramon and Jurgens 1983). In the rodent, biochemical stimulation of the rostral ACC (see also ▶ Cingulate Cortex, Nociceptive Processing, Behavioral Studies in Animals, ▶ Pain Processing in the Cingulate Cortex, Behavioral

Studies in Humans, ▶ Cingulate Cortex, Functional Imaging) produced conditioned avoidance, and biochemical suppression of the same site impairs conditioned avoidance learning (Johansen and Fields 2004). Overall, the results suggest a negative affective state that must be associated with noxious stimuli for the normal experience of pain.

Electrophysiology

Several evoked potential and single-cell recording studies show that noxious stimuli evoke neuronal responses within the rostral and mid-anterior sectors of the ACC of humans and experimental animals. In the human, nociceptive responses appear within 300 ms of a noxious laser stimulus (Inui et al. 2003; Kakigi et al. 2000; Lenz et al. 1998). Neurophysiological studies show that the ACC becomes active nearly simultaneously with the S1 and S2 cortices, indicating that spatiotemporal and intensity analysis begins in parallel with the processing of affective-related information (Ploner et al. 2002; Schnitzler and Ploner 2000). The early parallel processing of affective and sensory information is consistent with the concept that noxious stimuli have an intrinsic, primary unpleasantness (Fields 1999; Melzack and Casey 1968; Price 2000).

The limited sample of cellular recordings from human subjects shows that some mid-anterior ACC neurons responded only to noxious thermal (heat or cold) stimuli but also responded in anticipation of noxious stimulation (Hutchison et al. 1999); the receptive field size could not be determined. In experimental animals, nociceptively responding cells had large receptive fields (Sikes and Vogt 1992; Yamamura et al. 1996). In the conscious monkey, neurons in the ACC and in the anatomically associated caudate nucleus responded during the anticipation of pain in a pain-avoidance task (Koyama et al. 1998, 2000, 2001).

Functional Imaging

Reviews of functional imaging studies show that either the rostral or mid-anterior or both sectors of the ACC are activated consistently during pain (Bushnell et al. 1999; Casey 1999; Derbyshire 2000, 2003). Davis and colleagues showed that a sector of the ACC rostral to that responding during pain is activated specifically by a pain independent attention-demanding task (Davis et al. 1997). Rainville and colleagues (1997) showed that an anterior, supragenual area of this cortex participates specifically in the affective coding of pain (Rainville 2002; Rainville et al. 1997). During prolonged, repetitive heat stimulation, the most rostral sector of the ACC is active only during the early phase, while activation of the more caudal part of the ACC appears during the later phase (Casey et al. 2001). This result is in accord with the observation that the most rostral sector of the ACC is associated with the anticipation of pain (Ploghaus et al. 1999).

Posterior Cingulate Cortex (PCC)

Nociceptive information arrives at the PCC within approximately 200–250 ms of the application of a noxious laser stimulus (Bentley et al. 2003; Bromm 2001), so, like the ACC, it is involved in the earliest stages of nociceptive analysis. As noted above in the general discussion of the cingulate cortex, there is neurophysiological and functional imaging (Gelnar et al. 1999) evidence that the PCC participates in both the sensory and behavioral aspects of pain processing.

Intermediate Cortical and Limbic Processing (Milliseconds-Minutes)

As the nociceptive identification process is sustained, it leads to the allocation of attentional resources and the distraction from competing stimuli. Ward and colleagues presented evidence that, in the visual system, the application of attentional mechanisms requires up to 500 ms (Ward et al. 1996). Following this very early stage, there is an elaboration of the identity of the stimulus so that its location, physical property, and affective qualities are recognized more clearly and begin to form the basis for further analysis and response. This early identification of the affective component of a noxious stimulus is probably identical to the "primary unpleasantness" of Fields (1999) and the "immediate pain unpleasantness" of Price (2000) and is closely tied to perceived stimulus intensity.

Somatic, visceral, and autonomic responses may be reflexive and preconscious at the earliest stages of cortical engagement but may be facilitated or prolonged by the action of specific cortical mechanisms. Immediate withdrawal from the stimulus and the facilitation of flexion reflexes is an example of an early somatomotor response that may be modified quickly to include more elaborate voluntary escape maneuvers. Early cognitive reactions at this stage may include the recognition of tissue injury, the experience of fear or anger, and the development of immediate defensive actions. Cortical connections with the amygdala are important for these affective and autonomic responses (LeDoux 2000; Neugebauer et al. 2004).

Inferior Parietal Lobule (B40)

Lesions involving the inferior parietal lobule (IPL) (see also ▶ PET and fMRI Imaging in Parietal Cortex (SI, SII, Inferior Parietal Cortex BA40)), particularly of the right hemisphere, are associated classically with hemibody neglect syndromes (Adams and Victor 1993; DeJong 1979; Mesulam 1990). In humans, lesions involving this cortical area, which includes the posterior parietal operculum and Brodmann's area 40, have clinically obvious deficits in detecting and responding to noxious stimuli (Bassetti et al. 1993). Neurons responding to noxious stimulation or the visual threat of noxious stimulation have been recorded from the posterior parietal (7b) cortex of monkeys, and a lesion in this same area was associated with contralateral hypalgesia (Dong et al. 1994, 1996). Functional imaging studies have frequently revealed activity in this lateral posterior parietal cortex during pain, particularly when the task involves attending specifically to the painful stimulation (Coghill et al. 2001; Duncan and Albanese 2003; Peyron et al. 1999, 2000; Svensson et al. 1997) or during simulated pathological pain states (Baron et al. 1999; Hsieh et al. 1994, 1995). However, (Karnath 2001; Karnath et al. 2001) has presented and summarized evidence that pure spatial neglect follows lesions of the superior temporal gyrus (STG) and that lesions involving the human

IPL are more likely to reflect deficits in organizing movements directed within extrapersonal, body-oriented hemispace. Downar and colleagues showed that activation of the cortex at the temporoparietal junction, which includes the inferior parietal lobule and superior temporal gyrus, is associated with the perceived salience of both painful and painless stimuli (Downar et al. 2002, 2003). In summary, there is substantial evidence that the cortex in the area of the temporoparietal junction, including the IPL (B40) and STG, participates in attentional mechanisms that are likely to be critical for the normal perception of and attention to injurious stimuli. This interpretation is of interest in view of the strong connections between this cortex and the premotor cortical areas (see below).

Premotor Cortex (B6)

The premotor cortex (see also ▶ Prefrontal Cortex, Effects on Pain-Related Behavior) is active during pain as shown in many functional imaging studies of normal human subjects (Casey 1999; Casey et al. 2001; Coghill et al. 1999; Ladabaum et al. 2000; Tracey et al. 2000), including those focused on the related sensation of itch (Drzezga et al. 2001; Hsieh et al. 1994). Fibromyalgia patients who show a high degree of catastrophizing about their pain have an increased response in the medial premotor cortex (B6) to somatic pressure stimuli (Gracely et al. 2004).

There is a strong anatomical and functional relationship between the parietal and frontal motor (see also ▶ Prefrontal Cortex, Effects on Pain-Related Behavior, ▶ PET and fMRI Imaging in Parietal Cortex (SI, SII, Inferior Parietal Cortex BA40)) areas of the monkey and human brain (Rizzolatti et al. 1998). The circuits connecting predominantly somatosensory posterior parietal association areas with the dorsal and ventral premotor cortex are important for nociceptive processing in the cortex. For example, a component of the dorsal premotor circuit is involved in motor planning based primarily on somatosensory information (Rizzolatti et al. 1998). A functional imaging study has now shown that both the ventral and dorsal premotor cortices participate in the development of the sense of ownership of a body part (Ehrsson et al. 2004). Therefore, it is possible that the premotor cortex contributes to stimulus recognition and identification in addition to participating in the development of a motor response to a stimulus.

Posterior Insula

Lesions

The recent observations by Greenspan and colleagues show that lesions involving the posterior insula may not attenuate pain threshold but are associated with a significant increase in pain tolerance as assessed by the cold pressor test (Greenspan et al. 1999). Clinical observations also show that lesions involving the deep parasylvian cortex are associated with significant hypalgesia (Bassetti et al. 1993; Davison and Schick 1935; Greenspan and Winfield 1992; Horiuchi et al. 1996; Schmahmann and Leifer 1992). Involvement of the insula also leads to the clinical phenomenon of pain asymbolia, in which the patient fails to recognize or respond to the threat of noxious stimuli (Berthier et al. 1988).

Stimulation

Ostrowsky and colleagues (Ostrowsky et al. 2000, 2002) explored the insular cortex in 43 patients undergoing evaluation for epilepsy surgery and were able to verify the location of the stimulating electrodes with MR imaging. Stimulation within the posterior insula evoked painful sensations in the upper posterior insular cortex at 17 of 93 (18.2 %) insular stimulation sites in 14 patients. The patients described the stimulus-evoked sensations as burning, stinging, or disabling, located either contralaterally or bilaterally and rarely outlasting the stimulation. Non-painful somatosensory sensations, described as warmth, cold, or tingling, were also evoked in 21 of these 93 insular sites (37.2 %). The pain-related region shows a striking overlap with the dorsal posterior insular site activated by heat pain (Craig et al. 2000). Although these electrical stimuli did not evoke after discharge, it is likely that other cortical areas were activated during the insular stimulation and may have participated in the elaboration of the sensory experiences.

Electrophysiology

A small number of neurons responding to noxious stimuli have been recorded from the posterior insula (Robinson and Burton 1980). Zhang and colleagues recorded single insular neurons that responded to both innocuous and noxious somatic stimuli; some of these cells were localized to the more posterior granular area and also responded to baroreceptor stimulation (Zhang et al. 1999). In addition, numerous electrophysiological studies in humans have shown insular responses to painful cutaneous stimulation (Kakigi et al. 2000, 2003; Treede et al. 2000).

Functional Imaging

The anterior insula is activated early in the course of repetitive stimulation with noxious heat, but this response is replaced by a shift of the peak activity caudally to the mid-posterior insula as the stimulation continues (Casey et al. 2001). The mid- and posterior insula are among the most regularly responsive regions found among a variety of functional imaging studies (Casey 1999; Craig et al. 2000; Peyron et al. 2000, 2002). Derbyshire (2003) has reviewed evidence showing that visceral distention activates both the anterior and posterior insula, but the intensity, timing, and duration of stimulation varied across studies (Derbyshire 2003).

Anterior and Posterior Cingulate Cortex

The more rostral perigenual sectors of the ACC are active only early in the course of prolonged repetitive heat stimulation. Such stimulation is followed by sustained activity more caudally within Brodmann area 24 (Casey et al. 2001). Both attentional mechanisms and the perception of pain activate adjacent but separate sectors of the ACC (Davis et al. 1997). Gelnar and colleagues showed that the posterior cingulate cortex, which receives input from thalamic targets of the spinothalamic tract (Apkarian and Shi 1997), responds specifically during painful heat stimulation (Gelnar et al. 1999); the PCC also participates in identifying the salience of prolonged (1 min) noxious stimuli (Downar et al. 2003).

Long-Term Cortical and Limbic Processes (Minutes-Years)

Unlike the preceding responses, which generally occur over periods of seconds or less, a more detailed evaluation of an injury proceeds over time periods ranging from minutes to years and, especially in the case of chronic pain, may change over time. The qualitative evaluation includes mnemonic processes such as a comparison of present and past experiences with injurious physical stimuli such as heat, mechanical distortions, and chemical changes in the tissue. Most critically, a global assessment of the environment and the context in which the noxious event occurred is an important determinant of the perceived severity of the noxious stimulation or injury. The affective component of the experience may vary over time but is fully developed at this stage; it is probably related to what Fields has called "secondary unpleasant-ness" (Fields 1999) and Price refers to as "secondary pain unpleasantness" (Price 2000). Cognitive mechanisms are engaged at this stage, and these may include mnemonic functions, adjustments to daily living, and planning for the future. Overall, this aspect of pain-related cortical function is closely related to the concept of pain as a "need state" (Wall 1979) or a homeostatic function (Craig 2002).

Anterior Cingulate Cortex

Patients with lesions specifically confined to the ACC have an attenuation of the affective component of pain that is sustained for many months and, where information is available, for years (Cohen et al. 1999; Hurt and Ballantine 1973). Unfortunately, these patients may also show impairment of sustained attention and spontaneous behavioral responses (Cohen et al. 1999). A patient with a surgical lesion confined to the right mid-anterior cingulate gyrus had impaired executive control of manual but not verbal responses; responses to noxious stimuli were not tested (Turken and Swick 1999). Surgical lesions within the more rostral, pregenual ("affective") sector of the ACC are associated with affective blunting and a reduction of concern about clinical pain, but the responses to acute experimentally

induced pain are generally not described in detail (Foltz and White 1962; Hornak et al. 2003; Hurt and Ballantine 1973); the same is true of an extensively analyzed case of medial prefrontal damage (Damasio et al. 1994). There is insufficient data to warrant comment on the long-term pain-related effects of posterior cingulate lesions in humans.

Medial Prefrontal Cortex (B9, 10, 11)

Although the medial prefrontal cortex participates in the elaboration of emotional and high-order cognitive states that are independent of noxious somatic and visceral stimuli (Ramnani and Owen 2004; Simpson et al. 2001; Wager et al. 2003), a significant minority of functional imaging studies have shown that this cortex is active during experimental somatic or visceral pain (Derbyshire 2003; Peyron et al. 2000; Wager et al. 2004) or simulated pathological states (Hsieh et al. 1995; Iadarola et al. 1998). Extensive lesions of the human medial prefrontal cortex extending rostral to the cingulate gyrus (see also ▶ Cingulate Cortex, Nociceptive Processing, Behavioral Studies in Animals, ▶ Pain Processing in the Cingulate Cortex, Behavioral Studies in Humans, ▶ Cingulate Cortex, Functional Imaging) are associated with profound and lasting neurological deficits ranging from blunting of emotional responses (abulia) to the syndrome of akinetic mutism and motor neglect (Damasio et al. 1994; Mochizuki and Saito 1990; Kumral et al. 2002; Minagar and David 1999). Although extensive damage to this cortex precludes an unbiased examination of pain sensations, it is likely that it participates in the long-term evaluation of the emotional impact of pain and the need to apply cognitive strategies to adapt accordingly.

Orbitofrontal Cortex (OFC)

The orbitofrontal cortex (see also ▶ Hypothalamus and Nociceptive Pathways, ▶ Prefrontal Cortex, Effects on Pain-Related Behavior) is the major cortical output to the hypothalamus (see also ▶ Hypothalamus and Nociceptive Pathways) and also has direct efferent connections to the amygdala and periaqueductal gray matter (Ongur et al. 1998). Functional imaging studies of pain show OFC activation during experimental pain studies of heat (Craig et al. 2000), visceral stimulation (Derbyshire 2003), and simulated pathological pain states (Lorenz et al. 2002). Other imaging investigations show that the OFC plays a critical role in assigning affective valence to sensory information and in establishing the rewarding or punishing value of experiences including pain (Rolls et al. 2003), which in turn guide behavioral responses (Rolls 2000). In studying 43 patients with a variety of prefrontal cortical lesions, Hornack and colleagues (2003) identified six patients with bilateral circumscribed surgical lesions involving the OFC. These patients had difficulty identifying emotions and impairments in their subjective emotional states. Therefore, the greatest long-term clinical significance of OFC participation in pain-related behavior may be a sustained impairment of the ability to recognize pain as a primary reinforcer, which, in turn, leads to adaptive behaviors and emotional responses to injury (Kringelbach and Rolls 2004).

Dorsolateral Prefrontal Cortex (DLPFC; B9, 46)

The DLPFC (see also ▶ Prefrontal Cortex, Effects on Pain-Related Behavior) is active during executive processes involving shifts of attention between or among tasks (Smith and Jonides 1999). Cortical activations that include the DLPFC are observed in a majority of neuroimaging studies of pain (Peyron et al. 2000). However, as Peyron and colleagues have shown, the activation appears to be due to the participation of attentional and executive processes involved in attempting to attend to or ignore the painful stimulation rather than an analysis of the sensory or affective dimensions of the stimulus (Peyron et al. 1999). Thus, during the pain of heat allodynia (Lorenz et al. 2002), a high level of activation intensity in the left DLPFC is correlated with reduced pain unpleasantness and reduced functional connectivity between the midbrain and medial thalamus; right DLPFC activity, however, is associated with a reduced correlation of anterior insular activity with both

pain unpleasantness and intensity (Lorenz et al. 2003). The DLPFC also appears to be an active agent in mediating placebo analgesia, because the intensity of DLPFC activity correlates with and predicts the intensity of expected pain relief in the placebo condition (Wager et al. 2004). The DLPFC thus participates in mediating acute pain, but the current author suggests that the sustained pain modulatory effects of DLPFC activity are most strongly engaged following the initial sensory and hedonic analysis of injuries and during chronic pain states, where the impairment of frontal lobe function would have clinically significant impact on the recruitment of pain coping strategies (Stuss and Benson 1986).

Hippocampus and Entorhinal Cortex

The hippocampus (see ▶ Nociceptive Processing in the Hippocampus and Entorhinal Cortex, Neurophysiology and Pharmacology) and its major input, the entorhinal cortex, are part of a cortical network, including the DLPFC and anatomically related frontal cortices, for the encoding, storage, and retrieval of polymodal sensory information emanating from parietal association areas (Sakai 2003; Simons and Spiers 2003). In the rodent, the synaptic excitability of hippocampal pyramidal neurons undergoes a prolonged, cholinergically dependent depression following noxious stimulation; this evoked depression shows a marked habituation to repeated stimulation (Khanna and Sinclair 1989, 1992). Noxious stimuli also activate immediate early genes in neurons within the same hippocampal sector (Khanna et al. 2004), further suggesting a role in mnemonic circuitry. Prolonged increases in neuronal activity are also seen in the rodent entorhinal cortex (Frank and Brown 2003). Hippocampal or entorhinal activation or deactivation is rarely seen in most neuroimaging studies, but appears in studies in which the stimulus intensity increases during the scan period (Derbyshire et al. 1997) or when painful, but not painless, stimulation is unexpected (Ploghaus et al. 2000). When the expectation of a noxious stimulus is manipulated so as to produce anxiety, there is an anxiety-related increase in pain and in the pain response in the entorhinal cortex (Ploghaus et al. 2001). Together, these findings show that these medial temporal lobe structures participate in elaborating the experience of pain based on emotional state, expectation, and past experience. The author suggests that this elaboration follows the earlier sensory identification and affective labeling of the noxious stimulation and that its sustained clinical impact is on the ability of patients to interpret the long-term significance of internal states, including clinically painful conditions. A well-studied example of this rare but important condition has been presented by Hebben and colleagues (Hebben et al. 1985).

References

Adams, R. D., & Victor, M. (1993). *Principles of neurology* (5th ed.). New York: McGraw-Hill.

Apkarian, A. V., & Shi, T. (1997). Thalamocortical connections of the cingulate and insula in relation to nociceptive inputs to the cortex. In S. Ayrapetian & A. V. Apkarian (Eds.), *Pain mechanisms and management* (pp. 212–220). Amsterdam: Ios Press.

Arendt-Nielsen, L., Yamasaki, H., Nielse, J., et al. (1999). Magnetoencephalographic responses to painful impact stimulation. *Brain Research, 839*, 203–208.

Augustine, J. R. (1996). Circuitry and functional aspects of the insular lobe in primates including humans. *Brain Research Reviews, 22*, 229–244.

Baron, R., Baron, Y., Disbrow, E., et al. (1999). Brain processing of capsaicin-induced secondary hyperalgesia: A functional MRI study. *Neurology, 53*, 548.

Bassetti, C., Bogousslavsky, J., & Regli, F. (1993). Sensory syndromes in parietal stroke. *Neurology, 43*, 1942–1949.

Bentley, D. E., Derbyshire, S. W. G., Youell, P. D., et al. (2003). Caudal cingulate cortex involvement in pain processing: An inter-individual laser evoked potential source localisation study using realistic head models. *Pain, 102*, 265–271.

Berthier, M., Starkstein, S., & Leiguarda, R. (1988). Asymbolia for pain: A sensory-limbic disconnection syndrome. *Annals of Neurology, 24*, 41–49.

Bromm, B. (2001). Brain images of pain. *News in Physiological Sciences, 16*, 244–249.

Burton, H., Videen, T. O., & Raichle, M. E. (1993). Tactile-vibration-activated foci in insular and parietal-opercular cortex studied with positron emission tomography: Mapping the second somatosensory area in humans. *Somatosensory and Motor Research, 10*, 297–308.

Burton, H., Fabri, M., & Alloway, K. (1995). Cortical areas within the lateral sulcus connected to cutaneous representations in areas 3b and 1: A revised interpretation of the second somatosensory area in macaque monkeys. *The Journal of Comparative Neurology, 355*, 539–562.

Bushnell, M. C., Duncan, G. H., Hofbauer, R. K., et al. (1999). Pain perception: Is there a role for primary somatosensory cortex? *Proceedings of the National Academy of Sciences of the United States of America, 96*, 7705–7709.

Casey, K. L. (1999). Forebrain mechanisms of nociception and pain: Analysis through imaging. *Proceedings of the National Academy of Sciences of the United States of America, 96*, 7668–7674.

Casey, K. L. (2000). The imaging of pain: Background and rationale. In K. L. Casey & M. C. Bushnel (Eds.), *Pain imaging* (pp. 1–29). Seattle: IASP Press.

Casey, K. L., Minoshima, S., Morrow, T. J., et al. (1996). Comparison of human cerebral activation patterns during cutaneous warmth, heat pain, and deep cold pain. *Journal of Neurophysiology, 76*, 571–581.

Casey, K. L., Morrow, T. J., Lorenz, J., et al. (2001). Temporal and spatial dynamics of human forebrain activity during heat pain: Analysis by positron emission tomography. *Journal of Neurophysiology, 85*, 951–959.

Coghill, R. C., Talbot, J. D., Evans, A. C., et al. (1994). Distributed processing of pain and vibration by the human brain. *Journal of Neuroscience, 14*, 4095–4108.

Coghill, R. C., Sang, C. N., Maisog, J. H., et al. (1999). Pain intensity processing within the human brain: A bilateral, distributed mechanism. *Journal of Neurophysiology, 82*, 1934–1943.

Coghill, R. C., Gilron, I., & Iadarola, M. J. (2001). Hemispheric lateralization of somatosensory processing. *Journal of Neurophysiology, 85*, 2602–2612.

Cohen, R. A., Kaplan, R. F., Moser, D. J., et al. (1999). Impairments of attention after cingulotomy. *Neurology, 53*, 819.

Craig, A. D. (2002). How do you feel? Interoception: The sense of the physiological condition of the body. *Nature Reviews Neuroscience, 3*, 655–666.

Craig, A. D., Chen, K., Bandy, D., et al. (2000). Thermosensory activation of insular cortex. *Nature Neuroscience, 3*, 184–190.

Damasio, H., Grabowski, T., Frank, R., et al. (1994). The return of Phineas Gage: Clues about the brain from the skull of a famous patient. *Science, 264*, 1102–1105.

Davis, K. D. (2000). The neural circuitry of pain as explored with functional MRI. *Neurological Research, 22*, 313–317.

Davis, K. D., Taylor, S. J., Crawley, A. P., et al. (1997). Functional MRI of pain- and attention-related activations in the human cingulate cortex. *Journal of Neurophysiology, 77*, 3370–3380.

Davis, K. D., Kwan, C. L., Crawley, A. P., et al. (1998). Functional MRI study of thalamic and cortical activations evoked by cutaneous heat, cold, and tactile stimuli. *Journal of Neurophysiology, 80*, 1533–1546.

Davison, C., & Schick, W. (1935). Spontaneous pain and other subjective sensory disturbances. *AMA Archives of Neurology and Psychiatry, 34*, 1204–1237.

DeJong, R. N. (1979). *The neurologic examination* (4th ed.). Hagerstown: Harper and Row.

Derbyshire, S. W. (2000). Exploring the pain "neuromatrix". *Current Review of Pain, 4*, 467–477.

Derbyshire, S. W. G. (2003). A systematic review of neuroimaging data during visceral stimulation. *American Journal of Gastroenterology, 98*, 12–20.

Derbyshire, S. W., Jones, A. K., Gyulai, F., et al. (1997). Pain processing during three levels of noxious stimulation produces differential patterns of central activity. *Pain, 73*, 431–445.

Dong, W. K., Chudler, E. H., Sugiyama, K., et al. (1994). Somatosensory, multisensory, and task-related neurons in cortical area 7b (PF) of unanesthetized monkeys. *Journal of Neurophysiology, 72*, 542–564.

Dong, W. K., Hayashi, T., Roberts, V. J., et al. (1996). Behavioral outcome of posterior parietal cortex injury in the monkey. *Pain, 64*, 579–587.

Downar, J. D., Crawley, A. P., Mikulis, D. J., et al. (2002). A cortical network sensitive to stimulus salience in a neutral behavioral context across multiple sensory modalities. *Journal of Neurophysiology, 87*, 615–620.

Downar, J., Mikulis, D. J., & Davis, K. D. (2003). Neural correlates of the prolonged salience of painful stimulation. *NeuroImage, 20*, 1540–1551.

Drzezga, A., Darsow, U., Treede, R. D., et al. (2001). Central activation by histamine-induced itch: Analogies to pain processing: A correlational analysis of O-15 H$_2$O positron emission tomography studies. *Pain, 92*, 295–305.

Dum, R. P., & Strick, P. L. (1993). Cingulate motor areas. In B. A. Vogt & M. Gabriel (Eds.), *Neurobiology of cingulate cortex and limbic thalamus: A comprehensive handbook*. Boston: Birkhauser.

Duncan, G. H., & Albanese, M. C. (2003). Is there a role for the parietal lobes in the perception of pain? *Advances in Neurology, 93*, 69–86.

Ehrsson, H. H., Spence, C., & Passingham, R. E. (2004). That's my hand! Activity in premotor cortex reflects feeling of ownership of a limb. *Science, 305*, 875–877.

Fields, H. L. (1999). Pain: An unpleasant topic. *Pain, 6*, 61–69.

Foltz, E. L., & White, L. E. (1962). Pain "relief" by frontal cingulotomy. *Journal of Neurosurgery, 19*, 89–100.

Frank, L. M., & Brown, E. N. (2003). Persistent activity and memory in the entorhinal cortex. *Trends in Neurosciences, 26*, 400–401.

Frot, M., & Mauguiere, F. (2003). Dual representation of pain in the operculo-insular cortex in humans. *Brain, 126*, 438–450.

Frot, M., Garcia-Larrea, L., Guenot, M., et al. (2001). Responses of the supra-sylvian (SII) cortex in humans to painful and innocuous stimuli: A study using intra-cerebral recordings. *Pain, 94*, 65–73.

Gabriel, M., Kubota, Y., Sparenborg, S., et al. (1991). Effects of cingulate cortical lesions on avoidance learning and training-induced unit activity in rabbits. *Experimental Brain Research, 86*, 585–600.

Gelnar, P. A., Krauss, B. R., Sheehe, P. R., et al. (1999). A comparative fMRI study of cortical representations for thermal painful, vibrotactile, and motor performance tasks. *NeuroImage, 10*, 460–482.

Gracely, R. H., Geisser, M. E., Giesecke, T., et al. (2004). Pain catastrophizing and neural responses to pain among persons with fibromyalgia. *Brain, 127*, 835–843.

Greenspan, J. D., & Winfield, J. A. (1992). Reversible pain and tactile deficits associated with a cerebral tumor compressing the posterior insula and parietal operculum. *Pain, 50*, 29–39.

Greenspan, J. D., Lee, R. R., & Lenz, F. A. (1999). Pain sensitivity alterations as a function of lesion location in the parasylvian cortex. *Pain, 81*, 273–282.

Hebben, N., Corkin, S., Eichenbaum, H., et al. (1985). Diminished ability to interpret and report internal states after bilateral medial temporal resection: Case H.M. *Behavioral Neuroscience, 99*, 1031–1039.

Hofbauer, R. K., Rainville, P., Duncan, G. H., et al. (2001). Cortical representation of the sensory dimension of pain. *Journal of Neurophysiology, 86*, 402–411.

Horiuchi, T., Unoki, T., Yokoh, A., et al. (1996). Pure sensory stroke caused by cortical infarction associated with the secondary somatosensory area. *Journal of Neurology, Neurosurgery & Psychiatry, 60*, 588–589.

Hornak, J., Bramham, J., Rolls, E. T., et al. (2003). Changes in emotion after circumscribed surgical lesions of the orbitofrontal and cingulate cortices. *Brain, 126*, 1691–1712.

Hsieh, J.-C., Hagermark, O., Stahle-Backdahl, M., et al. (1994). Urge to scratch represented in the human cerebral cortex during itch. *Journal of Neurophysiology, 72*, 3004–3008.

Hsieh, J.-C., Belfrage, M., Stone-Elander, S., et al. (1995). Central representation of chronic ongoing neuropathic pain studied by positron emission tomography. *Pain, 63*, 225–336.

Hurt, R. W., & Ballantine, H. T. (1973). Stereotactic anterior cingulate lesions for persistent pain: A report on 68 cases. *Clinical Neurosurgery, 21*, 334–351.

Hutchins, K. D., Martino, A. M., & Strick, P. L. (1988). Corticospinal projections from the medial wall of the hemisphere. *Experimental Brain Research, 71*, 667–672.

Hutchison, W. D., Davis, K. D., Lozano, A. M., et al. (1999). Pain-related neurons in the human cingulate cortex. *Nature Neuroscience, 2*, 403–405.

Iadarola, M. J., Berman, K. F., Zeffiro, T. A., et al. (1998). Neural activation during acute capsaicin-evoked pain and allodynia assessed with PET. *Brain, 121*, 931–947.

Inui, K., Tran, T. D., Qiu, Y., et al. (2003). A comparative magnetoencephalographic study of cortical activations evoked by noxious and innocuous somatosensory stimulations. *Neuroscience, 120*, 235–248.

Johansen, J. P., & Fields, H. L. (2004). Glutamatergic activation of anterior cingulate cortex produces an aversive teaching signal. *Nature Neuroscience, 7*, 398–403.

Johansen, J. P., Fields, H. L., & Manning, B. H. (2001). The affective component of pain in rodents: Direct evidence for a contribution of the anterior cingulate cortex. *Proceedings of the National Academy of Sciences of the United States of America, 98*, 8077–8082.

Kaas, J. H. (1993). The functional organization of somatosensory cortex in primates. *Annals of Anatomy, 175*, 509–518.

Kaas, J. H., Nelson, R. J., Sur, M., et al. (1979). Multiple representations of the body within the primary somatosensory cortex of primates. *Science, 204*, 521–523.

Kaas, J. H., Florence, S. L., & Neeraj, J. (1997). Reorganization of sensory systems of primates after injury. *The Neuroscientist, 3*, 123–130.

Kakigi, R., Koyama, S., Hoshiyama, M., et al. (1995). Pain-related magnetic fields following painful CO_2 laser stimulation in man. *Neuroscience Letters, 192*, 45–48.

Kakigi, R., Watanabe, S., & Yamasaki, H. (2000). Pain-related somatosensory evoked potentials. *Journal of Clinical Neurophysiology, 17*, 295–308.

Kakigi, R., Tran, T. D., Qiu, Y., et al. (2003). Cerebral responses following stimulation of unmyelinated C-fibers in humans: Electro- and magneto-encephalographic study. *Neuroscience Research, 45*, 255–275.

Kanda, M., Nagamine, T., Ikeda, A., et al. (2000). Primary somatosensory cortex is actively involved in pain processing in human. *Brain Research, 853*, 282–289.

Karnath, H. O. (2001). New insights into the functions of the superior temporal cortex. *Nature Reviews Neuroscience, 2*, 568–576.

Karnath, H. O., Ferber, S., & Himmelbach, M. (2001). Spatial awareness is a function of the temporal not the posterior parietal lobe. *Nature, 411*, 950–953.

Kenshalo, D. R., & Isensee, O. (1983). Responses of primate S1 cortical neurons to noxious stimuli. *Journal of Neurophysiology, 50*, 1479–1496.

Kenshalo, D. R., Chudler, E. H., Anton, F., et al. (1988). SI nociceptive neurons participate in the encoding process by which monkeys perceive the intensity of noxious thermal stimulation. *Brain Research, 454*, 378–382.

Kenshalo, D. R., Iwata, K., Sholas, M., et al. (2000). Response properties and organization of nociceptive neurons in Area 1 of monkey primary somatosensory cortex. *Journal of Neurophysiology, 84*, 719–729.

Khanna, S., & Sinclair, J. G. (1989). Noxious stimuli produce prolonged changes in the CA1 region f the rat hippocampus. *Pain, 39*, 337–343.

Khanna, S., & Sinclair, J. G. (1992). Responses in the CA1 region of the rat hippocampus to a noxious stimulus. *Experimental Neurology, 117*, 28–35.

Khanna, S., Seong Chang, L., Jiang, F., et al. (2004). Nociception-driven decreased induction of Fos protein in ventral hippocampus field CA1 of the rat. *Brain Research, 1004*, 167–176.

Koyama, T., Tanaka, Y. Z., & Mikami, A. (1998). Nociceptive neurons in the macaque anterior cingulate activate during anticipation of pain. *Neuroreport, 9*, 2663–2667.

Koyama, T., Kato, K., & Mikami, A. (2000). During pain-avoidance neurons activated in the macaque anterior cingulate and caudate. *Neuroscience Letters, 283*, 17–20.

Koyama, T., Kato, K., Tanaka, Y. Z., et al. (2001). Anterior cingulate activity during pain-avoidance and rewftasks in monkeys. *Neuroscience Research, 39*, 421–430.

Kringelbach, M. L., & Rolls, E. T. (2004). The functional neuroanatomy of the human orbitofrontal cortex: Evidence from neuroimaging and neuropsychology. *Progress in Neurobiology, 72*, 341–372.

Kumral, E., Bayulkem, G., Evyapan, D., et al. (2002). Spectrum of anterior cerebral artery territory infarction: Clinical and MRI findings. *European Journal of Neurology, 9*, 615–624.

Ladabaum, U., Minoshima, S., & Owyang, C. (2000). Pathobiology of visceral pain: Molecular mechanisms and therapeutic implications v. Central nervous system processing of somatic and visceral sensory signals. *American Journal of Physiology. Gastrointestinal and Liver Physiology, 279*, 1–6.

Laitinen, L. V. (1979). Emotional responses to subcortical electrical stimulation in psychiatric patients. *Clinical Neurology and Neurosurgery, 81*, 148–157.

LeDoux, J. E. (2000). Emotion circuits in the brain. *Annual Review of Neuroscience, 23*, 155–184.

Lenz, F. A., Rios, M., Zirh, A., et al. (1998). Painful stimuli evoke potentials recorded over the human anterior cingulate gyrus. *Journal of Neurophysiology, 79*, 2231–2234.

Lighthall, J. W. (1988). Controlled cortical impact: A new experimental brain injury model. *Journal of Neurotrauma, 5*, 1–15.

Lighthall, J. W., Dixon, C. E., & Anderson, T. E. (1989). Experimental models of brain injury. *Journal of Neurotrauma, 6*, 83–97.

Lorenz, J., Cross, D., Minoshima, S., et al. (2002). A unique representation of heat allodynia in the human brain. *Neuron, 35*, 383–393.

Lorenz, J., Minoshima, S., & Casey, K. L. (2003). Keeping pain out of mind: The role of the dorsolateral prefrontal cortex in pain modulation. *Brain, 126*, 1079–1091.

MacLean, P. D. (1955). The limbic system ("visceral brain") in relation to the central gray and reticulum of the brain stem. *Psychosomatic Medicine, 17*, 355–366.

Marshall, J. (1951). Sensory disturbances in cortical wounds with special reference to pain. *Journal of Neurology, Neurosurgery, and Psychiatry, 14,* 187–204.

Melzack, R., & Casey, K. L. (1968). Sensory, motivational and central control determinants of pain. In D. R. Kenshalo (Ed.), *The skin senses* (pp. 423–439). Springfield: CC Thomas.

Merskey, H., & Bogduk, N. (1994). *Classification of chronic pain: Descriptions of chronic pain syndromes and definitions of pain terms* (2nd ed.). Seattle: IASP Press.

Mesulam, M. M. (1990). Large-scale neurocognitive networks and distributed processing for attention, language, and memory. *Annals of Neurology, 28,* 597–613.

Mesulam, M. M., & Mufson, E. J. (1982a). Insula of the old world monkey II. Afferent cortical input and comments on the claustrum. *The Journal of Comparative Neurology, 212,* 23–37.

Mesulam, M. M., & Mufson, E. J. (1982b). Insula of the old world monkey: III efferent cortical input and comments on function. *The Journal of Comparative Neurology, 212,* 38–52.

Minagar, A., & David, N. J. (1999). Bilateral infarction in the territory of the anterior cerebral arteries. *Neurology, 52,* 886–888.

Mochizuki, H., & Saito, H. (1990). Mesial frontal lobe syndromes: Correlations between neurological deficits and radiological localizations. *The Tohoku Journal of Experimental Medicine, 161,* 231–239.

Neugebauer, V., Li, W., Bird, G. C., et al. (2004). The amygdala and persistent pain. *The Neuroscientist, 10,* 221–234.

Nii, Y., Uematsu, S., Lesser, R. P., et al. (1996). Does the central sulcus divide motor and sensory functions? Cortical mapping of human hand areas as revealed by electrical stimulation through subdural grid electrodes. *Neurology, 46,* 360–367.

Ongur, D., An, X., & Price, J. L. (1998). Prefrontal cortical projections to the hypothalamus in macaque monkeys. *The Journal of Comparative Neurology, 401,* 480–505.

Opsommer, E., Weiss, T., Plaghki, L., et al. (2001). Dipole analysis of ultralate (C-fibres) evoked potentials after laser stimulation of tiny cutaneous surface areas in humans. *Neuroscience Letters, 298,* 41–44.

Ostrowsky, K., Isnard, J., Ryvlin, P., et al. (2000). Functional mapping of the insular cortex: Clinical implication in temporal lobe epilepsy. *Epilepsia, 41,* 681–686.

Ostrowsky, K., Magnin, M., Ryvlin, P., et al. (2002). Representation of pain and somatic sensation in the human insula: A study of responses to direct electrical cortical stimulation. *Cerebral Cortex, 12,* 376–385.

Papez, J. W. (1937). A proposed mechanism of emotion. *Archives of Neurology and Psychiatry, 38,* 725–743.

Pastoriza, L. N., Morrow, T. J., & Casey, K. L. (1996). Medial frontal cortex lesions selectively attenuate the hot plate response: Possible nocifensive apraxia in the rat. *Pain, 64,* 11–17.

Peyron, R., Garcia-Larrea, L., Gregoire, M. C., et al. (1999). Haemodynamic brain responses to acute pain in humans: Sensory and attentional networks. *Brain, 122,* 1765–1780.

Peyron, R., Laurent, B., & Garcia-Larrea, L. (2000). Functional imaging of brain responses to pain. A review and meta-analysis. *Neurophysiologie Clinique, 30,* 263–288.

Peyron, R., Frot, M., Schneider, F., et al. (2002). Role of operculoinsular cortices in human pain processing: Converging evidence from PET, fMRI, dipole modeling, and intracerebral recordings of evoked potentials. *NeuroImage, 17,* 1336–1346.

Ploghaus, A., Tracey, I., Gati, J. S., et al. (1999). Dissociating pain from its anticipation in the human brain. *Science, 284,* 1979–1981.

Ploghaus, A., Trace, I., Clare, S., et al. (2000). Learning about pain: The neural substrate of the prediction error for aversive events. *Proceedings of the National Academy of Sciences of the United States of America, 97,* 9281–9286.

Ploghaus, A., Narain, C., Beckmann, C. F., et al. (2001). Exacerbation of pain by anxiety is associated with activity in a hippocampal network. *Journal of Neuroscience, 21,* 9896–9903.

Ploner, M., Schmitz, F., Freund, H. J., et al. (1999). Parallel activation of primary and secondary somatosensory cortices in human pain processing. *Journal of Neurophysiology, 81,* 3100–3104.

Ploner, M., Gross, J., Timmermann, L., et al. (2002). Cortical representation of first and second pain sensation in humans. *Proceedings of the National Academy of Sciences of the United States of America, 99,* 12444–12448.

Porro, C. A., Baraldi, P., Pagnoni, G., et al. (2002). Does anticipation of pain affect cortical nociceptive systems? *Journal of Neuroscience, 22,* 3206–3214.

Price, D. D. (1988). *Psychological and neural mechanisms of pain.* New York: Raven.

Price, D. D. (1999). *Psychological mechanisms of pain and analgesia.* Seattle: IASP Press.

Price, D. D. (2000). Psychological and neural mechanisms of the affective dimension of pain. *Science, 288,* 1769–1772.

Rainville, P. (2002). Brain mechanisms of pain affect and pain modulation. *Current Opinion in Neurobiology, 12,* 195–204.

Rainville, P., Duncan, G. H., Price, D. D., et al. (1997). Pain affect encoded in human anterior cingulate but not somatosensory cortex. *Science, 277,* 968–971.

Ramnani, N., & Owen, A. M. (2004). Anterior prefrontal cortex: Insights into function from anatomy and neuroimaging. *Nature Reviews Neuroscience, 5,* 184–194.

Rizzolatti, G., Luppino, G., & Matelli, M. (1998). The organization of the cortical motor system: New concepts. *Electroencephalography and Clinical Neurophysiology, 106*, 283–296.

Robinson, C. J., & Burton, H. (1980). Organization of somatosensory receptive fields in cortical areas 7b, retroinsula, postauditory and granular insula of M. fascicularis. *The Journal of Comparative Neurology, 192*, 69–92.

Rolls, E. T. (2000). The orbitofrontal cortex and reward. *Cerebral Cortex, 10*, 284–294.

Rolls, E. T., O'Doherty, J., Kringelbach, M. L., et al. (2003). Representations of pleasant and painful touch in the human orbitofrontal and cingulate cortices. *Cerebral Cortex, 13*, 308–317.

Ruben, J., Schwiemann, J., Deuchert, M., et al. (2001). Somatotopic organization of human secondary somatosensory cortex. *Cerebral Cortex, 11*, 463–473.

Russell, W. R. (1945). Transient disturbances following gunshot wounds of the head. *Brain, 68*, 79–97.

Sakai, K. (2003). Reactivation of memory: Role of medial temporal lobe and prefrontal cortex. *Reviews in the Neurosciences, 14*, 241–252.

Schmahmann, J. D., & Leifer, D. (1992). Parietal pseudothalamic pain syndrome: Clinical features and anatomic correlates. *Archives of Neurology, 49*, 1032–1037.

Schneider, R. J., Friedman, D. P., & Mishkin, M. (1993). A modality-specific somatosensory area within the insula of the rhesus monkey. *Brain Research, 621*, 116–120.

Schnitzler, A., & Ploner, M. (2000). Neurophysiology and functional neuroanatomy of pain perception. *Journal of Clinical Neurophysiology, 17*, 592–603.

Sikes, R. W., & Vogt, B. A. (1992). Nociceptive neurons in area 24 of rabbit cingulate cortex. *Journal of Neurophysiology, 68*, 1720–1732.

Simons, J. S., & Spiers, H. J. (2003). Prefrontal and medial temporal lobe interactions in long-term memory. *Nature Reviews Neuroscience, 4*, 637–648.

Simpson, J. R., Jr., Drevets, W. C., Snyder, A. Z., et al. (2001). Emotion-induced changes in human medial prefrontal cortex: II. During anticipatory anxiety. *Proceedings of the National Academy of Sciences of the United States of America, 98*, 688–693.

Smith, E. E., & Jonides, J. (1999). Storage and executive processes in the frontal lobes. *Science, 283*, 1657–1661.

Spiller, W. G., & Martin, E. (1912). The treatment of persistent pain of organic origin in the lower part of the body by division of the anterolateral column of the spinal cord. *Journal of the American Medical Association, 58*, 1489–1492.

Stevens, R. T., London, S. M., & Apkarian, A. V. (1993). Spinothalamocortical projections to the secondary somatosensory cortex (SII) in squirrel monkey. *Brain Research, 631*, 241–246.

Stuss, D. T., & Benson, D. F. (1986). *The frontal lobes.* New York: Raven.

Svensson, P., Minoshima, S., Beydoun, A., et al. (1997). Cerebral processing of acute skin and muscle pain in humans. *Journal of Neurophysiology, 78*, 450–460.

Sweet, W. H. (1982). Cerebral localization of pain. In R. A. Thompson & J. R. Green (Eds.), *New perspectives in cerebral localization* (pp. 205–242). New York: Raven.

Talairach, J., Bancaud, J., Geier, S., et al. (1973). The cingulate gyrus and human behaviour. *Electroencephalography and Clinical Neurophysiology, 34*, 45–52.

Tommerdahl, M., Delemos, K. A., Vierck, C. J., Jr., et al. (1996). Anterior parietal cortical response to tactile and skin-heating stimuli applied to the same skin site. *Journal of Neurophysiology, 75*, 2662–2670.

Tracey, I., Becerra, L., Chang, I., et al. (2000). Noxious hot and cold stimulation produce common patterns of brain activation in humans: A functional magnetic resonance imaging study. *Neuroscience Letters, 288*, 159–162.

Tran, T. D., Inui, K., Hoshiyama, M., et al. (2002). Cerebral activation by the signals ascending through unmyelinated C-fibers in humans: A magnetoencephalographic study. *Neuroscience, 113*, 375–386.

Treede, R. D., Kenshalo, D. R., Gracely, R. H., et al. (1999). The cortical representation of pain. *Pain, 79*, 105–111.

Treede, R. D., Apkarian, A. V., Bromm, B., et al. (2000). Cortical representation of pain: Functional characterization of nociceptive areas near the lateral sulcus. *Pain, 87*, 113–119.

Turken, A. U., & Swick, D. (1999). Response selection in the human anterior cingulate cortex. *Nature Neuroscience, 2*, 920–924.

Vogt, B. A., Finch, D. M., & Olson, C. R. (1992). Functional heterogeneity in cingulate cortex: The anterior executive and posterior evaluative regions. *Cerebral Cortex, 2*, 435–443.

Vogt, B. A., Derbyshire, S., Jones, A. K. P., et al. (1996). Pain processing in four regions of human cingulate cortex localized with co-registered PET and MR imaging. *European Journal of Neuroscience, 8*, 1461–1473.

von Cramon, D., & Jurgens, U. (1983). The anterior cingulate cortex and the phonatory control in monkey and man. *Neuroscience and Biobehavioral Reviews, 7*, 423–425.

Wager, T. D., Phan, K. L., Liberzon, I., et al. (2003). Valence, gender, and lateralization of functional brain anatomy in emotion: A meta-analysis of findings from neuroimaging. *NeuroImage, 19*, 513–531.

Wager, T. D., Rilling, J. K., Smith, E. E., et al. (2004). Placebo-induced changes in fMRI in the anticipation and experience of pain. *Science, 303*, 1162–1167.

Wall, P. D. (1979). On the relation of injury to pain the John J. Bonica lecture. *Pain, 6*, 253–264.

Ward, R., Duncan, J., & Shapiro, K. (1996). The slow time-course of visual attention. *Cognitive Psychology, 30*, 79–109.

Whitsel, B. L., Dreyer, D. A., & Roppolo, J. R. (1971). Determinants of body representation in postcentral gyrus of macaques. *Journal of Neurophysiology, 34*, 1018–1034.

Yamamura, H., Iwata, K., Tsuboi, Y., et al. (1996). Morphological and electrophysiological properties of ACC nociceptive neurons in rats. *Brain Research, 735*, 83–92.

Zhang, Z. H., Dougherty, P. M., & Oppenheimer, S. M. (1999). Monkey insular cortex neurons respond to baroreceptive and somatosensory convergent inputs. *Neuroscience, 94*, 351–360.

Cortical Feedback

Definition

Back projections from cortical regions that receive forward projections from the thalamus.

Cross-References

▶ Thalamic Plasticity and Chronic Pain

Cortical Information Flow

▶ Information Processing in Thalamocortical Circuits

Cortical Plasticity

Definition

Flexible organization of the cortical textures and circuits adapting to input variability and regulated by "experience," i.e., preferential use increase/reduction of definite pathways. At the synaptic level, strengthening or weakening of the synaptic connections depending on a correlation principle.

Cross-References

▶ Deafferentation Pain

Cortical Projections

▶ Corticothalamic and Thalamocortical Interactions

Cortical Reorganization

Definition

A functional or structural change in the primary sensorimotor areas of the cortex that can also occur in the adult nervous system but was long thought possible only in the developing organism.

Cross-References

▶ Phantom Limb Pain, Treatment

Cortical Retraining

▶ Brain Based Interventions

Cortical Spreading Depression

Definition

A wave of activation followed by deactivation that moves across the surface of the brain typically at 2–6 mm/min. It is widely considered to be the animal equivalent of migraine aura.

Cross-References

▶ Migraine, Pathophysiology

Cortical Stimulation for Relief of Pain

▶ Pain Treatment, Motor Cortex Stimulation

Corticocortical Pathways

▶ Information Processing in Thalamocortical Circuits

Corticolimbic Circuits

Definition

Pathway through parietal opercular cortex, including SII, and insula cortex to the medial temporal lobe including amygdala and hippocampus.

Cross-References

▶ Angina Pectoris, Neurophysiology and Psychophysics

Corticospinal Tract

Definition

A motor pathway originating in the motor cortex and terminating in brain stem and spinal motor nuclei. Injury of this pathway can lead to paresis of the ipsilateral lower limb.

Corticosteroid Injections

▶ Steroid Injections

Corticothalamic and Thalamocortical Interactions

Edward Jones
Center for Neuroscience, University of California, Davies, CA, USA

Synonyms

40 Hz oscillations; Bursting activity; Cortical projections; Reticular nucleus inputs from cortex; Thalamic projections; Thalamocortical and corticothalamic interactions

Definition

Cortical projections to dorsal thalamic nuclei and to thalamic reticular nucleus, together with projections from the thalamus to the cortex, form the basic thalamocortical-corticothalamic loop. Neuronal connectivity in these loops, involved neurotransmitters, and synaptic mechanisms define distinct dynamical states of this network.

Characteristics

▶ Corticothalamic fibers arise in every area of the cerebral cortex and every dorsal thalamic nucleus receives the terminations of corticothalamic fibers. The cells of origin of these fibers form two distinct classes. Both are a form of ▶ pyramidal neuron, with spiny dendrites and utilize glutamate as the transmitter. One class, with soma located in layer VI, is characterized by small size, a thin ascending apical dendrite that devolves into a tuft of branches in layer IV, where it receives monosynaptic inputs from thalamic fibers, and strongly recurrent, columnar axon collaterals (Fig. 1). The thin primary axon of these cells returns to the thalamic nucleus from which the cortical area in which it lies receives thalamic input, giving off collateral branches to the ▶ thalamic reticular nucleus *en route* (Fig. 2). A second class, with soma located in layer V, is larger, with an apical dendrite that ascends to layer I, ending there in a tuft of spiny branches, and a set of lengthy horizontal axon collaterals that connect extensive stretches of layers III and V (Fig. 1). The relatively thick primary axon of these cells descends and sends branches to multiple subcortical sites, such as the spinal cord, brainstem, tectum, basal ganglia, and thalamus. The thalamic branches traverse the reticular nucleus without giving off collaterals and end primarily in nuclei other than the primary sensory nuclei, especially in those characterized by a high

Corticothalamic and Thalamocortical Interactions, Fig. 1 The two principal classes of corticothalamic neurons. Neurons in layer VI are small with a tuft of dendrites in layer IV and strongly recurrent axon collaterals. Neurons in layer V are larger with a tuft of dendrites in layer I and lengthy horizontal axon collaterals. Bars: 1 mm (From Jones (2007))

Cerebral cortex   Striatum   Reticular   Relay   Intralaminar
                             nucleus   nucleus nuclei

I II-III IV V VI   GPe GPi To brainstem

Corticothalamic and Thalamocortical Interactions, Fig. 2 Schematic view of the circuitry connecting thalamus and cerebral cortex. Axons of layer VI corticothalamic neurons send collaterals to the reticular nucleus and terminate in the underlying dorsal thalamus. Axons of layer V corticothalamic neurons project to several subcortical sites, with collaterals to the dorsal thalamus only (From Jones 2007)

density of matrix cells (see ▶ Spinothalamic Terminations, Core and Matrix) (Fig. 2).

The reticular nucleus, which is innervated by the collaterals of the layer VI corticothalamic cells, occupies a key place in the circuitry connecting thalamus and cortex; it is also innervated by collaterals of thalamocortical axons

arising in the underlying dorsal thalamus. Because of the differences in the diameters of the corticothalamic and ▶ thalamocortical fibers that provide collateral inputs to the reticular nucleus, it is possible physiologically to identify the two kinds of collateral synapse. A brief electrical stimulus applied to the cerebral cortex or underlying white matter elicits short latency ▶ EPSCs in reticular nucleus cells due to ▶ antidromic invasion of thalamocortical collaterals and longer latency EPSCs due to ▶ orthodromic activation of the slower corticothalamic fibers (Golshani et al. 2001; Liu et al. 2001). EPSCs attributable to collateral corticothalamic synapses have a consistent amplitude, reflecting their small size and a single vesicle release site, but a wide range of rise times, reflecting their wide distribution over the dendritic tree of reticular nucleus cells (Liu et al. 2001). Unitary EPSCs attributable to collateral thalamocortical synapses tend to have large, although more variable amplitudes, reflecting their large size and multiple release sites, but very consistent rise times, reflecting their proximal location on the dendritic tree.

In the dorsal thalamus, more than 40 % of the synapses on a thalamic relay cell are derived from layer VI corticothalamic fibers and are concentrated on secondary and especially on tertiary dendrites (Liu et al. 1995a). Less than 20 % of the synapses are derived from medial lemniscal, optic tract, or other subcortical fibers and are concentrated on proximal dendrites. These terminals tend to have multiple points of synaptic contact, including on dendritic protrusions and their parent dendritic shafts. The remaining 40 % of synapses are inhibitory and tend to be concentrated on proximal and second-order dendrites and on the soma. The majority of these terminals are derived from axons of the reticular nucleus (Liu et al. 1995b). A lesser number are derived from the presynaptic dendrites of intrinsic interneurons in animals that possess these neurons. The terminals of layer V-originating corticothalamic fibers end in large terminals concentrated proximally, in numbers that have not yet been quantified.

The glutamatergic nature of corticothalamic synapses is evidenced by the ability to record ▶ NMDA-, ▶ AMPA-, and ▶ metabotropic glutamate receptor-based EPSCs in relay cells and in reticular nucleus cells when corticothalamic fibers are stimulated under appropriate conditions (McCormick and Von Krosigk 1992; Kao and Coulter 1997; Turner and Salt 1998, 1999;

Golshani et al. 1998; Zhang and Jones 2004). However, corticothalamic stimulation can also engender a powerful disynaptic, feed-forward inhibition of relay cells, resulting from co-activation of the reticular nucleus. The importance of this corticothalamic-induced inhibition of the relay cells is that it drives them toward the burst-firing mode. As they recover from this inhibition, the low threshold calcium conductance, I_T, is de-inactivated and the cells fire a burst of action potentials. This has the effect of re-exciting, via the collaterals of thalamocortical fibers, the reticular nucleus cells, which then fire a new burst of action potentials. These re-inhibit the relay cells, which burst again on recovering and so the cycle continues at 7–14 Hz, the spindle frequency. Recordings made simultaneously from reticular nucleus and relay cells in the underlying dorsal thalamus clearly demonstrate synchrony of their discharges at spindle frequencies, a result of the synaptic interplay between the two sets of cells.

Apart from causing repetitive burst firing in relay cells, a reticular nucleus cell, by reason of the widespread terminations of its axon within the dorsal thalamus, has the effect of distributing the disynaptic inhibitory effects of corticothalamic stimulation across many relay cells, thus helping to promote synchrony throughout the whole thalamo-cortico-thalamic network. Spread of corticothalamic effects across many reticular nucleus cells and recruitment of others by collateral inputs from the thalamocortical axons of bursting relay cells will also serve to spread spindle oscillations across most of the thalamus and cortex. The simultaneous onset of spindles throughout cortex and thalamus implies rapid diffusion of reticular nucleus effects on relay cells and equally rapid collateral excitation of widespread sections of the reticular nucleus by bursting relay cells (Steriade and Amzica 1996). Although the corticothalamic system is particularly powerful in inducing spindle oscillations, it is the reticular nucleus that is the prime mover in synchronizing the oscillations of virtually all cells in the network. Its capacity to do this is enhanced when the weak inhibitory effects of one reticular nucleus cell on another are removed (Sohal et al. 2000).

Corticothalamic and Thalamocortical Interactions, Fig. 3 Scheme of re-entrant thalamocortical and corticothalamic circuitry promoting intra- and inter- columnar and inter-areal synchronous activity (Based on Llinás and Paré (1997) and Jones (2007))

The power of the corticothalamic projection to induce ► low frequency oscillations in the spindle or lower frequency ranges clearly depends upon the capacity of the disynaptic inhibitory effect of the reticular nucleus to overcome the direct, monosynaptic excitatory effect of corticothalamic fibers upon the relay cells of the dorsal thalamus. To effect this, the strength of the corticothalamic input to the reticular nucleus is stronger than that to the relay cells. Corticothalamic EPSCs in the reticular nucleus cells are nearly three times larger than in relay cells (Golshani et al. 2001). The basis for this difference in AMPA-receptor-based synaptic strength depends upon the presence of nearly three times as many GluR$_4$ receptor subunits at the corticothalamic synapses on the reticular nucleus cells than at corticothalamic synapses on relay cells (Liu et al. 2001). Increases in channel opening time consequent upon this enrichment should account for the larger EPSCs in the reticular nucleus cells.

Unlike the corticothalamic synapses at which GluR$_4$ receptor subunits are enriched in comparison with other AMPA-receptor subunits, notably GluR$_3$, the larger collateral thalamocortical synapses in the reticular nucleus possess GluR$_4$ and GluR$_3$ subunits in equal proportions. The collateral thalamocortical synapses provide the capacity for the powerful ► re-entrant excitation of the reticular nucleus cells by bursts of action potentials in relay cells during the course of spindle oscillations.

When thalamic relay cells are relatively depolarized, as in the alert attentive state, they tend to display intrinsic membrane oscillations at 20–50 Hz (Pedroarena and Llinás 1997). Under these circumstances, with the disynaptic inhibition of the reticular nucleus rendered less effective, corticothalamic stimulation tends to promote oscillatory activity at ~40 Hz in the connected thalamo-cortico-thalamic network. ► 40 Hz Oscillations are an accompaniment of attention, perception, and higher cognitive states.

In cooperation with the corticothalamic system, calbindin-expressing cells of the thalamic matrix form a basis for dispersion of activity across larger areas of cortex than the parvalbumin-expressing cells of the thalamic core with their focused projections to an individual area (Jones 2007). Within a cortical area, the terminations of matrix cell axons on distal dendrites in superficial layers and of matrix cell axons on more proximal dendrites in middle layers should serve as a coincidence detection circuit leading to a high degree of temporal integration (Llinás and Paré 1997) (Fig. 3). This, in turn, should promote synchronous activity in the cells of individual cortical columns and in a group of columns activated by the same stimulus. Oscillatory activity in these cortical columns should be fed back by layer VI corticothalamic cells to the thalamic nucleus from which they receive input, serving to reinforce the synchrony. Synchronous activity would be spread across other cortical columns in the same cortical area and in adjacent cortical areas by the diffuse projections of matrix cells in the thalamic nucleus. However, other thalamic nuclei and, through their matrix cells, other cortical areas should also be recruited into large scale coherent activity by the diffuse intracortical and corticothalamic projections of layer V corticothalamic neurons. This is thought to be a key to the binding together of the various elements of a cognitive event.

References

Golshani, P., Warren, R. A., & Jones, E. G. (1998). Progression of change in NMDA, non-NMDA, and metabotropic glutamate receptor function at the developing corticothalamic synapse. *Journal of Neurophysiology, 80*, 143–154.

Golshani, P., Liu, X.-B., & Jones, E. G. (2001). Differences in quantal amplitude reflect GluR4- subunit number at corticothalamic synapses on two populations of thalamic neurons. *Proceedings of the National Academy of Sciences USA, 98*, 4172–4177.

Jones, E. G. (2007). *The thalamus* (2nd ed.). Cambridge: Cambridge University Press.

Kao, C. Q., & Coulter, D. A. (1997). Physiology and pharmacology of corticothalamic stimulation-evoked responses in rat somatosensory thalamic neurons in vitro. *Journal of Neurophysiology, 77*, 2661–2676.

Liu, X.-B., Bolea, S., Golshani, P., et al. (2001). Differentiation of corticothalamic and thalamocortical collateral synapses on mouse reticular nucleus neurons by EPSC amplitude and AMPA receptor subunit composition. *Thalamus Related Systems, 1*, 15–29.

Liu, X.-B., Honda, C. N., & Jones, E. G. (1995b). Distribution of four types of synapse on physiologically identified relay neurons in the ventral posterior thalamic nucleus of the cat. *The Journal of Comparative Neurology, 352*, 69–91.

Liu, X.-B., Warren, R. A., & Jones, E. G. (1995a). Synaptic distribution of afferents from reticular nucleus in ventroposterior nucleus of cat thalamus. *The Journal of Comparative Neurology, 352*, 187–202.

Llinás, R., & Paré, D. (1997). Coherent oscillations in specific and non-specific thalamocortical networks and their role in cognition. In M. Steriade, E. G. Jones, & D. A. McCormick (Eds.), *Thalamus* (Experimental and clinical aspects, Vol. II, pp. 501–516). Amsterdam: Elsevier.

McCormick, D. A., & von Krosigk, M. (1992). Corticothalamic activation modulates thalamic firing through glutamate "metabotropic" receptors. *Proceedings of the National Academy of Sciences USA, 89*, 2774–2778.

Pedroarena, C., & Llinás, R. (1997). Dendritic calcium conductances generate high-frequency oscillation in thalamocortical neurons. *Proceedings of the National Academy of Sciences USA, 94*, 724–728.

Sohal, V. S., Huntsman, M. M., & Huguenard, J. R. (2000). Reciprocal inhibitory connections regulate the spatiotemporal properties of intrathalamic oscillations. *Journal of Neuroscience, 20*, 1735–1745.

Steriade, M., & Amzica, F. (1996). Intracortical and corticothalamic coherency of fast spontaneous oscillations. *Proceedings of the National Academy of Sciences USA, 93*, 2533–2538.

Turner, J. P., & Salt, T. E. (1998). Characterization of sensory and corticothalamic excitatory inputs to rat thalamocortical neurones in vitro. *The Journal of Physiology, 510*, 829–843.

Turner, J. P., & Salt, T. E. (1999). Group III metabotropic glutamate receptors control corticothalamic synaptic transmission in the rat thalamus in vitro. *The Journal of Physiology, 519*, 481–491.

Zhang, L., & Jones, E. G. (2004). Corticothalamic inhibition in the thalamic reticular nucleus. *Journal of Neurophysiology, 91*, 759–766.

Corticothalamic Fibers

Definition

Axons with cell bodies located in the cortex with terminations in the thalamus.

Cross-References

▶ Corticothalamic and Thalamocortical Interactions

Cortisol

Definition

Hormone released by the adrenal glands. It is regulated by many endogenous substances, e.g., the serotonin 5-HT1 receptors, which induce a biphasic response. In fact, 5-HT1 agonists first induce a cortisol increase, followed by a decrease.

Cross-References

▶ Placebo Analgesia and Descending Opioid Modulation

Cost Shifting

Definition

Cost shifting can be said to occur in healthcare when changes in the reimbursement for the delivery of health services, or alterations in the parameters of health benefits displace part of the cost of healthcare expenditures from one sphere of healthcare to another, or from healthcare institutions to patients and/or their families. Examples of this would include attempts by state governments to try to restrict reimbursement for medications to cheaper nonbrand name medications, or to limit access to categories of medications such as the Cox B anti-inflammatory medications. This cost is displaced onto underinsured consumers. Often as a result of inability to absorb such costs, patients are hospitalized or placed in skilled nursing facilities.

Cross-References

▶ Cancer Pain Management, Undertreatment and Clinician-Related Barriers

Costoclavicular Syndrome

▶ Thoracic Outlet Syndrome

Co-transmission

Definition

Many neurons synthesize two different types of neurotransmitters. Small-molecule neurotransmitters such as glutamate and acetylcholine are recycled or synthesized at the level of the nerve terminal and stored in small synaptic vesicles. Peptidergic transmitters are synthesized as part of large precursor proteins on the rough endoplasmic reticulum in the cell body and transported into the Golgi apparatus where they are packaged into large, dense-core vesicles that are transported to the nerve terminal. Transmitters released from small clear vesicles generate rapid synaptic responses because small vesicles are located closest to the voltage-dependent Ca^{2+} channels that provide the influx of Ca^{2+} that activates the release process, and small-molecule transmitters in synaptic vesicles activate ionotropic receptors. As large, dense-core vesicles are located at a greater distance from the site of Ca^{2+} influx, higher firing frequencies are required to initiate release of transmitter from these vesicles. Peptidergic transmitters activate metabotropic receptors that couple to intracellular signaling pathways. Thus, peptidergic transmission is capable of modulating neurotransmission of small-molecule transmitters.

Cross-References

▶ Spinothalamic Tract Neurons, Peptidergic Input

Cotrimoxazol

Definition

Treatment of choice in the limited stage of WG (combination of 2 × 800 mg sulfamethoxazole and 2 × 160 mg trimethoprim).

Cross-References

▶ Headache Due to Arteritis

Cotunnius Disease

▶ Sciatica

Counterirritation

Definition

The mechanisms by which any kind of intense and noxious stimulus elicits an analgesic effect distant from the site of the pain producing effect.

Cross-References

▶ Acupuncture Mechanisms
▶ Tourniquet Test

Coupling Between Sympathetic Postganglionic Neurons and Primary Afferent Neurons (In The Dorsal Root Ganglion)

▶ Sympathetic-Afferent Coupling in the Dorsal Root Ganglion, Neurophysiological Experiments

Coupling Media of Ultrasound

Definition

Coupling media are mineral oil or several commercially available coupling gels. They have similar transmissivities to the reference standard of distilled degassed water.

Cross-References

▶ Ultrasound Therapy of Pain from the Musculoskeletal System

Coupling of Sympathetic Postganglionic Neurons onto Primary Afferent Nerve Fibers

▶ Sympathetic-Afferent Coupling in the Afferent Nerve Fiber, Neurophysiological Experiments

COX

▶ Cyclooxygenases
▶ Cyclooxygenases in Biology and Disease
▶ NSAIDs, COX-Independent Actions
▶ NSAIDs, Mode of Action
▶ Prostaglandins, Spinal Effects

COX Isozymes

Definition

Are subsets of the cyclooxygenase enzymes that catalyze the conversion of arachidonic acid to prostaglandins and other products involved in pain signaling.

Cross-References

▶ Drugs with Mixed Action and Combinations: Emphasis on Tramadol
▶ Pharmacology of Cyclooxygenase Inhibitors

COX-1

▶ COX-1 and COX-2 in Pain

COX-1 and COX-2 in Pain

Irmgard Tegeder
Pharmazentrum frankfurt, Institute of Clinical
Pharmacology, Goethe-University Hospital,
Frankfurt, Germany

Synonyms

COX-1; COX-2; Cyclooxygenase-1; Cyclooxy-
genase-2; PGHS-1; PGHS-2; Prostaglandin
endoperoxide synthase 1; Prostaglandin endoperoxide synthase 2; Ptgs1; Ptgs2

Definition

The isoenzymes ▶ cyclooxygenases-1 and −2
(COX-1 and COX-2) catalyze the conversion of
arachidonic acid (AA) to prostaglandin H2 (PGH2)
which is the rate-limiting step in the biosynthesis of
various ▶ prostaglandins and thromboxanes. The
subsequent specific step is accomplished by various prostaglandin synthases which convert PGH2
to PGE2, PGD2, PGF2alpha, PGI2, and TXA2.
Prostaglandin E2 (PGE2) plays a particularly
important role in pain signaling (Fig. 1).

Characteristics

COX-1 and COX-2 are the key enzymes in the
prostaglandin and thromboxane biosynthetic pathway. Although the structure and enzymatic activity
of both enzymes are very similar, COX-1 and COX-
2 perform different tasks which is allowed for
a different localization and regulation. COX-1 is
expressed in all tissues in certain cell types and
produces prostaglandins and thromboxanes that
are needed for the maintenance of physiological
functions such as regulation of blood flow, platelet
aggregation, and mucus production in the stomach.
On the other hand, COX-2 is not normally expressed
in "healthy" tissue but only occurs following
adequate stimulation, which may be any kind of
tissue damage such as trauma, ischemia, infection,

or inflammation. Such stimuli result in activation of
COX-2 regulating ▶ transcription factors including
nuclear factor kappa B (NFkappaB), activator pro-
tein-1 (AP-1), CCAAT enhancer-binding protein-1
(C/EBPbeta), and cAMP response element-binding
protein (CREB), which translate the stimulus into
COX-2 ▶ upregulation and excessive prostaglandin
production. PGE2 is particularly involved in the
activation and ▶ sensitization of the nociceptive
system. This effect is mediated by stimulation of
▶ nociceptors in the periphery, and direct ▶ depolarization of postsynaptic nociceptive neurons in the
spinal cord, which is mediated through prostaglandin E (EP2) receptors (Baba et al. 2001). In addition,
PGE2 causes a dys-inhibition of nociceptive neurons by suppressing inhibitory glycinergic synaptic
transmission in the spinal cord (Ahmadi et al. 2002).
Enhanced or mal-controlled PGE2 release results in
spontaneous pain and ▶ hyperalgesia, that is,
increased sensitivity to painful stimuli. The functional disparity of COX-1 and COX-2 has encouraged the development of COX-2 selective
inhibitors. These drugs are supposed to inhibit
only the "pain"-related PG synthesis without affecting physiological prostaglandins. Multiple clinical
trials have demonstrated that COX-2 selective
inhibitors cause less gastrointestinal toxicity than
unselective agents and are equally effective in the
treatment of arthritis (pain and function) and postoperative pain in clinical studies. However, there are
some exceptions from the COX-1/COX-2 rule, in
that COX-2 is also constitutively expressed in some
tissues, particularly neurons in the spinal cord and
brain, and COX-1 contributes to pain-associated
PG release.

Contribution of COX-1-Derived Prostaglandins to Nociception

Some experimental studies suggest that COX-1 is
involved in the "injury"-induced immediate
release of PGE2 in the spinal cord and periphery
and, thereby, contributes to the activation and
sensitization of nociceptive neurons (Tegeder
et al. 2001a; Wallace et al. 1998). However, this
is contradicted by others (Ghilardi et al. 2004;
Yaksh et al. 2001). The debate is primarily fuelled
by a controversy about the exact localization of
constitutive COX-2 and the time needed for its

Membrane phospholipids

Phospholipase activity:
cPLA2, sPLA2

Arachidonic acid

Cyclooxygenase activity
COX-1

Cyclooxygenase activity
COX-2

PGG2

Peroxidase activity
COX-1

Peroxidase activity
COX-2

PGH2   PGH2

PGFS   TXS   cPGES   mPGES   PGIS   PGDS   Prostanoid Synthases

PGF2β   TXA2   PGE2   PGE2   PGI2   PGD2   Metabolites

Pain   TXB2   Stomach   Pain   PGF1β   Δ15dPGJ2
Inflammation   ↓   Kidney   Inflammation
        Platelets   Arteries   Adipose tissue
                Veins   Healing

COX-1 and COX-2 in Pain, Fig. 1 Synthesis of prostaglandins and thromboxanes

upregulation. For example, constitutive COX-2 expression was observed in motor neurons of the ventral horn, but neither in dorsal horn sensory neurons nor glial cells (Maihofner et al. 2000). In line with the lack of baseline expression of COX-2 in nociceptive neurons, celecoxib (COX-2 inhibitor, injected i.p.) had no antinociceptive effect in a model of acute nociception, whereas a COX-1 inhibitor was effective (Tegeder et al. 2001a), which however may be mediated through COX-1-independent mechanisms (Brenneis et al. 2006). Another study, however, reported a constitutive expression of COX-2 in dorsal horn sensory neurons and radial glial cells (Ghilardi et al. 2004). In this study, the COX-2 inhibitor reduced acute nociceptive behavior and the COX-1 inhibitor was ineffective (Ghilardi et al. 2004). In this study, the drugs were delivered onto

the lumbar spinal cord via a chronically implanted spinal catheter. The catheter may by itself cause an activation of radial glia and COX-2 upregulation in the dorsal horn. Hence, the nociceptive system might have been pre-activated in this study, which would explain the antinociceptive effect of the COX-2 inhibitor. The studies generally agree upon the COX-2 upregulation in dorsal horn neurons following inflammatory or nociceptive stimulation (Beiche et al. 1996; Dolan et al. 2003; Ghilardi et al. 2004; Samad et al. 2001; Tegeder et al. 2001b).

COX-1 and COX-2 in Clinical Pain
In clinical studies, COX-2 inhibitors were generally as effective as traditional NSAIDs in the treatment of arthritis, postoperative pain, gout, migraine, dental pain, dental surgery, and

dysmenorrhea. Thus, selective COX-2 inhibitors not only reduce chronic inflammatory pain caused, for example, by chronic arthritis but also "acute" pain, although COX-2 is not constitutively expressed in most peripheral human tissues or in human sensory neurons in the spinal cord (Maihofner et al. 2003). However, some cortical neurons show constitutive COX-2 expression, and the acute effects of COX-2 inhibitors may be mediated through cortical actions. In clinical settings, "acute" pain (such as postoperative pain, acute gout, dental pain, dysmenorrhea) that requires treatment with non-opioid analgesics lasts for at least several hours, and hence, COX-2 is going to be upregulated in the course of the affection and contributes to PG production. The onset of effects will probably depend more on pharmacokinetic issues than on the speed of COX-2 upregulation.

COX-2 in Higher Brain Regions

Results of some clinical studies, however, cannot be satisfactorily explained by COX-2 upregulation. For example, preemptive treatment with a COX-2 inhibitor before starting surgery resulted in a reduction of postoperative pain, although the time from the skin incision to the first pain assessment was very short and no pain was reported before surgery. This suggests that prostaglandin release from constitutively expressed COX-2 in higher brain regions also contributes to nociception. This idea is supported by experimental data showing that COX-2-derived prostaglandins in the preoptic hypothalamus mediate lipopolysaccharide (LPS)-induced hyperalgesia in rats (Abe et al. 2001). An experimental pain study in humans also suggests an involvement of constitutive brain COX-2 (Koppert et al. 2004). This study employed electrically evoked noninflammatory pain, where the electrical current directly stimulated axons without affecting the nociceptive nerve terminals. Intravenous parecoxib (prodrug of valdecoxib) reduced secondary hyperalgesia and allodynia within 30 min after starting the stimulation so that COX-2 upregulation at any site was highly unlikely. The analgesic effects are therefore probably caused by inhibition of constitutive COX-2 in

higher brain regions such as hypothalamus or cortex. Alternatively, effects might be mediated through interaction with the cannabinoid system recently suggested for parecoxib and valdecoxib (Schröder et al. 2011).

Peripheral Versus Central Effects in Humans

In light of the growing evidence for central effects of COX inhibitors, dissection of peripheral versus central effects is challenging, particularly because NSAID gels and creams that only provide COX inhibition in the skin and directly underlying tissue are very popular in some countries. A recent clinical study has addressed this question, employing local and systemic diclofenac in the ► freeze lesion model (Burian et al. 2003). In this study, diclofenac gel significantly reduced mechanical hyperalgesia. However, overall pain relief with systemic diclofenac was stronger with equal tissue concentrations. Thus, both peripheral and central effects likely contribute to the analgesic efficacy in this inflammatory model. The relative contribution of peripheral versus central effects, however, may vary depending on the type and source of pain.

Neuropathic Pain

The role of prostaglandins in ► neuropathic pain is highly controversial and depends on the model used. It has recently been suggested that COX-1 plays a significant role because an increase of its expression in the spinal cord was observed. COX-1 is expressed in microglial cells in the spinal cord. These cells proliferate in response to a peripheral nerve injury. The observed COX-1 "upregulation" therefore likely reflects the activation and proliferation of microglial cells which contribute to the development of neuropathic pain through the release of pro-inflammatory mediators (Tsuda et al. 2005). The exact role of COX-1, however, is unknown. In addition to prostaglandin-mediated effects, cyclooxygenases are involved in the oxidative metabolism of the endogenous cannabinoid, anandamide, and may therefore modify this endogenous pain defense system. COX-2 is not or only weakly upregulated at mRNA levels in the spinal cord in response to a peripheral nerve injury, suggesting that its contribution to neuropathic pain is of minor importance. In support, inhibitors of cyclooxygenase activity (traditional NSAIDs and selective COX-2 inhibitors) are generally considered to be ineffective in reducing neuropathic pain in clinical practice.

References

Abe, M., Oka, T., Hori, T., et al. (2001). Prostanoids in the preoptic hypothalamus mediate systemic lipopolysaccharide-induced hyperalgesia in rats. *Brain Research, 916*, 41–49.

Ahmadi, S., Lippross, S., Neuhuber, W. L., et al. (2002). PGE(2) selectively blocks inhibitory glycinergic neurotransmission onto rat superficial dorsal horn neurons. *Nature Neuroscience, 5*, 34–40.

Baba, H., Kohno, T., Moore, K. A., et al. (2001). Direct activation of rat spinal dorsal horn neurons by prostaglandin E2. *Journal of Neuroscience, 21*, 1750–1756.

Beiche, F., Scheuerer, S., Brune, K., et al. (1996). Up-regulation of cyclooxygenase-2 mRNA in the rat spinal cord following peripheral inflammation. *FEBS Letters, 390*, 165–169.

Brenneis, C., Maier, T. J., Schmidt, H., et al. (2006). Inhibition of prostaglandin E2 synthesis by SC-560 is independent of cyclooxygenase 1 inhibition. *The FASEB Journal, 20*(9), 1352–1360.

Burian, M., Tegeder, I., Seegel, M., et al. (2003). Peripheral and central antihyperalgesic effects of diclofenac in a model of human inflammatory pain. *Clinical Pharmacology and Therapeutics, 74*, 113–120.

Dolan, S., Kelly, J. G., Huan, M., et al. (2003). Transient up-regulation of spinal cyclooxygenase-2 and neuronal nitric oxide synthase following surgical inflammation. *Anesthesiology, 98*, 170–180.

Ghilardi, J. R., Svensson, C. I., Rogers, S. D., et al. (2004). Constitutive spinal cyclooxygenase-2 participates in the initiation of tissue injury-induced hyperalgesia. *Journal of Neuroscience, 24*, 2727–2732.

Koppert, W., Wehrfritz, A., Korber, N., et al. (2004). The cyclooxygenase isozyme inhibitors parecoxib and paracetamol reduce central hyperalgesia in humans. *Pain, 108*, 148–153.

Maihofner, C., Tegeder, I., Euchenhofer, C., et al. (2000). Localization and regulation of cyclooxygenase-1 and −2 and neuronal nitric oxide synthase in mouse spinal cord. *Neuroscience, 101*, 1093–1108.

Maihofner, C., Probst-Cousin, S., Bergmann, M., et al. (2003). Expression and localization of cyclooxygenase-1 and −2 in human sporadic amyotrophic lateral sclerosis. *European Journal of Neuroscience, 18*, 1527–1534.

Samad, T. A., Moore, K. A., Sapirstein, A., et al. (2001). Interleukin-1beta-mediated induction of Cox-2 in the CNS contributes to inflammatory pain hypersensitivity. *Nature, 410*, 471–475.

Schröder, H., Höllt, V., Becker, A., et al. (2011). Parecoxib and its metabolite valdecoxib directly interact with cannabinoid binding sites in CB1-expressing HEK 293 cells and rat brain tissue. *Neurochemistry International, 58*(1), 9–13.

Tegeder, I., Niederberger, E., Vetter, G., et al. (2001a). Effects of selective COX-1 and −2 inhibition on formalin-evoked nociceptive behaviour and prostaglandin E(2) release in the spinal cord. *Journal of Neurochemistry, 79*, 777–786.

Tegeder, I., Niederberger, E., Israr, E., et al. (2001b). Inhibition of NF-{kappa}B and AP-1 activation by R- and S-flurbiprofen. *The FASEB Journal, 15*, 2–4.

Tsuda, M., Inoue, K., Salter, M. W., et al. (2005). Neuropathic pain and spinal microglia: A big problem from molecules in "small" glia. *Trends in Neurosciences, 28*(2), 101–107.

Wallace, J. L., Bak, A., McKnight, W., et al. (1998). Cyclooxygenase-1 contributes to inflammatory responses in rats and mice: Implications for gastrointestinal toxicity. *Gastroenterology, 115*, 101–109.

Yaksh, T. L., Dirig, D. M., Conway, C. M., et al. (2001). The acute antihyperalgesic action of nonsteroidal, anti-inflammatory drugs and release of spinal prostaglandin E2 is mediated by the inhibition of constitutive spinal cyclooxygenase-2 (COX-2) but not COX-1. *Journal of Neuroscience, 21*, 5847–5853.

COX-1 Inhibitor

Definition

A drug that inhibits type 1 of the enzyme cyclooxygenase, which takes part in the production of prostaglandins that cause inflammation and pain, cause blood platelets to become sticky and stop bleeding, and protect the gastric mucosal cells (and also have many other physiologic functions).

Cross-References

▶ COX-1 and COX-2 in Pain
▶ Pharmacology of Cyclooxygenase Inhibitors
▶ Postoperative Pain, Acute Pain Management Principles

COX-2

▶ COX-1 and COX-2 in Pain

COX-2 Inhibitor

Synonyms

Coxib; NSAIDs

Definition

Nonsteroidal anti-inflammatory drugs (NSAIDs) are increasingly used for postoperative analgesia. While lacking some of the troublesome adverse effects of opioids, nonselective NSAIDs may cause bleeding, gastric ulceration, and renal injury as a result of their inhibitory effects on cyclooxygenase-1 (COX-1). Cyclooxygenase-2 (COX-2) is normally present in small concentrations but is induced peripherally under conditions of inflammation. Cyclooxygenase-2 (COX-2) is constitutively expressed in the brain and spinal cord and is further upregulated after persistent noxious inputs. Spinal COX-2 inhibition may be an important mechanism for reducing post-injury hyperalgesia. For this reason, COX-2-selective inhibitors (COXIBS) could uncouple the therapeutic and adverse effects of the nonselective NSAIDs. COX-2-selective inhibitors are effective for the treatment of preoperative and postoperative pain and reduce postsurgical requirements for opioids. They have similar analgesic efficacy to the nonselective NSAIDs. Data from large multicenter, multidose comparative studies are needed to establish whether COX-2-selective inhibitors are more efficacious, cost-effective, and safe compared to the nonselective NSAIDs with respect to gastric, renal, and coagulation problems, and whether COX-2-selective inhibitors confer greater cardiovascular risk in the perioperative setting.

Cross-References

▶ Multimodal Analgesia in Postoperative Pain
▶ Pharmacology of Cyclooxygenase Inhibitors
▶ Postoperative Pain, Acute Pain Management Principles
▶ Postoperative Pain, Cox-2 Inhibitors

COX-2 Inhibitors

▶ Postoperative Pain, COX-2 Inhibitors

Coxib

▶ COX-2 Inhibitor

Coxibs

Definition

Structural heterogenous class of compounds inhibiting COX-2, more or less selectively.

Cross-References

▶ COX-1 and COX-2 in Pain
▶ COX-2 Inhibitor
▶ Coxibs and Novel Compounds, Chemistry
▶ NSAIDs, Chemical Structure and Molecular Mode of Action
▶ Pharmacology of Cyclooxygenase Inhibitors
▶ Postoperative Pain, COX-2 Inhibitors

Coxibs and Novel Compounds, Chemistry

Stefan A. Laufer
Institute of Pharmacy, Eberhard-Karls University of Tuebingen, Tuebingen, Germany

Synonyms

Cyclooxygenase inhibitors, chemistry; NSAIDs and coxibs

Definition

As most ► NSAIDs are believed to act via a substrate analogue of ► arachidonic acid mechanism at the active site of the molecular target, ► cyclooxygenases (COX), chemical structure, and biological activity are closely linked. Acidic groups and π-electron systems of NSAIDs mimic key structural features of the genuine substrate, arachidonic acid. COX-1/COX-2 selectivity is mainly determined by different conformational requirements rather than specific interactions.

Characteristics

Unselective COX Inhibitors

Structure and Metabolic Function of COX-1
COX-1 is a 70 kD enzyme, catalyzing the reaction of arachidonic acid to PGG_2 (cyclooxygenase reaction) and consecutively PGG_2 to PGH_2 (peroxidase reaction) as outlined in Fig. 1.

There are distinct active sites for the cyclooxygenase and peroxidase reactions (Fig. 2).

Coxibs and Novel Compounds, Chemistry, Fig. 1 Reactions catalyzed by COX enzymes

Coxibs and Novel Compounds, Chemistry, Fig. 2 A ribbon representation of the Co^{3+}-oPGHS-1 monomer with AA bound in the COX channel. The EGF domain, MBD, and catalytic domain are shown in *green, orange, and blue,* respectively; Co^{3+}-protoporphyrin IX is depicted in *red,* disulfide bonds (Cys36-Cys47, Cys37-Cys159, Cys41-Cys57, Cys59-Cys69, and Cys569-Cys575) in *dark blue* and side chain atoms for COX channel residues Arg120, Tyr355, and Tyr385 in *magenta* (From Malkowski et al. (2000))

Coxibs and Novel Compounds, Chemistry, Fig. 3 Mechanistic sequence for converting AA to PGG$_2$. Abstraction of the 13-proS hydrogen by the tyrosyl radical leads to the migration of the radical to C-11 on AA. An attack of molecular oxygen, coming from the base of the COX channel, occurs on the side interfacial to the hydrogen abstraction. As the 11R-peroxyl radical swings over C-8 for an R-side attack on C-9 to form the endoperoxide bridge, C-12 is brought closer to C-8 via rotation about the C-10/C-11 bond, allowing the formation of the cyclopentane ring. The movement of C-12 also positions C-15 optimally for addition of a second molecule of oxygen, formation of PGG$_2$, and the migration of the radical back to Tyr385 (From Malkowski et al. (2000))

Inhibitors

Different chemical classes can provide the structural features necessary to mimic arachidonic acid at the active site. The substrate arachidonic acid is a C_{20} carboxylic acid with four isolated double bonds at positions 5, 8, 11, and 14. For the enzyme reaction, arachidonic acid must adapt to a "folded" conformation, allowing the oxygen to insert between C_9 and C_{11} and the ring closure between C_8 and C_{12} (Fig. 3). To fix arachidonic acid in such a conformation, several interactions to the active site of the enzyme are necessary, e.g., ionic interaction (a salt bridge) between the carboxylic group and arginine 120, π-π interactions between the double bonds of arachidonic acid and aromatic amino acids, and numerous hydrophobic interactions (Fig. 4).

All these structural features can be identified in many NSAIDs. Most acidic NSAIDs are therefore believed to mimic arachidonic acid in its folded conformation at the active site of COX. The structure activity relationship follows these structural constrictions closely. The activity against COX-1 clearly correlates with torsion angles around the π-electron systems and the overall lipophilicity of the molecule (Moser et al. 1990).

Two classes of compounds, however, have a distinctly different molecular mode of action:
- ASS irreversibly acetylates serine 530 at the active site of the enzyme.
- The oxicames are believed to interfere with the peroxidase active site, which also explains their structural difference.

Coxibs and Novel Compounds, Chemistry, Fig. 4 A schematic of interactions between AA and COX channel residues. Carbon atoms of AA are *yellow*, oxygen atoms *red*, and the 13proS hydrogen *blue*. All *dashed lines* represent interactions within 4.0 Å between the specific side chain atom of the protein and AA (From Malkowski et al. (2000))

Most of the currently used NSAIDs, including diclofenac, ibuprofen, naproxen, piroxicam, and indomethacin, for instance, may produce full inhibition of both COX-1 and COX-2 with relatively poor selectivity under therapeutic conditions (Warner et al. 1999). Acidic NSAIDs like diclofenac accumulate particularly in blood, liver, milt, and bone marrow but also in tissues with acidic extracellular pH values. Such tissues are mainly inflamed tissues, such as gastric tissue and the manifolds of the kidney. In inflamed tissue, NSAIDs inhibit the pathological overproduction of prostaglandins. In contrast, neutral NSAIDs (paracetamol) and weakly acidic NSAIDs (metamizol) distribute themselves quickly and homogeneously in the organism. They also penetrate the blood–brain barrier.

Fenamate Group

The core structure is 2-aminobenzoic acid (anthranilic acid). The 2-amino group is substituted with aromatic residues.

- Flufenamic acid
- Mefenamic acid
- Meclofenamic acid
- Niflumic acid (core structure, 2-amino-pyridyl-3-carboxylic acid). For topical application, the carboxylic acid group is esterified with diethylene glycol
- Etofenamate (Fig. 5)

Coxibs and Novel Compounds, Chemistry, Fig. 5 Chemical structures of fenamate group NSAIDs

Fenamate core structure

Etofenamate

Flufenamic acid

Mefenamic acid

Meclofenamic acid

Nifluminic acid

Coxibs and Novel Compounds, Chemistry, Fig. 6 Chemical structures of fenac group NSAIDs

Fenac core structure

Diclofenac

Felbinac

Fenac Group

The core structure is 2-aminophenylacetic acid. The 2-amino group is substituted with aromatic residues.

- Diclofenac
- Felbinac (only used topically) (Fig. 6)

Heteroaryl Acetic Acid Group

- Indomethacin
- Acemetacine
- Proglumetacine
- Tolmetin (and its ring-closed analog ketorolac)
- Lonazolac (Fig. 7)

Profen Group

Core structure: 2-arylpropionic acid

- Ibuprofen
- Ketoprofen
- Thiaprofen
- Naproxen

- Ketorolac (can be seen formally as a ring-closed profene) (Fig. 8)

Oxicam Group

Core structure: 1,2-benzothiazine or 1,2 thienothiazine

- Piroxicam
- Tenoxicam
- Lornoxicam
- Droxicam
- Cinnoxicam
- Sudoxicam
- Meloxicam (Fig. 9)

Pyrazolone Group

The mode of action of the pyrazolones remains unclear. It is thought that they may not be involved in the inhibition of COX-1 or COX-2. The compounds of the pyrazol-3-on series at least are neutral molecules with no acidity. A central mode of action is suggested. They also act

Coxibs and Novel Compounds, Chemistry, Fig. 7 Chemical structures of heteroaryl acetic acid group NSAIDs

Coxibs and Novel Compounds, Chemistry, Fig. 8 Chemical structures of profen group NSAIDs

antispasmodically, and they are effective against visceral pain. In the past, pyrazolones were among the nonsteroidal anti-inflammatory drugs used very frequently. They show a high plasma protein binding and therefore have a high rate of interaction with other pharmaceuticals. Agranulocytosis is a rare but severe side effect.

The core structure is *3H*-pyrazol-3-one

- Propyphenazone
- Metamizole-Na
- Phenazone (Fig. 10)

Pyrazolidinedione Group
The core structure is pyrazolidin-3,5-dione
- Phenylbutazone
- Mofebutazone (Fig. 11)

Coxibs and Novel Compounds, Chemistry, Fig. 9 Chemical structures of oxicam group NSAIDs

Coxibs and Novel Compounds, Chemistry, Fig. 10 Chemical structures of pyrazolone group NSAIDs

Pyrazolone core structure

Propyphenazone

Metamizole-Na

Phenazone

Coxibs and Novel Compounds, Chemistry, Fig. 11 Chemical structures of pyrazolidinedione group NSAIDs

Pyrazolidindione core structure

Phenylbutazone

Mofebutazone

Coxibs and Novel Compounds, Chemistry, Fig. 12 Comparison of the active sites of COX-1 (PGHS-1) and COX-2 (PGHS-2) (From Wong et al. (1997))

COX-2-Selective Inhibitors

Isoform 2 of the COX enzyme catalyzes the identical reaction, AA to PGG_2; the active site, however, is slightly different from that of COX-1 (Fig. 12).

Isoleucine 523 is replaced by valine 509, making the active site of COX-2 more "spacious." This difference can be used to generate COX-2-selective inhibitors, as this active site tolerates more bulky molecules. Celecoxib is capable of producing full inhibition of COX-1 and COX-2. However, it shows a preferential selectivity toward COX-2 (>fivefold). The newer coxibs like rofecoxib strongly inhibit COX-2 with only weak activity against COX-1 (Warner et al. 1999).

A common pharmacophore cannot be identified; however, vicinal diaryl systems (celecoxib, rofecoxib, valdecoxib) and sulfone or sulfonamide groups seem to be advantageous (Dannhardt and Laufer 2000). Lumiracoxib is an excellent example of the fact that spatial demanding substituents (bulky groups) alone are sufficient to generate selectivity, even with a diclofenac-like pharmacophore.

Coxibs and Novel Compounds, Chemistry, Fig. 13 Chemical structures of coxibs

Structural Features of Selective COX-2 Inhibitors

Sulfonamide Structure
- Celecoxib
- Valdecoxib

Methylsulfone Structure
- Rofecoxib
- Etoricoxib

Aryl Acetic Acid
- Lumiracoxib

Others:
- Parecoxib (water-soluble prodrug for parenteral application, rapidly metabolized to valdecoxib) (Fig. 13)

References

Dannhardt, G., & Laufer, S. (2000). Structural approaches to explain the selectivity of COX-2 inhibitors: Is there a common pharmacophore? *Current Medicinal Chemistry, 7*, 1101–1112.

Malkowski, M., Ginell, S. L., Smith, W. L., & Garavito, R. M. (2000). The productive conformation of arachidonic acid bound to prostaglandin synthase. *Science, 289*, 1933–1937.

Moser, P., Sallmann, A., & Wiesenberg, L. (1990). Synthesis and quantitative structure-activity relationships of diclofenac analogues. *Journal of Medicinal Chemistry, 33*, 2358–2368.

Warner, T., Giuliano, F., Vojnovic, I., et al. (1999). Nonsteroid drug selectivities for cyclo-oxygenase-1 rather than cyclo-oxygenase 2 are associated with human gastrointestinal toxicity: A full in vitro analysis. *Proceedings of the National Academy of Sciences of the United States of America, 96*, 7563–7568.

Wong, E., Bayly, E., Waterman, H. L., et al. (1997). Conversion of prostaglandin G/H synthase-1 into an enzyme sensitive to PGHS-2-selective inhibitors by a double His[513]→Arg and Ile[523]→Val mutation. *Journal of Biological Chemistry, 272*, 9280–9286.

CPH

▶ Chronic Paroxysmal Hemicrania

CPSP

▶ Central Poststroke Pain
▶ Chronic Postsurgical Pain Syndromes

Cranial Autonomic Symptoms

▶ Hemicrania Continua (HC)

Cranial Nerve Neuralgia

Definition

A family of chronic neuropathic pain states associated with injury or dysfunction of a cranial nerve, most often the trigeminal or glossopharyngeal nerve. Ongoing pain and/or hypersensibility is felt primarily in the distribution of the injured nerve.

Cross-References

▶ Trigeminal and Cranial Nerve Neuralgias

Cranial Nerves

Definition

The 12 pairs of nerves originating from the brain that provide motor, sensory, and autonomic functions to the structures of the head and neck. The N. vagus (X) also innervates the viscera.

Cross-References

▶ Diabetic Neuropathies

Craniomandibular Disorders (CMDs)

▶ Temporomandibular Joint Disorders

Cranium

Definition

The skull and its associated soft tissues.

Cross-References

▶ Orofacial Pain: Taxonomy/Classification

CRD

▶ Colorectal Distension

CREB

▶ Cyclic Adenosine Monophosphate-Responsive Element-Binding Protein

Credibility Finding

Definition

An adjudicative finding of the extent to which a person's statements about the intensity, persistence, and functional effects of his or her symptoms can be believed and accepted as true. The credibility finding takes into consideration the objective medical evidence and the other evidence in the case record, to include the person's statements, the statements and other evidence provided by physicians and psychologists who have treated or examined the person, evidence from other health-care professionals, and statements from other persons, including the person's spouse, family members, neighbors, friends, coworkers, etc.

Cross-References

▶ Disability Evaluation in the Social Security Administration

Credibility, Assessment

Kenneth D. Craig and Samantha R. Fashler
Department of Psychology, University of British Columbia, Vancouver, BC, Canada

Synonyms

Deception; Malingering

Definition

Pain expression and disability are invariably subjected to scrutiny to establish their veracity, with context, self-report, nonverbal behavior, the reports of significant others and medical information of variable quality in contributing to judgments.

Introduction

Pain enmeshes people in complex social transactions, with the credibility of complaints invariably subjected to careful appraisal by others. A substantial range of caregiving resources may be available, whether motivated by sympathy, professional responsibility, contractual obligation (e.g., disability insurance), legislated entitlement

(e.g., workers' compensation benefits), or court order. In all instances, optimal caregiving and prudent stewardship of resources require judgments about the legitimacy of the representations of the person in distress. In health-care settings, norms of trust often do not prevail and patients must negotiate care by persuading health-care practitioners of the legitimacy of their symptoms. Patients are likely to be scrutinized to discover whether financial incentives, efforts to obtain illicit drugs, or avoidance of onerous responsibilities are involved. Widespread misuse and diversion of prescription opioids have intensified efforts to detect inappropriate complaints, placing those in legitimate need at a disadvantage. Patients sometimes warrant and often encounter distrust, particularly when physical pathology is not manifestly obvious. Werner and Malterud (2003) describe women with medically unexplained disorders encountering skepticism, lack of comprehension, rejection, being blamed for their condition and suffering feelings of being ignored or belittled.

Characteristics

The real incidence and costs of various forms of deception are unknown, because people work hard to avoid discovery of dishonesty (Vrij 2011). For example, Aguerrevere et al. (2009) suggested that rates might be as high as 20–50 % in chronic pain patients referred for psychological evaluation. However, the consensus is that the general incidence is lower, perhaps 8 % in medical or psychiatric cases involved in personal injury claims (Mittenberg et al. 2002). Denial of pain to avoid painful treatments or imposition of sick roles is a widespread form of deception, but it is typically ignored because it does not inflict immediate demands on others, despite the potential for collapse of personal health and long-term medical costs. Masedo and Esteve (2007) examined the impact of suppression in an experimental short-term pain context and found that suppression of pain was linked to greater pain and distress. Opportunistic feigning or exaggeration of minor pain is relatively common and may be approved by others (e.g., sick days), despite substantial costs (e.g., to employers). Flagrant, planned fraud is generally estimated to be rare; nevertheless, costs to employers, health-care providers, insurance companies, and social services could mount dramatically.

Decisions that another is misrepresenting pain and disability or malingering require a judgment of intentional production of false symptoms to obtain financial or other privileges. Many clinicians eschew providing reports on credibility because they perceive them inconsistent with the role of patient advocate. Challenging patients is also fraught with substantial risk of angry, emotional confrontations, and reprisal. Patients reasonably perceive serious consequences, including denial of treatment, loss of income, or criminal investigations.

Regrettably, universally acceptable and valid measures of neither pain nor deception are available, and judgments require inferences based upon information with demonstrable limitations. Efforts to detect deception of pain generally appear not to be particularly effective, although in everyday settings, people do reasonably well at detecting the serious lies of others. There are considerable individual differences in detecting deception. Hill and Craig (2004) reported that success rates in differentiating genuine and faked or suppressed facial expressions of pain varied across individuals from 18 % to 63 %. However, individual differences may be negligible if success is improbable, for example, with spoken lies (Bond and DePaulo 2008).

Strategies for Assessing Credibility

Useful information often emerges through interview accounts of symptoms and personal histories, data from standardized questionnaires, observation of nonverbal expression of distress and functional capability, and physical examination and study of archival documentation and reports of significant others. Examiners typically look for inconsistencies between their expectations as to how one should behave under the circumstances and the actual behavior of the individual.

We know more about cues that have "illusory correlations" with deception than about sensitive and specific cues to honest or dishonest representations. Erroneous cues to deception, not unique to pain, include signs of nervousness, clearing the throat, faltering speech, and biting the lips (Porter et al. 2002). Concerning pain, people who respond to placebos have been misconstrued as not experiencing "real" pain, despite clear evidence that placebos can dramatically reduce pain instigated by injury and surgery (Pollo and Benedetti 2011).

Contextual Analysis

The contexts in which people present themselves as in pain are important. Sometimes a discrete, precipitating event may be required to legitimize work-related injuries, or engineering reports must describe substantial physical forces in motor vehicle accidents, but these expectations are controversial as pain onset can occur without trauma.

Analyses of the settings in which people lead their lives may disclose circumstances that may encourage deception, for example, work dissatisfaction or domestic distress. Blair et al. (2010) provide studies in which detection of lying was improved when a judge's attention was directed to situational contexts. Of course, disincentives for assuming the life of a person with persistent pain also must be considered (e.g., loss of employment or impaired family and friendly relationships). The American Psychiatric Association diagnostic manual (2000) implicates malingering when there is a medicolegal context, a strong discrepancy between claimed stress and disability and objective findings, limited cooperation with assessment or lack of compliance in treatment. However, these alleged cues are commonplace and their use can cast inappropriate aspersions on people.

Use of Self-Report

Capitalizing on the unique human capacity to use language to describe subjective states, self-report has been favored as a modality for expression of pain. Assessment guidelines often assert "Pain is what the patient says it is."

But self-report is also a prime medium for dissembling pain. Distrust of self-report is widespread. Scarry (1985) captures the suspicion in the observation that, "To have great pain is to have certainty, to hear that another person has pain is to have doubt." Nevertheless, verbal statements can disclose useful information. Mendelson and Mendelson (2004) reviewed questionnaire studies contrasting people genuinely in pain with people coached to pretend they were in pain, finding that simulators exaggerated the genuine pattern of response. However, cutting scores yielded substantial numbers of misidentified false positives and negatives. A review of the MMPI-2 (Arbisi and Butcher 2004) concluded that it was useful in detecting inconsistent and inaccurate self-report responding, uncooperative responding, unusually positive or virtuous self-presentation, and general defensiveness.

Biomedical Approaches

The biomedical approach to pain requires lesions to account for pain complaints. The model is adequate for painful conditions arising from serious injury, surgery, or disease, provided there is allowance for individual differences and situational influence on pain expression. In some instances, controlled diagnostic blocks serve as useful diagnostic tools when a specific source of pain can be posited and is subject to being anesthetized (e.g., spinal pain) (Bogduk 2004). There also have been attempts to identify the underlying brain substrates related to feeling pain and suppressing pain (Apkarian et al. 2005; Freund et al. 2009). However, medical diagnosis often is unsatisfactorily reliable (Hunt et al. 2001). The reality is that medically unexplained pain is common and a serious health-care problem. Large numbers of patients with functional disorders (Mayer and Bushnell 2009) are marginalized from meaningful professional care and treatment when physicians cannot find evidence of physical pathology. The biopsychosocial approach to understanding pain implicates peripheral and central neuroplasticity as capable of producing persistent pain (Melzack et al. 2001) and

psychological and social factors also are major determinants of pain experience and expression (Hadjistavropoulos et al. 2011).

Nonverbal Cues

Nonverbal information weighs heavily when people make judgments about others' reports of pain, including their credibility (Schiavenato and Craig 2010). Painful distress can be manifest in (a) reflexive, automatic expression, for example, facial grimaces, escape actions, qualities of speech, and protective postures or movements, and (b) complex actions capable of being intentional or deliberate, for example, self-report of pain, use of medication, time off from work, and seeking medical care (Craig et al. 2010). Functional capacity examinations are of this latter type and are recognized as vulnerable to motivational and situational influence. Involuntary expression during pain is more capable of instigating reflexive empathetic reactions in observers (Goubert et al. 2005) than complex, conscious expression which is ambiguous as to its source in painful experience or efforts to control the situation (McCrystal et al. 2011). The dynamics and structure of facial displays permit differentiation of faked and genuine expressions of pain (Hill and Craig 2002). The differences are difficult to detect by the human observer (Hill and Craig 2004); nevertheless, automated, computer recognition systems offer promise of discriminating faked and genuine facial displays (Littlewort et al. 2009). Clinicians are often unable to observe chronic pain during acute exacerbations and are dependent upon self-report. Uncertainty as to the origins of symptomatic complaints of pain can bias judgments and lead to disagreement with patient reports, contributing greater influence on judgments of patient gender or ethnic status and clinician empathy and experience (Tait and Chibnall 2008).

Illustrations of inappropriate use of cues presumed indicative of deception are available. Evidence that pain behavior can be controlled by social consequences (e.g., release from responsibilities, potent medication, or sympathy) has been taken to suggest, incorrectly, that the behavior represented intentional dishonesty or efforts to secure benefits. However, environmental contingencies can operate on behavior outside consciousness (Gatzounis et al. 2012). There have been similar misinterpretations of diagnostic signs related to low back pain that could not have a basis in physical pathology, as we understand current anatomy and physiology (e.g., nonanatomic or superficial tenderness). Main and Waddell (1998) and Fishbain et al. (2004) concluded that there was little evidence of an association between these signs and secondary gain or malingering. Main and Waddell (1998) warn that these signs "are not on their own a test of credibility or faking" and consider inferences based on that supposition to be misuse, both in clinical and medicolegal contexts.

Other measures arise from observations that people who sincerely exercise effort in strength testing appear less likely to be misrepresenting themselves. Displays of weakness and decreased range and velocity of motion have been interpreted as conscious attempts to deceive. Investigations of a variety of indices of this type led Robinson and Dannecker (2004) to conclude "that the current 'state of the art,' while promising in some instances, does not warrant the clinical application of muscle testing as a means of sincerity of effort." Problems arise because performance also is influenced by fear of injury, pain, medications, and other motivational and cognitive factors, and discriminating maximal and submaximal effort remains a problem.

Reports of Significant Others
Reports of family members, friends, colleagues, and others can provide valuable collateral information able to confirm or disconfirm patient self-report (Kappesser and Williams 2008). Collusion would be possible, but contradiction seems more likely, unless there has been very careful grooming.

Surveillance
Unobtrusive behavioral observation can be undertaken using covert surveillance to provide photographic or video records of people claiming to have severely disabling pain engaging in

vigorous physical activity. While this may appear to represent flagrant abuse, careful analysis often suggests the need for income, or commitments to family and others can lead people to engage in vigorous activity, despite severe immediate or delayed pain. Perceptions of credibility can have an important impact on injury claims, as demonstrated by Le Page et al. (2008); credibility of court participants was significantly correlated with the amount of award granted for fibromyalgia claims.

Summary

Assessment of credibility is an evolving process. Those who would deceive can be well motivated and well tutored by prior personal experience, or instructional models in their families. In turn, those interested in assessing credibility must use a broad range of information sources that will hopefully improve in sensitivity and specificity to deceptive behavior. Sequential, progressive evaluation seems the best strategy, using broad screening approaches to trigger suspicion, followed by progressively more rigorous methods.

References

Aguerrevere, L. E., Greve, K. W., Bianchini, K. J., & Meyers, J. E. (2009). Validation of the Malingering Pain-Related Disability Criteria. A poster presented at the NAN's 26th Annual Conference & Workshops. New Orleans, La.

American Psychiatric Association. (2000). *Diagnostic and statistical manual of mental disorders* (4th ed. Text Rev.). Washington, DC: American Psychiatric Association.

Apkarian, A. V., Bushnell, M. C., Treede, R. D., et al. (2005). Human brain mechanisms of pain perception and regulation in health and disease. *European Journal of Pain, 9*, 463–484.

Arbisi, P. A., & Butcher, J. N. (2004). Psychometric perspectives on detection of malingering of pain: Use of the MMPI-2. *The Clinical Journal of Pain, 20*, 383–391.

Blair, J. P., Levine, T. R., & Shaw, A. S. (2010). Content in context improves deception detection accuracy. *Human Communication Research, 36*, 423–442.

Bogduk, N. (2004). Diagnostic blocks: A truth serum for malingering. *The Clinical Journal of Pain, 20*, 409–414.

Bond, C. F., & DePaulo, B. M. (2008). Individual differences in judging deception: Accuracy and bias. *Psychological Bulletin, 134*, 477–492.

Craig, K. D., Versloot, J., Goubert, L., Vervoort, T., & Crombez, G. (2010). Perceiving others in pain: Automatic and controlled mechanisms. *The Journal of Pain, 11*, 101–108.

Fishbain, D. A., Cutler, R. B., Rosomoff, H. L., & Rosomoff, R. S. (2004). Is there a relationship between non-organic physical findings (Waddell signs) and secondary gain/malingering? *The Clinical Journal of Pain, 20*, 399–408.

Freund, W., Klug, R., Weber, F., et al. (2009). Perception and suppression of thermally induced pain: A fMRI study. *Somatosensory and Motor Research, 261*, 1–10.

Gatzounis, R., Schrooten, M. G. S., Crombez, G., & Vlaeyen, J. W. S. (2012). Operant learning theory in pain and chronic pain rehabilitation. *Current Pain and Headache Reports, 13*. doi:10.1007/s11916-012-0247-1.

Goubert, L., Craig, K. D., Vervoort, T., Morley, S., Sullivan, M. J. L., de C Williams, A. C., Cano, A., & Crombez, G. (2005). Facing others in pain: The effects of empathy. *Pain, 118*, 286–288.

Hadjistavropoulos, T., Craig, K. D., Duck, S., Cano, A. M., Goubert, L., Jackson, P., Mogil, J., Rainville, P., Sullivan, M. J. L., de C Williams, A. C., Vervoort, T., & Dever Fitzgerald, T. (2011). A biopsychosocial formulation of pain communication. *Psychological Bulletin, 137*, 910–939.

Hill, M. L., & Craig, K. D. (2002). Detecting deception in pain expressions: The structure of genuine and deceptive facial displays. *Pain, 98*, 134–144.

Hill, M. L., & Craig, K. D. (2004). Detecting deception in facial expressions of pain: Accuracy and training. *The Clinical Journal of Pain, 20*, 415–422.

Hunt, D. G., Zuberbier, O. A., Kozlowski, A., Robinson, J., Berkowitz, J., Schultz, I. Z., Milner, R. A., Crook, J. M., & Turk, D. C. (2001). Reliability of the lumbar flexion, lumbar extension and passive straight leg raise test in normal populations embedded within a complete physical examination. *Spine, 26*, 2714–2718.

Kappesser, J., & de C Williams, A. C. (2008). Pain judgements of patients' relatives: Examining the use of social contract theory as theoretical framework. *Journal of Behavioral Medicine, 31*, 309–317.

Le Page, J. A., Iverson, G. L., & Collins, P. (2008). The impact of judges' perceptions of credibility in fibromyalgia claims. *International Journal of Law and Psychiatry, 31*, 30–40.

Littlewort, G. C., Bartlett, M. S., & Lee, K. (2009). Automatic coding of facial expressions displayed during posed and genuine pain. *Image and Vision Computing, 27*, 1797–1803.

Main, C. J., & Waddell, G. (1998). Behavioral responses to examination: A reappraisal of the interpretation of "Nonorganic Signs". *Spine, 23*, 2367–2371.

Masedo, A. I., & Esteve, M. R. (2007). Effects of suppression, acceptance and spontaneous coping on pain tolerance, pain intensity and distress. *Behaviour Research and Therapy, 45*, 199–209.

Mayer, E. A., & Bushnell, M. C. (2009). *Functional pain syndromes: Presentation and pathophysiology.* Seattle: IASP Press.

McCrystal, K. N., Craig, K. D., Versloot, J., Fashler, S. R., & Jones, D. M. (2011). Perceiving pain in others: Validation of a dual processing model. *Pain, 152*, 1083–1089.

Melzack, R., Coderre, T. J., Katz, J., & Vaccarino, J. L. (2001). Central neuroplasticity and pathological pain. *Annals of the New York Academy of Sciences, 933*, 157–174.

Mendelson, G., & Mendelson, D. (2004). Malingering pain in the medico-legal context. *The Clinical Journal of Pain, 20*, 423–432.

Mittenberg, W., Patton, C., Canyock, E. M., et al. (2002). Base rates of malingering and symptom exaggeration. *Journal of Clinical and Experimental Neuropsychology, 24*, 1094–1102.

Pollo, A., & Benedetti, F. (2011). Placebo/Nocebo: A two-sided coin in the clinician's hand. In M. E. Lynch, K. D. Craig, & P. W. H. Peng (Eds.), *Clinical pain management: A practical guide* (pp. 42–48). Chichester: Blackwell.

Porter, S., Campbell, M. A., Stapleton, J., & Birt, A. R. (2002). The influence of judge, target, and stimulus characteristics on the accuracy of detecting deceit. *Canadian Journal of Behavioural Science, 34*, 172–185.

Robinson, M. E., & Dannecker, E. A. (2004). Critical issues in the use of muscle testing for the determination of sincerity of effort. *The Clinical Journal of Pain, 20*, 392–398.

Scarry, E. (1985). *The body in pain.* New York: Oxford University Press.

Schiavenato, M., & Craig, K. D. (2010). Pain assessment as a social transaction: Beyond the "Gold Standard". *The Clinical Journal of Pain, 26*, 667–676.

Tait, R. C., & Chibnall, J. T. (2008). Provider judgments of patients in pain: Seeking symptom certainty. *Pain Medicine, 10*(1), 11–34.

Vrij, A. (2011). *Detecting lies and deceits: Pitfalls and opportunities* (2nd ed.). Oxford: Wiley.

Werner, A., & Malterud, K. (2003). It is hard work behaving as a credible patient: encounters between women with chronic pain and their doctors. *Social Science and Medicine, 57*(8), 1409–19.

Creep

Definition

Creep is a physiologic property of collagenous tissue. In response to the application of a constant force, collagen starts to deform (stretch). However, this deformation is not directly proportional to time. Initially, the deformation is small, but the longer the constant force is applied, the more rapidly collagen will deform (and weaken), even if the magnitude of the force remains constant. This phenomenon is defined as creep and may lead to complete failure of the collagen fiber.

Cross-References

▶ Ergonomic Counseling

Criterion-Referenced Assessment

Definition

In criterion-referenced assessment, the critical aspects of a job are analyzed to determine the standards essential for the worker's adequate performance of a job. The results are stated in terms of skills mastered, for example, Method Time Methods (MTM) (the time it would take for an average, well-trained worker to complete the work sample tasks if they were carried out over the 8-h work day in a typical industrial context), and the Worker Qualification Profiles given in Jist's Enhanced Dictionary of Occupational Titles.

Cross-References

▶ Vocational Counseling

Crohn's Disease

▶ Animal Models of Inflammatory Bowel Disease

Cross Excitation

Synonyms

Crossed afterdischarge; Ephaptic cross talk; Nonsynaptic functional coupling; Nonsynaptic neuron-to-neuron coupling; Volume conduction

Definition

An electrophysiological phenomenon in which impulse activity in a neuron, or a population of neurons, excites activity in a neighboring neuron or population of neurons in a manner that is not mediated by synaptic action. This may be mediated by the direct passage of electrical current among the coupled neurons ("ephaptic cross excitation"). Alternatively, it may be mediated by neurotransmitter molecules (or ions) that are released nonsynaptically into the extracellular space from the active neurons and then diffuse toward the neighboring passive neurons to excite them ("chemically mediated cross excitation" or "crossed afterdischarge"). Cross excitation occurs within sensory ganglia and at sites of nerve injury and may contribute to neuropathic pain.

Cross-References

▶ Pain Paroxysms

Crossed Afterdischarge

▶ Cross Excitation

Cross-Excitatory Discharge

Definition

See the definition for ▶ Cross Excitation.

Cross-References

▶ Ectopia, Spontaneous

Cross-Sectional Study

Definition

A research methodology that examines relationships between variables at one point in time.

Cross-References

▶ Pain in the Workplace, Risk Factors for Chronicity, Job Demands

Cross-System Viscero-Visceral or Viscero-Somatic Convergence

Definition

A connective situation in which neurons in the CNS receive converging input from different organs (viscero-visceral), sometimes in different functional systems (e.g., reproductive and urinary), and from different types of structures such as skin, muscle, and internal organs (viscero-somatic).

Cross-References

▶ Gynecological Pain, Neural Mechanisms

CRPS

▶ Complex Regional Pain Syndrome
▶ Complex Regional Pain Syndromes, Clinical Aspects
▶ Sympathetically Maintained Pain, Clinical Pharmacological Tests

CRPS (Currently Accepted Taxonomy)

▶ CRPS-1 in Children

CRPS I (Complex Regional Pain Syndrome Type I): Reflex Sympathetic Dystrophy

▶ Sympathetically Maintained Pain in CRPS I, Human Experimentation

CRPS II

▶ Sympathetically Maintained Pain in CRPS II, Human Experimentation

CRPS, Evidence-Based Treatment

Anne Louise Oaklander
Nerve Injury Unit, Departments of Neurology and Pathology, Massachusetts General Hospital Harvard Medical School, Boston, MA, USA

Synonyms

Algodystrophy (most common designation in Europe); Causalgia (from Greek, causos = heat; algia = pain, Oxford English Dictionary first use: S. Weir Mitchell 1872); Complex regional pain syndrome (term designated by International Association for the Study of Pain (IASP) in 1994); Major or minor causalgia; Postinfarctional sclerodactylia; Posttraumatic dystrophy; Pourfour du Petit syndrome; Reflex sympathetic dystrophy (RSD) (term first used by Evans in 1946); Shoulder-hand syndrome (refers specifically to CRPS originating in hand, spreading to shoulder); Sudeck's atrophy (1901); Traumatic angiospasm

Definition

Allodynia	is when pain is induced by a stimulus that does not pain in most instances.
Dependence	means that if a medication is abruptly discontinued, illness or discomfort will ensure. Although usually associated with opioids, this is common with many classes of medications including anti-hypertensives and anti-convulsants.
Hyperalgesia	is an excessive response to noxious stimulation.
Hyperhidrosis	means increased sweating. This is considered characteristic of CRPS as it is far more often noticed than hypohidrosis, which is often overlooked. Areas of hypohidrosis caused by nerve injuries can provoke compensatory hyperhidrosis in nearby skin; this can be mistakenly interpreted as the primary problem.
Neuralgia	is pain initiated or caused by a primary lesion or dysfunction in the nervous system.
Neuropathic pain	is pain initiated or caused by a primary lesion or dysfunction in the nervous system.

IASP Diagnostic Criteria for CRPS-I (Reflex Sympathetic Dystrophy)

1. The presence of an initiating noxious event or a cause of immobilization.

2. Continuing pain, ▶ allodynia, or ▶ hyperalgesia with which the pain is disproportionate to any inciting event.

3. Evidence at some time of edema, changes in skin blood flow, or abnormal sudomotor activity in the region of the pain.

4. This diagnosis is excluded by the existence of conditions that would otherwise account for the degree of pain and dysfunction.

Note: Criteria 2–4 must be satisfied (Merskey and Bogduk 1994)

IASP Diagnostic Criteria for CRPS-II (Causalgia)

1. The presence of continuing pain, allodynia, or hyperalgesia after a nerve injury, not necessarily limited to the distribution of the injured nerve.

2. Evidence at some time of edema, changes in skin blood flow, or abnormal sudomotor activity in the region of the pain.

3. This diagnosis is excluded by the existence of conditions that would otherwise account for the degree of pain and dysfunction.

Note: All three criteria must be satisfied (Merskey and Bogduk 1994)

Introduction

The biological causes of CRPS are only now being identified, and the definition, which lags behind the science, is overdue for revision. The definition is still phenotypic based on the presence of various symptoms, none of which are specific. The loose definition has led to diagnostic chaos, with the term CRPS sometimes indiscriminately applied to any ill-defined type of chronic pain. For research, many investigators create their own definitions, hampering meta-analysis. The International Association of Pain formulated consensus definitions in 1994 (see below) (Merskey and Bogduk 1994). "Type I" classifies patients without known nerve injury; "type II" classifies those with neural injuries. The recent discovery in at least three independent laboratories of nerve damage in patients categorized as CRPS-I essentially obviates any distinction. The dichotomy appears mostly to reflect how carefully the patient has been examined. Few CRPS patients are referred for a specialist's examination to look for nerve injuries, and even fewer are sent for electrodiagnostic studies, neurodiagnostic skin biopsy, or other tests that can detect the presence of nerve injuries. Although flawed, use of IASP criteria is encouraged until a better definition is developed.

More rigorous definitions have been proposed for research use. Most of these require visual confirmation of tissue edema or color change. This improves specificity at the cost of reduced sensitivity. Also, many patients report that their symptoms fluctuate, for instance, with edema of a dependent limb worsening with standing and improving after a night's recumbency. Similarly, color changes may be triggered by ambient temperature fluctuations. The concern about rigorous diagnostic criteria is that they may disproportionately favor inclusion of the most severely affected patients, who do not represent the majority of patients, and are least likely to improve in clinical trials. This author suggests adopting the hierarchy of definitions used for other neurologic research (definite CRPS, probable CRPS, possible CRPS) to permit greater or less sensitivity and specificity for use in individual research projects.

Characteristics

Complex regional pain syndrome (CRPS) is a "pain-plus" syndrome comprising regional pain plus additional symptoms. Most cases follow trauma, but the post-injury symptoms persist after apparent healing of injuries. Autonomic dysregulation (e.g., hyperhidrosis, vasomotor instability causing edema, abnormal skin color, or temperature) is part of the definition. Motor abnormalities are common and increasingly recognized. Some patients have mild or transient symptoms; others are chronically disabled. CRPS was first characterized in Weir Mitchell's astute descriptions of Civil War soldiers with *causalgia* after penetrating wounds. It is not known why patients develop the CRPS phenotype when countless others who sustain identical injuries do not. Modern epidemiological study has begun to identify the risk factors. The fundamental lesion in CRPS-I appears to involve specific peripheral axons: the small-diameter unmyelinated and thinly myelinated axons that subserve pain and some autonomic functions (van der Laan et al. 1998; Oaklander et al. 2000; Albrecht et al. 2006). These axons innervate the blood vessels of the

microcirculation, whose dysregulation is an integral part of CRPS (Oaklander and Fields 2009). In some cases, relative sparing of motor and large-fiber sensory axons preserves enough function to give patients a misleadingly normal appearance.

Treatments for Recent-Onset CRPS

Mild or very early cases may require only supportive efforts to mobilize the affected limb and to restore its function. Several trials evaluate treatments for CRPS symptoms present for less than 6 months. Inflammation and remodeling from the original trauma may still be present, and the final effects of injury on the CNS may not yet have occurred, leaving the possibility for treatments to modify disease course (e.g., provide permanent cure or improvement) rather than merely alleviate symptoms. It has not been examined whether these treatments are also effective for chronic CRPS, and their mechanisms are completely unknown. Two small unblinded trials (Christensen et al. 1982; Braus et al. 1994) support use of oral corticosteroids. Systematic meta-analysis (Perez et al. 2001) of several placebo-controlled randomized controlled trials (RCTs) of intranasal calcitonin 100–400 IU supports its use. Calcitonin is a neuropeptide with direct anti-hyperalgesic effects independent of effects on bone. Use of bisphosphonates, which also inhibit osteoclast-mediated bone resorption, is supported by several small RCT (Adami et al. 1997; Varenna et al. 2000). One study found pamidronate 30 mg daily for 3 days effective for 35 patients with acute or chronic CRPS (Maillefert et al. 1999). A single RCT supports efficacy of topical dimethyl sulfoxide in CRPS-I (Zuurmond et al. 1996) and vitamin C (Zollinger et al. 1999). The alpha-adrenergic antagonist phenoxybenzamine was effective in small trials in early CRPS (Muizelaar et al. 1997) and various ▶ neuralgias including CRPS-II (Ghostine et al. 1984). Several small studies suggest that topical application of the free-radical scavenger dimethyl sulfoxide (DMSO) can reduce pain (reviewed in (Perez et al. 2010)).

Conservative Treatments for Established CRPS

Rehabilitation is essential. In addition to preventing secondary problems including deconditioning, contractures, weight gain, and depression, functional brain imaging suggests that abnormal cortical plasticity related to disuse may be ameliorated by aggressive restoration of function (Maihöfner et al. 2004). A recent review opines that current physiotherapy guidelines should be interpreted cautiously as there is little evidence of efficacy (Daly and Bialocerkowski 2009). Only graded motor imagery is supported by good-quality evidence (Moseley 2006). The mechanisms are unknown but likely cortical and related to "neglect" phenomena more often associated with stroke. Although use of mirror therapy is intriguing, the quality of studies is also weak (Moseley et al. 2008). Psychotherapy and/or use of psychiatric medications are indicated in many patients who develop reactive depression or other psychological difficulties in response to their chronic pain and disability. However, the former view that CRPS is a primarily psychiatric syndrome has been discredited by the myriad of evidence of objective abnormalities in these patients. There is no evidence of efficacy for acupuncture, magnets, or other alternative treatments.

Medical Treatments for Established CRPS

Pharmacotherapy becomes the major treatment for patients who do not improve despite aggressive rehabilitation. Few pharmaceutical companies consider CRPS suitable for clinical trials, and so, there are few highly powered trials. Clinicians currently extrapolate from the results of well-designed and conclusive trials for other related conditions (primarily painful diabetic neuropathy and postherpetic neuralgia (PHN)). No medication has a US Food and Drug Administration indication for CRPS. Despite this, most chronic CRPS patients will take medications at some point, of necessity "off label."

For nerve-injury pain, randomized, controlled clinical trials have established efficacy and safety for four classes of medications. There are not enough data to support use of one class of medication over another, and information on efficacy much be tempered with consideration of possible adverse effects and cost. Standard principles of pharmacotherapy should be applied, namely, selection of medication based not only on the patient's symptoms but their overall medical, social, and financial situation. Each choice should be tried in an adequate dose and for long enough (1–2 months for tricyclics). If ineffective or ill tolerated, a medication should be tapered, discontinued, and replaced with the next choice. Multiple medications are indicated only when each provides added benefit and do not have adverse interactions. Usually, these medications will be from different classes. Polytherapy is common in CRPS because pain relief is often elusive. There are no good studies that evaluate efficacy of acetaminophen, paracetamol, or nonsteroidal anti-inflammatory drugs (Perez et al. 2010).

Emerging Treatment Options

Ketamine, a NMDA antagonist has achieved widespread interest for use in refractory CRPS. High doses (up to 100 mg for 4 h daily for 10 days) were reported efficacious for pain relief in a randomized double-blind placebo-controlled study (Schwartzman et al. 2009). Anesthetic doses given for 5 days were also efficacious in an open-label phase II study of 20 patients. Complete remission occurred at 1 month in all patients, at 3 months in 17 and at 6 months in 16 patients, although serious complications also ensued (Kiefer et al. 2008). Increasing evidence of autoimmunity in CRPS (Kohr et al. 2009) has prompted consideration of new types of therapies. A single well-designed randomized clinical trial finds modest efficacy for intravenous immunoglobulin (IVIG) in established refractory CRPS (Goebel et al. 2010). Given the high cost of treatment, the most important impact may be the additional evidence linking CRPS to autoimmunity.

Topical Local Anesthetics

The most important characteristic of medications applied to the skin, whose active ingredient remains in the skin, is their low rate of adverse effects. They are available in various forms including gels, creams, and sprays. The 5 % lidocaine patch is probably most commonly prescribed for CRPS. The patch itself protects allodynic skin from contact and the local anesthetics within reduce hyperalgesia. An RCT of 40 patients with focal ▶ neuropathic pain including CRPS documents efficacy (Meier et al. 2003), as does an uncontrolled open-label add-on study containing 6 patients with CRPS (Devers and Galer 2000).

Tricyclic Antidepressants (TCAs)

TCAs that augment noradrenergic neurotransmission are perhaps the most effective and best-studied treatments. Generic forms are available. TCAs have several mechanisms that contribute to efficacy, including potentiation of descending inhibitory pathway to decrease dorsal horn hyperactivity, mu opioid activity, and cation channel blockade. There are no studies specifically documenting efficacy of TCAs in CRPS, but efficacy in related conditions including diabetic neuropathy and PHN is better documented than for any other medications. TCAs that increase norepinephrine are effective; those with serotonergic-only effects are not, those that increase both may work best. Adverse effects often limit TCA use. These are more common with amitriptyline than for nortriptyline or desipramine. CRPS patients are often younger than other neuralgia patients and thus may tolerate TCAs better.

Antiepileptic Drugs (AEDs)

AEDs decrease central neuronal hyperexcitability to provide efficacy against pain as well as seizures. Gabapentin, well studied in PHN and diabetic neuropathy, is a primary option. It binds to the alpha2delta subunit of voltage-gated calcium channels and decreases calcium-mediated release of excitatory neurotransmitters. Gabapentin has no drug-drug interaction, and serum levels do not usually need to be monitored. Serious adverse effects appear to be less common than after use of TCAs, though minor ones may not. In CRPS, gabapentin was useful in an open-label uncontrolled study of nine adult patients given 600 mg/day (a lower dose than now used) (Mellick and Mellick 1997) and

in a single pediatric case (Wheeler et al. 2000). Carbamazepine, FDA-approved for trigeminal neuralgia, is supported by a single small uncontrolled study adding it to spinal cord stimulation (Harke et al. 2001).

Opioids

Reluctance to prescribe is common, though a number of CRPS patients do find benefit and are maintained on stable doses. Like other options, opioids help some but not all and usually provide only partial relief. Physical ▶ dependence or tolerance sometimes but not always develops and can usually be managed by avoiding sudden dose increases or decreases. There is increasing recognition that opioid doses should not be escalated indefinitely. A recent study reported a relationship between prescribed opioid dose and overdose events, with nine-fold increased risk of overdose at doses exceeding 100 mg/day morphine equivalent dose compared to doses below 20 mg/day in patients with chronic, noncancer pain (Dunn et al. 2010). There are inadequate data to recommend any particular opioid, although methadone is attractive because of very low cost. Direct evidence from CRPS patients is almost nil; rationale for use comes largely from well-designed studies in other neuralgic conditions. An add-on study of morphine in CRPS patients already treated with spinal cord stimulators did not demonstrate efficacy, but the study group was too unique to generalize its conclusions (Harke et al. 2001). Current or former substance abuse is the major contraindication to use, although opioids can sometimes be prescribed for abusers with severe pain if patients are rigorously monitored.

Other Medications to Consider

Systemic local anesthetics are sometimes efficacious but require invasive administration. Such use was supported by the results of a small RCT in 16 patients with CRPS-I and -II (Raja et al. 2002) and a case series (Linchitz and Raheb 1999). Administration must be carefully monitored to prevent seizures or cardiac arrest. The evidence supporting clonidine is weak. A case series of four CRPS patients screened by regional and systematic sympathetic blocks found that anti-hyperalgesic effect was restricted to skin directly under the clonidine patch (Davis et al. 1991). Thalidomide, which inhibits release of TNF-alpha, is under evaluation.

Surgical Treatments for Established CRPS
If medications have not provided significant relief after adequate trials, two surgical options deserve consideration, and one does not, namely, cutting nerves or neural structures transmitting pain sensation from the affected areas. Ablative neurosurgery has a simplistic appeal and occasionally provides a pain-free interval to terminal patients; it is not indicated for the bulk of CRPS patients who have largely preserved motor function and long life expectancy. Limb amputation is still occasionally performed and may very rarely be indicated for serious complications such as gangrene, but it should never be performed for pain control. CRPS pain is likely maintained by CNS abnormalities not addressed by amputation or nerve cutting.

In contrast, consideration should more often be given to whether the patient's CRPS might be caused by nerve entrapment or compression. Surgical decompression can provide a dramatic cure in this situation (Thimineur and Saberski 1996). Occasionally, unexpected causes of CRPS such as leprosy, tumors, or vascular malformations are diagnosable only by exploratory surgery. The single most effective treatment for severe established CRPS is implantation of bipolar neural stimulator. For properly selected patients, dramatic pain relief is possible with discontinuation of medications and functional gains. Experienced clinicians have noted that, unlike medications, stimulators can have disease-modifying effects, as documented by pain relief that persists for longer and longer periods after the stimulator is turned off. Similar properties have also been documented after CNS stimulation for other conditions. Stimulation of the dorsal column is the most common procedure because the electrode can be placed by epidural needle. Although device trials are difficult to control for, there is at least one well-designed study (Kemler et al. 2000) and several case series that demonstrate

efficacy in about half of implanted CRPS patients. Expense and not-infrequent complications are the major limitations. For patients whose CRPS is associated with damage to a single peripheral nerve, stimulation of the injured nerve (or - proximal to the site of damage, using surgical incision) or - offers even higher probability of significant pain relief, about 80 % in some series (Campbell and Long 1976). For both types of stimulators, a trial of temporary implantation is customary.

References

Adami, S., Fossaluzza, V., Gatti, D., Fracassi, E., & Braga, V. (1997). Bisphosphonate therapy of reflex sympathetic dystrophy syndrome. *Annals of the Rheumatic Diseases, 56*(3), 201–204.

Albrecht, P. J., Hines, S., Eisenberg, E., Pud, D., Finlay, D. R., Connolly, M. K., et al. (2006). Pathologic alterations of cutaneous innervation and vasculature in affected limbs from patients with complex regional pain syndrome. *Pain, 120*(3), 244–266.

Braus, D. F., Krauss, J. K., & Strobel, J. (1994). The shoulder-hand syndrome after stroke: A prospective clinical trial. *Annals of Neurology, 36*(5), 728–733.

Campbell, J. N., & Long, D. M. (1976). Peripheral nerve stimulation in the treatment of intractable pain. *Journal of Neurosurgery, 45*, 692–699.

Christensen, K., Jensen, E. M., & Noer, I. (1982). The reflex dystrophy syndrome response to treatment with systemic corticosteroids. *Acta Chirurgica Scandinavica, 148*(8), 653–655.

Daly, A. E., & Bialocerkowski, A. E. (2009). Does evidence support physiotherapy management of adult Complex Regional Pain Syndrome Type One? A systematic review. *European Journal of Pain, 13*(4), 339–353.

Davis, K. D., Treede, R. D., Raja, S. N., Meyer, R. A., & Campbell, J. N. (1991). Topical application of clonidine relieves hyperalgesia in patients with sympathetically maintained pain. *Pain, 47*(3), 309–317.

Devers, A., & Galer, B. S. (2000). Topical lidocaine patch relieves a variety of neuropathic pain conditions: An open-label study. *The Clinical Journal of Pain, 16*(3), 205–208.

Dunn, K. M., Saunders, K. W., Rutter, C. M., Banta-Green, C. J., Merrill, J. O., Sullivan, M. D., et al. (2010). Opioid prescriptions for chronic pain and overdose: A cohort study. *Annals of Internal Medicine, 152*(2), 85–92.

Ghostine, S. Y., Comair, Y. G., Turner, D. M., Kassell, N. F., & Azar, C. G. (1984). Phenoxybenzamine in the treatment of causalgia. Report of 40 cases. *Journal of Neurosurgery, 60*(6), 1263–1268.

Goebel, A., Baranowski, A., Maurer, K., Ghiai, A., McCabe, C., & Ambler, G. (2010). Intravenous immunoglobulin treatment of the complex regional pain syndrome: A randomized trial. *Annals of Internal Medicine, 152*(3), 152–158.

Harke, H., Gretenkort, P., Ladleif, H. U., Rahman, S., & Harke, O. (2001). The response of neuropathic pain and pain in complex regional pain syndrome I to carbamazepine and sustained-release morphine in patients pretreated with spinal cord stimulation: A double-blinded randomized study. *Anesthesia and Analgesia, 92*(2), 488–495.

Kemler, M. A., Barendse, G. A., van Kleef, M., de Vet, H. C., Rijks, C. P., Furnee, C. A., et al. (2000). Spinal cord stimulation in patients with chronic reflex sympathetic dystrophy. *The New England Journal of Medicine, 343*, 618–624.

Kiefer, R. T., Rohr, P., Ploppa, A., Dieterich, H. J., Grothusen, J., Koffler, S., et al. (2008). Efficacy of ketamine in anesthetic dosage for the treatment of refractory complex regional pain syndrome: An open-label phase II study. *Pain Medicine, 9*(8), 1173–1201.

Kohr, D., Tschernatsch, M., Schmitz, K., Singh, P., Kaps, M., Schafer, K. H., et al. (2009). Autoantibodies in complex regional pain syndrome bind to a differentiation-dependent neuronal surface autoantigen. *Pain, 143*(3), 246–251.

Linchitz, R. M., & Raheb, J. C. (1999). Subcutaneous infusion of lidocaine provides effective pain relief for CRPS patients. *The Clinical Journal of Pain, 15*(1), 67–72.

Maihöfner, C., Handwerker, H. O., Neundorfer, B., & Birklein, F. (2004). Cortical reorganization during recovery from complex regional pain syndrome. *Neurology, 63*(4), 693–701.

Maillefert, J. F., Cortet, B., & Aho, S. (1999). Pooled results from 2 trials evaluating biphosphonates in reflex sympathetic dystrophy. *Journal of Rheumatology, 26*(8), 1856–1857.

Meier, T., Wasner, G., Faust, M., Kuntzer, T., Ochsner, F., Hueppe, M., et al. (2003). Efficacy of lidocaine patch 5 % in the treatment of focal peripheral neuropathic pain syndromes: A randomized, double-blind, placebo-controlled study. *Pain, 106*(1–2), 151–158.

Mellick, G. A., & Mellick, L. B. (1997). Reflex sympathetic dystrophy treated with gabapentin. *Archives of Physical Medicine and Rehabilitation, 78*(1), 98–105.

Merskey, H., & Bogduk, N. (1994). *Classification of chronic pain: Descriptions of chronic pain syndromes and definitions of pain terms* (2nd ed.). Seattle: IASP Press.

Moseley, G. L. (2006). Graded motor imagery for pathologic pain: A randomized controlled trial. *Neurology, 67*(12), 2129–2134.

Moseley, G. L., Gallace, A., & Spence, C. (2008). Is mirror therapy all it is cracked up to be? Current evidence and future directions. *Pain, 138*(1), 7–10.

Muizelaar, J. P., Kleyer, M., Hertogs, I. A., & DeLange, D. C. (1997). Complex regional pain syndrome (reflex sympathetic dystrophy and causalgia): Management with the calcium channel blocker nifedipine and/or the alpha-sympathetic blocker phenoxybenzamine in 59 patients. *Clinical Neurology and Neurosurgery, 99*(1), 26–30.

Oaklander, A. L., & Fields, H. L. (2009). Is reflex sympathetic dystrophy/complex regional pain syndrome type I a small-fiber neuropathy? *Annals of Neurology, 65*(6), 629–638.

Oaklander, A. L., Rissmiller, J. G., & Yang, Y. (2000). Skin biopsies provide objective evidence of injury to nociceptors in patients with Complex Regional Pain Syndrome (Reflex Sympathetic Dystrophy) (abstract). *Neurology, 48*, 430A.

Perez, R. S., Kwakkel, G., Zuurmond, W. W., & de Lange, J. J. (2001). Treatment of reflex sympathetic dystrophy (CRPS type 1): A research synthesis of 21 randomized clinical trials. *Journal of Pain and Symptom Management, 21*(6), 511–526.

Perez, R., Zollinger, P., Dijkstra, P., Thomassen-Hilgersom, I., Zuurmond, W., Rosenbrand, K., et al. (2010). Evidence based guidelines for complex regional pain syndrome type 1. *BMC Neurology, 10*(1), 20.

Raja, S. N., Haythornthwaite, J. A., Pappagallo, M., Clark, M. R., Travison, T. G., Sabeen, S., et al. (2002). Opioids versus antidepressants in postherpetic neuralgia: A randomized, placebo-controlled trial. *Neurology, 59*(7), 1015–1021.

Schwartzman, R. J., Alexander, G. M., Grothusen, J. R., Paylor, T., Reichenberger, E., & Perreault, M. (2009). Outpatient intravenous ketamine for the treatment of complex regional pain syndrome: A double-blind placebo controlled study. *Pain, 147*, 107–115.

Thimineur, M. A., & Saberski, L. (1996). Complex regional pain sydrome type I (RSD) or peripheral mononeuropathy: A discussion of three cases. *The Clinical Journal of Pain, 12*, 145–150.

van der Laan, L., ter Laak, H. J., Gabreels-Festen, A., Gabreels, F., & Goris, R. J. (1998). Complex regional pain syndrome type I (RSD): Pathology of skeletal muscle and peripheral nerve. *Neurology, 51*(1), 20–25.

Varenna, M., Zucchi, F., Ghiringhelli, D., Binelli, L., Bevilacqua, M., Bettica, P., et al. (2000). Intravenous clodronate in the treatment of reflex sympathetic dystrophy syndrome. A randomized, double blind, placebo controlled study. *Journal of Rheumatology, 27*(6), 1477–1483.

Wheeler, D. S., Vaux, K. K., & Tam, D. A. (2000). Use of gabapentin in the treatment of childhood reflex sympathetic dystrophy. *Pediatric Neurology, 22*(3), 220–221.

Zollinger, P. E., Tuinebreijer, W. E., Kreis, R. W., & Breederveld, R. S. (1999). Effect of vitamin C on frequency of reflex sympathetic dystrophy in wrist fractures: A randomised trial. *Lancet, 354*(9195), 2025–2028.

Zuurmond, W. W., Langendijk, P. N., Bezemer, P. D., Brink, H. E., de Lange, J. J., & van Loenen, A. C. (1996). Treatment of acute reflex sympathetic dystrophy with DMSO 50 % in a fatty cream. *Acta Anaesthesiologica Scandinavica, 40*(3), 364–367.

CRPS-1 in Children

Elliot Krane
Anesthesia and Pediatrics, Stanford University School of Medicine Lucile Packard Children's Hospital at Stanford, Stanford, CA, USA

Synonyms

Complex Regional Pain Syndrome Type 1; CRPS (currently accepted taxonomy); Reflex Neurovascular Dystrophy (unconventional taxonomy used by a few clinicians); Reflex Sympathetic Dystrophy; RSD (obsolete taxonomy); Suddick's Atrophy (historical name)

Definition

Well-defined diagnostic criteria for ▶ CRPS-1 were established in 1995 (Stanton-Hicks et al. 1995), but the criteria for diagnosis are evolving and are periodically revised. Stanton-Hicks defined CRPS-1 as follows:

- Preceded by an eliciting event, but with no demonstrable injury to a peripheral nerve
- Pain, spontaneous or evoked, often with evidence of ▶ allodynia in a limb
- Signs of ▶ dysautonomia
- No alternative explanations for the pain

More recent and expanded well-validated criteria (the Budapest Criteria), for the diagnosis of CRPS were developed in an IASP consensus conference, are as follows (Harden et al. 2010):

- Continuing pain, which is disproportionate to any inciting event
- Must report at least one symptom in three of the four following categories:

- Sensory: reports of hyperesthesia and/or allodynia
- Vasomotor: reports of temperature asymmetry and/or skin color changes and/or skin color asymmetry
- Sudomotor/edema: reports of edema and/or sweating changes and/or sweating asymmetry
- Motor/trophic: reports of decreased range of motion and/or motor dysfunction (weakness, tremor, dystonia) and/or trophic changes (hair, nail, skin)

- Must display at least one sign at time of evaluation in two or more of the following categories:

- Sensory: evidence of hyperalgesia (to pinprick) and/or allodynia (to light touch and/or deep somatic pressure and/or joint movement)
- Vasomotor: evidence of temperature asymmetry and/or skin color changes and/or asymmetry
- Sudomotor/edema: evidence of edema and/or sweating changes and/or sweating asymmetry
- Motor/trophic: evidence of decreased range of motion and/or motor dysfunction (weakness, tremor, dystonia) and/or trophic changes (hair, nail, skin)

- There is no other diagnosis that better explains the signs and symptoms

The Budapest Criteria were developed and validated for adults with CRPS-1 and have been found to have good diagnostic specificity and sensitivity; however, there are no studies that have defined diagnostic criteria for CRPS-1 in the pediatric population. Hence, clinicians caring for children must extrapolate criteria from these adult case series.

Introduction

As in adults, CRPS-1 occurring during childhood is seen more frequently in females than in males, but unlike the case in adults, it involves the lower extremity more frequently than the upper. The most common age of onset is 10–14 years of age, and while reported, it is very infrequently, indeed rarely seen before the age of 8.

In both children and adults, the cause of CRPS-1 is unknown. Throughout the twentieth century, as the nomenclature RSD suggested, the primary abnormality was thought to reside in the function of the sympathetic nervous system, and chemical or surgical sympathectomy was believed to be the primary treatment (Coderre 2011). Today our understanding of the pathophysiology remains incomplete, but it is more sophisticated. Current hypotheses invoke subtle damage to small peripheral nerves during the antecedent injury or event; limb disuse and immobilization; peripheral neuroinflammation; central neuroinflammation; and spinal cord glial cell and astrocyte activation in the etiology, development, and maintenance of the chronic pain state (Bruehl 2010).

Whatever the etiology, it is curious that CRPS-1 is seldom seen in early childhood before the onset of pre-adolescence and the occurrence of early secondary sexual development, suggesting that biological changes are occurring in the peripheral and/or central nervous systems, perhaps triggered by hormonal changes, and that coincident with these developmental changes are those that establish the neurobiology permissive of the creation and maintenance of this neuropathic pain state. Furthermore, the very strong female to male predominance in both children and adults also implies a hormonal role in the neurobiology of CRPS-1.

An association of certain psychological profiles and behaviors, as well as antecedent psychosocial stressors, has been suggested in the past to play a causative role in the development of CRPS-1. Whether these factors are etiologic, represent true comorbidities, or are simply phenomena that establish a behavioral baseline that furthers the disability behaviors remains a topic of discussion. For example, it is frequently observed that children with CRPS-1 are overachievers, however this behavioral pattern may simply lead children to being more prone to sports injuries, therefore creating the opportunity for CRPS to occur, rather than biologically contributing to it. Nevertheless, and unlike observations made in adult patients, the coincidence of psychopathology is higher (87 %) in the childhood population with (severe) CRPS-1 than would be expected in other chronic severe pain populations (Krane et al. 2009), and is higher than the incidence of psychopathology associated with CRPS in adults (Beerthuizen et al. 2009).

Characteristics

Demographics

As is the case of adults, a large majority of children with CRPS are female, with a preponderance of greater than 8:1 (Sherry et al. 1999; Logan et al. 2012; Wilder et al. 1992). The reason for the female predominance in this disease is unknown at this time. The average age at presentation is in the early teen years, and typically the youngest patients seen in a busy pediatric pain management center are no younger than 8 years of age. There is one case report of CRPS-1 in a 3-year-old boy (Bukhalo and Mullin 2004), but a careful reading of this case report shows that this child did not meet today's established criteria for CRPS, and that his presentation was in most ways atypical of CRPS (MRI evidence of a joint effusion, warm and red extremity, and rapid complete resolution of symptoms in 2 weeks with nonsteroidal anti-inflammatory medication and TENS, neither of which would be expected to be of value in CRPS). Therefore, in spite of this case report, it is probably correct to say that rarely, if ever, does CRPS-1 present in preschool-age children, and is most common in the teen years.

CRPS is also less common, in our experience and that of other pediatric pain centers, in non-Caucasian children.

CRPS-1 typically, but not exclusively, affects the lower extremities in children. This probably does not represent unique pathophysiology as much as it is a reflection that CRPS typically follows an otherwise trivial or moderate injury, and that athletically active children injure their feet and ankles more frequently than they do their upper extremities, whereas falls on outstretched arms and wrist fractures are more common in adults who are not engaged in athletic activities. As is the case in adults (Maleki et al. 2000), some children with CRPS-1 may have their neuropathic pain syndrome jump to other previously unaffected and uninjured limbs, or, may have the CRPS spread proximally from the distal extremity to the proximal extremity, and even to the shoulder or hip girdle and from there to their spine. The propensity of some patients to generalize their CRPS to distant or to contiguous body areas has been associated with a unique genotype.

Characteristics and Physical Findings

There are no formal diagnostic criteria for CRPS-1 specific to childhood presentation. The IARS criteria (Budapest Criteria, Harden et al. 2010) are generally applicable to children, taking into account that in clinical experience, children differ from adults in the following features (Stanton-Hicks 2010):

1. Children generally come to medical attention earlier and have had established CRPS for shorter durations of time; therefore, trophic skin and nail changes, and joint contractures are decidedly less common in the pediatric age range.
2. Autonomic changes are more protean: in the adult, limbs are generally cyanotic and cold, but in children blood flow is more variable and indeed may change during the course of the day, from hyperemic, to mottled, and to frank cyanosis.
3. Motor dysfunction and dystonias seem to be less common in children, but do indeed occur.

Autonomic Changes

The limb affected by CRPS invariably shows alteration in blood flow, although the clinician may not observe such in the clinic, the child and parent will report pallor, cyanosis, or mottling associated with coldness of the limb during the course of almost every day. In its earliest stages, the dysautonomia of CRPS may manifest as a predominance of hyperemia and warmth, confusing the appearance of the limb with one afflicted by an inflammatory disorder, but this phase is usually short lived. Triple phase bone scans of a pediatric limb with CRPS typically shows decreased blood flow compared with the opposite normal limb, and occasionally the radiologic report of a bone scan will describe increased blood flow of the contralateral, normal limb, when in fact the correct interpretation is hypoperfusion of the affected limb. The clinical and prognostic significance of autonomic changes is unknown, but suggests an indication for a trial sympathectomy to produce analgesia.

Trophic Changes

Most children with CRPS come to appropriate medical attention within 6 months of onset of their pain syndromes, at which time trophic changes are not common, with the exception of disuse muscle atrophy and increased hair growth. Over time, however, one may see thickening and drying of the skin and nails, but rarely does one see advanced CRPS-1 in children with associated atrophic, shiny skin.

Figure 1 illustrates the typical appearance of CRPS-1 of 4 months' duration in a 13-year-old girl, demonstrating cyanosis and edema, while Fig. 2 illustrates longer standing CRPS-1 of 9 months' duration in a 16-year-old girl, demonstrating more severe cyanosis and edema, as well as thickening of the skin of the toes of the affected limb.

Sensory Alterations

The pain of CRPS does not adhere to a dermatomal or anatomic distribution but is usually stocking-glove in pattern. As in adults, CRPS-1 produces severe spontaneous and movement-induced pain that is out of proportion to the inciting injury. The most distressing aspect of CRPS-1 pain is light touch and thermal allodynia, especially cold allodynia. Occasionally, CRPS-1 is associated with areas of anesthesia and dysesthesia, but when these patterns exist it raises the question of whether there is an underlying peripheral nerve injury, and the CRPS is better categorized as Type 2 (anesthesia dolorosa).

Motor Function

As in adult CRPS, pediatric CRPS-1 is frequently accompanied by motor dysfunction best

CRPS-1 in Children, Fig. 1 CRPS-1 of the right foot and leg in a 13-year-old girl

CRPS-1 in Children, Fig. 2 Severe CRPS-1 of the left foot in a 16-year-old girl

categorized as dystonia. Less frequently, there may be gross motor tremors either at rest or with movement. It is unclear of the inability to move the digits or large joints of a limb affected by CRPS is a learned behavior due to the accentuation of pain, or whether there is truly a motor neuropathy. The fact that EMG and NCV testing is typically normal in CRPS suggests a behavioral, though subconscious behavioral, component if not etiology. In some cases of adult CRPS-1, and infrequently in children with CRPS-1, the motor function may be manifested as a dystonia with tonic contracture of the flexors of the foot or hand, or arm. These dystonias typically persist during sleep, suggesting a nonbehavioral etiology.

Psychological Characteristics

Contrary to conventional wisdom, there is no typical personality profile of children with CRPS-1. However, there is a clinical impression

that anxiety disorders are particularly common, as well as other DSM-IV diagnoses, and this impression was borne out by a case series from our practice that was reported at the 9th International Symposium on Pediatric Pain in 2009. In that sample of children, who failed outpatient management of CRPS and were hospitalized for intensive rehabilitation supported by continuous nerve blocks, DSM-IV psychiatric diagnoses were identified in 78 % of children, with most of these disorders predating the onset of CRPS (Krane et al. 2009). Diagnoses included anxiety disorders, ADHD, clinical depression, and obsessive-compulsive disorder. In adults with CRPS-1, there seems not to be an association with primary psychopathology (Beerthuizen et al. 2009).

Diagnosis

CRPS-1 remains a clinical diagnosis, and the Budapest criteria seem to apply equally well to children as to adults (Harden et al. 2010;

Stanton-Hicks 2010). Clinical laboratory testing and diagnostic imaging are of no value, except as far as ruling out other possible diagnoses in ambiguous situations, such as inflammatory and rheumatologic disorders, fractures, etc. While at one time triple phase bone scans were thought to be useful (Kozin et al. 1977), particularly for staging CRPS-1, the utility of staging the disease has now been dismissed, and bone scans are recognized to be highly variable and nonspecific in their appearance (Sarikaya et al. 2001).

Treatment

There are no randomized clinical trials of treatment of CRPS-1 in children, and relatively few case series or case reports to constitute an evidence base for treatment. The current state of treatment is based largely upon what seems to be a reasonable extrapolation from the growing treatment evidence base in adults (Baron et al. 2002), and expert opinion of pediatric pain management experts (Baron et al. 2002; Perez et al. 2001; Stanton-Hicks 2010; Wilder 2006).

Physical and Occupational Therapy

There is a broad consensus among pediatric pain management physicians that the mainstay of treatment of CRPS-1 is physical and occupational therapy directed at limb remobilization, strengthening, and progressive desensitization, and a large experience in some centers of success using only these modalities to the exclusion of medication and nerve blocks (Sherry et al. 1999), whereas there is a large clinical experience that medication and/or nerve blocks alone are rarely, if ever, curative. However, physical therapy alone is most often too painful and stressful for children to endure without amelioration of pain by analgesic medication or nerve blocks with supportive psychotherapy (see below) (Brooke and Janselewitz 2012; Lee et al. 2002; Logan et al. 2012). Mirror box therapy has been described in the management of CRPS in adults, but there is no experience using this modality in children.

Medication Management

Conventional analgesics, consisting of acetaminophen, nonsteroidal anti-inflammatory drugs, and opioids, have little or no use in the management of the pain of CRPS-1. They are generally quite ineffective. Unconventional analgesics (antiepileptic drugs, AEDs, and tricyclic antidepressants, TCAs) are frequently used for treating a variety of neuropathic pain conditions, including CRPS-1, and there is a long track record of their safe use in children for other indications, although they are associated with many adverse effects, and there are no randomized controlled trials determining efficacy or safety in CRPS-1. Gabapentin is the most widely prescribed AED, typically starting at low dose and titrating over weeks to approximately 30 mg/kg/day in three divided doses. Gabapentin is associated with emotional lability and worsening depressive symptoms, suicidal thinking, ataxia, dizziness, urinary retention, and cognitive dysfunction. Other AEDs used, particularly when gabapentin is poorly tolerated, are pregabalin and topiramate. The most frequently prescribed TCA is amitriptyline, with a target dose of 0.5–1 mg/kg administered at bedtime. TCAs are associated with anticholinergic side effects (drowsiness, dry mouth, tachycardia, urinary retention) as well as postural hypotension and prolongation of the QT interval of the electrocardiogram. Nortriptyline and desipramine have better side effect profiles when amitriptyline is poorly tolerated. TCAs, as all antidepressants, may paradoxically worsen depression and increase suicidal risk during the first month or two of therapy. The selective serotonin and norepinephrine reuptake inhibitors (SSNRIs) duloxetine and venlafaxine have analgesic and antidepressant properties and are useful in cases of CRPS-1 in adolescents who have depression.

Low–dose intravenous ketamine administration has been reported to be effective in reversing refractory CRPS-1 in adults (Schwartzman et al. 2009). There are no published cases of its use in the pediatric age group, but a growing anecdotal experience demonstrating its utility. Its use in children is therefore not established and its use should be considered to be conjectural at this time.

Interventional Procedures

The role and safety of interventional procedures (Nelson and Brett 2006) have not been well established in children, and their use has been described in the literature only in single case reports or small case series. Sympathetic nerve blocks have been conventionally used in the management of CRPS-1 in adults and their use has been extrapolated to children without an evidence base. Indeed, the evidence base for their use in adults is weak; a Cochrane review found scant evidence of short-term efficacy and no evidence of long-term efficacy (Cepeda et al. 2005). These blocks include the stellate ganglion block for CRPS of the upper extremity, lumbar sympathetic blocks and epidural blockade for the lower extremity, and intravenous regional anesthesia (IVRA) utilizing an alpha-adrenergic antagonist for both upper and lower extremities. Typically, sympathetic nerve blocks are reserved for patients with significant autonomic overactivity in the affected limb (cyanosis and coldness), and provide transient relief, if any, measured in days. They are not curative, but may permit the institution of physical therapy. In adults, sympathetic blocks are often repeated at intervals of several days many times, but this is less practical in children who will generally require either significant pharmacologic sedation or general anesthesia to perform the block, thus increasing risk and cost.

More recently, especially since the introduction of ultrasound to facilitate regional nerve blocks, continuous peripheral nerve blocks have been used with good effect to permit the institution of physical therapy (Dadure et al. 2005; Carayannopoulos et al. 2009). They now represent the most common interventional treatment modality to facilitate rehabilitation.

Psychological Interventions

In addition to a psychotherapist addressing any underlying and comorbid psychopathology, individual psychotherapy to improve coping strategies and to reduce perceptions of pain is used and thought to be a valuable adjunct to physical therapy, though there are no randomized controlled studies in children (Brooke and Janselewitz 2012; Lee et al. 2012; Wilder 2006). Appropriate modalities include cognitive behavioral therapy, hypnosis, and biofeedback.

Neuromodulation

Neuromodulation refers specifically to either spinal cord stimulation (SCS) or continuous administration of intrathecal opioids via implanted pump. The use of SCS is generally reserved for those patients refractory to less invasive and expensive treatment for CRPS-1. While often utilized, and known to provide short-term relief of spontaneous pain and allodynia, the evidence for their long-term efficacy is the subject of debate (Kemler et al. 2006). The utility of SCS, its long-term effects, particularly on the developing nervous system, and its efficacy is still less defined in children. Nevertheless, clinicians are faced from time to time with refractory cases of CRPS that defy all current management strategies, and in these "dead-end" situations, few alternatives exist to a trial of neuromodulation, knowing the benefit may not be sustained. Olsson et al. (2008) published the only series of this modality in children, reporting a group of 7 girls aged 11–14, in which 5 of the 7 were reported to have substantial or complete pain relief of a duration of 1–8 years. This very positive response rate exceeds that which would be expected in the adult population. Confirmation of SCS efficacy and benefit will be required before its routine use in children, but the Olsson report is encouraging.

Controversies

A few international centers, principally in Germany and Mexico, advocate prolonged multiday high-dose (anesthetic level) ketamine, the so-called ketamine coma, in the management of severe CRPS. This practice carries significant risk and requires prolonged endotracheal intubation and nursing care in an intensive care environment for several days to weeks. It is not recognized as a legitimate therapeutic option at this time, and in the view of the author should never be considered for children until studies demonstrate efficacy and safety in the adult population. Such studies have not been done nor have case series been published in the peer-reviewed literature.

Cases of children "total body CRPS" occasionally present to pain clinics. Patients with

CRPS-1 in Children, Table 1 Outcomes of children treated in an intensive inpatient program with nerve blocks, psychotherapy, and physical rehabilitation

Gender	n	Episodes	Age	CRPS duration (months)	Outcome	Recurrences
Male	9	11	12.7	9	Good: 8	2
					Partial: 3	
					Poor: 0	
Female	14	24	13.3	12	Good: 12	10*
					Partial: 5	
					Poor: 7*	

Outcome key: Good = No pain + Full function; Partial = Pain + Full function; Poor = Pain + Disability; * signifies $p < 0.05$ male vs. female

this diagnosis report generalized total body allodynia and profound disability. This diagnosis is usually self-assigned by patients or families based upon Internet descriptions of this entity, and the entity does not meet current diagnostic criteria for CRPS and is not a recognized diagnosis by any of the governing pain management societies.

Prognosis and Outcomes

There is a perception among pediatric pain management physicians that CRPS-1 in children is more amenable to therapy and carries a better long-term prognosis than in adults (Bruehl 2009), but the data suggest that indeed there is a high rate of refractory pain, and recurrent CRPS, leading to persistent disabilities. Olsson et al. (1990) reported that 25 % of children persisted in pain 5 years after diagnosis and treatment, Murray et al. (2000) reported a relapse rate of 27 % in children, and Wilder et al. (1992) reported 50 % of 70 children with CRPS reported persistent pain on long-term follow-up. We described the outcome of 35 episodes of inpatient CRPS-1 treatment in 23 children with a mean age of 13, and found that there was a statistically worse outcome in girls, who had 12 of 24 partial or poor outcomes and 10 recurrences in 14 patients, than in boys, who had only 2 of 9 poor outcomes and 2 recurrences in 9 patients (Table 1; Krane et al. 2009).

Therefore, aside from gender, at this time there are no known risk factors for the spread of CRPS symptoms to contiguous or remote body parts, or that may identify those children at risk for recurrence, and therefore we have no information on how we may prevent recurrences.

Summary

CRPS-1 is the most common neuropathic pain disorder in children. As the case in adults, the etiology of the disease is not well defined, but clues to its biology may be found in the fact that rarely, if ever, does CRPS-1 occur before the development of secondary sexual characteristics, and in the strong female preponderance of cases. Presently the Budapest Criteria, defined for the uniform diagnosis of CRPS-1 in adults, is generally applicable to children as well.

The evidence base for management of childhood CRPS-1 is poor, with no high quality randomized controlled trials to guide therapy. However, there is a broad consensus that the mainstay of therapy is physical rehabilitation and reactivation, supported by active psychotherapy, both to improve the child's coping mechanisms, but also to address underlying psychopathology, which is not uncommon in children with CRPS-1. There is probably a role for medication management and interventional nerve blocks, but only inasmuch as they support the child's ability to participate in physical and occupational therapy.

The prognosis of childhood CRPS-1 is favorable in terms of functional recovery; however, persistent pain is not uncommon, and recurrence in the same or other limbs often occurs, particularly in females. The management of refractory and disseminated CRPS-1 is not established, but may include neuromodulation, trials of low–dose continuous ketamine infusions, or both.

References

Baron, R., Fields, H. L., Jänig, W., Kitt, C., & Levine, J. D. (2002). National Institutes of Health Workshop: reflex sympathetic dystrophy/complex regional pain syndromes-state-of-the-science. *Anesth. Analg., 95,* 1812–1816.

Beerthuizen, A., van 't Spijker, A., Huygen, F. J., Klein, J., & de Wit, R. (2009). Is there an association between psychological factors and the Complex Regional Pain Syndrome type 1 (CRPS1) in adults? A systematic review. *Pain, 145,* 52–59.

Brooke, V., & Janselewitz, S. (2012). Outcomes of children with complex regional pain syndrome after intensive inpatient rehabilitation. *PM&R, 4,* 349–354.

Bruehl, S. (2009). Complex regional pain syndrome: outcomes and subtypes. *Clin. J. Pain, 25,* 598–599.

Bruehl, S. (2010). An update on the pathophysiology of complex regional pain syndrome. *Anesthesiology, 113* (713–725), 710.

Bukhalo, Y., & Mullin, V. (2004). Presentation and treatment of complex regional pain syndrome type 1 in a 3 year old. *Anesthesiology, 101,* 542–543.

Carayannopoulos, A. G., Cravero, J. P., Stinson, M. T., & Sites, B. D. (2009). Use of regional blockade to facilitate inpatient rehabilitation of recalcitrant complex regional pain syndrome. *PM R, 1,* 194–198.

Cepeda, M.S., Carr, D.B., & Lau, J. (2005). Local anesthetic sympathetic blockade for complex regional pain syndrome. Cochrane Database Syst Rev CD004598.

Coderre, T. J. (2011). Complex regional pain syndrome: What's in a name? *J. Pain, 12,* 2–12.

Dadure, C., Motais, F., Ricard, C., Raux, O., Troncin, R., & Capdevila, X. (2005). Continuous peripheral nerve blocks at home for treatment of recurrent complex regional pain syndrome I in children. *Anesthesiology, 102,* 387–391.

Harden, R. N., Bruehl, S., Perez, R. S. G. M., Birklein, F., Marinus, J., Maihofner, C., & Vatine, J. J. (2010). Validation of proposed diagnostic criteria (The Budapest Criteria) for complex regional pain syndrome. *Pain, 150,* 268–274.

Kemler, M. A., de Vet, H. C., Barendse, G. A., van den Wildenberg, F. A., & van Kleef, M. (2006). Spinal cord stimulation for chronic reflex sympathetic dystrophy–five-year follow-up. *N. Engl. J. Med., 354,* 2394–2396.

Kozin, F., Haughton, V., & Ryan, L. (1977). The reflex sympathetic dystrophy syndrome in a child. *J. Pediatr., 90,* 417–419.

Krane, E., Golianu, B., Larkin, K., Almgren, C., Sentivany-Collins, S., Leon Guerrero, A., & Bhandrari, R. (2009). Inpatient interventional management of complex regional pain syndromes in children and adolescents. Presented at the 8th International Symposium on Pediatric Pain, Acapulco, Mexico.

Lee, B. H., Scharff, L., Sethna, N. F., McCarthy, C. F., Scott-Sutherland, J., Shea, A. M., & Berde, C. B. (2002). Physical therapy and cognitive-behavioral treatment for complex regional pain syndromes. *J. Pediatr., 141,* 135–140.

Logan, D. E., Capino, E. A., cCHiang, G., Condon, N., Firn, E., Gaughan, V. J., Berde, C. B. (2012). A day-hospital approach to treatment of pediatric complex regional pain syndrome: initial functional outcomes. Clin J Pain, in press.

Maleki, J., LeBel, A. A., Bennett, G. J., & Schwartzman, R. J. (2000). Patterns of spread in complex regional pain syndrome, type I (reflex sympathetic dystrophy). *Pain, 88,* 259–266.

Murray, C. S., Cohen, A., Perkins, T., Davidson, J. E., & Sills, J. A. (2000). Morbidity in reflex sympathetic dystrophy. *Arch. Dis. Child., 82,* 231–233.

Nelson, D. V., & Brett, B. R. (2006). Interventional therapies in the management of complex regional pain syndrome. [Interventional therapies in the management of complex regional pain syndrome]. *Clin. J. Pain, 22,* 438–442.

Olsson, G. L., Meyerson, B. A., & Linderoth, B. (2008). Spinal cord stimulation in adolescents with complex regional pain syndrome type I (CRPS-I). *Eur. J. Pain, 12,* 53–59.

Olsson, G. L., Arner, S., & Hirtsch, G. (1990). Reflex sympathetic dystrophy in children. In: Tyler, D. C., Krane, E. J. (eds), Advances in Pain Research Therapy. New York: Raven Press, pp 323–331.

Perez, R. S. G. M., Kwakkel, G., Zuurmond, W. W. A., & de Lange, J. J. (2001). Treatment of reflex sympathetic dystrophy (CRPS Type 1): A research synthesis of 21 randomized clinical trials. *J. Pain Symptom Manage., 21,* 511–526.

Sarikaya, A., Sarikaya, I., Pekindil, G., Firat, M. F., & Pekindil, Y. (2001). Technetium-99 m sestamibi limb scintigraphy in post-traumatic reflex sympathetic dystrophy: preliminary results. *Eur. J. Nucl. Med., 28,* 1517–1522.

Schwartzman, R. J., Alexander, G. M., Grothusen, J. R., Paylor, T., Reichenberger, E., & Perreault, M. (2009). Outpatient intravenous ketamine for the treatment of complex regional pain syndrome: a double-blind placebo controlled study. *Pain, 147,* 107–115.

Sherry, D. D., Wallace, C. A., Kelley, C., Kidder, M., & Sapp, L. (1999). Short- and long-term outcomes of children with complex regional pain syndrome type I treated with exercise therapy. *Clin. J. Pain, 15,* 218–223.

Stanton-Hicks, M. (2010). Plasticity of complex regional pain syndrome (CRPS) in children. *Pain Med., 11,* 1216–1223.

Stanton-Hicks, M., Janig, W., Hassenbusch, S., Haddox, J. D., Boas, R., & Wilson, P. (1995). Reflex sympathetic dystrophy: changing concepts and taxonomy. *Pain, 63,* 127–133.

Wilder, R. T. (2006). Management of pediatric patients with complex regional pain syndrome. *Clin. J. Pain, 22,* 443–448.

Wilder, R. T., Berde, C. B., Wolohan, M., Vieyra, M. A., Masek, B. J., & Micheli, L. J. (1992). Reflex sympathetic dystrophy in children. Clinical characteristics and follow-up of seventy patients. *J. Bone Joint Surg. Am., 74,* 910–919.

CRPS2

▶ Causalgia

CRPS-I Neuropathy Model

▶ Neuropathic Pain Models, CRPS-I Neuropathy Model

Crura of Clitoris

Definition

The crura or bulbs extend posteriorly and horizontally as two cylinders of erectile tissue arising from the body of the clitoris where it meets the pubic bone for up to 9 cm.

Cross-References

▶ Clitoral Pain

Cryoablation

Definition

The use of sub-zero temperatures to destroy tissue.

Cross-References

▶ Cancer Pain Management, Neurosurgical Interventions

Cryoglobulinemia

Definition

Condition wherein cryoglobulins, a class of immunoglobulins that precipitate when cooled and redissolve when heated, are present in the blood.

Cross-References

▶ Viral Neuropathies

Cryotherapy

Definition

Application of cold to body surfaces.

Cross-References

▶ Chronic Pain in Children, Physical Medicine and Rehabilitation

CSF

▶ Cerebrospinal Fluid

CSF Examination

Definition

Shows pleocytosis, elevation of protein, or oligoclonal banding in neuro-Behcet, collagen vascular disease, systemic vasculitides, and the isolated angiitis of the central nervous system.

Cross-References

▶ Headache Due to Arteritis

CSF Leak

▶ Spontaneous Cerebrospinal Fluid (CSF) Leak

CT Myelography

Definition

Imaging of the spinal cord using computerized analysis of cross-sectional scans (tomograms) after injection of a contrast agent into the spinal fluid.

Cross-References

▶ Chronic Low Back Pain: Definitions and Diagnosis

CT Scanning

Peter Lau
Department of Clinical Research, Royal
Newcastle Hospital, University of Newcastle,
Newcastle, NSW, Australia

Synonyms

CAT Scan; Computerized axial tomography; Helical CT; Multichannel CT; Spiral CT

Definition

Computerized axial tomography (CAT or CT) is a means by which images of the internal structure of the body can be obtained in selected planes and at selected depths.

Characteristics

Principles
In CT scanning, a computer is used to synthesize images based on X-ray absorption data. The object to be scanned is placed at the center of a circular gantry, around which an X-ray camera and its receiver rotate. As the camera rotates, X-rays are beamed through the object along multiple, selected diameters. Along each diameter the X-ray shadows (see ▶ Plain Radiography) of the object and its internal structure are transmitted to the receiver and stored by a computer. Using an iterative program, the computer then analyzes the stored information to generate a synthetic image of what the internal, three-dimensional structure of the object must be, in order to account for all of the different patterns of X-ray absorption obtained along each of the diameters across which the object was viewed.

The plane across which the object is studied is typically transverse to the long axis of the object; but this plane can be varied by tilting the gantry. Also, the object can be translated relative to the gantry so that it can be studied across multiple, selected sections, also referred to as "slices." By scanning across its entire length, images of the entire object can be synthesized. By using special computer programs, images can be synthesized to view the object from any angle or perspective, at any depth. The images can be planar, i.e., depicting the appearance of a single, selected slice through the object (Fig. 1a), or they can be simulated, three-dimensional reconstructions of the internal structure of the object (Fig. 1b).

Capabilities
The first-generation scanner took about 7 min to perform a single scan. Since then, the technology involved has been improved greatly, through at least five generations of development. Today, with multi-slice or multichannel, slip-ring CT systems, which allow continuous scanner gantry motion, coupled with synchronous table feed provide the basis for so-called helical or spiral CT scanning (Fox et al. 1998). Ultrafast scanners are capable of performing as many as 32 scans in half a second which is the time taken per rotation of the gantry. This is achieved by unique design of detectors, isotropic scanning, and fast reconstruction. Submillimeter thin and ultrathin slices can be obtained. There is now commercially available CT scanner capable of more than 64 slices per half a second.

In the past, using slow scanners, CT provided images of all but the smallest of soft tissues. Today with submillimeter, microvoxel scanning

CT Scanning,
Fig. 1 Sagittal CT scans of
the cervical spine in which
the vertebrae C4 to C7 have
been fused with interbody
grafts and an anterior metal
plate. (**a**) Midline section.
(**b**) 3-D reconstruction
(Reproduced courtesy of
Toshiba Medical)

(Pretorius and Fishman 1999), the smallest structures in the body, particularly small bony structures such as those of the middle and inner ear, can be well demonstrated (Fig. 2). The cardinal advantage of CT is that it allows observers to "see inside" cavities such as the chest, abdomen, skull, and vertebral canal (Fig. 3). The ability to demonstrate soft tissues allows CT to detect displacement and distortion of these structures by tumors, infiltration, infection, and inflammation (Fig. 4). Of particular relevance to pain medicine is the ability of CT to demonstrate intracranial pathology in patients with headaches and the relationship between spinal nerves and the vertebral column in patients with radicular pain.

However, CT still relies on detecting the X-ray shadows of internal structures. Consequently, although CT can demonstrate the location, shape, and density of tissues, it cannot reveal their internal architecture, although hollow or permeable structures can be indirectly visualized by injection or infusion of radiopaque agents, which localize in targeted tissues. Thus, injection of contrast medium will outline the lumen of hollow organs and blood vessels, joint cavities, and the subarachnoid space (Fig. 5). The internal structure of the

intervertebral disc can be visualized by injection of contrast into the nucleus (Fig. 6), and intravenous infusion of a rapidly excreted contrast medium will enhance the structure of the kidney and urinary tract.

Multichannel techniques have had a significant impact on the style and quality of spine CT. Coverage is no longer limited by practical problems. Lumbar studies routinely extend from T12 to S1 and cervical spine studies from C2 to T2. With a 64-channel scanner, the full spine scan may be obtained in the duration of one breath-hold, which is easier for patients to tolerate than MRI scans.

Applications
Microvoxel volumetric acquisition techniques provide a significant boost in spatial and contrast resolution. They can be used to create exquisite sagittal images of the intervertebral foramina in patients with radicular pain, especially in those with persisting pain after surgery, when missed foraminal stenosis is a major source of chronic pain.

With the advent of fine, submillimeter technology, CT is becoming the modality of choice for the assessment of failed back surgery syndrome. Artifacts from intervertebral metal

CT Scanning, Fig. 2 A CT scan through the temporal bone, showing the external ear and contents of the middle and internal ear. (**a**) Transverse section showing the ossicles of the middle ear. (**b**) Transverse section showing the cochlea and semicircular canals

cages or pedicular screws can be minimized, which allows for assessment of failure of spinal fixation hardware by loosening or metal fracture and fatigue.

CT can demonstrate the progress of healing in the skeletal system and can detect pseudoarthrosis in the spine after attempted arthrodesis. Similar techniques can be used for the assessment of joint replacement in the hip, shoulder, and knee. Microvoxel scan techniques minimize partial-volume artifact and allow evaluation of the state of bone surrounding hardware implants, as well as evaluation of possible loosening and migration of the prosthesis as a chronic source of pain.

With the advent of 32 and 64 slice scanners, dynamic, real-time imaging of joint movements and spinal movement in order to assess biomechanical abnormalities is possible.

Real-time fluoroscopy CT is used to perform pain interventional procedures. It enables drugs to be injected accurately and safely. It enables the accurate and safe delivery of ablation devices, such as radio-frequency electrodes, to deep sites

that were considered difficult, unsafe, and inaccurate in the past. This opens up a new arena of oncological pain management by using radio-frequency ablation to destroy pain-producing tumors that have invaded nerve plexuses, e.g., metastatic carcinoma of the rectum with invasion of the lumbosacral plexus. Ablation of painful bone tumors, such as osteoid osteoma, is another avenue of application of CT in pain management.

High-contrast resolution and the seamless multi-planar capability of the latest multichannel CT imagers have their greatest application in musculoskeletal pain after trauma. Fractures are now delineated easily in any plane, while the patient is situated comfortably on the table. Positioning during the scan is not critical, as any plane can be created from the volumetric data. Trauma series are typically performed in less than 30 s, minimizing the likelihood of patient motion (Tanenbaum 2003). Treatment can be better planned when CT demonstrates the presence and extent of subtle cortical fractures, fracture fragments, intra-articular loose bodies, and

CT Scanning, Fig. 3 Sagittal CT scan of the abdomen showing internal viscera and vessels. *a* aorta, *l* liver, *m* superior mesenteric artery, *c* colon, *s* small intestine (Reproduced courtesy of Toshiba Medical)

CT Scanning, Fig. 4 Axial CT scan of the thoracic spine showing osteomyelitis of a vertebral body

articular surface offset/depression; this was not possible by plain radiography (Fig. 7).

Whereas magnetic resonance imaging is superior for the delineation of the extent of

tumors in medullary bone, CT is better able to delineate cortical bone, calcifications in the matrix of lesions, and periosteal new bone formation.

Utility

The utility of CT emerges from a tension between what can be seen with CT and what is relevant and valid. Although CT can provide detailed images of the internal structure of the body, abnormalities demonstrated by CT are not necessarily the cause of a patient's pain. In the context of pain medicine, CT is useful as a diagnostic test only if and when it reliably shows the cause of pain.

Reliability

For major and obvious lesions, the reliability of radiologists using CT has not been questioned. Such lesions, however, are rarely pertinent to pain medicine. Of greater relevance are minor abnormalities that are sometimes reported in scans of patients with pain. Conspicuously, reliability data are largely lacking in the radiology literature. There has been a tradition to assume that all radiologists agree on what they report. When this has been studied, however, the converse emerges.

In reading CT scans of the lumbar spine, radiologists fail to agree on the presence of such features as spinal stenosis, degenerative joint disease, and even herniated nucleus pulposus (Wiesel et al. 1986). Unfortunately, data have not been presented in a manner from which kappa scores can be calculated.

Validity

In the joints of the appendicular skeleton, CT can show abnormalities of joint space, subchondral bone, and even internal and external ligaments. However, there are no data that incriminate these abnormalities as the cause of pain (Bogduk 2003).

In CT scans of the lumbar spine, abnormalities, such as degenerative joint disease, disc bulges, and disc herniations, are the most commonly encountered. These abnormalities, however, also frequently occur in asymptomatic

CT Scanning, Fig. 5 Parasagittal and sagittal CT scans of the entire spine. Contrast medium has been injected into the subarachnoid space, which appears *white* (sa), and outlines the spinal cord (*sc*), which appears *dark* (Reproduced courtesy of Toshiba Medical)

individuals (Wiesel et al. 1986). Indeed, some 20–30 % of asymptomatic individuals exhibit herniated lumbar intervertebral discs. These figures warn that finding such abnormalities in symptomatic patients does not necessarily prove that they are the cause of pain.

Tumors and infections are assumed to be valid causes of pain, but they are rare causes of either acute or chronic pain. The same applies for conditions, such as osteonecrosis. For most patients with pain, there is nothing demonstrable by CT that has actually been proven to be a cause of pain. Even pseudoarthrosis following spine surgery has not been validated as a cause of pain.

Future Trends

Improvements in the technology of CT are likely to continue. There is an urgent need, however, for clinical science to catch up with the technology. Urgently needed are studies in which the presence and absence of morphological abnormalities are correlated with the presence and absence of pain. Such studies are necessary lest

CT Scanning, Fig. 6 An axial CT scan of a lumbar intervertebral disc. Contrast medium fills the nucleus pulposus (*NP*), a left posterolateral radial fissure (*1*) that extends into the annulus fibrosus and a right posterolateral fissure (*2*) that has a circumferential extension (*3*)

CT Scanning, Fig. 7 A sagittal CT scan of the ankle, showing an oblique fracture (*arrow*) through the posterior tibia, extending into the talocrural joint (Reproduced courtesy of Toshiba Medical)

abnormalities simply be assumed to be relevant to the diagnosis of pain.

Another opportunity lies in the fusion of morphological CT technology with physiological imaging techniques such ▶ as positron emission tomography (PET) (Erickson 2003). Already

PET-CT can produce the physiological properties of certain neoplasms combined with anatomical imaging. In the same way, it may become possible to demonstrate the physiology of lesions, seen on CT, that are suspected of being causes of pain.

References

Bogduk, N. (2003). Diagnostic procedures in chronic pain. In T. S. Jensen, P. R. Wilson, & A. S. C. Rice (Eds.), *Clinical pain management: Chronic pain* (pp. 125–144). London: Arnold.

Erickson, J. E. (2003). TRIP 3-possible future technologies. *Paper presented at the 89th annual meeting of the Radiological Society of North America*. Chicago, Illinois, November 2003.

Fox, S. H., Tanenbaum, L. N., Ackelsberg, S., et al. (1998). Future directions in CT technology. *Neuroimaging Clinics of North America, 8*, 497–513.

Pretorius, E. S., & Fishman, E. K. (1999). Volume-rendered three-dimensional spiral CT: Musculoskeletal application. *Radiographics, 19*, 1143–1160.

Tanenbaum, L. N. (2003). Multichannel helical CT of the musculoskeletal system. *Applied Radiology, 32*, 15–22.

Wiesel, S. W., Tsourmas, N., Feffer, H. L., et al. (1986). A study of computer-assisted tomography. 1. The incidence of positive CAT scans in an asymptomatic group of patients. *Spine, 9*, 549–551.

CT Scans

Definition

CT (computed tomography), sometimes called CAT scan (computed axial tomography), uses X-ray images taken at different angles around the body and then by computer processing of the information produces two- or three-dimensional slice images.

Cross-References

▶ Motor Cortex, Effect on Pain-Related Behavior

CTDs

▶ Disability, Upper Extremity

CTrP

▶ Central Trigger Point

Cuban Neuropathy

▶ Metabolic and Nutritional Neuropathies

Cumulative Trauma Disorders

Definition

A descriptive term used to label pain associated with repetitive activities.

Cross-References

▶ Disability, Upper Extremity

Cuneate Nucleus

Definition

A nucleus at the upper cervical spinal cord where the first synapse of the dorsal columns occurs. Discriminative sensory information from cutaneous regions from the upper half of the body is relayed to this nucleus.

Cross-References

▶ Brainstem Subnucleus Reticularis Dorsalis Neuron
▶ Postsynaptic Dorsal Column Projection, Anatomical Organization

Cutaneous

Definition

Relating to, or affecting, the skin.

Cross-References

▶ Psychiatric Aspects of Visceral Pain

Cutaneous Allodynia

Definition

Allodynia means "another form of stimulation." The term has been used for characterizing touch-evoked pain (i.e., pain induced by activation of sensitive mechanoreceptors) which can be seen after capsaicin treatment of the skin and in some forms of neuropathies. Nowadays, the term is used more loosely to characterize pain elicited by stimuli which are below the normal pain threshold. In this form, the term "allodynia" becomes synonymous to hyperalgesia, when it refers, e.g., to pain induced by sensitized heat nociceptors.

Cross-References

▶ Migraine Without Aura

Cutaneous Field Stimulation

Jens Schouenborg
Neuronano Research Center, Lund University, Lund, Sweden

Synonyms

CFS

Definition

Cutaneous field stimulation (CFS) allows topographically restricted and tolerable electrical stimulation of thin (Aδ and C) cutaneous fibers for the symptomatic relief of ▶ itch and pain.

Characteristics

The CFS technology is based on the organization of endogenous intermodality interactions and allows tolerable stimulation of cutaneous Aδ and C afferent fibers (McMahon and Koltzenburg 1992; Ward et al. 1996). For example, mechanical stimulation inhibits ongoing pain (Bromm et al. 1995; Sjölund 1990; Wall and Cronly-Dillon 1960; Ward et al. 1996). In fact, particularly strong interactions are found between submodalities of the nociceptive system (here including itch), e.g., low-frequency electrical stimulation of Aδ fibers may cause a durable depression of nociceptive C-fiber transmission both in vivo (Sjölund 1985, 1988) and in vitro spinal preparations (Sandkühler et al. 1997), and noxious mechanical stimulation, such as scratching, reduces itch. These interactions occur at several levels in the somatosensory system, the dorsal horn of the spinal cord (Cervero et al. 1979; Melzack and Wall 1965), the dorsal column nuclei (Saade et al. 1985), and the thalamus (Olausson et al. 1994) and are often topographically well organized. To use these interactions for the symptomatic relief of itch and pain, a new technique, termed cutaneous field stimulation (CFS), was introduced (Nilsson et al. 1997). CFS allows topographically restricted and tolerable electrical stimulation of thin (Aδ and C) cutaneous fibers.

CFS uses a flexible rubber plate with multiarray needle-like electrodes regularly fixed at 2 cm intervals (Fig. 1). Each electrode is surrounded by a "stop device" 2.0 mm in diameter that protrudes 2.0 mm from the plate. The electrode tip protrudes 0.3 mm from the stop device. When the electrode plate is gently pressed against the skin, the electrode tips are introduced

close to the receptors in the epidermis and the superficial part of dermis. Since the electrodes traverse the electrically isolating horny layer of the epidermis and the current density is high near the sharp electrode tips, the voltage and current necessary to stimulate cutaneous nerve fibers are small, typically less than 10 V and up to 0.8 mA, respectively. As the current density decreases rapidly with distance, localized stimulation is achieved. The electrodes are stimulated consecutively with a constant current stimulator (64 pulses/s), each electrode with a frequency of 1–10 Hz (pulse duration 1.0 ms). Treatment duration is 5–45 min. A self-adhesive surface (TENS) electrode serves as anode and is placed about 5–30 cm away from the needle electrode plate.

Several pieces of evidence demonstrate that CFS acts via stimulation of nociceptive Aδ and C-fibers (Nilsson et al. 1997, 2003): (1) The needle-like electrodes of the CFS plate are introduced close to the dermo-epidermal junction, known to be richly innervated by thin afferent fibers (Fundin et al. 1994; Kruger et al. 1985); (2) CFS evokes a sensation of pricking and slightly burning pain of weak/moderate intensity, indicating an activation of nociceptive Aδ- and/or C-fibers, respectively (Bromm and Treede 1987); (3) during a selective blockade of the impulse conduction in A fibers, CFS still produces a burning sensation, known to be due to heat sensitive nociceptive C-fibers; and (4) a flare, known to be caused by nociceptive C-fibers (Kenins 1981), develops around the CFS electrodes. It should be noted that while Aβ-fibers are presumably also activated by CFS due to their low electrical threshold, the stimulation frequencies used for CFS are very low for Aβ fibers (but relatively high for C-fibers).

CFS acts preferentially on nociceptive senses, since homo- but not heterotopical (with respect to testing sites) CFS, abolishes itch evoked by histamine in normal humans, and reduces CO_2-laser-evoked Aδ- and C-fiber-mediated heat pain and pinch-evoked pain (Nilsson and Schouenborg 1999), but has no effect on fabric-evoked prickle (assumed to be mediated

a

b

Cutaneous Field Stimulation, Fig. 1 (**a**) Schematic of CFS apparatus with electrodes. (**b**) The needle-like electrodes are inserted into the dermo-epidermal junction, permitting stimulation of thin nerve fibers with low electrical voltage and current

by high-threshold mechanoreceptive Aδ- and C-fibers; Garnsworthy et al. 1988) (Fig. 2). The relative potency of CFS on nociceptive skin senses is thus, in a descending order, itch, secondary heat pain, primary heat pain, pinch-evoked pain, and prickle. Regarding innocuous senses, CFS increases the warm and cold thresholds somewhat but has no effect on tactile sensibility.

CFS causes a dramatic reduction in histamine-induced itch. In fact, all subjects tested have reported a complete or near-complete inhibition of itch evoked by histamine. This inhibitory effect lasts 4–8 h and extends 10–20 cm along the same dermatome, but appears to be more restricted in a direction across dermatomal borders (Nilsson et al. 1997). Maximal effects are reached already after 8–10 min, and more

Cutaneous Field Stimulation, Fig. 2 Differential effect on different sensory qualities of nociception after CFS (n = 10 subjects except for itch where n = 7 subjects) and lack of effect after conventional high-frequency TENS (n = 10 subjects except for itch where n = 7 subjects). The conditioning stimulations and all sensory testings were performed on the right volar forearm. The mean magnitude of sensation after CFS (*black bars*) or TENS (*hatched bars*) is indicated as percentage of control (*unfilled bars*). Top diagram shows effects after CFS; bottom diagram shows effects after TENS. Statistical comparisons (Wilcoxons signed-ranks test, two-tailed) were made both between control values and values obtained after conditioning stimulation and also between values obtained after CFS and TENS, respectively. Normalized visual analogue scale (VAS) on the vertical axes. * $p < 0.05$, ** $p < 0.01$, *** $p < 0.001$; *ns* nonsignificant. Error bars indicate + S.E.M (From Nilsson et al. 1997; Nilsson and Schouenborg 1999)

prolonged CFS stimulation (up to 45 min tested) does not appear to add to the inhibitory effect (Nilsson et al. 2003). The optimal stimulation frequency appears to be around 4 Hz per

Cutaneous Field Stimulation, Fig. 3 Effects of CFS (*top diagram*) and TENS (*bottom diagram*) on chronic itch due to atopic dermatitis. For comparison, the mean VAS values were normalized with respect to control (*unfilled bars*). The mean sensory intensities after CFS (*black bars*, n = 15) and TENS (*hatched bars*, n = 12) are shown. ANOVA repeated measures with Dunnet's post hoc test was applied for matched comparisons within a group, whereas the Mann–Whitney U test was used when comparing effects of the two treatments; *ns* nonsignificant, * $p < 0.05$, ** $p < 0.01$. Error bars indicate + S.E.M (From Nilsson et al. 2004)

electrode (Nilsson et al. 2003). Notably, the effective stimulation parameters are similar to those known to cause long-term depression (LTD) (Sandkühler et al. 1997) – a memory-like mechanism – in the spinal cord and elsewhere.

Figure 3 shows the results from a randomized controlled clinical study on chronic itch due to localized atopic dermatitis (n = 27;

Nilsson et al. 2004). As can be seen, CFS depresses itch significantly for more than 7 h. Peak inhibitory effect (down to about 25 % of control) is reached between 1 and 5 h postconditioning. Similar results have recently been reached for chronic itch due to neurodermatitis (n = 30; Bäck et al. in preparation). In the latter single-blind study, CFS was compared to conventional TENS over a treatment period of 2 weeks (two treatments a day). It was found that CFS provided effective relief of itch in practically all patients and also markedly improved the skin condition. The improvement of the skin condition may be due to a reduced urge to scratch the skin (or care not to scratch the treated skin), thus stopping the vicious itch – scratch circle that may sustain this type of condition. Other workers have reported similar findings on the relief of chronic itch by CFS in various dermatoses (Wallengren 2002; Wallengren and Sundler 2001). In addition, a normalized innervation of the skin may result from CFS (Wallengren and Sundler 2001). No adverse effects of CFS have been reported (Wallengren 2002).

CFS affects not only cutaneous nociception but also musculoskeletal nociception. In a recent study on 142 TENS-resistant patients with chronic musculoskeletal pain (i.e., a group of patients that received no pain relief from established treatments of electrostimulation), 30 % received clinically relevant pain relief from CFS (Martinsson and Sjölund, in preparation). The group studied consisted of 91 patients with nociceptive pain and 51 patients with neuropathic pain according to established criteria. 33/91 patients with nociceptive pain and 10/51 with neuropathic pain got clinically relevant pain relief from CFS for more than 3 months.

In conclusion, CFS preferentially affects nociceptive senses via endogenous inhibitory memory-like mechanisms and has no significant adverse side effects. The findings that CFS causes a strong inhibitory effect on itch and that skin conditions improve considerably and the lack of adverse side effects demonstrate that CFS is a useful itch therapy. Moreover, CFS is clearly worth trying for the treatment of chronic musculoskeletal pain.

References

Bromm, B., & Treede, R.-D. (1987). Human cerebral potentials evoked by CO_2 laser stimuli causing pain. *Experimental Brain Research, 67*, 153–162.

Bromm, B., Scharein, E., Darsow, U., et al. (1995). Effects of menthol and cold on histamine-induced itch and skin reactions in man. *Neuroscience Letters, 187*, 157–160.

Cervero, F., Iggo, A., & Molony, V. (1979). An electrophysiological study of neurones in the substantia gelatinosa rolandi of the cat's spinal cord. *Quarterly Journal of Experimental Physiology, 64*, 297–314.

Fundin, B. T., Rice, F. L., Pfaller, K., et al. (1994). The innervation of the mystacial pad in the adult rat studied by anterograde transport of HRP-conjugates. *Experimental Brain Research, 99*, 233–246.

Garnsworthy, R. K., Gully, P. L., Kenins, P., et al. (1988). Identification of the physical stimulus and the neural basis of fabric-evoked prickle. *Journal of Neurophysiology, 59*, 1083–1097.

Kenins, P. (1981). Identification of the unmyelinated sensory nerves which evoke plasma extravasation in response to antidromic stimulation. *Neuroscience Letters, 25*, 137–141.

Kruger, L., Sampogna, S. L., Rodin, B. E., et al. (1985). Thin-fiber cutaneous innervation and its intraepidermal contribution studied by labelling methods and neurotoxin treatment in rats. *Somatosensory Research, 2*, 335–356.

McMahon, S. B., & Koltzenburg, M. (1992). Itching for an explanation. *Trends in Neurosciences, 15*(12), 497–501.

Melzack, R., & Wall, P. D. (1965). Pain mechanisms: A new theory. A gate control system modulates sensory input from the skin before it evokes pain perception and response. *Science, 150*, 971–979.

Nilsson, H.-J., & Schouenborg, J. (1999). Differential inhibitory effect on human nociceptive skin senses induced by local stimulation of thin cutaneous fibers. *Pain, 80*, 103–112.

Nilsson, H.-J., Levinsson, A., & Schouenborg, J. (1997). Cutaneous Field Stimulation (CFS) – a new powerful method to combat itch. *Pain, 71*, 49–55.

Nilsson, H.-J., Psouni, E., & Schouenborg, J. (2003). Long term depression of human nociceptive skin senses induced by thin fibre stimulation. *European Journal of Pain, 7*, 225–233.

Nilsson, H.-J., Psouni, E., Carstam, R., et al. (2004). Profound inhibition of chronic itch induced by stimulation of thin cutaneous nerve fibres. *Journal of the European Academy of Dermatology and Venereology, 18*, 37–43.

Olausson, B., Xu, Z.-Q., & Shyu, B.-C. (1994). Dorsal column inhibition of nociceptive thalamic cells mediated by gamma-aminobutyric acid mechanisms in the cat. *Acta Physiologica Scandinavica, 152*, 239–247.

Saade, N. E., Dajani, B. M., Atweh, S. F., et al. (1985). Inhibition of dorsal column nuclei by stimulation of trigeminal afferents in decerebrate-decerebellate cats. *Brain Research, 348*, 405–407.

Sandkühler, J., Chen, J. G., Cheng, G., et al. (1997). Low-frequency stimulation of afferent AÎ'-fibers induces long-term depression at primary afferent synapses with substantia gelatinosa neurons in the rat. *Journal of Neuroscience, 17*, 6483–6491.

Sjölund, B. H. (1985). Peripheral nerve stimulation suppression of C-fiber-evoked flexion reflex in rats. Part 1. Parameters of continuous stimulation. *Journal of Neurosurgery, 63*, 612–616.

Sjölund, B. H. (1988). Peripheral nerve stimulation suppression of C-fiber-evoked flexion reflex in rats. Part 2: Parameters of low-rate train stimulation of skin and muscle afferent nerves. *Journal of Neurosurgery, 68*, 279–283.

Sjölund, B. H. (1990). Transcutaneous and implanted electric stimulation of peripheral nerves. In J. Bonica (Ed.), *Management of pain* (2nd ed., pp. 1852–1861). Philadelphia: Lea & Fegiber.

Wall, P. D., & Cronly-Dillon, J. R. (1960). Pain, itch, and vibration, A.M.A. *Archives of Neurology, 2*, 365–375.

Wallengren, J. (2002). Cutaneous field stimulation of sensory nerve fibers reduces itch without affecting contact dermatitis. *Allergy, 57*, 1195–1199.

Wallengren, J., & Sundler, F. (2001). Cutaneous field stimulation in the treatment of severe itch. *Archives of Dermatology, 137*, 1323–1325.

Ward, L., Wright, E., & McMahon, S. B. (1996). A comparison of the effects of noxious and innocuous counterstimuli on experimentally induced itch and pain. *Pain, 64*, 129–138.

Cutaneous Freeze Trauma

▶ Freezing Model of Cutaneous Hyperalgesia

Cutaneous Hyperalgesia

▶ Freezing Model of Cutaneous Hyperalgesia

Cutaneous Mechanical Sensation

▶ Postsynaptic Dorsal Column Projection, Anatomical Organization

Cutaneous Mechanoreception

Definition

Neuronal responses to mechanical stimuli applied to the skin are transduced by specialized receptive endings of afferent nerve fibers. The responsible nerve fibers are mainly fast conducting A beta fibers. However, there are also low-threshold A delta and C fibers which can maintain mechanical receptivity after a loss of A beta fibers.

Cross-References

▶ Postsynaptic Dorsal Column Projection, Anatomical Organization

Cutaneous Neuroma

▶ Painful Scars

Cutaneous Pain

▶ Functional Imaging of Cutaneous Pain

CWP

▶ Chronic Widespread Pain

Cyclic Adenosine Monophosphate-Responsive Element-Binding Protein

Synonyms

CREB

Definition

CREB is a transcription factor that is activated by the cAMP signal transduction pathway. It is phosphorylated by protein kinase A, enters the nucleus of a neuron, and can alter gene expression.

Cross-References

▶ Spinothalamic Tract Neurons, Role of Nitric Oxide

Cyclic Alternating Pattern

Definition

Repetition of micro-arousal every 20–60 s as a sentinel during sleep that allows for a "reset" of physiological functions (e.g., heart rate, respiration, muscle tone), or prepares the body for an appropriate response if an event could be potentially disrupting.

Cross-References

▶ Orofacial Pain, Sleep Disturbance

Cyclic Pain

Definition

Cyclic pain occurs with definite association to the menstrual cycle, generally 1–2 days prior to the onset of or during menses.

Cross-References

▶ Dyspareunia and Vaginismus

Cyclooxygenase

▶ Cyclooxygenases in Biology and Disease
▶ NSAIDs, COX-Independent Actions
▶ NSAIDs, Mode of Action

Cyclooxygenase (COX) Inhibitors

▶ Pharmacology of Cyclooxygenase Inhibitors

Cyclooxygenase Inhibitors

▶ NSAIDs, Adverse Effects
▶ NSAIDs and Cancer
▶ Postoperative Pain, Nonsteroidal Anti-Inflammatory Drugs

Cyclooxygenase Inhibitors, Chemistry

▶ Coxibs and Novel Compounds, Chemistry

Cyclooxygenase-1

▶ COX-1 and COX-2 in Pain

Cyclooxygenase-2

▶ COX-1 and COX-2 in Pain

Cyclooxygenase-2 Inhibitors

Definition

A class of drugs related to nonsteroidal anti-inflammatory drugs (NSAIDs) that prevent the synthesis of prostaglandins by the enzyme cyclooxygenase-2.

Cross-References

Cyclooxygenases

Synonyms

COX

Definition

COX is a 70–72 kD enzyme, catalyzing the reaction of arachidonic acid to PGG_2 (cyclooxygenase reaction) and consecutively PGG_2 to PGH_2 (peroxidase reaction). Prostaglandin H2 (PGH_2) is further converted into various prostaglandins and thromboxanes. There is a distinct active site for both cyclooxygenase and peroxidase reactions. There are at least three isoforms, COX1, COX2, and COX3, which are also slightly different at the active site, explaining particular selectivities of inhibitors. Prostaglandins are involved in pain, inflammation, and fever but also in cytoprotection in the stomach and blood flow regulation in the kidneys. NSAIDS (nonsteroidal anti-inflammatory drugs) are therapeutic agents that block this enzyme and thus act as painkillers, antipyretic, and anti-inflammatory drugs.

Cross-References

Cyclooxygenases in Biology and Disease

Regina M. Botting

London, UK

Synonyms

COX; Cyclooxygenase; Prostaglandin endoperoxide synthase; Ptgs

Definition

Cyclooxygenases are enzymes which synthesize prostaglandins.

Characteristics

Two isoenzymes of cyclooxygenase have been characterized: cyclooxygenase-1 (COX-1; Ptgs-1) and cyclooxygenase-2 (COX-2; Ptgs-2).

Cyclooxygenase-1

In the 1930s, Goldblatt in England (Goldblatt 1935) and von Euler in Sweden (von Euler 1934) found that seminal fluid contained activity that contracted uterine smooth muscle, and also caused a fall in blood pressure. Von Euler identified the active principle as a lipid-soluble acid, which he named "prostaglandin" (PG) because he thought it originated from the prostate gland. In the 1960s, technical advances allowed the characterization of the PGs as a family of lipid compounds with a unique structure. They proved to be 20-carbon unsaturated carboxylic acids with a cyclopentane ring, and in 1964, PGE_2 was synthesized using ▶ arachidonic acid and an enzyme preparation from ram seminal vesicles (Bergström et al. 1964). Prostaglandins and related compounds are some of the most prevalent of autacoids and can be released from every tissue except red blood cells. As local hormones, they produce, in minute concentrations, an incredibly broad

Cyclooxygenases in Biology and Disease, Fig. 1 The arachidonic acid cascade. Cyclooxygenase-1 (COX-1) and cyclooxygenase-2 (COX-2) utilize arachidonic acid to form the unstable endoperoxide intermediates, prostaglandin G_2 (PGG$_2$) and prostaglandin H_2 (PGH$_2$). From these intermediates, the more stable prostaglandins are synthesized by individual synthases. Prostaglandin E_2 (PGE$_2$) is mostly involved in pain, fever, and inflammation; prostaglandin $F_{2\alpha}$ (PGF$_{2\alpha}$) in labor and parturition; and prostaglandin D_2 (PGD$_2$) is a mediator in brain and mast cells. Prostacyclin (PGI$_2$) with a half-life of 3 min protects the stomach mucosa from damage and prevents aggregation of blood platelets. It breaks down spontaneously to inactive 6-keto-PGF$_1$ $_\alpha$. Thromboxane A_2 is mainly found in blood platelets and promotes clotting of blood by inducing platelet aggregation. It has a half-life of 30 s and breaks down to inactive thromboxane B_2. The activity of COX-1 and COX-2 is inhibited by aspirin and similar drugs, the nonsteroid anti-inflammatory drugs (NSAIDs), thereby preventing formation of prostanoids

spectrum of effects that modulate almost every biological function. They derive mostly from the 20-carbon fatty acid, arachidonic acid (C20:4ω6), and almost 100 different derivatives have been identified, including lipoxins and isoprostanes. Prostaglandins are important to inflammatory processes, as evidenced by the anti-inflammatory effects of drugs that interfere with their synthesis, such as the steroid and nonsteroid anti-inflammatory drugs (NSAIDs). The discovery 22 years ago of a second cyclooxygenase (COX-2), inducible by cytokines, has led to an important clarification of the roles of PGs as mediators of inflammation. COX-2 is induced in inflammation and makes PGs locally, whereas COX-1 is a "housekeeping" enzyme involved in the modulation of physiological events.

Cyclooxygenase had long been studied in preparations from ram seminal vesicles and a homogeneous, enzymatically active prostaglandin endoperoxide synthase was isolated in 1976 (Hemler et al. 1976) and cloned in 1988 (DeWitt and Smith 1988). This membrane-bound hemo- and glycoprotein, with a molecular weight of 71kDA, is found in greatest amounts in the endoplasmic reticulum of prostanoid-forming cells. It both cyclizes arachidonic acid and adds the 15-hydroperoxy group to form PGG$_2$ (Fig. 1). The hydroperoxy group of PGG$_2$ is reduced to the hydroxy group of PGH$_2$, by a peroxidase (in the same enzyme protein) that utilizes a wide variety of compounds to provide the requisite pair of electrons. The ▶ hydroperoxides also drive the cyclooxygenase reaction by maintaining a "hydroperoxide tone." The three-dimensional

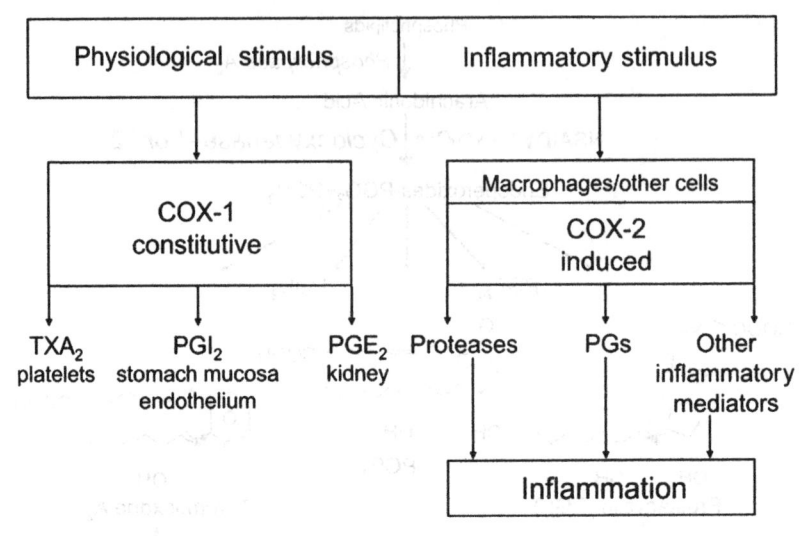

Cyclooxygenases in Biology and Disease, Fig. 2 Physiological and pathological functions of COX-1 and COX-2. Cyclooxygenase-1 (COX-1) functions as a constitutive enzyme and produces prostaglandins which are involved in physiological processes. For example, thromboxane A_2 (TXA$_2$) made by platelets promotes the clotting of blood when required, prostacyclin (PGI$_2$) made by the stomach mucosa protects it from damage by gastric acid, and PGE$_2$ made in kidney cells is involved in kidney function. PGI$_2$ made by COX-1 in vascular endothelial cells inhibits aggregation of platelets and the clotting of blood. PGI$_2$ is also formed by COX-2 induced in endothelial cells by shear stress produced by laminar blood flow. Cyclooxygenase-2 is mainly induced in inflammation and by mitogens and growth factors. It is induced in macrophages and other inflammatory cells by bacterial lipopolysaccharide and cytokines. It forms mainly PGE$_2$, which with other inflammatory mediators causes pain and fever. Some constitutive COX-2 is also expressed, particularly in the brain. In pain, PGE$_2$ is hyperalgesic, sensitizing peripheral sensory nerve terminals to nociceptive stimuli

structure of COX-1 was determined in 1994 (Picot et al. 1994). This enzyme consists of three independent folding units: an epidermal growth factor-like domain, a membrane-binding section, and an enzymatic domain. The sites for peroxidase and cyclooxygenase activity are adjacent but spatially distinct. Three of the helices of the COX-1 structure form an entrance channel to the active site, which is a long, hydrophobic channel. Since the enzyme integrates into only a single leaflet of the membrane lipid bilayer, the position of the cyclooxygenase channel allows arachidonic acid to gain access to the active site from the interior of the bilayer.

Most NSAIDs compete with arachidonic acid for binding to the active site (Fig. 1). ▶ Flurbiprofen, for example, inhibits COX-1 by excluding arachidonic acid from the upper portion of the channel and blocking its access to tyrosine (Tyr) 385 and serine (Ser) 530 at the apex of the long active site. Uniquely, aspirin irreversibly inhibits COX-1 by acetylation of

Ser 530, thereby excluding access for the arachidonic acid to Tyr 385 by steric hindrance.

Before 1971, many biochemical effects of the NSAIDs had been reported, and there were many hypotheses about the mechanism of action of these drugs. However, in 1971, John Vane elegantly demonstrated that aspirin, salicylate, and indomethacin inhibited COX of guinea pig lung homogenates, and thus prevented the formation of PGs (Vane 1971). Two other reports from the same laboratory, that aspirin prevented the release of prostanoids from aggregating human platelets and that NSAIDs blocked PG release from the perfused isolated spleen of the dog, lent support to and extended his finding.

COX-1 is constitutively expressed in most tissues and performs a "housekeeping" function to synthesize PGs which regulate normal cell activity (Fig. 2). The PGs produced are therefore protective to the organism. The concentration of the enzyme remains largely stable, but small two- to fourfold increases can occur in response to stimulation with hormones or growth factors. COX-1 has clear physiological functions (Vane and Botting 2001). Its activation leads, for instance, to the production of prostacyclin (PGI_2), which when released by endothelial cells is antithrombogenic and when released by the gastric mucosa is cytoprotective. However, it is likely that in humans, endothelial PGI_2 is also synthesized by COX-2, induced by ► shear stress on endothelial cells. Release of PGI_2 and PGE_2 made by COX-1 also sensitizes nociceptors on peripheral sensory nerve terminals and thus acutely amplifies painful stimulation. Since COX-2 is induced by inflammatory stimuli and by cytokines in migratory and other cells, it appears likely that the anti-inflammatory actions of NSAIDs are due to the inhibition of COX-2, whereas the unwanted side effects such as irritation of the stomach lining are due to the inhibition of the constitutive enzyme, COX-1.

Cyclooxygenase-2

The cloning of COX-1 from sheep seminal vesicles provided useful probes for the study of the relationship between PG production and COX mRNA induction. Many of these studies revealed a lack of correlation between COX activity and the induction of COX mRNA. For example, the levels of 2.8 kb COX-1 mRNA in epithelial cells of sheep trachea stimulated with growth factors did not reflect the increase in enzyme activity. However, a second mRNA (4.0 kb) recognized by the COX-1 cDNA probes and formed from a separate gene increased in parallel with PG production. Dexamethasone inhibited only the inducible COX activity in human IL-1 treated dermal fibroblasts and in human bacterial lipopolysaccharide (LPS)-stimulated monocytes (Masferrer et al. 1990). Basal levels of COX were not affected by dexamethasone.

The presence of the gene encoding for a second COX enzyme was discovered by Xie et al. while characterizing "immediate early" genes, switched on by expression of the v-src oncogene in chicken embryo fibroblasts (Xie et al. 1991). This new COX isoenzyme was also induced by serum, tetradecanoylphorbol acetate (TPA) or forskolin in established murine fibroblast cell lines. The chicken and mouse COX-2 gene, at 8.3 kb, is similar to the COX-2 gene of human but smaller than the 22 kb human COX-1 gene. The gene products also differ, with the size of the mRNA for COX-2 being approximately 4.5 kb and that for COX-1 being 2.8 kb. At the amino acid level, the protein sizes of COX-1 and COX-2 enzymes from different sources are approximately 80–90 % identical, with just over 600 amino acids, of which 63 % are in an identical sequence. Both enzymes have a molecular weight of 71 kDA and similar Km and Vmax values for metabolism of arachidonic acid. Both enzymes also have similar active sites for the attachment of arachidonic acid or NSAIDs, although the active site of COX-2 is larger than that of COX-1 and can accept a wider range of structures as substrates.

The residues that form the substrate-binding channel, the catalytic sites, and the residues immediately adjacent are all identical in COX-1 and COX-2 except for two small variations. The primary sequence difference in the active site itself is the valine 523 side chain in COX-2, which is smaller by a single methyl group than the isoleucine side chain that it replaces in COX-1. This opens up access to a side pocket off the main substrate channel in COX-2, allowing an enzyme inhibitor to react with arginine 513, which replaces histidine in COX-1. Another sequence difference is situated outside, but close to, the active site and can indirectly influence its conformation. The large residue, phenylalanine 503, in COX-1 is replaced by the smaller leucine in COX-2. This allows leucine 384, which borders the active site, to reorientate its methyl side chain out of the active site and thus leave more space available for a larger inhibitor molecule in COX-2 than in COX-1.

The importance of the discovery of the inducible COX-2 is highlighted by the differences in pharmacology of the two enzymes (Warner et al. 1999). Aspirin, indomethacin, and ibuprofen are much less active against COX-2 than against COX-1. Indeed, the strongest inhibitors of COX-1 such as aspirin, indomethacin, and piroxicam are the NSAIDs which cause the most damage to the stomach. The spectrum of activities of some ten standard NSAIDs against the two enzymes ranges from a high selectivity towards COX-1 (166-fold for aspirin) through to equiactivity on both (e.g., diclofenac). The range of activities of NSAIDs against COX-1 compared with COX-2 explains the variations in the side effects of NSAIDs at their anti-inflammatory doses. Drugs which have a high potency against COX-2 and a low COX-2/COX-1 activity ratio will have potent anti-inflammatory activity with few side effects on the stomach and kidney. Published epidemiological data on the side effects of NSAIDs reported that piroxicam and indomethacin in anti-inflammatory doses showed high gastrointestinal toxicity. These drugs have a much higher potency against COX-1 than against COX-2.

Inhibition of Cyclooxygenases

The inhibition of PG synthesis by the traditional NSAIDs (including aspirin, indomethacin, ibuprofen, diclofenac, piroxicam, flurbiprofen) explains all the pharmacological actions of these drugs. They prevent the pathological overproduction of PGs by COX-2 which contribute to the inflammatory process (therapeutic effects) and prevent the physiological formation of prostanoids by COX-1 (side effects). For instance, the ulcerogenic activity of aspirin arises from the inhibition of prostacyclin production, which has an important cytoprotective function in the gastric mucosa. Administration of various PGs reverses or prevents experimental gastric ulcers, and some of these PG derivatives are available for clinical use. In addition, the inhibition of COX-1 by NSAIDs correlates with their capacity to erode the gastric mucosa. NSAIDs also inhibit the aggregation of blood platelets, by preventing the formation of the potent pro-aggregatory and vasoconstrictor eicosanoid, thromboxane A_2 (TXA$_2$), synthesized by COX-1 in platelets. Aspirin irreversibly inhibits COX-1 in platelets by acetylation of the enzyme, thus completely preventing synthesis of TXA$_2$ until new platelets are produced after 8–11 days.

The inhibition of PG synthesis by NSAIDs has been demonstrated in a wide variety of cell types and tissues, ranging from whole animals and humans to microsomal enzyme preparations. For example, the concentration of PGE$_2$ is about 20 ng/ml in the synovial fluid of patients with rheumatoid arthritis. This decreases to zero in patients taking aspirin, a good demonstration of the effect of this drug on PG synthesis clinically. Several classes of NSAIDs have been identified, and at least 12 major chemical series are known to affect PG production by COX-1 and COX-2.

The discovery of COX-2, induced by inflammatory stimuli and cytokines in migratory and other cells, stimulated several laboratories to develop highly selective inhibitors of this enzyme (Fig. 2). Monsanto/Searle (now Pfizer) is marketing celecoxib and its successor, ▶ valdecoxib, whereas Merck and Company developed the selective COX-2 inhibitors, rofecoxib and etoricoxib. Early trials of celecoxib (Silverstein et al. 2000) and of rofecoxib (Bombardier et al. 2000) demonstrated good analgesia and anti-inflammatory activity, with less adverse effects on the stomach mucosa than comparator drugs. Millions of arthritic patients were then prescribed these drugs. However, the rofecoxib trial revealed that four times as many patients receiving rofecoxib suffered a myocardial infarction than those treated with naproxen. This was interpreted as a protective effect of naproxen, which would inhibit aggregation of platelets and therefore prevent heart attacks. Following these initial trials, Merck and Pfizer began large-scale trials of rofecoxib and celecoxib in patients with recurrent neoplastic ▶ polyps of the large bowel, a precancerous condition preceding colorectal neoplasia. These trials were instituted since cancer cells have high levels of COX-2, while surrounding healthy colon cells have none. In these anticancer trials, the effects of rofecoxib and celecoxib were compared with patients receiving placebo treatment. Both trials were terminated after 18 months, when it became clear that the incidence of adverse cardiovascular events, primarily myocardial infarctions and ischemic cerebrovascular events, was greater in the patients receiving selective COX-2 inhibitors. At this stage, Merck decided to withdraw rofecoxib from the market. The most plausible explanation for this increase in heart attacks is that selective COX-2 inhibitors prevent synthesis of anti-aggregatory prostacyclin in the vascular endothelium. Since these drugs do not affect TXA_2 synthesis by COX-1 in blood platelets, the aggregatory action of TXA_2 progresses unopposed and thrombotic emboli form to cause myocardial infarctions (Grosser et al. 2006).

Pfizer withdrew valdecoxib after it was rejected by the FDA because of an increase in cardiovascular events and serious skin reactions during its clinical trials. Parecoxib, the prodrug for valdecoxib, administered by intravenous injection, mainly for postoperative pain, was not granted registration either but is available in Europe. More recently, Novartis have made lumiracoxib, which is a weak acid and an analogue of diclofenac. It is 400 times more selective for COX-2 than for COX-1 and in clinical trials produced fewer gastric ulcer complications and no more cases of myocardial infarction than comparator traditional NSAIDs. However, lumiracoxib was withdrawn by the European Medicines Agency because of a high risk of serious liver toxicity. Thus, the only selective COX-2 inhibitors prescribed for musculoskeletal pain and inflammation in the USA are celecoxib and meloxicam, an inhibitor with low selectivity for COX-2. However, in Europe, patients can be prescribed celecoxib, parecoxib (for injection only), etoricoxib, meloxicam, or nimesulide.

▶ Paracetamol (acetaminophen) has potent analgesic and antipyretic actions but very little anti-inflammatory activity. Unlike aspirin, it does not inhibit aggregation of platelets or cause damage to the gastric mucosa. Its mechanism of action has remained a mystery for many years, since it is only a weak inhibitor of PG biosynthesis by either COX-1 or COX-2 in vitro but inhibits PG biosynthesis in vivo. Reduction of PG biosynthesis by paracetamol varies in different tissues, and COX activity in brain is particularly sensitive to inhibition (Flower and Vane 1972). In human whole blood assays, paracetamol inhibited COX-2 selectively in preference to COX-1 (Hinz et al. 2008). However, a selective COX-2 action does not explain the lack of anti-inflammatory activity of paracetamol.

In 2002, Simmons and colleagues cloned, characterized, and expressed a splice variant of COX-1 (COX-1v or COX-1b) from canine brain, which was sensitive to inhibition by low concentrations of paracetamol (Chandrasekharan et al. 2002). However, only the mRNA of COX-1v was found in the human brain. Later, Quin et al. (2005) identified an active splice variant of COX-1 in human brain, which was no more sensitive to paracetamol than COX-1 or COX-2. It has also been postulated that paracetamol is a COX inhibitor at the peroxidase active site and its activity is reduced by high concentrations of peroxides, arachidonic acid, or COX enzyme in the cell (Aronoff et al. 2006). These properties of paracetamol may explain its selective action on cells which are involved in fever and pain such as endothelial cells and brain neurons.

References

Aronoff, D. M., Oates, J. A., & Boutaud, O. (2006). New insights into the mechanism of action of acetaminophen: Its clinical characteristics reflect its inhibition of the two prostaglandin H_2 synthases. *Clinical Pharmacology and Therapeutics, 79*, 9–19.

Bergström, S., Danielsson, H., & Samuelsson, B. (1964). The enzymatic formation of prostaglandin E2 from arachidonic acid. Prostaglandins and related factors 32. *Biochimica et Biophysica Acta, 90*, 207–210.

Bombardier, C., Laine, L., Reicin, A., Shapiro, D., Burgos-Vargas, R., Davis, B., et al. (2000). Comparison of upper gastrointestinal toxicity of rofecoxib and naproxen in patients with rheumatoid arthritis. *The New England Journal of Medicine, 343*, 1520–1528.

Chandrasekharan, N. V., Dai, H., Roos, K. L. T., Evanson, N. K., Tomsik, J., Elton, T. S., & Simmons, D. L. (2002). COX-3, a cyclooxygenase-1 variant inhibited by acetaminophen and other analgesic/antipyretic drugs: Cloning, structure and expression. *Proceedings of the National Academy of Sciences, 99*, 13926–13931.

DeWitt, D. L., & Smith, W. L. (1988). Primary structure of prostaglandin G/H synthase from sheep vesicular gland determined from the complementary DNA sequence. *Proceedings of the National Academy of Sciences, 85*, 1412–1416.

Flower, R. J., & Vane, J. R. (1972). Inhibition of prosta-glandin synthetase in brain explains the antipyretic activity of paracetamol (4-acetamidophenol). *Nature*, *240*, 410–411.

Goldblatt, M. W. (1935). Properties of human seminal plasma. *Journal of Physiology London*, *84*, 208–218.

Grosser, T., Fries, S., & FitzGerald, G. A. (2006). Biological basis for the cardiovascular consequences of COX-2 inhibition: Therapeutic challenges and opportunities. *The Journal of Clinical Investigation*, *116*, 4–15.

Hemler, M., Lands, W. E. M., & Smith, W. L. (1976). Purification of the cycloxygenase that forms prostaglandins: Demonstration of two forms of iron in the holoenzyme. *Journal of Biological Chemistry*, *251*, 5575–5579.

Hinz, B., Cheremina, O., & Brune, K. (2008). Acet-aminophen (paracetamol) is a selective cyclooxy-genase-2 inhibitor in man. *The FASEB Journal*, *22*, 383–390.

Masferrer, J. L., Zweifel, B. S., Seibert, K., & Needleman, P. (1990). Selective regulation of cellular cyclooxygenase and endotoxin in mice. *The Journal of Clinical Investigation*, *86*, 1375–1379.

Picot, D., Loll, P. J., & Garavito, R. M. (1994). The X-ray crystal structure of the membrane protein prostaglandin H2 synthase-1. *Nature*, *367*, 243–249.

Quin, N., Zhang, S. P., Reitz, T. L., Mei, J. M., & Flores, C. M. (2005). Cloning, expression and func-tional characterization of human cyclooxygenase-1 splicing variants: evidence for intron 1 retention. *Journal of Pharmacology and Experimental Therapeutics*, *315*, 1298–1305.

Silverstein, F. E., Faich, G., Goldstein, J. L., Simon, L. S., Pincus, T., Whelton, A., et al. (2000). Gastrointestinal toxicity with celecoxib vs. nonsteroidal anti-inflammatory drugs for osteoarthritis and rheumatoid arthritis: The CLASS study: A randomised controlled trial. Celecoxib long-term arthritis safety study. *Journal of the American Medical Association*, *284*, 1247–1255.

Vane, J. R. (1971). Inhibition of prostaglandin synthesis as a mechanism of action for the aspirin-like drugs. *Nature*, *231*, 232–235.

Vane, J. R., & Botting, R. M. (2001). Formation and actions of prostaglandins and inhibition of their synthesis. In J. R. Vane & R. M. Botting (Eds.), *Therapeutic roles of selective COX-2 inhibitors* (pp. 1–47). London: William Harvey Press.

von Euler, U. S. (1934). Zur Kenntnis der Pharmako-logischen Wirkung von Nativsekreten und Extracten Männlicher Accessorischer Geschlechtsdrüsen. *Arch Exp Path Pharmak*, *175*, 78–84.

Warner, T. D., Giuliano, F., Vojnovic, I., Bukasa, A., Mitchell, J. A., & Vane, J. R. (1999). Nonsteroid drug selectivities for cyclooxygenase-1 rather than cyclooxygenase-2 are associated with human gastrointestinal toxicity: A full in vitro analysis. *Proceedings of the National Academy of Sciences*, *96*, 7563–7568.

Xie, W., Chipman, J. G., Robertson, D. L., Erikson, R. L., & Simmons, D. L. (1991). Expression of a mitogen-responsive gene encoding prostaglandin synthase is regulated by mRNA splicing. *Proceedings of the National Academy of Sciences*, *88*, 2692–2696.

Cyclophosphamide

Definition

An alkylating agent frequently used as immunosup-pressant. The treatment of choice, in combination with steroids, in systemic vasculitides like WG.

Cross-References

▶ Headache Due to Arteritis
▶ Vascular Neuropathies

Cyclosporin, Methotrexate

Definition

Drugs used to suppress immune functions in organ-transplanted patients and in patients with diseases with disturbances in immune functions, e.g., rheumatoid arthritis and psoriasis.

Cross-References

▶ Postoperative Pain, Acute Pain Management Principles

Cystitis Models

▶ Visceral Pain Model, Urinary Bladder Pain (Irritants or Distension)

Cytoarchitectural

Definition

Pertaining to the cellular composition of bodily structure.

Cross-References

▶ Pain Processing in the Cingulate Cortex, Behavioral Studies in Humans

Cytochrome Oxidase Staining

Synonyms

CO staining

Definition

Type of staining that reflects mitochondrial activity in neurons.

Cross-References

▶ Spinothalamic Terminations, Core and Matrix
▶ Thalamus, Receptive Fields, Projected Fields, Human

Cytokine Modulation of Opioid Action

Vasudeva Raghavendra[1] and Joyce DeLeo[2]
[1]Algos Therapeutics Inc., Saint Paul, MN, USA
[2]Dartmouth Hitchcock Medical Center, Dartmouth Medical School, Hanover, NH, USA

Synonyms

Opioid modulation by cytokines

Definition

A conceptual framework has recently emerged to support the hypothesis that an ▶ opioid may, in fact, be broadly defined as a ▶ cytokine. This is due to the presence of opioid receptors on immune cells and the ability of opioids to mediate interactions between cells and regulate processes in the extracellular environment. In addition, it has been demonstrated that a bidirectional interaction exists among endogenous opioids and classically defined cytokines. In the central nervous system (CNS), cytokines appear to be modulated by endogenous opioids and the neural effects of opioids altered by cytokines. In particular, cytokines modulate opioid regulation of neuronal activity, sensory motor function, body temperature, food intake, and regulation of hypothalamic-pituitary-adrenocortical axis functions. Recently, the role of cytokines in modulating opioid ▶ analgesia, tolerance, and ▶ hyperalgesia has also been reported.

Characteristics

Opioids are a class of the most effective analgesics for treating many forms of acute and chronic pain. In addition to the known adverse effects, the clinical utility of opioid analgesics is often hampered by the development of analgesic tolerance, hyperalgesia, and physical dependence. A wide range of neurotransmitters and neuromodulators play a role in the development of tolerance and dependence to opioids. Recent studies have demonstrated that cytokines are also capable of modulating opioid-induced analgesia and play a role in the development of opioid tolerance and withdrawal-induced hyperalgesia.

Cytokines

Cytokines, e.g., IL-1β, IL-6, IL-10, interferons, and chemokines, have the ability to modulate the analgesic action of opioids. For example, the proinflammatory cytokine, IL-1β, directly attenuates opioid analgesia (Gul et al. 2000). IL-1β also plays a role in the anti-analgesic activity of dynorphin and ligands of the peripheral benzodiazepine receptor (Laughlin et al. 2000; Rady and Fujimoto 2001). The role of IL-6 in modulating opioid analgesia is more complicated. Although IL-6 is known to induce hyperalgesia in animals, IL-6 knockout mice showed reduced morphine analgesia and an early development of morphine tolerance (Bianchi et al. 1999). We have shown that neutralizing or blocking IL-6, IL-1β, and TNF-α spared the analgesic action of morphine in rats following a peripheral nerve injury (Raghavendra et al. 2002). A sickness-inducing agent such as lipopolysaccharide or lithium chloride, which induces proinflammatory cytokine release, reduces morphine analgesia (Johnston and Westbrook 2003). Proinflammatory cytokines also play a role in the development of morphine tolerance and withdrawal-induced hyperalgesia. Chronic administration of heroin or morphine to mice or rats resulted in increased peripheral and central expression of proinflammatory cytokines (Raghavendra et al. 2002; Holan et al. 2003). The central expression of proinflammatory cytokines after chronic morphine treatment exhibited a temporal correlation with the development of analgesic tolerance and withdrawal-induced hyperalgesia in rats (Raghavendra et al. 2002). Conversely, IL-10, an anti-inflammatory cytokine, spared the analgesic actions of opioids in dynorphin-treated rats (Laughlin et al. 2000). Interferon-alpha (IFN-α), which exhibits structural and functional similarities to endorphins, induces opioid receptor-mediated analgesia and prevents the development of tolerance to opioids (Dafny 1984).

Chemokines

▶ Chemokines (chemotactic cytokines) also influence the perception of pain by interacting with G-protein-coupled opioid receptors. Recent studies have shown that cross-desensitization between opioid and chemokine receptors has significant implications in the regulation of leukocyte trafficking, progression of inflammatory disease, and regulation of opioid receptor function in the CNS (Steele et al. 2002). The direct administration of chemokines, particularly CXCL12 and CCL5,

into the periaqueductal gray matter, inhibited opioid-induced analgesia by cross-desensitizing μ-opioid receptors (Szabo et al. 2002). Opioid enhancement of chemokine upregulation is also responsible for enhanced HIV infection and progression to AIDS in opioid abusers (Peterson et al. 1998).

Glial Component

Major sources of proinflammatory cytokines and chemokines in the CNS are activated ▶ glial cells. The presence of opioid receptors on microglia and astrocytes is well documented, and priming these cells with opioids induces enhanced secretion of proinflammatory cytokines and chemokines (Peterson et al. 1998). Activation of spinal proinflammatory cytokines by chronic opioid treatment may be caused by its interaction with glial cells (Raghavendra et al. 2002). Attenuation of glial activation using the methylxanthine, propentofylline, restored opioid analgesia and prevented the development of morphine tolerance and morphine withdrawal-induced hyperalgesia and opioid-induced neuroimmune activation (Raghavendra and DeLeo 2004). This reduction in neuroimmune activation by propentofylline included decreased microglial, astrocytic activation, and cytokine expression in the lumbar spinal cord following morphine administration in a peripheral nerve injury in a rat model of neuropathy.

Intracellular Signaling Cascades

Mitogen-activated protein (MAP) kinase or protein kinase C (PKC) pathways may be involved in opioid-induced activation of proinflammatory immune responses. MAP kinase and PKC are intriguing, because they are key players in the intracellular signaling cascade leading to the development of morphine tolerance and production of proinflammatory immune activation (Raghavendra et al. 2002). Since proinflammatory cytokines and chemokines induce hyperalgesia in animals, it is possible that activation of pro-nociceptive pathways by these mediators could counteract opioid analgesia.

Cellular Adhesion Molecules

The immune system has the capacity to modulate pain directly at the site of tissue injury by liberating opioid peptides from immune cells. In an inflammatory pain model, the peripheral analgesic activity of opioids is modulated by ▶ selectins and intracellular adhesion molecule-1 (ICAM-1). ICAM-1, expressed on vascular endothelium, recruits immunocytes containing opioids to promote the local control of inflammatory pain. Anti-selectin or anti-ICAM-1 treatment strongly reduces endogenous peripheral opioid analgesia, apparently by blocking the extravasation of immune cells containing β-endorphin and by the consequent decrease of the β-endorphin level in the inflamed tissue (Machelska et al. 1998, 2002). Therefore, the immune control of opioid action may demonstrate a dichotomous action in the periphery as compared with the central nervous system.

Clinical Relevance

One of the reasons for decreased opioid analgesia in neuropathic pain conditions may be an enhanced expression of proinflammatory cytokines and chemokines in the CNS, which is evoked following peripheral nerve injury. It is well established that exaggerated pain states occur as a part of an inflammatory stress reaction. Preclinical studies demonstrated that neutralizing proinflammatory cytokines restored the analgesic action of opioids in neuropathic conditions, induced either by experimental diabetes or by L5 nerve transection in rats (Gul et al. 2000; Raghavendra et al. 2002). Pathological pain is the result of a dynamic interplay between hyperalgesic and analgesic mediators. During tissue injury or inflammation, both hyperalgesic and anti-hyperalgesic/analgesic mediators are produced by activated immune cells. Hyperalgesic mediators, such as proinflammatory cytokines and chemokines, and anti-hyperalgesic mediators, such as anti-inflammatory cytokines and adhesion molecules, modulate opioid analgesia. Drugs such as immunosuppressants that influence this interplay may also modulate endogenous or exogenously administered opioid analgesia.

Conclusions

Of particular clinical interest, the glial modulating agent, propentofylline, was shown to prevent opioid tolerance and hyperalgesia in a rat model of neuropathy (Raghavendra and DeLeo 2004; Raghavendra et al. 2003). Together, these data offer a novel clinical therapy for the prevention or delay of opioid tolerance and hyperalgesia: the coadministration of cytokine inhibitors and/or glial modulators with opioids. Recognizing that opioids remain the gold standard for the treatment of acute pain, improvement of their untoward side-effect profile would enhance their clinical efficacy.

References

Bianchi, M., Maggi, R., Pimpinelli, F., Rubino, T., Parolaro, D., Poli, V., et al. (1999). Presence of a reduced opioid response in interleukin-6 knockout mice. *European Journal of Neuroscience, 11*, 1501–1507.

Dafny, N. (1984). Interferon: A candidate as the endogenous substance preventing tolerance dependence to brain opioids. *Progress in Neuro-Psychopharmacology, 8*, 351–357.

Gul, H., Yildiz, O., Dogrul, A., Yesilyurt, O., & Isimer, A. (2000). The interaction between IL-1β and morphine: Possible mechanism of the deficiency of morphine-induced analgesia in diabetic mice. *Pain, 89*, 39–45.

Holan, V., Zajikova, A., Krulova, M., Blahoutova, V., & Wilczek, H. (2003). Augmented production of proinflammatory cytokines and accelerated allotransplantation reaction in heroin-treated mice. *Clinical and Experimental Immunology, 132*, 40–45.

Johnston, I. N., & Westbrook, R. F. (2003). Acute and conditioned sickness reduces morphine analgesia. *Behavioural Brain Research, 142*, 89–97.

Laughlin, T. M., Bethea, J. R., Yezierski, R. P., & Wilcox, G. L. (2000). Cytokine involvement in dynorphin-induced allodynia. *Pain, 84*, 159–167.

Machelska, H., Cabot, P. J., Mousa, S. A., Zhang, Q., & Stein, C. (1998). Pain control in inflammation governed by selectins. *Nature Medicine, 4*, 1425–1428.

Machelska, H., Mousa, S. A., Brack, A., Schopohl, J. K., Rittner, H. L., Schafer, M., et al. (2002). Opioid control of inflammatory pain regulated by intracellular adhesion molecule-1. *Journal of Neuroscience, 22*, 5588–5596.

Peterson, P. K., Molitor, T. W., & Chao, C. C. (1998). The opioid-cytokine connection. *Journal of Neuroimmunology, 83*, 63–69.

Rady, J. J., & Fujimoto, J. M. (2001). Confluence of antialgesic action of diverse agents through brain Interleukin$_1$ $_i^2$ in mice. *Journal of Pharmacology and Experimental Therapeutics, 299*, 659–665.

Raghavendra, V., & DeLeo, J. A. (2004). The role of astrocytes and microglia in persistent pain. In L. Hertz (Ed.), *Non-neuronal cells in the nervous system. Function and dysfunction* (pp. 951 966). Amster dam: Elsevier.

Raghavendra, V., Rutkowski, M. D., & DeLeo, J. A. (2002). The role of spinal neuroimmune activation in morphine tolerance/hyperalgesia in neuropathic and sham operated rats. *Journal of Neuroscience, 22*, 9980–9989.

Raghavendra, V., Tanga, F., Rutkowski, M. D., & DeLeo, J. A. (2003). Anti-hyperalgesic and morphine-sparing actions of propentofylline following peripheral nerve injury in rats: Mechanistic implications of spinal glia and proinflammatory cytokines. *Pain, 104*, 657–665.

Steele, A. D., Szabo, I., Bednar, F., & Rogers, T. J. (2002). Interaction between opioid and chemokine receptors: Heterologous desensitization. *Cytokine & Growth Factor Reviews, 13*, 209–222.

Szabo, I., Chen, X. H., Xin, L., Adler, M. W., Howard, O. M., Oppenheim, J. J., et al. (2002). Heterologous desensitization of opioid receptors by chemokines inhibits chemotaxis and enhances the perception of pain. *Proceedings of the National Academy of Sciences of the United States of America, 99*, 10276–10281.

Cytokines

▶ Proinflammatory Cytokines

Cytokines and Pain

Definition

Any of several small regulatory proteins (25 kDa), such as the interleukins and lymphokines, that are released by cells of the immune system and act as intercellular mediators in the generation of immune and inflammatory responses. They can act in an autocrine, paracrine, and endocrine manner. Interleukins, interferons, and tumor necrosis factor (TNF) alpha are typical members of the cytokine family.

Cross-References

▶ Adjuvant Analgesics in Management of Cancer-Related Bone Pain
▶ Animal Models of Inflammatory Bowel Disease
▶ Cytokine Modulation of Opioid Action
▶ Cytokines as Targets in the Treatment of Neuropathic Pain
▶ Cytokines, Regulation in Inflammation
▶ Inflammatory Neuritis
▶ Muscle Pain in Systemic Inflammation (Polymyalgia Rheumatica, Giant Cell Arteritis, Rheumatoid Arthritis)
▶ Neutrophils in Inflammatory Pain
▶ Vascular Neuropathies
▶ Wallerian Degeneration

Cytokines as Targets in the Treatment of Neuropathic Pain

Claudia Sommer[1] and Linda S. Sorkin[2]
[1]Department of Neurology, University of Würzburg, Würzburg, Germany
[2]Occupational Therapy and Rehabilitation, Anesthesia Research Laboratories, University of California, San Diego, CA, USA

Synonyms

Chemokines; Interleukins; Lymphokines

Definition

Extracellular signaling proteins within the immune system and between the immune system and nervous system.

Characteristics

Cytokines

▶ Cytokines are a heterogeneous group of proteins that were originally found to mediate

activation of the immune system and inflammatory responses. They are produced by white blood
cells and a variety of cells including neurons,
Schwann cells, and other glial cells. Most cytokines act on multiple tissue types, including those
of the peripheral and central nervous systems.
Cytokines are extracellular signaling proteins
that form part of a bidirectional circuit between
the immune system and the nervous system,
acting at hormonal concentrations through
high-affinity receptors, and producing endocrine,
paracrine, and autocrine effects. In contrast to
circulating hormones, they exert their effects
over short distances onto nearby cells. In vivo
concentrations are in the range of a few petagram
to nanogram per milliliter. Due to local cytokine
effects at low concentrations, serum levels may
not reliably reflect activity. Cytokines are called
"pleiotropic," due to a broad range of redundant,
frequently overlapping functions. Their activation or dysregulation is implied in a variety of
disease states like sepsis, rheumatoid arthritis,
Crohn's disease, multiple sclerosis, skin diseases,
and many more. Some cytokines are labeled
"pro-inflammatory" or "Th1," and others "anti-
inflammatory" or "Th2," depending on their
effects on immune cells, in particular on
lymphocytes.

Cytokines and Pain
Evidence from Experimental Studies
Numerous experimental studies provide evidence that pro-inflammatory cytokines induce
or facilitate inflammatory as well as neuropathic pain and hyperalgesia. Direct receptor-
mediated actions of cytokines on afferent nerve
fibers have been reported, as well as cytokine
effects involving downstream mediators. The
first indications of a hyperalgesic effect of cytokines came from studies using intraplantar cytokine injections in the rat; interleukin-1β (IL-1β)
and ▶ tumor necrosis factor alpha (TNFα)
reduced mechanical nociceptive thresholds in
a cyclooxygenase-dependent process. Cytokine
antagonists reduced hyperalgesia in this
animal model, indicating that the cytokines
were activated in a classical sequence
(TNF ⇒ IL-1 ⇒ ▶ IL-6), and that activation

of this pro-inflammatory cytokine cascade is
an important step in the development of inflammatory pain (for review see (Poole et al. 1999)).
A second parallel pathway involved in inflammatory pain is also triggered by TNF and
involves an IL-8 sympathetic efferent fiber linkage. It is unknown to what extent this second
adrenergic-dependent pathway participates in
neuropathic pain.

After nerve injury, cytokine production in
the injured nerve is upregulated, and blockade
of the pro-inflammatory cytokines TNFα and
IL-1β reduces behavioral signs of hyperalgesia
in experimental animals (for review see
(Sommer 2001)). In addition to its indirect
actions involving secondary production of
nitric oxide, bradykinin, or prostaglandins,
IL-1β has a direct excitatory action on nociceptive fibers, which are activated within 1 min of
its injection (Fukuoka et al. 1994). In a skin-
nerve in vitro preparation, brief exposure of the
corium to IL-1β resulted in a facilitation of
heat-evoked calcitonin gene-related peptide
release from peptidergic neurons (Opree and
Kress 2000). The short response latency and
the absence of the neuronal cell soma in the
preparation indicate that the heat sensitization
is independent of changes in gene expression or
receptor upregulation. Further experiments
showed that IL-1β can act directly on sensory
neurons to increase their sensitivity to noxious
heat via a mechanism involving IL-1 receptor I,
tyrosine kinase, and protein kinase C (Obreja
et al. 2002).

TNFα has been shown to lower mechanical
activation thresholds in C nociceptors of the rat
sural nerve when injected subcutaneously
(Junger and Sorkin 2000). While mechanisms of
this action are unknown, the acute TNF-induced
decrease in K^+ conductance may play an important role (Liu et al. 2008). In vitro perfusion of
TNFα onto dorsal root ganglia (DRG) elicits neuronal discharges in both A and C-fibers. Firing
frequency is markedly higher and the discharge
longer lasting after nerve injury, indicating an
increased sensitivity of injured afferent neurons
to TNFα (Schäfers et al. 2003). Furthermore,
DRG neurons with injured afferents, and

neighboring neurons attached to intact afferent fibers running within the same peripheral nerve, have an increased immunoreactivity to TNFα (probable increased TNFα protein), and both display increased sensitivity to TNFα (Schäfers et al. 2003). Interestingly, the effects on the neighboring neurons, but not the injured ones, seem to be mediated predominantly by the TNFR2 receptor. Injection or perfusion of TNFα into/onto rat DRGs in vivo induces allodynia. Subthreshold quantities of TNFα, injected into a DRG at the same time that its spinal nerve was ligated resulted in a synergy that manifests as faster onset of allodynia and increased spontaneous pain behavior (Schäfers et al. 2003). Thus, there is strong in vivo and in vitro evidence that injury results in increased endogenous TNFα, and that injured nerve fibers are sensitized to the excitatory effects of TNFα. Downstream mediators of this excitation at the DRG level include prostaglandins, PKA, p38 and changes in the excitability of both Na+ and K+ channels (See (Schäfers and Sorkin 2008) for review).

▶ Nucleus pulposus, the material within the vertebral disks, is highly enriched with a variety of pro-inflammatory cytokines. Subsequent to disk herniation, it is likely that this material comes into contact with the dorsal roots. Application of autologous nucleus pulposus to dorsal roots, or experimental disk herniation in animals, results in pain behavior as well as both ongoing and enhanced evoked activity in spinal nociceptive neurons (Onda et al. 2003). Animals treated either with neutralizing antibodies to TNFα directly on the nerve root or with systemic TNFα antagonists showed a marked reduction of both the neuronal activity and the pain behavior, implicating a role for TNFα in the process.

In particular, interleukin-6 (IL-6) seems to be a pivotal cytokine in conditions like muscle pain (Manjavachi et al. 2010) and chronic widespread pain, as in the fibromyalgia syndrome (Üçeyler et al. 2011). As another example, mice lacking the IL-6 receptor gp 130 in nociceptors are protected from tumor pain (Andratsch et al. 2009).

In addition to their sensitizing and directly activating effects on peripheral nociceptors, mechanisms of central sensitization by pro-inflammatory cytokines have been identified. For example, TNFα, IL-1, and IL-6 regulate synaptic activity in the superficial spinal cord by enhancing transmission via excitatory neurotransmitters, while concurrently decreasing inhibitory transmission (Kawasaki et al. 2008; Zhang et al. 2010). Peripheral inflammation induces TNF-dependent movement of the glutamate receptor 1 into neuronal membranes in the spinal cord (Choi et al. 2010).

The chemokine (C-C motif) ligand 2 (CCL2) also known as monocyte chemotactic protein-1 (MCP-1) is an exception in several ways. In particular, it can be released from nociceptors like a neurotransmitter and is thus involved in nociceptive signal processing in the spinal cord (Van Steenwinckel et al. 2011). Furthermore, CCL-2 may directly excite primary sensory neurons by increasing Nav1.8 current density and decreasing inactivation (Belkouch et al. 2011). Moreover, CCL-1 is involved in opening the blood-spinal cord barrier after peripheral nerve injury, which contributes to local inflammation in the spinal cord (Echeverry et al. 2011). Importantly, this enhanced permeability is countered by administration of anti-inflammatory cytokines. Both the increased permeability and its blockade hold significant implications for the development of neuropathic pain and efficacy of systemically administered therapeutic agents (Echeverry et al. 2011).

Anti-inflammatory cytokines like ▶ IL-4, IL-13, IL-10, and perhaps erythropoietin (Epo) seem to have anti-hyperalgesic actions in animal models of pain (Li et al. 2005; Wagner et al. 1998). IL-10 pretreatment reduces the hyperalgesic responses to intraplantar injections of carrageenan, IL-1β, IL-6, and TNFα (Poole et al. 1999). Systemic IL-10 downregulates local levels of IL-1β, TNFα, and nerve growth factor (NGF) after endotoxin injection into the hindpaw, and reduces thermal and mechanical hyperalgesia. Epo has similar actions reducing TNFα and IL-6 expression as well as inflammatory infiltrates and increasing the rate of recovery after nerve injury (Campana 2007). IL-4 and IL-10, delivered by a viral vector, reduce behavioral signs of pain in an animal model of neuropathic pain. Spinal administration of gene therapy to increase endogenous production of anti-inflammatory cytokines is also highly effective in not only reducing behavioral signs of pain, but also blocking injury-induced increases in DRG pro-inflammatory cytokine expression (Ledeboer et al. 2007).

Like in other compartments, in the skin, a multi-protein complex named an "inflammasome" can regulate cytokine release, which may be an important step in the pathophysiology of complex regional pain syndrome (CRPS) (Li et al. 2009). Another main switch seems to be the purinergic system. For example, the P2X7 receptor regulates microglia activation and release of IL-1ß in the spinal cord dorsal horn (Clark et al. 2010).

Clinical Applications

The most striking correlation between cytokine levels and neuropathic pain comes from leprosy, where a subgroup of patients has elevated serum levels of TNFα and IL-1; these patients suffer from excruciating pain. Treatment with thalidomide reduces TNFα secretion in peripheral blood mononuclear cells by >90 % and greatly reduces pain in the patients (Barnes et al. 1992). In other human neuropathies, preliminary data also point to a correlation between cytokine expression and pain. Not all patients with elevated cytokine levels in their sural nerve biopsy have pain, but pro-inflammatory cytokine levels (IL-1β, IL-6, and TNFα) were increased more often in patients with painful neuropathies (Lindenlaub and Sommer 2003), indicating that at least a subgroup of patients with painful neuropathies might benefit from cytokine inhibition. Several studies have shown changes of various pro-inflammatory cytokines in plasma of pain patients, but specific patterns vary with pain condition, whether mRNA or protein was measured and technique employed. Perhaps the most interesting is a study that looked at a variety of patients with neuropathies divided

into those that were painful and those not associated with pain (Üçeyler et al. 2007). Levels of TNFα and IL-2 were elevated in the patients with pain, compared to control and nonpainful disease, while the anti-inflammatory cytokines IL-4 and IL-10 were increased in the pain-free population. It has been posited that the loss of normal balance between the pro- and anti-inflammatory cytokines predisposes the patient to develop chronic pain. Occasionally, inflammatory neuropathies have been treated with TNFα inhibitors, but data on pain are frequently lacking. One remarkable case report describes remission of long-standing complex regional pain syndrome (CRPS) after treatment with thalidomide for Behçet's disease (Ching et al. 2003).

Anti-TNFα strategies have been used in various non-neuropathic painful conditions like AIDS-associated proctitis, rheumatoid arthritis, and IIIV-associated aphthous ulcers. Many other reports have followed since the advent of etanercept and ▶ infliximab: for example, a controlled trial showing a beneficial effect of etanercept in ankylosing spondylitis, and a prospective study with historical controls using infliximab in low back pain.

Several studies have reported increased levels of cytokines in the vicinity of herniated disks and a correlation of their presence to sciatica (Brisby et al. 2002). A recent double-blind, placebo-controlled, albeit small, study for pain associated with sciatica demonstrated reduction in pain and a positive global perceived effect with epidural etanercept (Cohen et al. 2009). It was postulated that the different outcomes between this study and that of the negative study of Korhonen et al. (Korhonen et al. 2005) were due to epidural rather than intravenous route of administration of the antagonist (infliximab) and the consequent higher drug levels at the nerve injury site. Previously, Dahl and Cohen (Dahl and Cohen 2008) published a series of case reports demonstrating efficacy of perineural etanercept for residual limb (stump) pain and neuroma formation. Importantly, the treatment was effective only for those patients whose legs had been amputated within the previous year.

Although inhibition of pro-inflammatory cytokines may have an analgesic effect, like TNF blockade in rheumatoid arthritis, care has to be taken when trying to use cytokine antagonists in neuropathic pain, since the same pro-inflammatory cytokines that are major mediators of pain, like IL-1β and TNFα, are needed for nerve regeneration (Nadeau et al. 2011).

Inhibitors of TNFα have been used successfully for patients with chronic nerve root pain, but up to now only data from case reports and uncontrolled studies have been available.

In summary, evidence from numerous preclinical studies and preliminary data in humans point to a possible beneficial role of cytokine inhibition in patients with painful neuropathy or radiculopathy. More randomized controlled trials are needed to determine the potential breadth of their role in the treatment of neuropathic pain.

References

Andratsch, M., Mair, N., Constantin, C. E., Scherbakov, N., Benetti, C., Quarta, S., Vogl, C., Sailer, C. A., Üçeyler, N., Brockhaus, J., Martini, R., Sommer, C., Zeilhofer, H. U., Müller, W., Kuner, R., Davis, J. B., Rose-John, S., & Kress, M. (2009). A key role for gp130 expressed on peripheral sensory nerves in pathological pain. *Journal of Neuroscience, 29*, 13473–13483.

Barnes, P. F., Chatterjee, D., Brennan, P. J., Rea, T. H., & Modlin, R. L. (1992). Tumor necrosis factor production in patients with leprosy. *Infection and Immunity, 60*, 1441–1446.

Belkouch, M., Dansereau, M. A., Reaux-Le Goazigo, A., Van Steenwinckel, J., Beaudet, N., Chraibi, A., Melik-Parsadaniantz, S., & Sarret, P. (2011). The chemokine CCL2 increases Nav1.8 sodium channel activity in primary sensory neurons through a Gbetagamma-dependent mechanism. *Journal of Neuroscience, 31*, 18381–18390.

Brisby, H., Olmarker, K., Larsson, K., Nutu, M., & Rydevik, B. (2002). Proinflammatory cytokines in cerebrospinal fluid and serum in patients with disc herniation and sciatica. *European Spine Journal, 11*, 62–66.

Campana, W. M. (2007). Schwann cells: Activated peripheral glia and their role in neuropathic pain. *Brain, Behavior, and Immunity, 21*, 522–527.

Ching, D. W. T., McClintock, A., & Beswick, F. (2003). Successful treatment with low-dose thalidomide in a patient with both Behcet's disease and complex regional pain syndrome type I. *Journal of Clinical Rheumatology, 9*, 96–98.

Choi, J. I., Svensson, C. I., Koehrn, F. J., Bhuskute, A., & Sorkin, L. S. (2010). Peripheral inflammation induces tumor necrosis factor dependent AMPA receptor trafficking and Akt phosphorylation in spinal cord in addition to pain behavior. *Pain, 149*, 243–253.

Clark, A. K., Staniland, A. A., Marchand, F., Kaan, T. K., McMahon, S. B., & Malcangio, M. (2010). P2X7-dependent release of interleukin-1beta and nociception in the spinal cord following lipopolysaccharide. *Journal of Neuroscience, 30*, 573–582.

Cohen, S. P., Bogduk, N., Dragovich, A., Buckenmaier, C. C., 3rd, Griffith, S., Kurihara, C., Raymond, J., Richter, P. J., Williams, N., & Yaksh, T. L. (2009). Randomized, double-blind, placebo-controlled, dose–response, and preclinical safety study of transforaminal epidural etanercept for the treatment of sciatica. *Anesthesiology, 110*, 1116–1126.

Dahl, E., & Cohen, S. P. (2008). Perineural injection of etanercept as a treatment for postamputation pain. *The Clinical Journal of Pain, 24*, 172–175.

Echeverry, S., Shi, X. Q., Rivest, S., & Zhang, J. (2011). Peripheral nerve injury alters blood-spinal cord barrier functional and molecular integrity through a selective inflammatory pathway. *Journal of Neuroscience, 31*, 10819–10828.

Fukuoka, H., Kawatani, M., Hisamitsu, T., & Takeshige, C. (1994). Cutaneous hyperalgesia induced by peripheral injection of interleukin-1ß in the rat. *Brain Research, 657*, 133–140.

Junger, H., & Sorkin, L. S. (2000). Nociceptive and inflammatory effects of subcutaneous TNFalpha. *Pain, 85*, 145–151.

Kawasaki, Y., Zhang, L., Cheng, J. K., & Ji, R. R. (2008). Cytokine mechanisms of central sensitization: Distinct and overlapping role of interleukin-1beta, interleukin-6, and tumor necrosis factor-alpha in regulating synaptic and neuronal activity in the superficial spinal cord. *Journal of Neuroscience, 28*, 5189–5194.

Korhonen, T., Karppinen, J., Paimela, L., Malmivaara, A., Lindgren, K. A., Jarvinen, S., Niinimaki, J., Veeger, N., Seitsalo, S., & Hurri, H. (2005). The treatment of disc herniation-induced sciatica with infliximab: Results of a randomized, controlled, 3-month follow-up study. *Spine (Phila Pa 1976), 30*, 2724–2728.

Ledeboer, A., Jekich, B. M., Sloane, E. M., Mahoney, J. H., Langer, S. J., Milligan, E. D., Martin, D., Maier, S. F., Johnson, K. W., Leinwand, L. A., Chavez, R. A., & Watkins, L. R. (2007). Intrathecal interleukin-10 gene therapy attenuates paclitaxel-induced mechanical allodynia and proinflammatory cytokine expression in dorsal root ganglia in rats. *Brain, Behavior, and Immunity, 21*, 686–698.

Li, X., Gonias, S. L., & Campana, W. M. (2005). Schwann cells express erythropoietin receptor and represent a major target for Epo in peripheral nerve injury. *Glia*, *51*, 254–265.

Li, W. W., Guo, T. Z., Liang, D., Shi, X., Wei, T., Kingery, W. S., & Clark, J. D. (2009). The NALP1 inflammasome controls cytokine production and nociception in a rat fracture model of complex regional pain syndrome. *Pain, 147*, 277–286.

Lindenlaub, T., & Sommer, C. (2003). Cytokines in sural nerve biopsies from inflammatory and non-inflammatory neuropathies. *Acta Neuropathologica, 105*, 593–602.

Liu, B. G., Dobretsov, M., Stimers, J. R., & Zhang, J. M. (2008). Tumor necrosis factor-alpha suppresses activation of sustained potassium currents in rat small diameter sensory neurons. *Open Pain Journal, 1*, 1.

Manjavachi, M. N., Motta, E. M., Marotta, D. M., Leite, D. F., & Calixto, J. B. (2010). Mechanisms involved in IL-6-induced muscular mechanical hyperalgesia in mice. *Pain, 151*, 345–355.

Nadeau, S., Filali, M., Zhang, J., Kerr, B. J., Rivest, S., Soulet, D., Iwakura, Y., de Rivero Vaccari, J. P., Keane, R. W., & Lacroix, S. (2011). Functional recovery after peripheral nerve injury is dependent on the pro-inflammatory cytokines IL-1beta and TNF: Implications for neuropathic pain. *Journal of Neuroscience, 31*, 12533–12542.

Obreja, O., Rathee, P. K., Lips, K. S., Distler, C., & Kress, M. (2002). IL-1 beta potentiates heat-activated currents in rat sensory neurons: Involvement of IL-1RI, tyrosine kinase, and protein kinase C. *The FASEB Journal, 16*, 1497–1503.

Onda, A., Yabuki, S., & Kikuchi, S. (2003). Effects of neutralizing antibodies to tumor necrosis factor-alpha on nucleus pulposus-induced abnormal nociresponses in rat dorsal horn neurons. *Spine, 28*, 967–972.

Opree, A., & Kress, M. (2000). Involvement of the proinflammatory cytokines tumor necrosis factor-alpha, IL-1beta, and IL-6 but not IL-8 in the development of heat hyperalgesia: Effects on heat-evoked calcitonin gene-related peptide release from rat skin. *Journal of Neuroscience, 20*, 6289–6293.

Poole, S., Cunha, F. Q., & Ferreira, S. H. (1999). Hyperalgesia from subcutaneous cytokines. In L. R. Watkins & S. F. Maier (Eds.), *Cytokines and pain* (pp. 59–87). Basel: Birkhäuser.

Schäfers, M., & Sorkin, L. (2008). Effect of cytokines on neuronal excitability. *Neuroscience Letters, 437*, 188–193.

Schäfers, M., Lee, D. H., Brors, D., Yaksh, T. L., & Sorkin, L. S. (2003). Increased sensitivity of injured and adjacent uninjured rat primary sensory neurons to exogenous tumor necrosis factor-alpha after spinal nerve ligation. *Journal of Neuroscience, 23*, 3028–3038.

Sommer, C. (2001). Cytokines and neuropathic pain. In P. Hansson, H. Fields, R. Hill, & P. Marchettini (Eds.), *Neuropathic pain: Pathophysiology and treatment* (pp. 37–62). Seattle: IASP Press.

Üçeyler, N., Rogausch, J. P., Toyka, K. V., & Sommer, C. (2007). Differential expression of cytokines in painful and painless neuropathies. *Neurology, 69*, 42–49.

Üçeyler, N., Häuser, W., & Sommer, C. (2011). Systematic review with meta-analysis: Cytokines in fibromyalgia syndrome. *BMC Musculoskeletal Disorders, 12*, 245.

Van Steenwinckel, J., Reaux-Le Goazigo, A., Pommier, B., Mauborgne, A., Dansereau, M. A., Kitabgi, P., Sarret, P., Pohl, M., & Melik, P. S. (2011). CCL2 released from neuronal synaptic vesicles in the spinal cord is a major mediator of local inflammation and pain after peripheral nerve injury. *Journal of Neuroscience, 31*, 5865–5875.

Wagner, R., Janjigian, M., & Myers, R. R. (1998). Anti-inflammatory interleukin-10 therapy in CCI neuropathy decreases thermal hyperalgesia, macrophage recruitment, and endoneurial TNF-alpha expression. *Pain, 74*, 35–42.

Zhang, H., Nei, H., & Dougherty, P. M. (2010). A p38 mitogen-activated protein kinase-dependent mechanism of disinhibition in spinal synaptic transmission induced by tumor necrosis factor-alpha. *Journal of Neuroscience, 30*, 12844–12855.

Cytokines, Effects on Nociceptors

Victoria C. J. Wallace
Department of Anaesthetics, Pain Medicine and Intensive Care Chelsea and Westminster Hospital Campus, London, UK

Synonyms

Immunocytokines

Definition

Cytokines are a family of growth factor proteins secreted primarily from ▶ leukocytes as part of the immune and inflammatory response.

Characteristics

Interactions with Peripheral Nociceptive Terminals

Proinflammatory cytokines, such as IL-1, TNF-α, and IL-6, as well as other inflammatory mediators

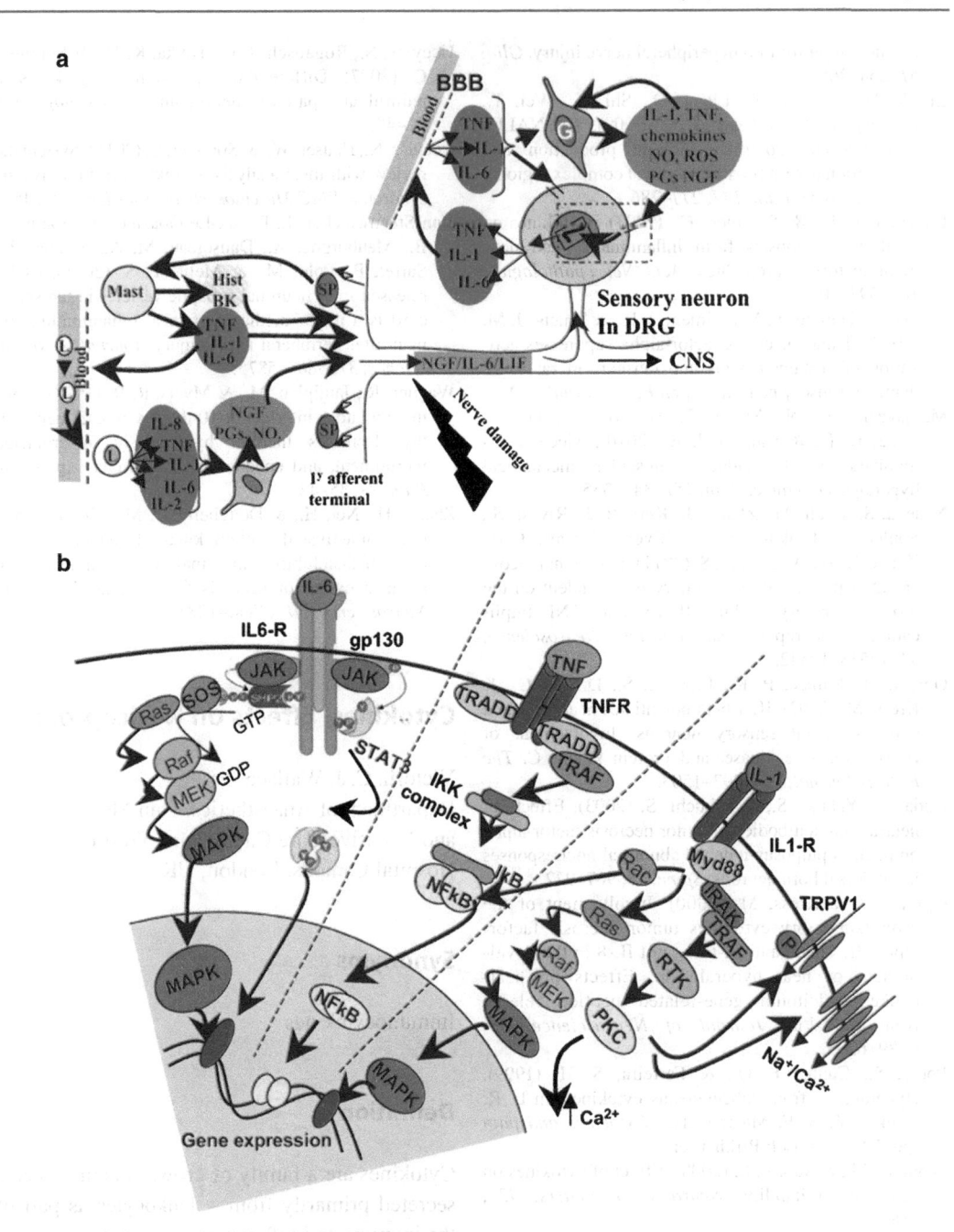

Cytokines, Effects on Nociceptors, Fig. 1 (a) Interactions of cytokines with nociceptors at their peripheral terminals and in the DRG. Injurious stimuli induce immune and inflammatory cells to release cytokines and chemokines, which stimulate further release of cytokines and non-cytokine inflammatory mediators such as prostaglandins (*PGs*), NGF, and nitric oxide (*NO*) from various tissue cells. Cytokines and various inflammatory mediators such as histamine (*Hist*) and bradykinin (*BK*) can activate nociceptor terminals to release neurotransmitters such as SP leading to neurogenic inflammation. Retrogradely transported signal proteins cause the release of cytokines and other factors from blood, neurones, and glia (*G*) into the DRG which can activate intracellular cascades (see 1b) regulating gene transcription, cell phenotype, and excitability and therefore sensory transmission to the spinal cord. (**b**) Highlighted region in (1a) (*dashed line*) representing examples of some of the intracellular

that cytokines influence as part of the inflammatory response, can directly stimulate and sensitize peripheral terminals of small diameter ▶ nociceptors (C-fibers) influencing transduction and transmission of the nociceptive signal contributing to ▶ peripheral sensitization (Fig. 1a). For example, a subcutaneous injection of IL-1, TNF-α, or IL-6 has been shown to directly excite peripheral nociceptive fibers (Poole et al. 1999). Also, TNF-α, which causes the subsequent release of IL-1, produces thermal hyperalgesia and mechanical allodynia, as well as endoneurial inflammation, demyelination, and axonal degeneration while lowering C-fiber thresholds and inducing ectopic activity in single primary afferent nociceptive fibers (Junger and Sorkin 2000). Such stimulation of nociceptors not only results in signalling to the spinal cord, leading to sensation/perception of pain, but also to a local axon reflex. Importantly, SP acts directly via surface-expressed NK1 receptors, to degranulate mast cells causing release of histamine, which further sensitize nociceptive terminals, and proinflammatory cytokines, thus inducing a SP-cytokine positive-feedback loop. Proinflammatory cytokines can also modulate nociceptors indirectly, by stimulating the release of a variety of pronociceptive inflammatory mediators such as the following: BK, which directly stimulates nociceptor terminals via BK receptors (Dray 1997), also upregulated by proinflammatory cytokines; prostaglandins, such as PGE2, which can be produced by almost all types of cells and is a well-known key player in the inflammatory cascade and inflammatory hyperalgesia (Samad et al. 2002); and ▶ chemokines (small

chemottractant cytokines) such as IL-8 (CXCL8) which in turn stimulate the production of sympathomimetic amines capable of directly sensitizing nociceptors.

Another major pronociceptive target of cytokine function is NGF, which is released by a variety of cells at the site of infection/inflammation in response to proinflammatory cytokines, and plays a role in exaggerated pain states (Woolf et al. 1994) via direct (binding to TrκA receptors on nociceptive fibers) and indirect (via cytokine-like actions, causing immune cells to accumulate and release their cellular contents exciting nociceptive fibers) actions on the nociceptive function.

Peripheral stimulation of nociceptive afferents can also produce an axon reflex, in which prodromic nerve impulses travelling up the sensory axon initiate ▶ antidromic impulses in neighboring axons. This results in a release of ▶ neuropeptides and ▶ neurogenic inflammation at sites distant from the original stimuli increasing the area of sensitivity.

Such cytokine-driven, neurogenic inflammatory loops stimulate and sensitize nociceptors in the absence of further peripheral stimulation, which could, in theory, create enough "drive" to create and maintain spinal cord sensitization, resulting in persisting pain.

Influences in the DRG

Cytokines can also have direct and indirect effects on nociceptor function, at the level of the DRG, resulting in excitability and phenotypic changes. This contributes to altered synaptic transmission in the spinal cord (central

Cytokines, Effects on Nociceptors, Fig. 1 (continued) mechanisms of proinflammatory cytokine-induced cell change. IL-1 via the IL-1R activates a number of protein kinase cascades including protein tyrosine kinase (*PTK*) and protein kinase C (*PKC*) inducing intracellular Ca^{++} increase and phosphorylation of ion channels such as TRPV1 and others inducing changes in Na^+ and Ca^{++} currents. Early events in all cascades involve MyD88, IL-1 receptor-associated kinase (*IRAK-1*), and a TNF receptor-associated factor (*TRAF*). Ras activation of p38 MAP kinase (*MAPK*) and Rac-mediated transactivation

of gene expression by NFkB regulation also occur. The IL-6/IL-6R complex activates JAK which recruits STAT3 to the phosphorylated receptor, gp130, and activates the Ras/MAP kinase pathway via the SHP-2 pathway ultimately affecting nuclear transcription factors such as CREB, c-fos, and c-jun. In contrast to such kinase signalling, the TNF/TNFR complex is regulated by intracellular adaptor proteins (TRADD and TRAF), via which it activates the IKK complex to phosphorylate IkB, liberating NFkB to the nucleus (see Hanada and Yoshimura 2002; Obreja et al. 2002; Heinrich et al. 2003)

sensitization) involved in the development of hyperalgesia and persistent pain (Fig. 1a). Levels of cytokines can increase in the DRG following inflammation and nerve damage in response to signal proteins produced at the site of damage, such as the cytokines LIF and IL-6 as well as cytokine-induced NGF, which are retrogradely transported by both intact and injured axons (Watkins and Maier 2000). Such signals can initiate the recruitment of immune cells to the DRG and induce proliferation and activation of satellite (glial) cells, resulting in the release of a variety of growth factors, including cytokines, into the extracellular fluid of the DRG where they modulate sensory neuron, and/or glial cell function via differentially expressed cytokine receptors. For example, the proinflammatory cytokines IL-1 and TNF-α bind directly to neuronally expressed receptors inducing rapid increases in sensory neuronal excitation, facilitating pain transmission to the spinal cord. On the other hand, the primary receptor for IL-6 (IL6-R) is predominantly expressed by glial cells, and therefore, IL-6, although produced by DRG neurons, likely affects nociceptor function indirectly (such as via Schwann cell release of IL-1 and TNF-α), although receptor expression is likely to change under pathological conditions (De Jongh et al. 2003). In fact, even in the absence of nerve damage or of any action at receptors expressed by peripheral nerve terminals, proinflammatory cytokines acting in the DRG can rapidly induce aberrant responses in peripheral nociceptors (Sorkin et al. 1997; Watkins and Maier 2000).

Retrogradely transported signals can also induce phenotypic changes in primary sensory neurons, such as increased expression of BDNF, TRPV1, NAv1.8, CGRP, SP, and galanin (Watkins and Maier 2000), all of which can change the properties of nociceptive neurons.

In the DRG, ▶ G protein-coupled, chemokine ▶ receptors are expressed by small diameter nociceptors, suggesting a direct involvement of chemokines in nociceptive signal transduction and the pathogenesis of pain. This is supported by the fact that endogenous chemokines such as

SDF1α/CXCL12, MDC/CCL22, and RANTES/ CCL5 or exogenous factors such as the HIV-1 coat protein gp120, which is a ligand at the chemokine receptors CXCR4 and CCR5, have been shown to produce powerful excitatory effects on sensory neurons via increased calcium mobilization and the lowering of the threshold for action potential generation (Oh et al. 2001). Chemokine receptors are also expressed by glial cells, the activation of which enhances the release of neuroactive substances, including NGF, NO, glutamate, and other chemokines, which can further affect nociceptor activity and even lead to neurotoxicity.

Intracellular Mechanisms

Cytokines modulate nociceptor activity contributing to ▶ neural plasticity and pain, via the initiation of intracellular cascades of phosphorylation of constitutively expressed signal proteins involved in nociception and pain such as PKC, MAP kinase, Ras/Raf c-jun, c-fos, and STAT (Fig. 1b). As a result, there are alterations to the transcription rate and/or posttranslational changes in proteins involved in the pain pathway, which ultimately lead to a modification of transduction, conduction, and transmission functionality of these neurons. For example, IL-1 binds to the specific cell surface receptors, IL-1RI and IL-1RII; members of the Toll-like receptor superfamily, IL-1R1, are expressed by nociceptive neurons. IL-1RI-mediated mechanisms, involving activation of tyrosine kinases and PKC-associated phosphorylation, lead to long-lasting increases in voltage-sensitive sodium and calcium channel conductance, increasing excitability of the cell as well as changes in specific receptor function (Obreja et al. 2002). In nociceptors, this may include phosphorylation of heat-transducing vanilloid receptors, TRPV1 or TRPVL-1, supported by evidence that IL-1β directly sensitizes rat sensory neurons to heat. Downstream pathways of IL-1Rs and TNF-α receptors (TNFRI and TFNRII) ultimately activate NFκB, which modulates the expression of a variety of proteins implicated in nociceptor sensitivity and pain, including growth factors and inducible enzymes such as cyclooxygenase 2 (COX2) and inducible nitric oxide synthase (iNOS) (Hanada and Yoshimura 2002). NFκB also promotes further proinflammatory cytokine and chemokine production, potentiating cytokine-mediated mechanisms of nociceptor excitability and signalling. In addition to receptor function, the TNF-α molecule itself may cause changes in neuronal excitability due to pH-facilitated insertion into lipid membranes, to form a central pore-like region forming voltage-dependent sodium channels, which may not only be associated with the generation of neuronal hyperexcitability but may interact with endogenous sodium and calcium channels to increase membrane conductance (Kagan et al. 1992). In contrast, IL-6 signalling occurs via binding to the α-subunit non-signalling receptor IL-6R (gp80), which leads to homodimerization of the transmembranous signalling transducer receptor gp130, which signals intracellularly via the JAK/STAT pathways to influence gene expression via MAPK (Heinrich et al. 2003). IL-6R and gp130 are expressed predominantly on glial cells; however, although peripheral nociceptors lack IL-6R subunits under normal physiological conditions, they do express gp130 molecules, and the expression patterns of each have been reported to change following nerve damage (De Jongh et al. 2003).

Antinociceptive Functions
In addition to the anti-inflammatory and antinociceptive function of cytokines, such as IL-10, IL-4, and IL-13, there is evidence of the direct interaction of cytokines with nociceptors in an antinociceptive manner. For example, IL-2 induces peripheral ► antinociception via the IL-2R specific receptor, which is expressed in small and medium (i.e., nociceptive) primary sensory neurons of DRG, and is transported to the peripheral axon terminals (Song et al. 2000). The mechanism underlying cytokine-induced antinociception may include inhibition of neurotransmitter release, due to Ca²⁺ channel-blocking properties decreasing nociceptor activity.

References

De Jongh, R. F., Vissers, K. C., Meert, T. F., Booij, L. H., De Deyne, C. S., & Heylen, R. J. (2003). The role of interleukin-6 in nociception and pain. *Anesthesia and Analgesia, 96*, 1096–1103.
Dray, A. (1997). Kinins and their receptors in hyperalgesia. *Canadian Journal of Physiology and Pharmacology, 75*, 704–712.
Hanada, T., & Yoshimura, A. (2002). Regulation of cytokine signaling and inflammation. *Cytokine & Growth Factor Reviews, 13*, 413–421.
Heinrich, P. C., Behrmann, I., Haan, S., Hermanns, H. M., Muller-Newen, G., & Schaper, F. (2003). Principles of interleukin (IL)-6-type cytokine signalling and its regulation. *Biochemical Journal, 374*, 1–20.
Junger, H., & Sorkin, L. S. (2000). Nociceptive and inflammatory effects of subcutaneous TNF-alpha. *Pain, 85*, 145–151.
Kagan, B. L., Baldwin, R. L., Munoz, D., & Wisnieski, B.J. (1992). Formation of ion-permeable channels by tumor necrosis factor-alpha. *Science, 255*, 1427–1430.
Obreja, O., Rathee, P. K., Lips, K. S., Distler, C., & Kress, M. (2002). IL-1 beta potentiates heat-activated currents in rat sensory neurons: Involvement of IL-1RI, tyrosine kinase and protein kinase C. *The FASEB Journal, 16*, 1497–1503.
Oh, S. B., Tran, P. B., Gillard, S. E., Hurley, R. W., Hammond, D. L., & Miller, R. J. (2001). Chemokines and glycoprotein120 produce pain hypersensitivity by directly exciting primary nociceptive neurons. *Journal of Neuroscience, 21*, 5027–5035.
Poole, S., Cunha, F. D. Q., & Ferreira, S. H. (1999). Hyperalgesia from subcutaneous cytokines. In L. R. Watkins & S. F. Maier (Eds.), *Cytokines and pain* (pp. 59–88). Basel: Birkhauser.
Samad, T. A., Sapirstein, A., & Woolf, C. J. (2002). Prostanoids and pain: Unraveling mechanisms and revealing therapeutic targets. *Trends in Molecular Medicine, 8*, 390–396.
Song, J., Jang, Y. Y., Shin, Y. K., Lee, C., & Chung, S. (2000). N-ethylmaleimide modulation of tetrodotoxin-sensitive and tetrodotoxin-resistant sodium channels in rat dorsal root ganglion neurons. *Brain Research, 855*, 267–273.
Sorkin, L. S., Xiao, W. H., Wagner, R., & Myers, R. R. (1997). Tumour necrosis factor-alpha induces ectopic activity in nociceptive primary afferent fibres. *Neuroscience, 81*, 255–262.
Watkins, L. R., & Maier, S. F. (2000). The pain of being sick: Implications of immune-to-brain communication for understanding pain. *Annual Review of Psychology, 51*, 29–57.
Watkins, L. R., & Maier, S. F. (2002). Beyond neurons: Evidence that immune and glial cells contribute to

pathological pain states. *Physiological Reviews, 82*, 981–1011. doi:10.1152/physrev.00011.2002.

Woolf, C. J., Safieh-Garabedian, B., Ma, Q. P., Crilly, P., & Winter, J. (1994). Nerve growth factor contributes to the generation of inflammatory sensory hypersensitivity. *Neuroscience, 62*, 327–331.

Cytokines, Regulation in Inflammation

Victoria C. J. Wallace
Department of Anaesthetics, Pain Medicine and Intensive Care Chelsea and Westminster Hospital Campus, London, UK

Synonyms

Immunocytokines

Definition

▶ Cytokines are a large family of growth factor proteins (MW 8–30 kD) secreted primarily from leukocytes (as well as a variety of cells) as part of the immune and inflammatory response.

Characteristics

Over 150 cytokines have been cloned, which may be grouped into various subfamilies, based upon functional properties and receptor utilization, although nomenclature is somewhat confusing and arbitrary (Vilcek 1998). Cytokines that are particularly involved in regulation of inflammation are termed the ▶ proinflammatory cytokines (which can also encompass the inflammatory chemokines) and the ▶ anti-inflammatory cytokines.

Proinflammatory Cytokines

The major proinflammatory cytokines include tumor necrosis factor-α (TNF-α), interleukin-1α and -β (IL-1α and β), and interleukin-6 (IL-6), which are secreted in a sequential cascade (TNF-α > IL-1 > IL-6), by a variety of immune and inflammatory cells, in response to inflammation. A wealth of literature supports a role for proinflammatory cytokines in the production of inflammation associated with pain and hyperalgesia. For example, pain associated with signs of inflammation after peripheral nerve injury in rodents correlates with the number of proinflammatory cytokine-producing cells at the injury site, and thermal hyperalgesia and mechanical allodynia can be reduced via the blockade of IL-1 or TNF-α (Watkins and Maier 2000). Furthermore, an intraplantar injection of TNF-α results in hyperalgesia, which lasts for several hours and can be associated with a cascade of cytokine involvement, including IL-1 followed by IL-6, as well as the activation of IL-8.

TNF-α, a member of the TNF family of cytokines, is produced primarily by macrophages and acts via the TNFRI and TNFRII receptors (Orlinick and Chao 1998). Binding of TNF-α triggers intracellular signalling cascades to induce changes in gene expression, mainly via the activation of NFκB. Effects can include the upregulation of proinflammatory cytokines, adhesion molecules, chemokines, growth factors, inducible enzymes such as cyclooxygenase 2 (COX2), and inducible nitric oxide synthase (iNOS), all of which can have effects on mechanisms of nociception. Importantly, TNF-α induces production of IL-1, a member of the interleukin family, which is produced in two molecular forms (α and β) by many cell types including mononuclear cells, fibroblasts, synoviocytes, macrophages, keratinocytes, mast cells, glial cells, and neurons (Bianchi et al. 1998). Additionally, the inactive precursor of IL-1 (pro-IL-1) is constantly produced by keratinocytes and fibroblasts in the skin, and in increasing amounts following tissue damage, during which degranulation of mast cells releases the enzyme, chymase, which cleaves the IL-1 precursor to its active form. IL-1 binds to the specific cell surface receptors, IL-1RI and IL-1RII, which are members of the Toll-like receptor super-family. Like TNF-α, the proinflammatory and pronociceptive actions of IL-1 are mediated via complex intracellular signalling cascades that ultimately activate NFκB. In the context of pain, IL-1 is likely to be the most important proinflammatory cytokine, with many outcomes of other proinflammatory cytokine actions occurring via IL-1 release. In most cell types, IL-1 and TNF-α induce the release of IL-6, a member of the neuropoietic cytokine family. IL-6 plays a minimal role in inducing inflammatory mediators or other inflammatory cytokines in tissues although may play a role in inducing chemokine production. IL-6 binds to the α-subunit non-signalling receptor, IL-6R (gap80), leading to homodimerization of the transmembranous signal transducer receptor, gp130, and an intracellular cascade of phosphorylation (De Jongh et al. 2003). This affects cell activity and initiates the release of many cell-specific ▶ neuroactive substances and inflammatory substances.

In general, proinflammatory cytokines act to increase vascular endothelial membrane permeability and induce upregulation of endothelial adhesion molecules and thus leukocyte adhesion, migration, and extravasation. Due to a self-promoting cytokine loop, there is a rapid accumulation of proinflammatory cytokines at the site of damage, ensuring a rapid onset of the inflammatory response (Fig. 1).

In the context of pain, which is a key component of the inflammatory response designed to prevent further insult to the damaged site, it is important to note that proinflammatory cytokines are released from and act upon neurones and glial cell, via which they signal to the brain that infection or injury has occurred (Watkins and Maier 2000; Watkins and Maier 2002). This ▶ neuroimmune interaction can influence mechanisms implicated in the development of hyperalgesia and persistent pain, which are associated with many inflammatory conditions. For example, cytokines released in damaged tissue can activate peripheral terminals of primary sensory neurones either directly, via neuronally expressed cytokine receptors, or indirectly via the stimulation of other substances that have relevance to the production of pain such as prostaglandins, nitric oxide, and nerve growth factor (NGF) (see ▶ Cytokines, Effects on Nociceptors). Such activation enhances the release of neuroactive transmitters, such as ▶ substance P (SP), enhancing neurogenic inflammation leading to sensitization and increased excitability of sensory primary afferents (Fig. 1) (Black 2002). Inflammation of, or damage to, the peripheral nerve itself causes the activation of recruited and resident macrophages, fibroblasts, mast cells, dendritic cells, and endothelial cells, all of which can release proinflammatory cytokines which, due to their induction of edema-associated disruption of the ▶ blood-nerve barrier, can lead to further immune invasion of the nervous system. Furthermore, activated Schwann cells, which myelinate peripheral nerves, release proinflammatory cytokines (as well as many other damaging substances such as NO, ROS, PGs, and ATP). This can damage myelin, with resulting demyelination and/or nerve degeneration (▶ Wallerian Degeneration), all of which can be associated with the development of pain.

Chemokines

Chemokines are a family of small chemoattractant cytokines (MW 8–10 kDa) that have potent leukocyte activation and/or chemotactic activity. Over 50 have been identified so far, the majority of which are classed into the CC group or CXC group based in their cysteine residues. Via actions on their specific G-protein-coupled receptors, chemokines are intimately involved in the orchestration of inflammatory responses, by inducing cell migration across a concentration gradient. Although most pain research related to neuroimmune function has focused on proinflammatory cytokines, the involvement of chemokines in inflammatory pain and hyperalgesia has become more apparent (Boddeke 2001). In addition to immune cells, many nervous system cells (neuronal and glial) are capable of synthesizing chemokines and express a variety of chemokine receptors,

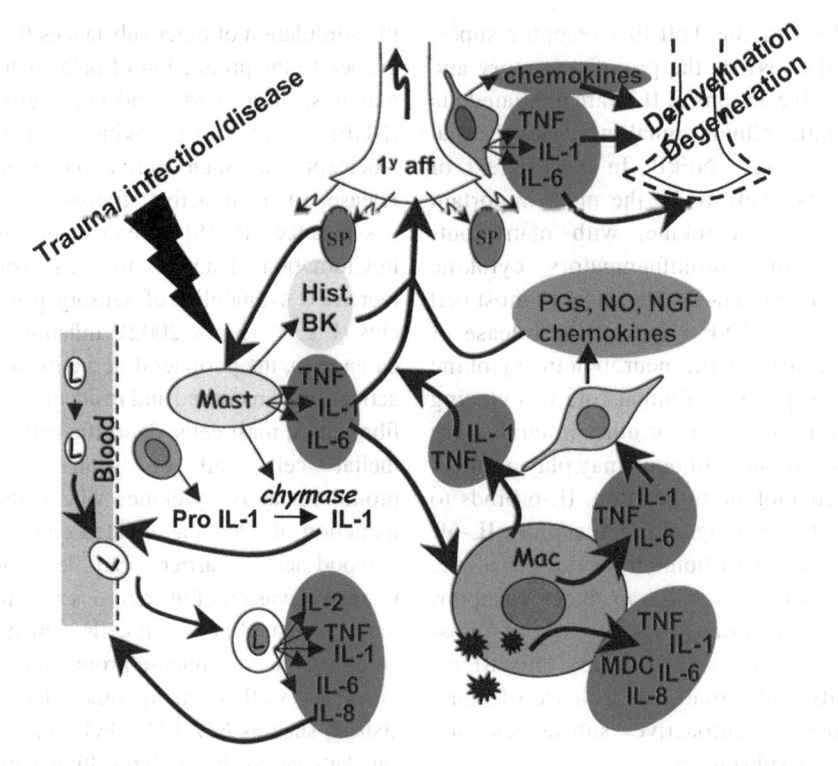

Cytokines, Regulation in Inflammation, Fig. 1 A representation of the network of cytokine production in tissue, following typical injurious stimuli. Degranulation of local and infiltrating mast cells releases proinflammatory cytokines which stimulate further cytokine release from macrophages (*Mac*) attracted to the site of injury to phagocytose pathogens (*), a process which in itself triggers cytokine production. Cytokines act to increase vascular permeability inducing leukocyte extravasation and further stimulating release of cytokines and non-cytokine inflammatory mediators such as PGs, growth factors, and NO from various tissue cells (*TC*). IL-1 is also formed from its precursor pro-IL11, and chemokines are secreted enhancing further leukocyte migration from the bloodstream. Cytokines and various inflammatory mediators such as histamine and BK can activate primary sensory neurones to release neurotransmitters such as SP. Activation of Schwann cells causes further cytokine and chemokine release which can cause demyelination and degeneration of 1y afferents

suggesting that chemokines act as messengers between peripheral immune cells and sensory afferent neurons from inflamed sites. Evidence supports a role for a number of chemokines in inflammation and pain such as IL-8/CXCL8 (Boddeke 2001), SDF1α/CXCL12, MDC/CCL22, and RANTES/CCL5 (Oh et al. 2001), which may act to increase intracellular calcium and, therefore, cell excitability and release.

Limiting Actions of Cytokines

Since cytokines are potent mediators of potentially damaging tissue responses, several mechanisms exist to ensure that the effects of these cytokines are restricted, thereby limiting the inflammatory response. For example, the IL-1 receptor antagonist (IL-1ra), a member of the interleukin cytokine family, is upregulated during inflammation and prevents binding of IL-1 to its receptors (Hopkins 2003). Additionally, the cellular receptors for both IL-1 and TNF-α exist in soluble forms and are able to bind and neutralize their cytokine ligands after being cleaved from the cell surface. Some cytokines, such as IL-10, IL-4 (Hamilton et al. 2002), and LIF (see Gadient and Patterson 1999), have anti-inflammatory activity and are able to downregulate expression of proinflammatory cytokines. The actual role of LIF is, however, controversial and has been implicated as a proinflammatory cytokine and as an anti-inflammatory and analgesic cytokine (Banner et al. 1998). It is likely that the increased activation of leukocytes and, therefore, release of proinflammatory cytokines, characteristic of chronic inflammation and associated persistent pain, involves a dysfunction of these negative regulatory mechanisms leading to sustained positive feedback cytokine-neural interactions.

References

Banner, L. R., Patterson, P. H., Allchorne, A., Poole, S., & Woolf, C. J. (1998). Leukemia inhibitory factor is an anti-inflammatory and analgesic cytokine. *Journal of Neuroscience, 18*, 5456–5462.

Bianchi, M., Dib, B., & Panerai, A. E. (1998). Interleukin-1 and nociception in the rat. *Journal of Neuroscience Research, 53*, 645–650.

Black, P. H. (2002). Stress and the inflammatory response: A review of neurogenic inflammation. *Brain Behavior and Immunity, 16*, 622–653.

Boddeke, E. W. (2001). Involvement of chemokines in pain. *European Journal of Pharmacology, 429*, 115–119.

De Jongh, R. F., Vissers, K. C., Meert, T. F., Booij, L. H., De Deyne, C. S., & Heylen, R. J. (2003). The role of interleukin-6 in nociception and pain. *Anesthesia and Analgesia, 96*, 1096–1103.

Gadient, R. A., & Patterson, P. H. (1999). Leukemia inhibitory factor, interleukin 6, and other cytokines using the GP130 transducing receptor: Roles in inflammation and injury. *Stem Cells, 17*, 127–137.

Hamilton, T. A., Ohmori, Y., & Tebo, J. (2002). Regulation of chemokine expression by anti-inflammatory cytokines. *Immunologic Research, 25*, 229–245.

Hanada, T., & Yoshimura, A. (2002). Regulation of cytokine signaling and inflammation. *Cytokine & Growth Factor Reviews, 13*, 413–421.

Hopkins, S. J. (2003). The pathophysiological role of cytokines. *Leg Med (Tokyo), 5*(Suppl 1), S45–S57.

Junger, H., & Sorkin, L. S. (2000). Nociceptive and inflammatory effects of subcutaneous TNFalpha. *Pain, 85*(1–2), 145–151.

Oh, S. B., Tran, P. B., Gillard, S. E., Hurley, R. W., Hammond, D. L., & Miller, R. J. (2001). Chemokines and glycoprotein120 produce pain hypersensitivity by directly exciting primary nociceptive neurons. *Journal of Neuroscience, 21*, 5027–5035.

Orlinick, J. R., & Chao, M. V. (1998). TNF-related ligands and their receptors. *Cellular Signalling, 10*, 543–551.

Vilcek, J. (1998). The cytokines: An overview. In A. Thomson (Ed.), *The cytokine handbook* (3rd ed., pp. 1–20). San Diego: Academic.

Watkins, L. R., & Maier, S. F. (2000). The pain of being sick: Implications of immune-to-brain communication for understanding pain. *Annual Review of Psychology, 51*, 29–57.

Watkins, L. R., & Maier, S. F. (2002). Beyond neurons: Evidence that immune and glial cells contribute to pathological pain states. *Physiological Reviews, 82*, 981–1011.

Cytokines, Upregulation in Inflammation Neuropathic Pain Model, Chronic Constriction Injury

▶ Wallerian Degeneration

Cytoprotection

Definition

As applied to the gastrointestinal tract, cytoprotection indicates the development of mechanisms that protect the tract from damage from the digestive enzymes and, in the stomach and duodenum, from the acidic contents.

Cross-References

▶ NSAIDs and Their Indications

Cytoskeleton

Definition

The scaffold structures of cells, also cytoskeletal system.

Cross-References

▶ Toxic Neuropathies

D

Many EMG experiments show repetition rates that seem to respond to analgesics when the primary stimulus (This nociceptive TMS is a [...] [...] level cognitive response or represents [...] [...] is spatially and temporally linked to nerve [...] [...] in nerve flow, and has been shown to be primarily due to neuronal inhibition.

(+/−)-6-Dimethylamino-4,4-Diphenylheptane-3-One-Hydrochloride

▶ Postoperative Pain, Methadone

Daily Persistent Headache

▶ New Daily Persistent Headache

DAMGO

Definition

DAMGO is a μ-opioid receptor-selective enkephalin derivative: [D-Ala 2, N-Me-Phe 4, Gly 5-ol]-enkephalin.

Cross-References

▶ Opioids and Inflammatory Pain

Dapsone

Definition

Dapsone is an antibacterial, belonging to a group of sulfonamide-like sulfones. It is used especially in the treatment of leprosy and administered orally.

Cross-References

▶ Hansen's Disease

Dark Disc Disease

▶ Discogenic Back Pain

Data

Definition

Data can be described in one of two of the following ways:
(i) Pieces of numerical or other information
(ii) A set of facts from which conclusions can be drawn

G.F. Gebhart, R.F. Schmidt (eds.), *Encyclopedia of Pain*, DOI 10.1007/978-3-642-28753-4,
© Springer-Verlag Berlin Heidelberg 2013

Cross-References

▶ Postoperative Pain, Data Gathering and Auditing

Data Auditing/Collection/Gathering/Management/Security

▶ Postoperative Pain, Data Gathering and Auditing

Data Collection

▶ Postoperative Pain, Data Gathering and Auditing

Data Gathering

▶ Postoperative Pain, Data Gathering and Auditing

Data Management

▶ Postoperative Pain, Data Gathering and Auditing

DBS

▶ Deep Brain Stimulation

DCS

▶ Dorsal Column Stimulators

Deactivation

Definition

Many FMRI experiments show regions of cortex that seem to respond in antiphase with the primary stimulus. This negative BOLD (blood oxygenated level dependent) response or deactivation is spatially and temporally linked to reductions in blood flow, and has been shown to be primarily due to neuronal inhibition.

Cross-References

▶ Hippocampus and Entorhinal Complex: Functional Imaging
▶ Nociceptor, Fatigue

Dead Sea

Definition

This region in Israel is a major spa area for patients with various types of arthritis.

Cross-References

▶ Spa Treatment

Deafferentation

Definition

Deafferentation means blocking peripheral afferent input from gaining access to the central nervous system (CNS). This can be accomplished by dorsal rhizotomy, root avulsion, or ganglionectomy. Deafferentation is frequently confused with "neurectomy" (peripheral nerve injury). In the latter, the afferent innervation of peripheral tissues is destroyed (i.e., the tissue is denervated). However, the nerve-end neuroma and the sensory ganglion cells remain connected with the CNS and can, in principle, deliver sensory input centrally.

Cross-References

▶ Anesthesia Dolorosa Model, Autotomy
▶ Atypical Facial Pain: Etiology, Pathogenesis, and Management
▶ Atypical Odontalgia
▶ Central Nervous System Stimulation for Pain
▶ Cordotomy Effects on Humans and Animal Models
▶ Dorsal Root Ganglionectomy and Dorsal Rhizotomy
▶ Hyperpathia, Assessment
▶ Peripheral Neuropathic Pain
▶ Plexus Injuries and Deafferentation Pain
▶ Thalamic Plasticity and Chronic Pain
▶ Thalamus, Dynamics of Nociception

Deafferentation Model of Neuropathic Pain

▶ Anesthesia Dolorosa Model, Autotomy

Deafferentation Pain

Definition

Pain resulting from deafferentation. This is a type of central neuropathic pain ("central pain") that frequently develops after avulsion of the brachial plexus, dorsal rhizotomy, or lesions of sensory ganglia (notably the trigeminal ganglion). Injury to nerves peripheral to the sensory ganglion may also result in neuropathic pain. This is termed peripheral neuropathic pain and probably has a different mechanism from deafferentation pain.

Cross-References

▶ Anesthesia Dolorosa Model, Autotomy
▶ Brachial Plexus Avulsion and Surgery in the Dorsal Root Entry Zone (DREZ)
▶ Central Pain, Diagnosis
▶ Dietary Variables in Neuropathic Pain
▶ Dorsal Root Ganglionectomy and Dorsal Rhizotomy

Deception

▶ Credibility, Assessment

Decerebrated

Definition

Cerebral brain function (in an animal) and its impact on lower brain and spinal cord can be eliminated experimentally by removing the cerebrum, cutting across the brain stem, or severing certain arteries in the brain stem.

Cross-References

▶ Postsynaptic Dorsal Column Projection, Anatomical Organization

Decidualization

Definition

Forming of the deciduas, which is the changed endometrium (the lining of the uterus), after the blastocyst (fertilized ovum after 4–9 days of development) is implanted onto the endometrium.

Cross-References

▶ NSAIDs, Adverse Effects

Deconditioned

Definition

Deconditioned refers to a state of poor physical conditioning brought on by inactivity. It is usually used to refer to a deterioration of physical conditioning, secondary to a reduction in an individual's level of activity. The deconditioning may involve loss of strength, loss of flexibility, loss of endurance, and reduced cardiovascular function.

Cross-References

▶ Disability, Effect of Physician Communication

Deep Brain Stimulation

I. Garonzik[1], Nirit Weiss[2], A. Samdani[1] and Frederick A. Lenz[3]
[1]Department of Neurosurgery, Johns Hopkins Medical Institutions, Baltimore, MD, USA
[2]Department of Neurosurgery, Mount Sinai Medical Center, New York, NY, USA
[3]Department of Neurosurgery, Johns Hopkins Hospital, Baltimore, MD, USA

Synonyms

Brain electrode; DBS

Definition

Electrical stimulation of subcortical structures in the brain, including periventricular gray, internal capsule, somatic sensory and intralaminar nuclei of thalamus, is an effective therapy for some patients with chronic pain.

Introduction

The electrical stimulation of subcortical brain structures for the treatment of chronic pain was first reported in the 1950s when hypothalamic nuclei were stimulated for pain control (Tulane University, School of Medicine, Dept. of Psychiatry and Neurology 1954; Pool et al. 1956). Over the next few decades, the sensory thalamus and periaqueductal gray (PAG) became the most frequent targets for deep brain stimulation (DBS) for the treatment of pain. This transition, which followed observations initially seen in animal studies, was subsequently confirmed in retrospective analyses of stimulation-evoked analgesia in humans (Hosobuchi et al. 1977; Richardson and Akil 1977). In addition to the sensory thalamus and PAG, a recent target of interest is the posteromedial hypothalamus for treatment of chronic cluster headache (Franzini et al. 2003, 2010). The mechanism by which posterior hypothalamic stimulation provides relief in cluster headaches is poorly understood.

Chronic pain is pain that occurs daily over a 6-month period. It can be divided into two categories: nociceptive and neuropathic. Nociceptive refers to pain arising from the activation of peripheral nociceptors and transmitted to the central nervous system through intact somatosensory pathways. Examples of nociceptive pain include pain of acute trauma and cancer pain secondary to invasion of bone. This pain responds well to opiates. Neuropathic pain refers to pain arising from injury to the nervous system either peripherally (deafferentation pain, e.g., diabetic neuropathy) or centrally (central pain, e.g., poststroke pain) (Portenoy 1989). It has been proposed that this type of pain does not respond to opiates (Arner and Meyerson 1988), although this proposal is not universally accepted (Reynolds 1969; Dellemijn 1999).

Characteristics

Mechanisms of Action

The mechanisms of action of deep brain stimulation of the PAG and sensory thalamus are not completely understood. Evidence supports the role of endogenous opioid release in PAG stimulation. Focal stimulation of the PAG prevents nociceptive responses to painful stimuli (Reynolds 1969), an effect which is reversed by opioid antagonists (Hosobuchi et al. 1977) and which is associated with increased endogenous opioid levels in the third ventricle (Akil et al. 1978; Hosobuchi et al. 1979). Several lines of evidence also suggest that the mechanism of analgesia in PAG stimulation involves spinal connections (Basbaum and Fields 1984; Maciewicz and Fields 1986). PAG neurons synapse upon neurons in the medullary nucleus raphe magnus (NRM) which, in turn, sends a strong serotonergic projection to the dorsal horn. This analgesic effect is abolished by cutting the descending pathways to the spinal cord. The PAG, medullary NRM, and dorsal horn all contain high levels of opiates and opiate receptors. Due to the important role apparently played by opioids, PAG stimulation is usually indicated in chronic nociceptive pain conditions.

The mechanism by which sensory thalamic stimulation provides pain relief is less clearly understood. The target nucleus for this type of stimulation is the principle somatosensory nucleus known as ventral caudal (Vc) in humans and ventral posterior (VP) in other species. Stimulation of rat VP inhibits responses to noxious stimulation through an opioid-independent pathway (Benabid et al. 1983). The effects of somatosensory thalamic stimulation on pain modulation may be due to an inhibitory effect on Rexed's laminae I to V spinothalamic tract (STT) neurons, likely through serotonergic pathways (Gerhart et al. 1984). Due to the relative independence from opioid pathways, somatosensory thalamic stimulation is indicated in cases of chronic neuropathic pain.

The characteristic circadian rhythm and alteration of hormones, such as melatonin and cortisol, during cluster headache periods, was thought to be evidence of a central origin (Leone and Bussone 1993). DBS was, therefore, investigated as a therapy (Franzini et al. 2003).

Patient Selection

The issue of patient selection for placement of DBS for the treatment of pain has been addressed in a number of published studies (Young and Rinaldi 1997; Levy et al. 1987; Hosobuchi 1986). These patients have experienced chronic pain unresponsive to medical or surgical therapies over several years, and have failed treatment administered during an admission to a pain clinic. Often, they have undergone psychosocial assessment, and are selected for stimulation if there is no evidence of psychological or secondary gain issues. In many published studies, patients have been subjected to intravenous morphine infusion tests, based on the hypothesis that nociceptive but not neuropathic pain responds to opioids. Reversibility of analgesia with naloxone was taken as further evidence of a nociceptive pain condition. If the patient's pain was likely to be nociceptive in origin, PAG stimulation was performed. If the pain appeared to be neuropathic, the patient underwent thalamic stimulation.

Intraoperative Targets

Localization of the surgical target for PAG has been determined using different means by different authors. All targets were determined relative to the anterior commissure-posterior commissure (AC-PC) line, a third ventricular radiologic landmark with a relatively fixed anatomical relationship to subcortical structures. Levy and coworkers targeted a point 1 mm below and 1 mm posterior to the posterior commissure and 3 mm lateral to the lateral wall of the third ventricle (Levy et al. 1987). Young and Rinaldi targeted a point 2–3 mm below the AC-PC line and 12–14 mm posterior to the midpoint (MC) of the AC-PC line and 2–3 mm lateral to the midline (Young and Rinaldi 1997). Hosobuchi targeted a site at the level of the opening of the aqueduct into the third ventricle and 3 mm lateral to the midline (Hosobuchi 1986). The correct target was identified by stimulation-evoked warm sensations, oscillopsia, loss of up-gaze, elevation of both heart rate and blood pressure, or pain relief. If the electrode is implanted more ventral and posterior, in PAG, then fear and anxiety are evoked by stimulation (Hosobuchi 1986). Periventricular gray (PVG), which is slightly rostral to PAG, has been targeted as an alternative site to PAG in some studies. The PVG is located 10 mm posterior to MC at the vertical level of the AC-PC line and 3–4 mm lateral to the midline (Young and Rinaldi 1997).

Levy and coworkers described sites of thalamic targets for the treatment of deafferentation pain (Young and Rinaldi 1997). Targets for facial deafferentation pain in the medial Vc (facial representation of Vc) were chosen 8 mm posterior to the MC point, 8 mm lateral to the midline, and 1–3 mm above the MC point. Deafferentation pain of the extremities was treated by stimulation of lateral Vc (representation of the extremities) defined as a point 9 mm posterior to MC, 10–12 mm lateral, and 2–3 mm above the MC point. In other studies, targets were calculated from standard atlas maps. All studies included a period of postoperative testing with the electrode attached to an externalized lead. Patients were stimulated at different electrode combinations, voltages, pulse widths, and frequencies until the optimal stimulation parameters for pain relief were determined. The results of this period of postoperative testing were used to decide whether or not to permanently implant the stimulator.

For the treatment of cluster headache a few separate targeting areas within the posteromedial hypothalamus have been suggested (Franzini et al. 2010). In the majority of cases, the area targeted was 2 mm lateral to the midline, 3 mm posterior to the MC point, and 5 mm inferior to the midcommissural plane.

Results

Three large studies have reported the results of deep brain stimulation for the treatment of pain. Young et al. retrospectively review 178 patients over 14 years who had DBS for chronic pain (Young and Rinaldi 1997). Early in their

experience, Young et al. relied on screening tests, such as response of pain to opiates and reversal of pain relief with naloxone, to determine the surgical target of DBS. Later in their experience, electrodes were implanted in both PAG and Vc in most patients, and determined the ideal stimulation parameters through postoperative DBS programming. Of the 89 patients with permanent implants placed, 62 % experienced long-term pain relief. Long-term pain relief was obtained in 70 % of patients with nociceptive pain, and in 50 % of patients with neuropathic pain.

Another study reported the results of DBS for chronic pain in 161 patients over 80 months. Patients with pain relief with morphine infusion and reversal with naloxone had PAG stimulation, those who had no response had Vc stimulation, and those with a mixed response had electrodes placed in both locations. Success was measured by regular use of the electrode at initial (6 weeks) and long-term (>6 weeks) follow-up. Overall, there was a 61 % initial success rate and a 30 % long-term success rate. Patients with nociceptive pain had a 56 % initial success rate and a 32 % long-term success rate, while patients with neuropathic pain experienced 20–50 % success rates (Levy et al. 1987). In a third study, 122 patients had thalamic DBS placed for chronic pain, with 68 % initial and 57 % long-term success rates (Hosobuchi 1986).

A meta-analysis of 13 studies (1,114 patients) evaluating DBS for the treatment of chronic pain reported that 50 % of all patients experienced long-term pain relief (Bendok and Levy 1998). Patients with nociceptive pain experienced a 60 % long-term relief of pain with PAG stimulation. Patients with neuropathic pain experienced a 56 % long-term success rate with Vc stimulation.

A recent database study of the *European Association of Neurology Panel on Neurostimulation for the Treatment of Pain* identified two recent studies using current standards of MRI target localization in thalamus or PAG/PVG (Cruccu et al. 2007). The first study described results in 15 patients with poststroke central pain(CPSP) who identified success (pain relief >30 %) in 67 % of patients at long term. The second included 21 patients with mixed neuropathic pain syndromes and concluded that only 24 % of patients had durable pain relief as defined by the use of DBS for over 5 years. Overall, they considered that there was weak positive evidence for use of DBS in peripheral neuropathic pain.

The most recent clinical study of DBS included 56 patients with different forms of neuropathic and mixed nociceptive/neuropathic pain with blinded, long-term follow-up (mean 3.5 years, range 1–8 years) (Rasche et al. 2006). Electrodes were routinely implanted in both the somatosensory thalamus and the PVG region. Before implantation of the stimulation device, a double-blinded evaluation tested the efficacy of stimulation alone and in combination. The best long-term results were attained in patients with chronic low-back and leg pain, who had greater than 50 % relief in 8/18 cases, and in those with Complex Regional Pain Syndrome (CRPS) Type II (4/6 cases). Least benefit was found in patients with central pain (spinal cord injury pain 0/12 cases, CPSP 1/11cases).

Complications

Hemorrhage is the most serious complication of DBS, occurring in 14/441 cases (3 %) among whom there were three deaths. This complication is much more frequent in older series using an electrode with a rougher outer profile than the current model (Krack et al. 2003; Rasche et al. 2006). Neurologic sequelae of stimulation were reported in 7 % of cases, including diplopia, vertical gaze paresis, blurred vision, oscillopsia, hemineglect, and hemiparesis. Persistent headache occurred in 50 % of cases in one study, but was not mentioned in others (Young and Rinaldi 1997).

Infections occurred in 5–12 % of cases (Young and Rinaldi 1997; Levy et al. 1987; Hosobuchi 1986). *Staphylococcus* species were the majority of causative organisms. Stitch and subgaleal infections were most common, and were usually adequately treated with antibiotics. Removal of the hardware system and intravenous antibiotics were successful in treating the hardware infections. Ventriculitis (*Propionibacterium*)

and subdural empyema (*Staphylococcus aureus*) rarely occurred.

Technical failures were reported in all studies, including electrode migration (2–10 %) and insulation fracture (3–4 %). Skin erosion was reported in 2 % of cases. The technical complication rate has decreased with the newly designed electrode.

References

Akil, H., Richardson, D. E., Hughes, J., & Barchas, J. D. (1978). Enkephalin-like material elevated in ventricular cerebrospinal fluid of pain patients after analgetic focal stimulation. *Science, 201*, 463–465.

Arner, S., & Meyerson, B. A. (1988). Lack of analgesic effect of opioids on neuropathic and idiopathic forms of pain. *Pain, 33*, 11–23.

Basbaum, A. I., & Fields, H. L. (1984). Endogenous pain control systems: Brainstem spinal pathways and endorphin circuitry. *Annual Review of Neuroscience, 7*, 309–338.

Benabid, A. L., Henriksen, S. J., McGinty, J. F., & Bloom, F. E. (1983). Thalamic nucleus ventro-postero-lateralis inhibits nucleus parafascicularis response to noxious stimuli through a non-opioid pathway. *Brain Research, 280*, 217–231.

Bendok, B., & Levy, R. M. (1998). Brain stimulation for persistent pain management. In P. L. Gildenberg & R. R. Tasker (Eds.), *Textbook of stereotactic and functional neurosurgery* (pp. 1539–1546). New York/St. Louis/San Francisco/Auckland/Bogotá/Caracas/Lisbon/London/Madrid/Mexico City/Milan/Montreal/New Delhi/San Juan/Singapore/Sydney/Tokyo/Toronto: McGraw-Hill.

Cruccu, G., Aziz, T. Z., Garcia-Larrea, L., Hansson, P., Jensen, T. S., Lefaucheur, J. P., Simpson, B. A., & Taylor, R. S. (2007). EFNS guidelines on neurostimulation therapy for neuropathic pain. *European Journal of Neurology, 14*, 952–970.

Dellemijn, P. (1999). Are opioids effective in relieving neuropathic pain? *Pain, 80*, 453–462.

Franzini, A., Ferroli, P., Leone, M., & Broggi, G. (2003). Stimulation of the posterior hypothalamus for treatment of chronic intractable cluster headaches: First reported series. *Neurosurgery, 52*, 1095–1099.

Franzini, A., Messina, G., Cordella, R., Marras, C., & Broggi, G. (2010). Deep brain stimulation of the posteromedial hypothalamus: Indications, long-term results, and neurophysiological considerations. *Neurosurgical Focus, 29*, E13.

Gerhart, K. D., Yezierski, R. P., Wilcox, T. K., & Willis, W. D. (1984). Inhibition of primate spinothalamic tract neurons by stimulation in periaqueductal gray or adjacent midbrain reticular formation. *Journal of Neurophysiology, 51*, 450–466.

Hosobuchi, Y. (1986). Subcortical electrical stimulation for control of intractable pain in humans. Report of 122 cases (1970–1984). *Journal of Neurosurgery, 64*, 543–553.

Hosobuchi, Y., Adams, J. E., & Linchitz, R. (1977). Pain relief by electrical stimulation of the central gray matter in humans and its reversal by naloxone. *Science, 197*, 183–186.

Hosobuchi, Y., Rossler, J., Bloom, F. E., & Guillemin, R. (1979). Stimulation of human periaqueductal gray for pain relief increases immunoreactive beta-endorphin in ventricular fluid. *Science, 203*, 279–281.

Krack, P., Batir, A., Van Blercom, N., Chabardes, S., Fraix, V., Ardouin, C., Koudsie, A., Limousin, P. D., Benazzouz, A., LeBas, J. F., Benabid, A. L., & Pollak, P. (2003). Five-year follow-up of bilateral stimulation of the subthalamic nucleus in advanced Parkinson's disease. *The New England Journal of Medicine, 349*, 1925–1934.

Leone, M., & Bussone, G. (1993). A review of hormonal findings in cluster headache. Evidence for hypothalamic involvement. *Cephalalgia, 13*, 309–317.

Levy, R. M., Lamb, S., & Adams, J. E. (1987). Treatment of chronic pain by deep brain stimulation: Long term follow-up and review of the literature. *Neurosurgery, 21*, 885–893.

Maciewicz, R., & Fields, H. L. (1986). Pain pathways. In A. K. Asbury, G. M. McKhann, & W. I. McDonald (Eds.), *Diseases of the nervous system. Clinical neurobiology* (pp. 930–940). Philadelphia/London: W.B. Saunders.

Pool, J. L., Clark, W. K., Hudson, P., & Lombardo, M. (1956). Hypothalamic-hypophysial dysfunction in man. laboratory and clinical assessment. In R. Guillemin, R. Guillemin, & C. A. Carton (Eds.), *Hypothalamic-hypophysial interrelationships* (pp. 114–124). Springfield: Thomas.

Portenoy, R. K. (1989). Mechanisms of clinical pain. Observations and speculations. *Neurologic Clinics, 7*, 205–230.

Rasche, D., Rinaldi, P. C., Young, R. F., & Tronnier, V. M. (2006). Deep brain stimulation for the treatment of various chronic pain syndromes. *Neurosurgical Focus, 21*, E8.

Reynolds, D. V. (1969). Surgery in the rat during electrical analgesia induced by focal brain stimulation. *Science, 164*, 444–445.

Richardson, D. E., & Akil, H. (1977). pain reduction by electrical brain stimulation in man: Part II: Chronic self-administration in the periaqueductal gray matter. *Journal of Neurosurgery, 47*, 184–194.

Tulane University, School of Medicine, Dept. of Psychiatry and Neurology. (1954). *Studies in schizophrenia. A multidisciplinary approach to mind-brain relationships*. Cambridge: Harvard University Press.

Young, R. F., & Rinaldi, P. C. (1997). Brain stimulation. In R. B. North & R. M. Levy (Eds.), *Neurosurgical management of pain* (pp. 283–301). New York/Berlin/Heidelberg: Springer.

Deep Dysparuenia

Definition

Deep dyspareunia is defined as pain perceived within the pelvis associated with penetrative sex.

Cross-References

▶ Gynecological Pain and Sexual Functioning

Deep Sedation

Definition

A drug-induced state of depressed consciousness during which patients are not easily aroused and may need airway and ventilatory assistance, although they purposefully respond to repeated or painful stimulation.

Cross-References

▶ Pain and Sedation of Children in the Emergency Setting

Deep Somatic Pain, Muscle and Joint Pain

▶ Spinal Dorsal Horn Pathways, Muscle and Joint

Deep Vein Thrombosis (DVT)

▶ Postoperative Pain, Venous Thromboembolism

Deep Venous Thrombosis

Synonyms

Blood clot; DVT

Definition

Deep venous thrombosis is a thrombotic event in the deep (and not only superficial) veins of the lower extremity. It commonly affects the lower leg and can present as a pain syndrome mimicking sciatica with primarily distal extremity pain. It can lead to life-threatening thrombembolism, in particular pulmonary embolism. It is often caused by immobility (e.g., bed rest, long distance flights), in particular in patients with genetic-, disease-, or lifestyle-related risk factors.

Cross-References

▶ Postoperative Pain, Venous Thromboembolism
▶ Sciatica

Default Value

Definition

Values attributed to a variable or data as a base value, which remains unless replaced by the user.

Cross-References

▶ Postoperative Pain, Data Gathering and Auditing

Definition of Disability (Adults)

Definition

Generically, "disability" refers to an inability to perform certain activities because of a medical disorder. An individual can be disabled in various spheres – for example, unable to perform activities of daily living (as seen in quadriplegia), or inability to manage finances (as might be seen following traumatic brain injury). Work disability refers to an inability to work because of a medical disorder. It is defined differently by various agencies that administer disability programs. The United States Social Security Administration defines work disability as the inability to engage in any substantial gainful activity by reason of any medically determinable physical or mental impairment or combination of impairments, which is expected to result in death or that has lasted or can be expected to last for a continuous period of not less than 12 months.

Cross-References

▶ Disability Evaluation in the Social Security Administration

Definition of Disability for Individuals under Age 18 (SSI Program Only)

Definition

A medically determinable physical or mental impairment or impairments that result in marked and severe functional limitations and that meet the duration requirement specifically for individuals under 18 years of age.

Cross-References

▶ Disability Evaluation in the Social Security Administration

Degenerative Disease

Definition

A biomechanical and pathologic condition caused by degeneration, inflammation, or infection.

Cross-References

▶ Chronic Back Pain and Spinal Instability

Degenerative Joint Disease

▶ Arthritis Model, Osteoarthritis

Dejerine-Roussy Syndrome

▶ Central Pain, Diagnosis
▶ Central Poststroke Pain

Dejerine-Sottas Neuropathy (DSN)

▶ Hereditary Neuropathies

Dejerine-Sottas Syndrome (Neuropathy)

Synonyms

DSN

Definition

A severe form of inherited neuropathy that is clinically detected during infancy.

Cross-References

▶ Hereditary Neuropathies

Delayed Muscle Soreness

▶ Delayed Onset Muscle Soreness

Delayed Onset Muscle Soreness

Jan Fridén
Department of Hand Surgery, Sahlgrenska
University Hospital, Gothenborg, Sweden

Synonyms

Delayed muscle soreness; DOMS; Exercise-induced muscle soreness

Definition

Delayed onset muscle soreness (DOMS) is the sensation of muscular discomfort and pain during active contractions that occur 24–48 h after strenuous exercise. The initial symptoms are most evident at the muscle-tendon junction and thereafter spread throughout the entire muscle. Muscle soreness and injury are associated with intense exercise and are more pronounced if the exercise performed is new to the individual.

Characteristics

Sore muscles after exercise are stiff, tender, and aching, symptoms that are aggravated by active muscle contractions. The symptoms of DOMS develop during the first 24–48 h, peak between 24 and 72 h, and usually disappear without intervention within 5–7 days (Armstrong 1984; Ebbeling and Clarkson 1989; Fridén et al. 1981).

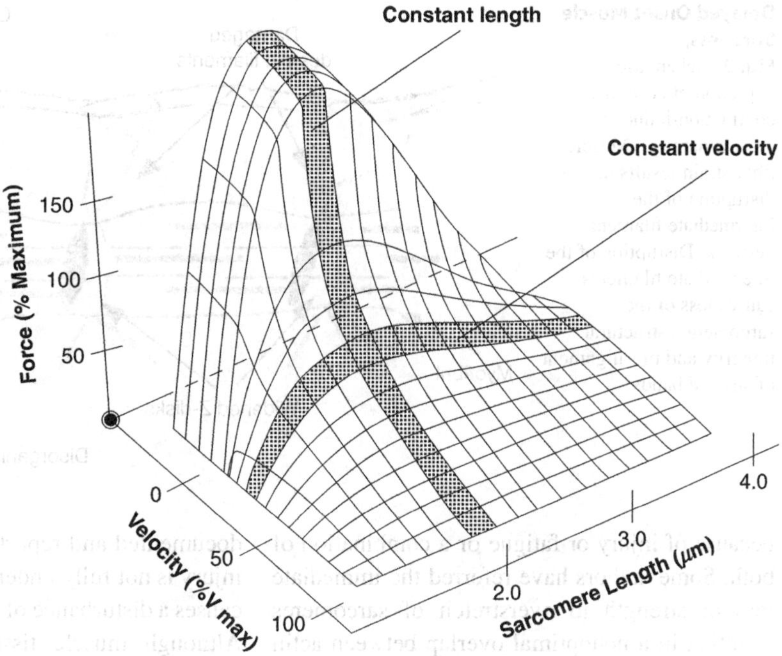

Delayed Onset Muscle Soreness,
Fig. 1 Generalized relationship between muscle length, force, and velocity. Note the very high forces occurring at "negative" velocities (From Fig. 6, Fridén and Lieber 1992, Med Sci Sports Exerc 24:521–530)

The soreness has been reported by some to be localized at the muscle-tendon junction, while others describe the pain as being diffusely spread throughout the muscle. Regardless of the exact location, palpation, passive stretching, and renewed activity aggravate the pain. Some controversy exists regarding the relation between maximum voluntary force and symptoms of soreness. Ebbeling and Clarkson suggested that there is no, or very little, relationship between the development of soreness and decrease of muscle strength (Ebbeling and Clarkson 1989). Return of maximum muscle strength to pre-exercise levels takes between 24 h and 2 weeks.

Bobbert and coworkers (Bobbert et al. 1986) reported an increase in limb volume 24–72 h after eccentric activity of the calf. Intramuscular edema of the biceps, as verified by magnetic resonance imaging, is maximized at 72 h after eccentric activity. The stiffness and swelling associated with DOMS is probably an effect due to edema occurring in the perimuscular connective tissue. Regardless of whether the swelling is intra- or extracellular, an increase in the intramuscular pressure is implicated. In addition to increased tenderness during

palpation, the examiner will find prolonged strength loss, a reduced range of motion, and elevated levels of creatine kinase in the blood. Frequently, muscle cramps accompany DOMS.

DOMS is primarily associated with the eccentric component of exercise (Asmussen 1952) During an eccentric action, a contracted muscle is forced to elongate while producing tension. Its counterpart, concentric action, produces tension during muscle shortening. The intermediate, isometric contraction produces tension, while the muscle remains essentially at the same length (Fig. 1). All three actions are common components of daily movement. The tension generated during eccentric action is higher than that for either of the other actions (Katz 1939) although fewer motor units are recruited (Bigland-Ritchie and Woods 1976).

Due to the nature of an eccentric action, the tension-generating mechanism can be expressed in two phases: the muscle contracts to generate tension, and then the muscle is passively stretched by an external force while generating additional tension.

It is still unknown whether the initial decline of strength after eccentric contractions (EC) is

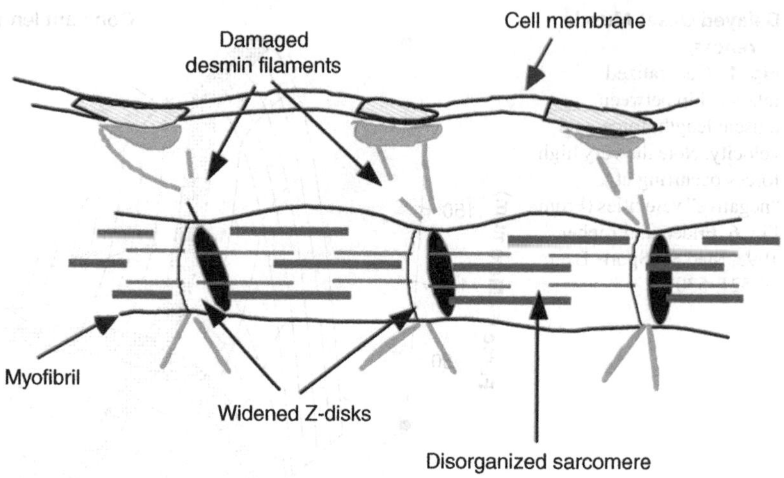

Delayed Onset Muscle Soreness,
Fig. 2 Schematic depiction of eccentric contraction-induced muscle damage. Muscle fiber strain results in disruption of the intermediate filament network. Disruption of the intermediate filaments causes loss of the sarcomere's structural integrity and misalignment of striated bands

because of injury or fatigue or a combination of both. Some authors have referred the immediate loss of strength to overstretch of sarcomeres resulting in a nonoptimal overlap between actin and myosin filaments. It has been suggested that during eccentric exercise, some sarcomeres are stretched beyond overlap and thereby injured, while others maintain their length. Despite the great force loss after eccentric contractions, the resting membrane potentials of isolated muscles fatigued by eccentric or isometric contractions were similar (Warren et al. 1993). Since there is no evidence of excitation failure, the ability to produce action potentials is expected to be unaffected. Lieber and Fridén showed that it is not high force per se which causes muscle damage following eccentric contraction but the magnitude of the active strain, i.e., strain during active lengthening (Lieber and Fridén 1993). The active strain hypothesis is described in terms of the interaction between the myofibrillar cytoskeleton, the sarcomere, and the sarcolemma.

It is generally agreed that the initial force decline is due to mechanical injury. A number of mechanical factors such as muscle length, force, and velocity seem to play roles in the consequences of eccentric contractions. Newham and colleagues found more pronounced strength loss after eccentric exercise at a long muscle length compared to a short muscle length (Newham et al. 1988). Although the nature of morphological injury to the muscle has been well documented and reported, the mechanism of the injury is not fully understood. Strenuous exercise causes a disturbance of the muscle's homeostasis. Although muscle tissue is extremely plastic, destructive changes in muscle ultrastructure may occur in response to unusual demands. DOMS and ultrastructural disruptions are selectively associated with exercise involving EC (Fridén 1984; Fridén et al. 1981) and have been reported in both animal and human models. Such changes may require several weeks to resolve and often lead to muscle hypertrophy and strengthening.

In previous studies, two pieces of mechanical evidence suggested that ▶ desmin plays a direct mechanical role in the force loss after EC (Fig. 2). Desmin immunostaining was rapidly lost, as fast as 15 min into a single bout of EC (Lieber et al. 1996), and a strong linear correlation was seen between the magnitude of desmin loss and loss of muscle force generated after EC. Recently, it has been hypothesized that there is a relationship between force loss, desmin immunostaining, transcriptional upregulation, and the ultimate increase in desmin content per sarcomere (Barash et al. 2002; Peters et al. 2003). Direct mechanical data demonstrated that reinforced sarcomeres are highly resistant to EC-induced injury.

In addition to injuries to muscle fibers, there is evidence of disturbance to muscle sense organs and of proprioception after eccentric exercise. The peripheral contribution to perturbations of force perception after eccentric exercise is, however, small, and the centrally derived sense of effort plays the most important role. Proske and Morgan made the common observation that a second period of exercise, for example, 1 week after the first, produces much less damage (Proske and Morgan 2001). One proposed mechanism for this adaptation may be an increase in sarcomere number in muscle fibers. It is postulated that the adaptation is likely to lead to a secondary shift in the muscle's optimum length for active tension. The authors conclude that the ability of muscle to adapt quickly after damage from a single bout of eccentric exercise implies the use in clinical applications of mild eccentric exercise for protecting a muscle against more major injuries.

Cross-References

▶ Nocifensive Behaviors, Muscle and Joint
▶ Stretching

References

Armstrong, R. B. (1984). Mechanisms of exercise-induced delayed onset muscular soreness: A brief review. *Medicine & Science in Sports & Exercise, 16*, 529–538.

Asmussen, E. (1952). Positive and negative muscular work. *Acta Physiologica Scandinavica, 28*, 364–382.

Barash, I. A., Peters, D., Fridén, J., et al. (2002). Desmin cytoskeletal modifications after a bout of eccentric exercise in the rat. *American Journal of Physiology, 283*, 958–563.

Bigland-Ritchie, B., & Woods, J. J. (1976). Integrated electromyogram and oxygen uptake during positive and negative work. *The Journal of Physiology, 260*, 267–277.

Bobbert, M. F., Hollander, A. P., & Huijing, P. A. (1986). Factors in delayed onset muscular soreness of man. *Medicine & Science in Sports & Exercise, 18*, 75–81.

Ebbeling, C. B., & Clarkson, P. M. (1989). Exercise-induced muscle damage and adaptation. *Sports Medicine, 7*, 207–234.

Fridén, J. (1984). Changes in human skeletal muscle induced by long term eccentric exercise. *Cell and Tissue Research, 236*, 365–372.

Fridén, J., Sjöström, M., & Ekblom, B. (1981). A morphological study of delayed muscle soreness. *Experientia, 37*, 506–507.

Katz, B. (1939). The relation between force and speed in muscular contraction. *The Journal of Physiology, 96*, 45–64.

Lieber, R. L., & Fridén, J. (1993). Muscle injury is not a function of muscle force but active muscle strain. *Journal of Applied Physiology, 74*, 520–526.

Lieber, R. L., Thornell, L. E., & Fridén, J. (1996). Muscle cytoskeletal disruption occurs within the first 15 min of cyclic eccentric contraction. *Journal of Applied Physiology, 80*, 278–284.

Newham, D. J., Jones, D. A., Ghosh, G., et al. (1988). Muscle fatigue and pain after eccentric contractions at long and short length. *Clinical Science, 74*, 553–557.

Peters, D., Barash, I., Burdi, M., et al. (2003). Asynchronous functional, cellular and transcriptional changes after a bout of eccentric exercise in the rat. *The Journal of Physiology, 553*, 947–957.

Proske, U., & Morgan, D. L. (2001). Muscle damage from eccentric exercise: mechanism, mechanical signs, adaptation and clinical applications. *The Journal of Physiology, 537*, 333–345.

Warren, G. W., Hayes, D., Lowe, D. A., et al. (1993). Mechanical factors in the initiation of eccentric contraction-induced injury in rat soleus muscle. *The Journal of Physiology, 464*, 457–475.

Delayed Respiratory Depression

Definition

Severe slowing or arrest of spontaneous breathing, usually caused by hydrophilic opioids (in particular morphine) administered close to the spinal cord (epidurally or intrathecally (directly into the cerebrospinal fluid)).

Cross-References

▶ Postoperative Pain, Acute Pain Team

References

Sultan, P., Gutierrez, M. C., & Carvalho, B. (2011). Neuraxial morphine and respiratory depression: Finding the right balance. *Drugs, 71*, 1807–19.

Delirium

Definition

An acute or subacute syndrome with waxing and waning levels of consciousness; global cognitive impairment; a reduced ability to focus attention, sustain attention, or to shift attention; and disorganized wake-sleep cycle.

Cross-References

▶ Cancer Pain, Assessment in the Cognitively Impaired

Delivery Route

▶ Opioid Therapy in Cancer Pain Management, Route of Administration

Delta(δ) Opioid Receptor(s)

Synonyms

DOP; OP1 receptors

Definition

Receptors that preferentially bind enkephalins and enkephalin-like drugs. It was originally named for its discovery in the mouse vas deferens. The receptor was cloned in 1992. They are present in the brain and spinal cord. Agonists are associated with antinociception in preclinical settings.

Cross-References

▶ Opiates During Development
▶ Opioid Receptors
▶ Postoperative Pain, Transition from Parenteral to Oral Drugs

Delta(δ) Sleep

Definition

Delta sleep refers to stages 3 and 4 of non-REM (rapid eye movement) sleep.

Cross-References

▶ Fibromyalgia

Demand Characteristics

Definition

This term refers to the demands of the situation that must be considered in the context of introducing hypnosis. Experimental subjects tend to guess at the experimenter's objectives and then behave accordingly in order to be socially compliant and cooperative. This tendency is clearly important in any therapeutic interaction where the patient may be trying to please the doctor.

Cross-References

▶ Therapy of Pain, Hypnosis

Dementia

Definition

A syndrome characterized by a decline in multiple cognitive functions occurring in clear consciousness.

Cross-References

▶ Cancer Pain, Assessment in the Cognitively Impaired

Demographic

Definition

Statistical characteristics of populations such as age, educational attainment, or income.

Cross-References

► Pain in the Workplace, Risk Factors for Chronicity, and Demographics

Demyelination

Matthew N. Rasband[1,2] and Peter Shrager[3]
[1]Department of Molecular and Cellular Biology, Baylor College of Medicine, Farmington, Houston TX, USA
[2]University of Connecticut Health Center, Farmington, CT, USA
[3]Department of Neurobiology and Anatomy, University of Rochester Medical Center, Rochester, NY, USA

Synonyms

Disease or injury; Loss of the myelin sheath

Definition

A large percentage of the axons in the nervous system are ensheathed by a lipid-rich membrane called myelin. This sheath is essential for rapid and efficient conduction of electrical signals along axons. Demyelination is the loss of the myelin sheath and results from disease or injury.

Characteristics

Proper nervous system function depends on the transmission of electrical signals (► Action Potential) along axons and over relatively long distances. In order to facilitate conduction in a rapid and efficient manner, vertebrates have developed a lipid-rich sheath, called ► myelin, which confers several passive electrical properties onto axons. These properties include high ► membrane resistance and low ► membrane capacitance. Myelin also plays several active roles that determine axonal excitability. For example, myelin-axon interactions restrict ► sodium (Na+) channels to regularly spaced gaps in the myelin sheath called nodes of Ranvier (Fig. 1a, arrow), and compact myelin regulates the kinds of Na^+ channels expressed in axons (for a more comprehensive review, see Salzer 2003). Together, these active and passive properties facilitate rapid and efficient action potential conduction by decreasing loss of electrical charge in regions covered by myelin and by regenerating the action potential at each node of Ranvier.

Myelin is made by glial cells: Schwann cells in the peripheral nervous system (PNS) and oligodendrocytes in the central nervous system (CNS). Demyelination, or loss of the myelin sheath, is most commonly associated with autoimmune diseases (e.g., multiple sclerosis (MS) in the CNS and Guillain-Barre syndrome in the PNS) or traumatic injury (e.g., spinal cord crush) but may also be a consequence of mutations in a variety of myelin proteins (e.g., various types of Charcot-Marie-Tooth disease). In most cases of demyelination in the PNS, Schwann cells are able to remyelinate axons. In contrast, in the CNS, oligodendrocytes are much less efficient at remyelination. This is thought to be a consequence of differences between the PNS and CNS extracellular environments as well as intrinsic differences between ► Schwann cells and oligodendrocytes.

Subsequent to demyelination, the passive electrical properties of axons are dramatically altered, resulting in failure to conduct action potentials through demyelinated zones. In particular, demyelination causes a striking increase in the axonal membrane capacitance. Thus, as electrical signals propagate down an axon, the current available to trigger the next group of Na^+ channels is reduced due to capacitive charging of the axonal membrane. Demyelination also results in dysregulation of myelin's "active" properties.

Demyelination, Fig. 1 Na$^+$ channel clusters in myelinated, demyelinated, and remyelinating peripheral axons. (a) A node of Ranvier (*arrow*) labeled for Na$^+$ channels. The outline of the myelin sheath can also be seen. (b) Cartoon illustrating that remyelinated axons have shorter and thinner sheaths with new nodes of Ranvier in formerly internodal zones. (c) A Na$^+$ channel cluster in a demyelinated axon. (d) A heminode with a Na$^+$ channel cluster (*arrow*) and "tail" of Na$^+$ channel immunoreactivity (*arrowheads*). (e) Two Na$^+$ channel clusters (*arrowheads*), located on the same remyelinating axon. (f) Two Na$^+$ channel clusters just before they fuse to form a new node of Ranvier. Scale bars = 10 μm

It may be that disruption of these "active" properties contributes to the generation of neuropathic pain following demyelination.

While alterations to the passive properties of axons after demyelination are well understood, the consequences for the active properties of myelin are mostly unknown. As a result, recent work has focused on how myelin-axon interactions regulate the excitable properties of axons. For example, a variety of experimental models have been used to investigate the consequences of demyelination and remyelination on Na^+ channel clustering and expression.

Acute Demyelination

Peripheral demyelination has been shown to dramatically influence the localization of ion channels. For example, injection of the weak detergent lysolecithin into a peripheral nerve causes a transient, focal demyelinated lesion about 1 week after the injection. This lesion can be remyelinated by Schwann cells during the second and third weeks of post-injection. However, the architecture of this new myelin is somewhat different than before: sheaths are thinner and shorter (Fig. 1b) (Dugandzija-Novakovic et al. 1995). Therefore, in order to support action potential conduction, new nodes of Ranvier with high densities of Na^+ channels must be formed in regions that were formerly covered by compact myelin and characterized by a low density of channels (Fig. 1b, arrowheads) (Shrager 1989). This model has permitted a careful analysis of Na^+ channel localization subsequent to acute demyelination and during remyelination (Dugandzija-Novakovic et al. 1995). Specifically, regions of demyelinated axon can be seen to retain Na^+ channel clusters. Some of these clusters even appear indistinguishable from those found in normal axons (Fig. 1c), suggesting that in the absence of overlying myelinating Schwann cells, neurons have intrinsic mechanisms that cause clusters of Na^+ channels to be retained. In other places, Na^+ channel clusters (Fig. 1d, arrow) can be seen to have a "tail" of immunoreactivity leading away from the main cluster (Fig. 1d, arrowheads). This situation is most often seen adjacent to heminodes (where

the myelin on one side of a node has been lost and the sheath on the other side remains intact). As remyelination occurs, single axons can be seen to have multiple Na^+ channel clusters between remyelinating Schwann cells (Fig. 1e; the two Na^+ channel clusters identified by the arrowheads are found in the same axon and are only about 25 μm apart). As remyelination progresses, these clusters appear to be "pushed" ahead of the myelinating Schwann cell until they finally fuse and form a new node of Ranvier. Figure 1f shows two Na^+ channel clusters (11 days after lysolecithin injection) just before they fuse at a nascent node of Ranvier.

Chronic Demyelination

One major difference between lysolecithin-mediated demyelination and many forms of disease or injury-related demyelination is that the latter forms are usually chronic. As a result, lessons learned from acute demyelination may not be generally applicable to chronic demyelination. To date, few studies have focused on the consequences of chronic demyelination for Na^+ channel expression and localization. Some of the first experiments relied on mutant dysmyelinating mice to assess these characteristics. For example, we showed that the localization of Na^+ channels was dramatically disrupted in axons of the hypomyelinated mouse mutant *shiverer* (Rasband et al. 1999). Further, Westenbroek et al. (1992) showed that the expression of one particular type of brain Na^+ channel (Nav1.2) was dramatically increased in these mice. However, subsequent studies demonstrated that during normal development, Nav1.2 is found in premyelinated and actively myelinating axons (Boiko et al. 2001). Therefore, the differences in Na^+ channel expression and localization seen in *shiverer* may be related to developmental defects rather than a consequence of demyelination.

To directly address whether chronic demyelination can influence the kinds of Na^+ channels present in CNS axons, Craner et al. (2003) examined the localization of Nav1.2 and Nav1.6 in an inflammatory mouse model of CNS demyelination [▶ Experimental Allergic Encephalitis

(EAE)], and Rasband et al. (2003) examined localization and expression levels of these same Na^+ channels in a genetic, adult onset model of chronic CNS demyelination. Each of these studies showed essentially the same result that chronic demyelination results in a loss of clustered Na^+ channels but a concomitant increase in Nav1.2 expression levels. Indeed, Rasband et al. (2003) showed a sixfold increase in their model of the amount of Nav1.2 found in chronically demyelinated optic nerve axons. These Nav1.2 channels were found diffusely distributed throughout the axon rather than in clusters. Together, these studies suggest that, in addition to altering the passive electrical properties of axons and the active clustering of Na^+ channels to nodes of Ranvier, myelin also regulates the kinds of channels present in CNS axons. The axon-glia signaling that mediates this regulation of channel expression, trafficking, and/or localization is an active area of investigation.

Demyelination and Neuropathic Pain

A very common symptom associated with MS is the presence of acute and/or chronic pain (e.g., ▶ Trigeminal Neuralgia). Similarly, some forms of Charcot-Marie-Tooth disease, characterized by ▶ segmental demyelination of PNS axons, also have as symptoms the development of ▶ hyperalgesia, spontaneous pain, and ▶ allodynia. Unfortunately, very little is known about the molecular and cellular mechanisms linking demyelination and neuropathic pain. Since the expression and localization of Na^+ channels appears to depend on interactions with myelinating glial cells and Na^+ channels are obvious candidates to modulate neuronal hyperexcitability, one intriguing possibility is that pain associated with demyelination is a consequence of loss of neuroglial interactions.

Recently, Wallace et al. (2003) showed that acute PNS demyelination by lysolecithin results in the development of both allodynia and hyperalgesia. The development of the neuropathic pain coincided with detection of Nav1.3 Na^+ channels in formerly myelinated axons – a surprising result since Nav1.3 is not normally found in myelinated nerve fibers or anywhere else

in the adult nervous system. Further, spontaneous action potentials were measured during the remyelination phase. These results suggest that peripheral demyelination may lead to aberrant Na^+ channel subtype expression and activity and the switching of receptor modality from proprioceptor or mechanoreceptor to nociceptor.

Although very little is known about the development of neuropathic pain as a consequence of demyelination, future studies designed to determine the molecular and cellular mechanisms underlying control of both Na^+ channel localization and expression may lead to the ability to directly modulate neuronal activity. However, one thing is clear: in the arena of neuropathic pain, myelinating cells such as Schwann cells and oligodendrocytes are not simply bystanders but instead play active roles by modulating the excitable properties of sensory neurons.

References

Boiko, T., Rasband, M. N., Levinson, S. R., et al. (2001). Compact myelin dictates the differential targeting of two sodium channel isoforms in the same axon. *Neuron, 30,* 91–104.

Craner, M. J., Lo, A. C., Black, J. A., et al. (2003). Abnormal sodium channel distribution in optic nerve axons in a model of inflammatory demyelination. *Brain, 126,* 1552–1561.

Dugandzija-Novakovic, S., Koszowski, A. G., Levinson, S. R., et al. (1995). Clustering of Na^+ channels and node of Ranvier formation in remyelinating axons. *Journal of Neuroscience, 15,* 492–503.

Rasband, M. N., Peles, E., Trimmer, J. S., et al. (1999). Dependence of nodal sodium channel clustering on paranodal axoglial contact in the developing CNS. *Journal of Neuroscience, 19,* 7516–7528.

Rasband, M. N., Kagawa, T., Park, E. W., et al. (2003). Dysregulation of axonal sodium channel isoforms after adult-onset chronic demyelination. *Journal of Neuroscience Research, 73,* 465–470.

Salzer, J. L. (2003). Polarized domains of myelinated axons. *Neuron, 40,* 297–318.

Shrager, P. (1989). Sodium channels in single demyelinated mammalian axons. *Brain Research, 483,* 149–154.

Wallace, V. C., Cottrell, D. F., Brophy, P. J., et al. (2003). Focal lysolecithin-induced demyelination of peripheral afferents results in neuropathic pain behavior that is attenuated by cannabinoids. *Journal of Neuroscience, 23,* 3221–3233.

Westenbroek, R. E., Noebels, J. L., & Catterall, W. A. (1992). Elevated expression of type II Na$^+$ channels in hypomyelinated axons of shiverer mouse brain. *Journal of Neuroscience, 12*, 2259–2267.

Dendrite

Definition

Dendrite refers to the post-synaptic element of a neuron.

Cross-References

▶ Opioid Receptor Trafficking in Pain States

Dendritic Spines

Definition

Small protrusions of the dendritic shafts specialized at the reception and processing of incoming signals at synapses. They most often look like pedicled knobs, but may also assume other configurations, such as sessile knobs, rose spines, or bunches of knobs connected to a sole pedicle.

Cross-References

▶ Spinothalamic Tract Neurons, Morphology

Dendritic Topography

Definition

Spatial distribution of the dendritic tree with respect to the cell body and gray matter regions located around it.

Cross-References

▶ Spinothalamic Tract Neurons, Morphology

Dendritic Tree

Definition

Ensemble of neuronal processes ramifying around the cell body and specialized to receive input (chemical or electrical signals) from other neurons.

Cross-References

▶ Spinothalamic Tract Neurons, Morphology

Dendrogram

Definition

A dendrogram hierarchically organizes stimulus objects, subclusters, clusters, etc. on the basis of their relative similarities.

Denervation

Definition

Situation where a limb (or an organ or part of it) is rendered insensitive to applied stimuli by severing sensory axons that innervate the limb. Sensory endings rapidly degenerate in the process of anterograde (Wallerian) degeneration.

Cross-References

▶ Anesthesia Dolorosa Model, Autotomy
▶ Facet Joint Pain

Denervation Model of Neuropathic Pain

▶ Anesthesia Dolorosa Model, Autotomy

Densocellular Subnucleus of the Mediodorsal Nucleus

Synonyms

MDdc

Definition

Envelops MD laterally and posteriorly. Its neurons resemble those of the central lateral nucleus (CL). MDdc receives input from many of the same subcortical sites as CL and projects to the striatum and to premotor cortical areas.

Cross-References

▶ Spinothalamic Terminations, Core and Matrix

Dental Pain, Etiology, Pathogenesis, and Management

Graham R. Holland
Department of Cariology, University of Michigan, Ann Arbor, MI, USA

Synonyms

Dentin sensitivity; Pulpalgia; Pulpitis; Toothache

Definition

A noxious experience that originates or appears to originate from a tooth.

Characteristics

Pain is not experienced from an entirely healthy tooth under normal physiological conditions, but can be induced by a cold stimulus at 0 °C or below. It is more readily felt in otherwise normal teeth in which the dentin beneath the enamel of the crown or beneath the cementum of the root is exposed. Then cold stimuli will induce pain and hot or osmotic stimuli *may* do so. This pain is sharp and lasts only for the duration of the stimulus. If the ▶ dental pulp, the soft tissue within the dentin and responsible for its production, is inflamed, the pain to an applied stimulus will be strong, dull, and throbbing. It may continue beyond the duration of the stimulus and may be present spontaneously. As the majority of nociceptors in the uninjured pulp are inactive under normal circumstances, they can be classified as "silent." No sensations other than pain can be experienced from the tooth pulp in response to nonelectrical stimuli, a property that has been utilized in a number of experimental studies. The stimuli capable of inducing pain from inflamed pulps would not be in the noxious range when applied to intact, healthy skin. It is difficult to elicit pain from a freshly cut cavity in ▶ dentin, but the tooth becomes much more responsive if bacteria are allowed to inhabit the cavity for 1 week allowing the underlying pulp to become inflamed (Anderson and Matthews 1967). The intensity of pain felt from a tooth with an inflamed pulp is highly variable, but it has been ranked using the ▶ McGill pain questionnaire as similar to the pain experienced from arthritis or a bone fracture but substantially less than that associated with childbirth (Melzack 1984). The descriptors most commonly applied to it are throbbing, boring, sharp, sickening, annoying, constant, and rhythmic (Melzack 1975).

Etiology

Dental caries is the predominant cause of pulpal inflammation and the pain associated with it. Toxins from the bacteria contained in a carious lesion permeate the dentin and induce inflammation in the pulp long before the bacteria themselves invade it. Pain from restored teeth without overt dental caries is usually the result of the micro-leakage of bacteria beneath the filling and the penetration of their toxins to the dental pulp. The pulps of all carious teeth are inflamed but not always painful; in almost one third of cases, the inflammation may progress to necrosis without the patient experiencing discomfort (Michaelson and Holland 2002). When the dental pulp is heavily inflamed or becomes necrotic, inflammatory mediators as well as bacteria and toxins may spread into the tissues around the apex of the tooth root. This tissue may be painful especially when the tooth is tapped but is often not noticed by the patient. If the inflammation around the tooth apex progresses to ► necrosis, ► abscess formation may occur and this can be acutely painful due to swelling, tissue distension, and the spread of inflammatory mediators. In contrast to pulpal pain, pain from the periodontal tissues is readily located.

The pulp has a dense afferent innervation of ► C fibers and Aδ fibers, which are a dedicated component of the pain system with a smaller number of Aβ fibers that may be recruited during inflammation. The Aδ fibers may be those which respond to cold stimuli applied to an uninjured tooth. The C fibers may be silent and only become active in inflammation, causing the dull throbbing pain characteristic of symptomatic pulpitis (Narhi et al. 1992). The stimuli which initiate dental pain when applied to the dentin, do so mainly by causing fluid movement in the narrow (approx. 1–2 μm in diameter) ► tubules which make up the tissue (Matthews and Vongsavan 1994). This movement rapidly deforms and activates the exposed axonal membranes of the nociceptors which are abundant in the dental pulp, and present to a limited extent within the tubules of the dentin. ATP activated P2X3, and P2X2/3 receptors are present on sensory afferents in tooth (Cook et al. 1997) and could mediate nociception perhaps via ATP released by the movement of cells or by inflammatory mediators. Vanilloid receptors (VR1) for which inflammatory mediators would be effective ligands are also present on these afferents (Renton et al. 2003).

It is also possible that some chemicals may diffuse slowly through the dentin and act directly on the nociceptors (Matthews and Vongsavan 1994) though this would be opposed by the outward flow of dentinal fluid especially during inflammation (Puapichartdumrong et al. 2003).

Once inflammation is initiated in the pulp, a large number of cytokines and inflammatory mediators are released some of which activate (e.g., ► bradykinin) or sensitize (► prostaglandins) nociceptive terminals. Inflammation also leads to axonal sprouting and increases in the neuropeptide content of pulpal nerves (Byers et al. 1990). The neuropeptide ► substance P is found within a subpopulation of nociceptive afferent nerve fibers. The expression of substance P within pulpal nerves undergoes dynamic changes during the inflammation induced by dental caries (Rodd and Boissonade 2000; Bowles et al. 2003). The dental pulp is encased in a rigid chamber of dentin where the interstitial fluid pressure is somewhat higher than in many other tissues and which rises in localized areas of inflammation (Heyeraas and Berggreen 1999). This may contribute to nociceptor activation.

The pulpal nociceptors are the terminals of afferent fibers carried in the mandibular and maxillary divisions of the trigeminal nerve with cell bodies in the trigeminal ganglion. They relay at various levels in the trigeminal nuclear complex and contribute to the central ascending pathways.

The variability in the presentation of pulpal pain may, in part, be due to variations in the extent of inflammation but peripheral and central mechanisms of pain suppression and hyperalgesia are also involved. For example, pulpal inflammation leads to gene product Fos expression (Chattipakorn et al. 2002) as well as central sensitization (Chiang et al. 1998) in the ► trigeminal subnucleus caudalis. Opioid peptides and receptors have been found in the dental pulp (Jaber et al. 2003) and increase during inflammation (Mudie and Holland 2006). The pulp also has a sympathetic innervation, and activity in this may modulate nociceptive output perhaps by neuropeptide release (Hargreaves et al. 2003).

Management

The diagnosis of dental pain is often complicated by its referral to other teeth and sometimes to extraoral structures. Central mechanisms and sensitization may explain this (Sessle et al. 1986). Conversely pain in extraoral structures, such as the masticatory apparatus and boney sinuses, may be referred to the teeth. Pain from cardiac muscle may sometimes be referred to the teeth. Thus, the first step in management is to determine that the pain is of dental origin and from which tooth it originates. This is done by using cold (ice or solid CO_2) and sometimes hot stimuli to test all the teeth in both upper and lower arches on the side indicated by the patient. The patient will experience greater pain when an affected tooth is stimulated. In some circumstances, ► selective anesthesia of individual teeth may help. If serious uncertainty exists, symptoms usually become clearer with time.

If a patient reports only occasional brief discomfort to hot and cold drinks and it is determined that the discomfort is due to exposed dentin in an otherwise healthy tooth, the sensitivity can be reduced or eliminated by blocking the dentinal tubules. Patent tubules are a characteristic of sensitive dentin (Yoshiyama et al. 1989). Rubbing the surface with a wood stick creates a smeared layer of dentin which occludes many tubules. Some salt solutions, such as potassium oxalate and sodium fluoride, induce precipitation within the tubules. These may be applied in a dental office setting. ► Desensitizing toothpastes for home use usually incorporate fluoride compounds and potassium nitrate. The potassium ions released from these pastes may permeate the dentin and inactivate nociceptors by altering the ionic composition of the extracellular fluid. Sometimes attaching an adhesive restoration to the surface of the dentin may, by occluding the dentinal tubules, be successful.

If the cause of the pain is pulpal inflammation resulting from dental caries, this must be dealt with by removing the infected dentin and replacing it with a filling material that does not by itself induce inflammation, and also has surface margins that prevent the migration of bacteria onto the dentinal surface. With the causative factor removed, the inflammation in the dental pulp should gradually subside and the symptoms with it. When the dentin is severely infected or bacteria themselves rather than just their toxins have reached the pulp, the inflammatory process may be irreversible and the only solution is to remove the pulp followed by ► root canal therapy or, alternatively, to extract the tooth. Unfortunately, it is difficult to diagnose the extent of inflammation in an affected pulp. No clear correlation has ever been established between symptoms and the inflammatory condition of the pulp though spontaneously painful teeth usually have extensively inflamed pulps. The most consistent finding in examining the pulps that have been removed from symptomatic teeth is that substance P is present in greater amounts in painful teeth than in normal teeth (Rodd and Boissonade 2000; Bowles et al. 2003). Currently, the most accepted standard for determining the presence of irreversible pulpal inflammation is the presence of spontaneous pain or pain that is prolonged well beyond the removal of a cold stimulus. Convincing evidence supporting this relationship is meager. In the acute management of dental pain, analgesics are of great value in relieving symptoms. As toothache is predominantly of inflammatory origin, ► nonsteroidal anti-inflammatory drugs (NSAIDs) are likely to be more effective than centrally acting analgesics. One presentation of toothache, which provides a special treatment challenge, is a condition often known as ► hot tooth syndrome. This presents as an intense and spontaneous pain in a tooth which is difficult to control with local anesthetics. When other signs of effective local anesthesia are present, the tooth remains painful. Regional block anesthesia and direct local infiltration may prove incompletely effective. Resolution is often only achieved by removing the pulp from the tooth with less than total anesthesia, a trial for both patient and practitioner. In these cases, long-standing inflammation and nociceptors stimulation has led to central sensitization (Hu et al. 1986; Chiang et al. 1998). The principal features of this are the recruitment of Aβ axons, hyperalgesia, allodynia, and spontaneous pain. A combination of peripherally acting anti-inflammatory drugs and centrally acting analgesics would best control symptoms. Drugs developed specifically to reduce or eliminate central sensitization may find application here. As central effects contribute substantially to these more serious toothaches the use of long acting local anesthetics may be helpful.

References

Anderson, D. J., & Matthews, B. (1967). Osmotic stimulation of human dentine and the distribution of dental pain thresholds. *Archives of Oral Biology, 12,* 417–426.

Bowles, W. R., Withrow, J. C., Lepinski, A. M., & Hargreaves, K. M. (2003). Tissue levels of immunoreactive substance P are increased in patients with irreversible pulpitis. *Journal of Endodontics, 29,* 265–267.

Byers, M. R., Taylor, P. E., Khayat, B. G., & Kimberly, C. L. (1990). Effects of injury and inflammation on pulpal and periapical nerves. *Journal of Endodontics, 16,* 78–84.

Chattipakorn, S. C., Sigurdsson, A., Light, A. R., Narhi, M., & Maixner, W. (2002). Trigeminal c-Fos expression and behavioral responses to pulpal inflammation in ferrets. *Pain, 99,* 61–69.

Chiang, C. Y., Park, S. J., Kwan, C. L., Hu, J. W., & Sessle, B. J. (1998). NMDA receptor mechanisms contribute to neuroplasticity induced in caudalis nociceptive neurons by tooth pulp stimulation. *Journal of Neurophysiology, 80,* 2621–2631.

Cook, S. P., Vulchanova, L., Hargreaves, K. M., Elde, R., & McCleskey, E. W. (1997). Distinct ATP receptors on pain-sensing and stretch-sensing neurons. *Nature, 387,* 505–508.

Hargreaves, K. M., Bowles, W. R., & Jackson, D. L. (2003). Intrinsic regulation of CGRP release by dental pulp sympathetic fibers. *Journal of Dental Research, 82,* 398–401.

Heyeraas, K. J., & Berggreen, E. (1999). Interstitial fluid pressure in normal and inflamed pulp. *Critical Reviews in Oral Biology and Medicine, 10,* 328–336.

Hu, J. W., Dostrovsky, J. O., Lenz, Y. E., Ball, G. J., & Sessle, B. J. (1986). Tooth pulp deafferentation is associated with functional alterations in the properties of neurons in the trigeminal spinal tract nucleus. *Journal of Neurophysiology, 56,* 1650–1668.

Jaber, L., Swaim, W. D., & Dionne, R. A. (2003). Immunohistochemical localization of mu-opioid receptors in human dental pulp. *Journal of Endodontics, 29,* 108–110.

Matthews, B., & Vongsavan, N. (1994). Interactions between neural and hydrodynamic mechanisms in dentine and pulp. *Archives of Oral Biology, 39,* 87S–95S.

Melzack, R. (1975). The McGill Pain Questionnaire: Major properties and scoring methods. *Pain, 1,* 277–299.

Melzack, R. (1984). The myth of painless childbirth (the John J. Bonica lecture). *Pain, 19,* 321–337.

Michaelson, P. L., & Holland, G. R. (2002). Is pulpitis painful? *International Endodontic Journal, 35,* 829–832.

Mudie, A. S., & Holland, G. R. (2006). Local opioids in the inflamed dental pulp. *Journal of Endodontics, 32,* 319–323.

Narhi, M., Jyvasjarvi, E., Virtanen, A., Huopaniemi, T., Ngassapa, D., & Hirvonen, T. (1992). Role of intradental A- and C-type nerve fibres in dental pain mechanisms. *Proceedings of the Finnish Dental Society, 88,* 507–516.

Puapichartdumrong, P., Ikeda, H., & Suda, H. (2003). Facilitation of iontophoretic drug delivery through intact and caries-affected dentine. *International Endodontic Journal, 36*(10):674–81.

Renton, T., Yiangou, Y., Baecker, P. A., Ford, A. P., & Anand, P. (2003). Capsaicin receptor VR1 and ATP purinoceptor P2X3 in painful and nonpainful human tooth pulp. *Journal of Orofacial Pain, 17,* 245–250.

Rodd, H. D., & Boissonade, F. M. (2000). Substance P expression in human tooth pulp in relation to caries and pain experience. *European Journal of Oral Sciences, 108,* 467–474.

Sessle, B. J., Hu, J. W., Amano, N., & Zhong, G. (1986). Convergence of cutaneous, tooth pulp, visceral, neck and muscle afferents onto nociceptive and non-nociceptive neurones in trigeminal subnucleus caudalis (medullary dorsal horn) and its implications for referred pain. *Pain, 27,* 219–235.

Yoshiyama, M., Masada, J., Uchida, A., & Ishida, H. (1989). Scanning electron microscopic characterization of sensitive vs. insensitive human radicular dentin. *Journal of Dental Research, 68,* 1498–1502.

Dental Pulp

Definition

The dental pulp is the connective enclosed by the dentin of the tooth. It contains a large number of afferent and sympathetic nerve fibers, and an extensive vascular plexus that supports the cells in the pulp that include those lining the dentin (the odontoblasts) that are responsible for dentin formation.

Cross-References

▶ Dental Pain, Etiology, Pathogenesis, and Management
▶ Nociceptors in the Dental Pulp
▶ Orofacial Pain: Taxonomy/Classification

Dentin

Definition

Dentin is a mineralized tissue similar in hardness to bone. It forms the bulk of a tooth, but is covered by the harder enamel (that forms the crown and visible part of the tooth) and the cementum that binds it to the bony socket. It is made up of many thousands of small (1–2 μm) tubules running from the dental pulp on its inside to its outer surface in contact with the enamel and cementum.

Cross-References

▶ Dental Pain, Etiology, Pathogenesis, and Management
▶ Nociceptors in the Dental Pulp

Dentin Sensitivity

▶ Dental Pain, Etiology, Pathogenesis, and Management

Dentinal Tubules

Definition

Fluid-filled thin tubules (diameter: 1–2 μm) in dentin, which extend all the way from the dental pulp to the dentin-enamel border. There are 30,000–40,000 tubules/mm^2 of dentin.

Cross-References

▶ Nociceptors in the Dental Pulp

Dependence

Definition

Dependence is a state in which an abstinence syndrome may occur following abrupt withdrawal, dose reduction, or administration of an antagonist. Although usually associated with opioids, this is common with many classes of medication including antihypertensives and anticonvulsants. It is seen in all subjects chronically given a dependence-inducing drug.

Cross-References

▶ CRPS, Evidence-Based Treatment
▶ Opioid Receptors
▶ Postoperative Pain, Opioids

Depolarization

Definition

A loss of polarity in general. Depolarization is a change in the electrical charge between the inside and the outside of a membrane. Specifically in cells, it is the response of a cell membrane to a stimulus, which leads to a loss of the negativity of the membrane potential beyond the resting potential, and thereby the generation of an action potential.

Cross-References

▶ COX-1 and COX-2 in Pain
▶ Mechanonociceptors
▶ Transition from Acute to Chronic Pain

Depression

Definition

A state of lowered mood, often accompanied by disturbances of sleep, energy, appetite, concentration, interests, and sexual drive. There are also associated negative thoughts like ideas of hopelessness, worthlessness, and helplessness. The term "depression" can be used to denote a mood state, a symptom, or a psychiatric disorder (Major Depressive Disorder).

Cross-References

▶ Depression and Pain
▶ Pain in the Workplace, Risk Factors for Chronicity, and Psychosocial Factors
▶ Psychiatric Aspects of the Management of Cancer Pain

Depression and Pain

Robert D. Kerns and Alicia A. Heapy
VA Connecticut Healthcare System, Westhaven, CT, USA

Synonyms

Depression; Major depressive disorder

Definition

▶ Depression is characterized by sadness and loss of interest or pleasure in previously enjoyed activities. The term depression can be used to denote a mood state, a symptom, or a disorder (major depressive disorder). Depression qualifies as a disorder when it is intense, frequent, or enduring enough to result in dysfunction and/or concern for the person, his or her significant others, or society, more generally. The formal

diagnosis of ▶ major depressive disorder (MDD), as defined by the *Diagnostic and Statistical Manual of Mental Disorders fourth edition* (DSM-IV) (American Psychiatric Association 1994), is the presence of a collection of symptoms that must include depressed mood or loss of interest or pleasure in most activities lasting at least 2 weeks. Additional symptoms include fatigue, feelings of excessive or inappropriate worthlessness or guilt nearly every day, significant weight loss or gain, insomnia or hypersomnia, diminished concentration or ability to make decisions, frequent thoughts of death, suicidal ideation, or suicide attempt. In order to meet criteria for MDD, symptoms must cause distress or impairment in functioning. ▶ Chronic pain is characterized as non-cancer pain lasting 6 or more months.

Characteristics

The presence of at least transient experiences of depressed mood among persons with chronic pain is thought to be understandable and even expected, and in fact, a high ▶ co-prevalence between the experience of chronic pain and ▶ depressive symptom severity has been noted, both clinically and in formal empirical investigations for many years. However, estimates of the prevalence of MDD among persons with chronic pain vary widely depending on the manner of assessment, population selection, and definition of depression that is used (Banks and Kerns 1996). MDD may be assessed with varying levels of stringency, and when standard diagnostic criteria are not used, persons with significant depressive symptoms, but not MDD, may be included in the depressed category, thereby providing a misleading picture of the true comorbidity of chronic pain and MDD. Further complicating the picture, several symptoms of MDD and pain overlap (e.g., sleep disturbance, loss of energy, change in appetite), which may inflate estimates of depression in populations with chronic pain. Use of standardized semi-structured interview methods, such as the

Structured Clinical Interview for DSM-III-R
(SCID, Spitzer 1992) and DSM-IV criteria, is
encouraged to ensure reliable and valid diagnosis
of MDD in persons with chronic pain. When
these criteria are used, a majority of studies report
▶ point prevalence rates ranging from 30 % to
54 % (Banks and Kerns 1996). Even when these
criteria are employed, however, MDD appears to
occur more frequently in chronic pain than in
other common, chronic medical conditions
(Banks and Kerns 1996). Studies also suggest
that the presence of chronic pain may interfere
with the detection of depression in primary care
settings (Bair et al. 2003). Patient emphasis on
somatic complaints in chronic pain conditions,
which may be exacerbated in the context of
depression, is thought to interfere with physician
detection of depression, particularly milder cases
of depression.

While depression and pain appear to
frequently co-occur, the temporal association
between the two and possible causal mechanisms
underlying the comorbidity have not been
identified. Three hypotheses have been asserted:
(1) pain precedes depression, (2) depression
precedes pain, and (3) the two occur simultaneously. Many proponents of the first hypothesis,
that pain precedes depression, argue that
depression is largely a psychological reaction to
pain, mediated by cognitive factors, behavioral
factors, or combination of both (e.g., Fordyce
1976; Rudy et al. 1988; Sullivan and D'Eon
1979). Those that assert that depression precedes
pain point to mood-induction studies that have
demonstrated increased self-focus of attention
and increased reports of aches and pain, as well
as decreased tolerance for laboratory-induced
pain in response to depressed mood (Ingram and
Smith 1984; Salovey and Birnbaum 1989).
Finally, evidence for the simultaneous
occurrence of depression and pain is rooted in
nervous system biogenic ▶ monoamines such as
serotonin and norepinephrine, which have been
implicated in both the development of depression
and the modulation of pain (Ward et al. 1982).
This theory hypothesizes that depression and pain
symptoms both use the descending pathways of

the central nervous system. Specifically, the
periaqueductal gray, the limbic structures, and
the rostral-ventromedial medulla modulate attention and affective response to peripheral pain
stimuli, typically suppressing these stimuli in
favor of attending to outside stimuli. However,
in the context of depression and its concomitant
depletion of serotonin and epinephrine, attention
and affective responses to these peripheral
signals are enhanced, providing a potential
explanation for the high comorbidity of pain
and depression and the effectiveness of
antidepressant medication for chronic pain complaints (Bair et al. 2003). Efforts to fully describe
the temporal development of pain and depression
have been hampered by lack of longitudinal
studies and other methodological shortcomings.
The reliability and validity of retrospective
patient reports of the temporal emergence of
pain and depression have been questioned due
to the subjective nature of pain and depression;
the difficulty identifying the exact onset of
depression, which is usually gradual; and the
difficulty in determining when acute pain can be
classified as chronic (Banks and Kerns 1996).

Several theories have been asserted to explain
the frequent co-occurrence of depression and
pain. These models are variations of major
▶ cognitive-behavioral theories and hypothesize
that an individual's pattern of cognitive assessment of pain and pain experience and their behavior in response to pain contribute to the
development and maintenance of depression.
Beck's cognitive distortion model (Beck 1976)
contends that certain individuals, by virtue of
preexisting negative cognitive schemas, are
particularly vulnerable to developing depression
and depressive symptomatology in response to
the challenges of living with chronic pain. Once
these negative schemas are activated, they serve
to elicit negative thoughts about the self, the
world, and the future. In turn, these negative
thoughts produce negative mood, which in turn
leads to other symptoms of depression such as
decreased activity and suicidal ideation. Learned
helplessness theory (Seligman 1975) asserts that
when individuals are exposed to uncontrollable

negative outcomes like chronic pain, they develop an attitude of helplessness or expectation that other future outcomes are also uncontrollable. These individuals do not attempt to alter future outcomes, even when those outcomes are amenable to their efforts, due to an erroneous belief in their own helplessness, thereby maintaining the circumstances that contribute to depression. Behavioral models posit that an individual with chronic pain faces greatly reduced opportunities for rewards and, consequently, is at higher risk of developing depression. Specifically, persons with chronic pain experience pain as a result of performing previously rewarding activities like work, social interactions, and physical activity, which leads to withdrawal from these activities. Additionally, fear of reinjury may also result in a reduction in activities and consequent reinforcement.

The preceding models, although widely accepted for explaining depression in general, are less able to account for the elevated levels of depression found in chronic pain compared with other chronic diseases or conditions. The cognitive and behavioral vulnerabilities that are thought to precipitate depression should be equally distributed across populations with other chronic illnesses, and the presence of a chronic illness could be conceptualized as a stressor capable of triggering the vulnerabilities necessary to set off depression. If these models are correct, the prevalence of depression would be approximately equal among persons with various chronic medical conditions. Banks and Kerns (1996) have asserted that pain has several unique qualities that make it a particularly potent stressor. First, chronic pain is a symptomatic condition, and the symptoms provide a constant reminder on the condition. Second, pain is typically a signal of danger and tissue damage and as a result often produces anxiety. Third, pain is often associated with a variety of losses resulting in impairment and disability. Secondary losses are especially prominent. Pain can interfere with employment, leisure activities, sexual function, and interpersonal relationships. Financial consequences can include high medical expenses, diminished capacity to work, and unemployment. The combination of the many losses experienced by persons with chronic pain can lead to feelings of social isolation, meaninglessness, uncertainty about the future, and loss of self-concept, self-esteem, and family role. Finally, due to the frequent discrepancy between structural pathology and pain severity, impairment, and disability, patients may experience a disconnection between their own experience and the messages they receive from the health-care system. Persons with chronic pain may be told that there is no medical basis for their pain or that there is a medical basis, but there is nothing more that can be done. The conflicting messages between what a person is told by the health-care system and their own experience can lead to self-doubt, distrust of health-care providers, confusion, frustration, and affective distress (e.g., Goldman 1991; Reid et al. 1991).

Negative mood states, including depressive symptoms and MDD, have been shown to hinder pain treatment (Haythornthwaite et al. 1991) and exacerbate chronic pain conditions. Somatic focus and amplification, autonomic arousal, and misinterpretation of symptoms are all mechanism through which negative mood states are thought to influence pain experience and outcomes (Dersh et al. 2002). The frequent correlation of pain and depression and the effect of depression on pain treatment argue for the importance of identification and treatment of depression for improving chronic pain treatment outcomes.

References

American Psychiatric Association. (1994). *Diagnostic and statistical manual of mental disorders* (4th ed.). Washington, DC: APA.

Bair, M. J., Robinson, R. L., Katon, W., et al. (2003). Depression and pain comorbidity: A literature review. *Archives of Internal Medicine, 163*, 2433–2455.

Banks, S. M., & Kerns, R. D. (1996). Explaining the high rates of depression in chronic pain: A diathesis stress framework. *Psychological Bulletin, 119*, 95–110.

Beck, A. T. (1976). *Cognitive therapy and the emotional disorders*. New York: International Universities Press.

Dersh, J., Polatin, P. B., Gatchel, R. J. (2002). Chronic pain and psychopathology: research findings and theoretical considerations. *Psychosomatic Medicine, 64,* 773–86.

Fordyce, W. E. (1976). *Behavioral methods for chronic pain and illness.* St. Louis: Mosby.

Goldman, B. (1991). Chronic pain patients must cope with chronic lack of physician understanding. *Canadian Medical Association Journal, 144,* 1492–1497.

Haythornthwaite, J. A., Sieber, W. J., & Kerns, R. D. (1991). Depression and the chronic pain experience. *Pain, 46,* 177–184.

Ingram, R. E., & Smith, T. S. (1984). Depression and internal versus external locus of attention. *Cognitive Therapy and Research, 8,* 139–152.

Reid, J., Ewan, C., & Lowy, E. (1991). Pilgrimage of pain: The illness experience of women with repetition strain injury and the search for credibility. *Social Science & Medicine, 32,* 601–612.

Rudy, T. E., Kerns, R. D., & Turk, D. C. (1988). Chronic pain and depression: Toward a cognitive-behavioral mediation model. *Pain, 35,* 129–140.

Salovey, P., & Birnbaum, D. (1989). Influence of mood on health-relevant cognitions. *Journal of Personality and Social Psychology, 57,* 539–551.

Seligman, M. E. P. (1975). *Helplessness: On depression, development, and death.* San Francisco: Freeman.

Spitzer, R. L. (1992). History and rationale and development of SCID for DSM-III-R. *Archives of General Psychiatry, 49,* 624–629.

Sullivan, M. J. L., & D'Eon, J. L. (1979). Relation between catastrophizing and depression in chronic pain patients. *Journal of Abnormal Psychology, 99,* 260–263.

Ward, N. G., Bloom, V. L., & Dworkin, S. (1982). Psychobiological markers in coexisting pain and depression: Toward a unified theory. *The Journal of Clinical Psychiatry, 43,* 32–41.

Depressive Symptom Severity

Definition

Depressive symptom severity refers to the presence, duration, and intensity of symptoms of depressive disorders such as dysphoria, anhedonia, insomnia, and hopelessness.

Cross-References

▶ Depression and Pain

Dermatomal Level

Definition

A dermatome is an area on the body supplied by a single nerve root. The dermatomal level is used to describe the highest level of anesthesia provided by either a spinal or epidural anesthetic. A thoracic 4 level of anesthesia is typically needed to provide pain relief for cesarean delivery.

Cross-References

▶ Analgesia During Labor and Delivery

Dermatomes

Definition

Dermatomes are the areas of skin that are innervated by single dorsal roots of the spinal cord or trigeminal nerve in the case of the oral-facial region. Dermatomes are labeled by the spinal segment with which they are associated, such as C_1, C_2, (cervical), T_1, T_2, (thoracic), and L_1, L_2 (lumbar).

Cross-References

▶ Central Nervous System Stimulation for Pain
▶ Cervical Transforaminal Injection of Steroids
▶ Dorsal Root Ganglionectomy and Dorsal Rhizotomy
▶ Pain in Humans, Psychophysical Law

Dermatomyositis

Definition

Immunogenic myositis, mainly with characteristic additional skin lesions. A paraneoplastic cause should be considered in adult cases. The development of muscle symptoms, especially in proximal muscles, is acute or subacute. Muscle pain is frequent. Muscle biopsies confirm the diagnosis.

Cross-References

▶ Myositis

Descending Anti-analgesic Systems

▶ Descending Facilitatory Systems

Descending Circuitry, Molecular Mechanisms of Activity-Dependent Plasticity

Ke Ren
Department of Neural and Pain Sciences,
University of Maryland Dental School,
Baltimore, MD, USA

Synonyms

Cellular and receptor mechanisms underlying the development of persistent pain; Pain modulatory circuitry and endogenous pain control

Definition

In the context of pain, the descending circuitry refers to a neural network between the brain and spinal cord that provides supraspinal modulation of nociceptive transmission at the spinal and trigeminal levels. Multiple brain sites and pathways are involved in descending pain modulation. One of the key pathways consists of midbrain periaqueductal gray (PAG), rostral ventromedial medulla (RVM), and spinal cord. Plastic changes occur after tissue or nerve injury within the descending pain modulatory circuitry. Such

changes involve transcriptional, translational, and posttranslational modulation of neurotransmitters and their receptors and increased activity of glial cells. The neuronal hyperexcitability, further facilitated by interactions between neurons and glia, leads to enhanced synaptic strength and altered net descending modulation and contributes to the development of persistent or chronic pain (Watkins et al. 1995; Wei et al. 2008; Roberts et al. 2009).

Characteristics

Changing Role of Descending Circuitry After Injury

The potency, efficacy, and net effect of descending circuitry are subject to modulation by functional status of the organism. The excitability of brain stem pain modulatory neurons is enhanced after injury (Heinricher et al. 2009). Rats with inflammatory hyperalgesia exhibit an increased sensitivity to opioid analgesics (Neil et al. 1986). Parallel changes in the net effect of descending modulation are observed at the spinal level where an enhanced descending inhibition of spinal nociceptive neurons has been demonstrated in animals with inflammation (Schaible et al. 1991; Ren and Dubner 2008, review). The source of the increased sensitivity to opioids and enhanced descending inhibition after inflammation can be traced to supraspinal structures. The antihyperalgesic and analgesic potency of mu- and delta-opioid receptor (MOR and DOR, respectively) agonists microinjected into the RVM (Fields et al. 2006) is progressively enhanced after adjuvant-induced hind paw inflammation (Hurley and Hammond 2000). Focal lesions of the RVM and locus coeruleus are followed by an increase in spinal Fos protein expression and hyperalgesia after inflammation (Tsuruoka and Willis 1996). Stimulation in the PAG produces an increased antinociception in the inflamed paw in rats (Morgan et al. 1991). Increased central (delta) opioid receptor trafficking to the cell membrane under persistent pain conditions provides a mechanism of increased opioid analgesia (Bie and Pan 2007).

The descending circuitry can also actively facilitate pain, particularly following peripheral tissue or nerve injury. The destruction of cells in the nucleus reticularis gigantocellularis in RVM with a soma-selective neurotoxin, ibotenic acid, leads to an attenuation of hyperalgesia and a reduction of inflammation-induced spinal Fos expression (see Ren and Dubner 2008). There is an enhancement of C-fiber windup and novel A-fiber windup during inflammation, which also depends on input from supraspinal circuitry (Herrero and Cervero 1996). A descending facilitatory drive contributes to the pathogenesis of secondary hyperalgesia and experimental neuropathic pain (Porreca et al. 2002, review). The tactile allodynia after nerve injury is dependent upon a tonic activation of net descending facilitation from supraspinal sites. Lesions of the RVM inhibit secondary hyperalgesia produced by topical application of mustard oil (Urban and Gebhart 1999). Hind paw formalin-induced hyperalgesia is prevented by RVM lesion (Wiertelak et al. 1997). Spinalization blocks mustard oil-produced secondary mechanical allodynia and mechanical hyperexcitability of spinal nociceptive neurons (Mansikka and Pertovaara 1997). In nerve injured rats, lesions of the dorsolateral fasciculus, local anesthetic block of the RVM, and lesions of RVM MOR-expressing cells do not prevent the onset but reverse the later maintenance of tactile and thermal hypersensitivity (Burgess et al. 2002). A selective role of RVM in the maintenance, but not expression, of persistent pain is also observed in a visceral pain model (Vera-Portocarrero et al. 2006). However, employing the same local anesthetic block strategy, RVM is found critical for both the initiation and maintenance of pain hypersensitivity after intramuscular injection of acidic saline (Tillu et al. 2008). In addition to RVM, the dorsal reticular nucleus in the caudal medulla is a major site to produce descending pain facilitation (Martins et al. 2010). From a physiological point of view, descending facilitation is necessary to maintain hyperalgesia as the tissue heals and to protect the injured tissue from further insult (Gebhart 2004). However,

under certain conditions, this central facilitatory drive may be key in transition to chronic pain that is no longer protective.

Activity-Dependent Cellular Changes Underlie Descending Pain Facilitation After Injury

Synaptic transmission within the descending circuitry involves multiple neurotransmitters and receptors. Correlated changes in gene transcription, protein translation, and receptor phosphorylation have been found in the descending circuitry after injury. For example, Williams and Beitz (1993) showed that hind paw inflammation increased neurotensin mRNA expression in the PAG and the lateral tegmental nuclei, from which neurotensinergic neurons project to the RVM. Extracellular signal-regulated kinase (ERK) phosphorylation is induced in the RVM and locus coeruleus in association with stress-induced hyperalgesia (Imbe et al. 2004). The sensitivity of excitatory amino acid receptors in the descending circuitry is changed during the development of inflammatory hyperalgesia, which is related to transcriptional and translational modulation of the receptors (Ren and Dubner 2008).

Glutamate Receptors

Excitatory amino acid receptors have been shown to activate pain modulatory neurons in the RVM and mediate morphine analgesia (Heinricher et al. 2001). Examination of mRNA expression of the GluN1 (NR1), GluN2A (NR2A), and GluN2B (NR2B) subunits of the NMDA receptor in the RVM reveals an upregulation that parallels the time course of the RVM excitability changes. This is accompanied by an increase in NMDA receptor subunit proteins. Protein phosphorylation is a major mechanism for regulation of receptor function. Phosphorylation of multiple sites in the cytoplasmic C-termini of the GluN1 and GluN2 subunits, including tyrosine, serine, and threonine residues, is known to modulate NMDA receptor activity and affect synaptic transmission. There is an increase in tyrosine phosphorylation of the NMDA receptor

GluN2A subunit in the RVM after inflammation as compared with naïve noninflamed rats. A time-dependent increase in GluN1 and GluA1 serine phosphorylation has also been identified in the RVM after hind paw inflammation. Consistent with a role for AMPA receptors, neuronal pentraxin 1 (NP1), a mediator of synaptic AMPA receptor recruitment, is induced in RVM after nerve injury (Zapata et al. 2012). Suppressing NP1 expression attenuates neuropathic pain, and rescue of RVM NP1 restores pain hypersensitivity in NP1 knockout mice. A common feature of the changes in glutamate receptor phosphorylation is that it occurs rapidly, as early as 10–30 min after inflammation, and persists for up to a week. The time course of changes in glutamate receptor phosphorylation in the RVM correlates well with changes in excitability in descending circuitry and behavioral hyperalgesia after inflammation (Ren and Dubner 2008).

Brain-Derived Neurotrophic Factors (BDNF)

A well-established regulator of synaptic plasticity, BDNF also plays a critical role in pain modulatory circuitry and contributes to the mechanisms of descending pain facilitation. After peripheral tissue injury, BDNFergic neurotransmission involving TrkB in the PAG-RVM-spinal cord circuitry is rapidly activated, evidenced by increased BDNF expression in PAG and upregulation and increased phosphorylation of TrkB receptors in the RVM. Through the intracellular signaling cascade that involves phospholipase C, IP_3, PKC, and non-receptor protein tyrosine kinases, NMDA receptors are phosphorylated and descending facilitatory drive is initiated (Guo et al. 2006).

Neurokinin 1 Tachykinin (NK1) Receptors

Injection of NK1 receptor antagonists into the RVM reverses inflammatory hyperalgesia but has no effect on ongoing neuronal activity and baseline nociception (Hamity et al. 2010; Khasabov et al. 2012; Pacharinsak et al. 2008), suggesting that NK1 receptors in RVM are not tonically active in normal nociception but

activated by injury. However, the source of substance P released into the RVM after injury is unclear. After inflammation, NK1 receptors exhibit a time-dependent upregulation in the RVM (LaGraize et al. 2010). Injection of an NK1 receptor antagonist into RVM attenuated the capsaicin-evoked enhanced responses of ON cells to noxious stimuli with less of an effect on the enhanced inhibitory responses of OFF cells, suggesting a selective localization of NK1 receptors in ON cells (Brink et al. 2012). Injection of substance P, the NK1 receptor agonist, into the RVM induces increased NKCC1 (the Na-K-Cl cotransporter) protein expression and converts postsynaptic GABAergic inhibition to excitation in the spinal dorsal horn, likely via an effect on the anionic equilibrium potential in dorsal horn neurons. The change in polarity of the effect of GABAergic transmission leads to spinal NMDA receptor phosphorylation that correlates with behavioral hyperalgesia (LaGraize et al. 2010).

Descending Serotonin

Serotonin (5-HT) is a major neurotransmitter in the RVM-spinal pathway, although 5-HT does not appear to be present in RVM pain modulatory neurons (Heinricher et al. 2009). The role of 5-HT in descending pain modulation is complicated by multiple subtypes of spinal 5-HT receptors (Millan 2002, review) and heterogeneity of 5-HT-containing RVM neurons (Zhang et al. 2006). Selective deletion of 5-HT by RNA interference of tryptophan hydroxylase-2, the rate-limiting enzyme for neuronal 5-HT synthesis, in RVM neurons reveals its role in descending pain facilitation after injury (Wei et al. 2010). After peripheral inflammation, about 70 % of RVM serotonin-containing neurons exhibit p38 mitogen-activated protein kinase, suggesting a role in RVM neuronal plasticity (Imbe et al. 2007). The pain facilitatory effect of descending 5-HT is mediated by the spinal/trigeminal $5-HT_3$ receptor, the lone ligand-gated cation channel in the 5-HT receptor family (Asante and Dickenson 2010; Chai et al. 2012; Dogrul et al. 2009; LaGraize et al. 2010; Zeitz et al. 2002).

Summary

The activity-induced plasticity in descending circuitry complements the activity-dependent neuronal plasticity in ascending pain transmission pathways. The spinal and brain stem plasticity is dependent upon increased activation of nociceptors at the site of injury, and its initiation and maintenance is dependent upon modulation of glutamatergic, opioidergic, neurotensinergic, and presumably GABAergic neurons. The activity-dependent plasticity at the spinal and brain stem levels shares similar but not identical and yet to be fully elucidated molecular mechanisms.

Acknowledgement The author's work is supported by NIH awards DE011964 and NS060735.

References

Asante, C. O., & Dickenson, A. H. (2010). Descending serotonergic facilitation mediated by spinal 5-HT3 receptors engages spinal rapamycin-sensitive pathways in the rat. *Neuroscience Letters, 484*(2), 108–112.

Bie, B., & Pan, Z. Z. (2007). Trafficking of central opioid receptors and descending pain inhibition. *Molecular Pain, 3*, 37.

Brink, T. S., Pacharinsak, C., Khasabov, S. G., Beitz, A. J., & Simone, D. A. (2012). Differential modulation of neurons in the rostral ventromedial medulla by neurokinin-1 receptors. *Journal of Neurophysiology, 107*, 1210–1221.

Burgess, S. E., Gardell, L. R., Ossipov, M. H., Malan, T. P., Jr., Vanderah, T. W., Lai, J., et al. (2002). Time-dependent descending facilitation from the rostral ventromedial medulla maintains, but does not initiate, neuropathic pain. *Journal of Neuroscience, 22*, 5129–5136.

Chai, B., Guo, W., Wei, F., Dubner, R., & Ren, K. (2012). Trigeminal-rostral ventromedial medulla circuitry is involved in orofacial hyperalgesia contralateral to tissue injury. *Molecular Pain, 23*(8), 78.

Dogrul, A., Ossipov, M. H., & Porreca, F. (2009). Differential mediation of descending pain facilitation and inhibition by spinal 5HT-3 and 5HT-7 receptors. *Brain Research, 1280*, 52–59.

Fields, H. L., Basbaum, A. I., & Heinricher, M. M. (2006). Central nervous system mechanisms of pain modulation. In S. B. McMahon & M. Koltzenburg (Eds.), *Wall and Melzack's textbook of pain* (5th ed., pp. 125–142). Philadelphia: Elsevier/Churchill Livingstone.

Gebhart, G. F. (2004). Descending modulation of pain. *Neuroscience and Biobehavioral Reviews, 27*, 729–737.

Guo, W., Robbins, M. T., Wei, F., Zou, S., Dubner, R., & Ren, K. (2006). Supraspinal brain-derived neurotrophic factor signaling: A novel mechanism for descending pain facilitation. *Journal of Neuroscience, 26*, 126–137.

Hamity, M. V., White, S. R., & Hammond, D. I. (2010). Effects of neurokinin-1 receptor agonism and antagonism in the rostral ventromedial medulla of rats with acute or persistent inflammatory nociception. *Neuroscience, 165*, 902–913.

Heinricher, M. M., Schouten, J. C., & Jobst, E. E. (2001). Activation of brainstem N-methyl-D-aspartate receptors is required for the analgesic actions of morphine given systemically. *Pain, 92*, 129–138.

Heinricher, M. M., Tavares, I., Leith, J. L., & Lumb, B. M. (2009). Descending control of nociception: Specificity, recruitment and plasticity. *Brain Research Reviews, 60*, 214–225.

Herrero, J. F., & Cervero, F. (1996). Supraspinal influences on the facilitation of rat nociceptive reflexes induced by carrageenan monoarthritis. *Neuroscience Letters, 209*, 21–24.

Hurley, R. W., & Hammond, D. L. (2000). The analgesic effects of supraspinal mu and delta opioid receptor agonists are potentiated during persistent inflammation. *Journal of Neuroscience, 20*, 1249–1259.

Imbe, H., Murakami, S., Okamoto, K., Iwai-Liao, Y., & Senba, E. (2004). The effects of acute and chronic restraint stress on activation of ERK in the rostral ventromedial medulla and locus coeruleus. *Pain, 112*, 361–371.

Imbe, H., Okamoto, K., Aikawa, F., Kimura, A., Donishi, T., Tamai, Y., Iwai-Liao, Y., & Senba, E. (2007). Effects of peripheral inflammation on activation of p38 mitogen-activated protein kinase in the rostral ventromedial medulla. *Brain Research, 1134*, 131–139.

Khasabov, S. G., Brink, T. S., Schupp, M., Noack, J., & Simone, D. A. (2012). Changes in response properties of rostral ventromedial medulla neurons during prolonged inflammation: Modulation by neurokinin-1 receptors. *Neuroscience, 224*, 235–248.

Lagraize, S. C., Guo, W., Yang, K., Wei, F., Ren, K., & Dubner, R. (2010). Spinal cord mechanisms mediating behavioral hyperalgesia induced by NK-1 tachykinin receptor activation in the rostral ventromedial medulla. *Neuroscience, 171*, 1341–1356.

Mansikka, H., & Pertovaara, A. (1997). Supraspinal influence on hindlimb withdrawal thresholds and mustard oil-induced secondary allodynia in rats. *Brain Research Bulletin, 42*, 359–365.

Martins, I., Costa-Araújo, S., Fadel, J., Wilson, S. P., Lima, D., & Tavares, I. (2010). Reversal of neuropathic pain by HSV-1-mediated decrease of noradrenaline in a pain facilitatory area of the brain. *Pain, 151*(1), 137–145.

Millan, M. J. (2002). Descending control of pain. *Progress in Neurobiology, 66*, 355–474.

Morgan, M. M., Gold, M. S., Liebeskind, J. C., & Stein, C. (1991). Periaqueductal gray stimulation produces a spinally mediated, opioid antinociception for the inflamed hindpaw of the rat. *Brain Research, 545*, 17–23.

Neil, A., Kayser, V., Gacel, G., Besson, J.-M., & Guilbaud, G. (1986). Opioid receptor types and antinociceptive activity in chronic inflammation: Both kappa and mu opiate agonistic effects are enhanced in arthritic rats. *European Journal of Pharmacology, 130*, 203–208.

Pacharinsak, C., Khasabov, S. G., Beitz, A. J., & Simone, D. A. (2008). NK-1 receptors in the rostral ventromedial medulla contribute to hyperalgesia produced by intraplantar injection of capsaicin. *Pain, 139*, 34–46.

Porreca, F., Ossipov, M. H., & Gebhart, G. F. (2002). Chronic pain and medullary descending facilitation. *Trends in Neurosciences, 25*, 319–325.

Ren, K., & Dubner, R. (2008). Descending control mechanisms. In A. I. Basbaum, A. Kaneko, G. M. Shepherd, & G. Westheimer (Eds.), *The senses: A comprehensive reference, Volume 5: Pain* (pp. 723–762). San Diego: Academic.

Roberts, J., Ossipov, M. H., & Porreca, F. (2009). Glial activation in the rostroventromedial medulla promotes descending facilitation to mediate inflammatory hypersensitivity. *European Journal of Neuroscience, 30*, 229–241.

Schaible, H.-G., Neugebauer, V., Cervero, F., & Schmidt, R. F. (1991). Changes in tonic descending inhibition of spinal neurons with articular input during the development of acute arthritis in the cat. *Journal of Neurophysiology, 66*, 1021–1032.

Tillu, D. V., Gebhart, G. F., & Sluka, K. A. (2008). Descending facilitatory pathways from the RVM initiate and maintain bilateral hyperalgesia after muscle insult. *Pain, 136*, 331–339.

Tsuruoka, M., & Willis, W. D. (1996). Bilateral lesions in the area of the nucleus locus coeruleus affect the development of hyperalgesia during carrageenan-induced inflammation. *Brain Research, 726*, 233–236.

Urban, M. O., & Gebhart, G. F. (1999). Supraspinal contributions to hyperalgesia. *Proceedings of the National Academy of Sciences of the United States of America, 96*, 7687–7692.

Vera-Portocarrero, L. P., Xie, J. Y., Kowal, J., Ossipov, M. H., King, T., & Porreca, F. (2006). Descending facilitation from the rostral ventromedial medulla maintains visceral pain in rats with experimental pancreatitis. *Gastroenterology, 130*, 2155–2164.

Watkins, L. R., Maier, S. F., & Goehler, L. E. (1995). Immune activation: The role of pro-inflammatory cytokines in inflammation, illness responses and pathological pain states. *Pain, 63*, 289–302.

Wei, F., Guo, W., Zou, S., Ren, K., & Dubner, R. (2008). Supraspinal glial-neuronal interactions contribute to descending pain facilitation. *Journal of Neuroscience, 28*, 10482–10495.

Wei, F., Dubner, R., Zou, S., Ren, K., Bai, G., Wei, D., et al. (2010). Molecular depletion of descending serotonin unmasks its novel facilitatory role in the development of persistent pain. *Journal of Neuroscience, 30*, 8624–8636.

Wiertelak, E. P., Roemer, B., Maier, S. F., & Watkins, L. R. (1997). Comparison of the effects of nucleus tractus solitarius and ventral medial medulla lesions on illness-induced and subcutaneous formalin-induced hyperalgesias. *Brain Research, 748*, 143–150.

Williams, F. G., & Beitz, A. J. (1993). Chronic pain increases brainstem proneurotensin/neuromedin-N mRNA expression: A hybridization-histochemical and immunohistochemical study using three different rat models for chronic nociception. *Brain Research, 611*, 87–102.

Zapata, A., Pontis, S., Schepers, R. J., Wang, R., Oh, E., Stein, A., Bäckman, C. M., Worley, P., Enguita, M., Abad, M. A., Trullas, R., & Shippenberg, T. S. (2012). Alleviation of neuropathic pain hypersensitivity by inhibiting neuronal pentraxin 1 in the rostral ventromedial medulla. *Journal of Neuroscience, 32*, 12431–12436.

Zeitz, K. P., Guy, N., Malmberg, A. B., Dirajlal, S., Martin, W. J., Sun, L., Bonhaus, D. W., Stucky, C. L., Julius, D., & Basbaum, A. I. (2002). The 5-HT3 subtype of serotonin receptor contributes to nociceptive processing via a novel subset of myelinated and unmyelinated nociceptors. *Journal of Neuroscience, 22*, 1010–1019.

Zhang, L., Sykes, K. T., Buhler, A. V., & Hammond, D. L. (2006). Electrophysiological heterogeneity of spinally projecting serotonergic and nonserotonergic neurons in the rostral ventromedial medulla. *Journal of Neurophysiology, 95*, 1853–1863.

Descending Circuitry, Opioids

Donna L. Hammond
Department of Anesthesia, University of Iowa,
Iowa City, IA, USA

Synonyms

Bulbospinal pathways; Descending modulation; Pain facilitatory pathways; Pain inhibitory pathways; Pain modulatory pathways

Definition

Descending circuitry is a generic term for populations of neurons in the medulla, pons, midbrain, or forebrain that modulate the evoked and spontaneous activity of neurons in the dorsal horn of the spinal cord. These neurons may project directly to the ▶ dorsal horn of the spinal cord, or they may project to brainstem nuclei that in turn project to and terminate in the dorsal horn. They are an important component of the neural circuitry by which opioid receptor agonists such as morphine produce ▶ analgesia or alleviate ▶ hyperalgesia or ▶ allodynia.

Introduction

The finding that direct injection of the ▶ opioid receptor agonist morphine into the cerebral ventricles of rats or mice produced analgesia provided some of the earliest evidence that opioid receptor agonists act centrally within the brain to produce analgesia. Mapping studies subsequently demonstrated that direct injection of small microgram doses of morphine into any one of a number of different nuclei in the midbrain, pons, or medulla was sufficient to produce analgesia in both laboratory rodents and primates (Yaksh and Rudy 1978). A central site of action was further supported by the finding that inactivation of these nuclei, by chemical or electrolytic lesions or by injection of an opioid receptor antagonist such as naloxone, could reduce or completely prevent the analgesic effects of systemically administered morphine. Coincidently, electrophysiological studies demonstrated that nuclei in the medulla, pons, and midbrain exert strong inhibitory and excitatory influences on the synaptic transmission of nociceptive information in the dorsal horn. Of particular interest, chemical or electrical stimulation of many of the same nuclei where microinjection of morphine-produced analgesia similarly suppressed the responses of dorsal horn neurons to noxious stimuli in anesthetized animals and produced analgesia in conscious animals. Collectively, these data

indicate that neurons in numerous supraspinal nuclei are involved in the modulation of nociception and in the production of analgesia by opioids. These nuclei include the dorsal and lateral reticular nucleus in the caudal lateral medulla, the nucleus raphe magnus and nucleus reticularis gigantocellularis pars α in the rostral ventromedial medulla, the locus coeruleus and A7 catecholamine cell group in the pons, the ventrolateral periaqueductal gray and cuneiform nucleus in the midbrain, and the hypothalamus, amygdala, and nucleus accumbens in the diencephalon. Neuroanatomical tract-tracing methods determined that only a few of these nuclei, those in the medulla and pons, project directly to the spinal cord and terminate in the dorsal horn. Other, more rostrally situated nuclei that do not project directly to the spinal cord are nonetheless able to modulate synaptic transmission in the spinal cord via their projections to the brainstem nuclei that do project to the dorsal horn. Hence, the term descending circuitry or bulbospinal pain modulatory pathways was coined as a generic term for a collection of supraspinal neurons that modulate synaptic transmission of sensory information in the spinal cord and that are implicated in the production of analgesia by opioid receptor agonists.

Characteristics

Redundant or Complementary Pathways to the Dorsal Horn

There are two pathways by which nuclei such as the periaqueductal gray, amygdala, or hypothalamus, which do not project directly to the dorsal horn, can influence the transmission of sensory information in the spinal cord. One pathway originates from the nucleus raphe magnus and nucleus reticularis gigantocellularis pars α on the midline of the medulla, collectively referred to as the ▶ rostral ventromedial medulla. Both nuclei contain serotonergic as well as nonserotonergic neurons, many of which project directly to the dorsal horn where they may presynaptically or postsynaptically modulate synaptic transmission in the dorsal horn. These neurons comprise a final common output pathway to the spinal cord. Not surprisingly, their function and particularly the role of the serotonergic neurons in these nuclei have been the subject of intense investigation (Ossipov et al. 2010; Heinricher and Ingram 2009). The second common output pathway is comprised of neurons in the more laterally situated ▶ locus coeruleus, which is nearly exclusively noradrenergic, and the ▶ dorsolateral pontine tegmentum including neurons in the A7 catecholamine cell group. For reasons that remain unclear, these nuclei have not been as extensively studied as those in the rostral ventromedial medulla (Pertovaara 2006). Yet, the literature reveals many instances in which experimental manipulations that produce analgesia, such as microinjection of opioid receptor agonists at supraspinal sites, are attenuated not only by ▶ intrathecal administration of serotonin receptor antagonists but also by noradrenergic receptor antagonists. Furthermore, coincident release of serotonin and norepinephrine in the spinal cord occurs after microinjection of morphine or electrical stimulation at supraspinal sites. Thus, the mechanisms by which opioid receptor agonists produce analgesia at supraspinal sites appear to involve several different bulbospinal pathways.

These two bulbospinal pathways may or may not function in concert, but their existence suggests complementary and possibly redundant means by which sensitivity to painful stimuli can be regulated and by which opioids can produce analgesia. Systemically administered opioids like morphine, meperidine, or fentanyl that cross the blood–brain barrier readily distribute throughout the brain and are therefore likely to act at multiple supraspinal sites to produce analgesia. As a consequence, spinally projecting neurons in the rostral ventromedial medulla and in the dorsolateral pontine tegmentum could be recruited independently and possibly in a redundant manner. However, it is also highly likely that these pathways are recruited in a concordant or coordinated manner. For example, neurons in the periaqueductal gray, an important site of opioid action, project to neurons in the

rostral ventromedial medulla and to neurons in the dorsolateral pontine tegmentum. Thus, both pathways are recruited by an action of opioids limited solely to the periaqueductal gray. Furthermore, neurons in the rostral ventromedial medulla project to and activate neurons in the locus coeruleus and A7 catecholamine cell group. Thus, even though the rostral ventromedial medulla does not contain noradrenergic neurons, microinjection of opioid receptor agonists in this region produces an analgesia that can be reversed by intrathecal administration of either serotonergic or noradrenergic receptor antagonists. The latter finding is concordant with a recruitment of spinally projecting noradrenergic neurons in the locus coeruleus or A7 catecholamine cell group.

Figure 1 summarizes the complementary pain modulatory pathways that arise from neurons in the rostral ventromedial medulla, comprised of the nucleus raphe magnus and nucleus reticularis gigantocellularis pars α and the dorsolateral pontine tegmentum, which includes the locus coeruleus and A7 catecholamine cell group. The strain and stock of Sprague–Dawley rat will determine whether the noradrenergic pathways in the dorsolateral pontine tegmentum that modulate nociceptive sensitivity arise from the locus coeruleus (e.g., Harlan Sprague–Dawley rats) or the A7 catecholamine cell group (Sasco or Charles River Sprague–Dawley rat) (Clark and Proudfit 1992). Neurons in the rostral ventromedial medulla and in the dorsolateral pontine tegmentum express several different types of opioid receptor although the μ opioid receptor predominates. Opioid receptor agonists can act directly in each nucleus to produce analgesia either by disinhibition of spinally projecting pain inhibitory neurons or by direct inhibition of pain facilitatory neurons. Figure 1 also illustrates the interconnections among these nuclei that can support concordant recruitment of these bulbospinal pain modulatory neurons even when the action of an opioid is limited to just one nucleus. Finally, Fig. 1 illustrates the efferent projections of the periaqueductal gray, which also contains opioid receptors, to neurons in both the rostral

Descending Circuitry, Opioids, Fig. 1 Schematic illustrating the interconnections among nuclei in the midbrain, pons, and medulla that mediate the analgesic, antiallodynic, and anti-hyperalgesic effects of opioid receptor agonists such as morphine. Neurons in these nuclei may function to either facilitate or inhibit (+/−) the synaptic transmission of sensory information in the dorsal horn of the spinal cord. Although not illustrated for the sake of clarity, more rostrally situated nuclei implicated in opioid-mediated antinociception, such as the hypothalamus and amygdala, send afferent projections to the periaqueductal gray. *PAG* periaqueductal gray, *DLPT* dorsolateral pontine tegmentum, *RVM* rostral ventromedial medulla

ventromedial medulla and the dorsolateral pontine tegmentum that can also support concordant recruitment of pain modulatory pathways in these nuclei. Although not illustrated, more rostrally situated nuclei, such as the hypothalamus and amygdala, which are implicated in nociception and opioid-mediated analgesia send afferent projections to the periaqueductal gray.

Evolving Concepts of Opioid Mechanisms of Action

The mechanisms by which the different classes of opioid receptor agonists (▶ μ, ▶ δ, and ▶ κ produce analgesia after acute administration or produce hyperalgesia upon the induction of tolerance or withdrawal have been the focus of much work. The results have greatly advanced our understanding of the cellular and subcellular mechanisms responsible. Just as importantly, they have also provided key insights into the physiology and pharmacology of the bulbospinal circuitry that modulates nociception.

It was long proposed that analgesia results from the excitation of brainstem neurons that then inhibit the transmission of nociceptive information in the dorsal horn. This proposal was predicated on the finding that local application of glutamate, an excitatory amino acid, or electrical stimulation of supraspinal nuclei produced analgesia. However, μ, δ, and κ opioid receptor agonists are almost exclusively inhibitory. They suppress synaptic transmission by acting at postsynaptic receptors to increase K^+ ▶ conductance resulting in ▶ hyperpolarization or by acting at presynaptic receptors to decrease Ca^{++} conductance thereby inhibiting neurotransmitter release. At first glance, the inhibitory effects of opioid receptor agonists are difficult to reconcile with the data indicating that excitation of neurons in brainstem nuclei produces analgesia. However, neurons in the periaqueductal gray, rostral ventromedial medulla, and the locus coeruleus and A7 catecholamine cell group express γ-aminobutyric acid (GABA)$_A$ and GABA$_B$ receptors and receive inputs from local GABAergic interneurons. Of note, the GABAergic neurons express opioid receptors. Thus, it was proposed that opioid receptor agonists produce analgesia by a process of ▶ disinhibition rather than by direct excitation. Specifically, opioid receptor agonists inhibit the activity of tonically active GABAergic neurons, thereby releasing spinally projecting neurons in the rostral ventromedial medulla or A7 neurons from inhibition and presumably resulting in excitation. The concept of disinhibition by opioids as a means by which spinally projecting brainstem neurons are activated to produce analgesia would dominate the literature for over a decade (Fields and Basbaum 1999). Less appreciated were other reports that stimulation of supraspinal neurons by microinjection of glutamate or electrical stimulation could also facilitate synaptic transmission of nociceptive information in the dorsal horn. The seminal report of Zhuo and Gebhart (1990) that glutamate or electrical stimulation of the rostral ventromedial medulla could, in a dose- or intensity-dependent manner, produce either hyperalgesia or analgesia firmly established that brainstem nuclei contain both pain facilitatory and pain inhibitory neurons and are capable of bidirectional control of nociception (Gebhart 2004). It is now well accepted that responsiveness and sensitivity to noxious stimuli reflect the net sum activity in these opposing bulbospinal pathways and that persistent pain states after injury reflect an imbalance in these pathways (Ren and Dubner 2009). With the recognition of spinally projecting pain facilitatory neurons came the proposal for a second mechanism by which opioid receptor agonists produce analgesia – direct inhibition of bulbospinal pain facilitatory neurons (Marinelli et al. 2002). These two mechanisms are not mutually exclusive, nor must they necessarily occur within the same nucleus.

The cellular actions of μ, δ, and κ opioid receptor agonists have helped shape several classification systems for neurons in the rostral ventromedial medulla and provided heuristic value to mechanistic models. One example is the cell classification scheme, based on extracellular recordings in lightly anesthetized rats, developed by Fields and further advanced by ▶ Heinricher and ▶ Mason. ON cells are identified by their increase in firing shortly before reflex withdrawal from a noxious stimulus. Opioid receptor agonists suppress the firing of ON cells, consistent with their identification as pain facilitatory neurons and the direct postsynaptic inhibitory actions of opioid receptor agonists, OFF cells are identified by the pause in firing that precedes the withdrawal reflex. Opioid receptor agonists eliminate the pause in firing, consistent with the identification of these cells as pain inhibitory neurons and the disinhibition hypothesis. Yet, another cell classification scheme has been proposed by Pan et al. (1990) and is based on whole-cell patch-clamp recordings from neurons in the in vitro brainstem slice preparation: primary and secondary cells. Primary cells are thought to be pain inhibitory neurons. Application of μ or δ opioid receptor agonists does not affect them, whereas κ opioid receptor agonists cause hyperpolarization. Secondary cells are hyperpolarized by μ or δ opioid receptor agonists but not by κ opioid receptor agonists. Secondary cells may represent the population of inhibitory neurons that are inhibited by opioids, resulting in disinhibition of primary cells and the production of analgesia. The ability of κ opioid receptor agonists to antagonize the actions of μ opioid receptor agonists is attributed to their direct inhibition of primary cells (Pan et al. 1997), while their ability to produce analgesia is due to their presynaptic inhibition of excitatory drive to secondary cells (Ackley et al. 2001).

Explicit in these hypotheses is the idea that the ongoing activity of the neurons in the RVM dictates the *state* of responsiveness to noxious stimuli. Increased discharge of ON cells is associated with or predisposing to hyperalgesia, while increased activity or the absence of a pause of OFF cells is associated with or predisposing to analgesia. However, the results of extracellular recordings in the rostral ventromedial medulla of conscious rats and mice may warrant a revision of our current hypotheses. Here again, the actions of opioid receptor agonists have been informative. Unlike lightly anesthetized rats, an analgesic dose of morphine in the conscious mouse does not alter the ongoing (tonic) discharge of either ON or OFF cells. It does, however, suppress the evoked (phasic) response of ON cells (i.e., burst increase in firing) and OFF cells (pause in firing) to noxious stimulation. Given that greater rates of ongoing discharge in OFF cells and greater silencing of OFF cells by the noxious stimulus correlate with an increased likelihood of withdrawal, this newer model proposes that OFF cells are in fact pronociceptive. In this model, opioid receptor agonists produce analgesia by suppressing the phasic responses of ON and OFF cells to a noxious stimulus, specifically the pause in firing of (pronociceptive) OFF cells and the burst in firing of the ON cell. In this model, a suppression of the burst in firing by ON cells is sufficient to produce analgesia, in the absence of any change in the activity of OFF cells (Mason 2012).

Given that nociceptive homeostasis reflects a balance of activity in bulbospinal pain facilitatory and pain inhibitory neurons, it should not be surprising that the apparent potency and efficacy of opioid receptor agonists also depends on this balance of activity. Behavioral data support the idea that opioid receptor agonists affect pain facilitatory and pain inhibitory neurons in concert and that the analgesia produced by opioids represents the sum of that activity. For example, microinjection of morphine in the A7 catecholamine cell group results in analgesia that is enhanced by antagonism of spinal α_1 noradrenergic receptors and diminished by antagonism of spinal α_2 noradrenergic receptors (Holden et al. 1999).

The relative balance of activity can also dictate the mechanisms that predominate. For example, neurons in the rostral ventromedial medulla and the dorsolateral pontine tegmentum can synapse directly on spinoreticular and spinothalamic tract neurons in the dorsal horn. In this instance, the functional effect of activation of the bulbospinal projection will be dictated by the postsynaptic receptor. For example, activation of α_1 noradrenergic receptors depolarizes dorsal horn neurons, whereas activation of α_2 noradrenergic receptors hyperpolarizes these neurons. Thus, activation of bulbospinal noradrenergic neurons of the locus coeruleus or the A7 catecholamine cell group can, respectively, facilitate or inhibit synaptic transmission of nociceptive information in the spinal cord. Here, the signal is dictated by the postsynaptic receptor on the dorsal horn neuron. Bulbospinal neurons may also synapse on interneurons in the dorsal horn that may be excitatory (i.e., contain substance P or glutamate) or inhibitory (i.e., contain ▶ enkephalin, glycine, or ▶ GABA). In this instance, the identity of the interneuron will dictate the functional consequence of activation or inhibition of the bulbospinal neuron.

For example, ▶ serotonin (5HT)$_3$ receptors are excitatory and cause depolarization of neurons. Serotonergic neurons of the rostral ventromedial medulla are postulated to inhibit synaptic transmission and produce analgesia via their projections to GABAergic interneurons in the dorsal horn that express a 5HT$_3$ receptor (Alhaider et al. 1991). The resultant excitation of the GABAergic interneurons in the dorsal horn in turn inhibits the activity of spinothalamic or spinoreticular tract neurons and produces analgesia. However, following peripheral nerve or inflammatory injury, serotonergic neurons in the rostral ventromedial medulla play a key role in maintaining the resulting hyperalgesia via their direct projections to and excitation of spinothalamic and spinoreticular neurons that also express the excitatory 5HT$_3$ receptor (Suzuki et al. 2004; Wei et al. 2010).

Plasticity in Opioid Actions

Until relatively recently, studies investigating the neural circuitry by which opioids modulate nociceptive sensitivity limited their investigation to naïve, uninjured rats and used noxious stimuli of brief duration. Work of the past decade has now made clear that the bulbospinal pain modulatory pathways exhibit substantial plasticity in terms of their physiological and pharmacological properties following peripheral injury (Ren and Dubner 2009; Vanegas and Schaible, 2004). Moreover, these changes have important ramifications for the actions of opioid receptor agonists among several ▶ other neurotransmitter systems. Specifically, the anti-hyperalgesic and analgesic effects of μ, δ, and κ opioid receptor agonists applied directly into the rostral ventromedial medulla are enhanced under conditions of persistent inflammatory nociception (Hurley and Hammond 2000; Schepers et al. 2008). In addition, persistent inflammation increases the levels of enkephalins, endogenous opioid peptides that have preferential affinity for δ opioid receptors, in the rostral ventromedial medulla. The enhancement of the analgesic effects of μ opioid receptor agonists appears to occur as a consequence of an additive or synergistic interaction of the μ opioid receptor agonist with the enhanced release of

▶ enkephalins that act preferentially at δ opioid receptors (Hurley and Hammond 2001; Sykes et al. 2007). These results suggest that endogenous opioids such as enkephalin can be recruited in persistent inflammatory pain states as part of a compensatory response on the part of brainstem pain modulatory pathways to mitigate the behavioral pain state.

Recent whole-cell ▶ patch-clamp recordings from spinally projecting neurons in the rostral ventromedial medulla have identified specific subpopulations of non-serotonergic and serotonergic neurons that exhibit an increased response to the μ opioid receptor agonist ▶ DAMGO following induction of a persistent inflammatory pain state. Interestingly, the same subpopulation of non-serotonergic neurons that show enhanced responsiveness to DAMGO also exhibits a more hyperpolarized resting membrane potential and a dramatic increase in their responsiveness to excitatory glutamatergic inputs. The population of serotonergic neurons, which were previously unresponsive to DAMGO, acquired opioid responsiveness under conditions of persistent inflammatory injury. The concordance of the populations of neurons whose passive membrane and action potential properties were altered as a consequence of peripheral inflammatory injury and those that exhibited increased responsiveness to DAMGO is intriguing. These two populations of neurons may have specific roles in the modulation of nociceptive sensitivity and merit further investigation (Zhang and Hammond 2009, 2010).

Summary

A principal challenge at present concerns the development of methods to reliably identify pain facilitatory and pain inhibitory neurons in the unanesthetized, awake behaving animal. Although a classification scheme has been developed whereby neurons in the rostral ventromedial medulla are characterized as ON (pain facilitatory), OFF (pain inhibitory), or neutral (no response) cells based on their response to noxious stimulation, the vast majority of this work has been conducted in anesthetized animals. Similar studies in conscious animals have suggested that this classification scheme and the function of these neurons may be more tightly linked to and reflective of behavioral state than originally envisioned (Mason 2001; Fields 2004) and that how we envision these neurons to function may require revision (Mason 2012). Advances in understanding the neural circuitry that mediates the analgesic, anti-allodynic, and anti-hyperalgesic effects of opioids will require a means to reliably identify pain inhibitory and pain facilitatory neurons on the basis of their neurotransmitter contents and receptor expression, their afferent and efferent projections, and their responses to a variety of nociceptive stimuli and pharmacological agents. Although definition of the neural circuitry that regulates pain sensitivity and mediates the actions of different drug classes requires a systems-based analysis, it is not complete without complementary investigations that address these same hypotheses using cellular and molecular approaches. Therefore, concordance is required so that classification schemes that are developed can be uniformly applied under both in vivo and in vitro conditions.

Acknowledgement This work was supported by R01DA006736 and R01DA023576.

References

Ackley, M. A., Hurley, R. W., Virnich, D. E., & Hammond, D. L. (2001). A cellular mechanism for the antinociceptive effect of a kappa opioid receptor agonist. *Pain, 91*, 377–388.

Alhaider, A. A., Lei, S., & Wilcox, G. L. (1991). Spinal 5-HT$_3$ receptor-mediated antinociception: Possible release of GABA. *Journal of Neuroscience, 11*, 1881–1888.

Clark, F. M., & Proudfit, H. K. (1992). Anatomical evidence for genetic differences in the innervation of the rat spinal cord by noradrenergic locus coeruleus neurons. *Brain Research, 591*, 44–53.

Fields, H. (2004). State-dependent opioid control of pain. *Nature Reviews Neuroscience, 5*, 565–575.

Fields, H. L., & Basbaum, A. I. (1999). Central nervous system mechanisms of pain modulation. In P. D. Wall & R. Melzack (Eds.), *The textbook of pain* (4th ed., pp. 309–329). Edinburgh: Churchill Livingstone.

Gebhart, G. F. (2004). Descending modulation of pain. *Neuroscience and Biobehavioral Reviews, 27,* 729–737.

Heinricher, M. M., & Ingram, S. L. (2009). The brainstem and nociceptive modulation. In A. I. Basbaum & M. C. Bushnell (Eds.), *Science of pain* (pp. 723–762). Amsterdam: Elsevier.

Holden, J. E., Schwartz, E. J., & Proudfit, H. K. (1999). Microinjection of morphine in the A7 catecholamine cell group produces opposing effects on nociception that are mediated by $\hat{I}\pm^{1-}$ and $\hat{I}\pm2$-adrenoceptors. *Journal of Neuroscience, 91,* 979–990.

Hurley, R. W., & Hammond, D. L. (2001). Contribution of endogenous enkephalins to the enhanced analgesic effects of supraspinal $\hat{I}\frac{1}{4}$ opioid receptor agonists after inflammatory injury. *Journal of Neuroscience, 21,* 2536–2545.

Hurley, R. W., & Hammond, D. L. (2000). The analgesic effects of supraspinal mu and delta opioid receptor agonists are potentiated during persistent inflammation. *The Journal of Neuroscience, 20,* 1249–1259.

Marinelli, S., Vaughan, C. W., Schnell, S. A., et al. (2002). Rostral ventromedial medulla neurons that project to the spinal cord express multiple opioid receptor phenotypes. *Journal of Neuroscience, 22,* 10847–10855.

Mason, P. (2001). Contributions of the medullary raphe and ventromedial reticular region to pain modulation and other homeostatic functions. *Annual Review of Neuroscience, 24,* 737–777.

Mason, P. (2012). Medullary circuits for nociceptive modulation. *Current Opinion in Neurobiology, 22,* 640–645.

Ossipov, M. H., Dussor, G. O., & Porreca, F. (2010). Central modulation of pain. *Journal of Clinical Investigation, 120,* 3779–3787.

Pan, Z. Z., Tershner, S. A., & Fields, H. L. (1997). Cellular mechanism for anti-analgesic action of agonists of the kappa-opioid receptor. *Nature, 389,* 382–385.

Pan, Z. Z., Williams, J. T., & Osborne, P. B. (1990). Opioid actions on single nucleus raphe magnus neurons from rat and guinea-pig in vitro. *The Journal of Physiology, 427,* 519–532.

Pertovaara, A. (2006). Noradrenergic pain modulation. *Progress in Neurobiology, 80,* 53–83.

Ren, K., & Dubner, R. (2009). Descending control mechanisms. In A. I. Basbaum & M. C. Bushnell (Eds.), *Science of pain* (pp. 723–762). Amsterdam: Elsevier.

Schepers, R. J., Mahoney, J. L., & Shippenberg, T. S. (2008). Inflammation-induced changes in rostral ventromedial medulla mu and kappa opioid receptor mediated antinociception. *Pain, 136,* 320–330.

Suzuki, R., Rygh, L. J., & Dickenson, A. H. (2004). Bad news from the brain: Descending 5-HT pathways that control spinal pain processing. *Trends in Pharmacological Sciences, 25,* 613–617.

Sykes, K. T., White, S. R., Hurley, R. W., Mizoguchi, H., Tseng, L. F., & Hammond, D. L. (2007). Mechanisms responsible for the enhanced antinociceptive effects of mu-opioid receptor agonists in the rostral ventromedial medulla of male rats with persistent inflammatory pain. *The Journal of Pharmacology and Experimental Therapeutics, 322,* 813–821.

Vanegas, H., & Schaible, H. G. (2004). Descending control of persistent pain: Inhibitory or facilitatory? *Brain Research. Brain Research Reviews, 46,* 295–309.

Wei, F., Dubner, R., Zou, S., Ren, K., Bai, G., Wei, D., & Guo, W. (2010). Molecular depletion of descending serotonin unmasks its novel facilitatory role in the development of persistent pain. *Journal of Neuroscience, 30,* 8624–8636.

Yaksh, T. L., & Rudy, T. A. (1978). Narcotic analgetics: CNS sites and mechanisms of action as revealed by intracerebral injection techniques. *Pain, 4,* 299–359.

Zhang, L., & Hammond, D. L. (2009). Substance P enhances excitatory synaptic transmission on spinally projecting neurons in the rostral ventromedial medulla after inflammatory injury. *Journal of Neurophysiology, 102,* 1139–1151.

Zhang, L., & Hammond, D. L. (2010). Cellular basis for opioid potentiation in the rostral ventromedial medulla of rats with persistent inflammatory nociception. *Pain, 149,* 107–116.

Zhuo, M., & Gebhart, G. F. (1990). Characterization of descending inhibition and facilitation of spinal nociceptive transmission from the nuclei reticularis gigantocellularis and gigantocellularis pars alpha in rat. *Pain, 42,* 337–350.

Descending Circuitry, Transmitters and Receptors

Alvin J. Beitz
Department of Veterinary and Biomedical Sciences, University of Minnesota, St. Paul, MN, USA

Synonyms

Receptors in the descending circuitry; Transmitters in the descending circuitry

Definition

▶ Neurotransmitters can be defined as chemicals released from nerve cells, which convey information from one neuron to another or to nonneuronal target cells such as muscle.

Nerve cells (i.e., neurons) communicate via a combination of electrical and chemical signals, and here the reference to neurotransmitters refers to chemical signaling. Following their release, transmitters typically affect postsynaptic neurons by binding to specialized proteins located in their neuronal membrane called receptors.

▶ Receptors can be defined as structures on the surface of a cell (or inside a cell) that selectively receives and binds a specific substance. There are many types of receptors associated with neurons including chemokine, cytokine, growth factor, hormone, and toll-like receptors; however, in the context of this entry, we will focus on neurotransmitter receptors. Once a neurotransmitter binds to its receptor, it either facilitates opening of an intrinsic ion channel, which can act quickly to depolarize or hyperpolarize the neuron or it activates a second messenger system, such as cyclic GMP, which ultimately leads to a response in the postsynaptic neuron. The descending circuitry associated with modulation of spinal nociceptive processing utilizes a wide variety of transmitters and receptors.

Introduction

It has been recognized for many years that there is a dissociation between injury and pain, such that soldiers severely injured during battle often experience little or no pain. Similarly, athletes who are injured on the playing field often do not experience the amount of pain that one might expect from their injury and they continue playing while injured. The existence of this type of dissociation implies that there is a system in the brain that can modulate pain. The idea of descending control systems dates back to 1970s and 1980s when the simplistic concept at the time was that structures in the brain stem send impulses that "descend" and inhibit pain transmission in the spinal cord. It is now well documented that an intrinsic pain modulation system does exist that can control the level of nociceptive information at various levels of the brain, although the major components do indeed descend to the spinal cord. The present of this

intrinsic pain modulation system provides a distinct survival advantage to all species, and we now know that it can both inhibit and enhance pain transmission at the level of the spinal cord and at several supraspinal sites. Like all neural systems in the brain, this system has a rich variety of neurotransmitters and neurotransmitter receptors that are critical to its function. Here, we review the major neurotransmitter and receptor components of this system.

Characteristics

Descending pain modulatory circuitry can either facilitate or inhibit nociceptive signaling from the spinal cord to higher brain regions by interacting with the terminals of primary afferent fibers, with projection neurons, intrinsic spinal cord interneurons, or terminals of other descending pathways (Wu et al. 2010). There is no absolute anatomical separation between brain regions that give rise to descending pathways that inhibit nociceptive transmission versus those that give rise to pathways that facilitate nociceptive transmission. The difference between whether a pathway facilitates or inhibits spinal nociceptive processing is dependent in most cases on the transmitters and ▶ receptors associated with that particular pathway (Milan 2002). For instance, neurons in the rostral ventromedial medulla, including the nucleus raphe magnus, give rise to descending ▶ serotonergic (5-HT) pathways, which can produce inhibition or facilitation of spinal cord dorsal horn neurons (Fields and Basbaum 1999; Godínez-Chaparro et al. 2012; Porreca et al. 2002), based in part on the type of 5-HT receptor that is activated upon 5-HT release. Thus, 5-HT activation of $5\text{-HT}_{1B, 1D, \text{ or } 3}$ receptors inhibits dorsal horn nociceptive transmission, while activation of $5\text{-HT}_{1A, 2A, 2C, 3, 4, 6 \text{ or } 7}$ receptors facilitates nociceptive transmission (Milan 2002). Similarly, descending dopaminergic, cholinergic, and noradrenergic pathways can facilitate or inhibit dorsal horn nociceptive transmission, by differential actions on dopaminergic, cholinergic, or noradrenergic receptor subtypes.

Descending Circuitry, Transmitters and Receptors, Table 1 Overview of the descending pathways that modulate nociception in the spinal cord dorsal horn

Brain structure	Primary transmitter	Other transmitters	Inhibits nociception	Facilitates nociception
Anterior cingulate gyrus	Glutamate	?	−	++
Frontal/parietal	Glutamate	?	++	−
Amygdala	GABA	Somatostatin, neurotensin, VIP, nociceptin, nocistatin	++	+
Paraventricular nucleus/ hypothalamus	Vasopressin, oxytocin, corticotrophin-releasing factor (CRF)	Dynorphin, enkephalin, neuropeptide FF, nitric oxide	$VP_{1a}/OTCRF_1K$-opioid	−
Posterior periventricular nucleus (A_{11})/ hypothalamus	Dopamine	RFamide-related peptides CGRP	D_2	D_1
Tuberomammillary nucleus/Hypothalamus	Histamine	GABA, galanin, enkephalin	−	H_1
Arcuate nucleus/ hypothalamus	β-endorphin, melanocortin, neuropeptide Y	CART	μ-opioid NPY_1	MC_4
Lateral hypothalamus	Orexin	Dynorphin, glutamate, CRF	Orexin-1 receptor	−
Periaqueductal gray	Glutamate, CCK/substance P	Serotonin, neurotensin, Enkephalin	NMDA5-HT_3	CCK_2/NK_1
Parabrachial nucleus	Glutamate, acetylcholine	Somatostatin, enkephalin, GABA	++	+
Pedunculopontine Nucleus	Acetylcholine	Dynorphin, vasopressin	++	−
A5, A6, A7 nuclei	Noradrenaline	GABA, glutamate, neuropeptide Y	α_{2A}, α_{2B}	α_{1A}, α_{1B}
Nucleus raphe magnus	Serotonin	GABA. glycine, glutamate, enkephalin, galanin, CCK	5-$HT_{1B, 1D, 3}$	5-HT $_{1A, 2A, 2C, 3, 4, 7}$
Rostral ventromedial medulla	Acetylcholine, GABA	Glycine, CCK, enkephalin, BDNF	Muscarinic, nicotinic	Nicotinic, TrkB
Caudal ventrolateral medulla	Angiotensin II			AT_1
Nucleus of the solitary tract	Glutamate	Somatostatin, orphanin FQ, CCK	++	+
Dorsal reticular nucleus	Glutamate	?	+	++

Since there are multiple pathways that descend from the brain into the spinal cord to modulate incoming nociceptive information, it is not surprising that there are a myriad of neurotransmitters and receptors associated with these pathways. The major descending pathways and their transmitters and receptors are summarized in Table 1. This list is by no means definitive, and the reader is referred to the comprehensive review by Milan (2002) for an in-depth summary of these pathways, transmitters and receptors.

A summary of the major descending pathways that synapse in the spinal cord dorsal horn and the transmitters and receptors associated with each pathway that inhibit or facilitate nociception. The primary transmitters of each pathway are listed in the second column, while additional transmitters (neuromodulators) are listed in the third column. The identity of other transmitters for corticospinal pathways and some reticulospinal pathways remains to be determined. The specific receptor subtypes involved in antinociceptive or pronociceptive actions are also listed. In cases where the specific receptors that facilitate or inhibit nociception have yet to be identified with certainty, the antinociceptive or pronociceptive effect of stimulating the pathway is indicated by pluses (++). A single "+" indicates that the evidence is controversial, and a lack of evidence is indicated by a minus sign. It is important to note that while the structures listed in this table have direct projections to the spinal cord, they may also influence nociception indirectly by activating other brain regions, particularly medullary and pontine nuclei, that in turn project directly to the spinal cord. Receptor abbreviations: α, adrenergic; CCK, cholecystokinin; CRF, corticotropin-releasing factor; D, dopamine; H, histamine; MC, melanocortin; NK, neurokinin; NMDA, N-methyl-D-aspartate; OT, oxytocin; 5-HT, serotonergic; VP, vasopressin.

Cortical and Subcortical Pathways

The anterior cingulate gyrus, agranular insular cortex, and the frontal/parietal cortical areas have direct projections to the spinal cord, but each of these regions also affects the spinal cord indirectly via relays to brain stem nuclei, including the ▶ rostral ventromedial medulla (RVM). All three pathways utilize ▶ glutamate as a major transmitter, but the anterior cingulate gyrus and the agranular insular cortex facilitate nociception, while the frontal/parietal cortex inhibits nociception, perhaps by acting on different glutamate receptor subtypes or by affecting different brain stem circuits or spinal interneurons. Based on these differential effects, the ▶ cerebral cortex has been proposed to modulate pain by acting on both pronociceptive and antinociceptive circuits, to change the set point of pain threshold (Jasmin et al. 2003). The ▶ amygdala is a subcortical structure that also projects directly to the ▶ spinal cord. It utilizes ▶ GABA and several peptides as neurotransmitters and has both an inhibitory and facilitatory effect on nociception at the spinal cord level. The amygdala participates together with the agranular ▶ insular cortex, and other cortical areas, in setting the pain threshold at the level of the spinal cord, and an imbalance in the output from the cortex and amygdala is likely to underlie some chronic pain states.

Hypothalamus

The ▶ hypothalamus has direct projections to the spinal cord but is also extensively linked to the ▶ periaqueductal gray (PAG), nucleus of the solitary tract, and RVM and thus can influence spinal cord nociception indirectly. For example, the paraventricular nucleus (PVN) of the hypothalamus provides the major source of oxytocin- and vasopressin-containing axons to the dorsal horn, and release of these peptides inhibits nociception by activation of VP_{1a} and oxytocin receptors on spinal cord inhibitory interneurons. A number of other peptides and transmitters have been identified in paraventricular hypothalamospinal projections, as summarized in Table 1. The posterior periventricular nucleus and the tuberomammillary nucleus are the major sources of ▶ dopamine and ▶ histamine, respectively, to the spinal cord. Dopamine acts at spinal cord D2 receptors to inhibit nociception and at D1 receptors to facilitate nociception, while histamine acts at H1 receptors to facilitate nociception. The hypothalamic arcuate nucleus contains a large number of neuroactive peptides, as indicated in Table 1, and sends projections directly to the spinal cord, where these neuroactive peptides are either antinociceptive or pronociceptive, depending on the neuropeptide receptors that are activated. Finally, the lateral hypothalamus provides a direct orexin projection to the spinal cord, which is thought to play a role in descending antinociceptive mechanisms, by activating inhibitory interneurons in the spinal cord dorsal horn (Grudt et al. 2002).

Brain Stem Nuclei

The periaqueductal gray (PAG) is a midbrain structure with a vast number of neurotransmitters and neuropeptides and has a rich history as a key component in the descending pain modulation system (Beitz 1995). While glutamate appears to be the key transmitter of descending projections from this region to the spinal cord and to other brain stem nuclei (Beitz 1995), a number of other transmitters including 5-HT, cholecystokinin, substance P, and neurotensin are present in descending PAG projections (Beitz 1982; Beitz et al. 1987). Neurotensin released from the PAG has a dual effect in the rostral ventromedial medulla (RVM): facilitation predominates at low (picomolar) doses of neurotensin injected into the RVM, whereas higher doses (nanomolar) produce antinociception (Smith et al. 1997). It is also worth noting that stimulation of μ-► opioid receptors in the PAG with ► morphine or other μ-opioid receptor agonists activates a descending antinociceptive circuit with a delta-opioid receptor link in the RVM (Hirakawa et al. 1999). It is also important to point out that the anatomical and physiological characteristics of the PAG and its descending projections to the RVM are sexually dimorphic with females having significantly more PAG-RVM output neurons along the rostrocaudal axis of the PAG compared to males, yet the number of PAG neurons activated by morphine is consistently and significantly higher in males compared to females (Loyd and Murphy 2009).

While a significant portion of the PAG's effect on spinal nociception is mediated via relays in other nuclei, like the RVM and A7 nucleus, direct descending pathways from this region can inhibit nociception at the spinal cord level via actions at NMDA receptors on spinal interneurons and facilitate nociception via activation of neurokinin and cholecystokinin receptors. The pons contains a number of noradrenergic cell groups including the locus coeruleus (A6), the A5 group in the ventrolateral brain stem, and the A7 group, whose axons project directly to the spinal cord dorsal horn. The bulbospinal axons arising from these cell groups release ► noradrenaline, which inhibits nociception by acting on α_2-adrenergic receptors on primary afferent terminals, to reduce the release of glutamate and substance P, and by increasing the release of inhibitory neurotransmitters from lamina II interneurons (Kawasaki et al. 2003). Noradrenaline released from these descending fibers can also facilitate nociception, by acting at spinal cord α_{1A}- and α_{1B}-adrenergic receptors. In addition to these noradrenergic cell groups, the pons provides a major cholinergic projection to the spinal cord, arising from the pedunculopontine nucleus and the parabrachial nucleus (PBN), as well as a glutamatergic projection arising from the parabrachial nucleus. The

pedunculopontine nucleus inhibits nociception via release of acetylcholine, dynorphin, and vasopressin, while the PBN can inhibit nociception via activation of ► glutamate receptors located on spinal inhibitory interneurons or via the direct release of GABA and enkephalin.

The medulla provides a rich contribution of descending projections to the spinal cord, which includes a major source of ► serotonin, glutamate, GABA, and neuropeptide input. The 5-HT innervation of the spinal cord dorsal horn arises in large part from the nucleus raphe magnus. This descending serotonergic projection can both inhibit and facilitate nociception via activation of different 5-HT receptors, as summarized in Table 1. It should be noted that descending 5-HT axons also contain a number of other neuropeptides and transmitters including glutamate (Hokfelt et al. 2000), and thus it is likely that additional neuroactive substances are co-released with 5-HT in the spinal cord dorsal horn, causing a wide spectrum of post- (and pre-) synaptic action. The rostral ventromedial medulla consists of several subnuclei, including the nucleus gigantocellularis pars alpha and the lateral paragigantocellular nucleus, which provide both a GABAergic and cholinergic projection to the spinal cord. This region can inhibit nociception, via activation of muscarinic and nicotinic cholinergic receptors, and can facilitate nociception via activation of nicotinic receptors. Descending projections from the RVM also contain glycine, CCK, enkephalin, and 5-HT (Milan 2002). These neurotransmitters also appear to modulate spinal nociception, but the precise action of each transmitter and the specific receptor subtypes activated within the dorsal horn remain to be adequately defined. It is important to note that rostral ventromedial medulla cells modulate nociceptive events on an on demand basis, only in concert with incoming nociceptive input (Mason 2012). It is also noteworthy that spinally projecting neurons in the RVM have mu-opioid and NK1 receptors on their surface and thus are modulated by opioid peptides and substance P released locally from afferent inputs or intrinsic neurons. The nucleus of the solitary tract (NTS) plays a crucial role in processing visceral information, and like the PBN, it serves as an interface between the autonomic and sensory branches of the nervous system. Stimulation of the NTS has both antinociceptive and pronociceptive effects, which appear to be mediated by glutamatergic bulbospinal projections, with contributions from a peptidergic component consisting of somatostatin, CCK, and orphanin FQ-containing axons. Since the NTS contains a myriad of neuropeptides and transmitters, it is likely that other NTS neuropeptides and transmitters may also contribute to nociceptive modulation at the spinal level. The dorsal reticular nucleus is located in the dorsal portion of the medulla, adjacent and lateral to the NTS, and stimulation of this region has been shown to enhance the responsiveness of spinal nociceptive neurons to peripheral stimulation (Dugast et al. 2003). Neurons in this region are activated by noxious stimuli, contain a high level of $GABA_B$ receptors, and send glutamatergic projections to the spinal cord to modulate nociception.

Given the information summarized above, it is clear that there are a large number of transmitters and receptors associated with the descending circuitry that modulates spinal nociceptive processing. An understanding of the role that each transmitter and receptor plays in pain modulation is complicated by the fact that a single transmitter and even a single receptor class can exert differential effects on nociception, at both spinal and supraspinal sites. Nonetheless, extensive progress has been made in defining the roles of many of the neuroactive substances listed above, and knowledge of the mechanisms by which descending pathways modulate nociception has increased dramatically in recent years, which should lead to improved management of pain.

Summary

Descending control of spinal nociception originates from many brain and brain stem areas and plays a critical role in determining the experience of both acute and chronic pain. While this descending control system was originally viewed as an endogenous "analgesia system," it is now thought to serve not only as a modulation system for pain, but in the context of integrated systems, it regulates pain input with respect to other competing behavioral needs and homeostatic demands. This system is quite complicated, and there are a large number of transmitters and receptors associated with the descending circuitry that modulates spinal nociceptive processing as outlined in this section. While many details remain to be elucidated, significant progress has been made in defining the roles of many of the neurotransmitters and neuropeptides that comprise the neurochemical backbone of this system, and this knowledge should provide important clues that will lead to improved therapeutic control of chronic pain.

References

Beitz, A. J. (1982). The sites of origin of brain stem neurotensin and serotonin projections to the rodent nucleus raphe Magnus. *Journal of Neuroscience, 2*, 829–842.

Beitz, A. J. (1995). Periaqueductal gray. In G. Paxinos (Ed.), *The rat nervous system* (2nd ed., pp. 173–182). San Diego: Academic.

Beitz, A. J., Clements, J. R., Ecklund, L. J., et al. (1987). The nuclei of origin of brainstem enkephalin and cholecystokinin projections to the spinal trigeminal nucleus of the rat. *Neuroscience, 1987*(20), 409–425.

Dugast, C., Almeida, A., & Lima, D. (2003). The medullary dorsal reticular nucleus enhances the responsiveness of spinal nociceptive neurons to peripheral stimulation in the rat. *European Journal of Neuroscience, 18*, 580–588.

Fields, H. L., & Basbaum, A. I. (1999). Central nervous system mechanisms of pain modulation. In P. D. Wall & R. Melzac (Eds.), *Textbook of pain* (4th ed., pp. 309–329). Edinburgh: Churchill Livingston.

Godínez-Chaparro, B., López-Santillán, F. J., Orduña, P., & Granados-Soto, V. (2012). Secondary mechanical allodynia and hyperalgesia depend on descending facilitation mediated by spinal 5-HT? 5-HT? and 5-HT? receptors. *Neuroscience, 222*, 379–391.

Grudt, T. J., van den Pol, A. N., & Perl, E. R. (2002). Hypocretin-2 (orexin-B) modulation of superficial dorsal horn activity in rat. *The Journal of Physiology, 538*(Pt 2), 517–525.

Hirakawa, N., Tershner, S. A., & Fields, H. L. (1999). Highly Í' selective antagonists in the RVM attenuate the antinociceptive effect of PAG DAMGO. *Neuroreport, 10*, 3125–3129.

Hokfelt, T., Arvidsson, U., Cullheim, S., et al. (2000). Multiple messengers in descending serotonin neurons: Localization and functional implications. *Journal of Chemical Neuroanatomy, 18*, 75–86.

Jasmin, L., Rabkin, S. D., Granato, A., et al. (2003). Analgesia and hyperalgesia from GABA-mediated modulation of the cerebral cortex. *Nature, 424*, 316–320.

Kawasaki, Y., Kumamoto, E., Furue, H., et al. (2003). Alpha 2 adrenoceptor-mediated presynaptic inhibition of primary afferent glutamatergic transmission in rat substantia gelatinosa neurons. *Anesthesiology, 98*, 682–689.

Loyd, D. R., & Murphy, A. Z. (2009). The Role of the Periaqueductal Gray in the Modulation of Pain in Males and Females: Are the Anatomy and Physiology Really that Different? *Neural Plasticity, 2009*, 12. Article ID 462879. doi:10.1155/2009/462879

Mason, P. (2012). Medullary circuits for nociceptive modulation. *Current Opinion in Neurobiology, 22*, 640–645.

Milan, M. J. (2002). Descending control of pain. *Progress in Neurobiology, 66*, 355–474.

Porreca, F., Ossipov, M. H., & Gebhart, G. F. (2002). Chronic pain and medullary descending facilitation. *Trends in Neurosciences, 25*, 319–325.

Smith, D. J., Hawranko, A. A., Monroe, P. J., et al. (1997). Dose-dependent pain-facilitatory and -inhibitory actions of neurotensin are revealed by SR 48692, a nonpeptide neurotensin antagonist: Influence on the antinociceptive effect of morphine. *Journal of Pharmacology and Experimental Therapeutics, 282*, 899–908.

Wu, S. X., Wang, W., Li, H., Wang, Y. Y., et al. (2010). The synaptic connectivity that underlies the noxious transmission and modulation within the superficial dorsal horn of the spinal cord. *Progress in Neurobiology, 91*, 38–54.

Descending Circuits in the Forebrain, Imaging

Elizabeth Roy Felix
The Miami Project to Cure Paralysis,
University of Miami Miller School of Medicine,
Miami, FL, USA

Synonyms

Cognitive modulation of pain; Functional imaging of descending modulation; Imaging descending circuits in the forebrain

Definition

Functional brain imaging techniques have been used as in vivo methods for examining which brain regions may be involved in or reflective of the modulation of pain. Current evidence using ► positron emission tomography (PET) and ► functional magnetic resonance imaging (fMRI) methodologies supports roles for the perigenual and mid-► anterior cingulate cortices (ACC) and portions of the ► prefrontal cortex in modulating input from nociceptive afferent pathways via descending projections.

Characteristics

Utilization of brain imaging techniques to study the descending modulation of pain is not straightforward and thus the results from these studies must be considered together with the results derived from other methodologies. One pitfall in the interpretation of functional imaging data is that increases in the oxygenation level of the blood (as is commonly measured in fMRI experiments) or metabolism (as measured with PET) can represent increased activity in either excitatory or inhibitory neurons (or both) within the area of interest, making it hard to distinguish the net neurophysiological outcome of the measured activity. A second caution when interpreting imaging results involves the distinction between changes that occur in each "active" area as a source of modulation and changes that occur as a result of such modulation (through the influence of other brain areas). Without supporting research from other methodologies, hypotheses based on results from imaging studies alone are speculative.

Given these caveats, the great advantage of functional brain imaging is that responsiveness in a number of brain regions can be looked at simultaneously to examine how these areas are affected by particular manipulations that occur within short experimental sessions and with relatively few subjects. This allows for a comprehensive assessment of the areas that may be involved in descending modulation of pain.

The regions most consistently activated by nociceptive stimulation alone, measured with PET and fMRI, have been well studied using various types of peripheral stimulation. The recognition that brain imaging techniques could bring special insight regarding which of these areas exhibited changes in activity that were correlated with changes in pain perception measured psychophysically has led to another level of investigation, one which examines the mechanisms of nociceptive modulation in the forebrain.

A first approach to understanding the mechanisms of pain modulation is the investigation of responsiveness to the presentation of the effective manipulation itself, in the absence of nociceptive stimulation. PET studies are especially useful in this regard and have provided a starting point for evaluating the possible mechanisms of descending modulation. Manipulations involving suggestions of pain relief (▶ placebo: Petrovic et al. 2002; hypnosis: Faymonville et al. 2000) result in increased regional cerebral blood flow (rCBF) in areas of the ACC compared to a resting condition. Administration of fentanyl (Adler et al. 1997; Casey et al. 2000) or remifentanil (Petrovic et al. 2002), both opioid agonists, also results in increased rCBF in the perigenual- and mid-ACC regions. Similarly, increased activations in prefrontal regions during placebo (Petrovic et al. 2002) or opioid administration (Firestone et al. 1996; Petrovic et al. 2002) support inclusion of prefrontal areas in the circuitry responsible for descending modulation of pain.

Perhaps a more direct way to implicate regional involvement in descending modulation of pain is to examine the nociceptive specific changes that occur in forebrain activity during manipulations aimed at altering pain perception. Indeed, several cortical and subcortical areas have been shown to modulate their level of responsiveness to painful stimuli in the presence of cognitive and pharmacological manipulations that modify the perception of such stimuli. One of the most frequently identified areas related to this type of modulation in human subjects is the ACC. Although the exact location and polarity of the changes in activity within the ACC during cognitive modulation of pain are not in full

agreement among all such studies, the most consistent effect in this area suggests that manipulations that cause decreases in pain perception (compared to equal intensity stimuli presented outside the cognitive modulation context) result in increased stimulus-induced activation here (Bantick et al. 2002; Frankenstein et al. 2001; Petrovic et al. 2002). In addition, a study that prompted increased pain perceptual ratings by increasing the anxiety associated with an impending noxious stimulus was correlated with decreased activation in this area of the ACC (Porro et al. 2002), suggesting that the direction of change in ACC modulatory activity can be reversed if the situation calls for increased vigilance.

Although the mechanisms underlying the relationship between the modulation of pain and activity in the perigenual/rostral ACC seen in brain imaging studies are far from fully elucidated, the evidence cited above demonstrates that this area is a likely candidate for the modulation of pain in certain contexts. Taken together, these studies suggest that increased activation in this area of the ACC serves to suppress incoming nociceptive information, causing decreases in pain perception. In fact, Petrovic and colleagues (2002) (see Fig. 1) have shown that, during administration of either remifentanil (an opioid receptor agonist) or placebo (administered with suggestions of pain relief), there is increased correlation between activation in rostral ACC and areas of the pons and the ▶ periaqueductal gray (PAG) in the brainstem (when compared with correlated stimulus-induced activation of these areas in the absence of pain modulation). A recent fMRI study (Bingel et al. 2006) supports this finding, identifying significant covariation of placebo-dependent activation between rostral ACC and the PAG and between rostral ACC and bilateral amygdalae, as well. These results implicate the ACC as a possible source of cortical nociceptive modulation, exerting control of activity in brainstem areas known to be involved in the descending modulation of pain. Despite the differences in the vehicle of modulation (administration of an opioid agonist or cognitive suggestions accompanying placebo administration), both means of manipulating pain perception

Descending Circuits in the Forebrain, Imaging, Fig. 1 Covariation of activity in brainstem regions with activity in the rACC (denoted by the *blue sphere*) in different pain conditions. (**a**) Activity in the rostral ACC covaried with activity in the PAG and in the lower pons/medulla during the pain + opioid condition. These covariations were significantly greater during the pain + opioid condition compared with the pain alone condition. (**b**) A similar covariation between the rostral ACC and the lower pons/medulla was observed during the pain + placebo condition. This covariation was significantly greater during pain + placebo as compared with pain alone. (**c**) No such regressions were observed during the pain alone condition. The activations are presented on an SPM99-template and a more detailed image of the brainstem indicating the approximate position of the PAG and the pons. The threshold of activation is at p = 0.005 (From Petrovic et al. 2002)

resulted in activation of similar brain circuits, suggesting at least some common mechanisms for modulation of pain at the cortical level. Additionally, fMRI has been used to further suggest the engagement of the PAG during modulation of pain via attentional mechanisms (Tracey et al. 2002).

Many prefrontal areas have also been implicated as possible sources of descending modulatory control. Changes similar to those seen in the perigenual/rostral ACC have been reported in the prefrontal cortex when the subjective perception of a noxious stimulus is decreased due to distraction or administration of pharmacological agents or when nociceptive specific increases in activation of prefrontal regions are evident (Adler et al. 1997; Bantick et al. 2002; Frankenstein et al. 2001; Petrovic et al. 2000). Administration of opioidergic agents in the absence of pain stimulation also increases activation in prefrontal areas (Firestone et al. 1996; Petrovic et al. 2002), further suggesting that such activation may serve to inhibit forthcoming nociceptive signals.

Experiments involving evoked increases in pain perception due to increased anticipatory anxiety however, report positive correlations between ratings of pain and activity in prefrontal areas (Hsieh et al. 1999; Porro et al. 2002). These results imply that the prefrontal cortex may work as a coding mechanism for pain intensity – simply a reflection of changes in pain perception – rather than a source of such modulation. Similarly, peripheral stimulation with capsaicin, which sets up conditions of ► allodynia and ► secondary hyperalgesia (mediated by central mechanisms), also results in increases in positive correlations between prefrontal activation and measures of pain intensity (Iadorola et al. 1998; Lorenz et al. 2002). The current evidence regarding the prefrontal cortex and its modulatory role in pain perception then suggests a complex relationship involving increases in activation of this area under contexts that cause reductions in pain (during distraction and opioid administration) and those that cause increases in pain sensitivity (during anxiety and stimulus-induced allodynia and hyperalgesia).

The insular cortex is another area that is consistently reported to show changes in its level of stimulus-induced activation under cognitive and pharmacological conditions that alter pain sensitivity. Manipulations resulting in increases in pain result in increases in activity in the insula (Hofbauer et al. 2001; Porro et al. 2002), and manipulations that result in decreases in pain perception correlate with similar decreases in this area (Brooks et al. 2002; Bantick et al. 2002; Petrovic et al. 2002). To date, there is little evidence from functional imaging studies that the insula has any direct influence over incoming nociceptive input. Instead, it may serve as a coding mechanism for the level of the sensory experience. Likewise, other areas involved in pain processing during baseline or control conditions (thalamus, primary and secondary somatosensory cortex) show changes in responsiveness that positively correlate with the direction of the perceptual changes (Bantick et al. 2002; Casey et al. 2000; Hofbauer et al. 2001; Lorenz et al. 2003; Petrovic et al. 2000) and are not implicated as sources of the cognitive or pharmacological modulation of pain.

Although the spatial and temporal resolution of PET and fMRI may not always be fine enough to resolve direct relationships between different brain structures known to exhibit descending pain modulation definitively, these methods do allow us access to the interrelated nature of the forebrain during such modulation. Functional brain imaging in human subjects has allowed the evaluation not only of pharmacological effects but also of complex psychological manipulations on changes in activation of particular brain areas that correspond with changes in pain perception. Current research, indicating increased activity levels in the ACC and in prefrontal regions during decreased pain perception due to experimental manipulation, points to these areas as the best candidates for exerting descending control over nociceptive signaling. Many other areas responsive to pain, including the insular cortex, the primary and secondary somatosensory cortices, and the thalamus, may also be found to be involved in the descending modulation of pain, but current functional imaging results do not strongly support such roles.

References

Adler, L. J., Gyulai, F. E., Diehl, D. J., et al. (1997). Regional brain activity changes associated with fentanyl analgesia elucidated by positron emission tomography. *Anesthesia and Analgesia, 84*, 120–126.

Bantick, S. J., Wise, R. G., Ploghaus, A., et al. (2002). Imaging how attention modulates pain in humans using functional MRI. *Brain, 125*, 310–319.

Bingel, U., Lorenz, J., Schoell, E., Weiller, C., & Büchel, C. (2006). Mechanisms of placebo analgesia: rACC recruitment of a subcortical antinociceptive network. *Pain, 120*, 8–15.

Brooks, J. C. W., Nurmikko, T. J., Bimson, W. E., et al. (2002). fMRI of thermal pain: Effects of stimulus laterality and attention. *NeuroImage, 15*, 293–301.

Casey, K. L., Svensson, P., Morrow, T. J., et al. (2000). Selective opiate modulation of nociceptive processing in the human brain. *Journal of Neurophysiology, 84*, 525–533.

Faymonville, M. E., Laureys, S., Degueldre, C., et al. (2000). Neural mechanisms of antinociceptive effects of hypnosis. *Anesthesiology, 92*, 1257–1267.

Firestone, L. L., Gyulai, F., Mintun, M., et al. (1996). Human brain activity response to fentanyl imaged by positron emission tomography. *Anesthesia and Analgesia, 82*, 1247–1251.

Frankenstein, U. N., Richter, W., McIntire, M. C., et al. (2001). Distraction modulates anterior cingulated gyrus activations during the cold pressor test. *NeuroImage, 14*, 827–836.

Hofbauer, R. K., Rainville, P., Duncan, G. H., et al. (2001). Cortical representation of the sensory dimension of pain. *Journal of Neurophysiology, 86*, 402–411.

Hsieh, J.-C., Stone-Elander, S., & Ingvar, M. (1999). Anticipatory coping of pain expressed in the human anterior cingulated cortex: A positron emission tomography study. *Neuroscience Letters, 262*, 61–64.

Iadorola, M. J., Berman, K. F., Zeffiro, T. A., et al. (1998). Neural activation during acute capsaicin-evoked pain and allodynia assessed with PET. *Brain, 121*, 931–947.

Lorenz, J., Cross, D. J., Minoshima, S., et al. (2002). A unique representation of heat allodynia in the human brain. *Neuron, 35*, 383–393.

Lorenz, J., Minoshima, S., & Casey, K. L. (2003). Keeping pain out of mind: The role of the dorsolateral prefrontal cortex in pain modulation. *Brain, 126*, 1079–1091.

Petrovic, P., Petersson, K. M., Ghatan, P. H., et al. (2000). Pain-related cerebral activation is altered by a distracting cognitive task. *Pain, 85*, 19–30.

Petrovic, P., Kalso, E., Petersson, K. M., et al. (2002). Placebo and opioid analgesia – Imaging a shared neuronal network. *Science, 295*, 1737–1740.

Porro, C. A., Baraldi, P., Pagnoni, G., et al. (2002). Does anticipation of pain affect cortical nociceptive systems? *The Journal of Neuroscience, 22*, 3206–3214.

Tracey, I., Ploghaus, A., Gati, J. S., et al. (2002). Imaging attentional modulation of pain in the periaqueductal gray in humans. *The Journal of Neuroscience, 22*, 2748–2752.

Descending Control

▶ Descending Modulation of Nociceptive Processing

Descending Control of Hyperalgesia and Allodynia

▶ Descending Modulation of Nociceptive Transmission During Persistent Damage to Peripheral Tissues

Descending Excitatory Modulation

▶ Descending Facilitatory Systems

Descending Facilitation

▶ Descending Modulation of Visceral Pain

Descending Facilitation and Inhibition in Neuropathic Pain

Michael H. Ossipov[1] and Frank Porreca[2]
[1]Department of Pharmacology, University of Arizona College of Medicine, Tucson, AZ, USA
[2]Department of Pharmacology, University of Arizona Health Sciences Center, Tucson, AZ, USA

Synonyms

Bulbospinal modulation of nociceptive inputs; Spinopetal modulation of pain; Supraspinal pain modulation

Definition

Neuropathic pain can occur following a lesion or dysfunction in peripheral nerves or in the central nervous system. Nerve injury may result from physical trauma including surgery, a disease state, cancer chemotherapy, or exposure to neurotoxins or medications. Neuropathic pain is often characterized by spontaneous or ongoing burning pain, paroxysmal episodes of pain. Some, but not all, patients suffer from pain in response to normally innocuous stimuli such as touch or cold (i.e., allodynia). Sensory gain can be demonstrated in neuropathic pain patients (e.g., hyperalgesia) and can paradoxically exist together with sensory loss. Neuropathic pain is difficult to treat with most patients achieving only partial pain relief.

Descending Modulation: Understanding of circuits mediating pain has evolved from a linear model of pain transmission to one of distributed signaling with multiple regions of the brain participating (i.e., the so-called pain matrix). Afferent nociceptive signals are modulated as a consequence of a "top-down" opioid-sensitive circuit. Modulation of afferent nociceptive inputs by the top-down circuit is an essential part of the human experience of pain and reflects context, emotionality, memories, attention, and expectation, an important component of the placebo response. The pain matrix includes important cortical sites (e.g., prefrontal cortex, somatosensory cortex, anterior cingulate cortex, insula), limbic sites (amygdala), thalamic nuclei, and midbrain and medullary sites. These regions ultimately lead to activation of discrete brain loci that send projections to the spinal or medullary dorsal horn and modulate nociceptive inputs (for reviews, see Ossipov et al. 2010). Since such influences descend from the brain to lower areas of the neuraxis, this process is referred to as descending modulation. Accordingly, electrical stimulation of discrete regions of the brain can produce robust antinociception in many species. Communications among these sites lead to activation of neurons within the rostroventromedial medulla (RVM), often called the final common relay in bulbospinal pain modulation

(for reviews, see Fields et al. (2005), Ossipov et al. (2010)). The RVM includes the midline nucleus raphe magnus (NRM) and the nucleus reticularis gigantocellularis and is defined anatomically as the region of the medulla between the pyramids, from the ventral surface to the top of the facial nucleus and extending rostrocaudally from the caudal aspect of the superior olive to the rostral aspect of the inferior olive. Projections from the RVM to the spinal cord course principally along the dorsolateral funiculus (DLF). Because of the prominent role of this region in pain modulation, the RVM is the best characterized node in descending pain modulation.

Characteristics

Descending Modulation of Pain

The idea that the brain can exert control over pain is strongly supported by studies showing attenuation of pain by placebo, stress, distraction, and hypnosis. Conversely, the expectation of increased pain enhances the pain experience and is termed "nocebo" (Benedetti et al. 2007). Importantly, placebo analgesia is blocked by naloxone, and brain regions that are activated by placebo analgesia, such as the ACC, insula, and prefrontal cortex, express the μ-opioid receptor (Zubieta et al. 2005). Convincing evidence that specific regions of the brain exert a control over nociception was demonstrated when focal electrical stimulation in the midbrain periaqueductal gray (PAG) of the awake rat produced analgesia that was sufficient to allow surgical procedures. Electrical stimulation of the PAG was used clinically to alleviate intractable pain in humans. Surprisingly, stimulation of the PAG also produced severe migraine-like headaches in some patients, even if they had not previously experience migraines. Electrical stimulation of other regions, including the thalamus, locus coeruleus, and RVM, also produces robust antinociception (for reviews, see Fields and Basbaum (2005)). These sites coincide with those that produce antinociception after the microinjection of morphine (Fields and Basbaum 2005). The PAG, locus coeruleus, and thalamus communicate with the RVM, suggesting that region functions as a critical brainstem relay in the modulation of descending nociceptive controls. This concept is supported by observations that interruption of the activity of the RVM by either lidocaine microinjections or lesions blocked antinociception elicited by stimulation of the PAG (Gebhart 1983). More recent studies show that the RVM may give rise to a facilitatory pathway that promotes pain, thus acting as a bidirectional pain modulatory system. For example, nociceptive inputs may activate pain facilitatory neurons of the RVM, thus promoting further nociception, setting the stage for the maintenance of a chronic pain state (Porreca et al. 2002). An illustration of this concept is observed when hyperalgesia induced by a variety of means, including naloxone-precipitated opiate withdrawal, inflammation, or nerve injury, is reversed by intra-RVM lidocaine or electrolytic lesion (Pertovaara 1998). Electrical stimulation of the RVM at low intensities has facilitated dorsal horn neuronal activity and the spinal nociceptive tail flick reflex, further demonstrating the existence of nociceptive facilitation arising from this region (Zhuo and Gebhart 1997).

In order to describe how the RVM may mediate apparently physiologically contradictory functions, it should be considered that the RVM is considered as containing three types of cells which have been extensively characterized by Fields and colleagues (Fields and Heinricher 1989). Based on response characteristics to nociception, these cells are described as "on" cells, "off" cells, and neutral cells. The physiologic function of the neutral cells is undetermined. The off-cells are tonically active and pause in their firing immediately before a withdrawal response occurs to nociceptive stimuli and are believed to be responsible for descending inhibition of nociceptive inputs (see Fields and Basbaum 2005). The on-cells accelerate firing immediately before the nociceptive reflex occurs and are likely to be the source of descending facilitation of nociceptive inputs and thus can mediate hyperalgesia (Fields and Basbaum 2005). Accordingly, manipulations

that increase nociceptive responsiveness, thus indicating facilitation, also increase on-cell activity. Naloxone-precipitated opioid withdrawal is associated with hyperalgesia and also with increased activity of RVM on-cells, which is in turn blocked by RVM lidocaine (for reviews, see Fields and Basbaum 2005; Porreca et al. 2002). Emerging evidence suggests that the functionality of the on-, off-, and neutral cells is not static but dynamic. Dickenson and colleagues (Sikandar and Dickenson 2011) identified RVM on-cells, off-cells, and neutral cells by application of thermal stimulus to the tail. However, when they applied a visceral stimulus in the form of colorectal distension, they found that there was no correlation between neuronal activity after somatic stimulation and that after visceral stimulation (Sikandar and Dickenson 2011). For example, on-cells identified somatically could appear as on-, off-, or neutral cells after visceral stimulation (Sikandar and Dickenson 2011).

It is now generally accepted that a spinal-bulbo-spinal loop, with increased pain facilitatory activity arising from the RVM, is important to the development and maintenance of neuropathic pain (Porreca et al. 2002; Suzuki and Dickenson 2005). Based on observations that somatically identified on-cells are directly hyperpolarized by μ-opioid receptors and activated by CCK_2 receptors, it was found that manipulation of on-cell activity could modulate behavioral signs of neuropathic pain. Behavioral signs of tactile hypersensitivity and thermal hyperalgesia were produced by CCK_8 injected into the RVM, and this effect was blocked by the CCK_2 receptor antagonist L365,260 (Kovelowski et al. 2000). Furthermore, microinjection of L365,260 or of lidocaine into the RVM blocked behavioral signs of neuropathic pain in rats with spinal nerve ligation (SNL) (Kovelowski et al. 2000). Selective ablation of RVM neurons expressing the μ-opioid or the CCK_2 receptor, and therefore presumed to be on-cells, by means of dermorphin or CCK conjugated to the cytotoxin saporin abolished behavioral signs of neuropathic pain (Zhang et al. 2009). Importantly, ablation of these putative pain facilitatory neurons did not block the initiation phase of neuropathic pain, likely driven by peripheral mechanisms, but blocked the maintenance phase, which is likely driven by descending facilitation (Zhang et al. 2009). Recent studies showed that blockade of descending modulation by microinjection of lidocaine into the RVM also abolished evidence of ongoing pain in rats with SNL (King et al. 2009). Similar results were obtained with lesions of the DLF made ipsilateral, but not those made contralateral, to SNL (Ossipov et al. 2000). Ablation of descending facilitation did not alter acute nociceptive thresholds (Ossipov et al. 2000; Zhang et al. 2009). Peripheral nerve injury resulted in increased pinch-evoked activity of RVM on-cells when compared to responses of those of sham-operated rats, consistent with a pronociceptive role of RVM on-cells (Goncalves et al. 2007). Studies with inflammatory pain corroborate these observations that facilitation from the RVM contributes to enhanced pain. Persistent nociceptive input from an injection of formalin into the hindpaw has facilitated the nocifensive tail flick reflex through an NMDA-mediated mechanism within the RVM (Wiertelak et al. 1994). RVM lidocaine or ibotenic acid also blocked hyperalgesia and increased activity of spinal wide-dynamic range neurons in rats with inflammation (Pertovaara 1998).

Several studies suggest a serotonergic component of this spinal-bulbal-spinal loop to maintain neuropathic pain. Although the NRM contains a large proportion of serotonergic neurons, the nature of the descending inhibitory and facilitatory neurons that modulate nociception is unknown. Aicher and colleagues found that more than 60 % of projection cells from the RVM are GABAergic (Winkler et al. 2006), and Mason and colleagues found that only neutral cells are serotonergic (Potrebic et al. 1994). It is unclear if these descending projections synapse with projection cells or interneurons in the spinal cord (Potrebic et al. 1994; Winkler et al. 2006). However, there is strong evidence indicating that the mechanisms of descending modulation at the level of the spinal cord are serotonergic, even though the nature of the descending projections remains to be clarified. Early studies showed blockade of stimulation-produced antinociception with serotonergic antagonists given intrathecally, and stimulation of the PAG or RVM was found to cause release of serotonin in the spinal cord (Cui et al. 1999). More recent studies indicate that the effect of spinal serotonin on nociception can be either inhibitory or facilitatory, depending on the receptor subtype activated. The spinal administration of a 5-HT$_7$ antagonist blocked the antinociceptive effect of morphine microinjected into the RVM, whereas that of a 5-HT$_3$ antagonist blocked hyperalgesia induced by CCK administered into the RVM (Dogrul et al. 2009), leading to conclusions of inhibitory and facilitatory activity through these receptors. Peripheral nerve injury results in sensitization of wide-dynamic range neurons of the deeper laminae of the spinal cord (Suzuki et al. 2004). The enhanced responses of these neurons to peripheral stimuli are abolished by spinal application of ondansetron, a 5-HT$_3$ receptor antagonist (Suzuki et al. 2004), as are behavioral signs of allodynia (Chen et al. 2009). Surprisingly, systemic ondansetron did not block ongoing pain or dynamic allodynia in patients with peripheral neuropathy (Tuveson et al. 2011). Although these observations indicate an important serotonergic role for pain modulation, the relevance of serotonergic mechanisms in descending pain modulation remains unclear.

In light of these considerations, it is believed that behavioral manifestations of neuropathic pain result, in part, from supraspinal neuroplastic changes in response to abnormal, persistent afferent inputs resulting from the nerve injury. These changes appear to induce the activation of medullospinal facilitatory pathway(s) that result in further increase in afferent input. These changes lead to a self-generating feed-forward mechanism that perpetuates sensations of neuropathic pain, even once the original injury has been resolved. This concept is illustrated by studies showing that peripheral nerve injury results in sensitization of spinal wide-dynamic range neurons. Electrophysiologic signs of sensitization are abolished by pretreatment with spinal substance P conjugated to saporin, which selectively lesions the projection neurons of the outer laminae of the spinal cord that express the NK1 receptor (Suzuki et al. 2004). Moreover, substance P-saporin treatment also abolishes the effect of ondansetron on WDR neurons, suggesting a loss of descending facilitation (Suzuki et al. 2004).

It has been proposed that descending facilitation results in upregulation of spinal dynorphin levels, which help maintain a sensitized state in the spinal cord. Spinal administration of antisera to dynorphin blocked behavioral signs of neuropathic pain in mice with SNL (Wang et al. 2001). Moreover, mice with selective deletion of the gene coding for prodynorphin developed initial neuropathic pain, which spontaneously resolved over the following 10 days (Wang et al. 2001). This parallels the loss of neuropathic pain after ablation of descending facilitation and coincides temporally with the elevation of spinal dynorphin observed after peripheral nerve injury (Wang et al. 2001). Thus, it is suggested that spinal dynorphin elevations may be a result of neuroplastic changes over a period of several days. This suggestion was supported by the observation that in mice tested within 2 days of SNL, tactile hypersensitivity was reversed by spinal injections of MK-801 but not by antiserum to dynorphin. In contrast, on the 10th day after SNL, both spinal injections of MK-801 and antiserum to dynorphin blocked signs of neuropathic pain (Wang et al. 2001). Recent evidence suggests that dynorphin may exert its pronociceptive effect through activation of spinal bradykinin receptors (Lai et al. 2006). Dynorphin evoked calcium influx in F-11 cells that express the B1 or B2 bradykinin receptors, and these effects are blocked by antagonists of these receptors (Lai et al. 2006). Moreover, intrathecal administration of bradykinin B1 or B2 receptor antagonists abolished signs of abnormal pain elicited by intrathecal injection of dynorphin or by peripheral nerve injury (Lai et al. 2006). These studies taken together suggest that elevated spinal dynorphin may not be critical to the initiation of increased pain but is a necessary component for the long-term maintenance of abnormal pain after nerve injury.

The relative activation of descending pain inhibitory and facilitatory systems may help explain the development of neuropathic pain in some individuals and not in others. A recent investigation found that peripheral nerve injury results in a loss of RVM off-cells and in descending projections to the spinal cord (Leong et al. 2011). It is suggested that injury can shift the balance toward less inhibition, or greater facilitation, to promote an enhanced pain state (Leong et al. 2011). It is therefore reasonable to expect that a shift toward greater descending inhibition would mitigate the enhanced facilitation and thus protect against development of neuropathic pain. A recent investigation found that, although a large (i.e., >85 %) proportion of Sprague-Dawley rats that receive SNL develop behavioral signs of neuropathic pain, nearly one-half of Holtzman rats, which are genetically close to the Sprague-Dawley rat, develop neither evoked nor ongoing neuropathic pain (De Felice et al. 2011). Microinjection of lidocaine into the RVM unmasked facilitation by provoking tactile allodynia in rats with SNL but not expressing this enhanced pain state (De Felice et al. 2011). Likewise, RVM lidocaine also unmasked evidence of ongoing spontaneous pain in these rats. Inhibition of off-cell activity by microinjection of the κ-opioid receptor agonist U69593 also unmasked an ongoing pain facilitation, as did the spinal administration of the α_2-adrenergic antagonist yohimbine (De Felice et al. 2011). Finally, animals not showing allodynia after SNL demonstrated significantly less on-cell activity in response to formalin injected into a hindpaw than did those expressing allodynia (De Felice et al. 2011). These studies indicate that engagement of descending inhibition is a critical factor that can prevent the transition from acute to chronic neuropathic pain. Thus, it is possible that individuals that develop neuropathic pain after injury may have a dysfunction in the endogenous pain inhibitory system. Conversely, therapeutics that mimic descending inhibition may provide a novel avenue for treatment of neuropathic pain.

In summary, initial nerve injury may result in increased peripheral neuronal hyperexcitability, along with sensitization of dorsal horn neurons. Increased afferent inputs to supraspinal centers elicit neuroplastic changes that ultimately result in enhanced descending pain facilitation from the RVM. These neuroplastic changes and enhanced descending facilitation lead to pathologic elevations in spinal dynorphin expression, which exert a pronociceptive effect through activation of spinal bradykinin receptors. Consequently, elevated spinal dynorphin further promotes nociceptive input and contributes to a positive feed-forward cycle that serves to maintain an abnormally sensitized pain state. Manipulations that interrupt this self-perpetuating cycle and restore a balance between pain facilitation and inhibition may be exploited as targets for rational novel therapies for the treatment of neuropathic pain states. These would include local anesthetic application at the site of injury, antagonists of excitatory neurotransmitters (e.g., CCK antagonists or NMDA antagonists), interruption of spinal neurotransmitter release or blockade of these receptors, and prevention of the upregulation of spinal dynorphin.

References

Benedetti, F., Lanotte, M., Lopiano, L., et al. (2007). When words are painful: Unraveling the mechanisms of the nocebo effect. *Neuroscience, 147*(2), 260–271.

Chen, Y., Oatway, M. A., & Weaver, L. C. (2009). Blockade of the 5-HT3 receptor for days causes sustained relief from mechanical allodynia following spinal cord injury. *Journal of Neuroscience Research, 87*(2), 418–424.

Cui, M., Feng, Y., McAdoo, D. J., et al. (1999). Periaqueductal gray stimulation-induced inhibition of nociceptive dorsal horn neurons in rats is associated with the release of norepinephrine, serotonin, and amino acids. *Journal of Pharmacology and Experimental Therapeutics, 289*(2), 868–876.

De Felice, M., Sanoja, R., Wang, R., et al. (2011). Engagement of descending inhibition from the rostral ventromedial medulla protects against chronic neuropathic pain. *Pain, 152*(12), 2701–2709.

Dogrul, A., Ossipov, M. H., & Porreca, F. (2009). Differential mediation of descending pain facilitation and inhibition by spinal 5HT-3 and 5HT-7 receptors. *Brain Research, 1280,* 52–59.

Fields, H. L., & Heinricher, M. M. (1989). Brainstem modulation of nociceptor-driven withdrawal reflexes. *Annals of the New York Academy of Sciences, 563,* 34–44.

Fields, H. L., Basbaum, A. I., & Heinricher, M. M. (2005). *Central nervous system mechanisms of pain modulation. Textbook of pain* (Vol. 5, pp. 125–142). Edinburgh: Churchill Livingstone.

Gebhart, G. F. (1983). Recent developments in the neurochemical bases of pain and analgesia. *NIDA Research Monograph, 45,* 19–35.

Goncalves, L., Almeida, A., & Pertovaara, A. (2007). Pronociceptive changes in response properties of rostroventromedial medullary neurons in a rat model of peripheral neuropathy. *The European Journal of Neuroscience, 26*(8), 2188–2195.

King, T., Vera-Portocarrero, L., Gutierrez, T., et al. (2009). Unmasking the tonic-aversive state in neuropathic pain. *Nature Neuroscience, 12*(11), 1364–1366.

Kovelowski, C. J., Ossipov, M. H., Sun, H., et al. (2000). Supraspinal cholecystokinin may drive tonic descending facilitation mechanisms to maintain neuropathic pain in the rat. *Pain, 87*(3), 265–273.

Lai, J., Luo, M. C., Chen, Q., et al. (2006). Dynorphin A activates bradykinin receptors to maintain neuropathic pain. *Nature Neuroscience, 9*(12), 1534–1540.

Leong, M. L., Gu, M., Speltz-Paiz, R., et al. (2011). Neuronal loss in the rostral ventromedial medulla in a rat model of neuropathic pain. *The Journal of Neuroscience: The Official Journal of the Society for Neuroscience, 31*(47), 17028–17039.

Ossipov, M. H., Hong Sun, T., Malan, P., Jr., et al. (2000). Mediation of spinal nerve injury induced tactile allodynia by descending facilitatory pathways in the dorsolateral funiculus in rats. *Neuroscience Letters, 290*(2), 129–132.

Ossipov, M. H., Dussor, G. O., & Porreca, F. (2010). Central modulation of pain. *The Journal of Clinical Investigation, 120*(11), 3779–3787.

Pertovaara, A. (1998). A neuronal correlate of secondary hyperalgesia in the rat spinal dorsal horn is submodality selective and facilitated by supraspinal influence. *Experimental Neurology, 149*(1), 193–202.

Porreca, F., Ossipov, M. H., & Gebhart, G. F. (2002). Chronic pain and medullary descending facilitation. *Trends in Neurosciences, 25*(6), 319–325.

Potrebic, S. B., Fields, H. L., & Mason, P. (1994). Serotonin immunoreactivity is contained in one physiological cell class in the rat rostral ventromedial medulla. *The Journal of Neuroscience: The Official Journal of the Society for Neuroscience, 14*(3 Pt 2), 1655–1665.

Sikandar, S., & Dickenson, A. H. (2011). Pregabalin modulation of spinal and brainstem visceral nociceptive processing. *Pain, 152*(10), 2312–2322.

Suzuki, R., & Dickenson, A. (2005). Spinal and supraspinal contributions to central sensitization in peripheral neuropathy. *Neurosignals, 14*(4), 175–181.

Suzuki, R., Rahman, W., Hunt, S. P., et al. (2004). Descending facilitatory control of mechanically evoked responses is enhanced in deep dorsal horn neurones following peripheral nerve injury. *Brain Research, 1019*(1–2), 68–76.

Tuveson, B., Leffler, A. S., & Hansson, P. (2011). Ondansetron, a 5HT3-antagonist, does not alter dynamic mechanical allodynia or spontaneous ongoing pain in peripheral neuropathy. *The Clinical Journal of Pain, 27*(4), 323–329.

Wang, Z., Gardell, L. R., Ossipov, M. H., et al. (2001). Pronociceptive actions of dynorphin maintain chronic neuropathic pain. *Journal of Neuroscience, 21*(5), 1779–1786.

Wiertelak, E. P., Furness, L. E., Horan, R., et al. (1994). Subcutaneous formalin produces centrifugal hyperalgesia at a non- injected site via the NMDA-nitric oxide cascade. *Brain Research, 649*(1–2), 19–26.

Winkler, C. W., Hermes, S. M., Chavkin, C. I., et al. (2006). Kappa opioid receptor (KOR) and GAD67 immunoreactivity are found in OFF and NEUTRAL cells in the rostral ventromedial medulla. *Journal of Neurophysiology, 96*(6), 3465–3473.

Zhang, W., Gardell, S., Zhang, D., et al. (2009). Neuropathic pain is maintained by brainstem neurons co-expressing opioid and cholecystokinin receptors. *Brain: A Journal of Neurology, 132*(Pt 3), 778–787.

Zhuo, M., & Gebhart, G. F. (1997). Biphasic modulation of spinal nociceptive transmission from the medullary raphe nuclei in the rat. *Journal of Neurophysiology, 78*(2), 746–758.

Zubieta, J. K., Bueller, J. A., Jackson, L. R., et al. (2005). Placebo effects mediated by endogenous opioid activity on mu-opioid receptors. *Journal of Neuroscience, 25*(34), 7754–7762.

Descending Facilitatory Systems

Min Zhuo
Department of Physiology, University of Toronto, Toronto, ON, Canada

Synonyms

Descending anti-analgesic systems; Descending excitatory modulation

Definition

Descending facilitation systems are the descending projection functional neuronal networks originating from supraspinal structures to the spinal cord. Activation of these systems enhances or increases evoked responses of spinal sensory neurons to peripheral sensory stimulation, including noxious and non-noxious stimuli. Reduction in the neuronal threshold to sensory stimuli contributes to the facilitatory effects. These systems are likely to be active in physiological conditions to enhance the detection, localization, and reaction to potentially dangerous or noxious stimuli. In pathological conditions, activation of descending facilitatory systems contributes to persistent pain related to inflammation or nerve injury.

Characteristics

Brain activity is able to affect sensory transmission in the spinal cord through descending modulatory systems (Basbaum and Fields 1984; Porreca et al. 2002; Gebhart 2004; Zhuo 2008). Descending modulation is biphasic, including descending inhibitory and facilitatory systems. Descending projection pathways derive directly or indirectly from central nuclei, including the ▶ anterior cingulate cortex (ACC), periaqueductal gray (PAG), and brainstem rostroventral ▶ medulla (RVM). As the last step of relay nuclei, neurons in the RVM play important roles in descending inhibition and facilitation of spinal sensory transmission. Biphasic modulation of spinal nociceptive transmission from the RVM offers fine regulation of spinal sensory thresholds and responses.

Integrative approaches have been used to investigate the mechanisms for descending facilitation, including electrophysiological, pharmacological, behavioral, and biochemical studies. The key evidence for descending facilitation is from electrophysiological recording of spinal sensory dorsal horn neurons (Zhuo and Gebhart 1991a, 1997, 2003; Zhuo et al. 2002). Through focally delivered electrical stimulation or ▶ glutamate into brain regions, it has been shown that responses of spinal dorsal horn neurons to peripheral sensory stimulation are enhanced or increased. No significant effects on spontaneous activity of spinal sensory neurons have been found. There are four major characteristics of descending facilitatory systems:

1. They affect spinal sensory transmission from both cutaneous and visceral inputs.
2. They reduce the neuronal threshold to nociceptive stimulation.
3. They are intensity dependent, and stimulation at lower intensities usually leads to facilitation.
4. They have longer latencies for the onset of descending facilitatory systems as compared to that of descending inhibitory systems. Descending inhibition from the RVM has a latency of 90 ms, while facilitation is 232 ms.

Descending facilitatory systems affect animals' behavioral withdrawal reflexive responses to noxious stimuli (Zhuo and Gebhart 1990, 1991b, 2002). In the spinal nociceptive tail-flick reflex, activation of descending facilitatory systems reduces the response latency to noxious heating of the tail. In visceromotor responses to colorectal distention (noxious visceral stimuli), descending facilitatory systems enhance responses to noxious visceral stimuli. These results consistently indicate that facilitation of spinal sensory neurons' responses leads to changes in behavioral reflexive responses in intact animals. Most 5-HT-containing nerve fibers in the spinal cord originate from the RVM. Pharmacological studies, using intrathecal drug administration, found that spinal 5-HT receptors mediate descending facilitatory modulation (Zhuo and Gebhart 1991b). It is likely that certain brain activity leads to the release of 5-HT at the level of the spinal cord; 5-HT binds to postsynaptic and presynaptic 5-HT receptors and enhances responses of spinal sensory neurons to peripheral sensory (noxious and non-noxious) stimuli. The enhanced neuronal activities in spinal dorsal horn neurons also contribute to the shortening of behavioral withdrawal response latencies by reducing the threshold.

It is important to point out that, in the case of cutaneous sensory transmission, descending facilitation is often small or difficult to elicit. This is because spinal sensory transmission receives tonic descending inhibitory modulation under normal physiological conditions. Removal of tonic descending inhibition (e.g., by lesioning the dorsolateral funiculi) often increases the chance of observing descending facilitation (Zhuo and Gebhart 1991b). Recent studies from the ACC show that descending facilitation may have stronger influences in pain perception at cortical levels. Activation of neurons in the ACC leads to pure facilitation of the spinal nociceptive tail-flick reflex in adult animals, without any significant inhibition (Calejesan et al. 2000). There are at least two different descending systems for the top-down facilitation from the cortex. The first one is the descending facilitatory system relayed through the RVM-spinal projection system. The second one is possible direct corticospinal projection from deeper layers of the ACC.

The cellular mechanism for descending facilitation is provided by in vitro spinal cord slice studies. Application of a low dose of 5-HT, or a selective 5-HT$_2$receptor agonist, induces facilitation of fast EPSCs in the lumbar spinal cord (Li and Zhuo 1998). 5-HT at low doses could facilitate fast EPSCs in the presence of an NMDA receptor antagonist AP-5, indicating that the facilitatory effect is NMDA receptor independent. Furthermore, the facilitatory effect induced by 5-HT at low doses persists during the washout of 5-HT. While the activation of 5-HT receptors is important for the induction of the facilitation, continuous activation of these receptors is not necessary for the expression of the facilitation. Application of methysergide after the 5-HT receptor agonist DOI fails to reverse the facilitatory effect (Li and Zhuo 1998).

One synaptic mechanism for the 5-HT produced facilitation is due to the recruitment of silent glutamatergic synapses. Application of 5-HT (5 μM) causes typically fast EPSCs to appear at synapses initially lacking AMPA/kainate receptor-mediated responses. 5-HT may affect spinal sensory transmission, by acting on presynaptic or postsynaptic receptors. Postsynaptic application of G protein inhibitors, introduced through the recording pipette, abolishes the effect of 5-HT to facilitate synaptic transmission, suggesting that postsynaptic 5-HT receptors are critical for the facilitatory effect. In support of this notion, postsynaptic Ca^{2+}-dependent processes are required for 5-HT-induced facilitation. In experiments with BAPTA (1,2-bis-(o-aminophenoxy)ethane-N,N,N',N'-tetraacetic acid) in the pipette solution, the facilitatory effect of 5-HT is abolished, indicating that an increase in postsynaptic Ca^{2+} is required. Additional evidence against a mechanism of 5-HT-induced synaptic facilitation, involving modulation of presynaptic glutamate release, comes from the observation that while 5-HT application clearly causes ▶ AMPA receptor-mediated EPSCs, NMDA receptor-mediated EPSCs are significantly decreased by 5-HT in the same neurons (Li and Zhuo 1998). This result suggests that postsynaptic enhancement of AMPA receptor-mediated currents by 5-HT is selective.

One possible mechanism for the recruitment of silent synapses is the interaction of glutamate AMPA receptors and proteins containing postsynaptic density-95/discs large/zona occludens-1 (PDZ) domains. GluR2 and GluR3 are widely expressed in sensory neurons in the superficial dorsal horn of the spinal cord. Glutamate receptor-interacting protein (GRIP), a protein with seven PDZ domains that bind specifically to the C-terminus of GluR2/GluR3, is also expressed in spinal dorsal horn neurons. A synthetic peptide corresponding to the last ten amino acids of GluR2 ("GluR2-SVKI": NVYGIESVKI), which disrupts binding of GluR2 to GRIP (Li et al. 1999), blocks the facilitatory effect of 5-HT. Experiments with different control peptides consistently indicate that the interaction between the c-terminus of GluR2/GluR3 and GRIP/ABP (or called GRIP1 and GRIP2) is important for 5-HT-induced facilitation.

Possible developmental factors have been raised, because silent experiments require a good space clamp in neurons, thus, most are performed in young neurons. In adult mouse spinal cord dorsal horn, some synaptic responses (26.3 % of a total of 38 experiments) between primary afferent fibers and dorsal horn neurons are almost completely mediated by NMDA receptors (Wang and Zhuo 2002). Dorsal root stimulation does not elicit any detectable AMPA/kainate receptor-mediated responses in these synapses. Unlike young spinal cord, 5-HT alone does not produce any long-lasting synaptic enhancement in adult spinal dorsal horn neurons. Activation of adenylyl cyclases by forskolin also does not significantly affect synaptic responses induced by dorsal root stimulation. However, co-application of 5-HT and forskolin produces long-lasting facilitation of synaptic responses. Possible contributors to the increases in the cAMP levels are the calcium-sensitive adenylyl cyclases. The facilitatory effect induced by 5-HT and forskolin is completely blocked in mice lacking AC1or AC8 (Wang and Zhuo 2002). One possible scenario for regulation of two different signaling pathways, under physiological or pathological conditions, is that postsynaptic increases in cAMP levels by sensory transmitters may favor 5-HT-induced facilitation.

Descending facilitatory systems play important roles in physiological and pathological conditions (Urban and Gebhart 1999; Porreca et al. 2002; Rahman et al. 2009; Tillu et al. 2008). One important physiological function for descending facilitation is that the activation of descending facilitatory systems increases the input of sensory information to enhance the detection, localization, and reaction to potentially dangerous or noxious stimuli. Imbalance between the descending facilitatory systems and the inhibitory systems contribute to chronic pain in pathological conditions. The recruitment of silent or ineffective synapses could significantly enhance spinal sensory transmission, including nociceptive transmission. AMPA receptors are expressed in the ▶ dorsal horn neurons both at synapses receiving high- and low-threshold inputs. Activation of silent synapses with low-threshold afferents might cause the dorsal horn neurons to exhibit an increased firing rate in response to non-noxious stimuli. Furthermore, the recruitment of silent synapses with high-threshold afferents might contribute to injury-related hyperalgesia, including the gain access of A fibers to C-fiber-driven spinal cord circuit (see Torsney 2011). Increases in postsynaptic AMPA receptor density and activation of silent synapses could contribute to plastic changes in the electrophysiological properties of dorsal horn sensory neurons, including ascending spinothalamic tract-projecting neurons. These

Descending Facilitatory Systems, Fig. 1 Descending facilitatory systems. (**a**) Example of facilitation of spinal visceral transmission produced by electrical stimulation and glutamate in the nucleus raphe magnus (*NRM*). Peristimulus time histograms (1s binwidth) and corresponding ocillographic records, in the absence (*top* histograms) and presence (*bottom* histograms) of, electrical stimulation (25 μA) and glutamate (5 nmoles) given in the same site in NRM. The intensity and duration of colorectal distension is illustrated below; the period of electrical stimulation (25 s) is indicated by the arrows. (**b**) Model of intensity-dependent biphasic modulation of spinal nociceptive transmission. (**c**) Model of different effects of facilitatory vs. inhibitory modulation on the SRFs of spinal dorsal horn neurons. (**d**) Model of the latency for descending facilitatory vs. inhibitory effects from the RVM. (**e**) Model of electrical stimulation in the RVM on the latencies of cardiovascular vs. neuronal effects

Descending Facilitatory Systems, Fig. 2 Cellular model for descending facilitation at the spinal cord dorsal horn. Peripheral tissue injury activates nociceptors and causes the release of glutamate (*filled circles*) as well as substance P, CGRP, and other putative transmitters (not shown) from the central terminals in the spinal dorsal horn. ▶ Serotonin released from descending projection terminals binds to its receptors and activates protein kinase C (*PKC*) by G protein-coupled receptors. Activation of PKC activated a series of intracellular signaling pathways and caused long-term facilitation of AMPA receptor mediated responses. The interaction between AMPA receptors and the PDZ protein glutamate receptor-interacting protein (*GRIP*) are likely to contribute to the recruitment of functional AMPA responses. In adult dorsal horn neurons, the activity of adenylyl cyclase subtype 1 and 8 is required for 5-HT produced facilitatory effects. Abbreviations: *Glu* glutamate, *AMPARs* α-amino-3-hydroxy-5-methyl-4-isoxalepropionate receptors, *5-HT* serotonin, *GRIP* glutamate receptor interacting protein, *AC1* and *AC8* adenylyl cyclase subtype 1 and 8

include increased receptive fields, enhanced responses to noxious and non-noxious stimuli, decreased firing thresholds, and increased background activity. There are two pieces of evidence that indirectly support the notion that alterations to descending serotonergic influences and spinal glutamate AMPA receptors may contribute to persistent pain: first, descending serotonergic systems from the RVM to the spinal cord have been implicated in different types of hyperalgesia after tissue injury (Calejesan et al. 1998; Urban and Gebhart 1999). Second, formalin-induced inflammation causes increases in activity of 5-HTdcontaining neurons in the RVM for up to 2 h (Robinson et al. 2002). Third, changes in the expression of AMPA receptors on spinal dorsal horn neurons after tissue injury has been reported and inhibiting the interaction of AMPA receptor and adapter proteins reduced chronic pain (Garry et al. 2003; Vikman et al. 2008;

Wang et al. 2011). Therefore, activation of silent synapses could serve as an important cellular mechanism underlying two features of chronic pain: hyperalgesia, where the intensity of responses to noxious stimuli is increased over baseline, and allodynia, where the nociceptive threshold is decreased and a normally non-noxious stimulus, such a gentle touch, can induce pain (Figs. 1 and 2).

References

Basbaum, A. I., & Fields, H. L. (1984). Endogenous pain control systems: brainstem spinal pathways and endorphin circuitry. *Annual Review of Neuroscience, 7*, 309–338.

Calejesan, A. A., Ch'ang, M. H., & Zhuo, M. (1998). Spinal serotonergic receptors mediate facilitation of a nociceptive reflex by subcutaneous formalin injection into the hindpaw in rats. *Brain Research, 798*, 46–54.

Calejesan, A. A., Kim, S. J., & Zhuo, M. (2000). Descending facilitatory modulation of a behavioral nociceptive response by stimulation in the adult rat anterior cingulate cortex. *Eur J Pain, 4*, 83–96.

Garry, E. M., Moss, A., Rosie, R., Delaney, A., Mitchell, R., & Fleetwood-Walker, S. M. (2003). Specific involvement in neuropathic pain of AMPA receptors and adapter proteins for the GluR2 subunit. *Mol Cell Neurosci, 24*, 10–22.

Gebhart, G. F. (2004). Descending modulation of pain. *Neuroscience and Biobehavioral Reviews, 27*, 729–737.

Li, P., & Zhuo, M. (1998). Silent glutamatergic synapses and nociception in mammalian spinal cord. *Nature, 393*, 695–698.

Li, P., et al. (1999). AMPA receptor-PDZ interactions in facilitation of spinal sensory synapses. *Nature Neuroscience, 2*, 972–977.

Porreca, F., Ossipov, M. H., & Gebhart, G. F. (2002). Chronic pain and medullary descending facilitation. *Trends in Neurosciences, 25*, 319–325.

Rahman, W., Bauer, C. S., Bannister, K., Vonsy, J. L., Dolphin, A. C., & Dickenson, A. H. (2009). Descending serotonergic facilitation and the antinociceptive effects of pregabalin in a rat model of osteoarthritic pain. *Molecular Pain, 5*, 45.

Robinson, D. A., Calejesan, A. A., & Zhuo, M. (2002). Long-lasting changes in rostral ventral medulla neuronal activity following inflammation in the adult rat. *The Journal of Pain, 3*, 292–300.

Tillu, D. V., Gebhart, G. F., & Sluka, K. A. (2008). Descending facilitatory pathways from the RVM initiate and maintain bilateral hyperalgesia after muscle insult. *Pain, 136*, 331–339.

Urban, M. O., & Gebhart, G. F. (1999). Supraspinal contributions to hyperalgesia. *Proceedings of the National Academy of Sciences of the United States of America, 96*, 7687–7692.

Vikman, K. S., Rycroft, B. K., & Christie, M. J. (2008). Switch to Ca2+-permeable AMPA and reduced NR2B NMDA receptor-mediated neurotransmission at dorsal horn nociceptive synapses during inflammatory pain in the rat. *The Journal of Physiology, 586*, 515–527.

Wang, G. D., & Zhuo, M. (2002). Synergistic enhancement of glutamate-mediated responses by serotonin and forskolin in adult mouse spinal dorsal horn neurons. *Journal of Neurophysiology, 87*, 732–739.

Wang, W., Petralia, R. S., Takamiya, K., Xia, J., Li, Y. Q., Huganir, R. L., et al. (2011). Preserved acute pain and impaired neuropathic pain in mice lacking protein interacting with C Kinase 1. *Molecular Pain, 7*, 11.

Zhuo, M. (2008). Cortical excitation and chronic pain. *Trends in Neurosciences, 31*, 199–207.

Zhuo, M., & Gebhart, G. F. (1990). Characterization of descending inhibition and facilitation from the nuclei reticularis gigantocellularis and gigantocellularis pars alpha in the rat. *Pain, 42*, 337–350.

Zhuo, M., & Gebhart, G. F. (1991a). Characterization of descending facilitation and inhibition of spinal nociceptive transmission from the nuclei reticularis gigantocellularis and gigantocellularis pars alpha in the rat. *Journal of Neurophysiology, 67*, 1599–1614.

Zhuo, M., & Gebhart, G. F. (1991b). Spinal serotonin receptors mediate descending facilitation of a nociceptive reflex from the nuclei reticularis gigantocellularis and gigantocellularis pars alpha in the rat. *Brain Research, 550*, 35–48.

Zhuo, M., & Gebhart, G. F. (1997). Biphasic modulation of spinal nociceptive transmission from the medullary raphe nuclei in the rat. *Journal of Neurophysiology, 78*, 746–758.

Zhuo, M., & Gebhart, G. F. (2002). Biphasic modulation of a visceral nociceptive reflex from the rostroventral medial medulla in the rat. *Gastroenterology, 122*, 1007–1019.

Zhuo, M., & Gebhart, G. F. (2003). Modulation of noxious and non-noxious spinal mechanical transmission from the rostroventral medial medulla in the rat. *Journal of Neurophysiology, 88*, 2928–2941.

Zhuo, M., Sengupta, J. N., & Gebhart, G. F. (2002). Biphasic modulation of spinal visceral nociceptive transmission from the rostroventral medial medulla in the rat. *Journal of Neurophysiology, 87*, 2225–2236.

Descending Inhibition

Definition

Inhibition of spinal nociceptive processing can be effected from numerous sites in the brain, including cortical, diencephalic, midbrain, and brainstem sites. Inhibition produced by stimulation in rostral sites in the brain is usually indirect and involves a relay in the brainstem, usually the rostral ventral medulla (RVM), including the serotonergic cell group in nucleus raphe magnus in the RVM. For example, stimulation in hypothalamic sites and the periaqueductal gray (PAG) relay in the RVM, which in turn activates neurons that send projections to the spinal dorsal horn as well as to other brainstem nuclei (most importantly to noradrenergic cell groups in the dorsolateral pons). Accordingly, inhibition at the level of the dorsal horn is effected by serotonin and noradrenaline released from axons descending from the brainstem. In addition, opioidergic and gabaergic contributions to spinal nociceptive inhibition have been documented as have contributions of neurotensin, cholecystokinin, and other atypical neurotransmitters/modulators.

Cross-References

- Descending Modulation of Visceral Pain
- Sex Differences in Descending Pain Modulatory Pathways
- TENS, Mechanisms of Action

Descending Inhibition/Facilitation

Definition

Descending inhibition refers to the ability of activation in various brain areas to attenuate spinal (or trigeminal) dorsal horn neuronal responses to peripheral stimuli, most commonly applied with respect to noxious stimuli. Descending facilitation refers to the ability of activation in various brain areas to increase spinal (or trigeminal) dorsal horn neuronal responses to peripheral stimuli, most commonly applied with respect to noxious stimuli. It should be appreciated that descending influences are not necessarily restricted to modulation of noxious input.

Cross-References

- Forebrain Modulation of the Periaqueductal Gray and Its Role in Pain

Descending Inhibitory Fibers

Definition

These fibers (axons), arising from cell bodies in the brainstem (principally), are contained in descending tracts in spinal dorsolateral and/or ventrolateral funiculi. See "▶ Centrifugal Control of Sensory Inputs" for sites in the brainstem from which these axons arise and

the neurotransmitters/modulators released from their terminals in the spinal cord.

Cross-References

- Descending Modulation and Persistent Pain
- Descending Modulation of Nociceptive Processing
- Infant Pain Mechanisms

Descending Inhibitory Mechanisms and GABA Mechanisms

- GABA Mechanisms and Descending Inhibitory Mechanisms

Descending Inhibitory Noradrenergic Pathways

Definition

Descending axons originating from noradrenergic cell groups in the brainstem dorsolateral pons that terminate in the spinal dorsal horn release noradrenaline and attenuate dorsal horn neuronal responses to noxious stimulation. Noradrenaline acts on alpha-2 adrenoceptors located on primary afferent spinal terminals and/or on cells postsynaptic to the noradrenaline-releasing descending axon.

Cross-References

- Alpha(α) 2-Adrenergic Agonists in Pain Treatment

Descending Inhibitory Pathway

- Sex Differences in Descending Pain Modulatory Pathways

Descending Inhibitory Pathway for Pain

Definition

A pain-modulating pathway originating in the brain and terminating in the dorsal horn of the spinal cord.

Cross-References

▶ Alternative Medicine in Neuropathic Pain

Descending Modulation

▶ Descending Circuitry, Opioids

Descending Modulation and Persistent Pain

Antti Pertovaara
Department of Physiology, Institute of
Biomedicine University of Helsinki, Helsinki,
Finland

Definition

Ascending nociceptive pathways carry information about potentially harmful peripheral events from the spinal cord to the brain, and these ascending signals may eventually evoke the sensation of pain. The magnitude of the ascending nociceptive signal and the consequent pain sensation can be greatly influenced by descending pathways originating in the brainstem and terminating in the spinal cord. In conditions that cause persistent pain, such as inflammation or injury, the function of descending pathways may change. This change may contribute to suppression or facilitation of pain sensitivity, depending among other things on the pathophysiological

condition, the brainstem-spinal pathway, time course of the injury, and the type of pain.

Characteristics

Multiple descending pathways originate in several different brainstem nuclei, release different neurotransmitters from their axonal endings at the spinal cord, and have different functional characteristics. Those descending pathways that terminate in the spinal dorsal horn, a critical relay for ascending nociceptive signals, contribute to pain modulation (Fields et al. 2006). Descending pain modulatory pathways are involved in mediating the ▶ top-down and ▶ feedback control of pain. This includes the optimal setting of gain for ascending signals according to various behavioral needs. In healthy animals, descending pathways predominantly suppress pain, as shown by increased spinal responses to noxious heat following inactivation of descending pathways (e.g., by a cold block at a midthoracic level; Fields et al. 2006). However, descending pathways may also facilitate pain, and the same brainstem site may be a source of descending facilitation as well as inhibition. This was shown by the finding that a low concentration of glutamate, or low electrical stimulus intensity in the ▶ rostroventromedial medulla (RVM), produced facilitation of spinal nociception in healthy animals, whereas a high glutamate concentration or high electrical stimulus intensity in the same brainstem site suppressed spinal nociception (Urban and Gebhart 1999).

Enhanced Inhibitory Controls

In conditions causing persistent pain, such as following injury or inflammation, the function of descending pathways may change. The first observations on the pathophysiological function of descending pain control systems indicated that the inhibitory effect of descending pathways on responses of spinal pain-relay neurons was stronger in animals with experimental arthritis than in controls (Schaible et al. 1991). Enhanced descending inhibitory controls have been observed in not only experimental arthritis, but also following paw inflammation or formalin treatment of the skin (Ren and Dubner 2002). The enhanced inhibitory controls required a few days to develop in inflamed animals. It was accompanied by an increased inhibitory efficacy of glutamate, via action on medial medullary ▶ N-methyl-D-aspartate (NMDA) and alpha-amino-3-hydroxy-5-methyl-4-isoxazole propionic acid (AMPA) receptors, and by a phenotypic switch in the response profile of RVM neurons (Ren and Dubner 2002). These results suggest that following inflammation, the RVM and its glutamatergic receptors exhibit activity-dependent plasticity, leading to enhanced inhibitory controls. Additionally, noradrenergic ▶ locus coeruleus nucleus in the dorsolateral pons have an important role in the enhanced inhibition, as shown by a stronger inflammation-induced decrease of a heat-evoked spinal withdrawal latency in animals with a bilateral lesion of the locus coeruleus than in sham-operated animals (Tsuruoka and Willis 1996). Enhanced descending inhibition provides a feedback control of pain, which helps to maintain the capacity to use an injured and painful body part for fight or flight in case of emergency, giving potential evolutionary benefits.

Enhanced Facilitatory Controls

An opposite effect to the enhanced descending inhibitory controls has also been described in conditions causing persistent pain. Namely, hypersensitivity induced by inflammation or nerve injury was attenuated following brainstem lesions or a blockade of descending pathways (Herrero and Cervero 1996; Pertovaara et al. 1996; Urban et al. 1996; Wiertelak et al. 1994). This finding indicates that descending spinal pathways may contribute to hypersensitivity by facilitating nociception at the spinal cord level. Thus, persistent pain may induce not only enhanced inhibitory controls, but also enhanced facilitatory controls, or concurrently both. There is evidence suggesting that the net effect of descending controls may critically depend on whether nociceptive signals originate in the site of injury (▶ primary hyperalgesia predominantly

due to peripheral mechanisms), or adjacent to it (▶ secondary hyperalgesia predominantly due to central mechanisms) (Urban and Gebhart 1999). Enhanced inhibitory controls were reported in experiments in which primary hyperalgesia was studied (Schaible et al. 1991), whereas descending facilitation was the net effect in experiments in which secondary hyperalgesia was studied (Pertovaara 1998; Urban et al. 1996; Wiertelak et al. 1994). The net effect of descending controls may also depend on the sub-modality of pain, since it was reported that particularly mechanical and cold hypersensitivity were dependent on intact medullospinal pathways (Kauppila et al. 1998). Furthermore, the net effect of descending controls may vary from facilitation to inhibition as time progresses from the start of the injury (Ren and Dubner 2002).

Neurocircuitry and Neurochemistry Underlying Descending Facilitation

The dorsal columns are at least partly involved in carrying the ascending injury signal to the brainstem level that maintains the descending facilitation, since a lesion of this ascending pathway selectively attenuated hypersensitivity in neuropathic animals (Porreca et al. 2002). The RVM, an important relay for many descending inhibitory pathways, is also a major link for mediating the facilitatory effect to the spinal cord. Attenuation of hypersensitivity following micro-administration of a local anesthetic into the RVM or following a lesion of the dorsolateral funiculus, a major conduit of descending projections from the RVM (Urban and Gebhart 1999), demonstrates this. Glutamatergic NMDA, neurotensin, and cholecystokinin-2 (CCK-2) receptors in the RVM are involved in the maintenance of descending facilitation, since micro-administration of antagonists of these receptors into the RVM reduced hypersensitivity induced by neurogenic inflammation or nerve injury (Porreca et al. 2002). In addition to neurons, glial cells in the RVM also may contribute to descending facilitation, as indicated by the antihypersensitivity produced following intramedullary administration of glial cell inhibitors in inflamed animals (Roberts et al. 2009). The finding that a single dose of an NMDA receptor antagonist microinjected into the RVM, prior to but not after a nerve injury, delayed the development of hypersensitivity suggests that medullary NMDA receptors trigger central changes that contribute to descending facilitation (Pertovaara 2000). However, attenuation of neuropathic symptoms by local anesthesia near the injury site indicates that central changes alone are not sufficient to maintain hypersensitivity, but continuous injury discharge from the periphery is of critical importance for the sustained facilitatory action of medullospinal pathways (Porreca et al. 2002). From the RVM, the facilitatory action may be mediated downwards by a class of mu-opioid receptor–expressing neurons called ON-cells, which typically discharge prior to a ▶ noxious stimulus–induced withdrawal, and are considered to have a role in the induction of protective (pain) reflexes. This is suggested by the finding that immunocytochemical destruction of these RVM neurons reduced hypersensitivity, and attenuated spinal upregulation of dynorphin, a pronociceptive substance, in neuropathic animals (Porreca et al. 2002). The contribution of increased ON-cell discharge to hypersensitivity is further supported by the finding that neurogenic inflammation-induced hyperreflexia was accompanied by an increase of ON-cell activity in the RVM, and these effects were reversed in parallel by an NMDA receptor antagonist (Heinricher et al. 2003). Unexpectedly, the association of response properties of medullary ON-neurons with behavioral hypersensitivity has varied with the model and duration of chronic ▶ neuropathy (Gonçalves et al. 2007; Pertovaara et al. 2001). However, a lack of change in response properties of medullary ON-neurons during the chronic phase of neuropathy does not exclude the possibility that the afferent barrage at the time of nerve injury would have caused long-term changes in the synaptic efficacy of the spinal terminals of the RVM neurons. This spinal change would not be revealed by characterization of neuronal response properties at the medullary level. The finding that intrathecal administration

Descending Modulation and Persistent Pain, Fig. 1 A schematic diagram showing one example of enhanced descending inhibitory controls and one of enhanced descending facilitatory controls in inflamed and/or nerve-injured conditions. *LC* noradrenergic locus coeruleus nucleus in the pons, *RVM* rostroventromedial medulla, − indicates inhibitory and + a facilitatory effect, *NA* noradrenaline acting on the spinal α2-adrenoceptor, *EAA* excitatory amino acid, *CCK* cholecystokinin acting on the CCK-2 receptor, *NT* neurotensin, *5-HT* serotonin acting on the spinal 5-HT3 receptor, *NO* nitric oxide. It should be noted that the RVM is also an important source of descending inhibitory controls

of a serotonin 5-HT3 receptor antagonist (Dogrul et al. 2009; Suzuki et al. 2002), a CCK-2 receptor antagonist (Urban et al. 1996), an NMDA receptor antagonist, and a nitric oxide synthesis inhibitor (Wiertelak et al. 1994) selectively attenuated hypersensitivity suggests that serotonin through action on 5-HT3 receptors, CCK through action on CCK-2 receptors, glutamate through action on NMDA receptors, and nitric oxide are involved in mediating the descending facilitatory action on spinal nociceptive neurons. Following a block of descending pathways, hypersensitivity in nociceptive spinal neurons was attenuated, but still significant (Pertovaara 1998). This finding indicates that medullary descending facilitation augments spinal segmental mechanisms, which alone may induce significant hyperalgesia following inflammation or injury. From the evolutional point of view, enhanced facilitation of pain might help the healing process by promoting immobilization and protection of the injured region. However, under human clinical conditions, this type of pain-enhancing loop may contribute to chronic hypersensitivity and persistent pain that serves no useful purpose. Further characterization of the neurochemical basis of this facilitatory circuitry may provide novel targets for selective suppression of chronic hyperalgesia, without influence on the physiological pain mechanisms that help to protect tissues (Fig. 1).

References

Dogrul, A., Ossipov, M. H., & Porreca, F. (2009). Differential mediation of descending pain facilitation and inhibition by spinal 5HT-3 and 5HT-7 receptors. *Brain Research, 1280*, 52–59.

Fields, H. L., Basbaum, A. I., & Heinricher, M. M. (2006). Central nervous system mechanisms of pain modulation. In S. B. McMahon & M. Koltzenburg (Eds.), *Wall and Melzack's textbook of pain* (5th ed., pp. 125–142). Hong Kong: Churchill Livingstone.

Gonçalves, L., Almeida, A., & Pertovaara, A. (2007). Pronociceptive changes in response properties of rostroventromedial medullary neurons in a Rat model of peripheral neuropathy. *European Journal of Neuroscience, 26*, 2188–2195.

Heinricher, M. M., Pertovaara, A., & Ossipov, M. H. (2003). Descending modulation after injury. *Program of Pain Research Management, 24*, 251–260.

Herrero, J. F., & Cervero, F. (1996). Supraspinal influences on the facilitation of Rat nociceptive reflexes induced by carrageenan monoarthritis. *Neuroscience Letters, 209*, 21–24.

Kauppila, T., Kontinen, V. K., & Pertovaara, A. (1998). Influence of spinalization on spinal withdrawal reflex responses varies depending on the submodality of the test stimulus and the experimental pathophysiological condition. *Brain Research, 797*, 234–242.

Pertovaara, A. (1998). A neuronal correlate of secondary hyperalgesia in the Rat spinal dorsal horn is submodality selective and facilitated by supraspinal influence. *Experimental Neurology, 149*, 193–202.

Pertovaara, A. (2000). Plasticity in descending pain modulatory systems. *Progress in Brain Research, 129*, 231–242.

Pertovaara, A., Wei, H., & Hämäläinen, M. M. (1996). Lidocaine in the rostroventromedial medulla and the periaqueductal gray attenuates allodynia in neuropathic rats. *Neuroscience Letters, 218*, 127–130.

Pertovaara, A., Keski-Vakkuri, U., Kalmari, J., et al. (2001). Response properties of neurons in the rostroventromedial medulla of neuropathic rats: attempted modulation of responses by [1DMe] NPYF, a neuropeptide FF analogue. *Neuroscience, 105*, 457–468.

Porreca, F., Ossipov, M. H., & Gebhart, G. F. (2002). Chronic pain and medullary descending facilitation. *Trends in Neurosciences, 25*, 319–325.

Ren, K., & Dubner, R. (2002). Descending modulation in persistent pain: An update. *Pain, 100*, 1–6.

Roberts, J., Ossipov, M. H., & Porreca, F. (2009). Glial activation in the rostroventromedial medulla promotes descending facilitation to mediate inflammatory hypersensitivity. *European Journal of Neuroscience, 30*, 229–241.

Schaible, H. G., Neugebauer, V., Cervero, F., et al. (1991). Changes in tonic descending inhibition of spinal neurons with articular input during the development of acute arthritis in the Cat. *Journal of Neurophysiology, 66*, 1021–1032.

Suzuki, R., Morcuende, S., Webber, M., Hunt, S. P., & Dickenson, A. H. (2002). Superficial NK1-expressing neurons control spinal excitability through activation of descending pathways. *Nature Neuroscience, 5*, 1319–1326.

Tsuruoka, M., & Willis, W. D. (1996). Bilateral lesions in the area of the nucleus locus coeruleus affect the development of hyperalgesia during carrageenan-induced inflammation. *Brain Research, 726*, 233–236.

Urban, M. O., & Gebhart, G. F. (1999). Supraspinal contributions to hyperalgesia. *Proceedings of the National Academy of Sciences of the United States, 96*, 7687–7692.

Urban, M. O., Jiang, M. C., & Gebhart, G. F. (1996). Participation of central descending nociceptive facilitatory systems in secondary hyperalgesia produced by mustard Oil. *Brain Research, 737*, 83–91.

Wiertelak, E. P., Furness, L. E., Horan, R., et al. (1994). Subcutaneous formalin produces centrifugal hyperalgesia at a non-injected site via the NMDA-nitric oxide cascade. *Brain Research, 649*, 19–26.

Descending Modulation of Nociceptive Processing

Ronald Dubner and Ke Ren
Department of Neural and Pain Sciences,
University of Maryland Dental School,
Baltimore, MD, USA

Synonyms

Descending control; Descending pain modulation; Endogenous analgesia systems; Pain facilitatory systems; Pain inhibitory systems; Pain modulation; Supraspinal pain control systems

Definition

Information about the condition of injured peripheral tissues enters the spinal cord and then ascends to higher centers in the brain stem, thalamus, and cerebral cortex, ultimately being elaborated as a complex sensory experience. What we perceive depends not only on the features of the stimulus, but is also related to its meaning, based on previous exposure to pain, its influence on our emotional state, and its relevance to our survival. Descending modulation refers to modulation of pain transmission at the spinal level by supraspinal sites, leading to changes in perceived pain. It is an important component of the sensory processing of pain because it provides the neural networks by which attention, motivation, and cognition modulate what we perceive.

Characteristics

Our knowledge of the existence of endogenous descending pain-modulatory systems (see ▶ Pain-Modulatory Systems, History of Discovery) spans at least three decades (for comprehensive reviews, see Fields and Basbaum 1999; Millan 2002). The first line of evidence to support endogenous pain control came from the study of Reynolds (1969) who demonstrated that focal brain stimulation in the periaqueductal gray (PAG) produced sufficient analgesia to permit abdominal surgery. Liebeskind and colleagues confirmed this finding and concluded that stimulation in the PAG activated a normal function of the brain, pain inhibition (Mayer et al. 1971; Mayer and Liebeskind 1974). They labeled the phenomenon stimulation-produced analgesia (SPA). These early studies found SPA to be specifically antinociceptive – producing no generalized sensory, attentional, emotional, or motoric deficits. They also showed that SPA could outlast the period of brain stimulation and that it occurred in a restricted peripheral field, such that noxious stimuli applied outside that field elicited normal defensive reactions. They reported that during stimulation, the sensation of

light touch was intact, there was no indication of seizure activity, and animals were still capable of eating. The analgesia produced by electrical stimulation in the PAG was of rapid onset and as potent as high doses of morphine – completely inhibiting withdrawal responses to even the most severe noxious stimuli.

An important final common descending modulatory site in the brain stem is the RVM or rostral ventromedial medulla, whose major component is the nucleus raphe magnus (NRM). The RVM receives signals directly from the PAG and indirectly from forebrain sites such as the prefrontal cortex, the amygdala, and the anterior cingulate cortex (Bandler and Shipley 1994). The descending PAG-RVM circuit is involved in the emotional, motivational, and cognitive factors that modulate the sensation of pain. Numerous anatomical studies have demonstrated that PAG afferents arise predominantly in the forebrain. PAG receives significant innervation from a number of cortical and subcortical sites involved in nociception. The forebrain projections terminate with a high degree of topographical specificity, forming sets of longitudinal input columns extending focally throughout the rostrocaudal extent of the PAG. These forebrain structures include the medial preoptic area (MPO), central nucleus of the amygdala (CNA), and certain medial prefrontal cortical fields. Thus, the PAG is ideally positioned to integrate information from sensory systems in the brainstem/spinal cord with information from the forebrain and is thought to modulate a number of homeostatic processes including fear, anxiety, cardiovascular tone, and vocalization. The descending nociceptive modulation from PAG appears to be normally under tonic gamma (γ)-aminobutyric acid (GABA) inhibitory control. Endogenous and exogenous opiates acting within the PAG probably produce analgesia via inhibition of inhibitory GABA-containing interneurons and thereby activate (disinhibit) PAG output neurons.

Besson and colleagues were the first to show that electrical stimulation of the nucleus raphe magnus in the RVM suppressed behavioral responses to a strong pinch of the tail or the limbs as well as modifying the threshold of the jaw opening reflex (Oliveras et al. 1975). Perhaps the most important finding concerning SPA was that analgesia induced by brain stimulation shared several characteristics with analgesia from opiate drugs. Areas of the brain from which SPA could be elicited were rich in opiate receptors and microinjection of morphine into these brain regions produced analgesia, indicating that common brain sites support both SPA and opiate analgesia. Tsou and Jang (1964) discovered that the most sensitive sites for the analgesic action of morphine were located in the PAG and the adjacent hypothalamic periventricular area. This led to the hypothesis that opiate drugs like morphine acted by binding to a receptor in the brain (Pert and Snyder 1973) and that there were probably endogenous ligands or chemical mediators whose actions were mimicked by opiates. These findings led to the discovery of the endogenous opioid peptides (Goldstein et al. 1979; Hughes et al. 1975) and subsequently to the cloning of the three major subtypes of opioid receptors, mu, delta, and kappa (Chen et al. 1993; Evans et al. 1992; Kieffer et al. 1992; Thompson et al. 1993).

Electric shock, restraint, rotation, forced swim, and intruder threat can all produce analgesia in laboratory animals without producing identifiable tissue damage (Hayes et al. 1978). Since the same environmental influences are known to produce stress and activate the hypothalamic-pituitary-adrenal (HPA) axis, this has been called "stress-induced analgesia" (SIA). Timing, severity, and location of a single stressor were all eventually reported to differentially produce opioid and nonopioid SIA (Terman et al. 1984). Some forms of opioid SIA, although depending on descending pain inhibitory pathways, appear to be activated entirely by the physical properties of the stressor and can be elicited even in the surgically anesthetized animal (Terman et al. 1984). On the other hand, other forms of opioid SIA depend on learning mechanisms and can be blocked by decerebration, muscarinic antagonists, amygdala lesions, and pentobarbital

anesthesia. Placing rats in a cage where foot shock was previously received, but is not currently administered, can also elicit analgesia ("conditioned analgesia"). Evidently, animals can not only activate their intrinsic pain inhibitory systems in the presence of stress, but they can also learn to activate them in anticipation of such stimuli. Unconditioned expectations also appear capable of activating opioid descending pain inhibitory substrates. The scent of an animal's natural predator, for instance, can induce analgesia.

Bidirectional Descending Control

Early studies established the presence of a descending inhibitory pain-modulatory circuit linking the brain stem PAG and RVM with the spinal cord (see Basbaum and Fields 1984; Fields and Basbaum 1999; Gebhart 1986; Millan 2002, for reviews). However, we now know that there are parallel descending facilitatory mechanisms (see Gebhart 2004; Ren and Dubner 2008a, for reviews). Brain stem descending pathways also facilitate nociceptive transmission at the spinal cord level. Excitation and inhibition of dorsal horn neurons can be produced by stimulation of the dorsolateral funiculus of the spinal cord (Dubuisson and Wall 1979; McMahon and Wall 1988), NRM (Dubuisson and Wall 1980), and nucleus reticularis gigantocellularis (NGC) (Haber et al. 1980). In the NGC, low intensity electrical stimulation or microinjection of a low dose of glutamate or neurotensin produce facilitation of behavioral and spinal dorsal horn neuronal responses to noxious stimulation (Thomas et al. 1995; Urban and Gebhart 1999; Zhuo et al. 2002; Zhuo and Gebhart 1992). RVM neurons may exert bidirectional control of nociception through descending serotonergic pathways and descending noradrenergic pathways (Holden et al. 1999; Zhuo and Gebhart 1991). Descending inhibitory influences from the RVM travel mainly in the dorsolateral funiculus whereas descending facilitatory effects reach the spinal cord via the ventral and ventrolateral funiculi (Zhuo et al. 2002).

Descending Modulation of Visceral Pain

Descending modulation of visceral pain processing at the spinal level is also under the control of both facilitatory and inhibitory influences. Activation of neurons in the RVM attenuates responses to inflammation of the viscera (Coutinho et al. 1998; Urban et al. 1999). Electrical stimulation or glutamate microinjection into the PAG, RVM, or the NGC inhibits the activity of dorsal horn neurons responsive to visceral afferent fiber stimulation or noxious colorectal distension (CRD) (Ammon et al. 1984; Cervero et al. 1985; Ness and Gebhart 1987). Other studies have shown that spinal visceral input is also subject to facilitation from descending brain stem sites (Zhuo et al. 2002). These descending effects are intensity dependent with facilitation produced by the lowest levels of electrical stimulation or drug doses and higher levels of stimulation accounting for descending inhibition at RVM and PAG sites. In addition, multiple transmitters may exert different actions depending on the type of neurons stimulated and the receptors activated (Saab et al. 2004). In addition, stimulus-dependent vagal afferent stimulation has also been shown to produce facilitation and inhibition of the nociceptive tail flick reflex as well as dorsal horn nociceptive neuronal activity (Randich and Gebhart 1992), and reduce pain in humans (Busch et al. 2012). Vagal afferents transmit signals from the esophagus, lower airways, heart, gastrointestinal tract, liver, gallbladder, and pancreas to the nucleus tractus solitarius in the medulla oblongata. Pharmacological activation of vagal afferents with intravenous administration of serotonin or low dose opioids also inhibits spinal nociceptive reflexes and the response of spinothalamic neurons to noxious stimuli.

Functional Properties of RVM Neurons

In the RVM, two types of cells have been identified by Fields and colleagues (Fields et al. 1991) as pain-modulatory neurons. On-cells are characterized by a sudden increase in activity before the initiation of a nocifensive behavior, in this case,

a tail flick to a transient noxious heat stimulus. Off-cells exhibit a pause in activity just prior to the initiation of the tail flick. While off-cells are usually associated with the inhibition of nocifensive behavior, on-cells are correlated with a facilitation of nocifensive behavior. A third type of cell, the neutral cell, was also identified, but its activity was not correlated with nocifensive behavior in response to transient stimuli.

The RVM also contains serotonergic cell types. Serotonergic cells comprise about 15 % of the cells in RVM. Most serotonergic RVM cells send unmyelinated axons through the dorsolateral funiculus into the spinal cord. Most serotonergic cells contain at least one and often several co-transmitters, including excitatory and inhibitory amino acids and a plethora of neuropeptides (Bowker et al. 1982). Since decreasing the availability of serotonin or antagonizing the activation of serotonergic receptors strongly attenuates antinociception, serotonergic RVM cells were originally thought to be the antinociceptive output cell of the RVM. However, serotonergic RVM cells are not activated by midbrain periaqueductal gray stimulation or opioid administration that evokes analgesia. Further, supraspinal opioid administration that causes analgesia does not consistently evoke serotonin release in the spinal cord. Serotonergic RVM cells fire at their highest rates when an animal is active and moving about, fire progressively more slowly as an animal proceeds from drowsy to slow wave sleep and are silent during rapid eye movement sleep. Within the dorsal horn, serotonin is likely to primarily function by presynaptically modulating the release of glutamate and other transmitters (Li and Zhuo 1998).

A second neurochemically distinct monoamine descending system involves noradrenergic neurons whose cell bodies are in A5, A6, and A7 cell groups in the brainstem. Several studies indicate that a descending norepinephrine (NE) system can mediate analgesia and dorsal horn inhibition and that the NE descending system is critical for opiate-induced analgesia. Blockade of

PAG SPA by intrathecal α2-adrenergic antago- nists (Graham et al. 1997) emphasizes the impor- tance of noradrenergic systems in descending nociceptive modulation.

Persistent Pain and Descending Modulation

Earlier studies of descending modulation mainly focused on responses to acute or transient stimuli. In contrast, recent studies have examined the effects of persistent pain on descending modula- tion following tissue damage or nerve injury. These persistent, or chronic, pain conditions are associated with prolonged functional changes in the nervous system, evidenced by the develop- ment of dorsal horn hyperexcitability and activ- ity-dependent plasticity, also commonly referred to as central sensitization (for reviews see Dubner and Ruda 1992; Woolf and Salter 2000; Ren and Dubner 2002, 2008a). There is evidence of enhanced net descending inhibition after inflam- mation at sites of primary hyperalgesia. Primary hyperalgesia involves increased sensitivity of pri- mary afferent neurons (peripheral sensitization) as well as central sensitization. Descending inhi- bition is greater in neurons with input from an inflamed knee as compared to a non-inflamed knee (Schaible et al. 1991). In rats with hind paw inflammation, spinal cord lidocaine block leads to an enhanced activity of dorsal horn noci- ceptive neurons that is greater in inflamed than that in non-inflamed rats (Ren and Dubner 1996). Similar findings are found using Fos protein expression (see ▶ c-Fos) as a marker of neuronal activation. There are more inflammation-induced Fos-immunoreactive neurons in the dorsal horn in spinally transected or dorsolateral funiculus lesioned rats as compared to sham-operated inflamed rats (Ren and Ruda 1996; Wei et al. 1998). Kauppila et al. (1998) showed that thermal but not mechanical nociceptive responses were further enhanced in hind paw inflamed and spinal nerve ligated rats after midthoracic spinalization. Finally, hyperalgesia is intensified in rats with lesions of the dorsal lateral quadrant of the spinal cord after inflammation or formalin injection (Abbott et al. 1996; Ren and Dubner 1996). These studies reveal the net descending

inhibitory effects of activation of multiple supraspinal sites. The findings suggest that injury-induced dorsal horn hyperexcitability and primary hyperalgesia can be dampened by descending pathways, due to enhancement of descending net inhibitory effects.

The source of the enhanced net inhibition can be found in the brain stem. Local anesthesia of the RVM results in a further increase in dorsal horn nociceptive neuronal activity in hind paw inflamed rats (Ren and Dubner 1996). Focal lesions of the RVM and locus coeruleus produce an increase in spinal Fos expression and hyperalgesia after inflammation (Tsuruoka and Willis 1996a, b; Wei et al. 1999b). It appears that both RVM and locus coeruleus descending pathways are major sources of enhanced net inhi- bition in inflamed animals.

After inflammation, descending facilitation parallels inhibition but can also dominate it, resulting in a net enhancement of activity or hyperalgesia. The selective destruction of the NGC with a soma selective neurotoxin, ibotenic acid, leads to an attenuation of hyperalgesia and a reduction of inflammation-induced spinal Fos expression (Wei et al. 1999a). A descending facilitatory effect may also originate from the medullary dorsal reticular nucleus (Lima and Almeida 2002) and other brain sites such as the anterior cingulate cortex (Calejesan et al. 2000).

A descending facilitatory drive contributes to the pathogenesis of certain types of persistent pain, particularly those associated with second- ary inflammatory hyperalgesia or nerve injury (Porreca et al. 2002; Ren and Dubner 2002). Spinalization blocks mustard oil–produced sec- ondary mechanical allodynia and hyperexcitability of spinal nociceptive neurons (Mansikka and Pertovaara 1997). Hind paw formalin-induced hyperalgesia is prevented by RVM lesions (Wiertelak et al. 1997). The rostral ventromedial medulla (RVM) in the brainstem activates a descending neuronal pathway to the medullary and spinal dorsal horns leading to enhanced excitability and behavioral hyperalgesia after injury (Porreca et al. 2002; Ren and Dubner 2008a). RVM lesions inhibit secondary hyperalgesia produced by topical application of mustard oil (Urban and Gebhart 1999). It also has been shown previously that the NGC plays a role in descending facilitation of nociceptive transmission after transient noxious stimuli (Zhuo and Gebhart 1992). The same phenomenon occurs in models of neuropathic pain. The tactile allodynia after nerve injury is dependent upon a tonic activation of net descending facilitation from supraspinal sites (Ossipov et al. 2000). In nerve injured rats, lesions of the dorsolateral funiculus, local anesthetic block of the RVM, and lesions of RVM mu-opioid receptor-expressing cells do not prevent the onset, but reverse the later maintenance of tactile and thermal hyperalgesia (Burgess et al. 2002; Porreca et al. 2001). These observations point to an ascending-descending loop that is activated in response to prolonged stimulation and resulting in facilitation at the spinal level. The descending projections are in part serotonergic (Green et al. 2000; Zeitz et al. 2002) and descending serotonin (5-HT) has a pronociceptive function in the spinal and medullary dorsal horns after peripheral inflammation (Suzuki et al 2002) and nerve injury (Wei et al. 2010; Chai et al. 2012). Descending 5-HT from the RVM and 5-HT3 receptors are involved in the central mechanisms underlying the enhanced neuronal excitability in dorsal horn nociceptive neurons and the behavioral long-term maintenance of secondary hyperalgesia (Suzuki et al. 2002; Wei et al. 2010; Chai et al. 2012). Both facilitatory and inhibitory circuitry may be activated by ascending input after injury (Gozariu et al. 1998; Herrero and Cervero 1996; Terayama et al. 2000). There is an increase in on- and off-cell activity after inflammation (Miki et al. 2002). What appears to be important is the balance between synaptic excitation and inhibition under different behavioral conditions.

Dynamic Shifts in Descending Modulation after Injury

Studies indicate that the enhancement of descending inhibition in response to tissue injury appears to build up gradually (Danziger et al.

1999; Dubner and Ren 1999; Hurley and Hammond 2000; Ren and Dubner 1996; Ren and Ruda 1996; Schaible et al. 1991). An effect of Freund's adjuvant on the spinal cord contralateral to tissue injury may also be subject to the inhibitory control of descending inputs, as suggested by a clear contralateral increase in Fos-positive cells in transected rats after inflammation (Ren and Ruda 1996). It appears that following inflammation, brainstem descending pathways can become progressively more involved in suppressing incoming nociceptive signals in primary hyperalgesic zones.

Persistent inflammation induces dramatic changes in the excitability of RVM pain modulating circuitry, suggesting that there are dynamic temporal changes in synaptic activation in the brain stem after inflammation (Guan et al. 2002; Terayama et al. 2000) and nerve injury (Burgess et al. 2002; LaGraize et al. 2010).

What are the cellular mechanisms that underlie these changes? Excitatory amino acids (EAAs) have previously been shown to mediate descending modulation in response to transient noxious stimulation and early inflammation (Heinricher et al. 1999; Urban and Gebhart 1999, for review) and they appear to be involved in the development of RVM excitability associated with inflammation and persistent pain (Da Silva et al. 2010; Guan et al. 2002; Miki et al. 2002; Terayama et al. 2000; Xu et al. 2007). The finding that a single dose of an NMDA receptor antagonist microinjected into the RVM, prior to but not after a nerve injury, delayed the development of hypersensitivity suggests that medullary NMDA receptors trigger central changes that contribute to descending facilitation (Pertovaara 2000). At 3 h post-inflammation, low doses of NMDA, a prototype NMDA receptor agonist, microinjected into the RVM produces facilitation of the response to noxious heat of the inflamed hind paw (Urban and Gebhart 1999). Higher doses of NMDA at 3 h post-inflammation only produce inhibition. At 24 h post-inflammation, NMDA produces only inhibition. AMPA, a selective AMPA receptor agonist, produces dose- and time-dependent inhibition at 3 and 24 h post-inflammation that is blocked by an AMPA receptor antagonist. The above findings indicate that there is an increase in the potency of the dose-response curves of NMDA- and AMPA-produced inhibition at 24 h post-inflammation as compared to 3 h. The results suggest that the time-dependent functional changes in descending modulation are mediated in part by enhanced EAA neurotransmission.

Serotonin (5-HT) is a major neurotransmitter in the RVM-spinal pathway, although 5-HT does not appear to be present in RVM pain-modulatory neurons (Heinricher et al. 2009). Selective deletion of 5-HT in RVM neurons by RNA interference of tryptophan hydroxylase-2, the rate-limiting enzyme for neuronal 5-HT synthesis, reveals its role in descending pain facilitation after injury (Wei et al. 2010). 5-HT axons terminating in the spinal and medullary dorsal horns originate mainly in the RVM and activate a number of heterogeneous receptors (Millan 2002; Wei et al. 2010). The pain facilitatory effect of descending 5-HT is mediated by the spinal/trigeminal 5-HT$_3$ receptor, the lone ligand-gated cation channel in the 5-HT receptor family (Chai et al. 2012; Dogrul et al. 2009; LaGraize et al. 2010; Suzuki et al. 2002; Wei et al. 2010; Zeitz et al. 2002).

A well-established regulator of synaptic plasticity, brain-derived neurotrophic factor (BDNF), also plays a critical role in pain-modulatory circuitry and contributes to the mechanisms of descending pain facilitation. After peripheral tissue injury, BDNFergic neurotransmission involving TrkB in the PAG-RVM-spinal cord circuitry is rapidly activated, evidenced by increased BDNF expression in PAG and upregulation and increased phosphorylation of TrkB receptors in the RVM. Through the intracellular signaling cascade that involves phospholipase C, IP$_3$, PKC, and non-receptor protein tyrosine kinases, NMDA receptors are phosphorylated and descending facilitatory drive is initiated (Guo et al. 2006).

Injection of neurokinin-1 (NK1) tachykinin receptor antagonists into the RVM reverses inflammatory hyperalgesia but has no effect on ongoing neuronal activity and baseline nociception (Hamity et al. 2010). After inflammation, NK1 receptors exhibit a time-dependent upregulation in the RVM (LaGraize et al. 2010). Injection of an NK1 receptor antagonist into RVM attenuated the capsaicin-evoked enhanced responses of on-cells to noxious stimuli with less of an effect on the enhanced inhibitory responses of off-cells, suggesting a selective localization of NK1 receptors in on-cells (Brink et al. 2012). Injection of substance P, the NK1 receptor agonist, into the RVM induces increased NKCC1 (the Na-K-Cl cotransporter) protein expression and reduces postsynaptic GABAergic inhibitory tone in the spinal dorsal horn, likely via an effect on the anionic equilibrium potential in dorsal horn neurons. The change in polarity of the effect of GABAergic transmission leads to spinal NMDA receptor phosphorylation that correlates with the enhancement of behavioral hyperalgesia (LaGraize et al. 2010).

Rats with inflammatory hyperalgesia exhibit an increased sensitivity to opioid analgesics (Neil et al. 1986). Typically, there is a leftward shift of the dose-response curve for opioids from the inflamed hyperalgesic paw when compared to the non-inflamed paw (Hylden et al. 1991). Kayser et al. (1991), suggesting that this increased opioid sensitivity in inflamed animals is related to a peripheral mechanism, as it is significantly attenuated after local injections of very low doses (0.5–1 µg) of naloxone. Recent observations indicate that the increased opioid sensitivity after inflammation may also reflect changes in central pain modulating pathways. Hurley and Hammond (2000, 2001) have demonstrated enhancement and plasticity of the descending inhibitory effects of mu and delta-2 opioid receptor agonists microinjected into RVM during the development and maintenance of inflammatory hyperalgesia. During persistent inflammation, an increased percentage of RVM neurons, both serotonergic and non-serotonergic, responded to a mu-opioid receptor agonist DAMGO (Zhang and Hammond 2010). It is likely that opioid peptide activation or GABA disinhibition (Fields and Basbaum 1999) is also important in the initiation and maintenance of RVM plasticity.

Neuron-Glial Interactions in Descending Circuitry

Glia, consisting of microglia, astrocytes and oligodendrocytes, are the major non-neuronal cells in the CNS. Microglia and astroglia, in particular, release inflammatory cytokines and function as immune cells in the brain. Neuron-glial interactions have emerged as critical mechanisms underlying chronic pain conditions (McMahon and Malcangio 2009; Ren and Dubner 2010). These interactions also occur in the brain pain-modulatory circuitry. A few studies have demonstrated injury-induced glial hyperactivity in the PAG-RVM circuitry, phenomena similar to that extensively reported at the spinal and trigeminal levels. Injury of the infraorbital nerve induces upregulation of CD11b (cluster of differentiation molecule 11b) and Iba1 (ionized calcium-binding adapter molecule), markers of microglia and astrocytes, respectively in the RVM (Wei et al. 2008). Raghavendra et al. (2004) showed that complete Freund's adjuvant-induced hind paw inflammation upregulated microglia and astroglia in brain stem samples. Roberts et al. (2009) further showed that upregulation of microglial and astroglial markers occurs in the RVM after carrageenan-induced hind paw inflammation. Associated with upregulated glial markers, inflammatory cytokines such as interleukin-1beta (IL-1b) and tumor necrosis factor-alpha (TNFa) are also increased (Raghavendra et al. 2004; Wei et al. 2008). The injury-related glial hyperactivity in the descending circuitry is correlated with behavioral hyperalgesia. Administration of glial inhibitors into the RVM attenuates or reverses behavioral hyperalgesia after nerve or tissue injury (Roberts et al. 2009; Wei et al. 2008). Microglial and astroglial inhibitors attenuated hyperalgesia at 3 and 14 days after infraorbital nerve injury, respectively, which correlates with an early and transient activation of microglia and prolonged activation of astroglia (Wei et al. 2008). Studies at the spinal and medullary levels also suggest a differential time-dependent coordinated involvement of microglia and astroglia in persistent pain (See Ren and Dubner 2008b, 2010).

Glial cells and associated cytokines in descending pathways impact neuronal activity. Upregulated TNFa and IL-1b are found in astroglia after infraorbital nerve injury, and receptors for TNFa and IL-1b colocalize with the GluN1 (NR1) subunit of the N-methyl-D-aspartate (NMDA) receptor in RVM neurons (Wei et al. 2008). Nerve injury increases the expression of TNF and IL-1 receptor levels and GluN1 phosphorylation in the RVM. These findings indicate the contribution of supraspinal astrocytes and associated cytokines as well as glia-neuronal interactions to descending facilitation of neuropathic pain.

Astroglia play a major role in maintaining homeostasis of glutamate, the major excitatory amino acid neurotransmitter in the nervous system, and actively prevent neuronal hyperexcitability and excitotoxic cascades. The glutamate transporter GLT1 (EAAT2), mainly expressed in astroglia, is crucial for glutamate removal from the synaptic cleft. Recent studies show that there are time-dependent biphasic changes in GLT1 expression in the RVM after tissue or nerve injury (Ren et al. 2011).

Endogenous opioid peptides and their receptors are distributed extensively in descending pathways and normally produce pain inhibition. However, a series of studies reveal that opioids may also act on glial cells through nonclassic sites (see review by Hutchinson et al. 2011). Instead of producing analgesia, the non-conventional action of opioids seems to be related to opioid tolerance and hyperalgesia. In mice with deletion of all three subtypes of opioid receptors, opioid-produced hyperalgesia remains, suggesting non-conventional opioid signaling (see Hutchinson et al. 2011).

Molecular Mechanisms of Plasticity in the RVM

What are the molecular mechanisms of this increased potency leading to enhanced synaptic activity and increases in descending net inhibition associated with primary hyperalgesia? Recent studies have examined whether transcriptional, translational, and posttranslational changes occur in the RVM after inflammation and may underlie the changes in EAA receptor sensitivity observed. Examination of the mRNA expression of the GluN1 (NR1), GluN2A (NR2A), and GluN2B (NR2B) subunits of the NMDA receptor in the RVM revealed an upregulation that parallels the time course of the RVM excitability changes (Miki et al. 2002). This is accompanied by an increase in NMDA receptor protein. There is also an increase in receptor phosphorylation of the GluN2A subunit of the NMDA receptor in the RVM after inflammation (Guo et al. 2006). Western blot analysis revealed a time-dependent increase in the AMPA receptor GluA1 (GluR1) subunit levels in the RVM at 5 h and 24 h post-inflammation as compared to naive animals (Guan et al. 2004). Western blots also demonstrated that GluR1 phosphoprotein levels were increased as early as 30 min and were time dependent, suggesting that posttranslational receptor phosphorylation may also contribute to the enhanced AMPA transmission (Guan et al. 2004). These findings support the hypothesis that activity-dependent plasticity takes place at the RVM level and involves both changes in excitatory amino acid receptor gene and protein expression and increased phosphorylation of these receptors.

This activity-induced plasticity in pain modulating circuitry after inflammation complements the activity-dependent neuronal plasticity in ascending pain transmission pathways (Dubner and Ruda 1992; Ren and Dubner 2002). Inflammation leads to peripheral sensitization of nociceptors and central sensitization or activity-dependent plasticity of spinal nociceptive neurons. The increased neuronal barrage at the spinal level activates spinal projection neurons, which leads to activation of glutamatergic, opioidergic, and GABAergic neurons at the brain stem level and a similar, but not identical, form of activity-dependent plasticity.

Phenotypic Changes in the RVM after Inflammation

The time-dependent plasticity in descending pain-modulatory circuitry also involves changes in the response profiles of RVM neurons. Miki et al. (2002) used paw withdrawal latency as a behavioral correlate to assess the relationship between nocifensive behavior and RVM neural activity after inflammation. Similar to the findings of Fields et al. (1991), who correlated tail flick responses to RVM neural activity after transient noxious thermal stimuli, Miki et al. (2002) observed on-like, off-like, and neutral-like cells based on the relationship of their responses to paw withdrawal behavior during the development of inflammation. They found that some neutral-like cells changed their response profile and were reclassified as on-like or off-like cells during continuous recordings of 5 h or more. The change in the response profile of RVM neurons correlated with the temporal changes in excitability in the RVM after inflammation (Terayama et al. 2000). This phenotypic switch of RVM neurons was verified in a population study which showed that there was a significant increase in the percentage of on-like and off-like cells, and a decrease in the neutral-like cell population 24 h after inflammation. There was also a greater increase in the magnitude of the responses of on-like cells after inflammation as compared to on-like responses in naive controls, suggesting an increase in facilitatory drive in the RVM.

Placebo Analgesia and Hypnosis: Role of Descending Effects

Placebo analgesia is that component of the reduction of pain due to an inert treatment (or the nonactive component of an active treatment) administered to a subject who expects that it will reduce pain. Thus, the context in which a placebo is administered and the expectation of the subject influence the level of analgesia produced. Conditioning stimulation can also influence the level of analgesia, but such conditioning effects appear to involve cognitive functions and are not merely reflex effects. The neural mechanisms of the placebo effect are unclear. There are numerous studies that suggest that the placebo effect, in part, involves endogenous opioid pathways and descending modulation. Studies using positron emission tomography (PET) have shown that the brain regions in the cerebral cortex and in the brainstem influenced by a placebo manipulation are also those activated by opioids, suggesting similar mechanisms in placebo-induced and opioid-induced analgesia (Petrovic et al. 2002). The areas activated include the rostral anterior cingulate cortex, the orbitofrontal cortex, and the brain stem. In addition, there are several pharmacological studies which show that placebo analgesia is attenuated or reversed by the opioid receptor antagonist naloxone (Amanzio and Benedetti 1999; Grevert et al. 1983; Levine et al. 1978, Levine and Gordon 1984), providing further evidence for the involvement of opioid mechanisms. It is important to note that placebo analgesia can also be insensitive to opioid receptor antagonists, suggesting that chemical mediators other than opioids may be involved (Amanzio and Benedetti 1999; Gracely et al. 1983).

Hypnotically induced reductions in pain are induced by suggestions and facilitated by alterations of consciousness (Hilgard and Hilgard 1983; Price and Barrell 1990; Rainville and Price 2003). Hypnosis can alter the sensory discriminative component of pain (see ▶ Sensory-Discriminative Aspects) as well as the affective dimension. The efficacy of hypnotic analgesia depends on several factors including the pain dimension that is measured, baseline pain intensity, the maintained presence of the hypnotist or hypnotic suggestions, and finally hypnotic ability (Price and Barber 1987). The neural mechanisms underlying hypnotic analgesia are poorly understood though it is thought that both ascending and descending nociceptive pathways are involved.

References

Abbott, F. V., Hong, Y., & Franklin, K. B. (1996). The effect of lesions of the dorsolateral funiculus on formalin pain and morphine analgesia: A dose-response analysis. *Pain, 65*, 17–23.

Amanzio, M., & Benedetti, F. (1999). Neuropharmacological dissection of placebo analgesia: Expectation-activated opioid systems versus conditioning-activated specific sub-systems. *Journal of Neuroscience, 19*, 484–494.

Ammon, W. S., Blair, R. W., & Foreman, R. D. (1984). Raphe magnus inhibition of primate T1-T2 spinothalamic cells with cardiopulmonary visceral input. *Pain, 20*, 247–260.

Bandler, R., & Shipley, M. T. (1994). Columnar organization in the midbrain periaqueductal gray: Modules for emotional expression? *Trends in Neurosciences, 17*, 379–389.

Basbaum, A. I., & Fields, H. L. (1984). Endogenous pain control systems: Brainstem spinal pathways and endorphin circuitry. *Annual Review of Neuroscience, 7*, 309–338.

Bowker, R. M., Westlund, K. N., Sullivan, M. C., Wilber, J. F., & Coulter, J. D. (1982). Transmitters of the raphe-spinal complex: Immunocytochemical studies. *Peptides, 3*, 291–298.

Burgess, S. E., Gardell, L. R., Ossipov, M. H., Malan, T. P., Jr., Vanderah, T. W., Lai, J., & Porreca, F. (2002). Time-dependent descending facilitation from the rostral ventromedial medulla maintains, but does not initiate, neuropathic pain. *Journal of Neuroscience, 22*, 5129–5136.

Busch, V., Zeman, F., Heckel, A., Menne, F., Ellrich, J., Eichhammer, P. (2012). The effect of transcutaneous vagus nerve stimulation on pain perception – An experimental study. *Brain Stimulation* [Epub ahead of print]

Calejesan, A. A., Kim, S. J., & Zhuo, M. (2000). Descending facilitatory modulation of a behavioral nociceptive response by stimulation in the adult rat anterior cingulate cortex. *European Journal of Pain, 4*, 83–96.

Cervero, F., Lumb, B. M., & Tattersall, J. E. H. (1985). Supraspinal loops that mediate visceral inputs to thoracic spinal cord neurons in the cat: Involvement of descending pathways from raphe and reticular formation. *Neuroscience Letters, 56*, 189–194.

Chai, B., Guo, W., Wei, F., Dubner, R., & Ren, K. (2012). Trigeminal-Rostral Ventromedial Medulla circuitry is involved in orofacial hyperalgesia contralateral to tissue injury. *Molecular Pain, 8*, 78.

Chen, Y., Mestek, A., Liu, J., Hurley, J. A., & Yu, L. (1993). Molecular cloning and functional expression of a mu-opioid receptor from rat brain. *Molecular Pharmacology, 44*, 8–12.

Coutinho, S. V., Urban, M. O., & Gebhart, G. F. (1998). Role of glutamate receptors and nitric oxide in the rostral ventromedial medulla in visceral hyperalgesia. *Pain, 78*, 59–69.

Da Silva, L. F., Desantana, J. M., & Sluka, K. A. (2010). Activation of NMDA receptors in the brainstem, rostral ventromedial medulla, and nucleus reticularis gigantocellularis mediates mechanical hyperalgesia produced by repeated intramuscular injections of acidic saline in rats. *The Journal of Pain, 11*, 378–387.

Danziger, N., Weil-Fugazza, J., Le Bars, D., & Bouhassira, D. (1999). Alteration of descending modulation of nociception during the course of monoarthritis in the rat. *Journal of Neuroscience, 19*, 2394–2400.

Dogrul, A., Ossipov, M. H., & Porreca, F. (2009). Differential mediation of descending pain facilitation and inhibition by spinal 5HT-3 and 5HT-7 receptors. *Brain Research, 1280,* 52–59.

Dubner, R., & Ren, K. (1999). Endogenous mechanisms of sensory modulation. *Pain, 6,* 45–53.

Dubner, R., & Ruda, M. A. (1992). Activity-dependent neuronal plasticity following tissue injury and inflammation. *Trends in Neurosciences, 15,* 96–103.

Dubuisson, D., & Wall, P. D. (1979). Medullary raphe influences on units in laminae 1 and 2 of cat spinal cord. *The Journal of Physiology, 300,* 33.

Dubuisson, D., & Wall, P. D. (1980). Descending influences on receptive fields and activity of single units recorded in laminae 1, 2 and 3 of cat spinal cord. *Brain Research, 199,* 283–298.

Evans, C. J., Keith, D. E., Jr., Morrison, H., Magendzo, K., & Edwards, R. H. (1992). Cloning of a delta opioid receptor by functional expression. *Science, 258,* 1952–1955.

Fields, H. (2004). State-dependent opioid control of pain. *Nature Reviews Neuroscience, 5,* 565–575.

Fields, H. L., & Basbaum, A. I. (1999). Central nervous system mechanisms of pain modulation. In P. D. Wall & R. Melzack (Eds.), *Textbook of pain* (pp. 309–329). London: Churchill Livingstone.

Fields, H. L., Heinricher, M. M., & Mason, P. (1991). Neurotransmitters in nociceptive modulatory circuits. *Annual Review of Neuroscience, 14,* 219–245.

Gebhart, G. F. (1986). Modulatory effects of descending systems on spinal dorsal horn neurons. In T. L. Yaksh (Ed.), *Spinal afferent processing* (pp. 391–416). New York: Plenum.

Gebhart, G. F. (2004). Descending modulation of pain. *Neuroscience and BioBehavioral Reviews, 27,* 729–737.

Goldstein, A., Tachibana, S., Lowney, L. I., Hunkapiller, M., & Hood, L. (1979). Dynorphin-(1-13), an extraordinarily potent opioid peptide. *Proceedings of the National Academy Science of the United States of America, 76,* 6666–6670.

Gozariu, M., Bouhassira, D., Willer, J. C., & Le Bars, D. (1998). The influence of temporal summation on a C-fibre reflex in the rat: Effects of lesions in the rostral ventromedial medulla (RVM). *Brain Research, 792,* 168–172.

Gracely, R. H., Dubner, R., Wolskee, P. J., & Deeter, W. R. (1983). Placebo and naloxone can alter postsurgical pain by separate mechanisms. *Nature, 306,* 264–265.

Graham, B. A., Hammond, D. L., & Proudfit, H. K. (1997). Differences in the antinociceptive effects of alpha-2 adrenoceptor agonists in two substrains of Sprague-Dawley rats. *The Journal of Pharmacology and Experimental Therapeutics, 283,* 511–519.

Grevert, P., Albert, L. H., & Goldstein, A. (1983). Partial antagonism of placebo analgesia by naloxone. *Pain, 16,* 129–143.

Guan, Y., Terayama, R., Dubner, R., & Ren, K. (2002). Plasticity in excitatory amino acid receptor-mediated descending pain modulation after inflammation. *The*

Journal of Pharmacology and Experimental Therapeutics, 300, 513–520.

Guan, Y., Guo, W., Robbins, M. T., Dubner, R., & Ren, K. (2004). Changes in AMPA receptor phosphorylation in the rostral ventromedial medulla after inflammatory hyperalgesia in rats. *Neuroscience Letters, 366,* 201–205.

Guo, W., Robbins, M. T., Wei, F., Zou, S., Dubner, R., & Ren, K. (2006). Supraspinal brain-derived neurotrophic factor signaling: A novel mechanism for descending pain facilitation. *Journal of Neuroscience, 26,* 126–137.

Haber, L. H., Martin, R. F., Chung, J. M., & Willis, W. D. (1980). Inhibition and excitation of primate spinothalamic tract neurons by stimulation in region of nucleus reticularis gigantocellularis. *Journal of Neurophysiology, 43,* 1578–1593.

Hamity, M. V., White, S. R., & Hammond, D. L. (2010). Effects of neurokinin-1 receptor agonism and antagonism in the rostral ventromedial medulla of rats with acute or persistent inflammatory nociception. *Neuroscience, 165,* 902–913.

Hayes, R. L., Bennett, G. J., Newlon, P. G., & Mayer, D. J. (1978). Behavioral and physiological studies of non-narcotic analgesia in the rat elicited by certain environmental stimuli. *Brain Research, 155,* 69–90.

Heinricher, M. M., McGaraughty, S., & Farr, D. A. (1999). The role of excitatory amino acid transmission within the rostral ventromedial medulla in the antinociceptive actions of systemically administered morphine. *Pain, 81,* 57–65.

Heinricher, M. M., Tavares, I., Leith, J. L., & Lumb, B. M. (2009). Descending control of nociception: Specificity, recruitment and plasticity. *Brain Research Reviews, 60,* 214–225.

Herrero, J. F., & Cervero, F. (1996). Supraspinal influences on the facilitation of rat nociceptive reflexes induced by carrageenan monoarthritis. *Neuroscience Letters, 209,* 21–24.

Hilgard, E. R., & Hilgard, J. R. (1983). *Hypnosis in the relief of pain* (p. 294). Los Altos: William Kaufmann.

Holden, J. E., Schwartz, E. J., & Proudfit, H. K. (1999). Microinjection of morphine in the A7 catecholamine cell group produces opposing effects on nociception that are mediated by alpha1- and alpha2-adrenoceptors. *Neuroscience, 91,* 979–990.

Hughes, J., Smith, T. W., Kosterlitz, H. W., Fothergill, L. A., Morgan, B. A., & Morris, H. R. (1975). Identification of two related pentapeptides from the brain with potent opiate agonist activity. *Nature, 258,* 577–580.

Hurley, R. W., & Hammond, D. L. (2000). The analgesic effects of supraspinal mu and delta opioid receptor agonists are potentiated during persistent inflammation. *Journal of Neuroscience, 20,* 1249–1259.

Hurley, R. W., & Hammond, D. L. (2001). Contribution of endogenous enkephalins to the enhanced analgesic effects of supraspinal mu opioid receptor agonists after inflammatory injury. *Journal of Neuroscience, 21*, 2536–2545.

Hutchinson, M. R., Shavit, Y., Grace, P. M., Rice, K. C., Maier, S. F., & Watkins, L. R. (2011). Exploring the neuroimmunopharmacology of opioids: An integrative review of mechanisms of central immune signaling and their implications for opioid analgesia. *Pharmacological Reviews, 63*, 772–810.

Hylden, J. L. K., Thomas, D. A., Iadarola, M. J., Nahin, R. L., & Dubner, R. (1991). Spinal opioid analgesic effects are enhanced in a model of unilateral inflammation / hyperalgesia: Possible involvement of noradrenergic mechanisms. *European Journal of Pharmacology, 194*, 135–143.

Kauppila, T., Kontinen, V. K., & Pertovaara, A. (1998). Influence of spinalization on spinal withdrawal reflex responses varies depending on the submodality of the test stimulus and the experimental pathophysiological condition in the rat. *Brain Research, 797*, 234–242.

Kayser, V., Chen, Y. L., & Guilbaud, G. (1991). Behavioural evidence for a peripheral component in the enhanced antinociceptive effect of a low dose of systemic morphine in carragenin-induced hyperalgesic rats. *Brain Research, 560*, 237–244.

Kieffer, B. L., Befort, K., Gaveriaux-Ruff, C., & Hirth, C. G. (1992). The delta-opioid receptor: Isolation of a cDNA by expression cloning and pharmacological characterization. *Proceedings of the National Academy Science of the United States of America, 89*, 12048–12052.

Levine, J. D., & Gordon, N. C. (1984). Influence of the method of drug administration on analgesic response. *Nature, 312*, 755–756.

Levine, J. D., Gordon, N. C., & Fields, H. L. (1978). The mechanisms of placebo analgesia. *Lancet, 2*, 654–657.

Li, P., & Zhuo, M. (1998). Silent glutamatergic synapses and nociception in mammalian spinal cord. *Nature, 393*, 695–698.

Lima, D., & Almeida, A. (2002). The medullary dorsal reticular nucleus as a pronociceptive centre of the pain control system. *Progress in Neurobiology, 66*, 81–108.

Mansikka, H., & Pertovaara, A. (1997). Supraspinal influence on hindlimb withdrawal thresholds and mustard oil-induced secondary allodynia in rats. *Brain Research Bulletin, 42*, 359–365.

Mayer, D. J., & Liebeskind, J. C. (1974). Pain reduction by focal electrical stimulation of the brain: An anatomical and behavioral analysis. *Brain Research, 68*, 73–93.

Mayer, D. J., Wolfle, T. L., Akil, H., Carder, B., & Liebeskind, J. C. (1971). Analgesia from electrical stimulation in the brainstem of the rat. *Science, 174*, 1351–1354.

McMahon, S. B., & Malcangio, M. (2009). Current challenges in glia-pain biology. *Neuron, 64*, 46–54.

McMahon, S. B., & Wall, P. D. (1988). Descending excitation and inhibition of spinal cord lamina I projection neurons. *Journal of Neurophysiology, 59*, 1204–1219.

Miki, K., Zhou, Q. Q., Guo, W., Guan, Y., Terayama, R., Dubner, R., & Ren, K. (2002). Changes in gene expression and neuronal phenotype in brain stem pain modulatory circuitry after inflammation. *Journal of Neurophysiology, 87*, 750–760.

Millan, M. J. (2002). Descending control of pain. *Progress in Neurobiology, 66*, 355–474.

Neil, A., Kayser, V., Gacel, G., Besson, J. M., & Guilbaud, G. (1986). Opioid receptor types and antinociceptive activity in chronic inflammation: Both kappa and mu opiate agonistic effects are enhanced in arthritic rats. *European Journal of Pharmacology, 130*, 203–208.

Ness, T. J., & Gebhart, G. F. (1987). Quantitative comparison of inhibition of visceral and cutaneous spinal nociceptive transmission from the midbrain and medulla in the rat. *Journal of Neurophysiology, 58*, 850–865.

Oliveras, J. L., Redjemi, F., Guilbaud, G., & Besson, J. M. (1975). Analgesia induced by electrical stimulation of the inferior centralis nucleus of the raphe in the cat. *Pain, 1*, 139–145.

Ossipov, M. H., Hong Sun, T., Malan, P., Jr., Lai, J., & Porreca, F. (2000). Mediation of spinal nerve injury induced tactile allodynia by descending facilitatory pathways in the dorsolateral funiculus in rats. *Neuroscience Letters, 290*, 129–132.

Pert, C. B., & Snyder, S. H. (1973). Opiate receptor: Demonstration in nervous tissue. *Science, 179*, 1011–1014.

Pertovaara, A. (2000). Plasticity in descending pain modulatory systems. *Progress in Brain Research, 129*, 231–242.

Petrovic, P., Kalso, E., Petersson, K. M., & Ingvar, M. (2002). Placebo and opioid analgesia -Imaging a shared neuronal network. *Science, 295*, 1737–1740.

Porreca, F., Burgess, S. E., Gardell, L. R., Vanderah, T. W., Malan, T. P., Jr., Ossipov, M. H., Lappi, D. A., & Lai, J. (2001). Inhibition of neuropathic pain by selective ablation of brainstem medullary cells expressing the micro-opioid receptor. *Journal of Neuroscience, 21*, 5281–5288.

Porreca, F., Ossipov, M. H., & Gebhart, G. F. (2002). Chronic pain and medullary descending facilitation. *Trends in Neurosciences, 25*, 319–325.

Price, D. D., & Barber, J. (1987). An analysis of factors that contribute to the efficacy of hypnotic analgesia. *Journal of Abnormal Psychology, 96*, 46–51.

Price, D. D., & Barrell, J. J. (1990). The structure of the hypnotic state: A self-directed experiential study. In J. J. Barrell (Ed.), *The experiential method: Exploring the human experience* (pp. 85–97). Massachusetts: Copely Publishing Group.

Raghavendra, V., Tanga, F. Y., & DeLeo, J. A. (2004). Complete Freunds adjuvant-induced peripheral inflammation evokes glial activation and proinflammatory cytokine expression in the CNS. *European Journal of Neuroscience, 20*, 467–473.

Rainville, P., & Price, D. D. (2003). Hypnosis phenomenology and the neurobiology of consciousness. *International Journal of Clinical and Experimental Hypnosis, 51*, 105–129.

Randich, A., & Gebhart, G. F. (1992). Vagal afferent modulation of nociception. *Brain Research Brain Research Reviews, 17*, 77–99.

Ren, K., & Dubner, R. (1996). Enhanced descending modulation of nociception in rats with persistent hindpaw inflammation. *Journal of Neurophysiology, 76*, 3025–3037.

Ren, K., & Dubner, R. (2002). Descending modulation in persistent pain: An update. *Pain, 100*, 1–6.

Ren, K., & Dubner, R. (2008a). Descending control mechanisms. In A. I. Basbaum, A. Kaneko, G. M. Shepherd, G. Westheimer, M. C. Bushnell, & A. I. Basbaum (Eds.), *The senses: A comprehensive reference* (pp. 723–762). San Diego: Academic Press.

Ren, K., & Dubner, R. (2008b). Neuron-glia crosstalk gets serious: Role in pain hypersensitivity. *Current Opinion in Anaesthesiology, 21*, 570–579.

Ren, K., & Dubner, R. (2010). Interactions between the immune and nervous systems in pain. *Nature Medicine, 16*, 1267–1276.

Ren, K., Guo, W., Gu, M., Zou, S.-P., Wie, F., & Dubner, R. (2011). Kappa B-motif binding phosphoprotein (KBBP), astroglial glutamate transporter GLT1 and behavioral hyperalgesia. *Glia, 59*(Suppl. 1), S113–S113.

Ren, K., & Ruda, M. A. (1996). Descending modulation of Fos expression after persistent peripheral inflammation. *Neuroreport, 7*, 2186–2190.

Reynolds, D. V. (1969). Surgery in the rat during electrical analgesia induced by focal brain stimulation. *Science, 164*, 444–445.

Roberts, J., Ossipov, M. H., & Porreca, F. (2009). Glial activation in the rostroventromedial medulla promotes descending facilitation to mediate inflammatory hypersensitivity. *European Journal of Neuroscience, 30*, 229–241.

Saab, C. Y., Park, Y. C., & Al-Chaer, E. D. (2004). Thalamic modulation of visceral nociceptive processing in adult rats with neonatal colon irritation. *Brain Research, 1008*, 186–192.

Schaible, H. G., Neugebauer, V., Cervero, F., & Schmidt, R. F. (1991). Changes in tonic descending inhibition of spinal neurons with articular input during the development of acute arthritis in the cat. *Journal of Neurophysiology, 66*, 1021–1032.

Suzuki, R., Morcuende, S., Webber, M., Hunt, S. P., & Dickenson, A. H. (2002). Superficial NK1-expressing neurons control spinal excitability through activation of descending pathways. *Nature Neuroscience, 5*, 1319–1326.

Terayama, R., Dubner, R., & Ren, K. (2000). The roles of NMDA receptor activation and nucleus reticularis gigantocellularis in the time-dependent changes in descending inhibition after inflammation. *Pain, 97*, 171–181.

Terman, G. W., Shavit, Y., Lewis, J. W., Cannon, J. T., & Liebeskind, J. C. (1984). Intrinsic mechanisms of pain inhibition: Activation by stress. *Science, 226*, 1270–1277.

Thomas, D. A., McGowan, M. K., & Hammond, D. L. (1995). Microinjection of baclofen in the ventromedial medulla of rats: Antinociception at low does and hyperalgesia at high doses. *The Journal of Pharmacology and Experimental Therapeutics, 275*, 274–284.

Thompson, R. C., Mansour, A., & Akil, W. S. J. (1993). Cloning and pharmacological characterization of a rat mu opioid receptor. *Neuron, 11*, 903–913.

Tsou, K., & Jang, C. S. (1964). Studies on the site of analgesic action of morphine by intracerebral microinjections. *Scientia Sinica, 13*, 1099–1109.

Tsuruoka, M., & Willis, W. D. (1996a). Bilateral lesions in the area of the nucleus locus coeruleus affect the development of hyperalgesia during carrageenan-induced inflammation. *Brain Research, 726*, 233–236.

Tsuruoka, M., & Willis, W. D. (1996b). Descending modulation from the region of the locus coeruleus or nociceptive sensitivity in a rat model of inflammatory hyperalgesia. *Brain Research, 743*, 86–92.

Urban, M. O., & Gebhart, G. F. (1999). Supraspinal contributions to hyperalgesia. *Proceedings of the National Academy Science of the United States of America, 96*, 7687–7692.

Urban, M. O., Coutinho, S. V., & Gebhart, G. F. (1999). Biphasic modulation of visceral nociception by neurotensin in rat rostral ventromedial medulla. *The Journal of Pharmacology and Experimental Therapeutics, 290*, 207–213.

Wei, F., Guo, W., Zou, S., Ren, K., & Dubner, R. (2008). Supraspinal glial-neuronal interactions contribute to descending pain facilitation. *Journal of Neuroscience, 28*, 10482–10495.

Wei, F., Ren, K., & Dubner, R. (1998). Inflammation-induced Fos protein expression in the rat spinal cord is enhanced following dorsolateral or ventrolateral funiculus lesions. *Brain Research, 782*, 136–141.

Wei, F., Dubner, R., & Ren, K. (1999a). Nucleus reticularis gigantocellularis and nucleus raphe magnus in the brain stem exert opposite effects on behavioral hyperalgesia and spinal Fos protein expression after peripheral inflammation. *Pain, 80*, 127–141.

Wei, F., Dubner, R., & Ren, K. (1999b). Laminar-selective noradrenergic and serotoninergic modulation includes spinoparabrachial cells after inflammation. *NeuroReport, 10*, 1757–1761.

Wiertelak, E. P., Roemer, B., Maier, S. F., & Watkins, L. R. (1997). Comparison of the effects of nucleus tractus solitarius and ventral medial medulla lesions on illness-induced and subcutaneous formalin-induced hyperalgesias. *Brain Research, 748*, 143–150.

Woolf, C. J., & Salter, M. W. (2000). Neuronal plasticity: Increasing the gain in pain. *Science, 288*, 1765–1768.

Xu, M., Kim, C. J., Neubert, M. J., & Heinricher, M. M. (2007). NMDA receptor-mediated activation of medullary pro-nociceptive neurons is required for secondary thermal hyperalgesia. *Pain, 127*, 253–262.

Zeitz, K. P., Guy, N., Malmberg, A. B., Dirajlal, S., Martin, W. J., Sun, L., Bonhaus, D. W., Stucky, C. L., Julius, D., & Basbaum, A. I. (2002).

The 5-HT3 subtype of serotonin receptor contributes to nociceptive processing via a novel subset of myelinated and unmyelinated nociceptors. *Journal of Neuroscience, 22*, 1010–1019.

Zhang, L., & Hammond, D. L. (2010). Cellular basis for opioid potentiation in the rostral ventromedial medulla of rats with persistent inflammatory nociception. *Pain, 149*, 107–116.

Zhuo, M., & Gebhart, G. F. (1991). Spinal serotonin receptors mediate descending facilitation of a nociceptive reflex from the nuclei reticularis gigantocellularis and gigantocellularis pars alpha in the rat. *Brain Research, 550*, 35–48.

Zhuo, M., & Gebhart, G. F. (1992). Characterization of descending facilitation and inhibition of spinal nociceptive transmission from the nuclei reticularis gigantocellularis and gigantocellularis pars alpha in the rat. *Journal of Neurophysiology, 67*, 1599–1614.

Zhuo, M., Sengupta, J. N., & Gebhart, G. F. (2002). Biphasic modulation of spinal visceral nociceptive transmission from the rostroventral medial medulla in the rat. *Journal of Neurophysiology, 87*, 2225–2236.

Descending Modulation of Nociceptive Transmission During Persistent Damage to Peripheral Tissues

Horacio Vanegas, Enrique Vazquez and Victor Tortorici
Instituto Venezolano de Investigaciones Cientificas (IVIC), Caracas, Venezuela

Synonyms

Descending control of hyperalgesia and allodynia; Descending pain control; Endogenous pain control; PAG; RVM

Definition

Brain stem structures exert inhibitory or facilitatory influences upon the spinal processing and transmission of pain messages that arise from inflamed peripheral tissues or damaged peripheral nerves. Animal models show that persistent nociception induces changes in the activity of brain stem structures and these in turn facilitate or inhibit spinal circuits responsible for hyperalgesia and allodynia.

Characteristics

Transmission of pain messages in the spinal cord is modulated by the descending pain-control system (Fields 2004; Heinricher et al. 2008; Porreca et al. 2002; Vanegas and Schaible 2004). In its simplest form, this system is pictured as follows: The gray matter around the Sylvian aqueduct (periaqueductal gray, PAG) collects information from vast areas of the central nervous system. PAG neurons project to the rostral ventromedial medulla (RVM), which includes the nucleus raphe magnus and other reticular nuclei. RVM neurons project to the dorsal horn of spinal cord and trigeminal nuclei. At these targets, the so-called on-cells of RVM increase, and the so-called off cells decrease, nociceptive transmission along segmental reflex circuits and along pathways that ascend towards the brain stem, the thalamus, and the cerebral cortex.

In animal models of inflammation or nerve damage, application of noxious stimuli to the affected body part elicits exaggerated nociceptive responses (primary hyperalgesia), and even previously innocuous stimuli can now elicit such aversive responses (primary allodynia). Also, stimulation of healthy body parts near the affected areas may elicit exaggerated responses (secondary hyperalgesia and allodynia). The descending pain-control system is modified by peripheral inflammation or nerve damage and in turn increases or decreases hyperalgesia and allodynia (Fields 2004; Heinricher et al. 2008; Porreca et al. 2002; Vanegas and Schaible 2004). Importantly, exogenous opiate and non-opiate analgesic drugs act upon PAG and RVM by imitating, or interacting with, endogenous opioids and endocannabinoids, and this contributes to their analgesic action during peripheral damage (Escobar et al. 2011; Vanegas and Tortorici 2002).

The present review mostly cites studies published after 2004. Citations for older studies can be found in Vanegas and Schaible (2004).

Inflammation of Peripheral Tissues

One model of inflammatory pain in rats involves the injection of antigenic substances, like complete Freund's adjuvant or carrageenan, into a hind paw or into an articular cavity. Another model consists of exciting C-fibers with irritants like capsaicin, formalin, or mustard oil and thus mimicking persistent peripheral damage (Vanegas and Schaible 2004). Acidic saline injections can be employed to induce muscle damage (Tillu et al. 2008).

Normally, there is a tonic descending inhibition of spinal nociceptive transmission. When inflammation of peripheral tissues begins, an increased impulse flow reaches the spinal cord (Vanegas and Schaible 2004). Here the neuronal pools that receive these impulses relay in turn an elevated number of impulses towards segmental and extrasegmental reflex circuits as well as towards brain stem, thalamic, and cortical targets. At the same time, the primary neuronal pools become hyperexcitable (central sensitization), and primary and secondary hyperalgesia set in. Normal tonic descending inhibition may hinder this process, and, also the hyperexcitable neuronal pools become more sensitive to descending noradrenergic and local opioidergic inhibition (Vanegas and Schaible 2004). Both factors tend to dampen the ensuing hyperalgesia. Particularly important are α_2-adrenergic receptors in primary afferent endings as well as intrinsic spinal neurons (Pertovaara 2006). The increased flow of ascending nociceptive impulses induces time-dependent changes in the inhibitory and facilitatory influences which simultaneously descend from RVM. In primary hyperalgesia, descending inhibition seems to predominate over facilitation, so that the resulting pain is not as intense as it could be (Vanegas and Schaible 2004). This is due to an increase in enkephalin synthesis in RVM and PAG as well as an increase in the expression of both NMDA and AMPA receptors in RVM, which results in an elevated opioidergic and glutamatergic drive on descending inhibitory mechanisms (Hurley and Hammond 2001; Ren and Dubner 2002; Schepers et al. 2008). Simultaneously, an increase in the activation of NMDA, AMPA, neurotensin, and cholecystokinin (CCK) receptors in RVM causes descending facilitation of secondary hyperalgesia (Ambriz-Tututi et al. 2011; Urban and Gebhart 1999; Vanegas and Schaible 2004). CCK and neurotensin are known to excite on-cells (Heinricher and Neubert 2004; Neubert et al. 2004), and secondary hyperalgesia requires an NMDA-dependent increase in the activity of on-cells (Kincaid et al. 2006; Xu et al. 2007).

It has recently been found that neurons in the PAG that express brain-derived neurotrophic factor (BDNF) project to RVM where, upon peripheral inflammation, they release BDNF, which activates its TrkB receptor. This leads to phosphorylation of the NR2 subunit of the NMDA receptor via IP$_3$, PKC, and Src and to descending facilitation of inflammatory hyperalgesia (Guo et al. 2006) (see below). Likewise, an upregulation of the NR1 subunit in RVM may contribute to hyperalgesia in muscle models of inflammatory pain (Da Silva et al. 2010). Descending facilitation of inflammatory hyperalgesia also depends on the activation of microglia and astroglia in RVM as well as on the phosphorylation of p38 MAPK in microglia and neurons (Roberts et al. 2009).

During peripheral inflammation, also activation of substance P's NK1 receptors in RVM contributes to hyperalgesia (Hamity et al. 2010), to the facilitation of on-cell excitability (Brink et al. 2012), and to the enhancement of NMDA-dependent excitation of spinally projecting non-serotonergic neurons (Zhang and Hammond 2009). Activation of spinal 5-HT$_3$ receptors by serotonergic projections from RVM is nevertheless heavily involved in hyperalgesia. Selective molecular depletion of serotonin in RVM neurons prevents the hyperalgesia induced by intra-RVM microinjection of BDNF or by peripheral inflammation or formalin injection (Gu et al. 2011; Wei et al. 2010). Activation of spinal 5-HT$_3$ receptors leads to spinal glial hyperactivity, neuronal hyperexcitability, and nociceptive hypersensitivity (Gu et al. 2011). This involves neuron-to-microglia signaling via the chemokine fractalkine, microglia-to-astrocyte signaling via interleukin-18, astrocyte-to-neuron signaling via interleukin-1β (IL-1β), and enhanced activation of NMDA receptors.

An interesting aspect of descending modulation during peripheral inflammation is the action of non-opioid analgesics upon the PAG. Aspirin and metamizole (dipyrone) are cyclooxygenase (COX) inhibitors and have an analgesic action not only on peripheral tissues but also at the PAG (Vanegas and Tortorici 2002, 2007). Microinjection of prostaglandin E_2 in PAG leads to activation of on-cells and depression of off-cells in RVM and thus to enhanced pain sensitivity (Heinricher et al. 2004). COX-1, COX-2, and prostaglandins are tonically active in the PAG (Breder et al. 1992). Aspirin and metamizole inhibit prostaglandin synthesis and thus, when acting upon the PAG, increase in RVM the activity of off-cells and decrease the activity of on-cells (Vanegas and Tortorici 2002) consequently leading to a reduction of the inflammation-induced hyperexcitability of spinal nociceptive neurons and spinal nocifensive reflexes (Vazquez et al. 2007). Apparently, the PAG output neurons responsible for descending inhibition of nociception are tonically inhibited by local GABAergic neurons (Moreau and Fields 1986; Park et al. 2010). Opioids directly increase the availability of arachidonic acid, and COX inhibitors do the same because they decrease the transformation of arachidonic acid into prostaglandins (Vaughan et al. 1997). Arachidonic acid derivatives and cannabinoids inhibit GABA release from presynaptic terminals (Vaughan 1998; Vaughan et al. 2000). Then, a depression of GABAergic synapses by opioids, COX inhibitors, or cannabinoids would free PAG output neurons from inhibition and thus trigger descending antinociception via the RVM neurons. Interestingly, COX inhibitors act in the PAG not only by inhibiting prostaglandin synthesis but also by way of, or in synergy with, endogenous opioids and endocannabinoids (Escobar et al. 2011; Vanegas and Tortorici 2002).

In sum, peripheral inflammation causes peripheral sensitization of nociceptors and central sensitization of spinal and trigeminal neurons, thus increasing the reflex and ascending nociceptive traffic. This in turn elicits a variety of interrelated changes in PAG and RVM, including activation of neuroglia; enhanced activity of BDNF, neurotensin, CCK, NK1, opioid, and glutamate receptors; and increase in descending inhibition and facilitation of nociception. Descending inhibition may predominate on primary hyperalgesia, probably due to an enhanced spinal α_2-adrenergic and opioidergic activity. Otherwise, descending facilitation predominates due to increased activity of RVM on-cells and serotonergic neurons. Activation of spinal 5-HT$_3$ receptors is critical for hyperactivity of spinal glia and neurons and for behavioral hyperalgesia. Endogenous and exogenous opioids and cannabinoids as well as COX-inhibiting drugs converge at the PAG to induce descending inhibition of nociception.

Damage to Peripheral Nerves

The role of the descending pain-control system has also been studied in animal models of nerve damage. These include ligation of the infraorbital nerve, spinal nerves L_5–L_6 or S_1 S_3, and various types of lesion of the sciatic nerve, all of which cause hyperalgesia and allodynia of the corresponding body part.

Making a lesion on the sciatic nerve elicits an immediate activation of spinal nociceptive neurons and, consequently, of RVM on-cells, which in turn facilitate the activity of spinal nociceptive neurons. This spinal-RVM-spinal positive feedback event contributes to the later development of the spinal neuronal hyperexcitability that underlies neuropathic pain (Sanoja et al. 2008; Vanegas and Schaible 2004; Wei and Pertovaara 1999). During the first few days after nerve damage, plasticity at both the damaged and neighboring nerve fibers as well as central sensitization at the spinal cord and trigeminal nuclei triggers the development of primary hyperalgesia and allodynia. Potentially, these disturbances may subside after a few days, but descending influences from RVM maintain the neuropathic syndrome for several weeks (Vanegas and Schaible 2004). Indeed, the primary neuronal pool at the spinal dorsal horn is under abnormal bombardment from peripheral fibers and sends exaggerated ascending impulses which eventually induce changes in PAG and RVM.

In the PAG, an activation of the astroglia can be detected some days after nerve lesion (Mor et al. 2010). Critical for the development of hyperalgesia and allodynia is the activation in the RVM of microglia in the early post-lesion days and astroglia thereafter, an elevation of tumor necrosis factor-α (TNF) and IL-1β (Wei et al. 2008) and, consequently, a phosphorylation of the NR1 subunit of NMDA receptors in neurons that express the receptors for TNF and IL-1β.

RVM neurons that express both μ-opioid and CCK receptors are essential for neuropathic pain (Vanegas and Schaible 2004; Zhang et al. 2009), and these neurons possess the properties of the electrophysiologically and pharmacologically defined on-cells (Heinricher and Neubert 2004). The ongoing and evoked activity of on-cells is elevated, among other factors, by the increased excitatory drive that ascends from spinal nociceptive neurons and the activation of their CCK receptors by increased CCK concentrations (Carlson et al. 2007; Goncalves et al. 2007; Marshall et al. 2012). Concomitantly, there is a decrease in off-cell activity. These and other mechanisms (see below) result in net facilitatory influences onto the spinal dorsal horn and thus close a spinal-RVM-spinal feedback circuit. This circuit is broken, and hyperalgesia and allodynia thus diminish, if the RVM is locally anesthetized, the neuroglia and the cytokines are blocked, the CCK receptors are antagonized, or the on-cells are destroyed with neurotoxins. Interestingly, the role of RVM neurons that express both μ-opioid and CCK receptors is limited to maintaining neuropathic pain (Vanegas and Schaible 2004; Zhang et al. 2009). Indeed, except for their plausible influence at the time of nerve lesion (Sanoja et al. 2008; Wei and Pertovaara 1999), activity of these neurons is not critical during the first few days of neuropathic pain development.

Interestingly, together with descending facilitation, during peripheral nerve damage, there are descending influences from PAG and RVM that hinder nociceptive responses. Descending facilitation is generally predominant, but descending inhibition of nociception may be found by means of clever experimental manipulations (De Felice et al. 2011) or outside the primary area of neuropathy (Vanegas and Schaible 2004).

Descending modulation of nociceptive transmission is ultimately exerted at the central terminals of nociceptive primary afferents and at the nociceptive neuronal circuits of the spinal cord and the trigeminal nuclei. One mechanism of spinal hyperexcitability during nerve damage is a loss of the neuronal inhibition normally caused by descending activation of α2-adrenoceptors (Rahman et al. 2008). Other mechanisms involve a direct increase in neuronal excitability. For example, the CCK-induced activation of RVM neurons causes an elevation of spinal COX activity and prostaglandin E2 release (Marshall et al. 2012). The CCK-induced activation also causes an elevation of spinal serotonin which, via de 5-HT3 receptor, contributes to hyperalgesia. Activation of spinal 5-HT3 receptors by serotonin from RVM neurons is a fundamental mechanism of descending facilitation during tissue or nerve damage (Wei et al. 2010). As mentioned above, activation of spinal 5-HT3 receptors leads to spinal glial hyperactivity, neuronal hyperexcitability, and nociceptive hypersensitivity as a result of glial-neuronal reciprocal interactions (Gu et al. 2011). 5-HT3 receptors are expressed at the spinal terminals of nociceptive primary afferents. These terminals also express the voltage-dependent calcium channels (VDCCs) involved in presynaptic neurotransmitter release. There is evidence that this relationship favors neurotransmitter release under descending serotonergic influences. The α2δ subunit of VDCCs is upregulated during nerve damage and is the molecular target of gabapentin and pregabalin. These gabapentinoids would thus alleviate neuropathic pain by depressing the enhanced spinal release of excitatory neurotransmitters induced by descending serotonergic influences during nerve damage (Suzuki et al. 2005).

In sum, damage to peripheral nociceptive fibers triggers an immediate activation of RVM on-cells that is important for the later development of neuropathic pain. Subsequent activation of RVM neuroglia, elevation of TNF and IL-1β, activation of on-cells by CCK, and phosphorylation of NMDA receptors in neurons are critical for descending facilitation of hyperalgesia and allodynia. A continued hyperactivity of presumed on-cells is not essential during the first few post-lesion days but is necessary for the subsequent maintenance of neuropathic pain. At the spinal cord level, hyperexcitability may be due to a loss of descending inhibition mediated by α_2-adrenergic receptors. On the other hand, active descending facilitation from RVM is caused by PGE_2 and 5-HT release. Activation of spinal 5-HT$_3$ receptors leads to glial and neuronal reciprocal excitation. Also, 5-HT$_3$ receptors are expressed together with VDCCs on the synaptic terminals of nociceptive afferents. An upregulation of the $\alpha_2\delta$ subunit of VDCCs converges with an overactivation of 5-HT$_3$ receptors by descending axons to increase spinal excitability and thus neuropathic pain.

References

Ambriz-Tututi, M., Cruz, S. L., Urquiza-Marin, H., & Granados-Soto, V. (2011). Formalin-induced long-term allodynia and hyperalgesia are maintained by descending facilitation. *Pharmacology Biochemistry and Behavior, 98,* 417–424.

Breder, C. D., Smith, W. L., Raz, A., Masferrer, J., Seibert, K., Needleman, P., & Saper, C. B. (1992). Distribution and characterization of cyclooxygenase immunoreactivity in the ovine brain. *The Journal of Comparative Neurology, 322,* 409–438.

Brink, T. S., Pacharinsak, C., Khasabov, S. G., Beitz, A. J., & Simone, D. A. (2012). Differential modulation of neurons in the rostral ventromedial medulla by neurokinin-1 receptors. *Journal of Neurophysiology, 107,* 1210–1221.

Carlson, J. D., Maire, J. J., Martenson, M. E., & Heinricher, M. M. (2007). Sensitization of pain-modulating neurons in the rostral ventromedial medulla after peripheral nerve injury. *Journal of Neuroscience, 27,* 13222–13231.

Da Silva, L. F. S., Walder, R. Y., Davidson, B. L., Wilson, S. P., & Sluka, K. A. (2010). Changes in expression of NMDA-NR1 receptor subunits in the rostral ventromedial medulla modulate pain behaviors. *Pain, 151,* 155–161.

De Felice, M., Sanoja, R., Wang, R., Vera-Portocarrero, L., Oyarzo, J., King, T., Ossipov, M. H., Vanderah, T. W., Lai, J., Dussor, G. O., Fields, H. L., Price, T. J., & Porreca, F. (2011). Engagement of descending inhibition from the rostral ventromedial medulla protects against chronic neuropathic pain. *Pain, 152,* 2701–2709.

Escobar, W., Ramirez, K., Avila, C., Limongi, R., Vanegas, H., & Vazquez, E. (2011). Metamizol, a non-opioid analgesic, acts via endocannabinoids in the PAG-RVM axis during inflammation in rats. *European Journal of Pain, 16*(5), 676–689. in press.

Fields, H. (2004). State-dependent opioid control of pain. *Nature Reviews. Neuroscience, 5,* 565–575.

Goncalves, L., Almeida, A., & Pertovaara, A. (2007). Pronociceptive changes in response properties of rostroventromedial medullary neurons in a rat model of neuropathy. *European Journal of Neuroscience, 26,* 2188–2195.

Gu, M., Miyoshi, K., Dubner, R., Guo, W., Zou, S., Ren, K., Noguchi, K., & Wei, F. (2011). Spinal 5-HT$_3$ receptor activation induces behavioral hypersensitivity via a neuronal-glial-neuronal signaling cascade. *Journal of Neuroscience, 31,* 12823–12836.

Guo, W., Robbins, M. T., Wei, F., Zou, S., Dubner, R., & Ren, K. (2006). Supraspinal brain-derived neurotrophic factor signaling: A novel mechanism for descending pain facilitation. *Journal of Neuroscience, 26,* 126–137.

Hamity, M. V., White, S. R., & Hammond, D. L. (2010). Effects of neurokinin-1 receptor agonism and antagonism in the rostral ventromedial medulla of rats with acute or persistent inflammatory nociception. *Neuroscience, 165,* 902–913.

Heinricher, M. M., & Neubert, M. J. (2004). Neural basis for the hyperalgesic action of cholecystokinin in the rostral ventromedial medulla. *Journal of Neurophysiology, 92,* 1982–1989.

Heinricher, M. M., Martenson, M. E., & Neubert, M. J. (2004). Prostaglandin E_2 in the midbrain periaqueductal gray produces hyperalgesia and activates pain-modulating circuitry in the rostral ventromedial medulla. *Pain, 110,* 419–426.

Heinricher, M. M., Tavares, I., Leith, J. L., & Lumb, B. M. (2008). Descending control of nociception: Specificity, recruitment and plasticity. *Brain Research Reviews, 60,* 214–225.

Hurley, R. W., & Hammond, D. L. (2001). Contribution of endogenous enkephalins to the enhanced analgesic effects of supraspinal m opioid receptor agonists after inflammatory injury. *Journal of Neuroscience, 21,* 2536–2545.

Kincaid, W., Neubert, M. J., Xu, M., Kim, C. J., & Heinricher, M. M. (2006). Role for medullary pain facilitating neurons in secondary thermal hyperalgesia. *Journal of Neurophysiology, 95,* 33–41.

Marshall, T. M., Herman, D. S., Largent-Milnes, T. M., Badghisi, H., Zuber, K., Holt, S. C., Lai, J., Porreca, F., & Vanderah, T. W. (2012). Activation of descending pain-facilitatory pathways from the rostral ventromedial medulla by cholecystokinin elicits release of prostaglandin-E_2 in the spinal cord. *Pain, 153,* 86–94.

Mor, D., Bembrick, A. L., Austin, P. J., Wyllie, P. M., Creber, N. J., Denyer, G. S., & Keay, K. A. (2010). Anatomically specific patterns of glial activation in the periaqueductal gray of the sub-population of rats showing pain and disability following chronic constriction injury of the sciatic nerve. *Neuroscience, 166*, 1167–1184.

Moreau, J.-L., & Fields, H. L. (1986). Evidence for GABA involvement in midbrain control of medullary neurons that modulate nociceptive transmission. *Brain Research, 397*, 37–46.

Neubert, M. J., Kincaid, W., & Heinricher, M. M. (2004). Nociceptive facilitating neurons in the rostral ventromedial medulla. *Pain, 110*, 158–165.

Park, C., Kim, J. H., Yoon, B. E., Choi, E. J., Lee, C. J., & Shin, H. S. (2010). T-type channels control the opioidergic descending analgesia at the low threshold-spiking GABAergic neurons in the periaqueductal gray. *Proceedings of the National Academy of Sciences of the United States of America, 107*, 14857–14862.

Pertovaara, A. (2006). Noradrenergic pain modulation. *Progress in Neurobiology, 80*, 53–83.

Porreca, F., Ossipov, M. H., & Gebhart, G. F. (2002). Chronic pain and medullary descending facilitation. *Trends in Neurosciences, 25*, 319–325.

Rahman, W., D'Mello, R., & Dickenson, A. H. (2008). Peripheral nerve injury-induced changes in spinal a_2-adrenoceptor-mediated modulation of mechanically evoked dorsal horn neuronal responses. *The Journal of Pain, 9*, 350–359.

Ren, K., & Dubner, R. (2002). Descending modulation in persistent pain: An update. *Pain, 100*, 1–6.

Roberts, J. M., Ossipov, M. H., & Porreca, F. (2009). Glial activation in the rostroventromedial medulla promotes descending facilitation to mediate inflammatory hypersensitivity. *European Journal of Neuroscience, 30*, 229–241.

Sanoja, R., Vanegas, H., & Tortorici, V. (2008). Critical role of the rostral ventromedial medulla in early spinal events leading to chronic constriction injury neuropathy in rats. *The Journal of Pain, 9*, 532–542.

Schepers, R. J.-F., Mahoney, J. L., & Shippenberg, T. S. (2008). Inflammation-induced changes in rostral ventromedial medulla mu and kappa opioid receptor mediated antinociception. *Pain, 136*, 320–330.

Suzuki, R., Rahman, W., Rygh, L. J., Webber, M., Hunt, S. P., & Dickenson, A. H. (2005). Spinal-supraspinal serotonergic circuits regulating neuropathic pain and its treatment with gabapentin. *Pain, 117*, 292–303.

Tillu, D. V., Gebhart, G. F., & Sluka, K. A. (2008). Descending facilitatory pathways from the RVM initiate and maintain bilateral hyperalgesia after muscle insult. *Pain, 135*, 331–339.

Urban, M. O., & Gebhart, G. F. (1999). Supraspinal contributions to hyperalgesia. *Proceedings of the National Academy of Sciences of the United States of America, 96*, 7687–7692.

Vanegas, H., & Schaible, H.-G. (2004). Descending control of persistent pain: Inhibitory or facilitatory? *Brain Research Reviews, 46*, 295–309.

Vanegas, H., & Tortorici, V. (2002). Opioidergic effects of non-opioid analgesics on the central nervous system. *Cellular and Molecular Neurobiology, 22*, 655–661.

Vanegas, H., & Tortorici, V. (2007). The periaqueductal gray as critical site for antinociception and tolerance induced by non-steroidal anti-inflammatory drugs. In S. Maione & V. Di Marzo (Eds.), *Neurotransmission in the antinociceptive descending pathway* (pp. 69–80). Trivandrum: Research Signpost.

Vaughan, C. W. (1998). Enhancement of opioid inhibition of GABAergic synaptic transmission by cyclooxygenase inhibitors in rat periaqueductal grey neurones. *British Journal of Pharmacology, 123*, 1479–1481.

Vaughan, C. W., Ingram, S. L., Connor, M. A., & Christie, M. J. (1997). How opioids inhibit GABA-mediated neurotransmission. *Nature, 390*, 611–614.

Vaughan, C. W., Connor, M., Bagley, E. E., & Christie, M. J. (2000). Actions of cannabinoids on membrane properties and synaptic transmission in rat periaqueductal gray neurons in vitro. *Molecular Pharmacology, 57*, 288–295.

Vazquez, E., Escobar, W., Ramirez, K., & Vanegas, H. (2007). A nonopioid analgesic acts upon the PAG-RVM axis to reverse inflammatory hyperalgesia. *European Journal of Neuroscience, 25*, 471–479.

Wei, H., & Pertovaara, A. (1999). MK-801, an NMDA receptor antagonist, in the rostroventral medulla attenuates development of neuropathic symptoms in the rat. *NeuroReport, 10*, 2933–2937.

Wei, F., Guo, W., Zou, S., Ren, K., & Dubner, R. (2008). Supraspinal glial-neuronal interactions contribute to descending pain facilitation. *Journal of Neuroscience, 28*, 10482–10495.

Wei, F., Dubner, R., Zou, S., Ren, K., Bai, G., Wei, D., & Guo, W. (2010). Molecular depletion of descending serotonin unmasks its novel facilitatory role in the development of persistent pain. *Journal of Neuroscience, 30*, 8624–8636.

Xu, M., Kim, C. J., Neubert, M. J., & Heinricher, M. M. (2007). NMDA receptor-mediated activation of medullary pro-nociceptive neurons is required for secondary thermal hyperalgesia. *Pain, 127*, 253–262.

Zhang, L., & Hammond, D. L. (2009). Substance P enhances excitatory synaptic transmission on spinally projecting neurons in the rostral ventromedial medulla after inflammatory injury. *Journal of Neurophysiology, 102*, 1139–1151.

Zhang, W., Gardell, S. E., Zhang, D., Xie, J. Y., Agnes, R. S., Badghisi, H., Hruby, V. J., Rance, N., Ossipov, M. H., Vanderah, T. W., Porreca, F., & Lai, J. (2009). Neuropathic pain is maintained by brainstem neurons co-expressing opioid and cholecystokinin receptors. *Brain, 132*, 778–787.

Descending Modulation of Pain

Definition

Descending pathways originating in the brain modulate pain, from spinal/trigeminal nociceptive transmission at the neuronal level to the integrated interpretation of and reaction to pain. Modulation can be either inhibitory or facilitatory, meaning that pain can either be attenuated or made worse. The modulation exerted is influenced by the presence/absence and interpretation of the nature of tissue injury, circumstances in which the pain arises, expectations, attention and motivations as well and cognitive factors.

Cross-References

▶ Descending Modulation and Persistent Pain
▶ Encoding of Noxious Information in the Spinal Cord
▶ Somatic Pain

Descending Modulation of Visceral Pain

Elie D. Al-Chaer
Center for Pain Research, University of Arkansas for Medical Sciences, Little Rock, AR, USA

Synonyms

Descending facilitation; Descending inhibition; Functional abdominal pain; IBS; Irritable bowel syndrome; Stress; Visceral hyperalgesia; Visceral hypersensitivity

Definition

Spinal visceral nociceptive transmission is subject to descending modulatory influences from supraspinal structures (e.g., ▶ Periaqueductal Gray [PAG], nucleus raphe magnus [NRM], nuclei reticularis gigantocellularis [NGC], and the ventrobasal complex of the thalamus [VBC]). This descending modulation can be either inhibitory, facilitatory, or both, depending on the context of the visceral stimulus or the intensity of the descending stimulation. The descending pathways originating in the brainstem and other cerebral structures play an important role in the modulation and integration of visceroceptive messages in the dorsal horn. Serotonergic, noradrenergic, and to a lesser extent dopaminergic networks comprise of major components of these descending mechanisms.

Introduction

Pain is continuously modulated by inhibitory and facilitatory mechanisms that may originate at any level of the nervous system. These can originate in the periphery, at the level of the spinal cord, within the brainstem, or in the brain and contribute separately or in concert to direct or indirect pain inhibition or facilitation. Pain modulation by descending inputs from the brain and brainstem has been observed both in humans and animals; however, our understanding of the mechanisms of descending controls originates largely from animal experiments.

Characteristics

Descending modulation of spinal visceral nociceptive processing is tonically active and includes both inhibitory and facilitatory influences. Correlative behavioral and neuronal studies have demonstrated that descending modulation influences behavior, physiology, and neuronal activity at spinal cord levels in acute and persistent pain; this modulation includes descending facilitatory as well as inhibitory influences that may impact on pain of visceral origin (Dubner and Ren 1999). Electrical stimulation or chemical activation of a number of supraspinal centers can modulate the neuronal and behavioral responses to visceral stimuli. For example, responses to an intraperitoneal injection of hypertonic saline in rats are attenuated by electrical stimulation in the PAG (Giesler and Liebskind 1976). Chemical activation of cell bodies in the rostroventral medulla (RVM) attenuated responses to visceral stimulation (Coutinho et al. 1998; Urban et al. 1999). Electrical stimulation or glutamate microinjection into the PAG, NRM, or NGC inhibit spinal dorsal horn neuron responses (including spinoreticular and spinothalamic tract neurons) to visceral afferent fiber stimulation or noxious ► colorectal distension (CRD) (Ammons et al. 1984; Cervero et al. 1985; Ness and Gebhart 1987). The descending inhibitory effects from the RVM have been shown to be primarily mediated by pathways traveling in the dorsolateral spinal cord (Zhuo and Gebhart 2002).

Similarly, electrical stimulation of the ventrobasal complex of the thalamus (VBC) inhibits the responses of dorsal horn neurons to colorectal distension in normal rats, an effect attenuated in part by dorsal column (DC) input. However, thalamic descending inhibition proved to be context specific. For example, in chronically sensitized adult rats that received neonatal colon irritation, thalamic stimulation caused a facilitation of neuronal responses to colorectal distension in more than 60 % of the neurons isolated. This facilitation was observed regardless of the intensity of electrical stimulation and was facilitated by DC input. Inhibition in response to thalamic stimulation was seen in less than 40 % of the neurons in chronically sensitized rats. This inhibition appeared to be controlled by DC input. This goes to show that descending pathways do not exclusively exert inhibitory actions in the dorsal horn. Indeed, individual transmitters may exert multiple actions in the dorsal horn as a function of the type of neuron targeted and the receptor activated (Park and Al-Chaer 2002; Saab et al. 2004).

Nevertheless, studies have revealed evidence of tonic descending inhibition under normal circumstances, whose removal produced increases in spontaneous activity and/or responses of spinal neurons to visceral stimuli. For example, reversible cold block of the cervical spinal cord

revealed that spinal viscerosomatic neurons are under tonic descending inhibitory influences from supraspinal structures (Akeyson et al. 1990; Cervero 1983; Tattersall et al. 1986).

On the other hand, a number of studies have shown that spinal visceral input is subject to facilitatory modulation, providing the basis of a mechanism that could enhance visceral perceptions in the absence of noxious visceral stimulation, which would explain in part the visceral hypersensitivity, often observed in cases of ► functional bowel disorders and irritable bowel syndrome. Tonic descending facilitatory influences have been documented. For example, 44 % of viscerosomatic neurons, recorded from thoracic segments of the cat spinal cord, gave no responses to splanchnic nerve stimulation during cervical cold block (Tattersall et al. 1986). Electrical stimulation in the RVM produced intensity-dependent biphasic modulation (intensity-dependent inhibition or facilitation) of the visceromotor response (VMR) to colorectal distention. Activation of glutamatergic receptors in the RVM also facilitated or inhibited the VMR to colorectal distention. The descending facilitatory effects were shown to be primarily mediated by pathways traveling in the ventral/ventrolateral spinal cord. Importantly, descending modulation, whether inhibitory or facilitatory, was shown to be linked to the stimulus; resting EMG activity was unaffected by either electrical stimulation or glutamate microinjection in the RVM (Zhuo and Gebhart 2002). Furthermore, ablation of putative RVM on-cells revealed a role of these cells in regulating basal visceral pain responses. The descending facilitation mediated by these cells is not necessary for the development of visceral hyperalgesia but rather for a shift in the balance of descending modulation from the brainstem, where enhanced descending facilitatory controls (or a reduction in inhibitory influences) produced by acute visceral stimuli produce a permissive physiological state of central excitability (Sikandar et al. 2012).

In addition to electrical and chemical manipulation of supraspinal centers, stressors – exteroceptive and interoceptive – have been shown to affect the visceral pain experience both in humans and experimental animals. In general, a dual visceral pain response – analgesia or hyperalgesia – is seen to exteroceptive stressors (psychosocial). By contrast, stress-induced visceral hyperalgesia appears to be the predominant outcome in response to interoceptive stressors such as inflammation and infection (Larauche et al. 2011). Regulation of physiological stress involves the hypothalamic-pituitary-adrenal axis; however, the specific pathway for the effect of stress on visceral pain remains the subject of study, given the diversity of targets affected and the variability in the final outcome.

References

Akeyson, E. W., Kneupfer, M. M., & Schramm, L. P. (1990). Splanchnic input to thoracic spinal neurons and its supraspinal modulation in the rat. *Brain Research, 536,* 30–40.

Ammons, W. S., Blair, R. W., & Foreman, R. D. (1984). Raphe magnus inhibition of primate T1-T2 spinothalamic cells with cardiopulmonary visceral input. *Pain, 20,* 247–260.

Cervero, F. (1983). Supraspinal connections of neurons in the thoracic spinal cord of the cat: Ascending projections and effects of descending impulses. *Brain Research, 275,* 251–261.

Cervero, F., Lumb, B. M., & Tattersall, J. E. H. (1985). Supraspinal loops that mediate visceral inputs to thoracic spinal cord neurons in the cat: Involvement of descending pathways from raphe and reticular formation. *Neuroscience Letters, 56,* 189–194.

Coutinho, S. V., Urban, M. O., & Gebhart, G. F. (1998). Role of glutamate receptors and nitric oxide in the rostral ventromedial medulla in visceral hyperalgesia. *Pain, 78,* 59–69.

Dubner, R., & Ren, K. (1999). Endogenous mechanisms of sensory modulation. *Pain,* (Suppl 6), S45–S53.

Giesler, G. J., & Liebskind, J. C. (1976). Inhibition of visceral pain by electrical stimulation in the periaqueductal gray matter. *Pain, 2,* 43–48.

Larauche, M., Mulak, A., & Taché, Y. (2011). Stress and visceral pain: From animal models to clinical therapies. *Experimental Neurology.*

Ness, T. J., & Gebhart, G. F. (1987). Quantitative comparison of inhibition of visceral and cutaneous spinal nociceptive transmission from the midbrain and medulla in the rat. *Journal of Neurophysiology, 58,* 850–865.

Park, Y. C., & Al-Chaer, E. D. (2002). Thalamic stimulation differentially modifies spinal neuronal responses to colorectal distension in rats with chronic visceral pain. *The Journal of Pain, 3*(Suppl 1), 31.

Saab, C. Y., Park, Y. C., & Al-Chaer, E. D. (2004). Thalamic modulation of visceral nociceptive processing in adult rats with neonatal colon irritation. *Brain Research, 1008,* 186–192.

Sikandar, S., Bannister, K., & Dickenson, A. H. (2012). *Neuroscience Letters, 519,* 31–36.

Tattersall, J. E. H., Cervero, F., & Lumb, B. M. (1986). Effects of reversible spinalization on the visceral input to viscerosomatic neurons in the lower thoracic spinal cord of the cat. *Journal of Neurophysiology, 56,* 785–796.

Urban, M. O., Coutinho, S. V., & Gebhart, G. F. (1999). Biphasic modulation of visceral nociception by neurotensin in rat rostral ventromedial medulla. *Journal of Pharmacology and Experimental Therapeutics, 290,* 207–213.

Zhuo, M., & Gebhart, G. F. (2002). Facilitation and attenuation of a visceral nociceptive reflex from the rostroventral medulla in the rat. *Gastroenterology, 122,* 1007–1019.

Descending Noradrenergic Pathways

Definition

Descending noradrenergic projections originating in dorsolateral pontine cell groups modulate spinal nociceptive transmission, inhibiting (see "▶ Descending Inhibitory Noradrenergic Pathways") or facilitating dorsal horn neuronal responses to noxious stimulation. Inhibition is typically mediated by alpha-2 adrenoceptors, whereas facilitation is mediated by alpha-1 adrenoceptors in the spinal dorsal horn.

Cross-References

▶ Nitrous Oxide Antinociception and Opioid Receptors

Descending Pain Control

▶ Descending Modulation of Nociceptive Transmission During Persistent Damage to Peripheral Tissues

Descending Pain Modulation

▶ Descending Modulation of Nociceptive Processing

Desensitization

▶ Nociceptor, Fatigue
▶ Psychology of Pain, Sensitization, Habituation, and Pain

Desensitizing Toothpastes

Definition

Desensitizing toothpastes apply medicaments to dentin that has become exposed and block fluid flow through the dentinal tubules and/or allow the diffusion of salts into the dental pulp that reduces the sensitivity of nociceptors.

Cross-References

▶ Dental Pain, Etiology, Pathogenesis, and Management

Desmin

Definition

The intermediate filament system (mainly desmin) of skeletal muscle is believed to be responsible for the mechanical integration of the myofibrillar lattice in both the longitudinal and radial directions.

Cross-References

▶ Delayed Onset Muscle Soreness

Destructive Metabolism

▶ Postoperative Pain, Pathophysiological Changes in Metabolism in Response to Acute Pain

Detection Threshold

Definition

The minimum amount of stimulus energy necessary to elicit a sensory response, defined as the lowest stimulus energy perceived by the subject in psychophysical experiments.

Cross-References

▶ Hyperaesthesia, Assessment
▶ Hyperpathia, Assessment
▶ Hypoesthesia, Assessment
▶ Pain in Humans, Electrical Stimulation (Skin, Muscle, and Viscera)

Development of Spinothalamic Tract Neurons

Steve Davidson
Pain Center & Department of Anesthesiology, Washington University School of Medicine, MO, USA

Synonyms

Ventrolateral system, ascending spinal projection, development

Definition

The spinothalamic tract (STT) consists of spinal neurons that send axon projections to the thalamus. The number and localization of these neurons in the spinal cord, and their projections to the thalamus, are similar to the adult form a day before birth in the mouse; however, the functionality of these cells during the perinatal period is uncertain.

Characteristics

The STT is part of a larger ventrolateral ascending system and a major transmission route for nociceptive information to the brain. Spinothalamic tract neurons are presumed to derive, like other dorsal horn neurons, from a population of precursor cells that migrate from the ventricular zone of the spinal cord. These neurons begin to differentiate in rats on embryonic day 12 (E12; Altman and Bayer 1984). By day E14, the majority of cell bodies of the neurons that will become the STT are already generated (Beal and Bice 1994; Bice and Deal 1997).

The axons of STT neurons are present in the thalamus at E18 in mice (Davidson et al. 2010). The number of STT neurons present at E18, about 3,500 projecting to each thalamic hemisphere, is not different from the number of STT neurons in adult. At E18, the locations of STT neurons in the spinal cord generally match the adult organization. STT neurons have a significantly smaller diameter than adult STT neurons until 1 week postnatal. Complex arborizations and terminals with boutons can be observed in the thalamus of P5 mouse pups.

It is not known specifically when STT neurons are able to respond to peripheral stimuli, however, unidentified dorsal horn neurons fire action potentials in response to A-fiber stimulation at P3, and to C-fiber stimulation at P10 (Fitzgerald 1988; Jennings and Fitzgerald 1998; Baccei et al. 2003).

References

Altman, J., & Bayer, S. A. (1984). The development of the rat spinal cord. *Advances in Anatomy, Embryology, and Cell Biology, 85*, 1–166.
Baccei, M. L., Bardoni, R., & Fitzgerald, M. (2003). Development of nociceptive synaptic inputs to the neonatal rat dorsal horn: Glutamate release by capsaicin and menthol. *The Journal of Physiology, 549*, 231–242.
Beal, J. A., & Bice, T. N. (1994). Neurogenesis of spinothalamic and spinocerebellar tract neurons in the lumbar spinal cord of the rat. *Brain Research Developmental Brain Research, 78*, 49–56.
Bice, T. N., & Beal, J. A. (1997). Quantitative and neurogenic analysis of neurons with supraspinal projections in the superficial dorsal horn of the rat lumbar spinal cord. *The Journal of Comparative Neurology, 388*, 565–574.
Davidson, S., Truong, H., & Giesler, G. J., Jr. (2010). Quantitative analysis of spinothalamic tract neurons in adult and developing mouse. *The Journal of Comparative Neurology, 518*, 3193–3204.
Fitzgerald, M. (1988). The development of activity evoked by fine diameter cutaneous fibres in the spinal cord of the newborn rat. *Neuroscience Letters, 86*, 161–166.
Jennings, E., & Fitzgerald, M. (1998). Postnatal changes in responses of rat dorsal horn cells to afferent stimulation: A fibre-induced sensitization. *The Journal of Physiology, 509*, 859–868.

Developmental Level

Definition

A period or phase in the life span. In children, foci include the development of cognition, language, affect, and motor skills.

Cross-References

▶ Cancer Pain, Assessment in Children

Dexamethasone

Definition

The long-acting parenteral or orally administered steroid dexamethasone prevents the increased expression of COX-2, which is stimulated by bacterial lipopolysaccharide, cytokines, or growth factors.

Cross-References

▶ Acute Pain in Children, Postoperative
▶ Cyclooxygenases in Biology and Disease

Diabetes Mellitus

Definition

The disease due to inadequate secretion of a pancreatic hormone (insulin) resulting in hyperglycemia (abnormally high glucose level in blood).

Cross-References

▶ Toxic Neuropathies

Diabetes Mellitus Type I

Definition

Diabetes brought on by an autoimmune attack on the insulin-producing beta cells of the pancreas. It is more common in the young, and onset of symptoms is usually abrupt. Patients are treated with a lifetime regimen of exogenous insulin.

Cross-References

▶ Diabetic Neuropathies

Diabetes Mellitus Type II

Definition

Diabetes in which cells develop difficulty uptaking insulin from the blood. Also known as adult-onset diabetes (it is more common in older individuals), type II is the most common type of diabetes in the Western world. Obesity is a major risk factor, and onset of symptoms is gradual.

Cross-References

▶ Diabetic Neuropathies

Diabetic Amyotrophy

▶ Diabetic Neuropathies

Diabetic Autonomic Polyneuropathy

▶ Diabetic Neuropathies

Diabetic Cranial Mononeuropathy

▶ Diabetic Neuropathies

Diabetic Mononeuritis

▶ Diabetic Neuropathies

Diabetic Neuropathies

Hsinlin Thomas Cheng[1] and
Anne Louise Oaklander[2]
[1]Department of Neurology, University of
Michigan, Ann Arbor, MI, USA
[2]Nerve Injury Unit, Departments of Neurology
and Pathology, Massachusetts General Hospital
Harvard Medical School, Boston, MA, USA

Synonyms

Acute painful diabetic neuropathy; Burning feet syndrome; Diabetic amyotrophy; Diabetic autonomic polyneuropathy; Diabetic cranial mononeuropathy; Diabetic mononeuritis; Diabetic sensorimotor polyneuropathy; Diabetic sensory neuropathy; Diabetic symmetrical proximal neuropathy; Diabetic truncal mononeuropathy; DSPN; Hyperglycemic neuropathy; Insulin neuropathy; Mixed forms of diabetic neuropathy; Proximal diabetic neuropathy; Vasculitic neuropathy

Definition

Diabetic neuropathy (DN) refers to several types of damage to various parts of the peripheral nervous system (PNS) caused by the increasingly common diseases type 1 (T1DM) and type 2 diabetes (T2DM). DN can be divided into diffuse and focal forms, which have different mechanisms. The diffuse diabetic polyneuropathies are usually distal-onset, bilaterally symmetric, and progressive. Any type of nerve-fiber, whether myelinated or unmyelinated, somatic or autonomic can be involved. The focal DNs include mononeuropathy, mononeuropathy multiplex, plexopathy, cranial neuropathy, and radiculopathy. These are less common, usually with acute onset, and often improve spontaneously.

Pain is a common symptom of DN and sometimes the onset of pain precedes awareness of the underlying diabetes. Patients with painful diabetic neuropathy (PDN) experience typical features of neuropathic pain including allodynia, hyperalgesia, and lancinating pain. These often are most bothersome at night when they preoccupy attention and can disturb sleep and cause significant distress. PDN is considered a major detriment to quality of life for patients with diabetes (Dworkin et al. 2005, 2007; Jensen et al. 2006).

Characteristics

Epidemiology

Diabetes and related metabolic disorders may be the most common cause of peripheral neuropathies in Western societies. The combined global prevalence of diabetes was estimated at 2.8 % in 2000, but with the rapidly climbing incidence of morbid obesity and type 2 diabetes mellitus (T2DM), prevalence of 4.4 % is predicted by

2030 (Wild et al. 2004). Although these trends are global, the United States has among the highest prevalences of obesity, with 65 % of Americans currently overweight and at least 30 % of them clinically obese (Hedley et al. 2004). In parallel, there are 25.8 million people (8.3 % of the population) in the United States with diabetes as of 2011 (www.cdc.org). Global health expenditure on diabetes was expected to total at least United States dollars (USD) 376 billion or International dollars (ID) 418 billion in 2010, and USD 490 billion or ID 561 billion in 2030 (Zhang et al. 2010).

DN affects about 26–47 % of patients with diabetes (Barrett et al. 2007). In the United States, DN is the leading cause of diabetes-related hospital admissions and non-traumatic amputations (Feldman et al. 1999a, b, 2001). In parallel, up to 27 % of diabetic patients will experience chronic pain from DN (Harris et al. 1993). However, this number is an underestimate as it does not include patients with painful neuropathy whose diabetes remains undiagnosed or those with impaired glucose tolerance (IGT) and impaired fasting glucose (IFG). There is evidence that these conditions alone may suffice to trigger painful DN, as several recent studies suggest that nearly one third of the patients with IGT seek medical attention for a pain syndrome identical to PDN (Singleton et al. 2003, 2005).

The Clinical Spectrum of Diabetic Nerve Damage

In diabetic polyneuropathy, sensory deficits and symptoms usually precede motor nerve dysfunction. These appear first in the distal portions of the extremities and progress proximally in a "stocking-glove" distribution with increasing duration or severity of diabetes. The first symptoms are due to small-fiber damage affecting C and A delta fibers, causing clinical symptoms of paresthesias, dysesthesias, and/or neuropathic pain sometimes accompanied by impairment in pain and temperature perception (numbness). Distal weakness occurs only at a later stage when the large myelinated fibers are affected as well. Diminished or absent deep-tendon reflexes, particularly of the Achilles tendon reflex that is most distal, often indicate mild and otherwise asymptomatic DPN. As DPN progresses, more advanced neuropathy may present with late complications such as ulceration or neuroarthropathy (Charcot's joint) of the foot.

Autonomic neuropathy is another feature of small-fiber polyneuropathy and significant sympathetic or parasympathetic autonomic dysfunction can develop early, presenting with difficulty with sexual function, gastric motility, or sweating. Dysfunction of autonomic innervation of the heart may contribute to excess cardiac disease in diabetics. Patients with diabetic autonomic nerve damage may no longer develop the warning signs of hypoglycemia (e.g., tachycardia, sweating), and thus can have clinically unapparent hypoglycemia that remains unnoticed and untreated until severe consequences such as syncope or seizures ensue. Similarly, loss of perception of anginal chest pain can lead to "silent" myocardial infarctions in diabetics. Although DAN is highly prevalent and associated with a markedly reduced quality of life and increased mortality, it is among the least recognized and most poorly understood complications of diabetes (Freeman 2005). Comprehensive autonomic function testing is the recommended diagnostic test.

The focal diabetic neuropathies reflect damage to single (mononeuropathy) or several individual peripheral nerves (mononeuropathy multiplex), cranial nerves, regions of the brachial or lumbosacral plexuses (plexopathy), or nerve roots (radiculopathy). Except for peripheral nerve mononeuropathy, the focal diabetic neuropathies are relatively uncommon, acute at onset, self-limiting, and tend to occur in older patients. The most common cranial neuropathy affects the oculomotor (III) nerve, producing unilateral headache, diplopia, and ptosis with pupillary sparing (diabetic ophthalmoplegia) (Vinik and Mehrabyan 2004). Isolated ophthalmoplegia from diabetic IV and VI nerve palsy is also not rare. However, other cranial nerves can be affected and diabetes should be tested for in patients presenting with otherwise-unexplained new cranial neuropathy.

Several forms of focal proximal nerve damage affect the spinal segments. Diabetes can cause isolated thoracic level radiculopathies that present with unilateral dermatomal pain and weakness. The differential diagnosis includes *zoster sine herpete* and structural causes including tumor, so such patients should be considered for MRI. Moving distally, diabetes can affect the lumbosacral, or less often the brachial plexus, to cause proximal limb pain and radicular weakness. Electrodiagnostic testing may be needed to localize the lesion and imaging can help exclude other causes of plexopathy including mass lesions.

Individual nerves can also be affected by focal diabetic neuropathy. Femoral neuropathy is well described in older male patients with T2DM. Although both motor and sensory fibers are involved, the typical picture is predominately motor. In contrast, the most common diabetic mononeuropathies, medial and ulnar neuropathy, are indistinguishable from entrapment neuropathies in nondiabetic patients, suggesting that the diabetic nerve has increased susceptibility to compression, perhaps from intraneural edema. Indeed, carpal tunnel syndrome can be demonstrated electrophysiologically in 20–30 % of diabetic patients, and presents as a clinically relevant problem in 5–10 % of patients with diabetes (Dyck et al. 1993).

Diagnosis of Diabetic Polyneuropathy

A consensus statement from the San Antonio Conference on Diabetic Neuropathy recommended that the diagnosis and classification of DPN for research and clinical trials be based on at least one standardized measure from each of the following categories: clinical symptoms, clinical examination, electrodiagnostic testing (EDX), quantitative sensory testing (QST), and autonomic function testing (AFT) (American Diabetes Association American Academy of Neurology 1988). This is a two-stage diagnosis involving, first, diagnosis of polyneuropathy and, second, establishing diabetes as the cause. The 2-h fasting glucose tolerance test remains the gold standard for diagnosis of diabetes but diagnosis of polyneuropathy is more complex.

Electrodiagnostic testing (EDX; electromyography and nerve-conduction testing) has long been the standard way to diagnose

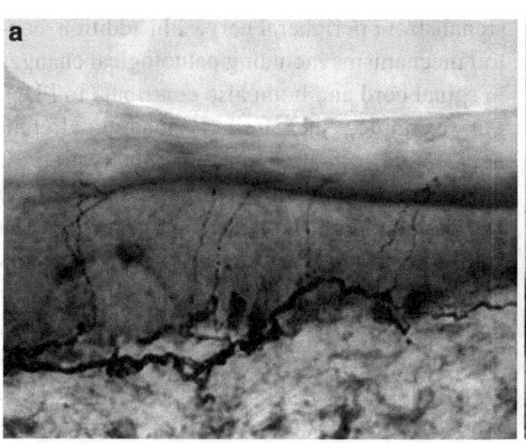

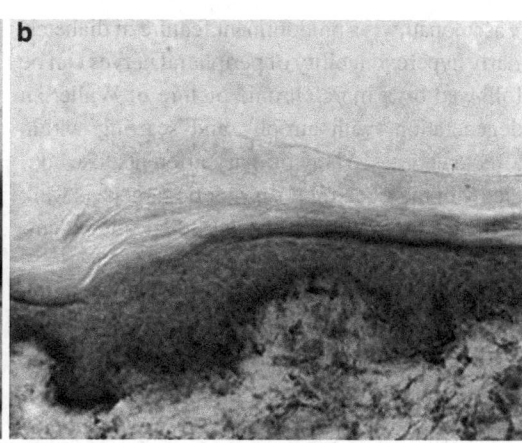

Diabetic Neuropathies, Fig. 1 PGP9.5-immunolabeled sensory nerve endings within punch skin biopsies. Representative labeled vertical skin-biopsy sections from subjects with and without diabetic neuropathy reveal distal loss of sensory nociceptive nerve endings. Both punches are from the standard biopsy site (10 cm above the lateral malleolus) used to diagnose peripheral neuropathy. (**a**) Skin biopsy from a normal subject is densely innervated, with a total of 320 epidermal neurites/mm^2 skin surface area; (**b**) biopsy from a patient with painful diabetic neuropathy shows near-total loss of innervation (two epidermal neurites/mm^2 skin surface area) (Immunohistochemistry and photography by Li Zheng)

polyneuropathy, but EDX is insensitive to small-fiber axonopathy, which predominates during early diabetic neuropathy. Patients whose polyneuropathy presents with pain and sensory complaints without weakness or loss of tendon reflexes often have normal EDX results. Nerve conduction studies (NCS) have great value for diagnosing patients with large fiber involvement in polyneuropathy, plexopathy, mononeuropathy, and mononeuropathy multiplex. These can also monitor severity and progression of DPN in patients with large fiber involvement.

The intensity and quality of the pain component is usually assessed subjectively by questionnaires, such as the regular or short-form McGill Pain Questionnaire (SF-MPQ) and the regular or short-form Neuropathic Pain Questionnaire. Mechanical and thermal thresholds can be measured by von Frey monofilaments and a radiant heat source, respectively. However, unlike the other methods discussed here, these are subjective tests wholly dependent on subject cooperation and attention (Freeman et al. 2003).

For diagnosing small-fiber-predominant polyneuropathy, the American Academy of Neurology currently recommends neurodiagnostic skin biopsy and comprehensive autonomic function testing (England et al. 2009). Sural nerve biopsy is not recommended for routine diagnosis of DN as it is invasive and cannot be repeated to monitor disease progress or the effects of therapy. Furthermore, it leaves about half of diabetics with persistent pain and sensory abnormalities (Dahlin et al. 1997). Fortunately, now clinicians can use intra-epidermal nerve fiber density (IENFD) measurement within ▶ punch skin biopsies, an easier, safer, and more sensitive diagnostic alternative (Kennedy et al. 1996), to diagnose, study, and monitor PDN (Fig. 1).

Reductions in density of epidermal nerve fibers occur early in the course of diabetes and are even seen in patients with impaired glucose tolerance (Sumner et al. 2003). However, the number of IENFs does not reflect the degree of PDN or predict the occurrence of PDN (Sorensen et al. 2006a, b).

Pathophysiology

The focal diabetic neuropathies in various locations are attributed to microvascular complications within nerves and nerve roots that cause hypoxia, inflammation, and, in severe cases, intraneural infarction with degeneration of distal axons.

Vasculopathy is a predominant feature of diabetes. Early hyperexcitability of peripheral nerves can be followed by a more chronic picture of Wallerian degeneration, with atrophy and scarring within affected nerves. Many primary afferents lose contact with their central postsynaptic targets as well, either from axonopathy or frank cell death, if damage is severe.

Diabetic sensory polyneuropathy, although more common, is less well understood. Over the past 25 years, various pathways have been linked to development of DPN; some have led to possible approaches to treatment. These include (1) glucose flux through the polyol pathway, (2) the hexosamine pathway, (3) excess/inappropriate activation of protein kinase C (PKC) isoforms, (4) accumulation of advanced glycation end products, (5) excessive activation of Poly (ADP-ribose) polymerase (PARP) pathway, (6) oxidative stress and apoptosis, (7) inflammation, and (8) lack of neurotrophic factors (Edwards et al. 2008). Most recently, mitochondrial-mediated oxidative stress in response to high glucose has been proposed as a primary cause of sensory neuron injury in DN. Hyperglycemia may trigger excessive mitochondrial fission that results in dysregulation of energy production, activation of caspase 3, and subsequent neuron injury (Vincent et al. 2010).

The molecular mechanisms of PDN are still under investigation and animal studies suggest multiple potential molecular mechanisms. Several peripheral nociceptive molecules, including ion channels, cell surface receptors for neurotrophic factors, and neurotransmitters have been demonstrated to play important roles in the development of PDN. Among them, the most studied are (1) pain-specific sodium channels (Craner et al. 2002; Hong et al. 2004), (2) purigenic receptors (Migita et al. 2009), (3) transient receptor potential (TRP) receptors (Pabbidi et al. 2008), (4) neurotrophic factors (Cheng et al. 2009), (5) neuropeptides (Cheng et al. 2009), and (6) nitric oxide (Cheng et al. 2010). These molecules are believed to be responsible for the enhanced afferent pain signals from peripheral nerves. In addition, central mechanisms including pathological changes in spinal cord and brain also contribute to PDN. For example, increased peripheral afferent nociceptive signals induce the activation of postsynaptic N-methyl-D-aspartate (NMDA) receptor on secondary nociceptive neurons in spinal cord dorsal horn. NMDA receptor is a glutamate receptor normally blocked by a magnesium ion during its quiescent state. The increased peripheral inputs from PDN hyperpolarize the postsynaptic membrane, removing magnesium blockade, and inducing calcium influx to further prolong the excitation of secondary nociceptive neurons (Cheng 2010). In addition, glial-mediated mechanisms within the spinal cord are known to facilitate prolonged painful symptoms in PDN. Potential molecules involved in glial reactions include nitric oxide, cytokines, chemokines, tumor necrosis factor-a, and purine receptors (Cheng 2010). Investigators are working vigorously to gain insight into these potential mechanisms associated with PDN.

Treatment Options

Two treatment categories exist with those that improve the underlying diabetes (disease-modifying treatments) of greatest importance. Weight loss is probably the single most important and safest therapy for T2DM. The second is optimal control of hyperglycemia, which has been convincingly demonstrated to delay the onset and slow the progression of DPN (The Diabetes Control and Complications Trial Research Group 1993). Second in importance are therapies that help preserve or restore injured neurons. A large trial of nerve growth factor (NGF), with the goal of increasing axonal regeneration in DPN, failed to show efficacy (Apfel et al. 1998), but methodological concerns were identified, and the efficacy of growth factors is not yet disproven. Trials of alternative treatments (e.g., myoinositol, alpha-lipoic acid, and aldose-reductase inhibitors) have been more helpful in animal models than in humans. Physical therapy to encourage motility and weight loss, as well as meticulous podiatric care to prevent and treat foot infections can go far in treating the underlying disease, as well as minimizing its complications. Decompressive neurosurgery is indicated only if entrapment contributes to clinically significant symptoms.

Symptomatic or palliative treatments are appropriate to consider for patients with moderate or severe symptoms, usually pain. For pain, randomized controlled clinical trials have established efficacy and safety for several classes of medications. According to the recommended guideline from the American Academy of Neurology, the American Association of Neuromuscular and Electrodiagnostic Medicine, and the American Academy of Physical Medicine and Rehabilitation, several classes of medication should be considered for treating PDN based on the existing evidence (Bril et al. 2011).

Pregabalin (300–600 mg/day, Level A) and gabapentin (900–3,600 mg/day, Level B) have been tested in large randomized trials for various neuropathic conditions. They are closely related molecules and selection can be based on price and convenience. Among antiepileptic medications, topiramate, oxcarbazepine, lamotrigine, and lacosamide have not been adequately tested to recommend consideration in PDN. There is limited experimental support for sodium valproate (500–1,200 mg/day, Level B) a second or third-line option.

Tricyclic compounds also constitute first-line therapy for DPN and related conditions. Amitriptyline (25–100 mg/day; Level B) has been most extensively tested, but as the secondary amine tricyclics nortriptyline and desipramine have been shown to have equal efficacy against other types of neuropathic pain to amitriptyline with fewer side effects, they are preferred. These compounds have the added benefit of efficacy against depression, a common complication of chronic pain, but their analgesic efficacy has been established in nondepressed patients, so their beneficial effects for neuropathic pain do not simply reflect antidepressant effects. Level B data supports the use of serotonin-norepinephrine reuptake inhibitors,

e.g., venlafaxine (75–225 mg/day), and duloxetine (60–120 mg/day). Level C evidence suggests that venlafaxine may be added to gabapentin for a better response. Serotonin-specific antidepressants, e.g., fluoxetine, have not been tested specifically in DPN but have been found ineffective for other neuropathic pain conditions so they are not indicated.

Opioid medications can be considered for treatment of DPN, including morphine sulfate (titrated to 120 mg/day), tramadol (210 mg/day), and oxycodone (up to 120 mg/day) (Level B). Data are insufficient to recommend one agent over the other.

Alternative Pharmacological Treatments

Several treatments have limited experimental support but have not found widespread clinical success and should be considered only after the above options. These include oral dextromethorphan (400 mg/day, Level B) and the topical lidocaine patch (5 %) (Level C). The major advantage of topical lidocaine is lack of systemic side effects but these small patches are impractical for numbing widespread areas of pain. Capsaicin cream (0.075 % qid) has limited experimental support (Level B) but is painful to apply and appears to worsen terminal axonal degeneration. Level B evidence argues against efficacy of clonidine, pentoxifylline, and mexiletine. There is insufficient evidence to support or refute the usefulness of vitamins and α-lipoic acid in the treatment of PDN (Level U).

Non-pharmacological Treatments

Percutaneous electrical nerve stimulation can be considered for the treatment of PDN (for 3–4 weeks, Level B). Electromagnetic field treatment, low-intensity laser treatment, and Reiki therapy should probably not be considered for the treatment of PDN (Level B). Noninvasive and invasive brain stimulation, though reported effective for other types of neuropathic pain, have not been adequately studied.

Acknowledgment The authors would like to thank Dr. Eva Feldman for revising the manuscript.

References

American Diabetes Association American Academy of Neurology. (1988). Consensus statement: Report and recommendations of the San Antonio conference on diabetic neuropathy. *Diabetes Care, 11*(7), 592–597.

Apfel, S. C., et al. (1998). Recombinant human nerve growth factor in the treatment of diabetic polyneuropathy. NGF Study Group. *Neurology, 51*(3), 695–702.

Barrett, A. M., et al. (2007). Epidemiology, public health burden, and treatment of diabetic peripheral neuropathic pain: A review. *Pain Medicine, 8*(Suppl. 2), S50–S62.

Bril, V., et al. (2011). Evidence-based guideline: Treatment of painful diabetic neuropathy: Report of the American Academy of Neurology, the American Association of Neuromuscular and Electrodiagnostic Medicine, and the American Academy of Physical Medicine and Rehabilitation. *Neurology, 76*(20), 1758–1765.

Cheng, H. (2010). Spinal cord mechanisms of chronic pain and clinical implications. *Current Headache and Pain Reports, 14*, 213–220.

Cheng, H. T., et al. (2009). Nerve growth factor mediates mechanical allodynia in a mouse model of type 2 diabetes. *Journal of Neuropathology and Experimental Neurology, 68*(11), 1229–1243.

Cheng, H. T., et al. (2010). p38 mediates mechanical allodynia in a mouse model of type 2 diabetes. *Molecular Pain, 6*, 28.

Craner, M. J., et al. (2002). Changes of sodium channel expression in experimental painful diabetic neuropathy. *Annals of Neurology, 52*(6), 786–792.

Dahlin, L. B., Eriksson, K. F., & Sundkvist, G. (1997). Persistent postoperative complaints after whole sural nerve biopsies in diabetic and non-diabetic subjects. *Diabetic Medicine, 14*(5), 353–356.

Dworkin, R. H., et al. (2005). Core outcome measures for chronic pain clinical trials: IMMPACT recommendations. *Pain, 113*, 9–19.

Dworkin, R. H., et al. (2007). Symptom profiles differ in patients with neuropathic versus non-neuropathic pain. *The Journal of Pain, 8*, 118–126.

Dyck, P. J., et al. (1993). The prevalence by staged severity of various types of diabetic neuropathy, retinopathy, and nephropathy in a population-based cohort: The Rochester Diabetic Neuropathy Study. *Neurology, 43*(4), 817–824.

Edwards, J. L., et al. (2008). Diabetic neuropathy: Mechanisms to management. *Pharmacology and Therapeutics, 120*(1), 1–34.

England, J. D., et al. (2009). Practice parameter: Evaluation of distal symmetric polyneuropathy: Role of autonomic testing, nerve biopsy, and skin biopsy (an evidence-based review). Report of the American Academy of Neurology, American Association of Neuromuscular and Electrodiagnostic Medicine, and American Academy of Physical Medicine and Rehabilitation. *Neurology, 72*(2), 177–184.

Feldman, E. L., et al. (1999a). Diabetic neuropathy. In S. I. Taylor (Ed.), *Current review of diabetes* (pp. 71–83). Philadelphia: Current Medicine Inc.

Feldman, E. L., et al. (1999b). Diabetic neuropathy. In *Diabetes in the new millennium* (pp. 387–402). Sydney: The Endocrinology and Diabetes Research Foundation of the University of Sydney.

Feldman, E. L., et al. (2001). Diabetic neuropathy. In: *Principles and practice of endocrinology and metabolism* (pp. 1391–1399). Philadelphia: Lippincott Williams & Wilkins.

Freeman, R. (2005). Autonomic peripheral neuropathy. *Lancet, 365*(9466), 1259–1270.

Freeman, R., Chase, K. P., & Risk, M. R. (2003). Quantitative sensory testing cannot differentiate simulated sensory loss from sensory neuropathy. *Neurology, 60*(3), 465–470.

Harris, M., Eastman, R., & Cowie, C. (1993). Symptoms of sensory neuropathy in adults with NIDDM in the U.S. population. *Diabetes Care, 16*(11), 1446–1452.

Hedley, A. A., et al. (2004). Prevalence of overweight and obesity among US children, adolescents, and adults, 1999–2002. *JAMA, 291*(23), 2847–2850.

Hong, S., et al. (2004). Early painful diabetic neuropathy is associated with differential changes in tetrodotoxin-sensitive and -resistant sodium channels in dorsal root ganglion neurons in the rat. *Journal of Biological Chemistry, 279*(28), 29341–29350.

Jensen, M. P., et al. (2006). Do pain qualities and spatial characteristics make independent contributions to interference with physical and emotional functioning? *The Journal of Pain, 7*, 644–653.

Kennedy, W. R., Wendelschafer-Crabb, G., & Johnson, T. (1996). Quantitation of epidermal nerves in diabetic neuropathy. *Neurology, 47*(4), 1042–1048.

Migita, K., et al. (2009). Modulation of P2X receptors in dorsal root ganglion neurons of streptozotocin-induced diabetic neuropathy. *Neuroscience Letters, 452*(2), 200–203.

Pabbidi, R. M., et al. (2008). Influence of TRPV1 on diabetes-induced alterations in thermal pain sensitivity. *Molecular Pain, 4*, 9.

Singleton, J. R., et al. (2003). Microvascular complications of impaired glucose tolerance. *Diabetes, 52*, 2867–2873.

Singleton, J. R., et al. (2005). Polyneuropathy with impaired glucose tolerance: Implications for diagnosis and therapy. *Current Treatment Options in Neurology, 7*, 33–42.

Sorensen, L., Molyneaux, L., & Yue, D. K. (2006a). The relationship among pain, sensory loss, and small nerve fibers in diabetes. *Diabetes Care, 29*(4), 883–887.

Sorensen, L., Molyneaux, L., & Yue, D. K. (2006b). The level of small nerve fiber dysfunction does not predict pain in diabetic neuropathy: A study using quantitative sensory testing. *The Clinical Journal of Pain, 22*(3), 261–265.

Sumner, C. J., et al. (2003). The spectrum of neuropathy in diabetes and impaired glucose tolerance. *Neurology, 60*(1), 108–111.

The Diabetes Control and Complications Trial Research Group. (1993). The effect of intensive treatment of diabetes on the development and progression of long-term complications in insulin-dependent diabetes mellitus. *The New England Journal of Medicine, 329*(14), 977–986.

Vincent, A. M., et al. (2010). Mitochondrial biogenesis and fission in axons in cell culture and animal models of diabetic neuropathy. *Acta Neuropathologica, 120*(4), 477–489.

Vinik, A. I., & Mehrabyan, A. (2004). Diabetic neuropathies. *The Medical Clinics of North America, 88*(4), 947–999, xi.

Wild, S., et al. (2004). Global prevalence of diabetes: Estimates for the year 2000 and projections for 2030. *Diabetes Care, 27*(5), 1047–1053.

Zhang, P., et al. (2010). Global healthcare expenditure on diabetes for 2010 and 2030. *Diabetes Research and Clinical Practice, 87*(3), 293–301.

Diabetic Neuropathy

▶ Ulceration, Prevention by Nerve Decompression

Diabetic Neuropathy Model

▶ Neuropathic Pain Model, Diabetic Neuropathy Model

Diabetic Neuropathy, Treatment

Anne K. Bertelsen[1] and John W. Griffin[2]
[1]Neurology, Aleris Hospital, Bergen, Norway
[2]Department of Neurology, Johns Hopkins University School of Medicine, Baltimore, MD, USA

Synonyms

Diabetic polyneuropathy; Symmetric diabetic neuropathy

Definition

Focal or generalized damage to the peripheral nerves is common with diabetes mellitus type I or II.

Characteristics

Most patients with diabetic neuropathy have the type designated diabetic sensory polyneuropathy (DSPN), and this section focuses on this form, emphasizing the aspects relevant to pain. The manifestations can include ▶ neuropathic pain, hypalgesia with loss of protective sensibility, a predilection for painless injuries, and autonomic dysfunction (Dyck and Dyck 1999). The autonomic manifestations can include cardiovascular (loss of heart period variability, loss of Valsalva response, orthostatic hypotension), genitourinary (impotence, retrograde ejaculation, incontinence), gastrointestinal (constipation, diarrhea, fecal incontinence), and sudomotor (loss of sweating) abnormalities. The hypalgesia is typically greatest in the feet and can contribute to the development of skin ulcers, painless fractures and joint damage (Charcot feet), and ultimately amputations.

Neuropathic pain is prevalent in diabetics, reaching 16.9 % of the diabetic population in a recent study, and is underreported to physicians (Daousi et al. 2004). ▶ Spontaneous pain can occur even in individuals with profound hypalgesia. The pain is usually greatest in the toes and feet. With progression, it moves into the fingers and can involve a "shield" over the sternum and parasternal areas, representing the most distal regions of innervation of the intercostals nerves. Patients use many descriptors for their pain, including "burning," "squeezing," and "throbbing." The particular adjective that they use does not have clinical or prognostic correlates. Sudden stabbing lightning pains are frequent. The clinical manifestations conform to those described in ▶ small fiber polyneuropathies.

Treatment of DSPN

The treatment can be divided into treatment of the underlying disease, treatment of neuropathic pain, protection from painless injuries, and currently topical unproved therapies.

Treatment of the Underlying Disease

The unequivocal message of large-scale, prospective, controlled studies is that tight glycemic control improves the prognosis for neuropathy (DCCT Research Group 1988). Fifteen years ago, this was a controversial point, and it is of fundamental significance in planning diabetic care. This concept can now be extended to an important group of individuals with impaired glucose tolerance but without frank diabetes. Such individuals, identified by glucose tolerance tests (GTT), are at substantial risk to progress to diabetes within a few years (Tuomilehto et al. 2001). Prospective studies have shown that progression to diabetes can be deferred or prevented by weight loss and exercise. This issue is relevant to neuropathy because it has recently been proved that a subgroup of individuals with impaired GTT develop painful neuropathies before frank diabetes (Singleton et al. 2001; Sumner et al. 2003) and because there is evidence of peripheral nerve damage in others who are asymptomatic (Smith et al. 2001).

Treatment of Neuropathic Pain

Education regarding the physiologic basis and meaning of neuropathic pain is an essential first step; to the extent patients understand that the pain sensation does not reflect ongoing damage to the feet or the tissues, they are better able to objectify the pain. Taking advantage of their observations about factors that exacerbate or improve pain is useful. For example, if they have cold allodynia, they may prefer to keep the feet warm. In contrast, many individuals are more comfortable with the feet cool and prefer to sleep without covers over the feet. Similarly, if there is mechanical hyperalgesia, the bedsheets are often painful at night; use of a foot cradle can eliminate that pain.

Drug treatments are adjuncts in management and are detailed below. The available agents include anticonvulsants (gabapentin, pregabalin, and newer analogs), antidepressants including SNRI, tricyclic, and probably SSRI representatives

(Sindrup and Jensen 2000; Sindrup et al. 2003), some non-opiate analgesics such as tramadol, and, if needed, opiates. There are special considerations in choice of agents in diabetics. Due to the frequent association with atherosclerotic cardiovascular disease and autonomic insufficiency, tricyclics require special caution. Also, due to gastrointestinal motility disturbances, the opiates, and even tramadol, may cause severe constipation. Some data support the use of α-lipoic acid, an antioxidant (Ametov et al. 2003), and adverse reactions have not been recognized.

Antiepileptic Drugs

Gabapentin has been demonstrated in two large clinical trials to result in significant pain relief in patients with painful DSPN (Backonja et al. 1998; Morello et al. 1999). Compared with amitriptyline, gabapentin has been shown to have equal efficacy; however, the side-effect profile of gabapentin has been shown to be more favorable, although dizziness and sedation are frequently reported (Dallocchio et al. 2000). The doses should therefore be titrated slowly to the maximum dose. It is recommended to start at 300 mg/day on day 1, increasing by 300 mg every day in 3 days to achieve a dose of 900 mg/day by day 3. Then, further dose increases by 300 mg/day up to 1,800 mg/day by day 14. Once this dose level is achieved, gabapentin can be titrated up to 3,600 mg/day as required over the following weeks to achieve a maximal response with good tolerability or until intolerable side effects occur.

Pregabalin is a follow-up drug to gabapentin, which also binds to the alpha$_2$delta subunit of voltage-dependent calcium channels but has a more linear bioavailability than gabapentin. Lower doses are therefore needed than for gabapentin, and the efficacy is more predictable. The therapeutic dose range for pregabalin is between 150 and 600 mg/day divided into two or three doses.

Oxcarbazepine may be effective in treating painful DSPN. Data from an open-label prospective study indicate that oxcarbazepine (mean effective dose 814 mg/day, range 150–1,200 mg/day) may significantly improve pain scores in patients with painful diabetic peripheral neuropathy.

Lamotrigine, a phenyltriazine derivate, inhibits the release of glutamate, possibly by stabilizing the neural membrane through encouraging inactivation of voltage-dependent sodium channels and thus suppressing abnormal ectopic discharges. To avoid side effects such as skin rash, the doses should be titrated slowly to the therapeutic dose. It is recommended to start at 25 mg/day on day 1, increasing the dose by 25 mg every 14 days until a dose of 400–600 mg/day divided on two doses has been achieved.

Antidepressant Drugs

Tricyclic antidepressants, like amitriptyline, doxepin, and desipramine, are strong sodium channel modulators in addition to their well-known effect on reuptake of serotonin and noradrenaline. The sodium channel-blocking effect is presumed to be the pain-relieving mechanism of these drugs in cases of painful diabetic neuropathy. The cumulative efficacy reported from trials with subjects with diabetic neuropathy suggests that about one third of patients will achieve a 50 % reduction in neuropathic pain; however, the benefits are often outweighed by side effects, especially among the elderly.

Selective serotonin reuptake inhibitors (SSRIs), like paroxetine and citalopram, differ from tricyclic antidepressants in that they selectively block serotonin reuptake. In clinical trials, SSRIs have shown to reduce the pain of diabetic neuropathy better than placebo, but not as effectively as tricyclic antidepressants.

Topical Analgesics

Lidocaine is an amide-type local anesthetic that stabilizes neural membranes by modulating voltage-sensitive sodium channels. It is the most frequently used local anesthetic in the treatment of neuropathic pain and has been shown to produce moderate reduction in pain in patients with diabetic neuropathy.

Capsaicin cream, (Zostrix) the active component in hot chili peppers, applied topically four to five times per day for 4 weeks is effective in treating painful diabetic peripheral neuropathy.

However, initial burning of the skin upon application is an unpleasant side effect that leads to treatment discontinuation among more than 30 % of patients. Capsaicin is thought to affect pain transmission by depleting the neural stores of substance P, a pain-modulating neuropeptide, in epidermal nerve fibers and temporarily decreasing the density of epidermal nerve fibers in treated regions. The capsaicin cream contains 0.025–0.075 % capsaicin and should be applied to intact skin only. Patients should wash their hands thoroughly after applying capsaicin cream in order to prevent inadvertent contact with other areas.

Opioid Analgesics and Opioid-Like Drugs

Although opioids are considered a last resort for chronic neuropathic pain control, limited data suggest efficacy and tolerability of oxycodone and other opioids for painful diabetic peripheral neuropathy. Tramadol, a centrally acting analgesic chemically unrelated to the opiates, has a low binding affinity at μ-opioid receptors and weak inhibition of norepinephrine and serotonin reuptake. Although its precise mechanism is not known, the facts that its analgesic effect is only partially antagonized by the opioid antagonist naloxone and that its (+) and (−) isomers (the former has opioid activity and the latter inhibits norepinephrine reuptake) interact synergistically suggest a mechanism involving an adrenergic-opioid interaction. Tramadol causes significantly less respiratory depression than morphine and, in contrast to morphine, does not stimulate the release of histamine. Tramadol does produce μ-opioid type dependence like codeine or dextropropoxyphene; however, there appears to be little potential for tolerance development or abuse. Tramadol is thus a useful drug in the treatment of painful diabetic peripheral neuropathy and is particularly effective in treating the nociceptive component of axonal pain such as that caused by ischemia. Its use is, however, limited in elderly patients or those with impaired hepatic and renal function, and it should be used with caution by patients taking other psychoactive drugs. The dose of tramadol should be in the range of between 200 and 400 mg/day. As pain treatment in diabetic neuropathy is a long-term undertaking, beginning with monotherapy and evaluating progress with a daily visual analog scale log can simplify decisions.

Protection from Painless Injuries

The problem of insensitive or anesthetic feet requires the individual to be constantly vigilant so that injuries to the feet can be prevented. It is particularly important to recognize early foot injuries and to treat all injuries promptly. Among the indications of failed foot care are ulcers in the skin and bony sequelae (Charcot joints) that result from unfelt injuries. In severe cases, amputation of digits, the foot, or even the lower leg may eventually be required. Repeated tiny injuries to the skin and bones of the feet, caused by loss of pain sensibility and loss of position sense, can be overcome by continual vigilance. Warning signs of danger are localized redness (erythema) or blisters on the feet, breaks in the skin, or hot points over the skin. The keystone is regular careful examination of the feet.

Shoes that fit well are essential. Callouses are a sign of long-standing excessive pressure on the skin and thus signal badly fitting shoes. Callouses are usually found on the balls of the feet, the sides of the feet, or the tops of the toes. A persistent or growing callous needs to be examined and treated by a medical professional.

Individuals with insensitive feet should not walk barefooted, as minor inadvertent injuries can cause major long-term problems. The feet should be bathed in lukewarm water, and the water temperature should be tested with a sensitive part of the body (the elbow or upper arm if the hands or wrists are insensitive). For some patients, heat feels good on the feet, but under no circumstances should a heating pad or hot water bottle be used, because of the potential for painless thermal injury. Insensitive fingers can be burned painlessly by smoking or during cooking. Gas ranges are preferable to electric ranges because it is apparent when they are "on."

Unproved Therapies

Patients can find a variety of devices that are advocated to provide relief in painful diabetic neuropathy, and the Internet provides a marketplace for such devices. Transcutaneous nerve stimulators are widely used, and it is reasonable to view their efficacy in terms of their effectiveness in other pain disorders. At present, no device specifically claimed to improve nerve function has been proved to be of value. Such claims have been made for a variety of bracelets, electrical devices, etc. Another unproved therapy is surgical release of putative nerve entrapments. Claims have been made that surgical release of multiple nerves in the legs and arms relieves pain and improves the neuropathic prognosis, including preventing ulcers, even in individuals in whom the only evidence of possible entrapment is a Tinel's sign to percussion over possible entrapment sites (Aszmann et al. 2004; Lee and Dellon 2004). Diabetics are at increased risk for nerve entrapments. In individuals with physiologic and clinical evidence of entrapment, surgical correction should be considered.

References

Ametov, A. S., Barinov, A., Dyck, P. J., et al. (2003). The sensory symptoms of diabetic polyneuropathy are improved with alpha-lipoic acid: The SYDNEY trial. *Diabetes Care, 26*, 770–776.

Aszmann, O., Tassler, P. L., & Dellon, A. L. (2004). Changing the natural history of diabetic neuropathy: Incidence of ulcer/amputation in the contralateral limb of patients with a unilateral nerve decompression procedure. *Annals of Plastic Surgery, 53*, 517–522.

Backonja, M., Beydoun, A., Edwards, K. R., et al. (1998). Gabapentin for the symptomatic treatment of painful neuropathy in patients with diabetes mellitus: A randomized controlled trial. *Journal of the American Medical Association, 280*, 1831–1836.

Dallocchio, C., Buffa, C., Mazzarello, P., et al. (2000). Gabapentin vs amitriptyline in painful diabetic neuropathy: An open-label pilot study. *Journal of Pain and Symptom Management, 20*, 280–285.

Daousi, C., MacFarlane, I. A., Woodward, A., et al. (2004). Chronic painful peripheral neuropathy in an urban community: A controlled comparison of people with and without diabetes. *Diabetic Medicine, 21*, 976–982.

DCCT Research Group. (1988). Factors in development of diabetic neuropathy: Baseline analysis of neuropathy in feasibility phase of diabetes control and complications trial (DCCT). *Diabetes, 37*, 476–481.

Dyck, P. J. B., & Dyck, P. J. (1999). Diabetic polyneuropathy. In P. J. Dyck & P. K. Thomas (Eds.), *Diabetic neuropathy* (pp. 255–278). Philadelphia: WB Saunders.

Lee, C. H., & Dellon, A. L. (2004). Prognostic ability of tinel's sign in determining outcome for decompression surgery in diabetic and nondiabetic neuropathy. *Annals of Plastic Surgery, 53*, 523–527.

Morello, C. M., Leckband, S. G., Stoner, C. P., et al. (1999). Randomized double-blind study comparing the efficacy of gabapentin with amitriptyline on diabetic peripheral neuropathy pain. *Archives of Internal Medicine, 159*, 1931–1937.

Sindrup, S. H., & Jensen, T. S. (2000). Pharmacologic treatment of pain in polyneuropathy. *Neurology, 55*, 915–920.

Sindrup, S. H., Bach, F. W., Madsen, C., et al. (2003). Venlafaxine versus imipramine in painful polyneuropathy: A randomized, controlled trial. *Neurology, 60*, 1284–1289.

Singleton, J. R., Smith, A. G., & Bromberg, M. B. (2001). Painful sensory polyneuropathy associated with impaired glucose tolerance. *Muscle & Nerve, 24*, 1225–1228.

Smith, A. G., Ramachandran, P., Tripp, S., et al. (2001). Epidermal nerve innervation in impaired glucose tolerance and diabetes-associated neuropathy. *Neurology, 57*, 1701–1704.

Sumner, C. J., Sheth, S., Griffin, J. W., et al. (2003). The spectrum of neuropathy in diabetes and impaired glucose tolerance. *Neurology, 60*, 108–111.

Tuomilehto, J., Lindstrom, J., Eriksson, J. G., et al. (2001). Prevention of type 2 diabetes mellitus by changes in lifestyle among subjects with impaired glucose tolerance. *The New England Journal of Medicine, 344*, 1343–1350.

Diabetic Polyneuropathy

Definition

Diabetic polyneuropathy refers to pathological changes to nerve function due to diabetes mellitus. This usually manifests as loss of sensation and burning pain (see also ▶ Diabetic Neuropathy).

Cross-References

▶ Diabetic Neuropathies
▶ Diabetic Neuropathy, Treatment
▶ Opioids in Geriatric Application

Diabetic Sensorimotor Polyneuropathy

▶ Diabetic Neuropathies

Diabetic Sensory Neuropathy

▶ Diabetic Neuropathies

Diabetic Symmetrical Proximal Neuropathy

▶ Diabetic Neuropathies

Diabetic Truncal Mononeuropathy

▶ Diabetic Neuropathies

Diagnosis

Definition

The term – diagnosis – is derived from the Greek words: δια (dia) meaning between and γνοσις (gnosis) meaning knowing. When used as a verb – to diagnose – it means to determine which of the two or more possible causes of a complaint is actually the cause, e.g., "the doctor will diagnose the cause of your chest pain." When used as an abstract noun, diagnosis means the act or process of determining the cause, e.g., "the patient was admitted to hospital for diagnosis of her chest pain." When used as a common noun, diagnosis means what is found as a result of that process and is a generic term for the name of the condition that is the cause of the complaint, e.g., "on the basis of the test results, the doctor provided a diagnosis of the patient's chest pain."

Classically, diagnosis requires establishing the pathology that is responsible for a patient's symptoms on the grounds that if the pathology is found, it can be rectified, and the patient's symptoms thereby are relieved. In pain medicine, this may not always be possible. Some sources of pain cannot be detected using currently available techniques of investigation.

In that event, physicians are nevertheless able to formulate an ▶ assessment of the patient and their problems. The assessment is not the same as a diagnosis, for a cause of the pain is not established. Nevertheless, the assessment may provide a sufficient basis for instituting treatment that may help the patient to reduce or eliminate their pain and associated problems.

Cross-References

▶ Taxonomy

Diagnosis and Assessment of Clinical Characteristics

▶ Diagnosis and Assessment of Clinical Characteristics of Central Pain

---

Diagnosis and Assessment of Clinical Characteristics of Central Pain

Ellen Jørum
The Laboratory of Clinical Neurophysiology,
Rikshospitalet University, Oslo, Norway

Synonyms

Central pain; Diagnosis and assessment of clinical characteristics

Definition

▶ Central pain is defined as "pain initiated or caused by a primary lesion or dysfunction in the central nervous system" (Merskey and Bogduk 1994). Some common causes of central pain are spinal cord injury (SCI), vascular brain lesions, arteriovenous malformations, syringomyelia, and multiple sclerosis (MS).

Characteristics

Clinical Characteristics

There is no major distinction between pain following a central or a peripheral nervous lesion. Chronic neuropathic pain in general may be ▶ stimulus independent (spontaneous) or ▶ stimulus dependent (evoked). The spontaneous pain may be ongoing and/or paroxysmal. The quality of the pain may be described in many terms, such as burning, pricking, stinging, aching, lancinating, throbbing, pressing (or a large variety of other adjectives) and many patients will experience two to four pain qualities. There is no single pain descriptor that characterizes the qualities of central pain. The onset of pain may be immediate or with a delay. Post-stroke pain was found to be delayed for at least 1 month in 37 % of the patients (Andersen et al. 1995). Similar results have been reported by Leijon et al. (1989). Of a total of 591 patients with pain and or dysesthesia following SCI, the onset of pain was within the first year in 58 % of the cases and pain was immediate from the moment of injury in 34 % of the cases (Störmer et al. 1997). The location of pain will depend upon which parts of the central nervous system are injured, but some general distinctions between pain of different etiologies exist. One example is post-stroke pain, where 75 % of the patients will suffer from a ▶ hemipain, situated contralateral to the lesion, while hemipain seldom occurs in MS (Boivie 2003). Ninety percent of MS patients will suffer from pain in the lower extremities and 36 % in the upper extremities (Boivie 2003). Neuropathic pain following SCI may occur in segments above, at, or below the level of injury and may be caused by factors other than the spinal cord lesion, such as compressive mononeuropathies and nerve root compression (Siddall et al. 1997, 2000). There is no correlation between the size of the lesion and the risk of developing central pain (Boivie 2003). The location of the lesion, however, plays a crucial part in the development of pain and of special importance is whether the spinothalamic tract system is involved in the lesion.

Clinical Diagnosis of Central Pain

The diagnosis of central pain is based on a thorough clinical and neurological examination of the patient. A thorough interview with the patient is mandatory, including detailed questions about spontaneous pain, both ongoing and paroxysmal as well as questions about stimulus-evoked pain. Pain drawings are recommended. But it is emphasized that a correct diagnosis cannot be made on evaluation of pain and sensory abnormalities alone. The investigation has to include examination of the motor system, reflexes, brain stem, and cerebellar functions in order to be able to pinpoint the extent and level of the central lesion. In patients with central pain, there will be a large variation in the presence of other neurological symptoms. No major difficulty will be encountered in making a diagnosis of a spinal cord injury or a vascular brain lesion, but the complete extent of neurological deficits needs to be evaluated. Typically, central pain is characterized by sensory abnormalities, such as a possible combination of hypoesthesia and ▶ allodynia/▶ hyperalgesia (or other combinations). Lesions involving the spinothalamic-thalamocortical projection system seem to be crucial for the development of central pain. The sensory abnormalities may be evaluated by simple means, such as cotton (for the evaluation of light touch), pins for pinprick and pain sensation and warm and cold water for the evaluation of the spinothalamic tract system. One should be aware, however, that in many cases, hypoesthesia to light touch may be masked by a concomitant allodynia to light touch and that a reduced sensitivity to painful pinprick may be masked by a hyperalgesia to pinprick. For the detection of sensory deficits in such cases, more sophisticated tools may be needed. The detailed sensory examination by ▶ quantitative sensory testing (QST) developed by Ulf Lindblom and colleagues has

major advantages for the detailed examination of sensory deficits, but is not, in general, crucial for making a diagnosis of central pain.

Quantitative Sensory Testing (QST)

QST is used to measure the intensity of stimuli needed to produce specific sensory perceptions. Tests are developed to determine sensory thresholds for tactile, vibratory, pressure, and temperature stimulation. All quantitative sensory tests are psychophysical tests requiring awake and alert patients who fully understand the instructions given and who are fully capable of cooperating. QST methods have been mainly used for research purposes for the classification of neuropathic pain, in studies of pathophysiological mechanisms and in pharmacological trials. For a purely clinical evaluation of the individual patient, the diagnosis of neuropathic pain may be made without using these time-consuming methods. The testing is restricted to a few laboratories and needs specially trained personnel. Through the use of QST, it has become evident that all patients with central post-stroke pain have lesions in the spinothalamic-cortical pathways, as seen by abnormal sensitivity to temperature and pain (Boivie et al. 1989). A high proportion of these patients (80 %) did not feel any temperature between 0 ° and 50 °C, indicating the severity of the sensory deficit. Many of these patients also had large deficits in the sensibility to touch and vibration, indicative of a disturbed function in the dorsal column-medial lemniscal pathway but, on the other hand, other patients had no sensory deficit related to these pathways. Similar results are obtained from studies of patients with MS, showing reduced sensibility to temperature and pain, but less prominent findings compared to patients with central post-stroke pain (Boivie 2003).

In studies of central pain following spinal cord injury, similar results have been obtained. In a study of 13 patients with below-level pain following SCI by Beric et al. (1988), 10 patients had absence of sensibility to thermal stimulation measured by QST, while the sensation of vibration and light touch was present, but reduced. These results suggest a relative preservation of dorsal column function with an abolition of spinothalamic system function. Results of such studies show how QST can help in differentiating between sensibility changes due to dorsal column and spinothalamic tract functions. No other method than QST will accurately assess an eventual allodynia and hyperalgesia to thermal stimuli, but thermal allodynia has scarcely been reported in patients with central pain (Boivie et al. 1989), mainly because of a substantial loss of thermal sensibility in patients with central pain. Since evoked pain is an important feature in central pain, diagnostic tools should include tests of central hyperexcitability.

Evaluation of Allodynia and Hyperalgesia

Patients with central pain may, like all patients with neuropathic pain, complain of evoked pain, pain that is evoked by stimulation of painful skin area. There are two types of allodynia/ hyperalgesia to tactile stimuli, one to light touch and one to punctate stimuli (the static type). Although both are caused by central sensitization mechanisms, they are mediated by different peripheral nerve fibers; allodynia to light touch is mediated by A β fibers, whereas the punctate type is mediated by A δ fibers. Mapping of the area may be recommended for both types. Punctate hyperalgesia may be mapped by the use of von Frey hairs, while dynamic allodynia is usually mapped by lightly brushing the skin.

Abnormal Temporal Summation

Temporal summation of neural impulses in nociceptive nerve fibers is a physiologically important mechanism that can intensify the sensation of pain. Repetitive stimulation of a painful skin area in a patient suffering from neuropathic pain may produce an intense and long-lasting pain in a phenomenon referred to as ► abnormal temporal summation (or "wind-up-like pain"). The abnormal temporal summation seen in patients may be assessed roughly by repetitive skin stimulation with the frequency of 2–3 Hz with a von Frey hair for up to 20–30s. If temporal summation is abnormal, the patient will report a sudden intense pain in the stimulated area, often occurring within a few seconds, with aftersensation

and radiation. An abnormal increase in temporal summation may be regarded as a sign of central hyperexcitability and can be quantified by measuring latency, duration of aftersensation, and area of radiation.

Evidence of Central Hyperexcitability in Central Pain Following SCI

In a study of 16 patients with traumatic SCI and below-level pain, somatosensory abnormalities in painful denervated skin areas below the level of the lesion as well as in nonpainful denervated skin areas at the level of the lesion were compared (Eide et al. 1996). In the painful denervated area, a highly significant reduction in sensibility to heat and cold, indicating reduced function in the spinothalamic tract system and supporting the results of Beric et al. was found. An important observation was that there was no significant difference in sensory thresholds in the painful denervated skin areas compared with the nonpainful denervated skin areas. Thus, in these patients, deafferentation of spinothalamic pathways was not a sufficient condition for the development of central neuropathic pain. One of the main findings was the demonstration of evoked pain, both allodynia to brush and an abnormal temporal summation in the painful denervated area alone. These data show that sensory loss must be coupled with abnormal pain responsiveness to reach a diagnosis of central neuropathic pain (Eide et al. 1996). Similar findings have more recently been reported by others (Finnerup et al. 2003). The main finding of their study was a higher intensity of brush-evoked dysesthesia and pinprick hyperalgesia at the lesion level in SCI patients with central pain as opposed to SCI patients without below-level pain, despite a similar degree of lost somatosensory function in the two groups. Brush-evoked pain or dysesthesia have also been reported to occur in patients with post-stroke pain (Boivie et al. 1989; Leijon et al. 1989; Andersen et al. 1995).

Supplementary Investigations

Supplementary investigations such as CT (computer tomography), MRI (magnetic resonance imaging), and electrodiagnostic evaluations can be performed to visualize and functionally assess the central lesions. In clinical neurophysiology, measurement of ▶ somatosensory-evoked potentials to electrical stimulation is routinely performed for investigating deficits in sensory pathways. It is important to be aware that electrical stimuli activate the dorsal column-lemniscal systems and that no information is obtained from the pain and temperature–mediating spinothalamic tract system. Laser-evoked late potentials, on the other hand, may evaluate function in the spinothalamic tract system, but few results have so far been published on central pain (Garcia-Larrea et al. 2002).

References

Andersen, G., Vestergaard, K., Ingeman-Nielsen, M., et al. (1995). Incidence of central post-stroke pain. *Pain, 61*, 187–193.

Beric, A., Dimitrijevic, M. R., & Lindblom, U. (1988). Central dysesthesia syndrome in spinal cord injury patients. *Pain, 34*, 109–116.

Boivie, J. (2003). Central pain and the role of quantitative sensory testing (QST) in research and diagnosis. *European Journal of Pain, 7*, 339–343.

Boivie, J., Leijon, G., & Johansson, I. (1989). Central post-stroke pain -a study of the mechanisms through analyses of the sensory abnormalities. *Pain, 37*, 173–185.

Eide, P. K., Jørum, E., & Stenehjem, A. E. (1996). Somatosensory findings in patients with spinal cord injury and central dysesthesia pain. *Journal of Neurology, Neurosurgery, and Psychiatry, 60*, 411–415.

Finnerup, N. B., Johannesen, I. L., Fuglsang-Frederiksen, A., et al. (2003). Sensory function in spinal cord injury patients with and without central pain. *Brain, 126*, 57–70.

Garcia-Larrea, L., Convers, P., Magnin, M., et al. (2002). Laser-evoked potential abnormalities in central pain patients: The influence of spontaneous and provoked pain. *Brain, 125*, 2766–2781.

Leijon, G., Boivie, J., & Johansson, I. (1989). Central post-stroke pain -neurological symptoms and pain characteristics. *Pain, 36*, 13–25.

Merskey, H., & Bogduk, N. (1994). *Classification of chronic pain: Descriptions of chronic pain syndromes and definitions of pain.* Seattle: IASP Press.

Siddall, P. J., Taylor, D. A., & Cousins, M. J. (1997). Classification of pain following spinal cord injury. *Spinal Cord, 35*, 69–75.

Siddall, P. J., Yezierski, R. P., & Loeser, J. D. (2000). Pain following spinal cord injury: Clinical features, prevalence and taxonomy. *IASP Newsletter, 3*, 3–7.

Störmer, S., Gerner, H. J., Grüninger, W., et al. (1997). Chronic pain/dysesthesia in spinal cord injury patients: Results of a multicentre study. *Spinal Cord, 35*, 446–455.

Diagnosis and General Pain Management

Nikolai Bogduk
University of Newcastle, Newcastle Bone & Joint Institute, Royal Newcastle Centre, Newcastle, NSW, Australia

Characteristics

Diagnosis

At their disposal, physicians have a variety of tools by which to pursue a diagnosis. These include taking a history (see ▶ Medical History) and performing a physical examination (see ▶ Musculoskeletal Examination). Depending on the nature of the complaint, the physician will use these tools to various extents. However, for the subsequent treatment to be effective, the diagnosis must be correct, and that requires that the tools used to make the diagnosis are both reliable and valid. Without these features, the diagnosis that is made will be false.

▶ Reliability is the extent to which two observers, who use the same test on the same group of patients, agree on the results of the test. If the agreement is high, when the test is positive, it is highly likely that the patient truly has the abnormality that has been detected. However, if the agreement is low, neither observer nor the patient can be certain whether or not the patient has the abnormality.

▶ Validity is the measure of the extent to which a test actually detects the presence and absence of the condition that it is designed to detect. Whereas reliability measures the ability of observers to perform the test correctly, validity measures the intrinsic ability of the test to detect what it is supposed to detect, when it is performed correctly.

Various subtypes of validity can apply. The most elementary is concept ▶ validity, which is no more than a statement of the purported, theoretical basis for the test. A test is considered to have concept validity if, in principle, it sounds as if it should work. All tests in pain medicine have concept validity. They do sound as if they should work; but having only concept validity does not guarantee, in a scientific sense, that the test does indeed work. Unfortunately, too many tests that have been used in pain medicine are based solely on concept validity. They have not been shown to have any greater degree of validity, or when subjected to scientific scrutiny, they have been shown to lack any greater degree of validity.

The next level of validity is content validity. This requires that the test be defined with strict criteria that must be satisfied before the result of the test can be considered positive. Having content validity ensures that physicians use the test consistently and do not differ from patient to patient in what they consider to constitute a positive result. In one sense, content validity serves to secure reliability in applying the test, but it also serves to ensure that the test is properly defined so as to identify specifically the particular condition that it is supposed to detect and not some other condition.

How well a test actually discriminates between a particular condition and some other condition or conditions defines its construct validity. Just because the result of a test is positive does not necessarily mean that the condition sought for is present. Other conditions might also cause the result to be positive. Such results are false positive results. Similarly, test results may be negative, but the condition is nevertheless present. Such results are false negative. Construct validity is the measure of the extent to which a test is compromised by false positive and false negative results. The greater the incidence of false results, the less accurate and less useful the test.

The cardinal measures of the validity of a diagnostic test are its ▶ sensitivity, ▶ specificity, and likelihood ratios. The sensitivity is the extent to which the test correctly detects the presence of a condition. The specificity is the extent

to which the test correctly detects the absence of the condition. The likelihood ratios indicate the extent to which a positive result is likely to be true positive and to which a negative result is likely to be true negative.

One of the means by which the reliability and validity of diagnosis in pain medicine is enhanced is by the use of taxonomies. A taxonomy is a catalogue that lists the conditions that can be a cause of pain and which stipulates for each condition the criteria that must be satisfied before the condition can be diagnosed. If the criteria are satisfied, the condition can be legitimately be diagnosed. If the criteria are not satisfied, the condition cannot be diagnosed as the cause of pain, irrespective of the intuition or conviction of the physician that it is the cause. Taxonomies serve to prevent physicians arbitrarily and incorrectly making diagnoses that may be wrong.

The International Association for the Study of Pain (IASP) has produced a taxonomy that lists the possible causes of chronic pain (Merskey and Bogduk 1994). The IASP ▶ taxonomy catalogues these conditions in terms of the region of the body affected by the pain, the anatomical location of its source (if known), and its pathology (if known). A similar taxonomy has been developed for headache, by the International Headache Society (IHS) (1988). The IHS ▶ taxonomy lists the criteria that must be satisfied before particular types of headache can be diagnosed. The American Psychiatric Association has produced the fourth edition of its taxonomy for psychiatric disorders, known as the *Diagnostic and Statistical Manual of Mental Disorders* (▶ DSM-IV) (American Psychiatric Association 2000). For certain conditions that may be associated with pain, the DSM-IV prescribes the criteria that must be satisfied before a patient can be considered to have a condition in which pain is due to or associated with a significant psychological disorder.

Despite the availability of taxonomies to assist diagnosis, it is often not possible to make a distinct, patho-anatomic, or even psychological diagnosis of a patient's pain. In such cases, the physician can formulate an ▶ assessment, instead of a diagnosis. The assessment typically describes the location and nature of the patient's pain, the associated psychological distress that the patient expresses, and the disabilities that both of these features seem to produce. Such a formulation allows the physician to treat various components of the patient's problems when they do not have access to the root cause of the pain. Diagnosis and assessment differ only with respect to the end point that is achieved. The former defines the cause of the symptoms; the latter carefully defines the symptoms when the cause cannot be found. Both processes rely on the same tools.

The ▶ pain history is a catalogue of all the features of the pain that a patient can describe. The catalogue is produced by the patient responding to enquiries from the physician into possible features that may or may not be present. A comprehensive pain history is obtained by enquiry into a set of prescribed domains (see below). These domains are based on traditional wisdom expressed by experts experienced in assessing patients with pain. Not all domains of enquiry will necessarily be relevant or fruitful in every patient, but the set of domains covers all possibilities. Following the set ensures that the physician does not neglect or forget domains that may be relevant. The domains of enquiry for a comprehensive pain history are:

- Length of illness
- Etiology
- Site
- Extent or spread
- Intensity
- Quality
- Frequency
- Duration
- Time of onset
- Mode of onset
- Precipitating factors
- Aggravating factors
- Relieving factors
- Associated features

By taking a comprehensive history, a physician will often be able to establish the nature of the pain and at least its possible causes. Most forms of ▶ headache can be diagnosed from the history alone.

Physical examination complements the history. By examining the patient, the physician looks for abnormalities of structure or function that can be detected by inspection, palpation, movement, or the application of simple tools at the bedside, such as a stethoscope or a patellar hammer and other neurological instruments. The examination can address particular body parts or body systems. A neurological examination (see ▶ Diagnosis of Pain, Neurological Examination) tests the integrity of the nervous system. A ▶ musculoskeletal examination tests the status and function of the muscles, bones, and joints. Other parts of the body can be examined according to the complaint with which the patient presents. These might include an examination of the ear, nose, and throat or the mouth, teeth, and jaws or the chest, abdomen, pelvis, or perineum.

However, whereas physical examination has been a traditional component of medical diagnosis, its reliability and validity are variable. Most reliable and valid is neurological examination. Any two observers will usually agree reasonably well on whether a patient has numbness or weakness, and the presence of such features accurately indicates particular causes. The same does not apply to musculoskeletal examination. Research studies have shown that observers have difficulties agreeing on the extent to which the patient has tenderness, stiffness, spasm, or impaired movement. Furthermore, even if the observers can agree on the presence of such abnormalities, data are missing as to what these abnormalities actually imply about the possible cause of pain. The reliability and validity of other forms of examination are not known, in the context of pain.

Because of these limitations, physical examination has limited utility in allowing a patho-anatomic diagnosis to be made. Nevertheless, it does serve as a screening test to exclude obvious and major abnormalities, and it serves for descriptive purposes to define the nature and extent of the physical abnormalities that the patient exhibits. In conjunction with the history, the results of the physical examination may serve to indicate what investigations may or may not be required.

Having obtained a history and having performed a physical examination, a physician should be able to establish, in broad terms, the nature of the patient's pain. The major types of pain are ▶ neuropathic pain, ▶ central pain, ▶ radicular pain, ▶ neuralgia, ▶ somatic pain, and ▶ visceral nociception and pain. Somatic pain and visceral pain may be associated with ▶ referred pain. Because the latter can be mistaken for radicular pain, recognition of its features is important. Failure to distinguish the two conditions will lead to inappropriate investigations and possibly inappropriate treatment. Recognition of these broad categories of pain is not tantamount to having made a diagnosis. Investigations may be required to establish the cause or source of pain or to define more precisely its mechanisms.

The investigations used in the assessment of patients with pain include, but are not necessarily limited to, blood tests, electrophysiological tests, and medical imaging. These tests, however, typically do not detect the pain but are designed to detect conditions that may be responsible for the pain. Their use is, therefore, indicated not by the presence of pain but by features obtained by the history or examination that suggest the possibility of a particular cause.

Blood tests can be diagnostic for conditions such as acute intermittent porphyria and systemic infection. The erythrocyte sedimentation rate is an important and inexpensive screening test for conditions such as infection and cancer. A raised sedimentation rate calls for further investigation of a possible serious cause of pain. A normal sedimentation rate is essentially reassuring in so far as further pursuit of serious conditions is not indicated unless and until other clinical features suggestive of a possible serious illness are manifest. In patients with features of a visceral disorder, specific blood tests may be obtained, such as serum amylase for suspected pancreatitis and prostate specific antigen levels for suspected prostatic cancer. Serum calcium levels and alkaline phosphatase levels can be tested in patients with suspected osseous metastases or hyperparathyroidism. Catecholamine levels will be diagnostic of headache caused by pheochromocytoma. Serological tests can be used to confirm rheumatoid arthritis and related inflammatory disorders. Serological tests are appropriate in patients with spinal pain in whom myeloma is a possibility.

In some circles, ▶ electromyography, i.e., EMG, and related electrophysiological tests such as ▶ conduction velocities are used in the assessment of patients with pain. Despite this practice however, the evidence challenges and even refutes their utility. Electrophysiological tests do not detect pain. They only show abnormalities in motor fibers and large diameter sensory fibers. Such abnormalities correlate poorly with pain. Systematic reviews and consensus statements have found no justification for electrophysiological tests for radicular pain that is otherwise clinically apparent (Andersson et al. 1996; Dvorak 1996, 1998). Only rarely might they be justified if the clinical features suggest peripheral neuropathy rather than radicular pain. Even for carpal tunnel syndrome, the utility of electrophysiological testing has been questioned because of the very high false positive rate (Hadler 1997; Atroshi et al. 1999).

Similarly, ▶ thermography has been advocated as a test for pain due to radiculopathy and to assess sympathetic function in complex regional pain syndromes, general aspects. For radiculopathy, however, thermography is essentially superfluous or redundant because the diagnosis can be made clinically and because other tests such as ▶ MRI establish the actual cause of the problem. Thermography may corroborate a diagnosis of complex regional pain syndrome but the operational criteria for this condition do not require thermography.

The mainstay of diagnosis in pain medicine is medical imaging. This may involve ▶ plain radiography, i.e., X-rays, ▶ CT scanning, ▶ ultrasound, ▶ magnetic resonance imaging, i.e., MRI, ▶ bone scan, or ▶ single-photon emission tomography, i.e., ▶ SPECT. Because of its limited sensitivity and specificity for disorders other than fractures, plain radiography is not suitable either as a diagnostic tool or a screening tool in patients with pain. It neither detects nor excludes serious disorders that are not otherwise manifest.

Meanwhile, tumors and infections are demonstrated with far greater sensitivity and specificity by other tests, such as MRI. For conditions such as rheumatoid arthritis, plain radiography can demonstrate the state of a joint but is not one of the diagnostic criteria for this condition. In patients with suspected osteoarthritis, features seen on plain radiography are barely more than coincidentally associated with pain. The indications for plain radiography are features possibly suggestive of fracture, namely, major trauma or minor trauma in elderly patients with osteoporosis, in patients on corticosteroids or with a risk factor for pathological fracture. It is also an appropriate screening test for patients suspected of having Paget's disease.

CT scanning is not an appropriate screening test for patients with pain. It is indicated only if clinical features, or other tests, implicate a serious disorder. In that context, CT is used to define better a lesion that has already been detected. Ultrasound can be used to demonstrate lesions in soft tissues. It is strongly diagnostic of visceral disorders that cause cystic lesions or abnormalities in hollow organs. For musculoskeletal disorders, its utility is questionable. Although it can show lesions such as tears in tendons, there is no evidence that these lesions are indeed the cause of the patient's pain.

Of all medical imaging tests, MRI has the highest sensitivity and highest specificity. It can detect lesions such as tumors, infections, and osteonecrosis, early in their evolution, and can allow these lesions to be specifically identified. No other imaging test has these properties. It defines soft tissue lesions as clearly as, or better than, ultrasound. However, specific lesions are uncommon in patients with pain. The utility of MRI is not so much to be diagnostic but to detect or exclude occult or cryptic lesions that do not manifest distinctive clinical features. Bone scan does not demonstrate specific lesions. It shows only hyperemia when this is a response to a lesion such as a stress, a fracture, an infection, or a tumor. Although classically used to detect such lesions, bone scan is largely being replaced by MRI because of its greater specificity in diagnosing these disorders.

SPECT scanning combines the virtues of bone scanning and CT. Its utility in pain medicine has not been established. For classical disorders of the musculoskeletal system, it does not reveal anything that is not otherwise evident on MRI. SPECT is used by some physicians to corroborate a diagnosis of complex regional pain syndrome, but it lacks sensitivity and specificity in this regard. Moreover, the diagnosis can be made clinically without recourse to tests such as SPECT.

More than any other branch of medicine, pain medicine is characterized by the use of ▶ diagnostic blocks. Injections of local anesthetics or other agents can be used to determine either the mechanism of pain or its source. In patients with neuropathic pain, diagnostic blocks can be used to determine whether or not a particular nerve is involved in mediating the patient's symptoms. In patients with somatic pain, diagnostic blocks can be used to locate the structure from which the pain seems to arise. In patients with complex regional pain syndromes, diagnostic blocks can be used to determine if the pain is sympathetically maintained.

Diagnostic blocks can be performed by injecting small amounts of local anesthetic onto peripheral nerves, in order to block conduction along that nerve. Relief of pain constitutes *prima facie* evidence that the nerve is responsible for mediating the pain and that the source of pain lies somewhere in the territory supplied by the nerve. Diagnostic blocks can be performed as ▶ intravenous infusions. If a particular region of the body is infused with a local anesthetic, relief of pain implies that the source of pain or its mechanism lies in the region infused. If a particular region is infused with a drug other than a local anesthetic agent, relief of pain implies a mechanism for pain that involves a process that is blocked by the agent used. In that regard, sympathoplegic agents may be infused to determine if a patient's pain involves activity of the sympathetic nervous system.

Local anesthetic agents may also be infused systemically. Systemic infusions are used to test if the patient has central pain. It is believed that systemic infusions suppress spontaneous activity in the central nervous system that is responsible for the pain. Local anesthetic agents and other drugs can be infused into the epidural or intrathecal space, as a diagnostic test. ▶ Epidural diagnostic blocks can be used when the actual source of pain is not evident. Relief of pain implies that the source lies somewhere in the periphery and that the pain is not central pain. Conversely, failure to relieve pain with an epidural block indicates that the source is not peripheral and strongly implicates some form of central cause.

For the diagnosis of spinal pain, diagnostic blocks have been refined to target particular and often small structures. They are performed under fluoroscopic control in order to deliver tiny amounts of local anesthetic into small joints of the spine or onto small nerves that innervate these joints. These procedures include ▶ cervical medial branch blocks for the investigations of neck pain, ▶ lumbar medial branch blocks and ▶ sacroiliac joint blocks for the investigation of low back pain, ▶ spinal nerve block for the investigation of lumbar radicular pain, ▶ thoracic medial branch blocks and intra-articular blocks for the investigation of thoracic spinal pain, and thoracic spinal nerve blocks for the investigation of thoracic radicular pain.

Analogous procedures do not involve the injection of local anesthetic agents to relieve pain but instead involve the injection of contrast medium to provoke the target structure with the objective of determining whether or not the patient's accustomed pain is reproduced or not. The singular examples of this form of test are ▶ cervical discography and ▶ lumbar discography.

An important requirement for any invasive test, be it a diagnostic block or a provocation test, is that the test must be controlled. Without some form of control observation, a positive response to the test is not compelling. A patient may report relief of pain for reasons other than the action of the local anesthetic injected. Such responses are known as ▶ placebo responses. Similarly, a patient may report provocation of their pain, not because the target structure is the source of their pain but because anything or everything in the target region hurts. Such false positive responses render the diagnosis incorrect. Performing control tests militates against false positive responses and incorrect diagnosis.

For diagnostic blocks, control tests can be performed using an inert agent or an active agent with a different duration of action. If an inert agent is used, the patient should report no relief of pain. If an alternative agent is used, the patient should report relief that lasts for the expected duration of action of the agent used. For provocation tests, structures other to the target one should also be tested. The target structure can be held to be symptomatic only if stimulating adjacent structures does not reproduce the pain.

Treatment

Having made a diagnosis or having formulated an assessment of the patient's pain, a physician can implement treatment for it. This treatment may be highly specific, if the cause of pain is known and can be resolved or removed. The treatment may be target specific but not curative, if the pain from a particular structure can be stopped but without resolving or removing its cause. Or the treatment may be palliative, in that the pain can be reduced, while the cause of pain takes its natural course. In addition, other dimensions of the patient's problem can be managed, without focusing or while focusing on their pain. Their distress can be managed using psychological interventions. Their disability can be managed using rehabilitation techniques.

For the relief of pain, physicians have at their disposal of a wide variety of tools, from which they can select ones that are appropriate for the patient's condition. These include ▶ pharmacotherapy, in which the physician may use ▶ analgesics, compound analgesics, ▶ nonsteroidal anti-inflammatory drugs (NSAIDS), ▶ opioids, ▶ tricyclic antidepressants, or ▶ muscle relaxants as systemic agents to relieve pain irrespective of its cause or source. They may use topical NSAIDS, topical local anesthetics, or topical capsaicin to relieve pain in particular regions. Agents known as ▶ nutraceuticals, which are naturally occurring substances that can be used as drugs, are increasingly available. They are used to treat pain from joints. These include ▸ glucosamine, ▸ chondroitin, and ▸ avocado-soya bean unsaponifiables. There is emerging evidence that certain vitamins are effective for joint pain. There is evidence that ▸ bisphosphonates may relieve complex regional pain.

For focal sources of pain, injection therapy may be indicated. Tender areas in muscles might be treated with local anesthetic injections, ▸ steroid injections, ▸ botulinum toxin, or ▸ prolotherapy. Radicular pain may be treated with ▸ epidural steroids or ▸ lumbar transforaminal steroids or ▸ cervical transforaminal steroids. Joint pain can be treated with ▸ intra-articular steroids or ▸ hyaluronan.

Regional pain problems might be treated with exercises, traction, manipulation, passive mobilization, massage, various physiotherapy modalities, such as ▸ shortwave diathermy, therapeutic ultrasound, or ▸ interferential therapy, or with acupuncture or ▸ transcutaneous electrical nerve stimulation. Apart from formal psychological interventions, the patient may be treated holistically with assurance and activation, coupled with patient education.

Specific entities may be treated with conventional or traditional neurosurgical procedures that include destructive and augmentative therapies. Certain forms of pain can be relieved by techniques using ▸ radiofrequency neurotomy. These include trigeminal neurotomy, ▸ lumbar medial branch neurotomy, ▸ cervical medial branch neurotomy, dorsal root ganglion lesioning, and sacroiliac neurotomy.

Critical to the selection of any intervention from this armamentarium is whether or not the treatment works and if it has any attributable effect beyond that of placebo. For this reason, physicians should be aware of the data concerning the efficacy and effectiveness of any procedure that they select and its effect size or ▸ number needed to treat (NNT). Not all the treatments available to physicians work as well as their reputations pretend. Systematic reviews have brought into question many hallowed interventions. Increasingly, through publications and the Internet, consumers are becoming aware of these data, allowing them to demand access to

treatments that have been shown to work and to question those treatments that do not.

References

American Psychiatric Association. (2000). *DSM-IV-TR. Diagnostic and statistical manual of mental disorders* (4th ed.). Washington, DC: American Psychiatric Association.

Andersson, G. B. J., Brown, M. D., Dvorak, J., et al. (1996). Consensus summary on the diagnosis and treatment of lumbar disc herniation. *Spine, 21*, 75–78.

Atroshi, I., Gummeson, C., Johnsson, R., Ornstein, E., Ranstam, J., & Rosen, I. (1999). Prevalence of carpal tunnel syndrome in a general population. *Journal of the American Medical Association, 282*, 153–158.

Dvorak, J. (1996). Neurophysiologic tests in diagnosis of nerve root compression caused by disc herniation. *Spine, 21*, 39–44.

Dvorak, J. (1998). Epidemiology, physical examination and neurodiagnostics. *Spine, 23*, 2663–2673.

Hadler, N. M. (1997). Carpal tunnel syndrome diagnostic conundrum. *Journal of Rheumatology, 24*, 417–418.

Headache Classification Committee of the International Headache Society. (1988). Classification and diagnostic criteria for headache disorders, cranial neuralgias and facial pain. *Cephalalgia, 8*, 1–96.

Merskey, H., & Bogduk, N. (Eds.). (1994). *Classification of chronic pain. Descriptions of chronic pain syndromes and definition of pain terms* (2nd ed.). Seattle: IASP Press.

Diagnosis of Pain, Epidural Blocks

Nikolai Bogduk
University of Newcastle, Newcastle Bone & Joint Institute, Royal Newcastle Centre, Newcastle, NSW, Australia

Synonyms

Epidural diagnostic blocks

Definition

▸ Epidural blocks are a diagnostic test in which a local anesthetic (or another drug) is injected into the epidural space in order to obtain information about a patient's pain.

Characteristics

Diagnostic epidural blocks should not be confused with epidural injections for therapeutic purposes. Classically, epidural injections of local anesthetic have been used to provide analgesia for childbirth and for certain operative procedures. Epidural injections of local anesthetic, morphine, clonidine, corticosteroids, and other agents, have been used to provide relief of pain for a variety of conditions such as CRPS, cancer pain, neuropathic pain, radicular pain, and back pain that has failed to be relieved by surgery. In these instances, epidural injections have been used to provide sustained relief of pain, as a treatment.

When used for diagnostic purposes, epidural blocks are used to obtain information. Relief of pain may or may not occur, but relief of pain is not the objective; it is one of the possible results of the test.

Since they block multiple nerves, epidural blocks do not provide information about the location of a patient's pain. Instead, they provide information about the nature of the patient's pain.

Technique

A needle is inserted into the epidural space, usually at a mid-lumbar or low-lumbar level, using an interlaminar approach. Through the needle, a plastic tube is threaded into the epidural space, and the needle is withdrawn. The tube is used to administer any of a selection of agents. A typical regimen requires the serial administration of normal saline, fentanyl, naloxone, and local anesthetic. After each drug is administered, the patient reports the intensity of their pain.

Application

Epidural blocks can be used to determine the nature of a patient's pain, if this cannot be determined from the ▶ medical history or other tests. In particular, they can be used to distinguish nociceptive pain (i.e., ▶ somatic pain) from ▶ neuropathic pain or ▶ central pain. Epidural injection of a local anesthetic should relieve nociceptive pain but will have a variable effect on neuropathic pain and no effect on central pain.

If a patient genuinely has nociceptive pain, they would obtain no relief on each occasion that normal saline is administered. Fentanyl should relieve their pain, but naloxone will subsequently reverse that effect. Finally, local anesthetic will relieve their pain. Thus, if a patient exhibits this pattern of response, under controlled conditions, the interpretation is that they have a peripheral source for their pain.

If they do not obtain relief, some other explanation is required. It may be that they have neuropathic pain, or central pain. That diagnosis might be pursued using ▶ intravenous infusions.

If the patient has inconsistent or paradoxical responses, such as relief after normal saline, or relief after naloxone but not after fentanyl, the response suggests that psychological factors, rather than nociceptive processes, are the major determinants of the patient's complaint. In that event, treatment can be focused on behavioral factors.

Utility

Diagnostic epidural blocks appear attractive in principle, as a means to not just distinguish between central and nociceptive pain but also to distinguish patients in whom nociception or psychological factors are the principal determinants of their presentation. Evidence in this regard, however, is meager. The application of epidural blocks has been described only in case series that illustrate their application.

The original series described several patients who had inconsistent or paradoxical responses (Cherry et al. 1985). These patients responded to behavioral therapy. Other patients had a clearly nociceptive response, but a cause for their pain was not evident on first assessment. The response to the epidural block prompted further investigation. In one patient, a previously unrecognized neoplasm was detected. Other patients responded well to spinal cord stimulation.

The second series reported how 40 patients with back pain could be classified into those who had a nociceptive response, those who had neuropathic pain, and others who were nonresponders (Sorensen et al. 1996). The authors argued that establishing these distinctions serves to channel patients appropriately into pharmacological, behavioral, or surgical treatment regimens.

References

Cherry, D. A., Gourlay, G. K., McLachlan, M., et al. (1985). Diagnostic epidural opioid blockade and chronic pain: Preliminary report. *Pain, 21*, 143–152.
Sorensen, J., Kalman, S., Tropp, H., et al. (1996). Can a pharmacological pain analysis be used in the assessment of chronic low back pain? *European Spine Journal, 5*, 236–242.

Diagnosis of Pain, Neurological Examination

Nikolai Bogduk
University of Newcastle, Newcastle Bone & Joint Institute, Royal Newcastle Centre, Newcastle, NSW, Australia

Definition

Neurological examination is a process in which the physician assesses the integrity of the nervous system by testing at the bedside, using hands and simple instruments, but not involving special devices.

Characteristics

Principles

Neurological examination involves testing the function of either individual nerves, or of large parts or entire divisions of the nervous system, such as the spinal cord, or the sympathetic nervous system. Abnormality is deemed to be present if the normal function that is expected is absent or reduced.

Disorders of sensory fibers in peripheral nerves, or of sensory tracts in the spinal cord, are indicated by loss of sensation from skin, muscles, bones, or joints. Disorders of motor nerves or of motor tracts are indicated by paralysis or weakness of movements. Disorders of the sympathetic nervous system are indicated by changes in temperature of the skin, discoloration, swelling, sweating, and by trophic changes in skin, muscles, and joints.

In certain neurological disorders, amplified or altered sensations can occur, when normal functions are disinhibited by injury to nearby or overlapping nerves. Hyperesthesia is a sensation of exaggerated touch. Hyperalgesia is an increased sensation of pain. Allodynia is the perception of pain in response to a stimulus that would normally produce a sensation of touch or brush.

Practice

Sensory function is tested by determining if the patient can feel selected stimuli, such as touch, brush, pinprick, cold, or heat, when applied to selected areas of skin. Position sensation is tested by asking if the patient can detect movements passively applied to selected joints. Other sensory functions can be tested by eliciting reflexes from the tendons, muscles, skin, or cornea.

Motor function is tested by having the patient execute movements, with and without resistance, to determine how strongly they can activate the muscles responsible for that movement. Motor function can also be tested indirectly by eliciting reflexes to selected muscles.

Function of sympathetic nerves is tested indirectly by examining for features of increased or decreased function in the form of temperature changes, discoloration, swelling in the skin, wasting of muscles, or stiffening of joints. For more direct testing of the sympathetic nervous system, laboratory tests are required, in which the responses of tissues are measured when sympathetic nerves are artificially stimulated or anesthetized.

Parasympathetic nerves are difficult to test directly, but disturbed function of parasympathetic nerves is inferred when features of that disturbance are evident. Typically, this occurs in the form of facial flushing or salivation in the case of cranial parasympathetic nerves, or bladder and ejaculatory dysfunction in the case of pelvic parasympathetic nerves.

Interpretation

By combining and correlating all the abnormalities detected in a neurological examination, a physician will usually be able to determine where in the nervous system an injury or disorder is located. When a peripheral nerve is affected, the abnormalities will all fall in the distribution of that nerve. When the disorder affects the spinal cord, the abnormalities will be distributed across greater or lesser regions of the body, according to which parts of the spinal cord are affected.

Reliability

Since it is a long-established, traditional part of medical practice, neurological examination at large has been assumed to be reliable. Consequently, it has not often been subject to formal study. When studied, however, it has typically been found to be reasonably reliable (Viikari-Juntura 1987; Waddell et al. 1982; McCombe et al. 1989; Nelson et al. 1979).

Validity

Neurological examination serves well to establish the location of a neurological disorder. In that regard, it is considered to be valid. However, neurological examination does not usually establish the cause of the abnormality. It is not designed to do so. If an injury has occurred, that should be evident from the patient's history (see ▶ medical history). Otherwise, the actual cause typically requires special investigations, such as ▶ electrodiagnostic testing or medical imaging.

Relevance to Pain Medicine

Only a minority of patients with pain has a neurological disorder or a neurological cause for their pain. Consequently, neurological examination is not relevant for most patients with pain, other than perhaps to establish that there is no abnormal neurological function and, therefore, no neurological disorder.

Neurological examination is specifically indicated when patients report neurological symptoms in association with their pain. They may have a neurological disorder that is responsible for their pain, or they may have a disorder that, on the one hand, causes pain and, on the other hand, causes the neurological features.

Some intrinsically painful neurological disorders lack any features detectable by neurological examination. Such disorders include ▶ migraine and ▶ trigeminal neuralgia. Other intrinsically painful neurological disorders are consistently associated with signs of reduced or exaggerated function. These include ▶ postherpetic neuralgia, peripheral neuropathies, syringomyelia, and some spinal cord tumors.

Some painful disorders do not arise in the nervous system but affect nearby nerves secondarily, usually by pressing on them. Such disorders include tumors, cysts, and infections of the vertebral column, and intervertebral disc herniations. In such cases, neurological examination does not establish the cause of pain but does point to its location.

When peripheral nerves have been injured, injury is the obvious cause of abnormalities. In such cases, neurological examination is not required to establish the cause, but it serves to establish the nature and extent of altered functions.

Neurological examination is also used to determine the accuracy of diagnostic blocks that might be performed in the course of managing a painful disorder. When somatic nerves are blocked, finding numbness in the appropriate territory indicates that the block has been delivered correctly to the target nerve. Similarly, when sympathetic nerves are blocked, finding increased temperature or a Horner's syndrome indicates that the target nerve has been successfully blocked.

References

McCombe, P. F., Fairbank, J. C. T., Cockersole, B. C., et al. (1989). Reproducibility of physical signs in low-back pain. *Spine, 14*, 908–918.

Nelson, M. A., Allen, P., Clamp, S. E., et al. (1979). Reliability and reproducibility of clinical findings in low-back pain. *Spine, 4*, 97–101.

Viikari-Juntura, E. (1987). Interexaminer reliability of observations in physical examinations of the neck. *Physical Therapy, 67*, 1526–1532.

Waddell, G., Main, C. J., Morris, E. W., et al. (1982). Normality and reliability in the clinical assessment of backache. *British Medical Journal, 284*, 1519–1523.

Diagnostic and Statistical Manual of Mental Disorders

Nikolai Bogduk
University of Newcastle, Newcastle Bone & Joint Institute, Royal Newcastle Centre, Newcastle, NSW, Australia

Synonyms

DSM, DSM-IV, DSM-IVR; DSM-IV; DSM-IVR; Mental disorders, diagnostics and statistics

Definition

The DSM is a taxonomy and schedule designed to secure the accurate and reliable use of diagnostic terms for mental disorders.

Developed by the American Psychiatric Association, the manual has appeared in several editions, the latest being the fourth (DSM-IV) (American Psychiatric Association 1994), and a revised (DSM-IV-TR) (American Psychiatric Association 2000) edition.

Characteristics

The DSM consists of two principal parts. A general section outlines how the assessment and description of patients with mental disorders should be conducted according to five axes. These are Clinical Disorders, Personality Disorders, General Medical Conditions, Psychosocial and Environmental Problems, and Global Assessment of Functioning. The remainder of the manual systematically lists and numbers all psychiatric disorders, provides descriptions of each, and specifies criteria that must be satisfied before a particular diagnosis can be accorded. It is this latter section that potentially pertains to pain.

Relevance to Pain

Pain can be a feature of certain psychiatric or behavioral disorders. The DSM indicates under which conditions a psychiatric diagnosis might be applied to a patient with pain. The Handbook of Differential Diagnosis (First et al. 2002), which accompanies the manual, recommends the algorithm depicted in Fig. 1 for classifying patients with pain.

The algorithm primarily invites the decision that the pain is no more than a symptom of a medical condition and requires no psychiatric diagnosis. For a psychiatric diagnosis to apply, the DSM prescribes explicit criteria that must be satisfied.

Other Mental Disorders

The critical criterion for determining a mental disorder is that pain is not the focus of attention. By definition, such patients, who present with features of the mental disorder, may admit pain as a symptom, but it is neither their primary concern nor the cardinal cause of disability. Most patients with chronic pain would not qualify for this diagnosis, as pain is their primary complaint and primary concern.

Conversion Disorder

Patients with pain are sometimes, if not commonly, regarded as having a conversion disorder. The essence of this diagnosis, however, is that the patient has a loss of motor or sensory function, such as paralysis, blindness, or numbness. The notes and diagnostic criteria for this rubric specifically preclude its use if pain is the sole symptom.

Hypochondriasis

The diagnostic criteria for hypochondriasis stipulate that the patient has a preoccupation with fears of having a serious disease. Although patients with chronic pain may appear preoccupied with their pain, to the extent of seeking

Diagnostic and Statistical Manual of Mental Disorders, Fig. 1 Decision tree for the psychiatric diagnosis of pain

relief, this differs from having a preoccupation with a fear of having pain, or fear of a serious disorder.

Somatization Disorder

When an organic explanation cannot be found for a patient's pain, physicians are often attracted to attributing the pain to a psychological origin. Somatization disorder is one of the available rubrics that they might invoke. The diagnostic criteria for this condition, however, stipulate that the patient must have pain in at least four different sites, as well as two gastrointestinal symptoms, one sexual symptom, and one pseudoneurological symptom, the latter not limited to pain. Furthermore, the symptoms are not intentional. These criteria specify a complex disorder that some patients with chronic pain may have; but by definition, the criteria do not apply to patients with pain in a single region, such as back pain, abdominal pain, or neck pain.

Undifferentiated Somatoform Disorder

The diagnostic criteria for this disorder are more liberal than those for somatization disorder, and some physicians may be attracted to attributing undiagnosed chronic pain to this disorder. The notes for this disorder, however, specify fatigue, loss of appetite, gastrointestinal, or urinary symptoms. Pain is conspicuously not mentioned, which suggests that this rubric was not designed to accommodate patients with pain for which there is no explanation.

Factitious Disorder

This condition is characterized by the intentional production or feigning of physical or psychological signs or symptoms in order to assume a sick role but where external incentives for the behavior are absent. A further qualification is that individuals with factitious disorder present their history with dramatic flair but are extremely vague and inconsistent when questioned in greater detail.

Pain may be the symptom produced, and distinction between a genuine and fabricated complaint may be difficult. Lack of external incentives distinguishes factitious disorder from malingering, and a patient who is consistent and lacks dramatic flair when describing their symptoms is unlikely to qualify as having a factitious disorder.

For a physician to apply a diagnosis of factitious disorder, they would need to be certain that the patient is intentionally complaining of a false symptom. This conviction should be based on objective evidence. It should not be a reflection of the physician's lack of knowledge about unusual or unfamiliar pain problems.

Malingering

Malingering is not a formal psychiatric diagnosis, and the DSM does not provide diagnostic criteria for it. It states only that malingering should be strongly suspected if any combination is noted of medicolegal context of presentation, marked discrepancy between the person's claimed distress or disability and the objective findings, lack of cooperation during the diagnostic evaluation and in complying with the prescribed treatment regimen, and presence of antisocial personality disorder.

For many patients with chronic pain, some of these features must be carefully interpreted lest they be unjustly incriminating. The criterion for medicolegal context pertains to overt and deliberate compensation-seeking behavior. It does not apply to injured patients covered by workers compensation who cannot avoid a medicolegal context. Some causes of pain may lack objective findings. In other instances, physical findings on musculoskeletal examination lack reliability and validity. Therefore, physicians will naturally fail to find a correlation between findings and symptoms. They should not misrepresent this natural feature of the disorder as malingering.

A systematic review found little scientific evidence to justify malingering as a diagnosis in patients with chronic pain (Fishbain et al. 1999; Osterwels 1987). Expert opinion considers it to be a rare condition (Fishbain et al. 1999; Osterwels 1987).

Adjustment Disorder

The essential feature for this diagnosis is a psychological response to an identifiable stressor that results in the development of clinically significant emotional or behavioral symptoms. Examples of stressors include termination of a relationship, business difficulties, living in a crime-ridden neighborhood, and natural disasters. Although pain is listed as a symptom by the handbook, it is conspicuously not mentioned by the manual. Rather the manual emphasizes behavioral symptoms such as anxiety and depression.

The use of this rubric may not be valid if the physician attributes pain to a stressor because superficially it is convenient to do so, instead of pursuing a diagnosis in a more rigorous manner.

Pain Disorder
The DSM provides an entry that explicitly accommodates patients with chronic pain. It provides three subtypes:
- Pain disorder associated with psychological factors
- Pain disorder associated with both psychological factors and a general medical condition
- Pain disorder associated with a general medical condition

For the first two subtypes, the critical criterion is that "psychological factors are judged to have an important role in the onset, severity, exacerbation, or maintenance of the pain." What the DSM does not provide are guidelines by which this judgment is to be made. Therefore, the judgment becomes a matter of the physician's opinion. For these rubrics to apply, psychological factors must overtly be responsible for the onset, severity, exacerbation, or maintenance of the pain. This is not the same as finding that patients have psychological features that are secondary to the persistence of pain.

The third subtype is not specified as a mental disorder. Having pain, for which psychological factors are judged not to be important, is no more than a medical condition. Unless there is explicit evidence to the contrary, most patients with chronic pain should fall into this category. It is this category that would accommodate patients with psychological features that are caused by the pain, not vice versa.

Utility
The cardinal utility of the DSM is that it stipulates rigorous criteria that must be satisfied before a psychiatric diagnosis can be applied to a patient with pain. Unless those criteria are expressly satisfied, a psychiatric diagnosis cannot be justified.

The pivotal weakness of the DSM is that "judgment" is required to determine whether or not psychological factors are associated with the disorder. This is a vexatious issue and may be more a measure of the physician's personal beliefs than a diagnosis of the patient.

References
American Psychiatric Association. (1994). *DSM-IV. Diagnostic and statistical manual of mental disorders* (4th ed.). Washington, DC: American Psychiatric Association.
American Psychiatric Association. (2000). *DSM-IV-TR. Diagnostic and statistical manual of mental disorders* (4th ed.). Washington, DC: American Psychiatric Association.
First, M. B., Frances, A., & Pincus, H. A. (2002). *DSM-IV-TR. Handbook of differential diagnosis*. Washington, DC: American Psychiatric Association.
Fishbain, D. A., Cutler, R., Rosomoff, H. L., et al. (1999). Chronic pain disability exaggeration/malingering and submaximal effort research. *The Clinical Journal of Pain, 5*, 244–274.
Osterwels, R., Kleinman, A., & Mechanic, D. (1987). *Institute of medicine committee on pain, disability, and chronic illness behavior: Pain and disability. Clinical behavioral and public policy perspectives* (p. 171). Washington, DC: National Academy Press.

Diagnostic Block

Definition

Relief of pain with local anesthetic injection.

Cross-References

- ▶ Chronic Low Back Pain: Definitions and Diagnosis
- ▶ Pain Treatment: Spinal Nerve Blocks
- ▶ Peripheral Nerve Blocks

Diagnostic Blocks

▶ Peripheral Nerve Blocks

Diagnostic Studies

▶ Low Back Pain Patients, Imaging

Diagnostic Subdivisions

▶ Orofacial Pain: Taxonomy/Classification

Diaphragmatic Breathing

Definition

A type of relaxation training in which the individual is taught to monitor his or her respiration and to take deep breaths using the muscles of the diaphragm. This technique is associated with increased oxygen perfusion and reduction in muscle tension.

Cross-References

▶ Coping and Pain
▶ Psychological Treatment in Acute Pain
▶ Relaxation in the Treatment of Pain

Diaries

Definition

Diaries are self-report devices that involve the collection of information about variables over time capture (i.e., when it occurs, such as a rating of pain at a specific predesignated interval, or whenever medication is taken), or gathering of information in relatively close proximity to an event of interest, rather than relying on retrospective reporting of an extended time period such as weeks or months (e.g., pain over the past month).

Cross-References

▶ Multiaxial Assessment of Pain

Diathermy

Definition

Literally "heating through." Originally the term was used to describe the deep heating effects of ultrasound, shortwaves, and microwaves. This usage remains correct, but the word is now often broadened to include the putative nonthermal effects of these agents as well.

Cross-References

▶ Therapeutic Heat, Microwaves, and Cold

Diathesis Stress

Definition

Diathesis stress is an understanding of the relationship between a stressor and an individual's response.

Cross-References

▶ Psychiatric Aspects of Pain and Dentistry

Diathesis-Stress Model of Chronic Pain

Definition

The diathesis-stress model views pain as a result of the interaction of predisposing and

Diathesis-Stress Model of Chronic Pain, Fig. 1   A diathesis-stress model delineating the main factors contributing to the development and maintenance of chronic pain

stress-related factors and assumes that social, psychological, and physiological factors are equally important in the understanding of chronic pain states Fig. 1.

Diazepam

Definition

Diazepam is a classical benzodiazepine facilitating the effect of GABA at $GABA_A$ receptors. It causes sedation, amnesia, anxiolysis, and (central) muscle relaxation and acts as an anticonvulsant.

Cross-References

▶ GABA and Glycine in Spinal Nociceptive Processing

Dictionary of Occupational Titles

Synonyms

DOT

Definition

The U.S. Department of Labor provides definitions of contents and demands (qualitative and quantitative) of occupations in order properly to match jobs and workers. These definitions are identified by electronic sources from the Occupational Information Network at (http://www.oalj.dol.gov/publicgov/public/dot/refrne/how2find.htm), or in Jist's Enhanced Dictionary of Occupational Titles at (http://jist.emcpublishingllc.com/enhanced-occupational-outlook-handbook-hardcover.html). The dictionary is organized under the headings:

Nine-digit Occupational Code identifying: (a) a particular occupational group; (b) worker functions concerning data, people, and things, and (c) differences between occupations.
Occupational title
Industry designation
The body of the occupational definition states:

Worker actions, their purpose, machines, tools, equipment, or work aids used by the worker, material used, products made, subject matter dealt with, or services rendered, instructions followed or judgments made.

The definition trailer presents worker traits, i.e., selected occupational characteristics on a General Educational Development Scale, whereas Reasoning, Mathematical, and Language Development Scales and the Specific Vocational Preparation Scale determine the requirements of a specific occupation.

Cross-References

▶ Vocational Counseling

Diencephalic Mast Cells

Daniel Menétrey
Biomédicale, Paris, France

Synonyms

Bone marrow; Mast cells; Paracrine immune cells

Definition

▶ Paracrine immune cells as part of the neuroimmune axis making a contribution to developing mature and degenerating nervous system.

Characteristics

Mast cells are immune system elements, mostly found at host-environment interfaces such as skin and gastrointestinal and respiratory tracts. They are known to derive from the ▶ bone marrow and enter the tissues as immature or precursor cells which then mature under microenvironmental conditions (Metcalfe et al. 1997). Mast cells are routinely identified by their ▶ proteoglycan contents, mostly heparin and chondroitin sulfate. Two main types of cells have been recognized, connective and mucosal cells, which can be distinguished by their proteolytic enzyme contents, proteases I and II, respectively. Several lines of evidence indicate that the functions of mast cells are linked to their capacity to respond to a broad array of mediators (neuropeptides, chemokines, etc.) and their ability to synthesize and release potent bioactive and vasoactive molecules (histamine, heparin, cytokines, nerve growth factor, nitric oxide, serotonin, etc. (Gordon et al. 1990)). In addition to its anticoagulant properties, heparin is related to regulation of angiogenesis, suppression of delayed hypersensitivity, and binding of proteins. Bioactive substances are stored in intracellular granules and released by extrusion. Mast cell activation may correspond to either intragranular changes accompanied with differential release of mediators or massive granule compound ▶ exocytosis (degranulation) as in ▶ anaphylactic reactions. Mast cells have been demonstrated to be fundamental cellular elements in the development of allergic reactions and inflammation (Wasserman 1979).

Mast cells also exist within the brain of several species, including mammals and humans (Olsson 1968; Persinger 1977; Silver et al. 1996, 2000). These include mast cells in ▶ dura mater, ▶ leptomeninges, and central nervous system (CNS), the last being exclusively located within the diencephalon. Evidence has been given that the diencephalic homing of mast cells can be triggered behaviorally and be influenced by physiological parameters (Theoharides 1990). The prevalent location of CNS mast cells within the dorsal thalamus, including nuclei with cortical projections, and their ability to react to physiological conditions raise the question of a potential role in sensory integrative processes. Diencephalic mast cell numbers vary with age, sex, and physiological state, as well as with environmental and hormonal factors. In small mammals, they increase in situations such as hibernation, exposure to cold or magnetic fields, fostering, parturition, estrus phase, subordination stress, and pharmacological insult. However, they decrease during social isolation or with daily handling. Diencephalic mast cells are more numerous in females than in males and show a left-sided predominance, especially in females. In doves, their number increases in aggressively behaving males and courted females as well as with steroid treatments. In all species, younger animals have more diencephalic mast cells. Mast cells release histamine by degranulation and may account for up to 50 % of this mediator in the adult rat brain (Sugimoto et al. 1995). Unlike dura mater cells, diencephalic mast cells store serotonin and heparin in different granules. Intravascular serotoninergic cells having mast cell features but lacking heparin also exist (Menétrey and Dubayle 2003).

Diencephalic mast cell functions still remain to be determined. It has been hypothesized that they could play a role in maintaining the normal physiology of vessel walls or local hemodynamics by regulating the opening of the blood–brain barrier (Zhuang et al. 1996; Esposito et al. 2001). A detoxifying action has also been envisaged since excessive numbers are often associated with infectious or degenerative processes. Their preferential functional relationships with ▶ astrocytes and their ability to provide delivery of neuromodulators to specific regions of the brain suggest neural-endocrine interactions at central level and the existence of a neuroimmune axis based on the cross talk between mast cells, astrocytes, and neurons (Purcell and Atterwill 1995).

References

Esposito, P., Gheorghe, D., Kandere, K., et al. (2001). Acute stress increases permeability of the blood-brain-barrier through activation of brain mast cells. *Brain Research, 888*, 117–127.

Gordon, J. R., Burd, P. R., & Galli, S. J. (1990). Mast cells as a source of multifunctional cytokines. *Immunology Today, 11*, 458–464.

Menétrey, D., & Dubayle, D. (2003). A one-step dual-labeling method for antigen detection in mast cells. *Histochemistry and Cell Biology, 120*, 435–442.

Metcalfe, D. D., Baram, D., & Mekori, Y. A. (1997). Mast cells. *Physiological Reviews, 77*, 1033–1079.

Olsson, Y. (1968). Mast cells in the nervous system. *International Review of Cytology, 24*, 27–70.

Persinger, M. A. (1977). Mast cells in the brain: Possibilities for physiological psychology. *Physiological Psychology, 5*, 166–176.

Purcell, W. M., & Atterwill, C. K. (1995). Mast cells in neuroimmune function: Neurotoxicological and neuropharmacological perspectives. *Neurochemical Research, 20*, 521–532.

Silver, R., Silverman, A.-J., Vitkovic, L., et al. (1996). Mast cells in the brain: Evidence and functional significance. *Trends in Neurosciences, 19*, 25–31.

Silverman, A.-J., Sutherland, A. K., Wilhelm, M., et al. (2000). Mast cells migrate from blood to brain. *Journal of Neuroscience, 20*, 401–408.

Sugimoto, K., Maeyama, K., Alam, K., et al. (1995). Brain histaminergic system in mast cell-deficient (Ws/Ws) rats: Histamine content, histidine decarboxylase activity, and effects of (S) alpha-fluoromethylhistidine. *Journal of Neurochemistry, 6*, 791–797.

Theoharides, T. C. (1990). Mast cells: The immune gate to the brain. *Life Sciences, 4*, 607–617.

Wasserman, S. I. (1979). The mast cell and the inflammatory response. In J. Pepys & A. M. Edwards (Eds.), *The mast cell – Its role in health and disease* (pp. 9–20). Turnbridge Wells: Pitman Medical.

Zhuang, X., Silverman, A.-J., & Silver, R. (1996). Brain mast cell degranulation regulates blood-brain barrier. *Journal of Neurobiology, 31*, 393–403.

Dietary Variables in Neuropathic Pain

Jasenka Borzan[1], Srinivasa N. Raja[1] and Yoram Shir[2]
[1]Department of Anesthesiology and Critical Care Medicine, Johns Hopkins University, Baltimore, USA
[2]The Alan Edwards Pain Management Unit, McGill University Health Centre, Montreal, QC, Canada

Definition

Foods and beverages are composed of various substances that can be helpful or harmful to human health. However, data on a possible analgesic role of dietary constituents in acute and chronic pain states are still limited. Recent in vitro and in vivo animal studies have indicated that dietary amino acids, proteins, and oils can be beneficial in neuropathic conditions. Conversely, carbohydrate consumption by individuals with untreated diabetes or by chronic alcohol abusers can lead to painful neuropathic conditions. Many of the published human studies lack proper control groups, but the results corroborate the idea that specific dietary choices or supplements have the potential to alleviate chronic neuropathic pain. These beneficial effects probably depend on both genetic factors and the specific type of the neuropathic condition. A large number of dietary constituents have yet to be explored for their potential benefits in the treatment of neuropathic pain.

Introduction

Some dietary components hold properties that may be beneficial in the treatment of neuropathic pain. Advances made in preclinical studies have enabled us to begin to understand the mechanisms by which certain compounds may attenuate pain or slow the progression of neuropathic disease. An estimated 35–63 % of chronic pain patients report using complementary and alternative approaches to manage pain, of which, approximately 57 % use vitamins or herbal remedies (Lee and Raja 2011; Brunelli and Gorson 2004). Unfortunately, published clinical trials to date often lack proper control groups, sufficient sample sizes, or randomization to convincingly demonstrate efficacy of specific dietary ingredients in neuropathic pain. We will review the most compelling basic science and clinical evidence available with a reminder that recommendations for or against the use of any compound should be made only after considering potential interactions with concomitant medications or contraindications with other pathological states.

Characteristics

Many peptides derived from foods have been identified to have opioid-like properties and hence have been named *exorphins* (in contrast to endogenous opioid peptides, or endorphins). Because exorphins show opioid receptor agonist or antagonist properties, they are an indication that diet selection is of potential importance in neuropathic and other painful conditions (for review of bioactive proteins and peptides including exorphins, see Kitts and Weiler 2003 (Kitts and Weiler 2003)). Other dietary constituents implicated as beneficial against neuropathic pain include soy isoflavones, polyunsaturated fatty acids, tryptophan, taurine, alpha lipoic acid, acetyl-L-carnitine, and vitamin E.

Preclinical Evidence

Only a small number of dietary constituents have been shown to possess analgesic properties in animal models of neuropathic pain. Diets enriched with the amino acids, tryptophan (Abbott and Young 1991) and taurine (Belfer et al. 1998), and diets based on casein protein (Shir et al. 1997) have been shown to suppress *autotomy* (self-mutilation) levels in a rat model of *deafferentation pain* that is produced by total peripheral neurectomy. Tryptophan, a precursor of serotonin, might exert its analgesic effects by activating spinal and supraspinal serotonergic *antinociceptive* mechanisms. In a placebo-controlled study, tryptophan supplements were shown to reduce neuropathic pain in human (as discussed in Abbott and Young (1991)). Taurine, an inhibitory amino acid found most abundantly in the central nervous system, is bioavailable primarily through ingestion of seafood and meat. Rats with a genetic tendency to develop high levels of autotomy (HA rats) expressed reduced levels of self-mutilation following total hind limb denervation when they consumed taurine-enriched water either pre- or post-nerve injury (Belfer et al. 1998). The authors of that study argued that taurine might protect the inhibitory neurons in the central nervous system from the damaging *excitotoxic* barrage that accompanies nerve injury.

The most comprehensive preclinical evidence for the analgesic effect of diet in neuropathic pain derives from the soy studies. In these studies, rats with *partial sciatic nerve ligation injury (the PSL*

Dietary Variables in Neuropathic Pain, Fig. 1 The effect of bread and cucumber (BC), casein-rich (CAS), or soy-rich (SOY) diet on average mechanical paw withdrawal thresholds before and 14 days after partial sciatic nerve ligation. Animals were fed one of the diets for 28 days (*left panel*: intact), or 14 days before and after the nerve injury (*right panel*: post-PSL). After PSL injury, rats fed soy-rich diet displayed significantly less mechanical allodynia than BC or CAS fed rats ($p = 0.04$ and $p = 0.001$ respectively) (Modified from Shir et al. 2001)

model) exhibited significantly less tactile and heat *allodynia* and hyperalgesia (Fig. 1) when fed soy-rich diets than when fed soy-deficient diets (Shir et al. 2001). It was further established that soy has a short-lasting prophylactic analgesic effect, but no palliative effect (Shir et al. 2001). *Phytoestrogens* (found abundantly in soy products) are the primary candidates for this preventive effect because they possess biological activities that could be relevant to pain (e.g., alteration of steroid hormone metabolism, inhibition of various protein kinases, antioxidant properties) (Shir et al. 2002). In the soy study, average plasma concentrations of phytoestrogens (approximately 400–900 nmol/l) were associated with reduced levels of tactile allodynia and mechanical hyperalgesia, but not with reduced heat allodynia, in rats with PSL injury. More recently, Valsecchi et al. (Valsecchi et al. 2008) reported that after chronic constriction injury of the sciatic nerve, administration of the phytoestrogen genistein dose dependently attenuates mechanical and heat allodynia in rats. This effect was attributed to immunomodulatory (e.g., inhibition of estrogen receptor β activation), anti-inflammatory (e.g., decrease in interleukin-1β and -6), and antioxidant (e.g., reduction in reactive oxygen species) properties of genistein.

Because highly variable levels of neuropathic pain among rats with nerve injury have also been attributed to genetic factors, investigators have studied the interaction between genetic and dietary factors in PSL-injured rats. Levels of heat hyperalgesia were found to be highly variable across eight different rat strains fed seven different diets, suggesting that genetic factors may influence the effects of diet on neuropathic pain (Shir and Seltzer 2001).

Polyunsaturated fatty acids, mainly the essential fatty acids linoleic acid (omega-6) and alpha-linolenic acid (omega-3), are dietary groups with biological activities that could be relevant to the suppression of chronic neuropathic pain (e.g., inhibition of voltage-gated sodium and calcium channels; inhibition of several protein kinases (Shapiro 2003)). Perez et al. (2004) have tested the role of dietary fat and fat/protein interactions in the development of tactile allodynia and heat hyperalgesia in PSL-injured rats. Their experiments showed that dietary fat is a significant independent predictor of neuropathic pain levels in rats and that this effect could be accentuated by dietary protein.

Clinical Evidence

Chronic neuropathic pain conditions in humans have a variety of etiologies, making the association between dietary constituents and neuropathic pain very complex. Clinical evidence for the effects of dietary components on the development or treatment of neuropathic pain exists mainly from studies in alcohol abusers and diabetic patients. Chronic alcohol abuse may result in a peripheral neuropathy in the lower extremities as a result of direct degenerative effects on both myelinated and unmyelinated nerve fibers, disruption of *axonal transport*, and enhancement of *glutamate neurotoxicity* in neurons (Koike et al. 2001). People with hyperglycemia, if untreated, often develop a painful neuropathic condition that most commonly affects the extremities (Duby et al. 2004). A diet high in carbohydrates can contribute to high intracellular glucose levels that can lead to pathological changes such as vasoconstriction and endothelial hyperplasia, increased *oxidative stress*, and activation of protein kinase C. These changes, in turn, promote *neuronal dysfunction* (e.g., decreased nerve blood flow resulting in neural hypoxia).

Vegan diet combined with exercise has been reported to effectively reduce symptoms of painful diabetic neuropathy, purportedly by helping to regulate blood viscosity and filterability above and beyond glycemic control (McCarty 2002). There is good evidence from randomized controlled trials in patients with diabetic neuropathy suggesting that oral or intravenous supplementation with alpha lipoic acid (600 mg daily) results in significant improvement of sensory symptoms (Lee and Raja 2011). Dietary supplements such as evening primrose oil and capsaicin also might have some role in decreasing pain symptoms in painful diabetic neuropathy without directly affecting blood glucose (Halat and Dennehy 2003). However, demonstration of efficacy will require additional studies. Another over-the-counter dietary supplement, acetyl-L-carnitine, has been shown to alleviate neuropathic pain in diabetic patients, particularly if treatment is begun soon after the onset of the disease (Evans et al. 2008). Some evidence supports a beneficial effect of acetyl-L-carnitine in the treatment of antiretroviral- or chemotherapy-induced neuropathic pain, but these claims need to be substantiated in large, randomized clinical trials (Lee and Raja 2011). Several randomized control trials also have reported that vitamin E treatment (400–600 mg/day) reduces neurotoxicity more than does placebo in patients undergoing chemotherapy.

Summary

It is reasonable to assume that multiple dietary constituents have biologically significant consequences, some of which may be directly related to neuropathic pain suppression. Preclinical evidence has supported the possibility that soy and some of its constituents promote antihyperalgesic effects in animal models of neuropathic pain. Clinically, the use of alpha lipoic acid and acetyl-L-carnitine has had promising results for complementary treatment of diabetic neuropathy. However, much careful work is still required to demonstrate a beneficial effect of specific nutrients for the treatment of neuropathic pain and to identify their active constituents. Future investigations to identify new effective dietary constituents and describe their underlying mechanisms in pain processing may provide novel therapeutic strategies in the treatment of neuropathic pain.

Acknowledgment We would like to thank Claire Levine for carefully editing this entry.

References

Abbott, F. V., & Young, S. N. (1991). The effect of tryptophan supplementation on autotomy induced by nerve lesions in rats. *Pharmacology, Biochemistry, and Behavior, 40*, 301–304.

Belfer, I., Davidson, E., Ratner, A., Beery, E., Shir, Y., & Seltzer, Z. (1998). Dietary supplementation with the inhibitory amino acid taurine suppresses autotomy in HA rats. *Neuroreport, 9*, 3103–3107.

Brunelli, B., & Gorson, K. C. (2004). The use of complementary and alternative medicines by patients with peripheral neuropathy. *Journal of Neurological Sciences, 218*, 59–66.

Duby, J. J., Campbell, R. K., Setter, S. M., White, J. R., & Rasmussen, K. A. (2004). Diabetic neuropathy: An intensive review. *American Journal of Health-System Pharmacy, 61*, 160–173.

Evans, J. D., Jacobs, T. F., & Evans, E. W. (2008). Role of acetyl-L-carnitine in the treatment of diabetic peripheral neuropathy. *The Annals of Pharmacotherapy, 42*, 1686–1691.

Halat, K. M., & Dennehy, C. E. (2003). Botanicals and dietary supplements in diabetic peripheral neuropathy. *The Journal of the American Board of Family Practice, 16*, 47–57.

Kitts, D. D., & Weiler, K. (2003). Bioactive proteins and peptides from food sources. Applications of bioprocesses used in isolation and recovery. *Current Pharmaceutical Design, 9*, 1309–1323.

Koike, H., Mori, K., Misu, K., Hattori, N., Ito, H, Hirayama, M., & Sobue, G. (2001). Painful alcoholic polyneuropathy with predominant small-fiber loss and normal thiamine status. *Neurology, 56*, 1727–1732.

Lee, F. H., & Raja, S. N. (2011). Complementary and alternative medicine in chronic pain. *Pain, 152*, 28–30.

McCarty, M. F. (2002). Favorable impact of a vegan diet with exercise on hemorheology: Implications for control of diabetic neuropathy. *Medical Hypotheses, 58*, 476–486.

Perez, J., Ware, M. A., Chevalier, S., Gougeon, R., Bennett, G. J., & Shir, Y. (2004). Dietary fat and protein interact in suppressing neuropathic pain-related disorders following a partial sciatic ligation injury in rats. *Pain, 111*, 297–305.

Shapiro, H. (2003). Could n-3 polyunsaturated fatty acids reduce pathological pain by direct actions on the nervous system? *Prostaglandins, Leukotrienes, and Essential Fatty Acids, 68*, 219–224.

Shir, Y., & Seltzer, Z. (2001). Heat hyperalgesia following partial sciatic ligation in rats: Interacting nature and nurture. *Neuroreport, 12*, 809–813.

Shir, Y., Campbell, J. N., Raja, S. N., & Seltzer, Z. (2002). The correlation between dietary soy phytoestrogens and neuropathic pain behavior in rats after partial denervation. *Anesthesia & Analgesia, 94*, 421–426, table.

Shir, Y., Raja, S. N., Weissman, C. S., Campbell, J. N., & Seltzer, Z. (2001). Consumption of soy diet before nerve injury preempts the development of neuropathic pain in rats. *Anesthesiology, 95*, 1238–1244.

Shir, Y., Ratner, A., & Seltzer, Z. (1997). Diet can modify autotomy behavior in rats following peripheral neurectomy. *Neuroscience Letters, 236*, 71–74.

Valsecchi, A. E., Franchi, S., Panerai, A. E., Sacerdote, P., Trovato, A. E., & Colleoni, M. (2008). Genistein, a natural phytoestrogen from soy, relieves neuropathic pain following chronic constriction sciatic nerve injury in mice: Anti-inflammatory and antioxidant activity. *Journal of Neurochemistry, 107*, 230–240.

Different Nociceptor Populations

▶ Nociceptors, Action Potentials, and Post-Firing Excitability Changes

Differentiation

Definition

The process cells undergo as they mature into normal cells. Differentiated cells have distinctive characteristics different from undifferentiated cells, perform specific functions, and are less likely to divide than undifferentiated cells.

Cross-References

▶ Cell Therapy in the Treatment of Central Pain

Difficulty to Disengage

Definition

Difficulty in switching attention towards neutral information once a threat stimulus has been detected and is being processed.

Cross-References

▶ Hypervigilance and Attention to Pain

Diffuse Noxious Inhibitory Controls

Synonyms

DNIC

Definition

The theoretical basis for understanding the inhibitory controls activated by noxious stimuli hypothesizes that noxious stimuli activate a surround inhibition that sharpens contrast between the stimulus zone and adjacent areas. This sharpened contrast has a net enhancing effect on the perceived intensity of the painful stimulus and an analgesic effect outside the stimulated zone. That is, inhibition within the spinal cord by noxious stimulation applied to one of one part of the body inhibits nociception in another part of the body. This surround analgesia may explain counterirritation and acupuncture analgesia.

Cross-References

▶ Acupuncture Mechanisms
▶ Opiates During Development
▶ Pain-Modulatory Systems, History of Discovery
▶ Tourniquet Test

Diffusion

Definition

The diffusion describes the effect of movement of particles from a region of higher concentration to a region of lower concentration.

Cross-References

▶ NSAIDs, Pharmacokinetics

Diffusion Tensor Imaging

Synonyms

DTI

Definition

Allows the mapping of white matter tracts in vivo.

Cross-References

▶ Human Thalamic Response to Experimental Pain (Neuroimaging)

Digital Nerve Blocks

Definition

Local anesthetic blockade of the finger or toe.

Cross-References

▶ Acute Pain in Children, Postoperative

Digital Radiography

▶ Plain Radiography

Dignity

▶ Ethics of Pain Control in Infants and Children

Dihydromorphinone

▶ Postoperative Pain, Hydromorphone

Dilaudid

▶ Postoperative Pain, Hydromorphone

Dimension

Definition

A dimension serves to define a point by its coordinates.

Cross-References

▶ Pain Measurement by Questionnaires, Psychophysical Procedures, and Multivariate Analysis

Dimethylsulfoxide

Synonyms

DMSO

Definition

Dimethylsulfoxide is a substance that has been used for pain treatment, e.g., as an ointment applied to inflamed joints. It has been given up because of side effects. It easily penetrates the skin and has been used in experimental settings to transport other agents through the skin.

Cross-References

▶ Opioid Modulation of Nociceptive Afferents In Vivo

Dioinguinal and Diohypogastric Nerve Block

Definition

Blockade of sensory innervation to the groin and scrotal sac on the ipsilateral side by infiltration of the abdominal wall in the region of the superior anterior iliac spine, or fan-shaped injection of 3–5 ml local anesthetic just beneath the internal oblique fascia.

Cross-References

▶ Acute Pain in Children, Postoperative

Direct Suggestion

Definition

Direct suggestion in the context of hypnosis refers to a suggestion that directly indicates what experience or behavior will occur (e.g., "your hand will feel progressively number and less painful").

Cross-References

▶ Hypnotic Analgesia

Disability

Definition

According to the WHO International Classification of Impairments, Disabilities, and Handicaps (ICIDH) it is defined as "any restriction or lack of ability to perform an activity in a manor or within the range considered normal for a human being." All in all the term disability serves as an umbrella concept for impairments, activity limitations, and participation restrictions. It can therefore be seen as the negative term of functioning. The interaction of disability and health condition works in both directions, because the presence of disability may even modify the health condition.

Cross-References

▶ Disability in Fibromyalgia Patients
▶ Disability, Upper Extremity
▶ Functioning and Disability Definitions
▶ Impairment, Pain-Related
▶ Oswestry Disability Index
▶ Pain as a Cause of Psychiatric Illness

Disability and Health (ICF) and ICF Core Sets

▶ WHO System on Impairment and Disability

Disability Assessment, Psychological/Psychiatric Evaluation

Robert J. Gatchel
Department of Psychology, The University of Texas, Airlington, TX, USA

Synonyms

Psychiatric evaluation; Psychological evaluation

Characteristics

The fundamental goal of the disability evaluation procedure is usually to ascertain whether a patient can or cannot work. However, such a determination is quite difficult in evaluating painful conditions because of the misguided assumption that impairment can be precisely and objectively measured and is closely linked to "mechanical failure" of an organ or body part. Often however, chronic pain patients will report activity restrictions that cannot be fully understood in terms of a specific "mechanical failure." There is often a low concordance between subjective reports of pain and objective data of impairment. This will introduce vagaries into the disability evaluation process. One disability evaluator may tend to ignore the patient's subjective reports of pain and disability and rely more heavily on any objective evidence of mechanical dysfunction that is available. Another evaluator may rely more exclusively on the subjective appraisals and activity restrictions reported by the patient, regardless of whether they can be objectively quantified in terms of any measurable mechanical failure or dysfunction. Still another evaluator may attempt to develop a composite of both the subjective and objective measures. Unfortunately, there is currently no totally-agreed-upon disability evaluation system that can be used (Dembe 2000; Gatchel et al. 2010; Robinson 2001; Schultz 2005).

A "▶ stepwise approach" should be used in the psychological/psychiatric evaluation process (Gatchel 2000, 2005). This is because most clinicians are under various time constraints, as well as billing constraints imposed by third-party payers, when considering the best method of evaluating possible disability in their patients with pain. Therefore, a frequently asked question is, "If I were to choose the most time- and cost-efficient assessment method, which one should I select?" However, one must not make the assumption that there is a single approach that can serve as the best assessment. For many patients, several methods and measures will be needed. Rather than asking which instrument and approach should be used, a better question is, "What sequence of testing should I consider to develop the best understanding of potential disability problems that might be encountered with this patient with reports of pain?" Therefore, psychosocial evaluation should be viewed as a stepwise process, proceeding from global indices of emotional distress and disturbance to more detailed evaluations of specific diagnoses of psychopathology. Thus, for example, an initial evaluation of disability may involve the administration of specific assessment tools that have been developed for this purpose, such as the Pain Disability Questionnaire (Anagnostis et al. 2004), the Million Visual Analog Scale (Million et al. 1982), or the Roland and Morris Disability Questionnaire (Roland and Morris 1983). A comprehensive review of many of these instruments is provided by Gatchel (2001, 2005) and Turk and Melzack (2001). In addition to one of these instruments, an initial screening process that can be done efficiently to flag obvious psychosocial distress might consist of the administration of the SF-36 (Ware et al. 1993), the SCL-90 (Derogatis 1983, pp. 115–132) and the Beck Depression Inventory (BDI;

Beck et al. 1961). Any pronounced scale elevations on these instruments would alert clinical staff to the degree of emotional distress or dysfunction in a pain patient, and would indicate the need for a more thorough evaluation. This may then lead to the administration of the MMPI-2 (Keller and Butcher 1991) or a structured interview for DSM diagnosis (SCID) to develop official American Psychiatric Association's Diagnostic and Statistical Manual, 4th edition, DSM-IV-based diagnoses of an Axis I or Axis II disorders (First et al. 1995). The Patient Health Questionnaire (PHQ) is a self-report alternative to the SCID that can be used for some Axis I diagnoses. The final step would be the administration of a psychosocial clinical interview, which is one of the most powerful assessment tools of the clinician. In addition to the traditional areas explored in a clinical history, there are other areas that would be explored that may pose potential barriers to recovery and could affect response to treatment. Some of these topics include: history of head injury, convulsions, or impairment of cognitive function; any stressful changes in lifestyle or marital status before or since the injury that precipitated the pain; any litigation pending for the patient's current medical or pain problem; and a work history, including explanation of job losses, changes, and dissatisfaction. The interview allows the clinician to contrast the patient's current psychosocial functioning with past functioning and to compare the testing data with the interview data. The clinician can then estimate the degree of disability reported by the patient and the potential for getting the patient to change behavior and possibly work toward rehabilitation.

Of course, the common denominator of all assessment methods is that they have adequate ▶ validity, reliability (reproducibility), and predictive value. Even when an instrument appears satisfactory on the basis of these psychometric criteria, ambiguity may exist regarding the relevance of findings in scientific reports to the circumstances in which the health care professional is using the instrument. Such ambiguities can be minimized by examining the match between the context in which a test was studied scientifically and the patient to whom it is applied in the clinical setting. Basically, it is answering the question of test validity by addressing the "valid for what purpose?" issue. Assessment methods may be valid for measuring specific biological states but have no validity in predicting, for example, disability or activities of daily living. Finally, it must be kept in mind that, when embracing a comprehensive biopsychosocial evaluation (see ▶ biopsychosocial perspective) model (Turk and Monarch 2002; Gatchel et al. 2007), it is important to consider each successive assessment measure in context with the other measures to be integrated. This will lead to the most comprehensive disability assessment of a patient that will then significantly contribute to the development of the most effective treatment regimen for dealing with the disability problem. A "stepwise approach" to assessment is recommended (Gatchel 2000, 2005), which proceeds from global indices of biopsychosocial concomitants of pain to more detailed evaluations of specific diagnoses.

References

Anagnostis, C., Gatchel, R. J., & Mayer, T. G. (2004). The pain disability questionnaire: A new psychometrically sound measure for chronic musculoskeletal disorders. *Spine, 29*, 2290–2302.

Beck, A. T., Ward, C. H., Mendelson, M. M., et al. (1961). An inventory for measuring depression. *Archives of General Psychiatry, 4*, 561–571.

Dembe, A. E. (2000). Pain, function, impairment and disability: Implications for workers' compensation and other disability insurance systems. In T. G. Mayer, J. Gatchel, & P. B. Polatin (Eds.), *Occupational musculoskeletal disorders: Function, outcomes and evidence*. Philadelphia: Lippincott Williams and Wilkins.

Derogatis, L. (1983). *The SCL90-R manual-II: Administration, scoring and procedures*. Baltimore: Clinical Psychometric Research.

First, M. B., Spitzer, R. L., Gibbon, M., et al. (1995). *Structured clinical interview for DSM-IV axis I disorders-nonpatient edition (SCID-I/NP, version 2.0)*. New York: New York State Psychiatric Institute.

Gatchel, R. J. (2000). How practitioners should evaluate personality to help manage patients with chronic pain. In R. J. Gatchel & J. N. Weisberg (Eds.),

Personality characteristics of patients with pain. Washington, DC: American Psychological Association.

Gatchel, R. J. (2001). *A compendium of outcome instruments for assessment and research of spinal disorders.* LaGrange: North American Spine Society.

Gatchel, R. J. (2005). *Clinical essentials of pain management.* Washington, DC: American Psychological Association.

Gatchel, R. J., Peng, Y., Peters, M. L., Fuchs, P. N., & Turk, D. C. (2007). The biopsychosocial approach to chronic pain: Scientific advances and future directions. *Psychological Bulletin, 133*, 581–624.

Gatchel, R. J., Kishino, N. D., & Minotti, D. E. (2010). The three major components of behavior used for assessing pain: Problems faced when there is discordance among the three. *Psychological Injury and Law, 3*, 212–219.

Keller, L. S., & Butcher, J. N. (1991). *Assessment of chronic pain patients with the MMPI-2*. Minneapolis: University of Minnesota Press.

Million, S., Hall, W., Haavik, N. K., et al. (1982). 1981 volvo award in clinical science: Assessment of the progress of the back-pain patient. *Spine, 7*, 204–212.

Robinson, R. C. (2001). Disability evaluation in painful conditions. In D. C. Turk & R. Melzack (Eds.), *Handbook of pain assessment* (2nd ed.). New York: Guilford.

Roland, M., & Morris, R. (1983). A study of the natural history of back pain. Part I: Development of a reliable and sensitive measure of disability and low back pain. *Spine, 8*, 141–144.

Schultz, I. Z. (2005). Impairment and occupational disability in research and practice. In I. Z. Schultz & R. J. Gatchel (Eds.), *Handbook of complex occupational disability claims: Early risk identification, intervention and prevention.* New York: Springer.

Turk, D. C., & Melzack, R. (2001). *Handbook of pain assessment* (2nd ed.). New York: Guilford.

Turk, D., & Monarch, E. S. (2002). Biopsychosocial perspective on chronic pain. In D. C. Turk & R. J. Gatchel (Eds.), *Psychological approaches to pain management: A practitioner's handbook* (2nd ed.). New York: Guilford.

Ware, J. E., Snow, K. K., Kosinski, M., et al. (1993). *SF-36 health survey: Manual and interpretation guide.* Boston: The Health Institute, New England Medical Center.

Disability Determination

▶ Disability Evaluation in the Social Security Administration

Disability Evaluation in the Social Security Administration

Victoria Dorf
Social Security Administration, Baltimore, MD, USA

Synonyms

Disability determination; Eligibility; Social security disability insurance (United States); Supplemental security income

Definition

Disability is the inability to engage in any ▶ substantial gainful activity by reason of any medically determinable physical or mental impairment that can be expected to result in death or that has lasted or can be expected to last for a continuous period of not less than 12 months.

Characteristics

The United States Social Security Administration (SSA) has responsibility for the administration of the Social Security ▶ Disability Insurance (SSDI) program provided for under title II of the Social Security Act (the Act) and the ▶ Supplemental Security Income (SSI) disability program provided for under title XVI of the Act. The SSDI program provides disability benefits to disabled workers insured under the Act, children of insured workers who become disabled before age 22 and to disabled widows or widowers and certain surviving divorced spouses of insured workers. The SSI program is a means-tested program that provides monthly payments to needy disabled adults and disabled children (individuals under age 18).

Both disability programs are national programs (United States), with the same ▶ definition of disability for adults regardless of the city or

State in which a person lives or where the disability determination is made. For adults, "disability" is defined as the inability to engage in any substantial gainful activity by reason of any medically determinable physical or mental impairment or combination of impairments that is expected to result in death or that has lasted or can be expected to last for a continuous period of not less than 12 months (1). Under SSI, "disability" for an individual under age 18 means a medically determinable physical or mental impairment or impairments that results in marked and severe functional limitations and that meets the ▶ duration requirement (2). For all individuals, a "medically determinable physical or mental impairment" is an impairment that results from anatomical, physiological, or psychological abnormalities which can be shown by medically acceptable clinical and laboratory diagnostic techniques. Likewise, the rules for considering ▶ symptoms, such as ▶ pain, in determining disability are essentially the same for both adults and children.

The SSA's policy for the evaluation of pain and other symptoms is set forth in regulations (3) and further clarified in agency instructions called Social Security Rulings (SSRs), SSR 96-3p (4), SSR 96-4p (5), and SSR 96-7p (6). A symptom, whether pain, fatigue, shortness of breath, nervousness, or any other symptom, is not a medically determinable impairment (5). Rather, SSA defines a symptom as a person's own perception or description of the impact of his or her physical or mental impairment(s) (7). Neither are symptoms alone enough to establish the existence of a physical or mental impairment (7) or to establish disability. This is so regardless of how many symptoms the person alleges or how genuine the person's report may appear to be (5, 6). There must be an underlying impairment (s), shown by medically acceptable clinical and laboratory diagnostic techniques (8), which could reasonably be expected to produce the symptom(s) (3).

To evaluate a person's symptoms the SSA first determines if there is a medically determinable physical or mental impairment that could reasonably be expected to produce the pain or other symptom(s) that the person alleges (3, 6). Thus, for example, an impairment of the lumbar spine could reasonably be expected to produce low back pain or pain radiating into a lower extremity, but could not reasonably be expected to produce arm pain. The threshold for finding that there is a medical basis for the alleged symptom(s) is low and does not involve a determination as to the intensity, persistence, or functionally limiting effects of the symptom(s), only whether the impairment(s) could reasonably be expected to cause the kind of symptom(s) the person alleges (3). If there is no medically determinable physical or mental impairment, or if there is a medically determinable physical or mental impairment(s) but the impairment(s) could not reasonably be expected to produce the person's pain or other symptoms, the symptom(s) cannot be found to affect the person's ability to do ▶ basic work activities (6). Basic work activities are the abilities and aptitudes necessary to do most jobs. Examples of such activities include physical functions such as walking, standing, sitting, lifting, pushing, pulling, reaching, and carrying or handling the capacities for seeing, hearing, and speaking; the mental abilities for understanding, carrying out and remembering simple instructions, use of judgment, and responding appropriately to supervision, coworkers, and usual work situations; and dealing with changes in a routine work setting (9). If there is a medically determinable impairment that could reasonably be expected to produce the types of symptoms a person reports, the SSA proceeds to consider the person's statements about the intensity, persistence, and functionally limiting effects of his or her impairment-related symptoms, unless a fully favorable determination can be made solely on the basis of the ▶ objective medical evidence. If the person's statements about the intensity, persistence, or functionally limiting effects of symptoms are not substantiated by the objective medical evidence, the adjudicator must consider all of the evidence in the case record and make a finding as to the credibility of the person's statements.

This ▶ credibility finding is a determination as to the degree to which the person's statements about his or her symptoms can be believed and accepted as true. To make the credibility finding, consideration is given to the objective medical evidence, the person's own statements about his or her symptoms, statements and other information provided by physicians and psychologists who have treated or examined the person and others about the person's symptoms and how they affect the person, and any other relevant evidence in the case record. The person's pain or other symptoms will be found to diminish his or her capacity to do basic work activities to the extent that the alleged symptom-related functional limitations and restrictions can reasonably be accepted as consistent with the objective medical evidence and other evidence.

A credibility finding is straightforward if the person's statements are consistent with what would be expected on the basis of the objective medical evidence. But the SSA does not require objective medical evidence to establish a direct cause and effect relationship between the impairment and the alleged intensity, persistence, or functional effects of the symptom(s). Rather, the SSA recognizes that even people with the same impairment may experience symptoms differently (10), that some people who experience pain may also report other symptoms related to the pain (11), and that a person's statements about symptoms may not always be consistent with what can be expected solely on the basis of the objective medical evidence and may, in fact, indicate a greater severity of impairment than suggested by that evidence.

Because of this, the SSA's adjudicators do not disregard the person's statements about his or her symptoms and their functional effects solely because the statements are not fully corroborated by the objective medical evidence. Rather, if the information in the case record is insufficient to assess the credibility of the person's statements, the SSA's adjudicators must make every reasonable effort to obtain additional information that could shed light on the credibility of the statements.

In assessing the credibility of the person's statements about his or her symptoms, the objective medical evidence must always be considered. While the SSA does not require objective medical evidence to corroborate the person's statements about his or her symptoms, the clinical and ► laboratory findings cannot be ignored, whether those findings are positive or negative. However, adjudicators are aware that the objective medical evidence is only one of many medical and nonmedical factors that must be considered.

SSA needs evidence from acceptable medical sources to establish the existence of a medically determinable impairment (12). Once the existence of a medically determinable physical or mental impairment(s) is established, the SSA considers information from all medical sources – not just physicians or psychologists, but also other medical sources, such as physical therapists, nurse practitioners, audiologists, and chiropractors – about the severity of the impairment(s) and how it affects the person's ability to work. Other sources, such as educational personnel and public and private social welfare agency personnel, can also provide information about the person and how his or her impairment(s) affects function. And, evidence from relatives, friends, neighbors, coworkers, clergy, and others can provide valuable information about the person's functioning on a day-to-day basis – information about what the person does and how well he or she does these things now and how the person functions now compared to how the person functioned in the past.

The SSA also considers the person's longitudinal medical history. Repeated attempts to seek or try medical treatment tend to lend credibility to the person's statements, but the failure to pursue regular medical treatment does not necessarily mean the person's statements about the intensity, persistence, and functionally limiting effects of symptoms are not credible without first considering any explanations that he or she may provide and any other information in the case record that may explain infrequent or irregular medical visits or the failure to seek medical treatment.

The adjudicator does not have to totally accept or totally reject the person's statements. Rather, in any given case, the adjudicator may determine that the evidence supports a finding that all of the person's statements are believable, that only some of those statements are believable, or that none of the statements are believable. Or some of the statements may be believable, but only to a certain degree.

Once the adjudicator has arrived at a credibility finding, he or she must provide specific reasons for the finding, supported by the evidence in the case record. The reasons must be sufficiently specific to make clear the weight that was given to the person's statements and the reasons for that weight.

Although the credibility finding is important in drawing conclusions as to the extent to which the person's symptoms affect his or her ability to perform basic work activities, it is not a determination as to whether the person is or is not disabled. Thus, a finding that all of the person's statements about the functionally limiting effects of pain or other symptoms are credible does not automatically mean that the person will be found disabled. Nor does a finding that only some or even none of the person's statements about pain or other symptoms are credible mean that the claim will be denied. Rather, the adjudicator must consider all of the evidence in the case record before a conclusion can be made about disability.

References

20 CFR 404.1513(a) and 416.913(a); http://www.socialsecurity.gov, under "Our Program Rules." Accessed February 2006.

20 CFR 404.1521(b) and 416.921(b); http://www.socialsecurity.gov, under "Our Program Rules." Accessed February 2006.

20 CFR 404.1528(a), 404.1529, 416.928(a) and 416.929; http://www.socialsecurity.gov, under "Our Program Rules." Accessed February 2006.

20 CFR 404.1545(e) and 416.945(e); http://www.socialsecurity.gov, under "Our Program Rules." Accessed February 2006.

Pain and Disability, Clinical, Behavioral and Public Policy Perspectives (1987) In: Osterweis M, Kleinman A, Mechanic D (eds) Institute of Medicine, p 116.

Section 1614(a)(3)(C) [42 U.S.C. 1382c] of the Social Security Act, as amended. Compilation of the Social Security Laws, U.S. GPO, Washington D.C. (2005); http://www.socialsecurity.gov under "Our Program Rules." Accessed February 2006.

Sections 223(d)(1)(A) and (2)(A) [42 U.S.C. 423] and 1614(a)(3)(A) and (B) [42 U.S.C. 1382c] of the Social Security Act, as amended. Compilation of the Social Security Laws, U.S. Government Printing Office, Washington D.C. (2005); http://www.socialsecurity.gov under "Our Program Rules." Accessed February 2006.

Sections 223(d)(3) [42 U.S.C. 423] and 1614(a)(3)(D) [42 U.S.C. 1382c] of the Social Security Ac, as amended; http://www.socialsecurity.gov, under "Our Program Rules." Accessed February 2006.

SSR 96-3p, "Titles II and XVI: Considering Allegations of Pain and Other Symptoms in Determining Whether a Medically Determinable Impairment is Severe," 61 FR 34468 (July 2, 1996); http://www.socialsecurity.gov, under "Our Program Rules." Accessed February 2006.

SSR 96-4p "Titles II and XVI: Symptoms, Medically Determinable Physical and Mental Impairments, and Exertional and Nonexertional Limitations" 61 FR 34488 (July 2, 1996); http://www.socialsecurity.gov, under "Our Program Rules." Accessed February 2006.

SSR 96-7p, "Titles II and XVI: Evaluation of Symptoms in Disability Claims" 61 FR 34483 (July 2, 1996) http://www.socialsecurity.gov, under "Our Program Rules." Accessed February 2006.

Title 20 of the Code of Federal Regulations (CFR), Parts 404.1529 and 416.929; Evaluation of Symptoms, Including Pain, 56 Federal Register (FR) 57941 and 56 FR 57944, effective November 14, 1991; www.gpoaccess.gov/nara ("GPO Access"). Accessed February 2006.

Disability in Fibromyalgia Patients

Robert Bennett
Department of Medicine, Oregon Health and Science University, Portland, USA

Synonyms

Disability; Handicap; Impairment

Definition

Fibromyalgia

A chronic pain disorder defined in terms of widespread pain and tenderness as per the American College of Rheumatology 1990 Classification criteria (Wolfe et al. 1990).

Disability

It involves the impact of impairment in societal or work functions, and notes the gap between what an individual can physically do and what the specific job or labor market requires. A disability evaluation thus takes into account the loss of function (impairment) *and* the patient's work requirements and home situation. Disability is usually decided by *nonphysician administrators*. Nevertheless, the clinician's determination of impairment, and inclusions of comments about ability to perform specific job tasks, the extent and permanence of the impairment, are important contributors to the administrative decision concerning disability.

Impairment

This is the loss of function, or derangement, of an organ or part of the body compared to what previously existed. This is a *medical* determination.

Handicap

An impairment that can be overcome by some type of reasonable accommodation to the impairment, i.e., glasses. New federal regulations mandate greater employer attention to "reasonable accommodation" to the needs of the permanently handicapped post-injury worker.

Characteristics

Prevalence of Disability

An analysis based on CDC the U.S. Census Bureau data from 2004 reported that the overall prevalence of disability in the United States was 22 % (CDC 2009). The top causes of disability were musculoskeletal disorders (8.6 million persons), spinal pain (7.6 million), and cardiac disorders (3 million). Age and sex were strongly associated with disability, with a 45.3 % prevalence in males aged >65 years compared to a 56.5 % prevalence in females aged >65 years. A survey of 1,604 patients with fibromyalgia over a course of 8 years in six U.S. academic medical centers reported that 64 % reported were able to work on all or most days and over 70 %

were employed or homemakers. There was considerable variation in disability rates from center to center, but overall 16 % of patients were receiving Social Security disability benefits (lower center-rate 6.3 % to highest center-rate 35.7 %) and 27 % were receiving at least one form of disability payment (4,108). A similar degree of the disability has been reported in Canada and the UK (White et al. 1999; Al-Allaf 2007).

Causes of Disability

Disability is the result of an impairment. An impairment is defined as an anatomical, physiological, or a psychological impediment. Impairment can relate to disorders of function at the organ level (e.g., a left-sided hemiplegia – an anatomic problem, epilepsy – a physiologic problem, schizophrenia – a psychological problem). Disability is an integrated concept that views impairment in a multidimensional context that embraces the following: age, sex, educational level, psychological profile, past attainments, job satisfaction, motivation, retraining prospects, social support systems, economic consequences, and potential for being competitive in the workforce. Factors affecting disability in FM are given in Table 1.

Disability in Fibromyalgia Patients, Table 1 Contributors to disability in fibromyalgia patients

Pain
Fatigue
Weakness
Psychological distress
Musculoskeletal stiffness
Mood disorders
Personality disorders
Medications
Cognitive impairment
Ongoing litigation
Poor social support
Job dissatisfaction
Lack of work autonomy
Low socioeconomic class
Heavy physical work
Associated disorders

Disability in fibromyalgia patients is a multidimensional problem (Grodman et al. 2011). Most FM patients report that chronic pain, stiffness, and fatigue adversely affect the quality of their life and negatively impact their ability to be competitively employed.

Pain and fatigability adversely affect motor performance in fibromyalgia patients, as everyday activities take longer. Two studies have reported that on self-assessment, FM patients have higher pain ratings and poorer functional status than patients with RA, OA, SLE, and scleroderma. In general, patients need more time to get going in the morning and usually require extra rest periods during the day. Although many fibromyalgia patients can tolerate work activity for short periods of time, the same tasks carried out for prolonged periods result in increased pain and fatigue. Heavy physical tasks, particularly involving repetitive activity, sustained muscle contraction, overhead work, and power grip demands, are especially difficult for fibromyalgia patients (Henriksson et al. 2005). Furthermore, psychological stress, as is often found in time censored work, aggravate fibromyalgia symptomatology. Cognitive dysfunction, particularly short-term memory and concentration, is increasingly recognized as being a common problem in fibromyalgia patients (Glass 2009) and has detrimental effects on most jobs, especially those that are intellectually demanding.

Disability is especially likely to occur in fibromyalgia patients with low socioeconomic status. These patients are more likely to have occupations involving manual labor, generally have poor flexibility as regards pacing their work and often labor in an environment lacking a positive social environment with appreciative supervisors and colleagues.

The fact that fibromyalgia patients usually look normal and can perform tasks effectively for short periods often generates pejorative attitudes in their immediate supervisors and coworkers who suspect that fibromyalgia is just an "excuse for not pulling their full weight." The resulting workplace disharmony causes further stress and aggravation of fibromyalgia symptomatology in a chronic downward spiral.

Assessment of Disability

The problems that physicians encounter in assessing the chronic pain patient are largely related to four issues: (1) Pain is a purely subjective sensation which is usually interpreted in an emotional context. (2) Chronic pain cannot be fully understood in terms of the classical model of disease that equates pathogenesis with tissue damage or dysfunction. (3) Many "non-sick" people have persistent pain but are not disabled. (4) Apparent impairment due to pain results from a complex interplay between past experiences, education, income level, work related self-esteem, motivation, psychological distress, fatigue, personal value systems, ethno-cultural background, and the availability of financial compensation.

In performing a disability evaluation, the physician should take a detailed medical, developmental, behavioral, and psychosocial history to assess a patient's current and premorbid level of functioning in an attempt to understand why the patient is now seeking disability.

There are no validated instruments for assessing disability in FM patients. A useful practical resource for the assessment of fibromyalgia related disability is the American Medical Association Guides to the Evaluation of Permanent Impairment (American Medical Association 1993). The (chapter 15) on chronic pain notes the following: (1) Pain evaluation does not lend itself to strict laboratory standards of accuracy. (2) The evaluation of chronic pain cannot be made on the basis of the degree of tissue damage – the classical medical model. (3) Pain evaluation requires a thorough understanding of a multifaceted biopsychosocial model of disease. (4) The physician's judgment of impairment represents a blend of the art and science of medicine, and judgment must be characterized not so much by scientific accuracy as by procedural regularity. It acknowledges that physicians are often uncomfortable in evaluating chronic pain states, but notes that they regularly make decisions on the basis of probabilities backed up by experience and stated in terms of reasonable medical probability.

There are several studies that have correlated higher scores on the Fibromyalgia Impact Questionnaire (FIQ) with disability (White et al. 2009). A structured observation based on Assessment of Motor and Process Skills (AMPS) appears to provide a reasonable assessment of functional ability in FM patients (Amris et al. 2011).

The various factors that seemingly influence the development of disability in fibromyalgia patients should also be considered:

Coping strategies. The lack of belief in one's ability to manage pain, and cope and function despite persistent pain is a significant predictor of the extent to which individuals with chronic pain become disabled and/or depressed. Therefore, therapy should target multiple goals, including, pain reduction, functional improvement, and the enhancement of self-efficacy beliefs.

Self-efficacy. The belief that one is capable of doing an activity is related to the development from an impairment to disability. The coping strategies questionnaire has been used to determine factors associated with disability. Coping attempts (reinterpreting pain, ignoring pain sensations, diverting attention, coping self-statements, and increased activity level) are associated with lower levels of psychosocial disability but higher levels of physical and total disability. Catastrophizing is highly associated with disability.

Fibromyalgia subgroups. Fibromyalgia patients are a very heterogeneous population and it is likely that different subgroups will display different levels of disability. The Yale multidimensional pain inventory has been used to classify fibromyalgia subjects into three groups. It was found that nearly 90 % of patients could be classified as dysfunctional, interpersonally distressed, or adaptive copers. The dysfunctional and interpersonally distressed groups had higher levels of disability. Observed physical function (cervical spine mobility) was positively correlated with patient perceived disability in the adaptive copers group but not in the other two groups (Turk et al. 1996).

Work conditions. Work is physically demanding, especially in adverse environmental conditions, it is strongly associated with disability in fibromyalgia patients. An inability or reluctance to modify work conditions and lack of work autonomy are associated with disability. A recent study from the Netherlands identified several features that were associated with continuation of being employed; they were: ability to pace work, a moderate workload, opportunities to recover from an FM flare, a good match between work and capabilities, opportunities for development and promotion, understanding and help from colleagues, a supportive and understanding employer, work agreements with the management (Bossema et al. 2012).

Associated conditions. Fibromyalgia is often associated with other disorders (e.g., irritable bowel syndrome, multiple chemical sensitivity, irritable bladder, restless leg syndrome, depression, orthostatic hypotension) that often tend to act synergistically in causing disability.

Rheumatic disorders. Fibromyalgia is commonly a complication of other rheumatic diseases, such as rheumatoid arthritis, systemic lupus erythematosus, and Sjogren's syndrome. In such instances, the impact of the central sensitization component of fibromyalgia amplifies the symptoms of the associated rheumatic and increases disability (Naranjo et al. 2002).

Depression. Mood disorders, especially depression, are a common accompaniment of chronic pain states. Depression by itself is, of course, a common cause of disability. When depression is accompanied by chronic pain, disability is magnified.

Prevention of Disability

Gainful employment is a powerful force in fulfilling one's obligations to society, maintaining self-esteem, and achieving financial security. Chronic pain invariably produces an existential crisis as it changes a person's perception of "self." This often results in varying degrees of anxiety, depression, and anger with loss of self-esteem, reduced self-efficacy, and functional decline. Enabling fibromyalgia patients do continue with productive employment must be based on two major strategies: (1) optimal management of fibromyalgia symptomatology and (2) a willingness to intercede in issues relating to job and workplace modifications (Liedberg & Henriksson 2002).

The optimal management of fibromyalgia symptomatology needs to be based on a multimodal approach to management (Spaeth & Briley 2009). At a minimum, this involves attention to the following issues: patient education, socioeconomic strata, pain, fatigue, sleep, psychological distress, deconditioning, and attention to commonly associated disorders (e.g., irritable bowel syndrome).

A readiness on the part of physicians to act in an administrative role in making recommendations to employers is an essential component of minimizing disability in fibromyalgia patients. In most cases, a careful workplace evaluation by an occupational therapist is invaluable tool for making appropriate recommendations. Lack of willingness to act on a patient's behalf often results in a downhill spiral of increasing workplace stressors, depressive symptoms, and sense of victimization.

Thus, the prevention of disability in fibromyalgia patients ultimately depends on both the expert management of their symptoms and a willingness to act proactively on their behalf in securing appropriate work place modifications. If work efficiency and productivity progressively decline over several years of observation, in a patient who has been cooperative in management recommendations, then the physician should recommend the patient seek disability and be prepared to help administratively in this process.

References

Al-Allaf, A. W. (2007). Work disability and health system utilization in patients with fibromyalgia syndrome. *Journal of Clinical Rheumatology, 13*(4), 199–201.

American Medical Association. (1993). *Guides to the evaluation of permanent impairment* (4th ed.). Chicago: American Medical Association.

Amris, K., Waehrens, E. E., Jespersen, A., Bliddal, H., & Danneskiold-Samsoe, B. (2011). Observation-based assessment of functional ability in patients with chronic widespread pain: A cross-sectional study. *Pain, 152*(11), 2470–2476.

Bossema, E. R., Kool, M. B., Cornet, D., Vermaas, P., de Jong, M., van Middendorp, H., et al. (2012). Characteristics of suitable work from the perspective of patients with fibromyalgia. *Rheumatology (Oxford), 51*(2), 311–318.

CDC. (2009). Prevalence and most common causes of disability among adults – United States, 2005. *MMWR Morbidity and Mortality Weekly Report, 58*(16), 421–426.

Glass, J. M. (2009). Review of cognitive dysfunction in fibromyalgia: A convergence on working memory and attentional control impairments. *Rheumatic Diseases Clinics of North America, 35*(2), 299–311.

Grodman, I., Buskila, D., Arnson, Y., Altaman, A., Amital, D., & Amital, H. (2011). Understanding fibromyalgia and its resultant disability. *The Israel Medical Association Journal, 13*(12), 769–772.

Henriksson, C. M., Liedberg, G. M., & Gerdle, B. (2005). Women with fibromyalgia: Work and rehabilitation. *Disability and Rehabilitation, 27*(12), 685–694.

Liedberg, G. M., & Henriksson, C. M. (2002). Factors of importance for work disability in women with fibromyalgia: An interview study. *Arthritis and Rheumatism, 47*(3), 266–274.

Naranjo, A., Ojeda, S., Francisco, F., Erausquin, C., Rua-Figueroa, I., & Rodriguez-Lozano, C. (2002). Fibromyalgia in patients with rheumatoid arthritis is associated with higher scores of disability. *Annals of the Rheumatic Diseases, 61*(7), 660–661.

Spaeth, M., & Briley, M. (2009). Fibromyalgia: A complex syndrome requiring a multidisciplinary approach. *Human Psychopharmacology, 24* (Suppl. 1), S3–S10.

Turk, D. C., Okifuji, A., Sinclair, J. D., & Starz, T. W. (1996). Pain, disability, and physical functioning in subgroups of patients with fibromyalgia. *Journal of Rheumatology, 23*, 1255–1262.

White, K. P., Speechley, M., Harth, M., & Ostbye, T. (1999). Comparing self-reported function and work disability in 100 community cases of fibromyalgia syndrome versus controls in London, Ontario: The London Fibromyalgia Epidemiology Study. *Arthritis and Rheumatism, 42*(1), 76–83.

White, D. H., Faull, K., & Jones, P. B. (2009). An exploratory study of long-term health outcomes following an in-patient multidisciplinary program for people with fibromyalgia syndrome. *International Journal of Rheumatic Diseases, 12*(1), 52–56.

Wolfe, F., Smythe, H. A., Yunus, M. B., Bennett, R. M.,
    Bombardier, C., Goldenberg, D. L., et al. (1990).
    The American College of Rheumatology 1990 criteria
    for the classification of fibromyalgia. Report of
    the Multicenter Criteria Committee. *Arthritis and
    Rheumatism, 33*(2), 160–172.

Disability Insurance Benefits

Definition

Social Security disability insurance benefits
are monthly benefits provided for under the
provisions of title II of the United States Social
Security Act. Entitlement to disability insurance
benefits (DIB) requires that an individual meet
both medical and legal requirements defined by
law and regulations.

Cross References

▶ Disability Evaluation in the Social Security
    Administration

Disability Management

▶ Disability, Effect of Physician Communication

Disability Management in Managed Care System

Bill McCarberg
Pain Services, Kaiser Permanente,
San Diego, USA

Synonyms

Capitated care; Managed care; Prepaid care

Definition

▶ "Managed care" does not have a precise definition. In general, it refers to techniques intended
to reduce costs and improve the quality outcomes. It also reflects systems of financing and
delivering care. Most often multiple providers
use a systematic approach to the delivery of
healthcare. The providers and the patients they
manage defines group health. Group health insurance provides members of a group at a discounted
rate medical insurance. Managed care systems
may be mandated by governmental agencies or
be established spontaneously by groups of physicians or health insurance companies. Sometimes,
the providers accept financial risk for the costs
associated with medical care. However, ▶ fee-
for-service medical treatment can also be provided according to a managed care plan. The
key feature of managed care is that a set of guidelines is established to determine how certain
medical conditions will be treated. Thus,
a managed care system can be contrasted with
a medical community in which several solo practitioners or small medical groups provide medical
care based on informal practice about when specialty referrals will be made, when certain diagnostics will be performed, and so on.

Characteristics

A major goal of managed care is to provide cost-
effective treatment with the use of evidence-
based guidelines. Rising medical costs in the
1980s and 1990s led to attempts by insurers to
move from traditional fee-for-service to managed
care plans. Workers' Compensation (WC) plans
did not join this push to managed care (MC) and
continued fee-for-service financing. As a result of
this different strategy, from 1985 to 1992, annual
medical expenditures for WC rose 45 % faster
than ▶ group health plans (Workers Comp Res
Inst 1997). WC cash indemnity and medical benefits payments increased (Worker's compensation costs 1992), despite nationally declining

prevalence rates of work disability (Prevalence of work disability-United States 1990) between 1980 and 1990.

In an attempt to control costs, many states made a concerted effort to embrace the tenets of cost-containment already proven effective in managed care group models (BNA's Workers' Compensation Report 1993). The use of managed care in workers' compensation has increased 150 % since 1991. With 75 % of employers who use managed WC plans finding them effective in controlling costs.

Differences Between Group Health and WC That Are Relevant to Managed Care

Returning patients to health and function in a cost-effective manner appears to be the goal of both group health and WC. However, there are fundamental differences that make use of group health managed care techniques problematic and difficult to implement in WC plans.

The state defines the level of coverage and medical services under WC, which are employer arranged under group health. There is salary replacement in WC, not in group health. Duration of coverage for a medical condition is generally longer for group health plans than for WC, because group health plans have an open ended commitment to a patient, whereas WC systems move toward claim closure when a person reaches maximal medical improvement. Employee productivity is all important. Since medical costs are just a part of expenses in WC, salary replacement costs must also be considered. Up to 60 % of WC plan costs are nonmedical (Ducatman 1987), arguing for more aggressive, earlier intervention to return employees to the workplace. More medical resources are used to treat injuries under WC plans than under group health plans (Johnson et al. 1996), expenditures that may be appropriate given the added cost of lost work time.

The principles of managed care do not easily translate in WC. Managed care describes a broad concept of active management and control over the use of medical services. Active oversight and control of medical services reduces costs and eliminates unnecessary expenditures under this model. Gate keeping is a fundamental tool of managed care, controlling access to providers and services. Many of the fundamental precepts used to control cost in managed care may not apply to the workers' compensation system.

The primary care physician is the gatekeeper in the group health model. All access to care must come through this provider. The primary care provider is less well trained in specific areas than specialists, but more familiar with the patient and his or her family. Since the primary care provider can adequately treat 90 % of medical conditions, this is the natural place for a gatekeeper. However, lack of familiarity with WC rules and procedures on the part of primary care providers can have deleterious consequences on disability and costs when these providers treat injured workers. Conversely, expedient and appropriate care by physicians familiar with the WC system has, in several studies, demonstrated reductions in disability and litigation, and improvement in patient satisfaction. Workers with chronic work-related conditions often require ongoing specialty care by musculoskeletal specialists, rendering the primary care provider less effective.

In group health, the parties involved are the patient and the primary care provider. In WC, besides these parties, the insurance carrier, the employer, and not infrequently a lawyer are also involved. Therefore, due to a more complicated treatment environment in WC, the primary occupational provider (POP) may be a better choice as gatekeeper, and is being employed in many WC managed care plans (Greenberg and Leopold 1998).

Cost sharing is frequently used in group health managed care. Co-payment for appointments or medications alerts the patient to the cost of services, and provides an incentive for patients to use fewer resources. WC laws explicitly prohibit patient cost sharing. States require employers to pay all medical expenses.

The incentive structure in WC favors continued disability. With salary replacement, full coverage of medical expenses and exemption of WC salary from most income taxes, return to work can be a problem. Generosity of WC benefits and duration of *injury* have been linked in a number of studies (Loeser et al. 1995). Managed care techniques do not alter this inherent incentive bias to remain disabled.

Provider risk sharing is another strategy frequently used in group health managed care. The reward system provides an incentive to providers to decrease medical expenditures. Capitation is an example of risk sharing, where providers receive a flat fee (often monthly) for each health plan enrollee. This payment is independent of medical resource utilization, thereby favoring the provider who manages the patient more efficiently. Less resource utilization produces higher profits. ▶ Case rate arrangements are another risk sharing technique. Under case rate agreements, the provider receives a set payment for a given diagnosis. The provider is incentivized to use fewer resources. Restricting care due to profit concerns may be effective in group health models where benign neglect produces favorable results in many medical conditions. In WC, where early return to work is the goal, waiting and watching the patient recover may defeat this goal. Capitation is losing favor in group health plans, and is rarely used in WC. Case rate arrangements are also less common in WC, since early and aggressive intervention may cost more yet return the patient to work sooner.

Opportunities for Effective Use of Managed Care in WC

Many strategies have been effectively utilized in managed care WC despite the differences listed above. Restricted provider networks allow for improved patient care, by steering patients to providers with expertise and competence in the management of occupational injury and illness. In so doing, managed care organizations can leverage their control over membership in provider networks to negotiate lower fees. Fewer providers allows for fewer billing sources, thereby decreasing administrative and management costs. A defined network of physicians also promotes standardized care based on treatment protocols, such as the ones that Aetna US Healthcare publishes in Coverage Policy Bulletins. With restricted provider networks, standardized care is possible resulting in significantly reduced medical expenditures (Cheadle et al. 1999).

The disadvantage to restricting choice is decreased patient satisfaction. Whether in group health models or WC, restricting the patient's ability to choose a provider leads to unhappy patients. Evaluations of Florida, Oregon and Washington State WC managed care plans, which use restricted networks, found that injured workers were significantly less satisfied than similar workers in traditional WC plans. These results were independent of treatment outcomes (Dembe 1998).

Case management is another managed care strategy that can be used in both group health and WC. It is frequently utilized for diseases that require a lot of patient compliance, are labor intensive and costly if not managed aggressively. Examples in group health settings include HIV, diabetes, and congestive heart failure. Examples in WC include work-related traumatic brain injuries and spinal cord injuries. Case managers are often nurses who work with physicians and other providers to maximize the coherence of a treatment plan. Case management is employed in WC due to the need for close follow-up, and the negative incentive for patients to return to work. Coordinating this care leads to better outcomes and reduced costs.

Something akin to case management can be used as a preventative strategy to reduce work-related injuries. A physician/nurse team assigned to an employer with specific repeated work-related injuries can risk manage, modify return to work policies, and be the first contact for the employer/employee in injury and illness (Feldstein and Breen 1998). This is an example of health maintenance through preventative services, long understood and practiced in managed care. More use of these strategies, through preventative safety engineering and ergonomic controls to abate workplace hazards and prevent injuries, are another managed care which needs to be used (Robinson 1998). Many of these techniques are difficult when individual providers offer care to the injured worker, but are ideal organizational fits for managed care and workers' compensation.

When Is a Medical Condition Work Related?

Separating work-related injuries from general medical health is often artificial. A patient with a chronic low back condition may worsen with repetitive work stresses. What part of the condition is related to work, what part is the underlying chronic condition? Attempts to integrate group health and WC into one managed care delivery system are problematic. The advantage of keeping all services in one seamless system of care is obvious and enticing. Unfortunately, the culture, structure, and financing of the WC system create barriers to such integration. In addition, healthcare providers, unfamiliar with management of work-related injuries and the WC system, often have difficulty providing effective care. Several pilot programs were started but never really developed fully due to these obstacles.

References

Cheadle, A., Wickizer, T. M., Franklin, G., et al. (1999). Evaluation of the Washington state workers' compensation managed care pilot project II. *Medical Care, 37*, 982–993.

Dembe, A. E. (1998). Evaluating the impact of managed health care in workers' compensation. *Occupational Medicine, 13*, 787–798.

Ducatman, A. M. (1987). The inevitable growth of workers' compensation costs. In *Workers compensation best's review* (pp. 50–52, 79–80). Oldwick: AM Best Company.

Feldstein, A., & Breen, V. (1998). Prevention of work-related disability. *American Journal of Preventive Medicine, 14*(3S), 33–39.

Greenberg, E. L., & Leopold, R. (1998). Performance measurement in workers' compensation managed care organizations. *Occupational Medicine, 13*, 755–772.

Johnson, W. G., Baldwin, M. L., & Burton, J. F., Jr. (1996). Why is the treatment of work-related injuries so costly? New evidence from California. *Inquiry, 33*, 53–65.

Loeser, J. D., Henderlite, S. E., & Conrad, D. A. (1995). Incentive effects of workers' compensation benefits: A literature synthesis. *Medical Care Research and Review, 52*, 34–59.

Managed care for workers' compensation up 150 %, new survey finds (pp. 181–182). (1993). *BNA's workers' compensation report*. Washington, DC: The Bureau of National Affairs.

Prevalence of work disability-United States. (1990). *MMWR, 42*(39).

Robinson, J. C. (1998). The rising long term trend in occupational injury rates. *American Journal of Public Health, 78*, 276–281.

Worker's compensation costs. (1992). *Medical benefits*, 1–2 Sept. New York: Panel Publishers.

Workers Comp Res Inst. (1997). *Managed care and medical cost containment in workers' compensation: A national inventory, 1997–1998* (Rep WC-97-6). Cambridge, MA: Workers Comp Res Inst.

Disability Payment

▶ Worker Compensation Benefits

Disability Prevention

Gary M. Franklin[1,4], Judith A. Turner[2,3], Thomas M. Wickizer[5], Deborah Fulton-Kehoe[1] and Robert D. Mootz[6]

[1]Department of Environmental and Occupational Health Sciences, University of Washington, Seattle, WA, USA
[2]Department of Psychiatry and Behavioral Sciences, University of Washington, Seattle, WA, USA
[3]Department of Rehabilitation Medicine, University of Washington, Seattle, WA, USA
[4]Department of Health Services, University of Washington, Seattle, WA, USA
[5]Health Services Management and Policy, Ohio State University, Colombus, OH, USA
[6]Washington State Department of Labor and Industries, Olympia, WA, USA

Synonyms

Secondary prevention

Definition

Disability prevention is the prevention of *long-term disability* (LTD) that may develop following a work-related injury when such disability would

Disability Prevention, Table 1 Disability prevention: Changing the paradigm

Primary prevention	Prevent workplace injuries and illnesses
Secondary prevention	Percent disability among workers with work-related injuries and illnesses
Tertiary prevention	Manage disability to reduce residual deficit and dysfunction

Disability Prevention, Fig. 1 Disability prevention is the key health policy issue (Adapted from Cheadle et al. (1994) Am J Publ Health 84:190–196). Inability to conduct any reasonably continuous work for at least 4 days

not normally be expected to result from the natural history of that injury. The term includes the *integrated actions* (healthcare, workplace, administrative) implemented to prevent long-term disability. Three types of prevention may be defined in the context of work-related injuries (Table 1): (1) *primary prevention* – prevention of the work-related injury (Frank et al. 1996a); (2) *secondary prevention* – prevention of long-term disability after an injury has occurred (disability prevention) (Frank et al. 1996b); and (3) *tertiary prevention* – prevention of additional clinical consequences once long-term disability has become established.

Long-term disability can be defined as having occurred if the worker is still totally disabled from gainful employment (temporary total disability) for a continuous or nearly continuous period of 3–6 months following injury. LTD can be said to be *established* if the worker is still disabled at 1 year.

Disability management, a common term in the US insurance industry, has been applied to a variety of case management interventions aimed at both secondary and tertiary prevention. This type of resource use (e.g., case management) has usually been applied to cases well after development of LTD and would therefore not be a strategic use of resources if prevention of LTD is the primary goal.

Characteristics

Disability Prevention as a Critical Public Health Issue
The development of disability after non-catastrophic work-related injuries is a critical public health problem. The indirect costs to

workers and employers, including loss of productive work life, are enormous (Fulton-Kehoe et al. 2000). In terms of direct costs, approximately 5–10 % of all work injuries result in long-term disability and these account for approximately 80 % of all work disability costs (Hashemi et al. 1997). This translates to billions of dollars nationally per year; in 2009, lost wage benefits totaled $29.4 billion in the USA (National Academy of Social Insurance 2011). Additional intangible costs of long-term work disability arise from its serious psychosocial consequences – depression, relationship difficulties, and social isolation are common.. For these reasons, *disability prevention* is of the utmost importance in achieving the goals of fiscal stability of the workers' compensation system and of preventing workers from falling into the disability chasm.

Is There a Natural History of Disabling Injury?
Most workers who are unable to work for a few days after a work-related injury return to work within 4–6 weeks (Fig. 1). (Cheadle et al. 1994) The slope of the disability curve changes abruptly, however, at 3 months. A worker who has 3 months of disability has a 50 % chance of remaining disabled at 1 year (Cheadle et al. 1994). By 1 year of disability, the curve is nearly flat and the likelihood of returning to work at reasonable wages is probably less than 15 %.

Five percent of all work-loss injuries result in LTD, but only 10 % of these could be considered medically severe (e.g., requiring hospitalization in the first 28 days) (Cheadle et al. 1994). Thus, the majority of LTD cases begin as routine injuries (e.g., back sprains) and it is likely most physicians who see these workers soon after their injuries are unaware of the potential for serious disability. Physicians' expectations of the natural history of most injuries are that they heal. How a common back sprain may actually worsen over time, so that a worker with a seemingly mild injury is totally disabled 3 years later, defies current medical understanding. However, it may be possible to identify, soon after injury, those workers at high risk for LTD by assessing known risk factors for poor outcomes. For example, high levels of physical disability and widespread pain a few weeks after a work-related back injury resulting in time off work were shown to be strong predictors of LTD (Turner et al. 2008).

The concept of "healing periods," expected times for an injury to resolve, may contribute to physicians' failure to identify patients at high risk for LTD soon after injury (Work-loss data institute. Official disability guidelines 2012). Data underlying healing periods are based upon injury-specific actuarial analyses similar to the disability curve in Fig. 1. These guidelines incorporate little clinical evidence but merely present statistical profiles that take no account of case-specific medical, employment, psychosocial or administrative barriers to recovery or return to work. Failure to address these barriers in the treatment of injured workers is a missed opportunity to prevent long-term disability.

Prediction of Long-Term Disability

At this time, there exists no widely implemented method for accurately predicting which injured workers will develop LTD. Systematic reviews of this literature indicate that a few factors, assessed soon after a work-related musculoskeletal injury, are strong and consistent predictors of chronic disability: greater pain, widespread pain, greater physical disability, and not receiving an early offer of work accommodation (Cats-Baril and Frymoyer 1991; Turner et al. 2000; Chou and Shekelle 2010). Other factors have been found in some studies to predict disability but have not been replicated in multiple studies. Such

variables include presence of radiating leg pain in association with lumbar injuries, smaller company size, various measures of job physical and psychosocial demands, certain types of heavy work (e.g., construction), catastrophizing and other psychosocial measures, and low patient-reported return to work expectations (Cheadle et al. 1994; Turner et al. 2000; Chou and Shekelle 2010).

Much of the early literature on this topic identified the principal culprit in LTD as the "accident process" itself (Behan and Hirschfeld 1963). This unproven theory assumes that it is the emotional/mental state of the worker at the time of the injury that primarily leads to subsequent disability. Unfortunately, much of this literature placed blame on the worker seeking undeserved benefits. Actual worker fraud in the workers' compensation system has never been shown to account for more than a few percent of LTD cases. Most workers would never choose to live the life of a disabled, socially isolated person; however, injured workers may not be aware of actions taken that might result in increased or decreased risk of long-term disability.

Which Predictors of Disability Are Most Likely Modifiable?

To achieve the goal of preventing chronic disability, it is logical to identify factors that are both predictive of chronic disability and modifiable. Table 2 outlines general categories of risk factors. There are opportunities for modifying important risk factors in the domains of health care, work, administrative processes, and selected psychosocial factors. In the domain of health care, the availability of occupational health resources and

Disability Prevention, Table 2 General categories of risk factors

What are the key modifiable factors leading to disability?	Medical[a]
	Work[a]
	Administrative[a]
	Psychosocial[a]
	Economic
	Demographic
	Legal

[a]Most modifiable

timely communication with the employer and insurer are likely critical for early return to work. The primary care provider is in the best position to identify and correct through education worker fears that activity is harmful and therefore should be avoided ("fear-avoidance") and to be certain not to promote the debunked clinical axiom that rest promotes healing. Primary care providers may also be able to identify and help reduce patient catastrophizing (magnification of the threat of, rumination about, and perceived inability to cope with pain) and low expectation for return to work (Chou and Shekelle 2010; Fritz et al. 2001; Turner et al. 2006; Turner et al. 2007). Patient education soon after injury holds the potential to modify these factors and thus to decrease risk of LTD (Godges et al. 2008; Brox et al. 2008; Rainville et al. 2011). In the employment category, no offer of accommodation (job modification, such as light duty or reduced hours) to allow the worker to return to work has consistently been found to be associated with lower rates of long-term disability (Turner et al. 2008; Chou and Shekelle 2010). Finally, avoidance of delays in claims administration decision making can mitigate this cause of worse worker outcomes (Herbert et al. 1999).

Other risk factors such as depression and other psychiatric disorders, unhappiness with one's job, and family issues that work against return to work may be associated with increased likelihood of chronic disability, but these individual issues are less easy to address in broadscale programmatic efforts. Although demographic factors, such as age, are not modifiable, they are important to identify in risk modeling and may be important to consider in selecting workers for early intervention programs designed to facilitate return to work.

Litigation or retention of attorneys to assist with claims, while much heralded as an important contributor to disability, is likely significantly overrated. Most workers retain attorneys either because they have received insufficient information regarding their right to benefits or because there has been an administrative decision that the worker may see as adverse to his/her interests. Furthermore, most workers do not retain attorneys until late in the first year or in the second year following injury, well beyond a secondary prevention timeframe (Wickizer et al. 2004).

Timing and Type of Interventions Likely to Prevent Disability

To have a substantial impact on disability prevention, an intervention would need to occur *prior* to 3 months of disability, and ideally, prior to 6 weeks of disability. However, targeting every worker for special management at an early stage would be unnecessary and not cost effective. A more cost-effective targeted approach is needed.

Screening of injured workers, at a population level, to identify workers at high risk for long-term disability and most likely to benefit from early disability prevention interventions, is a promising approach, although more work is needed to develop brief, feasible, and highly accurate screening tools (Turner et al. 2008; Fulton-Kehoe et al. 2008; DuBois and Donceel 2008; Hill et al. 2008; Mellow et al. 2009; Hill et al. 2011). Of critical importance is the development of early interventions that address the identified risks and barriers to return to work and are effective in preventing long-term disability (Loisel et al. 1997; Wickizer et al. 2001).

Reorganization of health care delivery, with an occupational medicine focus, holds considerable promise as a cost-effective means to prevent chronic disability (Wickizer et al. 2001; Wickizer et al. 2011). This is because workers' compensation laws in most states rely on attending doctors' assessments of injured workers' ability to return to work; the actions of these providers are thus critical to disability prevention. A related area with substantial potential for secondary prevention is the availability of expertise and resources to assist the return to work effort. Opportunity for workers to return to modified work is key. In one randomized trial, for example, an employer-based labor-management team assisted health care providers to achieve important degrees of disability prevention compared to routine medical care (Loisel et al. 1997). Finally, it seems plausible that improved timeliness and accuracy of claims administration decision making would be critical, but not sufficient, to achieve disability prevention. For example, we conducted a field-based LTD prevention trial randomizing workers by employer to either an elite, resource-rich claims unit or to usual claims administration (Washington State Department of Labor and Industries. Long Term Disability Prevention Pilots and Legislature 1997). Although all parties were more satisfied with claims management response times and service in the novel intervention, significant disability prevention did not occur. Claims administration changes alone, in the absence of significant health care delivery and employer-based interventions, are not likely to substantially prevent LTD. Similarly, insurance-based interventions using computerized algorithms to identify those at risk early after injury and linked to subsequent claims adjudicators actions have also not proven fruitful (Battie et al. 2002).

Conclusions

Systematic strategies aimed at work-related injury cases prior to 3 months of disability are needed to achieve disability prevention. Such efforts are likely to benefit from the following:

1. Development of brief screening instruments that are feasible to administer at a population level and that can identify accurately workers at greatest risk of long-term disability and most likely to benefit from specific early intervention efforts to prevent LTD
2. Development of electronic systems to facilitate real-time population-based worker screening and tracking such that (a) workers at high risk of long-term disability can be identified soon after injury, (b) their outcomes can be tracked over time, and (c) care can be modified based on worker outcomes
3. Reorganization of health care delivery with enhanced capacity to deliver timely, occupationally focused care in the first 6 weeks after injury
4. Reassessment of the current incentives and resources available to employers and workers that are most likely to enhance early return to work
5. Reorganization of claims management to focus on disability prevention as a strategy

distinct from disability management, and with greater integration in assisting the delivery of appropriate health care services

Until more complete models of disability prediction are developed and broad systems changes are implemented, it is recommended that clinicians closely monitor work status and elapsed time since injury in their patients with work-related injuries. If a worker has gotten to 4 weeks of lost work time from a non-catastrophic injury, and return to work is not imminent, it may be advisable to seek occupational health expertise to assist recovery and return to work.

References

Battie, M. C., Fulton-Kehoe, D., & Franklin, G. (2002). The effects of a medical care utilization review program on back and neck injury claims. *Journal of Occupational and Environmental Medicine, 44*, 365–371.

Behan, R. C., & Hirschfeld, A. H. (1963). The accident process. II. Toward more rational treatment of industrial injuries. *Journal of the American Medical Association, 186*, 300–306.

Brox, J. I., Storheim, K., Grotle, M., Tveito, T. H., Indahl, A., & Eriksen, H. R. (2008). Systematic review of back schools, brief education, and fear-avoidance training for chronic low back pain. *The Spine Journal, 8*, 948–958.

Cats-Baril, W. L., & Frymoyer, J. W. (1991). Identifying patients at risk of becoming disabled because of low-back pain. The vermont rehabilitation engineering center predictive model. *Spine, 16*, 605–607.

Cheadle, A., Franklin, G., Wolfhagen, C., et al. (1994). Factors influencing the duration of work-related disability: A population-based study of Washington state workers' compensation. *American Journal of Public Health, 84*, 190–196.

Chou, R., & Shekelle, P. (2010). Will this patient develop persistent disabling low back pain? *Journal of the American Medical Association, 303*, 1295–1302.

DuBois, M., & Donceel, P. (2008). A screening questionnaire to predict no return to work within 3 months for low back calimants. *European Spine Journal, 17*, 380–385.

Frank, J. W., Kerr, M. S., Brooker, A. S., et al. (1996a). Disability resulting from occupational low back pain. Part I: What do we know about primary prevention? A review of the scientific evidence on prevention before disability begins. *Spine, 21*, 2908–2917.

Frank, J. W., Brooker, A. S., DeMaio, S. E., et al. (1996b). Disability resulting from occupational low back pain. Part II: What do we know about secondary prevention?

A review of the scientific evidence on prevention before disability begins. *Spine, 21*, 2918–2929.

Fritz, J. M., George, S. Z., & DeLitto, A. (2001). The role of fear-avoidance beliefs in acute low back pain: Relationships with current and future disability and work status. *Pain, 94*, 7–15.

Fulton-Kehoe, D., Franklin, G., Weaver, M., & Cheadle, A. (2000). Years of productivity lost among injured workers in Washington state: Modeling disability burden in workers' compensation. *American Journal of Industrial Medicine, 37*, 656–662.

Fulton-Kehoe, D., Stover, B. D., Turner, J. A., et al. (2008). Development of a brief questionnaire to predict long-term disability. *Journal of Occupational and Environmental Medicine, 50*, 1042–1052.

Godges, J. J., Anger, M. A., Zimmerman, G., & Delitto, A. (2008). Effects of education on return-to-work status for people with fear-avoidance beliefs and acute low back pain. *Physical Therapy, 88*, 231–239.

Hashemi, L., Webster, B. S., Clancy, E. A., & Volinn, E. (1997). Length of disability and cost of workers' compensation low back pain claims. *Journal of Occupational and Environmental Medicine, 39*, 937–945.

Herbert, R., Janeway, K., & Schechter, C. (1999). Carpal tunnel syndrome and workers' compensation among an occupational clinic population in New York State. *American Journal of Industrial Medicine, 35*, 335–342.

Hill, J. C., Dunn, K. M., Lewis, M., et al. (2008). A primary care back pain screening tool: Identifying patient subgroups for initial treatment. *Arthritis & Rheumatism, 59*, 632–641.

Hill, J. C., Whitehurst, D. G. T., Lewis, M., et al. (2011). Comparison of stratified primary care management for low back pain with current best practice (STarT Back): A randomised controlled trial. *Lancet, 378*, 1560–1571.

Loisel, P., Abenhaim, L., Durand, P., et al. (1997). A population-based, randomized clinical trial on back pain management. *Spine, 22*, 2911–2918.

Mellow, M., Elfering, A., Egli, P. C., et al. (2009). Identification of prognostic factors for chronicity in patients with low back pain: A review of screening instruments. *International Orthopaedics, 33*, 301–313.

National Academy of Social Insurance. (2011). Workers' compensation: Benefits, coverage, and costs, 2009. URL: http://www.nasi.org/sites/default/files/research/Workers_Comp_Report_2009.pdf. Accessed 7Aug2012.

Rainville, J., Smeets, R. J., Bendix, T., Tveito, T. H., Poiraudeau, S., & Indahl, A. J. (2011). Fear-avoidance beliefs and pain avoidance in low back pain–translating research into clinical practice. *The Spine Journal, 11*, 895–903.

Turner, J. A., Franklin, G., & Turk, D. C. (2000). Predictors of chronic disability in injured workers: A systematic literature synthesis. *American Journal of Industrial Medicine, 38*, 707–722.

Turner, J. A., Franklin, G., Fulton-Kehoe, D., et al. (2006). Worker recovery expectations and fear avoidance predict work disability in a population-based workers' compensation back pain sample. *Spine, 31*, 682–689.

Turner, J. A., Franklin, G., Fulton-Kehoe, D., et al. (2007). Early predictors of chronic work disability associated with carpal tunnel syndrome: A longitudinal workers' compensation cohort study. *American Journal of Industrial Medicine, 50*, 489–500.

Turner, J. A., Franklin, G. M., Fulton-Kehoe, D., et al. (2008). ISSLS prize winner: Early predictors fo chronic work disability: A prospective, population-based study of workers with back injuries. *Spine, 33*, 2809–2818.

Washington State Department of Labor and Industries. Long Term Disability Prevention Pilots, Annual Report to the Legislature, 1997.

Wickizer, T. M., Franklin, G., Plaeger-Brockway, R., & Mootz, R. D. (2001). Improving the quality of workers' compensation health care delivery: The Washington state occupational health services project. *The Milbank Quarterly, 79*, 5–33.

Wickizer, T. M., Franklin, G. M., Turner, J. A., et al. (2004). Use of attorneys in appeal filing in the Washington State workers' compensation program: Does patient satisfaction matter? *Journal of Occupational and Environmental Medicine, 46*, 331–339.

Wickizer, T. M., Franklin, G., Fulton-Kehoe, D., et al. (2011). Improving quality, preventing disability and reducing costs in workers' compensation healthcare: A population-based intervention study. *Medical Care, 49*, 1105–1111.

Work-loss data institute. Official disability guidelines. 2012, (17th ed.).

Disability Rating

▶ Rating Impairment Due to Pain in a Workers' Compensation System

Disability, Effect of Physician Communication

James P. Robinson
Department of Anesthesiology & Pain Medicine, University of Washington, Seattle, WA, USA

Synonyms

Disability management; Secondary prevention

Definition

Physicians engage in a wide range of behaviors to help injured workers return to work. The most obvious interventions are medical or surgical treatments that promote an injured worker's recovery from injury. But physicians also perform activities that facilitate return to work without directly altering the biology of an injury. These activities can collectively be called *disability management activities*. They include interacting with a worker's employer or workers' compensation claims manager (Pransky et al. 2004). They also include *communication with a patient* about the significance of his or her injury, and, in particular, its implications for the patient's ability to work.

Characteristics

Extensive research demonstrates that the manner in which physicians communicate with patients affects encounters between them. Effective communication by physicians promotes more open disclosure by patients, increased patient satisfaction with their medical encounters, and increased patient adherence with medical regimens (Fuertes et al. 2009; Chou et al. 2007). Much of the research in this area has addressed *how* physicians communicate with patients. In contrast, this entry is focused on the content of *what* physicians communicate to patients with work injuries. In particular, it will focus on communication that addresses common fears among patients who have sustained work-related low back injuries. The entry assumes that a physician is communicating with a worker who has localized back pain from a relatively minor work injury. That is, it assumes that the worker has not sustained major trauma such as a spinal fracture, does not have a lumbar radiculopathy, and does not have any "red flags" suggestive of nonmechanical sources of low back pain (LBP) such as neoplasm (Turner et al. 2000).

Abundant literature supports the conclusion that the outcomes of work injuries are affected psychosocial factors as well as biomedical ones (Vlaeyen and Linton 2000). The relevant psychosocial factors are diverse. They include pre-injury problems such as substance abuse or major depression. However, there is also abundant evidence that workers' responses to low back injuries are affected by a variety of factors that become evident only after injury, and are more appropriately viewed as expected reactions to injury than as mental disorders. In particular, recent research indicates fear of pain and/or reinjury acts as a major impediment to return to normal activities for individuals who have sustained back injuries (Boersma et al. 2004).

Research suggests that various kinds of psychological interventions can lead to reductions in patients' fears regarding their pain conditions, and that clinical improvement following therapies is mediated by reductions in fear (Woby et al. 2004; Burns et al. 2003; Anema et al. 2002). While these studies used psychologically based treatments such as in vivo exposure, there is a consensus among experts (Pransky et al. 2004; Deyo 1988; Catchlove and Cohen 1982; Liddle et al. 2007) and some empirical support (Du Bois and Donceel 2012) for the hypothesis that communications by physicians and other health care providers can allay the fears of low back pain patients, and hasten their return to normal activities.

Relatively little research has been performed on the question of whether communications by a physician are effective in reducing work disability on the part of patients with low back pain, as opposed to improving other kinds of outcomes. The results are somewhat mixed, but on balance support the conclusion that when physicians tell patients about the benign nature of most back injuries and the importance of early return to normal activities, the patients tend to return to work relatively early (Hall et al. 1994; Dasinger et al. 2001; Hazard et al. 1997; Cheadle et al. 1994).

The general conclusion that communications by physicians can facilitate resolution of work disability among patients with low back pain begs the question: What exactly should the physician say? As Liddle et al. (Du Bois and Donceel 2012) have noted, the exact interventions made by physicians in studies on the effectiveness of physician communication have varied widely. In the absence of specific guidance from research, informal experience and expert opinion suggest several common concerns among injured workers with back problems, and messages from physicians that might allay these concerns.

Concern #1 – "Low back injuries are disabling, and potentially catastrophic. I might end up in a wheelchair."

Response: The physician can appropriately point out that LBP is an extremely common problem in modern societies, that it usually occurs in self-limited episodes, and that it only rarely causes protracted disability. Thus, although the severity of pain associated with acute LBP is often distressing, evidence about the natural history of LBP provides ground for optimism that the symptoms will subside (Hagen et al. 2004).

Concern #2 – "I should reduce my activity to avoid pain in my back, because pain indicates that I am reinjuring myself."

Response: Patients with acute LBP often have enough pain that they have no choice but to reduce their activity levels for a short period of time. It is understandable that when people experience significant pain when they try to return to normal activities, they become concerned that they are making their problem worse. But research supports the conclusion that pain associated with moderate activity is unlikely to make a back problem worse. That is, the "hurt" that patients may feel during activity usually does not mean that they are harming their back, or delaying their recovery. Moreover, prolonged inactivity can create new problems because patients become deconditioned. In general, research supports the strategy of early activation (Rainville et al. 2004), and indicates that even fairly vigorous exercise is more likely to facilitate recovery than to impede it (Von Korff and Saunders 1996).

Concern #3 – "I need to let my back heal. I should not return to work or to normal activities until then." It makes sense for patients with acute LBP to reduce their activity for a short period of time. But it is unrealistic and usually counterproductive for people to restrict their activities severely until all their symptoms are gone. Most people return to normal activity – and

in particular, to work – before their pain has resolved completely (White and Gordon 1982). Conversely, people who take the position that they must remain inactive until they are completely free of symptoms run the risk of adverse effects of deconditioning.

Concern #4 – "My pain implies that I have an injury that needs to be fixed. I expect doctors to find the specific cause of my pain (the pain generator), and to fix the problem. Only treatment by a skilled doctor will allow me to recover from my back injury – there is nothing that I can do about the problem."

Response: This perspective rests on the assumptions that back pain is a result of a specific structural lesion, that physicians have the skill to identify and fix the lesion, and that it is essentially impossible for an individual to return to normal functioning until the lesion has been fixed. All of these assumptions are incorrect for most patients with non-radicular LBP. In addressing this belief system, it is important to note that physicians are often unable to identify a precise pain generator for patients with non-radicular LBP (Robinson et al. 2005), and that it is often doubtful that a single pain generator even exists (Rhodes et al. 1999). Perhaps because of the difficulty in determining the anatomic basis of LBP, definitive interventions to "fix" the back are often only modestly helpful. Conversely, the vast majority of people who hurt their backs rely on self-management, and recover with only minimal professional help. Thus, patients' interests are best served by taking an active role in managing their problem, rather than by waiting for a professional to fix it.

Concern #5 – "My MRI scan shows a bulging disk. I am convinced that it is the cause of my pain, and that I would not get better until someone fixes my disk."

Response: This is a variant of the conceptual model outlined in #4 above. It embodies the enormous significance that LBP patients tend to place on imaging studies that demonstrate anatomic abnormalities (Boos et al. 2000). Many patients conclude that since they have a bad disk, they cannot possibly recover unless the disk is fixed. Patients with this perspective need to realize that research demonstrates that MRI findings in non-radicular LBP tend to be nonspecific – e.g., that abnormalities such as bulging disks are commonly found among individuals with no current back pain or history of back pain (Chatterjee 2006).

Barriers to Effective Communication

Physicians who attempt to allay the fears of their patients, to motivate them to take a vigorous role in rehabilitating their backs, and to encourage them to return to normal activities soon after injury face an uphill battle for many reasons. One is that a rehabilitative approach that makes heavy demands on the patient conflicts with prevalent myths in advanced societies about the magical power of medical/surgical treatment, and the sense that patients are entitled to rapid, effortless relief of suffering. Second, a physician who attempts to steer patients away from invasive therapies and promote rapid return to work may be viewed as an agent of the employer or insurance company. Third, there are ongoing differences of opinion among spine physicians and surgeons about the benefits of alternatives to a rehabilitative approach that places a great deal of responsibility on patients. Claims about new interventional therapies create ambiguity in the medical community about how to approach patients with LBP. Moreover, regardless of their benefits as demonstrated in well-controlled clinical trials, interventional therapies have the allure of promising patients a "fix" for their problem based on the magic of modern technology. In comparison to new interventional therapies, recommendations for self-management and return to activities in the absence of a total cure may be perceived by patients as outdated and unimaginative.

Summary

Research to date suggests that physicians can influence the course of recovery following work-related back injuries by their communications with injured workers and with other interested parties (e.g., employers) (Pransky et al. 2004). This entry has focused on a specific area of physician communication – messages that address the belief systems and fears that patients frequently harbor regarding their back injuries. Expert opinion and a limited body of research support the conclusion that physicians can promote return to normal activity in injured workers with LBP by addressing these concerns.

References

Anema, J. R., Van Der Giezen, A. M., Buijs, P. C., & Van Mechelen, W. (2002). Ineffective disability management by doctors is an obstacle for return-to-work: A cohort study on low back pain patients sicklisted for 3–4 months. *Occupational and Environmental Medicine, 59*(11), 729–733.

Boersma, K., Linton, S., Overmeer, T., Jansson, M., Vlaeyen, J., & de Jong, J. (2004). Lowering fear-avoidance and enhancing function through exposure in vivo. A multiple baseline study across six patients with back pain. *Pain, 108*(1–2), 8–16.

Boos, N., Semmer, N., Elfering, A., Schade, V., et al. (2000). Natural history of individuals with asymptomatic disc abnormalities in magnetic resonance imaging. *Spine, 25*, 1482–1492.

Burns, J. W., Glenn, B., Bruehl, S., Harden, R. N., & Lofland, K. (2003). Cognitive factors influence outcome following multidisciplinary chronic pain treatment: A replication and extension of a cross-lagged panel analysis. *Behaviour Research and Therapy, 41*(10), 1163–1182.

Catchlove, R., & Cohen, K. (1982). Effects of a directive return to work approach in the treatment of workman's compensation patients with chronic pain. *Pain, 14*(2), 181–191.

Chatterjee, J. S. (2006). From compliance to concordance in diabetes. *Journal of Medical Ethics, 32*(9), 507–510. Review. PubMed PMID: 16943329; PubMed Central PMCID:PMC2563412.

Cheadle, A., Franklin, G., Wolfhagen, C., Savarino, J., Liu, P. Y., Salley, C., & Weaver, M. (1994). Factors influencing the duration of work-related disability: A population-based study of Washington State workers' compensation. *American Journal of Public Health, 84*(2), 190–196.

Chou, R., Qaseem, A., Snow, V., Casey, D., Cross, J. T., Jr., Shekelle, P., & Owens, D. K. (2007). Diagnosis and treatment of low back pain: A joint clinical practice guideline from the American College of Physicians and the American Pain Society. *Annals of Internal Medicine, 147*(7), 478–491.

Dasinger, L. K., Krause, N., Thompson, P. J., Brand, R. J., & Rudolph, L. (2001). Doctor proactive communication, return-to-work recommendation, and duration of disability after a workers' compensation low back injury. *Journal of Occupational and Environmental Medicine, 43*(6), 515–525.

Deyo, R. A. (1988). The role of the primary care physician in reducing work absenteeism and costs due to back pain. *Occupational Medicine, 3*, 17–30.

Du Bois, M. G., & Donceel, P. (2012). Guiding low back claimants to work a randomized controlled trial. *Spine (Phila Pa 1976), 37*, 1425–1431.

Fuertes, J. N., Boylan, L. S., & Fontanella, J. A. (2009). Behavioral indices in medical care outcome: The working alliance, adherence, and related factors. *Journal of General Internal Medicine, 24*(1), 80–85.

Hagen, K. B., Hilde, G., Jamtvedt, G., & Winnem, M. (2004). Bed rest for acute low-back pain and sciatica (Cochrane Review). In *The Cochrane Library*, Issue 3. Chichester: Wiley.

Hall, H., McIntosh, G., Melles, T., Holowachuk, B., & Wai, E. (1994). Effect of discharge recommendations on outcome. *Spine, 19*(18), 2033–2037.

Hazard, R. G., Haugh, L. D., Reid, S., McFarlane, G., & MacDonald, L. (1997). Early physician notification of patient disability risk and clinical guidelines after low back injury: A randomized, controlled trial. *Spine, 22*(24), 2951–2958.

Liddle, S. D., Gracey, J. H., & Baxter, G. D. (2007). Advice for the management of low back pain: A systematic review of randomized controlled trials. *Manual Therapy, 12*, 310–327.

Pransky, G. S., Shaw, W. S., Franche, R., & Clarke, A. (2004). Disability prevention and communication among workers, physicians, employers, and insurers – Current models and opportunities for improvement. *Disability and Rehabilitation, 26*(11), 625–634.

Rainville, J., et al. (2004). Exercise as a treatment for chronic low back pain. *The Spine Journal, 4*, 106–115.

Rhodes, L. A., McPhillips-Tangum, C. A., Markham, C., & Klenk, R. (1999). The power of the visible: The meaning of diagnostic tests in chronic back pain. *Social Science & Medicine, 48*(9), 1189–1203.

Robinson, J. P., Ricketts, D., & Hanscom. D. A. (2005). Musculoskeletal pain. In H. Merskey, JD Loeser and R Dubner (Eds), The Paths of Pain (pp 1975–2005). Seattle: IASP Press.

Turner, J. A., Franklin, G., & Turk, D. C. (2000). Predictors of chronic disability in injured workers: A systematic literature synthesis. *American Journal of Industrial Medicine, 38*, 707–722.

Vlaeyen, J. W., & Linton, S. J. (2000). Review. Fear-avoidance and its consequences in chronic musculoskeletal pain: a state of the art. *Pain, 85*(3), 317–332.

Von Korff, M., & Saunders, K. (1996). The course of back pain in primary care. *Spine, 21*(24), 2833–2837.

White, A. A., & Gordon, S. L. (1982). *Symposium on idiopathic low back pain*. St. Louis: Mosby.

Woby, S. R., Watson, P. J., Roach, N. K., & Urmston, M. (2004). Are changes in fear-avoidance beliefs, catastrophizing, and appraisals of control, predictive of changes in chronic low back pain and disability? *European Journal of Pain, 8*(3), 201–210.

Disability, Fear of Movement

Stéphanie Volders[1], Maaike Leeuw[2], Johan W. S. Vlaeyen[3,4] and Geert Crombez[5]
[1]Research Group on Health Psychology, Leuven, Belgium
[2]Department of Medical, Clinical and Experimental Psychology, Maastricht University, Maastricht, The Netherlands
[3]Research Group Health Psychology, Catholic University of Leuven, Leuven, Belgium
[4]Department of Clinical Psychological Science, Maastricht University, Maastricht, The Netherlands
[5]Department of Experimental Clinical and Health Psychology, Ghent University, Ghent, Belgium

Synonyms

Fear of pain; Fear-avoidance; Kinesiophobia; Pain-related fear

Definition

▶ Fear of movement/(re)injury is a specific form of pain-related fear and refers to the fear that certain movements and physical activities will cause (re)injury. It is especially prominent in patients with (chronic) musculoskeletal and low back pain in particular.

Kinesiophobia is defined as "an excessive, irrational, and debilitating fear of physical movement and activity resulting from a feeling of vulnerability to painful injury or (re)injury" (Kori et al.. 1990). It is associated with avoidance of fear-eliciting activities. Both fear of movement/(re)injury and ▶ avoidance behavior are postulated to significantly contribute to the development and maintenance of (chronic) musculoskeletal pain and low back pain in particular (e.g., Kori et al. 1990; Leeuw et al. 2007a; Vlaeyen et al. 1995; Vlaeyen and Linton 2000).

Characteristics

The experience of pain can be characterized by psychophysiological (e.g., muscle activity), cognitive (e.g., anticipation), and behavioral (e.g., escape and avoidance) responses. The role of fear of movement/(re)injury in the transition from acute to chronic low back pain, and in the maintenance of chronic low back, has been well documented. In other pain syndromes, such as chronic fatigue syndrome, whiplash, headache, and fibromyalgia, the evidence on the importance of fear of movement/(re)injury in the process of chronicity is still preliminary. Fear of movement/(re)injury is strongly related to escape and avoidance behaviors. Some patients believe that performance of certain activities may promote pain and (re)injury. This belief leads to fear of movement/(re)injury and consequently to the avoidance of these activities, although there are no medical explanations for this behavioral pattern of avoidance. In most instances, these fear-avoidance beliefs take the form of erroneous "if......then" cognitions, for example, "If I lift this shopping bag, I will damage a nerve and end up paralysed."

Despite the fact that in acute pain the avoidance of daily activities may be adaptive in facilitating healing and recovery, avoidance behavior is no longer necessary for recovery in chronic pain. The long-term maintenance and enhancement of avoidance behavior can be established by the short-term effect of reduced pain and suffering as proposed by Fordyce (1982), and by certain beliefs and expectations that confrontation with particular activities will promote pain and (re)injury (e.g., Asmundson et al. 1999; Kori et al. 1990; Philips 1987; Vlaeyen et al. 1995; Vlaeyen and Linton 2000).

Several studies have shown that fear of movement/(re)injury is associated with poor physical performance on behavioral tasks (Crombez et al. 1999; Vlaeyen and Crombez 1999). The effect of pain-related fear on physical performance also seems to generalize to daily life situations. For example, fear-avoidance beliefs about work are strongly related to disability of daily living and work lost in the previous year, even more so than pain severity or other pain variables (Waddell et al. 1993). Moreover, self-reported fear of movement/(re)injury is a better predictor of self-reported disability than biomedical symptoms and pain severity (Crombez et al. 1999; Vlaeyen

et al. 1995, 2004; Vlaeyen and Crombez 1999; Waddell et al. 1993). As fear of movement/(re)injury and avoidance behavior are important determinants of disability, Waddell et al. (1993) even concluded that "fear of pain and what we do about it may be more disabling than pain itself." Yet, recent evidence showed that pain intensity is able to explain a significant portion of the variance in disability (Sieben et al. 2005), which means that the relation between pain intensity and disability may be more important than previously has been suggested (Leeuw et al. 2007a).

A Cognitive-Behavioral Model of Chronic Low Back Pain

Building upon the work of Lethem et al. (1983), the ▶ cognitive-behavioral model of fear of movement/(re)injury was developed to make sense of the development of chronic suffering in nonspecific low back pain (Vlaeyen et al 1995). According to the model, two opposing behavioral responses may occur in response to acute pain: "confrontation" and "avoidance." A gradual confrontation and resumption of daily activities despite pain is considered an adaptive response that eventually leads to the reduction of fear, the encouragement of physical recovery and functional rehabilitation. In contrast, a catastrophic interpretation of pain is considered a maladaptive response, which initiates a vicious circle in which fear of movement/(re)injury, and the subsequent avoidance of activities, augment functional disability and the pain experience by means of hypervigilance, depression, and disuse (Fig. 1).

This cognitive-behavioral model has several presumptions (Vlaeyen and Linton 2000):

1. ▶ Pain catastrophizing, an exaggerated negative orientation toward actual or anticipated pain experiences, is assumed to be a precursor of pain-related fear (Vlaeyen and Crombez 1999; Vlaeyen and Linton 2000). Studies have indeed found a strong association between fear of pain and catastrophizing in chronic pain patients, as well as an association between pain catastrophizing and a higher pain intensity in various pain problems (for an overview, see Leeuw et al. 2007a), although most research is cross-correlational

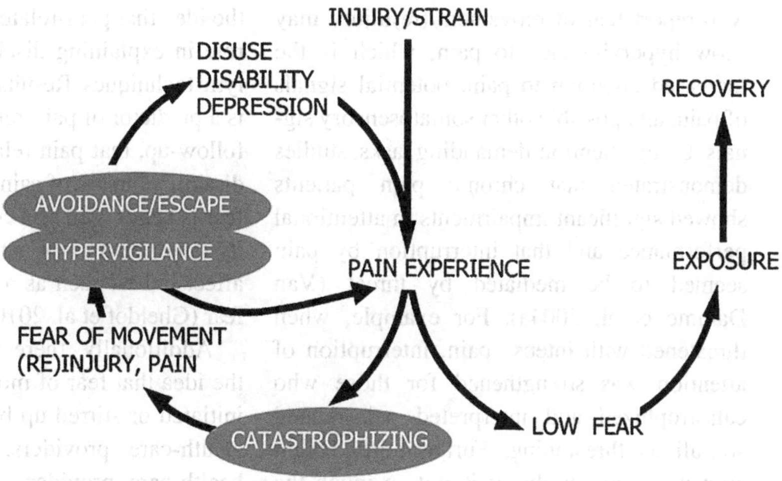

Disability, Fear of Movement,
Fig. 1 A cognitive behavioral model of pain-related fear (From: Vlaeyen, J. W. S. (2002). Fear in musculoskeletal pain. In J. O. Dostrovsky, D. B. Carr, & M. Koltzenburg (Eds.), Proceedings of the 10th world congress on pain (pp. 631–650). Seattle: IASP press. Reprinted with permission

in nature. For example, chronic pain patients who catastrophized reported higher pain intensities, felt more disabled, and experienced more psychological distress (Vlaeyen et al. 2002b). Laboratory and longitudinal studies allow investigation as to whether catastrophizing may actually precede pain-related fear (Leeuw et al. 2007b; Vlaeyen et al. 2004). For example, pain-free individuals who catastrophized about pain became more fearful when threatened with an intense pain experience than did students who reported low catastrophizing (Crombez et al. 1998). Furthermore, pain catastrophizing appeared to be the most powerful predictor of back pain chronicity 1 year after the acute onset (Burton et al. 1995). Both studies indicate that catastrophizing is a precursor of fear of movement/(re)injury.

2. Fear is characterized by escape and avoidance behaviors, leading to an impaired ability to accomplish daily living tasks and subsequent functional disability. Although studies have not been able to confirm the mediating role of avoidance behavior in disability (Leeuw et al. 2007a), several studies did show that patients with chronic low back pain, who associate pain with damage, tend to avoid activities that they assume will promote pain (Asmundson et al. 1999; Kori et al. 1990; Philips 1987; Vlaeyen and Crombez 1999; Vlaeyen and Linton 2000). Avoidance

behaviors can occur as early in the anticipation of pain rather than as a response to pain. Therefore, opportunities are excluded for correction of (erroneous) catastrophic cognitions and beliefs about the consequences of activities. By this means, avoidance is likely to expand and become more persistent.

3. Excessive and long-lasting avoidance of physical activities may have detrimental physical and psychological consequences. A decrease of mobility, decreased muscle strength, and loss of aerobic fitness can occur, possibly resulting in a "▶ disuse syndrome." Avoidance can also result in loss of self-esteem, deprivation of reinforcers, depression, and worrying. Both disuse and depression are known to decrease pain tolerance and, can thereby, augment the pain experience (Crombez et al. 1999). However, studies provide ambiguous evidence with regard to what extent patients actually suffer from a loss of aerobic fitness and are less physically active (Bousema et al. 2007; Verbunt et al. 2003; 2010). In studies showing a relationship between disability and lower levels of aerobic fitness, it remains inconclusive to what extent the latter is related to pain-related fear (Smeets et al. 2009).

4. In accordance with other forms of fear, pain-related fear interferes with cognitive functioning, such as attention (Eccleston and Crombez 1999; Vlaeyen and Linton 2000). Pain patients

who report fear of movement/(re)injury may show hypervigilance to pain, which is the increased attention to pain, potential signals of pain, and possible other somatosensory signals. Using attention-demanding tasks, studies demonstrated that chronic pain patients showed significant impairments in attentional performance and that interruption by pain seemed to be mediated by threat (Van Damme et al. 2004a). For example, when threatened with intense pain, interruption of attention was strengthened for those who catastrophized and interpreted pain-related stimuli as threatening. Furthermore, recent evidence suggests that it is not so much the shift of attention toward painful stimuli that has disabling effects, rather than it is the inability of disengaging attention from painful stimuli and redirecting it toward other relevant stimuli in the situation (Roelofs et al. 2005). This effect is strongest in patients with high levels of pain catastrophizing (Van Damme et al. 2002, 2004b). It seems that hypervigilance in fearful chronic pain patients appears at the expense of other tasks, such as everyday activities or pain coping strategies.

5. Pain-related fear is associated with increased psychophysiological reactivity when the individual is confronted with situations that are evaluated as "dangerous." Results showed that contextual fear caused by the experimental setting resulted in muscular reactivity, to which nonfearful patients readily habituated over time, while fearful patients did not (Vlaeyen et al. 1999).

There is substantial support for the cognitive-behavioral model in explaining the development of chronic nonspecific low back pain. Most findings stem from cross-sectional studies, but recently, prospective studies have also been conducted that are able to confirm the dynamic relationships among the variables of this model, which further validate it (Gheldof et al. 2010). One study found that fear-avoidance beliefs are a strong predictor of sickness absence one year later among female health-care workers, with the same level of low back pain intensity (Jensen et al. 2010). Another study was able to confirm the idea that pain-related fear plays an important role in explaining disability by using path analytic techniques. Results showed that pain severity is a predictor of pain-related fear and disability at follow-up, that pain-related fear is a predictor of disability but not of pain severity, that pain-related fear is better seen as a consequence of pain severity rather than as an antecedent and that negative affect is best seen as a precursor of pain-related fear (Gheldof et al. 2010).

Additionally, there is increasing support for the idea that fear of movement/(re)injury can be initiated or stirred up by beliefs and attitudes of health-care providers. Besides the fact that health-care providers on average hold beliefs that are consistent with current evidence, many would still advise patients to avoid painful movements. Others still believe that pain-reduction is a necessary requirement for returning to work, or that sick leave is an adequate treatment for back pain. The fear-avoidance beliefs of certain practitioners can, therefore, reinforce the beliefs and avoidance behaviors of patients (Linton et al. 2002b), although not all studies support this hypothesis (Sieben et al. 2009). In a related vein, a lot of attention is paid to the social context of pain and pain behavior. For example, research has shown that both over- and underestimations of pain by an observer (e.g., a partner, a therapist) can moderate the level of displayed pain and distress (Goubert et al. 2005).

Cognitive-Behavioral Therapy with Exposure In Vivo

Evidence is accumulating that the advice to restrict activity or to rest in case of back pain is counterproductive, delays recovery, and may lead to chronic disability and complicated rehabilitation. Substantial evidence favors the recommendation to stay active and continue usual daily activities (e.g., Waddell 1998; Verbunt et al. 2010).

One of the clinical implications of the idea that chronic suffering and disability is caused by fear of movement and avoidance, is that treatment of CLBP patients should directly target these perpetuating and exacerbating factors. Therefore, in analogy with phobia and anxiety disorders, exposure in vivo has been proposed as a potentially effective treatment (Kori et al. 1990; Philips 1987; Vlaeyen et al. 2002a). Philips (1987) was the first to argue that graded exposure should be systematically applied to realize disconfirmations between the erroneous expectations of harm, pain, or other consequences and the actual pain of patients who are fearful of movement.

Exposure in vivo does not aim to reduce pain levels but to restore the functional abilities despite pain. The treatment is embedded in a cognitive-behavioral treatment approach and is comprised of three stages (e.g., Vlaeyen et al. 2004). The treatment starts with a cognitive-behavioral assessment, consisting of an extensive interview regarding cognitive, behavioral, and psychophysiological aspects of the symptoms, and estimates the magnitude of the influence of pain-related fear on the pain problem. An indication of the level of fear of movement can be obtained, for example, by the Tampa Scale for ▶ Kinesiophobia (For an overview of measures for fear of movement/(re)injury, see, e.g., Vlaeyen et al. 2002a; McCracken et al. 1996). Furthermore, insight is gained into avoidance behavior and catastrophizing cognitions concerning the relationship between activities and pain and (re)injury. Besides defining feasible treatment goals, the cognitive-behavioral assessment is concluded with the establishment of a personal graded hierarchy of activities that the person is actually afraid of, starting with activities that produce only slight distress and concluding with activities that are beyond the present capabilities of the patient. A valuable tool to assess this is the Photograph Series of Daily Activities (PHODA; Kugler et al. 1999; short electronic version: PHODA -SeV; Leeuw et al. 2007c) because it is able to provide information about which *specific* stimuli evoke the least and the most fear. Following this first stage of exposure in vivo, education is provided, aimed to motivate the patient to participate in previously avoided fear-provoking activities. During education, the pain problem is conceptualized as a common condition that can be self-managed, rather than a serious disease that needs careful protection. Additionally, a careful explanation of the cognitive-behavioral model is provided, in which the patients' idiosyncratic symptoms, cognitions, and behaviors are discussed. The third stage of the treatment proceeds with the implementation of behavioral experiments, during which the patient is systematically exposed to fear-provoking activities, leading to disconfirmation of harm beliefs and reduction of fear, thereby promoting recovery of activities and functional abilities (Vlaeyen et al. 2004).

Effectiveness of Cognitive-Behavioral Therapy with Exposure In Vivo

The effectiveness of graded exposure in vivo has been investigated in single case experimental designs, in which exposure in vivo is compared to usual graded activity in reducing pain-related fear, catastrophizing, and disability in CLBP patients reporting fear of movement/(re)injury (Vlaeyen et al. 2002b). Results show that remarkable improvements were observed on self-report measures of pain-related fear, catastrophizing, and disability, and in objective physical activity measurements whenever exposure in vivo was initiated. Two Swedish single case studies (Boersma et al. 2004; Linton et al. 2002a) also found evidence for the effectiveness of exposure in vivo, which demonstrated generalization to other therapists and treatment settings, adding support for the external validity of the treatment. Recently, randomized controlled trials have also been executed. Although these show beneficial effects of other treatment strategies like graded activity, results are usually slightly in favor of exposure in vivo (e.g., Leeuw et al. 2008; Linton et al. 2008; Woods and Asmundson 2008). An extensive overview of 17 studies compared exposure in vivo with three other treatment protocols (graded activity, Acceptance and Commitment Therapy, and mixed protocols). It revealed positive effects for all treatments, but exposure in vivo therapy turned out to be somewhat better in reducing pain intensity and pain catastrophizing (Bailey et al. 2010). Additionally, one study suggests that exposure in vivo is cost-effective as it diminishes health-care use, work absenteeism, and productivity loss as well as it reduces pain-related disability and increases quality of life (Goossens et al. 2009).

References

Asmundson, G. J. G., Norton, P. J., & Norton, G. R. (1999). Beyond pain: The role of fear and avoidance in chronicity. *Clinical Psychology Review, 19*, 97–119.

Bailey, K. M., Carleton, R. N., Vlaeyen, J. W., & Asmundson, G. J. (2010). Treatments addressing pain-related fear and anxiety in patients with chronic musculoskeletal pain: a preliminary review. *Cognitive Behaviour Therapy, 39*(1), 46–63.

Boersma, K., Linton, S. J., Overmeer, T., Jansson, M., Vlaeyen, J. W. S., & de Jong, J. R. (2004). Lowering fear-avoidance and enhancing function through exposure in vivo: a multiple baseline study across six patients with back pain. *Pain, 108*, 8–16.

Bousema, E. J., Verbunt, J. A., Seelen, H. A. M., Vlaeyen, J. W. S., & Knottnerus, J. A. (2007). Disuse and physical deconditioning in the first year after the onset of back pain. *Pain, 130*(3), 279–286.

Burton, A. K., Tillotson, K. M., Main, C. J., & Hollis, S. (1995). Psychosocial predictors of outcome in acute and subchronic low back trouble. *Spine, 20*, 722–728.

Crombez, G., Eccleston, C., Baeyens, F., & Elen, P. (1998). When somatic information threatens, catastrophic thinking enhances attentional interference. *Pain, 75*, 187–198.

Crombez, G., Vlaeyen, J. W. S., Heuts, P. H. T. G., & Lysens, R. (1999). Pain-related fear is more disabling than pain itself: Evidence on the role of pain-related fear in chronic back pain disability. *Pain, 80*, 329–339.

Eccleston, C., & Crombez, G. (1999). Pain demands attention: A cognitive-affective model of interruptive function of pain. *Psychological Bulletin, 3*, 356–366.

Fordyce, W. E., Shelton, J. L., & Dundore, D. E. (1982). The modification of avoidance learning in pain behaviors. *Journal of Behavioral Medicine, 5*, 405–414.

Gheldof, E. L. M., Crombez, G., Van den Bussche, E., Vinck, J., Van Nieuwenhuyse, A., Moens, G., Mairiaux, P., & Vlaeyen, J. W. S. (2010). Pain-related fear predicts disability, but not pain severity: A path analytic approach of the fear-avoidance model. *European Journal of Pain, 14*, 870–879.

Goossens, M., Vlaeyen, J. W. S., Leeuw, M., de Jong, J., & Ruijgrok, J. (2009). Is graded exposure cost-effective? An economic evaluation study in chronic low back pain. Paper presented at the VI 6th Congress of the European Federation of IASP® Chapters (EFIC).

Goubert, L., Craig, K. D., Vervoort, T., Morley, S., Sullivan, M. J. L., De C Williams, A. C., et al. (2005). Facing others in pain: The effects of empathy. *Pain, 118*, 285–288.

Jensen, J. N., Karpatschof, B., Labriola, M., & Albertsen, K. (2010). Do fear-avoidance beliefs play a role on the association between low back pain and sickness absence? A prospective cohort study among female health care workers. *Journal of Occupation and Environment Medicine, 52*, 85–90.

Kori, S. H., Miller, R. P., & Todd, D. D. (1990). Kinesiophobia: A new view of chronic pain behavior. *Pain Management, 3*, 35–43.

Kugler, K., Wijn, J., Geilen, M., de Jong, J. R., & Vlaeyen, J. W. S. (1999). The photograph series of daily activities (PHODA). [CD-rom version 1.0].

Leeuw, M., Goossens, M., Linton, S., Crombez, G., Boersma, K., & Vlaeyen, J. (2007a). The fear-avoidance model of musculoskeletal pain: Current state of scientific evidence. *Journal of Behavioral Medicine, 30*, 77–94.

Leeuw, M., Houben, R. M. A., Severeijns, R., Picavet, H. S. J., Schouten, E. G. W., & Vlaeyen, J. W. S. (2007b). Pain-related fear in low back pain: A prospective study in the general population. *European Journal of Pain, 11*, 256–266.

Leeuw, M., Goossens, M. E. J. B., van Breukelen, G. J. P., Boersma, K., & Vlaeyen, J. W. S. (2007c). Measuring perceived harmfulness of physical activities in patients with chronic low back pain: The photograph series of daily activities–short electronic version. *The Journal of Pain, 8*, 840–849.

Leeuw, M., Goossens, M. E. J. B., van Breukelen, G. J. P., de Jong, J. R., Heuts, P. H. T. G., Smeets, R. J. E. M., et al. (2008). Exposure in vivo versus operant graded activity in chronic low back pain patients: Results of a randomized controlled trial. *Pain, 138*, 192–207.

Lethem, J., Slade, P. D., Troup, J. D., & Bentley, G. (1983). Outline of a fear-avoidance model of exaggerated pain perception: I. *Behaviour Research and Therapy, 21*, 401–408.

Linton, S. J., Overmeer, T., Janson, M., Vlaeyen, J. W. S., & de Jong, J. R. (2002a). Graded in vivo exposure treatment for fear-avoidant pain patients with functional disability: A case study. *Cognitive Behaviour Therapy, 31*, 49–58.

Linton, S. J., Vlaeyen, J. W. S., & Ostelo, R. W. (2002b). The back pain beliefs of health care providers: Are we fear-avoidant? *The Journal of Occupational Rehabilitation, 12*, 223–232.

Linton, S. J., Boersma, K., Jansson, M., Overmeer, T., Lindblom, K., & Vlaeyen, J. W. S. (2008). A randomized controlled trial of exposure in vivo for patients with spinal pain reporting fear of work-related activities. *European Journal of Pain, 12*, 722–730.

McCracken, L. M., Gross, R. T., Aikens, J., & Carnrike, C. L. M., jr. (1996). The assessment of anxiety and fear in persons with chronic pain: A comparison of instruments. *Behaviour Research and Therapy, 34*, 927–933.

Philips, H. C. (1987). Avoidance behaviour and its role in sustaining chronic pain. *Behaviour Research and Therapy, 25*, 273–279.

Roelofs, J., Peters, M. L., Fassaert, T., & Vlaeyen, J. W. S. (2005). The role of fear of movement and injury in selective attentional processing in patients with chronic low back pain: A dot-probe evaluation. *The Journal of Pain, 6*, 294–300.

Sieben, J. M., Portegijs, P. J. M., Vlaeyen, J. W. S., & Knottnerus, J. A. (2005). Pain-related fear at the start of a new low back pain episode. *European Journal of Pain, 9*, 635–641.

Sieben, J. M., Vlaeyen, J. W. S., Portegijs, P. J. M., Warmenhoven, F. C., Sint, A. G., Dautzenberg, N., et al. (2009). General practitioners' treatment orientations towards low back pain: Influence on treatment behaviour and patient outcome. *European Journal of Pain, 13*, 412–418.

Smeets, R. J., van Geel, K. D., & Verbunt, J. A. (2009). Is the fear avoidance model associated with the reduced level of aerobic fitness in patients with chronic low back pain? *Archives of Physical Medicine and Rehabilitation, 90*, 109–117.

Van Damme, S., Crombez, G., & Eccleston, C. (2002). Retarded disengagement from pain cues: The effects of pain catastrophizing and pain expectancy. *Pain, 100*, 111–118.

Van Damme, S., Crombez, G., Eccleston, C., & Roelofs, J. (2004a). The role of hypervigilance in the experience of pain. In G. J. G. Asmundson, J. W. S. Vlaeyen, & G. Crombez (Eds.), *Understanding and treating fear of pain*. Oxford University press, Inc.

Van Damme, S., Crombez, G., & Eccleston, C. (2004b). Disengagement from pain: The role of catastrophic thinking about pain. *Pain, 107*, 70–76.

Verbunt, J. A., Seelen, H. A., Vlaeyen, J. W., van der Heijden, G. J., & Knottnerus, J. A. (2003). Fear of injury and physical deconditioning in patients with chronic low back pain. *Archives of Physical Medicine and Rehabilitation, 84*(8), 1227–1232.

Verbunt, J. A., Smeets, R. J., & Wittink, H. M. (2010). Cause or effect? Deconditioning and chronic low back pain. *Pain, 149*(3), 428–430.

Vlaeyen, J. W. S., & Crombez, G. (1999). Fear of movement/(Re)injury, avoidance and pain disability in chronic Low back pain patients. *Manual Therapy, 4*, 187–195.

Vlaeyen, J. W. S., & Linton, S. J. (2000). Fear-avoidance and its consequences in chronic musculoskeletal pain: A state of the Art. *Pain, 85*, 317–332.

Vlaeyen, J. W. S., Kole-Snijders, A. M. J., Rotteveel, A. M., Ruesink, R., & Heuts, P. H. T. G. (1995). The role of fear of movement/(Re)injury in pain disability. *The Journal of Occupational Rehabilitation, 5*, 235–252.

Vlaeyen, J. W. S., Seelen, H. A. M., Peters, M., de Jong, P., Aretz, E., Beisiegel, E., & Weber, W. (1999). Fear of movement/ (re)injury and muscular reactivity in chronic low back pain patients: An experimental investigation. *Pain, 82*, 297–304.

Vlaeyen, J. W. S., de Jong, J. R., Sieben, J. M., & Crombez, G. (2002a). Graded exposure *In vivo* for pain-related fear. In D. C. Turk & R. J. Gatchel (Eds.), *Psychological approaches to pain management: A practitioner's handbook* (pp. 210–233). New York: The Guilford Press.

Vlaeyen, J. W. S., de Jong, J. R., Geilen, M., Heuts, P. H. T. G., & van Beukelen, G. (2002b). The treatment of fear of movement/(re) injury in chronic lowback pain: Further evidence on the effectiveness of exposure in vivo. *The Clinical Journal of Pain, 18*, 251–261.

Vlaeyen, J. W. S., de Jong, J. R., Leeuw, M., & Crombez, G. (2004a). Fear reduction in chronic pain: Graded exposure *In vivo* with behavioral experiments. In G. J. G. Asmundson, J. W. S. Vlaeyen, & G. Crombez (Eds.), *Understanding and treating fear of pain*. Oxford University Press.

Vlaeyen, J. W. S., Timmermans, C., Rodriguez, L. M., Crombez, G., van Horne, W., Ayers, G. M., et al. (2004b). Catastrophic thinking about pain increases discomfort during internal atrial cardioversion. *Journal of Psychosomatic Research, 56*(1), 139–144.

Waddell, G. (1998). *The back pain revolution*. London: Churchill Livingstone.

Waddell, G., Newton, M., Henderson, I., Somerville, D., & Main, C. J. (1993). A fear avoidance beliefs questionnaire (FABQ) and the role of fear-avoidance beliefs in chronic low back pain and disability. *Pain, 52*, 157–168.

Woods, M. P., & Asmundson, G. J. G. (2008). Evaluating the efficacy of graded in vivo exposure for the treatment of fear in patients with chronic back pain: A randomized controlled clinical trial. *Pain, 136*(3), 271–280.

Disability, Functional Capacity Evaluations

Douglas P. Gross[1] and Michele C. Battie[2]
[1]Department of Physical Therapy, University of Alberta WCB-Alberta Millard Health, Edmonton, AB, Canada
[2]Department of Physical Therapy, University of Alberta, Edmonton, AB, Canada

Synonyms

Functional abilities evaluation; Functional capacity assessment; Functional capacity battery; Functional capacity evaluation; Job capacity evaluation; Occupational capacity evaluation; Physical capacity evaluation; Vocational assessment; Work capacity evaluation; Work fitness test; Work performance evaluation; Work sample; Work-related assessment

Definition

Functional capacity evaluations (FCEs) are standardized protocols of performance tests used to determine an individual's ability for work-related activity, readiness to return-to-work, or need for modified work duties (Soer et al. 2008h). FCEs typically involve a series of assessment items including manual handling (lifting and carrying), position tolerance testing, and mobility and coordination tests. To make return-to-work decisions, clinicians compare demonstrated ability to required job demands. A subject whose performance meets or exceeds job demands is deemed suitable for return-to-work.

Introduction

Context and Uses of Functional Capacity Evaluation

▶ Musculoskeletal disorders, including regional pain conditions such as neck and back pain, continue to be some of the most common health problems and leading causes of work disability. Improved management strategies are needed, especially strategies aimed at promoting sustainable return-to-work. Many common musculoskeletal conditions are nonspecific in nature (i.e., back pain, neck pain, cumulative activity–related disorders), making it difficult to rely on clinical exam findings or diagnostic imaging results to determine tissue healing or readiness for return to activity. Additionally, only a modest correlation exists between severity of injury or bodily impairment and functional ability for real-world tasks. Individuals with more "minor" medical conditions such as back strain are often very disabled whereas those with "major" conditions such as limb amputation are at times capable of performing strenuous work at high levels of physical demand. This has led to the development of functional capacity assessments aimed at determining work ability, with the intent of determining readiness to return-to-work or facilitating job placement. Decisions based on these assessments have considerable consequences for those undergoing testing as well as their employers. Thus,

results need to be accurate and trustworthy (Pransky and Dempsey 2004).

Functional capacity evaluations (FCE) are frequently relied on for assessing readiness for "safe" return-to-work, particularly for workers' ▶ compensation and other insurance systems. The term "▶ functional capacity evaluation" was coined by ▶ rehabilitation professionals, whose goals typically focus on assisting injured or ill individuals to optimize ability to perform desired work and leisure activities, and assessing such capacities. The goal of functional capacity testing is to directly measure capacity for specific tasks, often in relation to the required physical demands of a particular activity.

As mentioned, FCEs are frequently used to assess readiness to return-to-work following injury, but they are also used to guide other clinical decisions. FCEs are sometimes used to establish baseline performance levels to assist in planning ▶ functional restoration and work hardening programs and to measure program outcomes. In addition, FCEs are used to guide retraining programs and vocational planning following disabling injuries. Another suggested application is in assessing the extent of disability to assist in permanent ▶ impairment judgments and determination of wage-earning potential in litigation cases.

Characteristics

Scientific Basis of Functional Capacity Evaluation

Return-to-work testing incorporating performance testing led by a trained therapist (i.e., actually having clients lift, carry, walk, etc., in the clinic) has traditionally been considered most useful, trustworthy, and "objective" or free of bias (Mayer et al. 1985; Menard and Hoens 1994; Rainville et al. 2007). However, some controversy exists over just what items and activities should be measured during functional capacity testing. This has resulted in numerous testing methods reportedly measuring functional capacity. A variety of proprietary protocols are available from various providers and have recently been reviewed in detail elsewhere (Genovese et al. 2009).

Due to the important decisions made based on FCE results and the major implications arising from such decisions, important FCE measurement properties for appropriate, ethical use include adequate reliability, or test consistency, validity, and responsiveness to change in status over time. The reliability and validity of functional capacity evaluation has been examined in individuals with a variety of work-related injuries (Gouttebarge et al. 2004). Reliability of therapists' ► judgments of safe, maximum work levels has been found to be good and some research indicates the FCE does in fact measure work-related functional ability. However, performance on FCE does not appear to be entirely "objective" as previously supposed. Subject performance on FCEs undertaken within insurance or compensation environments is not only determined by physiological capacity, but also appears to be influenced by pain intensity, beliefs about disability, self-efficacy or confidence in abilities, and testing context, among other factors (Asante et al. 2007; Gross 2006; Gross and Battie 2005; Reneman, et al. 2006). More importantly, results of assessment are only modestly predictive of future return-to-work (Gouttebarge et al. 2009; Gross et al. 2004; Streibelt et al. 2009) and do not accurately predict whether or not subjects will sustain a recurrent injury after returning to work (Mahmud et al. 2010). Responsiveness of FCE for detecting change over time also appears inadequate (Durand et al. 2008; Kuijer et al. 2006).

As mentioned, performance-based FCEs have typically been considered more "► objective" than other forms of assessment such as questionnaires. However, FCE testing can be time-consuming, require specialized equipment, and be as expensive as advanced diagnostic imaging. Since testing is often strenuous, it is frequently associated with increased pain reports (Soer et al. 2008a). Because of this, some research has compared FCE to work assessments that do not require strenuous physical activity such as interviews or questionnaires. Typically interviews and questionnaires have been found to provide lower ability estimates than performance testing, with

estimates of lifting ability approximately 5 kg lower on average (Asante et al. 2007). The clinical implication of this lower estimation is unknown, but potentially could lead to delayed return-to-work or reduced ability recommendations (i.e., return to light duty as opposed to unrestricted work). However, this has not been studied and whether FCE leads to superior clinical outcomes needs to be determined.

Based on previous findings of modest FCE predictive ability, body region-specific short-form FCE protocols have been developed and tested in an attempt to reduce test burden and the likelihood of pain exacerbation (Gross et al. 2006). In a randomized controlled trial, return-to-work outcomes were comparable regardless of whether subjects underwent testing with the short-form or full-length protocol (Gross et al. 2007). Claimant satisfaction with the assessment process was similar between groups. The short-form FCE was also found to enhance prediction of future return-to-work, with percentage of explained variation in outcomes increased from <10 % to between 18 % and 27 % (Branton et al. 2010). In these studies, functional assessment was performed within a broader continuum of care decision-making framework (Stephens and Gross 2007). Therefore, it is unknown how much FCE actually contributed to the positive outcomes beyond the case management decision and further research is needed.

FCE Performance as an Indicator of Sincerity of Effort

Considering the context in which FCEs are often performed and the fact that many of the individuals undergoing testing have pain-mediated or nonspecific disability conditions, questions have arisen as to whether functional capacity testing is measuring true physical capacity or other constructs. As mentioned, performance during FCE or on any physical test can be influenced by multiple factors including, but not limited to, physiological limitations, pain intensity levels, motivation to perform, fear of movement or pain, and wellness on the day of the assessment. Several approaches have been used in an attempt to judge sincerity of effort during FCE or whether performance during testing is being limited by physical capacity or by some other factor (Lechner et al. 1998). These approaches include observations of inconsistency between findings of musculoskeletal pathology and demonstrated functional performance, inconsistent performance on related functional tasks in an assessment, and other measures of inconsistent effort, such as the coefficient of variation applied to maximal strength tests. However, the validity of the various tests of sincerity of effort has either not been thoroughly evaluated, or found deficient in some cases (Genovese et al. 2009; Jay et al. 2000; Shechtman 2001). More importantly, the reason why individual subjects do not perform to maximum physical ability levels is typically unknown. Poor or submaximal performance on FCE appears to be related more to ▶ psychological factors, such as depression or high perceived disability, than to true physical limitations (Gibson and Strong 1998). In fact, given the important influence of psychological and environmental factors on performance during such testing, some authors have recommended that FCEs be viewed not merely as maximum physical ability tests, but as ▶ behavioral assessments that must be interpreted within each subject's unique personal and environmental context (Gross 2004; Rudy et al. 1996).

Summary

FCEs are commonly used to identify the ability of injured workers to perform physical work activities and to inform decisions about safety for return-to-work. Research has supported FCEs as reliable measures of work-related functional ability, providing modest information for predicting return-to-work. Work ability estimates appear higher when FCE is used as compared to results from self-report questionnaires, but whether this actually leads to better clinical outcomes is unknown. Abbreviated or short-form FCE protocols appear as effective as longer protocols and have a reduced test burden and enhanced predictive ability. Given the important influence of psychological and other contextual factors on

subject performance during FCE, these tests may be more accurately viewed as behavioral assessments rather than evaluations of maximum physical ability. Accordingly, FCE results must be interpreted within each subject's unique personal and environmental context when return-to-work or related decisions are being made.

References

Asante, A. K., Brintnell, E. S., & Gross, D. P. (2007). Functional self-efficacy beliefs influence functional capacity evaluation. *Journal of Occupational Rehabilitation, 17*, 73–82.

Branton, E. N., Arnold, K. M., Appelt, S. R., Hodges, M. M., Battie, M. C., & Gross, D. P. (2010). A short-form functional capacity evaluation predicts time to recovery but not sustained return-to-work. *Journal of Occupational Rehabilitation, 20*, 387–393.

Durand, M. J., Brassard, B., Hong, Q. N., Lemaire, J., & Loisel, P. (2008). Responsiveness of the physical work performance evaluation, a functional capacity evaluation, in patients with low back pain. *Journal of Occupational Rehabilitation, 18*(1), 58–67.

Genovese, E., Galper, J. S., & American Medical Association. (2009). *Guide to the evaluation of functional ability: How to request, interpret, and apply functional capacity evaluations.* Chicago: American Medical Association.

Gibson, L., & Strong, J. (1998). Assessment of psychosocial factors in functional capacity evaluation of clients with chronic back pain. *British Journal of Occupational Therapy, 61*(9), 399–404.

Gouttebarge, V., Wind, H., Kuijer, P. P., & Frings-Dresen, M. H. (2004). Reliability and validity of functional capacity evaluation methods: A systematic review with reference to blankenship system, ergos work simulator, ergo-kit and isernhagen work system. *International Archives of Occupational and Environmental Health, 77*(8), 527–537.

Gouttebarge, V., Kuijer, P. P., Wind, H., van Duivenbooden, C., Sluiter, J. K., & Frings-Dresen, M. H. (2009). Criterion-related validity of functional capacity evaluation lifting tests on future work disability risk and return to work in the construction industry. *Occupational and Environmental Medicine, 66*(10), 657–663.

Gross, D. P. (2004). Measurement properties of performance-based assessment of functional capacity. *Journal of Occupational Rehabilitation, 14*(3), 165–174.

Gross, D. P. (2006). Are functional capacity evaluations affected by the patient's pain? *Current Pain and Headache Reports, 10*(2), 107–113.

Gross, D. P., & Battie, M. C. (2005). Factors influencing results of functional capacity evaluations in workers' compensation claimants with low back pain. *Physical Therapy, 85*(4), 315–322.

Gross, D. P., Battie, M. C., & Cassidy, J. D. (2004). The prognostic value of functional capacity evaluation in patients with chronic low back pain: Part 1: Timely return to work. *Spine (Phila Pa 1976), 29*(8), 914–919.

Gross, D. P., Battie, M. C., & Asante, A. (2006). Development and validation of a short-form functional capacity evaluation for use in claimants with low back disorders. *Journal of Occupational Rehabilitation, 16*(1), 53–62.

Gross, D. P., Battie, M. C., & Asante, A. K. (2007). Evaluation of a short-form functional capacity evaluation: Less may be best. *Journal of Occupational Rehabilitation, 17*(3), 422–435.

Jay, M. A., Lamb, J. M., Watson, R. L., Young, I. A., Fearon, F. J., Alday, J. M., et al. (2000). Sensitivity and specificity of the indicators of sincere effort of the EPIC lift capacity test on a previously injured population. *Spine, 25*(11), 1405–1412.

Kuijer, W., Brouwer, S., & Reneman, M. F. (2006). Determining responsiveness of FCEs, mission impossible? *The Clinical Journal of Pain, 22*(7), 664–665.

Lechner, D. E., Bradbury, S. F., & Bradley, L. A. (1998). Detecting sincerity of effort: A summary of methods and approaches. *Physical Therapy, 78*(8), 867–888.

Mahmud, N., Schonstein, E., Schaafsma, F., Lehtola, M. M., Fassier, J. B., Verbeek, J. H., et al. (2010). Functional capacity evaluations for the prevention of occupational re-injuries in injured workers. *Cochrane Database of Systematic Reviews, 7*, CD007290.

Mayer, T. G., Gatchel, R. J., Kishino, N., Keeley, J., Capra, P., Mayer, H., et al. (1985). Objective assessment of spine function following industrial injury. A prospective study with comparison group and one-year follow-up. *Spine (Phila Pa 1976), 10*(6), 482–493.

Menard, M. R., & Hoens, A. M. (1994). Objective evaluation of functional capacity: Medical, occupational, and legal settings. *Journal of Orthopaedic and Sports Physical Therapy, 19*(5), 249–260.

Pransky, G. S., & Dempsey, P. G. (2004). Practical aspects of functional capacity evaluations. *Journal of Occupational Rehabilitation, 14*(3), 217–229.

Rainville, J., Kim, R. S., & Katz, J. N. (2007). A review of 1985 volvo award winner in clinical science: Objective assessment of spine function following industrial injury: A prospective study with comparison group and 1-year follow-up. *Spine (Phila Pa 1976), 32*(18), 2031–2034.

Reneman, M. F., Kool, J., Oesch, P., Geertzen, J. H., Battie, M. C., & Gross, D. P. (2006). Material handling performance of patients with chronic low back pain during functional capacity evaluation: A comparison between three countries. *Disability and Rehabilitation, 28*(18), 1143–1149.

Rudy, T. E., Dieber, S. J., & Boston, J. R. (1996). Functional capacity assessment: Influence of behavioural and environmental factors. *Journal of Back and Musculoskeletal Rehabilitation, 6*, 277–288.

Shechtman, O. (2001). The coefficient of variation as a measure of sincerity of effort of grip strength, Part II: Sensitivity and specificity. *Journal of Hand Therapy, 14*(3), 188–194.

Soer, R., Groothoff, J. W., Geertzen, J. H., van der Schans, C. P., Reesink, D. D., & Reneman, M. F. (2008a). Pain response of healthy workers following a functional capacity evaluation and implications for clinical interpretation. *Journal of Occupational Rehabilitation, 18*(3), 290–298.

Soer, R., van der Schans, C. P., Groothoff, J. W., Geertzen, J. H., & Reneman, M. F. (2008b). Towards consensus in operational definitions in functional capacity evaluation: A delphi survey. *Journal of Occupational Rehabilitation, 18*(4), 389–400.

Stephens, B., & Gross, D. P. (2007). The influence of a continuum of care model on the rehabilitation of compensation claimants with soft tissue disorders. *Spine, 32*(25), 2898–2904.

Streibelt, M., Blume, C., Thren, K., Reneman, M. F., & Mueller-Fahrnow, W. (2009). Value of functional capacity evaluation information in a clinical setting for predicting return to work. *Archives of Physical Medicine and Rehabilitation, 90*(3), 429–434.

Disability, Upper Extremity

J. Mark Melhorn
Section of Orthopaedics, Department of Surgery, University of Kansas School of Medicine, Wichita, KS, USA

Synonyms

CTDs; Cumulative trauma disorders; Disability; Impairment, function loss; MSDs/MSPs; MSPs; Musculoskeletal disorders; Musculoskeletal pain; Repetitive strain injuries; RSIs; Work-related upper-extremity disorders; WRUEDs

Definition

Disability arises out of an individual's inability to perform a task successfully because of an insufficiency in one or more areas of functional capability which include physical function, mental function, agility, dexterity, coordination, strength, endurance, knowledge, skill, intellectual ability, or experience. Disability (the inability to perform a task) has to be related to a specific health impairment or medical condition (e.g., aging). However, a medical condition or impairment may cause or contribute to disability (e.g., loss of an arm). This point is often overlooked since most "disability agencies" (e.g., Social Security Administration) require evidence of impairment in order to determine that an individual is disabled, and therefore "entitled" to financial compensation. This is an example of the differences between "medical" impairment and "legal" disability (Melhorn 2011a).

Disability requires a conceptual definition and is context specific. Disability is the gap between what a person can do and what the person needs or wants to do in a specific environment and may be temporary or permanent and is defined by the American Medical Association Guides to the Evaluation of Permanent Impairment sixth edition (AMA Guides) as an umbrella term for activity limitations and/or participation restrictions in an individual with a health condition, disorder, or disease (AMA 2009). This definition is somewhat different than the AMA Guides fifth edition, which is an alteration in an individual's capacity to meet personal, social, or occupational demands or statutory or regulatory requirements, because of an impairment.

The AMA Guide's sixth edition definition of impairment as a significant deviation, loss, or loss of use of any body structure or function in an individual with a health condition, disorder, or disease (AMA 2008) is similar to the AMA Guide's fifth definition of loss of a physiological function or of an anatomical structure. The evaluation of impairment in the AMA Guide's sixth edition is the acquisition, recording, assessment, and reporting of medical evidence, using a standard method such as described in the Guides, to determine permanent impairment associated with a physical or mental condition, which is similar to the fifth edition.

The change in definitions reflects the adoption of the International Classification of Functioning, Disability, and Health (ICF) as developed by the World Health Organization (WHO 2001). The current ICF model of disablement blurs the traditional lines separating impairment from disability and the consideration of handicapped.

While disability is determined in an operational setting, such as the workplace or in a structured environment, impairment is the functional loss of the body part regardless of subsequent impact on the ability or capacity to carry out particular tasks or perform specified functions. Thus, an impaired person is not necessarily disabled.

Function, disability, health, and impairment become important when you consider the impact pain has on the musculoskeletal system. The ability to perform a task is dependent on three factors: capacity, tolerance, and risk (Talmage and Melhorn 2005). Musculoskeletal capacity is defined as what the body is physically capable of doing. Tolerance is what the individual is willing to do or not do based on the level of pain or biopsychosocial considerations. Pain can impact or alter capacity and tolerance, while the risk from a specific activity is relatively constant for most individuals over periods of time (Melhorn 2010). However, risk for the individual from a specific activity can change as one ages or if a musculoskeletal condition results in an impairment (a functional loss), such as the individual with the arm amputation. In other words, an individual's functional capacity changes with time. Individuals are dynamic – we are constantly changing. Therefore, most would agree that as we age, our functional capacity diminishes which implies that disabilities usually increase with age (Melhorn 2011b). Risk is defined as the likelihood of experiencing pain or injury from a specific activity. Pain is an unpleasant perception associated with actual or potential cellular damage (Turk and Okifuji 2001). It is a concept, not an entity. People with pain experience the unpleasant effects of nociceptive stimulation or they suffer in ways that they associate with pain. Recognition of this point is basic to understanding pain and suffering. If pain consisted solely of a hardwired stimulus response capability, there would be no role for learning or other cognitive behavior processes that can impact pain behavior. That is, if a hardwired stimulus response capability were the defining characteristic of pain, neuroanatomical and pathophysiological parameters would suffice in dealing with pain, which, obviously, they do not (Melhorn 2002b). Persistent pain in the absence of continued tissue trauma is pathological and is influenced by learned behaviors, which can then impact or alter the individual's capacity and tolerance (Melhorn 2012).

Nociception is the response to an unpleasant (noxious) stimulus that produces pain in human subjects under normal circumstances (Turk and Okifuji 2001). Acute pain is elicited by the injury of body tissues and activation of nociceptive transducers at the site of local tissue damage. The local injury alters the response characteristics of the nociceptors and perhaps their central connections and the autonomic nervous system in the region. In general, the state of acute pain lasts for a relatively limited time and generally remits when the underlying pathology resolves. This type of pain is often a reason to seek care and occurs after trauma, surgery, and some disease processes (Melhorn 2002b).

Chronic pain is elicited by an injury but may be perpetuated by factors that are both pathogenetically and physically remote from the originating cause. Chronic pain extends for a long period of time and represents low levels of underlying pathology that does not adequately explain the presence and/or extent of the pain. This type of pain frequently prompts patients to seek health care and is rarely treated effectively by medication alone (Melhorn 2012). Because the pain persists, it is likely that environmental and affective factors eventually interact with the tissue damage, contributing to the persistence of pain and illness behaviors. Additionally, just as the brain is modified by experiences especially in early life, it may also alter the way noxious information is processed to augment or reduce its effect on subjective awareness (Melhorn 2002b). Functional restoration programs based on congenital behavioral therapy appear to be the most effective treatment approach (Melhorn 2012).

Musculoskeletal pain (MSP) is any pain that may involve the muscles, nerves, tendons, ligaments, bones, or joints. This pain can be the result of physiological injury or non-physiological stimulus. "Musculoskeletal disorders" (MSDs)

is a term used to describe musculoskeletal pain that may include indeterminate or specific ICD-9 diagnoses. Other terms, such as cumulative trauma disorders (CTDs), repetitive stress injury (RSI), and repetitive motion injury (RMI), mean roughly the same thing as MSDs. However, RSI and RMI are arguably inaccurate, because these terms imply that repetition is the primary risk factor, which may or may not be the case. Again, MSDs are not a medical diagnosis but a descriptive term for musculoskeletal pain (Melhorn 2002a).

Work related or work compensable is the term used to describe or assign (financial) responsibility. Musculoskeletal disorders that are determined to be associated with the workplace may be considered work related or work compensable based on each state's legislative requirements (Melhorn et al. 2008). Work-related musculoskeletal disorders (WRMSDs) require two elements, an individual and a job. Each element has unique risk factors. WRMSDs often become work compensable and fall under the workers' compensation system (BLS 1996). Unfortunately, "compensation status" negatively impacts outcomes from treatment and surgery (Harris et al. 2005; Sperka et al. 2008).

Workers compensation is designed to be a no-fault and exclusive remedy system. Although described as a system, it is not; each state, territory, and federal employees have different and separate workers compensation laws and regulations. The workers and their dependents are not required to prove fault for personal injuries, diseases, or deaths arising out of and in the course of employment. The employer agrees to provide rapid payment to the injured worker for lost wages and medical care costs in exchange for limiting or eliminating the employer's potential liability for said occupational illness, injuries, and death, and, thereby, the possibility of large tort verdicts.

Introduction

Incidence rates and costs for disabilities of the upper extremity are difficult to determine.

Four different reports are provided to consideration. In 1997, 52.6 million people or 19.7 % of the population had some level of disability and 33.0 million or 12.3 % were defined as with a severe disability (McHeil 2001). A 2007 study reported more than 40 million people in the United States have a physical or mental impairment that significantly affects life activities and work performance. The total annual costs of disability are currently estimated at $300 billion (Zeller et al. 2007). A 2009 study found that the Federal Employees' Compensation Act (FECA) program, administered by the Department of Labor (DOL), covered over 2.7 million federal employees in more than 70 different agencies. FECA costs rose from $1.4 billion in 1990 to $2.6 in 2006, while the federal workforce remained essentially unchanged. Additionally, federal civilian employees represent only 2.1 % of all workers eligible for workers' compensation benefits while federal programs account for 6 % of the benefits paid (Ladou 2009). A 2010 study reported that workers' compensation is a far more significant expense to the US economy than is commonly recognized with the total annual cost of the health care and disability benefits in the United States at least $300 billion. Furthermore, the health care costs shifted by employers to Medicare/Medicaid and the disability costs shifted to the Social Security system far exceed the total costs of all the state Workers' Compensation programs (Ladou 2010). Finally, in 2005, studies found that a 20-year-old worker had a 3 in 10 chance of becoming disabled before reaching retirement age (Melhorn et al. 2005).

Characteristics

Work-related upper extremity disorders are an increasingly common cause of work-related musculoskeletal pain and disability. Although most upper extremity disorders are acute and self-limited, a small percentage of workers develop chronic nonmalignant musculoskeletal pain which can progress to permanent disability and this group accounts for the majority of costs

associated with WRUEDs. The progression from symptoms to disability and how the disability might be prevented requires an understanding of the multiple factors involved.

Depending upon one's definition of disability, between 35 and 46 million Americans can be labeled as disabled. The definition of disability and the determination of who is disabled continue to challenge governments and adjudicating bodies and therefore expand and contract more along political and ideological lines than according to any clear physical determinations. If the cost of exclusion from the workplace, medical care, legal services, and earning replacements are summed, the 1980 estimate was $177 billion or approximately 6.5 % of the gross domestic product (Demeter et al. 2001).

At any given time, up to 45 % of currently employed workers could file work-related injury or disability claims, but most do not, choosing instead to carry out their job responsibilities, accepting some discomfort as part of living (Biddle et al. 1998). This ability to tolerate discomfort is determined by the level of the biological stimulus (pain), existing psychosocial or biosocial issues, and previous learned behaviors (Melhorn 2002b). Whenever one of these elements exceeds a personal toleration level, health care is sought.

Individual risk factors for the development of disability from the biological stimulus include age, gender, and inherited health characteristics, from psychosocial and/or biosocial issues depression, current substance abuse, somatoform pain disorders, longer duration of symptoms, higher association of anxiety disorders, higher levels of stress in life events, and lower levels of lifestyle organization (goal directedness, performance focus and efficiency, timeliness of task completion, and organization of physical space). Risk factors from previously learned behavior include less time on the job, more surgeries, a higher frequency of acute antecedent trauma, indeterminate musculoskeletal diagnoses, self-reported higher pain levels, more anger with their employer, a greater psychological response or reactivity to pain, having an attorney, and being involved in litigation with their employer (Himmelstein et al. 1995; Sikorski et al. 1989).

Workplace risk factors for the development of musculoskeletal pain include any part of the production process (the manufacturing of a product). The production process usually includes input (raw materials), the production (methods, materials, machines, environment, physical stressors – such as repetition, force, posture, vibration, cold temperatures, contact stress and static muscle loads, unaccustomed activities, and combinations thereof), and the output (the finished product) (Melhorn 1998). As discussed before, the individual risk factors become contributors, moderators, and buffers as to how the workplace may affect the individual for the development of MSDs.

Preventing disability and musculoskeletal pain is challenging. Physicians cannot prove or disprove the existence of pain clinically. With traumatic injuries occurring at a rate of 7.1 per 100 equivalent full-time employees in the private business sector at an estimated cost of over $1.25 trillion, there is a need for better disability management.

Since the risk for developing musculoskeletal pain and the associated disability is 65 % for individual risk factors and 35 % for the workplace, the solution will require intervention for both (Melhorn et al. 2001). Rest is often overprescribed (Deyo et al. 1986). Healing is usually rapid and is often promoted by motion (Nachemson 1992). Rest, excessively or carelessly prescribed, may inhibit or interfere with healing while also encouraging deactivation. The problem is compounded if the task of determining whether healing has occurred and activity is now appropriate is assigned to the untrained patient, as by saying to the patient: "Let pain be your guide" to determine when to terminate rest and resume activity (Fordyce et al. 1986). Often, the result of this arrangement is that the patient, on moving, experiences some discomfort and interprets it to mean that healing has not occurred and that continued movement may impair healing. The patient then moves less, thereby increasing the adverse effects of disuse and

making it more probable that future movement of the involved body part will be painful. Thus, begins a vicious circle. Each painful movement is interpreted to reaffirm that healing has not occurred and encourages greater disuse and more pain. Pain with movement may also become paired to an environment, which through "conditioning" can create environmental cues capable of eliciting more discomfort. In sum, the pain problem may persist and worsen despite adequate healing because the consequences of use and movement were misinterpreted secondary to ambiguous clinician guidance. Behavior that avoids or postpones an anticipated aversive consequence, known as avoidance learning, has been studied in clinical pain trials (Fordyce et al. 1982). Therefore, physicians should discourage patients from prolonging disability beyond medical necessity. This will require a more aggressive approach to pain control; prevention of unnecessary surgery, directed efforts to improve patients' abilities to manage residual pain and distress, and attention to employer-employee conflicts may be important in preventing the development of prolonged work disability in this population.

Summary

Musculoskeletal impairment and disability have become increasingly important and more individuals are claiming work-compensable injuries. As the number of claims has increased so has the cost, financially to the employer and healthwise to the individual. Legislatures have passed new definitions and thresholds for workplace causation (Melhorn 2011c). The future is reduction of costs and improvement of quality of life is through prevention.

References

AMA. (2008). Conceptual foundations and philosophy. In R.D. Rondinelli (Ed.), *Guides to the evaluation of permanent impairment* (6th ed., pp. 1–18). Chicago, IL: American Medical Association.

AMA. (2009). Glossary. In R. D. Rondinelli (Ed.), *Guides to the evaluation of permanent impairment* (6th ed., pp. 609–616). Chicago, IL: American Medical Association.

Biddle, J., Roberts, K., Rosenman, K. D., et al. (1998). What percentage of workers with work-related illnesses receive workers' compensation benefits? *Journal of Occupational and Environmental Medicine, 40*, 325–331.

Demeter, S. L., Andersson, G. B. J., & Smith, G. M. (2001). *Disability evaluation*. St. Louis, MO: Mosby.

Deyo, R. A., Diehl, A. K., & Rosenthal, M. (1986). How many days of bed rest for acute low back pain? A randomized clinical trial. *The New England Journal of Medicine, 315*, 1064–1070.

Fordyce, W. E., Brockway, J. A., Bergman, J. A., et al. (1986). Acute back pain: a control-group comparison of behavioral vs traditional management methods. *Journal of Behavioral Medicine, 9*, 127–140.

Fordyce, W. E., Shelton, J. L., & Dundore, D. E. (1982). The modification of avoidance learning pain behaviors. *Journal of Behavioral Medicine, 5*, 405–414.

Harris, I., Mulford, J., Solomon, M., van Gelder, J. M., & Young, J. (2005). Association between compensation status and outcome after surgery: A meta-analysis. *JAMA: The Journal of the American Medical Association, 293*, 1644–1652.

Himmelstein, J. S., Feuerstein, M., Stanek, E. J., et al. (1995). Work related upper extremity disorders and work disability: Clinical and psychosocial presentation. *Journal of Occupational and Environmental Medicine, 37*, 1278–1286.

Ladou, J. (2009). Federal Employees' Compensation Act. *International Journal of Occupational and Environmental Health, 15*, 180–194.

Ladou, J. (2010). Workers' compensation in the United States: Cost shifting and inequities in a dysfunctional system. *New Solutions, 20*, 291–302.

McHeil, M. (2001). *In Americans with disabilities. Household economic studies* (pp. 70–73). Washington, DC: United States Department of Commerce.

Melhorn, J. M. (1998). *Management of work related upper extremity musculoskeletal disorders*. Kansas Case Managers annual meeting (pp. 16–25). Wichita, KS: Wesley Rehabilitation Hospital.

Melhorn, J. M. (2002a). Cumulative Trauma Disorders (CTDs): The science, myths and folklore. In *16th annual scientific session*. Chicago, IL: American Academy of Disability Evaluating Physicians.

Melhorn, J. M. (2002b). Understanding and managing chronic non-malignant pain for the occupational orthopaedists. In J. M. Melhorn & R. E. Strain Jr. (Eds.), *Occupational orthopaedics and workers' compensation: A multidisciplinary perspective*. Rosemont, IL: American Academy of Orthopaedic Surgeons.

Melhorn, J. M. (2010). Causation – Understanding why. In: *65th Annual Workers' Compensation Educational Conference and 22nd Annual Safety & Health Conference* (pp. 45–55). Orlando, FL: Florida Workers Compensation Institute.

Melhorn, J. M. (2011a). Unnecessary disability – Why can't I work? In *Impairment without disability*. Rochester, MN: Mayo Clinic.

Melhorn, J. M. (2011b). Causation: Factually evaluating work-relatedness and injury-relatedness. In *Advances skills course*. Chicago, IL: American Academy of Disability Evaluating Physicians.

Melhorn, J. M. (2011c). Workers' compensation: What's your threshold? *AAOS Now, 5*, 11.

Melhorn, J. M. (2012). Chronic pain treatment – What works? In *Effectively working within your workers' compensation system*. Sacramento, CA: California Orthopaedic Association.

Melhorn, J. M., Wilkinson, L. K., & O'Malley, M. D. (2001). Successful management of musculoskeletal disorders. *J Hum Ecolog Risk Assessment, 7*, 1801–1810.

Melhorn, J. M., Ackerman, W. E., Glass, L. S., & Deitz, D. C. (2008). Understanding work-relatedness. In J. M. Melhorn & W. E. Ackerman (Eds.), *Guides to the evaluation of disease and injury causation* (pp. 13–32). Chicago, IL: AMA Press.

Melhorn, J. M., Lazarovic, J., & Roehl, W. K. (2005). Do we have a disability epidemic? In R. J. Gatchel & I. Z. Schultz (Eds.), *Handbook of complex occupational disability claims: Early risk identification, intervention and prevention*. New York: Springer.

Nachemson, A. L. (1992). Newest knowledge of low back pain. A critical look. *Clinical Orthopaedics, 13*, 8–20.

Sikorski, J. M., Molan, R. R., & Askin, G. N. (1989). Orthopaedic basis for occupationally related arm and neck pain. *Australian and New Zealand Journal of Surgery, 59*, 471–478.

Sperka, P., Cherry, N., Burnham, R., & Beach, J. (2008). Impact of compensation on work outcome of carpal tunnel syndrome. *Occupational Medicine (London), 58*, 490–495.

Talmage, J. B., & Melhorn, J. M. (2005). How to think about work ability and work restrictions: Risk capacity, tolerance. In J. B. Talmage & M. JM (Eds.), *A physician's guide to return to work* (pp. 7–19). Chicago, IL: AMA Press.

Turk, D. C., & Okifuji, A. (2001). Pain terms and taxonomies of pain. In *Bonica's management of pain* (pp. 17–25). Philadelphia: Lippincott Williams & Wilkins.

United States Bureau of Labor Statistics. (1996). *Survey of occupational injuries and illnesses in 1994*. Washington, DC: United States Government Printing Office.

World Health Organization. (2001). *International classification of functioning, disability and health: ICF*. New York, NY: Unites Nations.

Zeller, J. L., Burke, A. E., & Glass, R. M. (2007). JAMA patient page. Assessing disability. *JAMA: The Journal of the American Medical Association, 298*, 2096.

Disability-Litigation System

Definition

A method of compensation that rewards injured workers for non-function or the inability to recover resulting from a work-related injury.

Cross-References

▶ Pain in the Workplace, Risk Factors for Chronicity, Job Demands

---

Discharge Frequency

Definition

The number of times a neuron produces an action potential in a given unit of time.

Cross-References

▶ Encoding of Noxious Information in the Spinal Cord

Discharge Pattern

Definition

Particular form in which action potentials are fired in a given period of time.

Cross-References

▶ Mechanonociceptors

Discogenic Back Pain

Edward C. Benzel[1] and Tiffany Grace Perry[2]
[1]Department of Neurosurgery, Center for Spine Health Neurological Institute Cleveland Clinic, Cleveland, OH, USA
[2]Center for Spine Health, The Cleveland Clinic Foundation, Cleveland, OH, USA

Synonyms

Black disc disease; Dark disc disease; Symptomatic annular tear

Definition

Low back pain is a nebulous term that describes many various types of pain and encompasses the endless etiologies for the pain. Many sources of low back pain (LBP) exist including but not limited to the facets, fractures, musculoskeletal, intervertebral discs (IVD), infections, tumors, and structural deformities. In 1994, Schwarzer et al. performed a study to evaluate the relative contribution of the disc and the facet joints to chronic LBP. They had 92 patients with chronic LBP who were all evaluated with discograms and facet blocks. He reported that the disc was responsible for LBP in 39 % of the 92 subjects with only 3 % of the patients having pain elicited from both the discogram and relief from the facet block. Schwarzer et al. reported that this implicated the disc as a leading contributor to chronic LBP and also showed that multifactorial LBP is much less common (Schwarzer et al. 1994).

Since that study, many studies have been done on the diagnostic reliability of discograms. Today, this test has lost much of its popularity due to its low sensitivity and specificity. The difficulty with diagnosing discogenic back pain is that there is no definable radiographic or diagnostic study that can assist in diagnosing pain originating in the disc as opposed to other potential sources for the pain. Another confounding variable is that commonly LBP is indeed multifactorial and may not be attributable to only one cause. Perhaps most importantly, as pertains to this entry, is the notion that "discogenic back pain" may not be an appropriate diagnosis at all. After all, how do we know the pain is originating in the "disc," and how do we define the entity termed "discogenic back pain"? Herein, we will leave this up to conjecture. However, as you read on, please be cognizant of the fact that the term discogenic back pain characterizes pain that originates within the confines of the intervertebral disc. Without proof of the existence of such a pain syndrome, we question the value of establishing such from a diagnostic and therapeutic perspective. Regardless, in the pages that follow, we discuss this nebulous entity and present current state of knowledge regarding management.

Characteristics

Pathophysiology
Luschka first described the sinuvertebral nerve (SVN) in the 1800s as a spinal nerve that receives input from the sympathetic chain. In 1959, Malinsky then demonstrated multiple types of nerve endings on the annulus fibrosus, now known to be innervated by the SVN. The SVN also supplies nerve endings to the dura mater, the posterior longitudinal ligament (PLL), and the annulus fibrosus. The SVNs arise from two roots, one from the ventral primary ramus and the other from the ramus communicans (Fig. 1). The SVN enters the neuroforamen and sends its major branch rostrally to the PLL and the other branch medially to the annulus. When the annulus is stimulated by an annular tear or by a herniated nucleus pulposus extending into the tear, pain from the disc ensues, causing low back pain. It may masquerade as a radiculopathy or buttock pain.

Pain Modulation
The nucleus pulposus is composed of collagen, mucopolysaccharides, and water and is void of

Discogenic Back Pain, Fig. 1 The SVN supplies nerve endings to the PLL, annulus fibrosus, and the dura mater

nerve endings. Substance P is increased in concentrations in degenerative discs and likely modulates the initially nociceptive pain signals in the dorsal spinal cord. Somatostatin in the dorsal gray matter is released from the dorsal root ganglion (DRG) after noxious stimulation or biomechanical forces acting on the disc. The release of somatostatin then stimulates the synthesis of inflammatory agents (cytokines and prostaglandins). The cytokines (IL-1, IL-6, IL-10, TNF-a) are produced in response to neural injury in the CNS and play a role in the neural hypersensitization contributing to the hyperalgesia. The prostaglandin molecules break into smaller molecules that bind less water and result in lower pH within the disc. Studies have found that in animal models, subjects with artificially lowered intradiscal pH exhibited more pain-related behavior and hyperalgesia. Nitric oxide (NO) is elevated in degenerated discs and inhibits prostaglandin synthesis in the nucleus pulposus, leading to decreased water content, causing further disc degeneration.

Differential Diagnosis

Low back pain has a multitude of etiologies. One of the difficulties in diagnosing and treating the pathology is that multiple pathologies may be the source for the patient's symptoms. Sources of low back pain other than the disc may include the facets, the facet capsules, ligaments, paraspinal muscles, fractures, infection, tumor, deformity, and instability. Some of these can be ruled out based on XR, CT, and/or MRI. Once the diagnosis may be narrowed down to the facets, ligaments, or discs, steroid or lidocaine injections may be utilized to do a diagnostic block to aid in determining the source of the pain (Fig. 2).

Diagnosis

The entity of discogenic back pain can be difficult to appropriately diagnose. MRI T2-weighted sequences may exhibit a high-intensity zone (HIZ) that is a hyperintense region around the annular tear (Figs. 3 and 4). Provocative discography is the injection of water-soluble contrast into the disc under fluoroscopy. A positive discogram is one in which the patient's symptoms are reproduced at the level of the injection. The reliability of the patient is key to the diagnostic validity of the discogram. Patients who have abnormal psychometric testing or multiple complaints are more prone to false-positive or nondiagnostic discograms. Provocative discography had at one point been the diagnostic test of choice for discogenic back pain. However, it has now become less popularized due to the low specificity and sensitivity. In 1999, Carragee et al. performed a study in which 24 discograms were performed on eight volunteers with no axial low back pain. The patients had undergone iliac crest autograft harvest in the few months preceding the discogram. Of the eight patients, 50 % of them had pain with discography upon pressurization of the disc that was concordant with their post-iliac crest harvest pain in the gluteal region. The purpose of the study was to illustrate that the ability of a patient to differentiate between spinal

Discogenic Back Pain, Fig. 2 A herniated disc may cause focal compression of the exiting nerve root. Diagnostic blocks (needle) may be performed to assist in narrowing the diagnosis

and non-spinal etiology of pain is difficult thereby questioning the accuracy of discography even in patients with no history of low back pain (Carragee et al. 1999). Therefore, discography may be able to diagnose pain originating from the intervertebral disc about 50 % of the time, a rate which is arguably unacceptable.

Pain

Mechanical low back pain is usually insidious in onset and is characterized by pain originating in the lower back and radiating to the buttock, hip, or proximal lower extremity. It is also characterized by pain that is typically relieved by lying down, off-loading the spine and made worse by axial loads such as sitting or standing. These patients have a tendency to shift positions frequently to alleviate the discomfort in their back. Mechanical back pain may be caused by any number of pathologies. Because it is most often multifactorial, the success of both surgical and nonsurgical interventions is nearly impossible to

Discogenic Back Pain, Fig. 3 T2-weighted MRI showing the sagittal and axial high-intensity zone (arrows) of an annular tear

Discogenic Back Pain, Fig. 4 (**a**) The IDET procedure uses thermal energy to denature the collagen of the annulus and to coagulate the annular nociceptors. (**b**) The herniated nucleus pulposus shrinks and the annulus scars at the site of IDET

predict. Mechanical low back pain, as discussed here is a distinct and well-defined entity compared to discogenic back pain. The quality of discogenic back pain has not yet been defined.

The origin of mechanical low back pain is a dysfunctional motion segment. The origin of discogenic back pain allegedly is the intervertebral disc.

Prevention

Overall healthy habits in lifestyle, diet, and exercise can aid in prevention of degenerative disc disease. The paraspinal muscles and ligaments that support the back and the core abdominal muscles should be strengthened to off-load the spine, including the bones, discs, and joints. Smoking should be avoided as it has been shown to have a strong correlation to disc degeneration.

Treatment

Once the preliminary diagnosis of discogenic back pain is made by a process of exclusion, the patient should explore nonsurgical means of treatment. Conservative modalities such as physical therapy, weight loss, and NSAIDs are the mainstay of the initial treatment of mechanical low back pain. Invasive nonsurgical measures include intradiscal electrothermal therapy (IDET), nucleoplasty, percutaneous intradiscal radiofrequency thermocoagulation (PIRFT), and laser discectomy. Surgical treatment whether minimal access or standard open approach is reserved for patients who have failed all conservative and less-invasive treatments. The following treatments have been purported to be of value for discogenic back pain. Their true clinical value is suspect at best. They are included herein to provide comprehensive coverage of the subject.

Annuloplasty

The term "disc annuloplasty" refers to any procedure that directly affects the annulus fibrosus to treat pain due to the nucleus pulposus causing an annular tear or disc herniation. There are two basic procedures to treat pain thought to be originating from the annulus fibrosus as described below.

- *Intradiscal Electrothermal Therapy (IDET)*
  This technique first described in 1997 involves heating the posterior annulus with a thermal resistive coil in an attempt to repair an annular tear and denervate the injured area. The thermal energy exposes the collagen to 60 °C, leading the denaturing of the collagen strands of the annulus. The passive heat dissipates along the outer annulus at a temperature of about 42 °C coagulating the nociceptors along the annulus.

Discogenic Back Pain, Fig. 5 Biaculoplasty incorporates a radiofrequency energy between the bilaterally inserted electrodes

- *Intervertebral Disc Biacuplasty (IDB)*
  IDB utilizes two small bipolar radiofrequency electrodes that insert bilaterally into the disc space terminating in the nucleus pulposus (Fig. 5). Once within the disc, a radiofrequency energy is delivered between the two electrodes heating the tissue and lesioning the SVNs to the annulus. The circulating water within the disc dissipates the heat thereby preventing any thermal injury to the dorsal root ganglion (DRG) or the nerve root. Case reports and small study cohorts have shown improvement in patients' Oswestry Disability Index and SF-36. Due to the relatively recent development of this procedure, to date, there have not been any prospective randomized trials (Kapural et al. 2008).

Nucleoplasty

A nucleoplasty defines any intervention that directly acts on the nucleus pulposus to decrease the disc water content and thereby reducing the volume, diminishing the effect of a herniated disc. This intervention was FDA-approved in 2001 and is indicated for herniated discs that are contained within the annulus fibrosus. A bipolar radiofrequency device uses a localized energy field that dissolves tissue without significant thermal energy release. The claim is that there is

Discogenic Back Pain, Fig. 6 (a) Nucleoplasty introduces a thermal energy that dissociates the water molecules of the nucleus pulposus to shrink the disc volume by about 10 %. (b) In nucleoplasty, the primary target is the nucleus pulposus, purportedly causing minimal scarring and denaturing of the annulus and nociceptors

minimal thermal injury to surrounding structures. The wand with a bipolar coil is directed to the center of the nucleus pulposus (Fig. 6a). The coil creates a plasma field dissolving the water content of the disc into hydrogen and oxygen ions. The volume of the disc is supposed to decrease by about 10 % facilitating decompression of the disc and decreased intradiscal pressure (Fig. 6b). Singh et al. (2002) reported about 80 % of patients obtained statistically significant improvement in their pain scales which remained stable over 1 year (Singh et al. 2002).

Percutaneous Intradiscal Radiofrequency Thermocoagulation (PIRFT)

Similar to nucleoplasty, a radiofrequency lesion is made in the nucleus pulposus. Thermal energy dissipates causing fibrosis of the disc and loss of nociceptive input on the surface of the annulus. The spread of the thermal energy to the outer annulus is difficult to measure.

Laser Discectomy

Today with the popularity of "laser spine surgery," spine surgeons frequently are asked by patients if their back pain can be alleviated by laser surgery. The laser discectomy is indicated in a very small portion of the population. The procedure involves using laser energy to vaporize a portion of the nucleus volume to decrease the intradiscal pressure and shrink the disc protrusion.

Surgical Options

Is there ever an indication to operate for chronic low back pain? The authors are of the opinion that there are few select patients who benefit from surgery for chronic low back pain. Patients who fail conservative measures may be offered surgical options at the discretion of their spine surgeon. For one-level mechanical back pain thought to be originating from the disc, an artificial disc replacement (ADR) may be a viable option for this certain subset of patients. There are currently two FDA-approved discs on the market (Charité and ProDisc-L). This procedure is an anterior approach that involves a complete discectomy of the affected level and is only approved for patients with one-level disc disease. ADRs are for patients with isolated disc degeneration without significant degeneration of the facets or other structures including adjacent levels that could contribute to their pain. Proponents of the ADR state that one of the benefits to having a disc replacement is that the patient's nociceptors to the painful disc are eliminated and the motion of the segment is preserved with the artificial disc.

Discogenic Back Pain, Fig. 7 The Charité ADR, manufactured by DePuy Synthes™, employed an ultra-high molecular weight polyethylene core affixed to 2 cobalt chrome alloy endplates. This device is no longer available in the United States

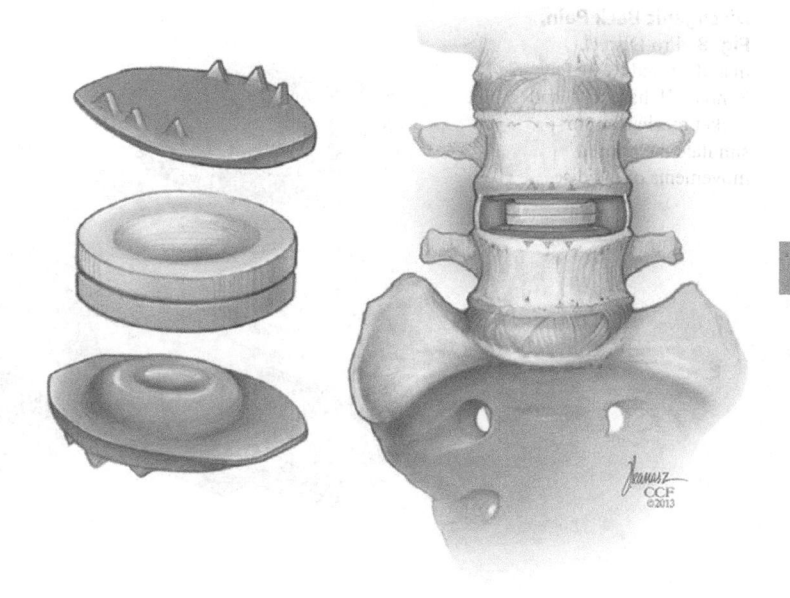

In 2009, the data was published for the first 5-year prospective randomized trial with follow-up of 160 patients who underwent the Charité ADR versus BAK cage placement (Fig. 7). There was no statistical difference in the clinical outcomes and satisfaction between the two groups in this study. The Charité patients did have a statistically greater rate of employment and lower rate of disability than the BAK patients. The range of motion (ROM) at the surgical level and the adjacent levels was not statistically different from the ROM seen at the 2-year follow-up (Guyer et al. 2009).

In October 2012, Zigler et al. reported their 5-year follow-up for 161 patients who underwent ADR with the ProDisc-L and 75 patients who underwent circumferential fusions (Fig. 8). This study basically showed that both procedures maintained a statistically significant improvement in the Oswestry Disability Index (ODI), visual analog score (VAS), and satisfaction of the subjects. The percentage of ADR patients who stated they would have the surgery again was 82 % compared to 68 % in the fusion group. This study demonstrates that both ADR and fusion are reasonable options for patients with mechanical LBP (Zigler and Delamarter 2012).

Keep in mind, though, that long-term efficacy of ADRs has not yet been established. Their utilization should be undertaken with such in mind.

As indicated in the above studies, spinal fusion for mechanical back pain has been studied widely encompassing various procedures for different pathology (Fig. 9). In 2009, the Spine Patient Outcomes Research Trial (SPORT) showed that fusion for degenerative lumbar spondylolisthesis has been shown to have success at alleviating LBP (Weinstein et al. 2009).

Xu et al. reported that patients in their synovial cyst study from 2010, spine fusion at lumbar segments with synovial cysts was more successful in preventing recurrent cysts and therefore recurrent LBP and radiculopathy (Xu et al. 2010).

Spondylolisthesis and synovial cysts are two examples of structural pathologies that may respond well to surgical fusion. Pain that might originate from the disc alone is extremely difficult, if not impossible, to diagnosis and therefore very difficult to treat successfully. The theory of a fusion for "discogenic back pain" is that by eliminating motion at the level of pain, the nociceptors of the disc are not going to be

Discogenic Back Pain,
Fig. 8 Pro-Disc L,
manufactured by DePuy
Synthes™, has a ball and
socket mechanism to
simulate the normal
movements of the disc

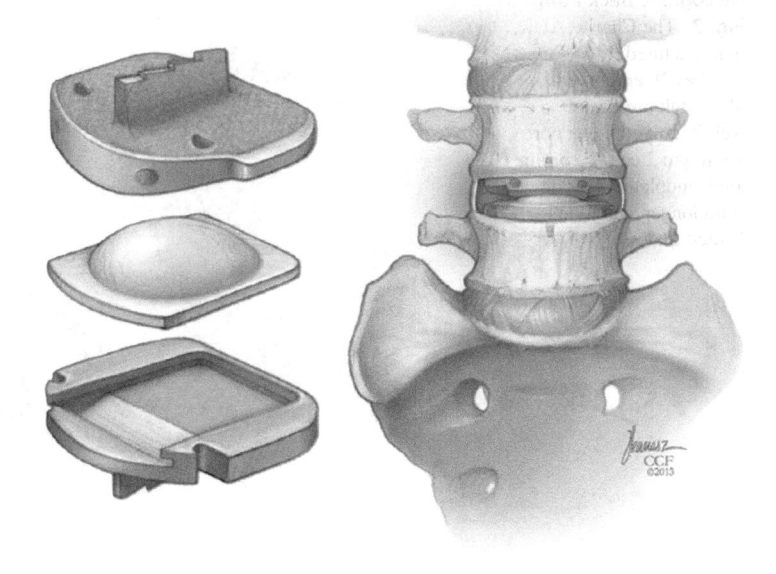

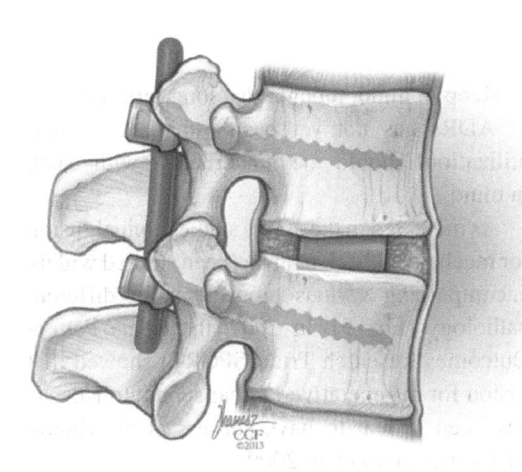

Discogenic Back Pain, Fig. 9 One-level posterior
instrumented fusion, as depicted, with or without an
interbody graft can be used in selective patients to diminish pain from an abnormal motion segment

stimulated. In fact, this is a theoretical hypothesis, one that has worked on paper, but not in
human subjects. Mechanical back pain is one of
the most physiologically and psychologically
complex entities to treat for medical and surgical
spine physicians. This entry just touches on the
obscurity of pain and the intricacies of the contributors to pain. The more we strive to understand it, the more convoluted it becomes.

Surgical management of low back pain,
including fusion and ADR, in the absence of
structural pathology, is associated with a
tarnished track record. If one critically assesses
the literature, the true value of fusion is less
than what one might expect. Regardless,
surgical intervention may be of clinical utility in
some cases.

Summary

If your impression and clinical opinion of
discogenic back pain has been clarified after
reading this entry, you are one step ahead of the
authors. The entity termed discogenic back pain
remains ill defined and is a diagnostic enigma.
Hopefully, we will all, at the very least, be stimulated to investigate our patients' history, examination, and diagnostic tests astutely and to utilize
sound clinical judgment for the management of
chronic low back pain.

References

Carragee, E. J., Tanner, C. M., Yang, B., Brito, J. L., &
    Truong, T. (1999). False-positive findings on lumbar
    discography. Reliability of subjective concordance

assessment during provocative disc injection. *Spine (Phila Pa 1976), 24*(23), 2542–2547.

Guyer, R. D., McAfee, P. C., Banco, R. J., Bitan, F. D., Cappuccino, A., Geisler, F. H., Hochschuler, S. H., Holt, R. T., Jenis, L. G., Majd, M. E., Regan, J. J., Tromanhauser, S. G., Wong, D. C., & Blumenthal, S. L. (2009). Prospective, randomized, multicenter Food and Drug Administration investigational device exemption study of lumbar total disc replacement with the CHARITE artificial disc versus lumbar fusion: Five-year follow-up. *The Spine Journal, 9*(5), 374–386. Epub 2008 Sep 19.

Herkowitz, H., Longley, M., Lenke, L., Emery, S., & Hu, S. S. (2009). Surgical compared with nonoperative treatment for lumbar degenerative spondylolisthesis. four-year results in the Spine Patient Outcomes Research Trial (SPORT) randomized and observational cohorts. *J Bone Joint Surg Am, 91*(6): 1295–304.

Kapural, L., Ng, A., Dalton, J., Mascha, E., Kapural, M., de la Garza, M., & Mekhail, N. (2008). Intervertebral disc biacuplasty for the treatment of lumbar discogenic pain: Results of a six-month follow-up. *Pain Medicine, 9*(1), 60–67.

Schwarzer, A. C., Aprill, C. N., Derby, R., Fortin, J., Kine, G., & Bogduk, N. (1994). The relative contributions of the disc and zygapophyseal joint in chronic low back pain. *Spine (Phila Pa 1976), 19*(7), 801–806.

Singh, V., Piryani, C., Liao, K., & Nieschulz, S. (2002). Percutaneous disc decompression using coblation (nucleoplasty) in the treatment of chronic discogenic pain. *Pain Physician, 5*(3), 250–259.

Weinstein, J. N., Lurie, J. D., Tosteson, T. D., Zhao, W., Blood, E. A., Tosteson, A. N., Birkmeyer, N., Herkowitz, H., Longley, M., Lenke, L., Emery, S., & Hu, S. S. (2009). Surgical compared with nonoperative treatment for lumbar degenerative spondylolisthesis four-year results in the Spine Patient Outcomes Research Trial (SPORT) randomized and observational cohorts. *The Journal of Bone and Joint Surgery. American Volume, 91*(6), 1295–1304.

Xu, R., McGirt, M. J., Parker, S. L., Bydon, M., Olivi, A., Wolinsky, J. P., Witham, T. F., Gokaslan, Z. L., & Bydon, A. (2010). Factors associated with recurrent back pain and cyst recurrence after surgical resection of one hundred ninety-five spinal synovial cysts: Analysis of one hundred sixty-seven consecutive cases. *Spine (Phila Pa 1976), 35*(10), 1044–1053.

Zigler, J. E., & Delamarter, R. B. (2012). Five-year results of the prospective, randomized, multicenter, Food and Drug Administration investigational device exemption study of the ProDisc-L total disc replacement versus circumferential arthrodesis for the treatment of single-level degenerative disc disease. *Journal of Neurosurgery. Spine, 17*(6), 493–501. doi:10.3171/2012.9. SPINE11498. Epub 2012 Oct 19.

Discogenic Pain

Definition

Pain due to an abnormality of the vertebral disc.

Cross-References

▶ Chronic Back Pain and Spinal Instability
▶ Chronic Low Back Pain: Definitions and Diagnosis

Discomfort

▶ Pain in Early Pregnancy

Discordant Illness Behavior

▶ Hypochondriasis, Somatoform Disorders, and Abnormal Illness Behavior

Discriminability

Definition

Discriminability reflects the capacity for detection of the presence of a stimulus or differences between stimuli, usually those of intensity.

Cross-References

▶ Pain in Humans, Psychophysical Law
▶ Pain Measurement by Questionnaires, Psychophysical Procedures, and Multivariate Analysis

Disease or Injury

▶ Demyelination

Disease-Modifying Antirheumatic Drugs

Synonyms

DMARDs

Definition

Disease-modifying antirheumatic drugs (DMARDs) are used to treat chronic inflammatory diseases such as rheumatoid arthritis: They target immune cells in order to inhibit the cellular inflammatory response. Examples are methotrexate, leflunomide, and cyclosporine.

Cross-References

▶ Neutrophils in Inflammatory Pain
▶ NSAIDs and Their Indications

Disinhibition

Definition

Excitation due to an inhibition of inhibitory processes.

Cross-References

▶ Stimulation-Produced Analgesia

Disinhibition of Nociceptive Neurons

Definition

A loss of inhibitory control of nociceptive neurons, which is normally exerted by GABAergic and glycinergic neurons in the spinal cord dorsal horn, is increasingly recognized as a major source of chronic pain. It can result from inhibition of glycine or GABA receptors, or reduced GABA or glycine release, apoptotic death of GABA and glycinergic neurons, and from changes in the chloride gradient, which renders GABAergic and glycinergic input less inhibitory.

Cross-References

▶ GABA and Glycine in Spinal Nociceptive Processing

Disk Displacement

Definition

The cartilaginous disk, interposed between the mandibular condyle and the fossa of the temporomandibular joint, may get displaced. Most often an anterior disk displacement occurs, which may give rise to clicking noises during mandibular movement. If the disk is displaced permanently, a locking of the joint occurs that is habitually accompanied by a limitation of mandibular movement.

Cross-References

▶ Orofacial Pain: Taxonomy/Classification

Displacement

▶ Social Dislocation and the Chronic Pain Patient

Disposition

▶ Personality and Pain

Disruption

▶ Social Dislocation and the Chronic Pain Patient

Dissection

Definition

Separation of (usually arterial) vessel wall layers by an intramural hemorrhage.

Cross-References

▶ Headache Due to Dissection

Dissociation

Definition

Separation or detachment from one's immediate environment or the compartmentalization of various components of conscious experience. Dissociation can be produced by severe psychological trauma, and can reoccur for years after the trauma. Hypnotic analgesia may be achieved, for example, by encouraging the subject to dissociate himself or herself from the procedure room or from the painful body part.

Cross-References

▶ Therapy of Pain, Hypnosis

Dissociative Imagery

Definition

Dissociative imagery is imagery that is disconnected from the felt sense of the body.

Cross-References

▶ Hypnotic Analgesia

Dissociative Sedation

Definition

A trance-like cataleptic state induced by the dissociative agent ketamine and characterized by profound analgesia and amnesia, usually with the retention of protective airway reflexes, spontaneous respirations, and cardiopulmonary stability.

Cross-References

▶ Pain and Sedation of Children in the Emergency Setting

Distal Axonopathy

Synonyms

Central-peripheral distal axonopathy; Dying-back neuropathy

Definition

Peripheral nerve disorder resulting from degeneration of the most terminal parts of either central or peripheral processes of sensory neurons. This is the major pathology of toxic neuropathies.

Cross-References

▶ Toxic Neuropathies

Distance Treatment

▶ Remote Management of Pediatric Pain

Distractibility

Definition

Tendency to give attention to any stimulus regardless of its relevance.

Cross-References

▶ Hypervigilance and Attention to Pain

Distracting Responding

▶ Spouse, Role in Chronic Pain

Distracting Responses

Definition

Distracting responses refer to cues from significant others intended to encourage alternative, presumably more adaptive, well behaviors

(e.g., increased activity, use of distraction to cope with pain) on the part of the person experiencing pain.

Cross-References

▶ Spouse, Role in Chronic Pain

Distraction

Definition

The use of materials to provide alternative sensory stimulation for the infant during a painful procedure (e.g., music, mobiles).

Cross-References

▶ Acute Pain Management in Infants
▶ Psychological Treatment in Acute Pain

Distraction Signs

▶ Lower Back Pain, Physical Examination

Distribution

Definition

The distribution characterizes the reversible transfer of a drug or a substance into regions within the body.

Cross-References

▶ NSAIDs, Pharmacokinetics

Disuse Syndrome

Definition

Decreased level of physical activities in daily life, in the long term leading to physical deconditioning.

Cross-References

▶ Disability, Fear of Movement
▶ Muscle Pain, Fear-Avoidance Model

Diurnal Variations of Pain in Humans

Gaston Labrecque
Faculty of Pharmacy, University of Laval,
Quebec City, QC, Canada

Synonyms

24-h variation; Biological rhythms; Circadian variations in pain level

Definition

Diurnal changes in pain intensity usually mean that the bouts of pain occur during the daytime. However, pain intensity fluctuates throughout the day and night, and patients often report that peak pain occurs at specific hours of the day. In this case, we will talk about ▶ biological rhythms or ▶ circadian variations (about 24 h) in pain level.

Characteristics

Pain is one of the most common symptoms for which patients seek advice and help from health professionals. This is a complex, subjective, and unpleasant phenomenon influenced by factors such as anxiety, fatigue, suggestions or emotions, and prior experience. Pain is rarely constant, and patients usually report bouts of pain throughout the 24-h period. Using the visual analog scale (VAS), many studies have indicated that the intensity of pain varied specifically during the day or the night spans. Clinicians must rely on the patient's evaluation of pain intensity to decide which and how much drug must be prescribed and when it must be taken by the patient. Information on biological rhythms of pain can be used to maximize the effect of analgesic drugs and/or to minimize their side effects. A recent literature review summarizes the information regarding rhythms, pain, and pain management (Labrecque and Vanier 2003).

Rhythms in Endogenous Opioid Peptide Levels

In the last 25 years, data obtained from laboratory animals indicate the existence of 24-h variations in plasma and brain concentrations of β-endorphin or enkephalin: peak values were obtained late during the resting period or at the beginning of the activity period. Similar data were obtained in healthy volunteers and pregnant women: the highest β-endorphin plasma levels occurred between 6 a.m. and 8 a.m., while the lowest levels were found between 8 p.m. and midnight. It is also interesting to point out that a twofold rise in the plasma endogenous opioid peptide was found in the last semester of the pregnancy. This time-dependent variation was not found on the fourth day after delivery (2). Finally, a circannual variation of endorphin levels was reported 20 years ago in the CSF of patients with chronic pain syndrome of psychogenic and organic etiology: highest concentrations were found in January and February, whereas lowest concentrations occurred in July and August (see Labrecque and Vanier 1997; Labrecque and Vanier 2003 for references). Thus, time-dependent variations in pain level and/or in the requirements for analgesia should be expected in patients with pain.

Diurnal Variations of Pain in Humans, Table 1 Biological rhythms of pain in patients (see Labrecque and Vanier (1997, 2003), for references)

Causes of pain	No. of patients	Hours of peak	Hours of trough
Anginal pain	7,788	6 a.m. to noon	Midnight to 6 a.m.
Unstable angina	2,586	8–10 a.m.	2–4 a.m.
Myocardial infraction	1,229	5–9 a.m.	
	703	5–10 a.m.	
Biliary colic	50	11 p.m. to 3 a.m.	9 a.m. to 1 p.m.
Cancer pain	130	6 p.m.	4–10 a.m.
Heavy burns	9	8 a.m. to 4 p.m.	Midnight to 8 a.m.
Intractable pain	41	10 p.m.	8 a.m.
Migraine	15	10 a.m.	Midnight
	117	8 a.m. to noon	Midnight
	114	4–8 a.m.	Noon
Osteoarthritis of the knee	20	10 p.m.	2–6 a.m.
	57	2–10 p.m.	7 a.m.
	4	7–11 p.m.	
Rheumatoid arthritis	19	6–8 a.m.	6 p.m.
Toothache	543	8 a.m.	3 p.m.

Circadian Rhythms in Human Pain

Patients often complain that pain intensity increases in the evening or at night. This phenomenon is usually explained by fatigue related to daily activity, or to the anxiety induced by the incoming sleeping period, or by the departure of the family visitors. Clinicians rarely consider that the pain level changes specifically during the day or night, but they forget that time-dependent changes have been documented in hospitalized and outpatients. In fact, these studies showed that the hour of highest and lowest pain level is specific for each painful stimulus.

Table 1 shows that pain related to the onset of cardiovascular events is found very early in the morning, while the peak in the frequency of migraine and toothache is largest in the morning. On the other hand, the peak of biliary colic, intractable, and low back pain was found in the evening. In cancer patients, most studies were carried out in patients receiving opioid analgesics, and they were mainly concerned with the pain relief produced by these drugs. For instance, Sittl et al. (1990) reported that cancer pain was largest by 6 p.m. Other studies are obviously needed in cancer patients, and it would be interesting to find whether the hour of pain varies with the causes and/or the severity of this disease.

The hour of arthritic pain deserves a special note, because the time of peak pain varies according to the cause of arthritis. In patients with rheumatoid arthritis (RA), the intensity of pain is highest at the beginning of the day, whereas it occurred late in the afternoon, at the end of the day, in patients with osteoarthritis (OA) of the knee (Bellamy et al. 2002; Kowanko et al. 1981; Lévi et al. 1985). It must be pointed out that interindividual differences in the hour of highest pain intensity were reported in OA patients. Studies by Lévi et al. (1985) and Bellamy et al. (2002) indicated that most OA patients reported peak pain between 4 p.m. and 11 p.m., but they also reported that morning pain was highest in the morning in 5 % of the OA patients, while about 10 % of these patients did not find any rhythmic changes in the intensity of their arthritic pain. These disease-related and interindividual differences in the hour of arthritic pain should be taken into account when prescribing antiarthritic medications.

Biological Rhythm and the Effect of Medications

The circadian variation in arthritic pain can be used to maximize the effect of nonsteroidal anti-inflammatory agents (NSAIDs). For instance, it was shown that RA patients must take an evening dose of flurbiprofen to control morning stiffness and pain (Kowanko et al. 1981). In OA patients, a multicenter study was carried out to answer the question often asked by patients: When should I take this once-daily NSAID? The optimal time of administration for each patient was related to the time of peak pain. The evening administration of indomethacin was most effective in patients with a predominantly nocturnal pain, while drug ingestion around noon was best for patients with peak pain occurring late in the afternoon or early in the evening. The analgesic effect of the NSAID was increased by 60 % when the medication was taken at the time preferred by the patients (Lévi et al. 1985). Furthermore, double-blind crossover trials indicated that the frequency of side effects of sustained-release indomethacin (Lévi et al. 1985) and ketoprofen (Boissier et al. 1990) was significantly larger when these drugs were taken at 8 a.m. than at 8 p.m. Thus, appropriate selection of the time of ingestion of the well-established NSAIDs can increase their effectiveness, and it may reduce the side effects of the well-established agents. Unfortunately, there is no data for the newer NSAIDs, such as celecoxib and rofecoxib, but it is expected that biological rhythms can also be used to maximize their effectiveness.

Very few investigators have studied the temporal variation in the effects of morphine and other opioids in patients with acute pain. In acute surgical pain, the demands for morphine (Mo) or hydromorphone (Hm) administered with a patient-controlled analgesia (PCA) device were largest in the morning than in the evening. For instance, Graves et al. (Graves et al. 1983) reported that the demands for Mo by patients with gastric bypass or abdominal surgery were 18 % larger at 9 a.m. than at 9 p.m. Similar data was obtained in postsurgical cancer patients (Auvril-Novak et al. 1990), but morning and evening peaks in opioid demands were reported by others (Labrecque and Vanier 1997; Labrecque and Vanier 2003).

Only 2 groups of investigators looked at the temporal changes in the effect of opioids in patients with chronic pain. In patients with chronic cancer pain and in patients with metastasis, the doses of Mo or Hm were significantly larger late in the afternoon or early evening (Vanier et al. 1992; Wilder-Smith and Wilder-Smith 1992). On the other hand, the Mo doses required to reduce the pain level of patients with heavy burns were significantly larger between 8 a.m. and 4 p.m., because this is the time of the day for daily personal health by nurses and/or physiotherapy treatment (see Labrecque and Vanier 2003 for reference). Finally, Bruera et al. (1992) reviewed the distribution of the extra doses of opioids received by 61 patients admitted to a palliative care unit at 4-h intervals over a 24-h period. The data indicated that 76 % of the patients received their extra doses between 10 a.m. and 10 p.m. than during the sleeping period; the number of extra doses during this period of day was 60 % larger during the night.

As pain can easily be altered by many factors such as anxiety, suggestions, and emotions, it is interesting to determine whether biological rhythms can be found in the effect of placebo. To our knowledge, Pöllmann (1987) is the only investigator who evaluated this effect of placebo on pain relief. When healthy individuals ingested a sugar-coated placebo tablet in the morning, the pain threshold was increased by 25–30 % between 9 a.m. and 9 p.m. When administered at night, the placebo did not produce any analgesic effect.

Guidelines for Using Rhythmic Changes in Clinical Situations Pertinent to Pain

The time-dependent changes in pain level and in the analgesic action of medications are relevant for the daily practice of health professionals. By selecting the most appropriate time for ingestion of analgesic agents, the clinicians can optimize and individualize drug treatment and also reduce the frequency of side effects of medications. To optimize and individualize the effect of analgesics, the clinicians should:

- Accept that pain intensity fluctuates during the 24-h period. Circadian variations are now well described in pain intensity.
- Determine when the pain level is highest and lowest during the 24-h period. Using a VAS, the practitioner can quickly determine when pain levels are highest and lowest. This approach also gives information on the interindividual variations in the intensity of pain.
- Be aware that the action and pharmacokinetics of analgesics and most medications are not constant throughout the day or night. The time-dependent variations were found with most drugs, even when the slow-release formulation was used.
- Administer analgesics to produce highest blood levels when pain is highest. In practice, drugs are administered at regular intervals such as 4 h after the last dose. This traditional approach does not take into account the time-dependent variations in pain intensity and the effect of analgesics; determinations of temporal changes in pain and in the effect of analgesics are the basis for the chronotherapeutic approach to pain management.
- Refrain from administering medications at time of highest toxicity or when the frequency of side effects is highest. Special attention must be given to the temporal changes in the pharmacokinetics of NSAIDs. The data indicate clearly that early morning doses should be avoided, because this is the time of day where the frequency of side effects is largest.

Conclusions

Pain is a very complex phenomenon influenced by anxiety, fatigue, suggestions or emotions, and prior experience. Human data now indicates that time of day is another factor that must be taken into account when prescribing pain medications. The data on biological rhythms suggest that inadequate pain management, which occurs frequently in clinical practice, may be reduced when time-dependent variations in pain and analgesic action are taken into account in daily practice.

References

Auvril-Novak, S. E., Novak, R. D., Smolensky, M. H., et al. (1990). Temporal variation in the self-administration of morphine sulfate via patient-controlled analgesia in post-operative gynecologic cancer patients. *Annual Review Chronopharmacol, 7*, 253–256.

Bellamy, N., Sothern, R. B., Campbell, J., et al. (2002). Rhythmic variations in pain, stiffness, and manual dexterity in hand of osteoarthritis. *Annals of the Rheumatic Diseases, 61*, 1075–1080.

Boissier, C., Decousus, H., Perpoint, B., et al. (1990). Timing optimizes sustained release ketoprofen treatment osteoarthritis. *Annual Review Chronopharmacol, 7*, 289–292.

Bruera, E., Macmilland, K., Huehn, N., et al. (1992). Circadian distribution of extra doses of narcotic analgesics in patients with cancer pain: A preliminary report. *Pain, 49*, 311–314.

Graves, B. A., Batenhorst, R. L., Bennett, J. G., et al. (1983). Morphine requirements using patient-controlled analgesia: Influence of diurnal variation and morbid obesity. *Clinical Pharmacy, 2*, 49–53.

Kowanko, I. C., Pownall, R., Knapp, M. S., et al. (1981). Circadian variations in the signs and symptoms of rheumatoid arthritis and in the therapeutic effectiveness of flurbiprofen at different times of the day. *British Journal of Clinical Pharmacology, 11*, 477–484.

Labrecque, G., & Vanier, M. C. (1997). Biological rhythms in pain and in analgesics. In P. H. Redfern & B. Lemmer (Eds.), *Physiology and pharmacology of biological rhythms* (pp. 619–649). Berlin: Springer.

Labrecque, G., & Vanier, M. C. (2003). Rhythms, pain and pain management. In P. H. Redfern (Ed.), *Biological clocks: Pharmaceutical and therapeutics applications*. London: Pharmaceutical Press.

Levi, F., LeLouarn, C., & Reinberg, A. (1985). Timing optimized sustained indomethacin treatment of osteoarthritis. *Clinical Pharmacology and Therapeutics, 37*, 77–84.

Pöllmann, L. (1987). Circadian variation of potency of placebo as analgesic. *Functional Neurology, 22*, 99–103.

Raisanen, I. (1988). Plasma levels and diurnal variation of beta-endorphin, beta-lipotropin and corticotropin during pregnancy and early puerperium. *European Journal of Obstetrics, Gynecology, and Reproductive Biology, 27*, 13–20.

Sittl, R., Kamp, H. D., & Knoll, R. (1990). Zirkadiane rhythmik des schmerzempfindens bei tumorpatienten. *Nervenheilkunde, 9*, 22–24.

Vanier, M. C., Labrecque, G., & Lepage-Savary, D. (1992). Temporal changes in the hydromorphone analgesia in cancer patients. *Proceedings of the 5th international conference biological rhythms and medications*, Amelia Island, FL. Abstract # XIII-8

Wilder-Smith, C. H., & Wilder-Smith, O. H. (1992). Diurnal patterns of pain in cancer patients during treatment with long-acting opioid analgesics. *Proceedings of the 5th conference biological rhythms and medications*, Amelia Island, FL. Abstract #XIII-7

Divalproex Sodium

Definition

Anticonvulsant medication.

Cross-References

▶ Preventive Migraine Therapy

Diver's Headache

▶ Primary Exertional Headache

Dizziness

Definition

A nonspecific term that describes an altered orientation in space, reflecting a discrepancy between internal sensation and external reality, creating sensory conflicts. The conflicts can be due to peripheral problems and occur between any of the vestibular, visual, or somatosensory systems, or it may be caused by central problems involving not one particular modality but rather the integration and weighting of the different modalities and their relation to memory. Words like light-headedness, faintness, giddiness, unsteadiness, imbalance, falling, waving, and floating may be used to describe dizziness.

Cross-References

▶ Coordination Exercises in the Treatment of Cervical Dizziness

DMARDs

▶ Disease-Modifying Antirheumatic Drugs

DMSO

▶ Dimethylsulfoxide

DNA Recombination

Definition

Biologically active deoxyribonucleic acid (DNA), which has been formed by the in vitro joining of segments of DNA from different sources.

Cross-References

▶ Cell Therapy in the Treatment of Central Pain

DNIC

▶ Diffuse Noxious Inhibitory Controls

Dobutamine

Definition

Dobutamine is an intravenously administered inotropic sympathomimetic medication, which acts on beta receptors of cardiac muscle to increase contractility, and is used to stress the heart in "stress tests" to detect myocardial ischemia.

Cross-References

▶ Thalamus and Visceral Pain Processing (Human Imaging)
▶ Thalamus, Clinical Visceral Pain, Human Imaging

Doctor-Patient Communication

▶ Chronic Gynecological Pain, Doctor-Patient Interaction

DOMS

▶ Delayed Onset Muscle Soreness

DOP

▶ Delta(δ) Opioid Receptor(s)

DOP Receptor

Definition

The term δ-opioid peptide receptor represents the G-protein-coupled receptor protein that responds selectively to a group of largely experimental opioid drugs and endogenous opioid peptides. It is homologous with the MOP receptor and is expressed in areas of the nervous system that moderately mediate analgesia with a side-effect profile distinct from μ-opioids. The DOP receptor protein is produced by a single gene. When activated, the DOP receptor predominantly transduces cellular actions via inhibitory G-proteins. The electrophysiological consequences of DOP receptor activation are usually inhibitory.

Cross-References

▶ Delta(δ) Opioid Receptor(s)
▶ Opioid Electrophysiology in PAG

Dopamine

Definition

Dopamine, a biogenic monoamine or catecholamine, is an endogenous transmitter synthesized from the amino acid tyrosine by actions of tyrosine hydroxylase (to produce DOPA, dihydroxyphenylalanine) and then DOPA decarboxylase to generate dopamine. Dopamine is a precursor to norepinephrine (noradrenaline) and thus to epinephrine (adrenaline) but also is a neurotransmitter in its own right (e.g., substantia nigra pars compacta neurons).

Cross-References

▶ Descending Circuitry, Transmitters and Receptors

DOR-1

Definition

DOR-1 refers to a clone that encodes a delta opioid receptor.

Cross-References

▶ Opioid Receptors

Dorsal Column

Definition

The dorsal (or posterior) columns constitute an ipsilateral ascending pathway in the dorsomedial spinal cord composed of myelinated axons associated with conveying information about fine touch, vibration, and proprioception. The dorsal columns include the fasciculus gracilis and fasciculus cuneatus associated with, respectively, sensory information from the lower and upper body.

Cross-References

▶ Angina Pectoris, Neurophysiology and Psychophysics
▶ Pain Treatment: Spinal Cord Stimulation
▶ Postsynaptic Dorsal Column Projection, Functional Characteristics
▶ Visceral Nociception and Pain
▶ Visceral Pain and Nociception

Dorsal Column Nuclei

Definition

The dorsal column nuclei include the nucleus cuneatus and nucleus gracilis in the dorsomedial caudal medulla. These nuclei contain the second-order neurons upon which axons ascending in the ipsilateral spinal dorsal columns terminate. Axons arising from neurons in dorsal column nuclei cross (decussate) at the level of the medulla to the opposite side and form the medial lemniscal pathway to the thalamus where they terminate on third-order neurons.

Cross-References

▶ Opioids in the Spinal Cord and Modulation of Ascending Pathways (N. gracilis)
▶ Postsynaptic Dorsal Column Projection, Anatomical Organization
▶ Postsynaptic Dorsal Column Projection, Functional Characteristics
▶ Spinothalamic Projections in Rat

Dorsal Column Stimulation (DCS)

▶ Pain Treatment: Spinal Cord Stimulation

Dorsal Column Stimulators

Synonyms

DCS

Definition

Dorsal column stimulators provide electrical pulses via electrodes implanted over the dorsal aspect of the spinal cord and are used as an alternative to or adjunct to other strategies for management of chronic pain. Dorsal column stimulation has been reported useful in treating failed back surgery syndrome, ischemic limb pain, and refractory angina pectoris by mechanisms not fully understood at present.

Cross-References

▶ Pain Treatment: Spinal Cord Stimulation
▶ Postherpetic Neuralgia, Pharmacological and Nonpharmacological Treatment Options

Dorsal Horn

Definition

This structure is that part of the spinal cord gray matter in which the cell bodies of second-order neurons that receive somatovisceral afferent input are located. These second-order neurons are of two general types: projection neurons and intrinsic neurons. The axons of projection neurons extend rostrally in the spinal white matter to convey information to sites in the brain (e.g., spinothalamic tract). The axons of intrinsic neurons remain within the spinal cord, either terminating locally or within other (typically nearby) spinal segments. The dorsal horn is organized into layers or laminae, numbered I to VI in a dorsal to ventral direction. Although its architecture is complex, with cells from deeper laminae sending dorsal dendrites to more superficial laminae, the principal targets of nociceptive primary afferent input are spinal dorsal horn neurons located in lamina I (the marginal layer) and lamina II (the substantia gelatinosa, further divided into II outer and II inner), collectively referred to as the superficial dorsal horn. Deeper dorsal horn laminae contain cells that are generally associated with non-nociceptive sensory processing. In addition to neurons and glia, the spinal dorsal horn also contains the axonal terminals of modulatory neurons that descend from the brainstem, including monoaminergic terminals that have been implicated in modulation of spinal nociceptive processing.

Cross-References

- ▶ Amygdala, Pain Processing and Behavior in Animals
- ▶ Forebrain Modulation of the Periaqueductal Gray and Its Role in Pain
- ▶ Infant Pain Mechanisms
- ▶ Opiates During Development
- ▶ Postsynaptic Dorsal Column Projection, Functional Characteristics
- ▶ Prostaglandins, Spinal Effects
- ▶ Responses of Spinothalamic Tract Neurons to Pruritic Stimuli
- ▶ Somatic Pain
- ▶ Spinothalamocortical Projections from SM

Dorsal Horn Neurons

Definition

They are neurons whose cell bodies lie in the dorsal horn of the spinal cord, which are of two general types: projection neurons and intrinsic neurons. The axons of projection neurons extend rostrally in the spinal white matter to convey information to sites in the brain

(e.g., spinothalamic tract). The axons of intrinsic neurons remain within the spinal cord, either terminating locally or within other (typically nearby) spinal segments. Spinal dorsal horn neurons receive input from somatovisceral primary afferent fibers as well as modulatory input from intrinsic neurons within the dorsal horn and axon terminals descending from the brain.

Cross-References

- ▶ Cancer Pain Model, Bone Cancer Pain Model
- ▶ Postsynaptic Dorsal Column Neurons, Responses to Visceral Input

Dorsal Horn Opiate Systems

Definition

The spinal dorsal horn contains endogenous opioid peptides (enkephalins, dynorphins, and endomorphins) as well as mu (μ), delta (δ), and kappa (κ) opioid receptors. Most opioid receptors are located on the spinal terminals of somatovisceral primary afferents, documented by their significant reduction after experimental dorsal root rhizotomy, but both spinal dorsal horn neurons and terminals of descending axons also express opioid receptors.

Cross-References

- ▶ Pain Treatment, Implantable Pumps for Drug Delivery

Dorsal Longitudinal Myelotomy

- ▶ Midline Myelotomy

Dorsal Rhizotomy

Definition

Dorsal rhizotomy is the transection of the dorsal roots of spinal nerves as they enter the spinal cord. The dorsal roots contain the central process of primary afferent fibers, including those of nociceptors, and thus prevent transmission of sensory information from the peripheral terminals of primary afferent fibers to the central nervous system.

Cross-References

► Cancer Pain Management, Neurosurgical Interventions
► Dorsal Root Ganglionectomy and Dorsal Rhizotomy
► Muscle Pain Model, Inflammatory Agents-Induced

Dorsal Root Entry Zone

Synonyms

DREZ

Definition

Dorsal Root Entry Zone (DREZ) includes: (1) the ventrolateral part of the central portion of the dorsal rootlets, where there is a lateral regrouping of fine fibers and (2) the medial part of the Lissauer's tract, where the small afferent enter and where they trifurcate to reach the dorsal horn, either directly or via pathways which ascend or descend several segments.

Cross-References

► Brachial Plexus Avulsion and Surgery in the Dorsal Root Entry Zone (DREZ)

Dorsal Root Entry Zone and Brachial Plexus Avulsion

► Brachial Plexus Avulsion and Surgery in the Dorsal Root Entry Zone (DREZ)

Dorsal Root Entry Zone Lesioning

Synonyms

DREZ lesioning

Definition

The dorsal root entry zone (DREZ) includes the central portion of the dorsal rootlets and the medial portion of Lissauer's tract (see Dorsal Root Entry Zone above), areas where afferent nociceptive fibers enter the spinal cord. Destroying this anatomical area, described initially by Sindou in 1972, interrupts the nociceptive pathway and can result in decreased pain from syndromes such as those resulting from plexus avulsion and spinal cord injuries. A modification involving a deeper destruction of the spinal dorsal horn, including laminae I – V (termed the "Nashold procedure"), is also a DREZ lesion procedure.

Dorsal Root Ganglia

► Dorsal Root Ganglion

Dorsal Root Ganglion

Synonyms

Dorsal root ganglia; DRG

Definition

The collection (ganglion) of pseudounipolar sensory neuron cell bodies in the vicinity of the spinal cord, with a peripheral process to the target organs and a central process to the spinal cord to terminate in the dorsal horn or the dorsal column nuclei. These cell bodies comprise of the nucleus as well as the cellular machinery for protein synthesis. Following their synthesis, the proteins have to be axonally transported to both the central and peripheral nerve terminals. The axons within the dorsal root mainly convey somatosensory information. Dorsal root ganglia also contain local glia cells (satellite cells).

Cross-References

▶ Central Pain, Diagnosis
▶ Cytokines, Effects on Nociceptors
▶ Dorsal Root Ganglion
▶ Dorsal Root Ganglionectomy and Dorsal Rhizotomy
▶ Inflammation, Role of Peripheral Glutamate Receptors
▶ Neuropathic Pain Model, Tail Nerve Transection Model
▶ Opioid Modulation of Nociceptive Afferents In Vivo
▶ Opioid Receptor Localization
▶ Opioids and Inflammatory Pain
▶ Prostaglandins, Spinal Effects
▶ Spinal Cord Nociception, Neurotrophins
▶ Substance P Regulation in Inflammation
▶ Toxic Neuropathies
▶ Visceral Pain Model; Esophageal Pain

Dorsal Root Ganglion Radiofrequency

Way Yin[1] and Nikolai Bogduk[2]
[1]Department of Anesthesiology, University of Washington, Seattle, WA, USA
[2]University of Newcastle, Newcastle Bone & Joint Institute, Royal Newcastle Centre, Newcastle, NSW, Australia

Synonyms

Partial dorsal rhizotomy; Partial dorsal root ganglion lesioning; Partial radiofrequency dorsal root ganglion lesion; RF-DRG

Definition

Partial radiofrequency (RF) dorsal root ganglion (DRG) lesion (RF-DRG) is a procedure in which a radiofrequency electrode is placed in the vicinity of the dorsal root ganglion of a spinal nerve and a radiofrequency current is passed through the electrode, for the purpose of creating a lesion in the nerve of sufficient magnitude sufficient to relieve pain but without actually damaging the nerve (see ▶ Radiofrequency Neurotomy, Electrophysiological Principles).

Characteristics

In principle, RF-DRG was born out of a need for a procedure that could treat spinal pain arising from a particular spinal segment that could not be treated by other methods, or which had not responded to other, more target-specific therapy. In practice, RF-DRG arose as a means of treating patients whose pain had not been relieved by medial branch neurotomy.

Rationale

RF-DRG is not a procedure intended to destroy the dorsal root ganglion. Indeed, a critical objective of the procedure is to preserve function in the target nerve and its dermatome. One stated rationale of the procedure is "to expose the dorsal root ganglion to temperatures that prevail in the peripheral part of an RF lesion to preserve the large myelinated fibers and to deactivate the small unmyelinated fibers" (van Kleef et al. 1996a).

Notwithstanding numerous theories, the mechanism by which RF-DRG is supposed to operate has not been demonstrated. A differential effect of heat lesions on myelinated and unmyelinated sensory fibers has been refuted (Zervas and Kuwayama 1972). In effect, the procedure advocated by some amounts to no more than placing an electrode sufficiently close to the target nerve in order to do something, but not so close as to actually damage the nerve. How close, or how far away, the electrode should be placed has not been defined. Other investigators have advocated placing the electrode tip within the DRG (Stolker et al. 1994b), but the one anatomical study that has been conducted demonstrated that electrode placement is variable in relation to the target ganglion and sometimes too far away for any lesion to have an effect on the nerve. Placement of the electrode tip within the DRG occurred in only 61 % of levels studied (Stolker et al. 1994a).

Notwithstanding these limitations to the rationale and mechanism of the treatment, RF-DRG has assumed popularity in various regions across the world, to various extents. That popularity, however, is dissonant with the quality of the available literature on the procedure and its results.

Efficacy

The popularity of RF-DRG has been sustained largely on the grounds of observational studies and word of mouth. Those studies claim some degree of effectiveness, and the procedure has gained a reputation of being something that works.

The first of the cervical studies (van Kleef et al. 1993) reported that 2/20 patients (10 %) were pain-free at 3 months; 4/20 were so at 6 months, and 2/17 (11 %) at 12 months. Eight patients had good relief at 3 months, but their numbers dropped to two at 6 months and two at 12 months. The authors did not conclude that their treatment was successful. They portrayed it as a "reversible procedure" that can be useful to provide a relatively pain-free period, which can be used to obtain the maximum benefit from conservative forms of treatment.

Despite the less than modest outcomes of this study, investigators from the same institution undertook a placebo-controlled trial (van Kleef et al. 1996b). Success was defined as a reduction by 2 points or more on a 10-point visual analogue scale. A significantly greater proportion, 8/9 patients, had a successful outcome following active treatment than those (2/11) who underwent sham treatment. Follow-up, however, was limited to only 8 weeks. At this time, the actively treated patients had reduced their mean pain scores from 6.4 to 3.3, whereas the sham-treated patients maintained the same scores (5.9, 6.0).

Although this study provided data to the effect that the initial effects of cervical RF-DRG were not due to placebo, they did not attest to a successful, lasting effect. Unlike its preceding study, the controlled study did not report on the number of patients completely relieved. Success was defined only as a 2-point decrease in pain scores. The duration of effect beyond 8 weeks was not measured. These data attest only to a modest and short-lived therapeutic effect. The data of the preceding observational study would predict that outcomes would attenuate substantially beyond 8 weeks.

The first of the lumbar studies (van Wijk et al. 2001) was a retrospective review of 361 patients, but results were available for only 279. At 2 months, 61 (17 %) patients were pain-free. This number became 23 (6 %) at a mean follow-up of 22.9 months, but the range of that period was 2–70 months. A further 103 (29 %) patients reported incomplete but greater than 50 % relief at 2 months. This number dropped to 73 (20 %) at longer-term follow-up.

Lumbar RF-DRG was eventually subjected to a rigorous placebo-controlled trial (Geurts et al. 2003). In all patients, the DRG was anesthetized and a radiofrequency electrode was placed into position. The active treatment was RF-DRG at 67 °C. The control treatment was no generation of current. At 3 months, 16 % of the 44 patients treated with RF-DRG had greater than 50 % reduction in pain. Meanwhile, 25 % of the 36 patients who had sham treatment experienced the same outcome. These proportions are not significantly different statistically.

This study, therefore, denied any attributable effect of the procedure. Patients whose DRG was anesthetized, without a lesion being produced, had the same outcomes as actively treated patients.

The first of the thoracic studies (Stolker et al. 1994b) announced astounding results. At 2 months, 30/45 patients (67 %) were pain-free, and a further 11 (24 %) had greater than 50 % reduction in pain. At long-term follow-up, ranging from 13 to 46 months, 20 patients were pain-free, and 15 had greater than 50 % relief of pain.

The second thoracic study (van Kleef et al. 1995) did not reproduce these outcomes. At 8 weeks, only 8 of 43 patients (18 %) had complete relief of pain, and 9 (21 %) had greater than 50 % relief. At follow-up beyond 36 weeks, only 5 patients (12 %) were pain-free and 8 (18 %) had greater than 50 % relief.

There is no obvious explanation for the discrepancy between these two studies. Patient selection may have been the source of difference. The first study had a large proportion (38 %) of patients with postsurgical pain (thoracotomy, mastectomy, abdominal scar), in whom good outcomes were achieved. Such patients were absent from the second study. Conversely, the second study had a large proportion (47 %) of patients with neuralgia. In the first study, patients with neuralgias had less than average outcomes. (The second study did not stratify its results according to diagnosis.) Another factor which may account for the discrepancies between the studies may have been technical. In the first study, the goal was to place the electrode within the DRG, whereas in the second study, the goal of electrode placement was next to the DRG.

The outcomes of RF-DRG are not consistent across cervical, lumbar, and thoracic levels. The outcomes of cervical RF-DRG are less than

modest, even in a controlled trial. For lumbar RF-DRG, a rigorous controlled trial has shown that sham therapy achieves at least equivalent outcomes to those of active therapy. Since it was conducted by authors of the foregoing observational study, that trial surely refutes lumbar RF-DRG as a valid treatment and, by extrapolation, casts doubt on the validity of RF-DRG in general.

Thoracic RF-DRG has not been subjected to a controlled trial, but it might be effective for certain types of thoracic pain, for which there is not an analogue at cervical and lumbar levels. The available data hint at the possibility that thoracic RF-DRG could be useful for postsurgical pain, although not for neuralgias, and placement of the electrode within the DRG itself may be an intensive as well as procedural prerequisite.

References

Geurts, J. W. M., can Wijk, M. A. W., Wunne, H., et al. (2003). Radiofrequency lesioning of dorsal root ganglia for chronic lumbosacral radicular pain: A randomized, double-blind, controlled trial. *Lancet, 361*, 21–26.

Stolker, R. J., Vervest, A. C., & Groen, G. J. (1994a). The treatment of chronic thoracic segmental pain by radiofrequency percutaneous partial rhizotomy. *Journal of Neurosurgery, 80*, 986–992.

Stolker, R. J., Vervest, A. C. M., Ramos, L. M. P., et al. (1994b). Electrode positioning in thoracic percutaneous partial rhizotomy: An anatomical study. *Pain, 57*, 241–251.

van Kleef, M., Spaans, F., Dingemans, W., et al. (1993). Effects and side effects of a percutaneous thermal lesion of the dorsal root ganglion in patients with cervical pain syndrome. *Pain, 52*, 49–53.

van Kleef, M., Barendse, G. A. M., Dingemans, W. A. A. M., et al. (1995). Effects of producing a radiofrequency lesion adjacent to the dorsal root ganglion in patients with thoracic segmental pain. *The Clinical Journal of Pain, 11*, 325–332.

van Kleef, M., Barendse, G., & Sluijter, M. (1996a). Response to invited commentary. Assessing a new procedure: Thoracic radiofrequency dorsal root ganglion lesions. *The Clinical Journal of Pain, 12*, 76–78.

van Kleef, M., Liem, L., Lousberg, R., et al. (1996b). Radiofrequency lesion adjacent to the dorsal root ganglion for cervicobrachial pain. A prospective double-blind study. *Neurosurgery, 38*, 1127–1132.

van Wijk, R. M. A. W., Geurts, J., & Wynne, H. J. (2001). Long-lasting analgesic effect of radiofrequency treatment of the lumbosacral dorsal root ganglion. *Journal of Neurosurgery (Spine 2), 94,* 227–231.

Zervas, N. T., & Kuwayama, A. (1972). Pathological characteristics of experimental thermal lesions. Comparison of induction heating and radiofrequency electrocoagulation. *Journal of Neurosurgery, 37,* 418–422.

Dorsal Root Ganglionectomy

▶ Dorsal Root Ganglionectomy and Dorsal Rhizotomy

Dorsal Root Ganglionectomy and Dorsal Rhizotomy

Allan J. Belzberg and Michael J. Dorsi
Department of Neurosurgery, Johns Hopkins School of Medicine, Baltimore, MD, USA

Synonyms

Dorsal rhizotomy; Dorsal root ganglionectomy; Sensory ganglionectomy; Sensory rhizotomy

Definition

Dorsal root ganglionectomy and dorsal rhizotomy are neuroablative procedures, which interrupt peripheral sensory pathways. Dorsal root ganglionectomy is the surgical removal of the ▶ dorsal root ganglion of a spinal nerve. ▶ Dorsal rhizotomy is the transection of the dorsal root of a spinal nerve.

Characteristics

Anatomy
The primary sensory afferents project proximally (via the dorsal root) to form synapses in the dorsal horn of the spinal cord or the dorsal column

nuclei, with pseudobipolar cell bodies of these axons being located in the dorsal root ganglion. The traditional belief, established by the "Law of Bell and Magendie" is that for a given spinal nerve, sensory and motor functions are segregated in the dorsal and ventral roots, respectively. However, it is now clear that afferent sensory fibers are also found in the ventral roots, with up to 29 % of fibers in human ventral roots being small unmyelinated (presumably afferent) fibers (Coggeshall et al. 1975). Some of these fibers course from the periphery into the ventral root and loop back into the dorsal root before entering the dorsal horn (Coggeshell 1979). Others bypass the dorsal root and enter the spinal cord directly through the ventral root (Yamamoto et al. 1977). In addition, cell bodies of sensory afferents are sometimes located outside the DRG in the dorsal root, ventral root, or along the nerve in the periphery.

Rationale
Neuroablative procedures, such as dorsal root ganglionectomy and dorsal rhizotomy, block the transmission afferent activity arising from ▶ nociceptors and so diminish pain evoked by experimental stimuli. Dorsal rhizotomy interrupts input to dorsal horn neurons from DRG cells that project centrally via the dorsal root. Removal of the dorsal root ganglion leads to ▶ Wallerian degeneration of afferent fibers in the periphery, dorsal root, and ventral root. Therefore, both procedures lead to deafferentation of the central nervous system.

It has been suggested that ganglionectomy is superior to dorsal rhizotomy, because it results in interruption of all afferent input at that spinal segment. One theoretical disadvantage of ganglionectomy is that the resulting Wallerian degeneration of peripheral afferents contributes to neuropathic pain in animal models (Li et al. 2000). Wallerian degeneration in the periphery or target tissue denervation may alter the function of intact afferents adjacent to degenerating axons (Li et al. 2000). This effect may lead to sensitization of afferents or neurons at higher levels in the pain-signaling pathway.

Indications

Neuroablation has been implemented in a diverse array of painful conditions including radiculopathy, failed back surgery syndrome, post-herpetic neuralgia, malignancy, and multiple sclerosis. Dorsal rhizotomy at the level of C-2 has been performed for treatment of occipital neuralgia and cervicogenic headache. In general, neuroablative procedures are carried out in patients who have failed physical therapy, medical treatment, and other nonsurgical therapies.

Neuroablative techniques are a plausible approach to pain treatment in cases in which there is a clearly identifiable pain generator. Determination of the spinal segmental level in which the pain occurs is complicated by preganglionic intersegmental anastomoses, ventral root afferents, and denervation hypersensitivity. Diagnostic testing including electromyographic, imaging studies, and nerve blocks is implemented to identify the painful segment.

Nerve blocks have not proven to be a reliable predictor of outcome, and the validity of peripheral nerve blocks has been brought into question (North et al. 1996). A positive block occurs when there is good pain relief with a small volume of local anesthetic injected in the ▶ neural foramen, and no pain relief achieved with placebo injection or injection at the nerve root above or below, performed in a blinded fashion.

Even when a single ▶ dermatome is identified as the pain generator, it is not clear how many roots may supply that distribution or how many segments should be denervated for pain relief. In primates, it is likely that at least three adjacent roots innervate each dermatome. It may be that one segment above and one below the target level should be included to achieve a clinical effect.

Outcome

Response rates for dorsal rhizotomy and ganglionectomy vary between 19 % and 69 % for rhizotomy and between 0 % and 100 % for ganglionectomy. The lack of consistent results can be attributed to multiple uncontrollable variables.

The results of dorsal rhizotomy performed in 51 patients with chronic lumbar radiculopathy were published by Wetzel et al. (1997). At 6 months after surgery, 55 % were believed to have a good or excellent outcome, while at 2–4 years, such outcomes were obtained in only 19 % of patients. Similar deterioration of outcomes has been observed in several series and has led many to favor ganglionectomy over rhizotomy. The most impressive results following ganglionectomy have been reported for treatment of thoracic and occipital pain, with some series reporting long-term success rates as high as 68 % for thoracic pain (Young 1996) and 80 % for occipital pain (Lozano et al. 1998). Results of ganglionectomy for ▶ failed back surgery syndrome (FBSS) are much less favorable. In one recent series, success was obtained in only 2 of 13 patients with FBSS at 2 years after surgery and none at 5.5 years (North et al. 1991), although another series reported treatment success in four of six patients with FBSS at greater than 2 years (Wilkinson and Chan 2001).

In summary, both dorsal root ganglionectomy and dorsal rhizotomy are neuroablative procedures implemented for treatment of chronic pain. These procedures should be considered carefully in the light of available augmentative procedures and should be limited to cases where the spinal segmental level of pain generation has been clearly defined.

Cross-References

▶ Wallerian Degeneration

References

Coggeshall, R. E., Applebaum, M. L., Fazen, M., Stubbs, T. B., 3rd, & Sykes, M. T. (1975). Unmyelinated axons in human ventral roots, a possible explanation for the failure of dorsal rhizotomy to relieve pain. *Brain, 98,* 157–166.

Coggeshell, R. E. (1979). Afferent fibers in the ventral root. *Neurosurgery, 4,* 443–448.

Li, Y., Dorsi, M. J., Meyer, R. A., & Belzberg, A. J. (2000). Mechanical hyperalgesia after an L5 spinal nerve lesion in the rat is not dependent on input from injured afferents. *Pain, 85*(3), 493–502.

Lozano, A. M., Vanderlinden, G., Bachoo, R., & Rothbart, P. (1998). Microsurgical C-2 ganglionectomy for chronic intractable occipital pain. *Journal of Neurosurgery, 89*, 359–365.

North, R. B., Kidd, D. H., Campbell, J. N., & Long, D. M. (1991). Dorsal root ganglionectomy for failed back surgery syndrome: A 5-year follow-up study. *Journal of Neurosurgery, 74*, 236–242.

North, R. B., Kidd, D. H., Zahurak, M., & Piantadosi, S. (1996). Specificity of diagnostic nerve blocks: A prospective, randomized study of sciatica due to lumbosacral spine disease. *Pain, 65*, 77–85.

Wetzel, F. T., Phillips, M., Aprill, C. N., Bernard, T. N., & LaRocca, H. S. (1997). Extradural sensory rhizotomy in the management of chronic lumbar radiculopathy a minimum 2-year follow-up study. *Spine, 22*, 2283–2292.

Wilkinson, H. A., & Chan, A. S. (2001). Sensory ganglionectomy: Theory, technical aspects, and clinical experience. *Journal of Neurosurgery, 95*, 61–66.

Yamamoto, T., Takahashi, K., Staomi, H., & Ise, H. (1977). Origins of primary afferent fibers in the ventral spinal roots in the cat as demonstrated by the horseradish peroxidase method. *Brain Research, 126*, 350–354.

Young, R. F. (1996). Dorsal rhizotomy and dorsal root ganglionectomy. In J. R. Youmans (Ed.), *Neurological surgery* (4th ed., pp. 3442–3451). Philadelphia: WB Saunders.

Dorsal Root Reflexes

Definition

If primary afferent depolarization becomes suprathreshold, it can elicit action potentials in the central terminals of primary afferent nociceptors, which can then travel retrogradely to the periphery, release proinflammatory neuropeptides, and support neurogenic inflammation.

Cross-References

▶ Arthritis Model, Kaolin-Carrageenan-Induced Arthritis (Knee)
▶ GABA and Glycine in Spinal Nociceptive Processing

Dorsal Root-Spinal Cord Junction

▶ Brachial Plexus Avulsion and Surgery in the Dorsal Root Entry Zone (DREZ)

Dorsolateral Fasciculus

Definition

The dorsolateral fasciculus (DLF) is a longitudinal bundle of axons traveling in the dorsolateral quadrant of the spinal cord. Descending axons from the rostroventral medial medulla are contained in the DLF, targeted transection of which is often used to investigate the contribution of descending pathways in spinal nociceptive transmission.

Cross-References

▶ Descending Circuitry, Molecular Mechanisms of Activity-Dependent Plasticity
▶ Stimulation-Produced Analgesia
▶ Vagal Input and Descending Modulation

Dorsolateral Pons

Definition

The dorsolateral pons contains groups of neurons that synthesize and release norepinephrine (noradrenalin). The noradrenergic cell groups are designated as A6 (locus coeruleus), A6sc (sub-coeruleus), and A7, all of which contribute axons to descending pathways that terminate at many levels of the spinal cord and contribute modulation of spinal nociceptive transmission. Other cell groups in the dorsolateral pons have been implicated in nociceptive processing (e.g., the parabrachial area).

Cross-References

▶ Spinothalamic Tract Neurons, Descending Control by Brain Stem Neurons

Dorsomedial Nucleus (DM)

Definition

The largest of the medial nuclei of the thalamus. It makes extensive connections with most of the other thalamic nuclei.

Cross-References

▶ Human Thalamic Response to Experimental Pain (Neuroimaging)

Dose Titration

Definition

Dose titration refers to an approach for achieving a therapeutic response by individualizing the dose that involves planned increases in dose, with assessment of the response at each dose level. Dose increases occur until a satisfactory therapeutic response is achieved or side effects prevent further dose increments.

Cross-References

▶ Opioid Rotation

Dosing Interval

Definition

The dosing interval is the time interval between the administered doses of a drug.

Cross-References

▶ Opioid Rotation

Dosing Regimen

Definition

Dosing regimen is the treatment taken by the patient, defined by the specific drug, the dose, and the dosing interval.

Cross-References

▶ Opioid Rotation

DOT

▶ Dictionary of Occupational Titles

Double Depression

Definition

Double Depression refers to a dual diagnosis of "Major Depression" and "Dysthymia."

Cross-References

▶ Psychiatric Aspects of the Epidemiology of Pain

Double-Blind

Definition

The patient and treating physician are both unaware which treatment the patient is receiving.

Cross-References

▶ Antidepressants in Neuropathic Pain
▶ Central Pain, Pharmacological Treatments

Downregulated

Definition

A state whereby a physiologic feedback loop causes a substance to reduce the production or action of another substance.

Cross-References

▶ Cancer Pain Management, Orthopedic Surgery

DREZ

▶ Dorsal Root Entry Zone

DREZ Lesion

▶ DREZ Procedures

DREZ Lesioning

▶ Dorsal Root Entry Zone Lesioning

DREZ Lesions

Synonyms

DREZotomy

Definition

DREZ lesions are destructive therapeutic lesions applied onto the dorsal root entry zone (DREZ) by techniques including microsurgical

ablation, Radio-Frequency-Thermocoagulation, Laser coagulation, and Ultrasonic lesioning.

Cross-References

▶ Anesthesia Dolorosa Model, Autotomy
▶ Brachial Plexus Avulsion and Surgery in the Dorsal Root Entry Zone (DREZ)
▶ Dorsal Root Entry Zone
▶ Dorsal Root Entry Zone Lesioning
▶ DREZ Procedures

DREZ Procedures

Allan H. Friedman and Kenneth M. Little
Division of Neurosurgery, Duke University
Medical Center, Durham, NC, USA

Synonyms

DREZ lesion; Junctional DREZ coagulation; Microsurgical DREZotomy; Nucleus caudalis DREZ

Definition

The dorsal root entry zone (DREZ) includes the central portion of the dorsal spinal rootlets, ▶ Lissauer's tract, and layers I through V of the dorsal horn. At the cervicomedullary junction, the dorsal horn is contiguous with the nucleus caudalis, an analogous structure within the spinal trigeminal nucleus. Surgical DREZ lesions may be accomplished with radiofrequency-induced heating, mechanical incision, bipolar coagulation, laser coagulation, or ultrasonic destruction to treat a variety of ▶ central pain syndromes.

Characteristics

Rationale
At the DREZ, fibers conveying nociceptive sensory information enter the spinal cord in the

ventrolateral aspect of the dorsal root. These relatively small axons, with sparse or absent myelination, enter the Lissauer's tract, ascend and descend up to four segments, and terminate in laminae I through VI (principally I, II, and V) of the ipsilateral dorsal horn. Within the dorsal horn, sensory information, including pain, is modulated through neurochemical signaling and inhibitory anatomical connections. Central pain syndromes (pain mediated by the central nervous system), caused by spinal cord injury or peripheral deafferentation, may result from aberrations of this system. In animal models, epileptiform activity has been observed within the dorsal horn after root avulsion, possibly because of regenerative sprouting with abnormal neuronal reorganization. The rationale for DREZ procedures is to ablate or interrupt the central structures and thus the abnormal physiological processes implicated in some pain syndromes.

Indications
Spinal and nucleus caudalis DREZ lesioning procedures can be an effective means of treating deafferentation pain syndromes, which are refractory to medical management or other operative interventions. The key to a successful outcome lies in careful patient selection.

Brachial Plexus Avulsion
Traumatic brachial plexus traction causing nerve root avulsions frequently result in a characteristic pain syndrome. The patient notes a constant burning or aching pain punctuated by paroxysms of crushing pain. Prior to proceeding with DREZ lesioning in these patients, root avulsion must be confirmed, as peripheral nerve injury pain does not respond to DREZ lesions. Care must be taken to identify and treat all of the painful segments, not just those identified as abnormal by radiographic studies or gross inspection.

Spinal Cord Injury
There are different types of pain associated with trauma to the cervical, thoracic, and lumbosacral spinal cord. The pain that responds best to DREZ

procedures is radicular or segmental, occurring in the partially deafferented levels adjacent to the level of injury. Diffuse pain occurring below the level of injury, especially constant burning pain in the sacral dermatomes, only occasionally responds to DREZ procedures.

Conus Medullaris and Cauda Equina Injury
This type of pain often occurs following trauma to the T12 to L1 levels, injuring elements at both the conus medullaris and cauda equina. Patients that respond best to DREZ procedures in this region are those with incomplete neurologic deficits, those with pain that is "electrical" in character, and those with injury due to blunt trauma.

Phantom Limb Pain
Patients suffering pain after limb amputation may experience stump pain, phantom limb pain, or both. Typically, phantom limb pain responds significantly better to DREZ procedures than does stump pain.

Cancer Pain
The type of cancer-related pain that responds best to DREZ procedures is topographically limited to a few spinal segments (as with Pancoast syndrome). Patients with pain from thoracic or abdominal wall invasion or lumbosacral root involvement may also respond well. Lumbosacral pain must be limited, however, to avoid the increased risk of lower extremity hypotonia or sphincter dystonia associated with bilateral or more extensive DREZ lesions.

Craniofacial Pain
The caudalis DREZ procedure may be indicated in patients with central craniofacial pain including anesthesia dolorosa, post-tic dysesthesias, atypical facial pain, postherpetic pain, pain related to neoplasms in the region of the Gasserian ganglion, and facial pain caused by brainstem lesions such as infarction, tumors, and multiple sclerosis. The caudalis DREZ procedure, however, is controversial due to the relatively high risk of postoperative deficits and the transience of pain relief.

Methods

Several DREZ lesioning methods have been described (Nashold and El-Naggar 1992; Iskandar and Nashold 1998; Sindou 2002). Laminectomies or, more elegantly, hemilaminectomies are performed over the spinal levels to be lesioned. In the case of ▶ brachial plexus avulsion injuries and spinal cord trauma, the pathologic segments are identified by a combination of gross inspection and impedance measurements. In cases in which there is no spinal cord pathology, the DREZ to be lesioned is identified by following nerve rootlets from their entry into the spine to their entry into the spinal cord or by recording evoked potentials from the DREZ while stimulating the affected dermatomes. Lesions are made along the intermediolateral sulcus, extending approximately 2 mm into the DREZ. They may be made with radiofrequency-generated heat to 80 °C at 1 mm intervals, mechanical incision, laser coagulation, bipolar coagulation, or ultrasonic destruction. It is important that the lesion extends into the DREZ above and below the affected dermatomes. For the caudalis DREZ procedure, a small suboccipital hemicraniectomy and bilateral C1-C2 laminectomy is performed (Iskandar and Nashold 1998; Nashold and El-Naggar 1992). Classically, a specialized 3 mm electrode (with the proximal 1 mm being insulated to protect the overlying spinocerebellar tract) is used to make two rows of RF lesions at the cervicomedullary junction. The first row begins at the dorsal rootlets of C2 extending rostrally to about 5 mm above the obex. The second row parallels the first, 1 mm dorsal to the DREZ. Recently, trigeminal evoked potentials, EMG, and SSEPs have been used during the caudalis DREZ procedure to target the symptomatic nucleus caudalis region specifically and to identify and protect the adjacent corticospinal tract and dorsal column (Husain et al. 2002).

Outcomes

In the larger series reported in the literature, adequate, long-term pain relief has been reported in about 60–90 % (Dreval 1993; Rath et al. 1997; Sindou et al. 2001). Variations in results can be attributed to differences between criteria for patient selection, outcome measures, times of follow-up, and techniques (Friedman et al. 1988; Nashold and El-Naggar 1992; Dreval 1993; Thomas and Kitchen 1994; Sampson et al. 1995; Iskandar and Nashold 1998; Sindou 2002; Spaic et al. 2002).

Complications

Following DREZ lesions for brachial plexus avulsion pain, 41 % experienced objective sensory deficits and 41 % objective motor deficits (Friedman et al. 1988). However, the majority of these deficits were mild, with only a single patient suffering a deficit sufficient to limit ambulation. In spinal cord and conus medullaris DREZ procedures, permanent sensory and motor deficits occur in about 13 % and 12 %, respectively. Non-neurologic complications such as infection, CSF leak, and epidural hematoma occur in about 7 %. The nucleus caudalis DREZ procedure is associated with more complications. Rates of postoperative ataxia have ranged from 39 % to 54 % and diplopia or corneal anesthesia in about 20 %. For this reason, the caudalis DREZ procedure is performed at a limited number of centers. In a recently reported series of nucleus caudalis DREZ procedures, aided by trigeminal evoked potentials, EMG, and SSEPs, fewer lesions were made resulting in no permanent neurological deficits and pain relief in 71 % of patients at 12 months (Husain et al. 2002).

References

Dreval, O. N. (1993). Ultrasonic DREZ-operations for treatment of pain due to brachial plexus avulsion. *Acta Neurochirurgica, 122,* 76–81.

Friedman, A. H., Nashold, B. S., & Bronec, P. R. (1988). Dorsal root entry zone lesions for the treatment of brachial plexus avulsion injuries: A follow-up study. *Journal of Neurosurgery, 22,* 369–373.

Husain, A. M., Elliott, S. L., & Gorecki, J. P. (2002). Neurophysiological monitoring for the nucleus caudalis dorsal root entry zone operation. *Neurosurgery, 50,* 822–827.

Iskandar, B. J., & Nashold, B. S. (1998). Spinal and trigeminal DREZ lesions. In P. L. Gildenberg & R. R. Tasker (Eds.), *Textbook of stereotactic and functional neurosurgery* (pp. 1573–1583). New York: McGraw-Hill, Health Professional Division.

Nashold, B. S., Jr., & El-Naggar, A. O. (1992). Dorsal root entry zone (DREZ) lesioning. In S. S. Rengachary & R. H. Wilkins (Eds.), *Neurosurgical operative atlas* (pp. 9–24). Baltimore: Williams & Wilkins. vol 2.

Rath, S. A., Seitz, K., Soliman, N., Hahamba, J. F., Antoniadis, G., & Richter, H. P. (1997). DREZ coagulations for deafferentation pain related to spinal and peripheral nerve lesions: Indication and results of 79 consecutive procedures. *Stereotactic and Functional Neurosurgery, 68*, 161–167.

Sampson, J. H., Cashman, R. E., Nashold, B. S., & Friedman, A. H. (1995). Dorsal root entry zone lesions for intractable pain after trauma to the conus medullaris and cauda equina. *Journal of Neurosurgery, 82*, 28–34.

Sindou, M. P. (2002). Dorsal root entry zone lesions. In Burchiel (Ed.), *Surgical management of pain* (pp. 701–713). New York: Thieme Medical.

Sindou, M., Mertens, P., & Wael, M. (2001). Microsurgical DREZotomy for pain due to spinal cord and/or cauda equina injuries: Long-term results in a series of 44 patients. *Pain, 92*, 159–171.

Spaic, M., Markovic, N., & Tadic, R. (2002). Microsurgical DREZotomy for pain of spinal cord and cauda equina injury origin: Clinical characteristics of pain and implications for surgery in a series of 26 patients. *Acta Neurochirurgica, 144*, 453–462.

Thomas, D. G., & Kitchen, N. D. (1994). Long-term follow-up of dorsal root entry zone lesions in brachial plexus avulsion. *Journal of Neurology, Neurosurgery & Psychiatry, 57*, 737–738.

DREZotomy

▶ DREZ Lesions

DRG

▶ Dorsal Root Ganglion

Drive

▶ Motivational Aspects of Pain

Drug Guidelines

▶ Analgesic Guidelines for Infants and Children

Drugs and Procedures to Treat Neuropathic Pain

George L. Wilcox
Department of Neuroscience, Pharmacology and Dermatology, University of Minnesota Medical School, Minneapolis, MN, USA

Characteristics

Disease Entities

Of the disease entities covered explicitly by entries, postherpetic neuralgia (PHN) (see ▶ Drugs Targeting Voltage-Gated Sodium and Calcium Channels) and diabetic neuropathy (DN) (see ▶ Diabetic Neuropathy, Treatment) are the two with the highest incidence, perhaps accounting for half of the overall incidence of neuropathic pain. These high incidence disorders also account for the vast majority of well-designed and conclusive trials; therefore, the evidence basis for the majority of several other disease entities is an extension of what has been learned from these two entities. The multifactorial causes of cancer pain, including both somatic and neuropathic components, make estimation of the incidence or contribution of neuropathic cancer pain difficult (see ▶ Cancer Pain Management, Treatment of Neuropathic Components). The combined incidence of complex regional pain syndrome (CRPS) (see ▶ CRPS, Evidence-Based Treatment) and fibromyalgia (FMS) (see ▶ Fibromyalgia, Mechanisms, and Treatment) is probably comparable to that of PHN or DN, alone. The other classifications of trigeminal neuralgia (TN) (see ▶ Trigeminal Neuralgia: Diagnosis and Treatment), phantom pain (see ▶ Phantom Limb Pain, Treatment), and movement-related neuropathic pain (see ▶ Evoked and Movement-Related

Neuropathic Pain) collectively represent a relatively small fraction of neuropathic pain conditions. Comparison of the common treatment modalities effective across these various types of neuropathic pain yields a homogeneity that both derives from the evidence basis mentioned above and suggests a commonality in underlying mechanisms. This commonality of mechanisms is highlighted by the medical therapies, anticonvulsants, antidepressants, and calcium channel blockers, described below.

Medical Versus Non-Medical Therapies

Five entries address medical therapies used most frequently in management of neuropathic pain disorders or a target of some of these therapies, descending modulatory systems. None of the therapies is curative, rather treating the symptoms of painful peripheral neuropathies. Antidepressants may be the most used medications, although their mechanisms of action are among the least well understood and perhaps the least target-selective (see ▶ Antidepressants in Neuropathic Pain). Anticonvulsants (▶ Drugs Targeting Voltage-Gated Sodium and Calcium Channels, and see Voltage-Gated Channels below) are perhaps becoming the most frequently used agents in neuropathic pain; most seem to target voltage-gated Na⁺ channels but some of the newer agents in this class seem to target a subunit of voltage-gated Ca⁺⁺ channels. Medical treatments targeting multiple receptors, a prototype for which is tramadol (see ▶ Drugs with Mixed Action and Combinations: Emphasis on Tramadol), are less widely used, but nonetheless represent a significant fraction of medical treatments. By comparison, drugs targeting adrenergic receptors have a very low prevalence of use, largely because of the necessity of spinal targeting by implanted catheter (Alpha(α) 2-Adrenergic Agonist). The spinal receptors for adrenergic agonists are also the target of descending inhibitory systems covered in ▶ Descending Facilitation and Inhibition in Neuropathic Pain; the descending terminals of these inhibitory systems may be targeted indirectly by antidepressants and tramadol (▶ Drugs with Mixed Action and Combinations: Emphasis on Tramadol). The entry ▶ Descending

Facilitation and Inhibition in Neuropathic Pain also addresses the potential for descending facilitatory influences to contribute to the neuroplastic changes thought to mediate formation of chronic neuropathic pain. These descending systems may be fruitful future targets for therapies.

Five entries address nonmedical therapies. The entry ▶ Central Nervous System Stimulation for Pain discusses the indications and possible mechanisms for spinal cord (SCS), deep brain stimulation (DBS), and motor cortex stimulation; the first is useful in a broad range of disorders, such as CRPS, DN, PHN, and spinal cord injury pain, but not back pain; DBS is indicated in central pain, ▶ anesthesia dolorosa, postcordotomy dysesthesias, and possibly cluster headaches; the cortical stimulation site has shown utility in pain following deep brain or spinal ischemia and some forms of deafferentation pain. However, the mechanisms and utility of CNS stimulation for neuropathic pain remain obscure and controversial. The entry ▶ Phantom Limb Pain, Treatment explores the therapies useful (and not useful) in phantom pain: Many treatments useful in other forms of neuropathic pain fail to exceed the efficacy of placebo in phantom pain; nonetheless, opioids, NMDA antagonists, gabapentin, TENS, and sensory discrimination training appear to be effective. Prevention may be a promising future area of development for controlling phantom pain, combining, for example, memantine and peripheral regional anesthesia. Four entries explore noninvasive, nonmedical therapies including complementary and alternative medicine (▶ Alternative Medicine in Neuropathic Pain), dietary variables (▶ Dietary Variables in Neuropathic Pain), the utility of exercise particularly with respect to FMS (▶ Fibromyalgia, Mechanisms, and Treatment), and movement-related pain (▶ Evoked and Movement-Related Neuropathic Pain). Although several studies document that many patients with neuropathic pain (20–50 % in various countries) use alternative and complementary medicine, only electroacupuncture has been validated by randomized controlled trials. Some aspects of diet probably contribute to a patient's predisposition to developing neuropathic pain, but research into pro-analgesic nutrients is in its infancy. The causes and mechanisms underlying fibromyalgia (FMS) remain obscure, though suggestive evidence implicates glutamate, neuropeptides, and nerve growth factor. Tricyclic antidepressants and tramadol have shown efficacy, but exercise remains the most effective therapy overall. This field of disorders – FMS and movement-related pain and therapeutic regimens, TENS, acupuncture and nonmedical therapies – seems to constitute a fruitful area for development of future therapies.

Voltage-Gated Channels

The dominance of diverse anticonvulsants as effective therapeutic agents in a broad range of neuropathic pain states (see ► Drugs Targeting Voltage-Gated Sodium and Calcium Channels) together with the commonality of their molecular targets strongly links voltage-gated sodium and calcium channels to mediation of pain accompanying neuropathy. The involvement of calcium channels derives largely from presumptive targeting of the alpha-2 delta subunit of ► voltage-gated calcium channels by gabapentin and newer pregabalin. Gabapentin's targets would presumably be the central terminals of primary afferent fibers where reduction in calcium-dependent release of excitatory transmitter could account for a pain-attenuating action. A very different recently introduced agent, ziconotide, targets with extreme selectivity N-type voltage-gated calcium channels, which are thought to be directly linked to transmitter release in axon terminals. Ziconotide is approved for intrathecal application in neuropathic pain patients, but several side effects may limit the prevalence of its ultimate use (Wallace 2002).

The important sites of action of agents targeting voltage-gated sodium channels have been variously assigned to peripheral and central targets, but are generally invoked as sources of ► ectopic discharges in hypersensitive zones of peripheral regenerating axon tips or in neuromas (Michaelis 2002). Axonal sensitization has been associated most often with injury-related or neurotrophin-imposed plasticity in expression of ► voltage-gated sodium channels (Black et al. 2002) and most recently with a genetic neuropathic pain–linked polymorphism in the $Na_v1.7$ subtype that lowers the threshold for spike initiation (Dib-Hajj et al. 2005). This finding may prove to be seminal in seeking axonal mediators of sensory neuron sensitization accompanying neuropathic pain. Erythromelalgia manifests with paroxysmal episodes of burning pain in distal extremities initiated by warming; as such, this disorder embodies many of the enigmatic characteristics of several neuropathic pain syndromes, including DN, Fabry disease, and idiopathic burning hands and feet. Only rarely has axonal sensitization after nerve injury been linked to inflammatory mediators released by mast cells (Zuo et al. 2003), in proximal regenerating nerve tips (Zochodone et al. 1997) or distally in the dermis (Marchand et al. 2005). Release of excitatory substances (cytokines, histamine, serotonin, noradrenaline, or ATP) has also been hypothesized proximally in DRG (Michaelis 2002) and distally in dermis (Marchand et al. 2005) as a promoter of ectopic activity. Site-specific study of cellular and molecular mediators of axonal or terminal sensitization would appear to be a fruitful area for future study.

Neurotransmitters

Glutamate, the most prominent fast excitatory neurotransmitter between nociceptors and spinal neurons, activates ligand-gated channels that admit monovalent (mostly Na^+ and K^+ for all AMPA- and kainate-operated channels) and divalent (Ca^{++} or Mg^{++}, some AMPA / kainate receptors, and all NMDA receptors) cations, depolarizing neural processes (Wilcox et al. 2005). Ca^{++} entry mediated by these channels can activate such signaling systems as calcium-calmodulin kinase (CaM-kinase II), which can set in motion numerous intracellular cascades contributing to long-lasting changes in synaptic strength. Both types of receptors are thought to be involved in initiation and maintenance of neuropathic pain states. Normally, NMDA receptors are more directly involved in responses of dorsal horn neurons to intense noxious stimuli than are AMPA / kainate receptors, but increased surface expression of AMPA receptors by spinal neurons may accompany neuropathic pain. NMDA receptors in spinal cord are subject to positive modulation by PKC, which accompanies strong activation by, among others, SP acting at NK_1 receptors (see below) and can be blocked by dissociative anesthetics, including phencyclidine (PCP), MK-801, ketamine, memantine, and the inactive opioid congener dextromethorphan. Only the latter three agents have been investigated clinically for use in neuropathic pain; the other agents like AMPA antagonists are not likely to be of clinical use due to numerous side effects. AMPA receptor activation produces strong depolarization that is not dependent upon concurrent activity, making this process "non-contingent." On the other hand, the NMDA receptor is blocked by Mg^{++} at resting potential, preventing ligand gating of the channel; depolarization of the plasma membrane removes this Mg^{++} block allowing subsequent ligand gating. Removal of this Mg^{++} block by prior depolarization makes NMDA-evoked depolarization "contingent" and may underlie the participation of this receptor in the "wind-up" phenomenon recruited by repetitive activation of C-fibers. NMDA receptors are critical participants in the induction of acute (e.g., formalin) and chronic (the chronic constriction injury neuropathic pain model, CCI) hyperalgesia.

The entry ▶ Peptides in Neuropathic Pain States explores five of the peptide neurotransmitters, vasoactive intestinal polypeptide (VIP and similar PACAP), dynorphin (DYN), cholecystokinin (CCK), and neuropeptide Y (NPY), most directly contributing to enhanced excitatory neurotransmission following peripheral nerve injury. VIP and PACAP, contained in small diameter fibers, dramatically increase in DRG after peripheral nerve injury and participate strongly in induction and maintenance of hyperalgesic states induced by these injuries (Kashiba et al. 1992). DYN expression is also dramatically increased after nerve injury and somehow facilitates excitatory transmission, but this occurs in spinal cord neurons rather than in

DRG. CCK is another central neuropeptide that is upregulated in spinal cord and in pain-relevant brain structures after peripheral nerve injury; CCK probably plays an important role in maintaining neuropathic pain at both the spinal and supraspinal levels. Moderate levels of NPY normally occur in large diameter primary afferent fibers, but these levels are markedly upregulated after peripheral nerve injury; this increased expression is an important component of hypersensitivity accompanying these injuries. These five peptides and their receptors represent likely targets for new drug development directed at neuropathic pain therapy.

Several other neuropeptides are altered after peripheral nerve injury and may participate in development or maintenance of hyperalgesic states. Galanin (GAL) levels in primary sensory neurons increase in DRG soon after nerve injury (Villar et al. 1991) or treatment with the chemotherapeutic agent vinblastine (Kashiba et al. 1992), apparently as a compensatory reaction via Gal_1 receptors countering hypersensitivity (Blakeman et al. 2003). However, GAL actions via its three receptor subtypes are complex, manifesting both inhibitory and excitatory effects depending on dose (Liu et al. 2001). Calcitonin gene–related peptide (CGRP) is found normally in small diameter primary afferent fibers and its levels in DRG decrease after axotomy (Dumoulin et al. 1992) or partial sciatic nerve ligation (Ma et al. 2003). Conversely, spinal nerve ligation increases capsaicin-evoked CGRP release in spinal cord dorsal horn coincident with hyperalgesia (Gardell et al. 2003). Evidently, CGRP participates in maintenance of persistent hyperalgesia following peripheral nerve injury. Substance P (SP), like CGRP, is also found in small diameter primary afferent fibers and its levels similarly decrease in DRG after axotomy (Nielsch et al. 1987); as is the case with CGRP, SP signaling via spinal neurokinin-1 (NK_1) receptors is increased after peripheral nerve injury and contributes to nerve injury–induced hyperalgesia (Cahill and Coderre 2002). Surprisingly, clinical trials of NK_1 antagonists in patients with painful diabetic neuropathy were negative (Goldstein et al. 2001). Somatostatin (SOM), another peptide contained in small diameter primary afferent fibers, decreases in dorsal horn and increases in ventral horn after partial sciatic nerve injury (Swamydas et al. 2004); the mechanical hyperalgesia following this injury is also reversed by systemically administered SOM receptor antagonists (Pinter et al. 2002). One or more of these peptides may also be fruitful targets for therapeutic drug development.

Inflammatory Mediators

Although immunocytes and cytokines are often invoked as factors contributing to neuropathic pain, the sites of their actions are rarely invoked. For example, a recent study of thalidomide in a rodent model of neuropathic pain opens with this sentence: "In almost every neuropathic pain state caused by peripheral nerve damage, whether due to trauma or disease, both structural damage and an inflammatory response exist" (Bennett 2000). The verity of this statement is accepted by most researchers in the field without question, but the design of this particular study, or indeed most such studies, rarely offers significant data or interpretation explicitly delineating the specific sites of the inflammatory response. Only within the past decade have cytokines and subsequently microglia been identified as necessary and sufficient spinal cord mediators of hyperalgesia in rodent models of neuropathic pain (see also ▶ Proinflammatory Cytokines). Presumably, intense activity on primary nociceptive afferent fibers is sufficient to activate spinal microglia, which in turn adopt an activated phenotype, upregulate synthesis of proinflammatory cytokines, and release excitatory neurotransmitters and neuromodulators. The substances released by spinal microglia are thought to include glutamate and nitric oxide, which contribute to the ongoing excitatory cascade thought to establish the spinal and spinal-supraspinal substrates of chronic pain.

Studies of peripheral neuroimmune action contemporary with the central microglial studies from Watkins's and Salter's groups have investigated the role of inflammatory cells at sites of experimental nerve injury. The chronic constriction injury developed by Bennett and Xie (1988) was found to rely on a local inflammatory reaction at the point of injury along the sciatic nerve. However, few recent studies have investigated inflammatory reactions taking place at the distal end of the afferent arm, that is in or near tissue hosting terminals of sensory nerves (e.g., epidermis) undergoing Wallerian degeneration. Do activated immunocytes near the terminal fields (e.g., subepidermal plexus) release ▶ Algogens capable of activating or sensitizing afferent axons?

References

Bennett, G. J. (2000). A neuroimmune interaction in painful peripheral neuropathy. *The Clinical Journal of Pain, 16*, 139–143.

Black, J. A., Cummins, T. R., Dib-Haaj, S. D., et al. (2002). Sodium channels and the molecular basis for pain. In A. B. Malmberg & S. R. Chaplan (Eds.), *Mechanisms and mediators of neuropathic pain* (pp. 23–50). Basel/Boston/Berlin: Birkhäuser Verlag.

Blakeman, K. H., Hao, J. X., Xu, X. J., et al. (2003). Hyperalgesia and increased neuropathic pain-like response in mice lacking galanin receptor 1 receptors. *Neuroscience, 117*, 221–227.

Cahill, C. M., & Coderre, T. J. (2002). Attenuation of hyperalgesia in a rat model of neuropathic pain after intrathecal pre- or post-treatment with a neurokinin-1 antagonist. *Pain, 95*, 277–285.

Dib-Hajj, S. D., Rush, A. M., Cummins, T. R., et al. (2005). Gain-of-function mutation in Nav1.7 In familial erythromelalgia induces bursting of sensory neurons. *Brain, 128*, 1847–1854.

Dumoulin, F. L., Raivich, G., Haas, C. A., et al. (1992). Calcitonin gene-related peptide and peripheral nerve regeneration. *Annals of the New York Academy of Sciences, 657*, 351–360.

Gardell, L. R., Vanderah, T. W., Gardell, S. E., et al. (2003). Enhanced evoked excitatory transmitter release in experimental neuropathy requires descending facilitation. *Journal of Neuroscience, 23*, 8370–8379.

Goldstein, D. J., Wang, O., Gitter, B. D., et al. (2001). Dose–response study of the analgesic effect of lanepitant in patients with painful diabetic neuropathy. *Clinical Neuropharmacology, 24*, 16–22.

Hilliges, M., Wang, L., & Johansson, O. (1995). Ultrastructural evidence for nerve fibers within all vital layers of the human epidermis. *The Journal of Investigative Dermatology, 104*, 134–137.

Hsieh, S. T., Choi, S., Lin, W. M., et al. (1996). Epidermal denervation and its effects on keratinocytes and langerhans cells. *Journal of Neurocytology, 25*, 513–524.

Hsieh, S. T., Chiang, H. Y., Lin, W. M., et al. (2000). Pathology of nerve terminal degeneration in the skin. *Journal of Neuropathology and Experimental Neurology, 59,* 297–307.

Karanth, S. S., Springall, D. R., Kuhn, D. M., et al. (1991). An imunocytochemical study of cutaneous innervation and the distribution of neuropeptides and protein gene product 9.5 in man and commonly employed laboratory animals. *The American Journal of Anatomy, 191,* 369–383.

Kashiba, H., Senba, E., Kawai, Y., et al. (1992). Axonal blockade induces the expression of vasoactive intestinal polypeptide and galanin in rat dorsal root ganglion neurons. *Brain Research, 577,* 19–28.

Kennedy, W. R., & Said, G. (1999). Sensory nerves in skin: Answers about painful feet? *Neurology, 53,* 1614–1615.

Kennedy, W. R., & Wendelschafer-Crabb, G. (1993). The innervation of human epidermis. *Journal of Neurological Sciences, 115,* 184–190.

Lindenlaub, T., & Sommer, C. (2002). Epidermal innervation density after partial sciatic nerve lesion and pain-related behavior in the rat. *Acta Neuropathologica, 104,* 137–143.

Liu, H. X., Brumovsky, P., Schmidt, R., et al. (2001). Receptor subtype-specific pronociceptive and analgesic actions of galanin in the spinal cord: Selective actions via GalR1 and GalR2 receptors. *Proceedings of the National Academy of Sciences of the United States of America, 98,* 9960–9964.

Ma, W., Chabot, J. G., Powell, K. J., et al. (2003). Localization and modulation of calcitonin gene-related peptide-receptor component protein-immunoreactive cells in the rat central and peripheral nervous systems. *Neuroscience, 120,* 677–694.

Marchand, F., Perretti, M., & McMahon, S. B. (2005). Role of the immune system in chronic pain. *Nature Reviews Neuroscience, 6,* 521–532.

Michaelis, M. (2002). Electrophysiological characteristics of injured peripheral nerves. In A. B. Malmberg & S. R. Chaplan (Eds.), *Mechanisms and mediators of neuropathic pain* (pp. 23–50). Basel/Boston/Berlin: Birkhäuser Verlag.

Nielsch, U., Bisby, M. A., & Keen, P. (1987). Effect of cutting or crushing the rat sciatic nerve on synthesis of substance P by isolated L5 dorsal root ganglia. *Neuropeptides, 10,* 137–145.

Pinter, E., Helyes, Z., Nemeth, J., et al. (2002). Pharmacological characterisation of the somatostatin analogue TT-232: Effects on neurogenic and non-neurogenic inflammation and neuropathic hyperalgesia. *Naunyn-Schmiedebergs Archives Pharmacology, 366,* 142–150.

Swamydas, M., Skoff, A. M., & Adler, J. E. (2004). Partial sciatic nerve transection causes redistribution of pain-related peptides and lowers withdrawal threshold. *Experimental Neurology, 188,* 444–451.

Villar, M. J., Wiesenfeld-Hallin, Z., Xu, X. J., et al. (1991). Further studies on galanin-, substance P-, and CGRP-like immunoreactivities in primary sensory

neurons and spinal cord: Effects of dorsal rhizotomies and sciatic nerve lesions. *Experimental Neurology, 112,* 29–39.

Wallace, M. (2002). In A. B. Malmberg & S. R. Chaplan (Eds.), *Mechanisms and mediators of neuropathic pain.* Basel/Boston/Berlin: Birkhäuser Verlag.

Wilcox, G. L., Stone, L. S., Ossipov, M. H., et al. (2005). Pharmacology of pain tranmission and modulation. I. Central mechanisms. In M. Pappagallo (Ed.), *The neurologic basis of pain* (pp. 31–52). New York: McGraw-Hill.

Zochodne, D. W., & Cheng, C. (2000). Neurotrophins and other growth factors in the regenerative milieu of proximal nerve stump tips. *Journal of Anatomy, 196,* 279–283.

Zochodone, D. W., Theriault, M., Cheng, C., et al. (1997). Peptides and neuromas. Peptides and neuromas: Calcitonin gene-related peptide, substance P and mast cells in a mechanosensitive human sural neuroma. *Muscle & Nerve, 7,* 875–880.

Zuo, Y., Perkins, N. M., Tracey, D. J., et al. (2003). Inflammation and hyperalgesia induced by nerve injury in the rat: A key role of mast cells. *Pain, 105,* 467–479.

Drugs Targeting Voltage-Gated Sodium and Calcium Channels

Anne K. Bertelsen[1] and Misha-Miroslav Backonja[2]
[1]Neurology, Aleris Hospital, Bergen, Norway
[2]Department of Neurology, University of Wisconsin-Madison, Madison, WI, USA

Synonyms

Antiarrhythmics; Anticonvulsants; Ion channel blockers; Membrane-stabilizing drugs; Voltage-gated channels

Definition

Drugs that attenuate inward sodium or calcium ion currents in nociceptive afferent neurons, thus exhorting a membrane-stabilizing action on these neurones.

Characteristics

Voltage-gated sodium channels (VGSCs) and voltage-gated calcium channels (VGCCs) have a fundamental role in the excitability of all neurons. Agents that block these channels or encourage inactivation, often referred to as membrane-stabilizing agents, reduce nerve excitability by blocking the initiation of action potentials.

The role of VGCCs in ▶ neuropathic pain is under intense investigation, but their role is far from clear. In contrast, VGSCs in sensory neurons are well studied and thought to play a crucial role in neuropathic pain caused by peripheral nerve injury. Alteration in VGSC expression or function has been shown to have profound effects on the firing pattern of primary afferent sensory neurons, as well as neurons in the central nervous system. After nerve injury, sodium channels may accumulate not only on the neuroma or sprouts of damaged peripheral nerve endings but also along the rest of the axons and on uninjured neighboring axons (Cummins and Waxman 1997; England et al. 1996). This accumulation of sodium channels dramatically lowers the depolarization threshold of the nerves, and the nerves fire more readily in response to low-threshold stimulation (Matzner and Devor 1994). Increased firing of primary sensory neurons results in increased release of glutamate and substance P from the central terminals of primary afferent fibers, leading to subsequent activation of the NMDA receptors and development of a state called ▶ central sensitization. Pathologically unstable membranes on primary afferent neurons are critical for induction of central sensitization and development of neuropathic pain after peripheral nerve injury. Patients with this condition will typically experience stimulus-independent pain and/or stimulus-dependent pain as ▶ hyperalgesia and ▶ allodynia.

There are at least nine VGSCs present in the nervous system of mammals that differ in their expression patterns within the nervous system, their kinetics, and their recovery from inactivation. There are, for example, certain VGSCs that are only expressed in the dorsal root ganglia, secondary to downregulation of certain sodium

channel subtypes and the upregulation of others (Cummins et al. 2000). Further, the sodium channels can be divided into tetrodotoxin-sensitive and tetrodotoxin-resistant channels, both of which play a physiological role in nociceptive impulse transmission (Brock et al. 1998). However, there is evidence to suggest that preferential expression and accumulation of tetrodotoxin-resistant sodium channels, especially $Na_v1.8$ channels, may be relevant to the mechanisms of peripheral nerve injury-induced neuropathic pain (Lai et al. 2004).

A number of drugs modulate sodium channels in the peripheral nervous system and are thought to provide relief from neuropathic pain by suppressing ▶ ectopic discharges originating in the injured ▶ nociceptors or at the level of the associated dorsal root ganglia. The fact that the available agents lack selectivity for sodium channel subtypes means that multiple channel subtypes are affected, resulting in frequent adverse side effects. The therapeutic effects of these sodium channel-modulating drugs result from their ability to prolong the refractory period following action potentials and to prevent the generation of spontaneous ectopic discharges at concentrations lower than those required to block or inhibit normal impulse generation and propagation.

Individual Ion Channel Modulators in Treatment of Neuropathic Pain

A number of drugs have been used for treatment of neuropathic pain, and this use was primarily empirically derived (Dworkin et al. 2003a). The drugs used for treatment of neuropathic pain belong to a few classes, and, because they modulate the activity of the nervous system to achieve this effect, they are sometimes referred to as neuromodulators. It should be noted that all of them provide only symptomatic pain relief, and none of them reverses underlying pathology.

Anticonvulsant Agents

Gabapentin has been used since 1994 for the management of partial epilepsy, and is now one of the most widely used drugs in the management of neuropathic pain. Structurally, it is an analog of gamma-aminobutyric acid, GABA, but pharmacologically gabapentin appears to have no direct effect on GABA uptake or metabolism. Its mechanisms of action appear to involve antagonism of non-NMDA receptors and binding to the alpha$_2$delta subunit of voltage-dependent calcium channels. The latter action could mediate reduced release of excitatory neurotransmitters. Although gabapentin is associated with few drug interactions and is well tolerated, the concentration range and value of therapeutic drug monitoring of this drug is unclear; therefore, the dosage should be adjusted based on efficacy and tolerability. Gabapentin seems to be most effective at or above doses of 1,800 (up to 3,600 mg/day) divided into three doses. Lower doses (<900 mg/day) have shown to have limited effect in the treatment of neuropathic pain (Backonja and Glanzman 2003). The dosage should be increased slowly, from the lowest recommended therapeutic daily dose up to the maximum recommended daily dose, or until adequate pain relief is achieved and/or the patient complains about side effects.

Pregabalin, a follow-up drug to gabapentin, also binds to alpha$_2$delta subunit of voltage-dependent calcium channels but has a more linear bioavailability than gabapentin. Lower doses are therefore needed for reducing pain and allodynia than gabapentin, and the efficacy is more predictable. The therapeutic dose range for pregabalin is between 150 and 600 mg/day divided in two or three doses (Dworkin et al. 2003b).

Carbamazepine binds preferentially to the inactivated state of voltage-gated sodium channels, thereby stabilizing axonal membranes. It is approved by the US Food and Drug Administration as the drug of first choice to relieve shock-like pain, such as that experienced with trigeminal neuralgia (TN). The pain relief in TN is so fast and effective with carbamazepine that it is sometimes used as a marker for determining whether a patient has TN or not. However, carbamazepine may not be as effective for atypical trigeminal neuralgia, and data with regard to other neuropathic pain disorders are negative, limited, or both. Dosages are adjusted for each person

individually, but usually vary between 200 and 1,200 mg/day. However, intolerance to the side effects and a need for laboratory monitoring limits its use, especially in the elderly. Phenytoin, next to carbamazepine, is another example of a drug that has negative clinical trial results. Theoretically, both of these drugs could be effective because of their specific effects on sodium channels, but their complex pharmacology limits their use.

Lamotrigine is a new sodium channel and presynaptic glutamate-blocking drug that may possess beneficial properties for pain relief by suppressing abnormal ectopic discharge. It has recently been reported that lamotrigine, at doses of 400–600 mg/day, results in moderate pain relief with tolerable side effects in trials involving patients with diabetic and human immunodeficiency virus (HIV)-associated neuropathy (Simpson et al. 2000).

Topiramate, tiagabine, and bupropion are other new-generation anticonvulsants that display an ability to modulate sodium channels, but randomized trials for topiramate have been negative, and the others still lack indication of their utility in neuropathic pain management.

Local Anesthetics/Antiarrhythmics
Local anesthetic drugs should be considered as alternative therapies in the treatment of chronic neuropathic pain, because they reportedly provide effective relief in different neuropathic pain conditions such as painful diabetic neuropathy, postherpetic neuralgia, lumbar radiculopathies, and complex regional pain syndrome (CRPS) types I and II and in traumatic peripheral nerve injuries. However, the usefulness of these agents is limited because of frequent association with numerous adverse effects, particularly CNS-related side effects, for example, nausea and emesis, dizziness, ataxia, and tinnitus, as well as cardiotoxicity, which severely limits its use, especially in the elderly.

Intravenous lidocaine (lignocaine) is the most frequently used local anesthetic in the treatment of neuropathic pain and has been shown to produce a moderate reduction in pain in patients with diabetic neuropathy (Kastrup et al. 1987).

It has been shown to be especially effective in pain states like trigeminal and post-herpetic neuralgia and painful diabetic neuropathy. However, intravenous administration poses a challenge for the long-term treatment because oral dosing is unavailable. Therefore, this treatment is administered periodically on a scheduled basis or for excruciating episodes of neuropathic pain. In addition to its acute effect, it also produces pain relief for several days, an effect far outlasting drug elimination from the plasma. The mechanisms related to this phenomenon remain unclear. Intravenous lidocaine/lignocaine is advocated as a diagnostic aid for the presence of pain associated with nerve injury and for its predictive value of potential analgesic efficacy of oral local anesthetic agents, such as mexiletine.

The orally administered "local anesthetics" include both tocainide and the type 1b antiarrhythmic agent mexiletine; both have been used with success either as monotherapies, or sequentially following an initial lignocaine infusion. However, there have been inconsistent results with the use of mexiletine. Patients with diabetic neuropathy obtained a beneficial effect from mexiletine (Oskarsson et al. 1997), and another demonstrated efficacy with regard to secondary outcomes but not with regard to global pain relief (Stracke et al. 1992); a fourth trial in patients with HIV-associated neuropathy failed to demonstrate a benefit (Kemper et al. 1998).

A critical aspect of the analgesic action of local anesthetics is their ability, at low subanesthetic doses, to block the spontaneous and/or evoked repetitive, ectopic impulse activity in afferent fibers apparently mediated by both TTX-S and slowly inactivating TTX-R sodium channels. Lidocaine/lignocaine can suppress the generation of this abnormal impulse traffic and restore a more normal firing rhythm by acting either directly at the site of origin or at distant sites. Injured nerves that fire repetitively at high frequency are specifically sensitive to concentrations of these drugs that have minimal impact on normal, somatosensory (i.e., nociceptive) neuronal function, and this sensitivity of injured neurons is the basis for the efficacy of these drugs.

Antidepressant Drugs

▶ Tricyclic antidepressants are the most well studied and most used agents for relief of neuropathic pain. Interestingly, amitriptyline, doxepin, and desipramine have been found to be strong sodium channel modulators (Pancrazio et al. 1998), in addition to their well-known effect on reuptake of serotonin and noradrenaline. This sodium channel-blocking effect is presumed to be the pain-relieving mechanism of these drugs in the case of neuropathic pain due to peripheral nerve injury, and they have efficacy on both spontaneous pain as well as for hyperalgesia. The accumulative efficacy reported from trials suggests that about one third of patients will achieve a 50 % reduction in neuropathic pain; however, the benefits are often outweighed by side effects, especially among the elderly.

References

Backonja, M., & Glanzman, R. L. (2003). Gabapentin dosing for neuropathic pain: Evidence from randomized, placebo-controlled clinical trials. Clinical Therapeutics, 25, 81–104.

Brock, J. A., McLachlan, E. M., & Belmonte, C. (1998). Tetrodotoxin-resistant impulses in single nociceptor nerve terminals in guinea-pig cornea. The Journal of Physiology, 512, 211–217.

Cummins, T. R., & Waxman, S. G. (1997). Downregulation of tetrodotoxin-resistant sodium currents and upregulation of a rapidly repriming tetrodotoxine-sensitive sodium current in small spinal sensory neurons after nerve injury. Journal of Neuroscience, 17, 3503–3514.

Cummins, T. R., Dib-Hajj, S. D., Black, J. A., et al. (2000). Sodium channels and the molecular pathophysiology of pain. Progress in Brain Research, 129, 3–19.

Dworkin, R. H., Backonja, M., Rowbotham, M. C., et al. (2003a). Advances in neuropathic pain: Diagnosis, mechanisms, and treatment recommendations. Archives of Neurology, 60, 1524–1534.

Dworkin, R. H., Corbin, A. E., Young, J. P., Jr., et al. (2003b). Pregabalin for the treatment of postherpetic neuralgia: A randomized, placebo-controlled trial. Neurology, 60, 1274–1283.

England, J. D., Happel, L. T., Kline, D. G., et al. (1996). Sodium channel accumulation in humans with painful neuromas. Neurology, 47, 272–276.

Kastrup, J., Petersen, P., Dejgård, A., et al. (1987). Intravenous lidocaine infusion – A new treatment of chronic painful diabetic neuropathy? Pain, 28, 69–75.

Kemper, C. A., Kent, G., Burton, S., et al. (1998). Mexiletine for HIV-infected patients with painful peripheral neuropathy; A double-blind, placebo-controlled, crossover treatment trial. *Journal of Acquired Immune Deficiency Syndromes and Human Retrovirology, 19*, 367–372.

Lai, J., Porreca, F., Hunter John, C., et al. (2004). Voltage-gated sodium channels and hyperalgesia. *Annual Review of Pharmacology and Toxicology, 44*, 371–397.

Matzner, O., & Devor, M. (1994). Hyperexcitability at sites of nerve injury depends on voltage-sensitive Na^+ channels. *Journal of Neurophysiology, 72*, 349–359.

Oskarsson, P., Lunggren, J. G., & Lins, P. E. (1997). Efficacy and safety of mexiletine in the treatment of painful diabetic neuropathy. *Diabetes Care, 20*, 1594–1597.

Pancrazio, J. J., Kamatchi, G. L., Roscoe, A. K., et al. (1998). Inhibition of neuronal Na^+ channels by antidepressant drugs. *The Journal of Pharmacology and Experimental Therapeutics, 284*, 208–214.

Simpson, D. M., Olney, R., McArthur, J. C., et al. (2000). A placebo-controlled trial of lamotrigine for painful HIV-associated peripheral neuropathy. *Neurology, 54*, 2115–2119.

Stracke, H., Meyer, U. E., Schumacher, H. E., et al. (1992). Mexiletine in the treatment of diabetic peripheral neuropathy. *Diabetes Care, 15*, 1550–1555.

Drugs with Mixed Action and Combinations: Emphasis on Tramadol

Robert B. Raffa
Department of Pharmaceutical Sciences,
Temple University School of Pharmacy,
Philadelphia, PA, USA

Synonyms

Mixed-acting analgesics; Multimodal analgesics

Definition

Analgesics for which a significant portion of their analgesic effect is attributable to the combined action of more than one analgesic mechanism. The different mechanisms can be contained within a single drug or by combination of separate drugs.

Characteristics

Because the nature of neuropathic pain is diverse in its etiology, neuro(chemical/logical) pathology, and clinical manifestations, it is highly likely that multiple diverse pain-signal-transmitting pathways are involved. It seems reasonable, then, that a strategy to treat this type of pain might incorporate the simultaneous combined activation of diverse analgesic pathways. This section considers the treatment of neuropathic pain with analgesic drugs that have multiple mechanisms of analgesic action. The strategy of combining analgesics that have a single mechanism of action is highlighted.

Individual Agents (Classes)

In treatment algorithms for pharmacologic management of neuropathic pain, such as that published by Namaka et al. (2009), four drug categories are commonly listed as first-line therapy: "antidepressants" (tricyclics), "antiepileptics" (including gabapentin), opioids and tramadol, and topicals (including lidocaine). From within this list, the majority of drugs can be considered to produce their analgesic effect through a single analgesic mechanism (i.e., they are mono-mechanistic), whereas the nonselective TCAs and tramadol can be considered to be multi-mechanistic analgesics.

Gabapentin and Pregabalin. The attributes and place for gabapentin in the management of neuropathic pain have been the subject of reviews such as that by Rosner and Diwan (2003). The exact mechanism of analgesic action of gabapentin or pregabalin in treating neuropathic pain is not known. Gabapentin does not have affinity for opioid receptors (μ, δ, or κ) and does not involve $GABA_A$ or $GABA_B$ binding sites (Hwang and Yaksh 1997). Each drug attenuates both cold- and touch-induced hyperesthesias (Hunter et al. 1997). The recent research suggests that an inhibitory action on N-type Ca^{2+} currents in DRG neurons is the analgesic mechanism (Ferreira et al. 2010). The analgesic efficacy of gabapentin and its adverse effect profile are summarized in Dworkin et al. (2003) and in Rosner and Diwan (2003).

Lidocaine. Synthesized in 1943 by Löfgren, lidocaine is an amide local anesthetic that blocks voltage-gated Na^+ channels. The uncharged form of lidocaine, a weak base, penetrates the axonal membrane. After reestablishment of steady state, the charged (cationic) form binds to the receptor accessible from the internal side of the membrane. Low plasma blood levels during lidocaine patch (5 %) (Rowbotham et al. 1996) application suggest a purely local effect.

Opioids. All of the opioids produce their analgesic effect through a well-known single mechanism: agonist action at seven-transmembrane G protein-coupled (7TM-GPCR) opioid receptors located in the spinal cord and higher CNS regions. Activation of opioid receptors on neurons has two well-established direct actions (Schumacher et al. 2004): (1) inhibition of voltage-gated Ca^{2+} currents on presynaptic neurons (thereby reducing the release of neurotransmitters) and (2) enhancement of postsynaptic neuronal K^+ efflux (thereby hyperpolarizing the neuron and inhibiting excess neuronal firing).

Tricyclics. The tricyclics (TCAs) produce an analgesic effect that is demonstrably independent of their well-known antidepressant action. The efficacy of TCAs against neuropathic pain has been documented in controlled clinical trials (for summary, see Dworkin et al. 2003). TCAs have varying degrees of selectivity for inhibiting the neuronal reuptake of norepinephrine and/or 5-HT (5-hydroxytryptamine; serotonin). They are not always well tolerated, but a systematic review concludes that they are effective in relieving at least some types of neuropathic pain (McQuay et al. 1996).

Tramadol. Tramadol hydrochloride (tramadol) is the only analgesic of these first-line medications that has been demonstrated to have at least two distinct mechanisms that contribute to its analgesic action. Tramadol, (1*RS*,2*RS*)-2-[(dimethylamino) methyl]-1-(3-methoxyphenyl)   cyclohexanol hydrochloride (Fig. 1), is a centrally acting synthetic analgesic. Based on animal and human studies, tramadol appears to produce its analgesic effect through two mechanisms (for summary, see Raffa and Friderichs 1996). One of the mechanisms appears to relate to a weak affinity of

Drugs with Mixed Action and Combinations: Emphasis on Tramadol, Fig. 1 Tramadol hydrochloride (a chiral center gives rise to enantiomers)

parent drug for μ-opioid receptors, estimated to be about 6,000-fold less than the affinity of morphine and about the same affinity as dextromethorphan. Its mono-*O*-desmethyl metabolite (designated M1) binds to μ-opioid receptors with greater affinity than the parent compound and is presumably responsible for the opioid component. However, in most animal tests (reviewed in Raffa and Friderichs 1996) and in human clinical tests (Collart et al. 1993), the antinociceptive and analgesic effect of tramadol is only partially blocked (< 50 %) by the opioid antagonist naloxone, tramadol retains a significant antinociceptive effect in μ-opioid receptor knockout mice (Ide et al. 2006), and tramadol-induced antinociception is attenuated by 5-HT antagonists (Yanarates et al. 2010), all of which suggest a significant contribution of a non-opioid mechanism. The apparent non-opioid component of tramadol's mechanism appears to be related to its ability to inhibit the neuronal reuptake of norepinephrine and 5-HT. Tramadol does not have a significant peripheral or anti-inflammatory mechanism of action, and a mixed agonist–antagonist action has been ruled out. In addition to the individual contributions made by the opioid and non-opioid components of tramadol's mechanism of action, it appears that there is a synergistic interaction between these two mechanisms. The (+) enantiomer of tramadol binds to μ-opioid receptors and inhibits neuronal reuptake of 5-HT more potently than does the (−) enantiomer, whereas the (−) enantiomer inhibits the neuronal reuptake of norepinephrine more potently than does the (+) enantiomer (Table 1). Each of the enantiomers individually produces centrally mediated antinociception in mice; however, in several tests, combination of enantiomers is more potent than either enantiomer alone (i.e., the interaction

Drugs with Mixed Action and Combinations: Emphasis on Tramadol, Table 1 In vitro activity (affinity) of tramadol, its enantiomers, and its major metabolite (mono-O-desmethyl tramadol; M1) at opioid μ, δ, and κ receptors and at norepinephrine (NE) and serotonin (5-HT) neuronal reuptake sites. Values are K_i in mM (Modified from Raffa and Friderichs (1996))

	μ	δ	κ	NE	5-HT
Tramadol	2.1	57.6	42.7	0.78	0.99
(+) Enantiomer	1.3	62.4	54.0	2.51	0.53
(−) Enantiomer	24.8	213	53.5	0.43	2.35
M1 metabolite	0.0121	0.911	0.242	1.52	5.18

is synergistic) (Raffa et al. 1993). Importantly, the enantiomers interact less than synergistically or even less than additively in several tests predictive of side-effect potential. Thus, the clinical profile of tramadol apparently results from fortuitous combinations and interactions of component parts. It is the proposed duality of mechanism of analgesic action that has been suggested to form the foundation for understanding tramadol's clinical attributes (e.g., Raffa and Friderichs 1996).

Tramadol is rapidly and almost completely absorbed after its oral administration, with an absolute bioavailability (100 mg dose) of about 75 % that is not significantly affected by food. The plasma protein binding of tramadol is relatively low (about 20 %). Tramadol's volume of distribution is about 2.7 L/kg. The analgesic effect of tramadol in humans begins within 1 h after its oral administration and usually peaks in about 2–3 h. In humans, about 30 % of tramadol is excreted unchanged in the urine, whereas animals metabolize tramadol much more extensively – only about 1–2 % is excreted unchanged. The major metabolic pathways involve N- and O-demethylation and glucuronidation or sulfation in the liver. The M1 (O-desmethyl tramadol) pathway is catalyzed by the CYP-2D6 isozyme of cytochrome P-450. Tramadol and its metabolites are excreted primarily via the kidney. The mean terminal plasma elimination half-life of tramadol is about 6.3 h, which increases slightly (about 1 h) upon multiple dosing.

The efficacy of tramadol for treating neuropathic pain has been reported from double-blind, placebo-controlled randomized clinical trials (for review see Dworkin et al. 2003). Pain was significantly reduced compared to placebo, and beneficial effects on allodynia and quality of life were reported.

Tapentadol. The recently approved dual-mechanism analgesic tapentadol has an open-chain structure instead of tramadol's cyclic structure, which imparts its distinct pharmacologic and clinical properties. Tapentadol combines relatively strong μ-opioid receptor agonism with the inhibition of neuronal norepinephrine reuptake (Tzschentke et al. 2007). Both of the mechanisms reside within the parent molecule, and tapentadol does not have a relevant functional neuronal serotonin reuptake inhibition. The two mechanisms of antinociceptive action of tapentadol interact in a synergistic manner (Schröder et al. 2011). Tapentadol undergoes direct O-glucuronidation to an analgesically inactive metabolite (Schröder et al. 2010; Schiene et al. 2011).

The efficacy and tolerability profile of tapentadol in the clinical setting has been reviewed (Riemsma et al. 2011; Daniels and Golf 2012; Hoy 2012). It has largely been established based on comparison with oxycodone (Merker et al. 2012) with a more favorable gastrointestinal tolerability (Daniels et al. 2009; Hartrick et al. 2009; Afilalo et al. 2010). An approximate equianalgesic ratio of 5:1 for long-acting formulations of tapentadol and oxycodone was estimated based on individually adjusted dose (Lange et al. 2010).

Combinations

Another approach to introducing a diversity of analgesic mechanisms into neuropathic pain therapy is the use of combinations of individual agents. Advantages and potential disadvantages of this strategy have been reviewed (e.g., Raffa 2001; Raffa et al. 2003). In relation to neuropathic pain, the major advantages could be facilitation of prescribing and enhanced compliance, broader coverage of diverse pain types, decreased side effects, and possible synergistic analgesic interaction. The major potential disadvantage is higher cost, but many combination products are priced at about the combined expense of the single entities. The most common combinations involve opioids or tramadol with nonsteroidal anti-inflammatory drugs (NSAIDs) or with acetaminophen (paracetamol).

The mechanism of action of the NSAIDs was delineated in the early 1970s and is now generally accepted to be via inhibition of cyclooxygenase (COX) isozymes.

The mechanism of action by which acetaminophen, N-acetyl-p-aminophenol, produces its analgesic action is not well understood. It has been shown that acetaminophen-induced antinociception is related to its plasma concentration. It is also known that the analgesia induced by acetaminophen is not attributable solely, or even to any significant extent, to an anti-inflammatory action. There is a significant amount of evidence that points to the CNS as the major site of analgesic action of acetaminophen (see reviews by Walker (1995) and by Björkman (1995)).

In the case of the combination of tramadol plus acetaminophen, in addition to expected advantages of the simple additive effect of the combination, the combination produces a synergistic effect in animals (Tallarida and Raffa 1996) and humans (Filitz et al. 2008).

For all analgesic combinations, it is important to assess possible enhancement of adverse effect end points in addition to possible enhancement of analgesic end points (Hanna et al. 2008).

References

Afilalo, M., Etropolski, M. S., Kuperwasser, B., Kelly, K., Okamoto, A., Van Hove, I., Steup, A., Lange, B., Rauschkolb, C., & Haeussler, J. (2010). Efficacy and safety of tapentadol extended release compared with oxycodone controlled release for the management of moderate to severe chronic pain related to osteoarthritis of the knee: A randomized, double-blind, placebo- and active-controlled phase III study. *Clinical Drug Investigation, 30*, 489–505.

Björkman, R. (1995). Central antinociceptive effects of Non-steroidal anti-inflammatory drugs and paracetamol. *Acta Anaesthesiologica Scandinavica, 39*, 2–44.

Collart, L., Luthy, C., & Dayer, P. (1993). Partial inhibition of tramadol antinociceptive effect by naloxone in man. *British Journal of Clinical Pharmacology, 35*, 73P.

Daniels, S. E., & Golf, M. (2012). Clinical efficacy and safety of tapentadol immediate release in the postoperative setting. *Journal of the American Podiatric Medical Association, 102*, 139–148.

Daniels, S. E., Upmalis, D., Okamoto, A., Lange, C., & Haeussler, J. (2009). A randomized, double-blind, phase III study comparing multiple doses of tapentadol IR, oxycodone IR, and placebo for postoperative (bunionectomy) pain. *Current Medical Research and Opinion, 25*, 765–776.

Dworkin, R. H., Backonja, M., Rowbotham, M. C., Allen, R. R., Argoff, C. R., Bennett, G. J., Bushnell, M. C., Farrar, J. T., Galer, B. S., Haythornthwaite, J. A., Hewitt, D. J., Loeser, J. D., Max, M. B., Saltarelli, M., Schmader, K. E., Stein, C., Thompson, D., Turk, D. C., Wallace, M. S., Watkins, L. R., & Weinstein, S. M. (2003). Advances in neuropathic pain: diagnosis, mechanisms, and treatment recommendations. *Archives of Neurology, 60*, 1524–1534.

Ferreira, S. H., Prado, W. A., & Ferrari, L. F. (2010). Potassium and calcium channels in pain pharmacology. In P. Beaulieu, D. Lussier, F. Porreca, & A. H. Dickenson (Eds.), *Pharmacology of pain* (pp. 163–184). Seattle: IASP Press.

Filitz, J., Ihmsen, H., Günther, W., Tröster, A., Schwilden, H., & Schüttler, K. W. (2008). Supra-additive effects of tramadol and acetaminophen in a human pain model. *Pain, 136*, 262–270.

Hanna, M., O'Brien, C., & Wilson, M. C. (2008). Prolonged-release oxycodone enhances the effects of existing gabapentin therapy in painful diabetic neuropathy patients. *European Journal of Pain, 12*, 804–813.

Hartrick, C., Van Hove, I., Stegmann, J.-U., Oh, C., & Upmalis, D. (2009). Efficacy and tolerability of tapentadol immediate release and oxycodone HCl immediate release in patients awaiting primary joint replacement surgery for end-stage joint disease: A 10-day, phase III, randomized, double-blind, active- and placebo-controlled study. *Clinical Therapeutics, 31*, 260–271.

Hoy, S. M. (2012). Tapentadol extended release: In adults with chronic pain. *Drugs, 72*, 375–393.

Hunter, J. C., Gogas, K. R., Hedley, L. R., et al. (1997). The effect of novel anti-epileptic drugs in Rat experimental models of acute and chronic pain. *European Journal of Pharmacology, 324*, 153–160.

Hwang, J. H., & Yaksh, T. L. (1997). Effect of subarachnoid gabapentin on tactile-evoked allodynia in a surgically induced neuropathic pain model in the rat. *Regional Anesthesia, 22*, 249–256.

Ide, S., Minami, M., Ishihara, K., Uhl, G. R., Sora, I., & Ikeda, K. (2006). Mu opioid receptor-dependent and independent components in effects of tramadol. *Neuropharmacology, 51*, 651–658.

Lange, B., Kuperwasser, B., Okamoto, A., Steup, A., Haufel, T., Ashworth, J., & Etropolski, M. (2010). Efficacy and safety of tapentadol prolonged release for chronic osteoarthritis pain and low back pain. *Advances in Therapy, 27*, 381–399.

McQuay, H. J., Tramèr, M., Nye, B. A., et al. (1996). A systematic review of antidepressants in neuropathic pain. *Pain, 68*, 217–227.

Merker, M., Dinges, G., Koch, T., Kranke, P., & Morin, A. M. (2012). Undesired side effects of tapentadol in comparison to oxycodone. A meta-analysis of randomized controlled comparative studies. *Der Schmerz, 26*, 16–26.

Namaka, M., Leong, C., Grossberndt, A., Klowak, M., Turcotte, D., Esfahani, F., Gomori, A., & Intrater, H. (2009). A treatment algorithm for neuropathic pain: An update. *The Consultant Pharmacist, 24*, 885–902.

Raffa, R. B. (2001). Pharmacology of oral combination analgesics: Rational therapy for pain. *Journal of Clinical Pharmacology & Therapeutics, 26*, 257–264.

Raffa, R. B., & Friderichs, E. (1996). The basic science aspect of tramadol hydrochloride. *Pain Reviews, 3*, 249–271.

Raffa, R. B., Friderichs, E., Reimann, W., et al. (1993). Complementary and synergistic antinociceptive interaction between the enantiomers of tramadol. *Journal of Pharmacology and Experimental Therapeutics, 267*, 331–340.

Raffa, R. B., Clark-Vetri, R., Tallarida, R. J., et al. (2003). Combination strategies for pain management. *Expert Opinion on Pharmacotherapy, 4*, 1697–1708.

Riemsma, R., Forbes, C., Harker, J., Worthy, G., Misso, K., Schafer, M., Kleijnen, J., & Sturzebecher, S. (2011). Systematic review of tapentadol in chronic severe pain. *Current Medical Research and Opinion, 27*, 1907–1930.

Rosner, H., & Diwan, S. (2003). The role of gabapentin in the management of neuropathic pain. In C. Bountra, R. Munglani, & W. K. Schmidt (Eds.), *Pain: Current understanding, emerging therapies, and novel approaches to drug discovery* (pp. 781–794). New York: Marcel Dekker.

Rowbotham, M. C., Davies, P. S., Verkempinck, C., et al. (1996). Lidocaine patch: A double-blind controlled study of a new treatment method for post-herpetic neuralgia. *Pain, 65*, 39–44.

Schiene, K., De Vry, J., & Tzschentke, T. M. (2011). Antinociceptive and antihyperalgesic effects of tapentadol in animal models of inflammatory pain. *Journal of Pharmacology and Experimental Therapeutics, 339*, 537–544.

Schröder, W., De Vry, J., Tzschentke, T. M., Jahnel, U., & Christoph, T. (2010). Differential contribution of opioid and noradrenergic mechanisms of tapentadol in rat models of nociceptive and neuropathic pain. *European Journal of Pain, 14*, 814–821.

Schröder, W., Tzschentke, T. M., Terlinden, R., De Vry, J., Jahnel, U., Christoph, T., & Tallarida, R. J. (2011). Synergistic interaction between the two mechanisms of action of tapentadol in analgesia. *Journal of Pharmacology and Experimental Therapeutics, 337*, 312–320.

Schumacher, M. A., Basbaum, A. I., & Way, W. L. (2004). Opioid analgesics & antagonists. In B. G. Katzung (Ed.), *Basic & clinical pharmacology* (pp. 497–516). New York: McGraw-Hill.

Tallarida, R. J., & Raffa, R. B. (1996). Testing for synergism over a range of fixed ratio drug combinations: Replacing the isobologram. *Life Sciences, 58*, 23–28.

Tzschentke, T. M., Christoph, T., Kögel, B., Schiene, K., Hennies, H. H., Englberger, W., Haurand, M., Jahnel, U., Cremers, T. I., Friderichs, E., & De Vry, J. (2007). (−)-(1R,2R)-3-(3-Dimethylamino-1-ethyl-2-methyl-propyl)-phenol hydrochloride (tapentadol HCl): a novel μ − opioid receptor agonist/norepinephrine reuptake inhibitor with broad-spectrum analgesic properties. *Journal of Pharmacology and Experimental Therapeutics, 323*, 265–276.

Walker, J. S. (1995). NSAID: An update on their analgesic effects. *Clinical and Experimental Pharmacology and Physiology, 22*, 855–860.

Yanarates, O., Dogrul, A., Yildirim, V., Sahin, A., Sizlan, A., Seyrek, M., Akgül, O., Kozak, O., Kurt, E., & Aypar, U. (2010). Spinal 5-HT7 receptors play an important role in the antinociceptive and antihyperalgesic effects of tramadol and its metabolite, O-desmethyltramadol, via activation of descending serotonergic pathways. *Anesthesiology, 112*, 696–710.

Dry Needling

Chang-Zern Hong
Department of Physical Medicine and Rehabilitation, University of California, Irvine, CA, USA

Synonyms

Intramuscular sensory nerve stimulation; Twitch-obtained intramuscular stimulation

Definition

▶ Dry needling is a procedure in which a small-tipped needle is inserted multiple times into a myofascial trigger point region or acupuncture point in order to stimulate the sensory nerve endings and relieve muscle pain.

Characteristics

Muscle pain can be caused by various conditions, including neurological lesions, muscular lesions, soft tissue lesions, skeletal lesions, and systemic diseases. In the cases of neurological lesions or systemic diseases, the muscle pain is usually diffused and without tender spots. More frequently, however, one or more tender spots can be identified in a muscle. These tender spots are usually myofascial trigger points. A ▶ myofascial trigger point (MTrP) is a hyperirritable spot that is well localized within a taut band of skeletal muscle fibers. Compression of an MTrP can cause pain (tenderness). An ▶ active myofascial trigger point is a spot with spontaneous pain or pain in response to movement, while a ▶ latent myofascial trigger point is a spot with pain or discomfort in response to compression only. Referred pain can also be elicited when the compression pressure is high enough. If the examiner compresses the MTrP with a fingertip, then follows with a snapping motion over the MTrP, in a direction perpendicular to the muscle fibers (this motion is known as snapping palpation), a brief contraction of the tense muscle fibers may occur. This is a polysynaptic spinal reflex known as a ▶ local twitch response (LTR).

Recent research studies on both human and animal subjects have determined that MTrPs are commonly found in the neuromuscular junction region (endplate zone). Latent MTrPs may be developed as early as 6 months after birth (Hong 2000, 2002). A latent MTrP is probably caused by minor peripheral nerve injury due to repetitive minor trauma early in life (Gunn 1996; Hong 2000). A latent MTrP may become active (painful) after the soft tissue or the lesions involving various structures sustain an acute injury or endure chronic repetitive trauma or even after the patient suffers emotional stress (Hong 1996; Hong and Simons 1998).

It is possible that the formation of active MTrPs is actually a defense mechanism in the body. If muscle mobility is limited, further damage to the existing lesion – and further injury to the already injured soft tissue – may be prevented (Hong 1996). In some cases though, chronic active MTrPs may persist even after the etiological lesion is controlled. If the pain becomes so intolerable that it interferes with daily life, the MTrP may need to be inactivated. Conservative therapy, such as manual therapy combined with thermotherapy and electrotherapy, can usually inactivate painful MTrPs. Other situations, however, might necessitate dry needling. Such cases include the following: (1) poor response to conservative therapy, (2) intolerable pain, (3) deep location of an MTrP, rendering it inaccessible by conservative manual therapy, (4) unavailable or inadequate time to accept the time-consuming conservative therapy, (5) persistent pain or discomfort after complete elimination of the underlying pathological lesions that activate MTrPs, or (6) personal preference.

Mechanisms

Although dry needling is one of the most effective methods for rapid inactivation of an active MTrP, the mechanism of pain relief is still unclear. Recently, clinical and basic science research studies of MTrPs have been summarized (Hong 1996, 2000, 2002; Hong and Simons 1998; Simons et al. 1999). The animal model developed by Hong and Torigoe in 1994 has facilitated understanding of the nature of MTrPs (Hong and Simons 1998; Hong and Torigoe 1994). Based on evidence from animal studies as well as clinical observations, a model of multiple small MTrP loci in an MTrP region has been proposed (Fig. 1) (Hong 1994a). According to this model, there are two components of an ▶ MTrP locus, the sensory component, known as the ▶ sensitive locus or local twitch response locus (LTR locus), and the motor component, known as the ▶ active locus or endplate noise locus (EPN locus).

A sensitive locus (LTR locus) is the site from which local and referred pain and LTR can be elicited by mechanical stimulation (Fig. 2). A weak stimulation to the sensitive locus can elicit pain and sometimes referred pain, but a strong stimulation (high pressure with a needle tip) is usually required to elicit an LTR. If the MTrP is extremely hyperirritable (painful), the patient may consistently experience referred

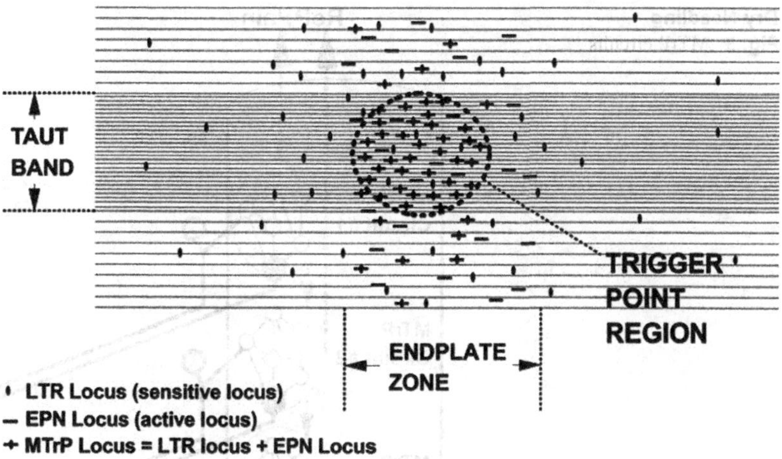

Dry Needling,
Fig. 1 Multiple MTrP loci
in an MTrP region

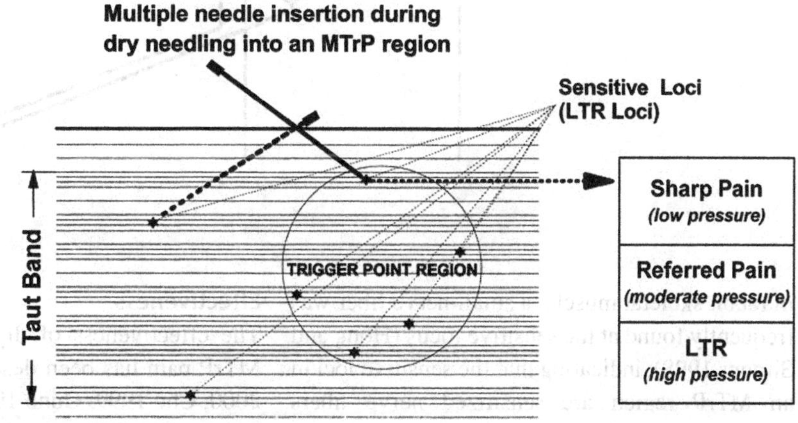

TAUT
BAND

TRIGGER
POINT
REGION

ENDPLATE
ZONE

• LTR Locus (sensitive locus)
— EPN Locus (active locus)
+ MTrP Locus = LTR locus + EPN Locus

Multiple needle insertion during
dry needling into an MTrP region

Sensitive Loci
(LTR Loci)

Taut Band

TRIGGER POINT REGION

Sharp Pain
(low pressure)

Referred Pain
(moderate pressure)

LTR
(high pressure)

Dry Needling,
Fig. 2 Characteristics of
a sensitive (LTR) locus

pain in addition to the local pain and LTRs can be elicited easily with a light snapping palpation. Based on the studies by Mense and Simons (1999), the ▶ referred muscle pain (from muscle to muscle) elicited following noxious stimulation to the sensitive loci in an MTrP region seems to be a consequence of central sensitization in the spinal cord. The consistent pattern of referred pain from the MTrP of a specific muscle indicates that the interneuronal connections among different dorsal horn neurons follow a fixed pattern, similar to the "meridian connections" of acupuncture points (Hong 2000). In both human and rabbit subjects, the EMG activity of an LTR can be recorded specifically from the muscle fibers of the taut band in response to stimulation

of the MTrP (Hong and Simons 1998; Hong and Torigoe 1994). The EMG activity of an LTR is diminished in a denervated human muscle or in a rabbit muscle following lidocaine block or transection of the innervating nerve. In the rabbit study, this activity disappeared temporarily after spinal cord transection during the spinal shock stage but was almost completely recovered following the spinal shock period, indicating that the LTR is mediated through the spinal cord. The ▶ MTrP circuit (Hong 2002) cannot only transfer the nociceptive impulses to the brain but may also control the referral pain patterns via connections to other MTrP circuits (Fig. 3). Furthermore, the occurrence of LTRs is mediated via the MTrP circuit. In a histological study of sensitive loci

Dry Needling,
Fig. 3 MTrP circuits

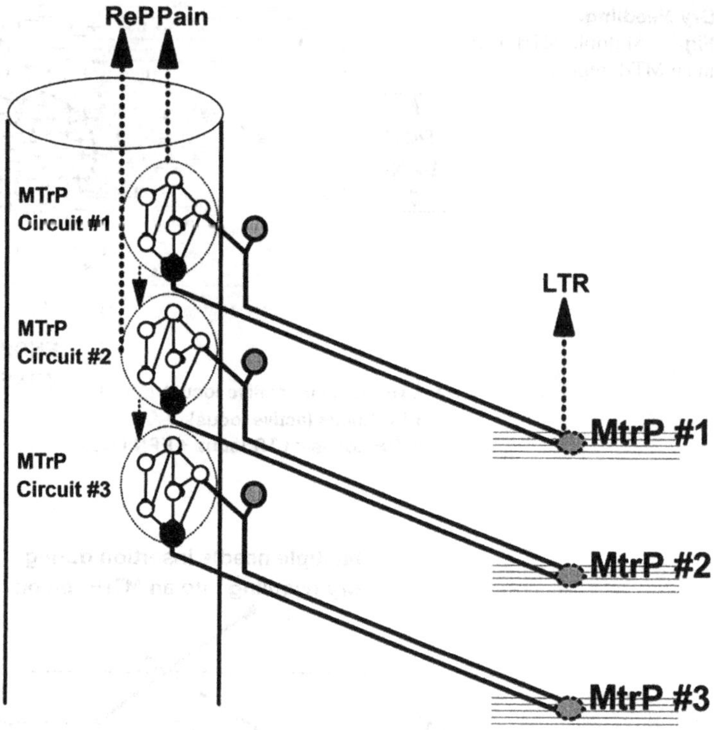

RePPain

MTrP
Circuit #1

MTrP
Circuit #2

MTrP
Circuit #3

LTR

MtrP #1

MtrP #2

MtrP #3

in rabbit skeletal muscle, a small nerve fiber was frequently found at the sensitive locus (Hong and Simons 1998), indicating that the sensitive loci in an MTrP region are sensitized nerve fibers (nociceptors).

An active locus (EPN locus) is the site from which spontaneous electrical activity, known as endplate noise (EPN), can be recorded. The active locus is in the immediate vicinity of a sensitive locus, with the MTrP locus comprising both active and sensitive loci. Simons has suggested that EPN is a consequence of excessive acetylcholine release, which may subsequently cause "taut band" formation (Fig. 4) (Simons et al. 1999). The contraction knots in an MTrP region appear to be directly responsible for the palpable nodule and the taut band of an MTrP (Simons et al. 1999). The resulting increase in energy consumption, combined with a reduction in the energy supply, may produce a local energy crisis evidenced by severe localized hypoxia (Fig. 4) (Hong and Simons 1998; Simons et al. 1999).

Effectiveness

The effectiveness of dry needling in relieving MTrP pain has been described in detail (Baldry 2000; Chu 1999; Gunn 1996; Hong 1994b, 2000; Lewit 1979; Melzack 1981; Simons et al. 1999), but no randomized controlled trials are available. By using a special technique of "fast-in, fast-out" needle movement (to generate high pressure) into multiple loci in an MTrP region, immediate pain relief can be achieved with local anesthetic injection or dry needling, but only if LTRs are elicited during needling (Hong 1994a, b). Since an MTrP can be inactivated without injection of any solution, the mechanical stimulation of the MTrP seems to be the most important factor for pain relief. It has been hypothesized that hyperstimulation analgesia "closes the gate" by disrupting reverberatory neural circuits (MTrP circuits) in the central nervous system and that this is the mechanism of pain relief for both dry needling and acupuncture (Hong 2002; Melzack 1981). Many authors have documented the similarity between acupuncture and dry needling in

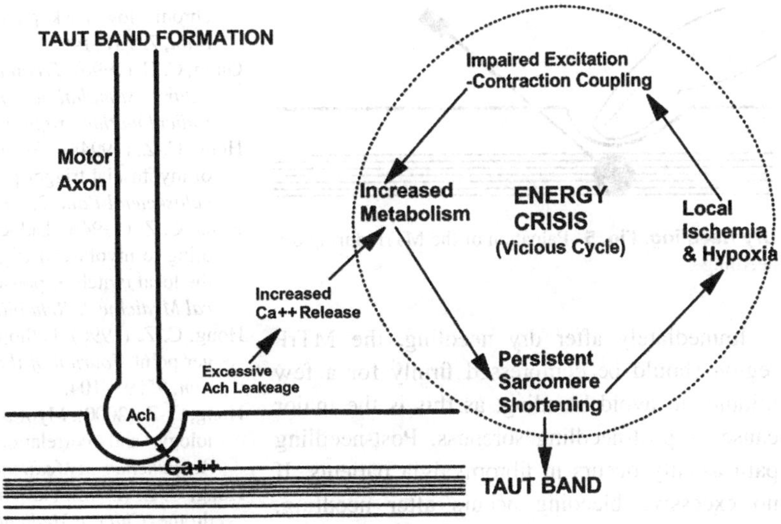

Dry Needling,
Fig. 4 Energy crisis in
a taut band

treating MTrP (Baldry 2000; Gunn 1996; Hong 1994a, 2000; Lewit 1979; Melzack 1981). The "Teh-Chi" effect, described by acupuncturists as an important sign for obtaining an optimal effect in acupuncture therapy, is similar to the sensation reported by patients when the needle tip approaches a sensitive locus in an MTrP region during MTrP injection. The patients being treated by dry needling or MTrP injection may also experience this sensation at the moment when an LTR is elicited. In fact, all MTrPs are acupuncture points, although many acupuncture points are not trigger points. Additionally, nociceptors are diffusely distributed in the subcutaneous tissues and may also connect to the MTrP circuit, although stimulation of subcutaneous nociceptors will not cause LTRs. This may explain the effectiveness of superficial dry needling (Baldry 2000).

Equipment

A hypodermal needle, an acupuncture needle, or an electromyographic (EMG) needle (Chu 1999) can be used for this procedure. The disadvantage of a hypodermal needle is the sharp cutting edge of the needle tip, which may actually damage muscle fibers. An acupuncture needle is too flexible and is difficult to manipulate during the multiple muscle insertions. An EMG needle is therefore the best selection, since it can avoid the disadvantages of the

other two. However, EMG needles (even disposable ones) are quite expensive.

Technique

A specific and very effective technique of dry needling has been recommended (Hong 1994b). This technique is similar to the procedure of MTrP injection. Gunn (1996) and Chu (1999) recommend a similar technique, but they fail to stress the importance of MTrP palpation during needling. Before dry needling, the exact spot (MTrP) that causes discomfort or pain should be carefully identified (pain recognition). Most of the time, patients can simply point to the painful spot with their finger. In severe cases, however, the pain may be diffuse and vague. Illustrations in certain textbooks may thus be needed to help locate MTrPs (Simons et al. 1999; Travell and Simons 1992).

During dry needling, it is important to identify the most painful region (MTrP site) by finger palpation with the hand not holding the needle (usually the non-dominant hand) (Fig. 5). Multiple rapid insertions of the needle into the MTrP site, without pulling the needle tip out from under the skin (i.e., the needle remains in the subcutaneous layer), are then performed. It is important to maintain an up-and-down motion of the needle, as a side-to-side motion can cut vessels or nerves. Moving the needle rapidly will also elicit LTRs more easily. LTRs should be elicited as often as possible by multiple needle insertions into the MTrP region.

Dry Needling, Fig. 5 Palpation of the MTrP during dry needling

Immediately after dry needling, the MTrP region should be compressed firmly for a few minutes to avoid bleeding, as this is the major cause of post-needling soreness. Post-needling pain usually occurs in fibromyalgia patients. If no excessive bleeding occurs after needling, application of thermotherapy after compression may help to reduce pain or soreness. A cold pack is usually unnecessary – and not recommended – unless the bleeding is massive. Cold temperatures can induce vasoconstriction, which may then lead to the impairment of local circulation and interfere with the resolution of any tissue damage caused by the needle.

Gunn uses an acupuncture needle (Intramuscular stimulation) (Gunn 1996), while Chu uses an EMG needle (▶ Twitch-Obtaining Intramuscular Stimulation) (Chu 1999) to perform dry needling. Both apply multiple insertions into the MTrP region. Baldry applies the needle into the subcutaneous, but not into muscle tissue (▶ Superficial dry needling) (Baldry 2000). Sometimes, the needle is inserted into a trigger point in a region other than the subcutaneous tissue or muscle (such as a ligament, tendon, and bursa).

Cross-References

▶ Acupuncture

References

Baldry, P. E. (2000). Superficial dry needling. In C. L. Chaitow (Ed.), *Fibromyalgia syndrome: A practitioner's guide to treatment*. Edinburgh: Churchill Livingston.
Chu, J. (1999). Twitch-obtaining intramuscular stimulation: Observation in the management of radiculopathic chronic low back pain. *Journal of Musculoskeletal Pain, 7*, 131–146.
Gunn, C. C. (1996). *Treatment of chronic pain. Intramuscular stimulation for myofascial pain of radiculopathic origin*. London: Churchill Livingston.
Hong, C.-Z. (1994a). Consideration and recommendation of myofascial trigger point injection. *Journal of Musculoskeletal Pain, 2*, 29–59.
Hong, C. Z, (1994b). Lidocaine injection versus dry needling to myofascial trigger point: The importance of the local twitch response. *American Journal of Physical Medicine & Rehabilitation, 73*, 256–263.
Hong, C.-Z. (1996). Pathophysiology of myofascial trigger point. *Journal of the Formosan Medical Association, 95*, 93–104.
Hong, C.-Z. (2000). Myofascial trigger points: Pathophysiology and correlation with acupuncture points. *Acupuncture in Medicine, 18*, 41–47.
Hong, C.-Z. (2002). New trends in myofascial pain syndrome. *Chinese Medical Journal, 65*, 501–512.
Hong, C.-Z., & Simons, D. G. (1998). Pathophysiologic and electrophysiologic mechanism of myofascial trigger points. *Archives of Physical Medicine and Rehabilitation, 79*, 863–872.
Hong, C.-Z., & Torigoe, Y. (1994). Electrophysiologic characteristics of localized twitch responses in responsive bands of rabbit skeletal muscle fibers. *Journal of Musculoskeletal Pain, 2*, 17–43.
Lewit, K. (1979). The needle effect in relief of myofascial pain. *Pain, 6*, 83–90.
Melzack, R. (1981). Myofascial trigger points: Relation to acupuncture and mechanism of pain. *Archives of Physical Medicine and Rehabilitation, 62*, 114–117.
Mense, S., & Simons, D. G. (1999). *Muscle pain: Understanding its nature, diagnosis, and treatment*. Baltimore: Williams & Wilkins.
Simons, D. G., Travell, J. G., & Simons, L. S. (1999). *Travell & Simons's myofascial pain and dysfunction: The trigger point manual* (2nd ed.). Baltimore: Williams & Wilkins. vol. I.
Travell, J. G., & Simons, G. (1992). *Myofascial pain and dysfunction: The trigger point manual* (Vol. II). Baltimore: Williams & Wilkins.

DSM, DSM-IV, DSM-IVR

▶ Diagnostic and Statistical Manual of Mental Disorders

DSM-IV

▶ Diagnostic and Statistical Manual of Mental Disorders

DSM-IVR

▶ Diagnostic and Statistical Manual of Mental Disorders

DSN

▶ Dejerine-Sottas Syndrome (Neuropathy)

DSPN

▶ Diabetic Neuropathies

DTI

▶ Diffusion Tensor Imaging

Dual-specificity Phosphatases (DUSP)

▶ Phosphatases and Glia

Ductus Arteriosus

Definition

Ductus arteriosus is a blood vessel in a fetus that bypasses pulmonary circulation by connecting the pulmonary artery directly to the ascending aorta. It is open in the fetus but normally closes after birth.

Cross-References

▶ NSAIDs, Adverse Effects
▶ NSAIDs and Their Indications

Dummy

▶ Placebo

Dura Mater

Definition

A tough, fibrous membrane that surrounds the brain and provides protection for the central nervous system.

Cross-References

▶ Diencephalic Mast Cells

Dural Puncture

Synonyms

Dural tap

Definition

A puncture of the dura mater is associated with a spinal nerve block and is a complication of epidural nerve block in about 1 in 100 patients. The consequence of a dural puncture can be postdural puncture headache (PDPH). The incidence of PDPH depends on a number of factors, including age, puncture technique, needle type, and size. As an example, the puncture hole resulting from an epidural needle of 16 gauge is much larger than after a deliberate puncture with a spinal needle of 25 or smaller gauge. PDPH requires sometimes treatment, which may involve an epidural blood patch.

Cross-References

▶ Postpartum Pain

References

Thew, M., & Paech, M. J. (2008). Management of postdural puncture headache in the obstetric patient. *Current Opinion in Anaesthesiology, 21*, 288–292.

Dural Sensory Receptors

▶ Nociceptors in the Orofacial Region (Meningeal/Cerebrovascular)

Dural Tap

▶ Dural Puncture

Duration of Ultrasound Treatment

Definition

The duration is 1–15 min. Usually, the duration should be at least 3 min.

Cross-References

▶ Ultrasound Therapy of Pain from the Musculoskeletal System

Duration Requirement

Definition

In order for a person to be eligible for disability benefits from the United States Social Security Administration, he or she must have a physical or mental impairment that is expected to result in death, or has lasted or can be expected to last for a continuous period of not less than 12 months.

Cross-References

▶ Disability Evaluation in the Social Security Administration

Durogesic®

Definition

Durogesic® is a transdermal patch system that continuously delivers fentanyl for 72 h.

Cross-References

▶ Postoperative Pain, Fentanyl

Durogesic® – Transdermal Patch

▶ Postoperative Pain, Fentanyl

DUSP1

▶ Phosphatases and Glia

DUSP6

▶ Phosphatases and Glia

DVT

▶ Deep Venous Thrombosis

Dying Child

▶ Cancer Pain, Palliative Care in Children

Dying-Back Neuropathy

▶ Distal Axonopathy

Dynamic Allodynia

Definition

Light brush evokes a sensation of pain.

Cross-References

▶ Allodynia (Clinical, Experimental)
▶ Hyperpathia, Assessment
▶ Neuropathic Pain Model, Tail Nerve Transection Model

Dynamic Ensemble

Definition

A conceptualization of pain mechanisms that suggests that the experience of pain is brought about by a variable network of interconnected regions, subject to dynamic influences of past history and reproductive status; compare with pain pathway.

Cross-References

▶ Gynecological Pain, Neural Mechanisms

Dynamic Mechanical Hyperalgesia

Definition

Unusually intense pain produced by moving a mechanical stimulus across the skin that normally produces only mild pain.

Cross-References

▶ Allodynia (Clinical, Experimental)
▶ Restless Legs Syndrome

Dynamic Pain

Definition

Pain provoked by movement, such as deep breathing or coughing, getting out of bed, or walking (during the postoperative period).

Cross-References

▶ Postoperative Pain, Acute Pain Management Principles
▶ Postoperative Pain, Acute Pain Team

Dynorphins

Definition

Peptides of several sizes, including dynorphin A and dynorphin B, which preferentially bind to KOP receptors. They are formed from prodynorphin and found through the central and peripheral nervous system, and may have a role in modulating pain sensation.

Cross-References

▶ Endogenous Opioid Peptides
▶ Nitrous Oxide Antinociception and Opioid Receptors
▶ Opiates During Development
▶ Peptides in Neuropathic Pain States

Dysaesthesia (Dysesthesia)

Definition

An unpleasant abnormal sensation, whether spontaneous or evoked

Cross-References

► Anesthesia Dolorosa Model, Autotomy
► Central Nervous System Stimulation for Pain
► Cervical Transforaminal Injection of Steroids
► Dorsal Root Ganglionectomy and Dorsal Rhizotomy
► Hyperpathia
► Hyperpathia, Assessment
► Metabolic and Nutritional Neuropathies
► Postherpetic Neuralgia, Etiology, Pathogenesis, and Management
► Postherpetic Neuralgia, Pharmacological and Nonpharmacological Treatment Options
► Satellite Cells
► Viral Neuropathies

Dysautonomic, Dysautonomia

Definition

Pertaining to pathophysiological changes of autonomic functions.

Cross-References

► CRPS-1 in Children

Dyschesia

► Pain Mechanisms in Endometriosis

Dyschezia

Definition

Difficult or painful evacuation of feces from the rectum generally associated with endometriosis of the rectovaginal septum.

Cross-References

Dyspareunia and Vaginismus

Dysesthesia, Assessment

Ellen Jørum
The Laboratory of Clinical Neurophysiology, Rikshospitalet University, Oslo, Norway

Synonyms

Abnormal sensation; Paresthesia; Unpleasant

Definition

An unpleasant abnormal sensation, whether spontaneous or evoked (Classification of chronic pain 1994).

Characteristics

Assessment of Dysesthesia
Assessment of Spontaneously Occurring Dysesthesia
Dysesthesia is frequently present in ► neuropathic pain (Lindblom and Verrillo 1979; Eide et al. 1996; Nasreddine and Saver 1997) but may also occur in non-neuropathic pain states (Aikens et al. 2003) and in ► epilepsy (Duchowny 1993). The mechanisms underlying dysesthesia are largely unknown. Since this chapter is dealing only with the assessment of dysesthesia, further

discussion of possible neurophysiological mechanisms will not be undertaken. The only possible way of assessing dysesthesia to ▶ spontaneous pain is to obtain thorough verbal information from the patient, where they are asked to describe in detail the various sensations they may experience. It may be difficult for some patients to decide whether a sensation is unpleasant or not and thereby sometimes difficult to decide whether one is dealing with dysesthesia or paresthesia. However, if patients are describing a purely painful sensation, dysesthesia will be the right term. The descriptors used by the patients may vary, probably even for the same experience, but since we are dealing with purely subjective sensations, a more objective evaluation is impossible. One may ask patients to describe the quality of the ▶ unpleasant sensation, or use one of a large variety of scales, where patients may choose from a list of pain quality descriptors. One frequently employed scale is the McGill Pain Questionnaire (MPQ), which is a well-validated and psychometrically reliable instrument consisting of 20 lists of pain descriptors (Melzack 1975). However, many other scales also exist. It may be worthwhile to assess not only the type of sensation but also the intensity of a dysesthetic sensation by, for example, using a visual analogue scale (VAS) from 0 to 10 cm or 0 to 100 mm, where 0 represents no pain and 10 or 100 the worst thinkable pain (or other intensity scales).

Assessment of Evoked Dysesthesia

One may assess evoked dysesthesia to ▶ tactile stimuli or ▶ thermal stimuli. Special cases of dysesthesia include hyperalgesia and allodynia, making the distinctions between the different phenomena difficult. There is normally no painful sensation to light touch. However, light touch may evoke a painful sensation in a patient with neuropathic pain. In daily life, patients may experience evoked dysesthesia in many situations, such as being touched by fellow passengers in a crowded bus and bed linen in bed. In many reports, there are no distinctions made between allodynia to light touch or to brush and dysesthesia (Jørum et al. 2003). However,

a dysesthetic sensation will always be an abnormal sensation, described by the patient as different from painful touch. There may be a large number of descriptors, which again may be assessed by asking the patient for a description or by letting the patient choose from various descriptors in one of many possible pain scales. The intensity of an evoked sensation may also be assessed by means of VAS (or another intensity scale). Dysesthesia to thermal stimuli, especially to cold stimuli, is a common finding in neuropathic pain (Jørum et al. 2003) and in more generalized pain like fibromyalgia (Berglund et al. 2001). In most studies where cold dysesthesia has been described, the sensation of cold is studied by quantitative sensory testing (QST), where the thresholds for cold pain are assessed (Berglund et al. 2001; Jørum et al. 2003). In the study by Berglund and coworkers, the perceived quality of each thermal stimulus was assessed by means of a list of preselected verbal descriptors. Patients with fibromyalgia not only had lower threshold of cold pain and cold tolerance (allodynia and hyperalgesia) but also used more pain-related descriptors than normal controls. A hyperfunction for heat pain and heat-pain tolerance was also found. Aberrant perceptions to cold stimulation were also reported (paradoxical heat, dysesthesia). In neuropathic pain, the sensation of cold is frequently reported as dysesthetic, as an unpleasant aberrant sensation, often a sensation of warmth (Jørum and Arendt-Nielsen 2003).

Dysesthesia is frequently reported in studies of neuropathic pain, but listed here are some examples of other conditions where the term may be encountered:

1. ▶ Vulvar dysesthesia is not an uncommon condition, frequently described in the literature and where spontaneous dysesthesia may occur. (Aikens et al. 2003; Reed et al. 2003) The dysesthesia is often characterized as a constant or intermittent burning sensation, but descriptors such as tender and sore have also been employed (Aikens et al. 2003).

2. Dysesthesia may occur in relation to epileptic seizures or as an adverse effect to vagal nerve stimulation (Akman et al. 2003).

In this report, the authors describe a patient who was treated for refractory complex partial epilepsy with vagal nerve stimulation, who reported complaints of intermittent and unpredictable "funny" feelings and tightness in his throat and stomach 4 months after the implantation of the vagal nerve stimulator. However, as the authors make clear, somatosensory symptoms, including various forms of dysesthesia, may occur in various forms of epilepsy, such as benign epilepsy of childhood with temporal spikes, where pharyngeal sensations of tingling, burning, constriction, or unpleasant sensations (dysesthesia) are described (Duchowny 1993).

References

Aikens, J. E., Reed, B., Gorenflo, D., et al. (2003). Depressive symptoms among women with vulvar dysesthesia. *American Journal of Obstetrics and Gynecology, 189*, 462–466.

Akman, C., Riviello, J. J., Madsen, J. R., et al. (2003). Pharyngeal dysesthesia in refractory complex partial epilepsy: New seizure or adverse effect of vagal nerve stimulation? *Epilepsia, 44*, 855–858.

Berglund, B., Harju, E.-L., Kosek, E., et al. (2001). Quantitative and qualitative perceptual analysis of cold dysesthesia and hyperalgesia in fibromyalgia. *Pain, 96*, 177–187.

Duchowny, M. (1993). Identification of the surgical candidates and timing of the operation: Overview. In E. Wyllie (Ed.), *The treatment of epilepsy: Principles and practice* (pp. 999–1008). Philadelphia: Lea & Fabiger.

Eide, P. K., Jørum, E., & Stenehjeme, A. (1996). Somatosensory findings in spinal cord injury patients with central dysesthesia pain. *Journal of Neurology, Neurosurgery, and Psychiatry, 60*, 411–415.

Jørum, E., & Arendt-Nielsen, L. (2003). Sensory testing and clinical neurophysiology. In Breivik, H., Campbell, W., & Eccleston, C. (Eds.), *Clinical Pain Management, Practical Applications and Procedures* (vol. 4, pp. 27–38). London: Hodder Arnold.

Jørum, E., Warncke, T., & Stubhaug, A. (2003). Cold allodynia and hyperalgesia in neuropathic pain: The effect of N-methyl- D aspartate (NMDA) receptor antagonist ketamine: A double-blind, cross-over comparison with alfentanil and placebo. *Pain, 101*, 229–235.

Lindblom, U., & Verrillo, R. T. (1979). Sensory functions in chronic neuralgia. *Journal of Neurology, Neurosurgery, and Psychiatry, 42*, 422–435.

Melzack, R. (1975). The McGill pain questionnaire: Major properties and scoring methods. *Pain, 23*, 345–356.

Nasreddine, Z. S., & Saver, J. L. (1997). Pain after thalamic stroke: Right diencephalic predominance and clinical features in 180 Patients. *Neurology, 48*, 1196–1199.

Reed, B. D., Haefner, H. K., & Cantor, L. (2003). Vulvar dysesthesia (vulvodynia). A follow-up study. *The Journal of Reproductive Medicine, 48*, 409–416.

Dysesthesias

Definition

An unpleasant abnormal sensation that may or may not be painful.

Cross-References

▶ Spinal Cord Injury Pain Model, Hemisection Model

Dysesthetic Vulvodynia

▶ Vulvodynia

Dysfunctional Pain and the International Classification of Function

Bengt H. Sjölund
Department of Community Medicine and Rehabilitation Medicine, Umea University, Umea, Sweden

Synonyms

Sensory impairment

Definition

Pain conditions due to disturbances of function in the nervous system. In a wider sense, all ongoing (chronic) pain can be considered as a sensory impairment, i.e., a disturbance of normal sensory nociceptive function (WHO 2001).

Characteristics

Often, there is not a traditional pathology in chronic pain conditions. For example, it was recently reported that 35 % of nearly 2,500 responders to a questionnaire in a southern rural/small town area in Sweden reported chronic musculoskeletal pain (Bergman et al. 2001) whereas thorough clinical screening of smaller cohorts showed that only 0.5 % fulfilled the criteria for rheumatoid arthritis (Simonsson et al. 1999) and 1.3 % those for fibromyalgia (Lindell et al. 2000).

How should we classify such "nonspecific" pain conditions? Several research groups have found that localized chronic musculoskeletal pain may be associated with sensory alterations in or close to the painful region, demonstrating a change of function in the sensory nervous system (Bendtsen et al. 1996; Graven-Nielsen et al. 2000; Kosek and Hansson 2002; Persson et al. 2003). It is not established at present whether such changes contribute to the pain but this may well be the case.

With the current official definition of neuropathic pain by the International Association for the Study of Pain (Merskey and Bogduk 1994), not only conditions due to primary lesions of the nervous system but also conditions due to dysfunction of the nervous system belong to this group. However, most pain management physicians would think of such pain conditions not as neuropathic in the sense secondary to structural nerve change but as disturbances in nervous system function. I have earlier suggested that pain conditions due to disturbances of function in the nervous system should be denoted dysfunctional pain (Sjölund 1994). These conditions do not really show "signs of organic disease" at all and may therefore be misinterpreted as "somatizing" but are analogous to, e.g., epilepsy.

To be able to interpret such chronic pain conditions, we should therefore include among the

Dysfunctional Pain and the International Classification of Function, Fig. 1 Sites of action of traditional pain therapies. Abbreviations: NSAIDs, nonsteroidal anti-inflammatory drugs; TENS, transcutaneous electrical nerve stimulation; SCS, spinal cord stimulation (Modified from Sjölund 2003)

nociceptive, neuropathic, and psychogenic pain categories a fourth main category, dysfunctional pain. The diagnosis of dysfunctional pain should be made whenever there are signs of central sensitization or central disinhibition, independent of type of afferent input. Likely examples of dysfunctional pain would be fibromyalgia (Graven-Nielsen et al. 2000), myofascial pain (Bendtsen et al. 1996), referred pain (Kosek and Hansson 2002), and centralization phenomena in neuropathic pain (Eide 2000; Flor et al. 2001). This proposal is in agreement with the new principle suggested for classification of pain syndromes, based on underlying pain mechanisms (Woolf et al. 1998).

Can we then adequately assess and manage a person in chronic pain after labeling the category of pain/the mechanism of pain? Most clinicians trying to manage chronic pain patients work very hard to temporarily prevent the nociceptive stimuli from reaching the conscious sensory level (sensorium; Fig. 1).

What about all those other factors important for that patient? How do we label the consequences of pain for that individual in a particular pain condition and in a particular setting? More specifically, is it possible with existing taxonomies to categorize the interpretation by the patient as well as the affective and behavioral consequences of that interpretation for a specific person in a certain context? As illustrated in Fig. 2, the awareness of painful stimuli immediately gives rise to an interpretative process where previous experience and environmental cues as well as affective aspects play

Dysfunctional Pain and the International Classification of Function, Fig. 2 A comprehensive model of chronic pain (Modified from Sjölund 2003)

Dysfunctional Pain and the International Classification of Function, Fig. 3 Pain versus health. The ICF components (*italics*) integrated into the comprehensive model of chronic pain (Modified from Sjölund 2003)

important roles. This interpretation thereby becomes heavily influenced by ▶ personal factors. These cognitive processes also form the basis for the resulting motor output, i.e., behaviors. In addition, the behaviors immediately influence the sensorium anew, since the occurrence and severity of pain is usually influenced by body movement, giving rise to phenomena like fear-avoidance. The present lack of a more generalized taxonomy for a person in chronic pain hampers the choice of treatment strategies, since the analysis may be limited to factors less relevant for the individual case.

A more general and suitable system to describe pain and related influences on health is now available in the form of the international classification of function (▶ ICF) (WHO 2001) that is basically a component of a health system and describes functioning on the personal level. The new classification has two parts: One is denoted functioning and disability, where functioning is an umbrella term encompassing all body functions, activities, and participation and disability serves as an umbrella term for impairments, ▶ activity limitations, and ▶ participation restriction. The other part, ▶ contextual factors forms the complete background of an individual. The term "body functions" refers to the physiological functions of the body systems including psychological function. Impairments are problems in body function and structure such as significant deviation or loss. Activity is the execution of a task or action by an individual, and activity limitations are difficulties an individual

may have in executing activities. Participation is involvement in a life situation and participation restrictions are problems an individual may experience in involvement in such situations. In addition, the ICF contextual factors are environmental factors, that make up the physical, social, and attitudinal environment in which people live and conduct their lives and personal factors that are internal influences and attributes, such as race, gender, lifestyle, education, path, and current experiences. These components have multiple interactions.

To facilitate use of the ICF for the taxonomy and management of chronic pain, I have proposed that long-standing pain in general (>3–6 months duration) should be classified as a significant deviation from normal function of the nociceptive transmission and thereby be seen as a severe sensory impairment (Sjölund 2003). This implementation enables us to use the ICF classifications to describe pain-related activity limitations and participation restrictions, both of which are cardinal features of the much used but officially denounced "chronic pain syndrome" (Sanders 1999) and of the much debated (somatoform) pain disorder (APA 1994). As can be seen in Fig. 3, impairments, activity limitations, and participation restrictions act in an integrated fashion to explain the complex resulting picture in the sensory/interpretative/behavioral clinical presentation of long-standing pain. It may be much more meaningful (and true!) to describe a person in chronic pain with a combination of one or several of the four main pain categories mentioned previously and the ICF, rather than speculating if a particular organ system is affected by an uncertain pathology in a painful region and adding "pain behaviors" (Sanders 1999), "somatization/pain disorder" (APA 1994), or an empirical psychological characterization (Rudy et al. 1995).

For therapeutic strategies, it is easier to define indications for pain management by deriving them from the ICF approach, assessing the pain category and its characteristics (the sensory impairment) as well as the activity limitations and participation restrictions in that individual in his/her context and to focus pain management on those persons who are distinctly limited or restricted by their pain. The planning and goal setting of pain management can be facilitated by directly asking, "How can we reduce the sensory impairment? Can we remove the cause of chronic pain? Can we achieve temporary or permanent pain relief? How do we compensate for activity limitations and focus on specific participation restrictions in our rehabilitation efforts? How do we best consider environmental and personal factors? Such a structured and individualized ICF-related assessment may allow pain management to be tested in a more meaningful scientific manner from now on.

References

APA. (1994). *Diagnostic and statistical manual of mental disorders*. Washington, DC: American Psychiatric Association.

Bendtsen, L., Jensen, R., & Olesen, J. (1996). Qualitatively altered nociception in chronic myofascial pain. *Pain, 65*, 259–264.

Bergman, S., Herrstrom, P., Hogstrom, K., et al. (2001). Chronic musculoskeletal pain, prevalence rates, and sociodemographic associations in a Swedish population study. *The Journal of Rheumatology, 28*, 1369–1377.

Eide, P. (2000). Wind-up and the NMDA receptor complex from a clinical perspective. *European Journal of Pain, 4*, 5–17.

Flor, H., Denke, C., Schäfer, M., et al. (2001). Sensory discrimination training alters both cortical reorganization and phantom limb pain. *Lancet, 357*, 1763–1764.

Graven-Nielsen, T., Aspegren Kendall, S., et al. (2000). Ketamine reduces muscle pain, temporal summation, and referred pain in fibromyalgia patients. *Pain, 85*, 483–491.

Kosek, E., & Hansson, P. (2002). The influence of experimental pain intensity in the local and referred pain area on somatosensory perception in the area of referred pain. *European Journal of Pain, 6*, 413–425.

Lindell, L., Bergman, S., Petersson, I. F., et al. (2000). Prevalence of fibromyalgia and chronic widespread pain. *Scandinavian Journal of Primary Health Care, 18*, 149–153.

Merskey, H., & Bogduk, N. (1994). *Classification of chronic pain. Descriptions of chronic pain syndromes and definitions of pain terms*. Seattle: IASP Press.

Persson, A. L., Hansson, G. A., Kalliomaki, A., et al. (2003). Increases in local pressure pain threshold after muscle exertion in women with chronic shoulder pain. *Archives of Physical Medicine and Rehabilitation, 84*, 1515–1522.

Rudy, T., Turk, D., Kubinski, J. A., et al. (1995). Differential treatment responses of TMD patients as a function of psychological characteristics. *Pain, 61*, 103–112.

Sanders, S. (1999). Clinical practise guidelines for chronic non-malignant pain syndromes. *Journal of Back and Musculoskeletal Rehabilitation, 5*, 115–120.

Simonsson, M., Bergman, S., Jacobsson, L. T., et al. (1999). The prevalence of rheumatoid arthritis in Sweden. *Scandinavian Journal of Rheumatology, 28*, 340–343.

Sjölund, B. H. (1994). Chronic pain in society -a case for chronic pain as a dysfunctional state? *Quality of Life Research, 3*, 5–9.

Sjölund, B. H. (2003). Current and future treatment strategies for chronic pain. In N. Soroker & H. Ring (Eds.), *Advances in physical and rehabilitation medicine* (pp. 307–313). Bologna: Monduzzi Editore.

WHO. (2001). *International classification of functioning. Disability and health* (pp. 1–209). Geneva: World Health Organization.

Woolf, C. J., Bennett, G. J., Doherty, M., et al. (1998). Towards a mechanism-based classification of pain? *Pain, 77*, 227–229.

Dysfunctional Segmental Motion

▶ Chronic Back Pain and Spinal Instability

Dysmenorrhea

▶ Gynecological Pain and Sexual Functioning
▶ Gynecological Pain, Neural Mechanisms
▶ Pain Mechanisms in Endometriosis

Dysnosognosia

▶ Hypochondriasis, Somatoform Disorders, and Abnormal Illness Behavior

Dyspareunia

Definition

Dyspareunia is defined as pain perceived within the pelvis associated with penetrative sex.

It tells us nothing about the mechanism and may be applied to the female or male. It is usually applied to penile penetration, but is often associated with pain during insertion of any object. It can also be associated with anal intercourse, which is called anodyspareunia. It is classically subdivided into superficial and deep.

Cross-References

▶ Dyspareunia and Vaginismus
▶ Gynecological Pain and Sexual Functioning
▶ Gynecological Pain, Neural Mechanisms
▶ Myalgia
▶ Pain Mechanisms in Endometriosis
▶ Visceral Pain Models, Female Reproductive Organ Pain

Dyspareunia and Vaginismus

Yitzchak M. Binik, Marie Faaborg-Andersen and Marie Andree Lahaie
McGill University, Montreal, QC, Canada

Synonyms

Chronic pelvic pain; Coital pain; Genito-pelvic pain/penetration disorder; Provoked vestibulodynia; Sexual pain disorder; Urogenital pain; Vulvodynia

Definition

The terms dyspareunia and vaginismus refer to diagnoses which were first proposed in the nineteenth century (Binik 2010a). This entry will explore recent research which challenges the reliability and validity of these diagnoses and has paved the way for a new diagnosis to appear in the DSM-5 as well as for new therapeutic interventions.

Introduction

Dyspareunia is defined in the DSM-IV as "recurring genital pain associated with sexual activity in either a male or a female" (APA 2000). Most classification systems and practitioners consider dyspareunia to be a sexual dysfunction. This classification, however, is problematic as it has resulted in a focus on the "association with sexual activity" rather than the main symptom of dyspareunia, the pain (Binik 2010a). While the majority of pain disorders are classified by the characteristics and mechanisms of the pain, dyspareunia is defined by the primary activity with which it interferes, sexual intercourse (Binik 2010a). The genital pain of dyspareunia can, however, also be experienced during nonsexual activities such as tampon insertion, gynecological examinations, urination, sitting, and sports and may precede first sexual experiences (e.g., tampon insertion in adolescence). As a result, Binik (2010a) proposed that dyspareunia be reclassified as a pain disorder, rather than as a sexual dysfunction or a simple manifestation of underlying physical pathology.

The previous usage of the term "dyspareunia" was inconsistent and vague and was often associated with a number of poorly defined urogenital pain syndromes (e.g., provoked vestibulodynia [PVD; previously termed vulvar vestibulitis], vulvodynia) and with pain symptoms presumably linked to physical pathology. In addition, dyspareunia was described in a variety of ways (e.g., superficial, deep, localized, generalized, postpartum). This failure of the DSM-IV and other classifications to differentiate between potentially different subtypes of dyspareunia resulted in women with vastly different symptoms being given identical diagnoses.

Traditionally, dyspareunia has been considered a woman's health problem. However, it has recently become apparent that recurrent genital pain also exists in males. In men, dyspareunic pain is currently referred to as urological chronic pelvic pain syndrome (UCCPS; Davis et al. 2011a). Men with UCCPS most often experience pain in their penis, perineum, and testicles, and the pain can be associated with both sexual (ejaculation) and nonsexual activities (sitting, sports; Davis et al., in press).

Characteristics

Dyspareunia

Epidemiology
The prevalence of dyspareunia in women varies depending on the definition used and the population sampled. A recent epidemiological study found that 9.3 % of women between the ages of 20 and 60 suffered from severe and prolonged genital pain during sexual activity, with women between the ages of 20 and 29 having the highest incidence at 13 % (Danielsson et al. 2003). This is likely, however, to be an overestimate of the number of women seeking treatment for dyspareunia, as the level of distress pertaining to dyspareunia was not taken into account. Despite the fact that dyspareunia appears to be highly prevalent across the life span, many women never seek treatment, and of those who do, many are never formally diagnosed (Harlow et al. 2001). Current studies examining the prevalence of male dyspareunia are preliminary but suggest that it is less prevalent in males than in females (Clemens et al. 2005).

Assessment
The primary step in evaluating patients with dyspareunia is conducting a thorough multidisciplinary assessment of the pain and its resulting interference with the individual's quality of life. Assessment of the pain component should include specific information on its history, location, quality, intensity, exacerbating and ameliorating factors, temporal pattern, interference, and potential underlying pathology. Evaluation of the impact of the pain on sexual functioning and on personal well-being is crucial to treatment planning. Additionally, the couple's relationship dynamic, such as partner responses to pain, should be evaluated when relevant (Rosen et al. 2012). A comprehensive gynecological examination can be useful to screen for biological causes, such as genital abnormalities or infections. Depending on the history and interview, this examination might include an inspection and cotton swab evaluation of the vulvar area; laboratory tests to investigate bacterial or viral infection; assessment for atrophic vaginitis; palpation and examination of the uterus, adnexa, urethra, bladder, and surrounding structures; evaluation of the pelvic floor musculature; colposcopy; and laparoscopy.

In general, the clinician should attempt to replicate the quality of the pain reported by the patient during the manual examination. It is crucial to keep in mind that the presence of physical pathology may or may not be closely related to the experienced dyspareunia. Additionally, the ability of clinicians to differentiate reliably between different subtypes of dyspareunia is questionable (Binik 2010b).

Physical Factors Associated with Dyspareunia
There is a wide range of underlying physical pathology associated with dyspareunia. Deep dyspareunia, characterized by diffuse pain or localized tenderness, is frequently associated with chronic pelvic pain and pelvic pathology. Some reported causes of deep dyspareunia are pelvic inflammatory disease, endometriosis, ovarian pathology, fixed uterine retroversion, and inflammatory bowel disease. It is often the case, however, that correcting the presumed underlying pathology does not alleviate the dyspareunia. Studies examining pelvic floor muscle function in women have found higher pain responsiveness and pelvic floor hypertonus in women with PVD as compared with controls (Rosenbaum and Owens 2008). Similar results have been found for male dyspareunia, as men

with UCCPS showed increased pelvic floor resting activity (Davis et al. 2011b). Superficial dyspareunia refers to pain localized at the entry of the vagina, which can be associated with vaginal infections, scar tissue in an episiotomy repair, vaginitis, vulvo-vaginal atrophy, PVD, interstitial cystitis, and urethral conditions. Additionally, both men and women with dyspareunia have been found to have lower tactile and pain thresholds on genital sites as well as on nongenital sites, such as the deltoid (Davis et al. 2011a; Pukall et al. 2005). However, most cases of dyspareunia in men and woman are not associated with physical pathology (Meana and Binik 2011).

Psychosocial Factors Associated with Dyspareunia

The psychosocial factors associated with dyspareunia vary depending on the age of onset, duration, course, and subtype of dyspareunia. Not surprisingly, the psychosocial profile of a 25-year-old single woman suffering from dyspareunia due to PVD will be different from that of a 55-year-old married woman who just started experiencing genital pain. Recent studies examining the psychosocial factors associated with both male and female dyspareunia found that people with recurrent genital pain show more psychological distress (e.g., increased interpersonal sensitivity, depression, anxiety, and phobic avoidance), sexual dysfunction, and relationship problems (Meana and Binik 2011; Gonen et al. 2005). Studies examining partner responses to PVD pain have found that varying partner responses are associated with either increased or reduced pain and sexual satisfaction (Rosen et al. 2012). In terms of sexual dysfunction, women with dyspareunia have lower frequencies of intercourse, lower levels of sexual desire and arousal, and greater difficulty achieving orgasms through oral stimulation and intercourse. Women suffering from PVD have been found to display hypervigilance for pain relevant information and to catastrophize about their pain, which may result in an increased attention to pain and a diversion away from sexual stimuli (Payne et al. 2005). Other factors that have been found to be associated with PVD include the following: depression,

physical abuse (Harlow and Stewart 2003), early puberty, pain with first tampon use (Harlow et al. 2001), and increased state and trait anxiety (Payne et al. 2005). Sexual abuse has been hypothesized to result in dyspareunia; however, current research has questioned this hypothesis (Reed 2009).

Treatment for Dyspareunia

Given that dyspareunia is often conceptualized as resulting from either a medical or psychosexual condition, most treatments have been unidisciplinary. A wide variety of psychological, medical, and surgical procedures have been used to treat dyspareunia, producing varied effects.

Even though the central complaint of women with dyspareunia is pain, the vast majority of present therapeutic strategies focus on presumed underlying physical pathologies or psychosexual mechanisms. For example, estrogen replacement therapy is a common medical treatment for postmenopausal women suffering from dyspareunia (Kao et al. 2008). For dyspareunia resulting from PVD, common treatments include corticosteroids, anesthetic topical creams, antidepressant medications, vestibulectomy, cognitive-behavioral pain management (CBT), and pelvic floor physical therapy including biofeedback. There have been five randomized controlled double-blind placebo studies examining the efficacy of pharmacological interventions in treating PVD. None of the drugs tested were found to be significantly more effective at alleviating pain than placebo (see Landry et al. 2008; Petersen et al. 2009; Foster et al. 2010).

Additionally, three studies evaluating the efficacy of CBT as a treatment for PVD have been conducted, and it has been shown to be effective in reducing pain (Weijmar Schultz et al. 1996; Bergeron et al. 2001; Masheb et al. 2009). In an RCT, Bergeron and colleagues (2008) found that vestibulectomy was, however, more effective at reducing pain than CBT. Based on available data, vestibulectomy is considered the most effective treatment for PVD. However, it has been proposed that multidisciplinary interventions including CBT, physical therapy, and sex therapy be attempted prior to surgical interventions (Bergeron et al. 2001).

Vaginismus

Definition of Vaginismus

Vaginismus has been defined as "recurrent or persistent involuntary muscle spasms of the outer third of the vagina which interfere with sexual intercourse" (APA 2000). This definition has recently been questioned (Binik 2010b) since only about 28 % of women with this diagnosis appear to experience spasm. Recent definitions (e.g., Basson et al. 2004) suggest that the inability to experience desired vaginal penetration during intercourse is the key feature. The reasons for this inability may include spasm or hypertension but also include fear of pain/penetration, actual experienced pain, or other emotional states including disgust. Women who suffer from vaginismus can have diverse clinical presentations, with some experiencing penetration difficulties only related to sexual intercourse and others demonstrating interference in all activities involving vaginal penetration (e.g., tampon insertion, gynecological examinations).

Epidemiology

Although vaginismus is a common problem reported to gynecologists and sex therapists, the population prevalence in Western societies is low. For example, Christensen et al. (2011) found that 0.4 % of a representative sample of the Danish population experience distress regarding their vaginismus (Christensen et al. 2011).

Assessment

Evaluating the presence of vaginismus is complicated by the lack of empirically supported diagnostic criteria. Assessment involves a multidisciplinary team to evaluate both physical and psychosocial factors. Physical factors should be evaluated by a gynecologist using an educational pelvic examination (EPE). An EPE involves preparing the patient for what to expect and providing her with relaxation mechanisms to reduce her anxiety both prior to and during the EPE. The ultimate goals of the EPE are to rule out any organic comorbidity and to educate the patient about her anatomy. A successful EPE can also lead to feelings of achievement and reassurance and can be a useful first step in

treating vaginismus. Psychologists should assess for couple conflict, untreated trauma, and any sexual difficulties in the partner as these factors can also interfere with treatment. Additionally, fear and anxiety concerning pain and penetration-related activities should be assessed. Finally, structured measurements, such as the Vaginal Penetration Cognition Questionnaire (Klaassen and Ter Kuile 2009) and the sex subscale of the Disgust Scale (Haidt et al. 1994), can be used to assess concepts such as genital incompatibility cognitions and disgust.

Physical Factors Associated with Vaginismus

While previous definitions identified vaginal spasms as a primary diagnostic criterion for vaginismus, recent findings (Reissing et al. 2004) have demonstrated that these spasms are not characteristic of vaginismus. Additionally, EMG studies do not support the idea of a vaginal spasm mechanism in vaginismus (Binik 2010b). Other suggested physical causes of vaginismus include hymenal abnormalities, PVD, vulvovaginal atrophy, endometriosis, infections, and vaginal lesions (Reissing 2009). However, a recent review suggests that somatic explanations for vaginismus exist only in up to 5 % of cases (Van Lankveld et al. 2010).

Psychosocial Factors Associated with Vaginismus

Numerous psychosocial factors (e.g., history of sexual abuse, religious orthodoxy, negative sexual attitudes, dyspareunia) have been implicated in the etiology of vaginismus. There is, however, little empirical evidence to support these factors. Reissing et al. (2003) found that more women with vaginismus reported a history of childhood sexual interference and less positive attitudes regarding sexuality compared to women with dyspareunia and controls; however, the differences were small and need to be replicated. Currently, fear of pain and vaginal penetration are believed to be key factors associated with vaginismus (Reissing 2009). These fears appear to result in an elevated level of "genital incompatibility" cognitions (e.g., I am afraid my vagina is too narrow for penetration) as compared to controls and women with dyspareunia

Ethnomethodology

(Klaassen and Ter Kuile 2009). Finally, recent studies have found elevated levels of disgust in women suffering from vaginismus, suggesting that vaginismus may be, in part, a disgust-induced defense response (Borg et al. 2010).

Treatment for Vaginismus

Currently, the most popular treatment for vaginismus is cognitive-behavioral therapy. This treatment strategy focuses on erroneous cognitive beliefs, conditioned vaginal spasms, Kegel exercises, and vaginal desensitization through the use of vaginal dilators. Although this treatment regimen is widely accepted as effective, only one randomized controlled treatment study has evaluated this approach. Van Lankveld and colleagues (2007) compared a CBT intervention to a wait-list control group of women and found CBT to result in 18 % of participants being able to engage in sexual intercourse posttreatment as compared to none of the wait-list controls. While an 18 % success rate reflects a clinically significant result, it does not solve the problem for most women. Ter Kuile and colleagues (2009) innovated an exposure-based treatment based on the fear of pain or vaginal penetration mechanism of vaginismus. This treatment consists of exposure to vaginal penetration focused specifically on the fear of penetration and consists of 3-h long sessions within a 2-week period taking place in a hospital. These exposure sessions are similar to the typical treatment for specific phobias, which require massed in vivo exposure (e.g., Barlow 2002). During these sessions, the patient and their partner, assisted by a therapist, engage in gradual penetration exercises, starting with the client simply touching their vulva, and moving towards the insertion of fingers and dilators, and the ultimate goal of penetration. Therapist-aided exposure is completely controlled by the patient; however, both the therapist and the partner play active roles in reducing anxiety by suggesting relaxation techniques and providing support when needed. The results of this intervention were impressive in a preliminary controlled study (Ter Kuile et al., 2009). These promising findings appear to have been replicated in a submitted RCT (Ter Kuile, unpublished).

Summary

A Reconceptualization of Dyspareunia and Vaginismus in the DSM-5.

The DSM-IV (APA 2000) classified dyspareunia and vaginismus as separate disorders. However, women suffering from dyspareunia and vaginismus have many common characteristics, such as pain during, fear of, and difficulty with vaginal penetration. In fact, the comorbidity of vaginismus and vulvar pain ranges from 40 to almost 100 % (Binik 2010b). Moreover, health practitioners (i.e., gynecologists, physical therapists, psychologists) cannot reliably differentiate vaginismus from dyspareunia (Reissing et al. 2004). This data led Binik (2010a, b) to propose that the diagnoses of vaginismus and dyspareunia be merged into a single, new diagnosis called genito-pelvic pain/penetration disorder. This new diagnosis was included in the DSM 5 (American Psychiatric Association, 2013). It is assessed based on the percentage of successful penetration attempts, the pain experienced during vaginal penetration or penetration attempts, the significant fear of penetration and/or of genito-pelvic pain during penetration, and pelvic floor muscle dysfunction. This new diagnosis attempts to eliminate the current difficulty in making differential diagnoses by focusing the clinician on the relevant symptom dimensions.

There is currently no comprehensive biopsychosocial model which can account for the phenomena of dyspareunia and vaginismus (genito-pelvic pain/penetration disorder). However, the fear-avoidance model (FAM; Vlaeyen and Linton 2000) does address the psychosocial and behavioral aspects, as well as the pain, and is consistent with the fear mechanism previously discussed. It suggests that people may have an initial experience of pain during sex, which can lead to catastrophizing about the "threatening" experience. This, in turn, will result in a fear of painful sex, which gives way to hypervigilance or avoidance of behaviors specific to the fearful thoughts. The hypervigilance will result in an exaggerated attention to the physical sensations of pain during intercourse and will reinforce and exacerbate the fear of pain. This chain of events, taken together, will result in the perceived inability to achieve penetration. While the FAM does present a plausible model of genito-pelvic pain/ penetration disorder, it does not take into consideration possible biological factors. Future research should attempt to incorporate these biological factors with a view to creating a truly biopsychosocial model.

References

American Psychiatric Association. (2000). *Diagnostic and statistical manual of mental disorders* (4th ed.). Washington, DC: American Psychiatric Association.

American Psychiatric Association. (2013). *Diagnostic and statistical manual of mental disorders* (5th ed.). Washington, DC: Author.

Barlow, D. H. (2002). *Anxiety and its disorders: The nature and treatment of anxiety and panic* (2nd ed.). New York: Guilford Press.

Basson, R., Leiblum, S., Brotto, L., Derogatis, L., Fourcroy, J., Fugl-Meyer, K., & Schultz, W. W. (2004). Revised definitions of women's sexual dysfunction. *The Journal of Sexual Medicine, 1*(1), 40–48.

Bergeron, S., Binik, Y. M., Khalifé, S., Pagidas, K., Glazer, H. I., Meana, M., & Amsel, R. (2001). A randomized comparison of group cognitive-behavioral therapy, surface electromyographic biofeedback, and vestibulectomy in the treatment of dyspareunia resulting from vulvar vestibulitis. *Pain, 91*(3), 297–306.

Bergeron, S., Khalife, S., Glazer, H. I., & Binik, Y. M. (2008). Surgical and behavioral treatments for vestibulodynia: Two-and-one-half year follow-up and predictors of outcome. *Obstetrics and Gynecology, 111*(1), 159–166.

Binik, Y. M. (2010a). The DSM diagnostic criteria for dyspareunia. *Archives of Sexual Behavior, 39*(2), 292–303.

Binik, Y. M. (2010b). The DSM diagnostic criteria for vaginismus. *Archives of Sexual Behavior, 39*(2), 278–291.

Borg, C., de Jong, P. J., & Weijmar Schultz, W. (2010). Vaginismus and dyspareunia: Automatic vs deliberate disgust responsivity. *The Journal of Sexual Medicine, 7*(6), 2149–2157.

Christensen, B. S., Grønbæk, M., Osler, M., Pedersen, B. V., Graugaard, C., & Frisch, M. (2011). Sexual dysfunctions and difficulties in Denmark: Prevalence and associated sociodemographic factors. *Archives of Sexual Behavior, 40*(1), 121–132.

Clemens, J. Q., Meenan, R. T., O'Keeffe Rosetti, M. C., Gao, S. Y., & Calhoun, E. A. (2005). Incidence and clinical characteristics of National Institutes of Health type III prostatitis in the community. *Journal of Urology, 174*(6), 2319–2322.

Danielsson, I., Sjoberg, I., Stenlund, H., & Wikman, M. (2003). Prevalence and incidence of prolonged and severe dyspareunia in women: Results from a population study. *Scandinavian Journal of Public Health, 31*(2), 113–118.

Davis, S. N. P., Binik, Y. M., Amsel, R., & Carrier, S. (in press). A subtypes based analysis of urological chronic pelvic pain syndrome in men. *Journal of Urology.*

Davis, S. N. P., Maykut, C. A., Binik, Y. M., Amsel, R., & Carrier, S. (2011a). Tenderness as measured by pressure pain thresholds extends beyond the pelvis in chronic pelvic pain syndrome in men. *The Journal of Sexual Medicine, 8*(1), 232–239.

Davis, S. N. P., Morin, M., Binik, Y. M., Khalife, S., & Carrier, S. (2011b). Use of pelvic floor ultrasound to assess pelvic floor muscle function in urological chronic pelvic pain syndrome in men. *Journal of Sexual Medicine, 8*(11), 3173–3180.

Foster, D. C., Kotok, M. B., Huang, L. S., Watts, A., Oakes, D., Howard, F. M., & Dworkin, R. H. (2010). Oral desipramine and topical lidocaine for vulvodynia: A randomized controlled trial. *Obstetrics and Gynecology, 116*(3), 583.

Gonen, M., Kalkan, M., Cenker, A., & Ozkardes, H. (2005). Prevalence of premature ejaculation in Turkish men with chronic pelvic pain syndrome. *Journal of Andrology, 26*(5), 601–603.

Haidt, J., McCauley, C., & Rozin, P. (1994). Individual differences in sensitivity to disgust: A scale sampling seven domains of disgust elicitors. *Personality and Individual Differences, 16*(5), 701–713.

Harlow, B. L., & Stewart, E. G. (2003). A population-based assessment of chronic unexplained vulvar pain: have we underestimated the prevalence of vulvodynia? *Journal of the American Medical Women's Association, 58*(2), 82–88.

Harlow, B. L., Wise, L. A., & Stewart, E. G. (2001). Prevalence and predictors of chronic lower genital tract discomfort. *American Journal of Obstetrics and Gynecology, 185*(3), 545–550.

Kao, A., Binik, Y. M., Kapuscinski, A., & Khalife, S. (2008). Dyspareunia in postmenopausal women: A critical review. *Pain Research & Management, 13*(3), 243–254.

Klaassen, M., & Ter Kuile, M. M. (2009). Development and initial validation of the vaginal penetration cognition questionnaire (VPCQ) in a sample of women with vaginismus and dyspareunia. *The Journal of Sexual Medicine, 6*(6), 1617–1627.

Landry, T., Bergeron, S., Dupuis, M. J., & Desrochers, G. (2008). The treatment of provoked vestibulodynia: A critical review. *The Clinical Journal of Pain, 24*(2), 155–171.

Masheb, R. M., Kerns, R. D., Lozano, C., Minkin, M. J., & Richman, S. (2009). A randomized clinical trial for women with vulvodynia: Cognitive-behavioral therapy vs. supportive psychotherapy. *Pain, 141,* 31–40.

Meana, M., & Binik, Y. M. (2011). Dyspareunia: Causes and treatments (Including Provoked Vestibulodynia). In P. Vercellini (Ed.), *Chronic pelvic pain* (1st ed., pp. 125–136). Oxford, UK: Blackwell.

Payne, K. A., Binik, Y. M., Amsel, R., & Khalifé, S. (2005). When sex hurts, anxiety and fear orient attention towards pain. *European Journal of Pain, 9*(4), 427–436.

Petersen, C. D., Giraldi, A., Lundvall, L., & Kristensen, E. (2009). Botulinum toxin type A-a novel treatment for provoked vestibulodynia? Results from a randomized, placebo controlled, double blinded study. *The Journal of Sexual Medicine, 6*(9), 2523–2537.

Pukall, C. F., Strigo, I. A., Binik, Y. M., Amsel, R., Khalifé, S., & Bushnell, M. C. (2005). Neural correlates of painful genital touch in women with vulvar vestibulitis syndrome. *Pain, 115*(1), 118–127.

Reed, B. D. (2009). Dyspareunia and sexual/physical abuse. In *Female sexual pain disorders: Evaluation and management* (pp. 213–217). Oxford: Blackwell.

Reissing, E. D. (2009). Vaginismus: Evaluation and management. In *Female sexual pain disorders: Evaluation and management* (pp. 229–234). Oxford: Blackwell.

Reissing, E. D., Binik, Y. M., Khalife, S., Cohen, D., & Amsel, R. (2003). Etiological correlates of vaginismus: Sexual and physical abuse, sexual knowledge sexual self-schema and relationship adjustment. *Journal of Sex & Marital Therapy, 29*, 47–59.

Reissing, E. D., Binik, Y. M., Khalifé, S., Cohen, D., & Amsel, R. (2004). Vaginal spasm, pain, and behavior: An empirical investigation of the diagnosis of vaginismus. *Archives of Sexual Behavior, 33*(1), 5–17.

Rosen, N. O., Bergeron, S., Lambert, B., & Steben, M. (2012). Provoked vestibulodynia: Partner and patient catastrophizing and dyadic adjustment as mediators of the associations between partner responses, pain and sexual satisfaction. *Archives of Sexual Behavior, 42*(1), 129–141.

Rosenbaum, T. Y., & Owens, A. (2008). Continuing medical education: The role of pelvic floor physical therapy in the treatment of pelvic and genital pain? Related sexual dysfunction (CME). *The Journal of Sexual Medicine, 5*(3), 513–523.

ter Kuile, M. M., Bulte, I., Weijenborg, P. T. M., Beekman, A., Melles, R., & Onghena, P. (2009). Therapist-aided exposure for women with lifelong vaginismus: A replicated single-case design. *Journal of Consulting and Clinical Psychology, 77*(1), 149–159.

van Lankveld, J., ter Kuile, M. M., de Groot, H. E., Melles, R., Nefs, J., & Zandbergen, M. (2007). Cognitive-behavioral therapy for women with lifelong vaginismus: A randomized waiting-list controlled trail of efficacy. *Journal of Consulting and Clinical Psychology, 74*(1), 168–178.

Van Lankveld, J. J., Granot, M., Weijmar Schultz, W., Binik, Y. M., Wesselmann, U., Pukall, C. F., & Achtrari, C. (2010). Women's sexual pain disorders. *The Journal of Sexual Medicine, 7*(1pt2), 615–631.

Vlaeyen, J. W., & Linton, S. J. (2000). Fear-avoidance and its consequences in chronic musculoskeletal pain: A state of the art. *Pain, 85*, 317–332.

Weijmar Schultz, W. C., Gianotten, W. L., van der Meijden, W. I., van de Wiel, H. B., Blindeman, L., Chadha, S., & Drogendijk, A. C. (1996). Behavioral approach with or without surgical intervention to the vulvar vestibulitis syndrome: A prospective randomized and non-randomized study. *Journal of Psychosomatic Obstetrics and Gynaecology, 17*(3), 143–148.

Dyspareunia Model

▶ Visceral Pain Models, Female Reproductive Organ Pain

Dyspepsia

Definition

Recurrent or persistent pain or discomfort in the upper abdomen and also used to describe sensations of nausea, bloating, and early satiety in relation to meals.

Cross-References

▶ Recurrent Abdominal Pain

Dysphagia

Definition

Dysphagia is defined as difficulty in swallowing.

Cross-References

▶ Opioid Therapy in Cancer Pain Management, Route of Administration

Dysraphism

Definition

Any failure of closure of the primary neural tube. This general category would include the disorder myelomeningocele.

Cross-References

▶ Lower Back Pain, Physical Examination

Dysthymia

Definition

Dysthymia is the term for chronic depression of a duration of 2 years or more that is not severe enough to qualify as chronic major depression.

Cross-References

▶ Psychiatric Aspects of the Epidemiology of Pain

Dysthymia, Chronic Depression

▶ Psychiatric Aspects of the Epidemiology of Pain

Dystonia

Definition

Involuntary tonic contraction of a muscle or a muscle group due to a central nervous or inflammatory lesion. Most cases of dystonia are painful. The reasons for the pain are assumed to be low tissue pH and ischemia.

Cross-References

▶ Sensitization of Muscular and Articular Nociceptors

Dysraphism

Definition

Any failure of closure of the primary neural tube. But generally, surgery would refer to the disorder involving the spine.

Cross-References

▶ Lower Back Pain: Physical Examination

Dysthymia

Definition

Dysthymia refers to a chronic depression, a duration of 2 years or more, failing to meet criteria to qualify as a major depression.

Cross-References

▶ Psychiatric Aspects of the Epidemiology of Pain

Dysthymia, Chronic Depression

▶ Psychiatric Aspects of the Relationships of Pain

Dyskinesia

Definition

Involuntary tonic contraction of a muscle or a muscle group, leading to a cramped posture or inflammatory lesion. Most cases of pain have no parallel. The reasons for the pain are resolved by the low tissue pH and ischemia.

Cross-References

▶ Sensitization of Muscular and Articular Nociceptors

E

EAE

▶ Experimental Allergic Encephalitis

Early Mobilization

Definition

Early mobilization is the early resumption of activities of daily living, usually within 24 h after surgery.

Cross-References

▶ Postoperative Pain, Importance of Mobilization

ECD

▶ Equivalent Current Dipole

Ectopia

▶ Ectopic Activity (Ectopic Discharge)

Ectopia, Spontaneous

Robert H. LaMotte
Department of Anesthesiology, Yale School of Medicine, New Haven, CT, USA

Synonyms

Spontaneous Ectopia

Definition

Ongoing nerve impulses produced by an atypical generator ("pacemaker"), at an abnormal (ectopic) locus in the primary sensory neuron, for example, at a site of injury. The ectopic discharges are spontaneous, i.e., endogenously generated by the neuron in the absence of an apparent external stimulus.

Characteristics

The types of injuries that produce spontaneous ectopia (SE) (see ▶ Ectopia, Spontaneous) include axotomy, mechanical trauma such as nerve compression, demyelination, and inflammation. However, after each kind of injury, only a subpopulation of neurons exhibits SE. The SE typically originates in a part of the neuron that is close to the site of injury, and/or within the ▶ dorsal root ganglion (DRG). In addition,

G.F. Gebhart, R.F. Schmidt (eds.), *Encyclopedia of Pain*, DOI 10.1007/978-3-642-28753-4,
© Springer-Verlag Berlin Heidelberg 2013

Ectopia, Spontaneous, Fig. 1 Typical patterns of injury-induced ectopic discharge recorded intracellularly from the intact DRG. (**a**) Tonic discharge recorded from an L5 DRG with a myelinated axon, transected 5 days earlier by a spinal nerve ligation (from Ma et al. 2003, Fig. 7). (**b**) Bursting discharge in a non-axotomized L4 with a myelinated axon, 5 days after transection of the adjacent L5 spinal nerve. Each burst follows an increase in the amplitude of oscillation of the resting membrane potential (Liu et al. 2000, Fig. 1C, action potentials have been truncated). (**c**) Irregular discharge in a non-axotomized neuron with an unmyelinated axon and a cell body in a ganglion that had been compressed for 6 days by a rod inserted into the intervertebral foramen (from Zhang et al. 1999, Fig. 2E). Voltage scale: 10 mV for (**a**, **b**), 20 mV for (**c**). Timescale: 100 ms for (**a**), 200 ms for (**b**), and 500 ms for (**c**)

SE sometimes also occurs in the neighboring uninjured neurons that presumably activated as a result of the injury (Wu et al. 2002; Ma et al. 2003). The SE in neurons with myelinated (A-fiber) axons typically has a regular pattern (with a more or less constant interval between nerve impulses, e.g., 35–65 ms) that, for some neurons, is continuous and, for others, periodically interrupted by periods of silence resulting in "bursts" (Fig. 1a, b). Other neurons with A-fibers, and most of those with unmyelinated (C-fiber) axons, exhibit an irregular pattern with longer intervals between impulses, e.g., 100–1,000 ms or more (Devor and Seltzer 1999) (Fig. 1c).

Although SE may appear to be endogenous to an electrophysiologically recorded hyperexcitable neuron, i.e., occurs in the absence of experimentally applied stimuli, it can in many instances be modulated (or in some cases produced in silent "injured" neurons) by externally applied mechanical, thermal, or chemical stimuli. Such stimuli might normally be present in vivo in the form of body temperature, movement of limbs, local ischemia, anoxia, and the presence of inflammatory mediators at the injury site. For example, there is a host of endogenous chemical factors that might modulate or induce SE in hyperexcitable neurons, including substances released from sympathetic neurons (e.g., norepinephrine), or from sensory neurons (neuropeptides), or inflammatory mediators normally present or released in injured or inflamed tissue (Devor and Seltzer 1999).

SE in Axotomized Sensory Neurons

If the axon of an adult DRG neuron is transected (axotomized), for example, by a cut, a severe compression, or as a result of disease, the end of the portion still connected to the cell body (soma) may begin to sprout within hours or days. After the transection of peripheral but not central (dorsal root) axons, some of these sprouts may regenerate back to the target tissue and reestablish functional connections. Certain other regenerating sprouts may become trapped and entangled forming a structure called a "▶ neuroma". A proportion of the injured neurons develop SE, which typically originates at the site of injury, but can also originate within the DRG.

After transection of the sciatic nerve, the incidence of SE is initially greater during the first 2 weeks for neurons with myelinated axons (A-neurons) than those with unmyelinated axons (C-neurons) (Devor and Seltzer 1999). Thereafter, the reverse is true, with SE being more common for C-neurons rather than for A-neurons. For A-neurons, the incidence of SE is greater for nerve lesions closer to the somata (transection of the spinal nerve as opposed to the sciatic nerve). However, the reverse is true for C-neurons, where there is virtually no SE after transection of the spinal nerve (Liu et al. 2000). The reasons for these observations must await identification, both of the chemical factors that alter the excitability of a neuron after axotomy, the cells that release the chemicals, and the events that lead to their release.

Several weeks after a sciatic nerve injury, some pairs of injured nerve fibers develop an abnormal, but stable, electrical ("ephaptic") interaction between certain nerve fibers, for example, between the endings of two fibers terminating in a neuroma or within the site of a crush injury. Ephaptic communication can also develop within pairs of fibers that are in close apposition, with patches of demyelination presumably where the glial insulation is absent (Devor and Seltzer 1999). One possible functional consequence of ephaptic connections is that activity in nociceptive afferent nerve fibers is evoked by activity in low-threshold mechanoreceptive, sympathetic, or motor nerve fibers.

Ephaptic connections do not form in the DRG. However, another type of cross-excitation can occur within the DRG between repetitively discharging neurons and their passive neighbors. This "cross-excitatory discharge" is mediated not electrically but by an unknown chemical mediator (Devor and Seltzer 1999).

SE in Intact Sensory Neurons

SE can also develop in intact neurons after either an axotomy or, in the absence of axotomy, after a local demyelination, inflammation, or compression of the nerve or the DRG. The SE can originate in the DRG, even in many instances when the site of injury or inflammation appears confined to the peripheral nerve.

After a transection of either a mixed (spinal) nerve or a ventral root (containing only motor nerve fibers) at L5, a low level of SE develops in intact sensory neurons with unmyelinated axons and cell bodies in the adjacent L4 ganglion (Wu et al. 2002). In addition, after a transection of the L5 spinal nerve, the various patterns of SE that develop in the axotomized L5 A-neurons are also found in intact A-neurons with cell bodies in the adjacent L4 DRG (Ma et al. 2003). Thus, the initiation of SE in the DRG may not be due to the absence of an axonally transported signal from the peripheral target, but rather to the retrograde delivery of a positive signal from the injury site to the DRG. Possible signals that might induce SE in the somata or distal axons of intact neurons include cytokines such as nerve growth factor (NGF), tumor necrosis factor-α (TNFα), interleukin-1α (IL-1α), IL-1β, and IL-6 released by reactive cells in the target tissue and/or, by Schwann cells surrounding nerve fibers undergoing Wallerian degeneration and by invading macrophages.

Certain neurodegenerative diseases such as multiple sclerosis or injuries produced by nerve entrapment may cause focal demyelination. A proportion of demyelinated axons do exhibit SE in the absence of axotomy or a loss of conduction (Kapoor et al. 1997).

Inflammatory processes per se can lead to SE in a subpopulation of peripheral sensory neurons. In an experimental model of immune mediated neuritis, topical application of Complete Freund's Adjuvant to a focal region of the sciatic nerve produced ipsilateral cutaneous ▶ hyperalgesia and SE described as irregular, occurring at a slow rate (about 1 Hz), and confined to a subpopulation of unmyelinated and thinly myelinated nerve fibers innervating subcutaneous musculoskeletal tissue (Bove et al. 2003).

A ▶ chronic constriction injury (CCI) produced in the rat by a loose ligation of the sciatic nerve is accompanied by a partial axotomy, local inflammation, and demyelination in the vicinity of the injury site and ipsilateral cutaneous hyperalgesia. SE in both C- and A-neurons originated at the injury site or within the DRG, and its rate increased by stimulation of sympathetic efferent neurons or by application of norepinephrine to the DRG or to the injury site. Most of the neurons with SE had transected axons, though a minority conducted through the injury site and thus could be considered as intact (Kajander and Bennett 1992).

A model of chronic compression of the DRG (CCD), which might occur with a laterally herniated disk or foraminal stenosis, was produced in the rat by the insertion of a rod into the intervertebral foramen, one at L4 and another at L5. This produced cutaneous hyperalgesia in the ipsilateral hind paw and SE, originating in the DRG, in neurons with intact, conducting axons. The SE occurred in both neurons with unmyelinated and myelinated axons, and was exhibited even after the formerly compressed DRGs were removed from the animal and recorded in vitro from the intact ganglion (Zhang et al. 1999).

Cellular Mechanisms of SE

There are similarities in the characteristics of somal hyperexcitability after seemingly different types of injury such as peripheral or spinal nerve axotomy, peripheral nerve constriction, or DRG compression without axotomy: These include SE that is regular, bursting, or irregular (Fig. 1), lower than normal current thresholds (a lesser magnitude of current injected into the soma required to elicit an action potential), decreased accommodation (increased firing to an injection of steady, suprathreshold current), and subthreshold membrane potential oscillations (Devor and Seltzer 1999;

Ma et al. 2003). These characteristics are exhibited by dissociated somata of the DRG neuron (after prior injury). The dissociated neurons can be labeled by a dye delivered peripherally prior to injury to determine whether the neuron innervated, for example, skin or muscle. For example, a subpopulation of A-neurons with medium size cell bodies, with narrow non-inflected action potentials and including some cutaneous but primarily muscle afferents, exhibited spontaneous subthreshold oscillations of their resting membrane potential or upon depolarization after peripheral or spinal nerve transection (Liu et al. 2002). A-neurons with similar oscillations and action potential properties were recorded intracellularly from the intact ganglion after spinal nerve transection (Liu et al. 2000) or after chronic compression of the ganglion (Zhang et al. 1999), and also from axons after experimentally induced demyelination (Kapoor et al. 1997). Blockers of sodium or potassium currents can respectively block or initiate an increase in the oscillations in these A-neurons (Kapoor et al. 1997). Of those DRG neurons exhibiting a bursting pattern of SE, the amplitude of ongoing membrane oscillation typically increased just prior to each burst, suggesting that the former triggered the latter (Fig. 1b).

Patch-clamp recordings of isolated currents in dissociated DRG somata provide an insight into the effects of different injurious/inflammatory conditions in altering ion channel properties. Dissociated neurons from lumbar DRGs ipsilateral to a transection of the spinal or sciatic nerve express a variety of changes in sodium, potassium, and calcium currents. These include an increase and faster repriming of a kinetically fast TTX-sensitive sodium current (Nav 1.3) and decreases in other currents including a slow TTX-resistant current (Waxman et al. 1999), a high-voltage-activated calcium current (Baccei and Kocsis 2000), and sustained and transient potassium currents (Everill and Kocsis 1999). TTX-R current in DRG somata also decreases after CCI (Dib-Hajj et al. 1999). The hyperpolarization cation current (Ih) increases after CCD (Yao et al. 2003). A decrease in potassium current and increase in sodium and/or Ih current could contribute to the decrease in current threshold and accommodation in excitable DRG neurons after nerve injury.

The responses of SE neurons to norepinephrine and inflammatory mediators such as serotonin, histamine, bradykinin, and prostaglandin E_2 may be intrinsic properties of the cell bodies (somata) of neurons after a nerve injury such as CCI, because they are recorded in vitro from the intact ganglion and from cells that have been acutely dissociated (Petersen et al. 1996).

Contribution of SE to Pain, Hyperalgesia, and Allodynia

In addition to producing pain and paresthesiae, it is likely that SE in appropriate primary sensory nociceptive neurons could increase the sensitivity of second-order neurons in the spinal dorsal horn ("central sensitization"). Central sensitization can occur in the absence of neuronal injury (Fig. 2a). For example, a local intradermal injection of capsaicin in the arm produces a wide area of hyperalgesia (enhanced pain to normally painful stimuli such as a pin prick), and allodynia (pain to normally non-painful stimuli, such as a touch) to mechanical stimulation of the skin surrounding the area exposed to the chemical. The basis for this is believed to be the lowered threshold responses and increased suprathreshold responses of spinal nociceptive neurons in the dorsal horn (LaMotte et al. 1991). Thus, it is possible that SE, resulting, for example, from

Ectopia, Spontaneous, Figure 2 Schematic of how pain might be evoked by normal activity in cutaneous dorsal root ganglion (DRG) neurons and by ectopic spontaneous activity after neuronal injury. (a) Normal activity in uninjured neurons. Pain and touch are normally elicited by respective activity in nociceptive (N1, N2) and low-threshold mechanoreceptive (M) neurons. N terminates onto a spinothalamic tract (STT) neuron (S) whose cutaneous receptive field is shown on the left (large, dashed oval). M projects into the dorsal columns (DC) but sends a collateral axon to S. After a localized injury or inflammation of the skin (such as a minor cut to the skin resulting in release of inflammatory chemicals, e.g., from blood vessels, mast cells), activity in N1 releases a neurotransmitter that sensitizes S ("central sensitization") such that its output in response to nociceptive mechanical stimuli (e.g., a pricking with a stiff hair that activates N2) is increased (contributing to hyperalgesia), and its response to innocuous touch (e.g., lightly stroking the skin that activates M) is enhanced, thereby contributing to touch-evoked allodynia both within and outside (larger dashed oval) the area of injury. (b) Spontaneous ectopic activity (SE) in injured neurons. SE can originate at the site of a local demyelination and/or inflammation (a), from the proximal ends of transected axons forming a neuroma (b), from the DRG of injured or inflamed neurons (c), or as a result of compression/inflammation of the ganglion itself. After axotomy, a loss of input to inhibitory neurons (d) might disinhibit activity in nociceptive STT neurons, thereby causing pain of central origin. In this example, chronic ectopic discharges in N1 could elicit, in addition to pain, central sensitization leading to chronic allodynia and hyperalgesia. SE in mechanoreceptive afferents mediating touch could elicit abnormal sensations (paresthesiae) and, in the presence of central sensitization, chronic pain

nerve injury, may produce and maintain a state of chronic central sensitization (Fig. 2b), thereby contributing to a chronic state of hyperalgesia and allodynia.

References

Baccei, M. L., & Kocsis, J. D. (2000). Voltage-gated calcium currents in axotomized adult rat cutaneous afferent neurons. *Journal of Neurophysiology, 83*, 2227–2238.

Bove, G. M., Ransil, B. J., Lin, H. C., et al. (2003). Inflammation induces ectopic mechanical sensitivity in axons of nociceptors innervating deep tissues. *Journal of Neurophysiology, 90*, 1949–1955.

Devor, M., & Seltzer, Z. (1999). Pathophysiology of damaged nerves in relation to chronic pain. In P. D. Wall & R. Melzack (Eds.), *Textbook of pain* (4th ed., pp. 129–164). London: Churchill Livingstone.

Dib-Hajj, S. D., Fjell, J., Cummins, R. R., et al. (1999). Plasticity of sodium channel expression in DRG neurons in the chronic constriction injury model of neuropathic pain. *Pain, 83*, 591–600.

Everill, B., & Kocsis, J. D. (1999). Reduction of potassium currents in identified cutaneous afferent dorsal root ganglion neurons after axotomy. *Journal of Neurophysiology, 82*, 700–708.

Kajander, K. C., & Bennett, G. J. (1992). Onset of a painful peripheral neuropathy in rat: A partial and differential deafferentation and spontaneous discharge in A beta and A delta primary afferent neurons. *Journal of Neurophysiology, 68*, 734–744.

Kapoor, R., Li, Y. G., & Smith, K. J. (1997). Slow sodium-dependent potential oscillations contribute to ectopic firing in mammalian demyelinated axons. *Brain, 120*, 647–652.

LaMotte, R. H., Shain, C. N., Simone, D. A., et al. (1991). Neurogenic hyperalgesia: Psychophysical studies of underlying mechanisms. *Journal of Neurophysiology, 66*, 190–211.

Liu, C. N., Michaelis, M., Amir, R., et al. (2000). Spinal nerve injury enhances subthreshold membrane potential oscillations in DRG neurons: Relation to neuropathic pain. *Journal of Neurophysiology, 84*, 205–215.

Liu, C. N., Devor, M., Waxman, S. G., et al. (2002). Subthreshold oscillations induced by spinal nerve injury in dissociated muscle and cutaneous afferents of mouse DRG. *Journal of Neurophysiology, 87*, 2009–2017.

Ma, C., Shu, Y., Zheng, Z., et al. (2003). Similar electrophysiological changes in axotomized and neighboring intact dorsal root ganglion neurons. *Journal of Neurophysiology, 89*, 1588–1602.

Petersen, M., Zhang, J., Zhang, J. M., et al. (1996). Abnormal spontaneous activity and responses to norepinephrine in dissociated dorsal root ganglion cells after chronic nerve constriction. *Pain, 67*, 391–397.

Waxman, S. G., Dib-Hajj, S., Cummins, T. R., et al. (1999). Sodium channels and pain. *Proceedings of the National Academy of Sciences of the United States of America, 96*, 7635–7639.

Wu, G., Ringkamp, M., Murinson, B. B., et al. (2002). Degeneration of myelinated efferent fibers induces spontaneous activity in uninjured C-fiber afferents. *The Journal of Neuroscience, 22*, 7746–7753.

Yao, H., Donnelly, D. F., Ma, C., et al. (2003). Upregulation of the hyperpolarization cation current after chronic compression of the dorsal root ganglion. *The Journal of Neuroscience, 23*, 2069–2074.

Zhang, J. M., Song, X. J., & LaMotte, R. H. (1999). Enhanced excitability of sensory neurons in rats with cutaneous hyperalgesia produced by chronic compression of the dorsal root ganglion. *Journal of Neurophysiology, 82*(6), 3359–3366.

Ectopic Activity (Ectopic Discharge)

Synonyms

Ectopia; Ectopic electrogenesis; Ectopic pacemaker activity

Definition

Ectopic activity (or ectopic discharge) refers to trains of ongoing (spontaneous) electrical nerve impulses, or impulses generated upon stimulation, for which the discharge originates at any location other than the normal location. The normal location for sensory impulse generation is the sensory ending of afferent neurons in the skin, viscera, and other organs. Common ectopic locations for sensory impulse generation are sites of nerve injury and associated dorsal root ganglia. Ectopic impulse activity is thought to be an important contributor to neuropathic pain.

Cross-References

▶ Drugs Targeting Voltage-Gated Sodium and Calcium Channels
▶ Ectopia, Spontaneous
▶ Neuroma Pain
▶ Neuropathic Pain, Joint and Muscle Origin
▶ Proinflammatory Cytokines
▶ Trigeminal Neuralgia: Diagnosis and Treatment

Ectopic Electrogenesis

▶ Ectopic Activity (Ectopic Discharge)

Ectopic Excitability

Synonyms

Ectopic hyperexcitability

Definition

Primary sensory neurons normally generate impulses at their sensory transducer ending, in the skin or other innervated tissues. After nerve injury, these neurons may begin to generate impulses spontaneously, or upon mechanical, thermal, or chemical stimulation, at abnormal (ectopic) locations. Examples are the nerve injury site and associated dorsal root ganglia. The acquired ability to general impulse activity at any location other than the normal one is ectopic excitability.

Cross-References

▶ Neuroma Pain

Ectopic Hyperexcitability

▶ Ectopic Excitability

Ectopic Mechanosensitivity

Definition

Mechanosensitive sensory neurons normally generate impulses upon application of mechanical force at their sensory transducer ending in the skin and other innervated tissues. Nerve fibers are not responsive to mechanical stimulation at other locations, along the course of a healthy nerve, for example. After nerve injury, sensory nerve fibers may begin to generate impulses upon mechanical stimulation at abnormal (ectopic) locations. Examples are nerve-end neuromas and other sites of nerve injury. The acquired ability to generate impulse activity upon mechanical stimulation at ectopic locations is ectopic mechanosensitivity.

Cross-References

▶ Neuroma Pain

Ectopic Nerve Impulses

Definition

Electrical nerve impulses that originate at any location other than the normal one, the sensory nerve ending for primary afferent neurons. Typical loci for ectopic impulse generation are the nerve injury site and associated DRGs. Ectopic impulse activity is thought to be an important contributor to neuropathic pain.

Cross-References

▶ Ectopia, Spontaneous
▶ Ectopic Activity (Ectopic Discharge)
▶ Painless Neuropathies

Ectopic Pacemaker Activity

▶ Ectopic Activity (Ectopic Discharge)

Edema

Definition

The accumulation of excessive fluid in the intercellular spaces of subcutaneous tissue. This can be caused by an increased permeability of the microvascular endothelium, resulting in leakage of vascular components into tissue.

Cross-References

▶ Inflammation, Modulation by Peripheral Cannabinoid Receptors

EDT

▶ Electrodiagnostic Testing (Studies)

Education

▶ Information and Psychoeducation in the Early Management of Persistent Pain

Education and Chronicity

▶ Pain in the Workplace, Risk Factors for Chronicity, and Demographics

EEG

▶ Electroencephalogram/ Electroencephalography

EEG (electroencephalography)

▶ Pain in Humans, EEG Documentation

Effect Size

Nikolai Bogduk
University of Newcastle, Newcastle Bone & Joint Institute, Royal Newcastle Centre, Newcastle, NSW, Australia

Synonyms

Attributable effect; Effectiveness; Efficacy

Definition

▶ Effect size is a measure of how effective a treatment is when applied to a group of patients. It can be used to measure by how much a group of patients improves after treatment, or by how much a particular treatment is better than another treatment to which it is compared.

Characteristics

For any outcome variable, a group of patients will typically exhibit a normal distribution of values (Fig. 1). That distribution will have a mean value (μ) and a standard deviation (sd). About 68 % of the patients will express a value between one standard deviation less than the mean value and one standard deviation greater. The outcome measure may be pain scores or any other outcome of interest.

If that group of patients undergoes a treatment, their outcome variables will change but will again assume a normal distribution (Fig. 2). The change in their scores will be reflected by the difference between the mean values before (μ_0) and after (μ_1) treatment.

Whether or not the change is a clinically meaningful one depends on the magnitude of the change with respect to the spread of values, before and after treatment. A given change in mean values may not be impressive if the patients exhibit a wide spread of values (Fig. 3a). Under

Effect Size, Fig. 1 In a sample of patients, the values of any outcome measure will assume a normal distribution. The mean value (μ) represents the average value. The spread of the values is reflected by the standard deviation (sd)

Effect Size, Fig. 2 Before treatment, a group of patients express values of an outcome measure that are distributed in a normal fashion around a mean value (μ_0). After treatment, they express values distributed around a new mean value (μ_1)

those conditions, the change is not impressive, for it is encompassed by the extent to which patients normally differ in their values. After treatment, many patients still have scores that others had before treatment, which means that they have not improved categorically. The treatment does not render patients different from patients before treatment.

The same change, however, may be more impressive if it is appreciably greater than the normal spread of values (Fig. 3b). Under those conditions, most patients have scores after

treatment, which are outside the range of scores before treatment. This attests to a definite categorical improvement. After treatment, most patients are distinctly better than patients before treatment.

The qualitative significance of a change can, therefore, be expressed as a ratio between the change in mean values and the distribution of values. Specifically, the ▶ effect size is the ratio between the difference in the mean value of a selected outcome measure and the standard deviation of that value (Cohen 1977). Some authorities recommend that the distribution of values be expressed as the pooled standard deviation of the samples before and after treatment, but the simplest method is to use the standard deviation of the sample before treatment. Under that definition:

$$\text{Effect size} = \frac{\mu_1 - \mu_0}{\text{sd}}$$

This equation yields a single number that indicates how effective the intervention has been. That number can be translated into a verbal description that gauges the effect size (Table 1) (Cohen 1977).

Effect sizes can be calculated to determine by how much a group of patients responds to a particular treatment. Separate effect sizes can be calculated for patients undergoing a particular treatment and patients undergoing a comparison or control treatment. Those effect sizes can be compared in order to determine the extent to which a treatment is better than control, or a relative effect size can be calculated by using the mean scores of the treatment group and the control group after treatment and the standard deviation of the control group.

Utility

The attraction of effect size is that it provides, in a single number, a measure of how much a group of patients improves after treatment, or by how much a treatment achieves outcomes better than those of a control treatment. Different treatments can be compared according to the magnitude of their effect sizes.

Effect Size, Fig. 3 Graphic representations of the relative significance of a given change in mean values, before and after treatment, with respect to the spread of values exhibited by a group of patients. (**a**) The change in mean values is not greater than the spread of values in the group.

Many patients after treatment have scores like those of many patients before treatment. (**b**)The same change in mean values is appreciably larger than the spread of values. The scores of most patients after treatment are outside the range of scores before treatment

Effect Size, Table 1 Verbal translations of selected effect sizes

Effect size	Descriptor
0.00	Nil
0.20	Small
0.50	Medium
0.80	Large

A disadvantage of effect size is that it only describes the effect of treatment on a group of patients as a whole. It does not relate to individual patients. It does not indicate what chances a given patient has of benefiting, or by how much.

A further disadvantage stems from a statistical idiosyncrasy. Since the calculation of effect size requires mean values and standard deviations, the outcome variables must be normally distributed. If outcomes are not normally distributed, a mean value and, therefore, effect size cannot be calculated.

Cross-References

▶ Meta-Analysis
▶ Oswestry Disability Index
▶ Psychological Aspects of Pain in Women
▶ Psychology of Pain, Efficacy of Cognitive-Behavioral Treatment

References

Cohen, J. (1977). *Statistical power analysis for the behavioural sciences* (pp. 20–23, 40). New York: Academic.

Effective Analgesia

Definition

Effective analgesia is the provision of adequate analgesia to enable the patient to perform "activities of daily living."

Cross-References

▶ Postoperative Pain, Importance of Mobilization

Effectiveness

Definition

The objective of any treatment is to relieve the problems that a patient suffers. For a patient with pain, the objective may be to relieve their pain, or it may be to relieve one or other of the other problems that they suffer, such as psychological distress or disabilities in everyday activities or work.

Effectiveness is a particular property of a treatment that indicates how well it achieves these objectives under average or typical conditions, e.g., when average practitioners perform the treatment on typical patients encountered in their practice. In this regard, effectiveness is distinguished from efficacy, which indicates how well the treatment works under ideal conditions.

Effectiveness is usually established by studies conducted after studies have determined the efficacy of a treatment. Efficacy studies pave the way for effectiveness studies by showing how well a treatment can work. Studies of effectiveness are undertaken to determine if the treatment works sufficiently well under average conditions to justify its wholesale application.

As a rule, the efficacy of a treatment will usually be greater than its effectiveness. In some instances, a treatment may well have efficacy, but it might lack effectiveness. Although it works under ideal conditions, the results of the treatment might be confounded by the other problems that patients in typical practice present, or average practitioners may not be as skilled in providing the treatment as experts are.

A treatment may not necessarily work for all domains of a patient's problems. It may work for one but not for others. The effectiveness of a treatment, therefore, should be qualified by the domain for which it works, e.g., efficacy for relief of pain, efficacy for depression, efficacy for return to work. Qualifying treatments in this manner avoids misinterpretation and misrepresentation, lest patients expect that because a treatment works for one symptom, it will work for another.

The assessment of effectiveness is based on measuring the changes of outcome measures after treatment, and comparing them with changes after control treatments. Efficacy can be expressed statistically in terms of number needed to treat (NNT) or effect-size.

Cross-References

▶ Attributable Effect and Number Needed to Treat
▶ Effect Size
▶ Efficacy

Effectiveness Measure

▶ Central Pain, Outcome Measures in Clinical Trials
▶ Central Pain: Outcome Measures in Preclinical Research

Efficacy

Definition

The objective of any treatment is to relieve the problems that a patient suffers. For a patient with pain, the objective may be to relieve their pain, or it may be to relieve one or other of the other problems that they suffer, such as psychological distress or disabilities in everyday activities or work.

Efficacy refers to whether a treatment achieves a specific objective. Tests of efficacy typically are performed under carefully controlled conditions, often those that are relatively optimal in terms of the specific effect. In this regard, efficacy is distinguished from ▶ effectiveness, which indicates how well the treatment works under "real life," typically less ideal conditions.

One purpose of determining the efficacy of a treatment is to set a benchmark, which indicates how well the treatment can work under ideal conditions. Under less than ideal conditions, however, that benchmark might not be achieved. Therefore, although practitioners might aspire to achieve the same results as those encountered in efficacy studies, they should not necessarily expect to achieve those results. The patients that they treat may not be as ideal as those recruited to a study.

Another, more subtle purpose of an efficacy study is to identify treatments that do not work. If a treatment does not work in expert hands on ideal patients, it is highly unlikely to work under normal or average conditions. In essence, if a treatment has no efficacy, it is unlikely to have effectiveness.

A treatment may not necessarily work for all domains of a patient's problems. It may work for one but not for others. If a treatment has efficacy, it should be qualified by the domain for which it works, e.g., efficacy for relief of pain, efficacy for depression, efficacy for return to work. Qualifying treatments in this manner avoids misinterpretation and misrepresentation, lest patients expect that because a treatment works for one symptom it will work for another.

The assessment of efficacy is based on measuring the changes of outcome measures after treatment, and comparing them with changes after control treatments. Efficacy can be expressed statistically in terms of number needed to treat (NNT) or effect size.

Cross-References

▶ Attributable Effect and Number Needed to Treat
▶ Effect Size
▶ Opioid Responsiveness in Cancer Pain Management

Efficacy of Drugs

Definition

It is possible to express the efficacy of a drug in terms of a fraction of the total receptor population (fractional receptor occupancy) that an agonist must occupy to yield a given effect. The number of receptors to be occupied is inversely proportional to the intrinsic activity. As the amount of the remaining unoccupied receptors depends on this property, the larger the receptor reserve, the greater the intrinsic efficacy. It has been

suggested that the degree of tolerance is inversely related to the reserve of spare opioid receptors.

Cross-References

▶ Opioid Responsiveness in Cancer Pain Management

Efficacy Study

Definition

Efficacy studies determine the effects of specific treatments. In order to do this, studies are carried out where as many possible likely confounding factors are controlled by the careful selection of patients and standardized implementation of the treatment. Specific treatments are compared with highly controlled alternative treatments including those designed as a placebo, i.e., treatments in which the presumed active ingredients have been excluded. The randomized controlled trial is the de facto standard for efficacy studies. In contrast, effectiveness studies aim to determine whether the treatment may be generalized to the real world across a range of patient populations, therapists, and delivery settings.

Cross-References

▶ Effect Size
▶ Psychology of Pain, Efficacy of Cognitive-Behavioral Treatment

Effleurage

▶ Massage and Pain Relief Prospects

Effort Headache

▶ Primary Exertional Headache

eHealth

► Remote Management of Pediatric Pain

Eicosanoids

Definition

Eicosanoids refer to any product derived from arachidonic acid, an unsaturated fatty acid found in the plasma membrane of neurons. Eicosanoids are lipids that include prostaglandins, prostacyclins, thromboxanes, and leukotrienes. The eicosanoids can collectively mediate almost every aspect of the inflammatory response.

Cross-References

► Immunocytochemistry of Nociceptors
► Prostaglandins, Spinal Effects

Electrical Stimulation Analgesia

► TENS, Mechanisms of Action

Electrical Stimulation Therapy (EST)

► Transcutaneous Electrical Nerve Stimulation Outcomes
► Transcutaneous Electrical Nerve Stimulation (TENS) in Treatment of Muscle Pain

Electrical Stimulation-induced Analgesia

► Stimulation-Produced Analgesia

Electrical Therapy

► Transcutaneous Electrical Nerve Stimulation

Electroacupuncture (EA)

Definition

Electrical stimulation through one or more pairs of acupuncture needles.

Cross-References

► Acupuncture Mechanisms

Electroanalgesia

► Transcutaneous Electrical Nerve Stimulation Outcomes
► Transcutaneous Electrical Nerve Stimulation (TENS) in Treatment of Muscle Pain

Electrodermal Activity (EDA)

► Sympathetic Skin Response

Electrodiagnosis and EMG

D. Paul Harries
Pain Treatment Center, Samaritan Hospital
Lexington, Lexington, KY, USA

Synonyms

Electromyography; Nerve conduction studies

Definition

Electrodiagnosis includes nerve conduction studies and electromyography (EMG). Nerve conduction studies are the measurement of electrical transmission within nerves as the result of an artificial, electrical, or magnetic stimulus. EMG is the measurement of electrical activity within a muscle for diagnostic purposes. EMG is unfortunately incorrectly regarded by many as being synonymous with electrodiagnostic studies.

Characteristics

Nerve Conduction Studies

Physiology

When a peripheral nerve is stimulated electrically, action potentials will be evoked in all the axons within that nerve. By recording from the nerve, the combined response of those axons can be obtained. Generally, however, only the signals from the larger, faster, myelinated fibers are recorded. In clinical practice, impulses from slow fibers do not contribute significantly to the electrodiagnostic results.

Two types of nerve injury occur: axonal and demyelinating. These usually occur together to some degree, but one is usually predominant. The pattern of injury helps in diagnosis.

In demyelinating nerve injuries, the nerve remains intact and continues to function, but the demyelination results in slowing of the conduction velocity. In a severe demyelinating nerve injury, conduction block appears. In conduction block, the impulses are slowed to such an extent that they cease to propagate, resulting in a drop in amplitude. A drop in amplitude of 50 % is normally required to diagnose conduction block. It should be noted that conduction block is a result of a focal lesion in the nerve membrane. Despite there being a drop in amplitude, there is no axonal injury.

In axonal injury, the nerve is damaged, and the affected axons cease to function. The degree of nerve damage is proportional to the number of its axons that are affected. Those axons that still function will generally conduct at a normal velocity, but there will be fewer of them. So, the amplitude of activity recorded from the nerve will be reduced.

If a nerve is transected, the effects depend on the location of the injury with respect to the cell bodies of the nerve. If sensory nerves are transected distal to the dorsal root ganglion, axons distal to the site of injury will degenerate and exhibit no function, and therefore, no conduction. Proximal to the lesion, however, the axons will remain intact and will conduct at normal velocity. If the lesion is in the dorsal root, the entire peripheral nerve remains connected to the dorsal root ganglion and will exhibit normal function throughout. For motor nerves, the cell bodies lie in the spinal cord. When transected, the axons distal to the lesion will degenerate and cease to function, while their proximal ends will remain intact.

Sensory

Sensory testing is performed by stimulating a nerve and then recording a response in the nerve, either distal or proximal to the site of stimulation. A waveform known as the SNAP (sensory nerve action potential) is obtained. From the waveform, the amplitude and peak latency are measured. The conduction velocity is defined as the distance between stimulator and recording electrode, divided by the time from stimulation to peak latency.

Motor

Motor testing is performed by stimulating a nerve and recording from the muscles that it supplies. A compound muscle action potential (CMAP) is obtained. This waveform is very different to the SNAP, as it represents the response of a muscle, not a nerve. The impulse is effectively magnified at least 100-fold by the neuromuscular junction. The latency is measured to the onset of the impulse. Due to the transit time across the neuromuscular junction, velocity cannot simply be calculated from distance/time. Instead, two readings have to be taken, typically by stimulating the nerve at two locations at least 10 cm apart and recording from the same location. The nerve conduction velocity is given by (distance between proximal and distal stimulation sites)/(time to onset latency from proximal stimulation time to onset latency from distal stimulation).

H-Reflex

When a muscle nerve is stimulated, impulses are propagated both distally and proximally from the site of stimulation. The orthodromic propagation occurs along motor axons and elicits a response in the muscle, called the M wave. The antidromic propagation occurs in all muscle afferent fibers. These travel to the spinal cord where they synapse on anterior horn cells, which are activated. Their impulses then pass orthodromically into the nerve and eventually reach the muscle, where they evoke a second response, called the H wave. This is the electrophysiological equivalent of the stretch reflex.

The H wave is best seen at low-stimulation voltages prior to the production of the M wave. The reason for this is that amplification of the stimulus occurs at the spinal cord. H reflexes can only be consistently obtained from the gastrocnemius, upon stimulation of the tibial nerve, and to a lesser degree the flexor carpi radialis, upon stimulation of the median nerve.

F Wave

Upon stimulation of a muscle nerve, antidromic activity also occurs in the motor axons. When these impulses reach the anterior horn cell, some can be "reflected" in the axon. They eventually reach the muscle and evoke a third response, called the F wave. A relatively high stimulus is required to produce these waves. They can, however, be obtained from any nerve.

Somatosensory Evoked Potentials (SSEP)

A peripheral nerve with sensory fibers is repeatedly stimulated. Small changes in voltage are detected as the impulse passes into the spinal cord and then onto the brain. SSEP is useful in determining where in the CNS a delay in conduction occurs. Traditionally, it has been used to help diagnose multiple sclerosis. Today it is often used during spinal surgery to warn a surgeon of compromise to neural structures.

Current Perception Threshold (CPT)

In this test, a small electrical current is applied to an extremity. The patient indicates when they can first feel the current. CPT has been used in an attempt to pick up neuropathy and also to determine which patients will be able to cooperate during sensory testing while performing radiofrequency neurotomy. CPT has been used on an experimental basis in evaluating radiculopathy and in CRPS (Yamashita 2002).

EMG

In order to study the activity of a muscle in detail, a fine needle is inserted into it, and the recorded waveforms are examined during three separate phases:
1. At rest
2. With the patient still, but moving the needle to "provoke" the muscle
3. During voluntary contraction of muscle

The results are interpreted in the context of responses from other muscles in the same or different myotome or peripheral nerve distribution.

The EMG provides information regarding the muscle itself, the neuromuscular junction, and the motor axon. Classically, in axonal motor injury, fibrillations and positive sharp waves will be seen 3 weeks after injury and will persist for many months. These both represent signs of muscle membrane instability. The muscle is easily provoked by the needle and responds by producing these waveforms. In chronic cases, fibrillations and positive sharp waves become less evident.

Indications

The indications for nerve conduction studies and EMG in pain medicine are contentious. Pain is mediated by small diameter afferents, but these are not sampled by nerve conduction studies or EMG. Any relationship to pain is circumstantial or surrogate.

Radiculopathy

The Taxonomy of Pain (Merskey and Bogduk 1994) defines a radiculopathy as "objective loss of sensory and or motor function as a result of conduction block in axons of a spinal nerve or its roots." This is synonymous with a demyelinating injury. The electromyographer defines radiculopathy as a lesion proximal to the dorsal root ganglion.

If the lesion is proximal to the dorsal root ganglion, the peripheral sensory nerve will have normal nerve conduction studies. If there is an axonal injury to the motor nerve, conduction studies may demonstrate reduced amplitude in motor nerves. If there is a pure demyelinating lesion of the motor root, then nerve conduction studies will be normal.

The H-reflex can in theory identify demyelinating sensory and motor root involvement. Unfortunately, the H-reflex is only of limited value in diagnosing S1 radiculopathy (American Academy of Electrodiagnostic Medicine 1998; Dumitru 1995).

EMG will detect a radiculopathy that has axonal motor involvement. EMG abnormalities, when present, are clearest 3–30 weeks after onset of the axonal injury. SSEP has been evaluated in radiculopathy and found not to be clinically useful (American Academy of Electrodiagnostic Medicine 1998).

It can be deduced from basic physiological principles that EMG and nerve conduction studies will not pick up pure sensory radiculopathies or pure demyelinating motor neuropathies. EMG can be invaluable in differentiating between radiculopathy and other causes of motor weakness. The AAEM mini monograph on radiculopathy claims that electrodiagnostic studies can pick up the vast majority of radiculopathies using EMG (American Academy of Electrodiagnostic Medicine 1998). Clearly this will not be the case in pure sensory radiculopathies.

The difficulty in determining validity of electrodiagnostic testing in radiculopathy stems from disagreement with regard to the definition of radiculopathy (Merskey and Bogduk 1994; American Academy of Electrodiagnostic Medicine 1998) and the absence of a criterion standard.

Peripheral Neuropathy

EMG and nerve conduction studies, together with nerve biopsy, are the diagnostic tests of choice for peripheral neuropathies. Unfortunately, the information gained does not in any way help with the management of pain, except in very rare instances where the cause of the neuropathy can be effectively treated.

Plexopathy

Differentiating radiculopathy from plexopathy can be done very elegantly using nerve conduction studies and EMG. Together with MRI, it is the test of choice.

Peripheral Nerve Injuries

In peripheral nerve injuries, nerve conduction studies and EMG can provide data that help determine prognosis, site of lesion, and need for surgical intervention.

Complex Regional Pain Syndrome

Electrodiagnosis can help differentiate between type 1 and type 2 CRPS. CPT studies are asymmetric in CRPS but do not correlate with pain (Yamashita 2002).

Utility

Electrodiagnostic studies are explicitly designed to study loss of function in sensory and motor nerves. However, they have no direct application for the investigation of pain. They might be used to obtain details in patients with a painful peripheral neuropathy or plexopathy, but such cases would be uncommon.

Although commonly used to investigate patients with spinal and ▶ radicular pain, nerve conduction studies and EMG have no proven utility in such patients (Bogduk and Govind 1999; Bogduk 1999). They may confirm certain features of radiculopathy, but radiculopathy and ▶ radicular pain are not synonymous.

References

American Academy of Electrodiagnostic Medicine. (1998). AAEM minimonograph 32: The electrodiagnostic examination in patients with radiculopathies. *Muscle Nerve, 21*, 1612–1631.

Bogduk, N. (1999). *Medical management of acute cervical radicular pain: An evidence-based approach* (pp. 67–69). Newcastle: Newcastle Bone and Joint Institute.

Bogduk, N., & Govind, J. (1999). *Medical management of acute lumbar radicular pain: An evidence-based approach* (pp. 53–58). Newcastle: Newcastle Bone and Joint Institute.

Dumitru, D. (1995). *Electrodiagnostic medicine*. Philadelphia: Hanley and Belfus.

Merskey, H., & Bogduk, N. (1994). *Classification of chronic pain* (2nd ed., p. 16). Seattle: IASP Press.

Yamashita, T. (2002). A quantitative analysis of sensory function in lumbar radiculopathy using current perception threshold testing. *Spine, 27*, 1567–1570.

Electrodiagnostic Testing (Studies)

Synonyms

EDT

Definition

Electrodiagnostic studies are those that involve the use of electrical stimulation.

Traditional evaluation of the peripheral nervous system uses electrical stimuli to measure nerve conduction velocity (NCV) and muscle function (electromyography, EMG). This testing can be painful, but is objective, requiring no cognitive input from the person being tested for a response to the stimuli.

Cross-References

▶ Carpal Tunnel Syndrome
▶ Causalgia: Assessment

Electroencephalogram/ Electroencephalography

Synonyms

EEG

Definition

Synchronized extracellular currents in a few square centimeters of cortex generate electrical potentials measurable with electrodes on the scalp. The signal is low-pass filtered to 50 Hz.

Cross-References

▶ Insular Cortex, Neurophysiology, and Functional Imaging of Nociceptive Processing
▶ Thalamotomy for Human Pain Relief

Electrolytic Lesion

Definition

Electrolytic lesion is a procedure that can produce death of neurons and nerve fibers in a nervous center by passing an electrical current through an electrode. The extent of the lesion depends on the intensity and duration of the applied current.

Cross-References

▶ Lateral Thalamic Lesions, Pain Behavior in Animals
▶ Poststroke Pain Model, Thalamic Pain (Lesion)
▶ Thalamotomy

Electromyography

Definition

This is a method of studying the electrical activity of a muscle by recording action potentials from the contracting muscle fibers and is used to assess the level of muscular arousal. In newborn infants, the usual way of doing this is through surface electrodes applied to the overlying skin. However, in adults, for diagnostic purposes, it is useful to employ concentric needle electrodes that are inserted through the skin and into the muscle fibers themselves. The analogue signal from the electrical activity is either recorded on an oscilloscope, or more usually digitized and displayed via computer, where more complex waveform analysis can be conducted.

Cross-References

► Electrodiagnosis and EMG
► Infant Pain Mechanisms
► Psychophysiological Assessment of Pain

Electron Microscopy

Definition

Electron microscope is a high-resolution imaging technique using electron beams.

Cross-References

► Toxic Neuropathies

Electrophysiological Mapping

Definition

Electrophysiological mapping is the recording of electrical responses in the brain in response to a stimulus that is applied to the body.

Cross-References

► Thalamic Nuclei Involved in Pain: Cat and Rat

Electrophysiology

Definition

Electrophysiology is the science and branch of physiology that pertains to the flow of ions in biological tissues and, in particular, to the electrical recording techniques that enable the measurement of this flow. It is useful for studying response properties of neurons.

Cross-References

► Amygdala, Pain Processing and Behavior in Animals
► Cancer Pain Model, Bone Cancer Pain Model

Electrotherapy

Definition

Electrotherapies are techniques, including TENS, CFS, and acupuncture, used to stimulate nerve fibers or receptors to achieve analgesia or itch relief.

Cross-References

► Cutaneous Field Stimulation
► Modalities

Eligibility

► Disability Evaluation in the Social Security Administration

Elimination

Definition

The elimination of a drug characterizes its irreversible loss from the body due to excretion or metabolism into another chemical molecule.

Cross-References

► NSAIDs, Pharmacokinetics

Elimination Half-Life (t1/2)

Definition

The elimination half-life ($t_{1/2}$) of a drug describes the time needed to reduce the drug concentration in blood, plasma, or serum to one-half. The elimination half-life may be influenced by a variation in urinary excretion (pH), intersubject variation (e.g., polymorphic enzymes), age, drug-drug interactions, and diseases (especially renal and liver diseases). Elimination means the annihilation of the administered drug, not its metabolites, from the body by biliary, urinary, or other pathways of excretion (e.g., lung, skin) or rather biotransformation through metabolism.

Cross-References

▶ NSAIDs, Pharmacokinetics
▶ Opioid Rotation

Elmiron®

Definition

Pentosan polysulfate sodium, a commonly prescribed medication for IC. Structure similar to glycosaminoglycan layer of bladder mucosa.

Cross-References

▶ Interstitial Cystitis and Chronic Pelvic Pain

Embolism

▶ Postoperative Pain, Venous Thromboembolism

Emergence

Definition

Emergence is the process by which a system of interacting elements spontaneously acquires a qualitatively new pattern and structure that is unpredictable from knowledge of the individual elements.

Cross-References

▶ Consciousness and Pain

EMG-Assisted Relaxation

Definition

EMG-assisted relaxation is a form of biofeedback that employs information about muscle tension levels in order to facilitate overall relaxation.

Cross-References

▶ Biofeedback in the Treatment of Pain

Emotional Factors

▶ Pain in the Workplace, Risk Factors for Chronicity, and Psychosocial Factors

Emotional Factors/Reactions

Definition

Strong and roughly appropriate reaction to a significant event, tagged by an affective label, in the environment of the species. Emotional components arise from motivational and autonomic changes with a modulation by individual experience and a subjective perception. In the case of pain: defensive/aversive behavior with a myriad of autonomic changes and the avoidance learning (fear) of the noxious event.

Cross-References

▶ Hypothalamus and Nociceptive Pathways
▶ Pain in the Workplace, Risk Factors for Chronicity, and Psychosocial Factors
▶ Parabrachial Hypothalamic and Amygdaloid Projections

Empathic

Definition

Making an effort to understand what the patient is thinking, feeling, and wanting. Being empathic often includes reflecting one's understanding of the patient back to the patient in the form of reflective listening.

Cross-References

▶ Chronic Pain, Patient-Therapist Interaction

Empathy

John D. Banja
Department of Rehabilitation Medicine, Center for Ethics Emory University, Atlanta, GA, USA

Definition

Definitions and Connotations

The word "empathy" originally appeared as the English translation of the German word *Einfuhlung* (literally meaning "feeling into") and derives from the Greek cognate *empatheia*,

which means an appreciation of another's feelings (Hojat 2007). Empathy is a common term in the clinical literature that is frequently used in conjunction with a health professional's humanity, sensitivity, or compassion. Nevertheless and despite a century of attempts, a consensually adopted definition of empathy does not exist, illustrating the word's theoretical complexity and multiple properties. For example, the following definitions or characterizations of empathy have been popularly advanced:

> An ability "to perceive the internal frame of reference of another with accuracy as if one were the other person but without ever losing the 'as if' condition." (Rogers 1959)
> "[T]he ability to put oneself into the mental shoes of another person to understand her emotions and feelings." (Goldman 1993)
> "A two-way street in which the distinct affective associations of individuals interact and mutually shape each other to yield new thought." (Halpern 2001)
> "An affective response that stems from the apprehension or comprehension of another's emotional state or condition, and that is similar to what the other person is feeling or would be expected to feel." (Eisenberg and Fabes 1990)
> "A complex form of psychological inference in which observation, memory, knowledge, and reasoning are combined to yield insights into the thoughts and feelings of others." (Ickes 1997)

The literature also discusses a variety of empathic ecologies or contexts, such as empathic concern, empathy for pain, empathic anger, empathic contagion, cognitive empathy, affective empathy, and clinical empathy (De Waal et al. 2006).

Despite these multiple definitions and characterizations, the above nevertheless suggest certain themes or commonalities that will dominate the remainder of this entry. For example, empathy is a decidedly relational phenomenon between an observer and another involving the observer's ability to leave his or her frame of self-reference and enter into the other's. Thus, empathy is more than sympathy or compassion since the one who empathizes seeks to gain a deep insight into the way the other's feelings and beliefs make sense out of experience. Such a perspective change in the observer requires both unconscious and affective processes as are present in sympathy, but as Halpern's and Ickes's characterizations of empathy suggest, the perspectival alteration also requires conscious, cognitive abilities at least involving understanding, memory, and inference. Considerable disagreement, however, exists over the degree of affective resonance versus conscious inference or reasoning that is necessary to practice empathy; the precise nature of the observer's empathic experience, especially as to whether it is possible or even desirable for him or her to sensorially "feel" rather than just infer, imagine, or understand the other's perspective; and, on a related point, in what empathic "attunement" or resonance precisely consists.

Introduction

Because the primary readership of this entry will likely be health professionals who might wish to improve the therapeutic possibilities of their relationships with patients, this entry will begin by examining some contemporary neuroscientific positions on empathy. Two will be of special interest: How empathy is understood to have occurred as a neuroevolutionary product, and how "empathy for pain," i.e., an observer's experience of another's pain behaviors or expressions, is characterized in contemporary neuroscience.

Characteristics

The Neuroevolution of Empathy

Although scholars, practitioners, and neuroscientists might argue about the essential properties of empathy, virtually all of them agree that empathy is a product of evolution's selective pressures bearing on reproduction and survival. Clearly, a species whose members can have some insight into or resonance with the experience of their conspecifics, especially when the latter are in distress, is more adaptively suited for survival than the species that cannot. Thus, primatologist Frans de Waal describes how infant rhesus monkeys will embrace and comfort another infant monkey who is screaming from having been

punished or rejected (De Waal et al. 2006), while Langford et al. (2006) and her colleagues from McGill University found that mice display pain behaviors like writhing when they observe the writhing behaviors of their cagemates (but not stranger mice) who are actually in pain.

The neuroevolutionary explanation of these phenomena looks to the discovery of "mirror neurons" which form neural circuits throughout the brain and are activated in observers when they witness another's behaviors (Decety and Jackson 2004). A good example is Dimburg's electromyographic findings of invisible, unconscious muscle contractions in humans' faces when they look at pictures of human facial expressions (Dimberg et al. 2000). Imaging studies of the activation patterns in their brains show them to be similar to the ones in the observed entity's brain, presumably causing the observer to have an experience that somewhat simulates the experience of the observed individual. These studies have identified structures in the anterior cingulate gyrus and the bilateral insula as especially prominent in empathy for pain – that is, in the observer's experience of another's pain (Fitzgibbon et al. 2010).

Evolutionary theory emphasizes the value of this kind of relational resonance, especially as it fosters nurturing and rearing offspring. Assisted by neurochemicals like oxytocin, opioids, and prolactin that foster parent–offspring attachment, the ability to sense and accommodate the needs of offspring is crucial to a species' reproductive success (Decety 2011). Furthermore, performing helping and comforting behaviors trigger a release of dopamine in the nucleus accumbens, causing pleasure sensations that are reinforcing. Nevertheless, the empathic properties being discussed here involve only neurologically primitive events that trigger largely affective responses in observers. Neuroscientists call this phenomenon "emotional contagion" wherein the experience of a sufferer like the infant rhesus monkey spreads to and is absorbed by other monkeys. In deploying soothing behaviors toward the suffering monkey, however, the infant monkeys are actually thought to be motivated by their need to soothe *their own* unpleasant feelings – as they are triggered by various nodes in their "mirroring" neural circuitry – not the sufferer's. As yet, these infant monkeys are neurologically unable to take the perspective of another, which is the essence of empathy (De Waal et al. 2006). Nevertheless and however narcissistically based these neural triggers subserving emotional contagion may be, they constitute an utterly essential first step in developing the more elaborate, clinically familiar empathic response that is at the heart of this entry. Absent the unconscious, gut-level feelings that mirror neurons enable, the more sophisticated expressions of clinical empathy would not occur. Also, the prominence of pain or other unpleasant feelings in the development of empathy is telling, as pain or distress not only alerts the sufferer to something wrong, but its public expression is communicated to others who can render help. Thus, "empathy for pain theory" is solidly in accord with the way evolutionary theory understands the evolving brain's early dependence on brain stem and limbic structures that use emotional and affective neural networks for survival information and strategies.

Empathy for Pain

According to mirror neuron theory, empathy for pain depends on the overlapping or sharing of certain neural circuits in both the person experiencing pain and the pain observer. Although neuroscientific findings are not unanimous, the circuitry in the bilateral anterior insula and the rostral anterior cingulate cortices seem consistently activated in both research subjects experiencing pain and in their observers. Disagreement exists, however, over the degree and quality of the observer's pain experience as some investigators like Singer et al (2004) have found only affective and not sensory components of the pain networks to be activated while other investigators claim that some observers actually feel the pain of the other (Lamm et al. 2011). Fitzgibbon et al. (2010) have pointed out, however, that discrepant findings like these might owe to the different research methodologies employed, as direct visual observation of people in pain seems more likely to activate sensory cortices among observers than nonvisual ones.

Also, studies vary regarding the intensity of the pain stimuli used, e.g., pin pricks versus Q-tip prods; the relationship differential between the observer and observed, e.g., close family members versus strangers; and the nature of the instructions given to the research subjects, e.g., whether research subjects receiving pain stimuli are instructed to suppress or evince pain behaviors. Consequently, these differing methodologic variables might account for disagreement over precisely what the pain observer experiences.

Bottom-Up and Top-Down Neural Networks

A well-accepted research finding in empathy for pain is that the pain experience in both the individual experiencing pain and the pain observer is governed by "bottom-up" neural circuitry that is automatic, unintentional, and reflexive, as well as by "top-down" networks that subserve intentional, controlled, purposive, thoughtful, and regulatory behavior (Craig et al. 2010). Automatic reactions among individuals experiencing pain might include screaming or facial expressions, while controlled reactions can take the form of self-reflective reports or purposeful regulations of motor activity, such as inhibiting facial grimacing. Automatic responses among humans observing pain behavior can include visceral, "gut-level," or reflexive reactions that sense or mimic the experience of the individual in pain, and top-down "reflective appraisal of the nature of the other person's situation, as well as attention to likely sources of the activity observed." (Craig et al. 2010) Importantly, bottom-up and top-down processes occurring in the pain observer are mediated by different neural networks, such that sophisticated, therapeutically intended empathic responses – which require effortful conscious activity – ultimately constitute a "choice" that is up to an observer or clinician to make.

The reflexive interplay of gut feelings and reflective behaviors, especially among pain observers, tokens a more sophisticated form of empathy that can include a curiosity as to why the other is experiencing pain, what events led to the sufferer's pain situation, and what an observer might do to resolve the situation, both for the individual in pain as well as for whatever pain threats might be presented to the observers. Consequently, although bottom-up processes will often activate top-down ones in humans, each one can regulate the other, ultimately resulting in modulating the nature and intensity of an empathic response (Goubert et al. 2005). This explains why human (and nonhuman primate) research on empathy for pain has found that the helping reactions of pain observers will vary according to whether the individual observed to be in pain is a family relative, a nonrelative, or a stranger. Indeed, investigations of physicians' empathy for the pain of their patients have found some evidence suggesting that certain doctors might intentionally diminish (or "down regulate") their gut-level feeling reactions to their patients, presumably because they believe the psychological costs of doing so are excessive, or that they want to conserve their attentional energy and only direct it toward treatment strategies (Shaw et al. 1994; Decety et al. 2010). Indeed, Kenny (2004) has discussed how a pain patient's complaint believability can influence the physician's empathic response, especially depending on the absence or presence of organic pathology.

This continuum of gut-level to cognitive to perspective taking empathic response recalls De Waal's analogizing empathy to a Russian Doll (De Waal et al. 2006). The Russian Doll model contains dolls within dolls, such that the innermost doll represents the neurologically primitive, limbic, unconscious response to one's own or another's pain. The next, larger doll would represent a more advanced, decidedly cognitive response that reflects on and assesses the situation and considers reasons that explain the other's situation. A third stage representing the largest doll would token the kind of empathic responses or strategies implicated in the definitions of empathy that begin this entry, especially involving "placing oneself in another's shoes" and exploring the way another's emotional associations "unify the details and nuances of the patient's life into an integrated experience." (Halpern 2001) The remainder of this entry will explore this last and most advanced form of empathy, which is called "clinical empathy."

The Affective Foundations of Clinical Empathy

The above shows that empathy evolved as a form of adaptive fitness that enhances the survival and reproductive success prospects of a species. Species with less neurologically evolved empathy will perform helping behaviors to soothe their feelings of distress caused by the distress signals of their conspecifics. But more advanced forms of empathy represent a conscious and intentional *other-regard*, where the observer attempts to assume the perspective of the other and perform helping measures for the other's benefit.

In clinical contexts involving health professionals' empathic responses to clients, empathic behaviors comprise a set of relation-building communicative skills. The empathic aim is not only to achieve a shared definition or characterization of the patient's experience that can inform interventional strategies, but also to develop an alliance that serves as a form of therapy in its own right.

Just as neurologically primitive empathic responses rest on feelings and emotions so, too, emotions and affects take center stage in the clinical empathy model. As numerous scholars have observed, affects and emotions play a critical role in "interpreting" or giving meaning to experiences, and in organizing and shaping the organism's behaviors relevant to those experiences (Westen and Blagov 2007; Halpern 2001). We are neurologically programmed to engage in behaviors whose outcomes are anticipated to feel gratifying, safe, pleasant, enjoyable, etc. and we are just as inclined to avoid behaviors and situations that are affectively anticipated to be distressing, painful, awkward, and humiliating. Thus, operational conditioning has been called a form of "affective forecasting," wherein such predictions inform and direct the organism's behaviors toward activities that would not jeopardize its safety and functional capacity (Westen and Blagov 2007). Consequently, affect theory maintains that gaining a deep comprehension of the meanings that people ascribe to their experience at least requires an imaginative entry into the *meaning constituting properties* of their feelings and emotions. In clinical contexts like pain medicine, wherein such understandings of the patient's experience might significantly contribute to if not determine clinical success, the clinician's ability to explore the patient's emotional and affective experience thus represents a powerful therapeutic tool.

Three Skills Bearing on Clinical Empathy

Curiosity and Checking

Clinical empathy fundamentally requires a willingness on the clinician's part to be curious about the patient's comprehension of and feelings about of his or her ailments or maladies. The empathic clinician will be disposed toward discerning the patient's perspective (or "stand in his or her shoes") and so will deploy curiosity-like responses toward the client such as exercising acute attention, probing the client's narrative, and subordinating his or her urge to formulate the next question or comment and just attend to the patient's narrative (Banja 2006; Matthews et al. 1993). The clinician will be conscious of his or her body language and nonverbal acknowledgement of the patient's narrative by sitting down, using good eye contact, and not interrupting. He or she will probe the client's narrative with directives or questions designed to gain a deeper insight into the patient's world-constitution such as "Tell me more about that," "How did that make you feel?" "What was that like?" or, simply, "Um-hmm, go ahead."

One of empathy's early champions, the famous therapist Carl Rogers (1959), encouraged clinicians to constantly check or recalibrate their understanding of the client's narrative. Here, the clinician might use empathic statements like "So, what you are saying is that..." or "Let me make sure I am understanding what you are saying. Your point is that ......" The aim here is not only to insure the reliability of the clinician's understanding of the patient but also to convey the clinician's valuing that information and, thus, showing his or her respect for the patient.

Unconditional Regard

Rogers's theory of "unconditional regard" also deserves comment since it is widely mentioned, if not universally embraced, as an important dimension of empathy. Rogers believed that successful therapy in counseling *only* consisted of (1) the persistent attempts of the therapist to explore and secure the perspective of the other's experience of the world "as if" he or she were the patient, and (2) an unconditional regard for the client, which Rogers characterized as "a warm acceptance of each aspect of the client's experience as being part of that client." (Rogers 1957) Kirschenbaum has further characterized Rogers's unconditional regard as creating "a psychological climate in which the client feels the kind of warmth, understanding and freedom from attack in which he may drop his defensiveness, and explore and reorganize his life style" (Kirschenbaum 1979). Also, Kirschenbaum (1979) noted that "whatever his [the client's] feelings and behaviors are or have been, no matter how troubling or frightening or socially disapproved of, he is still accepted as a worthy human being by the therapist."

Regardless of how far a pain specialist follows Rogers on these empathic strategies, it clearly accrues to both the clinician's and the patient's benefit to develop a relational atmosphere wherein a patient feels at liberty to discuss his or her innermost struggles, frustrations, disappointments, fears, and humiliations. Furthermore, some therapists have asserted that the more the clinician can enter into and understand the patient's perspective or world, the more the patient becomes humanized and the less his or her behaviors and decisions seem bizarre or upsetting (Halpern 2001).

Resonance and Attunement

According to virtually all accounts of clinical empathy, an important goal is reached when the clinician and patient achieve a resonance and attunement with one another that is characterized by a mutual sense of trust, understanding, strength, and commitment. Havens (2004) has written particularly sensitive accounts of this phenomenon in his discussions of relationship building. He writes about the clinician's needing "to find the patient, and the patient to find the clinician." (Havens 2004) He comes very close to Rogerian sensibilities in encouraging the clinician to identify and emphasize the patient's strengths, insure that the patient feels heard, and reflect back re-descriptions or reframings of the patient's experiences to achieve a shared understanding of the patient's therapeutic and existential experience.

Havens's account of empathy may be of particular value for any clinician treating pain patients in its concentrating on (1) the challenges in forming a therapeutic alliance with individuals who are disappointed with prior attempts to resolve their problems and (2) narrowing the relational distance between a healthy, reasonably adequate and successful clinician versus a patient who feels helpless, worthless, hateful, etc. Wassan and his colleagues (2005) observe how "[p]ain patients have a tendency to be angry, argumentative, mistrustful, anxious, and depressed...and...frequently present with comorbid emotional liabilities such as rage, inflexibility, and entitled behavior...They can also display destructive behaviors, such as threats of suicide, self-mutilation, extreme noncompliance with treatment, and opiod misuse." Havens (2004) wants to equip the clinician with empathic skills that can overcome these relational obstacles and that can mitigate the "collision" between a clinician's experience of the world and the patient's.

Havens focuses his empathic approach on deepening the patient's belief that "the clinician is on my side" – which does not imply that a clinician will support a patient's self-destructive habits like drug-seeking but rather that the clinician will not abandon or disrespect the patient and will "not shake what little faith the patient has in himself." (Havens 2004) Much like Rogers, Havens (2004) believes that "relationship building is the treatment" because "it makes other tasks possible...and because so many patients enter treatment precisely because they have relationship difficulties." Havens encourages clinicians to identify and emphasize their patients' strengths; refrain from blaming them; and acknowledge that the patient's navigational strategies, however maladaptive they might be, nevertheless seem reasonable to him or her.

Psychiatrist Jodi Halpern, however, believes that if resonance or attunement only implies a "merging" of egos, then, clinicians must go beyond them. She argues for empathic practice culminating in a kind of "emotional reasoning" wherein the clinician comes to understand the comprehensive feeling-thinking psyche of the patient. For Halpern, empathic practice lays the groundwork for developing an "interpretive context" wherein the clinician discerns how those meanings link up with the patient's other feelings and meaning constructs that inform the patient's world experience. Thus, Halpern understands attunement or reasonance more cognitively or imaginatively such that they enable better diagnoses and result in patients feeling more autonomy and therapeutic support (Halpern 2001).

Empathy's Clinical Value and Its Challenge

If this physiological and social isolation is a significant cause of human distress, illness exacerbates its unpleasantness and the onset of chronic pain can make it unbearable. The evolutionary purpose of pain sensory signaling is to alert the organism that something is wrong and to prompt remedial action (Decety 2011). However, chronic pain can also exacerbate the sufferer's sense of isolation such that it leaves him or her with no energy or vitality for anything else. It is hardly surprising, then, that some chronic pain patients present with the psychological symptoms that Wasan et al. (2005) describe. In so many instances, their pain incapacitates their ability to navigate their environments and triggers their feeling hopeless and embittered. The enduring experience of chronic pain and its libido-depleting impact can encourage a profoundly negative identity change, wherein patients come to understand themselves as inadequate, incapable, cursed, abandoned, and worthless.

The therapeutic value of empathy is thus understood as a mechanism that counters these relational and existential disappointments. An empathic approach enables the patient and the clinician to arrive at a mutually agreed-upon narrative that identifies the problems at issue, delineates a treatment course with shared expectations and responsibilities, and projects a shared vision of therapeutic success (Halpern 2001). Especially here, one sees how empathy's reach exceeds mere expressions of sympathy or compassion. The constant probing, inquiring, rechecking, and re-stating of the clinician's evolving, empathic insight requires both the patient and the clinician to refine whatever is at issue with increasingly greater clarity. The therapeutic dividends are that patients will often feel immensely respected and valued by the clinician's persistence and interest in gaining an understanding of what it is like to be them. Furthermore, by requiring the patient to confirm the clinician's empathic inferences and hunches, the patient is forced to reflect on his or her situation rather than just feel its oppressiveness and brutality. When done skillfully, empathic practices evolve the narrative of a therapeutic game plan to which both clinician and patient become committed and to which they can devote their hopes and energy in a coherent way (Halpern 2001).

The Challenge and Choice of Empathy

Pain patients will frequently "project" or share their feelings and beliefs with their clinicians who, after all, are holding themselves out as helping agents (Havens et al. 2001). The phenomenon of projection recalls the observations made above about clinical empathy as a choice or decision. The patient's presenting condition and sharing of painful psychic materials can easily trigger feelings of distress in the clinician that the clinician may be tempted to ignore, resist, or deny (Halpern 2003). At that moment, "top-down" executive neural processing can dispose the clinician to self-protective behaviors, whose purpose is to diminish or eliminate the discomfort triggered by the patient's distress. Empathic behaviors might seem ill-advised to some clinicians, who fear that taking the patient's perspective will prove too emotionally burdensome or too time consuming (Shaw et al. 1994). For example, Decety et al. (2010) have recorded event-related potentials of physicians observing pictures of body parts being pricked by a needle or touched by a Q-tip and compared their reactions to a nonphysician group. They noted that the physicians they studied tended to suppress their affective (bottom-up) response to the visual

stimuli in contrast to the nonphysician comparison group. While this might have professional survival value in reducing burnout and "compassion fatigue," it obviously presents a challenge to maintaining empathy (Banja 2008).

Other instances wherein clinicians might find their empathy severely challenged involve what psychiatrists call "projective identification." Here, patients who might have been victims of abuse and neglect unconsciously re-enact masochistic behaviors that aim to trigger a sadistic response from their "authority figure" clinicians (Passik et al. 2007). While the clinician may be entirely unaware of these patients' psychohistories, such patients are re-enacting scenarios with authority figures that do not feel right if they are not being abused. Less dramatic but common nevertheless are those pain patients who simply do not respond to the clinician's interventions and who can exhaust the most empathic clinician's patience. Thus, Cohen and his colleagues (2011) have written about "negative empathy," wherein the clinician's feelings of helplessness and inadequacy in the face of clinical failure result in his or her blaming or stigmatizing the patient for lack of progress. This entry will conclude with a discussion on how empathic practices can be learned and maintained without their inviting an unreasonable degree of psychological strain or burden on the clinician.

Empathy Training and Education

If a clinician (or any professional) wishes to develop empathic skills, he or she might learn much from three sources: the psychological aspects or character traits of empathic people; education or training in empathic techniques; and organizational factors that support or discourage the development of empathy.

The Empathic Personality

Perhaps the fundamental characteristics of empathic individuals involve their willingness to "be" with a distressed person along with their ability to explore that distress in considerable existential depth. As just discussed, many health professionals resist doing so for fear that it requires excessive emotional work, is psychologically threatening, or that it is unreasonably time consuming. The literature suggests, however, that empathic clinicians experience less stress or "burnout"; that the chances of developing a treatment plan to which the patient will commit are improved by an empathic relationship; and that the clinician's own humanity is often deepened (Halpern 2001). Using the revised NEO Personality Inventory, Hojat et al. (1999) found that physicians who were considered role models of empathic patient care scored considerably lower on psychological measures of anger, hostility, depression, self-consciousness, and vulnerability than comparison groups of internal medicine residents and the adult US population.

The empathic personality can be nicely described by contrasting it with its opposite formation, the pathological narcissist. Decety (2011) has noted that people who are skilled at empathy often exhibit feelings and behaviors demonstrating "attachment security" that facilitates their feeling safe and secure in the world. This observation may reflect the parenting they received, as observations on "good enough" parenting theorize that capable or healthy parents communicate a love to children bordering on the "unconditional" (Winnicott 1965). Alternatively, personality theory proposes that certain individuals with narcissistic traits or inclinations suffered, as children, from cold and distant parents who only loved "conditionally," i.e., only as long as the child accommodated parental expectations. These children experienced attachment insecurity as they always felt threatened and very anxious by the withdrawal of their parents' affection. Personality theorists have proposed that these children learn to cultivate a rigid set of beliefs and values that shape a series of lifestyle decisions – always reflecting parentally imposed lovability conditions – that subsequently compromises their ability to assume the perspective of others. Narcissists are notable for the intractability of their beliefs, attitudes, and world view, largely because those psychical materials inform who they are and maintain their always fragile self-esteem. Assuming the perspective of others is not only foreign to the narcissist's way of understanding the world but

would provoke his or her anxiety over trying out other world views (Banja 2005).

An alternative psychogenic account that contrasts the empathic personality with the narcissistic one looks to parents whose love for their child was suffocating and all-consuming. These children, who would otherwise be labeled "spoiled rotten," learn very early that any and all of their beliefs, inclinations, and choices are correct and privileged because their parents made them feel utterly precious and special. Consequently, their learned reaction toward anyone who disagrees with them or who holds contrary views is to label that person stupid or ridiculous. Just like his or her narcissistic counterpart who was deprived of healthy parental love, this "hyper-loved" narcissist would regard the idea of taking the perspective of another as foolish and a waste of time (Banja 2006).

Another salient contrast between the narcissistic and empathic personalities is the latter's practicing or espousing a humility marked by an acceptance of life's limitations, unfairness, disappointments, and tragedies. Unlike the narcissist who lives in an imagined (and false) world of idealization wherein others are always to blame for the narcissist's failures, the empathic clinician accepts the unpredictability and knowledge limitations of any clinical relationship. These persons would have given up a need to appear wondrous and glorious in their patients' eyes and instead acknowledge that occasional failure is inevitable and that much of what occurs during and after their client relationship is and always will be beyond their control (Banja 2011).

Developing Empathic Skills

Regardless of the degree of an individual's narcissism, virtually anyone can at least learn rudimentary empathic skills. This skill development generally takes two forms that involve cultivating humanistic insight and learning communication techniques that build rapport.

Developing insight whereby the clinician better discerns another's perspective might be enhanced by the study of narrative accounts of illness, where clinicians analyze the meaning behind the stories of others (Charon 2001);

group sessions like those originated and popularized by the psychiatrist Michael Balint (Novack et al. 1997), where clinicians recount challenging cases and look to one another for advice; mindfulness and meditation exercises, where clinicians become more attentive to psychological details of patient care and feel less distress in the midst of patients who exhibit misery, anger, or frustration; and self-awareness exercises, where clinicians practice exerting greater insight into their own defensive or coping styles (Novack et al. 1997). For these exercises to succeed, however, it is important to create a safe environment for clinicians to discuss their anxieties which, in turn, might encourage them to give up the idea that they must always exhibit a narcissistic-like self-confidence around their peers and patients.

Perhaps the most accessible dimension of empathy training, however, is simply learning a rudimentary set of empathic responses. A list of the most common ones is offered in Table 1. Even when the clinician only learns these empathic techniques (as a kind of empathy "lite"), the therapeutic impact on patients can be considerable. Note that these responses are inquisitive, nonjudgmental, and respectful, giving patients an impression of the clinician's concern. Although a familiar criticism of these utterances is that they might sometimes sound insincere and artificial, the more the clinician uses and becomes comfortable with them, the more authentic or real his or her empathy can become. Another opportunity for empathic skill development is for clinicians to study any of the countless videos of eminent psychotherapists practicing empathy, or to practice empathic techniques themselves with standardized patients and get their feedback.

Organizational Factors Affecting Empathy

Many clinicians learn to embrace or reject empathy from observing their role models. Branch et al. (2001) have noted that organizations wanting to encourage the use of empathy among clinicians should establish a climate of humanism, where clinicians consistently demonstrate interest in and respect for patients; where staff role model humility – such as in hospital settings where clinicians can adjust a patient's pillows,

Empathy, Table 1 Some empathic phrases and responses

Curiosity Statements
"So this must have caused/must be causing you a lot of . . . (stress, confusion, etc.)."
"I wonder what you are feeling right now."
"What is it about that that . . . (frightens, angers, perturbs, etc.) . . . you?"
"What is it about talking about that that . . . (upsets, saddens, peeves, etc.) . . . you?"
"What would you like to have happen from this?"
"Anything else?"
"I have a hunch that you are . . . (holding something back, feeling worse than you are letting on). . ."
Checking/clarifying statements
"This must be . . . (alarming, horrible, exasperating, etc.). . . for you to hear."
"This is obviously making you feel very . . ."
"Now, let me make sure I am understanding you. You are asking me . . . Is that correct?"
Probing statements
"Tell me more about that."
"And what was that like?"
"It must be hard for you to be here when you are so disappointed with . . . (me, how things have)".
Empowering statements
"That is an important question."
"That is amazing. And you got through it!"
"I hear you."

ask permission to sit, pull curtains for privacy, etc.; and where clinicians can receive constructive feedback on their relational skills. These organizations also recommend and take steps to have clinician behaviors observed and evaluated, provide feedback to staff, and encourage staff to step outside themselves and watch themselves be themselves.

Empathic excellence probably best develops from input or mentoring from others (Branch et al. 2001). It seems also true that cultivating superior empathic technique can require a lifetime of study and practice. However, when one considers the emotional work involved in practicing health care, the importance of the patient's attitude toward treatment, and the way empathy can augment the clinician's therapeutic skill set, gaining a reasonable degree of competence in empathic practice seems to be work well worth doing.

References

Banja, J. (2005). *Medical errors and medical narcissism.* Sudbury: Jones and Bartlett.

Banja, J. D. (2006). Empathy in the physician's pain practice: Benefits, barriers, and recommendations. *Pain Medicine, 7,* 265–275.

Banja, J. (2008). Toward a more empathic relationship in pain medicine. *Pain Medicine, 9,* 1125–1129.

Banja, J. (2011). Stigmatization, empathy, and the ego depletion hypothesis. *Pain Medicine, 12,* 1579–1580.

Branch, W. T., Kern, D., Haidet, P., et al. (2001). Teaching the human dimensions of care in clinical settings. *Journal of the American Medical Association, 286,* 1067–1074.

Charon, R. (2001). Narrative medicine: A model for empathy, reflection, profession, and trust. *Journal of the American Medical Association, 286,* 1897–1902.

Cohen, M., Quintner, J., Buchanan, D., Nielsen, M., Guy, L. (2011). Stigmatization of patients with chronic pain. *Pain Medicine, 12,* 1637–1643.

Craig, K. D., Versloot, J., Goubert, L., Vervoort, T., & Crombez, G. (2010). Perceiving pain in others: Automatic and controlled mechanisms. *The Journal of Pain, 11,* 101–108.

De Waal, F., Wright, R., Korsgaard, C. M., Kitcher, P., & Singer, P. (2006). *Primates and philosophers.* Princeton: Princeton University Press.

Decety, J. (2011). The neuroevolution of empathy. *Annals of the New York Academy of Sciences, 1231,* 35–45.

Decety, J., & Jackson, P. L. (2004). The functional architecture of human empathy. *Behavioral and Cognitive Neuroscience Reviews, 3,* 71–100.

Decety, J., Yang, C., & Cheng, Y. (2010). Physicians down-regulate their pain empathy response. *NeuroImage, 50,* 1676–1682.

Dimberg, U., Thunberg, M., & Elmehed, K. (2000). Unconscious facial reactions to emotional facial expressions. *Psychological Science, 11,* 86–89.

Eisenberg, N., & Fabes, R. A. (1990). Empathy: Conceptualization, measurement, and relation to prosocial behaivor. *Motivation and Emotion, 14,* 131–149.

Fitzgibbon, B. M., Giummarra, M. J., Georgiou-Karistianis, N., Enticott, P. G., & Bradshaw, J. L. (2010). Shared pain: From empathy to synaesthesia. *Neuroscience and Biobehavioral Reviews, 34,* 500–512.

Goldman, A. (1993). Ethics and cognitive science. *Ethics, 103,* 337–360.

Goubert, L., Craig, K. D., Vervoort, T., Morley, S., Sullivan, M. J. L., et al. (2005). Facing others in pain: The effects of empathy. *Pain, 118,* 285–288.

Halpern, J. (2001). From detached concern to empathy: *Humanizing Medical Practice.* Oxford, UK: Oxford University Press.

Halpern, J. (2003). What is clinical empathy? *Journal of General Internal Medicine, 18,* 670–674.

Havens, L. (2004). The best kept secret: How to form an effective alliance. *Harvard Review of Psychiatry, 12,* 56–62.

Havens, L., Vaillant, G., Price, B. H., Goldstein, M., & Kim, D. (2001). Soundings: A psychological equivalent of medical percussion. *Harvard Review of Psychiatry, 9,* 147–157.

Hojat, M. (2007). *Empathy in patient care: Antecedents, development, measurement, and outcomes.* New York: Springer.

Hojat, M., Naca, T. J., Maagee, M., et al. (1999). A comparison of the personality profiles of internal medicine residents, physician role models, and the general population. *Academic Medicine, 12,* 1327–1333.

Ickes, W. (1997). *Empathetic accuracy.* New York: The Guilford Press.

Kenny, D. T. (2004). Constructions of chronic pain in doctor-patient relationships: Bridging the communication chasm. *Patient Education and Counseling, 52,* 297–305.

Kirschenbaum, H. (1979). *On becoming Carl Rogers.* New York: Delta/Dell.

Lamm, C., Decety, J., & Singer, T. (2011). Meta-analytic evidence for common and distinct neural networks associated with directly experienced pain and empathy for pain. *NeuroImage, 54,* 2492–2502.

Langford, D. J., Crager, S. E., Shehzad, Z., Smith, S. B., Sotocinal, S. G., Levenstadt, J. S., Chanda, M. L., Levitin, D. J., & Mogil, J. S. (2006). Social modulation of pain as evidence for empathy in mice. *Science, 312,* 1967–1970.

Matthews, D. A., Suchman, A. L., & Branch, W. T. (1993). Making 'connexions': Enhancing the therapeutic potential of patient-client relationships. *Annals of Internal Medicine, 15,* 973–977.

Novack, D. H., Suchman, A. L., Clark, S., et al. (1997). Calibrating the physician: Personal awareness and effective patient care. *Journal of the American Medical Association, 278,* 502–509.

Passik, S. D., Byers, K., & Kirsh, K. L. (2007). Empathy and the failure to treat pain. *Palliative & Supportive Care, 5,* 167–172.

Rogers, C. (1957). The necessity and sufficient conditions of therapeutic personality change. *Journal of Consulting Psychology, 21,* 95–103.

Rogers, C. (1959). A theory of therapy, personality and interpersonal relationships, as developed in the client-centered framework. In S. Koch (Ed.), *A study of science. Study 1. Conceptual and systematic* (Formulations of the person and the social context, Vol. 3, pp. 184–256). New York: McGraw-Hill.

Shaw, L. L., Batson, C. D., & Todd, R. M. (1994). Empathy avoidance: Forestalling feeling for another in order to escape the motivational consequences. *Journal of Personality and Social Psychology, 67,* 879–887.

Singer, T., Seymour, B., O'Doherty, J., Kaube, H., Dolan, R. J., & Frith, C. D. (2004). Empathy for pain involves the affective but not sensory components of pain. *Science, 303,* 1157–1162.

Wasan, A. D., Wootton, J., & Jamison, R. N. (2005). Dealing with difficult patients in your pain practice. *Regional Anesthesia and Pain Medicine, 30,* 184–192.

Westen, D., & Blagov, P. (2007). A clinical-empirical model of emotion regulation: From defense and motivated reasoning to emotional constraint satisfaction. In J. J. Gross (Ed.), *Handbook of emotion regulation* (pp. 373–392). New York: Guilford Press.

Winnicott, D. W. (1965). *The maturational processes and the facilitating environment*. New York: International Universities Press.

Empathy and Pain

Liesbet Goubert[1], Tine Vervoort[1] and Kenneth D. Craig[2]
[1]Department of Experimental Clinical and Health Psychology, Ghent University, Ghent, Belgium
[2]Department of Psychology, University of British Columbia, Vancouver, BC, Canada

Synonyms

Clinical decision making; Judgment; Pain communication; Social and interpersonal features of pain

Definition

The term "empathy" has been defined in various ways, generally featuring the capacity to understand and respond to the unique affective experiences of another person (Decety and Jackson 2006). In the context of pain, a consensus definition of empathy has been advanced (see Goubert et al. 2005): "Empathy is best construed as a sense of knowing the experience of another person, with cognitive, affective and behavioral components."

Introduction

Biological and psychological determinants of pain, pain-related distress, and ▶ disability are increasingly well understood. Only in recent years have the social dimensions of pain received the clinical and research attention required by comprehensive models recognizing pain as a ▶ biopsychosocial phenomenon (Hadjistavropoulos et al. 2011). Of major importance has been increasing insight into how and when observers attend to, decode, interpret, emotionally react, and behave in response to another's pain. These advances have enhanced our understanding of pain as a social phenomenon. The last decade has brought increased attention to understanding the nature and mechanisms engaged by pain empathy. Insight into these determinants of observers' reactions to another's pain has important clinical consequences, as understanding why parents/partners or health-care professionals engage in particular actions contributes useful information to providing optimal care. As well, empathy for pain has implications for the observers' own experience of pain and pain-related ▶ coping. Empathy appears to be prerequisite to ▶ observational learning about pain; people learn about potentially dangerous situations, their impact, and pain-related ▶ coping by observing others (Goubert et al. 2011). Furthermore, insight into dysregulation of the empathic system may further our understanding of some "medically unexplained" pain conditions. For instance, cases have been reported in which (predominantly phantom) pain patients experience pain themselves by simply observing another person in pain, hypothesized to be caused by a dysfunctioning empathic system (Fitzgibbon et al. 2010).

We provide an overview of existing models and research on empathy as they relate to an extended model of pain empathy and discuss directions for future research.

Characteristics

Current Models of Empathy
A growing body of evidence indicates that observing other people in pain activates neural structures that overlap those involved in the direct experience of pain, including the anterior ▶ insula and the medial/anterior cingulate cortex (Jackson et al. 2006b; Lamm et al. 2011). This evidence is consistent with the "shared representations" account of empathy, proposed by Preston and de Waal (2002). In their "perception-action

model," shared representations between self and another person allow observers to come to understand the other's inner experience through activation of brain areas producing a similar state in the observer. Brain imaging studies have demonstrated that brain regions tapping into the ▶ affective-motivational properties of pain are activated when seeing someone else in pain (e.g., Singer et al. 2004). Moreover, some brain imaging studies have found evidence for the activation of brain regions subserving the sensory-discriminative properties of pain, suggesting that empathic responses to others in pain may also involve mimicry of sensory qualities of pain (Bufalari et al. 2007).

A meta-analysis by Lamm et al. (2011) has suggested that the differences found across studies in the activation of particular brain areas (e.g., somatosensory or affective areas) were due to differences in experimental paradigms, as the studies varied in the types of cues signaling pain to which research participants were exposed. The core neural network of empathy typically activated is the bilateral anterior insular cortex and medial/anterior cingulated cortex. This network could be co-activated with distinct brain regions dependent on whether the experimental methodology directed attention to specific features of the experience of pain and the situational context. Showing pictures of body parts in painful situations (e.g., Jackson et al. 2006a) simultaneously recruited core features as well as brain areas underpinning action understanding (inferior parietal/ventral premotor cortices). Showing abstract visual information about the other's affective state (e.g., by providing signals that another is receiving painful stimulation, e.g., Singer et al. 2004) engaged areas associated with inferring and representing mental states of self and other (precuneus, ventral medial prefrontal cortex, superior temporal cortex, and temporoparietal junction). Only in studies using pictures of body parts in painful situations were somatosensory brain sites also activated in observers. Based upon these meta-analytic findings, Lamm et al. (2011) suggested that somatosensory activation reflects nonspecific co-activation elicited by viewing body parts being touched rather than

a specific nociceptive somatosensory state. Further, there is evidence from meta-analyses that distinct and associated neural bases underlie not only the affective-perceptual features but also the cognitive-evaluative dimensions of empathy (Fan et al. 2011). Finally, in contrast to the common assumption that the mirror neuron system participates in pain empathy, Fan et al. (2011) showed it was not consistently activated in empathy, suggesting that this system is not central to empathy (see also Decety 2010).

Based upon the perception-action model, two models bearing upon pain empathy have emerged. A first model of empathy was developed by Goubert et al. (2005, 2009a), specifically in the context of pain. Within this model, empathy has been defined as a sense of knowing the experience of another person, including ▶ cognitive, ▶ affective, and ▶ behavioral components. The cognitive component constitutes estimation of the nature, meaning, and severity of the other's pain. On an affective level, two types of emotions may be evoked, those oriented to the self and those oriented to the other person. These may resemble the person in pain's affect but would not necessarily be isomorphic with this experience. Self-oriented affective responses involve feelings of distress for oneself, including apprehension of personal danger, whereas other-oriented affective responses refer to feelings that address the well-being of the other person, including concern or sympathy (see Goubert et al. 2005, 2009a). These two types of emotional reactions are likely to have divergent motivational and behavioral consequences. Self-oriented feelings of distress may motivate behavior aimed at relieving personal distress, in particular, seeking safety; other-oriented feelings of sympathy would motivate toward attending to the needs of the other and helping behavior (Batson 2009; Caes et al. 2011). The model also proposes that empathic responses are influenced by three types of determinants: (1) characteristics of the person in pain (i.e., "bottom-up" determinants, such as the ▶ pain behaviors of the person in pain; see Williams 2002; Craig et al. 2010), (2) characteristics of the observer (i.e., "top-down" determinants, such as the observers' learning experiences, goals, and beliefs about pain), and (3) contextual or relational determinants, for example, kinship between the person in pain and the observer. Since 2005, this model has been broadly used as a framework for empathy research in the context of pain.

A second model of empathy applied in the context of pain is the neurocognitive model outlined by Decety and Jackson (2006). This model emphasizes characteristics of "adequate" empathy and extends the understanding of shared representations as a necessary component for empathic appreciation of another's subjective experiences. It argues that a sense of self-awareness, or the observers' ability to differentiate between one's own and another's emotional responses, plays a pivotal role in the functional consequences of empathy. They conceptualized "adequate" empathy as the ability to take the perspective of another person, without confusing it with one's own interests. In other words, individuals need to disentangle their own experiences from others' experiences, a process which requires emotion self-regulation capabilities. "Adequate" empathy should thus involve higher brain activation in the right temporoparietal function, a brain region involved in self-other distinction (Spengler et al. 2009). The concept also accommodates our understanding that total concordance of personal experience with another's painful experience is improbable (and possibly undesirable) – rather, it is likely that most people operate in terms of "good enough" correspondence between the pain of another and their personal empathic reaction for the other (Goubert et al. 2009a).

An Extended Model of Empathy

Both models agree that empathy involves (a) a capacity to take the cognitive perspective of another person and (b) an affective response to the other person, which often, but not always, entails (partial) sharing of that person's emotional state. Both models also have in common an emphasis on bottom-up and top-down processing, as well as contextual factors. The Decety and Jackson (2006) model has a greater focus than the Goubert et al. (2005) formulation on "adequate" or functional forms of empathy.

The latter represents a general framework for understanding the processes involved in empathic responding to another's pain that should accommodate features of the Decety and Jackson model. It is noteworthy that Goubert and colleagues (2005) consider behaviors ensuing from decoding and emotionally responding to another's pain as a core component of empathy.

Based upon current evidence, this entry proposes an extended model of the Goubert et al. (2005) empathy model in the context of pain, incorporating elements from other models, including the Decety and Jackson (2006) model. First, it is argued that the cognitive component of empathy should be decomposed by examining different features. Understanding another's pain engages different, yet related, elements, that is, the detection and discrimination of available pain information, which implies attentional processing of another's pain-expressive behaviors, and the process of meaning attribution or interpretation of those bottom-up cues (Hadjistavropoulos et al. 2011). Thus, the "sense of knowing another's pain," as described in the Goubert et al. (2005) model, is made more explicit by examining constituent processes in the extended model. Second, the role of emotion regulation as it relates to the self-other distinction is examined as a crucial determinant of ensuing effective helping behaviors and is explicitly incorporated in the extended model of empathy. The extension of the Goubert et al. (2005) model also emphasizes the value of detailed analyses of how cognitive, affective, and behavioral components of empathy are elicited and linkages among the three components. The intent is to stimulate further research. Figure 1 provides an overview of the extended model of empathy, with the elements elaborated below.

The process of decoding another's pain provides the basis for the cognitive component of empathy and associated emotional reactions. A broad range of "bottom-up" information may be available to the observer, including the ▶ specific actions of the person in pain and information concerning the painful events inflicted upon the individual (e.g., a cut, burn, or sprained joint). The behavior of the person in pain may be conceptualized as primarily automatic or

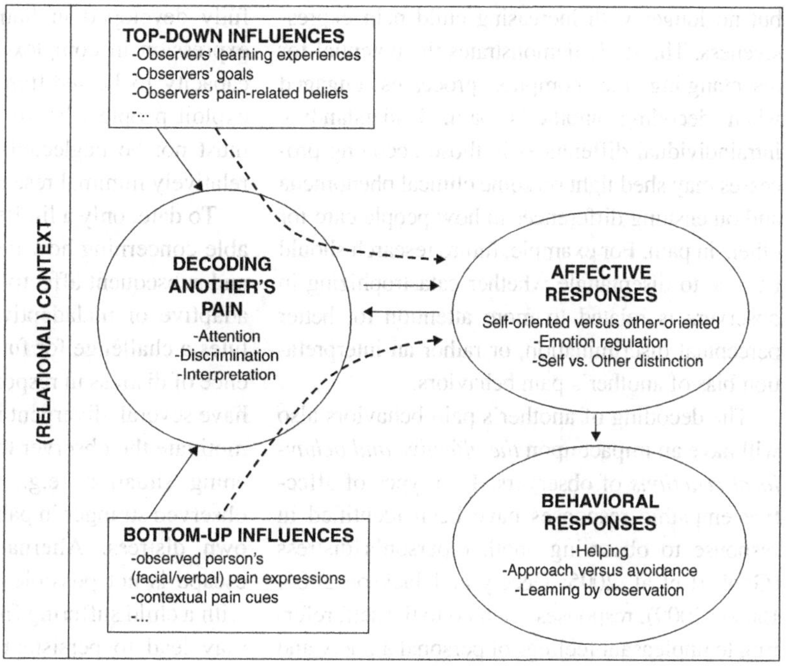

Empathy and Pain,
Fig. 1 An extended model
of empathy for another's
pain

controlled behavior. The former includes reflex-ive withdrawal from a source of pain, facial gri-maces, changes of paralinguistic qualities of voice, etc., whereas the latter includes self-report and organized behavior, such as use of medica-tion or accessing health care (McCrystal et al. 2011). Automatic behaviors instigate the peripheral autonomic and central limbic reactiv-ity associated with emotional reactions to the other's distress characterized as "gut level" reac-tions in observers, whereas controlled behaviors elicit considered appraisal of the nature of the person's distress (Craig et al. 2010).

Considerable research has focused on how the pain of others is estimated by observers, with the vast majority of studies showing underestimation of another's pain (see Kappesser and Williams 2010 and Prkachin et al. 2007, for an overview). Perfect agreement, however, would seem unlikely and perhaps a coincidence as an observer and the observed person in pain can be expected to use different cues to come to an estimate of pain (Kappesser and Williams 2010). In particu-lar, the person in pain has somatosensory experi-ence commanding attention. Fine-grained analyses of the processes underlying the estimate

of another person's pain would seem necessary to achieve a good understanding of how pain-expressive behavior/cues are processed and perceived by others.

Attentional processing of another's pain may underlie various empathic reactions, including pain estimations (Goubert et al. 2009b). Before decoding or other empathic processes can occur, attention is likely to be deployed to another's pain (cues). A recent study (Vervoort et al. 2011) examined parental attention to children's facial displays, ordinarily a commanding source of information. Findings indicated that, although high- and not low-▶ catastrophizing parents increasingly attended away from low pain faces, parental attention to child pain faces was facili-tated by increasing levels of child facial pain expressiveness, particularly when parental threat perceptions regarding child pain (i.e., catastrophizing thoughts) were high. The authors suggested that for high-catastrophizing parents, it may be that automatic orienting to children's pain faces instigates urges to escape or avoid paying attention to a child's pain in order to alleviate the aversive mood elicited by viewing child pain. This may be successful for low pain expression, but no longer with increasing child pain expressiveness. The study demonstrates the potential for disentangling the complex processes engaged when decoding another's pain. Understanding intraindividual differences in those decoding processes may shed light on some clinical phenomena and on ensuing differences in how people care for others in pain. For example, future research should be able to disentangle whether catastrophizing in observers is related to more attention to, better perceptual discrimination, or rather an interpretation bias of another's pain behaviors.

The decoding of another's pain behaviors also will have an impact upon *the affective and behavioral reactions* of observers. Two types of affective empathic responses have been identified in response to observing another person's distress (Goubert et al. 2005; Decety and Jackson 2006; Batson 2009): responses oriented to the self, referring to unpleasant feelings of personal anxiety and discomfort, sometimes including apprehension of personal danger, and responses oriented to the other, comprising feelings of sympathy and compassion. Research has shown that various bottom-up and top-down factors influence the type and extent of emotional reactions. For example, facial displays of pain (bottom-up) have been found to impact observers' distress (Williams 2002). Regarding observers' characteristics (top-down), interpreting another's pain as particularly threatening (i.e., catastrophizing thoughts) provokes a more intense experience of distress in partners (Leonard and Cano 2006) and parents (Caes et al. 2011; Goubert et al. 2006). On the other hand, psychopathic characteristics diminish affective responding to another's pain (Caes et al. 2012).

Emotions include action tendencies, with different emotions capable of motivating divergent behavioral reactions when observing another's pain. The personal experience of distress instigates an egoistic or self-oriented motivation to avoid or escape personal threat, a capacity probably originating in primordial ancestors. Feelings of sympathy on the other hand have been associated with an altruistic or other-oriented motivation, that is, the behavioral tendency to help another person (see Batson 2009). From an evolutionary perspective, this capacity would be most

fully developed in humans and sees its fullest expression in complex health-care systems. The capacity to be indifferent, or to be cruel, or to exploit people who are in pain or suffering also must not be neglected, yet these have received relatively minimal research attention.

To date, only a limited understanding is available concerning how decoding of another's pain and subsequent affective responses translate into adaptive or maladaptive behavior. This constitutes a challenge for future research. The experience of distress in response to another's pain may have several, divergent implications. First, it may motivate the observer to avoid/escape the threatening situation (e.g., walking away from an observed stranger in pain) in an attempt to reduce own distress. Alternatively, when avoidance/escape is not possible (e.g., in case of a parent with a child suffering from chronic pain), distress may lead to persistent (potentially ineffective) efforts to provide pain relief or (over)protective behaviors (Caes et al. 2011). Third, it may also result in learning experiences impacting upon behavior over an extended time frame, for example, a child observing another child acting fearful in response to needle prick may result in a child learning pain-related fear and wanting to avoid the next needle prick (Goubert et al. 2011).

Although our understanding of the action consequences of pain empathy is preliminary, studies typically have indicated that observers react automatically with distress and are predisposed to avoidance tendencies when seeing another individual in pain (e.g., Olsson and Phelps 2007; Avenanti et al. 2009; Caes et al. 2012). One study demonstrated that observers demonstrated heightened defensive reflexes when observing another person undergoing a painful test (Caes et al. 2012). Another study demonstrated that observation of painful stimulation delivered to another's hand automatically induced freezing and avoidance responses in the observers' corticospinal system (Avenanti et al. 2009). However, such defensive tendencies would seem to conflict with the often-observed emergence of other-oriented emotions (e.g., sympathy) and associated approach tendencies (e.g., comforting behavior) when viewing others in pain. How are observers inclined to approach and help others in pain when they themselves feel distressed? This refers to an intriguing paradox. It has been suggested that observers should be able to regulate their self-oriented distress, allowing other-oriented emotions to prevail thereby motivating approach (Goubert et al. 2009b; Caes et al. 2011). A key factor here may be observers' ability to monitor and regulate emotional processes to prevent confusion between self and other, that is, emotional self-regulation (Decety and Jackson 2006; Goubert et al. 2009b). Observers who are not able to disentangle their own feelings as evoked by observing the other's pain from the other's feelings may experience high levels of distress or anxiety. It remains to be investigated whether the capabilities of emotional self-regulation and making the self-other distinction allows other-oriented emotions and approach behaviors to prevail when being confronted with another in pain. One informative study by Decety et al. (2010) showed that physicians were better able to downregulate automatic distress reactions in response to observing pictures showing needle pricks to another's body parts than control participants. It was suggested that, in a clinical context, downregulation of distress in response to their patients' pain expression may have beneficial consequences such as an enhanced capacity to respond to patients' needs (Decety et al. 2010). This is corroborated by research in the developmental literature, showing that children with better effortful control exhibit greater empathic concern, as they are able to regulate negative vicarious emotion and maintain an optimal level of emotional arousal (Eisenberg and Eggum 2009). Alternatively, enhanced levels of personal distress may result in avoidance of the threatening event (e.g., patients' expressive pain behaviors) in an attempt to downgrade distress, which might give rise to ineffective management of patients' pain.

Summary

In recent years, accumulating research has shown that observing others in pain impacts others, attesting to the importance of the social dimensions in the conceptualization of pain as a biopsychosocial phenomenon. Specifically, observing pain-expressive behaviors activates a decoding process in observers and a variety of affective and behavioral responses. A systematic investigation of pain empathy is expected to add to a comprehensive model of empathy and guide clinical intervention strategies to relieve pain and suffering. Although knowledge on pain empathy is growing, the different components of the decoding process (i.e., attention, discrimination, interpretation) need further study. Also, further research should investigate the implications of (in)accurate decoding and the impact of (in)effective emotional self-regulation on (professional) caregiver behaviors.

Acknowledgment Tine Vervoort is postdoctoral researcher of the Fund for Scientific Research, Flanders. This manuscript was supported by grant no. G.0299.09 from the Fund for Scientific Research (FWO) Flanders.

References

Avenanti, A., Minio-Paluello, I., Sforza, A., & Aglioti, S. M. (2009). Freezing or escaping? Opposite modulations of empathic reactivity to the pain of others. *Cortex, 45*(9), 1072–1077.

Batson, C. D. (2009). These things called empathy: Eight related but distinct phenomena. In W. Ickes & J. Decety (Eds.), *The social neuroscience of empathy* (pp. 3–16). Cambridge, MA: The MIT Press.

Bufalari, I., Aprile, T., Avenanti, A., Di Russo, F., & Aglioti, S. M. (2007). Empathy for pain and touch in the human somatosensory cortex. *Cerebral Cortex, 17*, 2553–2561.

Caes, L., Uzieblo, K., Crombez, G., De Ruddere, L., Vervoort, T., & Goubert, L. (2012). Negative emotional responses elicited by the anticipation of pain in others: psychophysiological evidence. *The Journal of Pain, 13*, 467–476.

Caes, L., Vervoort, T., Eccleston, C., Vandenhende, M., & Goubert, L. (2011). Parental catastrophizing about child's pain and its relationship with activity restriction: The mediating role of parental distress. *Pain, 152*, 212–222.

Craig, K. D., Versloot, J., Goubert, L., Vervoort, T., & Crombez, G. (2010). Perceiving pain in others: Automatic and controlled mechanisms. *The Journal of Pain, 11*(2), 101–108.

deC Williams, A. C. (2002). Facial expression of pain: An evolutionary account. *The Behavioral and Brain Sciences, 25*(4), 439–488.

Decety, J. (2010). To what extent is the experience of empathy mediated by shared neural circuits? *Emotion Review, 2*, 204–207.

Decety, J., & Jackson, P. L. (2006). A social-neuroscience perspective on empathy. *Current Directions in Psychological Science, 15*, 54–58.

Decety, J., Yang, C. Y., & Cheng, Y. W. (2010). Physicians down-regulate their pain empathy response: An event-related brain potential study. *NeuroImage, 50*(4), 1676–1682.

Eisenberg, N., & Eggum, N. D. (2009). Empathic responding: Sympathy and personal distress. In W. Ickes & J. Decety (Eds.), *The social neuroscience of empathy* (pp. 71–83). Cambridge, MA: The Mitt Press.

Fan, Y., Duncan, N. W., de Greck, M., & Northoff, G. (2011). Is there a core neural network in empathy? An fMRI based quantitative meta-analysis. *Neuroscience and Biobehavioral Reviews, 35*(3), 903–911.

Fitzgibbon, B. M., Giummarra, M. J., Georgiou-Karistianis, N., Enticott, P. G., & Bradshow, J. L. (2010). Shared pain: From empathy to synaesthesia. *Neuroscience and Biobehavioral Reviews, 34*, 500–512.

Goubert, L., Craig, K. D., Vervoort, T., Morley, S., Sullivan, M. J. L., deC Williams, A. C., Cano, A., & Crombez, G. (2005). Facing others in pain: The effects of empathy. *Pain, 118*, 285–288.

Goubert, L., Craig, K. D., & Buysse, A. (2009a). Perceiving others in pain: Experimental and clinical evidence on the role of empathy. In W. Ickes & J. Decety (Eds.), *The social neuroscience of empathy* (pp. 153–165). Cambridge, MA: The MIT Press.

Goubert, L., Eccleston, C., Vervoort, T., Jordan, A., & Crombez, G. (2006). Parental catastrophizing about their child's pain. The parent version of the pain catastrophizing scale (PCS-P): A preliminary validation. *Pain, 123*(3), 254–263.

Goubert, L., Vervoort, T., & Crombez, G. (2009b). Pain demands attention from others: The approach/avoidance paradox. *Pain, 143*, 5–6.

Goubert, L., Vlaeyen, J. W. S., Crombez, G., & Craig, K. D. (2011). Learning about pain from others: An observational learning account. *The Journal of Pain, 12*, 167–174.

Hadjistavropoulos, T., Craig, K. D., Duck, S., Cano, A., Goubert, L., Jackson, P. L., Mogil, J. S., Rainville, P., Sullivan, M., de C. Williams, A. C., Vervoort, T., & Fitzgerald, T. D. (2011). A biopsychosocial formulation of pain communication. *Psychological Bulletin, 137*, 910–939.

Jackson, P. L., Brunet, E., Meltzoff, A. N., & Decety, J. (2006a). Empathy examined through the neural mechanisms involved in imagining how I feel versus how you feel pain. *Neuropsychologia, 44*(5), 752–761.

Jackson, P. L., Rainville, P., & Decety, J. (2006b). To what extent do we share the pain of others? Insight from the neural bases of pain empathy. *Pain, 125*(1–2), 5–9.

Kappesser, J., & Williams, A. C. D. (2010). Pain estimation: Asking the right questions. *Pain, 148*(2), 184–187.

Lamm, C., Decety, J., & Singer, T. (2011). Meta-analytic evidence for common and distinct neural networks associated with directly experienced pain and empathy for pain. *NeuroImage, 54*, 2492–2502.

Leonard, M. T., & Cano, A. (2006). Pain affects spouses too: Personal experience with pain and catastrophizing as correlates of spouse distress. *Pain, 126*(1–3), 139–146.

McCrystal, K. N., Craig, K. D., Versloot, J., Fashler, S. R., & Jones, D. N. (2011). Perceiving pain in others: Validation of a dual processing model. *Pain, 152*, 1083–1089.

Olsson, A., & Phelps, E. A. (2007). Social learning of fear. *Nature Neuroscience, 10*, 1095–1103.

Preston, S. D., & de Waal, F. B. M. (2002). Empathy: Its ultimate and proximate bases. *The Behavioral and Brain Sciences, 25*, 1–71.

Prkachin, K. M., Solomon, P. E., & Ross, J. (2007). Underestimation of pain by health-care providers: towards a model of the process of inferring pain in others. *CJNR (Canadian Journal of Nursing Research), 39*, 88–106.

Singer, T., Dolan, R. J., & Frith, C. D. (2004). Empathy for pain involves the affective but not sensory components of pain. *Science, 303*, 1157–1162.

Spengler, S., von Cramon, D. Y., & Brass, M. (2009). Was it me or was it you? How the sense of agency originates from ideomotor learning revealed by fMRI. *NeuroImage, 46*, 290–298.

Vervoort, T., Caes, L., Crombez, G., Koster, E. H. W., Van Damme, S., Dewitte, M., & Goubert, L. (2011). Parental catastrophizing about children's pain and selective attention to varying levels of facial expression of pain in children: A dot-probe study. *Pain, 152*, 1751–1757.

Employment Assessment

▶ Vocational Assessment in Chronic Pain

ENaC/DEG

Definition

ENaC/DEG is a family of ion channel proteins, several gated by low pH, for example, the ASIC channel.

Cross-References

▶ Species Differences in Skin Nociception

Enamel

Definition

Enamel is the hardest biological tissue known and forms the crown of the tooth. It overlies the softer and more flexible dentin.

Cross-References

▶ Dental Pain, Etiology, Pathogenesis, and Management

Enantiomer of Tramadol

Definition

Enantiomer of tramadol refers to one of the two stereoisomers that comprise tramadol, a racemate.

Cross-References

▶ Drugs with Mixed Action and Combinations: Emphasis on Tramadol

Encephalopathy

Definition

Encephalopathy is a frequent symptom in cerebral vasculitis and is a manifestation of systemic vasculitides and the isolated angiitis of the central nervous system.

Cross-References

▶ Headache Due to Arteritis

Encoding of Noxious Information in the Spinal Cord

Robert Coghill
Department of Neurobiology and Anatomy, Wake Forest University School of Medicine, Winston-Salem, NC, USA

Synonyms

Encoding of noxious stimuli; Nociceptive processing; Spinal cord nociception

Definition

The encoding of noxious information by the spinal cord is the process by which information from ▶ primary afferents/neurons is assimilated, integrated, and prepared for transmission to central nervous system regions important in nociception and nociceptive reflex modulation. Information from ▶ descending modulation systems of the brain can strongly shape afferent information processing at the spinal level and contributes substantially to spinal nociceptive processing.

Characteristics

The subjective sensory experience of pain is multidimensional and encompasses a conscious appreciation of the intensity, location, quality, and temporal aspects of a stimulus that damages (or holds the potential to damage) the integrity of the body. Given that the experience of pain evoked by a noxious stimulus is largely built upon sensory information ascending from the spinal cord to the brain, information about the subjectively available sensory features of a noxious stimulus must necessarily be encoded and transmitted by neurons within the spinal cord.

The processing of nociceptive information within the spinal cord gray matter is accomplished by neurons with two broad classes of response properties, ▶ wide dynamic range neurons (WDR) and ▶ nociceptive specific neurons (NS). WDR neurons were the first class of nociceptive neurons to be identified (Wall and Cronly-Dillon 1960) and were named on the basis of their responsiveness to a wide range (i.e., noxious and innocuous) of stimulus intensities (Mendell 1966). NS neurons were identified several years later in 1970 (Christensen and Perl 1970). As their name implies, NS neurons respond exclusively to noxious stimulus intensities. Both classes of neurons are interspersed through laminae I, II, IV, V, VI, VII, and X of the spinal cord gray matter. However, NS neurons tend to predominate in the marginal layers of the dorsal horn, while WDR neurons predominate in the deeper lamina. Both classes of neurons project supraspinally to brain regions involved in pain and/or pain modulation. Since their discovery, debate has raged over the functional significance of both NS and WDR neurons.

Much of the current understanding of spinal nociceptive processing is significantly limited by the failure to explore and appreciate how populations of spinal neurons work together to encode pain. In particular, the contribution of WDR neurons to the encoding of noxious stimuli cannot be understood if the responses of single WDR neurons are considered in isolation. Single WDR neurons respond with equal vigor to both noxious and innocuous stimuli and are, therefore, incapable of encoding a distinction between a noxious and innocuous stimulus. However, these cells have complex ▶ receptive field properties that result in population responses being markedly different from the responses of single neurons within that population. WDR neurons have a small, central receptive field zone in which both noxious and innocuous stimuli evoke increases in discharge frequencies (see ▶ Discharge Frequency). This central receptive field zone is surrounded by a larger peripheral zone in which only noxious stimuli are sufficient to elicit increases in discharge frequency. Both receptive field zones have gradients of sensitivity, such that progressively greater discharge frequencies are evoked as the noxious

stimulus is applied progressively closer to the central receptive field zone. If the stimulated region of the body surface is sufficiently large to encompass multiple, overlapping peripheral receptive field zones of WDR neurons, then a noxious stimulus would recruit far more WDR neurons than an innocuous stimulus of equal size. Thus, the total output of the WDR population would be far greater for the noxious stimulus. Functional imaging studies of the spinal cord have confirmed that noxious stimuli produce far greater rostrocaudal recruitment of neuronal activity than innocuous stimuli (Coghill et al. 1993a) and strongly suggest that the output of populations of WDR neurons is sufficient to support a subjective distinction between noxious and innocuous stimuli. Importantly, both NS and WDR neurons are activated by noxious stimuli under most conditions and both probably work together to encode distinctions between noxious and innocuous stimuli.

To date, the neural mechanisms which encode the intensity of a noxious stimulus represent the best understood dimension of spinal nociceptive processing. Both NS and WDR neurons exhibit monotonic increases in discharge frequencies as the intensity of noxious stimulation increases, so both classes of neurons are capable of contributing to the subjective experience of pain intensity. However, WDR neurons are more sensitive to smaller changes in stimulus intensity in that they have steeper stimulus-response curves than NS neurons (Price et al. 1978). Furthermore, the stimulus-response functions obtained from WDR neurons closely parallel those obtained from human psychophysical studies (Price et al. 1978). Finally, WDR neurons encode sufficiently small distinctions in ▶ noxious stimulus intensity to support behavioral discriminative capacity (Maixner et al. 1986). Population recruitment also appears to be a critical dimension in the encoding of stimulus intensity. Progressive increases in noxious stimulus intensity recruit a progressively greater rostrocaudal distribution of neuronal activity (Coghill et al. 1991).

Mechanisms supporting the encoding of stimulus location by spinal cord neurons remain poorly understood (see ▶ Noxious Stimulus Location). There is, however, a clear somatotopic organization, with caudal body regions being represented in the caudal spinal cord and rostral body regions being represented in the rostral spinal cord. There is also a mediolateral organization, where distal structures are represented in the medial aspect of the dorsal horn and proximal structures are represented in the lateral aspect of the dorsal horn. Clearly, a significant degree of spatial processing and modulation occurs at the spinal level. In spinal cord-transected animals, spinally mediated reflex withdrawal responses are appropriate for the body site stimulated. Furthermore, these responses can be modulated in a complex and elegant fashion when multiple body sites are stimulated simultaneously (Le Bars et al. 1979; Morgan 1999). Such interactions probably serve to generate the most appropriate withdrawal response to complex, multifocal painful stimuli and underscore the fact that the spinal cord is equipped to both process and utilize spatial information. It remains unknown if this spinally processed spatial information can become subjectively available. Both NS and WDR neurons may participate in the encoding of spatial information. NS neurons appear to be ideally suited for the encoding of stimulus location based on their relatively small and well-localized receptive fields. The complex receptive fields of WDR neurons could also yield population responses with a high degree of spatial information, but no existing data have confirmed such population-based encoding of stimulus location. However, population-based mechanisms may be significantly involved in the perceptual radiation of pain. Since recruitment of spatially remote neuronal populations occurs during intense noxious stimulation, such recruitment may contribute substantially to the experienced spatial distribution of a painful stimulus (Price et al. 1978; Coghill et al. 1991).

The encoding of ▶ stimulus quality (i.e., the subjective experience of burning, stinging, crushing, pinching) has been minimally explored. A subclass of NS neuron, however, receives selective input from A-δ afferents that respond solely to noxious mechanical information. Such

neurons would be well positioned to encode a distinction between noxious mechanical and noxious thermal information. WDR neurons receive input from a number of different classes of primary afferents and appear unlikely to make a clear contribution to the encoding of stimulus quality.

The experience of pain has multiple temporal aspects derived from interactions between differing populations of primary afferents and spinal cord neurons. In particular, when a noxious stimulus is applied to the distal portion of an extremity, the distinction between information carried by rapidly conducting (Aδ) nociceptors and slowly conducting (C) nociceptors becomes sufficiently amplified to become subjectively available. The sharp, well-localized pain sensation that is first perceived (▶ First Pain) is thought to arise largely from information transmitted by Aδ nociceptors, while the later, more diffuse burning sensation (▶ Second Pain Assessment) is thought to reflect information arising from C nociceptors. Both NS and WDR neurons have convergent input from Aδ and C fibers and exhibit responses that would be consistent with their ability to support first and second pain. In the case of prolonged, repetitive noxious stimulation, C afferents undergo progressive decreases in their discharge frequencies. However, if interstimulus intervals are sufficiently brief, the discharge frequencies of spinal nociceptive neurons actually increase and perceived pain increases over time. This progressive ▶ temporal summation has been termed "▶ wind-up" (Mendell and Wall 1965). Neurons in both the deep and superficial laminae exhibit such temporal summation and could be sufficient to subserve temporal summation of pain and to overcome progressive diminution of C afferent input. In the case of maintained pain, NS and WDR neurons exhibit differing responses over the course of prolonged nociceptive stimuli. NS neurons exhibit prolonged responses to mechanical stimuli but exhibit significant adaptation to prolonged thermal stimuli. In contrast, WDR neurons exhibit prolonged responses to both heat and capsaicin pain. As such, either class of neuron appears sufficient to encode the duration of prolonged nociceptive stimuli of the appropriate modality (Coghill et al. 1993b).

The processing and encoding of nociceptive information by spinal cord neurons is both complex and elaborate. Clearly multiple spinal cord regions interact during the processing of noxious information. Neurons in the superficial dorsal horn can modulate the responses of neurons in deeper laminae (Suzuki et al. 2002). Spatially remote stimuli can modulate responses of ongoing stimuli (Le Bars et al. 1979; Morgan 1999). Significant gaps in our understanding of these processes represent substantial obstacles to the better understanding and treatment of pain. Although much work as been done at the level of the single neuron, much more work is needed to understand how populations of nociceptive neurons encode the magnitude, spatial, and temporal features of noxious stimuli.

References

Christensen, B. N., & Perl, E. R. (1970). Spinal neurons specifically excited by noxious or thermal stimuli: Marginal zone of the dorsal horn. *Journal of Neurophysiology, 33,* 292–307.

Coghill, R. C., Price, D. D., Hayes, R. L., et al. (1991). Spatial distribution of nociceptive processing in the rat spinal cord. *Journal of Neurophysiology, 65,* 133–140.

Coghill, R. C., Mayer, D. J., & Price, D. D. (1993a). The roles of spatial recruitment and discharge frequency in spinal cord coding of pain: A combined electrophysiological and imaging investigation. *Pain, 53,* 295–309.

Coghill, R. C., Mayer, D. J., & Price, D. D. (1993b). Wide dynamic range but not nociceptive specific neurons encode multidimensional features of prolonged repetitive heat pain. *Journal of Neurophysiology, 69,* 703–716.

Le Bars, D., Dickenson, A. H., & Besson, J. M. (1979). Diffuse noxious inhibitory controls (DNIC). I. Effects on dorsal horn convergent neurones in the rat. *Pain, 6,* 283–304.

Maixner, W., Dubner, R., Bushnell, M. C., et al. (1986). Wide-dynamic-range dorsal horn neurons participate in the encoding process by which monkeys perceive the intensity of noxious heat stimuli. *Brain Research, 374,* 385–388.

Mendell, L. M. (1966). Physiological properties of unmyelinated fiber projection to the spinal cord. *Experimental Neurology, 16,* 316–332.

Mendell, L. M., & Wall, P. D. (1965). Response of single dorsal cord cells to peripheral cutaneous unmyelinated fibres. *Nature, 206,* 97–99.

Morgan, M. M. (1999). Paradoxical inhibition of nociceptive neurons in the dorsal horn of the rat spinal cord during a nociceptive hindlimb reflex. *Neuroscience, 88,* 489–498.

Price, D. D., Hayes, R. L., Ruda, M., et al. (1978). Spatial and temporal transformations of input to spinothalamic tract neurons and their relation to somatic sensations. *Journal of Neurophysiology, 41,* 933–947.

Suzuki, R., Morcuende, S., Webber, M., et al. (2002). Superficial NK1-expressing neurons control spinal excitability through activation of descending pathways. *Nature Neuroscience, 5,* 1319–1326.

Wall, P. D., & Cronly-Dillon, J. R. (1960). Pain, itch, and vibration. *Archives of Neurology, 2,* 365–375.

Encoding of Noxious Stimuli

Martin Schmelz
Department of Anesthesiology in Mannheim, University of Heidelberg, Mannheim, Germany

Definition

External and internal signals interacting with the body are transformed by sensory nerve terminals into electrical signals, such that the frequency of action potentials in their axons encodes the intensity, duration, and rate of change of the stimulus. Classically, the spike initiation site, separating receptive nerve terminal and conductive axon, is the crucial point at which a stimulus interacting with a sensory nerve terminal is encoded to trains of action potentials, which are then propagated centrally.

Characteristics

Electrophysiological investigation of nociceptive nerve endings is hampered by their small size. Results from extracellular recordings in nociceptive terminals in the guinea pig cornea suggest, that the generator potential and action potentials can be measured by superficially placed patch clamp pipettes (Brock et al. 2001; Carr et al. 2002, 2003). Spike generation in the

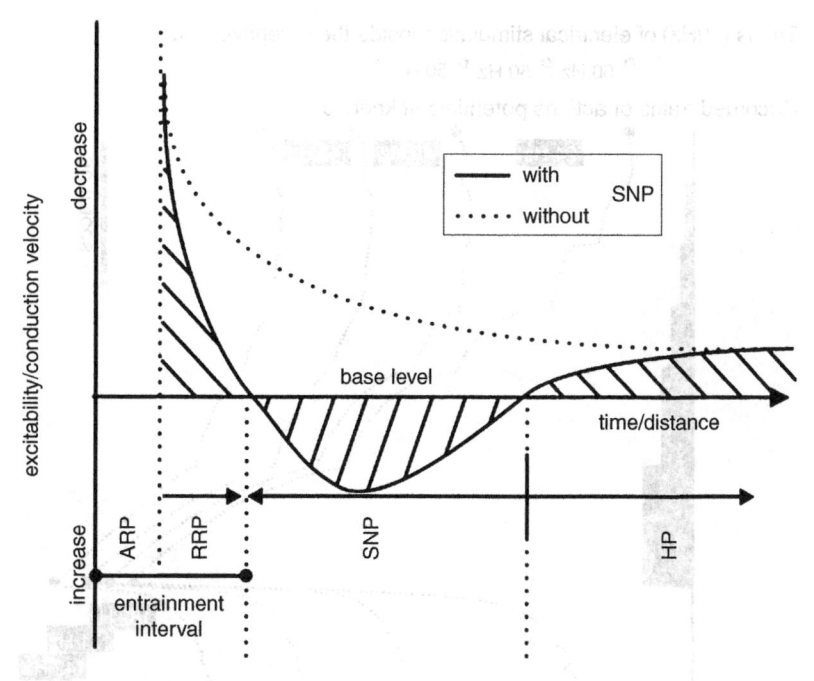

Encoding of Noxious Stimuli, Fig. 1 Post-excitatory membrane potential changes following the action potential. *ARP* absolute refractory period, during which no second *AP* can be induced (frog A-fiber 1.5–2 ms). RRP, relative refractory period that lasts from the end of APR to the intersection of threshold with the control level (frog A-fiber 1–2.5 ms, ARP + RRP human C-fiber > 5–10 ms). SNP, facultative supernormal period that extends from the end of RPR to the second intersection with the control level (frog A-fiber ~500 ms, human Aβ-fiber 15–20 ms, human Aα-fiber 4 s, human C-fiber ~500 ms). *HP* the hypoexcitable period that lasts from the end of SNP to normalization of threshold (frog A-fiber and human C-fiber minutes) (Modified from Weidner et al. 2000b)

terminals was not sensitive to tetrodotoxin (Brock et al. 1998) but blocked by lidocaine (Brock et al. 2001) implying the importance of TTX-resistant sodium channels for spike initiation (Black and Waxman 2002). The exact location of the spike initiation site is still unknown, but in recordings from the guinea pig cornea it has been suggested that it might shift distally upon hyperpolarization (Carr et al. 2009). Recently, even direct recordings from sensory nerve endings succeeded showing the expression of both, TTX-sensitive and TTX-insensitive sodium channels and a membrane potential of about -60mV (Vasylyev and Waxman 2012).

Post-excitatory Modulation of Excitability
The axon has been regarded mainly as passively propagating action potentials; however, along their axonal path action potential frequencies can be massively modulated. During the time period immediately following an action potential (AP), successive periods of post-excitatory conduction and excitability modulation are known (Weidner et al. 2000b). The absolute refractory period is directly followed by the relative refractory period, during which action potentials are conducted slower (Fig. 1). After the relative refractory period, a supernormal period (SNP) with increased conduction velocity and excitability may follow in some nerve fibers. After the short-lasting SNP, a long-lasting period of conduction velocity slowing and reduced excitability can be regularly observed (hypoexcitable period). In humans (Serra et al. 1999; Weidner et al. 1999; Bostock et al. 2003) and in animals (Gee et al. 1996; Thalhammer et al. 1994; Raymond et al. 1990), this lasting slowing of conduction has been

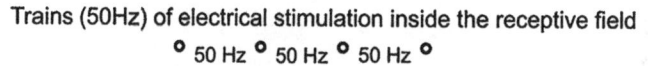

Trains (50Hz) of electrical stimulation inside the receptive field

° 50 Hz ° 50 Hz ° 50 Hz °

Recorded trains of actions potentials at knee level:

Encoding of Noxious Stimuli, Fig. 2 Changes in instantaneous frequency with repetitive impulse firing. Repetitive trains of 4 pulses were applied to the innervation territory in the foot at 50 Hz, and responses of a C nociceptor were recorded by microneurography in the peroneal nerve at knee level. Responses of the nociceptor are depicted in one trace and subsequent stimulations are shown from *top* to *bottom*. Interspike intervals of the first response were longer as compared to the interstimulus interval, leading to instantaneous frequencies of 24–40 Hz (*white letters* in *black box*). When the repetition frequency of the trains was increased stepwise (*left panel*), instantaneous frequencies of the evoked response gradually increased, although the interstimulus interval was kept constant at 20 ms (50 Hz). With increasing repetition, frequency activity–dependent hyperpolarization becomes more intense and supernormal conduction becomes more pronounced. Thus, gradually higher instantaneous frequencies of the evoked responses are observed far above the stimulation frequency in the periphery of 50 Hz. Maximum frequencies (entrainment) of more than 180 Hz were recorded in this example (Modified from Weidner et al. 2000b)

shown to correlate with the receptive properties of the nerve fibers. Increased slowing was found to reduce mechanical responsiveness in rat skin (Taguchi et al. 2010) and dura mater (Decol et al. 2012).

Axonal Modulation of Discharge Frequency

The activity–induced conduction velocity slowing typically leads to a reduction of the conduction velocity of a subsequent action potential, i.e., the interval between two successive action potentials increases along the axonal propagation, and consequently the impulse frequency decreases with axonal length (Fig. 2). However, should the subsequent action potential fall into the supernormal period, it will be conducted faster than the preceding one, and, i.e., the instantaneous frequency will increase with axonal length. The supernormal period in human C-fibers increases with the magnitude of the activity-dependent hyperpolarization. Therefore, the more action potentials an axon had conducted before, the more pronounced the activity-dependent hyperpolarization and the supernormal period will be. Thus, instantaneous frequencies will gradually increase along the axon, eventually exceeding the original stimulation frequency in the peripheral innervation territory. Remarkably, with a high degree of hyperpolarization, the instantaneous frequencies can reach the maximum frequency of C axon ("entrainment frequency"), which is about 200 Hz. Thereby, a peripheral stimulation at a frequency of 50 Hz can provoke spike trains that will increase in instantaneous frequency along the axon, and may reach the spinal cord at a frequency that is 3–4 × higher than the original frequency generated in the periphery (Weidner et al. 2002).

The most interesting part of this axonal modulation of the discharge frequency is given by the fact that modulation of instantaneous frequency is determined by the degree of hyperpolarization of the axon, which depends on the history of the stimulated fiber: The more active a unit has been in the time preceding a stimulus, the more prone it will be to accelerate the subsequent axon potentials. The discharge frequency reaching the spinal cord can, under these conditions, be increased by a factor of 3–4. This mechanism might, therefore, compensate in part for fatigue or adaptation of the peripheral endings when confronted to sustained or repetitive stimulation. Moreover, it might contribute to windup phenomena seen with repetitive electrical stimulation, which is usually attributed to postsynaptic processing in the spinal cord.

The acceleration of a subsequent action potential is restricted to the supernormal period after an action potential, which lasts from about 15–300 ms in human C-fibers. While instantaneous frequencies can be increased in this time, actions potential arising later will be slowed down. Thus, this modulation will also act as some kind of contrast enhancement mechanism.

Modulation of Action Potential Shape

Repetitive discharge is known to prolong the duration of action potentials. As action potentials of longer duration provoke higher calcium influx, the synaptic strength will be increased (Geiger and Jonas 2000). The modulatory effects of axonal propagation, including propagation failures, axo-axonal coupling, and reflected propagation, are extensively discussed in a recent review (Debanne 2004).

Unidirectional Block

Excitation of one peripheral ending in the axonal tree will generate action potentials which are conducted centrally, but at the branching points they will also propagate antidromically. Assuming that at each branching point the incoming action potential would be conducted in both directions, the entire axonal tree would be depolarized (see also axon reflex, neurogenic inflammation). Should two or more endings be active simultaneously, only the action potential reaching the common branching point first is expected to reach the central nervous system, whereas the latter will collide with the retrogradely invading action potential. For a given stimulus that maximally activates sensory terminals of a single nociceptor simultaneously, this mechanism would limit the maximum discharge frequency of the parent axon to the one of the fastest small branch. On one hand this provides a large safety margin, as only a minority of sensory endings are needed to produce the maximum discharge in the parent axon, but maximum discharge frequency in the parent axon would be limited by the maximum frequencies tolerated by the small branches. It is well known that the safety factor for propagation at branching points is low, especially when a thin axon enters an axon of larger diameter (Segev and Schneidman 1999).

Thus, it is not surprising that axonal propagation can be blocked branching points. If the action potential only propagates centrally, but does not invade the daughter branch antidromically (unidirectional block), action potentials generated in the non-invaded branches will not collide and can also reach the central nervous system (Fig. 3). It is

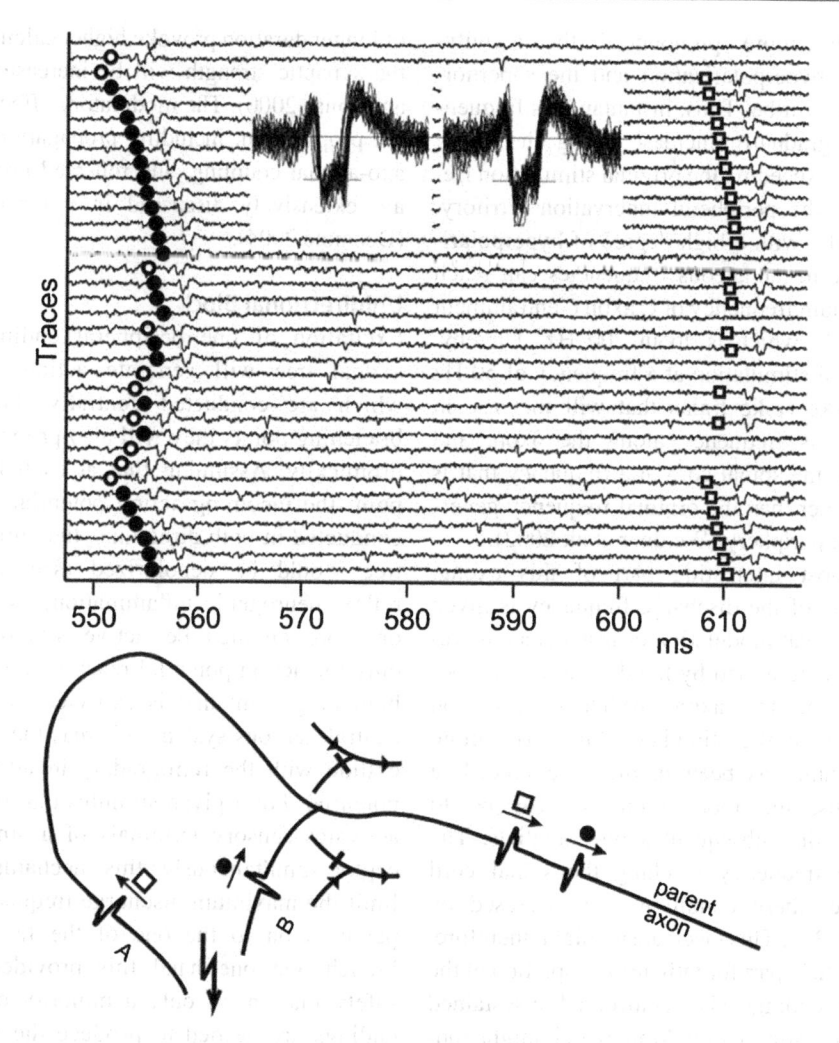

Encoding of Noxious Stimuli, Fig. 3 Unidirectional block of a human CMiHi unit. (**a**) Responses of a human C-fiber to electrical stimulation in the receptive field in the dorsum of the foot are shown in successive traces from *top* to *bottom*. The fiber was recorded by microneurography at knee level. Although only one electrical stimulus was applied in each trace, in most of the traces the unit conducts two action potentials. Following each double response, conduction velocity slowing can be observed in both branches. In contrast, recovery follows the end of the double activation period. A single double pulse (*open dot*) intermittently leads to latency increase of the following single pulse immediately. In the *inset*, the action potentials of all first (*left*) and all second (*right*) responses during the unidirectional block depicted are superposed to show their identical AP shape. (**b**) The scheme of a branched axon is shown to illustrate the mechanism of the unidirectional block (Modified from Weidner et al. 2003)

interesting to note that the unidirectional block in the periphery will enhance discharge frequency in the parent axon, whereas unidirectional blocks of its central endings in the spinal cord will reduce the sensory input (Weidner et al. 2000a). Unidirectional blocks have been discussed as a mechanism of pain sensitization (Serra et al. 2011), and their occurrence was found to be increased in patients with neuropathic pain (Schmidt et al. 2012).

Cross-References

▶ Encoding of Noxious Information in the Spinal Cord

References

Black, J. A., & Waxman, S. G. (2002). Molecular identities of two tetrodotoxin-resistant sodium channels in corneal axons. *Experimental Eye Research, 75*, 193–199.

Bostock, H., Campero, M., Serra, J., & Ochoa, J. (2003). Velocity recovery cycles of C fibres innervating human skin. *The Journal of Physiology, 553*, 649–63.

Brock, J. A., Mclachlan, E. M., & Belmonte, C. (1998). Tetrodotoxin-resistant impulses in single nociceptor nerve terminals in guinea-pig cornea. *The Journal of Physiology, 512*, 211–217.

Brock, J. A., Pianova, S., & Belmonte, C. (2001). Differences between nerve terminal impulses of polymodal nociceptors and cold sensory receptors of the guinea-pig cornea. *The Journal of Physiology, 533*, 493–501.

Carr, R. W., Pianova, S., & Brock, J. A. (2002). The effects of polarizing current on nerve terminal impulses recorded from polymodal and cold receptors in the guinea-pig cornea. *Journal of General Physiology, 120*, 395–405.

Carr, R. W., Pianova, S., Fernandez, J., Fallon, J. B., Belmonte, C., & Brock, J. A. (2003). Effects of heating and cooling on nerve terminal impulses recorded from cold-sensitive receptors in the guinea-pig cornea. *Journal of General Physiology, 121*, 427–439.

Carr, R. W., Pianova, S., McKemy, D. D., & Brock, J. A. (2009). Action potential initiation in the peripheral terminals of cold-sensitive neurones innervating the guinea-pig cornea. *The Journal of Physiology, 587*, 1249–1264.

Debanne, D. (2004). Information processing in the axon. *Nature Reviews. Neuroscience, 5*, 304–316.

Decol, R., Messlinger, K., & Carr, R. W. (2012). Repetitive activity slows axonal conduction velocity and concomitantly increases mechanical activation threshold in single axons of the rat cranial dura. *The Journal of Physiology, 590*, 725–736.

Gee, M. D., Lynn, B., & Cotsell, B. (1996). Activity-dependent slowing of conduction velocity provides a method for identifying different functional classes of C-fibre in the rat saphenous nerve. *Neuroscience, 73*, 667–675.

Geiger, J. R. P., & Jonas, P. (2000). Dynamic control of presynaptic Ca^{2+} inflow by fast-inactivating K + channels in hippocampal mossy fiber boutons. *Neuron, 28*, 927–939.

Raymond, S. A., Thalhammer, J. G., Popitz, B. F., & Strichartz, G. R. (1990). Changes in axonal impulse conduction correlate with sensory modality in primary afferent fibers in the rat. *Brain Research, 526*, 318–321.

Schmidt, R., Kleggetveit, I. P., Namer, B., et al. (2012). Double spikes to single electrical stimulation correlates to spontaneous activity of nociceptors in painful neuropathy patients. *Pain, 153*, 391–398.

Segev, I., & Schneidman, E. (1999). Axons as computing devices: Basic insights gained from models. *Journal of Physiology, Paris, 93*, 263–270.

Serra, J., Campero, M., Ochoa, J., & Bostock, H. (1999). Activity-dependent slowing of conduction differentiates functional subtypes of C fibres innervating human skin. *The Journal of Physiology, 515*, 799–811.

Serra, J., Sola, R., Aleu, J., Quiles, C., Navarro, X., & Bostock, H. (2011). Double and triple spikes in C-nociceptors in neuropathic pain states: An additional peripheral mechanism of hyperalgesia. *Pain, 152*, 343–353.

Taguchi, T., Ota, H., Matsuda, T., Murase, S., & Mizumura, K. (2010). Cutaneous C-fiber nociceptor responses and nociceptive behaviors in aged Sprague-Dawley rats. *Pain, 151*, 771–782.

Thalhammer, J. G., Raymond, S. A., Popitz Bergez, F. A., & Strichartz, G. R. (1994). Modality-dependent modulation of conduction by impulse activity in functionally characterized single cutaneous afferents in the rat. *Somatosensory and Motor Research, 11*, 243–257.

Vasylyev, D. V., & Waxman, S. G. (2012). Membrane properties and electrogenesis in the distal axons of small dorsal root ganglion neurons in vitro. *Journal of Neurophysiology, 108*(3), 729–740.

Weidner, C., Schmelz, M., Schmidt, R., Hansson, B., Handwerker, H. O., & Torebjörk, H. E. (1999). Functional attributes discriminating mechano-insensitive and mechano-responsive C nociceptors in human skin. *Journal of Neuroscience, 19*, 10184–10190.

Weidner, C., Schmelz, M., Schmidt, R., Handwerker, H. O., & Torebjörk, H. E. (2000a). Unidirectional conduction block at branching points of human nociceptive C-afferents: A peripheral mechanism for pain amplification. In M. Devor, M. Rowbotham, & Z. Wiesenfeld-Hallin (Eds.), *Progress in pain research and management* (pp. 233–240). Seattle: IASP Press.

Weidner, C., Schmidt, R., Schmelz, M., Hilliges, M., Handwerker, H. O., & Torebjörk, H. E. (2000b). Time course of post-excitatory effects separates afferent human C fibre classes. *The Journal of Physiology, 527*, 185–191.

Weidner, C., Schmelz, M., Schmidt, R., et al. (2002). Neural signal processing: The underestimated contribution of peripheral human C-fibers. *Journal of Neuroscience, 22*, 6704–6712.

Weidner, C., Schmidt, R., Schmelz, M., Torebjörk, H. E., & Handwerker, H. O. (2003). Action potential conduction in the terminal arborisation of nociceptive C-fibre afferents. *The Journal of Physiology, 547*, 931–940.

Endocannabinoids

Definition

Endocannabinoids are the endogenous ligands for the known cannabiniod CB1 and CB2 receptors. Endogenous cannabinoids have been demonstrated to have analgesic actions in pre-clinical models. They participate in the regulation of diverse body functions such as vasodilation, neuronal activity, immune functions, and pain.

Cross-References

▶ Acute Pain Mechanisms
▶ Inflammation, Modulation by Peripheral Cannabinoid Receptors
▶ NSAIDs, Adverse Effects

End-of-life Care

▶ Cancer Pain, Palliative Care in Children
▶ Pain Treatment in Pediatric Palliative Care

Endogenous

Definition

Endogenous means to originate internally.

Cross-References

▶ Endogenous Analgesia System
▶ Stimulation-Produced Analgesia

Endogenous Analgesia System

Definition

Endogenous analgesia system refers to pathways originating in the brainstem and terminating in the spinal cord (and in trigeminal sensory nuclei) that inhibit spinal/trigeminal nociceptive processing. Neurotransmitters released by this system include endogenous opioids and mono-amines (e.g., norepinephrine, serotonin).

Cross-References

▶ Descending Modulation of Nociceptive Processing
▶ GABA Mechanisms and Descending Inhibitory Mechanisms
▶ Spinothalamic Tract Neurons, Descending Control by Brain Stem Neurons

Endogenous Analgesia Systems

▶ Descending Modulation of Nociceptive Processing

Endogenous Descending Pain Modulatory Pathways

▶ Sex Differences in Descending Pain Modulatory Pathways

Endogenous Opiate/Opioid System

Definition

Endogenous opiate system refers to that component of the endogenous analgesia system that produces inhibition of spinal/trigeminal nociceptive transmission through release of endogenous opioids (e.g., enkephalins).

Cross-References

▶ Deep Brain Stimulation

Endogenous Opioid Peptides

Definition

Endogenous opioid peptides include endorphins (e.g., β-endorphin), enkephalins, and dynorphins derived from three corresponding genes: prepro-opiomelanocortin (POMC), preproenkephalin, and preprodynorphin, respectively. Opioid peptides share a common amino terminal sequence of Tyr–Gly–Gly–Phe and either Met or Leu. They function commonly like other neuropeptide transmitters, are heterogeneously distributed throughout the body, and exert their actions through opioid receptors.

Cross-References

▶ Nitrous Oxide Antinociception and Opioid Receptors
▶ Opioids and Inflammatory Pain
▶ Pain Treatment, Implantable Pumps for Drug Delivery
▶ Placebo Analgesia and Descending Opioid Modulation
▶ Trigeminal Brain Stem Nuclear Complex, Immunohistochemistry, and Neurochemistry

Endogenous Oplold Receptors

Definition

Opioid receptors are 7-transmembrane-spanning, G-protein-coupled receptors that are the targets of exogenous opioids (e.g., mophine, fentanly, pentazocine) and endogenous opioid peptides. Classically, three families of opioid receptors were initially described – mu (μ), delta (δ), and kappa (κ), of which several subtypes have since been discovered (e.g., μ_1, μ_2, μ_3; δ_1, δ_2; κ_1, κ_2, κ_3). A nonclassical opioid receptor, nociceptin, has also been described. Opioid receptors are heterogeneously distributed throughout the body and different effects are exerted by activation of different receptors/receptor subtypes, although analgesia can be produced by activation of μ, δ, or κ receptors. A new IUPHAR terminology recommends that μ, δ, κ, and nociceptin receptors be designated MOP, DOP, KOP, and NOP, respectively.

Cross-References

▶ Nitrous Oxide Antinociception and Opioid Receptors

Endogenous Opioids/Opiates

Definition

See "▶ Endogenous Opioid Peptides"

Cross-References

▶ Opioid Electrophysiology in PAG

Endogenous Pain Control

▶ Descending Modulation of Nociceptive Transmission During Persistent Damage to Peripheral Tissues

Endogenous Pain Control Pathway

Definition

Endogenous pain control pathway is synonymous with "endogenous analgesia system" and refers to pathways (not a single pathway) originating in the brainstem and terminating in the spinal cord (and in trigeminal sensory nuclei) that inhibit spinal/trigeminal nociceptive processing. Neurotransmitters released by these pathways include endogenous opioids and monoamines (e.g., norepinephrine, serotonin).

Cross-References

▶ Primary Stabbing Headache

Endometriosis

Definition

The presence and growth of endometrial glands and stroma, similar to the lining of the uterus (endometrium) but located outside the uterus, usually on the pelvic peritoneum, ovaries, or rectovaginal septum, and other pelvic viscera such as bowel and bladder, but rarely seen on more distant structures, such as diaphragm, lungs, or brain. The pathogenesis of the disease, as well as the pain and infertility, is the subject of extensive research. The most widely accepted theory is that the disorder originates from retrograde menstruation, with passage and then implantation of endometrial tissue within the peritoneal cavity. It is uncertain whether in women with this hormonally mediated disease there is an underlying endometrial abnormality, unusual menstrual uterine contractile pattern, local peritoneal immunologic abnormalities, or other aberrant angiogenic and neuropathic factors. Associated symptoms can include dyspareunia, dyschezia, irregular uterine spotting, and infertility. Pelvic exam often reveals a fixed retroverted uterus, focal tenderness, or nodularity of the uterosacral ligaments and ovarian masses, which on ultrasound usually has the characteristics of an endometrioma or "chocolate" cyst.

Cross-References

► Dyspareunia and Vaginismus
► Gynecological Pain, Neural Mechanisms
► Pain Mechanisms in Endometriosis
► Visceral Pain Models, Female Reproductive Organ Pain

Endometriosis Externa

► Chronic Pelvic Pain, Endometriosis

Endometriosis Model

► Visceral Pain Models, Female Reproductive Organ Pain

Endometritis Model

► Visceral Pain Models, Female Reproductive Organ Pain

Endomorphin 1 and Endomorphin 2

Definition

Endomorphin 1 and endomorphin 2 are opioid tetrapeptides with a high affinity and selectivity to μ opioid receptors. Endomorphin 1 is more widely distributed throughout the brain, and endomorphin 2 is more prevalent in the spinal cord. Precursor proteins of endomorphin 1 and 2 are not yet known.

Cross-References

► Opiates During Development
► Opioids and Inflammatory Pain

Endone®

► Postoperative Pain, Oxycodone

Endorphins

Definition

Endorphins are endogenous opioid peptides that are produced in the body and have an affinity and activity at opioid receptors.

Cross-References

► Alternative Medicine in Neuropathic Pain
► Endogenous Opioid Peptides
► Nitrous Oxide Antinociception and Opioid Receptors

Endoscopic Sympathectomy

Definition

Endoscopic sympathectomy is a minimally invasive technique for resecting components of the sympathetic chain.

Cross-References

▶ Complex Regional Pain Syndrome and the Sympathetic Nervous System

Endothelins

Definition

Endothelins are a family of three peptides that are released from endothelial cells and some tumor cells that can activate nociceptors, mount an inflammatory response, and stimulate angiogenesis and growth of tumor cells.

Cross-References

▶ Cancer Pain, Animal Models

Endovanilloid

Definition

Endovanilloids are endogenous metabolites that activate the capsaicin receptor by binding to the capsaicin binding site.

Cross-References

▶ Capsaicin Receptor
▶ Thalamus, Clinical Pain, Human Imaging

Endplate Noise

Synonyms

EPN

Definition

Complex noise-like potentials generated by a large increase (up to 1,000 times) in spontaneously released acetylcholine packets resulting in subsynaptic miniature end-plate potentials. Many electromyographers have been taught that end-plate noise represents normal function of the motor end plate.

Cross-References

▶ Myofascial Trigger Points

Endurance Exercise

▶ Exercise

Enforced Mobilization

▶ Postoperative Pain, Importance of Mobilization

Enkephalin

Definition

Enkephalin is one of two pentapeptides, comprised of five amino acids, which differ only at the last one (leu-enkephalin; met-enkephalin) that preferentially binds to DOP receptor. They are formed from proenkephalin, prodynorphin, or proopiomelanocortin. They are found throughout the central and enteric nervous systems and in the adrenal medulla.

Cross-References

▶ Endogenous Opioid Peptides
▶ Nitrous Oxide Antinociception and Opioid Receptors
▶ Opiates During Development
▶ Opioid Receptors

Enteric Coating

Definition

A coating applied to tablet granules in capsules. The enteric coating is insoluble in the acid contents of the stomach, but breaks down in the approximately neutral contents of the small intestine where the drug is released.

Cross-References

▶ NSAIDs and Their Indications

Enterohepatic Recirculation (Biliary Recycling)

Definition

The enterohepatic recirculation (biliary recycling) describes the effect, where drugs are excreted via bile into the small intestine, but can be reabsorbed from the distal intestinal lumen.

Cross-References

▶ NSAIDs, Pharmacokinetics

Enthesis

Definition

Enthesis is the periosteal attachment of a ligament or tendon.

Cross-References

▶ Myofascial Trigger Points

Enthesopathy

Definition

Enthesopathy is a disease process at musculotendinous or musculo-osseous junctions.

Cross-References

▶ Myofascial Trigger Points

Entorhinal Cortex and Hippocampus

▶ Hippocampus and Entorhinal Complex: Functional Imaging

Entrapment Neuropathies

Definition

Entrapment neuropathies are abnormal peripheral nerve functions arising from compression by an anatomical structure.

Cross-References

▶ Spinal Cord Injury Pain

Entrapment Neuropathies, Carpal Tunnel Syndrome

▶ Carpal Tunnel Syndrome

Environmental Factors

Definition

Environmental factors refer to all aspects of the external or extrinsic world that form the context of an individual's life and, as such, have an impact on that person's functioning.

Cross-References

▶ Functioning and Disability Definitions

Ephaptic Coupling

Definition

This term, based on the idea of an "electrical synapse," refers to the passage of an electrical current from one neuron to a closely apposed neighbor, in the absence of the apparatus, delay, and pharmacology associated with a conventional chemical synapse. Current flow in ephaptically coupling cells may be due to the presence of gap junctions, or it may be due to a sufficient area of close membrane apposition without cytoplasmic continuity between the coupled cells. It is usually assumed to be an abnormal but stable, electrical interaction between two nerve fibers, for example, between the endings of two fibers terminating within a neuroma or a site of a crush injury whereby nerve impulses in one fiber trigger impulses in the other.

Cross-References

▶ Ectopia, Spontaneous
▶ Pain Paroxysms

▶ Trigeminal and Cranial Nerve Neuralgias
▶ Trigeminal Neuralgia: Diagnosis and Treatment

Ephaptic Cross Talk

▶ Cross Excitation

Epicondylitis

Definition

Inflammation of the tendon attachments for muscles on the inside (medial) or outside (lateral) of the elbow joint.

Cross-References

▶ Ergonomics Essay

Epicritic

Definition

Relating to the perception of slight differences in the intensity of stimuli, especially touch, temperature, vibration, and limb position in space. Cutaneous discriminative perception is typically attributed to the larger myelinated sensory nerve fibers. Epicritic pain refers to pains that convey precise information about location, duration, and intensity. For example, pricking pain from needle penetration is a form of epicritic pain.

Cross-References

▶ Postsynaptic Dorsal Column Projection, Functional Characteristics

Epidemiology

Definition

Although classically epidemiology has been understood to be the study of the characteristics of larger populations, such as in a census or transcultural studies, epidemiological methods are applied to a great variety of scientific inquiries including the study of natural history of illnesses, the comparison of clinical or social policy interventions or natural events that affect health outcomes, studies of risk factors predicting new illnesses, factors affecting prognosis, and other applications.

Cross-References

▶ Low Back Pain, Epidemiology
▶ Prevalence of Chronic Pain Disorders in Children
▶ Psychiatric Aspects of the Epidemiology of Pain
▶ Psychological Aspects of Pain in Women

Epidemiology of Chronic Pelvic Pain

Abigail Broderick[1], Stephen Kennedy[1] and Krina Zondervan[2]
[1]Nuffield Department of Obstetrics and Gynaecology, University of Oxford, Oxford, UK
[2]Genetic and Genomic Epidemiology Unit, Nuffield Dept of Obstetrics and Gynaecology, Wellcome Trust Centre for Human Genetics, Oxford, UK

Synonyms

Chronic pelvic pain, epidemiology; Pelvic pain syndrome; Recurrent pelvic pain

Definition

The most commonly used definition of chronic pelvic pain (CPP) for research purposes is "Recurrent or constant pain in the lower abdominal region that has lasted for at least 6 months." The International Association for the Study of Pain (1986) defines CPP without obvious pathology (CPPWOP) as "chronic or recurrent pelvic pain that cannot be sufficiently explained by an apparent physical cause." The definition assumes a causal link between pelvic pathology and pain, which clinical experience and the published literature would suggest is not always the case.

Since then, two further definitions have been suggested. The American College of Obstetricians and Gynecologists (2004) has proposed a definition of CPP as "non-cyclical pain of at least 6 months' duration that appears in locations such as the pelvis, anterior abdominal wall, lower back, or buttocks, and that is serious enough to cause disability or lead to medical care." The Royal College of Obstetricians and Gynaecologists in the United Kingdom (2005) further defines the pain as "intermittent or constant pain in the lower abdomen or pelvis of at least 6 months' duration, not occurring exclusively with menstruation or intercourse and not associated with pregnancy." These definitions solely consider the location and duration of the pain; no assumptions are made about its cause (Campbell and Collett 1994), emphasizing the entity of CPP as a symptom and not a diagnosis.

Although there are common elements to these definitions, there is no consensus regarding an absolute definition of CPP. In fact, in a MEDLINE survey of the definitions used in papers published from 1966 to 2001 (Williams et al. 2004), there was no mention of the following: duration of pain (44 %), restriction by pathology (74 %), location of pain (93 %), or restriction by comorbidity (95 %).

Introduction

This entry examines purely the epidemiology of CPP, not exploring any of the specific pathology underlying the diagnosis.

Characteristics

The lack of an unambiguous definition makes the study of the epidemiology of CPP difficult.

Early studies focused mainly on the frequency with which pelvic pathology was found at laparoscopy as an explanation for CPP and on attempts to explain the symptoms when no such pathology was found.

In 1998, a systematic review was conducted of the prevalence and incidence of CPP in the UK general population (Zondervan et al. 1998). It was clear that few attempts had been made to define an appropriate study population with an adequately powered sample size, which is essential if results are to be generalized to the wider population. The only study that investigated the epidemiology of CPP in any setting was a hospital-based survey of 559 pathology-free women who had undergone a laparoscopy for sterilization or investigation of infertility (Mahmood et al. 1991). The prevalence of CPP, defined as "recurrent pelvic pain unrelated to menstruation or coitus," was 39 %.

Worldwide, the only truly community-based prevalence estimate at that time was 14.7 % (95 % confidence interval: 12.7–16.7 %), based upon a telephone survey of 5,263 women, aged 18–50, randomly selected from the general US population (Mathias et al. 1996). CPP was defined as "pelvic pain of at least 6 months' duration and with pain having occurred in the past 3 months." In another US study, prevalence was assessed in 581 patients and their female companions aged 18–45 in waiting rooms of gynecology and family medicine practices (Jamieson and Steege 1996). Thirty-nine percent of women reported having some degree of pelvic pain, while 20 % reported pelvic pain of more than 1 year's duration.

The lack of epidemiological data on CPP in the UK outside a hospital setting led to a study investigating its prevalence and incidence in primary care (Zondervan et al. 1999a). For this purpose, the MediPlus UK Primary Care Database (UKPCD) was used, which contains anonymized clinical and prescribing data from 1991 onwards on approximately 1,700,000 patients. The most common definition of CPP was used, and cases were identified from a denominator of 278,509 women, on the basis of contacts for pelvic pain with the practices contributing information to the database between 1991 and 1995. The annual prevalence of CPP in women aged 15–73 was 38/1,000. This figure was comparable to those reported elsewhere for asthma (37/1,000) and back pain (41/1,000). The annual prevalence for women aged 18–50 was similar, at 37/1,000. CPP prevalence was found to vary with age, from 18/1,000 in 15–20-year-olds to 28/1,000 in women older than 60.

Although the prevalence was high, the monthly incidence was only 1.6/1,000: one possible explanation for this combination would be long symptom duration. Therefore, CPP duration was estimated in a cohort of 5,051 incident cases from MediPlus UKPCD, who were followed for 3–4 years from their first pelvic pain contact in primary care (Zondervan et al. 1999b). The median symptom duration was 15 months, with a third of women having persistent symptoms after 2 years. These results were probably influenced by the actual duration of symptoms, as well as by health-care-seeking behavior. A quarter of incident CPP cases received no diagnostic label during the 3–4-year follow-up, and only 40 % were referred to a hospital specialist. The majority of women only received one diagnosis, the most common being irritable bowel syndrome (IBS) and cystitis in all age groups. In addition, pelvic inflammatory disease (PID) was common in women aged up to 40, and other gastrointestinal diagnoses in women above 50. The likelihood of receiving a diagnosis varied with age: women under 20 and older than 60 were less likely than others to receive a diagnosis. The referral rate varied similarly with age: women in the 31–40 age group were twice as likely to have been referred as the youngest or oldest women in the cohort.

As the true community prevalence could not be estimated from MediPlus UKPCD, because it only provided information on women with CPP seeking health care, a postal questionnaire survey was conducted among 4,000 women, randomly selected from 141,400 women aged 18–49 on the Oxfordshire Health Authority (OHA) register (Zondervan et al. 2001a). The most common definition of CPP was used: pain in the lower abdominal region, not exclusively related to periods, intercourse, or pregnancy. The response rate among those who received the questionnaire was 74 %. The study group consisted of 2,016 women, after exclusion of those who had been pregnant in the last 12 months. CPP in the last 3 months was reported by 24.0 % of the women (95 % confidence interval: 22.1–25.8). This prevalence estimate was higher than the 14.7 % reported by Mathias et al. (1996) in the USA. However, their study excluded women with ovulation-related pain; exclusion of those able to report ovulation-related pain would have reduced the Oxford estimate to a similar figure of 16.9 %. The prevalence varied slightly with age: the lowest rate (20 %) was found in the age groups 18–25 and 31–35 and the highest (28 %) in 36–40-year-olds. Non-Caucasian women had a much lower prevalence (10 %) than Caucasian women (25 %; $p = 0.003$); a risk ratio (RR) of 0.4 (95 % confidence interval: 0.2–0.8). Adjustment for age and social class did not affect this result. Prevalence did not vary with social class, marital status, or employment status. A third of women with CPP reported that their pain had started more than 5 years ago. Women with CPP who had consulted for their symptoms more than 12 months previously (27 % of cases) appeared slightly less affected by their symptoms than recent consulters. However, a substantial number (43 %) reported that the pain restricted their activities. The reasons why these women discontinued seeking medical advice despite ongoing symptoms were unclear. Some women may have been controlling their pain after medical advice or treatment. It is also possible, however, that some may simply have been dissatisfied with their medical care. Women with CPP who had never consulted for pelvic pain (41 % of cases) were similar in terms of general health, pain severity, use of health care, and measures of pain-related functioning to women with dysmenorrhea alone. However, the reported effect of dysmenorrhea on women's lives when they had the pain was not negligible: three-quarters used medication for symptom relief, and a third reported that the pain restricted their activities. Thus, non-consulting women with CPP appear to perceive their symptoms as no greater a burden than having painful periods. Another important observation in the study was that a third of women with CPP reported they were anxious about their pain, particularly its cause. This symptom-related anxiety was common in non-consulters as well as consulters.

Subsequently, a worldwide systematic review was conducted by the World Health Organization (WHO) (Latthe et al. 2006a). They reviewed literature investigating the prevalence of dysmenorrhea, dyspareunia, and noncyclical pain from 1887 to 2004. Of the 178 studies in 148 articles, 18 were on noncyclical pain, giving a prevalence of 4–43.4 %. They identified only three high-quality studies with representative samples, which still provided prevalence rates that varied widely from 2.1 % to 24 %. The rate of dysmenorrhea was 16.8–81 % and dyspareunia 8–21.8 %. However, they found significant heterogeneity across studies, and the rates of pain identified appeared to vary with study features, including design, sample size, and sampling stratification. Unfortunately, there were few good-quality studies assessing the burden of CPP in less developed countries, and any observed variation in rates of CPP worldwide may be explained by variation in the quality of published studies.

Since this review, a further study was undertaken in New Zealand examining the prevalence, pain severity, diagnoses, and use of health services (Grace and Zondervan 2004). They utilized a postal questionnaire, based on the Oxford questionnaire and piloted in New Zealand, and reported similar findings to the UK study (Zondervan et al. 2001a). With a response rate of 66 %, they had a sample of 1,160 respondents aged 18–50 years, with a 3-month prevalence of 25.4 % (95 % confidence interval 22.8–27.9 %), including mid-cycle pain. Among the 817 women who had periods and were sexually active, the prevalence was 26.2 %. However, their sample was skewed towards women with a higher personal income and slightly skewed towards older women and those with higher qualifications. Interestingly, the prevalence of CPP varied with ethnicity, with a prevalence of 27.4 % among those of New Zealand European descent, 18.8 % among Maori women, and 14.1 % in other ethnicities. There were also significant differences between age groups. The highest rate was among those aged 26–30 years, with a prevalence of 31.8 %, closely followed by women aged 31–35 years at 31.0 %. The lowest rate of 19.7 % was found in the lowest age group (18–25 years). Of those women with CPP, 51.9 % reported it as mild in severity, 30.4 % had never consulted health services, 31.2 % had consulted recently (within the last 12 months), and 38.3 % in the past (more than 12 months ago). Those who had consulted were more likely to have experienced pain in the last month, describe longer pain episodes, and have more episodes during the last 12 months. Of those women reporting CPP, 47.7 % did not have a diagnosis, and of those that did, irritable bowel syndrome (IBS) was the most common diagnosis at 26.2 %. Other diagnoses included ovarian cysts or polycystic ovarian syndrome, constipation, stress, endometriosis, cystitis, and back problems.

In Australia, the prevalence of CPP was examined as part of a broader national study of health and relationships (Pitts et al. 2008). Telephone interviews were conducted with 4,366 women aged 16–64 years, with 18 of the 200 questions related to pelvic pain. Of the whole sample, 1983 women aged 16–49 years were still menstruating and sexually active, thus suitable for data analysis. They looked at pain experienced during the previous 12 months, defining pelvic pain as "any type of pain in the lower part of your belly – that is, from your belly button down." "Other CPP" was further defined as "pelvic pain not occurring with periods or intercourse, either on and off or constantly," although no duration was specified. They found an overall prevalence of 21.5 % for other CPP. Of these women, severe pain was experienced by 20 % (95 % confidence interval 16.1–24.6 %). These prevalence figures correlate well with the UK and New Zealand studies. Additionally, unlike previous studies they found no significant demographic differences between women reporting other CPP, including age, educational status, region of residence, and language spoken at home.

Most recently, a population-based study of IBS and CPP was undertaken in Olmsted County,

Minnesota (Choung et al. 2010). The Talley bowel disease questionnaire was posted to a random sample of 1,031 residents aged 30–64 years, with a response rate of 69 %, and 52 % of responders female, giving a total sample of 339 women. Of these, 20 % reported some pelvic pain, defined as "pain in your pelvic region separate from pain in your stomach or belly in the last year," again with no statement of duration.

With information on the burden of CPP in less developed countries lacking, a cross-sectional population-based study was conducted in Hermosilla, Mexico (Garcia-Perez et al. 2010). They conducted face-to-face interviews on a random sample of 1,307 women aged 25–54 years utilizing a Spanish translation of a shortened version of the Oxfordshire Women's Health Study questionnaire. Of the 1,201 nonpregnant women, 6 % reported CPP, defined as "constant or intermittent pain below the belly button or in the female organs over a period of 6 months or longer, excluding pain associated with menstruation, sexual intercourse, or pregnancy." Four percent of these women described their pain as moderate or severe. They further demonstrated an increased risk in age groups around 30 years, an adjusted odds ratio of 3.13 (95 % confidence interval 1.90–5.13) for ages 25–34 years compared to ≥35 years. Of those with CPP, only 40.3 % had consulted a physician.

These studies demonstrate the variation in the prevalence of CPP depending on the population and definition used. Early studies looking at a women undergoing laparoscopy found higher rates of 39 %, while in primary care, the prevalence was much lower at 3.8 %. Later studies in the UK, the USA, New Zealand, and Australia utilizing population-based questionnaires demonstrated more consistent results, with a prevalence of 20–25.4 % including mid-cycle pain and 14.7–16.9 % excluding mid-cycle pain. They also consistently found a lower rate among non-Caucasian ethnicities. Looking at variation across age groups, all studies found lower prevalence in younger groups aged 18–25 years. There is less consensus about the age group with the highest rate, although both the New Zealand and Mexican studies found it to be most common around the age of 30 years.

Symptom Complexity and Diagnoses

CPP is difficult to diagnose and treat, mainly due to the wide range of possible causes with overlapping symptomatology, e.g., endometriosis, chronic PID, adhesions, IBS, interstitial cystitis, and the urethral syndrome. In gynecology clinics, high IBS prevalence rates (50–79 %) have been reported in women referred for CPP (Prior et al. 1989). Similarly, high prevalence rates of dyspareunia (42 %) and urinary symptoms (61 %) have been reported in IBS patients (Whorwell et al. 1986). In the second part of the Oxfordshire Women's Health Study, the symptom overlap was investigated between CPP, IBS, genitourinary symptoms, dysmenorrhea, and dyspareunia in the community, and associated investigations and diagnoses (Zondervan et al. 2001b). The diagnoses were based only on the woman's recall and were not validated using their medical records. A substantial overlap was found between CPP, GU symptoms, and IBS. Approximately half of the 483 women with CPP had at least one additional symptom; 39 % had IBS and 24 % had GU symptoms. Prevalence rates of dysmenorrhea and dyspareunia were much higher among women with CPP (81 % and 41 %, respectively) than among those without CPP (58 % and 14 %). These rates were high in all subgroups of CPP, irrespective of the presence of additional GU symptoms or IBS. Moreover, the study found that women with CPP and additional GU symptoms, and IBS in particular, were most affected in terms of pain severity and episode duration. This group also reported having received the widest range of diagnoses (Zondervan et al. 2001b).

Two further studies demonstrated an overlap between different pelvic pain symptoms, although the proportions experiencing multiple symptoms were less that in the Oxford study. In Australia, Pitts et al. (2008) demonstrated rates of dysmenorrhea, dyspareunia, and all three symptoms among those with CPP of 56 %, 2 %, and 27 %, respectively. In New Zealand, Grace et al. (2004) found that of those with CPP, 83 % complained of dysmenorrhea and/or dyspareunia, 42.5 % had dysmenorrhea alone, 4.7 % dyspareunia alone, and 36 % all three symptoms. In the same study, 47.7 % remained undiagnosed, although only 25.4 % of those who had consulted a medical practitioner. The most common diagnosis offered by medical practitioners was IBS in 26.2 % of patients, lower than in the Oxford study. Other diagnoses included ovarian cysts and PCOS (23.5 %), constipation (20.8 %), stress (19.5 %), endometriosis (14.8 %), cystitis (14.1 %), and back problems (13.4 %).

Choung and colleagues (Choung et al. 2010) specifically addressed the potential overlap of IBS and CPP in their study in the USA. Using three different criteria for IBS (Manning, Rome I and modified Rome III), they demonstrated that of the 67 participants with CPP, 40 % met one or more sets of IBS criteria. This represents an overlap rate approximately two times higher than expected and significantly higher than would be expected by chance ($p < 0.001$). Additionally, they identified a number of risk factors that were associated with reporting both IBS and CPP as opposed to either condition alone. They conclude that there may be a subset of IBS that shares a common pathophysiology with CPP, and somatization may represent a key explanation.

Risk Factors for CPP

Studying risk factors for CPP provides a major challenge to the epidemiologist. Gradually emerging knowledge indicates that a host of somatic, psychological, and socio-environmental factors may act and interact in its causation, together forming a multidimensional "biopsychosocial" model of disease etiology. Women with CPP may suffer from various underlying somatic conditions, which may all have different or even conflicting risk factors. Furthermore, psychological factors (e.g., traumatic events, depression, illness beliefs) and socio-environmental factors (e.g., the role of "significant others") may play a role in pain etiology or maintenance. Risk of CPP may be influenced by a variety of these factors, some of which can be difficult to measure. Furthermore, different populations are likely to exhibit different frequencies of underlying somatic and psychosocial conditions, thus limiting comparability of results from one country to another.

The situation may be even more complex when CPP cases are identified from primary instead of secondary/tertiary care settings. Associations between CPP and risk factors in these settings may be very different, because of differences in health-care-seeking behavior and referral patterns. For instance, if pelvic pain were common in young women, but if they were less inclined to seek medical advice than older women, they would be less likely to be identified as CPP cases, and risk of CPP in this group would be seen as low. Differences in referral patterns from primary to secondary/tertiary care between certain demographic groups may result in an even further distorted picture of CPP risk. Therefore, risk-factor analysis of CPP is at present only of some value using cases identified at community level, and interpretation of the results remains complicated due to its multi-causality.

The information from cross-sectional studies is very limited. The community survey in the USA (Mathias et al. 1996), and the semi-community study of women and their companions in primary care clinics (Jamieson and Steege 1996), described the association between certain demographic factors and CPP. In a logistic regression model adjusted for age, ethnicity, education, and marital status, Mathias et al. (1996) found a slightly decreased risk of CPP in women older than 35 compared to younger women (OR: 0.7, 95 % CI: 0.6–0.8). Jamieson and Steege (1996) reported unadjusted results for the association between pelvic pain (defined as any current pain not associated with menstruation or intercourse) and age, with rates varying significantly from 49 % in 26–30-year-olds to 22 % in 36–40-year-olds. Mathias et al. (1996) found a reduced risk in women of African-American origin compared to Caucasian women (OR: 0.7, 95 % CI: 0.5–0.95), whereas Jamieson and Steege (1996) found an increased risk (53 % vs. 35 %, respectively). Neither study found associations with educational status, income, or parity.

In the Oxfordshire Women's Health Study (Zondervan 1999), significant differences in risk were found only with age, ethnicity, height, condom use, length of bleeding, and subfertility. Compared to women aged 18–25, ORs of other age categories varied from 0.72 (95 % confidence interval: 0.44–1.18) for 46–49-year-olds to

1.30 (95 % confidence interval: 0.88–1.90) for 26–30-year-olds. Non-Caucasian women had a much lower risk of CPP compared to Caucasian women (OR = 0.35, 95 % confidence interval: 0.16–0.78). However, although this result was consistent with the Mathias et al. (1996) findings, it was based on only 80 non-Caucasian women participating in the study, which reflects the small percentage of non-Caucasian women in the Oxfordshire population. Surprisingly, tall women (≥1.70 m) appeared to have an increased risk of CPP, even after adjustment for possible confounding factors such as social class or ethnicity (OR = 1.37, 95 % confidence interval 1.00–1.88). Another unexpected result was the association of condom use (but not the diaphragm) with increased risk of CPP, but this effect was limited only to use more than 12 months earlier (OR = 1.55, 95 % confidence interval 1.17–2.05). A tentative explanation for this elevated risk could be that condom users may not have used this method consistently and that they were less likely to be long-term users of oral contraceptives. In terms of menstrual patterns, longer bleeding periods (7+ days) were associated with an increased risk of CPP (OR = 1.34, 95 % confidence interval 1.00–1.80). Subfertility was also found to be positively associated with CPP (OR = 1.50, 95 % confidence interval 1.08–2.07). Since both pelvic pain and infertility are common in women with endometriosis, this association was also not surprising.

The later population-based study in New Zealand (Grace et al. 2004) also demonstrated a lower rate in the youngest age group of 18–25 years (19.7 %) and highest rate among the 26–30-year age group (31.8 %) and 31–35 years (31.0 %). The findings regarding ethnicity also demonstrated differences with Maori women (18.8 %) and non-NZ/European, non-Maori women (14.1 %) reporting a lower prevalence than NZ/European women (27.4 %), p = 0.01. Women with higher (university) qualifications were significantly more likely to report CPP (33 %) than women with no qualifications (23.2 %) or school qualifications (21.5 %) (p = 0.02). However, the Australian study demonstrated no significant demographic differences between those reporting CPP and controls including age, educational or employment status, region of residency, or language spoken at home. Additionally there was no correlation between gravidity, parity, miscarriage, or termination status (Pitts et al. 2008).

In a recent systematic review of risk factors predisposing women to CPP (Latthe et al. 2006b), 40 studies were examined evaluating 48 risk factors for noncyclical pelvic pain. Of these, 28 met three or more of their quality criteria, which were prospective study design, consecutive or random recruitment of subjects, validated ascertainment of risk factors and outcome, temporal relationship between exposure and outcome, and control for confounding factors. They concluded that drug or alcohol abuse, miscarriage, heavy menstrual flow, pelvic inflammatory disease, previous caesarean section, pelvic pathology, abuse, and psychological comorbidity were associated with an increased risk of CPP, acknowledging there could be a significant overlap between cause and effect with these factors. However, a response to this paper (Zondervan et al. 2006) expressed concerns regarding the accuracy of performing a meta-analysis on such a heterogeneous study population, including a lack of standardized definition of CPP across study groups. They also questioned the methods regarding the need for appropriate controls given the greatly variable medical, personal, and social characteristics of women investigated for CPP in secondary care, as well as the fact that these may change during the course of their symptoms. They feel that a prospective, long-term, women's health study is required to ascertain risk factors for CPP.

A recent study involving a large Australian twin cohort sampled from the community supported the multifaceted origin of CPP in terms of genetic background (Zondervan et al. 2005). It was found that, although CPP showed a moderate heritability of 41 %, there was no independent genetic factor underlying CPP causation. Instead, CPP heritability could be completely attributed to genetic susceptibility for endometriosis, fibroids, dysmenorrhea, and somatic anxiety (the latter a potential marker for increased nociception).

In a longitudinal study looking at the clinical course of CPP, Weijenborg et al. (2007) followed a cohort of 72 women who had initially visited a multidisciplinary team over a mean follow-up period of 3.4 years (range 1–6 years). They found that a quarter of women reported recovery from CPP (pelvic pain for less than 3 months per year), with 11 % reporting complete recovery, i.e., no pain at all. No patients reported relapse of symptoms once recovered. They found no association between clinical course and any intervention undertaken, or demographic, clinical, or pain-related variable at baseline.

It is evident that while it is difficult to categorize rates of CPP according to a consistent definition, it represents a major health-care problem and a burden to both primary and secondary care services worldwide, as well as having a significant impact on those women who experience it. A prospective long-term women's health study would not only allow further identification of risk factors but would also further knowledge of this complex problem, assisting in tailoring services to provide diagnoses and appropriate management where possible.

References

American College of Obstetricians and Gynecologists. (2004). ACOG Practice Bulletin No.51. Chronic pelvic pain. *Obstetrics and Gynecology, 103*, 589–605.

Campbell, F., & Collett, B. J. (1994). Chronic pelvic pain. *British Journal of Anaesthesia, 73*(5), 571–573.

Choung, R. S., Herrick, L. M., Locke, G. R., Zinsmeister, A. R., & Talley, N. J. (2010). Irritable bowel syndrome and chronic pelvic pain a population-based study. *Journal of Clinical Gastroenterology, 44*(10), 696–701.

Garcia-Perez, H., Harlow, S. D., Erdmann, C. A., & Denman, C. (2010). Pelvic pain and associated characteristics among women in Northern Mexico. *International Perspectives on Sexual and Reproductive Health, 36*(2), 90–98.

Grace, V. M., & Zondervan, K. T. (2004). Chronic pelvic pain in New Zealand: Prevalence, pain severity diagnoses and use of health services. *Australian and New Zealand Journal of Public Health, 28*(4), 369–375.

International Association for the Study of Pain. (1986). Classification of chronic pain. Definitions of chronic pain syndromes and definition of pain terms. *Pain,* (Suppl 3), S1–S225.

Jamieson, D. J., & Steege, J. F. (1996). The prevalence of dysmenorrhea, dyspareunia, pelvic pain, and irritable bowel syndrome in primary care practices. *Obstetrics and Gynecology, 1*, 55–59.

Latthe, P., Latthe, M., Say, L., Gulmezoglu, M., & Khan, K. S. (2006a). WHO systematic review of prevalence of chronic pelvic pain: A neglected reproductive health morbidity. *BMC Public Health, 6*, 177.

Latthe, P., Mignini, L., Gray, R., Hills, R., & Khan, K. (2006b). Factors predisposing women to chronic pelvic pain: Systematic review. *BMJ, 332*(7544), 749–755.

Mahmood, T. A., Templeton, A. A., Thomson, L., & Fraser, C. (1991). Menstrual symptoms in women with pelvic endometriosis. *British Journal of Obstetrics and Gynaecology, 98*, 558–563.

Mathias, S. D., Kuppermann, M., Liberman, R. F., Lipschutz, R. C., & Steege, J. F. (1996). Chronic pelvic pain: Prevalence, health-related quality of life, and economic correlates. *Obstetrics and Gynecology, 87*, 321–327.

Pitts, M. K., Ferris, J. A., Smith, A. M. A., Shelley, J. M., & Richters, J. (2008). Prevalence and correlates of three types of pelvic pain in a nationally representative sample of Australian women. *The Medical Journal of Australia, 189*(3), 138–143.

Prior, A., Wilson, K., Whorwell, P. J., & Faragher, E. B. (1989). Irritable bowel syndrome in the Gynecological clinic. Survey of 798 new referrals. *Digestive Diseases and Sciences, 34*(12), 1820–1824.

Royal College of Obstetricians and Gynecologists (RCOG). (2005). *The initial management of chronic pelvic pain. Green-top Guideline 41.* London: RCOG.

Weijenborg, P. T., Greeven, A., Dekker, F. W., Peters, A. A., & Ter Kuile, M. M. (2007). Clinical course of chronic pelvic pain in women. *Pain, 132*(Suppl. 1), S117–S123.

Whorwell, P. J., McCallum, M., Creed, F. H., & Roberts, C. T. (1986). Non-colonic features of irritable bowel syndrome. *Gut, 27*, 37–40.

Williams, R. E., Hartmann, K. E., & Steege, J. F. (2004). Documenting the current definitions of chronic pelvic pain: Implications for research. *Obstetrics and Gynecology, 103*(4), 686–691.

Zondervan, K. T. (1999). Epidemiology of chronic pelvic pain. Oxford University DPhil thesis.

Zondervan, K. T., Yudkin, P. L., Vessey, M. P., Dawes, M. G., Barlow, D. H., & Kennedy, S. H. (1998). The prevalence of chronic pelvic pain in women in the United Kingdom: A systematic review. *British Journal of Obstetrics and Gynaecology, 105*, 93–99.

Zondervan, K. T., Yudkin, P. L., Vessey, M. P., Dawes, M. G., Barlow, D. H., & Kennedy, S. H. (1999a). Prevalence and incidence of chronic pelvic pain in primary care: Evidence from a national general practice database. *British Journal of Obstetrics and Gynaecology, 106*, 1149–1155.

Zondervan, K. T., Yudkin, P. L., Vessey, M. P., Dawes, M. G., Barlow, D. H., & Kennedy, S. H. (1999b).

Patterns of diagnosis and referral in women consulting for chronic pelvic pain in UK primary care. *British Journal of Obstetrics and Gynaecology, 106*, 1156–1161.

Zondervan, K. T., Yudkin, P. L., Vessey, M. P., Jenkinson, C. P., Dawes, M. G., Barlow, D. H., & Kennedy, S. H. (2001a). The community prevalence of chronic pelvic pain in women and associated illness behaviour. *British Journal of General Practice, 51*, 541–547.

Zondervan, K. T., Yudkin, P. L., Vessey, M. P., Jenkinson, C. P., Dawes, M. G., Barlow, D. H., & Kennedy, S. H. (2001b). Chronic pelvic pain in the community: Symptoms, investigations and diagnoses. *American Journal of Obstetrics and Gynecology, 184*(6), 1149–1155.

Zondervan, K. T., Cardon, L. R., Kennedy, S. H., Martin, N. G., & Treloar, S. E. (2005). Multivariate genetic analysis of chronic pelvic pain and associated phenotypes. *Behavior Genetics, 35*(2), 177–188.

Zondervan, K. T., Moore, J., Kennedy, S. H. (2006). Issues in investigating the aetiology of chronic pelvic pain. Response to Latthe P, Mignini L, Gray R, Hills R, Khan K. Factors predisposing women to chronic pelvic pain: Systematic review. *BMJ, 332*(7544), 749–755.

Epidermal Nerves

Definition

Epidermal nerves are free nerve endings in the most superficial layer of the skin (epidermis), also termed intraepidermal nerve fibers.

Cross-References

▶ Toxic Neuropathies

Epidural

Synonyms

Peridural

Definition

Superficial to the dura, which covers the brain and spinal cord. The epidural space is a potential space lying just outside the dura mater. Both anesthesia and analgesia can be administered by placing local anesthetics and adjuvants in the epidural space, either by a single injection through a needle or longer term through a catheter.

Cross-References

▶ Cancer Pain Management, Anesthesiologic Interventions
▶ Epidural Space
▶ Pain Treatment, Implantable Pumps for Drug Delivery
▶ Postherpetic Neuralgia, Etiology, Pathogenesis, and Management
▶ Postoperative Pain, Appropriate Management

Epidural Analgesia

Synonyms

Peridural analgesia

Definition

Administration of local anesthetics, opioids, and other adjuvants into the epidural space, commonly by infusion through a catheter, to provide analgesia. It is used widely in labor pain but also highly effective in postoperative and posttraumatic pain.

Cross-References

▶ Analgesia During Labor and Delivery
▶ Cancer Pain Management, Anesthesiologic Interventions, and Neural Blockade
▶ Postoperative Pain, Acute Pain Management Principles
▶ Postoperative Pain, Acute Pain Team

Epidural Block

Synonyms

Peridural block

Definition

Regional anesthesia technique to block the nerve roots crossing the epidural space, to provide anesthesia and analgesia.

Cross-References

▶ Pain Treatment: Spinal Nerve Blocks

Epidural Diagnostic Blocks

▶ Diagnosis of Pain, Epidural Blocks

Epidural Drug Pumps

▶ Pain Treatment, Implantable Pumps for Drug Delivery

Epidural Hematoma: Acute Traumatic Epidural Hematoma

▶ Headache Due to Intracranial Bleeding

Epidural Infusions in Acute Pain

Anne Jaumees
Royal North Shore Hospital, St Leonards, NSW, Australia

Synonyms

Extradural infusions; Perispinal infusions; Peridural infusions

Definition

Epidural infusions are ▶ infusions of medications/drugs into the ▶ epidural space. They may be used in acute pain, chronic pain, and cancer pain. They may be used in all age groups from neonates to the elderly.

They have a role in acute pain in the pre-, intra-, and postoperative periods. They may be used in the management of pain during labor.

Characteristics

History

The epidural space has been used since 1901 for provision of analgesia. It was originally accessed through the ▶ sacral hiatus (now called ▶ caudal analgesia). It became obvious that higher access would be needed for procedures in the thorax and abdomen; this was done via the thoracic and lumbar spine. The use of a continuous infusion via an epidural catheter was described in 1949, but it was not until the latter third of the twentieth century that it was used with any frequency. Initially local anesthetics were the only drugs used, but with the report of long-lasting analgesia from intrathecal morphine in 1979, there was an increased use of epidural or intrathecal opioids (Miller 2004).

Many other drugs have been used. The most common include alpha agonists, ketamine and benzodiazepines. But various other drugs including neostigmine nonsteroidal anti-inflammatory drugs (NSAIDs) and droperidol have been trialed (Walker et al. 2002). Drugs are used either alone or in combination, reflecting clinical need and local practice.

Theoretical Basis

There are multiple receptors at spinal cord and brain levels that affect pain processing both in the ascending and descending pathways. These include, among others, sodium channels (site of action for local anesthetics), alpha-adrenergic receptors, opioid receptors, and NMDA receptors. The delivery of drugs into the epidural space allows diffusion across the dura to act on these receptors on the spinal cord and nerve roots. The proximity of the drug to the effect site means a smaller amount of drug is given for a particular effect and this reduces the risk of side effects. It also allows regional analgesia/anesthesia with reduced systemic effects, e.g., sedation.

Anatomy

The spinal cord is surrounded by three membranes called the meninges. The inner is the pia mater and is adherent to the spinal cord and the middle (the arachnoid) and the outer (the dura mater) lie close together and form the dural sac. The epidural space is a potential space that lies between the dura mater and the bone and ligaments of the spinal canal. It contains fat, connective tissue, blood vessels, and nerve roots. Deeper than the arachnoid mater is the intrathecal or subarachnoid space, which contains cerebrospinal fluid and the spinal cord or cauda equina.

When to Use Epidural Infusions

Preoperative

Epidurals may be used to obtain a good pain relief preoperatively, e.g., ischemic leg pain prior to amputation. Preemptive analgesia is the concept wherein by providing good analgesia prior to the painful stimulus (e.g., preoperatively), pain in the postoperative period is reduced. It is on this basis that epidurals have previously been promoted or proposed preoperatively. There have been good retrospective trials to support this, but prospective trials have been equivocal (Rodgers et al. 2000; Rigg et al. 2002). Therefore, it is not usually current practice to place an epidural except in the immediate preoperative phase.

Intraoperative

Previous meta-analyses have demonstrated the benefit of epidurals over systemic analgesia on multiple levels including postoperative morbidity and mortality, reduced pulmonary complications, lower incidence of deep vein thrombosis (DVT) and myocardial infarctions, and improved bowel recovery. However, two large randomized controlled trials showed no major advantages with regard to outcomes except in aortic surgery or the reduction of pulmonary complications in high risk surgery (Rigg et al. 2002). Both meta-analyses and prospective trials show improved pain relief (Rigg et al. 2002; Block et al. 2003). Because of the conflicting evidence, the role of epidurals in the perioperative phase is being reviewed constantly and use reflects patient and procedure indications, anesthetist preference, and hospital setup (trained nursing staff, etc.).

Postoperative
Infusions may be continued into the postoperative phase. The duration depends on factors such as clinical indication and risk of infection.

Other Indications Include
• Obstetric analgesia for labor
• Trauma, e.g., chest trauma with fractured ribs

Epidural Drugs
Local Anesthetics
Action Action is on nerve roots and spinal cord by crossing the dura.

Drugs The most commonly used local anesthetic (LA) is bupivacaine but various LAs may be used including ropivacaine and lignocaine. LAs are generally used in combination with an opioid that increases the efficacy of the pain relief.

Side Effects These include motor and sensory block, hypotension due to sympathetic block, bradycardia with high block, and urinary retention.

Opioids
Action Epidural-administered opioids produce analgesia by blocking opioid receptors in the dorsal horn of the spinal cord and by systemic absorption. Lipid solubility influences the onset and duration of the effect and side-effect profile. Very lipid-soluble drugs (e.g., fentanyl) have a more rapid onset and offset with more specific segmental analgesic effects and therefore require a more precise positioning of the epidural catheter. Less soluble drugs such as morphine have slower onset and are cleared more slowly from the CSF and may therefore spread more rostrally

and have a longer duration of action. Epidural opioids include morphine, hydromorphone, diamorphine (heroin), pethidine (meperidine), fentanyl, and sufentanil.

Side Effects These include sedation, respiratory depression (unusual), nausea and vomiting, pruritus, and urinary retention. They also show gastrointestinal motility but to a lesser extent than the equivalent systemic dose.

Alpha Agonists
Clonidine is the most commonly used alpha agonist. It is analgesic alone or in combination with other analgesics for postoperative pain. It is more effective with fewer side effects in combination with other drugs. It is also useful in the treatment of complex regional pain syndrome (CRPS) as an epidural infusion.

Side Effects Most common side effects include hypotension, bradycardia, and sedation.

Combinations
Combinations of analgesics are often additive in effect. The most commonly used combination is local anesthetic ± opioid. However, common combinations may include alpha agonists. Drugs such as neostigmine, midazolam, and ketamine have been shown to have some analgesic effect both alone and in combination, but there is a lack of large trials to fully evaluate analgesic efficacy and safety (Walker et al. 2002).

Contraindications
Staff Factors
Unskilled/untrained staff: Because of the risk of both more common but less severe side effects or less common but potentially devastating side effects and complications, it is important that epidural infusions are only used with staff trained in their use and in their monitoring.

Patient Factors
Absolute
• Patient refusal
• Local sepsis
• Coagulopathy

Relative/Controversial
- Some neurological diseases, e.g., multiple sclerosis
- Generalized sepsis
- Hypovolemia
- Presence of dural puncture
- Concurrent anticoagulation (dependent on drug) (Macintyre and Ready 2001)

Complications
Post-dural Puncture Headache
This occurs when the dura is punctured and leakage of CSF occurs, resulting in a decrease in CSF pressure and tension on meninges and blood vessels. The risk is $\sim < 1$ % of epidurals.

Treatment Treatment is simple analgesia, hydration, and an epidural blood patch.

Subarachnoid Injection (Subarachnoid Anesthesia)
This may result in a total spinal anesthetic with loss of consciousness $\pm$ cardiovascular and respiratory changes which represents an anesthetic emergency. The changes will resolve with time but require supportive care.

Subdural Injection
This may result in a higher than expected block for the dose given. CVS and respiratory changes may also occur depending on the height of the block. Close monitoring and urgent management as appropriate are needed.

Neurological Injury
Injury to either spinal cord or nerves may be either due to direct trauma or as a result of hypotension with secondary infarction.
   The risk hard to estimate but the complication is rare.

Epidural Hematoma
The risk is unknown but is likely to be <1/150,000.

Diagnosis It is important to have a high index of suspicion as signs and symptoms may be sudden in onset. Trained staff should regularly monitor

for these. Symptoms include back pain or nerve root pain and evidence of neurological deficit due to nerve compression including muscle weakness and sensory, bladder, or bowel dysfunction. MRI is the investigation of choice and should be arranged urgently.

Treatment Surgical decompression within 8–12 h will allow the best chance of full recovery.

Epidural Abscess
This may result from direct contamination on insertion of the catheter or from the infusion of contaminated fluids. It may also occur spontaneously.

Diagnosis Symptoms and signs may be similar to those of epidural hematoma but with an onset that is later and slower. The most frequent presenting symptoms are increasing and persistent back pain, back tenderness, and signs of infection. The imaging of choice is MRI and blood investigations (↑WCC, CRP); lumbar puncture may show evidence of infection.

Treatment Urgent neurosurgical assessment should be sought. Conservative treatment with antibiotics may be appropriate if there is no evidence of neurological compromise; however, decompression within 8–12 h of onset of neurological signs gives the best chance of recovery.

Intravascular Injection
Local Anesthetic Toxicity This can affect both the central nervous system (CNS) and the cardiovascular system (CVS). CNS toxicity occurs at lower doses than CVS toxicity and is initially excitatory due to the initial inhibition of inhibitory pathways. At higher doses, there is also inhibition of excitatory pathways and the general effect is one of CNS depression that may even result in death.

Anticoagulation and Epidurals
Anticoagulants are commonly used drugs in the community and in hospital. While it is generally contraindicated to use epidurals with a coagulation disorder, it is less clear when

people are on anticoagulants and there must be a risk/benefit assessment for each patient.

Nonsteroidal Anti-inflammatory Agents (NSAIDs)/Aspirin

These have not been shown to increase the risk of epidural hematoma.

Oral Anticoagulants

The risk of epidural hematoma in these patients is unknown. If warfarin is started postoperatively, the best time to remove the catheters is unclear and it should be done in conjunction with monitoring of coagulation status.

Heparin

Unfractionated Heparin The epidural should be placed at least 1 h before the dose. Catheters should be removed at least 6 h post-dose and a further 1–2 h before another dose is given.

Low Molecular Weight Heparin (LMWH) Concerns about the combination of LMWH and epidural or spinal anesthesia causing epidural hematoma arose after a series of cases in the USA where patients received LMWH in a different dosing regimen from that used in many other parts of the world. Patients received twice-daily doses and had a total higher daily dose.

Currently, the use of concurrent LMWH and epidurals varies depending on local prescribing practice (e.g., od or bd dosing). There is no clear evidence for any particular prescribing practice but a general guideline could be that insertion may not be within 24 h of the last dose, with epidural catheter removal at least 12–18 h following the last dose and subsequent doses not being given for 6 h.

Summary

Epidural infusions clearly have a role in addressing acute, postoperative pain in infants and children. Research is expanding rapidly to better define safety and outcome parameters, as well as the potential for broader use.

References

Block, B. M., Liu, S. S., Rowlingson, A. J., et al. (2003). Efficacy of postoperative epidural analgesia: A meta-analysis. *Journal of the American Medical Association, 290*, 2455–2463.

Macintyre, P. E., & Ready, L. B. (2001). *Acute pain management: A practical guide* (2nd ed.). London: WB Saunders.

Miller, R. D. (2004). *Anesthesia* (6th ed.). Edinburgh: Elsevier/Church Livingstone.

Rigg, J. R., Jamrozik, K., Myles, P. S., et al. (2002). Epidural anaesthesia and analgesia and outcome of major surgery: A randomised trial. *Lancet, 359*, 1276–1282.

Rodgers, A., Walker, N., Schug, S., et al. (2000). Reduction of post-operative mortality and morbidity with epidural or spinal anaesthesia: Results from overview of randomized trials. *BMJ, 321*, 1493–1497.

Walker, S. M., Goudas, L., Cousins, M. J., et al. (2002). Combination spinal analgesic chemotherapy: A systematic review. *Anesthésie et Analgésie, 95*, 674–715.

Epidural Injection of Corticosteroids

▶ Epidural Steroid Injections

Epidural Space

Synonyms

Extradural space; Peridural space

Definition

The epidural space surrounds the dural mater sac and is also called the extradural or peridural space. Anteriorly, it is bound by the posterior longitudinal ligament; posteriorly by the ligamenta flava and the periosteum of the laminae; and laterally by the pedicles and the intervertebral foramina with their neural roots. Cranially, the epidural space is closed at the foramen magnum where the spinal dura attaches with the endosteal dura of the cranium. Caudally, the epidural space ends at the sacral hiatus that is closed by the sacrococcygeal ligament. The epidural space contains loose areolar connective tissue, fat, lymphatics, arteries, a plexus of veins, and the spinal nerve roots as they leave the dural sac and pass through the intervertebral foramina. The epidural space communicates freely with the paravertebral space through the intervertebral foramina.

Cross-References

▶ Acute Pain in Children, Postoperative
▶ Epidural Infusions in Acute Pain

References

Newell, R. L. (1999). The spinal epidural space. *Clinical Anatomy*, *12*, 375–379.

Epidural Spinal Electrical Stimulation (ESES)

▶ Pain Treatment: Spinal Cord Stimulation

Epidural Steroid Injections

Nikolai Bogduk
University of Newcastle, Newcastle Bone & Joint Institute, Royal Newcastle Centre, Newcastle, NSW, Australia

Synonyms

Caudal epidural steroids; Epidural injection of corticosteroids; Lumbar epidural steroids; Transforaminal injection of steroids

Definition

Epidural injection of steroids is a treatment for radicular pain in which corticosteroid preparations are injected into the epidural space, using an interlaminar, caudal, or transforaminal route.

Characteristics

Epidural injections of steroids are a common treatment that has been in use since 1952. The popularity of the treatment is supported by a large body of descriptive literature (Bogduk et al. 1993).

Technique

Common to all techniques is placement of a needle into, or near, the epidural space at lumbar or sacral levels, so that material can be injected into that space. Access to the epidural space can be obtained through the ligamentum flavum, the sacral hiatus, or the intervertebral foramina. These techniques are known as the lumbar or interlaminar route, the caudal route, and the transforaminal route, respectively (Bogduk et al. 1993).

Classically, lumbar and caudal injections have been performed blind, i.e., without radiological guidance, although increasingly commentators have called for fluoroscopic guidance to be adopted. Transforaminal injections have, since their inception, been performed under fluoroscopic guidance.

Rationale

The rationale for epidural injection of steroids is that radicular pain is caused by inflammation of the lumbar or sacral nerve roots and that corticosteroids should suppress this inflammation and, thereby, provide relief of pain. In the past, this rationale had been presumptive. More recently, considerable evidence from animal experiments and clinical studies has established an inflammatory basis for lumbar radiculopathy (Bogduk and Govind 1999).

Indications

Epidural steroids are a possible treatment for lumbar radicular pain. They are not indicated for low back pain. No literature supports their indiscriminate use for low back pain.

Efficacy

The evidence concerning the efficacy of epidural steroids differs in quantity, strength, and conclusion, for each of the different techniques.

Lumbar Injections

Early reviews were divided concerning the efficacy of lumbar epidural injections of steroids (Benzon 1986; Kepes and Duncalf 1985). An Australian government report found no evidence to either refute or endorse the practice (Bogduk et al. 1993). Later reviews concluded that there was no evidence of efficacy for lumbar epidural injections of steroids (Koes et al. 1995, 1999; Nelemans et al. 2001).

Since those reviews, two placebo-controlled studies have refuted the efficacy of lumbar epidural steroids. One found no differences in pain or disability, at 3 months after treatment with either epidural steroids or placebo (Carette et al. 1997). The other found no differences in outcome at 35 days (Valat et al. 2003). For lumbar epidural steroids, the ▶ number needed to treat (NNT) was 100.

Caudal Injections

A meta-analysis found that the pooled data suggested that epidural steroids did have an attributable effect (Watts and Silagy 1995). For lumbar epidural steroids, that conclusion has now been refuted (Carette et al. 1997; Valat et al. 2003). For injections by the caudal route, the conclusion is still open. It essentially rests on the results of two studies, each of which alone lacked sufficient statistical power to provide a conclusion.

One study suggested that steroids were reasonably effective but that control injections of saline were markedly ineffective (Breivik et al. 1976). The small sample size, however, failed to reach statistical significance. Similarly, the second study showed a trend in favor of steroids at 4 weeks although not at 1 year (Bush and Hillier 1991), but the differences in outcome were not statistically significant from control.

Transforaminal Injections

Transforaminal injections of steroids are only modestly better than control treatment for the relief of pain (Vad et al. 2002). However, they are very effective at sparing patients from surgery (Riew et al. 2000; Weiner and Fraser 1997). An open study found that 47 % of patients, previously listed for surgery, no longer required it at an average of 3 years after treatment with transforaminal injections (Weiner and Fraser 1997). This effect was corroborated in a controlled study, in which 29 % of patients treated with steroids still required surgery, compared with 67 % who were treated with transforaminal injections of bupivacaine alone (Riew et al. 2000).

Discussion

The evidence from controlled trials and systematic reviews does not support the popular use of lumbar epidural injections of steroids. These injections have proven to be no more effective than placebo therapy. For caudal injections, the evidence only hints at these injections being effective.

With respect to relief of pain, transforaminal injections of steroids are barely better than the control treatments against which they have been compared. However, transforaminal injections do have a significant effect in saving patients from surgery for lumbar radicular pain.

Of all the techniques available for epidural injection of steroids, the transforaminal appears to be the most effective. The available literature on transforaminal injections, however, is limited to patients with chronic radicular pain, awaiting surgery. Their efficacy for acute radicular pain has not been reported.

References

Benzon, H. T. (1986). Epidural steroid injections for low back pain and lumbosacral radiculopathy. *Pain, 24*, 277–905.

Bogduk, N., Christophidis, N., Cherry, D., et al. (1993). *Epidural is of steroids in the management of back pain and sciatica of spinal origin. Report of the Working party on epidural use of steroids in the management of back pain.* Canberra: National Health and Medical Research Council.

Bogduk, N., & Govind, J. (1999). *Medical management of acute lumbar radicular pain: An evidence-based approach.* Newcastle: Newcastle Bone and Joint Institute.

Breivik, H., Hesla, P. E., Molnar, I., et al. (1976). Treatment of chronic low back pain and sciatica. Comparison of caudal epidural injections of bupivacaine and methylprednisolone with bupivacaine followed by Saline. In J. J. Bonica & D. Albe-Fessard (Eds.), *Advances in pain research and therapy* (Vol. 1, pp. 927–932). New York: Raven.

Bush, K., & Hillier, S. (1991). A controlled study of caudal epidural injections of triamcinolone plus pro-caine for the management of intractable sciatica. *Spine, 16*, 572–575.

Carette, S., LeClaire, R., Marcoux, S., et al. (1997). Epidural corticosteroid injections for sciatica due to herniated nucleus pulposus. *The New England Journal of Medicine, 336*, 1634–1640.

Kepes, E. R., & Duncalf, D. (1985). Treatment of back-ache with spinal injections of local anesthetics, spinal and systemic steroids: A review. *Pain, 22*, 33–47.

Koes, B. W., Scholten, R. J. P. M., Mens, J. M. A., et al. (1995). Efficacy of epidural steroid injections for low-back pain and sciatica: A systematic review of randomized clinical trials. *Pain, 63*, 279–288.

Koes, B. W., Scholten, R. J. P. M., Mens, J. M. A., et al. (1999) Epidural steroid injections for low back pain and sciatica: An updated systematic review of random-ized clinical trials. *Pain Digest, 9*, 241–247.

Nelemans, P. J., de Bie, R. A., de Vet, H. C. W., et al. (2001). Injection therapy for subacute and chronic benign low back pain. *Spine, 26*, 501–515.

Riew, K. D., Yin, Y., Gilula, L., et al. (2000). The effect of nerve-root injections on the need for operative treat-ment of lumbar radicular pain. A prospective, random-ized, controlled, double-blind study. *Journal of Bone & Joint Surgery, 82A*, 1589–1593.

Vad, V. B., Bhat, A. L., Lutz, G. E., et al. (2002). Transforaminal epidural steroid injections in lumbosa-cral radiculopathy. *Spine, 27*, 11–16.

Valat, J. P., Giraudeau, B., Rozenberg, S., et al. (2003). Epidural corticosteroid injections for sciatica: A randomised, double-blind, controlled clinical trial. *Annals of the Rheumatic Diseases, 62*, 639–643.

Watts, R. W., & Silagy, C. A. (1995). A meta-analysis on the efficacy of epidural corticosteroids in the treatment of sciatica. *Anaesthesia and Intensive Care, 23*, 564–569.

Weiner, B. K., & Fraser, R. D. (1997). Foraminal injection for lateral lumbar disc herniation. *Journal of Bone & Joint Surgery, 79B*, 804–807.

Epidural Steroid Injections for Chronic Back Pain

Katie Nixdorf and Brett R. Stacey
Comprehensive Pain Center, Oregon Health and Science University, Portland, OR, USA

Synonyms

Caudal injection; Interlaminar epidural steroid injection; Midline epidural steroid injection; Nerve block; Selective nerve root block; Selective nerve root injection; Transforaminal epidural steroid injection; Translaminar epidural steroid injection

Definition

"Epidural steroid injection" is a minimally invasive procedure, in which corticosteroids are placed in the epidural space. Epidural steroid injections (ESIs) were initially reported decades ago and continue to be utilized for pain management.

Epidural steroid injections are typically used in patients whose pain has not responded to con-servative therapies, although in selected patients ESIs may be indicated earlier in the course of treatment. The most accepted use of ESIs is to treat lumbar radicular pain associated with herni-ated intervertebral disc(s), although they may also be employed for other causes of radicular pain including spondylosis, degenerative disc disease, and spinal stenosis. Epidural steroid injections have been used for non-radicular pain such as axial low back pain or spinal stenosis, compression fractures, or other conditions as well; however, there is limited evidence for non-radicular pain. ESIs are also used in the cervical spine, typically to treat cervical radiculopathy.

Radicular pain is thought to result from sev-eral pathophysiologic factors. Compression of the nerve can occur due to a herniated disc or other material pressing on the nerve, with direct pressure leading to structural changes and vascu-lar compromise (Lipetz 2002). Swelling and edema at the site can also cause nerve irritation. In addition to simple compression, the nucleus pulposus may be a source of proinflammatory substances which lower the threshold for noci-ceptive fibers to fire, or may cause direct nerve root irritation and subsequent pain. The proinflammatory changes can involve the dorsal root ganglion as well (Kobayashi et al. 2004).

Steroids placed within the epidural space are thought to work by decreasing neurogenic inflammation and stabilizing membranes (Lee et al. 1998). These effects take place over a few days, with the ultimate effect of pain relief. The addition of local anesthetic provides pain relief through an alternative mechanism of sodium channel blockade, which provides more immediate, but short-acting, pain relief. However, there is some evidence that local anesthetics such as lidocaine have anti-inflammatory properties as well (Yabuki et al. 1998). Alternatively, the steroid can be delivered in a preservative-free normal saline vehicle.

Characteristics

Accessing the Epidural Space

In the lumbar spine, there are three primary approaches by which the epidural space can be entered. For each of these approaches, there is wide variability in practice techniques (needle, use of imaging, injectate details, frequency of repeating the procedure, use of sedation, setting in which the procedure is performed, concomitant treatments, etc.) with no clear standard (Cluff et al. 2002). Such variability makes comparisons difficult. Interlaminar and caudal approaches without imaging were widely utilized and standard of practice for years. The transition to transforaminal injections has increased with the growing popularity of fluoroscopy for procedures. Evidence also suggests that using fluoroscopy or other imaging increases the likelihood of delivering medication to the targeted area of the spine and may confer increased safety (Fredman et al. 1999; Bartynski et al. 2005; Stretanski and Chopko 2005; Alemo 2010). Fredman demonstrated in a prospective study a high failure rate of 53 % in placing an epidural needle at the desired level based on surface anatomy alone, and only a 26 % success rate in delivering injectate to the level of pathology. In the cervical and thoracic spine, the epidural space can be approached by both the interlaminar and transforaminal approaches, although nearby anatomic structures add complexity and risk to transforaminal injections.

Interlaminar Approach

The interlaminar (or translaminar) approach enters the epidural space posteriorly, in the space between adjacent vertebral lamina. This approach is similar to the technique utilized for epidural perioperative anesthesia and analgesia, and thus is familiar to many practitioners. This approach can be done with a midline, paramedian, or parasagittal technique. Traditionally the interlaminar approach was completed blindly using bony landmarks as reference, but radiographic guidance such as fluoroscopy is increasingly utilized to identify specific spinal levels, confirm the location of the needle, and evaluate for the appropriate spread of radio-opaque injectate. A study evaluating contrast spread of the parasagittal technique showed anterior epidural spread, which allows for injectate to be located closer to disc pathology (Candido et al. 2008).

The interlaminar approach typically utilizes a specialized epidural needle and a loss of resistance syringe, which helps to identify the moment when the ligamentum flavum is traversed and the needle tip has entered the epidural space, located just anterior to the ligamentum flavum. After a loss of resistance, aspiration of the syringe is intended to help decrease, but not completely eliminate the risk of inadvertent intrathecal or intravascular injection. When completed using radiographic guidance, monitoring the spread of contrast confirms the location in the epidural space and movement of the medication towards the intended target.

Typically, injectate volumes ranging from 5 - 10 ml are deposited along the posterior aspect of the epidural space, usually with less ability to target a location such as a specific nerve root or dorsal root ganglion. This injectate includes steroid in a vehicle of saline and/or local anesthetic.

Transforaminal Approach

The transforaminal approach gains access to the epidural space laterally, via the intervertebral foramen. This location allows for placement of medication at a level targeting a specific spinal nerve and dorsal root ganglion. In addition, like the parasagittal interlaminar approach, the transforaminal approach is theoretically closer to the anterior epidural space, where most disc pathology occurs (Candido et al. 2008).

Techniques for the transforaminal approach focus on entering the foramen in a "safe" location, to avoid the nerve and vascular structures. This approach is done under radiographic guidance (typically fluoroscopy) to assist with proper placement of a smaller needle (typically 22–25 gauge) in the superior aspect of the neural foramen. Multiplanar fluoroscopy aids in positioning the needle and verifying radio-opaque contrast spread medially into the epidural space and laterally along the targeted nerve.

A "selective nerve root block" is completed in a similar manner to the transforaminal approach, although the intended target is slightly more lateral, to avoid epidural spread. The ability to "selectively" block a nerve root has been questioned (Furman and O'Brien 2000). Yeom et al. completed a study to determine the efficacy of lumbar selective nerve root blocks. In patients with monoradicular pain determined by clinical, radiographic, and MR imaging findings, selective nerve root blocks were completed on affected and non-affected levels. On the basis of at least 70 % pain relief at the correct level of pathology, they found a sensitivity of 57 %, a specificity of 86 %, and an accuracy of 73 % (Yeom et al. 2008).

The injectate for a transforaminal epidural steroid injection is typically steroid mixed with a small volume of local anesthetic and/or saline. The total volume of this injection is typically 2–4 ml, less volume than its interlaminar counterpart, as the injected area is intended to be more precise.

Caudal Approach
The caudal approach enters the epidural space through the sacral hiatus, a normally occurring gap in the lower portion of the sacrum typically closed by the sacrococcygeal ligament. Due to the distance from the sacral hiatus to the intended targets more cephalad in the spine, either a larger volume than other approaches must be injected, or a catheter is threaded through the needle towards the intended target in the lumbar spine. Utilizing either a fiberoptic scope or a stiff, steerable catheter under fluoroscopic guidance, lysis of adhesions or neuroplasty can be accomplished (Igarashi et al. 2004).

Procedural Considerations
Contraindications
Contraindication to epidural steroid injection includes systemic infection or local infection at the location of injection, bleeding disorder or anti-coagulated state, allergic reaction to injectate, or acute spinal cord compression. Caution should be utilized in patients who are pregnant, have poorly controlled diabetes, or are immunosuppressed. ESIs can also lead to transient blood pressure changes.

Complications
A review of the literature of lumbar ESIs done by Chou et al. found that the reporting of serious adverse events was rare in randomized trials (Chou et al. 2009a). When reported, adverse events were typically transient and minor. This was true for both interlaminar and transforaminal approaches. Serious events have been reported in case reports, including paralysis, infection, and hematoma. More common adverse events reported were post-injection headache, nausea, and post-dural puncture headache.

Complications and adverse events in the thoracic spine are commonly reported in the context of epidural catheters for perioperative analgesia, but there is limited literature available for thoracic epidural steroid injections. A retrospective review of 39 thoracic interlaminar ESIs showed a transient adverse event rate of 20.5 %, but no major complications (Botwin et al. 2006). Adverse events were described as transient symptoms including injection site pain, facial flushing and headache.

In cervical epidural steroid injections, the true frequency of complications has been difficult to determine as well, due to the limited number of studies that evaluated the procedure and the rate of complication (Abbasi et al. 2007). While several studies have stated a 0 % complication rate, other studies reported complications as frequently as 16.8 %. Many of the case reports discuss complications which are similar to those seen in the lumbar spine, but rare complications with devastating neurologic consequences and death have been published for interlaminar cervical ESIs, including epidural hematoma, spinal cord injury, persistent paresthesias, subdural injections and hematoma, and death. Controversy over the use of transforaminal injections in the cervical spine is fueled by several case reports of potentially catastrophic complications including brain and spinal cord infarction, vertebral artery dissection, spinal cord trauma, seizure, and epidural hematomas (Malhotra et al. 2009). Adding to this controversy is the limited availability of definitive prospective randomized trials in support of this technique.

In a review of the American Society of Anesthesiologists closed-claims database from 2005 to 2008, claims associated with cervical procedures represented 22 % of the claims for chronic pain treatments (Rathmell et al. 2011). Of those who underwent cervical spinal cord procedures, 58 % had spinal cord injuries compared to 11 % of patients with other pain complaints having spinal cord injury.

Particulate Versus Non-particulate Steroids
Particulate steroids have been implicated in cases of spinal cord and brain infarction after ESI, thought related to intra-arterial injection. An in vivo study evaluating the impact of intra-arterial injection of steroids in the rat brain found that Depo-Medrol (a particulate formulation of methylprednisolone), Solu-Medrol (a non-particulate formulation of methylprednisolone), and the Depo-Medrol carrier (non-particulate in nature) could all produce significant tissue damage, whereas the non-particulate steroid Decadron (dexamethasone) did not produce any injury (Dawley et al. 2009). They theorize that these findings suggest that the cause of tissue injury is related to more than just particulate-related embolism.

A few studies have evaluated the differences in pain relief between particulate and non-particulate steroid injections. In a study by Park, there was a statistically significant advantage to using triamcinolone (particulate steroid) over dexamethasone (non-particulate) in a lumbar transforaminal ESI (Park et al. 2010). Other studies done with cervical transforaminal ESI found no statistical difference between the steroid types (Lee et al. 2009c; Dreyfuss et al. 2006).

Other Unanswered Questions
Several other questions remain unanswered in regard to the use of ESIs. Some practitioners commonly complete a series of injections in the attempt to improve outcomes. Review of the literature showed there was no significant evidence to support this commonly used practice (Novak 2008). There is variability in practice of the volume of solution injected. Rabinovitch et al. completed a systemic review of literature to assess for a correlation between the volume injected and amount of pain relief (Rabinovitch et al. 2009). These results suggested a positive correlation between larger volumes and improved pain relief in radicular leg pain. These studies evaluate just a portion of the practices that are commonly utilized in ESIs that are routinely done, but not rooted in strong evidence.

Summary

Efficacy of Epidural Steroid Injections
A review of the literature regarding the use of ESIs for chronic pain shows both evidence supporting their use as well as evidence disputing their efficacy. Much of the available literature is difficult to utilize for developing evidence-based practice guidelines, as it is presented in unblinded, retrospective, or smaller studies with flawed designs. It also becomes difficult to compare data between studies when there is often variability in patient selection, indication for ESI, technique used, use of fluoroscopy, solution injected, and the timing and frequency of injections. Older studies are less likely to involve radiographic guidance or comprehensive outcome assessment while newer studies tend to incorporate such radiographic guidance and are more likely to include functional assessments of outcome.

Lumbar Radiculopathy
The majority of strong evidence for ESIs has been completed in lumbar radiculopathy due to disc herniation. A study done by Ackerman evaluated the different approaches for epidural steroid injection in treating a patient with lumbar

radiculopathy due to disc herniation (Ackerman 2007). Ninety patients with symptoms consistent with an S1 radiculopathy were randomized to one of the three approaches to ESI. At 24 weeks, complete pain relief was present in 9 of 30 in the transforaminal route, 3 of 30 in the interlaminar group, and 1 of 30 in the caudal group. In a controlled case series by Schlaufele, interlaminar ESI was compared to transforaminal in treating radicular pain from herniated disc (Schaufele et al. 2006). Patients receiving transforaminal injections had improved short-term pain control, and lower rates of surgical interventions. A study comparing transforaminal injection to blind interlaminar injection at 6 months showed statistically significant improvements favoring the transforaminal group in regard to pain relief, daily living, leisure and work activities, and anxiety and depression (Thomas et al. 2003).

Comprehensive reviews of the literature have also been completed, including by larger medical societies as they create practice guidelines. These reviews and guidelines will be outlined later in this entry.

Lumbar Spinal Stenosis

The evidence behind use of ESIs in lumbar spinal stenosis is limited. Specific studies have compared the different techniques for ESI in treatment of spinal stenosis. A retrospective study of 19 patients comparing results of interlaminar and transforaminal injections found both procedures to lead to pain relief, with no statistical difference between the two techniques (Smith et al. 2010). A retrospective study comparing all three techniques for up to 2 months after injection found higher success rates in pain control and patient satisfaction in interlaminar and transforaminal compared to caudal ESIs, with the largest reductions in the transforaminal group when compared to the caudal approach (Lee et al. 2009b). This author then went on to complete a prospective randomized trial comparing interlaminar injection to bilateral transforaminal injection for spinal stenosis up to 4 months after injection (Lee et al. 2009a). The transforaminal group showed more significant benefits in pain reduction and

patient satisfaction compared to the interlaminar group. They found that 61.4 % of patients in the transforaminal group had "successful outcomes" compared to 35.7 % in the interlaminar group. Their study did use higher volumes in the transforaminal approach than previous studies, placing 4.5 ml per side.

Axial Low Back Pain

Limited evidence is available for the efficacy of ESIs for axial low back pain without radicular symptoms. In an observational study of 81 patients with axial low back pain without radicular symptoms and no imaging evidence of nerve root compression, interlaminar ESIs were evaluated (Lee et al. 2010). Sixty-three of 81 patients had relief with interlaminar ESI, with a median symptom-free interval of 154 days, and 37 % of patients had relief for greater than 6 months.

Cervical Radiculopathy

A retrospective review of 76 patients with neck pain or cervical radiculopathy was completed to determine the efficacy of cervical interlaminar ESI at 2 weeks (Kwon et al. 2007). Seventy-two percent of patients had effective pain relief, with those having pain from radiculopathy having better results than those having pain from spinal stenosis. In a study of 58 patients who underwent interlaminar ESI, 41.4 % of patients had greater than 90 % pain relief lasting 6 months, and 29 % of patients had greater than 50 % of pain relief lasting at least 6 weeks (Cicala et al. 1989).

Prospective, randomized studies involving transforaminal cervical ESIs are limited. An observational study of 21 patients with cervical radiculopathy scheduled for surgery, and felt to have limited potential for spontaneous recovery, evaluated whether a series of two transforaminal ESIs would assist with pain control and delay or eliminate need for surgery (Kolstad et al. 2005). Five of 21 patients had "excellent" pain relief lasting at 4 month follow-up and cancelled their planned surgery, leaving 16 of 21 patients with moderate or no pain relief. Those who responded had improvement in both neck and radicular pain. Evaluation of the responders showed pre-procedure pain scores that were lower than the mean. A prospective, randomized study evaluating transforaminal cervical ESIs in patient with unilateral radicular pain with corresponding MRI pathology, who had previously had a positive response to a selective nerve root block, compared a steroid injection to a local anesthetic injection (Anderberg et al. 2007). At 3 week follow-up, both the steroid and the control local anesthetic group had improvement in their symptoms. There was no difference between patients receiving a steroid/local anesthetic injection versus a saline/local anesthetic injection, raising the question of whether local anesthetic alone can provide long-term pain relief.

Guidelines for Epidural Steroid Injections

As the amount and quality of the available literature is difficult to fully decipher, various professional societies have created consensus statements or clinical practice guidelines to assist practitioners in the appropriate use of ESIs. Most of these recommendations focus on lumbar radicular pain.

The American Pain Society reviewed the evidence for ESIs prior to the publication of their clinical guidelines in 2009. For lumbar radicular pain, they identified six studies in which epidural injections were compared to soft tissue injection, and 5 of 6 studies showed short-term benefit for ESI, while only 2 of 11 studies comparing ESIs to epidural saline or local anesthetic injections were reported to be of benefit (Chou et al. 2009a). Based on their review of the literature, the American Pain Society stated in their guidelines that the level of evidence was "fair," showing a moderate net benefit for short-term outcomes only (Chou et al. 2009b).

The American Society of Anesthesiologists and American Society of Regional Anesthesia combined to create a set of guidelines for chronic pain management (American Society of Anesthesiologists 2010). These guidelines only address the use of ESIs for radicular pain, where members strongly supported the use of ESIs with the utilization of imaging. These recommendations did not decipher between short-term versus long-term pain relief.

The American Academy of Neurology (AAN) created guidelines for the use of ESIs for lumbar radicular pain, stating these injections can provide short-term pain relief, but the AAN did not feel the evidence supported impact on average impairment of function, on need for surgery, or on long-term pain relief beyond 3 months (Armon et al. 2007). Therefore, the recommendation stated that routine use of ESI for these indications is not recommended.

The American Society of Interventional Pain Physicians (ASIPP) published guidelines in 2010 that looked at the individual approaches to ESIs and provided individual recommendations (Manchikanti et al. 2009). The evidence quality ranged from low to high quality, but overall they gave strong recommendations for use of transforaminal lumbar ESI for short- and long-term pain relief in radicular pain, interlaminar lumbar ESI for short-term pain control, and caudal ESIs for both short and long-term pain control.

Guidelines for the use of ESIs in spinal stenosis are limited. In the American Pain Society guidelines, the authors felt the evidence was poor, and they were unable to estimate any net benefit of ESIs (Chou 2009). In the ASIPP guidelines, a strong recommendation was given for use of caudal ESI for spinal stenosis, with low- to moderate-quality evidence (Manchikanti et al. 2009). They gave a weak recommendation for use of interlaminar ESIs for spinal stenosis, based on low-quality evidence.

The guidelines published by ASIPP gave a strong recommendation for cervical interlaminar epidural steroid injections, based on low-quality evidence (Manchikanti et al. 2009). The other guidelines noted above focused on low back pain, and did not address cervical procedures.

Despite these guidelines, based on the available evidence, there continues to be differences amongst societies and individual practitioners regarding the use of ESIs in subacute and chronic pain.

Conclusions

In conclusion, ESIs remain a potentially valuable treatment option for providing short-term pain relief for patients with radicular leg pain.

Controversy persists for assertions regarding long-term pain relief, improvement in function or neurologic status, avoidance of more invasive management such as surgery, or their use for other conditions and in nonlumbosacral locations. ESIs are commonly used for the diagnosis of radicular pain associated with a herniated intervertebral disc. ESIs are also used to treat pain associated with symptomatic spinal stenosis, intervertebral disc disease, and spondylosis.

The techniques for performing these injections vary in multiple ways, including patient selection, indication for treatment, technique used, use of fluoroscopy, steroid injected, use of local anesthetic, and the timing and frequency of injections, making comparisons of studies difficult. It is likely that image guidance enhances the safety and efficacy of ESIs. Large, randomized, placebo-controlled studies are needed to clarify several of these unresolved questions.

References

Abbasi, A., Malhotra, G., Malanga, G., Elovic, E. P., & Kahn, S. (2007). Complications of interlaminar cervical epidural steroid injections: A review of the literature. *Spine, 32*(19), 2144–2151.

Ackerman, W. E., 3rd, & Ahmad, M. (2007). The efficacy of lumbar epidural steroid injections in patients with lumbar disc herniations. *Anesthesia and Analgesia, 104*(5), 1217–1222.

Alemo, S., & Sayadipour, A. (2010). Observational study of the use of an epidurogram in interlaminar lumbar epidural steroid injection. *British Journal of Anaesthesia, 104*(5), 665–666.

American Society of Anesthesiologists Task Force on Chronic Pain Management, & American Society of Regional Anesthesia and Pain Medicine. (2010). Practice guidelines for chronic pain management: An updated report by the American society of anesthesiologists task force on chronic pain management and the American society of regional anesthesia and pain medicine. *Anesthesiology, 112*(4), 810–833.

Anderberg, L., Annertz, M., Persson, L., Brandt, L., & Saveland, H. (2007). Transforaminal steroid injections for the treatment of cervical radiculopathy: A prospective and randomised study. *European Spine Journal, 16*(3), 321–328.

Armon, C., Argoff, C. E., Samuels, J., & Backonja, M. M. (2007). Therapeutics and technology assessment subcommittee of the American academy of neurology, assessment: Use of epidural steroid injections to treat radicular lumbosacral pain: Report of the therapeutics and technology assessment subcommittee of the American academy of neurology. *Neurology, 68*(10), 723–729.

Bartynski, W. S., Grahovac, S. Z., & Rothfus, W. E. (2005). Incorrect needle position during lumbar epidural steroid administration: Inaccuracy of loss of air pressure resistance and requirement of fluoroscopy and epidurography during needle insertion. *AJNR. American Journal of Neuroradiology, 26*(3), 502–505.

Botwin, K. P., Baskin, M., & Rao, S. (2006). Adverse effects of fluoroscopically guided interlaminar thoracic epidural steroid injections. *American Journal of Physical Medicine & Rehabilitation, 85*(1), 14–23.

Candido, K. D., Raghavendra, M. S., Chinthagada, M., Badiee, S., & Trepashko, D. W. (2008). A prospective evaluation of iodinated contrast flow patterns with fluoroscopically guided lumbar epidural steroid injections: The lateral parasagittal interlaminar epidural approach versus the transforaminal epidural approach. *Anesthesia and Analgesia, 106*(2), 638–644.

Chou, R., Atlas, S. J., Stanos, S. P., & Rosenquist, R. W. (2009a). Nonsurgical interventional therapies for low back pain: A review of the evidence for an American pain society clinical practice guideline. *Spine, 34*(10), 1078–1093.

Chou, R., Loeser, J. D., Owens, D. K., Rosenquist, R. W., Atlas, S. J., Baisden, J., et al. (2009b). Interventional therapies, surgery, and interdisciplinary rehabilitation for low back pain: An evidence-based clinical practice guideline from the American pain society. *Spine, 34*(10), 1066–1077.

Cicala, R. S., Thoni, K., & Angel, J. J. (1989). Long-term results of cervical epidural steroid injections. *The Clinical Journal of Pain, 5*(2), 143–145.

Cluff, R., Mehio, A. K., Cohen, S. P., Chang, Y., Sang, C. N., & Stojanovic, M. P. (2002). The technical aspects of epidural steroid injections: A national survey. *Anesthesia and Analgesia, 95*(2), 403–408.

Dawley, J. D., Moeller-Bertram, T., Wallace, M. S., & Patel, P. M. (2009). Intra-arterial injection in the rat brain: Evaluation of steroids used for transforaminal epidurals. *Spine, 34*(16), 1638–1643.

Dreyfuss, P., Baker, R., & Bogduk, N. (2006). Comparative effectiveness of cervical transforaminal injections with particulate and nonparticulate corticosteroid preparations for cervical radicular pain. *Pain Medicine, 7*(3), 237–242.

Fredman, B., Nun, M. B., Zohar, E., Iraqi, G., Shapiro, M., Gepstein, R., et al. (1999). Epidural steroids for treating "failed back surgery syndrome": Is fluoroscopy really necessary? *Anesthesia and Analgesia, 88*(2), 367–372.

Furman, M. B., & O'Brien, E. M. (2000). Is it really possible to do a selective nerve root block? *Pain, 85*(3), 526.

Igarashi, T., Hirabayashi, Y., Seo, N., Saitoh, K., Fukuda, H., & Suzuki, H. (2004). Lysis of adhesions and epidural injection of steroid/local anaesthetic during epiduroscopy potentially alleviate low back and leg pain in elderly patients with lumbar spinal stenosis. *British Journal of Anaesthesia, 93*(2), 181–187.

Kobayashi, S., Yoshizawa, H., & Yamada, S. (2004). Pathology of lumbar nerve root compression. Part 2: Morphological and immunohistochemical changes of dorsal root ganglion. *Journal of Orthopaedic Research, 22*(1), 180–188.

Kolstad, F., Leivseth, G., & Nygaard, O. P. (2005). Transforaminal steroid injections in the treatment of cervical radiculopathy. A prospective outcome study. *Acta Neurochirurgica, 147*(10), 1065–1070.

Kwon, J. W., Lee, J. W., Kim, S. H., Choi, J. Y., Yeom, J. S., Kim, H. J., et al. (2007). Cervical interlaminar epidural steroid injection for neck pain and cervical radiculopathy: Effect and prognostic factors. *Skeletal Radiology, 36*(5), 431–436.

Lee, H. M., Weinstein, J. N., Meller, S. T., Hayashi, N., Spratt, K. F., & Gebhart, G. F. (1998). The role of steroids and their effects on phospholipase A2. An animal model of radiculopathy. *Spine, 23*(11), 1191–1196.

Lee, J. H., An, J. H., & Lee, S. H. (2009a). Comparison of the effectiveness of interlaminar and bilateral transforaminal epidural steroid injections in treatment of patients with lumbosacral disc herniation and spinal stenosis. *The Clinical Journal of Pain, 25*(3), 206–210.

Lee, J. H., Moon, J., & Lee, S. H. (2009b). Comparison of effectiveness according to different approaches of epidural steroid injection in lumbosacral herniated disk and spinal stenosis. *Journal of Back and Musculoskeletal Rehabilitation, 22*(2), 83–89.

Lee, J. W., Park, K. W., Chung, S. K., Yeom, J. S., Kim, K. J., Kim, H. J., et al. (2009c). Cervical transforaminal epidural steroid injection for the management of cervical radiculopathy: A comparative study of particulate versus non-particulate steroids. *Skeletal Radiology, 38*(11), 1077–1082.

Lee, J. W., Shin, H. I., Park, S. Y., Lee, G. Y., & Kang, H. S. (2010). Therapeutic trial of fluoroscopic interlaminar epidural steroid injection for axial low back pain: Effectiveness and outcome predictors. *AJNR. American Journal of Neuroradiology, 31*(10), 1817–1823.

Lipetz, J. S. (2002). Pathophysiology of inflammatory, degenerative, and compressive radiculopathies. *Physical Medicine and Rehabilitation Clinics of North America, 13*(3), 439–449.

Malhotra, G., Abbasi, A., & Rhee, M. (2009). Complications of transforaminal cervical epidural steroid injections. *Spine, 34*(7), 731–739.

Manchikanti, L., Boswell, M. V., Singh, V., Benyamin, R. M., Fellows, B., Abdi, S., et al. (2009). Comprehensive evidence-based guidelines for interventional techniques in the management of chronic spinal pain. *Pain Physician, 12*(4), 699–802.

Novak, S., & Nemeth, W. C. (2008). The basis for recommending repeating epidural steroid injections for radicular low back pain: A literature review. *Archives of Physical Medicine and Rehabilitation, 89*(3), 543–552.

Park, C. H., Lee, S. H., & Kim, B. I. (2010). Comparison of the effectiveness of lumbar transforaminal epidural injection with particulate and nonparticulate corticosteroids in lumbar radiating pain. *Pain Medicine, 11*(11), 1654–1658.

Rabinovitch, D. L., Peliowski, A., & Furlan, A. D. (2009). Influence of lumbar epidural injection volume on pain relief for radicular leg pain and/or low back pain. *Spine Journal: Official Journal of the North American Spine Society, 9*(6), 509–517.

Rathmell, J. P., Michna, E., Fitzgibbon, D. R., Stephens, L. S., Posner, K. L., & Domino, K. B. (2011). Injury and liability associated with cervical procedures for chronic pain. *Anesthesiology, 114*(4), 918–926.

Schaufele, M. K., Hatch, L., & Jones, W. (2006). Interlaminar versus transforaminal epidural injections for the treatment of symptomatic lumbar intervertebral disc herniations. *Pain Physician, 9*(4), 361–366.

Smith, C. C., Booker, T., Schaufele, M. K., & Weiss, P. (2010). Interlaminar versus transforaminal epidural steroid injections for the treatment of symptomatic lumbar spinal stenosis. *Pain Medicine, 11*(10), 1511–1515.

Stretanski, M. F., & Chopko, B. (2005). Unintentional vascular uptake in fluoroscopically guided, contrast-confirmed spinal injections: A 1-yr clinical experience and discussion of findings. *American Journal of Physical Medicine & Rehabilitation, 84*(1), 30–35.

Thomas, E., Cyteval, C., Abiad, L., Picot, M. C., Taourel, P., & Blotman, F. (2003). Efficacy of transforaminal versus interspinous corticosteroid injection in discal radiculalgia – a prospective, randomised, double-blind study. *Clinical Rheumatology, 22*(4–5), 299–304.

Yabuki, S., Kawaguchi, Y., Nordborg, C., Kikuchi, S., Rydevik, B., & Olmarker, K. (1998). Effects of lidocaine on nucleus pulposus-induced nerve root injury. A neurophysiologic and histologic study of the pig cauda equina. *Spine, 23*(22), 2383–2389.

Yeom, J. S., Lee, J. W., Park, K. W., Chang, B. S., Lee, C. K., Buchowski, J. M., et al. (2008). Value of diagnostic lumbar selective nerve root block: A prospective controlled study. *AJNR. American Journal of Neuroradiology, 29*(5), 1017–1023.

Epigastric

Definition

Epigastric is pertaining to the region overlying the upper abdomen that can be a site of pain associated with digestive system components such as the stomach and pancreas.

Cross-References

▶ Animal Models and Experimental Tests to Study Nociception and Pain
▶ Visceral Pain Model, Pancreatic Pain

Epigastric Pain

▶ Visceral Pain Model, Pancreatic Pain

Epilepsy

Definition

Epilepsy is the term given for syndromes of epileptic seizures, a disorder of the nervous system in which abnormal electrical activity in the brain causes seizures (sudden uncontrolled waves of electrical activity in the brain, causing involuntary movement or loss of consciousness).

Cross-References

▶ Central Pain, Outcome Measures in Clinical Trials
▶ Dysesthesia, Assessment

Episode Tension-Type Headache

▶ Headache, Episodic Tension-Type

Episodic Pain

Definition

Episodic pain is a synonymous term for breakthrough pain that is used in other countries.

Cross-References

▶ Evoked and Movement-Related Neuropathic Pain

Episodic Paroxysmal Hemicrania

Definition

As defined in the International Classification of Headache Disorders – second edition – attacks of severe unilateral had pain lasting from 2 to 30 min accompanied by cranial autonomic features, such as lacrimation, conjunctival injection, and nasal symptoms, with a typical daily frequency of ten attacks occurring in 6–8 week bouts once or twice a year. Indomethacin completely abolishes attacks.

Cross-References

▶ Paroxysmal Hemicrania

Episodic Tension-Type Headache (Frequent and Infrequent)

Definition

This distinction was introduced with the second edition of the International Headache Classification Committee. Infrequent episodic tension-type headache (<12 days/year) is often described as a "normal" headache, although normal is what is routine for an individual, with marginal influence on daily activities and no impact on quality of life.

Cross-References

▶ Headache, Episodic Tension-Type

EPN

▶ Endplate Noise

EPN Locus

▶ ActiveLocus

EPSC

▶ Excitatory Postsynaptic Current

Equianalgesic Dose

Definition

Equianalgesic refers to the dose of one analgesic that produces the same analgesic response as another. The equianalgesic dose is estimated from a relative potency determination that compares graded doses of the two analgesics.

Cross-References

▶ Cancer Pain Management, Principles of Opioid Therapy, Drug Selection
▶ Opioid Rotation

Equianalgesic Dose Ratios

Definition

Equianalgesic dose ratios are estimates of the dose ratio of drug A and B that provide equivalent analgesia. If incomplete cross tolerance is observed between A and B, then the equianalgesic dose ratios of A/B will vary depending on the length of administration of A before B is substituted in opioid rotation.

Cross-References

▶ Opioid Rotation

Equianalgesic Dose Table

Definition

Equianalgesic dose table is a dosing table that describes the relative potencies for different opioids and routes of administration. All values correspond to intravenous morphine 10 mg.

Cross-References

▶ Opioid Rotation in Cancer Pain Management

Equivalent Current Dipole

Synonyms

ECD

Definition

A summation of currents of many neurons with the same positive-negative direction can be mimicked as one strong dipole. This is termed the equivalent current dipole (ECD).

Cross-References

▶ Magnetoencephalography in Assessment of Pain in Humans

Ergonomic Approach

Definition

A scientific approach aiming at reducing the physical load associated with work-related. In addition to a curative effect on potential tissue damage, a proper ergonomic approach has a preventive impact on symptoms and will also improve performance, which in a working environment implies an increase in turnover and/or quality.

Cross-References

▶ Ergonomic Counseling

Ergonomic Counseling

Christophe Maes[1], Guy Vanderstraeten[2] and John De Clercq[2]
[1]Manager Innovatie AZ Alma, Belgium
[2]Department of Rehabilitation Sciences and Physiotherapy, Ghent University Hospital, Ghent, Belgium

Synonyms

Work-related musculoskeletal disorders

Definition

An ergonomic approach to a ▶ work-related musculoskeletal disorder (WMSD) is directed at decreasing the impact of the ▶ load, since the risk for the development of a WMSD is dependent on the relation between the imposed load and the physical resistance (▶ load-bearing capacity) of the tissue.

Characteristics

Work-related musculoskeletal disorders have become a major problem in many industrialized countries and are among the most prevalent lost-time injuries in almost every industry (Bernard 1997). Consequently, the prevention of WMSDs is an important topic, not only for the welfare of the employees, but also for the employer. However, workers adapt themselves relatively easily to difficult work situations and accept high levels of effort and/or discomfort as an inevitable "part of the job" (Pope et al. 1984). The principles of prevention are therefore a part of the total treatment of the complaint rather than an effective prevention tool. Ergonomic counseling should be a part of the treatment, not only to promote the healing process, but also to prevent recurrences.

General Principles of Treatment

The aim of treatment is to restore compromised tissues to normal function. When a WMSD has been diagnosed, standard conservative treatment consists of medical treatment and physiotherapy, combined or not. The goal of traditional medical treatment with NSAIDs is to diminish the inflammation, which will alleviate the pain and break the vicious circle of pain, immobilization, dysfunction, inflammation, and pain. The treatment with drugs can also be focused on blocking the pain stimulus, resulting in pain relief. Traditional physiotherapy with mobilization, stretching, muscle strengthening, and electrotherapy aims at normalizing the load-bearing capacity of the tissue.

The risk for the development of a WMSD is determined by the relation between the imposed load and the physical resistance (load-bearing capacity) of the tissue. When the imposed load exceeds the load-bearing capacity, the tissue will be damaged and a WMSD may develop. The prevention as well as the treatment of WMSDs is focused on these two factors, with the aim of normalizing the relationship between them.

Impact on Force

The load itself is determined by multiple factors. The best-known biomechanical factor is the impact of force. While certain loads can be tolerated, all soft tissues, including muscle, tendon, ligament, fascia, synovia, cartilage, intervertebral disc, and nerve, fail when subjected to sufficient force (Adams and Dolan 1995). Data from cadaveric studies provide ranges within which such failure occurs. However, even at force levels clearly below the failure level, scientific evidence shows that tissue response to deformation may produce inflammation and injury at microscopic levels and also muscle fatigue (▶ Tissue Fatigue).

An important factor in the ergonomic approach is to diminish the force, to decrease the loads, and to adapt them to the load-bearing capacity of the tissue. For instance, decreasing the weight of cement sacks (load) from 50 to 25 kg has a positive effect on the prevalence of WMSDs, especially on low back pain. This also applies to other WMSDs. For instance, decreasing the required pinch force on a pair of tongs will diminish the occurrence of carpal tunnel syndrome. The lower the force, the lower is the risk of exceeding the tissue resistance and of developing a WMSD. Consequently, it is essential to have an idea of all activities and the extent to which the affected structure will be (over) used. An ergonomic assessment is necessary. The intensity of the exertion is not only related to the magnitude of the load but also to the resistance, the drag, the inertia, the accelerations, and the reaction forces of the work objects.

Impact on Posture

Biomechanical modeling of load-handling activities shows that ▶ posture also plays an important role in the structure of the load and its effect on the origin of WMSDs. Posture is defined as a position or attitude of the body or of a body part. Some positions require more effort than others or result in compression or stretching of tissue in or around the joints. Consideration of both the position and the length of time in the position are critical. In general, the postural requirements for the upper extremity (minimal load) should include keeping the elbows close to the body, shoulder (abduction) angles below 20–30, and forward arm flexion below 25–30° combined with no pronation or supination. The neutral position of the forearm should be held with a minimal strength of pinch grip. Posture is always related to static force. The magnitude of this static force is in its turn determined by the posture itself.

This static force has an influence on the blood flow in the tissue and on ▶ creep. If a static force is imposed, the blood flow in the concerned tissue will decrease. Consequently, fewer metabolites will be washed away, possibly leading to complaints. In addition, this static force has an impact on the origin of the "creep" phenomenon. Creep

is the change that occurs in a ligament (collagen) when a constant force is imposed. As a consequence, disc herniations can theoretically be explained as being associated with work activities involving flexed and torsed postures. This hypothesis, however, remains to be confirmed epidemiologically in occupational settings (Grandjean 1980). In a study by Boos et al. (1995), the presence of symptoms was related to nerve compromise and psychosocial aspects of work but not to exposure to physical stressors. The ergonomic (treatment) approach will be focused on the avoidance of critical postures. The need to adapt the working postures will have an influence on the workspace. For instance, the height of worktables must be adapted to the worker to avoid a burdened posture.

Repetitiveness

A third important biomechanical factor is ▶ repetitiveness of movement. Inflammatory muscle responses have been documented in humans subjected to repetitive loading. Repetitive loading also negatively affects ligaments and tendons (Armstrong 1994).

Health-care providers tend to think of repetitiveness in terms of joint motions, whereas ergonomists consider the task cycle or standard task completion. Consequently, repetition has been characterized in terms of the number of exertions per unit of time or frequency. An exertion is defined as the contraction of the muscle to produce force or movement. Other investigators have characterized repetition as the number of parts produced per unit of time or cycle time (Mish and Gilman 1991). The cycle time can easily be obtained from production records and interviews with the workers and supervisors. In general, ergonomists define a repetitive task as one with a task cycle shorter than 30 s. Consequently, it becomes easy to identify the persons at risk. However, in some cases, this approach will fail to prevent WMSDs. For instance, in keyboard work, the task cycle will be much longer than 30 s, but it is clear that the number of exertions is very high. Several studies have shown a clear relation between visual display work and the prevalence of WMSDs (Melhorn 1998).

Thus, it is important in the ▶ ergonomic approach to avoid repetitiveness, taking into account both definitions. WMSDs caused by repetitiveness can be prevented not only by lowering the frequency or increasing the cycle time but also by implementing a different muscle activity. The implementation of a different muscle activity, in which no or very little activity of the strained muscles is required, will provide more recovery time for the strained muscles.

Vibration

Another underestimated biomechanical stressor is cyclic loading, i.e., ▶ vibration. The effect of vibration of major mechanical significance concerns the duration of the fatigue life of tissues. Fatigue, which is defined as a loss of strength resulting from intermittent stressors over time, can lead to tissue failure. This can occur with relatively small stressors compared to those required for static stress failure (Hertzberg and Manson 1980). A tissue's fatigue life (the number of loading cycles it can withstand) depends on the range of stress imposed. A ligament is a biological connective tissue and consists of collagen composed of polypeptide chains. A study of Fung (1981) showed that after cyclic loading, the ultimate strength of the ligament was significantly less than that of its unstressed counterpart. The amount of softening in the ligament was related to the cyclic stress. Vibration of the tissue induces so-called vibrocreep, which has been defined as the acceleration of creep under vibrations (Kazarian 1972). Vibrocreep has been demonstrated in cadaveric specimens, and it presumably also occurs in living subjects. The ergonomic approach has to determine which vibrations resonate with the body vibrations and must be minimized to avoid creep.

Organization

In addition to biomechanical factors, organizational components may also give rise to WMSDs (Henrick and Brown 1984). The most important ones are work pressure, work rhythm, breaks, and time of exposure. Some less important factors that may also have an effect on the development of WMSDs are internal stress, monotony of work, and environmental factors like temperature, light, and noise. Some of these organizational components, e.g., the environmental factors, can "easily" be influenced by ergonomic counseling. Although the direct influence of these environmental factors on the prevalence of WMSDs seems small, it is important to try to optimize these components as well.

The role of stress in the development and exacerbation of chronic conditions has been well documented (Feuerstein et al. 1985; Holmes and Rahe 1967; Liker et al. 1984). Lampe et al. (1998) studied chronic back pain patients whose pain had a clearly organic etiology and patients with pain of unknown origin (idiopathic group). Significantly more members of the idiopathic group had experienced at least one highly stressful event preceding the most recent episode of pain exacerbation. The type of stress associated with "overload" occurring when the level of demand exceeds one's coping capacity has clearly a negative effect on the course of WMSDs (Niemcryk et al. 1987). Within the context of the work environment, it is important to avoid all factors that contribute to overwhelming stress levels in the employees, regardless of how small the impact may seem at first sight.

The way in which a factor (stressor) is experienced as stressful by the employee (patient) can only be determined by the employee himself. It is thus extremely important to involve the employee (patient) in the ergonomic study, the so-called participated ergonomics. The goal of such an ergonomic study is to determine all stressors that are stressful for the individual and to minimize or remove them at a later stage. For instance, the 8 h working day is an organizational component, which can be a psychological and muscular stressor. The introduction of a "work rotation system," 4 h job A and 4 h job B, decreases muscular stress if the muscles that have been working in job A are different from those active during the execution of job B. These kinds of muscular pauses are called macro-pauses. They have a positive effect on the prevention of WMSDs because they reduce the total load on the tissue.

During a job task, the musculoskeletal tissue is not loaded continuously. Because of movement, the muscles work alternately. When the agonist works, the antagonist is inactive. These types of pauses are called micro-pauses. They have been shown to have a positive effect on the prevention of WMSDs, especially in static muscle-work situations. The blood flow, lowered during the static effort, will normalize during the micro-pause. The ergonomic measurement has to be focused on the increase in the amount and duration of the micro-pauses.

Conclusion

The biomechanical and organizational factors discussed earlier all have an influence on the magnitude of the load. As explained, a WMSD occurs when the load-bearing capacity of the tissue (tissue resistance) is exceeded. This capacity mainly depends on personal characteristics like age, gender, physical fitness, and medical history. All the risk factors that have an influence on the load are interrelated. The ergonomic approach must therefore study the whole "human work environment system." Only then can all the risks be identified and evaluated. In a third step of the ergonomic approach, a risk qualification can be done, so that the most important risk factors can be modified to reduce their impact on the pain.

References

Adams, M. A., & Dolan, P. (1995). Recent advances in lumbar spinal mechanics and their clinical significance. *Clinical Biomechanics, 10*, 3–19.

Armstrong, T. J. (1994). Ergonomics and cumulative trauma disorders. In M. L. Kasdan (Ed.), *Occupational injuries* (pp. 553–567). Philadelphia: WB Saunders.

Bernard, B. P. (1997). *A critical review of epidemiologic evidence for work-related musculoskeletal disorders of the neck, upper extremity, and low back*. US Department of Health and Human Services.

Boos, N., Reider, V., Schade, K., et al. (1995). The diagnostic accuracy of magnetic resonance imaging, work perception and psychosocial factors in identifying symptomatic disc herniations. *Spine, 20*, 2613–2625.

Feuerstein, M., Sult, S., & Houle, M. (1985). Environmental stressors and chronic low back pain: Life events, family and work environment. *Pain, 22*, 295–307.

Fung, Y. C. (1981). *Mechanical properties of living tissues*. New York: Springer.

Grandjean, E. (1980). *Fitting the task to the human: An ergonomic approach*. London: Taylor & Francis.

Henrick, H., & Brown, O. Jr. (Eds.). (1984). *Human factors in organizational design and management*. (pp. 99–155). Amsterdam: North-Holland.

Hertzberg, R. W., & Manson, J. A. (1980). *Fatigue of engineering plastics* (p. 64). New York: Academic Press.

Holmes, T. H., & Rahe, R. H. (1967). The social readjustment rating scale. *Journal of Psychosomatic Research, 11*, 213–218.

Kazarian, L. E. (1972). Dynamic response characteristics of the human vertebral column. An experimental study on human autopsy specimens. *Acta Orthopaedica Scandinavica Supplement, 146*, 1–186.

Lampe, A., Sollner, W., Krismer, M., et al. (1998). The impact of stressful life events on exacerbation of chronic low back pain. *Journal of Psychosomatic Research, 44*, 555–563.

Liker, J. K., Joseph, B. S., & Armstrong, T. J. (1984). From ergonomic theory to practice: Organizational factors affecting the utilization of ergonomic knowledge. In J. Henrick & A. Brown (Eds.), *Human factors in organizational design and management* (pp. 1–256). New York: Wiley.

Melhorn, J. M. (1998). Cumulative trauma disorders and repetitive strain injuries: The future. *Clinical Orthopaedics, 351*, 107–126.

Mish, F. C., & Gilman, E. W. (1991). *Webster's ninth new collegiate dictionary* (pp. 87–88). Springfield: Merriam-Webster.

Niemcryk, S., Jenkins, C. D., Rose, R. M., et al. (1987). The prospective impact of psychosocial variables on rates of illness and injury in professional employees. *Journal of Occupational Medicine, 29*, 645–652.

Pope, M. H., Frymoyer, J. W., & Andersson, G. (1984). *Occupational low back pain*. New York: Praeger.

Ergonomics

▶ Pain in the Workplace, Risk Factors for Chronicity, Job Demands

Ergonomics Essay

Peregrin O. Spielholz
Washington State Department of Labor and Industries, SHARP Program, Olympia, WA, USA

Synonyms

Human factors engineering; Workplace design

Definition

Ergonomics has become a buzz word in product marketing for everything from chairs and automobiles to power tools. Many people first think of keyboarding and working at their desk when they hear the word. The roots of the word, ergo meaning "work" and -nomics meaning "the study of," lead us toward the true roots of this discipline in manufacturing around the turn of the century. The field of Ergonomics incorporates much more than the design and evaluation of workstations.

Ergonomics, or Human Factors Engineering, is a field of study that incorporates the disciplines of biomechanics, industrial engineering, psychology, and physiology to design jobs, products, or tools to be compatible with human capabilities. Ergonomists are employed for tasks that include evaluation of injury risks to workers and job-redesign, usability testing and software or product design, accident investigation, sports and recreation, and engineering of complex display-control systems such as airplane cockpits. The field has its roots in work done by Frank and Lillian Gilbreth and Frederick Taylor between the 1890s and 1920s. The studies of these scientists focused on determining the most efficient way to design and perform industrial work. They found that raising and tilting parts bins toward the worker makes it easier to retrieve parts, shoveling lighter loads more frequently can be more efficient and less strenuous, and that work is better distributed to include both hands. These findings are design principles used regularly today.

Characteristics

History
The formal founding of the field of Ergonomics and Human Factors came directly out of World War II. Experimental psychologists were instrumental in the design of instruments for ranging weapons and the design of aircraft cockpit displays both during and after the war. The success of this work led to increased interest in human

performance and capabilities, and the founding of the Human Factors (and Ergonomics) Society in 1957.

Ergonomics has received much attention in the United States, due to actions by government agencies using the word to describe regulations that require companies to evaluate the risk of work-related musculoskeletal disorders (WMSDs) and to reduce hazardous levels of physical risk factors. These regulations on both the federal and state level have become a lightning rod for both some business groups, symbolizing what they perceive to be far-reaching government regulation. Labor groups, on the other hand, have supported regulations to curb the large number of debilitating musculoskeletal injuries in the workforce.

Most large corporations in the United States have been addressing ergonomics as part of their safety and health programs without the threat of regulations. Companies and government agencies are doing these activities to protect workers; however, the economic consequences of WMSDs are a strong driving force. Over 700,000 workers a year are off the job due to a WMSD, which accounts for about one-third of all reported occupational injuries and illnesses. These injuries cost business and the US economy at least $20 billion annually in lost productivity, health costs, and direct costs, not to mention the cost to society and the injured worker (AFL-CIO 1997).

Injuries occur on the job due to a variety of factors that may include physical work requirements, work organization, psychosocial issues, and individual differences (NRC 1999). These factors lead to the development of disorders of the soft tissues that include ► tendinitis, ► epicondylitis (tennis or golfer's elbow), ► tenosynovitis, disc herniation or ► sciatica, ► carpal tunnel syndrome, rotator cuff tendinitis, and ► Raynaud's Vibration White Finger Syndrome.

The current body of research has shown that many of these disorders are likely to be a product of a combination of factors. Work-related factors have been shown by most epidemiology studies to be the strongest predictor of soft tissue musculoskeletal disorders (NIOSH 1997). However, the underlying health conditions of an individual must be evaluated by a physician in any case, as the development of a disorder can also be influenced by other conditions such as diabetes, obesity, thyroid conditions, or pregnancy to name a few.

Several reviews of the epidemiologic evidence for work-related musculoskeletal disorders and effect of company ergonomics programs were performed in the late 1990s (GAO 1997; NIOSH 1997; NRC 1999). The conclusions of the National Academy of Sciences review (NRC 1999) were that: (1) there is strong biological plausibility to the relationship between the incidence of musculoskeletal disorders and exposure risk factors in high-exposure jobs; (2) there is a higher incidence of disorders in workers in high-exposure jobs versus low-exposure jobs; and (3) research clearly shows that the rate of musculoskeletal disorders can be reduced by specific interventions in high-exposure jobs. The General Accounting Office (GAO 1997), in an evaluation of five different companies, also found that ergonomics programs yielded positive results when designed for the needs of the individual workplace.

The most comprehensive review of the epidemiologic literature was done by the National Institute for Occupational Safety and Health (NIOSH 1997). Over 2,000 relevant studies were identified and over 600 studies were reviewed in detail. Four physical work factors were identified as being significant contributors to WMSDs of the neck/shoulder, elbow, and hand/wrist: (1) Repetition, (2) Force, (3) Posture, and (4) Vibration. The following four risk factors showed evidence of causing back disorders: (1) Lifting and forceful movements, (2) Awkward postures, (3) Heavy physical work, and (4) Whole-body vibration.

Representative Research on Ergonomic Factors

Carpal tunnel syndrome has received a large amount of attention in the press and scientific literature. Two epidemiologic studies of carpal tunnel syndrome were identified by NIOSH (1997), which separated force as an exposure variable and met all the review inclusion criteria regarding participation rate, health outcome definition, blinding, and exposure assessment (Chiang et al. 1993; Silverstein et al. 1987). Chiang et al. (1993) found a statistically significant increase in carpal tunnel syndrome (CTS) symptoms (OR 1.8, 95 % CI 1.1–2.9) in jobs with an average hand force of 30 N. Silverstein et al. (1987) found an OR of 2.9 for jobs with an adjusted hand force of 58 N compared to a group with a hand force < 58 N.

NIOSH (1997) also reviewed epidemiologic studies of the hand/wrist which addressed repetitiveness. Five studies met the inclusion criteria of 70 % participation rate, health outcome defined by symptoms and physical examination, blinding to health or exposure status, and independent exposure assessment. Several studies (Osorio et al. 1994; Silverstein et al. 1987) have found increased odds ratios ranging from 1.9 to 5.5 for carpal tunnel syndrome symptoms in jobs classified as highly repetitive. While the definition is not widely agreed upon, Silverstein et al. (1987) defined high hand/wrist repetitiveness as a cycle time less than 30 s, or greater than 50 % of the cycle time performing the same kind of fundamental motions. Results from Silverstein et al. (1987) showed that high repetition and high force may act in a multiplicative fashion, raising the OR to 15.5 in jobs with high force and repetition.

NIOSH (1997) in its review of epidemiologic evidence of risk factors for low-back musculoskeletal disorders found strong evidence for relationships with work-related lifting, forceful movements, and whole-body vibration. The review also found evidence that awkward postures and heavy physical work were associated with low-back disorders. Punnett et al. (1991) conducted a case-control study examining back disorders and work-related exposures in auto assembly workers. Mild or severe back flexion (OR 8.09, CI 1.4–44) and lifting (OR 2.16, CI 1.0–4.7) were associated with back disorders.

Marras et al. (1993, 1995) studied the relationship between back disorders and spinal loading during work-related lifting, taking dynamic variables such as velocity and acceleration into account. Spinal loading while lifting was associated with back injury reports (OR 10.7, CI 4.9–23.6) when accounting for lift frequency, velocity, sagittal angle, and load weight. Dynamic variables were generally shown to increase the predicted spinal loading over static models.

Dynamic motion and other physical risk factors have also been studied for their relationship with work-related musculoskeletal disorders. Velocity and acceleration (peak and mean) of wrist motions have been demonstrated as potential risk factors in their relation to musculoskeletal disorders (Marras and Schoenmarklin 1993). Other possible risk factors include: localized mechanical stress, short recovery times, low temperatures, and segmental vibration (Hagberg et al. 1992; Hammarskjold et al. 1992; Sundelin and Hagberg 1989). Some research has also shown personal attributes such as age, gender, physical condition, or carpal tunnel cross-sectional area may play a role in the development of disorders along with occupational factors (Nathan et al. 1992). However, NIOSH (1997) concluded that, while individual risk factors may influence the development of disorders, they do not act synergistically with physical risk factors.

The Role of Psychosocial Factors

Physical demands of jobs are not the only factors that put workers at risk for musculoskeletal disorders. Psychosocial and work organization factors have also been shown to play a significant role in musculoskeletal disorder development, either as a confounder or direct contributor (NIOSH 1997; NRC 1999). Studies have found psychosocial issues such as job dissatisfaction within each worker and within the workplace as being a contributor to the development of work-related musculoskeletal disorders (Niemcryk et al. 1987). A study by Marras et al. (2000) found that physical loading may increase as a result of external psychological stress. Other studies also relate evidence that psychosocial stress may influence central nervous system function and promote disease development, retard the repair of damaged tissue, or alter an individual's perception of pain (Clauw and Williams 2002). These nonphysical factors could play a significant role in WMSD development by aiding disorder development, hindering healing processes, or increasing physical exposure.

Summary

The scientific evidence is very strong for a relationship between musculoskeletal disorder development and physical factors of posture, force, repetition, heavy lifting, and vibration. However, other work-related and nonwork factors may also play a role in the development of a disorder. The work done over the last several decades in this field has also demonstrated that a multitude of successful intervention strategies are available to reduce a person's chances of suffering a musculoskeletal condition. With further work in this area and continuing workplace changes we can hope to reduce the presence of these disorders for people in every environment.

References

AFL-CIO. (1997). *Stop the pain.* Washington, DC: AFL-CIO.

Chiang, H., Ko, Y., Chen, S., et al. (1993). Prevalence of shoulder and upper-limb disorders among workers in the fish-processing industry. *Scandinavian Journal of Work, Environment & Health, 19,* 126–131.

Clauw, D. J., & Williams, D. A. (2002). Relationship between stress and pain in work-related upper extremity disorders: The hidden role of chronic multisymptom illnesses. *American Journal of Industrial Medicine, 41,* 370–382.

General Accounting Office (GAO). (1997). Worker protection: Private sector ergonomics programs yield positive results. United States General Accounting Office, Report to Congressional Requesters, publication #GAO/HEHS, pp. 97–163.

Hagberg, M., Morgenstern, H., & Kelsh, M. (1992). Impact of occupations and job tasks on the prevalence of carpal tunnel syndrome: A review. *Scandinavian Journal of Work, Environment & Health, 12,* 277–279.

Hammarskjold, E., Harms-Ringdahl, K., & Ekholm, J. (1992). Reproducibility of carpenters work after cold exposure. *International Journal of Epidemiology, 9,* 195–204.

Marras, W., & Schoenmarklin, R. (1993). Wrist motions in industry. *Ergonomics, 36,* 341–351.

Marras, W., Lavender, S., Leurgans, S., et al. (1993). The role of dynamic three-dimensional trunk motion in occupationally-related low back disorders: The effects of workplace factors, trunk position, and trunk motion characteristics on risk of injury. *Spine*, *18*, 617–628.

Marras, W., Lavender, S., Leurgans, S., et al. (1995). Biomechanical risk factors for occupationally-related low back disorders. *Ergonomics, 38*, 377–410.

Marras, W. S., Davis, K. G., Heaney, C. A., et al. (2000). The influence of psychosocial stress, gender, and personality on mechanical loading of the lumbar spine. *Spine, 25*, 3045–3054.

Nathan, P. A., Keniston, R. C., Myers, L. D., et al. (1992). Longitudinal study of median nerve sensory conduction in industry: Relationship to age, gender, hand dominance, occupational hand use and clinical diagnosis. *The Journal of Hand Surgery, 17A*, 850–857.

National Academy of Sciences: National Research Council (NRC). (1999). *Work-related musculoskeletal disorders: A review of the evidence, steering committee for the workshop on work-related musculoskeletal injuries. The research base, committee on human factors*. Washington, DC: National Academy Press.

National Institute for Occupational Safety and Health (NIOSH). (1997). Musculoskeletal disorders and workplace factors: A critical eview of epidemiologic evidence for work-related musculoskeletal disorders of the neck, upper extremity, and low back. DHHS (NIOSH) Publication No. 97–141.

Niemcryk, S. J., Jenkins, C. D., Rose, R. M., et al. (1987). The prospective impact of psychosocial variables on rates of illness and injury in professional employees. *Journal of Occupational Medicine, 29*, 645–652.

Osorio, A. M., Ames, R. G., Jones, J., et al. (1994). Carpal tunnel syndrome among grocery store workers. *American Journal of Industrial Medicine, 25*, 229–245.

Punnett, L., Fine, L. J., Keyserling, W. M., et al. (1991). Back disorders and nonneutral trunk postures of automobile assembly workers. *Scandinavian Journal of Work, Environment & Health, 17*, 337–346.

Silverstein, B. A., Fine, L. J., & Armstrong, T. J. (1987). Occupational factors and carpal tunnel syndrome. *American Journal of Industrial Medicine, 11*, 343–358.

Sundelin, G., & Hagberg, M. (1989). The effects of different pause types on neck and shoulder EMG activity during VDU work. *Ergonomics, 32*, 527–537.

ERK

▶ Extracellular Signal-Regulated Protein Kinase

ERK Regulation in Sensory Neurons During Inflammation

Koichi Obata, Yi Dai and Koichi Noguchi
Department of Anatomy and Neuroscience, Hyogo College of Medicine, Hyogo, Japan

Synonyms

MAPK family

Definition

Extracellular signal-regulated protein kinase (ERK) is involved in cell proliferation and differentiation and in neuronal plasticity, including long-term potentiation, learning, and memory. The phosphorylation of ERK in primary afferent neurons occurs in response to acute noxious stimulation, such as with capsaicin or inflammatory mediators. Furthermore, the activation of ERK occurs after peripheral inflammation induced by complete Freund's adjuvant (CFA) and contributes to persistent inflammatory pain, via transcriptional regulation of key gene products. Thus, activation of ERK in sensory neurons during inflammation may participate in generating pain hypersensitivity through transcription-dependent and transcription-independent means.

Characteristics

▶ Mitogen-activated protein kinases (MAPK) are a family of serine/threonine protein kinases that transduce extracellular stimuli into intracellular posttranslational and transcriptional responses (Widmann et al. 1999). The MAPK family includes ▶ extracellular signal-regulated protein kinase (ERK), ▶ p38 MAPK, c-*Jun* N-terminal kinase/stress-activated protein kinase (JNK/SAPK), and ERK5. ERK is activated by membrane depolarization and calcium influx; is activated by an upstream kinase, MAPK/ERK kinase (MEK); and is known to be one of the intracellular signaling pathways involved in neuronal plasticity, such as long-term potentiation, learning, and memory. Physiological and pathological activity-dependent activation of ERK occurs in the CNS, especially in the hippocampus. Several studies have reported ERK phosphorylation in the nociceptive pathway; for example, acute noxious stimulation, such as that produced by formalin or capsaicin, induces ERK phosphorylation in spinal dorsal horn neurons (Ji et al. 1999; Karim et al. 2001). A MEK inhibitor PD 98059 reduces acute pain behavior after subcutaneous formalin injection, suggesting a role for ERK in acute nociceptive processing by a nontranscriptional mechanism (Ji et al. 1999; Karim et al. 2001). On the other hand, the activation of ERK in dorsal horn neurons contributes to persistent inflammatory pain via transcriptional regulation of prodynorphin and neurokinin-1 (Ji et al. 2002a). Neuronal plasticity occurs in primary afferent neurons in the dorsal root ganglion (DRG), as well as in spinal dorsal horn neurons.

Inflammatory mediators, such as prostaglandin E_2, serotonin, epinephrine, and ▶ nerve growth factor (NGF), produce hyperalgesia through activation of ▶ protein kinase A (PKA) or ▶ protein kinase C (PKC) in primary afferent neurons. It has been shown that the ERK cascade acts in epinephrine-induced hyperalgesia; also, the Ras-MEK-ERK pathway is activated independently of PKA or PKC (Aley et al. 2001; Dina et al. 2003). ▶ Bradykinin (BK) injected into the peripheral tissue also increases phosphorylated-ERK (p-ERK) labeling in DRG neurons (Rashid et al. 2004). On the other hand, ERK pathway involvement in neurotrophin-dependent survival and differentiation of developing peripheral neurons has been characterized in detail. For example, the high-affinity receptor for NGF ▶ tyrosine kinase A (trkA) can signal through at least six different pathways, a major one of which is a MAPK pathway (i.e., the ERK pathway). In this pathway, activated receptors induce GTP loading and activation of the small G-protein Ras. In turn, Ras-GTP recruits

a three-tiered enzyme cascade in which a MAPK kinase kinase (Raf) phosphorylates MEK, which phosphorylates and activates ERK. It has been reported that p-ERK is prominent in a few trkA-containing DRG neurons and in satellite glial cells and is increased by NGF treatment (Averill et al. 2001; Delcroix et al. 2003). Furthermore, this p-ERK undergoes fast anterograde and retrograde axonal transport, indicated by accumulation at a sciatic nerve ligature, and NGF reduces the level of retrograde p-ERK transport (Averill et al. 2001).

Phosphorylation of ERK in primary afferent neurons occurs in response to noxious stimulation of the peripheral tissue or electrical stimulation to the peripheral nerve, that is, activity-dependent activation of ERK in DRG neurons (Dai et al. 2002). For example, the electrical stimulation of Aδ-fibers induces p-ERK, primarily in neurons with myelinated fibers, whereas c-fiber activation by capsaicin injection induced p-ERK in small neurons with unmyelinated fibers containing the transient receptor potential ion channel ▶ TRPV1, formerly known as the vanilloid receptor-1 (Fig. 1). In addition, noxious heat stimulation, as well as electrical stimulation, induces p-ERK in primary afferents in a stimulus intensity-dependent manner. On the other hand, capsaicin injection into the skin also increases p-ERK labeling significantly in peripheral fibers and terminals in the skin and a MEK inhibitor U0126 dose-dependently attenuates thermal hyperalgesia after capsaicin injection. All of these results suggest that the activation of ERK pathways in primary afferents is involved in peripheral sensitization in acute pain conditions. The phosphorylation of ERK in DRG neurons after noxious stimulation might be useful for examining the activation state of each neuron that contains various pain-related molecules (Dai et al. 2002). For example, ▶ P2X receptors in primary afferent neurons increase their activity with enhanced sensitivity of the intracellular ERK signaling pathway during inflammation and then contribute to the hypersensitivity to mechanical noxious stimulation in the inflammatory state (Dai et al. 2004).

ERK Regulation in Sensory Neurons During Inflammation, Fig. 1 Stimulus-evoked ERK phosphorylation in rat DRG neurons. (a) L4 DRG section immunostained for p-ERK in the naïve rat. (b) p-ERK labeling in many small L4 DRG neurons 2 min after intraplantar injection of capsaicin. (c) Time course of capsaicin-evoked p-ERK expression in L4/5 DRG neurons. $*p < 0.01$; $**p < 0.05$ compared with the naive control. (d) p-ERK-IR (*green*) and TRPV1-IR (*red*) in the ipsilateral L4 DRG after capsaicin injection are shown. Double labeling of neurons with p-ERK and TRPV1 was observed (*yellow*). Scale bars, 100 µm (Modified from Dai et al. 2002)

The ERK in the nervous system produces not only short-term functional (nontranscriptional) changes by phosphorylating kinases, receptors, and ion channels but also long-term adaptive changes by activating transcriptional factors. For example, activated ERK translocates from the cytoplasm to the nucleus and activates Rsk2, which then phosphorylates the transcription factor cAMP response element-binding protein (CREB) on serine 133. The activated CREB then binds to the activating cAMP response element (CRE) sites on the promoter regions of the DNA and initiates the transcription of genes, such as c-*fos*, dynorphin, and ▶ brain-derived neurotrophic factor (BDNF). The activation of ERK regulates gene expression of BDNF in primary afferent neurons after peripheral inflammation induced by CFA (Obata et al. 2003). Peripheral inflammation induces an increase in the phosphorylation of ERK, mainly in trkA-containing small-to-medium diameter DRG neurons (Fig. 2). Treatment with the MEK inhibitor, U0126, reverses the pain hypersensitivity and the increase in p-ERK and BDNF in DRG neurons. Furthermore, the intrathecal application of NGF induced an increase in the number of p-ERK- and BDNF-labeled cells, mainly small neurons. These findings suggest that the activation of ERK in the primary afferents occurs in DRG neurons after peripheral inflammation through alterations in the target-derived NGF

ERK Regulation in Sensory Neurons During Inflammation, Fig. 2 (**a, b**) Photomicrographs showing the p-ERK-IR in the contralateral (**a**) and ipsilateral (**b**) L4/5 DRG 1 day after peripheral inflammation. There was an increase in the number of p-ERK-IR neurons in the ipsilateral DRG (*arrows*). (**c**) Time course of the mean percentages of p-ERK-IR neurons relative to the total number of neurons in the L4/5 DRG. The mean percentages of p-ERK-IR neurons on the ipsilateral side significantly increased at 1 day after CFA injection compared to the naive control rats. $*p < 0.05$ compared with the naive control. (**d**) p-ERK-IR (*green*) and trkA-IR (*red*) in the ipsilateral L4/5 DRG after the CFA are shown. Double labeling of neurons with p-ERK and trkA was observed (*yellow*). Scale bars, 100 µm (Modified from Obata et al. 2003)

and contributes to persistent inflammation via transcriptional regulation of BDNF expression (Obata et al. 2003).

It has been demonstrated that TRPV1 is regulated by NGF-induced activation of the ERK/MAPK pathway in DRG neurons in vitro (Bron et al. 2003), whereas Ji and colleagues showed that p38 MAPK activation in the DRG is required for NGF-induced increases in TRPV1 expression and contributes to the maintenance of inflammatory pain hypersensitivity (Ji et al. 2002b). Therefore, in addition to ERK, other MAPK pathways, such as the p38 or JNK/SAPK or ERK5 pathways, also may be activated by inflammation and play an important role in the generation of pain hypersensitivity (Obata and Noguchi 2004). Activation of MAPK clearly has a substantial role in the establishment and maintenance of nociceptive-induced plasticity in DRG neurons, not only by posttranslational modifications of target proteins but also by increasing gene transcription. These attractive targets for study will give us new approaches for understanding the cellular/molecular mechanisms underlying pain hypersensitivity. Furthermore, inhibition of MAPK signaling in the primary afferents, as well as in the spinal cord, may provide a fruitful strategy for the development of novel analgesics.

References

Aley, K. O., Martin, A., McMahon, T., et al. (2001). Nociceptor sensitization by extracellular signal-regulated kinases. *The Journal of Neuroscience, 21*, 6933–6939.

Averill, S., Delcroix, J. D., Michael, G. J., et al. (2001). Nerve growth factor modulates the activation status and fast axonal transport of erk 1/2 in adult nociceptive neurones. *Molecular and Cellular Neuroscience, 18*, 183–196.

Bron, R., Klesse, L. J., Shah, K., et al. (2003). Activation of Ras is necessary and sufficient for upregulation of vanilloid receptor type 1 in sensory neurons by neurotrophic factors. *Molecular and Cellular Neuroscience, 22*, 118–132.

Dai, Y., Fukuoka, T., Wang, H., et al. (2002). Phosphorylation of extracellular signal-regulated kinase in primary afferent neurons by noxious stimuli and its involvement in peripheral sensitization. *The Journal of Neuroscience, 22*, 7737–7745.

Dai, Y., Iwata, K., Fukuoka, T., et al. (2004). Contribution of sensitized P2X receptors in inflamed tissue to the mechanical hypersensitivity revealed by phosphorylated ERK in DRG neurons. *Pain, 108*, 258–66.

Delcroix, J. D., Valletta, J. S., Wu, C., et al. (2003). NGF signaling in sensory neurons: Evidence that early endosomes carry NGF retrograde signals. *Neuron, 39*, 69–84.

Dina, O. A., McCarter, G. C., de Coupade, C., et al. (2003). Role of the sensory neuron cytoskeleton in second messenger signaling for inflammatory pain. *Neuron, 39*, 613–624.

Ji, R. R., Baba, H., Brenner, G. J., et al. (1999). Nociceptive-specific activation of ERK in spinal neurons contributes to pain hypersensitivity. *Nature Neuroscience, 2*, 1114–1119.

Ji, R. R., Befort, K., Brenner, G. J., et al. (2002a). ERK MAP kinase activation in superficial spinal cord neurons induces prodynorphin and NK-1 upregulation and contributes to persistent inflammatory pain hypersensitivity. *The Journal of Neuroscience, 22*, 478–485.

Ji, R. R., Samad, T. A., Jin, S. X., et al. (2002b). p38 MAPK activation by NGF in primary sensory neurons after inflammation increases TRPV1 levels and maintains heat hyperalgesia. *Neuron, 36*, 57–68.

Karim, F., Wang, C. C., & Gereau, R. W., 4th. (2001). Metabotropic glutamate receptor subtypes 1 and 5 are activators of extracellular signal-regulated kinase signaling required for inflammatory pain in mice. *The Journal of Neuroscience, 21*, 3771–3779.

Obata, K., & Noguchi, K. (2004). MAPK activation in nociceptive neurons and pain hypersensitivity. *Life Sciences, 74*(21), 2643–2653.

Obata, K., Yamanaka, H., Dai, Y., et al. (2003). Differential activation of extracellular signal-regulated protein kinase in primary afferent neurons regulates brain-derived neurotrophic factor expression after peripheral inflammation and nerve injury. *The Journal of Neuroscience, 23*, 4117–4126.

Rashid, M. H., Inoue, M., Matsumoto, M., et al. (2004). Switching of bradykinin-mediated nociception following partial sciatic nerve injury in mice. *The Journal of Pharmacology and Experimental Therapeutics, 308*, 1158–1164.

Widmann, C., Gibson, S., Jarpe, M. B., et al. (1999). Mitogen-activated protein kinase: Conservation of a three-kinase module from yeast to human. *Physiological Reviews, 79*, 143–180.

ERP

▶ Phosphatases and Glia

Error of Measurement

Definition

As one aspect of the reliability of an outcome instrument, it is a measure of the precision of the instrument in detection of score change. It is calculated as a range of score units with 95 % probability limits, taking into account the standard deviation of the score change and the test-retest reliability coefficient.

Cross-References

▶ Oswestry Disability Index

Erythema

Definition

Erythema is local redness or flushing of the skin due to dilatation of capillaries in the dermis.

Cross-References

▶ UV Erythema, a Model for Inducing Hyperalgesias
▶ UV-Induced Erythema

Erythema Nodosum

Definition

Erythema nodosum is a frequent symptom in cerebral vasculitis.

Cross-References

▶ Headache Due to Arteritis

Erythema Nodosum Leprosum

▶ Type 2 Reaction (Leprosy)

Erythropoietin

Definition

Erythropoietin is a substance that is naturally produced by the kidneys and that stimulates the bone marrow to make red blood cells. When erythropoietin is made in the laboratory, it is called epoetin alfa or epoetin beta.

Cross-References

▶ Cancer Pain Management, Chemotherapy

Escape Response

Definition

Escape response is a response made to terminate a stimulus that is usually noxious.

Cross-References

▶ Visceral Pain Models, Female Reproductive Organ Pain

Esophageal Pain, Noncardiac Chest Pain

▶ Visceral Pain Model; Esophageal Pain

Essential Vulvodynia

▶ Vulvodynia

Estrous Cycle

Definition

Estrous cycle is a mammalian reproductive cycle, consisting of four stages of varying lengths, which are associated with changes in ovarian hormones and reproductive behaviors. The four stages include proestrus, estrus, metestrus, and diestrus.

Cross-References

▶ Visceral Pain Models, Female Reproductive Organ Pain

Ethics

▶ Ethics of Pain Control in Infants and Children

Ethics of Pain Control in Infants and Children

Maureen Kelley[1,3] and Gary A. Walco[2]
[1]Department of Pediatrics, Bioethics Division,
University of Washington School of Medicine,
WA, USA
[2]Department of Anesthesiology and Pain
Medicine, University of Washington School of
Medicine, Seattle Children's Hospital, Seattle,
WA, USA
[3]Seattle Children's Hospital, Seattle, WA, USA

Synonyms

Dignity; Ethics; Harm principle; Human rights

Characteristics

Balancing Harm and Benefit in Pediatric Pain Management

The ethical pain management in infants and children is guided by the same principles of clinical ethics that shape other areas of clinical practice. The central challenge for managing pain in any patient is to balance the benefits of pain therapy against potential harms of interventions, including the harm of unmanaged pain. For pediatric patients, such decisions are made in partnership between parents and providers and against the backdrop of pragmatic, contextual, and cultural considerations.

Ethical problems arise when there is disagreement or ambiguity in our assessment of this balance, whether among clinicians and pain experts or between parents and providers. Rapidly changing data on the effectiveness and side effects of pain interventions sometimes requires clinicians to make decisions under uncertainty. Cultural beliefs about pain, suffering, and quality of life can also complicate decisions about pain management, particularly in the palliative care context. However, rarely is the disagreement over the prevention of pain and suffering a fundamental disagreement. Since failure to address pain in any patient would constitute a straightforward violation of the harm principle, "first, do no harm," failure to do so on behalf of the youngest and most vulnerable persons, such as infants and children, is viewed as especially egregious.

This view, and the move to correct the historic undertreatment of pain in children, is reflected in a much stronger view, namely, that the failure to address pain violates important and basic human right to dignity and prevention of suffering (James 1993). Framing the question of pain management as a question of dignity and rights offers a stronger argument for expanding access to pain treatment to all children, regardless of background, socioeconomic status, or geographical location (Southall et al. 2000). This view is reflected in the position of the World Health Organization on pain in children (World Health Organization 1998) and the recent International Association for the Study of Pain, the Declaration of Montreal (Cousins and Lynch 2011), which states:

> ...recognizing the intrinsic dignity of all persons and that withholding of pain treatment is profoundly wrong, leading to unnecessary suffering which is harmful; we declare that the following human rights must be recognized throughout the world:
>
> *Article 1.* The right of all people to have access to pain management without discrimination.
> *Article 2.* The right of people in pain to acknowledgment of their pain and to be informed about how it can be assessed and managed.
> *Article 3.* The right of all people with pain to have access to appropriate assessment and treatment of the pain by adequately trained health care professionals.

Similarly, the IASP SIG on Pain in Childhood supported the Bellagio Declaration (ChildKind International 2010), calls for all health-care facilities to commit to the developmentally appropriate prevention, assessment, and management of pain in children and adolescents aged 0–18 years. These statements jointly recognize the universality of pain among children being treated in health-care facilities, the short- and long-term negative effects of poorly treated pain in children, and the wrongfulness of failures to adequately address pediatric pain.

Clinical Ethics at the Bedside

The seemingly straightforward ethical weighing of risks and benefits in pain management in infants and children can be much more complicated in clinical decisions at the bedside (Walco and Kelley 2013; Lantos and Meadow 2007). The balance of harm and benefit in alleviating pain is an evidence-based decision, but it occurs against the backdrop of evolving evidence and changing or inconsistent standards of practice (Walco et al. 1994; Olmstead et al. 2010). Pediatric pain management is also still influenced by a relatively recent history of pervasive under-recognition of pain in infants and children. Until about four decades ago, the general belief was that young children, in particular, did not experience pain and certainly had no lasting memory of it if they did. Beginning in the late 1970s, there was increasing recognition that pain was undertreated in children compared to adults and that even very young premature infants have nociceptive systems in place and thus perceive pain. The work of developmental physiologists provided even greater insights, and by the turn of the century, data clearly showed that untreated or undertreated pain in young children was associated with lasting harmful effects (Fitzgerald 2005). Because of this relatively recent shift in the recognition of harm and some persistence in the undertreatment of pediatric pain, it is helpful to label and respond to the possible objections to pain treatment in infants and children (Walco et al. 1994). Three potential justifications for not adequately treating pain in children have shaped that discussion and still arise in clinical decision-making.

Revisionist Justification

Medical decision-making often relies on objective data. Pain is a subjective experience, however, and therefore direct quantitative assessment is difficult. This difficulty is compounded in young children in whom articulate verbal expressions of their pain and discomfort are either impossible or of questionable reliability, simply due to developmental factors. The fact that pain is so often undertreated in the young raises the possibility that available signs of pain, as well as children's reports of pain, are in some fashion "revised" in the context of observers' biases, so that evaluations of pain severity are minimized.

In this view, there may be a "correct" or "reasonable" amount of pain that is associated with a given procedure or pathophysiological state, but this completely ignores the developmental factors related to pain systems (Andrews 2003), the individual differences in pain response that may be seen throughout the lifespan (Walco and Harkins 1999), and the complexity of factors that modulate the relationship between nociceptive stimulation and pain ratings in children and adolescents (Ilowite et al. 1992). It is often assumed that adults are more reliable reporters of pain than are children, and undue credence is put on parental or caregiver reports, rather than reports or expressions by children themselves. This assumption is not always valid, however, as parents may project their own concerns onto the situation and they rely principally on behavioral cues to ascertain pain levels, which are not always consistent with children's internal states (Walco et al. 2005).

Incorrect notions about pain in children, especially neonates, are slowly reversing. For example, there was a prevailing view that neonates did not experience pain and if they did, the pain would not be remembered or it would have no lasting effect. Data are now clear that even premature infants, as young as 23 weeks gestation, experience pain. Furthermore, there is evidence to show that if untreated or undertreated, such pain experiences in the young may have significant lasting effects on the subsequent development of pain networks and may lead to hyperalgesic responses in the future, all in addition to the potentially detrimental hormonal and metabolic responses during the acute phases of the pain (Anand and Hickey 1987; Fitzgerald and de Lima 2001; Goldschneider and Anand 2003).

Finally, many questioned the reliability and value of pain assessment strategies in the young.

There is now a plethora of literature on assessment strategies that may be used even in the smallest premature infants, through childhood and into adolescence. These instruments have been shown to be reliable, valid, and clinically sensitive enough to detect changes in pain states, especially as a function of intervention (Stevens and Franck 2001). Thus, it is clear that even very young infants communicate their pain, and it is the caregiver's job to utilize available strategies to make valid assessments. Denial of relief from the pain that is proportionate to the expressed need for such relief must be judged an unjustified harm, unless such deprivation serves a substantially greater good.

Comparative Justification

Once it has been ascertained that a child is in pain, one must weigh the benefits and risks of unrelieved pain against those of pain relief. In other words, while it always seems beneficial to alleviate pain, the "cost" of pain relief may be too high and thus imprudent to be invoked. In this arena, however, the reasons cited for withholding pain medications, for example, are often exaggerated or incorrect. The modal intervention for moderate to severe pain in children involves the use of potent analgesic medications, such as opioids. These medications have problematic side effects, and they have been associated with addiction. Thus, if one is concerned about respiratory depression (a known side effect of opioids), one may withhold these medications as it is presumed better to have a child in pain and breathing than it is to relieve pain and induce respiratory arrest.

Similar arguments are made for addiction, as it would be better for a child to endure acute pain for a defined period if it spared opioid addiction in the future. However, data make it clear that with appropriate dosing and adequate monitoring, the likelihood of a clinically significant respiratory depression related to opioids in otherwise healthy children is quite low (Yaster et al. 2003). Likewise, the probability of a young child with no prior difficulties with substance abuse becoming addicted to opioid medication administered for pain is close to zero (Yaster et al. 2003). Based on these data, denying children adequate analgesia

due to concerns about these extremely low probability events is way out of line with the usual level of risk physicians assume in prescribing treatment (Angell 1982).

Pragmatic Justification

The pragmatic justification acknowledges that unrelieved pain is bad but asserts that such pain may be necessary to achieve a greater goal. Specifically, pain may be useful to monitor a clinical condition, especially if the diagnosis or extent of disease is not known or to assess the efficacy of some other treatment. Masking pain in these circumstances may be harmful to the patient, and thus, one must weigh the benefit of immediate pain relief against the benefit of long-term recovery.

In circumstances when residual or unrelieved pain is pragmatically justified, certain important questions must be considered. First, is the pain useful? Is it the means through which a goal may be attained? Second, is the pain necessary? If the same diagnostic or therapeutic goal may be met without the presence of pain, that should be considered as the choice alternative. Third, is the pain at the lowest possible level? If there is a therapeutic benefit to be derived from a child being in pain, one must be exquisitely economical with it, as the cost is the child's suffering.

Socioeconomic and Cultural Considerations

Patient populations are increasingly culturally and socioeconomically diverse. As such, ethical decisions surrounding pain management sometimes invoke parents' or family's cultural beliefs about pain or issues surrounding access to adequate pain treatment. Regarding access, if untreated pain is an unjust suffering, then it is so universally for all children, regardless of their socioeconomic or political status. Such a position, as articulated in the human rights statements cited above, provides a strong case for improving universal access to pain treatment within and across borders. Practical realities may constrain available treatment options, but there will remain a strong ethical imperative to do everything possible to overcome barriers to access for all children.

Cultural beliefs regarding pain, such as "boys need to experience pain to be strong," can be addressed respectfully in conversations between providers and parents but does not alter the basic moral obligation to treat a child or infant's pain. While more research is needed, some data suggest that differences in pain and responses to pain to be insignificant across ethnic groups; an important counter to assuming relativism in the experience of pain (Edwards et al. 2005). In approaches to pain management and end of life care at the bedside, it is important to distinguish cultural beliefs and practices of a particular family from lack of knowledge about the nature of pain in a child or unfamiliarity with available treatments (Zalon et al. 2008; Walco and Kelley 2013). This can provide impetus for educating parents to better address their child's pain.

Final Considerations

There are now published guidelines that represent a consensus on the standard of care for pain assessment and management in children (American Academy of Pediatrics 2001; American Pain Society 2003), both of which emphasize the need for the regular assessment and appropriately aggressive treatment of acute pain in children. All health professionals should provide care that is consistent with the technological growth of the field and reflects current standards of practice. The assessment and treatment of pain in children are important parts of pediatric practice, and failure to provide adequate control of pain amounts to substandard and unethical medical practice. Through ongoing educational efforts, it is hoped that more and more practitioners will provide care that meets stated standards and that those standards will change over time to reflect advances in the field.

References

American Academy of Pediatrics Committee on Psychosocial Aspects of Child and Family Health, American Pain Society Task Force on Pain in Infants, Children, and Adolescents. (2001). Policy statement: The assessment and management of acute pain in infants, children and adolescents (0793). *Pediatrics, 108,* 793–797.

American Pain Society. (2003). *Principle of analgesic use in the treatment of acute pain and cancer pain* (5th ed.). Glenview: American Pain Society.

Anand, K. J., & Hickey, P. R. (1987). Pain and its effects in the human neonate and fetus. *The New England Journal of Medicine, 317,* 1321–1329.

Andrews, K. A. (2003). The human developmental neurophysiology of pediatric pain. In N. L. Schechter, C. B. Berde, & M. Yaster (Eds.), *Pain in infants, children, and adolescents* (2nd ed., pp. 19–42). Philadelphia: Lippincott Williams & Wilkins.

Angell, M. (1982). The quality of mercy. *The New England Journal of Medicine, 306,* 98–99.

ChildKind International. (2010). The ChildKind Initiative: A program to reduce pain in child health facilities worldwide [Homepage of International Association for the Study of Pain], [Online]. Retrieved 1 Feb 2012 from http://www.iasp-pain.org/PainSummit/ChildKind_Initiative2010.pdf.

Edwards, R. R., Moric, M., Husfeldt, B., Buvanendran, A., & Ivankovich, O. (2005). Ethnic similarities and differences in the chronic pain experience: A comparison of african american, Hispanic, and white patients. *Pain Medicine, 6,* 88–98.

Fitzgerald, M. (2005). The development of nociceptive circuits. *Nature Reviews Neuroscience, 6,* 507–520.

Fitzgerald, M. F., & de Lima, J. (2001). Hyperalgesia and allodynia in infants. In G. A. Finley & P. M. McGrath (Eds.), *Progress in pain research and management* (pp. 1–12). Seattle: IASP Press. Vol 20.

Goldschneider, K. R., & Anand, K. S. (2003). Long-term consequences of pain in neonates. In N. L. Schechter, C. B. Berde, & M. Yaster (Eds.), *Pain in infants, children, and adolescents* (2nd ed., pp. 58–70). Philadelphia: Lippincott Williams & Wilkins.

Ilowite, N. T., Walco, G. A., & Pochaczevsky, R. (1992). Pain assessment in juvenile rheumatoid arthritis: Relation between pain intensity and degree of joint inflammation. *Annals of the Rheumatic Diseases, 51,* 343–346.

James, A. (1993). Painless human right. *Lancet, 342,* 567–568.

Lantos, J., & Meadow, W. (2007). Ethical issues in the treatment of neonatal and infant pain. In K. J. S. Anand, B. J. Stevens, & P. J. McGrath (Eds.), *Pain in neonates and infants* (3rd ed., pp. 211–217). Edinburgh, Scotland: Elsevier.

Olmstead, D. L., Scott, S. D., & Austin, W. J. (2010). Unresolved pain in children: a relational ethics perspective. *Nursing Ethics, 17,* 695–704.

Southall, D. P., Burr, S., Smith, R. D., Bull, D. N., Radford, A., Williams, A., Nicholson, S. (2000). The Child-Friendly Healthcare Initiative (CFHI): Healthcare provision in accordance with the UN Convention on the Rights of the Child. Child Advocacy International. Department of Child and Adolescent Health and Development of the World Health

Organization (WHO). Royal College of Nursing (UK). Royal College of Paediatrics and Child Health (UK). United Nations Children's Fund (UNICEF). *Pediatrics, 106*, 1054–1064.

Stevens, B. J., & Franck, L. S. (2001). Assessment and management of pain in neonates. *Paediatric Drugs, 3*, 539–558.

Walco, G. A., & Harkins, S. (1999). Life-span developmental approaches to pain. In R. J. Gatchel & D. C. Turk (Eds.), *Psychosocial factors in pain: Critical perspectives* (pp. 107–117). New York: Guilford Publications.

Walco, G. A., Cassidy, R. C., & Schechter, N. L. (1994). Pain, hurt, and harm: The ethical issue of pediatric pain control. *The New England Journal of Medicine, 331*, 541–544.

Walco, G. A., Conte, P. M., Labay, L. E., et al. (2005). Procedural distress in children with cancer: Self-report, behavioral observations, and physiological parameters. *The Clinical Journal of Pain, 21*, 484–490.

World Health Organization. (1998). *Pain in children with cancer: The World Health Organization-IASP guidelines*. Geneva, Switzerland: World Health Organization.

Yaster, M., Kost-Byerly, S., & Maxwell, L. G. (2003). Opioid agonists and antagonists. In N. L. Schechter, C. B. Berde, & M. Yaster (Eds.), *Pain in infants, children, and adolescents* (2nd ed., pp. 181–224). Philadelphia: Lippincott Williams & Wilkins.

Zalon, M. L., Constantino, R. E., & Andrews, K. L. (2008). The right to pain treatment: A reminder for nurses. *Dimensions of Critical Care Nursing, 27*, 93–101.

Ethics of Pain in the Newborn Human

Carlo Valerio Bellieni
Neonatal Intensive Care Unit, University Hospital, Siena, Italy

Introduction

Physiological, behavioral, and hormonal responses indicate that neonates possess a functional ▶ pain system, which was recently widely recognized among the medical profession. Until 20 years ago, neonatal pain was largely underestimated. Opioids were rarely used even in surgery (De Lima et al. 1996), and many doctors maintained that neonates do not feel pain (Beyer et al. 1983). The neonatal brain was thought to be insufficiently developed to recognize noxious stimuli. Also the current definitions of pain do not really apply to the neonate.

In 1991, the International Association for the Study of Pain (IASP) proposed the following definition of *pain*: "A sensory and emotional experience based on actual or potential tissue damage or described in terms of such damage" (Merskey 1979). This definition implies introspection and expression by language not present in the neonate. The limits of this definition for the human neonate are now recognized (Anand and Craig 1996), and new definitions of pain and ▶ suffering have been proposed to encompass preverbal experience:

"Pain is an inherent quality of life itself, expressed by all viable living organisms and while influenced by life's events, it does not require prior experience in the first instance." (Anand and Craig 1996).

"Although pain and suffering are closely identified in the medical literature, they are phenomenologically distinct: pain, is a fundamentally "physical" phenomenon, the clash arising from an attack to our physical integrity, whereas suffering is something broader with pain as one of its sources and desire as its condition: we can define it the clash arising from an attack to our person's integrity" (Bellieni 2005).

Definition

Recent reports show that neonatal pain is undertreated. Many newborns undergo painful treatments such as chest air drainage or circumcision without analgesia (Simons et al. 2003; Lago et al. 2005; Debillon et al. 2002). Some reasons lay at the basis of this behavior: the awareness of impunity, the lack of time, or the lack of awareness of newborns' capability to feel pain. Here we highlight this last point, showing that newborns not only feel pain, but can also suffer; then, we will focus on non-pharmacologic pain treatment, on the risks preverbal babies undergo when enrolled in clinical trials, and the importance of taking care even of all caregivers to achieve an effective newborns' pain treatment.

Characteristics

Principles in the Management of Neonatal Pain

In the last few years, analgesic drugs were used systemically and topically with success in neonatal care. In addition, non-pharmacological methods, such as non-nutritional sucking (Blass and Watt 1999) and the instillation of sucrose on the newborn's tongue (Balon-Perin et al. 1991), were proposed to interfere with pain reactions in minor invasive procedures. Nevertheless, newborns are, in a certain sense, treated differently from other children or adults during potentially painful procedures; an older baby is reassured before, distracted during, and soothed after blood sampling or injections. This rarely happens in dealing with neonates, are there physiological or psychological reasons for this? According to Als and Gilkerson (1997), preterm infants are neurobiologically social and the fact that they have needs and fears and feel pleasure should oblige us to reassure, distract, and soothe them.

Studies on Sensory Saturation

In a recent study (Bellieni et al. 2001), we exploited newborns' social behavior and sensoriality to stem pain. We studied premature babies during blood sampling using a technique of sensory stimulation and comforting. This technique was termed ▶ sensory saturation (SS) because competing sensory activation may result in saturation of central sensory pathways, which thus may prevent pain signals from reaching a response level. It consists in distracting and comforting babies by massaging them, speaking to them, establishing eye contact, and instilling a sweet liquid to the tongue before and during heel-pricks.

In short, it follows a "3Ts" rule: using touch, taste and talk to distract the baby and antagonize pain. We recorded in the SS group a reduction in crying time and pain score with respect to a control group and with respect to groups in which only oral sugar, only sucking, or a combination of the two was used. We repeated the experiment with term newborns and obtained similar results (Bellieni et al. 2002, 2007).

We attributed these results to the gate control theory of pain (Melzack and Wall 1965), but as this theory involves affective-motivational dimensions of pain experience, some further observation is necessary. We showed that giving sugar 2 min before heel lancing or mechanically putting a pacifier in the baby's mouth before taking a blood sample produces no sufficient analgesia; analgesia is more effective when newborns feel a nearby presence. This happens during breastfeeding or with SS, because a more intense social relationship is established and babies do not feel alone and abandoned during the painful experience. These findings demonstrate that treating newborns with the respect due to an individual person, trying to comfort them during a stressful event is not only ethical, but also analgesic.

Newborns Not Only Feel Pain: They also Suffer

Newborns not only feel pain, they also experience suffering and anxiety (Als et al. 1997) and anxiety calls for the presence and involvement of a person of reference, not only analgesia. Infants, families, and staff can be expected to grow in an environment where strengths are emphasized and vulnerability partnered. Unfortunately, in many cases, this partnership is often reduced to fulfillment of a task (Cunningham et al. 1984). Caregivers should provide adequate relationship-based care and help couples to become parents, to say "you" to the little preterm baby they see through the incubator window in the first hours of life (Bellieni 2003), recognizing its wishes, fears, and suffering.

Schopenhauer defined suffering as the gap between what we demand or expect from life and what we actually get and his definition fits well with neonates. In fact, we can affirm that newborns suffer, because they are persons who try to express their needs (Bellieni 2005), sometimes in a desperate way. We are right when we say that a newborn, though immature, is a person, because he/she has the features to be one: being alive, being an individual, and being human.

Ethical Recommendations for Analgesia in Clinical Trials

It should be good practice to compare every new analgesic drug or treatment with others of well-known efficacy, avoiding pure placebo or no treatment. Minor procedures (e.g., injections, heel-pricks) are rarely painful for adults, but are very painful for newborns, who lack full inhibitory systems. Moreover, the pain produced by a simple heel-prick can provoke a series of consequences, such as a dramatic increase in intracranial pressure and in free radical release (Bellieni et al. 2009). Unfortunately, most studies on neonatal procedural pain are carried out with no concern about these evidences. A recent review about neonatal "minor" procedural pain research (Axelin and Salantera 2008) reported that 75 % of the studies reviewed did not provide adequate pain management to the infants enrolled. Similar data emerge from the analysis of similar studies published in 2009: 24 in 29 studies on painful procedures were performed without offering analgesia to control infants (Bellieni and Buonocore 2010a). Even for circumcision, analgesic treatments are seldom compared with placebos (Van Howe and Svoboda 2008) This is why recently developed guidelines for an ethical approach to neonatal research, recommend that "[w]hen research studies evaluate an analgesic treatment, the control group should receive a generally accepted analgesia and not only a placebo," and "[s]tudies that were likely to have provoked avoidable pain in neonates should be rejected as a general journal policy"(Bellieni and Buonocore 2010b).

Medical and Social Synergy to Prevent Pain

We strongly promote the position that medical intervention should not be limited to drugs and technical procedures. What is required and needed is "presence" (Bellieni et al. 2003). Nurses and neonatologists should provide this presence and they should lead parents to become an active presence too, not being disappointed by their babies' apparent fragility, immaturity, and lack of reactivity, which make relief and suffering difficult to decode. Neonatal suffering can be expressed in many ways, as hinted consolation seeking, sometimes as a desperate need to sense the presence of the caregiver. Our task is to recognize this and to take care of it. Nevertheless, measures should be taken to avoid unnecessary pain, including specific teachings in medical schools, and legal sanctions for those who provoke pain to babies during clinical procedures, when time and tools to prevent it are available.

The Therapeutic Tripod

Neonatal pain treatment cannot be limited to administer analgesia to the newborns, but should include the care of the other subjects of newborn's development, namely, parents and caregivers, as a part of a therapeutic tripod. In facts, stress and burnout are very frequent among these subjects (Henning et al. 2009; Scheurer et al. 2009), and can influence babies' treatment and well-being (Dessy 2009). Conversely, babies' pain can influence adults' behavior. Moreover, caregivers' pain can influence parents' and vice versa. This is why we advocate a psychological support of adults involved in babies' intensive care: This is not only useful for them, but it is an absolute newborns' right.

References

Als, H., & Gilkerson, L. (1997). The role of relationship-based developmentally supportive newborn intensive care in strengthening outcome of preterm infants. *Seminars in Perinatology, 21*, 178–189.

Anand, K. J. S., & Craig, K. D. (1996). New perspectives on the definition of pain. *Pain, 67*, 3–6.

Axelin, A., & Salantera, S. (2008). Ethics in neonatal pain research. *Nursing Ethics, 15*, 492–499.

Balon-Perin, S., Kolanowski, J., Berbinschi, A., et al. (1991). The effects of glucose ingestion and fasting on plasma immunoreactive beta-endorphin, adrenocorticotropic hormone and cortisol in obese subjects. *Journal of Endocrinological Investigation, 14*, 919–925.

Bellieni, C. V. (2003). Withholding and withdrawing neonatal therapy: An alternative glance. *Ethics in Medicine, 19,* 99–102.

Bellieni, C. (2005). Pain definitions revised: Newborns not only feel pain, they also suffer. *Ethics in Medicine, 21,* 5–9.

Bellieni, C. V., & Buonocore, G. (2010a). No analgesia to the control group: Is it acceptable? *Pediatrics, 125*(3), e709.

Bellieni, C. V., & Buonocore, G. (2010b). Recommendations for an ethical treatment of newborns involved in clinical trials. *Acta Paediatrica, 99*(1), 30–32.

Bellieni, C. V., Buonocore, G., Nenci, A., et al. (2001). Sensorial saturation: An effective tool for heel-prick in preterm infants. *Biology of the Neonate, 80,* 15–18.

Bellieni, C. V., Bagnoli, F., Perrone, S., et al. (2002). Effect of multi-sensory stimulation on analgesia in term neonates: A randomized controlled trial. *Pediatric Research, 51,* 460–463.

Bellieni, C. V., Bagnoli, F., & Buonocore, G. (2003). Alone no more: Pain in premature children. *Ethics in Medicine, 19,* 5–9.

Bellieni, C. V., Cordelli, D. M., Marchi, S., Ceccarelli, S., Perrone, S., Maffei, M., & Buonocore, G. (2007). Sensorial saturation for neonatal analgesia. *The Clinical Journal of Pain, 23*(3), 219–221.

Bellieni, C. V., Iantorno, L., Perrone, S., Rodriguez, A., Longini, M., Capitani, S., & Buonocore, G. (2009). Even routine painful procedures can be harmful for the newborn. *Pain, 147*(1–3), 128–131.

Beyer, J. E., De Good, D. E., Ashley, L. C., et al. (1983). Patterns of postoperative analgesic use with adults and children following cardiac surgery. *Pain, 17,* 71–81.

Blass, E. M., & Watt, L. B. (1999). Suckling and sucrose-induced analgesia in human newborns. *Pain, 83,* 611–623.

Cunningham, C. C., Morgan, P. A., & McGucken, R. B. (1984). Down's Syndrome: Is dissatisfaction with disclosure of diagnosis inevitable? *Developmental Medicine and Child Neurology, 26,* 33–39.

De Lima, J., Lloyd-Thomas, A. R., Howard, R. F., et al. (1996). Infant and neonatal pain: Anaesthetists' Perceptions and prescribing patterns. *BMJ, 313,* 787.

Debillon, T., Bureau, V., Savagner, C., Zupan-Simunek, V., & Carbajal, R. (2002). French national federation of neonatologists. Pain management in French neonatal intensive care units. *Acta Paediatrica, 91*(7), 822–826.

Dessy, E. (2009). Effective communication in difficult situations: Preventing stress and burnout in the NICU. *Early Human Development, 85*(10 Suppl), S39–S41.

Henning, M. A., Hawken, S. J., & Hill, A. G. (2009). The quality of life of New Zealand doctors and medical students: What can be done to avoid burnout? *The New Zealand Medical Journal, 122*(1307), 102–110.

Lago, P., Guadagni, A., Merazzi, D., Ancora, G., Bellieni, C. V., & Cavazza, A. (2005). Pain study group of the Italian society of neonatology. Pain management in the neonatal intensive care unit: A national survey in Italy. *Paediatric Anaesthesia, 15*(11), 925–931.

Melzack, R., & Wall, P. D. (1965). Pain mechanisms: A new theory. *Science, 150,* 971–979.

Merskey, H. (1979). Pain terms: A list with definitions and notes on usage. *Pain, 6,* 249–252.

Robin, M. (2000). Premier regards, premiers échanges. In E. Herbinet & M.-C. Busnel (Eds.), *L'aube des* (sensth ed., pp. 55–66). Paris: Stock.

Scheurer, D., McKean, S., Miller, J., & Wetterneck, T. (2009). U.S. Physician satisfaction: A systematic review. *Journal of Hospital Medicine, 4*(9), 560–568.

Simons, S. H., van Dijk, M., Anand, K. S., Roofthooft, D., van Lingen, R. A., & Tibboel, D. (2003). Do we still hurt newborn babies? A prospective study of procedural pain and analgesia in neonates. *Archives of Pediatrics & Adolescent Medicine, 157*(11), 1058–1064.

Thirion, M. (1994). *Les compétences du nouveau-né* (pp. 121–126). Ramsay: Albin Michel SA.

Van Howe, R. S., & Svoboda, J. S. (2008). Neonatal pain relief and the Helsinki declaration. *The Journal of Law, Medicine & Ethics, 36*(4), 803–823.

Ethics of Pain, Culture, and Ethnicity

David B. Morris
University of Virginia, Charlottesville, VA, USA

Synonyms

Biopsychosocial; Principalism; Race

Definition

Ethnicity refers in everyday speech to groups having "common racial, cultural, religious, or linguistic characteristics" (Oxford English Dictionary). Anthropologists debate its meaning; "instrumentalists" view ethnicity as fluid, dynamic, and situational – a collective strategy for pursuing economic and political interests – while "primordialists" see ethnicity as a stable, ineffable, potentially coercive bond reinforced by culture. Culture is defined by primatologists as "social learning." Human cultures are the constructed environments of material objects, ideas, institutions, social practices, and interpersonal relationships in which we live. Ethics is the study of moral knowledge. It concerns principles of conduct – both personal and professional – as well as the wider field of individual and communal values.

Characteristics

Ethnicity

Ethnicity is sometimes employed as a soft synonym for race, with an implication that ethnicity and race are both genetically determined. Some ethnic groups show physiological and morphological distinctness, such as variations in drug metabolism and in muscle enzyme levels after exercise. Such genetic variations, however, do not confirm a genetic basis for ethnicity or explain what ethnicity is. Patterns of gene frequencies resulting from marriage within a cultural group are "not so much the cause of ethnicity as the outcome" (Macbeth 2001).

Most scientists view skin color, eye color, hair type, and other outward signs associated with race as phenotypic variations insignificant at the level of the genotype. Human migration, intermarriage, and genetic polymorphism mean that populations are rarely homogeneous. Africans, Caucasians, and Asians (the most common racial designations) show wider genetic differences within groups than across groups. It is not race but racism that accounts for most documented differences in health status between minority and majority groups.

Ethnicity is more dubious than race as a biological category. Government agencies in the USA and UK increasingly prefer subjective criteria for establishing ethnicity. The term Hispanic offers insights into ethnic fluidity because Latin America has seen widespread intermarriage among European, Asian, African, and native peoples. Health and census data show wide variation in racial self-identification across Hispanic groups.

Like race, ethnicity is a powerful social fact, a source of intense political and personal identification as well as a source of inequalities in health care. In medicine, ethnicity matters greatly when questions turn on the adequate treatment of minorities and on culturally sensitive medical practices that respect the traditions and beliefs of minority groups. Such practice must recognize the limits of ethnic stereotypes; any social group includes members who resist its practices or reject its standards. In immigrant families, ethnicity is especially unstable across generations. Recognizing such difficulties, Crews and Bindon (1991) suggest that researchers state how they propose to use ethnicity, what their chosen categories imply biologically and sociologically, and why their particular analyses are needed.

Pain in its relations to ethnicity has been the subject of significant confusion in the design and interpretation of research studies (Morris 2001). Strong preliminary evidence suggests the existence of distinctive ethnic variations in pain (Bates et al. 1993; Greenwald 1991; Wolff 1985). Future studies, however, need to proceed with care in stipulating their assumptions. Ethnicity is shaped even by the methods of data collection that researchers employ to measure it. Senior and Bhopal (1994) argue that ethnicity implies one or more of three conditions: (1) a common language or religious tradition, (2) shared origins or social background, and (3) shared culture and traditions that are distinctive, maintained between generations, and conducive to a sense of identity and group. This formula, while not trouble-free, avoids genetic or biological markers and suggests that ethnicity is inseparable from shared social experience or culture.

Culture

Culture, as the constructed environment in which we live, plays no role in models of pain that emphasize the transmission of nociceptive impulses from the peripheral nervous system to the brain. The International Association for the Study of Pain, however, emphasizes that ▶ nociception is not pain. Pain, always psychological and always a subjective state, is inseparably linked to culture through the complex ▶ neuromatrix of thoughts, memories, and emotions centered in the brain, which is, as neurosurgeon John Loeser writes, the organ responsible for all pain. A ▶ biopsychosocial model of pain implicitly recognizes the role of cultures. Cultures, in concert with the necessary biological processes, crucially shape and frame the human experience of pain.

Different historical cultures have framed the experience of pain within very different systems of thought (Morris 1991). The English word *pain* derives from the Latin word for punishment (*poena*), and Western societies until the nineteenth century extended the ancient tradition that employed pain as a public method of legal penalty, whipping, branding, and crushing. Nineteenth-century legal reforms replaced pain with incarceration, but they did not remove pain from the public realm. From Babylon to Homer, the ancient world had understood pain as a punishment sent by the gods. From the mid-nineteenth century onwards, the modern world has understood pain as caused by a lesion. How people understand pain, as recent studies in ▶ pain beliefs demonstrate, correlates with pain intensity and with treatment outcomes.

The study of pain beliefs has focused in general on how individuals understand the cause, duration, and outcome of their pain (Williams and Keefe 1991). ▶ Catastrophizing – the patient's mental-emotional conviction of a disastrous outcome – is both a barrier to successful treatment and a means of intensifying pain. Research shows that fear intensifies pain, but the fear in catastrophizing is more than a mere feeling. It is the product of beliefs that circulate within a culture long before they circulate within an individual brain. The pain associated with specific medical practices, such as the surgery without anesthesia until recently practiced on newborns, is no less the product of a culture or medical subculture. As for adolescents who cut their skin or who endure violent initiation rites, pain is a malleable experience that means different things in different contexts. Culture is the context through which pain connects nervous systems with the realm of social practices and individual meanings.

Children and patients with AIDS demonstrate how culture influences pain. Recently, pediatric pain units have made special efforts to overcome inherent limitations in language skills and in experience that typify children, as well as to address the distinctive fears and emotions that children feel. McGrath (1993) has emphasized the particular ways in which meaning and emotion enter into the child's experience of pain. The pain of patients with HIV/AIDS is complicated by social stigma associated with sexual deviance and illicit drug use. Heterosexual women, who make up an increasing percentage of AIDS patients, face the isolating cultural beliefs that AIDS is mainly a disease of homosexual males. The culturally charged categories of gender and race are associated with pain intensity among ambulatory AIDS patients in New York City (Breitbart et al. 1996). The role of multinational pharmaceutical companies in pricing drugs has a significant impact on the pain of AIDS patients in sub-Saharan Africa. Individual meanings, social practices, and cultural assumptions shape, for better or for worse, the pain that patients experience.

Ethics
Ethics is a cultural construction with ancient roots. By the 1970s, advances in medical technologies had so far outstripped traditional moral knowledge that a new subfield emerged to address the gap, bioethics. Bioethics addresses moral issues that arise in medical practice, and in particular, it seeks to help doctors, hospitals, and the modern corporate enterprise of medicine treat patients with respect and concern for individual rights. Such rights are widely understood to include the right to effective and appropriate treatment for pain.

The still dominant Western system of bioethics is known as principalism. Ethical decisions are considered under four main principles, autonomy, beneficence, justice, and nonmaleficence. Principalism, however, is under pressure to explain what standard underlies the selection of principles. Enlightenment assumptions about reason as a universal and absolute basis for moral judgments are no longer widely shared in a postcolonial, postmodern world that continually tests the limits of what is considered reasonable or ethical. Bioethical principles – especially when they are understood not as absolutes but as based in "common morality" and as "prima facie binding" (Beauchamp and Childress 1994) – always require interpretation and application by fallible human agents in specific cases where details are unique and outcomes uncertain. Pain specialists thus have no legitimate substitute for active, regular discussion of ethical issues.

The 1970s brought very practical dilemmas. Pain research using animals was in effect unregulated and ran the risk of exceeding both what is humane science and what is tolerable to the nonscientific public. Public attitudes were changing. The year 1975 saw publication of philosopher Peter Singer's groundbreaking Animal Liberation: A New Ethic for Our Treatment of Animals. In 1983, the International Association for the Study of Pain published its guidelines on experimental pain in conscious animals. The IASP report begins with the claim that research producing experimental pain in conscious animals is "essential" for new, clinically relevant knowledge about the mechanisms of pain. It also stipulates that researchers take measures to provide "reasonable assurance" that animals are exposed to "the minimal pain necessary for the purposes of the experiment." Yet, who defines what is reasonable or minimal? Are the purposes of an experiment ever other than human purposes? Human pain, with the far more complicated questions it evokes, did not receive official IASP research guidelines until 1995.

The usefulness of ethical guidelines is not in question. Guidelines, however, do not preempt the need for ongoing discussion, especially in the shift from a right-based ethics to an ethics based on values (Morris 2003). A study of oncology nurses in the USA concluded that the most frequently cited ethical dilemma involves the management of pain (Raines 2000). A pioneer of bioethics asserts that failure to relieve pain optimally is "tantamount to moral and legal malpractice" (Pellegrino 1998). Undermedication for pain, more pronounced in

children than in adults, is not generally recognized as an ethical issue but treated as a technical problem of medical education, licensing, and drug regulation. Its broader ethical implications are mostly unaddressed. Improved professional awareness of the ethical dilemmas surrounding pain is needed.

References

Bates, M. S., Edwards, W. T., & Anderson, K. O. (1993). Ethnocultural influences on variation in chronic pain perception. *Pain, 52,* 101–112.
Beauchamp, T. L., & Childress, J. F. (1994). *Principles of biomedical ethics* (4th ed.). New York: Oxford University Press.
Breitbart, W., Rosenfeld, B. D., Passik, S. D., et al. (1996). The undertreatment of pain in ambulatory AIDS patients. *Pain, 65,* 243–249.
Crews, D. E., & Bindon, J. R. (1991). Ethnicity as a taxonomic tool in biomedical and biosocial research. *Ethnicity & Disease, 1,* 42–49.
Greenwald, H. P. (1991). Interethnic differences in pain perception. *Pain, 44,* 157–163.
Macbeth, H. (2001). Ethnicity and race as epidemiological variables: Centrality of purpose and context. In H. Mabeth & P. Shetty (Eds.), *Health and ethnicity* (pp. 10–20). London: Taylor & Francis.
McGrath, P. A. (1993). Psychological aspects of pain perception. In N. L. Schechter, C. B. Berde, & M. Yaster (Eds.), *Pain in infants, children, and adolescents* (pp. 39–63). Baltimore: Williams & Wilkins.
Morris, D. B. (1991). *The culture of pain.* Berkeley: University of California Press.
Morris, D. B. (2001). Ethnicity and pain. *Pain: Clinical Updates, 9,* 1–4.
Morris, D. B. (2003). Ethics beyond guidelines: Culture, pain, and conflict. In J. O. Dostrovsky, D. B. Carr, M. Koltzenburg (Eds.), *Proceedings of the 10th World Congress on pain* (pp. 37–48). Seattle: IASP Press.
Pellegrino, E. D. (1998). Emerging ethical issues in palliative care. *Journal of the American Medical Association, 279,* 1521–1522.
Raines, M. L. (2000). Ethical decision making in nurses: Relationships among moral reasoning, coping style, and ethical stress. *JONA's Healthcare Law, Ethics and Regulation, 2,* 29–41.
Senior, P. A., & Bhopal, R. (1994). Ethnicity as a variable in epidemiological research. *British Medical Journal, 309,* 327–330.
Williams, D. A., & Keefe, F. J. (1991). Pain beliefs and the use of cognitive-behavioral coping strategies. *Pain, 46,* 185–190.
Wolff, B. B. (1985). Ethnocultural factors influencing pain and illness behavior. *The Clinical Journal of Pain, 1,* 23–30.

Ethics of Pain, Human Dignity, and the Ethical Management of Pain and Suffering

Daryl Pullman
Medical Ethics, Memorial University of
Newfoundland, St. John's, NL, Canada

Definition

Dignity is a complex value concept that speaks to the worth human beings place upon themselves both individually and corporately. The evaluative content of the notion is determined to some degree by the context in which it is used. In some cases, the term "dignity" plays a largely formal role and is used to describe the intrinsic or inherent moral worth of every human being. Simply being a member of the human species affords one this dignity. One does nothing to earn it, and nothing can take it away (Kant 1785/1987). It is this sense of dignity that underwrites our general moral obligation to respect our fellow human beings and to treat them accordingly (Human Rights and Bioethics 1999). Our moral duty to respond to the pain and suffering of our fellow human beings is anchored in the intrinsic or basic dignity we all share as human beings.

There is another sense of dignity, however, that is more individualistic, social, and transient in nature. This dignity is a socially constructed notion that is tied both to individual goals and social circumstances. Such personal dignity is related to an individual's sense of how he or she is perceived in the social world by both self and others. Personal dignity can be either enhanced or diminished depending upon a variety of contingent factors that may be largely outside the control of the person involved.

Characteristics

Unmitigated pain and suffering are sometimes said to rob people of their dignity. When pain is not managed appropriately either through ignorance, indifference, or neglect, the goals and projects that together serve to define a person as an individual can be overwhelmed. Inasmuch as a person's dignity is to some extent dependent upon the ability to make and execute plans that provide meaning and purpose, poorly managed pain can undermine an individual's sense of dignity and self-worth.

Our moral obligation to respond appropriately to the pain and suffering of other human beings is linked to our general moral obligation to recognize, honor, and otherwise preserve the basic dignity of our fellow human beings. Notice, however, that the dignity referred to in this paragraph is distinct from the dignity of the former paragraph. Here, we are speaking of the basic dignity possessed by every human being. Pain and suffering cannot undermine this dignity because our intrinsic moral worth is immune to the vicissitudes of life. When we speak of the manner in which unmitigated pain might undermine the dignity of an individual, then it is personal dignity that is in view. The "death with dignity" movement in particular trades upon this more individualistic notion (Quill 1991; Pullman 1996).

A clearer understanding of the nature and extent of our moral obligation to respond appropriately to human pain and suffering requires a fuller explication of some of the terminology employed. In particular, the relationship between human pain and suffering must be explored (Pullman 2002).

Although the terms "pain" and "suffering" are often coupled in the literature on pain and symptom management, the terms are phenomenologically distinct. Pain does not necessarily involve suffering, and it is possible to suffer without being in pain. An individual afflicted with a congenital insensitivity to pain, for example, does not feel pain, but might nevertheless suffer from the effects of injuries sustained because of this insensitivity. Conversely, the pain of childbirth might be experienced as severe but rewarding.

Human beings are meaning conferring creatures. Our capacity for reflective self-consciousness enables us to make sense of the world as we experience it. It is this capacity that enables us to distinguish between ourselves and our bodies. Intense pain, however, can overwhelm this reflective capacity, such that it becomes impossible for the individual to distinguish between the self as a person and the body in pain. Many have experienced this phenomenon for a relatively brief duration when stubbing a toe or striking a thumb with a hammer. The more problematic occurrences, however, are those that involve acute or chronic pain occasioned by illness or disease that occurs over a sustained period of time. Intractable or poorly managed pain can inhibit the ability for reflective awareness and the capacity to make life plans and to engage in other meaning conferring projects that contribute to a sense of dignity and self-worth.

Though phenomenologically distinct, pain and suffering are nevertheless closely related. Suffering is experienced by persons, not bodies (Cassell 1991). The manner in which a person suffers depends to some degree on their understanding of their pain. Hence, it is possible to relieve a person's suffering in the presence of continuing pain by identifying the source of the pain, by changing its meaning, by demonstrating that the pain can be controlled, or by indicating that an end is in sight (Cassell 1982). The key observation here is that the primary link between pain and suffering is epistemic. That is, suffering occurs when the person in pain is unable to make sense of an experience that appears to undermine his or her integrity as a person. When the epistemic link is broken, the individual's sense of personal dignity is compromised.

However, suffering is not confined only to those who experience pain. Inasmuch as suffering includes a significant epistemic component, others who share in or identify with the life project of the one in pain may suffer as well, as they strive corporately to make sense of their loved one's pain experience. "The loss that accompanies illness," writes Frank, "begins in the body, as pain does, then moves out until it affects the relationships connecting that body with others" (Frank 1991, p. 36). We use terms like "sympathy" and "compassion" to describe the process of "suffering with." Clearly, sympathy

and compassion serve an important existential function as a means of psychological and social identification and support. However, they can also be viewed as part of an epistemic process by which we strive together to make sense of the pain endured by those whose lives we share. Ideally, "making sense" of such experiences involves an attempt to weave them into a meaningful narrative that can be shared, both by those who are actually in pain and by the sympathetic and compassionate ones who suffer with them. It is a process and a project that involves reference to our shared dignity.

The foregoing has several immediate implications for the ethical management of pain and suffering. First, health-care providers who attend to the physical needs of patients in pain must be aware that their patient's suffering is not only physical and emotional, but to a large extent, it is epistemic. While it is essential that the source of the pain is identified and that appropriate analgesics are provided if available, it is equally important to discuss the nature of the pain with the patient, to explain possible sequelae, and to attempt to help the patient make some sense of the pain experience. Second, ethical management of pain and suffering extends beyond the person who experiences physical or psychical pain, to include those compassionate and sympathetic loved ones who share the illness narrative with the patient in pain. Inasmuch as we are social beings, our sense of personal dignity is tied up in the life stories of those we hold most dear. Hence, when our loved ones are in pain, we suffer. Thus, ethical pain management must extend to the community of care. Finally, the human resources necessary to provide this kind of ethical intervention are generally shared across a health-care team rather than found in any single caregiver. Palliative care teams that include a full range of professionals that cover the gamut of needs from the physiological, to the psychosocial, to the spiritual are especially adept at providing this kind of intervention. Such teams could well serve as a model for other care providers who are striving for the highest ethical standards in the dignified management of human pain and suffering.

References

Cassell, E. (1982). The nature of suffering and the goals of medicine. *The New England Journal of Medicine, 306*(11), 639–645.

Cassell, E. (1991). *The nature of suffering.* New York: Oxford University Press.

Frank, A. W. (1991). *At the will of the body: Reflections on illness.* Mifflin: Houghlin.

Human Rights and Bioethics. (1999). C.H.R. res. 1999/63, U.N. Doc. E/CN.4/RES/1999/63.

Kant, I. (1785/1987). *Fundamental principles of the metaphysics of morals* (T. K. Abbott, Trans.). Buffalo: Prometheus.

Pullman, D. (1996). Dying with dignity and the death of dignity. *Health Law Journal, 4*, 197–219.

Pullman, D. (2002). Human dignity and the ethics and aesthetics of pain and suffering. *Theoretical Medicine and Bioethics, 23*, 75–94.

Quill, T. E. (1991). Death and dignity. *The New England Journal of Medicine, 324*, 691–694.

Ethics of Pain-Related Disability Evaluations

Mark D. Sullivan
Psychiatry and Behavioral Sciences, University of Washington, Seattle, WA, USA

Synonyms

Exaggerated pain; Impairment rating; Malingering

Definition

The standards used for evaluating disability due to pain are primarily moral, rather than scientific, in nature. Criteria for legitimate entry to the sick role have evolved with society, with only modern industrial society placing heavy emphasis on tissue damage demonstrated on medical tests (Fabrega 1996). The highly variable relation between clinical pain and tissue damage poses a serious challenge to this strategy of illness behavior validation. The amount of objective tissue damage is a poor proxy for the amount of subjective pain. The strategy of validating pain complaints by matching them with something else is fundamentally flawed. Disability cannot be invalidated by matching public pain behavior with private pain experience. If this pain experience is truly private, it is not available to scientific investigation. Rather, pain behavior is judged as appropriate or exaggerated through complex assessments of the function of this behavior in its social context. It will remain necessary to triage suffering presented to health care providers into that which should be addressed in the medical setting and that which is better addressed elsewhere. However, current strategies of pain validation and triage are pseudoscientific and distort the clinical care of those with pain.

Characteristics

Malingering exists, though it is very rare in most clinical settings (Fishbain et al. 1999). The incentives for dissimulation in jails, emergency rooms, and disability ratings may make it more common in these settings. Much more commonly, in clinical settings such as primary care or pain clinics, one hears about patients "exaggerating" their pain or reporting pain "out of proportion" with their physical findings. We need to inquire into what we mean by these claims of exaggeration, taking malingering as a special case of exaggeration. Specifically, we need to inquire into the standards we use for exaggeration. These are primarily moral, rather than scientific standards. This is because there is no scientific procedure by which we can determine whether private pain and public pain behaviors are "equal," because private pain is not accessible to scientific observation. Rather, we determine whether someone is being a good sport and playing by the rules of the game established by the group.

Assessments of illness behavior and disability ratings are part of the very general problem of accommodating the weak, the ill, the old, and the young in any social group. It is one of the core ethical challenges of civilization. Though care of the sick is a universal feature of human societies, it is often morally ambiguous

(Wolff and Langley 1977). This ambiguity cannot be resolved by sorting the ill into accurate and inaccurate reporters of pain. The quest to identify and extinguish all "secondary gain" is futile.

Validating Pain Behavior

Matching Pain Behavior and Pain Experience at a Personal Level

Consider our own ability to match our pain behavior with our pain experience. We certainly know if we are being insincere about the magnitude of our pain. The intention to lie about pain is apparent on introspection, but how could we be sure we were being sincere? How could we verify that our sensation was equivalent to our expression of pain? How could we verify that we were being "completely transparent" about the nature and extent of our pain? We do not simply rely upon accurate self-observation; checking and negotiating with others is necessary. Sincerity may not be the gold standard for "true" pain behavior that we think it is. What constitutes "too much" pain behavior is judged by social, not scientific, standards. Knowledge of pain severity, like knowledge of one's personal location, requires a context, not just a private experience. This context provides us with a metric to measure our experience and our behavior (Wittgenstein 1953).

Interpreting the Pain Behavior of Another Person

A similar point can be made concerning our knowledge of others' pain. I do not validate pain behavior by confirming its equality with the private pain experience of the person I am evaluating. By definition, I have no access to the other person's private experience. I validate pain behavior if it has a comprehensible place within a wider social context. We think that our personal experience with pain must allow us to judge whether others are exaggerating or simply lying about their pain, for these judgments are made in clinical settings every day. Yet, it is the concept of "pain," rather than our personal experience of the sensation of pain, which provides a framework within which the appropriateness of pain behavior can be judged. Judgments about exaggerating or faking pain, whether

these are about others or about oneself, are not made on the basis of a simple encounter with a sensation. These judgments must be made on the basis of established criteria for appropriate behavior. One cannot establish meaningful criteria entirely by oneself (Wittgenstein 1953).

Interpreting Pain Behavior on the Basis of Physical Performance

The bulk of recent scientific literature about exaggeration and malingering among patients with chronic pain uses measures of physical function rather than measures of pain itself (Lechner et al. 1998). While this might appear to be a different issue, it is liable to the same critique. Such physical testing relies on the notion of honest or "maximal" effort. Various devices and strategies (e.g., isometric or isokinetic testing) have been devised to detect "submaximal" effort. Most have shown poor ability to detect malingerers in experimental settings (Lechner et al. 1998), but this is actually beside the point. The problem lies in defining what constitutes honest, sincere, or maximal effort. Maximal voluntary effort is determined, not just by the individual and his pain – as is implied by this type of physical capacity testing. Maximal effort is also defined by the social context, including the standards and incentives it provides. It is not possible for even the individual being tested to determine what a maximal effort is without reference to this context. Does maximal effort occur in the presence of verbal encouragement, a $1,000 reward, or a loaded gun pointed at the head?

Interpreting Pain Behavior by Comparison to Tissue Damage

Another variant of the matching of pain with pain behavior is matching pain with tissue damage. The validation of pain complaint through correlation with tissue damage remains common in clinical and medicolegal settings (Rondinelli and Katz 2000) but is inconsistent with modern pain research. The relationship between the amount of tissue damage and the amount of pain experienced is highly variable. Physiological mechanisms whereby the organism increases (e.g., central sensitization) or decreases (e.g., pain modulation with endogenous opioids) the pain experienced from a given amount of tissue damage have been identified (Fields 2000). The inference that a patient's pain is "psychogenic" because a peripheral lesion "adequate" to explain the pain cannot be identified is not justified by current science.

Interpreting Pain Behavior by Comparing Experimental Pain to Clinical Pain

An enormous amount of research has been done on experimentally induced pain in normal people. Some researchers have asked subjects with ongoing, clinically significant pain to compare their experience of this pain to their experience of experimentally induced pain (Koelbaek et al. 1999). Research on comparisons between clinically significant and experimental pain is new enough that its relevance to the broad issue of interpreting pain cannot yet be discerned. However, it is doubtful that such research will resolve the dilemmas associated with interpreting clinical pain. The social context and the behavioral incentives differ markedly between experimental and clinical pain, making it very difficult to infer from the former from the latter.

Summary

New standards of disability evaluation are necessary if we are to fairly evaluate the patient claiming disability due to pain. The ethical component of these standards should be acknowledged and publicly debated.

References

Fabrega, H. (1996). *The evolution of sickness and healing* (p. 62). Berkeley: University of California Press.
Fields, H. L. (2000). Pain modulation: Expectation, opioid analgesia and virtual pain. *Progress in Brain Research, 122*, 245–253.
Fishbain, D. A., Cutler, R., Rosomoff, H. L., et al. (1999). Chronic pain disability exaggeration/malingering and submaximal effort research. *The Clinical Journal of Pain, 15*, 244–274.
Koelbaek, M., Graven-Nielsen, T., Olesen, A. S., et al. (1999). Generalised muscular hyperalgesia in chronic whiplash syndrome. *Pain, 83*, 229–234.
Lechner, D. E., Bradbury, S. F., & Bradley, L. A. (1998). Detecting sincerity of effort: A summary of methods and approaches. *Physical Therapy, 78*, 867–888.
Rondinelli, R. D., & Katz, R. T. (2000). *Impairment rating and disability evaluation*. Philadelphia: WB Saunders.
Wittgenstein, L. (1953). *Philosophical investigations* (G. E. M. Anscombe, Trans., p. 253). Oxford: Blackwell.
Wolff, B. B., & Langley, S. (1977). Cultural factors and the response to pain. In D. Landy (Ed.), *Culture and disease and healing* (pp. 313–319). New York: MacMillan.

Ethnic Group

Definition

A self-identified cultural group that resides within another society and has beliefs, values, and rules that permit appropriate interactive behavior among its members.

Cross-References

▶ Cancer Pain, Assessment of Cultural Aspects

Ethnicity

Definition

An individual's self-identification with a group or pertaining to a group of people recognized as a class on the basis of certain distinctive characteristics such as religion, language ancestry, culture, or national origin.

Cross-References

▶ Cancer Pain, Assessment of Cultural Aspects

Ethyl 1-Methyl-4-Phenylpiperidine-4-Carboxylate Hydrochloride

▶ Postoperative Pain, Pethidine

Etomidate

Definition

Etomidate is an intravenous anesthetic acting at least partly via $GABA_A$ receptors.

Cross-References

▶ GABA and Glycine in Spinal Nociceptive Processing

Evaluation

▶ Psychological [Behavioral Health] Assessment of Pain

Evaluation of Pain in Humans

Lars Arendt-Nielsen
Center for Sensory-Motor Interaction,
Department of Health Science and Technology,
School of Medicine, Aalborg University,
Aalborg, Denmark

Characteristics

Assessment of Ongoing Clinical Pain and Treatment Effects

Assessment of pain in a systematic and consistent manner is an important instrument for promoting identification of unrelieved pain at the individual patient care level. This requires some collaboration from the patient and special considerations/ precautions should be taken when pain is assessed, for example, in neonates, children, the elderly, or cognitively impaired persons (see ▶ Pain Assessment in Children, ▶ Pain Assessment in Neonates, ▶ Pain Assessment in the Elderly). Pain assessment scales and methods used for neonates, children, and in the elderly have received special attention.

Factors known to aggravate or relieve pain may initially be recorded and qualitative information about the interaction with daily living, including work and recreational activities, diurnal variation, sleep disturbances, mobility, appetite, sexual functioning, and mood, should be gathered. These data are normally not quantitative but may still be recorded in databases for later analysis. The psychosocial assessment should emphasize the effect of pain on patients and their families, as well as patients' preferences among pain management methods. Technological developments have provided small hand-held devices (data logger), which can be used as pain diaries.

Some of the quantitative methods used to assess ongoing clinical pain may also be used to assess experimentally induced pain, whereas others are exclusively for clinical purposes. An assessment of pain intensity should include an evaluation of not only the present pain intensity but also pain at its least and worst and when aggravated by various internal or external factors (activity, weather, stress, hormonal level, etc.).

The clinical pain assessment techniques can roughly be divided into five categories of which only 1–3 are dealt with here:
- Quality of life
- Sensory component
  - ▶ Sensory-discriminative aspect (intensity)
  - Quality
  - Location
  - Temporal profile
- Psychophysiological methods
- Affective components
- Cognitive component

Quality of Life

The pain inventory SF-36 (see ▶ SF-36 Health Status Questionnaire) is a measure of various quality-of-life dimensions, and the scores are given within eight individual items related to physical functioning, physical problems, bodily pain, general health perspectives, vitality, social function, emotional problems, and mental health. Specific self-report questionnaires have been developed in, for example, physical therapy (Oswestry disability questionnaire, Roland and Morris questionnaire) and cancer pain research (functional living index cancer scale, 36-item European organization for research and treatment of cancer quality of life questionnaire for cancer).

The ▸ West Haven Yale multidimensional pain inventory consists of 12 scales designed to measure the impact of pain on a patient's activities of daily living and is a comprehensive instrument developed to evaluate psychosocial and cognitive-behavioral factors related to pain. The symptom checklist 90-revised (SCL-90-R) is used for measuring the psychological symptomatology associated with chronic pain.

Sensory Components
Sensory-Discriminative Aspect (Intensity)
The intensity and/or unpleasantness of ongoing or evoked pain can in principle be assessed by:
• Visual analogue scale (VAS)
• Numerical rating scale (NRS)
• Verbal rating scale (VRS)

The use of two separate psychophysical scales (mainly visual analogue scales or verbal rating/descriptor scales) for the assessment of pain intensity and pain unpleasantness has become a kind of a standard. The two dimensions have been proven to be sufficiently independent to justify separate assessments.

The visual analogue scale (VAS) can be a horizontal or a vertical line of, for example, 10 cm length anchored at the beginning like "no pain" and at the end like "worst pain imaginable." The subject is required to express the pain intensity by indicating a line length subjectively proportional to the perception. The pain intensity can also be rated on an electronic device where the scores are logged by a computer for later analysis or as part of a pain diary. The scales can also be defined as thermometers or have different shapes to compensate for the possible nonlinearity – an increased rating from 1 to 2 – many not in relative magnitude correspond to an increase from 7 to 8. Generally, it is not recommended to apply words or numbers along the line as this may lead to uneven distribution (clustering) along the line. The cognitive

demands in using a VAS may be too high for some patients (e.g., cognitively impaired and elderly). The direct scaling methods are said to produce scales with interval or even ratio qualities. Such metric qualities are, however, not unalterable features because they are dependent on the rating behavior of the given population under investigation.

The numerical rating scales (NRS) use verbal anchors such as "no pain" and "worst pain imaginable" at the starting and end points of the scale to define the scale range. Only numbers represent the categories themselves. It is tempting to assume that the numbers guarantee the equidistance between categories, which is not normally the case. Numerical scales are like verbal scale easy to explained and understood.

A verbal rating scale (VRS) is normally divided into several categories, which are labeled, for example, with "no pain," "slight pain," "moderate pain," and "strong pain." A subject is asked to express the pain perception by use of these categories. The scores are quantified afterward in case of, for example, a 4-point numerical rating scale by assigning a number from 1 to 4 according to the category chosen. The advantage of this approach is that it is easily understood by the subject and does not require a skilled investigator. However, the assumed equidistance between categories has not been proven for many of these scales. Therefore, ordinal scale properties ought to be taken for granted instead of interval scale properties. These limitations can be avoided by calibrating the psychometric distance between categories in advance.

Different pain assessment tools have been developed for different disciplines, for example, the most frequently used multi-item measure of pain intensity in cancer pain research is the pain intensity scale of the brief pain inventory (BPI) combining four NRS (current pain, worst, least, and averaged) into a single pain intensity score. Variants of the methods of direct scaling are magnitude estimation and cross-modality matching. Magnitude estimation requires the subjects to assign numbers proportionally to the intensity of pain perception. At the beginning, the numbers can either be chosen completely arbitrarily by the subjects or the investigator can introduce certain values. Thus, the subject is instructed to assign a 10 to the first stimulus or a 50 to a "barely painful stimulus." Furthermore, certain stimulus intensities can be used as references, which are assigned certain scale values. Such preset scale values are called a modulus and are thought to homogenize the use of a scale. The metric properties of such scales can be outstanding as long as a trained and skilled subject is the rater. However, the high cognitive demands associated with such scales prevent the use in populations with low education and cognitive disabilities. Cross-modality matching requires the subject to express the intensity of pain perception in a second sensory modality. A well-known example is the handgrip method, in which a pressure sensitive grip is to be compressed proportionally to the pain intensity.

The scales can be combined in various ways, for example, numbers and descriptors. The scales (VAS, NRS, and VRS) can also be used to assess pain relief. There is no evidence that one is better than another. However, practical considerations might suggest that VRS of pain relief (e.g., no relief, slight relief, moderate relief, lots of relief, and complete relief) may help the chances that the patient will not confuse the relief rating with pain intensity ratings because NRS and VAS pain intensity measures can look very similar to NRS and VAS measures of pain relief.

Quality

A multidimensional approach to assess pain qualities is the ▶ McGill pain questionnaire (MPQ) where a total of 78 pain descriptors are classified into 20 categories of pain to assess four major dimensions, sensory, affective, evaluative, and "miscellaneous" pain as well as the total severity score. The questionnaire has been translated into many languages and validated independently. A short form MPQ (SF-MPQ) consists of a subset of 15 descriptors drawn from the sensory and affective categories. Whereas many clinical pain conditions have been quantified by means of the MPQ, it has infrequently been used to assess experimental pain.

A further approach consists of the determination of the dimensionality of pain itself without using an a priori set of dimensions. The method of multidimensional scaling allows statistically finding the dimensionality of a given pain by using similarity estimates for series of perceptions of this pain under systemically varied conditions (e.g., different intensities, different durations, and different areas). The number of distinct dimensions of pain is by this, not a prerequisite, but a result. The resulting dimensions are, however, not necessarily straightforward in their physiological or psychological meaning. In case of neurological disorders, ▶ paresthesia may occur and here sensory qualities like numbness, tingling, burning, and prickling should be assessed.

Location

For painful muscular and visceral pain conditions, the referred pain patterns are of clinical importance and represent many years of empirical descriptions. Only more recently, basic studies have tried to understand the underlying mechanisms of referred pain, and there is a general consensus that it represents a central phenomenon and that central sensitization may enlarge the referred pain areas. Referred pain can still be projected to an area completely anesthetized or deafferentiated – indicating the role of the central nervous system.

The areas of pain (▶ pain drawings) or abnormal sensory perception can be assessed by a pain drawing or a pain site checklist. A pain drawing consists of an outline drawing of the human body, and patients are asked to indicate on the drawing specific sites of pain, other sensations, or abnormal sensitivity to various standardized stimuli (e.g., mechanical and thermal). The areas can be quantified and used for later quantitative comparisons.

Temporal Profile

The temporal aspects of pain, such as its variability, frequency, and duration as well as its pattern across time (minutes, hours, days, and months) can be assessed. Attempts have been made to use VRS to quantify how frequently the pain occurs over the day (ranging from "not at all" to "all day"). An important temporal aspect in relation to, for example, neurogenic pain is whether stimulus-independent pain may be continuous or ▶ paroxysmal or whether the stimulus-evoked pain may show, for example, facilitated ▶ wind-up-like pain with long after-sensation.

Another aspect is the temporal response to a standardized experimental stimulus, for example, a brief 51° stimulus. Such a stimulus will cause a well-localized, sharp, pricking first pain sensation (predominantly Aδ-fiber-mediated response) followed by a less well-localized, diffuse, burning second pain sensation (predominantly C-fiber-mediated response). This perceptual and temporal pattern may be markedly changed in patients suffering from, for example, neuropathic pain. Under normal conditions, it may be that pain evoked by even a cutaneous stimulus extends beyond the boundaries of the actual stimulus location (radiation). This radiation is more prominent in many types of clinical pain (e.g., assessment of ▶ neuralgia and ▶ causalgia).

Psychophysiological Methods

Psychophysiological measures can be muscular or autonomic indicators. Muscle tension as assessed by use of surface electrodes over the muscles of interest (surface EMG) can be measured. The surface EMG methods aim at assessing the increased tension of certain muscles regionally related to the myalgic complaints, in rest or under physical or psychological stress, the slowing of muscle relaxation after use, or the static or dynamic asymmetries in certain muscles groups responsible for posture and balance. Tension of many muscles can be scanned or that of a few muscles can be analyzed quantitatively in detail. Simple linear relationships between muscle tension and subjective pain have been little observed. Attempts have also been made to measure firing from single motor units in relation to painful musculoskeletal disorders and stressful demanding work operations.

The regional blood flow through body areas affected by painful conditions and the blood stream in certain vessels can be assessed by use of, for example, photo-plethysmography, laser Doppler flowmetry, reflectance spectroscopy, thermography, and thermistors. Blood pressure, heart rate, and skin conductance level are nonspecific correlates of nociception and may reflect physiological arousal. They are not capable of differentiating painful states precisely because the stimulus–response relationships are very variable and loosely determined.

Experimentally Induced Pain

Standardized activation/stimulation of the nociceptive and nonnociceptive pathways is important for quantitative sensory testing in basic experiments or for diagnostic purposes (Fig. 1). From a mechanism-based point of view, experimental tools are important for acquiring differentiated information about the different mechanisms/pathways involved/affected (Fig. 2).

The perception of pain cannot, however, be validly examined by experimental tools due to the fact that the pain stimulus used in the experiments is an artificial one. Moreover, the experimental measurements of pain have, as a rule, been developed on healthy individuals. Indeed, the experimental pain stimulus is usually short, harmless, predictable, easily tolerated, and hardly emotionally stressful and therefore exhibits marked differences from clinical pain. Despite these objections, various clinical pain conditions have been increasingly analyzed using experimental techniques.

The main advantages of an experimental approach are as follows:

- Stimulus intensity, duration, and modality are controlled and can be repeated over time.
- Differentiated responses to activation of different structures (skin, muscle, viscera) can be assessed (multitissue sensory testing approach).
- Differentiated responses to different stimulus modalities can be assessed (multimodal sensory testing approach).
- The physiological and psychophysical responses can be assessed quantitatively and compared over time.
- Pain sensitivity can be compared quantitatively between various normal/affected regions.

Evaluation of Pain in Humans, Fig. 1 Assessing of pain can be performed in patients and healthy volunteers. In patients the ongoing pain and clinical manifestations are assesed by pain scales and questionnaires. It is also possible to apply experimental pain stimuli to pain patients and assess quantitatively the evoked response. In both cases the effect of pharmacological and non-pharmacological modulation can be monitored. In basic studies on healthy volunteers experimental pain can be provoked under normal conditions or when the pain system has been changed experimentally into a hyperexcitable stage (proxy for clinical mainfestations i.e., allodynia or hyperalgesia) and the effects of various modulations can be assessed quantitatively

Evaluation of Pain in Humans, Fig. 2 In chronic pain patients the ongoing pain is composed by a variety of underlying mechanisms (A, B, and/or C) but is is not known which mechanism is the best target for intervention. By applying a variety of mechanistic experimental pain stimuli to various tissues it is theoretical possible to tease out which mechanism(s) (e.g., A) are the most attenuated in the patient and hence target this mechanism by the pharmacological interventions. This mechanism based therapeutic approach is only possible if sufficiently advanced pain assessment tools are available

The requirements for an ideal stimulator are as follows:

1. Can be applied to all body parts without causing damage
2. Activates selective receptor populations
3. Can induce temporal and spatial summation
4. Has a causal relationship between stimulus intensity and pain intensity
5. Elicits different and discriminatory pain intensities and qualities
6. Delivers reproducible stimuli
7. Is easy to apply and control in laboratory and clinical settings

Unfortunately, the ideal stimulator does not exist, and hence, the most adequate for a particular study should be selected according to pros and cons. The probes or electrodes used for delivering the stimuli will be developed particularly for skin, muscle, or visceral stimulation and, as such, not generally versatile.

Stimuli can have phasic (short lasting, milliseconds to a few seconds) or tonic properties (long lasting, many seconds to minutes). The various pain induction techniques are summarized in the list below and, as can be seen, most methods have been developed for cutaneous applications. Most of the phasic stimuli can be applied (1) as a single stimulus or as a series of repeated stimuli (to evoke temporal summation) and (2) as a stimulus activating a small or large/multiple area(s) for the study of spatial summation.

As pain is a multidimensional perception, the reaction to a single standardized experimental stimulus of a given modality can obviously only represent a very limited fraction of the entire pain experience. Therefore, it is mandatory to combine different stimulation and assessment approaches to gain advanced differentiated information about the nociceptive system under normal and pathophysiological conditions.

Pain Induction Modalities and Methods

Electrical
- Trans-cutaneous
- Intra-cutaneous
- Tooth-pulp
  - Intra-muscular
  - Trans-mucosal (oral, viscera)

Thermal
- Heat
  - Radiant (laser, light, infrared) (skin)
  - Contact thermode (skin)
  - Circulating hot water (skin, muscle, viscera)
- Cold
  - Cold pressor test (skin)
  - Contact thermode (skin)
  - Evaporation of gas (skin)
  - Menthol (skin)
  - Circulating cold water (muscle, viscera)

Mechanical
- Brushing/stroking (skin, in case of allodynia)
  - Pin prick (skin)
- Pinch (skin)
- Impact stimuli (skin)
  - Pressure (skin, muscle)
  - Tourniquet/(skin, muscle)
  - ▶ Ischemia (skin, muscle)
  - Distension (viscera)

Chemical
- ▶ Capsaicin (skin, muscle, viscera)
- Mustard oil (skin)
- Melittin (skin)
- Hypertonic saline (skin, muscles)
- Bradykinin, serotonin, substance P and other algogenic substances (muscle)
- CO_2 (nasal mucosa)
  - Glutamate (muscle)
- Nerve growth factor (muscle)
- Post-exercise soreness (muscle)
- Glycerol (viscera)
- Hydrochloride acid (viscera)

Stimulus Modalities

Selectivity is a major problem for experimental pain stimuli. Most studies have used electrical stimulation, but this technique is nonselective and bypasses the receptors by depolarizing the afferent nerve fiber. Many attempts have been made to refine electrodes to activate the pain fibers selectively but so far without success.

The most selective heat stimulator is probably a ▶ laser that can emit concentrated light or heat

Evaluation of Pain in Humans, Fig. 3 When a Peltier element for thermal stimulation is used, it is possible to heat and cool with fixed temperature rise (e.g., 2/s) and fall times and determine the various psychophysical thresholds (cold pain, cold detection, warm, heat pain, and heat pain tolerance). Normally, a given threshold is measured several times due to variability and the average is calculated

radiation. Many types have been applied to the skin (argon (488–515 nm), copper vapor lasers (510–577 nm), semiconductor lasers (e.g., 980 nm), Nd-YAG (1,064 nm), thulium-YAG (1,800 nm), and CO_2 (10,600 nm)). The advantage is that the laser light can be applied without touching the skin and hence does not contaminate the stimulation by mechanosensitive input. The stimuli can be short (e.g., 50 ms) and be used for both psychophysical and electrophysiological (evoked potentials, reflexes) evaluations.

The Peltier contact thermode, which can either warm or cool depending on the direction of the current (Fig. 3), is easy to use and is often used in the clinic. For more tonic cold stimulation, the so-called cold pressor test (immerse the hand into ice water) has been used. During this test, the person continuously scores the pain intensity/unpleasantness. Depending on the duration, for which the person can endure the pain, they are classified as pain tolerant or pain sensitive.

Depending on the conditions, ▶ mechanical stimuli can activate all fibers – ranging from Aβ-fibers (brushing/stroking used to assess ▶ allodynia, tactile (▶ von Frey hair) to assess hyper-, hypo-, or dysaesthesia (dysesthesia)), via activation of both Aβ- and Aδ-fibers (pin prick,

often used to assess hyper- or hypo-algesia) to full activation of Aβ-, Aδ-, and C-fibers (pressure or pinch). For visceral stimulation, different distension techniques have been used, where the most advanced methods also include the assessment of the cross sectional area, which can be used to calculate the actual strain, applied to the wall of a hollow organ.

Chemical stimuli are of nature tonic and difficult to repeat over time as some of them induce peripheral and central sensitization (e.g., capsaicin). Stimulation of muscle tissue has in particular utilized chemical stimulation, as, for example, intramuscular hypertonic saline is useful to generate local and referred pain phenomena in healthy volunteers or in patients. An adequate chemical method to elicit pain from the nasal mucosa is by an air stream with high CO_2 concentration.

Temporal and Spatial Summation

The phenomenon that a single painful or even subpain threshold stimulus by repetition causes exaggerated perceptions of pain in man is called temporal summation (Fig. 4).

Temporal summation can be induced from the skin, muscles, and viscera by repeated thermal, electrical, mechanical, or chemical stimuli.

Evaluation of Pain in Humans, Fig. 4 *Left panel.* If consecutive painful stimuli are applied with an interval of 4 s, the individual stimuli are theoretically perceived to be of equal intensity (e.g., just exceeding the pain threshold (PT)). If the interval between consecutive stimuli is reduced to, for example, 1 s, the pain intensity increases during the series, although the stimulus intensity is the same – this is temporal summation. *Right panel.* A given stimulus is applied to one probe, and the perceived intensity is just exceeding the pain threshold (PT). If the same stimulus is now applied simultaneously to more probes, the pain intensity increases as the number of probes is increased – this is spatial summation

Temporal summation can be measured by either psychophysical (pain summation threshold) or electrophysiological responses (facilitation of the ► nociceptive withdrawal reflex). Temporal summation is stimulus intensity and frequency dependent – and can, for high stimulus intensities, be elicited down to 0.3 Hz. Summation up to 20 Hz is reported. Temporal summation is a very potent mechanism, which is difficult to block pharmacologically and is assumed to be most probably an important mechanism in neuropathic pain. Spatial summation is the phenomenon which occurs when a given stimulus intensity is applied to an area and when the same stimulus is applied to a larger area, the pain is perceived as stronger. Spatial summation has predominantly been studied with thermal stimuli (Fig. 4).

Assessment of Experimentally Induced Pain: Basic and Clinical Applications

As there are various experimental ways of pain induction, there are also various experimental methods for assessing pain and its physiological correlates. In addition to the methods available to assess clinical ongoing pain, a number of psychophysical, electrophysiological, and imaging procedures have been specifically developed to assess experimentally induced pain (see list below).

Assessment of Experimentally Induced Pain

Psychophysics
- VAS/VRS/NRS
  - Intensity/Unpleasantness
  - Stimulus–response function
- Cross-modality matching
- Pain detection and tolerance thresholds
- McGill Pain Questionnaire (MPQ)
- Signal detection theory (SDT)

Electrophysiological
- Microneurography
- Excitatory or inhibitory reflexes
- ► Evoked potentials

- ▶ Electroencephalogram/
  Electroencephalography
- ▶ Magnetoencephalogram

Psychophysiological

- Cutaneous temperature (thermography)
  - Cutaneous blood flow (laser Doppler flowmetry)
- Blood pressure
- Heart rate
- Respiration
  - Pupillary responses
- Galvanic skin resistance

Imaging techniques

- PET
- FMRI
- SPECT

Psychophysical Methods

The psychophysical methods are developed on basis of the ▶ psychophysical laws and can be divided into response- and stimulus-dependent methods (Fig. 5). The response-dependent methods rely on how the person evaluates the stimulus intensity or unpleasantness on a given

scale (VAS/VRS/NRS) or as a cross-modality match (Fig. 5).

The stimulus-dependent methods are based on adjustment of the stimulus intensity until a predefined response, typically a threshold, is reached. The stimulus intensity required to reach the threshold is in physical units, and therefore, the use of scales is avoided (Fig. 5). Stimulus–response functions are more informative than a threshold determination as superthreshold response characteristics can be derived from the data. All quantitative sensory tests are psychophysical tests, which require conscious and alert patients, who fully understand the instructions given. Stimulus–response functions are valuable to assess hyperpathia to various stimulus modalities.

The determination of pain thresholds (pain detection and pain tolerance thresholds) has most often been used in experimental pain research. Thresholds are of great value because they are reliably, easily, and simply assessed. There are several ways of assessing these thresholds: the method of constant stimuli, the method

Evaluation of Pain in Humans, Fig. 5 Psychophysical assessment of pain can be done by stimulus (*top*) or response (*two lower panels*) dependent methods. Using a stimulus-dependent method, the stimulus intensity is gradually increased until a given threshold (e.g., pain detection or pain tolerance) is reached. For response-dependent methods, fixed stimulus intensities (*middle panel*) are delivered and the volunteers/patients rate the perceived pain intensities (*lower panel*) on a given scale (e.g., VAS). Based on those responses, e.g., stimulus–response curves can be constructed

of adjustment, the method of limits, method of levels, forced choice procedure, staircase method, etc. They all look for the least amount of physical energy necessary to elicit pain in the case of pain threshold or for the most amount of physical energy still tolerable for an individual in the case of pain tolerance threshold. Thresholds only demonstrate where the pain range starts (pain threshold), and where it ends (pain detection and pain tolerance thresholds). There is no information about pain perception between these range delimiters as obtained using stimulus–response functions.

The multidimensional scaling using, for example, MPQ is not often used in experimental studies, as a standardized stimulus is less multidimensional than ongoing clinical pain. Sensory decision/detection theory (SDT) yields two measures of perceptual performance (1) the accuracy with which a person judges whether, for example, a strong painful or a weak painful event occurs and (2) the subject's response bias which quantifies a general tendency to report one of the events as occurring more frequently than the other. These parameters can be presented and interpreted using the receiver operating characteristics (ROC) curves.

The SDT paradigms ask for long series of pain stimuli under constant conditions, which constitutes a prerequisite that is difficult to meet. The study of patients does not often allow for sessions of long duration. Sensitization, adaptation, or habituation prevents constant conditions in long series of stimuli. Such limitations have led to a decline in the use of SDT in experimental pain research.

Before new scales or new configurations are applied in the laboratory or clinic, they need to be validated and the reliability evaluated. Although it seems simple to rate a given pain on a scale, there are many endogenous and exogenous confounding factors – training, instruction/introduction, sex of investigator versus sex of patient/volunteer, experimental environment (e.g., is the spouse present or not), educational level of patient/volunteers, if the scores are done by experimenter/clinician/nurse or self-reported, etc. Other situational and individual factors

known to influence pain are, for example, diurnal variations (time of day), menstrual cycle, gender, race, and ethnicity.

Electrophysiological Methods

The electrophysiological assessment methods have mainly been applied in basic studies as they are complicated to use in the clinic. Microneurography is the direct recording of the activity of nerve fibers following peripheral stimulation and has been used to characterize firing. This method may identify activation patterns of nociceptive of Aδ- and C-fibers and of different classes of nociceptors.

The nociceptive withdrawal reflex is a spinal reflex of the lower limb that is elicited by painful somatic stimulus. The reflex can be used for assessment in two ways, either as a reflex threshold or as an amplitude to a fixed suprathreshold stimulus intensity. The reflex can be used to assess the response to both a single stimulus and a repeated stimulus (temporal summation). Under some conditions, the reflex threshold and the pain detection threshold coincide. The generation of a withdrawal reflex is initiated by the nociceptive input, but an extensive processing takes place within the spinal cord. The neural connection from the primary sensory neurons to the motor neurons is a polysynaptic pathway, and other afferent input, descending activity, and the excitability of the neurons in this pathway modulate the generation of the spinal nociceptive reflex. The reflex threshold shows substantial inter- and intraindividual variation, and hence, the method has exclusively been used in experimental studies. Under some conditions, a nociceptive stimulus may also cause an inhibitory reflex (e.g., silent period, exterosuppression period) of a contracting muscle (e.g., the jaw muscle). The inhibitory reflexes are not nociceptive specific.

The evoked potentials to painful stimulation and in particular the late event related vertex potential have been widely used in pain research due to the relation between amplitude and pain intensity. The vertex potential is modulated by many cognitive factors such as attention, habituation, vigilance, and interstimulus interval and may, therefore, exhibit large inter- and intra-individual variation.

In more recent years, it has been possible, on the basis of multichannel (e.g., 64–128 channels) electroencephalographic or magnetoencephalographic recordings, to calculate the scalp topography of the evoked potentials elicited by painful stimuli. This has also provided the possibility of estimating the various electrical brain sources (dipoles) responsible for a given topographic pattern. Since the estimation procedures for source analysis are becoming more and more advanced, it may be a reliable technique in the future.

The vertex potential can be elicited by any abrupt change in a sensory modality – including pain perception. The vertex potential elicited by an unspecific stimulus such as electrical stimulation may not correlate with the pain intensity, whereas a specific nociceptive stimulus (Aδ-fiber activation) from a laser may generate potentials that are a result of traffic along the nociceptive pathways. As such the laser-evoked potential has gained some clinical impact. Both the electro- and magnetoencephalographic responses have been used. Quantifying the C-fiber-mediated laser-evoked potential (latency >1,000 ms) has also been attempted. However, due to strongly varying latencies, the results have rarely been reliable.

Since the brain potentials to painful stimuli mirror only the cortical electro- or magnetoencephalographic activity elicited by very brief painful stimuli, attempts have been made to use the spontaneous EEG for assessing the cerebral activity accompanying experimental pain over long periods of time. Some of the spectral components of the EEG have shown a causal relationship to the presence of pain and to the intensity of pain. Furthermore, some of these spectral changes occur over specific cortical areas with relation to the somatic structure stimulated. Studies on EEG and ongoing clinical pain have not yet revealed better understanding of the cortical structures involved in the processing of pain.

To disclose aspects of cortical processing of nociceptive information, a change in the regional blood flow can be detected and represented by various brain imaging techniques. Positron emission tomography (PET) and single photon emission tomography (SPECT) require the injection of water or glucose labeled by a radioisotope into the blood stream to the brain. The radioisotopes with a short half-life accumulate for a brief period of time in the active areas of brain and can there be localized by scanners sensitive to the transient increase in radioactivity. The temporal resolution and spatial resolution are not as good as with functional magnetic resonance imaging (fMRI, see below), but PET and SPECT allow in addition the assessment of the concentrations of ligands and receptors with relevance for pain transmission in the brain.

fMRI is based on the different magnetic properties of the tissues in the brain, which can be made obvious in very strong magnetic fields. Red blood cells loaded with oxygen (oxygenated hemoglobin) present with different magnetic properties from unloaded ones (desoxygenated hemoglobin). Active areas in the brain have higher levels of oxygenated hemoglobin. This blood-oxygenation-level-dependent signal allows the detection of active areas with good spatial and temporal resolution by magnetic field scanners. The brain imaging techniques have disclosed some of the many structures involved in pain processing including, for example, somatosensory cortices I and II, the thalamus, the anterior part of the gyrus cinguli, the cerebellum, and the basal ganglia.

Induction and Assessment of Hyperexcitability

Peripheral and/or central hyperexcitability (hyperalgesia, allodynia) can be induced experimentally. This approach can be used to study neuroplastic changes important for various pain syndromes (e.g., neuropathic pain).

Experimental hyperexcitability has mainly been studies following sensitization of the skin, but a few models exist for the induction of muscular and visceral sensitization.

The manifestations of cutaneous hyperexcitability are flare reaction (neurogenic inflammation) and primary and secondary hyperalgesia. The flare reaction can be assessed by laser Doppler flowmetry, reflectance spectroscopy and thermography, and by visual inspection (mark the area). Primary hyperalgesia occurs in the skin underlying the actual site of stimulation (e.g., topical application) and is mainly a result of peripheral sensitization of the nociceptors. Normally, thermal stimuli are used to quantify primary hyperalgesia. A pain threshold can drop from approximately 44 °C to 33 °C and responses to suprathreshold stimulation increase.

Secondary hyperalgesia can be separated into an area of brush-evoked hyperalgesia (dynamic hyperalgesia, allodynia) and a slightly larger area of pinprick (punctuate) hyperalgesia (static hyperalgesia). The secondary hyperalgesia is predominantly a central phenomenon reflecting neuroplastic changes in spinal cord neurons and ongoing pain seems essential to maintain the secondary manifestations. Allodynia is normally assessed by stroking the skin with a cotton swab in a standardized way, and the area where the sensation changes from touch to pain is marked. Punctuate hyperalgesia is normally determined by a nylon filament (e.g., von Frey hair, bending force of, e.g., 70 g). The stimuli are applied along, for example, 6 vectors toward the center of the injury in steps of a few millimeters. The points where the pinprick sensation changes into stronger pain are marked for every vector and the area estimated. Electrical stimulation can also be used, whereas the existence of secondary thermal hyperalgesia is still controversial.

Cutaneous hyperalgesia can be induced by chemical, thermal, and mechanical stimuli. Intradermal injection (e.g., 100 µg of capsaicin in a 20 µl volume) or topical (e.g., 1 % capsaicin, 1.5 g cream applied to 4 cm^2 for 40 min) application of capsaicin is the most commonly used model. Intradermal capsaicin elicits a severe stinging/burning pain lasting for approximately a minute leaving well-characterized areas of primary and secondary hyperalgesia, which lasts up to 24 h. Topical application is better tolerated, but the manifestation and duration of hyperalgesia are less as compared to i.d. application. Cutaneous hyperexcitability can also be induced by noxious cold (e.g., −28 °C for 10 s), heat (e.g., 47 °C

for 7 min), or strong, prolonged, or repeated mechanical pressure stimuli (e.g., 8 N repeated pinching for 2 min).

Few possibilities exist to induce experimental hyperexcitability from deep tissue. Muscle sensitization can be transiently induced by intramuscular injection of capsaicin or algogenic substances, for example, bradykinin, serotonin, substance P, and nerve growth factor. Visceral sensitization can be elicited by capsaicin, glycol, or acid. The manifestations of deep pain hyperexcitability are increased in response to pressure/distension (most likely peripheral sensitization) and enlarged referred pain areas to standardized muscle/visceral experimental stimuli. The enlargement of the referred area is a central effect and is also found in patients with chronic musculoskeletal or visceral pain.

References

Angst, M. S., Brose, W. G., & Dyck, J. B. (1999). The relationship between the visual analog pain intensity and pain relief scale changes during analgesic drug studies in chronic pain patients. *Anesthesiology, 91*, 34–41.

Arendt-Nielsen, L. (1994). Characteristics, detection and modulation of laser-evoked vertex potentials. *Acta Anaesthesiologica Scandinavica, 38*, 1–44.

Arendt-Nielsen, L. (1997). Induction and assessment of experimental pain from human skin, muscle and viscera. In T. S. Jensen, J. A. Turner, & Z. Wiesenfeld-Hallin (Eds.), *Proceedings of the 8th world congress on pain* (pp. 393–425). Seattle: IASP.

Arendt-Nielsen, L., & Bjerring, P. (1988). Sensory and pain threshold characteristics to laser stimuli. *Journal of Neurology, Neurosurgery, and Psychiatry, 51*, 35–42.

Arendt-Nielsen, L., & Svensson, P. (1997). Referred muscle pain: Basic and clinical findings. *The Clinical Journal of Pain, 17*, 11–19.

Arendt-Nielsen, L., Brennum, J., Sindrup, S., et al. (1994). Electrophysiological and psychophysical quantification of central temporal summation of the human nociceptive system. *European Journal of Applied Physiology, 68*, 266–273.

Backonja, M. M., & Galer, B. S. (1998). Pain assessment and evaluation of patients who have neuropathic pain. *Neurologic Clinics, 16*, 775–790.

Boivie, J., Hansson, P., & Lindblom, U. (1994). *Touch, temperature, and pain in health and disease: Mechanisms and assessment* (Progress in pain research and management, Vol. 3, p. 548). Seattle: IASP Press.

Bromm, B. (1984). *Pain measurement in man.* Amsterdam: Elsevier.

Bromm, B., & Desmedt, J. (1995). *Pain and the brain: From nociception to cognition* (Advances in pain research and therapy, Vol. 22). New York: Raven Press.

Chang, P. F., Arendt-Nielsen, L., Graven-Nielsen, T., et al. (2003). Psychophysical and EEG responses to repeated experimental muscle pain in humans: Pain intensity encodes EEG activity. *Brain Research Bulletin, 15*, 533–543.

Chapman, C. R., & Loeser, J. D. (1989). *Issues in pain measurement* (Advances in pain research and therapy, Vol. 12). New York: Raven Press.

Chapman, C. R., Casey, K. L., Dubner, R., et al. (1985). Pain measurement: An overview. *Pain, 22*, 1–31.

Collins, S. L., Moore, R. A., & McQuay, H. J. (1997). The visual analogue pain intensity scale: What is moderate pain in millimetres? *Pain, 72*(1–2), 95–97.

Edwards, C. L., Fillingim, R. B., & Keefe, F. (2001). Race, ethnicity and pain. *Pain, 94*, 133–137.

Finley, G. A., & McGrath, P. J. (1998). *Measurement of pain in infants and children* (Progress in pain research and management, Vol. 10, p. 224). Seattle: IASP.

Gebhart, G. F. (1995). *Visceral pain, progress in pain research and management* (Vol. 5). Seattle: IASP.

Gibson, S. J., & Helme, R. D. (2001). Age-related differences in pain perception and report. *Clinics in Geriatric Medicine, 17*, 433–456.

Gracely, R. H. (1994). Studies of pain in normal man. In P. D. Wall & R. Melzack (Eds.), *Textbook of pain* (pp. 315–336). London: Churchill Livingstone.

Gracely, R. H., Lota, L., Walter, D. J., et al. (1988). A multiple random staircase method of psychophysical pain assessment. *Pain, 32*, 55–63.

Graven-Nielsen, T., Arendt-Nielsen, L., Svensson, P., et al. (1997). Quantification of local and referred muscle pain in humans after sequential i.m. Injections of hypertonic saline. *Pain, 69*, 111–117.

Handwerker, H. O., & Kobal, G. (1993). Psychophysiology of experimentally induced pain. *Physiological Reviews, 73*, 639–671.

Kems, R. D., Turk, D. C., & Rudy, T. E. (1985). The West Haven-Yale multidimensional pain inventory (WHYMPI). *Pain, 23*, 345–356.

Lautenbacher, S., & Rollman, G. B. (1993). Sex differences in responsiveness to painful and non-painful stimuli are dependent upon the stimulation method. *Pain, 53*, 255–264.

Lindblom, U. (1994). Analysis of abnormal touch, pain, and temperature sensation in patients. In J. Boivie, P. Hansson, & U. Lindblom (Eds.), *Touch, temperature, and pain in health and disease: Mechanisms and assessments* (Progress in pain research and management, Vol. 3, pp. 63–84). Seattle: IASP.

Melzack, R. (1975). The McGill pain questionnaire: Major properties and scoring methods. *Pain, 1*, 277–299.

Melzack, R., & Katz, J. (1994). Pain measurement in persons in pain. In P. D. Wall & R. Melzack (Eds.),

Textbook of pain (pp. 337–356). London: Churchill Livingstone.

Ness, T. J., & Gebhart, G. F. (1990). Visceral pain: A review of experimental studies. *Pain, 41*, 167–234.

Opsommer, E., Weiss, T., Miltner, W. H., et al. (2001). Scalp topography of ultralate (C-fibres) evoked potentials following thulium YAG laser stimuli to tiny skin surface areas in humans. *Clinical Neurophysiology, 112*, 1868–1874.

Petersen-Felix, S., Arendt-Nielsen, L., Bak, P., et al. (1996). The effects of isoflurane on repeated nociceptive stimuli (central temporal summation). *Pain, 64*, 277–281.

Price, D. D. (1999). *Psychological mechanisms of pain and analgesia* (Progress in pain research and management, Vol. 15). Seattle: IASP.

Riley, J. L., III, Robinson, M. E., Wise, E. A., et al. (1998). Sex differences in the perception of noxious experimental stimuli: A meta-analysis. *Pain, 74*, 181–187.

Riley, J. L., III, Robinson, M. E., Wise, E. A., et al. (1999). A meta-analytic review of pain perception across the menstrual cycle. *Pain, 8*, 225–235.

Rollman, G. B. (1977). Signal detection theory measurement of pain: A review and critique. *Pain, 3*, 187–211.

Rollman, G. B., & Lautenbacher, S. (2001). Sex differences in musculoskeletal pain. *The Clinical Journal of Pain, 17*, 20–24.

Turk, D. C., & Melzack, R. (1992). *Handbook of pain assessment.* New York: Guilford Press.

Vecchiet, L., AlbeFessard, D., Lindblom, U., et al. (1993). *New trends in referred pain and hyperalgesia.* Amsterdam: Elsevier.

Willis, W. (1992). *Hyperalgesia and allodynia.* New York: Raven.

Woolf, C. J., Bennett, G. J., Doherty, M., et al. (1998). Towards a mechanism-based classification of pain? *Pain, 77*(3), 227–229.

Yarnitsky, D., Sprecher, E., Zaslansky, R., et al. (1995). Heat pain thresholds: Normative data and repeatability. *Pain, 60*, 329–332.

Evaluation of Pain-Related Thought

▶ Psychology of Pain, Assessment of Cognitive Variables

Evaluation of Permanent Impairment

▶ Rating Impairment Due to Pain in a Workers' Compensation System

Event-Related Potential

Definition

Event-related potential is a cortical response to stimulation consisting of changes in electrical and magnetic fields (the so-called late near-field potentials are generated in the cortex cerebri).

Cross-References

▶ Nociception in Nose and Oral Mucosa

Evidence-Based Treatments

Definition

Treatment effect evidenced by randomized, double-blind, controlled clinical trials.

Cross-References

▶ Antidepressants in Neuropathic Pain

Evoked Activity

Definition

Activity of neurons elicited by stimulation of its receptive field.

Cross-References

▶ Molecular Contributions to the Mechanism of Central Pain

Evoked and Movement-Related Neuropathic Pain

Lois J. Kehl
Department of Anesthesiology, University of Minnesota, Minneapolis, MN, USA

Synonyms

Breakthrough pain; Episodic pain; Incident pain; Pain flares

Definition

Evoked and movement-related neuropathic pains are intermittent episodes of pain that occur in response to nerve injury, persist after the acute injury, and are elicited by external stimuli or movement. The term ▶ breakthrough pain is often used to refer to evoked and ▶ movement-related pain. Breakthrough pain has been defined as "Intermittent exacerbations of pain that can occur spontaneously or in relation to specific activity; pain that increases above the level of pain addressed by the ongoing analgesic; includes incident pain and end-of-dose failure" (American Pain Society 2002). Note that different definitions of breakthrough pain may be used in different countries, with some defining breakthrough pain as occurring only after background pain is controlled (Zeppetella et al. 2001). Breakthrough pain occurring as the result of normal voluntary or involuntary movement is referred to as ▶ incident pain. Unfortunately, much of the literature does not differentiate between incident pain and breakthrough pain. To address this problem, a Breakthrough Pain Questionnaire has been developed (Portenoy and Frager 1999). This instrument is not yet validated.

Characteristics

Our understanding of mechanisms contributing to neuropathic pain has grown tremendously since the development of appropriate animal models. This increased knowledge is reflected in advances made in clinical management of this pain condition since the conception of these models. Unfortunately, one important aspect of neuropathic pain that is still poorly understood and evades effective management is pain associated with movement. Movement-evoked pain is a significant clinical problem because of its high incidence, the distress it causes affected individuals, and its resistance to currently available treatments. Lack of adequate pain control with movement can interfere with normal function and rehabilitation efforts following trauma, surgery, or malignant disease.

Much of the literature supports a multifactorial etiology for movement-related pain in clinical conditions that include a neuropathic pain component, such as cancer pain and ► low back pain. This discussion will highlight these two prevalent clinical problems because they are perhaps the best characterized with respect to mechanisms that mediate evoked and movement-related neuropathic pain.

Pathobiology Associated with Movement-Evoked Neuropathic Pain Disorders

Neuropathic Pain Associated with Bone Cancer
Clinical and preclinical research studies indicate that cancer pain shares clinical and neuropathological characteristics with neuropathic pain. These similarities are best characterized in studies using animal models of bone cancer pain. More complete descriptions of this work are included elsewhere in this volume. This entry will focus on what is known about incident pain during the phase of cancer-induced bone pain that most closely resembles neuropathic pain.

Von Frey fiber testing is commonly used in studies of neuropathic pain. This method directly assesses cutaneous sensitivity to innocuous and noxious tactile stimuli. The effect of nerve damage on movement-related hyperalgesia can only be inferred indirectly from these results. A few investigators have directly evaluated the effect of sensory nerve injury on movement-related hyperalgesia using observation of ambulatory behaviors, performance on the rotarod, and

measurement of forelimb grip force. The movement made during grip force measurement is thought to activate musculoskeletal nociceptors via the innocuous pressure placed on muscle, joint, and deep tissue afferents during movement (Kehl et al. 2000). These dependent measures model incident pain by approximating clinical situations in which bone cancer pain is aggravated by movement and weight bearing.

Characteristics consistent with neuropathic pain begin to manifest between 9 and 16 days following implantation of ► osteolytic fibrosarcoma cells into bone in one murine model of bone cancer pain (Cain et al. 2001). These include spontaneous activity in C-fibers near the tumor and a significant reduction in epidermal nerve fibers (Cain et al. 2001). In this model, the level of movement-related hyperalgesia, measured using grip force, more than doubled from 7 to 10 days following implantation of tumor cells (Wacnik et al. 2003). At this time point, systemically administered morphine attenuated tumor-induced movement-related hyperalgesia in a dose-dependent manner. Morphine was 3.5 times less potent (ED_{50} = 23.9 mg/kg (11.4–50.1)) for tumor-induced hyperalgesia than for carrageenan-induced hyperalgesia (ED_{50} = 6.9 mg/kg (4.8–9.7)) measured concomitantly in the same study (Wacnik et al. 2003). Furthermore, morphine was less efficacious 10 days post-implantation than at 7 days post-implantation and when administered to carrageenan-injected mice. These results were mirrored in a parallel study evaluating the capacity of the nonselective ► cannabinoid agonist WIN55,212−2 (i.p.) to reverse movement-evoked hyperalgesia in the same two animal models (i.e., tumor and carrageenan). WIN55,212−2 was less efficacious and approximately four times less potent in reversing movement-evoked hyperalgesia following tumor implantation (ED_{50} = 23.3 mg/kg (13.6–40.0)) compared to i.m. carrageenan (ED_{50} = 5.6 mg/kg (3.4–9.6)) (Kehl et al. 2003). These findings parallel what is observed clinically with respect to the reduced potency and efficacy of opioids in managing movement-related cancer pain.

A second group of investigators reported changes consistent with neuropathic pain in a model of bone cancer pain involving implantation of mammary gland carcinoma cells (Donovan-Rodriguez et al. 2004). By the 11th day post-implantation, lamina I wide dynamic range neurons exhibited significant increases in electrically evoked C-fiber and post-discharge responses. Between days 15 and 17, Aβ-fiber-evoked responses also increased significantly. These spinal cord changes are similar to those seen in neuropathic pain models. The electrophysiological changes coincided with behavioral findings consistent with incident pain, including significant reductions in weight bearing ipsilateral to the tumor, general activity, and latency to fall on the rotarod test (Donovan-Rodriguez et al. 2004).

Although some characteristic elements of neuropathic pain exist in bone cancer pain models, some differences exist. For example, dorsal horn hyperexcitability (Urch et al. 2003) and spinal immunohistochemical profiles (Honore et al. 2000) seen in these models do not closely resemble changes typically seen in animal models of neuropathic pain. These findings suggest that cancer pain may be a unique pain state that shares some characteristics of neuropathic pain. Urch and colleagues propose that differences in the degree of C-fiber denervation between neuropathic and bone cancer pain models may help to explain these differences (Urch et al. 2003).

Neuropathic Pain Associated with Intervertebral Disc Herniation

For many years, mechanical compression of spinal nerve roots has been considered the primary cause for low back pain (LBP) and its associated neurological symptoms. Compression of spinal nerve roots or dorsal root ganglia (DRG), such as that which may occur secondary to intervertebral disc (IVD) herniation or swelling, produces spontaneous activity in the peripheral sensory nerve fibers of laboratory animals. This mechanical compression is thought to be responsible for the associated edema, ischemia, and demyelination that occur in DRG and nerve roots in these situations. Findings such as these provide support the view that mechanical compression leads to the pain and neurological symptoms associated with LBP.

In addition to dorsal nerve root and DRG compression, the density of pain-sensing nerve fibers increases in degenerated joint structures. For example, Freemont and colleagues reported that deep nerve growth into the inner third of lumbar IVDs (normally only the outer third of the IVD is innervated) was present in 57 % of patients with chronic LBP compared to 25 % of discs harvested post-mortem from subjects without a history of back pain (Freemont et al. 1997). Diseased IVD nerve fibers associated with blood vessels were immunopositive for substance P (SP) and growth-associated protein (GAP-43) (Freemont et al. 1997), a protein expressed in areas of axonal sprouting following nerve injury and remodeling of the nervous system. Nociceptive nerve ingrowth into painful IVDs was correlated with local production of NGF by blood vessels growing into IVDs from adjacent vertebral bodies (Freemont et al. 1997). TrkA receptors were expressed in the nerve terminals and the cytoplasm of chondrocytes in painful IVDs but not in controls (Freemont et al. 2002).

In addition, evidence also exists for immune system involvement in the development of LBP. Various cell types in herniated IVDs produce pro-nociceptive compounds such as cyclo-oxygenase-2, prostaglandin E2, proinflammatory cytokines, leukotrienes, nitric oxide, and matrix metalloproteinases. Indeed, it has been shown that application of ▶ nucleus pulposus or some of its constituents (i.e., tumor necrosis factor-α, interleukin-1β) to the epidural space produces increases in spinal nerve root excitability and sodium channel density as well as mechanical allodynia in the ipsilateral hind limb.

Clinical Presentation

Breakthrough Pain Associated with Cancer

At present, the phenomenon of movement-evoked pain is best characterized in studies of cancer pain. In this context, breakthrough pain is thought to occur secondarily to tumor progression, cancer treatments such as chemotherapy or radiation, or factors unrelated to the cancer.

Estimates for the prevalence of breakthrough pain in patients with malignancy range from 46 % to 93 % (Portenoy and Hagen 1990). Banning and colleagues (Banning et al. 1991) reported that of 184 cancer pain patients, 93 % had movement-evoked pain and 78 % had pain at rest. One to 2 weeks after analgesic regimens were begun, 63 % of patients still had movement-evoked pain. A study of hospice patients with cancer reported breakthrough pain in 89 % of subjects; 10 % were considered to be neuropathic in origin (Zeppetella et al. 2000). The number of daily breakthrough pain episodes of all types ranged from 1 to 14; 38 % of all breakthrough pain episodes were reported to be severe or excruciating. ▶ Rescue analgesic agents were not prescribed to more than 40 % of patients on long-acting opioids, even though these patients reported frequent episodes of breakthrough pain. These findings provide evidence that breakthrough pain is a substantial problem for most cancer pain patients and that it remains a major problem for half of those who are undergoing an established analgesic regimen.

Typically, neuropathic breakthrough pains associated with malignancy are for the most part brief (91 % have ≤30 min duration) when compared to breakthrough pain of somatic or visceral origin (69 % and 62 %, respectively, have ≤30 min duration) (Zeppetella et al. 2000). The implications of this pattern for successful treatment are addressed in the next section.

Neuropathic Breakthrough Pain Associated with Nonmalignant Pathology

Although many studies have investigated the clinical and pathobiological characteristics of neuropathic pain, the breakthrough and incident pain component of this type of pain is much less well studied in patients with nonmalignant pain. However, some information has been reported. In one such study, breakthrough pain was characterized in a group of hospice patients with terminal nonmalignant disease (Zeppetella et al. 2001). Sixty-three percent of patients evaluated reported breakthrough pain. Of these, 25 % were classified as neuropathic, 46 % as somatic, 14 % as visceral, and 15 % were of mixed etiology. Note that a larger percentage of nonmalignant pain patients were classified with neuropathic breakthrough pain (i.e., 25 % vs. 10 %, respectively). These

subjects (nonmalignant pain patients) also reported a larger percentage of severe or excruciating breakthrough pain (60 % vs. 38 %, respectively) compared to the previous study of malignant pain. Both groups reported a similar number (1–13/day) and duration of daily breakthrough pain episodes (75 % were ≤30 min duration).

Treatment

Breakthrough pain is typically managed using oral or parenteral supplemental rescue medications that are administered in addition to regularly scheduled analgesics. As indicated in the previous section, breakthrough pains of neuropathic origin are typically brief but occur frequently. Consequently, effective management requires that rescue analgesics should be quickly absorbed and have a rapid onset of action. Because opioids (i.e., morphine, hydromorphone, oxycodone) used conventionally in the past for breakthrough pain are not very lipophilic, their slower absorption reduces their effectiveness for managing breakthrough pain. As a result, short-acting oral immediate release or transmucosal opioids on an as needed basis are becoming more widely used. Intranasal morphine formulations are also under evaluation.

The management of neuropathic breakthrough pain associated with cancer presents a somewhat difficult situation in that this type of pain does not respond well to opioids. Furthermore, in most patients neuropathic breakthrough pains last 30 min or less, but short-acting oral morphine preparations may require up to an hour for onset and then last for 4 h. Consequently, adjuvant analgesics used to treat nonmalignant neuropathic pain, such as anticonvulsants and antidepressants, may more effectively manage this component of breakthrough cancer pain (Zeppetella et al. 2000).

References

American Pain Society. (2002). *Guideline for the management of pain in osteoarthritis, rheumatoid arthritis and juvenile chronic arthritis.* Glenview, Ill: American Pain Society.

Banning, A., Sjogren, P., & Henriksen, H. (1991). Treatment outcome in a multidisciplinary cancer pain clinic. *Pain, 47*(129–134), 127–128.

Cain, D. M., Wacnik, P. W., Turner, M., et al. (2001). Functional interactions between tumor and peripheral nerve: Changes in excitability and morphology of primary afferent fibers in a murine model of cancer pain. *Journal of Neuroscience, 21*, 9367–9376.

Donovan-Rodriguez, T., Dickenson, A. H., & Urch, C. E. (2004). Superficial dorsal horn neuronal responses and the emergence of behavioural hyperalgesia in a rat model of cancer-induced bone pain. *Neuroscience Letters, 360*, 29–32.

Freemont, A. J., Peacock, T. E., Goupille, P., et al. (1997). Nerve ingrowth into diseased intervertebral disc in chronic back pain. *Lancet, 350*, 178–181.

Freemont, A. J., Watkins, A., Le Maitre, C., et al. (2002). Nerve growth factor expression and innervation of the painful intervertebral disc. *The Journal of Pathology, 197*, 286–292.

Honore, P., Rogers, S. D., Schwei, M. J., et al. (2000). Murine models of inflammatory, neuropathic and cancer pain each generates a unique set of neurochemical changes in the spinal cord and sensory neurons. *Neuroscience, 98*, 585–598.

Kehl, L. J., Trempe, T. M., & Hargreaves, K. M. (2000). A new animal model for assessing mechanisms and management of muscle hyperalgesia. *Pain, 85*, 333–343.

Kehl, L. J., Hamamoto, D. T., Wacnik, P. W., et al. (2003). A cannabinoid agonist differentially attenuates deep tissue hyperalgesia in animal models of cancer and inflammatory muscle pain. *Pain, 103*, 175–186.

Portenoy, R. K., & Frager, G. (1999). Pain management: Pharmacological approaches. *Cancer Treatment and Research, 100*, 1–29.

Portenoy, R. K., & Hagen, N. A. (1990). Breakthrough pain: Definition, prevalence and characteristics. *Pain, 41*, 273–281.

Urch, C. E., Donovan-Rodriguez, T., & Dickenson, A. H. (2003). Alterations in dorsal horn neurones in a rat model of cancer-induced bone pain. *Pain, 106*, 347–356.

Wacnik, P. W., Kehl, L. J., Trempe, T. M., et al. (2003). Tumor implantation in mouse humerus evokes movement-related hyperalgesia exceeding that evoked by intramuscular carrageenan. *Pain, 101*, 175–186.

Zeppetella, G., O'Doherty, C. A., & Collins, S. (2000). Prevalence and characteristics of breakthrough pain in cancer patients admitted to a hospice. *Journal of Pain and Symptom Management, 20*, 87–92.

Zeppetella, G., O'Doherty, C. A., & Collins, S. (2001). Prevalence and characteristics of breakthrough pain in patients with nonmalignant terminal disease admitted to a hospice. *Palliative Medicine, 15*, 243–246.

Evoked Pain

Definition

Evoked pain is pain due to stimulation.

Cross-References

▶ Dysesthesia, Assessment

Evoked Pain and Morphine

▶ Opioids, Effects of Systemic Morphine on Evoked Pain

Evoked Potentials

Definition

Brain activity elicited by a sensory stimulation and extracted from the brain electroencephalographic background activity by averaging multiple responses during a window time-locked to the stimulus.

Cross-References

▶ Migraine Without Aura

Evolution of Pediatric Pain Treatment

Patrick McGrath
Departments of Psychology, Pediatrics and Psychiatry, Dalhousie University, Canada

Synonyms

Pediatric pain treatment, evolution; Undertreated pain in children; Unrecognized pain in children

Definition

Pain treatment in children has changed dramatically over the past 30 years. Pain management has evolved from being essentially nonexistent to being considered an essential aspect of humane medical care of children.

Characteristics

Prior to 1970, the few references in the literature to pain in children were anecdotal and reflected two prevailing biases (1) that children's pain as a symptom is not worthy of independent treatment because once the underlying illness is addressed, pain will dissipate naturally and (2) that children probably do not experience pain as intensely as adults because their nervous systems are immature and therefore they are even less worthy of treatment. Such attitudes are reflected in the work of Swafford and Allen who justify the negligible use of opioids postoperatively in their intensive care unit by stating, "pediatric patients seldom need medication for relief of pain. They tolerate discomfort well" (Swafford and Allen 1968).

Beginning in the 1970s and continuing on through the mid-1980s, a series of papers emerged which specifically examined pain treatment in children. Eland's seminal study in 1974 (Eland 1974) revealed that only half of the postoperative children on her unit received any analgesia during their hospital stay and that adults were 25× more likely to receive analgesia than children. An outpouring of papers subsequently replicated and expanded on her findings. In general, this research identified the under treatment of pain in essentially all domains of pediatric care (postoperative, procedure related, newborn, acute illness) and revealed significant differences in the way that children and adults with similar problems were addressed (Miser et al. 1987; Purcell-Jones et al. 1988; Schechter et al. 1986). Children's postoperative pain was rarely treated, and if treated, the pain control drugs and doses used were often inadequate. Sedation to lessen distress during painful procedures was rarely used even in children with cancer who required repeated noxious procedures such as bone marrow aspirations. When sedation was considered, the agents used often were inappropriate (such as benzodiazepines or chloral hydrate, neither of which are analgesic). Pain in newborns likewise was essentially ignored even in neonatal intensive care units where children would be subjected to almost continuous discomfort secondary to repeated procedures (chest tubes, blood sampling) as well as mechanical ventilation.

Changing Clinical Practice

Anand and colleagues (1987) in the mid-1980s, documented that premature infants undergoing ductal ligation surgery with minimal or no anesthesia (standard practice at that time) exhibited profound surgical stress responses and, in fact, experienced a much higher morbidity and mortality rate than infants who received anesthesia. This finding prompted a series of editorials in major medical journals demanding a change in surgical practice (Berry and Gregory 1987; Fletcher 1987), leading to a perceptible change in the attitude and practice of clinicians in the early 1990s. New research on how to treat pain in infants and children began to seep into textbooks and post-graduate education sessions. Recent studies suggest that, although far from perfect, we have come a very long way (Broome et al. 1996; Tesler et al. 1994). Postoperative pain management is much improved with the advent of better assessment techniques and the development of new tools such as patient controlled analgesia and regional anesthetic techniques. Sedation is now standard for painful procedures and although not uniform, it has been well studied and more reasonable choices are available to clinicians. Pain management in the newborn and premature has improved dramatically with sedation for ventilation, postoperative analgesia, and local anesthetic usage being the standard of care.

Historical Reasons for Undertreatment

It seems logical to ask how it could have occurred that children's pain was so uniformly ignored given the strong parental instinct to protect children from suffering coupled with the fact that pediatric providers tend to be caring compassionate individuals. Several factors coalesced to dampen interest in this area and limit the research needed to generate new knowledge. Without the ability to counter the prevailing mythology with knowledge, misinformation persisted which further undermined the sense of urgency to address pain and allowed the status quo to persist. It is helpful to examine some of the historical reasons for undertreatment so that we can gain additional insight into overcoming those barriers that continue to prevent adequate pain management in children.

The Multifactorial Nature of Pain

Children's pain is a complex phenomenon with neurological, psychological, social, and developmental contributions. Because pain is so ubiquitous, associated with almost every disease process and its nature is so complex, no one discipline in the past has had "ownership" of the symptom of pain. Pain was perceived as a fellow traveler with disease and the elimination of disease was the focus, not the elimination of pain. As a result, investigation of pain as a symptom was limited and narrow in scope. It was not until the development of pediatric multidisciplinary pain services in the 1990s, which were capable of addressing the wide-ranging factors that both amplified and muffled it, that the symptom of pain could be adequately addressed.

Difficulty with Pain Assessment in Infants and Children

As discussed in other entries on pain assessment, adequate assessment is the cornerstone of pain treatment and the individual's self-report of his or her discomfort is the foundation of assessment (Finley and McGrath 1998). However, clinicians often assumed that children could not rate their pain in a valid manner so that they failed to document children's pain or verify that a treatment had lessened their pain. Over the past 15 years, new pain measures have been developed for infants, toddlers, children, and adolescents, as well as for neurologically handicapped individuals. This not only allows for dramatically improved clinical care but also serves as a critical component in the further understanding of pain and its treatment.

Societal Attitudes About Children's Pain

Several sociocultural attitudes impact on the vigor with which pain in children is treated. In eras where disease or circumstance has devalued individual life, concerns about quality-of-life issues like pain management are less likely to be emphasized (McGrath and Unruh 1987). Until recently, the high incidence of child mortality limited the importance placed on child comfort. There was little interest in pain control when survival was at stake. This has changed dramatically in developed countries in the twentieth century. Similarly with the enormous improvement in neonatal and pediatric cancer survival over the past 20 years, quality of life concerns are now perceived as legitimate areas for research investigation.

Broader societal attitudes toward pain also affect the expression and response to pain (Bernstein and Pachter 2003). In some cultures, the ability to endure pain is both valued and promoted as character building while in others, pain expression is reinforced. It is clear that if pain is perceived as growth promoting, there will be little enthusiasm for vigorously treating it. These societal attitudes influence the intellectual and financial investment that a society is willing to make in pain relief.

Disciplinary Biases About Children's Pain

In addition to cultural and societal values, there are distinct differences in the way that varying disciplines within medicine understand and treat pain. There are multiple cultures within the medical and health-care community, each of which may have unique attitudes regarding pain and its implications. It is quite clear in surveys of physician attitudes toward pain in children, that there is now increasing uniformity in the belief that children are capable of experiencing pain at birth and that pain should be aggressively treated regardless of its origin. Such values represent a major shift from attitudes held merely 20 years ago.

Lack of Interest by the Pharmaceutical Industry

In the past, there was essentially no research by the pharmaceutical industry aimed at developing drugs that reduce pain in children. Almost no analgesic drugs had formally approved pediatric indications. There are many reasons for this lack of interest. Although mild to moderate pain is a common occurrence in children, severe pain necessitating pharmacological intervention is far less frequent in children than in adults. Thus, the potential market for these agents was quite limited. As a result, there were few financial incentives to entice the pharmaceutical industry to invest in research on pediatric pain. Ethical constraints on pediatric pain research coupled with the complexities of pain assessment further reduced whatever enthusiasm still persisted. As a result, even now, most drugs used to treat pain in children have not been fully researched and their dose and indications are often extrapolated from adult literature. This is most unfortunate because not only do children have different pharmacokinetics from adults, they often have different types of pain problems. In some countries, various legislative initiatives are underway to remedy this situation by providing additional incentives for the pharmaceutical industry to study children in addition to adults when developing new drugs.

Lack of Recognition of Long-Term Effects of Persistent Pain

Although it has long been perceived that pain treatment is necessary for humanitarian reasons, it has only recently been recognized that there are important physiological and psychological consequences that may stem from inadequately treating pain (see ▶ Long-Term Effects of Pain in Infants). The biological alteration of the nervous system due to repeated or persistent pain may sometimes be ameliorated through the use of adequate anesthesia and analgesia, thus advancing the case for aggressive pain control.

Research has also suggested that there are significant psychological ramifications of inadequately treated pain. Children who have had bone marrow aspirations with inadequate analgesia report more pain on subsequent bone marrow aspirations even when adequate analgesia is used, than do children who have had appropriate analgesia (Weisman et al. 1998). The recognition that there may be long-term consequences to inadequately treated pain has clearly impacted practice. Most neonatal units now aggressively treat pain. The American Academy of Pediatrics has recently changed its policy on circumcisions based on this data and now endorses analgesia/anesthesia for them. Sedation is now viewed as essential during painful procedures such as bone marrow aspirations.

Lack of Systemic Institutional Approach

Because pain is a symptom that is associated with almost every medical or surgical condition, most physicians perceive they have developed some expertise in pain management. In the past, their focus however, was typically on alleviation of the underlying condition producing the pain and they brought far less sophistication to the treatment of the symptom itself. In addition, most of the research on pain management is not published in journals typically perused by practicing clinicians but in pain-oriented journals. Unfortunately, as a result, many clinicians have not been aware of the broad advances that have occurred in pain management and these advances are not brought to the bedside.

The emergence of multidisciplinary pediatric pain services has significantly improved this situation (Berde and Solodiuk 2003). Pain services not only provide care for the small subset of children who are referred to them but they create an institutional environment in which pain is more likely to be considered and addressed in all patients. Institutions with pain services are more likely to offer educational programs in pain management and to insist on better documentation of pain ratings. Pain services often develop treatment protocols that will create a far more uniform approach to pain control in the institution and make it less subject to the variations of knowledge of each clinician.

Summary

Pain management in children has been a long ignored dimension of the medical care of children. Until the mid-1980s, there was essentially no research and concomitantly no information on the management of pain in children. Children received minimal analgesia postoperatively and were subjected to medical diagnostic and treatment procedures without sedation using only physical restraint. Pain control was rarely considered in the care of fragile newborns, who were frequently subjected to noxious treatments.

Pain management for children has improved dramatically in the past 20 years. Multidisciplinary pediatric pain services, the gradual development of uniform approaches to pain control for all children and an increased awareness of the problem of undertreatment of children have created a cultural shift in clinical care. The development of better pain assessment techniques has allowed for improved recognition and monitoring. Advances in understanding the developmental biology of pain transmission have emphasized the importance of the adequate treatment of pain to prevent long-term negative consequences, as well as providing additional clinical insights. Research has documented the importance of psychological and physical as well as pharmacological strategies for pain relief. We now recognize that treating pain in children is not only humane but also medically prudent and, for the overwhelming majority of problems, we have the tools available to accomplish this important task.

References

Anand, K. J. S., Sippell, W. G., & Aynsley-Green, A. (1987). Randomized trial of fentanyl anesthesia in preterm babies undergoing surgery: Effects on stress response. *Lancet, 1*, 243–248.

Berde, C. B., & Solodiuk, J. (2003). Multidisciplinary programs for the management of acute and chronic pain. In N. L. Schechter, C. B. Berde, & M. Yaster (Eds.), *Pain in infants, children, and adolescents* (2nd ed., pp. 471–489). Philadelphia: Lippincott Williams and Wilkins.

Bernstein, B., & Pachter, L. (2003). Cultural consideration in children's pain. In N. L. Schechter, C. B. Berde, & M. Yaster (Eds.), *Pain in infants, children, and adolescents* (2nd ed., pp. 142–157). Philadelphia: Lippincott Williams and Wilkins.

Berry, F. A., & Gregory, G. A. (1987). Do premature infants require anesthesia for surgery? *Anesthesiology, 67*, 291–293.

Broome, M. E., Richtsmeier, A., Maikler, V., et al. (1996). Pediatric pain practices: A national survey of health professionals. *Journal of Pain and Symptom Management, 11*, 312–320.

Eland, J. M. (1974). *Children's communication of pain (thesis)*. Iowa City: University of Iowa.

Finley, G. A., & McGrath, P. J. (1998). *Measurement of pain in infants and children. Progress in pain research and measurement*. Seattle: IASP Press.

Fletcher, A. B. (1987). Pain in the neonate. *New England Journal of Medicine, 317*, 1347–1348.

McGrath, P. J., & Unruh, A. (1987). *Pain in infants and children*. Amsterdam: Elsevier.

Miser, A. W., Dothage, J. A., Wesley, R. A., et al. (1987). The prevalence of pain in pediatric and young adult cancer population. *Pain, 29*, 73–83.

Purcell-Jones, G., Dorman, F., & Sumner, E. (1988). Paediatric anaesthetists perception of neonatal pain and infant pain. *Pain, 33*, 181–187.

Schechter, N. L., Allen, D. A., & Hanson, K. (1986). Status of pediatric pain control: A comparison of hospital analgesic use in adults and children. *Pediatrics, 77*, 11–15.

Swafford, L., & Allen, D. (1968). Relief in pediatric patients. *Medical Clinics of North America, 52*, 131–136.

Tesler, M. D., Wilkie, D. J., Holzemer, W. F., et al (1994). Postoperative analgesics for children and adolescents: Prescription and administration. *Journal of Pain and Symptom Management, 9*, 85–94.

Weisman, S. J., Bernstein, B., & Schechter, N. L. (1998). The consequences of inadequate analgesia during painful procedures in children. *Archives of Pediatrics and Adolescent Medicine, 152*, 147–149.

Exaggerated Pain

► Ethics of Pain-Related Disability Evaluations

Excitability

► Sensitization of Visceral Nociceptors

Excitatory Amino Acids

Definition

Amino acid neurotransmitters that depolarize neurons and mediate excitatory synaptic transmission. Potential candidates include gluta-mate, aspartate, cysteate, and homocysteate. Excitatory amino acids are used as transmitters for synapses in ascending and descending pain pathways.

Cross-References

▶ Descending Circuitry, Molecular Mechanisms of Activity-Dependent Plasticity
▶ Fibromyalgia, Mechanisms, and Treatment

Excitatory Postsynaptic Current

Synonyms

EPSC

Definition

Generates the excitatory postsynaptic potential, EPSP.

Cross-References

▶ Corticothalamic and Thalamocortical Interactions

Excitotoxic

Definition

Excitotoxic effects are those that initially excite and then destroy cells and tissues. For example, glutamate is an excitatory amino acid that in excessively high concentrations can cause neuro-nal death by its stimulatory effects.

Cross-References

▶ Dietary Variables in Neuropathic Pain
▶ Glutamate Homeostasis and Opioid Tolerance

Excitotoxic Lesion

Definition

An excitotoxic lesion is produced by infusing a chemical that overexcites the local neurons. This results in the preferential destruction of cells but not axons of passage.

Cross-References

▶ Lateral Thalamic Lesions, Pain Behavior in Animals

Excitotoxic Model

▶ Spinal Cord Injury, Excitotoxic Model

Excruciating Pain

Definition

Excruciating pain refers to extremely severe, atrocious, pain.

Cross-References

▶ Sunct Syndrome

Exercise

Susan Mercer
School of Biomedical Sciences, The University of Queensland, Brisbane, Australia

Synonyms

Aerobic exercise; Endurance exercise; Extension exercises; Flexion exercise; Physical therapy; Stabilizing exercise; Strengthening exercise

Definition

Exercise is an activity undertaken by an individual that is characterized by the deliberate use of muscles either to produce movement or to maximize muscle activity by resisting the movement that the muscles would produce. There are no explicit synonyms for exercise, but various adjectives may be used to define a particular form of exercise, according to the purpose of the exercise or the type of movement that it produces. Examples include aerobic exercise, strengthening exercise, endurance exercise, flexion exercise, extension exercises, and stabilizing exercise. Some other exercises are named after the individual who invented them.

Characteristics

Mechanism

The use of exercise in pain management is advocated for a variety of reasons. It is used to strengthen target muscles, increase joint mobility, improve motor control, stretch tight tissues, improve endurance, improve function, and improve general fitness. Notwithstanding the intrinsic merits of these objectives, in most cases the association between the biological rationale of exercise and the relief of pain is only conjectural.

The classical role of exercise has been to strengthen the target muscle. That this can be achieved is not in doubt, but doubts arise when exercise is prescribed to treat pain, for in that event, there is no known relationship between improvements in muscle strength and the mechanisms of pain or its relief. On the other hand, exercises might be used simply to encourage or increase mobility and function, despite pain. In that event, exercises are used to treat the effects of pain, not the pain itself.

A more sophisticated application of exercise requires training patients to coordinate their muscles differently, for example, to remember to co-contract their transversus abdominis and multifidus in order to enhance the stability of the lumbar spine (Richardson et al. 2000). With respect to rationale, the cardinal limitation to this model of intervention is that its physiological basis has not been elaborated. The supposed instability has not been defined nor has it been shown how the instability relates to the production of pain and how it should be identified clinically. The model hinges on the perplexing demonstration that in certain patients with chronic low back pain, the onset of activation of transversus abdominis is delayed during the performance of, for example, upper limb tasks (Hodges and Richardson 1996). This delay has loosely been taken to imply impaired stability of the lumbar spine through lack of action of the transversus abdominis on the thoracolumbar fascia (Hodges and Richardson 1996). What the model ignores is that even at maximum contraction, the transversus abdominis contributes barely more than 5 Nm to the moments acting on the lumbar spine (Macintosh et al. 1987), which is no more than 2 % of the moment required during lifting. The model, therefore, rests on the significance of a delayed onset of a less than trivial influence on the lumbar spine. In essence, this is a model with an ambiguous physiological basis. Nevertheless, it has attracted adherents and has been tested clinically.

Applications

In pain medicine, exercises have not been used prominently in a therapeutic sense for conditions such as cancer pain, neuropathic pain, headache, or postoperative pain. They have principally been used in the treatment of musculoskeletal pain and of complex regional pain syndromes.

Efficacy

Systematic reviews have provided data on the efficacy of exercises for several common, regional pain conditions. They provide mixed conclusions.

Acute Low Back Pain

A Cochrane review found that exercise therapy is not more effective than inactive treatments or other active treatments for acute low back pain (van Tulder et al. 2000). It also found that flexion and extension exercises are not effective.

Chronic Back Pain

Exercise therapy is more effective than usual care by a general practitioner and better than back school, but the evidence is conflicting on whether or not exercise is more effective than an inactive, sham treatment (Bogduk 2004). Strengthening exercises are not more effective than other types of exercises (Bogduk 2004). Specific stabilizing exercises have been shown to be more effective than usual care for patients with spondylosis or spondylolisthesis (O'Sullivan et al. 1997). They improve pain and function by a factor of 50 %, and the effects are enduring. It is not evident from this study, however, whether the therapeutic benefit is due to a biomechanically specific stabilizing effect on the lumbar spine or the intensity of intervention and attention during the 10-week treatment program is the operant factor.

Acute Neck Pain

Simple exercises designed to encourage and maintain mobility of the neck appear to be the single most effective intervention (Australian Acute Musculoskeletal Pain Guidelines Group 2003). Of all interventions that have been tested, exercises have produced the most enduring outcomes. Compared to manual therapy, exercise therapy is no less effective in the short term but achieves a significantly greater proportion of patients free of pain 2 years after treatment.

Chronic Neck Pain

Exercise is the only conservative therapy that has been shown to provide any benefit. Depending on the study, and depending on the type of exercises, reductions in pain range between 25 % and 75 % (Bogduk and McGuirk 2006). Strengthening exercises are no more effective than endurance exercises, but intensive exercises are more effective than light exercises, although not necessarily more effective than ordinary activity

(Bogduk and McGuirk 2006). Special stabilizing exercises are not more effective than home exercises and are barely more effective than heat treatment (Bogduk and McGuirk 2006).

Shoulder Pain

The evidence is limited and confounded by difficulties concerning accurate diagnosis of various conditions. For so-called rotator cuff disease, exercise seems to be of benefit in both the short and longer terms (Green et al. 2002). For other entities, evidence is lacking. For chronic impingement syndrome, exercises are superior to placebo therapy and are as effective as arthroscopic surgery, both in the short term (Brox et al. 1993) and at follow-up 2.5 years later (Brox et al. 1999). However, about one in 4 patients treated by exercise ultimately turns to surgery (Brox et al. 1999).

Anterior Knee Pain

Alone, or in combination with other interventions, exercises are the only intervention for anterior knee pain for which there is evidence of efficacy (Crossley et al. 2001). Eccentric quadriceps exercises are more effective, particularly in relation to functional outcomes, than standard quadriceps strengthening exercises.

Complex Regional Pain Syndrome

In conjunction with other interventions, such as analgesics and therapeutic blocks, exercises are recommended and used to mobilize the affected part in complex regional pain syndrome (Harden 2000; Stanton-Hicks et al. 1998). This prescription serves to prevent or reduce atrophy and contractures and thereby preserve the affected tissues. However, the benefit achieved has not been demonstrated in controlled studies, and any efficacy of exercises in relieving the pain of complex regional pain syndrome has not been established.

Side Effects

A particular virtue of exercises is that they are conspicuously free of any deleterious side effects.

Cross-References

▶ Training by Quotas

References

Australian Acute Musculoskeletal Pain Guidelines Group. (2003). Evidence-based management of acute musculoskeletal pain. Australian Academic Press, Brisbane Online. Available at http://www.nhmrc.gov.au

Bogduk, N. (2004). Management of chronic low back pain. *The Medical Journal of Australia, 180*, 79–83.

Bogduk, N., McGuirk, B. (2006). Medical management of acute and chronic neck pain. An evidence-based approach. Amsterdam: Elsevier (in press).

Brox, I., Staff, P. H., Ljunggren, A. E., et al. (1993). Arthroscopic surgery compared with supervised exercises in patients with rotator cuff disease (stage II impingement syndrome). *BMJ, 307*, 899–903.

Brox, J. I., Gjengedal, E., Uppheim, G., et al. (1999). Arthroscopic surgery versus supervised exercises in patients with rotator cuff disease (stage II impingement syndrome). *Journal of Shoulder and Elbow Surgery, 8*, 102–111.

Crossley, K., Bennell, K., Green, S., et al. (2001). A systematic review of physical interventions for patellofemoral pain syndrome. *Clinical Journal of Sport Medicine, 11*, 103–110.

Green, S., Buchbinder, R., Glazier, R., et al. (2002). Interventions for shoulder pain. *The Cochrane Library, (2)*. Update Software, Oxford.

Harden, R. N. (2000). A clinical approach to complex regional pain syndrome. *The Clinical Journal of Pain, 16*, 26–32.

Hodges, P. W., & Richardson, C. (1996). Inefficient muscular stabilization of the lumbar spine associated with low back pain. A motor control evaluation of transversus abdominis. *Spine, 21*, 2650–2650.

Macintosh, J. E., Bogduk, N., & Gracovetsky, S. (1987). The biomechanics of the thoracolumbar fascia. *Clinical biomechanics, 2*, 78–83.

O'Sullivan, P. B., Twomey, L. T., & Allison, G. T. (1997). Evaluation of specific stabilizing exercise in the treatment of chronic low back pain with radiologic diagnosis of spondylolysis or spondylolisthesis. *Spine, 22*, 2959–2967.

Richardson, C. A., Jull, G. A., & Hides, J. A. (2000). A new clinical model of the muscle dysfunction linked to the disturbance of spinal stability: Implications for treatment of low back pain. In L. T. Twomey & J. R. Taylor (Eds.), *Physical therapy of the low back* (3rd ed., pp. 249–267). New York: Churchill Livingstone.

Stanton-Hicks, M., Baron, R., Boas, R., et al. (1998). Complex regional pain syndromes: Guidelines for therapy. *The Clinical Journal of Pain, 14*, 155–166.

van Tulder, M. M., Malmivaara, A., Esmail, R. et al. (2000). Exercise therapy for low back pain (Cochrane Review). *The Cochrane Library, (2)*. Update Software, Oxford.

Exercise-induced Muscle Soreness

▶ Delayed Onset Muscle Soreness

Exertional Activity

Definition

One of the primary physical activities (sitting, standing, walking, lifting, carrying, pushing, and pulling) defining a level of work. The Social Security definition is the same as that used by the United States Department of Labor to classify occupations by strength levels. Any job requirement that is not exertional (as defined above by the primary strength activities) is considered as nonexertional.

Cross-References

▶ Disability Evaluation in the Social Security Administration

Exertional Capability

Definition

Exertional capability is the ability to perform any of the primary strength activities, i.e., sitting, standing, walking, lifting, carrying, pushing, and pulling.

Cross-References

▶ Disability Evaluation in the Social Security Administration

Exertional Limitations and Restrictions

Definition

Limitations or restrictions that affect the capability to perform an exertional (primary strength) activity – sitting, standing, walking, lifting, carrying, pushing, and pulling.

Cross-References

▶ Disability Evaluation in the Social Security Administration

Existential Distress

Definition

Concerns regarding survival.

Cross-References

▶ Cancer Pain Management, Interface Between Cancer Pain Management and Palliative Care

Exocytosis

Synonyms

Extrusion

Definition

Exocytosis is the process of exporting material in vesicles that is used to release substances, such as hormones or neurotransmitters, from a cell.

Cross-References

▶ Diencephalic Mast Cells

Exogenous Muscle Pain

Thomas Graven-Nielsen
Laboratory for Musculoskeletal Pain and Motor Control, Center for Sensory-Motor Interaction, Aalborg University, Aalborg, Denmark

Synonyms

Experimental muscle pain

Definition

Exogenous muscle pain is muscle pain provoked by external interventions, for example, electrical stimulation of muscle afferents, injection of algesic substances, or strong mechanical pressure. In contrast, endogenous muscle pain is caused by natural stimuli, for example, by ischemia or by exercise.

Characteristics

Sensory manifestations of muscle pain are seen as a diffuse aching pain in the muscle, pain referred to distant somatic structures, and modifications in the superficial and deep sensitivity in the painful areas (Graven-Nielsen 2006; Graven-Nielsen and Arendt-Nielsen 2008). These manifestations are different from cutaneous pain, which is normally superficial and localized around the injury, with a burning and sharp quality. Kellgren (1938) was one of the pioneers to study experimentally the diffuse characteristics of exogenous muscle pain and the actual locations of ▶ referred pain to selective activation of specific muscle groups. The sensation of exogenous muscle pain is the result of activation of group III (Aδ-fiber) and group IV (C-fiber) polymodal muscle nociceptors (Mense 1993). The nociceptors can be sensitized by release of neuropeptides from the nerve endings. This may eventually lead to ▶ hyperalgesia and ▶ central sensitization of dorsal horn neurons manifested as prolonged neuronal discharges, increased responses to defined noxious stimuli, response to non-noxious stimuli, and expansion of the receptive field (Mense and Hoheisel 2008).

Electrical

Intramuscular electrical stimulation (Fig. 1) is used to assess the sensitivity of muscles (Vecchiet et al. 1988) and to study basic aspects of muscle and referred pain (Arendt-Nielsen et al. 1997). Intramuscular electrical stimulation is a tissue-specific (although receptor unspecific) and reliable model to study sensory manifestations of exogenous muscle pain, such as referred pain (Laursen et al. 1999) and ▶ temporal summation of muscle pain (Arendt-Nielsen et al. 1997) but is confounded by concurrent activated muscle twitches. Electrical stimulation offers a unique possibility to compare both muscle and cutaneous tissues with the same stimulus modality. Intraneural microstimulation of muscle nociceptive afferents causes a muscle pain sensation (area and intensity), which is dependent on the stimulation time (temporal summation) and the number of stimulated afferents (▶ spatial summation).

Mechanical

Mechanical painful stimulation (Fig. 1) can be achieved with pressure algometers. The most widely used technique is manual pressure algometry (Fischer 1998). It is important to recognize that pressure stimulates both the skin and muscle. Methodological concerns like short-term and long-term reproducibility, influence of pressure rates, muscle contraction levels, and examiner expectancy have all been carefully addressed.

An alternative to pressure algometry, with the inherent variability related to manual application, is computer-controlled pressure algometry, where the rate and peak pressure can be predefined and automatically controlled. This method allows estimation of the stimulus–response function between pressure and pain intensities. Pressure algometry assesses a relatively small volume of tissue. However, a larger volume can be assessed using the computer-controlled cuff-algometry technique. In short, the pain intensity related to inflation of

a tourniquet applied around an extremity is used to establish stimulus–response curves. After intramuscular injections of lidocaine, the stimulus–response curve between the tourniquet pressure and pain intensity was shifted right, indicating the ability to assess the sensitivity of muscles (Polianskis et al. 2002).

Chemical

Intramuscular injections of algesic substances have been used to induce exogenous muscle pain (Fig. 1) (for specific references, see Graven-Nielsen 2006). The experimental method that has been used extensively is an IM injection of hypertonic saline, as the quality of the exogenous muscle pain is comparable to acute clinical muscle pain with localized and referred pain. The work of Kellgren and Lewis in the late 1930s (1938) initiated the method of saline-induced muscle pain, and the safety of the technique is illustrated by there being no reports of side effects after more than 1,000 IM infusions. A major advantage of the hypertonic saline model is that a detailed description of sensory and motor effects can be obtained, as the pain lasts for minutes. Furthermore, the model is reliable for studying referred pain from musculoskeletal structures, due to the longer-lasting pain. A systematic evaluation of the relation between infusion parameters (infusion concentration, volume, rate, and tissue) and pain intensity, quality, and local and referred pain patterns has been described.

The sensitization of muscle nociceptors is the best-established peripheral mechanism for the subjective tenderness and pain during movement of a damaged muscle. The sensitized nociceptors not only have a lowered mechanical excitation threshold but also exhibit larger responses to noxious stimuli. If the muscle lesion is extensive, high amounts of endogenous algesic agents will be released, which could lead to direct excitation of nociceptors, resulting in spontaneous pain. In humans, this has been seen as decreased pressure pain thresholds after intramuscular injections of ▶ capsaicin (Witting et al. 2000). Intra-arterial injections of serotonin, bradykinin, and prostaglandin have been found to be effective in sensitizing animal nociceptors (Mense 1993). In humans, a decrease in the pressure pain threshold

Exogenous Muscle Pain, Fig. 1 Stimulus modalities for exogenous muscle pain (*top*). Induction of exogenous muscle pain in the tibialis anterior muscle by electrical and mechanical stimulation and infusion of an algesic substance (*bottom*)

a

chemical:
- hypertonic saline
- potassium
- acidic saline
- capsaicin
- bradykinin
- serotonin
- calcitonin gene-related peptide
- glutamate

mechanical:
- pressure
- tourniquet pressure

electrical:
- intramuscular
- intraneural

miscellaneous:
- heated isotonic saline
- focused ultrasound

muscle nociceptor

exogenous muscle pain

b

electrical
chemical

mechanical

after combined intramuscular injections of serotonin and bradykinin was found (Babenko et al. 1999). Moreover, intramuscular coinjection of serotonin and a serotonin receptor antagonist, granisetron, reduced the spontaneous pain evoked by injection of serotonin and prevented allodynia/hyperalgesia to mechanical pressure stimuli (Ernberg et al. 2000). Thus, peripheral serotonergic receptors could be involved in the regulation of musculoskeletal pain disorders.

Glutamate receptors (ionotropic and metabotropic) are other receptor types that are potentially involved in muscle hyperalgesia. Intramuscular injections of glutamate produce pain and muscle hyperalgesia to pressure stimuli in humans (Svensson et al. 2003). The glutamate-induced muscle pain is attenuated by an N-methyl-D-aspartate (NMDA) receptor antagonist (ketamine) when coinjected with glutamate in humans, or decrease the afferent activity recorded in animals (Cairns et al. 2003). This indicates that activation of peripheral NMDA receptors may contribute to exogenous muscle pain. Interestingly, injections of glutamate in women induced significantly more muscle pain than similar injections in men (Cairns et al. 2001). This might be explained by greater glutamate-evoked afferent fiber activity in female rats, compared to male rats (Cairns et al. 2001). The gender difference is particularly important in relation to the high dominance of women with musculoskeletal pain syndromes.

Miscellaneous

From animal studies, afferent recordings have shown that a subgroup of muscle nociceptors responds to thermal stimulation (Mense 1993). It was also found that injections of heated isotonic saline induced muscle pain in contrast to isotonic saline at room temperature (Graven-Nielsen et al. 2002). It may be the thermal-induced muscle pain that is involved in conditions of widespread muscle pain observed in fever conditions.

Focused ultrasound has been used to induce muscle, joint, and skin pain (Wright et al. 2002). The transducer for ultrasound stimulation is located externally, but as a result of focusing the beam, the energy can be applied maximally to the deeper tissues, thereby selectively activating nociceptors in deep structures. By appropriate adjustment, it is possible to focus the main ultrasound energy to muscles and other subcutaneous structures. Repeated focused ultrasound stimulation of muscle-induced pain, which was progressively increasing to a larger extent than skin stimulation, indicates that temporal summation of muscle pain is more pronounced than skin pain (Wright et al. 2002).

References

Arendt-Nielsen, L., Graven-Nielsen, T., Svensson, P., & Jensen, T. S. (1997). Temporal summation in muscles and referred pain areas: An experimental human study. *Muscle & Nerve, 20*, 1311–1313.

Babenko, V., Graven-Nielsen, T., Svensson, P., Drewes, A. M., Jensen, T. S., & Arendt-Nielsen, L. (1999). Experimental human muscle pain and muscular hyperalgesia induced by combinations of serotonin and bradykinin. *Pain, 82*, 1–8.

Cairns, B. E., Hu, J. W., Arendt-Nielsen, L., Sessle, B. J., & Svensson, P. (2001). Sex-related differences in human pain and rat afferent discharge evoked by injection of glutamate into the masseter muscle. *Journal of Neurophysiology, 86*, 782–791.

Cairns, B. E., Svensson, P., Wang, K., Hupfeld, S., Graven-Nielsen, T., Sessle, B. J., Berde, C. B., & Arendt-Nielsen, L. (2003). Activation of peripheral NMDA receptors contributes to human pain and rat afferent discharges evoked by injection of glutamate into the masseter muscle. *Journal of Neurophysiology, 90*, 2098–2105.

Ernberg, M., Lundeberg, T., & Kopp, S. (2000). Effect of propranolol and granisetron on experimentally induced pain and allodynia/hyperalgesia by intramuscular injection of serotonin into the human masseter muscle. *Pain, 84*, 339–346.

Fischer, A. A. (1998). *Muscle pain syndromes and fibromyalgia. Pressure algometry for quantification of diagnosis and treatment outcome.* New York: Haworth Medical Press.

Graven-Nielsen, T. (2006). Fundamentals of muscle pain, referred pain, and deep tissue hyperalgesia. *Scandinavian Journal of Rheumatology, 35*(Suppl. 122)(5):1–43.

Graven-Nielsen, T., & Arendt-Nielsen, L. (2008). Human models and clinical manifestations of musculoskeletal pain and pain-motor interactions. In T. Graven-Nielsen, L. Arendt-Nielsen, & S. Mense (Eds.), *Fundamentals of musculoskeletal pain* (pp. 155–187). Seattle: IASP Press.

Graven-Nielsen, T., Arendt-Nielsen, L., & Mense, S. (2002). Thermosensitivity of muscle: High-intensity thermal stimulation of muscle tissue induces muscle pain in humans. *The Journal of Physiology, 540*, 647–656.

Kellgren, J. H. (1938). Observations on referred pain arising from muscle. *Clinical Science, 3*, 175–190.

Laursen, R. J., Graven-Nielsen, T., Jensen, T. S., & Arendt-Nielsen, L. (1999). The effect of compression and regional anaesthetic block on referred pain intensity in humans. *Pain, 80*, 257–263.

Mense, S. (1993). Nociception from skeletal muscle in relation to clinical muscle pain. *Pain, 54*, 241–289.

Mense, S., & Hoheisel, U. (2008). Mechanisms of central nervous hyperexcitability due to activation of muscle nociceptors. In T. Graven-Nielsen, L. Arendt-Nielsen, & S. Mense (Eds.), *Fundamentals of musculoskeletal pain* (pp. 61–73). Seattle: IASP Press.

Polianskis, R., Graven-Nielsen, T., & Arendt-Nielsen, L. (2002). Pressure-pain function in desensitized and hypersensitized muscle and skin assessed by cuff algometry. *The Journal of Pain, 3*, 28–37.

Svensson, P., Cairns, B. E., Wang, K., Hu, J. W., Graven-Nielsen, T., Arendt-Nielsen, L., & Sessle, B. J. (2003). Glutamate-evoked pain and mechanical allodynia in the human masseter muscle. *Pain, 101*, 221–227.

Vecchiet, L., Galletti, R., Giamberardino, M. A., Dragani, L., & Marini, F. (1988). Modifications of cutaneous, subcutaneous and muscular sensory and pain thresholds after the induction of an experimental algogenic focus in the skeletal muscle. *The Clinical Journal of Pain, 4*, 55–59.

Witting, N., Svensson, P., Gottrup, H., Arendt-Nielsen, L., & Jensen, T. S. (2000). Intramuscular and intradermal injection of capsaicin: A comparison of local and referred pain. *Pain, 84*, 407–412.

Wright, A., Graven-Nielsen, T., Davies, I., & Arendt-Nielsen, L. (2002). Temporal summation of pain from skin, muscle and joint following nociceptive ultrasonic stimulation in humans. *Experimental Brain Research, 144*, 475–482.

Exon

Definition

An exon comprises of the protein-coding region in the DNA sequence of a gene. Hence, it reflects that part of a gene, whose sequence is present in the mature mRNA.

Cross-References

▶ NSAIDs, Pharmacogenetics

Expectancies

Definition

Expectancies are the perceived probability of occurrence of a particular outcome.

Cross-References

▶ Psychology of Pain, Assessment of Cognitive Variables

Expectation

Definition

Expectation is the anticipation of an event. According to expectation theories, expecting an outcome affects the outcome itself.

Cross-References

▶ Placebo Analgesia and Descending Opioid Modulation

Experimental Allergic Encephalitis

Synonyms

EAE

Definition

Experimental allergic encephalitis (EAE) is an autoimmune disease that is initiated by immunizing experimental animals with myelin proteins. EAE is used as a model for CNS demyelinating diseases such as multiple sclerosis.

Cross-References

▶ Demyelination

Experimental Diabetic Neuropathy

▶ Neuropathic Pain Model, Diabetic Neuropathy Model

Experimental Endpoint

Definition

Experimental endpoint is a biological effect used as an index of the effect of a chemical or other manipulation on an organism.

Cross-References

▶ Amygdala, Pain Processing and Behavior in Animals

Experimental Jaw Muscle Pain

Definition

Numerous techniques are available to evoke a deep, diffuse pain in jaw muscles of healthy volunteers, for example, by injection of algogenic substances such as hypertonic saline, excitatory amino acids, capsaicin, prostaglandins, and serotonin. Electrical and mechanical stimuli can also be used.

Cross-References

▶ Orofacial Pain, Movement Disorders

Experimental Muscle Pain

▶ Exogenous Muscle Pain

Experimental Pain

Definition

Stimuli that artificially replicate a painful condition in humans.

Cross-References

▶ Experimental Pain in Children
▶ Human Thalamic Response to Experimental Pain (Neuroimaging)

Experimental Pain in Children

Jennie C. I. Tsao, Lonnie K. Zeltzer and Laura C. Seidman
Pediatric Pain Program, Department of Pediatrics David Geffen School of Medicine at UCLA, Los Angeles, CA, USA

Synonyms

Children; Cold-pressor task; Experimental pain; Laboratory pain

Definition

Experimental pain involves the use of standardized tasks within a controlled environment to test specific hypotheses regarding pain responsivity in children that would be difficult to test outside the laboratory. These tasks are designed to induce safely mild to moderate pain sensations in a reliable fashion across research participants. The nature of the pain stimulation (e.g., cold, pressure) and the types of pain tasks vary and have typically been based on procedures developed and refined in adults. Although some have questioned the utility of experimental pain studies in children, it is increasingly recognized that such work can assist researchers in understanding individual differences in pain responses while posing minimal harm to children. In adults, experimental pain methods have often been used to investigate questions concerning clinical pain, to teach ▶ pain coping skills and to evaluate the effectiveness of ▶ psychosocial treatments for pain. Such applications have found increased popularity in studies with children. The reproducibility of experimental pain procedures allows the investigation of pain responses without confounding variables (e.g., variations in intensity and/or duration) inherent to clinical pain episodes and painful "real world" medical procedures.

Characteristics

The most commonly used experimental pain tasks in children are described below.

Cold-Pressor Task

The cold-pressor task (CPT) involves placing a hand or forearm in cold water, a stimulus that produces a slowly mounting pain of mild to moderate intensity terminated by voluntary withdrawal of the limb. Since the first reported use of the CPT with children in 1937, the procedure has been used in more than 24 published studies involving over 1,700 children without reported adverse effects (for review, see (von Baeyer et al. 2005)) The CPT has been used with both healthy children and children with a variety of clinical conditions including ▶ juvenile rheumatoid arthritis (JRA), ▶ recurrent abdominal pain and recurrent headache. Thus, it is the most widely used experimental pain task in studies of children. Because the CPT is a painful stimulus lasting up to 3 or 4 min (i.e., the typical ceiling at which participants are instructed to withdraw from the apparatus if they have not already done so voluntarily), it is likely to be most comparable with acute somatic clinical pains lasting from a few minutes to several hours (e.g., postoperative pain), rather than visceral or chronic pain (von Baeyer et al. 2005). The majority of cold-pressor studies have used a custom-made apparatus to maintain constant water temperature and to achieve a flow of water over the hand. More recently, investigators have used commercial devices such as the Techne TE-10D Thermoregulator, B-8 Bath, and RU-200 Dip Cooler (Techne, Burlington, NJ) (see Fig. 1). The bath unit is separated into two compartments by a metal screen – one side of the bath contains the refrigeration and circulating pump machinery, which keeps the water at the designated temperature. Holes in the metal screen allow water to flow between the two sections, preventing the development of a microenvironment of warm water around the child's hand. Children are instructed to place their hand on the other side of the bath unit with their palm facing up, toward the ceiling, and the back of their hand laying flat against the bottom of the cooler. This results in submersion of the entire hand and an additional few inches of the arm above the wrist. Although varied cold temperatures are used, von Baeyer and colleagues (2005) recommend a continuously circulating water bath at a temperature of $10 \pm 1\,°C$ for children.

Assessment of Pain Response

The CPT provides a context in which a variety of pediatric pain outcome measures may be considered, including pain threshold, pain tolerance, pain intensity ratings, and distress ratings, as well as stress hormonal and autonomic/cardiovascular responses. Some studies have also used observational measures of facial, verbal, and bodily expression of pain during the procedure. The CPT provides a useful measure of pain tolerance which is typically defined as the duration of immersion (in seconds) from the time the hand is placed in the water to the time it is voluntarily withdrawn. Alternatively, cold-pressor tolerance may be defined as the duration of immersion time following pain threshold (i.e., the point at which the sensation is considered by the child to be painful). However, younger children may be less reliable in reporting pain threshold (Miller et al. 1994), perhaps because they focus on the procedure and forget to indicate when they first feel pain. Hence, the former definition of tolerance is recommended for use with children when conducting the CPT. Self-report ratings of pain can be obtained concurrently with the immersion by verbal numerical rating scales (0–10), or with mechanical ▶ visual analog scales (VAS) or faces scales employing the non-immersed hand. Self-report pain scales should be explained to the child prior to the CPT in order to minimize distractions during the task. However, any rating procedure may entail reactive effects of measurement, that is, changes produced by the measurement procedure itself (von Baeyer 1994). These reactive effects may operate to decrease tolerance by drawing the participant's attention to pain (LeBaron et al. 1989; Zeltzer et al. 1989; Fanurik et al. 1993), or to increase tolerance because of the inherent distraction of interacting with the experimenter (Mikail et al. 1986). Retrospective ratings immediately upon limb withdrawal from the water can be obtained using any validated self-report method (Champion et al. 1998).

Practical Considerations

Although there have been no documented adverse physical or psychological effects arising from use of the CPT in children, it is possible that a rare participant may react to the pain with a stress response, for example, increased heart rate and, in extreme instances, fainting associated with a vasovagal response. As noted by von Baeyer et al. (2005), this risk can be eliminated by: (a) excluding children with a history of: cardiovascular disease, fainting or seizures, frostbite

Experimental Pain in Children, Fig. 1 Example of a cold-pressor task (CPT) apparatus.

COLD PRESSOR

APPARATUS

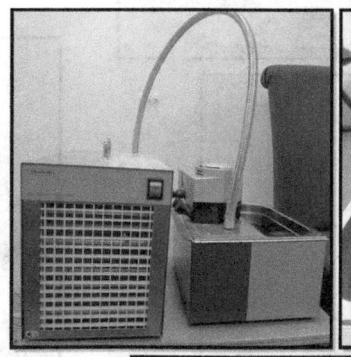

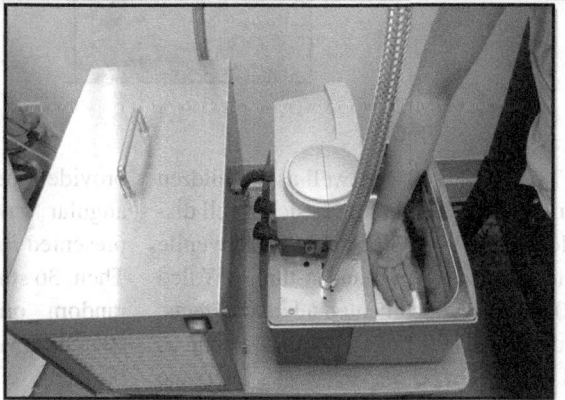

or ▶ Raynaud's syndrome and (b) giving children fruit juice to drink beforehand to ensure that they are hydrated before the experiment. This procedure can reduce the risk of vasovagal stress responses in children who have not had anything to eat or drink all day at the time of the experiment. Children should also be given time to habituate to the laboratory setting before starting the CPT. That time can be occupied with completing pretest measures.

Pressure Pain

A few studies have examined laboratory pressure-pain responsivity in children. The most commonly used device, the Forgione-Barber

focal pressure stimulator, in which a dull Lucite edge applies continuous pressure to the finger, produces pressure sensations that gradually build to a dull aching pain (Forgione and Barber 1971). The rate at which the sensations become painful may be varied by applying different weights to the apparatus, thus changing the intensity of the pressure. The stimulator is considered to be reliable and relatively unmodified by extraneous physiological events (e.g., vasomotor activity) (Forgione and Barber 1971). Pressure-pain reactivity to sphygmomanometer (blood pressure cuff) inflation has also been examined in children (Walco et al. 1990). These pressure-pain tasks have been administered to healthy

Experimental Pain in Children, Fig. 2 Evoked pressure apparatus

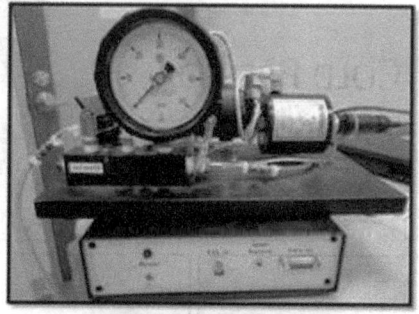

EVOKED PRESSURE

APPARATUS

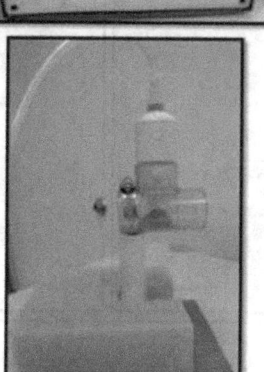

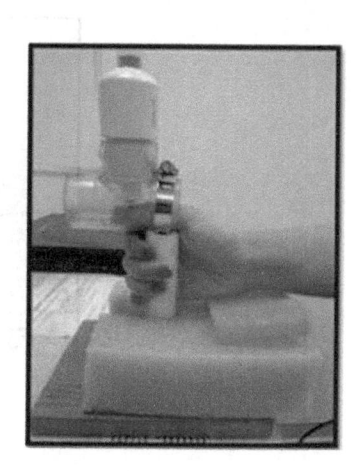

children (Tsao et al. 2004) as well as to children with chronic illnesses including ▶ sickle cell disease (Walco et al. 1990; Gil et al. 1997), juvenile rheumatoid arthritis (JRA), and asthma (Walco et al. 1990), without documented adverse effects. These studies have included children as young as 5 years of age (Walco et al. 1990). The tolerance tasks described above provide information regarding behavioral responses to different modalities of pain, essentially measuring the child's willingness to endure a painful stimulus. Judgments of discrete supra-threshold stimuli presented in a random (or seemingly random) order are considered optimal to assess pain sensitivity while minimizing the influence of expectancy effects and affective factors (Gracely et al. 2003; Petzke et al. 2005). Recently, investigators have begun testing a custom-made apparatus developed by Dr. Richard Gracely to assess children's sensitivity to evoked pressure (See Fig. 2). Discrete 5-s pressure stimuli are applied to the thumbnail of the dominant hand with a 1 × 1 cm hard rubber probe. The rubber probe is attached to a hydraulic piston, which is controlled by a computer-activated pump to provide repeatable pressure-pain stimuli of rectangular waveform. First, a series of stimuli are presented in a predictable, "ascending" manner. Then, 36 stimuli are delivered at 20-s intervals in random order, using the multiple random-staircase (MRS) pressure-pain sensitivity method (Gracely et al. 1988). The MRS method is response dependent, that is, it determines the stimulus intensity needed to elicit a specified response. Two thresholds are used: "moderate" pain (5–6/10 pain scale) and "mild pain" (2–3/10). A major advantage of the MRS technique is that a minimum of stimuli are delivered above the highest threshold. Such studies have been reliably conducted in adults and are currently being piloted in children.

Assessment of Pain Response

In the study by Walco and colleagues (1990), pressure applied by the Forgione-Barber stimulator was increased by approximately 250 g/cm^2 at 2-s intervals, although the initial amount of force applied was not specified. For the sphygmomanometer, the cuff was inflated rapidly to 40-mmHg and then increased in increments of

Experimental Pain in Children, Fig. 3 Modified Forgione-Barber pressure stimulator

FOCAL PRESSURE

STIMULATOR

10-mmHg/s. For both tasks, children were instructed to state when the sensations were first perceived as painful, thus providing a measure of pain threshold. Pain tolerance was not examined. Results indicated that threshold assessments for both tasks were reliable across trials. Gil and colleagues (1997) assessed pressure-pain responses to four stimulus intensities (force = 2.83–8.76) by asking children with sickle cell disease to rate each of 32 trials using a 10-point scale of verbal descriptors (1 = not noticeable to 10 = worst possible pain). Children were asked for pain ratings at 15 s or when the child did not want to tolerate the trial any longer and withdrew from the apparatus. Tolerance times were not reported. Tsao and colleagues (2004) using a modified Forgione-Barber device (see Fig. 3), exposed healthy children to four trials of pressure pain at two weight levels (322.5 g and 465 g). The trials were conducted with an uninformed ceiling of 3 min. Tolerance, defined as time in seconds elapsed from the onset of the pain stimulus to participants' withdrawal from the stimulus (M = 41.6 s, SD = 45.4), as well as pain intensity, rated on a 10-point VAS, were assessed for each trial.

Thermal Heat Pain and Vibration Pain

Quantitative sensory testing (QST), a noninvasive computer-based method, has been used to assess thermal sensations and vibration sensation. Although its use in children has been limited, QST has been used extensively in adults to screen for ▶ peripheral neuropathies with loss of sensation and increased cutaneous sensitivity and to monitor disease progression and responsiveness to therapy. In children, Hilz and colleagues have reported normative ranges for thermal and vibration perception using a commercially available apparatus (Somedic, Stockholm, Sweden) (Hilz et al. 1998a; Hilz et al. 1998b). Using a different device (Medoc Ltd. Advanced Medical Systems, Ramat Yishai, Israel), Meier and colleagues (2001) reported normative data on thermal and vibration detection thresholds in 101 healthy children aged 6–17 years. Recently, investigators conducted QST encompassing all somatosensory modalities using a TSA 2001 II (Medoc, Israel) in 176 children (aged 6–16 years) to establish reference values according to age, gender, and body site (Blankenburg et al. 2010). The findings from these investigators support the use of QST in documenting and monitoring the clinical course of sensory abnormalities in children with neurological disorders or neuropathic pain. These above studies examined threshold to detect sensations and did not assess pain tolerance or intensity. In their study of healthy children, Tsao et al. (2004), using a commercially available device (Ugo Basile Biological Research Apparatus #7360 Unit, Comerio, Italy) (see Fig. 4), administered four trials of infrared radiant heat 2" proximal to the wrist and 3" distal to the elbow

THERMAL PAIN
APPARATUS

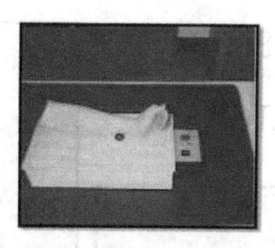

Experimental Pain in Children, Fig. 4 Device for
assessment of thermal heat pain reactivity

on both volar forearms, with an uninformed ceiling of 20 s. Thermal pain tolerance was electronically measured with an accuracy of 0.1 s
($M = 9.4$ s, $SD = 5.0$).

In summary, given their novelty, experimental
pain tasks have the advantage of being relatively
free from the influence of potentially
confounding factors such as anxiety engendered
by prior pain experiences (e.g., as with painful
medical procedures). Experimental pain methods
allow greater control over details of the stimulus
location, duration, and intensity than is possible
with clinical pain, thereby facilitating the testing
of specific hypotheses regarding individual difference variables and potential moderators of
pain reactivity such as age, sex, and ethnicity.

References

Blankenburg, M., Boekens, H., Hechler, T., Maier, C.,
    Krumova, E., Scherens, A., Magerl, W., Aksu, F., &
    Zernikow, B. (2010). Reference values for quantitative
    sensory testing in children and adolescents:
    developmental   and   gender   differences   of
    somatosensory perception. *Pain, 149*, 76–88.
Champion, G. D., Goodenough, B., von Baeyer, C. L., &
    Thomas, W. (1998). Measurement of pain by self-
    report. In G. A. Finley & P. J. McGrath (Eds.),
    Measurement of Pain in Infants and Children
    (pp. 123–160). Seattle, WA: IASP Press.
Fanurik, D., Zeltzer, L. K., Roberts, M. C., & Blount, R. L.
    (1993). The relationship between children's coping
    styles and psychological interventions for cold pressor
    pain. *Pain, 52*, 255–257.
Forgione, A. G., & Barber, T. X. (1971). A strain gauge
    pain stimulator. *Psychophysiology, 8*, 102–106.
Gil, K. M., Edens, J. L., Wilson, J. J., Raezer, L. B.,
    Kinney, T. R., Schultz, W. H., & Daeschner, C.
    (1997). Coping strategies and laboratory pain in
    children with sickle cell disease. *Annals of Behavioral
    Medicine, 19*.

Gracely, R. H., Grant, M. A., & Giesecke, T. (2003).
    Evoked pain measures in fibromyalgia. *Best Practice
    & Research. Clinical Rheumatology, 17*, 593–609.
Gracely, R. H., Lota, L., Walter, D. J., & Dubner, R.
    (1988). A multiple random staircase method of
    psychophysical pain assessment. *Pain, 32*, 55–63.
Hilz, M. J., Axelrod, F. B., Hermann, K., Haertl, U.,
    Duetsch, M., & Neundorfer, B. (1998). Normative
    values of vibratory perception in 530 children,
    juveniles and adults aged 3–79 years. *Journal of the
    Neurological Sciences, 159*, 219–225.
Hilz, M. J., Stemper, B., Schweibold, G., Neuner, I.,
    Grahmann, F., & Kolodny, E. H. (1998). Quantitative
    thermal perception testing in 225 children and juve-
    niles. *Journal of Clinical Neurophysiology, 15*, 529–
    534.
LeBaron, S., Zeltzer, L. K., & Fanurik, D. (1989). An
    investigation of cold pressor pain in children Part I.
    Pain, 37, 161–171.
Meier, P. M., Berde, C. B., DiCanzio, J., Zurakowski, D.,
    & Sethna, N. F. (2001). Quantitative assessment of
    cutaneous thermal and vibration sensation and thermal
    pain detection thresholds in healthy children and ado-
    lescents. *Muscle & Nerve, 24*, 1339–1345.
Mikail, S. F., Vandeursen, J., & von Baeyer, C. L. (1986).
    Rating pain or rating serenity: Effects on cold-pressor
    pain tolerance. *Canadian Journal of Behavioral Sci-
    ence, 18*, 126–132.
Miller, A., Barr, R. G., & Young, S. N. (1994). The cold
    pressor test in children: methodological aspects and
    the analgesic effect of intraoral sucrose. *Pain, 56*,
    175–183.
Petzke, F., Harris, R. E., Williams, D. A., Clauw, D. J., &
    Gracely, R. H. (2005). Differences in unpleasantness
    induced by experimental pressure pain between
    patients with fibromyalgia and healthy controls. *Euro-
    pean Journal of Pain, 9*, 325–335.
Tsao, J. C. I., Myers, C. D., Craske, M. G., Bursch, B.,
    Kim, S. C., & Zeltzer, L. K. (2004). Role of anticipa-
    tory anxiety and anxiety sensitivity in children's and
    adolescents' laboratory pain responses. *Journal of
    Pediatric Psychology, 29*, 379–388.
von Baeyer, C. L. (1994). Reactive effects of measure-
    ment of pain. *The Clinical Journal of Pain, 10*,
    18–21.
von Baeyer, C. L., Piira, T., Chambers, C. T., Trapanotto,
    M., & Zeltzer, L. K. (2005). Guidelines for the cold
    pressor task as an experimental pain stimulus for use
    with children. *The Journal of Pain, 6*, 218–227.
Walco, G. A., Dampier, C. D., Hartstein, G., Djordjevic,
    D., & Miller, L. (1990). The relationship between
    recurrent clinical pain and pain threshold in children.
    In D. C. Tyler & E. J. Krane (Eds.), *Advances in Pain
    Research Therapy* (pp. 333–340). New York: Raven
    Press, Ltd.
Zeltzer, L. K., Fanurik, D., & LeBaron, S. (1989).
    The cold pressor pain paradigm in children:
    feasibility of an intervention model Part II. *Pain, 37*,
    305–313.

Expert Patient

Definition

Patients with trigeminal neuralgia need to gather information about the condition in order to be able to make informed decisions about their management and improve their pain control.

Exposure In Vivo

Definition

Cognitive-behavioral treatment during which chronic pain patients are gradually exposed to fear-eliciting stimuli in order to disconfirm erroneous cognitions and to extinguish fear, thereby restoring activities and decreasing disability levels. Exposure in vivo treatment has proven to be effective in anxiety disorders and in chronic low back pain patients with fear of movement/(re)injury.

Cross-References

▶ Disability, Fear of Movement
▶ Fear and Pain
▶ Fear Reduction Through Exposure In Vivo
▶ Muscle Pain, Fear-Avoidance Model

Exposure Techniques for Reducing Activity Avoidance

▶ Behavioral Therapies to Reduce Disability

Exposure Treatment

▶ Fear Reduction Through Exposure In Vivo

Extension Exercises

▶ Exercise

Extensive First-Pass Metabolism

Definition

Morphine, a weak base (pKa – 8.0), undergoes extensive first-pass metabolism, and its oral bioavailability is 25–30 %.

Cross-References

▶ Postoperative Pain, Morphine

Exteroception

Definition

The term "exteroception" is antonymous to the term "interoception," and defines the sense for outside stimuli such as vision, hearing, taste, smell, and touch.

Cross-References

▶ Functional Imaging of Cutaneous Pain

Exteroceptive

Definition

See "exteroception."

Cross-References

▶ Trigeminal Brainstem Nuclear Complex, Anatomy

Exteroceptive Suppression

▶ Jaw-Muscle Silent Periods (Exteroceptive Suppression)

Extinction

▶ Fear Reduction Through Exposure In Vivo

Extinction and Extinction Training

Definition

Extinction is a process by which a conditioned emotional reaction is eliminated, when the conditioned stimuli are no longer paired with the stimuli that had previously elicited the occurrence of that behavior.

Cross-References

▶ Fear Reduction through Exposure in Vivo

Extracellular Signal-Regulated Protein Kinase

Synonyms

ERK

Definition

ERK is activated by an upstream kinase, MAPK (MEK), which produces not only short-term functional changes in the nervous system by post-translational modifications of target proteins but also long-term adaptive changes by increasing gene transcription. The activation of ERK in dorsal horn neurons by noxious stimulation or after peripheral inflammation has been associated with to pain hypersensitivity.

Cross-References

▶ ERK Regulation in Sensory Neurons during Inflammation
▶ Spinal Cord Nociception, Neurotrophins

Extradural Infusions

▶ Epidural Infusions in Acute Pain
▶ Postoperative Pain, Epidural Infusions

Extradural Space

▶ Epidural Space

Extralemniscal Myelotomy

▶ Midline Myelotomy

Extrasegmental Analgesia

Definition

Extrasegmental analgesia is the analgesic effect outside the stimulated segment.

Cross-References

▶ Transcutaneous Electrical Nerve Stimulation Outcomes

Extrasegmental Antinociceptive Mechanisms

Synonyms

Supraspinal antinociceptive mechanisms

Definition

Neural circuitry that originates above the spinal cord and when active prevents the onward transmission of noxious information en route to higher centers in the brain.

Cross-References

▶ Transcutaneous Electrical Nerve Stimulation (TENS) in Treatment of Muscle Pain

Extrusion

▶ Exocytosis

Eye Nociceptors

▶ Ocular Nociceptors

Eye Pain Receptors

▶ Ocular Nociceptors

F

F2 Intercross

Definition

The second-generation descendents of a cross of two contrasting populations (e.g., inbred strains). The offspring of the cross (i.e., F_1 hybrids) are sib-mated to produce F_2 hybrids, in which homologous recombination has "shuffled" the genomes of the progenitors in a unique manner. F_2 hybrids from inbred strain progenitors are useful for quantitative trait locus (QTL) mapping.

Cross-References

▶ Quantitative Trait Locus Mapping

Fabry's Disease

▶ Metabolic and Nutritional Neuropathies

Faces Pain Scale

Definition

Visual pain scale of seven faces now tested in older adults as well as children.

Cross-References

▶ Cancer Pain, Assessment in the Cognitively Impaired

Facet Denervation

▶ Facet Joint Procedures for Chronic Back Pain

Facet Joint

Definition

Facet is a flat, platelike surface that acts as part of a joint, as seen in the vertebrae of the spine and in the subtalar joint of the ankle. Each vertebra has two superior and two inferior facets. Facet joints are small stabilizing synovial joints located between and behind adjacent vertebrae.

Cross-References

▶ Chronic Back Pain and Spinal Instability
▶ Chronic Low Back Pain: Definitions and Diagnosis
▶ Facet Joint Pain
▶ Pain Treatment: Spinal Nerve Blocks

G.F. Gebhart, R.F. Schmidt (eds.), *Encyclopedia of Pain*, DOI 10.1007/978-3-642-28753-4,
© Springer-Verlag Berlin Heidelberg 2013

Facet Joint Injection

▶ Facet Joint Procedures for Chronic Back Pain

Facet Joint Pain

Jenna E. Zauk[1], Alfred T. Ogden[2] and
Christopher J. Winfree[2]
[1]St. George's University, School of Medicine,
Grenada, West Indies
[2]Department of Neurological Surgery,
Neurological Institute Columbia University,
New York, NY, USA

Synonyms

Facet syndrome; Sciatica; Zygapophysial joint
pain

Definition

Although the intervertebral joint has long been
known to be a common generator of low back
pain and leg pain, the facet joint has been
proposed as another potential focus of degenerative pathology that can produce similar symptoms. Structurally analogous to joints in the
extremities, the facet joint has the capacity to
degenerate over time and cause pain through the
same mechanisms that are at play in osteoarthritis
of the hip or the knee. This pain can be "felt" by
the patient in the area of the facet, producing low
back pain, or it can potentially be referred to other
parts of the body through the activation of
adjacent nociceptive fibers. Radiofrequency,
cryoneuroablation, and chemical neurolysis are
all possible treatment modalities for lumbar facet
syndrome (Wheeler and May 2010). Although
a "facet syndrome" including a component of
sciatica was initially proposed, stimulation of
the nerve supply to facet joints in live subjects
has shown reproducible patterns of pain that are
limited to the low back, flank, abdomen, and

buttock. Many studies have claimed a benefit
from treatment of low back pain through interventions aimed at disrupting the primary afferents supplying the facet joint, but the handful of
randomized clinical trials of these interventions
has generally failed to demonstrate any benefit
beyond placebo (Carette et al. 1991; Leclaire
et al. 2001; Lilius et al. 1989, Slipman 2003).
These studies have been criticized, however,
and their generally negative results may stem
from poor patient selection and improper selection of therapeutic target. In 2009, a panel of
experts in interventional therapies and surgery
recommended against the use of facet joint
steroid injections as a treatment for lower back
pain. Their decision was based upon insufficient
evidence from previously conducted studies
(Chou et al. 2009b). Although rates of use for
facet joint injections increased by 231 % from
1994 to 2001, these therapies have shown to only
improve short-term pain management (Friedly
et al. 2007; Chou et al. 2009a).

Characteristics

Anatomical Considerations

Ghormley is credited with coining the phrase the
"facet syndrome" in 1933 (Ghormley 1933).
In his seminal article, he noted that facet joints
were the only "true" joints in the spinal column,
meaning that they contain a complete joint capsule with a clear synovial membrane and hyaline
cartilage at the articular surface. Such true joints
are termed zygapophysial joints and exist in all
the typical weight-bearing places affected by
osteoarthritis such as the hip, the knee, and the
ankle. He conjectured and offered histopathological evidence that these joints degenerate over
time like their counterparts in the extremities.
He theorized that this process could produce
a syndrome of low back pain, scoliosis, and
sciatica, perhaps induced by rotatory strain of
the lumbosacral region.

Detailed anatomical studies of the nerve supply of facet joints have provided the blueprint for
testing and treating the "facet syndrome."
Each joint is innervated by spinal nerves from

the adjacent and superior vertebral level on the ipsilateral side (Bogduk and Long 1979; Maldjian et al. 1998; Mooney and Robertson 1976). Arising from the dorsal root ganglion, the medial branch of the posterior ramus passes through a notch at the base of the transverse process. Twigs are given off to the facet joint at the same level before the nerve descends inferiorly, giving off muscular and cutaneous branches, as well as a branch to the superior aspect of the facet joint one vertebral level below. Bogduk and Long published the most detailed anatomical study of these relationships, and proposed calling the branches comprising this dual nerve supply the proximal and distal zygapophysial nerves (Bogduk and Long 1979). These same authors recognized variability at L5–S1 mandated by the absence of transverse processes at this level, and noted the medial branch of the posterior ramus of L5 passes through a groove between the sacral ala and the root of the superior articular process of the sacrum. The key points from this anatomical study were that denervation of a facet joint would require a lesion of the medial branch of the posterior ramus at the same vertebral level and the level above, and that a denervation procedure directed at the facet joint proper might result in an incomplete lesion.

Provocative Studies

The first report of nervous stimulation of facet joints that induced back and leg pain was by Hirsch et al. in 1963 (Hirsch et al. 1963). The first systematic analysis of referred lumbar facet joint pain was written by Mooney and Robertson in 1975 (Mooney and Robertson 1976). These authors studied five normal subjects and 15 patients with low back pain, and reported a nonspecific pattern of pain referral to the flank, buttock, and the leg upon injection of 5 % hypertonic saline into the facet joint. An important discrepancy was noted between normal and low back pain patients in that the latter complained of more frequent and more widespread patterns of pain referral. The only reports of induced pain in a sciatic distribution came from the low back pain

cohort. These authors also related their results with fluoroscopically guided anesthetic injections, and offered a detailed description of the procedure, which served as a model for future studies. Due to difficulty in locating the precise vertebral level of the pain generator, injections at three lumbar levels were advocated and injections were directed into the facet joint proper. The facet joint, as the target of treatment, has been used by many clinicians and in many studies ever since, but this approach has come under criticism by some, who maintain that a more efficacious target of anesthetization or neurotomy is the medial branch of the posterior ramus (Bogduk and Long 1979).

The technique of Mooney and Robertson was applied (McCall et al. 1979) to six healthy individuals who received fluoroscopically guided injections of 6 % hypertonic saline into lumbar facet joints of L1–2 and L4–5. They noted several important findings. L1–2 injections reproduced pain to the adjacent lower back, flank, and groin. L4–5 injections produced pain in the adjacent lower back, buttock, groin, and lateral thigh. There was significant overlap in the distribution of the patterns of referred pain, even though the facet joints selected were three levels apart, and there was no patient that complained of pain below the mid-thigh.

Provocative studies have shown fairly clearly that stimulation of the sensory nerves around facet joints can induce pain, and that this stimulation can reproducibly generate pain that is referred to other parts of the body. It is still unclear to what extent this induced pain has a relationship to the clinical entity of low back pain in the general population. The paucity of provocative evidence for the induction of referred leg upon stimulation of facet joints in control populations casts some doubt on the sciatic component of the alleged "facet syndrome." The lack of dermatomal specificity for referred pain from facet joint stimulation may reflect the dual level innervation, and contribute to the difficulty in localization of the potential pain-generating level.

Randomized Clinical Trials

There have been multiple studies of treatments for facet joint pain including both anesthetic and steroid injections and radiofrequency nerve ablations. These studies vary greatly in their selection criteria and reported results. For example, reported efficacy of facet joint steroid injection ranges from 10 % to 63 % (Carette et al. 1991). The vast majority of these studies are retrospective case series. These divergent results likely stem from variable selection criteria, variable technique and target selection, variable follow-up, and variable criteria for a successful treatment. As such, they are very difficult to interpret. One review of these studies found "sparse evidence" to support the use of interventional techniques in the treatment of facet joint pain, and called for more randomized clinical trials (Slipman 2003).

Four peer-reviewed randomized clinical trials for the treatment of low back pain through facet joint procedures have been published, and the conclusion from three out of four of these studies was that no treatment has demonstrated any benefit beyond placebo (Carette et al. 1991; Leclaire et al. 2001; Lilius et al. 1989; van Kleef et al. 1999). There are criticisms of each study, however, which could have significantly affected results, and these criticisms are discussed below.

In 1989, Lilius et al. (Lilius 1989) randomized 109 patients with low back pain to receive injections of cortisone, local anesthetic, or saline into two facet joints. Patients were examined at 1 hour, 2 weeks, and 6 weeks, and also filled out a questionnaire regarding work performance and pain level at 3 months. Seventy percent of patients experienced initial pain relief and 36 % of patients reported continued benefit at 3 months. These results were irrespective of the contents of the injection. The major flaw in this study was that there were no selection criteria beyond low back pain to ensure that the facet joints were the pain generators in these patients. Another criticism was that the facet joint was used as the therapeutic target and not the medial branch of the posterior ramus.

In 1991, Carette et al. (Carette 1991) randomized 97 patients who reported >50 % immediate relief from low back pain following local anesthetic injection into facet joints, both at L4–5 and at L5–S1, to receive either steroid injections or saline injections. They followed the technique described by Mooney and Robertson (i.e., fluoroscopic guidance using contrast to localize the facet joint and injection into the facet joint proper). Patients were assessed immediately following the procedure and at 1-, 3-, and 6-month follow-up intervals. They found that 11 patients in the steroid group and 5 patients in the saline group had prolonged relief from the injections. The difference was not statistically significant. A post-hoc analysis of the subgroup of patients who claimed >90 % relief from the initial anesthetic injections yielded similar results. As mentioned previously, the target selection according to the technique of Mooney and Robertson has been criticized. It is also notable that their results approached significance ($p = 0.19$), begging the question of whether their study was simply underpowered.

In 1999, van Kleef et al. (van Kleef 1999) randomized 31 patients selected for >50 % relief from facet nerve block to receive radiofrequency nerve ablation or sham treatment. These authors targeted the medial branch of the posterior ramus according to the description of Bogduk and Long. Patients were assessed immediately after the procedure and at 1-, 3-, 6-, and 12-month intervals. Although initial analysis of their patient population showed no statistically significant benefit of radiofrequency ablation over placebo, a post-hoc analysis of the patients that reported the most relief from screening anesthetic injections demonstrated a benefit from the procedure. The authors concluded that this subpopulation of patients were the true sufferers of facet joint pain and that these patients, when properly selected, would benefit from radiofrequency neurotomy.

In 2001, Leclaire et al. (Leclaire 2001) randomized 70 patients selected for "significant" relief of low back pain after two level anesthetic injections. The target selected was the facet joint itself. Patients were assessed at 4 weeks and at

12 weeks using two measures of functionalability and one pain scale. Although one of the functional assessments showed a small but statistically significant improvement in the treatment group at 4 weeks, there were no statistically significant differences in the other functional assessment or the pain assessment at 4 weeks, or in any of the outcome measures at 12 weeks. The authors concluded that beyond a mild transient reduction in functional disability, radiofrequency facet joint neurotomy had no proven benefit in the treatment of low back pain. No post-hoc analysis of the patients who were most relieved by the selecting anesthetic injections was done. This study was criticized for its vague selection criteria and for its use of the facet joint as a target (Dreyfuss et al. 2002).

In 2005, Fuchs et al. (Fuchs 2005) randomized 60 patients suffering from chronic lumbar pain in a blind-observer clinical trial to evaluate the therapeutic results of intra-articular sodium hyaluronate (SII) injections versus glucocorticoid (TA) injections. The patients were divided into two groups to be treated with either 10-mg SH or 10-mg TA. The injections were delivered per facet joint bilaterally (from L3-S1) on a weekly basis. Level of pain was determined by the Visual Analog Scale (VAS), and life quality was determined by the Roland Morris questionnaire, Oswestry Disability questionnaire, Low Back Outcome Score and the Short Form 36 (SF-36) questionnaire. Throughout the 180-day trial period, both groups of patients demonstrated increased positive results in pain relief and quality of life. As for long-term results (>6 months), the sodium hyaluronate injections were potentially more beneficial for patients when compared to the glucocorticoid injection therapy (Fuchs et al. 2005).

While the existence of a "facet syndrome" that includes sciatica seems unlikely, a syndrome of low back pain caused by degenerative changes in the facet joints seems plausible. Provocative studies of sensory nerves to facet joints, as well as the close anatomical association between the nervous supply to the facet joint and the dorsal root ganglion, provide evidence of a pattern of referred pain to areas as distant as the buttocks and inguinal region. The prevalence of this entity within the vast population of patients with low back pain remains unknown.

Randomized clinical trials have failed to demonstrate convincing data to justify facet joint steroid injections or radiofrequency neurotomy within the populations of patients studied, but these results could easily be the result of improper patient selection. In the absence of a reliable radiographic diagnostic tool, more stringent screening criteria are required before these procedures should be dismissed. The cut-off of >50 % pain relief after a single session of anesthetic injections used by the studies reviewed may be too liberal and/or too unreliable. One interesting study probed this issue. Starting with 176 patients with low back pain, 47 were selected that reported a "definite" or "complete" response after facet block with a short-acting anesthetic. When this cohort was brought back for a confirmatory block 2 weeks later, only 15 % reported >50 % response (Schwarzer 1994). Facet joint degeneration may thus be a relatively rare cause of low back pain. Perhaps, anesthetic injections are simply not a reliable screening tool. Also, the difficulty to locate the exact site of the pain is a common problem in injection therapy (Fuchs et al. 2005). Another explanation for the negative results from clinical trials may lie in target selection. It has yet to be determined whether the facet capsule or the medial branch of the posterior ramus is preferred. Only randomized trials of steroid injections or ablation procedures that use more stringent selection criteria and compare results using different therapeutic targets will answer these questions.

References

Bogduk, N., & Long, D. M. (1979). The anatomy of the so-called "Articular Nerves" and their relationship to facet denervation in the treatment of low-back pain. *Journal of Neurosurgery, 51*, 172–177.

Carette, S., Marcoux, S., Truchon, R., et al. (1991). A controlled trial of corticosteroid injections into facet joints for chronic low back pain. *The New England Journal of Medicine, 325*, 1002–1007.

Chou, R., Atlas, S., Stanos, S., & Rosenquist, R. (2009a). Nonsurgical interventional therapies for low back pain, a review of the evidence for an American pain society clinical practice guideline. *Spine, 34*(10), 1078–1093.

Chou, R., Loeser, J., Owens, D., Rosenquist, R., Atlas, S., Baisden, J., Carragee, E., Grabois, M., Murphy, D., Resnick, D., Stanos, S., Shaffer, W., & Wall, E. (2009b). Interventional therapies, surgery, and interdisciplinary rehabilitation for low back pain, an evidence-based clinical practice guideline from the American pain society. *Spine, 34*(10), 1066–1077.

Dreyfuss, P., Baker, R., & Leclaire, R. (2002). Radiofrequency facet joint denervation in the treatment of low back pain: A placebo-controlled clinical trial to assess efficacy. *Spine, 27*, 556–557.

Friedly, J., Chan, L., & Deyo, R. (2007). Increase in lumbosacral injections in the medicare population. *Spine, 32*, 1754–1760.

Fuchs, S., Erbe, T., Fischer, H. L., & Tibesku, C. O. (2005). Intraarticular hyaluronic acid versus glucocorticoid injections for nonradicular pain in the lumbar spine. *Journal of Vascular and Interventional Radiology, 16*(11), 1493–1498.

Ghormley, R. K. (1933). Low back pain: With special reference to the articular facets. *Journal of the American Medical Association, 101*, 1773–1777.

Hirsch, D., Ingelmark, B., & Miller, M. (1963). The anatomical basis for low back pain. *Acta Orthopaedica Scandinavica, 33*, 1.

Leclaire, R., Fortin, L., Lambert, R., Bergeron, Y. M., & Rossignol, M. (2001). Radiofrequency facet joint denervation in the treatment of low back pain: A placebo-controlled clinical trial to assess efficacy. *Spine, 26*, 1411–1417.

Lilius, G., Laasonen, E. M., Myllynen, P., Harilainen, A., & Gronlund, G. (1989). Lumbar facet joint syndrome. A randomised clinical trial. *Journal of Bone and Joint Surgery, 71*, 681–684.

Maldjian, C., Mesgarzadeh, M., & Tehranzadeh, J. (1998). Diagnostic and therapeutic features of facet and sacroiliac joint injection. Anatomy, pathophysiology, and technique. *Radiologic Clinics of North America, 36*, 497–508.

McCall, I. W., Park, W. M., & O'Brien, J. P. (1979). Induced pain referral from posterior lumbar elements in normal subjects. *Spine, 4*, 441–446.

Mooney, V., & Robertson, J. (1976). The facet syndrome. *Clinic Orthopaedics and Related Research, 1976, 115*, 149–156.

van Kleef, M., Barendse, G. A., Kessels, A., Voets, H. M., Weber, W. E., & de Lange, S. (1999). Randomized trial of radiofrequency lumbar facet denervation for chronic low back pain. *Spine, 24*, 1937–1942.

Wheeler, A. H. May 2010 – http://emedicine.medscape.com/article/1144130-overview

Facet Joint Procedures for Chronic Back Pain

David M. Sibell
Comprehensive Pain Center/Spine Center, Oregon Health and Science University, Portland, OR, USA

Synonyms

Facet denervation; Facet joint injection; Facet rhizolysis; Medial branch block; Median branch block; Radiofrequency ablation; Zygapophyseal joint injection; Zygapophysial joint injection

Characteristics

Zygapophyseal (facet) joints are synovial diarthroses and are present from C1 to S1, inclusive. These joints allow for articular motion in the posterior spinal column and are innervated by medial branch of the primary posterior ramus of the segmental spinal nerves. Each articular process receives innervation from a spinal nerve, so each joint, comprised of two articular processes, is innervated by two medial branches (Fig. 1).

The medial branch is primarily sensory to the joint and surrounding structures and is innervated richly with nociceptive fibers. Numerous pain-mediating neurotransmitters (e.g., bradykinin, substance P, and neuropeptide Y) are also found in these neurons (Morinaga et al. 1996). These nerves are also motor to the multifidus muscles, and multifidus EMG studies have been used to validate the results of radiofrequency medial branch denervation (Dreyfuss et al. 2000).

Initially, the approach to treating facet arthropathy-related pain was limited to surgical excision and/or stabilization. It is difficult to assess the results of the surgical approaches to facet arthropathy, as patients do not uniformly have diagnostic procedures first, and the surgical treatment is almost always a part of another surgical procedure (e.g., fusion, laminectomy).

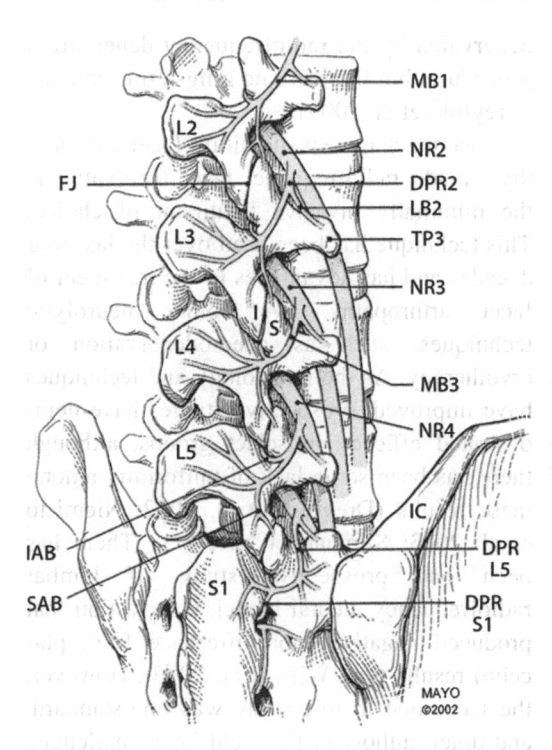

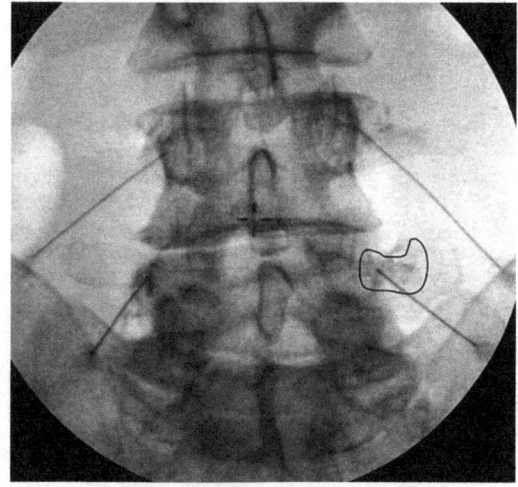

Facet Joint Procedures for Chronic Back Pain, Fig. 2 Lumbar medial branch block AP view with localized spread of contrast medium

Facet Joint Procedures for Chronic Back Pain, Fig. 1 Illustration of right posterior view of lumbosacral spine showing key right posterior neural structures. L2 through S1 spinous processes labeled. *Right*: *MB1* medial branch of L1 dorsal primary ramus, *NR2* L2 nerve root, *DPR2* L2 dorsal primary ramus, *LB2* lateral branch of L2 dorsal primary ramus, *TP3* L3 transverse process, *NR3* L3 nerve root, *MB* medial branch of L3 dorsal primary ramus that extends around the base of the right superior articular process (S) of L4 and innervates portions of the right L3–4 and L4–5 facet joint capsules, *NR4* L4 nerve root, *IC* iliac crest, *DPRL5* L5 dorsal primary ramus, *DPRS1* S1 dorsal primary ramus, *I* inferior articular process L3, *S* superior articular process of L4. *Left*: *FJ* L2–3 facet (zygapophysial) joint, which is innervated by branches of L1 and L2 medial branch nerves, *IAB* inferior articular branches from medial branch of L4 dorsal primary ramus, *SAB* superior articular branches from medial branch of L4 dorsal primary ramus (From Czervionke and Fenton 2003)

Joint injections with local anesthetic and steroid are still popular in many practices, but these injections have not been demonstrated to be reliably diagnostic (due to potential epidural spread of injectate) or of any prolonged therapeutic value (Dreyfuss and Dreyer 2003). Numerous prospective, double-blinded, randomized controlled trials have shown these injections to be no better than placebo in the treatment of chronic back and neck pain (Barnsley et al. 1994;

Carette et al. 1991). There have been no significant studies since these indicate that there is a role for intraarticular steroid injections.

Fluoroscopically guided diagnostic medial branch blocks anesthetize the facet joint selectively and are used to provide prognostic information for radiofrequency medial branch denervation. They are not intended for prolonged analgesia. Generally, a two-block paradigm is used: one injection of short-acting local anesthetic and one of long-acting anesthetic (Lord et al. 1995). When performed correctly, these blocks have high specificity for anesthetizing the facet joint (Dreyfuss et al. 1997) (Figs. 2 and 3). Some authors have attempted to use ultrasound as guidance for this procedure, and in ideal settings, it is possible that there may be a role. However, in clinical practice, body habitus and the inability to detect intravascular injections or perform injections at more than one site at a time will remain significant barriers (Rauch et al. 2009).

There has been debate regarding the exact interpretation of these blocks, fueled by the problems inherent in attempting to make an objective diagnosis in a subjective disorder (i.e., pain). Much of this debate has focused on the test characteristics of the procedure and uses terms such as "placebo response" and "false positive"

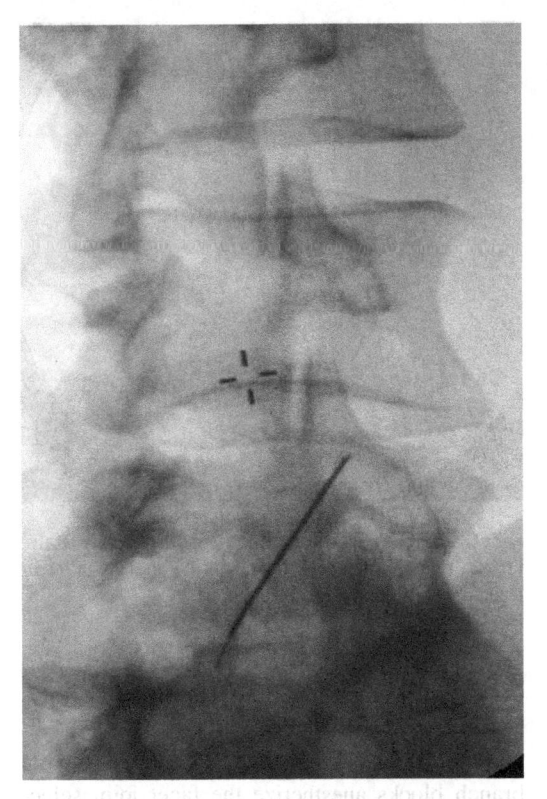

Facet Joint Procedures for Chronic Back Pain, Fig. 3 Lumbar medial branch block lateral oblique view

(Barnsley et al. 1993). Unfortunately, these terms are misleading in this sense. A subject may have an unanticipated response to an injection, but if an active treatment is used, by definition, that response is not a placebo response. Moreover, it is inappropriate to use the term "false positive" in this situation, as there is no gold standard test with which to compare the results.

These studies do not take into account the analgesic effect that simultaneously anesthetizing the multifidus muscle has, which may account for the prolonged duration of some subjects' responses. Therefore, since this field deals with subjective responses, most operators use a somewhat more liberal interpretation of the results of diagnostic medial branch blocks and allow for prolonged concordant responses (i.e., both responses more prolonged than would be expected solely due to the local anesthetic but of duration proportional to the anticipated duration). In addition, multifidus is

denervated by the radiofrequency denervation procedure, but there are no long-term sequelae (Dreyfuss et al. 2009).

Once the diagnosis of painful facet arthropathy is made, radiofrequency facet denervation is the minimally invasive treatment of choice. This technique has been used over the last four decades and has advantages in the treatment of facet arthropathy over other neurolytic techniques, such as chemodenervation or cryotherapy. As the technology and techniques have improved, prospective studies have demonstrated efficacy in select groups, although there has been some lack of uniformity among these results (Dreyfuss et al. 2002; Niemisto et al. 2003; Slipman et al. 2003). There has been one prospective study of lumbar radiofrequency medial branch denervation that produced negative (non-difference from placebo) results (van Wijk et al. 2005). However, the technique in this study was non-standard, and other authors in this field have challenged the results, as the technique was likely to miss the target nerve (Bogduk 2006). Since then, several studies have demonstrated positive results, though there is still a need for a large, multicenter, prospective, randomized trial (Nath et al. 2008; Burnham et al. 2009).

The technique involved in radiofrequency facet denervation is similar to that of medial branch blocks, inasmuch as the instrument is placed in proximity to the medial branches innervating a joint, as opposed to entering the joint itself (Lau et al. 2004). However, instead of using plain needles, special cannulae are used. These are coated with heat-shrink Polyethylene Terephthalate (PET) tubing in order to insulate most of the needle. This focuses the release of radiofrequency energy on the active tip, which leads to a focused, reproducible lesion. When positioned appropriately, this lesion includes the medial branch, while limiting collateral damage to surrounding structures. As a result of this precision, the risk of adverse events is exceedingly low (Kornick et al. 2004). The safety profile is one of the features that make this procedure an attractive alternative in the treatment of this common disorder.

The desired outcome in this procedure is the focal denervation of the joints in which the patient's back pain was relieved upon performance of diagnostic medial branch blocks. This does not treat the underlying arthropathy but reduces the painful limitation to mobility that it causes. The nature of radiofrequency denervation does allow for regrowth of the medial branch nerve. Therefore, the procedure may require repetitive treatments over time. Longer-term cohorts and studies of repeated intervention indicate that the clinical results of the procedure last approximately 12 months, which is consistent with a period of denervation, followed by regrowth of the medial branch nerve (Tomé-Bermejo et al. 2011; Rambaransingh et al. 2010). Since the denervation procedure does not address comorbidities, such as myofascial pain, post-denervation physical therapy may be used to extend the benefits of the procedure, to include relief of myofascial pain and associated loss of range of motion.

Due to the increasing use of this procedure in the United States (Friedly et al. 2007), insurers have sought to limit its availability by placing stringent requirements on patients with facetogenic pain prior to approving radiofrequency medial branch denervation (Noridian Local Coverage Determination; United Health Care Medical Policy). While the data supporting this procedure are imperfect, there is only one significant negative study, which used a nonstandard technique. The weight of evidence for the positive symptomatic and functional effects of radiofrequency medial branch denervation is far greater than for most analgesic interventional procedures. It remains to be seen whether this useful, safe procedure will continue to be available to those who most benefit from its use.

References

Barnsley, L., Lord, S., Wallis, B., et al. (1993). False-positive rates of cervical zygapophysial joint blocks. *The Clinical Journal of Pain, 9*, 124–130.

Barnsley, L., Lord, S. M., Wallis, B. J., et al. (1994). Lack of effect of intraarticular corticosteroids for chronic pain in the cervical zygapophyseal joints. *The New England Journal of Medicine, 330*, 1047–1050.

Bogduk, N. (2006). Lumbar radiofrequency neurotomy. *The Clinical Journal of Pain, 22*(4), 409.

Burnham, R. S. MSc, MD, FRCPC, Holitski, S. BScPT, & Dinu, I. PhD. (2009). A prospective outcome study on the effects of facet joint radiofrequency denervation on pain, analgesic intake, disability, satisfaction, cost, and employment. *Archives of Physical Medicine and Rehabilitation, 90*, 201–205.

Carette, S., Marcoux, S., Truchon, R., et al. (1991). A controlled trial of corticosteroid injections into facet joints for chronic low back pain. *The New England Journal of Medicine, 325*, 1002–1007.

Czervionke, L. F., & Fenton, D. S. (2003). Facet joint injection and medial branch block. In L. F. Czervionke & D. S. Fenton (Eds.), *Image-guided spine intervention*. Philadelphia: WB Saunders.

Dreyfuss, P., & Dreyer, S. J. (2003). Lumbar zygapophysial (facet) joint injections. *The Spine Journal, 3*, 50S–59S.

Dreyfuss, P., Schwarzer, A. C., Lau, P., et al. (1997). Specificity of lumbar medial branch and L5 dorsal ramus blocks. A computed tomography study. *Spine, 22*, 895–902.

Dreyfuss, P., Halbrook, B., Pauza, K., et al. (2000). Efficacy and validity of radiofrequency neurotomy for chronic lumbar zygapophysial joint pain. *Spine, 25*, 1270–1277.

Dreyfuss, P., Baker, R., Leclaire, R., et al. (2002). Radiofrequency facet joint denervation in the treatment of low back pain: A placebo-controlled clinical trial to assess efficacy. *Spine, 27*, 556–557.

Dreyfuss, P. MD, Stout, A. DO, April, C. MD, Pollei, S. MD, Johnson, B. MD, & Bogduk, N. MD, PhD. (2009). The significance of multifidus atrophy after radiofrequency neurotomy for low back pain successful radiofrequency neurotomy for low back pain. *Physical Medicine and Rehabilitation, 1*, 719–722.

Friedly, J. MD, Chan, L. MD, MPH, & Deyo, R. MD, MPH. (2007). Increases in lumbosacral injections in the medicare population 1994 to 2001. *Spine, 32*(16), 1754–1760.

Kornick, C., Kramarich, S. S., Lamer, T. J., et al. (2004). Complications of lumbar facet radiofrequency denervation. *Spine, 29*, 1352–1354.

Lau, P., Mercer, S., Govind, J., et al. (2004). The surgical anatomy of lumbar medial branch neurotomy (facet denervation). *Pain Medicine, 5*, 289–298.

Local Coverage Determination (LCD) for paravertebral facet nerve blockade (L30813). www.cms.gov. Accessed 1 Feb 2013.

Lord, S. M., Barnsley, L., & Bogduk, N. (1995). The utility of comparative local anesthetic blocks versus placebo-controlled blocks for the diagnosis of cervical zygapophysial joint pain. *The Clinical Journal of Pain, 11*, 208–213.

Morinaga, T., Takahashi, K., Yamagata, M., et al. (1996). Sensory innervation to the anterior portion of lumbar intervertebral disc. *Spine, 21*, 1848–1851.

Nath, S. MD, FRCA, Nath, C. A. SRN, & Pettersson, K. MD, PhD. (2008). Percutaneous lumbar zygapophysial (facet) joint neurotomy using radiofrequency current, in the management of chronic low back pain a randomized double-blind trial. *Spine, 33*(12), 1291–1297.

Niemisto, L., Kalso, E., Malmivaara, A., et al. (2003). Radiofrequency denervation for neck and back pain: A systematic review within the framework of the Cochrane Collaboration Back Review Group. *Spine, 28*, 1877–1888.

Rambaransingh, B. BSc, MD, Stanford, G. DC†, & Burnham, R. MSc, MD, FRCPC. (2010). The effect of repeated zygapophysial joint radiofrequency neurotomy on pain, disability, and improvement duration. *Pain Medicine, 11*, 1343–1347.

Rauch, S. MD, Kasuya, Y. MD, Turan, A. MD, Neamtu, A. MD, Vinayakan, A. MD, Sessler, D. I. MD, et al. (2009). Ultrasound-guided lumbar medial branch block in obese patients a fluoroscopically confirmed clinical feasibility study. *Regional Anesthesia and Pain Medicine, 34*(4), 340–342.

Slipman, C. W., Bhat, A. L., Gilchrist, R. V., et al. (2003). A critical review of the evidence for the use of zygapophysial injections and radiofrequency denervation in the treatment of low back pain. *The Spine Journal, 3*, 310–316.

Tomé-Bermejo, F. MD, Barriga-Martin, A. PhD, & Martin, J. L. R. PhD. (2011). Identifying patients with chronic low back pain likely to benefit from lumbar facet radiofrequency denervation. *Journal of Spinal Disorders & Techniques, 24*, 69–75.

United Healthcare Medical Policy. Ablative treatment for spinal pain policy number: 2013T0107M. https://www.unitedhealthcareonline.com. Accessed 8 May 2013.

van Kleef, M., Barendse, G. A. M., Kessels, A., et al. (1999). Randomized trial of radiofrequency lumbar facet denervation for chronic low back pain. *Spine, 24*, 1937–1942.

Van Wijk, R. M. A. W. MD, PhD, Geurts, J. W. M. MD, PhD, Wynne, H. J. PhD, Hammink, E. MD, Buskens, E. MD, PhD, Lousberg, R. PhD, et al. (2005). Radiofrequency denervation of lumbar facet joints in the treatment of chronic low back pain a randomized, double-blind, sham lesion-controlled trial. *The Clinical Journal of Pain, 21*(4), 335–344.

Facet Rhizolysis

▶ Facet Joint Procedures for Chronic Back Pain

Facet Syndrome

▶ Facet Joint Pain

Facial Ganglion Neuralgia

▶ Geniculate Neuralgia

Facial Pain

Definition

Facial pain is identified by its location, usually excluding tic douloureux.

Cross-References

▶ Pain Treatment, Motor Cortex Stimulation

Facial Pain Associated with Disorders of the Cranium

▶ Headache from Cranial Bone

Facilitative Tucking

Definition

A caregiver uses his/her hands to swaddle an infant by placing a hand on the infant's head and feet while providing flexion and containment.

Cross-References

▶ Acute Pain Management in Infants

Factor Loading

Definition

Factor analysis is a statistical procedure that groups together variables that share common variance. Variables that "load" on the same factor are presumed to reflect a similar underlying process.

Cross-References

► Psychology of Pain, Self-Efficacy

Factors Associated with Low Back Pain

► Low Back Pain, Epidemiology

Failed Back

Definition

Clinical syndrome characterized by back or lower extremity pain or both following surgery for decompression of neural elements in the lower back.

Cross-References

► Pain Treatment: Spinal Cord Stimulation

Failed Back Surgery Syndrome

Definition

Failed back surgery syndrome is axial or radicular pain persisting after surgical approaches to

relieve the pain. It is also known as failed back syndrome.

Cross-References

► Central Nervous System Stimulation for Pain
► Dorsal Root Ganglionectomy and Dorsal Rhizotomy

False Affirmative Rate

Definition

False affirmative rate is the probability of response "A" when event B has occurred.

Familial Adenomatous Polyposis

Synonyms

FAP

Definition

An inherited disease which is characterized by the formation of numerous polyps on the inside walls of the colon and rectum. The FAP disease is associated with a 100 % risk for developing colorectal cancer.

Cross-References

► NSAIDs and Cancer

Familial Amyloid Polyneuropathy (FAP)

► Hereditary Neuropathies

Familial Factors

▶ Impact of Familial Factors on Children's Chronic Pain

Familial Hemiplegic Migraine

Definition

Familial hemiplegic migraine is an inherited form of migraine with aura in which patients experience weakness and other neurological disturbances as their aura.

Cross-References

▶ Migraine, Pathophysiology

Familial Polyposis Coli

Definition

People with this syndrome have massive numbers of colonic polyps and almost invariably develop cancer of the colon.

Cross-References

▶ NSAIDs and Their Indications

Family Environment

▶ Impact of Familial Factors on Children's Chronic Pain

Family-Centered Care

Lisa M. Peters
Department of Anesthesiology and Pain Medicine, Seattle Children's Hospital, Seattle, WA, USA

Synonyms

Family-centered theory; Patient-centered care

Definition

What Is Family-Centered Care?
The practice of pediatrics focuses on the health of a child. The experience of pediatric care, however, is not isolated to the child; the experience impacts the whole family. In fact, a bidirectional relationship exists whereby efforts to provide holistic care to a child necessarily involve accounting for relevant contextual factors. Context includes elements internal to the child, such as developmental level, as well as external to the child, such as familial influences. The method and degree to which this relationship is embraced in the care setting is the focus of an approach called family-centered care.

While the central tenants of family-centered care may be practiced to varying degrees around the world, most of the published literature on this topic has originated from the United States and United Kingdom in the last two decades. Definitions articulate an approach to decision making and information sharing, captured most powerfully in one word, partnership. Together, healthcare practitioners, patient, and family navigate each step along the care pathway with intention and respect.

A framework for family-centered care in pediatrics, originally described by Shelton in 1987, includes the following nine elements:
• Recognizing the family as a constant in the child's life

- Facilitating parent-professional collaboration at all levels of health care
- Honoring the racial, ethnic, cultural, and socioeconomic diversity of families
- Recognizing family strengths and individuality and respecting different methods of coping
- Sharing complete and unbiased information with families on a continuous basis
- Encouraging and facilitating family-to-family support and networking
- Responding to child and family developmental needs as part of healthcare practices
- Adopting policies and practices that provide families with emotional and financial support
- Designing health care that is flexible, culturally competent, and responsive to family needs

According to the Institute for Family-Centered Care, established in 1992 in the USA, the summary themes include: respect and dignity; information sharing; participation; and collaboration (Committee on Hospital Care. American Academy of Pediatrics & Institute for Patient- and Family-Centered Care, 2003).

Characteristics

Pediatric Pain and Family-Centered Care

Pain in children is complex. Yet, the fundamental human right to pain care drives the field to further scientific study and the individual practitioner to prevent and alleviate children's pain. From parents' perspective, data show that the top two (among 36) issues of priority for their children admitted to an acute care setting were "Find out what is wrong with my child" and "taking care of my child's pain if it is relevant." And yet, pain presented the greatest discrepancy between priority and level of satisfaction (Ammentorp 2005). Through the alignment of science and humanity, the greatest possible impact on the short- and long-term consequences of poorly treated or untreated pain in children can be realized. Therein lays the

synergy of a family-centered care approach to improve a child's experience with pain; a partnership of evidence-based health care and the expertise of self-report or parents' knowledge of their child.

Clinical: A Call to Action

A commitment to partner along the clinical care continuum particularly illustrates the powerful opportunity, including assessment, planning, intervention, and evaluation. First, the cornerstone of effective pain care is assessment. Evolution of study has led to an iterative set of developmental-age specific tools to assess the intensity of children's pain. However, healthcare practitioners are not treating an intensity number, they are treating a person. Thus, it becomes essential to engage a parent to share and observe behavioral cues noticed in their child, or honor an adolescent's self-report of the quality, location, duration, and aggravating and relieving factors (Reid et al. 1995). From the outset, it must be noted that communication is foundational and continuous (Garland and Kenny 2006). Thus, eliminating language disparity via routine and readily accessible interpreter services is critical (Guerrero et al. 2010). This exchange informs planning for pain care and draws upon the critical component of respect. The well-recognized, positive effects of an interdisciplinary approach to pain care match the collaborative model of family-centered care (Huth et al. 2003; Shields et al. 2006; Simons et al. 2001). This is an opportunity to welcome and optimize a family's strengths and an individual child's coping style. The importance of optimizing patient and parent participation and coping has been clearly demonstrated, such as in a cohort of adolescents with sickle cell disease (Mitchell 2007). A partnership to share information allows for unbiased presentation of evidence-based interventions and nonjudgmental discussion. It becomes essential to recognize that all parties may hold different elements of the care at a premium, including evaluation of the risk and benefit of the pain experience itself, as well as the interventions and goals.

Education: Empowering Advocates

Efforts to enhance awareness and knowledge of children's pain through education can give new voice to all. Education offers a link to close the gap between the prevalence of undertreated or untreated pain in children and the lack of timely, effective interventions to prevent or alleviate that pain. In its 2011 publication titled "Relieving Pain in America," the Institute of Medicine cites education as a foundational endeavor to transform care (IOM 2011). The aim of educational activities may focus on expectations and current evidence of pain care to inform a shared mental model, yet importantly also focus on the benefit of each member's individual expertise. In other words, to recognize that a healthcare provider has the responsibility to contribute expertise in current evidence-based treatment options, while a parent has the responsibility to contribute expertise in their child's context and current experience. Such messaging again emphasizes communication as key; there is support for anyone to speak up and advocate on behalf of a child, including truly creating space for and hearing the voice of a parent.

Research: Data to Improve Outcomes

In recent years, further collaboration between the scientific community and the healthcare consumer offers another example of, and opportunity for, a powerful partnership (MacKean et al. 2005). For example, the Patient-Centered Outcomes Research Institute (PCORI) is an independent, nonprofit organization in the USA created to contribute specifically to shared, informed decision making by patients, families, and healthcare providers. Patients play a central role to define research priorities, contribute to the direction of clinical queries, and gain access to results. Another example is the Patient Reported Outcomes Measurement Information System (PROMIS), which is funded by the US National Institutes of Health (NIH). These measures become the tools of scientific study to inform the effectiveness of interventions and have already

invested in pain as a focal area. Specifically in pediatrics, this entity has made available validated tools to assess pain interference and intensity, as well as focusing on areas of physical function, emotional distress, fatigue, and peer relationships. Thus, research offers an arena not only for patient and parent participation, but also for collaboration to define priorities for study and new forums in which to share results.

What Do the Data Show with Regard to Family-Centered Care?

A systematic review (Shields et al. 2008) underscored the paucity of solid quantitative research on the effectiveness of family-centered care. Qualitative studies, and subsequent quantitative efforts, have shown the impact of family-centered care to (Bamm and Rosenbaum 2008; Kuo 2012):

• Improve patient outcomes
• Improve patient and family satisfaction
• Improve healthcare professionals' satisfaction
• Decrease cost
• Improve utilization of resources in health care

In line with these results, leading professional and safety organizations have endorsed family-centered care. In the USA, for example, the Institute of Medicine (IOM) identified family-centered care as one of the six attributes of high-quality health care (IOM 1999, 2001), and the Joint Commission incorporated a bill of rights that included patient comfort and that active involvement of patients and families is a strong strategy to ensure patient safety.

Summary

The synergy between family-centered care and the practice of pediatric pain medicine is undeniable. The data available assert its value to the extent that it is sanctioned as a primary safety and quality strategy. Yet, the challenge remains in the translation of a philosophy in approach to bedside practice. For some healthcare settings this would present a complete interdisciplinary paradigm shift, while at the same time offer a tremendous opportunity for culture change in treating pain in children. It is possible. First, the system must define the professional standards, set expectations for behavior, and create the space for a rich partnership. Then, the system must maintain accountability to that model, including individual practitioner responsibility. In time, a culture will emerge in which the partnership of family-centered care will transform the experience of a child in pain.

References

Ammentorp, J., Mainz, J., & Sabroe, S. (2005). Parents' priorities and satisfaction with acute pediatric care. *Archives of Pediatric and Adolescent Medicine, 159*(2), 127–131.

Bamm, E. L., & Rosenbaum, P. (2008). Family-centered theory: Origins, development, barriers, and supports to implementation in rehabilitation medicine. *Archives of Physical Medicine and Rehabilitation, 89*(8), 1618–1624.

Committee on Hospital Care. American Academy of Pediatrics & Institute for Patient- and Family-Centered Care. (2003). Family-centered care and the pediatrician's role. *Pediatrics, 112*(3 Pt 1), 691–697.

Garland, L., & Kenny, G. (2006). Family nursing and the management of pain in children. *Paediatric Nursing, 18*(6), 18–21.

Guerrero, A. D., et al. (2010). Racial and ethnic disparities in pediatric experiences of family centered care. *Medical Care, 48*(4), 388–393.

Huth, M. M., Broome, M. E., Mussatto, K. A., & Morgan, S. W. (2003). A study of the effectiveness of a pain management education booklet for parents of children having cardiac surgery. *Pain Management Nursing, 4*(1), 31–39.

Institute for Family-Centered Care. Available at: www.familycenteredcare.org. Accessed 11 Apr 2012.

Institute of Medicine. (1999). *To err is human: Building a safer health system*. Washington, DC: National Academies Press.

Institute of Medicine. (2001). *Crossing the quality chasm: A new health system for the 21st Century*. Washington, DC: National AcademiesPress.

Institute of Medicine. (2011). *Relieving pain in America: A blueprint for transforming prevention, care, education, and research*. Washington, DC: National AcademiesPress.

Kuo, D. (2012). Family-centered care: Current applications and future directions in pediatric health care. *Maternal and child health journal*, (1092–7875), 16 (2), 297.

MacKean, G. L., Thurston, W. E., & Scott, C. M. (2005). Bridging the divide between families and health professionals' perspectives on family-centered care. *Health Expectations, 8*, 74–85.

Mitchell, M. J., Lemanek, K., Palermo, T. M., Crosby, L. E., Nichols, A., & Powers, S. W. (2007). Parent perspectives on pain management, coping, and family functioning in pediatric sickle cell disease. *Clinical Pediatrics, 46*(4), 311–319.

Patient Reported Outcomes Measurement Information System. Available at: www.nihpromis.org. Accessed 11 Apr 2012.

Patient-Centered Outcomes Research Institute. Available at: www.pcori.org. Accessed 11 Apr 2012.

Reid, G. J., Hebb, J. P., McGrath, P. J., Finley, G. A., & Forward, S. P. (1995). Cues parents use to assess postoperative pain in their children. *The Clinical Journal of Pain, 11*(3), 229–235.

Shields, L., Pratt, J., & Hunter, J. (2006). Family-centred care: A review of qualitative studies. *Journal of Clinical Nursing, 15*(10), 1317–1323.

Shields, L., Pratt, J., Davis, L. M., & Hunter, J. (2008). Family-centred care for children in hospital. *Cochrane Database of Systematic Reviews*, (1), CD004811.

Simons, J., Franck, L., & Roberson, E. (2001). Parent involvement in children's pain care: Views of parents and nurses. *Journal of Advanced Nursing, 36*(4), 591–599.

Family-Centered Care, Approach and Basis

Definition

The American Academy of Pediatrics Committee on Hospital Care and the Institute for Family-Centered Care issued a joint policy statement in which they defined terms as follows: "Family-centered care is an approach to health care that shapes health care policies, programs, facility design, and day-to-day interactions among patients, families, physicians, and other health care professionals." Further, "Family-centered care in pediatrics is based on the understanding that the family is the child's primary source of strength and support and that the child's and family's perspectives and information are important in clinical decision making."

References

American Academy of Pediatrics. (2003). *Pediatrics, 112*, 691–696 (http://www.umassmed.edu/uploadedFiles/shriver/education/LEND/Courses/Pediatrics_FCC_and_Peds_Role_2003.pdf)

Family-Centered Care, Objective

Definition

Care directed at improving the health and well-being of the family and its members by assessing the family health needs and identifying potential obstacles.

Cross-References

▶ Chronic Pain in Children, Physical Medicine and Rehabilitation

Family-centered Theory

▶ Family-Centered Care

FAP

▶ Familial Adenomatous Polyposis

Fascia Iliaca Compartment Block

Definition

Injection via needle or catheter of local anesthetic deep to the fascia lata and iliaca medial to the anterior superior iliac spine and inferior to the inguinal ligament.

Cross-References

▶ Acute Pain in Children, Postoperative

Fasciculus Cuneatus

Definition

The lateral bundle of nerves in the dorsal column referred to as the cuneate fasciculus, which terminates in the cuneate nucleus just off the dorsal midline in the caudal medulla.

Cross-References

▶ Postsynaptic Dorsal Column Projection, Anatomical Organization

Fasciculus Gracilis

Definition

The medial bundle nerves in the dorsal column referred to as the fasciculus gracilis, which terminates in the gracile nucleus in the dorsal midline of the caudal medulla.

Cross-References

▶ Postsynaptic Dorsal Column Projection, Anatomical Organization

Fast-Track Surgery

▶ Postoperative Pain, Importance of Mobilization

Fatigue

Definition

Fatigue is a decrement of response seen with repeated stimulation and is a prominent attribute of nociceptors and other primary afferents.

Cross-References

▶ Pain in Humans, Electrical Stimulation (Skin, Muscle, and Viscera)
▶ Polymodal Nociceptors, Heat Transduction

FCA

▶ Freund's Complete Adjuvant

FCA-Induced Arthritis

Definition

This is the same as CFA (Complete Freund's Adjuvant)-induced Arthritis. An experimental model of inducing rheumatoid arthritis by injecting a suspension of killed mycobacteria (Freund's complete adjuvant, FCA) into various tissues in rats. The classical model involves injection of high doses into the tail base which produces multiple joint arthritis (polyarthritis) accompanied by wide-spread lesions of skin and other organs. A modified method uses low-dose local injections around a joint to produce unilateral single joint arthritis.

Cross-References

▶ Opioids and Inflammatory Pain

FCE

▶ Functional Capacity Evaluation

Fear and Pain

Kim Helsen[1,3], Maaike Leeuw[2] and Johan W. S. Vlaeyen[4,5]
[1]Research Group on Health Psychology, KU Leuven Katholieke Universiteit Leuven, Leuven, Belgium
[2]Department of Medical, Clinical and Experimental Psychology, Maastricht University, Maastricht, The Netherlands
[3]Department of Experimental–Clinical and Health Psychology, Ghent University, Ghent, Belgium
[4]Research Group Health Psychology, Catholic University of Leuven, Leuven, Belgium
[5]Department of Clinical Psychological Science, Maastricht University, Maastricht, The Netherlands

Synonyms

Fear of movement/(re)injury; Kinesiophobia; Pain-related anxiety; Pain-related fear

Definition

▶ Fear of pain is a general term used to describe several forms of fear with respect to pain. Depending on the anticipated source of threat, the content of fear of pain varies considerably. For example, fear of pain can be directed toward the occurrence or continuation of pain, toward physical activity, or toward the induction of (re) injury or physical harm. A more specific fear of pain concerns ▶ fear of movement/(re)injury, which is the specific fear that physical activity will cause (re)injury. Synonymously, ▶ kinesiophobia is defined as "an excessive, irrational, and debilitating fear of physical movement and activity resulting from a feeling of vulnerability to painful injury or re-injury" (Kori et al. 1990; Lundberg et al. 2006).

Characteristics

In recent years, chronic pain has no longer been conceptualized as purely a medical problem but rather as a complex biopsychosocial phenomenon in which the relationship among impairments, pain, and disability is weak. In chronic pain patients, anxiety disorders frequently co-occur, indicating that patients with persistent musculoskeletal pain fear a variety of situations that are not essentially related to pain (Asmundson and Katz 2009; Asmundson et al. 2004; Gatchel et al. 2007). Besides the finding that chronic pain patients seem to suffer more frequently from anxiety symptoms, fear and anxiety are often an integral part of the chronic pain problem. The experience of pain can be characterized by psychophysiological (e.g., muscle reactivity), cognitive (e.g., worry), and behavioral (e.g., escape and avoidance) responses, showing similarities with responses regarding fear and anxiety (Leeuw et al. 2007; Vlaeyen and Linton 2000).

There are multiple pathways by which pain-related fear mediates disability, namely, through escape and avoidance behaviors, through interference with cognitive functioning, through reduced opportunities to correct the erroneous underlying cognitions guiding the avoidance behaviors, and through detrimental effects of long-lasting avoidance on various physiological systems (Crombez et al. 1999; Gheldof et al. 2006; Turk and Wilson 2010). Empirical findings support the notion that fear of pain is a significant contributor to the chronification and maintenance of chronic pain syndromes (Hirsh et al. 2008; Turk and Wilson 2010; Vlaeyen and Linton 2000).

Fear and Anxiety
In the literature describing fear of pain, the concepts of fear and anxiety are often used interchangeably. Despite the fact that these concepts

are substantially related, some differences can be distinguished (Asmundson et al. 2004; Leeuw et al. 2007; Sylvers et al. 2011).

Fear is the emotional expression of the fight-flight response, which is the immediate readiness of the body to respond to an event that is perceived as dangerous or threatening. Fear is therefore a present-oriented state that is designed to protect the individual from the perceived immediate threat. ▶ Anxiety, however, is a cognitive-affective state that is rather future oriented. It tends to occur in the anticipation of a dangerous or threatening event and is therefore more indefinite and uncertain in nature. Instead of initiating the fight-flight response as in case of fear, the state of anxiety seems to facilitate and stimulate the fight-flight response only in case the threatening event occurs. Both in fear and anxiety, cognitive, physiological, and behavioral dimensions of responses can be distinguished. Physiologically, fear and anxiety responses are characterized by the activation of the sympathetic nervous system, designed to increase the likelihood of survival by promoting escape from or protection against the perceived threat. In anxiety, these physiological responses are less present than in fear. The cognitive element, relatively more present in anxiety, is more narrowed in anxiety and directed in such a way that a source of threat, when present, will be detected. Fear, on the other hand, comprises thoughts of danger, threat, or death, through which the attention toward the threat is advanced, while irrelevant distracters are ignored and the initiation of action is stimulated. On a behavioral level, anxiety guides motivation to engage in preventative and avoidance behaviors, while fear motivates to engage in defensive behaviors. Despite the definition of pain-related fear, both fear and anxiety are distinct processes that contribute significantly to chronic pain (Asmundson et al. 2004).

Hierarchy of Fear
Besides the important distinction between fear and anxiety in chronic pain, understanding about the hierarchical nature of fear and anxiety is also an important consideration. The hierarchical structure of anxiety is displayed in Fig. 1.

Traits		Trait anxiety	
Fundamental fears	Anxiety Sensitivity	Injury/Illness sensitivity	Social Evaluation Sensitivity
Common fears		Fear of pain	
Specific fears		Fear of movement/(re)injury	

Fear and Pain, Fig. 1 Hierarchical structure of fear of pain

Fundamental Fears

According to the expectancy theory (Reiss and McNally 1985), most fears can be derived from one of three more fundamental fears or sensitivities: (1) fear of anxiety symptoms (anxiety sensitivity), (2) fear of negative evaluation (social evaluation sensitivity), and (3) fear of illness/injury (injury/illness sensitivity). Anxiety sensitivity refers to the fear of anxiety symptoms arising from the belief that anxiety has harmful somatic, psychological, and social consequences. Social evaluation sensitivity reflects anxiety and distress that is associated with expectations that others will have negative evaluations about oneself and with avoidance of evaluative situations. Finally, injury/illness sensitivity refers to fear concerning injury, illness, or death. These fundamental fears are quite distinct from each other and comprise stimuli that are considered to be essentially aversive to most people (Asmundson et al. 2000; Vlaeyen 2003).

Common Fears: Fear of Pain

Common fears (such as spider phobia, agoraphobia, fear of pain) arise as the result of an interaction between the fundamental fears and learning experiences and can thus be logically derived from these three fundamental fears. In contrast to fundamental fears, they do not refer to a wide variety of stimuli and are not essentially considered to be aversive to most people. Due to fear of anxiety, causing one to fear the symptoms that are associated with anxiety, a common fear about

a particular situation easily arises when in the concerning situation anxiety symptoms are expected or likely to be experienced. In essence, anxiety sensitivity can be considered as a vulnerability factor that exacerbates the development and maintenance of common fears, the same holding for injury sensitivity and social evaluation sensitivity (Asmundson et al. 2000).

In chronic low back pain, fear of pain is a common fear that can be derived from the fundamental fear of anxiety symptoms (anxiety sensitivity). When someone who is highly anxiety sensitive expects to encounter anxiety symptoms during the experience of pain, fear of pain will likely develop (Asmundson et al. 2000; Ocañez et al. 2010). However, Keogh and Asmundson (2004) argue that it is more reasonable to assume that fear of pain is related to injury/illness sensitivity, which is also supported by Van Cleef et al. (2006). Fear of pain is still a relatively general construct. Fear of pain can be directed at pain sensations as well as activities and situations that are associated with pain. In chronic low back pain patients, one of the more specific forms of fear of pain is fear of movement/(re)injury, which is the specific fear that physical activity will cause (re)injury (Vlaeyen and Crombez 1999).

Specific Fears: Fear of Movement/(Re)Injury

A number of CLBP patients believe that performance of certain activities may induce or promote pain and (re)injury. Beliefs concerning harmful consequences of activities lead to fear

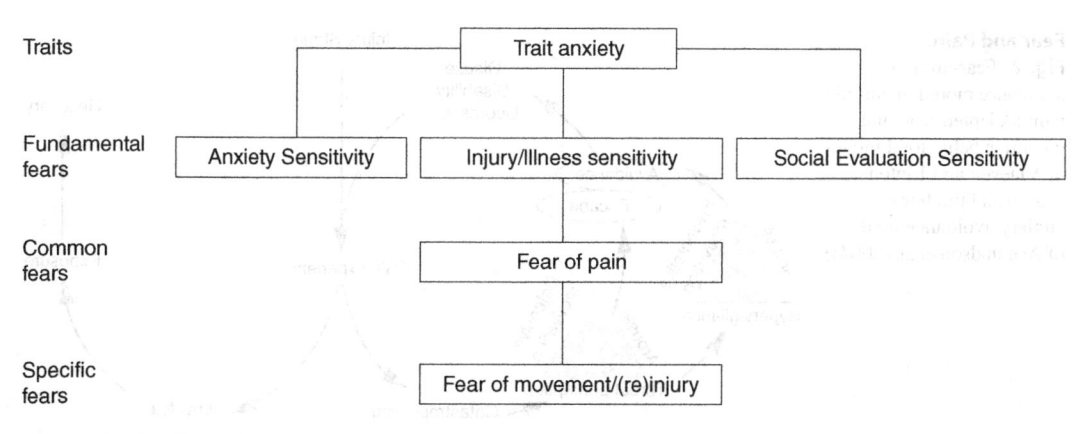

Fear and Pain,
Fig. 2 Fear-anxiety-
avoidance model of chronic
pain (Adapted from the
cognitive behavioral model
of Vlaeyen and Linton
(2000) and the fear-
anxiety-avoidance model
of Asmundson et al. (2004))

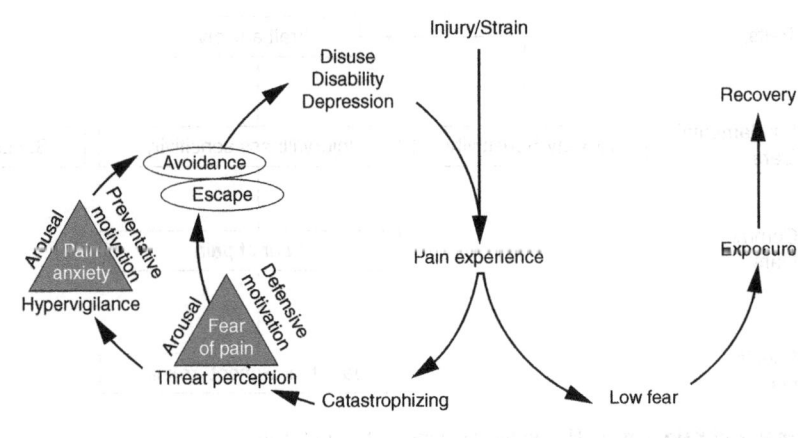

of movement/(re)injury and consequently to the avoidance of these activities, although medical indications for this behavioral pattern of avoidance are lacking. Despite the fact that in acute pain the avoidance of daily activities may be adaptive in facilitating healing and recovery, avoidance behavior is no longer necessary for recovery in chronic pain (Leeuw et al. 2007; Vlaeyen and Linton 2000).

Cognitive Behavioral Models

A cognitive behavioral model of chronic low back pain has been proposed, which emphasizes the crucial importance of the role of fear of movement/(re)injury and avoidance behavior in chronic low back pain patients (Vlaeyen 2003; Vlaeyen and Linton 2000). According to the model, two opposing behavioral responses may occur in response to acute pain: "confrontation" and "avoidance." A gradual confrontation and resumption of daily activities despite pain is considered as an adaptive response that eventually leads to the reduction of fear, the encouragement of physical recovery, and functional rehabilitation. In contrast, a catastrophic interpretation of pain is considered to be a maladaptive response, which initiates a vicious circle in which fear of movement/(re)injury and the subsequent avoidance of activities augment functional disability and the pain experience by means of hypervigilance, depression, and disuse. Substantial support for this cognitive behavioral

model and the role of the specific fear of movement/(re)injury has been found (Leeuw et al. 2007; Vlaeyen and Linton 2000).

In addition to this cognitive behavioral model, Asmundson et al. (2004) propose to update the model by integrating the concept of anxiety in addition to fear, referring to this as the fear-anxiety-avoidance model (Fig. 2).

This model states that catastrophizing about pain produces fear of pain, designed to protect the individual from the perceived immediate threat. This fear of pain in turn might promote pain-related anxiety. Pain-producing stimuli result through pain-related fear in escape and protecting behaviors aimed at reducing pain intensity. These behaviors in turn strengthen erroneous beliefs about pain, increase catastrophizing, and further enhance pain-related fear. The addition of an anxiety-related pathway to the pathway of pain-related fear provides a more accurate explanation for the fact that chronic pain interferes with one's daily life. In the anticipation, rather than in the presence of pain and/or injury, anxiety is evoked, leading to an increased attention (hypervigilance) for evidence of potential pain or injury. This hypervigilance and psychical responses may interact with memories and may promote misinterpretations of harmless stimuli as impending danger of pain or injury. Behaviorally, anxiety results in avoidance and preventative behaviors, increasing disability and disuse (Asmundson et al. 2004).

Recently, the sequential relationships between changes in pain catastrophizing, subsequent fear and avoidance, and functional outcomes such as disability or return to work were examined using a prospective design (Wideman et al. 2009). Although overall changes in pain catastrophizing and fear of movement were predictive for return to work, no evidence was found that early changes in catastrophizing were associated with late changes in fear of movement. These findings suggest that both components might independently influence disability. At present, new cognitive behavioral models regarding pain-related fear are evolving, focussing on behavior in context. Fear of pain might result in diverse behaviors depending on the current goal context, and likewise, seemingly similar behaviors may be driven by dissimilar motivational strategies (Van Damme et al. 2008; Vlaeyen et al. 2009).

Other Objects of Fear in Pain

Morley and Eccleston (2004) propose the existence of a range of "feared objects" in chronic pain, because of the overwhelming threat value of pain and three associated capacities to (1) interrupt, (2) interfere, and (3) impact on one's identity. Many potential fears arise because of the ability of pain to threaten the whole range of a person's existence. Interruption is established because the immediate pain experience interrupts behavior and influences the person's cognitive functioning (e.g., thoughts about possible harm). Interference is visible in the diminished accomplishment of daily functional activities. Finally, when repeated interference occurs to a degree that it concerns major goals, a threat to the identity is instigated. As chronic pain interferes with current tasks, plans, and goals, the person's perspective of oneself is changed, both with respect to the future and the past. Fear and anxiety are likely to occur when the goals and the identity of a person are threatened (Morley 2008).

Assessment of Fear of Pain

Several measurements of fear of pain are available (for an overview see McNeil and Vowles 2004).

Anxiety sensitivity can be measured by the 16-item Anxiety Sensitivity Index (ASI) (Peterson and Reiss 1987), measuring the degree to which people are concerned about the possible negative consequences of anxiety symptoms. Injury/illness sensitivity can be measured with the corresponding subscale of the sensitivity index (Taylor 1993). Fear of pain can be measured by, for example, the Pain Anxiety Symptoms Scale (PASS) (McCracken and Dhingra 2002; McCracken et al. 1993), designed to assess pain-specific fearful appraisals, cognitive symptoms of anxiety, physiological symptoms of anxiety, and escape and avoidance behavior. Another way to assess pain-related fear is by means of the Fear of Pain Questionnaire (FPQ), a self-report measure utilized in clinical as well as in nonclinical populations (McNeil and Rainwater 1998). Fear of movement/(re)injury can best be measured with the 17-item Tampa Scale for Kinesiophobia (TSK; Kori et al. 1990).

Treatment Implications

Due to the inextricable binding between fear and (chronic) pain, treatment of chronic pain should aim to focus on these perpetuating factors. In graded exposure in vivo, a hierarchy of fearful activities is established, which leads to disconfirmation of pain beliefs and reduction of fear, thereby promoting recovery of activities and functional abilities (Vlaeyen et al. 2004; Woods and Asmundson 2008). Fear- and anxiety-focused treatments seem to provide promising results in chronic low back pain patients (Bailey et al. 2010; Lohnberg 2007).

References

Asmundson, G. J. G., & Katz, J. (2009). Understanding the co-occurrence of anxiety disorders and chronic pain: State-of-the-art. *Depression and Anxiety, 26*(10), 888–901.

Asmundson, G. J., Wright, K. D., & Hadjistavropoulos, H. D. (2000). Anxiety sensitivity and disabling chronic health conditions: State of the arts and future directions. *Scandinavian Journal of Behaviour Therapy, 29*, 100–117.

Asmundson, G. J., Norton, P. J., & Vlaeyen, J. W. S. (2004). Fear-avoidance models of chronic pain: An overview. In G. J. G. Asmundson, J. W. S. Vlaeyen,

& G. Crombez (Eds.), *Understanding and treating fear of pain*. Oxford: Oxford University Press.

Bailey, K. M., Carleton, R. N., Vlaeyen, J. W. S., & Asmundson, G. J. G. (2010). Treatments addressing pain-related fear and anxiety in patients with chronic musculoskeletal pain: A preliminary review. *Cognitive Behaviour Therapy, 39*(1), 46–63.

Crombez, G., Vlaeyen, J. W. S., Heuts, P. H. T. G., et al. (1999). Pain-related fear is more disabling than pain itself: Evidence on the role of pain-related fear in chronic back pain disability. *Pain, 80*, 329–339.

Gatchel, R. G., Peng, Y. B., Peters, M. L., Fuchs, P. N., & Turk, D. C. (2007). The biopsychosocial approach to chronic pain: Scientific advances and future directions. *Psychological Bulletin, 133*(4), 581–624.

Gheldof, E. L. M., Vinck, J., Van den Bussche, E., Vlaeyen, J. W. S., Hidding, A., & Crombez, G. (2006). Pain and pain-related fear are associated with functional and social disability in an occupational setting: Evidence of mediation by pain-related fear. *European Journal of Pain, 10*(6), 513–525.

Hirsh, A. T., George, S. Z., Bialosky, J. E., & Robinson, M. E. (2008). Fear of pain, pain catastrophizing, and acute pain perception: Relative prediction and timing of assessment. *The Journal of Pain, 9*(9), 806–812.

Keogh, E., & Asmundson, G. J. G. (2004). Negative affectivity, catastrophizing, and anxiety sensitivity. In G. J. G. Asmundson, J. W. S. Vlaeyen, & G. Crombez (Eds.), *Understanding and treating fear of pain*. Oxford: Oxford University Press.

Kori, S. H., Miller, R. P., & Todd, D. D. (1990). Kinesiophobia: A new view of chronic pain behavior. *Pain Manage*, 35–43.

Leeuw, M., Goossens, M. E. J. B., Linton, S. J., Crombez, G., Boersma, K., & Vlaeyen, J. W. S. (2007). The fear-avoidance model of musculoskeletal pain: Current state of scientific evidence. *Journal of Behavioral Medicine, 30*(1), 77–94.

Lohnberg, J. A. (2007). A review of outcome studies on cognitive behavioral therapy for reducing fear-avoidance beliefs among individuals with chronic pain. *Journal of Clinical Psychology in Medical Settings, 14*, 113–122.

Lundberg, M., Larsson, M., Östlund, H., & Styf, J. (2006). Kinesiophobia among patients with musculoskeletal pain in primary healthcare. *Journal of Rehabilitation Medicine, 38*, 37–43.

McCracken, L. M., & Dhingra, L. (2002). A short version of the pain anxiety symptoms scale (PASS-20): Preliminary development and validity. *Pain Research & Management, 7*(1), 45–50.

McCracken, L. M., Zayfert, C., & Gross, R. T. (1993). The pain anxiety symptoms scale (PASS): A multidimensional measure of pain-specific anxiety symptoms. *Behavior Therapist, 16*, 183–184.

McNeil, D. W., & Rainwater, A. J. (1998). Development of the fear of pain questionnaire-III. *Journal of Behavioral Medicine, 21*(4), 389–410.

McNeil, D. W., & Vowles, K. E. (2004). Assessment of fear and anxiety associated with pain: Conceptualisation, methods, and measures. In G. J. G. Asmundson, J. W. S. Vlaeyen, & G. Crombez (Eds.), *Understanding and treating fear of pain*. Oxford: Oxford University Press.

Morley, S. (2008). Psychology of pain. *British Journal of Anaesthesia, 101*(1), 25–31.

Morley, S., & Eccleston, C. (2004). The object of fear in pain. In G. J. G. Asmundson, J. W. S. Vlaeyen, & G. Crombez (Eds.), *Understanding and treating fear of pain*. Oxford: Oxford University Press.

Ocañez, K. L. S., McHugh, R. K., & Otto, M. W. (2010). A meta-analytic review of the association between anxiety sensitivity and pain. *Depression and Anxiety, 27*, 760–767.

Peterson, R. A., & Reiss, S. (1987). *Anxiety sensitivity index manual*. Palos Heights: International Diagnostic Systems.

Reiss, S., & McNally, R. J. (1985). The expectancy model of fear. In S. Reis & R. R. Bootzin (Eds.), *Theoretical issues in behavior therapy*. New York: Academic.

Sylvers, P., Lilienfeld, S. O., & LaPrairie, J. L. (2011). Differences between trait fear and trait anxiety: Implications for psychopathology. *Clinical Psychology Review, 31*(1), 122–137.

Taylor, S. (1993). The structure of fundamental fears. *Journal of Behavior Therapy and Experimental Psychiatry, 24*, 289–299.

Turk, D. C., & Wilson, H. D. (2010). Fear of pain as a prognostic factor in chronic pain: Conceptual models, assessment, and treatment implications. *Current Pain and Headache Reports, 14*(2), 88–95.

Van Cleef, L. M. G., Peters, M. L., Roelofs, J., & Asmundson, G. J. G. (2006). Do fundamental fears differentially contribute to pain-related fear and pain catastrophizing? An evaluation of the sensitivity index. *European Journal of Pain, 10*(6), 527–536.

Van Damme, S., Crombez, G., & Eccleston, C. (2008). Coping with pain: A motivational perspective. *Pain, 139*, 1–4.

Vlaeyen, J. W. S. (2003). Fear in musculoskeletal pain. In O. Dostrovsky, D. B. Carr, & M. Koltzenburg (Eds.), *Proceedings of the 10th World Congress on pain, progress in pain research and management* (Vol. 24, pp. 631–650). Seattle: IASP Press.

Vlaeyen, J. W. S., & Crombez, G. (1999). Fear of movement/(re)injury, avoidance and pain disability in chronic low back pain patients. *Manual Therapy, 4*(4), 187–195.

Vlaeyen, J. W. S., & Linton, S. J. (2000). Fear-avoidance and its consequences in chronic musculoskeletal pain: A state of the art. *Pain, 85*, 317–332.

Vlaeyen, J. W. S., de Jong, J. R., Leeuw, M., & Crombez, G. (2004). Fear reduction in chronic pain: Graded exposure in vivo with behavioral experiments. In G. J. G. Asmundson, P. J. Norton, G. R. Norton,

J. W. S. Vlaeyen, & G. Crombez (Eds.), *Understanding and treating fear of pain* (pp. 313–343). Oxford: Oxford University Press.

Vlaeyen, J. W. S., Hanssen, M., Goubert, L., Vervoort, T., Peters, M. L., Van Breukelen, G., et al. (2009). Threat of pain influences social context effects on verbal pain report and facial expression. *Behaviour Research and Therapy, 47*(9), 774–782.

Wideman, T. H., Adams, H., & Sullivan, M. J. L. (2009). A prospective sequential analysis of the fear-avoidance model of pain. *Pain, 145*, 45–51.

Woods, M. P., & Asmundson, G. J. G. (2008). Evaluating the efficacy of graded in vivo exposure for the treatment of fear in patients with chronic back pain: A randomized controlled clinical trial. *Pain, 136*, 271–280.

Fear Avoidance

Definition

Fear avoidance is the avoidance of activities in order to prevent injury, reinjury, or exacerbation of any injury or pain. It is a learned behavior to reduce negative pain-related emotions. In this way, pain-related fear can lead to disability, catastrophizing beliefs, hypervigilance to bodily signals, and avoidance behavior.

Cross-References

► Disability, Fear of Movement
► Pain in the Workplace, Risk Factors for Chronicity, and Psychosocial Factors

Fear Avoidance Beliefs

Definition

Fear avoidance beliefs are cognitions often found in individuals with pain, which support avoidance behavior and make participation in physical activity programs more difficult. Patients believe that physical activity initiated their pain and that physical activity is bound to aggravate the pain in the long run.

Cross-References

► Disability
► Fear Avoidance
► Pain
► Psychological Treatment of Pain in Older Populations

Fear of Movement/(Re)Injury

Definition

Fear related to movement and physical activity, inextricably associated with the fear that physical activity will cause pain and (re)injury. Fear of movement is most prominent in patients with chronic pain syndromes.

Cross-References

► Disability, Fear of Movement
► Fear and Pain

Fear of Pain

Definition

Pain-related fear is a general term to describe several forms of fear with respect to pain. Depending on the anticipated source of threat, the content of fear of pain varies considerably. Fear of pain can be directed toward the occurrence or continuation of pain, toward physical activity, or toward the induction of (re)injury or physical harm.

Cross-References

► Disability, Fear of Movement
► Fear and Pain
► Muscle Pain, Fear-Avoidance Model

Fear Reduction Through Exposure In Vivo

Jeroen R. De Jong[1] and Johan W. S. Vlaeyen[2,3]
[1]Department of Rehabilitation, University
Hospital Maastricht, Maastricht,
The Netherlands
[2]Research Group Health Psychology, Catholic
University of Leuven, Leuven, Belgium
[3]Department of Clinical Psychological Science,
Maastricht University, Maastricht,
The Netherlands

Synonyms

Exposure in vivo; Exposure treatment; Extinction; Graded exposure; Graded exposure in vivo with behavioral experiments

Definition

Exposure in vivo, originally based on ▶ extinction of ▶ Pavlovian conditioning (Bouton 1988), is currently viewed as a cognitive process during which fear is activated and catastrophic expectations are being challenged and disconfirmed, resulting in reduction of the threat value of the originally fearful stimuli. During graded exposure, special attention goes to the establishment of an individual hierarchy of the ▶ pain-related fear stimuli. Exposure in vivo includes activities that are selected based on the fear hierarchy and the idiosyncratic aspects of the fear stimuli.

Characteristics

In order to produce disconfirmations between expectations of pain and harm, the actual pain, and the other consequences of the activity or movement, Philips (1987) was one of the first to argue for repeated, graded, and controlled exposures to such situations. Experimental support for this idea is provided by the match-mismatch model of pain (Rachman and Arntz 1991), which states that people

initially tend to overpredict how much pain they will experience, but after some exposures these predictions tend to be corrected to match with the actual experience. Crombez et al. (2002) and Goubert et al. (2002) found a similar pattern in a sample of chronic low back pain patients, who were requested to perform certain physical activities. As predicted, the chronic low back pain patients initially overpredicted pain, but after repetition of the activity, the overprediction was readily corrected. Overpredictions of pain and the negative consequences of pain are more pronounced in individuals reporting increased fear of pain. A number of studies examined the effectiveness of a graded ▶ exposure in vivo treatment in reducing pain-related fear, ▶ pain catastrophizing, and pain disability in chronic pain patients who were referred for outpatient behavioral rehabilitation (Vlaeyen et al. 2002a, b, c; Vlaeyen et al. 2001; de Jong et al. 2005a, b, 2008, 2012; Leeuw et al. 2008; Linton et al. 2008; Woods and Asmundson 2008). Theses showed improvements in pain-related fear, pain catastrophizing, and pain disability whenever the graded exposure was initiated. Measured with ambulatory activity monitors, the improvements also generalized to increases in physical activity in the home situation (Vlaeyen et al. 2002a). Besides behavioral and cognitive changes, one study in patients with ▶ Complex Regional Pain Syndrome (CRPS) even found observed positive changes in edema, skin color, excessive sweating, and motor function disturbances (de Jong et al. 2005a). It seems that graded exposure to painful activities, movements, and/or situations that were previously avoided may indeed be a successful treatment approach for pain patients reporting substantial pain-related fear.

Graded Exposure In Vivo
We suggest that the intervention generally be designed in three steps: cognitive-behavioral assessment, education, and exposure in vivo with ▶ behavioral experiments.

Cognitive-Behavioral Assessment
Specific Questionnaires A basic question that may be asked is as follows: "What is the nature of the perceived threat?" The answer is not as simple

as it seems. Patients may not view their problem as involving fear at all and may simply see inability in performing certain activities due to pain. In addition, the specific nature of pain-related fear varies considerably, making an idiosyncratic approach almost indispensable. Patients may not fear pain itself, but the impending (re)injury it signals: pain is seen as a warning signal for a seriously threatening situation. The list below outlines pain-related fear questionnaires, including sample items:

Pain and Impairment Relationship Scale (*PAIRS*: Riley et al. 1988)

"I have to be careful not to do anything that might make my pain worse."

"All of my problems would be solved if my pain would go away."

Survey of Pain Attitudes (*SOPA* [*Subscale Harm*]: Jensen et al., 1992)

Measures of Pain Catastrophizing

Pain Catastrophizing Scale (*PCS*: Sullivan et al. 1995)

Rumination: "I keep thinking about how much it hurt."

Amplification: "I grow afraid that the pain will get worse."

Helplessness: "There's nothing I can do to reduce the intensity of the pain."

Pain Cognition List (*PCL* [*Subscale catastrophizing*]: Van Breukelen and Vlaeyen 2005)

Cognitive Error Questionnaire (Lefebvre 1981)

Vignettes about catastrophizing, overgeneralization, personalization, and selective abstraction: e.g., "You have a painful back problem but have continued to work. Although you got quite a bit done today, you quit work a little early because your back was really hurting. You think to yourself, 'What a terrible day, it seems like I can't get anything done'."

Coping Strategies Questionnaire (*CSQ* [*Subscale catastrophizing*]: Rosenstiel and Keefe 1983)

"I feel I can't stand it anymore."

Pain-Related Fear and Anxiety

Tampa Scale for Kinesiophobia (*TSK*: Miller et al. 1991)

Harm:

"My body is telling me I have something dangerously wrong."

Avoidance of activity:

"Pain lets me know to stop exercising so that I don't injure myself."

Pain Anxiety Symptoms Scale (*PASS*: McCracken et al. 1992)

Cognitive anxiety:

"I can't think straight when in pain."

Escape/avoidance:

"I will stop any activity as soon as I sense pain coming on."

Fear:

"When I feel pain, I am afraid that something terrible will happen."

Physiological anxiety:

"I begin trembling when engaged in an activity that increases pain."

Fear-Avoidance Beliefs Questionnaire (*FABQ*: Waddell et al. 1993)

Fear-avoidance beliefs about work:

"My work might harm my back."

Fear-avoidance beliefs about physical activity:

"My pain was caused by physical activity."

Interview The semi structured interview is an additional tool to better estimate the role of pain-related fear in the pain problem. It includes information about the antecedents (situational and/or internal), catastrophic (mis)interpretations, and consequences of the pain-related fear. Information is gathered about the assumptions patients make of the association between activities, pain, and (re) injury. Factors that often seem to be associated with the development of fear are the characteristics of pain onset and the ambiguity surrounding the presence or absence of positive findings on medico-diagnostics. Reports about misconceptions and misinterpretations of information can later be used during the educational part of the intervention. Finally, the interview should also clarify whether other problems such as major depression, marital conflicts, or disability claims warrant specific attention before or after treatment.

Graded Hierarchies What is the patient actually afraid of? In addition to checklists of daily activities, the presentation of visual materials, such as pictures of stressing activities and movements reflecting the full range of situations avoided by the patient, can be quite helpful in the development of graded hierarchies. They start with activities or situations that provoke only mild discomfort and end with those that are beyond the patient's present abilities. The Photograph Series of Daily Activities uses photographs representing various physical daily activities to be placed along a fear thermometer (Dubbers et al. 2003; Jelinek et al. 2003; Kugler et al. 1999; Leeuw et al. 2007).

Behavioral Tests Sometimes, patients find it hard to really estimate the harmfulness of an activity when it has been avoided extensively. For cases in which even pictorial methods (PHODA) do not work, behavioral tests can be introduced. These consist of performing an activity that has been avoided previously, while performance indices (e.g., time, distance, number of repetitions) are measured. Target behaviors can be derived from PHODA items, and in most cases the behavioral tests can be considered as a variant of the exercise tolerance test described by Fordyce (Fordyce 1976).

Education
One of the major goals of the educational section is to increase the willingness of patients to finally engage in activities they avoided for a long time. The aim is to correct the misinterpretations and misconceptions that occurred early on during the development of the pain-related fear. The educational section is more than just reassuring that there are no specific physical abnormalities. Unambiguously educating the patient in a way that the patient views his or her pain as a common condition that can be self-managed, rather than a serious disease or condition that needs careful protection, is a useful first step. It can be explained to patients that they may have probably overestimated the value of diagnostic tests and that in symptom-free people similar abnormalities can also be found.

Graded Exposure In Vivo with Behavioral Experiments
Graded Exposure In Vivo As firsthand evidence of actually experiencing oneselves

behaving differently is far more convincing than rational argument, the most essential step consists of graded exposure to the situations the fearful patient has identified as "dangerous" or "threatening." The patient is encouraged to engage in these fearful activities as much as possible, until disconfirmation has occurred and anxiety levels have decreased. If the rating has decreased significantly, the therapist may consider moving on to the next item of the hierarchy. Each activity or movement is first modeled by the therapist. His presence, initially acting as a safety signal to promote more exposures, is gradually withdrawn to facilitate independence and to create contexts that mimic those of the home situation (Vlaeyen et al. 2012).

Behavioral Experiments The graded exposure to fear-eliciting activities can be carried out in the form of a behavioral experiment in which a collaborative empiricism is the bottom line. The essence of a behavioral experiment is that the patient performs an activity to challenge the validity of his catastrophic assumptions and misinterpretations. These assumptions take the form of "If ... then ..." statements and are empirically tested in such a behavioral experiment.

Generalization and Maintenance of Change
Exposure to physical activities is not likely to result in a fundamental change in the belief of the pain patient that certain movements are harmful or painful (Goubert et al. 2002). More likely, the patient will learn that the movements involved in the exposure treatment are less harmful or painful than anticipated. In other words, during successful exposure, exceptions to the rule are learned, rather than there being a fundamental change of that rule. Generalization and maintenance can be enhanced by the following measures. First, exposure to the full spectrum of contexts and natural settings in which fear has been experienced is required. Second, during the exposure, it is best that stimuli be varied as much as possible. Third, expanded-spaced, rather than a massed exposure, is preferred (Vlaeyen et al. 2012).

References

Bouton, M. E. (1988). Context and ambiguity in the extinction of emotional learning: Implications for exposure therapy. *Behaviour Research and Therapy, 26*, 137–149.

Crombez, G., Eccleston, C., Vlaeyen, J. W., Vansteenwegen, D., Lysens, R., & Eelen, P. (2002). Exposure to physical movements in low back pain patients: Restricted effects of generalization. *Health Psychology, 21*, 573–578.

de Jong, J. R., Vlaeyen, J. W. S., Onghena, P., Cuypers, C., den Hollander, M., & Ruygrok, J. (2005a). Reduction of pain-related fear in complex regional pain syndrome type I: The application of graded exposure in vivo. *Pain, 116*, 264–275.

de Jong, J. R., Vlaeyen, J. W., Onghena, P., Goossens, M. E., Geilen, M., & Mulder, M. (2005b). Fear of movement/(Re) injury in chronic low back pain: Education or exposure *In Vivo* as mediator to fear reduction? *The Clinical Journal of Pain, 21*, 9–17.

de Jong, J. R., Vangronsveld, K., Peters, M. L., Goossens, M. E., Onghena, P., Bulte, I., et al. (2008). Reduction of pain-related fear and disability in post-traumatic neck pain: A replicated single-case experimental study of exposure in vivo. *The Journal of Pain, 9*(12), 1123–1134.

de Jong, J. R., Vlaeyen, J. W. S., van Eijsden, M., Loo, C., & Onghena, P. (2012). Reduction of pain-related fear and increased function and participation in work-related upper extremity pain (WRUEP): Effects of exposure in vivo. *Pain, 10*, 2109–2118.

Dubbers, A. T., Vikström, M. H., & De Jong, J. R. (2003). *The Photograph series of Daily Activities (PHODA): Cervical spine and shoulder* [CD-ROM version 1.2]. The Netherlands: Hogeschool Zuyd, University Maastricht and Institute for Rehabilitation Research (iRv).

Fordyce, W. E. (1976). *Behavioral methods for chronic pain and illness*. St. Louis: Mosby.

Goubert, L., Francken, G., Crombez, G., Vansteenwegen, D., & Lysens, R. (2002). Exposure to physical movement in chronic back pain patients: No evidence for generalization across different movements. *Behaviour Research and Therapy, 40*, 415–429.

Jelinek, S., Germes, D., Leyckes, N., & De Jong, J. R. (2003). *The Photograph Series of Daily Activities (PHODA): Low extremities* [CD-ROM Version 1.2]. The Netherlands: Hogeschool Zuyd, University Maastricht and Institute for Rehabilitation Research (iRv).

Jensen, M. P., & Karoly, P. (1992). Pain-specific beliefs, perceived symptom severity, and adjustment to chronic pain. *The Clinical Journal of Pain, 8*, 123–130.

Kugler, K., Wijn, J., Geilen, M., De Jong, J. R., & Vlaeyen, J. W. S. (1999). *The Photograph series of Daily Activities (PHODA)* [CD-ROM version 1.0]. The Netherlands: Institute for Rehabilitation Research and School for Physiotherapy Heerlen.

Leeuw, M., Goossens, M. E. J. B., van Breukelen, G., Boersma, K., & Vlaeyen, J. W. S. (2007). Measuring perceived harmfulness of physical activities in patients with chronic low back pain: The photograph series of daily activities [Short electronic version]. *The Journal of Pain, 8*, 840–849.

Leeuw, M., Goossens, M. E. J. B., van Breukelen, G. J. P., de Jong, J. R., Heuts, P. H. T. G., Smeets, R. J. E. M., et al. (2008). Exposure in vivo versus operant graded activity in chronic low back pain patients: Results of a randomized controlled trial. *Pain, 138*, 192–207.

Lefebvre, M. F. (1981). Cognitive distortion and cognitive errors in depressed psychiatric and low back pain patients. *Journal of Consulting and Clinical Psychology, 49*, 517–525.

Linton, S. J., Boersma, K., Jansson, M., Overmeer, T., Lindblom, K., & Vlaeyen, J. W. (2008). A randomized controlled trial of exposure in vivo for patients with spinal pain reporting fear of work-related activities. *European Journal of Pain, 12*, 722–730.

McCracken, L. M., Zayfert, C., & Gross, R. T. (1992). The Pain Anxiety Symptoms Scale: Development validation of a scale to measure fear of pain. *Pain, 50*, 67–73.

Miller, R. P., Kori, S. H., & Todd, D. D. (1991). *The Tampa Scale for Kinisophobia*. Unpublished Report, Tampa, Fl.

Philips, H. C. (1987). Avoidance behaviour and its role in sustaining chronic pain. *Behaviour Research and Therapy, 25*, 273–279.

Rachman, S., & Arntz, A. R. (1991). The overprediction and underprediction of pain. *Clinical Psychology Review, 11*, 339–355.

Riley, J. F., Ahern, D. K., & Follick, M. J. (1988). Chronic pain and functional impairment: Assessing beliefs about their relationship. *Archives of Physical Medicine and Rehabilitation, 69*, 579–582.

Rosenstiel, A. K., & Keefe, F. J. (1983). The use of coping strategies in chronic low back pain. *Pain, 17*, 33–44.

Sullivan, M. J., Bishop, S. R., & Pivik, J. (1995). The pain catastrophizing scale: Development and validation. *Psychological Assessment, 7*, 524–532.

Van Breukelen, G. J., & Vlaeyen, J. W. (2005). Norming clinical questionnaires with multiple regression: The pain cognition list. *Psychological Assessment, 17*, 336–444.

Vlaeyen, J. W. S., De Jong, J. R., Geilen, M., Heuts, P. H. T. G., & van Breukelen, G. (2001). Graded exposure in vivo in the treatment of pain-related fear: A replicated single-case experimental design in four patients with chronic low back pain. *Behaviour Research and Therapy, 39*, 151–166.

Vlaeyen, J. W., de Jong, J. R., Geilen, M., Heuts, P. H., & van Breukelen, G. (2002a). The treatment of fear of movement/(Re) injury in chronic low back pain: Further evidence on the effectiveness of exposure *in vivo*. *The Clinical Journal of Pain, 18*, 251–261.

Vlaeyen, J. W., de Jong, J. R., Onghena, P., Kerckhoffs-Hanssen, M., & Kole-Snijders, A. M. (2002b).

Can pain-related fear be reduced? The application of cognitive-behavioral exposure *in vivo*. *Pain Research & Management, 7,* 144–153.

Vlaeyen, J. W., de Jong, J. R., Sieben, J. M., & Crombez, G. (2002c). Graded exposure *in vivo* for pain-related fear. In D. C. Turk & R. J. Gatchel (Eds.), *Psychological approaches to pain management a practitioner's handbook* (pp. 210–233). New York: Guilford.

Vlaeyen, J. W., Morley, S. J., Linton, S. J., Boersma, K., & de Jong, J. R. (2012). *Pain-related fear: Exposure-based treatment of chronic pain.* Seattle: IASP Press.

Waddell, G., Newton, M., Henderson, I., Somerville, D., & Main, C. J. (1993). A Fear-Avoidance Beliefs Questionnaire (FABQ) and the role of fear-avoidance beliefs in chronic low back pain and disability. *Pain, 52,* 157–168.

Woods, M. P., & Asmundson, G. J. G. (2008). Evaluating the efficacy of graded in vivo exposure for the treatment of fear in patients with chronic back pain: A randomized controlled clinical trial. *Pain, 136,* 271–280.

Fear-Anxiety-Avoidance Model

Definition

The Fear-Anxiety-Avoidance Model states that catastrophic misinterpretations in response to acute pain can lead to fear of pain and subsequently to pain-related anxiety. Fear urges the escape from the pain stimulus, while anxiety in anticipation of pain or threat urges avoidance of those situations. As a result of this, a self-perpetuating cycle develops in which subsequent avoidance of activities augments functional disability and the pain experience by means of hypervigilance, depression, and decreased physical fitness, thereby further advancing chronicity.

Cross-References

▶ Fear and Pain

Fear-Avoidance

▶ Disability, Fear of Movement

Fear-Avoidance Model

Definition

The basic tenet of the fear-avoidance model is that catastrophic misinterpretations in response to acute pain lead to pain-related fear, which subsequently urges the escape from the painful stimuli, and to selectively attend to bodily sensations. As a result, a self-perpetuating cycle develops in which subsequent avoidance of activities augments functional disability, depression, and decreased physical fitness.

Cross-References

▶ Fear Reduction through Exposure in Vivo
▶ Muscle Pain, Fear-Avoidance Model

Feasible

Definition

The simple, economic, and easy application of a measure

Cross-References

▶ Pain Assessment in Neonates

Feedback Control of Pain

Definition

Ascending nociceptive signals activate supraspinal mechanisms (e.g., an inhibitory brainstem-spinal pathway) that suppress spinal/trigeminal nociceptive transmission.

Cross-References

▶ Descending Modulation and Persistent Pain

Fee-for-Service

Definition

Patients or insurance pay for medical care as it is needed and according to the service provided.

Cross-References

▶ Disability Management in Managed Care System

Female Reproductive Organ Pain Model

▶ Visceral Pain Models, Female Reproductive Organ Pain

Female Sex Steroid Hormones and Pain

Jennifer Brawn[1] and Katy Vincent[2]
[1]P.A.I.N. Group, Center for Pain and the Brain, Boston Children's Hospital, Harvard Medical School, Waltham, MA, USA
[2]Nuffield Department of Obstetrics and Gynaecology, The Women's Centre, John Radcliffe Hospital, University of Oxford, Oxford, UK

Introduction

It widely understood that females are disproportionately affected by pain syndromes when compared to their male counterparts. This clinical skew is notable in conditions such as migraine

headache, temporomandibular joint disorder (TMJD), irritable bowel syndrome (IBS), and fibromyalgia (Unruh 1996). In addition to the higher prevalence of females with pain conditions, there have been reports of cyclical patterns of pain onset and severity (Martin 2009). This disparity in pain experience has resulted in a significant body of research comparing gender-specific physiological differences. Most experimental pain studies suggest that females are more sensitive to pain than males, although there are inconsistencies in the literature (Martin 2009; Fillingim et al. 2009). Consequently, researchers have focused on female biology, specifically the menstrual cycle, to untangle this important clinical and biological conundrum. The two main female sex steroid hormones, progesterone and estradiol, are thought to influence pain processing through alteration of a variety of neural mechanisms.

Characteristics

Pain Perception in Healthy Females over the Menstrual Cycle

Animal models have allowed for improved mechanistic understanding of the central effects of progesterone and estradiol; however, differences in the physiology of the ovarian cycle between species causes difficulty in extrapolating some of these findings to humans. In rodents, these hormones have been implicated in a variety of pain-related pathways including the gamma-aminobutyric acid (GABA), glutamate, opioid, and dopamine systems. Thus, estradiol appears to mediate inflammatory responses, increase opioidergic tone, and promote reuptake of glutamate, while progesterone is thought to alter GABAergic inhibitory activity by increasing binding affinity through its metabolite allopregnanolone. Certainly, in rodents there is a notable cyclical trend toward greater pain sensitivity in the proestrus phase, a time that is characterized by dramatic fluctuations of progesterone and estradiol. However, researchers have also observed increased sensitivity at other points including estrus, metestrus, and diestrus (Reviewed in Martin 2009). In humans, results are more difficult to interpret. A number of factors contribute to this problem, not least the different definitions and methods of verification of the menstrual cycle phases (Sherman and LeResche 2006).

Evaluating Menstrual Cycle Changes in Pain Perception

In a meta-analysis performed by Riley et al. 1999, 16 studies were categorized by stimuli and assessed (Fillingim et al. 2009). For the cold pressor test, pressure, heat, and ischemic pain, sensitivity was lowest in the follicular phase, characterized by low hormone levels with estradiol increasing in the latter half. Electrical pain sensitivity, however, was lowest in the luteal and highest in the periovulatory phases. Only one of the included studies obtained hormonal measures. In a subsequent methodological review, researchers were skeptical of this analysis. They found inconsistent cyclical patterns in thermal, mechanical, and ischemic pain perception. The only notable pattern involved electrical pain with studies reporting decreased sensitivity in the early luteal phase (Sherman and LeResche 2006).

In a recent narrative review of cyclical pain response in healthy females, Martin 2009 analyzed the results of 19 human studies. To make findings comparable, the author redefined the reported cycle phases based on the actual test dates. Observed trends were summarized using three categories. The author noted that six studies reported no cyclical changes in pain sensitivity. However, five studies observed that pain sensitivity increased in phases characterized by fluctuating estradiol and increasing progesterone (late follicular and early luteal). Seven studies found that pain sensitivity increased in phases characterized by low or decreasing estradiol and progesterone (late luteal or early follicular). It should be noted that only four of the studies analyzed had blood progesterone and estradiol levels to verify menstrual cycle phase. An additional three studies confirmed ovulation through biological measures.

Hormonal Measures and Pain Perception

Hormonal assays are key to untangling the complex relationship between progesterone, estradiol, and pain. While countless studies have evaluated changes in pain perception over the menstrual cycle, only a small number have obtained either plasma or serum hormone levels. These measures allow for analysis of pain changes independent of experimental sessions or somewhat arbitrary divisions of the menstrual cycle (Sherman and LeResche 2006). Fillingim et al. 1997 conducted the first study that used blood-derived estrogen and progesterone levels to evaluate changes in pain sensitivity, specifically pain onset and threshold, over the menstrual cycle. A significant correlation was observed between elevated plasma estrogen levels and low thermal pain onset and threshold measures. In a study that examined diffuse noxious inhibitory control (DNIC), within the ovulatory phase, pain-induced analgesia was positively correlated with progesterone levels (Tousignant-LaFlamme and Marchand 2009). Ring et al. 2009 observed that within the luteal phase, progesterone, estradiol, and pain ratings from needle and catheter insertion were positively correlated. While Stening et al. 2007 used the cold pressor test to measure transition from mild to moderate pain (activation time) and noted that shorter activation times were significantly correlated with high progesterone levels; as were maximal pain ratings. Estradiol, however, was not significantly correlated with any of the results. Stening and colleagues (2007) also analyzed the combined impact of estradiol and progesterone on pain perception. When present together at high levels, estradiol and progesterone appeared to decrease pain sensitivity, with estradiol mediating progesterone's apparent pronociceptive influence. Alternatively, several studies have not observed any correlations between hormone levels and pain measures (Craft 2007; Klatzkin et al. 2010; Kowalczyk et al. 2006).

Menstrual Cycle Disorders

Due to their close correlation with hormonal changes, several pain studies have evaluated females with cyclical medical conditions.

However, a major difficulty with this strategy is disentangling hormonal influences on pain perception per se from influences on the pathology causing the symptoms. Dysmenorrhea is a menstrual cycle condition characterized by extreme abdominal and back pain during menses. It has been speculated that low or decreasing progesterone levels may trigger dysmenorrheic symptoms (Martin 2009). In a clinical study, Granot et al. 2001 compared heat and laser pain thresholds in dysmenorrheic women and healthy females over four experimental sessions. Dysmenorrheic women were more sensitive to pain from both stimuli. For the laser stimuli, both groups appeared to have the lowest pain sensitivity in the luteal phase (high plasma progesterone and estradiol) and the highest during the follicular phase (increasing plasma estradiol). Cycle phase, however, did not affect dysmenorrheic women's response to the heat stimuli. Similarly, Vincent and colleagues (2011) demonstrated increased sensitivity to noxious heat in dysmenorrheic women compared to pain-free controls, but found no influence of cycle phase in either group. In this study, the timing of the three experimental sessions was chosen to maximize endogenous variation in levels of estradiol and progesterone.

Premenstrual dysphoric disorder (PMDD) is a luteal phase condition characterized by extreme physical and emotional changes. While the underlying pathology of the condition remains unclear, it has been postulated that women with PMDD may have abnormal levels of sex steroid hormones and/or altered neurotransmitter regulation by these hormones. In support of this, two studies have shown luteal phase estradiol and progesterone levels to be higher in women with PMDD than controls (Straneva et al. 2002; Epperson et al. 2002). Interestingly, although there was no difference in cortical GABA levels between the groups, GABA levels decreased across the cycle in healthy women but increased in those with PMDD, with the steroid hormones appearing to exert opposing effects in the two groups (e.g., a positive correlation between GABA and both estradiol and progesterone in the women with PMDD and a negative correlation in the control women) (Epperson et al. 2002).

PMDD sufferers demonstrated higher pain and unpleasantness ratings in response to noxious stimuli in the luteal compared to the follicular phase, and significantly lower pain tolerance and threshold measures than healthy controls in all menstrual cycle phases (Straneva et al. 2002).

Chronic Pain Conditions and Hormonal Treatment

Given the prevalence of pain conditions in the female population, studies have focused on both evaluating differences in experimental pain response and deconstructing patterns of increased pain. Chronic pain conditions such as fibromyalgia, temporomandibular joint disorder (TMJD), and migraine appear to worsen at times of low or rapidly decreasing estradiol (LeResche et al. 2003), while symptoms from IBS, interstitial cystitis/painful bladder syndrome, and migraine as well as a variety of gynecological pain conditions can be improved by pharmacological manipulation of hormonal status (Mathias et al. 1994; Lentz et al. 2002; Silberstein and Goldberg 2007). Once again, effects of hormones on the underlying pathologies cannot be excluded; however, Sherman et al. 2005 did find that females with TMJD were more sensitive to experimental pain during the mid-luteal and menstrual phases.

Neuroimaging Studies

Neuroimaging technology has added an interesting dimension to the field (reviewed by Vincent and Tracey 2010). Studies of pain perception in healthy subjects suggest alterations in the cerebral response to noxious stimuli in differing hormonal states, although the findings from individual studies are not entirely consistent (Choi et al. 2006; de Leeuw et al. 2006). Unfortunately, to date, no studies have used neuroimaging techniques to investigate the relationship between steroid hormones and the cerebral response to pain in patients with a chronic pain condition. However, studies in healthy subjects further strengthen the evidence for interactions between sex hormones and endogenous pain modulatory systems; most strikingly for estradiol and the mu-opioid system

(Smith et al. 2006), but also with GABA as discussed earlier and potentially other systems including the descending pain inhibitory system.

Summary

Conclusion

Despite the sheer number of studies, much work is needed to develop a cohesive understanding of individual and combined effects of estradiol and progesterone on pain perception. As noted by many reviews, reproducible, standardized methodologies are greatly needed (Martin 2009; Sherman and LeResche 2006; Riley et al. 1999). The field of pain and hormones is complex, thus it is important to minimize as much research variation as possible.

References

Choi, J. C., Park, S. K., Kim, Y. H., Shin, Y. W., Kwon, J. S., Kim, J. S., et al. (2006). Different brain activation patterns to pain and pain-related unpleasantness during the menstrual cycle. *Anesthesiology, 105*(1), 120–127.

Craft, R. M. (2007). Modulation of pain by estrogens. *Pain, 132*(Suppl. 1), S3–S12.

de Leeuw, R., Albuquerque, R. J., Andersen, A. H., & Carlson, C. R. (2006). Influence of estrogen on brain activation during stimulation with painful heat. *Journal of Oral and Maxillofacial Surgery: Official Journal of the American Association of Oral and Maxillofacial Surgeons, 64*(2), 158–166.

Epperson, C. N., Haga, K., et al. (2002). Cortical gamma-aminobutyric acid levels across the menstrual cycle in healthy women and those with premenstrual dysphoric disorder: a proton magnetic resonance spectroscopy study. *Archives of General Psychiatry, 59*(9), 851–858.

Fillingim, R. B., Maixner, W., Girdler, S. S., Light, K. C., Harris, M. B., Sheps, D. S., et al. (1997). Ischemic but not thermal pain sensitivity varies across the menstrual cycle. *Psychosomatic Medicine, 59*(5), 512–520.

Fillingim, R. B., King, C. D., Ribeiro-Dasilva, M. C., Rahim-Williams, B., & III Riley, J. L. (2009). Sex, gender, and pain: A review of recent clinical and experimental findings. *The Journal of Pain, 10*(5), 447–485.

Granot, M., Yarnitsky, D., Itskovitz-Eldor, J., Granovsky, Y., Peer, E., & Zimmer, E. Z. (2001). Pain perception in women with dysmenorrhea. *Obstetrics and Gynecology, 98*(3), 407–411.

Klatzkin, R. R., Mechlin, B., & Girdler, S. S. (2010). Menstrual cycle phase does not influence gender differences in experimental pain sensitivity. *European Journal of Pain, 14*(1), 77–82.

Kowalczyk, W. J., Evans, S. M., Bisaga, A. M., Sullivan, M. A., & Comer, S. D. (2006). Sex differences and hormonal influences on response to cold pressor pain in humans. *The Journal of Pain: Official Journal of the American Pain Society, 7*(3), 151–160.

Lentz, G. M., Bavendam, T., Stenchever, M. A., Miller, J. L., & Smalldridge, J. (2002). Hormonal manipulation in women with chronic, cyclic irritable bladder symptoms and pelvic pain. *American Journal of Obstetrics and Gynecology, 186*(6), 1268–1271; discussion 1271–1263.

LeResche, L., Mancl, L., Sherman, J. J., Gandara, B., & Dworkin, S. F. (2003). Changes in temporomandibular pain and other symptoms across the menstrual cycle. *Pain, 106*(3), 253–261.

Martin, V. T. (2009). Ovarian hormones and pain response: A review of clinical and basic science studies. *Gender Medicine, 6*(Suppl. 2), 168–192.

Mathias, J. R., Clench, M. H., Roberts, P. H., & Reeves-Darby, V. G. (1994). Effect of leuprolide acetate in patients with functional bowel disease. Long-term follow-up after double-blind, placebo-controlled study. *Digestive Diseases and Sciences, 39*(6), 1163–1170.

Riley, J. L., 3rd, Robinson, M. E., Wise, E. A., & Price, D. D. (1999). A meta-analytic review of pain perception across the menstrual cycle. *Pain, 81*(3), 225–235.

Ring, C., Veldhuijzen van Zanten, J. J., & Kavussanu, M. (2009). Effects of sex, phase of the menstrual cycle and gonadal hormones on pain in healthy humans. *Biological Psychology, 81*(3), 189–191.

Sherman, J. J., & LeResche, L. (2006). Does experimental pain response vary across the menstrual cycle? A methodological review. *American Journal of Physiology. Regulatory, Integrative and Comparative Physiology, 291*(2), R245–R256. doi:10.1152/ajpregu.00920.2005.

Sherman, J. J., LeResche, L., Mancl, L. A., Huggins, K., Sage, J. C., & Dworkin, S. F. (2005). Cyclic effects on experimental pain response in women with temporomandibular disorders. *Journal of Orofacial Pain, 19*(2), 133–143.

Silberstein, S. D., & Goldberg, J. (2007). Menstrually related migraine: Breaking the cycle in your clinical practice. *The Journal of Reproductive Medicine, 52*(10), 888–895.

Smith, Y. R., Stohler, C. S., Nichols, T. E., Bueller, J. A., Koeppe, R. A., & Zubieta, J. K. (2006). Pronociceptive and antinociceptive effects of estradiol through endogenous opioid neurotransmission in women. *The Journal of Neuroscience: The Official Journal of the Society for Neuroscience, 26*(21), 5777–5785.

Stening, K., Eriksson, O., Wahren, L., Berg, G., Hammar, M., & Blomqvist, A. (2007). Pain sensations to the cold pressor test in normally menstruating women: Comparison with men and relation to menstrual phase and serum sex steroid levels. *American*

Journal of Physiology. Regulatory, Integrative and Comparative Physiology, 293(4), R1711–R1716.

Straneva, P. A., Maixner, W., Light, K. C., Pedersen, C. A., Costello, N. L., & Girdler, S. S. (2002). Menstrual cycle, beta-endorphins, and pain sensitivity in premenstrual dysphoric disorder. *Health Psychology: Official Journal of the Division of Health Psychology, American Psychological Association, 21*(4), 358–367.

Tousignant-Laflamme, Y., & Marchand, S. (2009). Excitatory and inhibitory pain mechanisms during the menstrual cycle in healthy women. *Pain, 146*(1–2), 47–55.

Unruh, A. M. (1996). Gender variations in clinical pain experience. *Pain, 65*(2–3), 123–167.

Vincent, K., & Tracey, I. (2010). Sex hormones and pain: The evidence from functional imaging. *Current Pain and Headache Reports, 14*(5), 396–403.

Vincent, K., Warnaby, C., Stagg, C. J., Moore, J., Kennedy, S., & Tracey, I. (2011). Dysmenorrhoea is associated with central changes in otherwise healthy women. *Pain, 152*(9), 1966–1975. doi:10.1016/j.pain.2011.03.029.

Femoral Nerve Block

Definition

Local anesthetic blockade of the femoral nerve that provides sensory innervation to the upper leg. It provides prompt analgesia and muscle relaxation in children with femoral shaft fractures.

Cross-References

▶ Acute Pain in Children, Postoperative

Fentanyl

Definition

Fentanyl is a synthetic opioid that is a phenylpiperidine derivative and structurally related to meperidine.

Cross-References

▶ Postoperative Pain, Fentanyl

Fetal Pain During Prenatal Surgery

Carlo Valerio Bellieni
Neonatal Intensive Care Unit, University Hospital, Siena, Italy

Introduction

The main argument in favor of fetal in utero analgesic treatment is that fetuses prematurely born are commonly entitled to receive routine analgesia, because newborns' pain is a widely accepted fact. This is substantiated by observational studies, and by the anatomical development of the pain system, that is increasingly mature with age, in prematurely born fetuses. Thus, if a premature is entitled to analgesia, the same should be due to a fetus with the same level of development.

Conversely, those arguing against fetal analgesic treatment say that until a certain fetal age, the absence of a cortico-thalamic link is an absolute obstacle to awareness and to pain sensation. Moreover, they affirm that the fetus is in a continuous state of sleep, with almost absence of awakeness that cannot be evocated neither by external stimuli, and therefore, it is in a continuous state of endogenous sedation.

Here, we will analyze three main data: (a) the development of pain pathways in the fetus; (b) fetus' behavior; (c) fetal analgesia.

Definition

Fetal surgery has recently made rapid progresses (Wilson et al. 2010). Now it is possible to treat some mechanical anomalies before birth, and this in some cases has brought clear advantages to the babies, who have not to wait birth to be cured. For instance, fetal surgery is used in fetuses with congenital diaphragmatic hernia, in whom lung growth is triggered by percutaneous tracheal occlusion. It can also be used to treat urinary obstructions. Many fetal interventions remain investigational, but randomized trials have established the role of in utero surgery for a number of conditions, making fetal surgery a clinical reality. The safety of fetal surgery is such that even nonlethal conditions, such as myelomeningocele repair, can be considered an indication. This leads to consider the opportunity of providing analgesia and anesthesia to the fetus. Of course, we have to wonder if a concern about fetal pain is justified, and this means to have an evidence-based opinion about the possibility for the fetus to feel pain.

Characteristics

Development of Pain Pathways

Anatomical and functional perception of pain develops throughout the gestational process. Peripheral receptors develop very early and may be seen by the ninth week of gestational age (GA) and are abundant by the 22nd week GA (Lowery et al. 2007). Four or five weeks after fertilization, pain receptors appear around the mouth, followed by nerve fibers, which carry stimuli to the brain. By 18 weeks GA, pain receptors have appeared throughout the body (Blackburn 2003). Connections between peripheral receptors and afferent fibers ending in the dorsal horn start as early as 8 weeks GA and the myelination of these fibers continues during development. Starting at 10–13 weeks GA, the afferent system located in the substantia gelatinosa of the spinal cord's dorsal horn is developing. Connections to the thalamus begin at 14 weeks and are completed by 20 weeks GA, and thalamocortical connections or are present from 20 weeks GA, and are more developed by 26–30 weeks (Kostovic and Goldman-Rakic 1983). The neurons of the cerebral cortex begin a migration from the periventricular zone at 8 weeks gestation, and by 24–30 weeks GA, the cortex has acquired a full complement of neurons (Rodeck and Whittle 2008). Electroencephalographic activity appears for the first time at 20 weeks, becomes synchronized at 26 weeks GA, and reveals wake-sleep cycles at 30 weeks GA. It is possible to measure evoked potentials from the cortex from 24 weeks GA, with an objective evidence that a peripheral stimulus can cause cortical activation (Wilken and Gortner 2000).

An increase in stress hormones happens during fetal potentially painful transfusions: Piercing the fetal abdomen to access the intrahepatic vein (IHV) for transfusion is associated with substantial rises in these hormones from as early as 18 weeks GA. The median increase in beta-endorphin levels was 590 %, in cortisol levels was 183 %, and median increase in noradrenaline levels was 196 % (Giannakoulopoulos et al. 1994, Giannakoulopoulos et al. 1999).

It has been hypothesized that during surgery, despite general analgesia with opioids, metabolic changes due to pain might be present, indicating a production of stress hormones independent of consciousness, (Fitzgerald et al. 2003) but several studies show that analgesia inhibits both consciousness and the production of stress hormones (Desborough 2000), thus, this increase seems due to actual stress.

Clinical evidence for perception mediated by subcortical centers comes from infants and children with hydranencephaly (Marín-Padilla 1997; Takada et al. 1989). Despite total or near-total absence of the cortex, these children clearly possess discriminative awareness. They seem to distinguish familiar from unfamiliar people and environments and are capable of some social interaction, visual orienting, musical preferences, appropriate affective responses, and associative learning (Shewmon et al. 1999). Recent studies highlight the possibility of a subcortical form of consciousness, that has been recently evidenced by several authors in adult healthy patients (Denton et al. 2009; Merker 2007; Johnson 2005; Ohman et al. 2007).

Fetal Behavioral States

Fetuses spend most gestational time in sleep: Nevertheless term fetuses spend 9 % of their daytime in a wake state (Pillai and James 1990). Some argue that this continuous sleep state is induced by warmth and adenosine in the blood, but fetal adenosine levels (Yoneyama et al. 1996) are not distinct from mothers' (Yoneyama et al. 2000), in whom it does not induce sleep or analgesia. Vibroacoustical stimulation, though little above the background noise, within the womb can induce awakening in fetuses (Leader and Fifer 1995). More conspicuous stimuli awaken the fetus at birth. Fetal sleep can increase pain threshold, but cannot annihilate the possibility of experiencing pain.

Fetal Analgesia

In the case of mayor surgery, some researchers recommend to administer 20 microg/kg of intramuscular fentanyl to the fetus prior to the procedure, to increase the effect of volatile anesthetics that arrive to it through mother's circulation (Schwarz and Galinkin 2003). Recent studies show that direct administration of 10 microg/kg fentanyl blunts the fetal stress response to intrauterine needling (Fisk et al. 2001): These authors showed that the magnitude of the beta-endorphin response to pain with these doses was halved, and the cerebral Doppler response was ablated. Comparison with control fetuses transfused without fentanyl indicated that the beta endorphin and cerebral Doppler response to intrahepatic vein transfusion with fentanyl approached that of nonstressful placental cord transfusions. Fentanyl significantly attenuated the endorphin and cerebrovascular response, but not the cortisol response (Fisk et al. 2001). This differential effect is not surprising. Using the same dose, intravenous fentanyl in preterm babies ablated most of the stress responses to surgery, but the reduction in the cortisol response after fentanyl failed to achieve statistical significance (Anand et al. 2001). It is worth highlighting that the amount of opioids that can arrive to the mother from the fetus is clearly low and relatively safe.

During open fetal surgery under maternal general anesthesia, inhalational agents given to the mother are still considered to provide adequate fetal anesthesia and produce uterine relaxation essential for successful surgery, so that additional analgesia for the fetus would be unnecessary. Nevertheless, several evidences against this approach should be considered: for instance after cesarean sections made with maternal general anesthesia, fetuses are born awake or are easily awakened at birth, providing evidence that maternal anesthetic drugs are not sufficiently anesthetic to the fetus. Endoscopic procedures performed directly on the fetus, such as tracheal occlusion or meningomyelocele repair, are usually performed under maternal local or regional anesthesia.

As these procedures do not require maternal general anesthesia, additional fetal anesthesia is desired and is usually done by direct administration of opioids and muscle relaxants to the fetus. Two possible routes of administration for these drugs are injection into the umbilical cord and intramuscular injection into the fetus. Sometimes direct fetal administration of fentanyl and pancuronium is sometimes limited to cases where the fetus moves during the procedure in other cases, it is given routinely (Sutton 2008).

Some researchers utilized intra-amniotic opioids for fetal analgesia on lamb fetuses: They showed that greater plasma concentrations were obtained in the fetal lamb as compared with the ewe, suggesting that this route might be utilizable for humans.

The presence of an active cortical subplate (Anand et al. 2001) and of spino-thalamic connections seem sufficient to transmit the painful stimulus during the second half of pregnancy. Thus, the use of fetal analgesia is at least a due precautionary procedure.

Summary

In the second trimester of gestation, the incomplete development of the cortex does not yet consent awareness, but this does not seem to be an absolute obstacle to pain from the 20th week of GA. Evidence is present in adult patients for skills that do not need an involvement of the cortex. Several stimuli are processed without the need of the cortex (Mulckhuyse and Theeuwes 2010); nevertheless, these stimuli give useful visual information to the individual (Sewards and Sewards 2000; Pasley et al. 2004), or provoke complex experiences such as fear (Ohman et al. 2007). A typical example of this is the perception of fearful stimuli or of unusual objects (for instance, during a mechanical skill like driving a car) that do not arrive to awareness, but that nevertheless can provoke complex reactions or behaviors: These stimuli do not arrive to awareness, but can affect consciousness. We can hypothesize a similar scenario for subcortical fetal processing of pain in this trimester: After the 20th week of GA, we should not exclude that pain can be felt by the fetus

(Mahieu-Caputo et al. 2000), with possible long-term consequences (Lowery et al. 2007).

In the second half of pregnancy, fetuses have a higher pain threshold due to their sleep state, but this does not absolutely rule out pain. The presence of thalamocortical pathways and of the cortical subplate point out the possibilities of pain at this stage of development.

References

Anand, K. J., & Sippel, W. G. (2001). Aynsley green a: Randomized trial of fentanyl anesthesia in preterm babies undergoing surgery: Effects on the stress response. *Lancet, 1*, 62–66.

Blackburn, S. T. (2003). *Maternal, fetal, and neonatal physiology* (2nd ed.). Philadelphia: WB Saunders.

Denton, D. A., McKinley, M. J., Farrell, M., & Egan, G. F. (2009). The role of primordial emotions in the evolutionary origin of consciousness. *Consciousness and Cognition, 18*(2), 500–514.

Desborough, J. P. (2000). The stress response to trauma and surgery. *British Journal of Anaesthesia, 85*, 109–117.

Fisk, N. M., Gitau, R., Teixeira, J. M., Giannakoulopoulos, X., Cameron, A. D., & Glover, V. A. (2001). Effect of direct fetal opioid analgesia on fetal hormonal and hemodynamic stress response to intrauterine needling. *Anesthesiology, 95*(4), 828–835.

Fitzgerald, R. D., Hieber, C., Schweitzer, E., Luo, A., Oczenski, W., & Lackner, F. X. (2003). Intraoperative catecholamine release in brain-dead organ donors is not suppressed by administration of fentanyl. *European Journal of Anaesthesiology, 20*, 952–956.

Giannakoulopoulos, X., Sepulveda, W., Kourtis, P., Glover, V., & Fisk, N. M. (1994). Fetal plasma cortisol and betaendorphin response to intrauterine needling. *Lancet, 344*, 77–81.

Giannakoulopoulos, X., Teixeira, J., Fisk, N., & Glover, V. (1999). Human fetal and maternal noradrenaline responses to invasive procedures. *Pediatric Research, 45*, 494–499.

Johnson, M. H. (2005). Subcortical face processing. *Nature Reviews Neuroscience, 6*(10), 766–774.

Kostovic, I., & Goldman-Rakic, P. S. (1983). Transient cholinesterase staining in the mediodorsal nucleus of the thalamus and its connections in the developing human and monkey brain. *The Journal of Comparative Neurology, 219*, 431–447.

Leader, L. R., & Fifer, W. P. (1995). The potential value of habituation in the prenate. In J. P. Lecanuet (Ed.), *Fetal development: A psychobiological perspective* (pp. 83–404). Hillsdale, NJ: Lawrence Erlbaum.

Lowery, C. L., Hardman, M. P., Manning, N., Hall, R. W., Anand, K. J., & Clancy, B. (2007). Neurodevelopmental changes of fetal pain. *Semin Perinatol, 31*(5), 275–82.

Mahieu-Caputo, D., Dommergues, M., Muller, F., & Dumez, Y. (2000). Fetal pain. *Presse Médicale, 29*(12), 663–669.

Marín-Padilla, M. (1997). Developmental neuropathology and impact of perinatal brain damage. II: White matter lesions of the neocortex. *Journal of Neuropathology and Experimental Neurology, 56*(3), 219–235.

Merker, B. (2007). Consciousness without a cerebral cortex: A challenge for neuroscience and medicine. *The Behavioral and Brain Sciences, 30*, 63–81.

Mulckhuyse, M., & Theeuwes, J. (2010). Unconscious attentional orienting to exogenous cues: A review of the literature. *Acta Psychologica, 134*(3), 299–309.

Ohman, A., Carlsson, K., Lundqvist, D., & Ingvar, M. (2007). On the unconscious subcortical origin of human fear. *Physiology & Behavior, 92*(1–2), 180–185.

Pasley, B. N., Mayes, L. C., & Schultz, R. T. (2004). Subcortical discrimination of unperceived objects during binocular rivalry. *Neuron, 42*(1), 163–172.

Pillai, M., & James, D. (1990). Behavioural states in normal mature human fetuses. *Archives of Disease in Childhood, 65*(1), 39–43. Spec No.

Rodeck, C. H., & Whittle, M. J. (2008). *Fetal medicine: Basic science and clinical practice*. London: Elsevier Health Sciences.

Schwarz, U., & Galinkin, J. L. (2003). Anesthesia for fetal surgery. *Seminars in Pediatric Surgery, 12*(3), 196–201.

Sewards, T. V., & Sewards, M. A. (2000). Visual awareness due to neuronal activities in subcortical structures: A proposal. *Consciousness and Cognition, 9*(1), 86–116.

Shewmon, D. A., Holmes, G. L., & Byrne, P. A. (1999). Consciousness in congenitally decorticate children: Developmental vegetative state as self-fulfilling prophecy. *Developmental Medicine and Child Neurology, 41*(6), 364–374.

Sutton, L. N. (2008). Fetal surgery for neural tube defects. *Best Practice & Research Clinical Obstetrics & Gynaecology, 22*(1), 175–188.

Takada, K., Shiota, M., Ando, M., Kimura, M., & Inoue, K. (1989). Porencephaly and hydranencephaly: A neuropathological study of four autopsy cases. *Brain & Development, 11*(1), 51–56.

Wilken, B., & Gortner, L. (2000). Early auditory evoked potentials in very small premature infants. *Zeitschrift für Geburtshilfe und Neonatologie, 204*(1), 14–19.

Wilson, R. D., Lemerand, K., Johnson, M. P., Flake, A. W., Bebbington, M., Hedrick, H. L., & Adzick, N. S. (2010). Reproductive outcomes in subsequent pregnancies after a pregnancy complicated by open maternal-fetal surgery (1996–2007). *American Journal of Obstetrics and Gynecology, 203*(3), 209 e1–209–e6.

Yoneyama, Y., Sawa, R., Suzuki, S., Shin, S., Power, G. G., & Araki, T. (1996). The relationship between uterine artery Doppler velocimetry and umbilical venous adenosine levels in pregnancies complicated by preeclampsia. *American Journal of Obstetrics and Gynecology, 174*(1Pt 1), 267–271.

Yoneyama, Y., Suzuki, S., Sawa, R., Takeuchi, T., Kobayashi, H., Takei, R., Kiyokawa, Y., Otsubo, Y., Hayashi, Z., & Araki, T. (2000). Changes in plasma adenosine concentrations during normal pregnancy. *Gynecologic and Obstetric Investigation, 50*(3), 145–148.

Fibroblast

Definition

Fibroblast is a connective tissue cell.

Cross-References

▶ Wallerian Degeneration

Fibrocartilage

Definition

Fibrocartilage is cartilage that is largely composed of fibers like those in ordinary connective tissue.

Cross-References

▶ Sacroiliac Joint Pain

Fibromyalgia

David Vivian[1] and Nikolai Bogduk[2]
[1]Metro Spinal Clinic, South Caulfield, VIC, Australia
[2]University of Newcastle, Newcastle Bone & Joint Institute, Royal Newcastle Centre, Newcastle, NSW, Australia

Synonyms

Chronic widespread pain; Fibrositis; Muscular rheumatism; Psychogenic rheumatism

Definition

Fibromyalgia is a condition characterized by chronic widespread pain, fatigue, unrefreshed sleep, and a tendency to experience generalized somatic symptoms.

Characteristics

Fibromyalgia is present in about 5 % of the population. It is a syndrome of symptoms that include chronic widespread pain, fatigue, disordered sleep, and increased stress in association with abnormal neurosensory processing and nonspecific endocrine and autonomic dysfunction.

The original American College of Rheumatology (ACR) classification criteria for fibromyalgia included tenderness on pressure palpation at 11 or more of 18 specific tender points (Wolfe et al. 1990). Subsequently, the recognition that examination for tender points lacks validity and reliability has lead to the deletion of tender point examination from the diagnostic formulation for fibromyalgia.

Fibromyalgia is now a symptom-based diagnosis with the primary feature being widespread pain present for at least 3 months in association with other symptoms including fatigue, non-restorative sleep (NRS), and "somatic" symptoms such as anxiety, irritable bowel and bladder syndrome, cognitive inefficiency, tinnitus, and nausea. Justification for the shift from the physical examination criterion has been elicited by large studies that have developed diagnostic classification criteria scales. The ACR has now combined the widespread pain index (WPI) and the symptom severity (SS) scale to produce "simple, practical criteria for the clinical diagnosis of fibromyalgia" (Wolfe et al. 2010).

The Symptom Intensity Scale (SIS) is another scoring system for fibromyalgia (Wilke 2009). It combines the number of areas of pain present in the previous 7 days with a subjective measurement of fatigue as indicated on a 10-cm line to produce an overall

SIS score, the level of which categorizes patients with possible fibromyalgia as having features "probably consistent with fibromyalgia" or "diagnostic."

Pathophysiology

The pathophysiology of fibromyalgia is unknown and remains in dispute. Patients with fibromyalgia exhibit increased pain sensitivity to pressure, heat, cold, and electrical stimulation, but these features are not unique to fibromyalgia (Gracely et al. 2003).

Other markers of the disorder have been explored, but the results have not been consistent or reproducible. Nor are the abnormalities, when evident, expressed by all patients with fibromyalgia. Nor are they unique to these patients.

NRS, which is defined variably but includes the feeling of being unrefreshed upon awakening in spite of normal sleep duration, is common in the general population but more prevalent in FM and chronic fatigue syndrome (CFS) (Staud 2010). Fragmented sleep or difficulty getting to sleep is also common in FM patients. Although not unique to NRS, ► alpha(α)-delta(δ) sleep is more common in subjects with NRS than in healthy controls. The same anomaly occurs in other conditions and in 15 % of healthy individuals (Carette 1995). Histological abnormalities have been reported in the muscles of patients with fibromyalgia, but no differences have been found in adequately controlled studies (Carette 1995). Nor have disturbances in muscle metabolism been verified (Carette 1995). Patients with fibromyalgia exhibit deficiencies in ► serotonin, increased levels of ► substance P in their cerebrospinal fluid, and abnormalities of the ► hypothalamic pituitary axis (Carette 1995), but these differences have not been shown to be unique to fibromyalgia.

One model that has been proposed is that fibromyalgia is due to an impairment of the diffuse noxious inhibitory control system (DNIC) (Gracely et al. 2003). The implication is that patients perceive spontaneous pain because of a lack of tonic inhibition of the central nociceptive pathways. The circumstantial evidence is that of conditioning, that painful stimuli produce analgesia in normal subjects but fail to do so in patients with fibromyalgia (Gracely et al. 2003).

Although this may be so, commentators have questioned whether the syndrome is due to altered central nociception or to hypervigilance, and if there is altered nociception, does it arise because of somatic factors or psychogenic influences (Cohen and Quintner 1998).

Nosology

The relegation of tenderness as a diagnostic criterion for diagnosis of fibromyalgia together with the development of symptom-based standards has provided clinicians with a more robust and reliable method for managing and studying patients with these symptoms. As a minimum, the diagnosis of fibromyalgia offers patients a diagnostic label that is more palatable than a diagnosis of widespread pain (Carette 1995).

Treatment

A variety of treatments have been applied to patients with fibromyalgia. Many treatments are based on hearsay or anecdote, and some of the conclusions from studies are based on previous diagnostic criteria for fibromyalgia. The EULAR report in 2008 found 146 studies eligible for review and made 9 recommendations for the management of fibromyalgia (Carville 2008). Exercise in general including heated pool exercise had some limited evidence and was included on the basis of expert opinion; tramadol had benefits over 13 weeks in a randomized controlled trial, but simple analgesics were supported and opioids refuted by expert opinion; antidepressants tended to reduce pain and improve function over a 12-week period but not over longer periods. A multidisciplinary approach to management was considered relevant, but this was again based on expert opinion rather than any evidence.

A meta-analysis of pregabalin found five high-quality trials (Straube 2010). Pregabalin seems to have similar efficacy to other medication treatments including duloxetine, combined paracetamol and tramadol, and amitriptyline 25 mg/day. Pregabalin enhances slow-wave sleep, and sodium oxybate (gamma-hydroxybutyrate) reduces alpha EEG sleep, and both may have a role in symptomatic control of FM for those reasons (Staud 2010).

Cross-References

▶ Chronic Low Back Pain: Definitions and Diagnosis
▶ Fibromyalgia, Mechanisms, and Treatment
▶ Human Thalamic Response to Experimental Pain (Neuroimaging)
▶ Muscle Pain, Fibromyalgia Syndrome (Primary, Secondary)
▶ Myalgia
▶ Nocifensive Behaviors, Muscle and Joint
▶ Opioids and Muscle Pain
▶ Physical Exercise
▶ Psychological Aspects of Pain in Women

References

Carette, S. (1995). Fibromyalgia 20 years later. What have we accomplished? *The Journal of Rheumatology, 22,* 590–594.
Carville, S. F., Arendt-Nielsen, S., Bliddal, H., et al. (2008). EULAR evidence-based recommendations for the management of fibromyalgia syndrome. *Annals of the Rheumatic Diseases, 67,* 536–541.
Cohen, M. L., & Quintner, J. L. (1998). Fibromyalgia syndrome and disability: A failed construct fails those in pain. *The Medical Journal of Australia, 168*(8), 402–404.
Gracely, R. H., Grant, M. A. B., & Giesecke, T. (2003). Evoked pain measures in fibromyalgia. *Best Practice & Research Clinical Rheumatology, 17,* 593–609.
Staud, S. (2010). Nonrestorative sleep in fibromyalgia: From assessment to therapy – focus on diagnosis. Medscape. http://cme.medscape.com.viewarticle/727292
Straube, S., Derry, S., Moore, R. A., et al. (2010). Pregabalin in fibromyalgia: Meta-analysis of efficacy and safety from company clinical trial reports. *Rheumatology, 49,* 706–715.
Wilke, W. S. (2009). New developments in the diagnosis of fibromyalgia syndrome: Say goodbye to tender points? *Cleveland Clinic Journal of Medicine, 76,* 345–352.
Wolfe, F., Smythe, H. A., Yunus, M. B., et al. (1990). The American college of rheumatology 1990 criteria for the classification of fibromyalgia. *Arthritis & Rheumatism, 33,* 160–172.
Wolfe, F., Clauw, D. J., Fitzcharles, M., et al. (2010). The American college of rheumatology preliminary diagnostic criteria for fibromyalgia and measurement of symptom severity. *Arthritis Care & Research, 62,* 600–610.

Fibromyalgia Syndrome

▶ Muscle Pain, Fibromyalgia Syndrome (Primary, Secondary)

Fibromyalgia, Mechanisms, and Treatment

Alice A. Larson and Katalin J. Kovacs
Department of Veterinary and Biomedical Sciences, University of Minnesota, St. Paul, MN, USA

Introduction

Historically, fibromyalgia syndrome (FMS) was referred to as fibrositis, a term coined by Sir Edward Gowers at the turn of the century to describe the inflammation responsible for the stiffness and pain experienced by a group of non-arthritic patients. Fibromyalgia syndrome was the name adopted in 1990 by the American College of Rheumatology (Wolfe et al. 1990), and the consensus document on fibromyalgia at the MYOPAIN conference held in Copenhagen, Denmark, in 1992 provided the original description of the syndrome, the diagnostic criteria, and its prevalence. More recently, the American College of Rheumatology has provisionally approved a revised set of criteria for the diagnosis of fibromyalgia that does not require a tender point assessment (Wolfe et al. 2010).

Definition

Many chronic pain conditions display symptoms that overlap with those of FMS. Those conditions must be diagnosed separately and eliminated as the first step in the diagnostic protocol. The diagnostic criteria include widespread pain in all four quadrants of the body for a minimum of 3 months, originally based on tenderness to digital palpation at 11 of the 18 anatomically defined tender points that are sensitive to pressure (mechanical pain). Wolfe et al. (2010) now recommend the diagnosis be based on the combination of a widespread pain index (WPI) and a symptom severity scale (SS) as follows: (WPI ≥ 7 and SS ≥ 5) or (WPI 3–6 and SS ≥ 9). Reductions in electrical and thermal pain thresholds have also been reported in patients with FMS, suggesting multimodal changes in pain sensitivity.

Characteristics

Prevalence

In North America, approximately 2 % of the population, or greater than 3.7 million people, are estimated to suffer from FMS. The majority of these patients, approximately three out of four, are female. While it is the second most common syndrome presented in rheumatology clinics, the treatment is typically unsatisfactory, resulting in disability and handicap. Studies in several countries and ethnicities suggest a poor prognosis over several years of treatment.

Comorbidity

Chronic widespread pain is the primary complaint bringing most patients with FMS into the clinic. In addition to pain, non-restorative sleep, fatigue, anxiety and depression, interstitial cystitis, and irritable bowel syndrome are symptoms frequently diagnosed in patients with FMS. A higher incidence of cold intolerance, restless leg syndrome, cognitive dysfunction, rhinitis, and multiple chemical sensitivities is also more common in FMS patients than in healthy normal controls, adding to the perplexing mosaic of the disease. The etiology of FMS is unknown, and the relationships between pain and the other symptoms of FMS are unclear. Acute pain usually results in increased activation of the pituitary-adrenal and sympathomedullary pathways as well as growth hormone production. Patients with FMS, in contrast, present with hypofunction of the HPA, thyroidal, and gonadal axis; diminished growth hormone production (reviewed by Bennett 1998); and abnormally low sympathetic

output (Clauw and Chrousos 1997). While exposure to stressful situations, including noise, lights, and weather, exacerbates symptoms of FMS, patients have an impaired ability to activate the hypothalamic-pituitary portion of the HPA axis as well as the sympathoadrenal system, leading to reduced adrenocorticotropic hormone (ACTH) and epinephrine responses. Yet many symptoms of FMS are sympathetically maintained, such as the neuropathic-type pain (reviewed by Martinez-Lavin 2007), abnormal heart rate variability, tilttable responses, and inappropriate responses to daily stressful situations.

Unique Location of Tender Points

While tender points are no longer necessary diagnostically, they offer insight into the etiology of pain. Tender points are generally distributed in areas whose primary afferent nerves project to spinal lumbar levels 3–5 and cervical levels 2–7 spinal cord segments. It may be of importance that these areas surround the thoracic 1 through lumbar 3 spinal cord segments that are involved with regulation of the sympathetic nervous system, whose function is significantly altered in patients with FMS. It is noteworthy that the pain of FMS is often in areas that receive a relatively low density of afferent innervation (trunk and proximal limbs) compared to areas normally considered "sensitive" by virtue of their dense innervation, e.g., the fingers, mouth, feet, and genitals. The sensitivity of these tender points does not, then, originate from a simple hyperactivity of all mechanically sensitive tissues but rather from an unknown, symmetrically distributed change in afferent input. Based on this and other psychophysical evidence of windup, the etiology of this syndrome is widely believed to include the CNS.

Excitatory Amino Acids

Excitatory amino acids, such as glutamate and aspartate, transmit pain signals, including those in patients with FMS. While the concentrations of most amino acids in the cerebrospinal fluid do not differ between subjects with FMS compared to healthy normal controls, a high degree of correspondence was found between specific amino acids and the degree of pain reported at the 18 tender points (tender point index, TPI). Excitatory amino acid activity is known to trigger synthesis of nitric oxide (NO), a gaseous-signaling compound that is critical for the development and expression of chronic pain. Enhanced synthesis of NO results from a variety of depolarizing events, including excitatory amino acid and substance P activity, both of which lead to influx of calcium and activation of the enzyme NO synthase. NMDA activity is also associated with the production of nerve growth factor (NGF), a compound that regulates the synthesis of peptides in primary afferent C-fibers such as substance P.

Substance P

The first documentation of a biochemical characteristic consistent with chronic pain in patients with FMS was the increased concentration of substance P in their cerebrospinal fluid (Vaeroy et al. 1988; Russell et al. 1994; Welin et al. 1995), similar to that in many other pain states. Substance P is a neuroactive peptide released from small-diameter, unmyelinated primary afferent fibers, called nociceptors. Substance P is upregulated in chronic, inflammatory pain conditions by increased concentrations of NGF. Because substance P does not mediate acute pain but supports hyperalgesia, enhanced substance P content in the cerebrospinal fluid of patients with FMS is consistent with a chronic hyperalgesic state.

Growth Factors

Synthesis of nerve growth factor (NGF), a neurotrophin, in inflamed tissue is responsible for the upregulation of SP synthesis during chronic pain and the development of mechanical hyperalgesia and allodynia. Intravenous administration of NGF in humans causes muscle pain in a dose-dependent manner, primarily in bulbar and truncal musculature, and affects women more than men. While there is no gross inflammation at tender points to account for a peripheral source of NGF, delivery of NGF to the spinal area of mice or rats is sufficient to cause hyperalgesia.

The concentration of NGF in the cerebrospinal fluid is normally low or unmeasurable in healthy individuals, but NGF is elevated in patients with primary but not secondary FMS (Giovengo et al. 1999). In addition, the concentration of brain-derived neurotrophic factor (BDNF), a mediator of pain in the peripheral nervous system, is elevated in the serum of patients with FMS (Laske et al. 2007). It is, therefore, possible that elevated concentrations of NGF and or BDNF are responsible for the hyperalgesia and allodynia associated with FMS, while secondary FMS results from conditions producing large areas of enhanced pain sensitivity.

FMS and Thalamic Activity

Positron emission tomography (PET) and functional magnetic resonance imaging (fMRI) studies in humans indicate that the thalamus as well as the anterior insula and S2 area are important in the perception of pain. Activation of the thalamus by pain normally initiates descending inhibitory activity that controls pain. Stimulation of the somatosensory thalamus has even been successfully used to treat extreme conditions of chronic pain in humans. Descending activity originating from the thalamus is sufficiently important that thalamotomy in rats results in both thermal as well as mechanical hyperalgesia. Although pain normally stimulates the thalamus in healthy individuals, a different pattern of activation occurs in patients with chronic pain. Regional blood flow in response to painful stimulation also differs in patients with FMS compared to that in normal control individuals (Risberg et al. 1995; Mountz et al. 1998; Petzke et al. 2000). These studies suggest that either the ascending spinothalamic nociceptive pathways are not fully functional or the thalamus responds abnormally to input from spinothalamic neurons. Because other parts of the CNS respond appropriately in patients with FMS, ascending activity appears sufficient. Rather, thalamic neurons likely fail to initiate sufficient descending inhibitory activity controlling nociceptive processing at the spinal cord level, a concept supported by altered nociceptive responses in patients with FMS (Staud et al. 2003).

Treatment

A variety of muscle relaxants, sedatives, analgesics, and drugs that promote sleep or combat fatigue are frequently used to treat symptoms of FMS. Unfortunately, even nonnarcotic analgesic drugs provide only temporary relief in most patients. When all else fails, narcotic analgesics are transiently effective in controlling flares, but not only does tolerance develop to their analgesic effect but long-term use can unmask an opioid-induced hyperalgesia. The efficacy and side effects of each compound vary greatly from one patient to another, but in general, their side effects are minimized and efficacy maximized by combination treatments using two drugs with distinct mechanisms of action. For example, gabapentin (Neurontin) is a compound used for neurogenic pain either alone or in combination with opioids. Gabapentin inhibits calcium channel activity, thereby minimizing the release of neurotransmitters.

Low doses of tricyclic antidepressants, such as amitriptyline, provide immediate but temporary analgesia, similar to their efficacy in other chronic pain conditions (O'Malley et al. 2000). The efficacy of this, and related drugs, is likely due to their action on nociceptive pathways and is not related to their antidepressant activity, which requires higher doses and longer periods of treatment to develop. Because tolerance to their analgesic activity develops, low doses (10 mg or less) are recommended in the beginning, gradually increasing by 10 mg as needed until the optimal dose of 70–80 mg is reached.

Only pregabalin (Lyrica), duloxetine (Cymbalta), and milnacipran (Savella) are FDA approved for the treatment of FMS symptoms. Duloxetine and milnacipran are antidepressants that inhibit serotonin and norepinephrine reuptake. Over the short term, this increases the synaptic concentration of these transmitters but, by doing so, influences their receptor sensitivity over the long term. In contrast, pregabalin binds to a voltage-dependent calcium channel in the central nervous system, reducing calcium influx into the nerve terminals and decreasing release of neurotransmitters such as glutamate, noradrenaline, and substance P. Pregabalin also increases the concentration of GABA, an inhibitory neurotransmitter, by increasing glutamic acid decarboxylase activity, the enzyme required for the synthesis of GABA.

Drug treatment to temporarily alleviate pain, including amitriptyline, NSAIDS, and even narcotic analgesics, should be geared to facilitate a new level of physical activity (reviewed by Clauw and Crofford 2003) as the most effective treatment for FMS is exercise (Sim and Adams 2002). Patients need to understand that exercise is the key to recovery, even if this means merely doubling their distance walked within the house from one day to the next. The exercise program should be initiated gradually, or it will temporarily exacerbate their symptoms. Heat therapy, in the form of a long hot bath immediately prior to bedtime, has been anecdotally reported to be effective for episodic bouts of intense pain. In contrast, cold exacerbates their condition and should be avoided.

Cross-References

▶ Chronic Low Back Pain: Definitions and Diagnosis
▶ Disability in Fibromyalgia Patients
▶ Fibromyalgia, Mechanisms, and Treatment
▶ Human Thalamic Response to Experimental Pain (Neuroimaging)
▶ Muscle Pain, Fibromyalgia Syndrome (Primary, Secondary)
▶ Myalgia
▶ Nocifensive Behaviors, Muscle and Joint
▶ Opioids and Muscle Pain
▶ Physical Exercise
▶ Psychological Aspects of Pain in Women

References

Bennett, R. M. (1998). Disordered growth hormone secretion in fibromyalgia: A review of recent findings and a hypothesized etiology. *Zeitschrift für Rheumatologie, 57*(Suppl 2), 72–762.
Clauw, D. J., & Chrousos, G. P. (1997). Chronic pain and fatigue syndromes: Overlapping clinical and

neuroendocrine features and potential pathogenic mechanisms. *Neuroimmunomodulation, 4*, 143–1533.

Clauw, D. J., & Crofford, L. J. (2003). Chronic widespread pain and fibromyalgia: What we know, and what we need to know. *Best Practice & Research Clinical Rheumatology, 17*, 685–7014.

Giovengo, S. L., Russell, I. J., & Larson, A. A. (1999). Increased concentrations of nerve growth factor in cerebrospinal fluid of patients with fibromyalgia. *Journal of Rheumatology, 26*, 1564–15695.

Laske, C., Stransky, E., Eschweiler, G. W., et al. (2007). Increased BDNF serum concentration in fibromyalgia with or without depression or antidepressants. *Journal of Psychiatric Research, 41*, 600–605.

Martinez-Lavin, M. (2007). Stress, the stress response system, and fibromyalgia. *Arthritis Research & Therapy, 9*, 216.

Mountz, J. M., Bradley, L. A., & Alarcon, G. S. (1998). Abnormal functional activity of the central nervous system in fibromyalgia syndrome. *The American Journal of the Medical Sciences, 315*, 385–3968.

O'Malley, P. G., Balden, E., Tomkins, G., et al. (2000). Treatment of fibromyalgia with antidepressants. *Journal of General Internal Medicine, 15*, 659–666.

Petzke, F., Clauw, D. J., Wolf, J. M., & Gracely, R. H. (2000). Pressure pain in fibromyalgia and healthy controls: Functional MRI subjective pain experience versus objective stimulus intensity. *Arthritis and Rheumatism, 43*, 200010.

Risberg, J. G., Rosenhall, U., Orndahl, G., et al. (1995). Cerebral dysfunction in fibromyalgia: Evidence from regional cerebral blood flow measurements, otoneurological tests and cerebrospinal fluid analysis. *Acta Psychiatrica Scandinavica, 91*, 86–9411.

Russell, I. J., Orr, M. D., Littman, B., et al. (1994). Elevated cerebrospinal fluid levels of substance P in patients with the fibromyalgia syndrome. *Arthritis and Rheumatism, 37*, 1593–160112.

Sim, J., & Adams, N. (2002). Systematic review of randomized controlled trials of nonpharmacological interventions for fibromyalgia. *The Clinical Journal of Pain, 18*, 324–33613.

Staud, R., Robinson, M. E., Vierck, C. J., et al. (2003). Diffuse noxious inhibitory controls (DNIC) attenuate temporal summation of second pain in normal males but not in normal females or fibromyalgia patients. *Pain, 101*, 167–17414.

Vaeroy, H., Helle, R., Forre, O., et al. (1988). Elevated CSF levels of substance P and high incidence of Raynaud's phenomenon in patients with fibromyalgia: New features for diagnosis. *Pain, 32*, 21–2615.

Welin, M., Bragee, B., Nyberg, F., et al. (1995). Elevated substance P levels are contrasted by a decrease in met-enkephalin-arg-phe levels in CSF from fibromyalgia patients. *Journal of Musculoskeletal Pain, 3*, 4.

Wolfe, F., Clauw, D. J., Fitzcharles, M.-A., et al. (2010). The American college of rheumatology preliminary diagnostic criteria for fibromyalgia and measurement of symptom severity. *Arthritis Care & Research, 62*, 600–610.

Wolfe, F., Smythe, H. A., Yunus, M. B., et al. (1990). The American college of rheumatology 1990 criteria for the classification of fibromyalgia. Report of the multi-center criteria committee. *Arthritis and Rheumatism, 33*, 160–172.

Fibrositis

▶ Fibromyalgia
▶ Myalgia
▶ Psychiatric Aspects of the Epidemiology of Pain

Fibrositis Syndrome

▶ Muscle Pain, Fibromyalgia Syndrome (Primary, Secondary)
▶ Myalgia

Field Block

Definition

Local anesthetic injected through intact skin adjacent to the surgical site to create a subcutaneous wall encompassing the injury.

Cross-References

▶ Acute Pain in Children, Postoperative

Fifth Lobe

▶ Insular Cortex, Neurophysiology, and Functional Imaging of Nociceptive Processing

Firing of Suburothelial Afferent Nerves

Definition

Firing of suburothelial afferent nerves and the threshold for bladder activation may be modified by both inhibitory (e.g., NO) and stimulatory (e.g., ATP, tachykinins, prostanoids) mediators. These mechanisms can be involved in the generation of detrusor overactivity causing urgency, frequency, and incontinence, but also bladder pain.

Cross-References

▶ Opioids and Bladder Pain/Function

First and Second Pain

Definition

When a needle is pierced into the skin of an extremity, the subject feels typically an immediate stinging pain followed after a short delay by a second wave of predominantly burning pain. These two pain phenomena are called "first" and "second" pain. They are attributed to the excitation of A-delta and C-fibers. Since the former have conduction velocities up to 30 m/s, the nerve impulses reach the central nervous system so quickly that no delay is experienced. C-fibers having conduction velocities around 1 m/s convey their impulses considerably slower. If the needle prick is applied to the foot of a grown-up person, the delay could be more than a second. If the sting is applied to the face, the two pains are not separated by a delay due to the short conductance distance.

Cross-References

▶ Encoding of Noxious Information in the Spinal Cord
▶ Opioids, Effects of Systemic Morphine on Evoked Pain

First and Second Pain Assessment (First Pain, Pricking Pain, Pin-Prick Pain, Second Pain, Burning Pain)

Donald D. Price
Oral Surgery and Neuroscience, McKnight Brain Institute University of Florida, Gainesville, FL, USA

Synonyms

Burning pain; First pain assessment; Pin-prick pain; Pricking pain; Second pain assessment

Definition

Prior to the development of methods for physiological and anatomical characterization of axons within peripheral nerves, Henry Head concluded that the skin was served by epicritic and protopathic afferent systems in which each gave rise to its own particular qualities of sensations (Head 1920). Epicritic pain, for example, was said to be accurately localized, to not outlast the stimulus, and to provide precise qualitative information about the nature of the stimulus. Thus, epicritic pain could be elicited by mild pricking of the skin with a needle, a form of pain known to almost everyone as pricking pain. In contrast, protopathic pain was described as less well localized, slow in onset, often outlasting the stimulus, and summating with repeated stimulus application. Protopathic pain was considered more difficult to endure and contained special feelings of unpleasantness or "feeling tone." The concept of "protopathic" has been applied to burning pain, aching pain, throbbing pain, and dull pain. Head based these ideas largely on observations made of his own experiences of pain, after nerve division and during nerve regeneration.

With the advent of modern electrophysiological and neuroanatomical techniques, it became clear that pain depended on two types of peripheral nerve axons. The first is that of thinly myelinated A-delta axons, whose conduction velocities range between 3 and 30 m/s, and the second is that of unmyelinated C axons, whose conduction velocities range between 0.5 and 2.0 m/s. Based on their differences in conduction velocity and reminiscent of the functional dichotomy proposed by Head, Zotterman proposed that A-delta and C afferent axons could account for first and second pain, which often occur in response to a brief intense stimulus to the hand or foot (Zotterman 1933). Landau and Bishop explicitly related first pain to epicritic pain and second pain to protopathic pain (Landau and Bishop 1953). Interestingly, they based their conclusions as a result of an approach similar to that of Head. They carefully observed and recorded their own pain experiences in response to "experimental" pain stimuli, before and after selective conduction block of A-delta or C axons of peripheral nerves. They used local injections of dilute solutions of procaine to selectively block C axons within small nerve branches in order to study pain from impulses in A-delta axons. They selectively blocked all myelinated axons in peripheral nerves of their lower arms by means of a 250 mmHg pressure cuff in order to assess the types of pains evoked by impulses in C axons. They applied various types of painful stimuli to skin, fascia, and periosteal surfaces, including bee stings, turpentine injections, intramuscular KCL injections, and application of deep and sharp pressure with mechanical probes. When they blocked C axons with procaine, well-localized brief-duration stinging sharp pains, such as those elicited by pinpricks (i.e., pricking pain or first pain), were preserved, but prolonged, deep, and diffuse burning pains evoked by inflammatory stimuli, such as the second pain from a bee sting, could no longer be elicited. When they blocked all myelinated (A) axons by means of the blood pressure cuff, the latter types of pain could be elicited and were more intense than before blockade of myelinated axons.

These general observations have since been corroborated in conventional studies, wherein investigators studied responses of volunteer participants (Price 1972, 1999; Collins et al. 1960). An important aspect of the study by Collins et al. was that compound action potentials were monitored (Collins et al. 1960). They placed stimulating and recording electrodes under the sural nerve of cancer patients undergoing anterolateral chordotomy for relief of pain. They found that stimulation of A-delta axons produced sharp pricking pain sensations that were accurately localized. When A-delta axons were blocked by cold, stimulation of C axons at a rate of 3/s resulted in summating unbearable diffuse burning pain that was not as well localized.

The association of first and second pain with impulses in A-delta and C axons and the relationship of these two types of pain to epicritic and protopathic pain have pivotal roles in the history of pain research. However, like all functional dichotomies, it is important not to overgeneralize their explanatory role, for example, to prematurely label different central pain-related pathways as "epicritic" or "protopathic." Even Head warned against this type of error, when he concluded that epicritic and protopathic systems recombined once they entered the dorsal horn (Head 1920). Thus, although zones of relatively pure epicritic or protopathic sensibility could be found after dorsal root or peripheral nerve lesions, he noted that such zones were never observed after lesions of the spinal cord or brain. More than 40 years before electrophysiological studies were carried out on dorsal horn nociceptive neurons, Head anticipated the synaptic convergence of two functional types of primary nociceptive afferents (known since as A-delta and C) onto neurons of the dorsal horn.

One must also take note of the fact that A-delta and C nociceptive afferents are rarely activated in isolation. Most acute pains are likely to reflect a combination of input from A-delta and C nociceptive afferents and, in most cases, from non-nociceptive afferents as well. Indeed, the composition of input from different types of nociceptive and non-nociceptive afferents undoubtedly contributes a lot to the diverse qualities of both painful and non-painful somatic sensation (Price 1999). However, many long-duration pains, especially those that are diffuse, spatially spreading, and especially unpleasant in their "feeling tone" may depend to a greater extent

First and Second Pain Intensity

Recording of Spinothalamic Neuron

Pain Ratings of Normal Controls

First and Second Pain Assessment (First Pain, Pricking Pain, Pin-Prick Pain, Second Pain, Burning Pain), Fig. 1 (*Left*) Double response of spinothalamic tract dorsal horn neuron to synchronous stimulation of A-delta and C axons by means of electrical shocks to a cutaneous nerve. The delay between the two

on tonic input from C nociceptive afferents than from input from A-delta nociceptive afferents. The initial pains from abrupt injuries (e.g., stepping on a tack) are likely to depend heavily on A-delta nociceptive afferents.

Characteristics

Temporal Characteristics of First and Second Pain and Their Relationships to Neural Mechanisms

First and second pains are often easily distinguished when a sudden tissue damaging or potentially tissue-damaging stimulus occurs on a distal part of the body, such as the hand or foot (Fig. 1).

The 0.5–1.5-s delay between the two pains occurs as a result of the fact that nerve impulses in C axons travel much slower (0.5–1.5 m/s) than those in thinly myelinated A axons (6–30 m/s). Lewis and Pochin independently mapped the body regions wherein they experienced both first and second pains (Lewis and Pochin 1938). The body maps of

postsynaptic impulse responses is related to the faster and slower conduction velocities of A and C axons, respectively. (*Right*) Subjects' mean ratings of intensities of first and second pains. Note similarity of profiles of neuron responses to pain ratings (Modified from Lewis and Pochin (1938))

both Lewis and Pochin were nearly identical. The maps showed that first and second pain could be perceived near the elbow but not the lower trunk, even though both sites were about the same distance from the brain. The reason for this difference is that C fibers that supply the trunk have a short conduction distance to the spinal cord, whereas C fibers that supply the skin near the elbow have a long conduction distance. Once both "trunk" and "elbow" C fibers reach the spinal cord, they synapse on nerve cells that have fast-conducting axons. As a result of differences in peripheral conduction distance and time, first and second pain can be discriminated at the elbow but not the trunk.

Later, studies using psychophysical methods replicated and extended the ones just described (Price 1972, 1999). These methods relied on delivering brief and well-controlled experimental stimuli to the distal part of an extremity, such as the hand or foot. These regions were chosen because they allowed for discrimination of first and second pain related to impulse conduction in A-delta and C axons, respectively. Reaction time

**First and Second Pain
Assessment (First Pain,
Pricking Pain, Pin-Prick
Pain, Second Pain,
Burning Pain),**
Fig. 2 Subjects' mean
ratings of second pain in
response to series of 9-mA
electric shocks. Note
temporal summation of
second pain at interstimulus
intervals of 3 s but not 5 s
and the fact that first pain
remains at the same
intensity throughout series
of shocks (From Price
(1999))

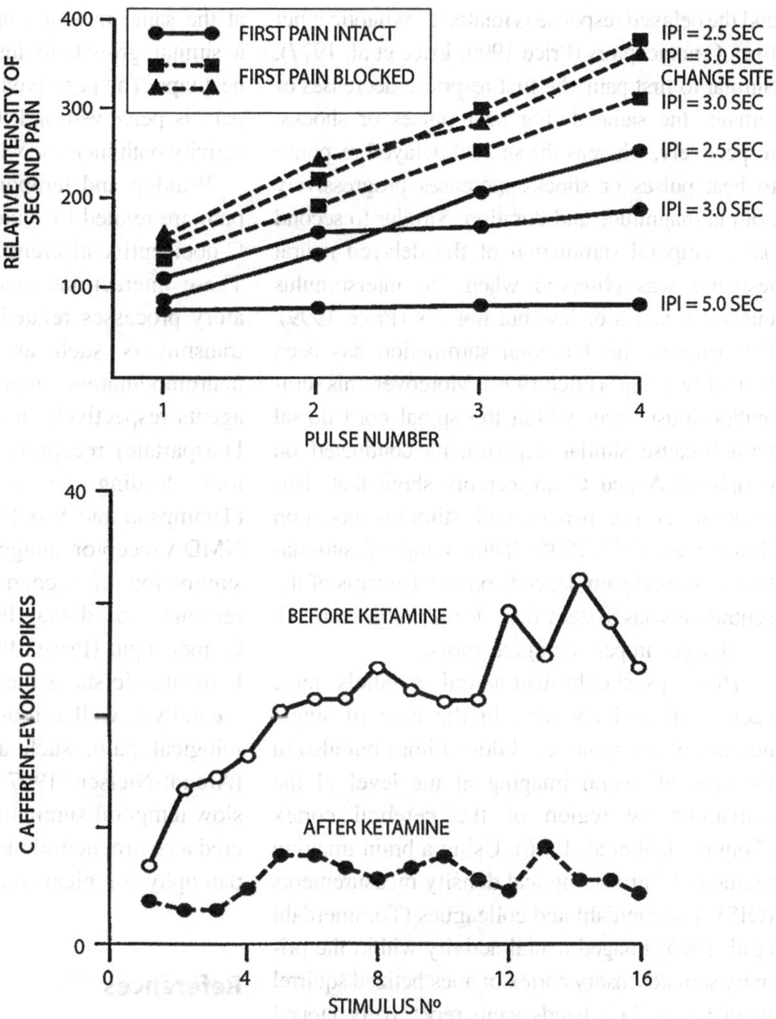

measurements were used to confirm that subjects
could indeed distinguish the two pains. Both
trained and untrained subjects reported qualities
of first pain as "pricking," "stinging," or "sharp"
(i.e., pin-prick pain), without provocation or suggestion that such qualities existed. Subjects were
trained to judge the perceived magnitudes of first
and second pain by squeezing a handgrip dynamometer in some experiments (Price et al. 1977)
or using a mechanical visual analogue scale in
others (Price et al. 1994). Mean psychophysical
ratings of intensities of first pain during trains of
computer-driven heat pulses (2.5 s duration, peak
temperature 52°C) decreased progressively
(Price 1999; Price et al. 1977, 1994) and stayed
the same in the case of 5–9 mA electric shocks

(Price 1972; Price et al. 1994). Unlike first pain,
second pain progressively increased in mean
intensity and duration throughout a series of
shocks or heat pulses, when the interstimulus
interval was less than 3 s but not when it was
5 s (Price 1972, 1999; Price et al. 1977, 1994)
(Fig. 2). Moreover, second pain became stronger,
more diffuse, and more unpleasant with repeated
heat pulses or repeated electrical shocks.

When the same types of heat pulses or electrical
shocks described above are applied to the skin of
monkeys, spinothalamic tract neurons within the
spinal cord dorsal horn respond with a double
response (i.e., two sets of impulse discharges)
(Price 1999), as shown in Fig. 1. The earlier of the
two is related to synaptic input from A-nociceptors,

and the delayed response is related to synaptic input from C nociceptors (Price 1999; Price et al. 1977). Similar to first pain, the first response decreases or remains the same during heat pulses or shocks, respectively, whereas the second delayed response to heat pulses or shocks increases progressively both in magnitude and duration. Similar to second pain, temporal summation of the delayed neural response was observed when the interstimulus interval was 3 s or less but not 5 s (Price 1999). For neurons, this temporal summation has been termed "windup" (Price 1999). Moreover, this summation must occur within the spinal cord dorsal horn because similar experiments conducted on peripheral A and C nociceptors show that their responses do not increase with stimulus repetition (Price et al. 1977, 1994). Thus, temporal summation of second pain depends on mechanisms of the central nervous system (i.e., dorsal horn neurons), not changes in peripheral receptors.

These psychophysical-neural parallels have been confirmed not only in the case of single neurons of the spinal cord dorsal horn but also in the case of neural imaging at the level of the somatosensory region of the cerebral cortex (Tommerdahl et al. 1996). Using a brain imaging method of intrinsic optical density measurements (OIS), Tommerdahl and colleagues (Tommerdahl et al. 1996) imaged neural activity within the primary somatosensory cortex of anesthetized squirrel monkeys as their hands were repetitively tapped with a heated thermode. These taps reliably evoke first and second pain in human subjects. Their method of neural imaging has a high degree of both spatial and temporal resolution, measuring local cortical neural activity within 50–100 μm and sampling neural activity that has occurred within a third of a second. Heat taps produced localized activity in two regions of the primary somatosensory cortex, termed 3a and 1. When a train of heat taps was presented at a rate of once per 3 s, each tap evoked delayed neural activity within these regions. The neural response to each tap grew progressively more intense and larger in area with each successive tap. This temporal summation of this neural response paralleled human psychophysical experiences of second pain in several distinct ways. Both types of responses summate

at the same rate of stimulus repetition and have a similar growth in intensity during a series of heat taps. The perceived skin area in which second pain is perceived and the area of cortical neural activity both increase with repeated heat taps.

Windup and temporal summation of second pain are related to synaptic interactions between C-nociceptive afferents and dorsal horn neurons. These interactions involve long-duration excitatory processes related to the release of neurotransmitters such as glutamate/aspartate and neuromodulators such as substance P. These agents respectively activate NMDA (N-methyl-D-aspartate) receptors and neurokinin 1 receptors, leading to prolonged depolarizations (Thompson and Woolf 1990, for review). Thus, NMDA receptor antagonists block both temporal summation of second pain and the "windup" responses of dorsal horn neurons to repeated C fiber input (Price 1999). Windup is related to hyperalgesic states that can be produced experimentally as well as those occurring in pathophysiological pain, such as postherpetic neuralgia (Arendt-Nielsen 1997; Dubner 1991). Indeed, slow temporal summation has long been considered a central neural mechanism that has a role in pathophysiological pain (Noordenbos 1959).

References

Arendt-Nielsen, L. (1997). Induction and assessment of experimental pain from human skin, muscle, and viscera. In T. S. Jensen, J. A. Turner, & Z. Wiesenfeld-Hallin (Eds.), Proceedings of the 8th World Congress on pain (pp 393–403). Seattle: IASP Press.
Collins, W. F., Nulsen, F. E., & Randt, C. T. (1960). Relation of peripheral nerve fiber size and sensation in man. Archive of Neurology (Chicago), 3, 381–385.
Dubner, R. (1991). Neuronal plasticity and pain following peripheral tissue inflammation or nerve injury. In M. Bond, E. Charlton, & C. J. Woolf (Eds.), Proceedings of Vth World Congress on pain. Pain research and clinical management (Vol. 5, pp. 263–276). Amsterdam: Elsevier.
Head, H. (1920). Studies in neurology. London: Oxford University Press.
Landau, W., & Bishop, G. H. (1953). Pain from dermal, periosteal, and fascial endings and from inflammation. Archives of Neurology and Psychiatry, 69, 490–504.
Lewis, T., & Pochin, E. E. (1938). The double response of the human skin to a single stimulus. Clinical Science, 3, 67–76.

Noordenbos, W. (1959). *Pain*. Amsterdam: Elsevier.

Price, D. D. (1972). Characteristics of second pain and flexion reflexes indicative of prolonged central summation. *Experimental Neurology, 37,* 371–387.

Price, D. D. (1999). *Psychological mechanisms of pain and analgesia* (p. 250). Seattle: IASP Press.

Price, D. D., Hu, J. W., Dubner, R., & Gracely, R. (1977). Peripheral suppression of first pain and central summation of second pain evoked by noxious heat pulses. *Pain, 3,* 57–68.

Price, D. D., Frenk, H., Mao, J., & Mayer, D. J. (1994). The NMDA receptor antagonist dextromethorphan selectively reduces temporal summation of second pain. *Pain, 59,* 165–174.

Thompson, S. W. N., & Woolf, C. J. (1990). Primary afferent-evoked prolonged potentials in the spinal cord and their central summation: Role of the NMDA receptor. In: M. R. Bond, J. Carlton, & C. J. Woolf (Eds.), *Proceedings of the VIth World Congress on pain* (pp. 291–298). Amsterdam: Elsevier.

Tommerdahl, M., Delemos, K. A., Vierck, C. J., Favorov, O. V., & Whitsel, B. L. (1996). Anterior parietal cortical response to tactile and skin heating stimulation. *Journal of Neurophysiology, 75*(6), 2662–2670.

Zotterman, Y. (1933). Studies in the peripheral nervous mechanisms of pain. *Acta Medica Scandinavica, 80,* 185–242.

First Pain

Definition

First pain is a rapid onset, sharp, pricking noxious sensation that is associated with the activation of Aδ nociceptors. When a noxious stimulus is sufficient to activate both Aδ and C nociceptors, first pain is perceived before second pain due to differences in nerve conduction velocity.

Cross-References

▶ Encoding of Noxious Information in the Spinal Cord

▶ First and Second Pain Assessment (First Pain, Pricking Pain, Pin-Prick Pain, Second Pain, Burning Pain)

▶ Opioids, Effects of Systemic Morphine on Evoked Pain

First Pain Assessment

▶ First and Second Pain Assessment (First Pain, Pricking Pain, Pin-Prick Pain, Second Pain, Burning Pain)

First-Generation Antidepressants

▶ Antidepressant Analgesics in Pain Management

First-Order Sensory Neurons

▶ Opioid Modulation of Nociceptive Afferents In Vivo

First-Order Subtype: Medial Thalamotomy

▶ Thalamotomy for Human Pain Relief

First-Pass Metabolism

Definition

The first-pass metabolism describes the transformation of the active drug after absorption prior to reaching the systemic circulation, i.e., pre-systemic elimination of the drug. It mainly occurs after oral or deep rectal administration and can be avoided by intravenous, intramuscular, sublingual, buccal, or topical administration.

Cross-References

▶ NSAIDs, Pharmacokinetics

Fitness Training

▶ Physical Exercise

Fits of Pain

▶ Pain Paroxysms

Flare

▶ Nociceptor, Axonal Branching

Flare, Flare Response

Definition

See also "neurogenic inflammation" and "axon reflex." Since the flare response is dependent on the extension of the terminal arborization of the C-fibers mediating this axon reflex response, i.e., their receptive fields, flare sizes vary in different species and depend also on the type of nerve fibers involved. Rodents do not show a visible flare response since the receptive fields of their C-afferents are small. They produce, however, a neurogenic plasma extravasation upon the stimulation of C-fibers. In humans, the flare responses following histamine stimulation are particularly large due to the large receptive fields of the histamine-sensitive C-fibers.

Cross-References

▶ Nociceptor, Axonal Branching
▶ Polymodal Nociceptors, Heat Transduction
▶ Quantitative Thermal Sensory Testing of Inflamed Skin

Flexion Exercise

▶ Exercise

Flexion Withdrawal Reflex

Definition

This is a protective reflex usually elicited in the lower limb and was originally described by Charles Sherrington as "the withdrawal of a limb from an offending stimulus." As originally characterized, it involved the activation of flexor muscles via a group of afferent nerve fibers called "flexor reflex afferents" with corresponding inhibition of extensor muscles. However, more recent research has revealed that the flexion withdrawal reflex has a more complex modular organization, its activation being contingent upon the area of skin being stimulated. Nevertheless, in the adult, the nociceptive (i.e., produced by noxious stimuli) flexion withdrawal reflex has proved invaluable as a model of nociceptive processing, and shows a clear correlation with pain perception in terms of threshold, peak intensity, and sensitivity to analgesics. However, in the newborn infant, the flexion withdrawal reflex can also be evoked with low-intensity mechanical stimuli to the foot, such as calibrated monofilaments (also called von Frey hairs), and has a much lower threshold than the nociceptive flexion withdrawal reflex in the adult.

Cross-References

▶ Infant Pain Mechanisms

Flinching

Definition

At its most vigorous, flinching of the paw and/or hindquarters is paw shaking, and when less vigorous, rapid paw lifting. It is usually seen as drawing the paw under the body and rapidly vibrating it, and this causes a shudder or rippling motion across the back which is easy to observe even when the paw is not visible. Each episode is recorded as a single flinch.

Cross-References

▶ Formalin Test

Flip-Flop Isoform of AMPA Receptors

Definition

AMPA receptors have four subunits named GluR1-4 (or GluRA-D), respectively. Each subunit of the AMPA receptor exists in two isoforms, so-called flip and flop due to alternative splicing of a 115-base pair region encoding 38 amino acid residues immediately preceding the predicted fourth membrane domain. The flip and flop isoforms of AMPA receptors may be differentially involved in pain transmission and response to injury.

Cross-References

▶ Descending Circuitry, Molecular Mechanisms of Activity-Dependent Plasticity

Flunarizine

Definition

Calcium channel, dopamine, and histamine antagonist used in the preventive treatment of migraine.

Cross-References

▶ Headache Due to Arteritis

Fluorocitrate

Definition

Fluorocitrate blocks the activation of glial cells, without directly affecting neurons, and functions to inhibit the activity of aconitase, an enzyme in the Krebs cycle of glia, but not neurons. Peri-spinal administration of fluorocitrate blocks exaggerated pain.

Cross-References

▶ Cord Glial Activation

Flurbiprofen

Definition

Flurbiprofen is a nonsteroid anti-inflammatory drug (NSAID), which inhibits the formation of prostaglandins by cyclooxygenase. It is selective for COX-1, thus inhibits COX-1 at lower concentrations than COX-2.

Cross-References

▶ Cyclooxygenases in Biology and Disease

fMRI

▶ Functional Magnetic Resonance Imaging

fMRI Imaging and PET in Parietal Cortex

▶ PET and fMRI Imaging in Parietal Cortex (SI, SII, Inferior Parietal Cortex BA40)

FMS

▶ Muscle Pain, Fibromyalgia Syndrome (Primary, Secondary)

Focal Pain

Definition

Focal pain is that which is experienced at one site, superficial to the underlying nociceptive lesion.

Cross-References

▶ Cancer Pain, Goals of a Comprehensive Assessment

Focused Analgesia

Definition

Focused analgesia is based on increased and directed attention to that part of the body for which suggestions of analgesia have been given. For example, suggestions for numbness in the hand may produce the experience of numbness as well as reduced pain sensation.

Cross-References

▶ Hypnotic Analgesia

Follicular Phase

Definition

The follicular phase is the time during which a single dominant ovarian follicle develops. The follicle should be mature at mid-cycle for ovulation. The average length of this phase is about 10–14 days. Variability in the length of this phase is responsible for variations in total cycle length.

Cross-References

▶ Premenstrual Syndrome

Forced-Choice Procedure

Definition

Forced-choice procedure is a statistical decision theory method in which a sensory decision is made after two or more stimulus presentations.

Forearm Ischemia Procedure

▶ Tourniquet Test

Forearm Occlusion Pain

▶ Tourniquet Test

Forebrain

Definition

Forebrain is the part of the brain including the cerebral cortex, limbic system, and hypothalamus.

Cross-References

▶ Forebrain Modulation of the Periaqueductal Gray and Its Role in Pain

Forebrain Modulation of the Periaqueductal Gray and Its Role in Pain

Dayna R. Loyd[1] and Anne Z. Murphy[2]
[1]US Army Institute of Surgical Research, Fort Sam Houston, TX, USA
[2]Neuroscience Institute, Georgia State University, Atlanta, GA, USA

Introduction

Forebrain Inputs to PAG

In the 1980s and 1990s, tract-tracing studies from a number of laboratories (e.g., Beitz 1982; Shipley et al. 1991) revealed an important feature of the PAG; afferent inputs to the PAG arise from a staggering number of cortical and subcortical forebrain sites. These inputs to the PAG are much heavier than previously suspected and terminate with a remarkable degree of topographic specificity. Major identified forebrain afferents to the PAG include the medial prefrontal cortex (MPF), the medial preoptic area (MPO), the central and medial nuclei of the amygdala (CeA, MeA), the ventromedial hypothalamus (VMH), and the paraventricular nucleus of the hypothalamus (PVN).

Definition

The midbrain ▶ periaqueductal gray (PAG) and its descending projections to the ▶ rostral ventromedial medulla (RVM) comprise a critical descending neuroanatomical pathway modulating pain and analgesia. The PAG projects heavily and directly to the RVM, which in turn projects to the ▶ dorsal horn of the spinal cord. Both pain and analgesia are modulated by emotional, motivational, and cognitive factors indicating higher-order cortical and subcortical ▶ forebrain modulation of the PAG-RVM-spinal cord pathway. Anatomical studies have demonstrated that the PAG predominantly receives afferents from the forebrain, including cortical and

subcortical sites involved in nociception. These forebrain projections terminate with a high degree of topographical specificity within the PAG, forming longitudinal input columns extending throughout the rostrocaudal extent of the PAG and activating output neurons projecting to the RVM. Major identified forebrain afferents to the PAG include the medial prefrontal cortex (MPF), the medial preoptic area (MPO), the central and medial nuclei of the ▶ amygdala (CeA, MeA), the ventromedial ▶ hypothalamus (VMH), and the paraventricular nucleus of the hypothalamus (PVN).

Characteristics

Medial Prefrontal Cortex (MPF)

The PAG receives dense inputs from large groups of neurons in the orbital, medial prefrontal, and lateral (insular and perirhinal) cortices (Beitz 1982; Shipley et al. 1991; Floyd et al. 2000). In total, inputs to the PAG arise from at least 8 distinct cytoarchitectonic fields in the medial prefrontal and lateral cortices (Shipley et al. 1991). ▶ Retrograde tracing demonstrates that projections to the PAG from the cerebral cortex arise from all fields of the medial prefrontal cortex infralimbic, prelimbic, anterior cingulate, and precentral medialis. Equally extensive projections arise from the lateral suprarhinal (insular cortex) and perirhinal cortical areas. Anterograde tracing reveals that the pattern of terminal labeling from each medial or lateral cortical field is highly organized and selectively targets discrete, columnar subregions of the PAG along its entire rostrocaudal axis. Inputs from different cortical fields terminate as different, largely complementary, longitudinal columns. Imaging studies in humans have shown that anterior cingulate and insular orbital cortices are consistently activated by the sensory-perceptual and affective (i.e., emotional or unpleasant) aspects of pain (Price 2000; Rolls et al. 2003). Analgesia, including opiate receptor-dependent analgesia, can be elicited from the anterior cingulate and insular cortices (see

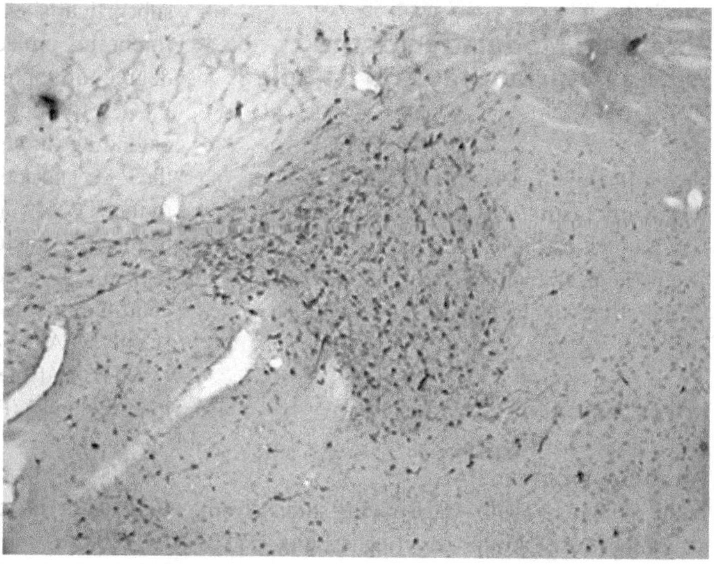

Forebrain Modulation of the Periaqueductal Gray and Its Role in Pain,
Fig. 1 Photomicrograph illustrating projection neurons in the CeA of the female rat as identified by retrograde tracing from the lateral/ventrolateral PAG and observed as a brown precipitate following immunohistochemical processing. Also shown is Fos expression to identify neural activity in CeA neurons and CeA-PAG projection neurons, observed as a *black* precipitate, following 24 h of hindpaw inflammation

Shipley et al. 1991; Burkey et al. 1996), and both areas are activated in humans following opioid or placebo treatment (Petrovic et al. 2002).

Central and Medial Nuclei of the Amygdala (CeA, MeA)

Projections from the amygdala to the PAG arise predominantly from the central and medial nuclei of the CeA. The CeA projects in a columnar manner to the PAG (Shipley et al. 1991). Additionally, a substantial population of PAG neurons project to the CeA. The CeA projection terminates as rostrocaudally oriented input columns that focally target different PAG subdivisions. The dorsomedial and lateral/ventrolateral subdivisions are especially heavily targeted (Shipley et al. 1991). Retrograde tract-tracing from the lateral/ventrolateral PAG reveals a dense population of projection neurons in the CeA of both male and female rats, and these projection neurons are activated by 24 h of hindpaw inflammation, observed as colocalization with Fos as a measure of neural activity (Fig. 1). There is also a dense projection of neurons from the MeA to the lateral/ventrolateral PAG that are also active during inflammatory pain (Fig. 2). The CeA is a key component of circuits involved in defense reactions, as well as

the mediation of both conditioned and innate fear-related responses. In this regard, it is noteworthy that conditioned and unconditioned fear and aversive responses are accompanied by antinociception (Helmstetter and Tershner 1994). Stimulation of the CeA produces analgesia (Shipley et al. 1991; Oliveira and Prado 2001), and the CeA is involved in analgesia elicited by systemically administered opiates (Manning and Mayer 1995). It is reasonable to speculate, therefore, that CeA projections to the PAG may mediate antinociceptive responses that accompany fear and aversive responses. In agreement with this hypothesis, lesions of the PAG and RVM attenuate analgesia associated with aversive conditioned responses (Helmstetter and Tershner 1994), and PAG lesions block CeA-stimulation-produced analgesia (Oliveira and Prado 2001).

Medial Preoptic Area (MPO)

The MPO-PAG projection is very dense and exhibits columnar organization, similar to the CeA-PAG projection. This projection arises from neurons in several cytoarchitectonically distinct subdivisions of the MPO, including the sexually dimorphic medial preoptic nucleus (Fig. 3). These neurons are also activated following 24 h of inflammation in the rat hindpaw.

Forebrain Modulation of the Periaqueductal Gray and Its Role in Pain, Fig. 2 Photomicrograph illustrating projection neurons in the MeA of the female rat as identified by retrograde tracing from the lateral/ventrolateral PAG and observed as a brown precipitate following immunohistochemical processing. Also shown is Fos expression to identify neural activity in MeA neurons and MeA-PAG projection neurons, observed as a *black* precipitate, following 24 h of hindpaw inflammation

Forebrain Modulation of the Periaqueductal Gray and Its Role in Pain, Fig. 3 Photomicrograph illustrating MPO-PAG projection neurons in the female rat labeled by retrograde tract-tracing from the lateral/ventrolateral PAG and observed as brown precipitate following immunohistochemistry. Also shown are Fos-positive MPO neurons and MPO-PAG projection neurons indicating neural activity following 24 h of inflammation in the rat hindpaw (observed as *black* precipitate)

Injections of anterograde tracers into the MPO label dense, highly organized, and topographically specific projections to the PAG. A hallmark of this projection, like other forebrain inputs studied to date, is that it forms longitudinally organized input columns that selectively target discrete subregions of the PAG (i.e., the dorsomedial and lateral/ventrolateral PAG) along its rostrocaudal axis. In particular, MPO inputs to the PAG selectively terminate within the lateral/ventrolateral caudal PAG, where output neurons to the RVM are selectively located (Murphy and Hoffman 2001). Activation of the MPO by a variety of physiological factors also activates longitudinally organized columns of PAG neurons, including those that project to the

Forebrain Modulation of the Periaqueductal Gray and Its Role in Pain, Fig. 4 Photomicrograph illustrating MPO-PAG projection neurons identified by retrograde tracing from the lateral/ventrolateral PAG and observed as brown precipitate following immunohistochemistry. Also shown is ERα-positive MPO neurons and colocalization with MPO-PAG neurons observed as *black* precipitate

ventral medulla (Murphy and Hoffman 2001; Normandin and Murphy 2008). The MPO is sexually dimorphic and plays a key role in neuroendocrine and steroidal regulation and maternal/reproductive behaviors. Nociceptive thresholds shift across the estrus cycle, with steroid hormone administration and during reproductive activities (see ▶ Sex Differences in Descending Pain Modulatory Pathways). Both the MPO and the PAG contain high levels of estrogen and androgen receptors (Murphy and Hoffman 2001; Loyd and Murphy 2008). The MPO-PAG projection neurons also coexpress the estrogen receptor ERα (Fig. 4). In this regard, it is interesting to note that there are sex differences in pain thresholds as well as sex differences in how inflammatory pain activates the PAG-RVM circuit (Loyd and Murphy 2006). Furthermore, the PAG-RVM pathway is anatomically and functionally sexually dimorphic (Loyd and Murphy 2006) and is essential for the observed sexual dimorphism in morphine analgesia (Loyd et al. 2008). The MPO, therefore, may provide a key link mediating hormonal influences on nociceptive thresholds and may also regulate descending nociception during maternal/reproductive behaviors (Murphy et al. 1999). Consistent with this hypothesis, stimulation of the MPO activates PAG neurons that project to the RVM (Rizvi et al. 1996), inhibits

dorsal horn spinal cord neuronal responses to nociceptive stimuli, and also elicits analgesia (see Shipley et al. 1991; Zhang and Ennis 2005).

Hypothalamic Projections to the PAG

There is also evidence of various hypothalamic nuclei projections to the lateral/ventrolateral PAG as observed by retrograde tract-tracing, including the ventromedial hypothalamus (VMH; Fig. 5) and the paraventricular nucleus of the hypothalamus (PVN; Fig. 6). These hypothalamic nuclei are known to modulate fear and stress responses, as well as sex and satiety behaviors. Hypothalamic projections to the PAG likely play a role in modulating these behavioral responses during pain and analgesia; however, little research has addressed these possibilities. Future research is warranted based on known anatomical connectivity between the hypothalamus and the descending pain inhibitory system.

PAG-Forebrain Afferents Coordinate a Multitude of Behavioral Responses

There has been a growing recognition that the PAG itself is far more complex than initially suspected and is clearly involved in many more physiological functions than nociception. For example, stimulation of different columnarly organized regions of the PAG produces

Forebrain Modulation of the Periaqueductal Gray and Its Role in Pain, Fig. 5 Photomicrograph illustrating VMH-PAG projection neurons in the female rat identified by retrograde tract-tracing from the lateral/ventrolateral PAG and observed as *brown* precipitate following immunohistochemistry. Also shown are Fos-positive VMH neurons and VMH-PAG output neurons indicating neural activity following 24 h of hindpaw inflammation

Forebrain Modulation of the Periaqueductal Gray and Its Role in Pain, Fig. 6 Photomicrograph illustrating PVN-PAG projection neurons in the female rat identified by retrograde tract-tracing from the lateral/ventrolateral PAG and observed as *brown* precipitate following immunohistochemistry. Also shown are Fos-positive PVN neurons and PVN-PAG output neurons indicating neural activity following 24 h of hindpaw inflammation

a number of distinctly different behavioral and physiological responses including vocalization, autonomic changes, sexual/reproductive behaviors, and fear and rage reactions (for review see Shipley et al. 1991; Murphy et al. 1994; Murphy and Hoffman 2001; Loyd and Murphy 2009, Loyd and Murphy 2008). Not surprisingly, many of the forebrain sites that project to the PAG are also known to regulate similar functions. Based on these findings, a more integrative conceptual framework has been adopted guided by the working hypothesis that the PAG is a structure that plays a central role in the production of certain stereotypical behaviors (e.g., reproduction, defense reactions, vocalization) essential to the animal's survival (Loyd and Murphy 2009). These behaviors require rapid, profound autonomic adjustments and simultaneously, significant alterations in pain thresholds. From this perspective, it is reasonable to consider that more highly elaborated forebrain structures interact with the PAG to coordinate antinociceptive, behavioral, and autonomic responses in concert with the dominant role of the forebrain in cognitive and emotional processing. Descending PAG output neurons also exhibit columnar organization, such that different output columns terminate with medial to lateral specificity in the ventral medulla in sites involved in nociception, autonomic responses, sexual/reproductive behaviors, and vocalization (Murphy and Hoffman 2001; Murphy et al. 1999; Loyd and Murphy 2008, Loyd and Murphy 2009, Loyd and Murphy 2006). Taken together, these findings suggest that forebrain sites may trigger or modulate specific nociceptive and autonomic adjustments via activation of discrete columns of PAG output neurons.

Summary

The PAG and the RVM are major nodes for bidirectional modulation of nociception and spinal cord neuronal responses to noxious sensory input. The PAG-RVM circuit is demonstrably central to analgesia elicited by activation of endogenous opioidergic systems as well as that resulting from exogenously administered opioids. Nociceptive regulation is but one aspect of this circuit as the PAG is clearly a highly organized structure that integrates defensive, fear/anxiety, and hormonal/reproductive behaviors in discrete columns that extend longitudinally through the structure. These columns receive dense and topographically specific input from cortical and subcortical forebrain areas centrally involved in these same functions. Some forebrain areas also project in parallel to the RVM, and the functional significance of such dual projections is not known. In humans, whose behavior is dominated by the massive and highly elaborated forebrain, the projections to the PAG and RVM are likely to regulate nociception in concert with forebrain-mediated cognitive and emotional processing of pain and its perceived impact.

Acknowledgement The authors would like to recognize the contributions of Dr. Matthew Ennis and Xijing Zhang on the first edition of this chapter.

References

Beitz, A. J. (1982). The organization of afferent projections to the periaqueductal gray of the rat. *Neuroscience, 7*, 133–159.

Burkey, A. R., Carstens, E., Wenniger, J. J., et al. (1996). An opioidergic cortical antinociception triggering site in the agranular insular cortex of the rat contributes to morphine antinociception. *The Journal of Neuroscience, 16*, 6612–6623.

Floyd, N. S., Price, J. L., Ferry, A. T., et al. (2000). Orbitomedial prefrontal cortical projections to distinct longitudinal columns of the periaqueductal gray in the rat. *The Journal of Comparative Neurology, 422*, 556–578.

Helmstetter, F. J., & Tershner, S. A. (1994). Lesions of the periaqueductal gray and rostral ventromedial medulla disrupt antinociceptive but not cardiovascular aversive conditional responses. *The Journal of Neuroscience, 14*, 3099–7108.

Loyd, D. R., & Murphy, A. Z. (2006). Sex differences in the anatomical and functional organization of the periaqueductal gray-rostral ventromedial medullary pathway in the rat: A potential circuit mediating the sexually dimorphic actions of morphine. *The Journal of Comparative Neurology, 496*, 723–738.

Loyd, D. R., & Murphy, A. Z. (2008). Androgen and estrogen (α) receptor localization on periaqueductal gray neurons projecting to the rostral ventromedial medulla in the male and female rat. *Journal of Chemical Neuroanatomy, 36*, 216–226.

Loyd, D. R., & Murphy, A. Z. (2009). The role of the periaqueductal gray in the modulation of pain in males and females: are the anatomy and physiology really that different? *Neural Plast.* doi: 10.1155/2009/462879.

Loyd, D. R., Wang, X., & Murphy, A. Z. (2008). Sex differences in mu opioid receptor expression in the rat midbrain periaqueductal gray are essential for eliciting sex differences in morphine analgesia. *The Journal of Neuroscience, 28*(52), 14007–14017.

Manning, B. H., & Mayer, D. J. (1995). The central nucleus of the amygdala contributes to the production of morphine antinociception in the rat tail flick test. *The Journal of Neuroscience, 15*, 8199–8213.

Murphy, A. Z., Ennis, M., Shipley, M. T., & Behbehani, M. M. (1994). Directionally specific changes in arterial pressure induce differential patterns of fos expression in discrete areas of the rat brainstem: a double-labeling study for Fos and catecholamines. *The Journal of Comparative Neurology, 349*(1), 36–50.

Murphy, A. Z., Rizvi, T. A., Ennis, M., et al. (1999). The organization of preoptic-medullary circuits in the male rat: Evidence for interconnectivity of neural structures involved in reproductive behavior, antinociception and cardiovascular regulation. *Neuroscience, 91*, 1103–1116.

Murphy, A. Z., & Hoffman, G. E. (2001). Distribution of gonadal steroid receptor-containing neurons in the preoptic-periaqueductal gray-brainstem pathway: a

potential circuit for the initiation of male sexual behavior. *The Journal of Comparative Neurology, 438*(2), 191–212.
Normandin, J. J., & Murphy, A. Z. (2008). Nucleus paragigantocellularis afferents in male and female rats: organization, gonadal steroid receptor expression, and activation during sexual behavior. *The Journal of Comparative Neurology, 508*(5), 771–794.
Oliveira, M. A., & Prado, W. A. (2001). Role of the PAG in the antinociception evoked from the medial or central amygdala in rats. *Brain Research Bulletin, 54*, 55–63.
Petrovic, P., Kalso, E., Petersson, K. M., et al. (2002). Placebo and opioid analgesia-imaging a shared neuronal network. *Science, 295*, 1737–1740.
Price, D. D. (2000). Psychological and neural mechanisms of the affective dimension of pain. *Science, 288*, 1769–1772.
Rizvi, T. A., Ennis, M., Murphy, A. Z., et al. (1996). Medial preoptic afferents to periaqueductal gray medullo-output neurons: A combined Fos and tract tracing study. *The Journal of Neuroscience, 16*, 333–344.
Rolls, E. T., O'Doherty, J., Kringelbach, M. L., et al. (2003). Representations of pleasant and painful touch in the human orbitofrontal and cingulate cortices. *Cerebral Cortex, 13*, 308–317.
Shipley, M. T., Ennis, M., Rizvi, T. A., et al. (1991). Topographical specificity of forebrain inputs to the midbrain periaqueductal gray: Evidence for discrete longitudinally organized input columns. In A. Depaulis & R. Bandler (Eds.), *The midbrain periaqueductal gray matter* (pp. 417–448). New York: Plenum Press.
Zhang, Y. H., & Ennis, M. (2007). Inactivation of the periaqueductal gray attenuates antinociception elicited by stimulation of the rat medial preoptic area. *Neurosci Lett. 429*(2–3), 105–110.

Formalin Test

Terence J. Coderre[1], Frances V. Abbott[2] and Jana Sawynok[3]
[1]Departments of Anesthesia, Neurology & Neurosurgery and Psychology, McGill University, Montreal, QC, Canada
[2]Departments of Psychiatry and Psychology, McGill University, Montreal, QC, Canada
[3]Department of Pharmacology, Dalhousie University, Halifax, Canada

Synonyms

Nociception induced by injection of dilute formaldehyde

Definition

The formalin test refers to the quantification of characteristic nociceptive behaviors that occur in response to subcutaneous (s.c.) or intradermal injection of a dilute solution of formaldehyde in 0.9 % saline, typically into the dorsal or plantar hindpaw of rodents.

Characteristics

The formalin test was originally described by Dubuisson and Dennis (1977) using 50 μl of 5 % formalin injected s.c. into the dorsal surface of one forepaw in rats and cats. "Five percent formalin" consisted of 1 ml of saturated formaldehyde (37 %) in water + 19 ml 0.9 % saline (i.e., 1.85 % formaldehyde). It is now more common to inject between 0.2 % and 5 % formalin into the dorsal or plantar hindpaw, using 20–50 μl in rats or 10–25 μl in mice. Another common site is the lateral aspect of the muzzle, or the temporomandibular joint, in rats, as a model of orofacial pain (Clavelou et al. 1995). The hindpaw has replaced the forepaw as a preferred site, because rats and mice frequently lick the forepaw in normal grooming. The formalin test has also been used in other species, including guinea pigs, rabbits, primates, crocodiles, domestic fowl, and octodon degus (Tjølsen et al. 1992).

Nociceptive behaviors increase with formalin concentration in rats and mice but reach a plateau between 2 % and 5 % formalin regardless of the scoring method used (see below) (Tjølsen et al. 1992; Coderre et al. 1993; Abbott et al. 1995; Sawynok and Liu 2003). With further increases in concentration, the magnitude and duration of the behavioral response are not increased; rather the animal's behavior becomes more disorganized, and the nociceptive scores may actually fall. Formalin concentrations of 1 % or less are useful for detecting the actions of weak analgesics, avoiding ceiling effects (Abbott et al. 1995; Sawynok and Liu 2003).

Factors that influence the magnitude of the response include the site of injection (plantar injections produce greater responses), age of the

Formalin Test, Fig. 1 Behavioral categories used in the weighted-scores technique of formalin test scoring devised by Dubuisson and Dennis (1977), illustrated for hindpaw injections (*0* injected paw is not favored, *1* injected paw has little or no weight on it, *2* injected paw is elevated and not in contact with any surface, *3* injected paw is licked, bitten, or shaken)

Formalin Test, Fig. 2 Time course of nociceptive responses after injection of 5 % formalin in rats. (**a**) Pain intensity ratings using the weighted-scores technique. (**b**) Number of flinches (*closed squares*) and time spent licking (*closed circles*). Note that all three measures illustrate the biphasic nature of nociceptive responses (Modified with permission from (**a**) Dubuisson and Dennis (1977) and (**b**) Wheeler-Aceto and Cowan (1991))

animal (responses are greater in infant rats), the strain, the degree to which the animal is habituated to handling, and the testing environment (unfamiliar environments can produce stress-induced analgesia), level of morbidity (e.g., time since surgery), and temperature of the animal colony and the testing environment (heat increases peripheral blood flow and the inflammatory response). Other environmental factors, such as sounds, odors, light, atmospheric pressure, or even activity of humans in the test room, can also influence the expression of nociceptive behaviors. Formalin concentrations should be decided on the basis of the scientific objectives of the study, with adjustments for the

response of animals under the conditions prevailing at the laboratory (Tjølsen et al. 1992; Sawynok and Liu 2003). For ethical reasons, the lowest possible concentration of formalin consistent with scientific objectives should be used.

Dubuisson and Dennis (1977) quantified formalin nociception using a ▶ weighted-scores technique (WST), which involves assigning weights to each behavioral category measured (paw favoring, paw elevation, or licking, biting, or shaking of the paw) (Figs. 1, 2a). The ordinality and validity of the category weights in the WST have been well established in rats (Coderre et al. 1993; Abbott et al. 1995). Others have used single behavioral scoring methods, including recording of the time spent licking/biting the injected paw (Hunskaar et al. 1985) or counting the number of flinches (Wheeler-Aceto and Cowan 1991) (Fig. 2b), or have used automated scoring techniques (Jourdan et al. 2001). Parametric analysis suggests that the WST is superior to any single nociceptive measure; however, it is clear that assessment of paw favoring adds little to the equation and may be omitted from the analysis (Abbott et al. 1995). Paw licking/biting scores are commonly used in the mouse formalin test, since these behaviors predominate in mice and are easily quantified. It has been argued that ▶ flinching is more robust and less influenced by treatments affecting other behaviors (e.g., motor function), while licking/biting is regarded as being more variable and subject to motor influences, stereotypy, and perhaps taste aversion following earlier licking episodes (Wheeler-Aceto and Cowan 1991). Flinching and licking/biting may reflect distinct neuronal mechanisms, as they can be differentially modulated by certain drugs and procedures (e.g., amitriptyline and naloxone decrease licking/biting behaviors while simultaneously increasing flinching behaviors) (Sawynok and Liu 2003).

Concerning validation, the formalin concentration-response relationship, and the analgesic effects of opioids, has been examined for most scoring methods. However, few studies have determined whether nociceptive scores are suppressed by drugs known not to be analgesic in humans. The latter is particularly important, because sedation and other toxic effects can appear as analgesia. When different behaviors compete (e.g., a rat cannot favor its paw at the same time as licking it), then the interpretation of the change depends on whether one behavior represents greater nociception than another. This is problematic for some scoring methods, because a decrease in a behavior considered to represent greater nociception may occur because of drug side effects (Abbott et al. 1995). On the other hand, the selection of scoring method may also depend on other factors, such as the drug administration method. For example, in some cases, the presence of a chronic intrathecal (I.T.) cannula can eliminate the expression of licking/biting behaviors, yet leave flinching unaltered. This may be the reason why studies using rats and chronic I.T. cannulas (PE-10 tubing) generally do not report licking/biting behaviors as an outcome, yet when drugs are given spinally by acute lumbar puncture, such behaviors are often reported (Sawynok and Reid 2003).

An important characteristic of the formalin test is that there are two distinct phases of nociceptive behavior, described as the early and late, or first and second, phases. The early phase lasts 5–10 min; this is followed by a quiescent interval of 5–10 min, and then a subsequent late phase of activity is observed up to 60–90 min. The biphasic response to formalin is prominent in rats and mice, regardless of the nociceptive scoring method (i.e., WST, flinches, licking/biting, or automated) (Fig. 2), but is less obvious in other species (Tjølsen et al. 1992). The biphasic nature of the formalin response is also observed for heart rate and blood pressure (Taylor et al. 1995), as well as the neuronal activity of Aδ- and C-fiber sensory afferent neurons (Puig and Sorkin 1995) or deep dorsal horn neurons of the spinal cord (Dickenson and Sullivan 1987) (Fig. 3). The two phases of nociceptive activity likely reflect an initial direct activation of nociceptive primary afferents (i.e., nociceptors) by formalin, followed by afferent activation produced by inflammatory mediators released following tissue injury (Dubuisson and Dennis 1977; Tjølsen et al. 1992). The late phase may also involve ▶ central sensitization (Coderre 2001). A quiescent interval in C-fiber and dorsal horn neuron activation (Puig and Sorkin 1995; Dickenson and Sullivan 1987) tends to support the notion that distinct mechanisms underlie the two phases of behavioral and electrophysiological activation. However, the "interphase" also reflects ▶ active inhibition initiated by processes activated in phase 1. For example, a second formalin injection, administered 20 min after the initial injection, produces inhibition of behavioral responses and electrophysiological activity, with a time frame that corresponds to the first interphase interval (Fig. 4) (Henry et al. 1999).

Formalin Test, Fig. 3 Time course of (**a**) mean C-fiber responses in sural nerve and (**b**) an individual dorsal horn (DH) convergent neuron response to hindpaw formalin injections of 5 % formalin in rats. Note the biphasic nature of neuronal responses in both C-fibers and DH neurons (Modified with permission from (**a**) Puig and Sorkin (1995) and (**b**) Dickenson and Sullivan (1987))

Formalin Test, Fig. 4 Effects of one or two injections (20-min interinjection interval) of 2.5 % formalin into the plantar surface of one hindpaw on nociceptive scores in the formalin test. *Filled triangles* show the typical biphasic nociceptive response to a single injection. *Open triangles* show the effects of two injections of formalin. Note the second inhibitory interphase after the second formalin injection in the group that had two injections (Modified with permission from Henry et al. (1999))

Electrophysiological and pharmacological studies suggest that the early-phase formalin response depends on both the direct activation of nociceptors and ► neurogenic inflammation generated by the release of bradykinin, 5-hydroxytryptamine, histamine, and adenosine triphosphate (Tjølsen et al. 1992; Sawynok and Liu 2003). Formalin might also disrupt the perineurium, and this could enhance the access of tissue mediators to the sensory nerve. The late-phase formalin response depends in part on neurogenic inflammation induced by the above mediators as well as inflammatory responses associated with the release of cytokines and breakdown of arachidonic acid (Tjølsen et al. 1992; Sawynok and Liu 2003). There is also evidence to suggest that substance P- and glutamate-mediated central sensitization, initiated during the early phase, contributes to nociception in the late phase of the formalin test. This evidence arose from observations that the spinal administration of local anesthetics, opioids, and substance P or glutamate (N-methyl-D-aspartate) receptor antagonists inhibits late responses when given prior to, but not after, the early phase (Dickenson and Sullivan 1987; Coderre 2001). These findings have prompted the extensive use of the formalin test as an animal model for examining the potential of treatments for producing ► preemptive analgesia. The mechanisms underlying the active inhibition during the interphase are not completely understood but probably depend on both spinal and supraspinal influences, since the inter-phase is still present in spinalized animals but is eliminated by brain transection at the mes-encephalic-diencephalic junction. A role for GABA$_A$ receptors is suggested because the interphase is reduced both by spinal adminis-tration of GABA$_A$ receptor antagonists and by systemic administration of low doses of barbi-turates (Sawynok and Liu 2003).

While release of inflammatory mediators con-tributes to formalin-induced nociception, inflam-mation alone is not sufficient to produce the nociceptive behaviors induced by formalin. Thus, injection of inflammatory agents such as yeast, carrageenan, or Complete Freund's Adju-vant (CFA) into the hindpaw produces profound edema but essentially no nociceptive shaking, licking, or flinching behaviors as produced by formalin. This implies that activation of nociceptors is a complex process that is to some extent dissociable from inflammatory processes (Sawynok and Liu 2003). Importantly, the inflammatory response to formalin, and the reli-ance of nociceptive response on inflammation, depends significantly on the concentration and volumes of formalin administered. Assuming that injection volumes are similar, low concen-trations of formalin (1–2.5 % formalin or less) produce very little ► plasma extravasation and edema, while greater concentrations (5 %) pro-duce significant hindpaw inflammation, lasting for as long as 1 week after injection. In addition, studies using adult pretreatment with capsaicin (to desensitize TRPV1-expressing nociceptors) and anti-inflammatory drugs (to target ► nonneurogenic inflammation) suggest that low concentrations of formalin produce effects that depend on neurogenic inflammation, whereas greater concentrations of formalin depend on nonneurogenic inflammation (Sawynok and Liu 2003; Coderre 2001). With greater concentrations of formalin, it is also more difficult to demonstrate a role for central sensitization in the late phase. Thus, nociceptive behaviors produced by 1–2.5 % formalin, but not those produced by 3.75–5 % formalin, are sensitive to preemptive effects of spinal adminis-tration of local anesthetics (Coderre 2001). The

evidence suggesting differing roles of neurogenic and nonneurogenic inflammation with low and high concentrations of formalin has been recog-nized in rats, but not mice. However, it should be noted that the relatively large volumes of forma-lin injections typically used in mice produce sig-nificant inflammation even at low concentrations of formalin (reflecting different injection to tissue volume ratios) and might not allow for such a clear distinction.

Recent studies using knockout mice provide further insights into molecular mechanisms involved in formalin activation of nociceptors. In vitro data indicates that formaldehyde acti-vates TRPA1 receptors, which is supported by complementary in vivo data that TRPA1-deficient ($-/-$) mice exhibit markedly reduced behavioral responses to formalin (Macpherson et al. 2007; McNamara et al. 2007). However, the attenuation of nociceptive responses is seen only at a low concentration of formalin (0.5 %), and there are no differences in responses at greater concentrations of formalin (2 % and 5 %) which are used more commonly in nociceptive behavioral tests (Bráz and Basbaum 2010). At the lower concentration, most formalin-responsive afferents are unmy-elinated, but greater concentrations recruit both unmyelinated and myelinated afferents (Bráz and Basbaum 2010). The involvement of myelinated afferents in greater concentra-tions of formalin-induced behaviors is further supported by observations that ablation of the vast majority of C-fiber nociceptors does not alter such responses (Shields et al. 2010). Nerve injury occurs following administration of formalin, even at 0.5 %, as all concentra-tions of formalin lead to induction of activat-ing transcription factor 3 (ATF3), a reliable marker of nerve injury (Bráz and Basbaum 2010). Hindpaw inflammation produced by CFA did not induce ATF3, and carrageenan led to only a slight effect; these observations suggest that nerve damage, not increased activ-ity in nociceptors, resulted in ATF3 induction (Bráz and Basbaum 2010).

Formalin injections also activate processes that lead to long-term changes in the central nervous system. Formalin injections lead to the spinal activation of various intracellular signaling molecules, such as cyclic-adenosine monophosphate (cAMP), protein kinase C, nitric oxide, cyclic-guanosine monophosphate (cGMP), mitogen-activated protein kinase (MAPK), cAMP-response element binding protein (CREB), as well as proto-oncogenes and their protein products, including c-*fos*, c-*jun*, Krox-24, Fos, Jun B, and Jun D. Formalin injections also produce delayed activation of microglia in spinal cord dorsal horn. These changes may underlie long-term changes in mechanical and thermal sensitivity that occur in sites adjacent to or remote from the injection site, which last from days to weeks after the injection (Sawynok and Liu 2003; Coderre 2001).

Cross-References

▶ Behavioral Studies in Animals
▶ Cingulate Cortex
▶ Nociceptive Processing in the Hippocampus and Entorhinal Cortex, Neurophysiology and Pharmacology

References

Abbott, F. V., Franklin, K. B. J., & Westbrook, R. F. (1995). The formalin test: Scoring properties of the first and second phases of the pain response in rats. *Pain, 60*, 91–102.

Bráz, J. M., & Basbaum, A. I. (2010). Differential ATF3 expression in dorsal root ganglion neurons reveals the profile of primary afferents engaged by diverse chemical stimuli. *Pain, 150*, 290–301.

Clavelou, P., Dallel, R., Orliaguet, T., Woda, A., & Raboisson, P. (1995). The orofacial formalin test in rats: Effects of different formalin concentrations. *Pain, 62*, 295–301.

Coderre, T. J. (2001). Noxious stimulus-induced plasticity in spinal cord dorsal horn: Evidence and insights on mechanisms obtained using the formalin test. In M. M. Patterson & J. W. Grau (Eds.), *Spinal cord plasticity: Alterations in reflex function* (pp. 163–183). Boston: Kluwer Academic.

Coderre, T. J., Fundytus, M. E., McKenna, J. E., Dalal, S., & Melzack, R. (1993). The formalin test: A validation

of the weighted-scored method of behavioral pain rating. *Pain, 54*, 43–50.

Dickenson, A. H., & Sullivan, A. F. (1987). Subcutaneous formalin-induced activity of dorsal horn neurones in the rat: Differential response to an intrathecal opioid administered pre or post formalin. *Pain, 30*, 349–360.

Dubuisson, D., & Dennis, S. G. (1977). The formalin test: A quantitative study of the analgesic effects of morphine, meperidine, and brain stem stimulation in rats and cats. *Pain, 4*, 161–174.

Henry, J. L., Yashpal, K., Pitcher, G. M., & Coderre, T. J. (1999). Physiological evidence that the "Interphase" in the formalin test is due to active inhibition. *Pain, 82*, 57–63.

Hunskaar, S., Fasmer, O. B., & Hole, K. (1985). Formalin test in mice, a useful technique for evaluating mild analgesics. *Journal of Neuroscience Methods, 14*, 69–76.

Jourdan, D., Ardid, D., & Eschalier, A. (2001). Automated behavioural analysis in animal pain studies. *Pharmacological Research, 43*, 103–110.

Macpherson, L. J., Xiao, B., Kwan, K. Y., Petrus, M. J., Dubin, A. E., Hwang, S. W., Cravatt, B., Corey, D. P., & Patapoutian, A. (2007). An ion channel essential for sensing chemical damage. *The Journal of Neuroscience, 27*, 11412–11415.

McNamara, C. R., Mandel-Brehm, J., Bautista, D. M., Siemens, J., Deranian, K. L., Zhao, M., Hayward, N. J., Chong, J. A., Julius, D., Moran, M. M., & Fanger, C. M. (2007). TRPA1 mediates formalin-induced pain. *Proceedings of National Academy of Sciences USA, 104*, 13525–13530.

Puig, S., & Sorkin, L. S. (1995). Formalin-evoked activity in identified primary afferent fibres: Systemic lidocaine suppresses phase-2 activity. *Pain, 64*, 345–355.

Sawynok, J., & Liu, X. J. (2003). The formalin test: Characteristics and usefulness of the model. *Reviews in Analgesia, 7*, 145–163.

Sawynok, J., & Reid, A. R. (2003). Chronic intrathecal cannulas inhibit some and potentiate other behaviors elicited by formalin injection. *Pain, 103*, 7–9.

Shields, S. D., Cavanaugh, D. J., Lee, H., Anderson, D. J., & Basbaum, A. I. (2010). Pain behavior in the formalin test persists after ablation of the great majority of C-fiber nociceptors. *Pain, 151*, 422–429.

Taylor, B. K., Peterson, M. A., & Basbaum, A. I. (1995). Persistent cardiovascular and behavioral nociceptive responses to subcutaneous formalin require peripheral nerve input. *The Journal of Neuroscience, 15*, 7575–7584.

Tjølsen, A., Berge, O. G., Hunskaar, S., Rosland, J. H., & Hole, K. (1992). The formalin test: An evaluation of the method. *Pain, 51*, 5–17.

Wheeler-Aceto, H., & Cowan, A. (1991). Standardization of the rat paw formalin test for the evaluation of analgesics. *Psychopharmacology, 104*, 35–44.

Forty Hz Oscillations

Definition

40 Hz activity is oscillatory activity described in the brain when the organism is actively engaged in a task.

Cross-References

▶ Corticothalamic and Thalamocortical Interactions

Fos Expression

Definition

The expression of Fos protein, the gene product of the c-Fos gene. Fos expression in a particular set of central nervous system neurons is often taken to indicate activity of that set of neurons in response to a noxious stimulus.

Cross-References

▶ Visceral Pain Model, Lower Gastrointestinal Tract Pain

Fos Protein

Definition

Fos is the protein that is expressed by the c-Fos gene in nociceptive neurons after noxious stimulation. Fos protein is found in the nucleus, where it serves as a transcription factor (third messenger).

Cross-References

▶ c-fos
▶ Spinal Cord Injury Pain Model, Contusion Injury Model
▶ Spinal Dorsal Horn Pathways, Dorsal Column (Visceral)
▶ Spinothalamic Tract Neurons, Role of Nitric Oxide

Fractalkine

Definition

Fractalkine is a cell-surface protein expressed by neurons in the spinal cord. When sufficiently activated, spinal neurons release fractalkine, which then binds to receptors expressed on microglia, thereby inducing the release of the proinflammatory cytokine interleukin (IL)-1. Peri-spinal administration of fractalkine, as well as IL-1, induces exaggerated pain responses. Importantly, inhibiting fractalkine activity blocks neuropathic pain, implying that endogenous fractalkine generates enhanced pain responses.

Cross-References

▶ Cord Glial Activation

Fractalkine (CX3CL1)

▶ Cathepsin and Microglia

Free from Bias

Definition

A personal attitude that may influence the evaluation of patients' pain conditions or the interpretation of study results. Pain measures should be bias-free in that children should use them in the same manner regardless of differences in how they wish to please adults.

Cross-References

▶ Pain Assessment in Children

Free Magnitude Estimation

Definition

Free magnitude estimation is the numerical rating of sensation magnitude without upper or lower boundaries.

Cross-References

▶ Opioids, Effects of Systemic Morphine on Evoked Pain

Free Nerve Endings

▶ Non-corpuscular Sensory Endings

Freeze Lesion

Definition

A brief punctual freeze lesion of human skin that causes moderate burning or itching sensations together with a reddening and a short-lived edema, after 24 h hyperalgesia to punctuate stimuli, hyperalgesia to blunt pressure and impact stimuli, and increased heat sensitivity. Brush-evoked hyperalgesia does not develop after freezing.

Cross-References

▶ Freezing Model of Cutaneous Hyperalgesia

Freezing Model of Cutaneous Hyperalgesia

Susanne K. Sauer
Department of Physiology and Pathophysiology,
University of Erlangen, Erlangen, Germany

Synonyms

Cold-induced hyperalgesia; Cutaneous freeze trauma; Cutaneous hyperalgesia; Inflammatory hyperalgesia by skin freezing

Definition

A brief ▶ freeze lesion of human skin is a viable model to induce a slowly developing and sustained pattern of ▶ hyperalgesia. During freezing, moderate burning or itching sensations together with a reddening and a short-lived edema of the lesioned skin site can be observed. No spontaneous pain or hyperalgesia arises during the first hours after freezing the skin. About 24 h after induction, a typical pattern of hyperalgesia develops: hyperalgesia to punctuate stimuli, hyperalgesia to blunt pressure and impact stimuli, and increased heat sensitivity. Brush-evoked hyperalgesia (▶ allodynia) does not usually develop after freezing.

Characteristics

Injury of the skin is often followed by augmented pain sensations to mechanical, thermal, or chemical stimuli. Since the mechanism of the induction and maintenance of the different types of hyperalgesia are areas of considerable interest, scientists have searched for adequate experimental models, e.g., standardized lesions of the skin. Freeze injury of the skin was first described by

Lewis and Love in 1926 (Lewis and Love 1926), maybe stirred by reports from expeditions to the North Pole in this decade, describing freeze injuries of their fingers and feet.

Experimental Procedures

Experimental inflammation by freezing the skin is normally induced on the upper leg, the forearm, or the back of the hand. For this purpose a cylindrical copper bar with a standardized diameter and weight (e.g., 15 mm and 290 g, respectively) is cooled to $-28\,^{\circ}$C (Kilo et al. 1994). This bar is placed on the skin, with its long axis perpendicular to the surface and a contact pressure provided by its weight. To improve thermal contact to the skin, saline-soaked filter paper is placed on the skin beneath the copper bar. The degree of developing inflammation is determined by the temperature of the copper bar and the thawing time of the lesioned skin site. The duration of freezing assessed that thawing duration normally account as to 20–40 s. Sensory testing for different forms of hyperalgesia follows 22–24 h after freezing the skin.

Early Skin Reactions and Activation of Primary Afferents

Volunteer subjects describe freezing of the skin as moderately painful, with sensations of electric prickling or slight itching. Intense burning sensations are reported rarely. At the lesioned skin site, a sharply delineated region of local reddening develops, due to both cold-induced vasodilatation and local edema which subside within 1–2 h (Lewis and Love 1926; Kilo et al. 1994). Freezing itself obviously provokes relatively little spontaneous activity in superficial ▶ nociceptors because they are rapidly inactivated by the decrease in skin temperature. Before inactivation by freezing Ad and C-nociceptors respond to noxious temperatures down to about $-10\,^{\circ}$C with graded responses. Units are recruited progressively with decreasing temperature and are thus suited to contribute to the sensation of cold pain (Simone and Kajander, 1996; Simone and Kajander, 1997; Campero et al. 1996)). The obviously brief activation of nociceptors is reflected by a very spatially limited region of

▶ neurogenic inflammation around the frozen skin, since no flare or wheal reaction can be observed. For transduction of cold stimuli, different mechanisms have been proposed: activation of transient receptor potential channels, namely, ▶ TRPA1 and ▶ TRPM8, or inhibition of background potassium channels (Belmonte et al., 2009; Foulkes and Wood, 2007). Further on also a TTX-resistant sodium channel ▶ NaV1.8, which does not inactivate while cooling, is particularly important to retain excitability while cooling (Zimmermann et al., 2007).

The destruction of the cells in the upper skin layers evokes a complex inflammatory process, responsible for the altered pain sensation developing within the first day after freezing the skin.

Activation and Sensitization of Nociceptive Spinal Cord Neurons

Nociceptive spinal dorsal horn neurons respond to freeze injury of the skin. Both ▶ wide dynamic range (WDR) and ▶ nociceptive specific neurons (NS) are stimulated when the skin temperature in their innervation territories is lowered to $-15\,^{\circ}$C (Khasabov et al. 2001). WDR and NS neurons respond with a high-frequency discharge at the onset of freezing and subsequent ongoing activity. Moreover, a cross-sensitization by freezing of heat and cold activation thresholds is observed. Although little is known about the specific subtypes of peripheral afferents that excite the WDR and NS neurons, it is known that both types of dorsal horn neurons contribute to the cold and heat hyperalgesia produced by freeze injury. In a rat model, freezing the skin of the hindpaw also leads to increased ▶ c-fos expression in the spinal cord (Abbadie et al. 1994; Doyle and Hunt 1999).

Heat Hyperalgesia

Heat hyperalgesia, for instance, that following ▶ capsaicin application, is restricted to the primary hyperalgesic zone, and the underlying mechanism responsible for this is a peripheral sensitization of polymodal, mechano-heat sensitive, mostly unmyelinated nociceptors (Treede et al. 2004; Ali et al. 1996). Similarly heat hyperalgesia is observed at the injured skin site after freezing (1° zone) and to a minor extent also in the 2° zone. The peripheral sensitization of heat sensitive ion channels such as TRPV1 – which is expressed in these nerve fibers – by locally released inflammatory mediators is responsible for the increased heat sensitivity after freeze lesion (Patapoutian et al. 2003). The sensitization to heat stimuli of spinal nociceptive neurons after freeze injury has also been demonstrated in rat and thus contributes to the development of heat hyperalgesia (Khasabov et al. 2001).

Patterns of ▶ Mechanical Hyperalgesia

After freezing the skin a characteristic pattern of mechanical hyperalgesia develops to punctuate stimulation (pin-prick hyperalgesia), to pressure stimulation, and to impact stimulation of the treated skin area. Increased sensitivity to non-noxious mechanical stimulation (brush-evoked; Allodynia) is not observed. The alterations of sensitivity to different types of mechanical stimuli at the site of the freeze injury (primary zone or 1°) and the area surrounding the site of injury (secondary zone or 2°) are described as follows:

Brush-Evoked Hyperalgesia

Brush-evoked hyperalgesia or allodynia (formerly also called hyperesthesia), normally occurring in the 1° and 2° zone, is not consistently found in the 2° zone and only rarely found in the 1° zone of the freeze model (Kilo et al. 1994). Mechanical allodynia is mediated by an enhanced central processing of normal sensory input coming from myelinated, mechano-sensitive A fibers (Ab fibers) that transmit non-painful, tactile sensations (Torebjork et al. 1992; Treede et al. 2004). Allodynia after capsaicin application is found in the primary and secondary hyperalgesic zone and results from central plasticity induced and sustained by persistent nociceptor activity from the primary zone. After freezing no psychophysiological correlate of Ab fiber activation is reported, suggesting that a lack of activity of Ab fibers together with the limited ongoing activity of nociceptors may explain the lack of allodynia.

Pin-Prick Hyperalgesia

Pin-prick hyperalgesia tested by punctuate stimulation of the skin develops in the primary and secondary zone after freezing the skin in a manner very similar to the pattern of pin-prick hyperalgesia observed after capsaicin application (Kilo et al. 1994). In the 1° zone a sensitization of C-nociceptors to mechanical stimulation leads to an increased input to the spinal cord. Astonishingly only a few studies have shown a peripheral sensitization of C or Ad fibers to suprathreshold mechanical stimuli, and mechanical activation thresholds remained unchanged (Cooper et al. 1991; Schmelz et al. 1996). However, the normally mechano-insensitive class of nociceptors is sensitized by repeated mechanical stimulation and could contribute to the primary mechanical hyperalgesia (Schmidt et al. 2000). Which fiber types contribute to the central sensitization resulting in the pin-prick hyperalgesia in the freeze model is not clear, but after triggering the central sensitization, no ongoing activity of these nociceptors is required. The hyperalgesia in the 2° zone surrounding the injured skin site is also caused by central sensitization following excitation of C or Ad nociceptors in the injured area that facilitate Ad fiber input from the unlesioned secondary zone.

Hyperalgesia to Impact Stimuli

C and Ad fiber nociceptors are able to encode impacts of different strengths and after sensitization these fibers mediate hyperalgesia to impact stimuli (Koltzenburg and Handwerker 1994). The freezing lesion, in contrast to capsaicin application, induced this special type of hyperalgesia in the primary zone (Kilo et al. 1994). Maybe due to a long lasting and severe inflammatory tissue damage, that again induces the sensitization of mechano- and/or polymodal nociceptors.

References

Abbadie, C., Honore, P., & Besson, J. M. (1994). Intense cold noxious stimulation of the rat hindpaw induces c-fos expression in lumbar spinal cord neurons. *Neuroscience, 59,* 457–468.

Ali, Z., Meyer, R. A., & Campbell, J. N. (1996). Secondary hyperalgesia to mechanical but not heat stimuli following a capsaicin injection in hairy skin. *Pain, 68,* 401–411.

Belmonte, C., Brock, J. A., & Viana, F. (2009). Converting cold into pain. *Experimental Brain Research, 196,* 13–30.

Campero, M., Serra, J., & Ochoa, J. L. (1996). C-polymodal nociceptors activated by noxious low temperature in human skin. *The Journal of Physiology, 497*(Pt 2), 565–572.

Cooper, B., Ahlquist, M., Friedman, R. M., Loughner, B., & Heft, M. (1991). Properties of high-threshold mechanoreceptors in the oral mucosa I. Responses to dynamic and static pressure. *Journal of Neurophysiology, 66*, 1272–1279.

Doyle, C. A., & Hunt, S. P. (1999). A role for spinal lamina I neurokinin-1-positive neurons in cold thermoreception in the rat. *Neuroscience, 91*, 723–732.

Foulkes, T., & Wood, J. N. (2007). Mechanisms of cold pain. *Channels (Austin, Tex.), 1*, 154–160.

Khasabov, S. G., Cain, D. M., Thong, D., Mantyh, P. W., & Simone, D. A. (2001). Enhanced responses of spinal dorsal horn neurons to heat and cold stimuli following mild freeze injury to the skin. *Journal of Neurophysiology, 86*, 986–996.

Kilo, S., Schmelz, M., Koltzenburg, M., & Handwerker, H. O. (1994). Different patterns of hyperalgesia induced by experimental inflammation in human skin. *Brain, 117*, 385–396.

Koltzenburg, M., & Handwerker, H. O. (1994). Differential ability of human cutaneous nociceptors to signal mechanical pain and to produce vasodilatation. *The Journal of Neuroscience, 14*, 1756–1765.

Lewis, T., & Love, W. S. (1926). Vascular reactions of the skin to injury. Part III.– Some effects of freezing, of cooling and of warming. *Heart, 13*, 27–60.

Patapoutian, A., Peier, A. M., Story, G. M., & Viswanath, V. (2003). ThermoTRP channels and beyond: Mechanisms of temperature sensation. *Nature Reviews Neuroscience, 4*, 529–539.

Schmelz, M., Schmidt, R., Ringkamp, M., Forster, C., Handwerker, H. O., & Torebjork, H. E. (1996). Limitation of sensitization to injured parts of receptive fields in human skin C-nociceptors. *Experimental Brain Research, 109*, 141–147.

Schmidt, R., Schmelz, M., Torebjork, H. E., & Handwerker, H. O. (2000). Mechano-insensitive nociceptors encode pain evoked by tonic pressure to human skin. *Neuroscience, 98*, 793–800.

Simone, D. A., & Kajander, K. C. (1996). Excitation of rat cutaneous nociceptors by noxious cold. *Neuroscience Letters, 213*, 53–56.

Simone, D. A., & Kajander, K. C. (1997). Responses of cutaneous a-fiber nociceptors to noxious cold. *Journal of Neurophysiology, 77*, 2049–2060.

Torebjork, H. E., Lundberg, L. E., & LaMotte, R. H. (1992). Central changes in processing of mechanoreceptive input in capsaicin-induced secondary hyperalgesia in humans. *The Journal of Physiology, 448*, 765–780.

Treede, R. D., Handwerker, H. O., Baumgärtner, U., Meyer, R. A., & Magerl, W. (2004). Hyperalgesia and allodynia: Taxonomy, assessment, and mechanisms. In K. Brune & H. Handwerker (Eds.), *Hyperalgesia: Molecular mechanisms and clinical implications*. Seattle: IASP Press.

Zimmermann, K., Leffler, A., Babes, A., Cendan, C. M., Carr, R. W., Kobayashi, J., Nau, C., Wood, J. N., & Reeh, P. W. (2007). Sensory neuron sodium channel Nav1.8 is essential for pain at low temperatures. *Nature, 447*, 855–858.

Frequency of Low Back Pain

▶ Low Back Pain, Epidemiology

Frequency of Ultrasound Treatment

Definition

The most commonly used frequencies are in the range of 0.8–1.1 MHz, although frequencies around 3.0 MHz are also fairly common.

Cross-References

▶ Ultrasound Therapy of Pain from the Musculoskeletal System

Frequency-dependent Nociceptive Facilitation

▶ Windup of Spinal Cord Neurons

Freund's Complete Adjuvant

Synonyms

FCA

Definition

Freund's complete adjuvant (FCA) is a suspension of heat-killed mycobacterium in neutral oil, usually paraffin or mineral oil. Suspension is often achieved by ultrasonication.

Cross-References

▶ Arthritis Model, Adjuvant-induced Arthritis

Friction

▶ Massage and Pain Relief Prospects

Friction Massage

▶ Massage, Basic Considerations

Frontal-Posterior Neck Electromyographic Sensor Placement

Definition

Frontal-posterior neck electromyographic sensor placement is an approach designed to assess muscle activity across a broad region, ranging from the back of the neck to the front of the head.

Cross-References

▶ Psychophysiological Assessment of Pain

Frustration

▶ Anger and Pain

Fulcrum

Definition

The fulcrum is a pivot about which a lever turns.

Cross-References

▶ Sacroiliac Joint Pain

Functional

Definition

Functional is a term used before a symptom or disease to describe a process where a symptom or disease cannot be explained by any lesion, change in structure, or derangement of an organ.

Cross-References

▶ Functional Abdominal Pain in Children

Functional Abdominal Pain

▶ Descending Modulation of Visceral Pain
▶ Visceral Pain Model, Irritable Bowel Syndrome Model

Functional Abdominal Pain in Children

Sara E. Williams
Division of Pediatric Gastroenterology, Medical College of Wisconsin, Milwaukee, WI, USA

Synonyms

Chronic abdominal pain; Recurrent abdominal pain (RAP)

Definition

The term "functional abdominal pain" (FAP) refers to chronic or recurrent abdominal pain without evidence of a pathologic or organic condition. FAP is included as a diagnostic category among the childhood functional gastrointestinal disorders (FGID) as defined by the Rome criteria (Rasquin et al. 2006). FAP is considered a diagnosis of exclusion, with criteria only requiring the presence of episodic or continuous abdominal pain that is not better defined by another FGID or related to an organic process (e.g., anatomic, metabolic, infectious, or inflammatory disorder). FAP Syndrome is also described in the Rome Criteria, which requires the additional presence of disability related to pain (e.g., school absences) and other somatic symptoms (e.g., headache). Within the Rome criteria, there are other pain-related FGIDs that are not defined as FAP because the presence of associated symptoms qualifies as a separate diagnosis. For example, Irritable Bowel Syndrome (IBS) is defined as abdominal pain associated with changes in and relief by bowel movements. As a result, clinically, FAP typically describes a small number of FGID patients with abdominal pain. However, in the research setting, it is more common for FAP to be used as a term to describe abdominal pain conditions that are not associated with organic disease in general.

Introduction

Abdominal pain is a common complaint of childhood. Originally identified in the pediatric literature in the 1950s, "recurrent abdominal pain" was defined as abdominal pain severe enough to interfere with normal activities (Apley and Naish 1958). It was described as a common occurrence among school-aged children, with a greater prevalence in females than males (Apley 1975). More recent population studies have found that it is common for children to experience abdominal pain at some point during development, with up to 43 % of school children reporting transient abdominal pain (Roth-Isigkeit et al. 2005) and 10–15 % of children reporting chronic abdominal pain problems (Liakopoulou-Kairis et al. 2002). Studies of FAP prevalence among pediatric

patients in the medical setting have shown that 2–4 % of all pediatric primary care patients are seen for abdominal pain complaints and up to 50 % of pediatric gastroenterology patients have a pain-related FGID (Nurko and Di Lorenzo 2008).

From a biological standpoint, FAP is understood to be a physiological disorder generated by a combination of disruptions in intestinal and central nervous system activity (Chiou and Nurko 2010). Changes in intestinal motility can result in painful muscular spasms and contractions. Visceral hypersensitivity, a result of nerves in the gastrointestinal (GI) tract becoming over sensitized, contributes to the production of pain or discomfort from normal stimuli. Finally, communication between the brain and gut becomes dysregulated, resulting in increased pain signaling and perception.

From a psychosocial standpoint, disturbances in mood and functioning amplify illness experience and adversely affect health status for patients with FAP. FAP is disruptive to normal childhood development. FAP is associated with anxiety, depression, and somatic symptoms; in addition, FAP is associated with high levels of functional disability, school absence, and health care utilization (Campo et al. 2004).

Because of the complex relationship between these factors that contribute to the presentation of FAP, it is widely agreed that FAP is best understood through a *biopsychosocial* model (Chiou and Nurko 2010). The biopsychosocial model "integrates biological science with the unique features of the individual and determines the degree to which biological and psychosocial factors interact to explain the disease, illness, and outcome" (Drossman 1998, p. 262). In addition to the biological process, psychological, social, environmental, and behavioral factors also are recognized as potential contributing and interacting factors in patients' experience of FAP. It is equally important to assess and treat FAP from a broad perspective. Research has identified several factors that may predispose a child to develop FAP. Genetics play a role in making it more likely for a child to develop physiological disruptions in the GI tract (Talley 2008). Early life events, such as premature birth or early medical procedures, have been identified as potential precursors to the development of visceral hypersensitivity (Miranda 2008). Parent and family factors also play an important role; there is a clustering of chronic pain in families that may occur through parental modeling of illness behavior and/or through shared genetics (Levy et al. 2000).

Characteristics

Many studies have been conducted to examine concurrent and long-term medical and psychosocial outcomes in children with FAP. Pediatric FAP patients demonstrate more anxious, depressive, and somatic symptoms, experience more stressful life events, and have more school absences than well children (e.g., Campo et al. 2004; Walker et al. 1993; Walker and Greene 1989). Furthermore, children with FAP have increased depression, more somatic symptoms, and greater functional impairment compared to children with organic gastrointestinal disease (Walker et al. 1991). A long-term follow-up study demonstrated that patients with FAP continued to have more abdominal pain episodes, greater disability, more functional impairment, and greater health service utilization 5 years after initial evaluation compared to a well sample (Walker et al. 1995). In the same study, examination of patients' subsequent medical diagnoses revealed that only 1 child out of 31 was later diagnosed with an organic condition that may have accounted for his or her symptoms. A recent study of adults who had FAP as children found that one third continued to have abdominal pain into adulthood, in addition to the development of other chronic pain conditions (Walker et al. 2010).

Studies of parents of children with FAP have revealed a high level of anxiety (Garber et al. 1990) as well as a strong history of FGIDs (Levy et al. 2000) compared to parents of well children. Parents of children with FAP typically report high levels of distress about their children's symptoms, such as describing themselves as helpless in dealing with their children's pain and identifying threat of serious disease as their most central concern (Van Tilburg et al. 2006). These factors may lead parents to reassure their children and protect them from harm by encouraging children to withdraw from activities. Although parents have good intentions, this type of protective parenting behavior has been associated with high levels of school absence and pain-related disability in pediatric FAP patients (e.g., Walker and Zeman 1992). In contrast, distracting or coping-promoting parenting behavior that draws children's attention away from symptoms and encourages participation in regular activities is associated with fewer symptom complaints and better psychosocial outcomes among children with FAP (e.g., Walker et al. 2006).

Treatment

Several interventions have been found to be effective for treating FAP. As stated, it is essential that FAP is assessed and treated from a biopsychosocial model to ensure that all factors that contribute to the FAP presentation are addressed.

Medical therapies typically include dietary changes to improve motility (e.g., increased fiber intake), medication from the tricyclic antidepressant or selective serotonin reuptake inhibitor classes to regulate visceral hypersensitivity and the brain-gut pathway (e.g., amitriptyline, citalopram), and other common GI therapies to regulate the digestive system as a whole (e.g., antacid or proton pump inhibitor therapies, supplements such as peppermint oil). Regulating the GI system from the biological standpoint can help decrease pain and discomfort and is one of the essential features of treating FAP (Nurko and Di Lorenzo 2008).

Psychological therapies have been found to have a promising evidence base for pediatric FAP. Treatment typically involves cognitive behavioral therapy (CBT) performed individually with the child, or with the child and family. The goal of CBT is for children to learn pain coping skills (e.g., relaxation techniques, positive thinking strategies) and adopt a functional pain coping approach (e.g., adopting active and distracting pain coping skills to continue with regular activities). Even if a child is seen for individual therapy, it is essential for the treating provider to work with the child's school and family to ensure that the functional pain coping approach is supported in the larger system. Another modality of psychological therapy is biofeedback, consisting of focused relaxation practice that helps children restore physiological balance to their bodies, which has also been associated with improvements in pain and functioning among children with FAP (Schurman et al. 2010). A recent review of psychological therapies for youth with chronic pain found that cognitive behavioral therapy and relaxation techniques are effective in reducing pain severity and frequency, as well as have a lasting effect for symptom improvement (Eccleston et al. 2009).

Summary

FAP is a common complaint of childhood that has been extensively studied in the literature. FAP is a physiological disruption of the GI tract in the absence of organic disease. Pediatric FAP patients experience high levels of disability and psychological distress. Furthermore, many patients who experience FAP as children report higher levels of pain, functional disability, and health service utilization as adults, and a significant proportion go on to develop other chronic pain problems. From a biopsychosocial perspective, the role of central sensitization of pain perception in the brain, in addition to ongoing psychological and social stressors, likely combine to influence maintenance or persistence of pain. These negative concurrent and long-term

emotional and behavioral outcomes underscore the significant potential impact of FAP on children's lives and futures, adding to the importance of improving the understanding and treatment of patients with these conditions through a biopsychosocial approach.

References

Apley, J. (1975). The child with abdominal pains. London: Blackwell.

Apley, J., & Naish, N. (1958). Recurrent abdominal pain: A field survey of 1,000 school children. Archives of Disease in Childhood, 33, 165–170.

Campo, J., Bridge, J., Ehmann, M., Altman, S., Lucas, A., Birmaher, B., et al. (2004). Recurrent abdominal pain, anxiety, and depression in primary care. Pediatrics, 113, 817–824.

Chiou, E., & Nurko, S. (2010). Management of functional abdominal pain and irritable bowel syndrome in children and adolescents. Expert Review of Gastroenterology & Hepatology, 4, 293–304.

Drossman, D. A. (1998). Gastrointestinal illness and the biopsychosocial model. Psychosomatic Medicine, 60, 258–267.

Eccleston, C., Palermo, T. M., Williams, A. C. D. C., Lewandowski, A., & Morley, S. (2009). Psychological therapies for the management of chronic and recurrent pain in children and adolescents. Cochrane Database of Systematic Reviews, (2):1–50.

Garber, J., Zeman, J., & Walker, L. S. (1990). Recurrent abdominal pain in children: Psychiatric diagnoses and parental psychopathology. Journal of the American Academy of Child and Adolescent Psychiatry, 29, 648–656.

Levy, R. L., Whitehead, W. E., Von Korff, M. R., & Feld, A. D. (2000). Intergenerational transmission of gastrointestinal illness behavior. The American Journal of Gastroenterology, 95, 451–456.

Liakopoulou-Kairis, M., Alifieraki, T., Protagora, D., Korpa, T., Kondyli, K., Dimosthenous, E., et al. (2002). Recurrent abdominal pain and headache. European Child & Adolescent Psychiatry, 11, 115–122.

Miranda, A. (2008). Early life events and the development of visceral hyperalgesia. Journal of Pediatric Gastroenterology and Nutrition, 47, 680–684.

Nurko, S., & Di Lorenzo, C. (2008). Functional abdominal pain: Time to get together and move forward. Journal of Pediatric Gastroenterology and Nutrition, 47, 679–680.

Rasquin, A., Di Lorenzo, C., Forbes, D., Guiraldes, E., Hyams, J. S., Staiano, A., et al. (2006). Childhood functional gastrointestinal disorders. Gastroenterology, 130, 1527–1537.

Roth-Isigkeit, A., Thyen, U., Stöven, H., Schwarzenberger, J., & Schmucker, P. (2005). Pain among children and adolescents: Restrictions in daily living and triggering factors. *Pediatrics, 115*, e152–e162.

Schurman, J. V., Wu, Y. P., Grayson, P., & Friesen, C. A. (2010). A pilot study to assess the efficacy of biofeedback-assisted relaxation training as an adjunct treatment for pediatric functional dyspepsia associated with duodenal eosinophilia. *Journal of Pediatric Psychology, 35*, 837–847.

Talley, N. J. (2008). Genetics and functional bowel disease. *Journal of Pediatric Gastroenterology and Nutrition, 47*, 680–682.

Van Tilburg, M. A. L., Venepalli, N., Ulshen, M., Freeman, K. L., Levy, R., & Whitehead, W. E. (2006). Parents' worries about recurrent abdominal pain in children. *Gastroenterology Nursing, 29*, 50–55.

Walker, L. S., & Greene, J. W. (1989). Children with recurrent abdominal pain and their parents: More somatic complaints, anxiety, and depression than other patient families? *Journal of Pediatric Psychology, 14*, 231–243.

Walker, L. S., & Zeman, J. L. (1992). Parental response to child illness behavior. *Journal of Pediatric Psychology, 17*, 49–71.

Walker, L. S., Garber, J., & Greene, J. W. (1991). Somatization symptoms in pediatric abdominal pain patients: Relation to chronicity of abdominal pain and parent somatization. *Journal of Abnormal Child Psychology, 19*, 379–394.

Walker, L. S., Garber, J., & Greene, J. (1993). Psychological correlates of recurrent abdominal pain: A comparison of pediatric patients with recurrent abdominal pain, organic illness, and psychiatric disorders. *Journal of Abnormal Psychology, 102*, 248–258.

Walker, L. S., Garber, J., Van Slyke, D. A., & Greene, J. W. (1995). Long-term health outcomes in patients with recurrent abdominal pain. *Journal of Pediatric Psychology, 20*, 233–245.

Walker, L. S., Williams, S. E., Smith, C. A., Garber, J., Van Slyke, D. A., & Lipani, T. A. (2006). Parent attention versus distraction: Impact on symptom complaints by children and adolescents with and without chronic functional abdominal pain. *Pain, 122*, 43–52.

Walker, L. S., Dengler-Crish, C. M., Rippel, S., & Bruehl, S. (2010). Functional abdominal pain in childhood and adolescence increases risk for chronic pain in adulthood. *Pain, 150*, 568–572.

Functional Abilities Evaluation

▶ Disability, Functional Capacity Evaluations

Functional Aspects of Visceral Pain

▶ Psychiatric Aspects of Visceral Pain

Functional Bowel Disorder

Definition

Functional bowel disorder is a disorder or disease where the primary abnormality is an altered physiological function (the way the body works) rather than an identifiable structural or biochemical cause. A functional disorder does not show any evidence of an organic or physical disease. It is a disorder that generally cannot be diagnosed in a traditional way, as an inflammatory, infectious, or structural abnormality that can be seen by commonly used examination, x-ray, or laboratory test.

Cross-References

▶ Descending Modulation of Visceral Pain
▶ Visceral Pain Model, Irritable Bowel Syndrome Model

Functional Brain Imaging

Definition

Functional brain imaging is a noninvasive neuroimaging technique that detects changes in brain metabolism or neuronal activity in response to sensory, motor or cognitive tasks, giving information regarding the function of specific brain regions; see also ▶ functional imaging and ▶ fMRI.

Cross-References

▶ Thalamus and Visceral Pain Processing (Human Imaging)

Functional Capacity

Definition

For United States Social Security purposes, functional capacity represents the measure of a person's ability to perform particular work-related physical and mental activities.

Cross-References

▶ Disability Evaluation in the Social Security Administration

Functional Capacity Assessment

▶ Disability, Functional Capacity Evaluations

Functional Capacity Battery

▶ Disability, Functional Capacity Evaluations

Functional Capacity Evaluation

Synonyms

FCE

Definition

Functional capacity evaluation systematically measures the worker's ability to carry out work tasks safely. It includes trait-oriented systems of norm-referenced assessment, e.g., the Baltimore Therapeutic Equipment (BTE) (http://www.bteco.com/). The results demonstrate the individual's physical functional capacity, e.g., lifting capacity, fine motor dexterity, work tolerance, and preparedness for returning to work.

FCE has been criticized for the lack of correspondence between the individual's functional capacity and a job's real requirements: It does *not* demonstrate the individual's opportunities of returning to, and retaining, a job. Synonyms: physical capacity evaluation; work capacity evaluation.

Cross-References

▶ Disability, Functional Capacity Evaluations
▶ Vocational Counseling

Functional Changes in Sensory Neurons Following Spinal Cord Injury

▶ Functional Changes in Sensory Neurons Following Spinal Cord Injury in Central Pain

Functional Changes in Sensory Neurons Following Spinal Cord Injury in Central Pain

Jing-Xia Hao[1] and Xiao-Jun Xu[2]
[1]Department of Physiology and Pharmacology, Karolinska Institutet, Stockholm, Sweden
[2]Department of Physiology and Pharmacology, Section of Integrative Pain Rsearch Karolinska Institutet, Stockholm, Sweden

Synonyms

Central pain; Functional changes in sensory neurons following spinal cord injury

Definition

Chronic pain is a common consequence of spinal cord injury (SCI), often mixed with

nociceptive, visceral, and neuropathic components. Neuropathic SCI pain is a form of central
pain that represents a major challenge for those
responsible for clinical pain management
(Siddall et al. 2002). Several models of SCI
have been developed in recent years that allow
the direct examination of neuronal mechanisms
mediating SCI-induced central pain (Vierck
et al. 2000).

▶ Central pain refers to neuropathic pain associated with a lesion of the central nervous system.
▶ Spinal cord injury pain (SCI pain) is
a particular form of central pain seen in patients
with spinal cord injury or diseases of the spinal
cord, e.g., syringomyelia and tumors. The dorsal
horn of the spinal cord is the first relay point for
somatosensory perception. Neurons in the dorsal
horn process sensory input and transmit this
information to higher brain centers, while initiating motor and autonomic reflexes. Sensory neurons in the dorsal horn can be classified into three
types with respect to responses to mechanical
stimulation: ▶ low-threshold (LT) neurons that
respond maximally to innocuous stimuli,
▶ wide dynamic range (WDR) neurons that give
graded responses to innocuous and noxious stimulation, and ▶ high-threshold (HT) neurons that
respond exclusively to noxious stimulation.

Introduction

Chronic pain is a common consequence of spinal
cord injury (SCI), often mixed with nociceptive,
visceral, and neuropathic components. Neuropathic SCI pain is a form of central pain that
represents a major challenge for those responsible for clinical pain management (Siddall et al.
2002). Several models of SCI have been
developed in recent years that allow the direct
examination of neuronal mechanisms mediating
SCI-induced central pain (Vierck et al. 2000).

Characteristics

In recent years, several studies have been
published characterizing the response patterns

of dorsal horn neurons in rats with spinal cord
injury and behaviors suggesting the presence of
SCI pain. The models of injury employed include
photochemically induced ischemic SCI (Hao
et al. 1992; Hao et al. 2004), excitoxicity associated with intraspinal injection of quisqualic acid
(Yezierski and Park 1993), contusion (Drew et al.
2001; Hoheisel et al. 2003; Hains et al. 2003a),
transection (Scheifer et al. 2002; Zhang et al.
2005), and hemisection (Hains et al. 2003b; Lu
et al. 2008). Below are described four functional
characteristics of dorsal horn neurons that have
been described as changing following spinal cord
injury.

Receptive Fields

The proportion of dorsal horn neurons without
a demonstrable mechanical receptive field (RF)
is increased after spinal cord injury (Hoheisel
et al. 2003; Hao et al. 2004). These neurons tend
to have high-frequency ongoing activity and are
clustered around the rostral end of the injured
spinal segment. This suggests that the normal
afferent input (either direct or indirect) to some
of these neurons has been interrupted by injury,
resulting in a lack of RF. The original RFs of
these neurons are likely to be on skin areas rendered anesthetic by the injury. Abnormal spontaneous discharges of these neurons, especially
those projecting to higher centers, may give rise
to dysesthesia or pain referred to the anesthetic
skin areas caudal to the lesion. This type of pain is
referred to as below level pain and is a major
component of the pain syndrome in patients
with SCI (Siddall et al. 2002). There are also
behavioral indications of below level pain in rat
models in the form of autotomy and excessive
scratching/grooming leading to skin damage
(Vierck et al. 2000).

Spontaneous Activity

In our analysis of activity of dorsal horn neurons
after SCI, we have described animals with segmental allodynia, and these rats have a larger
proportion of neurons with high rates of spontaneous activity (SA) (Hao et al. 2004). These
neurons were located near the edge of the
lesion site. This is similar to the results of

Hoheisel et al. (2003) and Zhang et al. (2005) who found that neurons close to a spinal contusion or transection exhibited increased levels of ongoing activity. Yezierski and Park (1993) and Drew et al. (2001) have also reported a higher level of such activity in neurons in spinally injured rats. In our sample, the level of spontaneous activity of HT neurons and neurons without receptive fields was significantly higher than that of LT or WDR neurons. Interspike interval analysis indicated that the discharges were irregular and burst-like in the majority of SA neurons. Since HT neurons receive input from nociceptors and are involved in pain signaling (Chung et al. 1986), the high rate of SA in HT neurons may give rise to spontaneous pain sensations in skin areas with sensory loss. Hoheisel et al. (2003) have also noted several forms of pathophysiological background activity that were not seen in normal animals, including bursting-like activity.

Changes in the Functional Type of Neurons Recorded

In normal rats, the proportion of different neuronal types recorded depends on several factors, including classification criteria used, type of preparations, anesthesia, etc. In our study, we used an intact rat preparation with urethane anesthesia and we recorded from neurons with response characteristics resembling those of LT, WDR, and HT neurons (Chung et al. 1986). The relative proportion of these neuronal types in a sample of normal rats was similar to that reported in most previous studies in the field. By contrast, considerably more WDR neurons were encountered in spinally injured rats two to three segments above the injury (Hao et al. 2004). Since neurons in normal and injured rats were recorded from different animals, it is impossible to know the original phenotype of these WDR neurons. We have speculated that the increased proportion of WDR neurons reflects either the appearance of novel innocuous input to HT neurons or increased responses of LT neurons to noxious input. Both possibilities imply that there is an increased excitability in sensory pathways in the spinal cord underlying behavioral allodynia. A decrease in inhibitory influences may also

contribute to these functional changes. Drew et al. (2001) have also compared the response properties of neurons in normal, spinally injured non-allodynic and allodynic rats. Interestingly, they observed that the proportion of LT neurons was increased and they became more common than WDR neurons rostrally, but not caudally, to the lesion site. The difference between our results and those of Drew et al. (2001) may be due to differences in classification criteria, since they did not have HT neurons in their sample.

Responses to Peripheral Stimulation

Mechanical ▶ allodynia is a common symptom in patients with SCI pain (Siddall et al. 2002), and similar behavior is consistently observed in rat models of SCI pain (Xu et al. 1992; Vierck et al. 2000). In support of the behavioral observations, electrophysiological recording of dorsal horn activity in these animals has revealed increased neuronal responsiveness to mechanical stimulation. This included increased responses of WDR neurons to brush, pinch, and graded von Frey hair stimulation, decreased von Frey threshold, increased response of HT neurons, and increases in afterdischarges (Hao et al. 1992; Yezierski and Park 1993; Drew et al. 2001; Hains et al. 2003; Hoheisel et al. 2003; Zhang et al. 2005). These changes are likely to underlie the mechanical allodynia observed following spinal injury. Similarly, we have also found that the responses of dorsal horn neurons to cold stimulation increased after SCI, corresponding to cold allodynia in these rats. A higher percentage of WDR and LT neurons, which normally react poorly to cold, react to cold stimulation in allodynic rats (Hao et al. 2004).

Conclusions

Abnormal electrophysiological properties of dorsal horn neurons can be documented in rats with chronic pain-related behaviors after SCI. Based on the similarities between the neuronal and behavioral responses, it is likely that neuronal changes in the dorsal horn around the level of injury are responsible for the behavioral manifestation of pain-like responses. Some of the abnormalities are most probably the result of deafferentation in the zone immediately rostral to the spinal lesion. The reduction of spinal GABAergic inhibition has also been shown to be involved in the mechanical hypersensitivity of dorsal horn neurons after SCI (Hao et al. 1992; Lu et al. 2008). Moreover, SCI induces complex neurochemical changes in areas rostral to the lesion, which may also contribute to the neuronal abnormalities. It is important to note that some of neuronal changes observed in spinally injured rats, noticeably spontaneous hyperactivity in the dorsal horn, have also been seen in patients with SCI pain and lesions of the dorsal root entry zone; targeting areas of spontaneous activity have been shown to alleviate pain (Falci et al. 2002).

References

Chung, J. M., Surmeier, D. J., Lee, K. H., et al. (1986). Classification of primate spinothalamic and somatosensory thalamic neurons based on cluster analysis. *Journal of Neurophysiology, 56*, 308–327.

Drew, G. M., Siddal, P. J., & Duggan, A. W. (2001). Responses of spinal neurons to cutaneous and dorsal root stimuli in rats with mechanical allodynia after contusive spinal cord injury. *Brain Research, 893*, 59–69.

Falci, S., Best, L., Bayles, R., et al. (2002). Dorsal root entry zone microcoagulation for spinal cord injury-related central pain: Operative intramedullary electrophysiological guidance and clinical outcome. *Journal of Neurosurgery, 97*, 193–200.

Hains, B. C., Klein, J. P., Saab, C. Y., et al. (2003a). Upregulation of sodium channel $Na_V1.3$ and functional involvement in neuronal hyperexcitability associated with central neuropathic pain after spinal cord injury. *Journal of Neuroscience, 23*, 8881–8892.

Hains, B. C., Willis, W. D., & Hulsebosch, C. E. (2003b). Serotonin receptors5-HT1A and 5-HT3 reduces hyperexcitability of dorsal horn neurons after spinal cord hemisection injury in rat. *Experimental Brain Research, 149*, 174–186.

Hao, J. X., Xu, X. J., Yu, Y. X., et al. (1992). Baclofen reverses the hypersensitivity of dorsal horn wide dynamic range neurons to low threshold mechanical stimuli after transient spinal cord ischemia: Implication for a tonic GABAergic inhibitory control upon myelinated fiber input. *Journal of Neurophysiology, 68*, 392–396.

Hao, J. X., Kupers, R., & Xu, X. J. (2004). Response characteristics of spinal cord dorsal horn neurons in chronic allodynic rats after spinal cord injury. *Journal of Neurophysiology, 92*, 1391–1399.

Hoheisel, U., Scheifer, C., Trudrung, P., et al. (2003). Pathophysiological activity in rat dorsal horn neurons in segments rostral to a chronic spinal cord injury. *Brain Research, 974*, 134–145.

Lu, Y., Zheng, J. H., Xiong, L. Z., Zimmermann, M., & Yang, J. (2008). Spinal cord injury-induced attenuation of GABAergic inhibition in spinal dorsal horn circuits is associated with down-regulation of the chloride transporter KCC2 in rat. *The Journal of Physiology, 586*, 5701–5715.

Scheifer, C., Hoheisel, U., Trudrung, P., et al. (2002). Rats with chronic spinal cord transection as a possible model for the at-level pain of paraplegic patients. *Neuroscience Letters, 323*, 117–120.

Siddall, P. J., Yezierski, R. P., & Loeser, J. D. (2002). Taxonomy and epidemiology of spinal cord injury pain. In R. P. Yezierski & K. J. Burchiel (Eds.), *Spinal cord injury pain: Assessment, mechanisms, management* (pp. 9–24). Seattle: IASP Press.

Vierck, C. J., Jr., Siddall, P., & Yezierski, R. P. (2000). Pain following spinal cord injury: Animal models and mechanistic studies. *Pain, 89*, 1–5.

Xu, X. J., Hao, J. X., Aldskogius, H., et al. (1992). Chronic pain-related syndrome in rats after ischemic spinal cord lesion: A possible animal model for pain in patients with spinal cord injury. *Pain, 48*, 279–290.

Yezierski, R. P., & Park, S. H. (1993). The mechanosensitivity of spinal sensory neurons following intraspinal injections of quisqualic acid in therat. *Neuroscience Letters, 157*, 115–119.

Zhang, H. J., Xie, W. R., & Xie, Y. K. (2005). Spinal cord injury triggers sensitization of wide dynamic range dorsal horn neurons in segments rostral to the injury. *Brain Research, 1055*, 103–110.

Functional Changes in Thalamus

▶ Thalamic Plasticity and Chronic Pain

Functional Gastrointestinal Disorders

Definition

Functional gastrointestinal disorders are a collection of symptom based gastrointestinal conditions for which no biomedical abnormality can be found to explain symptoms. Similar to other functional conditions such as fibromyalgia.

Cross-References

Functional Imaging

Definition

Functional imaging is a general term used to describe methodologies that allow function to be located either spatially or temporally within the brain (and other organs). The methods allow detection of molecular signals that indicate the presence of biochemical activity and changes, such as cell activity or death. They are generally noninvasive and used for human studies. The term "neuroimaging" is often used when applied specifically to brain studies. Methods include: Functional Magnetic Resonance Imaging (FMRI), Positron Emission Tomography (PET), Magneto-Encephalography (MEG), and Electro-Encephalography (EEG).

Cross-References

Functional Imaging of Cutaneous Pain

Jürgen Lorenz
Faculty of Life Science, Competence Center
Gesundheit, University of Applied Science
Hamburg, Hamburg, Germany

Synonyms

Cutaneous pain; Functional imaging

Definition

This chapter reviews methodologies that monitor the spatial distribution of cerebral metabolic, hemodynamic, electrical, or magnetic reactions by which areas of the brain are identified that are active during the processing and perception of experimentally induced or clinical pain originating in the human skin.

Characteristics

Functional imaging techniques applied for the study of cutaneous pain are ▶ positron emission tomography (PET), ▶ functional magnetic resonance imaging (fMRI), multi-lead electroencephalography (EEG), and ▶ magnetoencephalography (MEG). PET measures cerebral blood flow, glucose metabolism, or neurotransmitter kinetics. A very small amount of a labeled compound (called the radiotracer) is intravenously injected into the patient or volunteer. During its uptake and decay in the brain, the radionuclide emits a positron, which, after traveling a short distance, "annihilates" with an electron from the surrounding environment. This event results in the emission of two gamma rays of 511 keV in opposite directions, the coincidence of which is detected by a ring of photomultipliers inside the scanner. In case of the most common use of O^{15}-water injection, counting and spatial reconstruction of these occurrences within the brain anatomy allow visualization of the regional cerebral blood flow response (rCBF) as an indicator of neuronal activity. Usually scans during painful stimulation are statistically compared with scans during the resting state or non-painful stimulation (blocked design) and plotted as 3-dimensional color-coded t- or Z-score maps.

FMRI images blood oxygenation using the so-called BOLD (blood oxygen level-dependent) technique which exploits the phenomenon that oxygenated and deoxygenated hemoglobin possess different magnetic properties. Both the rCBF using O^{15}-water PET and the BOLD technique rely on neurovascular coupling mechanisms that are not yet fully understood, but which overcompensate local oxygen consumption, thus causing a flow of oxygenated blood into neuronally active brain areas in excess of that utilized.

EEG and MEG are noninvasive neurophysiological techniques that measure the respective electrical potentials and magnetic fields generated by neuronal activity of the brain and propagated to the surface of the skull where they are picked up with EEG-electrodes or, in the case of its magnetic counterpart, received by SQUID (supra conducting quantum interference device) sensors located outside the skull. Compared with PET and fMRI, EEG and MEG are direct indicators of neuronal activity and yield a higher temporal resolution of the investigated brain function. The spatial distributions of EEG potentials and MEG fields at characteristic time points following noxious stimulation (► noxious stimulus) are analyzed using an inverse mathematical modeling approach called equivalent current dipole (ECD) reconstruction. An ECD evoked by painful stimuli hence represents a source model of pain-relevant activity within the brain. The spatial acuity of MEG is higher than that of EEG because the latter measures the extracellular volume currents that are distorted by the differentially conducting tissues such as gray and white matter, cerebrospinal fluid, durae, and bone. In contrast, MEG measures the magnetic field perpendicular to the intracellular currents undistorted by the surrounding tissue. Given the different geometry of electrical potentials and magnetic fields, MEG is predominantly sensitive to dipoles oriented tangentially to the head convexity, whereas EEG depends primarily on radial but also on tangential dipoles. Functional imaging using PET, fMRI, EEG, and MEG has substantially increased the knowledge about the cerebral representation of pain (Casey 1999) and its modulation by psychological phenomena (Porro 2003).

A bone of contention in the research field dealing with the cerebral representation of pain is the involvement of primary and secondary somatosensory cortex. When using contact heat to activate ► nociceptors, multiple areas of the brain such as the thalamus, primary (SI) and secondary (SII) somatosensory cortices, posterior

and anterior parts of the insula, and mid-caudal parts of the anterior cingulate cortex (ACC) respond to this input in a correlated manner with perceived intensity (Coghill et al. 1999). However, the touch of the probe or attentional effects inherent in block designs could explain SI and SII cortex activity being unrelated to pain. Yet, if infrared laser stimuli that lack a concomitant tactile component are applied over a range of randomly presented intensities in event-related designs by MEG (Timmermann et al. 2001) or fMRI (Bornhoevd et al. 2002), SI activity shows a steady increase with laser heat intensity, whereas the SII and insula respond robustly to painful but not or only slightly to non-painful intensities. Using BOLD responses of FMRI, Moulton et al. (2012) recently compared different degrees of non-painful and painful contact heat intensities and found best correlation with heat rather than pain intensity in contralateral SI and SII cortices. Discrepancy between laser and contact heat may be due to the fact that laser stimuli involve a very narrow range of pre-pain intensities. These stimulus–response functions fit with the behavior of distinct wide dynamic range (WDR) and nociceptive-specific (NS) neurons of the dorsal horn, thalamus, and SI cortex (Price et al. 2003). These features and the importance of a relay of afferent activity in lateral thalamic nuclei substantiate the key role of SI and SII cortices as part of a lateral pain system subserving the sensory-discriminative function of pain (Melzack and Casey 1968; Price 2000). Under conditions of specific psychological interventions such as hypnosis (Rainville et al. 1997), pain anticipation (Ploghaus et al. 1999), or ► placebo cognitions (Petrovic et al. 2002), frontal brain regions such as ACC and the anterior insula appear to subserve more specifically the affective-motivational and cognitive component of pain (Melzack and Casey 1968; Price 2000). They represent the major targets of the medial pain system given a major afferent input to these brain areas from medial thalamic nuclei. Other areas such as ► basal ganglia, ► cerebellum, and various structures within the prefrontal cortex yielded activation by experimental pain in several studies; their role in pain processing, however, remained elusive and is mainly deduced from their importance in other cognitive, motor, or behavioral functions.

Craig (2003) challenged the concept of a sensory-discriminative function of pain. An important component of his hypothesis is the assumption that pain is primarily an interoceptive perception like hunger, thirst, or itch, originating in specific spinothalamic tract (STT) neurons of the superficial dorsal horn (lamina I). As "labeled lines," these STT neurons impinge upon specific thalamic nuclei, such as the posterior part of the ventromedial nucleus (VMpo) and the ventral caudal part of the mediodorsal nucleus (MDvc), which relay afferent input to dorsal posterior insula and caudal ACC, respectively. These two pathways are regarded as important elements of a hierarchical system subserving ▶ homeostasis, linking the sense of the physiological condition of the body (▶ interoception) with subjective feelings and emotion. Although the interoceptive function of pain is well conceivable, negation of an exteroceptive function of pain is neither intuitive nor supported by functional imaging data (see above). Word descriptors such as those compiled in the McGill Pain Questionnaire, e.g., cutting, pinching, stinging, squeezing or crushing, and many more, express how a stimulus from outside the body causes pain and are often used by pain patients to describe their pain, although there is no exteroceptive stimulus. In contrast, there is clearly less richness of exteroceptive sensory descriptors for hunger or thirst, even though the latter are much more everyday sensations or feelings than pain. Lenz et al. (2004) electrically stimulated thalamic termination areas of STT neurons in conscious patients undergoing stereotactic procedures for the treatment of movement disorders and chronic pain. Patients used mainly exteroceptive words, rarely internal or emotional phenomena, to describe their subjective responses to the stimuli. Furthermore, two groups of responses were observed following stimuli applied to distinct thalamic locations, a binary response signaling pain, but no non-painful sensation, and an analog response covering graded intensities across non-painful and painful sensations. This result is consistent with a sensory-discriminative or exteroceptive pain function through pathways that convey alarm according to an all-or-nothing stimulus–response relationship and others that indicate how strong and where the stimulus is.

The ambiguity of pain as exteroceptive or interoceptive phenomenon likely relates to the fact that many painful experiences throughout the entire life span of an individual derive from the skin (burns, cuttings, etc.). The human skin forms the interface between the body and the outer world. Cutaneous sensitivity is therefore primarily reflecting the capability of the organism to receive signals from outside the body in the form of touch, temperature, and pain sensations. As a consequence, cutaneous sensitivity primarily subserves an exteroceptive function. However, pathological changes of the skin due to inflammation or injury lead to pain that informs the individual about the body rather than a stimulus from the outer world and, therefore, serves an interoceptive function. Often the term interoception is attributed to visceral sensation, or homeostatic feelings such as hunger and thirst. Yet, inflammatory and injury pain of the skin are indeed good indications that a change in the bodily state can also involve a shift from primarily exteroceptive to interoceptive cutaneous pain function. Sound support for the hypothesis that cutaneous pain inherits both exteroceptive and interoceptive functions is provided by a PET imaging study conducted by Lorenz et al. (2002). These authors addressed the question of whether nociceptive activity resulting from two equally painful contact heat stimuli applied to normal skin or the same skin sensitized by a topical solution of capsaicin would yield different functional imaging results. They tested whether the effect of skin sensitization would be similar to that of pain intensity observed by the same group in an earlier study when painful heat was compared with non-painful warmth (Casey et al. 2001). Thus, whereas the critical aim of the Lorenz et al. experiment was to match subjective intensities of a slowly ramped and continuously applied contact heat stimulus across normal and sensitized skin conditions to minimize a confound

Functional Imaging of Cutaneous Pain, Fig. 1 Stimulus paradigms to differentially manipulate body-related (interoceptive) and stimulus-related (exteroceptive) pain processes with positron emission tomography (*PET*). The top section shows the temporal profile of two slowly ramped heat stimuli reaching different plateaus equated for subjective pain intensity. One stimulus is applied on the normal skin of the left volar forearm to approximately 47 °C (*solid black*), which is 2 °C above the average heat pain threshold (*dotted black*) and felt as a clear, but tolerable pain sensation. Another stimulus is applied to approximately 43 °C on the same skin, when it had been sensitized by a topical solution of capsaicin (*solid red*). This treatment caused a drop in the heat pain threshold by approximately 4 °C (*dotted red*), rendering the heat stimulus as painful as the 4 °C more intense stimulus applied on normal skin. The subtraction of PET scans between the two conditions (capsaicin-treated minus normal skin) is shown at the bottom of section A. Surface-rendered images and median sagittal cuts from both sides, a horizontal slice 11 mm above the AC-PC line, and a superior view are displayed. The images demonstrate activity of the midbrain, medial thalamus, anterior insula, perigenual cingulate, and prefrontal cortex representing the change in the physiological status of the skin (sensitized *vs.* normal). The bottom section shows the temporal profile of two heat stimuli that were set to a constant temperature of either 50 °C or 40 °C in different blocks and repeatedly applied to different spots of the normal skin, each contact lasting 5 s. Images comparing these two stimuli illustrate activity in the lateral thalamus, lenticular nucleus, mid-posterior insula/SII, mid-anterior cingulated, cerebellum, and premotor cortex, the latter extending into the MI/SI region

with perceived intensity and test the importance of the physiological status of the tissue (interoception), the Casey et al. study tested the importance of different applied and perceived intensities of repeated rapidly ramped contact heat stimuli without changing the tissue status (► exteroception). Their results illustrated in Fig. 1 show that interoceptive and exteroceptive manipulations of burning pain engage clearly different forebrain structures. Key structures responding to manipulating the exteroceptive stimulus condition are the lateral thalamus, lenticular nucleus/putamen, mid-posterior insula/SII, SI/MI, caudal ACC, premotor cortex, and cerebellum. In contrast, key structures responding to manipulating the interoceptive stimulus condition are the medial thalamus, ventral caudate/nucleus accumbens, anterior insula, dorsolateral prefrontal and orbitofrontal cortices, and the perigenual ACC. Notably, although the perceived intensities were equated between the two skin conditions in the Lorenz et al. study, pain on sensitized skin (heat allodynia) yielded greater negative affect according to ratings subjects made using both a visual analog scale of "unpleasantness" and a short form of the McGill Pain Questionnaire at the end of each scan. This result is consistent with the close relationship of interoception with emotion giving rise to intrinsically stronger affective pain experiences during pathological tissue states. Furthermore, the involvement of cerebral motor systems differs between exteroceptive and interoceptive pain. Whereas exteroceptive pain recruits brain structures such as the putamen, motor and premotor cortex, and cerebellum suited to govern an immediate and spatially guided defense or withdrawal due to their somatotopic organization (Bingel et al. 2004a, b), interoceptive pain engages the ventral caudate and nucleus accumbens, which are part of a limbic basal ganglia loop relevant for motivational drive of behavior rather than motor execution. Overall, these results substantiate suggestions that different projection systems originating in the dorsal horn of the spinal cord mediate normal pain and pain from sensitized nociceptors (Hunt and Mantyh 2001). In differentiating different pain types by their origins from either outside or inside the body, the brain may engage different behavioral adaptations according to the meaning of the pain in relation to the physiological status of the body.

References

Bingel, U., Gläscher, J., Weiller, C., & Büchel, C. (2004a). Somatotopic representation of nociceptive information in the putamen: An event related fMRI study. *Cerebral Cortex, 14*, 1340–1345.

Bingel, U., Lorenz, J., Glauche, V., Knab, R., Gläscher, J., Weiller, C., & Büchel, C. (2004b). Somatotopic organization of human somatosensory cortices for pain: A single trial fMRI study. *NeuroImage, 23*, 224–232.

Bornhoevd, K., Quante, M., Glauche, V., et al. (2002). Painful stimuli evoke different stimulus–response functions in the amygdala, prefrontal, insula, and somatosensory cortex: A single trial fMRI study. *Brain, 123*, 601–619.

Casey, K. L. (1999). Forebrain mechanisms of nociception and pain: Analysis through imaging. *Proceedings of the National Academy of Sciences of the United States of America, 96*, 7668–7674.

Casey, K. L., Morrow, T. J., Lorenz, J., et al. (2001). Temporal and spatial dynamics of human cerebral activation pattern during heat pain: Analysis of positron emission tomography. *Journal of Neurophysiology, 85*, 951–959.

Coghill, R. C., Sang, C. N., Maisog, J. M., et al. (1999). Pain intensity processing in the human brain: A bilateral, distributed mechanisms. *Journal of Neurophysiology, 82*, 1934–1943.

Craig, A. D. (2003). A new view of pain as a homeostatic emotion. *Trends in Neurosciences, 6*, 303–307.

Hunt, S. P., & Mantyh, P. W. (2001). The molecular dynamics of pain control. *Nature Neuroscience, 2*, 83–91.

Lenz, F. A., Ohara, S., Gracely, R. H., et al. (2004). Pain encoding in the human forebrain: Binary and analog exteroceptive channels. *The Journal of Neuroscience, 24*, 6540–6544.

Lorenz, J., Cross, D. J., Minoshima, S., et al. (2002). A unique representation of heat allodynia in the human brain. *Neuron, 35*, 383–393.

Melzack, R., & Casey, K. L. (1968). Sensory, motivational, and central control determinants of pain. In D. R. Kenshalo (Ed.), *The skin senses* (pp. 423–443). Springfield Illinois: Thomas.

Moulton, E. A., Pendse, G., Becerra, L. R., & Borsook, D. (2012). BOLD responses in somatosensory cortices better reflect heat sensation than pain. *The Journal of Neuroscience, 32*(17), 6024–6031.

Petrovic, P., Kalso, E., Petersson, K. M., et al. (2002). Placebo and opioid analgesia – imaging a shared neuronal network. *Science, 295*, 1737–1740.

Ploghaus, A., Tracey, I., Gati, J. S., Clare, S., Menon, R. S., Matthews, P. M., et al. (1999). Dissociating pain from its anticipation in the human brain. *Science, 284*, 1979–1981.

Porro, C. A. (2003). Functional imaging and pain: Behavior, perception, and modulation. *The Neuroscientist, 9*, 354–369.

Price, D. D. (2000). Psychological and neural mechanisms of the affective dimension of pain. *Science, 288*, 1769–1772.

Price, D. D., Greenspan, J. D., & Dubner, R. (2003). Neurons involved in the exteroceptive function of pain. *Pain, 106*, 215–219.

Rainville, P., Duncan, G. H., Price, D. D., et al. (1997). Pain affect encoded in human anterior cingulate but not somatosensory cortex. *Science, 277*, 968–971.

Timmermann, L., Ploner, M., Haucke, K., et al. (2001). Differential coding of pain in the human primary and secondary somatosensory cortex. *Journal of Neurophysiology, 86*, 1499–1503.

Functional Imaging of Descending Modulation

▶ Descending Circuits in the Forebrain, Imaging

Functional Loss

Definition

Functional loss is a decrease or loss of physiological function.

Cross-References

▶ Disability, Upper Extremity

Functional Magnetic Resonance Imaging

Synonyms

fMRI

Definition

Functional Magnetic Resonance Imaging (or fMRI) is the use of MRI to learn which regions of the brain are active in a specific cognitive task or during sensory stimulation, e.g. during a pain experiment. It is one of the most recently developed forms of brain imaging, and measures hemodynamic signals related to neural activity in the brain or spinal cord of humans or other animals. As nerve cells "fire" impulses, they metabolize oxygen from the surrounding blood. Approximately 6 s after a burst of neural activity, a hemodynamic response occurs and that region of the brain is infused with oxygen-rich blood. As oxygenated hemoglobin is diamagnetic, while deoxygenated blood is paramagnetic, MRI is able to detect a small difference (a signal of the order of 3 %) between the two. This is called a blood-oxygen-level-dependent, or "BOLD," signal. The precise nature of the relationship between neural activity and the BOLD signal is a subject of current research.

Cross-References

▶ Amygdala, Functional Imaging
▶ Cingulate Cortex, Functional Imaging
▶ Descending Circuits in the Forebrain, Imaging
▶ Human Thalamic Response to Experimental Pain (Neuroimaging)
▶ Nociceptive Processing in the Nucleus Accumbens, Neurophysiology and Behavioral Studies
▶ Pain Processing in the Cingulate Cortex, Behavioral Studies in Humans
▶ PET and fMRI Imaging in Parietal Cortex (SI, SII, Inferior Parietal Cortex BA40)
▶ Thalamus and Visceral Pain Processing (Human Imaging)
▶ Thalamus, Clinical Pain, Human Imaging
▶ Thalamus, Clinical Visceral Pain, Human Imaging

Functional Magnetic Resonance Imaging (fMRI)

▶ Amygdala, Functional Imaging

Functional Restoration

Definition

Functional restoration is a pain management approach geared specifically for chronic low back pain patients. This approach places a strong emphasis on function, and combines a quantitatively directed exercise progression with disability management and psychosocial interventions such as individual and group therapy.

Cross-References

▶ Psychological Treatment of Chronic Pain, Prediction of Outcome

Functional Restoration Program

▶ Multidisciplinary Pain Centers, Rehabilitation

Functional Restoration/Exercise Programs

▶ Physical Conditioning Programs

Functioning

▶ Functioning and Disability Definitions
▶ Physical Exercise

Functioning and Disability Definitions

Reuben Escorpizo and Gerold Stucki
Swiss Paraplegic Research, Nottwil, Switzerland
Department of Health Sciences and Health Policy, University of Lucerne, Lucerne, Switzerland
ICF Research Branch in cooperation with the World Health Organization (WHO) Collaborating Centre for the Family of International Classifications in Germany, Nottwil, Switzerland

Synonyms

Classification; Disability; Functioning; ICD; ICF

Definition

Functioning; Disability; ICF; Classification; ICD

Introduction

The International Classification of Functioning, Disability and Health (ICF) (WHO 2001) was intended by the WHO to be the common reference framework to describe the full spectrum of human functioning and disability. As a conceptual model and a classification system, the ICF can be applied in various ways in clinical care, health and social policy, and research, and for epidemiological and population surveys. The ICF can be used to understand health and health-related domains and compare data across regions and countries, and as a system of coding (WHO 2001). The ICF can be applied regardless of the setting, culture, and context.

As a conceptual model, the ICF illustrates the interrelationship between a health condition and its impact on the individual's body (body functions and body structure), and activities and

Health condition
(disorder or disease)

Body Functions and Structures ← → Activities ← → Participation

Environmental Factors | Personal Factors

Functioning and Disability Definitions, Fig. 1 Interactions between the components of ICF

participation (societal impact). This relationship may be influenced by environmental factors and personal factors (Fig. 1).

Characteristics

The current framework of functioning and disability is the ICF (WHO 2001), in which disability is determined by impairment of body structures and body functions, activity limitations, and participation restrictions. Historically, there have been two major conceptual frameworks in the field of disability: the International Classification of Impairment, Disability and Handicap (ICIDH), which is the precursor to the ICF, and the "functional limitation," or Nagi, framework (Nagi 1964). In the ICIDH (WHO 1980), the four concepts were disease, impairment, disability, and handicap. In the Nagi framework, the four concepts were pathology, impairment, functional limitation, and disability. Different from the ICIDH, the Nagi framework was not accompanied by a classification. Building on the conceptual frameworks of the ICIDH and Nagi, the US Committee on a National Agenda for the Prevention of Disabilities developed a model emphasizing the interaction between the disabling process, quality of life, and individual risk factors (Pope and Tarlov 1991).

The ICF attempts to achieve a synthesis and integration of the different perspectives of health from biological, psychological, and social perspectives (i.e., biopsychosocial approach). It emphasizes health-related "building blocks" or domains to form specific disability models in the context of health. The ICF has addressed many of the criticisms of prior conceptual frameworks (e.g., lack of environmental consideration, purely biomedical perspective), and has been developed in a worldwide and comprehensive expert consensus process. For all these reasons, the ICF is the generally accepted conceptual framework and classification to describe a persons' level of functioning and disability in light of his or her health condition. *Disability* serves as an umbrella term for impairments, activity limitations, and participation restrictions. So it can be seen as the negative term of functioning. The interaction of disability and health condition works in both directions, because the presence of disability may even modify the health condition. The level of disability may be indicated for body functions, body structures, activities and participation, and environmental by using the generic ICF qualifiers of the WHO (WHO 2001). More information on the qualifiers is provided later.

It is important to recognize that the ICF can serve as a common language (Stucki et al. 2002), in that the texts that can be created with it (ICF) can be used by different users, within their intended purpose, and their scientific orientation. The understanding of bidirectional interactions and relationships between the components of the ICF is shown in Fig. 1.

Health condition refers to any kind of disorder, illness, or disease. It may include information about pathogeneses and/or etiology. There are (possible) interactions with all or any of the components of functioning, namely, ▶ body functions and body structures, activities, and participation (WHO 2001).

Body functions are the physiological (and psychological) functions of body systems. ▶ Body structures are anatomical parts of the body. Problems in body functions and body structures are called impairments, which are defined as a significant deviation or loss (e.g., deformity) of structures (e.g., joints), and/or functions (e.g., reduced range of motion (ROM), muscle weakness, pain, or fatigue).

Activity is described as the execution of a task or action by an individual. Difficulties an individual may have in executing activities are called activity limitations (e.g., limitations in mobility such as walking, climbing steps, grasping, or carrying).

Participation is described as involvement in a life situation (i.e., societal involvement). It represents the societal perspective of functioning. Problems an individual may experience in involvement in life situations are called participation restrictions. Limitations and restrictions are assessed against a generally accepted population standard. "Participation" records discordance between the observed performance and the expected performance. The expected performance is the population norm, which represents the experience of people without the specific health condition (e.g., with associated restrictions in community life, recreation, and leisure).

A person's functioning (and disability) is conceived as a dynamic interaction between health conditions (diseases, disorders, injuries, traumas, etc.) and contextual factors. Likewise, there are (possible) interactions with all components of functioning and contextual factors.

Contextual factors are factors that constitute the context of an individual's life, and in particular, the backgrounds against which health states are classified in the ICF. There are two sets of contextual factors: ▶ Environmental factors and personal factors.

Environmental factors refer to all aspects of the external or extrinsic world that form the context of an individual's life and, as such, may have an impact on that person's functioning. Environmental factors include the physical world and its features, the human-made physical world, other people in different relationships and roles, attitudes and values, social systems and services, and governing policies, rules, and laws.

ICF qualifiers can be assigned to rate the problem for each domain under the ICF components on body functions, body structures, activities and participation, and environmental factors (Table 1).

Functioning and Disability Definitions, Table 1 ICF qualifiers

ICF qualifier	Equivalent percentage
Body functions, body structures, and activities and participation	
0 NO problem (none, absent, negligible, etc.)	0–4 %
1 MILD problem (slight, low, etc.)	5–24 %
2 MODERATE problem (medium, fair, etc.)	25–49 %
3 SEVERE problem (high, extreme, etc.)	50–95 %
4 COMPLETE problem (total, etc.)	96–100 %
Environmental factors	
+4 Complete facilitator	96–100 %
+3 Substantial facilitator	50–95 %
+2 Moderate facilitator	25–49 %
+1 Mild facilitator	5–24 %
0 Neither barrier nor facilitator	0–4 %
1 Mild barrier	5–24 %
2 Moderate barrier	25–49 %
3 Severe barrier	50–95 %
4 Complete barrier	96–100 %

In the case of environmental factors, the extent of being a barrier or a facilitator is used.

Personal factors are contextual factors that relate to the individual such as age, gender, social status, life experiences, lifestyle, habits, coping styles, education, behavior, and other personal characteristics. Risk factors could be described under personal factors (e.g., lifestyle, coping mechanism) or environmental factors (e.g., living and work conditions, architectural barriers) that are associated with conditions such as pain. Bidirectional arrows in Fig. 1 indicate the possibility of feed forward and feedback process. Risk factors also affect the progression of disability and, may include, depending on the stage, treatment, rehabilitation, age of onset, financial resources, healthcare provision, expectations, and environmental barriers. Personal factors are defined and included in the ICF model, but not yet classified (hence, it has no categories nor qualifiers) into a coding system like the rest of the components.

The ICF is intended for use in multiple sectors that include health, education, insurance, labor, health and disability policy, etc. In the clinical

Functioning and Disability Definitions, Table 2 ICF categories of the comprehensive ICF Core Set for chronic widespread pain (brief core set marked with *)

ICF category (code)	Title
Body functions	
Global mental functions	
b122	Global psychosocial functions
b126	Temperament and personality functions
b130*	Energy and drive functions
b134*	Sleep functions
Specific mental functions	
b140	Attention functions
b147*	Psychomotor functions
b152*	Emotional functions
b1602*	Content of thought
b164	Higher-level cognitive functions
b180	Experience of self and time function
Additional sensory functions	
b260	Proprioceptive function
b265	Touch function
b270	Sensory function related to temperature and other stimuli
b280*	Sensation of pain
Functions of the haematological and immunological systems	
b430	Haematological system functions
Additional functions and sensations of the cardiovascular and respiratory systems	
b455*	Exercise tolerance functions
Genital and reproductive functions	
b640	Sexual functions
Functions of the joints and bones	
b710	Mobility of joint functions
Muscle functions	
b730*	Muscle power functions
b735	Muscle tone functions
b740	Muscle endurance functions
Movement functions	
b760*	Control of voluntary movement functions
b780	Sensations related to muscles and movement functions
Body structures	
s770	Additional musculoskeletal structures related to movement
Activities and participation	
Applying knowledge	
d160	Focusing attention
d175*	Solving problems

(continued)

Functioning and Disability Definitions, Table 2 (continued)

ICF category (code)	Title
General tasks and demands	
d220	Undertaking multiple tasks
d230*	Carrying out daily routine
d240*	Handling stress and other psychological demands
Changing and maintaining body position	
d410	Changing basic body position
d415	Maintaining a body position
Carrying, moving, and handling objects	
d430*	Lifting and carrying objects
Walking and moving	
d450*	Walking
d455	Moving around
Moving around using transportation	
d470	Using transportation
d475	Driving
Self-care	
d510	Washing oneself
d540	Dressing
d570	Looking after one's health
Acquisition of necessities	
d620	Acquisition of goods and services
Household tasks	
d640*	Doing housework
Caring for household objects and assisting others	
d650	Caring for household objects
d660	Assisting others
General interpersonal interactions	
d720	Complex interpersonal interactions
Particular interpersonal interactions	
d760*	Family relationships
d770*	Intimate relationships
Work and employment	
d845	Acquiring, keeping and terminating a job
d850*	Remunerative employment
d855	Non-remunerative employment
Community, social and civic life	
d910	Community life
d920*	Recreation and leisure
Environmental factors	
Products and technology	
e1101*	Drugs

(continued)

Functioning and Disability Definitions, Table 2 (continued)

ICF category (code)	Title
Support and relationships	
e310*	Immediate family
e325	Acquaintances, peers, colleagues, neighbors, and community members
e355*	Health professionals
Attitudes	
e410*	Individual attitudes of immediate family members
e420	Individual attitudes of friends
e425	Individual attitudes of acquaintances, peers, colleagues, neighbors, and community members
e430	Individual attitudes of people in positions of authority
e450	Individual attitudes of health professionals
e455	Attitudes of other professionals
e460	Societal attitudes
e465	Social norms, practices, and ideologies
Services, systems, and policies	
e570*	Social security services, systems, and policies
e575	General social support services, systems, and policies
e580	Health services, systems, and policies
e590	Labor and employment services, systems, and policies

context, it is intended for use in needs assessment and matching interventions to specific health states, rehabilitation program, and outcome evaluation. The joint use of the ICF and the International Classification of Diseases (ICD) needs to be addressed when applying the ICF in medicine in general (Kostanjsek et al. 2011a, b) and pain rehabilitation in particular. The WHO considers the ICF and the ICD to be distinct but complementary classifications. According to this view, patient functioning and health are associated with a condition or disease, but are not merely and necessarily a consequence of it. However, with the hundreds of codes within the ICF, the ICF needs to be tailored in order to be practical. Hence, for practical purposes, short lists of ICF categories called ICF Core Sets have been developed (www.icf-research-branch.org/home.html) and are relevant to a number of health conditions with a high burden, including chronic widespread pain, i.e., pain in several regions of the body (Cieza et al. 2004a). ICF Core Sets have also been developed for several other conditions that may have pain as a common denominator, including low back pain (Cieza et al. 2004b), ankylosing spondylitis (Boonen et al. 2010), osteoarthritis (Dreinhöfer et al. 2004), and rheumatoid arthritis (Stucki et al. 2004). Table 2 shows the ICF Core Set for chronic widespread pain (Cieza et al. 2004a). A *comprehensive version* of the ICF Core Set includes the ICF categories (units of domains) considered relevant for a comprehensive multidisciplinary assessment such as inhospital settings. The *brief version* of an ICF Core Set, marked with an asterisk (*) in the table, includes those ICF categories to be assessed in every clinical study and clinical encounter to allow for a meaningful description of a patient's functioning and health. A practical demonstration of how to use the ICF is shown in Fig. 2.

Current Measures in Pain Can Be Linked to the ICF

There are measures available, generic or condition specific, that can be linked to the ICF according to the domains that these measures cover. The concepts contained in such measures can be linked to the ICF using defined linkage rules (Cieza et al. 2005). For example, low back pain can be linked to the body function domain of the ICF, and other measures that assess pain and depression in relation to activities of daily living can be linked to the activities and participation domains of the ICF. The linking can be made more specific to the category level (e.g., b280 for *sensation of pain*).

It is important to recognize that different measures address different components of the ICF, and it seems that the concepts contained in these measures are generally represented in the ICF (Weigl et al. 2003). Condition-specific measures typically focus on body functions and activities, while generic measures tend to cover activities and participation including aspects of physical, mental, and social health. Contextual factors are hardly covered by any of these instruments.

Functioning and Disability Definitions, Fig. 2 Illustration of how the ICF components can be used to structure patient problems (listed in the *upper* section "patient perspective") as well as findings, and observations by the rehabilitation team (listed in the *lower* section "health professional perspective"). *Lines* between the selected target problems from the patient's perspective (*circled* in the *upper* section, i.e., "walking/standing"), impaired body functions, and structures as well as the given personal and environmental factors (in the *lower* section) denote their hypothesized associative relationship. Note that the wording denotes patient's words or special medical terms which can be linked back to specific domains of the ICF

Nonetheless, there is a need to address all components when assessing functioning and health in patients with chronic health conditions (Ewert et al. 2004).

Finally, it is important to note that the perspective of functioning, disability, and health is different when viewed from a medical versus functioning-oriented perspective, e.g., in rehabilitation. From the medical perspective, functioning and health are seen primarily as a consequence of a disease or condition. Accordingly, medical interventions are targeted toward mitigation or elimination of the disease process. From a broader view, the measurement of functioning, disability, and health is required to evaluate patient-relevant outcomes of an intervention. Hence, from a functioning-oriented or rehabilitation perspective as depicted in the ICF, patients' functioning and health are *associated* with, but not merely a consequence of, a condition or disease. Furthermore, functioning and health are not only seen in association with a condition, but also in association with personal and environmental factors. Therefore, the measurement of functioning, disability, and health is relevant to evaluate intervention outcomes and for the diagnosis (assessment) and overall clinical care. For the management of patients with pain conditions, consideration of the individual's functioning is becoming more and more critical than before.

Summary

In summary, the ICF can be used as a conceptual framework and classification system to examine and understand the impact and relation of pain with regard to the functioning and disability of an individual. The ICF provides a holistic biopsychosocial perspective that is beyond the traditional biomedical model and, hence, can aid in the thought process to guide clinical decision making and clinical care in the context of pain and pain-related conditions.

Acknowledgment Previous edition of this entry was written by Gerold Stucki and Thomas Ewert.

References

Boonen, A., Braun, J., van der Horst Bruinsma, I. E., Huang, F., Maksymowych, W., Kostanjsek, N., Cieza, A., Stucki, G., & van der Heijde, D. (2010). ASAS/WHO ICF Core Sets for ankylosing spondylitis (AS): How to classify the impact of AS on functioning and health. *Annals of the Rheumatic Diseases, 69*(1), 102–107.

Cieza, A., Stucki, G., Weigl, M., Kullmann, L., Stoll, T., Kamen, L., Kostanjsek, N., & Walsh, N. (2004a). ICF Core Sets for chronic widespread pain. *Journal of Rehabilitation Medicine*, (44 Suppl.), 63–68.

Cieza, A., Stucki, G., Weigl, M., Disler, P., Jäckel, W., van der Linden, S., Kostanjsek, N., & de Bie, R. (2004b). ICF Core Sets for low back pain. *Journal of Rehabilitation Medicine*, (44 Suppl.), 69–74.

Cieza, A., Geyh, S., Chatterji, S., Kostanjsek, N., Ustün, B., & Stucki, G. (2005). ICF linking rules: An update based on lessons learned. *Journal of Rehabilitation Medicine, 37*(4), 212–218.

Dreinhöfer, K., Stucki, G., Ewert, T., Huber, E., Ebenbichler, G., Gutenbrunner, C., Kostanjsek, N., & Cieza, A. (2004). ICF Core Sets for osteoarthritis. *Journal of Rehabilitation Medicine*, (44 Suppl.), 75–80.

Ewert, T., Fuessl, M., Cieza, A., et al. (2004). Identification of the most common patient problems in patients with chronic conditions using the ICF checklist. *Journal of Rehabilitation Medicine, 36*, 186–188.

Kostanjsek, N., Escorpizo, R., Boonen, A., Walsh, N. E., Ustün, T. B., & Stucki, G. (2011a). Assessing the impact of musculoskeletal health conditions using the international classification of functioning, disability and health. *Disability and Rehabilitation, 33*(13–14), 1281–1297.

Kostanjsek, N., Rubinelli, S., Escorpizo, R., Cieza, A., Kennedy, C., Selb, M., Stucki, G., & Ustün, T. B. (2011b). Assessing the impact of health conditions using the ICF. *Disability and Rehabilitation, 33*(15–16), 1475–1482.

Nagi, S. Z. (1964). A study in the evaluation of disability and rehabilitation potential: Concepts, methods, and procedures. *The American Journal of Public Health, 54*, 1568–1579.

Pope, A. M., & Tarlov, A. R. (1991). *Disability in America: Toward a national agenda for prevention.* Washington, DC: National Academy Press.

Stucki, G., Ewert, T., & Cieza, A. (2002). Value and application of the ICF in rehabilitation medicine. *Disability and Rehabilitation, 24*, 932–938.

Stucki, G., Cieza, A., Geyh, S., Battistella, L., Lloyd, J., Symmons, D., Kostanjsek, N., & Schouten, J. (2004). ICF Core Sets for rheumatoid arthritis. *Journal of Rehabilitation Medicine*, (44 Suppl.), 87–93.

Weigl, M., Cieza, A., Harder, M., et al. (2003). Linking osteoarthritis specific health status measures to the international classification of functioning, disability and health (ICF). *Osteoarthritis and Cartilage, 11*, 1–5.

WHO. (1980). *ICIDH. International Classification of Impairments, Disabilities and Handicaps.* Geneva: WHO.

WHO. (2001). *International classification of functioning, disability and health: ICF.* Geneva: WHO.

Funiculus

Definition

Longitudinal subdivisions of the spinal white matter that are named according to their location within the spinal cord.

Cross-References

▶ Spinothalamic Projections in Rat

Encyclopedia of Pain

Encyclopedia of Pain

Gerald F. Gebhart • Robert F. Schmidt
Editors

Encyclopedia of Pain

Second Edition

Volume 3

G–M

With 795 Figures and 217 Tables

Springer Reference

Editors

Gerald F. Gebhart
Department of Anesthesiology
Center for Pain Research
University of Pittsburgh
Pittsburgh, PA, USA

Robert F. Schmidt
Physiologisches Institut der
Universität Würzburg
Würzburg, Germany

ISBN 978-3-642-28752-7   ISBN 978-3-642-28753-4 (eBook)
ISBN 978-3-642-28754-1 (print and electronic bundle)
DOI 10.1007/978-3-642-28753-4
Springer Heidelberg New York Dordrecht London

Library of Congress Control Number: 2013951551

Printed on acid-free paper

Springer is part of Springer Science+Business Media (www.springer.com)

Preface to the Second Edition

As was pointed out in the Preface to the first edition, pain is the principal reason individuals seek medical and dental attention. Fortunately, acute tissue insults associated with pain – and typically inflammation – readily resolve with appropriate treatment and care. Chronic pain states, in contrast, are notoriously difficult to manage and are commonly associated with comorbidities (e.g., depression, cross-organ sensitization) and reduced quality of life. Indeed, chronic pain has come to be considered by some as a disease in its own right. This edition of the *Encyclopedia of Pain* updates the content of the 1st edition and introduces new topics consistent with advances in our knowledge of underlying mechanisms of pain. Thus, new content on channelopathies, expanded and updated material on the role of glia and immune-competent cell interactions with nociceptive neurons, and advances in imaging and changes in brain structure in chronic pain states are included in this edition.

The cardinal objective of research on pain is the translation of knowledge between the laboratory and the clinic. The present prevailing refrain is "bench to bedside," but in fact many ideas for research in pain derive from clinical observations and the absence of knowledge to explain clinical pain states. In the recent past, basic and clinical science research has seen the application of imaging techniques to visualize areas of the brain that are involved in both the sensory and affective aspects of pain, cloning of pain-related channels and receptors, and the application of elegant genetic manipulations that selectively "label" functionally distinct dorsal root ganglion sensory and spinal neurons, two-photon imaging, and optogenetic approaches.

Accordingly, much is new in this second edition of the *Encyclopedia of Pain*. The number of authors who have contributed to this edition is close to 800. In addition to a print edition, the publisher, Springer-Verlag, will produce an electronic version that will make comprehensive searches easy to manage. The electronic version, to be made available on the online publishing platform SpringerReference, can be updated regularly to ensure that the content remains current.

October 2013
Gerald F. Gebhart
Pittsburgh, PA, USA

Robert F. Schmidt
Würzburg, Germany

Preface to the First Edition

As all medical students know, pain is the most common reason for a person to consult a physician. Under ordinary circumstances, acute pain has a useful, protective function. It discourages the individual from activities that aggravate the pain, allowing faster recovery from tissue damage. The physician can often tell from the nature of the pain what its source is. In most cases, treatment of the underlying condition resolves the pain. By contrast, children born with congenital insensitivity to pain suffer repeated physical damage and die young (see Sweet WH (1981) Pain 10:275).

Pain resulting from difficult to treat or untreatable conditions can become persistent. Chronic pain "never has a biologic function but is a malefic force that often imposes severe emotional, physical, economic, and social stresses on the patient and on the family…." (Bonica JJ (1990) The Management of Pain, vol 1, 2nd edn. Lea & Febiger, Philadelphia, p 19). Chronic pain can be considered a disease in its own right.

Pain is a complex phenomenon. It has been defined by the Taxonomy Committee of the International Association for the Study of Pain as "An unpleasant emotional and sensory experience associated with actual or potential tissue damage, or described in terms of such damage" (Merskey H and Bogduk N (1994) Classification of Chronic Pain, 2nd edn. IASP Press, Seattle). It is often ongoing, but in some cases it may be evoked by stimuli. Hyperalgesia occurs when there is an increase in pain intensity in response to stimuli that are normally painful. Allodynia is pain that is evoked by stimuli that are normally non-painful.

Acute pain is generally attributable to the activation of primary efferent neurons called nociceptors (Sherrington CS (1906) The Integrative Action of the Nervous System. Yale University Press, New Haven; 2nd edn, 1947). These sensory nerve fibers have high thresholds and respond to strong stimuli that threaten or cause injury to tissues of the body. Chronic pain may result from continuous or repeated activation of nociceptors, as in some forms of cancer or in chronic inflammatory states, such as arthritis.

However, chronic pain can also be produced by damage to nervous tissue. If peripheral nerves are injured, peripheral neuropathic pain may develop. Damage to certain parts of the central nervous system may result in central neuropathic pain. Examples of conditions that can cause central neuropathic pain include spinal cord injury, cerebrovascular accidents, and multiple sclerosis.

Research on pain in humans has been an important clinical topic for many years. Basic science studies were relatively few in number until experimental work on pain accelerated following detailed descriptions of peripheral nociceptors and central nociceptive neurons that were made in the 1960s and 1970s, by the discovery of the endogenous opioid compounds and the descending pain control systems in the 1970s and the application of modern imaging techniques to visualize areas of the brain that are affected by pain in the 1990s. Accompanying these advances has been the development of a number of animal models of human pain states, with the goal of using these to examine pain mechanisms and also to test analgesic drugs or non-pharmacologic interventions that might prove useful for the treatment of pain in humans. Basic research on pain now emphasizes multidisciplinary approaches, including behavioral testing, electrophysiology, and the application of many of the techniques of modern cell and molecular biology, including the use of transgenic animals.

The *Encyclopedia of Pain* is meant to provide a source of information that spans contemporary basic and clinical research on pain and pain therapy. It should be useful not only to researchers in these fields but also to practicing physicians and other health care professionals and to health care educators and administrators. The work is subdivided into 35 fields, and the Field Editor of each of these describes the areas covered in each in a brief review entry. The topics included in a field are the subject of a series of short essays, accompanied by key words, definitions, illustrations, and a list of significant references. The number of authors who have contributed to the encyclopedia exceeds 550. The plan of the publisher, Springer-Verlag, is to produce both print and electronic versions of this encyclopedia. Numerous links within the electronic version should make comprehensive searches easy to manage. The electronic version will be updated at sufficiently short intervals to ensure that the content remains current.

The editors thank the staff at Springer-Verlag who have provided oversight for this project, including Rolf Lange, Thomas Mager, Claudia Lange, Natasja Sheriff, and Michaela Bilic. Working with these outstanding individuals has been a pleasure.

July 2006
ROBERT F. SCHMIDT   WILLIAM D. WILLIS
Würzburg, Germany   Galveston, Texas, USA

About the Editors

Gerald F. Gebhart received his PhD from The University of Iowa at Iowa City, Iowa, USA, after which he completed a 2-year fellowship at the Université de Montréal, Montréal, Canada, under the direction of Herbert H. Jasper. He began his academic career in the Department of Pharmacology at The University of Iowa in 1973, where he developed a productive research program and led a Pain Interest Group. His career includes a sabbatical year in Heidelberg, Germany (1981–1982), and leadership as head of the Department of Pharmacology from 1996 to 2006. In 2006, Dr. Gebhart was named director of the Center for Pain Research, Department of Anesthesiology, the University of Pittsburgh at Pittsburgh, Pennsylvania, USA. Dr. Gebhart has served the pain community throughout his career. He was elected to the Board of Directors of the American Pain Society (1991–1994) and subsequently elected president of the APS (1997). In 2000, he became the founding editor of *The Journal of Pain*, the official journal of the APS, serving as editor-in-chief until 2010. Dr. Gebhart also served on the Council and later as president-elect, president, and past president of the International Association for the Study of Pain (2005–2012).

Dr. Gebhart is best known for his research contributions in descending modulation of pain and mechanisms of visceral pain, and was designated in 2004 by Thomson ISI as a Highly Cited Researcher (Neuroscience). His scientific contributions have been recognized with several awards, including a 5-year Bristol Myers-Squibb Award for Excellence in Pain Research (1989–1994), a 10-year MERIT Award from the National Institutes of Health (1993–2003),

the Frederick W.L. Kerr Award from the American Pain Society (1994), the Purdue Pharma Prize for Pain Research (2004), the Janssen Award in Gastroenterology (2005), the Founders Award from the American Academy of Pain Medicine (2006), and the Patrick D. Wall Award from the British Pain Society (2012).

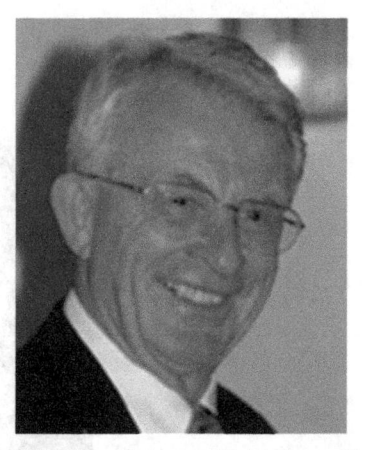

Robert F. Schmidt studied medicine at Heidelberg University (Germany). He graduated in 1959 and was promoted to Dr.med. in the same year. After residencies in clinical disciplines, Dr. Schmidt went to the John Curtin School of Medical Research of the Australian National University in Canberra, ACT, where he worked under the head of the Department of Neurophysiology, Sir John C. Eccles. He received his Ph.D. in 1963, the same year Sir John was awarded the Nobel Prize for Physiology and Medicine.

Returning to Heidelberg Medical School, Dr. Schmidt continued his academic career. He completed his habilitation in 1964, becoming a tenured lecturer (Wissenschaftlicher Rat) in 1966 and an associate professor in 1970. After an 8-month sabbatical with Sir John at the State University of New York, Dr. Schmidt became professor and director of the Department of Physiology at the University of Kiel, Germany. In 1982, he moved to the University of Würzburg to take the same position in the Department of Physiology. In 2000, Dr. Schmidt became professor emeritus at this department.

The research work of Dr. Schmidt and his associates in both Kiel and Würzburg focused on the neurophysiology of fine sensory afferents and their connections and functions in the spinal cord. In Würzburg, he concentrated on the processes leading to the changes in response characteristics of nociceptive afferents (i.e., peripheral sensitization) during inflammation of joints, discovering in the process silent nociceptors, which have subsequently been described in all vertebrates studied, including humans. In addition, Dr. Schmidt has edited, coedited, authored, and coauthored numerous textbooks and monographs, most of them having several editions and translations.

In recognition of his research contributions, Dr. Schmidt received many prizes, awards, and honors, including a D.Sc. honoris causa from the University of New South Wales, Sydney, Australia, and an Honorary Professorship from the University of Tübingen, Germany. He was elected to membership in the Akademie der Wissenschaften und der Literatur, Mainz, Germany. Dr. Schmidt is an Honorary Member of the Colombian Association for the Study of Pain, the Japan Physiological Society, the German Society for the Study of Pain, the German Pain Association, the International Association for the Study of Pain (IASP), and the German Physiological Society (in which he served as president in 1979). In 2000, he was awarded the Bundesverdienstkreuz 1st Class (Order of Merit of the German Federal Republic).

Section Editors

Ebtesam Ahmed College of Pharmacy and Health Sciences, St. John's University, Department of Pain Medicine and Palliative Care, Beth Israel Medical Center, New York, NY, USA

A. Vania Apkarian Department of Physiology, Northwestern University Feinberg School of Medicine, Chicago, IL, USA

Lars Arendt-Nielsen Center for Sensory-Motor Interaction, Department of Health Science and Technology, School of Medicine, Aalborg University, Aalborg, Denmark

Carlos Belmonte Instituto de Neurociencias, Universidad Miguel Hernandez-CSIC, San Juan de Alicante, Spain

Niels Birbaumer Institute of Medical Psychology and Behavioral Neurobiology, University of Tübingen, Tübingen, Germany

Nikolai Bogduk University of Newcastle, Newcastle Bone & Joint Institute, Royal Newcastle Centre, Newcastle, NSW, Australia

Jin Mo Chung Department of Neuroscience and Cell Biology, University Chair in Neuroscience, University of Texas Medical Branch, Galveston, TX, USA

Joyce DeLeo Dartmouth Hitchcock Medical Center, Dartmouth Medical School, Hanover, NH, USA

Marshall Devor Department of Cell & Developmental Biology, Institute of Life Sciences and Center for Research on Pain, The Hebrew University of Jerusalem, Jerusalem, Givat Ram, Israel

Hans-Christoph Diener Department of Neurology, University Hospital Essen, University Duisburg-Essen, Essen, Germany

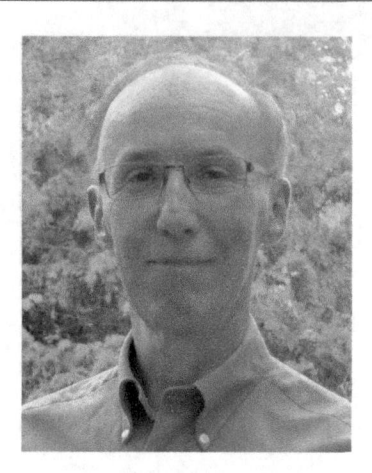

Jonathan O. Dostrovsky Department of Physiology, University of Toronto, Toronto, ON, Canada

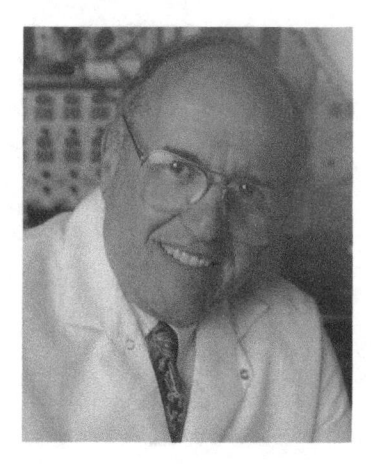

Ronald Dubner Department of Neural and Pain Sciences, University of Maryland Dental School, Baltimore, MD, USA

Michel Y. Dubois NYU Medical School, NYU Langone Medical Centers, Center for Research & Treatment of Pain, NYU Ambulatory Care Center, New York, NY, USA

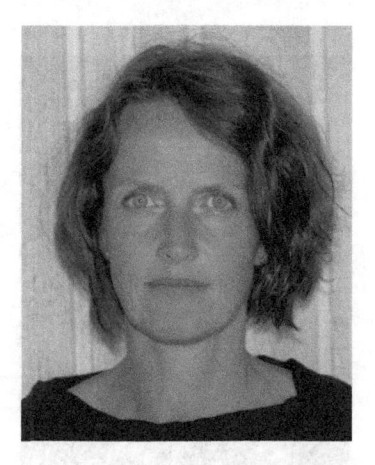

Nanna Brix Finnerup Aarhus University Hospital, Danish Pain Research Center, Aarhus, Denmark

Herta Flor Department of Cognitive and Clinical Neuroscience, Central Institute of Mental Health, Medical Faculty Mannheim, Heidelberg University, Mannheim, Germany

Gerald F. Gebhart Department of Anesthesiology, Center for Pain Research, University of Pittsburgh, Pittsburgh, PA, USA

Maria Adele Giamberardino Pathophysiology of Pain Laboratory and Centre for Fibromyalgia and Musculoskeletal Pain, Department of Medicine and Science of Aging, University of Chieti, Chieti, Italy

Glenn J. Giesler Department of Neuroscience, University of Minnesota, Minneapolis, MN, USA

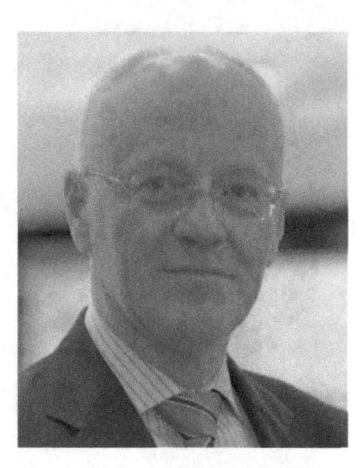

Peter Goadsby Headache Center, Department of Neurology, University of California, San Francisco, San Francisco, CA, USA

Michael S. Gold Department of Anesthesiology, Center for Pain Research, University of Pittsburgh, Pittsburgh, PA, USA

Hermann O. Handwerker Department of Physiology and Pathophysiology, University of Erlangen/Nürnberg, Erlangen, Germany

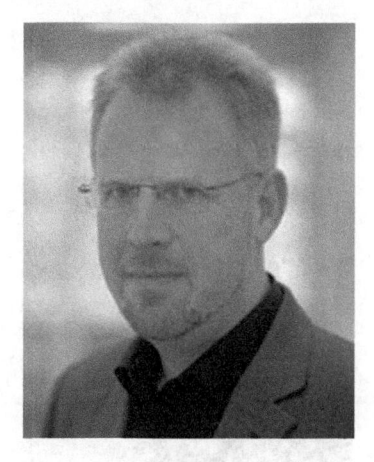

Burkhard Hinz Institute of Toxicology and Pharmacology, University of Rostock, Rostock, Germany

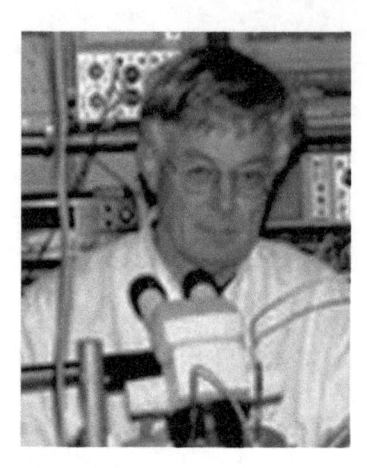

Wilfrid Jaenig Physiologisches Institut, Christian-Albrechts-Universität zu
Kiel, Kiel, Germany

Michaela Kress Department für Physiologie und Medizinische Physik,
Medizinische Universität Innsbruck, Innsbruck, Austria

Frederick A. Lenz Department of Neurosurgery, Johns Hopkins Hospital, Baltimore, MD, USA

Tonya M. Palermo Seattle Children's Hospital Research Institute, CW8-6 - Child Health, Behavior and Development, University of Washington, Seattle, WA, USA

Frank Porreca Department of Pharmacology, University of Arizona Health
Sciences Center, Tucson, AZ, USA

Russell K. Portenoy Department of Pain Medicine and Palliative Care, Beth
Israel Medical Center, New York, NY, USA

James P. Robinson Department of Anesthesiology & Pain Medicine, University of Washington, Seattle, WA, USA

Hans-Georg Schaible Department of Physiology - Neurophysiology, Friedrich-Schiller-University of Jena, Jena, Germany

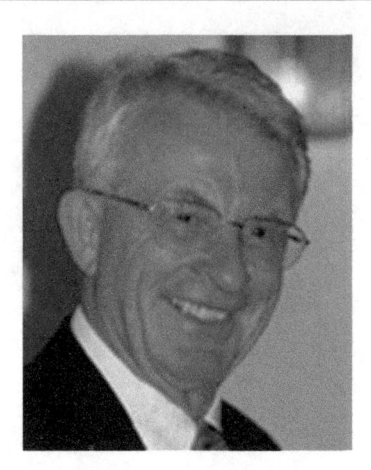

Robert F. Schmidt Physiologisches Institut der Universität Würzburg, Würzburg, Germany

Stephan A. Schug School of Medicine and Pharmacology, Royal Perth Hospital, Faculty of Medicine, Dentistry & Health Sciences, The University of Western Australia, UWA Anaesthesiology, Perth, Australia

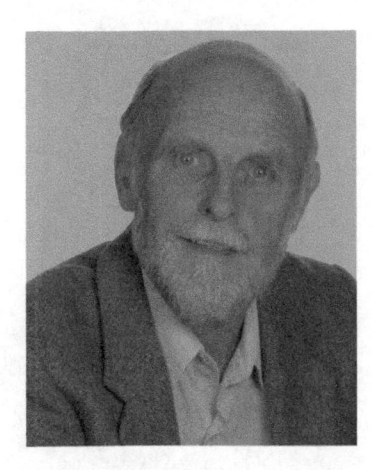

Barry J. Sessle Faculty of Dentistry, University of Toronto, Toronto, ON, Canada

Bengt H. Sjölund Department of Community Medicine and Rehabilitation Medicine, Umea University, Umea, Sweden

Mark D. Sullivan Psychiatry and Behavioral Sciences, University of Washington, Seattle, WA, USA

Irene Tracey Pain Imaging Neuroscience (PaIN) Group, Department of Physiology, Anatomy and Genetics, Oxford University, Oxford, UK

Dennis C. Turk Department of Anesthesiology & Pain Research, University of Washington, Seattle, WA, USA

Félix Viana Instituto de Neurociencias, Universidad Miguel Hernandez-CSIC, San Juan de Alicante, Alicante, Spain

Katy Vincent Nuffield Department of Obstetrics and Gynaecology, The Women's Centre, John Radcliffe Hospital, University of Oxford, Oxford, UK

Gary A. Walco Department of Anesthesiology and Pain Medicine, University of Washington School of Medicine, Seattle Children's Hospital, Seattle, WA, USA

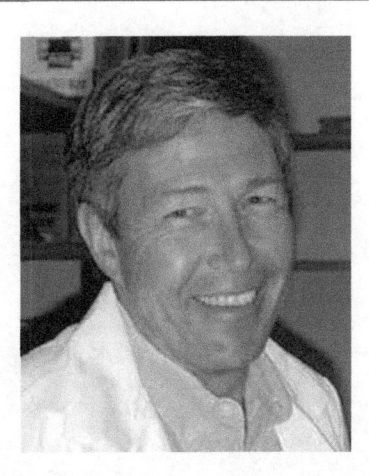

George L. Wilcox Department of Neuroscience, Pharmacology and Dermatology, University of Minnesota Medical School, Minneapolis, MN, USA

Katja Wiech Nuffield Department of Clinical Neurosciences, John Radcliffe Hospital, Oxford, UK

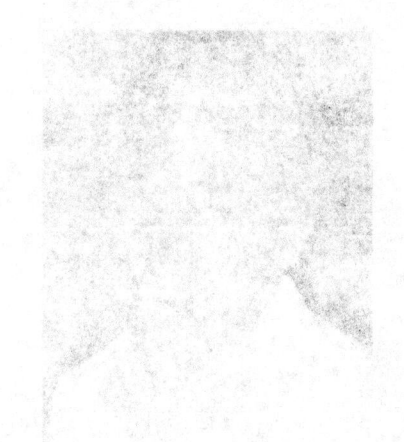

George E. Wilcox Department of Neuroscience, Pharmacology and Dermatology, University of Minnesota Medical School, Minneapolis, MN, USA

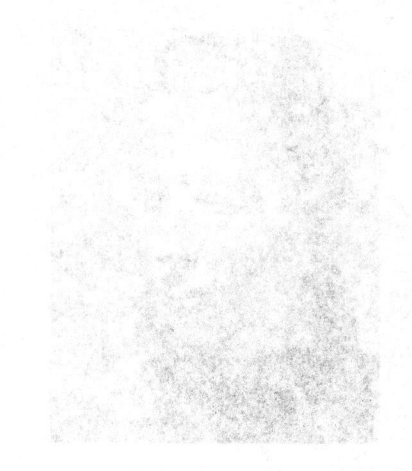

Katja Wiech Nuffield Department of Clinical Neurosciences, with Radcliffe Hospital Oxford UK

Contributors

Tipu Aamir The Auckland Regional Pain Service, Auckland Hospital, Auckland, New Zealand

Catherine Abbadie Department of Pharmacology, Merck Research Laboratories, Rahway, NJ, USA

Frances V. Abbott Departments of Psychiatry and Psychology, McGill University, Montreal, QC, Canada

Heather Adams Department of Surgery, McGill University Health Centre, Montreal, QC, Canada

Mary O. Adedoyin Department of Anatomy, University of California San Francisco, San Francisco, CA, USA

Rolf H. Adler University of Berne Medical School, Kehrsatz, Switzerland

Claire D. Advokat Department of Psychology, Louisiana State University, Baton Rouge, LA, USA

Reto M. Agosti Headache Center Hirlsanden, Zurich, Switzerland

Ebtesam Ahmed College of Pharmacy and Health Sciences, St. John's University, Department of Pain Medicine and Palliative Care, Beth Israel Medical Center, New York, NY, USA

Marie-Claire Albanese Department of Psychology, McGill University, Montreal, QC, Canada

Kathryn M. Albers Department of Neurobiology, University of Pittsburgh School of Medicine, Pittsburgh, PA, USA

Phillip J. Albrecht Center for Neuropharmacology and Neuroscience, Albany Medical College, Albany, NY, USA

Elie D. Al-Chaer Center for Pain Research, University of Arkansas for Medical Sciences, Little Rock, AR, USA

Terry Altilio Department of Pain Medicine and Palliative Care, Beth Israel Medical Center, New York, USA

Rainer Amann Medical University Graz, Graz, Austria

Ron Amir Institute of Life Sciences, The Hebrew University of Jerusalem, Jerusalem, Israel

Praveen Anand Peripheral Neuropathy Unit, Imperial College London, London, UK

Karen O. Anderson The University of Texas MD Anderson Cancer Center, Houston, TX, USA

Ole K. Andersen Department of Health Science & Technology, Aalborg University, Aalborg, Denmark

Stanley W. Anderson Department of Neurosurgery, Johns Hopkins Hospital, Johns Hopkins Medical Institutions, Johns Hopkins University, Baltimore, MD, USA

Trevor Anderson Fairview Pain & Palliative Care, University of Minnesota Medical Center, Minneapolis, MN, USA

Karl-Erik Andersson Wake Forest Institute for Regenerative Medicine, Wake Forest University School of Medicine, Winston Salem, USA

Frank Andrasik Department of Psychology, Applied Psychophysiology and Biofeedback, The University of Memphis, Memphis, TN, USA

A. Vania Apkarian Department of Physiology, Northwestern University Feinberg School of Medicine, Chicago, IL, USA

Lars Arendt-Nielsen Center for Sensory-Motor Interaction, Department of Health Science and Technology, School of Medicine, Aalborg University, Aalborg, Denmark

Kirsten Arndt CNS Pharmacology, Boehringer Ingelheim Pharma GmbH & Co. KG, Biberach, Germany

Charles-Daniel Arreto Université Paris Descartes, INSERM UMR 894, France

Erik Askenasy Neurobiology and Anatomy, Health Science Center, University of Texas, Houston, TX, USA

J. H. Atkinson Psychiatry Service, VA San Diego Healthcare System and Department of Psychiatry, University of California, San Diego, CA, USA

Nadine Attal Centre d'Evaluation et de Traitement de la Douleur, Hôpital Ambroise Paré, INSERM U 987 APHP, Boulogne-Billancourt, France

Christoph Aufenberg Functional Neurosurgery, Neurosurgical Clinic University Hospital, Zürich, Switzerland

Beate Averbeck Sanofi-Aventis Deutschland GmbH, Frankfurt, Germany

Qasim Aziz Section of GI Sciences, Hope Hospital University of Manchester, Salford, UK

Cathrine Baastrup Danish Pain Research Center, Aarhus University Aarhus University Hospital, Aarhus, Denmark

F. W. Bach Danish Pain Research Center, Department of Neurology, Aarhus University Hospital, Aarhus, Denmark

S. K. Back Neuroscience Research Institute and Department of Physiology, Korea University, College of Medicine, Seoul, Republic of Korea

Misha-Miroslav Backonja Department of Neurology, University of Wisconsin-Madison, Madison, WI, USA

Banani Banerjee Division of Gastroenterology and Hepatology and Pediatric Gastroenterology, Medical College of Wisconsin, Milwaukee, WI, USA

John D. Banja Department of Rehabilitation Medicine, Center for Ethics Emory University, Atlanta, GA, USA

Carsten Bantel Department of Surgery & Cancer? Anaesthetics Section, Imperial College of Science and Medicine Chelsea and Westminster Campus, London, UK

Andrew Paul Baranowski The Pain Management Centre, National Hospital for Neurology and Neurosurgery, UCL Hospitals NHS Foundation Trust, London, UK

Alex Barling Department of Neurology, Birmingham Muscle and Nerve Centre Queen Elizabeth Hospital, Birmingham, UK

Ralf Baron Department of Neurological Pain Research and Therapy, Christian Albrechts University Kiel, Kiel, Germany

Gordon A. Barr Department of Anesthesiology and Critical Care Medicine, Children's Hospital of Philadelphia University of Pennsylvania, Philadelphia, PA, USA

Emily J. Bartley Department of Psychology, The University of Tulsa, Tulsa, OK, USA

Jeffrey R. Basford Department of Physical Medicine and Rehabilitation, Mayo Clinic and Mayo Foundation, Rochester, NY, USA

Michele C. Battie Department of Physical Therapy, University of Alberta, Edmonton, AB, Canada

Ulf Baumgartner Chair of Neurophysiology, Center for Biomedicine and Medical Technology Mannheim, Heidelberg University, Mannheim, Germany

Nicola Beale Nuffield Department of Anaesthetics, Oxford University Hospitals NHS Trust, UK

Alain Beaudet Montreal Neurological Institute, McGill University, Montreal, QC, Canada

Lino Becerra P.A.I.N. Group, Department of Anesthesiology, Perioperative and Pain Medicine, Boston Children's Hospital Department of Radiology Massachusetts General Hospital Center for Pain and the Brain, Harvard Medicals, Boston, MA, USA

Yaakov Beilin Department of Anesthesiology, and Obstetrics, Mount Sinai School of Medicine, New York, NY, USA

Alvin J. Beitz Department of Veterinary and Biomedical Sciences, University of Minnesota, St. Paul, MN, USA

Miles Belgrade Fairview Pain Management Center, University of Minnesota Medical Center, Minneapolis, MN, USA

Carlo Valerio Bellieni Neonatal Intensive Care Unit, University Hospital, Siena, Italy

Carlos Belmonte Instituto de Neurociencias, Universidad Miguel Hernandez-CSIC, San Juan de Alicante, Spain

Allan J. Belzberg Department of Neurosurgery, Johns Hopkins School of Medicine, Baltimore, MD, USA

Matt Bender Department of Neurosurgery, Johns Hopkins Hospital, Baltimore, MA, USA

Fabrizio Benedetti Department of Neuroscience, Clinical and Applied Physiology Program, University of Turin Medical School, Turin, Italy

David L. H. Bennett Wolfson CARD, Kings College London, London, UK

Nuffield Department of Clinical Neurosciences, John Radclife Hospital, The University of Oxford, Oxford, UK

Robert Bennett Department of Medicine, Oregon Health and Science University, Portland, USA

Edward C. Benzel Department of Neurosurgery, Center for Spine Health Neurological Institute Cleveland Clinic, Cleveland, OH, USA

David A. Bereiter Department of Diagnostic and Biological Sciences, University of Minnesota School of Dentistry, Minneapolis, MN, USA

Elmar Berendes Klinik für Anästhesiologie, Operative Intensivmedizin und Schmerztherapie, Klinikum Krefeld, Krefeld, Germany

Karen J. Berkley Program in Neuroscience, Florida State University, Tallahassee, FL, USA

Peter Berlit Department of Neurology, Alfried Krupp Hospital, Essen, Germany

Jean-François Bernard INSERM UMR-894, UPMC Site Pitié-Salpêtrière, Paris, France

Anne K. Bertelsen Neurology, Aleris Hospital, Bergen, Norway

Jennifer Bickel Integrative Pain Management, Children's Mercy Hospitals and Clinics, University of Missouri-Kansas City School of Medicine, Kansas City, MO, USA

Marcelo E. Bigal Department of Neurology and Montefiore Headache Unit, Albert Einstein College of Medicine, Bronx, NY, USA

Yitzchak M. Binik McGill University, Montreal, QC, Canada

Niels Birbaumer Institute of Medical Psychology and Behavioral Neurobiology, University of Tübingen, Tübingen, Germany

Frank Birklein Department of Neurology, University of Mainz, Mainz, Germany

Kathryn A. Birnie Department of Psychology, Dalhousie University, Halifax, NS, Canada

Thomas Bishop Centre for Neuroscience, King's College London, London, UK

Henning Bliddal The Parker Institute, Frederiksberg Hospital, Copenhagen, Denmark

Nikolai Bogduk University of Newcastle, Newcastle Bone & Joint Institute, Royal Newcastle Centre, Newcastle, NSW, Australia

Jorgen Boivie Department of Neurology, University Hospital, Linköping, Sweden

Michael R. Bond University of Glasgow, Glasgow, UK

Richard L. Boortz-Marx Division of Pain Medicine, Department of Anesthesiology, University of North Carolina, Chapel Hill, North Carolina, USA

James M. Borowczyk Department of Orthopaedics and Musculoskeletal Medicine, University of Otago, Christchurch, Christchurch, New Zealand

David Borsook P.A.I.N. Group, Department of Anesthesiology, Perioperative and Pain Medicine, Boston Children's Hospital Department of Radiology Massachusetts General Hospital Center for Pain and the Brain, Harvard Medicals, Boston, MA, USA

George S. Borszcz Department of Psychology, Wayne State University, Detroit, USA

Jasenka Borzan Department of Anesthesiology and Critical Care Medicine, Johns Hopkins University, Baltimore, USA

Regina M. Botting London, UK

Didier Bouhassira Centre d'Evaluation et de Traitement de la Douleur, Hôpital Ambroise Paré, INSERM U 987 APHP, Boulogne-Billancourt, France

Laurence Bourgeais INSERM UMR 894, Centre de Psychiatrie et Neuro-sciences, Paris, France

David Bowsher Pain Research Institute, University Hospital Aintree, Liverpool, UK

Emma L. Brandon Anaesthesiology and Pain Medicine, University of Western Australia Royal Perth Hospital, Perth, WA, Australia

Jennifer Brawn P.A.I.N. Group, Center for Pain and the Brain, Boston Children's Hospital, Harvard Medical School, Waltham, MA, USA

Harald Breivik Department of Pain Management and Research, Oslo University Hospital, Rikshospitalet and University of Oslo, Oslo, Norway

Kori L. Brewer Department of Emergency Medicine, Brody School of Medicine East Carolina University, Greenville, NC, USA

Christopher R. Brigham Brigham and Associates, Inc., Portland, USA

Abigail Broderick Nuffield Department of Obstetrics and Gynaecology, University of Oxford, Oxford, UK

Jeffrey A. Brown Department of Neurological Surgery, Wayne State University School of Medicine, Detroit, MI, USA

Stephen C. Brown Department of Anesthesia, The Hospital for Sick Children, Toronto, ON, Canada

Eduardo Bruera Department of Palliative Care and Rehabilitation Medicine, The University of Texas M. D. Anderson Cancer Center, Houston, TX, USA

Pablo Brumovsky Faculty of Biomedical Sciences, Austral University, Buenos Aires, Argentina

Consejo Nacional de Investigaciones Científicas y Técnicas (CONICET), Buenos Aires, Argentina

Kay Brune Department of Experimental and Clinical Pharmacology and Toxicology, Friedrich Alexander University Erlangen-Nürnberg, Erlangen, Germany

Maria Burian Department of General, Visceral and Transplantation Surgery, University of Heidelberg, Heidelberg, Germany

Dan Buskila Rheumatic Disease Unit, Soroka Medical Center, Beer Sheva, Israel

Catherine M. Cahill Anesthesiology & Perioperative Care, University of California, Irvine, Irvine, CA, USA

Brian E. Cairns Faculty of Pharmaceutical Sciences, University of British Columbia, Vancouver, BC, Canada

Nigel A. Calcutt Department of Pathology, University of California San Diego, La Jolla, CA, USA

Margarita Calvo Wolfson CARD, Kings College London, London, UK

James N. Campbell Johns Hopkins University School of Medicine, Baltimore, MD, USA

Lauren Campbell Department of Psychology, York University, Toronto, ON, Canada

Ling Cao Department of Biomedical Sciences, College of Osteopathic Medicine, University of New England, ME, USA

Adam Carinci Department of Anesthesia, Critical Care and Pain Medicine, Massachusetts General Hospital Harvard Medical School, Boston, MA, USA

Giancarlo Carli Department of Physiology, University of Siena, Siena, Italy

Christer P. O. Carlsson Rehabilitation Clinic, Lunds University Hospital, Lund, Sweden

Susan M. Carlton Neuroscience and Cell Biology, University of Texas Medical Branch, Galveston, TX, USA

Richard Carr Clinic for Anaesthesiology, Heidelberg University, Mannheim, Baden-Wuerttemberg, Germany

Benjamin S. Carson Department of Neurosurgery, Johns Hopkins Hospital, Baltimore, MA, USA

Brian Carter University of Missouri at Kansas City Children's Mercy Hospital and Bioethics Center, Kansas City, MO, USA

Kenneth L. Casey Department of Neurology, University of Michigan Medical Center, Ann Arbor, MI, USA

Christine T. Chambers Departments of Pediatrics and Psychology, Dalhousie University, Halifax, NS, Canada

Sandra Chaplan Janssen Research and Development, LLC, San Diego, CA, USA

C. Richard Chapman Department of Anesthesiology, University of Utah School of Medicine, Salt Lake City, UT, USA

Santosh K. Chaturvedi Psychiatry, National Institute of Mental Health & Neurosciences, Bangalore, India

Andrew C. N. Chen Human Brain Mapping and Cortical Imaging Laboratory, Aalborg University, Aalborg, Denmark

Hsinlin Thomas Cheng Department of Neurology, University of Michigan, Ann Arbor, MI, USA

Pavel Cherkas Faculty of Dentistry, University of Toronto, Toronto, ON, Canada

Andrea Cheville Department of Physical Medicine and Rehabilitation, Mayo Clinic, Rochester, MN, USA

Chen Yu Chiang Faculty of Dentistry, University of Toronto, Toronto, ON, Canada

N. L. Chillingworth School of Physiology and Pharmacology, University of Bristol, Bristol, UK

Edward Chow Department of Radiation Oncology, Sunnybrook Health Sciences Centre Odette Cancer Centre (T2-182), Toronto, ON, Canada

Julie A. Christianson Department of Anatomy and Cell Biology, University of Kansas Medical Center, Kansas City, KS, USA

MacDonald J. Christie Discipline of Pharmacology, University of Sydney, Sydney, Australia

Jin Mo Chung Department of Neuroscience and Cell Biology, University Chair in Neuroscience, University of Texas Medical Branch, Galveston, TX, USA

Kyungsoon Chung Department of Neuroscience and Cell Biology, University of Texas Medical Branch, Galveston, TX, USA

Anna K. Clark Wolfson Centre for Age Related Diseases, King's College London, London, UK

W. Crawford Clark College of Physicians and Surgeons, Columbia University, New York, USA

James F. Cleary Pain & Policy Studies Group, UW Carbone Cancer Center, Madison, WI, USA

John De Clercq Department of Rehabilitation Sciences and Physiotherapy, Ghent University Hospital, Ghent, Belgium

Jacqui Clinch Royal National Hospital for Rheumatic Diseases NHS Foundation Trust, Bath, UK

Terence J. Coderre Departments of Anesthesia, Neurology & Neurosurgery and Psychology, McGill University, Montreal, QC, Canada

Robert Coghill Department of Neurobiology and Anatomy, Wake Forest University School of Medicine, Winston-Salem, NC, USA

Lindsey Cohen Psychology, Georgia State University, Atlanta, GA, USA

Alan L. Colledge Utah Labor Commission, International Association of Industrial Accident Boards and Commissions Central Utah Medical Clinic, South Salt Lake City, UT, USA

Beverly J. Collett University Hospitals of Leicester, Leicester, UK

Lesley A. Colvin Department of Anaesthesia, Critical Care and Pain Medicine Western General Hospital, Edinburgh, UK

Kevin C. Conlon Trinity College Dublin, University of Dublin, Dublin, Ireland

Mark Connelly Integrative Pain Management, Children's Mercy Hospitals and Clinics, University of Missouri-Kansas City School of Medicine, Kansas City, MO, USA

Brian Y. Cooper Department of Oral and Maxillofacial Surgery, University of Florida, Gainesville, FL, USA

Christine Cordle University Hospitals of Leicester, Leicester, UK

Laura A. Cousins Psychology, Georgia State University, Atlanta, GA, USA

Michael J. Cousins Department of Anesthesia and Pain Management, Royal North Shore Hospital University of Sydney, Sydney, NSW, Australia

Verne C. Cox Department of Psychology, University of Texas at Arlington, Arlington, TX, USA

George A. Crabill Department of Neurosurgery, Johns Hopkins Hospital, Baltimore, MA, USA

Rebecca M. Craft Psychology, Washington State University, Pullman, WA, USA

Kenneth D. Craig Department of Psychology, University of British Columbia, Vancouver, BC, Canada

Richard Crevenna Medizinische Universität Wien, Wien, Austria

Geert Crombez Department of Experimental Clinical and Health Psychology, Ghent University, Ghent, Belgium

Joan Crook School of Nursing, Faculty of Health Sciences, McMaster University, Hamilton, ON, Canada

Giorgio Cruccu Department of Neurological Sciences, La Sapienza University, Rome, Italy

Michele Curatolo Department of Anesthesiology, Division of Pain Therapy, University Hospital of Bern Inselspital, Bern, Switzerland

Nachum Dafny Neurobiology and Anatomy, Health Science Center, University of Texas, Houston, TX, USA

Jorgen B. Dahl Department of Anaesthesiology, Glostrup University Hospital, Herlev, Denmark

Yi Dai Department of Anatomy and Neuroscience, Hyogo College of Medicine, Hyogo, Japan

Stefaan Van Damme Department of Experimental Clinical and Health Psychology, Ghent University, Ghent, Belgium

Alexandre F. M. DaSilva Biologic & Materials Sciences, University of Michigan Dental School, MI, USA

Steve Davidson Pain Center & Department of Anesthesiology, Washington University School of Medicine, MO, USA

Brian Davis Department of Neurobiology, University of Pittsburgh School of Medicine, Pittsburgh, PA, USA

Karen D. Davis Brain, Imaging and Behaviour – Systems Neuroscience, Toronto Western Research Institute, Toronto, ON, Canada

Department of Surgery, University of Toronto, Toronto, ON, Canada

Richard O. Day Department of Clinical Pharmacology and Toxicology, St Vincent's Hospital, Sydney, NSW, Australia

Isabelle Decosterd Pain Center, University Hospital Center and University of Lausanne, Lausanne, Switzerland

Ruth Defrin Physical Therapy, Faculty of Medicine, Tel-Aviv University, Tel-Aviv, Israel

Antoon F. C. De Laat Department of Oral/Maxillofacial Surgery, Catholic University of Leuven, Leuven, Belgium

Joyce DeLeo Dartmouth Hitchcock Medical Center, Dartmouth Medical School, Hanover, NH, USA

A. Lee Dellon Department of Plastic Surgery and Neurosurgery, Johns Hopkins University, Baltimore, USA

Dellon Institutes for Peripheral Nerve Surgery, Baltimore, MD, USA

Marshall Devor Department of Cell & Developmental Biology, Institute of Life Sciences and Center for Research on Pain, The Hebrew University of Jerusalem, Jerusalem, Givat Ram, Israel

Perry Dhaliwal Department of Neurosurgery, Cleveland Clinic Spine Institute, Cleveland, OH, USA

David Diamant Neurological and Spinal Surgery, Lincoln, NE, USA

Anthony H. Dickenson Department of Pharmacology, University College London, London, UK

Hans-Christoph Diener Department of Neurology, University Hospital Essen, University Duisburg-Essen, Essen, Germany

Natalia Dmitrieva Program in Neuroscience, Florida State University, Tallahassee, FL, USA

L. F. Donaldson School of Life Sciences, University of Nottingham, Nottingham, UK

Henri Doods CNS Pharmacology, Boehringer Ingelheim Pharma GmbH & Co. KG, Biberach, Germany

Victoria Dorf Social Security Administration, Baltimore, MD, USA

Michael J. Dorsi Department of Neurosurgery, Johns Hopkins School of Medicine, Baltimore, MD, USA

Jonathan O. Dostrovsky Department of Physiology, University of Toronto, Toronto, ON, Canada
Faculty of Dentistry, University of Toronto, Toronto, ON, Canada

Robyn Drach Fairview Pain Management Center, Minneapolis, MN, USA

Ron Dressler Utah Labor Commission, International Association of Industrial Accident Boards and Commissions Central Utah Medical Clinic, South Salt Lake City, UT, USA

Paul Dreyfuss Department of Rehabilitation Medicine, University of Washington, Seattle, USA

Peter D. Drummond School of Psychology, Murdoch University, Perth, Australia

Ronald Dubner Department of Neural and Pain Sciences, University of Maryland Dental School, Baltimore, MD, USA

Anne Ducros Headache Emergency Department, INSERM, Paris, France

Gary H. Duncan Département de stomatologie, Faculté de médecine dentaire, Université de Montréal, Montreal, QC, Canada

Department of Neurology and Neurosurgery, McGill University, Montreal, QC, Canada

Mary J. Eaton Research Service (151), Miami VA Health Care System, Miami VAMC, Miami, FL, USA

Andrea Ebersberger Department of Physiology, Friedrich Schiller University of Jena, Jena, Germany

Christopher Eccleston Pain Management Unit, University of Bath, Bath, UK

John Edmeads Sunnybrook and Woman's College Health Sciences Centre, University of Toronto, Toronto, Canada

Edzard Ernst Complementary Medicine, Peninsula Medical School Universities of Exeter and Plymouth, Exeter, UK

Reuben Escorpizo Swiss Paraplegic Research, Nottwil, Switzerland

Department of Health Sciences and Health Policy, University of Lucerne, Lucerne, Switzerland

ICF Research Branch in cooperation with the World Health Organization (WHO) Collaborating Centre for the Family of International Classifications in Germany, Nottwil, Switzerland

Schonstein Eva RehabCo, Fyshwick, ACT, Australia

Joshua C. Eyer Department of Psychology, University of Alabama, Tuscaloosa, AL, USA

Marie Faaborg-Andersen McGill University, Montreal, QC, Canada

Keri L. Fakata College of Pharmacy and Pain Management Center, University of Utah, Salt Lake City, USA

Marie Fallon Department of Oncology, Edinburgh Cancer Center University of Edinburgh, Edinburgh, UK

Melissa Farmer Department of Physiology / Feinberg School of Medicine, Northwestern University, Chicago, IL, USA

Samantha R. Fashler Department of Psychology, University of British Columbia, Vancouver, BC, Canada

Jean-Baptiste Fassier Department of Occupational Health and Medicine, UMRESTTE-Université Claude Bernard Lyon 1, Lyon, France

Charlotte Feinmann Eastman Dental Institute London, University College London, London, UK

Elizabeth Roy Felix The Miami Project to Cure Paralysis, University of Miami Miller School of Medicine, Miami, FL, USA

Bin Feng Center for Pain Research, Department of Anesthesiology, School of Medicine, University of Pittsburgh, Pittsburgh, Pennsylvania, USA

Yi Feng Department of Anesthesiology, Peking University People's Hospital, Peking, China

Andres Fernandez Department of Neurosurgery, Johns Hopkins Hospital, Baltimore, MD, USA

Antonio Vicente Ferrer-Montiel Institute of Molecular and Cell Biology, University Miguel Hernández, Elche, Alicante, Spain

Perry G. Fine Pain Research Center, School of Medicine, University of Utah, Salt Lake City, UT, USA

David J. Fink Department of Neurology, University of Michigan and VA Ann Arbor Healthcare System, Ann Arbor, MI, USA

Nanna Brix Finnerup Aarhus University Hospital, Danish Pain Research Center, Aarhus, Denmark

David A. Fishbain Department of Psychiatry, University of Miami, Miami, FL, USA

Maria Fitzgerald Neuroscience, Physiology & Pharmacology, University College London, London, UK

Per Flisberg Department of Anaesthesia and Intensive Care, Lund University Hospital, Lund, Sweden

Herta Flor Department of Cognitive and Clinical Neuroscience, Central Institute of Mental Health, Medical Faculty Mannheim, Heidelberg University, Mannheim, Germany

Christopher M. Flores Drug Discovery, Johnson and Johnson Pharmaceutical Research and Development, Spring House, PA, USA

Wilbert E. Fordyce Department of Rehabilitation, University of Washington School of Medicine, Seattle, WA, USA

Robert D. Foreman Department of Physiology, College of Medicine, University of Oklahoma Health Sciences Center, Oklahoma City, OK, USA

Gary M. Franklin Department of Environmental and Occupational Health Sciences, University of Washington, Seattle, WA, USA

Department of Health Services, University of Washington, Seattle, WA, USA

Jan Fridén Department of Hand Surgery, Sahlgrenska University Hospital, Gothenborg, Sweden

Justin Friedlander The Arthur Smith Institute for Urology, North Shore-LIJ Healthcare System, Hofstra North Shore LIJ School of Medicine, New Hyde Park, NY, USA

Allan H. Friedman Division of Neurosurgery, Duke University Medical Center, Durham, NC, USA

Perry N. Fuchs Department of Psychology, University of Texas at Arlington, Arlington, TX, USA

Deborah Fulton-Kehoe Department of Environmental and Occupational Health Sciences, University of Washington, Seattle, WA, USA

Andrea D. Furlan Institute for Work & Health, Toronto, ON, Canada

Vasco Galhardo Departamento de Biologia Experimental, Faculdade de Medicina, Universidade do Porto, Portugal

Andreas R. Gantenbein RehaClinic Bad Zurzach, Bad Zurzach, Switzerland

I. Garonzik Department of Neurosurgery, Johns Hopkins Medical Institutions, Baltimore, MD, USA

Robert Gassin Musculoskeletal Medicine Clinic, Frankston, VIC, Australia

Robert J. Gatchel Department of Psychology, The University of Texas, Arlington, TX, USA

Gerald F. Gebhart Department of Anesthesiology, Center for Pain Research, University of Pittsburgh, Pittsburgh, PA, USA

Gerd Geisslinger Department of Clinical Pharmacology, University Frankfurt am Main, Frankfurt, Germany

Robert D. Gerwin Department of Neurology, Johns Hopkins University, Bethesda, MD, USA

Maria Adele Giamberardino Pathophysiology of Pain Laboratory and Centre for Fibromyalgia and Musculoskeletal Pain, Department of Medicine and Science of Aging, University of Chieti, Chieti, Italy

Stephen Gibson National Ageing Research Institute Department of Medicine, University of Melbourne, Parkville, VIC, Australia

Glenn J. Giesler Department of Neuroscience, University of Minnesota, Minneapolis, MN, USA

Philip L. Gildenberg Baylor Medical College, Houston, TX, USA

Myra Glajchen Department of Pain Medicine and Palliative Care, Beth Israel Medical Center Albert Einstein College of Medicine, New York, NY, USA

Hartmut Göbel Kiel Migraine and Headache Center, Kiel Pain Center, Kiel, Germany

Peter Goadsby Headache Center, Department of Neurology, University of California, San Francisco, San Francisco, CA, USA

Philippe Goffaux Faculty of Medicine, University of Sherbrooke, Sherbrooke, QC, Canada

Ana Gomis Instituto de Neurociencias de Alicante, Universidad Miguel Hernández-CSIC, San Juan de Alicante, Spain

Gilbert R. Gonzales Department of Neurology, Memorial Sloan-Kettering Cancer Center, New York, USA

Allan Gordon Wasser Pain Management Centre, Mount Sinai Hospital University of Toronto, Toronto, ON, Canada

Liesbet Goubert Department of Experimental Clinical and Health Psychology, Ghent University, Ghent, Belgium

Jayantilal Govind Pain Clinic, Department of Anaesthesia, University of New South Wales, Sydney, NSW, Australia

Richard H. Gracely Departments of Medicine-Rheumatology and Neurology, University of Michigan Health System, Ann Arbor, MI, USA

Garry G. Graham Department of Pharmacology, School of Medical Sciences, University of New South Wales, Sydney, NSW, Australia

David K. Grandy Department of Physiology and Pharmacology, Oregon Health and Science University, Portland, OR, USA

Thomas Graven-Nielsen Laboratory for Musculoskeletal Pain and Motor Control, Center for Sensory-Motor Interaction, Aalborg University, Aalborg, Denmark

Barry Green Pierce Laboratory, Yale School of Medicine, New Haven, CT, USA

Paul G. Green Oral & Maxillofacial Surgery, University of California San Francisco, San Francisco, CA, USA

Joel D. Greenspan Department of Neural and Pain Sciences, University of Maryland School of Dentistry, Baltimore, MD, USA

John W. Griffin Department of Neurology, Johns Hopkins University School of Medicine, Baltimore, MD, USA

Cornelius Groenwald Department of Anesthesiology and Pain Medicine, Seattle Children's Hospital, University of Washington School of Medicine, Seattle, WA, USA

Douglas P. Gross Department of Physical Therapy, University of Alberta WCB-Alberta Millard Health, Edmonton, AB, Canada

Sabine Grösch Institute of Clinical Pharmacology, Clinical Center of the Goethe University, Frankfurt, Germany

Blair D. Grubb Department of Cell Physiology and Pharmacology, University of Leicester, Leicester, UK

Ruth Eckstein Grunau Pediatrics Dept, University of British Columbia Developmental Neurosciences & Child Health, Child & Family Research Institute, Vancouver, BC, Canada

Christoph Gutenbrunner Department of Physical Medicine and Rehabilitation, Hanover Medical School, Hanover, Germany

Maija Haanpää Department of Neurosurgery, Helsinki University Central Hospital, Helsinki, Finland

Sherri Haas Fairview Pain Management Center, Minneapolis, MN, USA

Ute Habel Department of Psychiatry and Psychotherapy, RWTH Aachen University, Aachen, Germany

Thomas Hachenberg Clinic for Anaesthesiology and Intensive Therapy, Otto-von-Guericke University, Magdeburg, Germany

Nouchine Hadjikhani Martinos Center for Biomedical Imaging, Massachusetts General Hospital Harvard Medical School, Charlestown, MA, USA

Ole Haegg Department of Orthopedic Surgery, Sahlgren University Hospital, Goteborg, Sweden

Bryan C. Hains Department of Neurology, Yale University School of Medicine, West Haven, CT, USA

Else K. B. Hals University of Oslo, Faculty of Dentistry, Oslo, Norway

Darryl T. Hamamoto Diagnostic and Biological Sciences, University of Minnesota, Minneapolis, MN, USA

Donna L. Hammond Department of Anesthesia, University of Iowa, Iowa City, IA, USA

Menachem Hanani Hadassah-Hebrew University Medical Center Hebrew University Medical School, Jerusalem, Israel

Hermann O. Handwerker Department of Physiology and Pathophysiology, University of Erlangen/Nürnberg, Erlangen, Germany

George M. Hanna Department of Anesthesia, Critical Care and Pain Medicine, Massachusetts General Hospital Harvard Medical School, Boston, MA, USA

Jing-Xia Hao Department of Physiology and Pharmacology, Karolinska Institutet, Stockholm, Sweden

Eleni G. Hapidou Department of Psychiatry and Behavioural Neurosciences, McMaster University and Hamilton Health Sciences Chronic Pain Management Unit, Chedoke Hospital, Hamilton, ON, Canada

Geoffrey Harding Sandgate Spinal Medicine Clinic, Sandgate, QLD, Australia

Louise Harding Hospitals NHS Foundation Trust, University College London, London, UK

Kenneth M. Hargreaves Department of Endodontics, University of Texas Health Science Center at San Antonio, San Antonio, USA

D. Paul Harries Pain Treatment Center, Samaritan Hospital Lexington, Lexington, KY, USA

Denise Harrison Nursing, Children's Hospital of Eastern Ontario (CHEO), University of Ottawa, Ottawa, ON, Canada

Steven Harte Anesthesiology and Internal Medicine/Rheumatology, University of Michigan VA Ann Arbor Healthcare System, Ann Arbor, MI, USA

Barbara A. Hastie Community Dentistry & Behavioral Science, University of Florida College of Dentistry, Gainesville, FL, USA

Jennifer A. Haythornthwaite Psychiatry & Behavioral Sciences, Johns Hopkins University, Baltimore, MD, USA

Heinz-Joachim Häbler FH Bonn-Rhein-Sieg, Rheinbach, Germany

John H. Healey Orthopaedic Service, Department of Surgery, Memorial Sloan-Kettering Cancer Center Weill Medical College of Cornell University, New York, NY, USA

Alicia A. Heapy VA Connecticut Healthcare System, Westhaven, CT, USA

Anne M. Heffernan The Adelaide and Meath Hospital Incorporating The National Children's Hospital, Dublin, Ireland

Geert Van der Heijden Julius Center, Department of Epidemiology, University Medical Center Utrecht, Utrecht, The Netherlands

Mary M. Heinricher Departments of Neurological Surgery and Behavioral Neuroscience, Oregon Health & Science University, Portland, USA

Robert D. Helme Department of Medicine, Royal Melbourne Hospital, University of Melbourne, Melbourne, Australia

Kim Helsen Research Group on Health Psychology, KU Leuven Katholieke Universiteit Leuven, Leuven, Belgium

Department of Experimental–Clinical and Health Psychology, Ghent University, Ghent, Belgium

Amin S. Herati The Arthur Smith Institute for Urology, North Shore-LIJ Healthcare System, Hofstra North Shore LIJ School of Medicine, New Hyde Park, NY, USA

Robert D. Herbert Neuroscience Research Australia, Sydney, Australia

Christiane Hermann Department of Clinical Psychology, Justus-Liebig University Giessen, Giessen, Germany

Juan F. Herrero Department of Physiology, Faculty of Medicine University of Alcala, Alcalá de Henares Madrid, Spain

Aki J. Hietaharju Department of Neurology and Rehabilitation, Pain Clinic Tampere University Hospital, Tampere, Finland

Karin Westlund High Department of Physiology, University of Kentucky, Lexington, KY, USA

Marita Hilliges Dalarna University, Halmstad, Sweden

Max J. Hilz University of Erlangen-Nuremberg, Erlangen, Germany

Burkhard Hinz Institute of Toxicology and Pharmacology, University of Rostock, Rostock, Germany

Anthony R. Hobson Section of GI Sciences, Hope Hospital University of Manchester, Salford, UK

Edward Högestätt Department of Laboratory Medicine, Lund University, Division of Clinical Chemistry and Pharmacology, Lund, Sweden

Tomas Hökfelt Department of Neuroscience, Karolinska Institute, Stockholm, Sweden

Sarah V. Holdridge Department of Pharmacology and Toxicology, Queen's University, Kinston, Canada

Kjell Hole University of Bergen, Bergen, Norway

Graham R. Holland Department of Cariology, University of Michigan, Ann Arbor, MI, USA

Chang-Zern Hong Department of Physical Medicine and Rehabilitation, University of California, Irvine, CA, USA

Minoru Hoshiyama Department of Integrative Physiology, National Institute for Physiological Sciences, Okazaki, Japan

Fred M. Howard University of Rochester, Rochester, NY, USA

Sung-Tsang Hsieh Department of Anatomy and Cell Biology, Department of Neurology, National Taiwan University Hospital, National Taiwan University, Taipei, Taiwan

James W. Hu Faculty of Dentistry, University of Toronto, Toronto, Canada

Claire E. Hulsebosch Spinal Cord Injury Research, University of Texas Medical Branch, Galveston, TX, USA

Thomas Hummel Department of Otorhinolaryngology, Technical University of Dresden Medical School, Dresden, Germany

Mark R. Hutchinson School of Medical Sciences, University of Adelaide, Adelaide, SA, Australia

Charles E. Inturrisi Department of Pharmacology, Weill Cornell Medical College, New York, USA

Koji Inui Department of Integrative Physiology, National Institute for Physiological Sciences, Okazaki, Japan

Koichi Iwata Department of Physiology, Nihon University School of Dentistry, Chiyoda-ku, Japan

Suhayl J. Jabbur Anatomy, Cell Biology and Physiology, Faculty of Medicine, American University of Beirut, Beirut, Lebanon

Neuroscience Program, Faculty of Medicine, American University of Beirut, Beirut, Lebanon

Nora Janjan University of Texas M. D. Anderson Cancer Center, Houston, TX, USA

Luc Jasmin Department of Anatomy, University of California San Francisco, San Francisco, CA, USA

Anne Jaumees Royal North Shore Hospital, St Leonards, NSW, Australia

Wilfrid Jaenig Physiologisches Institut, Christian-Albrechts-Universität zu Kiel, Kiel, Germany

Daniel Jeanmonod Functional Neurosurgery, The Center of Ultrasound Functional Neurosurgery, Solothurn, Switzerland

Andrew Jeffreys Director Department of Anaesthesia & Pain Medicine, Western Health, Melbourne, VIC, Australia

Mark P. Jensen Department of Rehabilitation Medicine, University of Washington, Seattle, WA, USA

Troels Staehelin Jensen Danish Pain Research Center, Department of Neurology, Aarhus University Hospital, Aarhus, Denmark

Kurt L. Johnson Department of Rehabilitation Medicine, University of Washington, Seattle, WA, USA

Mark I. Johnson Faculty of Health and Social Sciences, Leeds Pallium Research Group, Leeds Metropolitan University, Leeds, UK

Mark Johnston Hibiscus Coast Highway, Orewa, New Zealand

Edward Jones Center for Neuroscience, University of California, Davies, CA, USA

Jeroen R. De Jong Department of Rehabilitation, University Hospital Maastricht, Maastricht, The Netherlands

Sven-Eric Jordt Department of Pharmacology, Yale University School of Medicine, New Haven, CT, USA

Ellen Jørum The Laboratory of Clinical Neurophysiology, Rikshospitalet University, Oslo, Norway

Stefan Just CNS Pharmacology, Boehringer Ingelheim Pharma GmbII & Co. KG, Biberach, Germany

Timothy K. Y. Kaan Department of Anatomy, University of California San Francisco, San Francisco, CA, USA

Noufissa Kabli Department of Pharmacology and Toxicology, Queen's University, Kinston, Canada

Ryusuke Kakigi Department of Integrative Physiology, National Institute for Physiological Sciences, Okazaki, Japan

Peter C. A. Kam Anaesthesia, Central Clinical School, The University of Sydney, Sydney, Australia

Giresh Kanji The Sports and Pain Clinic, Wellington, New Zealand

Joel Katz Department of Psychology, McGill University, Montreal, QC, Canada

Valerie Kayser NSERM U677 91Bd de l'Hôpital, Paris, France

Francis J. Keefe Pain Prevention and Treatment Research, Department of Psychiatry and Behavioral Sciences, Duke University Medical Center, Durham, NC, USA

Lois J. Kehl Department of Anesthesiology, University of Minnesota, Minneapolis, MN, USA

Maureen Kelley Department of Pediatrics, Bioethics Division, University of Washington School of Medicine, WA, USA

Seattle Children's Hospital, Seattle, WA, USA

Stephen Kennedy Nuffield Department of Obstetrics and Gynaecology, University of Oxford, Oxford, UK

Edmund Keogh Department of Psychology, University of Bath, Bath, UK

Robert D. Kerns VA Connecticut Healthcare System, Westhaven, CT, USA

Sanjay Khanna Department of Physiology, National University of Singapore, Singapore

Kok Eng Khor Department of Pain Management, Prince of Wales Hospital University of New South Wales, Randwick, NSW, Australia

Mina Khoshnejad Department of Neuroscience, University of Montreal, Montreal, QC, Canada

H. J. Kim Department of Life Science, Yonsei University, Wonju, Republic of Korea

Sara King Faculty of Education (School Psychology), Mount Saint Vincent University, Halifax, NS, Canada

Wade King Department of Clinical Research, Royal Newcastle Centre University of Newcastle, Newcastle, NSW, Australia

K. L. Kirsh Behavioral Medicine, The Pain Treatment Center of the Blue grass, Lexington, KY, USA

Henriette Klit Danish Pain Research Center, Aarhus University Hospital, Aarhus C, Denmark

Ricardo J. Komotar Department of Neurological Surgery, Neurological Institute Columbia University, New York, NY, USA

Sungtae Koo Division of Meridian and Structural Medicine, School of Korean Medicine, Pusan National University, Yangsan, Korea

Rhonda K. Kotarinos Oakbrook Terrace, IL, USA

Katalin J. Kovacs Department of Veterinary and Biomedical Sciences, University of Minnesota, St. Paul, MN, USA

Malgorzata Krajnik Chair of Palliative Care, Nicolaus Copernicus University in Torun, Collegium Medicum in Bydgoszcz, Bydgoszcz, Poland

Elliot Krane Anesthesia and Pediatrics, Stanford University School of Medicine Lucile Packard Children's Hospital at Stanford, Stanford, CA, USA

Ajit Krishnaney Department of Neurosurgery, Cleveland Clinic, Cleveland, OH, USA

Greg Krohm International Association of Industrial Accident Boards and Commissions (IAIABC), Utah Valley University, USA

Lawrence Kruger UCLA Geffen School of Medicine, Los Angeles, CA, USA

M. M. ter Kuile Department of Gynecology, Leiden University Medical Center, Leiden, The Netherlands

Promil Kukreja Department of Anesthesiology, University of Alabama at Birmingham, Birmingham, AL, USA

Alice Kvåle Department of Public Health and Primary Health Care, Faculty of Medicine and Dentistry University of Bergen, Bergen, Norway

Gunnvald Kvarstein Department of Pain Management and Pain Research, Oslo University Hospital, Rikshospitalet, Oslo, Norway

Jean-Jacques Labat Neurology and Rehabilitation, Clinique Urologique, Nantes, France

Gaston Labrecque Faculty of Pharmacy, University of Laval, Quebec City, QC, Canada

Marco Lacerenza Pain Medicine Center, Casa di Cura S. Pio X, Opera San Camillo Foundation, Milan, Italy

Uri Ladabaum Gastroenterology & Hepatology, Stanford Cancer Institute, Redwood City, CA, USA

Marie Andree Lahaie McGill University, Montreal, QC, Canada

Miguel J. A. Lainez-Andres Department of Neurology, University of Valencia, Valencia, Spain

Robert H. LaMotte Department of Anesthesiology, Yale School of Medicine, New Haven, CT, USA

James W. Lance Institute of Neurological Sciences, University of New South Wales, Sydney, Australia

Robert G. Large The Auckland Regional Pain Service, Auckland Hospital, Auckland, New Zealand

Alice A. Larson Department of Veterinary and Biomedical Sciences, University of Minnesota, St. Paul, MN, USA

Roberta Lattanzi Department of Physiology and Pharmacology, Sapienza University of Rome, Rome, Italy

Peter Lau Department of Clinical Research, Royal Newcastle Hospital, University of Newcastle, Newcastle, NSW, Australia

Stefan A. Laufer Institute of Pharmacy, Eberhard-Karls University of Tuebingen, Tuebingen, Germany

Stefan Lautenbacher Physiologische Psychologie, Otto-Friedrich-Universität, Bamberg, Germany

Gilles J. Lavigne Facultés de Médecine Dentaire et de Médecine, Université de Montréal, Hopital Sacre Coeur de Montreal. Surgery-trauma dept, Montréal, QC, Canada

Emily Law Center for Child Health, Behavior, and Development, Seattle Children's Research Institute, Seattle, WA, USA

Peter G. Lawlor Our Lady's Hospice, Medical Department, Dublin, Ireland

Michel Lazdunski Institut de Pharmacologie Moléculaire et Cellulaire (IPMC) CNRS/UNS UMR7275, Valbonne-Sophia Antipolis, France

Vincenzo Di Lazzaro Institute of Neurology, Campus Biomedico University, Rome, Italy

Alyssa Lebel Boston Children's Hospital, Boston, MA, USA

Frank Lee Johns Hopkins University, Baltimore, MD, USA

Maaike Leeuw Department of Medical, Clinical and Experimental Psychology, Maastricht University, Maastricht, The Netherlands

Siri Graff Leknes Research Fellow, Psykologisk institutt, Oslo, Norway

Frederick A. Lenz Department of Neurosurgery, Johns Hopkins Hospital, Baltimore, MD, USA

Guillaume Léonard Faculty of Medicine, University of Sherbrooke, Sherbrooke, QC, Canada

Pauline Lesage Department of Pain Medicine and Palliative Care, Beth Israel Medical Center, New York, USA

Isobel Lever Neural Plasticity Unit, Institute of Child Health University College London, London, UK

Khan W. Li Department of Neurosurgery, Johns Hopkins Hospital, Baltimore, MD, USA

Alan R. Light Department of Anesthesiology, University of Utah, Salt Lake City, UT, USA

Michael Lim Department of Neurosurgery and Department of Neurology, Johns Hopkins Hospital, Baltimore, MA, USA

Deolinda Lima Department of Experimental Biology, Faculdade de Medicina da Universidade do Porto, Porto, Portugal

Volker Limmroth Department of Neurology, University of Essen, Essen, Germany

Eric Lingueglia Institut de Pharmacologie Moléculaire et Cellulaire (IPMC) CNRS/UNS UMR7275, Valbonne-Sophia Antipolis, France

Steven J. Linton Department of Occupational and Environmental Medicine, Orebro Medical Center, Orebro, Sweden

Arthur G. Lipman Departments of Pharmscotherpay and Anesthesiology and Pain Management Center, University of Utah, Salt Lake City, UT, USA

Richard B. Lipton Departments of Neurology, Epidemiology and Population Health, Albert Einstein College of Medicine, Bronx, NY, USA

Diana Lisi Department of Psychology, York University, Toronto, ON, Canada

Kenneth M. Little Division of Neurosurgery, Duke University Medical Center, Durham, NC, USA

Edward Littleton Queen Elizabeth Hospital, Birmingham, UK

Spencer S. Liu Virginia Mason Medical Center, Seattle, WA, USA

Anne Elisabeth Ljunggren Section for Physiotherapy Science, Department of Public Health and Primary Health, Faculty of Medicine, University of Bergen, Bergen, Norway

John D. Loeser Department of Neurological Surgery, University of Washington, Seattle, WA, USA

Joern Loetsch Institute for Clinical Pharmacology, Pharmaceutical Center Frankfurt, Johann Wolfgang Goethe University, Frankfurt, Germany

Patrick Loisel Disability Prevention Research and Training Centre, Université de Sherbrooke, Longueuil, Canada

Elisa Lopez-Dolado Servicio de Rehabilitacion, National Hospital of Paraplèjicos SESCAM, Toledo, Spain

Jürgen Lorenz Faculty of Life Science, Competence Center Gesundheit, University of Applied Science Hamburg, Hamburg, Germany

Dayna R. Loyd US Army Institute of Surgical Research, Fort Sam Houston, TX, USA

Daniel Lubelski Cleveland Clinic Lerner College of Medicine, Cleveland, OH, USA

Stephen Lutz Blanchard Valley Regional Cancer Center, Findlay, OH, USA

Bruce Lynn Department of Physiology, University College London, London, UK

Ross MacPherson Department of Anesthesia and Pain Management, Royal North Shore Hospital University of Sydney, Sydney, NSW, Australia

Christophe Maes Manager Innovatie AZ Alma, Belgium

Michel Magnin INSERM, Neurological Hospital, Lyon, France

Bryan Mahony Department of Anesthesiology, Ohio State University Medical Center, Mount Sinai Hospital, New York, NY, USA

Steven F. Maier Department of Psychology and Neuroscience, and The Center for Neuroscience, University of Colorado at Boulder, Boulder, CO, USA

Thorsten J. Maier Institute of Pharmaceutical Chemistry, Goethe University, Frankfurt, Germany

Chris J. Main Keele University, Keele, Staffordshire, UK

Marzia Malcangio Wolfson Centre for Age Related Diseases, King's College London, London, UK

Paolo L. Manfredi Department of Neurology, Memorial Sloan-Kettering Cancer Center, New York, USA

Kaisa Mannerkorpi Department of Rheumatology and Inflammation Research, Institute of Medicine, University of Gothenburg, Göteborg, Sweden

Barton H. Manning Amgen Inc., Thousand Oaks, CA, USA

Patrick W. Mantyh Department of Pharmacology, The University of Arizona, AZ, USA

Jianren Mao Department of Anesthesia and Critical Care, Harvard Medical School, Boston, MA, USA

Serge Marchand Faculty of Medicine, University of Sherbrooke, Sherbrooke, QC, Canada

Paolo Marchettini Pain Medicine Center, Scientific Institute San Raffaele, Milan, Italy

Sarah R. Martin Psychology, Georgia State University, Atlanta, GA, USA

Peggy Mason Department of Neurobiology, Pharmacology and Physiology, University of Chicago, Chicago, IL, USA

Scott Masters Caloundra Spinal and Sports Medicine Centre, Caloundra, QLD, Australia

Laurence E. Mather Department of Anaesthesia and Pain Management, University of Sydney at Royal North Shore Hospital, St. Leonards, Sydney, NSW, Australia

Bill McCarberg Pain Services, Kaiser Permanente, San Diego, USA

Lance McCracken Health Psychology/Psychology Department, King's College London, London, UK

Kathleen McGinn Anesthesiology & Pain Medicine, Seattle Children's Hospital, Seattle, WA, USA

Patricia A. McGrath The Hospital for Sick Children Pain Research Center, Toronto, ON, Canada

Patrick McGrath Departments of Psychology, Pediatrics and Psychiatry, Dalhousie University, Canada

Greg McIntosh Canadian Back Institute, Toronto, ON, Canada

Peter McIntyre Department of Pharmacology, University of Melbourne, Melbourne, Australia

Elspeth M. McLachlan Neuroscience Research Australia, University of New South Wales, Randwick, NSW, Australia

University of Glasgow, Glasgow, UK

Peter A. McNaughton Department of Pharmacology, University of Cambridge, Cambridge, UK

J. Mark Melhorn Section of Orthopaedics, Department of Surgery, University of Kansas School of Medicine, Wichita, KS, USA

Ronald Melzack Department of Psychology, McGill University, Montreal, QC, Canada

Lorne M. Mendell Department of Neurobiology and Behavior, Stony Brook University, Stony Brook, NY, USA

George Mendelson Department of Psychological Medicine, Monash University, Caulfield, VIC, Australia

Siegfried Mense Neuroanatomy, Heidelberg University, Medical Faculty Mannheim, CBTM, Mannheim, Germany

Daniel Menétrey Biomédicale, Paris, France

Sebastiano Mercadante Pain Relief & Palliative Care, La Maddalena Cancer Center University of Palermo, Palermo, Italy

Susan Mercer School of Biomedical Sciences, The University of Queensland, Brisbane, Australia

Karl Messlinger University of Erlangen-Nürnberg, Erlangen, Germany

Richard A. Meyer Department of Neurosurgergy, School of Medicine Johns Hopkins University, Baltimore, MD, USA

Walter J. Meyer III Shriners Hospitals for Children, Shriners Burns Hospital, Galveston, TX, USA

Department of Psychiatry, University of Texas Medical Branch, Galveston, TX, USA

Björn A. Meyerson Department of Clinical Neuroscience, Section of Neurosurgery, Karolinska Institute/University Hospital, Stockholm, Sweden

Martin Michaelis Sanofi-Aventis Deutschland GmbH, Frankurt am Main, Germany

Joseph Mignogna Michael E. DeBakey VA Medical Center, Houston VA HSR&D Center of Excellence (152), Houston, TX, USA

Judith A. Mikacich Department of Obstetrics and Gynecology, University of California Los Angeles Medical Center, Los Angeles, CA, USA

Kensaku Miki Department of Integrative Physiology, National Institute for Physiological Sciences, Okazaki, Japan

Vjekoslav Miletic School of Medicine and Public Health, University of Wisconsin, Madison, WI, USA

Scott R. Miller Memorial Sloan-Kettering Cancer Center, New York, USA

Erin D. Milligan Department of Neurosciences, University of New Mexico, Albuquerque, NM, USA

Michael S. Minett Molecular Nociception Group, Wolfson Institute for Biomedical Research, University College London, London, WC, UK

J. Mocco Department of Neurological Surgery, Neurological Institute Columbia University, New York, NY, USA

Jeffrey S. Mogil Department of Psychology, McGill University, Montreal, QC, Canada

Harvey Moldofsky Sleep Disorders Clinic of the Centre for Sleep and Chronobiology, Toronto, ON, Canada

Robert M. Moldwin The Arthur Smith Institute for Urology, North Shore-LIJ Healthcare System, Hofstra North Shore LIJ School of Medicine, New Hyde Park, NY, USA

Derek C. Molliver Department of Neurobiology, University of Pittsburgh School of Medicine, Pittsburgh, PA, USA

Robert D. Mootz Washington State Department of Labor and Industries, Olympia, WA, USA

Laura Mordillo-Mateos FENNSI Group, Hospital Nacional de Parapléjicos, Toledo, Spain

Anne Morel Laboratory for Functional Neurosurgery, Neurosurgical Clinic University Hospital Zürich, Zürich, Switzerland

Anne Morinville Montreal Neurological Institute, McGill University, Montreal, QC, Canada

Stephen Morley Academic Unit of Psychiatry and Behavioral Sciences, University of Leeds, Leeds, UK

David B. Morris University of Virginia, Charlottesville, VA, USA

G. Lorimer Moseley Sansom Institute for Health Research, University of South Australia Neuroscience Research Australia, Adelaide, Australia

Joachim Mossner Medical Clinic and Policlinic II, University Clinical Center Leipzig AöR, Leipzig, Germany

Eric A. Moulton P.A.I.N. Group, Anesthesiology, Boston Children's Hospital, Harvard Medical School, Waltham, MA, USA

Francis Githae Muriithi Department of Obstetrics and Gynaecology, Aga Khan University, Nairobi, Kenya

Anne Z. Murphy Neuroscience Institute, Georgia State University, Atlanta, GA, USA

Paul M. Murphy Department of Anaesthesia and Pain Management, Royal North Shore Hospital, St. Leonards, NSW, Australia

Alison Murray Foothills Medical Center, Calgary, AB, Canada

H. S. Na Neuroscience Research Institute and Department of Physiology, Korea University, College of Medicine, Seoul, Republic of Korea

Daisuke Naka Department of Integrative Physiology, National Institute for Physiological Sciences, Okazaki, Japan

Yoshio Nakamura Pain Research Center, Department of Anesthesiology University of Utah School of Medicine, Salt Lake City, UT, USA

Barbara Namer Department of Physiology & Pathophysiology, FAU, Erlangen/Nuremberg, Bavaria, Germany

Matti V. O. Närhi Department of Biomedicine/Physiology, Department of Dentistry, University of Eastern Finland, Institute of Medicine, Kuopio, Finland

H. J. W. Nauta University of Texas Medical Branch, Galveston, TX, USA

Lucia Negri Department of Physiology and Pharmacology, Sapienza University of Rome, Rome, Italy

Gary Nesbit Department of Radiology, Oregon Health and Science University, Portland, OR, USA

Timothy Ness Department of Anesthesiology, University of Alabama at Birmingham, Birmingham, AL, USA

Volker Neugebauer Department of Neuroscience & Cell Biology, University of Texas Medical Branch, Galveston, TX, USA

Lawrence C. Newman Albert Einstein College of Medicine, New York, NY, USA

Michael K. Nicholas Royal North Shore Hospital, University of Sydney, St. Leonards, NSW, Australia

Ellen Niederberger Pharmazentrum frankfurt, Institute of Clinical Pharmacology, Goethe-University Hospital, Frankfurt, Germany

Lone Nikolajsen Department of Anaesthesiology, Danish Pain Research Center Aarhus University Hospital, Aarhus, Denmark

Katie Nixdorf Comprehensive Pain Center, Oregon Health and Science University, Portland, OR, USA

Donald R. Nixdorf Department of Diagnostic and Biological Sciences, School of Dentistry University of Minnesota, Minneapolis, MN, USA

Carl E. Noe University of Texas Southwestern Medical Center, Lubbock, TX, USA

Koichi Noguchi Department of Anatomy and Neuroscience, Hyogo College of Medicine, Hyogo, Japan

Richard B. North The Johns Hopkins University, Baltimore, MD, USA

Turo J. Nurmikko Department of Neurological Science, University of Liverpool, and Pain Research Institute, The Walton Centre NHS Foundation Trust, Liverpool, UK

Anne Louise Oaklander Nerve Injury Unit, Departments of Neurology and Pathology, Massachusetts General Hospital Harvard Medical School, Boston, MA, USA

Koichi Obata Department of Anatomy and Neuroscience, Hyogo College of Medicine, Hyogo, Japan

Tim F. Oberlander Division of Developmental Pediatrics, University of British Columbia, Vancouver, BC, Canada

Bruno Oertel Institute for Clinical Pharmacology, Pharmaceutical Center Frankfurt, Johann Wolfgang Goethe University, Frankfurt, Germany

Andrea C. Ogbonna Wolfson Centre for Age Related Diseases, King's College London, London, UK

Alfred T. Ogden Department of Neurological Surgery, Neurological Institute Columbia University, New York, NY, USA

S. Ohara Department of Neurosurgery, Johns Hopkins Hospital, Johns Hopkins Medical Institutions, Johns Hopkins University, Baltimore, MD, USA

Peter T. Ohara Department of Anatomy, University of California San Francisco, San Francisco, CA, USA

Akiko Okifuji Department of Anesthesiology, University of Utah, Salt Lake City, UT, USA

Jean-Louis Oliveras Inserm, Paris, France

Antonio Oliviero FENNSI Group, National Hospital of Parapléjicos, Toledo, Spain

Sean O'Mahony Palliative medicine, Rush University Medical Center, Chicago, IL, USA

Peregrine B. Osborne Dept Anatomy & Neuroscience, University of Melbourne, Melbourne, VIC, Australia

Michael H. Ossipov Department of Pharmacology, University of Arizona College of Medicine, Tucson, AZ, USA

Tonya M. Palermo Seattle Children's Hospital Research Institute, CW8-6 - Child Health, Behavior and Development, University of Washington, Seattle, WA, USA

Shreela Palit Department of Psychology, The University of Tulsa, Tulsa, OK, USA

Juan A. Pareja Department of Neurology, Fundación Hospital Alcorcón, Madrid, Spain

Steven D. Passik Psychiatry and Anesthesiology, Vanderbilt University, Nashville, TN, USA

Gavril W. Pasternak Laboratory of Molecular Neuropharmacology, Memorial Sloan-Kettering Cancer Center, New York, USA

S. H. Patel Department of Neurosurgery, Johns Hopkins Hospital, Baltimore, MD, USA

Gavin Pattullo Department of Anesthesia and Pain Management, Royal North Shore Hospital, University of Sydney, St. Leonards, NSW, Australia

Kevin Pauza Tyler Spine and Joint Hospital, Tyler, TX, USA

Eric M. Pearlman Pediatrics, Mercer University School of Medicine, The Children's Hospital at Memorial University Medical Center, Savannah, USA

Elvira De la Pena Instituto de Neurociencias, Universidad Miguel Hernández-CSIC, Alicante, Spain

Yuan Bo Peng Department of Psychology, University of Texas at Arlington, Arlington, TX, USA

Jose Pereira Foothills Medical Center, Calgary, AB, Canada

Edward Perl Department of Cell and Molecular Physiology, School of Medicine University of North Carolina, Chapel Hill, NC, USA

Tiffany Grace Perry Center for Spine Health, The Cleveland Clinic Foundation, Cleveland, OH, USA

Marie Pertin Pain Center, University Hospital Center and University of Lausanne, Lausanne, Switzerland

Antti Pertovaara Department of Physiology, Institute of Biomedicine University of Helsinki, Helsinki, Finland

Peter Peter Department of Neurobiology and Anatomy, University of Rochester Medical Center, Rochester, NY, USA

Lisa M. Peters Department of Anesthesiology and Pain Medicine, Seattle Children's Hospital, Seattle, WA, USA

Pramit Phal Department of Radiology, Oregon Health and Science University, Portland, OR, USA

Rebecca Pillai Riddell Department of Psychology, York University, Toronto, ON, Canada

Department of Psychiatry, Hospital for Sick Children, Toronto, ON, Canada

Issy Pilowsky University of Adelaide, Adelaide, Australia

Leon Plaghki Institute of Neuroscience (IoNS), Université Catholique de Louvain, Brussels, Belgium

Markus Ploner Department of Neurology, Technische Universität München, Munich, Germany

Esther M. Pogatzki-Zahn Department of Anesthesiology and Intensive Care Medicine, University Muenster, Muenster, Germany

Michael Polydefkis Department of Neurology, Johns Hopkins University School of Medicine, Baltimore, MD, USA

James D. Pomonis Algos Preclinical Services, Roseville, MN, USA

Dieter Pongratz Friedrich Baur Institute, Ludwig Maximilians University, Munich, Germany

Hayley Poole Hadley House, Leicester General Hospital, Leicester, UK

Frank Porreca Department of Pharmacology, University of Arizona Health Sciences Center, Tucson, AZ, USA

Russell K. Portenoy Department of Pain Medicine and Palliative Care, Beth Israel Medical Center, New York, NY, USA

Ian Power Critical Care and Pain Medicine, The University of Edinburgh, Edinburgh, UK

Donald D. Price Oral Surgery and Neuroscience, McKnight Brain Institute University of Florida, Gainesville, FL, USA

Daryl Pullman Medical Ethics, Memorial University of Newfoundland, St. John's, NL, Canada

Yunhai Qiu Department of Integrative Physiology, National Institute for Physiological Sciences, Okazaki, Japan

Michael Quittan Department of Physical Medicine and Rehabilitation, Kaiser-Franz-Joseph Hospital, Vienna, Austria

Raymond M. Quock Department of Pharmaceutical Sciences, Washington State University College of Pharmacy, Pullman, WA, USA

Jennifer Rabbitts Department of Anesthesiology and Pain Medicine, Seattle Children's Hospital, University of Washington School of Medicine, Seattle, WA, USA

Nicole Racine Psychology, York University, Toronto, ON, Canada

Gabor B. Racz Texas Tech University Health Sciences Center, Lubbock, TX, USA

Lukas Radbruch Department of Palliative Medicine, University of Bonn, Bonn, Germany

Robert B. Raffa Department of Pharmaceutical Sciences, Temple University School of Pharmacy, Philadelphia, PA, USA

Vasudeva Raghavendra Algos Therapeutics Inc., Saint Paul, MN, USA

Pierre Rainville Department of Stomatology, Faculty of Dental Medicine, University of Montreal, Montreal, QC, Canada

Srinivasa N. Raja Department of Anesthesiology and Critical Care Medicine, Johns Hopkins University, Baltimore, USA

Alan Randich University of Alabama, Birmingham, AL, USA

Andrea J. Rapkin Department of Obstetrics and Gynecology, University of California Los Angeles Medical Center, Los Angeles, CA, USA

Z. Harry Rappaport Department of Neurosurgery, Rabin Medical Center Tel-Aviv University, Petah Tiqva, Israel

Matthew N. Rasband Department of Molecular and Cellular Biology, Baylor College of Medicine, Farmington, Houston TX, USA

University of Connecticut Health Center, Farmington, CT, USA

Douglas Rasmusson Department of Physiology and Biophysics, Dalhousie University, Halifax, Canada

James P. Rathmell Department of Anesthesiology, University of Vermont College of Medicine, Burlington, USA

K. Ravishankar The Headache and Migraine Clinic, Jaslok Hospital and Research Centre Lilavati Hospital and Research Centre, Mumbai, India

Ke Ren Department of Neural and Pain Sciences, University of Maryland Dental School, Baltimore, MD, USA

Seth Resnick Silver Hill Hospital, New York University School of Medicine, USA

Cielito Reyes-Gibby The University of Texas MD Anderson Cancer Center, Houston, TX, USA

Jamie L. Rhudy Department of Psychology, The University of Tulsa, Tulsa, OK, USA

Alfredo Ribeiro-da-Silva Department of Pharmacology and Therapeutics, McGill University, Montreal, QC, Canada

Frank L. Rice Center for Neuropharmacology and Neuroscience, Albany Medical College, Albany, NY, USA

Matthias Ringkamp Department of Neurosurgergy, School of Medicine Johns Hopkins University, Baltimore, MD, USA

Pierre J. M. Rivière Ferring Research Institute, San Diego, CA, USA

Claude Robert Université Paris Descartes, INSERM UMR 894, France

Roger Robert Neurosurgery, Hotel Dieu, Nantes, CHU, France

James P. Robinson Department of Anesthesiology & Pain Medicine, University of Washington, Seattle, WA, USA

John Robinson Pain Management Centre, Burwood Hospital, Christchurch, New Zealand

Alfonso Romero-Sandoval Department of Pharmaceuticals and Administrative Sciences, Presbyterian College School of Pharmacy, Clinton, SC, USA

David Roselt Aberdovy Clinic, Bundaberg, QLD, Australia

Jackie Ross Early Pregnancy Unit, King's College Hospital Early Pregnancy Unit, London, UK

Shlomo Rotshenker Department of Medical Neurobiology, IMRIC, Faculty of Medicine Hebrew University of Jerusalem, Jerusalem, Israel

Paul C. Rousseau Medical University of South Carolina, Charleston, SC, USA

Ranjan Roy Department of Clinical Health Psychology, Faculty of Medicine, University of Manitoba, Winnipeg, MB, Canada

Carolina Roza Biología de Sistemas (Unidad Fisiología), Universidad de Alcalá, Alcalá de Henares, Madrid, Spain

Todd D. Rozen Department of Neurology, Geisinger Health System, Wilkes-Barre, PA, USA

Roman Rukwied Department of Anaesthesiology and Intensive Care Medicine, Faculty of Clinical Medicine Mannheim University of Heidelberg, Heidelberg, Germany

Irwin Jon Russell Division of Clinical Immunology and Rheumatology, Department of Medicine, The University of Texas Health Science Center, San Antonio, TX, USA

Nayef E. Saadé Anatomy, Cell Biology and Physiology, Faculty of Medicine, American University of Beirut, Beirut, Lebanon

Mary Ann C. Sabino Oral and Maxillofacial Surgery, Hennepin County Medical Center University of Minnesota, Minneapolis, MN, USA

Thomas E. Salt Institute of Ophthalmology, University College London, London, UK

A. Samdani Department of Neurosurgery, Johns Hopkins Medical Institutions, Baltimore, MD, USA

Yasmin Sana King's College Hospital, London, UK

Margarita Sanchez-del-Rio Neurology Department, Hospital Ruber Internacional, Madrid, Spain

Jürgen Sandkühler Department of Neurophysiology, Center for Brain Research Medical University of Vienna, Vienna, Austria

Peter S. Sándor RehaClinic, Cantonal Hospital Baden, Baden, Switzerland

Johannes Sarnthein Functional Neurosurgery, Neurosurgical Clinic University Hospital, Zürich, Switzerland

Susanne K. Sauer Department of Physiology and Pathophysiology, University of Erlangen, Erlangen, Germany

Steven Savvas Biomedical Division, National Ageing Research Institute, Melbourne, VIC, Australia

Jana Sawynok Department of Pharmacology, Dalhousie University, Halifax, Canada

John W. Scadding The National Hospital for Neurology and Neurosurgery, London, UK

Frederieke Geraldine Schaafsma Department of Public and Occupational Health, EMGO+ Institute/VU Medical Center, Amsterdam, MB, The Netherlands

Hans-Georg Schaible Department of Physiology - Neurophysiology, Friedrich-Schiller-University of Jena, Jena, Germany

Steven S. Scherer Department of Neurology, The Perelman School of Medicine at the University of Pennsylvania, Philadelphia, PA, USA

Michael Schäfer Department of Anesthesiology and Intensive Care Medicine, Charité University, Berlin, Germany

Martin Schmelz Department of Anesthesiology in Mannheim, University of Heidelberg, Mannheim, Germany

Robert F. Schmidt Physiologisches Institut der Universität Würzburg, Würzburg, Germany

Suzan Schneeweiss The Hospital for Sick Children, Toronto, ON, Canada

Frank Schneider Department of Psychiatry and Psychotherapy, RWTH Aachen University, Aachen, Germany

Alfons Schnitzler Department of Neurology, Heinrich Heine University, Düsseldorf, Germany

Jean Schoenen Department of Neurology, University of Liège, CHR Citadelle, Belgium

Department of Neuroanatomy, University of Liège, CHR Citadelle, Belgium

Jens Schouenborg Neuronano Research Center, Lund University, Lund, Sweden

Stephan A. Schug School of Medicine and Pharmacology, Royal Perth Hospital, Faculty of Medicine, Dentistry & Health Sciences, The University of Western Australia, UWA Anaesthesiology, Perth, Australia

Patrick Schuss Department of Neurosurgery, University-Hospital Bonn, Bonn, Germany

Anthony C. Schwarzer Faculty of Medicine and Health Sciences, The University of Newcastle, Newcastle, Australia

Whitney Scott Department of Psychology, McGill University, Montreal, QC, Canada

Laura C. Seidman Pediatric Pain Program, Department of Pediatrics David Geffen School of Medicine at UCLA, Los Angeles, CA, USA

Ze'ev Seltzer Faculty of Dentistry, University of Toronto Centre for the Study of Pain, Toronto, ON, Canada

Peter Selwyn Albert Einstein College of Medicine, Bronx, NY, USA

Patrick B. Senatus Department of Neurological Surgery, Neurological Institute Columbia University, New York, NY, USA

Jyoti N. Sengupta Division of Gastroenterology and Hepatology and Pediatric Gastroenterology, Medical College of Wisconsin, Milwaukee, WI, USA

Mariano Serrao Department of Neurology and Otorinolaringoiatry, University of Rome La Sapienza, Rome, Italy

Barry J. Sessle Faculty of Dentistry, University of Toronto, Toronto, ON, Canada

Faculty of Dentistry, University of Toronto, Toronto, ON, Canada

Virginia S. Seybold Department of Neuroscience, University of Minnesota, Minneapolis, MN, USA

T. Shah Royal Perth Hospital and University of Western Australia, Crawley, WA, Australia

Yair Sharav Department of Oral Medicine, School of Dental Medicine, Hebrew University-Hadassah, Jerusalem, Israel

S. Murray Sherman Department of Neurobiology, University of Chicago, Chicago, IL, USA

Yoshio Shigenaga Department of Oral Anatomy and Neurobiology, Osaka University, Osaka, Japan

Susan Shin Neurology, Mount Sinai School of Medicine, New York, NY, USA

Edward A. Shipton Department of Anaesthesia, Christchurch School of Medicine and Health Sciences, University of Otago, Christchurch, New Zealand

Yoram Shir The Alan Edwards Pain Management Unit, McGill University Health Centre, Montreal, QC, Canada

Peter Shrager Department of Neurobiology and Anatomy, University of Rochester Medical Center, Rochester, NY, USA

Marc J. Shulman VA Connecticut Healthcare System, Yale University, New Haven, CT, USA

David M. Sibell Comprehensive Pain Center/Spine Center, Oregon Health and Science University, Portland, OR, USA

Philip J. Siddall Department of Pain Management, Greenwich Hospital, University of Sydney, Sydney, NSW, Australia

Stephen D. Silberstein Jefferson Medical College, Thomas Jefferson University and Jefferson Headache Center, Thomas Jefferson University Hospital, Philadelphia, PA, USA

Donald A. Simone Diagnostic and Biological Sciences, University of Minnesota, Minneapolis, MN, USA

David G. Simons Department of Rehabilitation Medicine, Emory University, Atlanta, GA, USA

David Simpson Clinical Neurophysiology Laboratories, Mount Sinai Medical Center, New York, NY, USA

Marc Sindou Department of Neurosurgery, Hopital Neurologique Pierre Wertheimer, University of Lyon 1, Lyon, France

Soeren H. Sindrup Department of Neurology, Odense University Hospital, Odense, Denmark

Bengt H. Sjölund Department of Community Medicine and Rehabilitation Medicine, Umea University, Umea, Sweden

Carsten Skarke Institute for Translational Medicine and Therapeutics (ITMAT), University of Pennsylvania Perelman School of Medicine, Philadelphia, PA, USA

Paul A. Sloan Department of Anesthesiology, University of Kentucky, Lexington, KY, USA

Kathleen A. Sluka Physical Therapy and Rehabilitation Science Graduate Program, University of Iowa, Iowa City, IA, USA

Seymour Solomon Montefiore Medical Center, Albert Einstein College of Medicine, Bronx, NY, USA

Claudia Sommer Department of Neurology, University of Würzburg, Würzburg, Germany

Linda S. Sorkin Occupational Therapy and Rehabilitation, Anesthesia Research Laboratories, University of California, San Diego, CA, USA

Ingrid Söderback Uppsala University, Uppsala, Sweden

Kari Sørensen Department of Pain Management and Research, Oslo University Hospital, Rikshospitalet, Oslo, Norway

Michael Spaeth Friedrich-Baur-Institute, University of Munich, Munich, Germany

Peregrin O. Spielholz Washington State Department of Labor and Industries, SHARP Program, Olympia, WA, USA

Peter S. Staats Division of Pain Medicine, The Johns Hopkins University, Baltimore, MD, USA

Brett R. Stacey Comprehensive Pain Center, Oregon Health and Science University, Portland, OR, USA

Wendy M. Stein San Diego Hospice and Palliative Care, UCSD, San Diego, CA, USA

Christoph Stein Department of Anesthesiology and Intensive Care Medicine, Freie Universitaet Berlin Charité Campus Benjamin Franklin, Berlin, Germany

Michael P. Steinmetz Department of Neurosurgery, MetroHealth Medical Center, Case Western Reserve University School of Medicine, Cleveland, OH, USA

Jair Stern Functional Neurosurgery, Neurosurgical Clinic University Hospital, Zürich, Switzerland

Bonnie J. Stevens University of Toronto and The Hospital for Sick Children, Toronto, ON, Canada

Carl-Olav Stiller Division of Clinical Pharmacology and Department of Medicine, Karolinska University Hospital, Stockholm, Sweden

Jennifer N. Stinson The Hospital for Sick Children, Toronto, ON, Canada

Henry Stockbridge University of Washington School of Public Health and Community Medicine, Seattle, WA, USA

Christian S. Stohler Baltimore College of Dental Surgery, University of Maryland, Baltimore, MD, USA

Laura Stone Faculty of Dentistry, Alan Edwards Centre for Research on Pain, Montreal, QC, Canada

William Stones Department of Obstetrics and Gynaecology, Aga Khan University, Nairobi, Kenya

Andreas Straube Department of Neurology, Ludwig Maximilians University, Munich, Germany

Jenny Strong University of Queensland, Brisbane, QLD, Australia

Audun Stubhaug Department of Pain Management and Research, Oslo University Hospital, Rikshospitalet and University of Oslo, Oslo, Norway

Gerold Stucki Swiss Paraplegic Research, Nottwil, Switzerland

Department of Health Sciences and Health Policy, University of Lucerne, Lucerne, Switzerland

ICF Research Branch in cooperation with the World Health Organization (WHO) Collaborating Centre for the Family of International Classifications in Germany, Nottwil, Switzerland

Cheryl L. Stucky Department of Cell Biology Neurobiology and Anatomy, Medical College of Wisconsin, Milwaukee, WI, USA

Mathias Sturzenegger Department of Neurology, University Hospital, Inselspital, University of Bern, Bern, Switzerland

Yasuo Sugiura School of Human Care Study, Nagoya University of Arts and Sciences, Nissin City, Aichi, Japan

Mark D. Sullivan Psychiatry and Behavioral Sciences, University of Washington, Seattle, WA, USA

Michael J. L. Sullivan Department of Psychology, McGill University, Montreal, QC, Canada

Rie Suzuki Department of Pharmacology, University College London, London, UK

Kristina B. Svendsen Danish Pain Research Center, Aarhus University Hospital, Aarhus, Denmark

Peter Svensson Clinical Oral Physiology, Aarhus University, Aarhus, Denmark

Peter Szucs Spinal Neuronal Networks, Instituto de Biologia Molecular e Celular - IBMC, Porto, Portugal

A. Taghva Department of Neurosurgery, Johns Hopkins Hospital, Baltimore, MD, USA

Kimberson Cochien Tanco Department of Palliative Care and Rehabilitation Medicine, The University of Texas M. D. Anderson Cancer Center, Houston, TX, USA

Kersi Taraporewalla Royal Brisbane and Womens' Hospital, University of Queensland, Brisbane, QLD, Australia

Irmgard Tegeder Pharmazentrum frankfurt, Institute of Clinical Pharmacology, Goethe-University Hospital, Frankfurt, Germany

Astrid Juhl Terkelsen Danish Pain Research Center, Department of Neurology, Aarhus University Hospital, Aarhus, Denmark

Gregory W. Terman Department of Anesthesiology and the Graduate Program in Neurobiology and Behavior, University of Washington, Seattle, WA, USA

Alison Thomas Lismore, NSW, Australia

Benjamin Thomas Chelsea & Westminster Hospital Foundation Trust, London, UK

Jacob Thomas School of Medical Sciences, University of Adelaide, Adelaide, SA, Australia

Miles Thompson Goldsmiths, University of London, London, UK

Beverly E. Thorn Department of Psychology, University of Alabama, Tuscaloosa, AL, USA

Cindy L. Thurston-Stanfield Department of Biomedical Sciences, University of South Alabama, Mobile, AL, USA

Arne Tjølsen Department of Bomedicine, University of Bergen, Haukeland University Hospital, Bergen, Norway

Andrew J. Todd Institute of Neuroscience and Psychology, College of Medical, Veterinary and Life Sciences, University of Glasgow, Glasgow, UK

Makoto Tominaga Department of Cellular and Molecular Physiology, Mie University School of Medicine, Tsu Mie, Japan

Victor Tortorici Instituto Venezolano de Investigaciones Cientificas (IVIC), Caracas, Venezuela

Irene Tracey Pain Imaging Neuroscience (PaIN) Group, Department of Physiology, Anatomy and Genetics, Oxford University, Oxford, UK

Diep Tuan Tran Department of Integrative Physiology, National Institute for Physiological Sciences, Okazaki, Japan

Rolf-Detlef Treede Institute of Physiology and Pathophysiology, Johannes Gutenberg University, Mainz, Germany

Jennie C. I. Tsao Pediatric Pain Program, Department of Pediatrics David Geffen School of Medicine at UCLA, Los Angeles, CA, USA

Lawrence Tsen Department of Anesthesiology, Perioperative and Pain Medicine, Brigham and Women's Hospital Harvard Medical School, Boston, MA, USA

Jeanna Tsenter Department of Physical Medicine and Rehabilitation, Hadassah University Hospital, Jerusalem, Israel

Maurits Van Tulder Institute for Research in Extramural Medicine, Free University Amsterdam, Amsterdam, The Netherlands

Eldon Tunks Hamilton Health Sciences Corporation, Hamilton, ON, Canada

Susan Tupper Pediatrics, University of Saskatchewan, Saskatoon, SK, Canada

Dennis C. Turk Department of Anesthesiology & Pain Research, University of Washington, Seattle, WA, USA

Judith A. Turner Department of Psychiatry and Behavioral Sciences, University of Washington, Seattle, WA, USA

Department of Rehabilitation Medicine, University of Washington, Seattle, WA, USA

Stephen P. Tyrer Royal Victoria Infirmary, University of Newcastle upon Tyne, Newcastle-on-Tyne, UK

Alexander Tzabazis Department of Anesthesia, Stanford University, Stanford, CA, USA

Kene Ugokwe Department of Neurosurgery, St Elizabeth Health Center, Youngstown, OH, USA

Lindsay S. Uman Department of Psychology, IWK Health Centre, Halifax, NS, Canada

Christiane Vahle-Hinz Department of Neurophysiology and Pathophysiology, University Medical Center Hamburg-Eppendorf, Hamburg, Germany

Muthukumar Vaidyaraman Division of Pain Medicine, The Johns Hopkins University, Baltimore, MD, USA

Todd W. Vanderah Department of Pharmacology, College of Medicine University of Arizona, Tucson, AZ, USA

Guy Vanderstraeten Department of Rehabilitation Sciences and Physiotherapy, Ghent University Hospital, Ghent, Belgium

Horacio Vanegas Instituto Venezolano de Investigaciones Cientificas (IVIC), Caracas, Venezuela

Jean-Jacques Vatine Outpatient and Research Division, Reuth Medical Center, Sackler Faculty of Medicine, Tel Aviv University, Tel Aviv, Israel

Leigh M. Vaughan Medical University of South Carolina, Charleston, SC, USA

Linda K. Vaughn Department of Biomedical Sciences, Marquette University, Milwaukee, WI, USA

Enrique Vazquez Instituto Venezolano de Investigaciones Cientificas (IVIC), Caracas, Venezuela

Louis Vera-Portocarrero Department of Pharmacology, College of Medicine University of Arizona, Tucson, AZ, USA

Paul Verrills Metropolitan Spinal Clinic, Prahran, VIC, Australia

Tine Vervoort Department of Experimental Clinical and Health Psychology, Ghent University, Ghent, Belgium

Félix Viana Instituto de Neurociencias, Universidad Miguel Hernández-CSIC, San Juan de Alicante, Alicante, Spain

Charles J. Vierck Jr. Department of Neuroscience and McKnight Brain Institute, University of Florida College of Medicine, Gainesville, FL, USA

Luis Villanueva INSERM UMR 894, Centre de Psychiatrie et Neurosciences, Paris, France

Marcelo Villar Faculty of Biomedical Sciences, Austral University, Buenos Aires, Argentina

Katy Vincent Nuffield Department of Obstetrics and Gynaecology, The Women's Centre, John Radcliffe Hospital, University of Oxford, Oxford, UK

David Vivian Metro Spinal Clinic, South Caulfield, VIC, Australia

Johan W. S. Vlaeyen Research Group Health Psychology, Catholic University of Leuven, Leuven, Belgium

Department of Clinical Psychological Science, Maastricht University, Maastricht, The Netherlands

Stéphanie Volders Research Group on Health Psychology, Leuven, Belgium

Ernest Volinn Pain Research Center, University of Utah, Salt Lake City, UT, USA

Lucy Vulchanova Department of Veterinary and Biomedical Sciences, University of Minnesota, St. Paul, MN, USA

Paul W. Wacnik Neuromodulation Research, Medtronic Inc., Minneapolis, MN, USA

Gordon Waddell University of Glasgow, Glasgow, UK

Gary A. Walco Department of Anesthesiology and Pain Medicine, University of Washington School of Medicine, Seattle Children's Hospital, Seattle, WA, USA

Suellen M. Walker University of Sydney Pain Management and Research Centre, Royal North Shore Hospital, NSW, Australia

Victoria C. J. Wallace Department of Anaesthetics, Pain Medicine and Intensive Care Chelsea and Westminster Hospital Campus, London, UK

Xiaohong Wang Department of Integrative Physiology, National Institute for Physiological Sciences, Okazaki, Japan

Gunnar Wasner Department of Neurological Pain Research and Therapy, Neurological Clinic, Kiel, Germany

Shoko Watanabe Department of Integrative Physiology, National Institute for Physiological Sciences, Okazaki, Japan

Linda R. Watkins Department of Psychology and Neuroscience, and The Center for Neuroscience, University of Colorado at Boulder, Boulder, CO, USA

Peter N. Watson Department of Medicine, University of Toronto, Toronto, ON, Canada

James Watt North Shore Hospital, Auckland, New Zealand

Stephen G. Waxman Department of Neurology, Yale School of Medicine, New Haven, CT, USA

Inge Wegner Julius Center, Department of Epidemiology, University Medical Center Utrecht, Utrecht, The Netherlands

Feng Wei Department of Neural and Pain Sciences, University of Maryland Dental School, Baltimore, MD, USA

Christian Weidner Department of Physiology and Experimental Pathophysiology, University Erlangen, Erlangen, Germany

P. T. M. Weijenborg Department of Gynecology, Leiden University Medical Center, Leiden, The Netherlands

Robin Weir School of Nursing, Faculty of Health Sciences, McMaster University, Hamilton, ON, Canada

Nirit Weiss Department of Neurosurgery, Mount Sinai Medical Center, New York, NY, USA

Ursula Wesselmann Departments of Neurology and Anesthesiology / Division of Pain Management, University of Alabama at Birmingham, Birmingham, AL, USA

Dagmar Westerling Department of Anesthesiology, Central Hospital of Kristianstad, Kristianstad, Sweden

Karin N. Westlund Department of Physiology, University of Kentucky, Lexington, KY, USA

Thomas M. Wickizer Health Services Management and Policy, Ohio State University, Colombus, OH, USA

Eva Widerström-Noga The Miami Project to Cure Paralysis, Department of Neurological Surgery, University of Miami Miller, School of Medicine, Miami, FL, USA

Julie Wieseler Department of Psychology and Neuroscience, and The Center for Neuroscience, University of Colorado at Boulder, Boulder, CO, USA

Zsuzsanna Wiesenfeld-Hallin Karolinska Institute, Stockholm, Sweden

Ann Wigham McCoull Clinic, Prudhoe Hospital, Northumberland, UK

George L. Wilcox Department of Neuroscience, Pharmacology and Dermatology, University of Minnesota Medical School, Minneapolis, MN, USA

Oliver H. G. Wilder-Smith Pain Centre, Department of Anaesthesiology, Radboud University Nijmegen Medical Centre, Nijmegen, The Netherlands

Kjersti Wilhelmsen Section for Physiotherapy Science, Department of Public Health and Primary Health, Faculty of Medicine, University of Bergen, Bergen, Norway

Victor Wilk Brighton Spinal Group, Brighton, VIC, Australia

W. A. Williams Royal Perth Hospital and University of Western Australia, Perth, WA, Australia

Sara E. Williams Division of Pediatric Gastroenterology, Medical College of Wisconsin, Milwaukee, WI, USA

William D. Willis Department of Neuroscience and Cell Biology, University of Texas Medical Branch, Galveston, TX, USA

John B. Winer Department of Neurology, Birmingham Muscle and Nerve Centre Queen Elizabeth Hospital, Birmingham, UK

Christopher J. Winfree Department of Neurological Surgery, Neurological Institute Columbia University, New York, NY, USA

Janet Winter Novartis Institute for Medical Science, London, UK

Alain Woda Faculté de Chirurgie Dentaire, University Clermont-Ferrand, Clermont-Ferrand, France

John N. Wood Molecular Nociception Group, Wolfson Institute for Biomedical Research, University College London, London, WC, UK

Lee Woodson Shriners Hospitals for Children, Shriners Burns Hospital, Galveston, TX, USA

Department of Anesthesia, University of Texas Medical Branch, Galveston, TX, USA

Anava A. Wren Pain Prevention and Treatment Research, Department of Psychiatry and Behavioral Sciences, Duke University Medical Center, Durham, NC, USA

Xiao-Jun Xu Department of Physiology and Pharmacology, Section of Integrative Pain Rsearch Karolinska Institutet, Stockholm, Sweden

Hiroshi Yamasaki Department of Integrative Physiology, National Institute for Physiological Sciences, Okazaki, Japan

Michael Yelland School of Medicine, Logan campus, Griffith University, Logan, Australia

Sriram Yennurajalingam Department of Palliative Care and Rehabilitation Medicine, The University of Texas M. D. Anderson Cancer Center, Houston, TX, USA

David C. Yeomans Department of Anesthesia, Stanford University, Stanford, CA, USA

Robert P. Yezierski Comprehensive Center for Pain Research and Department of Neuroscience, The McKnight Brain Institute University of Florida, Gainesville, FL, USA

Way Yin Department of Anesthesiology, University of Washington, Seattle, WA, USA

Atsushi Yoshida Department of Oral Anatomy and Neurobiology, Osaka University, Osaka, Japan

Joanna M. Zakrzewska Eastman Dental Hospital UCLH NHS Foundation Trust, London, UK

Jenna E. Zauk St. George's University, School of Medicine, Grenada, West Indies

Hanns Ulrich Zeilhofer Institute for Pharmacology and Toxicology, University of Zurich, and Swiss Federal Institute of Technology (ETH) Zurich, Zürich, Switzerland

Lonnie K. Zeltzer Pediatric Pain Program, Department of Pediatrics David Geffen School of Medicine at UCLA, Los Angeles, CA, USA

Giovambattista Zeppetella St. Clare Hospice, Hastingwood, UK

Xijing Zhang Department of Neuroscience, University of Minnesota, Minneapolis, MN, USA

Min Zhuo Department of Physiology, University of Toronto, Toronto, ON, Canada

C. Zöllner Department of Anesthesiology and Intensive Care Medicine, Charité University, Berlin, Germany

Krina Zondervan Genetic and Genomic Epidemiology Unit, Nuffield Dept of Obstetrics and Gynaecology, Wellcome Trust Centre for Human Genetics, Oxford, UK

Peter Zygmunt Department of Laboratory Medicine, Lund University, Division of Clinical Chemistry and Pharmacology, Lund, Sweden

Zbigniew Zylicz Hildegard Hospiz, Basel Switzerland, Basel, Switzerland

David C. Yeomans Department of Anesthesia, Stanford University, Stanford, CA, USA

Robert P. Yezierski Comprehensive Center for Pain Research and Department of Neuroscience, The McKnight Brain Institute, University of Florida, Gainesville, FL, USA

Way Yin Department of Anesthesiology, University of Washington, Seattle, WA, USA

Atsushi Yoshida Department of Oral Anatomy and Neurobiology, Osaka University, Osaka, Japan

Joanna M. Zakrzewska Eastman Dental Hospital, UCLH NHS Foundation Trust, London, UK

Senna L. Zaak St. George's University, School of Medicine, Grenada, West Indies

Hanns Ulrich Zeilhofer Institute of Pharmacology and Toxicology, University of Zurich, and Swiss Federal Institute of Technology (ETH) Zürich, Zürich, Switzerland

Lonnie K. Zeltzer Pediatric Pain Program, Department of Pediatrics, David Geffen School of Medicine at UCLA, Los Angeles, CA, USA

(Gwenabaihisn Zuppella) St. Clare Hospice, Hastingwood, UK

Yiting Zhang Department of Neuroscience, University of Minnesota, Minneapolis, MN, USA

Min Zhuo Department of Physiology, University of Toronto, Toronto, ON, Canada

C. Zöllner Department of Anaesthesiology and Intensive Care Medicine, Charité University, Berlin, Germany

Katja Zadravec Gamete and Gynecology, Department of Obstetrics and Gynaecology, Wellcome Trust Centre for Human Genetics, Oxford, UK

Peter Zygmunt Department of Pharmacology, Skåne University Hospital, Division of Clinical Chemistry and Pharmacology, Lund, Sweden

Zbigniew Zylicz Hildegard Hospital, Basel, Switzerland, Basel, Switzerland

G

GABA

▶ Gamma(γ)-Aminobutyric Acid

GABA and Glycine in Spinal Nociceptive Processing

Hanns Ulrich Zeilhofer[1] and William D. Willis[2]
[1]Institute for Pharmacology and Toxicology, University of Zurich, and Swiss Federal Institute of Technology (ETH) Zurich, Zürich, Switzerland
[2]Department of Neuroscience and Cell Biology, University of Texas Medical Branch, Galveston, TX, USA

Synonyms

Inhibitory synaptic transmission

Definition

Fast inhibitory neurotransmission in the mammalian CNS is mediated by the amino acids γ-aminobutyric acid (GABA) and glycine. GABA and glycine open ligand-gated anion channels, designated ▶ GABA$_A$ receptors, and inhibitory (▶ strychnine-sensitive) glycine receptors, respectively. GABA, but not glycine, acts also on G-protein-coupled GABA$_B$ receptors, which inhibit neuronal excitability through activation of K$^+$ (potassium) channels and synaptic transmitter release through inhibition of presynaptic Ca^{2+} channels. In most parts of the brain, GABA is the only fast inhibitory neurotransmitter, while GABA and glycine act in concert in the spinal cord, the brain stem, and selected areas of the brain, such as the cerebellum.

Introduction

Cellular Functions

Upon activation, GABA$_A$ and strychnine-sensitive glycine receptors permit the permeation of chloride (and to a lesser extent of bicarbonate) ions through the plasma membrane. In most parts of the CNS, this increase in anion conductance inhibits neuronal activity through two major mechanisms. First, in most neurons chloride flux is inwardly directed at the physiological resting potential and hyperpolarizes neurons. Second, activation of GABA$_A$ and glycine receptors causes a "shunting" conductance and thereby impairs the propagation of excitatory postsynaptic currents along the dendrite and of action potentials along axons.

Molecular Structure and Pharmacology
GABA Receptors

GABA$_A$ receptors and strychnine-sensitive glycine receptors are pentameric protein complexes and belong to the family Cys-loop ion channels

G.F. Gebhart, R.F. Schmidt (eds.), *Encyclopedia of Pain*, DOI 10.1007/978-3-642-28753-4,
© Springer-Verlag Berlin Heidelberg 2013

GABA and Glycine in Spinal Nociceptive Processing, Fig. 1 Molecular structure of GABA$_A$ and glycine receptors. (**a**) GABA$_A$ and glycine receptors are Cys-loop ion channels with a disulfide bond between two cysteine residues (C) in the N terminus and with four transmembrane domains (1–4). (**b**) Chemical structure of GABA. (**c**) Five subunits form one functional GABA$_A$ receptor channel. (**d**) Dendrogram depicting the phylogenetic tree of GABA$_A$ receptor subunits. (**e**) Chemical structure of glycine. (**f**) Five subunits form one functional glycine receptor channel. (**g**) Dendrogram depicting the phylogenetic tree of glycine receptor subunits (modified from Zeilhofer et al., 2012a)

(Fig. 1a). Mammalian GABA$_A$ receptors (Fig. 1b–d) are assembled from a repertoire of 19 subunits ($\alpha1$–$\alpha6$, $\beta1$–$\beta3$, $\gamma1$–$\gamma3$, δ, ϵ, ϑ, π, $\rho1$–$\rho3$), each of which is encoded by a separate gene. The most prevalent form of the GABA$_A$ receptor in the brain is $\alpha1$-$\beta2$-$\gamma2$. $\alpha2$, $\alpha3$, and $\beta3$ receptors are more prevalent in the spinal cord. GABA$_A$ receptors containing $\alpha1$, $\alpha2$, $\alpha3$, or $\alpha5$ subunits and a $\gamma2$ subunit form benzodiazepine-sensitive GABA$_A$ receptors which are potentiated by classical benzodiazepines including ▶ diazepam, while $\alpha4$ and $\alpha6$ subunits cannot form benzodiazepine-insensitive GABA$_A$ receptors. The β subunits bind barbiturates and intravenous anesthetics including ▶ propofol and ▶ etomidate, which also facilitate the activation of GABA$_A$ receptors. ▶ Bicuculline-insensitive ionotropic

GABA receptors have been termed GABA$_C$ receptors. They contain ρ subunits, which can form functional homomeric receptor channels.

GABA$_B$ receptors are G-protein-coupled (or 7 transmembrane) receptors. They are obligatory heterodimers composed of a GABA$_{B1}$ and a GABA$_{B2}$ subunit. They couple to pertussis toxin-sensitive (inhibitory) G-proteins, activate G-protein-coupled K$^+$ channels, inhibit Ca^{2+} channels, and reduce the formation of c-AMP.

Glycine Receptors
Glycine receptors (GlyRs) show much less diversity than GABA$_A$ receptors. Four α subunits, $\alpha1$ through $\alpha4$, are known and one β subunit, each encoded by a separate gene (Fig. 1 e–g). The α subunits bind glycine and are capable of

forming functional homomeric or heteromeric ion channels, while the β subunits alone do not form functional glycine receptors. They contribute to the formation of the glycine binding site and confer postsynaptic clustering through an interaction with the intracellular protein gephyrin. Heteropentameric channels composed of GlyRα1 and GlyRβ subunits constitute the most prevalent adult strychnine-sensitive glycine receptor isoform. GlyRα2 subunits are mainly expressed during early development where they form homomeric receptors. In some tissues, they remain however expressed into adulthood. GlyRα3 is another adult GlyR subunit with a rather restricted expression pattern. Due to the presence of a premature stop codon, GlyRα4 is a pseudogene in humans.

Although glycine is the major, most likely the only, endogenous activator of synaptic glycine receptors, extrasynaptic glycine receptors may also be activated endogenously by β-alanine and taurine.

Glycine serves a dual role in spinal neurotransmission. It binds not only to inhibitory glycine receptors but also to a strychnine-insensitive binding site at excitatory glutamate receptors of the N-methyl-D-aspartate (▶ NMDA) subtype. These receptors are primarily gated by glutamate but require glycine or d-serine as an obligatory coagonist. Increasing evidence indicates that the glycine binding site of NMDA receptors is not saturated under resting conditions, opening the possibility that spinal NMDA receptor activity is modulated by changes in extracellular glycine.

Synthesis, Storage, and Reuptake of GABA and Glycine

GABA is synthesized through two isoforms of ▶ glutamic acid decarboxylase (GAD-65, *gad2*, and GAD-67, *gad1*), which both exist in the dorsal horn (Fig. 2). Unlike GABA, glycine is a proteogenic amino acid and therefore ubiquitously present. In glycinergic neurons, it is however enriched through specific uptake from the extracellular space by the glycine transporter GlyT2. The transport of GABA and glycine into the presynaptic storage vesicles is mediated by the ▶ vesicular inhibitory amino acid transporter

(VIAAT or VGAT; *slc32a1*). VIAAT/VGAT has a rather low affinity for glycine, and the presence of GlyT2 is required to achieve glycine concentrations sufficiently high for glycine transport into presynaptic vesicles. Following synaptic release, glycine and GABA are removed from the synaptic cleft by specific membrane-associated transporters, which belong to the family of Na^+-Cl^--dependent neurotransmitter transporters. Two forms of ▶ glycine transporters (▶ GlyT-1, *slc6a9*, and ▶ GlyT-2, *slc6a5*) exist. GlyT2 is primarily expressed in glycinergic neurons and hence restricted mainly to the spinal cord, brain stem, and cerebellum. GlyT1 is found both in glial cells (astrocytes) and neurons and expressed more widely in the CNS. Data obtained in knockout mice suggest that GlyT1 mediates fast removal of glycine from the synaptic cleft, whereas GlyT2 enables the recycling of glycine in glycinergic neurons. Experiments with pharmacological inhibitors suggest less distinct functions and suggest that both transporters shape the time course of the postsynaptic signal. Pharmacological blockade of these transporters induces analgesia. Five types of ▶ GABA transporters (GAT1, GAT2, GAT3, BGT1, TAUT) have been identified. Their contribution to spinal nociceptive processing is less clear.

Characteristics

Functional Aspects of GABAergic and Glycinergic Pain Control in the Dorsal Horn

GABA and glycine reduce the excitability of dorsal horn neurons and thereby exert a negative control on the transmission of nociceptive signals to higher CNS areas (Sandkühler 2009; Todd 2010; Zeilhofer et al. 2012a). They constitute the "gate" in the gate-control theory of pain of Melzack and Wall (Fig. 3). Many inhibitory neurons in the spinal cord release GABA and glycine from the same synaptic vesicles. Removal of this inhibitory tone promotes nociception and pain. More specifically, blockage of GABA_A receptors (with, e.g., ▶ bicuculline) and of GABA_B receptors (with, e.g., phaclofen) or of glycine receptors in the rodent spinal cord leads to behavioral signs

GABA and Glycine in Spinal Nociceptive Processing, Fig. 2 GABAergic (*left*) and glycinergic (*right*) axon terminal. GABA is synthesized from L-glutamate by GAD-65 and GAD-67 and transported into presynaptic storage vesicles by VGAT/VIAAT. Degradation of GABA occurs through GABA transaminase (GABA-T) to form succinic semialdehyde (SSA) which is further metabolized in the Krebs cycle. Glycine is formed from serine by serine methyl transferase and taken up from extracellular space by GlyT2. Transport into presynaptic vesicles occurs again through VGAT/VIAAT. GABA acts at postsynaptic $GABA_A$ receptors and presynaptic and postsynaptic $GABA_B$ receptors. Some neurons, in particular primary sensory afferent neurons, carry $GABA_A$ receptors also on their presynaptic terminals. Glycine acts at postsynaptic glycine receptors (modified from Zeilhofer et al., 2012a)

GABA and Glycine in Spinal Nociceptive Processing, Fig. 3 Gate-control theory of pain. Inhibitory interneurons in the substantia gelatinosa (SG) of the superficial dorsal horn control the relay of sensory and nociceptive signals to higher CNS areas (modified from Melzack and Wall, 1965)

of hyperalgesia, allodynia, and spontaneous discomfort, while application of GABA or glycine diminishes pain. The consequences of $GABA_A$ and glycine receptor blockade thus resemble most of the symptoms of chronic pathological pain in patients. On a cellular level, activation of $GABA_A$ and glycine receptors limits the excitability of dorsal horn projection neurons, including that of nociceptive projection neurons; prevents the excitation of normally nociceptive-specific neurons by input from non-nociceptive sensory fibers; and inhibits spontaneous discharges of dorsal horn neurons in the absence of sensory input.

GABA$_A$ receptors are also expressed on the presynaptic (spinal) terminals of primary sensory neurons. In these cells, the intracellular chloride concentration is higher than in most adult neurons of the CNS, and its equilibrium potential is more depolarized than the resting membrane potential rendering GABA$_A$ receptors depolarizing rather than hyperpolarizing. Primary sensory neurons achieve this high intracellular chloride concentration due to the expression of a ▶ Na$^+$K$^+$Cl$^-$ cotransporter (*slc12a2*), which accumulates Cl$^-$ inside cells, and the lack of K$^+$Cl$^-$ cotransporters, which extrude Cl$^-$ and K$^+$ from cells. The high intracellular chloride concentration causes GABAergic input to depolarize the central terminals of primary sensory neurons and causes GABA$_A$ receptor-mediated primary afferent depolarization (PDA). It is generally accepted that this depolarization is still inhibitory under physiological conditions (▶ presynaptic inhibition) as it reduces glutamate release from sensory nerve terminals possibly through voltage-dependent inactivation of Na$^+$ and Ca^{2+} channels in the terminal. Under pathological conditions, primary afferent depolarization can become suprathreshold and evoke action potentials giving rise to so-called ▶ dorsal root reflexes, which can enhance pain and contribute to neurogenic inflammation through the release of neuropeptides from the peripheral sensory nerve endings. While classical postsynaptic inhibition in the dorsal horn is mediated by both GABA and glycine, primary afferent depolarization and presynaptic inhibition of primary afferent terminals are mediated by GABA but not by glycine because primary sensory neurons lack functional glycine receptors.

Distribution of GABAergic and Glycinergic Neurons and of GABA$_A$ and Glycine Receptors

Inhibitory interneurons are found throughout the spinal dorsal horn (Todd 2010). GABAergic interneurons are densely packed in lamina II of the dorsal

horn where glycinergic neurons are largely missing. The latter are most abundant in lamina III and in deeper laminae. The majority of GABAergic neurons are also glycinergic. Although cell bodies of glycinergic neurons are largely absent from lamina II, lamina II is still densely innervated by glycinergic axon terminals which likely is the basis of the glycinergic control of spinal nociception.

The subunit composition of GABA$_A$ and glycine receptors varies between laminae (Figs. 4 and 5). In laminae I and II, where mainly nociceptive fibers terminate, GABA$_A$ receptor α2 and α3 subunits are densely expressed, while α1 and α5 are almost absent. Deeper dorsal horn laminae show less specific subunit expression and contain also α1 and α5 GABA$_A$ receptor subunits. A significant portion of the α2 and α3 GABA$_A$ receptor subunits found in laminae I and II are located on the terminals of primary nociceptive afferents.

Glycine receptor subunits also show a lamina-specific expression pattern (Fig. 5). α3 glycine receptor subunits are specifically expressed in lamina II, mainly lamina II inner, while α1 subunits are expressed throughout the dorsal horn. Besides α1 and α3 subunits, α2 subunits are found in the dorsal horn.

Diminished GABAergic and Glycinergic Pain Control in Pathological Pain States

GABAergic and glycinergic neurotransmission is the target of several processes leading to central sensitization in inflammatory and neuropathic pain conditions (Sandkühler 2009; Zeilhofer et al. 2012b) (Fig. 6):

- Peripheral nerve injury causes a transsynaptic decrease in the expression of a potassium chloride transporter (KCC2). This inhibition reduces the chloride gradient of superficial dorsal horn neurons and in turn renders GABAergic and glycinergic input less inhibitory. The contributing signal transduction steps involve recruitment and activation of dorsal horn microglia by the chemokine CCL and ATP through CCR2 and P2X4/7 receptors, which triggers the release of BDNF from microglia and the activation of trk-B receptors in dorsal horn neurons.

GABA and Glycine in Spinal Nociceptive Processing, Fig. 4 Expression of $GABA_A$ receptor alpha subunit in the mouse lumbar spinal cord. Cold colors (*blue*) indicate absent or low expression; warm colors (*yellow* and *red*) indicate high levels of expression (Modified from Paul et al. (2012))

- Peripheral inflammation diminished specifically glycinergic inhibition through the spinal production of PGE_2, the subsequent activation of EP2 receptors, and a protein kinase A-dependent phosphorylation of $GlyR\alpha3$. Mice deficient in the $GlyR\alpha3$ subunit not only lack PGE_2-mediated inhibition of glycine receptors but also show reduced inflammatory pain sensitization.
- Both peripheral nerve injury and inflammation cause an epigenetic downregulation of GAD-65 in the spinal dorsal horn and thus impair GABAergic pain control.
- Furthermore, a large number of neuromodulators (e.g., endocannabinoids, neurosteroids, noradrenaline, purines, opioid peptides, serotonin) are released in the spinal dorsal horn from neurons and/or glia cells to modulate either the excitability of inhibitory neurons, to interfere with the release of GABA and glycine or with the function of GABA and glycine receptors. Some of these modulators, in particular serotonin and noradrenaline, likely originate from fiber tracts descending from the brain stem of midbrain.

Therapeutic Interventions Targeting Spinal GABA and Glycine Receptors

The role of glycine and GABA in spinal nociceptive processing is increasingly recognized, but no

GABA and Glycine in Spinal Nociceptive Processing, Fig. 5 Distribution of glycine receptor α1 and α3 subunits in the mouse lumbar spinal dorsal horn. (**a**) GlyRα3 (yellow) is restricted to the superficial dorsal horn. Gephyrin (red) illustrates the distribution of inhibitory synapses throughout the different laminae of the spinal cord. (**b**) GlyRα1 (red) is found throughout the gray matter of the spinal cord, while GlyRα3 (blue) is distinctly expressed in lamina II inner below the termination area of peptidergic nociceptors identified by the presence of immunoreactivity against calcitonin gene-related peptide (CGRP) (Modified from Harvey et al. (2004))

GABA and Glycine in Spinal Nociceptive Processing, Fig. 6 Schematic diagram illustrating mechanisms leading to synaptic disinhibition in inflammatory and neuropathic pain conditions (modified from Zeilhofer et al., 2012a)

analgesic drugs, perhaps apart from the $GABA_B$ receptor agonist baclofen, target these transmitter systems.

Although potentiation of $GABA_A$ receptor-mediated inhibition by classical benzodiazepines does not evoke clinically relevant analgesia in humans, more recently developed subtype-selective benzodiazepine-site agonists evoke analgesia in rodent models of inflammatory and neuropathic pain. Experimental drugs which spare $\alpha1$-$GABA_A$ receptors are largely devoid of sedative effects but show pronounced antihyperalgesic and antiallodynic effects, which mainly originate from their activity at $\alpha2$-$GABA_A$ receptors. Such experimental compounds include L-838417, NS11394, and HZ166. Gaboxadol, which targets specifically $\alpha4$-$GABA_A$ receptors, also exhibits analgesic actions (Zeilhofer et al. 2012b).

Few drugs specifically target glycine receptors. Nevertheless, a considerable number of compounds do act as allosteric modulators of glycine receptors. In particular compounds structurally related to cannabinoids have recently been shown to facilitate the activation of glycine receptors and to exert analgesic properties in rodents (Yévenes and Zeilhofer 2011).

Summary

GABAergic and glycinergic neurons exert an inhibitory control on spinal sensory processing through presynaptic and postsynaptic inhibition. They limit the excitability of spinal projection neurons by nociceptive input, prevent their excitation by input from non-nociceptive fibers, and inhibit spontaneous activity of dorsal horn neurons. Several signal transduction pathways triggered by peripheral inflammation or nerve damage impair GABAergic or glycinergic pain control and thereby contribute to central (spinal) sensitization.

Acknowledgement The research of the author is supported by grants from the Swiss National Science Foundation, the European Research Council, and the Deutsche Forschungsgemeinschaft.

References

Harvey et al. (2004). GlyR 3α: an essential target for spinal PGE2-mediated inflammatory pain sensitization. *Science, 304*, 884–887.

Melzack, R., & Wall, P. D. (1965). Pain mechanisms: a new theory. *Science, 150*, 971–979.

Paul et al. (2012). Selective distribution of GABAA receptor subtypes in mouse spinal dorsal horn neurons and primary afferents. *Journal of Comparative Neurology, 520*, 3895–3911.

Sandkühler, J. (2009). Models and mechanisms of hyperalgesia and allodynia. *Physiological Reviews, 89*, 707–758.

Todd, A. J. (2010). Neuronal circuitry for pain processing in the dorsal horn. *Nature Reviews Neuroscience, 11*, 823–836.

Yévenes, G. E., & Zeilhofer, H. U. (2011). Allosteric modulation of glycine receptors. *British Journal of Pharmacology, 164*, 224–236.

Zeilhofer, H. U., Wildner, H., & Yévenes, G. E. (2012a). Fast synaptic inhibition in spinal sensory processing and pain control. *Physiological Reviews, 92*, 193–235.

Zeilhofer, H. U., Benke, D., & Yévenes, G. E. (2012b). Chronic pain states: Pharmacological strategies to restore diminished inhibitory spinal pain control. *Annual Review of Pharmacology and Toxicology, 52*, 111–133.

GABA Mechanisms and Descending Inhibitory Mechanisms

Ronald Dubner
Department of Neural and Pain Sciences,
University of Maryland Dental School,
Baltimore, MD, USA

Synonyms

Centrifugal control of nociceptive processing; Descending inhibitory mechanisms and GABA mechanisms; Endogenous analgesia system; GABAergic inhibition; Supraspinal regulation

Definition

Pathways that originate in the brain can inhibit nociceptive neurons in the dorsal horn, including spinothalamic tract (STT) cells. Several different inhibitory neurotransmitters are used by the endogenous analgesia system. One of these is gamma-aminobutyric acid (GABA). GABA can be released either directly by the axons of brainstem neurons that descend to the spinal cord or indirectly by the excitation of GABAergic inhibitory interneurons in the spinal cord through release of excitatory transmitters from descending axons. There are several mechanisms for GABAergic inhibitory actions. These include pre- and postsynaptic inhibition following actions of GABA on $GABA_A$ or $GABA_B$ receptors.

Characteristics

▶ Gamma(γ)-aminobutyric acid (GABA) is a major inhibitory neurotransmitter in the spinal cord, especially in the dorsal horn (Willis and Coggeshall 2004). Its synthetic enzyme is ▶ glutamic acid decarboxylase (GAD). There are several forms of GAD. The sources of GABAergic terminals in the spinal cord include GABAergic spinal interneurons (see ▶ GABAergic Cells (Inhibitory Interneurons)) (Carlton and Hayes 1990) and axons that descend from the rostral ventral medulla (Millhorn et al. 1987; Reichling and Basbaum 1990). GABA-containing synapses have been demonstrated on primate spinothalamic tract cells in an electron microscopic study (Carlton et al. 1992). GABA is also contained in presynaptic contacts with primary afferent terminals (Willis and Coggeshall 2004).

There are at least 3 types of GABA receptors: $GABA_A$, $GABA_B$, and $GABA_C$ receptors. Emphasis here will be on the first two of these. $GABA_A$ receptors are ionotropic receptors and cause the opening of chloride channels (Willis and Coggeshall 2004). This can result in either a hyperpolarization or a depolarization, depending on where Cl^- is concentrated. The concentration of Cl^- depends on the type of ▶ Cl^- transporter that is present in the neuronal membranes (Willis and Coggeshall 2004; Willis 1999). In the case of presynaptic endings, $GABA_A$ receptor activation results in primary afferent depolarization and ▶ presynaptic

GABA Mechanisms and Descending Inhibitory Mechanisms, Fig. 1 Inhibition of the activity of a primate spinothalamic tract (STT) neuron by iontophoretic release of GABA. (**a**) shows the responses of an STT cell to pulsed release of glutamate. The iontophoretic currents lasted for 5 s and were repeated at 10 s intervals. At the times indicated by the horizontal bars, GABA was also released, using the indicated currents. (**b**) shows the inhibitory effects of GABA release on the responses of the same STT neuron to continuous noxious pinch of the skin in the receptive field. The iontophoretic currents are indicated (From Willcockson et al. 1984)

inhibition (Willis 1999). In the case of postsynaptic neurons, $GABA_A$ receptors cause hyperpolarization and ▶ postsynaptic inhibition (Willis and Coggeshall 2004).

$GABA_B$ receptors are metabotropic G-protein-coupled receptors (see ▶ Metabotropic Glutamate Receptors) (Willis and Coggeshall 2004). They are found both pre- and postsynaptically. Their activation can also cause pre- or postsynaptic inhibition. However, presynaptic inhibition in this case is not accompanied by primary afferent depolarization. Instead, it is due to a reduction in the Ca^{++} current that is necessary for release of transmitter from presynaptic terminals.

GABA Mechanisms and Descending Inhibitory Mechanisms, Fig. 2 Inhibition of the activity of a primate STT cell by spinal cord microdialysis administration of baclofen and the antagonistic effect of phaclofen. (a) Left histogram shows the background discharge of an STT cell. During the time indicated by the horizontal bar, baclofen was infused into the spinal cord by microdialysis. (a) Right histogram shows the antagonistic action of phaclofen when this agent was coadministered with baclofen. (b) shows the inhibitory effects of baclofen on the responses of the neuron to brush and pinch stimuli and to noxious heat. The stimuli were applied at the times indicated by the horizontal bars or by the temperature monitor. The upper row of histograms shows the control responses, the middle row the inhibited responses during baclofen administration, and the lower row the responses when phaclofen was coadministered with baclofen. In (c), baclofen is shown to block orthodromic but not antidromic responses of the STT cell. The effect was reversed by phaclofen (From Lin et al. 1996a)

Postsynaptic inhibition that is mediated by GABA$_B$ receptors results from an increased conductance for K$^+$ ions.

Experiments in which drugs were released by ▶ microiontophoresis near primate STT cells have shown that the excitation of these neurons by the pulsed release of glutamate or by noxious compression of the skin can be reduced by GABA (Fig. 1) (Willcockson et al. 1984). The iontophoretic release of the GABA$_A$ receptor agonist, muscimol, also inhibited the activity of all of the STT cells tested (Lin et al. 1996a). However, the GABA$_B$

GABA Mechanisms and Descending Inhibitory Mechanisms, Fig. 3 Reduction in the PAG inhibition of nociceptive dorsal horn neurons in the rat spinal cord following microdialysis administration of the $GABA_A$ receptor antagonist, bicuculline. The upper row of histograms in (a) shows repeated periods of inhibition of the responses of a dorsal horn neuron to brush, press, and pinch stimuli due to four periods of stimulation in the periaqueductal gray (PAG). The stimulus monitor pulses in the lowest row of records indicate the timing of PAG stimulation. The second row of histograms shows that the PAG inhibition was blocked during the microdialysis administration of bicuculline into the spinal cord. The third row of histograms shows recovery from the bicuculline. (b) shows the grouped results for 19 dorsal horn neurons. Bicuculline infusion resulted in a significant reduction in the PAG inhibition (From Peng et al. 1996)

receptor agonist, baclofen, produced inhibition in only 17 % of STT cells examined. On the other hand, microdialysis administration of baclofen into the dorsal horn resulted in a strong inhibition of STT cells (Fig. 2). This was counteracted by coadministration of the GABA$_B$ receptor antagonist, phaclofen. ▶ Microdialysis administration of the GABA$_A$ antagonist, bicuculline, or the GABA$_B$ receptor antagonist, phaclofen, enhanced the responses of STT cells (Lin et al. 1996a). This evidence suggests that GABA$_B$ receptors are likely to be more important for presynaptic inhibition than for postsynaptic inhibition.

Stimulation in the midbrain ▶ periaqueductal gray (PAG) can produce a strong inhibition of the responses of nociceptive dorsal horn neurons, including STT cells (Hayes et al. 1979). This inhibition is at least partly due to the effects of the release of GABA in the spinal cord (Lin et al. 1996a; Peng et al. 1996). Evidence for this was obtained by administration of antagonists of GABA receptors into the spinal cord by microdialysis. The antagonists reduced the amount of inhibition produced by PAG stimulation. GABA$_A$ receptors are activated by PAG stimulation, since PAG inhibition is partially blocked by the GABA$_A$ antagonist, bicuculline (Fig. 3). GABA$_B$ receptors are less involved in PAG inhibition of STT cells, since administration of the GABA$_B$ antagonist, phaclofen, reduced the inhibition produced by PAG stimulation in only 22 % of the STT cells tested. Thus, GABA, as well as opioids and monoamines, such as serotonin and norepinephrine, is one of the inhibitory neurotransmitters utilized by the ▶ endogenous analgesia system (Willis and Coggeshall 2004).

The inhibitory action of GABA has been shown to have an important antinociceptive action (see ▶ Antinociception) in the spinal cord that is mediated by GABA$_A$ receptors (Aanonsen and Wilcox 1989). Release of GABA has been offered as an explanation for the antinociceptive effects of spinal cord stimulation (Linderoth et al. 1994). In contrast, antagonism of GABA$_A$ receptors in rats produces a profound state of ▶ mechanical allodynia (Sivilotti and Woolf 1994). Consistent with this is the observation that the inhibition of primate STT cells produced by iontophoretic

application of GABA or muscimol is greatly reduced during the ▶ central sensitization that follows intradermal injection of capsaicin (Lin et al. 1996b). Central sensitization is likely to be due to both an increase in the responsiveness of excitatory amino acid receptors and a decrease in the responsiveness of inhibitory amino acid receptors (Willis and Coggeshall 2004).

References

Aanonsen, L. M., & Wilcox, G. L. (1989). Muscimol, γ-aminobutyric acid$_A$ receptors and excitatory amino acids in the mouse spinal cord. *Journal of Pharmacology and Experimental Therapeutics, 248*, 1034–1038.

Carlton, S. M., & Hayes, E. S. (1990). Light microscopic and ultrastructural analysis of GABA-immunoreactive profiles in the monkey spinal cord. *The Journal of Comparative Neurology, 300*, 162–182.

Carlton, S. M., Westlund, K. N., Zhang, D., et al. (1992). GABA-immunoreactive terminals synapse on primate spinothalamic tract cells. *The Journal of Comparative Neurology, 322*, 528–537.

Hayes, R. L., Price, D. D., Ruda, M. A., et al. (1979). Suppression of nociceptive responses in the primate by electrical stimulation of the brain or morphine administration: Behavioral and electrophysiological comparisons. *Brain Research, 167*, 417–421.

Lin, Q., Peng, Y. B., & Willis, W. D. (1996a). Role of GABA receptor subtypes in inhibition of primate spinothalamic tract neurons: Difference between spinal and periaqueductal gray inhibition. *Journal of Neurophysiology, 75*, 109–123.

Lin, Q., Peng, Y. B., & Willis, W. D. (1996b). Inhibition of primate spinothalamic tract neurons by spinal glycine and GABA is reduced during central sensitization. *Journal of Neurophysiology, 76*, 1005–1014.

Linderoth, B., Stiller, C. O., Gunasekera, L., et al. (1994). Gamma-aminobutyric acid is released in the dorsal horn by electrical spinal cord stimulation: An in vivo microdialysis study in the rat. *Neurosurgery, 34*, 484–488.

Millhorn, D. E., Håkfelt, T., Seroogy, K., et al. (1987). Immunohistochemical evidence for colocalization of γ-aminobutyric acid and serotonin in neurons of the ventral medulla oblongata projecting to the spinal cord. *Brain Research, 410*, 179–185.

Peng, Y. B., Lin, Q., & Willis, W. D. (1996). Effects of GABA and glycine receptor antagonists on the activity and PAG-induced inhibition of rat dorsal horn neurons. *Brain Research, 736*, 189–201.

Reichling, D. B., & Basbaum, A. I. (1990). Contribution of brainstem GABAergic circuitry to descending antinociceptive control. I. GABA-immunoreactive projection neurons in the periaqueductal gray and nucleus raphe magnus. *The Journal of Comparative Neurology, 302*, 370–377.

Sivilotti, L., & Woolf, C. J. (1994). The contribution of GABA_A and glycine receptors to central sensitization: disinhibition and touch-evoked allodynia in the spinal cord. *Journal of Neurophysiology, 782*, 169–179.

Willcockson, W. S., Chung, J. M., Hori, Y., et al. (1984). Effects of iontophoretically released amino acids and amines on primate spinothalamic tract cells. *Journal of Neuroscience, 4*, 732–740.

Willis, W. D. (1999). Dorsal root potentials and dorsal root reflexes: A double-edged sword. *Experimental Brain Research, 124*, 395–421.

Willis, W. D., & Coggeshall, R. E. (2004). *Sensory mechanisms of the spinal cord* (3rd ed.). New York: Kluwer Academic/Plenum Publishers.

GABA Transporter

Definition

Plasma membrane transporters, which transport GABA into neurons and glial cells. These are distinct from the transporters that load GABA into vesicles.

Cross-References

▶ GABA and Glycine in Spinal Nociceptive Processing

GABAA Receptors

Definition

An ionotropic receptor activated by γ-amino butyric acid (GABA). These are pentomeric anion channels permeable to both Cl^- and $HCO3^-$ at a ratio of ~4:1. They are members of the cis-loop family of ionotropic receptors that also includes glycine receptors, nicotinic acetylcholine receptors, and serotonin type 3 receptors. Subunit composition determines receptor pharmacology, biophysical properties, and cellular distribution. GABAA receptors a primarily responsible for fast inhibitory synaptic transmission in the central nervous system.

Cross-References

▶ GABA and Glycine in Spinal Nociceptive Processing

GABAB Receptors

Definition

GABA_B receptors are metabotropic receptors for γ-amino butyric acid (GABA). The "B" is used to distinguish the metabotropic receptors from the ionotropic GABA_A receptors. GABA_B receptors are members of the G-protein coupled receptor (GPCR) family characterized by the presence of seven membrane spanning alpha helix domains that are also responsible for G-protein coupling. In contrast to many GPCR family members, GABA_B receptors only function as a heterodimer. The B1 subunit binds GABA and other allosteric modulators, but is unable to bind and activate G-proteins and is unable to reach the cell membrane by itself. The B2 subunit is unable to bind GABA, but enables the B1-B2 complex to reach the cell surface and is responsible for G-protein signaling. GABA_B receptors are typically coupled to inhibitory G-proteins, Gi and Go.

Cross-References

▶ GABA Mechanisms and Descending Inhibitory Mechanisms
▶ Thalamic Plasticity and Chronic Pain

GABAergic

Definition

Synaptic transmission at which the γ-amino acid GABA (γ-aminobutyric acid) is used as a neurotransmitter.

Cross-References

▶ Opioid Receptors at Postsynaptic Sites

GABAergic Cells (Inhibitory Interneurons)

Definition

GABAergic cells are neurons that use gamma-aminobutyric acid (GABA), an inhibitory neurotransmitter, as their neurotransmitter. GABAergic inhibitory interneurons are local circuit inhibitory neurons that use GABA as their neurotransmitter.

Cross-References

▶ GABA Mechanisms and Descending Inhibitory Mechanisms
▶ Information Processing in Thalamocortical Circuits
▶ Thalamic Nuclei Involved in Pain: Cat and Rat

GABAergic Inhibition

▶ GABA Mechanisms and Descending Inhibitory Mechanisms

Gabapentin

Definition

Gabapentin is an antiepileptic drug that has some efficacy for the treatment neuropathic pain conditions such as postherpetic neuralgia and diabetic neuropathy. While originally developed as a novel GABA agonist, the compound has no activity at GABA receptors. Rather, its primary binding target is the $\alpha2\delta1$ subunit of voltage-gated Ca^{2+} channels.

Cross-References

▶ Postoperative Pain, Gabapentin
▶ Postoperative Pain, Postamputation Pain, Treatment, and Prevention
▶ Preventive Migraine Therapy

GAD

▶ Glutamic Acid Decarboxylase

Gainful Work Activity

Definition

Gainful work activity is that which is done for pay or profit. Work activity is gainful if it is the kind of work usually done for pay or profit, whether or not a profit is realized.

Cross-References

▶ Disability Evaluation in the Social Security Administration

Galactorrhea

Definition

Galactorrhea is the normal production and flow of milk after pregnancy. This condition is considered abnormal in the absence of recent pregnancy.

Cross-References

▶ Cancer Pain Management, Opioid Side Effects, Endocrine Changes, and Sexual Dysfunction

Galanin

Definition

Galanin is a 29-amino acid peptide (30 in humans), which was purified from porcine intestine. It is widely distributed in the nervous system and has inhibitory effects on its target cells. Galanin-containing dorsal root ganglion neurons seem to play a role in pain processing, particularly following nerve injury. Roles in the central control of feeding, body weight, and affect are also discussed.

Galvanic Skin Response (GSR)

▶ Sympathetic Skin Response

Gamma Knife

Definition

A Gamma Knife is a highly specific and focused, noninvasive, gamma radiation device used as a surgical unit.

Cross-References

▶ Cancer Pain Management, Anesthesiologic Interventions

Gamma(γ)-Aminobutyric Acid

Synonyms

GABA

Definition

GABA is a biologically active amino acid found in plants as well as in the brain and other animal

tissues. It is one of the principal inhibitory amino acid neurotransmitters in the central nervous system. GABA acts as an agonist at two receptors, the $GABA_A$ receptor, which is an anion (primarily) selective ionotropic receptor, and the $GABA_B$ receptor, which is a G-protein-linked receptor. While GABA is found in some projection neurons, it is found primarily in inhibitory interneurons throughout the neuraxis.

Cross-References

▶ GABA and Glycine in Spinal Nociceptive Processing
▶ GABA Mechanisms and Descending Inhibitory Mechanisms
▶ Molecular Contributions to the Mechanism of Central Pain
▶ Nociceptors in the Orofacial Region (Temporomandibular Joint and Masseter Muscle)
▶ Somatic Pain
▶ Spinal Cord Nociception, Neurotrophins
▶ Stimulation-Produced Analgesia
▶ Thalamic Neurotransmitters and Neuromodulators
▶ Thalamic Plasticity and Chronic Pain

Ganglionopathies

John W. Griffin
Department of Neurology, Johns Hopkins
University School of Medicine, Baltimore,
MD, USA

Synonyms

Idiopathic ataxic neuropathy; Sensory ganglionitis; Sensory neuronopathy

Definition

A group of disorders of the peripheral nervous system characterized by loss of primary sensory neurons in the dorsal root ganglia, with or without concomitant loss of autonomic neurons from the peripheral ganglia.

Characteristics

The primary sensory neurons lie in the dorsal root ganglia and cranial nerve ganglia. In a group of disorders of the peripheral nervous system (PNS), they degenerate and are lost, along with their peripheral and central axons. There are several etiologies that can produce such sensory ganglionopathies, including inflammatory, toxic, and infectious causes. Most types of sensory ganglionopathies affect large sensory neurons and so produce loss of proprioception, joint position sensibility, and kinesthesia (Denny-Brown 1948; Griffin et al. 1990; Kuntzer et al. 2004; Windebank et al. 1990). When the legs are affected, there is sensory ataxia, characterized by ataxia associated with the inability to stand with the eyes closed (Romberg's sign). When the arms are affected, there is typically drift of the outstretched arms with the eyes closed, associated with "piano-playing" involuntary movements in the fingers (pseudoathetosis). Pseudoathetosis is easiest to see with the arms outstretched, but in severe cases, it can be recognized in the hands at rest. A useful test is to ask the patient to find the thumb of one hand with the index finger of the other without visual guidance. In normal individuals this is a prompt, secure movement. In patients with loss of kinesthesia, the moving hand must search for the thumb. In the most severe cases, so much touch sensibility is lost that the patient may not recognize when contact is made.

These features reflect loss of sensation from relatively proximal levels of the arms and legs. Most nerve diseases produce length-dependent loss of function, so that sensation is lost in the toes and feet first, and only with advanced disease would gait ataxia, drift of the arms, and pseudoathetosis develop. Thus, a characteristic feature of the ganglionopathies is the loss of large fiber sensory functions, which is not length-dependent in fashion, so that short as well as long nerves are affected.

Spontaneous sensations (tingling paresthesias or burning pain) may be present in affected regions. Reflecting the nonlength-dependent pathology, the affected regions frequently include the face and the trunk (Denny-Brown 1948; Griffin et al. 1990; Windebank et al. 1990), regions rarely affected in length-dependent axonal degenerations. A characteristic electrodiagnostic finding is loss of sensory nerve action potential (SNAP) amplitudes in a nonlength-dependent fashion – the SNAP amplitudes from the ulnar, median, or radial nerves may be reduced at times when the responses from the sural nerves in the legs are still elicitable or all SNAP responses may be lost at the same time (Lauria et al. 2003). In length-dependent axonal degenerations, the sural SNAP amplitude is reduced or lost well before the SNAP amplitudes in the arms. There are two other indications that the neuronal loss in ganglionopathies is nonlength-dependent. First, magnetic resonance imaging (MRI) of the spinal cord shows evidence of fiber loss in the dorsal column at all levels of the spinal cord, reflecting loss of the central processes of large primary afferent neurons (Lauria et al. 2000). In length-dependent axonal degenerations, loss of fibers in the dorsal columns occurs first in the rostral gracile tract. Second, skin biopsies immunostained for nerve fibers show loss of nerve fibers from proximal sites such as the thigh, back, and chest, as well as distal sites (Lauria et al. 2001). In length-dependent axonal degeneration fibers are lost first from the distal leg.

Immune-Mediated Ganglionopathies

Three disorders are included under the designation sensory ganglionitis: carcinomatous sensory neuropathy, ataxic neuropathy associated with Sjogren's syndrome, and idiopathic sensory neuronopathy. All share the pathologic features of lymphocytic infiltration of dorsal root ganglia and destruction of sensory neurons. As indicated in Table 1, there are clinical differences that allow suspicion of the correct diagnosis in the clinic.

Carcinomatous Sensory Neuropathy

This disorder was the first recognized sensory ganglionopathy (Denny-Brown 1948), and its

Ganglionopathies, Table 1 Sensory ganglionitis syndromes

Feature	Sicca syndrome	Idiopathic	Carcinomatous
Female predilection	Marked	Modest	Absent
Course	Variable, acute to chronic	Variable, acute to chronic	Subacute
Progression	Variable, may stabilize or improve	Variable, may stabilize or improve	Progressive
Fiber predilection	Large fiber, kinesthetic loss	Large fiber, kinesthetic loss	More global
Associated central nervous system involvement	Usually none	None	Cerebellar involvement
Serologic studies	ANA$^+$, elevated IgG	Normal	Anti-Hu in many
Antineuronal nuclear antibody	−	−	+
Cerebrospinal fluid	Normal	Normal	Some cells, protein
Nerve biopsy	Inflammation, large fiber loss	Large fiber loss	More global fiber loss

diagnosis has special urgency because it may be the presenting manifestation of an underlying malignancy. Sensory ganglionopathy can be associated with underlying lung, breast, ovary, or other carcinomas. The pathology is an intense lymphocytic infiltration of the dorsal root ganglia, often with autoaggressive T cells invading the sensory nerve cells (Denny-Brown 1948). The initial symptom is often neuropathic pain that involves the hands, trunk, and/or face as well as the feet. The sensory loss is global, as reflected in nerve biopsies that show loss of small myelinated and unmyelinated fibers as well as large myelinated fibers. Some individuals have evidence of CNS involvement such as dementia, cerebellar dysfunction, or myelopathy. The spinal fluid often has a mild lymphocytic pleocytosis and elevated protein. A key diagnostic test is a search for antineuronal antibodies associated with sensory ganglionopathies. These antibodies include anti-Hu antibodies, also called ANNA-1 antibodies, directed against a 37 kD nuclear antigen. Other antibodies include anti-ampiphysin antibodies, ANNA-2, and ANNA-3. Patients who present with subacute sensory ganglionopathy require meticulous examination for underlying carcinoma, especially when one of these antibodies is detected.

The course is usually inexorable, although occasionally the disease stabilizes. Early discovery of the neoplasm can be life-saving and occasionally results in stabilization of the ganglionopathy. Immunosuppressive therapies have had disappointing results. Control of the neuropathic pain is often difficult and may require opiates.

Ataxic Ganglionopathy with Features of Sjogren's Syndrome

Sjogren's syndrome is an autoimmune rheumatologic disorder that includes dry eyes, dry mouth, and serologic abnormalities that often include antinuclear antibodies. The dry mouth and dry eyes reflect lymphocytic infiltration of the salivary and lacrimal glands, respectively. Testing for tear production (Shirmer test) and lip biopsy (minor salivary gland biopsy) and looking for lymphocytic inflammation are adjuncts to the diagnosis. Several types of neuropathy can be associated with Sjogren's syndrome (Griffin et al. 1990; Mellgren et al. 1989). In typical Sjogren's syndrome, it has been shown that ataxic neuropathy is rare, whereas multiple mononeuropathies and axonal sensorimotor neuropathies are frequently encountered. In the ganglionopathy, the ocular and other features of Sjogren's syndrome are often minor, and the neuropathy is usually the presenting manifestation (Griffin et al. 1990). The ataxic neuropathy patients thus form a distinct subgroup of the Sjogren's patients.

Ataxic neuropathy associated with features of Sjogren's syndrome is a syndrome that can be recognized by clinical and laboratory testing. The laboratory features useful in recognizing this syndrome are positive Schirmer and rose bengal tests for dry eyes and keratitis, respectively, inflammation of the minor salivary glands on lip biopsy, and a markedly abnormal antinuclear antibody titer, with or without the Ro or La reactivity often associated with Sjogren's syndrome.

The neurologic examination is similar to the other ganglionopathies, with sensory ataxia and loss of kinesthesia in the arms (Griffin et al. 1990). Autonomic dysfunction, reflected in orthostatic hypotension and loss of heart period variability, is common. A characteristic abnormality is the development of Adie's pupils, unilaterally or bilaterally (Griffin et al. 1990). Adie's pupils are mid-position pupils with a slow reaction to changes in illumination, prompt reconstriction to accommodation, and vermicular movements under a slit lamp. Bilateral Adie's pupils are sufficiently rare in other disorders that they suggest the possibility of Sjogren's ganglionopathy.

The histologic appearance is remarkable for a variable degree of neuronal loss and marked lymphocytic infiltration of the ganglia and dorsal roots (Griffin et al. 1990). In cutaneous sensory nerves, large myelinated fibers are lost. Small myelinated and unmyelinated fiber densities are relatively preserved. Several nerve biopsy specimens have had small perivascular inflammatory cuffs around epineurial vessels.

In general, attempts at immunotherapy have proved disappointing. In most patients, oral and intravenous corticosteroids, cyclophosphamide, and azathioprine have had no obvious effect on the course of the disorder, although a slowing of the progression cannot be excluded. Most patients received their initial therapy at a time when SNAPs were markedly reduced or absent, making the likelihood of recovery low. Rare patients with relatively preserved SNAPs stabilize or improve after treatment with intravenous methylprednisolone and oral azathioprine.

Idiopathic Sensory Neuronopathy

This category, one of the most frequent causes of sensory ganglionopathy, includes patients with acute, subacute, and chronic disease courses (Windebank et al. 1990). In the acute form, devastating sensory loss develops over a few days. The spinal fluid protein value may be normal or elevated. Results of other laboratory tests are normal and useful principally in excluding Sjogren's syndrome and occult cancer. Pathologic studies have been rare, but the results have been similar to the findings in Sjogren's syndrome with sensory ganglionitis.

The prognosis is highly variable, but in time, the majority of patients with this disorder are able to return to their previous career (Windebank et al. 1990). The role of therapy is uncertain; most patients have received corticosteroids at some point, and such therapy may minimize progression (Windebank et al. 1990). More important are reassurance, gait training, and safety instruction, as described below.

The Fisher Syndrome

The Fisher syndrome is a variant of the Guillain Barre syndrome (*link to chapter on Guillain Barre, Scadding*) that is characterized by ataxia, loss of reflexes, and inability to move the eyes (ophthalmoparesis) associated with pupils that do not react to light or looking at near objects. Like the other forms of the Guillain Barre syndrome, the Fisher syndrome is an acute monophasic autoimmune disorder that can follow infections. Like the other forms, it is likely that the immune response to antigens on infectious agents result in an immune attack on similar moieties within peripheral nerve molecular mimicry. In the Fisher syndrome, serologic studies have found the presence of antibodies against the ganglioside GQ1b (Chiba et al. 1993; Willison et al. 1993) and experimental studies have shown that exposure to strains of *Campylobacter jejuni*, that bear related epitopes with their lipopolysaccharides, can produce pathogenic anti-GQ1b antibodies (Plomp et al. 1999). In the inflammatory demyelinating form of Guillain Barre syndrome, treatment with plasmapheresis – the removal of the immunoglobulin fraction of plasma by exchange for albumin – or the infusion of large amounts of human immunoglobulin, a procedure that ameliorates many immune disorders, speeds recovery. Although no data applies specifically to the Fisher syndrome, it is reasonable to infer that these treatments would be efficacious in this disorder as well.

Whether the ataxic represents a ganglionopathy is unresolved. Immunization of rabbits with GD1b produces an inflammatory ganglionopathy and ataxia (Kusunoki et al. 1999). However, the reversibility of the ataxia in the Fisher syndrome suggests that the ganglion cells need not be destroyed.

Toxic Causes

Several pharmacologic agents can produce ataxic neuropathies. Whether these are truly sensory ganglionopathies is questionable. Some agents, such as pyridoxine, can produce loss of large DRG neurons experimentally, but the reversibility of the effects of intoxication by large doses of pyridoxine in man suggests that it at least begins as a length-dependent axonal degeneration. Regeneration can occur when the agent is discontinued. Several can produce neuropathic pain, and experimental models of taxane-induced painful neuropathies have been developed.

References

Chiba, A., Kusunoki, S., Obata, H., Machinami, R., & Kanazawa, I. (1993). Serum anti-GQ1b antibody is associated with ophthalmoplegia in Miller Fisher syndrome and Guillain-Barre syndrome: Clinical and immunohistochemical studies. *Neurology, 43*, 1911–1917.

Denny-Brown, D. (1948). Primary sensory neuropathy with muscular changes associated with carcinoma. *Journal of Neurology, Neurosurgery, and Psychiatry, 11*, 73–87.

Griffin, J. W., Cornblath, D. R., Alexander, E., Campbell, J., Low, P. A., Bird, S., & Feldman, E. L. (1990). Ataxic sensory neuropathy and dorsal root ganglionitis associated with Sjogren's syndrome. *Annals of Neurology, 27*, 304–315.

Kuntzer, T., Antoine, J. C., & Steck, A. J. (2004). Clinical features and pathophysiological basis of sensory neuronopathies (ganglionopathies). *Muscle & Nerve, 30*, 255–268.

Kusunoki, S., Hitoshi, S., Kaida, K., Arita, M., & Kanazawa, I. (1999). Monospecific anti-GD1b IgG is required to induce rabbit ataxic neuropathy. *Annals of Neurology, 45*, 400–403.

Lauria, G., Pareyson, D., Grisoli, M., & Sghirlanzoni, A. (2000). Clinical and magnetic resonance imaging findings in chronic sensory ganglionopathies. *Annals of Neurology, 47*, 104–109.

Lauria, G., Sghirlanzoni, A., Lombardi, R., & Pareyson, D. (2001). Epidermal nerve fiber density in sensory ganglionopathies: Clinical and neurophysiologic correlations. *Muscle & Nerve, 24*, 1034–1039.

Lauria, G., Pareyson, D., & Sghirlanzoni, A. (2003). Neurophysiological diagnosis of acquired sensory ganglionopathies. *European Neurology, 50*, 146–152.

Mellgren, S. I., Conn, D. L., Stevens, J. C., & Dyck, P. J. (1989). Peripheral neuropathy in Sjogren syndrome. *Neurology, 39*, 390–394.

Plomp, J. J., Molenaar, P. C., O'Hanlon, G. M., Jacobs, B. C., Veitch, J., Daha, M. R., vanDoorn, P. A., van der Meche, F. G. A., Vincent, A., Morgan, B. P., & Willison, H. J. (1999). Miller Fisher anti-GQ1b antibodies: Î±-latrotoxin-like effects on motor end plates. *Annals of Neurology, 45*, 189–199.

Willison, H. J., Veitch, J., Patterson, G., & Kennedy, P. G. E. (1993). Miller Fisher syndrome is associated with serum antibodies to GQ1b ganglioside. *Journal of Neurology, Neurosurgery, and Psychiatry, 56*, 204–206.

Windebank, A. J., Blexrud, M. D., Dyck, P. J., Daube, J. R., & Karnes, J. L. (1990). The syndrome of acute sensory neuropathy: Clinical features and electrophysiologic and pathologic changes. *Neurology, 40*, 584–591.

Gap Junctions

Definition

Gap junctions are intercellular channels established between cells through which small molecules can pass. Within the spinal cord, the majority of gap junctions are found between astrocytes. Gap junctions allow for fast communication between cells over long distances. Due to this, gap junction communication may be salient to extraterritorial/mirror-image pain.

Cross-References

▶ Cord Glial Activation

Gasserian Ganglion

▶ Trigeminal Ganglion

Gastroesophageal Reflux Disease

▶ GERD

Gastroesophageal Reflux Disease (GERD)

▶ Visceral Pain Model; Esophageal Pain

Gastrointestinal Tract

▶ Nocifensive Behaviors, Gastrointestinal Tract

GAT 1, GAT 2, GAT 3

Definition

Plasma membrane GABA transporter, isoforms 1–3. GAT 1 is encoded by the gene SLC6A1, GAT 2 by SLC6A12, and GAT 3 by SLC6A11.

Cross-References

▶ GABA and Glycine in Spinal Nociceptive Processing

Gate Control Theory

Definition

The Gate Control Theory was devised by Melzack and Wall in 1965. It proposed an explanation (later falsified) on how innocuous stimulation inhibits pain via a presynaptic inhibitory mechanism. It was claimed that innocuous stimulation, such as produced by rubbing your skin, activates large sensory nerve fibers and inhibits nociceptive neurons in the spinal cord. If small-fiber nociceptive primary afferents are simultaneously being activated by a noxious stimulus such as a bee sting, less pain is felt because the "pain gate" is closed.

Cross-References

▶ Central Nervous System Stimulation for Pain
▶ Pain Treatment: Spinal Cord Stimulation

Gazelius Model

▶ Retrograde Cellular Changes After Nerve Injury

GDNF

▶ Glial Cell Line-Derived Neurotrophic Factor

GDNF-dependent Neurons

▶ IB4-Positive Neurons, Role in Inflammatory Pain

Gender

Definition

Gender is the psychosocial identity in males and females such as masculinity and femininity. It is both the person's representation as male or female, and how that person is responded to by social institutions on the basis of the individual's gender presentation. Gender refers to the social, political, and psychological aspects of what it means to live as male or female in a given society.

Cross-References

▶ Gender and Pain
▶ Psychological Aspects of Pain in Women

Gender and Pain

Edmund Keogh
Department of Psychology, University of Bath, Bath, UK

Definition

The perception and experience of pain varies between individuals, and the search to identify and explain such differences is critical to developing effective treatment interventions. One individual difference factor that seems to play a general role in moderating pain is the sex of the person, i.e., whether they are male or female. Given that pain itself is believed to be due to complex interaction between biological, psychological, and social processes, the similarities and differences between males and females are best considered within a biopsychosocial framework. Indeed, alongside biological factors, psychosocial influences are also thought to be important, and it is for this reason that "gender" is also used in this context. Gender includes constructs such as masculinity and femininity, and reflects a more fluid construct where perceptions and beliefs are socially learned and variable, incorporating notions such as gender identity. However, the terms "sex" and "gender" are sometimes used interchangeably, which can cause problems and confusion. For the purposes of this entry, the term "sex" is used when considering the general differences between males and females, incorporating physiological (e.g., anatomy, genetics) and well as psychological and social factors. Gender

will be used when specifically referring to the psychosocial construction of being male or female, and it incorporates a wider psychological and social view of what is it to be a man or a woman.

Introduction

There is extensive evidence to support the notion that there are important differences in the perception and experience of pain between males and females (Fillingim 2000). Studies have generally shown that for a range of different types of pain, women seem to suffer from more pain, with greater frequency, when compared to men. There are numerous sources of information which collectively support this view, and this section will provide a general overview as to the nature and extent of such differences. It will draw on clinical, as well as laboratory-based studies, as well as consider the biopsychosocial mechanisms that may be involved in explaining why such differences occur.

Characteristics

Sex Differences in Clinical Pain

One of the main sources of evidence that supports the view that there are important sex and gender differences in pain comes from epidemiology and investigations into sex-specific incidence rates for clinical conditions (Berkley 1997; Fillingim et al. 2009; LeResche 1999; Unruh 1996). Generally, there seems to be a sex-related bias, with greater incidence rates in females for a number of common painful conditions, including musculoskeletal pain, osteoarthritis, headache/migraine, temporomandibular joint disorder, abdominal pain, and fibromyalgia. Furthermore, when one considers the number of conditions that show this sex-related bias, there seem to be more that show a female-prevalent bias than a male-prevalent bias (Berkley 1997). These differences are not limited to chronic pain, as women also report more frequent acute and everyday pain, as well as suffer greater postoperative pain. Indeed, women report more pain, in more body areas and with greater frequency than men seem to do (Unruh 1996). Women are also more likely to be users of healthcare for pain and are greater users of both over-the-counter and prescription-based analgesics. Reasons for this are varied, and may in part reflect different ways in which males and females are socialized into using healthcare systems. These sex differences in pain experience seem to occur across all phases of the life cycle, in that female-prevalent biases have been reported in both child/adolescent groups, as well as within older adults. However, there is variation, in that for a number of painful conditions these sex-related biases seem to become most apparent in females of reproductive age (i.e., menarche to menopause), e.g., migraine headache, temporomandibular joint disorder (LeResche 1999). Thus, the general pattern that emerges is that while females report more pain and suffer more pain-related conditions across all phases of life, they are particularly pronounced during the reproductive years. Taken together, these studies indicate a clear sex-related bias in the incidence and experience of both acute and chronic pain states.

Sex Differences in Experimentally Induced Pain

Sex differences have also been examined within the laboratory, where it is possible to make use of pain induction procedures (Fillingim et al. 2009; Riley et al. 1998). These pain induction methods often make use of Quantitative Sensory Testing (QST) techniques and involve well-controlled methods such as thermal and electrical stimulation, the cold pressor task, ischemia-inducing tourniquets, pressure pain algometers, and chemical application such as capsaicin. These approaches allow for systematic comparisons between men and women in terms of pain threshold and tolerance levels within thermal, mechanical, and chemical domains, as well as the investigation of centrally mediated processes, including temporal summation and diffuse noxious inhibitory control. The pattern that emerges from these studies is that on average, when compared to males, females show greater sensitivity to a wide range of experimentally induced pain sensations (Fillingim et al. 2009). This difference has been found for pain threshold and tolerance levels, as well as suggestions that there are centrally mediated sex differences as well (Popescu et al. 2010). However, there is also variation in these effects, in that there are differences within the sexes as well as between. This variation may partly be due to the type of pain induction method used, in that some methods seem to show stronger sex effects than others (Fillingim et al. 2009; Riley et al. 1998). For example, ischemic and thermal pain inductions seem to result in weaker sex differences than pressure or electrical stimulation methods. However, these sex differences in pain induction responses are also influenced by a range of psychosocial factors, including coping strategies (e.g., avoid vs. focus), cognitive and emotional processes, as well as social context/interactions. Thus, psychosocial factors may also account for some of the variation within, and between, the sexes. While it seems as if there are consistent sex and gender differences in experimental pain, with females showing greater sensitivity than males, there is variation that needs to be accounted for.

Sex Differences in Responses to Pain Interventions

Besides differences between men and women in the perception and experience of pain, sex and gender have also been considered within the context of pain interventions. Most of this research has focused on whether there are sex differences in response to pain medication (i.e., analgesic effectiveness), using both experimental pain induction procedures as well as clinical pain models, e.g., postsurgical pain. The pattern is somewhat mixed, in that some studies indicate that females exhibit greater analgesic effectiveness than males, whereas others report contradictory findings (Fillingim and Gear 2004; Fillingim et al. 2009; Niesters et al. 2010). Part of the problem is that there are a wide range of study parameters and so these effects may depend on specific factors, such as the type of analgesia involved. Despite this variability, some patterns are emerging under certain conditions. For example, women seem to exhibit greater analgesia from morphine in experimental pain conditions, as well as in clinical patient-controlled analgesia studies (Niesters et al. 2010). There is also the added complication that there may be sex differences in the side effects associated with analgesics (Ciccone and Holdcroft 1999). Indeed, it is possible that reduced opioid consumption may partly be due to avoidance of side effects rather than due to effective pain relief. There is also emerging evidence that there may be sex differences in placebo (and nocebo) analgesia. This implies that there are wider contextual influences that need to be considered when interpreting these treatment-related differences.

Alongside analgesic drug studies, there are also examples of sex and gender differences in interventions that involve psychosocial approaches, including multidisciplinary approaches for chronic pain. Some experimental treatment analogue studies have been conducted, in which attentional coping strategies have been manipulated through simple instructions, e.g., distraction vs. focus, sensory vs. emotional focusing. There seem to be differences in the effectiveness of such approaches, in that men seem to benefit more from attentional focusing than women. There are also investigations as to whether men and women differ in responses to multidisciplinary/ multimodal interventions (Keogh et al. 2005). Here the pattern is somewhat mixed, and like pharmacological studies suffers from inconsistencies and a wide range of differences between studies. There are examples suggesting that there may be sex and gender differences in how patients respond to these treatments, but it is too early to say under which parameters and for which treatments this may occur.

One of the reasons why there is somewhat limited and inconsistent evidence for sex and gender differences in treatment responses is a general failure within clinical studies to report whether sex differences exist (Greenspan et al. 2007). Indeed, despite the good evidence to suggest that there are sex differences in pain and analgesia, there still seems to be a reluctance to directly seek out and/or report such differences in treatment intervention studies. This is a fairly widespread problem, in that biases exist in recruitment of both sexes into clinical trials, and even when the data does exist, the possibility for reporting sex differences is ignored or even statistically controlled for. As a response to this, a consensus report and guidelines were published by the Sex, Gender and Pain Special Interest Group of the International Association for the Study of Pain to try and change this study bias in pain research (Greenspan et al. 2007).

Biological Explanations for Sex Differences in Pain

Explanations for why there may be differences between males and females in terms of pain and analgesia are multifaceted and reflect a range of different biopsychosocial mechanisms (Fillingim 2000). Physiological differences between the sexes are apparent (Holdcroft and Berkley 2005) and include genetics, sex hormones, as well as cardiovascular and immune functions. There have been advances in our knowledge about the genetic and neural mechanisms that may be involved in pain and analgesia, which may also act in a sex-specific way. For example, it has been shown that certain genes may not only show a sex-specific influence on pain sensitivity but may also account for some of the within-sex variations as well (Fillingim et al. 2005; Mogil et al. 2003). Additionally, the possibility that there are different brain mechanisms in male and female pain processing has been explored using imaging methods. Sex differences in areas of brain activation to experimental pain stimuli have been found in both healthy individuals, as well as patient groups. However, there is variability in the methods and results found, meaning that while there is some evidence for neural differences, the actual pattern is not clear.

Progress has been made into the role that sex hormones have in explaining the differences between male and female pain. Here it is suggested that estrogen serves as a vulnerability factor to pain and is the reason for an increased vulnerability within women (Craft et al. 2004; Sherman and LeResche 2006). Evidence supporting the hormonal hypothesis stems from observations that sex differences in pain typically occur during the reproductive years when estrogen levels are more pronounced. Furthermore, it also seems as if fluctuations in sex hormones may be related to changes in pain sensitivity. For example, not only can pain sensitivity decrease over the course of pregnancy, but some chronic conditions, such as temporomandibular joint disorder and rheumatoid arthritis, can change during this period. Postmenopause hormone-replacement theory has been implicated in an increase in headaches in some women, and at least one study that has shown that gender differences in pain sensitivity may change following gender-reassignment surgery, where there are accompanying changes in hormonal profile (Aloisi et al. 2007). Perhaps the main line of evidence implementing sex hormones in human pain sensitivity stems from studies examining pain across the menstrual cycle (Riley et al. 1999; Sherman and LeResche 2006). Both healthy adult studies utilizing pain induction methods, as well as clinical pain studies, suggest that pain sensitivity varies according to phase of cycle. For example, experimental pain studies have shown that there may be greater pain sensitivity during the luteal phase, and at least one study has shown that the effectiveness of some analgesics may vary as a function of the menstrual cycle; analgesia from morphine and side effects were more likely to occur during the follicular phase (Ribeiro-Dasilva et al. 2011). Some clinical pain conditions also seem to vary across the menstrual cycle (e.g., temporomandibular joint disorder), with suggestions that it is the change in sex hormone levels that are important. There are, however, examples where inconsistent effects are found, with suggestions that patterns may depend on methods employed, e.g., type of induction procedure, phase of cycle, etc. Few studies have examined the proposed protective effect of testosterone on pain.

Psychosocial Explanations for Sex Differences in Pain

Alongside these physiological mechanisms, it also seems as if there are important psychosocial factors that may help to explain why there are sex

and gender differences in pain. There seem to be sex differences in pain behaviors, with women being more likely to engage in a range of coping strategies than men. Women also show specific differences in the coping strategies used. For example, women report a greater use of social support systems, as well as being more likely to engage in pain-related catastrophic thinking. Catastrophizing has also been found to explain why there may be sex differences in pain behavior, in both healthy experimental and clinical studies (Sullivan et al. 2001). Interestingly, sex differences have also been found in child and adolescent groups, suggesting that sex differences in coping behaviors are learnt and established early in life.

Despite evidence that there are general differences between the sexes in terms of anxiety and depression, there does not seem to be consistent evidence that men and women differ in terms of these pain-related emotions (Fillingim et al. 2009). However, sex-specific differences are found in the way in which anxiety and depression are related to clinical pain outcomes. Pain-related anxiety seems to be more strongly related to pain outcomes in men, whereas depression has been more strongly related to pain in women. There are exceptions, and these relationships seem to depend on the precise measure under investigation. For example, anxiety sensitivity, which is the fear of bodily sensations, seems closely related to pain in women more than men.

Although pain is a subjective and personal experience, this does not mean that it occurs in isolation. Indeed, wider social factors and context also play an important role in understanding why there are sex differences in pain (Bernardes et al. 2008). For example, it has been shown that the presence of another person can influence the amount of pain that is reported, and that this can occur in a sex-specific way. In pain induction studies it has been shown that if the experimenter is female, then males will report less pain than if the experimenter is male (Levine and Desimone 1991). Part of the reason for this could be due to the person in pain being more or less expressive, but similarly, it could also be due to the behaviors of those observing. For example, the type of supportive behaviors exhibited by spouses to their partner's pain may vary according to whether they are male or female, and parents can differ in their treatment of male and female offspring. The assumption here is that both the expression and detection of pain reflect important social processes and that these pain behaviors can be influenced by those around them (Craig 2009). Recently, there have also been attempts to see whether nonverbal pain behaviors (e.g., facial expressions) are encoded (expressed) and decoded (recognized) in a gender-specific way.

Such differences may be related to the expectations and beliefs that people have about how males and females deal with pain (Bernardes et al. 2008). These beliefs form part of a wider gender identity which develops during childhood and is acquired through the observation and learning of gender-related norms and behaviors from family, friends, teachers, and wider society. Generally, it seems as if the typical man is perceived as stoic and less likely to seek out help for pain, whereas the typical women is viewed as more likely to be sensitive to pain. It has also been shown that there are gender expectations about how the typical man and woman behave when in pain. For example, with respect to pain coping strategies, the typical woman is believed to catastrophize, use distraction, and pray more than the typical man, whereas the typical man is believed to be more likely to ignore pain (Keogh and Denford 2009).

Gender-role expectations also help explain why there are differences between men and women in terms of their pain, as well as highlight that these beliefs influence how observers appraise another person's pain. For example, gender-role expectations have been found to mediate sex differences to experimental pain. It has also been shown that when watching patients in pain, differences in ratings of males and female patients can be explained by observers' gender-role pain beliefs. Such appraisals have also been explored within healthcare professionals using experimental methods (e.g., vignettes, virtual reality patients) where a patient's gender and/or other characteristics can be manipulated. These studies demonstrate that decisions about levels of pain and pain behaviors are influenced by whether the person being observed is male or female. Such biases are thought to have an effect on treatment decisions, although there are only a few published examples that have explored such biases in actual healthcare setting, and even when they are reported there are methodological inconsistencies to consider (LeResche 2011).

Summary

In summary, this entry clearly implicates sex and gender as important in the perception, experience, and response to pain. There are clear and consistent differences between men and women in pain from a wide range of sources, including epidemiology, clinical investigations, and experimental laboratory-based studies on healthy groups. Generally, women seem to show a greater sensitivity to pain and seem to be more vulnerable to painful conditions. While this bias seems to occur at all ages, there seems to be a particular vulnerability during the reproductive years. There is, however, considerable variation within the sexes, and these differences depend on a range of other factors linked to methodology, biology, psychology, and social context. Explanations for sex differences in pain are complex and multidimensional, and we are only just beginning to piece together some of the aspects that help us understand how and why such differences occur. These similarities and differences are not just linked to underlying biology (e.g., sex hormones) but are also influenced by what we think, feel, and behave when in pain. A question that often arises is whether we should start tailoring pain interventions in a sex-specific manner. The answer is unknown at present and likely to be a long way off. This is partly because there is a reluctance to report and investigate sex and gender-related differences in pain research. We need to see clinical trials routinely reporting results for males and females, irrespective of outcome. Once we have this information, we might be in a better position to decide the best approach to effectively manage male and female pain.

References

Aloisi, A. M., Bachiocco, V., Costantino, A., Stefani, R., Ceccarelli, I., Bertaccini, A., et al. (2007). Cross-sex hormone administration changes pain in transsexual women and men. *Pain, 132*(Suppl. 1), 60–67.

Berkley, K. J. (1997). Sex differences in pain. *The Behavioral and Brain Sciences, 20*(3), 371–380.

Bernardes, S. F., Keogh, E., & Lima, M. L. (2008). Bridging the gap between pain and gender research: A selective literature review. *European Journal of Pain, 12*(4), 427–440.

Ciccone, G. K., & Holdcroft, A. (1999). Drugs and sex differences: A review of drugs relating to anaesthesia. *British Journal of Anaesthesia, 82*(2), 255–265.

Craft, R. M., Mogil, J. S., & Aloisi, A. M. (2004). Sex differences in pain and analgesia: The role of gonadal hormones. *European Journal of Pain, 8*(5), 397–411.

Craig, K. D. (2009). The social communication model of pain. *Canadian Psychology, 50*, 22–32.

Fillingim, R. B. (Ed.). (2000). *Sex, gender, and pain* (Vol. 17). Seattle: IASP Press.

Fillingim, R. B., & Gear, R. W. (2004). Sex differences in opioid analgesia: Clinical and experimental findings. *European Journal of Pain, 8*(5), 413–425.

Fillingim, R. B., Kaplan, L., Staud, R., Ness, T. J., Glover, T. L., Campbell, C. M., et al. (2005). The A118G single nucleotide polymorphism of the mu-opioid receptor gene (OPRM1) is associated with pressure pain sensitivity in humans. *The Journal of Pain, 6*(3), 159–167.

Fillingim, R. B., King, C. D., Ribeiro-Dasilva, M. C., Rahim-Williams, B., & Riley, J. L., 3rd. (2009). Sex, gender, and pain: A review of recent clinical and experimental findings. *The Journal of Pain, 10*(5), 447–485.

Greenspan, J. D., Craft, R. M., LeResche, L., Arendt-Nielsen, L., Berkley, K. J., Fillingim, R., et al. (2007). Studying sex and gender differences in pain and analgesia: A consensus report. *Pain, 132*(1), S26–S45.

Holdcroft, A., & Berkley, K. J. (2005). Sex and gender differences in pain and its relief. In S. B. McMahon, M. Koltzenburg, P. D. Wall, & R. Melzack (Eds.), *Wall and Melzack's textbook of pain* (5th ed., pp. 1181–1197). Edinburgh: Elsevier Churchill Livingstone.

Keogh, E., & Denford, S. (2009). Sex differences in perceptions of pain coping strategy usage. *European Journal of Pain, 13*, 629–634.

Keogh, E., McCracken, L. M., & Eccleston, C. (2005). Do men and women differ in their response to interdisciplinary chronic pain management? *Pain, 114*(1–2), 37–46.

LeResche, L. (1999). Gender considerations in the epidemiology of chronic pain. In I. K. Crombie, P. R. Croft, S. J. Linton, L. LeResche, & M. Von Korff (Eds.), *Epidemiology of pain* (pp. 43–52). Seattle: IASP Press.

LeResche, L. (2011). Defining gender disparities in pain management. *Clinical Orthopaedics and Related Research, 469*(7), 1871–1877.

Levine, F. M., & Desimone, L. L. (1991). The effects of experimenter gender on pain report in male and female subjects. *Pain, 44*(1), 69–72.

Mogil, J. S., Wilson, S. G., Chesler, E. J., Rankin, A. L., Nemmani, K. V., Lariviere, W. R., et al. (2003). The melanocortin-1 receptor gene mediates female-specific mechanisms of analgesia in mice and humans. *Proceedings of the National Academy of Sciences of the United States of America, 100*(8), 4867–4872.

Niesters, M., Dahan, A., Kest, B., Zacny, J., Stijnen, T., Aarts, L., et al. (2010). Do sex differences exist in opioid analgesia? A systematic review and meta-analysis of human experimental and clinical studies. *Pain, 151*(1), 61–68.

Popescu, A., LeResche, L., Truelove, E. L., & Drangsholt, M. T. (2010). Gender differences in pain modulation by diffuse noxious inhibitory controls: A systematic review. *Pain, 150*(2), 309–318.

Ribeiro-Dasilva, M. C., Shinal, R. M., Glover, T., Williams, R. S., Staud, R., Riley, J. L., 3rd, et al. (2011). Evaluation of menstrual cycle effects on morphine and pentazocine analgesia. *Pain, 152*(3), 614–622.

Riley, J. L., Robinson, M. E., Wise, E. A., Myers, C. D., & Fillingim, R. B. (1998). Sex differences in the perception of noxious experimental stimuli: A meta-analysis. *Pain, 74*(2–3), 181–187.

Riley, J. L., Robinson, M. E., Wise, E. A., & Price, D. D. (1999). A meta-analytic review of pain perception across the menstrual cycle. *Pain, 81*(3), 225–235.

Sherman, J. J., & LeResche, L. (2006). Does experimental pain response vary across the menstrual cycle? A methodological review. *American Journal of Physiology - Regulatory, Integrative and Comparative Physiology, 291*(2), R245–R256.

Sullivan, M. J., Thorn, B., Haythornthwaite, J. A., Keefe, F., Martin, M., Bradley, L. A., et al. (2001). Theoretical perspectives on the relation between catastrophizing and pain. *The Clinical Journal of Pain, 17*(1), 52–64.

Unruh, A. M. (1996). Gender variations in clinical pain experience. *Pain, 65*(2–3), 123–167.

Gender Differences in Opioid Analgesia

▶ Sex Differences in Opioid Analgesia

Gender Role Expectation of Pain Scale

Synonyms

GREP scale

Definition

The Gender Role Expectation of Pain Scale measures gender-related stereotypic attributions of pain sensitivity, endurance, and willingness to report pain.

Cross-References

▶ Psychological Aspects of Pain in Women

Gene

Definition

A gene contains hereditary information encoded in the form of DNA and is located at a specific position on a chromosome in a cell's nucleus. Genes individually determine many aspects of physiological functions by controlling the production of proteins.

Cross-References

▶ NSAIDs, Pharmacogenetics

Gene Array

Definition

Nucleic acid arrays work by hybridization of labeled RNA or DNA in solution to DNA molecules attached at specific locations on a surface. The hybridization of a sample to an array is, in effect, a highly parallel search by each molecule for a matching partner on an "affinity matrix" with the eventual pairings of molecules on the surface determined by the rules of molecular recognition.

Cross-References

▶ Retrograde Cellular Changes after Nerve Injury

Gene Therapy and Opioids

▶ Opioids and Gene Therapy

Gene Transcription

Definition

Gene transcription is the process of constructing a messenger RNA molecule using a DNA molecule as a template, with resulting transfer of genetic information to the messenger RNA.

Cross-References

▶ NSAIDs, Pharmacogenetics

General Adaptation Syndrome

▶ Postoperative Pain, Pathophysiological Changes in Neuroendocrine Function in Response to Acute Pain

General Anesthesia

Definition

General anesthesia is a drug-induced loss of consciousness during which patients are not arousable and often have impaired cardiorespiratory function needing support.

Cross-References

▶ Pain and Sedation of Children in the Emergency Setting

Generator Current

▶ Nociceptor Generator Potential

Generator Currents

Definition

Membrane currents originated at the membrane of sensory receptor endings when transduction channels are open or closed by the stimulus. These currents spread passively from the stimulated membrane patch to neighbor sites (electrotonic propagation) according to the spatial and temporal cable properties of the axon.

Cross-References

▶ Nociceptor Generator Potential

Generic Carbamazepine

▶ Tegretol

Genetic Correlation

Definition

Genetic correlation is a mediation by similar sets of genes, suggestive of overlapping physiological mediation. Genetic correlation of two traits can be inferred by a significant correlation of inbred strain means on each trait.

Genetic Linkage

Cross-References

▶ Quantitative Trait Locus Mapping

Geniculate Ganglionitis

▶ Geniculate Neuralgia

Geniculate Neuralgia

Synonyms

Facial ganglion neuralgia; Geniculate ganglionitis; Nervus intermedius neuralgia

Definition

Geniculate neuralgia is a rare, severe chronic pain condition in which stabbing or electric shock-like pain paroxysms, sometimes accompanied by ongoing burning pain, are felt in the distribution of the sensory component of the facial (seventh cranial) nerve (the outer ear, ear canal, mastoid, or infrequently in the region of the eye). It is similar to the other cranial nerve neuralgias, trigeminal neuralgia, and glossopharyngeal neuralgia. Pain attacks may occur spontaneously, or be triggered by sensory stimuli or by swallowing.

Cross-References

▶ Neuralgia
▶ Neuralgia, Assessment
▶ Trigeminal and Cranial Nerve Neuralgias
▶ Trigeminal, Glossopharyngeal, and Geniculate Neuralgias

Geniculate neuralgia: *Nervus Intermedius*, Primary Otalgia

▶ Trigeminal, Glossopharyngeal, and Geniculate Neuralgias

Genital Mucosa, Nociception

▶ Nociception in Mucosa of Sexual Organs

Genito-Pelvic Pain/Penetration Disorder

▶ Dyspareunia and Vaginismus

Genome

Definition

A genome is the total set of genes carried by an individual or a cell. The genome determines, in part, the final morphology or body or form of the individual human or cell.

Cross-References

▶ Cell Therapy in the Treatment of Central Pain

Genotype

Definition

The genotype describes the genetic makeup of an individual organism, determined by the full complement of genes that the organism possesses.

Cross-References

▶ NSAIDs, Pharmacogenetics

GERD

Synonyms

Gastroesophageal reflux disease

Definition

Frequent reflux of gastric acid into the esophagus during transient lower esophageal relaxation. Acid reflux produces heartburn and in chronic cases it produces erosive esophagitis.

Cross-References

▶ Visceral Pain Model; Esophageal Pain

Geriatric Medicine

Definition

Geriatric medicine is a discipline dedicated to the study, assessment, and treatment of diseases related to older populations.

Cross-References

▶ Psychological Treatment of Pain in Older Populations

Giant Cell Arteritis (Arthritis)

Definition

Giant cell arteritis is an autoimmune disease with vasculitis, with preference for extracranial branches of the arteria carotis, including the temporal and arteries to the retina and the optical nerve. The symptoms include muscle pain as in polymyalgia.

Cross-References

▶ Muscle Pain in Systemic Inflammation (Polymyalgia Rheumatica, Giant Cell Arteritis, Rheumatoid Arthritis)

Giant Cell Arthritis

▶ Muscle Pain in Systemic Inflammation (Polymyalgia Rheumatica, Giant Cell Arteritis, Rheumatoid Arthritis)

Gigantocellular Reticular Nucleus

Definition

Gigantocellular reticular nucleus is the medial zone of the rostral medullary reticular formation containing large cell bodies.

Cross-References

▶ Spinothalamocortical Projections to Ventromedial and Parafascicular Nuclei

GIRK Channel

Definition

GIRK stands for G-protein-coupled inwardly rectifying potassium (K) channel. As the name implies, these channels are activated by G-protein coupled receptors, where the $\beta\gamma$ subunit of the G-protein is ultimately responsible for channel activation. The channels are described as inward rectifying because they have a higher conductance for potassium entering than leaving the cell as a result of Mg2+ or polyamine block of the channel from the inside which limits the amount of K+ that can flow out. These channels are encoded by four genes (KCNJ3, 5, 6, and 9) which encode GIRK 1, 4, 2, and 3, respectively. These channels are also known as K_{ir}3.1–3.4. One of the primary mechanisms of opioid-induced inhibition of neural activity in the central nervous system is via the activation of GIRK channels which will drive membrane hyperpolarization.

Cross-References

▶ Opioid Electrophysiology in PAG

Glabrous

Definition

Glabrous means smooth and non-hairy, for example, glabrous skin.

Cross-References

▶ Substance P Regulation in Inflammation

Glabrous Skin

Definition

The glabrous skin is the completely hairless skin areas, including the palmar surface of the hand and the plantar surface of the foot.

Cross-References

▶ Pain in Humans, Thresholds

Glans Clitoris

Definition

Glans is the very sensitive and visible part of the clitoris, made up entirely of erectile tissue, soft to the touch even when aroused, and engorged with blood land, usually averaging 4–5 mm in diameter.

Cross-References

▶ Clitoral Pain

Glia

Definition

Glia are cells that surround and support neurons in the central and peripheral nervous systems; glial and neural cells together compose the tissue of the central nervous system. There are three types of CNS glial cells: astrocytes (part of the blood brain barrier), oligodendrocytes (myelin-producing cells of the CNS), and microglia (CNS macrophages). In the periphery, glial cells comprise the ▶ satellite cells and Schwann cells.

Cross-References

- ▶ Cord Glial Activation
- ▶ Cytokine Modulation of Opioid Action
- ▶ Glia and Neuropathic Pain
- ▶ Glial Cells in Orofacial Pathological Pain Mechanisms

Glia and Neuropathic Pain

Julie Wieseler, Steven F. Maier and Linda R. Watkins
Department of Psychology and Neuroscience, and The Center for Neuroscience, University of Colorado at Boulder, Boulder, CO, USA

Definition

It is currently well accepted that glia, the major nonneuronal, immunocompetent cells of the central nervous system, specifically astrocytes and microglia, play a significant role in the development and maintenance of neuropathic pain states. They do this at least in part through their release of proinflammatory, neuroexcitatory signals. This relationship has been identified and characterized in multiple exaggerated pain rodent models, with evidence for the relationship also supported by clinical data (Backonja et al. 2008).

Introduction

Glia, specifically astrocytes, are intimately entangled with neurons insofar as they encapsulate neuronal synapses, as well as wrap their processes around neuronal somas, axons, and dendrites (Araque et al. 1999). This proximity puts astrocytes in a unique position to modulate the activity of nocisponsive neurons. Garrison et al. were the first to investigate the relationship between exaggerated pain and ▶ glial activation (Garrison et al. 1994). Using the sciatic chronic constriction injury (CCI) model, they compared the expression of CCI-induced hyperalgesia and immunohistochemical markers of astrocyte activation (Garrison et al. 1991). In addition, they evaluated whether CCI-induced astrocyte activation would be affected by an ▶ N-methyl-D-aspartate (NMDA) receptor antagonist (MK801), known to inhibit CCI-induced hyperalgesia (Garrison et al. 1994). They found that (1) hyperalgesia induced by CCI is associated with astrocyte activation, and (2) this glial activation (like CCI-induced hyperalgesia) is inhibited by MK801. Thus, these studies revealed a predictive relationship between glial activation and exaggerated pain. Since then, positive correlations between exaggerated pain and activation of spinal cord microglia (see caption of Fig. 1 for discussion) and astrocytes have been revealed by diverse animal models of exaggerated pain, including those induced by inflammation and/or trauma to tissues, peripheral nerves, spinal roots, and/or spinal cord underlying clinical pathologies including peripheral nerve injury, spinal cord injury, and possibly migraine (DeLeo and Colburn 1999; Hulsebosch et al. 2009; Milligan and Watkins 2009; Watkins et al. 2001). Such data naturally raise the question of whether glial activation is necessary and/or sufficient for induction and maintenance of exaggerated pain states.

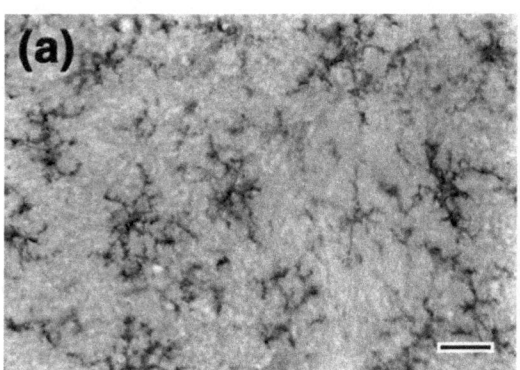

Glia and Neuropathic Pain, Fig. 1 Example of microglial activation in response to a stimulus that creates exaggerated pain states. Perispinal administration of HIV-1 envelope glycoprotein gp120 creates exaggerated response to both thermal and touch/pressure stimuli. Disruption of glial activation abolishes the pain changes. Furthermore, these pain changes are correlated with anatomical evidence of activation of both astrocytes and microglia. An example of such microglial activation is illustrated here. *Panel A* illustrates the dorsal horn of a rat injected perispinally with vehicle. *Panel B* is identical except that the rat received perispinal HIV-1 gp120 at a dose that creates exaggerated pain states. These photomicrographs are from the tissues collected for analysis reported in Milligan et al. (2001). Activation of microglia induces these cells to upregulate their expression of complement type 3 receptors. Thus, enzyme labeled antibodies directed against complement receptor type 3 (OX-42 monoclonal antibodies) can be used to detect microglial activation by light microscopy. In the photomicrographs shown here, the bound enzyme labeled antibodies catalyzed a reaction leading to precipitation of a colored reaction product. Thus, more antibody bound (and hence more colored reaction product produced) reflects upregulation of the complement receptor type 3 microglial activation marker. By this method, activated microglia appear darker and more densely stained in accordance with their increased expression of OX-42 antibody bound receptors. Scale bar, 25mm in (**a**) and (**b**) (Reproduced from Milligan et al. (2001) with permission)

That glial activation plays a prominent role in the induction of exaggerated pain has been demonstrated using glial inhibitors. ▶ Fluorocitrate selectively blocks glial metabolic activity by inhibiting aconitase, a Krebs cycle enzyme (Hassel et al. 1992; Paulsen et al. 1987). Blocking the activation of spinal cord glial cells with fluorocitrate blocks exaggerated pain states, including thermal and mechanical hyperalgesia and mechanical allodynia (Meller et al. 1994; Milligan et al. 2000). A similar pattern has also been observed with the putative microglial inhibitor, ▶ minocycline. Minocycline selectively inhibits the release of several neuroexcitatory substances from microglia, including inflammatory mediators. Administration of minocycline attenuates the development of mechanical allodynia and thermal hyperalgesia in response to spinal nerve transection, peripheral nerve inflammation, or spinal cord inflammation (Raghavendra et al. 2003). Thus, microglial activation plays a significant role in the development of exaggerated pain. Intriguingly, minocycline failed to reverse, or minimally reversed, these exaggerated pain states once they were fully expressed. These data, supported by parallel analyses of glial activation markers, revealing upregulation of GFAP (astrocyte activation marker), suggest that microglial activation may lead to astrocyte activation. Astrocytes may then play a more significant role in the maintenance of the pain facilitatory process. Taken together, these data suggest that glial activation is important for both the induction and maintenance of exaggerated pain. Additional glial inhibitors including propentofylline (Sweitzer and De Leo 2011) and ibudilast (Rolan et al. 2009) are also found to attenuate the development of exaggerated pain states, validating these earlier data.

It is important to note another key finding from these early studies with glial inhibitors. That is, glia are not involved in normal pain processing (Milligan et al. 2000). Normally, glia are quiescent, releasing nothing that enhances pain. However, in situations leading to exaggerated pain states, glia become activated. Once activated, they begin releasing substances that contribute to the amplification of pain. Therefore, the involvement of glia in pain is determined by their state of activation.

The data discussed above was the first to show that glial activation is important for the expression of exaggerated pain, and was followed up the question: is glial activation sufficient for the induction of pain? To address this question, glia alone needed to be directly activated. While glia can be activated by classical pain neurotransmitters, including EAA, NO, prostaglandins, and extracellular ATP, these substances also activate neurons. To activate spinal cord glia without activating neurons, the portions of bacteria and viruses that bind to and activate the immunocompetent glial cells have been used. Administration of these substances indeed induces mechanical allodynia and thermal hyperalgesia (Meller et al. 1994; Milligan et al. 2000). Thus, taken together, spinal cord glial activation is necessary and sufficient in both the creation and maintenance of exaggerated pain states (see also ▶ Glial Cells in Orofacial Pathological Pain Mechanisms).

Characteristics

Activation of Glia via Neuron-Glia Communication

The discussion above describes glia as integral to the creation and maintenance of exaggerated pain. The question remains as to what signal(s) induces glial activation in the dorsal horn of the spinal cord, thereby creating exaggerated pain. As described, one mechanism is via the release of classical pain neurotransmitters/ neuromodulators including EAA, substance P, prostaglandins, ATP, and NO among others, by incoming sensory neurons and/or by intrinsic dorsal horn neurons in proximity to glia. While each of these substances excites spinal cord neurons in the pain pathway, they also activate astrocytes and microglia. As such, these traditional neuronal mediators may exert their pain modulating effects, at least in part, via glial activation.

There are a number of signals that may potentially trigger glial activation, such as altered neuronal activity, transported molecules from sensory neurons, or the removal of tonic neuronal inhibitory signals. Current signals being considered include neuronal chemokines. Fractalkine is considered to be a putative neuron-to-glia signal. Fractalkine is a protein expressed on the extracellular surface of sensory neurons as well as dorsal horn neurons. Fractalkine is tethered to neuronal membranes by a mucin stalk (Chapman et al. 2000). When neurons are sufficiently activated, the mucin stalks break, releasing fractalkine into the extracellular fluid. Within the spinal cord, only microglia express receptors for fractalkine, supporting the concept of fractalkine as a neuron-to-glia signal (Fig. 2). Another chemokine playing a significant role in neuropathic pain and glial activation is monocyte chemoattractant protein-1 (MCP-1; also known as CCL2, binds to CCR2). MCP-1 is upregulated in the dorsal root ganglia following nerve trauma, and is known to activate glia (White and Miller 2010). These neuron-to-glia signals point to a means by which sensory nocisponsive neurons and dorsal horn neurons communicate to glia, thereby inducing activation. In response to neuron-to-glia signals, glia release substances that then excite other glia and neurons. For example, microglial binding of fractalkine rapidly induces the release of the neuro- and glial-excitatory substances (including interleukin-1beta and nitric oxide) from dorsal spinal cord (Watkins and Maier 2003). The release of these excitatory substances induces exaggerated pain.

Once glia are activated, the question becomes what signal(s) maintains the persistent glial activation. One possible answer currently being considered studied in the field is that the signal is something released from dying or degenerating neurons associated with some models of chronic pain (Coggeshall et al. 2001; Moore et al. 2002; Sugimoto et al. 1990). Neurons modulate glial activation via cell-to-cell contact beyond that described above (Gao and Ji 2010; Milligan and Watkins 2009). It is well known that neuronal injury and degeneration, often a component of peripheral or central neural trauma, activate glia

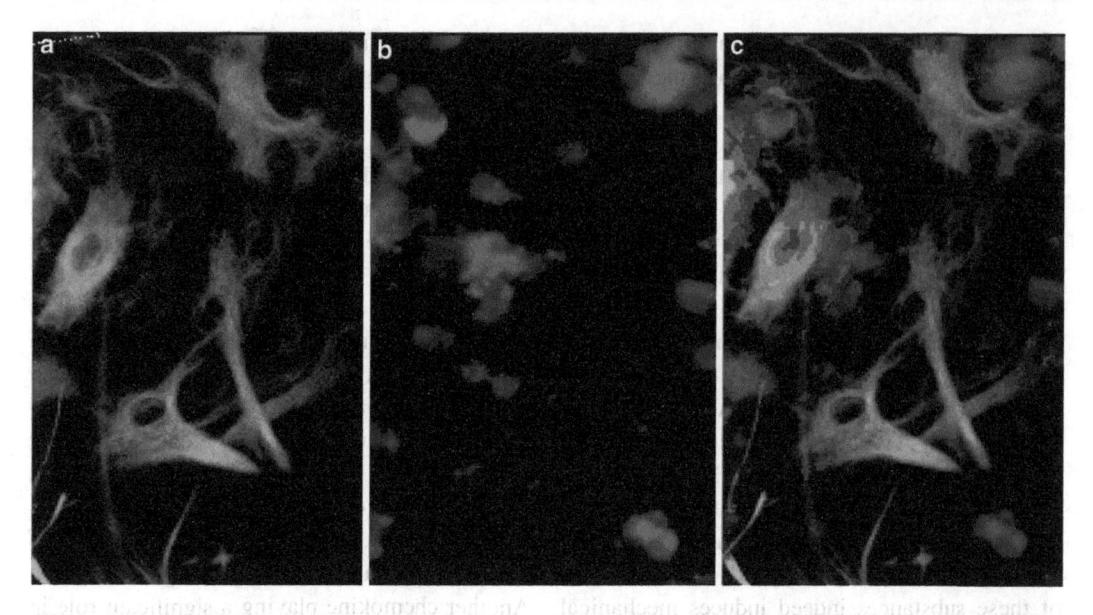

Glia and Neuropathic Pain, Fig. 2 Mixed astrocyte and microglia culture from neonatal rat pups showing selectivity of fractalkine for microglia. *Panel A* reveals astrocytes labeled with GFAP (glial fibrillary acidic protein) visualized by a fluorescent antibody conjugated to Alexa 488. GFAP is a cytoskeletal protein that is highly specific to astrocytes. Note that, despite these astrocytes being in a mixed culture containing microglia, no microglia are labeled for GFAP. *Panel B* shows microglia labeled with fractalkine, which is directly conjugated to the Cy3b fluorochrome. In the dorsal spinal cord, fractalkine receptors are found exclusively on microglia. Note that, despite these microglia being in a culture containing astrocytes, no astrocytes are labeled by fractalkine. *Panel C* presents an overlay of *Panels A* and *B* to provide the true picture of the mixed culture

via different pathways. Stressed, damaged, or dead neurons have porous membranes through which endogenous danger signals, referred to as "alarmins," are released. These "alarmins" include heat-shock proteins, HMGB1 (high mobility group box protein 1), and ATP (adenosine triphosphate) are released (Watkins et al. 2009), and all are known to activate glia. Most current evidence supports that the "alarmins" are binding toll-like receptors (TLRs), such as TLR4. Thus, when a neuron dies, the cell-to-cell contact is disrupted, removing the inhibitory signal and as well as releasing the "alarmins," thereby leading to glial activation. Thus, it seems possible that persistent glial activation may be part of a positive feedback loop in that glia are activated in response to the neuronal death cascade, in turn causing further cell death of local neurons.

As ▶ immunocompetent cells, activated glia release classical immune products in response to neuronal signals. These include proinflammatory cytokines such as interleukin-1beta, interleukin-6,

and ▶ tumor necrosis factor alpha (TNF-α) reduced. In their role as classical immune signals, these cytokines are "proinflammatory" because they orchestrate the early immune response to infection and damage by recruiting immune cells to the site and activating them. In spinal cord, these proinflammatory cytokines are signal molecules that activate neurons as well as glia, via binding to specific receptors on these cells. This is one means by which glia cannot only signal neurons, but can further activate neighboring glia. Proinflammatory cytokines are not constitutively released in the spinal cord, and pharmacologically blocking their activity does not alter normal pain responsiveness. However, pain is enhanced upon intrathecal delivery, and blocking their spinal actions blocks allodynia in diverse pain models (reviewed in (Milligan and Watkins 2009; Watkins et al. 2007; Wieseler-Frank et al. 2005)). Moreover, in conditions leading to exaggerated pain, including chronic constriction injury (CCI), sciatic inflammatory neuropathy (SIN), and spinal

Glia and Neuropathic Pain, Fig. 3 Illustration of the spread of a calcium wave among astrocytes. Astrocytes are electrically coupled by gap junctions into vast networks. These electrical couplings allow astrocytes to rapidly communicate with one another over widespread areas. These images show a sequence of time lapse photographic images, illustrating the rapid spread of nerve transection, blocking proinflammatory cytokine activity in spinal cord blocks mechanical allodynia and thermal hyperalgesia (reviewed in (DeLeo and Colburn 1999)). Proinflammatory cytokines are therefore necessary for these exaggerated pain states. Mechanical allodynia and mechanical and thermal hyperalgesia are induced by spinal administration of proinflammatory cytokines (reviewed in (Milligan and Watkins 2009)). Thus, proinflammatory cytokines are not only sufficient for exaggerated pain, but likely potentiate glia activation.

In addition to released signals acting on glia, astrocytes are also able to communicate with other astrocytes via ▶ gap junctions. This spread of excitation leads to the activation of distant glia (Fig. 3) and their release of neuroexcitatory substances (Araque et al. 1999).

The importance of this type of astrocyte-to-astrocyte communication, causing excitation of excitation that occurs after the activation of a single glial cell at time 0. These newly activated, distant cells can then release pain-enhancing substances, thereby activating nearby pain neurons. Such gap junctional spread of excitation has recently been linked to pain facilitation, especially to mirror-image pain (Reproduced from Araque et al. (1999) with permission)

distant glia, raises the question of whether gap junctions may potentially provide an explanation for the spread of human pain beyond the original site of trauma; that is, mirror image pain and extraterritorial pain. Such a communication network could, theoretically at least, lead to release of pain-enhancing substances by newly activated glia distant from the initial site of gap junctional activation. While no data yet exist in humans, the spread of pain in animal models is indeed blocked by spinal administration of the gap junction decoupler ▶ carbenoxolone. Carbenoxolone blocks the spread of mechanical allodynia in peripheral neuropathic pain models, CCI and SIN (Spataro et al. 2004), and in the central neuropathic pain model, spinal cord hemisection (Roh et al. 2010). Control studies, using glycyrrhetinic acid, verify that carbenoxolone-induced disruption of CCI and SIN-induced mirror image pains are indeed due to its disruption of gap junctions (Spataro et al. 2004). Thus, gap junctional communication among astrocytes provides an intriguing explanation for pain that arises in otherwise healthy body parts.

Summary

In summary, spinal cord glial activation is integrally involved in diverse exaggerated pain phenomena. The powerful role that glia play in pain facilitation suggests that drugs targeting this nonneuronal cell type may hold promise for effective clinical pain control.

Acknowledgment This work was supported by National Institutes of Health Grants D A024044, DA 023132, and DE021966.

References

Araque, A., Parpura, V., Sanzgiri, R. P., & Haydon, P. G. (1999). Tripartite synapses: Glia, the unacknowledged partner. *Trends in Neurosciences, 22*, 208–215.

Backonja, M. M., Coe, C. L., Muller, D. A., & Schell, K. (2008). Altered cytokine levels in the blood and cerebrospinal fluid of chronic pain patients. *Journal of Neuroimmunology, 195*(1–2), 157–163.

Chapman, G. A., Moores, K., Harrison, D., Campbell, C. A., Stewart, B. R., & Strijbos P. J. (2000). Fractalkine cleavage from neuronal membranes represents an acute event in the inflammatory response to excitotoxic brain damage. *Journal of Neuroscience, 20*(15), RC87. http://www.ncbi.nlm.nih.gov/entrez/query.fcgi?cmd=Retrieve&db=PubMed&dopt=Citation&list_uids=10899174

Coggeshall, R. E., Lekan, H. A., White, F. A., & Woolf, C. J. (2001). A-fiber sensory input induces neuronal cell death in the dorsal horn of the adult rat spinal cord. *The Journal of Comparative Neurology, 435*(3), 276–282.

DeLeo, J. A., & Colburn, R. W. (1999). Proinflammatory cytokines and glial cells: Their role in neuropathic pain. In L. Watkins (Ed.), *Cytokines and pain* (pp. 159–182). Basel: Birkhauser.

Gao, Y. J., & Ji, R. R. (2010). Chemokines, neuronal-glial interactions, and central processing of neuropathic pain. *Pharmacology and Therapeutics, 126*(1), 56–68.

Garrison, C. J., Dougherty, P. M., Kajander, K. C., & Carlton, S. M. (1991). Staining of glial fibrillary acidic protein (GFAP) in lumbar spinal cord increases following a sciatic nerve constriction injury. *Brain Research, 565*(1), 1–7.

Garrison, C. J., Dougherty, P. M., & Carlton, S. M. (1994). GFAP expression in lumbar spinal cord of naive and neuropathic rats treated with MK-801. *Experimental Neurology, 129*, 237–243.

Hassel, B., Paulsen, R. E., Johnson, A., & Fonnum, F. (1992). Selective inhibition of glial cell metabolism by fluorocitrate. *Brain Research, 249*, 120–124.

Hulsebosch, C. E., Hains, B. C., Crown, E. D., & Carlton, S. M. (2009). Mechanisms of chronic central neuropathic pain after spinal cord injury. *Brain Research Reviews, 60*(1), 202–213.

Meller, S. T., Dykstra, C., Grzybycki, D., Murphy, S., & Gebhart, G. F. (1994). The possible role of glia in nociceptive processing and hyperalgesia in the spinal cord of the rat. *Neuropharmacology, 33*(11), 1471–1478.

Milligan, E. D., & Watkins, L. R. (2009). Pathological and protective roles of glia in chronic pain. *Nature Reviews Neuroscience, 10*(1), 23–36.

Milligan, E. D., Mehmert, K. K., Hinde, J. L., Harvey, L. O., Martin, D., Tracey, K. J., et al. (2000). Thermal hyperalgesia and mechanical allodynia produced by intrathecal administration of the human immunodeficiency virus-1 (HIV-1) envelope glycoprotein, gp120. *Brain Research, 861*(1), 105–116.

Milligan, E. D., O'Connor, K. A., Nguyen, K. T., Armstrong, C. B., Twining, C., Gaykema, R. P., Holguin, A., Martin, D., Maier, S. F., & Watkins, L. R. (2001). Intrathecal HIV-1 envelope glycoprotein gp120 induces enhanced pain states mediated by spinal cord proinflammatory cytokines. *Journal of Neuroscience, 21*(8), 2808–2819. http://www.ncbi.nlm.nih.gov/entrez/query.fcgi?cmd=Retrieve&db=PubMed&dopt=Citation&list_uids=11306633

Moore, K. A., Kohno, T., Karchewski, L. A., Scholz, J., Baba, H., & Woolf, C. J. (2002). Partial peripheral nerve injury promotes a selective loss of GABAergic inhibition in the superficial dorsal horn of the spinal cord. *Journal of Neuroscience, 22*(15), 6724–6731.

Paulsen, R. E., Contestabile, A., Villani, L., & Fonnum, F. (1987). An in vivo model for studying function of brain tissue temporarily devoid of glial cell metabolism: The use of fluorocitrate. *Journal of Neurochemistry, 48*, 1377–1385.

Raghavendra, V., Tanga, F., & DeLeo, J. A. (2003). Inhibition of microglial activation attenuates the development but not existing hypersensitivity in a rat model of neuropathy. *Journal of Pharmacology and Experimental Therapeutics, 306*(2), 624–630.

Roh, D. H., Yoon, S. Y., Seo, H. S., Kang, S. Y., Han, H. J., Beitz, A. J., et al. (2010). Intrathecal injection of carbenoxolone, a gap junction decoupler, attenuates the induction of below-level neuropathic pain after spinal cord injury in rats. *Experimental Neurology, 224*(1), 123–132.

Rolan, P., Hutchinson, M., & Johnson, K. (2009). Ibudilast: A review of its pharmacology, efficacy and safety in respiratory and neurological disease. *Expert Opinion on Pharmacotherapy, 10*(17), 2897–2904.

Spataro, L. E., Sloane, E. M., Milligan, E. D., Wieseler-Frank, J., Schoeniger, D., Jekich, B. M., Barrientos, R. M., Maier, S. F., & Watkins, L. R. (2004). Spinal gap junctions: potential involvement in pain facilitation. *Journal of Pain, 5*(7), 392–405. http://www.ncbi. nlm.nih.gov/entrez/query.fcgi?cmd=Retrieve&db= PubMed&dopt=Citation&list_uids=15501197

Sugimoto, T., Bennett, G. J., & Kajander, K. C. (1990). Transsynaptic degeneration in the superficial dorsal horn after sciatic nerve injury: Effects of a chronic constriction injury, transection, and strychnine. *Pain, 42*(2), 205–213.

Sweitzer, S., & De Leo, J. (2011). Propentofylline: Glial modulation, neuroprotection, and alleviation of chronic pain. *Handbook of Experimental Pharmacology, 200,* 235–250.

Watkins, L. R., & Maier, S. F. (2003). Glia: A novel drug discovery target for clinical pain. *Nature Reviews. Drug Discovery, 2,* 973–985.

Watkins, L. R., Milligan, E. D., & Maier, S. F. (2001). Glial activation: A driving force for pathological pain. *Trends in Neurosciences, 24,* 450–455.

Watkins, L. R., Hutchinson, M. R., Milligan, E. D., & Maier, S. F. (2007). "Listening" and "talking" to neurons: Implications of immune activation for pain control and increasing the efficacy of opioids. *Brain Research Reviews, 56*(1), 148–169.

Watkins, L. R., Hutchinson, M. R., Rice, K. C., & Maier, S. F. (2009). The "toll" of opioid-induced glial activation: Improving the clinical efficacy of opioids by targeting glia. *Trends in Pharmacological Sciences, 30*(11), 581–591.

White, F. A., & Miller, R. J. (2010). Insights into the regulation of chemokine receptors by molecular signaling pathways: Functional roles in neuropathic pain. *Brain, Behavior, and Immunity, 24*(6), 859–865.

Wieseler-Frank, J., Maier, S. F., & Watkins, L. R. (2005). Central proinflammatory cytokines and pain enhancement. *Neurosignals, 14*(4), 166–174.

Glial Activation

Definition

Glia (microglia, astrocytes) exit their basal states and become activated in response to inflammation, damage, infection, and various neuron-to-glia signals (fractalkine, various neurotransmitters, etc.). The basal state of astrocytes is active but not activated. That is, astrocytes regulate the extracellular ion and chemical environment, among other functions. The basal state of microglia is quiescent surveillance. Upon activation, both types of glia can release a variety of neuroexcitatory substances (activation markers) such as proinflammatory cytokines, nitric oxide, and prostaglandins. Glial activation is associated with exaggerated pain states.

Cross-References

▶ Cord Glial Activation
▶ Proinflammatory Cytokines

Glial Cell Line-Derived Neurotrophic Factor

Synonyms

GDNF

Definition

Glial cell line-derived neurotrophic factor (GDNF) was the first described member of a novel family of trophic factors that also includes neurturin, persephin, and artemin. In addition to effects in the CNS, GDNF, neurturin, and artemin promote the in vitro survival of many peripheral neurons, including enteric, sympathetic, and sensory neurons. Members of the GDNF family exert their effects via a multicomponent receptor complex consisting of RET, a tyrosine kinase receptor acting as a signal-transducing domain, in combination with a member of the GFRα family of GPI-linked receptors (GFRα, GFRα4) acting as ligand-binding domains. Either GFRα1 or GFRα2 in conjunction with RET can mediate GDNF signaling, although GDNF is thought to bind preferentially to GFRα1. GFRα1 and GFRα2 overlap extensively with the IB4-positive population of sensory neurons.

Cross-References

▶ IB4-Positive Neurons, Role in Inflammatory Pain
▶ Immunocytochemistry of Nociceptors

Glial Cells in Orofacial Pathological Pain Mechanisms

Koichi Iwata[1], Chen Yu Chiang[2] and Pavel Cherkas[2]
[1]Department of Physiology, Nihon University School of Dentistry, Chiyoda-ku, Japan
[2]Faculty of Dentistry, University of Toronto, Toronto, ON, Canada

Definition

▶ Glia play an important role in the pathogenesis of both acute and chronic pain conditions. In the periphery, the ▶ satellite glial cells (SGCs) in the ▶ trigeminal ganglion (TG) have an important function in modulating the activity of TG neurons and undergo changes following tissue inflammation and nerve injury which contribute to orofacial pain conditions. Likewise, in the trigeminal ▶ subnucleus caudalis (Vc) both ▶ microglia and ▶ astroglia have been shown to be crucially involved in the mechanisms underlying the development of inflammatory and neuropathic orofacial pain.

Introduction

▶ Glial cells far outnumber neurons in the central nervous system (CNS), and include several different subtypes such as ▶ astroglia (astrocytes), resident ▶ microglia, perivascular (migrated) microglia, and oligodendrocytes and each has distinct functions in the CNS. Glial cells also play an important role in the peripheral nervous system where they are present as ▶ satellite glial cells (SGCs) in the sensory and autonomic ganglia and as ▶ Schwann cells surrounding the axons. It has long been known that glial cells nurture neurons and enhance their function, maintaining the chemical environment around neurons, protecting and assisting in the repair and regeneration of neurons, and scavenging dead cells and bacteria in instances of injury, inflammation, or infection (for review, see Haydon and Carmignoto 2006). Although these cells do not fire electrical impulses, they can release several neurotransmitters and growth factors that can influence the activity of neurons or that can act on other cells (e.g., of the immune system). Furthermore, it is now becoming clear that these cells play an important role in the pathogenesis of both acute and chronic pain conditions and may offer new targets for management of these conditions (for review, see Tsuda et al. 2005; Ren and Dubner 2010). Several other entries describe the role of glia in spinal dorsal horn and dorsal root ganglia (DRG) pain mechanisms, and this entry focuses on SGCs in the ▶ trigeminal ganglion (TG) and astrocytes and microglia in the ▶ trigeminal subnucleus caudalis (Vc). See also entries on "▶ Glial Dysregulation of Opioids" and "▶ Glia and Neuropathic Pain."

Characteristics

SGCs in the TG

The SGCs share many properties with the astrocytes in the CNS, including expression of glutamine synthetase and transporters of amino acid transmitters and, like astrocytes, are likely involved in removing extracellular K^+ and maintenance of metabolic requirements. The anatomical and physiological characteristics of SGCs in the TG are similar to those in the DRG and are described more fully in "Satellite Cells and Inflammatory Pain."

Communication Among Cells in the TG

SGCs communicate with each other and with neurons utilizing many different neuroactive agents as well as via ▶ gap junctions. Recent studies in the TG have focused on ▶ purinergic receptors on TG neurons and SGCs (Ceruti et al. 2011; Suadicani et al. 2010; Villa et al. 2010), and suggest that ATP is important for signal transmission between cells in the TG. Additionally, electrical activation of neurons in vitro releases a purinergic agonist that binds to P2Y and P2X receptors of SGCs, producing intracellular Ca^{2+} elevation, a response blocked by the purinergic antagonist suramin (Suadicani et al. 2010). Additionally, ▶ calcitonin gene-related peptide (CGRP) has been shown in vitro to function as a mediator between TG neurons and SGCs (Ceruti et al. 2011). As in the DRG, SGC gap junctions (Thalakoti et al. 2007; Garrett and Durham 2008; Ohara et al. 2008) as well as paracrine processes appear to be important in spreading excitation between SGCs and to TG neurons, e.g., activated TG SGCs may release mediators (e.g., ▶ interleukin 1-beta) which modulate TG neuronal excitability (Takeda et al. 2007).

Pathological Changes in SGCs in the TG Following Inflammation

There are several reports on pathological changes in SGCs in the TG that result from tissue inflammation. For example, peripheral inflammation triggers the production of IL-1β in SGCs that leads to neuronal depolarization (Takeda et al. 2007), and also triggers SGCs surrounding axotomized and "healthy" TG neurons to increase their expression of glial fibrillary acidic protein (GFAP) (Stephenson and Byers 1995; Takeda et al. 2007; Villa et al. 2010). In addition, peripheral orofacial inflammation triggers functional changes in inward rectifying K^+ currents and sensitivity of purinergic receptors in TG SGCs (Kushnir et al. 2011; Takeda et al. 2011). It has been proposed that post-inflammatory activation of SGCs is associated with their developing hypersensitivity to ATP, supporting SGCs' role in peripheral pain mechanisms (Kushnir et al. 2011). Additionally, an increase in expression of gap junction proteins has been reported in TG following acute or chronic temporomandibular joint inflammation (Garrett and Durham 2008). Also noteworthy is that inflammation in one trigeminal division may produce intracellular changes (e.g., ▶ P38 MAPK, ▶ ERK, and phosphatases) in TG neurons innervating all three trigeminal divisions, likely involving SGCs and thus possibly contributing peripheral mechanisms to the spread of pain within the orofacial region (Chiang et al. 2011).

Pathological Changes in SGCs in the TG Following Nerve Injury

Trigeminal nerve damage-induced changes in trigeminal primary afferent activity are associated with a variety of changes in SGCs in the TG (for

review, see Bender 2008; Ohara et al. 2009), including increased coupling among SGCs (Cherkas et al. 2004) and a rise in the extracellular K^+ concentration. The augmented coupling might enable SGCs to redistribute K^+ ions (and other harmful substances) more effectively, a view supported by the fact that silencing of the K^+ channel Kir4.1 in SGCs, which are responsible for K^+ buffering in the TG, leads to facial pain-like behavior (Vit et al. 2008). The augmented glial coupling may also contribute to the symptoms of neuropathic pain, possibly by permitting greater intercellular communication through movement of ions and signaling molecules between the SGCs enveloping adjacent TG neurons (Hanani 2005). Satellite glial cell-TG neuron communication also has an important role in inducing ▶ ectopic discharges following nerve injury. For example, ATP (e.g., released from TG neurons) may bind to SGCs, producing increased Ca^{2+} influx and K^+ outflow and thereby SGC activation. Furthermore, such SGC-TG neuronal mechanisms may contribute to findings that injury of nerve fibers in one trigeminal division may be associated with functional changes in TG neurons of another trigeminal division and thus be involved in pain spread, along the lines of the effects of inflammation as noted above (for review, see Chiang et al. 2011; Iwata et al. 2004).

All these findings suggest that SGCs have an important function in modulating the activity of TG neurons and underscore the role that ion channels, paracrine processes involving neuropeptide, purinergic and other receptors, intracellular signaling, and SGC/TG neuron communication and interactions may play as peripheral ▶ mechanisms in orofacial pain conditions (for review, see Chiang et al. 2011).

Role of Vc Glial Cells in Orofacial Pain

Microglia

After trigeminal nerve injury in rats, in parallel with the development of orofacial allodynia, ▶ microglial activation (evaluated by increase in OX-42 expression) in the superficial laminae of Vc is initiated within 1 day, is maximal at 3 days, and maintained for 14 days. Experimental tooth movement-induced masseter muscle allodynia is associated with increased OX-42 expression in the Vc as early as day 1 after surgery, and the microglial inhibitor ▶ minocycline can attenuate the allodynia as well as the OX-42 expression (Liu et al. 2009). Recent findings provide in vivo electrophysiological evidence that microglia, like astroglia (see below), are indeed rapidly (within 20 min) involved in Vc central sensitization associated with acute inflammatory pain. In an acute inflammatory model involving mustard oil (MO) application to a single molar tooth pulp in the rat or mouse, ▶ central sensitization can be detected electrophysiologically within 20 min of MO application but can be preemptively blocked by prior (<30 min) medullary application of SB230580, an inhibitor of microglial p38MAPK (Xie et al. 2007) or minocycline (Itoh et al. 2011). Furthermore, the effects of microglial inhibitors do not affect the baseline nociceptive receptive field (RF) and response properties of the nociceptive neurons but are limited to their *hyper*excitable state of central sensitization, as reflected in an increased RF size and responses to noxious stimuli and a decreased activation threshold, consistent with the behavioral literature (for review, see Chiang et al. 2011).

It is well known that purinergic P2X and P2Y receptors are involved in glial cell activation as well as in modulation of neurons under pathological conditions. In vivo electrophysiological studies in the acute pulp inflammatory pain model also have demonstrated a role for P2X mechanisms that may involve microglia in Vc central sensitization (Chiang et al. 2005; Xie et al. 2007). In extremely abnormal conditions, P2X7 receptor activation and its pore formation may promote microglia to produce and to release excessive amounts of pro-inflammatory ▶ cytokines and ATP and also trigger astroglia to release chemokines, cytokines, and glutamate, and lead to persistent pathological changes and eventually to ▶ apoptosis and gliosis. These mechanisms may explain the multiple signals released from microglia (and astroglia). The presence of the positive feedback may lead to a progressive enhancement of neuron-glial interactions and increase nociceptive neuronal activity. These

various findings indicate an important role in central sensitization of P2X receptors and the likely involvement also of microglia, since P2X4 and P2X7 receptors in the Vc (and spinal dorsal horn) are located primarily on microglia (for review, see Chiang et al. 2011).

Astroglia

Reactive astroglia, identified by their increased GFAP expression, have been found in Vc as early as 1 h after acute tooth pulp inflammation and within 1 day after induction of trigeminal chronic inflammation and trigeminal nerve injury, although astroglial activation (based on increases in GFAP expression) appears most intense at 7 and 14 days following trigeminal nerve injury (Piao et al. 2006; Guo et al. 2007; Liu et al. 2009; Okada-Ogawa et al. 2009; Tsuboi et al. 2011, for review, see Chiang et al. 2011). Furthermore, the mRNA levels of GFAP are increased by day 1 and last for 1 week after trigeminal nerve injury in rats (Latrémolière et al. 2008). The involvement of Vc astroglia in trigeminal neuropathic pain is also indicated from findings that at 7 days following trigeminal nerve injury, medullary application of the astroglial metabolic inhibitor fluoroacetate (FA) can temporarily (30–60 min) disrupt the sustained central sensitization of Vc nociceptive neurons as well as increase the reduced head withdrawal latency to noxious heating that characterizes this chronic neuropathic pain model (Okada-Ogawa et al. 2009).

The i.t. administration of the astroglial glutamine synthetase inhibitor methionine sulfoximine at 7 days after tooth pulp damage can temporarily relieve the mechanical behavioral hypersensitivity and reduce the Vc nociceptive neuronal hyperexcitability reflecting a maintained central sensitization in this model of chronic inflammatory pain (Tsuboi et al. 2011). Medullary application of methylaminoisobutyric acid, which specifically inhibits uptake of extracellular glutamine into presynaptic terminals, also potently suppresses the MO-induced central sensitization in Vc (Chiang et al. 2008). As with the effects of microglial inhibitors, these astroglial inhibitors do not affect the neuronal baseline nociceptive properties but are limited to their *hyper*excitable state, consistent with the behavioral literature (for review, see Chiang et al. 2011).

Also noteworthy is astroglial involvement in the ▶ descending modulation of trigeminal nociceptive processing. This is suggested by the prolonged elevations of pro-inflammatory cytokines ▶ TNFα and IL-1β expressed in astroglia in the ▶ rostral ventromedial medulla (RVM) at 14 days after trigeminal nerve injury, and by findings that intra-RVM injection of the astrocytic inhibitor ▶ fluorocitrate temporarily attenuates the mechanical hyperalgesia and allodynia at day 14 but not day 3 (Ren and Dubner 2010; Wei et al. 2008). These findings suggest that hyperactivated astroglia are a primary source of the increased TNFα and IL-1β in the RVM and are involved in the maintenance of neuropathic pain after nerve injury.

Glial cells are known to have functional interactions with each other and with neurons via gap junctions under pathological conditions. It has also been reported that the medullary administration of the astroglial gap junction blocker ▶ carbenoxolone can completely block the initiation of the Vc central sensitization in the acute tooth pulp inflammatory model (Chiang et al. 2010). These findings suggest that gap junctions in glial cells are involved in the exaggerated spatiotemporal features of Vc central sensitization.

References

Bender, S. D. (2008). Neuron-glia signaling in the trigeminal ganglion. *Headache, 48,* 299–300.

Ceruti, S., Villa, G., Fumagalli, M., Colombo, L., Magni, G., Zanardelli, M., Fabbretti, E., Verderio, C., van den Maagdenberg, A. M., Nistri, A., & Abbracchio, M. P. (2011). Calcitonin gene-related peptide-mediated enhancement of purinergic neuron/glia communication by the algogenic factor bradykinin in mouse trigeminal ganglia from wild-type and R192Q Cav2.1 Knock-in mice: Implications for basic mechanisms of migraine pain. *Journal of Neuroscience, 31,* 3638–3649.

Cherkas, P. S., Huang, T. Y., Pannicke, T., Tal, M., Reichenbach, A., & Hanani, M. (2004). The effects of axotomy on neurons and satellite glial cells in mouse trigeminal ganglion. *Pain, 110,* 290–298.

Chiang, C. Y., Dostrovsky, J. O., Iwata, K., & Sessle, B. J. (2011). Role of glia in orofacial pain. *The Neuroscientist, 17,* 303–320.

Chiang, C. Y., Zhang, S., Xie, Y. F., Hu, J. W., Dostrovsky, J. O., & Sessle, B. J. (2005). Endogenous ATP involvement in mustard oil-induced central sensitization in trigeminal subnucleus caudalis (medullary dorsal horn). *Journal of Neurophysiology, 94,* 1751–1760.

Chiang, C. Y., Li, Z., Dostrovsky, J. O., Hu, J. W., & Sessle, B. J. (2008). Glutamine uptake contributes to central sensitization in the medullary dorsal horn. *Neuroreport, 19,* 1151–1154. http://www.ncbi.nlm.nih.gov/pubmed/18596618.

Chiang, C. Y., Li, Z., Dostrovsky, J. O., & Sessle, B. J. (2010). Central sensitization in medullary dorsal horn involves gap junctions and hemichannels. *Neuroreport, 21,* 233–237.

Garrett, F. G., & Durham, P. L. (2008). Differential expression of connexins in trigeminal ganglion neurons and satellite glial cells in response to chronic or acute joint inflammation. *Neuron Glia Biology, 4,* 295–306.

Guo, W., Wang, H., Watanabe, M., Shimizu, K., Zou, S., LaGraize, S. C., Wei, F., Dubner, R., & Ren, K. (2007). Glial cytokine-neuronal interactions underlying the mechanisms of persistent pain. *Journal of Neuroscience, 27,* 6006–6018.

Hanani, M. (2005). Satellite glial cells in sensory ganglia: From form to function. *Brain Research. Brain Research Reviews, 48,* 457–476.

Haydon, P. G., & Carmignoto, G. (2006). Astrocyte control of synaptic transmission and neurovascular coupling. *Physiological Reviews, 86,* 1009–1031.

Itoh, K., Chiang, C. Y., Li, Z., Lee, J. C., Dostrovsky, J. O., & Sessle, B. J. (2011). Central sensitization of nociceptive neurons in rat medullary dorsal horn involves purinergic P2X7 receptors. *Neuroscience, 192,* 721–731.

Iwata, K., Tsuboi, Y., Shima, A., Harada, T., Ren, K., Kanda, K., & Kitagawa, J. (2004). Central neuronal changes after nerve injury: Neuroplastic influences of injury and aging. *Journal of Orofacial Pain, 18,* 293–298.

Kushnir, R., Cherkas, P. S., & Hanani, M. (2011). Peripheral inflammation upregulates P2X receptor expression in satellite glial cells of mouse trigeminal ganglia: A calcium imaging study. *Neuropharmacology, 61,* 739–746.

Latrémolière, A., Mauborgne, A., Masson, J., Bourgoin, S., Kayser, V., Hamon, M., & Pohl, M. (2008). Differential implication of proinflammatory cytokine interleukin-6 in the development of cephalic versus extracephalic neuropathic pain in rats. *Journal of Neuroscience, 28,* 8489–8501.

Liu, X. D., Wang, J. J., Sun, L., Chen, L. W., Rao, Z. R., Duan, L., Cao, R., & Wang, M. Q. (2009). Involvement of medullary dorsal horn glial cell activation in mediation of masseter mechanical allodynia induced by experimental tooth movement. *Archives of Oral Biology, 54*, 1143–1150.

Ohara, P. T., Vit, J. P., Bhargava, A., & Jasmin, L. (2008). Evidence for a role of connexin 43 in trigeminal pain using RNA interference in vivo. *Journal of Neurophysiology, 100*, 3064–3073.

Ohara, P. T., Vit, J.-P., Bhargava, A., Romero, M., Sundberg, C., Charles, A., & Jasmin, L. (2009). Gliopathic pain: When satellite glial cells go bad. *The Neuroscientist, 15*, 450–463.

Okada-Ogawa, A., Suzuki, I., Sessle, B. J., Chiang, C. Y., Salter, M. W., Dostrovsky, J. O., Tsuboi, Y., Kondo, M., Kitagawa, J., Kobayashi, A., Noma, N., Imamura, Y., & Iwata, K. (2009). Astroglia in medullary dorsal horn (trigeminal spinal subnucleus caudalis) are involved in trigeminal neuropathic pain mechanisms. *Journal of Neuroscience, 29*, 11161–11171.

Piao, Z. G., Cho, I. H., Park, C. K., Hong, J. P., Choi, S. Y., Lee, S. J., Lee, S., Park, K., Kim, J. S., & Oh, S. B. (2006). Activation of glia and microglial p38 MAPK in medullary dorsal horn contributes to tactile hypersensitivity following trigeminal sensory nerve injury. *Pain, 121*, 219–231.

Ren, K., & Dubner, R. (2010). Interactions between the immune and nervous systems in pain. *Nature Medicine, 16*, 1267–1276.

Stephenson, J. L., & Byers, M. R. (1995). GFAP immunoreactivity in trigeminal ganglion satellite cells after tooth injury in rats. *Experimental Neurology, 131*, 11–22.

Suadicani, S. O., Cherkas, P. S., Zuckerman, J., Smith, D. N., Spray, D. C., & Hanani, M. (2010). Bidirectional calcium signaling between satellite glial cells and neurons in cultured mouse trigeminal ganglia. *Neuron Glia Biology, 6*, 43–51.

Takeda, M., Tanimoto, T., Kadoi, J., Nasu, M., Takahashi, M., Kitagawa, J., & Matsumoto, S. (2007). Enhanced excitability of nociceptive trigeminal ganglion neurons by satellite glial cytokine following peripheral inflammation. *Pain, 129*, 155–166.

Takeda, M., Takahashi, M., Nasu, M., & Matsumoto, S. (2011). Peripheral inflammation suppresses inward rectifying potassium currents of satellite glial cells in the trigeminal ganglia. *Pain, 152*, 2147–2156.

Thalakoti, S., Patil, V. V., Damodaram, S., Vause, C. V., Langford, L. E., Freeman, S. E., & Durham, P. L. (2007). Neuron-glia signaling in trigeminal ganglion: Implications for migraine pathology. *Headache, 47*, 1008–1023; discussion 1024–1005.

Tsuboi, Y., Iwata, K., Dostrovsky, J. O., Chiang, C. Y., Sessle, B. J., & Hu, J. W. (2011). Modulation of astroglial glutamine synthetase activity affects nociceptive behaviour and central sensitization of medullary dorsal horn nociceptive neurons in a rat model of chronic pulpitis. *European Journal of Neuroscience, 34*, 292–302.

Tsuda, M., Inoue, K., & Salter, M. W. (2005). Neuropathic pain and spinal microglia: A big problem from molecules in "small" glia. *Trends in Neurosciences, 28*, 101–107.

Villa, G., Fumagalli, M., Verderio, C., Abbracchio, M. P., & Ceruti, S. (2010). Expression and contribution of satellite glial cells purinoceptors to pain transmission in sensory ganglia: An update. *Neuron Glia Biology, 6*, 31–42.

Vit, J. P., Ohara, P. T., Bhargava, A., Kelley, K., & Jasmin, L. (2008). Silencing the Kir4.1 potassium channel subunit in satellite glial cells of the rat trigeminal ganglion results in pain-like behavior in the absence of nerve injury. *Journal of Neuroscience, 28*, 4161–4171.

Wei, F., Guo, W., Zou, S., Ren, K., & Dubner, R. (2008). Supraspinal glial-neuronal interactions contribute to descending pain facilitation. *Journal of Neuroscience, 28*, 10482–10495.

Xie, Y. F., Zhang, S., Chiang, C. Y., Hu, J. W., Dostrovsky, J. O., & Sessle, B. J. (2007). Involvement of glia in central sensitization in trigeminal subnucleus caudalis (medullary dorsal horn). *Brain, Behavior, and Immunity, 21*, 634–641.

Glial Dysregulation of Opioids

Jacob Thomas[1], Linda R. Watkins[2] and Mark R. Hutchinson[3]
[1]School of Medical Sciences, University of Adelaide, Adelaide, SA, Australia
[2]Department of Psychology and Neuroscience, and The Center for Neuroscience, University of Colorado at Boulder, Boulder, CO, USA
[3]School of Medical Sciences, University of Adelaide, Adelaide, SA, Australia

Introduction

Opioids not only suppress pain, they also activate endogenous counter-regulatory mechanisms that, for example, actively oppose opioid-induced pain suppression, enhance analgesic tolerance wherein repeated opioids lose their ability to suppress pain, and enhance dependence wherein organisms require continued opioid exposure to stave off drug withdrawal. Despite the continual clinical use of opioids over several millennia, and intense scientific research in the past century, the importance of certain pharmacological actions of opioids has not yet been fully appreciated. Several systems have been shown to contribute to opioid tolerance and addiction, including hypertrophy of the cyclic AMP system, enhancement of N-methyl-D-aspartate (NMDA) receptor activity, upregulation of P-glycoprotein, and hetero-dimerization and trafficking of μ-opioid/δ-opioid receptor. However, immunocompetent cells within the central nervous system (CNS), and their immune signaling, have shown to play a large role in opioid pharmacology. Opioid-induced "central immune signaling" profoundly affects all types of cells within the CNS, contributing to the development of negative side effects such as toler-ance and enhanced pain states.

Glia are intimately entangled with neurons in what is termed a tetrapartite synapse where microglia and astrocytes encapsulate pre- and post- neuronal synapses, as well as wrap their processes around neuronal somata, axons, and dendrites (Goldstein et al. 1971). This proximity puts astrocytes in a unique position to modulate the activity of nociceptive neurons. Song and Zhao (2001) were the first to investigate the relationship of glia and morphine tolerance. They demon-strated that chronic systemic morphine increased astrocyte activation in the spinal cord but, impor-tantly, co-administration with fluorocitrate (selec-tively blocks glial metabolic activity by inhibiting aconitase, a Krebs cycle enzyme) significantly attenuated not only glial activation but also the development of tolerance. Since then, a similar pattern has also been observed with microglial attenuators minocycline or ibudilast, suggesting that glial activation plays a prominent role in the induction of opioid-induced tolerance and hyperalgesia. Further work has demonstrated that the administration of morphine induces the upregulation of markers of both microglial and astrocytic activation while also upregulating and/or causing the release of proinflammatory cyto-kines, further confirming the involvement of glia in opposing opioid analgesia.

Characteristics

Soluble Factors Release by Glia

As outlined above, immunocompetent cells of the brain and CNS release soluble factors that contribute to opioid-induced pain enhancement.

Here, we highlight the evidence that acute opioid analgesia is substantially modified by the rapid opioid-induced initiation of central immune signaling and that, upon repeated opioid exposure, continued central immune signaling leads to anal-gesic tolerance and enhanced pain states.

Cytokines

Within the CNS, astrocytes, microglia, oligodendrocytes, and endothelial cells are robust producers of cytokines, in particular, IL-1b, IL-6, and TNF-a. Rodent models have demonstrated that by blocking the key inflammatory cytokine IL-1b by administering exogenous IL-1 receptor antagonist (IL-1ra), a nearly eightfold increase in morphine analgesic potency is seen, as demon-strated by a profound leftward shift in the mor-phine dose–response function in the presence of IL-1ra (Kawasaki et al. 2008). The involvement of IL-1b in opioid analgesia has also been confirmed using three separate genetically modi-fied strains: a transgenic knock-in of IL-1ra such that endogenous IL-1ra is over-expressed; IL-1 receptor knockout, leaving IL-1b without its cog-nate receptor; and an IL-1 receptor accessory protein knockout preventing IL-1 receptor signaling due to the lack of an intracellular link to the associated Toll/IL-1 receptor signaling cascade. In each case, morphine analgesia was significantly potentiated and prolonged (Shavit et al. 2005). This proinflammatory-mediated anti-analgesic effect is not limited to IL-1b, as unmasking and/or potentiation of morphine analgesia is also observed by blocking the action of IL-6 and TNF-α (Hutchinson et al. 2008; Johnston et al. 2004). This induction of anti-analgesic central immune signaling is not a phenomenon limited to morphine, as oxyco-done analgesia was also potentiated threefold by co-administration with ibudilast (Hutchinson et al. 2009). In addition, the activation of endog-enous anti-inflammatory systems that results in elevations of IL-10 or administration of exoge-nous IL-10 was also capable of potentiating acute morphine analgesia (Johnston et al. 2004). Thus, acute opioid-induced proinflammatory central immune signaling can be pharmacologically modified to enhance acute opioid analgesia.

Proinflammatory Cytokine-Mediated Neuronal Excitation

Opioid-induced release of proinflammatory products from glia actively opposes the beneficial analgesic actions of opioids and is thought to do so via enhanced neuronal excitation in the dorsal horn of the spinal cord (Kawasaki et al. 2008).

IL-1b release has been shown to induce the phosphorylation of NMDA receptors on neurons, increasing calcium conductivity. TNF-α increases α-amino-3-hydroxy-5-methyl-4-isoxazolepropionic acid (AMPA) receptor conductivity while also increasing spontaneous neurotransmitter release from neuronal presynaptic terminals. IL-1b and TNF-α synergistically upregulate neuronal cell surface expression of both NMDA and AMPA receptors while downregulating gamma-aminobutyric acid (GABA) receptors. TNF-α also enhances neuroexcitability in response to glutamate, while IL-1b induces the release of the neuroexcitant ATP via an NMDA-mediated mechanism. Beyond these actions, proinflammatory cytokines also lead to the release of a host of neuroexcitatory substances, including more proinflammatory cytokines, nitric oxide, prostaglandins, nerve growth factors, reactive oxygen species, proinflammatory chemokines (e.g., CCL2/MCP-1, CXCL8/IL-8, CXCL10/IP-10), and BDNF. Proinflammatory cytokines can also indirectly lead to elevations in extracellular glutamate levels via downregulation of glial and neuronal glutamate transporters (Kawasaki et al. 2008; Milligan and Watkins 2009). Thus, taken together, proinflammatory cytokines exert multiple effects with the end result of increased neuronal excitatory tone, which is in part the basis behind enhancing the development of hyperalgesia and tolerance.

How Do Opioids Activate Glia?

In the infancy of opioid research, attention was focused directly toward the stereoselective receptors that were discovered to be critical for opioid analgesic responses. Classical opioid receptors are stereoselective, as they bind (−)-opioid isomers but not (+)-opioid isomers.

Intriguingly, nonclassical opioid actions were observed in some of the first studies that used synthesized unnatural opioid-receptor inactive stereoisomers but were considered to be experimental "noise" in the assays (Goldstein et al. 1971; Takagi et al. 1960). To highlight the involvement of nonclassical opioid receptors in opposing opioid analgesia, continuous infusion of morphine, oxymorphone, or fentanyl administered to triple-opioid-receptor knockout mice initiated immediate and steady declines in nociceptive thresholds, culminating in several days of unremitting hyperalgesia (Juni et al. 2007; Waxman et al. 2009). Not only did this suggest that opioids could bind to nonclassical opioid receptors but that they could also activate endogenous counter-regulatory mechanisms that actively oppose opioid-induced pain suppression, enhance analgesic tolerance, and enhance dependence independent of opioid receptors.

Non-stereoselective Activation of Central Immune Cells

Recent work by Hutchinson and coworkers demonstrated that at least some of the non-stereoselective binding can be attributed to the innate immune toll-like receptor 4 (TLR4) (Hutchinson et al. 2010; Wang et al. 2012). In vivo, in vitro, and in silico approaches provided converging lines of evidence that members of each structural class of opioids activate TLR4 (some non-stereoselectively) and that opioid antagonists such as naloxone and naltrexone non-stereoselectively block TLR4 signaling (Hutchinson et al. 2010). It was demonstrated that acute blockade of TLR4, genetic knockout of TLR4, or blockade of TLR4 downstream signaling each lead to a marked potentiation of the magnitude and duration of opioid analgesia, with TLR4 modulation of opioid actions in wild-type animals occurring within minutes (Hutchinson et al. 2010). Furthermore, it was demonstrated (Wang et al. 2012) that morphine binds the human TLR4 accessory protein, MD-2, inducing TLR4 oligomerization activating TLR4 signaling. Within the CNS, TLR4 is predominantly expressed by microglia and astrocytes, but expression has also been demonstrated on other nonneuronal cells such as endothelial cells (Hutchinson et al. 2011).

Summary

It is now thought that exclusively considering neuronal activity provides an incomplete understanding of the initiation and maintenance of opioid tolerance, hyperalgesia, and dependence. Here we highlight the notion that the activation of nonneuronal cells, in particular glia, within the CNS can profoundly affect the neuronal homeostatic environment, leading to significant alterations in neuronal firing. The current work on central immune signaling complements existing work on neuron-mediated opioid side effects, such as receptor internalization and recycling. Importantly, the negative side effects of opioids can be separated from the beneficial actions by targeting opioid-induced glial activation using blood–brain barrier permeable pharmacotherapies such as ibudilast or (+)-opioid antagonists. There is immense clinical utility for such non-opioid treatments for chronic pain and opioid dependence, as it would enable co-administration and the possibility of greater efficacy with decreased side effects.

References

Goldstein, A., Lowney, L. I., & Pal, B. K. (1971). Stereospecific and nonspecific interactions of the morphine congener levorphanol in subcellular fractions of mouse brain. *Proceedings of the National Academy of Sciences of the United States of America, 68*(8), 1742–1747.

Hutchinson, M. R., Coats, B. D., Lewis, S. S., Zhang, Y., Sprunger, D. B., Rezvani, N., Baker, E. M., et al. (2008). Proinflammatory cytokines oppose opioid-induced acute and chronic analgesia. *Brain, Behavior, and Immunity, 22*(8), 1178–1189.

Hutchinson, M. R., Lewis, S. S., Coats, B. D., Skyba, D. A., Crysdale, N. Y., Berkelhammer, D. L., Brzeski, A., et al. (2009). Reduction of opioid withdrawal and potentiation of acute opioid analgesia by systemic AV411 (ibudilast). *Brain, Behavior, and Immunity, 23*(2), 240–250.

Hutchinson, M. R., Zhang, Y., Shridhar, M., Evans, J. H., Buchanan, M. M., Zhao, T. X., Slivka, P. F., et al. (2010). Evidence that opioids may have toll-like receptor 4 and MD-2 effects. *Brain, Behavior, and Immunity, 24*(1), 83–95.

Hutchinson, M. R., Shavit, Y., Grace, P. M., Rice, K. C., Maier, S. F., & Watkins, L. R. (2011). Exploring the neuroimmunopharmacology of opioids: An integrative review of mechanisms of central immune signaling and their implications for opioid analgesia. *Pharmacological Reviews, 63*(3), 772–810.

Johnston, I. N., Milligan, E. D., Wieseler-Frank, J., Frank, M. G., Zapata, V., Campisi, J., Langer, S., et al. (2004). A role for proinflammatory cytokines and fractalkine in analgesia, tolerance, and subsequent pain facilitation induced by chronic intrathecal morphine. *Journal of Neuroscience, 24*(33), 7353–7365.

Juni, A., Klein, G., Pintar, J. E., & Kest, B. (2007). Nociception increases during opioid infusion in opioid receptor triple knock-out mice. *Neuroscience, 147*(2), 439–444.

Kawasaki, Y., Zhang, L., Cheng, J.-K., & Ji, R.-R. (2008). Cytokine mechanisms of central sensitization: Distinct and overlapping role of interleukin-1beta, interleukin-6, and tumor necrosis factor-alpha in regulating synaptic and neuronal activity in the superficial spinal cord. *Journal of Neuroscience, 28*(20), 5189–5194.

Milligan, E. D., & Watkins, L. R. (2009). Pathological and protective roles of glia in chronic pain. *Nature Reviews Neuroscience, 10*(1), 23–36.

Shavit, Y., Wolf, G., Goshen, I., Livshits, D., & Yirmiya, R. (2005). Interleukin-1 antagonizes morphine analgesia and underlies morphine tolerance. *Pain, 115*(1–2), 50–59.

Song, P., & Zhao, Z. Q. (2001). The involvement of glial cells in the development of morphine tolerance. *Neuroscience Research, 39*(3), 281–286.

Takagi, K., Fukuda, H., & Watanabe, M. (1960). Studies on antitussives. III. (+)-Morphine. *Yakugaku Zasshi, 80*, 1506–1509.

Wang, X., Loram, L. C., Ramos, K., de Jesus, A. J., Thomas, J., Cheng, K., Reddy, A., et al. (2012). Morphine activates neuroinflammation in a manner parallel to endotoxin. *Proceedings of the National Academy of Sciences of the United States of America, 109*(16), 6325–6330.

Waxman, A. R., Arout, C., Caldwell, M., Dahan, A., & Kest, B. (2009). Acute and chronic fentanyl administration causes hyperalgesia independently of opioid receptor activity in mice. *Neuroscience Letters, 462*(1), 68–72.

Glial Proliferation

Definition

Following a peripheral nerve injury, the satellite glia around large diameter sensory neurons within dorsal root ganglia that project into the damaged nerve start to proliferate. They form an onion-like structure of many layers around these somata. If the neuron is able to regenerate, the extent of proliferation is readily reduced so that the usual glial investment is recovered.

Cross-References

▶ Glial Activation
▶ Glial Cells in Orofacial Pathological Pain Mechanisms

Glomerulations

Definition

Glomerulations are punctate hemorrhages often seen in the bladder of IC patients and often noted upon hydrodistention of bladder.

Cross-References

▶ Interstitial Cystitis and Chronic Pelvic Pain

Glossopharyngeal Neuralgia

Definition

Glossopharyngeal neuralgia is a chronic neuropathic pain state associated with injury or dysfunction of the ninth cranial nerve (glossopharyngeal nerve) or its ganglion and felt in the distribution of this nerve. It is described as sharp, jabbing, electric, or shock-like pain located deep in the throat on one side.

Cross-References

▶ Neuralgia, Assessment
▶ Trigeminal and Cranial Nerve Neuralgias
▶ Trigeminal, Glossopharyngeal, and Geniculate Neuralgias

Glossopharyngeal Neuralgia: Vagoglossopharyngeal Neuralgia

▶ Trigeminal, Glossopharyngeal, and Geniculate Neuralgias

Glove Anesthesia

Definition

Glove anesthesia is a common clinical technique in which analgesia/anesthesia is induced in a hand by hypnotic suggestion. A variety of approaches is used, including imagining a protective glove being placed on the hand. Once good anesthesia is achieved and validated, this may be transferred in imagination to the painful area.

Cross-References

▶ Therapy of Pain, Hypnosis

Glucocorticosteroid

Definition

Glucocorticosteroids are a class of drugs related to the endogenous hormone cortisol from the adrenal glands. They have potent anti-inflammatory effects and relieve acute pain effectively.

Cross-References

▶ Postoperative Pain, Acute Pain Management Principles

Glucosamine

▶ Nutraceuticals

Glucose Intolerance

Definition

Glucose intolerance is a difficulty in removing glucose from the extracellular fluids for intracellular use or storage and can be an early sign of impending diabetes.

Cross-References

▶ Diabetic Neuropathies

Glutamate

Definition

L-glutamate is the principal excitatory amino acid neurotransmitter in the nervous system. Glutamate activates excitatory ionotropic amino acid receptors, which are nonselective cation channels that permit the movement of sodium, potassium, and, in some cases, calcium ions across the neuronal membrane, as well as metabotropic, G-protein-coupled glutamate receptors that link with intracellular second messenger pathways. Ionotropic glutamate receptors include NMDA (N-methyl-D-aspartate), KA (kainic acid), and AMPA (α-amino-3-hydroxyl-5-methyl-4-isoxazole propionic acid), named after the agonists that specifically bind at them; metabotropic glutamate receptors are termed mGluRs (mGluR1-mGluR6).

Cross-References

▶ Descending Circuitry, Transmitters and
  Receptors
▶ Glutamate Homeostasis and Opioid Tolerance
▶ Glutamate Neurotoxicity
▶ Nociceptors in the Orofacial Region
  (Temporomandibular Joint and Masseter
  Muscle)

▶ Opioid Modulation of Nociceptive Afferents
  In Vivo
▶ Somatic Pain
▶ Spinothalamic Tract Neurons, Glutamatergic
  Input

Glutamate Homeostasis

Definition

A balanced glutamate system for maintaining glutamate-mediated physiological functions.

Cross-References

▶ Glutamate Homeostasis and Opioid Tolerance

Glutamate Homeostasis and Opioid Tolerance

Jianren Mao
Department of Anesthesia and Critical Care,
Harvard Medical School, Boston, MA, USA

Synonyms

Glutamate homeostasis; Glutamate regulation; Glutamate transporter; Morphine tolerance; Opioid tolerance and glutamate homeostasis

Definition

Regulation of endogenous ligands of ▶ glutamate receptors such as glutamate, through a highly efficient ▶ glutamate transporter system, may play a significant role in ▶ opioid tolerance a pharmacological phenomenon related to chronic opioid administration. Regulation of glutamate transporters is also implicated in the mechanisms of opioid-induced neuronal apoptosis and increased pain sensitivity associated with the development of opioid tolerance. Modulation of GT activity and expression with pharmacological agents has been shown to regulate the development of morphine tolerance, suggesting a new strategy for improving opioid analgesic efficacy in pain management through regulating regional ▶ glutamate homeostasis.

Characteristics

Glutamate Transporters and Glutamate Regulation

▶ Glutamate is a major excitatory amino acid neurotransmitter in the central nervous system (CNS), which participates in the maintenance of important physiological functions such as synaptic plasticity and cognitive awareness. Maintaining a low extracellular glutamate concentration is key to preventing glutamate overexcitation and neurotoxicity that could occur under many pathological conditions. Regulating extracellular glutamate is primarily carried out by an efficient, high-capacity glutamate transporter (GT) system within the CNS, because clearance of extracellular glutamate via glutamate metabolism or diffusion is virtually negligible. To date, at least five cell membrane GT proteins have been identified and cloned (Robinson and Dowd 1997; Danbolt 2001). In general, GTs are labeled by a common name "excitatory amino acid transporter" (e.g., EAAT1 to EAAT5). Among these cell membrane GTs, EAAT1 (GLAST), EAAT2 (GLT-1), and EAAT3 (EAAC1) are particularly relevant to the regulation of glutamate uptake in broad CNS regions. EAAT4 is likely to be associated with Purkinje cells in the cerebellum, and the exact location of EAAT5 (a likely retinal GT) in the mammalian system remains unclear. In addition, there have been increasing reports of vesicular GTs, but their role and regulations remain to be determined.

Cellular Distribution of GTs

Among the five membrane GTs, EAAC1 is generally considered as a neuronal GT, whereas GLAST and GLT-1 are primarily astroglial GTs, although both GLAST and GLT-1 have also been demonstrated in neuronal cells during the developmental but not adult stage (Robinson and Dowd 1997; Danbolt 2001). GTs are primarily located in the CNS with a sporadic extra-CNS presence in the heart, kidney, and gastrointestinal system. There is evidence for the existence of GTs in glutamatergic nerve endings (Robinson and Dowd 1997; Danbolt 2001), indicating the capability by GTs of tightly regulating glutamate uptake at glutamatergic synapses besides glial cells. Subcellularly, GTs are located in plasma membranes, mitochondria, and synaptic vesicles, with the vast majority of GTs being associated with plasma membranes of both neuronal and glial cells (Robinson and Dowd 1997; Danbolt 2001). GTs participate in regulating the uptake of L-glutamate as well as L- or D-aspartate in a Na^+- and K^+-dependent manner.

The exact stoichiometry of glutamate uptake by GTs in relation to Na^+ and K^+ ions remains unclear. In general, GTs transport glutamate from the low-concentration extracellular compartment to the high-concentration intracellular compartment, at the cost of both Na^+ and K^+ ion gradients. Under certain circumstances such as global CNS ischemia, reversed uptake (from intracellular to extracellular compartment) could take place secondary to a weakened driving force from decreased transmembrane electrochemical gradients (Robinson and Dowd 1997; Danbolt 2001).

Role of GTs in Broad Neurological Disorders

Glutamate plays a dual role both as a major excitatory neurotransmitter essential for physiological functions and as a neurotoxic mediator contributory to pathological processes. Given the critical role of GTs in maintaining the homeostasis of extracellular glutamate, an imbalance in such a crucial regulatory system could become a fundamental cause of many neurological disorders. Of significance is that although inhibition of GT activity may not significantly prolong a single stimulus-induced excitatory postsynaptic glutamate current, it does so if the stimulus is repetitive and excessive (Overstreet et al. 1999), a condition that can be encountered under many pathological circumstances including peripheral nerve injury. Reduced GT function leads to accumulation of extracellular glutamate, causing excessive activation of glutamate receptors and initiating processes of glutamate-mediated neuronal over-excitation and ▶ excitotoxicity. To date, a large number of studies have shown the detrimental effects of reduced GT function on the pathogenesis of a variety of neurological disorders including brain ischemia, epilepsy, spinal cord injury, amyotrophic lateral sclerosis, AIDS neuropathy, and Alzheimer's disease.

Role of GTs in Nociceptive Processing

GTs have been shown to be involved in the spinal nociceptive processing in response to the hindpaw formalin injection or exogenous NMDA or prostaglandins (Minami et al. 2001; Niederberger et al. 2003). A series of recent experiments have demonstrated that both expression and uptake activity of spinal GTs changed following peripheral nerve injury and contributed to neuropathic pain behaviors in rats (Sung et al. 2003). Intrathecal administration of the tyrosine kinase receptor inhibitor K252a and the mitogen-activated protein kinase inhibitor PD98059 reduced and nearly abolished the increase in GT expression, respectively. Moreover, peripheral nerve injury significantly reduced spinal GT uptake activity, which was prevented by riluzole (a positive GT activity regulator). Riluzole also effectively attenuated and gradually reversed neuropathic pain behaviors. These results indicate that spinal GTs may play a critical role in both induction and maintenance of neuropathic pain following nerve injury via regulating regional glutamate homeostasis. The involvement of GTs in the mechanism of ▶ neuropathic pain is of particular interest, because compelling evidence has indicated that neuropathic pain and opioid tolerance may have much in common in terms of their neural mechanisms (Mao et al. 1995a).

Role of GTs in Opioid Tolerance

In animal models of ▶ morphine tolerance, subcutaneous injection of a proposed GT activator MS-153 diminished the development of morphine tolerance (Nakagawa et al. 2001). More recently, chronic morphine administered through either intrathecal boluses or continuous infusion has been shown to induce a dose-dependent downregulation of GTs (EAAC1 and GLAST) in the rat's superficial spinal cord dorsal horn (Mao et al. 2002b). This GT downregulation was mediated through opioid receptors because naloxone blocked such GT changes. Morphine-induced GT downregulation reduced the ability to maintain in vivo glutamate homeostasis at the spinal level, since the hyperalgesic response to exogenous glutamate was enhanced, including an increased magnitude and a prolonged time course, in morphine-treated rats with reduced spinal GTs. Moreover, the downregulation of spinal GTs exhibited a temporal correlation with the development of morphine tolerance. Consistently, the GT inhibitor PDC potentiated, whereas the positive GT regulator riluzole reduced, the development of morphine tolerance. The effects from regulating spinal GT activity by PDC were at least in part mediated through activation of the N-methyl-D-aspartate receptor (NMDAR), since the noncompetitive NMDAR antagonist MK-801 blocked morphine tolerance that was potentiated by PDC. These results indicate that spinal GTs may contribute to the neural mechanisms of morphine tolerance by means of regulating regional glutamate homeostasis.

Role of GTs in Opioid Dependence

Recent evidence has suggested that GTs may also play a role in opioid dependence. Firstly, changes in the GLT-1 mRNA level occurred following naloxone-precipitated morphine withdrawal (Oazwa et al. 2001). Secondly, during the withdrawal period from a sustained morphine treatment, glutamate uptake activity at hippocampal synapses was substantially increased, accompanied by an increase in the expression of GLT-1 (Xu et al. 2003). These results indicate that there may be a compensatory change in GT activity and expression, a response that is likely to serve as a buffer system to minimize the impact of a glutamate surge accompanying opioid withdrawal.

Role of GTs in Opioid-Induced Apoptosis and Pain Sensitivity

Both preclinical and clinical studies have indicated that the development of morphine tolerance is associated with an increased pain sensitivity, which may contribute to the manifestation of opioid tolerance (Mao et al. 1995a, b; Ossipov et al. 1995; Celerier et al. 2001). In addition, neurotoxic events in the form of neuronal ▶ apoptosis have been demonstrated in association with the development of morphine tolerance (Mao et al. 2002a). These apoptotic cells were predominantly located in the superficial spinal cord dorsal horn, and most apoptotic cells also expressed GAD, a key enzyme for the synthesis of the inhibitory neurotransmitter GABA. In addition, increased nociceptive sensitivity to heat stimulation was observed in these same rats, and modulation of GT activity regulated the occurrence of both opioid-induced neuronal apoptosis and increased pain sensitivity (Mao et al. 2002a). These results are consistent with a role of the spinal glutamatergic system in both opioid tolerance and neuropathic pain and provide new insights into interactions between the cellular mechanisms underlying both opioid tolerance and pain hypersensitivity.

Possible Mechanisms of GT Actions

The cellular mechanisms of GT regulation and actions in response to chronic opioid administration remain to be investigated. There are at least two possible regulatory mechanisms of GT actions. The GT expression could be regulated by extracellular glutamate, as suggested by the observations that downregulation of GLT-1 and GLAST occurs in the rat's brain regions following an impaired cortical glutamatergic connection and, conversely, that an increase in extracellular glutamate upregulates GLT-1 in astroglial cultures. Another possibility is that opioids could regulate GTs via opioid receptor-mediated intracellular changes such as cAMP, because cAMP has been shown to regulate the expression of GLT-1 and GLAST in cell cultures. These mechanisms are under current investigation, and each may play a role in opioid-induced GT regulation.

Clinical Implications

A functional role for GTs in the development of opioid tolerance suggests a new strategy for preventing opioid tolerance, opioid-induced neuronal apoptosis, and pain sensitivity, by regulating regional glutamate homeostasis using a GT regulator such as riluzole. Extensive investigation is under way to further explore such possibilities. Further, studies on GT regulation and opioid tolerance may provide new insights into the neural mechanisms of ▶ substance abuse, which may be particularly relevant to the mechanisms of heroin addiction, since heroin metabolites (6-monoacetylmorphine or morphine) interact with opioid receptors.

Acknowledgement This work was supported in part by PHS grants DA08835 and NS42661.

References

Celerier, E., Laulin, J. P., Corcuff, J. B., Le Moal, M., & Simonnet, G. (2001). Progressive enhancement of delayed hyperalgesia induced by repeated heroin administration: A sensitization process. *Journal of Neuroscience, 21*, 4074–4080.

Danbolt, N. C. (2001). Glutamate uptake. *Progress in Neurobiology, 65*, 1–105.

Mao, J., Price, D. D., & Mayer, D. J. (1995a). Mechanisms of hyperalgesia and opiate tolerance: A current view of their possible interactions. *Pain, 62*, 259–274.

Mao, J., Price, D. D., & Mayer, D. J. (1995b). Experimental mononeuropathy reduces the antinociceptive effects of morphine: Implications for common intracellular mechanisms involved in morphine tolerance and neuropathic pain. *Pain, 61*, 353–364.

Mao, J., Sung, B., Ji, R. R., & Lim, G. (2002a). Neuronal apoptosis associated with morphine tolerance: Evidence for an opioid-induced neurotoxic mechanism. *Journal of Neuroscience, 22*, 7650–7661.

Mao, J., Sung, B., Ji, R. R., & Lim, G. (2002b). Chronic morphine induces downregulation of spinal glutamate transporters: Implications in morphine tolerance and abnormal pain sensitivity. *Journal of Neuroscience, 22*, 8312–8323.

Minami, T., Matsumura, S., Okuda-Ashitaka, E., Shimamoto, K., Sakimura, K., Mishina, M., et al. (2001). Characterization of the glutamatergic system for induction and maintenance of allodynia. *Brain Research, 895*, 178–185.

Nakagawa, T., Ozawa, T., Shige, K., Yamamoto, R., Minami, M., & Satoh, M. (2001). Inhibition of morphine tolerance and dependence by MS-153, a glutamate transporter activator. *European Journal of Pharmacology, 419*, 39–45.

Niederberger, E., Schmidtko, A., Rothstein, J. D., Geisslinger, G., & Tegeder, I. (2003). Modulation of spinal nociceptive processing through the glutamate transporter GLT-1. *Neuroscience, 116*, 81–87.

Oazwa, T., Nakagawa, T., Shige, K., Minami, M., & Satoh, M. (2001). Changes in the expression of glial glutamate transporters in the rat brain accompanied with morphine dependence and naloxone-precipitated withdrawal. *Brain Research, 905*, 254–258.

Ossipov, M. H., Lopez, Y., Nichols, M. L., Bian, D., & Porreca, F. (1995). The loss of antinociceptive efficacy of spinal morphine in rats with nerve ligation injury is prevented by reducing spinal afferent drive. *Neuroscience Letters, 199*, 87–90.

Overstreet, L. S., Kinney, G. A., Liu, Y. B., Billups, D., & Slate, N. T. (1999). Glutamate transporters contribute to the time course of synaptic transmission in cerebellar granule cells. *Journal of Neuroscience, 19*, 9663–9673.

Robinson, M. B., & Dowd, L. A. (1997). Heterogeneity and functional properties of subtypes of sodium-dependent glutamate transporters in the mammalian central nervous system. *Advances in Pharmacology, 37*, 69–115.

Sung, B., Lim, G., & Mao, J. (2003). Altered expression and uptake activity of spinal glutamate transporters following peripheral nerve injury contributes to the pathogenesis of neuropathic pain in rats. *Journal of Neuroscience, 23*, 2899–2910.

Xu, N. J., Bao, L., Fan, H. P., Bao, G. B., Pu, L., Lu, Y. L., et al. (2003). Morphine withdrawal increases glutamate uptake and surface expression of glutamate transporter GLT-1 at hippocampal synapses. *Journal of Neuroscience, 23*, 4775–4784.

Glutamate Neurotoxicity

Definition

Glutamate neurotoxicity is neuronal death caused by excessive activation of glutamate receptors. Repeated activation of glutamate receptors results in increased intracellular calcium ion levels, which initiate a cascade of events producing free radicals and causing cell death.

Cross-References

▶ Dietary Variables in Neuropathic Pain

Glutamate Receptor (Metabotropic)

▶ Metabotropic Glutamate Receptors in Spinal Nociceptive Processing

Glutamate Receptors

Definition

Glutamate receptors are membrane receptors for the amino acid transmitter L-glutamate. These receptors may also be acted upon by other endogenous substances, e.g., L-aspartate, L-homocysteate, and N-acetyl-aspartyl-glutamate. They consist of ionotropic and metabotropic subclasses. The ionotropic glutamate receptors are further divided into AMPA, kainate, and NMDA subtypes, defined by ligands able to selectively activate them. Ionotropic receptors are composed of heteromeric combinations of four subunits where AMPA receptors are composed of GluR1-4 subunits. Kainate receptors are composed of GluR5-7 (GluK1-3) and/or KA1-2 (GluK4-5) subunits. NMDA receptors are composed of a combination of NR1, NR2A-D, and/or NR3A-B subunits. The ionotropic receptors are responsible for fast excitatory neurotransmission.

The metabotropic glutamate receptors are G-protein-coupled receptors and have eight subtypes that can be divided into three groups based on sequence identity, pharmacological profile, and signal transduction. Group 1 receptors include mGluR1 and mGluR5 and are mainly coupled Gq proteins which activate phospholipase C, resulting in the release of intracellular Ca^{2+}. Group 2 receptors include mGluR2 and mGluR3, both of which are generally associated with the inhibition of adenylyl cyclase, and are therefore inhibitory. Group 3 receptors include mGluR4, 6, 7, and 8, and are also coupled to Gi/Go resulting in the inhibition of adenylyl cyclase, or other inhibitory processes. Metabotropic glutamate receptors function as homodimers (a pair of the same receptor subunits) cross-linked by a disulfide bridge.

Glutamatergic synapses may have a combination of different subtypes of glutamate receptors that interact with each other and facilitate synaptic responses. For example, post-synaptic metabotropic glutamate receptors may enhance ionotropic NMDA receptor phosphorylation and contribute to spinal dorsal horn hyperexcitability. All receptor subtypes are highly expressed in the spinal cord dorsal horn and at nociceptive synapses.

Cross-References

▶ Descending Circuitry, Molecular Mechanisms of Activity-Dependent Plasticity
▶ Descending Circuitry, Transmitters and Receptors
▶ Glutamate Homeostasis and Opioid Tolerance
▶ Metabotropic Glutamate Receptors in the Thalamus
▶ NMDA Receptors in Spinal Nociceptive Processing
▶ Thalamus, Nociceptive Neurotransmission

Glutamate Regulation

▶ Glutamate Homeostasis and Opioid Tolerance

Glutamate Transporter

Definition

A high capacity system responsible for glutamate uptake from either extracellular space or cytoplasm. Transporters in the plasma membrane are referred to as excitatory amino-acid trans-porters (EAATs), and five of these have been identified that are differentially distributed in glial and endothelial cells (EAAT1-2), neurons (EAAT3-4), or the retina (EAAT5). GLAST, GLT1, and EAAC1 are the rodent orthologs of EAAT1-3.

Cross-References

▶ Glutamate Homeostasis and Opioid Tolerance

Glutamatergic

Definition

Synaptic transmission at which the amino acid glutamate is used as a neurotransmitter. Glutamate acting at ionotropic receptors is responsible for the majority of fast excitatory synaptic transmission in the nervous system. Glutamate acting at metabotropic receptors may be either excitatory or inhibitory depending on the second messenger pathways engaged following receptor activation.

Cross-References

▶ Opioid Receptors at Postsynaptic Sites

Glutamic Acid Decarboxylase

Synonyms

GAD

Definition

An enzyme mediating the synthesis of GABA from glutamic acid. Two isoforms exist (GAD-65 and GAD-67).

Cross-References

▶ GABA and Glycine in Spinal Nociceptive Processing
▶ GABA Mechanisms and Descending Inhibitory Mechanisms
▶ Thalamic Neurotransmitters and Neuromodulators

Gluteus Medius

Definition

Gluteus medius is the middle of three gluteal muscles.

Cross-References

▶ Sacroiliac Joint Pain

Glycerol Rhizotomy

Definition

Glycerol rhizotomy is the treatment of TN by a mild injury produced by instillation of glycerol into the space around the sensory (Gasserian) ganglion containing the cell bodies of the sensory fibers in the trigeminal nerve.

Cross-References

▶ Trigeminal, Glossopharyngeal, and Geniculate Neuralgias

Glycine Transporter

Definition

A plasma membrane protein that transports glycine into neurons and/or astrocytes, together with $3Na^+$ and $1Cl^-$ ion (GlyT2), or together with $2Na^+$ and $1Cl^-$ ion (GlyT-1).

Cross-References

▶ GABA and Glycine in Spinal Nociceptive Processing

Glycolysis

Definition

Glycolysis is a ubiquitous metabolic pathway in cells that allows for the production of energy by the breakdown of sugars. This complex pathway involves nine different enzymes and can function in the presence or absence of oxygen.

Cross-References

▶ Arthritis Model, Osteoarthritis

Glycosaminoglycan

Definition

Glycosaminoglycan is a type of long, unbranched polysaccharide molecule. They are major structural components of cartilage and are also found in the cornea of the eye.

Cross-References

▶ Perireceptor Elements

Glycyrrhizic Acid

Definition

Glycyrrhizic acid inhibits 1 beta-hydroxysteroid dehydrogenase, as does carbenoxolone, but without disrupting gap junctions. Due to the specific activity of glycyrrhizic acid, it can be used in carbenoxolone experiments to assess nonspecific effects of carbenoxolone (i.e., effects other than gap junction decoupling). In studies completed to date, glycyrrhizic acid does not affect basal nociceptive responses, territorial pain, or mirror-image pain.

Cross-References

▶ Cord Glial Activation

GlyT-1

Definition

Glyt-1 is a plasma membrane glycine transporter isoform found on astrocytes and neurons. It mediates fast reuptake of glycine after synaptic release. It is encoded by gene Slc6a9.

Cross-References

▶ GABA and Glycine in Spinal Nociceptive Processing

GlyT-2

Definition

Glyt-2 is a plasma membrane glycine transporter isoform found in glycinergic neurons. It mediates accumulation of glycine in glycinergic neurons and recycles synaptically released glycine. It is encoded by gene Slc6a5.

Cross-References

▶ GABA and Glycine in Spinal Nociceptive Processing

Goals for Pain Treatment in the Elderly

Definition

Complete pain relief is not a realistic goal for pain treatment in the elderly. Apart from a better

control of the pain, the treatment aims primarily at the maintenance or restitution of functional independence as a precondition of social participation.

Cross-References

▶ Psychological Treatment of Pain in Older Populations

gp120

Definition

gp120 is a glycoprotein found on the outer surface of the human immunodeficiency virus (HIV-1). Immunocompetent cells including glial cells recognize gp120. Peri-spinal administration of gp120 induces exaggerated pain responses.

Cross-References

▶ Cord Glial Activation

G-Protein-Coupled Glutamate Receptor

▶ Metabotropic Glutamate Receptors in the Thalamus

Gracile Nucleus

Definition

Discriminative sensory information from cutaneous regions in the lower half of the body is relayed to the thalamus through the gracile nucleus in the dorsal midline of the caudal medulla.

Cross-References

▶ Cuneate Nucleus
▶ Postsynaptic Dorsal Column Projection, Anatomical Organization

Graded Activity Approaches to Chronic Pain

▶ Behavioral Therapies to Reduce Disability

Graded Exposure

▶ Fear Reduction Through Exposure In Vivo

Graded Exposure In vivo with Behavioral Experiments

▶ Fear Reduction Through Exposure In Vivo

Gradiometer

▶ Magnetoencephalography in Assessment of Pain in Humans

Granulocyte Colony-Stimulating Factor

Definition

Granulocyte colony-stimulating factor is a naturally occurring protein that stimulates the production of certain white blood cell precursors.

Cross-References

▶ Cancer Pain Management, Orthopedic Surgery

Granulocytopenia

Definition

Granulocytopenia is a deficiency in the number of granulocytes, a type of white blood cell, predisposed to infection.

Cross-References

▶ Cancer Pain Management, Chemotherapy

Granulomas

Definition

Granulomas of the nose and the paranasal sinuses, in the limited stage of WG, causes headache, compression of cranial nerves, diabetes insipidus, or exophthalmos. For granulomatous arteritis of the nervous system (GANS).

Cross-References

▶ Angiitis of the CNS
▶ Headache Due to Arteritis

Gravity-assisted Traction

▶ Traction

Gray Matter Density

▶ Thalamus, Clinical Pain, Human Imaging

GREP Scale

▶ Gender Role Expectation of Pain Scale

Grip Force

Definition

Grip force is the force produced by grasping that can be measured using a strain gauge.

Cross-References

▶ Cancer Pain Model, Bone Cancer Pain Model
▶ Muscle Pain Model, Inflammatory Agents-Induced

Group Health

Definition

Group health is the care provided by a network of providers, including physicians and physician extenders, with the goal of a coordinated practice, primarily in one or more group practice facilities, often sharing common overhead expenses.

Cross-References

▶ Disability Management in Managed Care System

Group III/Group IV Afferent Fibers

Definition

Group III fibers are small-diameter myelinated (afferent) fibers, which are also called A-delta fibers. The conduction velocity is between 2.5 and 30 m/s depending on the species. Group IV fibers are unmyelinated afferent nerve fibers which are also called C-fibers. The conduction velocity is in the range from 0.5 to 2.5 m/s.

Cross-References

▶ Afferent Fiber/Afferent Neuron
▶ Exogenous Muscle Pain
▶ Sensitization of Muscular and Articular Nociceptors

Group Stimulus Space

Definition

Group stimulus space is a configuration of points (stimulus objects) along dimensions in continuous space or as clusters in discrete space.

Cross-References

▶ Pain Measurement by Questionnaires, Psychophysical Procedures, and Multivariate Analysis

Group-oriented Practice

▶ Physical Medicine and Rehabilitation, Team-Oriented Approach

Growth Factor

Definition

Growth factor is a substance that affects the growth of a cell or an organism.

Cross-References

▶ Animal Models of Inflammatory Bowel Disease

Growth Hormone

Definition

Growth hormone is released by the anterior pituitary gland. It is regulated by many endogenous substances, e.g., the serotonin 5-HT1 receptors, which increase its secretion.

Cross-References

▶ Placebo Analgesia and Descending Opioid Modulation

Guarding

▶ Nocifensive Behaviors, Muscle and Joint

Guided Imagery

▶ Relaxation in the Treatment of Pain

Guided Imagery/Guided Mental Imagery

Definition

Guided imagery is a relaxation technique that is often combined with diaphragmatic breathing and progressive muscle relaxation. It involves the use of mental imagery to produce calming cognitive and physical effects. It is most effective in individuals who report good visualization skills.

Cross-References

▶ Coping and Pain
▶ Psychological Treatment in Acute Pain
▶ Relaxation in the Treatment of Pain

Guillain-Barré Syndrome

John W. Scadding
The National Hospital for Neurology and
Neurosurgery, London, UK

Synonyms

Acute idiopathic demyelinating polyneuropathy; Acute idiopathic demyelinating polyradiculoneuropathy; Acute postinfectious polyradiculoneuropathy; Landry-Guillain-Barré-Strohl syndrome; Postinfective polyneuritis

Definition

The terms Guillain-Barré syndrome (GBS) and acute idiopathic demyelinating polyneuropathy are used interchangeably. Other terms less frequently used are listed above. The essential diagnostic features were described by Guillain, Barré, and Strohl in 1916. The pathology is acute multifocal, diffuse demyelination, affecting spinal nerve roots, peripheral nerves, and, to a less often and to a lesser extent, cranial nerves. GBS includes two major types of polyneuropathy: acute idiopathic demyelinating polyradiculoneuropathy (AIDP) and acute motor axonal neuropathy (AMAN), the latter due to an axonal immune-mediated attack, producing a clinically similar illness. A similar pathology affecting motor and sensory roots, acute motor and sensory axonal neuropathy (AMSAN) is also described. These conditions generally produce a symmetrical neurological deficit. By contrast, multifocal acquired demyelinating sensory and motor neuropathy (MADSAM, Lewis-Sumner syndrome) presents with multifocal root and peripheral nerve lesions but less acutely, over weeks or months, sometimes with a fluctuating course. This more chronic condition is not considered further here.

GBS variants include the Miller Fisher syndrome (ophthalmoplegia, ataxia, and areflexia, with mild or absent limb weakness), accounting for about 5 % of GBS, and a rare acute autonomic neuropathy (acute pandysautonomia).

Neuropathy: Disease or damage of the peripheral nerves. Mononeuropathy refers to affection of a single nerve, mononeuritis multiplex to several nerves, and polyneuropathy to diffuse involvement of the peripheral nerves.

Polyradiculoneuropathy: Describes conditions that affect spinal nerve roots and peripheral nerves.

Paresthesiae: An abnormal sensation, whether spontaneous or provoked. (Dysesthesia refers to abnormal sensations that are painful).

Sciatica: Pain in the distribution of the sciatic nerve, due either to disease of the sciatic nerve itself or the nerve roots that contribute to the nerve.

Neuropathic Pain: Pain arising as a direct consequence of a lesion affecting the somatosensory system.

Nociceptive Pain: Pain that originates in nonneural somatic tissues and is signaled and sensed by an intact nervous system.

Myalgia: Pain of muscular origin

Characteristics

Epidemiology
The mean annual incidence in different countries ranges from 0.6 to 1.9 per 100,000 of the population. The mean age of onset is 40 years, but there is a bimodal distribution with peaks in the third and sixth decades. However, GBS can occur at any age. Familial cases have been described, indicating a possible genetic susceptibility mechanism in a small minority of cases.

Diagnostic Criteria
The diagnostic criteria are summarized in Table 1.

Clinical Features
Antecedent Illness
There is a history of a preceding illness or other event in about two thirds of patients with GBS, the commonest being either a respiratory or gastrointestinal infection. The organisms most often identified are Campylobacter jejuni (approximately 26 % of patients, particularly those with the axonal form of GBS) and Mycoplasma pneumoniae (10 %). Other infective agents include cytomegalovirus, HIV, Epstein-Barr virus, varicella zoster virus, and hepatitis A and B. Widespread swine influenza vaccination in the United States in 1976 led to an increased incidence of 1 per 100,000 of the population. Preceding surgery may predispose to the development of GBS.

Campylobacter jejuni infection shows a particularly strong association with the axonal form of GBS; in northern China, evidence of infection with this microorganism is found in 76 % of patients with AMAN and in 42 % of patients with AIDP.

Cytomegalovirus infection causes GBS characterized by younger age of onset, high incidence of respiratory weakness, and more marked sensory and cranial nerve involvement.

Guillain-Barré Syndrome, Table 1 Diagnostic criteria for Guillain-Barré syndrome

Required clinical features	Clinical features supportive of diagnosis	Laboratory features supportive of diagnosis
Progressive weakness of arms and legs	Progression over days, up to 4 weeks	Raised CSF protein concentration
Areflexia	Mild cutaneous sensory loss; more marked vibration and position sense loss	<10 white cells/μl CSF
	Symmetry of deficit	Slowing of nerve conduction and/or conduction block
	Cranial nerve involvement, particularly bifacial palsies	Electrodiagnostic features of axonopathy in AMSAN/AMAN
	Autonomic dysfunction	

Adapted from Asbury and Cornblath (1990)

GBS has been reported in immunocompromised patients, either due to underlying disease, such as Hodgkin's lymphoma, or to therapeutic immunosuppression.

Pain and Other Sensory Symptoms

Although paralysis (often severe and generalized) is the dominant problem in GBS, pain is an early and often the first symptom, being troublesome in up to 85 % of patients at presentation. Sites include the interscapular and lumbar regions, buttocks, thighs, and calves. Pain quality is typically deep aching and sometimes burning. This early pain is usually transient but may be severe enough to require treatment with opiates. Associated tingling paresthesiae, with numbness, usually in a distal neuropathic distribution are common, and some patients complain of an unpleasant tightness in the legs. Similar painful symptoms and paresthesiae occur in the arms and around the shoulder girdle but are less frequent and usually milder than the lumbar and lower limb symptoms. Pain may be asymmetric and occasionally focal. Pain mimicking ▶ sciatica is recognized, and because pain in GBS may precede the development of weakness by several days, spinal imaging may occasionally be needed to exclude alternative, particularly compressive, pathology.

Pain of myalgic type is less common and is not always easy to distinguish clinically from pain of neuropathic type.

Pain in GBS tends to subside over the first 2–3 weeks of the illness but in a small proportion may persist for longer.

Paralysis

Weakness is the leading and most severe neurological symptom in the great majority of patients. Weakness of the limbs and trunk follows the pain within hours or days and progresses over a variable period. Oropharyngeal and respiratory weakness is common; ventilation is required in up to 23 % of patients. Severe weakness can develop within 12–24 h, but the average time from onset of weakness to its maximum severity is between 7 and 14 days. By definition, weakness progressing for longer than 4 weeks is no longer

AIDP; subacute inflammatory demyelinating polyradiculoneuropathy refers to patients in whom weakness progresses for up to 10 weeks, and progression after this time becomes chronic inflammatory demyelinating polyradiculoneuropathy (CIDP). CIDP may take the form of a chronic progressive or a chronic relapsing illness.

Weakness in AIDP may be predominantly distal, or global, involving limbs and trunk. It is usually more severe in the legs than the arms. Cranial nerve involvement occurs in approximately 30 % of patients, particularly unilateral or bilateral facial palsies.

Autonomic Disturbances

Autonomic disturbances develop in up to 65 % of patients with GBS (excluding acute autonomic neuropathy). There may be either reduced or increased autonomic activity. Signs of decreased sympathetic activity include orthostatic hypotension and anhidrosis; reduced parasympathetic activity leads to urinary retention, intestinal ileus, and iridoplegia. Increased sympathetic activity may cause hypertension, sinus tachycardia, or cardiac arrhythmias. Excessive parasympathetic vagal activity can lead to sinus bradycardia, heart block, and even asystole. Electrocardiographic abnormalities are common in GBS, often asymptomatic. These autonomic disturbances are more common in patients with severe weakness and particularly those with respiratory paralysis.

Other Features

Raised intracranial pressure, with ▶ papilledema is a rare complication in GBS. It is often related to a high cerebrospinal fluid (CSF) protein concentration, but it has also been described in patients with normal CSF protein concentrations.

Physical Examination

Examination reveals reduced muscle tone, weakness, and reduced or absent tendon reflexes. Despite the often prominent early sensory symptoms of pain and paresthesiae, objective and glove and stocking distribution cutaneous sensory loss is usually mild. Loss of vibration and

Guillain-Barré Syndrome, Table 2 Differential diagnosis of Guillain-Barré syndrome

Brain stem and spinal cord disorders	Radiculopathies	Peripheral neuropathies	Neuromuscular junction	Myopathies
Basilar artery thrombosis	Cytomegalovirus	Diphtheria	Myasthenia gravis	Rhabdomyolysis
Brain stem encephalitis	Lyme borreliosis	Porphyria	Botulinum toxin	Polymyositis
Poliomyelitis		Toxins: organophosphates, thallium, arsenic, lead, shellfish poisons		Hypokalemic paralysis
Rabies		Tick paralysis		
Transverse myelitis		Critical illness polyneuropathy		
		Vasculitic polyneuropathy		

joint position sensation is frequently more marked than the cutaneous sensory loss.

Straight leg raising may exacerbate lumbar and leg pain, reflecting the prominent spinal root pathology.

Autonomic disturbances, as described above, may be present.

It is mandatory to measure the vital capacity (VC) at presentation and to monitor this frequently during the period of progressive weakness. A presenting VC of 1 l is an indication that artificial ventilation will be required.

Investigation
In the early stages, peripheral nerve conduction may be normal, as the pathology is often predominantly in the spinal roots. It may be several days after the onset of weakness before peripheral electrophysiological abnormalities are found. Prolonged F wave latencies or absent F waves are early signs. Later, there is motor conduction slowing, and conduction block, manifest by greater than 30 % reduction in the compound motor action potentials (CMAP). A CMAP amplitude of less than 20 % of the lower limit of normal is associated with a poor outcome in GBS.

Cerebrospinal fluid (CSF) characteristically shows a raised total protein content, with a normal cell count or a mild pleocytosis (<10 cells/ul). Higher cell counts should raise the possibility of Borrelia infection (Lyme disease). It is often several days before the CSF protein level rises.

In patients with GBS and preceding Campylobacter infection, particularly those presenting with AMAN, serum anti-ganglioside antibodies are found in raised titers, notably anti-GM1 and anti-GD1a. However, this is not specific, these antibodies also occurring on other types of GBS. There is a strong association of anti-GQ 1b ganglioside antibodies with Miller Fisher syndrome.

In view of the risk of autonomic involvement, the electrocardiogram should be monitored, particularly in patients with rapidly progressive weakness.

Differential Diagnosis
The diagnosis of GBS is often straightforward, but presentation with a long prodrome of pain without weakness, asymmetric or focal onset weakness, weakness confined to the legs, or prominent autonomic features early in the course of the illness can lead to diagnostic difficulties. Table 2 lists the wide differential diagnosis.

Pathogenesis of GBS
GBS is an organ-specific autoimmune disorder mediated by autoreactive T cells and humoral antibodies. The nature of the antigens remains uncertain. A preceding infection, immunization, or other insult may provoke a reaction in which an immune response is mounted against antigens shared with the host's peripheral nerves.

Pathogenesis of Pain in GBS
Demyelination and axonal degeneration are likely to lead to ectopic impulse generation, causing ▶ neuropathic pain and paresthesiae. Pain may also result from root swelling and

entrapment. A third component may be a nociceptive type of pain, arising from acute inflammatory changes in the nerve roots, signaled by the nervi nervorum (nerve trunk pain).

Treatment of GBS

High-dose intravenous human immunoglobulin (IVIG) and plasma exchange have been found to be equally effective in GBS. Due to ease of administration, IVIG is the treatment of choice. Corticosteroids are ineffective.

General supportive measures, and in particular, elective ventilation, together with these disease-specific treatments, are vital and have reduced mortality from around 30 % to 5 %. Patients with a falling VC or cardiovascular instability should be monitored in a high-dependency care setting, where full resuscitation facilities are available. Respiratory and cardiac function and blood pressure require close monitoring. Tracheal suctioning may provoke episodes of arrhythmia or hypotension. Nasogastric feeding is needed in patients with oropharyngeal weakness; early assessment and frequent monitoring of swallowing is essential.

Pain Relief

Simple analgesia or weak opiates are sufficient in many patients. Stronger opiates may be needed for the early, usually transient, trunk and limb pains referred to above. However, opiates may further impair already reduced respiratory function and should be used with extreme caution in patients with moderate to severe weakness. For more severe neuropathic pain, ▶ tricyclic antidepressants may be indicated, but tricyclics are best avoided in patients with disturbances of autonomic function. Gabapentin and pregabalin are alternatives, though neither has been shown in clinical trials to relieve neuropathic pain in GBS.

Course and Prognosis

Following the progressive phase of the illness, there is a period of 2–4 weeks on average, during which the deficit remains static. Recovery then begins. Up to 5 % of patients die during the acute stage. 82 % recover completely within 2 years. There is a recurrence rate of about 3 %. Factors associated with a poorer outcome include

CMAPs less than 20 % of the lower limit of normal, age greater than 60 years, the need for ventilation, and rapid progression of weakness (maximum at less than 7 days).

Summary

GBS is one of the commonest types of polyneuropathy, characterized by acute onset of weakness, progression over a relatively short period, and then usually complete recovery. Although pain and painful paresthesiae are generally not severe, these can be troublesome symptoms, requiring careful assessment and specialist management. Severe ongoing pain, lasting longer than a few weeks, is uncommon.

References

Asbury, A. K., & Cornblath, D. R. (1990). Assessment of current diagnostic criteria for Guillain-Barré syndrome. *Annals of Neurology, 27*(Suppl.), S21–S24.

Gynecological Cancer

▶ Gynecological Pain and Sexual Functioning

Gynecological Pain and Sexual Functioning

P. T. M. Weijenborg and M. M. ter Kuile
Department of Gynecology, Leiden University Medical Center, Leiden, The Netherlands

Synonyms

Acute pelvic pain; Chronic pelvic pain; Chronic vulvar pain; Dysmenorrhea; Dyspareunia; Gynecological cancer; Sexual dysfunctions; (Un)provoked vestibulodynia; Vulvar vestibulitis syndrome; Vulvodynia

Characteristics

Gynecological Pain

Gynecological pain might be located in the pelvis and/or the vulvovaginal region of the body and can be divided into acute vs. chronic (intermittent, cyclic, or continuous) pain. Radiation to, for instance, the upper abdomen, groins, back, or legs is possible. Report of pelvic pain for some days each month, specifically associated with the menstrual period (i.e., dysmenorrhea), is a common complaint that can affect as many as 50 % of all premenopausal women (Dawood 1990).

Sexual Functioning

The definitions of the most common sexual dysfunctions presented in medical practice are given in Table 1 (APA 2000). All problems can occur as a primary, secondary, situational, or general condition. As part of the diagnostic criteria for "dysfunction," the sexual problem has to cause personal distress (Shifren et al. 2008). A higher level of sexually related distress is one of the factors associated with help-seeking for sexual problems in women (Shifren et al. 2009).

To assess the impact of gynecological pain on sexual functioning, the health-care professional has to raise the issue of sexual health and sexual functioning of women suffering pain explicitly (Berman et al. 2003). The diagnosis is made after taking the medical and sexual history of the patient preferably together with her partner (see Table 2) followed by a gynecological examination and/or laboratory tests for endocrinological analyses, if appropriate.

A history of sexual abuse is addressed before doing the exam. Traumatic memories may be aroused by the intimacy of a gynecological examination. For instance, the lithotomy position during the pelvic examination or the insertion of instruments, such as a speculum or fingers, into a woman's vagina, may be regarded and experienced as an abusive situation (Chalmers 1998).

The impact of different gynecological pain conditions on sexual functioning depends on various factors like the duration and location of the pain, the quality of sexual functioning before the onset of pain, and relational as well as psychosocial factors.

Gynecological Pain and Sexual Functioning, Table 1 Sexual problems, definition

302.71 Hypoactive sexual desire disorder	Persistently or recurrently deficient (or absent) sexual fantasies and desire for sexual activity. The judgment of deficiency or absence is made by the clinician, taking into account factors that affect sexual functioning, such as age and the context of the person's life
302.72 Female sexual arousal disorder	Persistent or recurrent inability to attain, or to maintain until completion of the sexual activity, an adequate lubrication-swelling response of sexual excitement
302.73 Female orgasmic disorder	Persistent or recurrent delay in, or absence of, orgasm following a normal sexual excitement phase. Women exhibit wide variability in the type or intensity of stimulation that triggers orgasm. The diagnosis of female orgasmic disorder should be based on the clinician's judgment that the woman's orgasmic capacity is less than would be reasonable for her age, sexual experience, and the adequacy of sexual stimulation she receives
302.76 Dyspareunia	Recurrent or persistent genital pain associated with sexual intercourse in either a male or a female
302.79 Sexual aversion disorder	Persistent or recurrent extreme aversion to, and avoidance of, all (or almost all) genital sexual contact with a sexual partner
306.51 Vaginismus	Recurrent or persistent involuntary spasm of the musculature of the outer third of the vagina that interferes with sexual intercourse

A revision of the APA Diagnostisc Classificaction DSM-IV-TR (APA 2000) is anticipated for 2012. The members of the Sexual and Gender Identity Disorders Work Group have published their reviews on female sexual dysfunctions. Each review focuses on a critical appraisal of the relevant diagnoses that appeared in the DSM-IV with proposed suggestions for reform and revision (Arch Sexual Behavior 2010)

Gynecological Pain and Sexual Functioning, Table 2 To assess the impact of the gynecological pain on sexual functioning

1	Pain and the consequences of pain on sexuality of the woman and her partner
2	Sexual functioning in the past, including sexual, physical, and emotional abuse
3	Relational factors
4	Psychosocial functioning (i.e., anxiety and depression)

Acute Gynecological Pain

Acute Pelvic Pain

There is a variety of gynecological diagnoses that cause acute *abdominal or pelvic* pain, for instance, extrauterine pregnancy, missed abortion, and pelvic inflammatory disease. Typically, pain and the accompanying symptoms of these conditions subside after adequate treatment. As a consequence, sexuality and sexual functioning of the woman and her partner will only be impaired during the short period of pain. Most women will avoid sexual contact.

Acute Vulvar Pain

Acute vulvar pain can be related to specific disorders (infectious, inflammatory, neoplasia, or neurologic) like, for instance, candidiasis, lichen planus, lichen sclerosis et atroficans vulvae (LSEAV), vulvar intraepithelial neoplasia (VIN), or herpes. Despite adequate diagnosis and treatment (anticandidal, antiherpetic, class IV- corticosteroids, laser evaporization, or local excision), vulvar pain might persist. The impact of these vulvar pain conditions on sexual functioning is understudied.

Chronic Gynecological Pain

According to the International Society for the Study of Vulvovaginal Diseases (ISSVD), ▶ vulvodynia is defined as "vulvar discomfort, most often described as burning pain, occurring in the absence of relevant visible findings or a specific clinically identifiable, neurologic disorder" (Moyal-Barracco and Lynch 2004). A distinction is made between localized and generalized vulvodynia, both subdivided in provoked, unprovoked, or mixed.

Intermittent or Localized Vulvar Pain

Provoked Vestibulodynia or Provoked Vulvodynia (PV) previously known as ▶ vulvar vestibulitis syndrome (VVS), is considered to be a common cause of ▶ superficial dyspareunia, affecting up to 12 % of premenopausal women from the general population (Harlow and Stewart 2003). According to Friedrich (1987), PV is characterized by (1) severe pain upon vestibular touch or attempted vaginal entry, (2) acute pain during cotton swab palpation of the vestibular area, and (3) vestibular erythema. However, the reliability of this last criterion has been disconfirmed (Bergeron et al. 2001; Masheb et al. 2004). The etiology of PV remains largely unknown. Women with PV demonstrate impaired sexual functioning like lower levels of sexual desire, arousal, and frequency of intercourse. Specific psychological states such as anxiety and depression, fear of pain, hypervigilance, and catastrophizing are more frequently reported by PV affected women (Van Lankveld et al. 2010). A wide variety of surgical, medical, cognitive-behavioral, and alternative treatments have been proposed. A critical review of published studies concerning the treatment of PV with an emphasis on pain and sexual function outcomes (Landry et al. 2008) showed that vestibulectomy (still) reached the best success rates. Given its degree of invasiveness and the potential risks and/or negative side effects (worsening of pain, recurrence of pain, and decreased lubrication), it cannot be recommended as a first-line intervention.

Vaginismus is commonly described as a persistent difficulty to allow vaginal entry of a penis, a finger, and/or any object despite the woman's expressed wish to do so. There is a variable involuntary pelvic muscle contraction, (phobic) avoidance and anticipation, and fear or experience of vulvar pain. In some cases, it is not possible to differentiate vaginismus from PV. Therefore, a multidimensional diagnosis termed "genito-pelvic pain/penetration disorder" is suggested (Binik 2010). Lifelong vaginismus occurs when a woman has never been able to have intercourse. The use of gradual exposure exercises, with gradual habituation to vaginal touch and penetration, and applied relaxation (systematic desensitization) has a relatively long history in the treatment of vaginismus (McGuire and Hawton 2001). The results of the first controlled studies are promising (Ter Kuile et al. 2010a).

Generalized Vulvar Pain

Nowadays, *unprovoked vulvodynia* (former dysesthetic, essential, or idiopathic ▶ vulvodynia) is a chronic pain condition without a clear skin disorder or other somatic explanation for the complaints. It seems logical that affected women avoid intercourse, being anxious about a worsening of the vulvar pain. However, studies on sexuality and chronic vulvar complaints are unknown.

Intermittent Pelvic Pain

Deep dyspareunia is defined as pain in the abdomen during intercourse, especially at the time of deep penetration and/or hours or days thereafter. Gynecological causes have to be ruled out like chronic pelvic inflammatory disease (PID), fixed retroverted uterus, enlarged uterus by myomata, endometriosis deposits in the cul-de-sac and/or the rectovaginal space, or an ovarian cyst and pelvic mass, irritable bowel syndrome, or other gastrointestinal diseases. When no pathology can be diagnosed, the woman and her partner are advised to try different positions during sexual intercourse aiming toward the avoidance of pain following penetration (see ▶ chronic pelvic pain).

Dysmenorrhea can be defined as recurrent cyclic pain, starting some days before or just at the menstrual period and lasting for some days (see ▶ chronic pelvic pain and dysmenorrhea). About 10 % of women suffering from this pain are incapacitated for 1–3 days each menstrual cycle (Dawood 1990). The complaints can result in physical and emotional distress with associated negative consequences for sexual functioning during pain. The first choice of treatment is the prescription of nonsteroidal anti-inflammatory drugs (NSAID) often in combination with oral contraceptives (Marjoribanks et al. 2010). With this regimen, pain relief will be established in the majority of affected women. The impact of dysmenorrhea on sexual functioning is not studied but will depend on the effectiveness of pain management.

Chronic Pelvic Pain

▶ Chronic pelvic pain (*CPP*) is described as pain in the lower abdomen or pelvis for at least 3 months, not exclusively associated with the menstrual cycle (dysmenorrhea) and/or sexual intercourse (deep dyspareunia).

CPP is a long-lasting and often disabling condition. Although the pain suggests pathology, still it is difficult to find an exclusive diagnosis and/or offer an effective treatment (Stones et al. 2005).

Problems with sexual functioning resulting from CPP have to be addressed and assessed by the health-care professional. As a result of the pain itself, extreme fatigue, and/or depressive mood, women may report all kinds of sexual problems ranging from decreased pleasure in and frequency of intercourse, deficient lubrication during sexual contact, superficial or deep dyspareunia, and/or problems reaching orgasm to a total aversion toward sexual intimacy.

Reports in the literature about the coincidental prevalence of sexual problems with CPP are scarce. In community based studies in the USA (Mathias et al. 1996), UK (Zondervan et al. 2001), New Zealand (Grace and Zondervan 2004), and Australia (Pitts et al. 2008), a substantially larger proportion of women with CPP report dyspareunia (between 29 % and 82 %) than women without CPP (between 11 % and 14 %). Also, within clinical/gynecological populations, patients with CPP report more sexual problems such as dyspareunia, decreased desire, arousal, and lubrication than patients without CPP. Moreover, patients with CPP tend to suffer more sexual problems than women with any other type of chronic pain problem (Collett et al. 1998). The available studies suggest that women with CPP report more frequently a history of sexual abuse (Roelofs and Spinhoven 2007) and show higher levels of anxiety and depression (McGowan et al. 1998) compared to controls. A history of sexual abuse as well as higher levels of anxiety and depression are both related to sexual problems. Recently, in a clinical sample, it was found that CPP women reported higher levels of sexual problems than healthy controls. These sexual problems were associated with anxiety, depression, and a sexual abuse history but not with somatic factors such as the level of pain and physical impairment. Anxiety as well as depression mediated the effect of CPP on sexual problems, irrespective of the report of sexual abuse. Sexual abuse turned out to be a general predictor of sexual problems in both women with and without CPP (Ter Kuile et al. 2010). These findings indicate that during the assessment of sexual problems in women with CPP, interest has also to be focused on current anxiety and depressive mood.

Gynecological Cancer and Sexuality

In the last decades, an increasing number of women survive gynecological cancer due to early detection and advances in treatment modalities. Sexual dysfunction is a potential long-term complication of cancer treatment and might impact quality of life (QOL). The cause of sexual dysfunction may be multifactorial and may result from direct or indirect effects of surgery, radiation and/or hormone therapy, lowered physical functioning, and an increase in symptoms of depression and anxiety. The introduction of, for instance, nerve-preserving surgical techniques to prevent nerve damage as well as psychophysiological studies to demonstrate the effect of these new techniques on sexual functioning shows the growing interest in this topic. All kinds of sexual dysfunction are mentioned during the first year of posttreatment for early stage gynecological cancer. Also, marked differences in rated distress are reported, depending on physical factors as treatment modalities and organ(s) involved (uterus, cervix, ovary, or vulva), as well as psychological and social factors (Weijmar Schultz and Van de Wiel 2003). Our knowledge about the effectiveness of medical and/or psychological interventions for sexual dysfunctions in women with gynecological cancer is hampered by a serious lack of recent studies of good methodological quality (Flynn et al. 2009).

References

American Psychiatric Association. (2000). *Diagnostic and statistical manual of mental disorders -text revision* (4th ed.). Washington, DC: American Psychiatric Association.

Arch Sex Behavior. (2010) Special section: Reports from the DSM-V work group on sexual and gender identity disorders. *39*, 221–513

Bergeron, S., Binik, Y. M., Khalifé, S., Pagidas, K., & Glazer, H. I. (2001). Vulvar vestibulitis syndrome: Reliability of diagnosis and evaluation of current diagnostic criteria. *The Obstetrician and Gynaecologist, 98*, 45–51.

Berman, L., Berman, J., Feder, S., Pollets, D., Chhabra, S., Miles, M., & Powell, J. A. (2003). Seeking help for sexual function complaints: What gynecologists need to know about the female patient's experience. *Fertility and Sterility, 79*, 572–576.

Binik, Y. M. (2010). The DSM diagnostic criteria for vaginismus. *Archives of Sexual Behavior, 39*, 278–291.

Chalmers, B. (1998). Psychosomatic obstetrics and gynecology in the next millennium: Some thoughts and observations. *Journal of Psychosomatic Obstetrics and Gynecology, 19*, 62–69.

Collett, B. J., Cordle, C. J., Stewart, C. R., & Jagger, C. (1998). A comparative study of women with chronic pelvic pain, chronic nonpelvic pain and those with no history of pain attending general practitioners. *British Journal of Obstetrics and Gynaecology, 105*, 87–92.

Dawood, M. Y. (1990). Dysmenorrhea. *Clinical Obstetrics and Gynecology, 33*, 168–178.

Flynn, P., Kew, F., & Kisely, S. R. (2009). Interventions for psychosexual dysfunction in women treated for gynaecological malignancy. *Cochrane Database of Systematic Reviews*, CD004708. doi:10.1002/14651858.CD004708.pub2.

Friedrich, E. G. (1987). Vulvar vestibulitis syndrome. *The Journal of Reproductive Medicine, 32*, 110–114.

Grace, V. M., & Zondervan, K. T. (2004). Chronic pelvic pain in New Zealand: Prevalence, pain severity, diagnoses and use of health services. *Australian and New Zealand Journal of Public Health, 28*, 369–375.

Harlow, B. L., & Stewart, E. G. (2003). A population-based assessment of chronic unexplained vulvar pain: Have we underestimated the prevalence of vulvodynia? *Journal of American Medical Women's Association, 58*, 82–88.

Landry, T., Bergeron, S., Dupuis, M., & Desrochers, G. (2008). The treatment of provoked vestibulodynia. A critical review. *The Clinical Journal of Pain, 24*, 155–171.

Marjoribanks, J., Proctor, M., Farquhar, C., & Derks, R. S. (2010). Nonsteroidal anti-inflammatory drugs for dysmenorrhoea. *Cochrane Database of Systematic Reviews*, CD001751. doi:10.1002/14651858.CD001751.pub2.

Masheb, R. M., Lozano, C., Richman, S., Minkin, M. J., & Kerns, R. D. (2004). On the reliability and validity of physician ratings for vulvodynia and the discriminant validity of its subtypes. *Pain Medicine, 5*, 349–358.

Mathias, S. D., Kuppermann, M., Liberman, R. F., Lipschutz, R. C., & Steege, J. F. (1996). Chronic pelvic pain: Prevalence, health-related quality of life, and economic correlates. *Obstetrics and Gynecology, 87*, 321–327.

McGowan, L. P. A., Clark-Carter, D. D., & Pitts, M. K. (1998). Chronic pelvic pain: A meta-analytic review. *Psychology & Health, 13*, 937–951.

McGuire, H., & Hawton, K. (2001). Interventions for vaginismus. *Cochrane Database of Systematic Reviews*, CD001760. doi:10.1002/14651858.CD001760.

Moyal-Barracco, M., & Lynch, P. J. (2004). 2003 ISSVD terminology and classification of vulvodynia: A historical perspective. *The Journal of Reproductive Medicine, 49*, 772–777.

Pitts, M. K., Ferris, J. A., Smith, A. M., Shelley, J. M., & Richters, J. (2008). Prevalence and correlates of three types of pelvic pain in a nationally representative sample of Australian women. *The Medical Journal of Australia, 189*, 138–143.

Roelofs, K., & Spinhoven, P. (2007). Trauma and medically unexplained symptoms: Towards an integration of cognitive and neuro-biological accounts. *Clinical Psychology Review, 27*, 798–820.

Shifren, J. L., Monz, B. U., Russo, P. A., Segreti, A., & Johannes, C. B. (2008). Sexual problems and distress in United States women. Prevalence and correlates. *Obstetrics Gynecology, 112*, 970–978.

Shifren, J. L., Johannes, C. B., Monz, B. U., Russo, P. A., Bennet, L., & Rosen, R. (2009). Help-seeking behavior of women with self-reported distressing sexual problems. *Journal of Women's Health, 18*, 461–468.

Stones, W., Cheong, Y. C., & Mountfield, J. (2005). Interventions for treating chronic pelvic pain in women. *Cochrane Database Systematic Reviews*, CD000387. doi:10.1002/14651858.CD000378.

Ter Kuile, M. M., Both, S., & Van Lankveld, J. J. D. M. (2010a). CBT for sexual dysfunction in women. *Psychiatric Clinics of North America, 33*, 595–610.

Ter Kuile, M. M., Weijenborg, P. T., & Spinhoven, P. (2010b). Sexual functioning in women with chronic pelvic pain: The role of anxiety and depression. *The Journal of Sexual Medicine, 7*, 1901–1910.

Van Lankveld, J. J. D. M., Granot, M., Weijmar Schultz, W. C. M., Binik, Y. M., Wesselmann, U., Pukall, C. F., Bohm-Starke, N., & Achtrari, C. (2010). Women's sexual pain disorders. *The Journal of Sexual Medicine, 7*, 615–631.

Weijmar Schultz, W. C. M., & Van de Wiel, H. B. M. (2003). Sexuality, intimacy and gynaecological cancer. *Journal of Sex & Marital Therapy, 29*(s), 121–128.

Zondervan, K. T., Yudkin, P. L., Vessey, M. P., Jenkinson, C. P., Dawes, M. G., Barlow, D., & Kennedy, S. H. (2001). The community prevalence of chronic pelvic pain in women and associated illness behaviour. *British Journal of General Practice, 51*, 541–547.

Gynecological Pain Model

▶ Visceral Pain Models, Female Reproductive Organ Pain

Gynecological Pain, Neural Mechanisms

Karen J. Berkley and Natalia Dmitrieva
Program in Neuroscience, Florida State University, Tallahassee, FL, USA

Synonyms

Adenomyosis; Chronic pelvic pain; Dysmenorrhea; Dyspareunia; Endometriosis; Labor pain; Mastitis; Mittelschmerz; Pelvic inflammatory disease; Referred pain; Vulvodynia

Definition

Gynecological pain is defined as pain associated with the female reproductive organs. It is usually considered in the context of pelvic pain as a subcategory of visceral pain, although pains associated with the breast are sometimes included. This entry is confined to pelvic pain in women.

Characteristics

Although gynecological pain is usually associated with specific disease conditions such as endometriosis, parametritis, salpingo-oophoritis, uterine leiomyoma, and ovarian cysts, the severity of the disease and the pain often do not correlate well. In fact, gynecological pain also occurs without obvious pathophysiology. Such pains can be chronic (e.g., primary dysmenorrhea, vulvodynia, chronic pelvic pain), and their severity can vary with reproductive status, i.e., puberty, ovarian cycle, birth control hormones, pregnancy, puerperium, parity, reproductive senescence, and hormone replacement (Stratton and Berkley 2011).

Pain associated with pathophysiology of internal reproductive organs is usually poorly localized internally and is often also "felt" in somatic structures (muscles, skin) of the affected organ's bodily segments, a phenomenon called "referred pain" (Giamberardino 2003). Sometimes the skin and particularly the muscles in the referral region are also tender, a phenomenon called "referred hyperalgesia." Referred muscle hyperalgesia remains for many months after the initial pathology has resolved (Giamberardino 2003). Furthermore, distressing chronic gynecological pains are surprisingly frequently found to co-occur with other painful disorders, mostly pelvic, such as irritable bowel syndrome and interstitial cystitis/painful bladder syndrome. Gynecological pain can also co-occur with non-pelvic painful conditions, such as migraine or temporomandibular joint disorder (Stratton and Berkley 2011).

Much of our current and still poor understanding of mechanisms of gynecological pain comes from work on rodents. The studies have shown that female reproductive organs are supplied by afferent and efferent fibers in a roughly rostro-caudal topographic fashion (Fig. 1a) (Papka and Traurig 1993; Berkley et al. 1993; Temple et al. 1999). A similar topographic pattern is also likely to exist for women (Wesselmann et al. 1997; Papka and Traurig 1993). The afferent fibers in these nerves convey modality- and organ-specific information to the spinal cord and brain about events occurring in relatively small and localized areas (Berkley et al. 1993).

Response sensitivity and other characteristics of peripheral afferent fibers supplying pelvic reproductive organs can be altered by reproductive status (puberty, ovarian cycle, pregnancy, lactation, reproductive senescence, hormone replacement; e.g., Tingåker and Irestedt 2010) as well as by pathophysiology (e.g., Berkley et al. 1993). Notably, however, despite limited evidence that the peripheral nerve fibers can branch to supply several different pelvic organs such as the bladder and colon (Ustinova et al. 2010), the responses of those nerve fibers remain organ-specific. For example, although injury to the uterus can increase sensitivity of hypogastric nerve fibers in response to stimulation to a small part of the uterus, and the receptive field may increase slightly, the sensory afferent fibers will still respond only to stimulation of the uterus, and fail to respond to stimulation of other nearby organs such as the vagina, bladder or colon (Berkley et al. 1993). Thus, although such changes in response sensitivity of peripheral nerve fibers are likely to contribute to changes in nociceptive/pain sensitivity of the innervated regions, they fail to explain the many and common gynecological pains mentioned above that do not correlate with peripheral pathophysiology, nor can they explain referred pain, referred hyperalgesia, and the co-occurrence of pain associated with other pelvic organs.

What is important, therefore, is how the central nervous system processes the information it receives to create pain. The most common conceptualization of this process, shown in Fig. 2a, involves a "pathway" mechanism in which there are parallel and separate tracks through the CNS whose activation gives rise to the different bodily perceptions of touch or pain. Regarding the pain pathway, the experience of pain is created when a noxious or pathophysiological stimulus in an organ (e.g., a reproductive organ) activates peripheral afferent fibers, which convey their information to the spinal cord, from where the information is relayed through a series of central neural structures to reach the cerebral cortex. This collection of structures can be referred to as the "pain matrix" (Tracey and Mantyh 2007). To account for variations in reported pain experiences despite similar peripheral noxious stimuli, an additional rheostat-like, pain-modulatory control mechanism exists, in which neurons in various parts of the brain influence, via descending fibers, the information arriving in the spinal cord.

One problem with this common view is that it requires a peripheral noxious stimulus for pain to occur and thereby fails to explain pain without injury as well as referred pain/hyperalgesia and co-occurrence of different types of pain syndromes. These problems are leading many

Gynecological Pain, Neural Mechanisms, Fig. 1 (a) Innervation of female rat reproductive tract and spinal entry segments (Adapted from Berkley et al. (1993), Temple et al. (1999)). Note topographic organization of the input; i.e., vagina/cervix via pelvic nerve to L6/S1 segments; cervix/uterus via hypogastric nerve to T13-L1 segments; ovary via superior ovarian and ovarian plexus nerves to T10-L1 segments. Although the fibers enter the spinal cord via dorsal root ganglia in these segments, the terminations of the afferents in the spinal cord extend to many segments rostrally and caudally from their entry. Note also that the entire tract is also innervated by the vagus nerve (Komisaruk and Sansone 2003). (b) Innervation of pelvic organs in women (This diagram is Fig. 2, reprinted with permission, from Wesselmann et al. (1997)). *CEL* celiac plexus, *DRG* dorsal root ganglion, *HGP* hypogastric plexus, *IHP* inferior hypogastric plexus, *PSN* pelvic splanchnic nerve, *PUD* pudendal nerve, *SA* short adrenergic projections, *SAC* sacral plexus, *SCG* sympathetic chain ganglion, *SHP* superior hypogastric plexus, *Vag.* vagina

toward a different conceptualization of pain mechanisms, a view, that takes into account the dynamic properties of neural processing (Fig. 2b) and that can give rise to unique and variable pain signatures (Tracey and Johns 2010; Vincent et al. 2011). With respect to pelvic reproductive organs, this dynamic concept derives from six features about how the CNS processes

Gynecological Pain, Neural Mechanisms, Fig. 2 Conceptualizations of how the transmission of stimulus information from bodily receptors in skin, muscle, and internal organs (including the reproductive organs) to and through various central nervous system regions might give rise to perceptions of pain and touch and modulate visceral function. (**a**) Represents the current traditional "pathway" view in which different pathways are invested with different perceptual functions.

(**b**) Represents a "dynamic distributed ensemble" view that takes into account connective features such as viscero-viscero-somatic convergence of input into the dorsal column nuclei (dcn), solitary nucleus, and spinal cord, widespread divergence and convergence in the brain and spinal cord, and dynamic processes such as central sensitization and hormonal modulation of CNS activity (Diagrams modified from Berkley (2001))

information arriving from these and other visceral structures.

The first feature is a connective "cross-talk" between different organ systems (Berkley 2001; Dmitrieva and Berkley 2002; Winnard et al. 2006; Ustinova et al. 2010). It involves the fact that organ-specific information provided by peripheral sensory afferents when conveyed to neurons in many segments in the spinal cord, as well as to the dorsal column and solitary nuclei in the brainstem, converges in those central regions with information arriving not only from skin and muscles in the same segments, but also from other organs such as the colon and bladder (Jänig 2006). The second feature is connective widespread divergence and convergence in the CNS. It involves the fact that information from

neurons in each of the three initial recipient regions (spinal cord, dorsal column, and solitary nuclei) is then conveyed to the other two regions, as well as to a large number of other sites in the brain which themselves interconnect (Jänig 2006). The third feature is the fact that information processing in the CNS is subject to both inhibitory and facilitatory influences, which involve interacting influences of endogenous opioids and other neuroactive agents such as serotonins (Dogrul et al. 2009) as well as, as recently discovered, endocannabinoids (Guindon and Hohmann 2009), which also act peripherally (Dmitrieva et al. 2010). The fourth feature is the dynamic process called "central sensitization" (Woolf 2011). When afferent fibers are sensitized (hyperactive) by noxious events in the periphery, this enhanced input can in turn sensitize recipient CNS neurons; this central sensitization can remain long after the noxious peripheral event (injury, disease) has resolved, producing, in effect, "injury memories" (Nathan 1985). The fifth feature involves CNS influences on autonomic output to internal organs (Jänig 2006), which in turn can influence functioning of peripheral nociceptive sensory afferent fibers, via a poorly understood mechanism referred to as sympathetic-sensory coupling (Jänig 2006). This process can be exaggerated under conditions of visceral injury or inflammation, which can give rise to sprouting of sympathetic fibers in the peripheral organ or on sensory neuronal cell bodies in dorsal root ganglia (Berkley et al. 2005; Xia et al. 2011). The sixth feature, also dynamic, is hormonal modulation. It involves the fact that the responses of recipient CNS neurons, regardless of whether or not they have been sensitized, are subject to modulation by circulating hormones (e.g., Bradshaw et al. 2000).

This dynamic conceptualization of central pain mechanisms (Fig. 2b), supported by these five features, clearly places the decision for the perception of pain within networks of CNS neurons, thereby allowing pain to depend as much if not more on the manipulation of information by CNS neurons than it does on peripheral nociceptive events. The connective features of this view also illustrate that the CNS provides a substrate by which events occurring in one organ can influence the events occurring in another organ, while the dynamic features provide a means by which these cross-talk influences can be modified by injury, pathophysiology, or reproductive status. In so doing, this view helps explain the poor correlation between pathophysiology of reproductive organs and pain, referred pain, referred hyperalgesia, and cross-system co-occurrence of painful conditions.

Perhaps even more interesting is that this dynamic view also provides a different way of assessing pain and its neural mechanisms, with important clinical implications. Here are three examples:

1. In rats, it has been found that vaginal nociception varies only moderately with estrous stage (rats tolerate more pressure in the vaginal canal when fertile than they do in other stages; Bradshaw et al. 1999). However, when the rats have been subjected to surgery to produce endometriosis, the estrous changes are exaggerated and modified (rats develop vaginal hyperalgesia when in their fertile period; Cason et al. 2003). A similar situation exists for women. Thus, the menstrual cyclicity of muscle and skin pain thresholds in women with dysmenorrhea is exaggerated when compared with women without dysmenorrhea (Giamberardino et al. 1997). These findings lend support to newly emerging and important brain imaging evidence that severe dysmenorrhea, like other chronic pain conditions, is associated with major changes in brain anatomy and function, as shown in Fig. 3, (Berkley and McAllister 2011; Tu et al. 2010; Vincent et al. 2011).

2. Using a rat model of surgically induced endometriosis, it has been found that this condition exacerbates pain behaviors and referred muscle hyperalgesia arising from a ureteral stone, but the surgical control procedure (partial unilateral hysterectomy) actually reduces those pain behaviors and the referred hyperalgesia, producing in effect a condition that has been named "silent stones" (Giamberardino et al. 2002). Similarly, in women who pass multiple ureteral stones, ureteral colic pain episodes and referred hyperalgesia are more frequent in those who suffer from endometriosis or severe dysmenorrhea than in women without those problems (Giamberardino et al. 2001). Importantly, this and other findings have led to the possibility that, when painful visceral conditions co-occur, treatment focused on one of the conditions, such as dysmenorrhea, can be effective in alleviating the pain symptoms associated with the other condition, such as irritable bowel syndrome (Giamberardino et al. 2010).

3. In rats, it has been found that the actions of a drug (a cannabinoid agonist) on uterine contractility are reduced when the bladder of the

Gynecological Pain, Neural Mechanisms, Fig. 3 Significant regional gray matter volume changes in women with severe primary dysmenorrhea. Morphological changes in these women were associated with increased gray matter volume in the (**a**) right posterior hippocampus; (**c**) anterior/dorsal posterior cingulate cortex (ACC/dPCC, BA 23/24), midbrain, and hypothalamus; (**d**) left ventral portion of precuneus (BA 31), (**e**) left superior/middle temporal gyrus (STG/MTG, BA 22). Decreased grey matter volumes were observed in the (**b**) right central portion of precuneus (BA 7) and medial prefrontal cortex (mPFC, BA 10); (**d**) right ventral portion of precuneus (BA 7/31); (**e**) bilateral secondary somatosensory cortex (SII)/posterior insula; (**f**) mid insula (13). *Red/blue* colors represent increased/decreased volume (This figure is Fig. 1 from Tu et al. (2010), reprinted here with permission)

rat has been inflamed, and this effect is strongly influenced by estrous stage (Dmitrieva and Berkley 2002). Cross-system *pharmacological* effects, in which pathophysiology in one system could interfere with the efficacy of a drug for another condition, remain unstudied in women.

In summary, conceptualizations of neural mechanisms of gynecological pain are now changing to take into account new data on cross-system viscero-visceral and viscerosomatic interactions, the widespread divergence and convergence and facilitation and inhibition of information flow throughout the CNS, sympathetic-sensory coupling, central sensitization (injury memories), and hormonal modulation. In so doing, such conceptualizations help explain puzzling features of gynecological pain that are common to pain associated with other internal organs, such as poor correlation of pathophysiology and pain severity, referred pain and referred hyperalgesia, and co-occurrence of painful conditions associated with different systems. In addition, this view of central pain mechanisms is also helping to improve our understanding of how central dynamic processing of information gives rise to the many variations of gynecological pain that a woman experiences throughout her life and the importance and value of therapeutic strategies that involve combining pharmaceutical, somatic, and situational treatments tailored to the individual (Stratton and Berkley 2011).

Acknowledgment Research from our laboratory discussed in this review was supported by NIH grant NS 11892.

References

Berkley, K. J. (2001). Multiple mechanisms of pelvic pain: Lessons from basic research. In A. MacLean, R. W. Stones, & S. Thornton (Eds.), *Pain in obstetrics and gynaecology* (pp. 26–38). London: RCOG Press.

Berkley, K. J., & McAllister, S. L. (2011). Don't dismiss dysmenorrhea! *Pain, 152*, 1940–1941.

Berkley, K. J., Robbins, A., & Sato, Y. (1993). Functional differences between afferent gibers in the hypogastric and pelvic nerves innervating female reproductive organs in the rat. *Journal of Neurophysiology, 69*, 533–544.

Berkley, K. J., Rapkin, A. J., & Papka, R. E. (2005). The pains of endometriosis. *Science, 10*(308), 1587–1589.

Bradshaw, H. B., & Berkley, K. J. (2000). Estrous variations in responses of neurons in the rat gracile nucleus to stimulation of hindquarter skin and pelvic viscera. *Journal of Neuroscience, 20*, 7722–7727.

Bradshaw, H. B., Temple, J. L., Wood, E., & Berkley, K. J. (1999). Estrous variations in behavioral responses to vaginal and uterine distention in the rat. *Pain, 82*, 187–197.

Cason, A., Samuelsen, C., & Berkley, K. J. (2003). Estrous changes in vaginal nociception in a rat model of endometriosis. *Hormones and Behavior, 44*, 123–131.

Dmitrieva, N., & Berkley, K. J. (2002). Contrasting effects of WIN 55212-2 on motility of the rat bladder and uterus. *Journal of Neuroscience, 22*, 7147–7153.

Dmitrieva, N., Nagabukuro, H., Resuehr, D., Zhang, G., McAllister, S. L., McGinty, K. A., Mackie, K., & Berkley, K. J. (2010). Endocannabinoid involvement in endometriosis. *Pain, 151*, 703–710.

Dogrul, A., Ossipov, M. H., & Porreca, F. (2009). Differential mediation of descending pain facilitation and inhibition by spinal 5HT-3 and 5HT-7 receptors. *Brain Research, 1280*, 52–59.

Giamberardino, M. A. (2003). Referred muscle pain/hyperalgesia and central sensitisation. *Journal of Rehabilitation Medicine, 41*(Suppl.), 85–88.

Giamberardino, M. A., Berkley, K. J., Iezzi, S., de Bigontina, P., & Vecchiet, L. (1997). Pain threshold variations in somatic wall tissues as a function of menstrual cycle, segmental site and tissue depth in non-dysmenorrheic women, dysmenorrheic women and men. *Pain, 71*, 187–197.

Giamberardino, M. A., De Laurentis, S., Affaitati, G., Lerza, R., Lapenna, D., & Vecchiet, L. (2001). Modulation of pain and hyperalgesia from the urinary tract by algogenic conditions of the reproductive organs in women. *Neuroscience Letters, 304*, 61–64.

Giamberardino, M. A., Berkley, K. J., Affaitati, G., Lerza, R., Centurione, L., Lapenna, D., & Vecchiet, L. (2002). Influence of endometriosis on pain behaviors and muscle hyperalgesia induced by ureteral calculosis in female rats. *Pain, 95*, 247–257.

Giamberardino, M. A., Costantini, R., Affaitati, G., Fabrizio, A., Lapenna, D., Tafuri, E., & Mezzetti, A. (2010). Viscero-visceral hyperalgesia: Characterization in different clinical models. *Pain, 151*, 307–322.

Guindon, J., & Hohmann, A. G. (2009). The endocannabinoid system and pain. *CNS & Neurological Disorders Drug Targets, 8*, 403–421.

Jänig, W. (2006). *Integrative action of the autonomic nervous system: Neurobiology of homeostasis*. New York: Cambridge University Press.

Komisaruk, B. R., & Sansone, G. (2003). Novel pathways mediating vaginal function: The vagus nerves and spinal cord oxytocin. *Scandinavian Journal of Psychology, 44*, 441–450.

Nathan, P. W. (1985). Pain and nociception in the clinical context. *Philosophical Transactions of the Royal Society B: Biological Sciences, 308*, 219–226.

Papka, R. E., & Traurig, H. H. (1993). Autonomic efferent and visceral sensory innervation of the female reproductive system: Special reference to neurochemical markers in nerves and ganglionic connections. In C. A. Maggi (Ed.), *Nervous control of the urogenital system* (pp. 423–466). Chur: Harwood Academic.

Stratton, P., & Berkley, K. J. (2011). Chronic pelvic pain and endometriosis: Translational evidence of the relationship and implications. *Human Reproduction Update, 17*, 327–346.

Temple, J., Bradshaw, H. B., Wood, E., & Berkley, K. J. (1999). Effects of hypogastric neurectomy on escape responses to uterine distention in the rat. *Pain*, (Suppl. 6), S13–S20.

Tingåker, B. K., & Irestedt, L. (2010). Changes in uterine innervation in pregnancy and during labour. *Current Opinion in Anaesthesiology, 23*, 300–303.

Tracey, I., & Johns, E. (2010). The pain matrix: Reloaded or reborn as we image tonic pain using arterial spin labelling. *Pain, 148*, 359–360.

Tracey, I., & Mantyh, P. W. (2007). The cerebral signature for pain perception and its modulation. *Neuron, 55*, 377–391.

Tu, C. H., Niddam, D. M., Chao, H. T., Chen, L. F., Chen, Y. S., Wu, Y. T., Yeh, T. C., Lirng, J. F., & Hsieh, J. C. (2010). Brain morphological changes associated with cyclic menstrual pain. *Pain, 150*, 462–468.

Ustinova, E. E., Fraser, M. O., & Pezzone, M. A. (2010). Cross-talk and sensitization of bladder afferent nerves. *Neurourology and Urodynamics, 29*, 77–81.

Vincent, K., Warnaby, C., Stagg, C. J., Moore, J., Kennedy, S., & Tracey, I. (2011). Dysmenorrhoea is associated with central changes in otherwise healthy women. *Pain, 152*, 1966–1975.

Wesselmann, U., Burnett, A. L., & Heinberg, L. J. (1997). The urogenital and rectal pain syndromes. *Pain, 73*, 269–294.

Winnard, K. P., Dmitrieva, N., & Berkley, K. J. (2006). Cross-organ interactions between reproductive, gastrointestinal, and urinary tracts: Modulation by estrous stage and involvement of the hypogastric nerve. *American Journal of Physiology – Regulatory, Integrative and Comparative Physiology, 291*, R1592–R1601.

Woolf, C. J. (2011). Central sensitization: Implications for the diagnosis and treatment of pain. *Pain, 152*, S2–S15.

Xia, C. M., Colomb, D. G., Jr., Akbarali, H. I., & Qiao, L. Y. (2011). Prolonged sympathetic innervation of sensory neurons in rat thoracolumbar dorsal root ganglia during chronic colitis. *Neurogastroenterology and Motility, 23*, 801–1339.

H

5-HT

▶ Serotonin

5-Hydroxytryptamine

▶ Serotonin

Habituation

Definition

Habituation is the reverse process of sensitization, consisting of a waning response to repeated stimulation that can be rapid or gradual. It involves a reduction in excitability and means that neurons cease to fire when stimulated. However, if the interval between stimuli is altered randomly and the strength of the stimulus is increased, this process can be reversed.

Cross-References

▶ Infant Pain Mechanisms
▶ Migraine Without Aura
▶ Psychology of Pain, Sensitization, Habituation, and Pain

Hamstring Muscle Strain

Definition

Hamstring muscle strain produces pain in the biceps femoris muscles in the back of the thigh. Stretching the muscle can produce pain, as is found with a straight leg-raising maneuver.

Cross-References

▶ Sciatica

Handicap

▶ Disability in Fibromyalgia Patients

Hansen's Disease

Aki J. Hietaharju[1] and Maija Haanpää[2]
[1]Department of Neurology and Rehabilitation, Pain Clinic Tampere University Hospital, Tampere, Finland
[2]Department of Neurosurgery, Helsinki University Central Hospital, Helsinki, Finland

Synonyms

Leprosy

G.F. Gebhart, R.F. Schmidt (eds.), *Encyclopedia of Pain*, DOI 10.1007/978-3-642-28753-4,
© Springer-Verlag Berlin Heidelberg 2013

Definition

Hansen's disease is a chronic granulomatous infection of the skin and peripheral nerves caused by the obligate intracellular organism ► Mycobacterium leprae. It is associated with marked disabilities which result from the impairment of both sensory and motor nerve function.

Characteristics

Hansen's disease used to be widely distributed all over the world, but now the major part of the global burden of the disease is represented by low-resource countries, in tropical and warm temperate regions. In 1985, there were an estimated 12 million people with ► leprosy worldwide, resulting in a prevalence of 12 per 10,000 (Britton and Lockwood 2004). In the twenty-first century, significant progression has been made in controlling leprosy. Globally, the annual detection of new cases has been steadily declining, from 620,638 cases in 2002 to 249,007 in 2008. At the beginning of 2009, the registered prevalence of Hansen's disease globally was 213,036. The registered prevalence at the start of 2009 was highest in the following six countries: India, Brazil, Indonesia, Nigeria, Democratic Republic of the Congo, Ethiopia, and Bangladesh (WHO 2009).

Mycobacterium leprae is an acid-fast Gram-positive bacillus, which is supposed to be transmitted mainly by aerosol spread of nasal secretions and uptake through nasal or respiratory mucosa (Noordeen 1994). The infection is not spread by touching because the bacterium cannot penetrate intact skin. *Mycobacterium leprae* has a peculiar ► tropism for macrophages and ► Schwann cells. After having invaded the Schwann cell, the leprosy bacilli replicate slowly over years. These bacilli show preference for growth in cooler regions of the body causing damage to superficial nerves. Peripheral nerves are affected in fibro-osseous tunnels near the surface of the skin, e.g., posterior tibial nerve near the medial malleolus (Britton and Lockwood 2004).

The dynamic nature of the immune response to *Mycobacterium leprae* often leads to spontaneous fluctuations in the clinical state, which are called ► leprosy reactions. They are divided into two types. ► Leprosy type 1 reaction or reversal reaction is caused by spontaneous increases in T-cell reactivity to mycobacterial antigens (Britton 1998). In type 1 reactions, especially borderline patients may present with reactions to nerve pain, sudden palsies, and many new skin lesions. ► Leprosy type 2 reaction or erythema nodosum leprosum is a systemic inflammatory response to the deposition of extravascular immune complexes. It occurs only in borderline lepromatous and lepromatous leprosy (Lockwood 1996).

Diagnosis of leprosy is clinical. Three cardinal diagnostic features include (1) the presence of skin lesions with definite sensory loss, (2) the presence of thickened peripheral nerves, and (3) demonstration of acid-fast bacilli on slit skin smears or tissue biopsy (Walker and Lockwood 2007). Leprosy is divided into five subtypes based on histologic and immunologic features: tuberculoid (► tuberculoid leprosy), borderline tuberculoid (► borderline leprosy), midborderline, borderline lepromatous, and ► lepromatous leprosy (Ridley and Jopling 1966). For assignment to treatment regimes, WHO is using a simplified operational classification based on the number of skin lesions. Leprosy is called paucibacillary if the number of skin lesions in a patient is between 1 and 5 and multibacillary if the number of skin lesions exceeds 5 (Walker and Lockwood 2007). According to WHO guidelines, the former is treated with ► dapsone 100 mg daily and rifampicin 600 mg monthly for 6 months. In multibacillary leprosy, duration of the treatment is a minimum 2 years, and ► clofazimine 50 mg daily and 300 mg monthly is added to paucibacillary regimen (Britton and Lockwood 2004). Early case finding and prompt treatment with the above-mentioned multidrug therapy (MDT) are the cornerstone of leprosy control programs.

The most devastating clinical consequence of the intracutaneous nerve damage is the total sensory loss of the extremities (Brand and Fritschi 1985). Pain and temperature sensation are most strikingly decreased in early cases, and later, tactile and pressure sense are also lost. Anesthesia of the extremities predisposes the patient to chronic ulcers and severe secondary deformities, and therefore, leprosy remains a significant cause of neurologic disability worldwide. Besides sensory nerve involvement, leprosy affects also motor and autonomic nerves. Peripheral motor nerve palsies usually occur during type 1 reactions. The most common manifestation of autonomic nerve dysfunction in leprosy patients is hypohidrosis. Sometimes deterioration of sensory and motor functions can occur insidiously without signs or symptoms of inflammation. This phenomenon is called silent neuropathy (van Brakel and Khawas 1994).

Since Hansen's disease causes severe sensory loss, it is assumed that pain is uncommon in leprosy. However, peripheral nerve pain, dysesthesias, and paresthesias may complicate leprosy, both during and after treatment. Data on consumption of analgesics by patients with neuropathic pain gives some indication of the extent of the problem. In a Malaysian study of 235 leprosy patients, neuritic pain was the main reason for consumption of analgesic preparations. In 46 patients (19.5 %), an overall total intake had been more than 2 kg of analgesics. The duration of intake ranged from 2 to more than 20 years (Segasothy et al. 1986).

Acute pain in one or several nerves may be the presenting feature in Hansen's disease. Pain is a familiar symptom of reactions and neuritis, due to entrapment of the edematous inflamed nerve in sites of predilection (Nations et al. 1998). Neuritis of cutaneous nerves may also be painful (Theuvenet et al. 1993). Peripheral nerve abscesses, which are often associated with severe acute pain, occur in all types of leprosy and a variety of nerve trunks and cutaneous nerves (Kumar et al. 1997). Leprosy-related acute pain can usually be relieved by steroids or other therapeutic measures, such as anti-inflammatory drugs and immobilization or surgical intervention.

Chronic neuropathic pain in Hansen's disease has received scant attention. Hietaharju et al. (2000) reported on moderate or severe chronic neuropathic pain in 16 patients with treated multibacillary leprosy. In 10 patients, the pain had a glove and stocking-like distribution, and in 2 patients, it followed the course of a specific nerve. The quality of pain was burning in 9, biting in 3, pricking in 3, cutting in 2, and electric shock-like in 2 patients. The occurrence of pain was continuous in 50 % of the patients. In an evaluation of 303 patients from a Brazilian referral center, 174 (57 %) patients complained of neuropathic pain (Stump et al. 2002). In 84 patients (48 %), pain manifested as bursts. Pain affected one or more peripheral nerve territories; ulnar nerve in 101 (58 %) patients and tibial nerve course in 48 (28 %). There was a polyneuropathic distribution as glove-like in 47 patients (27 %) and sock-like in another 47 patients. At the time of evaluation, pain was present in 47 (27 %) patients.

There is little data on the occurrence of sensory disturbances such as dysesthesias, paresthesias, or allodynia in patients with leprosy. In a study by Hietaharju et al. (2000), four patients complained of a tingling sensation, which was considered as unpleasant and painful, i.e., they had dysesthesia. Dysesthesia followed glove and stocking distribution in two patients, the course of femoral cutaneous nerve in one patient, and was located in both legs below mid-thigh in one patient. Allodynia, pain due to a stimulus that does not normally provoke pain, was noticed in two patients. In both of these patients, enlargement and tenderness of the nerves (cutaneous femoral, common peroneal, and posterior tibial) without abscess formation was discovered in clinical examination.

The most typical sensory abnormalities in leprosy patients are severely impaired perception of tactile stimuli and mechanical and thermal pain, indicating damage of $A\beta$, $A\delta$, and C fibers at the painful site (Hietaharju et al. 2000). The cases with sensory loss associated with pain suggest peripheral deafferentation, i.e., pain due to loss of sensory input into the central nervous system, as occurs with different types of lesions of peripheral nerves. However, in a considerable proportion of the patients with pain, the sensory function may be quite well preserved, suggesting other pathophysiological mechanisms. Early involvement of small fibers due to mycobacterial invasion can cause dysfunction and damage leading to paresthesia and pain (Lund et al. 2007). Other possible explanations include the impact of previous episodes of reactions, neuritis, and inflammation, which may leave the nerve fibrosed and at risk of entrapment (Negesse 1996). Some patients may have a chronic ongoing neuritis manifesting clinically with pain (Haanpaa et al. 2004). Inflammation along nerve trunks has been shown to produce ectopic activity in nerves, and therefore, past or present inflammatory conditions represent a source for central sensitization, which may manifest as chronic neuropathic pain. In a study of 17 leprosy patients, who had completed MDT and were suffering from chronic neuropathic pain, intraneural acid-fast bacilli were detected in 5 nerve biopsies (Lund et al. 2007). It is possible that, despite adequate therapy, in some leprosy patients, bacteria are persisting in the nerve, and this chronic infection may give cause to nerve pathology and neuropathic pain.

To date, there are no controlled studies on the efficacy of any drug in leprosy-related neuropathic pain.

References

Brand, P. W., & Fritschi, E. P. (1985). Rehabilitation in leprosy. In R. C. Hastings (Ed.), *Leprosy* (1st ed., pp. 287–319). Edinburgh: Churchill-Livingstone.

Britton, W. J. (1998). The management of leprosy reversal reactions. *Leprosy Review, 69*, 225–234.

Britton, W. J., & Lockwood, D. N. (2004). Leprosy. *Lancet, 363*, 1209–1219.

Haanpaa, M., Lockwood, D. N., & Hietaharju, A. (2004). Neuropathic pain in leprosy. *Leprosy Review, 75*, 7–18.

Hietaharju, A., Croft, R., Alam, R., Birch, P., Mong, A., & Haanpää, M. (2000). The existence of chronic neuropathic pain in treated leprosy. *Lancet, 356*, 1080–1081.

Kumar, P., Saxena, R., Mohan, L., Thacker, A. K., & Mukhija, R. D. (1997). Peripheral nerve abscess in leprosy: Report of twenty cases. *Indian Journal of Leprosy, 69*, 143–147.

Lockwood, D. N. (1996). The management of erythema nodosum leprosum: Current and future options. *Leprosy Review, 67*, 253–259.

Lund, C., Koskinen, M., Suneetha, S., Lockwood, D. N. J., Haanpää, M., Haapasalo, H., & Hietaharju, A. (2007). Histopathological and clinical findings in leprosy patients with chronic neuropathic pain: A study from Hyderabad, India. *Leprosy Review, 78*, 369–380.

Nations, S. P., Katz, J. S., Lyde, C. B., & Barohn, R. J. (1998). Leprous neuropathy: An American perspective. *Seminars in Neurology, 18*, 113–124.

Negesse, Y. (1996). Comment: 'Silently arising clinical neuropathy' and extended indication of steroid therapy in leprosy neuropathy. *Leprosy Review, 67*, 230–231.

Noordeen, S. K. (1994). The epidemiology of leprosy. In R. C. Hastings (Ed.), *Leprosy* (2nd ed., pp. 29–48). Edinburgh: Churchill-Livingstone.

Ridley, D. S., & Jopling, W. H. (1966). Classification of leprosy according to immunity: A five-group system. *International Journal of Leprosy, 54*, 255–273.

Segasothy, M., Muhaya, H. M., Musa, A., Rajagopalan, K., Lim, K. J., Fatimah, Y., Kamal, A., & Ahmad, K. S. (1986). Analgesic use by leprosy patients. *International Journal of Leprosy and Other Mycobacterial Diseases, 54*, 399–402.

Stump, P., Baccarelli, R., Marciano, L., Lauris, J., Ura, S., Teixeira, M., & Virmond, M. (2002). Prevalence and characteristics of neuropathic pain and consequences of the sensory loss in 303 patients with leprosy. In: *Abstracts of the 10th world congress on pain* (p. 93), August 17–22, 2002, San Diego, CA. Seattle: IASP Press.

Theuvenet, W. J., Finlay, K., Roche, P., Soares, D., & Kauer, J. M. G. (1993). Neuritis of the lateral femoral cutaneous nerve in leprosy. *International Journal of Leprosy, 61*, 592–596.

van Brakel, W. H., & Khawas, I. B. (1994). Silent neuropathy in leprosy: An epidemiological description. *Leprosy Review, 65*, 350–360.

Walker, S. L., & Lockwood, D. N. J. (2007). Leprosy. *Clinics in Dermatology, 25*, 165–172.

WHO. (2009). Weekly epidemiological record. No. 33, 84: 333–340.

Hargreaves Test

▶ Thermal Nociception Test

Harm Principle

▶ Ethics of Pain Control in Infants and Children

Head Pain

▶ Headache

Headache

Nikolai Bogduk
University of Newcastle, Newcastle Bone & Joint
Institute, Royal Newcastle Centre, Newcastle,
NSW, Australia

Synonyms

Cephalalgia; Head pain

Definition

Headache is pain perceived in the head. Specifically, in order to constitute headache, the pain must be perceived in the occipital, temporal, parietal, or frontal regions of the head or in some combination of these regions. Pain from these regions may extend to encompass the orbital region, and some forms of headache may particularly affect the orbit. Pain localized to the eye, however, is not generally regarded as headache and is better described as eye pain. Similarly, pain in the face does not conventionally constitute headache; it is regarded separately as facial pain.

Characteristics

There are many varieties of headache, with many possible causes (Headache Classification Subcommittee of the International Headache Society 2004; Olesen et al. 2000). For the most common types of headache, the actual causes are not known. Although several theories are available, they relate rather to the mechanism of pain production and do not explain the fundamental reason why the headache occurs. That remains unknown.

Headaches are distinguished and defined largely on the basis of their clinical features. These can be described systematically under the categories of enquiry recommended for taking

a ► history of a pain problem (see ► Medical History):
- Length of illness
- Site
- Radiation
- Intensity
- Quality
- Frequency
- Duration
- Time of onset
- Mode of onset
- Precipitating features
- Aggravating features
- Relieving features
- Associated features

Length of Illness

This domain pertains to whether or not this is the first episode of headache that the patient has suffered. Headache for the first time is the cardinal clue for a small set of serious headaches, such as those caused by aneurysm, subarachnoid hemorrhage, meningitis, or sudden, severe hypertension.

Site

The site in which pain is felt is of no diagnostic significance, other than to establish that the complaint is one of headache. However, whether the pain is unilateral or bilateral does bear on the diagnosis of some forms of headache. For example, tension-type headache typically affects the entire head, whereas most other forms of headache are typically, although not always, unilateral.

Pain of a neuralgic quality (see below), in the distribution of the nerve affected, is diagnostic of trigeminal neuralgia, glossopharyngeal or vagal neuralgia, and C2 neuralgia.

Radiation

Noting the areas *to which* pain radiates does not help in diagnosis. Different forms of headache may have the same pattern of radiation. However, it can be helpful to recognize sites *from which* pain appears to be referred. Although the pain may be perceived in the forehead, if it appears to have spread from the occiput or neck, a possible cervical source should be considered.

Intensity

All forms of headache can be mild, moderate, or severe in intensity. So, intensity alone does not serve to discriminate different types of headache. However, certain types of headache are characterized by severe headache of sudden onset, sometimes described as "thunderclap" headache. Possible causes include subarachnoid hemorrhage, meningitis, and phaeochromocytoma.

Quality

Most headaches will be dull, aching, or throbbing in quality. These features do not help in making a diagnosis. On the other hand, a lancinating quality of pain establishes that the pain is neuralgic and is one of the defining features of trigeminal neuralgia, glossopharyngeal or vagal neuralgia, and of C2 neuralgia. Stabbing pain, or jabs of pain, is characteristic of cluster headache, although other features more strongly define this condition.

Frequency

Of all pain problems, headache is the one condition in which frequency is a cardinal diagnostic feature. Short, repeated jabs of pain recurring in bouts over several minutes are what characterize cluster headache. Periods of pain lasting half a day, or up to 3 or 4 days, interspersed with periods free of pain, are what characterizes migraine. Other types of headache occur in paroxysmal bouts, i.e., sustained periods of repeated jabs of intense pain that then switch off. These include chronic paroxysmal hemicrania (CPH) and SUNCT (sudden, unilateral, neuralgiform headache with conjunctival injection and tearing).

Duration

Duration of pain is often inextricably linked to frequency. In cluster headache and its congeners, the frequency of jabs of pain is high, but the duration of each jab is short, i.e., seconds. In migraine, the headache is established and remains constant, such that its duration is measured in hours or days, but then a pain-free interval appears.

Time of Onset

This is probably an obsolete category of enquiry for diagnostic purposes. Its heritage is that early morning headache was once regarded as pathognomonic of hypertension headache, but this has been disproved.

Nevertheless, sometimes the time of onset can provide clues to the cause of headache. For example, headache caused by exposure to chemicals or allergens may occur at only particular times of the day, particular days of the week, or particular seasons of the year. Synchrony with menstrual cycle strongly suggests menstrual migraine.

Mode of Onset

Severe headaches of sudden onset suggest subarachnoid hemorrhage, meningitis, or hypertension as the causes. Otherwise, most headaches come on gradually or in an unremarkable fashion. However, some forms of migraine can have a prodrome. A prodrome of neurological symptoms is virtually diagnostic of classical migraine (now known as migraine with prodrome) (Headache Classification Subcommittee of the International Headache Society 2004). Some patients with migraine will suffer cravings for certain types of foods, before the onset of headache. This fits with serotonin mechanisms that on the one hand are involved with pain and on the other hand are involved with satiety.

Precipitating Factors

Some forms of recurrent headache can be precipitated, inadvertently or consciously, by certain actions. In some patients, the pain of trigeminal neuralgia can be precipitated by touching trigger spots on the face or in the mouth. Headache precipitated by sexual activity is referred to as "sex headache" and appears to be related to a rapid rise in blood pressure. A rare but distinctive entity is colloid cyst of the third ventricle, in which headache can be precipitated by extension of the head, which causes the cyst to occlude the cerebral aqueduct and precipitate a sudden rise in cerebrospinal fluid pressure. "Ice-pick headache" is the term accorded to headache precipitated by exposure to cold foods or liquids.

For headaches of recent onset, not experienced before, an antecedent event may indicate the possible or likely cause. A classical example is post-lumbar puncture headache. A vexatious issue is trauma. Patients may report an injury that apparently caused the headache. However, a direct link between trauma and headache may be difficult to prove and is sometimes contentious. Nevertheless, a history of trauma may be the only defining feature of some forms of headache, on which grounds, the entity of "post-traumatic headache" is recognized.

Some headaches can be caused by exposure to drugs such as alcohol. Some headaches, paradoxically, can be caused by excessive consumption of analgesics. Withdrawal of analgesics can cause rebound headache.

Aggravating Factors

Few features that aggravate headache help in establishing a diagnosis, for many different forms of headache may be aggravated by activities such as turning the head, or exertion. However, certain features that appear to aggravate the pain are better classified as associated features (see below).

Relieving Factors

Many patients with headache resort to lying down. So, lying down per se is not a discriminating feature. However, when lying down promptly relieves the headache and when resumption of the upright posture restores it, the leading diagnosis is low-pressure cerebrospinal fluid, which can be idiopathic or secondary to lumbar puncture.

Associated Features

It is in the domain of associated features that most headaches can be distinguished. Photophobia, nausea, and vomiting are the cardinal diagnostic features of migraine. Lacrimation, rhinorrhoea, and conjunctival injection are reflex parasympathetic effects that occur with a family of headaches. Classically, they are the associated features of cluster headache, but they also occur in paroxysmal hemicrania and SUNCT syndrome.

Papilloedema and focal neurological signs are the classic features of space-occupying lesion of the cranium. Other important features are neck stiffness and Kernig's sign, which are virtually diagnostic of spread of infection or hemorrhage into the cervical subarachnoid space. Neurological signs affecting the III, IV, and VI cranial nerves are diagnostic of Tolosa-Hunt syndrome, i.e., granuloma of the cavernous sinus.

Age is an important feature. New headache in an elderly patient may be the only warning feature of temporal arteritis.

The diagnosis of acute herpes zoster can be made immediately once the eruption of vesicles occurs, but the pain may precede the eruption by up to 3 days.

Diagnosis

Figure 1 illustrates how taking a systematic history can allow many types of headache to be diagnosed on the basis of certain clinical features, singly or in combination. Migraine is diagnosed on the basis of periodic pain associated with photophobia, nausea, or vomiting. Cluster headache is defined by paroxysmal pain associated with lacrimation, rhinorrhoea, and conjunctival injection. Its relatives – CPH and SUNCT – only differ essentially with respect to the periodicity and duration of the headache. Intracranial lesions are diagnosed on the basis of associated neurological signs.

Certain entities, however, cannot be recognized clinically because they do not have any distinctive features. Those entities are benign intracranial hypertension, sphenoid sinusitis, cervicogenic headache, and tension-type headache.

The first three of these entities require investigations. Benign intracranial hypertension requires a CT scan. Sphenoid sinusitis is perhaps the most "impalpable" headache. It exhibits nothing but pain, felt somewhere deep in the center of the head. The diagnosis is established eventually by medical imaging. The diagnosis of cervicogenic headache requires the establishment of a cervical source of pain, by medical imaging or by diagnostic blocks of cervical structures or nerves.

Headache, Fig. 1 The differential diagnosis of headache by clinical history and examination

Tension-type headache is notable because there are no positive diagnostic criteria for this entity. It is a diagnosis by exclusion of other possible causes.

Other ill-defined entities include so-called vascular headache, whose cardinal feature is throbbing pain, but which does not exhibit any of the diagnostic features of migraine.

Cross-References

▶ Chronic Daily Headache in Children

References

Headache Classification Subcommittee of the International Headache Society. (2004). The international classification of headache disorders, 2nd edn. *Cephalalgia, 24*(Suppl. 1), 1–160.

Olesen, J., Tfelt-Hansen, P., & Welch, K. M. (2000). *The headaches* (2nd ed.). Philadelphia: Lippincott Williams & Wilkins.

Headache Associated with Disorders of the Cranium

▶ Headache from Cranial Bone

Headache Associated with Psychotic Disorder

▶ Headache Due to Somatoform Disorder

Headache Associated with Somatization Disorder

▶ Headache Due to Somatoform Disorder

Headache Attributed to a Substance or Its Withdrawal

Stephen D. Silberstein
Jefferson Medical College, Thomas Jefferson University and Jefferson Headache Center, Thomas Jefferson University Hospital, Philadelphia, PA, USA

Synonyms

Headaches associated with substances or their withdrawal; Medication-induced headaches

Definition

The International Headache Society (IHS) previously grouped medication-induced headaches under the rubric "headaches associated with substances or their withdrawal" (Headache Classification Committee of the International Headache Society 1988). The ICHD-2 and new ICHD-3 classification (Headache Classification Committee 2003, 2013) now calls these headaches "headaches attributed to a substance or its withdrawal" (Monteiro and Dahlof 2000).

Food, chemical, and drug ingestion or exposure can be both a cause of and a trigger for headache (Silberstein 1998). Their association is often based on reports of adverse drug reactions and anecdotal data and does not prove causality. When a new headache occurs for the first time in close temporal relation to substance exposure, it is coded as a secondary headache attributed to the substance. When a preexisting primary headache is made worse in close temporal relation to substance exposure, the patient is given the diagnosis of both the preexisting primary or secondary headache and the diagnosis of headache attributed to the substance (Headache Classification Committee 2013).

Headache Attributed to Acute Substance Use or Exposure (Headache Classification Committee 2013)

Headache attributed to acute substance use or exposure may start instantly or within hours or days. This group of headache disorders can be caused by (1) an unwanted effect of a toxic substance, (2) an unwanted effect of a substance in normal therapeutic use, and (3) a substance used in an experimental study. Headache as a side effect has been recorded with many drugs, often as just a reflection of the very high prevalence of headache.

Characteristics

Alcohol, food and food additives, and chemical and drug ingestion and withdrawal have all been reported to provoke or activate migraine in susceptible individuals. Since headache is a complaint often attributed to placebo, substance-related headache may arise as a result of expectation. The association between a headache and an exposure may be coincidental (occurring just on the basis of chance) or due to a concomitant illness or a direct or indirect effect of the drug and may depend on the condition being treated. Headache can be a symptom of a systemic disease, and drugs given to treat such a condition will be associated with headache. Some disorders may predispose to substance-related headache. Alone, neither the drug nor the condition would produce headache. Nonsteroidal anti-inflammatory drugs may produce headache by inducing aseptic meningitis in susceptible individuals. The possible relationships between drugs and headache are outlined below (Silberstein 1998).

Drug- and Substance-Related Headache
1. Coincidental
2. Reverse causality
3. Interaction headache
4. Causal
Acute: Primary effect and secondary effect

Acute Drug-Induced Headache

Whether or not a drug triggers a headache often depends on the presence or absence of an underlying headache disorder. Headaches are usually similar to the preexisting headache. The drugs most commonly associated with acute headache can be divided into several classes (Monteiro and Dahlof 2000).

Vasodilators

Headache is a frequent side effect of antihypertensive drugs. It has been reported with the beta-blockers, ▶ calcium channel blockers (especially nifedipine), ACE inhibitors, and methyldopa. Nicotinic acid, dipyridamole, and hydralazine have also been associated with headache. The headache mechanism is uncertain (Thomson Healthcare 2003).

Nitric Oxide Donor-Induced Headache

Headache is a well-known side effect of therapeutic use of nitroglycerin (GTN) and other ▶ nitric oxide (NO) donors. They may cause headache by activating the trigeminal vascular pathway. There is an immediate NO donor-induced headache (GTN headache), which develops within 10 min after NO donor absorption and resolves within 1 h after release of NO has ended. There is also a delayed NO donor-induced headache, which develops after NO is cleared from the blood and resolves within 72 h after single exposure (Ashina et al. 2000).

Phosphodiesterase Inhibitor-Induced Headache

Phosphodiesterases (PDEs) are a large family of enzymes that break down cyclic ▶ nucleotides (cGMP and cAMP). PDE-5 inhibitors include sildenafil and dipyridamole. The headache, unlike GTN-induced headache, is monophasic. In normal volunteers it has the characteristics of TTH, but in migraine sufferers it has the characteristics of migraine without aura (Headache Classification Committee 2003).

Histamine-Induced Headache

Histamine causes an immediate headache in non-headache sufferers and an immediate as well as a delayed headache in migraine sufferers.

The mechanism is primarily mediated via the H_1 receptor because it is almost completely blocked by mepyramine. The immediate histamine-induced headache develops within 10 min and resolves within 1 h after absorption of histamine has ceased. The delayed histamine-induced headache develops after histamine is cleared from the blood and resolves within 72 h (Krabbe and Olesen 1980).

Nonsteroidal Anti-inflammatory Drugs
The nonsteroidal anti-inflammatory drugs, especially indomethacin, have been associated with headache. Mechanisms include aseptic meningitis (especially with ibuprofen) and reverse causality.

Serotonin Agonists
M-chlorophenylpiperazine, a metabolite of the antidepressant trazodone, can trigger headache by activating the serotonin (5-hydroxytryptamine [HT]) 2B and 2C receptors (Brewerton et al. 1988). This may be the mechanism of headache induction during early treatment with selective serotonin reuptake inhibitors.

Foods and Natural Products (Headache Induced by Food Components and Additives)
Chocolate, alcohol, citrus fruits, cheese, and dairy products are the foods that patients most commonly believe trigger their migraine, but the evidence is not persuasive.

Amino Acids
Monosodium glutamate (MSG) (Schamburg et al. 1969) and aspartame, the active ingredient of "NutraSweet," may cause headache in susceptible individuals (Schiffmann et al. 1987). Phenylethylamine, tyramine, and aspartame have been incriminated, but their headache-inducing potential is not sufficiently validated.

Monosodium Glutamate-Induced Headache (Chinese Restaurant Syndrome)
MSG can induce headache and the Chinese restaurant syndrome in susceptible individuals. The headache is typically dull or burning and non-pulsating, but may be pulsating in migraine sufferers. It is commonly associated with other symptoms, including pressure in the chest; pressure and/or tightness in the face; burning sensations in the chest, neck, or shoulders; flushing of the face; dizziness; and abdominal discomfort (Schamburg et al. 1969).

Aspartame
Aspartame, a sugar substitute, is an O-methyl ester of the dipeptide L-α-aspartyl-L-phenylalanine that blocks the increase in brain tryptophan, 5-HT, and 5-hydroxyindolacetic acid normally seen after carbohydrate consumption (Schiffmann et al. 1987). It produced headache in two controlled studies but not in the third (Silberstein 1998).

Tyramine
Tyramine is a biogenic amine that is present in mature cheeses. It is probably not a migraine trigger (Silberstein 1998).

Phenylethylamine
Chocolate contains large amounts of β-phenylethylamine, a vasoactive amine that is, in part, metabolized by monoamine oxidase. The evidence to support it as a trigger is weak (Silberstein 1998).

Ethanol
Alone or in combination with ► congeners (wine), ethanol can induce headache in susceptible individuals. The attacks often occur within hours after ingestion. In the United Kingdom, red wine is more likely to trigger migraine than white, while in France and Italy white wine is more likely to produce headache than red. Headaches are more likely to develop in response to white wine if red coloring matter has been added. Migraineurs who believed that red wine (but not alcohol) provoked their headaches were challenged either with red wine or with a vodka mixture of equivalent alcoholic content. The red wine provoked migraine in 9/11 subjects, the vodka in 0/11. Neither provoked headache in other migraine subjects or controls (Littlewood et al. 1988). It is not known which component of red wine triggers headache, and the study may not have been blinded to oenophiles.

The susceptibility to hangover headache has not been determined. Migraineurs can suffer a migraine the day following only modest alcoholic intake, while non-migraineurs usually need a high intake of alcoholic beverages to develop hangover headache. A few subjects develop headache due to a direct effect of alcohol or alcoholic beverages (cocktail headache). This is much rarer than delayed alcohol-induced headache (hangover headache).

Lactose Intolerance
Lactose intolerance is a common genetic disorder, occurring in over two-thirds of African-Americans, Native Americans, and Ashkenazi Jews and in 10 % of individuals of Scandinavian ancestry. The most common symptoms are abdominal cramps and flatulence. How lactose intolerance triggers migraine is uncertain (Silberstein 1998).

Chocolate
Chocolate is the food most frequently believed to trigger headache, but the evidence supporting this belief is inconsistent (Scharff and Marcus 1999). Chocolate is probably not a migraine trigger, despite the fact that many migraineurs believe that it triggers their headache. It is the most commonly craved food in the United States. Women are more likely than men to have migraine, and they crave chocolate more than men. Sweet craving is a premonitory symptom of migraine, and menses are often associated with an increase in carbohydrate and chocolate craving.

Chemotherapeutic Drugs
▶ Intrathecal methotrexate and diaziquone can produce aseptic meningitis and headache. Methyldichlorophen, interferon-β, and interleukin-2 are all associated with headache (Boogerd 1995).

Immunomodulating Drugs
Cyclosporine, FK-506, thalidomide, and antithymocyte globulin have been associated with headache (Shah and Lisak 1995).

Antimicrobial and Antimalarial Drugs
Amphotericin, griseofulvin, tetracycline, and sulfonamides have been associated with headache. Chloroquine and ethionamide are also associated with headache.

Other Substances
Carbon Monoxide-Induced Headache (Warehouse Workers' Headache)
Typically this is a mild headache without associated symptoms with carboxyhemoglobin levels of 10–20 %, a moderate pulsating headache and irritability with levels of 20–30 %, and a severe headache with nausea, vomiting, and blurred vision with levels of 30–40 %. When carboxyhemoglobin levels are higher than 40 %, headache is not usually a complaint because of changes in consciousness.

Cocaine-Induced Headache
Headache is common, develops immediately or within 1 h after use, and is not associated with other symptoms unless there is concomitant stroke or TIA (Dhopesh et al. 1991).

Cannabis-Induced Headache
Cannabis use is reported to cause headache associated with dryness of the mouth, paresthesias, feelings of warmth, and suffusion of the conjunctivae (elMallakh 1987).

References

Ashina, M., Bendtsen, L., Jensen, R., et al. (2000). Nitric oxide-induced headache in patients with chronic tension-type headache. *Brain, 123*, 1830–1837.

Bix, K. J., Pearson, D. J., & Bentley, S. J. (1984). A psychiatric study of patients with supposed food allergy. *The British Journal of Psychiatry, 145*, 121–126.

Boogerd, W. (1995). Neurological complications of chemotherapy. In F. A. DeWolff (Ed.), *Handbook of clinical neurology* (p. 527). Amsterdam: Elsevier.

Brewerton, T. D., Murphy, D. L., Mueller, E. A., et al. (1988). Induction of migraine-like headaches by the serotonin agonist m-chlorophenylpiperazine. *Clinical Pharmacology and Therapeutics, 43*, 605–609.

Dhopesh, V., Maany, I., & Herring, C. (1991). The relationship of cocaine to headache in polysubstance abusers. *Headache, 31*, 17–19.

elMallakh, R. S. (1987). Marijuana and migraine. *Headache, 27*, 442–443.

Headache Classification Committee. (2004a). The international classification of headache disorders 2nd edition. *Cephalalgia, 24*(Suppl 1), 1–160.

Headache Classification Committee. (2004b). The international classification of headache disorders 3rd edition. *Cephalalgia* (in press).

Headache Classification Committee of the International Headache Society. (1988). Classification and diagnostic criteria for headache disorders, cranial neuralgia, and facial pain. *Cephalalgia, 8*, 1–96.

Krabbe, A. A., & Olesen, J. (1980). Headache provocation by continuous intravenous infusion of histamine, clinical results and receptor mechanisms. *Pain, 8*, 253–259.

Littlewood, J. T., Glover, V., Davies, P. T., et al. (1988). Red wine as a cause of migraine. *Lancet, 559*.

Monteiro, J. M., & Dahlof, C. G. (2000). Single use of substances. In J. Olesen, P. Tfelt-Hansen, & K. M. Welch (Eds.), *The headaches* (pp. 861–869). Philadelphia: Lippincott Williams & Wilkins.

Rose, F. C. (1997). Food and headache. *Headache Quarterly, 8*, 319–329.

Schamburg, H. H., Byck, R., Gerstl, R., et al. (1969). Monosodium L-glutamate: Its pharmacology and role in the Chinese restaurant syndrome. *Science, 163*, 826–828.

Scharff, L., & Marcus, D. A. (1999). The association between chocolate and migraine: A review. *Headache Quarterly, 10*, 199–205.

Schiffmann, S. S., Buckley, C. E., Sampson, H. A., et al. (1987). Aspartame and susceptibility to headache. *The New England Journal of Medicine, 317*, 1181–1185.

Shah, A. K., & Lisak, R. (1995). Neurological complications of immunomodulating therapy. In F. A. DeWolff (Ed.), *Handbook of clinical neurology* (p. 547). Amsterdam: Elsevier.

Silberstein, S. D. (1998). Drug-induced headache. *Neurological Clinics of North America, 16*, 107–123.

Thomson Healthcare. (2003). *Physicians' desk reference*. Montvale: Thomson PDR.

VanDenEeden, S. K., Koepsell, T. D., Longstreth, W. T., et al. (1994). Aspartame ingestion and headaches: A randomized crossover trial. *Neurology, 44*, 1787–1793.

Headache Due to Arteritis

Peter Berlit
Department of Neurology, Alfried Krupp Hospital, Essen, Germany

Synonyms

Angiitis of the CNS; Vasculitis

Definition

Headache is the most common complaint in cranial or ▶ temporal arteritis. The major symptoms of central nervous system arteritis are multifocal neurological symptoms following ▶ stroke, in combination with headache and some degree of ▶ encephalopathy, with and without ▶ seizures. CNS-vasculitis may be part of a systemic autoimmune disease or the only manifestation of angiitis (isolated or primary ▶ angiitis of the central nervous system – PACNS).

Characteristics

Temporal Arteritis

Temporal arteritis (cranial arteritis, giant cell arteritis) is an autoimmune disease of elderly people, affecting women more frequently than men (3:1). Mean age at the beginning of the disorder is 65 years or more; the disease rarely appears before the age of 50. The incidence in Europe is 18/100,000 per year; there is a frequent association with HLA-DRB1. The diagnosis is confirmed by the histological examination of a ▶ biopsy specimen from the temporal artery, demonstrating the arteritis with necrosis of the media and a granulomatous inflammatory exudate containing lymphocytes, leukocytes, and giant cells.

Headache is the most common complaint in temporal arteritis, associated with a markedly elevated sedimentation rate. The patient develops an increasingly intense head pain, usually unilateral, sometimes bilateral. It has a nonpulsating often sharp and stabbing character, sometimes with a temporal pronunciation. But the localization of the headache may be frontal, occipital, or even nuchal (Pradalier and Le Quellec 2000). The pain increases during the night hours and persists throughout the day. Due to ischemia of the masseter muscles during mastication, ▶ jaw claudication may appear (Kraemer et al. 2010). The superficial temporal artery may be thickened and tender without pulsation – "cord-sign." Ultrasound examinations of the temporal artery reveal the halo-sign; the inflammation of the vessel wall may be demonstrated by contrast enhanced MRI.

Diagnostic Criteria of Temporal Arteritis

- Age 50 years or more
- Newly developed headache

Headache Due to Arteritis, Table 1 Frequency of signs and symptoms with temporal arteritis (Adapted from Caselli and Hunder 1993)

Symptom	All (%)	Initial symptom (%)
Headache	72	33
Polymyalgia rheumatica	58	25
Malaise, weight loss	56	20
Jaw claudication	40	4
Fever	35	11
Cough	17	8
Neuropathies (mono-, or multiplex)	14	0
Disorders of swallowing	11	2
Amaurosis fugax	10	2
Permanent loss of vision	8	3
Claudication of limbs (legs)	8	0
Stroke	7	0
Neuro-otologic disorders	7	0
Flimmer scotoma	5	0
Pain of the tongue	4	0
Depression	3	0.6
Diplopia	2	0
Myelopathy	0.6	0

- Tenderness of the superficial temporal artery
- Elevated sedimentation rate, at least 50 mm/h
- Giant cell arteritis in a biopsy specimen from the temporal artery

Besides the headache, there may be severe pain, aching, and symmetrical stiffness in proximal muscles of the limbs (polymyalgia rheumatica) in as many as 50 % of patients. Many patients present the symptoms of a cryptogenic neoplasm, anorexia, loss of weight, anemia, malaise, and low-grade fever.

Sudden blindness results from involvement of the posterior ciliary arteries, and blindness of one eye may be followed by the other. Other complications include the affection of intracranial or spinal vessels, necrosis of the scalp or tongue, and generalization of the arteritis affecting the coronary arteries, the aorta, or the intestines (Table 1).

The treatment of choice at the earliest suspicion of cranial arteritis is prednisone 60–80 mg/day. If ischemic complications are present, a steroid pulse-therapy for 3 to 5 days with at least 500 mg prednisone i.v. is recommended. Patients respond quickly and often very impressively to steroids. The start of this therapy should not be delayed for the biopsy. Depending on the clinical symptoms and the sedimentation rate, steroids are gradually reduced. In the majority of patients, steroid treatment is necessary for at least 20 months; therefore, a biopsy is mandatory in all cases. During the long-term course, the CRP is more helpful in the prediction of relapses than the sedimentation rate (Berlit 1997). In order to avoid side effects of the treatment pantoprazole, aspirin, vitamin D, and calcium are recommended. If necessary, ▶ methotrexate may be used as steroid sparing agent.

Systemic Lupus Erythematosus (SLE)

▶ Systemic lupus erythematosus (SLE) is the most frequent systemic autoimmune disease, incidence 7/100,000 (Ruiz-Irastorza et al. 2001); the prevalence in Europe and the USA is 10 to 60/100,000 per year, women:men = 10:1. The most common age of manifestation is 15–30 years. Both migraine and tension-type headaches (see ▶ migraine) are frequent. SLE is characterized by a disturbed regulation of T- and B-cell immunity with antinuclear antibodies and autoreactivity against other autoantigens in the progressive relapsing course of the disease. The multilocular manifestations are caused by a thrombotic vasopathy or antibodies interacting with cell membrane functions; a true vasculitis is rare. Antinuclear antibodies are present in 95 %, ds-DNA-antibodies in 80 %. ▶ Photosensitivity of the skin with a ▶ butterfly erythema of the face is a typical symptom of SLE. Arthritis and serositis with pulmonary and cardiac manifestations are frequent. Neurological symptoms are present in about 50 % of the patients, encephalopathy (60 %), seizures (60 %), and stroke (40 %). In SLE, strokes are frequently caused by a secondary ▶ antiphospholipid syndrome (25 % of all SLE patients). This diagnosis is made with the detection of lupus anticoagulant and IgG-anticardiolipin antibodies. Stroke may also be caused by cardiogenic embolism with Libmann-Sacks endocarditis or by thrombotic thrombocytopenic purpura. Some autoantibodies (ab) are associated with certain clinical manifestations: ribosomal P – psychosis, Jo 1 – polymyositis, antineuronal – epilepsy and encephalopathy. A classification of the neuropsychiatric SLE manifestations including headache has been given by the ACR Ad Hoc Committee on Neuropsychiatric Lupus in 1999. In case-control studies, there was no difference between SLE patients and the general population regarding the prevalence and incidence of migraine or tension-type headache (Sfikakis et al. 1998). In SLE patients with tension-type headache, there was an association with personality changes, emotional conflicts, and depression (Omdal et al. 2001). Most of these patients have higher disease activity scores (Amit et al. 1999). There was no association between anticardiolipin antibodies and migraine in a prospective study (Vazquez-Cruz et al. 1990). If a SLE patient develops a new headache, a neurological examination including ▶ MRI and lumbar puncture is mandatory. The association with a ▶ pseudotumor cerebri should be excluded. Treatment of idiopathic headache syndromes in SLE is the same as in the general population. A headache as the sole neurological symptom of SLE should not alter the immunosuppressive strategy in the individual patient.

Sjögren's Syndrome

▶ Sjögren's syndrome is clinically characterized by keratoconjunctivitis sicca and symptomatic xerostomia (the sicca-syndrome) and associated with the detection of anti-Ro (SSA-97 %) and anti-La (SSB-78 %) autoantibodies. In addition to multifocal CNS symptoms with encephalopathy, depression, or headache, a polyneuropathy and myopathy occur frequently. Whenever possible, the diagnosis should be verified with a salivary gland biopsy. The incidence of migraine is higher in patients with a sicca syndrome or Raynaud phenomenon (Pal et al. 1989). ▶ Flunarizine may be helpful for prophylaxis in rheumatologic patients with migraine (Mazagri and Shuaib 1992).

Granulomatosis with Polyangiitis (GPA)

GPA ▶ Wegener's granulomatosis is a rare autoimmune disease (1 per 100,000) associated with antineutrophil cytoplasmic antibodies (c-ANCA); men are affected twice as often as women. In the limited stage of the disease, necrotizing granulomas of the nose and the paranasal sinuses may lead to compression of neighborhood structures with cranial nerve lesions, diabetes insipidus, or exophthalmus. With generalization, the systemic necrotizing vasculitis involving small arteries and veins leads to affections of the lung and kidney.

In the limited stage of GPA, headaches are frequent and often caused by sinusitis, nonseptic meningitis, or local granulomas (Lim et al. 2002). MRI may show enhancement of the basal meninges especially of the tentorium (Specks et al. 2000); the development of an occlusive or communicating hydrocephalus is possible (Scarrow et al. 1998) and must be excluded.

Prednisone and ▶ cyclophosphamide are the treatment of choice in generalized GPA. Alternatively, Rituximab may be given in the acute phase; once the patient is in remission Methotrexate or Azathioprine are less toxic drugs. Headaches are treated symptomatically with paracetamol or nonsteroidal antiphlogistics.

Behçet's Disease

▶ Behçet's disease presents with the triad of iridocyclitis and oral and genital ulcers. The underlying systemic vasculitis of especially the veins may lead to an ▶ erythema nodosum, a thrombophlebitis, polyarthritis, or ulcerative colitis. Behçet's syndrome is rare in the USA and Germany (incidence 1/500,000), but frequent in Turkey (300/100,000); men are affected twice as often as women, usually between the ages of 20 and 40. There is an association with HLA-B5. Neurological manifestations occur in approximately 30 % (neuro-Behçet), either as ▶ meningoencephalitis of the brain stem and cerebellum (80 %) or as a ▶ sinus thrombosis, which presents often as pseudotumor cerebri (20 %; Akman-Demir et al. 1999). Headaches are the most common complaint in ▶ neuro-Behçet (87 % of patients). The holocephal stabbing severe pain does not usually respond to conventional analgetics but resolves with steroid treatment. MRI and lumbar puncture are diagnostic. Steroids and immunosuppressants like azathioprine are the treatment of choice. In sinus thrombosis, anticoagulants must be given in addition.

Primary Angiitis of the Central Nervous System (PACNS) (Isolated Arteritis of the Nervous System: IAN)

Primary angiitis of the central nervous system (PACNS) is an idiopathic medium and small vessel vasculitis affecting exclusively CNS vessels of the brain or spinal cord. About 500 cases have been documented worldwide (Berlit 2009, 2010). The major symptoms of PACNS are multifocal neurological symptoms following stroke, in combination with headache and some degree of encephalopathy, with or without seizures, cranial nerve palsies, or ▶ myelopathy.

The encephalopathy occurs in 40–80 %, subacute or chronic headaches in 40–60 %, focal symptoms in 40–70 %, and seizures are present in 30 %. An acute beginning of PACNS has been described in only 11 %; most patients develop the symptoms slowly and progressively. Systemic signs of inflammation (fever, ESR, CRP) are rare (10–20 %). On the other hand, there are usually signs of inflammation in the CSF (pleocytosis, elevation of protein, oligoclonal banding). The specificities of cerebral ▶ angiography or MRI are below 30 % (Schmidley 2000). For definitive diagnosis of PACNS, a combined leptomeningeal and parenchymal biopsy is necessary, especially in order to exclude infections (endocarditis; Berlit 2009) or tumors (lymphoma). Before the treatment of choice with prednisone and cyclophosphamide is established, a systemic inflammation or infection must be excluded and leptomeningeal and parenchymal biopsies must demonstrate the vascular inflammation. Without histological verification of the diagnosis, blind treatment is dangerous and possibly harmful for the patient and must be strictly avoided. With immunosuppressive therapy, the headaches resolve completely within a few weeks.

References

Ad Hoc Committee on Neuropsychiatric Lupus Nomenclature. (1999). The American College of Rheumatology nomenclature and case definitions for neuropsychiatric lupus. *Arthritis and Rheumatism, 42,* 599–608.

Akman-Demir, G., Serdaroglu, P., & Tasci, B. (1999). Clinical patterns of neurological involvement in Behçet's disease: evaluation of 200 patients. The Neuro-Behçet Study Group. *Brain, 122,* 2171–2182.

Amit, M., Molad, Y., Levy, O., et al. (1999). Headache in systemic lupus erythematosus and its relation to other disease manifestations. *Clinical and Experimental Rheumatology, 17,* 467–470.

Berlit, P. (1997). Giant Cell Arteritis. In R. Lechtenberg & H. S. Schutta (Eds.), *Practice guidelines for neurologic therapy.* New York: Marcel Dekker.

Berlit, P. (2009). The differential diagnosis of isolated angiitis of the CNS and bacterial endocarditis - similarities and differences. *Journal of Neurology, 256,* 792–795.

Berlit, P. (2010). Diagnosis and treatment of cerebral vasculitis. *Therapeutic Advances in Neurological Disorders, 3,* 29–42.

Caselli, R. J., Hunder, G. G. (1993). Neurologic aspects of giant cell arteritis. *Rheumatic Disease Clinics of North America, 19,* 941–953.

Kraemer M, Herold M, Venker C, Metz A, Berlit P (2011) Reduction of jaw opening: A neglected symptom pointing to temporal arteritis. *Rheumatology International, 31,* 1521–1523.

Lim, I. G., Spira, P. J., & McNeil, H. P. (2002). Headache as the initial presentation of Wegener's granulomatosis. *Annals of the Rheumatic Diseases, 61,* 571–572.

Mazagri, R., & Shuaib, A. (1992). Flunarizine is effective in prophylaxis of headache associated with scleroderma. *Headache, 32,* 298–299.

Omdal, R., Waterloo, K., Koldingsnes, W., et al. (2001). Somatic and psychological features of headache in systemic lupus erythematosus. *Journal of Rheumatology, 28,* 772–779.

Pal, B., Gibson, C., Passmore, J., et al. (1989). A study of headaches and migraine in Sjogren's syndrome and other rheumatic disorders. *Annals of the Rheumatic Diseases, 48,* 312–316.

Pradalier, A., & Le Quellec, A. (2000). Headache due to temporal arteritis. *Pathol Biol, 48,* 700–706.

Ruiz-Irastorza, G., Khamashta, M. A., Castellino, G., et al. (2001). Systemic lupus erythematosus. *Lancet, 357,* 1027–1032.

Scarrow, A. M., Segal, R., Medsger, T. A., Jr., et al. (1998). Communicating hydrocephalus secondary to diffuse meningeal spread of Wegener's granulomatosis: case report and literature review. *Neurosurgery, 43,* 1470–1473.

Schmidley, J. W. (2000). *Central nervous system angiitis.* Boston: Butterworth-Heinemann.

Sfikakis, P. P., Mitsikostas, D. D., Manoussakis, M. N., et al. (1998). Headache in systemic lupus erythematosus: a controlled study. *Br J Rheumatol, 37,* 300–303.

Specks, U., Moder, K. G., & McDonald, T. J. (2000). Meningeal involvement in Wegener granulomatosis. *Mayo Clinic Proceedings, 75,* 856–859.

Vazquez-Cruz, J., Traboulssi, H., Rodriquez-De la Serna, A., et al. (1990). A prospective study of chronic or recurrent headache in systemic lupus erythematosus. *Headache, 30,* 232–235.

Headache Due to Brain Metastases

Definition

Intracranial metastases are found in about 25 % of all patients who have died of cancer. Some of these tumors are silent, but the majority cause the syndrome consisting of headache, nausea and vomiting, mental change, confusion, seizures, and neurological deficit. Some tumors frequently produce brain secondary's (e.g., cancers of the lung and breast as well as melanoma), some only infrequently (e.g., cancer of ovary). Primary tumors, although relatively rare, can produce the same syndrome. Headache may arise from an expanding mass within the skull and distension of meninges. The treatment of choice is skull irradiation accompanied by the use of dexamethasone.

Cross-References

▶ Cancer Pain

Headache Due to Dissection

Mathias Sturzenegger
Department of Neurology, University Hospital, Inselspital, University of Bern, Bern, Switzerland

Synonyms

Carotidynia; Horner's syndrome; Tolosa-Hunt syndrome (painful ophthalmoplegia)

Definition

As already indicated by the title, the headaches are defined by their underlying pathology, i.e., dissection of the arteries. Since we are talking about headache, it is evident that we talk about dissections of cervico-cranial arteries; it is exceptional that dissection of the subclavian artery or aorta produces headache.

Pathogenesis

We have however to keep in mind that pain is usually a symptom of arterial dissections anywhere in the body, e.g., also of the aorta and renal or coronary arteries. The question rises why dissections are painful. Other pathologies of arterial walls may also be painful such as arteritis (e.g., giant cell arteritis), whereas atherosclerosis is usually painless.

We know that the walls of extracranial and also basal intracranial arteries are densely supplied with nociceptive, mainly trigeminal nerve fibers (Norregaard and Moskowitz 1985). These fibers are sensitive to inflammatory stimuli such as in vasculitis or to distension of the vessel wall that may take place during balloon dilatation or as a consequence of intramural hemorrhage such as in dissection. In atherosclerosis, although usually considered as an inflammatory process too, the inflammatory activity is probably simply too low to cause nociceptor discharge. It is controversial whether the irritation of the perivascular sympathetic nerve plexus, existing around the carotid as well as the vertebral arteries, is another explanation or a contributing etiological factor of dissection-associated pain (De Marinis et al. 1991). In my personal experience, pain may be equally intense in dissections with definite vessel diameter extension as in those dissections without enlargement but merely vessel stenosis or occlusion. Furthermore, also from merely personal experience, pain is not more frequent in patients with Horner's syndrome compared to those without.

Dissection-associated pain is usually reported with internal carotid artery (ICA) or vertebral artery (VA) dissections and their branches. We do not have data regarding dissection of external carotid arteries or subclavian arteries and their branches, nor whether such pathologies exist and how frequently nor whether they may cause any pain.

We know that dissections may take place without causing pain. It seems to be rare, but we usually detect dissections because of their consequences such as pain or cerebral ischemia. That means that asymptomatic dissections may simply go undetected, and painless dissections with other complications, such as lower cranial nerve palsy or cerebral ischemia may go unrecognized, since adequate diagnostic methods to detect dissections are not performed. In patients with "painful Horner's syndrome," many physicians have learned to think of ipsilateral internal carotid artery dissection (ICAD), but in what percentage of painless Horner's syndrome is ICAD the cause or is ICAD searched for? To detect ICAD as a cause, the MRI with special sequences must be performed of the neck region – and not (only) of the head!

The larger the affected vessel (carotid versus vertebral arteries), the more easily dissections are detected, i.e., can be imaged. Yet, without applying fat-saturated T1 MRI sequences and, furthermore, that specific method to the appropriate vessel segment (e.g., the high cervical retromandibular ICA segment), painful ICAD without causing vessel stenosis may not be detected even when performing Doppler- or Duplex examinations, MRA and/or conventional MRI. In the VA, it is well known that for various reasons MRI, as well as Doppler/duplex examination, is much less sensitive to dissections as compared to angiography, an invasive procedure not completely without risks.

To summarize: pain may herald dissections early on, absence of pain does not exclude dissections, the frequency of painless and asymptomatic dissections is not known, but they certainly exist.

Clinical Relevance

The most important aspect of headaches caused by dissections of cervico-cerebral arteries is the fact that they usually herald the onset of dissection and allow early recognition of the underlying pathology. Paying adequate attention to these warning symptoms enables the aversion of the often life-threatening sequelae of cerebral ischemia. Fifty to eighty percent of patients with a dissection of the cervico-cerebral arteries suffer a subsequent stroke; dissections are responsible for 20–30 % of all strokes in young (<45 years) persons; warning headaches preceding stroke have been noted in 47–74 % of patients with ICAD and in 33–85 % of patients with VAD (Fisher 1982; Silbert et al. 1995; Sturzenegger 1994, 1995).

Differential Diagnostic Aspects

The following three pathophysiologically poorly defined and most probably heterogeneous clinical syndromes with a diagnostic eponym may well be caused by dissection (Sturzenegger 1995).

Raeder's paratrigeminal syndrome (combination of pain, ipsilateral oculosympathetic defect and ipsilateral trigeminal dysfunction) may be due to spontaneous internal carotid artery dissection (ICAD) (Selky 1995). According to several studies, ICAD presents in 15 to 25 % as a painful Horner's syndrome (Sturzenegger 1995; Biousse 1994).

Carotidynia is a poorly defined syndrome with unilateral anterolateral cervical pain and tenderness. It is good advice to rule out underlying carotid dissection first, since most reports of this entity date from decades ago and the patients' carotid arteries have not been properly studied (no ultrasound, MRI, MRA, or angiographic evaluation) (Biousse and Bousser 1994).

The clinical features of Tolosa-Hunt syndrome ("painful ophthalmoplegia"; variable combination of periorbital pain, ipsilateral oculomotor nerve palsies, oculosympathetic palsy and trigeminal sensory loss) localise the pathological process to the region of the cavernous sinus. The causes may be traumatic, neoplastic, vascular, or inflammatory. Within the inflammatory category, there is a specific subset of patients with a steroid-responsive relapsing and remitting course – Tolosa-Hunt syndrome in the strict sense. The comprehensive patient evaluation is essential in establishing the correct diagnosis (Kline and Hoyt 2001).

Characteristics

Headaches caused by dissections have some typical although eventually unspecific features, which are not necessarily present in all cases. Independent of the affected vessel, these are as follows: high pain intensity, pain quality not experienced before, continuous more frequent than fluctuating pain over days, constant localization, sharp quality, and tenderness of the painful head, face, or neck area. Headache onset may be acute or even "thunderclap"-like suggesting subarachnoid hemorrhage, which indeed may be a complication of dissections of intracranial, especially vertebral artery, segments.

Additional characteristics are dependent on the vessel segment affected by the dissection. In the literature there are usually two broad categories, the traumatic and the spontaneous (nontraumatic) dissections. This distinction is somewhat arbitrary since in many ▶ spontaneous dissections, one will find some kind of so-called "trivial" trauma such as neck thrusting, a fall, or certain sports or other violent physical activities with questionable significance.

From the literature one gets the impression that traumatic dissections are more frequently painless, yet this may simply be an assessment bias since in traumatic dissections there are additional injuries readily explaining pain or the health state of the patients is too serious to worry about pain or to make pain assessment possible. In the spontaneous dissection subgroup, the literature reports four major categories: extracranial carotid dissections, intracranial carotid dissections and their branches, extracranial vertebral artery dissections, and intracranial vertebral artery dissections and their branches. It may however well be that these categories are human constructions, just for educational reasons, without reflecting the reality of, e.g., dissections affecting several segments of one artery or even several arteries at the same time. The vessel segments affected by dissections obviously show regional or probably ethnical variations with, e.g., dissections of intracranial vertebral artery segments and their branches predominantly reported from Asia (Japan).

Spontaneous Internal Carotid Artery Dissection (sICAD)

The most typical clinical syndrome and the most frequent dissection is that of the extracranial segment of the internal carotid artery. Usually the most distal (high cervical, retromandibular) carotid segment just before entering the petrous canal is affected.

Spontaneous Dissection of Extracranial Internal Carotid Artery (seICAD)

Headache is reported in 55–95 % of seICAD and was the first symptom in 47–68 % (Biousse et al. 1994; Fisher 1982; Schievink 2001; Silbert et al. 1995; Sturzenegger 1995). Headache and facial or orbital pain may be the sole symptom of dissection, probably more frequently than reported so far (5 %) and poses a diagnostic challenge (Maruyama et al. 2012). Local neurological manifestations (Horner's sign (35–48 %), lower cranial nerve palsies (~10 %), or pulsatile tinnitus (up to 30 %)) are found in 30–48 % of cases (Sturzenegger 1995; Sturzenegger and Huber 1993). Up to one third may complain of unilateral scalp tenderness and hair hypersensitivity. Ischemic cerebral events occur in 86 % (stroke in 60 % and TIA in 20 %) and retinal events in 20 % (Biousse et al. 1994; Schievink 2001; Sturzenegger 1995). Headache location is unilateral (79–90 %), ipsilateral to the side of dissection (almost all), in the forehead (~70 %), temple (~75 %), eye or periorbital (~60 %; ~10 % isolated), or ear (~20 %; ~10 % isolated). The headache quality is steady (~75 %), pulsating (25–40 %), of severe intensity in 85 %, thunderclap-like (14 %, mimicking SAB), severe periorbital (10 %, mimicking cluster headache), or unique and never experienced before (65 %). Headache duration is less than 1 week in 90 % (range, hours to years). Anterolateral neck pain is reported by 26–60 %, usually located in the upper neck behind the angle of the jaw.

Since migraine is a frequent disease and reported in up to 40 % of patients with carotid dissection and even considered a risk factor for dissection (D'Anglejean Chatillon et al. 1989), it is important not to confound migraine headaches with dissection headaches. The patient can usually distinguish these two headache types; dissection headache is a pain he/she never experienced before, is a continuous and not episodic pain, is not associated with general vegetative symptoms (nausea, vomiting, photophobia), and is usually constant not throbbing. Before assuming a so-called migrainous infarct, one should exclude underlying carotid dissection as the cause of the pain and (embolic) brain ischemia. Thus, if a patient with a history of migraine reports any change in the headache pattern (e.g., unique quality, long-lasting) or clinical characteristic, which he has not experienced before, ICAD should be considered and the appropriate investigation (MRI, MRA) performed soon.

The distinction from cluster headache is usually possible taking into account the duration (repetitive short attacks for cluster) and the autonomic symptoms (hyperhidrosis in cluster, anhidrosis in ICAD; tearing and stuffed or running ipsilateral nose in cluster).

Spontaneous Dissection of the Intracranial Internal Carotid Artery (siICAD)

Intracranial carotid artery dissection affecting the supraclinoid portion of the ICA and or the middle and anterior cerebral arteries is very rare, especially when compared with dissections of the extracranial ICA. Whether it represents a unique entity, different from the more common extracranial variant, is unclear. Diagnosis is more difficult and usually needs angiography or high-quality MRA. Thus the real incidence may be underestimated. According to the literature, it preferentially affects very young patients (between 15 and 25 years) without any vascular risk factors. The clinical presentation comprises severe unilateral retroorbital and temporal headache followed by contralateral hemiparesis usually immediately after headache onset (Chaves et al. 2002).

Spontaneous Vertebral Artery Dissection (sVAD)

Dissections of the VA most frequently affect the mobile and easily distorted V3 segment. The distal extension is frequently difficult to assess and distinction between extracranial and intracranial dissection more difficult in the vertebrobasilar territory than in the carotid.

Yet it could probably be of relevance, since anticoagulation of intracranial dissections, more frequently producing pseudoaneurysms, bears a significant risk of subarachnoid hemorrhage, which may accompany intracranial VAD even without anticoagulant treatment.

Headache is reported in 69–85 % and was the first symptom in 33–75 % (Silbert et al. 1995; Sturzenegger 1994). Headache location is ipsilateral to the side of dissection (almost always), usually in the occiput (∼80 %). Pain always started suddenly and was of sharp quality and severe intensity and different from any previously experienced headache. Headache was steady in about 60 % and pulsating in about 40 %. The time course of pain was monophasic with gradual remission of a persistent headache lasting 1–3 weeks.

Posterior neck pain is reported by 46–80 % and may be the only symptom (no associated headache). A delay between onset of head and neck pain heralding onset of dissection and neurological dysfunction is frequent (33–85 %) and may be of variable duration (hours to 3 weeks). Report of this distinct type of headache should raise suspicion of an underlying dissection of a vertebral artery. Its early diagnosis and immediate anticoagulation if confined to the extracranial segments may help prevent vertebrobasilar ischemic strokes, with frequently severe neurologic deficits. Presenting clinical features of VAD are extremely variable and include locked-in syndrome; Wallenberg syndrome, which represents the most frequently encountered type of neurological dysfunction; cerebellar syndrome; vestibular syndrome; transient amnesia; tinnitus; and hemianopia. Vertebral artery dissection may also occur silently, even without headache, and is detected by chance. This seems to happen predominantly in the case of multiple dissections of cervical arteries.

Vertebral artery dissection may be caused by neck manipulation (Williams and Biller 2003). If neck pain is the sole indication for such a treatment, especially in young people who never experienced such a pain before, one should be aware that VAD may be the cause and manipulation therapy might be fatal.

References

Biousse, V., & Bousser, M.-G. (1994). The myth of carotidynia. *Neurology, 44*, 993–995.

Biousse, V., D'Anglejan-Chatillon, J., Massiou, H., et al. (1994). Head pain in non-traumatic carotid artery dissection: A series of 65 patients. *Cephalalgia, 14*, 33–36.

Chaves, C., Estol, C., Esnaola, M. M., et al. (2002). Spontaneous intracranial internal carotid artery dissection. Report of 10 patients. *Archives of Neurology, 59*, 977–981.

D'Anglejan Chatillon, J., Ribeiro, V., Mas, J. L., et al. (1989). Migraine – a risk factor for dissection of cervical arteries. *Headache, 29*, 560–561.

De Marinis, M., Zaccaria, A., Faraglia, V., et al. (1991). Post-endarterectomy headache and the role of the oculosympathetic system. *Journal of Neurology, Neurosurgery, and Psychiatry, 54*, 314–317.

Fisher, C. M. (1982). The headache and pain of spontaneous carotid dissection. *Headache, 22*, 60–65.

Kline, L. B., & Hoyt, W. F. (2001). The tolosa-hunt syndrome. *Journal of Neurology, Neurosurgery, and Psychiatry, 71*, 577–582.

Maruyama, H., Nagoya, H., Kato, Y., et al. (2012). Spontaneous cervicocephalic arterial dissection with headache and neck pain as the sole symptom. *The Journal of Headache and Pain, 13*(3), 247–253.

Norregaard, T. V., & Moskowitz, M. A. (1985). Substance P and the sensory innervation of intracranial and extracranial feline cephalic arteries. *Brain, 108*, 517–533.

Schievink, W. I. (2001). Spontaneous dissection of the carotid and vertebral arteries. *The New England Journal of Medicine, 344*, 899–906.

Selky, A. K., & Pascuzzi, R. (1995). Raeder's paratrigeminal syndrome due to spontaneous dissection of the cervical and petrous internal carotid artery. *Headache, 35*, 432–434.

Silbert, P. L., Mokri, B., & Schievink, W. I. (1995). Headache and neck pain in spontaneous internal carotid and vertebral artery dissections. *Neurology, 45*, 1517–1522.

Sturzenegger, M. (1994). Headache and neck pain: The warning symptoms of vertebral artery dissection. *Headache, 34*, 187–193.

Sturzenegger, M. (1995). Spontaneous internal carotid artery dissection: Early diagnosis and management in 44 patients. *Journal of Neurology, 242*, 231–238.

Sturzenegger, M., & Huber, P. (1993). Cranial nerve palsies in spontaneous carotid artery dissection. *Journal of Neurology, Neurosurgery, and Psychiatry, 56*, 1191–1199.

Williams, L. S., & Biller, J. (2003). Vertebrobasilar dissection and cervical spine manipulation. A complex pain in the neck. *Neurology, 60*, 1408–1409.

Headache Due to Hypertension

John Edmeads
Sunnybrook and Woman's College Health
Sciences Centre, University of Toronto, Toronto,
Canada

Synonyms

Hypertensive encephalopathy; Hypertensive headaches; Reversible posterior leukoencephalopathy syndrome

Definition

Headaches and hypertension have in common the characteristics that both are prevalent, both are caused by multiple factors, and both can have acute and chronic phases. The relationships between them are thus multiple and complex.

1. They most often exist coincidentally, with no causal relation between high blood pressure and headache. It is a common lay misconception that chronic hypertension frequently causes headaches. Epidemiological studies show that the prevalence of headache is no higher in patients with mild or moderate hypertension than it is in age-matched normotensive populations (Badran et al. 1970) and, conversely, hypertension is no more common in headache populations than in those without headaches (Waters 1971). When patients with mild to moderate hypertension (diastolic below 120) have headaches, they are likely to be the same migraine and tension-type headaches that bedevil their normotensive brethren. However, severe hypertension (diastolic above 120) may cause headaches by a number of mechanisms.

2. Chronic severe hypertension may cause characteristic recurrent early morning headaches (see section on "Characteristics" for clinical details). It is believed that in chronic severe hypertension the ability of the cerebral circulation to autoregulate itself – that is, to

vasoconstrict in order to prevent an increased cerebral blood volume, increased brain capillary hydrostatic pressure, and cerebral edema – is impaired and that this, in combination with the effects of the head low position during sleep and perhaps with the vasodilating effects of CO_2 retention during sleep, causes increased intracranial pressure and headache.

3. As a rare complication of chronic hypertension, ▸ acute hypertensive encephalopathy may occur (see section on "Characteristics" for clinical details). In this condition there is segmental failure of protective constriction by some of the brain arterioles with the formation of pockets of paralytic vasodilatation. There is transudation of fluid and perhaps extravasation of blood in these regions, with the production of a pathological picture of multiple foci of cerebral edema with or without hemorrhages and of a clinical picture of increased intracranial pressure with headache, papilledema, obtundation, seizures, and/or multifocal deficits. When this vasogenic edema occurs mostly in the distribution of the posterior (vertebrobasilar) circulation, an MRI (magnetic resonance imaging) picture of reversible posterior ▸ leukoencephalopathy may result (see section on "Characteristics" for clinical details).

4. Abrupt marked rises in blood pressure (acute ▸ Paroxysmal Hypertension) may cause severe paroxysmal headaches, with or without other neurological symptoms (see section on "Characteristics" for clinical details). Here again there is failure of autoregulation with its protective vasoconstriction and headache results from sudden vasodilatation of the intracranial vessels, with or without an element of vasogenic cerebral edema and increased intracranial pressure. Such paroxysms of severe hypertension may occur in response to ingestion of exogenous pressor substances such as amphetamines or cocaine or to the taking of tyramine-containing foods or sympathomimetic medications with monoamine oxidase inhibitors. Endogenous pressor amine secretion, as in sexual intercourse or with pheochromocytoma, may produce ▸ paroxysmal hypertensive headaches. Failure of neurogenic regulation of blood pressure, as in paraplegia or in the Landry-Guillain-Barré syndrome, may lead to acute hypertensive headaches, as may the hypertension of preeclampsia and eclampsia.

5. On a more banal note, medications prescribed for the control of hypertension may themselves produce headaches, usually through the mechanism of cranial vasodilatation. These include some calcium channel blockers, enalapril, hydralazine, methyldopa, and some beta-blockers (Edmeads 2000).

Characteristics

The morning headache of severe hypertension either awakens the patient early in the morning or is present on spontaneous awakening. It is dull, often vaguely throbbing, and is maximal in the posterior part of the head, though on occasion it may be mostly bifrontal. Its characteristic feature is that as the patient gets up and about, the headache begins to abate and within a few hours it is gone – until the next morning. The patient may also complain of feeling "dull" or "muzzy" for the first few hours of the morning, and this too clears as the day wears on. The diastolic blood pressure is usually above 120. Often the patient is overweight or is known to snore – factors that predispose to nocturnal CO_2 retention. Otherwise, there are no characteristic findings on examination. The treatment of this headache is that for the high blood pressure.

Acute hypertensive encephalopathy is a rare but dreaded complication of chronic severe hypertension of any etiology. While the diastolic blood pressure (BP) is typically greater (often much greater) than 120, acute hypertensive encephalopathy has been reported with lower blood pressures, such as 150/100, particularly in children whose BP may normally be 90/50 and especially in children whose BP has been rapidly elevated by substances which also impair autoregulation, such as cancer chemotherapies or immunosuppressants (Pavlakis et al. 1999).

The patient acutely develops severe generalized headache, sometimes more marked bi-occipitally. Nausea is frequent; vomiting may occur. Visual symptoms are sometimes prominent, perhaps because the parieto-occipital white matter may be particularly edematous (see below) or because of papilledema; this visual impairment may range from nonspecific blurring through transient deficits to blindness. Papilledema, though often present, is no longer considered a prerequisite for the diagnosis. Some degree of confusion or mental obtundation is nearly always present. The foregoing symptoms are consistent with the cerebral edema and increased intracranial pressure that are the basis of this syndrome. Also, there may be focal neurological features such as seizures or hemiparesis.

Stroke (cerebral infarction or hemorrhage) is the major differential diagnosis of acute hypertensive encephalopathy, and distinguishing between the two can be difficult. Computerized tomographic (CT) scanning shows multifocal or diffuse cerebral edema, sometimes with scattered small hemorrhages or microinfarcts. Magnetic resonance (MR) imaging may show the appearance of a "reversible posterior leukoencephalopathy" (Hinchey et al. 1996), characterized by increased signal on T_2-weighted and FLAIR sequences found mostly in the posterior parts of the cerebrum, especially the parieto-occipital white matter. The frontal white matter, cerebellum, and brainstem may also be involved; there are rare cases of these changes being confined to the brainstem (Gondim and Cruz-Flores 2001). Diffusion-weighted imaging (DWI) can be normal or may demonstrate increased diffusion characteristic of vasogenic edema. This MRI picture is believed to be due to failure of autoregulation, particularly in the posterior (vertebrobasilar) circulation, which has been shown to have a relative paucity of sympathetic vasomotor innervation. Typically these MRI changes reverse when the hypertension is treated and any offending drugs eliminated.

While reduction of blood pressure is the mainstay of treatment for acute hypertensive encephalopathy, it is crucial to avoid precipitous lowering of BP (Weinberger 2003). With autoregulation severely impaired, overly rapid or excessive reduction of perfusion pressure can produce cerebral ischemia and infarction. If systolic blood pressure is greater than 200 mmHg on presentation, it must not be reduced below 160; if it is less than 200 on presentation, it must not be reduced below 140. In severe hypertensive encephalopathy (obtundation, seizures, visual impairment, or major focal deficits), this reduction should be effected over a period of about 15 min. The most controllable regimen is a slow intravenous infusion of sodium nitroprusside with constant BP monitoring, preferably through an arterial line. Alternative regimens include repeated intravenous boluses of labetalol, enalapril, or diltiazem; all require continuous BP monitoring for safe use. In patients with lesser degrees of acute hypertensive encephalopathy (awake, with no seizures, visual disturbances, or major neurological deficits), more slowly acting (30–45 min) medications can be given by mouth, with intermittent cuff pressures rather than continuous intra-arterial line monitoring. These oral medications include calcium channel blockers, ACE inhibitors, and angiotensin receptor blockers.

Acute paroxysmal severe hypertension can produce acute paroxysmal severe headache. The causes of such sudden marked rises are noted above (see section on "Definition"). These headaches are diffuse, severe, often pounding, and worse with movement and may be associated with nausea and vomiting. They resemble a very severe migraine headache but, unlike migraine, they do not usually present as multiple recurrent episodes – and, of course, they are accompanied by severe hypertension.

The International Headache Society (IHS) has provided diagnostic criteria for some of these acute hypertensive headaches (Headache Classification Subcommittee of the International Headache Society 2004). For a diagnosis of headache caused by an acute pressor response to an exogenous substance, the requirements are evidence of an appropriate toxin or medication, that the headache be accompanied by an acute rise in BP, and that the headache clears within 24 h of normalization of BP. For a diagnosis of headache caused by a pheochromocytoma, the IHS requires that the headache be accompanied by an abrupt rise in BP and by at least one adrenergic symptom such as sweating, palpitations, pallor, or anxiety; that there be demonstration of the pheochromocytoma by biochemical or imaging procedures and/or at surgery; and that the headache clears within 1 h of normalization of BP. For a headache to be attributed to preeclampsia or eclampsia, the IHS requires that the headache occurs during pregnancy (or the puerperium), that there be clinical features of preeclampsia or eclampsia (hypertension at least 140/90 and proteinuria), that appropriate investigation rules out other causes of hypertensive headaches such as medications, and that the headache clears within 7 days of normalization of BP.

The treatment of these acute paroxysmal hypertensive headaches is that of the underlying causes – removal of exogenous toxins and medications and treatment of exogenous conditions such as pheochromocytoma and preeclampsia/eclampsia. Where the cause cannot be identified, treatment is difficult, because the hypertension that causes the headaches is paroxysmal and this makes chronic treatment with hypotensive agents problematic.

References

Badran, R. H., Weir, R. J., & McGuinness, J. B. (1970). Hypertension and headache. *Scottish Medical Journal, 15*, 48–51.

Edmeads, J. (2000). Headache in the elderly. In J. Olesen, P. Tfelt-Hansen, & K. M. A. Welch (Eds.), *The headaches* (2nd ed., pp. 947–951). Philadelphia: Lippincott Williams & Wilkins.

Gondim, F., & Cruz-Flores. (2001). Two cases of brainstem variant of hypertensive encephalopathy with distinct outcomes. *Neurology, 56*, 430–431.

Headache Classification Subcommittee of the International Headache Society. (2004). The international classification of headache disorders, 2nd edn. *Cephalalgia, 24*(Suppl 1), 1–160.

Hinchey, J., Chaves, C., Appignani, B., et al. (1996). A reversible posterior leucoencephalopathy syndrome. *The New England Journal of Medicine, 334*, 494–500.

Pavlakis, S. G., Frank, Y., & Chusid, R. (1999). Hypertensive encephalopathy, reversible occipitoparietal encephalopathy, or reversible posterior leucoencephalopathy: Three names for an old syndrome. *Journal of Child Neurology, 14*, 277–281.

Waters, W. E. (1971). Headache and blood pressure in the community. *British Medical Journal, 1*, 142–143.

Weinberger, M. H. (2003). Hypertensive encephalopathy. In J. Noseworthy (Ed.), *Neurological therapeutics: Principles and practice* (pp. 592–595). London: Martin Dunitz.

Headache Due to Intracranial Bleeding

Seymour Solomon
Montefiore Medical Center, Albert Einstein College of Medicine, Bronx, NY, USA

Synonyms

Blood clot in brain; Epidural hematoma: acute traumatic epidural hematoma; Hemorrhagic stroke; Intracerebral hematoma: apoplexy: Subdural hematoma: posttraumatic subdural hemorrhage

Definition

Intracerebral Hematoma

Bleeding within the brain due to rupture of a blood vessel.

Subdural Hematoma

Hemorrhage between the brain and the dura mater, the covering of the brain, usually due to trauma.

Epidural Hematoma

Bleeding between the outer layer of the dura, the covering of the brain and the skull, due to trauma that fractures the skull and tears the artery within the bone.

Characteristics

Intracerebral Hematoma

Primary intracerebral hemorrhages are the third most common cause of stroke, after cerebral arterial thrombosis and embolism. (Hemorrhage may also be secondary to trauma, tumor, or hemorrhagic diseases). Cerebral infarcts may become hemorrhagic particularly those caused by embolism. In recent years, hemorrhagic transformation of a cerebral infarct may be a complication of early thrombolysis therapy (Fiorelli et al. 1999).

A spontaneous brain hemorrhage is usually the result of long-standing hypertension and associated degenerative changes in cerebral arteries (Fang et al. 2001). In recent decades, with widespread understanding of the need to control blood pressure, the incidence of intracerebral hemorrhage has decreased. In the vast majority of cases, the hemorrhage/stroke occurs while the person is up and about rather than during sleep (Caplan 1993). The neurological symptoms and signs are dependent on the location and size of the hemorrhage. About 50 % of hemorrhages are deep in the brain. Seepage or rupture of blood into the ventricular system is common with resultant bloody cerebrospinal fluid. Less common sites of hemorrhage are the cerebral hemispheres and the cerebellar lobes. If the hematoma is in the cerebellar hemisphere, the focal signs will be appropriate to the function of the cerebral site. Typically there is the sudden onset of headache and vomiting. If the cerebral hemorrhage is large and deep, there is depression of consciousness, hemiplegia, coma, and death.

If the hemorrhage is in the cerebellum, the initial symptoms of headache may be subacute or sudden. Inability to stand and walk may be the only signs and the diagnosis may be missed if the patient is examined only in bed or stretcher (Ropper and Brown 2005). As the hematoma grows, brain stem compression results in depressed consciousness, paralysis of gaze, other brain stem signs, small unequal pupils, and finally decerebrate posture, coma, and death. Computerized tomography will allow almost immediate visualization of the blood. Prompt surgical evacuation of the cerebellar hematoma is often lifesaving but the value of surgery for cerebral hemorrhages is more problematic. Surgery is almost always futile once the patient has become comatose (Rabinstein et al. 2002).

High plasma levels of proinflammatory biochemicals within 24 h of the hemorrhage are predictive of a poor outcome (Castillo et al. 2002). The prognosis for large and deep hemorrhages is grave (Arboix et al. 2002). About one-third of patients will die during the first few days to 1 month. In those who survive there may be a surprising degree of recovery since, contrary to cerebral infarction, the hemorrhage stretches brain tissue rather than destroys it. Nevertheless only one-third of patients regain independent functional status after 3 months (Weimar et al. 2003).

Headache is not invariable in patients with intracerebral hematoma. The frequency ranges from one-third to two-thirds of patients (Mitsias and Jensen 2006). The frequency of headache is highest in cerebellar and occipital hemorrhages and when the hematoma is large. The pain is presumably due to the stretching and stimulation of pain-sensitive structures by the mass effect of the hematoma and the irritation of blood (if subarachnoid bleeding is associated). The site of the headache often overlies the cerebral hematoma; occipital headache often is associated with cerebellar or occipital hematoma.

Subdural Hematoma

The course of neurological events may be acute or chronic, more or less corresponding to the degree of trauma. The location of the hematoma is usually over one or both cerebral hemispheres but rarely it may form in the posterior fossa.

In people who develop an acute subdural hematoma, the symptoms may be similar to those of acute epidural hematoma and the two conditions often occur together. Headache or loss of consciousness may be immediate, due to cerebral concussion following the blow to the head. Recovery from concussion is often associated with a lucid interval, but soon the effects of the mass of blood compressing the brain become evident with increasing headache and decreasing consciousness. Computerized tomography is used to visualize the hemorrhage. Prompt surgical evacuation of the hematoma is essential (Koc et al. 1997).

Chronic subdural hematoma is less clearly associated with head trauma (Iantosca and Simon 2000). The slowly developing blood clot is over one or sometimes both cerebral hemispheres. The trauma may be trivial or may have been forgotten. This is particularly true in the elderly when the brain shrinks and the veins bridging the skull and brain traverse a longer distance. For this reason the veins are more easily sheared by slight trauma. The symptoms develop slowly over weeks or months.

Headache is most common along with depression of mentation, drowsiness, inattentiveness, and confusion; focal signs are usually minor or absent. The initial symptoms are often subtle and may be mistaken for depression, Alzheimer's disease, drug intoxication, or brain tumor. Symptoms often fluctuate in severity, sometimes suggesting transient ischemic events. The hematoma eventually becomes encysted by a fibrous membrane (LaBadie and Glover 1976). As blood contents dissipate, computerized tomography may no longer reveal the striking density of blood, rather the diagnosis is made by the space-occupying effect of the mass. Contrast material used during the study will show the surrounding fibrous membrane. The hematoma may spontaneously reabsorb. But if it continues to grow, surgical drainage is necessary.

Subdural hematoma causes headache in about two-thirds of patients and is related more to the location in the head and irritation of the meninges than to the volume of the hematomas (Melo et al. 1996). Because the hematoma is due to venous bleeding, the symptoms as well as the course may range from acute to chronic. The time between injury and headache may range from hours to months. The headache is almost always ipsilateral to the hematoma but the qualities of the headache are not specific.

Epidural Hematoma

Because of arterial bleeding the course of epidural hemorrhage is acute (Castillo et al. 2002). Typically the blow to the head causes loss of consciousness due to cerebral concussion. Recovery of consciousness after a few minutes is followed by a lucid interval of minutes or hours. The epidural hematoma, rapidly expanding against the brain, causes headache, vomiting, drowsiness, coma, and (if not treated) death. Compression of the brain may cause hemiparesis and eventual compression of the brain stem causes generalized spasticity of the limbs and a dilated pupil on the side of the hematoma. Fracture of the skull is seen on standard roentgenograms and computerized tomograms reveal a lens-shaped clot. Usually the middle meningeal artery is sheared within the fractured temporal bone. Surgical drainage of the hematoma is the only lifesaving treatment.

If loss of consciousness is not immediate, head pain will be associated with trauma. Headache first on the side of the trauma, progresses in intensity as the clinical course progresses. The qualities of the headache are not specific.

References

Arboix, A., Comes, E., Garcia-Eroles, I., et al. (2002). Site of bleeding and early outcome in primary intracerebral hemorrhage. *Acta Neurologica Scandinavica, 105,* 282–288.

Caplan, L. R. (1993). *Stroke: A clinical approach* (2nd ed., pp. 425–467). Boston: Butterworth-Heinemann.

Castillo, J., Davalos, A., Alvarez-Sabin, J., et al. (2002). Molecular signatures of brain injury after intracerebral hemorrhage. *Neurology, 58,* 624–629.

Fang, X. H., Longstreth, W. T., Jr., Li, S. C., et al. (2001). Longitudinal study of blood pressure and stroke in over 37,000 people in China. *Cerebrovascular Diseases, 11,* 225–229.

Fiorelli, M., Bastianello, S., von Kummer, R., et al. (1999). Hemorrhagic transformation within 36 hours of a cerebral infarct: Relationships with early clinical deterioration and 3-month outcome in the European cooperative acute stroke study I (ECASSI) cohort. *Stroke, 11,* 2280–2284.

Iantosca, M. R., & Simon, R. I. I. (2000). Chronic subdural hematoma in adult and elderly patients. *Neurosurgery Clinics of North America, 11,* 447–454.

Koc, R. K., Akdemir, I. I., Oktem, I. S., et al. (1997). Acute subdural hematoma: Outcome and outcome prediction. *Neurosurgical Review, 20,* 239–244.

LaBadie, E., & Glover, D. (1976). Physiopathogenesis of subdural hematomas. *Journal of Neurosurgery, 45,* 382–393.

Melo, T. P., Pinto, A. N., & Ferro, J. M. (1996). Headache in intracerebral hematomas. *Neurology, 47,* 494–500.

Mitsias, P., & Jensen, T. J. (2006). Headache associated with ischemic stroke and intracranial hematoma. In J. Olesen, P. J. Goadsby, N. M. Ramadam, P. Tfelt-Hansen, & K. M. A. Welch (Eds.), *The headaches* (3rd ed., pp. 885–892). Philadelphia: Lippincott William and Wilkins.

Rabinstein, A. A., Atkinson, J. L., & Wijdicks, E. F. M. (2002). Emergency craniotomy in patients worsening due to expanded cerebral hematoma. To what purpose? *Neurology, 58,* 1367–1372.

Ropper, A. H., & Brown, R. H. (2005). Cerebrovascular diseases. In *Adams and Victor's principles of neurology* (8th ed., pp. 821–924). New York: McGraw-Hill.

Weimar, C., Weber, C., Wagner, M., et al. (2003). Management patterns and health care use after intracerebral hemorrhage. A cost of illness study from a societal perspective in Germany. *Cerebrovascular Diseases, 15,* 29–36.

Headache Due to Low Cerebrospinal Fluid Pressure

Peter S. Sándor[1] and Andreas R. Gantenbein[2]
[1]RehaClinic, Cantonal Hospital Baden, Baden, Switzerland
[2]RehaClinic Bad Zurzach, Bad Zurzach, Switzerland

Synonyms

Hypotension of spinal fluid; Low-intracranial-pressure headache; Post-lumbar puncture headache; Spontaneous aliquorrhea; Spontaneous intracranial hypotension; Symptomatic intracranial hypotension; Ventricular collapse

Definition

There are 3 types of headache attributed to low ▶ cerebrospinal fluid (CSF) pressure in the new classification of the International Headache Society (IHS, Headache Classification Committee of the International Headache Society 2004): post-dural puncture headache (7.2.1), CSF fistula headache (7.2.2), and headache related to spontaneous low CSF pressure (7.3.3). They have in common an ▶ orthostatic component, as the headache usually begins within 15 min of standing or sitting up. The headache mainly improves in the recumbent position; however, this is a diagnostic criterion only for post-▶ lumbar puncture in the IHS classification. The headache is associated with at least one of the following symptoms: neck stiffness, tinnitus, hypacusis, photophobia, or nausea. The etiology is different for the 3 types. See Table 1 for compared classification criteria (Headache Classification Committee of the International Headache Society 2004).

Characteristics

In 1938, Schaltenbrand, a German neurologist, wrote about two conditions regarding cerebrospinal fluid:

Headache Due to Low Cerebrospinal Fluid Pressure, Table 1 Compared diagnostic criteria of the International Headache Classification 2nd edition regarding headache attributed to low CSF pressure

Post-dural puncture headache	CSF fistula headache	Spontaneous low CSF
Headache within 15 min of sitting or standing		
Resolving within 15 min after lying		
Associated symptoms (1 of the following):		
Neck stiffness		
Tinnitus		
Hypacusis		
Photophobia		
Nausea		
Etiology		
Dural puncture	CSF leakage:	
	MRI evidence (pachymeningeal enhancement)	
	Conventional or CT myelography, cisternography	
	OP <60 mm CSF in sitting position	
Onset		
Within 5 days after lumbar puncture	Close relation to CSF leakage	No lumbar puncture or leakage
Resolving		
Spontaneously within 1 week or within 48 h after treatment	Within 1 week of sealing leak	Within 72 h after blood patch

1. "Liquorrhea" involving headache and ▶ papilledema, which later became known as pseudotumor cerebri.
2. "Spontaneous aliquorrhea," presenting with orthostatic headaches and features of intracranial hypotension. He explained the syndrome of low CSF pressure by three possible pathological mechanisms: decreased production, increased absorption, or leakage, e.g., after a lumbar puncture (Schaltenbrand 1938).

The brain is "swimming" in the CSF. The average weight of 1,500 g is reduced to 50 g by the intracranial pressure. The remaining weight is held by the meningeal blood vessels, the outgoing cranial nerves, and microstructures. If the CSF pressure decreases, there is traction on the supporting structures of the brain. Recent MRI studies have even shown the "descending" brain (Pannullo et al. 1993). It is thought that traction on the cranial nerves (V, IX, X), on the three upper cervical nerves, and on bridging veins which are pain-sensitive structures causes the headache and its associated features. However, there are contradictory reports to this so-called sagging theory (Levine and Rapalino 2001). As long as magnetic imaging cannot be performed in the erect position, it will probably be difficult to bring an end to that discussion.

Magnetic imaging of the head and spine has revolutionized the knowledge and the detection of this disorder (Fishman and Dillon 1993; Mokri 2001; Sable and Ramadan 1991). It was not until the 1990s that investigators demonstrated that the production of about 500 ml per day is relatively constant and therefore is rarely a cause for the problems. The CSF volume is estimated to be between 150 and 210 ml; this means that the total volume is renewed 2–3 times per day. CSF volume is smaller in women, younger persons, and obese persons. Most of the CSF is absorbed via the arachnoid villi into the venous sinuses and cerebral veins, and only a very small part is absorbed through simple diffusion. On this background, the most obvious and common reason for low intracranial pressure is CSF leakage. The spontaneous leaks are mostly located on the thoracic or cervical level (Sencakova et al. 2001).

Post-lumbar puncture headache happens in up to one-third of patients with lumbar puncture (Adams et al. 2002). Patients with preceding headaches and young women with low BMI may be at higher risk of developing headaches. Patients with postural headaches should be imaged before lumbar puncture, and if there are MRI signs of ▶ pachymeningeal enhancement, a lumbar puncture should not be performed.

Symptoms

Low CSF pressure usually causes orthostatic headaches, which develop in the upright position and improve when lying down (recumbency).

The onset of the headache is usually sudden or gradual. The character of the pain is often described as severe and throbbing or dull, and it can be diffuse or focal, with a frontal or occipital localization. The headache typically has orthostatic features in the beginning (e.g., onset within 30 min after standing up); however, these features may blur with chronicity and result in a chronic daily headache which is worse when the patient is in an upright position and improves when lying down. Other exaggerating factors include movements of the head, sneezing, coughing, straining, and jugular venous compression. In general pain killers do not sufficiently improve the headache. Recumbency is often the only measure which can relieve the pain, usually within 10–15 min. Associated features can be manifold: anorexia, nausea, vomiting, vertigo, dizziness, neck stiffness, blurred vision, and even photophobia are commonly described. Tinnitus, bilateral hyp(er)acusis, unsteadiness, staggering gait, diplopia, transient visual obscuration, hiccups, and dysgeusia have been reported.

Examination

The neurological examination is typically normal. However, mild neck stiffness is frequently noted.

The CSF opening pressure is typically below 70 mm; however, it can be low normally. Fluid is typically clear and colorless, occasionally ▶ xanthochromic. The CSF protein level is usually normal but may be high, mostly still below 100 mg/dl. Cell counts give variable results: erythrocytes and leukocytes may be normal or elevated. Cytologic and microbiologic tests are always negative, and glucose rate CSF/plasma is always between 0 and 1.

The standard diagnostic investigation for low CSF pressure and CSF leaks is MR imaging with gadolinium. The most common abnormality is diffuse pachymeningeal enhancement (Mokri 2004). According to the Monro-Kellie doctrine (brain volume + CSF + intracranial blood = constant), the CSF loss is compensated by venous hyperemia. Whereas the leptomeninges have blood–brain barriers, the pachymeninges (dura mater) do not and therefore accumulate the contrast medium. The enhancement is typically linear, thick and uninterrupted, and diffuse, including supra- and infratentorial meninges. Furthermore, there is commonly sinking or sagging of the brain, which can sometimes mimic Chiari I malformation, subdural fluid collections (they may be unilateral), and decrease in size of ventricles. Less common abnormalities include pituitary enlargement, engorged venous sinuses, and elongation of the brain stem. MRI of the spine can show spinal pachymeningeal enhancement, engorgement of venous plexus, and extra-arachnoidal fluid but only rarely reveal the site of the leak. The most accurate technique to find the exact site of CSF leaks is CT myelography. It is to be mentioned that different leaks and diverticles of different sizes can be found in the same patient. An exact identification of the site of the leak, however, is only necessary when surgical intervention is needed.

A CT scan of the head is usually unremarkable and therefore not very useful. Older diagnostic techniques include radioisotope cisternography with indium-111, myelography without CT, and meningeal biopsy.

Differential Diagnosis

Most patients present with a new onset daily headache following a lumbar puncture or another dural trauma or, if developing spontaneously, present as new daily persistent headache, which would act as a working diagnosis, unless low-CSF-pressure headache is diagnostically classified using imaging techniques (► MRI, CT myelography).

The orthostatic component is a salient feature. It is therefore hard to understand that in a recent study (Schievink 2003) spontaneous intracranial hypotension (SIH) was misdiagnosed in 94 % of the reviewed cases, with a mean diagnostic delay of 13 months (median 5 weeks, range 4 days to 13 years). Sometimes associated features, such as nausea or photophobia, can mimic migraine. Especially when there is a personal or family history of headaches, the picture can be diluted. Furthermore, the orthostatic feature becomes less prominent with time. Obviously, the diagnosis of low CSF pressure headache is easier when the patient is seen at the beginning of the problem and when there is a close temporal relationship to a lumbar puncture or another trauma affecting the spine.

The differential diagnosis to the other orthostatic headache due to raised intracranial pressure should be fairly easy. Interestingly, almost all patients with low CSF pressure develop headaches, but only 30–80 % of the patients with increased intracranial pressure do so (Mokri 2001).

Whereas both low- and high-pressure headaches can be aggravated by coughing or straining, intracranial hypertension typically develops when lying down, especially in the morning, and is mostly present with transient visual loss or papilledema in the neurological examination.

Management and Treatment

Fortunately, many low-pressure headaches dissolve spontaneously within days. Treatments vary for the three different types of low-CSF-pressure headache. Conservative strategies include bed rest, fluid intake, and an abdominal binder. Caffeine (250–500 mg i.v.), theophylline, and to a lesser part steroids can be effective. When conservative treatments give no sufficient pain relief within 24 h, an epidural ► blood patch (10–15 ml of autologous blood into the epidural space) would be indicated. Blood patches seem to have not only an immediate effect, through simple volume replacement, but also a delayed sealing of the leak. Post-lumbar puncture headaches are often relieved after the first (rarely the second) blood patch, while patients with spontaneous CSF leaks may need up to 4 or more. Instead of a second or third blood patch, an epidural saline infusion could be attempted, using a catheter placed at the L2–3 level and a flow rate of 20 ml/h for 72 h. If a leak is clearly located with imaging techniques and the headache is treatment refractory, a surgical closure may be considered.

References

Adams, M. G., Romanowski, C. A., & Wrench, I. J. (2002). Spontaneous intracranial hypotension – lessons to be learned for the investigation of post dural puncture headache. *International Journal of Obstetric Anesthesia, 11*, 65–67.

Fishman, R. A., & Dillon, W. P. (1993). Dural enhancement and cerebral displacement secondary to intracranial hypotension. *Neurology, 43*, 609–611.

Headache Classification Committee of the International Headache Society. (2004). The international classification of headache disorders. *Cephalalgia, 24,* 9–160. 2nd edn.

Levine, D. N., & Rapalino, O. (2001). The pathophysiology of lumbar puncture headache. *Journal of Neurological Sciences, 192,* 1–8.

Mokri, B. (2001). Low cerebrospinal fluid pressure headache. In S. D. Silberstein, R. B. Lipton, & D. J. Dalessio (Eds.), *Wolff's headache and other pain* (7th ed., pp. 417–433). New York: Oxford University Press.

Mokri, B. (2004). Spontaneous low cerebrospinal pressure/volume headaches. *Current Neurology and Neuroscience Reports, 4,* 117–124.

Pannullo, S. C., Reich, J. B., Krol, G., et al. (1993). MRI changes in intracranial hypotension. *Neurology, 43,* 919–926.

Sable, S. G., & Ramadan, N. M. (1991). Meningeal enhancement and low CSF pressure headache. An MRI study. *Cephalalgia, 11,* 275–276.

Schaltenbrand, G. (1938). Neuere anschauungen zur pathophysiologie der liquorzirkulation. *Zentralblatt für Neurochirurgie, 3,* 290–300.

Schievink, W. I. (2003). Misdiagnosis of spontaneous intracranial hypotension. *Archives of Neurology, 60,* 1713–1718.

Sencakova, D., Mokri, B., & McClelland, R. L. (2001). The efficacy of epidural blood patch in spontaneous CSF leaks. *Neurology, 57,* 1921–1923.

Headache Due to Somatoform Disorder

Reto M. Agosti
Headache Center Hirlsanden, Zurich, Switzerland

Synonyms

Headache associated with psychotic disorder; Headache associated with somatization disorder

Definition

Headaches of no typical characterization (such as migraine or cluster headaches) with close temporal association with undifferentiated somatoform disorder (as defined by DSM IV).

Characteristics

This type of headache does not have any characteristic symptoms that are unique to these types of headaches. Any other headache type, primary (such as migraine and cluster headaches) or secondary, must be excluded. By definition, there must be a close temporal relationship with the multiple symptoms of an undifferentiated somatoform disorder as defined by DSM-IV: (a) A physical complaint, plus headache, that, after appropriate investigation, cannot be fully explained by a known general medical condition, or by the direct effects of a substance or medication, or, when there is a related medical condition, that complaint or impairment is in excess of what would be expected from the history, examination, and/or laboratory findings; and (b) The physical complaint and headaches cause distress or impairment and last at least 6 months. The headache occurs exclusively during the course of the other physical complaint and resolves after the undifferentiated somatoform disorder remits. A similar condition with clearly more stringent criteria regarding the somatoform symptoms and complaints are headaches associated with somatization disorder, for which DSM-IV requires a minimum of eight somatoform symptoms or complaints and age of onset under 30. Both types of headaches (associated with somatoform and somatization disorder) only entered the IHS classification of headaches in 2004 and are highly debated regarding their existence as proper diagnoses, with the persistent lack of a biological marker for primary headaches as one of the major obstacles. The diagnosis is fully based on phenomenology and the treatment symptomatic towards treating the headaches and or the underlying psychiatric disorder. A causal relationship in any direction is under debate. Association of headaches with other psychiatric disturbances such as depression and phobias are classified separately. Headaches associated with somatization disorder are rare, and headaches associated with somatoform disorder are more frequent. So far only single cases and a few small series of cases have been published.

Headache Due to Venous Sinus Thrombosis

Volker Limmroth[1] and Hans-Christoph Diener[2]
[1]Department of Neurology, University of Essen, Essen, Germany
[2]Department of Neurology, University Hospital Essen, University Duisburg-Essen, Essen, Germany

Synonyms

Central venous thrombosis; Cerebral venous thrombosis; Intracranial venous thrombosis; Sinus-venous thrombosis (SVT); Venous sinus thrombosis

Definition

Cerebral venous sinus thrombosis (CVST) is a rare but challenging condition and is therefore often unrecognized. Its clinical presentation may vary significantly from case to case. Headache, however, is often the very first and leading symptom. The headache is mostly described as dull holocephalic pain of increasing intensity and can easily be mistaken for tension-type headache, migraine, or other disorders such as idiopathic intracranial hypertension (pseudotumor cerebri). Along with headache, additional symptoms typical to increasing intracranial pressure such as papilledema, nausea, vomiting, and cognitive decline may be present. Further symptoms are focal deficit and seizures. The headache does not typically respond to classical antiheadache drugs, which should be taken as an important sign that further evaluation is necessary. Typical patients with CVST are young females with risk factors such as oral contraceptive pill use, nicotine abuse, being overweight, or during pregnancy, but all age groups can be affected, and CVST can evolve secondary to an adjacent infectious process: dehydration, hypercoagulable state, inflammatory disorders, malignancies, or head traumas. The diagnosis can be easily confirmed by MRI with venography or modern

spiral CAT scan. The treatment of choice is intravenous heparin followed by oral anticoagulation for 3–6 months. The prognosis is good if treatment is initiated early but can be fatal when the condition is overlooked. Despite its low incidence, CSVT is, therefore, one of the most important differential diagnosis clinicians must bear in mind when evaluating patients with headache.

Characteristics

Pathophysiology

Venous blood drains through small cerebral veins into larger veins that empty into dural sinuses and eventually into the internal jugular veins. Preexisting anastomoses between cortical veins allow the development of collateral circulation in the event of an occlusion. The main cerebral venous sinuses affected by CVST are the superior sagittal sinus (72 %) and the lateral sinuses (70 %). In about one-third of cases more than one sinus is affected; in a further 30–40 % both sinuses and cerebral or cerebellar veins are involved (Ameri and Bousser 1992; Bousser and Barnett 1992; Villringer et al. 1994). In contrast to arterial thrombus, a venous thrombus evolves slowly, due good collateralization of the venous vessels, which probably explains the usually gradual onset of symptoms, frequently over weeks and months. Sudden onset, however, may occur and may then cause predominating focal deficits rather than headache. Hemorrhagic infarction occurs in approximately 10–50 % of cases, principally affecting the cortex and adjacent white matter (Bousser et al. 1985; de Bruijn et al. 1996; Buonanno et al. 1982; Provenzale et al. 1998). This is thought to be primarily due to elevated venous and capillary pressure caused by the persistence of thrombosis.

Predisposing Factors

An overview of predisposing factors is given below. In most of the cases one or more of these factors can be identified. In general, a distinction can be made between infective and noninfective causes or, as suggested by Bousser and Barnett, between local and systemic causes (Bousser and

Barnett 1992). Within recent decades, infective causes have declined and are now responsible for less than 10 % of cases and are mostly caused by staphylococcal infection of the face. Among the noninfective causes, systemic conditions such as connective tissue diseases, other granulomatous or inflammatory disorders, and malignancies are the most common. Other risk factors in otherwise healthy subjects are overweight, hormonal therapy, smoking, and underlying – mostly unknown – clotting disorders. In many cases several of these factors can be found.

Local Causes
- Penetrating head injury
- Intracranial infection
- Regional infection
- Stroke and hemorrhage
- Space-occupying lesions
- Neurosurgery

Systemic Causes
- Severe dehydration
- Hormonal and endocrine causes
- Cardiac disease
- Red blood cell disorders
- Thrombocythemia
- Coagulation disorders (acquired or hereditary)
- Infusions via central venous catheter
- Surgery with immobilization
- Malignancies
- Inflammatory bowel disease
- Connective tissue diseases
- Behcet's disease
- Sarcoidosis
- Nephrotic syndrome
- Drugs (L-asparaginase, epsilonaminocaproic acid, ecstasy)
- Sepsis and systemic infection

Clinical Presentation
Depending on the sinus involved and the extent of the venous thrombus, CVST presents with a wide spectrum of symptoms and signs. Headache is the leading symptom and present in 70–90 % of cases (Uzar et al. 2012; Bousser and Ferro 2007). Other important symptoms are focal deficits such as hemiparesis and hemisensory disturbance, seizures,

impairment of level of consciousness, and papilledema (Ameri and Bousser 1992; Bousser et al. 1985). The onset may also vary a great deal from acute, subacute, or insidious, but most patients develop symptoms over days or weeks.

In a series of 110 cases, Ameri and Bousser (1992) found several typical clinical constellations: up to 75 % of cases are characterized by a focal neurological deficit and headache, 30–50 % may present with seizures often followed by a Todd's paresis, and 18–38 % of cases present with a syndrome resembling benign intracranial hypertension with headache, papilledema, and visual disturbances. As indicated above, symptoms also depend on the location of the thrombus. The (isolated) thrombosis of the superior sagittal sinus (which occurs in less than 5 % of the cases) presents with bilateral or alternating deficits, particularly in the lower limbs and/or seizures, the (isolated) thrombosis of the cavernous sinus (3 % of the cases) with chemosis, proptosis, and painful ophthalmoplegia. Patients with lateral sinus thrombosis may present with a pseudotumor cerebri-like syndrome. Recently, Farb et al. (2003), using a technique called auto-triggered elliptic-centric-ordered 3-dimensional gadolinium-enhanced MR venography, found that 27 of 29 patients with idiopathic intracranial hypertension suffered from a bilateral sinovenous stenosis which was only seen in 4 of 59 control subjects. Severe cases with the involvement of the superior sagittal sinus, the cavernous sinus, and the lateral sinus, however, may present with a rapidly progressive condition including headache, nausea, pyramidal signs, and deepening coma.

CVST appears to be slightly more frequent in women with a suggested female-to-male ratio of 1.29:1. Interestingly, while 61 % of women with CVST were aged 20–35 years, a uniform age distribution has been suggested for men with CVST. The most likely explanation for this specific age distribution in women is the use of oral contraceptives and fact that CVST is frequently observed during pregnancies.

Diagnosis
Patients with a suspected CVST must undergo specific cranial imaging immediately. Magnetic

resonance imaging (MRI) combined with magnetic resonance venography (MRV) has largely replaced invasive cerebral angiography and conventional computed tomography (CT). Modern subsecond spiral CT and multidetector-row CT (MDCT) scanners, however, are now able to obtain whole-brain CT venograms in less than a minute. Unlike the conventional CT scanner, MDCT scanners have sufficient speed for high-resolution images of the entire brain and all dural sinuses during the peak venous enhancement (Casey et al. 1996; Wetzel et al. 1999). The technique can therefore – if available – be used as a first-line diagnostic tool since the procedure is cheaper and faster than MRI/MRV. Moreover, using a CT scanner of the latest generation, a recent study has been shown that CT venography may be superior to MRV in visualizing sinuses or smaller cerebral veins or cortical veins with low flow. However, it goes without saying that MRI/MRV are the imaging techniques of choice for pregnant women. Some doubtful cases may still require cerebral angiography. One of the common problems is the absence or hypoplasia of the anterior portion of the superior sagittal sinus, a normal variant that can simulate thrombosis on MRV (Provenzale et al. 1998; Wang 1997). Also, contrast enhancement along the edge of the thrombus can be mistaken for normal contrast material accumulating within a patient's sinus. Aside from confirming the diagnosis by cranial imaging, it is mandatory to search for the underlying causes including the search for local infection, head injury, malignancies, and connective tissue diseases with inflammatory markers, autoantibodies, and markers of coagulation disorders such as factor V Leiden mutation if resistance to activated protein C is abnormal, activities of proteins C and S, antithrombin III, plasminogen, fibrinogen, and anticardiolipin antibodies (de Bruijn et al. 1998; Deschiens et al. 1996; Kellett et al. 1998). All these investigations should probably be performed twice, i.e., before starting anticoagulation and 6 months later after finishing since the acute status of the disease may influence the expression of these parameters.

Treatment

Only a few therapeutic trials have evaluated potential therapeutic agents in CVST. Antithrombotic treatment modalities include heparin, thrombolysis, and oral anticoagulants. Einhäupl et al. (1991) in a randomized and placebo-controlled trial, demonstrated the benefits of heparin in a series of 20 patients. There was a significant difference in favor of intravenous heparin with respect to neurological recovery and mortality compared to placebo. Interestingly, in an additional retrospective analysis on 102 patients with CVST, the same authors suggested heparin to be beneficial, even in those patients who had an intracranial hemorrhage prior to treatment initiation. A few years later, de Bruijn et al. (1999) compared low-molecular-weight heparin followed by warfarin, or placebo. A significant difference between the groups could not be detected in this study (de Bruijn et al. 1999).

Several groups (Frey et al. 1999; Horowitz et al. 1995; Kim and Suh 1997; Smith et al. 1994) addressed the question of whether additional benefit could be achieved by thrombolysis via selective catheterization of the occluded sinus. Although all studies included a small number of patients (n = between 7 and 12 patients per study), all studies suggested that the majority of patients undergoing catheterization and thrombolysis with urokinase recovered well, and only a few patients suffered from an additional cerebral hemorrhage. Since there is no direct comparative trial between heparin and thrombolysis, the question if this approach provides an additional benefit and an acceptable benefit-to-risk ratio when compared to i.v. heparin is not answered. The disadvantage of catheterization in patients with CVST is the significant logistic effort and expertise necessary to have this intervention always available.

There is a general agreement that oral anticoagulants should follow as treatment of the acute phase for 3–6 months. In patients with known prothrombotic conditions anticoagulation may be a lifelong requirement. No agreement has been reached regarding the question whether patients who present with seizures should undergo antiepileptic treatment after the acute phase. This decision remains to be made from case to case and under the consideration of the individual circumstances.

Taken together, intravenous heparin is the first-line treatment in a dosage sufficient to increase the apTT to 2–3 times of the control value. Several authors suggest a start with a heparin bolus of 5,000 U and to continue according to the apTT elevation, which mostly requires dosages between 1,000 and 1,600 U/h for adults. Heparin is the first-line treatment, even in the presence of hemorrhagic infarction (Bousser 1999). In case of clinical deterioration despite adequate heparinization, selective local thrombolysis should be considered, in spite of the increased hemorrhagic risk.

Prognosis

Mortality in untreated cases of venous thrombosis has been reported to range from 13.8 % to 48 % (Preter et al. 1996). A recent Portuguese study suggested a morbidity of around 8 % despite adequate treatment in a group of 91 prospectively analyzed consecutively admitted patients with a mean 1-year follow-up interval (Ferro et al. 2004). Interestingly, 82 % of the patients recovered completely, but 59 % developed thrombotic events during the follow-up; 10 % had seizures, and 11 % complained of severe headaches. Buccino et al. (2003) found a good overall outcome in a series of 34 patients with CSVT. Still, ten patients (30 %) suffered from episodic headaches, three patients (8.8 %) from seizures, four patients (11.7 %) from pyramidal signs, two patients (5.9 %) from visual deficits, and six patients (17.6 %) from working memory deficit and depression. In a series of 100 patients (Gameiro et al. 2012), the outcome of cerebral venous thrombosis with isolated headache diagnosed early or late was favorable. All these studies clearly emphasize that CSVT is a treatable condition in the majority of cases and that early diagnosis and immediate initiation of heparin treatment are the key components for a good overall outcome.

References

Ameri, A., & Bousser, M. G. (1992). Cerebral venous thrombosis. *Neurologic Clinics, 10*, 87–111.

Bousser, M. G. (1999). Cerebral venous thrombosis: Nothing, heparin, or local thrombolysis? *Stroke, 30*, 481–483.

Bousser, M. G., & Barnett, H. J. M. (1992). *Cerebral venous thrombosis. Stroke: Pathophysiology, diagnosis and management* (2nd ed., pp. 517–537). New York: Churchill-Livingstone.

Bousser, M. G., & Ferro, J. M. (2007). Cerebral venous thrombosis: An update. *Lancet Neurology, 6*(2), 162–170. Epub 2007/01/24.

Bousser, M. G., Chiras, J., Bories, J., et al. (1985). Cerebral venous thrombosis – a review of 38 cases. *Stroke, 16*, 199–213.

Buccino, G., Scoditti, U., Patteri, I., et al. (2003). Neurological and cognitive long-term outcome in patients with cerebral venous sinus thrombosis. *Acta Neurologica Scandinavica, 107*, 330–335.

Buonanno, F. S., Moody, D. M., & Ball, T. L. M. (1982). CT scan findings in cerebral sinus venous occlusion. *Neurology, 12*, 288–292.

Casey, S. O., Alberico, R. A., Patel, M., et al. (1996). Cerebral CT venography. *Radiology, 198*, 163–170.

de Bruijn, S. F., Stam, J., & Kapelle, L. J. (1996). Thunderclap headache as first symptom of cerebral venous sinus thrombosis. CVST Study Group. *Lancet, 348*, 1623–1625.

de Bruijn, S. F., Stam, J., Koopman, M. M., et al. (1998). Case–control study of risk of cerebral sinus thrombosis in oral contraceptive users who are cautious of hereditary prothrombotic conditions. *BMJ, 316*, 589–592.

de Bruijn, S. F., Stam, J., & for the CVST Study Group. (1999). Randomised, placebo-controlled trial of anticoagulant treatment with low-molecular-weight heparin for cerebral sinus thrombosis. *Stroke, 30*, 484–488.

Deschiens, M. A., Conard, J., Horellou, M. H., et al. (1996). Coagulation studies, factor V Leiden and anticardiolipin antibodies in 40 cases of cerebral venous thrombosis. *Stroke, 27*, 1724–1730.

Einhäupl, K. M., Villringer, A., Meister, W., et al. (1991). Heparin treatment in sinus venous thrombosis. *Lancet, 338*, 597–600.

Farb, R. I., Scott, J. N., Willinsky, R. A., et al. (2003). Intracranial venous system: Gadolinium-enhanced three-dimensional MR venography with auto-triggered elliptic centric-ordered sequence -initial experience. *Radiology, 226*, 203–209.

Ferro, J. M., Canhao, P., Stam, J., ISCVT Investigators, et al. (2004). Prognosis of cerebral vein and dural sinus thrombosis: Results of the international study on cerebral vein and dural sinus thrombosis (ISCVT). *Stroke, 35*, 664–670.

Frey, J. L., Muro, G. J., McDougall, C. G., et al. (1999). Cerebral venous thrombosis: Combined intrathrombus rtPA and intravenous heparin. *Stroke, 30*, 489–494.

Gameiro, J., Ferro, J. M., Canhao, P., Stam, J., Barinagar-rementeria, F., Lindgren, A., et al. (2012). Prognosis of cerebral vein thrombosis presenting as isolated headache: Early vs. late diagnosis. *Cephalalgia, 32*(5), 407–412. Epub 2012/03/13.

Horowitz, M., Purdy, P., Unwin, H., et al. (1995). Treatment of dural sinus thrombosis using selective catheterisation and urokinase. *Annals of Neurology, 38*, 58–67.

Kellett, M. W., Martin, P. J., Enevoldson, T. P., et al. (1998). Cerebral venous sinus thrombosis connected with 20210, a mutation of the prothanbil gene. *Journal of Neurology, Neurosurgery, and Psychiatry, 65*, 611–612.

Kim, S. Y., & Suh, J. H. (1997). Direct endovascular thrombolytic therapy for dural sinus thrombosis: Infusion of alteplase. *American Journal of Neuroradiology, 18*, 639–664.

Preter, M., Tzourio, C., Ameri, A., et al. (1996). Long-term prognosis in cerebral venous thrombosis: Follow-up of 77 patients. *Stroke, 27*, 243–246.

Provenzale, J. M., Joseph, G. J., & Barboriak, D. P. (1998). Dural sinus thrombosis: Findings on CT and MRI imaging and diagnostic pitfalls. *American Journal of Roentgenology, 170*, 777–783.

Smith, T. P., Higashida, R. T., Barnwell, S. L., et al. (1994). Treatment of dural sinus thrombosis by urokinase infusion. *American Journal of Neuroradiology, 15*, 801–807.

Uzar, E., Ekici, F., Acar, A., Yucel, Y., Bakir, S., Tekbas, G., et al. (2012). Cerebral venous sinus thrombosis: An analyses of 47 patients. *European Review for Medical and Pharmacological Sciences, 16*(11), 1499–1505. Epub 2012/11/01.

Villringer, A., Mehraen, S., & Einhäupl, K. M. (1994). Pathophysiological aspects of cerebral sinus venous thrombosis. *Journal of Neuroradiology, 21*, 72–80.

Wang, A. M. (1997). MRA of venous sinus thrombosis. *Clinical Neuroscience, 4*, 158–164.

Wetzel, S. G., Kirsch, E., Stock, K. W., et al. (1999). Cerebral veins: Comparative study of CT venography with intraarterial digital subtraction angiography. *AJNR. American Journal of Neuroradiology, 20*, 249–255.

Headache from Cranial Bone

Hartmut Göbel
Kiel Migraine and Headache Center, Kiel Pain Center, Kiel, Germany

Synonyms

Facial pain associated with disorders of the cranium; Headache associated with disorders of the cranium

Definition

Pain in the head or face caused by a lesion within the cranial bone.

Characteristics

Most disorders of the skull (e.g., congenital abnormalities, fractures, tumors, metastases) are not usually accompanied by headache (Göbel and Edmeads 2000; International Headache Society Classification Subcommittee 2004). Exceptions of importance are osteomyelitis, multiple myeloma, and Paget's disease (Georgalas et al. 2011; Göbel 2012; Kazan et al. 2012; Lu and Yao 2011; Voorhies and Sundaresan 1985). Headache may also be caused by lesions of the mastoid and by petrositis. No epidemiological data are available on headaches due to lesions of the cranial bone.

The bone of the skull has limited sensitivity to pain because only a few nerve fibers enter it from the overlying periosteum. The periosteum is more pain sensitive, and skull lesions therefore produce headache, chiefly by involving it. The lesions of the skull most likely to do this are those that are rapidly expansile, aggressively osteoclastic, or have an inflammatory component.

Most skull lesions are asymptomatic and are discovered as incidental findings on roentgenograms or other imaging procedures done to investigate unrelated complaints, including fibrous dysplasia, osteomas, epidermoid cysts, metastatic cancers, hemangiomas, eosinophilic granulomas, and Paget's disease of the skull. Some of these lesions, notably hemangiomas and eosinophilic granulomas and the rare aneurysmal bone cysts, may present with a tender swelling on the calvarium but not with spontaneous headache.

Relatively few skull lesions produce headache. Multiple myeloma often presents with bone pain anywhere in the body, and skull deposits are sometimes a source of such pain. The multiplicity of the deposits, and the proclivity of the myeloma cells to produce osteoclast activating factor, are likely to account for the production of head pain by this particular bone tumor. Osteomyelitis produces spontaneous head pain because of its rapid evolution and its inflammatory component. Although most cases of Paget's disease of the skull are asymptomatic, remodeling of bone, by producing basilar invagination, may cause headache either through traction on the upper cervical nerve roots or by the production of cerebrospinal fluid pathway distortion with hydrocephalus.

Skull lesions as a cause of headache are infrequent, but usually require neurosurgical treatment. If necessary, surgical excision can serve to confirm the diagnosis and retard the progression of neurological dysfunction and head pain. Apart from specific medication, non-opioid and opioid analgesics may be used for pain relief.

References

Georgalas, C., Goudakos, J., & Fokkens, W. J. (2011). Osteoma of the skull base and sinuses. *Otolaryngologic Clinics of North America, 44*(4), 875–890. Aug.

Göbel, H. (2012). *Die Kopfschmerzen. Ursachen, Mechanismen, Diagnostik und Therapie in der Praxis* (pp. 1–950). Berlin: Springer Verlag.

Göbel, H., & Edmeads, J. G. (2000). Disorders of the skull and cervical spine. In J. Olesen, P. Tfelt-Hansen, & K. M. A. Welch (Eds.), *The headaches* (2nd ed., pp. 891–898). Philadelphia: Lippincott Williams & Wilkins.

International Headache Society Classification Subcommittee. (2004). International classification of headache disorders, 2nd edition. *Cephalalgia, 24* (suppl 1), 1–160. www.ihs-klassification.org.

Kazan, S., Ucar, T., & Turgut, U. (2012). Sutural diastasis caused by pseudotumor cerebri. *Turkish Neurosurgery, 22*(4), 458–460.

Lu, Z. H., & Yao, Z. W. (2011). Giant cell tumour of the posterior cranial fossa: A case report. *British Journal of Radiology, 84*(1007), e206–e209. Nov.

Voorhies, R. M., & Sundaresan, N. (1985). Tumors of the skull. In R. H. Wilkins & S. S. Rengachary (Eds.), *Neurosurgery* (pp. 984–1001). New York: McGraw-Hill.

Headache in Aseptic Meningitis

K. Ravishankar
The Headache and Migraine Clinic, Jaslok Hospital and Research Centre Lilavati Hospital and Research Centre, Mumbai, India

Synonyms

Abacterial meningitis; Aseptic meningitis; Serous meningitis; Viral meningitis

Definition

▶ Aseptic meningitis is the term applied to an acute clinical syndrome that comprises headache, fever, signs of meningeal inflammation, and a predominantly lymphocytic pleocytosis with normal glucose and normal to elevated proteins in the cerebrospinal fluid (CSF).

Historically, the word "aseptic" was introduced to denote the nonbacterial etiology of this syndrome and included forms of infective meningitis (viral and fungal) that were negative on routine bacteriologic stains and culture. With the introduction of polymerase chain reaction (PCR)-based investigations and improved diagnostic techniques, the yield has improved, and the list of conditions that can present with a clinical picture like aseptic meningitis has expanded considerably. Although often used interchangeably, this term is therefore no longer synonymous with ▶ viral meningitis.

Characteristics

Introduction

Both infective and noninfective conditions may present with a picture that fits the definition of aseptic meningitis. Infective causes (Table 1) are mostly viral in origin and less commonly of fungal, parasitic, nonpyogenic bacterial,

Headache in Aseptic Meningitis, Table 1 Viral conditions that may present with aseptic meningitis

Infectious etiologies	Noninfectious causes Drugs
Enteroviruses, polio, coxsackievirus, echovirus	NSAIDs
HSV types 1 and 2	Trimethoprim
Varicella zoster virus	Azathioprine
Adenovirus	Intravenous immunoglobulin
Epstein-Barr virus	Isoniazid
LCMV	Intrathecal
HIV	Methotrexate
Influenza A and B	Vaccines
	Allopurinol

Headache in Aseptic Meningitis, Table 2 Non-viral conditions that may present with aseptic meningitis

Infectious etiologies	Noninfectious causes
Bacteria	**Other diseases**
M. tuberculosis	Sarcoidosis
Borrelia burgdorferi	Leptomeningeal carcinoma
Treponema pallidum	SLE
Brucella	CNS vasculitis
Mycopl. pneumoniae	Behcet disease
Fungi	Vogt-Koyanagi-Harada syndrome
Crypto. neoformans	Migraine
Histo. capsulatum	
Coccidioides immitis	
Blasto. dermatitides	
Parasites	
Toxoplasma gondii	
Taenia solium	

rickettsial, or mycoplasmal origin; noninfective causes (Table 2) include tumors of the central nervous system, carcinomas, leukemias, sarcoidosis, systemic lupus erythematosus (SLE), rheumatoid arthritis, certain drugs, vaccines, immunoglobulins, intrathecal agents, and rarely some disorders of unproven etiology like Behcet's syndrome and Vogt-Koyanagi-Harada syndrome.

Aseptic meningitis is common and seen more often in children and young adults, especially during the summer months. Except in the

neonatal period, the mortality and morbidity rates are low (Norris et al. 1999; Cherry 1998). Most patients with aseptic meningitis due to viral causes have a benign course and spontaneously improve, while others may run a complicated course unless specifically treated. Worldwide prevalence varies depending on geographic factors, seasonal influence, epidemiologic patterns of diseases, and vaccination policies.

Clinical Features

Aseptic meningitis is characterized by abrupt onset of headache, fever, and neck stiffness. Additional clinical symptomatology may vary depending on the underlying cause. Focal signs and seizures are rarely seen in aseptic meningitis, but mumps, certain arboviruses, and lymphocytic choriomeningitis virus may cause a meningoencephalitis (Rice 2001).

The headache of aseptic meningitis has no typical characteristics. It is severe, most often bilateral, and may be associated with fever and vomiting. Lamonte et al. (1995), in their retrospective review of 41 patients with aseptic meningitis, noted that headache was present in all, started or worsened abruptly in 24; in 39 the headache was severe; and in 6 it was the worst headache. There was no consistent pattern of location or type of pain. In all cases, the headache was different from the usual headache. Systemic prodromal symptoms preceded the onset of headache in 19 patients. Nausea, vomiting, cognitive changes, back pain, blurred vision, phonophobia, photophobia, and tinnitus were the associated symptoms seen in their series (Lamonte et al. 1995).

Migraine headache may mimic aseptic meningitis, but if a patient presents acutely with fever and headache that is bilateral, throbbing not relieved with analgesics and different from their earlier headaches, then aseptic meningitis needs ruling out. Rarely migraine itself can cause aseptic meningitis. Bartleson et al. (1981) reported a series of patients with complicated migraine and CSF pleocytosis preceded by a viral-like illness (Gomez-Aranda et al. 1997). Other causes of similar headache that may confuse include subarachnoid hemorrhage and other acute headaches.

The cell count in aseptic meningitis is usually less than 1,000 per cu. mm, and there may be an early predominance of polymorphonuclear leukocytes. Repeated lumbar puncture in 8–12 h frequently shows a change from neutrophil to lymphocyte predominance. CSF glucose levels are normal and CSF proteins may be normal or elevated. CSF culture for viruses and PCR studies help in further confirming the diagnosis.

Differential Diagnosis

Viruses are the most common causative agents, but even when all viral diagnostic facilities are available, the causal agent may be difficult to identify in a good proportion of cases. Viral pathogens may enter the CNS through the hematogenous or neural route. Neural penetration is limited to herpes viruses (HSV-1, HSV-2, and varicella zoster virus) and some enteroviruses. Exposure to mosquito or tick vectors is a risk factor for transmission (Adams and Victor 2001). Over 80 % of aseptic meningitis are caused by enteroviruses (coxsackie A or B, enterovirus 68–71, echovirus and poliovirus), followed by the mumps virus, HSVû2, HIV, and less commonly HSV-1, varicella zoster virus (VZV), Epstein-Barr virus, and cytomegalovirus (CMV). Rarely arbovirus, lymphocytic choriomeningitis virus (LCMV), and adenovirus may be responsible for similar symptoms. Influenzal and parainfluenzal illnesses can also cause aseptic meningitis. The incidence of polio and mumps in the vaccination era has decreased significantly in developed countries. In younger people, measles virus may cause aseptic meningitis that is associated with a rash (Waisman et al. 1999).

Human immunodeficiency virus (HIV) infection may present with aseptic meningitis, particularly at the time of seroconversion (Levy et al. 1990). Patients may present with CSF pleocytosis, elevated protein level, and high intracranial pressure. Besides the usual meningeal signs, patients with HIV infection may have neurological deficits and may need imaging. Adenovirus may be a major cause of meningitis in patients with HIV infections. Varicella zoster virus can affect the immunocompromised.

Arbovirus accounts for approximately 5 % of cases of aseptic meningitis in North America, and the incidence varies depending on the life cycle of arthropod vectors, animal reservoirs, and their contact with humans. Some of the important viruses include Eastern and Western equine encephalitis viruses, St. Louis Encephalitis virus, West Nile virus, Japanese B virus, and Colorado tick fever. LCMV affects those at risk who come in contact with rodents or their excreta (Nelsen et al. 1993).

The immediate concern in practice should not be aimed at establishing a particular virus as the cause of the illness, but more importantly to exclude the few conditions with aseptic meningitis-like picture, but having another underlying non-viral cause warranting specific management. In every patient with aseptic meningitis, one has to look beyond viruses as the causative factor.

Non-viral causes have a more complicated course but can be managed with specific treatment. Tuberculous, fungal, syphilitic, spirochetal, rickettsial, parasitic, and other mycoplasmal infections can cause aseptic meningitis, which should be suspected in the appropriate clinical setting. In the early stages, tuberculous meningitis may appear like aseptic meningitis and can be difficult to diagnose. The glucose levels are reduced only in the later stages and the organism is difficult to find. CSF features of aseptic meningitis, but without fever, may be seen with acute syphilitic meningitis. Cryptococcal infections, other fungal infections, and some rare conditions like mycoplasma pneumonia, brucellosis, and Q fever can also present like aseptic meningitis. Brucellosis is common in specific geographic locations.

Conjunctival suffusion with transient erythema, severe leg and back pain, pulmonary infiltrates, and aseptic meningitis should suggest leptospiral infection. Infection is acquired by contact with soil or water contaminated by the urine of rats, dogs, or cattle. Lyme borreliosis is a common spirochetal cause of aseptic meningitis and meningoencephalitis. The spirochete is tick borne, common in northeastern United States from May to July (Eppes et al. 1999).

Leukemias in children and lymphomas in adults are common sources of meningeal reactions with aseptic meningitis like CSF picture. In these disorders, and in meningeal carcinomatosis, neoplastic cells are found throughout the leptomeninges with additional root involvement. Features of the aseptic meningitis syndrome can also be caused by brain abscess, parameningeal infections, and partially treated bacterial meningitis, when it may be mistakenly diagnosed as viral aseptic meningitis. A careful history of previous antibiotic administration must therefore be obtained in all patients with meningitis.

Sarcoidosis, Behcet's syndrome, vasculitis, and granulomatous angiitis can present with aseptic meningitis syndrome by infiltrating the leptomeninges. These conditions, however, rarely present with a clinical picture of meningitis alone, more often they are seen with other neurological accompaniments (Gullapalli and Phillips 2002; Nelsen et al. 1993). Some chronic diseases like systemic lupus erythematosus, serum sickness, and Vogt-Koyanagi-Harada syndrome may present with aseptic meningitis (Adams and Victor 2001).

Drug-induced aseptic meningitis (DIAM), either by (1) direct irritation of the meninges with intrathecal administration or by (2) immunological hypersensitivity to the drug, has been reported as an uncommon adverse reaction with numerous agents (Chaudhry and Cunha 1991). The major categories of causative agents are nonsteroidal anti-inflammatory drugs (NSAIDs), antimicrobials, intravenous immunoglobulins, isoniazid, allopurinol, and vaccines for measles, mumps, and rubella. In addition to headache, there may be signs of a hypersensitivity reaction. Trimethoprim-sulfamethoxazole, azathioprine, and intrathecal injections can result in the clinical findings of aseptic meningitis. The association between SLE and ibuprofen as a cause of DIAM is important to recognize. A high index of suspicion is necessary to make the diagnosis. Treatment is to withhold the drug. There are no long-term sequelae of DIAM.

Besides the typical CSF picture, it is essential to isolate the virus in CSF, stool, saliva, and throat swabs using PCR and other serologic tests (Jeffery et al. 1997). It is important to inquire about a past history of infectious disease, immunizations, contact with animals, insect bites, recent respiratory or gastrointestinal infection, and recent travel. The season during which the illness occurs and the geographical location are helpful pointers.

Recurrent aseptic meningitis is also known as Mollaret's meningitis and can be a diagnostic dilemma. There is spontaneous remission and no causative agent has been consistently found. It is difficult to identify the virus in the CSF. These patients need detailed investigations with repeat lumbar punctures, cytology or CSF bacterial cultures, PCR, HIV testing, and MRI with contrast if necessary. Recurrence in a few cases is caused by HSV-1 and HSV-2 infections (Cohen et al. 1994).

Conclusion

Most patients with aseptic meningitis need only supportive care. It may be prudent to start antibiotics until cultures are shown to be negative or a second examination of CSF shows a more typical picture. Most patients recover completely and rapidly when the etiology is viral, unless there is an associated encephalitic component. Precautions should be taken when specific viruses are identified. Effective antiviral therapy is available against HSV-1, varicella, and CMV. For HSV-2, acyclovir is the drug of choice. Other causes need appropriate management. Rarely, patients may have persistent headache, mild mental impairment, incoordination, or weakness that lasts for months. Although aseptic meningitis is an acute illness, most patients eventually improve.

References

Adams, R. D., & Victor, M. (2001). Viral infections in the nervous system. In R. D. Adams & M. Victor (Eds.), *Principles of neurology* (7th ed.). New York: McGraw-Hill.

Bartleson, J. D., Swanson, J. W., & Whisnant, J. P. (1981). A migrainous syndrome with cerebrospinal fluid pleocytosis. *Neurology, 1*, 1257–1262.

Chaudhry, H. J., & Cunha, B. A. (1991). Drug-induced aseptic meningitis: Diagnosis leads to quick resolution. *Postgraduate Medicine, 90*, 65–70.

Cherry, J. D. (1998). Aseptic meningitis and viral meningitis. In R. D. Feigin & J. D. Cherry (Eds.), *Textbook of pediatric infectious diseases, Vol. 2* (4th ed., pp. 450–457). Philadelphia: Saunders.

Cohen, B. A., Rowley, A. H., & Long, C. M. (1994). Herpes simplex type 2 in a patient with Mollaret's meningitis: Demonstration by polymerase chain reaction. *Annals of Neurology, 35*, 112–116.

Eppes, S. C., Nelson, D. K., Lewis, L. L., & Klein, J. D. (1999). Characterization of lyme meningitis and comparison with viral meningitis in children. *Journal of Pediatrics, 103*, 957–960. May.

Gomez-Aranda, F., Canadillas, F., Marti-Masso, J. F., et al. (1997). Pseudomigraine with temporary neurological symptoms and lymphocytic pleocytosis. A report of 50 cases. *Brain, 120*, 1105–1113.

Gullapalli, D., & Phillips, L. H. (2002). Neurologic manifestations of sarcoidosis. *Neurologic Clinics, 20*, 59–83.

Jeffery, K. J., Read, S. J., Petro, T. E., et al. (1997). Diagnosis of viral infections of the central nervous system: Clinical interpretation of PCR results. *Lancet, 349*, 313–317.

Lamonte, M., Silberstein, S. D., & Marcelis, J. F. (1995). Headache associated with aseptic meningitis. *Headache, 35*, 520–526.

Levy, R. M., Bredesen, D. E., & Rosenblum, M. L. (1990). Neurologic complications of HIV infection. *American Family Physician, 41*, 517–536.

Nelsen, S., Sealy, D. P., & Schneider, E. F. (1993). The aseptic meningitis syndrome. *American Family Physician, 48*, 809–815.

Norris, C. M., Danis, P. G., & Gardner, T. D. (1999). Aseptic meningitis in the newborn and young infant. *American Family Physician, 59*, 2761–2770.

Rice, P. (2001). Viral meningitis and encephalitis. *Medicine, 29*, 54–57.

Waisman, Y., Lotem, Y., Hemmo, M., et al. (1999). Management of children with aseptic meningitis in the emergency department. *Pediatric Emergency Care, 15*, 314–317.

Headache, Acute Posttraumatic

Miguel J. A. Lainez-Andres
Department of Neurology, University
of Valencia, Valencia, Spain

Synonyms

Posttraumatic headache; PTHA

Definition

Posttraumatic headache (PTHA) is usually one of the several symptoms of the "posttraumatic syndrome" and therefore may be accompanied by somatic, psychological, or cognitive disturbances (Solomon 2001). A variety of pain patterns may develop after head injury and may closely resemble primary headache disorders. Common headache pathways have been described for primary and posttraumatic headaches, but the pathogenesis of PTHA is still not well known (Martelli 1999).

Characteristics

Tension type is the most common variety of PTHA (more than 80 % of the patients suffered a tension-type headache after head or neck trauma), followed by cervicogenic headache. Exacerbations of migraine and cluster-like headaches also occur. Posttraumatic migraine (PTMA) represents approximately 8–10 % of PTHA. This is usually a migraine without aura, often found in children, adolescents, and young adults with familial history of migraine. Migraine with visual aura has been described in only a few patients (Hachinski 2000).

Mild, moderate, and severe head injuries can be associated with a PTHA. Clinical quantification of traumatic brain injury patients should be based on the Glasgow Coma Scale (GCS) score, duration of loss of consciousness (LOC), and presence of posttraumatic amnesia (PTA). In addition, a short practicable neuropsychological test may be useful in detecting minor memory and attentional deficits. Paradoxically, mild head injury is often accompanied by headache and additional symptoms, more frequently than moderate or severe head traumas.

To differentiate between a primary and a posttraumatic headache can be difficult in some cases. Patients who develop a new form of headache in close temporal relation to head or neck trauma should be coded as having a secondary headache. Patients in whom this type of headache was preexisting but significantly worsened in close temporal relation to trauma, without evidence of a causal relationship between the primary headache and the other disorder, receive only the primary headache diagnosis. However, if there is both a very close temporal relation to the trauma and other good evidence that the particular kind of trauma has aggravated the primary headache, that is, if trauma in scientific studies of good quality has been shown to aggravate the primary headache disorder, the patient receives the primary and the secondary headache diagnoses. In many cases of secondary headache, the diagnosis is definite only when the headache resolves or greatly improves within a specified time after effective treatment or spontaneous remission of the causative disorder. In such cases, this temporal relation is an essential part of the evidence of causation.

It is easy to establish the relationship between a headache and head or neck trauma when the headache develops immediately or in the first days after trauma has occurred. On the other hand, it is very difficult when a headache develops weeks or even months after trauma, especially when the majority of these headaches have the pattern of tension-type headache and the prevalence of this type of headache in the population is very high. Such late onset posttraumatic headaches have been described in anecdotal reports but not in case–control studies. In accordance with new IHS Classification that will soon be published, acute PTHA develops within 7 days after head trauma or regaining consciousness following head trauma and resolves within 3 months.

New Diagnostic Criteria for Acute Posttraumatic Headache

Acute Posttraumatic Headache with Moderate or Severe Head Injury

Diagnostic Criteria

1. Headache, no typical characteristics known, fulfilling criteria C and D.
2. Head trauma with at least one of the following:
   1. Loss of consciousness for >30 min
   2. Glasgow Coma Scale (GCS) <13
   3. Posttraumatic amnesia for >48 h
   4. Imaging demonstration of a traumatic brain lesion (cerebral hematoma, intracerebral and/or subarachnoid hemorrhage, brain contusion, and/or skull fracture)

3. Headache develops within 7 days after head trauma or after regaining consciousness following head trauma.
4. One or other of the following:
   1. Headache resolves within 3 months after head trauma.
   2. Headache persists but 3 months have not yet passed since head trauma.

Acute Posttraumatic Headache with Mild Head Injury

Diagnostic Criteria

1. Headache, no typical characteristics known, fulfilling criteria C and D.
2. Head trauma with all of the following:
   1. Either no loss of consciousness or loss of consciousness of <30 min duration
   2. Glasgow Coma Scale (GCS) >13
   3. Symptoms and/or signs diagnostic of concussion
3. Headache develops within 7 days after head trauma.
4. One or other of the following:
   1. Headache resolves within 3 months after head trauma.
   2. Headache persists but 3 months have not yet passed since head trauma.

Before new diagnosis criteria, acute PTHA might begin less than 14 days after head or neck trauma and continue for up to 8 weeks post-injury (Headache Classification Committee of IHS 1988). Headache that develops longer than 14 days after head injury has been termed "delayed PTHA or late-acquired headache." If such headaches persist beyond the first 3 months post-injury, they are subsequently referred to as chronic PTHA.

New Diagnostic Criteria for Chronic Posttraumatic Headache

Chronic Posttraumatic Headache with Moderate or Severe Head Injury

Diagnostic Criteria

1. Headache, no typical characteristics known, fulfilling criteria C and D.
2. Head trauma with at least one of the following:
   1. Loss of consciousness >30 min
   2. Glasgow Coma Scale (GCS) <13
   3. Posttraumatic amnesia >48 h

    4. Imaging demonstration of a traumatic brain lesion (cerebral hematoma, intracerebral and/or subarachnoid hemorrhage, brain contusion, and/or skull fracture)
3. Headache develops within 7 days after head trauma or after regaining consciousness following head trauma.
4. Headache persists for >3 months after head trauma.

Chronic Posttraumatic Headache with Mild Head Injury

Diagnostic Criteria

1. Headache, no typical characteristics known, fulfilling criteria C and D.
2. Head trauma with all of the following:
   1. Either no loss of consciousness or loss of consciousness of <30 min duration
   2. Glasgow Coma Scale (GCS) >13
   3. Symptoms and/or signs diagnostic of concussion
3. Headache develops within 7 days after head trauma.
4. Headache persists for >3 months after head trauma.

After mild head trauma, laboratory and ► neuroimaging investigations are not habitually needed. When the GCS score is less than 13 in the emergency room after head or neck trauma, LOC is longer than 30 min; there is PTA, neurological deficits, or personality disturbances; and neuroimaging studies (computer tomography scan, CT, or magnetic resonance imaging, MRI) are indicated. MRI (using at least T1-weighted, T2-weighted, proton density, and gradient-echo sequence images) is much more sensitive than CT in detecting and classifying brain lesions. Within 1 week of a head injury, MRI can identify cortical contusions and lesions in the deep white matter of the cerebral hemispheres underdiagnosed by CT. MRI thus provides a sounder basis for diagnosis and treatment in patients suffering from late sequelae of cranial injuries (Voller 2001).

Complementary studies (neuroimaging, EEG, evoked potentials, CSF examination, vestibular function tests) should also be considered for patients with ongoing posttraumatic headaches. The relationship between severity of the injury and severity of the posttraumatic syndrome has not been conclusively established. Moreover, there are some controversial data. Most studies suggest that PTHA is less frequent when the head injury is more severe. Differential diagnosis may include a symptomatic headache, secondary to structural lesions and simulation. There is no evidence that an abnormality in the complementary explorations changes the ► prognosis or contributes to treatment. Special complementary studies should be considered on a case-by-case basis or for research purposes.

After several months, some patients developed a daily headache. In the majority of patients with episodic headaches after head injury, this condition is self-limited, but a minority of individuals may develop persistent headaches. Neurological factors have been implicated in the initial phase, psychological and legal factors (litigation and expectations for compensation) in the maintenance of them. Premorbid personality can contribute to development of chronic symptoms, affecting adjustment to injury and treatment outcome. Surprisingly, the risk of developing chronic disturbances seems to be greater for mild to moderate head injury.

Age; gender; certain mechanical factors; a low intellectual, educational, and socioeconomic level; previous history of headache or alcohol abuse; and long duration of unconsciousness or neurological deficits after the head or neck injury are recognized ► risk factors for a poor outcome. Women have higher risk of PTHA, and increasing age is associated with a less rapid and less complete recovery. Mechanical impact factors, such as an abnormal position of the head (rotation or inclined), increase the risk of PTHA. Other predictor factors are presence of skull fracture, reduced value of Glasgow Coma Scale, elevated serum protein S-100B, and dizziness, headache, and nausea in the emergency room (De Krujik 2002).

The role of litigation in the persistence of headache is still discussed. The relationship between legal settlements and the temporal profile of chronic PTHA is not clearly established, but it is important to carefully assess patients who may be malingering and/or seeking enhanced compensation. In general, medicolegal issues should be solved as soon as possible.

Pathophysiology of PTHA

Pathophysiology of posttraumatic headaches is still not well understood, but biological, psychological, and social factors are included. In the pathogenesis, common headache pathways with primary headaches have been proposed.

During typical migraine, cerebral cortical and brain stem changes occur. The activation of the brainstem monoaminergic nuclei has been demonstrated with functional imaging studies (Bahra 2001). Disturbed neuronal calcium influx and/or hemostatic alterations have also been involved. However, these events have not been included for PTMA yet.

In recent years, several pieces of research have implicated similar neurochemical changes in both typical migraine and experimental traumatic brain injury, excessive release of excitatory amino acids, alterations in serotonin, abnormalities in catecholamines and endogenous opioids, decline in magnesium levels, abnormalities in nitric oxide formation, and alterations in neuropeptides (Packard 1997). Whether these changes are determining, contributing, or precipitating factors for headache in each patient is still unknown. In addition, in patients with late PTMA, a sensitization phenomenon is possible. In some patients without previous migraine and history of a recent mild head injury, trigeminal neuron sensitization could be a central cause in relation to focal lesions. Central and peripheral sensitizations have been proposed before by other authors (Malick 2000; Packard 2002).

Further researches are still necessary to clarify the relationship between chronic symptoms after mild head trauma and neuroimaging abnormalities. These abnormalities could provide a pathological basis for long-term neurological disability in patients with post-concussive syndrome. New techniques of MRI (especially diffusion tensor imaging and magnetization transfer ratio) are useful for the detection of small parenchymal brain lesions, diffuse axonal injury secondary to disruption of axonal membranes, or delayed cerebral atrophy (Hofman 2002). In normal-appearing white matter, magnetic resonance spectroscopy studies detect metabolic brain changes (an early reduction in N-acetyl aspartate and an increase in choline compounds), which correlate with head injury severity (Garnett 2000). Positron-emission tomography (PET), single-photon emission computed tomography (SPECT), and xenon-133 CT may provide evidence of brain perfusion abnormalities after mild head trauma and in the presence of chronic posttraumatic symptoms (Aumile 2002).

Management Strategies

Trauma-induced headaches are usually heterogeneous in nature, including both tension-type and intermittent migraine attacks. Over time, PTHA may take on a pattern of daily occurrence, although if aggressive treatment is initiated early, PTHA is less likely to become a permanent problem. Adequate treatment typically requires both "central" and "peripheral" measures. Delayed recovery from PTHA may be a result of inadequately aggressive or ineffective treatment, overuse of analgesic medications resulting in analgesia rebound phenomena, or comorbid psychiatric disorders (posttraumatic stress disorder, insomnia, substance abuse, depression, or anxiety) (Lane 2002).

In general, treatment strategies are based upon studies of non-traumatic headache types. Acute PTHA may be treated with analgesics, anti-inflammatory agents, and physiotherapy. PTMA may be also treated with ergotamine or triptans. Chronic PTHA needs prophylactic medication, chronic PTMA-specific antimigraine medications. Previously amitriptyline or propranolol used alone or in combination and verapamil have been demonstrated to improve all symptoms of post-concussive syndrome, especially the migraine. Recently, Packard has published very good results with divalproex sodium as a preventive option in the treatment of PTMA (Packard 2000). Additional physical therapy, psychotherapy (biofeedback), and appropriate educational support can be supplied, especially in patients with risk factors for poor prognosis. Explanation of the headache's nature can also improve the patient's evolution. In some cases, when a posttraumatic lesion is identified as a peripheral triggering factor for headache, specific treatment of the triggering lesion can resolve the pain. PTMA poorly treated will affect family life, recreation, and employment. There is no good evidence that litigation and economical expectation is associated with prolongation of headaches; however, litigation should be solved as soon as is possible.

Conclusions

Trauma-induced headache and headache attributed to whiplash should be treated early, or associated complications will appear (daily occurrence of headache, overuse of analgesic medications, and comorbid psychiatric disorders). Preventive and symptomatic treatments may be prescribed according to the clinical pattern of the headache (tension-type, migraine, cluster, or cervicogenic headaches) as a primary headache. Physiotherapy, psychotherapy, and resolution of litigation can be contributing factors to recovery.

References

Aumile, E. M., Sandel, M. E., Alavi, A., et al. (2002). Dynamic imaging in mild traumatic brain injury support for the theory of medial temporal vulnerability. *Archives of Physical Medicine and Rehabilitation, 83*, 1506–1513.

Bahra, A., Matharu, M. S., Buchel, C., et al. (2001). Brainstem activation specific to migraine headache. *Lancet, 357*, 1016–1017.

De Kruijk, J. R., Leffers, P., Menheere, P. P., et al. (2002). Prediction of posttraumatic complaints after mild traumatic brain injury: Early symptoms and biochemical markers. *Journal of Neurology, Neurosurgery, and Psychiatry, 73*, 727–732.

Garnett, M. R., Blamire, A. M., Rajagopalau, B., et al. (2000). Evidence for cellular damage in normal-appearing white matter correlates with injury severity in patients following traumatic brain injury: A magnetic resonance spectroscopy study. *Brain, 123*, 1043–1049.

Hachinski, W. (2000). Posttraumatic headache. *Archives of Neurology, 57*, 1780.

Headache Classification Committee of the International Headache Society. (1988). Classification and diagnostic criteria for headache disorders, cranial neuralgias and facial pain. *Cephalalgia, 8*, 1–96.

Hofman, P. A., Verhey, F. R., Wilmink, J. T., et al. (2002). Brain lesions in patients visiting a memory clinic with postconcussional sequelaes after mild to moderate brain injury. *The Journal of Neuropsychiatry and Clinical Neurosciences, 14*, 176–178.

Lane, J., & Arciniegas, D. B. (2002). Post-traumatic headache. *Current Treatment Options in Neurology, 4*, 89–104.

Malick, A., & Burstein, R. (2000). Peripheral and central sensitization during migraine. *Functional Neurology, 15*, 28–35.

Martelli, M. F., Grayson, R. L., & Zasler, N. D. (1999). Post-traumatic headache: Neuropsychological and psychological effects and treatment implications. *The Journal of Head Trauma Rehabilitation, 14*, 49–69.

Packard, R. C. (2000). Treatment of chronic daily posttraumatic headache with divalproex sodium. *Headache, 40*, 736–739.

Packard, R. C. (2002). The relationship of neck injury and posttraumatic headache. *Current Pain and Headache Reports, 6*, 30131–30137.

Packard, R. C., & Haw, C. P. (1997). Pathogenesis of PTH and migraine: A common headache pathway? *Headache, 37*, 142–152.

Solomon, S. (2001). Post-traumatic headache. *Medical Clinics of North America, 85*, 987–996.

Voller, B., Auff, E., Schnider, P., et al. (2001). To do or not to do MRI in mild traumatic brain injury? *Brain Injury, 15*, 107–115.

Headache, Episodic Tension-Type

Andreas R. Gantenbein[1] and Peter S. Sándor[2]
[1]RehaClinic Bad Zurzach, Bad Zurzach, Switzerland
[2]RehaClinic, Cantonal Hospital Baden, Baden, Switzerland

Synonyms

Episode tension-type headache; Idiopathic headache; Muscle contraction headache; Ordinary headache; Psychogenic headache; Psychomyogenic headache; Tension headache

Definition

The new classification of the International Headache Society (IHS) distinguishes an infrequent (less than 1 day per month) and a frequent form (at least 1 day but less than 15 days per month) of episodic tension-type

Headache, Episodic Tension-Type, Table 1 Compared diagnostic criteria frequent episodic tension-type headache and migraine without aura from the International Headache Classification, 2nd edn

Diagnosis	Frequent episodic tension-type headache	Migraine without aura
Number of episodes	At least 10	At least 5
Number of days with such headache	≥1 day and <15 days per month (for at least 3 months)	<15 days/month (untreated or unsuccessfully treated)
Duration of the headache	30 min to 7 days	4–72 h
Pain characteristics	At least two of the following:	At least two of the following:
	Pressing/tightening (non-pulsating) quality	Pulsating quality
	Mild or moderate intensity	Moderate or severe pain intensity
	Bilateral location	Unilateral location
	No aggravation by walking on stairs or similar routine physical activity	Aggravation by or causing avoidance of routine physical activity
Accompanying symptoms	Both of the following:	At least one of the following:
	No nausea or vomiting (anorexia may occur)	Nausea and/or vomiting
	Photophobia or phonophobia or none	Photophobia and phonophobia

headache. Duration varies from minutes to days. The pain is typically bilateral, of mild to moderate intensity, and has a ▶ pressing/tightening character. There is no worsening with routine physical activity. There is no nausea, but ▶ photophobia or ▶ phonophobia may be present. Both infrequent and frequent types can be subdivided according to the presence or absence of pericranial ▶ tenderness (jaw, scalp, and neck muscles). See Table 1 for the classification criteria (Headache Classification Subcommittee of the International Headache Society 2004).

Characteristics

With a lifetime prevalence of 30–78 % in general population, tension-type headache is the most common primary headache and has a high socioeconomic impact. The male to female ratio is 1:1.5. The prevalence in childhood ranges from 0.5 to 12 % (Anttila et al. 2002; Rasmussen et al. 1991; Rasmussen 2001; Schwartz et al. 1998). Tension headache was first defined by the IHS classification committee in 1988. This type of headache previously had a psychological label and was thought to be caused exclusively by mental conflicts, stress, tension, or emotional overload. There was exciting little interest from research and pharmaceutical companies. However, more recently a number of

studies have investigated neurobiological mechanisms. Peripheral pain mechanisms, such as myofascial tenderness, hyperalgesia, and muscle hardness, have been implicated in the episodic type and dysfunction of central sensitization in the chronic type. Overall tension-type headache appears to be a central disinhibitory phenomenon, probably with involved neurotransmitter changes, defective nociceptive control, increased sensitivity to both myofascial and vascular input, and associated personality traits (Jensen and Olesen 2000). Whether tension-type headache and migraine are separate entities, as suggested by epidemiological data, or rather represent a continuum with shared pathophysiology remains controversial (Rasmussen 1996; Ulrich et al. 1996).

Symptoms

The headache can be described in simple terms as pain in the head without associated symptoms. Unlike migraine there is no sensory hypersensitivity, e.g., to sound, light, or movements. Unlike cluster headache, autonomic features (tearing, redness of the eye, and blocked nose) are not present. Due to the usually mild and short-lasting character, patients with less frequent tension headache often view their symptoms as a nuisance and rarely seek the advice of a specialist. Accordingly, these patients often treat their headache with standard over-the-counter pain killers (often containing caffeine) or with non-pharmacological treatments, such as hot or cold packs or massage. If the headache is more frequent, it may well become distressing and interfere with daily life. This may be associated with regular intake of nonspecific analgesics and can lead to further problems, including chronification of the headache.

Episodic tension-type headache has a high intra- and interindividual variability with respect to frequency and intensity. Additionally, the duration of each attack may range from 30 min to 7 days. It usually has a diffuse pressing character, often described by patients as a "tight band around the head." The pain is dull, persistent, and often diurnal. The headache is bilateral in 80–90 % of cases. Most commonly intensity is mild to moderate and may interfere with (though not usually prevent) performance of daily activities. Characteristically it is not aggravated by routine physical activity. Nausea and vomiting are absent. Patients may complain of mild intolerance to loud sound or bright light, though true photophobia or phonophobia is rare and strongly suggests migraine (certainly when both are present). There is no blurred vision and no focal neurological disturbance. Patients may complain of a feeling of giddiness or lightheadedness, sometimes as a consequence of hyperventilation in association with anxiety. Many patients report difficulties in concentrating and lack of interest in work and hobbies. With age, tension-type headache can increase in frequency and duration, and there tends to be more variability of localization and rarely nausea may develop (Wober-Bingol et al. 1996).

Examination

Tension-type-headache patients require thorough neurological examination, including inspection and palpation of ▶ pericranial muscles. Pericranial tenderness is easily recorded by small rotating movements and a firm pressure with two fingers on the frontal, temporal, masseter, pterygoid, sternocleidomastoid, splenius, and trapezius muscles. A local tenderness score from 0 to 3 on each muscle can be summated to a total tenderness score for each individual. The use of a palpometer (pressure-sensitive device) can improve validity and reproducibility. Palpation is also a useful guide for treatment strategy and adds value and credibility to the explanations given to the patient.

Differential Diagnosis

An accurate diagnosis is essential, and migraine, as well as secondary headache, should be excluded. Tension-type headache is sometimes difficult to distinguish from migraine in patients who have both tension headache and migraine with or without aura. It is important to educate patients in the differentiation between these headaches, as the right treatment for the right headache can be administered and medication-overuse headache can be avoided. A diagnostic headache diary can be helpful to identify different patterns, since patients often describe only the characteristics of recent or the most severe attacks. In favor of tension-type headache is a highly variable temporal profile and pain improvement with exercise. Unsuccessful treatment with ergotamines or triptans for acute attacks, or with beta-blockers or flunarizine for prevention, also suggests a diagnosis of tension-type headache (Kaniecki 2002). See Table 1 for the comparison of classification criteria for migraine without aura and tension-type headache.

If headache is new (particularly over the age of 50), has a sudden onset, changes significantly in established pattern or characteristics, or does not fit a classical scheme, then secondary causes need to be excluded. Head trauma, vascular disorder, nonvascular intracranial disorder, substance abuse, noncephalic infection, metabolic disorder, and cranial structure defects can sometimes imitate tension-type headache. If the neurological examination is normal and the headache has no worrisome characteristics, there is no need for further investigations such as neuroimaging or lumbar puncture, and the patient can be reassured.

A number of precipitating factors have been described, including oromandibular dysfunction, nonpsychological motor stress, local myofascial release of irritants, sleep deprivation, and coexisting migraine (Spierings et al. 2001). More controversial is the role of psychological factors, although the triggering of attacks by psychological stress is recognized. Up to one third of tension-type-headache patients show associated symptoms of depression or anxiety, though surveys of personality profiles have not demonstrated significant abnormalities (Holroyd 2002; Merikangas et al. 1994; Mitsikostas and Thomas 1999).

Management and Treatment

As yet there is no specific treatment for tension-type headache. Episodic and mild headaches are often successfully treated with nonspecific analgesics, without the involvement of specialists. Drugs with evidence-based benefit for acute treatment include aspirin, paracetamol, and NSAIDs (Table 2). Compound analgesics should be used with caution, as repeated self-medication can yield to dependency, rebound headache, and chronification. The use of drugs for more than 2–3 days per week (>10 days per month) with associated chronic headache suggests an additional medication-overuse headache (IHS classification). The following rules apply for the acute treatment of ▶ episodic tension-type headache:

1. The analgesics should be taken at relatively high dose!
2. The intake should be as early as possible! (Cave: balance!)
3. Drugs should not be taken on more than 2 days a week!(Cave: balance!)
4. The use of compound analgesics (codeine, caffeine, etc.) should be avoided, or at least limited and carefully monitored!

As a preventative treatment for frequent tension-type headaches, a typical first choice is a tricyclic antidepressant, such as amitriptyline. High-dose magnesium may be effective. Combination with a non-pharmacological treatment, such as cognitive behavioral therapy, progressive muscle relaxation, or psychological counselling, may be useful. In addition, advice may be also needed about the mechanisms of hyperventilation. Management must include elimination of exacerbating factors, such as dental pathology, sinus disease, depressive disorders, unphysiological working conditions, and disturbed

Headache, Episodic Tension-Type, Table 2 Acute treatment options in episodic tension-type headache

List of effective acute drugs			
Paracetamol/ acetaminophen	1,000 mg		
Aspirin	1,000 mg	Steiner et al.	Cephalalgia (2003)
Ibuprofen	400 mg	Packman et al.	Headache (2000)
Ketoprofen	25 mg	Steiner et al.	Cephalalgia (1998)
Naproxen	750 mg	Autret et al.	Cephalalgia (1997)
Diclofenac	12.5–25 mg	Kubitzek et al.	EurJPain (2003)
Metamizol	1,000 mg	Martinez et al.	Cephalalgia (2001)
Medications for children			
Ibuprofen	10 mg/Kg		
Paracetamol	15 mg/Kg		

sleep patterns. Physiotherapy, physical treatment (hot and cold packs), ultrasound, electrical stimulation, posture improvement, relaxation, and exercise programs are helpful in certain cases. Some patients report beneficial effects of muscle relaxants, Tiger Balm, and peppermint oil (Stillman 2002).

References

Anttila, P., Metsahonkala, L., Aromaa, M., et al. (2002). Determinants of tension-type headache in children. *Cephalalgia, 22,* 401–408.

Autret, A., Unger, P. H., & Lesaichot, J. L. (1997). Naproxen sodium versus ibuprofen in episodic tension headache. *Cephalalgia, 17,* 446.

Headache Classification Subcommittee of the International Headache Society. (2004). *International classification of headache disorders, cephalalgia* (2nd ed.). Oxford: Blackwell.

Holroyd, K. A. (2002). Behavioral and psychologic aspects of the pathophysiology and management of tension-type headache. *Current Pain and Headache Reports, 6,* 401–407.

Jensen, R., & Olesen, J. (2000). Tension-type headache: An update on mechanisms and treatment. *Current Opinion in Neurology, 13,* 285–289.

Kaniecki, R. G. (2002). Migraine and tension-type headache: An assessment of challenges in diagnosis. *Neurology, 58,* 15–20.

Kubitzek, F., Ziegler, G., Gold, M. S., Liu, J. M., & Ionescu, E. (2003). Low-dose diclofenac potassium in the treatment of episodic tension-type headache. *Journal of European Pain, 7*(2), 155–62.

Martínez-Martín, P., Raffaelli, E., Jr, Titus, F., Despuig, J., Fragoso YD, Díez-Tejedor, E., Liaño, H., Leira, R., Cornet, M. E., van Toor, B. S., Cámara, J., Peil, H., Vix, J. M., & Ortiz, P. (2001). Co-operative Study Group. Efficacy and safety of metamizol vs. acetylsalicylic acid in patients with moderate episodic tension-type headache: A randomized, double-blind, placebo- and active-controlled, multicentre study. *Cephalalgia, 21*(5), 604–610.

Merikangas, K. R., Stevens, D. E., & Angst, J. (1994). Psychopathology and headache syndromes in the community. *Headache, 34*, 17–22.

Mitsikostas, D. D., & Thomas, A. M. (1999). Comorbidity of headache and depressive disorders. *Cephalalgia, 19*, 211–217.

Packman, B., Packman, E., Doyle, G., Cooper, S., Ashraf, E., Koronkiewicz, K., & Jayawardena, S. (2000). Solubilized ibuprofen: evaluation of onset, relief, and safety of a novel formulation in the treatment of episodic tension-type headache. *Headache, 40*(7), 561–567.

Rasmussen, B. K. (1996). Migraine and tension-type headache are separate disorders. *Cephalalgia, 16*, 217–220.

Rasmussen, B. K. (2001). Epidemiology of headache. *Cephalalgia, 21*, 774–777.

Rasmussen, B. K., Jensen, R., Schroll, M., et al. (1991). Epidemiology of headache in a general population: A prevalence study. *Journal of Clinical Epidemiology, 44*, 1147–1157.

Schwartz, B. S., Stewart, W. F., Simon, D., et al. (1998). Epidemiology of tension-type headache. *Journal of the American Medical Association, 279*, 381–383.

Spierings, E. L., Ranke, A. H., & Honkoop, P. C. (2001). Precipitating and aggravating factors of migraine versus tension-type headache. *Headache, 41*, 554–558.

Steiner, T. J., & Lange, R. (1998). Ketoprofen (25 mg) in the symptomatic treatment of episodic tension-type headache: Double-blind placebo-controlled comparison with acetaminophen (1000 mg). *Cephalalgia, 18*(1), 38–43.

Steiner, T. J., Lange, R., & Voelker, M. (2003). Aspirin in episodic tension-type headache: Placebo-controlled dose-ranging comparison with paracetamol. *Cephalalgia, 23*(1), 59–66.

Stillman, M. J. (2002). Pharmacotherapy of tension-type headaches. *Current Pain and Headache Reports, 6*, 408–413.

Ulrich, V., Russell, M. B., Jensen, R., et al. (1996). A comparison of tension-type headache in migraineurs and in non-migraineurs: A population-based study. *Pain, 67*, 501–506.

Wober-Bingol, C., Wober, C., Karwautz, A., et al. (1996). Tension-type headache in different age groups at two headache centers. *Pain, 67*, 53–58.

Headaches Associated with Substances or their Withdrawal

▶ Headache Attributed to a Substance or Its Withdrawal

Health Informatics

▶ Information and Psychoeducation in the Early Management of Persistent Pain

Heart Pain

▶ Visceral Pain Model, Angina Pain

Heat Hyperalgesia

Definition

Heat hyperalgesia is increased pain produced by a normally painful heat stimulus.

Cross-References

▶ Neuropathic Pain Model, Partial Sciatic Nerve Ligation Model
▶ Sympathetically Maintained Pain and Inflammation, Human Experimentation

Heat Lesion

▶ Radiofrequency Neurotomy, Electrophysiological Principles

Heat Sensor

▶ Capsaicin Receptor

Heightened Attention

▶ Hypervigilance and Attention to Pain

Helical CT

▶ CT Scanning

Helicobacter Pylori

Definition

Helicobacter pylori are bacteria that cause inflammation and ulcers in the stomach.

Cross-References

▶ NSAIDs, Adverse Effects

Heliotherapy

Definition

Heliotherapy is the exposure to sunrays and ultraviolet rays.

Cross-References

▶ Spa Treatment

Helplessness

Definition

Helplessness is a belief in one's inability to adequately manage or cope with a stressful situation and to exert any control over one's circumstances, symptoms, and life.

Cross-References

▶ Catastrophizing
▶ Cognitive-Behavioral Perspective of Pain

Hemianesthesia

Definition

Hemianesthesia is the sensory loss in the left or right side of the body.

Cross-References

▶ Central Nervous System Stimulation for Pain

Hemibody Radiation

Definition

Hemibody radiation is an external beam of radiation administered to half of the body, i.e., above or below the diaphragm, for systemic metastatic disease.

Cross-References

▶ Adjuvant Analgesics in Management of
  Cancer-Related Bone Pain

Hemicrania Continua

Definition

Hemicrania continua is a continuous (always present) but fluctuating unilateral headache, often accompanied by cranial autonomic symptoms, such as conjunctival injection, lacrimation, nasal congestion, rhinorrhea, ptosis, or eyelid edema. It is completely responsive to indomethacin.

Cross-References

Hemicrania Continua (HC)

Lawrence C. Newman
Albert Einstein College of Medicine,
New York, NY, USA

Synonyms

Cranial autonomic symptoms; SUNCT syndrome; Unilateral headache

Definition

An under-recognized, primary headache disorder that is characterized by a constant, one-sided headache with fluctuating intensity. In general, the headache is present as a persistent background discomfort of mild to moderate intensity, but exacerbations of more severe pain, superimposed upon the baseline pain, occur periodically. During these painful flare-ups, patients experience one or more symptoms on the side of the headache. These symptoms include drooping of the eyelid, reddening or tearing of the eye, constriction of the pupil, and stuffiness or dripping of the nostril. Recognition of the disorder is important, because the headache responds dramatically to treatment with the anti-inflammatory medication indomethacin.

Characteristics

Hemicrania continua (HC) is an under-recognized primary headache disorder. Initially, HC was believed to be a very rare disorder; however, in headache subspecialty practices, HC is a common cause of refractory, ▶ unilateral, chronic daily headache (Peres et al. 2001). Sjaastad and Spierings initially described the disorder in two patients with continuous headaches from onset (Sjaastad and Spierings 1984). Since that initial description, approximately 150 cases have been described in the literature.

Hemicrania continua demonstrates a marked female preponderance, with a female to male ratio of approximately 2:1. The condition most often begins during adulthood. The age of onset ranges from 5 to 67 years (mean 28 years) (Peres et al. 2001; Matharu et al. 2003).

Most sufferers describe strictly unilateral pain, without side-shift. Rarely, bilateral pain (Pasquier et al. 1987; Iordanidis and Sjaastad 1989; Trucco et al. 1992), or pain that alternated sides, has been described (Newman et al. 1992, 2004). The maximal pain is experienced in the eye, temple, and cheek regions. On occasion, the pain may radiate into the ▶ ipsilateral occiput, neck, and retro-orbital areas.

The pain is usually described as a steady ache or throbbing pain. Superimposed upon the continuous baseline low-level discomfort, the majority of patients report exacerbations of more intense pain lasting from 20 min to several days. Although significantly more intense than the usual background discomfort, the painful exacerbations never reach the level experienced by ▶ cluster headache sufferers. These exacerbations may occur at any time of the day or night and frequently awaken the patient from sleep. Migraine-like associated symptoms such as nausea, vomiting, ▶ photophobia, and ▶ phonophobia often accompany these exacerbations. Rarely, painful exacerbations may be preceded by a migrainous visual aura (Peres et al. 2002). ▶ Autonomic features of cluster headache, including ipsilateral ▶ ptosis, ▶ conjunctival injection, ▶ lacrimation, and nasal congestion, often accompany exacerbations of pain. When present, however, these associated features are usually much less pronounced than those seen in cluster headaches. Painful exacerbations are also associated with a sensation of ocular discomfort, often likened to a foreign body in the eye (typically reported as sand or hair). Concurrent ▶ primary stabbing headaches ("jabs and jolts") are reported by many patients, occasionally occurring only in association with the painful exacerbations. During exacerbations of pain, patients assume the pacing activity usually seen with cluster headaches. The International Headache Society (IHS) diagnostic criteria for HC are as follows:

Diagnostic Criteria
1. Headache for >3 months fulfilling criteria B-D
2. All of the following characteristics:
   (a) Unilateral pain without side-shift
   (b) Daily and continuous, without pain-free periods
   (c) Moderate intensity but with exacerbations of severe pain

1. At least one of the following autonomic features occurs during exacerbations and ipsilateral to the side of pain:
   (a) Conjunctival injection and/or ▶ lacrimation
   (b) Nasal congestion and/or ▶ rhinorrhea
   (c) Ptosis and/or miosis

1. Complete response to therapeutic doses of indomethacin
2. Not attributed to another disorder

Three temporal profiles of HC have been reported (Newman et al. 1994, Goadsby and Lipton 1997): a chronic form in which headaches persist unabated for years, an episodic form in which distinct headache phases are separated by periods of pain-free remissions, and an initially episodic form that over time evolves into the chronic, unremitting form. HC is chronic from onset in 53 %, chronic evolved from episodic in 35 %, and episodic in 12 % of sufferers (Matharu et al. 2003). There are also individual case reports of atypical presentations; one patient initially experienced the chronic form that over time became

episodic (Pareja 1995), and another patient with the episodic form experienced headaches with a clear seasonal pattern (Peres et al. 2001).

Organic mimics of HC have been reported to occur in association with brain tumors involving the bones of the skull and skull base (Matharu et al. 2003). HC has been reported to occur in a patient diagnosed with HIV, although a causal relationship was not definitively established (Brilla et al. 1998). Rarely, the diagnosis of HC is masked by a concurrent medication rebound headache. In these instances, discontinuation of the overused analgesic is not associated with headache cessation, and the diagnosis of HC is made by exclusion (Matharu et al. 2003). In rare instances, HC followed head trauma (Lay and Newman 1999).

Hemicrania continua is often misdiagnosed. Although it is not a true cluster headache variant, HC may be mistaken for cluster if the physician focuses on the painful flare-ups with associated autonomic features. A careful history should reveal the presence of the continuous, low-level baseline discomfort in addition to the more disabling exacerbations. Additionally, the autonomic features of HC, when present, tend to be much less pronounced than those of cluster. Similarly, the associated nausea, vomiting, photophobia, and phonophobia that accompany exacerbations of pain may be misdiagnosed as chronic migraine headaches. HC is distinguished from migraine by the presence of the persistent dull background discomfort.

Like all primary headache disorders, HC is diagnosed based on the patients' history and medical and neurological examinations. As it is a relatively uncommon headache disorder and because there have been serious disorders that mimic HC, all patients with features of HC should undergo an MRI scan of the brain prior to initiating therapy.

The treatment of HC is with the medication ▶ indomethacin. In fact, the diagnosis of HC is predicated on response to treatment with indomethacin. The initial dosage is 25 mg, three times daily. If clinical response is not seen within 1–2 weeks, the dosage should be increased to 50–75 mg, three times daily. Complete response to treatment with indomethacin is prompt, usually within 1–2 days of reaching the effective dose. The typical maintenance dose ranges from 25 to 100 mg, daily. Skipping or delaying the dose often results in headache recurrence. An intramuscular injection of indomethacin, 50–100 mg (the "indotest"), has been proposed as a diagnostic procedure for HC (Antonaci et al. 1998). Total resolution of the pain of HC was reported to occur within 2 h of the injection. Injectable indomethacin is not available in the United States.

Patients suffering with the episodic form should be instructed to continue the medication for 1–2 weeks longer than their typical headache phase and then gradually taper the dose. For those patients with the chronic form, medication tapering should be attempted every 6 months. Patients requiring long-term indomethacin therapy should be given medications such as antacids, misoprostol, histamine H_2 blockers, or proton pump inhibitors to mitigate the gastrointestinal side effects of this agent.

In patients who do not respond to treatment with adequate doses of indomethacin, another diagnosis should be considered. Other agents, which may have partial success in the treatment of HC, include naproxen and paracetamol, paracetamol in combination with caffeine, ibuprofen, piroxicam, and rofecoxib (Matharu et al. 2003). Six patients who met the clinical criteria for HC, yet failed to respond to treatment with indomethacin, have been reported (Matharu et al. 2003). Nonetheless, the IHS clinical criteria for HC specify that indomethacin responsiveness is necessary for the diagnosis.

References

Antonaci, F., Pareja, J. A., Caminero, A. B., et al. (1998). Chronic paroxysmal hemicrania and hemicrania continua: Parenteral indomethacin: The "indotest". *Headache, 38*, 122–128.

Brilla, R., Evers, S., Soros, P., et al. (1998). Hemicrania continua in an HIV-infected outpatient. *Cephalalgia, 18*, 287–288.

Goadsby, P. J., & Lipton, R. B. (1997). A review of paroxysmal hemicranias, SUNCT syndrome and other short-lasting headaches with autonomic feature, including new cases. *Brain, 120*, 193–209.

Iordanidis, T., & Sjaastad, O. (1989). Hemicrania continua: A case report. *Cephalalgia, 9*, 301–303.

Lay, C., & Newman, L. C. (1999). Posttraumatic hemicrania continua. *Headache, 39*, 275–279.

Matharu, M. S., Boes, C. J., & Goadsby, P. J. (2003). Management of trigeminal autonomic cephalgias and hemicrania continua. *Drugs, 63*, 1–42.

Newman, L. C., Lipton, R. B., Russell, M., et al. (1992). Hemicrania continua: Attacks may alternate sides. *Headache, 32*, 237–238.

Newman, L. C., Lipton, R. B., & Solomon, S. (1994). Hemicrania continua: Ten new cases and a review of the literature. *Neurology, 44*, 2111–2114.

Newman, L. C., Spears, R. C., & Lay, C. L. (2004). Hemicrania continua: A third case in which attacks alternate sides. *Headache, 44*, 821–823.

Pareja, J. A. (1995). Hemicrania continua: Remitting stage evolved from the chronic form. *Headache, 35*, 161–162.

Pasquier, F., Leys, D., & Petit, H. (1987). Hemicrania continua: The first bilateral case. *Cephalalgia, 7*, 169–170.

Peres, M. F. P., Silberstein, S. D., Nahmias, S., et al. (2001). Hemicrania continua is not that rare. *Neurology, 57*, 948–951.

Peres, M. F. P., Siow, H. C., & Rozen, T. D. (2002). Hemicrania continua with aura. *Cephalalgia, 22*, 246–248.

Sjaastad, S., & Spierings, E. L. (1984). Hemicrania continua: Another headache absolutely responsive to indomethacin. *Cephalalgia, 4*, 65–70.

Trucco, M., Antonaci, F., & Sandrini, G. (1992). Hemicrania continua: A case responsive to piroxicam-beta-cyclodextrin. *Headache, 32*, 39–40.

Hemicrania Simplex

▶ Migraine Without Aura

Hemipain

Definition

Hemipain is pain that is situated in one half of the body.

Cross-References

▶ Diagnosis and Assessment of Clinical Characteristics of Central Pain

Hemisphere

Definition

The hemisphere is either half of the cerebrum or brain; the human brain has a left and a right hemisphere.

Cross-References

▶ PET and fMRI Imaging in Parietal Cortex (SI, SII, Inferior Parietal Cortex BA40)

Hemorrhagic Stroke

▶ Headache Due to Intracranial Bleeding

Hereditary Motor and Sensory Neuropathy

Definition

Hereditary motor and sensory neuropathy is an alternative name for Charcot-Marie-Tooth disease.

Cross-References

▶ Hereditary Neuropathies

Hereditary Motor and Sensory Neuropathy (HMSN)

▶ Hereditary Neuropathies

Hereditary Neuropathies

Steven S. Scherer
Department of Neurology, The Perelman School of Medicine at the University of Pennsylvania, Philadelphia, PA, USA

Synonyms

Charcot-Marie-Tooth disease (CMT); Dejerine-Sottas Neuropathy (DSN); Familial Amyloid Polyneuropathy (FAP); Hereditary Motor and Sensory Neuropathy (HMSN); Hereditary Neuropathy with liability to Pressure Palsies (HNPP); Hereditary Sensory and Autonomic Neuropathy (HSAN); Hereditary Sensory Neuropathy (HSN)

Definition

Hereditary neuropathies are inherited diseases that injure axons in peripheral nerves.

Characteristics

The Biology of Myelinated Axons

Axons are the longest cells in the body and are vulnerable to defects in axonal transport; this is the basis for the long-standing doctrine that neuropathies are length dependent. Genetic evidence strongly supports this doctrine, as mutations in the genes that encode several axonal cytoskeleton and motor proteins cause axonal neuropathies. Neurofilaments are the main components of the axonal cytoskeleton, and are composed of three subunits, termed heavy, medium, and light. Dominant mutations in the gene encoding the light subunit (*NEFL*) cause an axonal neuropathy (CMT2E), and recessive mutations of *NEFL* also cause a neuropathy in which axons lack neurofilaments (Table 1). Microtubules are the tracks for axonal transport, and are comprised of a polymer comprised of

Hereditary Neuropathies, Table 1 Non-syndromic inherited neuropathies with a genetically identified cause

Disease (MIM)	Mutated gene/linkage	Clinical features
Autosomal or X-linked dominant demyelinating neuropathies		
HNPP (162500)	Usually deletion of one *PMP22* allele	Episodic mononeuropathies at typical sites of compression; also mild demyelinating neuropathy
CMT1A (118220)	Usually duplication of one *PMP22* allele	Onset 1st–2nd decade; weakness, atrophy, sensory loss; beginning in the feet and progressing proximally
CMT1B (118200)	*MPZ*	Similar to CMT1A; severity varies according to mutation (from "mild" to "severe" CTM1)
CMT1C (601098)	*LITAF/SIMPLE*	Similar to CMT1A; motor NCVs about 20 m/s
CMT1D (607687)	*EGR2*	Similar to CMT1A; severity varies according to mutation (from "mild" to "severe" CTM1)
CMT1X (302800)	*GJB1*	Similar to CMT1A, but distal atrophy more pronounced; men are more affected than are women
Dominant Intermediate CMT (DI-CMT; autosomal dominant)		
DI-CMTA (606483)	10q24.1-25.1	Onset of weakness in before 10 years, and median nerve motor conduction velocities range from 25 to 45 m/s
DI-CMTB (696482)	*DNM2*	Progressive, distal weakness and sensory loss; some families have associated neutropenia
DI-CMTC (608323)	*YARS*	Progressive distal weakness and sensory loss. Median motor responses range between 30 and 40 m/s
Autosomal dominant axonal neuropathies		
CMT2A (609260)	*MFN2*	Most have childhood onset with progression to severe proximal and distal weakness
CMT2B (600882)	*RAB7*	Onset 1st–3rd decade; severe sensory loss with distal ulcerations; also length-dependent weakness
CMT2C (606071)	*TRPV4*	Variable weakness, can be severe and proximal, with vocal cord and diaphragmatic weakness
CMT2D (601472)	*GARS*	Arm more than leg weakness; onset of weakness 2nd–3rd decade; motor axons more affected than sensory axons
CMT2E (162280)	*NEFL*	Variable onset and severity; ranging from DSS-like to CMT2 phenotype; pain sensation may be diminished
CMT2F (606595) allelic to HMN-II	*HSPB1*	Onset weakness 2nd–4th decade, followed by weakness in the distal muscles of arms
CMT2G (608591)		Late onset (average 30 years), progressing to mild, sensory, and motor neuropathy
CMT2I, CMT2J, CMT2-P₀ (118200)	*MPZ*	Late onset (30y or older); but progressive neuropathy; pain; hearing loss; abnormally reactive pupils
CMT2K (607831)	*GDAP1*	Onset ranges from infancy to childhood, and weakness and sensory loss worsen with time but typically milder than recessive *GDAP1* mutations
CMT2L (608673) allelic to HMN-II	*HSPB8*	Onset of weakness ranged from 15 to 33 years; distal legs are affected before the distal arms
CMT2N (613287)	*AARS*	Variable onset (10–54 years)
CMT2O (614228)	*DYNC1H1*	Childhood onset of motor > sensory axonal neuropathy
CMT2P (604484)		Adult onset proximal weakness with cramping and distal sensory loss
CFEOM3 (600638)	*TUBB3*	Patients with D417N and E410K can develop lower extremity weakness and sensory loss in the second to third decade, in the absence of congenital contractures

(continued)

Hereditary Neuropathies, Table 1 (continued)

Disease (MIM)	Mutated gene/linkage	Clinical features
Hereditary neuralgic amyotrophy (162100)	SEPT9	Attacks of severe pain in the neck, arm(s), and/or shoulder(s), followed by weakness and sensory loss in the distribution of the affected part of the brachial plexus
Hereditary thermosensitive neuropathy (602107)		Reversible episodes of ascending weakness, paresthesiae, and areflexia triggered by elevated body temperature
Severe demyelinating neuropathies (autosomal dominant or recessive; "CMT3 or HMSN III")		
Dejerine-Sottas neuropathy (145900)	dominant (PMP22; MPZ; GJB1; EGR2; NEFL) and recessive (MTMR2; PRX) mutations	Delayed motor development before 3y; severe weakness and atrophy; severe sensory loss particularly of modalities subserved by large myelinated axons; motor NCVs less than 10 m/s; dysmyelination on nerve biopsies
Congenital hypomyelinating neuropathy (605253)	dominant (EGR2; PMP22; MPZ) & recessive (EGR2) mutations	Clinical picture often similar to that of Dejerine-Sottas neuropathy but hypotonic at birth
Autosomal recessive demyelinating neuropathies ("CMT4")		
CMT4A (214400)	GDAP1	Early childhood onset; progressing to wheelchair-dependency; mixed demyelinating and axonal features
CMT4B1 (601382)	MTMR2	Early childhood onset; may progress to wheelchair-dependency; focally folded myelin sheaths
CMT4B2 (604563)	MTMR13	Childhood onset; progression to assistive devices for walking; focally folded myelin sheaths; glaucoma
CMT4C (601596)	SH3TC2/KIAA1985	Childhood onset typical; often progressing to wheelchair-dependency; scoliosis; severe to moderate NCV slowing
CMT4D (601455)	NDRG1	Childhood onset; progression to severe disability by 50 y; hearing loss and dysmorphic features
CMT4E (605253)	EGR2	Infantile onset; progressing to wheelchair-dependency
CMT4F (605260)	PRX	Childhood onset; usually progression to severe disability; prominent sensory loss
CMT4G (605285)	HK1	Also called HMSN Russe. Childhood onset of severe motor and sensory axonal neuropathy.
CMT4H (609311)	FDG4	Onset infancy to 1st decade, with severe distal weakness and sensory loss; slow progression
CMT4J (611228)	FIG4	Abrupt declines of strength; electrophysiologically a motor neuronopathy; minimal sensory involvement
Autosomal recessive axonal neuropathies ("AR-CMT2" or "CMT 2B")		
AR-CMT2A (605588)	LMNA	Onset of neuropathy in 2nd decade; progresses to severe weakness and atrophy in distal muscles
AR-CMT2B (605589)	MED25	Onset 28–42 y, weakness and atrophy in the distal muscles of the arms and legs, as well as distal sensory loss
AR-CMT2 (608340)	GDAP1	Onset typically by age 2, progressive neuropathy that results in severe proximal and distal weakness, and may cause inability to walk and vocal cord paresis
AR-CMT2 (no OMIM)	MFN2	The clinical onset is by age 3, and the neuropathy progresses to cause profound distal weakness, atrophy, and sensory loss
AR-CMT2 (no OMIM)	NEFL	Onset before age 2, with hypotonia and delayed motor milestones; progressive motor and sensory loss progress lead to severe distal and even proximal weakness

(continued)

Hereditary Neuropathies, Table 1 (continued)

Disease (MIM)	Mutated gene/linkage	Clinical features
AR-CMT2 (no OMIM)	*LRSAM1*	Early adult onset of weakness and sensory loss
Hereditary Sensory (and Autonomic) Neuropathies (HSN or HSAN)		
HSAN1A (162400)	Dominant *SPTLC1* mutations	Onset 2nd–3rd decade (often with phase of lacinating pain); severe sensory loss (including nociception) with distal ulcerations; also length-dependent weakness.
HSN1B (608088)	Dominant; 3p22-p24	Adult onset cough and sensory neuropathy; with sensory loss; painless injuries; and/or lacinating pains
HSN1C (616640)	Dominant *SPTLC2* mutations	Progressive distal sensory loss and distal muscle weakness in the lower limbs beginning in the 30s
HSN1D (613708)	Dominant *ATL1* mutations	Early adult onset of what becomes severe sensory > motor axonal neuropathy. Some patients also have a myelopathy.
HSN1E (126375)	Dominant *DNMT1* mutations	Sensorineural deafness and sensory neuropathy by the age of 20–35 years; cognitive and behavioral declines in fourth decade
HSN2A (201300)	Recessive *WNK1* mutations	Childhood onset of progressive loss of sensation, including pain, may culminate in ulcers, osteomyelitis, and amputation; no overt autonomic dysfunction
HSN2B (613115)	Recessive *FAM134B* mutations	Impaired sensation, mutilating ulcers, and arthropathy begins in childhood
HSN2C (614213)	Recessive *KIF1A* mutations	Clinical onset of an ulcero-mutilative neuropathy by the age of 10 years, leading to distal amputations.
HSN3 (Riley-Day syndrome; 223900)	Recessive *IKBKAP* mutations	Congenital onset; dysautonomic crises; decreased pain sensation; absent fungiform papilla; overflow tears
HSAN4 (CIPA; 256800)	Recessive *NTRKA* mutations	Congenital onset of dysautonomia, loss of pain sensation, no sweating; unheeded pain leads to development of Charcot joints; deceased sensation to multiple modalities
HSAN5 (608654)	Recessive *NGFB* mutations	Congenital onset of dysautonomia, loss of pain sensation, no sweating; unheeded pain leads to development of Charcot joints; deceased sensation to multiple modalities

The neuropathies are classified by OMIM (http://www.ncbi.nlm.nih.gov/Omim/); the references for the individual mutations are compiled in the CMT mutation database (http://molgen-www.uia.ac.be/CMTMutations/DataSource/MutByGene.cfm). **Bolded diseases** have pronounced affects on pain

α- and β-tubulin heterodimers. Dominant mutations in *TUBB3*, which encodes β3-tubulin, cause an axonal neuropathy, usually as part of a syndrome. Kinesins are the molecular motors that move cargo from the cell body to distal terminals. Recessive mutations in *KIF1A*, the gene encoding one of these kinesins, cause an axonal neuropathy (HSN2C). Conversely, mutations that affect components of the retrograde motor, a complex of dynactin and dynein, also cause neuropathies: A dominant mutation in *DYNC1H1*, which encodes a subunit of dynein, causes an axonal neuropathy (CMT2O), and a dominant mutation in the p150 subunit of dynactin causes a hereditary motor neuropathy (HMN7B). Defective axonal transport has been implicated in a host of other inherited neurological diseases, including the inherited spastic paraplegias, many of which can be regarded as length-dependent axonopathies of CNS neurons (Blackstone et al. 2011).

The structure and function of myelinating Schwann cells is the basis for understanding how mutations cause demyelinating neuropathies (Arroyo and Scherer 2000). The myelin sheath itself can be divided into two domains, compact

Hereditary Neuropathies, Fig. 1 The architecture of the myelinated axon in the PNS. In (**a**) one myelinating Schwann cell has been "unrolled" to reveal the regions forming compact myelin, as well as paranodes and incisures, regions of non-compact myelin. In (**b**) note that P_0, PMP22, and MBP and are found in compact myelin, whereas Cx32, MAG and E-cadherin are localized in non-compact myelin. Modified from (Kleopa and Scherer 2002), with permission of Elsevier Science

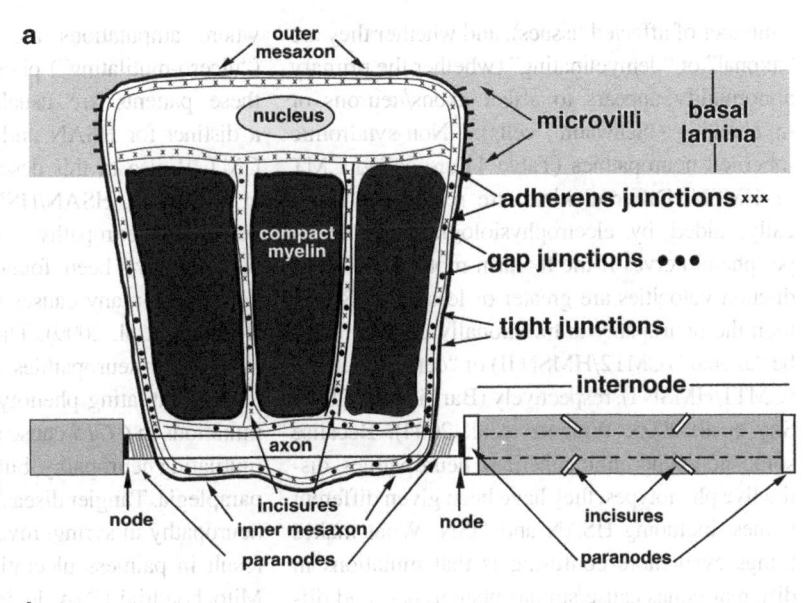

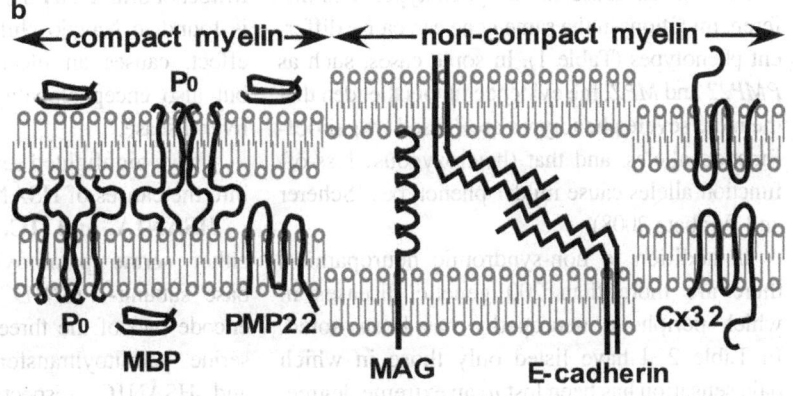

and non-compact myelin, each of which contains a nonoverlapping set of proteins (Fig. 1). Compact myelin forms the bulk of the myelin sheath. It is largely composed of lipids, mainly cholesterol and sphingolipids, including galactocerebroside and sulfatide, and three proteins – MPZ/P_0, PMP22, and myelin basic protein (MBP). Non-compact myelin is found in the paranodes and incisures, and contains tight junctions, gap junctions, and adherens junctions. In most cell types, these junctions join adjacent cells, whereas in Schwann cells, they are found between adjacent layers of non-compact myelin. Gap junctions formed by connexin32 may form a radial pathway, directly across the layers of the myelin sheath; this would be advantageous as it provides a much shorter pathway (up to 300-fold)

than a circumferential route. Homozygous mutations that cause loss-of-function in genes encoding proteins that play an essential function in myelinating Schwann cells would be expected to cause demyelination. Similarly, mutations of genes that are expressed by myelinating Schwann cells, that cause a toxic gain-of-function, would also be expected to cause demyelination (Scherer and Wrabetz 2008). Although Schwann cells are quite adept at remyelinating axons, if the remyelinating Schwann cells also express the causative mutation, then this does not result in long-lasting remyelination.

Classification of Inherited Neuropathies
Inherited neuropathies can be separated according to whether they are syndromic (i.e., one of

a number of affected tissues), and whether they are "axonal" or "demyelinating" (whether the primary abnormality appears to affect axons/neurons or myelinating Schwann cells). Non-syndromic inherited neuropathies (Table 1) are called CMT or HMSN. Different kinds are recognized clinically, aided by electrophysiological testing of peripheral nerves If the forearm motor nerve conduction velocities are greater or less than 38 m/s, then the neuropathy is traditionally considered to be "axonal" (CMT2/HMSN II) or "demyelinating" (CMT1/HMSN I), respectively (Barisic et al. 2008; Shy et al. 2005; Wrabetz et al. 2004). Because some non-syndromic inherited neuropathies distinctive phenotypes, they have been given different names, including HSAN and HSN. What makes things even more confusing is that mutations in different genes cause similar phenotypes, and different mutations in the same gene can cause different phenotypes (Table 1). In some cases, such as *PMP22* and *MPZ*, the evidence favors the idea that the more severe phenotypes are caused by gain-of-function alleles, and that (heterozygous) loss-of-function alleles cause milder phenotypes (Scherer and Wrabetz 2008).

In addition to non-syndromic neuropathies, there are more than 200 genetic diseases in which peripheral neuropathy has been noted. In Table 2, I have listed only those in which pain sensation has been lost to an extreme degree, and those with pronounced neuropathic pain.

Inherited Neuropathies Associated with Decreased Nociception

With the exception of some of the hereditary motor neuropathies, patients who have a hereditary neuropathy have diminished nociception. For hereditary axonal neuropathies – CMT2, AR-CMT2, HSAN, and HSN (Table 1) – this is expected, as these neuropathies affect unmyelinated axons, albeit to varying degrees. The diminished pinprick sensation that one finds in patients with inherited demyelinating neuropathies is probably due to the involvement of A-delta sensory axons, which are thinly myelinated. In some inherited neuropathies, however, the appreciation of pain can become so diminished that patients develop Charcot joints, ulcers, and infections, to the point

where amputations are required. This distinct ("ulcero-mutilating") phenotype is the reason that these patients are usually classified as having a distinct for HSAN and HSN, but patients with CMT2B also fit this description (Auer-Grumbach et al. 2006). HSAN/HSN is the rarest kind of hereditary neuropathy – only ~20 % of affected patients have been found to mutations in these genes – so many causes remain to be discovered (Rotthier et al. 2009). Three recessively inherited syndromic neuropathies are associated with an ulcero-mutilating phenotype (Table 2). Recessive mutations in *CCT5* cause a severe sensory, ulcero-mutilating neuropathy, but also a prominent spastic paraplegia. Tangier disease causes a severe sensory neuropathy in syringomyelia-like pattern, and can result in painless ulcerations and acromutilation. Mitochondrial DNA depletion syndrome 6, which is found in Navajo children owing to a founder effect, causes an ulcero-mutilating neuropathy, but also encephalopathy, myelopathy, and fatal liver disease.

I have highlighted several important advances into the causes of HSAN/HSN below.

HSAN1A and 1C. Missense mutations in either serine palmitoyltransferase, long chain base subunit-1 (*SPTLC1*) or *SPTLC2* – which encode two of the three subunits of the enzyme serine palmitoyltransferase – cause HSAN1A and HSAN1C, respectively. Serine palmitoyltransferase catalyzes the condensation of serine and palmitoyl-CoA, which is the first and rate-limiting step in the de novo synthesis of ceramide, a signaling molecule itself, and a precursor for complex sphingolipids. In addition to reducing enzymatic activity, the dominant *SPTLC1* and *SPTLC2* mutations associated with HSAN1 reduce the substrate specificity of serine palmitoyltransferase, allowing the incorporation of L-alanine and L-glycine, and subsequently the formation of two atypical deoxysphingolipids (Gable et al. 2010; Penno et al. 2010). Because they can neither be converted into complex sphingolipids nor degraded by the classical catabolic pathways, deoxysphingolipids accumulate in cells. Supplementation with L-serine reverses the accumulation of these abnormal lipids, making this a possible therapy for these diseases (Garofalo et al. 2011).

Hereditary Neuropathies, Table 2 Selected syndromic inherited neuropathies that prominently affect pain

Disease (MIM)	Mutated gene/ linkage	Clinical features
Recessively inherited, syndromic demyelinating neuropathies that prominently affect pain		
Metachromatic leuko-dystrophy (250100)	ARSA	Demyelinating neuropathy; optic atrophy; mental retardation; hypotonia; phase of neuropathic pain
Dominantly inherited, syndromic axonal neuropathies that prominently affect pain		
FAP 1 and 2 (176300)	TTR	Painful axonal neuropathy with prominent involvement of small axons; other organs involved; FAP 2 also has carpal tunnel syndrome
FAP 3/"Iowa" type (107680)	APOA1	Painful axonal neuropathy; renal and hepatic disease
Fabry disease (301500)	GLA	X-linked; painful neuropathy even painful crises; cardiomyopathy; renal failure; angiokeratoma
Acute intermittent porphyria (176000)	HMBS	Acute neuropathy follows crises of abdominal pain; psychosis; depression; dementia; seizures
Coproporphyria (121300)	CPOX	Skin photosensitivity; psychosis; crises of acute neuropathy (and abdominal pain) are rare
Variegate porphyria (176200)	PPOX	South Africa: founder effect; symptoms similar to those in acute intermittent porphyria
Erythropoietic proto-porphyria (177000)	FECH	Dermatitis; photosensitivity; liver disease; acute neuropathy rare
Recessively inherited; syndromic conditions that cause neuropathy and prominently affect pain		
Tangier disease (205400)	ABCA1	Atherosclerosis and/or peripheral neuropathy; syringomyelia-like loss of pain sensation can result in painless ulcerations and acromutilation
HSN with spastic paraplegia (256840)	CCT5	Infantile onset of spasticity, followed by severe sensory > motor axonal neuropathy
Hereditary tyrosinemia type 1 (276700)	FAH	Hepatic and renal disease; cardiomyopathy; crises of acute neuropathy and abdominal pain similar to those in porphyrias (but in infancy/childhood)
Mitochondrial DNA depletion syndrome 6 (256810)	MPV17	Affects Navajo infants/children; encephalopathy; myelopathy; neuropathy resulting in painless ulcerations and acromutilation; fatal liver disease
Recessively inherited conditions that prominently affect pain in which neuropathy is not documented		
Cold-induced sweating 1 (272430)	CRLF1	Poor sucking in infancy; cold-induced sweating; diminished pain caused by cold/hot/mechanical stimuli
Cold-induced sweating 2 (610313)	CLCF1	Poor sucking in infancy; unexplained fevers; cold-induced sweating; diminished pain caused by cold/hot/mechanical stimuli
Stüve-Wiedemann/Schwartz-Jampel type 2 syndrome (601559)	LIFR	Osteodysplasia with similar findings to HSN-3/familial dysautonomia: lack of corneal reflex, lack of fungiform papillae, tongue ulceration; also cold-induced sweating
Congenital indifference to pain (253000)	SCN9A	Congenital onset of inability to feel any kind of pain.
Dominantly inherited conditions that prominently affect pain in which neuropathy is not documented		
Familial episodic pain syndrome (no OMIM)	TRPA1	Debilitating upper body pain starting in infancy; skin biopsies and quantitative sensory testing is normal
Primary erythromelagia (133020)	SCN9A	Onset in childhood or adolescence of recurrent attacks of red, warm, and painful hands and/or feet
Paroxysmal extreme pain disorder (167400)	SCN9A	Neonatal or infantile onset of attacks of pain accompanied by autonomic signs, often triggered by defecation; skin biopsies show no consistent abnormalities

For references, see the following websites: http://www.ncbi.nlm.nih.gov/Omim/; http://molgen-www.uia.ac.be/CMTMutations/DataSource/MutByGene.cfm; and http://www.neuro.wustl.edu/neuromuscular/

HSAN2A and HSN2C. Recessive mutations in *WNK1* cause HSAN2. *WNK1* is a member of the family of With No Lysine Kinases, and regulates the function of ion channels and transporters in a variety of cell types, and presumably in neurons/axons, too. Although a global loss of *WNK1* function is probably lethal (Zambrowicz et al. 2003), the mutations that cause HSAN2A reside in exon 11, which is mainly expressed by PNS neurons (Shekarabi et al. 2008). The sequence that is encoded by exon 11 interacts with the portion of KIF1A (one on the main orthograde motors in axons) that is encoded by exon 25b, which is highly expressed in sensory neurons. Remarkably, a homozygous frameshift mutation in exon 25b causes an axonal neuropathy, providing strong support for the functional importance of this interaction (Riviere et al. 2011).

HSAN3/Familial Dysautonomia/Riley-Day Syndrome. Recessive mutations in *IKBKAP* cause HSAN3. Most patients are homozygous for a mutation in a donor splice site; this causes a cell type–specific reduction in the levels of IKBKAP protein (Cuajungco et al. 2003). IKBKAP is a component of the elongator complex, which has diverse cellular functions, including the acetylation of microtubules (which increases their stability) and neuronal maturation (Creppe et al. 2009).

HSAN4/CIPA syndrome and HSAN5. Recessive mutations in *NTRK1* and *NGFB* cause HSAN4 and HSAN5, respectively. *NGFB* and *NTRK1* encode nerve growth factor (NGF), and its receptor, TrkA, respectively. Based on the neurobiology of *Ngfb*- and *Ntrk1*-null mice, autonomic and small sensory neurons likely die in utero, so that HSAN4 and HSAN5 are likely to be congenital neuronopathies (Bibel and Barde 1999). HSAN4 is much more common than HSAN5, but has a similar phenotype (Carvalho et al. 2011; Indo 2001). As predicted from the knockout mouse model, myelinated axons (C-fibers) and thinly myelinated axons (A-delta fibers) that subserve nociception are globally and congenitally absent in HSAN4, whereas larger myelinated afferent axons (A-alpha and A-beta), which subserve several kinds of mechanoreceptors (associated with Merkel cells, hair follicles, Pacinian corpuscles, Meissner corpuscles, Golgi tendon organs,

and muscle spindles) appear to be spared. Biopsies of skin and sweat glands are largely devoid of axons (Nolano et al. 2000).

Inherited Neuropathies Associated with Prominent Neuropathic Pain
Neuropathic pain is seldom a prominent feature of CMT. It is typical of patients who have CMT2-P_0 (Gemignani et al. 2004), particularly the Thr124Met mutation. Several families have been found to have an adult onset neuropathy with painful lacinations and hearing loss. Nerve biopsies from clinically affected patients show axonal loss, clusters of regenerated axons, and some thinly myelinated axons. In spite of a late onset, many patients progress relatively rapidly to the point of using a wheelchair. Neuropathic pain, however, is not a prominent symptom in most patients with *MPZ* mutations; they have a demyelinating neuropathy (CMT1B or DSN).

Recurrent episodes of painful brachial plexus lesions are the hallmark of hereditary neuralgic amyotrophy (Windebank 1993). This is a dominantly inherited disorder caused by mutations in *SEPT9*. Individual episodes are similar to those in idiopathic neuralgic amyotrophy; both kinds are heralded by severe pain, followed by weakness within days, and recovery over weeks to months. Episodes may be triggered by immunization and childbirth, and perivascular inflammation and Wallerian degeneration are characteristic lesions (Klein et al. 2002). Subtle dismorphic features in affected patients with the inherited form indicate that this is a syndromic disorder.

Pain is a prominent feature of most of the syndromic neuropathies listed in Table 2. FAP types 1, 2, and 3 are caused by dominant mutations in transthyretin or apolipoprotein A1 that result in amyloid deposition in peripheral nerves and other tissues. Fabry disease is an X-linked disease caused by the lack of α-galactosidase, and is associated with a painful peripheral neuropathy and even painful crises. Several of the porphyrias, especially acute intermittent porphyria, as well as hereditary tyrosinemia, are associated with acute, painful, axonal neuropathies. Finally, abnormal pain is a feature of several conditions in which neuropathy has not been well documented – familial episodic pain syndrome, primary erythromelagia, paroxysmal extreme pain disorder, cold-induced sweating 1 and 2, and Stüve-Wiedemann/Schwartz-Jampel type 2 syndrome.

Summary

Almost all neuropathies result in diminished nociception; when this is profound, an ulcero-mutilating neuropathy is the result. The *Ntrka-* and *Ngfb*-null mice show how animal models can illuminate the clinical problem; other, genetically authentic animal models of these diseases are needed for further study. Finding the genes that are essential for nociception will potentially lead to novel targets for treating neuropathic pain, as exemplified by the search for compounds that selective block Nav1.7 (the Nav channel encoded by SCN9A). Except for the dominant mutations of *SCN9A* and *TRPA1*, both of which encode ion channels, how mutations result in enhanced neuropathic pain remains unexplained at a molecular level.

Acknowledgment I thank the NIH, the Muscular Dystrophy Association, and the National Multiple Sclerosis Society for supporting my work, and my colleagues for many informative discussions over the years.

References

Arroyo, E. J., & Scherer, S. S. (2000). On the molecular architecture of myelinated fibers. *Histochemistry and Cell Biology, 113*, 1–18.

Auer-Grumbach, M., Mauko, B., Surt-Grumbach, P., & Pieber, T. R. (2006). Molecular genetics of hereditary sensory neuropathies. *NeuroMolecular Medicine, 8*, 147–158.

Barisic, N., Claeys, K. G., Sirotkovic-Skerlev, M., Lofgren, A., Nelis, E., De Jonghe, P., & Timmerman, V. (2008). Charcot-Marie-Tooth disease: A clinicogenetic confrontation. *Annals of Human Genetics, 72*, 416–441.

Bibel, M., & Barde, Y.-A. (1999). Neurotrophins: Key regulators of cell fate and cell shape in the vertebrate nervous system. *Genes & Development, 14*, 2919–2937.

Blackstone, C., O'Kane, C. J., & Reid, E. (2011). Hereditary spastic paraplegias: Membrane traffic and the motor pathway. *Nature Reviews Neuroscience, 12*, 31–42.

Carvalho, O. P., Thornton, G. K., Hertecant, J., Houlden, H., Nicholas, A. K., Cox, J. J., Rielly, M., Al-Gazali, L., & Woods, C. G. (2011). A novel NGF mutation clarifies the molecular mechanism and extends the phenotypic spectrum of the HSAN5 neuropathy. *Journal of Medical Genetics, 48*, 131–135.

Creppe, C., Malinouskaya, L., Volvert, M.-L., Gillard, M., Close, P., Malaise, O., Laguesse, S., Cornez, I., Rahmouni, S., Ormenese, S., Belachew, S., Malgrange, B., Chapelle, J.-P., Siebenlist, U., Moonen, G., Chariot, A., & Nguyen, L. (2009). Elongator controls the migration and differentiation of cortical neurons through acetylation of α-tubulin. *Cell, 136*, 551–564.

Cuajungco, M. P., Leyne, M., Mull, J., Gill, S. P., Lu, W., Zagzag, D., Axelrod, F. B., Maayan, C., Gusella, J. F., & Slaugenhaupt, S. A. (2003). Tissue-specific reduction in splicing efficiency of IKBKAP due to the major mutation associated with familal dysautonomia. *American Journal of Human Genetics, 72*, 749–758.

Gable, K., Gupta, S. D., Han, G., Niranjanakumari, S., Harmon, J. M., & Dunn, T. M. (2010). A disease-causing mutation in the active site of serine palmitoyltransferase causes catalytic promiscuity. *Journal of Biological Chemistry, 285*, 22846–22852.

Garofalo, E., Penno, A., Schmidt, B. P., Lee, H. J., Frosch, M. P., von Echardstein, A., Brown, R. H., Hornemann, T., & Eichler, F. S. (2011). Oral L-serine supplementation in mice and humans with hereditary sensory autonomic neuropathy type 1. *J. Clin. Invest, 121*(12), 4735–4745.

Gemignani, F., Melli, G., Alfieri, S., Inglese, C., & Marbini, A. (2004). Sensory manifestations in Charcot-Marie-Tooth disease. *Journal of the Peripheral Nervous System, 9*, 7–14.

Indo, Y. (2001). Molecular basis of congenital insensitivity to pain with anhidrosis (CIPA): Mutations and polymorphisms in TRKA (NTRK1) gene encoding the receptor tyrosine kinase for nerve growth factor. *Human Mutation, 18*, 462–471.

Klein, C. J., Dyck, P. J. B., Friedenberg, S. M., Burns, T. M., Windebank, A. J., & Dyck, P. J. (2002). Inflammation and neuropathic attacks in hereditary brachial plexus neuropathy. *Journal of Neurology Neurosurgery and Psychiatry, 73*, 45–50.

Kleopa, K. A., & Scherer, S. S. (2002). Inherited neuropathies. *Neurologic Clinics of North America, 20*, 679–709.

Nolano, M., Crisci, C., Santoro, L., Barbieri, F., Casale, R., Kennedy, W. R., Wendelschafer-Crabb, G., Provitera, V., DiLorenzo, N., & Caruso, G. (2000). Absent innervation of skin and sweat glands in congenital insensitivity to pain with anhidrosis. *Clinical Neurophysiology, 111*, 1596–1601.

Penno, A., Reilly, M. M., Houlden, H., Laura, M., Rentsch, K., Niederkofler, V., Stoeckli, E. T., Nicholson, G., Eichler, F., Brown, R. H., von Eckardstein, A., & Hornemann, T. (2010). Hereditary sensory neuropathy type 1 is caused by the accumulation of two neurotoxic sphingolipids. *Journal of Biological Chemistry, 285*, 11178–11187.

Riviere, J.-B., Ramalingam, S., Lavastre, V., Shekarabi, M., Holbert, S., Lafontaine, J., Srour, M., Merner, N. D., Rochefort, D., Hince, P., Gaudet, R., Mes-Masson, A.-M., Baets, J., Houlden, H., Brais, B., Nicholson, G. A., Van Esch, H., Nafissi, S., De Jonge, P., Reilly, M. M., Timmerman, V., Dion, P., & Rouleau, G. A. (2011). *KIF1A*, an axonal transporter of synaptic vesicles, is mutated in hereditary sensory and autonomic neuropathy type 2. *American Journal of Human Genetics, 89*, 219–230.

Rotthier, A., Baets, J., De Vriendt, E., Jacobs, A., Auer-Grumbach, M., Levy, N., Bonello-Palot, N., Kilic, S. S., Weis, J., Nascimento, A., Swinkels, M., Kruyt, M. C., Jordanova, A., De Jonghe, P., & Timmerman, V. (2009). Genes for hereditary sensory and autonomic neuropathies: A genotype-phenotype correlation. *Brain, 132*, 2699–2711.

Scherer, S. S., & Wrabetz, L. (2008). Molecular mechanisms of inherited demyelinating neuropathies. *Glia, 56*, 1578–1589.

Shekarabi, M., Girard, N., Riviere, J.-B., Dion, P., Houle, M., Toulouse, A., Lafreniere, R. G., Vercauteren, F., Hince, P., Laganiere, J., Rochefort, D., Faivre, L., Samuels, M., & Rouleau, G. (2008). Mutations in the nervous system-specific HSN2 exon of WNK1 cause hereditary sensory neuropathy type II. *The Journal of Clinical Investigation, 118*, 2496–2505.

Shy, M. E., Lupski, J. R., Chance, P. F., Klein, C. J., & Dyck, P. J. (2005). Hereditary motor and sensory neuropathies: An overview of clinical, genetic, electrophysiologic, and pathologic features. In P. J. Dyck & P. K. Thomas (Eds.), *Peripheral neuropathy* (Vol. 2, pp. 1623–1658). Philadelphia: Saunders.

Windebank, T. (1993). Inherited recurrent focal neuropathies. In P. J. Dyck, P. K. Thomas, J. W. Griffin, P. A. Low, & J. F. Poduslo (Eds.), *Peripheral neuropathy* (pp. 1137–1148). Philadelphia: W.B. Saunders.

Wrabetz, L., Feltri, M. L., Kleopa, K. A., & Scherer, S. S. (2004). Inherited neuropathies: Clinical, genetic, and biological features. In R. A. Lazzarini (Ed.), *Myelin biology and disorders* (Vol. 2, pp. 905–951). San Diego: Elsevier Academic Press.

Zambrowicz, B. P., Abuin, A., Ramirez-Solis, R., Richter, L. J., Piggott, J., BeltrandelRio, H., Buxton, E. C., Edwards, J., Finch, R. A., Friddle, C. J., et al. (2003). Wnk1 kinase deficiency lowers blood pressure in mice: A gene-trap screen to identify potential targets for therapeutic intervention. *Proceedings of the National Academy of Sciences USA, 100*, 14109–14114.

Hereditary Neuropathy with Liability to Pressure Palsies (HNPP)

▶ Hereditary Neuropathies

Hereditary Sensory and Autonomic Neuropathy (HSAN)

▶ Hereditary Neuropathies

Hereditary Sensory and Autonomic Neuropathy Type IV, HSAN IV, HSAN 4

▶ Congenital Insensitivity to Pain with Anhidrosis

Hereditary Sensory Neuropathy

Definition

Hereditary sensory neuropathy is an inherited neuropathy that mainly affects sensory axons and/or sensory neurons.

Cross-References

▶ Hereditary Neuropathies

Hereditary Sensory Neuropathy (HSN)

▶ Hereditary Neuropathies

Heritable

Definition

A heritable trait is one which is passed on through generations (i.e., "runs in families"), such that offspring tend to resemble their parents. The strong implication is that inherited genetic factors are responsible, although non-genomic transmission has been demonstrated. Heritability is best established in humans using *twin studies*, in which the similarity of pairs of monozygotic and dizygotic twins is compared. In animals, heritability is best established by successful selective breeding for the trait.

Cross-References

▶ Opioid Analgesia, Strain Differences
▶ Twin Studies

Herpes Simplex Virus Vectors

Definition

Herpes simplex virus (HSV) is a human pathogen that causes the common cold sore and infections of the conjunctiva and is a double-stranded DNA virus with a capsid and surrounding tegument and envelope. The 152 kB genome can potentially accommodate up to approximately 44 kB of foreign DNA. The propagation of replication-incompetent HSV vectors in cultured cells is accomplished using cell lines that complement essential gene products that have been removed from the vector genome. HSV vector genomes do not integrate but remain as episomes in the nucleus of transduced cells. HSV vectors, like the parental virus, efficiently target to sensory neurons from the skin and can establish a lifelong latent state in those neurons.

Cross-References

▶ Opioids and Gene Therapy

Herpes Zoster

Definition

Herpes zoster is an infection of the nervous system caused by the varicella zoster virus (VZV), the same virus that causes chickenpox. VZV can remain dormant in sensory ganglia for decades after an infection. Herpes zoster results when the dormant virus in these nerve cell bodies is reactivated, often as a result of decline in cellular immunity to VZV with aging or immunosuppression.

Cross-References

▶ Cancer Pain, Assessment in the Cognitively Impaired
▶ Postherpetic Neuralgia
▶ Postherpetic Neuralgia, Pharmacological and Nonpharmacological Treatment Options

---

Herpes Zoster Pain

▶ Postherpetic Neuralgia, Etiology, Pathogenesis, and Management

---

Heterocyclic Antidepressants

▶ Antidepressant Analgesics in Pain Management

Heteromeric Channels

Definition

Heteromeric channels are protein complexes that form pores in the cell membrane. Typically, channels are made up of several subunits, which may be identical, resulting in homomeric channels, or different, resulting in heteromeric channels.

Cross-References

▶ GABAA Receptors
▶ Nicotinic Receptors
▶ Purine Receptor Targets in the Treatment of Neuropathic Pain

Heterotopic Ossification

Definition

Heterotopic ossification is the appearance of bony tissue elements in what are normally soft tissue structures.

Cross-References

▶ Spinal Cord Injury Pain

Heterozygosity

Definition

Heterozygosity is a state in which the maternal and paternal alleles of a gene are not the same. In this common situation, the expression of that gene will depend on dominance of the alleles. In inbred strains, all heterozygosity is lost, and every gene is fixed in a homozygous state.

Cross-References

▶ Alleles
▶ Inbred Strains
▶ Opioid Analgesia, Strain Differences

Heterozygous Carriers

Definition

Heterozygous carriers refer to the state of possessing two different alleles of a particular gene, one inherited from each parent.

Cross-References

▶ NSAIDs, Pharmacogenetics

Hidden Triggers

Definition

Hidden triggers are internal precipitating mechanisms.

Cross-References

▶ SUNCT Syndrome

High Dependency or Intensive Care Units

Definition

High dependency or intensive care units are specialized wards where one or two highly trained nurses take care of each patient.

Cross-References

▶ Postoperative Pain, Acute Pain Team

High Thoracic Epidural Anesthesia

Definition

High thoracic epidural anesthesia leads to a reversible cardiac sympathectomy blocking the segments T1–T5. The epidural catheters are inserted at levels C7–T1 or at levels T1–T2 by the median approach and with hanging drop technique.

Cross-References

▶ Postoperative Pain, Thoracic and Cardiac Surgery

High Threshold Neurons

Synonyms

HT neurons

Definition

HT neurons respond fairly selectively to noxious mechanical stimuli. They may have a minimal response (a few action potentials) to innocuous mechanical stimuli, but they are essentially tuned for strong stimuli. They may also respond to noxious thermal and chemical stimuli. Sometimes HT cells are referred to as nociceptive-specific neurons.

Cross-References

▶ Functional Changes in Sensory Neurons Following Spinal Cord Injury in Central Pain
▶ Nick Model of Cutaneous Pain and Hyperalgesia
▶ Spinothalamic Input: Cells of Origin (Monkey)
▶ Thalamus, Nociceptive Cells in VPI, Cat and Rat

High-threshold Calcium Channels

▶ Calcium Channels in the Spinal Processing of Nociceptive Input

High-Threshold Mechanoreceptor

▶ Nociceptors in the Orofacial Region (Skin/Mucosa)

High-Threshold Mechanoreceptors

▶ Mechanonociceptors

High-Threshold Mechanosensitive Muscle Receptors

Definition

In experiments employing recordings from single muscle receptors with unmyelinated or thin myelinated afferent fibers, many units exhibit a high mechanical threshold when tested with local pressure stimuli (e.g., using a forceps with broadened tips on the exposed muscle). These receptors do not respond to passively stretching the muscle or aerobic active contractions, but require pressure stimulation of tissue-threatening and subjectively painful intensity for activation. The receptors are also typically responsive to stimulation with algesic substances (e.g., bradykinin). The general interpretation is that these receptors are nociceptors and induce muscle pain when activated.

Cross-References

▶ Muscle Pain Model, Ischemia-Induced and Hypertonic Saline-Induced

High-Threshold Neurons

▶ Thalamus, Nociceptive Cells in VPI, Cat and Rat

High-threshold VDCCs

▶ Calcium Channels in the Spinal Processing of Nociceptive Input

High-Velocity Thrust Manipulation (HVTM)

▶ Spinal Manipulation, Pain Management

High-voltage Calcium Channels

▶ Calcium Channels in the Spinal Processing of Nociceptive Input

Hindlimb Flexor Reflex

▶ Opioids and Reflexes

Hippocampal Formation or Hippocampal Region

Definition

The hippocampus (dentate gyrus and pyramidal cell fields CA1–3) and the subiculum are together referred to as the hippocampal formations. Perhaps the most extensively studied structure in the brain, the hippocampal region has most often been implicated in memory processing.

Cross-References

▶ Hippocampus
▶ Hippocampus and Entorhinal Complex: Functional Imaging

Hippocampus

Definition

Brain structure comprising the dentate gyrus and the pyramidal cell fields of the hippocampus. There are three different pyramidal cell fields: CA1, CA2, and CA3. These subregions differ in their cellular organization and connectivity. The hippocampus is primarily organized as a unidirectional circuit. Information from the entorhinal cortex converges on the dentate gyrus, which in turn projects to field CA3, which sends projections to field CA1.

The circuit is completed as CA1 projects to the subiculum, the major output region of the hippocampus. Strictly speaking, the subiculum does not form part of the hippocampus, but together the two structures make up the hippocampal formation.

Cross-References

▶ Hippocampus and Entorhinal Complex: Functional Imaging

Hippocampus and Entorhinal Complex: Functional Imaging

Irene Tracey[1] and Siri Graff Leknes[2]
[1]Pain Imaging Neuroscience (PaIN) Group, Department of Physiology, Anatomy and Genetics, Oxford University, Oxford, UK
[2]Research Fellow, Psykologisk institutt, Oslo, Norway

Synonyms

Entorhinal cortex and hippocampus; Functional imaging; Neuroimaging; Parahippocampal region

Definition

The ▶ hippocampus is comprised of the dentate gyrus and the CA1, CA2, and CA3 pyramidal cell fields. The ▶ hippocampal formation consists of the hippocampus and the subiculum. The adjacent entorhinal, perirhinal, and parahippocampal cortices comprise the ▶ parahippocampal region (Fig. 1). These limbic subregions differ in their cellular organization and connectivity but are commonly implicated in memory and emotion processing.

The hippocampus lies at the end of a cortical processing hierarchy, and the entorhinal cortex is the major source of its cortical projections. Much of the cortical input to the entorhinal cortex originates in the adjacent perirhinal and

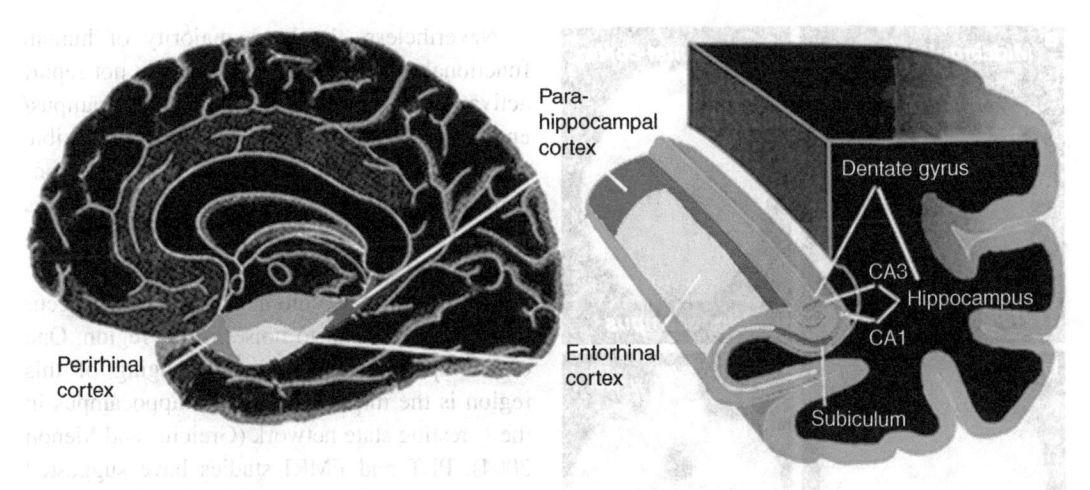

Hippocampus and Entorhinal Complex: Functional Imaging, Fig. 1 (*Left*) Medial view of the human brain outlining the perirhinal cortex (*orange*); parahippocampal cortex (*red*); and entorhinal cortex (*yellow*). (*Right*) Section of the temporal lobe showing the components of the hippocampal/entorhinal complex in some detail: the

dentate gyrus (*pale green*); the CA1 and CA3 hippocampal fields (*green*) that make up the hippocampus proper; the subiculum (*pink*); the perirhinal cortex (*orange*); parahippocampal cortex (*red*); and entorhinal cortex (*yellow*)

parahippocampal cortices, which in turn receive widespread projections from sensory and association areas in the frontal, temporal, and parietal lobes (Squire et al. 2004).

▶ Functional imaging is a general term used to describe methodologies that allow function to be located either spatially or temporally within the brain (and other organs). The methods are generally noninvasive and used for human studies; the term neuroimaging is often used when applied specifically to brain studies. Methods include functional magnetic resonance imaging (FMRI), positron-emission tomography (PET), magnetoencephalography (MEG), and electroencephalography (EEG). Unless otherwise stated, the studies discussed in this article are FMRI or PET studies of the brain.

Characteristics

Melzack and Casey (1968) proposed that the hippocampus and associated cortices participate in mediating the aversive drives and affective characteristics of pain perception. A wide range of animal studies support the notion that pain processing is a primary function of the hippocampal complex.

Importantly, Dutar and colleagues (1985) demonstrated that septo-hippocampal neurons in rats respond directly to noxious peripheral stimulation. Similarly, functional imaging studies of pain perception have repeatedly reported a direct implication of areas within the hippocampus in the processing of nociceptive stimuli. Since nociceptive information is typically novel and of high priority, a direct role for the hippocampus in nociceptive processing is consistent with comparator theories of hippocampal function (e.g., McNaughton and Gray 2000). Comparator theory maintains that the hippocampus is involved in novelty detection and that its function is to compare actual and expected stimuli (i.e., stimuli registered in memory).

In an early PET study, Derbyshire and colleagues (1997) found hippocampal activation in response to mildly and moderately painful heat stimuli, when contrasted with warm, non-painful stimulation. Using very specific nociceptive stimuli (laser stimulation of A-δ fibers only) to subjects' left and right hands, Bingel and colleagues (2002) found bilateral activation of the amygdala and hippocampal complex. The receptive fields of hippocampal neurons are predominantly large and bilateral (Dutar et al. 1985). As other pain-related

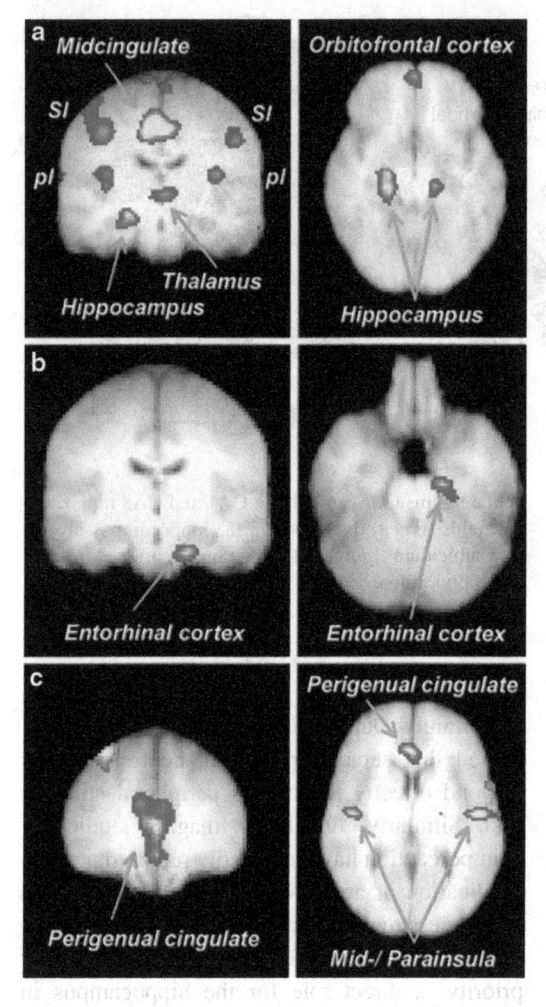

Hippocampus and Entorhinal Complex: Functional Imaging, Fig. 2 (a) Temperature-related activation increases in perceived pain: Bilateral S1, dorsal margin of posterior insula, thalamus, midcingulate and right hippocampus. (b) Anxiety-related activation increases in perceived pain associated with significant activation in left entorhinal cortex. (c) Activity in the perigenual cingulated and the mid-/para insula was significantly correlated with entorhinal FMRI signal during pain modulation by anxiety (Reproduced with permission from Ploghaus et al. (2001))

Nevertheless, the large majority of human functional imaging studies of pain do not report activation of regions within the hippocampus/ entorhinal complex. There are several possible explanations for this discrepancy. The first concerns the signal to noise ratio. As the complex is a relatively small structure, the spatial resolution of conventional whole-brain imaging paradigms means that partial volume effects might occur and decrease signal to noise in this region. One caveat specific to functional imaging of this region is the implication of the hippocampus in the ▶ resting state network (Greicius and Menon 2004). PET and FMRI studies have suggested that the resting brain has a default mode of internal processing in which the hippocampus is a central component. In the neuroimaging of pain perception, nociceptive processing is commonly compared with baseline (rest) conditions. The hippocampus' involvement in resting state/ baseline processing may mask out the activation of this region in such task-baseline comparisons if increased baseline activity reduces subsequent stimulation evoked responses and therefore could yield a false-negative result. Another factor that may mask out activation of regions within the medial temporal lobe is the registration of individual brains onto a standard template for group comparison. Traditional techniques that optimize whole-brain alignment (e.g., aligning to the atlas of Talairach & Tournoux) do not adequately account for variations in location and shape of medial temporal lobe structures (see Squire et al. 2004 for review).

Regions within the hippocampus/parahippocampal complex have been more consistently activated in studies where pain perception has been modulated by expectation and/or anxiety. It is clear that memory (which influences expectation) modulates pain perception. While certain expectation is associated with fear, uncertain expectation is associated with anxiety. For instance, a rat experiences fear when it must enter a space where a cat is present. Anxiety, on the other hand, corresponds to the state a rat is in when it must enter a space where a cat may or may not be present. While fear facilitates rapid reactions (fight or flight) and causes distraction

activation was lateralized, the authors suggested that the hippocampal activity reflected direct nociceptive projections to the hippocampus, perhaps revealing novelty detection (Bingel et al. 2002). Further, Ploghaus and colleagues (2001) found that pain modulation by drying stimulus temperature caused activation of a region of the hippocampus proper, consistent with a role of the hippocampus in pain intensity encoding (Fig. 2a).

and analgesia from the pain, anxiety is characterized by risk assessment behavior or behavioral inhibition (the rat hesitates to enter the space where a cat might be). This behavior is associated with increased somatic and environmental attention, which can lead to anxiety-driven hyperalgesia (McNaughton and Gray 2000).

Using FMRI to investigate the effects of expectation on pain perception, Ploghaus and colleagues (2000) found that areas in the hippocampal complex were activated during mismatches between expected and actual pain. Consistent with comparator theory, the same ► hippocampal regions were implicated in three different types of mismatch: when no pain was expected (novelty), when the nociceptive stimulus differed from expectation, and when the painful stimulus was unexpectedly omitted. In a subsequent study, Ploghaus and colleagues (2001) manipulated the certainty of expectation about impending nociceptive stimulation, to investigate its modulation on pain perception. This study examined the neural mechanism by which anxiety (uncertain expectation) causes increased pain perception (hyperalgesia) and contrasted it with the process by which a heightened nociceptive stimulation causes increased pain perception. The Gray-McNaughton theory proposes that the hippocampal formation responds to aversive events such as pain whenever they form part of a behavioral conflict, e.g., a conflict caused by uncertain expectation of pain. This conflict induces anxiety. Output from the comparator has two effects that underpin anxiety and behavioral inhibition. First, it tends to suppress both of the currently conflicting responses. Second, it increases the valence of the affectively negative associations of each of the conflicting goals (McNaughton and Gray 2000).

As predicted from theory, Ploghaus and colleagues reported activation of the entorhinal cortex during anxiety-driven hyperalgesia but not during increased pain perception caused by augmented nociceptive input (Fig. 2b; Ploghaus et al. 2001). Studies of other (not anxiety-related) types of hyperalgesia typically report no significant activation of the hippocampus/parahippocampal region (e.g., Zambreanu et al. 2005). One exception

is a recent FMRI study of drug modulation during pain (Borras et al. 2004). Naloxone, a predominantly μ opioid antagonist, was administered to naïve subjects in low doses. During rest (baseline) conditions where no stimulation was applied, regions in the hippocampal/entorhinal complex were activated more in the drug condition than during placebo. According to the Gray-McNaughton theory, the entorhinal cortex primes responses that are adaptive to an aversive input, such as the motor response necessary for escape from a threatening environment. Enhanced activation in this region after naloxone infusion indicates a change in basal activity, potentially lowering the threshold for activation of adaptive responses.

In line with this argument, differences between naloxone and placebo conditions during nociceptive processing were found in several areas within the hippocampus/parahippocampal region. When pain ratings were matched across conditions, an area within the posterior parahippocampal gyrus was significantly more activated in the naloxone condition. Activation of the hippocampus proper to nociceptive stimulation in the drug condition compared to the placebo condition was found only when subjects rated the pain intensity higher in the naloxone condition (nociceptive stimuli were of equal intensity across conditions). This result adds further support for the role of the hippocampus proper in pain intensity encoding. In their study of anxiety-driven hyperalgesia, Ploghaus and colleagues (2001) found that the entorhinal cortex activation was predictive of activity in the perigenual cingulate and mid-insula (Fig. 2c). Corresponding regions of the cingulate and insular cortices were also implicated in naloxone-induced increases in pain perception (Borras et al. 2004). The authors concluded that the regions where activation by noxious heat was modulated by naloxone were the sites of action of endogenous opioid pathways involved in regulating the central nervous system response to aversive stimuli.

Some support for the involvement of the hippocampus/parahippocampal region in opioid regulation of the brain's response to nociceptive

Hippocampus and Entorhinal Complex: Functional Imaging, Fig. 3 Visualization of slice positioning in a high-resolution, partial-coverage study of hippocampus/entorhinal complex function. By only covering a section of the brain, resolution can be improved significantly, and it may be possible to begin disentangling the function of small subregions within the hippocampus/parahippocampal complex for nociceptive processing and pain perception

input comes from functional imaging studies of acupuncture. Several studies investigating brain responses to acupuncture in healthy, pain-free volunteers have reported ▶ deactivation of regions within the hippocampus/entorhinal complex (e.g., Napadow et al. 2005). A recent study examining the effects of acupuncture in chronic pain patients does not report involvement of the hippocampus or parahippocampal areas (Pariente et al. 2005), but this study did not include a contrast for deactivation of specific brain regions.

There can be little doubt that the role of the hippocampus/entorhinal complex in nociceptive processing and the generation of pain perception demands further investigation in both healthy volunteers and in clinical pain patients. So far, the functional imaging studies of pain reporting hippocampus/entorhinal complex activation have been whole-brain studies examining the effects of nociceptive stimulation on all regions of the brain. This contrasts with the neuroimaging

literature on the role of the hippocampal complex in memory, where researchers have been able to focus solely on this narrow region of cortex, improving spatial resolution and avoiding registration caveats, e.g., by employing partial-coverage imaging techniques (Fig. 3) (see also Squire et al. 2004). To disentangle, the roles of the subregions within the hippocampus/entorhinal complex in nociceptive processing and pain perception, high-resolution studies of this region during pain, employing similar measures, are needed. Care must also be taken to optimize study design in order to avoid the masking out of nociceptive-related hippocampal activations by processing of the resting state network.

The role of hippocampus/entorhinal complex in clinical pain is still largely unknown. A study of patients suffering from irritable bowel syndrome (IBS) has recently shown involvement of hippocampus in pain processing in patients compared to healthy controls (Wilder-Smith et al. 2004). Given the known involvement of anxiety in irritable bowel syndrome, this result lends further support to the postulated involvement of the hippocampus/entorhinal complex in anxiety-driven increases of pain perception. Further, the hippocampus may form part of a system of central involvement that drives the visceral hypersensitivity of these patients. More studies of anxiety and hippocampus/entorhinal complex function in clinical pain should shed light on the importance of centrally generated pain and hyperalgesia.

In conclusion, converging evidence from human neuroimaging and animal studies points to a direct role for the hippocampus in the processing of nociceptive information such as pain intensity encoding. Areas within the hippocampus/entorhinal complex are involved in the comparison between actual and expected nociceptive stimuli and play a role in anxiety-driven hyperalgesia. The increases in pain perception caused by uncertain expectation may be due to a modulation of the opiate system, as hinted at by a study investigating the effects of the μ opioid antagonist naloxone (Borras et al. 2004) (Table 1).

Hippocampus and Entorhinal Complex: Functional Imaging, Table 1 Summary of functional imaging studies outlined here, listing stimulus type, neuroimaging technique, and activations/deactivations in hippocampal/parahippocampal regions

Authors	Stimulus type		Hippocampus proper	Parahippocampal region
Derbyshire et al. (1997)	Laser (heat nociception or warm)	PET	Nociceptive encoding	–
Ploghaus et al. (2000)	Thermal (heat nociception or warm)	FMRI	Expectation related	Expectation related
Ploghaus et al. (2001)	Thermal (heat nociception)	FMRI	Nociceptive encoding	Expectation related
Bingel et al. (2002)	A-δ-specific laser	FMRI	Nociceptive encoding	–
Wilder-Smith et al. (2004)	Rectal balloon distension and thermal (cold nociception)	FMRI	Patients more than controls	Patients more than controls
Borras et al. (2004)	Thermal (heat nociception)	FMRI	Nociceptive encoding	May be related to shift in threshold for adaptive response
Greicius and Menon (2004)	Visual (resting state examined)	FMRI	Resting state network	Resting state network
Napadow et al. (2005)	Acupuncture in pain-free controls	FMRI	Deactivation	–

References

Bingel, U., Quante, M., Knab, R., et al. (2002). Subcortical structures involved in pain processing: Evidence from single-trial fMRI. *Pain, 99*, 313–321.

Borras, M. C., Becerra, L., Ploghaus, A., et al. (2004). fMRI measurement of CNS responses to naloxone infusion and subsequent mild noxious thermal stimuli in healthy volunteers. *Journal of Neurophysiology, 91*, 2723–2733.

Derbyshire, S. W. G., Jones, A. K. P., Gyulai, F., et al. (1997). Pain processing during three levels of noxious stimulation produces differential patterns of central activity. *Pain, 73*, 431–445.

Dutar, P., Lamour, Y., & Jobert, A. (1985). Activation of identified septo-hippocampal neurons by noxious peripheral stimulation. *Brain Research, 328*, 15–21.

Greicius, M. D., & Menon, V. (2004). Default-mode activity during a passive sensory task: Uncoupled from deactivation but impacting activation. *Journal of Cognitive Neuroscience, 16*, 1484–1492.

McNaughton, N., & Gray, J. A. (2000). Anxiolytic action on the behavioural inhibition system implies multiple types of arousal contribute to anxiety. *Journal of Affective Disorders, 61*, 161–176.

Melzack, R., & Casey, K. L. (1968). Sensory, motivational, and central control determinants of pain. In D. R. Kenshalo (Ed.), *The skin senses* (pp. 423–439). Springfield: Thomas.

Napadow, V., Makris, N., Liu, J., et al. (2005). Effects of electroacupuncture versus manual acupuncture on the human brain as measured by fMRI. *Human Brain Mapping, 24*, 193–205.

Pariente, J., White, P., Frackowiak, R. S. J., et al. (2005). Expectancy and belief modulate the neuronal substrates of pain treated by acupuncture. *NeuroImage, 25*, 1161–1167.

Ploghaus, A., Tracey, I., Clare, S., et al. (2000). Learning about pain: The neural substrate of the prediction error for aversive events. *PNAS, 97*, 9281–9286.

Ploghaus, A., Narain, C., Beckmann, C. F., et al. (2001). Exacerbation of pain by anxiety is associated with activity in a hippocampal network. *Journal of Neuroscience, 21*, 9896–9903.

Squire, L. R., Stark, C. E. L., & Clark, R. E. (2004). The medial temporal lobe. *Annual Review of Neuroscience, 27*, 279–306.

Wilder-Smith, C. H., Schindler, D., Lovblad, K., et al. (2004). Brain functional magnetic resonance imaging of rectal pain and activation of endogenous inhibitory mechanisms in irritable bowel syndrome patient subgroups and healthy controls. *Gut, 53*, 1595–1601.

Zambreanu, L., Wise, R. G., Brooks, J. C. W., et al. (2005). A role for the brainstem in central sensitisation in humans. Evidence from functional magnetic resonance imaging. *Pain, 114*, 397–407.

Histamine

Definition

Histamine (2-(4-imidazolyl)-ethylamine), a biogenic amine, is synthesized by decarboxylation of the amino acid histidine mostly in mast cells. It is a hydrophilic vasoactive amine, a potent vasodilator, and also a neuromodulator in the central nervous system. As a strongly vasodilating substance, histamine plays a key role in many allergic reactions. In the superficial layers of the skin, histamine released from mast cells is strongly pruritogenic, by binding to H1 receptors in the terminal membranes of a special subgroup of C-afferent fibers.

Cross-References

▶ Nociceptor, Categorization

Histology of Nociceptors

Frank L. Rice[1], Lawrence Kruger[2] and Phillip J. Albrecht[1]
[1]Center for Neuropharmacology and Neuroscience, Albany Medical College, Albany, NY, USA
[2]UCLA Geffen School of Medicine, Los Angeles, CA, USA

Synonyms

Chemo-nociceptors; Mechano-nociceptors; Polymodal nociceptors; Silent nociceptors

Characteristics

Extremes of stress and damage to most tissues of the body elicit a very salient unpleasant or aversive sensation, broadly termed "pain." Pain serves an essential survival component to all species, and the loss of pain perception often results in dysfunction and untimely death of the organism. Sensory neurons located among the dorsal root and cranial nerve ganglia with thin, slowly conducting axons innervate all tissues from which a sensory perception of pain can be elicited, and the generation of pain sensations

has been attributed to activation of these "small-caliber" primary sensory neuron terminals, defined as ▶ nociceptors (Latin "nocere" meaning "to injure"). Axons supplying nociceptors are either thin, unmyelinated C axons or slightly thicker, lightly myelinated Aδ axons; the latter thought to account for the initial "fast, pricking pain" associated with noxious insults, while C axons account for the subsequent "slow, burning pain" following acute injury and particularly inflammation (Albrecht and Rice 2010; Rice and Albrecht 2008).

Electrophysiologic recording techniques have defined subsets of small-caliber axons that preferentially respond to specific modalities of painful stimuli such as noxious substances (*chemo-nociceptors*), extreme temperatures (*thermo-nociceptors*), intense mechanical stimuli (*mechano-nociceptors*), as well as various modality combinations (*polymodal nociceptors*). Most small-caliber axons are generally unresponsive to the broad variety of test stimuli used to elicit electrophysiological responses, but often, these same neurons can become responsive to noxious or even non-noxious stimuli under experimentally induced, prolonged pain conditions such as inflammation and nerve injury and, thus, have been inferred as *silent nociceptors*. However, it is more likely that the molecular signals involved in driving these small-caliber axons are more subtle and complex than the rather simplistic combinations of standard stimulus paradigms (Albrecht and Rice 2010; Rice and Albrecht 2008).

Most tissues, including skin, receive a rich variety of sensory innervation which includes large-caliber, myelinated Aβ axons that supply low-threshold mechanoreceptors, as well as small-caliber C and Aδ axons. Therefore, the identification of specific nociceptors was focused on tissues supplied principally by C and Aδ fibers where nearly any stimulation readily elicits a pain perception or withdrawal reflexes. Thus, most nociceptor histology was first elucidated in epithelial tissues such as cornea, inner tympanic membrane, tooth pulp, or highly vascularized connective tissues such as pericardium, testes, and joints (Kruger et al. 1981; Yeh and Kruger 1984). The nociceptive endings in such sites consistently lacked the distinctive unique morphologies and encapsulations associated with low-threshold mechanoreceptive endings supplied by Aβ fibers (Kruger et al. 1981; Kruger 1988). These non-corpuscular nociceptive endings have been referred to as "free nerve endings" (FNE), and the electrophysiologically characterized and accurately mapped testicular nociceptive "spots" enabled fine-structural characterization based on serial reconstruction (Kruger et al. 2003a).

With the initial development of antibodies selective for neuropeptides, most of C and Aδ fibers first identified by immunohistochemistry contained substance P (SP) and calcitonin gene-related peptide (CGRP) (i.e., peptidergic). The endings of these peptidergic fibers in various tissues were more numerous and marked than previously known terminations of thin-caliber axons best seen at the time with reduced silver stains (Kruger et al. 1989). Importantly, many of these newly revealed endings were located in the epidermis, which was thought to lack innervation, and around blood vessels that were only thought to have sympathetic innervation. In dorsal root and cranial nerve ganglia, SP and CGRP were detected in many small neurons generally accepted as the source of the thin-caliber nociceptor axons. However, a large proportion of small-caliber neurons did not express peptides but were revealed by fluoride-resistant acid phosphatase (FRAP) histochemistry, which only labeled the cell bodies and, thus, did not reveal their axons (Zylka et al. 2005). Additionally, most of these non-peptidergic neurons and their axons, as well as some containing peptides, were found to bind the B_4 fragment of plant isolectins in rodents (Silverman and Kruger 1988).

Perhaps the densest thin-fiber peptidergic innervation accessible to external stimuli is in the dentinal tubules underlying the grinding surface of molar teeth (Silverman and Kruger 1987). This led to the concept of a "noceffector" system that would not subserve sensory detection but rather efferent maintenance of peripheral tissues. While damage inflicted to a tooth can result in severe pain, such a dense innervation of relatively protected dentinal tubules far exceeds that of the exposed superficial cornea from which pain can be elicited with stimuli that are innocuous elsewhere.

As such, innervation density (axons/unit area) and pain do not appear to be linearly related. Similarly, thin-fiber innervation of glands, placodal epithelia (including the cochlear, vestibular, olfactory and vomeronasal, and gustatory organs), vascular and lymphoid tissue, as well as select secretory epithelia and most visceral organs is only more recently being explained in relation to nociception and pain. Examples of sheetlike distributions have enabled electron microscopic characterization of endings in the inner surface of the tympanic membrane (Yeh and Kruger 1984) and the tunica vasculosa of the testis, where 3-D reconstruction revealed clusters of "synaptic" vesicles at successive sites of a physiologically characterized nociceptive receptive field (Kruger et al. 2003a, b). Therefore, it is clearly evident that thin unmyelinated fibers are not invariably nociceptors, and it cannot be inferred from their distribution or simple histology that they respond to noxious stimuli with impulse production or that they are usually concerned with pain (Light and Perl 2003).

Importantly, not all thin-caliber neurons are nociceptors. In fact, a wide variety of non-nociceptive sensory perceptions including warmth, cold, tickle, itch, light touch, and stroking are associated with small to medium size C and/or Aδ axons which can be responsive innocuous thermal stimuli (thermoreceptors), molecules involved in normal tissue signaling (chemoreceptors), products of normal metabolism (metaboreceptors), and light mechanical stimuli (low-threshold mechanoreceptors) (Light and Perl 2003). Additionally, major sets of thin-caliber peptidergic C and Aδ fibers supply cutaneous blood vessels, where release of CGRP and SP has effector functions, including vasodilation and plasma extravasation, which are now known to mediate the process of "neurogenic inflammation" as described by Bayliss over 100 years ago (Bayliss 1901; Brain and Williams 1985, 1988; Brain et al. 1985). This innervation is likely involved in vasodilation normally related to thermal regulation and in meeting normal local tissue metabolic demands exacerbated by function, growth, or injury. Altogether, these lines of evidence raise questions about the functional identity of nociceptive sensory endings and evoked doubts about whether specific "nociceptors" even exist (Melzack and Wall 1962).

Subsequently, a major breakthrough occurred with the development of a specific antibody directed against protein gene product 9.5 (PGP9.5)/ubiquitin c-terminal hydrolase 1 (UCHL-1), a cytoplasmic enzyme important in protein regulation and enriched in neuroendocrine cells from all mammals and many nonmammalian species. Utilizing anti-PGP9.5 as an immunolabel distinctly and specifically marks peripheral nerves and was found to reveal far more extensive thin-caliber innervation and endings than were previously known. Under normal tissue and immunolabeling procedures, the PGP9.5 antibody consistently reveals the total innervation seen with all other antibodies against other neuronal antigens (Albrecht and Rice 2010; Rice and Albrecht 2008; Fundin et al. 1997).

Functional Characteristics of Nociceptors

Currently, a wide range of functional characteristics defined by molecular properties are implicated in the transduction and/or modulation of various types of nociceptive stimuli. The functions of a particular molecule are typically determined using a combination of in vitro and in vivo pharmacologic, electrophysiologic, and behavioral assessments involving specific agonists and antagonists, thermal and mechanical stimuli, and gain or loss of function mutations. The immunochemical and/or molecular identification of functional proteins expressed within presumptive small-caliber nociceptors includes specific neurotransmitters, cytokines, growth factors, amino acids, protons, ligand- and voltage-gated ion channels, and ligand-activated G-protein-coupled receptors (GPCR) (Albrecht and Rice 2010; Rice and Albrecht 2008) [see Stucky (2013) entry "Immunocytochemistry of Nociceptors" in this Encyclopedia].

In general, nociception is thought to be driven by direct activation of the nociceptor sensory potential by a specific noxious stimulus (i.e., thermal, chemical, mechanical) acting on an identified molecular receptor. More recently it has been demonstrated that nociceptor function may also be modulated indirectly via cells of the terminal tissue (i.e., keratinocytes, smooth muscle, endothelium, immune mediators). Furthermore, it is appreciated that the molecular signaling can lead to increased action potential generation (e.g., excitatory, pro-nociceptive, implicated in algesia) or decreased action potential generation (e.g., inhibitory, anti-nociceptive, implicated in analgesia) (Albrecht and Rice 2010; Rice and Albrecht 2008). For example, GPCR-activated second messenger signaling that opens sodium channels is implicated in pro-nociception, while signaling that activates inward-rectifying potassium channels (GIRKS) is usually implicated in anti-nociception.

Currently, there are several identified pro-nociceptive cellular signaling mechanisms which enhance small-fiber nociceptor activity either directly or indirectly, including, but not limited to:

1. *Neuropeptide signaling* (e.g., CGRP activation of RAMP1/CRLR complexes, substance P activation of NK1 receptors).
2. *Purinergic signaling* (e.g., ATP activation of multiple P2X receptors, ADP activation of multiple P2Y receptors) (Dussor et al. 2009).
3. *Endothelin-1 signaling* is mediated through two differentially expressed receptors (i.e., ETA, ETB); specific activation of ETA on nociceptors leads to painful responses (Balonov et al. 2006; Gokin et al. 2001).
4. *Biogenic amine signaling* (i.e., histamine activation of multiple H receptors) (Cannon et al. 2007).
5. *Cytokine signaling* (i.e., bradykinin activation of multiple B receptors).
6. *Proton activation* of multiple acid-sensing ion channels (ASICS), including even specificity for types of acids (i.e., lactic acid activation of ASIC3) (Molliver et al. 2005).
7. *MAS-related G-protein-coupled receptors* (*Mrgpr*) *activation*, particularly the epidermal fiber-targeted MrgprD, but which are also demonstrating genetic differentiation of non-noxious perceptions (i.e., itch, gentle stroking) (Zylka et al. 2005; Liu et al. 2008, 2009; Vrontou et al. 2013).
8. *Voltage-gated ion channel activation* directing largely sodium, calcium, and chloride movements, as exampled by the expression of Nav1.7 on nociceptors, for which a gain-of-function mutation results in hyperactivity of nociceptors and congenital painful

conditions (familial erythromelagia) (Estacion et al. 2008; Sheets et al. 2007; Dib-Hajj et al. 2005; Cho et al. 2012).

Additionally, of particular interest in pronociceptive mechanisms is the multimodal transient receptor potential (TRP) ligand-gated cation channel family, comprised of at least seven distinct subclasses, several of which are expressed robustly among small-caliber sensory neurons. The prototypical vanilloid family member (TRPV1) can be activated by noxious heat, protons, and exogenous chemicals (capsaicin) and is modulated by endogenous cannabinoid ligands (anandamide) (Caterina and Julius 2001). The prototypical ankyrin family member (TRPA1) is activated by multiple modalities including pungent exogenous chemicals (mustard oil) and endogenous (ankyrin) ligands but is also implicated in thermal perceptions of noxious cold and mechanical transduction (Kwan et al. 2009; Karashima et al. 2009). The prototypical melastatin-like family member (TRPM8) is also activated by noxious cold temperatures, exogenous chemicals (menthol, icilin), and endogenous ligands (lysophospholipids from PLA2 activity) (McCoy et al. 2011).

Conversely, there are some identified antinociceptive cellular signaling mechanisms which inhibit small-fiber nociceptor activity either directly or indirectly, including, but not limited to:

1. *Opioid signaling* mediated through several classes of opioid receptor (OR) and driven by numerous exogenous substances (e.g., morphine) and endogenous ligands (e.g., β-endorphin); particularly µOR activation leads to hyperpolarization of the nociceptors and inhibition of activity (Khodorova et al. 2003).

2. *Cannabinoid signaling* mediated through the cannabinoid receptors (CB1, CB2) driven by exogenous substances (THC) and endogenous phospholipid mediators (anandamide, 2-AG). Several lines of evidence suggest both indirect and direct modulation of nociceptor activity can be mediated through CB1 and/or CB2 activation, as well as CB ligand and receptor interactions with the TRP channels A1 and V1 (Ibrahim et al. 2005; Potenzieri et al. 2008; Akopian et al. 2008; Fioravanti et al. 2008).

3. *Voltage and ligand-gated ion channel activation* directing largely potassium movements, as exampled by activation of ATP-sensitive and various voltage-gated potassium channels in nociceptors, which directly inhibit activation and are modulated by a wide range of substances (Nicol et al. 1997; Rasband et al. 2001; Chi and Nicol 2007; Du et al. 2011; Tsantoulas et al. 2012; Passmore et al. 2012) and for which autoimmune-mediated loss of function leads to idiopathic pain conditions (Klein et al. 2012).

4. *Endothelin-1 signaling* is mediated through two differentially expressed receptors (i.e., ETA, ETB); specific activation of ETB leads to nociceptor inhibition directly or indirectly via keratinocyte-mediated opiate release and µOR activation (Khodorova et al. 2002, 2003). Importantly, ETB activation can have analgesic or algesic effects depending on the site where it is expressed and the timing of its activation following injury (Raffa et al. 1996; Werner et al. 2010).

The continued development of specific antibodies directed against many of these molecules, and the use of conditional promoter gene expression of fluorescent reporter molecules, has enabled the microscopic visualization (histology) of neurons that express these functional molecules, including neuronal cell bodies, axons, peripheral sensory endings, and central terminations. Structurally, the small-caliber sensory endings most often terminate with FNE morphology, although the location of the endings that express a particular immunochemical property may be specific to a particular target or multiple targets. For example, endings that express MrgprD only terminate in the epidermis, whereas endings that express CGRP terminate in numerous tissues, including the skin, where they can be located in the epidermis, and dermis around blood vessels, hair follicles, and sweat glands (Albrecht and Rice 2010; Zylka et al. 2005). Such CGRP expressing endings presumably serve different functions in these disparate locations (Rice and Albrecht 2008). The fine structure of physiologically marked FNE reveals nearly complete Schwann cell process envelopment and sequential zones of clustered vesicles (Kruger et al. 2003a), and the dense-core vesicles have been labeled with colloidal gold, revealing peptide localization within the dense core.

The multi-molecular profiles of sensory endings with particular morphologies and tissue locations have been elucidated through the use of multiple combinations of immunolabels and reporter gene expressions applied to alternating sections from the same and similar tissue specimens, as well as on sensory ganglia where mRNA expression can also be assessed by in situ hybridization. The combined molecular properties can then be used to infer the potential functional properties of a particular sensory ending. Continued small-caliber axon research, particularly with the sensory neuron-specific Mrgpr receptor family, is beginning to uncover molecular expression signatures that are consistent with the functional varieties of nociceptors defined by electrophysiology. Indeed, the use of multi-molecular morphological strategies has confirmed that sensory ending expression of certain molecules (i.e., TrpV1) correlates with electrophysiological properties (i.e., thermal nociception).

Critically, these comprehensive immunolabeling studies demonstrate that all sensory axon endings, including those which are likely nociceptors, express complex molecular properties responsible for integrating multiple stimulus modalities. Therefore, molecularly, all sensory axons appear to be capable of polymodal functioning. Certain structural and enzymatic properties of sensory neuron axons are consistently co-expressed. For example, co-expression of the pan-neuronal enzyme PGP (which reveals all innervation), the 200 kD neurofilament protein (NF), and the myelin basic protein (MBP) has been consistently revealed in species ranging from mice, mole rats, manatees, monkeys, to humans. Utilizing that preparation, the PGP-positive thin-caliber fibers which lack NF or MBP immunolabeling represent the C fiber populations, whereas the NF/MBP-positive population represents Aδ fiber populations. Further immunolabel combinations have revealed additional detailed classification breakouts, including that CGRP is expressed in subsets of C fibers and Aδ fibers,

while the P2X3 receptor is exclusively expressed in C fibers that nearly all lack CGRP. Thus, thin-caliber expression of P2X3 and CGRP appears to be mutually exclusive. Moreover, the μOR receptor is almost exclusively expressed by C fibers that express CGRP, but not C fibers that express P2X3, and rarely on Aδ fibers that express CGRP. Therefore, it can be inferred that peripheral opiate analgesia involving the μOR is likely acting on CGRP expressing C fibers. By contrast, TrpV1 expression appears in C fibers that express CGRP or P2X3 and can be found with varying proportions in different species. The myriad permutations of multi-molecular characteristics among the small-caliber sensory neuron populations are gradually being elucidated, and the functional implications of these markers revealed. Currently, several important observations can be highlighted from these comprehensive multi-molecular immunochemical assessments of small-caliber (nociceptor) peripheral innervation histology:

1. Dense concentrations of molecularly complex C and Aδ fiber-free nerve endings with nociceptive immunochemical labeling properties are located at sites which rarely receive nociceptive inputs (e.g., dental pulp, bone marrow) but which may be responsible for sensing increased stress of these hard structures prior to overt damage.

2. Thin-caliber C fibers endings with nociceptive immunochemical labeling properties terminate individually in some anatomical sites (i.e., epidermis, dermal vasculature) and also terminate in close association with certain types of Aβ low-threshold mechanoreceptors in dermal papilla of glabrous skin (i.e., Meissner corpuscles) and surrounding hair follicles (i.e., piloneural complexes). The Aβ fibers in both locations are purportedly rapidly adapting low-threshold mechanoreceptors that contribute to detecting vibration across glabrous skin and hairs. Some of the C fibers contain CGRP and SP, while the Aβ endings express neuropeptide receptors, allowing for stimulus integration at the site of sensory transduction. Moreover, the Aβ fiber endings in these locations can also express various molecular properties implicated in nociception, suggesting that these

low-threshold mechanoreceptors may impart functional signals as part of the pain perception (Pare et al. 2001). Indeed, patients suffering from complex regional pain syndrome type 1 (CRPS I) with severe hairy skin allodynia have profound neuropathology associated with the organization of both large- and small-caliber hair follicle endings (Albrecht et al. 2006).

3. Dense concentrations of C and Aδ fibers with nociceptive immunochemical labeling properties terminate within the adventitia of cutaneous and visceral arteries, arterioles, and arteriole-venule shunts. Subsets of these visceral fibers express the lactic acid receptor ASIC3, while others express the TrpV1 receptor, even though that vascular anatomical location would be unlikely challenged by noxious heat stimuli. Most of this vascular sensory innervation expresses CGRP and substance P, which are potent vasodilators, and implicated in inflammatory pain. As such, these thin-caliber sensory fibers presumably function in monitoring peripheral vascular status while contributing an essential effector role in vasodilatation.

4. Numerous studies have now demonstrated significant and essential interactions of thin-caliber axons and mediators derived from target tissues (i.e., epidermal keratinocytes and immune cells, vascular smooth muscle and endothelium). For example, a variety of neural signaling molecules are expressed in differentially stratified distributions across the deep to superficial layers of keratinocytes, involving mediators implicated in both analgesic (ETB, CB2) and algesic (Nav, CGRP, ATP/P2X, TrpV1) mechanisms (Albrecht and Rice 2010; Rice and Albrecht 2008; Ibrahim et al. 2005; Khodorova et al. 2002; Hou et al. 2011; Lumpkin and Caterina 2007; Zhao et al. 2008).

Importantly, protein immunocytochemistry has ushered in an era of labeling for antigens that have physiological properties implicated in eliciting or modulating nociception. These include neuropeptides, neurotransmitters, immune neuromodulators, G-protein-coupled receptors, endothelin and purinergic receptors, and a variety of ion channels. Concomitant molecular advances continue to indicate that specific molecular expressions can be linked to specific functional varieties of thin-axon nociceptors but importantly also to thin-caliber non-nociceptive neurons, such as those now implicated in itch (Liu et al. 2009) and even pleasurable perceptions such as gentle skin stroking (Vrontou et al. 2013), potentially demonstrating the existence of a small-caliber "volupticeptor" population.

Distinctive morphological patterns (Zylka et al. 2005), complex protein immunochemistry (Rice and Albrecht 2008), and microneurographic electrophysiological evidence (Campero et al. 2009; Obreja et al. 2010; Orstavik et al. 2006) are mounting to suggest that many thin-caliber sensory axon and associated small sensory neurons have an inherent capacity to be polymodal at the molecular/protein expression level (e.g., immunolabeling). However, these same classes remain somewhat singularly definable in electrophysiologic functionality testing (e.g., the nociceptor, or pruritic specific, or as noted above, the idea of a pleasure specific population) based from experimental stimuli. Thus, most (if not all) thin-caliber sensory neurons may have a spectrum of functional capabilities, potentially related to nociception, but that most likely also subserve homeostatic tissue maintenance mechanisms. Since peptide release may constitute a principal constitutive function of peptidergic sensory neurons, the afferent/efferent functional duality led to the term "noceffector" (Kruger 1988, 1996; Kruger et al. 1989) for those thin-caliber neurons whose peptides and target receptor-binding sites had been identified. This definition also serves as a compromise with the potentially misleading and erroneous nociceptor designation when evidence for nociception cannot be invoked or is exceedingly rare, as in teeth, testis, ducts, sphincters, etc. Certainly, the specific sensory neuron molecular identities and histologic characterizations subserving various nociceptive/pain perceptions are far from resolved.

References

Akopian, A. N., Ruparel, N. B., Patwardhan, A., et al. (2008). Cannabinoids desensitize capsaicin and mustard oil responses in sensory neurons via TRPA1 activation. *The Journal of Neuroscience, 28*(5), 1064–1075.

Albrecht, P. J., & Rice, F. L. (2010). Role of small-fiber afferents in pain mechanisms with implications on diagnosis and treatment. *Current Pain and Headache Reports, 14*(3), 179–188.

Albrecht, P. J., Hines, S., Eisenberg, E., et al. (2006). Pathologic alterations of cutaneous innervation and vasculature in affected limbs from patients with complex regional pain syndrome. *Pain, 120*(3), 244–266.

Balonov, K., Khodorova, A., & Strichartz, G. R. (2006). Tactile allodynia initiated by local subcutaneous endothelin-1 is prolonged by activation of TRPV-1 receptors. *Experimental Biology and Medicine (Maywood), 231*(6), 1165–1170.

Bayliss, W. (1901). On the origin of the vasodilator fibres of the hind-limb, and on the nature of these fibres. *The Journal of Physiology, 26*, 173–209.

Brain, S. D., & Williams, T. J. (1985). Inflammatory oedema induced by synergism between calcitonin gene-related peptide (CGRP) and mediators of increased vascular permeability. *British Journal of Pharmacology, 86*(4), 855–860.

Brain, S. D., & Williams, T. J. (1988). Substance P regulates the vasodilator activity of calcitonin gene-related peptide. *Nature, 335*(6185), 73–75.

Brain, S. D., Williams, T. J., Tippins, J. R., et al. (1985). Calcitonin gene-related peptide is a potent vasodilator. *Nature, 313*(5997), 54–56.

Campero, M., Baumann, T. K., Bostock, H., et al. (2009). Human cutaneous C fibres activated by cooling, heating and menthol. *The Journal of Physiology, 587*(23), 5633–5652.

Cannon, K. E., Chazot, P. L., Hann, V., et al. (2007). Immunohistochemical localization of histamine H3 receptors in rodent skin, dorsal root ganglia, superior cervical ganglia, and spinal cord: Potential antinociceptive targets. *Pain, 129*(1–2), 76–92.

Caterina, M. J., & Julius, D. (2001). The vanilloid receptor: A molecular gateway to the pain pathway. *Annual Review of Neuroscience, 24*, 487–517.

Chi, X. X., & Nicol, G. D. (2007). Manipulation of the potassium channel Kv1.1 and its effect on neuronal excitability in rat sensory neurons. *Journal of Neurophysiology, 98*(5), 2683–2692.

Cho, H., Yang, Y. D., Lee, J., et al. (2012). The calcium-activated chloride channel anoctamin 1 acts as a heat sensor in nociceptive neurons. *Nature Neuroscience, 15*(7), 1015–1021.

Dib-Hajj, S. D., Rush, A. M., Cummins, T. R., et al. (2005). Gain-of-function mutation in Nav1.7 in familial erythromelalgia induces bursting of sensory neurons. *Brain, 128*(8), 1847–1854.

Du, X., Wang, C., & Zhang, H. (2011). Activation of ATP-sensitive potassium channels antagonize nociceptive behavior and hyperexcitability of DRG neurons from rats. *Molecular Pain, 7*, 35.

Dussor, G., Koerber, H. R., Oaklander, A. L., et al. (2009). Nucleotide signaling and cutaneous mechanisms of pain transduction. *Brain Research Reviews, 60*(1), 24–35.

Estacion, M., Dib-Hajj, S. D., Benke, P. J., et al. (2008). NaV1.7 gain-of-function mutations as a continuum: A1632E displays physiological changes associated with erythromelalgia and paroxysmal extreme pain disorder mutations and produces symptoms of both disorders. *The Journal of Neuroscience, 28*(43), 11079–11088.

Fioravanti, B., De Felice, M., Stucky, C. L., et al. (2008). Constitutive activity at the cannabinoid CB1 receptor is required for behavioral response to noxious chemical stimulation of TRPV1: Antinociceptive actions of CB1 inverse agonists. *The Journal of Neuroscience, 28*(45), 11593–11602.

Fundin, B. T., Arvidsson, J., Aldskogius, H., et al. (1997). Comprehensive immunofluorescence and lectin binding analysis of intervibrissal fur innervation in the mystacial pad of the rat. *The Journal of Comparative Neurology, 385*(2), 185–206.

Gokin, A. P., Fareed, M. U., Pan, H. L., et al. (2001). Local injection of endothelin-1 produces pain-like behavior and excitation of nociceptors in rats. *The Journal of Neuroscience, 21*(14), 5358–5366.

Hou, Q., Barr, T., Gee, L., et al. (2011). Keratinocyte expression of calcitonin gene-related peptide beta: Implications for neuropathic and inflammatory pain mechanisms. *Pain, 152*(9), 2036–2051.

Ibrahim, M. M., Porreca, F., Lai, J., et al. (2005). CB2 cannabinoid receptor activation produces antinociception by stimulating peripheral release of endogenous opioids. *Proceedings of the National Academy of Sciences of the United States of America, 102*(8), 3093–3098.

Karashima, Y., Talavera, K., Everaerts, W., et al. (2009). TRPA1 acts as a cold sensor in vitro and in vivo. *Proceedings of the National Academy of Sciences of the United States of America, 106*(4), 1273–1278.

Khodorova, A., Fareed, M. U., Gokin, A., et al. (2002). Local injection of a selective endothelin-B receptor agonist inhibits endothelin-1-induced pain-like behavior and excitation of nociceptors in a naloxone-sensitive manner. *The Journal of Neuroscience, 22*(17), 7788–7796.

Khodorova, A., Navarro, B., Jouaville, L. S., et al. (2003). Endothelin-B receptor activation triggers an endogenous analgesic cascade at sites of peripheral injury. *Nature Medicine, 9*(8), 1055–1061.

Klein, C. J., Lennon, V. A., Aston, P. A., et al. (2012). Chronic pain as a manifestation of potassium channel-complex autoimmunity. *Neurology, 79*(11), 1136–1144.

Kruger, L. (1988). Morphological features of thin sensory afferent fibers: A new interpretation of 'nociceptor' function. *Progress in Brain Research, 74*, 253–257.

Kruger, L. (1996). The functional morphology of thin sensory axons: Some principles and problems. *Progress in Brain Research, 113*, 255–272.

Kruger, L., Perl, E. R., & Sedivec, M. J. (1981). Fine structure of myelinated mechanical nociceptor endings in cat hairy skin. *The Journal of Comparative Neurology, 198*(1), 137–154.

Kruger, L., Silverman, J. D., Mantyh, P. W., et al. (1989). Peripheral patterns of calcitonin-gene-related peptide general somatic sensory innervation: Cutaneous and deep terminations. *The Journal of Comparative Neurology, 280*(2), 291–302.

Kruger, L., Kavookjian, A. M., Kumazawa, T., et al. (2003a). Nociceptor structural specialization in canine and rodent testicular "free" nerve endings. *The Journal of Comparative Neurology, 463*(2), 197–211.

Kruger, L., Light, A. R., & Schweizer, F. E. (2003b). Axonal terminals of sensory neurons and their morphological diversity. *Journal of Neurocytology, 32*(3), 205–216.

Kwan, K. Y., Glazer, J. M., Corey, D. P., et al. (2009). TRPA1 modulates mechanotransduction in cutaneous sensory neurons. *The Journal of Neuroscience, 29*(15), 4808–4819.

Light, A. R., & Perl, E. R. (2003). Unmyelinated afferent fibers are not only for pain anymore. *The Journal of Comparative Neurology, 461*(2), 137–139.

Liu, Y., Yang, F. C., Okuda, T., et al. (2008). Mechanisms of compartmentalized expression of Mrg class G-protein-coupled sensory receptors. *The Journal of Neuroscience, 28*(1), 125–132.

Liu, Q., Tang, Z., Surdenikova, L., et al. (2009). Sensory neuron-specific GPCR Mrgprs are itch receptors mediating chloroquine-induced pruritus. *Cell, 139*(7), 1353–1365.

Lumpkin, E. A., & Caterina, M. J. (2007). Mechanisms of sensory transduction in the skin. *Nature, 445*(7130), 858–865.

McCoy, D. D., Knowlton, W. M., & McKemy, D. D. (2011). Scraping through the ice: Uncovering the role of TRPM8 in cold transduction. *American Journal of Physiology. Regulatory, Integrative and Comparative Physiology, 300*(6), R1278–R1287.

Melzack, R., & Wall, P. D. (1962). On the nature of cutaneous sensory mechanisms. *Brain, 85*, 331–356.

Molliver, D. C., Immke, D. C., Fierro, L., et al. (2005). ASIC3, an acid-sensing ion channel, is expressed in metaboreceptive sensory neurons. *Molecular Pain, 1*, 35.

Nicol, G. D., Vasko, M. R., & Evans, A. R. (1997). Prostaglandins suppress an outward potassium current in embryonic rat sensory neurons. *Journal of Neurophysiology, 77*(1), 167–176.

Obreja, O., Ringkamp, M., Namer, B., et al. (2010). Patterns of activity-dependent conduction velocity changes differentiate classes of unmyelinated mechano-insensitive afferents including cold nociceptors, in pig and in human. *Pain, 148*(1), 59–69.

Orstavik, K., Namer, B., Schmidt, R., et al. (2006). Abnormal function of C-fibers in patients with diabetic neuropathy. *The Journal of Neuroscience, 26*(44), 11287–11294.

Pare, M., Elde, R., Mazurkiewicz, J. E., et al. (2001). The Meissner corpuscle revised: A multiafferented mechanoreceptor with nociceptor immunochemical properties. *The Journal of Neuroscience, 21*(18), 7236–7246.

Passmore, G. M., Reilly, J. M., Thakur, M., et al. (2012). Functional significance of M-type potassium channels in nociceptive cutaneous sensory endings. *Frontiers in Molecular Neuroscience, 5*, 63.

Potenzieri, C., Brink, T. S., Pacharinsak, C., et al. (2008). Cannabinoid modulation of cutaneous Adelta nociceptors during inflammation. *Journal of Neurophysiology, 100*(5), 2794–2806.

Raffa, R. B., Schupsky, J. J., & Jacoby, H. I. (1996). Endothelin-induced nociception in mice: Mediation by ETA and ETB receptors. *The Journal of Pharmacology and Experimental Therapeutics, 276*(2), 647–651.

Rasband, M. N., Park, E. W., Vanderah, T. W., et al. (2001). Distinct potassium channels on pain-sensing neurons. *Proceedings of the National Academy of Sciences of the United States of America, 98*(23), 13373–13378.

Rice, F., & Albrecht, P. (2008). Cutaneous mechanisms of tactile perception: Morphological and chemical organization of the innervation to the skin. In A. Basbaum et al. (Eds.), *The senses: A comprehensive reference* (Somatosensation, Vol. 6, pp. 1–32). San Diego: Academic.

Sheets, P. L., Jackson, J. O., 2nd, Waxman, S. G., et al. (2007). A Nav1.7 channel mutation associated with hereditary erythromelalgia contributes to neuronal hyperexcitability and displays reduced lidocaine sensitivity. *The Journal of Physiology, 581*(Pt 3), 1019–1031.

Silverman, J. D., & Kruger, L. (1987). An interpretation of dental innervation based upon the pattern of calcitonin gene-related peptide (CGRP)-immunoreactive thin sensory axons. *Somatosensory Research, 5*(2), 157–175.

Silverman, J. D., & Kruger, L. (1988). Lectin and neuropeptide labeling of separate populations of dorsal root ganglion neurons and associated "nociceptor" thin axons in rat testis and cornea whole-mount preparations. *Somatosensory Research, 5*(3), 259–267.

Tsantoulas, C., Zhu, L., Shaifta, Y., et al. (2012). Sensory neuron downregulation of the Kv9.1 potassium channel subunit mediates neuropathic pain following nerve injury. *The Journal of Neuroscience, 32*(48), 17502–17513.

Vrontou, S., Wong, A. M., Rau, K. K., et al. (2013). Genetic identification of C fibres that detect massage-like stroking of hairy skin in vivo. *Nature, 493*(7434), 669–673.

Werner, M. F., Trevisani, M., Campi, B., et al. (2010). Contribution of peripheral endothelin ETA and ETB receptors in neuropathic pain induced by spinal nerve ligation in rats. *European Journal of Pain, 14*(9), 911–917.

Yeh, Y., & Kruger, L. (1984). Fine-structural characterization of the somatic innervation of the tympanic membrane in normal, sympathectomized, and neurotoxin-denervated rats. *Somatosensory Research, 1*(4), 359–378.

Zhao, P., Barr, T. P., Hou, Q., et al. (2008). Voltage-gated sodium channel expression in rat and human epidermal keratinocytes: Evidence for a role in pain. *Pain, 139*(1), 90–105.

Zylka, M. J., Rice, F. L., & Anderson, D. J. (2005). Topographically distinct epidermal nociceptive circuits revealed by axonal tracers targeted to Mrgprd. *Neuron, 45*(1), 17–25.

History

▶ History of Analgesics

History of Analgesics

Kay Brune[1] and Burkhard Hinz[2]
[1]Department of Experimental and Clinical Pharmacology and Toxicology, Friedrich Alexander University Erlangen-Nürnberg, Erlangen, Germany
[2]Institute of Toxicology and Pharmacology, University of Rostock, Rostock, Germany

Synonyms

Analgesics; History

Definition

Attempts to relieve pain are probably as old as mankind. Dioscourides, a Greek physician, prescribed extracts of willow bark against joint pain, while Hildegard von Bingen and the Reverend Stone, in his famous letter to the Royal Society of Medicine in London, suggested the same therapy (Brune 1997; Rainsford 1984). Local inflammation often goes along with "general inflammation" manifested by fever and malaise. The reasons for this were recently uncovered: the release of pyrogenic cytokines such as TNFα and IL-1. Fever along with malaise was treated on the basis of the Hippocratic concept by purgation, sweating, and bloodletting (Brune 1984).

Such practices were continued until the nineteenth century (Williams 1975) – probably without success. It was only recently that the inhibition of the cytokine effect has become feasible (Smolen et al. 2000).

Characteristics

A scientific approach to pain therapy became possible in the nineteenth century, with substances isolated from plants including the willow tree (salicylic acid esters) and then the description of the complete synthesis by Kolbe (Marburg) (Brune 1997; Rainsford 1984). To provide sufficient amounts, the first "scale-up" of a synthetic process was invented and the first drug factory built (Salicylic Acid Works founded by von Heyden in 1874; 6). Salicylic acid was found to be active against fever (Buss, Switzerland) and rheumatoid arthritis (Stricker, Berlin; Mac Lagan, Dundee) (Brune 1997; Rainsford 1984; Sneader 1985).

Earlier (1806), a pharmacist in Einbeck, Sertürner, had isolated morphine, the main analgesic ingredient of the opium resin. He checked extracts from opium for sedative activity in his pack of dogs and ended up with a pure substance (morphine) (Sertürner 1806; Sneader 1985). With morphine, for the first time, a pure (crystalline) drug was available. Death due to overdose or lack of effect could now be avoided by exact dosing (Bender 1966).

New Chemicals
The next step was taken by chemists who tried to compensate for an impaired supply of opium, china bark (quinine), and others by chemical synthesis. It was made possible by E. Fischer's discovery of phenylhydrazine, which allowed the synthesis of nitrogen-containing ring systems. His scholar, L. Knorr, tried to synthesize quinine, but produced phenazone (Fig. 1) (Brune 1997), which proved to be active against fever. The patent for this compound (Antipyrine®) was bought by a dye factory in Hoechst. This was the start of the pharmaceutical company Hoechst (Brune 1997). Another chemist (F. Hoffmann)

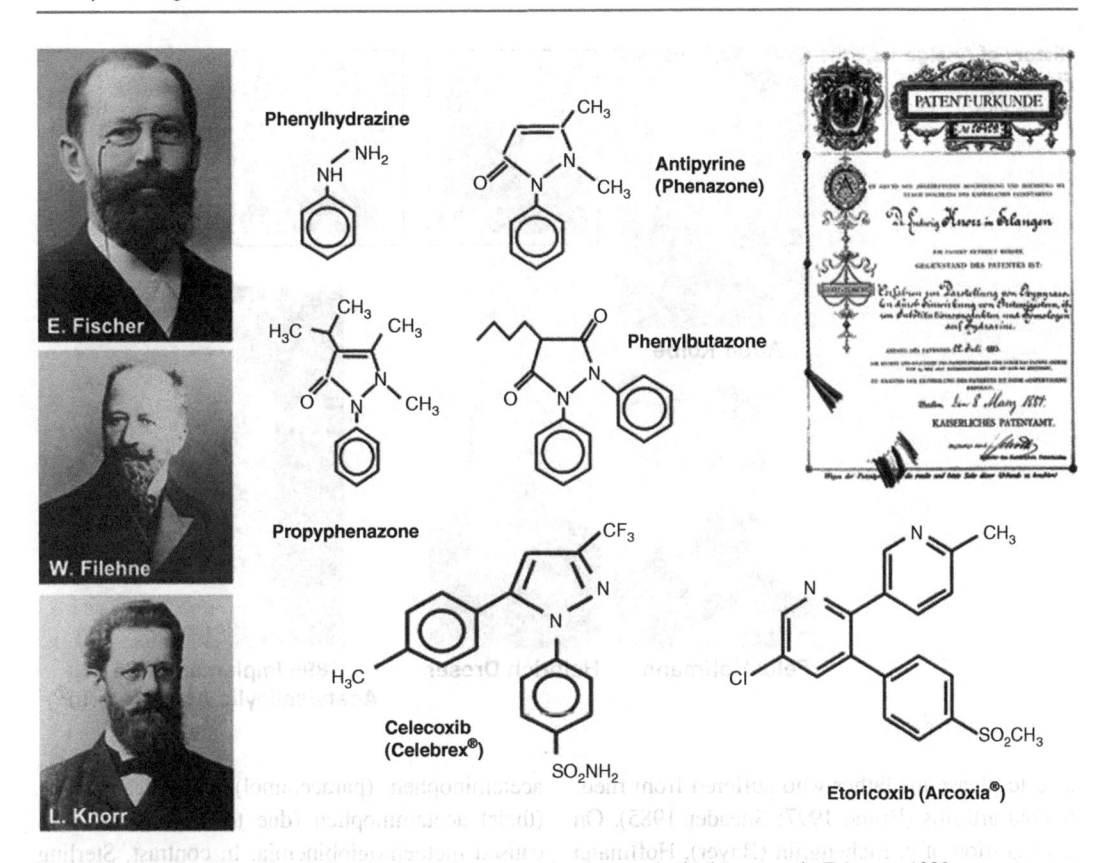

History of Analgesics, Fig. 1 Synthesis of phenazone, the first synthetic drug ever, in Erlangen 1882

esterified salicylic acid with acetate and (re-)discovered Aspirin. This synthesis was done in another dye factory, namely, Bayer (Rinsema 1999). The new science of chemistry helped to transform the dye industry by providing both synthetic dyes and new synthetic drugs.

Pain therapy was aided by another accidental discovery. In Strasbourg, two physicians, Cahn and Hepp, attempted to eradicate intestinal worms. The worms survived, but the fever resolved (Cahn and Hepp 1886). An analysis revealed that the pharmacy had provided acetanilide rather than naphthalene. This led to the discovery of acetanilide, which was marketed by another dye factory (Kalle) under the name Antifebrin® (Brune 1997; Sneader 1985). Bayer further investigated acetanilide and found that a by-product of aniline dye production, namely, "acetophenetidin," was equally effective. It was marketed as Phenacetin® (Sneader 1985). These discoveries constituted, as Tainter phrased it

(Tainter 1948), "[...] the beginning of the famous German drug industry and ushered in Germany's 40-year dominance of the synthetic drug and chemical field." Thus, by the end of the nineteenth century, four prototype substances were available for the treatment of pain: morphine, salicylic acid, phenazone, and phenacetin.

Chemical Modifications of Analgesics

Salicylic acid, phenazone, and phenacetin were widely used, and physicians soon recognized the disadvantages of these drugs. They were of low potency and had to be taken in gram quantities (spoon-wise). Sodium salicylate had an unpleasant taste. Taking several grams of phenacetin led to methemoglobinemia, while phenazone often caused allergic reactions. Consequently, the expanding drug industry set their chemists into action to produce improved derivatives.

F. Hoffman, a young chemist at the Bayer company, attempted to improve the taste of salicylic

**History of Analgesics,
Fig. 2** Synthesis of
acetylsalicylic acid in 1897

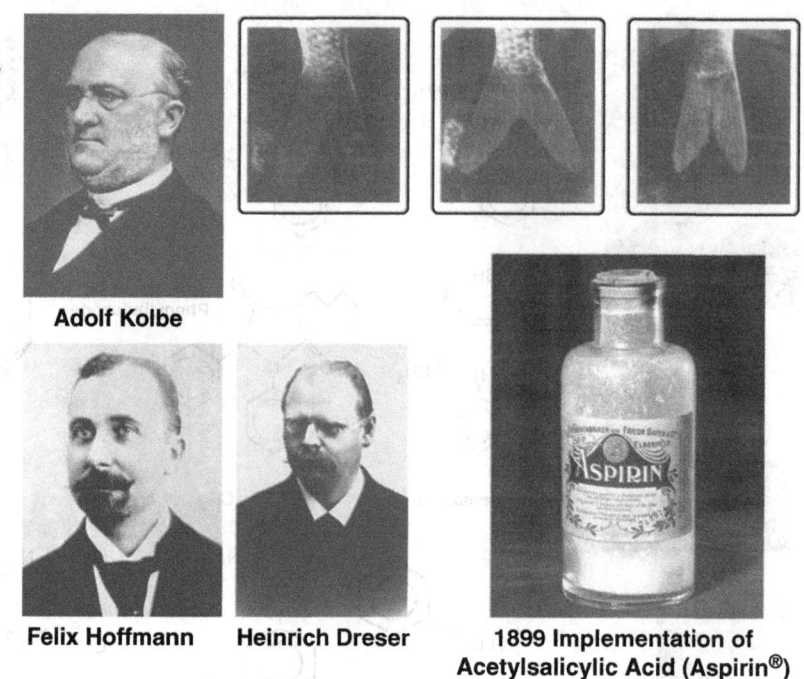

Adolf Kolbe

Felix Hoffmann   **Heinrich Dreser**   **1899 Implementation of
Acetylsalicylic Acid (Aspirin®)**

acid to please his father who suffered from rheumatoid arthritis (Brune 1997; Sneader 1985). On a suggestion of v. Eichengrün (Bayer), Hoffmann produced acetylsalicylic acid, which his father preferred (Brune 1997; Sneader 1985). Acetylsalicylic acid proved difficult to handle due to its instability. Bayer, therefore, took a patent on the water-free production process invented by Hoffmann and secured the name Aspirin® (derived from acetyl and the plant Spiraea ulmaria). H. Dreser, the first pharmacologist at Bayer, tried to demonstrate the reduced toxicity of aspirin as compared to salicylic acid. He employed a goldfish model, believing that the "mucosa" of their fins comprised an analogue of human intestinal mucosa. Dipping the fins of goldfish into solutions of either salicylic acid or aspirin, he observed that higher concentrations of aspirin were necessary to "cloudy" the fins (Fig. 2). He concluded that this was proof of better gastrointestinal tolerability (Dreser 1899). Later, Heinrich Dreser himself recognized that he did not measure a "gastrotoxic effect," but rather "acidity," and salicylic acid is more acidic than aspirin (Dreser 1907).

To further improve the tolerability of phenacetin, Bayer investigated a metabolite of phenacetin,

acetaminophen (paracetamol). It appeared that (their) acetaminophen (due to impurities?) also caused methemoglobinemia. In contrast, Sterling (UK) found acetaminophen free of methemoglobinemia and marketed it as Panadol® (Sneader 1985).

At Hoechst, the structure of phenazone was modified. The resulting compounds amidopyrine, melubrin, and dipyrone proved to be somewhat more active (Brune 1997). Roche substituted an amino group of phenazone with isopropyl. The resulting propyphenazone is still in use. It is relatively free of toxicity, i.e., it lacks kidney, gastrointestinal, and bone marrow toxicity (Kaufman et al. 1991). Finally, several companies combined two active principles, e.g., by producing salts of aspirin with amidopyrine or esters between acetaminophen and salicylic acid (Benorylate®). Moreover, analgesics were mixed and supplemented, e.g., with caffeine (APC powder) and with vitamins, minerals, and other partly obscure ingredients. This diversity of "drugs" pleased the consumer but was without major medical benefit – it may rather have led to abuse and kidney toxicity (Dubach et al. 1983).

History of Analgesics,
Fig. 3 Synthesis of
phenylbutazone in 1949

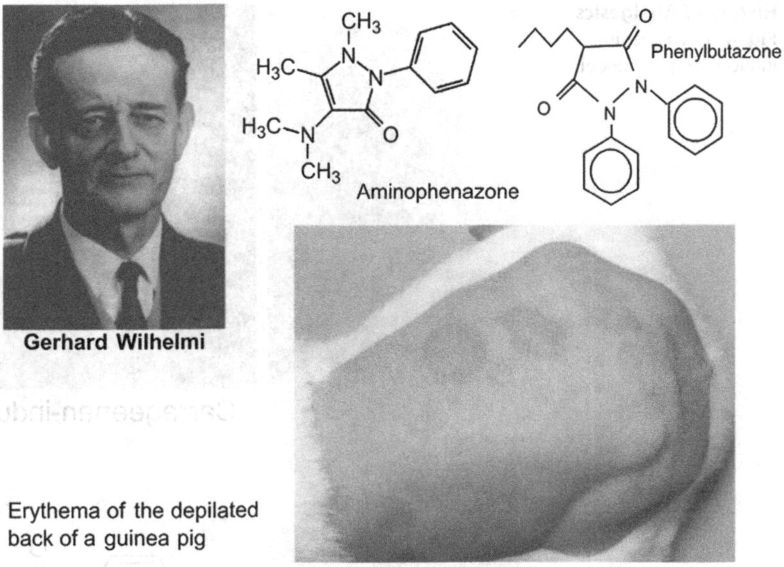

Gerhard Wilhelmi

Aminophenazone

Phenylbutazone

Erythema of the depilated
back of a guinea pig

New Compounds: Pharmacology Comes into Play

In 1949, an unexpected observation once again paved the way for new analgesics. Hoping to reduce toxicity and increase effectiveness of aminophenazone, Geigy (Basel) produced an injection containing the salt of the basic aminophenazone with an acidic derivative – later named phenylbutazone (Fig. 3). This salt was found to be very active, particularly in rheumatoid pain (Brune 1997; Sneader 1985). Burns and Brodie related this effect to phenylbutazone, which was present for much longer periods of time than aminophenazone (Domenjoz 1960). The conclusion was that the "salt forming" partner of aminophenazone was the dominant active ingredient. To further investigate this clinical observation, G. Wilhelmi (Geigy) developed novel models of inflammation (Wilhelmi 1949). Phenylbutazone turned out to be particularly active in reducing the UV erythema elicited in the skin of guinea pigs (Fig. 3) (Wilhelmi 1949). It was one of the first pharmacological models of inflammation, with which several phenylbutazone analogues were found.

In the USA, C. Winter, at Merck (MSD) and later at Parke Davis, developed his models of inflammatory pain. He introduced the cotton string granuloma and the carrageenin-induced rat paw model (Shen 1984). These assays turned out to be especially useful for measuring anti-inflammatory activity (Winter et al. 1962) (Fig. 4). A similar model was employed by Randall and Selitto for detecting analgesic activity (Randall and Selitto 1957). Using these models led to the discovery of several chemical classes of analgesics. Merck identified indols (including indomethacin and sulindac, T.Y. Shen) (Shen 1984); Boots found propionic acid derivatives (ibuprofen and flurbiprofen, S. Adams; Adams 1992); Parke Davis developed fenamates (e.g., mefenamic acid) (Shen 1984); Geigy was successful with new aryl-acetic acids, e.g., diclofenac (Shen 1984), and Rhone Poulenc with Bayer introduced ketoprofen (Shen 1984); and, finally, Lombardino at Pfizer rediscovered the ketoenolic acids (phenylbutazone). The advantage of these compounds is that all pharmacokinetic parameters can be tailored by minor changes in the molecular structure (Lombardino 1974). Pfizer's piroxicam (Otterness et al. 1982) was soon followed by tenoxicam (Roche) and meloxicam (Boehringer). All of these differ in their potency and in pharmacokinetic parameters including their metabolism and drug interactions, although their mode of action is basically the same. Most were identified using animal models before the mode of action of "aspirin-like" drugs – as these substances were

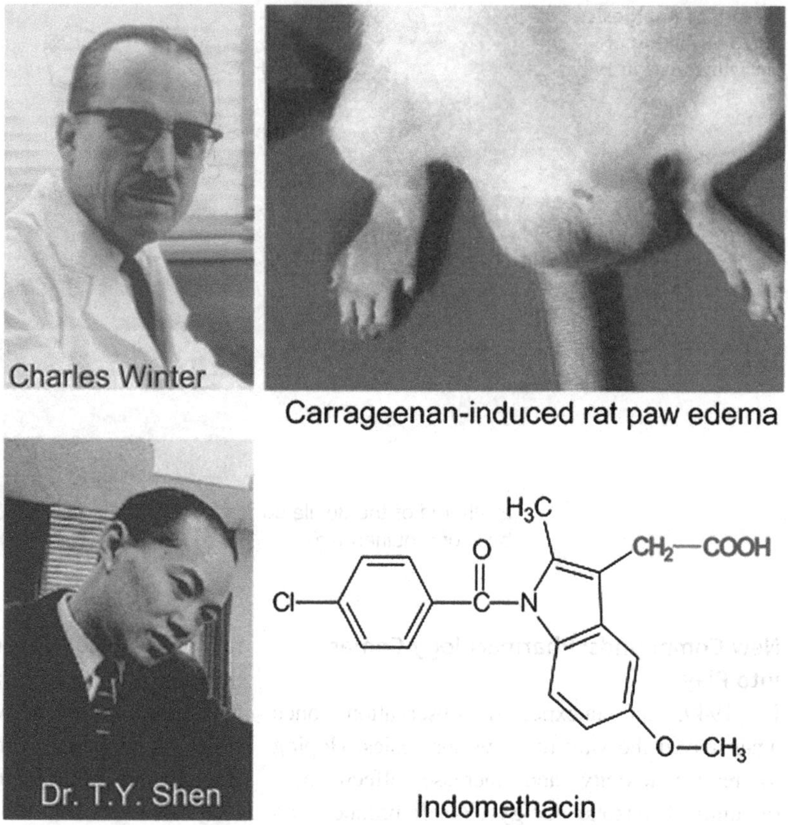

History of Analgesics,
Fig. 4 Carrageenin-
induced rat paw model

Charles Winter

Carrageenan-induced rat paw edema

Dr. T.Y. Shen

Indomethacin

formerly named – was determined. It was 70 years after the synthesis of aspirin when John Vane's group could demonstrate that these compounds were inhibitors of prostaglandin synthesis (Vane 1971).

This discovery, however, did not answer the question of why many of the old compounds (found by serendipity – such as phenazone, propyphenazone, phenacetin, paracetamol) were nonacidic chemicals that barely inhibited cyclooxygenases, while all the compounds developed in animal models of inflammation and pain were acidic and potent inhibitors (Brune 1974). All pharmacological models inflict an acute inflammation elicited by local prostaglandin production. Consequently, drugs that work by blocking cyclooxygenases in the inflamed tissue excel in these models. Acidic compounds (comprising pKa values of around 4, ~99 % protein binding, and amphiphilic structures) reach long-lasting high concentrations in inflamed tissue but

also relatively high concentrations in liver, kidney, and the stomach wall (Brune and Lanz 1985). This skewed distribution causes complete inhibition of prostaglandin synthesis in these locations resulting in superior anti-inflammatory activity but also liver, kidney, and stomach toxicity (Brune and Lanz 1985). This distributional selectivity may have reduced some of the side effects including CNS toxicity and increased the anti-inflammatory effects. Nonacidic compounds such as phenazone or paracetamol distribute homogeneously throughout the body. Their inhibition of prostaglandin production in inflamed tissue is small. Consequently, they are used to curb mild pain, but not inflammation.

The discovery of the existence of two cyclooxygenases, COX-1 and COX-2 (Flower 2003), has changed the landscape again. It provided a new dimension of selectivity, not limited to differences of tissue distribution, but based on enzyme selectivity.

Analgesics in the Age of Molecular Pharmacology

The discovery of prostaglandins and the inhibition of prostaglandin production by aspirin-like drugs caused the investigation of the effects of anti-inflammatory steroids on prostaglandin production. Many researchers observed that steroids can reduce prostaglandin production along with anti-inflammatory activity, but do not block it completely (e.g., Brune and Wagner 1979). Only P. Needleman came up with a molecular explanation that proposed two different enzymes, one being regulated by steroids (Fu et al. 1990; Masferrer et al. 1990). They were soon characterized (Kujubu et al. 1991). For the first time in the history of pharmacology, two molecular drug targets, COX-1 and COX-2 (the expression of COX-2 in the inflamed tissue is blocked by steroids), were identified before the biological role of the enzymes was fully known. It was soon clear that it might be advantageous to have drugs that block only COX-2, because this enzyme appeared not to be involved in the production of gastrointestinal-protective prostaglandins (for a review, see, e.g., Brune and Hinz 2004; Hinz and Brune 2002). Diclofenac and meloxicam were found to exert some but not sufficient selectivity to warrant gastrointestinal tolerance (Tegeder et al. 1999). Selective COX-2 inhibitors currently used for the treatment of osteoarthrosis and rheumatoid arthritis are the sulfonamide celecoxib and the methylsulfone etoricoxib. Others were withdrawn due to cardiovascular (rofecoxib) or cardiovascular and cutaneous (valdecoxib) side effects. However, comparable rates of thrombotic cardiovascular events have been subsequently reported for etoricoxib and diclofenac (Cannon et al. 2006), suggesting that there is presently no rationale for a differentiation of selective COX-2 inhibitors and nonselective COX inhibitors in terms of cardiovascular safety (for a review, see, e.g., Hinz et al. 2007; Brune et al. 2010). Another selective COX-2 inhibitor, lumiracoxib, is a relative of diclofenac and, like diclofenac, is sequestered into inflamed tissue. It combines COX-2 selectivity with selective tissue distribution (Feret 2003). However, the compound was withdrawn due to hepatotoxic effects.

Conclusion

After more than 120 years of development of pure analgesics, we have made some progress. Serendipity, as well as targeted research, has provided clinicians with many useful drugs that differ in many pharmacological and clinical aspects. Knowing a little of the history of their discovery and development may provide a perspective to better understand their effects and side effects. A humble acknowledgment of the role of serendipity may change our attitude towards research and marketing claims. But then serendipity is not all, as E. Kästner, a German poet, phrased it:

Irrtümer sind ganz gut, Jedoch nur hier und da. Nicht jeder, der nach Indien fährt, entdeckt AMERIKA.

Errors are fine, but only sometime(s).

Not everyone heading for India discovers AMERICA.

Acknowledgement The helpful discussion with many scientists, in particular I. Otterness, A. Sallmann, T.Y. Shen (†), and G. Wilhelmi, is gratefully acknowledged.

References

Adams, S. S. (1992). The propionic acids: A personal perspective. *Journal of Clinical Pharmacology, 32*, 317–323.
Bender, G. A. (1966). *Great moments in pharmacy*. Parke: Davis & Company.
Brune, K. (1974). How aspirin might work: A pharmacokinetic approach. *Agents and Actions, 4*, 230–232.
Brune, K. (1984). The concept of inflammatory mediators. In M. J. Parnham & J. Bruinvels (Eds.), *Discoveries in pharmacology* (Haemodynamics, hormones and inflammation, Vol. 2, pp. 487–498). Amsterdam: Elsevier.
Brune, K. (1997). The early history of non-opioid analgesics. *Acute Pain, 1*, 33–40.
Brune, K., & Hinz, B. (2004a). Selective cyclooxygenase-2 inhibitors: Similarities and differences. *Scandinavian Journal of Rheumatology, 33*, 1–6.
Brune, K., & Hinz, B. (2004b). The discovery and development of antiinflammatory drugs. *Arthritis and Rheumatism, 50*, 2391–2399.
Brune, K., & Lanz, R. (1985). Pharmacokinetics of nonsteroidal anti-inflammatory drugs. In I. L. Bonta, M. A. Bray, & M. J. Parnham (Eds.), *Handbook of inflammation* (Vol. 5, pp. 413–449). Amsterdam: Elsevier.

Brune, K., & Wagner, K. (1979). The effect of protein/ nucleic acid synthesis inhibitors on the inhibition of prostaglandin release from macrophages by dexamethasone. In K. Brune & M. Baggiolini (Eds.), *Arachidonic acid metabolism in inflammation and thrombosis* (pp. 73–77). Stuttgart: Birkhäuser. Agents Actions (Suppl 4), Basel, Boston.

Brune, K., Renner, B., & Hinz, B. (2010). Using pharmacokinetic principles to optimize pain therapy. *Nature Reviews Rheumatology, 6*, 589–598.

Cahn, A., & Hepp, P. (1886). Das antifebrin, ein neues fiebermittel. *Centralbl Klinicheskaia Meditsina, 7*, 561–564.

Cannon, C. P., Curtis, S. P., FitzGerald, G. A., Krum, H., Kaur, A., Bolognese, J. A., et al. (2006). Cardiovascular outcomes with etoricoxib and diclofenac in patients with osteoarthritis and rheumatoid arthritis in the multinational etoricoxib and diclofenac arthritis long-term (MEDAL) programme: A randomized comparison. *Lancet, 368*, 1771–1781.

Domenjoz, R. (1960). The pharmacology of phenylbutazone analogues. *Annals of the New York Academy of Sciences, 86*, 263.

Dreser, H. (1899). Pharmacologisches über aspirin (acetylsalicysäure). *Pflügers Archives Gesellschaft Physics, 76*, 306–318.

Dreser, H. (1907). Ueber modifizierte salizylsäuren. *Medizinische Klinik, 14*, 390–393.

Dubach, U. C., Rosner, B., & Pfister, E. (1983). Epidemiologic study of analgesics containing phenacetin – Renal morbidity and mortality (1968–1979). *The New England Journal of Medicine, 308*, 357–362.

Feret, B. (2003). Lumiracoxib: A Cox-2 inhibitor for the treatment of arthritis and acute pain. *Formulary, 38*, 529–537.

Flower, R. J. (2003). The development of Cox-2 inhibitors. *Nature Reviews, 2*, 179–191.

Fu, J. Y., Masferrer, J. L., Seibert, K., et al. (1990). The induction and suppression of prostaglandin H_2 synthase (cyclooxygenase) in human monocytes. *Journal of Biological Chemistry, 265*, 16737–16740.

Hinz, B., & Brune, K. (2002). Cyclooxygenase-2–10 years later. *Journal of Pharmacology and Experimental Therapeutics, 300*, 367–375.

Hinz, B., Renner, B., & Brune, K. (2007). Drug insight: Cyclo-oxygenase-2 inhibitors-a critical appraisal. *Nature Clinical Practice Rheumatology, 3*, 552–560.

Kaufman, D. W., Kelly, J. P., Levy, M., et al. (1991). *The drug etiology of agranulocytosis and aplastic anemia.* New York: Oxford University Press.

Kujubu, D. A., Fletcher, B. S., Varnum, B. C., et al. (1991). TIS10, A phorbol ester tumor promoter-inducible mRNA from Swiss 3 T3 cells, encodes a novel prostaglandin synthase/cyclooxygenase homologue. *Journal of Biological Chemistry, 266*, 12866–12872.

Lombardino, J. G. (1974). Enolic acids with anti-inflammatory activity. In R. A. Scherrer & M. W. Whitehouse (Eds.), *Medical chemistry: A series of monographs* (pp. 129–157). New York: Academic.

Masferrer, J. L., Zweifel, B. S., Seibert, K., et al. (1990). Selective regulation of cellular cyclooxygenase by dexamethasone and endotoxin in mice. *The Journal of Clinical Investigation, 86*, 1375–1379.

Otterness, I. G., Larson, D. L., & Lombardino, J. G. (1982). An analysis of piroxicam in rodent models of arthritis. *Agents and Actions, 12*, 308–312.

Rainsford, K. D. (1984). *Aspirin and the salicylates.* London: Butterworth.

Randall, L. O., & Selitto, J. J. (1957). A method for measurement of analgesic activity on inflamed tissue. *Archives International Pharmacodynamics, 111*, 409–419.

Rinsema, T. J. (1999). One hundred years of aspirin. *Medical History, 43*, 502–507.

Sertürner, F. W. (1806). Darstellung der reinen Mohnsäure (Opiumsäure) nebst einer chemischen Untersuchung des Opiums. *Journal der Pharmacie, 14*, 47–93.

Shen, T. Y. (1984). The proliferation of non-steroidal anti-inflammatory drugs (NSAIDS). In M. J. Parnham & J. Bruinvels (Eds.), *Discoveries in pharmacology* (Haemodynamics, hormones and inflammation, Vol. 2, pp. 523–553). Amsterdam: Elsevier.

Smolen, J. S., Breedveld, F. C., Burmester, G. R., et al. (2000). Consensus statement on the initiation and continuation of tumour necrosis factor blocking therapies in rheumatoid arthritis. *Annals of the Rheumatic Diseases, 59*, 504–505.

Sneader, W. (1985). *Drug discovery: The devolution of modern medicines.* Chichester: Wiley.

Tainter, M. L. (1948). Pain. *Annals of the New York Academy of Sciences, 51*, 3–11.

Tegeder, I., Lotsch, J., Krebs, S., et al. (1999). Comparison of inhibitory effects of meloxicam and diclofenac on human thromboxane biosynthesis after single doses and at steady state. *Pharmacology and Therapeutics, 65*, 533–544.

Vane, J. R. (1971). Inhibition of prostaglandin synthesis as a mechanism of action of aspirin-like drugs. *Nature New Biology, 231*, 232–235.

Wilhelmi, G. (1949). Guinea-Pig ultraviolet erythema test. *Schweizerische Medizinische Wochenschrift, 79*, 577–580.

Williams, G. (1975). *The age of agony. The art of healing 1700–1800.* London: Constable.

Winter, C. A., Risley, E. A., & Nuss, G. W. (1962). Carrageenin-induced edema in hindpaw of the rat as an assay for anti-inflammatory drugs. *Proceedings of the Society for Experimental Biology, 111*, 544–547.

Hit Rate or Sensitivity

Definition

Hit rate or sensitivity is the probability of response "A" when event A has occurred.

HIV and Pain

▶ Pain in Human Immunodeficiency Virus Infection and Acquired Immune Deficiency Syndrome

HIV/AIDS-related Pain

▶ Cancer Pain and Pain in HIV/AIDS

HIV-associated Neuropathy

▶ Pain in Human Immunodeficiency Virus Infection and Acquired Immune Deficiency Syndrome

HMSN

Definition

The acronym for hereditary motor and sensory neuropathy.

Cross-References

▶ Hereditary Neuropathies

Hoffmann-Tinel Sign

▶ Tinel Sign

Holistic Medicine

▶ Alternative Medicine in Neuropathic Pain

Homeopathy

Definition

Homeopathy is a system of medicine developed by Samuel Hahnemann in the nineteenth century based on a concept of vital energy inherent in all matter, which increases in potency with repeated dilution, and on the idea that substances can be used to treat conditions that mimic their toxicity.

Cross-References

▶ Alternative Medicine in Neuropathic Pain

Homeostasis

Definition

Homeostasis is a basic biological function associated with the maintenance of an internal environment that guarantees survival through adjusting important biological parameters (water, salt, glucose, temperature, acidity, etc.).

Cross-References

▶ Clinical Migraine with Aura
▶ Functional Imaging of Cutaneous Pain

Homeostatic Adaptations

Definition

Physiological responses or behavioral actions which maintain or restore normal levels of biological function (e.g., maintain or restore normal body temperature).

Cross-References

▶ Opioids, Effects of Systemic Morphine on Evoked Pain

Homework

Definition

Activities that a patient is asked to complete or practice, usually outside of the hospital or clinic, are referred to as homework.

Cross-References

▶ Multidisciplinary Pain Centers, Rehabilitation

Homologous Gene

Definition

A homologous gene has a similar, though often far from identical, sequence to another gene.

Cross-References

▶ Species Differences in Skin Nociception

Homomeric Channels

Definition

Channels are protein complexes, which form pores in the cell membrane. Typically, channels are made up of several subunits, which may be identical, resulting in homomeric channels, or different, resulting in heteromeric channels.

Cross-References

▶ Purine Receptor Targets in the Treatment of Neuropathic Pain

Homozygous Carriers

Definition

Homozygous carriers refer to the state of possessing two identical alleles of a particular gene, one inherited from each parent.

Cross-References

▶ NSAIDs, Pharmacogenetics

Horner's Syndrome

▶ Headache Due to Dissection

Horsley-Clarke Apparatus or Stereotaxic Frame

Definition

Horsley-Clarke apparatus or stereotaxic frame is a solid metallic frame made of two horizontal graduated bars fixed perpendicularly to a metal plate, holding a device for the fixation of the head of the animal through its upper jaw and orbits. The horizontal bars hold at mid-distance a device to fix two bars introduced into the ears (external auditory meatus).

Cross-References

▶ Poststroke Pain Model, Thalamic Pain (Lesion)

Hospice Care

Definition

The term "hospice care" can refer to the care provided in a special health-care system (as in the USA) or it can be used to denote end-of-life care or palliative care generally (in countries other than the USA). In the USA, hospice is a government-supported health-care system established more than 25 years ago to provide specialist palliative care for patients with far advanced illness. The core of this system is a home care program offering a range of services to the patient and family.

Cross-References

▶ Cancer Pain Management, Interface Between Cancer Pain Management and Palliative Care

Hostility

Definition

Hostility refers to "a set of negative attitudes, beliefs and appraisals concerning others." Can be inwardly directed toward oneself ("intrapunitiveness") or directed toward others ("extrapunitiveness"), Smith (1992).

Cross-References

▶ Anger and Pain
▶ Chronic Gynecological Pain, Doctor-Patient Interaction

References

Smith, T. W. (1992). Hostility and health: current status of a psychosomatic hypothesis. *Health Psychology*, 11(3), 139–150.

Hot Plate Test (Assay)

Definition

The hot plate test involves placement of a rat or mouse on a heated surface (usually $\geq 50\ ^{\circ}C$) and measuring the latency (in seconds) to licking of a hindpaw or jumping. The hot plate test, which employs a fixed temperature, has largely been supplanted by what is termed the "Hargreaves Test" in which a rat or mouse is placed on a glass surface and the animal's hindpaw is heated from below by a radiant heat source. Latency from stimulus to withdrawal of the paw is measured in this test.

Cross-References

▶ Thalamotomy, Pain Behavior in Animals

Hot Tooth Syndrome

Definition

A tooth is sometimes described as "hot" when it is very painful and difficult to anesthetize even with regional block anesthesia. The tooth is usually spontaneously painful, tender to touch, and difficult to treat.

Cross-References

▶ Dental Pain, Etiology, Pathogenesis, and Management

Household Income and Chronicity

▶ Pain in the Workplace, Risk Factors for Chronicity, and Demographics

HPA Axis

Definition

The hypothalamus-pituitary-adrenal axis forms the basic response triad regulating endogenous glucocorticoid concentrations in the circulation.

Cross-References

▶ Fibromyalgia, Mechanisms, and Treatment

HT Neurons

▶ High Threshold Neurons

Human Factors Engineering

▶ Ergonomics Essay

Human Models

▶ Human Models of Inflammatory Pain

Human Models of Inflammatory Pain

Louise Harding
Hospitals NHS Foundation Trust, University College London, London, UK

Synonyms

Human models; Inflammatory pain

Definition

Research tools used to investigate the mechanisms and pharmacology of inflammatory pain and neuronal sensitization.

Characteristics

Inflammation is a response of the body tissues to injury or irritation. Its most prominent features are pain, swelling, redness, and heat. Through activation and sensitization of nociceptors, inflammatory mediators also cause peripheral and central sensitization of the somatosensory system, altering the way we perceive mechanical and thermal stimuli in and around inflamed skin. Studying these changes can provide information on the underlying mechanisms and, when combined with drug studies, on the pharmacology of inflammation and neuronal sensitization. A number of experimental models have been developed for this purpose. In each model, inflammation is evoked by a different insult or injury and different models have specific characteristics. Table 1 compares the key features of the most common models which are summarized below.

The Capsaicin Model
▶ Capsaicin is the chemical component of chili peppers that gives them their "hot" quality. It directly activates ▶ TRPV1, a heat sensitive cationic ion channel expressed on cutaneous nociceptors, resulting in pain and inflammation. Capsaicin can either be applied topically, typically at 1 %, or injected intradermally (doses of 25–250 μg). Intradermal injection is associated with a quick hit of intense pain lasting 1–2 min compared to the mild-moderate pain of topical application that develops slowly over 10–30 min. Both methods produce neurogenic inflammation and similar changes in somatosensory function (LaMotte et al. 1991). At the primary zone, i.e., the area of inflammation, heat pain thresholds are reduced in the capsaicin model due to sensitization of TRPV1. Very high concentrations of capsaicin desensitize the heat responsive ion channel. This is sometimes evident following

Human Models of Inflammatory Pain, Table 1 Comparison of somatosensory changes produced by different human models of inflammatory pain

	Capsaicin	Burn	Heat/Capsaicin	Mustard oil	Electrical	UVB	Freeze
1° Heat pain[a]	Yes	Yes	Yes	Yes	Yes	Yes	Yes
2° Punctate[b]	Yes	Yes	Yes	Yes	Yes	Yes	Yes
2° Dynamic[c]	Yes	Yes	Yes	Yes	Yes	No	No
Dynamic duration[d]	<1 h	<1 h	~4 h	<1 h	>2 h	–	–
2° Punctate onset[e]	<1 h	<1 h	<1 h	<1 h	<1 h	>~4 h	>~4 h

[a]Decreased heat pain threshold in the inflamed site
[b]Area of secondary punctate hyperalgesia
[c]Area of secondary dynamic mechanical allodynia
[d]Duration of dynamic mechanical allodynia
[e]Time to onset of secondary punctate hyperalgesia. See text on individual models for data references

intradermal delivery and is characterized by an increase in heat pain thresholds in a 1–3 mm area around the injection site. Surrounding the primary zone, two discrete areas of ► secondary hyperalgesia develop: an area of dynamic ► mechanical allodynia and an area of ► punctate hyperalgesia. These two areas differ in development time, size, pharmacological sensitivity, and duration. The area of dynamic mechanical allodynia is maintained by ongoing afferent input from excited nociceptors and fades within an hour of capsaicin delivery as its concentration in the skin fades. In contrast, the area of punctate hyperalgesia, once established, appears independent of afferent input and may remain for 24 h.

The Burn Model

In this model, heat is used to produce a first degree burn on the skin. CO_2 lasers and electronically coupled thermodes are typically used to induce the burn, by heating the skin to approximately 47 °C for 7 min (Pedersen and Kehlet 1998). The burn stimulus is moderately painful during its application; however, the pain quickly subsides once the heat stimulus is stopped. The injury produces a flare response similar to the capsaicin model. Evoked somatosensory changes in the primary zone are heat pain sensitization (reduced heat pain threshold), together with a mild hypoesthesia (loss of sensation) to warming and cooling. A secondary area of punctate hyperalgesia develops around the primary zone, and dynamic mechanical allodynia can also develop, but this depends on

experimental conditions. Thermode size, location of skin stimulated, temperature, and duration of burn stimulus shape the intensity of the burn. If the burn is very mild, insufficient afferent drive is sustained to maintain dynamic mechanical allodynia once the burn stimulus is removed.

The Heat/Capsaicin Model

This model, as it suggests, uses both heat and capsaicin to produce inflammatory pain and hyperalgesia. A heat stimulus of 45 °C is applied to the skin for 5 min. followed by a 30 min. application of low-dose (0.075 %) topical capsaicin (Petersen and Rowbotham 1999). This produces areas of primary and secondary hyperalgesia comparable to the capsaicin model. Like the capsaicin model, the area of dynamic mechanical allodynia starts to fade after approximately 20 min, but in this model the area can be rekindled by restimulating the treated site with a heat stimulus of 40 °C for 5 min. This rekindling can be repeated every 20 min for up to 4 h, providing a much longer opportunity to study the mechanisms of dynamic mechanical allodynia than the capsaicin and heat models alone.

The Mustard Oil Model

The irritant mustard oil, allyl isothiocyanate, produces characteristics of inflammation and somatosensory changes comparable to the capsaicin model, i.e., sensitization to heat in the primary zone and secondary areas of dynamic mechanical allodynia and punctate hyperalgesia

(Koltzenburg et al. 1992). Applied topically for 5 min, either at 100 % or diluted for a lesser effect, mustard oil produces moderate to severe pain and neurogenic inflammation. Its mechanism of action is essentially unknown. Allyl isothiocyanate has recently been shown to be an agonist of the ▶ TRPA1 receptor (previously known as ANKTM1) expressed in nociceptors (Jordt et al. 2004), and this receptor may be key to its inflammatory effects. Prolonged application of mustard oil however causes blistering, which suggests the inflammation process in this model may also involve tissue damage pathways.

The Electrical Stimulation Model

As discussed, in experimental models of pain, dynamic mechanical allodynia is maintained by ongoing afferent input from excited C nociceptors. The electrical stimulation model uses continuous electrical activation of ▶ C fibers to evoke and maintain a stable area of dynamic allodynia throughout the experimental period. Current is injected at a frequency of 5 Hz and adjusted until the subject reports a pain intensity of 5/10 on a numerical pain intensity rating scale (mean current: 67 mA) (Koppert et al. 2001). This method produces an inflammatory pain response with stable dynamic allodynia for study periods of up to 2 h. Other characteristics of this model are the reduced heat pain thresholds in the primary zone and secondary area of punctate hyperalgesia common to most established models of inflammation.

The UVB/Sunburn Model

This model has two essential differences to those discussed so far. Firstly, there is a prolonged delay period of 6–12 h between the inflammatory stimulus and the development of erythema and hyperalgesia. Secondly, the stimulus event used to create inflammation is not in itself painful (Bickel et al. 1998). This model is particularly interesting, therefore, as the mechanisms of inflammation and hyperalgesia may differ somewhat to those evoked by direct activation of nociceptors. In this model, inflammation is produced by irradiating the skin with ultraviolet light in the UVB wavelength range (290–320 nm),

typically over an area of approximately 5 cm diameter. There is considerable intersubject variability in the dose of radiation required to produce inflammation; consequently, subjects are assessed prior to the experimental period to establish the minimum dose of UVB required. For studies of ▶ primary and secondary hyperalgesia, three times the minimum dose required to produce ▶ erythema is used for experimentation. The UV model produces primary hyperalgesia to heat and secondary hyperalgesia to punctate mechanical stimuli, but not dynamic mechanical allodynia. Both primary and secondary events have a delayed onset and are typically studied 20 h after irradiation. This model is therefore relatively demanding, compared to other models, as subjects are required on three consecutive days. An advantage of this model, however, is that the sensory changes are stable for 10 h, giving a long window for detailed study.

The Freeze Lesion Model

Delayed onset hyperalgesia is also a characteristic of the freeze lesion model. Freeze lesions can be created using a 1.5 cm diameter copper rod cooled to −28 °C and held perpendicularly against the skin for 10 s (Kilo et al. 1994). This produces mild to moderate sharp prickling pain, vasodilation of the stimulated and surrounding area, and a local edema. Pain, edema, and flushing outside the contact area subside within 2 h; however, a discrete erythema at the contact area remains for a number of days. No primary or secondary hyperalgesia can be detected in the first hours following the injury, but both are developed by the subsequent day. This model does not produce dynamic mechanical allodynia, and the area of punctate hyperalgesia produced by the freeze lesion model is typically much smaller than that produced by other models.

In addition to the models described above, inflammatory pain and hyperalgesia have been reported following administration of a number of other inflammatory stimuli. This is not an exhaustive list, but for reference includes melatonin from bee venom (Sumikura et al. 2003), acidic phosphate buffered solution (Steen and Reeh 1993), complete Freunds adjuvant (Gould 2000), and bradykinin (Manning et al. 1991).

References

Bickel, A., Dorfs, S., Schmelz, M., et al. (1998). Effects of antihyperalgesics on experimentally induced hyperalgesia in man. *Pain, 76,* 317–325.

Gould, H. J. (2000). Complete Freund's adjuvant-induced hyperalgesia: A human perception. *Pain, 85,* 301–303.

Jordt, S. V., Bautista, D. M., Chuang, H. H., et al. (2004). Mustard oils and cannabinoids excite sensory nerve fibres through the TRP channel ANKTM1. *Nature, 427,* 260–265.

Kilo, S., Schmelz, M., Koltzenburg, M., et al. (1994). Different patterns of hyperalgesia induced by experimental inflammation in human skin. *Brain, 117,* 385–396.

Koltzenburg, M., Lundberg, L. E. R., & Torebjork, H. E. (1992). Dynamic and static components of mechanical hyperalgesia in human hairy skin. *Pain, 51,* 207–219.

Koppert, W., Dern, S. K., Sittl, R., et al. (2001). A new model of electrically evoked pain and hyperalgesia in human skin. *Anesthesiology, 95,* 395–402.

LaMotte, R. H., Shain, C. N., Simone, D. A., et al. (1991). Tsai EFP neurogenic hyperalgesia: Psychophysical studies of underlying mechanisms. *Journal of Neurophysiology, 66,* 190–211.

Manning, D. C., Raja, S. N., Meyer, R. A., & Campbell, J. N. (1991). Pain and hyperalgesia after intradermal injection of bradykinin in humans. *Clinical Pharmacology and Therapeutics, 50,* 721–729.

Pedersen, J. L., & Kehlet, H. (1998). Hyperalgesia in a human model of acute inflammatory pain: A methodological study. *Pain, 74,* 139–151.

Petersen, K. L., & Rowbotham, M. C. (1999). A new human experimental pain model: The heat/capsaicin sensitization model. *Neuroreport, 10,* 1511–1516.

Steen, K. H., & Reeh, P. W. (1993). Sustained graded pain and hyperalgesia from harmless experimental tissue acidosis in human skin. *Neuroscience Letters, 154,* 113–116.

Sumikura, H., Andersen, O. K., Drewes, A. M., et al. (2003). A comparison of hyperalgesia and neurogenic inflammation induced by melittin and capsaicin in humans. *Neuroscience Letters, 337,* 147–150.

Human Rights

▶ Ethics of Pain Control in Infants and Children

Human Thalamic Nociceptive Neurons

Karen D. Davis[1,2] and Jonathan O. Dostrovsky[3,4]
[1]Brain, Imaging and Behaviour – Systems Neuroscience, Toronto Western Research Institute, Toronto, ON, Canada
[2]Department of Surgery, University of Toronto, Toronto, ON, Canada
[3]Department of Physiology, University of Toronto, Toronto, ON, Canada
[4]Faculty of Dentistry, University of Toronto, Toronto, ON, Canada

Definition

Central nervous system neurons whose cell bodies are located within the human thalamus (diencephalon) and that have a preferential or exclusive response to ▶ noxious stimuli. Such neurons are generally classified as ▶ wide dynamic range (WDR) and ▶ nociceptive specific (NS).

Introduction

It is possible to directly study human thalamic nociceptive neurons during the electrophysiological mapping used by some neurosurgical teams as part of functional neurosurgical procedures for treating chronic pain, Parkinson's disease, or other movement disorders (Lenz et al. 1988; Tasker and Kiss 1995). During these mapping procedures, microelectrodes are inserted into the thalamus to record the electrophysiological properties of individual thalamic neurons. The human thalamus, which is very similar to the monkey thalamus, includes several regions where neurons responding specifically or preferentially to nociceptive stimuli are found. However, in view of the very limited opportunities available to search for such neurons in the human and perform extensive testing on them, our knowledge concerning their properties and locations is extremely limited. See also ▶ Lateral Thalamic Pain-Related Cells in Humans and ▶ Central Pain, Human Studies of Physiology.

Characteristics

The unique opportunity afforded by functional stereotactic surgery to record and stimulate in the thalamus of awake patients has provided some interesting findings and validation of subhuman primate studies related to thalamic function in pain. Unfortunately, the inherent limitations of these studies (time constraints, ethical considerations, and lack of histological confirmation) limit the interpretation of the findings. The studies have attempted to address the following questions:

1. Where can one record nociceptive and thermoreceptive neurons?
2. What are the properties of nociceptive and thermoreceptive neurons?
3. What are the perceptual consequences of microstimulation in the regions containing nociceptive and thermoreceptive neurons?
4. Where can one evoke painful and temperature sensations by stimulating in the thalamus and what are the qualities of the sensations?
5. Are there any alterations in neuronal firing characteristics, receptive fields, or stimulation-evoked sensations in chronic pain patients?

This section briefly summarizes the findings pertaining to these questions.

Nociceptive Neurons in Lateral Thalamus

The existence of nociceptive neurons in Vc (ventrocaudal nucleus often termed VP or ventroposterior nucleus) and adjacent regions has been reported by Lenz and colleagues (for review see Lenz and Dougherty 1997). The vast majority of Vc neurons are classified as non-nociceptive tactile neurons, since they respond to light touch of a distinct area of skin (i.e., the neuron's receptive field). However, there have been a few reports of some nociceptive neurons in Vc. Approximately 5–10 % of Vc neurons have been classified as nociceptive, based on their responses to noxious thermal stimuli (Lenz et al. 1993a, 1994). A larger proportion of Vc neurons, up to 25 %, were found to respond selectively or preferentially to noxious mechanical stimuli (Lee et al. 1999; Lenz et al. 1994). These neurons were primarily located in the posterior-inferior portion of Vc. Interestingly, in the adjoining posterior-inferior area, which includes VMpo (Blomqvist et al. 2000), they identified NS neurons that responded to noxious heat, and none of the neurons in this area responded to innocuous tactile stimuli (Lenz et al. 1993a). In a more recent study by Lenz and colleagues (Kobayashi et al. 2009), a painful laser stimulus was used and evoked responses in 1/3 of the neurons tested; many responded with early and/or late latency peaks of activity, consistent with activity arising from Aδ and C fibers. Many of the activated neurons were located posterior and inferior to Vc.

The true proportion of thalamic nociceptive neurons may be underestimated in these studies for a variety of technical, physiological, and ethical reasons. First, there are very few opportunities to test for nociceptive responses in awake human subjects, and the small body of data that has been obtained derives from patients with either movement disorders or a chronic pain condition. Second, extensive testing for nociceptive responses (both in terms of the number of neurons tested and the skin area tested) is limited due to the painful nature of the stimulus. Third, the regions explored are limited in extent and location and are determined by the target location for the surgery. Fourth, it is not clear whether there is any selection bias in the ability of microelectrodes to record from nociceptive versus tactile neurons (e.g., based on cell size, spontaneous activity).

Medial Thalamus

Much less is known regarding the role of the medial thalamus compared to the lateral thalamus in human pain, largely due to the fact that there are few opportunities to record and stimulate in this region during functional stereotactic surgery. There are some discrepancies in the incidence of medial thalamic nociceptive responses across the few published studies. One group (Ishijima et al. 1975) reported a similar proportion of mechanically and thermally responsive nociceptive neurons in the CM-Pf region, as compared to the findings of Lenz and colleagues in lateral thalamus. However, another group found only 2 of 318 medial thalamic neurons that responded to noxious stimuli

(Jeanmonod et al. 1993). It is, however, difficult to evaluate these findings as few details were provided by the authors, and more recent studies have failed to replicate the findings (see Lenz and Dougherty (1997) for references).

Stimulation-Induced Pain

One of the unique aspects of electrophysiological studies in human patients is the ability to question the patient about sensations evoked by electrical stimulation within the brain. Electrical stimulation within Vc and adjacent regions of the thalamus usually evokes innocuous paresthesia. However, several early studies documented that stimulation in the area posterior-inferior to Vc elicited reports of painful sensations in some patients (Halliday and Logue 1972; Hassler and Riechert 1959; Tasker 1984). Recent studies have examined the effects of stimulation in much greater detail (Davis et al. 1996; Dostrovsky et al. 2000; Lenz et al. 1993b), and these show that pain and innocuous thermal sensations can be evoked from a region at the posterior-inferior border of Vc and extending several millimeters posterior, inferior, and medial. Microstimulation applied at the Vc sites of confirmed nociceptive neuronal responses rarely evokes pain but rather produces a non-painful tingling sensation (Lee et al. 1999; Lenz et al. 1993a, b, 1994). A greater incidence of stimulation-evoked pain in Vc and the ventroposterior region has been reported in patients with a history of visceral pain, phantom pain, or poststroke pain (Davis et al. 1995, 1996, 1998; Lenz et al. 1995).

The incidence of evoked pain/thermal sensations is much higher in the posterior-inferior area than within Vc proper. Unlike the paresthetic (tingling and "electric shock") sensations evoked in Vc, the pain/thermal sensations are usually reported as quite natural. They are always perceived on the contralateral side of the body, and the projected fields can be quite small. The painful sensations are frequently described as burning pain. In a few cases, sensations of pain referred to deep and visceral sites have been elicited (Lenz et al. 1994; Davis et al. 1995). Lenz and colleagues have reported that microstimulation within Vc (at sites where WDR neurons responding to noxious mechanical stimuli were found) rarely results in

pain, whereas at the sites in the region posterior-inferior to Vc where microstimulation evoked pain, there was a high likelihood of finding nociceptive neurons (Lenz and Dougherty 1997). Histological confirmation of these stimulation and recording sites has of course not been obtained in such patients, but it seems likely that the physiologically localized region posterior-inferior to Vc corresponds anatomically to VMpo.

A few studies have reported that stimulation in the posterior aspect of medial thalamus can evoke pain (Jeanmonod et al. 1993; Sano 1979), but in most cases, large-tipped electrodes and high intensities were used for stimulation, so current spread is an issue. More recent studies have failed to replicate these findings.

Innocuous Cool Neurons and Sensations

Cells responding to innocuous thermal stimuli are also of great interest and highly relevant, due to the well-known association of the pain and temperature pathways. Cooling-specific neurons are only found in lamina I of the spinal and trigeminal dorsal horns and have been shown to project to VMpo in the monkey (Dostrovsky and Craig 1996). In animal studies, cooling neurons in the thalamus have only been reported in VMpo (monkey) and medial VPM (cat). Cooling-specific neurons in human thalamus were located in the region medial and posterior-inferior to Vc that likely corresponds to the human VMpo (Davis et al. 1999). Of particular interest was the finding that stimulation at such sites evoked cooling sensations that were graded with stimulus intensity and that were referred to the same cutaneous region as the receptive fields of the cooling-specific neurons recorded at the site. Stimulation in this posterior-inferior region can also elicit pain (see above) and, as shown by Lenz and colleagues (1993a, b), this region also contains nociceptive-specific neurons.

References

Blomqvist, A., Zhang, E. T., & Craig, A. D. (2000). Cytoarchitectonic and immunohistochemical characterization of a specific pain and temperature relay, the posterior portion of the ventral medical nucleus, in the human thalamus. *Brain, 123*, 601–619.

Davis, K. D., Tasker, R. R., Kiss, Z. H. T., et al. (1995). Visceral pain evoked by thalamic microstimulation in humans. *NeuroReport, 6,* 369–374.

Davis, K. D., Kiss, Z. H. T., et al. (1996). Thalamic stimulation-evoked sensations in chronic pain patients and in non-pain (movement disorder) patients. *Journal of Neurophysiology, 75,* 1026–1037.

Davis, K. D., Kiss, Z. H. T., Luo, L., et al. (1998). Phantom sensations generated by thalamic microstimulation. *Nature, 391,* 385–387.

Davis, K. D., Lozano, R. M., et al. (1999). Thalamic relay site for cold perception in humans. *Journal of Neurophysiology, 81,* 1970–1973.

Dostrovsky, J. O., & Craig, A. D. (1996). Cooling-specific spinothalamic neurons in the monkey. *Journal of Neurophysiology, 76,* 3656–3665.

Dostrovsky, J. O., Manduch, M., et al. (2000). Thalamic stimulation-evoked pain and temperature sites in pain and non-pain patients. In M. Devor, M. Rowbotham, & Z. Wiesenfeld-Hallin (Eds.), *Proceedings of the 9th World Congress on pain* (Vol. 7, pp. 419–425). Seattle: IASP Press.

Halliday, A. M., & Logue, V. (1972). Painful sensations evoked by electrical stimulation in the thalamus. In G. G. Somjen (Ed.), *Neurophysiology studied man* (pp. 221–230). Amsterdam: Excerpta Medica.

Hassler, R., & Riechert, T. (1959). Clinical and anatomical findings in stereotactic pain operations on the thalamus. *Archiv für Psychiatrie und Nervenkrankheiten, vereinigt mit Zeitschrift für die gesamte Neurologie und Psychiatrie, 200,* 93–122.

Ishijima, B., Yoshimasu, N., Fukushima, T., et al. (1975). Nociceptive neurons in the human thalamus. *Confinia neurologica, 37,* 99–106.

Jeanmonod, D., Magnin, M., & Morel, A. (1993). Thalamus and neurogenic pain: Physiological, anatomical and clinical data. *NeuroReport, 4,* 475–478.

Kobayashi, K., Winberry, J., Liu, C. C., Treede, R. D., & Lenz, F. A. (2009). A painful cutaneous laser stimulus evokes responses from single neurons in the human thalamic principal somatic sensory nucleus ventral caudal (Vc). *Journal of Neurophysiology, 101,* 2210–2217.

Lee, J., Dougherty, P. M., Antezana, D., et al. (1999). Responses of neurons in the region of human thalamic principal somatic sensory nucleus to mechanical and thermal stimuli graded into the painful range. *The Journal of Comparative Neurology, 410,* 541–555.

Lenz, F. A., & Dougherty, P. M. (1997). Pain processing in the human thalamus. In M. Steriade, E. G. Jones, & D. A. McCormick (Eds.), *Thalamus* (Experimental and clinical aspects, Vol. II, pp. 617–652). Amsterdam: Elsevier.

Lenz, F. A., Dostrovsky, J. O., Kwan, H. C., et al. (1988). Methods for microstimulation and recording of single neurons and evoked potentials in the human central nervous system. *Journal of Neurosurgery, 68,* 630–634.

Lenz, F. A., Gracely, R. H., Rowland, L. H., et al. (1994). A population of cells in the human thalamic principal sensory nucleus respond to painful mechanical stimuli. *Neuroscience Letters, 180,* 46–50.

Lenz, F. A., Gracely, R. H., Romanoski, A. J., et al. (1995). Stimulation in the human somatosensory thalamus can reproduce both the affective and sensory dimensions of previously experienced pain. *Nature Medicine, 1,* 910–913.

Lenz, F. A., Seike, M., Lin, Y. C., et al. (1993a). Neurons in the area of human thalamic nucleus ventralis caudalis respond to painful heat stimuli. *Brain Research, 623,* 235–240.

Lenz, F. A., Seike, M., Richardson, R. T., et al. (1993b). Thermal and pain sensations evoked by microstimulation in the area of human ventrocaudal nucleus. *Journal of Neurophysiology, 70,* 200–212.

Sano, K. (1979). Stereotaxic thalamolaminotomy and posteromedial hypothalamotomy for the relief of intractable pain. In J.J. Bonica & V. Ventrafridda (Eds.), *Advances in Pain Research and Therapy,* (Vol. 2, pp. 475–485). New York: Raven Press.

Tasker, R. R. (1984). Stereotaxic surgery. In P. D. Wall & R. Melzack (Eds.), *Textbook of pain* (pp. 639–655). Edinburgh: Churchill Livingstone.

Tasker, R. R., & Kiss, Z. H. T. (1995). The role of the thalamus in functional neurosurgery. *Neurosurgery Clinics of North America, 6,* 73–104.

Human Thalamic Response to Experimental Pain (Neuroimaging)

Alexandre F. M. DaSilva[1] and Nouchine Hadjikhani[2]
[1]Biologic & Materials Sciences, University of Michigan Dental School, MI, USA
[2]Martinos Center for Biomedical Imaging, Massachusetts General Hospital Harvard Medical School, Charlestown, MA, USA

Synonyms

Thalamic response to experimental pain in humans

Definition

The thalamus is the major relay structure in the forebrain for noxious and non-noxious sensory inputs. In the case of noxious stimuli, the thalamus distributes the incoming information to other specific cortical areas for proper processing of their discriminative, cognitive, and affective components. Recent neuroimaging techniques can effectively detect transient thalamic neuronal activation following the application of experimental stimuli that artificially replicate painful conditions in humans.

Characteristics

Thalamic neuronal activation is frequently observed in functional neuroimaging studies following ▶ experimental pain. Through the use of neuroimaging techniques, the role of the thalamus has been gradually dissected in the nociceptive CNS network. Under those studies, experimental pain resultant of different noxious stimuli has revealed a pattern of thalamic activation that depends on the type of stimuli (e.g., thermal), area of application, and conditions inherent to the subject or patient, such as attention, or the presence of a chronic pain disorder.

Techniques

Of the neuroimaging technologies available, ▶ functional magnetic resonance imaging (fMRI) and ▶ positron emission tomography (PET) have greatly expanded our knowledge of human thalamic response to pain. Both indirectly measure the neuronal activity based on changes in the metabolism during a particular transient task (e.g., experimental pain) compared to a baseline state (e.g., no-pain state). The specific contrast for fMRI is the blood oxygenation level-dependent (BOLD) contrast, which does not require any tracer agent but relies on blood volume and blood flow, whereas radioactive labeled tracers are used to measure changes in cerebral blood flow and metabolism in PET.

Thalamic Nuclear Function

There are 14 major thalamic nuclei identified, but this number diverges depending on the histological technique applied. Some of them, or subdivisions, have specific roles in the thalamic nuclear configuration for pain processing. Activations of the

ventroposterior nuclei of the ventrobasal complex (lateral), and other more medial nuclei of the thalamus, have been consistently described in neuroimaging studies. These studies confirm previous animal experiments that noxious and innocuous discriminative input from cranial and the body parts is respectively processed by the ventroposterior medial nucleus (VPM) and the ▶ ventroposterior lateral nucleus (VPL), and afterward projected to the somatosensory cortex. The lateral nuclear activation has a clear somatotopic configuration for different kinds of sensory input, while the medial thalamus, such as the ▶ dorsomedial nucleus (DM), has particular thermoreceptive functions. Noxious thermal stimulation to the facial skin of each trigeminal division in healthy human volunteers activates the contralateral VPM, while the same noxious stimulation applied to the palmar surface of the thumb activates the VPL (DaSilva et al. 2002). In both cases, during trigeminal and thumb noxious thermal stimulation, the contralateral DM nucleus of the thalamus shows activation (Fig. 1).

A specific thalamic nuclear pathway is involved in interoceptive mechanism of homeostasis: the basal part of the ventromedial nucleus (VMb) and the ▶ posterior part of the ventromedial nucleus (VMpo) play an important role in thermal nociceptive inflow through main direct projections to the insular cortex (Craig 2002). With the future improvement of the spatial resolution (2 mm for PET and <1 mm for fMRI) and signal noise rate in the neuroimaging studies, as well as the superposition of structural ▶ diffusion tensor imaging (DTI) maps to delineate nuclear architecture under the activations (Fig. 2), we will be able to precisely define the pattern of thalamic activation following painful stimulus in human (DaSilva et al. 2003; Wiegell et al. 2003).

Experimental Noxious Stimuli

Most of the noxious stimuli used in neuroimaging studies are thermal in nature, applying temperatures higher than 45 °C for heat pain and usually lower than 6 °C for cold pain, enough to activate nociceptive fibers (C- and A-delta). The noxious thermal stimuli are delivered by nonmagnetic contact probes, water immersion, and laser

Human Thalamic Response to Experimental Pain (Neuroimaging), Fig. 1 Activation in the thalamus. (a) Activation in the thalamus contralateral to a noxious thermal stimulus to the V3 (mandibular trigeminal branch) region of the face. (b) Activation in the thalamus following contralateral stimulation to the face and hand. The *white areas* show regions of common activation following noxious thermal stimulation to the V1 (ophthalmic), V2 (maxillary), and V3 (mandibular) distributions of the face, in regions defined as the dorsomedial (*DM*) and ventroposteromedial (*VPM*) nuclei. Activation of the thumb is mapped onto the same anatomical section (*purple circle*) and corresponds to the ventroposterolateral (VPL) nucleus. The regions are defined anatomically using the Talairach atlas. Time courses of activation for each area are shown in inserts. Percent signal change is shown in arbitrary units (a.u.); numbers in *bottom corners* indicate the Talairach coordinate in the rostrocaudal (z) axis (Da Silva et al. 2002)

(heat) to a particular part of the body in an alternating fashion with a non-noxious state (e.g., neutral 32 °C x noxious 46 °C). Similar noxious sensations have also been produced by interlaced application of non-noxious warm and cold temperatures, which is known as the thermal-grill illusion of pain. Other noxious stimuli that have been used in pain studies are mechanical (e.g., tonic pressure), electrical (e.g., intramuscular electrical stimulation), and chemical (e.g., subcutaneous injection of ► capsaicin or ascorbic acid).

Neuroimaging studies applying experimental stimuli produce thalamic activation. Noxious and innocuous thermal stimuli (cold and heat) activate the medial and lateral thalamic nuclei, with predominant contralateral activation, while innocuous mechanical stimuli mostly activate the lateral thalamus. Noxious mechanical stimuli (tonic pressure) elicit inconsistent contralateral thalamic activation (Creac'h et al. 2000), as tonic pain (long duration) elicits less thalamic activation than phasic pain (short duration). In addition, the amount of thalamic activation observed depends on the size of the somatotopical representation of the body part being stimulated (the face has, e.g., a much bigger cortical representation than the foot).

If the experimental noxious stimulus is applied to the same region but in different tissues, the thalamic activation pattern can also be distinct, as in the case of experimental skin and

Human Thalamic Response to Experimental Pain (Neuroimaging), Fig. 2 Color-coded DTI map superimposed on high-resolution anatomical image. Comparison of an axial (**a**) MPRAGE structural image and the corresponding (**b**) color-coded DTI map. The ROI is taken from the *yellow box* shown in the whole slice image at *bottom left*. On the MPRAGE, the thalamus appears homogeneous, whereas the DTI map shows significant substructure. The thalamic nuclei have been labeled according to their anatomical position and fiber orientation. The color coding depicts the local fiber orientation (i.e., the principal eigenvector of the diffusion tensor) with *red* indicating mediolateral, *green* anteroposterior, and *blue* superoinferior. The color coding is also indicated by the *red-green-blue sphere* shown at *top center*. Abbreviations: *VA* ventral anterior, *VL* ventrolateral, *MD* mediodorsal, *VPL* ventral posterior lateral, *VPM* ventral posterior medial, *Pu* pulvinar (Da Silva et al. 2003)

muscular pain (Svensson et al. 1997). Although noxious intramuscular electrical stimulation and cutaneous pain, elicited by CO_2 laser in the left brachioradialis area, produce equal positive correlation between increases in regional cerebral flow (rCBF) in thalamus and anterior insula, only the cutaneous noxious stimulation shows a negative relationship in rCBF changes between thalamus and contralateral primary sensorimotor cortex, indicating a possible inhibitory mechanism between both structures.

Chemical experimental pain using capsaicin has been used in neuroimaging studies in two different ways: to induce acute and/or allodynic pain. Capsaicin is a hot pepper-derived substance that induces consistent ongoing pain, with a response including midline thalamic nuclei such as the DM nucleus (Iadarola et al. 1998). The cutaneous area treated with capsaicin, injected or topically applied, also develops secondary ▶ allodynia. Allodynia is a reversible state of painful sensitivity to non-noxious stimuli, such as brush and warm stimuli that replicates a clinical phenomenon common in ▶ neuropathic pain, burn lesions, and ▶ migraine patients. Capsaicin-allodynia to non-noxious heat

activates the medial thalamus simultaneously with the frontal cortex, orbital, and dorsolateral prefrontal (DLPFC), suggesting a greater affective and cognitive response, which correlates with the higher unpleasantness rating compared to normal heat pain rating (Lorenz et al. 2003).

Conditions inherent to the subjects also affect the thalamic response to experimental pain. There is an indication that gender differences in pain perception influence thalamus function. For the same thermal noxious pain, females show a higher rating for pain intensity than males, translated into higher activation in the contralateral thalamus, as well as in the prefrontal cortex and anterior insula (Paulson et al. 1998). Male subjects demonstrate higher μ-opioid system activation than female subjects in the anterior thalamus, ventral basal ganglia, and amygdala during sustained deep muscular pain (Zubieta et al. 2002). Pain perception is also altered by attention, hypnosis, or pharmacologic effect through a modulation of the pain system involving mainly the thalamus and cingulated cortex. Distraction tasks presented to subjects during thermal pain correlate with decreased perception of pain and consequently lower medial thalamic activation (Bantick et al. 2002). In a hypnotic state, the patient's reduced pain perception correlates with high functional modulation between the midcingulate cortex, the thalamus, and the midbrain (Faymonville et al. 2003). Under the influence of fentanyl, a μ-opioid receptor agonist, there is a strong attenuation of responses to noxious cold stimulation in the contralateral thalamus and primary somatosensory cortex (Casey et al. 2000).

Acute and chronic pain can alter the pattern of thalamic and cortical activation. Patients suffering acute post-dental extraction show increased response to heat pain applied to the ipsilateral hand in the somatosensory pathway, including thalamus and S1 (Derbyshire et al. 1999). This increased level of rCBF does not occur when the same noxious stimulus is applied to the hand contralateral to the dental extraction. This fact can be explained by the ongoing postsurgical inflammatory process, and its repercussions in the CNS awareness, amplifying any further sensory input from the ipsilateral areas, surrounding or distant (for safeguard) from the injury. Chronic pain disorders have shown a central distinct neuroplastic mechanism in response to the persistent pain input overflow. Instead of thalamic increase activation to painful stimulation, there is attenuation of the response and even a decrease of the rCBF in the thalamus. This is the case for ► fibromyalgia and neuropathic patients, where chronic thalamic activation following their persistent evoked and ongoing clinical pain attenuates or decreases its response after time (Gracely et al. 2002; Hsieh et al. 1995; Kwiatek et al. 2000). Patients suffering from ► cluster headache, a primary headache disorder, also show similar results, with significantly lower rCBF changes during the headache-free period compared to control subjects in the contralateral thalamus and S1 after ipsilateral tonic cold pain stimulation (Di Piero et al. 1997).

Can We Modulate the Thalamic Response to Pain?

The motor cortex is a reliable target to modulate the sensory and motor subthalamic activity associated with pain, independent of the type of stimulation applied (Garcia-Larrea et al. 1997, 1999; Strafella et al. 2004), which also affects other pain-matrix structures. This occurs directly and indirectly, because of the multiple connections between the corticospinal tract and thalamus. Using a forward modeling analysis for primary motor cortex (M1) modulation with transcranial direct current stimulation (tDCS), researchers predicted current flow in multiple areas of cortex, as well as several subcortical regions implicated in pain perception and modulation, especially the thalamus (DaSilva et al. 2012). In fact, M1-tDCS protocol can decrease pain in chronic migraine patients following repetitive sessions (DaSilva et al. 2011). There is an immediate reduction in μ-opioid receptor binding in response to an acute M1 neuromodulation. Levels of μ-opioid receptor-binding potential in a chronic trigeminal pain patient (post-herpetic neuralgia) during a single M1-tDCS application induced significant decrease in μ-opioid receptor binding in the thalamus (– below) and other pain-matrix structures, including nucleus accumbens (NAcc), ACC, and insula. Hence, analgesic effect of tDCS is possibly due to acute increase of endogenous opioids release, by direct and indirect effect of M1 stimulation on the thalamus and pain-matrix (DosSantos et al. 2012). Recently, it was reported that acute tDCS modulates functional connectivity depending on its polarity (Polania et al. 2012). Anodal stimulation over M1 with contralateral supracortical (SO) cathode placement (our protocol) immediately increases functional coupling between ipsilateral M1 and thalamus. On the contrary, cathodal tDCS over M1 decreases functional coupling between ipsilateral M1 and contralateral putamen.

Summary

Although it is clear that neuroimaging research can contribute to the understanding of the thalamic neuronal activation and its modulation regarding experimental and clinical pain, its nuclear specificity is yet to be completely defined. Technical improvement of imaging tools will provide better anatomical and functional nuclear maps of the thalamus and, consequently, of its correlation with each intrinsic aspect of a noxious event.

References

Bantick, S. J., Wise, R. G., Ploghaus, A., et al. (2002). Imaging how attention modulates pain in humans using functional MRI. *Brain, 125*, 310–319.

Casey, K. L., Svensson, P., Morrow, T. J., et al. (2000). Selective opiate modulation of nociceptive processing in the human brain. *Journal of Neurophysiology, 84*, 525–533.

Craig, A. D. (2002). How do you feel? Interoception: The sense of the physiological condition of the body. *Nature Reviews. Neuroscience, 3*, 655–666.

Creac'h, C., Henry, P., Caille, J. M., et al. (2000). Functional MR imaging analysis of pain-related brain activation after acute mechanical stimulation. *AJNR. American Journal of Neuroradiology, 21*, 1402–1406.

DaSilva, A. F., Becerra, L., Makris, N., et al. (2002). Somatotopic activation in the human trigeminal pain pathway. *The Journal of Neuroscience, 22*, 8183–8192.

DaSilva, A. F., Tuch, D. S., Wiegell, M. R., et al. (2003). Diffusion tensor imaging – A primer on diffusion tensor imaging of anatomical substructure. *Neurosurgical Focus, 15*, 1–4.

DaSilva, A. F., Volz, M. S., Bikson, M., & Fregni, F. (2011). Electrode positioning and montage in transcranial direct current stimulation. *Journal of Visualized Experiments, 51*, e2744.

DaSilva, A. F., Mendonca, M. E., Zaghi, S., Lopes, M., Dos Santos, M. F., Spierings, E. L., et al. (2012). tDCS-induced analgesia and electrical fields in pain-related neural networks in chronic migraine. *Headache, 52*, 1283–1295. doi:10.1111/j.1526 4610.2012.02196.x. (see http://www.ncbi.nlm.nih.gov/pubmed/22512348).

Derbyshire, S. W., Jones, A. K., Collins, M., et al. (1999). Cerebral responses to pain in patients suffering acute post-dental extraction pain measured by positron emission tomography (PET). *European Journal of Pain, 3*, 103–113.

Di Piero, V., Fiacco, F., Tombari, D., et al. (1997). Tonic pain: A SPET study in normal subjects and cluster headache patients. *Pain, 70*, 185–191.

DosSantos, M. F., Love, T., Martikainen, I. K., Nascimento, T. D., Fregni, F., Cummiford, C. M., et al. (2012). Immediate effect of tDCS on the μ-opioid system of a chronic pain patient. *Front Psychiatry, 3*, 93.

Faymonville, M. E., Roediger, L., Del Fiore, G., et al. (2003). Increased cerebral functional connectivity underlying the antinociceptive effects of hypnosis. *Brain Research. Cognitive Brain Research, 17*, 255–262.

Garcia-Larrea, L., Brousolle, E., Cezanne-Bert, G., Maugiere, F., et al. (1997). P 300 and frontal executive functions: application of a dual-task paradigm in normal subjects and patients with Parkinson's disease and progressive supranuclear palsy. *Electroencephalography and Clinical Neurophysiology, 103*, 148.

Garcia-Larrea, L., Peyron, R., Mertens, P., Gregoire, M. C., Lavenne, F., Le Bars, D., Convers, P., Mauguiere, F., Sindou, M., Laurent, B., et al. (1999). Electrical stimulation of motor cortex for pain control: a combined PET-scan and electrophysiological study. *Pain, 83*, 259–273.

Gracely, R. H., Petzke, F., Wolf, J. M., et al. (2002). Functional magnetic resonance imaging evidence of augmented pain processing in fibromyalgia. *Arthritis and Rheumatism, 46*, 1333–1343.

Hsieh, J. C., Belfrage, M., Stone-Elander, S., et al. (1995). Central representation of chronic ongoing neuropathic pain studied by positron emission tomography. *Pain, 63*, 225–236.

Iadarola, M. J., Berman, K. F., Zeffiro, T. A., et al. (1998). Neural activation during acute capsaicin-evoked pain and allodynia assessed with PET. *Brain, 121*, 931–947.

Kwiatek, R., Barnden, L., Tedman, R., et al. (2000). Regional cerebral blood flow in fibromyalgia: Single-photon-emission computed tomography evidence of reduction in the pontine tegmentum and thalami. *Arthritis and Rheumatism, 43*, 2823–2833.

Lorenz, J., Minoshima, S., & Casey, K. L. (2003). Keeping pain out of mind: The role of the dorsolateral prefrontal cortex in pain modulation. *Brain, 126*, 1079–1091.

Paulson, P. E., Minoshima, S., Morrow, T. J., et al. (1998). Gender differences in pain perception and patterns of cerebral activation during noxious heat stimulation in humans. *Pain, 76*, 223–229.

Polania, R., Paulus, W., & Nitsche, M. A. (2012). Modulating cortico-striatal and thalamo-cortical functional connectivity with transcranial direct current stimulation. *Human Brain Mapping, 33*(10), 2499–2508.

Strafella, A. P., Vanderwerf, Y., & Sadikot, A. F. (2004). Transcranial magnetic stimulation of the human motor cortex influences the neuronal activity of the subthalamic nucleus. *European Journal of Neuroscience, 20*, 2245–2249.

Svensson, P., Minoshima, S., Beydoun, A., et al. (1997). Cerebral processing of acute skin and muscle pain in humans. *Journal of Neurophysiology, 78*, 450–460.

Wiegell, M. R., Tuch, D. S., Larsson, H. B., et al. (2003). Automatic segmentation of thalamic nuclei from diffusion tensor magnetic resonance imaging. *NeuroImage, 19*, 391–401.

Zubieta, J. K., Smith, Y. R., Bueller, J. A., et al. (2002). mu-Opioid receptor-mediated antinociceptive responses differ in men and women. *The Journal of Neuroscience, 22*, 5100–5107.

Hunner's Ulcer

Definition

Hunner's ulcer is a focal inflammatory lesion of the bladder wall in chronic interstitial cystitis; its surface may crack and bleed with bladder distension.

Cross-References

▶ Interstitial Cystitis and Chronic Pelvic Pain

HVCCs

▶ Calcium Channels in the Spinal Processing of Nociceptive Input

Hyaline Cartilage

Definition

Hyaline cartilage is a translucent cartilage that is common in joints and the respiratory passages.

Cross-References

▶ Sacroiliac Joint Pain

Hyaluronan

Nikolai Bogduk
University of Newcastle, Newcastle Bone & Joint Institute, Royal Newcastle Centre, Newcastle, NSW, Australia

Synonyms

Hyaluronic acid; Viscosupplementation

Definition

Hyaluronic acid is a naturally occurring glycosaminoglycan, consisting of a repeating dimer of glucuronic acid and N-acetyl-glucosamine (Weissman and Meyer 1954). The proprietary form is known as hyaluronan.

This agent is administered by intra-articular injection, as a treatment for osteoarthritis.

Characteristics

Hyaluronic acid is a widely distributed polysaccharide, which plays an important role in all mammalian connective tissues, due to its peculiar physicochemical and biological properties. By nature of its propensity to form highly hydrated and viscous matrices, hyaluronic acid imparts stiffness, resilience, and lubrication to various tissues. The unique biophysical properties of hyaluronic acid are manifested in its mechanical function in the synovial fluid, the vitreous humor of the eye, and the ability of connective tissues to resist compressive forces (Laurent 1998).

In normal human synovial fluid, hyaluronic acid has a high molecular weight and acts in a viscoelastic manner. Due to its hyaluronic acid content, joint fluid acts as a viscous lubricant during slow movement of the joint, as in walking, and as an elastic shock absorber during rapid movement, as in running.

In osteoarthritis, both the concentration and molecular weight of hyaluronic acid in the synovial fluid are reduced (Marshall 1998; George 1998), which impacts on its biophysical properties. It was this finding that gave rise to the concept of *viscosupplementation*, in which injection of exogenous hyaluronic acid into the joint space is presumed to augment the functions of endogenous hyaluronic acid.

Mechanism

The mechanism by which intra-articular hyaluronic acid works in patients with osteoarthritis remains unknown. Although restoration of the elasto-viscous properties of synovial fluid seems to be the most logical explanation, other mechanisms must exist. The actual period that the injected hyaluronic acid product stays within the joint space is in the order of hours to days, but the time of clinical efficacy is often in the order of months (Cohen 1998; Balazs and Denlinger 1993). Possible explanations include stimulation of endogenous production of hyaluronic acid, inhibition of inflammatory mediators such as cytokines and prostaglandins, stimulation of cartilage matrix synthesis as well as inhibition of cartilage degradation, and a direct protective action on nociceptive nerve endings.

Technique

Hyaluronan is injected into the joint to be treated using a strict, no-touch, aseptic technique. If an effusion is present, aspiration of the joint is recommended before the injection, in order to prevent dilution of the injectate. Excessive weight-bearing physical activity should be avoided for 1–2 days.

Applications

The US Food & Drug Administration has approved the use of hyaluronan for patients with osteoarthritis of the knee, whose joint pain has not responded to non-medicinal measures and analgesic drugs. The guidelines for osteoarthritis from the American College of Rheumatology (American College of Rheumatology Subcommittee on Osteoarthritis Guidelines 2000) state that it may be "especially advantageous in patients in whom nonselective ▶ NSAIDs and Cox-2 specific inhibitors are contra-indicated, or in whom they have been associated either with a lack of efficacy or with adverse events." Intra-articular hyaluronic acid is generally used after non-pharmacologic treatments, analgesics, and a trial of several NSAIDs.

Efficacy

Since the 1970s, many studies have been carried out to evaluate the efficacy of hyaluronan. Despite a number of randomized, controlled trials having been carried out, the results and their interpretation remain conflicting. Whereas the earliest studies suggested benefits, more recent double-blind placebo-controlled trials did not show any benefit over placebo. In other studies, hyaluronan has been suggested to have an overall benefit over placebo.

An extensive review on intra-articular administration of hyaluronan, published by Brandt et al. (2000), concludes that "although several clinical trials indicate that intraarticular injection of [hyaluronan] results in relief of joint pain in patients with knee [osteoarthritis], and that this effect may last for months, similar results are seen with placebo, and it is not clear that the difference between [hyaluronan] and placebo, even if statistically significant, is clinically significant."

In response, Miller, in correspondence to the Journal of American Academy of Orthopaedic Surgeons (Miller 2001), argued that the decrease in the total number of knee replacements performed in the USA has occurred as a direct result of the use of viscosupplementation, citing a number of studies that formed the basis of the presentation to the FDA for its approval of hyaluronan as a treatment for osteoarthritis.

Side Effects

Transient localized pain and/or effusion is the most commonly reported side effect, albeit occurring in a low (0–3) percentage of patients, based on the majority of clinical trials conducted to date (Puttick et al. 1995). These resolve spontaneously within a short period. Several cases of pseudogout have been confirmed (Luzar and Altawil 1998). Long-term side effects have not been identified.

References

American College of Rheumatology Subcommittee on Osteoarthritis Guidelines. (2000). Recommendations for the medical management of osteoarthritis of the hip and knee. *Arthritis and Rheumatism, 43,* 1905–1915.

Balazs, E., & Denlinger, J. (1993). Viscosupplementation: A new concept in the treatment of osteoarthritis. *Journal of Rheumatology, 20*(Suppl.), 3–9.

Brandt, K., Smith, G., & Simon, L. (2000). Intraarticular injection of hyaluronan as treatment for knee osteoarthritis: What is the evidence? *Arthritis and Rheumatism, 43,* 1192–1203.

Cohen, M. (1998). Hyaluronic acid treatment (viscosupplementation) for OA of the knee. *Bulletin on the Rheumatic Diseases, 47,* 4–7.

George, E. (1998). Intra-articular hyaluronan treatment for osteoarthritis. *Annals of the Rheumatic Diseases, 57,* 637–640.

Laurent, T. C. (Ed.). (1998). *The chemistry, biology and medical applications of hyaluronan and its derivatives* (Wenner-Gren international series, Vol. 72). London: Portland Press.

Luzar, M., & Altawil, B. (1998). Pseudogout following intraarticular injection of sodium hyaluronate. *Arthritis & Rheumatism, 41,* 939–940.

Marshall, K. W. (1998). Viscosupplementation for osteoarthritis: Current status, unresolved issues and future directions. *Journal of Rheumatology, 25,* 2056–2058.

Miller, E. (2001). Correspondence. Viscosupplementation: Therapeutic mechanisms and clinical potential in osteoarthritis of the knee. *Journal of the American Academy of Orthopaedic Surgeons, 9,* 146–147.

Puttick, M., Wade, J., Chalmers, A., Connell, D., & Rangno, K. (1995). Local reactions after intra-articular hylan for osteoarthritis of the knee. *Journal of Rheumatology, 22,* 1311–1314.

Weissman, B., & Meyer, K. (1954). The structure of hyalbiuronic acid and of hyaluronic acid from umbilical cord. *Journal of the American Chemical Society, 76,* 1753–1757.

Hyaluronic Acid

▶ Hyaluronan

Hyaluronic Acid (HA)

Definition

Investigational drug for the treatment of IC; appears to temporarily replace defective mucosa.

Cross-References

▶ Hyaluronan
▶ Interstitial Cystitis and Chronic Pelvic Pain

Hydrodistention

Definition

Hydrodistention is the filling of bladder under anesthesia, to assess for mucosal tears, glomerulations, and bladder capacity; part of diagnostic work-up as well as therapy for IC.

Cross-References

▶ Interstitial Cystitis and Chronic Pelvic Pain

Hydrogen Magnetic Resonance Spectroscopy

▶ MRS

Hydromorphone

▶ Postoperative Pain, Hydromorphone

Hydroperoxides

Definition

Hydroperoxides such as PGG_2 are required to initiate the conversion of arachidonic acid into prostaglandins.

Cross-References

▶ Cyclooxygenases in Biology and Disease

Hydrotherapy

Definition

Hydrotherapy is the external application of water, e.g., the immersion of the body in thermal water.

Cross-References

▶ Chronic Pain in Children, Physical Medicine and Rehabilitation
▶ Spa Treatment

Hydroxy-7,8-Dihydrocodeinone

▶ Oxycodone

Hypalgia

▶ Hypoalgesia, Assessment

Hyperaesthesia, Assessment

Kristina B. Svendsen[1], Troels Staehelin Jensen[2]
and F. W. Bach[2]
[1]Danish Pain Research Center, Aarhus
University Hospital, Aarhus, Denmark
[2]Danish Pain Research Center, Department of
Neurology, Aarhus University Hospital, Aarhus,
Denmark

Hyperaesthesia, Assessment, Fig. 1 Hyperaesthesia refers to both lowered thresholds and to increased response to suprathreshold stimuli. Decreased pain threshold is called allodynia. Increased response to normally painful stimuli is called allodynia. *DT* detection threshold, *PT* pain threshold

Definition

Hyperaesthesia is increased sensitivity to stimulation, excluding the special senses (Merskey and Bogduk 1994). ▶ Allodynia and ▶ hyperalgesia are included in the definition.

Characteristics

Hyperaesthesia refers to both the finding of a lowered threshold to a nonnoxious or a noxious stimulus and to an increased response to suprathreshold stimuli (Merskey and Bogduk 1994). It can best be described as a leftward shift of the stimulus-response curve, which relates the response to the stimulus intensity (Fig. 1). Hyperaesthesia has to be distinguished from hyperpathia, which is a classical feature of neuropathic pain and easily demonstrated in skin territories innervated by damaged nerve fibers (Jensen et al. 2001; Jensen and Baron 2003).

Hyperaesthesia may occur after traumatic or inflammatory injury to the skin (Treede et al. 1992) or in the undamaged skin in neuropathic pain conditions (Boivie 1999; Woolf and Mannion 1999). Hyperaesthesia may also be found in the skin area of referred muscle (Svensson et al. 1998)

and visceral pain (Hardy et al. 1950; Stawowy et al. 2002). In tissue injury, increased sensibility to stimuli may be found in both the injured area (normally called primary hyperalgesia) and in surrounding noninjured skin area (secondary hyperalgesia) (Treede and Magerl 2000).

Induction and Assessment

Hyperaesthesia may be induced by different stimulus modalities including mechanical, thermal, and chemical stimuli (Treede et al. 1992; Jensen et al. 2001; Woolf and Mannion 1999).

Hyperaesthesia can be assessed by determining ▶ detection thresholds for a given stimulus. In the case of noxious stimuli, pain detection and pain tolerance thresholds can be used. Assessment of hyperalgesia includes stimulus-response curves, where noxious stimuli of different intensities (e.g., thermal stimuli or pressure) are applied in a random order and the pain sensation/intensity is assessed for each stimulus.

Hyperaesthesia is present when the detection and/or ▶ pain threshold for a given stimulus is decreased (Fig. 1) or the response to suprathreshold stimuli is increased. In the case where pain is induced by a normally nonpainful stimulus, the term allodynia is used. Increased response to normally painful stimuli, e.g., evaluated by the stimulus-response curve, is termed hyperalgesia.

Hyperaesthesia, Assessment, Table 1 Sensory testing

	Stimulus	Method	Sensation
Bedside examination	Mechanical stimuli – Dynamic touch – Static touch – Punctate stimuli	Stroking the skin with a paintbrush/cotton swab	Increased sensation/ pain = hyperaesthesia
		Gentle pressure with fingertip	
		Pinprick	
	Thermal stimuli – Cold – Warm	Metallic thermal roller kept at 20 °C	
		Acetone/menthol	
		Metallic thermal roller kept at 40 °C	
Quantitative sensory testing	Mechanical stimuli – Tactile detection threshold – Tactile pain threshold – Pressure pain threshold – Pressure pain threshold	von Frey hair	Decreased threshold = hyperaesthesia
		Pressure algometry	
	Thermal stimuli – Cold detection threshold – Warm detection threshold – Cold pain threshold – Heat pain threshold – Heat tolerance threshold	Thermotest	

Clinical Examination/Studies

Bedside sensory screening may be useful in the evaluation of the anatomical distribution and the qualitative characterization of sensory abnormalities of the skin (Hansson and Lindblom 1992). Bedside examination includes mechanical stimuli (cotton wool, paintbrush, pressure with fingertip, pinprick), thermal stimuli (thermal rollers kept at 20 °C and 40 °C, acetone drop), and vibration sense (tuning fork) (Table 1).

Sensory examination is normally done in the area with maximal pain and compared with the contralateral site of the body (Andersen et al. 1995; Jensen et al. 2001) or the adjacent body area not involved in disease. Hyperaesthesia is present in the case of increased sensation/pain to a nonpainful (hyperaesthesia/allodynia) or a painful stimulus (hyperalgesia).

Quantitative assessment of hyperaesthesia is performed using quantitative sensory testing (QST). QST includes mechanical (► Von Frey Hair, pressure algometry) and thermal stimuli (Thermotest) (Table 1). The results of QST from the affected site of the body are normally compared with results from an unaffected contralateral body site. However, when the contralateral site is also affected by disease, values from healthy subjects/general population may be used

(Kemler et al. 2000). For standardized regions such as feet, hands, and face, several laboratories have established normative data for thermal and mechanical stimuli. Hyperaesthesia is present in the case of lowered detection and/or pain thresholds. Pain detection and pain tolerance thresholds (see ► Pain Detection and Pain Tolerance Thresholds) indicate hyperalgesia.

The qualitative aspect of pain can be assessed by various questionnaires such as McGill pain questionnaire (Melzack 1975), verbal rating scales, visual analogue scales, and numerical rating scales (Turk and Melzack 1992).

Patient Example

A 54-year-old man with peripheral neuropathic pain following a trauma located to the right antebrachium. The sensory function of the right hand was assessed by QST, and the results were compared with the healthy contralateral site. The patient had signs of hyperaesthesia with decreased tactile detection threshold, allodynia with decreased tactile pain threshold, decreased pressure pain threshold, decreased heat and cold pain thresholds, and hyperalgesia with decreased pressure tolerance threshold (see Fig. 2). In addition, he had cold allodynia evoked by acetone drop.

Hyperaesthesia, Assessment, Fig. 2 Quantitative sensory testing in a patient with nerve lesion of the right antebrachium. *CDT* cold detection threshold, *HDT* heat detection threshold, *CPT* cold pain threshold, *HPT* heat pain threshold

Experimental Studies

Human

Hyperaesthesia is found in various human pain models:

Burn injury of the skin (a model of cutaneous injury) is followed by heat and mechanical allodynia in both injured and the adjacent noninjured surrounding skin (Pedersen and Kehlet 1998).

Capsaicin application of the skin produces allodynia with decreased heat pain threshold at the site of injection and pain induced by a light normally nonpainful mechanical stimulus in an area surrounding the injection site (Treede et al. 1992). Burn injury has also been combined with capsaicin in a heat-capsaicin sensitization model (Petersen et al. 2001).

Intramuscular injections of hypertonic saline, capsaicin, glutamate, and other excitatory or algogenic substances have been used as a model of localized and referred muscular pain (Graven-Nielsen and Arendt-Nielsen 2003). In these muscle pain models, decreased pressure pain thresholds have been found. Hypertonic saline may induce mechanical hyperaesthesia located to the overlying or adjacent skin (Svensson et al. 1998).

Animal

Strictly speaking, hyperaesthesia including allodynia and hyperalgesia with increased sensitivity to specific sensory stimulation cannot be determined in experimental animal models. Nevertheless, it is generally accepted that increased motor responses to mechanical (von Frey hair), thermal (cold bath, hot plate, acetone, focal heat), and chemical (capsaicin) stimuli in animal models of nerve injury, inflammation, or diabetes reflect a hypersensitivity of the animal to the pertinent stimulus (Scholz and Woolf 2002).

References

Andersen, G., Vestergaard, K., Ingeman-Nielsen, M., & Jensen, T. S. (1995). Incidence of central post-stroke pain. *Pain, 61*, 187–193.

Boivie, J. (1999). Central pain. In P. D. Wall & R. Melzack (Eds.), *Textbook of pain* (pp. 879–914). New York: Churchill Livingstone.

Graven-Nielsen, T., & Arendt-Nielsen, L. (2003). Induction and assessment of muscle pain, referred pain, and muscular hyperalgesia. *Current Pain and Headache Reports, 7*, 443–451.

Hansson, P., & Lindblom, U. (1992). Hyperalgesia assessed with quantitative sensory testing in patients with neurogenic pain. In W. D. Willis (Ed.), *Hyperalgesia and allodynia* (pp. 335–343). New York: Raven.

Hardy, J. D., Wolff, H. G., & Goodell, H. (1950). Experimental evidence on the nature of cutaneous hyperalgesia. *The Journal of Clinical Investigation, 29*, 115–140.

Jensen, T. S., & Baron, R. (2003). Translation of symptoms and signs into mechanisms in neuropathic pain. *Pain, 102*, 18.

Jensen, T. S., Gottrup, H., Sindrup, S. H., & Bach, F. W. (2001). The clinical picture of neuropathic pain. *European Journal of Pharmacology, 429*, 1–11.

Kemler, M. A., Schouten, H. J., & Gracely, R. H. (2000). Diagnosing sensory abnormalities with either normal values or values from contralateral skin: Comparison of two approaches in complex regional pain syndrome I. *Anesthesiology, 93*, 718–727.

Melzack, R. (1975). The MCGILL pain questionnaire: Major properties and scoring methods. *Pain, 1*, 277–299.

Merskey, H., & Bogduk, N. (1994). *Classification of chronic pain: Descriptions of chronic pain syndromes and definitions of pain terms, prepared by the international association for the study of pain, task force of taxonomy*. Seattle: IASP Press.

Pedersen, J. L., & Kehlet, H. (1998). Secondary hyperalgesia to heat stimuli after burn injury in man. *Pain, 76*, 377–384.

Petersen, K. L., Jones, B., Segredo, V., Dahl, J. B., & Rowbotham, M. C. (2001). Effect of remifentanil on pain and secondary hyperalgesia associated with the heat-capsaicin sensitization model in healthy volunteers. *Anesthesiology, 94*, 15–20.

Scholz, J., & Woolf, C. J. (2002). Can we conquer pain? *Nature Neuroscience, 5*(suppl), 1062–1067.

Stawowy, M., Rossel, P., Bluhme, C., Funch-Jensen, P., Arendt-Nielsen, L., & Drewes, A. M. (2002). Somatosensory changes in the referred pain area following acute inflammation of the appendix. *European Journal of Gastroenterology and Hepatology, 14*, 1079–1084.

Svensson, P., Graven-Nielsen, T., & Arendt-Nielsen, L. (1998). Mechanical hyperesthesia of human facial skin induced by tonic painful stimulation of jaw muscles. *Pain, 74*, 93–100.

Treede, R. D., & Magerl, W. (2000). Multiple mechanisms of secondary hyperalgesia. *Progress in Brain Research, 129*, 331–341.

Treede, R. D., Meyer, R. A., Raja, S. N., & Campbell, J. N. (1992). Peripheral and central mechanisms of cutaneous hyperalgesia. *Progress in Neurobiology, 38*, 397–421.

Turk, D. C., & Melzack, R. (1992). *Handbook of pain assessment*. New York: The Guilford Press.

Woolf, C. J., & Mannion, R. J. (1999). Neuropathic pain: Aetiology, symptoms, mechanisms, and management. *Lancet, 353*, 1959–1964.

Hyperalgesia

Rolf-Detlef Treede
Institute of Physiology and Pathophysiology, Johannes Gutenberg University, Mainz, Germany

Synonyms

Algesia; Hyperesthesia; Primary hyperalgesia; Secondary hyperalgesia

Definition

Increased pain sensitivity. Antonym: ► hypoalgesia (decreased pain sensitivity). Increased pain sensitivity at a site of tissue damage is called primary hyperalgesia. Increased pain sensitivity in normal skin surrounding a site of tissue damage is called secondary hyperalgesia.

Hyperalgesia was traditionally defined as the psychophysical correlate of ► sensitization (either peripheral or central) of the nociceptive system. As such, it is characterized by a decreased pain threshold and increased pain to suprathreshold

Hyperalgesia, Table 1 Types of hyperalgesia and their likely mechanisms

Test stimulus	Occurrence	Afferents	Sensitization
Heat	Primary zone	Type I and II AMH, CMH	Peripheral
Blunt pressure	Primary zone	MIA, (type I AMH?)	Peripheral
Impact	Primary zone	MIA, (type I AMH?)	Peripheral
Punctate	Neuropathic	Type I AMH	Central
	Secondary zone	Type I AMH	Central
	Primary zone	Type I AMH, MIA	Peripheral/central?
Stroking	Neuropathic	Aβ-LTM	Central
	Secondary zone	Aβ-LTM	Central
	Primary zone	Aβ-LTM	Central
Cold	Neuropathic pain	?	Central?
	Secondary zone?	?	Central?
Chemical	Inflammation	Type II AMH, CMH, MIA?	Peripheral?

Abbreviations: Aβ-LTM Aβ-fiber low-threshold mechanoreceptor ("touch receptor"), probably rapidly adapting subtype (Meissner corpuscle); type I AMH A-fiber nociceptor with slow high-threshold heat response (no TRPV1), probably equivalent to A-fiber high-threshold mechanoreceptor; type II AMH A-fiber nociceptor with rapid low-threshold heat response (TRPV1); CMH C-fiber mechano-heat nociceptor (TRPV1); MIA mechanically insensitive (silent) nociceptive afferent (From Treede et al. (2004))

stimuli. The current definition by the International Association for the Study of Pain (IASP) refers only to the latter phenomenon ("increased pain to a stimulus that is normally painful"). A decreased pain threshold would operationally fulfill the IASP definition of ▶ allodynia ("pain induced by stimuli that are not normally painful"). This narrow definition has proved to be counterproductive for two reasons: (1) all known mechanisms of sensitization lead to changes in both threshold and suprathreshold response and (2) the extended use of the term allodynia has distracted from its initial clinical meaning and has hampered the transfer of knowledge from animal research to the clinic. Therefore, this essay uses the traditional definition of hyperalgesia as the psychophysical correlate of sensitization, which will probably be adopted by IASP in the near future.

Characteristics

Increased pain sensitivity (hyperalgesia) can be differentiated according to the test stimulus that is perceived as more painful, mechanical hyperalgesia, heat hyperalgesia, cold hyperalgesia, and chemical hyperalgesia (Table 1). Mechanical

hyperalgesia can be further differentiated according to the size of the object contacting the skin (punctate or blunt) and the temporal dynamics of its application (static or dynamic). The underlying mechanisms are sensitization either in the periphery or in the central nervous system or both. Hyperalgesia at a site of tissue damage is called primary hyperalgesia; hyperalgesia surrounding this site is called secondary hyperalgesia. The sensory characteristics of primary and secondary hyperalgesia differ considerably (Fig. 1). Whereas primary hyperalgesia encompasses increased sensitivity to both mechanical and heat stimuli, secondary hyperalgesia is relatively specific for mechanical stimuli (Treede et al. 1992).

Primary hyperalgesia to heat stimuli is fully accounted for by ▶ peripheral sensitization of the terminals of primary nociceptive afferents (Raja et al. 1999). Peripheral sensitization shifts the stimulus response function for heat stimuli to the left. This leftward shift is associated with a decreased threshold, increased responses to suprathreshold stimuli, and spontaneous activity (Fig. 2). Primary nociceptive afferents express the heat-sensitive ion channel TRPV1 (Caterina and Julius 2001). This channel can be sensitized by inflammatory mediators, and the ensuing

Hyperalgesia, Fig. 1 Primary and secondary hyperalgesia in humans. Primary hyperalgesia is defined as increased pain sensitivity at a site of tissue damage. It is characterized by hyperalgesia to both heat and mechanical test stimuli. Secondary hyperalgesia is defined as increased pain sensitivity in normal skin surrounding a site of tissue damage. It is characterized by hyperalgesia to mechanical test stimuli only (From Treede and Magerl (1995))

Hyperalgesia, Fig. 2 Peripheral sensitization of nociceptive afferents by a burn injury in monkey. The stimulus response function relating the discharge rate of nociceptive C- (**a**) and A-fiber nociceptors (**b**) is shifted to the left following injury to the receptive field. This shift is characterized by a drop in threshold, increased responses to suprathreshold stimuli, and by spontaneous activity. Spontaneous discharges occur when the heat threshold is below body temperature. Peripheral sensitization is restricted to the injured part of the receptive field (From Treede et al. (1992))

drop in heat threshold turns normal body temperature into a suprathreshold stimulus (Liang et al. 2001). Thus, primary hyperalgesia to heat can also explain ongoing pain of inflammatory origin.

Secondary hyperalgesia to mechanical stimuli is not associated with any change in peripheral coding (Baumann et al. 1991) but can be explained by enhanced synaptic responses of second-order neurons in the spinal cord to their normal afferent input (▶ central sensitization). These neurons also exhibit a drop in threshold and an increase in suprathreshold responses (Simone et al. 1991). In addition, expansion of the ▶ receptive field is a prominent feature of central sensitization. The molecular mechanisms of central sensitization resemble those of long-term potentiation of synaptic efficacy (LTP). LTP has been demonstrated for neurons in isolated spinal cord slices, in intact animals, and on a perceptual level in human subjects (Klein et al. 2004; Sandkühler 2000; Treede and Magerl 1995). As a cellular correlate of learning and memory, LTP in the nociceptive system is a phylogenetically old mechanism, present even in invertebrates (Woolf and Walters 1991).

Although not characterized in as much detail, descending supraspinal mechanisms may contribute to both primary and secondary hyperalgesia, via reduced descending inhibition or via enhanced descending facilitation (Millan 2002; Porreca et al. 2002). Moreover, central sensitization may also occur at the thalamic or cortical level.

The mechanisms of cold hyperalgesia, which is a frequent finding in some ▶ neuropathic pain states, are still enigmatic (Wasner et al. 2004). Peripheral sensitization of nociceptive afferents cannot be ruled out because the peripheral encoding of noxious cold stimuli has not been investigated sufficiently (Raja et al. 1999). Some evidence supports the concept of central disinhibition by selective loss of a sensory channel specific for non-noxious cold that exerts a tonic inhibition on nociceptive channels (Craig and Bushnell 1994). Central sensitization, similar to mechanical hyperalgesia, is another possibility.

Clinical Implications

Primary and secondary hyperalgesia occur transiently after each injury and are hence part of the normal clinical picture of postoperative pain. Chronic inflammatory hyperalgesia resembles primary hyperalgesia. Hyperalgesia in neuropathic pain and referred hyperalgesia in visceral pain resemble secondary hyperalgesia (Treede et al. 1992). Cancer pain and musculoskeletal pain states including low-back pain may also be accompanied by hyperalgesia. Parallel to the definition of sensitization, hyperalgesia is characterized by a decrease in pain threshold, increased pain to suprathreshold stimuli, and spontaneous pain.

Hyperalgesia Versus Allodynia

The current IASP taxonomy has restricted the term "hyperalgesia" to increases in pain to suprathreshold stimuli (Merskey and Bogduk 1994). But are threshold changes and suprathreshold changes two independent phenomena needing two separate terms? This question can be addressed in a clinical example, the increased pain sensitivity to punctate mechanical stimuli in patients suffering from neuropathic pain (Baumgärtner et al. 2002). Figure 3 illustrates that hyperalgesia to calibrated pinpricks in these patients is characterized by both an increase in pain to suprathreshold stimuli and a decrease in pain threshold. According to the IASP taxonomy, the threshold decrease would be labeled "allodynia," whereas the increase in pain to suprathreshold stimuli would be labeled "hyperalgesia" (Fig. 3). Consistent use of the IASP taxonomy is obviously awkward in this case because these observations reflect two aspects of the same phenomenon and the same data, i.e., a dramatic leftward shift of the psychometric function and upward shift of the stimulus response function of pain to the same set of test stimuli. The traditional usage of the term "hyperalgesia" as an umbrella term for all phenomena of increased pain sensitivity describes hyperalgesia to punctate mechanical stimuli more adequately (Treede et al. 2004).

Hyperalgesia, Fig. 3 Hyperalgesia to punctate mechanical stimuli in neuropathic pain. Averaged data from a group of six patients with neuropathic pain were plotted in two different ways: as incidence (*left*) and as intensity (*right*) of pain sensation in neuropathic pain skin areas (*filled circles*) compared to normal skin (*open circles*). Stimuli were graded punctate probes (diameter 0.2 mm) of seven intensities (8–512 mN). *Left panel*: reduced threshold (intersection with *dotted line* at 50 %) implies pain due to a stimulus, which does not normally evoke pain ("allodynia?"). *Right panel*: Increased pain response to a stimulus, which is normally painful ("hyperalgesia?"). Note that both graphs are different aspects (pain incidence and pain intensity) plotted from the same data set. *Arrows*: leftward shift of pain incidence and upward shift of pain intensity. *VRS* verbal rating scale. Mean ± SEM across subjects. Post hoc least significant differences tests: ** $p < 0.01$; *** $p < 0.001$ (From Treede et al. (2004))

Cross-References

▶ Polymodal Nociceptors, Heat Transduction
▶ Postherpetic Neuralgia, Etiology, Pathogenesis, and Management
▶ Postherpetic Neuralgia, Pharmacological and Nonpharmacological Treatment Options
▶ Poststroke Pain Model, Thalamic Pain (Lesion)
▶ Psychiatric Aspects of Visceral Pain
▶ Psychology of Pain, Sensitization, Habituation, and Pain
▶ Sensitization of Visceral Nociceptors
▶ Spinal Cord Injury Pain Model, Contusion Injury Model
▶ Spinothalamic Tract Neurons, Central Sensitization
▶ Sympathetically Maintained Pain in CRPS II, Human Experimentation
▶ TENS, Mechanisms of Action
▶ Thalamotomy, Pain Behavior in Animals
▶ Thalamus, Clinical Pain, Human Imaging
▶ Thalamus, Dynamics of Nociception
▶ Transition from Acute to Chronic Pain
▶ Vagal Input and Descending Modulation
▶ Visceral Nociception and Pain
▶ Visceral Pain and Nociception

References

Baumann, T. K., Simone, D. A., Shain, C. N., et al. (1991). Neurogenic hyperalgesia: The search for the primary cutaneous afferent fibers that contribute to capsaicin-induced pain and hyperalgesia. *Journal of Neurophysiology, 66*, 212–227.

Baumgärtner, U., Magerl, W., Klein, T., et al. (2002). Neurogenic hyperalgesia versus painful hypoalgesia: Two distinct mechanisms of neuropathic pain. *Pain, 96*, 141–151.

Caterina, M. J., & Julius, D. (2001). The vanilloid receptor: A molecular gateway to the pain pathway. *Annual Review of Neuroscience, 24*, 487–517.

Craig, A. D., & Bushnell, M. C. (1994). The thermal grill illusion: Unmasking the burn of cold pain. *Science, 265*, 252–255.

Klein, T., Magerl, W., Hopf, H. C., et al. (2004). Perceptual correlates of nociceptive long-term potentiation and long-term depression in humans. *Journal of Neuroscience, 24*, 964–971.

Liang, Y. F., Haake, B., & Reeh, P. T. (2001). Sustained sensitization and recruitment of rat cutaneous nociceptors by bradykinin and a novel theory of its excitatory action. *The Journal of Physiology, 532*, 229–239.

Merskey, H., & Bogduk, N. (1994). *Classification of chronic pain* (p. 240). Seattle: IASP Press.

Millan, M. J. (2002). Descending control of pain. *Progress in Neurobiology, 66*, 355–474.

Porreca, F., Ossipov, M. H., & Gebhart, G. F. (2002). Chronic pain and medullary descending facilitation. *Trends in neurosciences, 25*, 319–325.

Raja, S. N., Meyer, R. A., Ringkamp, M., et al. (1999). Peripheral neural mechanisms of nociception. In P. D. Wall & R. Melzack (Eds.), *Textbook of pain* (4th ed., pp. 11–57). Edinburgh: Churchill Livingstone.

Sandkühler, J. (2000). Learning and memory in pain pathways. *Pain, 88*, 113–118.

Simone, D. A., Sorkin, L. S., Oh, U., et al. (1991). Neurogenic hyperalgesia: Central neural correlates in responses of spinothalamic tract neurons. *Journal of Neurophysiology, 66*, 228–246.

Treede, R. D., & Magerl, W. (1995). Modern concepts of pain and hyperalgesia: Beyond the polymodal C-nociceptor. *News in Physiological Sciences, 10*, 216–228.

Treede, R.-D., Meyer, R. A., Raja, S. N., et al. (1992). Peripheral and central mechanisms of cutaneous hyperalgesia. *Progress in Neurobiology, 38*, 397–421.

Treede, R.-D., Handwerker, H. O., Baumgärtner, U., et al. (2004). Hyperalgesia and allodynia: Taxonomy, assessment, and mechanisms. In K. Brune & H. O. Handwerker (Eds.), *Hyperalgesia: Molecular mechanisms and clinical implications* (pp. 1–15). Seattle: IASP Press.

Wasner, G., Schattschneider, J., Binder, A., & Baron, R. (2004). Topical menthol – A human model for cold pain by activation and sensitization of C nociceptors. *Brain, 127*, 1159–1171.

Woolf, C. J., & Walters, E. T. (1991). Common patterns of plasticity contributing to nociceptive sensitization in mammals and Aplysia. *Trends in neurosciences, 14*, 74–78.

Hyperalgesia, Primary and Secondary

Definition

Primary hyperalgesia is increased pain sensitivity at a site of tissue damage. Secondary hyperalgesia is increased pain sensitivity in normal skin surrounding a site of tissue damage. It is characterized by hyperalgesia to mechanical test stimuli.

Cross-References

▶ Allodynia (Clinical, Experimental)
▶ Hyperalgesia

Hyperemia

Definition

Increased blood flow or an excess of blood in a body parties known as hyperemia.

Cross-References

▶ Clinical Migraine with Aura

Hyperesthesia

▶ Hyperalgesia

Hyperexcitability

Definition

From Greek/Latin: extremely excitable. Besides being an awkward mixed Greek/Latin word, the term is not well defined. Basically, it can label afferent nerve fibers or neurons which are more excitable than usual. In the case of peripheral afferents, this state is usually characterized as "sensitized". However, the label "hyperexcitable" is most often used not for sensitized nociceptors, but for fast conducting mechanoreceptors which have become spontaneously active, and apparently more excitable, as a consequence of a nerve lesion.

Cross-References

▶ Sympathetic and Sensory Neurons after Nerve Lesions, Structural Basis for Interactions

Hyperglycemic Neuropathy

▶ Diabetic Neuropathies

Hyperhidrosis

Definition

Hyperhidrosis means increased sweating.

Cross-References

▶ CRPS, Evidence-Based Treatment

Hyperknesis

Definition

Hyperknesis is the abnormal pruriceptive state in which a normally pruritic stimulus (such as a fine diameter hair which can elicit a prickle sensation followed by an itch) elicits a greater than normal duration and/or magnitude of itch.

Cross-References

▶ Allodynia and Alloknesis

Hyperpathia

John W. Scadding
The National Hospital for Neurology and Neurosurgery, London, UK

Definition

The IASP, in its Classification of Chronic Pain (1994), defines hyperpathia thus:

> Hyperpathia is a painful syndrome characterized by an abnormally painful reaction to a stimulus, as well as an increased threshold.

The following note is added:

It may occur with ▶ allodynia, ▶ hyperesthesia, ▶ hyperalgesia, or ▶ dysesthesia. Faulty identification and localization of the stimulus, delay, radiating sensation, and after-sensation may be present, and the pain is often explosive in character. The changes in this note [*from the previous edition of the IASP Taxonomy*] are the specification of allodynia and the inclusion of hyperalgesia explicitly.

Allodynia: Allodynia is pain due to a stimulus which does not normally provoke pain.

Hyperalgesia: Hyperalgesia is an increased response to a stimulus which is normally painful.

Hyperesthesia: Hyperesthesia is an increased sensitivity to stimulation, excluding the special senses.

Dysesthesia: Dysesthesia is an unpleasant abnormal sensation, whether spontaneous or evoked.

Neuropathic Pain: Neuropathic pain is arising as a direct consequence of a lesion affecting the somatosensory system.

Characteristics

Introduction
Use of the term hyperpathia varies in the current scientific literature, and it is avoided by many. For example, in Wall and Melzack's Textbook of Pain (McMahon and Koltzenburg 2005), hyperpathia receives a single listing in the index and is not discussed a specific and complete entity in any chapter in the book. Likewise, several influential studies and reviews, tackling the elusive problem of linking individual symptoms and signs to underlying pathophysiological mechanisms, avoid use of the word hyperpathia altogether or make only passing reference to it (Woolf et al. 1998; Woolf and Mannion 1999; Otto et al. 2003; Jensen and Baron 2003).

The reason is that hyperpathia describes a complex sensory experience, with various components, occurring in the context of ▶ neuropathic pain. This complex can be broken down into component parts, each of which may be experienced by patients independent from the other constituent properties of hyperpathia.

However, the delay in perception of a stimulus, the summation of pain with repetitive stimulation, and aftersensation characteristic of the hyperpathic response (see below) are usually seen in combination with allodynia and hyperalgesia, when hyperpathia occurs in patients with neuropathic pain.

Historical Aspects
A brief historical examination reveals the variable usage of the term hyperpathia. Foerster (1927) suggested a lengthy and all-inclusive definition and description of hyperpathia, which most would agree comprehensively encapsulates the properties of stimulus-evoked painful sensations in patients suffering from neuropathic pain. He proposed the term hyperpathia be used when the following symptoms could be elicited from a regenerating area: A relative elevation of threshold when the duration of the stimulus or summation of stimuli become important, a latent period, an intensive explosive outbreak of pain of abnormal unpleasant character accompanied by strong withdrawal movements, vasomotor and vegetative reactions, lack of or insufficient relationship between the strength of the stimulus and the strength of the sensation, a long after-reaction of the pain when the stimulus has ceased, irradiation, faulty localization, and the inability to identify the nature of the stimulus which causes the pain.

Livingston (1943) equated hyperpathia with hyperalgesia:

Any injury that directly or indirectly involves the sensory nerves may lead to the development of an abnormal sensitiveness of the skin. All sensory experiences derived from the skin may be altered in this condition, so that it is frequently called a "hyperesthesia" or a "hyperpathia." However, since the principal alteration in sensibility is an intensification of pain sensation it is more commonly referred to as a "hyperalgesia." In this state the tissues are unduly sensitive and they tend to react to the most innocuous stimuli with explosive sensations of pain accompanied by withdrawal reflexes.

Finally, Noordenbos (1959) suggests

Hyperpathia is present when the response to noxious or non-noxious stimuli presents the following features: delay, overshooting and after-reaction.

Definition or Description

This brief historical survey serves to emphasize three important points. The first is the general point that a definition should include those characteristics that are the minimum necessary to categorize a condition, item, or state as separate and identifiably distinct. Secondly, in relation to the definition of diseases and clinical syndromes/ states, a definition must have clinical relevance and usefulness.

And thirdly, with reference specifically to the definition of hyperpathia, the definition is based on a collection of symptoms and signs. As is evident throughout this encyclopedic reference, the last few decades have witnessed enormous progress in the basic neuroscience of pain. However, it is still not yet possible to tightly define conditions and terms such as neuropathic pain, hyperalgesia, hyperesthesia, allodynia, and hyperpathia on the basis of pathophysiological mechanisms, though there are, of course, numerous candidate mechanisms, and it is likely that each of the symptoms and signs of painful states may be produced by more than one underlying pathophysiological property.

For the moment, however, we are forced to rely on frustratingly imprecise clinical syndromal definitions.

Symptoms and Signs Comprising Hyperpathia

It is notable that the current IASP definition of hyperpathia, quoted above, includes little detail, and it is left to the accompanying note to elaborate the symptoms. Most would agree that there are four main clinical features to hyperpathia:

1. An increased threshold to stimulation.
2. An abnormal delay in perception of a stimulus.
3. Summation, by which is meant increasingly painful sensation to a repetitive stimulus of steady intensity. Summation may take the form of an explosive, unbearable increase in pain, and it leads to brisk withdrawal from the provoking stimulus.
4. Aftersensation. This is a perception by the sufferer that the stimulus evoking the pain continues after the stimulus has in fact ceased. Painful aftersensations may persist for seconds, minutes, or even hours, following even brief periods of stimulation lasting only a few seconds.

Conditions in Which Hyperpathia Occurs

It is clear from numerous published accounts that hyperpathia may accompany (or, perhaps more accurately, be a part of) neuropathic pain, due to lesions at any level in the peripheral or central nervous system sensory pathways. This includes painful cutaneous scars; peripheral sensory or mixed peripheral neuropathies; brachial or lumbar plexopathies; spinal sensory radiculopathies; myelopathies; and brain stem, thalamic, subcortical, and, very occasionally, cortical lesions. In other words, all of the many causes of neuropathic pain may be associated with hyperpathia, and multiple etiologies are involved (Scadding 2009).

Noordenbos (1959) described in detail six patients with peripheral and central lesions, all of whom had severe hyperpathia, specifically to illustrate the occurrence of hyperpathia. Other classical accounts are to be found in Weir Mitchell et al. (1864), Riddoch (1938), and Livingston (1943).

Is Hyperpathia a Clinically Relevant and Useful Term?

Hyperpathia is very common and troublesome to patients, despite the impression one might get from perusal of the recent basic and clinical scientific literature on pain, which, as discussed above, tends to consider the component properties of hyperpathia rather than addressing hyperpathia as a whole. It is certainly highly relevant to patients. For example, a patient suffering from postherpetic neuralgia (PHN) in a mid-thoracic dermatome, with an accompanying hyperpathic response to normally innocuous stimulation, may find the gentle rubbing of clothes on the affected area of skin quite intolerable. Indeed, for patients with PHN, it is often hyperpathia, much more than ongoing pain, which is the major component of their suffering and immobilization. Hyperpathia at other sites has the same devastating effect on the lives of numerous patients.

Although tremendous advances have been made in the measurement of pain and particularly in the various attributes of neuropathic pain, hyperpathia is difficult to quantify and so has

Hyperpathia, Table. 1 Possible pathophysiological substrates for hyperpathia

Symptom	Mechanism
1. Increased threshold	Reduced input due to sensory lesion
2. Delay in perception	Reduced large fiber input
3. Summation	Afterdischarge in damaged sensory neurons
	Crossed afterdischarge in sensory neurons
	Ephaptic transmission in nerve lesion?
	Central sensitization
4. Aftersensation	Crossed afterdischarge in sensory neurons
	DRG ectopic firing
	Central disinhibition

DRG dorsal root ganglion
Sources: Woolf and Salter (2005), Devor (2005), Jensen and Baron (2003)

tended to be underestimated in published studies (routine quantitative sensory testing does not accurately assess this).

Pathophysiology

Table 1 lists some possible pathophysiological substrates for the development of hyperpathia.

Cross-References

▶ Cancer Pain
▶ Causalgia: Assessment
▶ Deafferentation Pain
▶ Hypoesthesia, Assessment
▶ Peripheral Neuropathic Pain

References

Devor, M. (2005). Response of nerves to injury in relation to neuropathic pain. In S. B. McMahon & M. Koltzenburg (Eds.), *Wall and melzack's textbook of pain* (5th ed., pp. 903–928). Edinburgh: Elsevier Churchill Livingstone. Chapter 58.
Foerster, O. (1927). *Die Leitungsbahnen des Schmerzgefühls*. Wien: Urban & Schwarzenberg.
International Association for the Study of Pain (IASP). (1994). Classification of chronic pain. In H. Merskey & N. Bogduk (Eds.), *Classification of chronic pain:*
Description of chronic pain syndromes and definitions of pain terms (2nd ed.). Seattle: IASP Press.
Jensen, T. S., & Baron, R. (2003). Translation of symptoms and signs into mechanisms in neuropathic pain. *Pain, 102*, 1–8.
Livingston, W. K. (1943). *Pain mechanisms. A physiologic interpretation of Causalgia and Its related states*. New York: Macmillan.
McMahon, S. B., & Koltzenburg, M. (Eds.). (2005). *Wall and melzack's textbook of pain* (5th ed.). Edinburgh: Churchill Livingstone.
Noordenbos, W. (1959). *Pain. Problems pertaining to the transmission of nerve impulses which give rise to pain.* Amsterdam: Elsevier.
Otto, M., Bak, S., Bach, F. W., Jensen, T. S., & Sindrup, S. H. (2003). Pain phenomena and possible mechanisms in patients with painful neuropathy. *Pain, 101*, 187–192.
Riddoch, G. (1938). Central pain. *Lancet, 1*, 1150–1156, 1205–1209.
Scadding, J. W. (2009). Neuropathic pain. In M. Donaghy (Ed.), *Brain's diseases of the nervous system* (12th ed., pp. 453–475). Oxford: Oxford University Press. Chapter 17.
Weir Mitchell, S., Morehouse, G. R., & Keen, W. W. (1864). *Gunshot wounds and other injuries of nerves.* Philadelphia: Lippincott.
Woolf, C. J., & Mannion, R. J. (1999). Neurogenic pain: Aetiology, symptoms, mechanisms, and management. *Lancet, 353*, 1959–1964.
Woolf, C. J., & Salter, M. W. (2005). Plasticity and pain: Role of the dorsal horn. In S. B. McMahon & M. Koltzenburg (Eds.), *Wall and melzack's textbook of pain* (5th ed., pp. 91–106). Edinburgh: Elsevier Churchill Livingstone. Chapter 5.
Woolf, C. J., Bennett, G. J., Doherty, M., Dubner, R., Kidd, B., Koltzenburg, M., Lipton, R., Loeser, J. D., Payne, R., & Torebjork, E. (1998). Towards a mechanism-based classification of pain? *Pain, 77*, 227–229.

Hyperpathia, Assessment

Astrid Juhl Terkelsen and Troels Staehelin Jensen
Danish Pain Research Center, Department of Neurology, Aarhus University Hospital, Aarhus, Denmark

Definition

The International Association for the Study of Pain (IASP) define hyperpathia as a painful syndrome characterized by an abnormally painful reaction to a stimulus, especially a repetitive stimulus, as well as an increased threshold.

Characteristics

Hyperpathia includes increased ▶ detection threshold. There is a steeper stimulus-response

Hyperpathia, Assessment, Table 1 Assessment of hyperpathia

Stimulus	Assessment of detection (DT) and pain (PT) threshold	Stimulus-response function following graded innocuous and/or noxious stimuli	Repetitive suprathreshold stimulation
Thermal stimulation:			
Screening with metallic cold and heat thermo rollers	–	–	–
Thermal quantitative sensory testing	DT, PT	Ass. of thermal allodynia/ hyperalgesia	Repetitive heat or cold pulses
Mechanical stimulation:			
Light touch: wisp of cotton or camel-hair brush	–	–	Brushing with a velocity >0.3 Hz
Punctuate stimuli: screening with safety pin. Testing with von Frey hair	DT, PT	Ass. of ▶ punctuate allodynia/hyperalgesia	Multiple >0.3 Hz pinprick stimuli
Static stimuli: pressure or skin fold	PT	Ass. of ▶ static allodynia/ hyperalgesia	+
Electrical stimulation	DT, PT	+	+
Vibrametry	DT, (PT)	+	+
Chemical stimulation:			
Topical capsaicin	Time before detection (DT) and pain (PT)	Pain increase as a function of time	–

Hyperpathia, Assessment, Fig. 1 Schematic description of different stimulus-response curves occurring with hyperpathia. Normal response (Normal) with normal detection threshold (DT) and normal pain threshold (PT). A: Hyperpathia with increased DT, decreased PT, allodynia, and steeper stimulus-response curve as compared to normal. B: Hyperpathia with increased DT, decreased PT, allodynia, hyperalgesia, and steeper stimulus-response curve as compared to normal. C: Hyperpathia with increased DT, normal PT, hyperalgesia, and steeper stimulus-response curve. D: Hyperpathia with increased DT, increased PT, hyperalgesia, and steeper stimulus-response curve. E: At repetitive stimulation, pain threshold decreases and the slope of the stimulus-response curve increases thereby unmasking hyperpathia

function than normal (Hansson and Lindblom 1992) with an intense, exaggerated, and explosive pain response to suprathreshold stimuli. Stimulus and response modality may be the same (▶ hyperalgesia) and/or different (allodynia) (Merskey and Bogduk 1994). The pain is often felt by a remarkable delay characterized by a time lag between the stimulus and the report of any sensory perception. The time lag can extend from 2–3 s to more than 10 s. There may be abnormal summation, aftersensations, and pain-radiating phenomena. There can be poor localization of the stimulus and faulty identification where the patient feels pain but not the specific modality of the stimulus (Fig. 2b) or with misnaming of the stimulus modality (Fig. 3). (Noordenbos 1979; Lindblom 1979; Merskey and Bogduk 1994; Bennett 1994).

Induction and Assessment

Thermal, mechanical, and chemical hyperpathia may exist singly or in any combination. Therefore, multiple different noxious and innocuous stimulus modalities have to be used to document or to exclude hyperpathia (Lindblom 1994) (Table 1). Hyperpathia is assessed by performing stimulus-response curves with assessment of detection- and pain thresholds, repetitive suprathreshold stimulation, and by asking the patient to report aftersensations, pain radiation, and coexistent phenomena (Table 1, Figs. 1, 2 and 3). Usually, the site of maximal pain reported by the patient is chosen as the test area. At unilateral involvement, the contralateral mirror image area is used as control. At bilateral involvement, data should preferably be compared with normal values from sex- and age-matched controls.

Temporal Summation

Temporal summation refers to an abnormally increasing painful sensation to repetitive stimulation, although the actual stimulus remains constant and is the clinical equivalent to ▶ wind-up (Mendell and Wall 1965; Price et al. 1992). Hyperpathia is most likely elicited when stimulus duration is increased or at repetitive stimulation. At repetitive stimulation above sensory detection threshold, hyperpathic subjects can report

Hyperpathia, Assessment, Fig. 2 Quantitative sensory testing. Schematic presentation of thermal quantitative sensory testing at control side (*empty box*) and affected side (*grey box*) in two patients (**a** and **b**). Warm (WT) and cold (CT) detection threshold and heat (HPT) and cold (CPT) pain threshold are expressed as average temperature of five consecutive stimuli. (**a**) A 46-year-old woman with pain following abdominal surgery. There is increased detection but normal pain threshold to warm and cold. The patient report increased cold and heat pain at threshold in the painful area as compared to the unaffected homologous region. Hyperpathia is present if there is steeper stimulus-response curve than normal (Fig. 1) or at exaggerated response to repetitive suprathreshold stimulation. (**b**) A 50-year-old man with pain following arm amputation after an explosion accident. The patient does not feel heat but pain induced by heat (*). Heat pain threshold is increased. Heat hyperpathia is present if there is steeper stimulus-response curve than normal (Fig. 1) or at exaggerated response to repetitive suprathreshold stimulation. Cold hyperpathia is present, as there is increased cold detection, decreased cold pain threshold, and the patient report exaggeration of pain following innocuous cold stimuli

a gradual change from a faint sensation to a mildly unpleasant sensation and then a sudden exaggerated response with unbearable pain (Lourie and King 1966). During this repetitive stimulation, pain threshold decreases and the slope of the stimulus-response curve increases (Fig. 1). This exaggerated response can be provoked by both noxious and innocuous stimuli (Table 1).

Aftersensations

Aftersensations refer to abnormal persistence of pain seconds to minutes after termination of the stimulation (Gottrup et al. 2003).

Hyperpathia, Assessment, Fig. 3 Stimulus-response function. Stimulus-response function for heat and cold sensation as a function of graded thermal stimulation in a 75-year-old man with pain in left part of face and right leg and arm following brain stem infarct. Stimulus-response curve at right affected forearm (Affected) shows hyperpathia with steeper stimulus-response curve than at control side (control) and faulty identification of stimulus modality with cold stimulation misnamed as heat stimulation. Control stimulus-response curve is assessed at contralateral mirror image area. (Modified from Vestergaard and coworkers (1995))

Pain Radiation

There may be a radiating sensation out from the point of stimulation to the cutaneous area around the stimulus or to wide adjacent areas (Bennett 1994).

Coexistent Phenomena

Hyperpathia is often accompanied by a general alerting response with strong withdrawal movements and vasomotor and vegetative reactions. It may occur with allodynia, ▶ hyperesthesia, hyperalgesia, or ▶ dysesthesia (Merskey and Bogduk 1994).

Clinical Examination

Pain history evaluates symptoms evoked by stimulation of the affected extremity like aftersensations, pain-radiating phenomena, and allodynia induced by movement, non-painful cold or heat, wind touching the extremity, contact with clothing or bed linen, etc.

A bedside screening for hyperpathia is performed with heated and cold thermo rollers kept at 20°C and 40°C, respectively; a wisp of cotton; and pinprick (Von Frey hair or safety pin), moving from the normal toward the painful area

(Jensen et al. 2001). This screening may detect areas with possible increased sensory detection and exaggerated pain responses. Hyperpathia is then assessed objectively as described in "induction and assesment".

During examination, patient's behavioral responses are observed, such as facial expression or withdrawal from stimulus.

Experimental Studies

Hyperpathia is a clinical phenomenon and cannot be induced in human or animal experimental conditions.

Cross-References

▶ Allodynia (Clinical, Experimental)
▶ Amygdala, Pain Processing and Behavior in Animals
▶ Cordotomy Effects on Humans and Animal Models
▶ Nerve Growth Factor Overexpressing Mice as Models of Inflammatory Pain
▶ Opioids, Effects of Systemic Morphine on Evoked Pain

References

Bennett, G. (1994). Neuropathic pain. In P. D. Wall & R. Melzack (Eds.), *Textbook of pain* (pp. 201–224). Amsterdam: Elsevier.

Gottrup, H., Kristensen, A. D., Bach, F. W., & Jensen, T. S. (2003). Aftersensations in experimental and clinical hypersensitivity. *Pain, 103,* 57–64.

Hansson, P., & Lindblom, U. (1992). Hyperalgesia assessed with quantitative sensory testing in patients with neurogenic pain. In W. D. Willis (Ed.), *Hyperalgesia and allodynia* (pp. 335–343). New York: Raven.

Jensen, T. S., Gottrup, H., Sindrup, S. H., & Bach, F. W. (2001). The clinical picture of neuropathic pain. *European Journal of Pharmacology, 19,* 1–11.

Lindblom, U. (1979). Sensory abnormalities in neuralgia. In J. Bonica (Ed.), *Advances in pain research and therapy* (pp. 111–120). New York: Raven.

Lindblom, U. (1994). Analysis of abnormal touch, pain, and temperature sensation in patients. In J. Boivie, P. Hansson, & U. Lindblom (Eds.), *Touch, temperature, and pain in health and disease: Mechanisms and assessments, progress in pain research and management* (pp. 63–84). Seattle: IASP Press.

Lourie, H., & King, R. B. (1966). Sensory and neurohistological correlates of cutaneous hyperpathia. *Archives of Neurology, 14,* 313–320.

Mendell, L. M., & Wall, P. D. (1965). Responses of single dorsal cord cells to peripheral cutaneous unmyelinated fibres. *Nature, 206,* 97–99.

Merskey, H., & Bogduk, N. (1994). *Classification of chronic pain: Descriptions of chronic pain syndromes and definitions of pain terms.* Seattle: IASP Press.

Noordenbos, W. (1979). Sensory findings in painful traumatic nerve lesions. In J. Bonica (Ed.), *Advances in pain research and therapy* (pp. 91–101). New York: Raven.

Price, D. D., Long, S., & Huitt, C. (1992). Sensory testing of pathophysiological mechanisms of pain in patients with reflex sympathetic dystrophy. *Pain, 49,* 163–173.

Vestergaard, K., Nielsen, J., Andersen, G., Ingeman-Nielsen, M., Arendt-Nielsen, L., & Jensen, T. S. (1995). Sensory abnormalities in consecutive, unselected patients with central post-stroke pain. *Pain, 61,* 177–186.

Hyperpolarization

Definition

Hyperpolarization is an increase in inside negativity of the transmembrane resting potential of an excitable cell, such as a neuron that can make a neuron less excitable and, if of sufficient magnitude, can prevent the occurrence of action potentials.

Cross-References

▶ Chronic Pain
▶ Descending Circuitry, Opioids
▶ Drugs with Mixed Action and Combinations: Emphasis on Tramadol
▶ Thalamic Bursting Activity

Hyperresponsiveness

Definition

Increased responsivity and improper frequency control of classes of sensory neurons in the central nervous system, e.g., originated by anomalous inputs.

Cross-References

▶ Deafferentation Pain

Hypersensitivity

Definition

Increased sensation of stimuli, or increased responsiveness of neuronal structures to stimuli.

Cross-References

▶ Chronic Pelvic Pain, Musculoskeletal Syndromes
▶ Deafferentation Pain
▶ Functional Abdominal Pain in Children
▶ Psychology of Pain, Sensitization, Habituation, and Pain
▶ Sensitization of Visceral Nociceptors

Hyperstimulation Analgesia

Definition

Hyperstimulation analgesia is mediated by a short but strong painful stimulation usually applied in the form of a train of electrical impulses which reduces another, background pain.

Cross-References

▶ Acupuncture Mechanisms

Hypertensive Encephalopathy

Definition

Hypertensive encephalopathy is a change in the brain caused by failure of autoregulation of the cerebral circulation in the presence of severe hypertension; characterized pathologically by vasogenic cerebral edema and, sometimes, microhemorrhages and microinfarcts; and characterized clinically by headache, obtundation, seizures, visual changes, and/or focal deficits.

Cross-References

▶ Headache Due to Hypertension

Hypertensive Headaches

▶ Headache Due to Hypertension

Hypertonic Saline

Definition

Hypertonic saline is a solution of greater than 155 mM sodium chloride. Sodium chloride solutions of 1.0 M can be injected into muscle tissue and produce pain, presumably due to their osmotic strength.

Cross-References

▶ Nociceptors in the Orofacial Region (Temporomandibular Joint and Masseter Muscle)

Hypervigilance

Definition

Hypervigilance is the excessive predisposition to attend to a certain class of events, or the excessive readiness to select and respond to a certain kind of stimulus from the external or internal environment. In the context of fear of movement, hypervigilance concerns the increased attention to pain, potential signals of pain, and other possible somatosensory signals. General hypervigilance is the tendency of highly anxious individuals to pay attention to other irrelevant (neutral) stimuli.

Cross-References

▶ Disability, Fear of Movement
▶ Dyspareunia and Vaginismus
▶ Fear and Pain
▶ Hypervigilance and Attention to Pain
▶ Muscle Pain, Fear-Avoidance Model

Hypervigilance and Attention to Pain

Stefaan Van Damme[1], Geert Crombez[1] and Christopher Eccleston[2]
[1]Department of Experimental Clinical and Health Psychology, Ghent University, Ghent, Belgium
[2]Pain Management Unit, University of Bath, Bath, UK

Synonyms

Heightened attention; Overalertness; Selective attention

Definition

Attention is a key theoretical and clinical construct in explaining amplified pain perception, disability, and distress. Patients with chronic pain are often thought to be hypervigilant, i.e., excessively attentive, for pain and/or somatic sensations (Crombez et al. 2005). Hypervigilance is best defined as the readiness to select pain-related information over other information from the environment (Eccleston and Crombez 1999). It should be noted that hypervigilance is conceptually different from other constructs that have been associated with pain hypersensitivity in chronic pain patients, such as hyperalgesia (increased pain response to a noxious stimulus) and allodynia (painful response to a normally innocuous stimulus) typically resulting from abnormalities in central nociceptive pathways. Hypervigilance may only be inferred when the involvement of attentional processes can be demonstrated (Crombez et al. 2005).

Characteristics

Hypervigilance to pain and somatic sensations may become manifest in a variety of ways. The following example may clarify this: Imagine a man, suffering from chronic low back pain, who is resuming his job after a period of pain-related work absence. Increases in pain intensity during work frequently attract his attention. Afraid of (re)injury due to back-related movements, he begins to scan his body for pain or for other potential signals of bodily harm, which may result in the rapid detection of any bodily sensation in his back. Once such bodily sensation is detected, the person may experience difficulties disengaging attention from it and to concentrate on his work.

Theoretical views emphasize that hypervigilance emerges as the working of normal attentional mechanisms in an abnormal situation, i.e., the chronic presence of threatening, high-intensity pain. In understanding hypervigilance, it is therefore important to consider "normal" attention to pain. Eccleston and Crombez (1999) were among the first to systematically investigate

"normal" attentional processes in the context of pain. In their cognitive-affective model of the interruptive function of pain, they argued that pain imposes an overriding priority for attentional engagement by activating a primitive defensive system that urges escape from bodily threat. In a more recent development of this model (Legrain et al. 2009), it has been argued that pain can be selected by the attentional system in two different ways: bottom-up and top-down.

First, bottom-up selection of pain is the unintentional, stimulus-driven capture of attention by pain because of its salience. The involuntary capture of attention by pain has been shown to be mediated by its sensory characteristics such as intensity and novelty (Eccleston and Crombez 1999; Legrain et al. 2009). Because chronic pain patients are characterized by sensitized nociceptive pathways and are continuously confronted with increases in pain intensity, there is frequent bottom-up selection of pain, leading to task interruptions, increased distractibility, and reduced cognitive performance (Moriarty et al. 2011).

Second, top-down selection of pain is an intentional and goal-directed process in which pain-related information is prioritized because of its relevance for current concerns or goals (Van Damme et al. 2010). It has been shown that the threat of bodily harm results in stronger interruption of attentional tasks by pain, makes one more rapidly aware of innocuous somatic sensations in the threatened body part, and results in more difficulty disengaging from that body part (Peters et al. 2000; Van Damme et al. 2007). Furthermore, the selection of pain-related information has been shown to be more pronounced when the goal to avoid or escape is strongly activated (Notebaert et al. 2011). In chronic pain patients, one may expect increased top-down selection of pain-related information because of the frequent presence of fear-avoidance beliefs, i.e., catastrophic misinterpretation of pain as a sign of serious injury or pathology over which one has little or no control (Vlaeyen and Linton 2000), and because the goal of pain relief or pain control is highly salient and often becomes chronically activated (Eccleston and Crombez 2007).

Implications

Our understanding of hypervigilance has a number of implications. First, it has often been assumed that hypervigilance is a maladaptive attentional mechanism resulting from abnormal personality characteristics such as negative affectivity (NA) or neuroticism. However, empirical evidence does not support the idea that hypervigilance in chronic pain patients emerges as a result of dysfunctional personality traits. For instance, Goubert et al. (2004) found that the key mediating variable in explaining hypervigilance to pain was not an abnormally high level of NA but the immediate threat value of pain. NA was rather conceived as a vulnerability factor, lowering the threshold at which pain is perceived as threatening.

Second, although hypervigilance primarily concerns attention to bodily sensations, the majority of studies in patients with chronic pain was conducted by means of visual attention paradigms using pain-related words (Van Damme et al. 2010). Such approach does, however, not allow firm conclusions about bodily hypervigilance. The development of experimental paradigms allowing the valid measurement of the attentional selection of pain and somatic sensations in clinical populations is therefore an important challenge for future research (Moore et al. 2012; Van Damme et al. 2010).

Third, as hypervigilance seems to be mediated by the threat value of pain, distraction is probably not an effective treatment technique in patients with a high level of catastrophic thinking about pain. This was confirmed in the study by Hadjistavropoulos et al. (2000), who found that distraction was not effective in chronic pain patients with a high level of health anxiety. Other techniques may be more suitable in those patients, such as the cultivation of mindful attention and nonjudgmental body awareness (Mehling et al. 2009).

Fourth, we have recast hypervigilance within a motivational perspective on pain and disability. Attentional distraction may be optimized by making the distraction task motivationally relevant. For example, it has been shown that the selection of pain-related information can be substantially reduced by the activation of a competing personal goal (Schrooten et al. 2012). However, techniques merely directed at reducing attention to pain might have limited value in these chronic pain patients in which hypervigilance has become part of persistent attempts to control a largely uncontrollable problem (Eccleston and Crombez 2007) and are probably best integrated within a broader cognitive-behavioral approach aimed at coping with disability and optimizing goal regulation.

References

Crombez, G., Van Damme, S., & Eccleston, C. (2005). Hypervigilance to pain: An experimental and clinical analysis. *Pain, 116,* 4–7.
Eccleston, C., & Crombez, G. (1999). Pain demands-attention: A cognitive-affective model of interruptive function of pain. *Psychological Bulletin, 3,* 356–366.
Eccleston, C., & Crombez, G. (2007). Worry and chronic pain: A misdirected problem solving model. *Pain, 132,* 233–236.
Goubert, L., Crombez, G., & Van Damme, S. (2004). The role of neuroticism, pain catastrophizing and pain-related fear in vigilance to pain: A structural equations approach. *Pain, 107,* 234–241.
Hadjistavropoulos, H. D., Hadjistavropoulos, T., & Quine, A. (2000). Health anxiety moderates the effects of distraction versus attention to pain. *Behaviour Research and Therapy, 38,* 425–438.
Legrain, V., Van Damme, S., Eccleston, C., Davis, K. D., Seminowicz, D. A., & Crombez, G. (2009). A neurocognitive model of attention to pain: Behavioral and neuroimaging evidence. *Pain, 144,* 230–232.
Mehling, W. E., Gopisetty, V., Daubenmier, J., Price, C. J., Hecht, F. M., & Stewart, A. (2009). Body awareness: Construct and self-report measures. *PloS One, 4,* e5614.
Moore, D. J., Keogh, E., & Eccleston, C. (2012). The interruptive effect of pain on attention. *Quarterly Journal of Experimental Psychology, 65,* 65–586.
Moriarty, O., McGuire, B. E., & Finn, D. P. (2011). The effect of pain on cognitive function: A review of clinical and preclinical research. *Progress in Neurobiology, 93,* 385–404.
Notebaert, L., Crombez, G., Vogt, J., De Houwer, J., Van Damme, S., & Theeuwes, J. (2011). Attempts to control pain prioritize attention towards signals of pain: An experimental study. *Pain, 152,* 1068–1073.
Peters, M. L., Vlaeyen, J. W. S., & van Drunen, C. (2000). Do fibromyalgia patients display hypervigilance for innocuous somatosensory stimuli? Application of a body scanning paradigm. *Pain, 86,* 283–292.
Schrooten, M., Van Damme, S., Crombez, G., Peters, M., Vogt, J., & Vlaeyen, J. W. S. (2012). Nonpain goal

pursuit inhibits attentional bias to pain. *Pain, 153*, 1180–1186.

Van Damme, S., Crombez, G., & Lorenz, J. (2007). Pain draws visual attention to its location: Experimental evidence for a threat-related bias. *The Journal of Pain, 8*, 976–982.

Van Damme, S., Legrain, V., Vogt, J., & Crombez, G. (2010). Keeping pain in mind: A motivational account of attention to pain. *Neuroscience and Biobehavioral Reviews, 34*, 204–213.

Vlaeyen, J. W. S., & Linton, S. J. (2000). Fear-avoidance and its consequences in chronic musculoskeletal pain: A state of the art. *Pain, 85*, 317–332.

Hypesthesia

Definition

Hypesthesia is a decreased sensitivity to stimulation, excluding special senses.

Cross-References

▶ Hyperaesthesia, Assessment
▶ Hypoaesthesia
▶ Hypoesthesia, Assessment

Hypnic "Alarm Clock" Headache Syndrome

▶ Hypnic Headache

Hypnic Headache

James W. Lance
Institute of Neurological Sciences, University of New South Wales, Sydney, Australia

Synonyms

Hypnic "alarm clock" headache syndrome

Definition

Headache awakening the subject from sleep, not occurring during waking hours, and usually lasting less than 180 min and not associated with autonomic features.

Characteristics

The headache is unilateral in about 40 % of patients but is bilateral in the remainder. It usually develops after the age of 50 years, recurs more than 15 times a month, and is not severe but persists for more than 15 min after waking. It may be accompanied by nausea, photophobia, or phonophobia, but not all three of these migrainous features. The bilateral site, mild intensity, and the lack of autonomic features distinguish it from cluster headache. It usually responds to the administration of caffeine or lithium taken on retiring to bed.

Clinical Reports and Pathophysiology

Raskin (1998) first drew attention to this uncommon syndrome. He reported six patients, five of whom were male, all aged 60 years or more who were waking up consistently with generalized headaches that persisted for 30–60 min. Two volunteered that they were always woken from a dream by these headaches. Three patients reported accompanying nausea. The headaches were not alleviated by amitriptyline or propranolol but responded to lithium 300 mg or 600 mg at night. Raskin attributed the condition to a disorder of the brain's ▶ "biological clock" in the hypothalamus, pointing out that cluster headache, cyclical migraine, and manic-depression disorder were also tied to bodily rhythms and responded to lithium.

Ten of the 19 patients described by Dodick et al. (1998) were awakened by headache at a consistent time, usually between 1:00 am and 3:00 am, giving rise to the term "alarm clock" headache. Three patients had infrequent but identical headaches during daytime naps. One described the headaches as developing during vivid dreams. Three patients mentioned infrequent nausea. It is not clear why one patient who had a severe unilateral headache with ipsilateral lacrimation and rhinorrhea was included in this series and not classified as cluster headache. An additional link with dreaming was provided by one of the three patients described by Morales-Asin et al. (1998).

In attempts to clarify this question, ▶ polysomnography has been carried out successfully in recording the onset of hypnic headache in six patients. Dodick (2000) found that an episode started during ▶ rapid eye movement (REM) sleep at a time of severe oxygen desaturation. Evers et al. (2003) reported two patients with onset during REM sleep, one of whom had periodic limb movements throughout the night. Oxygen desaturation did not exceed 85 % at any time. Pinessi et al. (2003) recorded four hypnic headaches in two patients, all emerging from the REM phase of sleep without any oxygen desaturation. These authors pointed out that a patient reported by Arjona et al. (2000) as being aroused by hypnic headache in stage 3 slow-wave sleep was being treated with venlafaxine, which may have altered her sleep pattern.

Cells that switch REM sleep cells off are found in the locus coeruleus and dorsal raphe nucleus and discharge regularly during waking hours, ceasing during REM sleep. Their action depends on noradrenergic and serotonergic transmission respectively. Since pathways from these areas form part of the body's endogenous pain control system their switching off could account for the onset of pain with REM sleep (Dodick et al. 2003; Pinessi et al. 2003). The sleep-wake cycle is controlled by the suprachiasmatic nucleus of the hypothalamus and reduced ▶ melatonin secretion is thought to play a part in the initiation of hypnic headache.

Martins and Gouveia (2001) reported the case of a patient in remission for 10 months after lithium therapy who flew from Portugal to Brazil over three time zones. Her hypnic headaches recurred each night for 10 days while away but ceased on her return.

Summary

Evers and Goadsby (2003) have reviewed the 71 cases of hypnic headache reported in the literature to date. There were 24 men and 41 women ranging in age from 26 to 83 years. The headache was bilateral in 61 % and unilateral in 39 %. It varied in frequency from one each week to six per night. It usually started 2–4 h after falling asleep, was moderate in intensity, and persisted for 15 min to 3 h.

Nausea was reported by 19.4 %. Mild photophobia, phonophobia, or both were experienced by 6.8 %. Some autonomic features such as lacrimation were recorded in six patients, two of whom developed ptosis. No relevant abnormality was found on CT, MRI, EEG, or carotid Doppler ultrasound studies.

Evers and Goadsby (2003) summarized the response to treatment in reported cases. Good results were achieved by lithium in 26/35 patients, caffeine in 6/16, indomethacin in 7/18, flunarizine in 4/5, melatonin in 3/7, and prednisone in the only two patients in whom it had been tried (Relja et al. 2002).

References

Arjona, J. A., Jimenez-Jimenez, F. J., Vela-Bueno, A., et al. (2000). Hypnic headache associated with stage 3 slow wave sleep. *Headache, 40*, 753–754.

Dodick, D. W. (2000). Polysomnography in hypnic headache syndrome. *Headache, 40*, 748–752.

Dodick, D. W., Mosek, A. C., & Campbell, J. K. (1998). The hypnic ("alarm clock") headache syndrome. *Cephalalgia, 18*, 152–156.

Dodick, D. W., Eross, E. J., & Parish, J. M. (2003). Clinical, anatomical and physiologic relationship between sleep and headache. *Headache, 43*, 282–292.

Evers, S., & Goadsby, P. J. (2003). Hypnic headache. Clinical features, pathophysiology and treatment. *Neurology, 60*, 905–909.

Evers, S., Rahmann, A., Schwaag, S., et al. (2003). Hypnic headache -the first German cases including polysomnography. *Cephalalgia, 23*, 20–23.

Martins, I. P., & Gouveia, R. G. (2001). Hypnic headache and travel across time zones: A case report. *Cephalalgia, 21*, 928–931.

Morales-Asin, F., Mauri, J. A., Iñiguez, C., et al. (1998). The hypnic headache syndrome: Report of three new cases. *Cephalalgia, 18*, 157–158.

Pinessi, L., Rainero, I., Cicolin, A., et al. (2003). Hypnic headache syndrome: Association of the attacks with REM sleep. *Cephalalgia, 23*, 150–154.

Raskin, N. H. (1998). The Hypnic headache syndrome. *Headache, 28*, 534–536.

Relja, G., Zarzon, M., Locetelli, L., et al. (2002). Hypnic headache: Rapid and long-lasting response to prednisone in two new cases. *Cephalalgia, 22*, 157–159.

Hypnosis

Definition

A process of focusing attention that typically produces deep relaxation and openness to verbal suggestions; it can be performed on oneself or by others by using a combination of relaxation and intensive guided imagery techniques. The resulting altered state of consciousness is known as a trance. Hypnosis is widely used in both adults and children and is broadly effective in the management of chronic and acute pain, especially cancer pain.

Cross-References

- ▶ Complex Chronic Pain in Children, Interdisciplinary Treatment
- ▶ Coping and Pain
- ▶ Hypnotic Analgesia
- ▶ Psychological Treatment in Acute Pain
- ▶ Relaxation in the Treatment of Pain

Hypnotherapy

- ▶ Therapy of Pain, Hypnosis

Hypnotic Analgesia

Donald D. Price[1] and Pierre Rainville[2]
[1]Oral Surgery and Neuroscience, Mcknight Brain Institute University of Florida, Gainesville, FL, USA
[2]Department of Stomatology, Faculty of Dental Medicine, University of Montreal, Montreal, QC, Canada

Definition

Psychological factors and interventions can sometimes powerfully modulate pain, and there is an emerging neurobiology of pain-modulatory mechanisms. Central neural mechanisms associated with such phenomena as placebo/nocebo, hypnotic suggestion (see ▶ Post-Hypnotic Suggestion), attention, distraction, and even ongoing emotions are now thought to modulate pain by decreasing or increasing neural activity within many of the brain structures shown in Fig. 1 (Rainville 2002). This modulation includes endogenous pain-inhibitory and pain-facilitation pathways that descend to spinal dorsal horn, the origin of ascending spinal pathways for pain as well as modulation, which takes place within corticolimbic circuits once nociceptive information has reached cortical levels (De Pascalis et al. 2001; Fields and Price 1997; Hofbauer et al. 2001; Porro et al. 2002; Rainville 2002). Hypnotically induced reduction in pain is based on changes in pain induced by suggestions and facilitated by an alteration of consciousness (Hilgard and Hilgard 1983; Price and Barrell 1990; Rainville and Price 2003). This alteration is accompanied by changes in brain activity involved in the regulation of consciousness (Rainville and Price 2003). Hypnotic changes in pain experience can consist of selective changes in the ▶ affective dimension (component) of pain or reductions in both sensory and affective dimensions, depending on the nature of the suggestions. Changes in affective and sensory components of pain are associated with corresponding changes in anterior cingulate cortical activity and somatosensory cortical activity, respectively (Rainville and Price 2003; Rainville 2002). Different hypnotic analgesic approaches are clinically useful.

Characteristics

What Are the Types of Hypnotic Suggestions for Analgesia

The suggestions for alteration of the experience of pain in studies of hypnotic analgesia relate closely to the dimensions of pain and to the psychological stages of pain processing. Thus, there are suggestions that specifically target the affective-motivational dimension of pain, as distinguished from the ▶ sensory-discriminative dimension (Rainville et al. 1999). These would

Hypnotic Analgesia, Fig. 1 Schematic of ascending pathways, subcortical structures, and cerebral cortical structures involved in processing pain. *PAG* periaqueductal gray, *PB* parabrachial nucleus of the dorsolateral pons, *VMpo* ventromedial part of the posterior nuclear complex, *MDvc* ventrocaudal part of the medial dorsal nucleus, *VPL* ventroposterior lateral nucleus, *ACC* anterior cingulate cortex, *PCC* posterior cingulate cortex, *HY* hypothalamus, *S-1* and *S-2* first and second somatosensory cortical areas, *PPC* posterior parietal complex, *SMA* supplementary motor area, *AMYG* amygdala, *PF* prefrontal cortex (Figure from Price, Science (2001))

include suggestions for reinterpreting sensations as neutral or pleasant rather than unpleasant, as well as suggestions for reducing or eliminating the implications of threat or harm from the sensations. Then, there are suggestions designed for specifically altering the quality and/or intensity of painful sensations so that they become less intense or absent altogether. There are three very different types of hypnotic suggestions for altering pain sensation intensity (De Pascalis et al. 1999, 2001). One type provides ▶ dissociative imagery by suggesting experiences that are disconnected from the felt sense of the body. An example would be a suggestion to imagine oneself "floating out of the body and up in the air" combined with the implicit or explicit suggestion that the pain belongs to the body and not to the one who experiences being somewhere else. Common to suggestions for dissociation is the intention of having subjects not feel parts of their bodies that would otherwise be painful and/or experience themselves in another location and context altogether. Another type is ▶ focused analgesia, which is intended to

replace sensations of pain with others, such as numbness or warmth or with the complete absence of sensation. In complete contrast to dissociative analgesia, focused analgesia requires increased attention to the body area wherein pain is present, combined with a replaced sensation in that body area. For example, focused analgesia might include suggestions to focus on sensations in the hand and to experience all sensations of the hand *as* if it were in a large glove. A third type of suggestion involves the reinterpretation of the meaning of the sensory experience. In this case, the significance of the experience for the integrity of the body is reduced or completely abolished, so that pain sensations are no longer associated with feelings of threat. Just as studies are needed to assess the role of hypnotic depth and individual components of hypnosis on pain, so there also need to be studies of differential effects of various types of suggestion on sensory and affective dimensions of pain experience. For example, what are the effects on pain of suggestions exclusively designed to reinterpret the meanings of the

Hypnotic Analgesia, Fig. 2 Pain sensory and distress ratings in response to noxious electrical stimulation delivered to the wrist in normal subjects with high (*Highs*), moderate (*Mids*), or low (*Lows*) hypnotic susceptibility. Both pain sensory and distress ratings decrease significantly in response to hypnotic suggestions for relaxation (Relax), dissociative imagery (Imag.), and focused analgesia (Analg) compared to the baseline wakefulness (Wak.) and placebo (Plac.) conditions. Larger pain reductions are observed in more susceptible subjects (*Highs*) and during focused analgesia. Also, note that there is no significant placebo analgesia observed for all three groups (De Pascalis et al. 2001)

sensations so that they are less threatening or unpleasant?

Which Types of Hypnotic Suggestions are Most Effective in Producing Analgesia

Very few hypnotic analgesia studies have directly compared effects from the different types of hypnotic suggestions described above. However, De Pascalis et al. conducted studies that compared analgesic effects produced by experimental conditions of deep relaxation, dissociated imagery, focused analgesia, and placebo in comparison to a waking control condition (De Pascalis et al. 1999, 2001). They compared these conditions across groups of high-, medium-, and low-hypnotizable participants and utilized several dependent pain-related measures. These included pain and distress ratings, pain threshold determinations, somatosensory event-related potentials (SERP), heart rate, and skin conductance responses (SCR). The experimental stimuli consisted of non-painful and painful levels of electrical pulses delivered to the right wrist.

Of the four experimental conditions, Deep Relaxation, Dissociated Imagery, and Focused Analgesia produced statistically significant reductions on all pain-related measures among all three groups of participants (i.e., low, mid, high).

However, these analgesic effects interacted with ▶ hypnotizability, as shown in Fig. 2. During focused analgesia, highly hypnotizable participants had larger reductions in pain ratings in comparison to low- and medium-hypnotizable participants. Furthermore, highly susceptible subjects had more pronounced reductions in distress ratings during focused analgesia and dissociated imagery in comparison to the other two groups. Focused analgesia produced the largest reductions in all dependent measures within highly hypnotizable participants. No significant placebo effects were obtained for any of the three groups. The combination of these results indicates several interesting features of hypnotic analgesia. First, hypnotic analgesia cannot simply be understood as a placebo effect and is more than just relaxation. Second, very different types of suggestions for analgesia are effective and are facilitated by hypnotizability. Third, hypnotic analgesia can affect physiological reflexive responses associated with pain (Hilgard and Hilgard 1983; Rainville 2002).

Each of the types of hypnotic suggestion discussed so far can be given directly or indirectly. A ▶ direct suggestion for analgesia would be "You will notice that the pain is less intense," whereas an ▶ indirect suggestion would be "I wonder if you will notice whether the sensation you once experienced as painful will be experienced as just warmth or pressure or perhaps even numbness." The latter is permissive, ambiguous, and refers to alternative experiences without the implication of a direct instruction. Resistance to hypnotic suggestions may be less in the case of permissive-indirect as compared to restrictive-direct suggestions, because one is not directly told what to experience. Furthermore, restrictive-direct suggestions may be perceived as unnecessarily authoritarian. One might expect that a larger proportion of people could benefit from a hypnotic approach that uses indirect suggestions, and there is some, albeit limited, evidence that this is so (Price and Barber 1987, 1990).

What are the Factors that Determine the Efficacy of Hypnotic Analgesia

The efficacy of hypnotic analgesia and its relationship to hypnotic susceptibility has been shown to depend on several factors (Price and Barber 1987). These include the pain dimension that is measured, baseline pain intensity, the maintained presence of the hypnotist or hypnotic suggestions, and finally hypnotic ability. Some of these factors are shown in Table 1. When suggestions were given for both reinterpreting the meaning of experimentally induced heat sensations and for experiencing them as less intense, pain sensation intensity was reduced by an average of about 50 %, and pain unpleasantness was reduced by 87 % in a group of 16 subjects. Thus, pain affect was more powerfully attenuated in comparison to pain sensation. Although hypnotic suggestions exerted a more powerful reduction of pain affect than pain sensation, it was also quite apparent that both dimensions were reduced, as has been amply demonstrated in several experimental laboratories (Barber and Mayer 1977; De Pascalis et al. 1999, 2001; Rainville et al. 1999, Rainville 2002). Reduction in pain sensation was statistically associated with hypnotic susceptibility, albeit at modest levels (Table 1). Therefore, the component of the hypnotic intervention that relied on hypnotic ability and a hypnotic state was the one most influential on pain sensation intensity. Interestingly, the association became stronger with increasing levels of pain intensity

Hypnotic Analgesia, Table 1 Hypnotic susceptibility and analgesia

Stimulus temperature	Sensory analgesia	Affective analgesia
	Spearman correlation coefficient	
44.5 °C	+0.04	−0.23
47.5 °C	+0.21	−0.11
49.5 °C	+0.43[*]	−0.08
51.5 °C	+0.56[*]	+0.10

[*]P < 0.05

(Table 1). It makes sense that the reduction in stronger pains requires more hypnotic ability than the reduction in weaker pains. A final factor was maintained contact between the hypnotist and the subject. Statistically significant analgesia developed in one group of subjects that had maintained contact with the hypnotist during the pain testing session and did not develop in the group that did not have maintained contact. Thus, multiple factors are involved in analgesia that results from a hypnotic intervention. These may include those that are unrelated to hypnotic susceptibility and perhaps even to a hypnotic state. Such potential multiple factors are closely related to different proposed mechanisms of hypnotic analgesia.

Rainville et al. further clarified the relationship between different types of hypnotic suggestions for analgesia and the dimensions of pain that are modulated by these suggestions (Rainville et al. 1999). This study conducted two types of experiments, one in which hypnotic suggestions were selectively targeted toward increasing or decreasing the sensory intensity of pain and the other in which hypnotic suggestions were targeted toward decreasing or increasing the affective dimension of pain. In both types of experiments, normal subjects who were trained in hypnosis rated pain intensity and pain unpleasantness produced by a tonic heat pain test (1-min immersion of the hand in 45.0–47.5 °C water). The results of the two experiments are illustrated in Fig. 3.

In the first experiment, suggestions to modulate pain sensation intensity resulted in significant changes in both pain sensation intensity ratings and pain unpleasantness ratings, that is, both

Hypnotic Analgesia, Fig. 3 Self-reports of the pain experienced during the immersion of the hand in hot water following hypnotic suggestions directed at the sensory and affective dimension of pain. Suggestions directed at the sensory aspect of pain (sensory modulation) produce parallel changes in self-reports of pain sensation intensity and unpleasantness. In contrast, suggestions for the reinterpretation of pain with decreased and increased sense of threat and discomfort (Affective modulation) produce specific changes in pain unpleasantness that largely exceed the changes in pain sensation intensity (Rainville et al. 1999)

dimensions were modulated in parallel. This was so, despite the fact that no suggestions were given about pain affect. In the second experiment, pain unpleasantness was significantly increased and decreased after suggestions were given for these changes, and these changes occurred without corresponding changes in pain sensation intensity. Hypnotic susceptibility (Stanford Hypnotic Susceptibility Scale Form A) was specifically associated with pain sensation intensity modulation in the first experiment (directed toward pain sensation; Spearman-r = 0.69) and with pain unpleasantness modulation in the second experiment (directed toward pain affect; Spearman-r = 0.43).

Thus, hypnotic changes in pain experience can consist of selective changes in the affective dimension of pain or reductions in both sensory and affective dimensions, depending on the nature of the suggestions. Selective changes in only affective components of pain are associated with corresponding changes in anterior cingulate cortical activity, and changes in sensory components are accompanied by corresponding changes in somatosensory cortical activity (Hofbauer et al. 2001; Rainville 2002). Reinterpretation of meanings of pain, dissociation, and focused analgesia reflect different psychological mechanisms

of hypnotic analgesia. These multiple mechanisms are likely to be associated with intracortical and descending brain-to-spinal cord mechanisms, to varying extents. Although there is some evidence that hypnotic analgesia has demonstrable clinical efficacy, there is a strong need for improvements in methodologies of clinical studies. In particular, there is a need to compare the efficacy of different hypnotic approaches and provide rigorous standardized outcome measures.

It is useful to consider how results of experiments by De Pascalis et al. (1999, 2001) and Rainville et al. (1999, Rainville 2002), described above, help identify the necessary and sufficient psychological factors for hypnotic analgesia. Hypnotic analgesia cannot work only by means of distraction, because suggestions for focused analgesia are among the most effective, particularly among highly hypnotizable participants. Focused analgesia requires greater not lesser attention to the body area wherein analgesia develops. Hypnotically induced changes in pain affect can occur directly through suggestions that alter the meaning of the experience of the stimulus or indirectly through suggestions that target the pain sensation. Hypnotic changes in the latter can also occur through suggestions for dissociation or through suggestions for changes in the way the sensory qualities are experienced (e.g., numbness versus burning). Hypnotic analgesia cannot only work by means of a placebo effect, because subjects are likely to experience placebo and hypnotic suggestions differently. Moreover, there is now good evidence that ► placebo analgesia, but not hypnotic analgesia, requires an endogenous opioid pain-inhibitory mechanism. Placebo analgesia is naloxone reversible in studies of experimental pain, whereas several studies have shown than hypnotic analgesia is not naloxone reversible (Barber and Mayer 1977; Goldstein and Hilgard 1975). Finally, placebo analgesia, unlike hypnotic analgesia, is not significantly associated with hypnotic susceptibility (Hilgard and Hilgard 1983).

Conclusions

The combination of anatomical, psychological, and neurophysiological approaches to understanding the brain mechanisms underlying sensory and affective dimensions of pain and its modulation by psychological interventions, such as hypnotic suggestions, has led to a vastly improved ability to answer questions that only 10 years ago were relatively impenetrable. In particular, studies that combine brain imaging with psychophysical methods and sophisticated experimental designs have led to the possibility of understanding complex mechanisms by which sensory and affective dimensions of pain are interrelated and how these dimensions can be modulated by cognitive factors. The brain networks for these mechanisms are extensive and involve both serial and parallel circuitry, which is itself under dynamic control from several brain regions.

References

Barber, J., & Mayer, D. (1977). Evaluation of the efficacy and neural mechanism of a hypnoticanalgesia procedure in experimental and clinical dental pain. *Pain, 4,* 41–48.
De Pascalis, V., Magurano, M. R., & Bellusci, A. (1999). Pain perception, somatosensory event-related potentials and skin conductance responses to painful stimuli in high, mid, and low hypnotizable subjects: Effects of differential pain reduction strategies. *Pain, 83,* 499–508.
De Pascalis, V., Magurano, M. R., Bellusci, A., et al. (2001). Somatosensory event-related potential and autonomic activity to varying pain reduction cognitive strategies in hypnosis. *Clinical Neurophysiology, 112,* 1475–1485.
Fields, H. L., & Price, D. (1997). Toward a neurobiology of placebo analgesia. In A. Harrington (Ed.), *The placebo effect* (pp. 93–115). Cambridge: Harvard University Press.
Goldstein, A., & Hilgard, E. R. (1975). Lack of influence of the morphine antagonist naloxone on hypnotic analgesia. *Proceedings of the National Academy of Science of the United States of America, 72,* 2041–2043.
Hilgard, E. R., & Hilgard, J. R. (1983). *Hypnosis in the relief of pain* (p. 294). Los Altos: William Kaufmann.
Hofbauer, R. K., Rainville, P., Duncan, G. H., et al. (2001). Cortical representation of the sensory dimension of pain. *Journal of Neurophysiology, 86,* 402–411.
Porro, C. A., Balraldi, P., Pagnoni, G., et al. (2002). Does anticipation of pain affect cortical nociceptive systems? *Journal of Neuroscience, 22,* 3206–3214.
Price, D. D., & Barber, J. (1987). An analysis of factors that contribute to the efficacy of hypnotic analgesia. *Journal of Abnormal Psychology, 96,* 46–51.
Price, D. D., & Barrell, J. J. (1990). The structure of the hypnotic state: A self-directed experiential study. In J. J. Barrell (Ed.), *The experiential method: Exploring the human experience* (pp. 85–97). Massachusetts: Copely Publishing Group.
Rainville, P. (2002). Brain mechanisms of pain affect and pain modulation. *Current Opinion in Neurobiology, 12,* 195–204.
Rainville, P., & Price, D. D. (2003). Hypnosis phenomenology and the neurobiology of consciousness. *International Journal of Clinical and Experimental Hypnosis, 51,* 105–129.
Rainville, P., Carrier, B., Hofbauer, R. K., et al. (1999). Dissociation of sensory and affective dimensions of pain using hypnotic modulation. *Pain, 82,* 159–171.

Hypnotic Relaxation

► Relaxation in the Treatment of Pain

Hypnotism

► Therapy of Pain, Hypnosis

Hypnotizability

Definition

Hypnotic susceptibility, hypnotic capacity, or hypnotic responding delineates a variable that determines the extent to which an individual is able to respond to hypnotic suggestion. Research has shown that hypnotizability can be measured with good reliability and is a remarkably stable trait in adults. It correlates with dissociative experiences and with measures of absorption. Highly hypnotizable individuals tend to have a high imaginative capacity.

Cross-References

▶ Hypnotic Analgesia
▶ Therapy of Pain, Hypnosis

Hypoaesthesia

▶ Hypoesthesia, Assessment

Hypoalgesia, Assessment

Misha-Miroslav Backonja
Department of Neurology, University of Wisconsin-Madison, Madison, WI, USA

Synonyms

Assessment of hypoalgesia; Hypalgia

Definition

IASP taxonomy (Merskey and Bogduk 1994) defines hypoalgesia as "decreased perception of noxious stimuli." Hypoalgesia could be in response to a wide variety of mechanical stimuli such as pinch, strong pressure, or punctuate and to thermal noxious stimuli of heat and cold, basically any physical force of sufficient intensity to disrupt or threaten the integrity or homeostasis of any tissue.

In other terms, hypoalgesia is diminished experience of pain in response to a normally painful stimulus. Hypoesthesia covers the case of diminished sensitivity to stimulation that is normally not painful.

Hypoalgesia is also defined as raised threshold to painful stimuli.

Characteristics

Hypoalgesia is a ▶ negative sensory phenomenon seen exclusively in patients with neurological disease or injury, including patients with ▶ neuropathic pain (Backonja and Galer 1998; Lindblom and Ochoa 1986; Backonja 2003). Hypoalgesia indicates a decrease or loss of function that comes as a result of neurological disease or injury affecting thermonociceptive pathways, anywhere from primary afferents to cerebral cortical structures. Distinction of hypoalgesia from hypoesthesia is based primarily on the type and intensity of the stimulus applied to the thermonociceptive sensory system.

Methods of Assessment and the Interpretation

Assessment of the sensory nervous system function is most commonly done at the bedside where testing is primarily qualitative in nature, while quantitative assessment, increasingly using computerized electronic equipment, is done in a quantitative sensory laboratory (Backonja and Galer 1998; Greenspan 2001). Either should be able to detect hypoalgesia, but the way these methods arrive to the conclusion about presence and severity of hypoalgesia is distinct, and that is also reflected in the definition. Qualitative bedside exam relies on patient report. Quantitative sensory testing arrives at its conclusion about hypoalgesia on the basis of the raised thresholds to painful stimuli.

Qualitative assessment is based on the subject's ability to compare and report quality of sensation from standard methods of stimulation, from the ▶ symptom affected areas, when it is compared to normal unaffected areas. Qualitative method is very convenient for bedside evaluations. Frequently utilized bedside methods include standard neurological examination tools such as safety pin, monofilament, and various metal objects that could be conveniently warmed or cooled in the clinical setting. A degree of quantification is possible and requires that the subject reports whether a decrease of pain from painful stimulation is mild, moderate, severe, or completely absent, when compared to a normal unaffected area. Since a qualitative method requires psychophysical interaction, this method can be used only in humans who can linguistically communicate with the examiner, and as such, it cannot be used in animal models of pain studies.

Quantitative assessment of sensory deficits requires a more sophisticated approach and frequently utilizes electronically controlled devices, although a number of psychophysical methods, especially mechanical stimuli, are used and all of them place much longer time demands on patients. Traditionally, this method is known as quantitative sensory testing (QST). The primary outcome of QST is determination of thresholds for specific modalities which are then compared to the established norms (Greenspan 2001; Getz et al. 2005). Increase in threshold to painful stimuli is then interpreted as hypoalgesia. QST methods could be used not only in human studies but also in animal models.

One of the main goals of neurological evaluation is to determine the site and level of ▶ neuraxis where pathological processes that produce symptoms, including pain, originate (Dyck and O'Brien 2003). In addition to establishing the nature of ▶ neurological deficit, such as hypoalgesia, to a specific modality, it is important to establish a special pattern of these abnormalities, since the pattern serves as the basis for the determination whether the lesion that is causing symptoms, including pain, involves specific peripheral nerve structures, such as peripheral nerves, plexus, or the nerve root, versus central nervous system structures, such as spinal cord, brain stem, or subcortical or cortical structures and pathways of the brain.

Caveats and Unresolved Issues

Difficulty of assessing hypoalgesia arises from the inherent difficulty of assessing negative sensory phenomenon. For example, conceptually it is easier to illustrate ▶ positive sensory phenomenon to subjects, such as pain with instruction that $0 =$ none and $10 =$ worst imaginable. In contrast, it is much harder conceptually to illustrate and request a rating of spontaneous loss of sensation because it is not possible to feel what one is not able to feel, in spite of the instruction that the patient is to imagine scaling between one end being a normal sensation and the other absence of sensation.

Another phenomenon that can result from painful stimulation and the one that is on the opposite end of the spectrum of sensory experience is ▶ hyperalgesia. The difficulty of assessing sensory abnormalities which are characterized by hypoalgesia to one sensory modality and hyperalgesia to another sensory modality in the same area frequently seen in patients with neuropathic pain leads to confusion not only for patients but also for inexperienced clinicians Depending on the way stimulation is conducted even when hypoalgesia is present, the outcome can be either hyperalgesia or hyperpathia. For example, in the case of partial hypoalgesia and when the stimulus is "strong enough," the outcome could be hyperalgesia, and in the case that stimulus is not "strong enough" with temporal and special summation, that could result in increased pain, which would become hyperpathia. Consequently, from the pain mechanisms prospective, the relationship between hypoalgesia and hyperalgesia still far from clear. In summary, hypoalgesia is a ▶ clinical sign of neurological injury or disease, which in some patients can lead to neuropathic pain. Methods for determining the presence of hypoalgesia are qualitative, such as in bedside exam or quantitative, such as QST. Mechanisms of hypoalgesia for specific modalities, and in particular its relationship to hyperalgesia and hyperpathia, are poorly understood.

Cross-References

▶ Allodynia (Clinical, Experimental)
▶ Amygdala, Pain Processing and Behavior in Animals
▶ Cordotomy Effects on Humans and Animal Models
▶ Nerve Growth Factor Overexpressing Mice as Models of Inflammatory Pain
▶ Opioids, Effects of Systemic Morphine on Evoked Pain

References

Backonja, M. (2003). Defining neuropathic pain. *Anesthesia and Analgesia, 97*, 785–790.

Backonja, M. M., & Galer, B. S. (1998). Pain assessment and evaluation of patients who have neuropathic pain. *Neurologic Clinics, 16*, 775–790.

Dyck, P. J., & O'Brien, P. C. (2003). Quantitative sensory testing: Report of the therapeutics and technology assessment subcommittee of the american academy of neurology. *Neurology, 61*, 1628–1630.

Getz, K. K., Cook, T., & Backonja, M. M. (2005). Pain ratings at the thresholds are necessary for interpretation of quantitative sensory testing. *Muscle & Nerve, 32*, 179–184.

Greenspan, J. D. (2001). Quantitative assessment of neuropathic pain. *Current Pain and Headache Reports, 5*, 107–113.

Lindblom, U., & Ochoa, J. (1986). Somatosensory function and dysfunction. In A. K. Asbury, G. M. McKhann, & W. I. McDonald (Eds.), *Diseases of the nervous system. Clinical neurobiology.* (pp. 283–298). Philadelphia: W.B. Saunders.

Merskey, H., & Bogduk, N. (1994). *Classification of chronic pain* (Vol. 2). Seattle: IASP Press.

Hypochondriaca

Definition

Hypochondriaca refers to a persistent conviction that one is or is likely to become ill, when a patient complains of symptoms that have no organic basis. These symptoms persist despite reassurance and medical evidence to the contrary.

Cross-References

▶ Psychiatric Aspects of Visceral Pain

Hypochondriasis

Definition

Hypochondriasis is a minimum 6-month preoccupation with fears of having a serious disease, based on misinterpretation of bodily symptoms (e.g., a sore throat is thought to be throat cancer), which persists in spite of medical evidence that the serious disease is not present.

Hypochondriasis, Somatoform Disorders, and Abnormal Illness Behavior

Issy Pilowsky
University of Adelaide, Adelaide, Australia

Synonyms

Abnormal illness behavior of the unconsciously motivated, somatically focussed type; Discordant illness behavior; Dysnosognosia; Somatoform disorders

Definition

In the fourth edition of the Diagnostic and Statistical Manual of the American Psychiatric Association (1994), hypochondriasis is defined according to the following criteria:

- Because of misinterpreting bodily symptoms, the patient becomes preoccupied with ideas or fears of having a serious illness.
- Appropriate medical investigation and reassurance do not relieve these ideas.
- These ideas are not delusional (as in delusional disorder) and are not restricted to concern about appearance (as in body dysmorphic disorder).

- They cause distress that is clinically important or impair work, social, or personal functioning.
- They have lasted 6 months or longer.
- These ideas are better explained by generalized anxiety disorder, major depressive episode, obsessive-compulsive disorder, panic disorder, separation anxiety, or a different somatoform disorder.

Specify when with poor insight: During most of this episode, the patient does not realize that the preoccupation is excessive or unreasonable.

It is of interest to compare this development with the earlier criteria for the diagnosis of hypochondriasis as listed in DSM-IIIR (the revised edition of DSM-III).

They are as follows:

1. Preoccupation with the fear of having, or the belief that one has, a serious disease, based on the person's interpretation of physical signs or sensations as evidence of physical illness.
2. Appropriate physical evaluation does not support the diagnosis of any physical disorder that can account for the physical signs or the person's unwarranted interpretation of them, and the symptoms in "A" are not just those of panic attacks.
3. The fear of having or belief that one has a disease persists despite medical reassurance.
4. Duration of the disturbance is at least 6 months.
5. The belief in A is not of delusional intensity, as in delusional disorder, somatic type (i.e., the person can acknowledge the possibility that the fear or the belief of having a serious illness is unfounded). [Comment: In which case, the psychopathological phenomenon could be labelled as an "abnormal preoccupation" or an "overvalued idea."]

Characteristics

Hypochondriasis is regarded as one of the somatoform disorders in both the DSM-IV and the tenth edition of the WHO Classification of Mental and Behavioral Disorders: (ICD-10): Clinical descriptions and diagnostic guidelines.

The somatoform disorders are defined in DSM-IV as essentially the presence of physical symptoms for which there are no demonstrable organic findings or known physiological mechanisms.

In both DSM-IV and ICD-10, a significant departure is made from the principle of classifying on the basis of phenomenological description only. Thus, in DSM-IV, we find the inclusion of the statement that "the symptoms are linked to psychological factors or conflicts."

Illness, the sick role, illness behavior, and abnormal illness behavior (Pilowsky 1969, 1978, 1997).

Illness is defined as any state of an organism which fulfils the requirements of a relevant reference group for admission to a sick role.

The Sick Role

As delineated by the sociologist Talcott Parsons (1964, 1978), the sick role is a partially and conditionally granted social role. The individual seeking this role is required to fulfil three obligations. These are the following: (a) accept that the role is "undesirable," one which should be relinquished a soon as possible; (b) cooperate with others so as to achieve "health"; and (c) utilize the services of those regarded by society as competent to diagnose and treat the condition. (In technologically advanced societies, this person is usually a formally registered doctor who is granted the authority to sign "sickness certificates").

If these obligations are met, the following privileges are granted: (a) the person is regarded as not "responsible" for the condition (i.e., he cannot produce or terminate it by an act of will and is not to be considered a malingerer), (b) the person is regarded as someone requiring care, and (c) the person is entitled to exemption from age-appropriate normal obligations.

All of these definitions demonstrate how central the role of the doctor is (in technologically advanced societies) when it comes to the allocation of healthcare resources. It also draws attention to the pressures on the doctor-patient relationship from without and their inevitable interaction with interpersonal and intrapersonal forces.

Abnormal Illness Behavior (AIB)

This is defined as an inappropriate or maladaptive mode of experiencing, evaluating, or acting in relation to one's own state of health, despite the fact that a doctor (or other recognized social agent) has offered accurate and reasonably lucid information concerning the person's health status and the appropriate course of management (if any), based on a thorough examination of all parameters of functioning (i.e., physical psychological, and social) taking into account the individuals age, educational, and sociocultural background.

A detailed analysis of this definition is found in Pilowsky (1997).

Clinical Characteristics

Pain is a common feature of hypochondriacal disorders. Since the patient tends to reject the presence of psychological problems, such individuals are not encountered in psychiatric settings but rather in medical, surgical, and, in the case of conversion disorders, neurological clinics.

Another relevant somatoform disorder is "conversion disorder," also known as "hysterical neurosis, conversion type." The difference between hypochondriasis and conversion is that the latter is defined as manifesting "a loss or alteration of physical functioning that suggests a physical disorder in the absence of physical signs on examination to support the presence of a physical disorder." However, "psychological factors are judged to be aetiologically related, because of a temporal relationship between a psychosocial stressor that is apparently related to a psychological conflict or need, and initiation or exacerbation of the symptom."

A feature often described in association with a conversion disorder is "la belle indifference," which refers not simply to an absence of concern but rather to a sort of positive serenity, clearly inappropriate to the apparent seriousness of the physical disability.

The condition named "somatoform pain disorder" is described in virtually the same terms as conversion disorder, except for the statement that it is primarily characterized by: "Pain which causes significant distress or impairment in functioning,

which cannot be fully explained by a physician. It must be judged to be related to psychological factors and cannot be better explained by another disorder."

Thus, the major difference between pain as a feature of hypochondriasis and pain as a conversion symptom is that in the former case there is concern and preoccupation as to what the pain may mean, in terms of specific illnesses such as cancer or heart disease, while in the latter the patient denies concern over any specific condition but is rather troubled by the experience of pain as a cause of disability and suffering.

Management

The key to management is the establishment of an alliance with the patient. This issue is of particular salience when the clinician is a psychologist or psychiatrist (has been discussed at length in Pilowsky (1997)) because patients often consider a referral to such a person to mean that the referring doctor believes "it is all in my mind," by which is meant that they are being accused of malingering. Achieving an alliance is not possible unless the acceptance of the reality of the symptoms is clearly conveyed to the patient by the attention paid to the alleviation of discomfort and prevention of further disability by appropriate supportive psychotherapeutic, physiotherapeutic, and psychopharmacological (e.g., antidepressants in low doses), and if necessary by psychological methods such as cognitive-behavioral therapy. In theory, this should be easiest at the initial presentation to the first doctor who sees the patient, especially as this is usually a nonpsychiatrist and most often a family doctor who should, ideally, be well acquainted with the patient and his circumstances, as well as, hopefully, his family.

How this doctor might manage the situation has been described and researched by Goldberg et al. (1989). They have developed a methodology whereby the doctor can help the patient to reattribute the physical symptoms to psychological causes. Once this has been achieved, it is reasonable to proceed with a problem-solving approach to any of the difficulties the patient is invariably experiencing in his life (Rost and Smith 1990; Wilkinson and Mynors-Wallis 1990; Scicchitano 2000).

When a multimodal approach is necessary, this is generally best provided by a multidisciplinary pain clinic, when it is available.

Some pain clinics have inpatient facilities with well-trained experienced staff that are able to provide program for patients manifesting severe invalidism and perhaps dependence on drugs such as opiates.

References

American Psychiatric Association. (1994). *Diagnostic and statistical manual of mental disorders* (3rd edn. Rev.). Washington, DC: American Psychiatric Association.

Goldberg, D. P., Gask, L., & O'Dowd, T. (1989). The treatment of somatization-teaching techniques of reattribution. *Journal of Psychosomatic Research, 33*, 689–696.

Parsons, T. (1964). *Social structure and personality*. London: Collier-Macmillan.

Parsons, T. (1978). *Action theory and the human condition*. New York: Free Press.

Pilowsky, I. (1969). Abnormal illness behaviour. *The British Journal of Medical Psychology, 42*, 347–351.

Pilowsky, I. (1978). A general classification of abnormal illness behaviours. *The British Journal of Medical Psychology, 51*, 131–137.

Pilowsky, I. (1997). *Abnormal illness behaviour*. Chichester: Wiley.

Rost, K. M., & Smith, G. R., Jr. (1990). Improving the effectiveness of routine care for somatization. *Journal of Psychosomatic Research, 43*, 463–465.

Scicchitano, J. P. (2000). *Identification and management of somatisation in the primary care setting in terms of illness behaviour and risk of psychiatric illness*. PhD thesis, University of Adelaide.

Wilkinson, P., & Mynors-Wallis, L. (1990). Problem-solving therapy in the treatment of unexplained physical symptoms in primary care. *Journal of Psychosomatic Research, 38*, 591–598.

Hypoesthesia, Assessment

Misha-Miroslav Backonja
Department of Neurology, University of Wisconsin-Madison, Madison, WI, USA

Synonyms

Hypesthesia; Hypoaesthesia

Definition

Hypoesthesia refers to decreased perception of innocuous stimuli, a condition where the body is much less sensitive than normal to stimulation that by its nature and intensity does not produce pain. Special senses are excluded (Merskey and Bogduk 1994). Hypoesthesia refers to diminished perception of a large range of mechanical stimuli such as touch, brush, pressure, and vibration and thermally innocuous stimuli of warm and cold. Stimulation and locus are specified. Hypoesthesia is also defined as a raised threshold to nonpainful stimuli and this definition is used as a criterion for hypoesthesia during quantitative sensory testing (QST). There are two phenomena that are the opposite of hypoesthesia, hyperesthesia and allodynia. Hyperesthesia is increased but not painful sensation from innocuous stimulation and allodynia is pain from innocuous stimulation. If stimulation is of nature or intensity to produce tissue damage and the subject perceives it as harmless, then the phenomenon is defined as hypoalgesia.

Characteristics

Hypoesthesia is a ▶ negative sensory phenomenon seen primarily in patients with neurological disease or injury, including patients with ▶ neuropathic pain (Lindblom and Ochoa 1986; Backonja and Galer 1998; Backonja 2003). Hypoesthesia indicates decrease or loss of function that arises as a result of neurological disease or injury affecting somatic sensory and thermal pathways, anywhere from primary afferents to cerebral cortical structures. Hypoesthesia is demonstrated by means of sensory examination during which standard methods of mechanical and thermal stimulus are applied with the goal of activating specific classes of receptors. The specificity of somatic sensory pathways in conducting particular somatic sensations is significantly altered by disease and injury of the ▶ somatosensory nervous system and hypoesthesia is probably the most sensitive and reliable indication of such injury. In contrast, positive sensory phenomena, such as allodynia and ▶ hyperalgesia, are relatively frequent components of neuropathic pain; the complexity of underlying mechanisms makes them much more difficult to interpret. Understanding the relationship between injury of the somatosensory neural structures and its manifestations, such as hypoesthesia, is relevant to pain mechanisms, because methods of testing and interpretations are based on the specificity of sensory modalities. Distinction of hypoesthesia from hypoalgesia is based primarily on the type and intensity of the stimulus applied to the thermonociceptive sensory system.

Methods of Assessment and Interpretation

Assessment of sensory nervous system function is most commonly done at the bedside, and under these circumstances, the testing is primarily qualitative in nature. Quantitative assessment increasingly utilizes computerized electronic equipment in the environment of a quantitative sensory laboratory (Backonja and Galer 1998; Greenspan 2001). Either approach should be able to detect hypoesthesia, but the ways in which these methods arrive at conclusions about the presence and severity of hypoesthesia are distinctly different and this distinction is also reflected in the definitions stated earlier. Qualitative bedside examination of the presence of hypoesthesia is based primarily on patient report that a stimulus is perceived as decreased. Quantitative sensory testing arrives at conclusions about hypoesthesia on the basis of raised thresholds to painful stimuli.

Qualitative somatosensory assessment is based on the subject's ability to compare and report quality of sensation resulting from standard methods of stimulation of affected areas compared to normal unaffected areas. The qualitative method is very convenient for bedside evaluations. Frequently utilized bedside methods include standard neurological examination tools such as cotton tips, monofilaments, tuning forks for testing of vibration, and various metal objects that can conveniently be warmed or cooled in the clinical setting. A degree of quantification is possible and requires that the subject report whether the decrease in perceived sensation from applied stimuli is mild, moderate, severe, or completely absent when compared to a normal unaffected area. Since the qualitative method requires psychophysical interaction, this

method can only be used in humans who can communicate linguistically with the examiner and hence cannot be used in infants or aphasic subjects or in animal models in somatosensory and pain research.

Quantitative assessment of sensory deficits requires a more sophisticated approach and frequently utilizes electronically controlled devices, though a number of psychophysical methods especially mechanical stimuli are used. Traditionally, this method is known as quantitative sensory testing (QST). All of the quantitative methods require much longer times for completion. The primary outcome of QST is determination of thresholds for specific modalities, which are then compared to the established norms (Greenspan 2001). Increases in the threshold to innocuous stimuli are interpreted as hypoesthesia. The QST method can be used not only in human studies but also in animal models.

A crucial step in the interpretation of QST is to obtain a pain rating at the threshold. In spite of the fact that the pain threshold is increased, the presence or absence of positive sensory phenomena, one of them being ► hyperpathia (increased threshold but even innocuous stimuli are perceived as painful) and consequently presence of a painful neuropathic disorder (Kelly Getz et al. 2005). The advantage of testing with innocuous stimuli and detecting hypoesthesia, especially for cold detection, is that it is one of the most sensitive methods of detecting somatic sensory deficits, which characterize neurological disorders, including neuropathic pain disorders (Dyck and O'Brien 2003).

One of the main goals of neurological evaluation is to determine the site and level of ► neuraxis where pathological processes that produce symptoms, including pain, originate (Dyck and O'Brien 2003). In addition to establishing the nature of the ► neurological deficit, such as hypoesthesia, to a specific modality, it is important to establish the special pattern of these abnormalities, since the pattern serves as the basis for the determination as to whether the lesion that is causing symptoms, including pain, involves specific peripheral nerve structures, such as peripheral nerves, plexuses, or nerve roots, or central nervous system structures, such as spinal cord, brain stem, subcortical or cortical structures, and pathways of the brain.

Caveats and Unresolved Issues

The difficulty of assessing hypoesthesia arises from the inherent difficulty of assessing a negative sensory phenomenon. For example, it is easier conceptually to illustrate to the subjects a ▶ positive sensory phenomenon, such as pain, with instructions that 0 = none and 10 = worst imaginable. In contrast, it is much harder conceptually to illustrate and request a rating of spontaneous loss of sensation, because it is not possible to feel what one is not able to feel, in spite of the instruction that the patient is to imagine scaling between normal sensation and absence of sensation.

Other phenomena that can result from innocuous stimulation and are on the opposite end of the spectrum of sensory experience are hyperesthesia, allodynia, or even hyperpathia. Confusion for the examiner as well as for the patient is caused by the difficulty of assessment that comes from the fact that pain disorders, most frequently neuropathic pain, are characterized by allodynia or hyperalgesia to one sensory testing modality but are also found to have evidence of hypoesthesia to another testing modality in the same area. Depending on the way stimulation is conducted, even when hypoesthesia is present, the outcome can be hyperpathia in case of repeated stimulation. For example, in the case of partial hypoesthesia, when the stimulus is administered in the way to produce temporal and spatial summation, it could result in increased threshold, but what is perceived is painful, which is then interpreted as hyperpathia. Consequently, from the pain mechanisms perspective, the relationship of hypoesthesia and hyperpathia is still far from clear.

In summary, hypoesthesia is a sensitive clinical sign of neurological injury or disease, which in some patients can lead to neuropathic pain. Methods for determining the presence of hypoesthesia are qualitative, as in bedside examination or quantitative, as with QST. Mechanisms of hypoesthesia for specific modalities and its relationship to hyperpathia are still not well understood.

References

Backonja, M. (2003). Defining neuropathic pain. *Anesthesia and Analgesia, 97,* 785–790.

Backonja, M. M., & Galer, B. S. (1998). Pain assessment and evaluation of patients who have neuropathic pain. *Neurologic Clinics, 16*(4), 775–790.

Dyck, P. J., & O'Brien, P. C. (2003). Quantitative sensory testing: Report of the Therapeutics and Technology Assessment Subcommittee of the American Academy of Neurology. *Neurology, 61,* 1628–1630.

Greenspan, J. D. (2001). Quantitative assessment of neuropathic pain. *Current Pain and Headache Reports, 5,* 107–113.

Kelly Getz, K., Cook, T., & Backonja, M. M. (2005). Pain ratings at the thresholds are necessary for interpretation of quantitative sensory testing. *Muscle and Nerve, 32,* 179–184.

Lindblom, U., & Ochoa, J. (1986). Somatosensory function and dysfunction. In A. K. Asbury, G. M. McKhann, & W. I. McDonald (Eds.), *Diseases of the nervous system -clinical neurobiology* (pp. 283–298). Philadelphia: W.B. Saunders.

Merskey, H., & Bogduk, N. (1994). *Classification of chronic pain* (Vol. 2). Seattle: IASP Press.

Hypogastric Neurectomy

Definition

Hypogastric neurectomy is a surgical resection of the hypogastric nerves.

Cross-References

▶ Visceral Pain Models, Female Reproductive Organ Pain

Hyponatremia

Definition

Hyponatremia represents less than normal levels of sodium ions, or salt, in the blood, which may result in cognitive impairment.

Cross-References

▶ Trigeminal Neuralgia: Diagnosis and Treatment

Hypophysectomy

Definition

Excision of the pituitary gland is known as hypophysectomy.

Cross-References

▶ Cancer Pain Management, Neurosurgical
Interventions

Hypotension of Spinal Fluid

▶ Headache Due to Low Cerebrospinal Fluid
Pressure

Hypothalamic Pituitary Axis

Definition

These organs combine with the gonads to play a critical role in the development and regulation of a number of the body's systems, such as the reproductive and immune systems.

Cross-References

▶ Fibromyalgia

Hypothalamic-Pituitary-Adrenal Axis Response

▶ Postoperative Pain, Pathophysiological Changes in Neuroendocrine Function in Response to Acute Pain

Hypothalamus

Definition

The hypothalamus is a very prominent group of neurons located below the thalamus at the base of the brain forming the ventral-most part of the diencephalon. It is divided into three lateral levels (medial, intermediate, and lateral) and five caudo-rostral levels (mammillary, posterior, intermediate, anterior, and preoptic). Its role includes the neuro-endocrine regulations (arcuate, paraventricular, and supraoptic nuclei), autonomic regulations (cardiorespiratory, thermoregulation, metabolic, digestive), and processing of motivational behaviors like sexual, feeding, drinking, waking/sleep state, aggressiveness, and illness feeling. It is also involved in modulating nociception.

Cross-References

▶ Descending Circuitry, Transmitters and
Receptors
▶ Parabrachial Hypothalamic and Amygdaloid
Projections

Hypothalamus and Nociceptive Pathways

Jean-François Bernard
INSERM UMR-894, UPMC Site
Pitié-Salpêtrière, Paris, France

Definition

The hypothalamus is a complex structure that occupies the ventral half of the diencephalon below the thalamus on either side of the third ventricle. It lies just above the ▶ pituitary gland responsible for neuroendocrine secretions.

The hypothalamus includes about 40 nuclei of very different shapes and sizes. For simplification, it is generally divided into three mediolateral zones, periventricular, medial, and lateral, and four caudo-rostral regions, mammillary, tuberal, anterior, and preoptic. Combination of zones and regions permitted the recognition of 12 hypothalamic areas (Simerly 1995).

Neurosecretory neurons are mainly located within the periventricular zone with a particularly high density in the paraventricular nucleus. In addition, another important group of neurosecretory neurons is located in the supraoptic nucleus, a well-individualized nucleus located in the lateral region, on the lateral border of the optic chiasm. The neurosecretory system is subdivided into two parts: (1) magnocellular neurosecretory neurons (oxytocin and vasopressin), which directly innervate the posterior pituitary gland, and (2) parvocellular neurosecretory neurons (corticotropin, gonadotropin, growth hormone, thyrotropin-releasing hormones, somatostatin, angiotensin II, and dopamine), which innervate the median eminence, the hypothalamic hormones being transported to the anterior pituitary gland via the hypophyseal portal system (Swanson 1987).

The medial and lateral zones of the hypothalamus are chiefly devoted to the control of ▶ autonomic functions (cardiovascular, respiratory, blood fluid balance, energy metabolism, thermoregulatory and digestive) and major basic instinctive behaviors (feeding, drinking, reproductive, flight, defensive, and aggressive) including the wakefulness-sleep cycles (Swanson 1987).

Characteristics

The hypothalamus is a fascinating region of the brain, which is much more than a control center for neuroendocrine secretion. Indeed, the hypothalamus is the upper center for autonomic functions and basic behaviors that assure the survival of both the individual and the species. It is easy to understand the role of the hypothalamus when it guarantees an adequate level of homeostasis for autonomic functions needed for survival. It is not so obvious to appreciate the importance of the myriad basic behaviors it generates. Thus, it is basically responsible for most of the motivations that govern our life, for example, hunger, the

pleasures of eating and satiety, sexual desire, aggressiveness, fear, drowsiness, alertness, and numerous other fundamental motivations of life.

Considering these functions, it seems that the hypothalamus should play an important role in the autonomic and ▶ motivational components of pain. All the same, the precise role of hypothalamus in different components of pain remains unclear. The only clearly accepted function of the hypothalamus in pain is the neuroendocrine corticotropin response.

In humans, imagery studies indicate that the acute traumatic pain comes with a noticeable activation of the hypothalamus (Hsieh et al. 1996). However, these studies provide neither information about the activation of different hypothalamic nuclei nor data about the role of hypothalamus in pain. In fact, until now, most evidence for an involvement of the hypothalamus in nociceptive processing comes from anatomical and c-fos data. Cross-checking these data with the known functions of hypothalamic nuclei, it becomes possible to make hypotheses about the involvement of the hypothalamus in pain.

Nociceptive Afferent Inputs to the Hypothalamus

The hypothalamus has three well-documented sources of nociceptive inputs, the spinal and trigeminal dorsal horn, the parabrachial area, and the ventrolateral medulla (Fig. 1).

• Spinal and trigeminal inputs – a number of spinal and trigeminal neurons are labeled after a large injection of retrograde axonal tracer within the hypothalamus. Labeled neurons are located in superficial and, above all, in deep laminae of the dorsal horn, that is, in regions known to be involved in nociceptive processing. Electrophysiological studies indicate that most spino-/trigemino-hypothalamic neurons respond to a variety of noxious stimuli (Burstein 1996). These data, which seem to indicate a major nociceptive input to the hypothalamus, are challenged by anterograde axonal tracing studies that show much lower spinal and trigeminal projection upon the hypothalamus (Gauriau and Bernard 2004). Comparative examination of all the studies

Hypothalamus and Nociceptive Pathways, Fig. 1 Schematic representation, in sagittal sections, of the three main hypothalamic nociceptive inputs: the PBl, the VLM/A1/C1 region, and the trigeminal and spinal dorsal horn (mainly the deep laminae). *Thick line*: extensive nociceptive projection; *thin line*: medium density nociceptive projection; *dotted line*: hypothetical nociceptive projection. Abbreviations: *A1* A1 noradrenaline cells, *C1* C1 adrenaline cells, *cc* corpus callosum, *Hyp* hypothalamus, *NTS* nucleus tractus solitarii, *PAG* periaqueductal gray matter, *PBl* lateral division of the parabrachial nucleus, *Pit* pituitary gland, *VLM* ventrolateral medulla

seems to point to at least a moderate but indisputable nociceptive projection, mainly to the lateral (Fig. 2) but also to the posterior and the paraventricular hypothalamic nuclei.

- Parabrachial inputs (see parabrachial hypothalamic and amygdaloid projections) – the lateral parabrachial area receives a heavy nociceptive input from spinal and trigeminal lamina I nociceptive neurons. The lateral parabrachial area projects heavily to the hypothalamic ventromedial nucleus and extensively to the retrochiasmatic, the median, and the ventrolateral preoptic hypothalamus. Although less extensive, a notable projection reaches the dorsomedial, the periventricular, the paraventricular, and the lateral nuclei (Fig. 2). Electrophysiological studies indicate that this strong afferent input to the hypothalamus from the parabrachial nucleus is primarily nociceptive (Bernard et al. 1996; Bester et al. 1997).

- Caudal ventrolateral medulla inputs – this reticular region includes the A1/C1 catecholaminergic neurons and receives nociceptive inputs from both the superficial and the deep laminae of the dorsal horn. The caudal ventrolateral medulla projects extensively to the paraventricular nucleus and, to a lesser extent, to the periventricular, the supraoptic, and the median preoptic hypothalamic nuclei (Fig. 2). Here again it was shown that this afferent input contains nociceptive neurons (Burstein 1996; Pan et al. 1999).

The nucleus of the solitary tract was also proposed as a nociceptive input for the hypothalamus. However, this nucleus is primarily a center for autonomic/visceral and gustatory information. The role and the importance of solitary tract neurons in conveying nociceptive messages from the spinal cord to the hypothalamus need to be confirmed.

To summarize, anatomical data indicate several hypothalamic subregions that appear to be more specifically involved in nociceptive processing:

1. The neuroendocrine group (the paraventricular nucleus and to a lesser extent the periventricular and supraoptic nuclei) that receives nociceptive messages from all the nociceptive sources described above.
2. The ventromedial nucleus, the perifornical, and the retrochiasmatic areas that receive a very prominent nociceptive input from the parabrachial area.

Hypothalamus and Nociceptive Pathways, Fig. 2 Summary diagram illustrating, in coronal sections, the location of nociceptive projections within the hypothalamus (*1–4*, caudal to rostral). The parabrachial "nociceptive" area projects primarily upon the VMH (*dark gray*) and extensively upon the DM and the perifornical area (1), the RCh and the Pe (2), the rostral Pe and the ventral AH (3), and the AVPO (4) hypothalamic nuclei (*gray*). Both the parabrachial nucleus and the A1/C1 group within the ventrolateral medulla project to the PVN (2), the SO (3), and the MnPO (4) (*black points*). Both the parabrachial area and the spinal and trigeminal dorsal horn project to the pLH (1) (*horizontal* hatching).

Abbreviations: *3V* third ventricle, *A1/C1* A1 noradrenaline cells, C1, adrenaline cells, *ac* anterior commissure, *AH* anterior hypothalamic area, *Arc* arcuate nucleus, *AVPO* anteroventral preoptic nucleus, *BST* bed nucleus of stria terminals, *DM* dorsal medial nucleus, *f* fornix, *ic* internal capsule, *LH* lateral hypothalamus, *MnPO* median preoptic nucleus, *MPA* medial preoptic area, *mt* mammillothalamic tract, *opt* optic tract, *ox* optic chiasm, *Pe* periventricular nucleus, *pLH* posterior portion of lateral hypothalamus, *PVN* paraventricular nucleus, *RCh* retrochiasmatic area, *SCh* suprachiasmatic nucleus, *SO* supraoptic nucleus, *sox* supraoptic decussation, *VMH* ventromedial hypothalamic nucleus

3. The median and ventrolateral preoptic area, the dorsomedial, the lateral, and the posterior hypothalamic region, which receive lower but yet substantial nociceptive inputs.

Corroborating the anatomical data closely, it was shown that various painful stimuli evoke c-fos expression in regions receiving nociceptive afferent projections. The strongest c-fos expression was observed in neuroendocrine neurons of the hypothalamus located in the paraventricular, the supraoptic, and the periventricular/arcuate nuclei. A substantial c-fos expression is evoked in the posterior, the ventromedial and the dorsomedial nuclei, and the retrochiasmatic, the lateral, and the anterior regions of the hypothalamus (Rodella et al. 1998; Snowball et al. 2000).

Role of Hypothalamus in Visceromotor Responses to Painful Stimuli
The anatomical data indicate that both parabrachial and A1/C1 projections to the paraventricular nucleus innervate more densely neurons containing corticotropin releasing hormone as well as ▶ magnocellular neurons (which contain vasopressin and oxytocin). Painful stimuli evoke specifically c-fos expression in neurons containing corticotropin releasing hormone, vasopressin, and oxytocin at the levels of paraventricular, arcuate, and supraoptic nuclei. This neuroendocrine response is specific for these neurohormones; it does not include gonadotropin, growth hormone, and thyrotropin-releasing hormones (Pan et al. 1996).

The neuroendocrine component of pain is indisputably under hypothalamic control with well-identified pain pathways to drive it. The role of these neurohormones in pain is not completely understood. It is likely that an increase in the corticotropin hormone axis is important to cope with the dangerous or traumatic situation that comes with pain (mobilization of metabolism and mental energy). Nonetheless, in the case of chronic pain, the stimulation of the corticotropin axis might become deleterious (anxiety, depression, decrease of immunity, neuronal loss). Vasopressin may accompany corticotropin secretion to increase or maintain blood pressure. The role and amount of oxytocin secretion in cases of acute pain remain yet poorly understood.

Importantly, psychological stress (immobilization, anxiety, and fear) acts on the corticotropin axis of the hypothalamus via limbic projections (bed nucleus of the stria terminalis, prefrontal cortex) different from those described for nociceptive stimuli (physical stress).

Numerous paraventricular neurons (chiefly in a dorsal position) are not neuroendocrine cells but provide descending projections to the brainstem and the spinal cord. Although the paraventricular nucleus provides the more extensive set of descending projections, other hypothalamic nuclei also receiving nociceptive messages send similar descending projections, namely, the periventricular, the retrochiasmatic area, the dorsomedial, the dorsal, the perifornical, and the lateral hypothalamic areas. These hypothalamic neurons project to the periaqueductal gray matter, the parabrachial area, the solitary tract, the motor vagus, the ambiguous nuclei, and the ventrolateral medulla in the brainstem. In the spinal cord, they project chiefly to the sympathetic preganglionic column (Saper 1995). These hypothalamic neurons are adequately placed to drive both the sympathetic and the ▶ parasympathetic components of pain. They might, in connection with brainstem neurons, increase or decrease blood pressure and cardiac frequency and modify circulatory territory, according to the nature of the painful stimuli.

Role of Hypothalamus in Behavioral Response to Painful Stimuli

Several hypothalamic nuclei, which receive an extensive nociceptive input, play an important role in motivational components of pain.

The first group, including the ventromedial and the dorsomedial nuclei, the perifornical and the retrochiasmatic areas, is markedly involved in defensive-aggressive behavior. This group of nuclei projects extensively to the periaqueductal gray matter, each nucleus targeting a specific quadrant. The periaqueductal gray matter appears to be a major hypothalamic-descending output to mediate aggressive-defensive behavior. Each nucleus of this hypothalamic group receives an extensive nociceptive input from the lateral parabrachial area, the ventromedial hypothalamic nucleus receiving the heaviest input. The ventromedial nucleus has been involved in aggressive-defensive behavior. Stimulation applied in this nucleus induces vocalization, attack, escape, piloerection, mydriasis, and micturition that resemble the pseudo-affective reactions induced by noxious stimuli (Bester et al. 1997; Swanson 1987). Recently, the dorsomedial portion of the ventromedial nucleus has been shown to be responsible for the vocalization induced by painful electrical shock applied to the tail (Borszcz 2002). The ventromedial nucleus has also been involved in feeding behavior (it has long been considered as the "satiety center") and regulation of energy metabolism (Swanson 1987). Recently, the ventromedial hypothalamic nucleus has thus been proposed to be responsible for the anorexia induced by migraine (Malick et al. 2001). Pain should act on appetite via a parabrachio-ventromedial CCKergic link. Leptin receptors, which are abundant in this hypothalamic nucleus, might also participate in the loss of appetite. Finally, stimulation applied within the ventromedial nuclei produces an analgesia, which is also probably mediated via the periaqueductal output. Thus, it appears that the medial zone of the tuberal (posterior) and the anterior hypothalamus is responsible for the defensive-aggressive and feeding motivational component of pain.

The second group, including the median and the anteroventral preoptic hypothalamic nuclei, is involved in osmotic/blood fluids balance regulation and sleep-promoting/thermoregulation functions. These nuclei also receive a substantial nociceptive input from the lateral parabrachial area. The influence of nociceptive input upon the neurons of this hypothalamic region is less clear. It might alter drinking behavior, vasopressin secretion, falling asleep, and the thermoregulation set point according to the nature of the nociceptive aggression (Saper et al. 2001; Swanson 1987).

The posterior portion of the lateral hypothalamus receives a diffuse but substantial nociceptive input directly from the deep laminae of the dorsal horn and indirectly via the internal lateral parabrachial nucleus. The role of the lateral hypothalamus in nociceptive processing remains obscure because this hypothalamic region was involved in a myriad functions, such as feeding behavior (it has long been considered to be a "feeding center"), drinking behavior, and cardiovascular and visceral regulation, as well as in wakefulness and anti-nociceptive and rewarding mechanisms. However, the recent discovery that ▶ narcolepsy can be induced by lack of orexin/hypocretin (a peptide located in the neurons of lateral hypothalamus) indicates that the lateral hypothalamus is probably markedly involved in the wakefulness mechanism (Saper et al. 2001). One role of nociceptive inputs upon neurons of the posterior lateral hypothalamus could be to trigger awakening.

Summary

Bringing together anatomical and functional data, the hypothalamus appears as a key center for most visceromotor (neuroendocrine, autonomic response) and motivational (aggressive-defensive reactions, ingestive behaviors, wakefulness, antinociception) components of pain. It yet remains to check experimentally the actual role of hypothalamic subregions and/or neuromodulators in the genesis of different components of pain. Anatomical data also indicate that hypothalamic functions are probably strongly modulated by the upper limbic structures (notably the extended amygdala and the cingulate/prefrontal cortex), which are also involved in the emotional appreciation of pain.

References

Bernard, J. F., Bester, H., & Besson, J. M. (1996). Involvement of the spino parabrachio-amygdaloid and -hypothalamic pathways in the autonomic and affective emotional aspects of pain. *Progress in Brain Research, 107*, 243–255.

Bester, H., Besson, J. M., & Bernard, J. F. (1997). Organization of efferent projections from the parabrachial area to the hypothalamus: A phaseolus vulgaris-leucoagglutinin study in the rat. *The Journal of Comparative Neurology, 383*, 245–281.

Borszcz, G. S. (2002). *The ventromedial hypothalamus contributes to generation of the affective dimension of pain in rats. Program N°. 653.5 2002 Abstract viewer/itinerary planner*. Washington, DC: Society for Neuroscience. Online.

Burstein, R. (1996). Somatosensory and visceral input to the hypothalamus and limbic system. *Progress in Brain Research, 107*, 257–267.

Gauriau, C., & Bernard, J. F. (2004). A comparative reappraisal of projections from the superficial laminae of the dorsal horn in the rat: The forebrain. *The Journal of Comparative Neurology, 468*, 24–56.

Hsieh, J. C., Stahle-Backdahl, M., Hagermark, O., et al. (1996). Traumatic nociceptive pain activates the hypothalamus and the periaqueductal gray: A positron emission tomography study. *Pain, 64*, 303–314.

Malick, A., Jakubowski, M., Elmquist, J. K., et al. (2001). A neurohistochemical blueprint for pain-induced loss of appetite. *Proceedings of the National Academy of Sciences of the United States of America, 98*, 9930–9935. Erratum in: Proc Natl Acad Sci USA 98:14186.

Pan, B., Castro-Lopes, J. M., & Coimbra, A. (1996). Activation of anterior lobe corticotrophs by electroacupuncture or noxious stimulation in the anaesthetized rat, as shown by colocalization of Fos protein with ACTH and beta-endorphin and increased hormone release. *Brain Research Bulletin, 40*, 175–182.

Pan, B., Castro-Lopes, J. M., & Coimbra, A. (1999). Central afferent pathways conveying nociceptive input to the hypothalamic paraventricular nucleus as revealed by a combination of retrograde labeling and c-fos activation. *The Journal of Comparative Neurology, 413*, 129–145.

Rodella, L., Rezzani, R., Gioia, M., et al. (1998). Expression of Fos immunoreactivity in the rat supraspinal regions following noxious visceral stimulation. *Brain Research Bulletin, 47*, 357–366.

Saper, C. B. (1995). Central autonomic system. In G. Paxinos (Ed.), *The rat nervous system* (2nd ed., pp. 107–135). San Diego: Academic Press.

Saper, C., Chou, T. C., & Scammell, T. E. (2001). The sleep switch: Hypothalamic control of sleep and wakefulness. *Trends in Neurosciences, 24*, 726–731.

Simerly, R. B. (1995). Anatomical substrates of hypothalamic integration. In G. Paxinos (Ed.), *The rat nervous system* (2nd ed., pp. 353–376). San Diego: Academic Press.

Snowball, R. K., Semenenko, F. M., & Lumb, B. M. (2000). Visceral inputs to neurons in the anterior hypothalamus including those that project to the periaqueductal gray: A functional anatomical and electrophysiological study. *Neuroscience, 99*, 351–361.

Swanson, L. W. (1987). The hypothalamus. In T. Hökfelt & L. W. Swanson (Eds.), *Handbook of chemical neuroanatomy, vol.5. Integrated systems of the CNS, part I* (pp. 1–124). Amsterdam: Elsevier.

Hypothalamus-Anterior Pituitary-Gonadal Axis

Definition

The hypothalamus controls endocrine function by direct release of neuropeptides or indirectly through the secretion of regulatory hormones to the anterior pituitary. These regulatory substances are secreted by the hypothalamus into the local portal plexus within the median eminence, which then drains into the blood vessels of the anterior pituitary. There are a wide number of substances released by the hypothalamus that either inhibit or stimulate the release of anterior pituitary hormones, including factors that affect the release of growth hormone, thyrotropin, and others. Related to sexual and reproductive function, the hypothalamus secretes prolactin-releasing factor (PRF), which stimulates the release of prolactin. Dopamine, also secreted by the hypothalamus, inhibits prolactin's release. Additionally, the hypothalamus secretes gonadotropin-releasing hormone (GnRH) to the pituitary gland, which triggers the secretion of luteinizing hormone (LH) from the pituitary gland. Luteinizing hormone then stimulates the Leydig cells of the testes to produce testosterone or the ovaries to produce progesterone.

Cross-References

▶ Cancer Pain Management, Opioid Side Effects, Endocrine Changes, and Sexual Dysfunction

Hypoxia

Definition

Hypoxia is a pathological condition in which the whole organism (*generalized hypoxia*) or only a region of the organism (*tissue hypoxia*) is deprived of adequate oxygen supply.

Cross-References

▶ NSAIDs and Cancer

Snowball R., ... Senneault ... F., M.? & Lamb, S.J.W. (2000). Visceral pain: ... lesions in ... anterior hypothalamus, including those that project to the periaqueductal grey. A functional anatomical and electrophysiological ... Neuroscience 99:3, 1–7.

Swanson L.W. (1987). The hypothalamus. In: T. Hökfelt & L.W. Swanson (Eds.) Handbook of chemical neuroanatomy: Integrated systems of the CNS, Part 1, pp1–124. Amsterdam: Elsevier.

Hypothalamus–Anterior Pituitary–Gonadal Axis

Definition

The hypothalamus controls endocrine function by the release of neuropeptides ... ... through the secretion of ... other hormones to the anterior pituitary. These regulatory substances are released by the hypothalamus into the hypophysial plexus within the median eminence, which is then ... into the blood vessels of the anterior pituitary. There are a wide number of substances released by the hypothalamus that either inhibit or stimulate the release of anterior pituitary hormones, including factors that affect the release of growth hormone, thyrotropin ... and others. Related to sexual and reproductive function, the hypothalamus secretes gonadotropin releasing factor (GnRH), which stimulates the release of ... Dopamine, also secreted by ...

---

the hypothalamus, inhibits prolactin's release. Additionally, the hypothalamus secretes gonadotropin-releasing hormone (GnRH) to the pituitary gland, which triggers the secretion of luteinizing hormone (LH) from the pituitary gland. Luteinizing hormone then stimulates the Leydig cells of the testes to produce testosterone or the ovaries to produce progesterone.

Cross-References

Cancer Pain Management, Gonadal Status
Bitters, Endocrine Changes and Sexual Dysfunction

Hypoxia

Definition

Hypoxia is a pathological condition in which the whole organism (generalized hypoxia) or only a region of the organism (tissue hypoxia) is deprived of adequate oxygen supply.

Cross-References

NSAIDs and Cancer

IAIABC System

▶ Impairment Rating, Ambiguity, IAIABC System

Iatrogenic Causes of Neuropathy

Paolo Marchettini
Pain Medicine Center, Scientific Institute San Raffaele, Milan, Italy

Synonyms

Chronic postoperative pain; Chronic postsurgical neuralgia; Postsurgical nerve injury

Definition

Iatros in Greek means physician; the term iatrogenic defines a pathological complication of medical care. ▶ Iatrogenic neuropathy is a nerve injury caused by surgical or pharmacological treatment or by its consequence.

It is estimated that postsurgical pain has a global incidence of 30 % (Macrae 2008) and iatrogenic neuropathy is increasingly recognized as its major cause (Perkins and Kehelet 2000; Birch 2008). Thus, the encompassing term "▶ Chronic Postsurgical Pain Syndromes," which is commonly used to cluster all the painful outcomes of surgery, should be abandoned. To properly highlight the pathophysiology operating in each case, postsurgical pain should be defined with the recognition of the eventual neuropathic nature of the pain and anatomical identification of the injured nerve structure. For the sake of clarity, the use of the widespread terms "failed" and "syndrome" are not useful when defining post-surgical pain conditions, as they do not determine the etiology of the pain. A proper attempt to diagnose the cause of chronic postsurgical pain permits appropriate steering of the clinical management.

Characteristics

Pain and other abnormal sensory symptoms replicate the typical description of sensory aberration in mono- or poly-neuropathy (Hansson et al. 2001; Treede and Baron 2008). Iatrogenic neuropathy, however, has more psychological complications, because the recognition of nerve injury is often missed or belated. The onset of neuropathic pain can be delayed for days, weeks or, rarely, months. Doctor and patient may erroneously attribute early onset neuropathic pain to the expected "normal" postoperative pain. The operation was technically successful, and there were no infections or other obvious complications. Delayed recognition may also be the consequence of immobilization by postoperative cast, bed rest, reduced activity, or sedation due to anesthetic or analgesic medication. The nerve injury may

G.F. Gebhart, R.F. Schmidt (eds.), *Encyclopedia of Pain*, DOI 10.1007/978-3-642-28753-4,
© Springer-Verlag Berlin Heidelberg 2013

selectively affect a sensory branch, hampering objective evidence of the lesion. The intensity of the pain is often unrelated to the severity of the nerve damage; indeed partial nerve injury may predispose to neuropathic pain to a greater extent than total injury. When the diagnosis is made, at times many years later, patients are frustrated by delayed treatment and by inadequate and unspecific clinical management (Horowitz 1984). The typical clinical picture of chronic pain, sensory disorder, insomnia, and depression (Langley et al. 2013) is complicated by frustration and disbelief toward physicians and other medical care providers, judged in general to be responsible for the prolonged and unnecessary suffering.

Symptoms
When a patient describes tingling (spontaneous discharges in large myelinated fibers), pins and needles (small myelinated fibers), and cramps and burning (unmyelinated fibers) in the region of the surgical procedure, iatrogenic nerve injury is high on the list of differential diagnoses. The diagnosis is made based on such typical neuropathic symptoms and on the identification of sensory disorder in the anatomical territory of the injured nerve(s). It is also fundamental to search for a Tinel sign to identify the site of the injury. A condition that may hamper the diagnosis is a partial injury in continuity. Painful iatrogenic neuropathy often originates from injury confined to a sensory nerve, without evidence of muscle wasting and weakness. At times, the nerve injury is rather selective, affecting small fascicular rami and may be hard to detect clinically and electrophysiologically. Command of peripheral nerve sub-anatomy helps early recognition. As all neuropathic pains, iatrogenic neuralgia also can be spontaneous and evoked by stimuli to the skin or perhaps just movement. At times, even minor stimuli, normally not painful, can evoke a burning and unpleasant tingling sensation or thermal sensations (dysesthesia-allodynia). Stimulus-evoked pain is so distressing that patients report sensory dysfunction beyond the anatomical distribution of hypoesthesia, in territories that are wider than those supplied by the injured nerve. However, sensory dysfunction is usually confined within a territory that surrounds the cutaneous distribution of the affected plexus, root, or peripheral nerve. Occasionally, the nerve injury is theconsequence of progressive entrapment in the surgical scar. Such entrapment, unlike anatomical entrapment syndromes, is rarely focal, usually being more homogeneously distributed along the scar. In the early phase, ectopic neural activity may exist in the absence of nerve conduction block, and neurophysiological signs may remain within normal values for a long time. The clinical equivalence is that the positive symptoms of pain, paresthesia, and dysesthesia may appear without striking negative phenomena (hypo-anesthesia).

Diagnosis
In clinical practice, recognizing a cutaneous nerve injury by the patient's description of tingling pins and needles and burning is relatively simple. Diagnosing an injury of nerves supplying deep structure is more demanding. In such cases the pain quality replicates the quality of a deep tissue injury, making the differential diagnosis between neuropathy and connective tissue disease not always so self-evident. Deep neuropathic pain is commonly more widespread and more difficult to localize; it is often referred at a distance from the injury site and might spread along an entire radicular territory. At times, entrapment can be related to or aggravated by movement or positioning. If a nerve injury is suspected, a thorough clinical examination including sensory hyperphenomena (identification and documentation of allodynia and hyperalgesia) as well as sensory loss and motor function is required. Nerve conduction studies and quantitative sensory testing may be helpful in supporting the clinical diagnosis. When a comparison with a contralateral nerve is possible, the sensitivity of such a test is improved. The clinical assessment should also take into account that any injury to the sensory component of a motor nerve may result in pain. The motor system should be examined properly through electromyography and motor nerve conduction studies. This comprehensive sensorimotor examination allows objective identification of the neurological dysfunction. Careful evaluation

means that precise follow-up can be ensured, providing a yardstick helpful in reassuring the patient that the nerve damage is not worsening and is possibly improving. Such evaluation also provides specific evidence in the event of medicolegal assessment. Assessment of the social and psychological aspects of the disease is also necessary, so that the full picture can be documented.

Epidemiology

Cox et al. (1974) found a 27 % incidence of neuropathic symptoms in the best-case scenario after saphenectomy. The majority of the injuries were symptomless. Only 12 % of the patients, less than half of those with a clinically identifiable injury, reported subjective sensory abnormalities, and only 5 % of all operated patients reported pain. In the USA, the estimated annual incidence of chronic postsurgical pain is 394,000–1,500,000 and in the UK is 41,000–103,000 (Perkins and Kehelet 2000). The most frequently injured nerves during medical interventions are:

1. Brachial plexus
2. Palmar cutaneous branch of the median nerve
3. Infrapatellar cutaneous branch of the saphenous nerve
4. Ilioinguinal, iliohypogastric, genitofemoral, and femoral nerves
5. Accessory and greater auricular nerves
6. Long thoracic nerve
7. Alveolar nerve

Our observation reiterates the homogeneous lists of Horowitz (1984), Dawson and Krarup (1989), and Sunderland (1991).

The surgical interventions most commonly associated with nerve injury are peripheral anesthetic nerve block, odontoiatric treatment, ENT surgery, general surgery, mastectomy, endoscopic cholecystectomy, and cardiovascular and thoracic surgery, i.e., thoracotomy, orthopedic surgery amputation, hip and knee replacement, and knee arthroscopy.

Predisposing Factors

Pain before surgery has been reported as major risk factor with a 62.7 incidence of persistent postsurgical pain (Fletcher et al. 2008).

Odontoiatric Treatment

In endodontic surgery, nerve damage may result from direct intrusion of instruments or implants into the mandibular canal or from introducing neurotoxic substances such as paraformaldehyde close to the inferior alveolar nerve.

ENT Surgery

Nasal sinus surgery may damage the maxillary branch of the trigeminal nerve, resulting in persistent pain in the region of the eye (Neuhaus 1990).

Cardiovascular Surgery

Varicose vein striping, particularly of the saphenous vein, may result in nerve injury (Offner and Loop 2012). After thoracotomy, a sensory disorder, in particular mechanical allodynia indicating a neuropathic component for the pain, has been found in 80 consecutive patients complaining of chronic pain after coronary artery bypass grafting (Eisenberg et al. 2001).

Orthopedic Surgery

Shoulder arthroscopy, as with all arthroscopic techniques, is becoming a more frequent cause of iatrogenesis, although reports vary widely (from 10 % to one case out of 439 (Stanish and Peterson 1995)).

Carpal tunnel release is a common surgical intervention that can be complicated by a lesion of the median and, more rarely, of the ulnar nerve. The reported incidence of iatrogenic nerve injury varies from 5 out of 83 cases to 0.8 % in a series of 3,035 hands reported in a review of 14 papers. Almost 10 % of patients complaining of complications following endoscopic carpal tunnel release have neuropathic pain (Kelly et al. 1994). Breast surgery is associated with chronic pain sequels in over 20 % of patients (Foley 1990). Harvesting of bone for grafting has been reported amongst the causes of injury to the lateral femoral cutaneous nerve (Hudson et al. 1979). Incision around the knee joint has been associated with lesions of the infrapatellar branch of the saphenous nerve. In 1983, Swanson and co-workers reported a 63.2 % incidence of prepatellar neuropathy in 87 patients immediately after open meniscectomy, while Sherman and co-workers noted only a 0.6 % incidence of reported symptoms following 2,640 arthroscopic procedures on the knee (Swanson 1983; Sherman et al. 1986). In 1995, Mochida and Kikuchi reported a 22.2 % incidence of sensory disturbances in 68 consecutive patients operated on between 1990 and 1991 (Mochida and Kikuchi 1995). Limb amputation is also frequently followed by chronic pain (the highest frequency reported is as high as 80 % of the amputees).

General Surgery

Lymph node biopsy in the neck is sometimes associated with injury to the accessory nerve and cutaneous nerves in the neck (Murphy 1983). Ilioinguinal nerve lesion is an important complication following inguinal hernioplasty (Heise and Starling 1998). Anterior hernioplasty, which requires the dissection of spermatic cord and sensory nerves, is a more common cause of this injury. Laparoscopic repair of the hernia is increasing simultaneously with the injuries to neural structures coursing through the groin. The femoral branches of the genitofemoral nerve and the lateral cutaneous nerve of the thigh, not visible during laparoscopic inguinal hernia repair, are more vulnerable. The symptoms might be confined to ejaculation (Aasvang et al. 2007). Abdominal rectopexy has been associated with femoral nerve injury in 6 patients out of a series of 24 patients, 21 of whom were operated on for rectal prolapse and 3 for rectorectal intussusception (Infantino et al. 1994).

Therapy

Treatment of painful iatrogenic neuropathy requires all means used in the management of any other neuropathic pain (Senegor 1991; Sindrup and Jensen 1999). In this medically provoked condition, it is even more fundamental to care for the psychological comorbidities. These should be prevented by early recognition of the neuropathic symptoms and the identification of the injured nerve. Early recognition, besides improving pain management, avoiding psychological overlay, and possibly preventing chronification, favors the reestablishment of a trustful patient doctor relationship.

References

Aasvang, E. K., Mohl, B., & Kehelet. (2007). Ejaculatory pain: A specific postherniotomy pain syndrome? *Anaesthesiology, 107*(2), 298–304.

Birch, R. (2008). Iatrogenous lesions of nerves and arteries in the leg and foot. *Foot and Ankle Surgery, 14*(3), 130–137.

Cox, S. J., Wellwood, J. M., & Martin, A. (1974). Saphenous nerve injury caused by stripping of the long saphenous vein. *British Medical Journal, 1*, 415–417.

Dawson, D. M., & Krarup, C. (1989). Perioperative nerve lesions. *Archives of Neurology, 46*, 1355–1360.

Eisenberg, E., Pultorak, Y., Pud, D., et al. (2001). Prevalence and characteristics of post coronary artery bypass graft surgery pain (PCP). *Pain, 92*, 11–17.

Fletcher, D., Fermanian, C., Mardaye, A., & Aegerter, P. (2008). A patient based national survey on post operative pain management in France reveals significant achievements and persistent challenges. *Pain, 137*(2), 441–451.

Foley, K. M. (1990). Brachial plexopathy in patients with breast cancer. In J. R. Harris, S. Hellman, I. C. Henderson, et al. (Eds.), *Breast diseases* (2nd ed., pp. 722–729). Philadelphia: Lippincott JB.

Hansson, P. T., Lacerenza, M., & Marchettini, P. (2001). Aspects of clinical and experimental neuropathic pain: The clinical perspective. In P. T. Hansson, H. L. Fields, et al. (Eds.), *Neuropathic pain: Pathophysiology and treatment. Progress in pain research and management* (pp. 1–18). Seattle: IASP Press.

Heise, C. P., & Starling, J. R. (1998). Mesh inguinodynia: A new clinical syndrome after inguinal herniorrhaphy? *Journal of the American College of Surgeons, 87*, 514–518.

Horowitz, S. H. (1984). Iatrogenic causalgia. Classification, clinical findings, and legal ramifications. *Archives of Neurology, 41*, 821–824.

Hudson, A. R., Hunter, G. A., & Waddell, J. P. (1979). Iatrogenic femoral nerve injuries. *Canadian Journal of Surgery, 22*, 62–66.

Infantino, A., Fardin, P., Pirone, E., et al. (1994). Femoral nerve damage after abdominal rectopexy. *International Journal of Colorectal Disease, 9*, 32–34.

Kelly, C. P., Pulisetti, D., & Jamieson, A. M. (1994). Early experience with endoscopic carpal tunnel release. *Journal of Hand Surgery (British), 19*, 18–21.

Langley, P. C., Van Litsenburg, C., Cappelleri, J. C., & Carroll, D. (2013). The burden associated with neuropathic pain in Western Europe. *Journal of Medical Economics, 16*(1), 85–95.

Macrae, W. A. (2008). Chronic post-surgical pain: 10 years on. *British Journal of Anaesthesia, 101*(1), 77–86.

Mochida, H., & Kikuchi, S. (1995). Injury to infrapatellar branch of saphenous nerve in arthroscopic knee surgery. *Clinical Orthopaedics and Related Research, 320*, 88–94.

Murphy, T. M. (1983). Complications of diagnostic and therapeutic nerve blocks. In F. K. Orkin &

L. H. Cooperman (Eds.), *Complications in anaesthesiology*. Philadelphia: Lippincott.

Neuhaus, R. W. (1990). Orbital complications secondary to endoscopic sinus surgery. *Ophthalmology, 97*, 1512–1518.

Offner, K. A., & Loop, T. (2012). Pathophysiology and epidemiology of pain in thoracic surgery. *Anästhesiologie, Intensivmedizin, Notfallmedizin, Schmerztherapie, 47*(9), 568–575.

Perkins, F. M., & Kehelet, H. (2000). Chronic pain as an outcome of surgery. A review of predictive factors. *Anaesthesiology, 93*(4), 1123–1133.

Senegor, M. (1991). Iatrogenic saphenous neuralgia: Successful therapy with neuroma resection. *Neurosurgery, 28*, 295–298.

Sherman, O. H., Fox, J. M., Snyder, S. J., et al. (1986). Arthroscopy-"no-problem surgery" an analysis of complications in two thousand six hundred and forty cases. *Journal of Bone and Joint Surgery, 68*, 256–265.

Sindrup, S. H., & Jensen, T. S. (1999). Efficacy of pharmacological treatments of neuropathic pain: An update and effect related to mechanism of drug action. *Pain, 83*, 389–400.

Stanish, W. D., & Peterson, D. C. (1995). Shoulder arthroscopy and nerve injury: Pitfalls and prevention. *Arthroscopy, 11*, 458–466.

Sunderland, S. (1991). Miscellaneous causes of nerve injury. In S. Sunderland (Ed.), *Nerve injuries and their repair. A critical appraisal* (pp. 197–199). Edinburgh: Churchill-Livingstone.

Swanson, A. J. (1983). The incidence of prepatellar neuropathy following medial meniscectomy. *Clinical Orthopaedics and Related Research, 181*, 151–153.

Treede, R. D., & Baron, R. (2008). How to detect a sensory abnormality. *European Journal of Pain, 12*(4), 395–396.

Iatrogenic Effect/Response

Definition

Iatrogenic effects/responses are outcomes inadvertently induced by a physician or surgeon or by medical treatment or diagnostic procedures.

Cross-References

▶ Acute Pain Management in Infants
▶ Cancer Pain, Assessment in Children
▶ Iatrogenic Causes of Neuropathy

Iatrogenic Neuropathy

Definition

Iatrogenic neuropathy is a nerve injury caused by surgical or pharmacological treatment or by its consequence.

Cross-References

▶ Iatrogenic Causes of Neuropathy

IB4-Binding Neurons

▶ IB4-Positive Neurons, Role in Inflammatory Pain

IB4-Positive Neurons, Role in Inflammatory Pain

Cheryl L. Stucky
Department of Cell Biology Neurobiology and Anatomy, Medical College of Wisconsin, Milwaukee, WI, USA

Synonyms

IB4-binding neurons; IB4-positive nociceptors; GDNF-dependent neurons

Definition

An IB4-positive neuron is a sensory neuron whose cell body lies in the dorsal root or trigeminal ganglia and whose membrane expresses surface carbohydrates (α-D-galactose groups of glycol conjugates) that bind the plant lectin isolectin B4 (IB4) from *Griffonia simplicifolia* I. IB4-positive neurons have small-sized cell bodies and primarily give rise to unmyelinated fibers, many of which are nociceptive.

Characteristics

IB4-positive neurons comprise one of two broad classes of small-diameter, C-fiber sensory neurons. The other class (IB4 negative) typically expresses neuropeptides such as substance P and calcitonin gene-related peptide (CGRP) and expresses ▶ trkA receptors for ▶ nerve growth factor (NGF). IB4-positive neurons express receptors for ▶ glial cell line-derived neurotrophic factor (GDNF) and depend on GDNF for survival after birth (Bennett et al. 1998). IB4-positive neurons selectively express the ▶ P2X3 Receptor for ATP. IB4-positive neurons are relatively poor in expression of neuropeptides or trkA receptors (Silverman and Kruger 1990; Averill et al. 1995; Molliver et al. 1995), but some studies show overlap between IB4 binding and neuropeptide or trkA expression (Wang et al. 1994; Kashiba et al. 2001). The central terminals of IB4-positive neurons terminate predominantly in the inner ▶ lamina II of the superficial dorsal horn of the spinal cord (Gerke and Plenderleith 2004), whereas IB4-negaitve neurons terminate mainly in lamina I and outer lamina II (Fig. 1).

The neurochemical differences between IB4-positive and IB4-negative neurons suggest that the two classes of small-diameter neurons may have different functional properties in conveying nociceptive information. Compared to IB4-negative neurons, IB4-positive neurons isolated from uninjured mice or rats have longer duration action potentials and higher densities of ▶ tetrodotoxin (TTX)-resistant sodium channel currents (Stucky and Lewin 1999), higher densities of N-type Ca^{2+} channel currents (Wu and Pan 2004), and higher densities of voltage-gated K^+ (Kv) channel currents (Vydyanathan et al. 2005). Furthermore, IB4-positive neurons from uninjured mice are significantly *less* responsive to noxious chemical stimuli, including capsaicin and protons, than IB4-negative neurons (Dirajlal et al. 2003).

Several authors have hypothesized that IB4-negative, but not IB4-positive, neurons contribute to inflammatory pain (Mantyh and Hunt 1998; Snider and McMahon 1998). To address this hypothesis, a recent study investigated whether peripheral inflammation in vivo alters the response properties of isolated IB4-positive or IB4-negative neurons to the noxious stimulus capsaicin, which activates the transient receptor potential vanilloid 1 (TRPV1) receptor (▶ TRPV1 receptor) (Breese et al. 2005). TRPV1 function was investigated because TRPV1 is a key heat transducer on nociceptors and mediates the heat hyperalgesia that accompanies peripheral inflammation (Caterina et al. 2000;

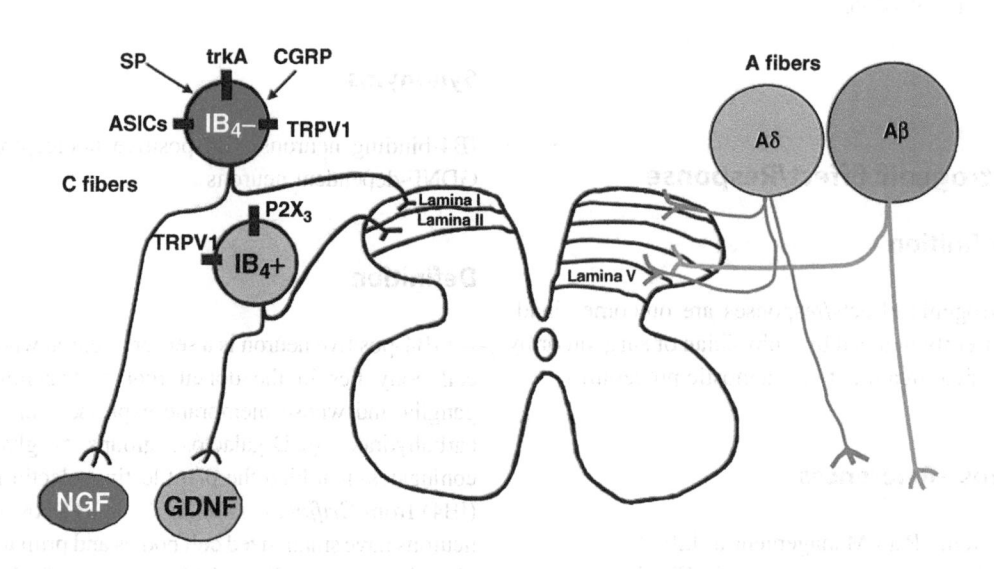

IB4-Positive Neurons, Role in Inflammatory Pain, Fig. 1 Schematic of cross-section of the lumbar spinal cord and innervation by the general subclasses of primary afferent neurons

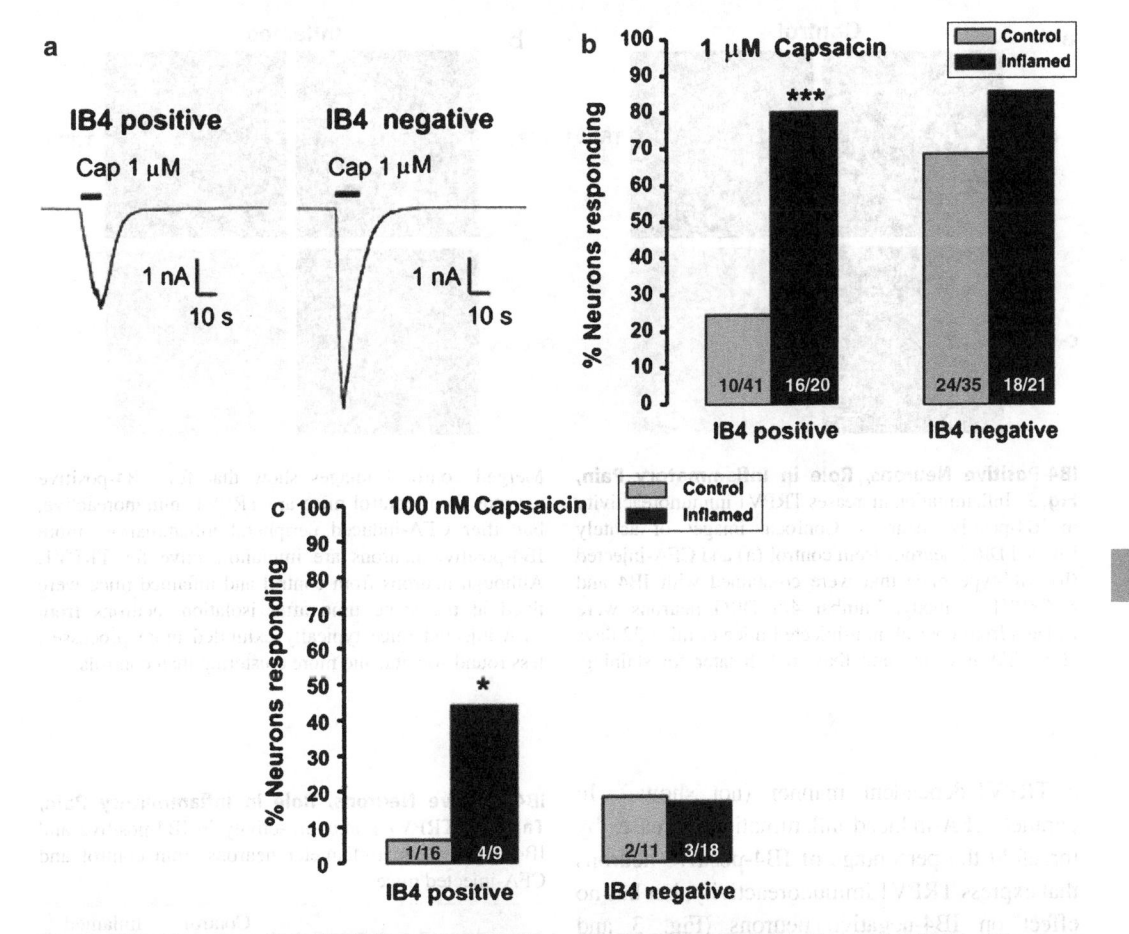

IB4-Positive Neurons, Role in Inflammatory Pain, Fig. 2 Inflammation sensitizes IB4-positive neurons to capsaicin. (**a**) Examples of whole-cell voltage clamp recordings from an IB4-positive and an IB4-negative small-diameter neuron ($\leq$26 µm) that responded to 1 µM capsaicin. (**b**) Percentage of IB4-positive and IB4-negative neurons from the L4/L5 DRG of control or CFA-injected mice that responded to a 10 s exposure to 1 µM capsaicin. Capsaicin-evoked currents in all cells tested were > 40 pA. *** indicates that significantly more IB4-positive neurons from inflamed mice responded

to capsaicin compared to the control IB4-positive group ($P < 0.0001$; Fisher's Exact test). (**c**) Percentage of IB4-positive and IB4-negative small-diameter neurons isolated from the L4/L5 DRG from control or CFA-injected mice that responded to 100 nM capsaicin. * indicates that significantly more IB4-positive neurons from inflamed mice responded to capsaicin compared to control IB4-positive neurons ($P < 0.05$; Fisher's Exact test). The percentage of IB4-negative neurons that responded to 100 nM capsaicin was unaltered by inflammation (Modified from Breese et al. 2005)

Davis et al. 2000). Peripheral inflammation was induced by injection of ▶ complete Freund's adjuvant (CFA) in the hind paw of mice. Two days later, the lumbar 4–5 ganglia, which contain the cell bodies of neurons that project to the hind paw, were isolated and dissociated and whole-cell patch clamp recordings were performed on small-diameter neurons. Figure 2 shows that the proportion of IB4-positive neurons that respond to capsaicin with an inward current is markedly

increased (threefold) after inflammation, whereas IB4-negative neurons are unaltered. The increase in capsaicin responsiveness in IB4-positive neurons is due to increased TRPV1 function because IB4-positive neurons from ▶ TRPV1-Null Mice with CFA inflammation are unresponsive to capsaicin. Similarly, inflammation increases the proportion of IB4-positive neurons, but not IB4-negative neurons, that respond to protons (pH 5.0) in

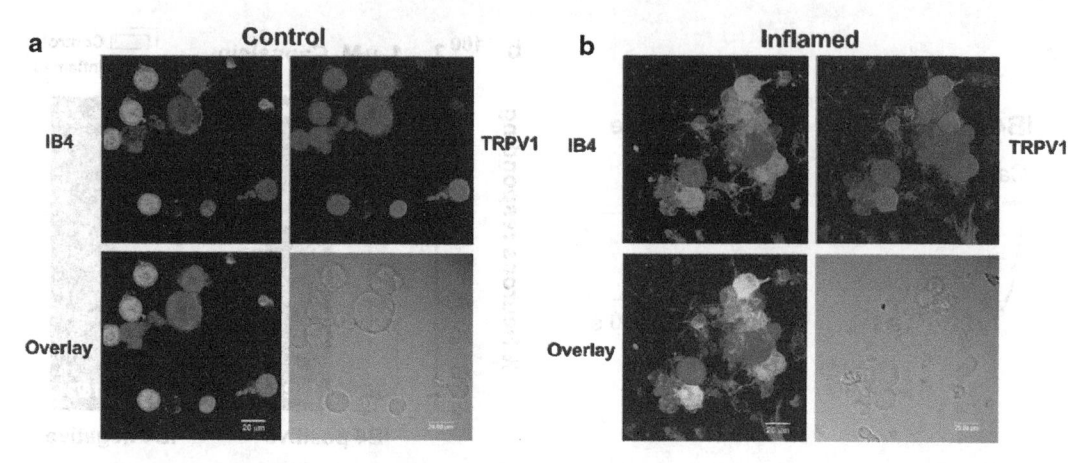

IB4-Positive Neurons, Role in Inflammatory Pain, Fig. 3 Inflammation increases TRPV1 immunoreactivity in IB4-positive neurons. Confocal images of acutely isolated DRG neurons from control (**a**) and CFA-injected (**b**) wild-type mice that were co-stained with IB4 and a TRPV1 antibody. Lumbar 4/5 DRG neurons were isolated from control, non-injected mice or mice 22 days after CFA injection and fixed 6–9 h later for staining. Merged confocal images show that few IB4-positive neurons from control mice are TRPV1-immunoreactive, but after CFA-induced peripheral inflammation, more IB4-positive neurons are immunoreactive for TRPV1. Although neurons from control and inflamed mice were fixed at the same time after isolation, neurons from CFA-injected mice typically exhibited more processes, less round somata, and more clustering than controls

a TRPV1-dependent manner (not shown). In parallel, CFA-induced inflammation increases by threefold the percentage of IB4-positive neurons that express TRPV1 immunoreactivity, but has no effect on IB4-negative neurons (Fig. 3 and Table 1).

Since the IB4-positive small-diameter neurons are selectively increased in TRPV1 function and expression during peripheral inflammation, they may play an important role in inflammatory pain. Natural stimuli for TRPV1 during inflammation may be noxious heat or endogenous TRPV1 ligands including protons, *N*-arachidonoyl-dopamine, anandamide or eicosanoids such as leukotriene B4, 12-(S)-HPETE or 15-(S)-HPETE molecules. An increase in the number of IB4-positive neurons that respond to TRPV1 stimuli including heat or endogenous TRPV1 inflammatory ligands could increase, by ▶ spatial summation, the amount of nociceptive information transmitted to the dorsal horn of the spinal cord and ultimately contribute to inflammatory pain and hyperalgesia. The IB4-positive neurons' capacity to become sensitized highlights them as a putative target for novel therapies for inflammatory pain.

IB4-Positive Neurons, Role in Inflammatory Pain, Table 1 TRPV1 immunoreactivity in IB4-positive and IB4-negative small-diameter neurons from control and CFA-injected mice

TRPV1-ir (%)	Control (%)	Inflamed (%)
Percentage total neurons	10.7 ± 4.0	25.0 ± 3.0[a]
Percentage of IB4-positive neurons	4.5 ± 2.1	15.5 ± 3.6[b]
Percentage of IB4-negative neurons	20.9 ± 12.2	34.7 ± 8.6

All data shown were collected on small-diameter (≤ 26 μm) L4/L5 DRG neurons co-stained for IB4 and TRPV1

[a]Indicates that the percentage of neurons that are TRPV1-immunoreactive (TRPV1-ir) was increased in CFA-injected mice compared to control mice ($P < 0.05$; two-tailed unpaired t-test)

[b]Indicates that the percentage of IB4-positive neurons that are TRPV1-ir was increased in CFA-injected mice compared to control mice ($P < 0.05$; two-tailed unpaired t-test). The percentage of TRPV1-ir IB4-negative neurons was not significantly altered in CFA-injected mice compared to control mice ($P > 0.3$; two-tailed unpaired t-test). n = 5 control mice (1,201 total neurons analyzed for TRPV1 and IB4 staining); n = 12 CFA-injected mice (1,232 total neurons analyzed for TRPV1 and IB4 staining) (Breese et al. 2005)

References

Averill, S., McMahon, S. B., Clary, D. O., et al. (1995). Immunocytochemical localization of trkA receptors in chemically identified subgroups of adult rat sensory neurons. *European Journal of Neuroscience, 7*, 1484–1494.

Bennett, D. L., Michael, G. J., Ramachandran, N., et al. (1998). A distinct subgroup of small DRG cells express GDNF receptor components and GDNF is protective for these neurons after nerve injury. *Journal of Neuroscience, 18*, 3059–3072.

Breese, N. M., George, A. C., Pauers, L. E., et al. (2005). Peripheral inflammation selectively increases TRPV1 function in IB4-positive sensory neurons from adult mouse. *Pain, 115*, 37–49.

Caterina, M. J., Leffler, A., Malmberg, A. B., et al. (2000). Impaired nociception and pain sensation in mice lacking the capsaicin receptor. *Science, 288*, 306–313.

Davis, J. B., Gray, J., Gunthrope, M. J., et al. (2000). Vanilloid receptor-1 is essential for inflammatory thermal hyperalgesia. *Nature, 405*, 183–187.

Dirajlal, S., Pauers, L. E., & Stucky, C. L. (2003). Differential response properties of IB(4)-positive and -negative unmyelinated sensory neurons to protons and capsaicin. *Journal of Neurophysiology, 69*, 1071–1081.

Gerke, M. B., & Plenderleith, M. B. (2004). Ultrastructural analysis of the central terminals of primary sensory neurones labeled by transganglionic transport of Bandeiraea simplicifolia I-isolectin B4. *Neuroscience, 127*, 165–175.

Kashiba, H., Uchida, Y., & Senba, E. (2001). Difference in binding by isolectin B4 to trkA and c-ret mRNA-expressing neurons in rat sensory ganglia. *Brain Research. Molecular Brain Research, 95*, 18–26.

Mantyh, P. W., & Hunt, S. P. (1998). Hot peppers and pain. *Neuron, 21*, 644–645.

Molliver, D. C., Radeke, M. J., Feinstein, S. C., et al. (1995). Presence or absence of TrkA protein distinguishes subsets of small sensory neurons with unique cytochemical characteristics and dorsal horn projections. *The Journal of Comparative Neurology, 361*, 404–416.

Silverman, J. D., & Kruger, L. (1990). Selective neuronal glycoconjugate expression in sensory and autonomic ganglia: Relation of lectin reactivity to peptide and enzyme markers. *Journal of Neuroscience, 23*, 789–801.

Snider, W. D., & McMahon, S. B. (1998). Tackling pain at the source: New ideas about nociceptors. *Neuron, 20*, 629–632.

Stucky, C. L., & Lewin, G. R. (1999). Isolectin B(4)-positive and -negative nociceptors are functionally distinct. *Journal of Neuroscience, 19*, 6497–6505.

Vydyanathan, A., Wu, Z. Z., Chen, S. R., et al. (2005). A-type voltage-gated K + currents influence firing properties of isolectin B4-positive but not isolectin B4-negative primary sensory neurons. *Journal of Neurophysiology, 93*, 3401–3409.

Wang, H., Rivero-Melian, C., Robertson, G., et al. (1994). Transganglionic transport and binding of the isolectin B4 from Griffonia simplicifolia I in rat primary sensory neurons. *Neuroscience, 62*, 539–551.

Wu, Z. Z., & Pan, H. L. (2004). High voltage-activated Ca^{2+} channel currents in isolectin B(4)-positive and -negative small dorsal root ganglion neurons of rats. *Neuroscience Letters, 368*, 96–101.

IB4-Positive Nociceptors

▶ IB4-Positive Neurons, Role in Inflammatory Pain

IBS

▶ Descending Modulation of Visceral Pain
▶ Irritable Bowel Syndrome
▶ Visceral Pain Model, Irritable Bowel Syndrome Model

ICD

▶ Functioning and Disability Definitions

Ice-Pick Pain

Definition

This describes a sharp jabbing pain in contrast to a persisting ache.

Cross-References

▶ Primary Stabbing Headache

Ice-Pick Pains

▶ Primary Stabbing Headache

Ice-Water Bucket Test

Definition

A test involving immersing an extremity into a bucket filled with ice and water, usually resulting in a temperature of 4°C. The test can be used as a test of pain sensitivity (e.g., by measuring how long the extremity is left immersed before pain becomes intolerable) or as a defined conditioning stimulus for testing the body's inhibitory responses to pain or nociceptive input. If the latter is intended, the extremity is left in the bucket for a defined length of time (or until a defined temperature is reached), with quantitative sensory testing (or other formal sensory testing) being performed before and after the test to quantitate the inhibitory response.

Cross-References

▶ Quantitative Sensory Testing

ICF

▶ Functioning and Disability Definitions

ICF, International Classification of Functional, Disability and Health

Definition

International Classification of Functioning, Disability and Health was developed by the World Health Organization. The ICF has addressed many of the criticisms of prior conceptual frameworks, and has been developed in a worldwide comprehensive consensus process over the few last years.

Cross-References

▶ Functioning and Disability Definitions

ICSI

Synonyms

O'Leary-Sant Interstitial Cystitis Symptom Index

Definition

ICSI stands for O'Leary-Sant Interstitial Cystitis Symptom Index, which is utilized for quantifying the degree of symptoms associated with interstitial cystitis.

Cross-References

▶ Interstitial Cystitis and Chronic Pelvic Pain

IDET

Synonyms

Intradiscal electrothermal therapy

Definition

IDET stands for Intradiscal Electrothermal Therapy – this technique is also called annuloplasty and involves treatment of posterior annulus with heat from a thermal resistive coil in an attempt to repair, denervate, and stabilize an annular tear.

Cross-References

▶ Discogenic Back Pain
▶ Intradiscal Electrothermal Therapy
▶ Spinal Fusion for Chronic Back Pain

Idiopathic

Definition

If a condition or disease is of unknown cause, it is described as idiopathic.

Cross-References

▶ Diabetic Neuropathies

Idiopathic Ataxic Neuropathy

▶ Ganglionopathies

Idiopathic Cramps

Definition

This type of cramp is the main symptom of a disease about which little is known, the cause of the cramp being obscure or speculative. Idiopathic cramps can be either sporadic or inherited and are usually not associated with any cognitive, pyramidal, cerebellar, or sensory abnormalities.

Cross-References

▶ Muscular Cramps

Idiopathic Headache

▶ Headache, Episodic Tension-Type

Idiopathic Myalgia

▶ Myalgia

Idiopathic Orofacial Pain

▶ Atypical Facial Pain: Etiology, Pathogenesis, and Management

Idiopathic Stabbing Headache (ISH)

▶ Primary Stabbing Headache

Idiopathic Vulvar Pain

▶ Vulvodynia

IEGs

▶ Immediate Early Genes

iGluRs

▶ Ionotropic Glutamate Receptors

Ignition Hypothesis

Definition

Hypothesis, proposed by Rappaport and Devor (1994), which attempts to explain paroxysmal pain in trigeminal neuralgia in terms of sustained afterdischarge and cross excitation, amplified by positive feedback among neighboring neurons in the trigeminal root and/or ganglion

Cross-References

▶ Pain Paroxysms

References

Rappaport, Z. H., & Devor, M., (1994). Trigeminal
    neuralgia : the role of self-sustaining discharge in the
    trigeminal ganglion. *Pain, 56*, 127–138.

IL-10

Definition

Interleukin-10 is an anti-inflammatory cytokine.

Cross-References

▶ Cytokines as Targets in the Treatment of
  Neuropathic Pain

IL-1beta

Definition

IL-1beta is a cytokine with many properties
similar to TNF. At higher endocrine concentrations, it is associated with fever and formation of
acute-phase plasma proteins in the liver.

Cross-References

▶ Cytokines as Targets in the Treatment of
  Neuropathic Pain

IL-4

Definition

Interleukin-4 is an anti-inflammatory cytokine.

Cross-References

▶ Cytokines as Targets in the Treatment of
  Neuropathic Pain

IL-6

Definition

IL-6 is a cytokine with mostly proinflammatory
and algesic actions. It is a member of the IL-6
cytokine family that includes leukemia inhibitory
factor (LIF) and ciliary neurotrophic factor
(CNTF).

Cross-References

▶ Cytokines as Targets in the Treatment of
  Neuropathic Pain

Ilium

Definition

The ilium is the upper portion of the hipbone.

Cross-References

▶ Sacroiliac Joint Pain

Illusory Cramp

Definition

Illusory cramp is a phenomenon in which the
sensation of cramping is experienced, but little
or no contraction of the muscle occurs.

Cross-References

▶ Muscular Cramps

Imagery

Definition

Representations in the mind of visual, auditory, tactile, olfactory, gustatory, or kinesthetic experiences. The subjective reality of these experiences may vary between and within individuals. Guided, directed, or elicited imagery can be a potent means of changing subjective experiences of pain.

Cross-References

▶ Therapy of Pain, Hypnosis

Imaging Descending Circuits in the Forebrain

▶ Descending Circuits in the Forebrain, Imaging

IME

▶ Independent Medical Examinations

Imitation Learning

▶ Modeling, Social Learning in Pain

Immediate Early Genes

Synonyms

IEGs

Definition

Immediate early genes (IEGs) are genes that are rapidly induced in the absence of *de novo* protein synthesis. The best characterized immediate early

genes are c-jun and c-fos; both genes encode transcription factors that bind to specific regulatory sequences and activate expression of responsive genes. c-jun has a leucine zipper motif, which forms a heterodimer with a variety of proteins including c-fos. The jun-fos protein complex, also known as the activator complex (AP)-1, binds to a number of cellular promoters through a common element (5'-TGACTCA-3'). IEGs are distinct from "late response" genes, which can only be activated later following the synthesis of early response gene products. IEGs have therefore been called the "gateway to the genomic response."

Cross-References

▶ Amygdala, Pain Processing and Behavior in Animals
▶ NGF, Regulation During Inflammation

Immortalization

Definition

Immortalization is to confer the property of continuous division by the addition or upregulation of a gene that when expressed in the cell, stimulates the cell to continuously divide until removed or downregulated.

Cross-References

▶ Cell Therapy in the Treatment of Central Pain

Immune Cell Recruitment

Definition

Inflammation induces immune cell migration from the circulation in multiple steps, including rolling, adhesion, and transmigration of immune cells through the vessel wall.

Cross-References

▶ Opioids and Inflammatory Pain

Immune Cells

Definition

B and T lymphocytes, macrophages, and monocytes are immune cells patrolling through the body. They can migrate to areas with inflammation (chemotaxis).

Cross-References

▶ Opioid Modulation of Nociceptive Afferents In Vivo

Immunocompetent Cells

Definition

Immunocompetent cells are able to respond to bacterial and viral stimuli. These cells respond by releasing classical immune mediators including proinflammatory cytokines. Within the CNS, these cells include astrocytes and microglia. Activation of these cells with immunogenic substances can induce exaggerated pain.

Cross-References

▶ Cord Glial Activation

Immunocytochemistry

Definition

Immunocytochemistry is a method for demonstrating the localization of compounds in tissues, based on the use of antibodies.

Cross-References

▶ Immunocytochemistry of Nociceptors
▶ Opioid Receptors at Postsynaptic Sites

Immunocytochemistry of Nociceptors

Cheryl L. Stucky
Department of Cell Biology Neurobiology and
Anatomy, Medical College of Wisconsin,
Milwaukee, WI, USA

Synonyms

Immunocytochemistry; Immunohistochemistry; Immunoreactivity; Immunostaining; Neurochemical Markers; Neurochemistry; Nociceptors

Definition

The study of molecules or proteins (antigens) found in the cytoplasm or membrane of nociceptors by using immunologic staining methods, such as the use of fluorescent antibodies or enzymes (e.g., horseradish peroxidase). The immunocytochemistry procedure is usually performed on sections of tissue that has been fixed by a cross-linking fixative like paraformaldehyde or glutaraldehyde. Typical tissues include the dorsal root or trigeminal ganglion, peripheral nerve, or target tissue such as skin. Most of these immunocytochemical markers label not only the cell bodies but also the axons and terminals of the sensory neurons.

Characteristics

Nociceptive sensory neurons express a large variety of molecules that are involved in neuronal communication, and many of these molecules are present in high enough concentrations to be detected by immunocytochemical methods.

For example, nociceptors contain neurotransmitters and neuropeptides which when released from central terminals act on spinal cord neurons or when released from peripheral terminals act on other cells in the skin or target tissues. Furthermore, nociceptors express ion channels and receptors in their plasma membrane that allow them to respond to external stimuli (heat, cold, mechanical force) or internal stimuli such as molecules released by other cell types. The goal of this entry is to highlight examples of the major immunocytochemical markers that are most frequently used to characterize subpopulations of nociceptors within the ▶ dorsal root ganglion (DRG) or trigeminal ganglion. However, because DRG and trigeminal neurons are exceptionally heterogeneous with respect to their expression of peptides, enzymes, and receptors, there are many molecules and subpopulations of neurons that are not described here.

Peptidergic Population: CGRP and Substance P

The most extensively characterized neuropeptides in nociceptive sensory neurons are ▶ calcitonin gene-related peptide (CGRP) and ▶ substance P. The CGRP- and substance P-expressing population of small-diameter neurons is frequently called the "peptidergic" population of small-diameter neurons because both of these neuropeptides are found in many small-diameter neurons that stain darkly with basic aniline dyes and are called "small dark cells." Small dark cells can be distinguished from large light cells, which stain clear or lightly, because they contain many neurofilaments that do not stain with the basic dyes.

Calcitonin Gene-Related Peptide

CGRP is expressed by more sensory neurons than other peptides and estimates range from 20 % to 80 % of DRG or trigeminal ganglion neurons expressing CGRP. The wide range among studies is due to differences between species, the location along the cervical to sacral neuraxis, the target tissue innervated, and whether or not colchicine, which blocks microtubule transport of neuropeptides, was used. Most studies agree that 35–50 %

of all rat lumbar DRG neurons express CGRP. Within the spinal cord, CGRP is localized exclusively in primary afferent neurons. Most CGRP-expressing sensory neurons are small or medium size in diameter (many are nociceptors), but a few are large diameter. This is consistent with the finding that the afferent fibers of CGRP-positive neurons are primarily unmyelinated C fibers and thinly myelinated Aδ fibers and a few are large myelinated Aβ fibers (McCarthy and Lawson 1990).

Substance P

Substance P (SP) is expressed in approximately 20 % of DRG neurons and most of these are small-diameter neurons, although a few are medium diameter. Consistent with this, individually identified SP-expressing DRG neurons have either C fiber or Aδ fiber axons and exhibit nociceptive response properties (Lawson et al. 1997). Substance P is highly colocalized with CGRP in that nearly all SP-expressing DRG neurons also contain CGRP, although only half of the CGRP-expressing neurons also contain SP.

Other Characteristics

The peptidergic population of small dark neurons overlaps extensively with expression of ▶ trkA, the high-affinity receptor for the neurotrophin ▶ nerve growth factor (NGF), and depends on NGF for survival during development (Averill et al. 1995). Because good antibodies are available for the ▶ trkA receptor, immunocytochemistry for trkA is frequently used as an alternative marker for the peptidergic population of small dark neurons. The central terminals of the peptidergic/trkA neurons are concentrated in lamina I and II_{outer} of the superficial dorsal horn as well as in lamina V and X of the spinal cord.

CGRP-positive and SP-positive neurons innervate all types of peripheral target tissue, including skin, muscle, joint, bone, and visceral organs. It is important to note that the expression pattern and levels of CGRP and SP in DRG neurons change with injury. Both CGRP and SP increase following persistent peripheral inflammation, and conversely, they decrease following peripheral nerve lesion (Donnerer et al. 1992;

Villar et al. 1991). This caveat must be carefully considered when using these markers to label and follow subpopulations of neurons in models of injury.

Other neuropeptides found in (typically smaller) subpopulations of mammalian DRG neurons that can be localized via antibody staining include somatostatin, vasoactive intestinal peptide, galanin, vasopressin, bombesin, dynorphin, enkephalin, neuropeptide Y, cholecystokinin, and endothelin-1. Their distribution and characteristics will not be discussed in detail here.

Non-peptidergic or "Peptide Poor" Population: Isolectin B4 and FRAP

The population of small dark C fiber neurons that is poor in expression of the neuropeptides CGRP or SP is typically characterized by labeling with the plant lectin, isolectin B4 (IB4). Isolectin B4 binds to surface carbohydrates, specifically the ▶ alpha(α)-D-galactose groups of glycol conjugates (Silverman and Kruger 1990). The IB4-binding technique is most frequently used today because it is very adaptable and easy to perform, and IB4 conjugated directly to fluorescein (IB4-FITC) or other fluorescent markers can be used to label live neurons within minutes after performing physiological experiments (Stucky and Lewin 1999). However, the "peptide poor" population can also be labeled by immunoreactivity to the antibody LA4, which recognizes the α-galactose oligosaccharides (Dodd and Jessell 1985), or by the presence of the fluoride-resistant acid phosphatase (FRAP) enzyme, an extralysosomal acid phosphatase that is resistant to fluoride ions (Silverman and Kruger 1988). The function of the α-D-galactose groups is not clear, but they have been hypothesized to play a role in the cell-cell interactions during the development of connections of primary afferent neurons to the dorsal horn of the spinal cord (Dodd and Jessell 1985). The function of FRAP is not known.

Other Characteristics

The IB4-positive/FRAP-positive population of small-diameter neurons also expresses receptors

for ▶ glial cell line-derived neurotrophic factor (GDNF) and depends on GDNF for survival during postnatal development (Molliver et al. 1997; Bennett et al. 1998). Receptors for GDNF include the ligand-binding domain GFRα1 or GFRα2 and the signal transducing, tyrosine kinase domain, RET. IB4-positive neurons primarily terminate in ▶ lamina II_{inner} of the superficial dorsal horn of the spinal cord, a region also known as the substantia gelatinosa.

Although a number of investigations have reported a clear separation between the peptidergic/trkA and the non-peptidergic/IB4-binding populations, other reports indicate that CGRP/SP/trkA and IB4/FRAP staining overlap extensively (Wang et al. 1994; Kashiba et al. 2001). The diverse findings may be due to differences in methods used or sensitivity of detection of the labels. It is likely that there is at least a subpopulation of neurons that expresses neuropeptides and binds IB4.

Importantly, IB4 binding decreases substantially following nerve injury (Bennett et al. 1998), and therefore, caution must be exercised when using this marker to label subpopulations of neurons in animal models of injury.

Neurofilament Markers: Antibodies RT97 and N52

The antibody clones RT97 and N52 both recognize the high molecular weight (200 kD) ▶ neurofilament protein NF200, which is found in sensory neurons that have myelinated axons. Both RT97 and N52 antibody labeling have been used to identify the large light population of DRG neurons. One difference between the two antibodies is that RT97 recognizes only the phosphorylated form of the 200 kD neurofilament protein whereas N52 recognizes both the phosphorylated and non-phosphorylated forms of the protein. Thus, there may be some differences in the populations of neurons labeled by these two markers. An advantage of RT97 is that a cell-by-cell correlation between RT97 staining and conduction velocities has been made for rat and all A fiber (Aδ and Aβ) DRG neurons were found to be RT97-positive whereas all C fiber neurons were RT97-negative (Lawson and Waddell 1991).

No such correlation between N52 and conduction velocity has yet been made.

There is almost no overlap between RT97 and IB4 staining or between N52 and IB4 staining, indicating that RT97/N52 and IB4 label distinct subpopulations of large light and small dark neurons, respectively. There is, however, significant overlap between RT97 and CGRP as approximately 30 % of RT97-positive neurons contain CGRP and these are medium/large size neurons (McCarthy and Lawson 1990).

Ion Channels: TRPV1 and P2X3

Subpopulations of nociceptors can also be identified by antibodies that recognize specialized transduction molecules on the plasma membrane. Examples of these include receptors for noxious heat (TRPV1 receptor) and for ATP (P2X$_3$ receptor).

Transient Receptor Potential Vanilloid 1 (TRPV1) Receptor

The TRPV1 receptor/ion channel is the receptor for capsaicin, the potent algogen found in "hot" chili peppers (Caterina et al. 1997). Formerly known as VR1, TRPV1 is a member of the ▶ transient receptor potential family of ion channels and, when activated, allows calcium and sodium to flow into the neuron, resulting in depolarization. Besides capsaicin, TRPV1 also responds to moderately noxious heat (>43 °C), acid (~pH 5.0), and a number of other endogenous ligands that may be present during inflammation or injury, including *N*-arachidonoyl-dopamine, anandamide, or ▶ eicosanoids such as leukotriene B4, 12-(S)-HPETE, or 15-(S)-HPETE molecules. Good antibodies to TRPV1 exist and TRPV1 immunoreactivity is found on many small-diameter and some medium-diameter sensory neurons (Caterina et al. 1997). Because capsaicin responsiveness is often used as a functional marker for nociceptors, TRPV1 immunoreactivity is frequently used as a neurochemical marker for nociceptors. Many TRPV1-immunoreactive neurons contain the neuropeptides CGRP or SP, whereas others bind IB4 (Tominaga et al. 1998). As is the case with neuropeptide immunoreactivity and IB4 binding, TRPV1 expression is also not stable after injury. The mRNA and protein for TRPV1 decreases in directly injured DRGs after ▶ axotomy or ▶ Spinal Nerve Ligation Model. Conversely, TRPV1 increases in adjacent DRGs after spinal nerve ligation and is reported to increase following peripheral inflammation (Hudson et al. 2001).

P2X$_3$ Receptor for ATP

Extracellular adenosine triphosphate (ATP) has been implicated in nociceptive signaling in normal and pathological pain conditions and ATP directly excites nociceptors. P2X$_3$ receptors are multimeric ion channels gated by ATP. They exist on native DRG neurons as either P2X$_3$ homomers or as a heteromeric combination with the P2X$_2$ receptor (P2X$_{2/3}$). Good antibodies against the P2X$_3$ receptor are available and studies document that P2X$_3$ receptors are selectively expressed on small-diameter DRG neurons. Under uninjured conditions, P2X$_3$ receptors are almost exclusively localized to the IB4-binding population and only a few P2X$_3$-positive neurons contain neuropeptides (Bradbury et al. 1998). Consistent with the pattern for IB4-binding neurons, the central terminals of P2X$_3$-expressing neurons project primarily to the lamina II$_{inner}$ of the superficial dorsal horn. The expression of P2X$_3$ receptors decreases following nerve injury and this decrease can be reversed by in vivo administration of GDNF (Bradbury et al. 1998).

References

Averill, S., McMahon, S. B., Clary, D. O., et al. (1995). Immunocytochemical localization of trkA receptors in chemically identified subgroups of adult rat sensory neurons. *European Journal of Neuroscience, 7*, 1484–1494.

Bennett, D. L., Michael, G. J., Ramachandran, N., et al. (1998). A distinct subgroup of small DRG cells express GDNF receptor components and GDNF is protective for these neurons after nerve injury. *Journal of Neuroscience, 18*, 3059–3072.

Bradbury, E. J., Burnstock, G., & McMahon, S. B. (1998). The expression of P2X3 purinoreceptors in sensory neurons: Effects of axotomy and glial-derived neurotrophic factor. *Molecular and Cellular Neurosciences, 12*, 256–268.

Caterina, M. J., Schumacher, M. A., Tominaga, M., Rosen, T. A., Levine, J. D., & Julius, D. (1997). The capsaicin receptor: A heat-activated ion channel in the pain pathway. *Nature, 389*, 816–824.

Dodd, J., & Jessell, T. M. (1985). Lactoseries carbohydrates specify subsets of dorsal root ganglion neurons projecting to the superficial dorsal horn of rat spinal cord. *Journal of Neuroscience, 5*, 3278–3294.

Donnerer, J., Schuligoi, R., & Stein, C. (1992). Increased content and transport of substance P and calcitonin gene-related peptide in sensory nerves innervating inflamed tissue: Evidence for a regulatory function of nerve growth factor in vivo. *Neuroscience, 49*, 693–698.

Hudson, L. J., Bevan, S., Wotherspoon, G., et al. (2001). VR1 Protein expression increases in undamaged DRG neurons after partial nerve injury. *European Journal of Neuroscience, 13*, 2105–2114.

Kashiba, H., Uchida, Y., & Senba, E. (2001). Difference in binding by isolectin B4 to trkA and c-ret mRNA-expressing neurons in rat sensory ganglia. *Brain Research. Molecular Brain Research, 95*, 18 – 26.

Lawson, S. N., & Waddell, P. J. (1991). Soma neurofilament immunoreactivity is related to cell size and fibre conduction velocity in rat primary sensory neurons. *The Journal of Physiology, 435*, 41–63.

Lawson, S. N., Crepps, B. A., & Perl, E. R. (1997). Relationship of substance P to afferent characteristics of dorsal root ganglion neurones in guinea-pig. *The Journal of Physiology, 505*, 177–191.

McCarthy, P. W., & Lawson, S. N. (1990). Cell type and conduction velocity of rat primary sensory neurons with calcitonin gene-related peptide-like immunoreactivity. *Neuroscience, 34*, 623–632.

Molliver, D. C., Wright, D. E., Leitner, M. J., et al. (1997). IB4-Binding DRG neurons switch from NGF to GDNF dependence in early postnatal life. *Neuron, 19*, 849–861.

Silverman, J. D., & Kruger, L. (1988). Lectin and neuropeptide labeling of separate populations of dorsal root ganglion neurons and associated "nociceptor" thin axons in rat testis and cornea whole-mount preparations. *Somatosensory Research, 5*, 259–267.

Silverman, J. D., & Kruger, L. (1990). Selective neuronal glycoconjugate expression in sensory and autonomic ganglia: Relation of lectin reactivity to peptide and enzyme markers. *Journal of Neurocytology, 19*, 789–801.

Stucky, C. L., & Lewin, G. R. (1999). Isolectin B(4)-positive and -negative nociceptors are functionally distinct. *Journal of Neuroscience, 19*, 6497–6505.

Tominaga, M., Caterina, M. J., Malmberg, A. B., et al. (1998). The cloned capsaicin receptor integrates multiple pain-producing stimuli. *Neuron, 21*, 531–543.

Villar, M. J., Wiesenfeld-Hallin, Z., Xu, X. J., et al. (1991). Further studies on galanin-, substance P- and CGRP-like immunoreactivities in primary sensory neurons and spinal cord: Effects of dorsal rhizotomies and sciatic nerve lesions. *Experimental Neurology, 112*, 29–39.

Wang, H., Rivero-Melian, C., Robertson, G., et al. (1994). Transganglionic transport and binding of the isolectin B4 from griffonia simplicifolia I in rat primary sensory neurons. *Neuroscience, 62*, 539–551.

Immunocytokines

▶ Cytokines, Effects on Nociceptors
▶ Cytokines, Regulation in Inflammation

Immunodeficient

Definition

An innate, acquired, or induced inability to develop a normal immune response.

Cross-References

▶ Animal Models of Inflammatory Bowel Disease

Immunoglobulin Therapy

Definition

Intravenous administration of immunoglobulin has been clinically shown to have anti-inflammatory immunomodulatory effects in GBS and CIDP. Doses required are higher than those required for immunodeficiency states.

Cross-References

▶ Inflammatory Neuritis

Immunohistochemistry

Definition

Immunohistochemistry is a staining method using the principle of antigen-antibody interactions to demonstrate a defined protein in tissue sections.

Cross-References

▶ Immunocytochemistry of Nociceptors
▶ Toxic Neuropathies

Immuno-Inflammatory Muscle Pain

▶ Muscle Pain in Systemic Inflammation (Polymyalgia Rheumatica, Giant Cell Arteritis, Rheumatoid Arthritis)

Immunoisolation

Definition

When supplying donor-grafted tissue or cells, immunoisolation allows the separation of that tissue from the host, for example, in an inert device that cannot be detected by the host immune system. Such a procedure is more likely to lead to long-term survival of donor tissue or cells.

Cross-References

▶ Cell Therapy in the Treatment of Central Pain

Immunoreactivity

▶ Immunocytochemistry of Nociceptors

Immunostaining

▶ Immunocytochemistry of Nociceptors

Immunosuppression

Definition

Inhibiting the activation and function of immune cells either by disease or by drugs.

Cross-References

▶ Proinflammatory Cytokines
▶ Vascular Neuropathies

Impact of Familial Factors on Children's Chronic Pain

Kathryn A. Birnie[1], Lindsay S. Uman[2] and Christine T. Chambers[3]
[1]Department of Psychology, Dalhousie University, Halifax, NS, Canada
[2]Department of Psychology, IWK Health Centre, Halifax, NS, Canada
[3]Departments of Pediatrics and Psychology, Dalhousie University, Halifax, NS, Canada

Synonyms

Familial factors; Family environment; Parental response; Psychosocial factors

Definition

Of the many psychosocial variables known to affect chronic pain in children, the family has emerged as one of the more prominent. Families are involved in all aspects of their children's pain, including assessment and management. It is now generally accepted that parents play an important role in influencing how their children learn to respond to pain. Having a child with chronic pain can also place a significant emotional and financial burden on families. This entry will provide a brief description of the major research areas examining the impact of familial factors on children's chronic pain. Several more comprehensive reviews of family factors and pain are also available elsewhere (e.g., Lewandowski et al. 2010; Evans et al. 2008; Palermo and Chambers 2005; Chambers 2003; Palermo 2000).

Introduction

It has long been observed that *chronic pain* tends to run in families, whereby children who report pain are likely to have parents who report pain and vice versa. Research by Saunders and colleagues (2007) observed a dose–response relationship, such that children were at increased risk of having multiple pain conditions the larger the number of pain sites reported by their mothers over the previous 6 months. Of concern, the majority of the studies examining aggregation of pain in families have been retrospective; however, a large-scale prospective study found considerable support for the association between pain and pain-related disability among children and their parents (Goodman et al. 1997).

Although genetic predisposition and shared environment may account in part for the tendency for pain to run in families, other psychological factors have also been revealed to influence the likelihood of childhood chronic pain. A recent prospective population-based study reported that both maternal and paternal anxiety in the first year of a child's life, as well as child temperament, were associated with later childhood recurrent abdominal pain (Ramchandani et al. 2006). Furthermore, a study by Bruehl and colleagues (2005) suggests that it is the child's recollection of the parental pain, not the confirmed parental pain, that predicts the child's own chronic pain status in adulthood. Additionally, psychological processes, particularly *social learning theory* and

operant conditioning, have emerged as important factors in understanding how children acquire information about how to respond to pain.

These theories indicate that children can learn how to cope with pain from their parents through various pathways. For example, children may learn through direct parental instruction or reprimands (i.e., being told to respond or not respond in a certain way to pain) as well as through observational or vicarious learning (i.e., by watching how the parent responds to his/her own pain; Craig 1986). Parents may also indirectly teach their children about how to express pain by how they react to their children's expressions of pain (e.g., by providing special time or exemptions when the child has pain). These childhood learning experiences may, in turn, provide the foundation for the development of pain-related behaviors and coping styles later in life. Fortunately, there is evidence that parental behavior and responses to pain can be modified successfully in order to produce improvements in pain and coping in children.

Characteristics

Parental Responses to Chronic Pain
The ways in which parents choose to respond to their children's pain complaints can have significant consequences for how children learn to express them (Walker et al. 1992; Walker et al. 2006). Principles of both positive and negative reinforcement can influence pain expression in children. *Positive reinforcement* may include providing tangible treats, such as presents or favorite foods, as well as non-tangible rewards, such as special privileges, sympathy, and increased attention for pain behaviors. *Negative reinforcement* may include allowing the child to stay home from school, as well as exemption from chores and other unpleasant or anxiety-provoking activities (e.g., music lessons, sport practices). Both positive and negative reinforcement may inadvertently promote pain behaviors and encourage their display, while simultaneously discouraging adaptive coping in the face of pain.

This combination of reinforcement can lead to problematic consequences, as children may learn to adopt a *sick role*. Children may be motivated to adopt the sick role because it provides an acceptable excuse for exemption from responsibilities as well as less than desirable performance or behavior (Walker et al. 1992). There is evidence that parents are more likely to excuse the misbehavior of children with physical illnesses when there is a medical explanation or known organic etiology for the pain compared to children with medically unexplained pain (Walker et al. 1995). Research suggests that parents of adolescents who report higher pain and pain-related functional disability are more likely to hold the belief that their adolescent's pain is only medical (i.e., without psychologic influence; Guite et al. 2009). Preliminary research suggests that parental beliefs about the cause of their child's chronic pain may impact the likelihood of the child's recovery. In a study by Crushell and colleagues (2003), more parents of children who had recovered from recurrent abdominal pain accepted a biopsychosocial model of illness as compared to parents of children with ongoing pain. Furthermore, the more the parents worry about their child's physical and emotional/behavioral well-being, the more likely they appear to engage in pain-promoting behaviors (Guite et al. 2009). Parental catastrophizing about their child's pain has been associated with more solicitous responses by parents (Hechler et al. 2011), greater parental emphasis on pain control over child activity engagement when pain is acute (Caes et al. 2012), as well as increased disability and lowered school attendance for the child independent of pain intensity (Goubert et al. 2006; Logan et al. 2012). Parental distress has mediated the relationship between parental pain catastrophizing and parents' desire to stop their child from continuing a painful activity (Caes et al. 2011). Taken together, parental worries and beliefs about their child's chronic pain appear to influence parents' behavior toward their child in potentially maladaptive ways, undermining the child's own ability to cope effectively over time.

Importantly, research suggests that parents can successfully learn and utilize more adaptive responses to their child's pain. As compared with adolescents whose parents had received no instruction, symptom complaints nearly doubled among both healthy adolescents and those with chronic pain whose parents engaged in attending talk during a laboratory-based pain induction task and were reduced by half among adolescents whose parents engaged in distraction talk (Walker et al. 2006). This effect was stronger among female pain patients than for male patients or healthy controls.

Recent research indicates that individual child characteristics are relevant when considering parent responses to child pain; for example, pain catastrophizing of children with chronic pain is associated with parent pain behaviors (Lynch-Jordan et al. 2010). A study by Williams and colleagues (2011) supported previous findings that parent symptom–related talk was associated with increased child symptom complaints; however, this relationship was greater for children who were high pain catastrophizers. Furthermore, increased parent non-symptom-related talk was associated with fewer child complaints for children with high pain threat appraisals. More research is needed to delineate the directionality of observed relationships between child characteristics and parent beliefs and behaviors.

The Adult Responses to Children's Symptoms (ARCS) was developed by Van Slyke and Walker (2006) to assess and quantify parental responses to child illness and pain behaviors. Parents rate each of the 29 items indicating how they respond when their child is ill or experiencing pain. The questions are designed to provide information about the extent to which parents engage in protective parental behaviors (i.e., give the child special attention or limit the child's normal activities), encouraging and monitoring responses (i.e., reassure and encourage to engage in activities, while continuing to monitor the child's pain behavior), or behaviors that minimize the pain (i.e., discount and criticize the child's pain as excessive). Research using the ARCS with children with mixed chronic pain conditions has shown that parental protective behaviors and their minimization of the pain were associated with negative child outcomes (Claar et al. 2010; Claar et al. 2008). The ARCS is a useful tool for elucidating and targeting maladaptive parental behaviors (for research and in clinical practice) which could be altered in order to promote better child health and pain management. Other measures assessing parental responses to child/adolescent pain include the Pain-related Parent Behavior Inventory (PPBI) (Hermann et al. 2008), the Inventory of Parent/Caregiver Responses to the Children's Pain Experience (IRPEDNA) (Huguet et al. 2008), and the Illness Behavior Encouragement Scale (IBES) (Walker and Zeman 1992) from which the ARCS was adapted and expanded.

Impact of Pain on the Family

While it is generally understood that families can significantly impact their children's pain, only recently has the reciprocal impact of pain on the family been acknowledged. Parenting a child with chronic pain typically involves extensive financial and emotional costs to the family, in the form of multiple medical appointments, missed time from work or school, disruptions in family activities, and increased stress (see Palermo and Eccleston 2009 for a brief review). Higher levels of anxiety and depression have been observed in mothers of children with chronic pain conditions as compared with mothers of healthy children (Campo et al. 2007). In a study examining the effect of chronic pain on the quality of life in adolescents and their families, it was found that more intense and frequent chronic pain in adolescents was associated with decreased self-reports of quality of life related to psychological functioning as well as physical and functional status (Hunfeld et al. 2001). Qualitative investigations with parents of children with chronic pain conditions describe a struggle to adapt to a different life than expected with feelings of uncertainty, loss, helplessness, fear, failure, and distress (Maciver et al. 2010; Jordan et al. 2007). Parents report modifying their representation of "good" parenting in order to better cope and develop more adaptive responses to their child's distress (Maciver et al. 2010).

Many children with chronic pain do not have an identified organic disease or underlying biological explanation for the pain, which can be an additional source of frustration for families who desire a definitive diagnosis with a time-limited treatment plan (Hunfeld et al. 2002).

Treatment of a child with chronic pain is typically an ongoing long-term process, involving a multidisciplinary team of health-care professionals. Given the numerous family resources (e.g., parental attention) devoted to caring for a child with chronic pain, it is possible that non-ill family members (e.g., siblings) may inadvertently feel neglected and resentful (Britton and Moore 2002). As a result, parents may differentially treat their other children in an effort to compensate. Thus, caring for a child with chronic pain is a process that affects all members of the family in some capacity. A significant gap in the literature is the impact of having a child with chronic pain on fathers (see Hechler et al. 2011 for an example). To date, studies have largely reported on mothers' experiences, even when both parents are eligible to participate. Although there are an increasing amount of investigations in the area, additional research is needed to further delineate the complex ways that pain can influence the health and well-being of children and their families.

Influence of Family System on Child Pain

Recent direction in understanding pediatric chronic pain and its management in children has moved toward a *family systems* approach. This approach understands the family as an emotional unit and takes a necessary framework to better describe the complex interactions within the family (Palermo and Chambers 2005). Results from investigations examining the moderating and mediating role of family characteristics on child pain and disability are largely mixed. A recent systematic review by Lewandowski and colleagues (2010) clarifies our understanding to date of the relationship between family functioning and child pain outcomes in families of children and adolescents with chronic pain. They observed that, in general, studies reported greater disturbances in family functioning, such as less cohesion and organization, and greater conflict, in families with children and adolescents with chronic pain as compared to families with healthy children. Furthermore, the majority of studies revealed that greater family dysfunction was related to increased pain-related disability for the child and vice versa (Lewandowski et al. 2010).

The parent–child relationship also appears relevant in the context of pediatric chronic pain conditions. A study conducted by Palermo and colleagues (2007) observed that lower adolescent autonomy was related to higher disability in adolescents with recurrent headaches. After controlling for pain severity, Lipani and Walker (2006) observed that mothers of children with chronic pain report greater worry and limitations on family activities when the child perceives their condition as serious and perceive themselves as having limited ability to cope with their pain. Taken together, this suggests that future research should include a developmental perspective as the relationship between children and their parents, as well as the child's capacity to cope with their condition, are expected to change over time.

Family Involvement in Interventions for Chronic Pain

Psychological interventions for chronic pain are typically *cognitive-behavioral* in nature and frequently involve training parents to encourage positive coping behaviors and take attention away from maladaptive behaviors in their children (Palermo et al. 2010). Interventions that involve parents have been shown to be more effective than those targeting the child alone, and are also more likely to result in maintained gains over time. For example, in a study examining the role of parent-mediated management of childhood migraines, it was found that the children undergoing biofeedback treatment whose parents also received pain management guidelines displayed significantly greater reductions in headache frequency and were more likely to be headache-free than children whose parents did not receive these guidelines (Allen and Shriver 1998). Parents were taught how their own behavior can influence the pain behavior of their children, and they were also trained to encourage

adaptive coping while discouraging maladaptive pain behaviors (e.g., being overly attentive to child pain behaviors, stopping normal activities during pain episodes).

Several studies have also examined the role of short-term cognitive-behavioral family interventions for pediatric recurrent abdominal pain (Sanders et al. 1989; Sanders et al. 1994). A study by Robins and colleagues (2005), in which parents were taught to be active partners in their children's pain management intervention, found that children whose families had been involved in their treatment reported significantly less abdominal pain and had fewer school absences as compared to children who received standard medical care. Including parents in the child's treatment likely empowers the parents in their own ability to cope. Indeed, improved anxiety, depression, and stress have been noted for parents attending an interdisciplinary cognitive-behavior therapy group with their children with chronic pain (Eccleston et al. 2003). Given the flexibility and increased access to care afforded by technology, new methods of intervention delivery are being explored. A preliminary study of an internet-based family cognitive-behavior therapy program suggests it is an acceptable and satisfactory intervention delivery method for both children with chronic pain and their parents (Palermo et al. 2009). Although reductions in pain intensity and activity limitations were reported by children with chronic pain in the intervention group as compared to a wait-list control, no changes were reported in parental protectiveness.

Alternative interventions to cognitive-behavioral therapy, which also involve parents, are being investigated with children with chronic pain with early success. A randomized controlled trial conducted by Wicksell and colleagues (2009) found that Acceptance and Commitment Therapy (ACT) was more effective than a multidisciplinary treatment including amitriptyline at improving perceived functional ability in relation to pain, fear of injury or kinesiophobia, pain intensity, and pain-related discomfort for adolescents with chronic pain. Both interventions had parental involvement.

Overall, most of the interventions involving parents have been based on behavioral principles. Research to date suggests that interventions involving the family in the management of chronic pain in children are advantageous. Future research should investigate the specific impact of including families in various interventions for pediatric chronic pain. Given the impact of pediatric chronic pain on the family unit more broadly, including parent-specific outcomes in intervention studies may be worthwhile.

Summary

It should be clear that familial factors are important to consider when evaluating and treating a child with chronic pain. These family influences are multifaceted and dynamic and can interact with various other biological, psychological, and social factors over time. Children learn how to cope with and express their pain from their parents. Parental responses to the child's pain can influence the child's ability to cope in both maladaptive and adaptive ways. Involving parents in interventions for chronic pain in children appears beneficial, and the financial and emotional burden related to caring for a child with chronic pain should also be considered. With some exceptions, the literature to date is largely based on retrospective and cross-sectional designs. Furthermore, there is substantial variability in measures used to assess family-related outcomes. Increased use of prospective or longitudinal investigations and greater consistency of outcome measures employed would largely benefit the field.

References

Allen, J. D., & Shriver, M. D. (1998). Role of parent-mediated pain behaviour management strategies in biofeedback treatment of childhood migraines. *Behaviour Therapy, 29*, 477–490.

Britton, C., & Moore, A. (2002). View from the inside, part 2: What the children with arthritis said, and the experiences of siblings, mothers, fathers and grandparents. *British Journal of Occupational Therapy, 65*, 413–419.

Bruehl, S., France, C. R., France, J., Harju, A., & Al'Absi M, M. (2005). How accurate are parental chronic pain histories provided by offspring? *Pain, 115*, 390–397.

Caes, L., Vervoort, T., Eccleston, C., Vandenhende, M., & Goubert, L. (2011). Parental catastrophizing about child's pain and its relationship with activity restriction: The mediating role of parental distress. *Pain, 152*, 212–222.

Caes, L., Vervoort, T., Eccleston, C., & Goubert, L. (2012). Parents who catastrophize about their child's pain prioritize attempts to control pain. *Pain, 153*, 1695–1701.

Campo, J. V., Bridge, J., Lucas, A., Savorelli, S., Walker, L., Di Lorenzo, C., Iyengar, S., & Brent, D. A. (2007). Physical and emotional health of mothers of youth with functional abdominal pain. *Archives of Pediatric and Adolescent Medicine, 161*, 131–137.

Chambers, C. T. (2003). The role of family factors in pediatric pain. In P. J. McGrath & G. A. Finley (Eds.), *Pediatric pain: Biological and social context, progress in pain research and management* (Vol. 26, pp. 99–130). Seattle: IASP Press.

Claar, R. L., Simons, L. E., & Logan, D. E. (2008). Parental response to children's pain: The moderating impact of children's emotional distress on symptoms and disability. *Pain, 138*, 172–179.

Claar, R. L., Guite, J. W., Kaczynski, K. J., & Logan, D. E. (2010). Factor structure of the adult responses to children's symptoms: Validation in children and adolescent with diverse chronic pain conditions. *The Clinical Journal of Pain, 26*, 410–417.

Craig, K. D. (1986). Social modeling influences: Pain in context. In R. A. Sternbach (Ed.), *The psychology of pain* (2nd ed., pp. 67–95). New York: Raven.

Crushell, E., Rowland, M., Doherty, M., Gormally, S., Harty, S., Bourke, B., & Drumm, B. (2003). Importance of parental conceptual model of illness in severe recurrent abdominal pain. *Pediatrics, 112*, 1368–1372.

Eccleston, C., Malleson, P. N., Clinch, J., Connell, H., & Sourbut, C. (2003). Chronic pain in adolescents: evaluation of a programme of interdisciplinary cognitive behaviour therapy. *Archives of Disease in Childhood, 88*, 881–885.

Evans, S., Tsao, J. C. I., Lu, Q., Myers, C., Suresh, J., & Zeltzer, L. (2008). Parent–child pain relationships from a psychosocial perspective: A review of the literature. *The Journal of Pain Management, 1*, 237–246.

Goodman, J. E., McGrath, P. J., & Forward, S. P. (1997). Aggregation of pain complaints and pain-related disability and handicap in a community sample of families. In T. S. Jensen, J. A. Turner, & Z. Wiesenfeld-Hallin (Eds.), *Proceedings of the 8th world congress on pain, progress in pain research and management* (Vol. 8, pp. 673–682). Seattle: IASP Press.

Goubert, L., Eccleston, C., Vervoort, T., Jordan, A., & Crombez, G. (2006). Parental catastrophizing about their child's pain. The parent version of the pain catastrophizing scale (PCS-P): A preliminary validation. *Pain, 123*, 254–263.

Guite, J. W., Logan, D. E., McCue, R., Sherry, D. D., & Rose, J. B. (2009). Parental beliefs and worries regarding adolescent chronic pain. *The Clinical Journal of Pain, 25*, 223–232.

Hechler, T., et al. (2011). Parental catastrophizing about their child's chronic pain: Are mothers and fathers different? *European Journal of Pain, 15*, 515.e1–515.e9.

Hermann, C., Zohsel, K., Hohmeister, J., & Flor, H. (2008). Dimensions of pain-related parent behavior: Development and psychometric evaluation of a new measure for children and their parents. *Pain, 137*, 689–699.

Huguet, A., Miro, J., & Nieto, R. (2008). The inventory of parent/caregiver responses to the children's pain experience (IRPEDNA): Development and preliminary validation. *Pain, 134*, 128–139.

Hunfeld, J. A. M., Perquin, C. W., Duivenvoorden, H. J., Hazebroek-Kampschreur, A. A. J. M., Passchier, J., van Suijlekom-Smit, L. W. A., & van der Wouden, J. C. (2001). Chronic pain and its impact on quality of life in adolescents and their families. *Journal of Pediatric Psychology, 26*, 145–153.

Hunfeld, J. A. M., Perquin, C. W., Hazebroek-Kampshreur, A. A. J. M., Passchier, J., van Suijlekom-Smit, L. W. A., & van der Wouden, J. C. (2002). Physically unexplained chronic pain and its impact on children and their families: The mother's perception. *Psychology and Psychotherapy: Theory, Research and Practice, 75*, 251–260.

Jordan, A. B., Eccleston, C., & Osborn, M. (2007). Being a parent of the adolescent with complex chronic pain: An interpretative phenomenological analysis. *European Journal of Pain, 11*, 49–56.

Lewandowski, A. S., Palermo, T. M., Stinson, J., Handley, S., & Chambers, C. T. (2010). Systematic review of family functioning in families of children and adolescents with chronic pain. *The Journal of Pain, 11*, 1027–1038.

Lipani, T. A., & Walker, L. S. (2006). Children's appraisal and coping with pain: Relation to maternal ratings of worry and restriction in family activities. *Journal of Pediatric Psychology, 31*, 667–673.

Logan, D. E., Simons, L. E., & Carpino, E. A. (2012). Too sick for school? Parent influences on school functioning among children with chronic pain. *Pain, 153*, 437–443.

Lynch-Jordan, A., Kashikar-Zuck, S., Flowers, S., Harding, K., Wolf, D., Paulford-Lecher, N., Desai, A., Szabova, A., & Goldschneider, K. (2010). The relationship between adolescent pain behaviors and catastrophizing on parent catastrophizing about their adolescents' pain. *The Journal of Pain, 11*, S18.

Maciver, D., Jones, D., & Nicol, M. (2010). Parents' experiences of caring for a child with chronic pain. *Qualitative Health Research, 20*, 1272–1282.

Palermo, T. (2000). Impact of recurrent and chronic pain on child and family daily functioning: A critical review of the literature. *Developmental and Behavioural Pediatrics, 21*, 58–69.

Palermo, T. M., & Chambers, C. T. (2005). Parent and family factors in pediatric chronic pain and disability: An integrative approach. *Pain, 119*, 1–4.

Palermo, T. M., & Eccleston, C. (2009). Parents of children and adolescents with chronic pain. *Pain, 146*, 15–17.

Palermo, T. M., Putnam, J., Armstrong, G., & Daily, S. (2007). Adolescent autonomy and family functioning are associated with headache-related disability. *The Clinical Journal of Pain, 23*, 458–465.

Palermo, T. M., Wilson, A. C., Peters, M., Lewandowski, A., & Somhegyi, H. (2009). Randomized controlled trial of an internet-delivered family cognitive-behavioral therapy intervention for children and adolescents with chronic pain. *Pain, 146*, 205–213.

Palermo, T. M., Eccleston, C., Lewandowski, A. S., Williams, A. C., & Morley, S. (2010). Randomized controlled trials of psychological therapies for management of chronic pain in children and adolescents: An updated meta-analytic review. *Pain, 148*, 387–397.

Ramchandani, P. G., Stein, A., Hotopf, M., Wiles, N. J., & The ALSPAC Study Team. (2006). Early parental and child predictors of recurrent abdominal pain at school age: Results of a large population-based study. *Journal of the American Academy of Child and Adolescent Psychiatry, 45*, 729–736.

Robins, P. M., Smith, S. M., Glutting, J. J., & Bishop, C. T. (2005). A randomized controlled trial of a cognitive-behavioural family intervention for pediatric recurrent abdominal pain. *Journal of Pediatric Psychology, 30*, 397–408.

Sanders, M. R., Rebgetz, M., Morrison, M., et al. (1989). Cognitive-behavioral treatment of recurrent nonspecific abdominal pain in children: An analysis of generalization, maintenance, and side effects. *Journal of Consulting and Clinical Psychology, 57*, 294–300.

Sanders, M. R., Shepherd, R. W., Cleghorn, G., & Woolford, H. (1994). The treatment of recurrent abdominal pain in children: A controlled comparison of cognitive-behavioral family intervention and standard pediatric care. *Journal of Consulting and Clinical Psychology, 62*, 306–314.

Saunders, K., Von Korff, M., LeResche, L., & Mancl, L. (2007). Relationship of common pain conditions in mother and children. *The Clinical Journal of Pain, 23*, 204–213.

Van Slyke, D. A., & Walker, L. S. (2006). Mothers' responses to children's pain. *The Clinical Journal of Pain, 22*, 387–391.

Walker, L. S., & Zeman, J. L. (1992). Parental response to child illness behaviour. *Journal of Pediatric Psychology, 20*, 329–345.

Walker, L. S., Garber, J., & Van Slyke, D. A. (1995). Do parents excuse the misbehaviour of children with physical or emotional symptoms? An investigation of the pediatric sick role. *Journal of Pediatric Psychology, 20*, 329–345.

Walker, L. S., Williams, S. E., Smith, C. A., Garber, J., Van Slyke, D. A., & Lipani, T. A. (2006). Parent attention versus distraction: Impact on symptom complaints by children with and without chronic functional abdominal pain. *Pain, 122*, 43–52.

Wicksell, R. K., Melin, L., Lekander, M., & Olsson, G. L. (2009). Evaluating the effectiveness of exposure and acceptance strategies to improve functioning and quality of life in longstanding pediatric pain – a randomized controlled trial. *Pain, 141*, 248–257.

Williams, S. E., Blount, R. L., & Walker, L. S. (2011). Children's pain threat appraisal and catastrophizing moderate the impact of parent verbal behaviours on children's symptom complaints. *Journal of Pediatric Psychology, 36*, 55–63.

Impact on Activities of Daily Living

▶ Rating Impairment Due to Pain in a Workers' Compensation System

Impairment

Definition

Impairment is a loss, loss of use, or derangement of any body part, organ system, or organ function as a direct consequence of illness.

Cross-References

▶ Disability in Fibromyalgia Patients
▶ Disability, Upper Extremity
▶ Functioning and Disability Definitions
▶ Pain as a Cause of Psychiatric Illness
▶ Physical Medicine and Rehabilitation, Team-Oriented Approach

Impairment Due to Pain

▶ Rating Impairment Due to Pain in a Workers' Compensation System

Impairment Evaluation

Definition

An evaluation performed by a physician to determine the presence and/or severity of a claimant's impairment. The evaluations are often commissioned by insurance companies or disability agencies.

Cross-References

▶ Impairment, Pain-Related

Impairment Rating

▶ Ethics of Pain-Related Disability Evaluations
▶ Rating Impairment Due to Pain in a Workers' Compensation System

Impairment Rating, Ambiguity

Alan L. Colledge[1], Ron Dressler[1] and Greg Krohm[2]
[1]Utah Labor Commission, International Association of Industrial Accident Boards and Commissions Central Utah Medical Clinic, South Salt Lake City, UT, USA
[2]International Association of Industrial Accident Boards and Commissions (IAIABC), Utah Valley University, USA

Synonyms

Ambiguity; Permanent Partial Disability; Permanent Partial Impairment; Workers' compensation

Definition

Permanent partial impairment – a permanent loss of or abnormality of psychological, physiological, or anatomical structure or function permanent.

Permanent partial disability – a permanently reduced ability to engage in substantial gainful employment by reason of any physical or mental impairment.

Introduction

Since the implementation of workers' compensation, accurately and consistently rating impairment has been a concern for the employee and employer, as well as rating physicians. In an attempt to standardize and classify impairments, the American Medical Association (AMA) publishes the AMA Guides ("Guides"), and recently published its 6th edition of the AMA Guides. Common critiques of the AMA Guides 6th edition are that they are too complex, lacking in evidence-based methods, and rarely yield consistent ratings. Many states mandate use of some edition of the AMA Guides, but many states are reluctant in adopting the current edition due to the increasing difficulty and frustration with their implementation. A clearer, simpler approach is needed.

Characteristics

For almost a century, employers in the United States have been required by jurisdictional State Laws to provide compensation to injured employees. This compensation is intended to cover the employees' loss of wages, earning capacity and medical expenses resulting from an on-the-job injury (Harger 2011). The basic concept of a workers' compensation is not new. The Babylonians had compensation systems 4,000 years ago, and ancient Greeks provided compensation due to military service (Williams 1991). Many examples of more primitive systems can be found throughout Europe and Asia, including pirates in the seventeenth century who had their own code of "compensation." These systems had intuitive means for cash awards for permanent injury, with amputation of extremities being the easiest cases to assess and assign specific benefits. Although a form of "workers' compensation"

existed officially in Europe as early as 1838 in Germany (Harger 2011), it was not until the early twentieth century that "Workers' Compensation" became legislated in the United States (Colledge et al. 2001a), with all 50 states adopting some form of workers' compensation legislation by 1949 (U.S. Chamber of Commerce 1997).

Basic Principles in US Workers' Compensation System
- Exclusive remedy for employee
- Employee gives up right to sue employer, except in extreme circumstances
- Limited liability for employer
- Employer enjoys immunity for civil action from employee
- Employer's liability is predictable
- Employee's right to sue third parties remains

In the United States, each individual state determines and administers its own workers' compensation system. These benefits include: (1) A statutory program. (2) Expeditious resolution of disputed issues. (3) Limited liability without fault: (Workers' compensation is a "no-fault" insurance program). (4) Automatic benefits, which include: (a) medical treatment coverage including: medical care, services, and supplies as necessary, such that the employee incurs no "out of pocket" medical expenses related to the injury. (b) Indemnity payments replacing wages, while the injured employee recovers to medical stability. (c) Death benefits, for the surviving spouse and dependent children in the case of employee death. (d) An impairment settlement giving compensation to an injured worker for permanent physical loss from a work-related injury, according to the state's defined compensation schedule.

Those injuries which are determined to result in a permanent loss qualify for an "impairment rating" *based on a predefined* scoring system, which varies from one jurisdiction to another, upon which the permanent partial and permanent total disability payments are determined for each case. Impairment ratings are a legal process whereby a physician or other qualified individual examines a patient and determines the permanent physical loss resulting from a workplace injury. The impairment numbers or percentages that are assigned are mostly consensus derived and have little medical evidence as to their origin. The rating percentage is given to administratively help understand the extent of an injured worker's residual limitations from injuries. It is a bridge between medical issues and legal determinations of fault, compensability, or benefit entitlements. Impairment ratings are to be performed after a worker attains "maximum medical improvement," a point at which medical recovery from injury has reached a plateau with no foreseeable significant medical improvement in the patient's future. Not all measured impairments are directly related to a specific event and an apportionment to other identified preexisting conditions is at times needed.

Measures of impairment are fundamentally different from measures in disability. Two workers with identical injuries will have the same impairment score, but may be assessed at drastically different levels of disability. For example, a professional piano player and an administrator may each lose their little finger. Their impairment rating could be identical; however, the professional piano player will be left with a significant impact on their earning capacity, while the administrator will probably see a minimal effect in their work performance. Impairment compensation relates solely to the effect of the injury on the body, while disability compensation includes the injury's specific effect on employment, social, and recreational performance. Physicians, whose expertise lies in the body's physical function, are only to assess impairment, while experts in the field of disability use their domain-specific expertise, including the impairment score from the physician, to assess disability. Currently, there are significant disparities of impairment ratings and post disability settlements raising serious questions about social justice with the disability determination processes (Burton 2008). The consistent objectification of a physical loss is impairment. When the impairment rating has been fairly and consistently established then the impact on life of that impairment can be discussed. It is the authors' opinion that only when a comprehensive and consistent impairment rating methodology has been developed to assess physical loss can

models of disability be adequately explored. The impairment rating is the beginning foundation for any disability discussion.

Permanent injury compensation is a large and growing component of workers' compensation system cost. In the United States, it represents about two thirds of all indemnity benefits. Moreover, the frequency and cost of permanent injury is increasing as a share of all workers' compensation claims. Citing data from the National Council on Compensation Insurance, *Durbin and Kish* state, "During the six-year period from 1988 to 1994, average PPD [permanent partial disability] costs increased 25 %, an increase of approximately 4 % per year, while the frequency increased almost 29 % or 4.4 % annually. These increasing average costs and frequencies have resulted in an increase in the permanent partial cost per worker of approximately 52 % (over 7 % annually) over the same six year period" (Durbin and Kish 1998).

Numerous wage-loss studies document that in virtually every state, and every classification of claimants, compensation awards are smaller than the actual or predicted wage losses that workers incur after an injury. Moreover, the correlation between impairment ratings and actual economic losses following injury is poor. For example, the statistical work by Durbin and Kish tries to measure the consistency of initial physician impairment ratings, disability ratings awarded, and final compensation given to injured workers across various US jurisdictions. They concluded that: *"The results show that impairment ratings are only one of a variety of factors that systematically influence the size of a final disability award. Specifically, even for cases with benefits awarded for non-economic loss, in addition to the treating physician's determination of physical impairment, the determination of the degree of permanent disability appears to take into account factors such as age, sex, pre-injury wage, weekly temporary total benefits, and whether an attorney is involved in the case. Moreover, even after these other factors are considered, a less than one-to-one relationship exists between impairment and final disability ratings, which might be expected."* (Durbin and Kish 1998)

Similar results were found by Park and Butler (2000). They found that degrees of permanent impairment assigned by physicians, even under Minnesota's relatively well administered guidelines, were not statistically related to the reduction in pre-injury wages after the injury. Even after adjusting permanent injury awards for age, occupation, and other economic factors, they found that the impairment ratings had a very poor statistical relation to the actual wage loss. This finding is consistent with mainstream belief by medical and non-medical researchers in workers' compensation.

Several factors account for this lack of correlation between permanent disability compensation and wage loss. First, most jurisdictions do not consciously or deliberately set out to provide benefit levels that match the future loss of earnings. The statutes and rules that establish permanent injury compensation do not equate impairment with disability, nor do they explicitly state what the PPD award is intended to compensate. Rather, benefit levels are set in a political arena. The employer's cost of workers' compensation, and how that affects the competitive position of one jurisdiction relative to its competitors, is a far more common metric in the political debate than statistical measures of wage loss from permanent injuries.

Second, differences across workers cause the permanent disability formulae to be relatively more generous to some injured workers and relatively prone to under compensate others. Especially in pure impairment states, the rules for scheduled injury benefits impose uniform awards for each degree of physical loss, e.g., a 5 % loss of the wrist is the same for a concert violinist and cement finisher. These uniform awards do not address the fact that relatively minor impairments often cause severe job limitations for construction workers, but minimal or no job limitations for office workers. In addition, younger workers tend to be under compensated for permanent injury relative to workers nearing retirement.

The focus of this entry is on the third reason for lack of predictability of impairment ratings: the physical measurements of bodily loss are not reliably and consistently measured by doctors.

Doctors do not set benefit levels for specific injuries, but the measurement of physical loss they provide translates directly into dollar benefits. This lack of consistency among rating physicians is widely observed (Colledge et al. 2009).

Much anecdotal evidence and a few formal studies suggest that a major problem with the system revolves around the consistency and defensibility of permanent impairment ratings made by physicians. Complaints in this area are routinely reported in the trade press. From the anecdotal evidence, the problem seems to exist in many jurisdictions.

Inconsistencies have been demonstrated when similar injury cases have been presented to multiple practitioners, and when the same practitioner has evaluated similar cases through time.

For example, in a 1999 study by the State of Texas, a significant number of the cases with multiple impairment ratings for the same injury showed disparities of 5 % or greater: Specifically, the study found that 24 % of injured workers with multiple impairment ratings had no difference between the first and last ratings, 29 % had a difference of 5–10 percentage points, and 14 % had a difference of 10 percentage points or more. One of the authors has documented similar inconsistencies in the State of Utah prior to that state's recodification of impairment rating guides (Colledge et al. 2001b).

Multiple factors contribute to these inconsistencies. A major source of the problem is that state laws and regulations related to impairment and disability evaluation are often poorly crafted and are inconsistent in awarding payment, how benefit levels are set, and the formulas and procedures for guiding physician impairment ratings. Likewise, the clarity of law on evaluating permanent injury helps control disputes (Victor and Boden 1991). Even if laws and regulations are consistent within states, there are often inconsistencies in methods and expectations across states. This is problematic for doctors who have multistate practices or attempt to deal with these issues from a national perspective. Another challenge for doctors is that state laws, regulations, and administrative law judges are sometimes out of step with the best available medical evidence.

Finally, injuries differ greatly in the ease with which they can be rated. A wide variety of common injuries are amenable to reasonably concrete and precise formulae, which convert measured losses to percentages of loss of use of a limb or the whole body. This type of injury is often classified as a scheduled injury. The rating of many scheduled injuries is not particularly difficult or ambiguous for a reasonably trained practitioner. For example, amputations or total loss of use of extremities are not a major source of error or inconsistency. They are objective and relatively easy to measure.

Problems can arise, though, in the evaluation of partial loss to an extremity. At what point does nerve, tendon, joint, or muscle damage render an arm or hand functionally useless? The rules on rating such partial disabilities are variable from jurisdiction to jurisdiction. A much greater problem arises in rating injuries to the spine, which are governed by much more general and subjective guidelines.

Psychological or mental injury and all the related behavioral and motivational consequences are another intractable problem. Psychological injury is real, but it is very difficult to differentiate from preexisting psychological illness not related to the workplace and it is difficult to rate or measure. For these reasons psychological injury is a lightning rod where it is being compensated by workers' compensation. New South Wales is one recent example: "...the rights of the people of NSW have been gravely compromised by the Government's subsequent decision to use the very flawed and unfair Psychiatric Impairment Rating Scale (PIRS) to measure such impairments, and by excluding psychologists from assessment of psychological and psychiatric impairment. Apparently the Government is attempting to save on compensation payouts at the expense of psychologically impaired workers. This is a dangerous election ploy." (Victor and Boden 1991)

Rating pain is the greatest problem of all. Some jurisdictions take the position that pain is not compensable in workers' compensation. Others hold that the pain must be directly linked to the loss of use of body part, e.g., acute pain

limiting the range of motion on a limb. Policy makers in such jurisdictions apparently hold the belief that workers' compensation is a no fault system that tries to get away from the problems of measuring pain and suffering that are so difficult to assign values to in civil tort cases. Other jurisdictions make some allowance for pain, if only in an indirect or implicit way. For example, some jurisdictions assign minimum impairment ratings for surgery even if the outcome was rated as 100 % successful in restoring function. This might implicitly be regarded as a reward to the injured worker for the uncertainty and trauma of the surgical procedure.

In the ideal situation, a skilled practitioner using a well-defined set of criteria might be able to fairly rate the intangibles of an injury, most importantly pain. However, there is ample evidence to suggest that pain is difficult to evaluate and subject to a host of psychosocial overlays that have nothing to do with the injury itself.

Inconsistencies develop not only over the severity of impairment associated with an injury, but also over causation of the impairment. Disagreements regarding causation are particularly likely to occur when workers sustain covert injuries (such as repetitive strain injuries of the upper extremity) rather than overt ones (such as an upper extremity fracture), when an injury is superimposed on a chronic musculoskeletal condition, workers present with symptoms that do not have a clear cause or etiology, e.g., sick building syndrome or stress claims.

The consequences of the inconsistencies described above are substantial. They tend to lead to "dueling doctors" (that is, differences of opinion between the treating physician and an expert hired by the payer of the claim), delays, and high cost litigation. In general, disputes over medical evidence are very expensive. The personnel and the support system needed for hearings tend to be among the most expensive parts of a workers' compensation agency budget. State agencies are continually experimenting with techniques to reduce case backlogs and speed decisions. Delays and other inefficiencies stimulate legislative inquiries, and as constituent frustration levels rise, lawmakers are inclined to "reform" the system. Some states go through a cycle of reform-dissatisfaction-reform.

For these reasons it is vital that permanent injury benefits be managed better in most states. The keys to success:

- Fixed conditions under which permanent injury benefits can be awarded
- A clear trigger for when permanent injury can be evaluated
- Well-defined responsibilities for the physician who is to make the legally required medical determinations
- Uniform procedures for the measurement and evaluation of the parameters of permanent impairment to the body
- An objective way to express the basis for the impairment rating

Whenever any one of these is lacking, doubt and mistrust by workers or their employers enters into the benefit award. Also, opportunists find ways to exploit the ambiguity to maximize gains by "gaming" the system. Gaming the system or adversarial disputes are signs of system failure in workers' compensation.

The State of Wisconsin presents an example of a very smoothly functioning impairment system for scheduled injuries. Wisconsin Administrative Rule 80.20 specifies quite clearly how impairments from scheduled injuries translate into percentages of body part loss. Loss of motion of fingers is a good example: the physician need only measure the loss of flexion and extension at each joint of the injured finger(s) to produce a precise measure of impairment under Wisconsin law. Even more serious/complex injuries to the knee have explicit standards. Finger and knee impairment ratings by treating physicians are almost never challenged by claims adjusters and virtually never litigated.

The role of physician or administrative judgment has been circumscribed by many jurisdictions. They have reacted to ambiguity in various ways: (1) eliminating the compensability of a class of injuries, (2) constraining the range of judgment about certain injuries, or (3) assigning a narrow range of estimates. These understandable responses to uncertainty have the undesirable consequence of introducing inequity. Some

workers are simply not compensated as much as they should be relative to other workers with more tangible and specific injuries. This is a political, not medical, issue.

Clearly, simple and direct rules work. They mete out compensation with efficiency and speed. Of course, some would object that "cookie cutter" justice is unfair. Yet, the very basis of workers' compensation is accepting administrative simplicity in benefit delivery for the individualistic tort based approach to equitable benefit determination.

In a companion entry (Impairment – Ambiguity, Part 2 – the IAIABC/Utah System), the work of the International Association of Industrial Accident Boards and Commissions (IAIABC along with the State of Utah) to improve the process of rating injuries is described.

Summary

In a companion entry (Impairment – Ambiguity, Part 2 – the IAIABC/Utah System), the work of the International Association of Industrial Accident Boards and Commissions (IAIABC along with the State of Utah) to improve the process of rating injuries is described.

References

Burton, J. F., Jr. (2008). The AMA guides and permanent partial disability benefits. *IAIABC Journal, 45,* 13–34.

Colledge, A., Sewell, J., & Hollbrook, B. (2001a). Impairment ratings in Utah: Reduction of variability and litigation within workers' compensation. *Disability Medicine, 1,* 16–27.

Colledge, A., Sewell, J., & Hollbrook, B. (2001). Impairment ratings in Utah, reduction of variability and litigation within workers' compensation. *Disability Medicine,* ABIME 1.

Colledge, A., Hunter, B., Bunkall, L., & Holmes, E. B. (2009). Impairment rating ambiguity in the United States: The Utah impairment guides for calculating workers' compensation impairments. *Journal of Korean Medical Science, 24*(Suppl. 2), S232–S441. The Korean Academy of Medical Sciences.

Durbin, D., & Kish, J. (1998). Factors affecting permanent partial disability ratings in workers' compensation. *The Journal of Risk and Insurance, 65,* 81–89. Mt. Vernon.

Harger, L. (2011). *Workers' compensation, a brief history. State of Florida division of worker' compensation.* Available from: http://www.fldfs.com/WC/history.html

Park, Y.-S., & Butler, R. (2000). Permanent partial disability awards and wage loss. *The Journal of Risk and Insurance, 67,* 331–350.

U.S. Chamber of Commerce. (1997). *Analysis of workers' compensation laws.* Washington, DC.

Victor, R., & Boden, L. (1991). *Model states show lawsuits can be prevented.* Chicago: National Underwriter.

Williams, C. A. (1991). *An international comparison of workers' compensation.* Boston: Kluwer.

Impairment Rating, Ambiguity, IAIABC System

Alan L. Colledge
Utah Labor Commission, International Association of Industrial Accident Boards and Commissions Central Utah Medical Clinic, South Salt Lake City, UT, USA

Synonyms

Ambiguity; IAIABC system; Permanent partial disability; Permanent partial impairment; Workers' compensation

Definition

Permanent partial impairment is defined as a permanent loss of or abnormality of psychological, physiological, or anatomical structure or function permanent.

Permanent partial disability is defined as a permanently reduced ability to engage in substantial gainful employment by reason of any physical or mental impairment.

Introduction

The Problem

As the long list of critical papers in the literature shows, the calculation of impairment is not an objective science and is based largely on consensus rather than on scientific evidence (Colledge and Krohm 2003; Gloss and Wardle 1982; Nitschke et al. 1999; Clark et al. 1988; Holmes). The purpose of worker's compensation is to deliver quick and elementary benefits to an injured worker, and it is not intended to be comprehensive.

A persistent problem in the current systems of impairment ratings is the lack of consistent and reliable metrics of the nature and extent of the loss of use of a body part or bodily system. This leads to variability in the impairment ratings themselves and has many negative, rippling effects, including frustration to patients, physicians, risk managers, state administrators, and payers (Spieler et al. 2000a). This entry will review specific methodology that jurisdictions can consider adopting for workers' compensation impairments.

Characteristics

A persistent problem in the current systems of impairment ratings is the lack of consistent and reliable metrics of the nature and extent of the loss of use of a body part or bodily system. This leads to variability in the impairment ratings themselves and has many negative, rippling effects, including frustration to patients, physicians, risk managers, state administrators, and payers (Spieler et al. 2000a). For example, studies have shown that the same patient can receive impairment ratings from three different physicians and receive a score of no impairment, one of moderate impairment, and one of SEVERE impairment. This variability causes disputes and is preventable. This variability and lack of reliable assessments of impairment often lead to unnecessary stress and cost to the injured worker. Likewise, disputes over impairment ratings are costly to the employer, insurer, and state regulator in the form of litigation. Doctors do not set the benefit levels for the injured workers, but the impairment measurements provided by the doctors of the permanent physical loss do directly impact the worker's benefit and dollar settlement. Studies have also shown that workers' compensation can also have an adverse effect upon patient recovery (Greenough and Fraser 1989; Hunter et al. 1998; Sander and Meyers 1986; Rainville et al. 1997), increase disability (Guest and Drummond 1992; Jamison et al. 1988), and decrease potential return to work (Leavitt 1992; Milhous et al. 1989; Fredrickson et al. 1988).

In the past, many jurisdictions, especially in North America, have deferred to the American Medical Association (AMA) Guides for rating impairment secondary to occupational injuries.

The AMA Guides ("Guides") were first published as a series of articles in the Journal of American Medical Association (JAMA) starting in 1958. The Guides were first published in book form (as a compilation of these articles) in 1971. The original book went through several publications until it was updated and republished in 1984. Since then, it has been reworked and published as the third edition (1988), third revised edition (1990), fourth edition (1993), fifth edition (2001), and now in 2008, the sixth edition. It is interesting to note that the first edition of the AMA Guides was published in 1971, was of 164 pages, and sold for $5.00. The sixth edition, published in 2008, is of 634 pages and sells for $189.00 to nonmembers. The goal of each edition was TO RESPOND to a public need for a standardized, objective approach to evaluating medical impairment. The stated reason for each revision is "to update the diagnostic criteria and evaluation processes used in impairment assessment, incorporating available scientific evidence and prevailing medical opinion" (Berkowitz and Burton 1987). Each of the six editions makes mention of the use of the latest science and, where lacking, consensus statements from individuals in each of the specialties are used.

The Guides have been helpful in taking steps to standardize and solidify impairment ratings across the country; however, consistency continues to BE an issue. Texas reported significant variation using the Guides for workers receiving more than one rating of impairment for the same condition, showing 25 % of these ratings

could differ BY more than 10 %, with 65 % of those who disputed their ratings, doing so on the methodology of how the rating was calculated ("Return to work" 1996). Some states have noted THAT disparities reflect a lack of consensus on the criteria physicians use, the objectivity of the criteria, and the understanding of how THE CRITERIA ARE to be applied. They also suggest that calculated ratings often take into account factors other than just clinical findings. These factors may include the competency and experience of the examiner, the patient's personality, and even financial motives.

With time the Guides have also become more and more complex, and increasingly difficult to interpret and apply in practice. There have been calls for major revisions of the Guides (Spieler et al. 2000b). The fifth edition of the Guides included impairments for pain and psychological effects of injuries. These are extremely hard to assess with any sort of consistency, and issues continually arise in relation to relative pain and the validity of qualitative claims. Currently, 38 states use some version of the AMA Guides, and it is estimated that over half of the states currently use the fifth edition. The sixth edition, since its release, has faced unprecedented resistance to its acceptance as a replacement for the fifth edition in workers' compensation. It is estimated that less than 15 states will adopt the new sixth edition of the AMA Guides (Brigham). New Hampshire, Vermont, and Rhode Island commissioned the National Council on Workers' Compensation Insurance (NCCI) to review the sixth edition for actuarial impact on their workers' compensation costs. These studies concluded: "NCCI is unable to provide an estimate of the potential impact on indemnity costs related to the adoption of the sixth edition of the AMA Guides. The ultimate impact from a change in average disability ratings under the sixth edition will flow through loss experience and will be reflected in future loss cost filings" (Hall 2008a, b, c).

In March 2008, Kentucky, one of the 16 states that use the "latest edition" of the Guides, passed a statute that delayed use of the sixth edition until at least July of 2009, pending a report from the

Office of Worker Claims. In the same month, Vermont's Department of Labor followed suit and indefinitely delayed implementation of the sixth edition.

In April 2008, Iowa's Labor Commissioner opted to maintain the applicability of the fifth edition for assessing impairment in that state's workers' compensation system until a task force had thoroughly reviewed the sixth edition and made recommendations to the state for adoption. To do this review Iowa's Labor Commission appointed an eight-member task force committee of various stakeholders to review testimony and make recommendations for adoption. Now completed, this is the most extensive non-biased review of the sixth edition to date. In the end, the committee recommended against adopting the AMA sixth edition (http://www.iowaworkforce.org/wc/amataskforce/2008amaguidesprocessreport.pdf). Rhode Island is delaying implementation of the sixth edition until it is fully reviewed by its Workers' Compensation Medical Advisory Board. In June 2008, New Hampshire's governor signed a bill replacing the language "most recent" with "fifth edition" in its compensation statute mandating use of the AMA guides. Other states have also acted through informal processes to delay or reject the sixth edition (Causey et al. 2008). At this time, there is a significant demand for a specific workers' compensation guide which is both current and straightforward, enabling physicians to give consistent, reliable impairment ratings.

IAIABC/Utah Impairment Guides

Continuing since Fall of 2001, the International Association of Industrial Accident Boards and Commissions (IAIABC) along with the state of Utah began to work together in developing a more specific workers' compensation impairment guide for the rating of permanent occupational injuries.

These supplemental guides contain the following:
- An introduction to practitioners on the nature of occupational impairment rating
- Definitions of key terms and concepts, such as maximum medical improvement

- Discussion of general issues, especially the measurement of pain
- Guides to rating surgical and nonsurgical injuries of the back
- Guides for rating the upper and lower extremities
- Guides for rating psychological loss
- Standardized reporting work sheets

An introduction to impairment rating is critical because many physicians who do ratings are unsure of their role, or have misunderstandings about the rating procedure in a given jurisdiction. This is particularly true of those who only do ratings on an occasional or infrequent basis, or who are confronted with a rating scheme from a jurisdiction outside their normal practice.

Methodology

The Impairment Rating Committee embraced some guiding principles in their development of the supplemental guides:
- They should be based on the best available empirical evidence on the reliability of tests, measurements, and correlations of measurements to biomechanical limits on the normal use and functioning of the body.
- They should be practical in their administration by physicians.
- They should be clearly explainable to practitioners.

These principles are challenging to implement. They involve trade-offs. For example, sophisticated measures that might be arguably more precise might be impractical and costly for a nonspecialist physician.

Also, the goals of objectivity and practicality often fly in the face of individual equity. The participants in the preparation of the draft IAIABC guides have elected to err on the side of simple consistent rules based on objective medical evidence. Pain is particularly a controversial issue. It is a real consequence of injury and surely affects postinjury reintegration into work and nonwork activities of daily living. Having said this, the committee could not find many reliable and practical tools for rating most injuries for residual pain. Some psychological tools may be able to consistently differentiate between simulated pain and "really experienced" pain. Further, such tools may be able to gauge the approximate degree of experienced pain. However, the tools available to the drafters appeared to fall short of being easily learned and applied by treating physicians who do not specialize in occupational injury.

Results

To date, Supplemental Guides have been produced. They provide specific guidance on the more common impairment ratings PERFORMED IN RELATION TO workers' compensation CLAIMS, muscle skeletal issues. THEY ALSO address issues not fully or clearly addressed in the AMA Guides.

The introduction of the Guides provides general background on the workers' compensation system and the role of the physician in awarding benefits for permanent work injuries. Much of this educational material is common knowledge for jurisdictional administrators, workers' compensation managers, and claims adjustors. Yet, it is surprising how few physicians who do not practice occupational medicine understand these fundamental concepts. The physician-rater must have an appreciation of their role in the benefit process for the system to work in accordance with the law.

This introduction also reviews some of the generic concepts of impairment rating, such as maximum medical improvement and apportionment of injury. The highly controversial subject of pain is also addressed.

These guides are now entirely independent of the AMA guides for the most common CONDITIONS THAT require impairment rating evaluations under workers' compensation. THESE include a complete impairment rating methodology for spine, upper extremity, lower extremity, peripheral neurologic loss, and psychological loss.

The current workers' compensation impairment guides can be reviewed at http://www.laborcommission.utah.gov/indacc/WC_Forms___Publications/2006UtahImpairmentGuides_may16,2006_.pdf.

Users to Date

To date, the best practical application of the Supplemental Guides is in the State of Utah. In many respects, the guides resemble administrative rules adopted by that state for the rating of permanent injuries.

The use of the Utah guides has had a dramatic reduction on litigation over impairment disputes. Less than one percent of claims with permanent disability are litigated since the revised impairment guides were adopted in 1997. This produced a dramatic cost savings to the Utah Labor Commission, since the direct cost to the agency is estimated to be $5,200 per case litigated. The chief administrator of the Utah Commission reports that the 1997 guides have been well received by attorneys, insurers, and worker representatives.

It is anticipated that the IAIABC/Utah impairment guides will eventually displace the need for the AMA guides with specific attention addressed to those concerns needed by workers' compensation jurisdictions.

Summary

Ambiguity and uncertainty over rating permanent injury produce higher administrative costs for workers' compensation systems, because (1) multiple examinations caused by disputed ratings are expensive, and (2) litigation over large discrepancies in ratings causes delay in benefits and further administrative cost.

One cure for the above problems of compensating for permanent injury is to establish clear and simple rules for awarding benefits. The downside to simplicity is a loss of individual equity and fairness among injured workers, that is, some are relatively better off, and some are disadvantaged by rules that do not take into consideration how their physical loss has affected their earnings potential.

This entry has argued for rating systems that stress objective medical evidence and consistent guidelines for injured workers. This methodology has been completed and validated for the most common injury types seen in injured workers, avoiding unnecessary disputes, challenges, and litigation, which many jurisdictions want to avoid. In effect, it is apparent that the IAIABC/Utah impairment guides subordinate the quest for better individual equity on claims to overall system savings.

We reemphasize that impairment rating does not predict disability or loss of earning capacity. Further, we take no position on whether a jurisdiction should use pure impairment ratings, adjustments for age/occupation, wage loss, or some hybrid system taking all of these factors into consideration. The purposes of this entry are to note that many impairment ratings using AMA guides are inconsistent and indefensible, and to describe steps taken by the IAIABC/Utah state government to develop supplemental workers' compensation impairment guides.

References

Berkowitz, M., & Burton, J. F., Jr. (1987). *Permanent disability benefits in workers' compensation.* Kalamazoo, MI: W.E. Upjohn Institute for Employment Research.

Brigham, C. Impairment resource discussions. Available from: http://www.impairment.com/use_of_AMA_guides_maps.htm. See http://www.ncbi.nlm.nih.gov/pmc/articles/PMC2690067. Accessed 6 June 2008.

Causey, J., McFarren, T., & Nimlos, J. (2008). The AMA 6th: Accelerating the demise of permanent disability in workers' compensation. *IAIABC Journal, 45*(2), 49–64.

Clark, W. L., Haldeman, S., Johnson, P., Morris, J., Schulenberger, C., Trauner, D., & White, A. (1988). Back impairment and disability determination, another attempt at objective, reliable rating. *Spine, 13*, 332–341.

Colledge, A., & Krohm, G. (2003). Rating of impairments from injury impairment. *IAIABC Journal, 4*, 32–48.

Fredrickson, B. E., Trief, P. M., Van Beveren, P., Yuan, H. A., & Baum, G. (1988). Rehabilitation of the patients with chronic back pain: A search for outcome predictors. *Spine, 13*, 351–353.

Gloss, D. S., & Wardle, M. G. (1982). Reliability and validity of American Medical Association's guide to ratings of permanent impairment. *Journal of the American Medical Association, 248*, 2292–2296.

Greenough, C. G., & Fraser, R. D. (1989). The effect of compensation of recovery from low back injury. *Spine, 14*, 947–955.

Guest, G. H., & Drummond, P. D. (1992). Effect of compensation on emotional state and disability in chronic back pain. *Pain, 48*, 125–130.

Hall, L. B. (2008a). *New Hampshire – Analysis of AMA guides* (6th ed.). Plainfield, VT: National Council on Compensation Insurance.

Hall, L. B. (2008b). *Vermont – Analysis of AMA guides* (6th ed.). Plainfield, VT: National Council on Compensation Insurance.

Hall, L. B. (2008c). *Rhode Island – Analysis of AMA guides* (6th ed.). Plainfield, VT: National Council on Compensation Insurance.

Holmes, E. B. Impairment rating and disability determination. Available from: http://www.emedicine.com/pmr/topic170.htm. Accessed 14 May 2008.

Hunter, S. J., Shaha, S., Flint, D. F., & Tracy, D. M. (1998). Predicting return to work, a long term follow-up study of railroad workers after low back injuries. *Spine, 23*, 2319–2328.

Jamison, R. N., Matt, D. A., & Parris, W. C. V. (1988). Effects of time, limited vs. unlimited compensation on pain behavior and treatment outcome in low back pain patients. *Journal of Psychosomatic Research, 32*, 277–283.

Leavitt, F. (1992). The physical exertion factor in compensable work injuries; a hidden flaw in previous research. *Spine, 17*, 307–310.

Milhous, R. L., Haugh, L. D., Frymoyer, J. W., et al. (1989). Determinants of vocational disability in patients with low back pain: A search for outcome predictors. *Archives of Physical Medicine and Rehabilitation, 70*, 589–593.

Nitschke, J., Nattrass, C., Disler, P., Chou, M., & Ooi, K. (1999). Reliability of the American Medical Association guides' model for measuring spinal range of motion: Its implication for whole-person impairment rating. *Spine, 24*, 262–268.

Rainville, J., Sobel, J. B., Hartigan, C., & Wright, A. (1997). The effect of compensation involvement on the reporting of pain and disability by patients referred from rehabilitation of chronic low back pain. *Spine, 22*, 2016–2024.

Return to work patterns for permanently impaired workers. (1996). *Texas Monitor, 1*, 7–8.

Sander, R. A., & Meyers, J. E. (1986). The relationship of disability to compensation status in railroad workers. *Spine, 11*, 141–143.

Spieler, E. A., Barth, P. S., Burton, J. F., Jr., Himmelstien, J., & Rudolph, L. (2000a). Recommendations to guide revision of the Guides to the Evaluation of Permanent Impairment. *JAMA: The Journal of the American Medical Association, 283*, 519–523.

Spieler, E. A., Barth, P. S., Burton, J. F., Jr., Himmelstien, J., & Rudolph, L. (2000b). Recommendations to guide revision of the Guides to the Evaluation of Permanent Impairment. *JAMA: The Journal of the American Medical Association, 283*, 519–523.

Impairment, Function Loss

▶ Disability, Upper Extremity

Impairment, Pain-Related

James P. Robinson
Department of Anesthesiology & Pain Medicine,
University of Washington, Seattle, WA, USA

Synonyms

Activity limitation; Disability; Organ or body part dysfunction; Pain-related impairment

Definition

"Impairment" does not have a unique definition. The American Medical Association (AMA) defines it as "a significant deviation, loss, loss of use of any body structure or body function in an individual with a health condition, disorder, or disease" (Rondinelli 2008, p. 5). The World Health Organization gives the following definition: "Impairments are problems in body function or structure as a significant deviation or loss" (p. 226) (World Health Organization (2001). The United States Social Security Administration defines impairments as "anatomical, physiological, or psychological abnormalities that can be shown by medically acceptable clinical and laboratory diagnostic techniques" (p. 3) (US Government Printing Office 1994). The present entry focuses on impairment as conceptualized by the AMA in *Guides to the Evaluation of Permanent Impairment*, Sixth Edition (Rondinelli 2008).

Characteristics

Although the above definitions of impairment differ somewhat, they all emphasize that impairments are biomedical abnormalities that can be analyzed at the level of organs or body parts. In fact, a critical distinction between impairment and disability is that they address limitations at different levels of analysis – impairment refers to a limitation in the function or structure of an organ or body part, whereas disability refers to a limitation in the behavior of a person. This distinction is reflected in the syntax used to describe impairments and disabilities. For example, one would say "Ms. Smith's right leg is weak because of her polio" to describe her impairment, and "Ms. Smith is unable to walk up stairs" to describe her consequent disability.

Since impairment refers to the function or structure of organs or body parts rather than to the behavior of people, observers often assume that physicians can perform impairment evaluations objectively and relatively independently of the persons to whom the organs/body parts belong. For example, a physician could assess renal impairment on the basis of creatinine clearance or cardiac impairment on the basis of ejection fraction. Thus, ▶ impairment evaluations are thought to bypass the complications that arise when examiners try to understand claimants at the "whole person" level, or to perform analyses that rely on claimants' self-reports and voluntary behavior.

The Significance of Impairment Evaluations

Agencies that administer disability programs need a method to determine whether people who seek benefits are actually disabled – i.e., whether they have medical conditions that significantly limit their ability to carry out certain activities. Among individuals who allege disability, ones who are truly disabled need to be distinguished from those who cannot perform activities because of nonmedical circumstances (e.g., inability to find work due to an economic depression) and from those who will not perform activities, even though they are capable of doing so. Physicians are often enlisted to make the above distinctions by objectively evaluating medical factors that may underlie a claimant's allegations of incapacitation. For at least the past 45 years, the AMA has persuasively argued that physicians can evaluate these medical factors by performing impairment evaluations. In fact, most physicians and disability adjudicators think of impairment evaluations as being synonymous with evaluations of medical factors that contribute to disability.

In practice, agencies that administer work disability programs often combine impairment data with nonmedical data (e.g., educational background and prior work history) to determine whether an individual is actually disabled from work. The logic of this approach can be summarized as follows:

1. Work disability results from a combination of medical and nonmedical factors.
2. Impairment evaluations measure the medical component of disability – i.e., the severity of medical factors that contribute to work disability.
3. Work disability awards, therefore, should be based on impairment evaluations supplemented by nonmedical data.

The Problem of Subjective Factors

However elegant the above syllogism appears to be, it leaves out an important category of information – the self-reports that disability applicants provide about their experiences as they try to engage in activities (Robinson et al. 2004, 2008). These reports provide a first-person perspective that, in principle, might be very important to the assessment of an applicant's disability.

In particular, applicants can provide subjective data regarding a variety of aversive experiences – such as anxiety, pain, fatigue, or subjective weakness – that make it difficult for them to engage in normal activities. Pain is the most common of the aversive experiences reported by applicants for work disability benefits.

The potential importance of subjective experiences is highlighted by Osterweis et al. in an Institute of Medicine monograph on the role of pain in disability awards by the Social Security Administration.

The notion that all impairments should be verifiable by objective evidence is administratively necessary for an entitlement program. Yet this notion is fundamentally at odds with a realistic understanding of how disease and injury operate to incapacitate people. Except for a very few conditions, such as the loss of a limb, blindness, deafness, paralysis, or coma, most diseases and injuries do not prevent people from

working by mechanical failure. Rather, people are incapacitated by a variety of unbearable sensations when they try to work (p. 28) (Osterweis et al. 1987).

Pain and other unbearable sensations cannot be incorporated into impairment evaluations, as conceptualized in the AMA system, for two reasons. First, since pain is inherently subjective, the methods used to assess it violate the tenet that impairment evaluations should be based on objective indices of the function of organs or body parts. Secondly, pain and its effects need to be analyzed at the level of the whole person, rather than at the level of a specific organ or body part. People with chronic pain typically attribute their pain and activity limitations to dysfunction of an organ or body part. However, these subjective reports are difficult to assess precisely because examination of the involved organ or body part often does not identify abnormalities that make the pain reports inevitable. It often appears to an observer that the affected organ or body part is capable of functioning, but that the claimant does not use it normally because of pain. The observer must consider the person as a whole in order to make sense of the situation. Thus, the assessment of incapacitation secondary to pain violates the tenet that impairment evaluations should assess the functioning of organs or body parts, rather than the functioning of an individual as a whole.

Attempts to Solve the Conundrum

At one level, the question of whether pain-related activity limitations can be construed as impairments is a simple one of semantics. In principle, it can be resolved in at least two ways. One is to expand the definition of "impairment," so that the concept includes limitations secondary to pain. A second strategy is to retain the interpretation of impairment as a problem at the level of organs or body parts, but to stipulate that in the medical assessment of a disability applicant, a physician should evaluate both impairment and subjective factors – such as pain – that contribute to the applicant's activity limitations.

Regardless of the semantic issue of whether or not pain is construed as a type of impairment, disability systems face a dilemma regarding the weight they place on pain and other subjective factors (such as fatigue or perceived weakness) during disability determinations. If a system equates impairment with the medical component of disability, it will ignore the subjective factors that often play a dominant role in preventing people from working. Such a system would be objective, but might also be somewhat irrelevant, since it would systematically exclude important factors that bear on the ability of an individual to work. Conversely, a system that gives undue weight to the subjective reports of claimants may inappropriately reward ▶ symptom magnifiers and ▶ malingerers. The challenge is to develop a disability assessment system that strikes a reasonable balance between the weight given to ▶ objective factors (such as amputations) and ▶ subjective factors (especially pain). No disability system has been able to resolve this challenge in a completely satisfactory way.

References

Osterweis, M., Kleinman, A., & Mechanic, D. (Eds.). (1987). *Pain and disability*. Washington, D.C: National Academy Press.

Robinson, J. P., Turk, D. C., & Loeser, J. D. (2004). Pain, impairment, and disability in the AMA Guides. *The Journal of Law, Medicine & Ethics, 32*(2), 315–326.

Robinson, J. P., Rathmell, J. P., & Gatchel, R. J. (2008). Pain-related impairment. In R. D. Rondinelli (Ed.), *Guides to the evaluation of permanent impairment* (6th ed.). Chicago: American Medical Association.

Rondinelli, R. D. (Ed.). (2008). *Guides to the evaluation of permanent impairment* (6th ed.). Chicago: American Medical Association.

U.S. Government Printing Office. (1994). *Disability evaluation under social security* (SSA Publication No. 64–039). Washington, DC: U.S. Government Printing Office.

World Health Organization. (2001). *International classification of functioning, disability and health.* World Health Organization: Geneva.

Impartial Medical Evaluation

▶ Independent Medical Examinations

In Vitro

Definition

Latin for "in glass"; in cells or tissues obtained from animals and maintained outside the living organism by artificial means; opposite to in vivo.

Cross-References

▶ Descending Circuitry, Opioids
▶ NSAIDs, COX-Independent Actions

In Vivo

Definition

Within a living organism.

Cross-References

▶ Animal Models of Inflammatory Bowel Disease
▶ Descending Circuitry, Opioids

Inactive Agent

▶ Placebo

Inappropriate Symptoms and Signs

▶ Nonorganic Symptoms and Signs

Inbred Strains

Definition

The result of repeated brother × sister matings over at least 20 generations. This breeding scheme eliminates heterozygosity, rendering each individual genetically identical (isogenic) to all others of the strain, barring rare de novo mutations. A large number of inbred mouse strains have been developed, all ancestors of European and East Asian "fancy mice" originally bred at the turn of the twentieth century. Due to their isogenicity, within-strain variation must be due to environmental factors, whereas between-strain variation is likely to be due to genetic factors.

Cross-References

▶ Heterozygosity
▶ Opioid Analgesia, Strain Differences

INCB

▶ International Narcotics Control Board

Incentive

▶ Motivational Aspects of Pain

Incidence

Definition

Incidence is the number of new events occurring in a defined population during a specified time period.

Cross-References

▶ Low Back Pain, Epidemiology
▶ Migraine Epidemiology
▶ Prevalence of Chronic Pain Disorders in Children

Incidence of Low Back Pain

▶ Low Back Pain, Epidemiology

Incident Pain

Synonyms

Pain flares

Definition

Incident pain is a subtype of breakthrough pain that occurs as the result of normal voluntary or involuntary movement but does not typically occur at rest. Incident pain often occurs predictably in response to identified triggers. Consequently, it may be possible to provide prophylactic pain control for these episodes.

Cross-References

▶ Breakthrough Pain
▶ Cancer Pain, Animal Models
▶ Cancer Pain, Goals of a Comprehensive Assessment
▶ Evoked and Movement-Related Neuropathic Pain
▶ Rest and Movement Pain

Incision Model for Postoperative Pain

▶ Nick Model of Cutaneous Pain and Hyperalgesia

Inclusion Body Myositis

Definition

Slowly progressive, immunogenic myositis especially in the elderly which involves proximal and distal muscles from the beginning. It is often asymmetric. Muscle wasting is nearly always pronounced in selected muscles. Muscle pain is absent. The diagnosis can be confirmed by muscle biopsy including electron microscopy.

Cross-References

▶ Myositis

Incomplete Spinal Cord Injury

Definition

Spinal cord injury where there is some preservation of distal function. The American Spinal Injury Association (ASIA) Classification defines a complete injury as ASIA impairment scale (AIS) A as a person with no motor or sensory function preserved in the sacral segments S4-S5. There are four categories of incomplete injuries AIS B-D. Please refer to http://www.asia-spinalinjury.org for more details.

Cross-References

▶ Spinal Cord Injury Pain

Inconsistent Symptoms and Signs

▶ Nonorganic Symptoms and Signs

Increased Extracellular Levels

▶ Opioid-Induced Release of CCK

Indemnity Costs

Definition

Indemnity costs are expenditures paid to an injured worker to cover the harm, loss, or damage associated with a work-related injury.

Cross-References

▶ Pain in the Workplace, Risk Factors for Chronicity, and Workplace Factors

Independent Medical Evaluation

▶ Independent Medical Examinations

Independent Medical Examinations

Christopher R. Brigham
Brigham and Associates, Inc., Portland, USA

Synonyms

Agreed medical examination; Binding medical examination; IME; Impartial medical evaluation; Independent medical evaluation; Neutral medical examination

Definition

Independent medical evaluations (IMEs) are examinations performed by a physician, or another healthcare provider, who is not involved in the person's care, to answer medical and case questions associated with a workers' compensation, personal injury, or disability claim (Brigham and Engelberg 1994).

Characteristics

Independent medical evaluations (IMEs) are performed to provide information for case management, and for evidence in hearings and other legal proceedings (Brigham et al. 1996). IMEs are a component of all workers' compensation statutes, although the specifics vary by state (Tompkins 1992). They are also widely used in automobile casualty, personal injury, and disability cases. IMEs are also an integral component of case management and are used universally by insurers, attorneys (defense and plaintiff), and others involved in these cases. They are performed at several stages during the cycle of injury/illness, treatment, rehabilitation, and return to work (Demeter et al. 1996). These evaluations are often performed to clarify physical, behavioral, psychological, psychosocial, vocational, and legal issues (Aronoff 1991; Turk et al. 1988).

Assessments may be requested because of a lack of medical information, or conflict on specific matters, such as the cause of a problem, the severity of the problem, the prognosis (the outlook for the future), the impairment (the loss of, loss of use, or derangement resulting from an injury), the functional abilities of the examinee, appropriateness of care, and recommendations. Due to the nature of these cases, most IMEs are done for pain-associated complaints.

These evaluations vary in complexity, dependent upon the specifics of the case and the issues involved. The process may require hours of physician time, dependent both on the case and on the effectiveness of the physician. Most evaluations are based on a single visit with the examining physician. Occasionally an evaluation may involve multiple physicians.

The examinee is not necessarily a willing participant, and may present with more dysfunctional behavior than is usually encountered in a typical treatment setting. The assessments, therefore, are more challenging and require greater patience. Depending on the complexity of the case and the tools used in the process, a comprehensive IME may require up to several hours of physician time, with approximately 2 h of that time spent with the examinee. The use of other staff, structured assessment protocols, and software will reduce the time needed to conduct a thorough evaluation.

The performance of these assessments requires specific skills in addition to clinical acumen. The physician must not only have a strong clinical background, but also have an appreciation of the biomedical, emotional, and vocational aspects of injury and illness. Credentials should be solid; this is not an arena for an inexperienced physician or someone who is unable to support conclusions if challenged. The specialty of the physician performing the evaluation is often dependent on the type of problem the examinee is experiencing, and/or the specialty of the treating physician. Specialties commonly involved in performing these evaluations are physical medicine and rehabilitation, orthopedic surgery, occupational medicine, neurology, neurosurgery, pain management, psychiatry, and psychology.

The examiner doing an independent medical evaluation needs to keep in mind the three elements of an IME: independent, medical, and examination. Evaluations are to be independent, impartial, and without bias toward the client. All physicians have biases based upon their experiences and perspective. The evaluator should strive not to allow these biases to distort the evaluation of a case. IMEs are medical evaluations. They involve the essential elements of a medical assessment including history, examination, and review of applicable diagnostic studies.

Independent medical evaluations are not done for the purpose of medical treatment. The sole purpose of an IME is to provide information to the client who requested the evaluation. There is no treating physician/patient relationship. The information presented by the individual being examined (i.e., the examinee) is presented in the report and may be conveyed to the client, as the examinee has waived the traditional physician/patient confidentiality relationship. The person being evaluated is known as the "examinee" and the person requesting the evaluation is known as the "client."

The product of an independent medical evaluation is the examiner's report to the client. The client requesting the IME relies upon the report and what is contained in it. The report should be considered a legal document, and therefore, should be a thorough, clear, easy-to-read report and address the issues requested by the client. The report reflects the quality of the evaluation.

Issues Addressed in IMEs

Issues associated with an IME often differ from clinical consultations in role and focus. The most common issues are diagnosis, causation, prognosis, maximal medical improvement, permanent impairment, functional ability (work capacity), appropriateness of care, and recommendations. Typically, the client requesting the examination will pose specific questions.

Pain, pathology, impairment, residual functional capacity, and disability are separate concepts. Pain, especially chronic pain, may not be associated with significant physical pathology (Loeser 1991). Pathology may be present without symptoms or dysfunction. Impairment is a measurable decrement in some physiologic function, whereas disability considers not only the physical or mental impairment but also the social, psychological, or vocational factors associated with a person's ability to work (Brena and Spektar 1993). Functional limitations are manifestations of impairment (Vasudevan 1992). Chronic pain, the most common problem seen in the performance of IMEs, may not be a symptom of an underlying acute somatic injury, but a multidimensional biopsychosocial phenomenon. A multiaxial assessment of biomedical, psychosocial, and behavioral-functional issues is necessary (Turk 1991; Turk et al. 1988; Turk and Melzack 1992).

Diagnoses, that is, clinical impressions, include the primary illness or injury and other significant conditions. They are commonly presented as a problem list of clinical diagnoses in relative order of significance. The significance of medical problems is discussed, such as the relationship between the extents of symptoms (subjective complaints) and signs (objective findings).

Causation is a critical issue in work-related and liability cases. A work-related problem is defined as one that "arose out of and during the course of employment" (Larson 1982). The physician must establish causation to a reasonable degree of medical probability, which implies that it is more probable than not (i.e., there is more than a 50 % probability) that a certain condition is secondary to an injury. Possibility implies less than 50 % likelihood. A medical condition may be the result of one or multiple factors. Apportionment refers to the extent to which a condition is related to each of the multiple factors. An aggravation implies a long-standing effect due to an event, resulting in a worsening, hastening, or deterioration of the condition. An exacerbation is a temporary increase in symptoms from the condition. Thus, an aggravation has an ongoing substantial impact on the physical condition, whereas an exacerbation results in a flare of symptoms, without a significant change in the underlying pathology.

The client may request that the physician identify the prognosis for a problem, that is, the extent and predicted time of recovery.

Maximal medical improvement is a phrase used to indicate when further recovery and restoration of function can no longer be anticipated to a reasonable degree of medical probability.

The client may request an impairment evaluation as part of the assessment. These are usually performed using the American Medical Association's Guides to the Evaluation of Permanent Impairment (American Medical Association 2000). The Guides distinguish between an impairment evaluation and a disability evaluation. According to the Guides, impairment is "the loss of, the loss of use of, or a derangement of any body part, system, or function" and disability is "the limiting loss of the capacity to meet personal, social, or occupational demands, or to meet statutory or regulatory requirements." Impairment is considered permanent when it has reached maximum medical improvement (MMI), meaning it is well stabilized and unlikely to change substantially in the next year with or without medical treatment.

Subjective complaints, such as fatigue or pain, when not accompanied by demonstrable organ dysfunction, clinical signs, or other independent, measurable abnormalities are generally not ratable with the AMA Guides. However, a significant change with the sixth edition is that pain may be ratable if there is an underlying organic cause, providing up to 3 % whole person permanent impairment in certain circumstances.

Functional ability, specifically work capacity, is a primary issue in many IMEs. Judgments of work capability are formed by the examinee's report, clinical condition, and measurements of functional performance. Disability durations that provide the length of disability typically associated with a diagnosis or disorder may also provide very valuable insight into how much and how long a problem may have an impact on functional ability (Reed 2001).

Clients may request a review of the appropriateness of care or recommendations. This review can include issues of unnecessary and/or omitted diagnostic evaluation, and inappropriate, excessive, and/or omitted treatment.

IME Process

The IME process consists of three phases: pre-evaluation, evaluation, and post-evaluation. The use of structured questionnaires, inventories, and software with report templates facilitates this procedure.

A request for an IME specifies issues to be addressed and background, demographic, clinical, and claims information. The examinee may be asked to complete a questionnaire and inventories, either as forms or computer-based, before the evaluation. Pain and functional inventories are particularly helpful in identifying behavioral and psychological components related to an illness or injury.

At the beginning of the visit, the physician explains the nature of the evaluation, that an independent evaluation will be conducted, and no treatment will be provided. There is no patient/physician relationship, and a report will be sent to the requesting client. The history usually commences with a detailed review of the injury or illness, including the reported mechanism, symptoms at the time, and events immediately thereafter. The history also includes relevant preexisting conditions and prior injuries. The chronology of events from the time of injury through to the present is determined. The current status is explored in detail, with attention directed to the examinee's primary concern. The person's perceived functional status is documented to clarify both

capacity and behavioral issues. The examinee's perceptions are useful in understanding the personal experience with the medical problem. The history includes a thorough occupational history, particularly if the individual is disabled, and psychosocial history. The history concludes with a traditional past medical history, which notes medical and surgical procedures, medications and allergies, review of systems, and family history.

The physical examination includes a behavioral assessment and detailed examination of the involved area(s). Regional examinations should ensure reliability and ideally reproducibility of positive, negative, and nonphysiologic findings (Waddell et al. 1992; Harris and Brigham 1990). The assessment should also look for nonphysiologic findings of symptom magnification.

Psychological and behavioral factors must be considered in a disability assessment (Osterweis et al. 1987). A number of self-report, interview, and behavioral inventories have been developed to assess pain and disability (Tait 1993). Numerous self-report instruments have been developed, using both comprehensive and brief formats. It is useful to include these in an IME to assess behavioral, personality, psychological, and psychosocial issues. Radiographic films brought by the examinee (or supplied by the client) are reviewed at the visit. The findings are compared with those of the reviewing radiologist.

Most clients prefer that the examiner does not discuss their findings directly with the examinee, nor send the examinee a copy of the report. Reports should be organized and detailed, and present the information obtained during the evaluation. The available medical records, the patient's behavior and quality as a historian, the individuals accompanying the examinee, and the context of the assessment should be described. The history, physical examination findings, results of pain inventories, and interpretation of radiographic studies should follow. Each of the issues and questions asked by the client should be answered in the discussion.

References

American Medical Association. (2000). *Guides to the evaluation of permanent impairment* (5th ed.). Chicago: American Medical Association.

Aronoff, G. M. (1991). Chronic pain and the disability epidemic. *The Clinical Journal of Pain, 7,* 330.

Brena, S. F., & Spektar, S. (1993). Systematic assessment of impairment and residual functional capacity in pain-impaired patients. *The Journal of Back and Musculoskeletal Rehabilitation, 3,* 6.

Brigham, C. R., & Engelberg, A. L. (1994). The disability/impairment evaluation. In R. J. McCunney (Ed.), *A practical approach to occupational and environmental medicine* (2nd ed., p. 85). Boston: Little, Brown.

Brigham, C. R., Mangraviti, J. J., & Babitsky, S. (1996). *The independent medical evaluation report: A step-by-step guide with models.* Falmouth: SEAK.

Demeter, S. L., Andersson, G. B. J., Smith, G. M., et al. (1996). *Disability evaluation.* St. Louis: Mosby.

Harris, J. S., & Brigham, C. R. (1990). Low back pain: Impact, causes, work relatedness, diagnosis and therapy. *OEM Reports, 4,* 84.

Larson, A. (1982). *The Law of Workmen's compensation.* New York: Matthew Bender.

Loeser, J. D. (1991). What is chronic pain? *Theoretical Medicine, 12,* 213.

Osterweis, M., Kleinman, A., Mechanic, D., & Institute of Medicine Committee on Pain, Disability and Chronic Illness Behavior. (Eds.). (1987). *Pain and disability: Clinical, behavioral, and public policy perspectives.* Washington, DC: National Academy Press.

Reed, P. (2001). *The medical disability advisor.* Boulder: Reed Group.

Tait, R. C. (1993). Psychological factors in the assessment of disability among patients with chronic pain. *The Journal of Back and Musculoskeletal Rehabilitation, 3,* 20.

Tompkins, N. (1992). Independent medical examination: The how, when, and why of this useful process. *OSHA Compliance Advisor, 215,* 7–12.

Turk, D. C. (1991). Evaluation of pain and disability. *Journal of Disability, 2,* 24.

Turk, D. C., & Melzack, R. (1992). *Handbook of pain assessment.* New York: Guilford Press.

Turk, D. C., Rudy, T. E., & Steig, R. L. (1988). The disability determination dilemma: Toward a multi-axial solution. *Pain, 3,* 217.

Vasudevan, S. V. (1992). Impairment, disability, and functional capacity assessment. In D. C. Turk & R. Melzack (Eds.), *Handbook of pain, assessment.* New York: Guilford Press.

Waddell, G., et al. (1992). Objective clinical evaluation of physical impairment in chronic low back pain. *Spine, 17,* 617.

Indirect Suggestion

Definition

Indirect suggestion, in the context of hypnosis, refers to a suggestion that is vague and/or permissive and offers several alternatives for experience and behavior (e.g., "I wonder if the sensations in your hand will feel unexpectedly different than before, perhaps more like numbness or tingling or perhaps nothing at all, with nothing to bother or disturb you").

Cross-References

▶ Hypnotic Analgesia

Individual Difference

▶ Personality and Pain

Individual Differences

Definition

Individuals differ markedly from one another in the pain they experience in response to virtually identical tissue trauma. Although this stems in part from genetic factors that influence nociceptive transduction and/or transmission, most interindividual variance reflects each individual's unique construction of the pain experience.

Cross-References

▶ Consciousness and Pain
▶ Personality and Pain

Indomethacin

Definition

A nonsteroidal, anti-inflammatory (NSAID) medication. These agents are usually used in the treatment of arthritis but are potent pain relievers and can be used in the treatment of many painful conditions. Several headache disorders are termed "indomethacin responsive" in that indomethacin, but none of the other NSAIDs, treats the pain and associated symptoms. Indomethacin-responsive headache disorders include hemicrania continua, the paroxysmal hemicranias, and primary stabbing headaches.

Cross-References

► Hemicrania Continua

Induration

Definition

Induration is an increase of fibrous elements in tissue commonly associated with inflammation.

Inert Agent

► Placebo

Infant

► Pain in Children

Infant Pain

► Long-Term Effects of Pain in Infants

Infant Pain Mechanisms

Maria Fitzgerald
Neuroscience, Physiology & Pharmacology,
University College London, London, UK

Synonyms

Neurobiological basis of developing pain pathways; Understanding the neural basis of pain in human infants

Definition

Infant pain mechanisms comprise the functional and structural processes involved in the transmission and processing of noxious stimuli within the developing nervous system in children under 1 year. An understanding of these processes is vital for appropriate pain assessment and management in this age group. While much remains to be discovered in this area, there have been important new advances in recent years.

Infant pain mechanisms can be understood through integration of information obtained in laboratory animal models and in clinical studies of hospitalized infants. In this way, it is possible to describe the developmental changes in nociceptor transduction and noxious information processing in spinal cord, brain stem, and cortical circuits over the first year of life and the influence of tissue injury upon those processes.

Introduction

Many infants experience events that we, as adults, would consider painful. For example, infants may need to undergo repeated invasive procedures in neonatal intensive care or require surgery in the first year of life or suffer from painful conditions with early life onset, such as early onset inflammatory joint disease. It is important to understand whether these infants process pain as in the same way as adults.

Furthermore, there is the additional possibility that, since their nervous system is still under construction, the normal development of pain pathways will be interrupted, changed, and even permanently altered by noxious stimulation and tissue damage in early life.

The challenge for the biologist working in this area is to dissect out how the normal process of nociception, which warns and protects the organism, develops over the first postnatal year and also how the immature pain system reacts and adapts in the presence of pathological tissue injury. The biologist also needs to help devise methods to accurately detect these changes in young patients, to provide a rational basis for the clinical treatment of pain and to help prevent maladaptive changes to the immature nervous system.

Characteristics

Properties of Infant Nociceptors

Nociceptors in peripheral organs that detect noxious mechanical, thermal, or chemical stimuli through a set of specific transduction molecules, such as the transient receptor potential (TRP) ion channel family, are functional in the infant. However, the full complement of TRP channels are not present in rat nociceptors until well into postnatal life (Hjerling-Leffler et al. 2007), and it is, therefore, likely that some forms of nociceptor transduction are less robust in infants. Conduction in immature afferent sensory nerves is also slower than in mature nerves as the axons are smaller and Aδ fibers are incompletely myelinated. As a result, nociceptive transmission will inevitably be slower in infants than adults (Fitzgerald 1987).

Increased pain sensitivity or hyperalgesia at the site of injury is evident from birth in both rat pups and human infants (Andrews and Fitzgerald 2002; Torsney and Fitzgerald 2002; Ririe et al. 2008). Nociceptors become sensitized to repeated noxious stimulation or inflammatory mediators in both young and adult animals (Constantinou et al. 1994) leading to increased action potential activity from a given noxious stimulus, and therefore, a greater barrage of nociceptive information reaching the central nervous system. Tissue injury also releases a lot of chemicals into the damaged region, particularly neurotrophins which play a key role in nociceptor sensitization. Levels are especially high in infants which may result in increased nerve density in the region of damage and therefore, increased sensitivity (Beggs et al. 2012a).

Infant Nociceptive Reflexes and Spinal Nociceptive Circuits

Noxious stimuli produce robust protective reflexes and these are especially strong in young infants. They can also be evoked by relatively low-intensity stimuli, especially in preterm infants and newborn rat pups, and the threshold increases with postnatal age. In addition, the spatial organization of these reflexes is immature, and they are often accompanied by generalized movements of all limbs or in some cases, poorly directed movements (Fitzgerald 2005; Waldenstrom et al. 2003). The dorsal horn of the spinal cord, the first site of central integration and processing of pain, undergoes a lot of change in the first year of life. The clear segregation of nociceptive C-fibers central terminals in lamina II and A fiber low-threshold tactile inputs in deeper laminae, observed in adults, is not present in the young dorsal horn. A fiber terminals are more diffuse and initially overlap with those of C fibers in lamina II and the gradual separation of the two types of sensory terminals is not complete until the rats are 3–4-weeks old (Beggs et al. 2002; Granmo et al. 2008). It is, therefore, perhaps not surprising that withdrawal reflexes do not discriminate between noxious and non-noxious stimuli as efficiently as in the adult.

When tissue is injured, nociceptor activity excites a mixed population of excitatory and inhibitory interneurons and some projection neurons, which together form what is commonly called a nociceptive circuit. This circuit undergoes considerable changes in early life (Fitzgerald 2005). Dorsal horn neuron excitatory receptive fields (the area of skin that, when stimulated, evokes activity) are relatively larger at birth and only gradually reduce in size with age (Torsney and Fitzgerald 2002; Fitzgerald 2005; Fitzgerald and Jennings 1999). Furthermore, dorsal horn neurons have lower thresholds and fire for longer durations than adult neurons and respond at longer and more variable latencies. It is likely that inhibitory interneurons in the circuit, in the case of adults, control evoked spike activity and regulate dorsal horn cell receptive field size, and mature more slowly than excitatory interneurons. The major inhibitory neurotransmitter, g-aminobutyric acid (GABA), can produce a similar functional inhibition of spinal nociceptive circuits to that seen in adults, but there is little activity from release of glycine, another important inhibitory neurotransmitter (Koch et al. 2012). This differing circuitry explains the exaggerated and excitable nature of protective spinal nociceptive reflexes in young animals and their relative lack of selectivity and spatial organization (Fig. 1).

Central Sensitization and Hyperalgesia in Infants

A feature of pain associated with tissue damage is hyperalgesia, (increased pain sensitivity) and allodynia, (pain from stimuli not normally felt to be painful). The neural pathways responsible for this 'central sensitization' to peripheral inputs are functional in infants but appear to differ in magnitude and time course from those in adults (Fitzgerald and Walker 2009). Tissue damage and inflammation in young animals is especially effective at modulating excitatory synapses in spinal nociceptive circuits. Surgical skin incision in the first weeks of life selectively increases excitatory synaptic activity in the dorsal horn 2–3 days after injury, without altering inhibitory synaptic activity, an effect that is not observed in older animals (Li et al. 2009). However, secondary hyperalgesia, that is, spread of pain to the area surrounding the injury, and referred hyperalgesia, pain in distant body areas, is weaker in very

Infant Pain Mechanisms, Fig. 1 Properties of dorsal horn nociceptive neurons in infant rats. Single dorsal horn neuronal activity recorded from anesthetized rats. Left panel shows that the neurons are more sensitive to mechanical stimulation of the paw in younger animals and the right panel shows that the receptive field sizes are larger in younger rats (see Fitzgerald and Jennings 1999)

young infants and increases with postnatal age (Walker et al. 2007).

Central sensitization involves glial cells and activation of neuroimmune pathways as well as neuronal circuits in the dorsal horn, and these play a major role in chronic pain states, especially those involving nerve damage. Interestingly, neither rat nor human infants develop neuropathic allodynia following nerve injury. Since acute and chronic inflammatory pain is evident from an early neonatal age, it appears that the mechanisms underlying neuropathic pain are differentially regulated over a prolonged postnatal period. Evidence suggests that the classic microglial and T cell response to nerve injury and C-fiber stimulation differs in infant and adult animals, and that infants may mount a protective anti-inflammatory response which is downregulated with maturity (Moss et al. 2007; Costigan et al. 2009), although more research is needed to understand this process.

Endogenous Pain Control: Descending Control of Nociceptive Circuits in Infants

An important aspect of central nervous system (CNS) pain processing is the system of endogenous control that modulates nociceptive activity through descending modulatory systems and endogenous opioids. In this way, a noxious stimulus or injury not only excites neuronal networks in the spinal cord and trigeminal system, but also activates brain stem descending inhibitory and excitatory activity, thus contributing to a homeostatic feedback mechanism of control. These descending and endogenous pain modulatory pathway controls are the mechanisms by which factors, such as attention and distraction, suggestion and expectation, stress and anxiety, context and past experience, influence pain responses. Recent research has revealed that the rostroventral medulla (RVM), which is the main output nucleus for brain stem descending control, undergoes a remarkable maturational switch in juvenile rats from being entirely facilitatory to inhibitory at older ages (Hathway et al. 2009). At an age equivalent to preadolescence, descending inhibition begins to dominate and becomes increasingly powerful with maturation.

Experimental evidence shows that the descending facilitation of spinal nociception in young animals is mediated by μ-opioid receptor pathways in the RVM. Furthermore, the developmental transition from descending facilitation to inhibition of pain is determined by activity in central opioid networks at a critical period of late childhood (Hathway et al. 2012). Thus in infancy, supraspinal centers have relatively little ability to dampen down nociceptive activity, but rather act to enhance it, suggesting that efforts to distract infants to diminish their pain may be less effective.

Pain Awareness and Pain Perception in Infancy: Infant Cortical Pain Networks

Although we frequently use the term 'pain' when discussing the mechanisms and responses at every level of this process, it is important to remember, especially when considering very young infants, that evidence of nociceptive activity at the level of the spinal cord or brain stem cannot be equated with pain experience. While input to the spinal cord and brain is necessary for perception of pain, it is not sufficient. Activities in the spinal cord, brain stem, and subcortical midbrain structures are sufficient to generate reflex behaviors and hormonal responses, but are not sufficient to support pain awareness.

Specialized output (projection) neurons, located in lamina I and V of the dorsal horn spinal cord send axons to several important nuclei in the brain stem, such as the parabrachial nucleus and to the thalamus. Thalamic neurons, in turn, send projections to the somatosensory cortex, cingulate cortex, amygdala, and insula which together form part of a complex 'pain matrix,' whose activity is proposed to underlie the awareness and experience of pain. Little is known about the maturation of the pain matrix over infancy and childhood in humans or laboratory animals, but recent studies show that in humans, nociceptive activity is processed at the level of the cortex from a very young age. Specific cortical hemodynamic and electrical (EEG) responses can be recorded from preterm and term infants in response to noxious heel lance (Slater et al. 2006; Slater et al. 2009), and while these

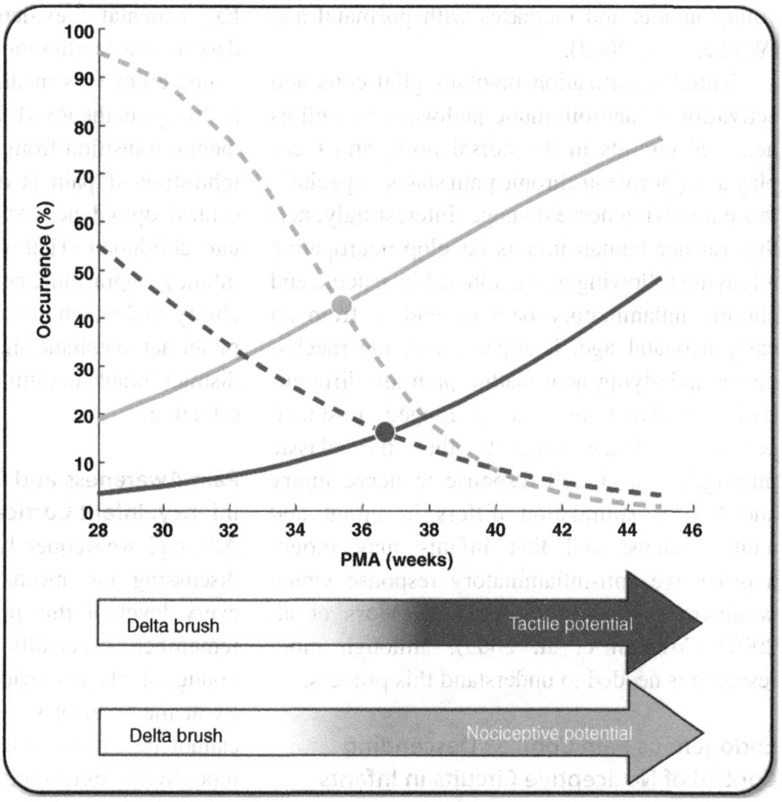

Infant Pain Mechanisms,
Fig. 2 The emergence of
touch and pain
discrimination in the
human infant brain. From
35 weeks gestation the
human brain is more likely
to generate touch- and pain-
specific activity (tactile and
pain potentials) rather than
generic neuronal bursts
(delta brushes). Data
obtained from EEG
recording of brain activity
in hospitalized infants in
response to touching the
heel or to a clinically
required heel lance
(Adapted from Fabrizi et al.
2011)

responses are often correlated with behavioral
and physiological responses, this is not always
the case (Slater et al. 2008). Infants who are born
prematurely and who have experienced at least
40 days of intensive or special care have
increased brain neuronal–evoked potential
responses to noxious stimuli compared to healthy
newborns at the same postmenstrual age (Slater
et al. 2010).

Furthermore, this type of research has also
revealed how noxious stimuli are distinguished
from innocuous stimuli in the human infant brain.
EEG recording of the brain neuronal activity in
response to time-locked touches and clinically
essential noxious lances of the heel in infants
aged 28–45 weeks gestation reveal a transition
in brain response following tactile and noxious
stimulation from nonspecific, evenly dispersed
neuronal bursts to modality-specific, localized,
evoked potentials. The results suggest that specific neural circuits necessary for discrimination

between touch and nociception emerge from
35–37 weeks gestation in the human brain
(Fabrizi et al. 2011) (Fig. 2).

Does Pain in Infancy Alter Pain Perception in Later Life?

The developing pain system, like other sensory
systems in the brain, requires a balanced sensory
input to develop normally. The natural twitching
that occurs in young infants during sleep has been
proposed as a major source of the sensory input
that drives the synaptic maturation of nociceptive
reflexes, and local anesthetic block of A fiber
touch receptor activity in young rats delays the
maturation of these reflexes (Waldenstrom et al.
2003). Activity can also arise spontaneously from
within the CNS: lamina I cells in young animals
display spontaneous, bursting "pacemaker" type
activity raising the possibility that intrinsic burst
firing could provide an endogenous drive to developing sensory networks (Li and Baccei 2011).

Infant Pain Mechanisms, Fig. 3 The late onset of neuropathic pain after infant nerve injury. Nerve injury in young rats (spared nerve injury, SNI) aged 10 days produces no pain initially on either the treated ipsilateral (Ipsi) or other contralateral (Contra) side and the sensitivity to mechanical stimulation of the foot does not differ in nerve injured (SNI) or sham operated (sham) rats. However, a gradual and significant mechanical hypersensitivity of the foot emerges on the treated side SNI-Ipsi as the rat grows up (from Vega-Avelaira et al. 2012)

There is evidence that altering the balance of sensory activity during development through tissue injury and hence excessive C fiber activity alters the subsequent development of somatosensory and pain processing with consequences that emerge later in life. Skin incision made during a critical stage in neonatal pups leads to a greater response to subsequent injury in later life (Beggs et al. 2012b; Walker et al. 2009). Thus, when an initial incision is performed in the first week of life, the degree of hyperalgesia following repeat incision six weeks later is greater than in animals having a single incision at the same age. Skin incisions in older animals do not have this lasting effect. The 'priming' in young animals is blocked by perioperative local anesthetic block, highlighting the important role of sensory activity at the time of the first injury in triggering this phenomenon. Prior neonatal injury also 'primes' the spinal microglial response to adult injury, resulting in an increased intensity, spatial distribution, and duration of microglial reactivity in the dorsal horn. Intrathecal minocycline at the time of adult injury selectively prevents both the hyperalgesia and early microglial reactivity associated with prior neonatal injury. A further contributing factor could be from the prolonged hyperinnervation of injured tissue that occurs in young animals. This long-lasting neonatal hyperinnervation is due to the release of neurotrophins into the wounded skin, in this case neurotrophin-3 (NT3), which is not normally expressed in skin but is released when skin is damaged. Long after healing, dorsal horn neurons have enlarged receptive fields in the previously injured skin (Beggs et al. 2012a) (Fig. 3).

It has been speculated that injury and pain in early life might produce the chronic pain patients of the future, and while there is no direct biological evidence to support this, it remains an intriguing possibility. Peripheral nerve injuries which trigger little or no neuropathic pain when they are sustained in infancy or early childhood may lead to a delayed pain that arises some considerable time after nerve damage. Recently this has also been shown in a young animal model of nerve injury where mechanical hypersensitivity only emerges later in life (Vega-Avelaira et al. 2012). This delayed adolescent onset in pain sensitivity is accompanied by neuroimmune activation and NMDA-dependent central sensitization of spinal nociceptive circuits. Furthermore, injury in infancy changes motivational behavior in adult rats (Low and Fitzgerald 2012).

Such novel findings may provide clues to understanding the long-term effects of early injury and the emergence of adolescent or adult chronic pain syndromes.

Summary

Infant nociceptors are capable of transducing and transmitting noxious information, and central pain processing in children is robust, but is likely to be less predictable and organized than in adults. Central sensitization can occur, leading to hyperalgesia and allodynia, but underlying mechanisms are likely to differ from the adult. Notably, peripheral nerve damage does not induce a strong neuroimmune response in children. Identifying the molecular mechanisms responsible for age-specific patterns of hyperalgesia will inform the rational development of novel therapeutic strategies for treating pediatric pain. New research suggests that early surgical injury has long-term consequences and primes the pain system such that responses to future noxious stimulation are enhanced. Furthermore, nerve damage, while producing no neuropathic pain in children, may cause alterations in pain sensitivity that only emerge at a later stage in life.

Infancy is a time when cortical connections required for pain perception are just beginning to form and the normal balance of descending brain stem controls are not yet established. A lack of balance or stability between inhibitory and excitatory supraspinal controls in early life may mean that infants are less able to mount effective endogenous control over noxious inputs compared to adults. Our understanding of the changing nature of infant and children's pain awareness and experience is poor. Nevertheless, quantifying the brain activity in infants evoked by noxious events using neurophysiological methods will allow us to study this further.

Acknowledgment The author gratefully acknowledges funding from the Medical Research Council and the Wellcome Trust, UK.

References

Andrews, K. A., & Fitzgerald, M. (2002). Wound sensitivity as a measure of analgesic effects following surgery in human neonates and infants. *Pain, 99*, 185–195.

Beggs, S., Torsney, C., Drew, L. J., & Fitzgerald, M. (2002). The postnatal reorganization of primary afferent input and dorsal horn cell receptive fields in the rat spinal cord is an activity-dependent process. *European Journal of Neuroscience, 16*, 1249–1258.

Beggs, S., Alvares, D., Moss, A., Currie, G., Middleton, J. & Fitzgerald, M. (2012). A role for NT-3 in the hyperinnervation of neonatally wounded skin. *Pain, 153*, 2133–2139.

Beggs, S., Currie, G., Salter, M. W., Fitzgerald, M., & Walker, S. M. (2012b). Priming of adult pain responses by neonatal pain experience: Maintenance by central neuroimmune activity. *Brain, 135*, 404–417.

Constantinou, J., Reynolds, M. L., Woolf, C. J., Safieh-garabedian, B., & Fitzgerald, M. (1994). Nerve growth factor levels in developing rat skin: Upregulation following skin wounding. *NeuroReport, 5*, 2281–2284.

Costigan, M., et al. (2009). T-cell infiltration and signaling in the adult dorsal spinal cord is a major contributor to neuropathic pain-like hypersensitivity. *Journal of Neuroscience, 29*, 14415–14422.

Fabrizi, L., Slater, R., Worley, A., Meek, J., Boyd, S., Olhede, S., & Fitzgerald, M. (2011). A shift in sensory processing that enables the developing human brain to discriminate touch from pain. *Current Biology, 21*, 1552–1558.

Fitzgerald, M. (1987). Cutaneous primary afferent properties in the hindlimb of the neonatal rat. *The Journal of Physiology, 383*, 79–92.

Fitzgerald, M. (2005). The development of nociceptive circuits. *Nature Reviews. Neuroscience, 6*, 507–520.

Fitzgerald, M., & Jennings, E. (1999). The postnatal development of spinal sensory processing. *Proceedings of the National Academy of Sciences of the United States of America, 96*, 7719–7722.

Fitzgerald, M., & Walker, S. (2009). Infant pain management: A developmental neurobiological approach. *Nature Clinical Practice. Neurology, 5*, 35–50.

Granmo, M., Petersson, P., & Schouenborg, J. (2008). Action-based body maps in the spinal cord emerge from a transitory floating organization. *Journal of Neuroscience, 28*, 5494–5503.

Hathway, G., Koch, S., Low, L., & Fitzgerald, M. (2009). The changing balance of brainstem-spinal cord modulation of pain processing over the first weeks of rat postnatal life. *The Journal of Physiology, 587*, 2927–2935.

Hathway, G., Vega-Avelaira, D., & Fitzgerald, M. (2012). A critical period in the supraspinal control of pain: Opioid-dependent changes in brainstem RVM function in preadolescence. *Pain, 153*, 775–783.

Hjerling-Leffler, J., Alqatari, M., Ernfors, P., & Koltzenburg, M. (2007). Emergence of functional sensory subtypes as defined by transient receptor potential channel expression. *Journal of Neuroscience, 27*, 2435–2443.

Koch, S., Hirschberg, S., Tochiki, K., & Fitzgerald, M. (2012). C- fibre activity-dependent maturation of glycinergic inhibition in the spinal dorsal horn of the postnatal rat. *Proceedings of the National Academy of Sciences of the United States of America, 109*, 12201–12206.

Li, J., & Baccei, M. L. (2011). Pacemaker neurons within newborn spinal pain circuits. *Journal of Neuroscience, 31*, 9010–9022.

Li, J., Walker, S. M., Fitzgerald, M., & Baccei, M. L. (2009). Activity-dependent modulation of glutamatergic signaling in the developing rat dorsal horn by early tissue injury. *Journal of Neurophysiology, 102*, 2208–2219.

Low, L. A., & Fitzgerald, M. (2012). Acute pain and a motivational pathway in adult rats: Influence of early life pain experience. *PLoS One, 7*, e34316.

Moss, A., et al. (2007). Spinal microglia and neuropathic pain in young rats. *Pain, 128*, 215–224.

Ririe, D., Bremner, L., & Fitzgerald, M. (2008). Comparison of the immediate effects of surgical incision on dorsal horn neuronal receptive field size and responses during postnatal development. *Anesthesiology, 109*, 698–706.

Slater, R., Worley, A., Fabrizi, L., Roberts, S., Meek, J., Boyd, S., & Fitzgerald, M. (2006). Cortical pain responses in human infants. *Journal of Neuroscience, 26*, 3662–3666.

Slater, R., Cantarella, A., Franck, L., Meek, J., & Fitzgerald, M. (2008). How well do clinical pain assessment tools reflect pain in infants? *PLoS Medicine, 5*, e129.

Slater, R., Worley, A., Fabrizi, L., Roberts, S., Meek, J., Boyd, S., & Fitzgerald, M. (2009). Evoked potentials generated by noxious stimulation in the human infant brain. *European Journal of Pain, 14*, 321–326.

Slater, R., Fabrizi, L., Worley, A., Meek, J., Boyd, S., & Fitzgerald, M. (2010). Premature infants display increased noxious-evoked neuronal activity in the brain compared to healthy age-matched term-born infants. *NeuroImage, 52*, 583–589.

Torsney, C., & Fitzgerald, M. (2002). Age-dependent effects of peripheral inflammation on the electrophysiological properties of neonatal rat dorsal horn neurons. *Journal of Neurophysiology, 87*, 1311–1317.

Vega-Avelaira, D., McKelvey, R., Hathway, G., & Fitzgerald, M. (2012). The emergence of adolescent onset pain hypersensitivity following neonatal nerve injury. *Molecular Pain, 8*, 30.

Waldenstrom, A., Thelin, J., Thimansson, E., Levinsson, A., & Schouenborg, J. (2003). Developmental learning in a pain-related system: Evidence for a cross-modality mechanism. *Journal of Neuroscience, 23*, 7719–7725.

Walker, S. M., Meredith-Middleton, J., Lickiss, T., Moss, A., & Fitzgerald, M. (2007). Primary and secondary hyperalgesia can be differentiated by postnatal age and ERK activation in the spinal dorsal horn of the rat pup. *Pain, 128*, 157–168.

Walker, S. M., Tochiki, K. K., & Fitzgerald, M. (2009). Hindpaw incision in early life increases the hyperalgesic response to repeat surgical injury: Critical period and dependence on initial afferent activity. *Pain, 147*, 99–106.

Infant Pain Reduction/Therapy/Treatment

▶ Acute Pain Management in Infants

Infant Pain Therapy

▶ Acute Pain Management in Infants

Infant Pain Treatment

▶ Acute Pain Management in Infants

Inflammation

Definition

Inflammation is a complex response to areas affected by injury or disease, which is meant to protect tissues, but is often a major source of damage. It involves the recruitment and activation of various cells (leukocytes and macrophages) and extracellular proteins and is induced and maintained by numerous mediators like cytokines, histamine, plasma proteases, arachidonic acid metabolites (prostaglandins, leukotrienes), growth factors, and ions (potassium, protons). The main signs of inflammation are dolor, rubor, calor, and tumor. Dolor, or pain, occurs due to activation of nociceptors that become sensitized with a lower threshold for activation. Rubor, or redness, is due to local vasodilatation with increased blood flow in the affected area. Calor, or heat, is due to a combination of increased local blood flow and an increase in cellular metabolic activity. Tumor or swelling occurs due to an increase in capillary permeability with plasma extravasation and localized edema. This may be exacerbated by restricted function in the affected area.

Cross-References

▶ Acid-Sensing Ion Channels
▶ COX-2 Inhibitors
▶ Lower Back Pain, Physical Examination
▶ Neuropeptide Release in Inflammation
▶ NGF, Regulation During Inflammation
▶ Opioids in the Periphery and Analgesia
▶ Pain-Modulatory Systems, History of Discovery
▶ Postoperative Pain, COX-2 Inhibitors

Inflammation, Modulation by Peripheral Cannabinoid Receptors

Isobel Lever
Neural Plasticity Unit, Institute of Child Health
University College London, London, UK

Definition

Cannabinoids are chemical compounds derived from the *cannabis sativa* plant and their synthetic analog. Pain responses resulting from inflammation can be reduced by cannabinoids, applied locally to peripheral tissues at systemically inactive doses. This analgesic mechanism involves the activation of peripheral cannabinoid receptors outside the CNS, making it distinct from the analgesic action of cannabinoids operating at receptors in the spinal cord and brain.

Characteristics

Two G-protein-coupled receptors that respond to cannabinoids (CBs) have been identified so far

and named CB_1 and CB_2. There is also evidence for additional cannabinoid receptor subtypes, which have yet to be characterized (Howlett et al. 2002). CB_1 receptors are found primarily in CNS neurons, and the majority of CB_2 receptors are expressed peripherally by cells with inflammatory and immune response functions. Both CB receptor subtypes have been detected in peripheral tissues from rat, mouse, and human (Howlett et al. 2002), with evidence for CB_1 expression in peripheral neurons and glia, as well as on immune cells (although expression levels are lower than for CB_2 receptors) (Howlett et al. 2002; Walter and Stella 2004). Richardson et al. (1998) first demonstrated modulation of inflammatory pain states by the activation of peripheral cannabinoid receptors in an animal model of inflammatory pain. Injection of the inflammatory agent carrageenan into rat hind paw skin was used to induce a localized inflammatory response (measured by paw edema due to ▶ plasma extravasation caused by increased microvasculature permeability) and ▶ hyperalgesia (measured by increased behavioral responses to noxious stimuli). After inflammatory injury, both the effects of paw edema and thermal hyperalgesia were attenuated by a local injection of a small dose of the cannabinoid anandamide (AEA) to the paw skin, but not when the same dose of AEA was applied systemically. Co-administration of a selective antagonist to the CB_1 receptor (SR141716), to the paw skin, reversed the effect of AEA on thermal hypersensitivity. It also attenuated the cannabinoid-mediated reduction of plasma extravasation (stimulated by a ▶ capsaicin-induced ▶ neurogenic inflammation), thus demonstrating both an anti-hyperalgesic and an anti-inflammatory action of cannabinoids operating via peripheral CB_1-type cannabinoid receptors.

Local application of other cannabinoids, WIN55, 212–2, HU-210, methanandamide, as well as AEA, in the formalin model of the rat hind paw skin inflammation, confirmed a reduction in pain behavior via the activation of peripheral CB_1 receptors (Calignano et al. 1998). The ▶ endocannabinoid ligand AEA was similarly effective in this model and could be detected in paw tissue, along with a related endogenous compound palmitoylethanolamide (PEA), which does not bind to any known cannabinoid receptor but the actions of which are reversed by the CB_2 receptor antagonist SR144528 (Howlett et al. 2002). Interestingly, when both of these compounds were injected into the paw together, they produced a synergistic reduction of pain behavior (Calignano et al. 1998). However, the anti-hyperalgesic effects of PEA were attenuated by local blockade of CB_2 but not CB_1 receptors in paw skin tissue. The contribution of CB_2 receptor signalling to the peripheral action of cannabinoids indicated by this study has since been confirmed, by using newly developed ligands that are selective for the CB_2 cannabinoid receptor subtype (Malan et al. 2003). Application of the selective CB_2 receptor agonist AM1241 to inflamed rat paw skin tissue can reduce both mechanical and thermal hyperalgesic responses, an action which can also be locally reversed by CB_2 receptor antagonists (Quartilho et al. 2003; Hohmann et al. 2004). Other CB_2 receptor agonists have also been shown to be effective against paw tissue edema when they were administered systemically (Clayton et al. 2002; Malan et al. 2003). Thus, cannabinoids have a dual anti-inflammatory and anti-hyperalgesic action mediated by both CB_1 and CB_2 peripheral cannabinoid receptor subtypes.

The Anti-hyperalgesic Action of Peripheral CB Receptors

Behavioral studies report that cannabinoid receptors can be exogenously activated in inflamed hind paw skin to produce a reduction in pain responses to noxious stimuli. Also, AEA and PEA (compounds that endogenously activate cannabinoid receptor signalling pathways) can be detected in this tissue (Calignano et al. 1998). It is therefore possible that these compounds interact with cannabinoid receptors on the peripheral endings of cutaneous ▶ nociceptive sensory neurons, in order to directly modulate the transmission of nociceptive signals to the CNS. As evidence of this mechanism, the local application of CB_1 and

CB_2 receptor agonists to carrageenan-inflamed skin can reduce the responses of spinal cord cells receiving noxious mechanical inputs from this area (Kelly et al. 2003; Sokal et al. 2003). CB_1 receptors can also inhibit ▶ neurosecretion of the vasodilatatory peptide CGRP from peripheral terminals of capsaicin-sensitive neurons in isolated paw skin (Richardson et al. 1998). This also suggests that cannabinoids might have a direct modulatory action on the signalling functions of nociceptive sensory neurons. In support of this mechanism, CB_1 mRNA and protein has been detected in the cell bodies of hind limb sensory nerves contained within dorsal root ganglia (DRG) (Bridges et al. 2003). However, the distribution of these receptors among the small, peptide-containing somas that are typical of nociceptive nerve fibers was found to be limited in favor of intermediate- to large-sized cells, which predominately have non-nociceptive sensory functions. It should be noted, however, that there is currently a discrepancy between the CB_1 receptor distribution reported in DRG tissue versus cultured DRG neurons (Bridges et al. 2003). CB_2 receptors have not been found on adult DRG neurons, although, as with CB_1, the presence of these receptors in functionally significant but as yet undetectable amounts in vivo cannot be entirely ruled out. Functional CB_1 receptors have been demonstrated on cultured sensory neurons, where they are reported to signal via inhibitory $G_{i/o}$ proteins, to inhibit N-type voltage-activated calcium currents. Crucially, the activity of these receptors has not been demonstrated on small-sized CGRP-containing sensory neurons in vitro (Khasabova et al. 2004).

The Anti-inflammatory Action of Peripheral CB Receptors

An alternative mechanism to explain the anti-hyperalgesic action of cannabinoids is linked to anti-inflammatory effects of these compounds in peripheral tissues. CB agonists may indirectly inhibit transmission in nociceptive afferents by reducing the release of by-products of inflammatory and excitotoxic processes, which serve to sensitize peripheral nociceptive fibers to mechanical and thermal stimuli. Cannabinoid receptors are present on cells that are functionally involved in immune responses to tissue injury, including B and T cells, natural killer cells, neutrophils, macrophages, and mast cells (Howlett et al. 2002; Samson et al. 2003; Walter and Stella 2004). Furthermore, mitogen activation of some immune cells, such as human lymphocytes, has been shown to stimulate an upregulation of cannabinoid receptors and also trigger the production of endocannabinoids (Howlett et al. 2002; Walter and Stella 2004). Taken together, this evidence suggests that cannabinoids are endogenous immunomodulators. Suppression of the pro-inflammatory functions of glial and immune cells (Walter and Stella 2004) is believed to mediate the inhibitory effects of cannabinoids on edema and plasma extravasation. However, much of the evidence for cell-specific regulatory actions of cannabinoids has been derived from studies with isolated cell lines.

Cannabinoids are reported to reduce the production and release of pro-inflammatory signalling molecules, including tumor necrosis factor, nitric oxide, and interleukin, and to enhance the release of anti-inflammatory cytokines like the IL-1 receptor antagonist, IL-4, and IL-10 (Molina-Holgado et al. 2003; Howlett et al. 2002; Walter and Stella 2004). Nerve growth factor (NGF) is an important mediator of inflammatory responses due to its ability to recruit and activate mast cells and also to directly sensitize the receptors involved in transducing noxious stimuli on sensory neurons. Intradermal injections of NGF are used to artificially produce inflammation and hyperalgesia, and AEA and PEA compounds can be used to alleviate both effects, operating via CB_1 and CB_2 receptors, respectively (Farquhar-Smith et al. 2003). The action of NGF in tissues is amplified by mast cell function, as these cells produce and release NGF locally, as well as being activated by it. They also synthesize and release other inflammatory mediators that increase local vascular permeability and others, like serotonin, which can sensitize nociceptors. The ability of CB_1 and CB_2 receptor signalling to reduce inflammatory responses has been linked to their regulation of mast cell function (Samson et al. 2003). CB_1

receptor activation on mast cell lines is linked to the suppression of serotonin secretory responses, whereas co-expressed CB_2 receptors regulate the activation of signalling pathways that regulate gene expression. Mast cells have also been reported to produce endocannabinoids and PEA, suggesting the existence of an autocrine regulatory signalling pathway involving cannabinoids in these cells (Samson et al. 2003). Cannabinoids are also reported to suppress immune cell proliferation and chemotactic processes, which contribute to the establishment of inflammatory sites in tissues (Howlett et al. 2002; Walter and Stella 2004). Specifically, PEA has been shown to reduce neutrophil accumulation in hind paw skin inflamed by NGF injections (Farquhar-Smith et al. 2003). This is speculated to involve peripheral CB_2 receptor signalling pathways, which act to reduce the production and/or release of mast cell-derived neutrophil chemotactic factors, such as Leukotriene B4 or even NGF itself.

PEA and endocannabinoids like AEA are likely by-products of the inflammatory and excitotoxic processes occurring after tissue injury, providing endogenous signalling molecules to activate CB_1 and CB_2 receptor pathways. There is evidence that CB receptors are endogenously activated in inflamed skin tissue, as administration of CB_2 receptor antagonists alone to noninflamed paw tissue increased the mechanical hypersensitivity and tissue edema caused by a subsequent carrageenan inflammation (Clayton et al. 2002). Also, the injection of both CB_1 and CB_2 receptor antagonists to hind paws before inflammation by formalin led to an enhancement of pain responses. This result similarly indicates that cannabinoid receptors are tonically activated in peripheral tissue (Calignano et al. 1998).

In summary, cannabinoids can act at CB_1 and CB_2 receptors in peripheral tissue to modulate immune and nociceptor functions: They suppress the pro-inflammatory actions of immune and glial cells leading to the reduction of tissue inflammation and reduce the hypersensitivity of sensory nerves responding to mechanical and thermal pain stimuli. These are two ways in which the action of cannabinoids at local receptors in peripheral tissue could be effective in altering pain responses to inflammatory injury. This peripheral action is of potential therapeutic importance because it presents a way of delivering the analgesic benefit of cannabinoid compounds, whilst circumventing their centrally mediated psychoactive side effects.

References

Bridges, D., Rice, A. S. C., Egertova, M., Elphick, M. R., Winter, J., & Michael, G. J. (2003). Localisation of cannabinoid receptor 1 in rat dorsal root ganglion using In situ hybridisation and immunohistochemistry. *Neurosci, 119*, 803–812.

Calignano, A., La Rana, G., Giuffrida, A., & Peiomelli, D. (1998). Control of pain initiation by endogenous cannabinoids. *Nature, 394*, 277–280.

Clayton, N., Marshall, F. H., Bountra, C., & O'Shaughnessy, C. T. (2002). CB1 and CB2 cannabinoid receptors are implicated in inflammatory pain. *Pain, 96*, 253–260.

Farquhar-Smith, W. P., & Rice, A. S. C. (2003). A novel neuroimmune mechanism in cannabinoid-mediated attenuation of nerve growth factor-induced hyperalgesia. *Anesthesiology, 99*, 1391–1401.

Hohmann, A. G., Farthing, J. N., Zvonok, A. M., & Makriyannis, A. (2004). Selective activation of cannabinoid CB2 receptors suppresses hyperalgesia evoked by intradermal capsaicin. *Journal of Pharmacology and Experimental Therapeutics, 308*, 446–453.

Howlett, A. C., Barth, F., Bonner, T. I., Cabral, G., Casellas, P., Devane, W. A., Felder, C. C., Herkenham, M., Mackie, K., Martin, B. R., Mechoulam, R., & Pertwee, R. G. (2002). International union of pharmacology. XXVII. Classification of cannabinoid receptors. *Pharmacological Reviews, 54*, 161–202.

Kelly, S., Jhaveri, D. M., Sagar, D. R., Kendall, D. A., & Chapman, V. (2003). Activation of peripheral cannabinoid CB1 receptors inhibits mechanically evoked responses of spinal neurons in Non-inflamed rats and rats with hind paw inflammation. *European Journal of Neuroscience, 18*, 2239–2243.

Khasabova, I. A., Harding-Rose, C., Simone, D. A., & Seybold, V. S. (2004). Differential effects of CB1 and opioid agonists on two populations of adult rat dorsal root ganglion neurons. *Journal of Neuroscience, 24*, 1744–1753.

Malan, T. P., Ibrahim, M. M., Lai, J., Vanderah, T. W., Makriyannis, A., & Porreca, F. (2003). CB2 Cannabinoid receptor agonists: Pain relief without psychoactive effects? *Current Opinion in Pharmacology, 3*, 62–67.

Molina-Holgado, F., Pinteaux, E., Moore, J. D., Molina-Holgado, E., Guaza, C., Gibson, R. M., & Rothwell, N. J. (2003). Endogenous interleukin-1 receptor

antagonist mediates anti-inflammatory and neuroprotective actions of cannabinoids in neurons and glia. *Journal of Neuroscience, 23*, 6470–6474.

Quartilho, A., Mata, H. P., Ibrahim, M. M., Vanderah, T. W., Porreca, F., Makriyannis, A., & Malan, T. P. (2003). Inhibition of inflammatory hyperalgesia by activation of peripheral CB2 cannabinoid receptors. *Anesthesiol, 99*, 955–960.

Richardson, J. D., Kilo, S., & Hargreaves, K. M. (1998). Cannabinoids reduce hyperalgesia and inflammation via interaction with peripheral CB1 receptors. *Pain, 75*, 111–119.

Walter, L., & Stella, N. (2004). Cannabinoids and neuroinflammation. *Brit J Pharmacol, 141*, 775–785.

Samson, M. T., Small-Howard, A., Shimoda, L. M. N., Koblan-Huberson, M., Stokes, A. J., & Turner, H. (2003). Differential roles of CB1 and CB2 cannabinoid receptors in mast cells. *Journal of Immunology, 170*, 4953–4962.

Sokal, D. M., Elmes, S. J. R., Kendall, D. A., & Chapman, V. (2003). Intraplantar injection of anandamide inhibits mechanically-evoked responses of spinal neurones via activation of CB2 receptors in anaesthetised rats. *Neuropharmacology, 45*, 404–411.

Inflammation, Role of Peripheral Glutamate Receptors

Susan M. Carlton
Neuroscience and Cell Biology, University of Texas Medical Branch, Galveston, TX, USA

Synonyms

Peripheral glutamate receptors; Role in inflammation

Definition

Inflammation and its accompanying pain is a complicated process. Recently it has been reported that peripheral glutamate and peripheral glutamate receptors contribute to inflammatory pain and these data are summarized here.

Introduction

There are several lines of evidence in animals and humans that glutamate receptors in the skin contribute to inflammatory pain. The data are derived from anatomical, behavioral, and physiological experiments (Carlton 2001).

Characteristics

Anatomical studies in the ▶ dorsal root ganglia (DRG) indicate that ▶ ionotropic glutamate receptors (iGluR) are expressed by significant populations of sensory neurons. N-methyl-D-aspartate receptor 1 (NMDAR1) receptors are localized in virtually all DRG cells; however, only subpopulations of DRG neurons express kainate and α-amino-3-hydroxy-5-methyl-4-isoxazolepropionic acid (AMPA) receptors (Sato et al. 1993). Immunohistochemistry at the electron microscopic level demonstrates that 48 ± 18 % of unmyelinated axons in the rat digital nerve are positively labeled for the NMDAR1 subunit, while 27 ± 3 % and 23 ± 8 % are labeled for subunits of the kainate and AMPA receptor (Coggeshall and Carlton 1998). Forty-eight hours after intraplantar injection of complete Freund's adjuvant (CFA), there is a significant increase in the number of digital axons expressing the iGluRs (Carlton and Coggeshall 1999). The percentage of NMDAR1-labeled fibers increases from 48 % to 61 %, for kainate the increase is from 27 % to 47 %, and for AMPA it is from 23 % to 40 %. These results suggest that there is an increased potential for activation of glutamate receptors in the inflamed state. ▶ Metabotropic glutamate receptors (mGluRs) have also been localized in primary sensory neurons (Bhave et al. 2001; Carlton et al. 2001; Carlton and Hargett 2007). Group III mGluRs have the highest expression with 75 % of DRG cells expressing mGluR8, followed by group II with 51 % expressing mGluR2/3, followed by group I with only 7 % expressing mGluR1α and 16 % expressing mGluR5 (Bhave et al. 2001; Carlton and Hargett 2007). At this time, it is unknown whether mGluR expression changes during inflammation.

In behavioral studies, activation of these peripheral receptors following intraplantar injection of glutamate (Carlton et al. 1995) or injection of more selective iGluR or mGluR agonists in normal skin results in increased mechanical sensitivity as well as thermal sensitivity (Zhou et al. 1996; Carlton et al. 1998; Bhave et al. 2001; Zhou et al. 2001). Intraplantar injection of 1 mM NMDA or kainate in normal animals produces a change in mechanical sensitivity equivalent to that seen 48 h post-CFA injection. Furthermore, concentrations of NMDA or kainate that have no effect in normal animals produce a significant reduction in mechanical thresholds in the inflamed hind paw (Du et al. 2003, 2005). This data suggests that subthreshold concentrations of glutamate may result in activation of glutamate receptors in the inflamed hind paw. Locally applied iGluR or group I mGluR antagonists and group II and III mGluR agonists can reduce nociceptive behaviors in several models of acute and chronic inflammation including those due to formalin, carrageenan, capsaicin, mustard oil, bee venom, and CFA (Carlton et al. 2001; Neugebauer and Carlton 2002; Ro 2003; Jin et al. 2009; Chen et al. 2010; Gangadharan et al. 2011).

Recording from identified nociceptors using an in vitro skin-nerve preparation demonstrates that normal nociceptors are excited and sensitized by application of glutamate (Du et al. 2001), NMDA (Du et al. 2003), or kainate (Du et al. 2005). Both Aδ and C nociceptors show a significant increase in activity compared to background when their receptive fields are exposed to 300 μM glutamate. Surprisingly, 1 mM glutamate results in lesser activity; however, it is possible that a desensitization of receptors occurs at this higher concentration. Furthermore, a 2 min exposure to 300 μM glutamate will increase unit responses to heat. This heat sensitization can occur whether the unit is excited by glutamate or not (Du et al. 2001). In 48 h CFA-inflamed skin, there is a slight shift to the left in the dose–response curve for NMDA or kainate-induced activation of nociceptors compared to normal skin, indicating that the amount of agonist required to induce nociceptor activation is reduced ten-fold in inflamed skin. Furthermore, the percentage of NMDA or kainate-activated nociceptors and NMDA or

Inflammation, Role of Peripheral Glutamate Receptors, Fig. 1 Glutamate-evoked activity in normal nociceptors. This schematic drawing depicts nociceptors terminating in the skin. In the normal state, many of the terminals contain glutamate in vesicles that are released upon stimulation. Through autocrine and/or paracrine routes, glutamate receptors (iGluRs, mGluRs) on nociceptors can be activated, modulating nociceptive transmission to the spinal cord

kainate-induced discharge rates in these nociceptors is significantly increased in inflamed compared to normal skin, and this activity can be blocked by co-application of the appropriate antagonists (Du et al. 2003, 2005). Activation of group II and group III mGluRs will depress responses of nociceptors to formalin or inflammatory soup (Du et al. 2008). Importantly, group II and III receptor activation can also significantly dampen TRPV1 responses (Carlton et al. 2009, 2011; Govea et al. 2012). Injection of glutamate into several craniofacial regions enhances nociceptor responses to capsaicin (Lam et al. 2009). TRPV1 receptors are major components producing the pain associated with inflammation so the ability of mGluRs to modulate TRP channels makes them important targets for the treatment of inflammatory pain. Thus, it is clear that glutamate receptors contribute to responses of nociceptors in the inflamed state.

A major source of ligand for activation of these peripheral receptors in normal and inflamed skin is the primary afferent terminals themselves since over 90 % of DRG cells contain glutamate at concentrations considered to be higher than that representing metabolic stores. Stimulation of primary afferent fibers in the non-noxious or noxious range results in release of glutamate from peripheral primary afferent terminals (deGroot et al. 2000; Jin et al. 2009). The endogenous glutamate could then initiate or enhance the activity of nociceptors in the microenvironment through paracrine or autocrine stimulation (Fig. 1). There are several lines of evidence that glutamate content increases in inflamed tissue in both animal and human studies. Nerves innervating inflamed knee joints show an increase in glutamate immunoreactivity suggesting enhanced glutamate content (Westlund et al. 1992). There is increased glutamate content in inflamed hind paws as measured by microdialysis (Omote et al. 1998). The macrophages and serum that infiltrate inflamed regions will also contribute to the glutamate content (Piani et al. 1991; McAdoo et al. 1997). In human studies, glutamate content increases in the synovial fluid of arthritic patients (McNearney et al. 2000). There is a ready source of ligand for glutamate receptors in inflamed regions (Fig. 2).

That glutamate receptors and ligand contribute to nociception and that both receptors and ligand increase during inflammation have clinical relevance. It has been demonstrated that iGluRs

Inflammation, Role of Peripheral Glutamate Receptors, Fig. 2 Glutamate-enhanced nociception in inflammation. In the inflamed state, there is an increase in glutamate content in terminals innervating the inflamed region, an increase in the number of glutamate receptor (GluR)-containing axons and thus an increased probability of GluR activation. In the inflamed region, other elements may contribute to an increase in glutamate content in the vicinity including mast cells and serum from leaky blood vessels. All of these factors will lead to enhanced nociceptive transmission

are present on unmyelinated axons in human skin (Kinkelin et al. 2000). Also, local treatment of the skin with ketamine, an NMDAR antagonist, reduces hyperalgesia associated with experimental burn injuries (Warncke et al. 1997). Injection of glutamate in the masseter muscle will result in sensitization to a subsequent capsaicin injection (Arendt-Nielsen et al. 2008).

Summary

Formulation of glutamate receptor antagonists that do not cross the blood brain barrier could be useful in reducing peripheral nociceptor activity while avoiding central side effects. Glutamate antagonists could be used in the periphery in combination with drugs that target the central nervous system to reduce both peripheral and central sensitization. These glutamate receptor antagonists would offer a non-opiate method for control of inflammatory pain.

Acknowledgement This work was supported by NIH NS027910, NS054765 and DA027460 to SMC.

References

Arendt-Nielsen, L., Svensson, P., Sessle, B. J., Cairns, B. E., & Wang, K. (2008). Interactions between glutamate and capsaicin in inducing muscle pain and sensitization in humans. *European Journal of Pain, 12*, 661–670.

Bhave, G., Karim, F., Carlton, S. M., et al. (2001). Peripheral group I metabotropic glutamate receptors modulate nociception in mice. *Nature Neuroscience, 4*, 417–423.

Carlton, S. M. (2001). Peripheral excitatory amino acids. *Current Opinion in Pharmacology, 1*, 52–56.

Carlton, S. M., & Coggeshall, R. E. (1999). Inflammation-induced changes in peripheral glutamate receptor populations. *Brain Research, 820*, 63–70.

Carlton, S. M., & Hargett, G. L. (2007). Colocalization of metabotropic glutamate receptors in rat dorsal root ganglion cells. *The Journal of Comparative Neurology, 501*, 780–789.

Carlton, S. M., Hargett, G. L., & Coggeshall, R. E. (1995). Localization and activation of glutamate receptors in unmyelinated axons of rat glabrous skin. *Neuroscience Letters, 197*, 25–28.

Carlton, S. M., Zhou, S., & Coggeshall, R. E. (1998). Evidence for the interaction of glutamate and NK1 receptors in the periphery. *Brain Research, 790*, 160–169.

Carlton, S. M., Hargett, G. L., & Coggeshall, R. E. (2001). Localization of metabotropic glutamate receptors 2/3 on primary afferent axons in the rat. *Neuroscience, 105*, 957–969.

Carlton, S. M., Du, J., & Zhou, S. (2009). Group II metabotropic glutamate receptor activation on peripheral nociceptors modulates TRPV1 function. *Brain Research, 1248*, 86–95.

Carlton, S. M., Zhou, S., Govea, R., & Du, J. (2011). Group II/III metabotropic glutamate receptors exert endogenous activity-dependent modulation of TRPV1 receptors on peripheral nociceptors. *Journal of Neuroscience, 31*, 12727–12737.

Chen, H. S., Qu, F., He, X., Kang, S. M., Liao, D., & Lu, S. J. (2010). Differential roles of peripheral metabotropic glutamate receptors in bee venom-induced nociception and inflammation in conscious rats. *The Journal of Pain, 11*, 321–329.

Coggeshall, R. E., & Carlton, S. M. (1998). Ultrastructural analysis of NMDA, AMPA and kainate receptors on unmyelinated and myelinated axons in the periphery. *The Journal of Comparative Neurology, 391*, 78–86.

deGroot, J. F., Zhou, S., & Carlton, S. M. (2000). Peripheral glutamate release in the hindpaw following low and high intensity sciatic stimulation. *NeuroReport, 11*, 497–502.

Du, J., Koltzenburg, M., & Carlton, S. M. (2001). Glutamate-induced excitation and sensitization of nociceptors in rat glabrous skin. *Pain, 89*, 187–198.

Du, J., Zhou, S., Coggeshall, R. E., et al. (2003). N-methyl-D-aspartate-induced excitation and sensitization of normal and inflamed nociceptors. *Neuroscience, 118*, 547–562.

Du, J., Zhou, S., & Carlton, S. M. (2005). Kainate-induced excitation and sensitization of nociceptors in normal and inflamed rat glabrous skin. *Neuroscience, 137*, 999–1013.

Du, J., Zhou, S., & Carlton, S. M. (2008). Group II metabotropic glutamate receptor activation attenuates peripheral sensitization in inflammatory states. *Neuroscience, 154*, 754–766.

Gangadharan, V., Wang, R., Ulzhofer, B., Luo, C., Bardoni, R., Bali, K. K., Agarwal, N., et al. (2011). Peripheral calcium-permeable AMPA receptors regulate chronic inflammatory pain in mice. *Journal of Clinical Investigation, 212*, 1608–1623.

Govea, R. M., Zhou, S., & Carlton, S. M. (2012). Group III metabotropic glutamate receptors and transient receptor potential vanilloid 1 co-localize and interact on nociceptors. *Neuroscience, 217*, 130–139.

Jin, Y. H., Yamaki, F., Takemura, M., Koike, Y., Furuyama, A., & Yonehara, N. (2009). Capsaicin-induced glutamate release is implicated in nociceptive processing through activation of ionotropic glutamate receptors and group I metabotropic glutamate receptors in primary afferent fibers. *Journal of Pharmacological Science, 109*, 233–241.

Kinkelin, I., Brocker, E.-B., Koltzenburg, M., et al. (2000). Localization of ionotropic glutamate receptors in peripheral axons of human skin. *Neuroscience Letters, 283*, 149–152.

Lam, D. K., Sessle, B. J., & Hu, J. W. (2009). Glutamate and capsaicin effects on trigeminal nociception I: Activation and peripheral sensitization of deep craniofacial nociceptive afferents. *Brain Research, 1251*, 130–139.

McAdoo, D. J., Hughes, M., Xu, G.-Y., et al. (1997). Microdialysis studies of the role of chemical agents in secondary damage upon spinal cord injury. *Journal of Neurotrauma, 14*, 507–515.

McNearney, T., Speegle, D., Lawand, N. B., et al. (2000). Excitatory amino acid profiles of synovial fluid from patients with arthritis. *Journal of Rheumatology, 27*, 739–745.

Neugebauer, E., & Carlton, S. M. (2002). Peripheral metabotropic glutamate receptors as drug targets for pain relief. *Expert Opinion on Therapeutic Targets, 6*, 349–361.

Omote, K., Kawamata, T., Kawamata, M., et al. (1998). Formalin-induced release of excitatory amino acids in the skin of the rat hindpaw. *Brain Research, 787*, 161–164.

Piani, D., Frei, K., Do, K. Q., et al. (1991). Murine brain macrophages induce NMDA receptor mediated neurotoxicity in vitro by secreting glutamate. *Neuroscience Letters, 133*, 159–162.

Ro, J. Y. (2003). Contribution of peripheral NMDA receptors in craniofacial muscle nociception and edema formation. *Brain Research, 979*, 78–84.

Sato, K., Kiyama, H., Park, H. T., et al. (1993). AMPA, KA and NMDA receptors are expressed in the rat DRG neurones. *NeuroReport, 4*, 1263–1265.

Warncke, T., Jorum, E., & Stubhaug, A. (1997). Local treatment with the N-methyl-D-aspartate receptor antagonist ketamine, inhibits development of secondary hyperalgesia in man by a peripheral action. *Neuroscience Letters, 227*, 1–4.

Westlund, K. N., Sun, Y. C., Sluka, K. A., et al. (1992). Neural changes in acute arthritis in monkeys. II. Increased glutamate immunoreactivity in the medial articular nerve. *Brain Research Reviews, 17*, 15–27.

Zhou, Z., Bonasera, L., & Carlton, S. M. (1996). Peripheral administration of NMDA, AMPA or KA results in pain behaviors in rats. *NeuroReport, 7*, 1–6.

Zhou, S., Komak, S., Du, J., et al. (2001). Metabotropic glutamate 1Î ± receptors on peripheral primary afferent fibers: Their role in nociception. *Brain Research, 913*, 18–26.

Inflammatory

Definition

Pertaining to the local response to an injury (usually including redness, warmth, and pain).

Cross-References

▶ Pain-Modulatory Systems, History of Discovery

Inflammatory Bowel Disease

▶ Animal Models of Inflammatory Bowel Disease

Inflammatory Cytokines

▶ Proinflammatory Cytokines

Inflammatory Hyperalgesia by Skin Freezing

▶ Freezing Model of Cutaneous Hyperalgesia

Inflammatory Mediators

Definition

Inflammatory mediators comprise of a variety of endogenous substances associated with the immune system including bradykinin, serotonin, histamine, prostaglandins, leukotrienes, amines, purines, cytokines, and chemokines that are blood-borne or released from resident immune cells, glia, sympathetic endings, and other sources in inflamed tissue. They act on primary afferent nociceptors to cause pain by either inducing activity in nociceptors (activation/excitation), or increasing nociceptor responses evoked by other stimuli (peripheral sensitization). In addition, inflammatory mediators may cause the local release of other mediators from leucocytes and phagocytotic cells and attract leucocytes to the site of inflammation. Inflammatory mediators are involved in the process of removal of injured tissue and repair of the injured site. Inflammatory mediators released in the spinal cord after tissue inflammation and peripheral nerve injury are thought to contribute to central sensitization.

Cross-References

▶ Neuropeptide Release in Inflammation
▶ Perireceptor Elements
▶ Quantitative Thermal Sensory Testing of Inflamed Skin

Inflammatory Myopathies

▶ Myositis

Inflammatory Neuritis

Linda S. Sorkin
Occupational Therapy and Rehabilitation, Anesthesia Research Laboratories, University of California, San Diego, CA, USA

Synonyms

Inflammatory neuropathy; Nerve inflammation; Neuritis

Definition

Inflammation of a nerve or nerves. Neuritis is considered to be a subset of neuropathy (Note: not to be used unless inflammation is thought to be present) (Linbloom et al. 1986). Neuritis must include or assume presence of neuroimmune reactions.

Characteristics

Inflammation of sensory nerves leading to pain is a common clinical occurrence. The etiology can range from responses to a local infection to neuroimmune reactions. In the latter case, antibodies generated in response to bacteria or viruses attack an axon or cell body of a sensory nerve or elements of local Schwann cells (usually myelin) because these structures share the

Inflammatory Neuritis, Fig. 1 This schematic illustrates the principle of molecular mimicry. If bacteria/viruses and Schwann cells/axons contain similar surface epitopes, antibodies generated as part of a normal immune response can cross-react with the neural elements and

epitope to which the antibody was originally directed. The autoimmune reaction resulting from this shared identity is termed "molecular mimicry" (Fig. 1). Alternatively, exogenous antibodies administered as part of a therapeutic regime can elicit similar responses resulting in an iatrogenic neuritis. In all cases, local immune reactions alter the microenvironments of the affected nerves. Under some conditions, the response seems to involve ► complement activation, and in others activation of T and/or B cells is implicated. Frank infiltration of T cells, macrophages, and neutrophils can occur, but perhaps due to the nonlinearity of the inflammation, they are not always seen in biopsy or autopsy specimens. Generation of prostaglandins, pro-inflammatory ► cytokines, oxygen-free radicals, and nitric oxide may all play a role in the immune cascade and help to break down the blood-nerve barrier. Antibody attack and/or the presence of pro-inflammatory agents within the local milieu can alter the activity of and/or destroy nerve fibers depending on the duration and severity of the response. When this includes A-delta and C nociceptive fibers, the disease is often characterized by painful dysesthesias during at least part of its course. Observed pathology can range from edema and axonal irritation, through localized demyelination with a resultant

trigger a local immune cascade around the nerve. The epitope can be part of a more complex antigen as long as it is exposed. Exogenously administered antibody can trigger similar immune responses

conduction slowing to irreversible axonal loss. Pain is frequently a prominent symptom of several stages of neuritis, but is not necessarily present throughout the full course of the disease.

Animal Models

Current animal models of neuritis with documented pain behavior frequently involve placement of pro-inflammatory substances on or around the nerve trunk (Eliav et al. 1999; Sorkin and Doom 2000; Gazda et al. 2001; Onda et al. 2001; Eliav et al. 2009). This elicits pain behavior referred to the innervated tissue, as well as a reversible endoneurial edema frequently accompanied by infiltration of both CD4 and CD8 staining T-lymphocytes (Eliav et al. 1999). Alternatively, resident peri-sciatic immune cells may release pro-inflammatory substances in the absence of infiltrating immune cells (Gazda et al. 2001).

Interestingly, while low doses/concentrations of inflammatory agents result in an ipsilateral response, increasing the severity of the insult (dose of the inflammatory agent) results in a bilateral response with "mirror image" pain behavior (Chacur et al. 2001; Eliav et al. 2009). In most cases, the pain behavior is restricted to mechanical, but not thermal sensitization. However, this might be a timing issue, as interleukin

1b applied to the nerve has recently been shown to induce thermal sensitivity starting several days after peri-sciatic application, and this is in contrast to the mechanical sensitization that develops within several hours following administration of TNF, IL-6, or GD_2 ganglioside antibody (an immunotherapy for several types of tumor). The pain behavior is reversed by spinal or systemic administration of glial inhibitors, antagonists to p38 MAP kinase or pro-inflammatory cytokines, NMDA receptor antagonists and κ-opioids, but not by morphine (Eliav et al. 1999; Milligan et al. 2003). Pretreatment with agents to antagonize complement receptor C5a or to prevent production of membrane attack complex substantially reduces GD_2-associated pain (Sorkin et al. 2010), and pain behavior resulting from zymosan applied to the nerve is prevented or reversed by blockade further up the alternative complement pathway at complement fragment C1 (Twining et al. 2005).

Significantly, epineural TNF, Complete Freund's adjuvant, IL-1β and IL-6 and systemic antibody to the GD_2 ganglioside elicit bursting activity in C and to a lesser extent in finely myelinated Aδ fibers (Sorkin et al. 1997; Ozaktay et al. 2002; Eliav et al. 2009). The TNF-induced axonal firing is enhanced under conditions of inflammation and/or nerve damage. Following damage to the selected sciatic nerve fibers, intermingled undamaged nerve fibers develop ongoing discharge and an increased response to TNF, mediated primarily through the TNFR2 receptor (Schafers et al. 2008). Low-dose intravenous lidocaine (1–2 mg/ml plasma levels) temporarily turns off this ectopic activity, indicating that sodium channel stabilizing agents may be an effective treatment of the symptomatology, but not the origins of the neuritis-associated pain. The TNF effect on peripheral nerve fibers is likely to include both PKA (Zhang et al. 2002) and p38 (Schafers et al. 2003) linkages. Activation of p38 may lead to potentiation of TTX-resistant sodium currents (Jin and Gereau 2006).

Animal models for multiple schlerosis (experimental autoimmune encephalomyelitis) have been used to study peripheral nerve injury, but until recently pain was not assessed in these animals. More recent studies have now demonstrated mechanical and thermal sensitivity in rats immunized with a myelin fragment (Aicher et al. 2004; Thibault et al. 2011) and have started to examine the pharmacology of the pain behavior.

Clinical Syndromes

Neuritis resulting from probable autoimmune pathogenesis has several common characteristics. There is a known or presumed antibody response against an epitope found within the peripheral nerve. The most common epitopes involved in molecular mimicry resulting in neuritis are the gangliosides and myelin-associated proteins (MAG) found in surface membranes. Neural membranes are enriched with sialic acid containing sphingolipids, including gangliosides. Detectable plasma levels of antibodies to various plasma proteins are associated with sensory ataxic neuropathy, Miller-Fisher syndrome, paraneoplastic neuropathy, and many variants of Guillain-Barré syndrome (GBS) (Quarles and Weiss 1999; Dalakas 1995). In the first month of the disease, 70–90 % of GBS patients complain of severe pain (Moulin et al. 1997). Acute GBS has been reported secondary to *Campylobacter jejuni*, Epstein-Barr virus, and cytomegalovirus (van der Meche 1995). More rarely it has occurred following therapeutic administration of ganglioside mixtures (Illa et al. 1995). Interestingly, systemic administration of antibodies to the GD_2 ganglioside as an immunotherapy results in a whole body allodynia; nerve biopsy in adult patients (who have more severe symptoms than pediatric patients) demonstrates demyelination and mononuclear infiltrates in both the endoneurial and perivascular spaces (Saleh et al. 1992). Thus, exogenous antibodies to nerve gangliosides can initiate a prominent neuritis. Recent animal studies indicated significant diminution of pain behavior when antibodies with a point mutation to decrease complement fixation were employed (Sorkin et al. 2010), and clinical trials with the mutated antibody are currently under way and thus far produce comparatively no pain behavior in children.

In both GBS and CIPD timely removal of
antibodies using plasma exchange or suppression
of their synthesis by ▶ immunoglobulin therapy
remains the most effective treatment. An early
study of diabetic patients with polyneuropathy
that meets the electrophysiological criteria for
CIPD suggests that immunoglobulin therapy
may elicit clinical improvement in this population indicating similar etiologies for both diseases
(Sharma et al. 2002).

References

Aicher, S. A., Silverman, M. B., et al. (2004).
    Hyperalgesia in an animal model of multiple sclerosis.
    Pain, 110(3), 560–570.
Chacur, M., Milligan, E. D., et al. (2001). A new model of
    sciatic inflammatory neuritis (SIN): Induction of uni-
    lateral and bilateral mechanical allodynia following
    acute unilateral peri-sciatic immune activation in
    rats. Pain, 94(3), 231–244.
Dalakas, M. C. (1995). Basic aspects of neuroimmunology
    as they relate to immunotherapeutic targets:
    Present and future prospects. Annals of Neurology,
    37(Suppl 1), S2–S13.
Eliav, E., Herzberg, U., et al. (1999). Neuropathic pain
    from an experimental neuritis of the rat sciatic nerve.
    Pain, 83, 169–182.
Eliav, E., Benoliel, R., et al. (2009). The role of IL-6 and
    IL-1beta in painful perineural inflammatory neuritis.
    Brain, Behavior, and Immunity, 23(4), 474–484.
Gazda, L. S., Milligan, E. D., et al. (2001). Sciatic inflam-
    matory neuritis (SIN): Behavioral allodynia is
    paralleled by peri-sciatic proinflammatory cytokine
    and superoxide production. Journal of the Peripheral
    Nervous System, 6(3), 111–129.
Illa, I., Ortiz, N., et al. (1995). Acute axonal Guillain-
    Barré syndrome with IgG antibodies against motor
    axons following parenteral gangliosides. Annals of
    Neurology, 38(2), 218–224.
Jin, X., & Gereau, R. W., 4th. (2006). Acute p38-mediated
    modulation of tetrodotoxin-resistant sodium channels
    in mouse sensory neurons by tumor necrosis factor-
    alpha. The Journal of Neuroscience, 26(1), 246–255.
Linbloom, U., Mersky, H., Mumford, J. M., Nathan, P. W.,
    Noordenbos, W., & Sunderland, S. (1986). Pain terms:
    a current list with definitions and usage. Pain, Supple-
    ment, 3, S215–S221.
Milligan, E. D., Twining, C., et al. (2003). Spinal glia and
    proinflammatory cytokines mediate mirror-image neu-
    ropathic pain in rats. The Journal of Neuroscience,
    23(3), 1026–1040.
Moulin, D. E., Hagen, N., et al. (1997). Pain in Guillain-
    Barré syndrome [see comments]. Neurology, 48(2),
    328–331.

Onda, A., Yabuki, S., et al. (2001). Effects of lidocaine on
    blood flow and endoneurial fluid pressure in a rat
    model of herniated nucleus pulposus. Spine (Phila Pa
    1976), 26(20), 2186–2191. Discussion 2191–2.
Ozaktay, A. C., Cavanaugh, J. M., et al. (2002). Dorsal
    root sensitivity to interleukin-1 beta, interleukin-6 and
    tumor necrosis factor in rats. European Spine Journal,
    11(5), 467–475.
Quarles, R. H., & Weiss, M. D. (1999). Autoantibodies
    associated with peripheral neuropathy. Muscle &
    Nerve, 22(7), 800–822.
Saleh, M. N., Khazaeli, M. B., et al. (1992). Phase I trial of
    the chimeric anti-GD2 monoclonal antibody ch14.18
    in patients with malignant melanoma. Human Anti-
    bodies and Hybridomas, 3(1), 19–24.
Schafers, M., Svensson, C. I., et al. (2003). Tumor necro-
    sis factor-alpha induces mechanical allodynia after
    spinal nerve ligation by activation of p38 MAPK in
    primary sensory neurons. The Journal of Neurosci-
    ence, 23(7), 2517–2521.
Schafers, M., Sommer, C., et al. (2008). Selective stimu-
    lation of either tumor necrosis factor receptor differ-
    entially induces pain behavior in vivo and ectopic
    activity in sensory neurons in vitro. Neuroscience,
    157(2), 414–423.
Sharma, K. R., Cross, J., Ayyar, D. R., Martinez-Arizala,
    A., & Bradley, W. G. (2002). Diabetic demyelinating
    polyneuropathy responsive to intravenous immuno-
    glubulin therapy. Archives of Neurology, 59, 751–757.
Sorkin, L., & Doom, C. (2000). Epineurial application of
    TNF elicits an acute mechanical hyperalgesia in the
    awake rat. Journal of the Peripheral Nervous System,
    5, 96–1000.
Sorkin, L. S., Xiao, W. H., et al. (1997). Tumour necrosis
    factor-alpha induces ectopic activity in
    nociceptive primary afferent fibres. Neuroscience,
    81(1), 255–262.
Sorkin, L. S., Otto, M., et al. (2010). Anti-GD(2) with an
    FC point mutation reduces complement fixation and
    decreases antibody-induced allodynia. Pain, 149(1),
    135–142.
Thibault, K., Calvino, B., et al. (2011). Characterisation of
    sensory abnormalities observed in an animal model of
    multiple sclerosis: A behavioural and pharmacological
    study. European Journal of Pain, 15, 231.e1–16.
Twining, C. M., Sloane, E. M., et al. (2005). Activation of
    the spinal cord complement cascade might contribute
    to mechanical allodynia induced by three animal
    models of spinal sensitization. The Journal of Pain,
    6(3), 174–183.
van der Meche, F. G., & van Doorn, P. A. (1995). Guillain-
    Barre syndrome and chronic inflammatory demyelin-
    ating polyneuropathy: Immune mechanisms and
    update on current therapies. Annals of Neurology, 37
    (Suppl 1), S14–S31.
Zhang, J. M., Li, H., et al. (2002). Acute topical applica-
    tion of tumor necrosis factor alpha evokes protein
    kinase A-dependent responses in rat sensory neurons.
    Journal of Neurophysiology, 88(3), 1387–1392.

Inflammatory Neuropathy

▶ Inflammatory Neuritis

Inflammatory Nociceptor Sensitization, Prostaglandins and Leukotrienes

Blair D. Grubb
Department of Cell Physiology and
Pharmacology, University of Leicester,
Leicester, UK

Synonyms

Leukotrienes in inflammatory nociceptor sensitization; Prostaglandins in inflammatory nociceptor sensitization

Definition

The term prostaglandins (PG) describes a group of oxygenated polyunsaturated fatty acids containing a cyclopentane ring and two alkyl side chains that are derived from arachidonic acid by the action of cyclooxygenase (cox) enzymes. The alphanumeric nomenclature (e.g., E_2, I_2, D_2, $F_{2\alpha}$) describes the nature and position of side groups and double bonds on the cyclopentane ring and of double bonds on the alkyl side groups. The leukotrienes (e.g., LTB_4, LTC_4, LTD_4) are a related group of conjugated trienes synthesized from arachidonic acid by the enzyme 5-lipoxygenase (Benedetto et al. 1987).

Characteristics

The link between PG synthesis and nociceptor sensitization was established over 30 years ago when it was discovered that nonsteroidal anti-inflammatory drugs (NSAIDs) produced analgesia by inhibiting PG synthesis (Vane 1971;

Ferreira 1972). PGs elicit nociceptor sensitization by lowering mechanical and thermal activation thresholds in nociceptive nerve endings, thus reducing the stimulus intensity required to elicit action potential firing in the afferent axon.

PGs are synthesized by cyclooxygenase (cox) enzymes from their natural precursor arachidonic acid (Fig. 1). Three cox isoforms have been cloned in mammalian species: cox-1, cox-2, and cox-3. Cox-1 is a constitutively expressed isoform present in most tissues and is responsible for PG synthesis associated with normal physiological processes. Cox-2 is constitutively expressed in some tissues, e.g., brain, spinal cord, and kidney, but in most tissues expression levels are low unless the tissues are damaged. Cox-2 gene expression is upregulated in somatic and neural tissues 3–12 h after tissue damage as part of the normal immune response. Since PGs are lipids and cannot be stored inside cells, tissue damage results in an increase in extracellular levels of PGs, which can contribute to the sensitization of nociceptors. Cox-3 was identified only recently and is a splice variant of cox-1 that is present in dog and appears to be the molecular target for acetaminophen (paracetamol) in this species. In humans, however, cox-3 protein is probably not expressed due to a frameshift mutation that results in an incomplete gene product.

Although several different PG species exist, only PGE_2 and PGI_2 (prostacyclin) produce pronounced nociceptor sensitization to mechanical and thermal stimuli (Schaible and Grubb 1993; Bley et al. 1998). It should be noted, however, that PGs can also directly excite some nociceptive nerve terminals suggesting that they might have more than one mode of action (Birrell et al. 1991; Schepelmann et al. 1992; Schmelz et al. 2003). There are many examples of PG sensitization of nociceptors skin, muscle, and joint in vivo. The sensitization of joint afferents has been studied extensively in studies where these compounds or their analogues have been applied by close arterial injections. PGE_2 primarily elicits sensitizing responses to mechanical stimuli, while PGI_2 and analogues produce both sensitization and excitation. Interestingly, PGs do not act independently to produce nociceptor

Inflammatory Nociceptor Sensitization, Prostaglandins and Leukotrienes, Fig. 1 A cartoon showing the possible membrane targets for prostaglandins and leukotrienes in nociceptors' peripheral terminals. Prostaglandins seem to mediate their effects through GPCRs located on the cell membrane, while leukotrienes can elicit either direct or indirect effects. *Solid lines* indicate pathways for which there is solid evidence, while *dotted lines* show possible pathways

sensitization but act in concert with other inflammatory mediators. A good example of this is PGE$_2$-mediated enhancement of bradykinin responses in articular nociceptors. Bradykinin alone is a potent algogen and electrophysiological recordings from naïve primary afferents show that it directly activates a proportion of nociceptors and also enhances responses to mechanical stimuli. In the presence of PGs however, these actions are significantly enhanced (Schaible and Schmidt 1988; Birrell et al. 1993). Interestingly, bradykinin-induced sensitization of nociceptors to some stimuli can be attenuated by NSAIDs, suggesting that a component of the bradykinin effect itself involves synthesis of PGs (Peth et al. 2001).

PGs elicit nociceptor sensitization by binding to G-protein-coupled PG receptors on nerve terminals and inducing changes in membrane excitability. PG receptors are named after the PG species that is the endogenous ligand (EP, IP, DP, and FP). There are four EP receptor subtypes, EP$_{1-4}$, and two of these, EP$_1$ and EP$_3$, can express varying numbers of splice variants in different species. The EP$_3$ receptor is most prolific in this respect having up 9 splice variants dependent on species. By contrast there are only single IP, DP, and FP receptor subtypes. This diversity of receptors and receptor subtypes mean that a number of signalling cascades are activated by PGs depending on the receptor subtype that is expressed in each tissue. In sensory neurons, mRNA for all EP receptors and for IP receptors has been identified (Oida et al. 1995; Donaldson et al. 2001) (Fig. 1). However, a lack of suitable antibodies has precluded an objective assessment of PG receptor distribution in sensory ganglia, and there is little known about the co-expression of individual prostaglandin receptors/subtypes with phenotypic markers of different class of primary afferent neurons, e.g., peptidergic nociceptors (CGRP expressing) and non-peptidergic (labelled by isolectin B4) and non-nociceptive fibers (labelled by microfilament protein antibody, RT-97).

EP$_1$ receptors couple primarily to the G$_{q/11}$ family of G proteins, and their activation results in an activation of protein kinase C (PKC) and elevated levels of inositol 1,4,5 trisphosphate (IP$_3$) and diacylglycerol. This results in an increase in intracellular calcium and activation of protein kinase C, which may underlie the regulation of ion channel conductances or mechanical or thermal transducer proteins. EP_2 and EP_4 receptors couple to the G_s family of G proteins, and ligand binding results in activation of adenylyl cyclase and an increase in intracellular concentrations of cAMP and protein kinase A (PKA). Like PKC, PKA can phosphorylate and regulate a number of different ion channels to change membrane excitability. The G-protein coupling of EP_3 receptors is less well understood since there is a multitude of similar but subtly different splice variants in different species. It is clear, however, that depending on the spice variant concerned, $G_{q/11}$, $G_{i/o}$ (or even G_s) proteins could be activated following ligand binding to EP_3 receptors. IP receptor coupling is primarily G_s linked, although some studies have suggested that there is a $G_{q/11}$-mediated activation of phosphoinositide turnover (Bley et al. 1998) (Fig. 1).

Primary afferent nociceptors are phenotypically distinct, expressing a number of ion channels that can be modulated by PGs. PGE_2 can regulate the mechanical and thermal transduction processes by modulating capsaicin/heat-sensitive TRPV1 currents and mechanosensitive currents in sensory neurons through PKA-dependent pathways (Lopshire and Nicol 1998; Cho et al. 2002). In addition to transducer channels, PGE_2 can modulate the activity of a number of channels that regulate action potential threshold and firing frequency (Fig. 2). In dorsal root ganglion neurons, for example, the activation kinetics (firing threshold) of the tetrodotoxin-resistant sodium current is altered by PGE_2 (England et al. 1996) through cAMP/PKA-dependent phosphorylation of $Na_{V1.8}$ (Fitzgerald et al. 1999) and PKC-dependent mechanisms (Gold et al. 1998).

Potassium channels that control action potential repolarization are also targets for PGs. PGE_2 and PGI_2 inhibit delayed rectifier potassium currents to increase cell excitability and action potential firing (Nicol et al. 1997; Jiang et al. 2003). In addition, calcium-activated potassium channels present in nociceptors are modulated by PGE_2 resulting in a decrease in spike frequency adaptation and an increase in action potential

Inflammatory Nociceptor Sensitization, Prostaglandins and Leukotrienes, Fig. 2 Pathways for prostaglandin and leukotriene synthesis showing the chemical species that have been shown to sensitize nociceptors in *bold text*. Those not implicated in nociceptor sensitization are in *grey*. Prostaglandins and leukotrienes mediate their actions through GPCRs (EP_{1-4}, IP, BLT1 and 2, $CysLT_{1/2}$) although leukotrienes and related polyunsaturated fatty acids (*dashed lines*) can directly gate the TRPV1 ion channel

firing frequency (Weinreich and Wonderlin 1987).

In comparison to PGs, the role of leukotrienes and their receptors in nociceptor sensitization has been neglected. There is good evidence showing that leukotrienes are present in inflammatory exudates but a limited number of studies where leukotrienes have been implicated in nociceptor sensitization (Martin et al. 1987; Madison et al. 1990). Furthermore, it is not yet known whether leukotriene receptors are present on nociceptor terminals nor if they mediate the sensitizing actions of leukotrienes (Figs. 1 and 2). Indeed there is much better evidence that leukotrienes can influence nociceptor sensitization through direct interactions with ion channel proteins. This has been demonstrated for LTB_4, which is structurally related to capsaicin and activates TRPV1, the heat-sensitive ion channel, by binding to the cytoplasmic domain of this protein (Hwang et al. 2000). Furthermore a number of other ion channels are known to be modulated by leukotrienes in different tissues, although it is not yet known whether this occurs in nociceptors.

References

Benedetto, C., McDonald-Gibson, R. G., Nigam, S., et al. (1987). *Prostaglandins and related substances: A practical approach.* Oxford/Washington, DC: IRL Press.

Birrell, G. J., McQueen, D. S., Iggo, A., et al. (1991). PGI2-induced sensitisation of articular mechanonociceptors. *Neuroscience Letters, 124*, 5–8.

Birrell, G. J., McQueen, D. S., Iggo, A., et al. (1993). Prostanoid-induced potentiation of the excitatory and sensitising effects of bradykinin on articular mechanonociceptors in the rat ankle joint. *Neuroscience, 54*, 531–544.

Bley, K. R., Hunter, J. C., Eglen, R. M., et al. (1998). The role of IP prostanoid receptors in inflammatory pain. *Trends in Pharmacological Sciences, 19*, 141–147.

Cho, H., Shin, J., Shin, C. Y., et al. (2002). Mechanosensitive ion channels in cultured sensory neurons of rats. *The Journal of Neuroscience, 22*, 1238–1247.

Donaldson, L. F., Humphrey, P. S., Oldfield, S., et al. (2001). Expression and regulation of prostaglandin E receptor subtype mRNAs in rat sensory ganglia and spinal cord in response to peripheral inflammation. *Prostaglandins & Other Lipid Mediators, 63*, 109–122.

England, S., Bevan, S., & Docherty, R. J. (1996). PGE$_2$ modulates the tetrodotoxin-resistant sodium current in neonatal rat dorsal root ganglion neurones via the cyclic AMP-protein kinase a cascade. *The Journal of Physiology, 495*, 429–440.

Ferreira, S. H. (1972). Prostaglandins, aspirin-like drugs and analgesia. *Nature: New Biology, 240*, 200–203.

Fitzgerald, E. M., Okuse, K., Wood, J. N., Dolphin, A. C., & Moss, S. J. (1999). cAMP-dependent phosphorylation of the tetrodotoxin-resistant voltage-dependent sodium channel SNS. *The Journal of Physiology, 516*(Pt 2), 433–446. doi:10.1111/j.1469-7793.1999.0433v.x.

Gold, M. S., Levine, J. D., & Correa, A. M. (1998). Modulation of TTX-R I$_{Na}$ by PKC and PKA and their role in PGE$_2$-induced sensitisation of rat sensory neurones in vitro. *The Journal of Neuroscience, 18*, 10345–10355.

Hwang, S. W., Cho, H., Kwak, J., et al. (2000). Direct activation of capsaicin receptors by products of lipoxygenases: Endogenous capsaicin-like substances. *PNAS, 97*, 6155–6160.

Jiang, X., Zhang, Y. H., Clark, J. D., et al. (2003). Prostaglandin E2 inhibits the potassium current in sensory neurones from hyperalgesic Kv1.1 knockout mice. *Neuroscience, 119*, 65–72.

Lopshire, J. C., & Nicol, G. D. (1998). The cAMP transduction cascade mediates the prostaglandin E2 enhancement of the capsaicin-elicited current in rat sensory neurons: Whole cell and single channel studies. *The Journal of Neuroscience, 18*, 6081–6092.

Madison, S., Whilsel, E. A., Suarez-Roca, H., et al. (1990). Sensitising effects of leukotriene B4 on intradental primary afferents. *Pain, 49*, 99–104.

Martin, H. A., Basbaum, A. I., Kwiat, G. C., et al. (1987). Leukotriene and prostaglandin sensitisation of cutaneous high threshold C- and A-delta mechanonociceptors in the hairy skin of the rat hindlimbs. *Neuroscience, 22*, 651–659.

Nicol, G. D., Vasko, M. R., & Evans, A. R. (1997). Prostaglandins suppress an outward potassium current in embryonic rat sensory neurones. *The Journal of Physiology, 77*, 167–176.

Oida, H., Namba, T., Sugimoto, Y., et al. (1995). In situ hybridization studies of prostacyclin receptor mRNA expression in various mouse organs. *British Journal of Pharmacology, 116*, 2828–2837.

Peth, Ã. G., Derow, A., & Reeh, P. W. (2001). Bradykinin-induced nociceptor sensitisation to heat is mediated by cyclooxygenase products in isolated rat skin. *European Journal of Neuroscience, 14*, 210–218.

Schaible, H. G., & Grubb, B. D. (1993). Afferent and spinal mechanisms of joint pain. *Pain, 55*, 5–54.

Schaible, H.-G., & Schmidt, R. F. (1988). Excitation and sensitisation of fine articular afferents from cat's knee joint by prostaglandin E$_2$. *The Journal of Physiology, 403*, 91–104.

Schepelmann, K., Messlinger, K., Schaible, H. G., et al. (1992). Inflammatory mediators and nociception in the joint: Excitation and sensitization of slowly conducting afferent fibres of the cat's knee by prostaglandin I$_2$. *Neuroscience, 50*, 237–247.

Schmelz, M., Schmidt, R., Weidner, C., et al. (2003). Chemical response pattern of different classes of c-nociceptors to pruritogens and algogens. *Journal of Neurophysiology, 89*, 2441–2448.

Vane, J. R. (1971). Inhibition of prostaglandin synthesis as a mechanism of action for aspirin-like drugs. *Nature: New Biology, 231*, 232–235.

Weinreich, D., & Wonderlin, W. F. (1987). Inhibition of calcium-dependent spike after-hyperpolarisation increases excitability of rabbit visceral sensory neurones. *The Journal of Physiology, 294*, 415–427.

Inflammatory Pain

Definition

Inflammatory pain is that which is associated with chronic inflammation (e.g., present in arthritis, back pain, or temporomandibular joint disorders).

Cross-References

▶ Human Models of Inflammatory Pain
▶ Neutrophils in Inflammatory Pain
▶ NSAIDs, Mode of Action

Inflammatory Pain and NGF

David L. H. Bennett
Wolfson CARD, Kings College London,
London, UK
Nuffield Department of Clinical Neurosciences,
John Radcliffe Hospital, The University of
Oxford, Oxford, UK

Definition

Inflammation is the body's reaction to injury in which part of the body becomes hot, red, swollen, and painful. At the cellular level initially neutrophils and subsequently macrophages infiltrate the affected area. Many chemical mediators are involved in this process, one of which is nerve growth factor (NGF). NGF is a secreted protein of molecular mass of 13 kD which exists as a homodimer. It is a member of the neurotrophin family, which also includes brain-derived neurotrophic factor (BDNF), neurotrophin-3 (NT3), and neurotrophin-4/5 (NT4/5). NGF binds to both a high-affinity tyrosine kinase receptor trkA and a low-affinity receptor p75.

Characteristics

NGF is the archetypal "target-derived neurotrophic factor" and was initially characterized for its neurodevelopmental effects. During the period of naturally occurring cell death, the survival of postganglionic sympathetic neurons and small-diameter nociceptive sensory neurons is critically dependent on limiting amounts of NGF produced in their target fields. This role of NGF is shown in people who have a mutation in the trkA-NGF receptor resulting in congenital insensitivity to pain and anhidrosis in which nociceptive neurons and sympathetic neurons fail to develop (Indo 2002). During development, virtually all small-diameter nociceptive neurons express trkA and are NGF dependent; however, this receptor is downregulated during the postnatal period (Bennett et al. 1996). In the adult trkA expression is restricted to those small-diameter DRG cells which express the neuropeptides substance P (SP) and calcitonin gene-related peptide (CGRP) (Fig. 1). There is a second group of small-diameter DRG cells which do not express neuropeptides but possess cell surface glycoconjugates that bind the lectin isolectin B4 (IB4). These cells do not possess trkA receptors but express receptors for and are sensitive to another trophic factor glial cell line-derived neurotrophic factor. NGF expression is required for the maintenance of a healthy neuronal phenotype in adult peptidergic DRG cells. NGF deprivation in vivo results in reduced thermal and chemical sensitivity in these neurons which is accompanied by a withdrawal of their terminals from their target fields (McMahon et al. 1995).

There is now strong evidence that NGF is an important mediator of inflammatory pain hypersensitivity: NGF levels increase following inflammation, NGF has been shown to sensitize nociceptive systems, and most importantly blocking NGF bioactivity reduces inflammatory pain. The expression of NGF increases rapidly following an inflammatory stimulus, and such increased levels have been shown in animals after inflammation induced by skin wounding, skin blistering, ultraviolet light, complete Freund's adjuvant, and carrageenan and in

Inflammatory Pain and NGF, Fig. 1 Pie chart relating trophic factor sensitivity to the different neurochemically defined classes of DRG cells. Large-diameter DRG cells (A-fibers) have myelinated axons and express the neurofilament NF200. These cells are responsive principally to the trophic factors NT-3 and GDNF. Small-diameter DRG cells (C-fibers) have unmyelinated axons and can be broadly divided into two groups. The first group expresses the neuropeptides SP and CGRP and is NGF sensitive. The second group is non-peptidergic and binds the lectin IB4, and these cells are GDNF sensitive

a model of cystitis. Increased NGF levels have also been demonstrated in the bladder of patients with cystitis and in the synovial fluid of patients with arthritis. Studies in cell culture indicate that many growth factors and cytokines can increase NGF mRNA levels and NGF secretion; these include interleukin-1 (IL-1), IL-4, IL-5, tumor necrosis factor-α (TNFα), transforming growth factor-β (TGFβ), platelet-derived growth factor (PDGF), acid and basic fibroblast growth factors (FGF-1, FGF-2), and epidermal growth factor (EGF). Both IL-1 and TNFα have been shown to be important in increasing NGF levels in vivo (Safieh-Garabedian et al. 1995; Woolf et al. 1997). Some agents have been shown to consistently reduce NGF mRNA levels and secretion; these include glucocorticoids and the interferons.

The administration of small doses of NGF to adult animals and man can produce pain and hyperalgesia. In rodents a thermal hyperalgesia develops within 30 min of systemic NGF administration, and both a thermal hyperalgesia and a mechanical hyperalgesia are present after a couple of hours (Lewin et al. 1993). Subcutaneous NGF produces a thermal and mechanical

hyperalgesia at the injection site (Andreev et al. 1995). In humans intravenous injections of very low doses of NGF produce widespread myalgia and hyperalgesia at the injection site. Injection site hyperalgesia was a prominent side effect in trials of the use of NGF in the treatment of diabetic neuropathy (Apfel et al. 2000).

The critical test of the hypothesis that NGF has a role in mediating inflammatory pain was to show that blocking NGF bioactivity could reduce the hyperalgesia induced by inflammation and this has now been demonstrated by multiple groups using different model systems. We have used a fusion protein of the ligand-binding domain of the trkA receptor and Fc region of human IgG (McMahon et al. 1995). This binds to and therefore blocks the activity of endogenous NGF. Application of trkA-IgG either locally or systemically could significantly reduce the thermal hyperalgesia resulting from intraplantar administration of carrageenan in the rodent (Fig. 2). This molecule also prevents the development of thermal hyperalgesia following skin incision and detrusor hyperexcitability in a rodent model of cystitis. Similar findings have been shown using function-blocking antibodies to NGF in inflammation evoked by complete Freund's adjuvant (Woolf et al. 1994; Safieh-Garabedian et al. 1995). Blocking NGF reduces thermal hyperalgesia and in some studies mechanical hyperalgesia, without altering many of the other aspects of the inflammatory process such as the degree of edema. There is now increasing understanding of how NGF contributes to inflammatory pain.

NGF administration causes a sensitization of nociceptors to both thermal and chemical stimuli. These effects are probably mediated by a combination of acute changes in second-messenger pathways within the nociceptor terminal itself as well as more chronic changes in gene expression within nociceptive neurons. The mechanisms by which NGF produces nociceptor sensitization are the subject of a separate chapter, NGF sensitization of nociceptors. The algesic actions of NGF are not only limited to peripheral actions, but in addition it has secondary effects on the spinal processing of nociceptive information. Firstly, the activation and sensitization of

Inflammatory Pain and NGF, Fig. 2 The NGF-sequestering molecule trkA-IgG was used to investigate the role of NGF in inflammation. Thermal hyperalgesia develops in rats within hours following intraplantar injection of carrageenan. Animals which are inflamed and concurrently treated with trkA-IgG fail to develop most of the thermal hyperalgesia (Adapted from McMahon et al. 1995)

primary afferent nociceptors may lead to sufficient afferent activity to trigger central changes. Peripheral NGF administration to some visceral structures results in somatotopically appropriate induction of *c-fos* in the dorsal horn. Secondly, NGF can modulate the expression of a number of neuropeptides and neuromodulators expressed within nociceptive neurons.

NGF administration has been shown to increase the expression of SP and CGRP within nociceptive neurons both in vitro and in vivo. Following NGF treatment, the release of SP within the dorsal horn is increased. The release of sensory neuropeptides is a well-known trigger for the induction of central sensitization. Several hours after systemic NGF treatment, C-fiber stimulation produces greater than normal amounts of central sensitization seen as windup of ventral root reflexes (Thompson et al. 1995). A-fibers also develop the novel ability to produce windup during inflammation, and this may be secondary to novel SP expression within A-fibers (Thompson et al. 1995; Neumann et al. 1996). NGF can also modulate the central processing of nociceptive information through a novel mechanism – by increasing the expression and release of a second neurotrophic factor BDNF. BDNF is normally expressed within the NGF-sensitive population of small-diameter DRG cells and is anterogradely transported to their terminals within the superficial laminae of the dorsal horn where it is present within dense-core vesicles

(Michael et al. 1997). NGF and inflammation increase the expression of BDNF within these cells. The elevated expression of BDNF is accompanied by increased release of this factor following C-fiber stimulation. Exogenous BDNF results in facilitation of the flexor reflex (Kerr et al. 1999). This facilitation is mediated through activation of ERK MAP kinase in dorsal horn neurons, phosphorylation of the NMDA receptor subunit 1, and facilitation of NMDA receptor-mediated responses (Lever et al. 2003). A trkB-IgG fusion protein has been used to sequester endogenous BDNF released within the dorsal horn and has demonstrated that BDNF has a role in the generation of hyperalgesia produced by inflammation or NGF administration (Kerr et al. 1999; Mannion et al. 1999).

In summary inflammation leads to a rapid induction of NGF expression, and this can lead to a sensitization of pain signalling systems via both peripheral and central mechanisms. There is now strong evidence that blocking NGF bioactivity reduces inflammatory hyperalgesia. NGF therefore provides a novel analgesic target, and indeed clinical trials of the use of function-blocking molecules in the treatment of inflammatory pain are under way.

Cross-References

▶ NGF, Sensitization of Nociceptors

References

Andreev, N. Y., Dimitrieva, N., Koltzenburg, M., et al. (1995). Peripheral administration of nerve growth factor in the adult rat produces a thermal hyperalgesia that requires the presence of sympathetic post-ganglionic neurones. *Pain, 63*, 109–115.

Apfel, S. C., Schwartz, S., Adornato, B. T., et al. (2000). Efficacy and safety of recombinant human nerve growth factor in patients with diabetic polyneuropathy: A randomized controlled trial. RhNGF clinical investigator group. *The Journal of the American Medical Association, 284*, 2215–2221.

Bennett, D. L., Averill, S., Clary, D. O., et al. (1996). Postnatal changes in the expression of the trkA high-affinity NGF receptor in primary sensory neurons. *The European Journal of Neuroscience, 8*, 2204–2208.

Indo, Y. (2002). Genetics of congenital insensitivity to pain with anhidrosis (CIPA) or hereditary sensory and autonomic neuropathy type IV. Clinical, biological and molecular aspects of mutations in TRKA (NTRK1) gene encoding the receptor tyrosine kinase for nerve growth factor. *Clinical Autonomic Research, 12*(Suppl 1), 120–132.

Kerr, B. J., Bradbury, E. J., Bennett, D. L., et al. (1999). Brain-derived neurotrophic factor modulates nociceptive sensory inputs and NMDA-evoked responses in the rat spinal cord. *The Journal of Neuroscience, 19*, 5138–5148.

Lever, I. J., Pezet, S., McMahon, S. B., et al. (2003). The signaling components of sensory fiber transmission involved in the activation of ERK MAP kinase in the mouse dorsal horn. *Molecular and Cellular Neurosciences, 24*, 259–270.

Lewin, G. R., Ritter, A. M., & Mendell, L. M. (1993). Nerve growth factor-induced hyperalgesia in the neonatal and adult rat. *The Journal of Neuroscience, 13*, 2136–2148.

Mannion, R. J., Costigan, M., Decosterd, I., et al. (1999). Neurotrophins: Peripherally and centrally acting modulators of tactile stimulus-induced inflammatory pain hypersensitivity. *Proceedings of the National Academy of Sciences USA, 96*, 9385–9390.

McMahon, S. B., Bennett, D. L., Priestley, J. V., et al. (1995). The biological effects of endogenous nerve growth factor on adult sensory neurons revealed by a trkA-IgG fusion molecule. *Nature Medicine, 1*, 774–780.

Michael, G. J., Averill, S., Nitkunan, A., et al. (1997). Nerve growth factor treatment increases brain-derived neurotrophic factor selectively in TrkA-expressing dorsal root ganglion cells and in their central terminations within the spinal cord. *The Journal of Neuroscience, 17*, 8476–8490.

Neumann, S., Doubell, T. P., Leslie, T., et al. (1996). Inflammatory pain hypersensitivity mediated by phenotypic switch in myelinated primary sensory neurons. *Nature, 384*, 360–364.

Safieh-Garabedian, B., Poole, S., Allchorne, A., et al. (1995). Contribution of interleukin-1 beta to the inflammation-induced increase in nerve growth factor levels and inflammatory hyperalgesia. *British Journal of Pharmacology, 115*, 1265–1275.

Thompson, S. W., Dray, A., McCarson, K. E., et al. (1995). Nerve growth factor induces mechanical allodynia associated with novel a fibre-evoked spinal reflex activity and enhanced neurokinin-1 receptor activation in the rat. *Pain, 62*, 219–231.

Woolf, C. J., Safieh-Garabedian, B., Ma, Q. P., et al. (1994). Nerve growth factor contributes to the generation of inflammatory sensory hypersensitivity. *Neuroscience, 62*, 327–331.

Woolf, C. J., Allchorne, A., Safieh-Garabedian, B., et al. (1997). Cytokines, nerve growth factor and inflammatory hyperalgesia: The contribution of tumour necrosis factor alpha. *British Journal of Pharmacology, 121*, 417–424.

Inflammatory Pain and Opioids

▶ Opioids and Inflammatory Pain

Inflammatory Pain Models

Definition

The inflammatory process can be divided into three phases: acute, subacute, and chronic. The acute and subacute phases are characterized by local vasodilatation, increased capillary permeability, and infiltration of phagocytic cells. The distinction is the duration of the inflammation. Chronic inflammation is long-term, hence the name. The above acute changes are usually characterized chronically by a proliferative phase, in which tissue degeneration and fibrosis occur. Models for testing acute and subacute somatic (skin, joints, etc.) inflammation include UV-induced cutaneous erythema in guinea pigs, or oxazolone- or croton-oil-induced ear edema in mice. A well-established model of chronic inflammation in rats is produced by intradermally injecting a suspension of *Mycobacterium butyricum* into both hindpaws (see FCA-induced arthritis). Visceral inflammation requires different models such as experimental colitis caused by intracolonic instillation of trinitrobenzene sulfonic acid, bladder inflammation or cystitis caused by repeated systemic administration of cyclophosphamide, or pancreatitis caused by repeated systemic administration of caerulein or caerulein in combination with ethanol.

Cross-References

▶ Neuropeptide Release in Inflammation

Inflammatory Syndrome

▶ Lower Back Pain, Physical Examination

Infliximab

Definition

Monoclonal antibodies to TNF-α that serve as a TNF-α inhibitor in human autoimmune disorders.

Cross-References

▶ Cytokines as Targets in the Treatment of Neuropathic Pain

Inflow Artifacts

Definition

Susceptibility artifacts. In MRI, susceptibility artifacts are caused, for example, by medical devices in or near the magnetic field or by implants of the patient. These materials with magnetic susceptibility distort the linear magnetic field gradients, which results in bright areas (misregistered signals) and dark areas (no signal) nearby the

magnetic material. Susceptibility artifacts also occur at air-sinus-bone interfaces (e.g., in the orbitofrontal cortex and amygdala).

Cross-References

▶ Amygdala, Functional Imaging

Information and Psychoeducation in the Early Management of Persistent Pain

Steven J. Linton
Department of Occupational and Environmental Medicine, Orebro Medical Center, Orebro, Sweden

Synonyms

Back schools; Education; Health informatics; Pain schools; Patient information; Psychoeducation

Definition

Educational and informational approaches are concerned with providing patients with knowledge about their painful illness that will help them to cope better with the problem. Psychoeducation refers to approaches that in particular utilize psychological knowledge and advice, often provided in a "study group" format. In part, the aim is to provide basic knowledge about pain and how it operates, in order to increase knowledge and decrease distress and uncertainty about the pain problem. Moreover, informational activities almost always strive to change the patient's behavior to enhance coping.

Characteristics

Modern approaches to back and neck pain stress self-management. In order to achieve this, patients need to understand their problem and how they can manage it and consequently change their behavior in accordance. For example, a patient may need to understand that the pain itself is not harmful and that activity will enhance recovery. Further, the patient may be asked to engage in exercises, to take pain relievers on a specific time basis in order to remain active, and to practice relaxation.

A lack of information or knowledge may therefore have grave consequences in the treatment of musculoskeletal pain for three reasons. First, lack of a correct explanation of the illness, its cause, course, and treatment may well lead to uncertainty and an increase in anxiety and psychological distress. In other words, patients may worry and suffer needlessly. Second, when not properly informed, patients may not feel engaged in the process. Decisions about assessment and treatment may be perceived as over the head of the patient, thus preempting active participation. Third, self-care by the patient may be limited if the patient does not understand its importance. For example, patients may be unclear as to what they should do to enhance recovery. Moreover, if instead the patient feels she is being "taken care of," this may reduce motivation to engage in self-care activities. This is particularly salient because many of the self-care activities such as exercise involve considerable time and effort. Thus, there is good reason to believe that educational interventions may serve an important purpose.

Since communication problems in medicine are common, several attempts have been made to provide patients with clear information. In the rehabilitation clinic, patient information is sometimes used as a basic building block in multidimensional treatments for chronic pain that include many components. It is difficult to ascertain the extent of the effect of such information in relation to the effects of the other components. However, including such information is probably not particularly effective because the information is often limited in scope and provided to patients with difficult problems. At best, the information may be necessary but not sufficient to produce significant improvements (Burton and Waddell 2002).

Another approach has been to provide general health information. Some health authorities, for example, have provided booklets designed to help patients self-manage minor illnesses such as colds, flu, and cuts. However, controlled scientific tests of their effectiveness have been quite disappointing, as these booklets have little or no effect in reducing the number of clinical consultations, for example (Heaney et al. 2001).

Written Materials

To be effective, written information needs to be read, understood, believed relevant, and acted upon. A review of these premises, however, shows that it is far from the usual case (Ley 1997). Indeed, written information as the only intervention appears to have small effect in changing behaviors. However, it might be incorporated into practices to enhance communication.

In the musculoskeletal pain area, several attempts have been made to provide patients with concrete educational materials in the form of pamphlets, books, videos, or Internet services. These materials reflect different approaches primarily to back pain. Historically, written material was based on an ergonomic approach to back pain, where patients were provided with information about the structure of the spine and how injuries may occur. The message was focused on avoiding movements that are not ergonomic. In addition, these booklets usually advised patients to contact their doctor if any of a host of pain situations occurred. Not surprisingly this type of information has not proved to be effective in lowering distress, absenteeism, or consultation (Burton and Waddell 2002).

There is now rather extensive theoretical and scientific knowledge on which to base potent patient information. As a result, the current trend is to combine the growing evidence about the role of psychosocial factors and the emergence of guidelines. It has become evident that psychosocial factors may create barriers for recovery and return to work. This idea is based on recent knowledge that psychological factors are strongly associated with the transition from acute to chronic pain (Linton 2000). At the same time, several authorities have developed guidelines for the treatment of back pain that are derived from the scientific literature. Although the guidelines vary (Koes et al. 2001), there are a number of key recommendations included in most. These include prescribing activity rather than rest and providing reassurance to reduce anxiety and fear. Written materials have been developed to provide patients with the key messages. Consequently, these pamphlets often provide clear messages aimed at reducing fear, avoidance of activities, and distress, for example (Symonds et al. 1995).

A review of the studies employing written materials indicates that under certain conditions they can be effective (Burton and Waddell 2002). While written information remains a rather weak intervention, a coordinated effort where the information is employed to enhance the messages and advice provided by clinicians may be of value. Carefully selected and presented information presented in an uncompromising style that is in line with current management guidelines can have a positive effect on the beliefs patients have, as well as on clinical outcomes.

Schools
A number of educational efforts have been initiated in the form of a "school" or study group; these include back schools, neck schools, and pain management schools. They assume that an important reason why people develop problems is a lack of knowledge. Back schools, for example, may include a wide range of topics, but typically focus on body mechanics and ergonomics. Other topics normally included are exercise, lifting, and stress. A trained professional such as a physical therapist almost always provides these educational efforts in groups. However, there is great variation in the content, number, and length of sessions.

Unfortunately, the school concept has demonstrated limited effects. In a review of the effects as an early, preventive intervention, it was concluded that neck and back schools were not

effective (Linton and van Tulder 2001). While other reviews find some effects on knowledge and correct back posture, there is considerable agreement that back schools, at best, have only slight effects on variables such as health-care utilization or absenteeism and virtually no effects on clinical variables such as pain intensity (Maier-Riehle and Harter 2001).

Psychoeducation: CBT Groups as an Early Intervention
These learning experiences focus on psychological aspects of pain, such as developing effective coping strategies, altering dysfunctional attitudes, or alleviating fears about the problem. These interventions are almost always provided in groups and are a method of providing a psychologically oriented therapy in an effective and inexpensive way. A key concept is promoting self-help.

A psychoeducational group program, based on cognitive behavioral therapy, was developed for patients with arthritis and then modified to be applicable for musculoskeletal pain problems (Lorig et al. 1993). In one study, chronic pain patients were randomly assigned either to the psychoeducational program or to a waiting list control (LeFort et al. 1998). The course offered six 2 h sessions designed to maximize group problem-solving and self-management skills. Results indicated that the educational intervention group made significant improvements in pain, vitality, functioning, satisfaction, and self-efficacy in comparison to the control group.

More recently, programs have also been adapted for early intervention aimed at preventing the development of chronic back or neck pain (Linton 2002). Although neck and back pain are very common, only a relatively small number of patients develop persistent problems restricting work capacity and requiring health care. However, these small numbers consume the majority of the resources. Consequently, identification and providing early, preventive interventions could be an effective strategy to reduce the problem. Psychological factors have been found to be an important link in the transition from acute to persistent back pain, and therefore psychologically oriented interventions seem to be needed to address the problem properly (Linton 2000). This is a particularly relevant idea because early interventions focus on changing the participant's behavior and lifestyle.

The effects of providing an early, cognitive behavioral group intervention were tested in a randomized study of 255 primary care back pain patients (Von Korff et al. 1998). The cognitive behavioral intervention consisted of four sessions focusing on problem-solving skills, activity management, and educational videos. Relative to the control group, those receiving the cognitive behavioral group intervention significantly reduced their worry and disability and increased their self-help skills. Similar results have been reported in other studies (Moore et al. 2000; Saunders et al. 1999).

A cognitive behavioral intervention was specifically developed for primary care patients found to be at risk on psychosocial variables. In this program patients meet six times in groups of about eight people for 2 h once a week. The course is led by a professional and is based on helping participants to change their beliefs and behaviors. Several steps are taken to enhance engagement, such as prompting discussion, homework, and "hands-on" skills training. Each session contains a review of homework, a short presentation of educational material, problem solving, coping skills acquirement, and individualized homework assignments. The end product for participants is a personalized coping program that they design themselves. The utility of this approach has been tested in four separate randomized controlled trials (Linton and Andersson 2000; Linton and Ryberg 2001; Linton et al. 2006). All of these studies have demonstrated that this intervention has a clear preventive effect on future sick absenteeism and function. Thus, there is support for the contention that a psychoeducational intervention designed to match the psychological needs of the participants may help prevent the development of persistent musculoskeletal pain.

References

Burton, A. K., & Waddell, G. (2002). Educational and informational approaches. In S. J. Linton (Ed.), *New avenues for the prevention of chronic musculoskeletal pain and disability* (pp. 245–258). Amsterdam: Elsevier.

Heaney, D., Wyke, S., Wilson, P., et al. (2001). Assessment of impact of information booklets on use of healthcare services: Randomized controlled trial. *BMJ, 322,* 1218.

Koes, B. W., van Tulder, M. W., Ostelo, R., et al. (2001). Clinical guidelines for the management of low back pain in primary care. *Spine, 26,* 2504–2514.

LeFort, S. M., Gray-Donald, K., Rowat, K. M., et al. (1998). Randomized controlled trial of a community-based psychoeducation program for the self-management of chronic pain. *Pain, 74,* 297–306.

Ley, P. (1997). Written communication. In A. Baum, S. Newman, J. Weinman, et al. (Eds.), *Cambridge handbook of psychology, health, and medicine* (pp. 331–334). Cambridge: Cambridge University Press.

Linton, S. J. (2000). A review of psychological risk factors in back and neck pain. *Spine, 25,* 1148–1156.

Linton, S. J. (Ed.). (2002). *New avenues for the prevention of chronic musculoskeletal pain and disability* (Vol. 1). Amsterdam: Elsevier.

Linton, S. J., & Andersson, T. (2000). Can chronic disability be prevented? A randomized trial of a cognitive-behavior intervention and two forms of information for patients with spinal pain. *Spine, 25,* 2825–2831.

Linton, S. J., & Ryberg, M. (2001). A cognitive-behavioral group intervention as prevention for persistent neck and back pain in a non-patient population: A randomized controlled trial. *Pain, 90,* 83–90.

Linton, S. J., & van Tulder, M. W. (2001). Preventive interventions for back and neck pain: What is the evidence? *Spine, 26,* 778–787.

Linton, S. J., Boersma, K., Jansson, M., et al. (2006). The effects of cognitive-behavioral and physical therapy preventive interventions on pain related sick leave: A randomized controlled trial. *The Clinical Journal of Pain, 21*(2), 109–119.

Lorig, K., Mazonson, P. D., & Holman, H. R. (1993). Evidence suggesting that health education for self-management in patients with chronic arthritis has sustained health benefits while reducing health care costs. *Arthritis and Rheumatism, 36,* 439–446.

Maier-Riehle, B., & Harter, M. (2001). The effects of back schools: A meta-analysis. *International Journal of Rehabilitation Medicine, 24,* 199–206.

Moore, J., Von Korff, M., Cherkin, D., et al. (2000). A randomized trial of a cognitive-behavioral program for enhancing back pain self care in a primary care setting. *Pain, 88,* 145–153.

Saunders, K. W., Von Korff, M., Pruitt, S. D., et al. (1999). Prediction of physician visits and prescription medicine use for back pain. *Pain, 83*, 369–377.

Symonds, T. L., Burton, A. K., Tillotson, K. M., et al. (1995). Absence resulting from low back trouble can be reduced by psychosocial intervention at the work place. *Spine, 20*, 2738–2745.

Von Korff, M., Moore, J. E., Lorig, K., et al. (1998). A randomized trial of a lay-led self-management group intervention for back pain patients in primary care. *Spine, 23*, 2608–2615.

Information Processing in Thalamocortical Circuits

S. Murray Sherman
Department of Neurobiology, University of Chicago, Chicago, IL, USA

Synonyms

Cortical information flow; Corticocortical pathways

Definition

Until recently, communication among related cortical areas (e.g., those for somatosensation and pain) was thought to involve only direct connections. We (Sherman and Guillery 2002, 2006; Guillery and Sherman 2002a, b; Sherman 2012) suggest a radically new view in which much corticocortical communication involves a transthalamic, cortico-thalamo-cortical route.

Characteristics

To understand how information is processed in a thalamocortical system, it is important to identify and follow the route of information transfer. A recent suggestion based on thalamic circuitry is that not all pathways are equivalent but instead can be divided into "drivers," which are the information bearing pathways, and "modulators,"

which serve to modulate the flow of information rather than transmitting it. A long-held notion is that information is carried in glutamatergic pathways, and all other inputs (e.g., cholinergic and noradrenergic) are modulatory. However, it is now clear that most glutamatergic inputs in both thalamus and cortex are also modulatory (Sherman and Guillery 2002, 2006; Guillery and Sherman 2002a, b; Sherman 2012), and it thus becomes important to identify the subset of driver glutamatergic inputs to better understand the functioning of thalamic and cortical circuits. Our discussion of drivers and modulators that follows is limited to a consideration of glutamatergic afferents.

Drivers and Modulators

Figure 1 shows the basic circuit of the thalamus, which varies only slightly among thalamic relays. As argued previously, the glutamatergic inputs to relay cells can be divided into two basic types: Class 1 (or *drivers*) and Class 2 (or *modulators*), and these differ on a number of morphological and functional grounds that are briefly summarized in Table 1 (for details, see Sherman and Guillery 2002, 2006; Guillery and Sherman 2002a, b; Sherman 2012). The drivers are the input that brings the information to be relayed. Examples are retinal input to the lateral geniculate nucleus, medial lemniscal input to the ventral posterior nucleus, and, as noted below for some thalamic relays, layer 5 input from cortex. The modulators are everything else, and their main function is to control the level and type of information relayed from drivers through thalamus to cortex. Examples are the local GABAergic cells (i.e., interneurons and cells of the thalamic reticular nucleus), feedback from cortical layer 6, and a projection from the brainstem reticular formation. Drivers represent relatively few of the synaptic inputs to relay cells (only about 5–10 %), but their synapses are relatively powerful. The other 90–95 % of synapses onto relay cells are divided roughly equally among modulatory inputs from local GABAergic cells, from cortical layer 6, and from the brainstem. The modulators require the vast majority of inputs for many subtle roles that affect the relay of driver inputs

Information Processing in Thalamocortical Circuits, Fig. 1 Schema of inputs to thalamic relay cells. Abbreviations: *5-HT* serotonin, *ACh* acetylcholine, *BRF* brainstem reticular formation, *GABA* gamma-aminobutyric acid, *Glu* glutamate, *LGN* lateral geniculate nucleus, *NA* noradrenalin, *TRN* thalamic reticular nucleus

Information Processing in Thalamocortical Circuits, Table 1 Types of glutamatergic input

Class 1 (driver)	Class 2 (modulator)
Activates only ionotropic receptors	Activates metabotropic receptors
Synapses show paired-pulse depression (high p)	Synapses show paired-pulse facilitation (low p)
Large EPSPs	Small EPSPs
Little or no convergence onto target	Much convergence onto target
Thick axons	Thin axons
Large terminals on proximal dendrites	Small terminals on distal dendrites
Dense, well-localized terminal arbors	Delicate terminal arbors

(see Sherman and Guillery 2002, 2006; Guillery and Sherman 2002a, b; Sherman 2012).

One main difference between thalamic relays is the origin of the driver input; the modulators are basically similar throughout thalamus, although there is some variation (see Sherman and Guillery 2006; Jones 2007).

The understanding that inputs to relay cells can be divided into drivers and modulators and that the former largely defines the function of a thalamic relay has implications that extend

beyond thalamus at least to cortical processing (see also below). Thus, the lateral geniculate nucleus is largely defined as a relay of retinal information. It is important to understand that consideration of anatomical information alone can obscure this. For the lateral geniculate nucleus, for instance, only 5–10 % of synapses onto relay cells derive from retina and roughly one third derive from brainstem. If we had only these anatomical data, most of us would conclude that the lateral geniculate nucleus relayed brainstem information and that retinal input provided some obscure, minor function. In other words, we would badly misconstrue this thalamic relay.

First- and Higher-Order Relays

Thus, as noted, identifying the driver is a major key in determining the role played by a thalamic relay. For instance, we define the role of the lateral geniculate nucleus based on its relay of retinal axons, that of the ventral posterior nucleus based on its relay of medial lemniscus axons and that of the ventral division of the medial geniculate nucleus based on its relay of axons from the inferior colliculus. However, until recently, the role played by many thalamic relays remained a mystery, because it was not clear what was

Information Processing in Thalamocortical Circuits, Fig. 2 First-order (*FO*; *left*) and higher-order (*HO*; *right*) thalamic relays. "Glomerulus" refers to a complex synaptic zone that is ubiquitous to thalamus and that is associated with driver input. Abbreviations as in Fig. 1, plus: *I* interneuron, *MGNv* or *MGNd* ventral or dorsal division of medial geniculate nucleus, *POm* posterior medial nucleus, *R* relay cell, *VPM/L* ventral posterior (medial/lateral) nucleus

being relayed. We used to think that the role of the thalamus was to relay subcortical information to cortex, and for large regions of thalamus, such as the pulvinar, it was not clear what was the subcortical source being relayed.

However, the recent realization that drivers for many thalamic relays originate in layer 5 of cortex led to a division of thalamus into "first-order" and "higher-order" relays, and this is summarized in Fig. 2 (Sherman and Guillery 2002, 2006; Guillery and Sherman 2002a, b; Sherman 2012). The term "relays" instead of "nuclei" is used, because some thalamic nuclei conventionally defined may contained a mixture of first order and higher order components. First-order relays transmit to cortex a particular type of information (e.g., retinal) for the first time, whereas higher-order relays are involved in further transmission of such information between cortical areas. The higher-order relay can be between a first-order and higher-order cortical area (as shown in Fig. 2) or between two higher-order cortical areas (see Fig. 3). Higher-order relays have been identified for the major sensory systems: the pulvinar for vision, the posterior medial nucleus for

somatosensation (and thus for pain), and the dorsal division of the medial geniculate nucleus for hearing. Other examples of higher-order relays have also been identified (see Sherman and Guillery 2002, 2006; Guillery and Sherman 2002a, b; Sherman 2012).

Several features from Fig. 2 bear further emphasis. All thalamic relays receive a modulatory input from layer 6 of cortex that is mainly feedback, whereas only the higher-order relays in addition receive a layer 5 cortical input, and this is feedforward. Note also that the driver afferents, both subcortical to first-order relays and from layer 5 for higher-order relays, are branches of axons that also innervate an extrathalamic target, which tends to be "motor" in nature; this is true for many and perhaps all driver inputs to thalamus (for details, see Sherman and Guillery 2002, 2006; Guillery and Sherman 2002a, b; Sherman 2012). For instance, many or all retinal afferents to the lateral geniculate nucleus branch also innervate midbrain structures associated with control of pupil size, eye movements, etc., and many layer 5 afferents to higher-order thalamic relays also innervate

Information Processing in Thalamocortical Circuits, Fig. 3 Involvement of higher-order thalamic relays in corticocortical communication. Note that cortical areas are connected in parallel by direct and transthalamic routes. For simplicity inputs from interneurons and cells of the thalamic reticular nucleus are omitted. Abbreviations as in Figs. 1 and 2

many levels of the brainstem, and some extend input to spinal levels. It is as if the information relayed to cortex through thalamus is a corollary of motor commands, and it is these motor commands that serve as the basis of perceptual information acted upon and further elaborated by cortex (Sherman and Guillery 2002, 2006; Guillery and Sherman 2002a, b; Sherman 2012).

It is also worth noting that, as sufficient information regarding various thalamic relays develops regarding the division into first order and higher order, the large majority of thalamus seems to be devoted to higher-order relays.

Role of Thalamus in Corticocortical Communication

Figure 3 summarizes the major implication of this division of thalamic relays into first order and higher order for cortical functioning. Information of a particular sort first reaches cortex via a first-order relay; this can apply to primary information about vision, sounds, pain, etc. Much further cortical processing of this primary information is based on cortico-thalamo-cortical, or transthalamic, pathways involving higher-order thalamic relays, although direct connections also exist to convey information between cortical areas. An important set of questions that remain unanswered are the following: What is different about the information sent directly and indirectly via thalamus between cortical areas? Why is one information pathway sent through thalamus? A small number of examples indicate that direct connections between cortical areas are paralleled by transthalamic ones, but how common is this arrangement?

To place this scheme in the proper perspective, it is important to appreciate that current dogma regarding functioning of cortical areas is based on direct connections between areas. For instance, the best studied is visual cortex, which is divided into more than 30 discrete areas in humans, and the detailed scheme of functional

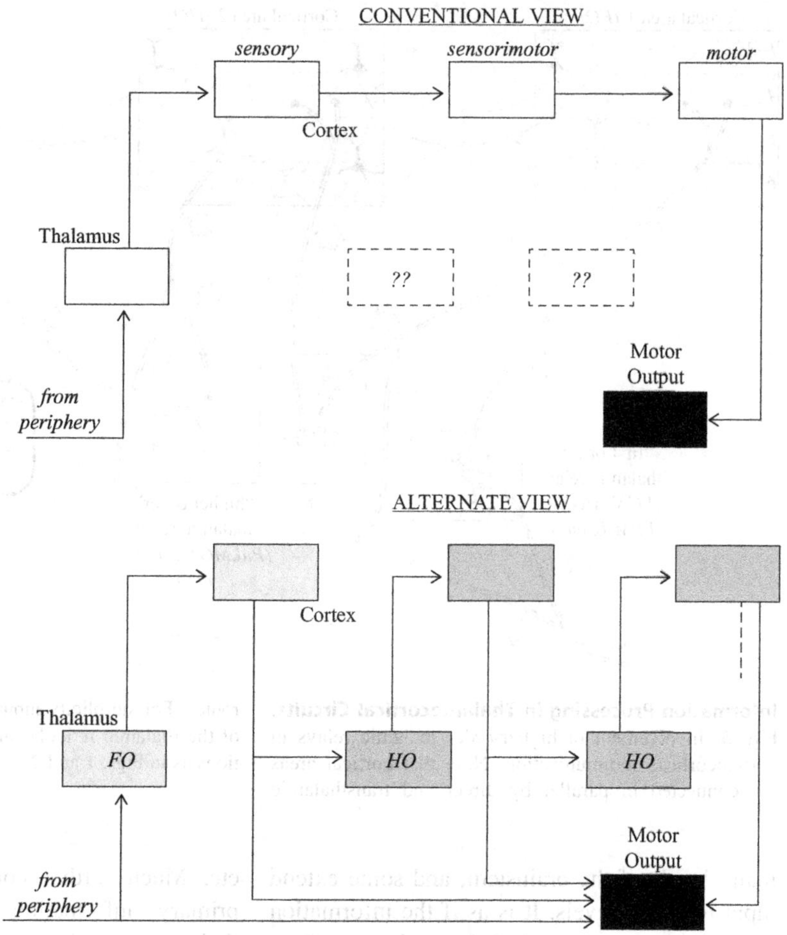

**Information Processing
in Thalamocortical
Circuits,**
Fig. 4 Conventional
(*upper*) versus alternate
(*lower*) views of cortical
processing

organization is based almost entirely on the pattern of direct corticocortical connections, with no place for thalamus (Felleman and Van Essen 1991; Van Essen et al. 1992; Kandel et al. 2000). A similar view dominates thinking about the organization of somatosensory cortical areas responsible for the cortical processing of pain. Understanding how cortical areas process information requires first identifying the routes of information, and since the driver/modulator distinction appears to hold for cortical pathways as it does in thalamus (Covic and Sherman 2011; DePasquale and Sherman 2011), then it becomes essential to distinguish among the direct corticocortical pathways those that are drivers from those that are modulators. As it happens, the current views of cortical organization are dominated by studies limited only to direct

corticocortical connections that have been identified almost entirely with anatomical techniques, and an implied assumption that needs to be made explicit is that all more or less contribute equally, in a sort of anatomical democracy, to information flow. This same logic applied to the thalamus would produce the misconception that the lateral geniculate nucleus relayed brainstem, and not retinal, inputs to cortex (see above).

Summary

To understand the implications of the proposal put forward here for the role of thalamus in corticocortical communication, it might be helpful to contrast it with the conventional view, and this is done in Fig. 4. In the conventional view

(Fig. 4, upper), sensory information is relayed from the periphery by thalamus to a primary sensory cortical area. From there, the information is processed strictly within cortex, eventually via sensorimotor areas to motor areas, and finally this leads to a motor output. Note that, in this view, the only role for thalamus is to get raw information to cortex in the first place and that most of thalamus, which we call higher-order relays, has no specific role to play. In the alternate view (Fig. 4, lower), offered here, from the very beginning, information relayed to cortex is corollary to motor commands, and further corticocortical processing involves higher-order thalamic relays of continuously elaborated and updated motor commands. Thus, thalamus not only gets information to cortex in the first place but also continues to play an essential role in corticocortical communication in some sort of as yet undefined partnership with direct corticocortical connections.

This has important implications for cortical functioning generally and also for cortical processing of pain information more specifically. That is, the higher-order thalamic relays involved in pain processing could be key. The best candidate for the higher-order thalamic relay of pain information would be the posterior medial nucleus, which lies mostly medial to the ventral posterior nucleus, most of which is the first-order somatosensory relay. We clearly need a better understanding of how pain is processed by somatosensory cortex, and the purpose of this entry is to provide a different theoretical framework that might fruitfully guide further research through this topic.

References

Covic, E. N., & Sherman, S. M. (2011). Synaptic properties of connections between the primary and secondary auditory cortices in mice. *Cerebral Cortex, 21,* 2425–2441.

DePasquale, R., & Sherman, S. M. (2011). Synaptic properties of corticocortical connections between the primary and secondary visual cortical areas in the mouse. *Journal of Neuroscience, 31,* 16494–16506.

Felleman, D. J., & Van Essen, D. C. (1991). Distributed hierarchical processing in the primate cerebral cortex. *Cerebral Cortex, 1,* 1–47.

Guillery, R. W., & Sherman, S. M. (2002a). Thalamic relay functions and their role in corticocortical communication: Generalizations from the visual system. *Neuron, 33,* 163–175.

Guillery, R. W., & Sherman, S. M. (2002b). The thalamus as a monitor of motor outputs. *Philosophical Transactions of the Royal Society B: Biological Sciences, 357,* 1809–1821.

Jones, E. G. (2007). *The thalamus* (2nd ed.). Cambridge, UK: Cambridge University Press.

Kandel, E. R., Schwartz, J. H., & Jessell, T. M. (2000). *Principles of neural science.* New York: McGraw Hill.

Sherman, S. M. (2012). Thalamocortical interactions. *Current Opinion in Neurobiology, 17,* 417–422.

Sherman, S. M., & Guillery, R. W. (2002). The role of thalamus in the flow of information to cortex. *Philosophical Transactions of the Royal Society B: Biological Sciences, 357,* 1695–1708.

Sherman, S. M., & Guillery, R. W. (2006). *Exploring the thalamus and its role in cortical function.* Cambridge, MA: MIT Press.

Van Essen, D. C., Anderson, C. H., & Felleman, D. J. (1992). Information processing in the primate visual system: An integrated systems perspective. *Science, 255,* 419–423.

Infraclavicular Block

Definition

A technique employed to block innervation by the brachial plexus. It is achieved by injection of local anesthetic inferior to and near the midpoint of the clavicle.

Cross-References

▶ Acute Pain in Children, Postoperative

Infusion

Definition

The therapeutic infusion of fluid.

Cross-References

▶ Epidural Infusions in Acute Pain

Inherited Factors Implicated in the Mechanisms of Migraine

▶ Migraine, Genetics

Inherited Variability of Drug Response

▶ NSAIDs, Pharmacogenetics

Inhibitory Synaptic Transmission

▶ GABA and Glycine in Spinal Nociceptive Processing

Injections of Corticosteroids

▶ Steroid Injections

Injections of Steroids

▶ Steroid Injections

Injury Discharge

Definition

When axons are cut across acutely, or neurons are punctured, they tend to produce a brief burst of impulses, usually lasting no more than a few seconds. This "injury discharge" is due to the intense depolarization caused by breaching the cell membrane.

Cross-References

▶ Neuropathic Pain Models, CRPS-I Neuropathy Model
▶ Peripheral Neuropathic Pain

Injury Memory

Definition

A process by which sensitized peripheral afferent fibers can produce a long-lasting sensitization of CNS neurons whose duration outlasts the peripheral sensitizing stimulus, in effect producing a memory of the injury by the brain.

Cross-References

▶ Gynecological Pain, Neural Mechanisms

Innervation of the Urinary Bladder

Definition

The bladder receives efferent sympathetic (hypogastric nerve), parasympathetic (pelvic nerve), and somatic-motor (pudendal nerve) innervation. Afferent (sensory) information is conveyed via the hypogastric, pelvic, and pudendal nerves.

Cross-References

▶ Opioids and Bladder Pain/Function

Insomnia

▶ Sleep and Pain

Instability

Definition

Instability is the inability to limit excessive or abnormal spinal displacement.

Cross-References

▶ Chronic Back Pain and Spinal Instability

Installation Block

Definition

Block achieved by pouring local anesthetic into the wound where it bathes injured tissue edges. It is removed 20–60 s later prior to wound closure.

Cross-References

▶ Acute Pain in Children, Postoperative

Instrumental, Operant Conditioning

Definition

Instrumental or Operant conditioning refers to a behavioral principle describing how the probability of occurrence of a specific behavior (e.g., limping) is influenced by the nature of its association with a particular outcome (e.g., attention). Thus, behavior may be maintained by reinforcement contingencies (i.e., positive or negative reinforcement, punishment).

Cross-References

▶ Behavioral Therapies to Reduce Disability

Insula

▶ Insular Cortex, Neurophysiology, and Functional Imaging of Nociceptive Processing

Insula/Insular Cortex

Synonyms

Insular area; Insular cortex

Definition

A part of the cerebral cortex which became infolded during embryonic development as a result of the deepening of the lateral cerebral fissures (of Sylvius). It can only be seen by separating the lips of the fissure or by cutting away the operculum which overhangs it. The insular cortex (or insula) is a cortical association area of whose function little is known.

Cross-References

▶ Amygdala, Pain Processing and Behavior in Animals
▶ Magnetoencephalography in Assessment of Pain in Humans
▶ Nociceptive Processing in the Secondary Somatosensory Cortex
▶ Secondary Somatosensory Cortex (S2) and Insula, Effect on Pain-Related Behavior in Animals and Humans

Insular Area

▶ Insula/Insular Cortex
▶ Insular Cortex, Neurophysiology, and Functional Imaging of Nociceptive Processing

Insular Cortex

► Insula/Insular Cortex
► Insular Cortex, Neurophysiology, and Functional Imaging of Nociceptive Processing

Insular Cortex, Neurophysiology, and Functional Imaging of Nociceptive Processing

Kenneth L. Casey[1] and Diep Tuan Tran[2]
[1]Department of Neurology, University of Michigan Medical Center, Ann Arbor, MI, USA
[2]Department of Integrative Physiology, National Institute for Physiological Sciences, Okazaki, Japan

Synonyms

Central lobe; Fifth lobe; Insula; Insular area; Insular cortex; Insular lobe; Island of reil

Definition

Insula in Latin means island. Despite its name, the insular cortex is not an island; it is an oval region of the cerebral cortex overlying the external capsule, lateral to the lenticular nucleus, buried in the ► sylvian fissure (lateral fissure) beneath the operculum, which forms the lips of the lateral fissure (Fig. 1, see also Fig. 1 Central lobe in ► Secondary Somatosensory Cortex (S2) and Insula, Effect on Pain-Related Behavior in Animals and Humans). The insula is buried simply because of the relatively greater growth of the rest of the hemisphere around it; therefore, it is not visible on the surface and can only be seen when the lips – frontoparietal and temporal ► operculum – are pulled apart or removed. The insular cortex has connections to widespread thalamic, limbic, and cortical regions related to all sensory modalities and autonomic and limbic functions (Augustine 1996).

Insular Cortex, Neurophysiology, and Functional Imaging of Nociceptive Processing, Fig. 1 The insula is not visible on the surface and can only be seen when the lips, frontoparietal and temporal operculum, are pulled apart or removed. *Ant. Ins* anterior insula (*orange*), *Pos. Ins* posterior insula (*blue*)

Characteristics

Anatomically, the insular cortex is mainly divided into three cytoarchitectural fields. In rostrocaudal order they are as follows: a rostroventral agranular field (Ia), a transitional dysgranular field (Id), and a posterior granular field (Ig). The anterior insula (AI) – Ia and Id – is predominantly connected with cortical areas related to limbic, olfactory, gustatory, and viscero-autonomic functions, whereas posterior insula (PI), Ig and part of Id, is predominantly connected with cortical areas related to somatosensory, auditory, and visual functions (Augustine 1996).

The evidence for a pain representation in the insular cortex is provided by the anatomical connectivity and electrophysiological properties of this area in monkeys (for review see Willis 1995). The ► spinothalamic tract (STT), which carries nociceptive information, terminates in thalamic nuclei that project to the insular cortex. In particular, the ventroposterior inferior nucleus (VPI), the suprageniculate-limitans nucleus (SG-Li), and the posterior complex (Po) project to the PI, and the posterior portion of the ventral medial nucleus (VMpo) projects to the AI cortex. In accord with these inputs, ► single-unit recordings reveal nociceptive responses in neurons of the PI and AI. These neurons have large ► receptive fields, either contralateral or bilateral.

In human lesion studies, brain lesions involving the posterior portion of the insula have led to an impairment of tactile and pain perception and changes in pain sensitivity (Greenspan et al. 1999). Large lesions that include the anterior portion of the insula apparently reduce pain affect and appropriate reactions to painful stimuli, the so-called asymbolia for pain (Berthier et al. 1988; Greenspan et al. 1999).

▶ Electroencephalography (EEG), ▶ nociceptive selective stimulation, and scalp or subdural ▶ laser-evoked potential (LEP) recordings reveal activity bilaterally in the SII-insular cortices (Lenz et al. 2000; Valeriani et al. 2000). ▶ Magnetoencephalography (MEG) with laser-evoked magnetic field (LEF) recordings and selective activation of ▶ Aδ fibers (Kakigi et al. 1996) and ▶ C fiber (Tran et al. 2002) show activity in SII bilaterally but not in the insula. This is because MEG can only detect tangential but not radial dipoles, which are believed to be generated in the insula. Since the SII and the insula are anatomically adjacent, there has long been a debate in dipole source modeling studies about whether the activity in the parasylvian area originates from two generators – SII and insula – or from only one generator in the insula. A meta-analysis of neuroimaging studies (PET, fMRI, scalp, and intracerebral LEP source analyses) shows that the spatial resolution of these techniques is not sufficient to segregate SII and insular activities (Peyron et al. 2002). However, using depth intracerebral electrodes in epileptic patients, a recent study has demonstrated a dual representation of pain in both SII and insular cortices (Frot and Mauguiere 2003). The high temporal resolution of LEP and LEF also reveals that the contralateral SII and insular cortices are activated earlier than the ipsilateral side by an average of 14–18 ms (Frot and Mauguiere 2003; Tran et al. 2002). Moreover, insular activity is delayed ~50 ms following SII activation (Frot and Mauguiere 2003). The delay may reflect either sequential activation from contralateral to ipsilateral through transcallosal transmission and from SII to insular cortices via intracortical connections or the direct activation of bilateral SII and insula through distinct thalamocortical projection fibers with different conduction times.

Human ▶ functional imaging studies reveal that pain-related activation is located in the anterior part of the insular cortex either contralateral or bilateral to the stimulation side (Casey et al. 1996; Coghill et al. 1994). The activity in AI is modulated by pain affect (Rainville et al. 1997), pain anticipation (Porro et al. 2002), pain attention (Brooks et al. 2002), pain illusion (Craig et al. 1996), pain anxiety (Ploghaus et al. 2001), and pain suggestion (Hofbauer et al. 2001). Neuroimaging studies also show another peak of pain-related activation in the contralateral posterior part of the insular cortex (Brooks et al. 2002; Casey et al. 2001; Craig et al. 2000). This activity in PI is correlated with thermal intensity and duration (Brooks et al. 2002; Casey et al. 2001; Craig et al. 2000). Moreover, stimulation studies show that electrical stimulation of the anterior part of the insula in conscious humans produces visceral sensory experiences and visceral motor responses, but not reports of pain, whereas stimulation of the posterior part of insula elicits somesthetic sensation including pain sensation in body areas contralateral to the stimulation site (Ostrowsky et al. 2002).

The imaging findings, neuronal projections, anatomical connections, and lesion and stimulation studies provide convergent evidence that insular activation is related to the affective-motivational and internal state of the body. However, PI activity may be more closely related to direct somatosensory processing. Averaged event-related activities (LEP, LEF) evoked by laser stimulation, which has less – if any – affective and emotional effect, often reveal pain-related dipole sources primarily in the parasylvian cortex, including the SII and the posterior part of the insula (Kakigi et al. 1996; Lenz et al. 2000; Tran et al. 2002; Valeriani et al. 2000). Neuroimaging studies, which reflect metabolic demands over several seconds, may however show both AI and PI activity in the course of repetitive noxious stimulation (Casey et al. 2001).

The insular cortex is in a special position to act as an interface between intrapersonal and extrapersonal information processes and between the lateral and medial pain systems as represented by the lateral and medial targets of the spinothalamic tract. The insula processes intrapersonal information and involves the affective-motivational components of pain in the medial system and somatosensory information related to the sensory-discriminative aspect of pain in the lateral system.

The circuitry of the insular cortex in primates including humans – with connections to widespread thalamic, limbic, and cortical regions – suggests a wide variety of functions related to all sensory modalities and to autonomic, neuroendocrine, and behavioral responses. The insular cortex may be an essential part of the neural substrate that integrates all external sensory sources and internal physiological conditions. Physiologically, the insula may be involved in autonomic reactions to a noxious stimulus, in sensory-discriminative and affective-motivational components of pain, and in pain-related memory and learning.

More studies are needed to explore the physiological functions of the AI and PI, the interaction of these two areas, and the temporal dynamics of the interaction. In addition, the anatomically adjacent SII and insular cortices require further investigations on their distinct roles in pain perception. The complementary use of neurophysiological techniques (LEP and LEF) with excellent temporal resolution in the order of milliseconds and functional imaging (PET and fMRI) with better spatial resolution is promising. The multimodal approach would certainly improve our understanding of normal pain perception and create a basis for studying pathological pain states and pain treatment.

Acknowledgment This work was supported by NIAMS AR46045 and the Dept. of Veteran's Affairs (KLC) and NINDS/WHO Fellowship (TDT).

References

Augustine, J. R. (1996). Circuitry and functional aspects of the insular lobe in primates including humans. *Brain Research. Brain Research Reviews, 22*, 229–244.

Berthier, M., Starkstein, S., & Leiguarda, R. (1988). Asymbolia for pain: A sensory-limbic disconnection syndrome. *Annals of Neurology, 24*, 41–49.

Brooks, J. C., Nurmikko, T. J., Bimson, W. E., et al. (2002). FMRI of thermal pain: Effects of stimulus laterality and attention. *NeuroImage, 15*, 293–301.

Casey, K. L., Minoshima, S., Morrow, T. J., et al. (1996). Comparison of human cerebral activation pattern during cutaneous warmth, heat pain, and deep cold pain. *Journal of Neurophysiology, 76*, 571–581.

Casey, K. L., Morrow, T. J., Lorenz, J., et al. (2001). Temporal and spatial dynamics of human forebrain activity during heat pain: Analysis by positron emission tomography. *Journal of Neurophysiology, 85*, 951–959.

Coghill, R. C., Talbot, J. D., Evans, A. C., et al. (1994). Distributed processing of pain and vibration by the human brain. *The Journal of Neuroscience, 14*, 4095–4108.

Craig, A. D., Reiman, E. M., Evans, A., et al. (1996). Functional imaging of an illusion of pain. *Nature, 384*, 258–260.

Craig, A. D., Chen, K., Bandy, D., et al. (2000). Thermosensory activation of insular cortex. *Nature Neuroscience, 3*, 184–190.

Frot, M., & Mauguiere, F. (2003). Dual representation of pain in the operculo-insular cortex in humans. *Brain, 126*, 438–450.

Greenspan, J. D., Lee, R. R., & Lenz, F. A. (1999). Pain sensitivity alterations as a function of lesion location in the parasylvian cortex. *Pain, 81*, 273–282.

Hofbauer, R. K., Rainville, P., Duncan, G. H., et al. (2001). Cortical representation of the sensory dimension of pain. *Journal of Neurophysiology, 86*, 402–411.

Kakigi, R., Koyama, S., Hoshiyama, M., et al. (1996). Pain-related brain responses following CO_2 laser stimulation: Magnetoencephalographic studies. *Electroencephalography and Clinical Neurophysiology, 47*, 111–120.

Lenz, F. A., Krauss, G., Treede, R. D., et al. (2000). Different generators in human temporal-parasylvian cortex account for subdural laser-evoked potentials, auditory-evoked potentials, and event-related potentials. *Neuroscience Letters, 279*, 153–156.

Ostrowsky, K., Magnin, M., Ryvlin, P., et al. (2002). Representation of pain and somatic sensation in the human insula: A study of responses to direct electrical cortical stimulation. *Cerebral Cortex, 12*, 376–385.

Peyron, R., Frot, M., Schneider, F., et al. (2002). Role of operculoinsular cortices in human pain processing: Converging evidence from PET, fMRI, dipole modeling, and intracerebral recordings of evoked potentials. *NeuroImage, 17*, 1336–1346.

Ploghaus, A., Narain, C., Beckmann, C. F., et al. (2001). Exacerbation of pain by anxiety is associated with activity in a hippocampal network. *The Journal of Neuroscience, 21*, 9896–9903.

Porro, C. A., Baraldi, P., Pagnoni, G., et al. (2002). Does anticipation of pain affect cortical nociceptive systems? *The Journal of Neuroscience, 22*, 3206–3214.

Rainville, P., Duncan, G. H., Price, D. D., et al. (1997). Pain affect encoded in human anterior cingulate but not somatosensory cortex. *Science, 277*, 968–971.

Tran, T. D., Inui, K., Hoshiyama, M., et al. (2002). Cerebral activation by the signals ascending through unmyelinated C-fibers in humans: A magnetoencephalographic study. *Neuroscience, 113*, 375–386.

Valeriani, M., Restuccia, D., Barba, C., et al. (2000). Sources of cortical responses to painful CO(2) laser skin stimulation of the hand and foot in the human brain. *Clinical Neurophysiology, 111*, 1103–1112.

Willis, W. D., Jr. (1995). From nociception to cortical activity. In B. Bromm & J. E. Desmedt (Eds.), *Pain and the brain* (pp. 1–19). New York: Raven.

Insular Lobe

▶ Insular Cortex, Neurophysiology, and Functional Imaging of Nociceptive Processing

Insulin Neuropathy

▶ Diabetic Neuropathies

Intensity of Pain

Definition

Intensity of pain is scored on a verbal points scale: 0 no pain; 1 mild pain, does not interfere with activities; 2 moderate pain, inhibits but does not prohibit activities; and 3 severe pain, prohibits activities. It may alternatively be measured in a visual analogue scale (VAS).

Cross-References

▶ Sunct Syndrome

Intensity of Ultrasound

Definition

The intensity is power per unit area.

Cross-References

▶ Ultrasound Therapy of Pain from the Musculoskeletal System

Intentionality

Definition

A complex adaptive system has intentionality when it exhibits directedness toward some future state or goal. Intent comprises of the endogenous initiation, construction and direction of perception, action, and goal-directed behavior.

Cross-References

▶ Consciousness and Pain

Intercostal Space

Definition

A potential space at the inferior margin of each rib through which the intercostal nerve, vein, and artery travel for each rib. The intercostal nerve provides sensory and motor function to the superficial chest and abdominal wall.

Cross-References

▶ Postoperative Pain, Regional Blocks

Interdisciplinary

Definition

Interdisciplinary refers to a team or collaborative process where members of different disciplines assess and treat patients jointly.

Cross-References

▶ Complex Chronic Pain in Children, Interdisciplinary Treatment
▶ Physical Medicine and Rehabilitation, Team-Oriented Approach

Interdisciplinary Assessment

▶ Multiaxial Assessment of Pain

Interdisciplinary Pain Management Programs

▶ Interdisciplinary Pain Rehabilitation

Interdisciplinary Pain Rehabilitation

Chris J. Main
Keele University, Keele, Staffordshire, UK

Synonyms

Cognitive-behavioral programs; Interdisciplinary pain management programs

Definition

Interdisciplinary pain management (IPM) does not offer cures for pain but comprises an integrated biopsychosocial approach to pain management from a team of health-care professionals with the aim of maximizing the patient's own ability to manage their pain and pain associated dysfunction. A range of cognitive and behavioral strategies focused on reactivation and minimization of unnecessary physical and psychosocial dysfunctions are used. Perhaps most important of all, as a defining characteristic of IPM, is the promulgation of self-help and personal responsibility. This is evident particularly in the emphasis on active rather than passive approaches to treatment.

Characteristics

The University of Washington in Seattle pioneered multidimensional treatment, research, education, and interdisciplinary working in the field of pain. From the perspective of interdisciplinary pain management, its fifth mission statement was critically important: "To enhance interaction and communication among all pain investigators at the University of Washington and to encourage cross-fertilization of ideas on pain research and therapy" (Bonica 1990). The "Seattle model" became not only an example of a practical implementation of a biopsychosocial model of pain but also an illustration of the mutual interdependence of clinical practice, teaching, training, and research.

More specifically, in the early days, the Seattle group promoted a view of illness that incorporated both physical and behavioral aspects. Although most pain management programs now would define themselves as cognitive-behavioral in their orientation, the power and incisiveness of the behavioral component (Fordyce 1976) in the original Seattle model was critical to the development of modern interdisciplinary pain management. During the last 30 years, variants of the Seattle model have appeared all over the world.

Modern pain management programs (PMPs) contain a number of discrete elements blended into a package of care. Each of these elements has its own pedigree as an approach to treatment and each element has its origins in a particular perspective on the nature of pain. Typically these will include rationalization of medication use, development of individually tailored reactivation programs, training in problem-solving techniques and enhancement of adaptive pain coping strategies including goal setting and pacing, with the objective of maximizing positive adaptation. Programs differ in the extent to which they have a specific work focus.

An appropriate educational component is a major ingredient of all PMPs, but it seems that education, although necessary, is not sufficient and it seems that unless increased understanding leads to significant behavioral change, it is of little value.

Specific therapeutic objectives usually include decreasing stiffness and immobility, recovery of fitness, development of strength, reversal of the "disuse syndrome," treatment of distress (including depression) and the modification of pain behavior. They differ from the "sports medicine approach," typified in functional restoration programs (Mayer and Gatchell 1988), in that they usually place less emphasis specifically on strength and mobility and, particularly outside N. America, have traditionally focused on clinical rather than on occupational outcomes.

It is not possible to overstress the importance of the behavioral perspective as an alternative to the disease/pathology model in offering a new and alternative understanding of pain. It has led to the focus on pain management as an alternative to cure, which is not available for many pain conditions. More recently incorporation of the cognitive perspective (Turk and Okifuli 2003; Flor and Turk 2011) has produced a psychologically oriented approach which, with a focus on reactivation and self-management, is frequently described as a cognitive-behavioral program and has become the treatment of choice for the dysfunctional chronic pain patient, with clear scientific evidence of its efficacy (Morley et al. 1999; Eccleston et al. 2009) over more narrowly focused traditional approaches. Newer variants such as contextual CBT, which are based on principles of acceptance and mindfulness are producing a sharper focus in therapy (Vowles and Thompson 2011).

Desirable Characteristics of PMPs

The British and Irish chapter of the International Association for the Study of Pain, therefore, inspired by the IASP report (IASP 1990) and by Sanders (1994) recently produced a set of "desirable characteristics" for PMPs (British Pain Society 2013). The key features are as follows:

- Behavioral rather than a disease perspective
- Focus on pain management rather than cure
- Blend of ingredients
- Interdisciplinary skill-mix
- Incorporation of group therapy
- Emphasis on active rather than passive approaches to treatment
- Promulgation of self-help and patient responsibility

The Context of Pain Management

There is a gradient of complexity in pain management, ranging from specific types of individualized intervention such as medication adjustment, spinal manipulation, or provision of educational material to "packages of care" involving not only multifaceted treatment but also intervention by a range of professionals using a team approach. Multidisciplinary treatment undoubtedly offers a much wider range of potential therapeutic options that can be offered by a single-handed practitioner, but there are also inherent dangers of iatrogenic confusion or distress if the patient is given inconsistent information or offered a package of care that is poorly integrated or even contradictory. Integrated care has become known as interdisciplinary pain management. The potential limitations of a multidisciplinary approach (in contrast to an interdisciplinary approach) are as follows:

- Poor communication leading to a lack of awareness of problem areas. (e.g., the physiotherapist may be unaware that a patient has an alcohol problem).
- Fragmentation of the approach to treatment and assessment in which the patient's

problems are tackled sequentially rather than in an integrated fashion. The patient then comes to view the different professionals as having overprescribed roles, making it difficult to tackle interdependent problems.

- There may be unnecessary duplication of certain parts of the assessment procedure. This in not only inefficient but can be irritating to patients.
- Different management strategies may be employed in dealing with the same problem.
- Patients and their relatives may receive conflicting information from different disciplines leading to confusion on their part about which approach should be adopted.
- In certain cases, manipulative patients can "play off" one team member against other members.
- Restricted communication within the multidisciplinary team may lead to a sense of isolation and lack of support among individual team members.

General Advantages of the Interdisciplinary Approach

In interdisciplinary pain management, a special effort is made to offer a consistent and coherent approach to both assessment and management. The structuring and delivery of IPM is dealt with in detail elsewhere (Main et al. 2008), but for present purposes a number of strategies merit consideration. In the interdisciplinary team, according to Melvin (1980) "...individuals not only require the skills of their own disciplines, but also have the added responsibility of the group effort on behalf of the activity or client involved. ...The group activity of an interdisciplinary team is synergistic, producing more than each individually and separately could accomplish" (p. 379).

Two differing patterns of cooperation within the interdisciplinary team have been described (Finset et al. 1995). In the coordinated interdisciplinary team, mutual goals are set and the individuals from each discipline attempt to work on these goals in their individual sessions. In the integrated interdisciplinary team, mutual

goals are worked on in joint treatment sessions with members of different disciplines participating in the sessions. The treatment could not be carried out by individual members of the team alone.

The key features of an interdisciplinary team are as follows:
- The team members share common assessment and treatment goals.
- The main task of all members of the team is to deliver treatments that are dependent on the patient's needs, not on the constraints of the individual disciplines involved.
- Unlike multidisciplinary teams, interdisciplinary teams are not necessarily led by a member of the medical profession. They are most likely to be led by an experienced clinician who has knowledge not only of working with patients with chronic pain but also a knowledge of working in a team.
- There are core areas of speciality unique to each discipline, but these are secondary to the goals of the team as a whole.
- Team members will have explicit knowledge of the skills of other disciplines involved in the team.
- Effective communication occurs between all team members.
- All team members have equal status within the team.
- Decisions are arrived at after considering input from all team members.

Recommendations for the development, establishment, and maintenance of the interdisciplinary team include the following:
- The philosophy of the team should be made articulate in an agreed policy document and made explicit to all members. This should be re-appraised at regular intervals and modified if necessary.
- There must be a system for staff induction and training, with a focus particularly on processes within groups, the development of core team skills required and education regarding the work of other disciplines.
- Communication is vital to the effective running of the team and a considerable amount of time should be spent on this area,

both formally and informally, including regular team meetings. However, team meetings are expensive; therefore, it is of importance that they are conducted efficiently and effectively.

- Meetings should have a clear agenda, with effective firm leadership to ensure focus, concluding with action points within a spirit of group cohesiveness.
- Rules of confidentiality concerning both staff and patients must be made explicit.

Team Maintenance

When different disciplines set up and work as an interdisciplinary team, this can give rise to problems in managerial, organizational, and interpersonal areas. Strategies to assist integration of external hospital managers might include involving management during the early stages of planning and development of the team, discussion with management of the policy document and associated specific working practices required, and developing specific and relevant outcome measures.

Ensure effective and efficient time management not only in patient assessment and treatment delivery but also in decision-making and audit. Clinic structuring and minimization of inappropriate referrals and failure to attend appointments lessen the dangers of fragmentation of the team and formation of "splinter groups" within the team. Ensure that good communication occurs between all team members by controlling the size of the team and encouraging cohesiveness by using team-building techniques. Any communication or interpersonal problems between team members need to be dealt with as early as possible. There should be time set aside specifically for team development and team building; this is particularly important in the early stages of forming an interdisciplinary team. Pay particular attention to the integration of new members to the team. A "probationary period" may be extremely useful. Ensure the availability of support systems to help team members cope with the emotional pressures that can occur when working with such a demanding group of patients.

Conclusion

The interdisciplinary clinic would seem to have a number of specific advantages over uni-professional or multidisciplinary clinics. The principal advantages include the following:

1. Greater range of professional expertise, permitting a wider range of treatment options
2. Decreased likelihood of inappropriate interventions as a consequence of failure to recognize (or take adequately into account) important obstacles to recovery
3. Minimizing the likelihood of producing further iatrogenic distress and confusion
4. Coherent and integrated treatment approach, with clear identification of goals and responsibilities

It could, I think, reasonably be argued that the framework, with its range of options and integration of perspectives, offers the potential for optimal clinical care. Clinicians who have genuinely embraced the interdisciplinary perspective seem to find it a professionally stimulating and satisfying approach to clinical management.

There are however potential difficulties with the approach. Some of the more frequent difficulties include patient and staff scheduling, negotiating adequate time for initial assessment, and failure among the staff to arrive at an agreed clinical formulation and treatment plan. Indeed over the last decade, there have been increasing difficulties in obtaining and securing funding for such clinics, particularly those run on an in-patient basis.

Nonetheless, as a way of managing patients, the interdisciplinary approach as developed in the early pain management programs has offered a powerful new way of managing patients and the philosophy and practice have had a powerful influence on the design and establishment of modern pain management, not only in tertiary health care, but also in secondary prevention and in occupational settings. Indeed, the general approach has been recommended specifically for low back pain patients reporting high disability and/or significant psychological distress and who have not derived significant benefit from at least one less intensive treatment (NICE 2009).

References

Bonica, J. J. (1990). Multidisciplinary / interdisciplinary pain programs. In J. J. Bonica (Ed.), *The management of pain* (2nd ed., pp. 197–208). Philadelphia: Lea & Febiger.

British Pain Society. (2013). *Recommended guidelines for pain management programmes for adults. A consensus statement prepared on behalf of the British Pain Society*. London: British Pain Society. (in press).

Eccleston, C., Williams, A. C. deC., & Morley, S. (2009). Psychological therapies for the management of chronic pain (excluding headache) in adults. *Cochrane Database of Systematic Reviews*, 1–101.

Finset, A., Krogstad, J. M., Hansen, H., et al. (1995). Team development and memory training in traumatic brain jury rehabilitation: Two birds with one stone. *Brain Injury, 9*, 495–507.

Flor, H., & Turk, D. C. (2011). *Chronic pain: an integrated biobehavioral approach*. Seattle: IASP Press.

Fordyce, W. E. (1976). *Behavioral methods for chronic pain and illness*. St. Louis: C.V. Mosby.

IASP. (1990). *Desirable characteristics for pain treatment facilities (Report of Task Force on Guidelines for Desirable Characteristics for Pain treatment Facilities)*. Seattle: IASP.

Main, C. J., Sullivan, M. J. L., & Watson, P. J. (2008). *Pain management: practical application of the biopsychosocial framework in clinical and occupational settings*. Edinburgh: Churchill-Livingstone.

Mayer, T. G., & Gatchel, R. (1988). *Functional restoration for spinal disorders*. Philadelphia: The sports medicine approach. Lea & Febiger.

Melvin, J. L. (1980). Interdisciplinary and multidisciplinary activities and the ACRM. *Archives of Physical Medicine and Rehabilitation, 61*, 379–380.

Morley, S., Eccleston, C., & Williams, A. (1999). A Systematic review and meta-analysis of randomised controlled trials of cognitive behaviour therapy and behaviour therapy for chronic pain in adults, excluding headache. *Pain, 80*, 1–13.

NICE. (2009). *LOW BACK PAIN: Early management of non-specific low back pain*. Nice Clinical Guideline 88. National Institute of Health and Clinical Excellence.

Sanders, S. H. (1994). An image problem for pain centers: Relevant factors and possible solutions. *APS Bulletin, Jan/Feb*, 17–18.

Turk, D. C., & Okifuli, A. (2003). In R. Melzack & P. D. Wall (Eds.), *Handbook of pain management: A clinical companion to wall and Melzack's textbook of pain* (pp. 533–541). Edinburgh: Churchill-Livingstone.

Vowles, K. E., & Thompson, M. (2011). Acceptance and commitment therapy for chronic pain. In L. M. McCracken (Ed.), *Mindfulness and acceptance in behavioral medicine: Current theory and practice* (pp. 31–60). Oakland: New Harbinger Press.

Interdisciplinary Pain Rehabilitation Programs

▶ Multidisciplinary Pain Centers, Rehabilitation

Interferential Therapy

▶ Modalities

Interlaminar Epidural Steroid Injection

▶ Epidural Steroid Injections for Chronic Back Pain

Interleukin(s) (IL)

Definition

A group of molecules within the cytokine family involved in signaling between cells participating in immune and inflammatory events; distinct interleukins are defined by different numbers and letters (e.g., IL-1α, IL-1β, and IL-6).

Cross-References

▶ Cytokines as Targets in the Treatment of Neuropathic Pain
▶ Wallerian Degeneration

Interleukins

▶ Cytokines as Targets in the Treatment of Neuropathic Pain

Intermittent

▶ Traction

Intermittent Claudication

Definition

Intermittent claudication is a condition that is characterized by pain and lameness during muscular exercise or – in the leg – during walking. The most frequent cause is narrowing of arteries in the lower leg and foot by arteriosclerosis.

Cross-References

▶ Muscle Pain Model, Ischemia-Induced and Hypertonic Saline-Induced

Internal Capsule

Definition

White matter pathway carrying fibers afferent to the cortex from the thalamus and efferent from the cortex to nuclei of the brain and to the spinal cord.

Cross-References

▶ Deep Brain Stimulation

Internal Disc Disruption

Definition

Ruptured vertebral disc at the center of the nucleus pulposus.

Cross-References

▶ Chronic Low Back Pain: Definitions and Diagnosis

Internal Neurolysis

Definition

Scarring related to a peripheral nerve can be external, causing adherence of the entire nerve to an adjacent structure, such as the roof of the carpal tunnel for the median nerve. When sufficient chronic compression or direct trauma creates scarring between the fascicles of the peripheral nerve, there is internal scarring. The microsurgical release of this interfascicular scarring is termed an "internal neurolysis."

Cross-References

▶ Carpal Tunnel Syndrome

Internal Validity

Definition

The treatment and comparison groups are selected and compared in such a manner that the observed differences between them on the dependent variables under study may only be attributed to the hypothesized intervention under investigation.

Cross-References

▶ Lumbar Traction

Internalization

Definition

Endocytosis, movement of proteins from the plasma membrane of a cell to intracellular compartments.

Cross-References

▶ Opioid Receptor Trafficking in Pain States

Internalization of Receptors

Definition

Endocytosis of surface membrane receptors after they bind a ligand. The receptors may later be recycled and reinserted into the membrane. When internalization occurs, it can be assumed that the ligand has been released in response to a stimulus. For example, when neurokinin-1 receptors are internalized, it is presumed that substance P (or a related neurokinin) has been released from nearby presynaptic terminals. Observation of this phenomenon is an indirect way of demonstrating the release of certain neurotransmitters.

Cross-References

▶ Visceral Pain Model, Lower Gastrointestinal Tract Pain

International Narcotics Control Board

Synonyms

INCB

Definition

An element of the United Nations drug control program. It is charged with prevention of illegal growth, manufacture, and distribution of narcotic and other controlled medications for nonmedical or nonscientific use. Its activities are set forth by the Single Convention on Narcotic Drugs and the Convention on Psychotropic Substances. It has recommended that governments facilitate the availability for the treatment of chronic pain including cancer pain. It collaborates closely with the World Health Organization.

Cross-References

▶ Cancer Pain Management, Undertreatment and Clinician-Related Barriers

Internet

▶ Remote Management of Pediatric Pain

Internet Therapy

▶ Psychological Treatment of Pain in Children

Interneuron

Definition

An interneuron is a neuron that communicates only to other neurons; they are often inhibitory.

Cross-References

▶ Information Processing in Thalamocortical Circuits
▶ Nociceptive Circuitry in the Spinal Cord
▶ Somatic Pain
▶ Visceral Nociception and Pain

Interoception

Definition

The term interoception is often used to define perceptual experiences that are not determined by "outside" stimuli. It refers to the sense of the physiological condition of the body and links sensations such as thirst, hunger, cold and warmth, and pain with emotional experience (feelings) and motivation, as it drives the individual towards adaptive behavior in order to establish homeostasis (e.g., by drinking, eating, and protection against environmental challenges).

Cross-References

▶ Angina Pectoris, Neurophysiology and Psychophysics
▶ Functional Imaging of Cutaneous Pain
▶ Spinothalamic Tract Neurons, in Deep Dorsal Horn
▶ Thalamus, Visceral Representation
▶ Trigeminal Brainstem Nuclear Complex, Anatomy

Interoceptive Sensation

▶ Angina Pectoris, Neurophysiology and Psychophysics

Interpolaris

Synonyms

Trigeminal subnucleus interpolaris, Vi, trigeminal nucleus interpolaris

Definition

The middle segment of the ▶ trigeminal sensory nuclear complex.

Cross-References

▶ Trigeminal Brainstem Nuclear Complex, Anatomy
▶ Trigeminal Brainstem Nuclear Complex, Physiology

Interprofessional Approach

Definition

Healthcare professionals co-working in an integrated process.

Cross-References

▶ Physical Medicine and Rehabilitation, Team-Oriented Approach

Interscalene

Definition

Pertaining to the potential fascial space between the anterior and meddle scalene muscles in the neck through which the brachial plexus traverses.

Cross-References

▶ Postoperative Pain, Regional Blocks

Interscalene Block

Definition

Injection via a needle or catheter of local anesthetic at the interscalene groove to block sensory and/or motor innervation of the brachial plexus.

Cross-References

▶ Acute Pain in Children, Postoperative

Interspike Interval

Synonyms

ISI

Definition

Interspike interval is the time interval (usually measured in milliseconds) between action potentials discharged by a neuron.

Cross-References

▶ Central Pain, Human Studies of Physiology
▶ Chronic Pain
▶ Thalamic Bursting Activity

Interstitial Cystitis

▶ Interstitial Cystitis and Chronic Pelvic Pain

Interstitial Cystitis and Chronic Pelvic Pain

Justin Friedlander, Amin S. Herati and Robert M. Moldwin
The Arthur Smith Institute for Urology, North Shore-LIJ Healthcare System, Hofstra North Shore LIJ School of Medicine, New Hyde Park, NY, USA

Synonyms

Bladder pain syndrome; Chronic pelvic pain; Interstitial cystitis

Definition

Interstitial cystitis/bladder pain syndrome (IC/BPS) is a chronic pelvic pain disorder characterized by the combination of pelvic pain (or pressure, discomfort) and irritative voiding symptoms, such as urinary urgency, frequency, and nocturia, *in the absence of an identifiable pathology*. Once thought to be a bladder specific disorder, recent findings suggest that other organ systems and central upregulation may play important roles in pathogenesis.

IC/BPS can be classified into "classic" and "nonclassic" types. The classic subtype, which accounts for only 5–10 % of patients, is characterized by the presence of Hunner's ulcers (or Hunner's patches). The much more common "nonclassical" subtype is typified by a normal cystoscopic examination (Peeker and Fall 2002).

Introduction

IC/BPS can present in children and in the elderly; however, patients most commonly seek medical attention from the third to fifth decade of life. The condition is 5–10 times more common in females than males. The lower estimates of male prevalence may be influenced by the clinical similarities between IC/BPS and chronic prostatitis/chronic pelvic pain syndrome.

The prevalence of IC/BPS in European countries ranges from 230 to 460/100,000 women (Leppilahti et al. 2005; Temml et al. 2007). A 2005 US population-based study found the prevalence in the Unites States to be 197/100,000 women and 41/100,000 men (Clemens et al. 2005). A more recent telephone survey of 131,000 US households estimated that 3.3–7.9 million women in the United States 18 years or older may suffer from IC/BPS symptoms (Berry et al. 2011).

Characteristics

Pathophysiology

The etiology of IC/BPS is poorly understood and thought to be multifaceted. Prevalent theories include infection, a defect in bladder permeability, local neurogenic and histamine-induced inflammation, and central neuropathic upregulation. Infectious etiologies such as bacteria, viruses, and fungi have also been investigated; however, no single organism has been implicated in the development of IC/BPS (Duncan and Schaeffer 1997). Defects in the urothelium, which normally provides an impermeable barrier to substances such as urinary electrolytes and metabolites, have also been identified in patients with IC/BPS. The effectiveness of this barrier can be compromised by inflammatory processes within the bladder, allowing irritative substances to permeate into the bladder interstitium (Negrete et al. 1996; Parsons et al. 1991). Leakage of these irritating metabolites can, in turn, release substance P and activate C fibers. Chronic activation of these fibers through chronic inflammation has been postulated to cause upregulation of neuroimmune pathways leading to the development of IC/BPS symptoms. Alternate theories involve the immune system, more specifically the hypersensitivity of mast cells, which release histamine when stimulated (Hofmeister et al. 1997; Filippou et al. 1999). Histamine induces local inflammatory changes, stimulates pain fibers, promotes angiogenesis, and allows the passage of white blood cells among other functions.

Decreased epithelial barrier integrity may also be due to the effects of antiproliferative factor (APF), a protein found in 94–97 % of patients with IC/BPS (Keay 2008). APF affects cell-signaling pathways, which can alter tight junction formation, cell differentiation, and proliferation of bladder of epithelial cells. APF has been demonstrated to induce changes in bladder epithelial cells that mimic changes seen in IC/BPS cells, both in vitro and in vivo.

Clinical and basic science research suggests that IC/BPS is a multiorgan disease process, especially given the high prevalence of comorbid conditions such as irritable bowel syndrome (IBS), chronic fatigue syndrome (CFS), fibromyalgia, vulvodynia, and neuropathic pain. In a questionnaire-based study comparing female patients with IC/BPS to controls without the condition (Nickel et al. 2010a), the prevalence of self-reported associated conditions diagnosed

with IC/BPS was 38.6 % vs. 5.2 % for IBS, 17.7 % vs. 2.6 % for fibromyalgia, and 9.5 % vs. 1.7 % for CFS, respectively. Of the 205 women in the IC/BPS cohort, 20.2 % reported multiple associated conditions. As the number of associated conditions increased, stress, depression, pain, and sleep disturbance increased, while sexual functioning, social support, and quality of life deteriorated.

Chronic pain creates time-dependent changes in cortical organization in patients with IC/BPS. These changes in cortical organization lead to the summation of all noxious afferent signals arising from the bladder, colon, uterus, and even pelvic floor musculature and erroneous interpretation of these pelvic signals as bladder in origin in a process termed "pelvic organ cross-talk" by Klumpp and Rudick (Klumpp and Rudick 2008). An important implication of this summation is a reduced threshold for the sensation of bladder pain.

Regardless of the mechanism, the result is a significant impact on quality of life (QOL). In an attempt to better understand psychosocial phenotypes in IC/BPS, a multi-institutional group had female patients with IC/BPS and controls without the condition complete a psychosocial phenotyping questionnaire battery (Nickel et al. 2010b). The group found that as compared to controls, patients with IC/BPS had significantly more pain (sensory, affective, and total), worse physical QOL, increased depression, sleep dysfunction, catastrophizing, stress, anxiety, and moderately more social/sexual function problems. Coping, suffering, and social parameters all correlated with the degree of general pain, whereas anxiety, depression, stress, and catastrophizing further correlated with IC/BPS-specific symptoms and with decreased QOL.

Diagnosis

In 1988, the National Institute of Diabetes & Digestive & Kidney Diseases set forth diagnostic criteria for IC/BPS; however, these criteria were intended for research purposes as a means of standardizing patients among research studies and were considered to be too restrictive, missing as many as 60 % of IC/BPS patients (Gillenwater and Wein 1998; Hanno et al. 1999).

In 2011, the American Urological Association (AUA) published guidelines for the diagnosis and management of IC/BPS (Hanno et al. 2011). The guidelines provide a framework to help determine whether or not a diagnosis of IC/BPS is appropriate. Diagnosis should include a thorough history and physical examination, followed by appropriate laboratory studies to exclude other commonly associated disorders.

Patients with IC/BPS typically present with a constellation of symptoms including pain (pressure or discomfort), low-volume urinary frequency, and urgency. Other common symptoms may include nocturia, dysuria, and dyspareunia. Patients often describe pain as "pressure," "aching," or "burning" that is worsened with bladder filling. Urinary frequency usually develops in an effort to maintain low bladder volumes, thus reducing pelvic pain. Voiding logs can be helpful to distinguish the various causes of urinary frequency.

The most frequent location for IC/BPS pain is midline suprapubic but may also include other areas of the pelvis such as urethra, vulva, vagina, and rectum (FitzGerald et al. 2006). Almost 90 % of patients with IC/BPS have evidence of pelvic floor myalgic pain which may give rise to varied extravesical symptoms (Peters et al. 2007). Extravesical sites such as the lower abdomen, inner thighs, and lower back are often described. Pain in these locations, along with symptoms such as urinary hesitancy, sensation of incomplete emptying, deep dyspareunia, and chronic constipation are often associated with this coexisting pelvic floor myalgia.

Various questionnaires are available to the clinician to monitor treatment response, including the O'Leary-Sant Interstitial Cystitis Symptom Index (IC/BPSSI) and pelvic pain, urgency, and frequency (PUF) patient symptom scale.

Physical examination is extremely important to exclude other well-defined causes for pelvic pain (such as vaginitis, urethral diverticulum, prostate pathology) and is often notable for midline suprapubic tenderness. Pelvic examination may demonstrate bladder tenderness, and, in

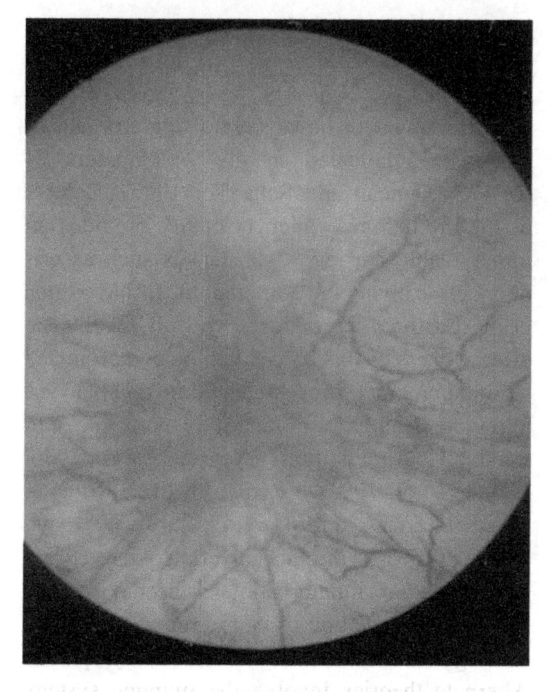

Interstitial Cystitis and Chronic Pelvic Pain, Fig. 1 Hunner's ulcer

most instances, trigger points of the pelvic floor musculature can be found.

Urinalysis is typically unremarkable. Urine cytology should be considered in older patients, particularly those with a smoking history. Hematuria, if found, should be investigated to exclude other pathologies such as calculus disease and cancer.

Cystoscopy and Hydrodistention of the Urinary Bladder. The nonclassical subtype of IC/BPS, which comprises 90–95 % of patients with IC/BPS will have grossly normal bladder mucosa on cystoscopic examination, whereas the remaining 5–10 % with the classic subtype will have inflammatory changes of the bladder wall, termed "Hunner's ulcers" (Fig. 1). Finding these ulcers is the only consistent cystoscopic finding that can lead to a diagnosis of IC/BPS. Glomerulations, submucosal petechial hemorrhages seen with bladder overdistention, are now known to be a nonspecific finding that can be seen in a variety of other conditions. In a retrospective study by Ottem and Teichman, patients with IC/BPS who received cystoscopy

Interstitial Cystitis and Chronic Pelvic Pain, Table 1 Diagnoses to be considered in the evaluation of IC/BPS

Category	Diagnosis		
	Urethral, ureteral, or bladder calculus	Urethral diverticulum	Prostadynia
Urologic	Intrinsic sphincter deficiency	Bladder outlet obstruction	Detrusor overactivity
	Urethrocele/cystocele	Radiation cystitis	Hemorrhagic cystitis
	Endometriosis	Pelvic inflammatory disease	
Gynecologic	Mittelschmerz	Vulvodynia	
	Ectopic pregnancy	Lichen sclerosis	
	Urethritis	Vulvovestibulitis	Schistosomiasis
Infectious	Infected Bartholin gland	Tuberculous cystitis	
	Infected Skene's gland	Vaginitis	
	Bladder cancer/carcinoma in situ	Urethral cancer	
Neoplastic	Uterine cancer		
	Ovarian cancer		
	Parkinson's disease	Spinal cord injury	
Neurologic	Multiple sclerosis	Spina bifida	
	Spinal stenosis	Diabetic cystopathy	

with hydrodistention were compared to those who had not undergone hydrodistention (Ottem and Teichman 2005). While 56 % of patients that had undergone hydrodistention reported improvement in symptoms, this was short lived (mean duration 2 months). Importantly, of the patients with and without improvement, no difference was found in mean age, duration of symptoms, anesthetic bladder capacity, or glomerulation grade. Cystoscopy with hydrodistention did not provide better information than what could be obtained from a thorough history and physical exam.

Potassium Sensitivity Testing (PST). The potassium sensitivity test was first described by Parsons et al. (1998) and is described as an intravesical challenge of a dilute solution containing potassium chloride. An increase in pelvic pain and irritative voiding symptoms following administration of potassium is considered a positive test and has been reported to elicit symptoms in approximately 75–78 % of patients who meet modified NIDDK diagnostic criteria for IC/BPS. Other conditions that may also elicit a painful response using this test include patients with prostatitis, acute bacterial cystitis, or urethritis (Parsons et al. 1998; Parsons and Albo 2002). Per the current guidelines, PST is no

longer recommended as it lacks the sensitivity or specificity to change clinical decision-making.

Urodynamics. Although not diagnostic of IC/BPS, urodynamics can be performed to rule out several competing conditions with similar symptom profiles including urge incontinence, detrusor sphincter dyssynergia (DSD), and bladder outlet obstruction among others. Patients with IC/BPS typically show reduced anesthetic bladder capacity, but again, this is a nonspecific finding (Evans and Sant 2007). Table 1 lists other conditions that may be relevant in the differential diagnosis of IC/BPS.

Treatment

The pathogenesis of IC/BPS is known to include epithelial dysfunction and inflammatory events at the local level and neuronal upregulation of pain perception. Because of the multifactorial nature of IC/BPS, a multimodal treatment approach, including dietary, behavioral, pharmacologic, intravesical instillations, and surgical treatments, tailored to the symptom profile of each patient is recommended. The guidelines suggest starting with conservative therapies before proceeding to more invasive treatments. Initial treatment should depend on patient preferences, symptom severity, and clinician judgment. Multiple, simultaneous treatments can also be considered if they are in the patient's best interest.

First-Line Treatments

First-line treatments include general relaxation strategies and stress management techniques, pain management, patient education, self-care practices, and behavioral modification. We highlight a few of these concepts.

Diet Fifty to sixty percent of IC/BPS patients are known to be influenced by ingested comestibles (Whitmore 1994). In a study performed by Shorter et al. (2007), the most bothersome comestibles were found to be caffeinated beverages followed closely by carbonated and alcoholic beverages, citrus fruits and beverages, artificial sweeteners, and spicy foods. In contrast, calcium glycerophosphate and baking soda were among comestibles that were reported to most consistently ameliorate symptoms. It is important to remember that not every IC/BPS patient is sensitive to the comestibles listed above and they should be advised to start with a bland diet and slowly add potentially irritating items. Patients should be encouraged to maintain a 3 day food and voiding diary to document the impact of each of the comestible challenges. Additionally, patients should be encouraged to maintain adequate hydration to dilute the noxious items that may be present in urine.

Bladder training, to gradually increase intervals between voids, may aid a minority of patients with IC/BPS, mostly those with more benign symptomatology. Some patients may benefit from bladder training programs, which include bladder relaxation and distraction techniques.

Second-Line Treatments

Second-line treatments are broken down into three categories: physical therapy techniques, oral therapies, and intravesical therapies.

Physical and Behavioral Therapy An article by Peters et al. (2007) investigated the prevalence of PFD in a small cohort of IC/BPS patients and reported it to be 87 %. Given the significant overlap in pelvic floor dysfunction symptoms, select IC/BPS patients with concomitant floor spasms may benefit from pelvic floor relaxation exercises and gentle stretching along with myofascial trigger point release as part of a comprehensive treatment plan.

Oral Therapies Pharmacologic therapies are available that target different pathophysiologic events involved in IC/BPS. Due to their different mechanisms of action, these drugs can be given in combination for synergistic effect. Patients with IC/BPS are often started on heparinoids, such as pentosan polysulfate sodium (PPS), which is believed to restore the normal protective barrier between the bladder and urine. PPS is the only Food and Drug Administration (FDA)-approved oral drug in the treatment of IC/BPS. PPS is routinely administered in 300 mg/day in three divided doses (AlTaweel et al. 2006).

Hydroxyzine, an antihistamine, is postulated to prevent the release of histamine from activated mast cells, cells characteristically abundant in the bladder urothelium of patients with IC/BPS. When used chronically, hydroxyzine suppresses mast cell degranulation. Hydroxyzine is started in doses of 10–25 mg daily, taken in the evening to minimize the sedative effect and increased to doses of 50–100 mg/day during the allergy season (Dimitrakov et al. 2007).

Another antihistamine that has shown some efficacy in treating IC/BPS is cimetidine. Dosing is typically 300–400 mg twice a day (Thilagarajah et al. 2001).

Significant upregulation of sensory nerves in the peripheral and central nervous system has been demonstrated in patients with long-standing IC/BPS. To inhibit the neuronal activation in these patients, a tricyclic antidepressant (TCA) such as amitriptyline or nortriptyline can be given. Amitriptyline has somnolent effects which is an added benefit given that 49 % of patients with IC/BPS have been shown to report insomnia and excessive daytime sleepiness (Koziol et al. 1993). Patients should be monitored for other side effects of TCAs including cardiac arrhythmias and acute urinary retention. Starting doses are usually 10 mg with increasing doses to 75 mg daily given at night.

Intravesical Therapies As previously mentioned, intravesical instillations of heparin sulfate (10,000 units in 10 ml sterile water) have been used for the treatment of IC/BPS. Like PPS, heparin's theoretical mechanism of action is an augmentation of the bladder's normal protective coating preventing noxious bladder solutes from stimulating the bladder surface. One study reported clinical improvement in 56 % of patients treated with intravesical heparin three times per week for 3 months and continued remission for nearly all patients treated for 6–12 months (Parsons et al. 1994).

Fifty percent dimethylsulfoxide (DMSO) is the only intravesical therapy FDA-approved for the treatment of IC/BPS and is proposed to work through a multifactorial mechanism of action with anti-inflammatory, analgesic, muscle relaxant, collagen degrading, and bacteriostatic properties. DMSO is administered with initial doses of 50 mL weekly for the first 6–8 weeks, followed by 50 mL every 2 weeks for the next 3–12 months. The combination of DMSO with heparin, sodium bicarbonate, and a steroid such as triamcinolone may prolong the beneficial response (Perez-Marrero et al. 1993). Patients should be warned that their symptoms may flare after the first several instillations. Additionally, the medication is partially excreted through a pulmonary route, giving patients a garlic-like odor to their breath that may last 8–12 h.

The use of intravesical lidocaine in severe IC/BPS was first reported in Sweden in 1989 by Asklin et al. (Asklin and Cassuto 1989), who reported symptomatic relief in one patient. However, the use of intravesical lidocaine failed to gain traction until 2005, when Parsons et al. (Parsons 2005) tested the efficacy of heparin combined with intravesical lidocaine in 47 newly diagnosed IC/BPS patients. Symptom improvements were seen in 75 % symptoms after one instillation and 94 % following a second instillation at higher concentrations of lidocaine.

Some IC/BPS patients have benefited from instillations of GAG, such as hyaluronic acid and chondroitin sulfate 0.2 %, which theoretically augment the bladder surface mucin. Hyaluronic acid, however, has not obtained approval in the United States presumably because of a lack of efficacy in placebo-controlled studies.

Third-Line Treatments
Third-line therapeutic options include treatment of Hunner's ulcers, if present, and cystoscopy with hydrodistention. Fulguration of Hunner's ulcers can be performed via laser ablation or with electrocautery, or alternatively injection of triamcinolone can be considered. Fulguration and excision of well-defined lesions can provide transient relief in the 5–10 % of patients who have the classic type of IC/BPS. Cystoscopy has three main functions: one, it allows for bladder inspection to rule out other pathologies; two, by determining anatomic bladder capacity, it identifies a subset of patients who have reduced bladder capacity due to fibrotic changes (which has therapeutic implications); and lastly, it may serve as treatment if hydrodistention is performed. Treatment cystoscopy under anesthesia should be short in duration, with low pressure used for hydrodistention.

Fourth-Line Treatments
Next in consideration is sacral neurostimulation. Several studies have shown this to be a safe and effective treatment option for patients with IC/BPS who have difficult-to-treat voiding dysfunction and perineal pain; however, most studies evaluating this modality have small sample sizes and lack long-term follow-up.

Fifth-Line Treatments
Fifth-line treatment options include cyclosporine A and intradetrusor injections of botulinum toxin A. Cyclosporine A is a calcineurin inhibitor typically used for prevention of graft rejection after organ transplantation. Lower doses are given to patients with IC/BPS as compared to transplant patients, with effect thought to be due to the drug's anti-inflammatory properties. Instillations of botulinum toxin have shown promising short-term

efficacy; however, longer follow-up is needed before recommendations regarding its use can be made (Giannantoni et al. 2008). At present, neither therapy is FDA-approved for the treatment of IC/BPS. It is recommended that use of either should be limited to experienced practitioners.

Sixth-Line Treatments
Surgery Surgery is often reserved for severe cases of IC/BPS that are refractory to the treatment options mentioned above. It is the last line of treatment outside of clinical trials. Surgical treatment options include substitution cystoplasty and urinary diversion with or without cystectomy. Subtotal cystectomy with bladder augmentation is reserved for patients whose chronic inflammation has resulted in severely diminished bladder capacities and unrelenting symptoms. Patients undergoing subtotal cystectomy with bladder augmentation are at high risk of postoperative urinary retention requiring lifelong clean intermittent catheterization (CIC/BPS). Total cystectomy and urinary diversion may be considered for those patients who have severe symptoms refractor to the other conservative modalities.

Treatments No Longer Offered
The updated guidelines highlight five specific treatments that should no longer be offered to patients based on lack of efficacy and/or unacceptable adverse event profiles. These include the following: long-term oral antibiotic administration; intravesical bacillus Calmette-Guerin (BCG); intravesical instillation of resiniferatoxin; high-pressure, long-duration hydrodistention; and systemic (oral) long-term glucocorticoid administration.

Summary

The diagnosis of interstitial cystitis remains one largely based upon presenting symptoms consistent with an irritative/painful bladder-based syndrome. Primary evaluation is dedicated to eliminating well-defined pathologies that might account for similar complaints. Treatment options range from lifestyle and behavioral modifications to oral and intravesical therapies, with surgery reserved for the most refractory patients. Continuous reevaluation of pain management is of great importance, with the goal to maximize function while minimizing pain and side effects. Although no cure is at hand, a combination of conservative modalities may afford significant symptom relief to the majority of IC/BPS patients.

References

AlTaweel, W., Hajebrahimi, S., & Corcos, J. (2006). Efficacy of interstitial cystitis treatments: A review. *EAU-EBU Update Series, 4*, 47–61.

Asklin, B., & Cassuto, J. (1989). Intravesical lidocaine in severe interstitial cystitis. Case report. *Scandinavian Journal of Urology and Nephrology, 23*, 311–312.

Berry, S. H., Elliot, M. N., Suttorp, M., et al. (2011). Prevalence of symptoms of bladder pain syndrome/interstitial cystitis among adult females in the United States. *Journal of Urology, 186*, 540–544.

Clemens, J. Q., Meenan, R. T., Rosetti, M. C., Gao, S. Y., & Calhoun, E. A. (2005). Prevalence and incidence of interstitial cystitis in a managed care population. *Journal of Urology, 173*, 98–102; discussion 102.

Dimitrakov, J., Kroenke, K., Steers, W. D., et al. (2007). Pharmacologic management of painful bladder syndrome/interstitial cystitis: A systematic review. *Archives of Internal Medicine, 167*, 1922–1929.

Duncan, J. L., & Schaeffer, A. J. (1997). Do infectious agents cause interstitial cystitis? *Urology, 49*, 48–51.

Evans, R. J., & Sant, G. R. (2007). Current diagnosis of interstitial cystitis: An evolving paradigm. *Urology, 69*, 64–72.

Filippou, A. S., Grant, G. S., & Theoharides, T. C. (1999). Increased expression of intercellular adhesion molecule 1 in relation to mast cells in the bladder of interstitial cystitis patients. *International Journal of Immunopathology and Pharmacology, 12*, 49–53.

FitzGerald, M. P., Brensinger, C., Brubaker, L., & Propert, K. (2006). What is the pain of interstitial cystitis like? *International Urogynecology Journal and Pelvic Floor Dysfunction, 17*, 69–72.

Giannantoni, A., Porena, M., Costantini, E., Zucchi, A., Mearini, L., & Mearini, E. (2008). Botulinum a toxin intravesical injection in patients with painful bladder syndrome: 1 year follow-up. *Journal of Urology, 179*, 1031–1034.

Gillenwater, J. Y., & Wein, A. J. (1998). Summary of the National Institute of Arthritis, Diabetes, Digestive and Kidney Diseases workshop on interstitial cystitis, National Institutes of Health, Bethesda, Maryland, August 28–29, 1987. *Journal of Urology, 140*, 203–206.

Hanno, P. M., Landis, J. R., Matthews-Cook, Y., Kusek, J., & Nyberg, L., Jr. (1999). The diagnosis of interstitial cystitis revisited: Lessons learned from the National Institutes of Health Interstitial Cystitis Database study. *Journal of Urology, 161*, 553–557.

Hanno, P. M., Burks, D. A., Clemens, J. Q., et al. (2011). AUA guideline for the diagnosis and treatment of interstitial cystitis/bladder pain syndrome. *Journal of Urology, 185*, 2162–2170.

Hofmeister, M. A., He, F., Ratliff, T. L., Mahoney, T., & Becich, M. J. (1997). Mast cells and nerve fibers in interstitial cystitis (IC): An algorithm for histologic diagnosis via quantitative image analysis and morphometry (QIAM). *Urology, 49*, 41–47.

Keay, S. (2008). Cell signaling in interstitial cystitis/painful bladder syndrome. *Cellular Signalling, 20*, 2174–2179.

Klumpp, D. J., & Rudick, C. N. (2008). Summation model of pelvic pain in interstitial cystitis. *Nature Clinical Practice Urology, 5*, 494–500.

Koziol, J. A., Clark, D. C., Gittes, R. F., & Tan, E. M. (1993). The natural history of interstitial cystitis: A survey of 374 patients. *Journal of Urology, 149*, 465–469.

Leppilahti, M., Sairanen, J., Tammela, T. L., Aaltomaa, S., Lehtoranta, K., & Auvinen, A. (2005). Prevalence of clinically confirmed interstitial cystitis in women: A population based study in Finland. *Journal of Urology, 174*, 581–583.

Negrete, H. O., Lavelle, J. P., Berg, J., Lewis, S. A., & Zeidel, M. L. (1996). Permeability properties of the intact mammalian bladder epithelium. *American Journal of Physiology, 271*, F886–F894.

Nickel, J. C., Tripp, D. A., Pontari, M., et al. (2010a). Interstitial cystitis/painful bladder syndrome and associated medical conditions with an emphasis on irritable bowel syndrome, fibromyalgia and chronic fatigue syndrome. *Journal of Urology, 184*, 1358–1363.

Nickel, J. C., Tripp, D. A., Pontari, M., et al. (2010b). Psychosocial phenotyping in women with interstitial cystitis/painful bladder syndrome: A case control study. *Journal of Urology, 183*(1), 167–172.

Ottem, D. P., & Teichman, J. M. (2005). What is the value of cystoscopy with hydrodistention for interstitial cystitis? *Urology, 66*, 494–499.

Parsons, C. L. (2005). Successful downregulation of bladder sensory nerves with combination of heparin and alkalinized lidocaine in patients with interstitial cystitis. *Urology, 65*, 45–48.

Parsons, C. L., & Albo, M. (2002). Intravesical potassium sensitivity in patients with prostatitis. *Journal of Urology, 168*, 1054–1057.

Parsons, C. L., Lilly, J. D., & Stein, P. (1991). Epithelial dysfunction in nonbacterial cystitis (interstitial cystitis). *Journal of Urology, 145*, 732–735.

Parsons, C. L., Housley, T., Schmidt, J. D., & Lebow, D. (1994). Treatment of interstitial cystitis with intravesical heparin. *British Journal of Urology, 73*, 504–507.

Parsons, C. L., Greenberger, M., Gabal, L., Bidair, M., & Barme, G. (1998). The role of urinary potassium in the pathogenesis and diagnosis of interstitial cystitis. *Journal of Urology, 159*, 1862–1866; discussion 1866–7.

Peeker, R., & Fall, M. (2002). Toward a precise definition of interstitial cystitis: Further evidence of differences in classic and nonulcer disease. *Journal of Urology, 167*, 2470–2472.

Perez-Marrero, R., Emerson, L. E., Maharajh, D. O., & Juma, S. (1993). Prolongation of response to DMSO by heparin maintenance. *Urology, 41*, 64–66.

Peters, K. M., Carrico, D. J., Kalinowski, S. E., Ibrahim, I. A., & Diokno, A. C. (2007). Prevalence of pelvic floor dysfunction in patients with interstitial cystitis. *Urology, 70*(1), 16–18.

Shorter, B., Lesser, M., Moldwin, R. M., & Kushner, L. (2007). Effect of comestibles on symptoms of interstitial cystitis. *Journal of Urology, 178*, 145–152.

Temml, C., Wehrberger, C., Riedl, C., Ponholzer, A., Marszalek, M., & Madersbacher, S. (2007). Prevalence and correlates for interstitial cystitis symptoms in women participating in a health screening project. *European Urology, 51*, 803–808; discussion 809.

Thilagarajah, R., Witherow, R. O., & Walker, M. M. (2001). Oral cimetidine gives effective symptom relief in painful bladder disease: A prospective, randomized, double-blind placebo-controlled trial. *BJU International, 87*, 207–212.

Whitmore, K. E. (1994). Self-care regimens for patients with interstitial cystitis. *The Urologic Clinics of North America, 21*, 121–130.

Interventional Therapies

Definition

In relation to low back pain, these typically encompass spinal surgery and a variety of injection therapies such as epidural steroid injections.

Cross-References

▶ Disability, Effect of Physician Communication

Intervertebral Disc

Definition

The portion between each vertebral body, composed of the annulus fibrosis and the nucleus pulposis.

Cross-References

▶ Chronic Low Back Pain: Definitions and Diagnosis

Intervertebral Foramen, Cervical

Definition

The cervical intervertebral foramen lies between adjacent cervical vertebrae and serves as the bony conduit through which the spinal nerve root exits the bony spinal canal. Its roof and floor are formed by the pedicles of consecutive vertebrae. Its posterolateral wall is formed largely by the superior articular process of the lower vertebra and in part by the inferior articular process of the upper vertebra and the capsule of the zygapophysial joint. The lower end of the upper vertebral body, the uncinate process of the lower vertebra, and the posterolateral corner of the intervertebral disc form the anteromedial wall.

Cross-References

▶ Cervical Transforaminal Injection of Steroids

Intra-Articular Blocks and Thoracic Medial Branch Blocks

▶ Thoracic Medial Branch Blocks and Intra-Articular Blocks

Intra-articular Corticosteroids

▶ Intra-articular Injections of Steroids

Intra-articular Injections of Steroids

Nikolai Bogduk
University of Newcastle, Newcastle Bone & Joint Institute, Royal Newcastle Centre, Newcastle, NSW, Australia

Synonyms

Intra-articular corticosteroids; Intra-articular steroid injections

Definition

Intra-articular injections of corticosteroids are a treatment for pain, ostensibly stemming from a synovial joint, in which a corticosteroid preparation is injected into the cavity of the painful joint. The injections may be blind or fluoroscopically guided.

Characteristics

Intra-articular injections of steroids were originally used as a treatment for overtly inflammatory joint diseases, such as rheumatoid arthritis. Their use for rheumatoid arthritis is not questioned. For that condition, intra-articular injections of steroids are held to be a useful adjunct to other therapy. They are not portrayed as a singular or curative treatment. They are used to suppress joint inflammation rapidly, while drug therapy is used to modify the disease process, long term.

The success of intra-articular injections of steroids for rheumatoid arthritis inspired their use for other painful conditions of joints, notably and most commonly osteoarthritis. In practice, any painful joint can be treated, and most joints of the body have attracted this form of treatment. The treatment has become very popular and is used not just by rheumatologists but also by orthopedic surgeons, anesthetists, physiatrists, and pain specialists. The literature on its efficacy, however, is sobering.

Rationale

The implicit rationale for the injection of steroids is that they relieve pain by suppressing inflammation. However, whereas this rationale is applicable to overtly inflammatory joint diseases, it is harder to sustain for osteoarthritis or undiagnosed joint pain. The evidence is weak or lacking that these latter conditions involve significant inflammation, if at all.

Rarely recognized in the literature is the fact that corticosteroids have a long-term local anesthetic effect (Johansson et al. 1990). This effect, rather than any anti-inflammatory action, may be the basis for the observed relief of pain following intra-articular injection.

Technique

The technique for intra-articular injection differs according to the joint targeted. Common to all techniques is the need to place a needle in the cavity of the target joint. Large joints, such as the knee are readily accessed, using palpation as a guide. The smaller the joint, and the deeper in the body that it lies, the more difficult it is to enter the joint accurately with a needle. Smaller and deeper joints can be accessed under fluoroscopic guidance.

Any of a number of corticosteroid preparations can be used. Those most commonly used contain betamethasone, dexamethasone, triamcinolone, or methylprednisolone. Often these corticosteroids are provided in a depot preparation which allows for sustained, slow release of the active agent.

Application

Although any joint can be targeted, the literature describes the use of intra-articular steroids for the larger joints of the body and some small ones. Most of the literature is anecdotal in nature, but there have been some controlled trials and systematic reviews.

Shoulder Joints

A systematic review (Buchbinder et al. 2003) assessed the literature on steroid injections for two common conditions of the shoulder. For rotator cuff disease, it found that subacromial steroid

injection had a small benefit over placebo in some trials, but no benefit over NSAIDs was demonstrated. For adhesive capsulitis, two trials suggested a possible early benefit of intra-articular steroid injection over placebo; and one trial suggested short-term benefit of intra-articular corticosteroid injection over physiotherapy in the short term.

Subsequent studies have shown that subacromial injections of steroids are no more effective than placebo for posttraumatic impingement of the shoulder (McInerney et al. 2003), and injection therapy is no more effective than physiotherapy for stiff shoulder (Hay et al. 2003). For adhesive capsulitis, fluoroscopically guided injections followed by exercises are more effective than injections alone or physiotherapy alone (Carette et al. 2003). For frozen shoulder, distension of the glenohumeral joint with steroids is more effective than sham treatment for 3 weeks, but thereafter the differences attenuate (Buchbinder et al. 2004).

Acromioclavicular Joint

Few studies have addressed treatment of pain stemming from the acromioclavicular joint. One controlled study found intra-articular steroids to be no more effective than placebo (Jacob and Sallay 1997).

Knee Joint

Reviews have concluded that intra-articular steroids for osteoarthritis of the knee are more effective than placebo, but they only have a short-term effect (3 weeks) (Arroll and Goodyear-Smith 2004; Creamer 1999; Raynauld et al. 2003; Towheed and Hochberg 1997). Steroid injections confer no long-term benefit for osteoarthritis of the knee (Creamer 1999).

Hip Joint

Although the hip joint has been less studied than the knee, the efficacy of intra-articular steroids seems to be the same. Injection of steroids is more effective than injection of local anesthetic alone but only for about 3 weeks (Kullenberg et al. 2004).

Thumb

The carpometacarpal joint of the thumb is another site affected by painful osteoarthritis and which attracts treatment with intra-articular injection of steroids. A controlled trial, however, found no ► attributable effect for this treatment (Meenagh et al. 2004).

Neck Pain

Injections of steroids into the cervical zygapophyseal joints have been used to treat chronic neck pain. A controlled study, however, found that injection of steroids was no more effective than injection of local anesthetic alone and that the beneficial effects of both agents rapidly disappeared in about 2 weeks (Barnsley et al. 1994).

Back Pain

Injection of steroids into the lumbar zygapophyseal joints has been a popular treatment for low back pain. A review of the observational studies and the controlled studies available found that injection of steroids is no more effective than sham therapy (Bogduk 2005).

Interpretation

The popularity of intra-articular injections of steroids for joint pain is not matched by the literature concerning their efficacy. For all joints that have been studied, the pattern is the same. Either the injections are not more effective than placebo or sham therapy or any pain-relieving effects last for only a matter of a few weeks.

References

Arroll, B., & Goodyear-Smith, F. (2004). Corticosteroid injections for osteoarthritis of the knee: Meta-analysis. *British Medical Journal, 328*, 869.

Barnsley, L., Lord, S. M., Wallis, B. J., et al. (1994). Lack of effect of intraarticular corticosteroids for chronic pain in the cervical zygapophyseal joints. *The New England Journal of Medicine, 330*, 1047–1050.

Bogduk, N. (2005). A narrative review of intra-articular corticosteroid injections for low back pain. *Pain Medicine, 6*, 297–298.

Buchbinder, R., Green, S., & Youd, J. M. (2003). Corticosteroid injections for shoulder pain. Cochrane Database Syst Rev, CD004016.

Buchbinder, R., Green, S., Forbes, A., et al. (2004). Arthrographic joint distension with saline and steroid improves function and reduces pain in patients with painful stiff shoulder: Results of a randomised, double blind, placebo controlled trial. *Annals of the Rheumatic Diseases, 63*, 302–309.

Carette, S., Moffet, H., Tardif, J., et al. (2003). Intraarticular corticosteroids, supervised physiotherapy, or a combination of the two in the treatment of adhesive capsulitis of the shoulder: A placebo-controlled trial. *Arthritis and Rheumatism, 48*, 829–838.

Creamer, P. (1999). Intra-articular corticosteroid treatment in osteoarthritis. *Current Opinion in Rheumatology, 11*, 417–421.

Hay, E. M., Thomas, E., Paterson, S. M., et al. (2003). A pragmatic randomised controlled trial of local corticosteroid injection and physiotherapy for the treatment of new episodes of unilateral shoulder pain in primary care. *Annals of the Rheumatic Diseases, 62*, 394–399.

Jacob, A. K., & Sallay, P. I. (1997). Therapeutic efficacy of corticosteroid injections in the acromioclavicular joint. *Biomedical Sciences Instrumentation, 34*, 380–385.

Johansson, A., Hao, J., & Sjolund, B. (1990). Local corticosteroid application blocks transmission in normal nociceptive C-fibres. *Acta Anaesthesiologica Scandinavica, 34*, 335–338.

Kullenberg, B., Runesson, R., Tuvhag, R., et al. (2004). Intraarticular corticosteroid injection: Pain relief in osteoarthritis of the hip? *The Journal of Rheumatology, 31*, 2265–2268.

McInerney, J. J., Dias, J., Durham, S., et al. (2003). Randomised controlled trial of single, subacromial injection of methylprednisolone in patients with persistent, post-traumatic impingement of the shoulder. *Emergency Medicine Journal, 20*, 218–221.

Meenagh, G. K., Patton, J., Kynes, C., et al. (2004). A randomised controlled trial of intra-articular corticosteroid injection of the carpometacarpal joint of the thumb in osteoarthritis. *Annals of the Rheumatic Diseases, 63*, 1260–1263.

Raynauld, J. P., Buckland-Wright, C., Ward, R., et al. (2003). Safety and efficacy of long-term intraarticular steroid injections in osteoarthritis of the knee: A randomized, double-blind, placebo-controlled trial. *Arthritis and Rheumatism, 48*, 370–377.

Towheed, T. E., & Hochberg, M. C. (1997). A systematic review of randomized controlled trials of pharmacological therapy in osteoarthritis of the knee, with an emphasis on trial methodology. *Seminars in Arthritis and Rheumatism, 26*, 755–770.

Intra-Articular Sacroiliac Joint Block

► Sacroiliac Joint Blocks

Intra-Articular Sacroiliac Joint Injection

▶ Sacroiliac Joint Blocks

Intra-articular Steroid Injections

▶ Intra-articular Injections of Steroids

Intracellular Labeling

Definition

Intracellular labeling is a technique that stains a single neuron by injecting a neural tracer intracellularly through a glass microelectrode containing the tracer.

Cross-References

▶ Trigeminal Brainstem Nuclear Complex, Anatomy

Intracerebral Hematoma: Apoplexy

▶ Headache Due to Intracranial Bleeding

Intracerebroventricular Drug Pumps

▶ Pain Treatment, Implantable Pumps for Drug Delivery

Intracerebroventricular, Intracerebral, and Intrathecal

Definition

Drugs are frequently administered directly into the brain of experimental animals because such drugs may not penetrate the blood-brain barrier or systemic administration of such drugs is prohibited by cost. Intracerebroventricular (i.c.v.) injections are made into the cerebral ventricle of the brain. Intracerebral (i.c.) injections are made into the brain tissue and generally require stereotaxic implantation of microinjection guide sleeves in order to deliver the drug to a specific locus in the brain. Intrathecal (i.t.) injections are made into the cerebrospinal fluid that bathes the spinal cord.

Cross-References

▶ Nitrous Oxide Antinociception and Opioid Receptors
▶ Proinflammatory Cytokines

Intracranial

Definition

Structures located within the skull.

Cross-References

▶ Pain Treatment, Intracranial Ablative Procedures

Intracranial Ablative Procedures

▶ Pain Treatment, Intracranial Ablative Procedures

Intracranial Nociceptors

▶ Nociceptors in the Orofacial Region (Meningeal/Cerebrovascular)

Intracranial Venous Thrombosis

▶ Headache Due to Venous Sinus Thrombosis

Intractable

Definition

Persistence of an abnormal or harmful function despite usual treatment. For instance, intractable pain that persists after the etiology is treated and defies usual pain treatments.

Cross-References

▶ Pain Treatment, Intracranial Ablative Procedures

Intradental Nociceptors

▶ Nociceptors in the Dental Pulp

Intradermal

Definition

Intradermal relates to areas between the layers of the dermis (skin).

Cross-References

▶ PET and fMRI Imaging in Parietal Cortex (SI, SII, Inferior Parietal Cortex BA40)

Intradiscal Electrothermal Annuloplasty

▶ Intradiscal Electrothermal Therapy

Intradiscal Electrothermal Therapy

Nikolai Bogduk[1] and Kevin Pauza[2]
[1]University of Newcastle, Newcastle Bone & Joint Institute, Royal Newcastle Centre, Newcastle, NSW, Australia
[2]Tyler Spine and Joint Hospital, Tyler, TX, USA

Synonyms

IDET; Intradiscal electrothermal annuloplasty

Definition

Intradiscal electrothermal therapy (IDET) is a treatment devised to relieve the pain of internal disc disruption. It involves threading into the painful disc a flexible wire electrode, which is used to heat the tissues of the posterior annulus fibrosus in an effort to relieve the pain.

Characteristics

IDET was born of the desire and need for a procedure for the treatment of discogenic low back pain, other than major surgical procedures such as fusion and disc replacement. Since its introduction into clinical practice, it has met with variable acceptance and considerable criticism. Much of the resistance, however, has been financial and social in nature, with insurers concerned about the costs of a new and potentially popular procedure and some critics expressing concerns that much of the literature on the efficacy of the procedure is being published by its inventors.

Indications
The procedure is expressly indicated for low back pain caused by internal disc disruption. Practitioners, however, differ in how rigorously they make this diagnosis. Some rely only on the results of lumbar discography. Others insist on more rigorous criteria (Karasek and Bogduk 2001)

that include demonstration of a radial fissure on post-discography CT (See ▶ CT Scans). Nevertheless, discography is common to both approaches and is essential for the diagnosis of discogenic pain.

Eligibility criteria for IDET (Karasek and Bogduk 2001):

- Chronic, intrusive low back pain for greater than 3 months
- Failure to achieve adequate improvement with comprehensive nonoperative treatment
- No red flag condition
- No medical contraindications
- No neurologic deficit
- Normal straight-leg raise
- Nondiagnostic MRI scan
- No evidence for segmental instability, spondylolisthesis at target level
- No irreversible psychological barriers to recovery
- Motivated patient with realistic expectations of outcome
- No greater than 25 % loss of disc height
- Criteria for IDD satisfied, namely:
  - Disc stimulation is positive at low pressures (<50 psi).

- Disc stimulation reproduces pain of intensity >6/10.
- Disc stimulation reproduces concordant pain.
- CT discography reveals a grade 3 or greater annular tear.
- Control disc stimulation is negative at one and preferably two adjacent levels.

Rationale

The professed rationale of IDET varies between proponents. None has been tested, and none has been validated. Among the conjectures is that heating the posterior annulus stiffens the collagen of the disc (Saal and Saal 2000a), denatures chemical exudates in radial and circumferential fissures, seals fissures, and destroys nociceptive nerve endings (Karasek and Bogduk 2001).

Technique

Practitioners differ in the manner in which they execute IDET. Indeed, the procedure has undergone various modifications since its original description.

Common to all variants is the introduction of a flexible electrode into the disc. A trochar is

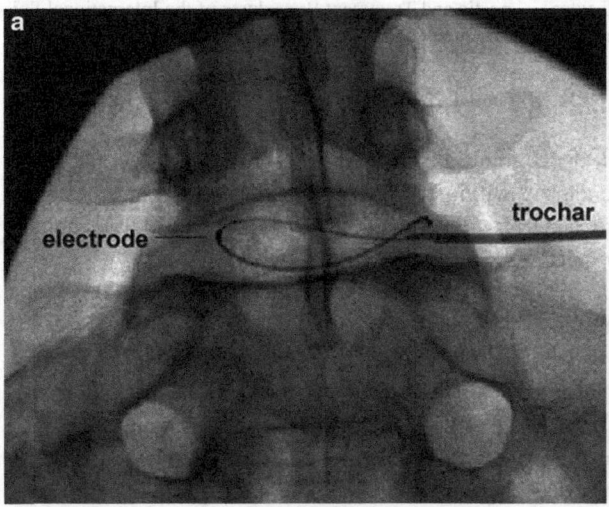
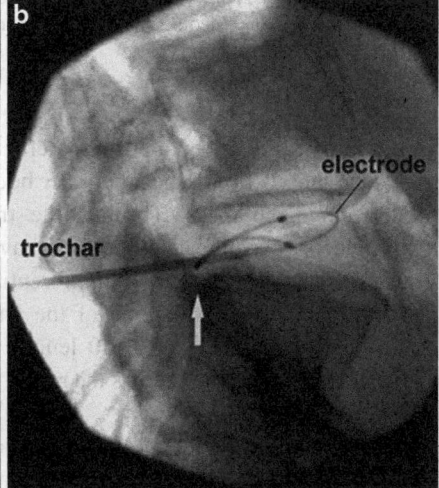

Intradiscal Electrothermal Therapy, Fig. 1 Radiographs of an IDET procedure on an L5-S1 disc. (**a**) Posteroanterior view. The electrode has been threaded through a trochar, into and around the disc. (**b**) Lateral view. The electrode has passed through the outer, posterior annulus (*arrow*) (Adapted from the Practice Guidelines for Spinal Diagnostic and Treatment Procedures of the International Spinal Intervention Society (Bogduk 2003))

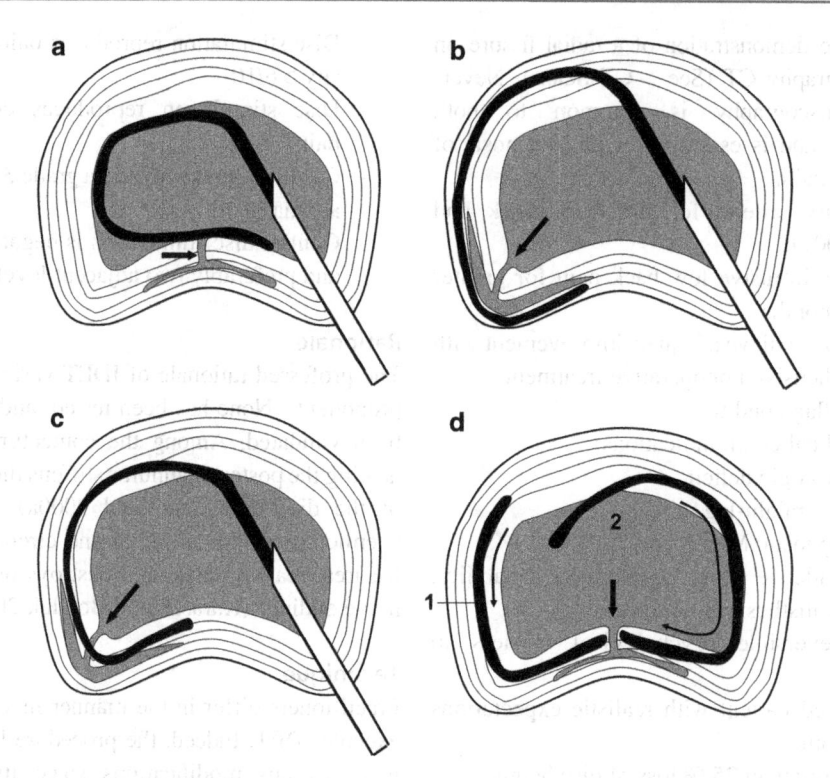

Intradiscal Electrothermal Therapy, Fig. 2 Variants of intradiscal electrothermal therapy. (**a**) The original description, in which the electrode is placed across the base of the radial fissure (*arrowed*) in the plane of the junction of the outer nucleus and inner annulus. (**b**) A variant in which the electrode is placed in the outermost annulus, parallel and external to the circumferential extension of the radial fissure (*arrowed*). (**c**) A variant in which the electrode is placed around the inner annulus, through the radial fissure (*arrowed*), but internal to its circumferential extension. (**d**) A variant in which the radial and circumferential fissures are attacked with two electrodes (1 and 2), introduced from opposites sides of the disc (Adapted from the Practice Guidelines for Spinal Diagnostic and Treatment Procedures of the International Spinal Intervention Society (Bogduk 2003))

passed into the disc along a posterolateral trajectory, such that its tip reaches the inner boundary of the annulus fibrosus. Through the trochar, a wire electrode is threaded and then navigated within the disc to assume a circumferential disposition, parallel to the lamina of the annulus. The objective is to place the 50 length of the active tip of the electrode across the radial fissure in the disc (Fig. 1).

The original description of the procedure required that the electrode be placed across the junction of the outer nucleus pulposus and the inner annulus fibrosus (Saal and Saal 2000a) (Fig. 2a). Subsequent variations have included the following: placing the electrode as far peripherally as possible in the outer annulus, attempting

multiple placements but at different heights within the annulus, and approaching the target fissure from both sides and at different heights (Fig. 2b, c, d).

Efficacy

In observational studies, practitioners of IDET have reported outcomes of various magnitudes for different durations of follow-up (Table 1). Most of these studies have reported clinically significant reductions in pain. Some have reported sustained relief for up to 2 years and beyond.

Others, however, have reported less than impressive outcomes. One retrospective audit found that 50 % of patients were dissatisfied

Intradiscal Electrothermal Therapy, Table 1 Outcomes of IDET based on observational studies

Study	N	Outcome Measures	Baseline	6 M	12 M	24 M
Saal and Saal (2000b, 2002)	58	VAS[a] mean	66		35	34
		Sd	19		23	20
		SF36: PF[b] mean	40		60	72
		Sd	25		22	23
Lutz and Lutz (2003)	33	VAS[a] mean	75		39[c]	
		Sd	24		30	
		RMDQ[d] mean	14		7[c]	
		Sd	5.9		7.2	
Lee et al. (2003)	62	VAS[a] mean	79			47
		Sd	13			30
		RMDQ[d] mean	15			8.8
		Sd	5.3			7.5
Cohen et al. (2003)	79	>50 % relief		0.48		
		VAS[a] mean	59	21		
		Sd	18	13		
		>90 % relief		0.10		
Gerszten et al. (2002)	27	SF36: BP[e]	27		38	
		SF36: PF[b]	32		47	
		ODI[f]	34		30	

[a]Visual analog pain scale (0–100)
[b]Physical functioning on the SF36 (0–100)
[c]Reported as 8–24 months
[d]Roland-Morris Disability Questionnaire (0–30)
[e]Bodily pain on the SF36
[f]Oswestry Disability Inventory (0–100)

with their outcome 1 year after treatment, and only 39 % had less pain (Davis et al. 2004). The other reported that 55 % of patients still required treatment after IDET; but 39 % resumed or remained at work (Webster et al. 2004).

Another study found that it was unable to achieve good outcomes in the same proportion of patients as reported in the more favorable outcome studies (Freedman et al. 2003). Nevertheless, it found some patients who had satisfying relief. The authors concluded that, although IDET was not a substitute for fusion, it nevertheless could be entertained as an option prior to undertaking fusion as a treatment.

Two controlled trials have produced favorable results (Table 2). In one (Bogduk and Karasek 2002), IDET was compared with the efficacy of a rehabilitation program. A significantly greater proportion of patients treated with IDET achieved relief of pain, which endured for up to 2 years. In particular, 20 % achieved complete relief of pain, and 57 % achieved greater than 50 % relief of pain. In the other study (Pauza et al. 2004), IDET was found to achieve a significantly greater reduction in pain than did sham treatment.

Another controlled study failed to find any difference in outcome between patients treated with IDET or sham therapy (Freeman et al. 2003). In that study, however, no patients benefited from either therapy. No placebo effect was encountered in either group.

Discussion

Both in observational studies and in the controlled trials, differences in outcome can potentially be attributed to a variety of factors: differences in patient selection, differences in diagnostic criteria, the particular technique that was used, and the rigor with which the technique was executed. None of these possible confounders has been formally tested. The one piece of evidence in this regard is that the best results reported to date have been in patients to whom the most rigorous diagnostic and selection criteria were applied and in whom the electrode was assiduously placed in the outer annulus. Implicitly, less stringent variants of the procedure are not as effective.

A conciliatory resolution of the current controversy concerning IDET is that, in its present form, IDET is only the forerunner of what might prove to be a better procedure. The electrode currently in use produces only a small lesion around its perimeter. That lesion may not be large enough for many of the intradiscal lesions

Intradiscal Electrothermal Therapy, Table 2 The outcomes of IDET based on controlled studies

Study		N	Outcome measures	Baseline	3 M	6 M	12 M	24 M
Bogduk and Karasek (2002)	Control	17	VAS[a] median	80	80		75	75
			IQR[b]		50–80	70–80	50–80	40–80
	IDET	35	VAS[a] median	80	35	30	30	30
			IQR[b]		70–90	10–50	10–60	10–70	10–70
			P value	0.70	0.00		0.01	0.03
	Control	17	50 % relief					0.22
	IDET	35	50 % relief					0.57
			P value					0.04
	Control	17	100 % relief					0.00
	IDET	35	100 % relief					0.20
			P value					0.047
Pauza et al. (2004)	Sham		ΔVAS[c] median		11			
			IQR[b]		26			
	IDET		ΔVAS[c] median		24			
			IQR[b]		24			
			P value		0.045			
	Sham		Δ ODI[d] median		4			
			IQR[b]		11			
	IDET		Δ ODI[d] median		11			
			IQR[b]		11			
			P value		0.05			
	Control		50 % relief		7			
			75 % relief		0			
			100 % relief		1			
	IDET		50 % relief		5			
			75 % relief		5			
			100 % relief		3			
			P value		0.03			

[a]Visual analog scale for pain (0–100)
[b]Interquartile range
[c]Change in visual analog score for pain (0–100)
[d]Change in Oswestry Disability Inventory (0–100)

suffered by patients. In effect, the IDET lesion undertreats these patients. This would explain the substantial failure rate of the procedure (50 %) even in the best hands. Improvements in the technology, such as producing a larger lesion, may provide for better outcomes of the procedure in a greater proportion of patients.

Cross-References

▶ IDET

References

Bogduk, N. (2003). *Practice guidelines for spinal diagnostic and treatment procedures. Intradiscal electrothermal therapy*. San Francisco: International Spinal Intervention Society.

Bogduk, N., & Karasek, M. (2002). Two-year follow-up of a controlled trial of intradiscal electrothermal annuloplasty for chronic low back pain resulting from internal disc disruption. *The Spine Journal, 2*, 343–350.

Cohen, S. P., Larkin, T., Abdi, S., et al. (2003). Risk factors for failure and complications of intradiscal electrothermal therapy: A pilot study. *Spine, 28*, 1142–1147.

Davis, T. T., Delamarter, R. B., Sra, P., et al. (2004). The IDET procedure for chronic discogenic low back pain. *Spine, 29*, 752–756.

Freedman, B. A., Cohen, S. P., Kuklo, T. R., et al. (2003). Intradiscal electrothermal therapy (IDET) for chronic low back pain in active-duty soldiers: 2-year follow-up. *The Spine Journal, 3*, 502–509.

Freeman, B. J. C., Fraser, R. D., Cain, C. M. J., et al. (2003). *A randomized double-blind controlled efficacy study: Intradiscal electrothermal therapy (IDET) versus placebo*. Presented at the 30th annual meeting of the international society for the study of the lumbar spine, Vancouver, May 13–15.

Gerszten, P. C., Welch, W. C., McGrath, P. M., et al. (2002). A prospective outcomes study of patients undergoing intradiscal electrothermy (IDET) for chronic low back pain. *Pain Physician, 5*, 360–364.

Karasek, M., & Bogduk, N. (2001). Intradiscal electrothermal annuloplasty: Percutaneous treatment of chronic discogenic low back pain. *Techniques in Regional Anesthesia and Pain Management, 5*, 130–135.

Lee, M. S., Cooper, G., Lutz, G. E., et al. (2003). Intradiscal electrothermal therapy (IDET) for treatment of chronic lumbar discogenic pain: A minimum 2-year clinical outcome study. *Pain Physician, 6*, 443–448.

Lutz, C., Lutz, G. E., & Cooke, P. M. (2003). Treatment of chronic lumbar diskogenic pain with intradiskal electrothermal therapy: A prospective outcome study. *Archives of Physical Medicine and Rehabilitation, 84*, 23–28.

Pauza, K. J., Howell, S., Dreyfuss, P., et al. (2004). A randomised, placebo-controlled trial of intradiscal electrothermal therapy for the treatment of discogenic low back pain. *The Spine Journal, 4*, 27–35.

Saal, J. S., & Saal, J. A. (2000a). Management of chronic discogenic low back pain with a thermal intradiscal catheter. A preliminary report. *Spine, 25*, 382–388.

Saal, J. A., & Saal, J. S. (2000b). Intradiscal electrothermal treatment for chronic discogenic low back pain. A prospective outcome study with minimum 1-year follow-up. *Spine, 25*, 2622–2627.

Saal, J. A., & Saal, J. S. (2002). Intradiscal electrothermal treatment for chronic discogenic low back pain. prospective outcome study with a minimum 2-year follow-up. *Spine, 27*, 966–974.

Webster, B. S., Verma, S., & Pransky, G. S. (2004). Outcomes of workers' compensation claimants with low back pain undergoing intradiscal electrothermal therapy. *Spine, 29*, 435–441.

Intralaminar Nuclei in Pain

▶ Parafascicular Nucleus, Pain Modulation

Intralaminar Thalamic Nuclei

Definition

A group of small nuclei including the centromedian, paracentral, parafascicular, and lateral and medial central nuclei, within the internal medullary lamina of the thalamus. The intralaminar thalamic nuclei are involved in pain processing as well as other behavioral and cognitive functions.

Cross-References

▶ Spinothalamic Terminations, Core and Matrix
▶ Thalamic Nuclei Involved in Pain, Human and Monkey
▶ Trigeminothalamic Tract Projections

Intramuscular Sensory Nerve Stimulation

▶ Dry Needling

Intraperitoneal

Definition

Within the peritoneal cavity, the area that contains the abdominal organs.

Cross-References

▶ Animal Models of Inflammatory Bowel Disease

Intrathecal

Definition

Administration of an agent directly into the cerebrospinal fluid (CSF) beneath the arachnoid membrane.

Cross-References

- ▶ Descending Circuitry, Opioids
- ▶ Headache Attributed to a Substance or its Withdrawal
- ▶ Opioid Receptor Trafficking in Pain States
- ▶ Opioids and Reflexes
- ▶ Opioids, Effects of Systemic Morphine on Evoked Pain
- ▶ Postoperative Pain, Appropriate Management

Intrathecal Drug Pumps

- ▶ Pain Treatment, Implantable Pumps for Drug Delivery

Intrathecal Injection

Synonyms

Spinal injection

Definition

Injection of drugs into the subarachnoid space, which contains cerebrospinal fluid. Placement of a catheter or needle into the subarachnoid space enables administration of drugs into the cerebrospinal fluid that bathes the spinal cord. This approach typically limits distribution of drugs to the spinal cord.

Cross-References

- ▶ Alpha(α) 2-Adrenergic Agonists in Pain Treatment
- ▶ Cancer Pain Management, Anesthesiologic Interventions
- ▶ Descending Circuitry, Opioids
- ▶ Intrathecal Space
- ▶ Pain Treatment, Implantable Pumps for Drug Delivery
- ▶ Postoperative Pain, Appropriate Management

Intrathecal Space

Definition

Deep to the arachnoid membrane and between the arachnoid mater and the pia mater lies the intrathecal or subarachnoid space. It contains cerebrospinal fluid, the spinal nerve roots, a trabecular network between the two membranes, blood vessels that supply the spinal cord, and the lateral extensions of the pia mater, the dentate ligaments.

Cross-References

- ▶ Postoperative Pain, Intrathecal Drug Administration

Intravenous Infusions, Regional and Systemic

James P. Robinson
Department of Anesthesiology & Pain Medicine,
University of Washington, Seattle, WA, USA

Synonyms

Intravenous regional analgesia; Intravenous regional block; IV block; IV infusion; Regional infusion

Definition

Intravenous (IV) infusions are a means of delivering a drug in order to determine if it relieves pain. The drug is administered through a needle or cannula inserted into a vein. The drug can be allowed to enter the circulation of the body, in which case the infusion becomes a systemic one. If a pressure cuff is applied to the limb proximal to the site of injection, the drug can be restricted to the circulation and tissue distal to the cuff, in which case the infusion becomes a regional one.

Characteristics

Principles
The objective of regional infusions is to relieve pain, either temporarily as a diagnostic test or for prolonged periods as a therapeutic intervention. Different drugs may be used, either to block nociceptive neurons in the periphery, or pain pathways in the central nervous system, or to block efferent sympathetic nervous activity that may be sensitizing nociceptive neurons.

Regional Infusions
Sympathetic Blocks
Intravenous regional sympathetic blocks (IRSBs) are designed to block efferent sympathetic activity. The drugs used are ones that either prevent the release of noradrenaline (norepinephrine) from the peripheral terminals of sympathetic nerves or block the receptors to noradrenaline in peripheral tissues.

For preventing the release of noradrenaline, the most commonly used agent is guanethidine (Jadad et al. 1995; Ramamurthy and Hoffman 1995). For blocking receptors phentolamine is used (Raja et al. 1991). Older drugs, whose use has been supplanted by these agents, include bretylium and reserpine.

Regional sympathetic blocks are typically used in the assessment and treatment of complex regional pain syndromes. As a diagnostic test, they are used to determine if the patient's pain or other symptoms are maintained by activity in efferent sympathetic nerves. The test is considered positive if administering the sympathetic blocking agent relieves the patient's pain. In that event, the pain is considered to be sympathetically maintained pain. Some practitioners use a positive response as an indication for sympathectomy to treat sympathetically maintained pain.

If pain is relieved for a prolonged and useful period, regional sympathetic blocks may be used as a treatment. In some patients, repeating the blocks progressively increases the duration of relief that ensues.

Local Anesthetic Blocks
Intravenous local anesthetic blocks involve the injection of a local anesthetic agent, typically lignocaine (lidocaine). They are used to anesthetize a larger area than might be possible with nerve blocks, or to anesthetize a region in which the source of pain is unknown and, therefore, not amenable to a specific nerve block. A typical dose is 30–40 ml of 0.5 % lignocaine, injected over 2–3 min (Buckley 2001).

Intravenous local anesthetic blocks are most commonly used in the management of complex regional pain syndrome. These blocks may provide prolonged periods of relief in their own right, but are usually used to provide analgesia so that other measures, such as physical therapy, can be instituted in patients who otherwise cannot bear to have the affected limb touched.

Systemic Infusions
Local Anesthetic
Systemic infusions of a local anesthetic agent can be used as a diagnostic test and as a treatment for ► central pain. It is believed that the agent circulates to the central nervous system where it decreases abnormal activity, either in central pain pathways or in the brain, to produce relief of pain without anesthesia. The agent can be delivered either as a bolus by slow injection from a syringe or as a slow infusion from a drip.

A positive response to a diagnostic infusion is taken as evidence that the patient has central pain. Treatment can be instituted either by prescribing oral local anesthetic agents or a slow infusion (Buckley 2001).

Phentolamine
Unlike other sympathetic blocking agents, phentolamine does not have significant side effects when administered systemically. Therefore, phentolamine does not need to be restricted to a regional infusion. It can be administered systemically in order to test for sympathetically maintained pain, in the manner in which regional sympathetic blocks are used.

Bisphosphonates

A recent innovation for the treatment of complex regional pain syndromes is the intravenous administration of ▶ bisphosphonates. These agents are believed to relieve pain by acting on descending pain modulatory pathways. Examples include alendronate (Adami et al. 1997), clodronate (Varenna et al. 2000), and pamidronate (Robinson 2002; Kingery 1997).

Other Agents

Other agents that have been used for systemic infusions, usually for the treatment of complex regional pain syndromes, include calcitonin (Kingery 1997), bretylium, clonidine, ketorolac, ketanserin, lysine acetylsalicylate, naftidrofuryl, methylprednisolone, labetalol, reserpine, hydralazine, thymoxamine (Shipton 1999) droperidol, and atropine (Galer et al. 2001).

Indications

The most common indications are complex regional pain syndromes or neuropathic pain for regional infusions, and central pain for systemic infusions. Intravenous methylprednisolone has been used successfully to treat acute cervical radicular pain after whiplash (Pettersson and Toolanen 1998).

Validity

When used as a diagnostic test, intravenous infusions have questionable validity. Although assumed to be valid, formal studies have rarely established their validity. Rather, placebo-controlled studies have shown that intravenous guanethidine and intravenous phentolamine have effects indistinguishable from those of normal saline, when administered under double-blind conditions (Robinson 2002; Kingery 1997). Consequently, in a given patient, responses to intravenous infusions cannot be considered to be positive unless accompanied by a negative response to placebo-controlled infusions.

Utility

When used as a treatment, intravenous infusions have been assumed to be effective. However, few agents have been submitted to placebo-controlled

studies, and even fewer have been shown to be more effective than placebo. The only agents proven to be effective are bretylium, ketanserin, and bisphosphonates (Robinson 2002; Kingery 1997). The evidence on phentolamine is conflicting, but the data favor that phentolamine does not produce analgesia greater than placebo or provides short-term relief in only a small subset of patients (Kingery 1997).

References

Adami, S., Fossaluzza, V., Gatti, D., Fracassi, E., & Braga, V. (1997). Bisphosphonate therapy for reflex sympathetic dystrophy syndrome. *Annals of the Rheumatic Diseases, 56*, 201–204.

Buckley, F. P. (2001). Regional anesthesia with local anesthetics. In J. D. Loeser (Ed.), *Bonica's management of pain* (pp. 1893–1952). Philadelphia: Lippincott, Williams & Wilkins.

Galer, B. S., Schwartz, L., & Allen, R. J. (2001). Complex regional pain syndromes – type I: Reflex sympathetic dystrophy, and type II: Causalgia. In J. D. Loeser (Ed.), *Bonica's management of pain* (pp. 388–411). Philadelphia: Lippincott, Williams & Wilkins.

Jadad, A. R., Carrol, D., Glynn, C. J., & McQuay, H. J. (1995). Intravenous regional sympathetic blockade in reflex sympathetic dystrophy: A systemic review and randomized crossover double blind study. *Journal of Pain and Symptom Management, 10*, 13–20.

Kingery, W. (1997). A critical review of controlled trials for peripheral neuropathic pain and complex regional pain syndromes. *Pain, 73*, 123–139.

Pettersson, K., & Toolanen, G. (1998). High-dose methylprednisolone prevents extensive sick leave after whiplash injury. A prospective, randomized, double-blind study. *Spine, 23*, 984–989.

Raja, A. N., Treed, R. D., Davis, K. D., & Campbell, J. N. (1991). Systemic alpha adrenergic blockade with phentolamine: A diagnostic test for sympathetically maintained pain. *Anaesthesiology, 746*, 691–698.

Ramamurthy, S., & Hoffman, J. (1995). Intravenous regional guanethidine in the treatment of RSD/causalgia: A randomized double blind study group. *Anaesthesia & Analgesia, 81*, 718–723.

Robinson, J. (2002). The treatment of complex regional pain syndrome type1. *Australasian Musculoskeletal Medicine, 7*, 101–105.

Shipton, E. A. (1999) (Ed.). Complex regional pain syndromes. In *Pain: acute and chronic* (pp. 210–231). London: Arnold.

Varenna, M., Zucchi, F., Ghiringhelli, D., Binelli, L., Bevilaqua, M., Bettica, P., & Sinigaglia, L. (2000). Intravenous clodronate in the treatment of reflex sympathetic dystrophy syndrome. *The Journal of Rheumatology, 27*, 1477–1483.

Intravenous Regional Analgesia

▶ Intravenous Infusions, Regional and Systemic

Intravenous Regional Analgesia/Block

Definition

Intravenous regional anesthesia is performed by occluding the circulation in a limb with a tourniquet and then injecting local anesthetic into a distal vein. Intravenous regional blocks (IVRB) using a range of agents have been utilized for the management of CRPS. Guanethidine depletes noradrenaline in the postganglionic adrenergic terminal, but its use in IVRBs has not been supported in controlled trials.

Cross-References

▶ Intravenous Infusions, Regional and Systemic
▶ Postoperative Pain, Acute Presentation of Complex Regional Pain Syndrome

Intravenous Regional Block

▶ Intravenous Infusions, Regional and Systemic

Intravenous Tramadol

Definition

Intravenous tramadol (2 mg/kg) provides similar antinociceptive effects to that of meperidine. Tramadol generally has no significant effects on gastrointestinal function. Tramadol has been shown to cause idiopathic seizures in clinical follow-ups, with an incidence

of <10 %. Tramadol is rapidly absorbed, with an oral bioavailability of 70 % after a single dose administration and ≃90 % after multiple doses (due to saturated first-pass hepatic metabolism).

Cross-References

▶ Postoperative Pain, Tramadol

Intraventricular

Definition

Within the cerebrospinal fluid-filled ventricles of the brain.

Cross-References

▶ Pain Treatment, Implantable Pumps for Drug Delivery

Intraventricular Drug Pumps

▶ Pain Treatment, Implantable Pumps for Drug Delivery

Intravesical

Definition

Within the bladder.

Cross-References

▶ Nocifensive Behaviors of the Urinary Bladder

Invasive

Definition

Procedures involving the introduction of instruments or other objects into the body or body cavities.

Cross-References

▶ Acute Pain in Children, Procedural

Inverse Agonist

Definition

Compound that stabilizes the inactive conformation of a receptor and inhibits the constitutive activity of the receptor.

Cross-References

▶ Metabotropic Glutamate Receptors in Spinal Nociceptive Processing

Inverse Problems

Definition

Problems in which the inputs or sources are estimated from the outputs or responses. This is in contrast to "direct" problems, in which outputs or responses are determined using a knowledge of the inputs or sources. (However, inverse problems are often ill-posed problems).

Cross-References

▶ Magnetoencephalography in Assessment of Pain in Humans

Ion Channel

Definition

An ion channel is a specialized membrane protein that acts as a pore and allows rapid diffusion of ions into and out of the cell through an aqueous pore within the molecule. It is usually selective for one ion (e.g., sodium, potassium, calcium) or a group of ions (e.g., all cations) and usually has some sort of gating mechanism (by voltage, binding a chemical, or sensory stimuli) that can switch the channel between open and closed states.

Cross-References

▶ Mechanonociceptors
▶ Nociceptors in the Orofacial Region (Skin/Mucosa)
▶ Nociceptors, Action Potentials, and Post-Firing Excitability Changes
▶ Painful Channelopathies
▶ Species Differences in Skin Nociception
▶ Trafficking and Localization of Ion Channels

Ion Channel Blockers

▶ Drugs Targeting Voltage-Gated Sodium and Calcium Channels

Ion Channel Trafficking

▶ Trafficking and Localization of Ion Channels

Ionotropic Glutamate Receptors

Synonyms

iGluRs

Definition

Family of ligand-gated ion channels that include N-methyl-D-aspartate (NMDA) receptors and non-NMDA (AMPA and kainate) receptors. They act quickly to depolarize the neuron and mediate fast synaptic events.

Cross-References

▶ Inflammation, Role of Peripheral Glutamate Receptors
▶ Metabotropic Glutamate Receptors in Spinal Nociceptive Processing
▶ Metabotropic Glutamate Receptors in the Thalamus
▶ Nociceptive Processing in the Amygdala: Neurophysiology and Neuropharmacology
▶ Thalamus, Nociceptive Neurotransmission

Ionotropic Receptor

Definition

Ionotropic receptor is a protein complex that forms an ion-permeable pore within the plasma membrane. Generally, two subunits of the complex have binding sites for a neurotransmitter. Binding of a neurotransmitter results in the opening of the ion channel. In the case of ionotropic receptors for glutamate, the ion channel is permeable to cations such as Na^+ and Ca^{2+}, resulting in depolarization of neurons. In the case of ionotropic receptors for GABA, the ion channel is permeable to Cl^-, generally resulting in hyperpolarization of the neuron.

Cross-References

▶ GABA Mechanisms and Descending Inhibitory Mechanisms
▶ NMDA Receptors in Spinal Nociceptive Processing
▶ Spinothalamic Tract Neurons, Peptidergic Input

Ionotropic Receptors

▶ NMDA Receptors in Spinal Nociceptive Processing

Ipsilateral

Definition

Situated or appearing on or affecting the same side of the body.

Cross-References

▶ Hemicrania Continua
▶ Opioid Receptor Trafficking in Pain States
▶ Spinothalamocortical Projections from SM

Ipsilateral Second Order Sensory Relay

▶ Postsynaptic Dorsal Column Projection, Anatomical Organization

Irritable Bowel Syndrome

Synonyms

IBS

Definition

Irritable bowel syndrome (IBS) is the most commonly reported functional gastrointestinal disorder affecting the large bowel and rectum. It is a syndrome characterized by abdominal pain or discomfort and altered defecation (such as diarrhea or constipation, with other common symptoms like bloating, passing mucus in the stools, or a sense that the bowels were not completely emptied) without abnormalities on conventional medical tests. It is one of the most commonly seen conditions by gastroenterologists in clinical practice. IBS is not an inflammatory bowel disease and should be diagnosed using the consensus "Rome" clinical criteria, that is, presence of recurrent abdominal pain or discomfort for at least 3 days/month in the last 3 months associated with two or more of the following: improvement with defecation, onset associated with a change in frequency of stool, onset associated with a change in form of stool. The criteria must be fulfilled for the past 3 months with symptom onset at least 6 months prior to diagnosis.

Cross-References

- ▶ Amygdala, Functional Imaging
- ▶ Amygdala, Pain Processing and Behavior in Animals
- ▶ Chronic Gynecological Pain, Doctor-Patient Interaction
- ▶ Descending Modulation of Visceral Pain
- ▶ Postsynaptic Dorsal Column Neurons, Responses to Visceral Input
- ▶ Psychological Aspects of Pain in Women
- ▶ Thalamus
- ▶ Thalamus and Visceral Pain Processing (Human Imaging)
- ▶ Thalamus, Clinical Visceral Pain, Human Imaging
- ▶ Visceral Pain Model, Irritable Bowel Syndrome Model

Irritable Bowel Syndrome Model

▶ Visceral Pain Model, Irritable Bowel Syndrome Model

Irritant

Definition

An irritant is a chemical used in prolotherapy solutions which acts by damaging cells directly or rendering them antigenic. They include phenol, guaiacol, and tannic acid.

Cross-References

- ▶ Prolotherapy

Ischemia

Definition

Insufficient or suppressed blood supply to an organ caused by blockage of a blood vessel (blood clot, atherosclerosis plaque, contractions, artery spasms). It prevents the tissue from receiving nutrients and oxygen and leads to accumulation of metabolic wastes like carbon dioxide and metabolic acids, and thus it induces local acidosis.

Cross-References

- ▶ Acid-Sensing Ion Channels
- ▶ TRPV1, Regulation by Protons

Ischemia Model

▶ Spinal Cord Injury Pain Model, Ischemia Model

Ischemia-Induced Muscle Pain

▶ Muscle Pain Model, Ischemia-Induced and Hypertonic Saline-Induced

Ischemic Heart Disease

Definition

Is a condition in which blood flow (and thus oxygen and nutrients) is reduced in areas of the heart where the diameters of the coronary arteries are significantly narrowed.

Cross-References

▶ Visceral Pain Model, Angina Pain

Ischemic Muscle Pain

▶ Muscle Pain Model, Ischemia-Induced and Hypertonic Saline-Induced

Ischemic Neuropathies

▶ Vascular Neuropathies

Ischemic Pain

▶ Tourniquet Test

Ischemic Pain/Test

Definition

Experimental ischemic pain from skeletal muscles can be easily produced by exercising a limb or an arm during a temporary block of the arterial blood supply.

Cross-References

▶ Tourniquet Test

Ischemic Test

▶ Tourniquet Test

ISI

▶ Interspike Interval

Island of Reil

▶ Insular Cortex, Neurophysiology, and Functional Imaging of Nociceptive Processing

Isolectin B4

Definition

The isolectin B4 of the plant lectin Griffonia simplicifolia recognizes a sugar moiety found selectively on the surface of small myelinated and unmyelinated nerve fibers, and is a useful marker for nociceptive neurons.

Cross-References

▶ Nociceptor, Categorization
▶ Trigeminal Brain Stem Nuclear Complex, Immunohistochemistry, and Neurochemistry

Itch

▶ Responses of Spinothalamic Tract Neurons to Pruritic Stimuli

Itch Man

Definition

A scale to quantify the severity of itching on a 10-point scale. It was developed and copyrighted by P. Blakeney and J. Marvin, Shriners Hospitals for Children, Galveston, Texas.

Cross-References

► Pain Control in Children with Burns

Itch/Itch Fibers

Martin Schmelz
Department of Anesthesiology in Mannheim,
University of Heidelberg, Mannheim, Germany

Synonyms

Pruritus

Definition

Itch (Latin: *pruritus*) obviously serves nociceptive functions, but it is clearly distinct from pain sensation. It is restricted to the skin and some adjoining mucosae. Whereas painful stimuli inflicted on the skin provoke withdrawal reflexes, itching stimuli provoke the very characteristic scratching reflex. This reflex pattern indicates that the neuronal apparatus for itch has developed as a nocifensive system for removal of irritating objects and agents affecting the skin. One might also describe scratching as a reflex pattern that is used in situations in which the noxious stimulus has already invaded the skin. In this situation, withdrawal would be useless; instead, localizing the injured site by scratching and a close inspection appear to be more adequate.

Characteristics

C-fibers, responding to histamine application in parallel to the itch ratings of subjects, have been discovered among the group of mechano-insensitive C-afferents (Schmelz et al. 1997) suggesting that there is a specific pathway for itch (Fig. 1). In contrast, the most common type of C-fibers, mechano-heat nociceptors (CMH or polymodal nociceptors) are either insensitive to histamine or only weakly activated by it (Schmelz et al. 2003b). This fiber type cannot account for the prolonged itch induced by the intradermal application of histamine.

The histamine-sensitive or "itch" fibers or pruriceptors are characterized by a particular low conduction velocity, large innervation territories, mechanical unresponsiveness, and high transcutaneous electrical thresholds (Schmelz et al. 2003b). In line with the large innervation territories of these fibers, two-point discrimination for histamine-induced itch is poor (15 cm in the upper arm).

The relative prevalence of the different C-fiber types has been estimated from recordings in the superficial peroneal nerve. About 80 % are polymodal nociceptors, which respond to mechanical, heat, and chemical stimuli. The remaining 20 % do not respond to mechanical stimulation. These fibers are mainly "mechano-insensitive nociceptors" (Schmidt et al. 1995), which are activated by chemical stimuli (Schmelz et al. 2000b), and can be sensitized to mechanical stimulation in the presence of inflammation (Schmelz et al. 2000; Schmidt et al. 1995). Among the mechano-insensitive afferent C-fibers, there is a subset of units that have a strong and sustained response to histamine. They comprise about 20 % of the mechano-heat-insensitive class of C-fibers, i.e., about 5 % of all C-fibers in the superficial peroneal nerve.

Nonhistaminergic itch is a histamine-independent itch that can be induced by Cowhage spicules inserted into human skin (LaMotte et al. 2009; Sikand et al. 2009). Interestingly, mechano-responsive "polymodal" C-fiber afferents are activated by cowhage in the cat (Tuckett and Wei 1987) and also in nonhuman primates (Johanek et al. 2007, 2008) and in human volunteers (Namer et al. 2008). These fibers are unresponsive to histamine and not involved in sustained axon reflex flare reactions (Schmelz et al. 2000a). This is consistent with the observation that cowhage-induced itch is not accompanied by a widespread axon reflex flare (Shelley and Arthur 1955, 1957; Johanek et al. 2007).

Itch/Itch Fibers, Fig. 1 In the *upper panel*, instantaneous discharge frequency of a mechano- and heat insensitive C-fiber (CMiHi) in the superficial peroneal nerve following histamine iontophoresis (20 mC; marked as *open box* in the diagram) is shown. In the *central panel*, activation of a spinal projection neuron following histamine iontophoresis is depicted. The *lower panel* shows average itch magnitude ratings of a group of 21 healthy volunteers after an identical histamine stimulus. Ratings at 10 s intervals on a visual analogue scale with the end points "no itch" – "unbearable itch" *Bars*: standard error of means (Modified from Schmelz et al. (1997) and Andrew and Craig (2001))

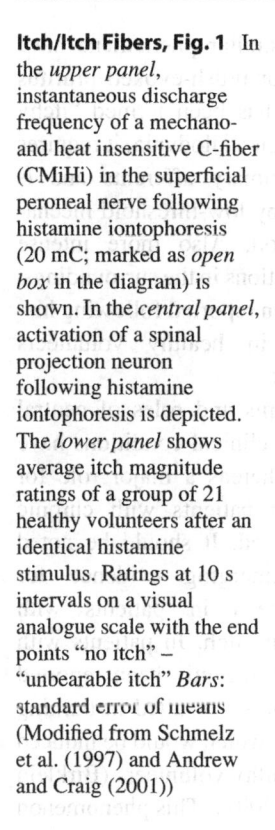

While in humans the segregation between histamine positive, mechano-insensitive fibers and cowhage positive mechanosensitive fibers is clear-cut, in monkey mechanosensitive C fibers also responded to histamine (Johanek et al. 2008).

Ad fibers responding to cowhage insertion for several minutes (Ringkamp et al. 2011) suggest an additional role of afferent input from myelinated fibers. Differential block of myelinated afferents does not reduce capsaicin-induced pain and only slightly reduces histamine-induced itch; however, it massively reduces cowhage-induced itch at least in part of the subjects (Ringkamp et al. 2011).

Specific Spinal Pruriceptive Neurons

The concept of dedicated pruriceptive neurons has now been complemented and extended by recordings from the cat spinal cord. A specific class of dorsal horn neurons projecting to the thalamus has been demonstrated, which respond strongly to histamine administered to the skin by iontophoresis (Andrew and Craig 2001). The time course of these responses was similar to that of itch in humans and matched the responses of the peripheral C-itch fibers (Fig. 1). These units were also unresponsive to mechanical stimulation and differed from the histamine–insensitive nociceptive units in lamina I of the spinal cord. In addition, their axons had a lower conduction velocity and anatomically distinct projections to the thalamus.

Interestingly, histamine- and cowhage-induced itch is not only processed separately in the primary afferent neurons, but the separation is maintained on the spinal level. Spinothalamic projection neurons in the dorsal horn could be separated into a histamine and cowhage-responsive population without overlap (Davidson et al. 2007). Moreover, the thalamic projections of the two subgroups also differed. The histamine-sensitive pathway can be assumed to be selective for itch albeit the recordings in monkey STT neurons also suggest some mechanical and capsaicin sensitivity. In contrast, the cowhage-sensitive pathway may be regarded as unspecific as activation of mechano-heat-responsive polymodal nociceptors are probably underlying the generation of cowhage-induced itch (Davidson and Giesler 2010). It is interesting to note that "specificity" is not only discussed for the neurons but also for mediators (Ross 2011): The classic algogen capsaicin generally provokes pain when applied to human skin. However, capsaicin induces itch when applied on the tip of an inactivated cowhage spicule (Sikand et al. 2009). This important finding indicates that the spatial characteristics of the application may be crucial and may functionally convert an algogenic mediator into a pruritic mediator.

Peripheral Sensitization

Increased intradermal nerve fiber density has been found in patients with chronic pruritus. In addition, increased epidermal levels of neurotrophin 4 (NT4) have been found in patients with atopic dermatitis and massively increased serum levels of NGF and SP have been found to correlate with the severity of the disease in such patients (Toyoda et al. 2002). Increased fiber density and higher local NGF concentrations were also found in patients with contact dermatitis. It is known that NGF and NT4 can sensitize nociceptors. These similarities between localized painful and pruritic lesions might suggest that similar mechanisms of nociceptor sprouting and sensitization exist on a peripheral level.

Central Sensitization

There is a remarkable similarity between the phenomena associated with central sensitization to pain and itch. Activity in chemo-nociceptors leads not only to acute pain but, in addition, can sensitize second-order neurons in the dorsal horn, thereby leading to increased sensitivity to pain (hyperalgesia).

In itch processing, similar phenomena have been described; touch or brush-evoked pruritus around an itching site has been termed "itchy skin" (Bickford 1938). Like allodynia, it requires ongoing activity in primary afferents and is most probably elicited by low-threshold mechanoreceptors (A-β fibers). Also more intense prick-induced itch sensations in the surroundings, "hyperknesis," have been reported following histamine iontophoresis in healthy volunteers (Atanassoff et al. 1999).

The exact mechanisms and roles of central sensitization for itch in clinical conditions have still to be explored, whereas a major role for central sensitization in patients with chronic pain is generally accepted. It should be noted that there is also emerging evidence for corresponding phenomena in patients with chronic pain and chronic itch. In patients with neuropathic pain, it has recently been reported that histamine iontophoresis resulted in burning pain instead of pure itch, which would be induced by this procedure in healthy volunteers (Birklein et al. 1997; Baron et al. 2001). This phenomenon is of special interest as it demonstrates spinal hypersensitivity to C-fiber input. Conversely, normally painful electrical, chemical, mechanical, and thermal stimulation is perceived as itching when applied in or close to lesional skin of atopic dermatitis patients (Ikoma et al. 2003).

Long-lasting activation of pruriceptors by histamine has been shown experimentally to induce central sensitization for itch in healthy volunteers (Ikoma et al. 2003); following the application of histamine via dermal microdialysis fibers, low pH stimulation of the skin close to the histamine site was perceived as itch instead of pain. Ongoing activity of pruriceptors, which might underlie the development of central sensitization for itch, has already been confirmed microneurographically in a patient with chronic pruritus (Schmelz et al. 2003a).

While there is obviously an antagonistic interaction between pain and itch under normal conditions, the patterns of spinal sensitization phenomena are surprisingly similar. It remains to be established whether this similarity will also include the underlying mechanism, which would also imply similar therapeutic approaches, such as gabapentin or clonidine, for the treatment of neuropathic itch.

References

Andrew, D., & Craig, A. D. (2001). Spinothalamic lamina 1 neurons selectively sensitive to histamine: A central neural pathway for itch. *Nature Neuroscience, 4*, 72–77.

Atanassoff, P. G., Brull, S. J., Zhang, J., Greenquist, K., Silverman, D. G., & LaMotte, R. H. (1999). Enhancement of experimental pruritus and mechanically evoked dysesthesiae with local anesthesia. *Somatosensory and Motor Research, 16*, 291–298.

Baron, R., Schwarz, K., Kleinert, A., Schattschneider, J., & Wasner, G. (2001). Histamine-induced itch converts into pain in neuropathic hyperalgesia. *NeuroReport, 12*, 3475–3478.

Bickford, R. G. L. (1938). Experiments relating to itch sensation, its peripheral mechanism and central pathways. *Clinical Science, 3*, 377–386.

Birklein, F., Claus, D., Riedl, B., Neundörfer, B., & Handwerker, H. O. (1997). Effects of cutaneous histamine application in patients with sympathetic reflex dystrophy. *Muscle & Nerve, 20*, 1389–1395.

Davidson, S., & Giesler, G. J. (2010). The multiple pathways for itch and their interactions with pain. *Trends in Neurosciences, 33*, 550–558.

Davidson, S., Zhang, X., Yoon, C. H., Khasabov, S. G., Simone, D. A., & Giesler, G. J., Jr. (2007). The itch-producing agents histamine and cowhage activate separate populations of primate spinothalamic tract neurons. *The Journal of Neuroscience, 27*, 10007–10014.

Ikoma, A., Rukwied, R., Stander, S., Steinhoff, M., Miyachi, Y., & Schmelz, M. (2003). Neuronal sensitization for histamine-induced itch in lesional skin of patients with atopic dermatitis. *Archives of Dermatology, 139*, 1455–1458.

Johanek, L. M., Meyer, R. A., Friedman, R. M., et al. (2008). A role for polymodal C-fiber afferents in nonhistaminergic itch. *The Journal of Neuroscience, 28*, 7659–7669.

Johanek, L. M., Meyer, R. A., Hartke, T., et al. (2007). Psychophysical and physiological evidence for parallel afferent pathways mediating the sensation of itch. *The Journal of Neuroscience, 27*, 7490–7497.

LaMotte, R. H., Shimada, S. G., Green, B. G., & Zelterman, D. (2009). Pruritic and nociceptive sensations and dysesthesias from a spicule of cowhage. *Journal of Neurophysiology, 101*, 1430–1443.

Namer, B., Carr, R., Johanek, L. M., Schmelz, M., Handwerker, H. O., & Ringkamp, M. (2008). Separate peripheral pathways for pruritus in man. *Journal of Neurophysiology, 100*, 2062–2069.

Ringkamp, M., Schepers, R. J., Shimada, S. G., et al. (2011). A role for nociceptive, myelinated nerve fibers in itch sensation. *The Journal of Neuroscience, 31*, 14841–14849.

Ross, S. E. (2011). Pain and itch: Insights into the neural circuits of aversive somatosensation in health and disease. *Current Opinion in Neurobiology, 21*(6), 880–7.

Schmelz, M., Hilliges, M., Schmidt, R., et al. (2003a). Active "itch fibers" in chronic pruritus. *Neurology, 61*, 564–566.

Schmelz, M., Schmidt, R., Weidner, C., Hilliges, M., Torebjörk, H. E., & Handwerker, H. O. (2003b). Chemical response pattern of different classes of C-nociceptors to pruritogens and algogens. *Journal of Neurophysiology, 89*, 2441–2448.

Schmelz, M., Schmidt, R., Bickel, A., Handwerker, H. O., & Torebjörk, H. E. (1997). Specific C-receptors for itch in human skin. *The Journal of Neuroscience, 17*, 8003–8008.

Schmelz, M., Michael, K., Weidner, C., Schmidt, R., Torebjörk, H. E., & Handwerker, H. O. (2000a). Which nerve fibers mediate the axon reflex flare in human skin? *NeuroReport, 11*, 645–648.

Schmelz, M., Schmidt, R., Handwerker, H. O., & Torebjörk, H. E. (2000b). Encoding of burning pain from capsaicin-treated human skin in two categories of unmyelinated nerve fibres. *Brain, 123*, 560–571.

Schmidt, R., Schmelz, M., Forster, C., Ringkamp, M., Torebjörk, H. E., & Handwerker, H. O. (1995). Novel classes of responsive and unresponsive C nociceptors in human skin. *The Journal of Neuroscience, 15*, 333–341.

Shelley, W. B., & Arthur, R. P. (1955). Mucunain, the active pruritogenic proteinase of cowhage. *Science, 122*, 469–470.

Shelley, W. B., & Arthur, R. P. (1957). The neurohistology and neurophysiology of the itch sensation in man. *AMA archives of dermatology, 76*, 296–323.

Sikand, P., Shimada, S. G., Green, B. G., & LaMotte, R. H. (2009). Similar itch and nociceptive sensations evoked by punctate cutaneous application of capsaicin, histamine and cowhage. *Pain, 144*, 66–75.

Toyoda, M., Nakamura, M., Makino, T., Hino, T., Kagoura, M., & Morohashi, M. (2002). Nerve growth factor and substance P are useful plasma markers of disease activity in atopic dermatitis. *British Journal of Dermatology, 147*, 71–79.

Tuckett, R. P., & Wei, J. Y. (1987). Response to an itch-producing substance in cat II. Cutaneous receptor population with unmyelinated axons. *Brain Research, 413*, 95–103.

Itchy Skin

Definition

The cutaneous areas of abnormally enhanced itch that are characterized by a greater than normal sensation of itch in response to a normally itchy stimulus (*hyperknesis*) and/or a sensation of itch evoked by stimuli that normally do not elicit itch (*alloknesis*). These states are sometimes accompanied by mild hyperalgesia in which a normally painful stimulus can evoke a greater than normal magnitude and duration of pain.

Cross-References

▶ Allodynia and Alloknesis

IV Block

▶ Intravenous Infusions, Regional and Systemic

IV Infusion

▶ Intravenous Infusions, Regional and Systemic

J

Jab-Like and Jolt-Like Headache

Definition

See ▶ Primary Stabbing Headache.

Cross-References

▶ Primary Exertional Headache

Jabs and Jolts Syndrome

▶ Primary Stabbing Headache

Jaw Claudication

Definition

Pain during mastication due to ischemia of the masseter muscles in temporal arteritis.

Cross-References

▶ Headache Due to Arteritis

Jaw-Muscle Silent Periods (Exteroceptive Suppression)

Giorgio Cruccu
Department of Neurological Sciences, La Sapienza University, Rome, Italy

Synonyms

Exteroceptive suppression; Masseter inhibitory reflex; Masseter silent periods

Definition

The jaw-muscle silent periods are trigemino-trigeminal inhibitory reflexes elicited by electrical, mechanical, or radiant heat stimuli delivered to the oral region (in the territory of the maxillary and mandibular trigeminal divisions); on electromyographic recordings from contracted jaw-closers, this reflex inhibition appears as an early phase and a late phase of suppression, also called ES1 and ES2 exteroceptive suppressions or SP1 and SP2 silent periods.

Characteristics

Whereas in the reflex control of jaw movement in lower mammals, the jaw-opening and jaw-closing muscles act in equilibrium, in humans, the jaw-openers (digastric and suprahyoid

G.F. Gebhart, R.F. Schmidt (eds.), *Encyclopedia of Pain*, DOI 10.1007/978-3-642-28753-4,
© Springer-Verlag Berlin Heidelberg 2013

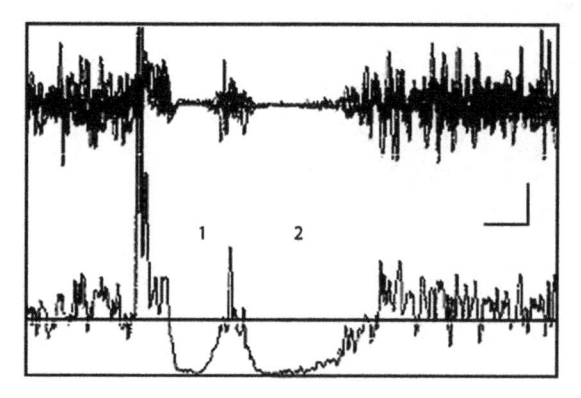 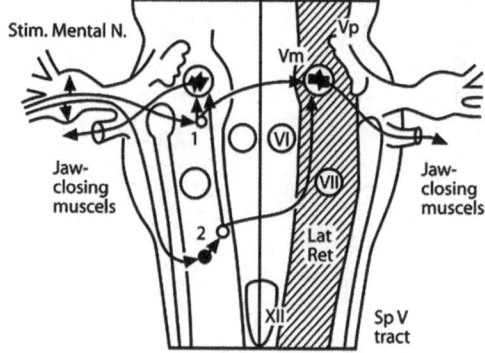

Jaw-Muscle Silent Periods (Exteroceptive Suppression), Fig. 1 Masseter inhibitory reflex and brainstem circuits. *Left*: first (*1*) and second (*2*) inhibitory periods. Recording from the right masseter muscle. Stimulation of the right mental nerve. *Upper traces*: 8 trials are superimposed. *Lower trace*: rectified and averaged signal. The horizontal line indicates 80 % of the background EMG level. Calibration 20 ms/200 μV. Right: schematic drawing of the reflex circuits. Afferents for the first inhibitory period connect with an inhibitory interneuron (*1*) located close to the ipsilateral trigeminal motor nucleus. The afferents for the second inhibitory period descend along the spinal trigeminal tract and connect with a multisynaptic chain of excitatory interneurons. The last interneuron is inhibitory (*2*) and projects bilaterally onto the jaw-closing motoneurons. Vm, trigeminal motor nucleus; Vp, trigeminal principal sensory nucleus; Sp V tract, spinal trigeminal tract; VI, abducens nucleus; VII, facial nucleus; XII, hypoglossal nucleus; Lat Ret, lateral reticular formation (From Cruccu and Deuschl 2000)

muscles) play a marginal role. The jaw-closers (masseter, temporalis, and pterygoid muscles) serve both functions: for reflex jaw closing, under normal circumstances (excitation) and for reflex jaw opening, when they undergo inhibition. The jaw-closers are excited by way of the Aα muscle spindle input and strongly inhibited by way of Aβ capsulated mechanoreceptors and Aδ free nerve endings. The powerful inhibition exerted by cutaneous and intraoral mechanoreceptors probably compensates for the unusual organization of the jaw-closing motoneurons, which undergo inhibitory control neither by reciprocal nor by recurrent inhibition.

Electrical or mechanical stimuli delivered to the oral region evoke a reflex inhibition of the jaw-closing muscles, the ▶ masseter inhibitory reflex (Fig. 1). On EMG recordings from contracted jaw-closers, this reflex inhibition appears as an early phase and a late phase of suppression, also called ES1 and ES2 exteroceptive suppressions (Godaux and Desmedt 1975), or SP1 and SP2 silent periods (Cruccu and Deuschl 2000). Probably because electrical stimuli yield a mixed – nociceptive and

non-nociceptive – input, whether the first or the second or both components are ▶ nociceptive reflexes remains controversial (Miles and Turker 1987). Innocuous mechanical stimuli will elicit both components, however, and indirect evidence supports the view that the afferents belong to the intermediately myelinated Aβ group. Afferent impulses reach the pons via the sensory mandibular or maxillary root of the trigeminal nerve. The first inhibitory period (10–13-ms latency) is probably mediated by one inhibitory interneuron, located close to the ipsilateral ▶ trigeminal motor nucleus. This interneuron projects onto jaw-closing motoneurons bilaterally. The whole circuit lies in the mid-pons. The afferents for the second inhibitory period (40–50-ms latency) descend in the spinal trigeminal complex and connect to a polysynaptic chain of excitatory interneurons, probably located in the ▶ medullary lateral reticular formation. The last interneuron of the chain is inhibitory and gives rise to ipsilateral and contralateral collaterals that ascend medially to the right and left spinal trigeminal complexes, to reach the trigeminal motoneurons (Cruccu et al. 2005; Ongerboer de Visser et al. 1990).

As shown by experiments with noxious, high-intensity laser stimuli directed to the perioral region, selective activation of Aδ afferents elicits a single, late (70-ms latency) silent period in jaw-closing muscles (Ellrich et al. 1997; Romaniello et al. 2002). This nociceptive reflex (► laser silent period, LSP) is supposedly mediated by the spinal trigeminal nucleus, pars caudalis.

Technical Requirements and Normal Values

The masseter inhibitory reflex is usually evoked by transcutaneous electrical stimulation of the mentalis territory (to test the mandibular division) or of the infraorbital territory (to test the maxillary division), by placing the cathode on the skin overlying the mental or the infraorbital foramen and the anode 2 cm laterally. A stimulus lasting 0.1 ms delivered at an intensity of about 3× the sensory threshold allows best visualization of the first and second inhibitory periods without causing excessive discomfort. The patient clenches the teeth at maximum strength, receives the stimulus, and is then allowed a few seconds' rest. At least 8 trials, but preferably 16, are repeated and superimposed. Quantitative studies of the excitability of the brainstem inhibitory interneurons require measurement of the size of responses (e.g., area of suppression); the level of background EMG activity must be kept constant and the signal is full-wave rectified and averaged (Fig. 1).

The onset latency should correspond to the beginning of the EMG suppression. This is usually taken at the intersection between the inhibitory shift and a line corresponding to 80 % of the mean background EMG activity. The size of the response can be evaluated by measuring the area of suppression or the duration, taking the end-latency at the point when EMG returns to 80 %. The onset latency is the most reliable measure for clinical applications. (Table 1 shows normal values.) The latency of the first inhibitory period (though not the second) has a relatively narrow range of variability, thus allowing comparisons between subjects. The intraindividual latency difference between sides is small (range 0–1.2 ms, mean 0.3 ms ± [SD] 0.37 ms in 100 normal subjects). A latency difference between sides larger than

Jaw-Muscle Silent Periods (Exteroceptive Suppression), Table 1 Masseter Inhibitory Reflex in 100 normal subjects aged 15–80 years

Latency (ms)	First inhibitory period (SP1 or ES1)	Second inhibitory period (SP2 or ES2)
Median	12	45
Mean	11.8	45.1
SD	0.8	5.2
Range	10–13.6	38–60
20-year-old subjects[a]	11.1	42
70-year-old subjects[a]	12.3	48

[a]Standard curve calculations for age-latency (from Cruccu and Deuschl 2000)

1.2 ms is abnormal. The second inhibitory period is considered abnormal if it is absent unilaterally or the latency difference between sides exceeds 8 ms. The second inhibitory period may be absent bilaterally in elderly patients or in patients with malocclusion (Cruccu et al. 1997).

Clinical Applications

Brainstem inhibitory reflexes cannot be tested by clinical procedures alone. In some patients, clinical examination discloses no signs of trigeminal impairment, yet testing the masseter inhibitory reflex reveals trigeminal or brainstem dysfunction. As in ► blink reflex studies, the pattern of abnormality (afferent, mixed, or efferent) provides information on the site of the lesion (Cruccu et al. 2005; Ongerboer de Visser et al. 1990). Nevertheless, except in rare conditions such as a purely motor trigeminal neuropathy and hemimasticatory spasm, the "efferent" type of abnormality (abnormal responses confined to the muscle on one side, regardless of the side of stimulation) is extremely uncommon.

The two inhibitory periods of the masseter inhibitory reflex have distinct EMG features and clinical applications. The first inhibitory period appears to be insensitive to peripheral conditioning and suprasegmental modulation, its latency varies little, and it is probably mediated by a small number of afferents. For these reasons, it is the best available response for assessing function of the maxillary and mandibular afferents in focal and in generalized diseases.

In patients with symptomatic ▶ trigeminal neuralgia or focal lesions within the pons, it has a diagnostic sensitivity similar to that of the R1 blink reflex (Cruccu and Deuschl 2000; Cruccu et al. 2005).

The second inhibitory period is far less sensitive than the first to lesions along the reflex arc. Being mediated by a multisynaptic chain of interneurons of the lateral reticular formation however, it is modulated by suprasegmental influences. Like the R2 blink reflex, the second inhibitory period shows a strongly enhanced recovery cycle in patients with extrapyramidal disorders and an increased habituation in hemiplegia (Cruccu and Deuschl 2000). The second inhibitory period (recorded from the temporalis muscle and called ES2) is a focus of research in several centers for patients with headache. Although the findings are still controversial, some data suggest that this response might help in differentiating tension-type headache from vasomotor headaches (Schoenen et al. 1987).

Unlike the electrically elicited masseter inhibitory reflex (Cruccu et al. 1997), the laser silent period, a purely nociceptive reflex, has been demonstrated to undergo modulation by experimental pain and found to be absent in patients with ▶ temporo-mandibular pain (Romaniello et al. 2002, 2003).

References

Cruccu, G., & Deuschl, G. (2000). The clinical use of brainstem reflexes and hand-muscle reflexes. *Clinical Neurophysiology, 111*, 371–387.

Cruccu, G., Frisardi, G., Pauletti, G., et al. (1997). Excitability of the central masticatory pathways in patients with painful temporo-mandibular disorders. *Pain, 73*, 447–454.

Cruccu, G., Iannetti, G. D., Marx, J. J., et al. (2005). Brainstem reflex circuits revisited. *Brain, 128*, 386–394.

Ellrich, J., Hopf, H. C., & Treede, R. D. (1997). Nociceptive masseter inhibitory reflexes evoked by laser radiant heat and electrical stimuli. *Brain Research, 764*, 214–220.

Godaux, E., & Desmedt, J. E. (1975). Exteroceptive suppression and motor control of the masseter and temporalis muscles in normal man. *Brain Research, 85*, 447–458.

Miles, T. S., & Turker, K. S. (1987). Reflex responses of motor units in human masseter muscle to electrical stimulation of the lip. *Experimental Brain Research, 65*, 331–336.

Ongerboer de Visser, B. W., Cruccu, G., Manfredi, M., et al. (1990). Effects of brainstem lesions on the masseter inhibitory reflex. Functional mechanisms of reflex pathways. *Brain, 113*, 781–792.

Romaniello, A., Arendt-Nielsen, L., Cruccu, G., et al. (2002). Modulation of trigeminal laser evoked potentials and laser silent periods by homotopical experimental pain. *Pain, 98*, 217–228.

Romaniello, A., Cruccu, G., Frisardi, G., et al. (2003). Assessment of nociceptive trigeminal pathways by laser-evoked potentials and laser silent periods in patients with painful temporomandibular disorders. *Pain, 103*, 31–39.

Schoenen, J., Jamart, B., Gerard, P., et al. (1987). Exteroceptive suppression of temporalis muscle activity in chronic headache. *Neurology, 37*, 1834–1836.

Job Analysis

Definition

Job analysis includes description of work tasks; methods, techniques, or processes involved and the work devices used, and its results; and worker (skills, knowledge, adaptations needed). Job analysis may expose environmental and organizational factors needed to accomplish the work tasks. The Revised Handbook of Analyzing Jobs (RHAJ) explores the procedures and techniques used to analyze jobs and to record the analyses. Such analyses underlie, and are congruent with, the occupational definitions of the Dictionary of Occupational Titles (DOT). A job analysis (according to RHAJ) addresses the worker's relationship to data, people, and things (i.e., Worker Functions); the methodologies and techniques employed (i.e., Work Fields); the machines, tools, equipment, and work aids used (MTEWA); the material, procedures, subject matter, or services (MPSMS); and what worker attributes contribute to successful job performance (Worker Characteristics).

Cross-References

▶ Vocational Counseling

Job Capacity Evaluation

▶ Disability, Functional Capacity Evaluations

Job Demands

Definition

Job demands are the mental and physical requirements necessary to fulfill obligations associated with specific jobs.

Cross-References

▶ Pain in the Workplace, Risk Factors for Chronicity, and Workplace Factors

Job Requirements

Definition

Job requirements are the demands an occupation or job task places on a worker for expected quality and quantity standard. The Dictionary of Occupational Titles (DOT) defines more than 20,000 occupations in the Labor Market in standards of Worker Functions; Work Fields, and Worker Characteristics.

Cross-References

▶ Vocational Counseling

Job Satisfaction

Definition

A measure of how happy or pleased someone is with different aspects of the job demands, work environment, or the occupation as a whole.

Cross-References

▶ Pain in the Workplace, Risk Factors for Chronicity, Job Demands

Job-Site Evaluation

▶ Situational Assessment

Joint Deformities

Definition

Joint deformities are changes from the typical size or shape of a particular joint.

Cross-References

▶ Chronic Pain in Children, Physical Medicine and Rehabilitation

Joint Nociceptors

▶ Articular Nociceptors

JRA

▶ Juvenile Rheumatoid Arthritis

Judgment

▶ Empathy and Pain

Junctional DREZ Coagulation

▶ DREZ Procedures

Just-Noticeable Difference

Definition

On a stimulus continuum a, is the smallest increment? (a) such that a+?(a) just noticeably exceeds a. This minimum increment of stimulus intensity is usually called the "just-noticeable difference" (JND or "difference threshold" or "difference limen").

Cross-References

► Pain Evaluation, Psychophysical Methods

Juvenile Rheumatoid Arthritis

Synonyms

JRA

Definition

A condition characterized by joint inflammation and stiffness for more than 6 weeks in children 16 years or younger. The inflammation causes redness, swelling, warmth, and soreness in the joints. Any joint can be affected and inflammation may limit the mobility of affected joints. JRA is a disease of the immune system. In JRA, the immune system attacks the body's own healthy cells, which causes inflammation in the lining and connective tissues of the joints.

Cross-References

► Experimental Pain in Children

Juxtaglomerular Apparatus

Definition

The functional entity in the kidney that consists of juxtaglomerular cells (epithelioid cells in the media of the afferent arterioles) and the macula densa (tubular epithelium at the region of afferent arteriole and efferent arteriole to the glomerulus) and is involved in the regulation of salt and water excretion and renal blood flow.

Cross-References

► NSAIDs, Adverse Effects

K

K/C arthritis

▶ Arthritis Model, Kaolin-Carrageenan-Induced Arthritis (Knee)

K⁺ Channel

Definition

K⁺ channel is a voltage-dependent permeation pathway for potassium ions.

Cross-References

▶ Ion Channel
▶ Ionotropic Receptor
▶ Trafficking and Localization of Ion Channels

Kainate Receptor

Definition

A type of ionotropic glutamate receptor that is activated by the agonist kainate. However, it should be noted that kainate will also activate other glutamate receptors and thus should not be regarded as a specific agonist. Kainate receptors comprise of several subunits (GluR5, GluR6, GluR7, KA1, KA2) that form a heteromeric receptor-ion-channel complex.

Cross-References

▶ Thalamus, Nociceptive Neurotransmission

Kainate-Induced Lesion

Definition

Neuronal death due to injection of the excitotoxin kainate.

Cross-References

▶ Lateral Thalamic Lesions, Pain Behavior in Animals

Kaolin

Definition

Hydrated aluminum silicate.

Cross-References

▶ Amygdala, Pain Processing and Behavior in Animals
▶ Arthritis Model, Kaolin-Carrageenan-Induced Arthritis (Knee)

G.F. Gebhart, R.F. Schmidt (eds.), *Encyclopedia of Pain*, DOI 10.1007/978-3-642-28753-4,
© Springer-Verlag Berlin Heidelberg 2013

Kaolin-Carrageenan-Induced Arthritis

Definition

Animal model of inflammatory pain arising from the knee joint, mimicking osteoarthritis in humans.

Cross-References

▶ Arthritis Model, Kaolin-Carrageenan-Induced Arthritis (Knee)
▶ Nociceptive Processing in the Amygdala: Neurophysiology and Neuropharmacology

Kapanol

Definition

Kapanol is a formulation of morphine prepared as polymer-coated sustained-release pellets contained in a capsule.

Cross-References

▶ Postoperative Pain, Morphine

Kapanol® Controlled-Release Pellets Within a Capsule

▶ Postoperative Pain, Morphine

Kappa Receptors and Visceral Pain

▶ Opioids, Kappa Receptors, and Visceral Pain

Kappa-Opioid Receptors

▶ Opioids, Kappa Receptors, and Visceral Pain

Keratinocytes

Definition

Keratinocytes are cells in skin that secrete keratin as well as neurotrophins. Basal keratinocytes release NGF, whereas suprabasal keratinocytes release NT-3.

Cross-References

▶ Nerve Growth Factor, Sensitizing Action on Nociceptors

Ketalar

▶ Postoperative Pain, Ketamine

Ketamine

▶ Postoperative Pain, Ketamine

Kinesiophobia

▶ Disability, Fear of Movement
▶ Fear and Pain
▶ Muscle Pain, Fear-Avoidance Model

Kinesophobia

Definition

Kinesophobia is an excessive, irrational, and debilitating fear of physical movement and activity resulting from a feeling of vulnerability to painful injury or (re)injury.

Cross-References

▶ Disability, Fear of Movement
▶ Fear and Pain
▶ Hypervigilance and Attention to Pain
▶ Muscle Pain, Fear-Avoidance Model

Kinesthesia

Definition

Conscious information about body position and movement.

Cross-References

▶ Postsynaptic Dorsal Column Neurons, Responses to Visceral Input

Knockout Mice

Definition

Mice in which a portion of a specified gene is disrupted or removed, eliminating the corresponding protein.

Cross-References

▶ Opioid Receptors

KOR

▶ Opioids, Kappa Receptors, and Visceral Pain

KOR-1

▶ Opioid Receptors

Kyphotic

Definition

Characteristic of or suffering from kyphosis, an abnormality of the vertebral column.

Cross-References

▶ Lower Back Pain, Physical Examination

Cross-References

- Disability, Fear of Movement
- Fear and Pain
- Hypervigilance and Attention to Pain
- Muscle Pain, Fear-Avoidance Model

Kinesthesis

Definition

Conscious information about body position and movement.

Cross-References

- Postsynaptic Dorsal Column Neurons, Responses to Visceral Input

Knockout Mice

Definition

Mice in which a portion of a specified gene is experimentally removed, eliminating the corresponding protein.

Cross-References

- Opioid Receptors

KOR

- Opioids, Kappa Receptors and Visceral Pain

KOR-1

- Opioid κ receptors

Kyphotic

Definition

Characteristic of or resulting from kyphosis, an abnormality of the vertebral column.

Cross-References

- Lower Back Pain, Physical Treatments

L

La Belle Indifference

Definition

Individuals with a health condition or pain problem who seem to appear unconcerned about the nature or implications of their condition. For children with certain complex pain conditions, either children or their parents may exhibit "la belle indifference." In some cases, children may be unable to walk during the physical examination, but parents are totally unaffected.

Cross-References

▶ Chronic Daily Headache in Children

Labor Pain

▶ Gynecological Pain, Neural Mechanisms

Labor Pain Model

▶ Visceral Pain Models, Female Reproductive Organ Pain

Laboratory Findings

Definition

Anatomical, physiological, or psychological phenomena that can be shown by the use of medically acceptable laboratory diagnostic techniques. Some of these diagnostic techniques include chemical tests, electrophysiological studies (electrocardiogram, electroencephalogram, etc.), medically acceptable imaging tests (X-rays, CAT scans, etc.), and psychological tests.

Cross-References

▶ Disability Evaluation in the Social Security Administration

Laboratory Pain

▶ Experimental Pain in Children

Lacrimation

Definition

Lacrimation is the tearing of an eye. During acute bouts of cluster headache and during exacerbations of hemicrania continua, the eye on the side of the pain often tears.

G.F. Gebhart, R.F. Schmidt (eds.), *Encyclopedia of Pain*, DOI 10.1007/978-3-642-28753-4,
© Springer-Verlag Berlin Heidelberg 2013

Cross-References

▶ Hemicrania Continua

Lamina II Outer and Lamina II Inner

Definition

The spinal cord in cross section has been divided into areas based on morphological characteristics (Rexed 1952). Laminae I and II comprise the most superficial aspect of the dorsal horn of the gray matter, and are known to receive the central terminal projections of many unmyelinated (C-fiber) and thinly-myelinated (Aδ-fiber) nociceptors. Lamina II is also known as the "substantia gelatinosa." The inner aspect of Lamina II receives the projections of many IB4-positive unmyelinated neurons. The outer aspects of Lamina II and Lamina I receive the terminals of many peptidergic unmyelinated neurons. Most neurons in Lamina II are interneurons.

Cross-References

▶ IB4-Positive Neurons, Role in Inflammatory Pain
▶ Immunocytochemistry of Nociceptors
▶ Morphology, Intraspinal Organization of Visceral Afferents
▶ Nociceptor, Categorization

Lamina Propria

Definition

A thin vascular layer of connective tissue beneath the epithelium of an organ.

Cross-References

▶ Animal Models of Inflammatory Bowel Disease

Laminae I and V Neurons

Definition

Laminae I and V neurons are neurons located in the superficial dorsal horn and deeper layers, respectively. Also known as the marginal layer of Waldeyer, lamina I is the most superficial lamina of the dorsal horn in Rexed's classification. It is a very thin layer of small neurons that often send long distance ascending projections to the brain. Numerous lamina I neurons are nociceptive (often nociceptive specific), with a smaller number being thermoreceptive or sensitive to itch inducing stimuli. Lamina V also contains many neurons which send long distance ascending projections to the brain. Both laminae I and V neurons often show convergent inputs from different tissues.

Cross-References

▶ Nociceptor, Categorization
▶ Opioids in the Spinal Cord and Central Sensitization
▶ Parabrachial Hypothalamic and Amygdaloid Projections
▶ Spinomesencephalic Tract
▶ Spinothalamocortical Projections from SM
▶ Thalamic Nuclei Involved in Pain: Cat and Rat

Laminae I/II Inputs to the Thalamus

▶ Spinothalamic Terminations, Core and Matrix

Laminae III–IV, X

Definition

The basal parts of the dorsal horn correspond to the nucleus proprius of the spinal cord. Lamina X is located around the central gray of the spinal cord. In particular, Lamina III receives most spinal terminations of non-nociceptive sensory afferents.

Cross-References

▶ Morphology, Intraspinal Organization of Visceral Afferents

Laminectomy

Definition

Laminectomy is the excision of the posterior arch of a vertebra.

Cross-References

▶ Chronic Back Pain and Spinal Instability

Lancinating Pain

▶ Pain Paroxysms

Landry-Guillain-Barré-Strohl Syndrome

▶ Guillain-Barré Syndrome

Langerhans Cells

Definition

Langerhans cells are dendritic cells in the epidermis. They have phagocytic properties and are responsible for antigen presentation in a variety of CD4-dependent immune responses. They are involved in the early stages of contact dermatitis, skin graft rejection, or HIV-1 infection.

Laparoscopic Pain Mapping

▶ Chronic Pelvic Pain, Laparoscopic Pain Mapping

Laparoscopy

Definition

Laparoscopy is a diagnostic tool designed to visualize the peritoneal cavity and the structures within by means of a laparoscope, which is a miniature telescope.

Cross-References

▶ Dyspareunia and Vaginismus

Laparoscopy (for Pain) Under Local Anesthesia

▶ Chronic Pelvic Pain, Laparoscopic Pain Mapping

Large Fibers

Definition

Large fibers is a collective term for large myelinated nerves, including motor nerves and the proprioceptive type of sensory nerves, and also large-diameter nerves.

Cross-References

▶ Toxic Neuropathies

Large-Fiber Neuropathy

Definition

Peripheral nerve disorders mainly affecting large myelinated nerves (large fibers).

Cross-References

▶ Toxic Neuropathies
▶ Ulceration, Prevention by Nerve Decompression

Laser

Bengt H. Sjölund
Department of Community Medicine and
Rehabilitation Medicine, Umea University,
Umea, Sweden

Synonyms

LLLT; Low-level laser therapy

Definition

Low-level laser therapy (LLLT) utilizes a pure light source of varying wavelength, intensity, and temporal characteristics to attempt to treat various types of pain. It does not generate heat or sensations, but may cause superficial photochemical reactions.

Characteristics

The use of laser (light) therapy originates in biophysical observations that fibroblast growth may be stimulated by ordinary light or laser light exposure. There is no proven link to pain-modulating mechanisms.

LLLT for pain was originally advocated by single authorities and by manufacturers of laser equipment, having little but case reports to substantiate their claims. Randomized controlled studies on patients with chronic low back pain have not found that low-energy laser stimulation plus exercise provides a significant advantage over exercise alone (Klein and Eek 1990). Similarly, Gur et al. (2003), in 75 patients with chronic low back pain exposed to a gallium arsenide laser + exercise, laser alone, or exercise alone, found significant improvements in all groups, but were not able to single out laser therapy as superior.

One possible cause of a treatment effect by low-power laser exposure in muscle pain conditions could be an increase in local microcirculation. In one study, the immediate effects on masseter muscle blood flow of low-power laser exposure in patients with chronic orofacial pain of muscular origin were examined in comparison to healthy individuals (Tullberg et al. 2003). Intramuscular laser Doppler flowmetry was performed unilaterally in the tenderest point (patients) or in a standardized point (healthy subjects) of the masseter muscle on 12 patients with myofascial pain of orofacial muscles and 12 age and gender-matched healthy individuals. The muscle was first exposed to a gallium aluminum arsenide laser or placebo laser for 2 min in a randomized and double-blind manner. After another 8 min, the muscle was treated with the other laser for 2 min, and the LDF recording continued for 8 min. Finally, the patients again assessed the pain intensity. The pain intensity was not affected by laser exposure, and the blood flow did not change significantly in the patients, but increased after active laser exposure and decreased after placebo exposure in the healthy individuals. Thus, the study did not support an effect of low-power laser exposure on masseter muscle microcirculation in patients with chronic orofacial pain.

There is conflicting evidence on the benefit of LLLT in the treatment of painful musculoskeletal conditions. Based on small trials, with different populations, LLLT doses, and comparison groups, there are insufficient data to either support or refute the effectiveness of LLLT for the treatment of LBP, as shown in a recent Cochrane review (Yousefi-Nooraie et al. 2011).

In another Cochrane review, Brosseau et al. (2009) found LLLT to provide short-term pain relief for patients with rheumatoid arthritis. A total of 222 patients were included in the five placebo-controlled trials, with 130 randomized to laser therapy. Relative to a separate control group, LLLT reduced pain by 1.10 points (95 % CI: 1.82, 0.39) on visual analogue scale relative to placebo, reduced morning stiffness duration by 27.5 min (95 % CI: 2.9–52 min), and increased tip to palm flexibility by 1.3 cm duration, dosage, and site of application.

Not uncommonly, a low-energy laser is applied in the form of laser acupuncture. In a recent controlled study on chronic tension-type headache (Ebneshahidi et al. 2005), the effects of laser acupuncture were examined in 50 patients, randomly allocated to low-energy laser acupuncture or placebo laser acupuncture (zero level) in selected acupuncture points for less than a minute in ten sessions, given three per week. Remarkably, there were significant differences between groups ($P < 0.001$) in changes from baseline in months 1, 2, and 3; in median score for headache intensity; in median duration of attacks; and in median number of days with headache per month.

Similarly, in a prospective, double-blind, randomized, controlled trial in patients with chronic myofascial pain syndrome in the neck to evaluate the effects of infrared low-level 904 nm gallium arsenide (Ga-As) laser therapy versus placebo laser on pain, disability, mood, and quality of life, statistically significant improvements were detected in all outcome measures compared with baseline ($P < 0.01$) in the active laser group, while in the placebo laser group, significant improvements were detected only in pain score at rest 1 week later. The score for self-assessed improvement of pain was significantly different between the active and placebo laser groups (63 % vs. 19 %) (Gur et al. 2004).

Painful symptoms of diabetic sensorimotor polyneuropathy (DSP) on the other hand were not relieved by LLLT in a randomized sham therapy-controlled clinical trial in 50 patients with painful DSP diagnosed with the Toronto clinical neuropathy score (Zinman et al. 2004). All patients received sham therapy over a 2-week baseline period and were then randomized to receive biweekly sessions of either sham or LLLT for 4 weeks. Both groups noted a decrease in weekly mean pain scores during sham treatment. After the 4-week intervention, the LLLT group had an additional reduction in weekly mean pain scores of -1.0 ± 0.4 compared with -0.0 ± 0.4 for the sham group, a difference that did not reach significance.

In summary, the present data do not allow the conclusion that LLLT can be recommended for the treatment of chronic pain other than possibly that from RA. It is interesting to speculate that a reason for this short-lasting effect may be a transient influence on superficial inflammatory processes.

References

Brosseau, L., Welch, V., Wells, G., et al. (2004). Low level laser therapy (classes I, II and III) for treating osteoarthritis. *The Cochrane Database of Systematic Reviews*, (3). Art. No. CD002046. doi: 10.1002/14651858.CD002046.pub2.

Brosseau, L., Welch, V., Wells, G., et al. (2009). Low level laser therapy (classes I, II and III) for treating rheumatoid arthritis. *The Cochrane Database of Systematic Reviews*, (4). Art. No. CD002049. doi: 10.1002/14651858.CD002049.

Ebneshahidi, N. S., Heshmatipour, M., Moghaddami, A., et al. (2005). The effects of laser acupuncture on chronic tension headache – A randomised controlled trial. *Acupuncture in Medicine, 23*, 13–18.

Gur, A., Karakoc, M., Cevik, R., et al. (2003). Efficacy of low power laser therapy and exercise on pain and functions in chronic low back pain. *Lasers in Surgery and Medicine, 32*, 233–238.

Gur, A., Sarac, A. J., Cevik, R., et al. (2004). Efficacy of 904 nm gallium arsenide low level laser therapy in the management of chronic myofascial pain in the neck: A double-blind and randomize-controlled trial. *Lasers in Surgery and Medicine, 35*, 229–235.

Klein, R. G., & Eek, B. C. (1990). Low-energy laser treatment and exercise for chronic low back pain: Double-blind controlled trial. *Archives of Physical Medicine and Rehabilitation, 71*, 34–37.

Tullberg, M., Alstergren, P. J., & Ernberg, M. M. (2003). Effects of low-power laser exposure on masseter muscle pain and microcirculation. *Pain, 105*, 89–96.

Yousefi-Nooraie, R., Schonstein, E., Heidari, K., Rashidian, A., Pennick, V., Akbari-Kamrani, M.,

Irani, S., Shakiba, B., Mortaz Hejri, S., Jonaidi, A. R., & Mortaz-Hedjri, S. (2008). Low level laser therapy for nonspecific low-back pain. *Cochrane Database of Systematic Reviews*, (2). Art. No.: CD005107. doi:10.1002/14651858.CD005107.pub4.

Zinman, L. H., Ngo, M., Ng, E. T., et al. (2004). Low-intensity laser therapy for painful symptoms of diabetic sensorimotor polyneuropathy: A controlled trial. *Diabetes Care, 27*, 921–924.

Laser Heat Stimulator

Definition

A laser (acronym for Light Amplification by Stimulated Emission of Radiation) is a very unique light source. In comparison to classical incandescent light heat stimulators which emit their radiative energy in all spatial directions and in a large spectrum of wavelengths, the laser energy is confined to a narrow beam of nearly parallel monochromatic electromagnetic waves. This results in a high power density (radiation per unit area). The combination of these characteristics makes the laser a light source with a spectral energy density (radiation per unit wavelength) several orders of magnitude higher than any known light source. This is an essential characteristic if fast and high transfer of radiation energy is needed.

Cross-References

▶ Pain in Humans, Thermal Stimulation (Skin, Muscle, Viscera), Laser, Peltier, Cold (Cold Pressure), Radiant, Contact

Laser Silent Period

Synonyms

LSP

Definition

Reflex inhibition of the contracted masseter muscle elicited by noxious, high-intensity laser stimuli in the trigeminal territory.

Cross-References

▶ Jaw-Muscle Silent Periods (Exteroceptive Suppression)

Laser Speckle Imaging

Definition

A blood flow technique that allows simultaneous measurement of vessel diameter both on the surface of the brain and within meningeal vessels.

Cross-References

▶ Clinical Migraine with Aura

Laser Stimulation

▶ Pain in Humans, Thermal Stimulation (Skin, Muscle, Viscera), Laser, Peltier, Cold (Cold Pressure), Radiant, Contact

Laser-Doppler Flowmetry

Definition

The laser-Doppler technique measures blood flow in the very small blood vessels of the microvasculature. Depending on the Doppler principle, low-power light from a laser is scattered by moving red blood cells and, as a consequence, is frequency broadened. The frequency-broadened light, together with laser light scattered from static tissue, is photodetected and the resulting photocurrent processed to provide a blood flow measurement.

Laser-Evoked Field

Synonyms

LEF

Definition

An event-related magnetic field, elicited by and time-locked to a laser stimulus.

Cross-References

▶ Insular Cortex, Neurophysiology, and Functional Imaging of Nociceptive Processing

Laser-Evoked Potential

Synonyms

LEP

Definition

An event-related potential, elicited by and time-locked to a laser stimulus.

Latent Myofascial Trigger Point

Definition

A latent myofascial trigger point is a sensitive spot with pain or discomfort in response to compression only but not causing a clinical pain complaint. Every person develops latent MTrPs as early as six months after birth. A latent MTrP is probably caused by minor peripheral nerve injury due to repetitive minor trauma early in life. A latent MTrP may become active (painful) after an acute injury or chronic repetitive minor trauma to the soft tissue or other structures, or even after an emotional stress.

Cross-References

▶ Dry Needling
▶ Myofascial Trigger Points
▶ Trigger Point

Lateral and Dorsal Funiculi

Definition

White mater adjacent to the dorsal horn dorsally and laterally, excluding Lissauer's tract.

Cross-References

▶ Morphology, Intraspinal Organization of Visceral Afferents

Lateral and Medial Thalamic Nociceptive System

▶ Thalamic Nuclei Involved in Pain: Cat and Rat

Lateral Cervical Nucleus

Synonyms

LCN

Definition

The lateral cervical nucleus is a collection of neurons adjacent to the spinal cord dorsal horn in the uppermost cervical segments. It receives input by way of the spinocervical tract, which originates from neurons in laminae III and IV on the ipsilateral side. The neurons of the lateral cervical nucleus project their axons into the contralateral medial lemniscus, and the axons synapse in the ventral posterior lateral thalamic nucleus. Some of the neurons in the spino-cervicothalamic pathway are nociceptive.

Cross-References

▶ Spinothalamic Input: Cells of Origin (Monkey)
▶ Spinothalamic Projections in Rat

Lateral Division (of the Spinal Dorsal Root)

Definition

The spinal dorsal roots of mammals separate into two major divisions as they join the spinal cord; the lateral division contains almost exclusively unmyelinated and thinly myelinated fibers.

Cross-References

▶ Nociceptor, Categorization

Lateral Geniculate Nucleus

Definition

The main visual information-transmitting nucleus in the thalamus.

Cross-References

▶ Information Processing in Thalamocortical Circuits

Lateral Pain System

Definition

The lateral pain system is a neural circuit consisting of laterally projecting spinothalamic and trigeminothalamic pathways that terminate in lateral thalamic nuclei (i.e., VPL, VPM), which in turn project to primary and secondary somatosensory cortices that process the sensory-discriminative dimension of the pain.

Cross-References

▶ Spinothalamocortical Projections to Ventromedial and Parafascicular Nuclei
▶ Thalamo-Amygdala Interactions and Pain

Lateral Spinal Nucleus

Synonyms

LSN

Definition

A spinal nucleus located caudally to the lateral cervical nucleus throughout the spinal cord.

Cross-References

▶ Spinothalamic Projections in Rat

Lateral Sulcus

▶ Sylvian Fissure

Lateral Thalamic Lesions, Pain Behavior in Animals

Perry N. Fuchs and Yuan Bo Peng
Department of Psychology, University of Texas at Arlington, Arlington, TX, USA

Synonyms

Thalamatomy, pain behaviors in animals; Thalamus lesion

Definition

Thalamotomy is a term that refers to damage which results in cellular loss to specific or numerous thalamic nuclei. Stereotaxic thalamotomy, using electrolytic and excitotoxic approaches, has been performed in animals as a method to ameliorate nociceptive conditions, or to simulate the human condition of central pain secondary to intracranial lesions.

Characteristics

Although chronic pain following thalamic lesion was first described 100 years ago (Dejerine and Egger 1903), it still remains a substantial clinical problem. The qualitative and quantitative experience of pain following thalamic lesion is variable, but is most often described as elevated sensitivity to touch, temperature, or pain in the affected body region that is also associated with burning or lacerating sensation (Fuchs et al. 2001). Central pain secondary to intracranial lesions is quite often associated with thalamic damage, but disturbances of the parietal lobe, basal ganglia, and spinothalamic tract distant from the cerebral hemispheres also cause central pain. In general, lesions anywhere along the spinothalamic/thalamocortical system can result in central pain. In particular, lesions involving the VPL in humans, due to trauma or infarct, or stereotactic electrolysis (Pagni 1976) often result in central pain syndrome.

There are a number of reasons why attention should be directed towards exploring the relationship between thalamotomy and pain behavior in animals. First, although significant advances have been made in identifying the underlying neural substrates that cause central pain when damaged, the physiological mechanisms associated with the peculiar symptoms of this syndrome remain poorly understood. Second, the large number of treatment options for central pain syndrome reflects the fact that there is no single pharmacological or surgical procedure that relieves pain in most cases (Fuchs et al. 2001). Third, the development of central pain syndrome following thalamotomy questions the traditional concept of the thalamus as a relay station for the transmission of sensory input.

Surprisingly, only a few studies have explored the relationship between the ▶ ventral-posterior lateral nucleus (VPL) and pain behavior in rats. In one study, responses to mechanical, thermal and acute inflammatory stimuli were tested following rather extensive ▶ electrolytic lesions of the thalamus (including medial, lateral and subtotal) (Saadé et al. 1999). In general, electrolytic thalamotomy produces a long-lasting enhancement of reactivity to normally noxious mechanical and thermal stimuli (▶ hyperalgesia). A more recent report measured the responses to mechanical and thermal stimuli before and up to 48 h after a selective ▶ excitotoxic lesion of the VPL induced by kainite microinjection (LaBuda et al. 2000) (Fig. 1). Similar to the outcome of the study performed by Saadé et al. (1999), ▶ kainate-induced lesion of the VPL results in a leftward shift in the force/response function, as revealed by a significant increase in response to normally innocuous mechanical and thermal stimuli. The mean frequency of responses at 24 h after the kainite injection more than doubled the baseline responses, and at 48 h, the responses were greater than three times the baseline response frequency (Fig. 2). The animal model described here mimics many of the peculiar symptoms of ▶ central pain

a Location of cannula tips

Bregma: -1.3 mm

Bregma: -4.3 mm

b Vehicle injection

c Kainate injection

Lateral Thalamic Lesions, Pain Behavior in Animals, Fig. 1 (a) Schematic representation of cannula tip locations for the Surgery (■), Vehicle injection (□), Kainate Miss (○) and Kainate Hit (·) groups based on plates from Paxinos and Watson (1986). (b) Photomicrograph of a subject that received vehicle injection within the region of the VPL. The arrow indicates the boundary of the cannula tip (dorsal aspect is right). (c) Photomicrograph of a subject that received kainate injection within the region of the VPL. The *arrow* indicates the boundary of the cannula tip and the *star* indicates the area of pronounced glial proliferation (dorsal aspect is right). All subjects included in the Kainate Hit group had cannula tip locations on target with the VPL and pronounced glial proliferation. Scale bar = 250 μM

syndrome, including delayed onset of pain and hyperalgesia to mechanical and thermal stimuli.

In addition to the possibility that the excitotoxic VPL lesion model can be used to study the underlying basic mechanisms associated with central pain syndrome, a number of additional general issues have arisen from the thalamotomy findings. First, the lateral thalamus is traditionally considered as a relay station, transmitting sensory input to the primary somatosensory cortex. Consequently, the initial expectation would be that a lesion of the VPL would result in contralateral loss of processing of somatosensory input, possibly reflected as a decrease in nociceptive threshold (▶ analgesia). However, the observation of pain behavior following thalamotomy in rats is an outcome that is opposite to the expectations provided by the traditional concept of the thalamus. This outcome highlights the complex nature of the somatosensory pathways responsible for transmitting nociceptive signals. Second, the experimental results following thalamotomy in rats have not addressed specific hypotheses that have been proposed for ▶ central pain syndrome. It remains to be determined if thalamotomy pain behavior in animals is based on: (1) irritable focus created at the site of injury, (2) sympathetic dysfunction,

Lateral Thalamic Lesions, Pain Behavior in Animals, Fig. 2 The symbol and line plots summarizes the effects of surgery alone (**a**), vehicle injection (**b**), injection of kainate confined to nuclei outside of the VPL (**c**) and injection of kainate localized to the VPL (**d**) on the mean percent of paw withdrawal response (± SEM) to punctate stimulation of the hindpaw. The responses to four intensities of punctate stimuli are shown at several time points including pre-surgery, pre-injection, 2, 8, 24 and 48 h post-injection. * = $p < 0.05$, ** = $p < 0.01$, *** = $p < 0.001$ compared to pre-surgery baseline

(3) hypothalamic dysfunction, (4) deafferentated central sensory nuclei, (5) hyperactivity of deafferentated nonspecific reticulothalamic pathways, and (6) deafferentation of cortical nociceptive pathways.

Thalamic stimulation has been used to relieve clinical pain syndromes (Duncan et al. 1998; Marchand et al. 2003; Mazars 1975). Paradoxically, thalamic stimulation has also been reported to induce pain (Lenz et al. 1995), an effect that is most likely due to the location of the stimulating electrode. Moderate increases in the function of the VPL, such as that produced by electrical stimulation or by microinjection of physiological concentrations of excitatory neurotransmitters or reuptake inhibitors, may be a suitable means of treatment for ▶ central pain syndrome.

References

Dejerine, J., & Egger, M. (1903). Le syndrome douloureaux thalamique. *Revista de Neurologia, 14,* 521–532.

Duncan, G., Kupers, R., Marchand, S., et al. (1998). Stimulation of human thalamus for pain relief. *Journal of Neurophysiology, 80*, 3326–3330.

Fuchs, P. N., Lee, J. I., & Lenz, F. A. (2001). Central pain secondary to intracranial lesions. In K. J. Burchiel (Ed.), *Pain surgery*. New York: Thieme.

LaBuda, C. J., Cutler, T. D., Dougherty, P. M., et al. (2000). Mechanical and thermal hypersensitivity develops following kainite lesion of the ventral posterior lateral thalamus in rats. *Neuroscience Letters, 290*, 79–83.

Lenz, F. A., Gracely, R. H., Romanoski, A. J., et al. (1995). Stimulation in the human somatosensory thalamus can reproduce both the affective and sensory dimensions of previously experienced pain. *Nature Medicine, 1*, 910–913.

Marchand, S., Kupers, R. C., Bushnell, M. C., et al. (2003). Analgesic and placebo effects of thalamic stimulation. *Pain, 105*, 481–488.

Mazars, G. J. (1975). Intermittent stimulation of nucleus ventralis posterolateralis for intractable pain. *Surgical Neurology, 4*, 93–95.

Pagni, C. A. (1976). Central pain and painful anesthesia. *Progress in Neurological Surgery, 8*, 132–257.

Paxinos, G., & Watson, C. (1986). *The rat brain in stereotaxic coordinates*. New York: Academic.

Saadé, N. E., Kafrouni, A. I., Saab, C. Y., et al. (1999). Chronic thalamotomy increases pain-related behavior in rats. *Pain, 83*, 401–409.

Lateral Thalamic Nuclei

Definition

The lateral thalamic nuclei include VPL, VPM, and other secondary nuclei. Activation of these neurons by the spinothalamic tract is important for sensory discrimination.

Cross-References

▶ Lateral Thalamic Lesions, Pain Behavior in Animals
▶ Parafascicular Nucleus, Pain Modulation
▶ Spinothalamic Tract Neurons, Visceral Input

Lateral Thalamic Nuclei and STT

▶ Spinothalamic Terminations, Core and Matrix

Lateral Thalamic Pain-Related Cells in Humans

Frederick A. Lenz[1], Andres Fernandez[1], S. Ohara[2], Nirit Weiss[3] and Stanley W. Anderson[2]
[1]Department of Neurosurgery, Johns Hopkins Hospital, Baltimore, MD, USA
[2]Department of Neurosurgery, Johns Hopkins Hospital, Johns Hopkins Medical Institutions, Johns Hopkins University, Baltimore, MD, USA
[3]Department of Neurosurgery, Mount Sinai Medical Center, New York, NY, USA

Synonyms

Mechanoreception; Pain system; Posterior nucleus; Thermoreception; Ventral posterior nucleus of thalamus

Definition

Neurons located in the human ▶ thalamus that respond selectively or differentially to painful stimuli.

Characteristics

Studies of patients at autopsy after lesions of the STT show that the human STT ascends to the thalamus medial to the medial geniculate (Mehler 1966) before terminating in the magnocellular medial geniculate, limitans and Vc portae nuclei, posterior to Vc (Mehler 1966). More anteriorly, the STT makes its most dense termination as irregular clusters in Vc (Mehler 1966). The STT terminations are concentrated in posterior inferior Vc and in dorsal Vc parvocellularis (Mehler 1966). Similarly, in monkeys, STT terminals occur as dense clusters in VPL (Boivie 1979; Apkarian and Hodge 1989). A more uniform, less dense termination is found in ventral posterior inferior – VPI (Apkarian and Hodge 1989), in the posterior nuclear group including posterior nucleus (Boivie 1979), pulvinar oralis, limitans, magnocellular medial geniculate, suprageniculate nuclei (Apkarian and Hodge 1989) and in the posterior division of the ventral medial nucleus (Blomqvist et al. 2000). Anatomic studies in patients following cordotomy demonstrate that nuclei where the STT terminates in humans are similar to those in monkeys (Mehler 1966). Finally, an area posterior, inferior and medial to monkey ventral posterior (▶ VP), corresponding to human Vc, is a proposed pain-related nucleus (ventral medial pars posterior – VMpo) (Blomqvist et al. 2000). Thus there is ample evidence of inputs from the STT to the region of Vc that could explain the occurrence of cellular responses to noxious and thermal stimuli.

Our physiological studies have demonstrated that cells in and posterior inferior to the human principal somatic sensory nucleus (ventral caudal – Vc) respond to painful mechanical stimuli (Lenz et al. 1994), painful heat stimuli (Lenz et al. 1993) and innocuous cool stimuli (Lenz and Dougherty 1998). The degree of convergence of thermal and mechanical modalities graded into the painful range has not previously been studied. Cells responding to both types of stimuli may explain both the sensation of pain, i.e., hyperalgesia and allodynia, evoked by normally nonpainful stimuli (Fruhstorfer and Lindblom 1984) and the alleviation of pain by thermal stimuli. The responses of human thalamic cells to painful and nonpainful thermal and mechanical stimuli in patients undergoing thalamic procedures for the treatment of movement disorders were examined.

The largest study of human cells responding to both thermal and mechanical stimuli graded into the painful range explored these neuronal responses in the region of Vc of 24 patients undergoing surgery for treatment of movement disorders (Lee et al. 1999). Preoperative somatic sensory testing was carried out with a series of thermal and somatic stimuli into the painful range on all patients. Intraoperative testing was carried out on 57 cells in the region where cells responded to innocuous cutaneous somatic sensory stimuli. Thermal stimuli consisted of contact cold or heat from 6 °C to 51 °C. The somatic series included stimulation with a camel hair brush and large, medium and small arterial clips. Preoperative somatic sensory testing established that both the mechanical and thermal series of stimuli spanned intensities extending into the painful range.

Of 57 cells tested, 15 had a graded response to mechanical stimuli extending into the painful range and thus were classified in the ▶ wide dynamic range (WDR) category. The mean stimulus–response function of cells in the WDR class, normalized to baseline, showed a fourfold mean increase in firing rate above baseline across the mechanical series of stimuli. Seven of these cells also responded to heat stimuli extending into the painful range (WDR-H) and 2 responded to cold stimuli (WDR-C). Twenty-five cells were in a class (multiple receptive – MR) that showed a response to both brush and compressive stimuli, although the responses were not graded into the painful range. Three of these cells (MR-H) had a response to heat stimuli and 5 cells responded to cold stimuli (MR-C). Nine cells responded to brushing without a response to the compressive stimuli (low threshold – LT). Although we have no direct anatomic evidence to confirm electrode location, the present results are consistent with monkey studies and suggest that cells differentially responsive to mechanical stimuli are located in Vc, Vcpc, Vcpor and anterior Po.

Brief stimuli spanning the range of the VAS (0–10) were used with preoperative training and intraoperative testing in the present study. This is unlike studies of awake monkeys where intense noxious stimuli were not used (Bushnell et al. 1993; Bushnell and Duncan 1987). The intense stimuli used in the present study may explain the large proportion of WDR cells in the region of Vc in this study. Another human study did not demonstrate WDR cells in the region of Vc (Tasker et al. 1997). The lack of such cells may be due to the stimuli used, although details of the methods have not been published for that study.

Microstimulation studies suggest that there is partitioning of thermal/pain sensations at different locations in the region of Vc. Stimulation sites where thermal/pain sensations are evoked are located near the posterior border of the core and within the posterior inferior region (Lenz et al. 1993). Microstimulation in the postero-inferior region evokes thermal sensations or pain often referred to large RFs and subcutaneous structures. Other reports identify sites where pain but not thermal sensations are evoked posterior and inferior to the core (Dostrovsky et al. 1991).

The largest study of micro-stimulation-evoked sensations in Vc reports explorations during stereotactic procedures for the treatment of tremor in 124 thalami and 116 patients. Core was defined as the area above the most inferior cell with a response to nonpainful cutaneous stimulation and anterior to the most posterior cell of this type. Warm sensations were evoked more frequently in the posterior region than in the core. The proportion of sites where microstimulation evoked cool and pain sensations did not differ between the core and the posterior region. In the posterior region, however, warm sensations were evoked more frequently in the lateral plane (10.8 %) than in the medial planes (3.9 %). No mediolateral difference was found for sites where pain and cool sensations were evoked. The presence of sites where stimulation evoked taste or where RFs and PFs were located on the pharynx were used as landmarks of a plane located as medial as VMpo. Microstimulation in this plane evoked cool, warm, and pain sensations. The results suggest that thermal and pain sensations are processed in the region of Vc as far medial as VMpo. Therefore, thermal and pain sensations seem to be mediated by neural elements in a region probably including the core of Vc, VMpo and other nuclei posterior and inferior to Vc.

Nociceptive neurons (see ▶ Human Thalamic Nociceptive Neurons) have been identified in the human medial thalamus. Ishijima et al. found that one quarter (20/80) of the cells they recorded from the central medial/parafascicularis complex of man responded to noxious pinprick (Ishijima et al. 1975). None of these cells responded to non-noxious cutaneous stimuli. One group of cells responded to painful stimuli with long latency and showed prolonged after-discharges while others had a time course similar to the stimulus. Another study (Tsubokawa and Moriyasu 1975) also found a relatively large number of nociceptive neurons which they localized to the central medial nucleus. Neither of these reports has been confirmed by more recent studies of patients with neuropathic pain (Rinaldi et al. 1991). Instead cells with very high rates of spontaneous bursting discharge activity were reported in the more recent studies.

Studies of nociceptive responsive cells in humans are consistent with the results reported in awake and anesthetized monkeys. In a study of responses of cells in VPM of awake monkeys to graded mechanical stimuli (Bushnell and Duncan 1987), 10 % of cells (9/89) are classified as WDR. These cells were clustered in ventral VPM. Another study reports 22 thermal responsive cells from a population of hundreds recorded in alert, trained, *Cynomologous* monkeys (Bushnell et al. 1993). Eighteen percent (4/22) of these cells responded to noxious heat only. Among cells with a WDR mechanical response pattern, those that also responded to heat stimuli graded into the noxious range comprised 27 % of cells (6/22) in that series.

Cells in VP, VPI and Po of anesthetized monkeys can respond to innocuous stimuli (Apkarian and Shi 1994). In a recent study 40 cells responded to noxious mechanical stimuli; of these 23 cells also responded to noxious heat and 9 responded to noxious cold. These cells were located in VPI and Po more commonly than in VP. Studies in awake squirrel monkeys have found that 8 % (3/36)–12 % (9/76) of cells in VP responded to noxious mechanical stimuli (Casey and Morrow 1983). These cells were widely distributed throughout VP. In another study, a smaller number of WDR and HT cells were found throughout VP in anesthetized rhesus monkeys (73 cells/thousands of cells) (Chandler et al. 1992). Overall, monkey studies suggest that cells responsive to innocuous and noxious inputs are located in VP, VPI and Po. Cells responding to cold stimuli have tentatively been located in the region proposed to correspond to human VMpo (Davis et al. 1999). Although we have no direct anatomic evidence, the present results are consistent with monkey studies and suggest that cells responsive to painful mechanical stimuli are located in Vc, Vcpc, Vcpor and Po.

Blockade of the activity of these cells by thalamic injection of local anesthetic significantly interferes with the monkey's ability to discriminate temperature in both the innocuous and noxious range (Duncan et al. 1993). These studies establish that cells in the region of the monkey principal somatic sensory nucleus are involved in pathways signaling cutaneous thermal sensations into the noxious range. Therefore, the region of Vc is a pain-signaling pathway, as demonstrated by the presence of afferent connections from the STT, of cells responding to noxious stimuli, of sites where stimulation evokes pain and of sites where lesions relieve pain (Duncan et al. 1993). This is strong evidence for a role of Vc and adjacent nuclei in the human pain system.

References

Apkarian, A. V., & Hodge, C. J. (1989). Primate spinothalamic pathways. III. Thalamic terminations of the dorsolateral and ventral spinothalamic pathways. *The Journal of Comparative Neurology, 288,* 493–511.

Apkarian, A. V., & Shi, T. (1994). Squirrel monkey lateral thalamus. I. Somatic nociresponsive neurons and their relation to spinothalamic terminals. *Journal of Neuroscience, 14,* 6779–6795.

Blomqvist, A., Zhang, E. T., & Craig, A. D. (2000). Cytoarchitectonic and immunohistochemical characterization of a specific pain and temperature relay, the posterior portion of the ventral medial nucleus, in the human thalamus. *Brain, 123,* 601–619.

Boivie, J. (1979). An anatomic reinvestigation of the termination of the spinothalamic tract in the monkey. *The Journal of Comparative Neurology, 186,* 343–369.

Bushnell, M. C., & Duncan, G. H. (1987). Mechanical response properties of ventroposterior medial thalamic neurons in the alert monkey. *Experimental Brain Research, 67,* 603–614.

Bushnell, M. C., Duncan, G. H., & Tremblay, N. (1993). Thalamic VPM nucleus in the behaving monkey. I. Multimodal and discriminative properties of thermosensitive neurons. *Journal of Neurophysiology, 69,* 739–752.

Casey, K. L., & Morrow, T. J. (1983). Ventral posterior thalamic neurons differentially responsive to noxious stimulation of the awake monkey. *Science, 221,* 675–677.

Chandler, M. J., Hobbs, S. F., Fu, Q.-G., et al. (1992). Responses of neurons in ventroposterolateral nucleus of primate thalamus to urinary bladder distension. *Brain Research, 571,* 26–34.

Davis, K. D., Lozano, A. M., Manduch, M., et al. (1999). Thalamic relay site for cold perception in humans. *Journal of Neurophysiology, 81,* 1970–1973.

Dostrovsky, J. O., Wells, F. E. B., & Tasker, R. R. (1991). Pain evoked by stimulation in human thalamus. In Y. Sjigenaga (Ed.), *International symposium on processing nociceptive information* (pp. 115–120). Amsterdam: Elsevier.

Duncan, G. H., Bushnell, M. C., Oliveras, J. L., et al. (1993). Thalamic VPM nucleus in the behaving monkey. III. Effects of reversible inactivation by lidocaine on thermal and mechanical discrimination. *Journal of Neurophysiology, 70,* 2086–2096.

Fruhstorfer, H., & Lindblom, U. (1984). Sensibility abnormalities in neuralgic patients studied by thermal and tactile pulse stimulation. In C. von Euler (Ed.), *Somatosensory mechanisms* (pp. 353–361). London: MacMillian.

Ishijima, B., Yoshimasu, N., Fukushima, T., et al. (1975). Nociceptive neurons in the human thalamus. *Confinia Neurologica, 37,* 99–106.

Lee, J.-I., Antezanna, D., Dougherty, P. M., et al. (1999). Responses of neurons in the region of the thalamic somatosensory nucleus to mechanical and thermal stimuli graded into the painful range. *The Journal of Comparative Neurology, 410,* 541–555.

Lenz, F. A., & Dougherty, P. M. (1998). Cells in the human principal thalamic sensory nucleus (Ventralis Caudalis -Vc) respond to innocuous mechanical and cool stimuli. *Journal of Neurophysiology, 79,* 2227–2230.

Lenz, F. A., Seike, M., Richardson, R. T., et al. (1993). Thermal and pain sensations evoked by microstimulation in the area of human ventrocaudal nucleus. *Journal of Neurophysiology, 70,* 200–212.

Lenz, F. A., Gracely, R. H., Rowland, L. H., et al. (1994). A population of cells in the human thalamic principal sensory nucleus respond to painful mechanical stimuli. *Neuroscience Letters, 180,* 46–50.

Mehler, W. R. (1966). The posterior thalamic region in man. *Confinia Neurologica, 27,* 18–29.

Rinaldi, P. C., Young, R. F., Albe-Fessard, D. G., et al. (1991). Spontaneous neuronal hyperactivity in the medial and intralaminar thalamic nuclei in patients with deafferentation pain. *Journal of Neurosurgery, 74,* 415–421.

Tasker, R. R., Davis, K. D., Hutchinson, W. D., et al. (1997). Subcortical and thalamic mapping in functional neurosurgery. In P. L. Gildenberg & R.R. Tasker (Eds.), *Stereotactic and functional neurosurgery* (pp. 883–923). New York: McGraw-Hill.

Tsubokawa, T., & Moriyasu, N. (1975). Follow-up results of centre median thalamotomy for relief of intractable pain. A method of evaluating the effectiveness during operation. *Confinia Neurologica, 37,* 280–284.

Lateral Thalamus Encodes Pain

▶ Thalamic Nuclei Involved in Pain, Human and Monkey

Laughing Gas

▶ Nitrous Oxide Antinociception and Opioid Receptors

Law of Bell and Magendie

Definition

The ventral root contains motor fibers and the dorsal root contains sensory fibers.

Cross-References

▶ Dorsal Root Ganglionectomy and Dorsal Rhizotomy

LCN

▶ Lateral Cervical Nucleus

Learned Helplessness

Definition

It is an experimental paradigm where animals learn to stop attempting to escape inescapable shock.

Cross-References

▶ Pain-Modulatory Systems, History of Discovery

LEF

▶ Laser-Evoked Field

Lemniscal Fibers

Definition

A band or bundle of ascending fibers from the secondary sensory nuclei to the ventral posterior part of the opposite thalamus. Lemniscal fibers conveying sensory-discriminative component of pain information.

Cross-References

▶ Trigeminothalamic Tract Projections

Lemniscus Trigeminalis

▶ Trigeminothalamic Tract Projections

LEP

▶ Laser-Evoked Potential

Lepromatous Leprosy

Definition

Lepromatous leprosy is the most malignant and infectious type of leprosy. It is characterized by widespread dissemination of leprosy bacilli in the tissues due to poor immune response to infection. Clinical features include widespread, symmetrical, and innumerable macules, which may progress to form nodules and infiltrations. The manifestations of nerve damage appear slowly.

Cross-References

▶ Hansen's Disease

Leprosy

Definition

Also called Hansen's disease. It is a slowly progressive, chronic infectious disease caused by *Mycobacterium leprae*. It is characterized by granulomatous or neurotrophic lesions in the skin, nerves, and viscera.

Cross-References

▶ Hansen's Disease

Leprosy Reaction

Definition

An acute or subacute hypersensitivity state occurring during the course of anti-leprosy treatment or in untreated leprosy. They are divided into two types: type 1 reaction and type 2 reaction.

Cross-References

▶ Hansen's Disease

Leptomeninges

Definition

A collective name for the arachnoid and pia mater membranes, the two innermost layers of the meninges, and between which the cerebrospinal fluid circulates.

Cross-References

▶ Diencephalic Mast Cells

Leptomeningitis

Definition

Inflammation of the leptomeninges (pia mater and arachnoid).

Cross-References

▶ Viral Neuropathies

Lesion

Definition

Selective controlled destruction or pathologic destruction of a structure within the brain.

Cross-References

▶ Pain Treatment, Intracranial Ablative Procedures

Leukocytes

Definition

White blood cells that are important in the induction of the immune response and host defense.

Cross-References

▶ Cytokines, Effects on Nociceptors

Leukoencephalopathy

Definition

Reversible posterior leukoencephalopathy is an MRI appearance seen in hypertensive encephalopathy and consists of T2-weighted and FLAIR changes in the cerebral and brainstem white matter, most prominent in the distribution of the vertebrobasilar circulation.

Cross-References

▶ Headache Due to Hypertension

Leukotrienes

Definition

Leukotrienes are lipid-soluble regulatory molecules synthesized from arachidonic acid by lipoxygenases. They participate in the regulation of diverse body functions such as bronchial constriction and allergic reactions, and are implicated in peripheral sensitization of afferent endings in inflamed tissues, and in ectopic discharge generated in injured nerves.

Cross-References

▶ NSAIDs, Adverse Effects

Leukotrienes in Inflammatory Nociceptor Sensitization

▶ Inflammatory Nociceptor Sensitization, Prostaglandins and Leukotrienes

Levator Ani Syndrome

Definition

A myofascial pain syndrome with painful, more or less permanent, spasms of the puborectal and levator ani muscles.

Cross-References

▶ Pudendal Neuralgia in Women

LFP

▶ Local Field Potential

Liberation

Definition

The liberation of a drug describes its release from the pharmaceutical product.

Cross-References

▶ NSAIDs, Pharmacokinetics

Libido

Definition

Libido, the desire for sexual intimacy, may be altered by emotional, metabolic, and physiologic phenomenon, as well as by many medications.

Cross-References

▶ Cancer Pain Management, Opioid Side Effects, Endocrine Changes, and Sexual Dysfunction

Lichen Sclerosis

Definition

Lichen sclerosis is a painful skin condition generally affecting the vulva (or penis) and anus. It is characterized by thinning and white patches of skin, itching and/or burning, painful sexual intercourse, and sores or lesions resulting from scratching. If left untreated, it can result in fusing of the skin, atrophy, and narrowing of the vagina.

Cross-References

▶ Clitoral Pain
▶ Dyspareunia and Vaginismus

Lidocaine

▶ Postoperative Pain, Lignocaine

Life-limiting Condition

▶ Pain Treatment in Pediatric Palliative Care

Life-threatening Condition

▶ Pain Treatment in Pediatric Palliative Care

Lifetime Prevalence

Definition

Lifetime prevalence is the total number of persons known to have had the disease or attribute for at least part of their life.

Cross-References

▶ Prevalence of Chronic Pain Disorders in Children

Ligands

Definition

Chemicals that have an affinity for a receptor.

Cross-References

▶ Alternative Medicine in Neuropathic Pain

Lightning Pain

▶ Pain Paroxysms

Likert Scale

Definition

A Likert scale presents a set of attitude statements and asks respondents to express agreement or disagreement on a numerical scale. Each degree of agreement is given a numerical value. Thus, a total numerical value can be calculated from all the responses.

Cross-References

▶ Pain Inventories

Limb Amputation

Definition

Limb amputation (removal of a limb) may be caused by surgery or trauma. In Western countries, limb amputation is most often performed because of medical disease. Many amputees experience phenomena such as phantom pain, phantom sensation, and stump pain.

Cross-References

▶ Postoperative Pain, Postamputation Pain, Treatment, and Prevention

Limb Pain

▶ Chronic Musculoskeletal Pain

Limbic Forebrain Matrix

▶ Thalamo-Amygdala Interactions and Pain

Limbic Forebrain/System

Definition

The limbic forebrain/system is a set of structures deep in the brain, including the amygdala, hypothalamus, and the parahippocampal and cingulate gyri, which are commonly grouped together as the limbic system. At the cortical level, it also includes the insular cortex and the cingulate cortex. These areas are phylogenetically older than the surrounding neocortex. The amygdala and the hippocampus form the central axis of the limbic system. The limbic structures are involved in the processing of emotion and motivation and are critical for normal human functioning.

Cross-References

▶ Amygdala, Functional Imaging
▶ Arthritis Model, Kaolin-Carrageenan-Induced Arthritis (Knee)
▶ Cingulate Cortex, Functional Imaging
▶ Hippocampus and Entorhinal Complex: Functional Imaging
▶ Pain Processing in the Cingulate Cortex, Behavioral Studies in Humans
▶ Pain Treatment, Intracranial Ablative Procedures
▶ Thalamo-Amygdala Interactions and Pain

Linkage Disequilibrium

Definition

Linkage disequilibrium is the condition in which the number of closely linked loci on a chromosome (haplotype frequencies) in a population deviates from the values they would have if the genes at each locus were combined at random.

Cross-References

▶ NSAIDs, Pharmacogenetics

Lipophilic

Definition

Chemically lipophilic or hydrophobic species are electrically neutral and nonpolar and, thus, prefer other neutral and nonpolar solvents or molecular environments such as lipids.

Cross-References

▶ Rest and Movement Pain

Lipophilic Substances

▶ NSAID-Induced Lesions of the Gastrointestinal Tract

Lissauer's Tract

Definition

Lissauer's tract is a pathway formed from the proximal end of small unmyelinated and poorly myelinated fibers in peripheral nerves, which enter at the lateral aspect of the dorsal horn and ascend and descend up to four segments, and terminate in Rexed's laminae I through to VI (principally I, II, and V) of the ipsilateral dorsal horn.

Cross-References

▶ Acute Pain Mechanisms
▶ DREZ Procedures
▶ Somatic Pain

List of Diagnoses and Their Definitions

▶ Taxonomy

LLLT

▶ Laser

Load

Definition

The total load imposed on a structure, consisting of the whole of stressors seeking to disrupt the initial integrity of the structure. Load is determined by various factors such as strength and repetitiveness.

Cross-References

▶ Ergonomic Counseling

Load-Bearing Capacity

Definition

The capacity of the structure to resist the destructive, negative impact of loading. Structures with

a high load-bearing capacity can resist more load before damage occurs than structures with a low load-bearing capacity.

Cross-References

▶ Ergonomic Counseling

"Local"

▶ Local Anesthetics

Local Analgesics

▶ Postoperative Pain, Local Anesthetics

Local Anesthetic Agents/Drugs

Definition

Local anesthetic agents are drugs that prevent the generation and transmission of action potentials by nerve cells. These drugs are sodium-channel blockers. They can be used topically by infiltration and by targeted injection/infusion beside peripheral nerves, nerve plexus and neuraxially. They provide anesthesia in a regional distribution, but affect not only pain fibers, but also other sensory fibers, motor fibers, and autonomic nerve function.

Cross-References

▶ Analgesia During Labor and Delivery
▶ Cancer Pain Management, Anesthesiologic Interventions, and Neural Blockade
▶ Pain Treatment: Spinal Nerve Blocks
▶ Postoperative Pain, Local Anesthetics

Local Anesthetic Motor Blockade

Definition

Skeletal muscles paralyzed by local anesthetic interruption of nerve impulses to muscle cells.

Cross-References

▶ Postoperative Pain, Acute Pain Team

Local Anesthetics

Nikolai Bogduk
University of Newcastle, Newcastle Bone & Joint Institute, Royal Newcastle Centre, Newcastle, NSW, Australia

Synonyms

"Local"

Definition

Local anesthetics are drugs with the unique property of being able to block conduction along peripheral nerves.

Characteristics

Local anesthetics are perhaps the most powerful and most valuable drug used in pain medicine. They are the only ones known to be able to stop pain completely. Their effect, however, is not lasting. They exert only a temporary effect, measured in hours. Their application, therefore, is limited to diagnostic tests.

Chemistry

Local anesthetics are drugs formed by aromatic and amino residues linked by either an amide or an ester (Fig. 1). The agents most commonly used in pain medicine are lignocaine and bupivacaine from among the amides and procaine from the esters. The differences in chemistry underlie the differences in duration of action and metabolism of these drugs.

Mechanisms

Local anesthetics act on sodium channels in nerve membranes to prevent or impede sodium flux, but their binding and duration of action depends on whether the channels are resting, open, closed, or inactivated (Butterworth and Strichartz 1990; Strichartz 1988). In turn, these states of the channels depend on whether the membrane is being depolarized and the frequency of depolarization. In nerves that are actively conducting, local anesthetics have a greater affinity for the channel than they have in resting nerves.

This difference in affinity underlies certain quantitative properties of local anesthetics that pertain to the interpretation of diagnostic blocks. The duration of action of local anesthetics has been determined by observation in subjects who were not experiencing ongoing or chronic pain. What was tested was cutaneous anesthesia in patients undergoing minor operative procedures or during childbirth. Accurate figures are difficult to ascertain because some investigators added adrenaline to the local anesthetic, whereas others did not. Nevertheless, representative figures for lignocaine would be mean duration of 2–4 h, with a standard deviation of 1–3 h (Moore et al. 1970a, b; Rubin and Lawson 1968; Watt et al. 1968). A duration of up to 7 h is not unusual, and periods longer than 10 h have been reported (Watt et al. 1968). For bupivacaine, a mean duration of 4–8 h with a standard deviation of 2–3 h would be representative (Moore et al. 1970a, b; Rubin and Lawson 1968; Watt et al. 1968). Periods of anesthesia up to 12 h are not unusual.

These figures should be understood to apply to individuals about to experience pain. They may not, and do not necessarily, apply to patients with persistent or chronic pain. Indeed, patients with neuropathic pain often exhibit extraordinarily prolonged responses to lignocaine (Arner et al. 1990), probably because the agent

Local Anesthetics, Fig. 1 The structural formulae of three common local anesthetics, illustrating how aromatic and amino residues are linked by either an amide or an ester

Local Anesthetics, Fig. 2 The relationship between the equilibrium equation of local anesthetics and their pharmacological properties

is able to bind strongly and persistently to open and changing sodium channels. What the "normal" duration of action of local anesthetics is in patients with chronic pain has not been determined.

Pharmacology

As local anesthetics are weak bases, they have an affinity to hydrogen ions but are subject to dissociation. The affinity and dissociation are described by the Henderson-Hasselbach equation. Different components of that equation

govern the various pharmacological properties of local anesthetics (Fig. 2).

The difference between the ambient pH and the equilibrium constant (Ka) determines the fraction of the local anesthetic available in the base form. That form is the active form. Being more lipid soluble, it is the base form that penetrates cell membranes to get to the active site. The lipid solubility of the base form determines the potency of the agent. Its protein binding determines its duration of action.

Meanwhile, the acid form of the local anesthetic is its soluble form. Local anesthetic agents

Local Anesthetics, Fig. 3 A flowchart showing the main steps of the metabolism of local anesthetic agents

METABOLISM	DISTRIBUTION	METABOLISM
(amides)	high vascularity	(esters)
LIVER ←	BLOOD ——→	PLASMA CHOLINESTERASE
	LIVER	
	LUNGS	
↓		↓
KIDNEY	HEART	KIDNEY
	BRAIN	
↓	↓	↓
EXCRETION	TOXICITY	EXCRETION

are prepared and stored in their soluble form. In order to permit storage in the soluble form, local anesthetic agents are stored at a low pH (e.g., 4.0). For most agents the pKa is about 7.9. In a pH environment of 4.0, the difference between pH and pKa is $4.0–7.9 = -3.9$. In which case, the fraction of the agent in the non-soluble form is $10^{-3.9}$, i.e., 1 in 10,000. That means that virtually all of the agent is in the soluble form.

When the agent is injected into tissues of pH 7.4, the difference between pH and pKa is $7.4–7.9 = -0.5$. Under those conditions, the fraction of the active, insoluble form is $10^{-0.5}$, i.e., 31 % is in the active form.

Metabolism

The amides and esters differ in how they are metabolized. Both are rapidly distributed through the blood stream and pass to tissues of high vascularity (Fig. 3). The amides are metabolized by the liver, after which they are excreted by the kidneys. The esters are metabolized in the blood by plasma cholinesterase. The products are excreted by the kidneys.

Once absorbed into the bloodstream, all local anesthetic agents reach the heart and the brain. At these sites, they can exert their toxic effects.

Toxicity

At toxic concentrations, local anesthetics can suppress electrical activity in the heart. In the central nervous system, they suppress inhibitory neurons, which results in excitation of disinhibited systems. This is manifest clinically as fitting.

The toxic doses required to produce these effects, however, are considerably higher than the amounts used for most clinical purposes in pain medicine. For bupivacaine, the total body dose that is typically toxic is about 150 mg; for lignocaine, the dose is 200 mg (Rosenberg et al. 2004).

Applications

In pain medicine, local anesthetics are used in three main ways. Most commonly, they are used to perform diagnostic blocks of peripheral nerves or of spinal nerves (see ► Peripheral Nerve Blocks). These blocks are used to identify either the source of pain or the nerves that are mediating the pain (Bogduk 2002). Local anesthetic agents can also be administered intravenously, for the diagnosis or treatment of neuropathic pain or central pain. Occasionally, they are administered orally for the treatment of neuropathic pain.

References

Arner, S., Lindblom, U., Meyerson, B. A., et al. (1990). Prolonged relief of neuralgia after regional anaesthetic blocks. A call for further experimental and systematic clinical studies. *Pain, 43*, 287–297.

Bogduk, N. (2002). Diagnostic nerve blocks in chronic pain. *Best Practice & Research. Clinical Anaesthesiology, 16*, 565–578.

Butterworth, J. F., & Strichartz, G. R. (1990). Molecular mechanisms of local anesthesia: A review. *Anesthesiology, 72*, 711–734.

Moore, D. C., Bridenbaugh, L. D., Bridenbaugh, P. O., et al. (1970a). Bupivacaine for peripheral nerve block: A comparison with mepivacaine, lidocaine, and tetracaine. *Anesthesiology, 32*, 460–463.

Moore, D. C., Bridenbaugh, L. D., Bridenbaugh, P. O., et al. (1970b). Bupivacaine: A review of 2,077 cases. *Journal of the American Medical Association, 214*, 713–718.

Rosenberg, P. H., Veering, B. T., & Urmey, W. F. (2004). Maximum recommended doses of local anesthetics: A multifactorial concept. *Regional Anesthesia and Pain Medicine, 29*, 564–575.

Rubin, A. P., & Lawson, D. I. F. (1968). A controlled trial of bupivacaine: A comparison with lignocaine. *Anaesthesia, 23*, 327–331.

Strichartz, G. R. (1988). Neural physiology and local anesthetic action. In M. J. Cousins & P. O. Bridenbaugh (Eds.), *Neural blockade in clinical anesthesia and management of pain* (2nd ed., pp. 25–45). Philadelphia: JB Lippincott.

Watt, M. J., Ross, D. M., & Atkinson, R. S. (1968). A double-blind trial of bupivacaine and lignocaine. *Anaesthesia, 23*, 331–337.

Local Anesthetics/Antiarrhythmics

Definition

Local anesthetics/antiarrhythmics are sodium-channel modulators that depress the action potential and enhance repolarization of primary afferent neurons.

Cross-References

▶ Drugs Targeting Voltage-Gated Sodium and Calcium Channels

Local Circuit Interneuron

Definition

Neuron with axonal arborization located in the area occupied by its own dendritic tree.

Cross-References

▶ Spinothalamic Tract Neurons, Morphology

Local Field Potential

Synonyms

LFP

Definition

Synchronized extracellular currents of a few hundred cells generate a LFP that reflects the average input to individual neurons. The LFP can be recorded with a microelectrode with impedance up to 0.5 MΩ. The signal is analyzed for frequencies up to 100 Hz.

Cross-References

▶ Thalamotomy for Human Pain Relief

Local Twitch Response

Synonyms

LTR

Definition

A transient, poorly synchronized, prolonged twitch-like contraction of the group of taut band fibers that is associated with a myofascial trigger point, MTrP. The response is a polysynaptic spinal reflex with its afferent limb commonly arising from mechanical stimulation of sensitized nociceptors in the trigger point, and the efferent limb begins as an activation of alpha motor neurons that supply the involved nerve terminals of taut band fibers in the same or sometimes in functionally related muscles. It can be easily elicited when the needle tip encounters a sensitive locus in the MTrP region during high-pressure stimulation.

Cross-References

▶ Dry Needling
▶ Myofascial Trigger Points

Local-Circuit Cells

Definition

Neurons in the brain are divided into two major groups, projection neurons and interneurons, local neurons, or local-circuit neurons.

Cross-References

▶ Trigeminal Brainstem Nuclear Complex, Anatomy

Localized Muscle Pain

▶ Myofascial Pain

Localized Provoked Vulvodynia

▶ Vulvodynia

Local-Twitch-Response Locus

▶ Sensitive Locus

Locus Coeruleus

Definition

A noradrenergic nucleus that is located in the pontomesencephalic junction and is involved in the regulation of vigilance and descending feedback control of pain. Many neurons in this region contain the neurotransmitter norepinephrine.

Cross-References

▶ Descending Modulation and Persistent Pain
▶ Vagal Input and Descending Modulation

Locus of Control

Definition

Beliefs about whether certain outcomes in life are a result of one's efforts (internal) or a result of luck, fate, or the actions of others (external).

Cross-References

▶ Psychological Treatment of Headache

Longitudinal Myelotomy

▶ Midline Myelotomy

Longitudinal Study

Synonyms

Cohort study

Definition

A cohort study is an epidemiological method identifying a study population by age, or by using other means or traits of grouping individuals for the purpose of research (Timmreck 2002) and following them over time.

Cross-References

Long-Lasting Acute Pain

Long-term Depression in the Spinal Cord

Long-term Effects

Long-Term Effects of Pain in Infants

Ruth Eckstein Grunau
Pediatrics Dept, University of British Columbia
Developmental Neurosciences & Child Health,
Child & Family Research Institute, Vancouver,
BC, Canada

Synonyms

Infant pain; Long-term effects

Definition

Historically, exposure to repeated or prolonged pain in the neonatal period was uncommon; however, this changed with the advent of neonatal intensive care. The tiniest infants born at very low gestational age (i.e., ≤ 32 weeks) are thereby physiologically immature, and their survival has increased significantly in recent years. Life-saving neonatal intensive care of infants born very prematurely or with serious medical complications involves prolonged exposure to procedures that result in ▶ stress and ▶ pain during the period of very rapid brain development and programming of stress systems.

Characteristics

Tissue injury induces numerous behavioral, physiological, and endocrine changes in neonates, which are consistent with pain responses in older children and adults under similar circumstances. Long-term functional and structural changes in ▶ nociception can occur, related to pain in the neonatal period (Fitzgerald 2005). Moreover, there is a growing body of evidence suggesting that cumulative neonatal pain and stress may contribute to long-term alterations of pain systems, and perhaps to multiple aspects of behavior and development in the most vulnerable infants (Anand et al. 2007; Grunau et al. 2006). A model for conceptualizing long-term effects of pain in human preterm neonates has been proposed (Grunau 2002). Long-term changes in pain processing of children exposed to substantial pain early in life are complex.

Neurobiology of Infant Pain

Pain perception is functional by mid-gestation in human infants (Fitzgerald 2005). Due to advances in knowledge of the developmental neurobiology of pain, it is clear that the immature nervous system responds differently to tissue damage and pain. In premature human infants and immature rat pups, reflex responses to touch show greater sensitivity compared to those of older infants and children (Fitzgerald and Beggs 2001). With repeated stimulation, even at innocuous levels, the thresholds for touch and pain responses drop even further, indicating that these immature infants become even more sensitive. This phenomenon of sensitization is greatest at 28–33 weeks post-conceptional age and is no longer evident by 42 weeks. Sensory neurons in the spinal cord and brain stem become hyperexcitable following inflammation, which is referred to as ▶ central sensitization. Activation of these central cells by repetitive pain results in hypersensitivity to low-level input of non-skin breaking handling. For example, due to greater tactile (touch) sensitivity in very immature mammals, frequent pain alters touch ▶ thresholds so that even noninvasive handling, such as diaper changes, elicits pain-like responses in extremely premature infants (see Grunau et al. 2006). In this way, repetitive pain may lay the groundwork for chronic pain and discomfort in the Neonatal Intensive Care Unit (NICU).

Using near-infrared spectroscopy studies have shown that clinical blood-taking in preterm infants evokes a specific localized hemodynamic response in the somatosensory cortex of the brain, suggesting that higher-level pain processing may occur as early as 25 weeks post-conceptional age (Slater et al. 2006; Bartocci et al. 2006). Furthermore, time-locked electroencephalography (EEG) provides evidence of noxious-evoked cortical neural activity in the preterm infant brain (Slater et al. 2010). Importantly, poorer development of brain microstructure from birth to term equivalent is associated with greater number of neonatal skin-breaking procedures, after adjusting for multiple medical confounders (Brummelte et al. 2012).

Neonatal Prolonged Pain and Reprogramming of Stress Systems

The process of maintaining physiological stability involves numerous behavioral, autonomic, and hormonal adjustments. The very immature developing organism is not yet capable of fine-tuning these multiple dimensions in a balanced fashion; thus, premature neonates born at extremely low gestational age are potentially more sensitive to external stress than infants born full-term. Of all the environmental input, the stress induced by invasive skin breaking procedures over a prolonged period is thought to be primary. While infants are still in the NICU, higher cumulative stress related to procedural pain since birth

is associated with lower capacity to mount a stress response both behaviorally and hormonally (Grunau et al. 2005). Of greater concern, however, are findings that cortisol (the main stress hormone in humans) is higher in extremely preterm infants much later in infancy, at 8 months "corrected age" (CA, age corrected for prematurity) (Grunau et al. 2004), persisting to 18 months CA (Grunau et al. 2007). These findings suggest a possible long-term "resetting" of stress regulation following prolonged cumulative stress and pain exposure associated with extreme prematurity.

In experimental animal studies, neonatal stress (e.g., due to maternal separation) can permanently reprogram stress responses in the hypothalamic-pituitary-adrenal (HPA) axis. In contrast, in studies of long-term effects of pain applied in the neonatal period, no changes were found to HPA responses subsequently in adult rats (Anand et al. 1999; Walker et al. 2003). Walker et al. showed that in rats, the frequency of maternal licking and grooming increased in those pups exposed to pain; this maternal care behavior appeared to prevent the reprogramming of the HPA axis. There is some evidence that mother-infant interaction in humans may moderate effects of neonatal pain. However, a great deal remains to be learned.

Animal Studies of Long-Term Effects of Neonatal Pain Exposure

Most of the evidence for long-term effects of neonatal pain on the developing nervous system is from experimental research on rats. Although we cannot extrapolate directly to human infants, these studies are useful because with humans there are multiple factors that cannot be controlled experimentally. Rats are most often studied because the central nervous system of rat pups at birth is very immature, providing an approximation to the central nervous system of a 24 week gestation, extremely premature human infant. In experimental animal studies, pain induced in the neonatal period can induce changes to nociceptive systems, which are not seen when the pain is applied at later ages. This "critical window" suggests that the long-term impact of pain is potentially greatest early in life.

Effects of early pain are complex and depend on multiple factors, such as age at pain stimulation, age at later testing, type of noxious stimulation (e.g., inflammation, pin prick, surgical cut), intensity, duration, and modality of input (see Fitzgerald 2005). For example, the direction of effects, that is, hypersensitivity (enhanced sensitivity) or hyposensitivity (reduced sensitivity), and the extent of effects vary depending on the type of pain stimulus and whether pain is short lasting or ongoing (see Ren et al. 2004).

Pain and Stress in the NICU Environment

The medical care of "high-risk" preterm infants in the NICU involves exposure to frequent invasive procedures. From birth to discharge from the NICU, very preterm infants undergo an average of about 113 procedures, which reflects approximately 2–10 procedures per day. However, some of the tiniest, sickest infants can have 300 procedures or more from admission through discharge, the majority of which are heel lances for blood tests. While infants are in the NICU, their tactile and pain responses are altered depending on the context, timing, and frequency of pain experience, gestational age at birth and postnatal age. Increased reactivity is typical in infants who are handled immediately before a procedure or who have had more skin breaking procedures the day before, consistent with sensitization. Conversely, however, cumulatively over time, those infants who have had more total exposure to prior pain since birth show decreased behavioral and stress hormone reactivity to invasive procedures later while they are still exposed to ongoing stress in the NICU. This is consistent with animal studies of more prolonged pain. Behavioral responses to pain may be dampened over time in both human infants and animal pups that have had more intense early pain exposure (see Grunau et al. 2006; Anand et al. 2007).

Pain Responsivity Later in Infancy and Childhood

Preterm Infants

Several studies have directly compared pain reactivity of premature infants with that of healthy infants born full-term (see Grunau and Tu 2007).

Contrary to expectation, after hospital discharge, preterm infants showed little difference in reactivity to pain at 4 months CA compared to full-term infants, except for a subset who had spent the longest time in the NICU. However, at 8 months CA, pain responses were more divergent in the preterm infants: They showed very brief initial behavioral hyper-reactivity to a finger lance, followed by greater behavioral and autonomic recovery. Recently, quantitative sensory testing of temperature threshold later in childhood (age 11–14 years) has shown both hyper- and hypo-reactivity (Hermann et al. 2006), consistent with the animal literature on long-term effects of early pain. Importantly, preterm infants who have also undergone neonatal surgery appear to be more affected (Walker et al. 2009). During nociceptive stimulation at age 11–16 years, children born preterm showed greater activation in the somatosensory cortex and several other brain regions on functional magnetic resonance imaging (fMRI), compared to children born full-term with or without early hospitalization (Hohmeister et al. 2010).

Parent-report indicated reduced pain sensitivity to everyday bumps and scrapes at 18 months CA in toddlers born extremely preterm, compared to more mature preterm and full-term infants. Interestingly, parents reported greater somatization (pain of no known medical cause) in preterm children at age 5 years; however, these differences were no longer evident at age 9 or 17 years (see Grunau and Tu 2007). In studies of self-reported pain, adolescents and young adults born preterm reported no differences in pain in daily life (Buskila et al. 2003; Schmelzle-Lubiecki et al. 2007; Saigal et al. 1996, 2006). There is wide variation in long-term outcomes of extreme prematurity for children exposed to comparable amounts of pain in the neonatal period.

Full-Term Infants

To evaluate long-term effects of pain in infants born full-term, children requiring early surgery or exposed to burns have been studied. Following early surgery, there is evidence of changes in later tactile sensitivity that vary depending on the type of surgery and whether testing is carried out in the region of surgery or elsewhere, suggesting both local and global effects. For example, children who had cardiac surgery in infancy displayed hyposensitivity to touch/cold/heat stimuli in the region of surgery, and globally to touch, compared to children with no surgery (Schmelzle-Lubiecki et al. 2007). Hypersensitivity to touch in the wound area and the intact contralateral abdominal side was found after abdominal surgery (Andrews and Fitzgerald 2002). In contrast, heart rate, cortisol and behavioral responses to vaccination did not differ in toddlers who had surgery early in life (Peters et al. 2003). Children who had suffered burns early in life (age 6–24 months) displayed altered mechanical or pain sensitivity later in childhood, that varied by severity of insult (Wollgarten-Hadamek et al. 2009). Taken together, it appears that early injury can induce changes in peripheral and central tactile and pain processing.

Developmental Outcomes in Preterm Infants

Learning, academic, and behavior problems are prevalent in children born extremely prematurely, persisting to late adolescence (e.g., Vohr 2000). Two mechanisms have been identified that may potentially link neonatal pain exposure to altered neurodevelopment in this vulnerable population. For the tiniest, most fragile infants, the period of pain exposure is prolonged, during the last trimester of "fetal" life, which is a period of very rapid and complex brain development. It is known that immature neurons are more sensitive to toxic influences, and brain volumes are smaller in preterm compared to full-term children (Bhutta and Anand 2002). At this time there is no direct evidence for a causal connection between neonatal pain and later developmental and behavioral difficulties in preterm infants. However, associations have been found between higher neonatal pain exposure and poorer motor and cognitive development to 18 months CA (Grunau et al. 2009), consistent with concerns that repeated neonatal pain may contribute to alterations in the developing preterm brain (Brummelte et al. 2012). Furthermore, prolonged exposure to repeated neonatal pain in preterm infants may contribute to altered programming

of hypothalamic-pituitary-adrenal (HPA) axis (Grunau et al. 2007), which has important implications for brain function.

Sex Differences

In animal studies, sex differences in vulnerability to early stress and effects on long-term pain reactivity have been reported. However, sex has rarely been examined in human infant studies of long-term effects of pain; therefore, no conclusions can be drawn at this point.

Maternal-Infant Factors

Multiple interacting intrinsic and extrinsic factors potentially ameliorate or exacerbate effects of neonatal experience on developmental trajectories. Later in childhood and as adults, pain perception and/or expression is affected by multiple interacting factors including social modeling, family variables, child temperament, culture, and sex. However, this work has been conducted almost entirely on older children born at term.

Maternal verbalizations that promoted coping in full-term human infants at age 6 months, but not general maternal sensitivity, were associated with altered infant behavior during immunization. Moreover, in former extremely preterm infants, child and family factors predicted ▶ somatization at age 4 1/2 years, and positive maternal responsivity was associated with normalized pain response in infancy (see Grunau 2003). Furthermore, in animal studies, it is very important to note that increased maternal behaviors appear to prevent negative effects of early pain on stress hormone response (Walker et al. 2003). Together these studies suggest that in mammals, alterations to pain systems may be modulated, at least to some extent, by ongoing caregiving.

Memory for Pain

Conscious awareness is not required for memory. There are different types of memory, other than declarative recall of events. Conditioned learning begins very early in life, which is an evidence of memory. Moreover, differing physiological changes after repeated pain experiences imply "biological memory." Thus, although infants cannot recall early experience, their central nervous system may retain a type of memory that is manifested later in altered responses to pain in new situations.

Summary

In human preterm infants, pain reactivity is altered while infants are in the NICU. There is now evidence of local and global changes in sensory processing that persist well into later childhood; however, effects are complex, consistent with the animal literature. Recent studies have shown similar effects of early injury in infants born full-term who are hospitalized due to major surgery or burns early in life.

Adaptation of an organism to the environment occurs through multiple processes during the prenatal and neonatal periods, infancy, and childhood. The mammalian brain is characterized by its plasticity, namely, the ability to adjust to changes in the internal or external environment, which especially applies to the developing nervous system. The extent to which human infants can compensate over time for the adverse early experiences of prolonged pain exposure in the NICU is unknown. The most long-lasting changes may be to generalized stress systems rather than specifically to pain. The extent to which sensitive and responsive parenting may modulate such effects is also unknown; however, this opens a promising new avenue for the development of interventions.

Acknowledgment Ruth Grunau's research program is supported by Grants from the Eunice Kennedy Shriver National Institute of Child Health and Human Development and the Canadian Institutes of Health Research; she receives a salary award from the Child and Family Research Institute.

References

Anand, K. J. S., Coskun, V., Thrivikraman, K. V., et al. (1999). Long-term behavioral effects of repetitive pain in neonatal rat pups. *Physiology & Behavior, 66,* 627–637.

Anand, K. J. S., Stevens, B. J., & McGrath, P. J. (2007). *Pain in neonates and infants* (Pain research and clinical management 3rd ed.). London: Elsevier.

Andrews, K., & Fitzgerald, M. (2002). Wound sensitivity as a measure of analgesic effects following surgery in human neonates and infants. *Pain, 99*(1–2), 185–195.

Bartocci, M., Bergqvist, L. L., Lagercrantz, H., & Anand, K. J. S. (2006). Pain activates cortical areas in the preterm newborn brain. *Pain, 122,* 109–117.

Bhutta, A. T., & Anand, K. J. S. (2002). Vulnerability of the developing brain: Neuronal mechanisms. *Clinics in Perinatology, 29,* 357–372.

Brummelte, S., Grunau, R.E., Chau, V., Poskitt, K.J., Brant, R., Vinall, J., Gover, A., Synnes, A., Miller, S. P. (2012). Procedural pain and brain development in premature newborns. *Annals of Neurology, 71*(3), 385–396.

Buskila, D., Neumann, L., Zmora, E., Feldman, M., Bolotin, A., & Press, J. (2003). Pain sensitivity in prematurely born adolescents. *Archives of Pediatrics & Adolescent Medicine, 157*(11), 1079–1082.

Fitzgerald, M. (2005). The development of nociceptive circuits. *Nature Reviews Neuroscience, 6,* 507–520.

Fitzgerald, M., & Beggs, S. (2001). The neurobiology of pain: Developmental aspects. *The Neuroscientist, 7,* 246–257.

Grunau, R. E. (2002). Early pain in preterm infants. A model of long-term effects. *Clinics in Perinatology, 29,* 373–394.

Grunau, R. E. (2003). Self-regulation and behavior in preterm children: Effects of early pain. In P. McGrath & G. A. Finley (Eds.), *Pediatric pain: Biological and social context* (pp. 23–51). Seattle: IASP Press.

Grunau, R. E., & Tu, M. T. (2007). Long-term consequences of pain in human neonates. In K. J. S. Anand, B. J. Stevens, & P. J. McGrath (Eds.), *Pain in neonates* (3rd ed., pp. 45–55). Amsterdam: Elsevier Science.

Grunau, R. E., Weinberg, J., & Whitfield, M. F. (2004). Neonatal procedural pain and preterm infant cortisol response to novelty at 8 months. *Pediatrics, 114*(1), e77–e84.

Grunau, R. E., Holsti, L., Haley, D. W., et al. (2005). Neonatal procedural pain exposure predicts lower cortisol and behavioral reactivity in preterm infants in the NICU. *Pain, 113,* 293–300.

Grunau, R. E, Holst, L, & Peters, J. W. (2006). Long-term consequences of pain in human neonates. In D Tibboel, R Bhat (Guest Eds.), *Pain Control and Sedation, Seminars in Fetal and Neonatal Medicine, 11*(4), 268–275.

Grunau, R. E., Haley, D. W., Whitfield, M. F., Weinberg, J., Yu, W., & Thiessen, P. (2007). Altered basal cortisol levels at 3, 6, 8 and 18 months in infants born at extremely low gestational age. *Journal of Pediatrics, 150*(2), 151–156.

Grunau, R. E., Whitfield, M. F., Petrie-Thomas, J., Synnes, A. R., Cepeda, I. L., Keidar, A., Rogers, M., Mackay, M., Hubber-Richard, P., & Johannesen, D. (2009). Neonatal pain, parenting stress and interaction, in relation to cognitive and motor development at 8 and 18 months in preterm infants. *Pain, 143*(1–2), 138–146.

Hermann, C., Hohmeister, J., Demirakça, S., Zohsel, K., & Flor, H. (2006). Long-term alteration of pain sensitivity in school-aged children with early pain experiences. *Pain, 125*(3), 278–285.

Hohmeister, J., Kroll, A., Wollgarten-Hadamek, I., Zohsel, K., Demirakça, S., Flor, H., & Hermann, C. (2010). Cerebral processing of pain in school-aged children with neonatal nociceptive input: An exploratory fMRI study. *Pain, 150*(2), 257–267.

Peters, J. W., Koot, H. M., de Boer, J. B., et al. (2003). Major surgery within the first 3 months of life and subsequent biobehavioral pain responses to immunization at later age: A case comparison study. *Pediatrics, 111*, 129–135.

Ren, K., Anseloni, V., Zou, S.-P., et al. (2004). Characterization of basal and re-inflammation-associated long-term alteration in pain responsivity following short-lasting neonatal local inflammatory insult. *Pain, 110*, 588–596.

Saigal, S., Feeny, D., Rosenbaum, P., Furlong, W., Burrows, E., & Stoskopf, B. (1996). Self-perceived health status and health-related quality of life of extremely low-birth-weight infants at adolescence. *Journal of the American Medical Association, 276*, 453–459.

Saigal, S., Stoskopf, B., Pinelli, J., Streiner, D., Hoult, L., Paneth, N., & Goddeeris, J. (2006). Self-perceived health-related quality of life of former extremely low birth weight infants at young adulthood. *Pediatrics, 118*(3), 1140–1148.

Schmelzle-Lubiecki, B. M., Campbell, K. A., Howard, R. H., Franck, L., & Fitzgerald, M. (2007). Long-term consequences of early infant injury and trauma upon somatosensory processing. *European Journal of Pain, 11*(7), 799–809.

Slater, R., Cantarella, A., Gallella, S., Worley, A., Boyd, S., Meek, J., & Fitzgerald, M. (2006). Cortical pain responses in human infants. *Journal of Neuroscience, 26*, 3662–3666.

Slater, R., Fabrizi, L., Worley, A., Meek, J., Boyd, S., & Fitzgerald, M. (2010). Premature infants display increased noxious-evoked neuronal activity in the brain compared to health age-matched term-born infants. *NeuroImage, 52*(2), 583–589.

Vohr, B. R. (2000). Guest Editor. Outcome of the very low-birth weight infancy. *Clinics in Perinatology, 27*(2). Philadelphia: Saunders.

Walker, C. D., Kudreikis, K., Sherrard, A., et al. (2003). Repeated neonatal pain influences maternal behavior, but not stress responsiveness in rat offspring. *Developmental Brain Research, 140*, 253–261.

Walker, S. M., Franck, L. S., Fitzgerald, M., Myles, J., Stocks, J., & Marlow, N. (2009). Long-term impact of neonatal intensive care and surgery on somatosensory perception in children born extremely preterm. *Pain, 141*(1–2), 79–87.

Wollgarten-Hadamek, I., Hohmeister, J., Demirakça, S., Zohsel, K., Flor, H., & Hermann, C. (2009). Do burn injuries during infancy affect pain and sensory sensitivity in later childhood? *Pain, 141*(1–2), 165–172.

Long-Term Potentiation

Synonyms

LTP

Definition

This is a long-lasting increase in synaptic efficacy resulting from repetitive activation of the synapse. This process was first described in hippocampus, whereby high-frequency stimulation of afferent pathways leads to a potentiated post-synaptic response that can last for hours to days. It is a form of activity-dependent plasticity. Some forms of LTP may increase the efficiency of pain transmission for weeks to months or longer.

Cross-References

▶ Alternative Medicine in Neuropathic Pain
▶ Opioids in the Spinal Cord and Central Sensitization
▶ Spinothalamic Tract Neurons, Role of Nitric Oxide

Long-Term Potentiation and Long-Term Depression in the Spinal Cord

Jürgen Sandkühler
Department of Neurophysiology, Center for Brain Research Medical University of Vienna, Vienna, Austria

Synonyms

Long-term depression in the spinal cord; Long-term potentiation in the spinal cord

Definitions

Pain amplification (hyperalgesia) may result from an acute noxious event such as a trauma, an inflammation, or a nerve injury. It may persist long after the primary cause for pain has disappeared. Hyperalgesia may also develop in the absence of any noxious event, e.g., during prolonged opioid application or upon abrupt opioid withdrawal. Altered processing of sensory information in the central nervous system contributes to these forms of hyperalgesia. Long-term potentiation of synaptic strength (LTP) in nociceptive pathways is a cellular model of pain amplification.

LTP and synaptic long-term depression (LTD) are defined as long-lasting changes in synaptic efficiency irrespective of the type and the location of chemical synapse. Enhanced synaptic strength which outlasts the conditioning stimulus by at least 30 min is usually classified as LTP. Shorter lasting forms of synaptic plasticity are short-term potentiation or -depression, post tetanic potentiation, and paired-pulse facilitation or -depression. LTP and LTD may be induced and maintained by pre- and/or postsynaptic mechanisms such as changes in transmitter release, receptor sensitivity, or density. LTP and LTD are divided into three phases: the induction phase, the consolidation phase, and the maintenance phase. The signal transduction pathways involved depend upon the induction protocol, type of synapse, types of pre- and postsynaptic neurons, polarity of synaptic plasticity, and developmental stage. It is now generally accepted the activation of glial cells contributes to the induction of LTP. While synaptic plasticity cannot directly be studied by recording action potential firing, polysynaptic responses, or behavioral variables, evaluation of these parameters is useful to show that synaptic plasticity is relevant to information processing downstream of the potentiated synapse in the nociceptive path.

Introduction

Synaptic Models for Learning and Memory in Nociceptive Pathways
LTP and LTD were first described at synapses in the hippocampus and are now considered major cellular models for learning and memory formation. The final proof is, however, still lacking, mainly since the neuronal elements and their activity patterns involved in cognitive or motor learning are largely unknown (Barnes 1995). Fortunately, in the nociceptive system the situation is considerably less vague: The primary afferent nerve fibers as well as their activity patterns that lead to pain are known. Second-order neurons in superficial spinal dorsal horn were identified that mediate hyperalgesia and allodynia. In particular, neurons in lamina I that express the NK1 receptor for substance P are essential for full expression of hyperalgesia in various animal models of inflammation and neuropathy (Nichols et al. 1999). These neurons are all nociceptive specific, most project to the parabrachial area and/or periaqueductal gray, and receive input from primary afferent, peptidergic C fibers.

Characteristics

Synaptic Long-Term Potentiation in Nociceptive Pathways
Conditioning stimulation of primary afferent C fibers leads to LTP at C-fiber synapses with these lamina I projection neurons, but not all lamina I neurons (Ikeda et al. 2003, 2006). Several independent and convergent lines of evidence suggest that synaptic LTP in superficial spinal dorsal horn is a cellular mechanism of afferent-induced hyperalgesia as reviewed in (Sandkühler 2009, 2012; Kuner 2010; Zhuo 2011):
1. Protocols that Induce LTP also Cause Hyperalgesia in Animals and Humans
    1. Continuous electrical stimulation of C fibers at low frequencies (Fig. 1a, LFS, 1–5 Hz for 2 or 3 min) or high frequency and burst-like stimulation (Fig. 1b, HFS, three to five 100 Hz bursts of 1 s duration) induce LTP in various spinal cord-dorsal root slice preparations under different recording conditions (Ikeda et al. 2006; Park et al. 2011; Luo et al. 2012) and in intact animal models (Drdla and Sandkühler 2008;

Long-Term Potentiation and Long-Term Depression in the Spinal Cord, Fig. 1 Illustrates the induction and the reversal of LTP at C-fiber synapses in the superficial spinal dorsal horn. In (**a**) LTP was assessed in a spinal cord-dorsal root slice preparation by measuring monosynaptically evoked excitatory postsynaptic currents (EPSCs). In all other panels (**b–f**) LTP was quantified in deeply anesthetized rats by amplitude changes of C-fiber-evoked field potentials in superficial lumbar spinal dorsal horn. LTP can be induced by conditioning low-frequency stimulation (**a**, LFS, 1 Hz for 2 min) or high-frequency stimulation (**b**, 100 Hz for 1 s given three times at 10 s intervals) of primary afferent nerve fibers at C-fiber intensity. LTP can also be induced

by selective activation of TRPV1-expressing primary afferents with subcutaneous capsaicin injections into the plantar surface of the ipsilateral hindpaw (**c**). In addition to these activity-dependent forms of LTP, abrupt withdrawal from a μ-opioid receptor agonist such as the ultrashort acting opioid remifentanil induces opioid withdrawal LTP in the absence of conditioning stimulation of sensory afferents (**d**). In contrast, when withdrawal from the opioid is tapered over a period of 30 min, opioid withdrawal LTP is not induced (data not shown). If an opioid is applied to an animal in which LTP has been induced previously by either conditioning HFS (**e**) or LFS (**f**), then C-fiber-evoked responses are acutely depressed during the opioid application. Upon withdrawal

Zhou et al. 2010). LTP at the first nociceptive synapse apparently affects downstream activity in nociceptive pathways. Action potential firing in deep dorsal horn, wide dynamic range neurons (Qu et al. 2009), and pain rating in human volunteers (Klein et al. 2004) are also potentiated by similar conditioning stimuli.

2. Natural patterns of afferent barrage during noxious stimuli [subcutaneous injections of capsaicin (Fig. 1c) or formalin] in intact (i.e., not spinalized) animals also induce LTP in spinal cord and hyperalgesia in behaving animals.

3. Abrupt withdrawal from opioids induces LTP at C-fiber synapses in spinal cord slices and in anesthetized animals (Fig. 1d) (Drdla et al. 2009).

2. Time Courses for LTP and Hyperalgesia Are Similar

1. LTP and hyperalgesia are induced within minutes after conditioning stimulation, excluding any time-consuming processes such as sprouting of nerve fibers. LTP and hyperalgesia may outlast the conditioning stimulation by hours.

2. LTP and hyperalgesia may spontaneously reverse within hours or days or may persist for longer periods depending upon induction protocols and the context of conditioning.

3. Shared Pharmacology and Signal Transduction Pathways for LTP and Hyperalgesia

1. Co-activation of NMDA, group I mGlu, and NK1 receptors is required.

2. Activation of voltage-gated calcium channels is required.

3. Ca^{2+}-dependent signal transduction pathways are involved.

4. Activation of protein kinase C, calcium-calmodulin-dependent protein kinase II,

and nitric oxide synthase (in some cases) is necessary.

5. Expression of TRPV1 receptors is required (Park et al. 2011).

6. Gliotransmitters including BDNF are involved in the induction of hyperalgesia and LTP at C-fiber synapses (Zhou et al. 2011).

4. LTP and Hyperalgesia Can Be Prevented by the Same Means

1. Activity in descending inhibitory pathways raises the threshold for induction of both LTP and hyperalgesia.

2. Opioids can preempt induction of LTP (Benrath et al. 2004) and hyperalgesia (Katz et al. 2003).

5. LTP and Hyperalgesia Can Be Reversed

1. Counter stimulation at Aδ-fiber intensity can reverse some forms of hyperalgesia and LTP.

2. Application of a brief, very high opioid dose can reverse both hyperalgesia and LTP (Drdla-Schutting et al. 2012).

3. Overlapping signalling pathways exist for the erasure of memory traces of pain and fear and reversal of LTP at the appropriate synaptic sites in the CNS (see Sandkühler and Lee 2013 for a review).

Not all details of signal transduction pathways that have been explored in one model have also been investigated in the other. All presently known cellular key elements are, however, shared by spinal LTP and some forms of centrally expressed hyperalgesia. This strongly suggests that LTP at synapses of primary afferent C fibers is a cellular mechanism of hyperalgesia. However, until now LTP has only been found at synapses of primary afferent C or Aδ fibers, but not at synapses of Aβ fibers. Thus, Aβ-fiber-induced allodynia can presently not be

Long-Term Potentiation and Long-Term Depression in the Spinal Cord, Fig. 1 (continued) the responses do, however, not return to predrug values (i.e., to the potentiated levels) but return to near 100 % control levels prior to conditioning stimulation. This demonstrates that either

abrupt (**e**) or tapered opioid withdrawal (*dots* in **f**) may depotentiate synaptic strength (Modified from Ikeda et al. (2006), Drdla et al. (2009), and Drdla-Schutting et al. (2012) and unpublished observations) explained by synaptic LTP in superficial spinal dorsal horn. Perhaps LTP at synapses of low-threshold mechanosensitive C-fibers (Seal et al. 2009) contributes to allodynia under some circumstances.

Synaptic Long-Term Depression in Nociceptive Pathways

Conditioning stimulation of primary afferent Aδ fibers but not Aβ fibers at low frequencies (1 Hz for 15 min) induces a homosynaptic LTD at Aδ-fiber synapses in vitro (Sandkühler et al. 1997; Chen and Sandkühler 2000; Randic et al. 1993) and a heterosynaptic LTD at synapses of C fibers in intact animals (Liu et al. 1998). Similar stimulation parameters lead to long-term depression of primary afferent-induced EPSCs in deep dorsal horn neurons (Garraway and Hochman 2001), of the jaw-opening reflex in mice (Ellrich 2004), and of human nociceptive skin senses (Nilsson et al. 2003; Klein et al. 2004). High-intensity, low-frequency forms of transcutaneous electrical nerve stimulation (TENS) can alleviate clinical pain in some human pain patients. (Electro-) Acupuncture, which leads to the "de-qi" sensation, probably also induces a low-frequency afferent barrage in Aδ fibers. When effective, pain relief outlasts the duration of these forms of TENS and acupuncture for hours or days. This is compatible with synaptic LTD in pain pathways if counterirritation is applied closely to the painful area. In contrast, TENS given at high frequencies, but at low (Aβ-fiber) intensity, does not lead to long-lasting analgesia and cannot be explained by synaptic LTD in pain pathways. This low-intensity-high-frequency form of TENS most probably involves excitation of spinal inhibitory interneurons as described in the "gate-control" theory [reviewed in (Sandkühler 2000)].

Summary

Convergent evidence suggest that long-term potentiation at or near the first central synapse in pain pathways is relevant to some forms of hyperalgesia (synaptic long-term potentiation)

and can also be used as a novel target to treat and perhaps prevent or even cure some forms of chronic pain (Ruscheweyh et al. 2011 for a review).

Acknowledgment The author's work is supported by grants from the FWF (Austrian Science Fund), the WWTF (Vienna Science and Technology Fund), and the Jubiläumsfonds der Oesterreichischen Nationalbank.

References

Barnes, C. A. (1995). Involvement of LTP in memory: are we "searching under the street light"? *Neuron 15*, 751–754.

Benrath, J., Brechtel, C., Martin, E., & Sandkühler, J. (2004). Low doses of fentanyl block central sensitization in the rat spinal cord in vivo. *Anesthesiology, 100*, 1545–1551.

Chen, J., & Sandkühler, J. (2000). Induction of homosynaptic long-term depression at spinal synapses of sensory Ad-fibers requires activation of metabotropic glutamate receptors. *Neuroscience, 98*, 141–148.

Drdla, R., & Sandkühler, J. (2008). Long-term potentiation at C-fibre synapses by low-level presynaptic activity in vivo. *Molecular Pain, 4*, 18.

Drdla, R., Gassner, M., Gingl, E., & Sandkühler, J. (2009). Induction of synaptic long-term potentiation after opioid withdrawal. *Science, 325*, 207–210.

Drdla-Schutting, R., Benrath, J., Wunderbaldinger, G., & Sandkühler, J. (2012). Erasure of a spinal memory trace of pain by a brief, high-dose opioid administration. *Science, 335*, 235–238.

Ellrich, J. (2004). Electric low-frequency stimulation of the tongue induces long-term depression of the jaw-opening reflex in anesthetized mice. *Journal of Neurophysiology, 1*, 1–2.

Garraway, S. M., & Hochman, S. (2001). Serotonin increases the incidence of primary afferent-evoked long-term depression in rat deep dorsal horn neurons. *Journal of Neurophysiology, 85*, 1864–1872.

Ikeda, H., Heinke, B., Ruscheweyh, R., & Sandkühler, J. (2003). Synaptic plasticity in spinal lamina I projection neurons that mediate hyperalgesia. *Science, 299*, 1237–1240.

Ikeda, H., Stark, J., Fischer, H., Wagner, M., Drdla, R., Jäger, T., et al. (2006). Synaptic amplifier of inflammatory pain in the spinal dorsal horn. *Science, 312*, 1659–1662.

Katz, J., Cohen, L., Schmid, R., Chan, V. W., & Wowk, A. (2003). Postoperative morphine use and hyperalgesia are reduced by preoperative but not intraoperative epidural analgesia: Implications for preemptive analgesia and the prevention of central sensitization. *Anesthesiology, 98*, 1449–1460.

Klein, T., Magerl, W., Hopf, H.-C., Sandkühler, J., & Treede, R.-D. (2004). Perceptual correlates of nociceptive long-term potentiation and long-term depression in humans. *Journal of Neuroscience, 24*, 964–971.

Kuner, R. (2010). Central mechanisms of pathological pain. *Nature Medicine, 16*, 1258–1266.

Liu, X.-G., Morton, C. R., Azkue, J. J., Zimmermann, M., & Sandkühler, J. (1998). Long-term depression of C-fibre-evoked spinal field potentials by stimulation of primary afferent Ad-fibres in the adult rat. *European Journal of Neuroscience, 10*, 3069–3075.

Luo, C., Gangadharan, V., Bali, K. K., Xie, R.-G., Agarwal, N., Kurejova, M., et al. (2012). Presynaptically localized cyclic GMP-dependent protein kinase 1 is a key determinant of spinal synaptic potentiation and pain hypersensitivity. *PLoS Biology, 10*, e1001283.

Nichols, M. L., Allen, B. J., Rogers, S. D., Ghilardi, J. R., Honoré, P., Luger, N. M., Finke, M. P., Li, J., Lappi, D. A., Simone, D. A., & Mantyh, P. W. (1999). Transmission of chronic nociception by spinal neurons expressing the substance P receptor. *Science 286*, 1558–1561.

Nilsson, H.-J., Psouni, E., & Schouenborg, J. (2003). Long term depression of human nociceptive skin senses induced by thin fibre stimulation. *European Journal of Pain, 7*, 225–233.

Park, C.-K., Lu, N., Xu, Z.-Z., Liu, T., Serhan, C. N., & Ji, R.-R. (2011). Resolving TRPV1- and TNF-α-mediated spinal cord synaptic plasticity and inflammatory pain with neuroprotectin D1. *Journal of Neuroscience, 31*, 15072–15085.

Qu, X.-X., Cai, J., Li, M.-J., Chi, Y.-N., Liao, F.-F., Liu, F.-Y., et al. (2009). Role of the spinal cord NR2B-containing NMDA receptors in the development of neuropathic pain. *Experimental Neurology, 215*, 298–307.

Randic, M., Jiang, M. C., & Cerne, R. (1993). Long-term potentiation and long-term depression of primary afferent neurotransmission in the rat spinal cord. *Journal of Neuroscience, 13*, 5228–5241.

Ruscheweyh, R., Wilder-Smith, O., Drdla, R., Liu, X.-G., & Sandkühler, J. (2011). Long-term potentiation in spinal nociceptive pathways as a novel target for pain therapy. *Molecular Pain, 7*, 20.

Sandkühler, J. (2000). Long-lasting analgesia following TENS and acupuncture: Spinal mechanisms beyond gate control. In M. Devor, M. C. Rowbotham, & Z. Wiesenfeld-Hallin (Eds.), *Proceedings of the 9th World Congress on pain* (pp. 359–369). Seattle: IASP Press.

Sandkühler, J. (2009). Models and mechanisms of hyperalgesia and allodynia. *Physiological Reviews, 89*, 707–758.

Sandkühler, J. (2012). Spinal plasticity and pain. In M. Koltzenburg & S. McMahon (Eds.), *Wall and Melzack's textbook of pain*. Churchill Livingstone: Elsevier. pp in the press.

Sandkühler, J. & Lee, J. (2013) How to erase memory traces of pain and fear. *Trends in Neurosciences*, in press.

Sandkühler, J., Chen, J. G., Cheng, G., & Randic, M. (1997). Low-frequency stimulation of afferent Ad-fibers induces long-term depression at primary afferent synapses with substantia gelatinosa neurons in the rat. *Journal of Neuroscience, 17*, 6483–6491.

Zhou, L.-J., Ren, W.-J., Zhong, Y., Yang, T., Wei, X.-H., Xin, W.-J., et al. (2010). Limited BDNF contributes to the failure of injury to skin afferents to produce a neuropathic pain condition. *Pain, 148*, 148–157.

Zhou, L.-J., Yang, T., Wei, X., Liu, Y., Xin, W.-J., Chen, Y., et al. (2011). Brain-derived neurotrophic factor contributes to spinal long-term potentiation and mechanical hypersensitivity by activation of spinal microglia in rat. *Brain, Behavior, and Immunity, 25*, 322–334.

Zhuo, M. (2011). Cortical plasticity as a new endpoint measurement for chronic pain. *Molecular Pain, 7*, 54.

Long-term Potentiation in the Spinal Cord

▶ Long-Term Potentiation and Long-Term Depression in the Spinal Cord

Loss of Consciousness Associated with Fentanyl

Definition

Loss of consciousness associated with fentanyl occurs at mean serum concentrations of 34 ng/ml.

Cross-References

▶ Postoperative Pain, Fentanyl

Loss of Olfactory Function

Definition

Also called dysosmia; anosmia describes the total loss of the sense of smell, and hyposmia describes a partial loss of olfactory function.

Cross-References

▶ Nociception in Nose and Oral Mucosa

Loss of the Myelin Sheath

▶ Demyelination

Low Back Pain

Definition

Low back pain affects almost everyone at some point in their lives. It can be acute or chronic in nature and is a primary cause of functional disability in the working population. Multiple causes exist for low back pain including degeneration of the lumbar spine and associated intervertebral discs; intervertebral disc herniation; strain of associated muscles, ligaments, and tendons; and referral from other organs and tissues.

Cross-References

▶ Chronic Low Back Pain: Definitions and Diagnosis
▶ Evoked and Movement-Related Neuropathic Pain
▶ Low Back Pain Patients, Imaging
▶ Low Back Pain, Epidemiology
▶ Spinal Fusion for Chronic Back Pain

Low Back Pain Patients, Imaging

Gary Nesbit and Pramit Phal
Department of Radiology, Oregon Health and
Science University, Portland, OR, USA

Synonyms

Diagnostic studies; Lumbar pain

Definition

Diagnostic evaluation of pain localized to the lower back, often associated with radiation in a radicular distribution.

Characteristics

The differential diagnosis for low back pain is broad, including mechanical, compressive, neoplastic, infectious, and referred visceral disease as causes. Only after thorough history and physical examination should one consider the use of imaging studies. For most adult patients under age 50 who do not exhibit signs or symptoms of systemic illness, conservative therapy without imaging is considered appropriate.

The choices available for imaging of back pain are also quite broad and often complimentary. They include the following: plain radiographs (Fig. 1), cross-sectional imaging such as computed tomography (CT) (see ▶ Single Photon Emission Computed Tomography) and ▶ magnetic resonance imaging (MRI), myelography, discography, and radionuclide bone scan. The selection of which modality is best suited for a given patient is based on what are the major considerations, such as bone or soft tissue, and what degree of detail is needed.

Plain Films

Plain radiographs are generally the first test performed, perhaps related to their ease of accessibility and low cost. A full lumbar spine series includes an anteroposterior (AP) and lateral projection, along with a spot lateral of the lumbosacral junction and bilateral oblique views (Fig. 1). AP and lateral views allow assessment of overall alignment, disk and vertebral body heights, and a gross assessment of bony density. Oblique views allow assessment of the pars interarticularis and facet joints, but are often dispensed with due to concerns of excessively irradiating gonadal tissue, particularly in females of reproductive age. If there are concerns of

Low Back Pain Patients, Imaging, Fig. 1 Lateral radiograph of the lumbar spine demonstrates a burst fracture of L1 with posterior spinal fusion with bilateral paraspinal rods and pedicle screws in T12 and L2. Severe degenerative disk changes are seen in the lower lumbar spine L3 through S1 with loss of disk height, subchondral sclerosis, and osteophyte formation

instability or ligamentous injury, lateral views in flexion and extension can be performed.

Plain radiographs have a limited role in the assessment of back pain. As many of the cases of back pain are due to mechanical causes, plain films suffer due to the inability to visualize soft tissue. Hence, disk protrusions and nerve root compression cannot be directly visualized. Only the secondary effects of disk degeneration on the adjacent vertebral body, namely, subchondral sclerosis and cysts, osteophytosis, disk space narrowing, and vacuum phenomenon, are observed. In spinal canal stenosis, only bony narrowing of the spinal canal can be seen. Plain films generally lack

sensitivity in diagnosing metastatic disease, as 50 % of the bony trabeculae must be destroyed before these lesions are visible (Sartoris et al. 1986).

The current role of plain films in assessment of lower back pain includes that of a screening test in trauma and limited dynamic assessment, by performing flexion and extension views. They are a useful screening test combined with lab tests for symptoms suggesting systemic illness and can demonstrate specific findings to complement the more sensitive but less specific studies such as MRI and radionuclide bone scan. Plain films have limited usefulness in radiculopathy, only showing bony narrowing of neural exit foramina. They do, however, remain useful in the postoperative patient, particularly in the presence of metallic hardware.

Cross-Sectional Imaging

With the advent of cross-sectional imaging, computed tomography (CT) and magnetic resonance imaging (MRI) have rapidly become mainstays in the imaging of chronic low back pain. CT utilizes x-rays and a series of detectors to measure density in an axial or helical cross-sectional envelope, whereas MRI utilizes magnetic fields and radiowaves to evaluate water concentrations and interactions within tissue (Fig. 2). Multi-planar capabilities using either technique provide the ability to interrogate the spine in sagittal, coronal, and even 3-dimensional methods. This, combined with improved contrast resolution over plain radiography, makes both techniques superior in evaluating the spinal canal, neural foramina, and paraspinal soft tissues. CT or MRI is usually reserved for those patients with lower back pain who are considering surgery, or in whom neoplastic disease is strongly suspected (Jarvik and Deyo 2002). Additional indications include those with progressive neurologic deficit, suspected cauda equina syndrome, or those who have failed to improve after 6–8 weeks of conservative management (Hicks et al. 2002). The indiscriminate use of MRI and CT may result in unnecessary intervention, which is, in part, due to the extremely high sensitivity of MRI. There is

Low Back Pain Patients, Imaging, Fig. 2 A sagittal T2-weighted MRI image of the lumbar spine. A bulge of the L4/5 disk encroaches on the central spinal canal. The relative decreased signal in the L4/5 and L5/S1 intervertebral disks is due to decreased water content, a finding of disk degeneration

a relatively high prevalence of disk bulges, protrusions, and facet arthropathy in the asymptomatic population, and the presence of these pathologies in patients with back pain may, in many cases, be coincidental (Gaensler 1999). These findings underscore the need for judicious use of diagnostic imaging when evaluating any patient with low back pain.

Pediatric patients, in contrast, rarely have degenerative disease. These patients usually have more serious underlying disease that should be assessed differently and often involves advanced imaging (Osborne 1994).

Determining whether to use CT or MRI in a given patient is usually by physician preference; however, there are situations when one modality is favored over another. Patients with

certain implanted devices or claustrophobia may prohibit the use of MRI, although metallic artifacts from spinal surgical implants are often worse on CT than MRI. CT provides greater bone detail than MRI and is excellent for evaluating neural foraminal and central canal stenosis (Gaensler 1999). The sensitivity and specificity of disk herniation is slightly higher with MRI than CT, but the two modalities are essentially equally effective in evaluating the degree of spinal stenosis (Jarvik and Deyo 2002). MRI has demonstrated superiority in evaluating for neoplasm, with sensitivity and specificity ranging from 0.83 to 0.93 and from 0.9 to 0.97, respectively (Jarvik and Deyo 2002). Infection is also best evaluated by MRI with sensitivity of 0.96 and specificity of 0.92 (Jarvik and Deyo 2002). In patients who have had prior surgery, MRI is the preferred modality. These patients should always be imaged with gadolinium contrast material to help distinguish scar tissue from recurrent disk herniation (Helms 1999). MRI has also been shown to be particularly useful as a screening tool in evaluating patients with symptomatic disks by identifying tears of the annulus fibrosus with high sensitivity; the relative low specificity, however, cannot replace the need for discography (Yoshida et al. 2002).

Myelography

When intrathecal contrast material is administered for myelography, the contents of the spinal canal and neural foramina are more clearly visualized (Fig. 3). When combined with CT, the additional anatomic information can provide a crucial supplement in the evaluation of nerve root impingement in the lateral recess as a cause for radiculopathy, because MRI tends to underestimate the degree of stenosis in this location (Bartynski and Lin 2003). Although myelography is invasive, and its use has decreased with the advent of MRI, it still remains an important technique when other modalities remain equivocal.

Nuclear Medicine

A radionuclide bone scan involves intravenous injection of a radionuclide taken up by bone,

Low Back Pain Patients, Imaging, Fig. 3 A frontal (**a**) and oblique (**b**) projection of the lumbar spine is shown following the intrathecal injection of contrast for a myelogram. The nerve roots of the cauda equina are well visualized. Surgical changes of posterior laminectomy and intervertebral disk prosthesis at L4–L5 are present. Extrinsic compression of the contrast-filled thecal sac is noted above the L4–L5 disk level. A CT scan of the same patient at L3 (**c**, *upper*) shows contrast outlining the normal nerve roots in the canal and L4 (**c**, *lower*); there are midline fracture, bilateral laminectomy, and the compression of the thecal sac. A sagittal reconstruction of the CT scan (**d**) delineates the disk prosthesis, thecal sac, and soft-tissue compression of the cauda equina

most commonly bound to a compound containing phosphate. Imaging is performed with a gamma camera and the use of cross-sectional techniques such as SPECT (single-photon emission computed tomography) can increase sensitivity and anatomic localization. Early imaging in the blood flow and blood pool phases detects areas of hyperemia. Delayed images demonstrate bony uptake. This imaging modality is most useful in conditions where there is increased bony turnover: osteoblastic metastases, infection, fractures, in particular in the diagnosis of stress fractures, and differentiating acute from chronic vertebral body compression fractures (Avrahami et al. 1989). While radionuclide bone scan has excellent sensitivity for detection of these conditions, it suffers from a lack of specificity and involves exposure to ionizing radiation.

Discography

Discography is a semi-invasive procedure that involves placement of a needle into the central nucleus pulposus of the disk and injection of iodinated contrast material directly into the nucleus.

Information gained from this procedure includes a morphological assessment of the disk. The procedure is most often done under fluoroscopy during which spot images are taken. CT of the lumbar spine is often performed after the procedure, including sagittal and coronal reformatted images. Degenerative change can range from fissures and clefts in the nucleus pulposus, annular tears with contrast extending to the disk periphery, and disk rupture with leakage of injected contrast (Anderson 2004; Sachs et al. 1987).

The patient's pain response during the procedure is also assessed. During the procedure the patient is asked to grade the severity pain (usually out of 5) and also the similarity of the pain to their usual symptoms.

Discography is very sensitive for detection of annular tears of the intervertebral disk, a common cause of back pain not confidently detected on MRI. Disparity between MRI and clinical presentation is a major indication for discography. Discography should only be performed on surgical candidates who have failed a period of conservative therapy of around 6 months. Other indications for discography include determining the symptomatic level in multilevel degenerative disease, preoperative planning, and evaluation of the postoperative patient.

Due to the semi-invasive nature of the procedure, potential risks of discography include infection, bleeding, neural damage, accentuation of pain, and postprocedural headache if the dura is traversed. The other main disadvantage of discography is its high false-positive rate, which is likely to be around 10 % (Saal 2002).

References

Anderson, M. W. (2004). Lumbar discography: An update. *Seminars in Roentgenology, 39*, 52–67.

Avrahami, E., Tadmor, R., Dally, O., et al. (1989). Early MR demonstration of spinal metastases in patients with normal radiographs and CT and radionuclide bone scans. *Journal of Computer Assisted Tomography, 13*, 598–602.

Bartynski, W. S., & Lin, L. (2003). Lumbar root compression in the lateral recess: MR imaging, conventional myelography, and CT myelography comparison with surgical confirmation. *American Journal of Neuroradiology, 24*, 348–360.

Gaensler, E. (1999). Nondegenerative diseases of the spine. In W. Brant (Ed.), *Fundamentals of diagnostic radiology* (pp. 233–280). Baltimore: Williams and Wilkins.

Helms, C. (1999). Lumbar spine disc disease and stenosis. In W. Brant (Ed.), *Fundamentals of diagnostic radiology* (pp. 281–288). Baltimore: Williams and Wilkins.

Hicks, G. S., Duddleston, D. N., Russell, L. D., et al. (2002). Low back pain. *The American Journal of the Medical Sciences, 324*, 207–211.

Jarvik, J. G., & Deyo, R. A. (2002). Diagnostic evaluation of low back pain with emphasis on imaging. *Annals of Internal Medicine, 137*, 586–597.

Osborne, A. (1994). *Nonneoplastic disorders of the spine and spinal cord. Neuroradiology* (pp. 820–875). St. Louis: Mosby.

Saal, J. S. (2002). General principles of diagnostic testing as related to painful lumbar spine disorders: A critical appraisal of current diagnostic techniques. *Spine, 27*, 2538–2546.

Sachs, B. L., Vanharanta, H., Spivey, M. A., et al. (1987). Dallas discogram description. A new classification of CT/discography in low-back disorders. *Spine, 12*, 287–294.

Sartoris, D. J., Clopton, P., Nemcek, A., et al. (1986). Vertebral-body collapse in focal and diffuse disease: Patterns of pathologic processes. *Radiology, 160*, 479–483.

Yoshida, H., Fujiwara, A., Tamai, K., et al. (2002). Diagnosis of symptomatic disc by magnetic resonance imaging: T2-weighted and gadolinium-DTPA-enhanced T^1-weighted magnetic resonance imaging. *Journal of Spinal Disorder Technology, 15*, 193–198.

Low Back Pain, Epidemiology

Jenna E. Zauk[1], J. Mocco[2] and
Christopher J. Winfree[2]
[1]St. George's University, School of Medicine,
Grenada, West Indies
[2]Department of Neurological Surgery,
Neurological Institute Columbia University,
New York, NY, USA

Synonyms

Factors associated with low back pain; Frequency
of low back pain; Incidence of low back pain;
Pervasiveness of low back pain; Prevalence of
low back pain

Definition

The word epidemiology comes from the Greek
words epi, "on or upon," demos, "the common
people," and logy, "study." Taken from these
roots, one can extrapolate epidemiology to be
the study of that which is upon the people.
According to the American College of
Epidemiology, epidemiology is "the study of
the distribution and determinants of disease
risk in human populations." Therefore, the
epidemiology of low back pain is the study of
the distribution and determinants of low back
pain.

Characteristics

The lumbar region of the spine (L1–L5) is
a common site for lower back pain because it
supports the weight of the upper body (National
Institute of Neurological Disorders and Stroke
2010). Disability from back pain is more common than any other cause of activity limitation in
adults of less than 45 years, and second only to
arthritis in people aged 45–65 years (Frank et al.
1996). Currently, it is the second leading

neurological illness of men and women in the
United States (National Institute of Neurological
Disorders and Stroke 2010). Not only is low back
pain (LBP) the leading cause of missed workdays, it is also becoming a bigger problem every
year. In fact, over the past 50 years, the number of
workdays missed secondary to LBP has increased
more than ten-fold (Clinical Standards Advisory
Group 1994). Nineteen percent of all disability
claims in the US work force are due to low back
pain. A recent review found that on a yearly basis
approximately 500,000 US employees are compensated for back injuries, and 19 % of all disability claims are specifically for low back pain
(Wheeler 2010). Therefore, efforts to better
understand the epidemiology of low back pain
are critical.

Prevalence is the number of people in
a defined population who have a specific disease
or condition at a particular time (Jekel 1996).
Unfortunately, because of the recurrent nature
and variable definitions of LBP that are used
from study to study, traditional epidemiological
concepts are difficult to apply to the experience
of low back pain. A recent comprehensive methodological review of the body of literature
concerning the prevalence of LBP found that
the 1-year prevalence ranged from 3.9 to 63 %
with a mean of 32 %, and the prevalence at
a single point in time ranged from 4.4 to 33 %
(Loney and Stratford 1999). If only the highest
quality studies were used, then the range
narrowed to 13.7–28.7 % (Loney and Stratford
1999).

Incidence is the occurrence of new cases of
disease in a candidate population over a specific
time period (Aschengrau 2005). True incidence is
essentially impossible to determine for LBP,
since onset is often difficult to determine and
recurrent episodes confound investigators" ability to identify true "new cases." The incidence of
LBP has been estimated to vary between 4 % and
15 % annually in most industrialized countries
(Leighton and Reilly 1995; Office of National
Statistics 1996).

After attempting to understand the general
distribution of LBP, the next step is to better
establish the determinants of LBP. In other

words, what factors are associated with the occurrence of LBP? There have been many studies undertaken to attempt to tease out the relevant factors associated with LBP in both patients" workplace and personal environments. Unfortunately, much of this literature is poorly designed and/or contains conflicting results.

In addition to injuries and degenerative conditions, there are also personal risk factors associated with LBP such as obesity, smoking, severe scoliosis, and depression (Devereaux 2003a; Devereaux 2003b). An individual's general physical fitness has not been shown to reduce the occurrence of acute LBP, but it has been shown to lower the incidence of chronic LBP and reduce the time needed to recover from episodes of acute LBP (Devereaux 2004). Despite some debate, the greater weight of evidence supports no association between gender and LBP (Devereaux 2004).

In the workplace, a number of characteristics have been identified as risk factors for the occurrence of LBP or a worse disease course. These can be divided into physical and psychosocial characteristics. Accepted physical risk factors include heavy physical labor, long static work postures, and excessive vibration (Devereaux 2003a, b). Psychosocial aspects of the workplace that affect LBP were reviewed in a systematic review of the literature, in which there was strong evidence for minimal workplace social support and low job satisfaction being associated with back pain. Additionally, financial and emotional stress can induce LBP. Psychological disturbances, such as depression and anxiety, may alter one's perception of pain and may also exacerbate the physiological systems that contribute to pain (Wheeler 2010). In contrast, other commonly held risk factors, such as a high work pace, job autonomy, and long periods of attentiveness, were not strongly associated with LBP, despite some weak supporting literature. Finally, a study found that monotonous work with few possibilities to learn new skills was associated with LBP in blue-collar workers (Leino and Hanninen 1995).

A final risk factor for LBP recurrence and conversion from acute to chronic status is the presence of litigation and/or workman's compensation claims. Litigation deserves special consideration as multiple studies have been performed to elucidate the association between LBP and litigation and/or workman's compensation claims, and the majority of these have found a positive association (Blake and Garrett 1997; Rainville et al. 1997; Valat et al. 1997). These include a well-designed prospective observational cohort study of 192 individuals, determining that patients with compensation involvement appear to have worse depression and disability before and after rehabilitation interventions (Rainville et al. 1997).

Two recent studies (a clinical trial, n = 501 and an observational study, n = 743) conducted by the Spinal Patient Outcomes Research Trial (SPORT) attempted to determine which mode of treatment for radicular lumbar disk pain, surgical or nonsurgical, provides a better outcome for patients. All of the subjects were suffering from pain>6 weeks prior to the study. Both operative and nonoperative care resulted in significant pain reduction. The authors concluded that there were no significant differences between the two treatment modalities after 1–2 years (Weinstein et al. 2006a, b).

The long-term effectiveness of invasive surgery or local injections for nonspecific chronic low back pain remains uncertain. Yet, other research demonstrates that bed rest is a very useful and appropriate remedy for acute back pain. It has been shown that two days versus seven days of bed rest provide the same therapeutic relief. This short recovery period (< 2 days) also allows for a quicker return to the workplace (Wheeler 2010; Deyo et al. 1986). However, these treatments do not resolve the psychosocial and environmental issues (as mentioned before) associated with lower back pain (Chou et al. 2009).

In conclusion, the study of the epidemiology of low back pain has determined that roughly a third of the adult population will suffer low back pain over the course of a year. Many of these individuals will be obese, smoke, suffer from depression, experience financial and emotional stress, or they will be employed in careers that expose them to physical labor, long static work postures, excessive vibration, low workplace social support, or low job satisfaction. Of those who suffer low back pain, those with either litigation or workers' compensation involvement are more likely to have a chronic course with worse outcomes and more workdays missed. Further research is necessary to understand the epidemiology of low back pain and provide appropriate treatment.

References

Aschengrau, S. (2005). *Essentials of epidemiology in public health*. Sudbury: Jones and Bartlett.

Blake, C., & Garrett, M. (1997). Impact of litigation on quality of life outcomes in patients with chronic low back pain. *Irish Journal of Medical Science, 166*, 124–126.

Chou, R., Loeser, J., Owens, D., Rosenquist, R., Atlas, S., Baisden, J., Carragee, E., Grabois, M., Murphy, D., Resnick, D., Stanos, S., Shaffer, W., & Wall, E. (2009). Interventional therapies, surgery, and interdisciplinary rehabilitation for low back pain, an evidence-based clinical practice guideline from the American pain society. *Spine, 34*, 1066–1077.

Clinical Standards Advisory Group. (1994). *Epidemiology review: The epidemiology and cost of back pain*. Clinical Standards Advisory Group. Back pain. London, UK: HMSO.

Devereaux, M. W. (2003a). *Approach to neck and low back disorders. ERW Saunders manual of neurologic practice* (pp. 745–751). Philadelphia: Elsevier.

Devereaux, M. W. (2003b). Neck and low back pain. *The Medical Clinics of North America, 87*, 643–662.

Devereaux, M. W. (2004). Low back pain. *Primary Care, 31*, 33–51.

Deyo, R. A., Diehl, A. K., & Rosenthal, M. (1986). How many days of bed rest for acute low back pain? A randomized clinical trial. *The New England Journal of Medicine, 315*(17), 1064–1070.

Frank, J. W., Kerr, M. S., Brooker, A., DeMaio, S. E., Maetzel, A., Shannon, H. S., Sullivan, T. J., Norman, R. W., & Wells, R. P. (1996). Disability resulting from occupational LBP: I: What do we know about primary prevention? A review of the scientific evidence on the prevention before disability begins. *Spine, 21*, 2908–2917.

Jekel, E. K. (1996). *Epidemiology, biostatistics, and preventative medicine* (pp. 20–22). Toronto, Canada: WB Saunders.

Leighton, D. J., & Reilly, T. (1995). Epidemiological aspects of back pain: The incidence and prevalence of back pain in nurses compared to the general population. *Occupational Medicine, 45*, 263–267.

Leino, P. I., & Hanninen, V. (1995). Psychosocial factors at work in relation to back and limb disorders. *Scandinavian Journal of Work, Environment & Health, 21*, 134–142.

Loney, P. L., & Stratford, P. W. (1999). The prevalence of low back pain in adults: A methodological review of the literature. *Physical Therapy, 79*, 384–396.

National Institute of Neurological Disorders and Stroke. June 14, 2010 http://www.ninds.nih.gov/disorders/backpain/detail_backpain.htm#157883102

Office of National Statistics. (1996). *Omnibus survey. Social survey division*. London, UK: Department of Health.

Rainville, J., Sobel, J. B., Hartigan, C., & Wright, A. (1997). The effect of compensation involvement on the reporting of pain and disability by patients referred for rehabilitation of chronic low back pain. *Spine, 22*, 2016–2024.

Valat, J. P., Goupille, P., & Vedere, V. (1997). Low back pain: Risk factors for chronicity. *Revue du Rhumatisme (English Ed.), 64*, 189–194.

Weinstein, J. N., Tosteson, T. D., Lurie, J. D., Tosteson, A. N., Hanscom, B., & Skinner, J. S. (2006a). Surgical vs nonoperative treatment for lumbar disk herniation: The spine patient outcomes research trial (SPORT): A randomized trial. *Journal of the American Medical Association, 296*(20), 2441–2450.

Weinstein, J. N., Lurie, J. D., Tosteson, T. D., Skinner, J. S., Hanscom, B., & Tosteson, A. N. (2006b). Surgical vs nonoperative treatment for lumbar disk herniation: The spine patient outcomes research trial (SPORT) observational cohort. *Journal of the American Medical Association, 296*(20), 2451–2459.

Wheeler AH. May 2010 – http://emedicine.medscape.com/article/1144130-overview

Low First-Pass Metabolism

Definition

Oral oxycodone undergoes low first-pass metabolism and is among those opioids possessing a short elimination half-life.

Cross-References

▶ Postoperative Pain, Oxycodone

Lower Back Pain

▶ Sacroiliac Joint Pain

Lower Back Pain, Acute

Stephan A. Schug[1] and T. Shah[2]
[1]School of Medicine and Pharmacology, Royal
Perth Hospital, Faculty of Medicine, Dentistry &
Health Sciences, The University of Western
Australia, UWA Anaesthesiology, Perth, WA,
Australia
[2]Royal Perth Hospital and University of
Western Australia, Crawley, WA, Australia

Synonyms

Acute backache; Acute lumbago; Acute sciatica;
Bad back

Definition

Acute lower back pain refers to spinal pain of
sudden onset originating from lumbar, sacral, or
lumbosacral areas. It may also be referred pain
from other areas such as the sacroiliac joints
(www.emia.com.au/MedicalProviders/Evidence
BasedMedicine/overview.html). Although often
used to mean lower back pain, sciatica spe
cifically refers to pressure on the sciatic nerve
causing pain passing down the posterior medial
aspect of the leg.

Characteristics

Lower back pain is extremely common, with
60–80 % of adults experiencing it at some time.
It is the most common reason cited for absence
from work, and its management, therefore,
has strong socioeconomic implications. Ninety
percent of acute lower back pain,
however, will resolve within 6 weeks (www.
emia.com.au/MedicalProviders/EvidenceBased-
Medicine/overview.html, http://www.acc.co.nz/
PRD_EXT_CSMP/groups/external_ip/documents/
internet/wcm002131.pdf, http://www.nhmrc.gov.
au/_files_nhmrc/publications/attachments/cp94.
pdf.). The underlying diagnosis is often
unknown. There is level I evidence that
common findings in patients with lower back
pain (e.g., osteoarthritis, spondylosis, or spinal
canal stenosis) may also be present in asymptomatic patients and do not necessarily represent
the cause of the pain (http://www.nhmrc.gov.au/
_files_nhmrc/publications/attachments/cp94.pdf).

Due to its frequent occurrence, it is essential to
identify early those rare causes of lower back pain
that may have an underlying serious pathology or
be non-biomechanical in origin. These should
prompt early referral to appropriate treatment
that might be urgent. A thorough history and simple examination largely identify these so-called
red flags. Red flags include neurological signs
and symptoms, a history of significant trauma,
weight loss and other signs of systemic illness,
and severe unrelenting pain, especially in those
at risk of pathological fractures. Conditions identified in such cases include cauda equina syndrome, fractures (traumatic or pathological),
tumors, and infection. In patients with no red
flags, imaging is largely unnecessary.

"Yellow flags" are also discussed with respect
to acute lower back pain. These are behavioral
and psychosocial factors that make progression to
chronic back pain more likely and that make
rehabilitation from acute back pain difficult and/
or undesirable. Problems such as depression,
poor job satisfaction, and a belief that physical
activity can cause further damage fall under this
heading. Early identification and intervention can
alter this course and ultimately reduce chronic
disability (www.emia.com.au/MedicalProviders/
EvidenceBasedMedicine/overview.html).

Management of simple mechanical lower back
pain has altered significantly in the last
few decades. There is now very good evidence
that previous recommendations of prolonged
bed rest are detrimental and that staying
active and an early return to work are beneficial
(www.emia.com.au/MedicalProviders/Evidence-
BasedMedicine/overview.html, http://www.acc.
co.nz/PRD_EXT_CSMP/groups/external_ip/docu-
ments/internet/wcm002131.pdf, http://www.nhmr
c.gov.au/_files_nhmrc/publications/attachments/
cp94.pdf.). Manipulative techniques in the first
4–6 weeks have been shown to improve outcome

(http://www.acc.co.nz/PRD_EXT_CSMP/groups/
external_ip/documents/internet/wcm002131.pdf,
Meade et al. 1995), as do simple analgesia (e.g.,
paracetamol) and heat wraps (http://www.nhmrc.
gov.au/_files_nhmrc/publications/attachments/
cp94.pdf.). The evidence for the use of a nonster
oidal anti-inflammatory drug is conflicting, and
side effects (in particular gastric mucosal ulcer
ation) are common with long-term use (http://
www.nhmrc.gov.au/_files_nhmrc/publications/
attachments/cp94.pdf.). Manipulation under
anesthesia and the use of muscle relaxants
(e.g., diazepam) have been shown to be
harmful (http://www.acc.co.nz/PRD_EXT_CSMP/
groups/external_ip/documents/internet/wcm002131.
pdf). Use of oral opioid analgesia may be necessary
to relieve pain in acute stages. If used, a short-acting
regular drug is preferable for as brief a period
as possible. The ongoing need for use should
prompt a reassessment of the patient (http://www.
nhmrc.gov.au/_files_nhmrc/publications/attach-
ments/cp94.pdf.).

Patients should have a time frame for their
progress and be educated about what to expect.
They should be aware that they play an active role
in their rehabilitation and should take this on as
their responsibility. Those not responding to
treatment at 4–12 weeks should be reassessed
and considered for specialist referral. They
might constitute a group of patients with
increased risk of chronification and subsequent
complex biopsychosocial consequences.

References

Evidence-Based Management of Acute Musculoskeletal
    Pain-Australian   Acute   Musculoskeletal   Pain
    Guideline   Group.   http://www.nhmrc.gov.au/
    _files_nhmrc/publications/attachments/cp94.pdf.
Meade, T. W., Dyer, S., Browne, W., et al. (1995).
    Randomised comparison of chiropractic and hospital
    outpatient management for low back pain: Results
    from extended follow-up. British Medical Journal,
    311, 349–351.
Overview of clinical guidelines for the management of acute
    lower back pain. www.emia.com.au/MedicalProviders/
    EvidenceBasedMedicine/overview.html.
The New Zealand acute lower back pain guide. http://
    www.acc.co.nz/PRD_EXT_CSMP/groups/external_
    ip/documents/internet/wcm002131.pdf

Lower Back Pain, Physical Examination

Jenna E. Zauk[1], Patrick B. Senatus[2] and
Christopher J. Winfree[2]
[1]St. George's University, School of Medicine,
Grenada, West Indies
[2]Department of Neurological Surgery,
Neurological Institute Columbia University,
New York, NY, USA

Synonyms

Distraction signs; Inflammatory syndrome;
Lumbago; Lumbosacral; Mechanical syndrome;
Myofascial syndrome; Neural compressive syndrome; Neurogenic claudication; Neuropathic
syndrome; Paravertebral muscle; Provocative
maneuver; Psychosocioeconomic syndrome;
Radiculopathy; Sacroiliac joint; Schrober's test;
Sciatic notch; Spondylolisthesis; Straight-leg-
raising sign

Definition

Component of the clinical evaluation focusing on
the region between L1 above and S1 below and
the paravertebral erector spinae muscles laterally
(Merskey and Bogduk 1994), relying on inspection palpation, and physical maneuvers to elicit
signs and reproduce symptoms in the diagnosis of
low back pathology.

Characteristics

Goals
The physical examination of the lower back is an
essential part of the evaluation of disorders stemming from this region, particularly lumbago and
sciatica. For sciatica, the pinching of the sciatic
nerve can be caused by a tumor, herniated or
ruptured disk, cyst, metastatic disease, or

degeneration of the nerve. In severe cases of sciatica, the patient experiences numbness and loses the ability to control leg movement (National Institute of Neurological Disorders and Stroke 2010). The diagnosis of back pain is complex and must distinguish among disparate entities such as myofascial syndrome, in which pain is derived from ligamentous and muscular soft tissue; mechanical syndrome, in which instability and abnormal motion across joints and ligaments causes pain; neural compressive syndrome, in which pain derives from pressure on individual nerve roots or the cauda equina; inflammatory syndrome, in which pain is derived from inflammatory diseases such as osteomyelitis, diskitis, or ankylosing spondylitis; osteogenic syndrome, in which pain is derived from bone itself such as with compression fractures and bony metastasis; neuropathic syndrome, in which pain is of neural origin arising from permanent injury to peripheral nerves, nerve roots, or spinal cord; and finally psychosocioeconomic syndrome, in which a major contributor to pain is personality dysfunction, depression, or secondary gain. Therefore, when presented with a patient for evaluation, the clinician must keep in mind the broad differential diagnosis from which to select the etiology of the patient's ailment (Kirkaldy-Willis 1983).

After a focused detailed history of present illness is obtained, a directed physical examination is undertaken. It should proceed in a fashion as to test, rule out, and distinguish among items on the differential diagnosis. Moreover, it should be used in conjunction with the history to guide and justify further examination including static and dynamic imaging, serological studies, and invasive provocative tests such as discography.

Inspection

The examination begins with a general inspection to determine the patient's overall state of comfort, pain, posture, and gait. The inability to achieve a stable comfortable position while standing or seated may speak to the intensity and nature of the cause of the disease.

A persistently flexed posture may be reflective of neural compressive syndrome, as that would tend to alleviate pressure on neural elements that would be exacerbated by hyperextension. Asymmetric gait, or favoring of one extremity, may point to lateralizing radiculopathy. Neurogenic claudication, manifesting as bilateral diffuse leg pain on standing or walking, is indicative of spinal stenosis.

From a lateral vantage point, the transition from the thoracic kyphotic curve to the lumbar lordotic curve must be assessed. Deviation from the expected degrees of curvature may be indicative of pathology. Loss of lumbar lordosis may be a sign of lumbar spondylopathy as the lumbar spine attempts to straighten itself to minimize neural compression and deleterious force vectors. Exaggerated lumbar lordosis may indicate spondylolisthesis. From a posterior vantage point, any lateral or scoliotic curve should be noted. The spinous processes for T1 down should lie along a straight line that should extrapolate through the gluteal cleft. Disparities with respect to shoulder level, the level of the iliac crests, and symmetry of gluteal folds may result from scoliosis or other vertebral abnormalities. Asymmetry of paraspinal musculature may indicate spasm or underlying scoliosis.

Tufts of hair, melanotic or vitiliginous patches, lipomas, skin dimples, or draining sinuses should raise the suspicion for occult spinal dysraphism that may require further investigation.

Inspection under dynamic conditions can yield important diagnostic data. As a patient rotates the shoulder with respect to the hips, bends forward, backward, and laterally, the range of motion, smoothness, symmetry, and changes in the lumbar curvature should be noted. Decreased mobility may point to ankylosing spondylitis, osteoarthritis, or paravertebral muscle spasm as causes of pain.

Schrober's test, in which the change in distance between two points (one 10 cm above S1 and one 5 cm below S1) during maximal forward bending, can be used to quantify restricted mobility. An increased distance of less than 5 cm is abnormal and may point to ankylosing spondylitis (Karnath 2003).

Exacerbation of symptoms on forward bending may point to myofascial syndrome as an etiology. Amelioration of symptoms on forward bending and exacerbation on hyperextension suggest spinal stenosis. Asymmetric exacerbation on lateral bending or rotation may point to paracentral or laterally herniated disk.

Palpation

Palpating the lower back in standing, sitting, and prone positions can be informative. Feeling the spines and spinous interspaces can help discern relationships between adjacent vertebrae. Unusual prominence of a spinous process may be a sign of spondylolisthesis or retrolisthesis of one body with respect to the other. Unusual prominence of an interspace may signal disruption or incompetence of posterior spinal elements. If palpation or gentle percussion of the spine causes pain, this raises the specter of fracture, malignancy, diskitis, or osteomyelitis, which may require further workup. Tenderness of the paravertebral muscle, spinous process, intervertebral joint, and sacroiliac notch may be indicative of intervertebral disk herniation.

Provocative Maneuvers

Provocative maneuvers are helpful in determining the etiology of low back pain. The ipsilateral and crossed straight-leg-raising (iSLR and cSLR) tests should be performed in patients in whom radiculopathy is suspected. With the patient lying supine, a positive iSLR sign is obtained if elevation of the ipsilateral lower extremity to between 30° and 60° reproduces sciatica. With the patient in the same position, a positive cSLR sign is obtained if elevation of the contralateral lower extremity reproduces sciatica in the ipsilateral lower extremity. Ipsilateral SLR sign has high sensitivity but low specificity whereas the crossed SLR has both high sensitivity and specificity. These tests can be performed in the seated position to elicit distraction signs used to rule out overlay.

Psychosocioeconomic Syndrome

Psychological conditions heavily influence the presence of chronic and acute back pain (Linton 2000). As depression, personality dysfunction, somatization, and secondary gain can affect the experience of disease and how it is expressed, the clinician should be cognizant of inconsistencies and non-physiologic findings that suggest supratentorial overlay or psychosocioeconomic syndrome. Overreaction during physical examination, inappropriate tenderness that is too generalized or too superficial, pain on simulated but not actual rotation of the spine or on simulated axial loading, regional impairment of strength or sensation that fails to conform to radicular or dermatomal distributions, and positive distraction signs, such as inconsistent performance during distracting maneuvers, can help the clinician identify psychosocioeconomic syndrome (Waddell et al. 1980).

References

Karnath, B. (2003). Clinical signs of low back pain. *Hospital Physician, 39*, 39–44.

Kirkaldy-Willis, W. H. (1983). The pathology and pathogenesis of low back pain. In W. H. Kirkaldy-Willis (Ed.), *Managing low back pain* (pp. 23–44). Philadelphia: Churchill Livingstone.

Linton, S. J. (2000). A review of psychological risk factors in back and neck pain. *Spine, 25*(9), 1148–1156.

Merskey, H., & Bogduk, N. (1994). *Classification of chronic pain: Descriptions of chronic pain syndromes and definitions of pain terms* (2nd ed., pp. 11–36). Seattle: IASP Press.

National Institute of Neurological Disorders and Stroke. June 14, 2010 http://www.ninds.nih.gov/disorders/backpain/detail_backpain.htm#157883102

Waddell, G., McCulloch, J. A., Kummel, E., & Venner, R. M. (1980). Nonorganic physical signs in low-back pain. *Spine, 5*, 117–125.

Lower Gastrointestinal Tract Pain Models

▶ Visceral Pain Model, Lower Gastrointestinal Tract Pain

Low-Frequency Oscillations

Definition

Oscillatory activity described in the brain when the organism is not attentive or drowsy.

Cross-References

▶ Corticothalamic and Thalamocortical Interactions

Low-Intracranial-Pressure Headache

▶ Headache Due to Low Cerebrospinal Fluid Pressure

Low-Level Laser Therapy

▶ Laser

Low-threshold Calcium Channels

▶ Calcium Channels in the Spinal Processing of Nociceptive Input

Low-Threshold Calcium Spike

Synonyms

LTS

Definition

A membrane depolarization resulting from an inflow of Ca^{++} ions through T-type channels. The LTS occurs when the membrane potential is sufficiently hyperpolarized to cause deinactivation of the T-channels. If the depolarization exceeds the threshold for generating action potentials, then the resulting short train of spikes, known as a calcium spike burst or LTS burst, has a characteristic pattern allowing it to be identified as a likely calcium spike-generated burst in extracellular recordings.

Cross-References

▶ Burst Activity in Thalamus and Pain

Low-threshold Calcium Spike (LTS) Burst

▶ Burst Activity in Thalamus and Pain

Low-threshold Calcium Spike Burst

Synonyms

LTS burst

Definition

Bursting discharges produced by thalamic and reticular cells, when their membranes are hyperpolarized by either disfacilitation or overinhibition. They are due to the deinactivation of calcium T-channels causing the appearance of a large calcium spike, itself inducing a burst of sodium action potentials.

Cross-References

▶ Burst Activity in Thalamus and Pain
▶ Thalamotomy for Human Pain Relief

Low-Threshold Calcium Spike Bursting (LTS)

▶ Burst Firing Mode

Low-Threshold Neuron

Synonyms

LT neurons

Definition

LT neurons are best activated by innocuous mechanical stimuli.

Cross-References

▶ Functional Changes in Sensory Neurons Following Spinal Cord Injury in Central Pain
▶ Postoperative Pain, COX-2 Inhibitors
▶ Spinothalamic Input: Cells of Origin (Monkey)

Low-threshold VDCCs

▶ Calcium Channels in the Spinal Processing of Nociceptive Input

Low-voltage Calcium Channels

▶ Calcium Channels in the Spinal Processing of Nociceptive Input

LSN

▶ Lateral Spinal Nucleus

LSP

▶ Laser Silent Period

LT Neurons

▶ Low-Threshold Neuron

LTP

▶ Long-Term Potentiation

LTR

▶ Local Twitch Response

LTR Locus

▶ Sensitive Locus

LTS

▶ Low-Threshold Calcium Spike

LTS Burst

▶ Low-Threshold Calcium Spike Burst

Lumbago

Definition

Pain in the back occasionally radiating into the buttocks.

Cross-References

▶ Chronic Back Pain and Spinal Instability
▶ Lower Back Pain, Physical Examination
▶ Radiculopathies

Lumbar Disc Stimulation

▶ Lumbar Discography

Lumbar Discogram

▶ Lumbar Discography

Lumbar Discography

David Diamant
Neurological and Spinal Surgery, Lincoln,
NE, USA

Synonyms

Lumbar disc stimulation; Lumbar discogram;
Provocation discography; Provocative
discography

Definition

Lumbar discography is a diagnostic procedure
designed to test if a patient's pain arises from
a lumbar intervertebral disc. It involves injecting
contrast medium into the disc in an effort to
reproduce the patient's pain.

Characteristics

Principles

Since they are innervated, the lumbar
intervertebral disc can be a source of low back
pain (Bogduk 1983). However, there are no conventional means, such as ▶ musculoskeletal
examination or medical imaging, whereby it can
be determined if a particular disc is painful or not.
For this reason, lumbar discography was adopted
and used as the only available means to test if
a disc was painful.

The test is analogous to palpation for tenderness, but because lumbar discs are inaccessible to
palpation, needles must be used. The needle is
used to stress the disc by injecting contrast
medium into its nucleus. Contrast medium will
outline the internal structure of the disc. This
process is described as discography. It verifies
that the disc has been correctly injected, but is
not the critical component of the test. Critical is
whether or not stressing the disc reproduces the
patient's accustomed pain. This phase is called
disc stimulation.

Technique

With the patient prone on the X-ray table and
their lumbar area prepped in sterile fashion,
a needle is inserted into the target disc via
a posterolateral approach through the skin and
muscles of the back. Once the needle has been
correctly placed (Fig. 1), a pressure transducer is
connected to the needle, and contrast dye, mixed
with antibiotic, is injected into the nucleus
pulposus in a relatively rate-controlled fashion
(Fig. 2). The flow of contrast medium, the pressure within the disc during injection, and the
patient's response to injection are all simultaneously monitored. The nucleus pulposus of
a typical lumbar disc may contain between 1.5
and 2.5 cc of such solution (Aprill 1991). Spread
of contrast medium from the nucleus pulposus
into the annulus fibrosus may be associated with
the onset of pain. It is the pain response, in accordance with a visualized annular spread of dye,
which is of critical importance when interpreting
the study. When diagnosing the lumbar disc as
the pain generator, injection of the disc must
provoke reproduction of the patient's typical pattern of pain (concordant pain). Stimulation of an
asymptomatic disc will typically not cause
such pain, or it will cause pain in an atypical
(non-concordant) pattern. Pain patterns may be

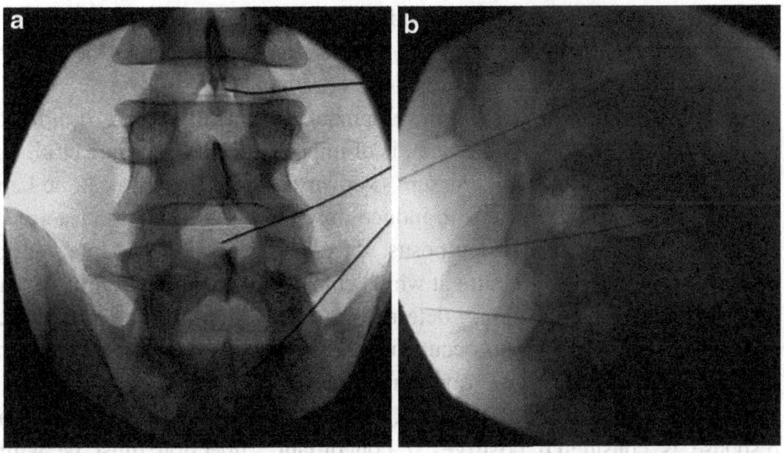

**Lumbar Discography,
Fig. 1** Radiographs of
needles placed for L3, L4,
and L5 discography. (**a**) AP
view. (**b**) Lateral view
(Reproduced courtesy of
the International Spinal
Intervention Society
(2004))

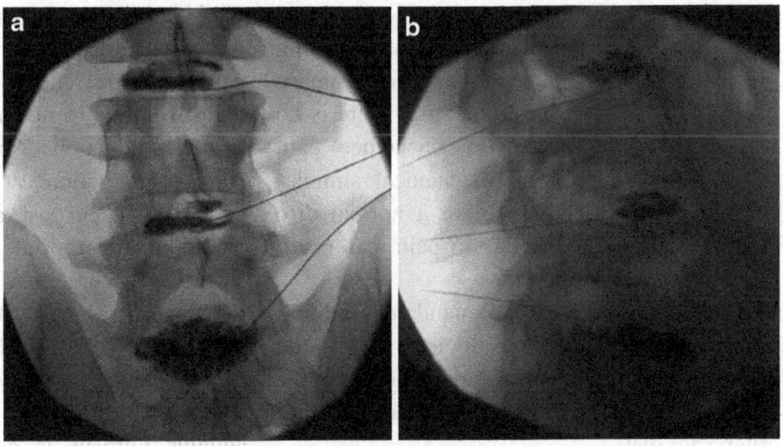

**Lumbar Discography,
Fig. 2** Radiographs of an
L3, L4, and L5 discogram,
after injection of contrast
medium. (**a**) AP view.
(**b**) Lateral view
(Reproduced courtesy of
the International Spinal
Intervention Society
(2004))

isolated to the lumbar area, or they may encompass areas that are topographically separated
from the site of pathology, such as the buttock
and/or lower limb. The manner in which contrast
medium spreads within the disc is salient for interpretation of the study. Although visualization of
such spread can be attained fluoroscopically, it is
best appreciated by computerized tomography
(CT) scan. In order to accept a disc as the pain
generator, not only must concordant pain have
been elicited during disc stimulation, but an internal tear extending at least to the outer one-third of
the posterior or posterolateral disc must be evident
on the CT scan. These features correlate significantly with the reproduction of pain (Moneta et al.
1994). Some 70 % of fissured discs are painful, and
some 70 % of painful discs exhibit radial fissures

(Vanharanta et al. 1987). If fissures are evident, the
diagnosis becomes one of internal disc disruption
(Merskey and Bogduk 1994; Bogduk and McGuirk
2002). However, further data is necessary from this
procedure in order to accept a particular disc as
a pain generator.

Although reproduction of concordant pain is
a necessary step in the diagnosis of discogenic
pain, discographic study of an adjacent disc or
discs is not ideally associated with concordant
pain reproduction. According to the standards of
the International Spinal Intervention Society
(2004), the diagnosis of lumbar discogenic pain
is most robust if stimulation of a single disc
reproduces concordant pain, while stimulation
of two adjacent discs does not. Furthermore, the
intensity of concordant pain must be of moderate

or severe magnitude in order to consider that disc a significant pain generator. In diagnosing lumbar discogenic pain, observation of the pressure within the disc, at the development of moderate-to-severe concordant pain, is of importance. A certain amount of pressure (known as opening pressure) is necessary to induce flow of contrast dye into the nucleus pulposus. Once this is ascertained, then the pressure at which concordant pain is reproduced is determined and recorded. If concordant pain occurs at relatively low pressure, e.g., 15 lb per square inch (psi) or less above the opening pressure, then the response is considered positive. If concordant pain occurs at a pressure of 15–50 psi above opening pressure, then the pain response is indeterminate, possibly being positive. If concordant pain occurs at greater than 50 psi over opening pressure, then such a concordant pain response cannot be considered positive, since there is no way to reliably differentiate whether pain occurred from significant mechanical stimulation, which may cause pain in a normal disc (Carragee et al. 2000), or from pain as a result of internal disc disruption.

Therefore, according to standard (International Spinal Intervention Society 2004), the intervertebral disc is deemed to be the generator of an individual's pain if the following criteria are met:

1. Stimulation of the target disc reproduces the patient's accustomed pain.
2. Stimulation of adjacent discs does not reproduce their pain.
3. Pain is produced at a pressure of injection of not greater than 50 psi and preferably at a pressure less than 15 psi.
4. The evoked pain has an intensity estimated by the patient as greater than 6, on a 10-point numerical pain rating scale.

Validity

In order for lumbar provocation discography to be valid, it must be demonstrated that the four criteria previously listed are adhered to. If these criteria are satisfied, the diagnosis becomes one of discogenic pain. False-positive results can occur due to technical errors, neurophysiological phenomena, and/or psychological factors. Some investigators have warned that discography can be false-positive, on the grounds that discs can be painful when stimulated in asymptomatic volunteers (Carragee et al. 2000). However, the false-positive rate can be kept to less than 10 % if the diagnostic criteria are strictly observed (International Spinal Intervention Society 2004).

Stimulation of the annulus fibrosus, either by injecting into it or by moving the needle within it, can be painful. Accepting a concordant pain response secondary to such an event would be inappropriate in the context of this procedure, as injection must be within the nucleus pulposus while the needle is not otherwise being manipulated. Acceptance of concordant pain that is of mild intensity does not necessarily implicate that disc as one's pain generator, as one of the principal diagnostic tenets for symptomatic internal disc disruption is moderate-to-severe concordant pain reproduction. Acceptance of concordant pain reproduction at manometric pressures greater than 50 psi above opening pressure would be inappropriate, as high intradiscal pressures have been noted to evoke pain in asymptomatic individuals (Carragee et al. 2000). Acceptance of concordant pain without significant annular disruption would violate the principles of this test, as only the outer one-third of the annulus fibrosus is definitively known to be innervated. Lastly, in those with somatization disorder, it has been noted that lumbar provocation discography may result in false-positive responses (Carragee et al. 2000).

Applications

The cardinal purpose of provocation discography is to establish if a patient's back pain is due to discogenic pain (internal disc disruption). The diagnostic utility is to provide a diagnosis and thus help the clinician in planning treatment of such pain. Treatment options include palliative care or surgical management, such as interbody fusion, disc replacement, or ► intradiscal electrothermal therapy. It has been shown that provocation discography in the lumbar spine can assist in predicting outcomes of surgical and nonsurgical management of symptomatic internal disc disruption (Derby et al. 1999).

References

Aprill, C. N. (1991). Diagnostic disc injection. In J. W. Frymoyer (Ed.), *The spine: Principle and practice* (pp. 45–84). New York: Raven.

Bogduk, N. (1983). The innervation of the lumbar spine. *Spine, 8*, 286–293.

Bogduk, N., & McGuirk, B. (2002). Causes and sources of chronic low back pain. In N. Bogduk & B. McGuirk (Eds.), *Medical management of acute and chronic low back pain. An evidence-based approach* (pp. 113–125). Amsterdam: Elsevier.

Carragee, E. J., Truong, T., Rossi, M., & Hagle, C. (2000). The rates of false-positive lumbar discography in select patients without low back symptoms. *Spine, 25*, 1373–1981.

Derby, R., Howard, M. W., Grant, J. M., Lettice, J. J., Van Peteghem, P. K., & Ryan, D. P. (1999). The ability of pressure-controlled discography to predict surgical and non-surgical outcomes. *Spine, 24*, 364–372.

International Spinal Intervention Society. (2004). Lumbar disc stimulation. In N. Bogduk (Ed.), *Practice guidelines for spinal diagnostic and treatment procedures.* San Francisco: International Spinal Intervention Society.

Merskey, H., & Bogduk, N. (Eds.). (1994). *Classification of chronic pain. Descriptions of chronic pain syndromes and definitions of pain terms* (2nd ed., p. 179). Seattle: IASP Press.

Moneta, G. B., Videman, T., Kaivanto, K., Aprill, C., Spivey, M., Vanharanta, H., Sachs, B. L., Guyer, R. D., Hochschuler, S. H., Raschbaum, R. F., & Mooney, V. (1994). Reported pain during lumbar discography as a function of anular ruptures and disc degeneration. A re-analysis of 833 discograms. *Spine, 17*, 1968–1974.

Vanharanta, H., Sachs, B. L., Spivey, M. A., Guyer, R. D., Hochschuler, S. H., Rashbaum, R. F., Johnson, R. G., Ohnmeiss, D., & Mooney, V. (1987). The relationship of pain provocation to lumbar disc deterioration as seen by CT/discography. *Spine, 12*, 295–298.

Lumbar Epidural Steroids

▶ Epidural Steroid Injections

Lumbar Facet Denervation

▶ Lumbar Medial Branch Neurotomy

Lumbar Facet Rhizolysis

▶ Lumbar Medial Branch Neurotomy

Lumbar Lordosis

Definition

The lumbar lordosis is the anterior concavity in the curvature of the lumbar spine as viewed from the side.

Cross-References

▶ Lumbar Traction

Lumbar Medial Branch Blocks

Paul Verrills
Metropolitan Spinal Clinic, Prahran, VIC, Australia

Synonyms

Lumbar zygapophysial joint; Musculoskeletal examination; Radiofrequency neurotomy

Definition

Lumbar medial branch blocks are a diagnostic procedure used to determine if a patient's pain arises from a lumbar zygapophysial joint. They involve anesthetizing the nerves that innervate the joint or joints suspected of being the source of pain.

Characteristics

The lumbar zygapophysial joints are a common source of chronic low back pain.

In younger-aged, injured workers with chronic low back pain, the prevalence of lumbar zygapophysial joint pain is about 15 % (Schwarzer et al. 1994). In older, non-injured patients, the prevalence may be as high as 40 % (Schwarzer et al. 1995). This pain, however, cannot be diagnosed by ▸ medical history, ▸ musculoskeletal examination, or medical imaging. Diagnostic blocks are the only means by which to establish the ▸ diagnosis.

Validity

Although lumbar zygapophysial joint pain can be diagnosed by intra-articular blocks, that type of block has not been validated. Nor have they been shown to have therapeutic utility. In contrast, ▸ lumbar medial branch blocks have been validated.

Of the structures innervated by the medial branches of the lumbar dorsal rami, the zygapophysial joints are the only ones known to be possible sources of chronic pain. There is no evidence to support the competing proposition that chronic pain can arise from the specific segments of the muscles innervated by individual medial branches (Bogduk et al. 1982).

Lumbar medial branch blocks are target specific, provided that precise target points are accurately accessed with needles introduced under fluoroscopic control (Dreyfuss et al. 1997). That is, structures other than the targeted nerves are not anesthetized by lumbar medial branch blocks.

In 89 % of cases, normal volunteers are protected from experimentally induced lumbar zygapophysial joint pain if the appropriate medial branches are anesthetized (Kaplan et al. 1998). The 11 % false-negative rate may be due local vascular puncture and repositioning of the needle (Kaplan et al. 1998).

Lumbar medial branch blocks are preferable to performing intra-articular injections, as they are easier to perform, safer, and more expedient. They are also more easily subjected to pharmacological controls.

Medial branch blocks do have therapeutic utility, for they lead to the target-specific treatment of ▸ radiofrequency neurotomy, which has

independently been validated as appropriate treatment for lumbar zygapophysial joint pain.

Control Blocks

Control blocks are essential to minimize false-positive responses. To this end, comparative local anesthetic blocks are used with a short-acting and a long-acting agent (Barnsley et al. 1993; Lord et al. 1995; International Spine Intervention Society 2004). In clinical practice, these would be lidocaine 2 % and bupivacaine 0.5 %.

A concordant response would be that the patient reports complete relief of pain for a shorter duration when the short-acting agent is used and a longer duration of relief when the long-acting agent is used (see ▸ Peripheral Nerve Blocks). These criteria provide for a highly specific diagnosis of zygapophysial joint pain, but may lack sensitivity. Due to the paradoxically prolonged durations of action of lignocaine, some patients with genuine zygapophysial joint pain may be excluded if these criteria are applied.

More generous criteria would require complete relief of pain on each of the two occasions that the nerves are blocked, irrespective of the duration of relief obtained. These criteria will result in a higher sensitivity but a lesser specificity, i.e., more patients will be diagnosed as having zygapophysial joint pain, but this will include false-positive cases. Since the treatment of lumbar zygapophysial joint pain is considered safe and otherwise relatively innocuous, false-positive responses are tolerable. Patients are not put at risk, for lack of a correct diagnosis, but may fail to benefit from the treatment.

Although placebo-controlled blocks would secure the most valid results, they require three blocks (Lord et al. 1995; International Spine Intervention Society 2004) and are not readily implemented in conventional practice. Meanwhile, comparative local anesthetic blocks have been shown to be cost-effective (Bogduk and Holmes 2000).

Patient Selection

Medial branch blocks are not indicated for acute pain. They are appropriate only for patients with persisting low back pain, for whom a diagnosis is required. A fundamental prerequisite is that serious possible causes of back pain, such as infection, tumors, vascular disease, and metabolic disease, have been excluded by careful and thorough ▶ medical history, ▶ musculoskeletal examination, ▶ neurological examination, laboratory tests, and medical imaging if necessary. The working diagnosis should be that of lumbar spinal pain of unknown origin.

Contraindications

Absolute contraindications include bacterial infection, either systemic or localized, in the region that blocks are to be performed, bleeding diatheses, or possible pregnancy. Relative contraindications include allergy to contrast media or local anesthetics.

Technique

The patient lies prone, and the target points are identified under a C-arm fluoroscope. The target points for the L1 to L4 medial branches are best approached from an oblique view. They lie where the target nerve crosses the junction of the superior articular process and the transverse process, midway between the superior border of the transverse process and the location of the mamillo-accessory notch (Fig. 1).

Correct placement of the needle is confirmed by obtaining a posterior-anterior view. A small dose of contrast medium is injected to ensure that there is no venous uptake. Following this, 0.3 ml of local anesthetic is injected onto the target nerve (Fig. 2).

At the L5 level, it is the dorsal ramus proper that is blocked. The target point lies opposite the middle of the base of the superior articular process and, hence, slightly below the silhouette of the top of the ala of the sacrum. An injection point higher or lower than this has an increased risk of foraminal or epidural spread (Dreyfuss et al. 1997) (Fig. 3).

Each zygapophysial joint is innervated from above and below by the medial branch of dorsal rami. Therefore, two nerves need to be blocked for any one joint. For example, an L4 medial branch and L5 dorsal ramus block anesthetizes the L5/S1 zygapophysial joint; an L3 and L4

Lumbar Medial Branch Blocks, Fig. 1 An oblique view of the lumbar spine, showing needles in correct position on the roots of the L4 and L5 transverse processes, to block the L3 and L4 medial branches (Reproduced, courtesy of the International Spinal Intervention Society (2004))

Lumbar Medial Branch Blocks, Fig. 2 A posteroanterior view of the lumbar spine, showing needles in correct position on the roots of the L4 and L5 transverse process, to block the L3 and L4 medial branches (Reproduced, courtesy of the International Spinal Intervention Society (2004))

Lumbar Medial Branch Blocks, Fig. 3 An oblique view of the lumbar spine, showing a needle in correct position on the ala of the sacrum, to block the L5 dorsal ramus (Reproduced, courtesy of the International Spinal Intervention Society (2004))

medial branch block anesthetizes the L4/5 zygapophysial joint.

Evaluation

The optimal means of reducing error and securing reliable diagnostic information is real-time assessment. The response to the diagnostic block is evaluated immediately after the block, and for some time afterwards, at the clinic at which the block was performed, by an independent observer using validated and objective instruments or tools.

Visual analogue scores are recorded before the block, immediately afterwards, at 30 min, and then hourly. The patient is instructed to monitor the extent and duration of any relief that ensues. Further, if relief occurs, the patient should carefully attempt movements and activities that are usually restricted by pain, to assess their response during the anesthetic phase.

If the block is negative, then zygapophysial joint pain can effectively be ruled out at the level tested. It is, therefore, common to perform an initial screening block at the L3, L4, and L5 levels, either unilaterally or bilaterally, depending on clinical pain patterns. If such

a screening block is negative, one procedure serves to rule out zygapophysial joint pain at the segmental levels most commonly affected.

If the screening block is positive, then further blocks can be undertaken to identify the particular segment or segments responsible. These blocks can target the L5-S1 joint, in the first instance, and subsequently L4-5, if necessary. If the patient has a concordant response for controlled local anesthetic medial branch blocks, then the putative diagnosis of zygapophysial joint pain is confirmed and the patient can be considered for ► radiofrequency neurotomy treatment (International Spine Intervention Society 2004).

References

Barnsley, L., Lord, S., & Bogduk, N. (1993). Comparative local anaesthetic blocks in the diagnosis of cervical zygapophysial joints pain. *Pain, 55*, 99–106.

Bogduk, N., & Holmes, S. (2000). Controlled zygapophysial joint blocks: The travesty of cost-effectiveness. *Pain Medicine, 1*, 25–34.

Bogduk, N., Wilson, A. S., & Tynan, W. (1982). The human lumbar dorsal rami. *Journal of Anatomy, 134*, 383–397.

Dreyfuss, P., Schwarzer, A. C., Lau, P., & Bogduk, N. (1997). Specificity of lumbar medial branch and L5 dorsal ramus blocks: A computed tomographic study. *Spine, 22*, 895–902.

International Spine Intervention Society. (2004). Lumbar medial branch blocks. In N. Bogduk (Ed.), *Practice guidelines for spinal diagnostic and treatment procedures*. San Francisco: International Spinal Intervention Society.

Kaplan, M., Dreyfuss, P., Halbrook, B., & Bogduk, N. (1998). The ability of lumbar medial branch blocks to anesthetize the zygapophysial joint. *Spine, 23*, 1847–1852.

Lord, S. M., Barnsley, L., & Bogduk, N. (1995). The utility of comparative local anaesthetic blocks versus placebo-controlled blocks for the diagnosis of cervical zygapophysial joint pain. *The Clinical Journal of Pain, 11*, 208–213.

Schwarzer, A. C., Aprill, C. N., Derby, R., Fortin, J., Kine, G., & Bogduk, N. (1994). Clinical features of patients with pain stemming from the lumbar zygapophysial joints. Is the lumbar facet syndrome a clinical entity? *Spine, 19*, 1132–1137.

Schwarzer, A. C., Wang, S., Bogduk, N., McNaught, P. J., & Laurent, R. (1995). Prevalence and clinical features of lumbar zygapophysial joint pain: A study in an Australian population with chronic low back pain. *Annals of the Rheumatic Diseases, 54*, 100–106.

Lumbar Medial Branch Neurotomy

Paul Dreyfuss
Department of Rehabilitation Medicine,
University of Washington, Seattle, USA

Synonyms

Lumbar facet denervation; Lumbar facet rhizolysis; Lumbar radiofrequency neurotomy

Definition

Lumbar medial branch neurotomy is a treatment for low back pain stemming from one or more of the zygapophysial joints of the lumbar spine. It involves coagulating the nerves that innervate the painful joint, or joints, with an electrode inserted through the skin. A generator delivers a high-frequency electrical current to the tip of the electrode, which causes the tissues immediately around the tip of the electrode to be heated and, thereby, denatured. Back pain is relieved because nerves incorporated into the heat lesion are made to be unable to conduct nociceptive information from the painful joint or joints (see ▶ Radiofrequency Neurotomy, Electrophysiological Principles).

Characteristics

Indications

Lumbar medial branch neurotomy is not a treatment for any form of back pain. It is explicitly and solely designed to relieve pain from the zygapophysial joints. Therefore, the singular indication for the procedure is complete, or near complete, relief of pain following controlled, ▶ diagnostic blocks of the nerves innervating the painful joint or joints, i.e., ▶ lumbar medial branch blocks. These blocks must be controlled, because the false-positive rate of uncontrolled blocks is such that responses to single blocks will be false in up to 60 % of patients, and those

patients will not benefit from the denervation procedure (Schwarzer et al. 1994a; Bogduk and Holmes 2000).

Setting

The procedure is conducted in a procedure room equipped with a fluoroscope and the necessary apparatus to generate the heat lesion. The procedure is performed under local anesthesia, and no premedication or sedation is required. The patient should be substantially alert in order to report any problems that might occur which threatens their safety. Sedation would be indicated only if the patient is particularly anxious and cannot be relaxed by explanation and assurance, but in that event, the patient must nevertheless always remain conscious.

Typically, the patient will lie facedown on a radiolucent table. The physician will cleanse the skin of the back. An x-ray will be taken, using the fluoroscope, in order to identify the target points at which the nerves will be coagulated.

Technique

Exactly what happens in the procedure depends on the type of electrode used. If a blunt electrode is used, the physician will first place a needle onto the target nerve as for a medial branch block. This is used to anesthetize the nerve and the surrounding tissues, so as to render painless the heating phase of the procedure. That needle is left in place in order to guide the insertion of the electrode. If a hollow electrode is used, this first step can be circumvented, for the electrode can be used to anesthetize the nerve prior to producing the heat lesion.

In order to maximize the effect on the nerve and to optimize the relief of pain, the electrodes should be inserted so as to lie parallel to the nerve. This requires inserting the electrode upwards and slightly medially through the skin and back muscles, until it reaches the target position. This lies on the root of the superior articular process (Fig. 1). Once the electrode has been correctly placed, the heating current is gradually increased by about 1 °C per second. Raising the temperature slowly provides time for both the patient and the physician to react if any untoward

Lumbar   Medial   Branch   Neurotomy, Fig. 1 Radiographs showing an electrode in position to coagulate an L4 medial branch. (**a**) Lateral view showing the course of the nerve depicted by a dotted line. (**b**) Oblique view. (**c**) Lateral view. (**d**) Anteroposterior view

sensations arise. This could occur if the electrode has dislodged, or the target site was not adequately anesthetized. The physician can respond before any injury occurs to the patient. Once a temperature of 80–85 °C has been achieved, it is maintained for about 90 s to ensure adequate coagulation of the nerve.

Depending on the patient's anatomy, the physician may readjust the electrode to accommodate variations in the possible location of the target nerve and will produce another lesion in that new location.

The procedure is repeated for all nerves that were anesthetized, in order to produce relief of pain during the prior conduct of diagnostic lumbar medial branch blocks.

Variants

The optimal technique requires that the electrode be placed parallel to the target nerves (Dreyfuss et al. 2000). Under this condition, a maximal length of the nerve is coagulated, which produces a lasting effect on the pain.

Some operators, however, use variants in which the electrode is not placed parallel to the nerve, but at some angle to it. This behavior can be traced to older versions of the procedure (Shealy 1975, 1976; Ogsbury et al. 1977; Mehta and Sluijter 1979; Sluijter and Mehta 1981; Gallagher et al. 1994), which were used before the nature of radiofrequency lesions was fully appreciated (Bogduk et al. 1987). These techniques are suboptimal, for they can fail to coagulate the target nerve or coagulate only a small section of it. In which case, the nerve regenerates rapidly and pain recurs.

Efficacy

A controlled trial has shown that the effects of lumbar medial branch neurotomy cannot be attributed to a ▶ placebo response (van Kleef et al. 1999). When correctly performed, the ▶ efficacy of lumbar medial branch neurotomy is genuine.

Provided that patients are correctly selected using controlled lumbar medial branch blocks, good outcome can be expected from lumbar medial branch neurotomy. Relief of pain is not permanent. In time, the coagulated nerves regenerate and may again transmit nociceptive information from the painful joint or joints. The time that it takes for this regeneration to occur depends on how accurately and how thoroughly the nerves were coagulated. If suboptimal techniques are used, little of the target nerve may be coagulated, and it recovers rapidly. If a substantial length of nerve is coagulated, relief of pain may last up to and beyond 12 months. Pain may return gradually; it may not return to its former intensity. In some patients the pain never returns.

If the optimal technique is used, 60 % of patients can expect at least 80 % relief of their pain, and some 80 % of patients can expect at least 60 % relief of their pain at 12 months after treatment (Dreyfuss et al. 2000).

If pain recurs and becomes sufficiently intense again as to warrant treatment, lumbar medial branch neurotomy can be repeated in order to reinstate relief. Patients have been successfully treated two, three, and more times, to reinstate and maintain prolonged relief of their pain (Schofferman in press).

Complications

Provided that the correct technique is used, no complications are associated with this procedure. Such complications that are known to be associated with lumbar medial branch neurotomy, incurred in medicolegal proceedings but not published in the literature, have been due to misplacement of the electrode and the use of general anesthesia, which prevented the patient from reporting the onset of complications.

Whereas it may be believed by some that denervating a joint will create a neuropathic joint (Charcot arthropathy), there is no evidence that this occurs, and there are no grounds for believing that it would occur. Charcot arthropathy occurs in limbs that have been completely denervated, in which potentially unstable joints are not protected by muscle activity. In contrast, the zygapophysial joints are intrinsically stable; they are stabilized further by the intervertebral disc, and most of the muscles that act on the affected segment remain functional.

Utility

Lumbar medial branch neurotomy is the singular means by which pain from lumbar zygapophysial joints can be eliminated. No other forms of treatment have been shown to be as effective for the treatment of proven lumbar zygapophysial joint pain. Given that the prevalence of lumbar zygapophysial joint pain is about 15 % in some populations (Schwarzer et al. 1994b) and as high as 40 % in others (Manchikanti et al. 1999; Schwarzer et al. 1995), this procedure has a possible application in a substantial proportion of patients with chronic low back pain.

References

Bogduk, N., & Holmes, S. (2000). Controlled zygapophysial joint blocks: The travesty of cost-effectiveness. *Pain Medicine, 1*, 25–34.

Bogduk, N., Macintosh, J., & Marsland, A. (1987). Technical limitations to the efficacy of radiofrequency neurotomy for spinal pain. *Neurosurgery, 20*, 529–535.

Dreyfuss, P., Halbrook, B., Pauza, K., Joshi, A., McLarty, J., & Bogduk, N. (2000). Efficacy and validity of radiofrequency neurotomy for chronic lumbar zygapophysial joint pain. *Spine, 25*, 1270–1277.

Gallagher, J., Petriccione di Valdo, P. L., Wedley, J. R., Hamann, W., Ryan, P., Chikanza, I., Kirkham, B., Price, R., Watson, M. S., Grahame, R., & Wood, S. (1994). Radiofrequency facet joint denervation in the treatment of low back pain: A prospective controlled double-blind study to assess its efficacy. *The Pain Clinic, 7*, 193–198.

Manchikanti, L., Pampati, V., Fellows, B., & Bakhit, C. E. (1999). Prevalence of lumbar facet joint pain in chronic low back pain. *Pain Physician, 2*, 59–64.

Mehta, M., & Sluijter, M. E. (1979). The treatment of chronic back pain. *Anaesthesia, 34*, 768–775.

Ogsbury, J. S., Simon, R. H., & Lehman, R. A. W. (1977). Facet denervation in the treatment of low back syndrome. *Pain, 3*, 257–263.

Schofferman, J. (2005). Psychotropic medications in chronic spinal disorders. *The Spine Journal (letter), 5,475*.

Schwarzer, A. C., Aprill, C. N., Derby, R., Fortin, J., Kine, G., & Bogduk, N. (1994a). The false-positive rate of uncontrolled diagnostic blocks of the lumbar zygapophysial joints. *Pain, 58*, 195–200.

Schwarzer, A. C., Aprill, C. N., Derby, R., Fortin, J., Kine, G., & Bogduk, N. (1994b). Clinical features of patients with pain stemming from the lumbar zygapophysial joints. Is the lumbar facet syndrome a clinical entity? *Spine, 19*, 1132–1137.

Schwarzer, A. C., Wang, S., Bogduk, N., McNaught, P. J., & Laurent, R. (1995). Prevalence and clinical features of lumbar zygapophysial joint pain: A study in an Australian population with chronic low back pain. *Annals of the Rheumatic Diseases, 54*, 100–106.

Shealy, C. N. (1975). Percutaneous radiofrequency denervation of spinal facets. *Journal of Neurosurgery, 43*, 448–451.

Shealy, C. N. (1976). Facet denervation in the management of back sciatic pain. *Clinical Orthopaedics, 115*, 157–164.

Sluijter, M. E., & Mehta, M. (1981). Treatment of chronic back and neck pain by percutaneous thermal lesions. In S. Lipton & J. Miles (Eds.), *Persistent pain. Modern methods of treatment* (Vol. 3, pp. 141–179). London: Academic.

van Kleef, M., Barendse, G. A. M., Kessels, A., Voets, H. M., Weber, W. E. J., & de Lange, S. (1999). Randomized trial of radiofrequency lumbar facet denervation for chronic low back pain. *Spine, 24*, 1937–1942.

Lumbar Pain

▶ Low Back Pain Patients, Imaging

Lumbar Plexus

Definition

A network of nerves originating from the spinal nerves of the second through fourth lumbar nerves. This network provides motor and sensory function to a large portion of the proximal leg, such as the knee.

Cross-References

▶ Postoperative Pain, Regional Blocks

Lumbar Puncture

Definition

An invasive diagnostic test, in which a needle is inserted into the spinal column at the level between the third and the fifth lumbar vertebra, to examine cerebrospinal fluid and measure intracranial pressure.

Cross-References

▶ Headache Due to Low Cerebrospinal Fluid Pressure

Lumbar Radiofrequency Neurotomy

▶ Lumbar Medial Branch Neurotomy

Lumbar Traction

Andrea D. Furlan[1], Inge Wegner[2],
Maurits Van Tulder[3] and Geert Van der Heijden[2]
[1]Institute for Work & Health, Toronto,
ON, Canada
[2]Julius Center, Department of Epidemiology,
University Medical Center Utrecht, Utrecht,
The Netherlands
[3]Institute for Research in Extramural Medicine,
Free University Amsterdam, Amsterdam,
The Netherlands

Definition

Traction is one of a wide variety of treatments used for low back pain (LBP). Lumbar traction is applied to stretch the lumbar segment of the spine, generally by putting harnesses around the lower rib cage and around the iliac crest. Traction can be applied through different techniques. The forces can be applied using a motorized pulley (mechanical traction) or by forces exerted by the therapist (manual traction). There are also variations depending on how the forces are applied; the patient may control the forces by grasping pulling bars (auto-traction), traction may be performed with the patient fixed perpendicularly in a deep pool and grasping a bar under the arms (underwater traction), or traction may be given in the bed rest position with forces exerted by a pulley and weights (gravitational traction). Duration and level of force exerted can be varied, and force can be applied in a continuous or intermittent mode. Only in mechanical and gravitational traction can the force be standardized.

Characteristics

It has been suggested that the mechanism by which traction works involves spinal elongation, decreasing lordosis and increasing intervertebral space, inhibiting nociceptive impulses, improving mobility, decreasing mechanical stress, reducing muscle spasm or spinal nerve root

compression, releasing ► luxation of a disc or capsule from the ► zygapophysial joint, and releasing adhesions around the zygapophysial joint and the ► annulus fibrosus (van der Heijden et al. 1995).

Proponents of traction claim that LBP patients with radiating symptoms are the ones most likely to benefit from traction (Krause et al. 2000). Some also advise lumbar traction for muscle spasm using appropriately applied traction of adequate force (Geiringer et al. 1993). There is discussion about what constitutes adequate technique and forces for lumbar traction. It has been suggested (Beurskens et al. 1995) that forces below 25 % of the body weight are ► placebo treatment, whereas others (Krause et al. 2000) argue that even low force traction can be expected to produce positive results.

Systematic Review of Lumbar Traction as a Treatment for LBP

A ► systematic review of ► randomized controlled trials (RCTs) involving traction was performed in order to determine whether this modality is effective for LBP with or without radiating symptoms. The review was an update of a previous Cochrane review (Clarke 2006) which was an update of an earlier review by van der Heijden (van der Heijden et al. 1995) and was conducted under the framework of the Cochrane Collaboration Back Review Group. Detailed information on the methods of the review and the specific RCTs is available in the Cochrane Library (Clarke et al. 2006).

The RCTs that were selected involved adults treated for ► nonspecific low back pain (acute, subacute, or chronic), with or without symptoms radiating below the knee (including radicular pain). Trials could involve any type of traction. The primary outcome measures were pain, global well-being, functional status, and return to work. Reported side effects were also considered.

The methodological quality of the RCTs was independently assessed by two reviewers, using Cochrane Collaboration guidelines (van Tulder et al. 2003). Because the studies did not provide sufficient data to enable statistical pooling, we performed a qualitative analysis, summarizing the results of the studies in terms of the strength of the scientific evidence. "Strong evidence" indicates consistent findings among multiple high-quality RCTs. "Moderate evidence" means consistent findings among multiple low-quality RCTs and/or one high-quality RCT. "Limited evidence" means only one low-quality RCT. "Conflicting evidence" means inconsistent findings among multiple RCTs. "No evidence" means no RCTs. High-quality studies were defined as RCTs that fulfilled five or more of the 11 risk of bias criteria.

The systematic review included 24 RCTs, 17 of these involved patients with radiating symptoms, and in the remaining eight there was a mix of patients with and without such symptoms. Seven studies included solely or primarily patients with chronic LBP (more than 12 weeks); in one study patients were all in the subacute range (4–12 weeks); in 12 studies the duration of LBP was a mix of acute, subacute, and chronic; in five studies duration was not specified. In general, the methodological quality of the RCTs was low. Only five studies were classified as "high-quality" studies.

For patients with radiating symptoms, the evidence indicated that continuous or intermittent lumbar traction is not better than placebo, ▶ sham, no treatment, or other treatments (e.g., spinal manipulation, exercise, corset, or infrared lamp). For mixed groups of patients (with and without radiating pain), the conclusions were the same as stated above; however the strength of the evidence was stronger due to more high-quality RCTs in this group. In the trials classifying their control groups as "sham traction," the force applied varied from 1.8 to 9.1 kg or from 10 % to 20 % of body weight. No significant differences were demonstrated in any of these trials.

Although it has been argued that some subgroups of patients may benefit from traction, such as patients with acute radicular pain with concomitant neurological deficit (Krause et al. 2000), the preponderance of evidence on the efficacy of traction comes from studies with methodological problems and potential ▶ bias, which do not distinguish between subgroups of patients (e.g., with differing pain duration and radicular symptoms).

Conclusion of the Systematic Review

Two recent high-quality studies have strengthened the findings that traction is not indicated in the treatment of LBP. Based on the RCTs currently available in the literature, there is moderate to strong evidence that traction is not an effective treatment for LBP patients, and some suggestion that patients receiving continuous traction are more likely to experience adverse effects, such as increased pain and subsequent surgery.

Regarding side effects, in our systematic review we found one study that mentioned there were no adverse effects, six studies reported some adverse effects (increased pain and more subsequent surgeries), and the remaining studies made no mention of this issue. However, most trials had small sample sizes that were not adequate to evaluate side effects. The few available case reports published in the literature suggest that there is some danger for nerve impingement in heavy traction (i.e., with forces exceeding 50 % of body weight). Other potential risks include respiratory constraints due to the traction harness or increased blood pressure during inverted positional traction.

References

Beurskens, A. J. H. M., van der Heijden, G. J. M. G., de Vet, H. C. W., et al. (1995). The efficacy of traction for lumbar back pain: Design of a randomized clinical trial. *Journal of Manipulative and Physiological Therapeutics, 18*, 141–147.

Clarke, J. A., van Tulder, M. W., Blomberg, S. E. I., et al. (2006). *Systematic Cochrane review of traction for low back pain with or without radiating symptoms.* Toronto: Cochrane Collaboration Back Group. submitted.

Geiringer, S. R., Kincaid, C. B., & Rechtien, J. R. (1993). Traction, manipulation, and massage. In J. A. DeLisa (Ed.), *Rehabilitation medicine: Principles and practice* (2nd ed., pp. 440–462). Philadelphia: JB Lippincott Company.

Krause, M., Refshauge, K. M., Dessen, M., et al. (2000). Lumbar spine traction: Evaluation of effects and recommended application for treatment. *Manual Therapy, 5*, 72–81.

van der Heijden, G. J. M. G., Beurskens, A. J. H. M., Koes, B. W., et al. (1995). The efficacy of traction for back and neck pain: A systematic, blinded review of randomized clinical trial methods. *Physical Therapy, 75*, 93–104.

van Tulder, M., Furlan, A., Bombardier, C., Editorial Board of the Cochrane Collaboration Back Review Group, et al. (2003). Updated method guidelines for systematic reviews in the Cochrane Collaboration Back Review Group. *Spine, 28*, 1290–1299.

Lumbar Transforaminal Injection of Steroids

James M. Borowczyk
Department of Orthopaedics and Musculoskeletal Medicine, University of Otago, Christchurch, Christchurch, New Zealand

Synonyms

Transforaminal steroids

Definition

Lumbar transforaminal injection (TFI) of corticosteroids is a procedure whereby an aliquot of a given corticosteroid preparation is delivered into the immediate vicinity of a lumbar spinal nerve and its respective roots, by way of the corresponding intervertebral foramen. The procedure is designed as a treatment for lumbar radicular pain.

Characteristics

Background

The first description of the "epidural" injection of corticosteroid was via the transforaminal route (Robecchi and Capra 1952). This became the standard route of administration in the 1950s and 1960s, but was superseded by the caudal and interlaminar routes in the United States and in Britain in the 1960s and later in Europe and Scandinavia. These latter routes became the standard for common practice.

In the late 1900s, the beginning of a reversion to the transforaminal route was prompted by a number of factors. These include, among others, reviews suggesting that caudal and interlaminar routes were not as effective as had previously been claimed, an increasing use of transforaminal nerve root injection for radicular pain diagnosis, and, subsequently, reports of successful outcomes for transforaminal injection in both observational studies and later in controlled trials (Vad et al. 2002).

Rationale

There is circumstantial evidence that inflammatory processes may play a major role in the production of at least some of the clinical symptoms experienced when a lumbar nerve root is compromised by an intervertebral disc herniation (McCarron et al. 1987; Olmarker et al. 1995; Kang et al. 1995). Corticosteroid delivery to this site of putative inflammation then becomes a logical intervention. This offers a theoretical advantage, in that it targets the site of such inflammation, and may reduce the supposed inflammatory process, presumably through inhibition of phospholipase-A and cellular inflammatory mechanisms (see ▶ Radicular Pain). A competing interpretation is that the steroid preparation may exert a long-lasting local anesthetic effect (Johansson et al. 1990) on the nerve root and the dura that surrounds it.

Efficacy

Observational studies have demonstrated statistical benefits from lumbar TFI. With respect to the conservative and nonsurgical management of radicular pain, several authors have reported a benefit from transforaminal epidural injection (Weiner and Fraser 1997; MacVicar et al. 2013). Prospective, randomized, controlled, double-blinded studies have also suggested a surgery-sparing effect for lumbar TFI, particularly for the corticosteroid component of the injectate (Riew et al. 2000).

Indications

Lumbar TFI is advocated for patients with lumbar radicular pain:

- Who require pain relief
- Who have not responded to other nonsurgical (conservative) interventions
- For whom other nonsurgical interventions are deemed inappropriate
- Whose pain may have an inflammatory basis

Patient Selection

According to the guidelines set out by the International Spinal Injection Society (2004), selected patients should have symptoms consistent with lumbar or sacral radicular pain that may be amenable to lumbar TFI. Specific signs and symptoms compatible with this include:

- Numbness or paresthesia in a dermatomal distribution
- Weakness in a myotomal distribution
- Inhibition of straight-leg raising to 30° or less

Additional confirmatory investigations should ideally include:

- Medical imaging demonstrating a cause of the radicular pain consistent with the clinical findings
- Medical imaging as above that would theoretically be amenable to lumbar TFI

Imaging findings which do not constitute intervention with LTFI would include:

- Tumor
- Angioma
- Cysts (with the possible exception of zygapophyseal synovial cysts)
- Arachnoiditis

Absolute Contraindications

- The (fully informed) patient is unable or unwilling to consent to the procedure.
- The use of contrast media is contraindicated.
- There is evidence of an untreated localized infection in the procedural field.
- The patient has a bleeding diathesis.

- The patient is unable to cooperate during the procedure.
- Adequate imaging of the target is difficult or impossible.

Relative Contraindications

- Allergy to any of the administrable drugs
- Concurrent use of anticoagulants, or relative anticoagulants
- Anatomical or surgical derangements within the procedural field
- Systemic infection
- Significant coexisting disease
- Immunosuppression
- Pregnancy

Technique

Full details of the technique are provided elsewhere (International Spinal Intervention Society 2004). In essence, the procedure requires placing the injectate as close as possible to the target nerve.

This is first facilitated by visualizing the appropriate intervertebral foramen under image intensification. Appropriate adjustments of the imaging apparatus are necessary to maximize the ease of approach to the injection site (International Spinal Intervention Society 2004). The point of needle placement should be in the so-called safe triangle (Bogduk et al. 1995a, b) (Fig. 1).

Upon successfully accessing the target point, needle placement is confirmed by visualization of the needle tip position against the posterior aspect of the vertebral body, in a lateral radiographic view. Proximity to the nerve is then confirmed in the AP projection by the injection of contrast medium so as to obtain a "radiculogram," and to confirm that there is no intravascular loss of the contrast medium in "real-time" imaging (Fig. 2). Having satisfied these criteria, the transforaminal corticosteroid injection may be placed.

Complications

Lumbar TFIs are subject to all the possible complications of transdermal injection procedures. These include infection, local bleeding, the puncture of adjacent structures, allergy to the various

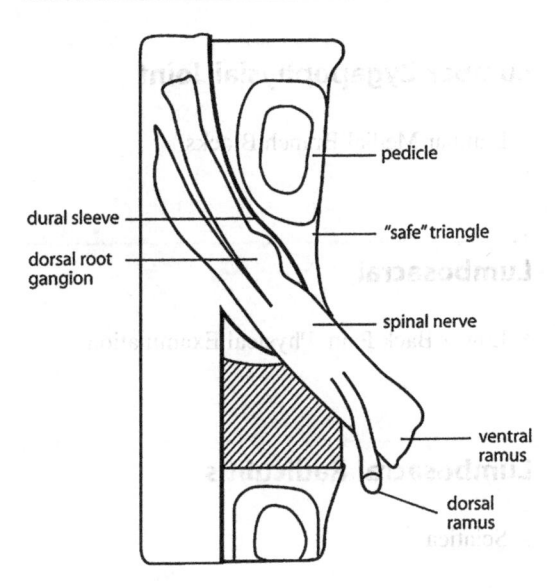

Lumbar Transforaminal Injection of Steroids, Fig. 1 A sketch of a lumbar intervertebral foramen, showing the "safe triangle" in relation to the spinal nerve (Reproduced, courtesy of the International Spinal Intervention Society (Riew et al. 2000))

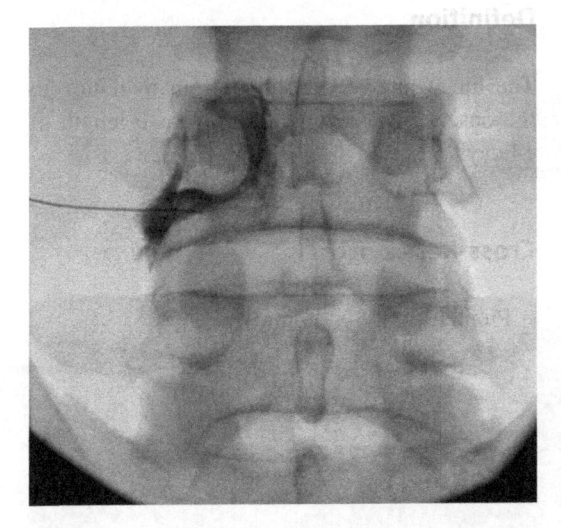

Lumbar Transforaminal Injection of Steroids, Fig. 2 An anteroposterior radiograph of the lumbar spine showing a transforaminal injection at L4-5. Contrast medium outlines the course of the L4 spinal nerve and its roots (Reproduced, courtesy of the International Spinal Intervention Society (Riew et al. 2000))

components of the injectate, and adverse vasovagal reactions of the injection recipient. With respect to the local anatomy, specific complications include epidural abscess, epidural hematoma, and puncture of the dural sac. Careful and aseptic technique will minimize these risks. Allergic reactions and induced hematomata remain a risk, even with careful patient screening and practiced technique.

Of the few studies reporting adverse reactions to the procedure, one cited non-positional headache in 4.8 % of patients, increased back pain in 4.8 %, increased leg pain in 1.6 %, and facial flushing (thought to be due to the effect of the corticosteroid component of the injectate) in a further 1.6 % (Botwin et al. 2000). Another report cited three patients who suffered paraplegia following the procedure, thought to be due to the inadvertent intravascular injection of colloid corticosteroid into a radicular artery supplying the distal end of the spinal cord (Houten and Errico 2002). Careful attention to the flow of contrast medium should detect inadvertent placement, or intravascular positioning of the needle, and should avoid such complications.

Utility

The injection of corticosteroid preparations next to the lumbar nerve root, using the transforaminal approach, has been shown to be a useful modality in the treatment of lumbar ▶ radicular pain. The technique is relatively noninvasive and carries few inherent dangers to the patients. While not universally effective, it may confer significant benefits to the patient in terms of relief of pain and return to normal function. Studies indicate that in an appropriately targeted population, the technique is more effective than, the previously more commonly used, caudal and interlaminar epidural injections. Lumbar TFI deserves consideration in all cases of lumbar radicular pain where conservative measures have failed, and surgery is the only alternative.

References

Bogduk, N., Aprill, C., & Derby, R. (1995a). Selective nerve root blocks. In D. J. Wilson (Ed.), *Interventional radiology of the musculoskeletal system* (pp. 121–132). London: Edward Arnold. Ch 10.
Bogduk, N., Aprill, C., & Derby, R. (1995b). Epidural steroids. In A. D. White (Ed.), *Spine care, vol 1:*

Diagnosis and conservative treatment (pp. 322–343). St Louis: Mosby. Ch 22.

Botwin, K. P., Gruber, R. D., Bouchlas, C. G., Torres-Ramos, F. M., Freeman, T. L., & Slaten, W. K. (2000). Complications of fluoroscopically guided transforaminal lumbar epidural injections. *Archives of Physical Medicine and Rehabilitation, 81,* 1045–1050.

Houten, J. K., & Errico, T. J. (2002). Paraplegia after lumbosacral nerve root block: Report of three cases. *The Spine Journal, 2,* 70–75.

International Spinal Intervention Society. (2004). Lumbar transforaminal injection of corticosteroids. In N. Bogduk (Ed.), *Practice guidelines for spinal diagnostic and treatment procedures.* San Francisco: International Spinal Intervention Society.

Johansson, A., Hao, J., & Sjolund, B. (1990). Local corticosteroid application blocks transmission in normal nociceptive C-fibres. *Acta Anaesthesiologica Scandinavica, 34,* 335–338.

Kang, J. D., Georgescu, H. I., McIntyre, L., et al. (1995). Herniated lumbar intervertebral discs make neutral metalloproteinases, nitric oxide, and interleukin-6. *Orthopaedic Transaction, 19,* 53.

Lutz, G. E., Vad, V. B., & Wisneski, R. J. (1998). Fluoroscopic transforaminal lumbar epidural steroids: An outcome study. *Archives of Physical Medicine and Rehabilitation, 79,* 1362–1366.

MacVicar, J., King, W., Landers, M., & Bogduk, N. (2013). The Effectiveness of Lumbar Transforaminal Injection of Steroids: A Comprehensive Review with Systematic Analysis of the Published Data. *Pain Medicine, 14,* 14–28.

McCarron, R. F., Wimpee, W. M., Hudkins, P. G., & Laros, G. S. (1987). The inflammatory effect of nucleus pulposus: A possible element in the pathogenesis of low back pain. *Spine, 12,* 760–764.

Olmarker, K., Blomquist, J., Stromberg, J., Nanmark, U., Thomsen, P., & Rydevik, B. (1995). Inflammatogenic properties of nucleus pulposus. *Spine, 20,* 665–669.

Riew, K. D., Yin, Y., Gilula, L., Bridwell, K. H., Lenke, L. G., Lauryssen, C., & Goette, K. (2000). The effect of nerve-root injections on the need for operative treatment of lumbar radicular pain. A prospective, randomized, controlled, double-blind study. *Journal of Bone and Joint Surgery, 82A,* 1589–1593.

Robecchi, A., & Capra, R. (1952). L'idrocortisone (composto F). Prime esperienze cliniche in campo reumatologico. *Minerva Medica, 98,* 1259–1263.

Vad, V. B., Bhat, A. L., Lutz, G. E., & Cammisa, F. (2002). Transforaminal epidural steroid injections in lumbosacral radiculopathy. *Spine, 27,* 11–16.

Weiner, B. K., & Fraser, R. (1997). Foraminal injection for lateral lumbar disc herniation. *Journal of Bone and Joint Surgery, 79B,* 804–807.

Lumbar Zygapophysial Joint

▶ Lumbar Medial Branch Blocks

Lumbosacral

▶ Lower Back Pain, Physical Examination

Lumbosacral Radiculitis

▶ Sciatica

Luteal Phase

Definition

The luteal phase is the time from ovulation to the onset of menses, with an average length of 14 days.

Cross-References

▶ Premenstrual Syndrome

Luxation

Definition

Dislocation.

Cross-References

▶ Lumbar Traction

LVCCs

▶ Calcium Channels in the Spinal Processing of Nociceptive Input

Lymphokines

▶ Cytokines as Targets in the Treatment of Neuropathic Pain

Lysis of Adhesions

Definition

An invasive procedure performed in the epidural space with a catheter. Solutions are injected to expand scar tissue in hopes of relieving pressure on nerve roots inside the spine.

Cross-References

▶ Sciatica

LVCCs

▶ Calcium Channels in the Basal Processing of
Nociceptive Input

Lymphokines

▶ Cytokines ... Freezes in the Treatment of
Neuropathic Pain

Lysis of Adhesions

Definition

An invasive procedure performed in the epidural
space with a catheter. Solutions are injected to
expand scar tissue in hopes of relieving pressure
on nerve roots inside the spine.

Cross-References

▶ Epiduroscopy

M

MAC

▶ Minimal Alveolar Concentration

Magnetic Resonance Imaging

Peter Lau
Department of Clinical Research, Royal
Newcastle Hospital, University of Newcastle,
Newcastle, NSW, Australia

Synonyms

MRI; NMR; Nuclear magnetic resonance

Definition

Magnetic resonance imaging (MRI) is a means of
obtaining images of the internal structure of the
body based on the detection of radio-frequency
waves emitted from the body under special
conditions when it has been placed in
a magnetic field.

Characteristics

Principles

Approximately 63 % of the atoms in the human
body are hydrogen atoms, largely in the forms of
fat or water. These hydrogen atoms consist of an
electron cloud and a nucleus containing
a single proton. The proton behaves as a tiny
magnet with a north and south pole. The proton
has a property called spin, which acts as a small
magnetic field. When placed in an external
magnetic field, the magnetic fields of the
protons are forced to align with the external
field. However, the alignment is not perfect.
There are protons lining up with and against the
external magnetic field. The distribution is
uneven as it takes less energy for a proton to
line up with than against the external magnetic
field. Slightly more protons therefore line up with
than against the external magnetic field, resulting
in a net magnetization of the human body.
This net magnetization is measured to make the
MR images.

However, the net magnetization cannot be
measured while it is lined up with the magnetic
field of the MR system. By tilting the net magnetization away from its alignment with the magnetic field of the MR system, the protons of the
body generate the signals that are used to make
the MR images.

The net magnetization of the protons behaves
like a spinning top on a table; it wobbles around
before finally falling over. This phenomenon is
called *Precession*. Precession is simply the wob-
bling motion of the net magnetization of a group
of the body's protons around the main magnetic
field of the MR system.

The speed at which magnetization precesses is
described by the Larmor equation

G.F. Gebhart, R.F. Schmidt (eds.), *Encyclopedia of Pain*, DOI 10.1007/978-3-642-28753-4,
© Springer-Verlag Berlin Heidelberg 2013

$$f = K \times B$$

where

f = precessional or Larmor frequency (in MHz)

K = the gyromagnetic ratio = 42.6 MHz per Tesla (for hydrogen nuclei)

B = magnetic field (in Tesla)

The net magnetization is tilted by an RF pulse called the excitation pulse. The RF energy tilts the net magnetization from the longitudinal plane to the transverse plane. After the RF pulse is turned off, the natural tendency of the net magnetization of the protons is to realign with the external magnetic field from the transverse plane back to the longitudinal plane. During this process, which is called *Relaxation*, energy that was absorbed by the excitation pulse is released back into the surrounding molecular environment in the form of radio waves, which are measured as RF signals.

The amount of energy released by the relaxation process of a decaying proton is small. Therefore, its emission is faint. However, if a radio-frequency excitation pulse is again transmitted into the body, it effectively bounces the proton's net magnetization back into the transverse plane, from which its precession once again decays. As it does so, it again emits the same amount of energy at the same rate as before. By repeating this process, the behavior of the proton can be studied repeatedly. Each faint signal is repeated and a cumulative consistent signal that is no longer faint can be harvested. This repetitive process is the basis of pulse sequences such as spin echo and fast spin-echo sequences.

In different molecules and in different tissues, the precessions of protons decay at different rates. Consequently, at different times after the commencement of decay, different tissues emit radio waves of different intensities. By sampling emissions at various selected times, a picture can be generated of different tissues according to the energy that they are emitting at that time. By exploiting the properties of frequency (how fast the net magnetization of the protons is precessing) and phase (the direction in which the net magnetization of the protons is pointing), the sampling of each RF excitation pulse forms

the basis of encoding anatomically where within a slice of the body, specific echo information originated. Furthermore, by exploiting variations in the gradient strengths of the external magnetic field in the X, Y, and Z axes, the net magnetization of the precessional frequency of the protons can be varied. This variation will allow sampling in different slice locations and thickness in all three planes (The MR CrossTrainer Study modules, Medical Imaging Consultants, Inc).

By recording the emissions across various diameters of the body, a computer can be used to synthesize a virtual anatomical image of the internal structure of the body, based on the inferred source of the emissions. These images can be printed as views at any depth across any plane, but sagittal, axial, and coronal or oblique views are typically used. By varying the sampling times and other variables such as the timing of radio-frequency pulses, the images obtained can focus on different types of tissues. The terms T1 and T2 refer to different time constants that apply to the net magnetization decay behavior of protons. Images based on these constants are referred to as T1 weighted and T2 weighted, respectively.

In general, T1-weighted signals generate images of solid structures, such as muscles and bones, and are used to assess the anatomical details of body structures. T2-weighted signals generate images of more fluid tissues and are used for tissue differentiation, particularly in pathological states. Fat and water have similar signal intensities under normal conditions, but a particular pulse sequence, known as fat saturation, can be used to suppress signals from fat, so that fatty tissue signals can be differentiated from water signals. This becomes relevant when signals from inflammatory exudates need to be differentiated from normal fatty tissue signals. In clinical practice, external magnets of low, mid and high field strengths can be used, ranging in strength from 0.3 to 1.5 T. Super-high field strength magnets are reserved for research purposes, but recently, 3 T units have become commercially available. The drive to higher field strength magnets is driven by higher signal to noise ratios and greater speed of imaging. As field strength increases however, safety issues

Magnetic Resonance Imaging, Fig. 1 A sagittal MRI scan showing a metastasis in a thoracic vertebral body, crossing the intervertebral disc into the next vertebral body. (a) T1-weighted image. (b) T1-weighted image with gadolinium enhancement. The gadolinium shows the increased vascular activity of the tumor as a white signal

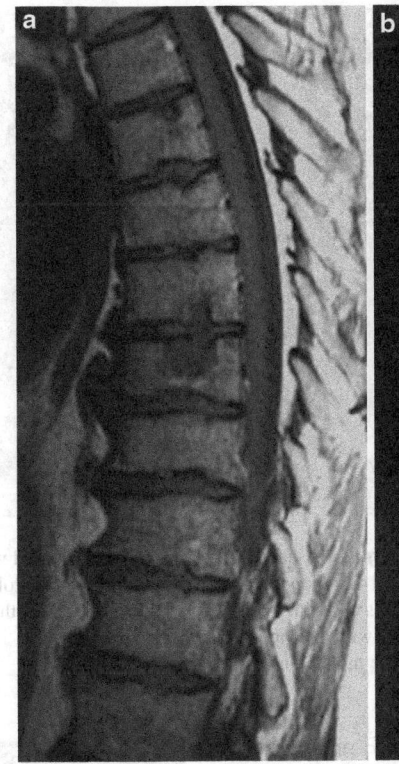
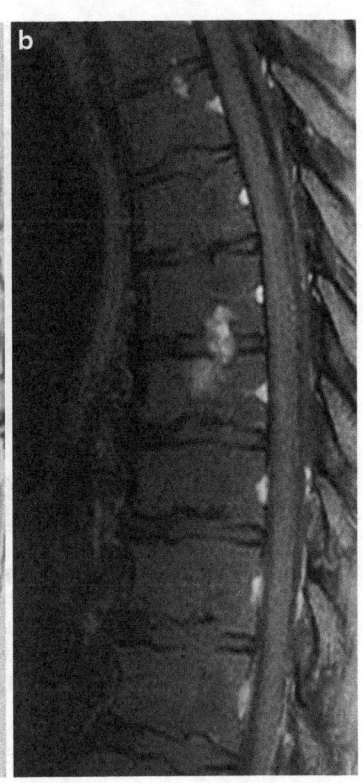

increase. These include not only the need for greater magnetic shielding but also possible adverse biological effects, although to date, there have been no significant reported clinical side effects. Body heating and possible superficial burns have been reported, but these have been attributed to faulty equipment such as surface coils and unsuspected internal metallic objects, rather than to the magnetic field itself (Shellock 2000).

Applications

Of all medical imaging tests, MRI has the highest sensitivity and highest specificity. It can detect lesions, such as tumors, infections, and osteonecrosis, early in their evolution (Figs. 1 and 2). Moreover, it allows these lesions to be specifically identified. No other imaging test has these properties. MRI defines soft tissue lesions as clearly as, or better than, ultrasound. For these reasons, MRI has assumed a paramount position in imaging for general medical conditions. Its relevance in pain medicine, however, is more limited.

The cardinal application of MRI in pain medicine is as a screening test to detect occult or cryptic lesions that do not manifest distinctive clinical features. Such lesions, however, are unusual and rare. Consequently, when MRI is used as a screening test, the default expectation is that it will not reveal lesions. Under those conditions, MRI is used to clear patients of serious disorders such as tumors, infection, or osteonecrosis.

In neurology, headache is a common clinical problem. Some 10–15 % of MR studies are undertaken for headache but fewer than 1 % of these investigations reveal significant pathology. Apart from tumors and infections, subarachnoid hemorrhage due to a leaking aneurysm in the circle of Willis can be demonstrated by MRI and by MR angiography. This has the advantage of being noninvasive and avoids the complications of a more invasive procedure such as four-vessel cerebral angiography. MRI can be used to diagnose headaches due to changes in pressure of cerebrospinal fluid (CSF), benign intracranial

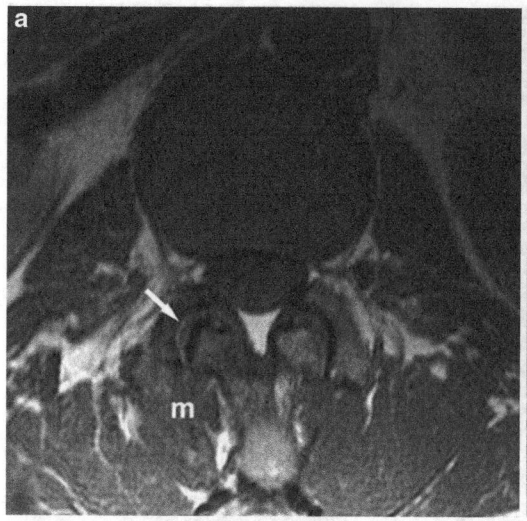

Magnetic Resonance Imaging, Fig. 2 Axial MRI scans of a septic arthritis of a right L1-2 zygapophysial joint. (**a**) The joint on the right exhibits a slightly bright signal in its cavity (*arrow*) compared with the opposite side. In addition, the architecture of the multifidus (m) is distorted compared with the other side. (**b**) Administered intravenously, gadolinium enhances the edema in the multifidus behind the infected joint, rendering it *bright white* (*arrow*)

hypertension, idiopathic low CSF pressure, and posttraumatic CSF leakage.

In musculoskeletal medicine, MRI is the best way to detect osteonecrosis (Fig. 3). For this condition, MRI is both sensitive and specific and is able to detect changes earlier than plain radiography or CT or even bone scan. MRI is also able to measure the size of the osteonecrotic bone fragment.

Because of its better resolution of nerves and its ability to provide coronal and sagittal images as well as axial images, MRI is the preferred means of investigating radiculopathy (Fig. 4). The dorsal root ganglion, the spinal nerve roots, and the nerve root sleeves can be clearly differentiated from the adjacent dura and the CSF in the spinal canal, making assessment of nerve root compression more reliable than is possible by CT.

Its ability to demonstrate the internal structure of intervertebral discs accords MRI a unique role in the investigation of chronic back pain. Some 30 % of patients with low back pain exhibit a high-intensity zone in one of their discs (April and Bogduk 1992). This zone is a bright spot located in the posterior anulus fibrosus (Fig. 5).

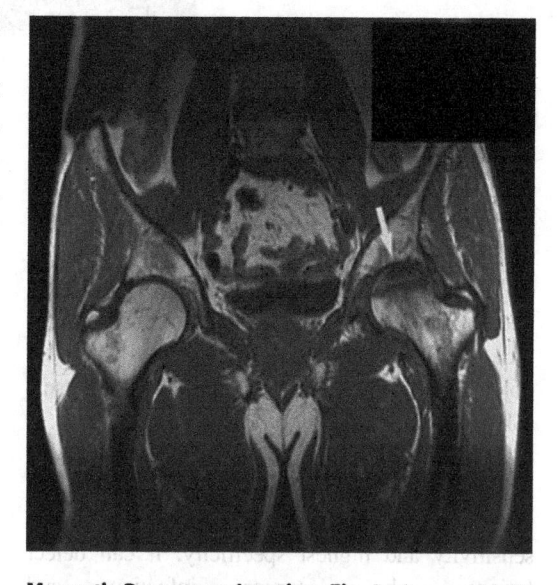

Magnetic Resonance Imaging, Fig. 3 A coronal MRI scan of the pelvis and hips, showing avascular necrosis of the head of the left femur. The necrosis appears as a *darkened area* (*arrow*)

The zone corresponds to a collection of fluid in a circumferential fissure crossing the posterior anulus and correlates strongly with that disc being the source of pain. When present, this sign

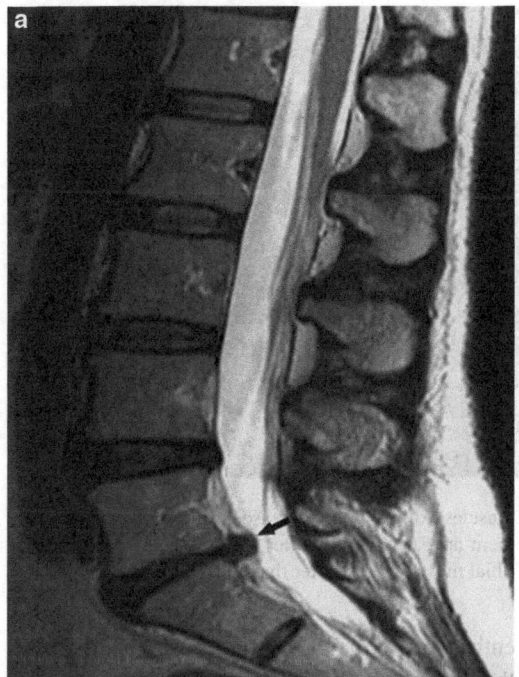

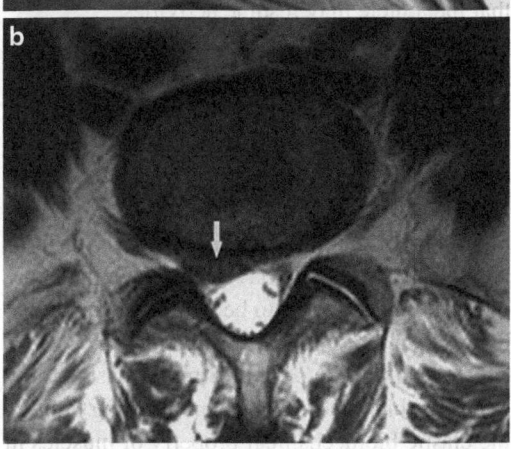

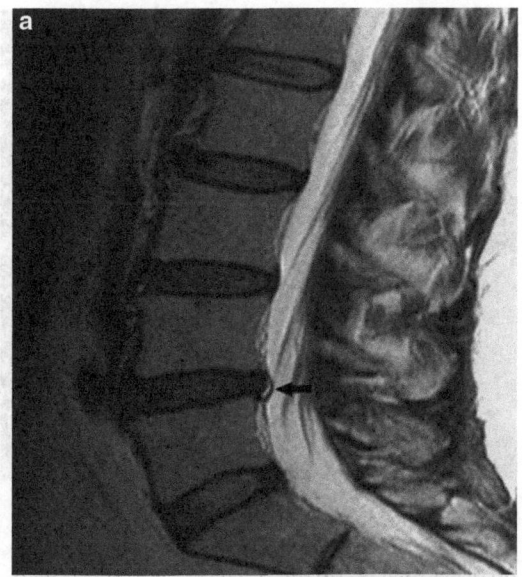

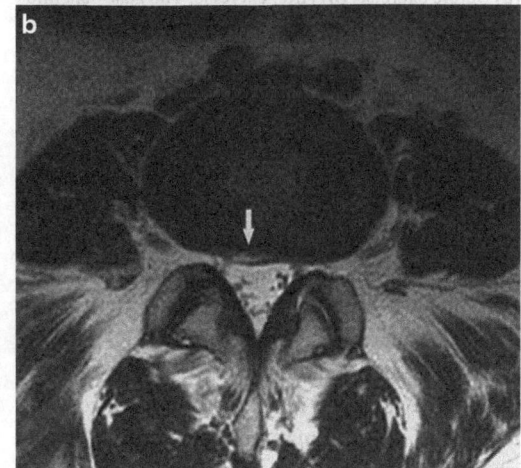

Magnetic Resonance Imaging, Fig. 4 MRI scans of a herniation of an L5-S1 intervertebral disc. (**a**) Sagittal scan. The disc herniation (*arrow*) protrudes beyond the *posterior margin* of the L5 vertebral body. (**b**) Axial scan. The herniation (*arrow*) protrudes *posteriorly to the right*, towards the zygapophysial joint

Magnetic Resonance Imaging, Fig. 5 MRI scans of a high-intensity zone (HIZ) in an L4-5 intervertebral disc. (**a**) Sagittal scan. In T2-weighted images, the HIZ appears as a bright signal in the posterior anulus fibrosus (*arrow*). (**b**) Axial scan. The HIZ corresponds to a circumferential tear (*arrow*), which appears as a bright crescentic signal in the posterior anulus

has positive likelihood ratio of about 6 for incriminating the affected disc as the source of pain (Bogduk and McGuirk 2002).

For other lesions and abnormalities affecting the spine, the utility of MRI is far more questionable. Lesions such as disc bulges, disc herniations, and degenerative disc disease occur very commonly in patients with no pain and increasingly with age (Jensen et al. 1994). Consequently, they are not diagnostic signs of back pain.

MR neurography can be used to enhance the appearance of nerves, such as the brachial plexus and lumbar plexus and large peripheral nerves such as the sciatic nerve and radial and ulnar nerves. However, unless there is gross lesion, such as tumor infiltration or neuroma, MR

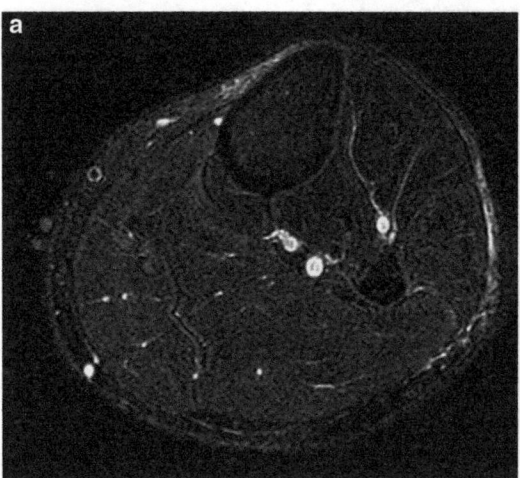
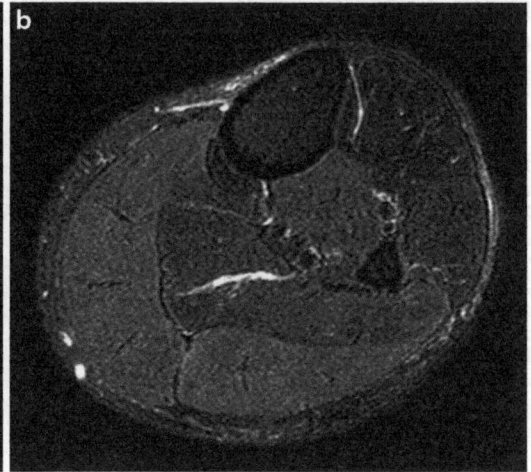

Magnetic Resonance Imaging, Fig. 6 Axial MRI scans of the calf showing the effects of strenuous exercise. (a) Before exercise. (b) After exercise. The gastrocnemius muscles exhibit a faint increase in signal intensity, making them appear slightly whiter than the soleus and anterior tibial muscles and whiter than they were before exercise

neurography has no proven role in detecting causes of pain arising in nerves.

The ability of MRI to resolve connective tissues makes it the premier means of assessing joints and periarticular structures. In some instances, MRI has an established and valid role. In others, the validity of various MR findings has not been established. In patients with chronic knee pain, MRI can detect meniscal tears and anterior cruciate or posterior cruciate ligament tears and is replacing direct arthroscopy as the preferred, primary diagnostic test. For the assessment of acute knee injury, a short MRI examination following radiography has been shown to save costs and improve quality of life after injury, in terms of time absent from work, additional investigations, and time required to recover (Nikken et al. 2003). Although MRI can demonstrate the structure of the shoulder in great detail, tears of the rotator cuff and other lesions occur in totally asymptomatic subjects and increasingly with age. Therefore, demonstrating these lesions by MRI is not diagnostic of the cause of shoulder pain. Pains in the elbow, wrist, ankle, or foot are common problems encountered by sportspeople, either after acute injury or as a result of chronic repeated stress. In such individuals, MRI is being used increasingly to detect bone bruises and chronic

enthesopathy or to rule out these conditions so that the athlete may resume activities.

Muscle pain and muscle imaging, however, remain an enigma. Signal changes, due to tissue disruption and edema, can be seen in patients with muscle strains or tears or with overuse and compartment syndromes (Fig. 6). However, no role for MRI has been established for more common entities such as myofascial pain and fibromyalgia. No features have been identified that allow these conditions to be diagnosed or studied by MRI.

There is work in progress on MR elastography (MRE), which may throw some imaging light on the elastic biomechanical property of muscles in health and disease. MRE is a noninvasive, phase-contrast MRI technique used to spatially map and measure displacement patterns in the muscle fibers in response to an external harmonic shear wave. Amplitude changes are detectable at the level of microns or less.

Cross-References

▶ Amygdala, Pain Processing and Behavior in Animals
▶ fMRI
▶ Headache Due to Arteritis

▶ Motor Cortex
▶ Secondary Somatosensory Cortex (S2) and Insula, Effect on Pain-Related Behavior in Animals and Humans

References

April, C., & Bogduk, N. (1992). High intensity zone: A diagnostic sign of painful lumbar disc on magnetic resonance imaging. *British Journal of Radiology, 65*, 361–369.
Bogduk, N., & McGuirk, B. (2002). *Medical management of acute and chronic low back pain. An evidence based approach* (p. 136). Amsterdam: Elsevier.
Jensen, M. C., Brant-Zawadzki, M. N., Modic, M. T., et al. (1994). Magnetic resonance imaging of the lumbar spine in people without back pain. *The New England Journal of Medicine, 331*, 69–73.
Nikken, J., Edwin, O., & Hunink, M. (2003). *A short low-field MRI examination in all patients with recent peripheral joint injury: Is it worth the costs?* 89th Annual Meeting of the Radiological Society of North America. Chicago, Nov 2003.
Shellock, F. G. (2000). Radiofrequency energy-induced heating during MR procedures: A review. *Journal of Magnetic Resonance Imaging, 12*, 30–36.

Magnetoencephalogram

▶ Magnetoencephalography in Assessment of Pain in Humans

Magnetoencephalography

Synonyms

MEG

Definition

Synchronized extracellular currents in a few square centimeters of cortex generate magnetic fields measurable with sensors on the surface of the scalp. The biggest advantage of MEG, as compared with electroencephalography (EEG),

is its high spatial resolution due to less of an effect of cerebrospinal fluid, skull, and skin, since magnetic fields are not affected by electric current conductivity.

Cross-References

▶ Insular Cortex, Neurophysiology, and Functional Imaging of Nociceptive Processing
▶ Magnetoencephalography in Assessment of Pain in Humans
▶ Thalamotomy for Human Pain Relief

Magnetoencephalography in Assessment of Pain in Humans

Ryusuke Kakigi, Minoru Hoshiyama, Koji Inui, Kensaku Miki, Daisuke Naka, Yunhai Qiu, Diep Tuan Tran, Xiaohong Wang, Shoko Watanabe and Hiroshi Yamasaki
Department of Integrative Physiology, National Institute for Physiological Sciences, Okazaki, Japan

Synonyms

Biomagnetometer; Gradiometer; Magnetoence-phalogram; Magnetometer; Source analysis; Superconducting quantum interference device; Topography

Definition

Cortical neurons are excited by the signals conducted through thalamocortical fibers from the thalamus. After signals are received, electric currents are conducted through apical dendrites of pyramidal cells of the cerebral cortex. The electric currents generate magnetic fields. The electric currents are recorded as electroencephalography (EEG), and magnetic fields are recorded as ▶ magnetoencephalography (MEG). There are two kinds of postsynaptic potentials (PSP), excitatory ones (EPSP) and inhibitory ones (IPSP). EPSP are considered to be the main generators for both EEG and MEG. There are two kinds of cellular currents, intracellular and extracellular. MEG mainly records intracellular currents. EEG records both but mainly extracellular currents.

To record clear MEG, at least 20,000 or 30,000 neurons must be activated simultaneously, which causes the same directed intracellular currents. A summation of the currents of many neurons with the same positive–negative direction can be mimicked as one strong dipole. Since it is easy and simple to imagine it present in the cortex, we hypothesized it by naming the ▶ equivalent current dipole (ECD).

Characteristics

At first, the advantages of MEG compared with EEG will be introduced. When electric currents generated in the cortex are recorded using scalp electrodes, there are effects of cerebrospinal fluid, skull, and skin, whose electric conductivities vary markedly. In contrast, since magnetic fields are not affected by current conductivity, the recorded MEG is theoretically unchanged. Therefore, the spatial resolution of MEG is higher than that of EEG.

The advantages and disadvantages of MEG compared with positron emission tomography (PET) and functional magnetic resonance imaging (fMRI) are listed below.

Advantages of MEG
1. Completely noninvasive.
2. Measures neuronal activity rather than blood flow changes or metabolic changes.
3. ▶ Temporal resolution is much larger, in the order of ms.
4. Stimulus-evoked or event-related responses can easily be measured in detail.
5. Frequency response (brain rhythm) analysis can be done, which means physiological changes can be analyzed spatiotemporally.
6. Results in an individual subject can be analyzed in detail, so that averaging results in

a number of subjects is not necessary. In other words, one can analyze interindividual differences.

Disadvantages of MEG
1. Spatial resolution is lower than for PET and fMRI, particularly when the ▶ signal-to-noise ratio is low.
2. An inverse problem solution program is necessary. In other words, measuring results are indirectly or artificially induced.
3. The quality of algorithms (solution programs) is not good enough at present when multiple areas are activated simultaneously. Therefore, it is sometimes difficult to use MEG to analyze long-latency components mainly relating to emotional and/or cognitive functions. However, this is an endless game, since users always want new and improved software.
4. It is difficult to detect magnetic fields generated in deep areas. Therefore, the smaller the distance between activated regions and detecting coils, the better.
5. It is impossible to record activities generated in white matter and pathways, since EPSP are not generated there.
6. It is difficult or impossible to record activities in some regions such as the thalamus showing a so-called physiological closed field.
7. Activated location measured by MEG must be overlaid on CT or MRI, but the location can change, though the change must be very small.

Considering the advantages and disadvantages of MEG outlined above, the temporal characteristics of primary components just after the period when signals ascending through ▶ Aδ fibers and ▶ C fiber reach the cortex are the best indication for MEG, and later activities relating to cognition are mainly analyzed by PET and fMRI.

Methods
Various methods are used to record pain-related SEP and SEF (see review by Kakigi et al. 2000a, b, 2003a, 2005). The first study was reported by Hari et al. based on dental pulp stimulation (Hari et al. 1983). Then, CO_2 gas was applied to the nasal mucosa, a painful impact stimulation (Arendt-Nielsen et al. 1999), and

epidermal electrical stimulation was applied (see review by Kakigi et al. 2000a, b, 2003a, 2005). Each method has its own advantages and disadvantages, but the ideal pain stimulation is pain specific, controllable, safe, and repeatable. At present, there are two main methods for recording pain-related SEP (SEF): (1) SEP (SEF) following high-intensity painful electrical stimulation (see review by Kakigi et al. 2000a, b) and (2) SEP (SEF) following painful CO_2 ► laser beam stimulation (see reviews by Bromm and Lorentz 1998; Kakigi et al. 2000a, b, 2003a, 2005). Since the latter method, which is usually called pain-related SEP (SEF), or laser-evoked potential (LEP), or magnetic field (LEF), has several advantages as described below, it is more popular.

1. Other methods causing pain or heat sensation such as needle stimulation of the skin activate not only nociceptive receptors but also mechanoreceptors. Therefore, for example, the SEP waveform recorded following needle stimulation is very similar to that following electrical stimulation. In contrast, since a CO_2 laser beam is light, it does not activate mechanoreceptors of the skin, that is, it is a purely noxious stimulation.

2. For analyzing temporal information in the order of msec, the time difference (lag) between the stimulus timing and the beginning of the sweep of the computer should be very stable and short, less than 1 ms. Figure 1 shows the procedure using our MEG device. A laser beam is applied to the subject's hand through optical fibers.

One interesting method reported recently is electrical stimulation of a very short needle, whose tip is located in the epidermis where only free nerve endings are present, named ES stimulation (Inui et al. 2003a, b). The biggest advantages of this method are as follows: (1) Only Aδ and C fibers are stimulated. (2) When the needle is inserted, subjects feel no uncomfortable painful feeling and show no bleeding. (3) Since a small intensity with a short duration is enough for recording Aδ fiber-related MEG, subjects feel only a tolerable tingling pain. (4) No special device is necessary except for a hand-made

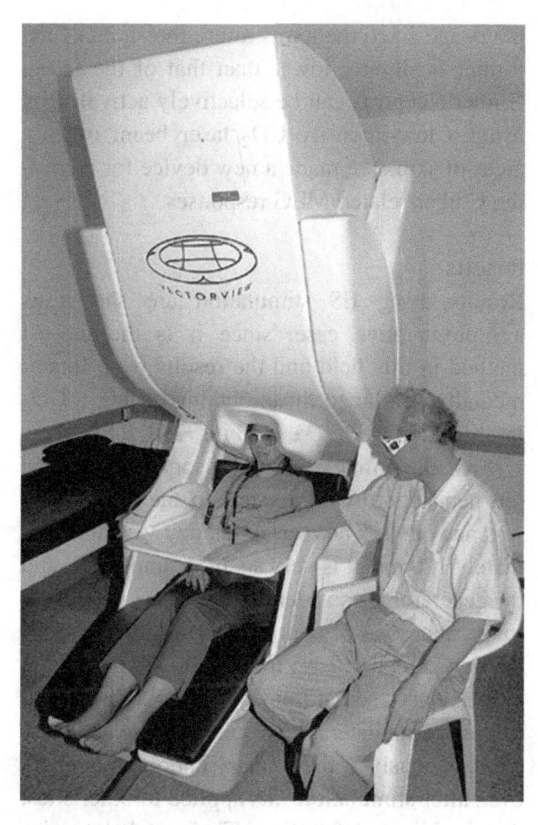

Magnetoencephalography in Assessment of Pain in Humans, Fig. 1 The procedure using our MEG device (VectorView 306 channels, Elekta Neuromag Oy, Helsinki). The laser stimulator is set outside the shielded room because of its large magnetic artifacts, and the laser beam is applied to the subject's hand through optical fibers. The laser beam can be applied to any part of the body except the eyes, so both the experimenter and the subject must wear special glasses or swimming goggles

short needle, which is easily made. Recently, we reported that ES stimulation with specific stimulus parameters selectively activates cutaneous C fibers (Otsuru et al. 2009).

One of the biggest recent topics in this field is the MEG response to the signals ascending through unmyelinated C fibers (see review by Kakigi et al. 2003a, 2005). Several methods have been reported to selectively stimulate C fibers, but they were difficult to record and the responses obtained were not consistently recorded. Our method was based on that reported by Brussels's group (Bragard et al. 1996). Since the number of polymodal receptors of C fibers relating to second pain is larger than that of Aδ fibers and since the temperature threshold of the former is slightly lower than that of the latter, C fiber receptors can be selectively activated by using a low-intensity CO_2 laser beam on tiny areas of skin. We made a new device for recording C fiber-related MEG responses.

Results

Results using ES stimulation are shown as a representative case, since it is the newest method in this field and the results were fundamentally similar to those obtained using a CO_2 laser (Inui et al. 2003a, b) (Figs. 2, 3).

First, the primary somatosensory cortex (SI) in the hemisphere contralateral to the stimulation was activated, whose peak latency was approximately 100 ms. It is very small in amplitude, and the generator is considered to be area 1 in SI. This long latency is, of course, due to the slow conduction velocity of $A\delta$ fibers (10–20 m/s). Then, ▶ secondary somatosensory cortex (SII) and insula in the bilateral hemispheres were activated simultaneously as the primary major component, even after stimulation was applied to other sites, that is, bilateral function. Their peak latencies were approximately 150 ms, but ipsilateral responses were significantly longer than contralateral ones, probably through the corpus callosum. SI component in the contralateral hemisphere was also recorded by laser stimulation (Kanda et al. 2000; Ploner et al. 1999). Then, cingulate cortex and mid-temporal regions around the amygdala in the bilateral hemispheres were activated; peak latencies of the first negative and second positive components were approximately 200 and 300 ms, respectively. We believe that MEG is the most appropriate method for a detailed spatiotemporal analysis of these early activities within 300–400 ms after stimulation, which PET and fMRI cannot do.

Factors That Affect the Waveforms of Pain

One frequently sustains an injury while playing sport but does not notice it until after the game. This clearly indicates that psychological conditions affect pain perception. These interesting findings were also confirmed by MEG (see reviews by Kakigi et al. 2000a, b, 2003b;

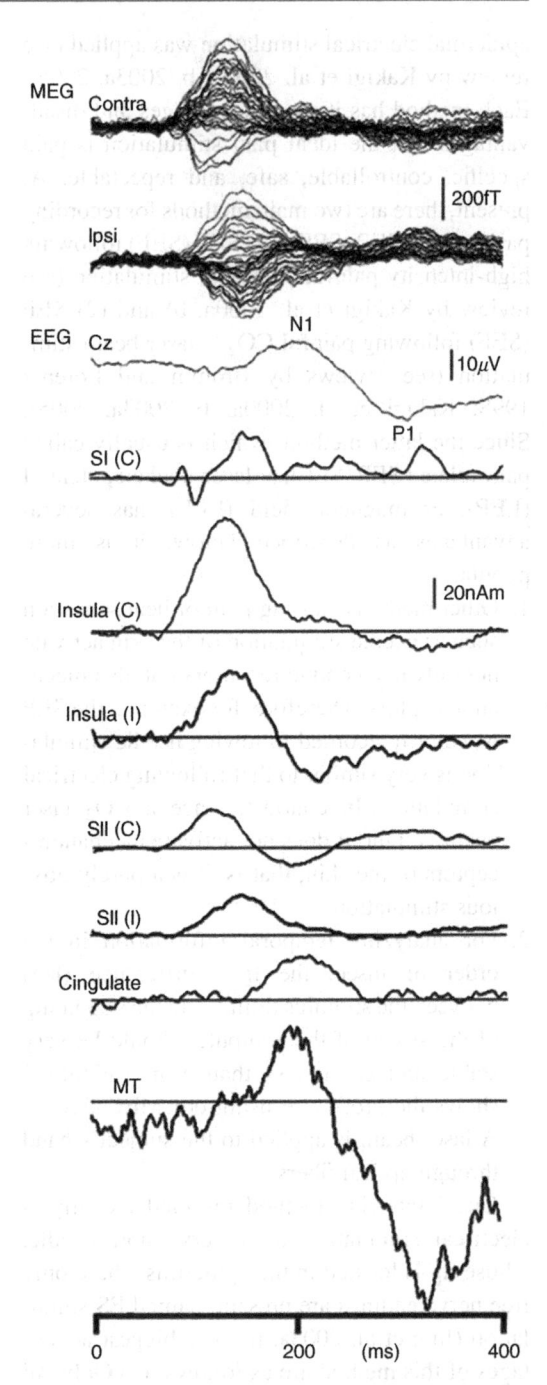

Magnetoencephalography in Assessment of Pain in Humans, Fig. 2 MEG waveforms measured using the ES method (intraepidermal electrical needle stimulation) for primary pain perception in humans by activating $A\delta$ fiber stimulation. *C* and *I*: Hemispheres contralateral and ipsilateral to the stimulation, respectively. *SI* and *SII*: Primary and secondary somatosensory cortices, respectively. *MT* mid-temporal region around amygdala (Applied from Inui et al. 2003a, b)

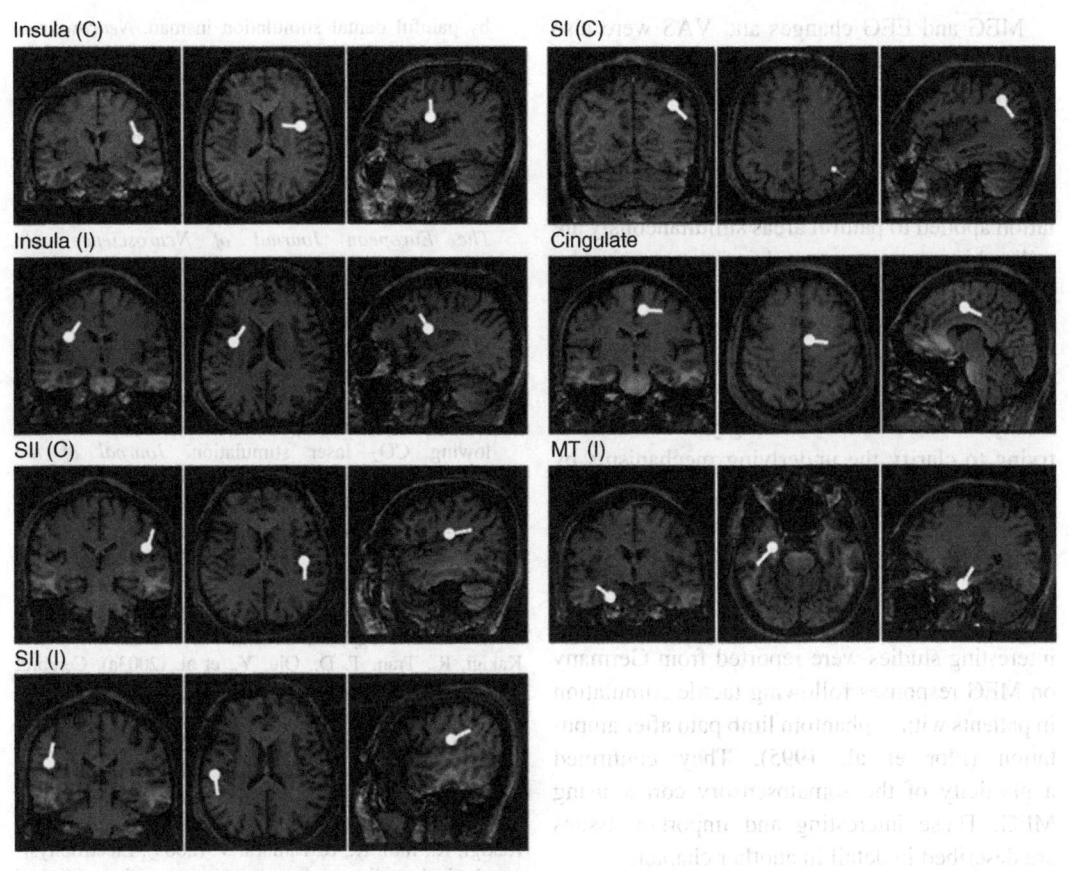

Magnetoencephalography in Assessment of Pain in Humans, Fig. 3 Activated regions using the ES method (intraepidermal electrical needle stimulation) for primary pain perception in humans by activating Aδ fiber stimulation. The calculated equivalent current dipole (ECD) of each region was overlaid on the MRI of this subject. See also legend of Fig. 2 (Applied from Inui et al. 2003)

Kakigi 2005; Kakigi and Watanabe 1996). By use of MEG, tactile effects on pain, for example (Inui et al. 2006), can be analyzed in detail. When subjects paid close attention to a mental activity such as a calculation or memorization (distraction task), MEG responses, particularly the later components generated in limbic systems, were much reduced in amplitude or disappeared, and a mental pain rating (▶ visual analogue scale, VAS) showed a positive correlation with the MEG changes. In contrast, early responses in SI and SII were reduced but still present. Therefore, early components seem important for primary functions such as the localization of stimulus points, but later components are very related to cognitive functions of pain perception. This pattern was more remarkable

after C fiber than Aδ stimulation. Therefore, second pain may be more related to cognitive functions than the first pain.

During sleep, we usually do not notice pain, unless it is very strong, so what happens in the brain while receiving painful sensations during sleep (see review by Kakigi et al. 2003b)? Most of the MEG responses, particularly later components, were much reduced in amplitude or disappeared. This finding is very similar to the changes in the distraction task but more remarkable. Interestingly, MEG responses during sleep to other modalities, such as touch, auditory, or visual stimuli, were enhanced, probably due to an inhibition of the inhibitory system while awake. Therefore, pain perception seems very different from other kinds of sensations.

MEG and EEG changes and VAS were also significantly reduced in amplitude by other kinds of sensations such as touch, vibration, cooling, and movements (see review by Kakigi et al. 2000a, b; 2003a, 2005; Kakigi and Watanabe 1996). The effects of pain relief by tactile stimulation applied to painful areas simultaneously are utilized in transcutaneous electric nerve stimulation (TENS). I think "movement" is most effective for pain relief, since we automatically move painful regions to reduce pain. For example, if we touch a very hot object with our hand, we involuntary shake that hand strongly. We are now trying to clarify the underlying mechanisms by MEG.

Clinical Application

Unfortunately, the clinical application of MEG for pain relief is not very popular at present. Most interesting studies were reported from Germany on MEG responses following tactile stimulation in patients with ▶ phantom limb pain after amputation (Flor et al. 1995). They confirmed a plasticity of the somatosensory cortex using MEG. These interesting and important issues are described in detail in another chapter.

The following three review articles are recommended on the basic physiological and clinical background of pain including MEG (Bromm and Lorentz 1998; Treede et al. 1999, 2000).

References

Arendt-Nielsen, L., Yamasaki, H., Nielsen, J., et al. (1999). Magnetoencephalographic responses to painful impact stimulation. *Brain Research, 839*, 203–208.

Bragard, D., Chen, A. C. N., & Plaghki, L. (1996). Direct isolation of ultra-late (C-fibre) evoked brain potentials by CO_2 laser stimulation of tiny cutaneous surface areas in man. *Neuroscience Letters, 209*, 81–84.

Bromm, B., & Lorentz, J. (1998). Neurophysiological evaluation of pain. *Electroencephalography and Clinical Neurophysiology, 107*, 227–253.

Flor, H., Elbert, T., Knecht, S., et al. (1995). Phantom-limb pain as a perceptual correlate of cortical reorganization following arm amputation. *Nature, 375*, 482–484.

Hari, R., Kaukoranta, E., Reinikainen, K., et al. (1983). Neuromagnetic localization of cortical activity evoked by painful dental stimulation in man. *Neuroscience Letters, 42*, 77–82.

Inui, K., Tran, T. D., Qiu, Y., et al. (2003a). A comparative magnetoencephalographic study of cortical activations evoked by noxious and innocuous somatosensory stimulations. *Neuroscience, 120*, 235–248.

Inui, K., Wang, X., Qiu, Y., et al. (2003b). Pain processing within the primary somatosensory cortex in humans. *The European Journal of Neuroscience, 18*, 2858–2866.

Inui, K., Tsuji, T., & Kakigi, R. (2006). Temporal analysis of cortical mechanisms for pain relief by tactile stimuli in humans. *Cerebral Cortex, 16*, 355–365.

Kakigi, R., & Watanabe, S. (1996). Pain relief by various kinds of interference stimulation applied to the peripheral skin in humans: pain-related brain potentials following CO_2 laser stimulation. *Journal of the Peripheral Nervous System, 1*, 189–198.

Kakigi, R., Watanabe, S., & Yamasaki, H. (2000a). The pain-related somatosensory evoked potentials. *Journal of Clinical Neurophysiology, 17*, 295–308.

Kakigi, R., Hoshiyama, M., Shimojo, M., et al. (2000b). The somatosensory evoked magnetic fields. *Progress in Neurobiology, 61*, 495–523.

Kakigi, R., Tran, T. D., Qiu, Y., et al. (2003a). Cerebral responses following stimulation of unmyelinated C-fibers in humans: Electro- and magneto-encephalographic study. *Neuroscience Research, 45*, 255–275.

Kakigi, R., Naka, D., Okusa, T., et al. (2003b). Sensory perception during sleep in humans: A magnetoencephalographic study. *Sleep Medicine, 4*, 493–507.

Kakigi, R., Inui, K., & Tamura, V. (2005). Electrophysiological studies on human pain perception. *Clinical Neurophysiology, 116*, 743–763.

Kanda, M., Nagamine, T., Ikeda, A., et al. (2000). Primary somatosensory cortex is actively involved in pain processing in human. *Brain Research, 853*, 282–289.

Otsuru, N., Inui, K., Yamashiro, K., et al. (2009). Selective stimulation of C fibers by an Intra-Epidermal needle electrode in humans. *The Open Pain Journal, 2*, 53–56.

Ploner, M., Schmitz, F., Freund, H. J., et al. (1999). Parallel activation of primary and secondary somatosensory cortices in human pain processing. *Journal of Neurophysiology, 81*, 3100–3104.

Treede, R. D., Kenshalo, D. R., Gracely, R. G., et al. (1999). The cortical representation of pain. *Pain, 79*, 105–111.

Treede, R. D., Apkarian, A. V., Bromm, B., et al. (2000). Cortical representation of pain: Functional characterization of nociceptive areas near the lateral sulcus. *Pain, 87*, 113–119.

Magnetometer

▶ Magnetoencephalography in Assessment of Pain in Humans

Magnification

Definition

The tendency to exaggerate the threat value associated with a particular symptom, situation, or outcome.

Cross-References

▶ Catastrophizing

Magnocellular Neurons

Definition

Magnocellular neurons are neurons with soma of a large size located in the paraventricular and supraoptic hypothalamic nuclei. These neurons synthesize vasopressin and oxytocin and transport them via their axons to the posterior neurohypophysis.

Cross-References

▶ Hypothalamus and Nociceptive Pathways

Maitland Mobilization

▶ Passive Spinal Mobilization

Major Analgesics

▶ Oral Opioids

Major Depressive Disorder

▶ Depression and Pain

Major or Minor Causalgia

▶ CRPS, Evidence-Based Treatment

Maladaptive Coping

▶ Catastrophizing

Maladaptive Thoughts

Definition

Cognitions that interfere with successful coping and adaptation to noxious stimuli or aversive events or contribute to inappropriate behavior.

Cross-References

▶ Multiaxial Assessment of Pain

Malalignment Syndromes

Definition

Malalignment syndromes are abnormal body alignments (e.g., an abnormally pronated rear foot) thought to predispose overuse syndromes.

Cross-References

▶ Stretching

Malignant Bone Pain

▶ Adjuvant Analgesics in Management of Cancer-Related Bone Pain

Malignant Bowel Obstruction

Definition

A mechanical obstruction of the gastrointestinal tract secondary to a malignant tumor that occurs in 5–43 % of all terminally ill patients. The obstruction may be partial or complete and at one site or multiple sites; however, the small bowel is more commonly involved than the large bowel. It is associated with colorectal, pancreatic, endometrial, gastric, mesothelial, breast, and prostatic cancers but may develop with any cancer predisposed to metastases to the gastrointestinal tract. Malignant bowel obstruction may occur secondary to extrinsic compression of the bowel lumen, intraluminal occlusion, intramural occlusion, intestinal motility disorders, and miscellaneous causes such as fecal impaction.

Cross-References

▶ Cancer Pain Management, Adjuvant Analgesics in Management of Pain Due To Bowel Obstruction

Malignant Pain

Definition

Pain associated with cancer.

Cross-References

▶ Cancer Pain
▶ Cancer Pain, Animal Models
▶ Pain Treatment, Implantable Pumps for Drug Delivery

Malingerer

Definition

A malingerer is an individual who consciously and deliberately feigns incapacitation. Malingering is a form of fraud.

Cross-References

▶ Impairment, Pain-Related

Malingering

Definition

A conscious decision to deliberately fake or lie about having pain in order to fool someone.

Cross-References

▶ Credibility, Assessment
▶ Ethics of Pain-Related Disability Evaluations
▶ Pain in the Workplace, Risk Factors for Chronicity, Job Demands

Malingering, Primary and Secondary Gain

David A. Fishbain
Department of Psychiatry, University of Miami, Miami, FL, USA

Synonyms

Reinforcers; Sick role

Definition

Primary gain (Fishbain 1994; Fishbain et al. 1995): A decrease in anxiety (gain) from an unconscious defensive operation, which then causes a physical or conversion symptom, e.g., an arm is voluntarily paralyzed because it was used to hurt somebody, thereby allaying guilt and anxiety.

Secondary gain (Fishbain 1994; Fishbain et al. 1995): The gain achieved from the physical or conversion symptom, which enables the patient to avoid a particularly noxious activity or which enables the patient to get support from the environment (gain) not otherwise forthcoming.

Malingering (Fishbain et al. 1999, 2002): It is the intentional production of false or grossly exaggerated physical or psychological symptoms motivated by external incentives (secondary gain), such as avoiding military duty or work or criminal prosecution and obtaining financial compensation or drugs.

Characteristics

Primary and Secondary Gain

Sigmund Freud was the first to define primary and secondary gain and apply these concepts to illness behavior, which was not medically explained. Since then, the concept of secondary gain has infiltrated the nomenclature of every medical specialty, usually being applied in case of medically unexplained symptoms and/or illness-affirming states. The following is a list of previously described secondary gains: gratification of dependency or revengeful strivings; fulfillment of needs for sympathy, concern, solicitousness, or attachment; maintenance of family status, love, or domination; desire for financial rewards and to prove entitlement for disability; avoidance of hazardous work conditions; permission to withdraw from an unsatisfactory life or socioemotional role; need for the ▶ sick role; acquisition of drugs; manipulation of spouse; and contraception. It is not clear whether secondary gains are the same as reinforcers.

Operationally, some are equivalent; the gain may be the reinforcer. Secondary gains, however, are the more unconscious motivation for the observed behaviors.

▶ Secondary losses has been described as follows: economic; inability to now relate to others through work; loss of family life, social support, and recreational activities; loss of community approval and resultant social stigma of being disabled; guilt over disability; negative sanctions from family and those in the helping roles (e.g., doctors); and loss of a clearly defined role. In general, the secondary losses far outweigh the secondary gains. Yet, patients act in spite of this economy. This problem with the economy of secondary gains and losses is a direct challenge to the integrity of the secondary gain concept and its importance to some of the illness-affirming states.

Scientifically, there are also major problems with the secondary gain concept. It is poorly defined and rests on psychoanalytic concepts, which are difficult to prove or disprove. Fishbain et al. (1995) recently reviewed the scientific evidence for this concept. They found that results of studies relating to this issue were in conflict, and many had methodological flaws relating to how the presence of secondary gain was established. However, the presence of a secondary gain agenda did change patient behaviors.

The secondary gain concept is also clinically abused. The presence of financial incentives, such as disability payments if associated with treatment failure, results in the accusation of secondary gain, which in turn is utilized as a rationalization for the treatment failure. Clinicians in this paradigm usually focus on the secondary gains and ignore the secondary losses. In addition, clinicians ignore two facts. The alleged presence of a secondary gain does not necessarily mean that the gain has had an etiological or reinforcing effect on the illness, and there is a significant amount of evidence that indicates that physician practices are influenced by secondary gain agendas (Fishbain et al. 1995).

Malingering

Malingering can be divided into the following types (Fishbain et al. 1999): Pure malingering is the feigning of disease or disability when it does not exist. Positive malingering (simulation) is the feigning of symptoms that do not exist. Partial malingering is the conscious exaggeration of symptoms that do exist. False imputation is the ascribing of actual symptoms to a cause consciously recognized to have no relationship to the symptoms. Dissimulation, also a form of malingering, is the concealment or minimization of symptoms for secondary gain reasons. Two other types of malingering may also occur: Genuine symptoms formerly present may cease to exist but may be fraudulently alleged to continue, or genuine symptoms may be fraudulently attributed to a cause other than the actual one.

There are no reliability studies on malingering, as these can only be generated if someone admits to being a malingerer. This is rarely the case. As such, three unique approaches for the study of malingering/disease simulation have evolved. In the first approach, a patient suspected of malingering is assigned a task that impacts on this alleged impairment or symptoms. This task is such that the patient is forced to make a choice (e.g., the patient complaining of sensory loss will be asked to close his or her eyes and guess which hand was being touched for a large number of trials). This is called a forced-choice technique. In this case, the patient has a 50 % chance of being correct if he has a sensory impairment. However, hysterical and malingering patients usually will perform at significantly below chance levels, indicating that they may be deliberately providing false answers.

The second approach involves a volunteer group performing a task such as completing a psychological test battery pretending to fake an illness (e.g., fake insanity). The resultant volunteer "faking scores" can be used in two ways. First, "the faking scores" can be compared with normal scores for differences and for the development of a "faking" profile. Second, the "faking scores" can be compared with authentic patient scores for differences and perhaps for the selection of similar faking profiles.

The third approach involves actual patients. In this case, a patient or groups of patients are asked to complete a task (e.g., a psychological battery) in such a way as to simulate an illness.

These faking patient scores may then be compared with scores from a new group of patients for identification of similar or different profiles. Fishbain et al. (1999) recently reviewed these different approaches in reference to pain. Their findings/conclusions were as follows: Malingering and dissimulation do occur within the chronic pain patient setting. Malingering may be present in 1.25–10.4 % of chronic pain patients. However, because of poor study quality, these prevalence percentages are not reliable. The study evidence also indicated that facial expression testing, questionnaire, sensory testing, or clinical examination were not reliable ways of identifying malingering. There was no acceptable scientific information on symptom magnification syndrome. Handgrip testing using the Jamar Dynamometer and other types of isometric strength testing did not reliably discriminate between a submaximal/malingering effort and a maximal/best effort. However, isokinetic strength testing appeared to have potential for discriminating between maximal and submaximal effort and between best and malingered efforts. Repetitive testing with the coefficient of variation was not a reliable method for discriminating a real/best effort from a malingered effort. It was concluded that as yet there is no reliable method for detecting malingering within chronic pain patients, although isokinetic testing shows promise.

A final issue relating to malingering is that of ▶ Waddell signs. These are eight physical signs frequently found in chronic pain patients. Historically, there has been disagreement on what these signs indicate. However, in a recent evidence-based structured review, Fishbain et al. (2003) concluded the following about Waddell signs:

- They do not discriminate organic from nonorganic problems.
- May represent an organic phenomenon.
- Are associated with poorer treatment outcome.
- Are associated with greater pain levels.

- Are not associated with secondary gain.
- As a group, studies demonstrate methodological problems.

In a further evidence-based review, specifically addressing the issue of whether Waddell signs indicate malingering, Fishbain et al. (2004) concluded the following: There is little or no evidence that Waddell signs are associated with malingering.

References

Fishbain, D. A. (1994). The secondary gain concept: Definition problems and its abuse in medical practice. *Journal of the American Pain Society, 3*, 264–273.

Fishbain, D. A., Rosomoff, H. L., Cutler, R. B., & Rosomoff, R. S. (1995). Secondary gain concept: A review of the scientific evidence. *The Clinical Journal of Pain, 11*, 6–21.

Fishbain, D. A., Cutler, R., Rosomoff, H. L., & Rosomoff, R. S. (1999). Chronic pain disability exaggeration/malingering and submaximal effort research. *The Clinical Journal of Pain, 15*, 244–274.

Fishbain, D. A., Cutler, R., Rosomoff, H. L., & Rosomoff, R. S. (2002). Does the conscious exaggeration scale detect deception within patients with chronic pain alleged to have secondary gain? *Pain Medicine, 3*, 39–46.

Fishbain, D. A., Cole, B., Cutler, R. B., Lewis, J., Rosomoff, H. L., & Rosomoff, R. S. (2003). A structured evidence-based review on the meaning of nonorganic physical signs: Waddell signs. *Pain Medicine, 4*, 141–181.

Fishbain, D. A., Cutler, R. B., Lewis, J., Cole, B., Rosomoff, R. S., & Rosomoff, H. L. (2004). Do Waddell signs indicate malingering? An evidence-based structured review. *The Clinical Journal of Pain, 20*, 399–408.

Malnutrition

▶ Metabolic and Nutritional Neuropathies

Managed Care

Definition

A board term used to describe various health-care payment systems that attempt to contain costs by controlling the type and level of services provided.

Cross-References

▶ Disability Management in Managed Care System

Management of Postoperative Pain

▶ Postoperative Pain, Appropriate Management

Mandibular Dysfunction

▶ Orofacial Pain, Movement Disorders

Manipulation Without Impulse

▶ Passive Spinal Mobilization

Manipulation Without Thrust

▶ Passive Spinal Mobilization

Manual or Continuous Traction

▶ Traction

Manual Therapy

Definition

Manipulation of body tissues to restore movement.

Cross-References

▶ Chronic Pain in Children, Physical Medicine and Rehabilitation

MAPK

▶ Mitogen-Activated Protein Kinase

MAPK Family

▶ ERK Regulation in Sensory Neurons During Inflammation

Marital Status and Chronicity

▶ Pain in the Workplace, Risk Factors for Chronicity, and Demographics

Marstock Method

Definition

The Marstock method is a threshold determination protocol for the thermal senses. This is a reaction time-dependent protocol that was originally developed by Drs. Heinrich Fruhstorfer and W. Schmidt of Marburg and Dr. Ulf Lindblom of Stockholm, hence the name.

Cross-References

▶ Threshold Determination Protocols

Massage and Pain Relief Prospects

Michael Quittan
Department of Physical Medicine and Rehabilitation, Kaiser-Franz-Joseph Hospital, Vienna, Austria

Synonyms

Effleurage; Friction; Petrissage; Tapotement; Vibration

Definition

The English word "massage" is derived from the Arabic word "mass'h" which means to press gently. At its most basic, massage is a simple way of easing pain, while at the same time aiding relaxation, promoting a feeling of well-being and a sense of receiving good care. Scientifically, massage may be defined as group of systematic and scientific manipulations of body tissues best performed with the hands, for the purpose of affecting the nervous and muscular system and the general circulation.

Although massage techniques may vary considerably between therapists, classical massage consists of a number of basic techniques that have remained essentially unchanged for centuries. Per Henrik Ling is considered the founder of modern massage, and subsequently these techniques are referred to as Swedish massage. Five elemental techniques are used in classical massage named effleurage, petrissage, friction, tapotement, and vibration. Specific purposes are attributed to each technique.

Effleurage
Effleurage (or stroking) is a deep stroking performed in the direction of venous or lymph flows. It is used to accustom the subject to the following massage procedures.

Petrissage
Petrissage (or kneading) is a deeper technique than effleurage and consists of repeated rolling, grasping, and lifting. The technique is directed towards muscles and connective tissue.

Friction
Friction (or rubbing) should be done in a slow elliptical or circular movement with the fingertips or the thenar eminence. It is a deep penetrating technique to relieve trigger points. There may be pain from the application of deep friction massage so this technique is optional. It should not be confused with deep transverse frictions.

Tapotement

Tapotement (or hacking) is a series of blows that may also be described as slapping, tapping, clapping, hacking, or cupping, depending on the positioning of the hands.

Vibration

Vibration (or shaking) consists of small tremulous movements with the hands and fingers and in some cases with an electromechanical device. The vibration is performed rhythmically in an attempt to enhance relaxation or stimulation.

Before initiating massage, the patient's individual preferences and responsiveness to touch and massage should be assessed. Cultural and social feelings towards massage should be considered, as well as the patient's ability to communicate concerns or express unwillingness to receive massage therapy. Deep transverse friction massage is a technique popularized by Dr. James Cyriax for pain and inflammation relief in musculoskeletal conditions (Cyriax 1975). It is a technique that attempts to reduce abnormal fibrous adhesions and makes scar tissue more mobile in subacute and chronic inflammatory conditions by realigning the normal soft tissue fibers. The technique has been advocated to enhance normal healing conditions by breaking crossbridges and preventing abnormal scarring. Its mechanical action causes hyperemia, which results in increased blood flow to the area.

Besides these classic techniques, there are numerous other procedures claiming specific effects. Reflexologists propose that there are reflex points on the feet corresponding to organs and structures of the body and that pain may be reduced by gentle manipulation or pressing certain parts of the foot. Pressure applied to the feet has been shown to result in an anesthetizing effect on other parts of the body (Ernst and Koeder 1997).

Characteristics

Analgesic Mechanisms

Classical massage is thought to improve physiological and clinical outcomes by offering the symptomatic relief of pain through different mechanisms. Massaging a particular area stimulates large-diameter nerve fibers, thus mechanisms described by the gate-control theory. Additionally, a moderate elevation of serum beta-endorphin levels lasting 1 h after cessation of massage has been demonstrated (Goats 1994). Motoneuron activity may be attenuated during application of massage, thus beneficially influencing the "pain-spasm-pain" cycle. It is noteworthy that massage both affects the motoneuron pool of the massaged muscles and lasts solely during application of massage (Dishman and Bulbulian 2001). It has recently been shown that intramuscular temperature in superficial layers of the m. vastus lateralis is increased after application of massage, mainly due to mechanical manipulation of the skin (Drust et al. 2003). Blood flow is increased, predominantly by dilatation of superficial blood vessels. Vigorous massage techniques enhance these effects by additional histamine release, which may then last as long as 1 h (Goats 1994). Finally, massage promotes physical and mental relaxation contributing to pain relief.

Indications and Outcomes

Despite definitions described above, there have been many variations, with differences in masseur expertise and time of application. This lack of standardization renders comparison of studies difficult. Nevertheless, massage has been increasingly investigated in the pain management area, for its potential to be a functional, nonpharmacological intervention for reducing chronic pain.

Back Pain

Massage therapy is among the most popular therapeutic strategies used by back pain patients. A survey of an urban rehabilitation medicine outpatient office in New York, addressing the use of alternative therapy and its perceived effectiveness, indicated that 29 % of the subjects had used one or more alternative medical therapies in the past 12 months and the most common therapy cited was massage (Wainapel et al. 1998). This evidence is confirmed by European data indicating that up to 87 % of back pain patients received massage as one form of treatment (Wiesinger et al. 1997).

Until recently, massage was included in RCTs only as a control treatment for various physical treatments. These massage control treatments were poorly described and often involved superficial massage techniques, brief treatment sessions of 10–15 min or few sessions (<5) (for reviews see Ernst (1999) and Furlan et al. (2002)). Recent published trials therefore address the effectiveness of massage therapy for back pain. Preyde randomly assigned 104 patients with low-back pain (lasting 1 week–8 months) to comprehensive massage therapy including stretching exercises, soft tissue manipulation alone, remedial exercise with posture education, and sham laser therapy. Over a 1-month period, all patients received six therapies. At 1-month follow-up, 63 % of subjects in the comprehensive massage therapy group reported no pain as compared with 27 % of the soft tissue manipulation group, 14 % of the remedial exercise group, and 0 % of the sham laser therapy group. Clinical significance was evident for the comprehensive massage therapy group and the soft tissue manipulation group on the measure of function (Preyde 2000).

Another RCT compared massage therapy with progressive muscle relaxation (Hernandez-Reif et al. 2001). Twenty-four adults with low-back pain of nociceptive origin with a duration of at least 6 months were randomly assigned to a massage therapy or a progressive muscle relaxation group. Sessions were 30 min long twice a week for 5 weeks. By the end of the study, the massage therapy group, as compared to the relaxation group, reported experiencing less pain, depression and anxiety, and improved sleep. They also showed improved trunk and pain flexion performance. Furthermore, their serotonin and dopamine levels were higher. The authors conclude that massage therapy is effective in reducing pain, stress hormones, and symptoms associated with chronic low-back pain. Finally, Cherkin et al. randomly assigned 262 patients with low-back pain to receive therapeutic massage compared to acupuncture or self-educational material. After the 10-week

treatment period, massage was found to be superior both to acupuncture (function) and self-education (function and symptoms). At 1-year follow-up, massage remained superior to acupuncture regarding function and symptoms (Cherkin et al. 2001).

In their updated review, Furlan, Imamura, and co-workers included 13 randomized studies with 1,596 participants. It is noteworthy that nine studies were published after the first version of the review in 2002. Only two studies compared massage to sham therapies and showed that massage was superior for pain and function on both short- and long-term observations.

The other studies compared massage to conservative therapies such as exercises, mobilization, relaxation, physical therapy, acupuncture, and self-care education. The overall result is that massage was found to be similar to or superior than the control conditions.

Therefore, massage is beneficial for patients with subacute and chronic nonspecific low-back pain, especially when combined with exercises and education. Additionally, present evidence suggests that acupuncture massage may be more effective than classic massage, but this should be confirmed. Furthermore, massage therapy may save money in health-care provider visits, pain medications, and costs of back care services (Furlan et al. 2009).

Neck Pain

In their evidence-based clinical practice guidelines, the authors of the Ottawa Panel demonstrated that therapeutic massage decreases pain and tenderness and improves range of motion for subacute and chronic neck pain. Despite long-term treatment effects were lacking, there is clear evidence for immediate posttreatment effects (Brosseau et al. 2012).

In a systematic review, Patel and co-workers found among 15 studies only 4 describing massage techniques appropriately. Despite major shortcomings in overall methodological quality of the studies, the conclusion is that massage as a stand-alone treatment was found to provide an immediate or short-term effectiveness or both in pain and tenderness (Patel et al. 2012).

Osteoarthritis

Perlman and co-workers subjected 68 patients with OA of the knee randomly to a massage treatment consisting of biweekly (4 weeks), then weekly (4 weeks) Swedish massage treatments or a waiting list serving as control condition. Massage therapy lasted for 1 h and comprised the therapist's choice of all classical massage techniques. Subjects receiving massage therapy demonstrated significant and clinically meaningful improvements in the Western Ontario and McMaster Universities Osteoarthritis Index (WOMAC) stiffness and physical functional disability domains and visual analogue pain scale. The benefits persisted up to 8 weeks following the cessation of massage (Perlman et al. 2006).

In order to identify the optimal dose of massage within an 8-week treatment regimen and to further examine durability of response, the same group performed a dose-finding trial with 30 or 60 min weekly or biweekly massage therapy. One hundred twenty-five patients with OA of the knee were randomly allocated to a five-arm study, containing weekly or biweekly massage 30 or 60 min each. The fifth group was a control group with usual care. Interestingly, massage therapy was not only focused on the affected joint but also comprised remote areas of the body like lower and upper back, head, neck, and chest, and if needed rib cage, flank, upper limbs, etc.

The authors published a dose–response curve plotting dose (total minutes over the course of 8 weeks of massage) (x-axis) vs. improvement (change in WOMAC Global scores after 8 weeks). Dose–response effects plateaued at 480 min, with no significant improvements noted in the 720-min group. Therefore, the authors conclude that a once-weekly protocol, cost savings, and consistency with a typical real-world massage protocol, the 60-min once-weekly dose was determined to be optimal (Perlman et al. 2012).

In conditions of unspecific affections of the shoulder joint soft tissue, massage was found effective for producing moderate improvements in active flexion and abduction range of motion, pain, and functional scores compared with no treatment, immediately after the cessation of treatment (van den Dolder et al. 2012).

Cancer-Related Pain

At present, evidence exists from multiple time series with or without the intervention that massage is beneficial in pain relief in cancer patients (Pan et al. 2000). In their review, Pan and co-workers identified two studies evaluating massage as single treatment for pain in palliative care patients and one study in combination with aromatherapy. In an unblinded randomized controlled trial, 28 patients (mean age 61.5 years) with cancer were assigned to either Swedish massage therapy or a visitor for 10 min. Men experienced immediate pain relief (VAS from 4.2 to 2.9, $P < 0.01$), but this effect subsided by 1 h after the massage. There was no significant benefit in women (Weinrich and Weinrich 1990). In a case series, nine male cancer patients (mean age 56.6 years) who received two consecutive 30-min evening massages reported significant reductions in pain as compared to baseline, lasting for 2 days. There was also a reduction in anxiety and enhanced feelings of relaxation (Ferrell-Torry and Glick 1993). In a case series involving 103 patients with cancer, a combination of massage and aromatherapy promoted pain relief in 33 % of patients who completed the study (47 %) (Wilkinson 1995).

In the past years, several reviews of the scientific literature have attributed numerous positive effects to massage, including improvements in the quality of patients' relaxation, sleep, and immune system responses and in the relief of their fatigue, pain, anxiety, and nausea. On the basis of these reviews, some large cancer centers in the United States have started to integrate massage therapy into conventional settings (Russell et al. 2008).

Critical Care and Geriatric Medicine

A review identified as many as 19 research articles evaluating the effect of back massage on pain, relaxation, and sleep in patients in various forms of critical care settings (Richards et al. 2000). The authors conclude that the most consistent effect of massage is a significantly decreased anxiety or perception of tension. Massage is therefore advocated as an effective treatment for promoting relaxation as indicated by significant changes in the expected direction in one or more physiological indicators. Additionally, massage was found to be effective in reducing pain. Further investigations of the beneficial effects of massage therapy comprise the incidence of chronic tension headache (Quinn et al. 2002) and post-exercise muscle pain (Weber et al. 1994).

As our population gets older, massage therapies get in the focus of geriatric medicine. Conditions such as pain, anxiety, and insomnia were once treated with massage. Therefore, Harris and Richards conducted a review to assess effects of massage in geriatric patients. All studies using slow-stroke back massage and hand massage showed statistically significant improvements on physiological or psychological indicators of relaxation. The most common protocols were 3-min slow-stroke back massage and 10-min hand massage. Physiological and psychological indicators suggest the effectiveness of slow-stroke back massage and hand massage in promoting relaxation in older people across all settings (Harris and Richards 2010).

Contraindications and Safety Considerations
Massage is recognized as a safe therapeutic modality, with little risk or adverse effects. However, there are contraindications such as applying massage over an area with acute inflammation, skin infection, non-consolidated fracture, burn, deep vein thrombosis, or over sites of active cancer tumor (Vickers and Zollman 1999). Caution should be used in patients with advanced osteoporosis regarding applying only a little pressure on the tissue and the underlying bones.

Complications
There are only few published complications of massage therapy. Among them are reports of nerve palsy (Wu et al. 2010), prolonged myopathy (Tanriover et al. 2009), carotid dissection (Wada et al. 2005), and ureteral stent displacement (Kerr 1997).

After a conventional relaxation massage performed in supine position, a transient weakness due to a spinal cervical cord injury was reported (Lee et al. 2011).

After a vigorous calf muscle massage, a massive pulmonary embolism was observed. It is believed that massage can dislodge an already established blood clot but can also predispose an individual to venous thrombosis and subsequent pulmonary embolism (Jabr 2007).

References

Brosseau, L., Wells, G. A., Tugwell, P., Casimiro, L., Novikov, M., Loew, L., et al. (2012). Ottawa panel evidence-based clinical practice guidelines on therapeutic massage for neck pain. *Journal of Bodywork and Movement Therapies, 16*, 300–325.

Cherkin, D. C., Eisenberg, D., Sherman, K. J., Barlow, W., Kaptchuk, T. J., Street, J., et al. (2001). Randomized trial comparing traditional Chinese medical acupuncture, therapeutic massage, and self-care education for chronic low back pain. *Archives of Internal Medicine, 161*, 1081–1088.

Cyriax, J. (1975). *Textbook of orthopaedic medicine* (Vol. 2, 9th ed.). Baltimore, MA: Williams and Wilkins.

Dishman, J. D., & Bulbulian, R. (2001). Comparison of effects of spinal manipulation and massage on motoneuron excitability. *Electroencephalography and Clinical Neurophysiology, 41*, 97–106.

Drust, B., Atkinson, G., Gregson, W., et al. (2003). The effects of massage on intra muscular temperature in the vastus lateralis in humans. *International Journal of Sports Medicine, 24*, 395–399.

Ernst, E. (1999). Massage therapy for low back pain: A systematic review. *Journal of Pain and Symptom Management, 17*, 65–69.

Ernst, E., & Koeder, K. (1997). An overview of reflexology. *European Journal of General Practice, 97*, 52–57.

Ferrell-Torry, A. T., & Glick, O. J. (1993). The use of therapeutic massage as a nursing intervention to modify anxiety and the perception of cancer pain. *Cancer Nursing, 16*, 93–101.

Furlan, A. D., Brosseau, L., Imamura, M., & Irvin, E. (2002). Massage for low-back pain: A systematic review within the framework of the Cochrane Collaboration Back Review Group. *Spine (Phila Pa 1976), 27*, 1896–1910.

Furlan, A. D., Imamura, M., Dryden, T., & Irvin, E. (2009). Massage for low back pain: An updated systematic review within the framework of the Cochrane Back Review Group. *Spine (Phila Pa 1976), 34*, 1669–1684.

Goats, G. C. (1994). Massage-the scientific basis of an ancient art: Part 2. Physiological and therapeutic effects. *British Journal of Sports Medicine, 28*, 153–156.

Harris, M., & Richards, K. C. (2010). The physiological and psychological effects of slow-stroke back massage and hand massage on relaxation in older people. *Journal of Clinical Nursing, 19*, 917–926.

Hernandez-Reif, M., Field, T., Krasnegor, J., & Theakston, H. (2001). Lower back pain is reduced and range of motion increased after massage therapy. *The International Journal of Neuroscience, 106*, 131–145.

Jabr, F. I. (2007). Massive pulmonary emboli after legs massage. *American Journal of Physical Medicine & Rehabilitation, 86*, 691.

Kerr, H. D. (1997). Ureteral stent displacement associated with deep massage. *Wisconsin Medical Journal, 96*, 57–58.

Lee, T. H., Chiu, J. W., & Chan, R. C. (2011). Cervical cord injury after massage. *American Journal of Physical Medicine & Rehabilitation, 90*, 856–859.

Pan, C. X., Morrison, R. S., Ness, J., et al. (2000). Complementary and alternative medicine in the management of pain, dyspnea, and nausea and vomiting near the end of life. A systematic review. *Journal of Pain and Symptom Management, 20*, 374–387.

Patel, K. C., Gross, A., Graham, N., Goldsmith, C. H., Ezzo, J., Morien, A., et al. (2012). Massage for mechanical neck disorders. *Cochrane Database of Systematic Reviews, 9*, CD004871.

Perlman, A. I., Sabina, A., Williams, A. L., Njike, V. Y., & Katz, D. L. (2006). Massage therapy for osteoarthritis of the knee: A randomized controlled trial. *Archives of Internal Medicine, 166*, 2533–2538.

Perlman, A. I., Ali, A., Njike, V. Y., Hom, D., Davidi, A., Gould-Fogerite, S., et al. (2012). Massage therapy for osteoarthritis of the knee: A randomized dose-finding trial. *PLoS One, 7*, e30248.

Preyde, M. (2000). Effectiveness of massage therapy for subacute low-back pain: A randomized controlled trial. *Canadian Medical Association Journal, 162*, 1815–1820.

Russell, N. C., Sumler, S. S., Beinhorn, C. M., & Frenkel, M. A. (2008). Role of massage therapy in cancer care. *Journal of Alternative and Complementary Medicine, 14*, 209–214.

Tanriover, M. D., Guven, G. S., & Topeli, A. (2009). An unusual complication: Prolonged myopathy due to an alternative medical therapy with heat and massage. *Southern Medical Journal, 102*, 966–968.

van den Dolder, P. A., Ferreira, P. H., & Refshauge, K. M. (2012). Effectiveness of soft tissue massage and exercise for the treatment of non-specific shoulder pain: A systematic review with meta-analysis. *British Journal of Sports Medicine*, July 26 [epub ahead of print].

Vickers, A., & Zollman, C. (1999). ABC of complementary medicine. The manipulative therapies: Osteopathy and chiropractic. *British Medical Journal, 319*, 1176–1179.

Wada, Y., Yanagihara, C., & Nishimura, Y. (2005). Internal jugular vein thrombosis associated with shiatsu massage of the neck. *Journal of Neurology, Neurosurgery, and Psychiatry, 76*, 142–143.

Wainapel, S. F., Thomas, A. D., & Kahan, B. S. (1998). Use of alternative therapies by rehabilitation outpatients. *Archives of Physical Medicine and Rehabilitation, 79*, 1003–1005.

Weber, M. D., Servedio, F. J., & Woodall, W. R. (1994). The effects of three modalities on delayed onset muscle soreness. *Journal of Orthopaedic & Sports Physical Therapy, 20*, 236–242.

Weinrich, S. P., & Weinrich, M. C. (1990). The effect of massage on pain in cancer patients. *Applied Nursing Research, 3*, 140–145.

Wiesinger, G. F., Quittan, M., Ebenbichler, G., et al. (1997). Benefit and costs of passive modalities in back pain outpatients: A descriptive study. *European Journal of Physical and Rehabilitation Medicine, 7/6*, 186.

Wilkinson, S. (1995). Aromatherapy and massage in palliative care. *International Journal of Palliative Nursing, 1*, 21–30.

Wu, Y. Y., Hsu, W. C., & Wang, H. C. (2010). Posterior interosseous nerve palsy as a complication of friction massage in tennis elbow. *American Journal of Physical Medicine & Rehabilitation, 89*, 668–671.

Massage, Basic Considerations   M

Susan Mercer
School of Biomedical Sciences, The University of Queensland, Brisbane, Australia

Synonyms

Classical massage; Friction massage; Soft tissue manipulation; Swedish massage

Definition

Massage is the application of force, typically by the hands, to the skin, fascia, and muscles of the body. Various techniques, e.g., stroking and gliding (effleurage), kneading (petrissage), tapping, clapping, percussion (tapotement), and deep friction that uses different rates and rhythm, direction, pressure, and duration of movements, are utilized in an effort to enhance circulation and remove waste products from the soft tissues of the body.

Characteristics

Mechanism

Massage therapy is believed to relieve pain by increasing local circulation; stimulating large diameter nerve fibers; stimulating the release of endorphins; decreasing muscle tone; normalizing the general mobility of muscles, tendons, ligaments, and fascia; reducing swelling; and relaxing the mind (Ernst 2003a, b; Furlan et al. 2002; Nicholson and Clendaniel 1989).

Applications

Massage is used for its psychological and physiological effects. Slow to evenly applied strokes of light to medium pressure are used in the direction of muscle fibers if general relaxation is desired. To stretch scarring in skin and subcutaneous tissue, moderate to heavy pressure in all directions is applied at a moderate, even rhythm. If the desired effect is to loosen or stretch connective tissue, a slightly less heavy pressure is applied more slowly. However, if adhesions in ligaments, tendons, and muscle are to be broken, a heavy pressure perpendicular to the direction of fibers applied at a moderate, even speed is advocated. Here the therapist's fingers and the patient's skin move as one. To reduce swelling, moderate to deep pressure is applied slowly and evenly in the direction of venous return. Tapping is used in desensitizing operative sites, e.g., following amputation.

Efficacy

Massage is widely used in the general community for the treatment of painful conditions. It can be provided by professional, remedial massage therapists. Physiotherapists and osteopaths may incorporate it into their treatment regimens. Lay persons, such as spouses, may provide it.

Despite the popularity of massage, the literature provides little evidence of efficacy. Reviews have established that massage has little or no effect in relieving pain (Ernst 2004), headaches (Jensen et al. 1990), postexercise muscle pain (Ernst 1998; Weber et al. 1994), or acute, subacute, or chronic low back pain (Ernst 1999; Furlan et al. 2002; Godfrey et al. 1984; Hoehler et al. 1981; Preyde 2000).

Side Effects

Incidences of adverse effects are unknown but are probably low. Reports include fracture of osteoporotic bone, hematoma, peripheral nerve damage, and hearing loss typically associated with the use of too much force (Ernst 2003a, b).

References

Ernst, E. (1998). Does post-exercise massage treatment reduce delayed onset muscle soreness? A systematic review. *British Journal of Sports Medicine, 32*, 212–214.

Ernst, E. (1999). Massage therapy for low back pain. A systematic review. *Journal of Pain and Symptom Management, 17*, 65–69.

Ernst, E. (2003a). Massage treatment for back pain. Evidence for symptomatic relief is encouraging but not compelling. *British Medical Journal, 326*, 562–563.

Ernst, E. (2003b). The safety of massage therapy. *Rheumatology, 42*, 1101. doi:10.1093.

Ernst, E. (2004). Manual therapies for pain control: Chiropractic and massage. *The Clinical Journal of Pain, 20*, 8–12.

Furlan, A. D., Brosseau, L., Imamura, M., et al. (2002). Massage for low-back pain: A systematic review within the framework of the Cochrane collaboration back review group. *Spine, 27*, 1896–1910.

Godfrey, C. M., Morgan, P. P., & Schatzker, J. (1984). A randomized trial of manipulation for low-back pain in a medical setting. *Spine, 9*, 301–304.

Hoehler, F. K., Tobis, J. S., & Buerger, A. A. (1981). Spinal manipulation for low back pain. *Journal of the American Medical Association, 245*, 1835–1838.

Jensen, O. K., Nielsen, F. F., & Vosmar, L. (1990). An open study comparing manual therapy with the use of cold packs in the treatment of post-traumatic headache. *Cephalalgia, 10*, 241–250.

Nicholson, G. G., & Clendaniel, R. A. (1989). Manual techniques. In R. M. Scully & M. R. Barnes (Eds.), *Physical therapy*. Philadelphia: JB Lippincott.

Preyde, M. (2000). Effectiveness of massage therapy for subacute low-back pain: A randomized controlled trial. *Canadian Medical Association Journal, 162*, 1815–1820.

Weber, M. D., Servedio, F. J., & Woodall, W. R. (1994). The effects of three modalities on delayed muscle soreness. *The Journal of Orthopaedic and Sports Physical Therapy, 20*, 236–242.

Masseter Inhibitory Reflex

Definition

Reflex inhibition of the jaw-closing muscles (masseter, temporalis, or medial pterygoid) that appears as a single or more often a double phase of electric silence during voluntary contraction and is elicited especially by electrical or mechanical stimuli in the intra- or perioral region.

Cross-References

▶ Jaw-Muscle Silent Periods (Exteroceptive Suppression)

Masseter Muscle

Definition

The jaw-closer muscle. This muscle elevates the mandible for mouth closing and during chewing.

Cross-References

▶ Nociceptors in the Orofacial Region (Temporomandibular Joint and Masseter Muscle)

Masseter Silent Periods

▶ Jaw-Muscle Silent Periods (Exteroceptive Suppression)

Mast Cells

Definition

Mast cells are immune-competent cells recruited to regions of injury including the skin. They are resident in most tissues and release proinflammatory mediators including NGF when activated.

Cross-References

▶ Diencephalic Mast Cells
▶ Nerve Growth Factor, Sensitizing Action on Nociceptors

Mastery Experience

Definition

Experiences that result in the elevation of an individual's assessment of his or her abilities in a specific skill domain.

Cross-References

▶ Psychology of Pain, Self-Efficacy

Masticatory Nucleus

▶ Trigeminal Motor Nucleus

Mastitis

▶ Gynecological Pain, Neural Mechanisms

Matrix Metalloproteinases

Synonyms

MMP

Definition

Matrix metalloproteinases (MMP) are a zinc- and calcium-dependent family of proteins that are collectively responsible for the degradation of the extracellular matrix tissues. MMP are involved in wound healing, angiogenesis, and tumor cell metastasis. An imbalance between the active enzymes and their natural inhibitors leads to the accelerated destruction of connective tissue, which is associated with the pathology of diseases such as arthritis, cancer, multiple sclerosis, and cardiovascular diseases.

Cross-References

▶ Neutrophils in Inflammatory Pain

McGill Pain Questionnaire

Ronald Melzack and Joel Katz
Department of Psychology, McGill University, Montreal, QC, Canada

Synonyms

MPQ; Pain evaluation; SF-MPQ; SF-MPQ-2; Subjective pain measurement

Definition

The ▶ McGill Pain Questionnaire (MPQ) consists of 78 words, obtained from pain-patient interviews and the clinical literature, which describe distinctly different aspects of the experience of pain (Melzack 1975). The words are categorized into three major classes:
1. Words that describe the sensory qualities of the experience in terms of temporal, spatial, pressure, thermal, and other properties
2. Words that describe ▶ affective qualities in terms of tension, fear, and autonomic properties that are part of the pain experience
3. Evaluative words that describe the subjective overall intensity of the total pain experience

Each class comprises several subclasses, which contain a group of words that are considered by most subjects to be qualitatively similar (Melzack and Torgerson 1971).

In addition to the list of pain descriptors, the questionnaire contains descriptors of the overall present pain intensity (PPI). The PPI is recorded as a number from 1 to 5, in which each number is associated with the following words: 1, mild; 2, discomforting; 3, distressing; 4, horrible; 5, excruciating. The mean scale values of these words, which were chosen from the evaluative category, are approximately equally far apart, so that they represent equal scale intervals and thereby provide "anchors" for the specification of the overall pain intensity.

Characteristics

The descriptor lists of the MPQ are read by (or to) a patient, with the explicit instruction that he or she choose only those words that describe his or her feelings and sensations currently, or during a specific period (such as "during the past week.") Four major indices are obtained:
1. The pain rating index (PRI) based on the rank values of the words. In this scoring system, the word in each subclass implying the least pain is given a value of 1, the next word is given a value of 2, and so on. The rank values of the words chosen by a patient are summed to obtain a score separately for the sensory (subclasses 1–10), affective (subclasses 11–15), evaluative (subclass 16), and miscellaneous (subclasses 17–20) words, in addition to providing a total score (subclasses 1–20).
2. The PRI-corrected scores, using empirically determined scale values to enhance discriminability.
3. The number of words chosen (NWC) in the word sets.
4. The PPI, the number-descriptor combination chosen as the indicator of the current overall pain intensity.

Since its introduction in 1975, the MPQ has been translated into more than 25 languages. As pain is a private, personal experience, it is impossible for us to know precisely what someone else's pain feels like. However, the MPQ

Short-Form McGill Pain Questionnaire-2 (SF-MPQ-2)

This questionnaire provides you with a list of words that describe some of the different qualities of pain and related symptoms. Please put an **X** through the numbers that best describe the intensity of each of the pain and related symptoms you felt during the past week. Use 0 if the word does not describe your pain or related symptoms.

Descriptor		0	1	2	3	4	5	6	7	8	9	10	
1. Throbbing pain	none	0	1	2	3	4	5	6	7	8	9	10	worst possible
2. Shooting pain	none	0	1	2	3	4	5	6	7	8	9	10	worst possible
3. Stabbing pain	none	0	1	2	3	4	5	6	7	8	9	10	worst possible
4. Sharp pain	none	0	1	2	3	4	5	6	7	8	9	10	worst possible
5. Cramping pain	none	0	1	2	3	4	5	6	7	8	9	10	worst possible
6. Gnawing pain	none	0	1	2	3	4	5	6	7	8	9	10	worst possible
7. Hot-burning pain	none	0	1	2	3	4	5	6	7	8	9	10	worst possible
8. Aching pain	none	0	1	2	3	4	5	6	7	8	9	10	worst possible
9. Heavy pain	none	0	1	2	3	4	5	6	7	8	9	10	worst possible
10. Tender	none	0	1	2	3	4	5	6	7	8	9	10	worst possible
11. Splitting pain	none	0	1	2	3	4	5	6	7	8	9	10	worst possible
12. Tiring-exhausting	none	0	1	2	3	4	5	6	7	8	9	10	worst possible
13. Sickening	none	0	1	2	3	4	5	6	7	8	9	10	worst possible
14. Fearful	none	0	1	2	3	4	5	6	7	8	9	10	worst possible
15. Punishing-cruel	none	0	1	2	3	4	5	6	7	8	9	10	worst possible
16. Electric-shock pain	none	0	1	2	3	4	5	6	7	8	9	10	worst possible
17. Cold-freezing pain	none	0	1	2	3	4	5	6	7	8	9	10	worst possible
18. Piercing	none	0	1	2	3	4	5	6	7	8	9	10	worst possible
19. Pain caused by light touch	none	0	1	2	3	4	5	6	7	8	9	10	worst possible
20. Itching	none	0	1	2	3	4	5	6	7	8	9	10	worst possible
21. Tingling or 'pins and needles'	none	0	1	2	3	4	5	6	7	8	9	10	worst possible
22. Numbness	none	0	1	2	3	4	5	6	7	8	9	10	worst possible

McGill Pain Questionnaire, Fig. 1 The short-form McGill Pain Questionnaire-2. The 22 descriptors comprise the following four subscales: continuous pain (items 1, 5, 6, 8–10), intermittent pain (items 2–4, 11, 16, 18), neuropathic pain (items 7, 17, 19–22), and affective descriptors (items 12–15). Each descriptor is rated on an 11-point numeric rating scale ranging from 0 = none to 10 = worst possible. Subscale scores are computed by calculating the mean of the ratings for subscale descriptors. Total score is the sum of the four subscale scores provides us with an insight into the qualities that are experienced. Recent studies indicate that each kind of pain is characterized by a distinctive constellation of words. There is a remarkable consistency in the choice of words used by patients suffering the same or similar pain syndromes, and there is strong evidence that the MPQ is a valid, reliable tool for ▶ pain measurement (Katz and Melzack 2011). The short-form McGill Pain Questionnaire (SF-MPQ) was developed for use in research settings when the time to obtain information from patients is limited and when more information is desired than that provided by intensity measures such as the VAS or PPI (Melzack 1987). The SF-MPQ consists of 15 representative words from the sensory ($n = 11$) and affective ($n = 4$) categories of the standard, long-form (LF) MPQ. The PPI and a VAS are included to provide indices of overall pain intensity. The 15 descriptors making up the SF-MPQ were selected based on their frequency of endorsement by patients with a variety of acute, intermittent, and chronic pains. Each descriptor is ranked by the patient on an intensity scale of 0 = none, 1 = mild, 2 = moderate, and 3 = severe. The SF-MPQ correlates very highly with the major PRI indices (sensory, affective, and total) of the LF-MPQ and is sensitive to traditional clinical therapies – analgesic drugs, ▶ epidural blocks, and ▶ transcutaneous electrical nerve stimulation. The SF-MPQ is now available in over 30 languages. A procedural model for testing the factorial validity of the SF-MPQ, which can be applied to translated versions, is provided by Wright et al. (2001).

Dworkin et al. (2009) developed the SF-MPQ-2 (Fig. 1), an expanded and revised version of the SF-MPQ, designed to measure of the qualities of both neuropathic and non-neuropathic pain in research and clinical settings. The SF-MPQ-2 (1) includes seven new descriptors relevant to neuropathic pain, (2) uses an 11-point numeric rating scale for each descriptor, (3) includes the qualifier "pain" associated with 13 descriptors, and (4) expands the instructions to take into account "different qualities of pain and related symptoms". Preliminary results suggest that SF-MPQ-2 is a reliable, valid, and sensitive measure of chronic pain that is capable of discriminating between neuropathic and non-neuropathic pain.

Cross-References

- ▶ Amygdala, Pain Processing and Behavior in Animals
- ▶ fMRI
- ▶ Headache Due to Arteritis
- ▶ Motor Cortex
- ▶ Secondary Somatosensory Cortex (S2) and Insula, Effect on Pain-Related Behavior in Animals and Humans

References

Dworkin, R. H., Turk, D. C., Revicki, D. A., Harding, G., Coyne, K. S., Peirce-Sandner, S., Bhagwat, D., Everton, D., Burke, L. B., Cowan, P., Farrar, J. T., Hertz, S., Max, M. B., Rappaport, B. A., & Melzack, R. (2009). Development and initial validation of an expanded and revised version of the short-form McGill pain questionnaire (SF-MPQ-2). *Pain, 144*, 35–42.

Katz, J., & Melzack, R. (2011). The McGill Pain Questionnaire: Development, psychometric properties, and usefulness of the long-form, short-form, and short-form-2. In D.C. Turk & R. Melzack (Eds.), *Handbook of pain assessment* (3rd ed., pp 45–66). New York: Guilford Press.

Melzack, R. (1975). The McGill pain questionnaire: Major properties and scoring methods. *Pain, 1*, 277–299.

Melzack, R. (1987). The short-form McGill pain questionnaire. *Pain, 30*, 191–197.

Melzack, R., & Torgerson, W. S. (1971). On the language of pain. *Anesthesiology, 34*, 50–59.

Wright, K. D., Asmundson, G. J., & McCreary, D. R. (2001). Factorial validity of the short-form McGill pain questionnaire (SF-MPQ). *European Journal of Pain, 5*, 279–284.

MCID

▶ Minimal Clinical Important Difference

Mcl-1

Definition

Human myeloid cell differentiation protein, which is involved in programming of differentiation and concomitant maintenance of viability but not of proliferation. Isoform 1 inhibits apoptosis.

Cross-References

▶ NSAIDs and Cancer

MDdc

▶ Densocellular Subnucleus of the Mediodorsal Nucleus

Measurement

▶ Pain Assessment in the Elderly

Measurement %20 Scales

Definition

Nominal scale: An arrangement of values of a categorical variable that has no meaningful order (such as hair color or occupation).

Ordinal scale: An order that can be imposed on the values of a variable in a subject, where the order ranges from the highest value (such as "very interested") to the lowest value (such as "not at all interested").

Interval scale: An interval scale allows for the classification and labeling of elements or objects into categories based on defined features that are numerically ranked or otherwise ordered with

respect to one another. In addition, equal differences between numbers reflect equal magnitude differences between the corresponding categories. Thus, this scale has nominal and ordinal properties, and in addition it incorporates a zero, but this zero value is not absolute (lacks true meaning). The lack of an absolute zero means that this scale cannot be used to calculate the ratio of two values (cannot say one level is twice as "painful" as another level). A common example of an interval scale is the (continuous) scale of a thermometer.

Ratio scale: The ordering of numeric values when zero is meaningful (such as money or weight). Ratio scales incorporate the properties of the 3 other scales, and because they make use of a meaningful 0, their values can be interpreted to mean a true difference between numbers, and the numbers reflect true ratios of magnitude.

Cross-References

▶ Pain Assessment in Neonates
▶ Pain Measurement by Questionnaires, Psychophysical Procedures, and Multivariate Analysis

Measuring Tools

▶ Outcome Measures

Mechanical Allodynia

Definition

Mechanical allodynia, also known as tactile allodynia, is pain evoked by mechanical (tactile) stimuli that are usually non-painful. It often occurs in inflamed tissue and in the presence of neuropathy.

Cross-References

► Cancer Pain
► GABA Mechanisms and Descending Inhibitory Mechanisms
► Neuropathic Pain Model, Chronic Constriction Injury
► Neuropathic Pain Model, Partial Sciatic Nerve Ligation Model
► Neuropathic Pain Model, Tail Nerve Transection Model
► Spinal Cord Injury Pain Model, Contusion Injury Model

Mechanical Allodynia Test

► Allodynia Test, Mechanical and Cold Allodynia

Mechanical Effects of Ultrasound

Definition

The mechanical effect on the target tissue results in an increased local metabolism, circulation, extensibility of connective tissue, tissue regeneration, and bone growth.

Cross-References

► Ultrasound Therapy of Pain from the Musculoskeletal System

Mechanical Hyperalgesia

Definition

A greater (stronger) response to a noxious/painful stimulus (e.g., pin prick), or pain in response to usually non-noxious/painful mechanical stimuli (see also Allodynia).

Cross-References

► Diabetic Neuropathies
► Freezing Model of Cutaneous Hyperalgesia
► Restless Legs Syndrome

Mechanical Low Back Pain

► Chronic Back Pain and Spinal Instability

Mechanical Nociceptors

► Mechanonociceptors

Mechanical Stimuli

Definition

Input produced by physical forces, such as touch, stretch, pressure, and vibration.

Cross-References

► Postsynaptic Dorsal Column Projection, Functional Characteristics

Mechanical Syndrome

► Lower Back Pain, Physical Examination

Mechanically Evoked Itch

Definition

Sensation of itch produced by innocuous mechanical stimulation of the skin.

Cross-References

▶ Spinothalamic Tract Neurons, Central Sensitization

Mechanism-Based Approaches to Pain and Nociception

Definition

A therapeutic approach to the treatment of pain and pain syndromes that is based upon treatment of the underlying mechanisms, as opposed to an empirical approach based on symptoms. The prerequisite to this approach is (1) knowledge of the mechanisms underlying the pain process or disorder in question and (2) knowledge of the mechanisms of action of the therapeutic intervention proposed.

Cross-References

▶ Quantitative Sensory Testing

Mechanoheat Nociceptor

▶ Nociceptors in the Orofacial Region (Skin/ Mucosa)
▶ Polymodal Nociceptors, Heat Transduction

Mechano-Insensitive C-Fibers, Biophysics

Christian Weidner
Department of Physiology and Experimental Pathophysiology, University Erlangen, Erlangen, Germany

Synonyms

Action potential conduction of C-fibers; Post-excitatory effects of C-fibers

Definition

Biophysical properties of nerve fibers comprise passive properties of the membrane and nerve (resistance, capacitance, and derived measures like ▶ time constant) and active properties of membrane channels. For unmyelinated (C-) fibers, i.e., nerve fibers with a functionally accessible receptive field in vivo and in vitro, these properties can only be derived from indirect measures, e.g., with the teased fiber technique in animals or ▶ microneurography in humans.

Characteristics

Directly after an action potential, several phases of aftereffects can be distinguished (see Fig. 1). An action potential is immediately followed by the absolute refractory period, which is determined by inactivity of most voltage-gated Na^+ channels and a repolarizing current through open voltage-gated K^+ channels. Thereafter, during the relative refractory period, action potentials can be induced at higher than normal stimulus strength and at lower than normal conduction velocity. The refractory periods together last at least 10 ms in all human C-fiber afferents without a difference between receptive classes. In other species they are shorter. After the refractory periods, a supernormal period might briefly turn a nerve fiber hyperexcitable and make action potentials traveling faster than initially. This supernormal period is not constantly observed but varies between species, fiber classes, and state of ▶ accommodation of nerve fibers. While an absolute supernormal period could be observed in different species, in human C-fibers the observed supernormality has always been relative. "Relative" means that cumulating long-term aftereffects must have established an accommodation that slows down all action potentials of the ongoing stimulus train. If a conditioning stimulus is interposed at an interstimulus interval adequate to induce supernormality in the conditioned action

Mechano-Insensitive C-Fibers, Biophysics, Fig. 1 Summary of aftereffects. Threshold (inversely correlated to conduction velocity) of this exemplified nociceptive nerve fiber starts at an unconditioned level A and increases to an infinite level during the absolute refractory period B followed by a subnormal level during the relative refractory period C. Several seconds after the excitation, there is still a remaining subnormality D which cumulates after a series of action potentials to a level E. Starting at level E, an action potential of the accommodated nerve fiber again increases threshold during absolute (F) and relative refractory period (G). Thereafter, a supernormal period H intermittently renders the nerve more excitable than before the respective action potential (level E) but never more than initially (level A). This relative supernormality is typical for human C-fibers, whereas other species can also develop absolute supernormality (exceeding level A). Again, a late subnormality (I) follows this activation

potential, this supernormality can speed up the conditioned action potential as compared to the conduction velocity it would have had without the conditioning stimulus (relative supernormality). The initial conduction velocity prior to accommodation cannot be exceeded. The supernormal period is directly followed by the subnormal periods. On a cellular basis several subtypes of ▶ afterhyperpolarization with different time constants have been suggested as a basis for the late subnormal period. For peripheral nerves, these types are intermingled, and a separation using a pharmacological access has not been reported. However, the cumulating long-lasting subnormality can most probably be equated with the slow AHP observed in central neurons.

Differences Between Human C-Fiber Classes

Biophysical and receptive properties of human afferent C-fibers are strongly correlated and separate two distinct fiber classes, the mechano-responsive (usually also heat responsive) fibers known as CMH fibers or ▶ polymodal nociceptors and the mechano-insensitive fibers, with two subclasses, heat responsive (CMi) and heat insensitive (CMiHi).

Unlike other species, mechano-insensitive fibers of humans do not even respond to the strongest mechanical stimuli but can gain mechanical responsiveness as a result of primary sensitization ("waking up" of a ▶ sleeping nociceptor). Mechano-insensitive fibers have larger ▶ receptive fields (Schmidt et al. 2002) and higher heat thresholds than CMH fibers (Weidner et al. 1999), sensitize to tonic pressure (Schmelz et al. 1997), respond longer to ▶ capsaicin injection (Schmelz et al. 2000b), and are responsible for ▶ neurogenic inflammation (Schmelz et al. 2000a). These properties make mechano-insensitive human afferent C-fibers prominent candidates for an outstanding role in the development of chronic and pathological forms of pain. Besides these receptive differences, the two fiber classes also differ with regard to their biophysical properties, in particular to the long-term effects described in the next paragraph.

Long-Term Aftereffects

Aftereffects with time constants in the range of seconds are studied with repetitive pulse protocols. These effects are well known from animal experiments, and good evidence is available that the threshold increase and conduction velocity decrease are equivalent measures (Raymond and Lettvin 1978). For experimental ease of assessment, all human studies use conduction velocity decrease as the measure. After the onset of a repetitive stimulation at a certain frequency or the increase to a higher frequency, the conduction latency increases until reaching a plateau. A frequency decrease leads to a velocity increase again. The time constants are in the range of 10–60 s. The amount of activity-dependent conduction velocity slowing but not the time constant clearly separates mechano-insensitive "sleeping" from mechano-responsive afferents (Serra et al. 1999; Weidner et al. 1999). Taking into account the difference in the unconditioned conduction velocities of mechano-insensitive (∼0.8 m/s) and mechano-responsive (∼1.0 m/s) afferents, the amount of conduction velocity slowing separates the two classes without overlap. A dissociation of these closely linked conductive and receptive properties in patients with pathological pain has been interpreted as "sleeping" nociceptors that "woke up," i.e., have been sensitized in the course of the disease and gained receptive properties like mechano-responsiveness, while their conduction velocity slowing remained typical for "sleeping" nociceptors (Orstavik et al. 2003).

Different mechanisms that could influence or account for the accommodation of nerve fibers have been suggested. The Na/K ATPase activity should lead to hyperpolarization of a nerve fiber and should be activated by Na ions that are brought into the cell by its activity.

Hyperpolarization in turn is thought to slow down action potential propagation by increasing the difference of resting membrane potential and the activation threshold of voltage-gated sodium channels. Therefore, electrotonic depolarization of the membrane in front of a fully depolarized (Na channels open) part of the axon will exceed the Na channel threshold over a shorter distance. The effect of ouabain, a blocker of the Na/K pump, on activity-dependent threshold increase makes its contribution likely but not necessarily the only factor (Raymond and Lettvin 1978). Changes of K conductance can also modulate the resting membrane potential and have an influence on conduction velocity and activation threshold. An ▶ apamin-insensitive Ca-dependent K current is responsible for the slow afterhyperpolarization (AHP) in central neurons (Wilson and Goldberg 2006). Furthermore, the AHP itself can activate the inward rectifying Ih current carried by hyperpolarization-activated cyclic nucleotide-gated channels (HCN) that have also been found in peripheral nerves (Takigawa et al. 1998). The Ih current counteracts the AHP and therefore limits the activity-dependent conduction velocity slowing. Most likely, however, activity-dependent decrease of sodium influx contributes largely to the observed threshold increase and conduction velocity decrease. DeCol et al. have shown that sodium channel inhibition mimics use-dependent effects (De Col et al. 2008); furthermore, the time course of slow inactivation of voltage-gated sodium channels parallels the time course of late aftereffects. Finally, geometrical reasons could also partly account for the observed differences in accommodation between mechano-responsive and mechano-insensitive C-fibers. The large receptive field of "sleeping" nociceptors is an indication of an extensively branched terminal arborization with long and thin terminals. In thin fibers with an increased surface to volume quotient (it is indirectly proportional to the radius), all surface-bound channels/pumps at a given density (parts per surface area) will have a higher effect on intracellular concentration (parts per volume).

Short-Term Aftereffects

Short-term aftereffects with time constants in the range of milliseconds cannot be assessed with the above-mentioned method. The delay between conditioning and conditioned action potential must be in the same range. Therefore, a protocol with a stable ongoing frequency and interposed conditioning action potentials at varying interstimulus intervals can be used to assess short-term aftereffects (Weidner et al. 2000; Weidner et al. 2002; Bostock et al. 2003). Extracellular accumulation of potassium was made responsible for the supernormal period and could explain relative and absolute supernormality. In human C-fibers, however, another mechanism seems to account for the relative supernormality. After an action potential, voltage-gated potassium channels repolarize the axon to the resting membrane potential. If accommodation has made the axon hyperpolarized, these potassium channels will only repolarize the fiber to the normal resting membrane potential but not to the hyperpolarized level prior to excitation. After closure of the repolarizing potassium channels, the remaining repolarization back to the accommodated hyperpolarized level is carried out by the "leakage" current, i.e., the current with all gated channels closed. It resembles the time constant to charge or discharge the membrane capacitor. No difference in the supernormal period and the underlying time constant of the axon could be observed for mechano-responsive or mechano-insensitive fibers with the provision that the same amount of accommodation is present. As described above, a higher stimulus frequency is needed for mechano-responsive fibers to yield the same accommodation as mechano-insensitive fibers.

Functional Implications

The most obvious implication of the long-term aftereffects is the use-dependent desensitization of nerve fibers. If the threshold to induce an action potential as tested by external electrical stimulation is increased, all receptor potentials will meet the same increased excitation threshold. Paradoxically the maximum frequency that can be reached by a high frequency burst of an accommodated nerve fiber exceeds that in an unused fiber. This is because an unused fiber will only have subnormal aftereffects delaying an action potential quickly following another. For an accommodated nerve fiber, the supernormal period can speed up a succeeding action potential. For a train of four action potentials at 50 Hz in the receptive field, this can lead to an "intra-burst" frequency of 160 Hz at the knee level. This is an effective contrast enhancement mechanism.

References

Bostock, H., Campero, M., Serra, J., et al. (2003). Velocity recovery cycles of C fibres innervating human skin. *Journal of Physiology (London), 553*, 649–663.

De Col, R., Messlinger, K., & Carr, R. W. (2008). Conduction velocity is regulated by sodium channel inactivation in unmyelinated axons innervating the rat cranial meninges. *The Journal of Physiology, 586*, 1089–1103.

Orstavik, K., Weidner, C., Schmidt, R., et al. (2003). Pathological C-fibres in patients with a chronic painful condition. *Brain, 126*, 567–578.

Raymond, S. A., & Lettvin, J. Y. (1978). Aftereffects of activity in peripheral axons as a clue to nervous coding. In S. G. Waxman (Ed.), *Physiology and pathobiology of axons* (pp. 203–225). New York: Raven.

Schmelz, M., Schmidt, R., Bickel, A., et al. (1997). Differential sensitivity of mechanosensitive and -insensitive C-fibers in human skin to tonic pressure and capsaicin. *Society of Neuroscience Abstract, 23*, 1004.

Schmelz, M., Michael, K., Weidner, C., et al. (2000a). Which nerve fibers mediate the axon reflex flare in human skin? *Neuroreport, 11*, 645–648.

Schmelz, M., Schmidt, R., Handwerker, H. O., et al. (2000b). Encoding of burning pain from capsaicin-treated human skin in two categories of unmyelinated nerve fibres. *Brain, 123*, 560–571.

Schmidt, R., Schmelz, M., Weidner, C., et al. (2002). Innervation territories of mechano-insensitive C nociceptors in human skin. *Journal of Neurophysiology, 88*, 1859–1866.

Serra, J., Campero, M., Ochoa, J., et al. (1999). Activity-dependent slowing of conduction differentiates functional subtypes of C fibres innervating human skin. *Journal of Physiology (London), 515*, 799–811.

Takigawa, T., Alzheimer, C., Quasthoff, S., et al. (1998). A special blocker reveals the presence and function of the hyperpolarization-activated cation current IH in peripheral mammalian nerve fibres. *Neuroscience, 82*, 631–634.

Weidner, C., Schmelz, M., Schmidt, R., et al. (1999). Functional attributes discriminating mechano-insensitive and mechano-responsive C nociceptors in human skin. *Journal of Neuroscience, 19,* 10184–10190.

Weidner, C., Schmidt, R., Schmelz, M., et al. (2000). Time course of post-excitatory effects separates afferent human C fibre classes. *Journal of Physiology (London), 527,* 185–191.

Weidner, C., Schmelz, M., Schmidt, R., et al. (2002). Neural signal processing: The underestimated contribution of peripheral human C-fibers. *Journal of Neuroscience, 22,* 6704–6712.

Wilson, C. J., & Goldberg, J. A. (2006). Origin of the slow afterhyperpolarization and slow rhythmic bursting in striatal cholinergic interneurons. *Journal of Neurophysiology, 95,* 196–204.

Mechanoinsensitive Nociceptor

▶ Silent Nociceptor

Mechanonociceptor

▶ Nociceptors in the Orofacial Region (Skin/Mucosa)

Mechanononociceptors

Elvira De la Pena
Instituto de Neurociencias, Universidad Miguel Hernández-CSIC, Alicante, Spain

Synonyms

Aδ-mechanoreceptor; C-mechanoreceptor; High-threshold mechanoreceptors; Mechanical nociceptors; Mechanosensory nociceptors

Definition

It refers to a subpopulation of sensory afferents activated only by strong mechanical stimulation, most effectively by sharp objects.

Characteristics

Mechanononociception has been observed in organisms of various evolutionary levels: paramecium, worm, insects, and mammals. In mammals, mechano- ▶ nociceptors are peripheral endings of primary sensory neurons that are activated only when harmful mechanical stimuli are applied to their ▶ receptive field, that is located in the skin, superficial mucosa, and cornea (Belmonte et al. 1991; Burgess and Perl 1967); in deep tissues, such as joints and muscles (Mense 1977; Schaible and Schmidt 1983); and in the viscera (Cerveró 1994). The force necessary to evoke a nerve impulse discharge in mechanononociceptors is several orders of magnitude higher than the force required for the activation of low-threshold mechanoreceptors, which are the sensory ▶ receptors detecting innocuous mechanical stimuli.

Mechanononociceptor neurons have peripheral axons of variable diameter and degree of ▶ myelination belonging either to the thin myelinated, Aδ (conduction velocity: 2–20 m/s) or to the unmyelinated, C (conduction velocity <2 m/s) fiber type that end as naked nerve endings. Their somas are located in the dorsal root and cephalic sensory ganglia, and are in general of small diameter.

Mechanononociceptors were first described by Burgess and Perl in 1967 in recordings of single ▶ primary afferents/neurons fibers innervating the hind limb of the cat. These authors reported the presence of fibers conducting between 6 and 37 m/s, which were classified as mechanononociceptors because they were activated only by damaging mechanical stimulation of the skin. Burgess and Perl observed that such fibers responded maximally to cutting or pinching the skin with serrated forceps, while noxious heat (50°), noxious cold (20°), acid applied to the receptive fields, and ▶ bradykinin injected into skin did not excite this class of sensory afferents. Typically, their receptive fields were 2–5-cm long by 1–2.5-cm wide, and consisted of responsive spots (under 1-mm diameter) separated by unresponsive areas. An example of the ▶ discharge pattern evoked by noxious mechanical

a

```
                                            ] 130
                                              70

                                            ]  70
                                               30
```

b

c

d

Mechanonociceptors, Fig. 1 Responses of a nociceptive fiber to mechanical stimuli. Fiber conduction velocity, 29 m/s. *Upper traces* in (**a**), (**b**), (**c**), and (**d**) show the output of a pulse circuit triggered by unitary action potentials. (**a**) Pressure is used to stimulate the receptive field. (**b**) Pinching with a needle used as stimuli. (**c**) Fold of skin in the *center* of the receptive field squeezed with serrated forceps used as stimuli. (**d**) Clip used to pinch the center of the receptive field used as stimuli. Calibrations on *right* in g (Modified from Burgess and Perl 1967)

forces in this type of nociceptor is shown in Fig. 1.

The classification of a peripheral sensory afferent as a mechanonociceptor, polymodal nociceptor, cold nociceptors, or "silent" (mechano-insensitive) nociceptor (see Nociceptor, Categorization) has been established experimentally by applying stimuli of different modality such as mechanical (pressure), thermal (heat or cold), and chemical (acid, inflammatory mediators) stimuli to the receptive field of the fiber. Often, the initial classification of a sensory afferent as nociceptive was defined by its response only to mechanical force of near injurious or injurious intensity. In most cases, strong heating was subsequently used to determine the ▶ polymodality of the nociceptive fiber. Nevertheless, in the skin, C- and A-delta fibers

responsive to high intensity mechanical stimuli but also to strong heating, named respectively CMHs and AMHs, have often been referred to as mechanonociceptors. This is incorrect, and only those units with high mechanical thresholds, absence of heat response, and no initial sensitivity to chemical stimuli (although these have been seldom tested systematically) can be considered pure mechanical nociceptors, sometimes also called "high-threshold mechanoreceptors." The designation as polymodal nociceptor is preferred for CMHs, AMHs, and other sensory afferent types that, in addition to their sensitivity to high intensity mechanical forces, also respond to other stimulus modalities (Lynn 1994).

Pure mechanonociceptors have a wide range of sensitivities to mechanical stimuli varying from near noxious to overtly noxious intensities. In this population, A fibers generally responded to equivalent stimuli at higher rates of firing than C fibers (Garell et al. 1996). No background discharges were usually present in mechanonociceptors, which develop partial inactivation after repetitive stimulation of their receptive field. The *adaptation* of mechanonociceptors to a sustained stimulus was intermediate, when compared with that of typical phasic or tonic mechanoreceptors (Perl 1968).

With the advent of microneurography in humans (Hallin and Torebjork 1970), mechanonociceptors have been also identified in the skin of the leg and foot of human subjects (Schmidt et al. 1997). Their receptive fields are not single sensitive spots as classical psychophysical studies had suggested, but cover a relatively large innervation territory.

The cellular and molecular mechanisms by which mechanonociceptors are sensitive only to mechanical stimuli of noxious or near-noxious intensity in contrast with the responsiveness to very light mechanical forces that characterizes low-threshold mechanoreceptors are still unknown.

Mechanotransduction is finally accomplished by opening or closing *ion channels* present in the cellular membrane. The identity of noxious mechanotransduction channels in sensory neurons remains elusive but new evidences indicates that transduction of noxious mechanical stimuli is carried out by more than one channel.

Many different types of mechanosensitive (MS) channels have been characterized in a wide variety of cell types. Also, a variety of extracellular and intracellular proteins have been implicated in the transmission/amplification/modulation of mechanical force to the MS channels present in the cellular membrane during the transduction process (see nociceptors, perireceptor events, and categorization).

Genetic screens for animals defective in touch sensation have revealed critical roles for genes encoding TRP channels and DEG/ENaCs in behavioral responses to mechanical inputs. These two families of channels are widely distributed in the sensory neurons of vertebrates and invertebrates. Today the proteins responsible for detecting mechanical stimuli have been identified only in a few mechanoreceptor neurons: md, multidendritic neurons in *Drosophila*, PVD and ASH neurons in *C. elegans*. These mechanonociceptors coexpress DEG/ECNaC and TRP channel proteins, both can form mechanosensitive channels, and both operate together to enable the proper sensory function. As in nematodes and flies, mechanoreceptor neurons in mice coexpress DEG/ECNaC and TRP channel proteins and are the principal channels implied in somatic sensations. Otherwise, they are evidence that different transient receptor potential channels (TRPA-1, TRPV-4, TRPC-1) have effects on the response of nociceptors to inflammation (for review see Geffeney and Goodman 2012).

Two-pore domain mechano-gated K^+ channels (K2P), ASIC-1,2,3, DEG/ENaC, TRPV-1, 4, and TRPA1 have been proposed as candidates to mediate mechanosensitive currents.

TREK-1, TREK-2, and TRAAK exhibit mechanosensitivity, but there is no experimental proof of their relationship with mechanonociception (Patel et al. 2001). TREK-1 was the first of these K^+ channels identified in sensory neurons, is expressed in a subset of C-fiber

nociceptors, and is activated by heat and pressure in cultured neurons. Mutant TREK-1 mice have an enhanced sensitivity to heat and mild mechanical stimuli, but there is no experimental proof of their relationship with mechanonociception (Alloui et al. 2006). TRESK, another K^+ channel related to K2P family, has been implied in mechanosensation, but it is not known whether TRESK is directly sensitive to mechanical stimulation or it may mediate responses to light touch and noxious mechanical stimuli (Bautista et al. 2008).

ASICs (acid-sensing ion channels), the mammalian homologues of MEC 4 and MEC-10 that mediate mechanotransduction in *Caenorhabditis elegans*, in mechanical sensibility, are expressed in peripheral mechanoreceptors and nociceptors (Garciía-Añoveros and Corey 1997; Hughes et al. 2007). ASICs were initially implicated in mechanotransduction, but several studies support a modulatory role for these channels, not a direct role in mechanotransduction (Drew et al. 2004).

TRPV-1 has been implicated in controlling the response of the bladder urothelium to stretch, but currents have not been recorded from urothelial cells in response to a mechanical stimulus (Birder et al. 2001, 2002).

TRPV-4, an *osmosensitive* channel in mammalian cells, has also been implicated in mechanotransduction (Alessandri-Haber et al. 2003). TRPV-4 is activated by swelling but requires fatty acid metabolites; for this reason, it cannot be considered as a mechanotransducer. Otherwise TRPV-4 is a crucial determinant in the response of nociceptive neurons to osmotic stress and to mechanical hyperalgesia during inflammation (Suzuki et al. 2003; Liedtke et al. 2000). It was recently shown that TRPV4 is required for mechanotransduction by high-threshold colonic afferents, but does not account for all aspects of visceral pain because there are residual mechanonociceptive responses in TRPV-4 knockout mice (Brierley et al. 2008; Cenac et al. 2008).

TRPA-1 is present in small diameter neurons of mammalian sensory ganglia, suggesting a role in mechanical pain sensation, but TRPA1−/− mice are not deficient in the detection of acute noxious mechanical stimulation applied to the extremities (Corey et al. 2004; Bautista et al. 2006). Otherwise, TRPA1 is required for normal mechanosensory function in specific subsets of vagal, splanchnic, and pelvic afferents. The behavioral responses to noxious colonic distension were substantially reduced in TRPA1−/− mice. TRPA1 is critical in the viscera for normal mechanosensory function, the signaling of noxious mechanical stimuli, and the alteration of mechanical responsiveness of visceral afferent endings by algesic chemical stimuli. (Brierley et al. 2009).

P2X3 receptors are present in small primary sensory neurons and respond to strong mechanical stimuli when expressed in oocytes of *Xenopus*. Thus, their implication in the mechanical sensitivity of nociceptor neurons has been also suggested (Nakamura and Strimatter 1996; Cook et al. 1997).

Other channels present in sensory neurons are not transducer channels, but they contribute to propagation and modulation of impulse discharge. Electrophysiological studies with knock-out mice have shown that voltage sodium channels Na_V 1.7 and Na_V 1.8 are not involved directly in the transduction of noxious mechanical stimuli in sensory neurons, but mechanonociception requires action potentials generated by Na_V 1.7 and Na_V 1.8. While Nav1.8 is the major contributor to actions potentials in nociceptors, the function of Nav1.7 is to amplify the subthreshold depolarization to facilitate the generation of action potentials (Raouf et al. 2012).

CaV3.2, a T-type calcium channel, seems to be required for normal function of D-hair mechanosensory neurons (the extremely low threshold mechanosensory receptors) but in Cav3.2 null mice mechano-nociception is not altered, suggesting that it does not play any role in high threshold mechanotransduction (Dubreuil et al. 2004; Shin et al. 2003; Wang and Lewin 2011).

The nature of the MS channels and of the extra- and intracellular matrix proteins involved in the different types of mechanotransduction in mammalian sensory neurons is still unsolved. It is ignored whether differences in mechanical threshold between low- and high-threshold mechanosensory neurons are due to the expression of different MS channels, to differences in channel density, or to the presence of different associated proteins. Cho et al. (2002) identified in isolated membrane patches of cultured neurons of the dorsal root ganglion two types of mechanosensitive (MS) channels, with different pressure thresholds as well as distinct biophysical properties; they named them as low-threshold mechanosensitive (LT-MS, 10–20-mmHg) and high-threshold mechanosensitive channels (HT-MS). The reversal potential of these MS channels was near zero, as occurs with nonselective cationic channels, and they were blocked by the MS channel–blocking agent Gd^{3+}. HT-MS channels were present only in small sensory neurons (10–17.5 mm), and were activated only by high pressures (>60 mmHg), and sensitized by PGE2, suggesting their implication in nociceptive transduction.

Neurons with low and high sensitivity to mechanical stimuli appear to have distinct electrophysiological properties. Viana et al. (2001), in cultured trigeminal neurons of the mouse, stimulated mechanically with hypoosmotic solution, described a small percent of neurons (12 %) that exhibited narrow action potentials and marked time-dependent inward rectification, which responded to hypoosmotic solution with fast and prominent elevations of intracellular calcium and were classified as low-threshold mechanosensitive neurons. In contrast, they identified another group of neurons (19 %) showing large and wide action potentials with an inflection in the falling phase and no rectification, which produced smaller and slower elevations of intracellular calcium in response to hypoosmotic stimuli and were classified as mechanonociceptor neurons.

Several studies establish that DRG neurons express four distinct types of mechanosensitive currents, according to the kinetic current decay in response to sustained

Mechanonociceptors, Fig. 2 Properties of mechanotransducer currents in mechanoreceptors and nociceptors sensory neurons. (**a–d**), Current-clamp responses evoked by mechanical stimuli in DRG neurons expressing I (**a**), I_{Int} (**b**), I_S (**c**), or I_{uS} (**d**). The *dotted line* indicates 0 mV level. Note that, although IR- and IInt-expressing neurons fire phasically in response to mechanical stimulus, they generate a train of action potentials in response to depolarizing current pulses. (**e**) *Frequency histograms* indicating the proportion of IR-, IInt-, IS-, and IuS-expressing neurons that express or not TTX-R Na + currents (*left*) or that are sensitive or not to capsaicin (*right*). Data collected from 110 cutaneous DRG neurons (Modified from Hao and Delmas 2010)

mechanical stimulation: I_R, rapid adapting currents (3–6 ms), I_{int}, intermediately adapting currents (15–30 ms), I_S, slowly adapting currents (200–300 ms), and Iu_S, ultra-slowly adapting currents (1000 ms). All of these currents were present in rat cutaneous DRG neurons. I_S and Iu_S were preferentially found in neurons expressing TTX-resistant Na^+ currents with sensitivity to capsaicin, properties of nociceptive neurons. Iu_S and Is are the dominant forms in mechanonociceptors (Hao and Delmas 2010; Rugiero et al. 2010) Fig. 2.

Recently, two additional classes of membrane proteins (Piezo and TMC, transmembrane channel-like proteins) have been linked to mechanotransduction. Piezo1 have been observed in skin and Piezo2 in DRG sensory neurons. Expression of Piezo1 and Piezo2 in heterologous systems induces two mechanosensitive nonselective cationic currents with different kinetics, I_S and I_R (Coste et al. 2010). The role of Piezo protein (Dmpiezo), in mechanical nociception, is essential in *Drosophila* for sensing noxious mechanical stimulus (Kim et al. 2012). Knocking down Dmpiezo in sensory neurons, which mediates nociception (multidendritic neurons), was sufficient to abolish responses to noxious mechanical stimuli. It is in discussion if piezo proteins are ion-conducting structures or are required for proper expression or modulation of other mechanosensory channels. Recently, Coste et al. (2012) have evidences to support the hypothesis that Piezo proteins are indeed ion channels.

On the other hand, it has been suggested by Kawashima et al. (2011) that transmembrane channel-like proteins (TMC) are necessary for hair cell mechanotransduction but are not related with mechanonociception.

In spite of the progress made in the last few years, the molecular and cellular bases of mechanotransduction in mammalian mechano- and polymodal nociceptors, and the molecular basis of their differences with low-threshold mechanoreceptors, require further research.

Cross-References

▶ Nociceptors in the Orofacial Region (Skin/Mucosa)
▶ Sensitization of Visceral Nociceptors
▶ Visceral Mechanoreceptors

References

Alessandri-Haber, N., Yeh, J. J., Boyd, A. E., Parada, C. A., Chen, X., Reichling, D. B., & Levine, J. D. (2003). Hypotonicity induces TRPV4-mediated nociception in rat. *Neuron, 39*(3), 497–511.

Alloui, A., Zimmermann, K., Mamet, J., Duprat, F., Noël, J., Chemin, J., Guy, N., Blondeau, N., Voilley, N., Rubat-Courdert, C., Borsotto, M., Romey, G., Heurteaux, C., Reeh, P., Eschalier, A., & Lazdunski, M. (2006). TREK-1 a K + channel involved in polymodal pain perception. *EMBO Journal, 25*, 2368–2376.

Bautista, D. M., Jordt, S. E., Nikai, T., Tsuruda, P. R., Read, A. J., Poblete, J., Yamoah, E. N., Basbaum, A. I., & Julius, D. (2006). TRPA1 mediates the inflammatory actions of environmental irritants and proalgesic agents. *Cell, 124*(6), 1269–1282.

Bautista, D. M., Sigal, Y. M., Milstein, A. D., Gariison, J. L., Zorn, J. A., Tsuruda, P. R., Nicoll, R. A., & Julius, D. (2008). Pungent agents from Szechuan peppers excite sensory neurons by inhibiting two-pore potassium channels. *Nature Neuroscience, 11*, 772–779.

Birder, L. A., Kanai, A. J., de Groat, W. C., Kiss, S., Nealen, M. L., Burke, N. E., Dineley, K. E., Watkins, S., Reynolds, I. J., & Caterina, M. J. (2001). Vanilloid receptor expression suggests a sensory role for urinary bladder epithelial cells. *Proceedings of the National Academy of Sciences of the United States of America, 98*, 13396–13401.

Birder, L. A., Nakamura, Y., Kiss, S., Nealen, M. L., Barrick, S., Kanai, A. J., Wang, E., Ruiz, G., De Groat, W. C., Apodaca, G., Watkins, S., & Caterina, M. J. (2002). Altered urinary bladder function in mice lacking the vanilloid receptor TRPV1. *Nature Neuroscience, 5*, 856–860.

Belmonte, C., Gallar, J., Pozo, M. A., & Rebollo, I. (1991). Excitation by irritant chemical substances of sensory afferent units in the cat's cornea. *The Journal of Physiology, 437*, 709–725.

Brierley, S. M., Hughes, P. A., Page, A. J., Kwan, K. Y., Martin, C. M., O'Donnell, T. A., Copper, N. J., Harrington, A. M., Adam, B., Liebregts, T., Holtmann, G., Corey, D. P., Rychkov, G. Y., & Blackshaw, L. A. (2009). The ion channel TRPA1 is required for normal mechanosensation and is modulated by algesic stimuli. *Gastroenterology, 137*, 2084–2095.

Brierley, S. M., Page, A. J., Hughes, P. A., Adam, B., Liebregts, T., Cooper, N. J., Holtmann, G., Liedtke, W., & Blackshaw, L. A. (2008). Selective role for TRPV4 ion channels in visceral sensory pathways. *Gastroenterology, 134*, 2059–2069.

Burgess, P. R., & Perl, E. R. (1967). Myelinated sfferent fibres responding specifically to noxious stimulation of the skin. *The Journal of Physiology, 190*, 541–562.

Cerveró, F. (1994). Sensory innervation of the viscera: Peripheral basis of visceral pain. *Physiological Reviews, 74*, 95–138.

Cenac, N., Altier, C., Chapman, K., Liedtke, W., Zamponi, G., & Vergnolle, N. (2008). Transient receptor potential vanilloid-4 has a major role in visceral hypersensitivity symptoms. *Gastroenterology, 135*, 937–946.

Cho, H., Shin, J., Shin, C. Y., Soos-youl, L., & Oh, U. (2002). Mechanosensitive ion channels in cultured sensory neurons of neonatal rats. *Journal of Neuroscience, 22*, 1238–1247.

Cook, S. P., Vulchanova, L., Hargreaves, K. M., Elde, R., & McCleskey, E. W. (1997). Distinct ATP receptors on pain-sensing and stretch-sensing neurons. *Nature, 387*, 505–508.

Corey, D. P., García-Añoveros, J., Holt, J. R., Kwan, K. Y., Lin, S. Y., Vollrath, M. A., Amalfitano, A., Cheung, E. L., Derfler, B. H., Duggan, A., Géléoc, G. S., Gray, P. A., Hoffman, M. P., Rehm, H. L., Tamasauskas, D., & Zhang, D. S. (2004). TRPA1 is a candidate for the mechanosensitive transduction channel of vertebrate hair cells. *Nature, 432*, 723–730.

Coste, B., Mathur, J., Schmidt, M., Earley, T. J., Ranade, S., Petrus, M. J., Dubin, A. E., & Patapoutian, A. (2010). Piezo1 and Piezo2 are essential components of distinct mechanically activated cation channels. *Science, 330*, 55–60.

Coste, B., Xiao, B., Santos, J. S., Syeda, R., Grandl, J., Spencer, K. S., Kim, S. E., Schmidt, M., Mathur, J., & Dubin, A. E. (2012). Piezo proteins are pore-forming subunits of mechanically activated channels. *Nature, 483*, 176–181.

Delmas, P., Hao, J., & Rodat-Despoix, L. (2011). Molecular mechanisms of mechanotransduction in mammalian sensory neurons. *Nature Reviews Neuroscience, 12*, 139–153.

Drew, L. J., Rohrer, D. K., Price, M. P., Blaver, K. E., Cockayne, D. A., Cesare, P., & Wood, J. N. (2004). Acid-sensing ion channels ASIC2 and ASIC3 do not contribute to mechanically activated currents in mammalian sensory neurones. *The Journal of Physiology, 556*, 691–710.

Dubreuil, A. S., Boukhaddaoui, H., Desmadryl, G., Martinez-Salgado, C., Moshourab, R., Lewin, G. R., Carroll, P., Valmier, J., & Scamps, F. (2004). Role of T-type calcium current in identified d-hair mechanoreceptor neurons studied in vitro. *Journal of Neuroscience, 24*, 8480–8484.

García-Añoveros, J., & Corey, D. P. (1997). The molecules of mechanosensation. *Annual Review of Neuroscience, 20*, 567–594.

Garell, P. C., Mcgillis, S. L., & Greenspan, J. D. (1996). Mechanical response properties of nociceptors innervating feline hairy skin. *Journal of Neurophysiology, 75*, 1177–1189.

Geffeney, S. L., & Goodman, M. B. (2012). How we feel: Ion channel partnerships that detect mechanical inputs and give rise to touch and pain perception. *Neuron, 74*, 609–619.

Hallin, R. G., & Torebjork, H. E. (1970). C-fibre components in electrically evoked compound potentials recorded from human median nerve fascicles in situ. A preliminary report. *Acta Societatis Medicorum Upsaliensis, 75*, 77–80.

Hao, J., & Delmas, P. (2010). Multiple desensitization mechanisms of mechanotransducer channels shape firing of mechanosensory neurons. *Journal of Neuroscience, 30*, 13384–13395.

Hughes, P. A., Brierley, S. M., Young, R. L., & Blackshaw, L. A. (2007). Localization and comparative analysis of acid-sensing ion channel (ASIC1, 2, and 3) mRNA expression in mouse colonic sensory neurons within thoracolumbar dorsal root ganglia. *The Journal of Comparative Neurology, 500*, 863–875.

Kawashima, Y., Géléoc, G. S. G., Kurima, K., Labay, V., Lelli, A., Asai, Y., Makishima, T., Wu, D. K., Della Santina, C. C., Holt, J. R., & Griffith, A. J. (2011). Mechanotransduction in mouse inner ear hair cells requires transmembrane channel-like genes 1 and 2*J. Journal of Clinical Investigation, 121*, 4796–4809.

Kim, S. E., Coste, B., Chadha, A., Cook, B., & Patapoutian, A. (2012). The role of drosophila piezo in mechanical nociception. *Nature, 483*, 209–212.

Liedtke, W., et al. (2000). Vanilloid receptor-related osmotically activated channel (VR-OAC), a candidate vertebrate osmoreceptor. *Cell, 103*, 525–535.

Lynn, B. (1994). The fibre composition of cutaneous nerves and the classifcation and response properties of cutaneous afferents, with particular reference to nociception. *Pain Reveiws, 1*, 172–183.

Mense, S. (1977). Muscular nociceptors. *Journal of Physiology, Paris, 73*, 233–240.

Nakamura, F., & Strimatter, S. M. (1996). P2Y1 Purinergic receptors in sensory neurons: Contribution to touch-induced impulse generation. *PNAS, 93*, 10465–10407.

Patel, A. J., Lazdunski, M., & Honoré, E. (2001). Lipid and mechano-gated 2P domain K + channels. *Current Opinion in Cell Biology, 13*, 422–427.

Perl, E. R. (1968). Myelinated afferent fibres innervating the primate skin and their response to noxious stimuli. *The Journal of Physiology, 197*, 593–615.

Raouf, R., Rugiero, F., Kiesewetter, H., Hatch, R., Hummler, E., Nassar, M. A., Wang, F., & Wood, J. N. (2012). Sodium channels and mammalian sensory mechanotransduction. *Molecular Pain, 26,* 8–21.

Rugiero, F., Drew, L. J., & Wood, J. N. (2010). Kinetic properties of mechanically activated currents in spinal sensory neurons. *The Journal of Physiology, 588,* 301–314.

Schaible, H. G., & Schmidt, R. F. (1983). Activation of groups III and IV sensory units in medial articular nerve by local mechanical stimulation of knee joint. *Journal of Neurophysiology, 49,* 35–44.

Schmidt, R. F., Schmelz, M., Ringkamp, M., Handwerker, H. O., & Torebjörk, H. E. (1997). Innervation territories of mechanically activated c nociceptor units in human skin. *Journal of Neurophysiology, 78,* 2641–2648.

Shin, J. B., Martinez-Salgado, C., Heppenstall, P. A., & Lewin, G. R. (2003). A T-type calcium channel required for normal function of a mammalian mechanoreceptor. *Nature Neuroscience, 6,* 724–730.

Suzuki, M., Mizuno, A., Kodaira, K., & Imai, M. (2003). Impaired pressure sensation in mice lacking TRPV4. *Journal of Biological Chemistry, 278,* 22664–22668.

Viana, F., de la Pena, E., Pecson, B., Schmidt, R. F., & Belmonte, C. (2001). Swelling-activated calcium signalling in cultured mouse primary sensory neurons. *European Journal of Neuroscience, 13,* 722–734.

Wang, R., & Lewin, G. R. (2011). The Cav3.2 T-type calcium channel regulates temporal coding in mouse mechanoreceptors. *The Journal of Physiology, 589*(9), 2229–2243.

Mechano-Nociceptors

▶ Histology of Nociceptors

Mechanoreception

▶ Lateral Thalamic Pain-Related Cells in Humans

Mechanoreceptive Endings

▶ Visceral Mechanoreceptors

Mechanoreceptive/ Mechanosensitive Visceral Receptors

Gerald F. Gebhart
Department of Anesthesiology, Center for Pain Research, University of Pittsburgh, Pittsburgh, PA, USA

Definition

Receptive endings in viscera that respond to mechanical stimuli. These receptive endings are not designated as nociceptors because most mechanoreceptive/mechanosensitive visceral receptors respond to low-intensity mechanical stimulation in the physiological range. A smaller proportion (~25 %) of mechanoreceptive/mechanosensitive visceral receptors, however, have response thresholds above the physiological range and thus likely function as nociceptors. In viscera (and in muscle), both low-threshold and high-threshold mechanoreceptive/mechanosensitive visceral receptors encode stimulus intensity into the noxious range and thus both may contribute to nociception.

In vitro organ-nerve preparations have led to broader functional characterization of mechanosensitive endings in the viscera. In all organs studied to date, including the esophagus, stomach, urinary bladder, and colon, three mechanosensitive endings – mucosal, tension (muscular), and serosal – have been characterized. Classification is based on responses to application of different mechanical stimuli rather than histological location (Fig. 1 below). In addition, receptive endings that respond to both mucosal stimulation and stretch (muscular-mucosal receptors) have been found in the stomach, colon, and urinary bladder, and mesenteric receptors have been identified in the innervation of the mouse colorectum (Brierley et al. 2004). The nature of these in vitro organ-nerve preparations allows for localization of receptive endings in the organ, which is not easily achieved using in vivo preparations, and also permits direct application of putative sensitizing mediators to the receptive

Mechanoreceptive/Mechanosensitive Visceral Receptors, Fig. 1 Responses of pelvic nerve mechanosensitive endings innervating the mouse colon. All mechanosensitive receptive endings respond to blunt probing (>1 g): endings in the serosa respond only to probing; muscular receptors also respond to circumferential stretch; mucosal receptors to stroking of the mucosa (10 mg; lines below the record); and muscular–mucosal afferents to both stroking of the mucosa and circumferential stretch. Instantaneous firing frequency is illustrated above each record

Mechanoreceptive/ Mechanosensitive Visceral Receptors, Fig. 2 Responses of a high threshold muscular bladder afferent are shown before and 5 min after application of an inflammatory soup (IS, pH 6.0) to its receptive field. The horizontal lines beneath recordings indicate the application of IS (1 min) or stretch (20 sec) and the loads applied. Instantaneous firing frequency is illustrated above each record. (Modified after Xu L and Gebhart GF J Neurophysiol 99:244-253, 2008.)

ending in the organ. Receptive fields in vitro are typically 1–4 mm^2 in area; multiple receptive endings are occasionally found. Figure 2 illustrates responses to stretch of a muscular mechanosensitive ending in the urinary bladder of the mouse and its sensitization following application of an inflammatory soup directly to the receptive ending in the urinary bladder.

Cross-References

▶ Sensitization of Visceral Nociceptors

References

Brierley, S. M., Jones, R. C., III, Gebhart, G. F., & Blackshaw, L. A. (2004). Splanchnic and pelvic mechanosensory afferents signal different qualities of colonic stimuli in mice. *Gastroenterology, 127,* 166–178.

Mechanoreceptor Afferents

Definition

Peripheral afferent fibers that respond to mechanical stimulation. They may begin in "specialized" endings to the skin (sensory receptors, e.g., Pacinian corpuscles) or as free nerve endings.

Cross-References

▶ Hypoaesthesia
▶ Nick Model of Cutaneous Pain and Hyperalgesia

Mechanosensitive

Definition

Mechanosensitive is the characteristic of nociceptors that are responsive to stimuli such as touch, stretch, vibration, and pressure.

Cross-References

▶ Postsynaptic Dorsal Column Projection, Functional Characteristics

Mechanosensitive Sympathetic Afferent Fibers

Definition

Afferent fibers that transmit information about mechanical changes that occur in the heart.

Cross-References

▶ Visceral Pain Model, Angina Pain

Mechanosensitivity

Definition

The ability to perceive input from mechanical stimulation of the skin or adjoining mucosa.

Cross-References

▶ Fibromyalgia, Mechanisms, and Treatment

Mechanosensory Nociceptors

▶ Mechanonociceptors

MED

Synonyms

Minimum erythema dose

Definition

MED is the abbreviation for minimum erythema dose. Ascending doses of UV irradiation determine the individual MED and enable the quantification of interindividually UV-evoked skin reactions. The lowest dose of UV radiation, just sufficient to induce a visible erythema that can be distinguished from normal and untreated skin, is defined as onefold MED. Increasing doses of irradiation are expressed as multiples of the MED. The MED is an accepted measure to assess the skin sensitivity of humans; it does not define the irradiance power or the duration of the UV irradiation.

Cross-References

► UV Erythema, a Model for Inducing Hyperalgesias

Medial Branch Block

► Facet Joint Procedures for Chronic Back Pain

Medial Lemniscus (ML)

Definition

The medial lemniscuses are the fiber bundle that originates from the dorsal column nuclei and projects to the thalamus.

Cross-References

► Information Processing in Thalamocortical Circuits
► Spinothalamic Projections in Rat

Medial Pain System

Definition

The medial pain system is the neural circuit consisting of medially projecting spinothalamic pathways that terminate in medial thalamic nuclei, which in turn project to limbic and cortical structures (anterior cingulate cortex, insula, prefrontal cortex, amygdala) that underlie the processing of the affective-motivational dimension of pain.

Cross-References

► Spinothalamocortical Projections to Ventromedial and Parafascicular Nuclei
► Thalamo-amygdala Interactions and Pain

Medial Part of the Posterior Complex

Synonyms

POm

Definition

The medial part of the posterior complex receives terminations from the spinothalamic tract and probably projects to the insula. The exact functional role remains unclear.

Cross-References

► Thalamus, Nociceptive Cells in VPI, Cat and Rat

Medial Thalamotomies

Definition

Medial thalamotomies are stereotactic operations that have been performed against both chronic nociceptive and neurogenic pain syndromes, and that targeted nuclei, such as the center median-parafascicular complex, central lateral nucleus, posterior complex, and medial pulvinar.

Cross-References

▶ Thalamotomy for Human Pain Relief

Medial Thalamus

Definition

The intralaminar and midline nuclei of the thalamus. This region of the thalamus most likely contributes to the affective components as well as to attention and arousal associated with pain.

Cross-References

▶ Spinothalamic Tract Neurons, Visceral Input
▶ Thalamo-Amygdala Interactions and Pain

Median Branch Block

▶ Facet Joint Procedures for Chronic Back Pain

Median Nerve Compression at the Wrist

▶ Carpal Tunnel Syndrome

Mediate

Definition

The mechanism or process through which one variable exerts its influence on a second variable.

Cross-References

▶ Psychology of Pain, Self-Efficacy

Medical Decision-Making Model

Definition

Medical decision-making model refers to the application of statistical decision theory to judgments concerning diagnostic decisions and treatment outcomes.

Medical Disorder

▶ Migraine Epidemiology

Medical History

Wade King
Department of Clinical Research, Royal Newcastle Centre University of Newcastle, Newcastle, NSW, Australia

Definition

Medical history is the account given by a patient of their symptoms, matters related to their current condition, and their general health.

Characteristics

The medical history helps the doctor appreciate subjective aspects of the patient's condition and circumstances. History-taking forms the basis of clinical assessment. It also helps establish the doctor-patient relationship that is essential for effective medical care.

One imperative in eliciting a history is for the doctor to be alert to so-called red flags,

Medical History, Table 1 Domains of a medical history

Identification
Presenting symptoms
History of the index condition
Concurrent conditions
Concurrent medical treatment
General medical history
Systems review
Psychological history
Social history

Medical History, Table 2 A checklist for obtaining the history of a patient's pain

Site
Distribution
Quality
Duration
Periodicity
Intensity
Aggravating factors
Relieving factors
Effects on activities of daily living
Associated symptoms
Onset
Previous similar symptoms
Previous treatment
Current treatment

i.e., clinical features known to be correlated with serious pathology like infection or neoplasm. Such possibilities must be explored at the initial assessment and at every subsequent contact.

History-taking needs to be systematic and thorough, so important issues are not overlooked. Taking a comprehensive history involves exploring each of nine domains listed in Table 1, and recording the relevant details of each.

Identification

The doctor should ask the patient's name, address, age, lateral dominance, and occupation. Age is a risk factor for some conditions. Lateral dominance and occupation relate to how the affected part has been used in the past, and activities the patient may undertake in the future.

Presenting Symptoms

All current symptoms should be listed and ranked in order of significance to the patient. The main one for which the patient is seeking treatment should be identified as the "index condition."

History of the Index Condition

The doctor should elicit the history of the index condition systematically using simple, open-ended questions to guide the patient so that they describe all relevant features. A suitable approach to a pain history is to follow the checklist provided in Table 2.

Site

The site of pain is the anatomical region in which the patient perceives their pain. That may or may not be the site of origin of the pain. The patient may be describing a region to which pain is referred. In recording the site of pain, the doctor should consider the conventional definition of the relevant region so that the record is interpreted correctly by any who might become engaged in the patient's management. For example, what a patient calls "back pain" may not be "lumbar spinal pain" as defined conventionally, and it should be distinguished from loin pain or buttock pain (Merskey and Bogduk 1994).

Distribution

The distribution of pain can provide clues to its source. Patterns called "pain maps" have been determined experimentally for pain from particular segments of the cervical spine (Dwyer et al. 1990; Aprill et al. 1990; Dreyfuss et al. 1994a; Fukui et al. 1996), thoracic spine (Dreyfuss et al. 1994b) and lumbar spine (Mooney and Robertson 1976; Fukui et al. 1997), from sacroiliac joints (Fortin et al. 1994), and from peripheral joints such as the sternoclavicular joint (Hassett and Barnsley 2001). In other cases, such as abdominal pain, the source can be deduced by considering structures with the same segmental innervation as the area of distribution.

Quality

The patient should be asked to describe the pain using adjectives related to its type (not its intensity). Patients often find it hard to describe pain quality, but it is important not to prompt them. The quality of pain provides clues to its source and mechanism. ► Somatic pain is typically dull, aching pain; such pain radiating from the spine to a limb suggests ► somatic referred pain of spinal origin. ► Radicular pain is typically sharp pain shooting from the spine into a limb. ► Neuropathic pain is often described as "burning." ► Visceral pain and ► visceral referred pain often begin as dull pain and later become sharp.

Duration

The time since the onset of pain establishes whether it is ► acute pain, ► subacute pain, or ► chronic pain. The distinction is important as it determines treatment.

Periodicity

Periodicity means the amounts of time a pain is present. Constant pain may be associated with continuing pathology. Intermittent pain, especially pain on movement, may be associated with impairment of parts that become painful under load. Some forms of neurogenic pain, particularly the ► neuralgias, are intrinsically periodic, i.e., occurring in bursts. Some forms of ► headache have characteristic periodicities.

Intensity

The intensity of pain should be measured using an instrument such as a visual analogue scale and the score recorded. If pain is not measured initially it cannot be said later to have been relieved (Huskisson 1974).

Aggravating Factors

Aggravating factors include stresses that load impaired tissues. Analysis of the biomechanical effects of such factors can assist in diagnosis. It is also useful in formulating management, providing guidance to potentially useful interventions and for advising the patient as to what they should not do, and what they might do in better ways despite the pain. Pain not aggravated by biomechanical loading or movement suggests the possibility of a remote origin, which should be pursued.

Relieving Factors

Relieving factors usually reduce stresses on the affected part, e.g., avoiding certain movements. Pain not relieved by rest raises the possibility of a serious cause that needs to be investigated.

Effects on Activities of Daily Living

The effects on activities of daily living (ADLs) denote ways in which the index condition disables or handicaps the patient. They comprise indices of the impact of the condition on the patient's lifestyle. If monitored during and after treatment, they provide guidance to any improvement.

Associated Symptoms

Associated symptoms are important clues to serious causes of pain like infection and neoplasia. Sooner or later, such serious causes will produce features other than pain and these features help identify the cause. Infection can be associated with fever and malaise; neoplasm with weight loss, malaise, and neurological symptoms; and vascular disorders with transient ischemic attacks. In the absence of associated symptoms, a serious cause of pain is unlikely.

Onset

The onset means the first appearance of the pain. The doctor should record the date of onset and the circumstances in which the pain started. Details of the onset can provide clues to the etiology. For example, if the pain arose from an incident resulting in injury, diagnosis may be assisted by knowing how forces were applied to the patient's body. If there is no history of trauma, the doctor should consider other causes such as an antecedent infection or neoplasm.

Previous Similar Symptoms

Previous related symptoms suggest a chronic or recurrent condition. In that event, persistent risk factors for recurrence should be explored and managed.

Previous Treatment

All measures used to treat the condition should be noted, with their outcomes and any unwanted effects. Interventions to which there have been favorable responses may be useful for further management. Treatment failures may provide clues to the cause of pain and also suggest treatments to avoid.

Current Treatment for the Index Condition

All current treatment should be recorded, including self-applied measures like local heat and all substances whether prescribed or otherwise, with the patient's appraisal of each.

Intercurrent Conditions

Any intercurrent problems should be noted, and consideration given to any "red flags."

Other Current Medical Treatments

All forms of treatment in use for other conditions should be considered for any effect they may have on the index condition or its treatment. Knowing all current treatments helps the doctor avoid deleterious interactions with further treatment that may be prescribed.

Past Medical History

Past history can provide clues to possible serious causes of pain. A past history of cancer warns of recurrence or metastases. Recent skin penetration, current infection, or immunological compromise warns of infection. Prolonged use of corticosteroids is a risk factor for peptic ulceration and for osteoporosis, resulting in stress factors. Renal disease, immunosuppression, and diabetes mellitus are risk factors for osteonecrosis.

Systems Review

Asking about past or present symptoms from each bodily system may yield clues to possible serious causes that may not have been evident otherwise.

Psychological History

The doctor should consider how the patient reacts to the condition mentally and try to identify any psychosocial risk factors, i.e., psychological and social issues correlated statistically with chronic disability. Sensitive exploration of a patient's psyche can enhance the doctor-patient relationship, enable development of empathy, and help the doctor care for them in the manner advocated by Cochrane (Cochrane 1977).

Aspects to be addressed include affect, in particular if anxious or depressed, understanding of the condition, any associated fears, relevant cognitions and beliefs, and coping strategies. The physician must decide if these factors are likely to influence the course of the condition (Lethem et al. 1983), and whether special psychological management is needed.

Social History

The social history helps the doctor understand the patient in their social environment. It should include their family, other close relationships, home, employment, sources of income, education, occupational qualifications, and leisure interests. These factors bear on what strategies may or may not be employed in the management plan. Some constitute risk factors for impeded recovery, in which case they need to be addressed.

References

Aprill, C., Dwyer, A., & Bogduk, N. (1990). Cervical zygapophyseal joint pain patterns: II. A clinical evaluation. *Spine, 15*, 458–461.

Cochrane, A. L. (1977). *Effectiveness and efficiency. Random reflections on health services* (p. 95). Cambridge, UK: Cambridge University Press.

Dreyfuss, P., Michaelsen, M., & Fletcher, D. (1994a). Atlanto-occipital and lateral atlanto-axial joint pain patterns. *Spine, 19*, 1125–1131.

Dreyfuss, P., Tibiletti, C., & Dreyer, S. J. (1994b). Thoracic zygapophyseal joint pain patterns. *Spine, 19*, 807–811.

Dwyer, A., Aprill, C., & Bogduk, N. (1990). Cervical zygapophyseal joint pain patterns: I. A study in normal volunteers. *Spine, 15*, 453–457.

Fortin, J. D., Dwyer, A. P., West, S., & Pier, J. (1994). Sacroiliac joint: Pain referral maps upon applying a new injection/arthrography technique; part 1: Asymptomatic volunteers. *Spine, 19*, 1475–1482.

Fukui, S., Ohseto, K., Shiotani, M., Ohno, K., Karasawa, H., Naganuma, Y., & Yuda, Y. (1996). Referred pain distribution of the cervical zygapophyseal joints and cervical dorsal rami. *Pain, 68*, 79–83.

Fukui, S., Ohseto, K., Shiotani, M., Ohno, K., Karasawa, H., & Naganuma, Y. (1997). Distribution of referred pain from the lumbar zygapophyseal joints and dorsal rami. *The Clinical Journal of Pain, 13*, 303–307.

Hassett, G., & Barnsley, L. (2001). Pain referral from the sternoclavicular joint: A study in normal volunteers. *Rheumatology, 40*, 859–862.

Huskisson, E. C. (1974). Measurement of pain. *Lancet, 2*, 1127–1131.

Lethem, J., Slade, P. D., Troup, J. D. G., & Bentley, G. (1983). Outline of a fear avoidance model of exaggerated pain perception – I. *Behaviour Research and Therapy, 21*, 401–408.

Merskey, H., & Bogduk, N. (Eds.). (1994). *Classification of chronic pain. Descriptions of chronic pain syndromes and definitions of pain terms* (2nd ed., p. 3, 11, 12). Seattle, WA: IASP Press.

Mooney, V., & Robertson, J. (1976). The facet syndrome. *Clinical Orthopaedics, 115*, 149–156.

Medical Hydrology

▶ Spa Treatment

Medical Misadventures

▶ Postoperative Pain, Adverse Events (Associated with Acute Pain Management)

Medical Mishaps

▶ Postoperative Pain, Adverse Events (Associated with Acute Pain Management)

Medical Outcomes Study 36-Item Short-Form Health Survey (SF-36)

Definition

The Medical Outcomes Study 36-Item Short-Form Health Survey (SF-36) is a 36-item general measure of perceived health status that is usually self-administered.

Cross-References

▶ Pain Inventories

Medical Signs

Definition

Anatomical, physiological, or psychological abnormalities that can be observed, apart from a person's statements (symptoms). Signs must be shown by medically acceptable clinical diagnostic techniques. Psychiatric signs are medically demonstrable phenomena that indicate specific psychological abnormalities, e.g., abnormalities of behavior, mood, thought, memory, orientation, development, or perception. They must also be shown by observable facts that can be medically described and evaluated.

Cross-References

▶ Disability Evaluation in the Social Security Administration

Medically Incongruent Symptoms and Signs

▶ Nonorganic Symptoms and Signs

Medication-induced Headaches

▶ Headache Attributed to a Substance or Its Withdrawal

Mediodorsal Nucleus (MD)

Definition

The main medial nucleus of the thalamus, limited by the intralaminar nuclei.

Cross-References

▶ Spinothalamic Terminations, Core and Matrix
▶ Thalamus, Visceral Representation

Mediolongitudinal Myelotomy

▶ Midline Myelotomy

Meditation

Definition

Meditation is a means of narrowing the focus of one's attention to a simple activity (e.g., awareness of the breath) or word (e.g., a mantra), thereby quieting and calming the mind.

Cross-References

▶ Relaxation in the Treatment of Pain

Medullary Dorsal Horn

Definition

The medullary dorsal horn is a caudal portion of the subnucleus caudalis of the spinal trigeminal nucleus, which has laminated cytoarchitecture analogous to the dorsal horn of the spinal cord. The medullary dorsal horn plays a major role in the processing of nociceptive information from the head, face, and mouth regions.

Cross-References

▶ DREZ Procedures
▶ Trigeminothalamic Tract Projections

Medullary Dorsal Horn (MDH)

▶ Subnucleus Caudalis

Medullary Lateral Reticular Formation

Definition

Lying dorsomedially to the spinal trigeminal nucleus, it contains the interneurons that mediate the long-latency trigeminal reflexes.

Cross-References

▶ Jaw-Muscle Silent Periods (Exteroceptive Suppression)

MEG

▶ Magnetoencephalography
▶ Pain in Humans, EEG Documentation

Meissner Receptor

Definition

Meissner receptor is an unprecise denomination of the meissner corpuscles in the skin. Meissner corpuscles are cell agglomerations of 30–150-μm length and 40–60-μm diameter in the stratum papillare of glabrous skin. They encapsulate the unmyelinated terminals of thick myelinated afferent nerve fibers. The whole complex serves as highly sensitive rapidly adapting mechanoreceptor. They encode the velocity of skin indentations. The molecules responsible for the mechanotransduction are probably in the nerve terminals. According to their response characteristics, these afferents have been named RA (rapidly adapting) or QA (quickly adapting) mechanoreceptors or mechanosensors. They are highly sensitive in particular for moving stimuli. Upon vibration at frequencies of 10–50 Hz, skin indentations of about 10 μm may be sufficient to excite them. There are no meissner corpuscles in the hairy skin where their role is played by hair follicle receptors.

Cross-References

▶ Perireceptor Elements

Melanotic

Definition

Melanotic refers to a high level of blackish pigmentation that produces a very dark or black color.

Cross-References

▶ Lower Back Pain, Physical Examination

Melatonin

Definition

Melatonin is a substance secreted by the pineal gland associated with circadian rhythms such as the sleep-wake cycle.

Cross-References

▶ Hypnic Headache

Member 1

▶ TRPV1 Receptor, Species Variability

Membrane Capacitance

Definition

The ability of a cell membrane to store ionic charge for a given transmembrane voltage. Membrane capacitance is proportional to total surface area. Together with the membrane resistance, it defines the basic, passive, electrical property of cellular membranes.

Cross-References

▶ Demyelination

Membrane Potential

Definition

The electric potential difference across the membrane of living cells at rest, due to a differential distribution of ions (particularly Na^+, K^+, and Cr) consecutive to the variable permeability of the membrane and the active transport of ions.

Cross-References

▶ Nociceptor Generator Potential

Membrane Resistance

Definition

A measurement of how easily ions cross the cell membrane.

Cross-References

▶ Demyelination

Membrane Stabilizers

Definition

Pharmacological agents that reduce the electrical excitability of neurons. Typical membrane-stabilizing drugs are local anesthetics and other Na^+ channel blockers.

Cross-References

▶ Pain Paroxysms
▶ Trigeminal and Cranial Nerve Neuralgias

Membrane-Stabilizing Agents

▶ Postoperative Pain, Membrane-Stabilizing Agents

Membrane-Stabilizing Drugs

▶ Drugs Targeting Voltage-Gated Sodium and Calcium Channels

Memory Decision Theory

Definition

Memory decision theory refers to the application of signal detection theory to judgments concerning whether a descriptor is "old," that is, previously used to describe a symptom, or "new," that is, not previously used.

Meningeal Afferents

▶ Nociceptors in the Orofacial Region (Meningeal/Cerebrovascular)

Meningeal Carcinomatous

Definition

This is widespread infiltration of the meninges with cancer. Symptoms and signs may reflect involvement of multiple sites in the central nervous system and may include altered mental status, headache, and pain. Carcinomatous meningitis occurs in 2–3 % of patients with adenocarcinoma of the lung and small cell lung cancer. The median survival of patients with carcinomatous meningitis is 2–4 months.

Meningeal Layer

Definition

The protective membranes that encase the central nervous system.

Cross-References

▶ Motor Cortex
▶ Secondary Somatosensory Cortex (S2) and Insula, Effect on Pain-Related Behavior in Animals and Humans

Meningeal Nociceptors

▶ Nociceptors in the Orofacial Region (Meningeal/Cerebrovascular)

Meningoencephalitis

Definition

Of the brain stem and cerebellum: neurologic manifestation of Behçet's disease.

Cross-References

▶ Headache Due to Arteritis

Meniscus

Definition

Flattened cartilaginous plate found in certain joints, including the knee. The meniscus serves to maintain the apposition of the opposed surfaces in their various motions, to ease the gliding movement of the joint, and to decrease the effects of pressure and sudden impact on the joint.

Cross-References

▶ Arthritis Model, Osteoarthritis

Mensendieck System

Definition

A pedagogically designed system of functional exercises based on an analysis of human anatomy, physiology, and biomechanics, developed by MD Bess M. Mensendieck, in the early 1900s. The aims are to better the individual's functional capacity in daily life by increasing the understanding of anatomically and physiologically correct movements and improving the ability to interpret body signals.

Cross-References

▶ Body Awareness Therapies

Menstrual Suppression

Definition

The use of hormonal or antihormonal medication such as combined hormonal contraception, high-dose progestins, androgens, or GnRH agonists for prolonged periods of time in order to obviate or minimize menstrual flow and pain.

Cross-References

▶ Dyspareunia and Vaginismus

Mental Disorders, Diagnostics and Statistics

▶ Diagnostic and Statistical Manual of Mental Disorders

Mental Pain

▶ Pain, Psychiatry, and Ethics

Menthol

Definition

Menthol is a natural substance (2-isopropyl-5-methylcyclohexanol) obtained from plant leaves of the genus *Mentha*, widely used in products such as common cold medications, toothpastes, confectionery, and cosmetics. l-Menthol evokes a sensation of coolness by stimulation of oral, nasal, and skin cold thermoreceptors.

Menthol can also act as an irritant involving capsaicin-sensitive pathways. A cold and menthol receptor, named TRPM8, was cloned from peripheral sensory neurons.

Cross-References

▶ Nociceptors, Cold Thermotransduction

Meperidine

▶ Postoperative Pain, Pethidine

Meralgia Paresthetica

Definition

Meralgia paresthetica is a painful mononeuropathy of the lateral femoral cutaneous nerve due to focal entrapment in the inguinal ligament. It results in pain in the lateral thigh.

Cross-References

▶ Neuralgia, Assessment
▶ Sciatica

Merkel Receptor

Definition

Like "Meissner Receptor," "Merkel Receptor" is an unprecise denomination of the Merkel cell-axon complex which serves as slowly adapting mechanosensor in the skin. Merkel cells form elongated structures in the basal layer of glabrous and hairy skin, oral and anal mucosa, external sexual organs, and mammary glands. They are contacted in synapse-like structures by the terminals of fast conducting myelinated afferent nerve fibers. The molecules responsible for the mechanotransduction are probably in the nerve terminals, and the coupling to the time courses of mechanical stimuli is too precise to allow for an intermittent synaptic process between Merkel cells and nerve terminals. The role of the Merkel cell for the mechanotransduction is still unclear. There are also Merkel cells without contact to nerves. The Merkel cell-axon complex serves as sensitive pressure transductor in the skin, and indentations of much less than 1 mm are precisely encoded in the spike frequency of the axon. Upon steady indentation the spike responses of these mechanosensors adapt but slowly. Therefore, they have been named SA (slowly adapting) mechanosensors.

Cross-References

▶ Perireceptor Elements

Mesencephalic Nucleus

Definition

Mesencephalic nucleus, containing the cell bodies of primary afferents innervating the jaw-muscle spindles or periodontal ligaments.

Cross-References

▶ Trigeminal Brainstem Nuclear Complex, Anatomy

Mesencephalic Tractotomy

▶ Pain Treatment, Intracranial Ablative Procedures

Mesencephalon

Definition

The narrow short part of the brain stem above the pons and below the thalamus. An operation to interrupt a pathway at the level of the mesencephalon is called a mesencephalotomy.

Cross-References

▶ Pain Treatment, Intracranial Ablative Procedures

Mesencephalotomy

▶ Pain Treatment, Intracranial Ablative Procedures

Mesmerism

▶ Therapy of Pain, Hypnosis

Meta-analysis

Definition

Meta-analysis is a statistical technique for combining the findings from independent studies. It is a systematic review that uses quantitative methods to summarize the results. It is most often used to assess the clinical effectiveness of healthcare interventions. Meta-analysis is an effective means of correcting both for bias and lack of power in individual randomized controlled studies. It provides a precise estimate of treatment effect, providing due weight to the size of the studies included. The validity of meta-analysis depends on the quality of the systematic review. A meta-analysis may also allow the investigator to explore whether features associated with studies influence the magnitude of the effects observed, e.g., do studies using skilled therapists obtain better effect than those using relatively unskilled ones?

Cross-References

▶ Operant Treatment of Chronic Pain
▶ Postoperative Pain, Preoperative Education
▶ Psychology of Pain, Efficacy of Cognitive-Behavioral Treatment
▶ Transition from Acute to Chronic Pain

Metabolic and Nutritional Neuropathies

John B. Winer[1], Edward Littleton[2] and Alex Barling[1]
[1]Department of Neurology, Birmingham Muscle and Nerve Centre Queen Elizabeth Hospital, Birmingham, UK
[2]Queen Elizabeth Hospital, Birmingham, UK

Synonyms

Alcoholism; Bariatric surgery; Beriberi disease; Cuban neuropathy; Fabry's disease; Malnutrition; Metabolic neuropathies; Nutritional neuropathies; Paraproteinemic neuropathies; Vitamin neuropathy

Definition

Neuropathy encompasses conditions occurring with systemic organ failure as well as inherited or acquired metabolic defects. Nutritional neuropathies are those related to malnutrition and resulting vitamin deficiency. Neuropathy related to alcoholism is also discussed in this section as it is frequently related to malnutrition, although a toxic metabolic cause has been suggested for the axonal injury. Axonal neuropathies usually result in damage to the nerve cell body and lead to a dying back of the longest axons.

Characteristics

Alcohol

Alcohol remains the second most common cause of neuropathy after diabetes mellitus and affects a third of alcoholics (Neundorfer 2001). It is unclear whether alcohol is toxic itself or simply associated with malnutrition that results in neuropathy. A study by Koike suggested a direct causal role (Koike 2003) in alcoholic Japanese subjects whose thiamine levels were normal and who did not display clinical features of other vitamin deficiencies. Coexistent alcoholism and thiamine deficiency was associated with a greater degree of motor involvement, large fiber involvement, and subperineurial edema on sural nerve biopsy than in subjects with "pure" alcoholic neuropathy.

The neuropathy typically manifests as burning feet with distal, symmetrical loss of pain sensation and ▶ allodynia in the lower limbs. As the neuropathy progresses, distal weakness occurs, initially in the legs and then the arms. These features can be accompanied by ataxia from joint position loss or cerebellar degeneration, Wernicke-Korsakoff encephalopathy, and stigmata of chronic liver disease. Impotence and postural hypotension from autonomic nerve involvement are often underdiagnosed (Monforte 1995), although the severity of coexistent somatic and autonomic neuropathy do not appear correlated (Nicolosi 2005).

There are no diagnostic investigations, but the clinical signs in conjunction with a history of alcohol consumption, typically greater than 3 l of beer or 300 ml of spirits per day for 3 years, are suggestive (Behse 1977). A red cell macrocytosis, elevated transaminase, and gamma glutamyltransferase are also pointers. Analysis of cerebrospinal fluid should be normal, although the protein may be minimally elevated. The nerve conduction studies (NCS) mirror the clinical findings and show distal, predominantly lower limb, symmetrically reduced sensory nerve action

Metabolic and Nutritional Neuropathies, Fig. 1 Teased fiber preparation of a peripheral nerve showing axonal degeneration as seen in alcoholic neuropathy

potentials. Histology demonstrates axonal loss in fibers of all sizes in the absence of inflammation (Fig. 1).

Treatment of alcoholic neuropathy requires abstinence and multivitamins including thiamine. Sensory symptoms may initially worsen on treatment.

Nutritional Neuropathies

Nutritional neuropathies occur predominantly in developing countries where crop failure and malnutrition are common. The last major outbreak was in Cuba in 1992 (Roman 1994). In developed countries, nutritional neuropathies are rare but still occur in the context of alcoholism, malignancy, chronic infection, and pregnancy. They are important to recognize, as prompt treatment can halt the progression and sometimes improve the neuropathy (Table 1).

Thiamine Deficiency, B₁

Thiamine deficiency causes the syndrome of beriberi, which encompasses peripheral neuropathy (dry beriberi) and congestive cardiac failure (wet beriberi). Initially recognized early in the nineteenth century, it was associated with a diet of processed rice or occurred during sea voyages.

Thiamine deficiency should be suspected in any malnourished patient. The symptoms and signs are similar to alcoholic neuropathy with burning ▶ dysesthesia in the legs. The condition progresses over weeks and months, with the

Metabolic and Nutritional Neuropathies, Table 1 Features of metabolic and nutritional neuropathy

Cause	Symptoms	Pain	Examination	Specific investigations	Pathology
Alcohol	Burning feet	+++	Distal, symmetrical, loss of pain and hyperesthesia to light touch	Raised red cell mean cell volume	Sensory axonal
				Raised transaminase	
				Raised gamma glutamyltransferase	
Thiamine deficiency	Burning feet	+++	Distal, symmetrical, loss of pain and hyperesthesia to light touch	Raised erythrocyte transketolase	Axonal
				Vitamin B_1 level	
Pyridoxine deficiency	Distal numbness and tingling		Distal, symmetrical sensory loss	Vitamin B_6 level	Sensory axonal
Cobalamin deficiency	Distal tingling and numbness		Distal loss of vibration followed by joint position sense	Low B_{12}, raised homocysteine and methylmalonic acid	Axonal
	+/− Ataxia and weakness		Absent ankle reflexes	Variable increase in hemoglobin and mean cell volume	
Vitamin E deficiency	Ataxia and weakness		Distal areflexia and loss of vibration +/− ataxia	Vitamin E level (in presence of normal lipid levels)	Axonal
Uremia	Distal paresthesia	+	Distal sensory loss	Raised urea and creatinine	Axonal
Hepatic dysfunction	Variable		Variable	Deranged liver function tests	Demyelinating or axonal
				+/− vitamin E deficiency	
Thyroid disease	Distal paresthesia	++	Distal sensory loss	TSH and T4	Sensory axonal
Porphyria	Distal paresthesia and proximal weakness	+	Distal paresthesia and proximal weakness	Urinary porphobilinogens	Axonal
			+/− Respiratory muscle weakness		
Amyloid	Distal paresthesia		Distal, symmetrical loss of pain and temperature	Congo red staining of nerve biopsy	Axonal
			+/− autonomic dysfunction		
Refsum's disease	Distal paresthesia	+	Distal, symmetrical loss of vibration and proprioception	Raised plasma phytanic acid and pipecolic acid	Demyelinating
			Mild distal weakness		
Fabry's disease	Burning hands and feet	+++	Progressive or episodic sensory loss	α-galactosidase levels in fibroblasts	Axonal

sensory symptoms spreading proximally followed by weakness. Less commonly, the cranial nerves can be affected, e.g., laryngeal weakness, facial weakness, and tongue deviation.

Laboratory tests for thiamine deficiency have in the past employed indirect functional methods rather than measuring thiamine directly. Numerous studies have looked at pyruvate which accumulates in thiamine deficiency; however, the sensitivity is poor. Assay of erythrocyte transketolase activity, a vitamin B_1-dependent enzyme, is more helpful if measured prior to any supplementation as it reflects B_1 stores. However, factors other than the vitamin B_1 level, including the presence of diabetes and liver disease, also influence this assay, and the assay has suffered from difficulty with standardization. More recently, high-performance liquid chromatographic methods for the direct measurement of thiamine diphosphate (pyrophosphate) in erythrocytes and in whole blood have been reported to compare favorably to erythrocyte transketolase assays in detecting the presence of thiamine deficiency (Talwar 2000). NCS are similar to that of alcoholic neuropathy with predominantly distal, symmetrical axonal loss. The pathology shows distal axonal degeneration.

The treatment of polyneuropathy is with oral thiamine, 300 mg per day. In the presence of heart failure or Wernicke's encephalopathy, thiamine should be administered parenterally (see ▶ Wernicke-Korsakoff syndrome). Supplementation with multivitamins is recommended as the neuropathy is often not the result of a single vitamin deficiency state.

Pyridoxine Deficiency, B6

Pyridoxine deficiency is rare and usually secondary to treatment with isoniazid or hydralazine. These drugs inhibit the phosphorylation of pyridoxine to the active coenzyme pyridoxal phosphate. Distal, symmetrical numbness with tingling is the common presentation. Discontinuing the offending drug usually reverses the symptoms. A daily oral dose of 50–100 mg of pyridoxine is generally advised. Doses above 180 g per day are toxic to nerves, leading to a pyridoxine-induced neuropathy.

Cobalamin Deficiency, B12

The most common cause of B_{12} deficiency is malabsorption secondary to pernicious anemia. In this condition, antibodies to intrinsic factor prevent absorption of B_{12} specifically. A deficiency state is also well documented in patients postgastrectomy or after resection of the terminal ileum. Less frequent causes of deficiency include tropical sprue, severe steatorrhea, drugs such as colchicine which reduce absorption, and rarely a strict vegan diet. Occasionally, general anesthesia or analgesia with nitrous oxide can precipitate myelopathy and neuropathy in patients with previously subclinical or borderline B_{12} deficiency; this is because nitrous oxide irreversibly oxidizes vitamin B_{12}, interfering with the B_{12}-dependent conversion of homocysteine to methionine. B_{12} is stored in the liver, and many years of deficiency are needed to deplete the stores.

B_{12} deficiency usually begins insidiously and progresses over weeks and months. Practically, it is very difficult to separate the symptoms of neuropathy from the myelopathy of subacute combined degeneration of the spinal cord as they usually coexist. In the early stages, ▶ paresthesia and numbness are the most common symptoms. Lower limb weakness and ataxia then ensue. Other symptoms include impotence, memory loss, and mental slowing.

Examination typically reveals loss of vibration sensation in the early stages, followed by loss of joint position sense and then light touch. The ankle reflexes are reduced or absent. The myelopathy leads to pyramidal motor dysfunction, hyperreflexia at the knee, and a positive Babinski sign.

Serum B_{12} levels can be measured, but the assay methods vary, and all are prone to inaccuracy, as B_{12} is predominantly protein bound. This means that a normal B_{12} concentration does not exclude a deficiency state. Also, falsely elevated levels are seen in myeloproliferative and hepatic disorders, and falsely low levels are seen in pregnancy. B_{12} deficiency is also associated with a macrocytic anemia. There is debate about whether significant neurological dysfunction from B_{12} deficiency can occur in the presence of a normal hemoglobin level. Significant reduction of the B_{12} concentration will cause an elevation in homocysteine and a reduction in methionine, due to impaired B_{12}-dependent remethylation of homocysteine to methionine, and an elevation of methylmalonic acid due to impaired B12-dependent metabolism of methylmalonyl-CoA to succinyl-CoA. Measuring the levels of these metabolites can be useful in cases of borderline deficiency. Of patients with B_{12} deficiency, with or without accompanying neurological symptoms, glossitis or anemia, 98 % have elevated levels of homocysteine, and 95 % have elevated methylmalonic acid (Savage 1994). NCS show reduced sural nerve action potentials resulting from presumed axonopathy. The pathological studies in this area are limited and inconclusive (Windebank 1993).

Vitamin B_{12} deficiency is treated with intramuscular injections. Partial neurological improvement is usually seen predominantly in the first 6 months, but recovery can continue for up to a year or more.

Vitamin E Deficiency

Vitamin E is a fat-soluble vitamin, and absorption is dependent on bile salts and pancreatic esterases in the small bowel. Causes of the deficiency include cystic fibrosis, bile salt deficiency from congenital cholestasis and small bowel resection, a rare inherited defect of chylomicrons and lipoprotein synthesis known as abetalipoproteinemia, and defects in alpha-tocopherol transfer protein genes which are inherited in an autosomal recessive manner.

There are significant reserves of vitamin E in adipose tissue; this means that deficiency states in adults take many years or decades to become clinically significant. Deficiency of vitamin E leads to spinocerebellar dysfunction with variable peripheral nerve involvement; affected patients usually complain of an unsteady gait and weakness. Examination reveals distal areflexia, prominent loss of vibration, and joint position sense with minimal weakness. Other sensory modalities such as light touch are seldom affected. Ataxia can be a feature of both joint position loss and cerebellar involvement. The clinical picture can be complicated by proximal muscle weakness and a positive Babinski sign.

Vitamin E assays are accurate in the presence of normal lipid levels as vitamin E is primarily located in lipoprotein; however, hyperlipidemic patients with normal vitamin E assays may still have a deficiency. In general, levels need to be undetectable to be able to attribute neurological disease to vitamin E deficiency. NCS demonstrate a reduction in sensory nerve action potentials with preservation of conduction velocity. Motor nerves are usually unaffected. ▶ Somatosensory evoked potentials and ▶ visual evoked potentials are often delayed due to spinal or optic nerve involvement. Electromyography can show mild denervation. CSF examination is normal.

The roles of vitamin E as an antioxidant and free radical scavenger are well documented, but the mechanism by which vitamin E deficiency affects the nervous system is not established.

Treatment of vitamin E deficiency involves large doses of oral vitamin E, 1–4 g/day (Sokol 1990). Every case should be carefully monitored and the dose individually tailored. In cases of severe damage to the nervous system, improvement is unlikely and treatment would aim to halt the progress of the disease. If conditions such as abetalipoproteinemia are diagnosed early enough, the condition may be prevented by supplementing vitamin E.

Neuropathy After Bariatric Surgery

Neurological complications are increasingly being recognized following bariatric surgery for morbid obesity. In a retrospective study of all patients undergoing bariatric surgery at the Mayo Clinic, Rochester, Thaisetthawatkul et al. found that mononeuropathy, polyneuropathy, or plexopathy occurred in 16 % subjects who underwent this type of procedure. In a retrospective study of cases referred to the neurology service following bariatric surgery, myelopathy and encephalopathy were also encountered (Juhasz-Pocsine 2007). The retrospective nature of these studies, with ascertainment bias, makes true estimates of the incidence and relative frequencies of the different types of neurological complication hard to establish. Studies do not consistently implicate deficiency of any one vitamin or mineral, and the causation may be multifactorial. Nerve biopsies have revealed axonal degeneration but also inflammatory infiltrates, raising the question of whether there is an autoimmune component (Thaisetthawatkul 2004).

Metabolic

Chronic uremia causes a progressive sensory motor, axonal neuropathy which can be painful (Wolfe 2002). The frequency has decreased over the last 10 years due to improved management with dialysis and renal transplant (Said 1987; Lagueny 2000). There are recordable changes in nerve excitability immediately before and after hemodialysis, with pre-dialysis peripheral nerve hyperpolarization presumably related to hyperkalemia (Krishnan 2005). It is unknown whether potassium or other dialyzable toxins determine the chronic neuropathy of end-stage renal failure.

Toxins can affect both the peripheral nerve and liver, but some neuropathies do occur as a consequence of hepatic dysfunction alone. Chronic hepatic failure can lead to an asymptomatic demyelinating neuropathy, and neuropathy is detectable in 93 % of patients pre-liver transplant (McDougall 2003). An acute demyelinating polyneuropathy similar to Guillain Barré syndrome can accompany viral hepatitis. A sensory neuropathy is associated with primary biliary cirrhosis, while childhood cholestatic liver disease can cause a sensory neuropathy secondary to vitamin E deficiency.

A mild chronic sensory motor neuropathy can be seen in both hyperthyroidism and hypothyroidism.

Acute intermittent porphyria can cause an axonal neuropathy which is preceded by the first attack of abdominal pain. Motor involvement is typically proximal and can affect the respiratory muscles; sensory involvement is mild, distal, and symmetrical. The neuropathy can also affect the autonomic system, leading to tachycardia, labile blood pressure, urinary retention, and diarrhea.

Amyloid neuropathy occurs in twenty percent of patients with light chain disease. This axonal neuropathy is predominantly sensory, distal, and symmetrical. Pain and temperature loss occur initially, but distal weakness can develop in the later stages. Involvement of the autonomic nerves leads to orthostatic hypotension, impotence, bladder dysfunction, hypohydrosis, and meiotic pupils. A multifocal neuropathy also occurs and frequently presents as carpal tunnel syndrome.

Refsum's disease is an autosomal recessive condition caused by a defect in phytanoyl-CoA-hydroxylase that causes ataxia, retinal damage, and deafness. The associated demyelinating neuropathy causes a slow, progressive, distal, symmetrical loss of vibration and proprioception. Weakness also occurs and progresses proximally. Painful paresthesia can be a feature.

Fabry's disease, caused by α-galactosidase deficiency, is an X-linked condition that affects the peripheral nervous system, the central nervous system, the heart, and the kidneys. The associated neuropathy is particularly painful. Patients describe burning in the hands and feet which can be episodic or chronic and progressive. Neurophysiology demonstrates an axonal neuropathy which affects the small fibers initially.

Paraproteinemic Neuropathies

Paraproteins are associated with ten percent of peripheral neuropathies, and they can be detected by immunoelectrophoresis or immunofixation of blood or urine. The identified proteins are usually the product of a single clone of plasma cells and are therefore termed monoclonal M proteins.

Paraproteinemic neuropathies can precede or be associated with a number of systemic disorders.

Monoclonal gammopathy of undetermined significance, MGUS, is the most common (Kelly 1981). In this condition, the paraprotein level is less that 3 g/dl in the serum, and the paraprotein usually possesses a kappa light chain component. MGUS is distinguished from myeloma by the lower level of paraprotein and the absence of systemic features. IgG is the most common (75 %) paraprotein found in patients with MGUS and no neuropathy, but when MGUS accompanies a neuropathy, the IgM isotype is more frequent (50 %). In half of cases with an IgM paraprotein, the IgM has antibody specificity against myelin-associated glycoprotein (MAG). Anti-MAG neuropathy typically affects men over the age of fifty and presents with distal numbness and paresthesia. Fifty percent of patients develop lancinating pains, dysesthesia, or aching discomfort in the limbs; light touch, joint position sense, and vibration sensation are also affected (Ropper 1998). In advanced cases, distal weakness can occur. Investigations usually reveal a demyelinating but occasionally an axonal neuropathy. Cerebrospinal fluid protein is typically elevated. Other neuropathy phenotypes associated with paraproteins include conditions clinically indistinguishable from chronic inflammatory

Metabolic and Nutritional Neuropathies, Table 2 Paraproteinemic neuropathies

Condition	Systemic symptoms	Neuropathy	Pain	Examination	Specific investigations	Pathology
MGUS	Nil	Distal numbness, paresthesia and pain with ataxia	+++	Predominantly sensory loss	IgG or IgM-K paraprotein <3 g/dl	Predominantly demyelinating
		Can resemble chronic inflammatory demyelinating polyneuropathy, CIDP		Weakness occurs late	Bone marrow biopsy <5 % plasma cells	
Multiple myeloma	Fatigue, bone pain	Distal weakness and dysesthesia		Distal, symmetrical weakness and numbness	IgM or IgG-K paraprotein >3 g/dl. Low hemoglobin	Predominantly axonal
					Raised calcium	
					Renal impairment	
					Bone marrow biopsy	
					Skeletal survey	
Waldenstrom's macroglobulin-anemia	Fatigue, weight loss, bleeding, and hyperviscosity	Numbness in the feet followed by foot drop	+	Distal, symmetrical paresthesia and numbness followed by distal weakness initially in the lower limbs	IgM-K paraprotein	Demyelinating
					Full blood count	
Amyloidosis (usually systemic)	Autonomic symptoms, e.g., postural hypotension, diarrhea and impotence	Numb feet initially progressing to lancinating pain and loss of temperature sensation	+++	Symmetrical, distal, predominantly sensory disturbance	IgG or IgA-λ paraprotein	Axonal
	Weight loss					
	Cardiac and renal dysfunction can occur				Biopsy of sural nerve, rectum or bone marrow with Congo red staining	
POEMS Syndrome (osteosclerotic myeloma)	Polyneuropathy	Progressive distal sensory disturbance and weakness	+/−	Symmetrical, distal allodynia, numbness, loss of joint position sensation and weakness	IgG or IgA-λ paraprotein	Demyelinating
	Organomegaly					
	Endocrinopathy					
	M protein				Skeletal survey	
	Skin changes					
	+/− Plasmacytoma					
Castleman's disease	Lymphadenopathy	Variable: can resemble motor neurone disease or CIDP	+/−	Variable	IgM or IgG paraprotein	Axonal
	+/− POEMS syndrome				Lymph node biopsy looking for angiofollicular hyperplasia	

(continued)

Metabolic and Nutritional Neuropathies, Table 2 (continued)

Condition	Systemic symptoms	Neuropathy	Pain	Examination	Specific investigations	Pathology
Cryoglobulin-anemia	Raynaud's phenomenon	Pain in the early stages	+++	Symmetric or asymmetric sensorimotor disturbance	IgM or IgG paraprotein but can be polyclonal	Axonal
	Evidence of lymphoproliferative disorders, collagen vascular or chronic inflammatory disease	Generalized or multifocal weakness			Precipitation of cryoglobulins from serum	
		Paresthesia can be precipitated by the cold				

demyelinating polyneuropathy. In these cases, the paraprotein is more likely to be IgG, and the pathogenic relevance of the paraprotein is doubtful.

The treatment of MGUS is not established, but intravenous immunoglobulin, plasma exchange, steroids, and immunosuppression have all been used with variable success. A number of recent small studies have reported beneficial effects treating MGUS with rituximab, a humanized monoclonal antibody directed against the CD20 B-lymphocyte surface epitope (Pestronk 2003; Renaud 2003; Dalakas 2009). The randomized placebo-controlled trial by Dalakas involved 13 patients with anti-MAG neuropathy randomized to rituximab and 13 randomized to placebo for 8 months. If one outlying patient was excluded, then the trial demonstrated a beneficial effect of treatment. Large randomized controlled trials are awaited to further assess the efficacy of rituximab. In addition to managing the neuropathy, it is important to monitor the paraprotein level in patients with MGUS, as 30 % will develop a malignant plasma-cell disorder within 25 years (Veneri 2004). The risk of progression of MGUS to malignancy is, on average, 1.5 % per year (Kyle 2003).

Other paraproteinemic disorders associated with a neuropathy are listed in Table 2. These conditions are important to recognize, as treatment can lead to remission of the condition and improvement in the neuropathy. Therefore, immunoelectrophoresis and immunofixation should be an essential part of the investigations of any unexplained neuropathy.

References

Behse, F., & Buchthal, F. (1977). Alcoholic neuropathy: Clinical, electrophysiological and biopsy findings. *Annals of Neurology, 2,* 95.

Dalakas, M. C., Rakocevic, G., et al. (2009). Placebo-controlled trial of rituximab in IgM anti-myelin-associated glycoprotein antibody demyelinating neuropathy. *Annals of Neurology, 65,* 286–293.

Juhasz-Pocsine, K., Rudnicki, S. A., Archer, R. L., & Harik, S. I. (2007). Neurologic complications of gastric bypass surgery for morbid obesity. *Neurology, 68,* 1843–1850.

Kelly, J. J., Kyle, R. A., O'Brien, P. C., et al. (1981). Prevalence of monoclonal protein in peripheral neuropathy. *Neurology, 31,* 1480–1483.

Koike, H., Iijima, M., Sugiura, M., et al. (2003). Alcoholic neuropathy is clinicopathologically distinct from thiamine-deficiency neuropathy. *Annals of Neurology, 54,* 19–29.

Krishnan, A. V., Phoon, R. K. S., et al. (2005). Altered motor nerve excitability in end-stage kidney disease. *Brain, 128,* 2164–2174.

Kyle, R. A., Therneau, T. M., Rajkumar, S. V., et al. (2003). Long-term follow-up of IgM monoclonal gammopathy of undetermined significance. *Seminars in Oncology, 30,* 169–171.

Lagueny, A. (2000). Metabolic and nutritional neuropathies. *La Revue du Praticien, 50,* 731–735.

McDougall, A. J., Davies, L., & McCaughan, G. W. (2003). Autonomic and peripheral neuropathy in end stage liver disease and following liver transplant. *Muscle & Nerve, 28,* 595–600.

Monforte, R., Estruch, R., Valls-Sole, J., et al. (1995). Autonomic and peripheral neuropathies in patients with chronic alcoholism. *Archives of Neurology, 52,* 45–51.

Neundorfer, B. (2001). Alcohol polyneuropathy. *Fortschritte der Neurologie-Psychiatrie, 69,* 341–345.

Nicolosi, C., Di Leo, R., et al. (2005). Is there a relationship between somatic and autonomic neuropathies in chronic alcoholics? *Journal of Neurological Sciences, 228,* 15–19.

Pestronk, A., Florence, J., Miller, T., et al. (2003). Treatment of IgM associated polyneuropathies using rituximab. *Journal of Neurology, Neurosurgery, and Psychiatry, 74*, 485–489.

Renaud, S., Gregor, M., Fuhr, P., et al. (2003). Rituximab in the treatment of anti-MAG associated polyneuropathy. *Muscle & Nerve, 27*, 611–615.

Roman, G. C. (1994). An epidemic in Cuba of optic neuropathy, sensorineural deafness, peripheral sensory neuropathy and dorsolateral myeloneuropathy. *Journal of Neurological Sciences, 127*, 11–28.

Ropper, A. H., & Gorson, K. C. (1998). Neuropathies associated with paraproteinemia. *The New England Journal of Medicine, 338*, 1601–1607.

Said, G. (1987). Acquired metabolic neuropathies (2): Kidney failure, hypothyroid and hypoglycaemia. *Revista de Neurologia, 143*, 785–790.

Savage, D. G., Lindenbaum, J., Stabler, S. P., & Allen, R. H. (1994). Sensitivity of serum methylmalonic acid and total homocysteine determinations for diagnosing cobalamin and folate deficiencies. *American Journal of Medicine, 96*, 239–246.

Sokol, R. J. (1990). Vitamin E and neurologic deficits. *Advances in Pediatrics, 37*, 119–148.

Talwar, D., Davidson, H., Cooney, J., & O'Reilly, D. S. J. (2000). Vitamin B_1 status assessed by direct measurement of thiamin pyrophosphate in erythrocytes or whole blood by HPLC: Comparison with erythrocyte transketolase activation assay. *Clinical Chemistry, 46*, 704–710.

Thaisetthawatkul, P., Collazo-Clavell, M. L., et al. (2004). A controlled study of peripheral neuropathy after bariatric surgery. *Neurology, 63*, 1462–1470.

Veneri, D., Aquel, H., Franchini, M., et al. (2004). Malignant evolution of monoclonal gammopathy of undetermined significance: analysis of 633 consecutive cases with long-term follow-up. *Haematologica, 89*, 876–878.

Windebank, A. (1993). Polyneuropathy due to nutritional deficiency and alcoholism. In *Peripheral neuropathy* (3rd ed., pp. 1310–1321). Philadelphia: W B Saunders.

Wolfe, G. I., & Barohn, R. J. (2002). Painful peripheralneuropathy. *Current Treatment Options in Neurology, 4*, 177–188.

Metabolic Neuropathies

▶ Metabolic and Nutritional Neuropathies

Metabolism

Definition

The metabolism of a drug describes its biotransformation into other substances due to chemical conversion.

Cross-References

▶ NSAIDs
▶ Pharmacokinetics

Metabotropic Glutamate Receptor

▶ Metabotropic Glutamate Receptors in the Thalamus

Metabotropic Glutamate Receptors

Synonyms

mGlu receptors; mGluRs

Definition

Family of G-protein-coupled glutamate receptors (mGluRs) that are coupled to intracellular second messenger (G protein) systems. There are eight known mGlu receptor subtypes (mGlu1–mGlu8), and some of these are known to form homodimers. The eight receptors can be placed into three groups (groups I, II, III) on the basis of sequence homology and pharmacology. They trigger long-lasting intracellular processes and mediate slow synaptic components (see also Glutamate Receptor).

Cross-References

▶ Inflammation, Role of Peripheral Glutamate Receptors
▶ Metabotropic Glutamate Receptors in the Thalamus
▶ Molecular Contributions to the Mechanism of Central Pain
▶ Nociceptive Processing in the Amygdala: Neurophysiology and Neuropharmacology
▶ Thalamus, Nociceptive Neurotransmission

Metabotropic Glutamate Receptors in Spinal Nociceptive Processing

Volker Neugebauer
Department of Neuroscience & Cell Biology,
University of Texas Medical Branch, Galveston,
TX, USA

Synonyms

Glutamate receptor (metabotropic); Nociceptive transmission; Spinal cord nociception

Definition

They are G-protein-coupled receptors that are activated by glutamate and linked to a variety of signal transduction pathways to regulate neuronal excitability and synaptic transmission in normal nervous system functions and in states of neuroplasticity associated with learning and memory processes as well as neuropsychiatric disorders. They also modulate nociceptive processing at different levels of the pain neuraxis, including the spinal dorsal horn.

Characteristics

Metabotropic glutamate receptors (mGluRs) (see ▶ Metabotropic Receptor) belong to family 3 of G-protein-coupled receptors, which can trigger long-lasting intracellular processes and "metabolic" changes that shape neurotransmission and synaptic plasticity. They are characterized by a seven transmembrane (heptahelical) domain topology and a large N-terminal extracellular domain, which contains important residues for ligand binding and forms two lobes that close like a Venus' flytrap upon ligand binding. The second (and possibly third) intracellular loop determines G-protein specificity and transduction mechanism. The intracellular C-terminal interacts directly with intracellular proteins such as Homer proteins. Eight mGluR subtypes have been cloned

to date and are classified into groups I (mGluRs 1 and 5), II (mGluRs 2 and 3), and III (mGluRs 4, 6, 7, and 8) based on their sequence homology, signal transduction mechanisms, and pharmacological profile (see Table 1). Several splice variants have been identified which may differ with regard to their pharmacology and G-protein coupling. For recent reviews, see Nicoletti et al. (2011) and Niswender and Conn (2010).

Signal Transduction

Group I mGluRs couple through $G_{q/11}$ proteins to the activation of phospholipase C (PLC), resulting in phosphoinositide (PI) hydrolysis and, subsequently, the formation of inositol-1,4,5-trisphosphate (IP3) and diacylglycerol (DAG) to release calcium from intracellular stores and activate protein kinase C (PKC), respectively (Ferraguti et al. 2008; Neugebauer 2007; Nicoletti et al. 2011; Niswender and Conn 2010). Group I mGluRs can also activate tyrosine kinases through mechanisms that may involve PLC or be independent of G-proteins (Ferraguti et al. 2008). The PKC- and tyrosine kinase–dependent pathways are two major signal transduction mechanisms that can activate the mitogen-activated protein kinases (MAPKs) such as the extracellular signal-regulated kinase 1/2 (ERK1/2) (Karim et al. 2001). mGluR1 can also couple to Gs and stimulate cAMP formation (Ferraguti et al. 2008). Group II and group III mGluRs are negatively coupled to adenylyl cyclase (AC) through G_i/G_o proteins, thereby inhibiting cyclic AMP (cAMP) formation and cAMP-dependent protein kinase (PKA) activation (Anwyl 1999; Neugebauer 2007; Nicoletti et al. 2011; Niswender and Conn 2010).

Modulation of Voltage- and Ligand-Gated Ion Channels

In general, the predominant effect of group I mGluR activation is enhanced neuronal excitability and synaptic transmission, whereas activation of groups II and III typically produces inhibitory effects. Exceptions exist however, and different subtypes within one group (e.g., mGluR1 and mGluR5) may exert opposing effects. mGluRs can regulate neuronal excitability through direct or

Metabotropic Glutamate Receptors in Spinal Nociceptive Processing, Table 1 Classification and pharmacology of mGluRs

Group	Group I	Group II	Group III
Subtype	mGluR1, 5	mGluR2, 3	mGluR4, 6, 7, 8
Activator			
Orthosteric agonist	S-DHPG (1,5)	NAAG (3)	LAP4 (4,6,8 > 7)
		2R, 4R-APDC(2,3)	LSOP
		LY354740 (2,3)	DCPH (8 > 4,6)
		LY379268 (2,3)	
Positive allosteric modulator (PAM)	Ro67-7476 (1)	LY487379 (2)	PHCCC (4)
	Ro67-4853 (1)	BINA (2)	AMN082 (7)
	VU71 (1)		
	CDPPB (5)		
	VU29 (5)		
	ADX47273 (5)		
Inhibitor			
Orthosteric antagonist	LY367385 (1)	LY341495 (2,3)	CPPG
			MAP4
			UPB1112
Negative allosteric modulator (NAM)	CPCCOEt (1)		MMPIP (7)
	BAY36-7620 (1)		
	JNJ16259685 (1)		
	SIB-1893 (5)		
	MPEP (5)		
	MTEP (5)		
	Fenobam (5)		
Effector	-Gq/11-protein	Gi/o-protein	Gi/o-protein
	PLC-DAG/IP3 $\uparrow$	AC-cAMP $\downarrow$	AC-cAMP $\downarrow$
	PKC $\uparrow$		
	MAPK-ERK $\uparrow$		
	(PLD $\uparrow$)		
	- (Gs-protein [mG1uR1a])		
	(AC $\uparrow$)		
	tyrosine kinase $\uparrow$		

indirect effects on a variety of voltage-sensitive ion channels, including high-voltage-gated Ca^{2+} channels, K^+ channels, and nonselective cationic channels (Anwyl 1999; Ferraguti et al. 2008; Neugebauer 2007; Nicoletti et al. 2011; Niswender and Conn 2010). Group I mGluRs activate L-type Ca^{2+} channels but inhibit N and P/Q type channels and certain K^+ channels such as Kv4.2 (Hu et al. 2007). Group II and III mGluRs inhibit Ca^{2+} channels but activate K^+ channels including G-protein-coupled inwardly rectifying potassium (GIRK) channels. The modulation of ligand-gated ion channels by mGluRs includes the group I mGluR-mediated enhancement of ionotropic glutamate receptor (see ▶ Ionotropic Receptor) function, likely through receptor phosphorylation. mGluRs can also modulate the release of transmitters by acting as autoreceptors (glutamate) or heteroreceptors (GABA, substance P, serotonin, dopamine, and acetylcholine) (Cartmell and Schoepp 2000).

Pharmacology and Modulation of mGluRs
Several potent and mGluR subgroup/subtype-selective compounds have been developed in recent years (see Table 1). A number of subtype-selective activators and inhibitors for group I mGluRs and some subtype-selective activators for group III are available now, but distinguishing between group II mGluR2 and mGluR3 remains a challenge. Many of the subtype-selective compounds are allosteric modulators rather than orthosteric agonists/antagonists, which means they bind to a site other than the glutamate binding site, typically within the heptahelical domain (Nicoletti et al. 2011; Niswender and Conn 2010).

The intracellular C-terminal is not only the critical site for G-protein coupling but also the target of interacting proteins, such as the Homer proteins, which can regulate receptor assembly, trafficking and subcellular localization, G-protein coupling, and constitutive (basal) receptor activity (Ferraguti et al. 2008; Nicoletti et al. 2011; Niswender and Conn 2010). Other interacting partners include calmodulin, β-tubulin and phosphatase1C. Unlike mGlu1 and mGlu5, the distal region of the C-terminal of mGluR7 does not interact with Homer proteins but binds the protein kinase C interacting protein (PICK1) that might be involved in the spatial organization of synaptic proteins. There is evidence to suggest that the interaction with PICK1 can be found in several members of group III mGluRs. Receptor phosphorylation is another important mechanism for modulating mGluR function, including PKC-mediated desensitization of group I mGluRs and uncoupling of groups II and III mGluRs from G-proteins by PKC and PKA (Ferraguti et al. 2008; Neugebauer 2007; Nicoletti et al. 2011; Niswender and Conn 2010).

Spinal Nociception
With the exception of mGluR6 and mGluR8, all mGluR subtypes are expressed in the spinal cord. The important role of group I mGluRs in spinal nociceptive processing, ► central sensitization, and spinally mediated pain behavior is now well established. The functions of groups II and III mGluRs are only beginning to emerge (Goudet et al. 2009; Neugebauer 2007; Nicoletti et al. 2011). The current focus of research is on the role of individual mGluR subtypes and signal transduction pathways.

Group I mGluRs
Both mGluR1 and mGluR5 are functionally expressed in the spinal dorsal horn. Anatomical and recent electrophysiological data further suggest that mGluR1 and mGluR5 are localized presynaptically as well as postsynaptically (Goudet et al. 2009; Neugebauer 2001, 2007; Nicoletti et al. 2011).

Effect of Activators
Activation of spinal group I mGluRs generally has pro-nociceptive effects in behavioral and electrophysiological assays, although mixed excitatory and inhibitory effects have been reported (Goudet et al. 2009; Neugebauer 2007; Nicoletti et al. 2011). Intrathecal administration of group I agonists such as S-DHPG produces or enhances spontaneous nociceptive behavior, thermal ► hyperalgesia (cold and heat), mechanical

hyperalgesia, mechanical ► allodynia, and formalin-induced nociception in the second phase (Fundytus et al. 1998; Hu et al. 2007; Karim et al. 2001; Neugebauer 2001, 2007). These effects are mediated through mGluR1 as well as mGluR5 since they were blocked with antagonists or antibodies for mGluR1 or mGluR5, where tested (Fundytus et al. 1998; Karim et al. 2001; Neugebauer 2001, 2007); the mGluR5-mediated effects involve Kv4.2-containing potassium channels (Hu et al. 2007; see next paragraph).

In electrophysiological studies in anesthetized animals in vivo, intraspinally administered group I mGluR agonists, including S-DHPG, had excitatory effects on spinal dorsal horn neurons and increased their responses to innocuous input/stimulus and, less consistently, to ► noxious mechanical stimulation of cutaneous or deep tissue (Neugebauer et al. 1999, 2001, 2007). In addition, mixed excitatory-inhibitory effects of group I mGluR activation have been observed in vivo as well as in spinal cord slices in vitro (Chen et al. 2000, Gerber et al. 2000a; Giles et al. 2007; Heinke and Sandkuehler 2007; Neugebauer 2001; Yang et al. 2011). In vivo, the facilitatory effects of group I agonists on spinothalamic tract cells can be blocked with antagonists for mGluR1 (CPCCOEt) or mGluR5 (MPEP) whereas the inhibitory effects are mimicked by an mGluR5 agonist (CHPG) (Neugebauer et al. 1999). In spinal cord slices in vitro, the facilitatory effect of DHPG on NMDA receptor-mediated excitatory transmission in dorsal horn neurons is blocked by an mGluR1 antagonist (AIDA; Yang et al. 2011); presynaptic facilitation of excitatory transmission by DHPG is inhibited by mGluR1 (CPCCOEt) and mGluR5 (MPEP) antagonists (Park et al. 2004); DHPG-induced increase of free cytosolic calcium concentration in lamina II neurons is blocked by antagonists for mGluR1 (S-4-CPG) and mGluR5 (MPEP) (Heinke and Sandkuehler 2007); DHPG-induced increase in excitability of dorsal horn neurons is prevented by an mGluR5 (MPEP), but not mGluR1 (LY367385), antagonist (Hu et al. 2007); mGluR5-driven increased neuronal excitability involves ERK-dependent inhibition of Kv4.2-containing potassium channels (Hu et al. 2007), particularly in NK1 and PKCγ positive excitatory neurons (Hu and Gereau 2011); finally, inhibitory effects of DHPG on excitatory transmission in lamina II neurons are blocked by an mGluR5 (MPEP), but not mGluR1 (LY367385), antagonist (Giles et al. 2007). The data suggest that mGluR1 and mGluR5 are involved in the excitatory modulation of spinal dorsal neurons whereas mGluR5 can also exert inhibitory effects.

Effect of Inhibitors

Behavioral and electrophysiological studies using antagonists, antibodies, or antisense oligonucleotides to block mGluR function have established the important contribution of spinal group I mGluRs to prolonged nociception and persistent pain states in models of inflammatory (second phase of *formalin test*; intradermal *capsaicin*; intraplantar carrageenan; *kaolin/carrageenan knee joint arthritis*; complete Freund's adjuvant-induced inflammation) and *neuropathic pain* (Goudet et al. 2009; Karim et al. 2001; Neugebauer et al. 1999, 2001, 2007; Soliman et al. 2005; Zhang et al. 2002). The role of group I mGluRs in brief and acute nociception is less clear.

Intrathecal administration of group I blockers, including subtype-selective antibodies and antagonists such as CPCCOEt (mGluR1) and MPEP (mGluR5), can inhibit nociceptive behavior in the second phase of the formalin test, thermal hyperalgesia and mechanical allodynia following intraplantar carrageenan, and capsaicin-induced mechanical, but not thermal, hypersensitivity (Fundytus et al. 1998; Karim et al. 2001; Neugebauer 2001, 2007; Soliman et al. 2005). There is also evidence for antinociceptive effects of group I blockade in models of neuropathic pain. In these studies, mGluR1 or mGluR5 antagonism or knockdown inhibited thermal hyperalgesia and mechanical allodynia (Dogrul et al. 2000; Fundytus et al. 1998; Neugebauer 2001, 2007). Others, however, found no effect of spinally administered mGluR5 antagonists (Urban et al. 2003). The direct comparison of systemically administered antagonists for mGluR1 (LY456236) and mGluR5 (MPEP and MTEP) suggests higher efficacy of the mGluR1 antagonist for inhibiting neuropathic pain behaviors in the spinal nerve ligation (SNL)

model, whereas mGluR1 and mGluR5 antagonists show similar antihyperalgesic effects in the second phase of formalin-induced pain model (Varty et al. 2005). Consistent with the contribution of spinal mGluR1 to neuropathic pain, bilateral upregulation of mGluR1 in postsynaptic densities in dorsal horn neurons has been reported in the SNL model of neuropathic pain (Aira et al. 2012). Group I mGluRs may also mediate brief and acute nociception because blockade of spinal group I mGluRs reduced noxious heat responses and the first phase of the formalin test; such effects, however, were not detected in other studies (see Neugebauer 2001, 2007). An mGluR5 antagonist (fenobam) also inhibited visceromotor responses (VMR) to bladder distention in normal animals in the absence of inflammation as well as in mice infected with a uropathogenic strain of *Escherichia coli* (UPEC) to develop bladder inflammation (Crock et al. 2012).

Electrophysiological studies of spinal dorsal horn neurons, including spinothalamic tract cells, in anesthetized animals in vivo show antinociceptive effects of spinally administered group I antagonists, including mGluR1 selective compounds (CPCCOEt), or antisense oligodeoxynucleotides directed against mGluR1, in pain-related central sensitization following intradermal capsaicin, topical application of mustard oil to the skin, or intra-articular kaolin/carrageenan injections (Neugebauer et al. 1999, 2001). In spinal cord slices in vitro, an mGluR1 antagonist (AIDA) inhibited enhanced NMDA receptor-mediated excitatory transmission in dorsal horn neurons in the CFA hindpaw inflammation model but had no effect on basal synaptic transmission in slices from normal animals (Yang et al. 2011). In this model, an antagonist for mGluR5 (MPEP), but not mGluR1 (LY367385), reduced Fos expression in the dorsal horn (Giles et al. 2007). In the streptozotocin model of diabetic neuropathy, an mGluR5 antagonist (MPEP) inhibited excitatory synaptic transmission in lamina II neurons more strongly than under control conditions, whereas an mGluR1 antagonist (CPCCOEt) had no effect under either condition (Li et al. 2010). An mGluR1 antagonist (S-4CPG) prevented the induction of LTP of field potentials in dorsal horn neurons without affecting basal C-fiber-evoked responses (Azkue et al. 2003).

In conclusion, there is good evidence for the contribution of mGluR1 and/or mGluR5 to pain-related spinal processing and behaviors in a number of models although a unifying pattern has yet to emerge. Their involvement in brief spinal nociception under normal circumstances is less consistent but mGluR5 seems to be a more likely contributor than mGluR1.

Groups II and III mGluRs

The roles of group II and group III mGluRs in spinal nociceptive processing are less clear. Both group II mGluR2/3 and group III mGluR4 and 7, but not mGluR6 and 8, are present in the spinal cord. Localization of group III mGluRs is predominantly presynaptic at the active zone of transmitter release, whereas presynaptic group II mGluRs are localized in the preterminal region away from the release site and mGluR3 is also found on postsynaptic sites and on glial cells. Thus, group II activation may require an excess of synaptic glutamate release or glutamate release from glial cells (Neugebauer 2001, 2007; Nicoletti et al. 2011).

Behavioral and electrophysiological studies suggest antinociceptive effects of spinal group II or group III mGluR activation (Neugebauer 2001, 2007; Nicoletti et al. 2011; Goudet et al. 2009). Intrathecal administration of a number of group II mGluR agonists has antinociceptive effects in models of inflammatory (intradermal/subdermal capsaicin; formalin) and neuropathic (SNL) pain, largely without affecting baseline mechanical or thermal sensitivity to acute stimuli; group II antagonists (EGLU or LY341495) block the inhibitory effect when tested (Dolan and Nolan 2002; Goudet et al. 2009; Soliman et al. 2005; Yamamoto et al. 2004; Zhou et al. 2011). Systemic administration of a selective group II agonist (LY379268) also had no significant effects on acute mechanical and thermal nociceptive function (Simmons et al. 2002). Intrathecal administered selective group III agonists such as LAP4 and a positive allosteric modulator for mGluR4 (PHCCC) have antinociceptive effects in several animal models

of inflammatory (carrageenan- or formalin-induced hindpaw inflammation; CFA ankle arthritis) and neuropathic (CCI, SNL, vincristine) pain; these effects are blocked with a selective antagonist (MAP4) when tested (Chen and Pan 2005; Goudet et al. 2009; Neugebauer 2007; Nicoletti et al. 2011; Soliman et al. 2005). Intrathecal administration of LAP4 had no effect in the absence of tissue injury (Chen and Pan 2005; Soliman et al. 2005).

Electrophysiological studies in anesthetized animals in vivo measured inhibitory effects of group II and group III agonists on spinal nociceptive processing (Neugebauer 2001, 2007; Goudet et al. 2009). Activation of spinal group II mGluRs inhibits electrically evoked C-fiber responses of nociceptive dorsal horn neurons in carrageenan-induced hind paw inflammation, whereas mixed effects (inhibition and facilitation) are observed in control rats. Intraspinal administration of a selective group II agonist (LY379268) also inhibits central sensitization of primate spinothalamic tract cells in the capsaicin pain model but has no effect on evoked responses of non-sensitized neurons (Neugebauer et al. 2000). Spinal application of a group III agonist (LAP4) inhibits capsaicin-induced central sensitization of spinothalamic tract cells but also their responses to brief noxious and innocuous mechanical cutaneous stimuli under control conditions (Neugebauer et al. 2000). LAP4 inhibits the evoked responses of dorsal horn projection neurons in neuropathic rats (SNL model) but not under control conditions although a selective antagonist (MAP4) potentiated the responses to brief noxious stimuli in normal animals in this study (Chen and Pan 2005).

In spinal cord slices in vitro, agonists for group II (DCG-IV) and group III (LAP4) inhibit excitatory synaptic transmission in lamina II neurons from neuropathic rats (SNL model) more strongly than in neurons from control rats. Both DCG-IV and LAP4 also inhibit inhibitory transmission and this effect is similar between the neuropathic model and control conditions (Zhang et al. 2009; Zhou et al. 2011). The inhibitory effects involve a presynaptic site of action. The data are in line with previous studies in slices from normal rats showing that agonists for group II (LCCG) and group III (LAP4) mGluRs inhibit excitatory and inhibitory transmission in lamina II neurons through a presynaptic mechanism; this effect is blocked with selective antagonists (Gerber et al. 2000b).

In conclusion, accumulating evidence suggests spinal groups II and III mGluRs as targets to inhibit nociceptive processing in models of inflammatory and neuropathic pain. Less clear is their modulatory function in baseline transmission and behavior.

References

Aira, Z., Buesa, I., Gallego, M., del Garcia, C. G., Mendiable, N., Mingo, J., et al. (2012). Time-dependent cross talk between spinal serotonin 5-HT2A receptor and mGluR1 subserves spinal hyperexcitability and neuropathic pain after nerve injury. *Journal of Neuroscience, 32*, 13568–13581.

Anwyl, R. (1999). Metabotropic glutamate receptors: Electrophysiological properties and role in plasticity. *Brain Research Brain Research Reviews, 29*, 83–120.

Azkue, J. J., Liu, X. G., Zimmermann, M., & Sandkuhler, J. (2003). Induction of long-term potentiation of C fibre-evoked spinal field potentials requires recruitment of group I, but not group II/III metabotropic glutamate receptors. *Pain, 106*, 373–379.

Cartmell, J., & Schoepp, D. D. (2000). Regulation of neurotransmitter release by metabotropic glutamate receptors. *Journal of Neurochemistry, 75*, 889–907.

Chen, J., Heinke, B., & Sandkuhler, J. (2000). Activation of group I metabotropic glutamate receptors induces long-term depression at sensory synapses in superficial spinal dorsal horn. *Neuropharmacology, 39*, 2231–2243.

Chen, S. R., & Pan, H. L. (2005). Distinct roles of group III metabotropic glutamate receptors in control of nociception and dorsal horn neurons in normal and nerve-injured rats. *Journal of Pharmacology and Experimental Therapeutics, 312*, 120–126.

Crock, L. W., Stemler, K. M., Song, D. G., Abbosh, P., Vogt, S. K., Qiu, C. S., et al. (2012). Metabotropic glutamate receptor 5 (mGluR5) regulates bladder nociception. *Molecular Pain, 8*, 20.

Dogrul, A., Ossipov, M. H., Lai, J., Malan, T. P., Jr., & Porreca, F. (2000). Peripheral and spinal antihyperalgesic activity of SIB-1757, a metabotropic glutamate receptor (mGLUR(5)) antagonist, in experimental neuropathic pain in rats. *Neuroscience Letters, 292*, 115–118.

Dolan, S., & Nolan, A. M. (2002). Behavioral evidence supporting a differential role for spinal group I and II metabotropic glutamate receptors in inflammatory hyperalgesia in sheep. *Neuropharmacology, 43*, 319–326.

Ferraguti, F., Crepaldi, L., & Nicoletti, F. (2008). Metabotropic glutamate 1 receptor: Current concepts and perspectives. *Pharmacological Reviews, 60*, 536–581.

Fundytus, M. E., Fisher, K., Dray, A., Henry, J. L., & Coderre, T. J. (1998). In vivo antinociceptive activity of anti-rat mGluR1 and mGluR5 antibodies in rats. *Neuroreproductive, 9*, 731–735.

Gerber, G., Youn, D. H., Hsu, C. H., Isaev, D., & Randic, M. (2000a). Spinal dorsal horn synaptic plasticity: Involvement of group I metabotropic glutamate receptors. *Progress in Brain Research, 129*, 115–134.

Gerber, G., Zhong, J., Youn, D., & Randic, M. (2000b). Group II and group III metabotropic glutamate receptor agonists depress synaptic transmission in the rat spinal cord dorsal horn. *Neuroscience, 100*, 393–406.

Giles, P. A., Trezise, D. J., & King, A. E. (2007). Differential activation of protein kinases in the dorsal horn in vitro of normal and inflamed rats by group I metabotropic glutamate receptor subtypes. *Neuropharmacology, 53*, 58–70.

Goudet, C., Magnaghi, V., Landry, M., Nagy, F., Gereau, R. W., & Pin, J. P. (2009). Metabotropic receptors for glutamate and GABA in pain. *Brain Research Reviews, 60*, 43–56.

Heinke, B., & Sandkuhler, J. (2007). Group I metabotropic glutamate receptor-induced Ca2 + -gradients in rat superficial spinal dorsal horn neurons. *Neuropharmacology, 52*, 1015–1023.

Hu, H. J., Alter, B. J., Carrasquillo, Y., Qiu, C. S., & Gereau, R. W., IV. (2007). Metabotropic glutamate receptor 5 modulates nociceptive plasticity via extracellular signal-regulated ERK4.2 Signaling in spinal cord dorsal horn neurons. *Journal of Neuroscience, 27*, 13181–13191.

Hu, H. J., & Gereau, R. W. (2011). Metabotropic glutamate receptor 5 regulates excitability and Kv4.2-containing K(+) channels primarily in excitatory neurons of the spinal dorsal horn. *Journal of Neurophysiology, 105*, 3010–3021.

Karim, F., Wang, C. C., & Gereau, R. W. (2001). Metabotropic glutamate receptor subtypes 1 and 5 are activators of extracellular signal-regulated kinase signaling required for inflammatory pain in mice. *Journal of Neuroscience, 21*, 3771–3779.

Li, J. Q., Chen, S. R., Chen, H., Cai, Y. Q., & Pan, H. L. (2010). Regulation of increased glutamatergic input to spinal dorsal horn neurons by mGluR5 in diabetic neuropathic pain. *Journal of Neurochemistry, 112*, 162–172.

Neugebauer, V. (2001). Metabotropic glutamate receptors: Novel targets for pain relief. *Expert Review of Neurotherapeutics, 1*(2), 207–224.

Neugebauer, V. (2007). Glutamate receptor ligands. *Handbook of Experimental Pharmacology, 177*, 217–249.

Neugebauer, V., Chen, P. S., & Willis, W. D. (1999). Role of metabotropic glutamate receptor subtype mGluR1 in brief nociception and central sensitization of primate STT cells. *Journal of Neurophysiology, 82*, 272–282.

Neugebauer, V., Chen, P.-S., & Willis, W. D. (2000). Groups II and III metabotropic glutamate receptors differentially modulate brief and prolonged nociception in primate STT cells. *Journal of Neurophysiology, 84*, 2998–3009.

Nicoletti, F., Bockaert, J., Collingridge, G. L., Conn, P. J., Ferraguti, F., Schoepp, D. D., et al. (2011). Metabotropic glutamate receptors: From the workbench to the bedside. *Neuropharmacology, 60*, 1017–1041.

Niswender, C. M., & Conn, P. J. (2010). Metabotropic glutamate receptors: Physiology, pharmacology, and disease. *Annual Review of Pharmacology and Toxicology, 50*, 295–322.

Park, Y. K., Galik, J., Ryu, P. D., & Randic, M. (2004). Activation of presynaptic group I metabotropic glutamate receptors enhances glutamate release in the rat spinal cord substantia gelatinosa. *Neuroscience Letters, 361*, 220–224.

Simmons, R. M., Webster, A. A., Kalra, A. B., & Iyengar, S. (2002). Group II mGluR receptor agonists are effective in persistent and neuropathic pain models in rats. *Pharmacology, Biochemistry, and Behavior, 73*, 419–427.

Soliman, A. C., Yu, J. S. C., & Coderre, T. J. (2005). mGlu and NMDA receptor contributions to capsaicin-induced thermal and mechanical hypersensitivity. *Neuropharmacology, 48*, 325–332.

Urban, M. O., Hama, A. T., Bradbury, M., Anderson, J., Varney, M. A., & Bristow, L. (2003). Role of metabotropic glutamate receptor subtype 5 (mGluR5) in the maintenance of cold hypersensitivity following a peripheral mononeuropathy in the rat. *Neuropharmacology, 44*, 983–993.

Varty, G. B., Grilli, M., Forlani, A., Fredduzzi, S., Grzelak, M. E., Guthrie, D. H., et al. (2005). The antinociceptive and anxiolytic-like effects of the metabotropic glutamate receptor 5 (mGluR5) antagonists, MPEP and MTEP, and the mGluR1 antagonist, LY456236, in rodents: A comparison of efficacy and side-effect profiles. *Psychopharmacology, 179*, 207–217.

Yamamoto, T., Hirasawa, S., Wroblewska, B., Grajkowska, E., Zhou, J., Kozikowski, A., et al. (2004). Antinociceptive effects of N-acetylaspartylglutamate (NAAG) peptidase inhibitors ZJ-11, ZJ-17 and ZJ-43 in the rat formalin test and in the rat neuropathic pain model. *European Journal of Neuroscience, 20*, 483–494.

Yang, K., Takeuchi, K., Wei, F., Dubner, R., & Ren, K. (2011). Activation of group I mGlu receptors contributes to facilitation of NMDA receptor membrane current in spinal dorsal horn neurons after hind paw inflammation in rats. *European Journal of Pharmacology, 670*, 509–518.

Zhang, H. M., Chen, S. R., & Pan, H. L. (2009). Effects of activation of group III metabotropic glutamate receptors on spinal synaptic transmission in a rat model of neuropathic pain. *Neuroscience, 158*, 875–884.

Zhang, L., Lu, Y., Chen, Y., & Westlund, K. N. (2002). Group I metabotropic glutamate receptor antagonists block secondary thermal hyperalgesia in rats with knee joint inflammation. *Journal of Pharmacology and Experimental Therapeutics, 300*, 149–156.

Zhou, H. Y., Chen, S. R., Chen, H., & Pan, H. L. (2011). Functional plasticity of group II metabotropic glutamate receptors in regulating spinal excitatory and inhibitory synaptic input in neuropathic pain. *Journal of Pharmacology and Experimental Therapeutics, 336*, 254–264.

Metabotropic Glutamate Receptors in the Thalamus

Thomas E. Salt
Institute of Ophthalmology, University College London, London, UK

Synonyms

G-protein-coupled glutamate receptor; Metabotropic glutamate receptor; mGluR; mGlu receptor; Thalamus metabotropic glutamate receptors

Definition

Metabotropic glutamate (mGlu) receptors are ▶ glutamate receptors that are coupled to intracellular second messenger (G protein) systems.

Characteristics

Eight metabotropic glutamate (mGlu) receptor subtypes (mGlu1–mGlu8) have been characterized to date. They can be placed into three groups (I, II, III) on the basis of their sequence homology, their pharmacological characteristics, and the types of intracellular transduction cascade to which they may couple in

Metabotropic Glutamate Receptors in the Thalamus, Table 1 Metabotropic glutamate receptor groups and subtypes with examples of ligands

Group	Receptor	Transduction mechanism	Orthosteric agonists		Orthosteric antagonists		Allosteric modulators	
			Subtype selective	Group selective	Subtype selective	Group selective	Positive	Negative
I	mGlu1	IP3/Ca^{++}cascade		DHPG	LY367385	4CPG	Ro67-4853	CPCCOEt
								JNJ16259685
	mGlu5		CHPG				CDDPB	MTEP
								MPEP
II	mGlu2	Inhibitory cAMP cascade	LY395756	LY354740		LY341495	BINA	Ro64-5229
	mGlu3			LY379268			LY487379	
				APDC	LY395756			
III	mGlu4	Inhibitory cAMP cascade	LSP4-2022	L-AP4			MAP4	VU0361737
	mGlu6							
	mGlu7			L-SOP			MPPG	MMPIP
	mGlu8		DCPG			CPPG		

APDC 2R,4R-4-aminopyrrolidine-2,4-decarboxylate, *BINA* 3′-[[(2-cyclopentyl-2,3-dihydro-6,7-dimethyl-1-oxo-1H-inden-5-yl)oxy]methyl]-[1,1′-biphenyl]-4-carboxylic acid, *CDDPB* 3-cyano-*N*-(1,3-diphenyl-1H-pyrazol-5-yl) benzamide, *CHPG* 2-chloro-5-hydroxyphenylglycine, *CPCCOEt* 7-(hydroxyimino)-cyclopropa[b]chromen-1a-carboxylate, *JNJ16259685* (3,4-Dihydro-2H-pyrano[2,3-b]quinolin-7-yl)-(cis-4-methoxycyclohexyl)-methanone, *4CPG* (*S*)-4-carboxyphenylglycine, *CPPG* alpha-cyclopropyl-4-phosphonophenylglycine, *DCPG* (S)-3,4-dicarboxyphenylglycine, *DHPG* 3,5-dihydroxyphenylglycine, *L-AP4* L-2-amino-4-phosphonobutyric acid, *L-SOP* L-serine-O-phosphate, *LSP4-2022* [((3S)-3-amino-3-carboxy)propyl][(4-(carboxymethoxy)phenyl) hydroxymethyl]phosphinic acid, *LY341495* 2S-2-amino-2 (1S,2S-2-carboxcyclopropyl-1-yl)-3-(xanth-9-yl)propanoic acid, *LY354740* (+)-2-aminobicyclo [3.1.0]hexane-2,6dicarboxylate, *LY367385* (+)-2-methyl-4-carboxyphenylglycine, *LY395756* (1S,2S,4R,5R,6S)-rel-2-amino-4-methylbicyclo[3.1.0]hexane-2,6-dicarboxylic acid, *LY487379* 2,2,2-trifluoro-*N*-[4-(2-methoxyphenoxy)phenyl]-*N*-(3-pyridinylmethyl)ethanesulfonamide, *MAP4* (S)-2-amino-2-methyl-4-phosphonobutanoic acid, *MMPIP* 6-(4-methoxyphenyl)-5-methyl-3-(4-pyridinyl)-isoxazolo[4,5-c]pyridin-4(5H)-one, *MPEP* 2-methyl-6-(phenylethynyl) -pyridine, *MPPG* (RS)-α-methyl-4-phosphonophenylglycine, *MTEP* 3-((2-methyl-1,3-thiazol-4-yl)ethynyl)pyridine, *Ro64-5229* (Z)-1-[2-cycloheptyloxy-2-(2,6-dichlorophenyl)ethenyl]-1H-1,2,4-triazole, *Ro67-4853* (9H-xanthen-9-ylcarbonyl)-carbamic acid butyl ester, *VU0361737* N-(4-chloro-3-methoxyphenyl)-2-pyridinecarboxamide

in vitro expression systems (Niswender and Conn 2010). In such expression systems, group I (mGlu1, mGlu5) receptors typically couple to postsynaptic inositol phosphate metabolism, while group II (mGlu2, mGlu3) and group III (mGlu4, mGlu6-8) receptors may couple to an inhibitory cyclic-AMP cascade. However it is becoming evident that these receptors can couple to alternative signaling pathways in a context-dependent manner, thus potentially adding complexity to the possible outcomes of mGlu receptor activation.

All of the mGlu receptors can be activated by L-glutamate with a variety of affinities. This amino acid is assumed to be the endogenous orthosteric ligand, but the receptors may also be activated by other endogenous ligands (e.g., sulfur-containing amino acids such as L-homocysteic acid or dipeptides such as N-acetyl-aspartyl-glutamate (NAAG)). The mGlu receptors form functional homodimers that appear to require binding of an orthosteric agonist to the ligand binding site of both dimers. There is also some suggestion that mGlu receptors may form heterodimers, but it is unknown whether this occurs in native systems (Niswender and Conn 2010).

A variety of synthetic orthosteric agonists and antagonists have been developed, some of which show selectivity for mGlu receptor groups or even subtypes (Table 1); however it appears that greater subunit selectivity can be achieved with positive or negative allosteric modulators (Niswender and Conn 2010). An additional advantage of positive allosteric modulators is that they allow/enhance activation of receptors in a more physiological manner as activation of the receptors by the orthosteric (physiological) ligand is a prerequisite for activation (Niswender and Conn 2010).

The group I receptors have been thought to predominantly mediate postsynaptic actions, whereas the group II receptors and group III receptors have been found to have presynaptic actions, regulating transmitter release. It is however becoming evident that the situation is more complex than this and that receptors of all three groups can have pre-, post-, or extra-synaptic actions (Niswender and Conn 2010). The complexity introduced by the variety of glutamate receptor types is compounded by their nonuniform distribution within the brain and within the neuropil, synapses, and extra-synaptic areas. Within the thalamus, the distribution of a considerable number of the various mGlu receptors has been described in some detail (Martin et al. 1992; Godwin et al. 1996; Petralia et al. 1996; Liu et al. 1998; Mineff and Valtschanoff 1999; Neto et al. 2000; Tamaru et al. 2001; Ngomba et al. 2011).

Group I Receptors

There is expression of the mRNA for the group I mGlu receptors (mGlu1 and mGlu5) in the thalamus (Martin et al. 1992; Ngomba et al. 2011), with both mGlu1 and mGlu5 receptor-like immunoreactivities being found on the distal dendrites of ▶ thalamocortical neurons opposed to the corticothalamic axon terminals (Martin et al. 1992; Godwin et al. 1996; Liu et al. 1998) and also on the dendrites of local circuit interneurons presynaptic to thalamocortical neurons (Godwin et al. 1996). A particular focus has been the function of mGlu1 receptors, as these have been localized postsynaptically predominantly beneath terminals of corticothalamic fibers (Martin et al. 1992; Godwin et al. 1996). Activation of mGlu receptors in thalamic relay neurons causes a slow depolarizing response associated with an increase in membrane resistance, as seen in many other parts of the brain, probably due a reduction in a potassium

conductance. This can be mediated specifically via mGlu1 receptors, which can be synaptically activated by stimulation of corticothalamic afferents (Turner and Salt 2000; Hughes et al. 2002; Reichova and Sherman 2004). A specific synaptic role for mGlu5 receptors in the thalamus is not so well defined, although these receptors do appear to be activated under physiological conditions (Salt and Binns 2000) and appear to be involved in the activation of glia by upon synaptic input to ventrobasal thalamus (VB) (Parri et al. 2010).

Group II Receptors

High levels of mRNA and protein for group II receptors have been found in the ▶ thalamic reticular nucleus (TRN), and much of this may be attributable to mGlu3 receptors, some of which may be localized in glial cells as well as neuronal bodies and dendrites (Petralia et al. 1996; Neto et al. 2000; Tamaru et al. 2001). Intriguingly, it has been shown that activation of these receptors can lead to an inhibition of TRN neuron activity (Cox and Sherman 1999). Ultrastructural studies suggest that group II receptors are localized in glial processes, some of which appear to be surrounding GABAergic terminals in VB (Liu et al. 1998; Mineff and Valtschanoff 1999), and mGlu3 receptors have been found to be concentrated on GABAergic axons in VB arising from TRN (Tamaru et al. 2001). Thus, group II receptors may modulate GABAergic transmission within VB, a notion supported by the finding that activation of group II receptors within VB results in a reduction of TRN-originating inhibition onto relay cells (Turner and Salt 2003; Copeland et al. 2012).

Group III Receptors

The mRNAs for mGlu4, mGlu7, and mGlu8 (but not mGlu6) receptors are expressed throughout the thalamus-TRN-cortex network (Ohishi et al. 1995; Saugstad et al. 1997; Neto et al. 2000), suggesting that these receptors may be involved in the control of transmission at corticothalamic and TRN-thalamic synapses. Thalamic relay neurons express mRNA for mGlu4 and mGlu7 receptors, whereas TRN neurons express mGlu8 and mGlu7 receptor mRNAs (Ohishi et al. 1995; Saugstad et al. 1997; Corti et al. 1998; Neto et al. 2000). mGlu4 and mGlu7 receptor protein is found in all thalamic relay nuclei, with less in the TRN (Bradley et al. 1998; Kinoshita et al. 1998; Corti et al. 2002). Ultrastructural localization of mGlu4 receptors has been shown presynaptically on cortical glutamatergic inputs in the TRN (Ngomba et al. 2008; Ngomba et al. 2008). Consistent with this are electrophysiological data, which suggest that group III receptors mediate a presynaptic reduction of the corticothalamic excitatory postsynaptic potential (Turner and Salt 1999) and a presynaptic reduction of the TRN-mediated inhibition of VB neuron responses to sensory stimuli in vivo and in vitro (Turner and Salt 2003).

Functional Considerations

There is little evidence to suggest that mGlu receptors are directly involved in ascending sensory transmission to the thalamic relay nuclei; rather it seems that such fast transmission is mediated via ▶ ionotropic glutamate receptors of the ▶ NMDA and ▶ AMPA variety (Salt and Eaton 1996). The characteristics of mGlu receptors are more suited to slow synaptic transmission or modulation or heterosynaptic modulation (Niswender and Conn 2010) and the roles in thalamic function that have been determined fit into this category.

The corticothalamic projection has been the subject of many studies. It has been speculated that the influence of the cortical input may operate via ▶ NMDA receptors or mGlu receptors (Sherman and Guillery 2000), largely because transmission via these receptor types would allow nonlinear amplification of excitatory inputs mediated via, for example, ▶ AMPA receptors. This is a particularly attractive hypothesis in the case of mGlu1 receptors, as these are restricted to corticothalamic synapses and

because NMDA receptor-mediated responses have been shown to be modulated by activation of group I (i.e., mGlu1/mGlu5) receptors in several brain areas. In the VB, activation of mGlu1 receptors potentiates responses mediated via either AMPA or NMDA receptors in vivo (Salt and Binns 2000). Thus, although the isolated corticothalamic synaptic potential which can be attributed to mGlu1 receptors in vitro appears to be rather small (Turner and Salt 2000; Reichova and Sherman 2004), it would be able to exert a large influence on ionotropic receptor-mediated responses, if the sensory stimulus was appropriate to recruit activity in the corticothalamic output. Recent studies with an mGlu1 positive allosteric modulator lend support to this idea (Salt et al. 2012). Conditions where this might come into play are, for example, during the processing of nociceptive information in the thalamus (Salt and Binns 2000) (Fig. 1).

The GABAergic inhibitory output from the TRN onto relay cells is a major contributor to the overall response profile of the relay cells, and thus the control of this inhibition by mGlu receptors is potentially of great functional significance. The location of group II and III receptors and the effects of their activation in this circuit suggest they play a pivotal role in these mechanisms (Fig. 1). All of these receptors are in locations removed from sites of synaptically released glutamate. Thus this suggests that these receptors are activated by glutamate which spills out of the conventional synaptic area, possibly under conditions of intense synaptic activity. This concept of "synaptic spillover" has been postulated on the basis of in vitro experiments from a number of non-thalamic brain areas (Kullmann 2000). This raises the possibility that the group II/III receptors may be activated by glutamate via a synaptic spillover mechanism. This glutamate could be released from terminals of sensory or cortical afferents or from astrocytes. This might occur under conditions of intense synaptic activation, such as seizure activity (Alexander and Godwin 2006; Ngomba et al. 2008, 2011), but perhaps also during normal sensory processing (Copeland et al. 2012).

Metabotropic Glutamate Receptors in the Thalamus, Fig. 1 Location of metabotropic glutamate receptors in the thalamus. Schematic illustration of some of the locations of glutamate receptors on different cellular elements in the thalamic circuitry (simplified and not to scale) and some possible synaptic interactions at glutamatergic and GABAergic synapses. See text for further details

References

Alexander, G. M., & Godwin, D. W. (2006). Metabotropic glutamate receptors as a strategic target for the treatment of epilepsy. *Epilepsy Research, 71*, 1–22.

Bradley, S. R., Rees, H. D., Yi, H., Levey, A. I., & Conn, P. J. (1998). Distribution and developmental regulation of metabotropic glutamate receptor 7a in rat brain. *Journal of Neurochemistry, 71*, 636–645.

Copeland, C. S., Neale, S. A., & Salt, T. E. (2012). Positive allosteric modulation reveals a specific role for mGlu2 receptors in sensory processing in the thalamus. *The Journal of Physiology, 590*, 937–951.

Corti, C., Restituito, S., Rimland, J. M., Brabet, I., Corsi, M., Pin, J. P., & Ferraguti, F. (1998). Cloning and characterization of alternative mRNA forms for the rat metabotropic glutamate receptors mGluR7 and mGluR8. *European Journal of Neuroscience, 10*, 3629–3641.

Corti, C., Aldegheri, L., Somogyi, P., & Ferraguti, F. (2002). Distribution and synaptic localisation of the metabotropic glutamate receptor 4 (mGluR4) in the rodent CNS. *Neuroscience, 110*, 403–420.

Cox, C. L., & Sherman, S. M. (1999). Glutamate inhibits thalamic reticular neurons. *Journal of Neuroscience, 19*, 6694–6699.

Godwin, D. W., Van Horn, S. C., Erisir, A., Sesma, M., Romano, C., & Sherman, S. M. (1996). Ultrastructural localization suggests that retinal and cortical inputs access different metabotropic glutamate receptors in the lateral geniculate nucleus. *Journal of Neuroscience, 16*, 8181–8192.

Hughes, S. W., Cope, D. W., Blethyn, K. L., & Crunelli, V. (2002). Cellular mechanisms of the slow (<1 Hz) oscillation in thalamocortical neurons in vitro. *Neuron, 33*, 947–958.

Kinoshita, A., Shigemoto, R., Ohishi, H., van der Putten, H., & Mizuno, N. (1998). Immunohistochemical localization of metabotropic glutamate receptors, mGluR7a and mGluR7b, in the central nervous system of the adult rat and mouse: A light and electron microscopic study. *The Journal of Comparative Neurology, 393*, 332–352.

Kullmann, D. M. (2000). Spillover and synaptic cross talk mediated by glutamate and GABA in the mammalian brain. *Progress in Brain Research, 125*, 339–351.

Liu, X. B., Munoz, A., & Jones, E. G. (1998). Changes in subcellular localization of metabotropic glutamate receptor subtypes during postnatal development of mouse thalamus. *The Journal of Comparative Neurology, 395*, 450–465.

Martin, L. J., Blackstone, C. D., Huganir, R. L., & Price, D. L. (1992). Cellular localization of a metabotropic glutamate receptor in rat brain. *Neuron, 9*, 259–270.

Mineff, E., & Valtschanoff, J. (1999). Metabotropic glutamate receptors 2 and 3 expressed by astrocytes in rat ventrobasal thalamus. *Neuroscience Letters, 270*, 95–98.

Neto, F. L., Schadrack, J., Berthele, A., Zieglgänsberger, W., Tölle, T. R., & Castro-Lopes, J. M. (2000). Differential distribution of metabotropic glutamate receptor subtype mRNAs in the thalamus of the rat. *Brain Research, 854*, 93–105.

Ngomba, R. T., Ferraguti, F., Badura, A., Citraro, R., Santolini, I., Battaglia, G., Bruno, V., De Sarro, G., Simonyi, A., van Luijtelaar, G., & Nicoletti, F. (2008). Positive allosteric modulation of metabotropic glutamate 4 (mGlu4) receptors enhances spontaneous and evoked absence seizures. *Neuropharmacology, 54*, 344–354.

Ngomba, R. T., Santolini, I., Salt, T. E., Ferraguti, F., Battaglia, G., Nicoletti, F., & van Luijtelaar, G. (2011). Metabotropic glutamate receptors in the thalamocortical network: Strategic targets for the treatment of absence epilepsy. *Epilepsia, 53*, 1211–1222.

Niswender, C. M., & Conn, P. J. (2010). Metabotropic glutamate receptors: Physiology, pharmacology, and disease. *Annual Review of Pharmacology, 50*, 295–322.

Ohishi, H., Akazawa, C., Shigemoto, R., Nakanishi, S., & Mizuno, N. (1995). Distributions of the mRNAs for L-2-amino-4-phosphonobutyrate-sensitive metabotropic glutamate receptors, mGluR4 and mGluR7, in the rat brain. *The Journal of Comparative Neurology, 360*, 555–570.

Parri, H. R., Gould, T. M., & Crunelli, V. (2010). Sensory and cortical activation of distinct glial cell subtypes in the somatosensory thalamus of young rats. *European Journal of Neuroscience, 32*, 29–40.

Petralia, R. S., Wang, Y. X., Niedzielski, A. S., & Wenthold, R. J. (1996). The metabotropic glutamate receptors, mGluR2 and mGluR3, show unique postsynaptic, presynaptic and glial localizations. *Neuroscience, 71*, 949–976.

Reichova, I., & Sherman, S. M. (2004). Somatosensory corticothalamic projections: Distinguishing drivers from modulators. *Journal of Neurophysiology, 92*, 2185–2197.

Salt, T. E., & Binns, K. E. (2000). Contributions of mGlu1 and mGlu5 receptors to interactions with N-methyl-D-aspartate receptor-mediated responses and nociceptive sensory responses of rat thalamic neurons. *Neuroscience, 100*, 375–380.

Salt, T. E., & Eaton, S. A. (1996). Functions of ionotropic and metabotropic glutamate receptors in sensory transmission in the mammalian thalamus. *Progress in Neurobiology, 48*, 55–72.

Salt, T. E., Jones, H. E., Andolina, I. M., Copeland, C. S., Clements, J. T. C., Knoflach, F., & Sillito, A. M. (2012). Potentiation of sensory responses in ventrobasal thalamus in vivo via selective modulation of mGlu1 receptors with a positive allosteric modulator. *Neuropharmacology, 62*, 1695–1699.

Saugstad, J. A., Kinzie, J. M., Shinohara, M. M., Segerson, T. P., & Westbrook, G. L. (1997). Cloning and expression of rat metabotropic glutamate receptor 8 reveals a distinct pharmacological profile. *Molecular Pharmacology, 51*, 119–125.

Sherman, S. M., & Guillery, R. W. (2000). *Exploring the thalamus*. New York: Academic.

Tamaru, Y., Nomura, S., Mizuno, N., & Shigemoto, R. (2001). Distribution of metabotropic glutamate receptor mGluR3 in the mouse CNS: Differential location relative to pre- and postsynaptic sites. *Neuroscience, 106*, 481–503.

Turner, J. P., & Salt, T. E. (1999). Group III metabotropic glutamate receptors control corticothalamic synaptic transmission in the rat thalamus in vitro. *The Journal of Physiology, 519*, 481–491.

Turner, J. P., & Salt, T. E. (2000). Synaptic activation of the group I metabotropic glutamate receptor mGlu1 on the thalamocortical neurons of the rat dorsal lateral geniculate nucleus in vitro. *Neuroscience, 100*, 493–505.

Turner, J. P., & Salt, T. E. (2003). Group II and III metabotropic glutamate receptors and the control of the TRN input to rat thalamocortical neurones in vitro. *Neuroscience, 122*, 459–469.

Metabotropic Receptor

Definition

A metabotropic receptor is a G-protein-coupled receptor. The term reflects the fact that transmitter binding results in the production of intracellular metabolites. Metabotropic receptors that couple to G-proteins are a complex of three proteins. Transmitter binding to the receptor results in a conformation change in the receptor, thereby activating the G-protein. One subunit of the G-protein may modulate ion channels; other subunits may increase or decrease the activity of enzymes that produce intracellular messengers that modulate the activity of kinases. Small-molecule neurotransmitters such as glutamate, acetylcholine, and serotonin activate metabotropic receptors as well as ionotropic receptors; mammalian peptides generally activate only metabotropic receptors.

Cross-References

▶ Amygdala, Pain Processing and Behavior in Animals
▶ GABA Mechanisms and Descending Inhibitory Mechanisms
▶ Metabotropic Glutamate Receptors in Spinal Nociceptive Processing
▶ Spinothalamic Tract Neurons, Peptidergic Input

Metalloproteases

Definition

Metalloproteases are peptide hydrolases using a metal in the catalytic mechanism and are implicated in many inflammatory processes.

Cross-References

▶ Vascular Neuropathies

Metastasis

Definition

Metastasis means the spread of cancer cells from the tissue of origin to distant organs. Cancer cells can break out of the primary tumor, penetrate into lymphatic and blood vessels, circulate through the bloodstream, and form a new focus (metastasize) within normal tissues elsewhere in the organism.

Cross-References

▶ NSAIDs and Cancer

Methadone

▶ Postoperative Pain, Methadone

Method of Adjustments

Definition

A psychophysical procedure used in a discrimination (or detection) experiment and in which the subjects adjust the value of the stimulus (e.g., by turning a dial) and set it to apparent equality with a standard or reference stimulus. Repeated applications of the procedure yield an empirical distribution of stimulus intensities that is used to estimate the just-noticeable difference (or threshold).

Cross-References

▶ Pain Evaluation, Psychophysical Methods

Method of Limits

Definition

It is a psychophysical procedure used in a discrimination (or detection) experiment, which consists in varying on each successive trial the intensity of the stimulus, in small ascending or descending steps. At each step the subject reports whether the stimulus appears smaller or larger than the reference stimulus (or is perceived or not). The values of the stimulus at which the subject's response shifts from one category to another are recorded, and the threshold is estimated by averaging these values.

Cross-References

▶ Pain Evaluation, Psychophysical Methods

Methotrexate

Definition

Methotrexate is a steroid-sparing agent in cranial arteritis.

Cross-References

▶ Headache Due to Arteritis

Methyl-2-(2,6-Xylyloxy)-Ethylamine-Hydrochloride

▶ Postoperative Pain, Mexiletine

Methylprednisolone

Definition

Steroid drug.

Cross-References

▶ Whiplash

Metoprolol

Definition

Beta-blocker.

Cross-References

▶ Preventive Migraine Therapy

Mexiletine/Mexitil

▶ Postoperative Pain, Mexiletine

Mexitil

▶ Postoperative Pain, Mexiletine

mGlu Receptor

▶ Metabotropic Glutamate Receptors in the Thalamus

mGlu Receptors

▶ Metabotropic Glutamate Receptors

mGluR

▶ Metabotropic Glutamate Receptors in the Thalamus

mGluRs

▶ Metabotropic Glutamate Receptors

MH, HTM, and PM

Definition

Nociceptors are often designated as mechanoheat (MH), high-threshold mechanoreceptive (HTM), or polymodal (PM). These designations refer to their capacity to respond to mechanical, thermal, or chemical stimuli. The designation HTM implies that the nociceptor responds only to mechanical stimuli. The designation MH implies that the nociceptor responds to both mechanical and heat stimuli. The designation PM implies that the nociceptor responds to mechanical, thermal, and at least one kind of chemical stimulus (e.g., bradykinin).

Cross-References

▶ Nociceptors in the Orofacial Region (Skin/ Mucosa)

Micro-arousal

Definition

An increase in EEG and heart rate frequency with a possible rise in muscle tone. It should last more than 3 s. but less than 10 s. The subject is unaware of such physiological activity.

Cross-References

▶ Orofacial Pain, Sleep Disturbance

Microdialysis

Definition

Microdialysis is a technique that allows the administration of drugs or sampling of substances in the extracellular fluid by means of a small dialysis fiber of the type used for renal dialysis. The semipermeable dialysis membrane has pores of a certain size, and molecules smaller than these pores diffuse across the dialysis membrane due to concentration gradients, either into the extracel- lular space or into the dialysate. The dialysis fluid is continually pumped through the fiber. Microdialysis allows minimal invasive insights into tissue metabolism. Drugs can be dissolved

in the dialysate for tissue administration, or samples of the dialysate can be taken for analysis of extracellular fluid concentrations of substances.

Cross-References

▶ Amygdala, Pain Processing and Behavior in Animals
▶ GABA Mechanisms and Descending Inhibitory Mechanisms
▶ Spinal Dorsal Horn Pathways, Dorsal Column (Visceral)
▶ Spinothalamic Tract Neurons, Role of Nitric Oxide
▶ Sympathetically maintained Pain and Inflammation, Human Experimentation

Microglia

Definition

Microglia are considered to be the resident immune cells of the CNS. These cells release classical immune proteins, respond to immuno- genic stimuli, and express cell surface receptors characteristic of peripheral phagocytic cells. Blocking microglial activation inhibits the onset of exaggerated pain.

Cross-References

▶ Cathepsin and Microglia
▶ Cord Glial Activation
▶ Glia
▶ Glial Cells in Orofacial Pathological Pain Mechanisms

Microglia Activation

▶ Cord Glial Activation

Microinjection

Definition

Microinjection is a common technique in behavioral neuroscience in which micro- or nanoliter amounts of a liquid are directly infused into a specific brain region.

Cross-References

▶ Cingulate Cortex, Nociceptive Processing, Behavioral Studies in Animals

Microiontophoresis

Definition

A technique used in electrophysiological studies of brain or spinal cord structures for releasing active agents near a neuron from which recordings are made. A multibarreled array of micropipettes is generally used. The central barrel records the activity of a neuron extracellularly or intracellularly, whereas the other barrels are filled with solutions that contain agonist or antagonist drugs or other substances. The dissolved agents are charged, and so they can be released from the micropipette by passing a current of the appropriate sign through the micropipette. A current of the opposite sign is generally used to restrain the agent when its release is not desired. The application is extracellularly close to the membrane of the recorded neuron.

Cross-References

▶ GABA Mechanisms and Descending Inhibitory Mechanisms

Microneurography and Intraneural Microstimulation

Barbara Namer
Department of Physiology & Pathophysiology, FAU, Erlangen/Nuremberg, Bavaria, Germany

Definition

Percutaneous single nerve fiber recordings of nociceptors in non-anesthetized humans with a needle electrode.

Introduction

History of Microneurography of Nociceptors

The first recordings of activity in peripheral nerves of humans were published 1960 by Hensel and Boman, who recorded activity from fine fascicles of an exposed nerve of a human volunteer (Hensel and Boman 1960). In 1967, two groups in Sweden could show that it is possible to insert microelectrodes percutaneously into a peripheral nerve of an awake human and record action potentials from this nerve (Knutsson and Widen 1967; Vallbo and Hagbarth 1968). The method of Hagbarth and Vallbo using a tungsten microelectrode adapted to recordings of C-fibers by their students Torebjörk and Hallin (Torebjork and Hallin 1974). The first attempt of Torebjörk and Hallin to publish the method of recording activity from single C-fibers in awake humans was rejected by the journal *Nature*, because the editors decided that "it is not possible." However over time, the method became accepted and unique and valuable facts could be found through microneurographic recordings of C-fibers in humans. Microneurography of single C-fibers for the study of pain is currently used only by very few groups in the world.

Characteristics

Method of Microneurography

The recording electrode is inserted into a peripheral nerve. When the needle is close to a C-fiber bundle, neuronal activity characteristic for unmyelinated afferents can be induced by mechanical scratching stimuli applied to the innervation territory of the nerve. Innervation territories of individual C-fibers are then located with transcutaneous electrical stimulation with a pointed electrode. C-fibers are identified by their low conduction velocity (<2 m/s). Following repetitive electrical stimulation at a fixed frequency from the skin, action potentials of individual C-fibers can be registered with the recording electrode at stable conduction latencies. An increase in conduction latency is observed when the stimulation frequency at the skin is increased or after the afferent fiber has been additionally activated, e.g., by natural stimuli ("marking") (Schmelz et al. 1995). Marking is due to activity-dependent slowing of conduction in C-fibers. Action potentials render the axonal membrane of afferent C-fibers less excitable for tens of seconds and thus slow down conduction velocity of subsequent action potentials.

These activity-dependent changes of conduction velocity form characteristic slowing patterns that segregate mechano-insensitive afferents (marked slowing), mechano-sensitive fibers (intermediate slowing), and cold fibers (little slowing) (Serra et al. 1999; Weidner et al. 1999).

Intraneural Microstimulation

Intraneural microstimulation (INMS) was developed from microneurography (Torebjork and Ochoa 1980; Vallbo 1981). During recordings, when fibers are characterized and counted, the same fibers as seen during recording can be stimulated electrically through the recording needle. Intraneural stimulation of nociceptors evokes pain sensations in the receptive field of the stimulated and recorded nociceptors. In most cases, a burning pain was described by the subjects. Also stinging and dull aching pain could arise. When increasing the stimulation strength, the subjective magnitude of pain and the projected area of pain increased due to additionally recruited nociceptive fibers (Jorum et al. 1989). By intraneural microstimulation of A- and C-fibers, it could be shown that mechanical allodynia induced by capsaicin injection into the skin has a central origin (Torebjork et al. 1992).

Nociceptor Classes Discovered by Microneurography

With microneurography, several classes of nociceptors have been described:

Mechano-sensitive C-fibers (CM, Class 1a fibers) are fibers that respond to probing the skin with a von Frey hair (>20 mN in most groups) and display moderate activity-dependent conduction velocity slowing. Most of these fibers exhibit also heat sensitivity with mean thresholds around 41 °C (Serra et al. 1999; Weidner et al. 1999). This fiber class is thought to transmit the time and spatial aspects of acute pain (Ruehle et al. 2006) and has been compared to the polymodal nociceptors found in cat and rodents. This fiber class seems to play a role in non-histaminergic itch (Namer et al. 2008) and is activated and sensitized by endothelin-1 and ATP (Hilliges et al. 2002; Namer et al. 2007).

Mechano-insensitive fibers (CMi, Class 1b fibers) form the other nociceptor class. These fibers do not respond under normal conditions to mechanical stimulation even when a cannula is inserted into the skin. Some of them are heat responsive with a threshold around 46 °C. They respond to most chemical irritants vigorously and thus can be sensitized to mechanical stimuli. Therefore, these fibers are often termed "sleeping nociceptors" since they can be "awakened" by chemical stimuli or tonic pressure (Schmelz et al. 2003, 2000b; Schmidt et al. 2000; Serra et al. 1998, 2004). Generally, these sleeping nociceptors seem to play a crucial role in states of hyperalgesia. Neuropeptide release from their terminals following noxious stimuli leads to the large flare reactions corresponding to their huge receptive fields in human skin (Schmelz et al. 2000a; Schmidt et al. 2002). A subgroup of the mechano-insensitive nociceptors mediates the histamine-dependent itch sensation and has been called "itch fibers" (Schmelz et al. 1997). The mechano-insensitive C-fibers have also distinct axonal properties. When stimulated electrically, they show a pronounced supernormal phase, i.e., a short time range after the action potential in which the next action potential is conducted faster than the first (Bostock et al. 2003; Weidner et al. 2000, 2002) In addition, they have much higher electrical thresholds than mechano-sensitive nociceptors.

Cold nociceptors and cold fibers responding to innocuous cold with low and high firing frequencies, respectively, have been described too (Campero et al. 1996, 2001, 2009). They have little activity-dependent conduction velocity slowing. Cold thresholds in cold fibers could be sensitized by menthol. Under pathological conditions of cold allodynia in a patient with neuropathic pain, even mechano-insensitive nociceptors can acquire menthol sensitivity (Serra et al. 2009).

Microneurography of Aging Nociceptors

In aged subjects, changes were found in the composition of the C-fiber population and in sensory and axonal properties. Although the relative incidence of afferent to efferent C-fibers remained almost constant, the ratio of mechano-responsive to mechano-insensitive nociceptors was 7:3 in aged subjects, while it was 8:2 in the young controls. In aged subjects, 13 % of the fibers showed atypical discharge characteristics, while those "atypical" fibers were extremely rare in young subjects. These atypical fibers were either degenerated fibers, which lost their responsiveness to mechanical stimuli or sensitized fibers, which resemble mechano-insensitive nociceptors which gained mechano-sensitivity. Astonishingly 2.5 % of the fibers in healthy aging subjects showed spontaneous activity, whereas this is virtually never seen in the skin of healthy young subjects (Namer et al. 2009).

Microneurography in Patients with Peripheral Neuropathy

In several microneurography studies on patients suffering from painful neuropathies, spontaneous activity in C-nociceptors has been found, which could underlay ongoing spontaneous pain (Ochoa et al. 2005; Orstavik et al. 2003, 2006). Also mechano-insensitive C-fibers, which were sensitized to mechanical stimuli, were described (Ochoa et al. 2005; Orstavik et al. 2003, 2006). Additionally as another mechanism for hyperalgesia and neuropathic pain, double and triple responses of an axon to single electrical stimulation were described as a mechanism of signal amplification in patients with neuropathic pain (Bostock et al. 2005; Serra et al. 2011). This double spiking due to unidirectional block in branching points of the axon occurs in healthy young subjects too, however, rarely (Weidner et al. 2003).

Not only signs of hyperexcitability were found. In a patient group with a lower pain level, mostly degenerated nociceptors were found, which had lost their heat and mechano-sensitivity (Orstavik et al. 2006). Thus, the data of group analyses obtained by microneurography in patients with neuropathy match the clinical presentation of loss of function and gain of function of these patient groups.

Acknowledgments I thank Prof. Handwerker for critical review of this manuscript. The German research council (DFG) supported the work by the grant NA 970 1–1.

References

Bostock, H., Campero, M., Serra, J., & Ochoa, J. (2003). Velocity recovery cycles of C fibres innervating human skin. *The Journal of Physiology, 553*, 649–663.

Bostock, H., Campero, M., Serra, J., & Ochoa, J. L. (2005). Temperature-dependent double spikes in C-nociceptors of neuropathic pain patients. *Brain, 128*, 2154–2163.

Campero, M., Serra, J., & Ochoa, J. L. (1996). C-polymodal nociceptors activated by noxious low temperature in human skin. *The Journal of Physiology, 497*(Pt 2), 565–572.

Campero, M., Serra, J., Bostock, H., & Ochoa, J. L. (2001). Slowly conducting afferents activated by innocuous low temperature in human skin. *The Journal of Physiology, 535*, 855–865.

Campero, M., Baumann, T. K., Bostock, H., & Ochoa, J. L. (2009). Human cutaneous C fibres activated by cooling, heating and menthol. *The Journal of Physiology, 587*, 5633–5652.

Hensel, H., & Boman, K. K. (1960). Afferent impulses in cutaneous sensory nerves in human subjects. *Journal of Neurophysiology, 23*, 564–578.

Hilliges, M., Weidner, C., Schmelz, M., Schmidt, R., Orstavik, K., Torebjork, E., & Handwerker, H. (2002). ATP responses in human C nociceptors. *Pain, 98*, 59–68.

Jorum, E., Lundberg, L. E., & Torebjork, H. E. (1989). Peripheral projections of nociceptive unmyelinated axons in the human peroneal nerve. *The Journal of Physiology, 416*, 291–301.

Knutsson, E., & Widen, L. (1967). Impulses from single nerve fibres recorded in man using microelectrodes. *Nature, 213*, 606–607.

Namer, B., Hilliges, M., Orstavik, K., Schmidt, R., Weidner, C., Torebjork, E., Handwerker, H., & Schmelz, M. (2007). Endothelin1 activates and sensitizes human C-nociceptors. *Pain, 137*(1), 41–49.

Namer, B., Carr, R., Johanek, L. M., Schmelz, M., Handwerker, H. O., & Ringkamp, M. (2008). Separate peripheral pathways for pruritus in man. *Journal of Neurophysiology, 100*, 2062–2069.

Namer, B., Barta, B., Orstavik, K., Schmidt, R., Carr, R., Schmelz, M., & Handwerker, H. O. (2009). Microneurographic assessment of C-fibre function in aged healthy subjects. *The Journal of Physiology, 587*, 419–428.

Ochoa, J. L., Campero, M., Serra, J., & Bostock, H. (2005). Hyperexcitable polymodal and insensitive nociceptors in painful human neuropathy. *Muscle & Nerve, 32*, 459–472.

Orstavik, K., Weidner, C., Schmidt, R., Schmelz, M., Hilliges, M., Jorum, E., Handwerker, H., & Torebjork, E. (2003). Pathological C-fibres in patients with a chronic painful condition. *Brain, 126*, 567–578.

Orstavik, K., Namer, B., Schmidt, R., Schmelz, M., Hilliges, M., Weidner, C., Carr, R. W., Handwerker, H., Jorum, E., & Torebjork, H. E. (2006). Abnormal function of C-fibers in patients with diabetic neuropathy. *Journal of Neuroscience, 26*, 11287–11294.

Ruehle, B. S., Handwerker, H. O., Lennerz, J. K., Ringler, R., & Forster, C. (2006). Brain activation during input from mechanoinsensitive versus polymodal C-nociceptors. *Journal of Neuroscience, 26*, 5492–5499.

Schmelz, M., Forster, C., Schmidt, R., Ringkamp, M., Handwerker, H. O., & Torebjork, H. E. (1995). Delayed responses to electrical stimuli reflect C-fiber responsiveness in human microneurography. *Experimental Brain Research, 104*, 331–336.

Schmelz, M., Schmidt, R., Bickel, A., Handwerker, H. O., & Torebjork, H. E. (1997). Specific C-receptors for itch in human skin. *Journal of Neuroscience, 17*, 8003–8008.

Schmelz, M., Michael, K., Weidner, C., Schmidt, R., Torebjork, H. E., & Handwerker, H. O. (2000a). Which nerve fibers mediate the axon reflex flare in human skin? *Neuroreport, 11*, 645–648.

Schmelz, M., Schmid, R., Handwerker, H. O., & Torebjork, H. E. (2000b). Encoding of burning pain from capsaicin-treated human skin in two categories of unmyelinated nerve fibres. *Brain, 123*(Pt 3), 560–571.

Schmelz, M., Schmidt, R., Weidner, C., Hilliges, M., Torebjork, H. E., & Handwerker, H. O. (2003). Chemical response pattern of different classes of C-nociceptors to pruritogens and algogens. *Journal of Neurophysiology, 89*, 2441–2448.

Schmidt, R., Schmelz, M., Torebjork, H. E., & Handwerker, H. O. (2000). Mechano-insensitive nociceptors encode pain evoked by tonic pressure to human skin. *Neuroscience, 98*, 793–800.

Schmidt, R., Schmelz, M., Weidner, C., Handwerker, H. O., & Torebjork, H. E. (2002). Innervation territories of mechano-insensitive C nociceptors in human skin. *Journal of Neurophysiology, 88*, 1859–1866.

Serra, J., Campero, M., & Ochoa, J. (1998). Flare and hyperalgesia after intradermal capsaicin injection in human skin. *Journal of Neurophysiology, 80*, 2801–2810.

Serra, J., Campero, M., Ochoa, J., & Bostock, H. (1999). Activity-dependent slowing of conduction differentiates functional subtypes of C fibres innervating human skin. *The Journal of Physiology, 515*(Pt 3), 799–811.

Serra, J., Campero, M., Bostock, H., & Ochoa, J. (2004). Two types of C nociceptors in human skin and their behavior in areas of capsaicin-induced secondary hyperalgesia. *Journal of Neurophysiology, 91*, 2770–2781.

Serra, J., Sola, R., Quiles, C., Casanova-Molla, J., Pascual, V., Bostock, H., & Valls-Sole, J. (2009). C-nociceptors sensitized to cold in a patient with small-fiber neuropathy and cold allodynia. *Pain, 147*, 46–53.

Serra, J., Sola, R., Aleu, J., Quiles, C., Navarro, X., & Bostock, H. (2011). Double and triple spikes in C-nociceptors in neuropathic pain states: An additional peripheral mechanism of hyperalgesia. *Pain, 152*, 343–353.

Torebjork, H. E., & Hallin, R. G. (1974). Identification of afferent C units in intact human skin nerves. *Brain Research, 67*, 387–403.

Torebjork, H. E., & Ochoa, J. L. (1980). Specific sensations evoked by activity in single identified sensory units in man. *Acta Physiologica Scandinavica, 110*, 445–447.

Torebjork, H. E., Lundberg, L. E., & LaMotte, R. H. (1992). Central changes in processing of mechanoreceptive input in capsaicin-induced secondary hyperalgesia in humans. *The Journal of Physiology, 448*, 765–780.

Vallbo, A. B. (1981). Sensations evoked from the glabrous skin of the human hand by electrical stimulation of unitary mechanosensitive afferents. *Brain Research, 215*, 359–363.

Vallbo, A. B., & Hagbarth, K. E. (1968). Activity from skin mechanoreceptors recorded percutaneously in awake human subjects. *Experimental Neurology, 21*, 270–289.

Weidner, C., Schmelz, M., Schmidt, R., Hansson, B., Handwerker, H. O., & Torebjork, H. E. (1999). Functional attributes discriminating mechano-insensitive and mechano-responsive C nociceptors in human skin. *Journal of Neuroscience, 19*, 10184–10190.

Weidner, C., Schmidt, R., Schmelz, M., Hilliges, M., Handwerker, H. O., & Torebjork, H. E. (2000). Time course of post-excitatory effects separates afferent human C fibre classes. *The Journal of Physiology, 527*(Pt 1), 185–191.

Weidner, C., Schmelz, M., Schmidt, R., Hammarberg, B., Orstavik, K., Hilliges, M., Torebjork, H. E., & Handwerker, H. O. (2002). Neural signal processing: the underestimated contribution of peripheral human C-fibers. *Journal of Neuroscience, 22,* 6704–6712.

Weidner, C., Schmidt, R., Schmelz, M., Torebjork, H. E., & Handwerker, H. O. (2003). Action potential conduction in the terminal arborisation of nociceptive C-fibre afferents. *The Journal of Physiology, 547,* 931–940.

Microneuroma

Definition

A neuroma is the structure that develops on the proximal cut end of a peripheral nerve branch or nerve fascicle. Severed axons form swollen terminal end bulbs, and there is usually initiation of sprouting. Regenerative sprouts are not able to elongate, they often form a tangled mass at the nerve end, a nerve end neuroma. If only a part of the cross-section of the nerve is cut, the cut axons form a neuroma-in-continuity adjacent to their intact neighbors. Transection of small groups of axons scattered throughout a nerve trunk, or of tiny nerve fascicles or tributaries yields microneuromas.

Cross-References

▶ Neuroma Pain
▶ Peripheral Neuropathic Pain

Microstimulation

Definition

See also "▶ Microneurography and Intraneural Microstimulation." This term has been used in different contexts to denominate stimulation of single axons, or single central neurons with fine microelectrodes which are also used for recording the action potentials of these or neighboring neurons.

Cross-References

▶ Central Pain, Human Studies of Physiology

Microsurgical DREZotomy

▶ Brachial Plexus Avulsion and Surgery in the Dorsal Root Entry Zone (DREZ)
▶ DREZ Procedures

Microswitch

Definition

A microswitch is a very small switch that acts by the movement of a small lever and is sensitive to minute motions.

Cross-References

▶ Assessment of Pain Behaviors

Microvascular Decompression

Synonyms

MVD

Definition

Neurosurgical procedures to relieve cranial nerve compression by intracranial arteries or veins as described under "vascular compression syndromes." The main indication is for pain relief in cranial nerve neuralgias.

Cross-References

▶ Pain Paroxysms
▶ Trigeminal and Cranial Nerve Neuralgias
▶ Trigeminal, Glossopharyngeal, and Geniculate Neuralgias

Microwaves

▶ Therapeutic Heat, Microwaves, and Cold

Midazolam

Definition

Midazolam is a benzodiazepine drug with an imidazole structure, commonly used as an anxiolytic, amnesic, and sedative/ hypnotic. Spinal administration has recently been advocated in the management of complex pain syndromes.

Cross-References

▶ Postoperative Pain, Appropriate Management

Midline Commissural Myelotomy

Definition

Dividing the spinal cord longitudinally in the midline to ablate nerve fibers at the point where they cross the spinal cord.

Cross-References

▶ Cancer Pain Management, Neurosurgical Interventions

Midline Epidural Steroid Injection

▶ Epidural Steroid Injections for Chronic Back Pain

Midline Myelotomy

H. J. W. Nauta
University of Texas Medical Branch, Galveston, TX, USA

Synonyms

Bischof myelotomy; Commissural myelotomy; Dorsal longitudinal myelotomy; Extralemniscal myelotomy; Longitudinal myelotomy; Mediolongitudinal myelotomy; Myelotomy; Pourpre myelotomy; Punctate midline myelotomy

Definition

Midline myelotomy refers to surgical procedures that involve an incision in the dorsal midline of the spinal cord. More recently, the term has come to include derived procedures that continue to include an intervention through the dorsal midline of the spinal cord.

Characteristics

Background
The spinal cord is commonly characterized by both its segmental organization and the long tracts that link the segments to each other and to the brain. Surgical interventions on the spinal cord similarly can be characterized by their effects on segmental components, long tracts, or both. The distinction is important. If the goal is to interrupt a long ascending tract, then a single lesion should suffice. If the goal is an intervention at the segmental level, then typically several spinal levels must be accessed, and an incision made in the spinal cord over several segments.

Myelotomy: Origin from Procedures Directed at Segmental Structures, not Long Tracts
The original myelotomy procedures were directed at the segments of the spinal cord for

treatment of pain or spasticity. The long spinal cord incision distinguished myelotomy from "cordotomy," where a single rostrocaudal level lesion is made, which interrupts longitudinal pathways. The distinction between myelotomy and cordotomy has blurred over time. It is now fairly safe to say that while the anterolateral cordotomy procedure continues to be understood as interrupting ascending tracts related to the conduction of pain, only the Bischof myelotomy (Bischof 1951, 1967), still occasionally used to treat painful spasticity, continues to be understood as a procedure directed at the segmental level and therefore requiring a long incision into the spinal cord. The original myelotomy procedures used to treat pain have mutated into much more concise procedures and are now understood to be effective through their interruption of ascending pain tracts near the midline of the dorsal columns (Nauta et al. 2002). Due to their history, however, these procedures continue to be described as myelotomies, although they no longer require a long incision into the spinal cord.

Impetus to Develop Myelotomy and Original Concepts of Benefit

Anterolateral cordotomy for interruption of the long tract system related to pain has evolved into an elegant fluoroscopic-guided percutaneous procedure at the C1–C2 level, which is effective for all levels below the lesion. The principal disadvantage of the anterolateral cordotomy was the unavoidable concomitant interruption of descending pathways in the anterolateral quadrant related to respiration and bladder control. The potential for the loss of automatic respiration (Ondine's curse) and development of incontinence required special strategies, if bilateral anterolateral cordotomy was contemplated for the all too common bilateral, midline, or visceral pain. These problems with bilateral cordotomy led to the proposal in 1926, by the neuropathologist Greenfield that a longitudinal midline incision of the spinal cord could interrupt the crossing fibers of the spinothalamic pathway bilaterally, near their segmental level of origin and before these fibers assemble into a long ascending tract. The longitudinal midline myelotomy required

definition of segmental levels of pain origin and a large enough exposure to incise the cord deeply over enough levels to bracket the levels of pain origin.

Clinical Observation at Odds with the Original Concept

The long midline myelotomy for pain was first performed by Armour in 1926 (Armour 1927), and while there were several early advocates, the procedure as originally conceived required a big operation which limited its wider acceptance. However, it became clear that the procedure was effective, even when a short myelotomy was performed at a level well above the segmental origin of the treated pain, and that pain relief extended to levels caudal to the zone of decreased pain sensation. Hirshberg et al. (1996) and Hitchcock (1974) originally attempting to treat neck and upper extremity pain made lesions of only limited rostrocaudal extent near the midline at C1 and observed relief of pain with analgesia extending into the legs. Such observations suggested that the midline myelotomy procedure was interrupting an ascending pain conducting system, perhaps a multisynaptic pathway separated from the spinothalamic tract, as proposed by Hitchcock (Hitchcock 1969, 1974) and others (Davis et al. 1929; Noordenbos 1959). Surprisingly, the midline myelotomy procedure remained effective even if a short midline incision was made above and not deep enough to reach either the commissural fiber systems or the central gray matter (Hirshberg et al. 1996).

Laboratory Evidence of a Dorsal Column Pain Pathway

The literature concerning myelotomy for pain shows an interesting progression, with late recognition that there is a pain pathway ascending near the midline of the dorsal columns (Hirshberg et al. 1996; Nauta et al. 2002). The old concept is that there are two pathways carrying somatic sensation to the brain: the spinothalamic system described as conducting crude or "protopathic" sensations such as pain and temperature and the dorsal column pathway conducting the "epicritic" sensations, such as vibration and proprioception. Recent evidence demonstrates that the dorsal columns do contain a postsynaptic pain pathway (Rustioni et al. 1979; Uddenberg 1968; Willis and Coggeshall 1991), but its role in somatic pain conduction was considered minor. Berkley and Hubscher (1995) gave evidence that neurons in the dorsal column nuclei can respond to innocuous and noxious stimulation of both pelvic viscera and skin. Finally, Al Chaer and others (1996b) demonstrated that the postsynaptic dorsal column pain pathway predominated in the conduction of visceral pain. Microdialysis fiber infusion of neurotransmitters or antagonists (morphine or CNQX) into the spinal cord suggested the presence of a synapse in the pathway (Al Chaer et al. 1996a, b). Retrograde tracer studies, depositing a marker in the nucleus gracilis (Christensen et al. 1996), revealed the likely cell bodies of origin of the pathway to reside at the medial base of the dorsal horn just above the central canal, an area known from earlier experiments to receive primary visceral fiber input. Injection of anterograde tracers into the same cell territory (Wang et al. 1999) resulted in fiber labeling ascending near the dorsal column midline, to end in the medial part of the nucleus gracilis. Further studies by this group defined a somatotopic organization within the postsynaptic dorsal column pathway, wherein the fibers originating in the lumbosacral cord ascend near the midline of the dorsal columns, while fibers originating in the thoracic cord terminate in the lateral part of nucleus gracilis and adjacent medial parts of nucleus cuneatus (Wang et al. 1999; Willis et al. 1999).

Subsequent Clinical Observations

Once the basis for the benefit from midline myelotomy was better understood, Nauta et al. (1997, 2000) made very small transverse (rather than sagittal longitudinal) midline incisions in the dorsal columns, well above the segmental level of pain origin, to treat medically intractable pelvic pain. The incisions, later crush, across the midline of the dorsal columns extended only 1 mm to either side of the exact midline and only to a depth of 5 mm. The effective lesion

was so small that it was dubbed "punctate myelotomy" since the original lesions were made with what amounted to a puncture across the spinal cord midline with a 16-gauge hypodermic needle. Pain relief from these tiny lesions confirmed that an ascending pathway in the dorsal column was being interrupted. The lesions were effective without reaching the commissures or the central gray, demonstrating that these latter structures were not essential to the benefit. Surprisingly, postoperative neurological examination in these patients by an independent neurologist failed to reveal any new deficit, emphasizing an important distinction between the consequences of small midline lesions of the dorsal columns and tabes dorsalis, in which a pathologic process effects the dorsal root ganglia in a widespread manner and results not only in visible changes in the dorsal columns but less visible changes in other sensory pathways as well. While Nauta et al. (1997, 2000) treated only pelvic origin visceral pain, more recent reports (Kim and Kwon 2000) suggest that more rostral visceral origin pain related to gastric cancer can also be treated effectively by punctate midline myelotomy at an upper thoracic level. Since it is now understood that the benefit depends on the interruption of an ascending pathway, there are good theoretical reasons to believe that it would be worthwhile to revive the percutaneous procedure described by Hitchcock (1970, 1974) and Schvarcz (1984), only modified to emphasize the medial dorsal columns rather than the subjacent central gray. Clearly the term "extralemniscal myelotomy" used by Schvarcz (Kanpolat 2002) would no longer apply to such an operation, because the dorsal columns are well known to contribute to signaling within the medial lemniscus.

The Bischof myelotomy and its variants can be used to treat severe, medically intractable, often painful, lower limb spasticity, usually in quadri- or paraplegic or paretic patients (Livshits et al. 2002; Putty and Shapiro 1991). The goal of the procedure is to interrupt the segmental reflex arcs underlying spasticity that pass between the dorsal horn and the motor neurons in the ventral horn. The goal is also to preserve as much residual function as possible and, above all, the anterior horn cells and their continuity with the muscles so that atrophy is minimized, and padding over boney prominences maintained to protect against the development of pressure sores. The latter can be a significant problem following ventral root section or neurectomy. The myelotomy method also offers an advantage over simple dorsal rhizotomy where recurrence of spasticity is more common. Since the availability of intrathecal baclofen infusion pumps (Penn and Kroin 1987), the procedure is performed less frequently but still remains a reasonable option for cases where severe problems with the pump are experienced or anticipated.

Since the Bischof myelotomy is directed at the segmental level, there is no alternative to an exposure of the spinal cord by laminectomy (typically, vertebral levels T10–L1) to gain access to those levels determined preoperatively to give rise to the worst spasticity (typically T12–S1). The required spinal cord incision is made in the longitudinal plane, perpendicular to the midsagittal plane. This plane passes through the central canal and dentate ligament of either side, thus filleting the spinal cord into a dorsal and ventral half. The incision along this plane was originally described from the lateral aspect of the cord (Bischof 1951) on either side and was typically performed with an angled triangular ophthalmology knife inserted just posterior to the dentate ligament and directed towards the central canal. Great care is required to minimize injury to the spinal cord vasculature. Pourpre (Pourpre 1960), who proposed a myelotomy en croix, in which the cord is entirely bisected into a right and left half (Bischof 1967), modified his original lateral approach making an upside down "T" incision instead, beginning with a long myelotomy in the midsagittal plane to a depth reaching the center of the cord but sparing the anterior white commissure. An angled triangular ophthalmology knife is then passed laterally along the depth of the midline incision aimed at the pia just above the dentate ligament of either side, thereby separating the dorsal horn from the ventral horn of each side. The lateral limbs of this myelotomy are intended to pass through the gray matter, but not far into the lateral funiculus containing whatever corticospinal fibers may still be functional.

Cross-References

▶ Postsynaptic Dorsal Column Neurons, Responses to Visceral Input
▶ Spinal Dorsal Horn Pathways, Dorsal Column (Visceral)

References

Al Chaer, E. D., Lawand, N. B., Westlund, K. N., et al. (1996a). Pelvic visceral input into the nucleus Gracilis is largely mediated by the postsynaptic dorsal column pathway. *Journal of Neurophysiology, 76*, 2675–2690.
Al Chaer, E. D., Lawand, N. B., Westlund, K. N., et al. (1996b). Visceral nociceptive input into the ventral posterolateral nucleus of the thalamus: A new function for the dorsal column pathway. *Journal of Neurophysiology, 76*, 2661–2674.
Armour, D. (1927). Surgery of the spinal cord. *Lancet, 2*, 691.
Berkley, K. J., & Hubscher, C. H. (1995). Are there separate central nervous system pathways for touch and pain? *Nature Medicine, 1*, 766–773.
Bischof, W. (1951). Die longitudinale myelotomie. *Zentralblatt für Neurochirurgie, 11*, 79–88.
Bischof, W. (1967). Zur dorsalen longitudinalen myelotomie. *Zentralblatt für Neurochirurgie, 28*, 123–126.
Christensen, M. D., Willis, W. D., & Westlund, K. N. (1996). Anatomical evidence for cells of origin of a postsynaptic dorsal column visceral pathway: Sacral spinal cord cells innervating the medial nucleus Gracilis. *Society for Neuroscience Abstract, 22*, 109.
Davis, L., Hart, J. T., & Crain, R. C. (1929). The pathway for visceral afferent impulses within the spinal cord. II. Experimental dilatation of the biliary ducts. *Surgery, Gynecology & Obstetrics, 48*, 647–651.
Hirshberg, R. M., Al Chaer, E. D., Lawand, N. B., et al. (1996). Is there a pathway in the posterior funiculus that signals visceral pain? *Pain, 67*, 291–305.
Hitchcock, E. (1969). Stereotaxic spinal surgery. A preliminary report. *Journal of Neurosurgery, 31*, 386–392.
Hitchcock, E. (1970). Stereotactic cervical myelotomy. *Journal of Neurology, Neurosurgery, and Psychiatry, 33*, 224230.
Hitchcock, E. (1974). Stereotactic myelotomy. *Proceedings of the Royal Society of Medicine, 67*, 771–772.

Kanpolat, Y. (2002). Percutaneous stereotactic pain procedures: Percutaneous cordotomy, extralemniscal myelotomy, trigeminal tractotomy-nucleotomy. In K. J. Burchiel (Ed.), *Surgical management of pain* (pp. 745–762). New York: Thieme.

Kim, Y. S., & Kwon, S. J. (2000). High thoracic midline dorsal column myelotomy for severe visceral pain due to advanced stomach cancer. *Neurosurgery, 46,* 85–90.

Livshits, A., Rappaport, Z. H., Livshits, V., et al. (2002). Surgical treatment of painful spasticity after spinal cord injury. *Spinal Cord, 40,* 161–166.

Nauta, H. J., Hewitt, E., Westlund, K. N., et al. (1997). Surgical interruption of a midline dorsal column visceral pain pathway. Case report and review of the literature. *Journal of Neurosurgery, 86,* 538–542.

Nauta, H. J., Soukup, V. M., Fabian, R. H., et al. (2000). Punctate midline myelotomy for the relief of visceral cancer pain. *Journal of Neurosurgery, 92,* 125–130.

Nauta, H. J. W., Westlund, K. N., & Willis, W. D. (2002). Midline myelotomy. In K. J. Burchiel (Ed.), *Surgical management of pain* (pp. 714–731). New York: Thieme.

Noordenbos, W. (1959). *Pain: Problems pertaining to the transmission of nerve impulses which give rise to pain.* Amsterdam: Elsevier.

Penn, R. D., & Kroin, J. S. (1987). Long-term intrathecal baclofen infusion for treatment of spasticity. *Journal of Neurosurgery, 66,* 181–185.

Pourpre, M. H. (1960). Traitement neuro-chirurgical des contractures chez les paraplegiques posttraumatiques. *Neurochirurgie, 6,* 229–236.

Putty, T. K., & Shapiro, S. A. (1991). Efficacy of dorsal longitudinal myelotomy in treating spinal spasticity: A review of 20 cases. *Journal of Neurosurgery, 75,* 397–401.

Rustioni, A., Hayes, N. L., & O'Neill, S. (1979). Dorsal column nuclei and ascending spinal afferents in macaques. *Brain, 102,* 95–125.

Schvarcz, J. R. (1984). Stereotactic high cervical extralemniscal myelotomy for pelvic cancer pain. *Acta Neurochirurgica, 33,* 431–435.

Uddenberg, N. (1968). Functional organization of long, second-order afferents in the dorsal funiculus. *Experimental Brain Research, 4,* 377–382.

Wang, C. C., Willis, W. D., & Westlund, K. N. (1999). Ascending projections from the area around the spinal cord central canal: A phaseolus vulgaris leucoagglutinin study in rats. *The Journal of Comparative Neurology, 415,* 341–367.

Willis, W. D., & Coggeshall, R. E. (1991). *Sensory mechanisms of the spinal cord.* New York: Plenum Press.

Willis, W. D., Al Chaer, E. D., Quast, M. J., et al. (1999). A visceral pain pathway in the dorsal column of the spinal cord. *Proceedings of the National Academy of Sciences of the United States of America, 96,* 7675–7679.

Migraine

Definition

Migraine is a common neurovascular headache disorder with two forms: episodic, in which there are less than 15 affected days a month, and chronic, in which 15 or more days a month are affected. Typical attacks are characterized by unilateral pulsating moderate-to-severe headache, typically lasting 4–72 h, of throbbing quality and with associated symptoms of light and sound sensitivity, nausea and vomiting, phono- and/or photophobia. The headache may be preceded by visual disturbances (aura).

Cross-References

▶ Calcitonin Gene-Related Peptide and Migraine Headaches
▶ Human Thalamic Response to Experimental Pain (Neuroimaging)
▶ New Daily Persistent Headache
▶ Preventive Migraine Therapy

Migraine Accompagnée

▶ Clinical Migraine with Aura

Migraine Epidemiology

Richard B. Lipton[1] and Marcelo E. Bigal[2]
[1]Departments of Neurology, Epidemiology and Population Health, Albert Einstein College of Medicine, Bronx, NY, USA
[2]Department of Neurology and Montefiore Headache Unit, Albert Einstein College of Medicine, Bronx, NY, USA

Synonyms

Incidence; Medical disorder; Prevalence

Definition

Migraine is a prevalent under-diagnosed and under-treated medical disorder, associated with a severe impact on the quality of life of the sufferers and their families as well as an enormous economic impact on society. This entry reviews the epidemiology and burden of migraine in population studies. Epidemiological studies often focus on the ► incidence and ► prevalence of disease in defined populations.

Characteristics

The Incidence of Migraine

The incidence of migraine has been investigated in a limited number of studies (Fig. 1). Stewart et al. found that in females, the incidence of migraine with aura peaked between ages 12 and 13 (14.1/1,000 person-years); migraine without aura peaked between ages 14 and 17 (18.9/1,000 person-years). In males, the incidence of migraine with aura peaked several years earlier, around 5 years of age at 6.6/1,000 person-years; the peak for migraine without aura was 10/1,000 person-years between 10 and 11 years. New onset of migraine was uncommon in men in their twenties. This study suggests that migraine begins earlier in males than in females and that migraine with aura begins earlier than migraine without aura (Stewart et al. 1993).

Three other studies assessed the incidence of migraine. A study performed in a random sample of young adults (21–30 years) found that the incidence of migraine was 5.0 per 1,000 person-years in males and 22.0 in females (Breslau et al. 1994), supporting the findings reported above (Stang et al. 1992). A second study using a linked medical records system showed a lower incidence (probably because many people with migraine do not consult doctors or receive a medical diagnosis) (Breslau et al. 1994). In this study, the average annual incidence rate per 1,000 person-years was 3.4 (4.8 in women and 1.9 in men). In women, incidence rates were low at the extremes of age and higher among those aged between 10 and 49 years, with

Migraine Epidemiology, Fig. 1 Incidence of migraine, by age and sex (From Stewart et al. 1993)

a striking peak at the age of 20–29 years. In this study, incidence also peaked later than in other studies, probably because medical diagnosis may occur long after the age of onset. Finally, in the Danish population, the annual incidence of migraine in those aged 25–64 years old was 8/1,000, being 15/1,000 in males and 3/1,000 in females (Lyngberg et al. 2003).

The Prevalence of Migraine

The published estimates of migraine prevalence have varied broadly, probably because of differences in the methodology. The studies presented herein primarily used the IHS definition.

Prevalence by Age

Before puberty, as suggested by the incidence data, migraine prevalence is higher in boys than in girls; as adolescence approaches, incidence and prevalence increase more rapidly. As a consequence, at all postpubertal ages, migraine is more common in girls than in boys. The prevalence increases throughout childhood and early adult life until approximately age 40, after which it declines (Fig. 2) (Scher et al. 1999). Overall, prevalence is highest from 25 to 55, the peak years of economic productivity. The gap between peak incidence in adolescence and peak prevalence in middle life indicates that migraine is a condition of long duration.

Despite suggestions to the contrary, the prevalence of migraine is probably not increasing.

Adjusted Prevalence of Migraine by Age

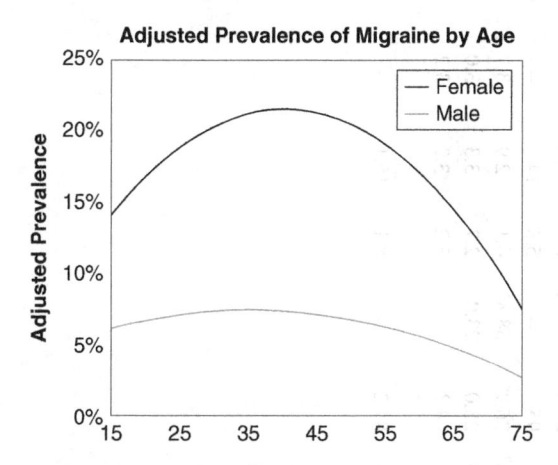

Migraine Epidemiology, Fig. 2 Adjusted prevalence of migraine by age from a meta-analysis of studies using IHS criteria (From Scher et al. 1999)

According to the Centers for Disease Control, self-diagnosed migraine prevalence in the USA increased by 60 %, from 25.8/1,000 to 41/1,000 persons, between 1981 and 1989 (MMWR 1991). Because this study relies on self-reported migraine, an increase in diagnosis or disease awareness could be mistaken for an increase in prevalence. Numerous population studies in the USA show that prevalence is stable, while consultation and diagnosis have increased (Lipton et al. 2001). It may be that these increases in medical consultation and diagnosis have caused an apparent rather than a real increase in migraine prevalence.

Prevalence in Children and Adolescents
The prevalence of headache in children, as investigated in a number of school- and population-based studies, shows that by age 3, headache occurs in 3–8 % of children. At age 5, 19.5 % have headache, and by age seven, 37–51.5 % have headaches. In 7–15-year-olds, headache prevalence ranges from 57 % to 82 %. The prevalence increases from ages 3 to 11 in both boys and girls, with higher headache prevalence in 3–5-year-old boys than in 3–5-year-old girls. Thus, the overall prevalence of headache increases from preschool age children to mid-adolescence, when examined using various cross-sectional studies (Bille 1989).

Recent studies report the prevalence of pediatric migraine in the Asian Middle East. The first one, performed in the southern Iran, evaluated a random sample of 1,868 teenage girls (aged 11–18 years). The overall prevalence rate for migraine was 6.1 % (95 % CI, 5.0–7.2). The second study evaluated 1,400 randomly selected Saudi children in grades one through nine. Overall, the headache prevalence was 49.8 %. The prevalence of migraine was 7.1 %. There was a sharp increase in the prevalence rate (from around 2 % to around 9 %) at ages 10–11, in both boys and girls. Age-adjusted prevalence for migraine between ages 6 and 15 was 6.2 % (for references on the original studies, the reader is referred to Bigal et al. 2004).

Another recent study evaluated the evolution of juvenile migraine without aura in adolescents over 5 years. Sixty-four subjects out of 80 previously selected were reevaluated. Thirty-two (50 %) had migraine without aura. After 5 years, migraine without aura persisted in 56.2 %, converted to migrainous disorder or non-classifiable headache respectively in 9.4 % and 3.1 % of cases, changed to episodic tension-type headache in 12.5 %, and remitted in 18.8 %.

Prevalence by Gender in Adults
Estimates of migraine prevalence range from 3.3 % to 21.9 % for women and 0.7–16.1 % for men. In the United States, the American Migraine Study II found that the prevalence of migraine was about 18 % in women and 6 % in men (Lipton et al. 2001). Table 1 summarizes several prevalence studies conducted in the last 12 years. The prevalence of migraine in different geographical locations, overall and by gender, is presented. The female to male gender ratio is about 3:1 in most places where it has been studied.

Prevalence of Migraine by Socioeconomic Status
The relationship between migraine prevalence and socioeconomic status is uncertain. In physician- and clinic-based studies, migraine appears to be associated with high intelligence and social class. In his studies of children, Bille did not find an association between migraine prevalence and intelligence (Bille 1989). Similarly, in adults, epidemiological studies do not support a relationship

Migraine Epidemiology, Table 1 Prevalence of headache and migraine by age in selected community- and school-based studies

Author(Y) country	Type of population	Sample size	Age range (years)	Time frame	Migraine definition	Headache prevalence			Migraine prevalence		
						Males	Females	Overall	Males	Females	Overall
Ayatollahi (2002) Iran	School teenage girls	1,868	11–18	?	IHS					6.1 %	
Al Jumah M (2002) Saudi Arabia	Schoolchildren	1,400	6–18	?	IHS				6.4 %	7.7 %	7.1 %
Abu-Arafeh (1994)	Schoolchildren	1,754	5–15	1 year	IHS[1]						10.6
Bille (1962) Sweden	Schoolchildren	8,993	7–15	Lifetime	Vahlquist[1]	58.0	59.3	–	3.3	4.4	–
Linet et al. (1989) USA	Community	10,132	12–29	1 year	2 of NV/U/VA	90	95	–	5.3	14	–
Mortimer et al. (1992) UK	General practice	1,083	3–11	1 year	IHS[1]	40.6[a]	36.9[a]	38.8[a]	4.1	2.9	3.7
Raielli (1995) Italy	Schoolchildren	1,445	11–14	1 year	IHS[1]	19.9	28.0	23.9	2.7	3.3	3.0
Sillanpaa (1976) Finland	Schoolchildren	4,825	3 / 7	? / ?	Vahlquist[1]		4.3		3.2	3.2	3.2
Sillanpaa (1983) Finland	Schoolchildren	3,784	13	1 year	Vahlquist[1]	79.8	84.2	–	8.1	15.1	–

N nausea, *U* unilateral, *V* vomiting, *VA* visual aura
[a] age adjusted

**Migraine Epidemiology,
Fig. 3** Adjusted
prevalence of migraine by
geographical area and
gender in a meta-analysis of
studies using IHS criteria
(From Scher et al. 1999)

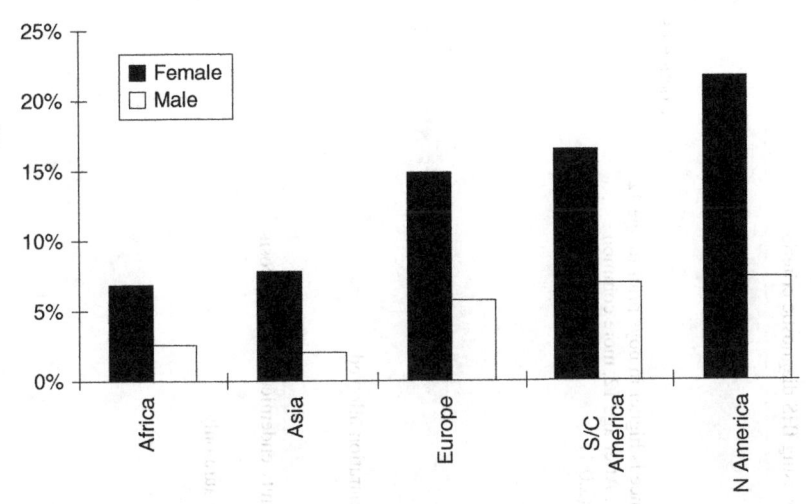

between occupation and migraine prevalence
(Lipton et al. 2001). In both the American
Migraine Studies I and II, migraine prevalence
was inversely related to household income (i.e.,
migraine prevalence fell as household income
increased) (Lipton et al. 2001). This inverse relationship between migraine and socioeconomic status was confirmed in another US study based on
members of a managed care organization and in
the National Health Interview Study. In Europe,
results are contradictory. While one large study
failed to demonstrate an association between
migraine and socioeconomic status (Launer et al.
1999), a second recent study in England showed
this relationship (Steiner et al. 2003).

The higher prevalence in the lower socioeconomic groups may be a consequence of
a circumstance associated with low income and
migraine, such as poor diet, poor medical care, or
stress. It may also reflect social selection, that is,
migraineurs may have lower incomes because
migraine interferes with educational and occupational function, causing a loss of income or the ability
to rise from a low-income group. The relationship of
migraine and socioeconomic status, especially in
children and adolescents, requires further study.

Prevalence of Migraine by Geographical Distribution

Migraine prevalence also varies by race and
geography. In the USA, it is highest in Caucasians, intermediate in African Americans, and

lowest in Asian Americans (Scher et al. 1999).
Similarly, a meta-analysis of prevalence studies
suggests that migraine is most common in North
and South America, similar in Europe, but lower
in Africa and often lowest in studies from Asia
(Fig. 3) (Scher et al. 1999). The data suggest that
race-related differences in genetic risk may
contribute.

The Burden of Migraine

The Impact of Migraine on the Individual

The burden of migraine is significant both to the
individual sufferer and to the society. In the
American Migraine Study II, 92 % of women
and 89 % of men with severe migraine had
some headache-related disability (Lipton
et al. 2001). About half were severely disabled
or needed bed rest. In addition to the attack-
related disability, many migraineurs live in fear,
knowing that at any time an attack could disrupt
their ability to work, care for their families, or
meet social obligations. Abundant evidence indicates that migraine reduces health-related quality
of life.

The World Health Organization (WHO) has
recently released a report on the burden of diseases (The World Health Report 2001). The
WHO report defines the "burden" of a disease to
include the economic and emotional difficulties
that a family experiences as a result of migraine,
as well as the lost opportunities – the adjustments
and compromises that prevent other family

Migraine Epidemiology, Table 2 Gender-specific prevalence estimates of migraine from 25 population-based studies using IHS diagnostic criteria

Author (year of publication)	Country	Source	Method	Sample size	Time frame	Age range	Migraine prevalence (%)			Comments
							Female	Male	Total	
Abu-Arafeh (1994)	Scotland	School	Clin interview	1,754	1 year	5–15	11.5	9.7	10.6	Prevalence is higher in boys prior to age 12 (1.14:1). After age 12, more common in girls (2.0:1)
Al Rajeh (1997)	Saudi Arabia	Community	Face to face/clin interview	22,630		All	6.8	3.2	5.0	
Alders (1996)	Malaysia	Community	Face to face	595	1 year	5+	11.3	6.7	9.0	
Arregui (1991)	Peru	Community	Clin interview	2,257		All	12.2	4.5	8.4	
Bánk et al. (2000)	Hungary	Community	Questionnaire	813	1 year	15–80			9.6	
Barea (1996)	Brazil	School	Clin interview	538	1 year	10–18	10.3	9.6	9.9	2–48-h duration allowed
Breslau (1991)	USA	Community	Face to face/telephone	1,007	1 year	21–30	12.9	3.4	9.2	
Cruz et al. (1995)	Ecuador	Community	Clin interview	2,723	Lifetime	All	7.9	5.6	6.9	Community endemic for cysticercosis
Cull et al. (1992)	UK	Community	Face to face	16,002		16+	11.0	4.3	7.8	Without aura only
Dahlöf et al. (2001)	Sweden	Community	Telephone interview	1,668	1 year	18–74	16.7	9.5	13.2	
Deleu et al. (2002)	Saudi Arabia	Community	Face to face	1,158	1 year	10+	5.6	4.5		
Göbel (1994)	Germany	Community	Mail SAQ	4,061	Lifetime	18+	15.0	7.0	11.0	
Hagen et al. (2000)	Norway	Community	Clin interview	51,833	1 year	20+	16.0	8.0	12.0	
Tekle Haimanot et al. (1995)	Ethiopia	Community	Face to face/clin interview	15,000	1 year	20+	4.2	1.7	3.0	
Henry et al. (1992)	France	Community	Face to face	4,204	1 year	15+	11.9	4.0	8.1	
Henry et al. (2002)	France	Community	Face to face	10,585	1 year	15+	11.2	4.0	7.9	

Study	Country	Setting	Method	Sample	Time period	Age	Female	Male	Overall	Comments
Abduljabbar et al. (1996)	Saudi Arabia	Community	Face to face	5,891	Lifetime	15+			8.0	
Jaillard et al. (1997)	Peru	Community	Clin interview	3,246	1 year	15+	7.8	2.3	5.3	
Kececi et al. (2002)	Turkey	Community	Face to face	1,320	1 year	All	17.1	7.9		
Lampl et al. (2003)	Sweden	Community	Face to face	997	1 year	15+	13.8	6.1	10.2	
Launer et al. (1999)	Netherlands	Community	Questionnaire and telephone	6,491	Lifetime / 1 year	18–65	33 / 25	13.3 / 7.5		
Lipton et al. (2001)	USA	Community	Telephone	29,727	1 year	12+	18.2	5.5		
Lipton et al. (2002)	USA	Community	Telephone	11,863	1 year	18–65	17.2	6		
Merikangas et al. (1993)	Switzerland	Community	Clin interview	379	1 year	28–29	32.6	16.1	24.5	Weighted prevalence
Michel et al. (1996)	France	Community	Mail SAQ	9,411	3 month	18+	18.0	8.0	13.0	
Miranda et al. (2003)	Puerto Rico	Community	Telephone	1,610	1 year	All	16.7	6	13.5	
O'Brien et al. (1994)	Canada	Community	Telephone	2,922	1 year	18+	21.9	7.4	15.2	
Raieli (1994)	Italy	School	Clin interview	1,445	1 year	11–14	3.3	2.7	3.0	
Rasmussen et al. (1992)	Denmark	Community	Clin interview	740	1 year	25–64	15.0	6.0	10.0	
Russell (1995)	Denmark	Community	Clin interview	3,471	Lifetime	40	23.7	11.7	17.7	
Sakai and Igarashi (1997)	Japan	Community	Mail SAQ	4,029	1 year	15+	12.9	3.6	8.4	Female to male prevalence ratio = 3.6. Regional differences
Steiner et al. (2003)	England	Community	CATI	4,007	1 year	18–65	18.3	7.6	14.3	
Stewart et al. (1992)	USA	Community	Mail SAQ	20,468	1 year	12–80	17.6	5.7	12.0	
Stewart et al. (1996)	USA	Community	Telephone	12,328	1 year	18–65	19.0	8.2	14.7	Racial differences

(continued)

Migraine Epidemiology, Table 2 (continued)

Author (year of publication)	Country	Source	Method	Sample size	Time frame	Age range	Migraine prevalence (%)			Comments
							Female	Male	Total	
Takeshima et al. (2004)	Japan	Community	Questionnaires and telephone	5,758	1 year	18+	9.1	2.3	6	Study done in the rural area of western Japan
Van Roijen et al. (1995)	Netherlands	Community	Face to face	10,480	1 year	1^2+	12.0	5.0	9.0	
Wang et al. (1997)	China	Community	Clin interview	1,533	1 year	65+	4.7	0.7	3.0	
Wong et al. (1995)	Hong Kong	Community	telephone	7,356	1 year	15+	1.5	0.6	1.0	
Zivadinov et al. (2001)	Croatia	Community	Telephone and face to face	5,173	Lifetime / 1 year	15–65	22.9 / 18	14.8 / 12.3	19	

members from achieving their full potential in work, social relationships, and leisure. The global burden of disease (GBD) is an analysis of the onset of disorders and the disability caused by them. Using the GBD methodology, migraine is estimated to account for 2.0 % years of life lost due to a disability in women of all ages. In both sexes of all ages, migraine is responsible for 1.4 % of total years of life lost due to a disability and ranks within the top 20 most disabled studied disorders (Table 2).

The Impact of Migraine on the Family

The impact of migraine extends to household partners and other family members. In a recent study, one half of the participants believed that, because of their migraine, they were more likely to argue with their partners (50 %) and children (52 %), while majorities (52–73 %) reported other adverse consequences for their relationships with their partner and children and at work. A third (36 %) believed they would be better partners but for their headaches. Participating partners partly confirmed these findings; 29 % felt that arguments were more common because of headaches and 20–60 % reported other negative effects on relationships at home. Compared with subjects who did not have migraine, regarding their work performance, a statistically significantly higher proportion of migraine partners were unsatisfied with work demands placed on them, with their level of responsibilities and duties and with their ability to perform (Lipton et al. 2003). Results from this study show that the impact of migraine extends to household partners and other family members (impact beyond the individual).

The Societal Impact of Migraine

Migraine has an enormous impact on society. Studies have evaluated the indirect costs of migraine as well as the direct costs (Lipton et al. 2001). Indirect costs include the aggregate effects of migraine on productivity at work (paid employment), for household work and in other roles. The largest components of indirect costs are the productivity losses caused by absenteeism and reduced productivity while at work. Hu et al.

Migraine Epidemiology, Table 3 Global burden of migraine. Leading causes of years of life lost due to a disability according to the global burden of disease initiative (Modified from Lipton et al. 2003)

	Females all ages	% Total		Both sexes all ages	% Total
1	Unipolar depressive disorders	1		Unipolar depressive disorders	11.9
2	Iron-deficiency anemia	2		Hearing loss, adult onset	4.6
3	Hearing loss, adult onset	3		Iron-deficiency anemia	4.5
4	Osteoarthritis	4		Chronic obstructive pulmonary disease	3.3
5	Chronic obstructive pulmonary disease	5		Alcohol use disorders	3.1
6	Schizophrenia	6		Osteoarthritis	3.0
7	Bipolar affective disorder	7		Schizophrenia	2.8
8	Falls	8		Falls	2.8
9	Alzheimer's and other dementias	9		Bipolar affective disorder	2.5
10	Obstructed labor	10		Asthma	2.1
11	Cataracts	11		Congenital abnormalities	2.1
12	Migraine	12		Perinatal conditions	2.0
13	Congenital abnormalities	13		Alzheimer's and other dementias	2.0
14	Asthma	14		Cataracts	1.9
15	Perinatal conditions	15		Road traffic accidents	1.8
16	Chlamydia	16		Protein-energy malnutrition	1.7
17	Cerebrovascular disease	17		Cerebrovascular disease	1.7
18	Protein-energy malnutrition	18		HIV/AIDS	1.5
19	Abortion	19		Migraine	1.4
20	Panic disorder	20		Diabetes mellitus	1.4

estimated that productivity losses due to migraine cost American employers 13 billion dollars per year (Hu et al. 1999). In a UK study, 5.7 working days were lost per year for every working person or student with migraine, although the most disabled 10 % accounted for 85 % of the total, projecting a lost 25 million days from work or school each year (Steiner et al. 2003).

Migraine's impact on healthcare utilization is marked as well. Studies show that 4 % of all visits to physicians' offices are for headache (Bigal et al. 2004). Migraine also results in major utilization of emergency rooms and urgent care centers (Bigal et al. 2004). Vast amounts of prescription and over-the-counter medications are taken for headache disorders. OTC sales of pain medication (for all conditions) were estimated to be 3.2 billion dollars in 1999 (USA) and headache accounts for about one third of OTC analgesic use (Bigal et al. 2004) (Table 3).

References

Abduljabbar, M., Ogunniyi, A., Al Balla, S., Alballaa, S., & Al-Dalaan, A. (1996). Prevalence of primary headache syndrome in adults in the Qassim region of Saudi Arabia. *Headache, 36*(6), 385–388.

Abu-Arafeh, I., & Russell, G. (1994 Sep 24). Prevalence of headache and migraine in schoolchildren. *BMJ, 309* (6957), 765–769.

Abu-Arafeh, I., & Russell, G. (1995 May). Prevalence and clinical features of abdominal migraine compared with those of migraine headache. *Archives of Disease in Childhood, 72*(5), 413–417.

Alders, E. E. A., Hentzen, A., & Tan, C. T. (1996). A community-based prevalence: Study on headache in Malaysia. *Headache: The Journal of Head and Face Pain, 36,* 379–384. doi:10.1046/j.1526-4610.1996.3606379.x.

Al Jumah, M., Awada, A., & Al, A. S. (2002 Apr). Headache syndromes amongst schoolchildren in Riyadh Saudi Arabia. *Headache., 42*(4), 281–286.

Al Rajeh, S. (1997). The prevalence of migraine and tension headache in Saudi Community; a community based study. *European Journal of Neurology, 4,* 502–506.

Arregui, A., Cabrera, J., Leon-Velarde, F., Paredes, S., Viscarra, D., & Arbaiza, D. (1991). High prevalence of migraine in a high-altitude population. *Neurology, 41,* 1668–1670.

Ayatollahi, S. M., Moradi, F., & Ayatollahi, S. A. (2002). Prevalence of migraine and tension type headache in for adolescent girls of shiraz (southern Iran). *Headache, 42,* 287–290.

Bánk, J., & Márton, S. (2000 Feb). Hungarian migraine epidemiology. *Headache, 40*(2), 164–169.

Barea, L. M., Tannhauser, M., & Rotta, N. T. (1996). An epidemiologic study of headache among children and adolescents of southern Brazil. *Cephalalgia, 16,* 545–549.

Bigal, M. E., Lipton, R. B., & Stewart, W. F. (2004). The epidemiology and impact of migraine. *Current Neurology and Neuroscience Reports, 4,* 98–104.

Bille, B. (1962). Migraine in school children. *Acta Paediatrica, 51,* 16–151.

Bille, B. (1989). Migraine in children: Prevalence, clinical features, and a 30-year follow-up. In M. D. Ferrari & X. Lataste (Eds.), *Migraine and other headaches.* New Jersey: Parthenon.

Breslau, N., Davis, G. C., & Andreski, P. (1991). Migraine, psychiatric disorders and suicide attempts: An epidemiologic study of young adults. *Psychiatry Research, 37,* 11–23.

Breslau, N., Davis, G. C., Schultz, L. R., et al. (1994). Joint 1994 Wolff award presentation. Migraine and major depression: A longitudinal study. *Headache, 34,* 387–393.

Cruz, M. E., Cruz, I., Preux, P. M., Schantz, P., & Dumas, M. (1995). Headache and cysticercosis in Ecuador, South America. *Headache, 35*(2), 93–97.

Cull, R. E., Wells, N. E. J., & Miocevich, M. L. (1992). The economic cost of migraine. *British Journal of Medical Economics, 2,* 103–115.

Dahlöf, C., & Linde, M. (2001). One-year prevalence of migraine in Sweden: A population-based study in adults. *Cephalalgia, 21*(6), 664–671.

Deleu, D., Khan, M. A., & Al Shehab, T. A. (2002). Prevalence and clinical characteristics of headache in a rural community in Oman. *Headache, 42*(10), 963–973.

Göbel, H., Petersen-Braun, M., & Soyka, D. (1994). The epidemiology of headache in Germany: A nationwide survey of a representative sample on the basis of headache classification of the International Headache Society. *Cephalalgia, 14,* 97–106.

Hagen, K., Zwart, J. A., Vatten, L., Stovner, L. J., & Bovim, G. (2000). Prevalence of migraine and nonmigrainous headache—head-HUNT, a large population-based study. *Cephalalgia, 20*(10), 900–906.

Henry, P., Michel, P., Brochet, B., et al. (1992). A nationwide survey of migraine in France: prevalence and clinical features in adults. *Cephalalgia, 12,* 229–237.

Hu, X. H., Markson, L. E., Lipton, R. B., et al. (1999). Burden of migraine in the United States: Disability and economic costs. *Archives of Internal Medicine, 159,* 813–818.

Jaillard, A. S., Mazetti, P., & Kala, E. (1997). Prevalence of migraine and headache in a high-altitude town of Peru: A population-based study. *Headache, 37*(2), 95–101.

Kececi, H., & Dener, S. (2002). Epidemiological and clinical characteristics of migraine in Sivas, Turkey. *Headache, 42*(4), 275–280.

Lampl, C., Buzath, A., Baumhackl, U., & Klingler, D. (2003). One-year prevalence of migraine in Austria: A nation-wide survey. *Cephalalgia, 23,* 280–286.

Launer, L. J., Terwindt, G. M., & Ferrari, M. D. (1999). The prevalence and characteristics of migraine in a population-based cohort: The GEM study. *Neurology, 3,* 537.

Linet, M. S., Stewart, W. F., Celentano, D. D., Ziegler, D., & Sprecher, M. (1989). An epidemiologic study of headache among adolescents and young adult. *Journal of the American Medical Association, 261*, 2211–2216.

Lipton, R. B., Stewart, W. F., Diamond, S., et al. (2001). Prevalence and burden of migraine in the United States: Data from the American migraine study II. *Headache, 41*(7), 646–657.

Lipton, R. B., Scher, A. I., Kolodner, K., et al. (2002). Migraine in the United States: epidemiology and patterns of health care use. *Neurology, 58*(6), 885–894.

Lipton, R. B., Bigal, M. E., Kolodner, K., et al. (2003). The family impact of migraine: Population-based studies in the USA and UK. *Cephalalgia, 23*, 429–440.

Lyngberg, A., Jensen, R., Rasmussen, B. K., et al. (2003). Incidence of migraine in a Danish population-based follow-up study. *Cephalalgia, 23*, 596. Abstract.

Merikangas, K. R., Whitaker, A. E., & Angst, J. (1993). Validation of diagnostic criteria for migraine in the Zürich longitudinal cohort study. *Cephalalgia, 13*, 47–53.

Michel, P., Pariente, P., Duru, G., et al. (1996). Mig access: A population-based, nationwide, comparative survey of access to care in migraine in France. *Cephalalgia, 16*, 30–55.

Miranda, H., Ortiz, G., Figueroa, S., Pena, D., & Guzman, J. (2003). Prevalence of headache in Puerto Rico. *Headache, 43*(7), 774–778.

Mortimer, M. J., Kay, J., & Jaron, A. (1992). Epidemiology of headache and childhood migraine in an urban general practice using Ad Hoc, Vahlquist and IHS criteria. *Developmental Medicine and Child Neurology, 34*(12), 1095–1101.

MMWR. (1991). Prevalence of chronic migraine headaches – United States, 1980-89. *Morbidity and Mortality Weekly Report, 40*, 331–338.

O'Brien, B., Goeree, R., & Streiner, D. (1994). Prevalence of migraine headache in Canada: A population-based survey. *International Journal of Epidemiology, 23*(5), 1020–1026.

Raieli, V., Raimondo, D., Cammalleri, R., & Camarda, R. (1995a). Migraine headache in adolescents: A student population-based study in Montreal. *Cephalalgia, 15*, 5–12.

Raieli, V., Raimondo, D., Cammalleri, R., & Camarda, R. (1995b). Migraine headaches in adolescents: A student population-based study in Monreale. *Cephalalgia, 15*(1), 5–12. doi:10.1046/j.1468-2982.1995.1501005.x.

Rasmussen, B. K., Jensen, R., & Olesen, J. (1992). Impact of headache on sickness absence and utilization of medical services: A Danish population study. *Journal of Epidemiology and Community Health, 46*, 443–446.

Sakai, F., & Igarashi, H. (1997). Prevalence of migraine in Japan: A nationwide survey. *Cephalalgia, 17*, 15–22.

Scher, A. I., Stewart, W. F., & Lipton, R. B. (1999). Migraine and headache: A meta-analytic approach. In I. K. Crombie (Ed.), *Epidemiology of pain* (pp. 159–170). Seattle: IASP Press.

Sillanpaa, M. (1976). Prevalence of migraine and other headaches in Finnish children starting school. *Headache, 15*, 288–290.

Sillanpaa, M. (1983). Prevalence of headache in prepuberty. *Headache, 23*, 10–14.

Stang, P. E., Yanagihara, T., Swanson, J. W., et al. (1992). Incidence of migraine headaches: A population-based study in Olmstead County, Minnesota. *Neurology, 42*, 1657–1662.

Steiner, T. J., Scher, A. I., Stewart, W. F., Kolodner, K., Liberman, J., & Lipton, R. B. (2003). The prevalence and disability burden of adult migraine in England and their relationships to age, gender and ethnicity. *Cephalalgia, 23*(7), 519–527.

Stewart, W. F., Lipton, R., Celentano, D. D., & Reed, M. L. (1992). Prevalence of migraine headache in the Unites States. *Journal of the American Medical Association, 267*, 64–69.

Stewart, W. F., Linet, M. S., Celentano, D. D., et al. (1993). Age and sex-specific incidence rates of migraine with and without visual aura. *American Journal of Epidemiology, 34*, 1111–1120.

Stewart, W. F., Lipton, R. B., & Simon, D. (1996). Work-related disability: Results from the American migraine study. *Cephalalgia, 16*(4), 231–238.

Takeshima, T., Ishizaki, K., Fukuhara, Y., Ijiri, T., Kusumi, M., Wakutani, Y., Mori, M., Kawashima, M., Kowa, H., Adachi, Y., Urakami, K., & Nakashima, K. (2004). Population-based door-to-door survey of migraine in Japan: The Daisen study. *Headache, 44*(1), 8–19.

Tekle Haimanot, R., Seraw, B., Forsgren, L., Ekbom, K., & Ekstedt, J. (1995). Migraine, chronic tension-type headache, and cluster headache in an Ethiopian rural community. *Cephalalgia, 15*(6), 482–488.

The World Health Report. (2001). Mental health, new understanding, new hope. Available at www.who.int/en/.

Van Roijen, L., Essink-Bot, M. L., Rutten, F. F. H., & Koopmanschap, M. A. (1995). Societal perspective on the burden of migraine. *Pharmacology & Economics, 7*, 170–179.

Wang, S. J., Liu, H. C., Fuh, J. L., et al. (1997). Prevalence of headaches in a Chinese elderly population in Kinmen: Age and gender effect and cross-cultural comparisons. *Neurology, 49*, 195–200.

Wong, T. W., Wong, K. S., Yu, T. S., & Kay, R. (1995). Prevalence of migraine and other headaches in Hong Kong. *Neuroepidemiology, 14*(2), 82–91.

Zivadinov, R., Willheim, K., Jurjevic, A., Sepic-Grahovac, D., Bucuk, M., & Zorzon, M. (2001). Prevalence of migraine in Croatia: A population-based survey. *Headache, 41*(8), 805–812.

Migraine Variants of Childhood

▶ Migraine, Childhood Syndromes

Migraine with Pleocytosis

▶ Transient Headache and CSF Lymphocytosis

Migraine Without Aura

Jean Schoenen
Department of Neurology, University of Liège,
CHR Citadelle, Belgium
Department of Neuroanatomy, University of
Liège, CHR Citadelle, Belgium

Synonyms

Clinical migraine without aura; Common migraine; Hemicrania simplex

Definition

Migraine without aura (MO) is a recurrent headache manifesting in attacks lasting between 4 h and 72 h. Typical features of this headache are unilateral location, pulsating quality, moderate or severe intensity, aggravation by routine physical activity, and association with nausea and/or photophobia and phonophobia (see diagnostic criteria according to the International Classification of Headache Disorders, 2nd edition or ICHD-II 2004).

Diagnostic Criteria of Migraine Without Aura
1. At least five attacks fulfilling criteria 2–4
2. Headache attacks lasting 4–72 h (untreated or unsuccessfully treated)
3. At least two of the following pain characteristics:
   • Unilateral location
   • Pulsating quality
   • Moderate or severe intensity
   • Aggravation by or causing avoidance of routine physical activity (e.g., walking or climbing stairs)
4. During headache at least one of the following:
   • Nausea and/or vomiting
   • Photophobia and phonophobia
5. Not attributed to another disorder

Characteristics

Epidemiology

The most recent population-based studies in adults, all using the diagnostic criteria of the IHS (International Headache Society), have reached very similar prevalence rates. Several European and American studies have reported somewhat congruent prevalence figures for migraine in adults (Steiner et al. 2003). The overall 1-year period prevalence of migraine with or without aura in adults is about 15 % (7.6 % among men and 19.1 % among women). The overall 1-year prevalence of migraine with aura is about 5.8 % (male 2.6 %, female 7.7 %). Studies in general populations agree that it is most common for migraineurs to have about one attack a month. In clinic samples the frequency is higher since high frequency may be a compelling reason for referral. A large number of headache sufferers with features of migraine fail to meet criteria for strict migraine but do meet criteria for probable migraine (attacks and/or headache missing one of the features 1–4 needed to fulfill all criteria for migraine without aura; see above). The 1-year prevalence for probable migraine was 14.6 % (15.9 % in women, 12.6 % in men) in a recent study (Patel et al. 2004).

Phenotype

The heterogeneity of the clinical phenotype of migraine is underestimated. Despite a common diagnostic denominator, some clinical features such as type of aura symptoms, pain intensity, presence of prodromes, coexistence of migraine with and without aura, or associated symptoms such as vertigo may characterize subgroups of patients bearing different underlying pathophysiological and genetic mechanisms. Pain intensity can help to distinguish migraine without aura from tension-type headache (TTH); indeed among subjects in the general population classified as TTH sufferers, clinical features suggestive of migraine, i.e., aggravation by routine physical exercise, pulsating quality, anorexia, photophobia, unilateral headache, and nausea, may occur in non-negligible proportions. Trigger factors are manifold and may vary between patients and during the disease course. The most common ones are stress, the perimenstrual period, and alcohol. Overuse of acute antimigraine drugs, in particular of combination analgesics and ergotamine, is another underestimated aggravating factor. There is a complex interrelation between migraine and depression, which are highly comorbid. It is not determined, however, whether the increased prevalence of depression in patients with frequent migraines is primary or secondary. Episodic vertigo without other signs of ▶ basilar migraine might belong to the migraine phenotype.

Genotype

The common migraine phenotypes appear to be complex genetic disorders, where additive genetic effects (susceptibility genes) and environmental factors are interrelated (Stewart et al. 1997). Migraine is characterized by recurrent attacks. Genetic load can be seen as determining, on the one hand, a critical threshold for migraine attacks. On the other hand, genetic abnormalities may induce incidental subclinical dysfunctions, such as, a reduced neuromuscular junction safety factor (Ambrosini et al. 2001) or subtle cerebellar hypermetria (Sandor et al. 2001). Various gene polymorphisms were found to be more prevalent in migraineurs than in controls (Freilinger et al. 2012). Their precise role remains to be determined; some of them may not be specific to migraine, but they could increase susceptibility to the disorder and induce endophenotypic vulnerability markers (see below).

Pathophysiology

The present consensus is that both neuronal and vascular components are relevant in migraine and most probably interrelated (Goadsby et al. 2002). The neuronal structures involved are the cerebral cortex, the brainstem (periaqueductal gray matter, aminergic nuclei), and the peripheral as well as the central components of the ▶ trigeminovascular system. The sequence of activation and the relative roles of these structures are still controversial.

Possible Role of the Brainstem

During attacks of migraine without aura, an area of increased blood flow has been identified contralaterally in the dorsolateral part of the brainstem in a positron emission tomography (PET) study (Weiller et al. 1995). This led to the hypothesis of a "migraine generator" in the region of the dorsal raphe nucleus, locus coeruleus, and periaqueductal gray which would suggest that migraine is a central pain disorder and explain recurrence after acute antimigraine drug treatment. Considering the well-known projections of second-order trigeminal nociceptors on the periaqueductal gray and the fact that activations were found in a similar location after other spontaneous and experimental pains, it cannot be excluded that the brain stem activations found in migraineurs reflect secondary modulation of pain.

▶ Cutaneous allodynia has been described during the attack, ipsilateral and contralateral to the hemicrania as well as in the forearm, indicating central ▶ sensitization at the level of 2nd- and thereafter of 3rd-order nociceptors (Burstein et al. 2000). This may justify the clinical rule that early treatment of the attack is more effective (Burstein et al. 2004).

Possible Role of the Cortex

During the headache-free interval, an abnormal functioning of the migrainous brain can be demonstrated by neurophysiological and metabolic studies. Neurophysiological methods show that cortical information processing in migraineurs is characterized by a deficient ▶ habituation during repetition of the stimulation (see Schoenen et al. 2003 for a review). This has been demonstrated for event-related and visual- ▶ evoked potentials (VEPs). Moreover, intensity dependence of auditory-evoked cortical potentials is increased in migraine patients compared to normal controls. The cortical abnormality is likely to be genetically determined. Although not confirmed in every study, it may represent an endophenotypic vulnerability marker (see above). It could be due to low activity in raphecortical serotonergic pathways, which would reduce preactivation levels of sensory cortices, allowing a wider range for suprathreshold corti-cal activation and thus enhancing intensity dependence and reducing habituation. During the attack, when brain stem activation has been found (see above), electrophysiological methods demonstrate a "normalization" of event-related and evoked potentials.

▶ Transcranial magnetic stimulation (TMS) studies to assess excitability of motor and visual cortices have yielded conflicting results, but taken together, they suggest normal or increased rather than decreased excitation thresholds between attacks (Schoenen et al. 2003). High-frequency repetitive TMS (rTMS), which acti-vates the underlying cortex, is able to normalize interictal VEPs in migraineurs, whereas low fre-quency rTMS, which has an inhibitory effect, induces in healthy volunteers a VEP potentiation similar to the one found in migraine (Bohotin et al. 2002). These rTMS results favor the hypoth-esis put forward to explain the habituation deficit found interictally in evoked potentials, i.e., that the preactivation level of sensory cortices is reduced in migraineurs.

Metabolic studies using nuclear magnetic res-onance spectroscopy (MRS) have shown a low mitochondrial phosphorylation potential (Welch et al. 1989) in brain and muscles of migraineurs. More recently, a 16 % decrease of absolute con-tent of ATP was measured in the occipital cortex of migraine without aura patients (Reyngoudt et al. 2011). Nevertheless, none of the known mutations of mitochondrial DNA has been found up to now in the common forms of migraine. It has been hypothesized that the con-junction of a decreased mitochondrial energy reserve and a deficit in habituation of cortical information processing, known to protect against overstimulation and lactate accumulation, might lead to activation of the trigeminovascular sys-tem (Schoenen 1994). It is still to be determined whether the mitochondrial abnormality is an independent pathophysiological component or rather a consequence of other functional deficits.

Role of the Trigeminovascular System

The trigeminovascular system is the major pain signaling structure of the visceral organ brain, but there is still no definite proof that an activation of its peripheral components is necessary to produce a migraine attack. Such activation is indeed suggested by the increase of CGRP (▶ Calcitonin-Gene-Related Peptide) in external jugular vein blood during the attack and by the effectiveness of most acute antimigraine medica-tions including ▶ triptans in the experimental neu-rogenic inflammation model in the rat. It is not known what activates the trigeminovascular noci-ceptive pathway in migraine. But a link between the migraine aura (probably due to a cortical ▶ Spreading Depression) and the headache is suggested by the experimental finding that cortical spreading depression is able to activate trigemino-vascular afferents and to evoke a series of cortical, meningeal, and brainstem events consistent with the development of headache (Bolay et al. 2002).

Role of Nitric Oxide

Nitric oxide (NO) donors such as glyceryl trinitrate (GTN) are able to induce attacks in patients suffering from migraine with or without aura. The underlying mechanisms are not well established, but there is clinical and experimental evidence that NO can favor central sensitization of trigeminal nociceptors.

Treatment of Migraine

Acute Treatment

During the last decade, the advent of highly effec-tive 5-HT$_{1B/1D}$ agonists, the triptans, has been a major breakthrough in the attack treatment. Triptans are able to act as vasoconstrictors via vascular 5-HT$_{1B}$ receptors and to inhibit neuro-transmitter release at the peripheral as well as at the central terminal of trigeminal nociceptors via 5-HT$_{1D/B}$ receptors. The site of action relevant for their efficacy in migraine is still a matter of controversy; possibly, their high-efficacy rate is due to their capacity to act at all three sites con-trary to other antimigraine drugs. Sumatriptan, the first triptan, was followed by several second-generation triptans (zolmi-, nara-, riza-, ele-, almo-, frovatriptan), which were thought to cor-rect some of the shortcomings of sumatriptan.

A large meta-analysis of a number of random-ized controlled trials performed with triptans

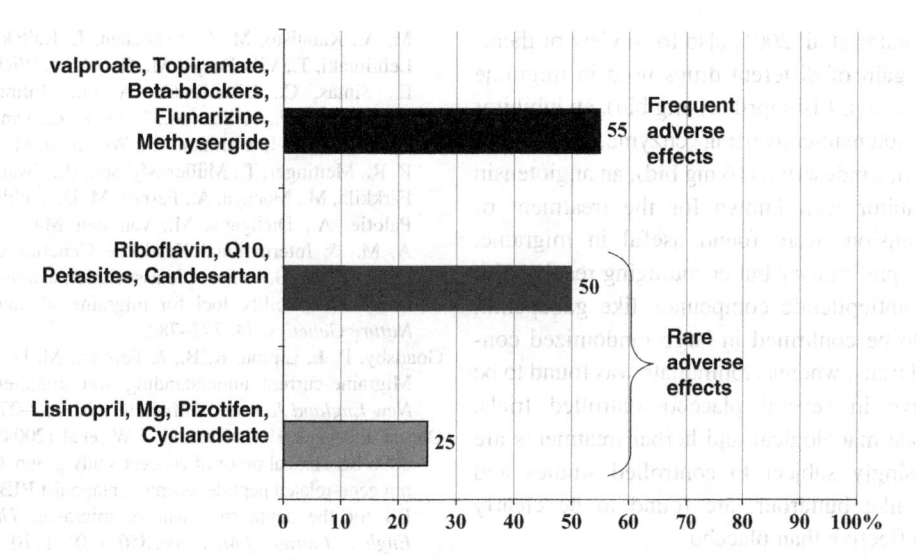

Migraine Without Aura, Fig. 1 Average absolute efficacy rates of the three major groups of preventive antimigraine drugs (estimated from available randomized controlled trials)

(Ferrari et al. 2001) confirms that the subcutaneous auto-injectable form of sumatriptan (6 mg) has the best efficacy, whatever outcome measure is considered. Differences between oral triptans do exist for some outcome measures, but in practice, each patient has to find the triptan that gives him the best satisfaction.

At present the major reason for not considering triptans as first-choice treatments for migraine attacks is their high cost and, in some patients, their cardiovascular side effects. However, stratifying care by prescribing a triptan to the most disabled patients has been proven cost-effective. In severely disabled migraineurs, the efficacy rate of injectable sumatriptan for pain-free at 2 h is twice that of ergot derivatives or NSAIDs taken at high oral doses and of IV acetylsalicylic acid lysinate. The ▶ therapeutic gain tends to be clearly lower for simple analgesics or NSAIDs such as acetaminophen 1,000 mg po, effervescent aspirin 1,000 mg, or ibuprofen 600 mg than for the oral triptans, when severe attacks are considered. Combining analgesics or NSAIDs to an antiemetic and/or to caffeine or administering them as suppositories increases their efficacy. As expected, the triptans have not solved the patients' problems. There is room for more efficient and safer oral acute migraine

treatments, among which CGRP receptor antagonists are at present most promising (Olesen et al. 2004).

Prophylactic Therapy in Migraine
Prophylactic antimigraine treatment has to be individually tailored to each patient taking into account the migraine subtype, the ensuing disability, the patient's history and demands, and the associated disorders. A major drawback of most classical prophylactics (beta-blockers devoid of intrinsic sympathomimetic activity, valproic acid, Ca^{2+} antagonists, antiserotoninergics, tricyclics), which have all on average a 50 % efficacy score, is the occurrence of side effects.

In recent years some new prophylactics with fewer side effects have been studied. High-dose magnesium or cyclandelate is well tolerated but poorly effective in comparison to the classical prophylactics (Fig. 1). A novel preventive treatment for migraine is high-dose (400 mg/day) riboflavin which has an excellent efficacy/side effect ratio and probably acts by improving the mitochondrial phosphorylation potential (see above). Coenzyme Q10 (100 mg tid.), another actor in the mitochondrial respiratory chain, is also effective in migraine prophylaxis (see Sandor et al. 2005, also for review of thera-peutic gain of different drugs used in migraine prophylaxis). Lisinopril (10 mg bid), an inhibitor of angiotensin-converting enzyme, and, even more so, candesartan (16 mg bid), an angiotensin II inhibitor well known for the treatment of hypertension, were found useful in migraine. Recent preliminary but encouraging results with novel antiepileptic compounds like gabapentin need to be confirmed in large randomized con-trolled trials, whereas topiramate was found to be effective in several placebo-controlled trials. Non-pharmacological and herbal treatments are increasingly subject to controlled studies and some, like butterbur, are found to be clearly more effective than placebo.

In chronic migraine where preventive drugs have very poor efficacy, onabotulinum toxin A (Botox) and percutaneous occipital nerve stim-ulation may be beneficial for a proportion of patients.

References

Ambrosini, A., Maertens de Noordhout, A., & Schoenen, J. (2001). Neuromuscular transmission in migraine: a single fiber EMG study in clinical subgroups. *Neurology, 56*, 1038–1043.

Bohotin, V., Fumal, A., Vandenheede, M., et al. (2002). Effects of repetitive transcranial magnetic stimulation on visual evoked potentials in migraine. *Brain, 125*, 912–922.

Bolay, H., Reuter, U., Dunn, A. K., et al. (2002). Intrinsic brain activity triggers trigeminal meningeal afferents in a migraine model. *Nature Medicine, 8*, 136–142.

Burstein, R., Cutrer, M. F., & Yarnitsky. (2000). The development of cutaneous allodynia during a migraine attack clinical evidence for the sequential recruitment of spinal and supraspinal nociceptive neu-rons in migraine. *Brain, 123*, 1703–1709.

Burstein, R., Jakubowski, M., & Collins, B. (2004). Defeating migraine pain with triptans: A race against the development of cutaneous allodynia. *Annals of Neurology, 19–26*.

Ferrarri, M. D., Roon, K. I., Lipton, R. B., et al. (2001). Oral triptans (serotonin 5-HT(1B/1D) agonists) in acute migraine treatment: A meta-analysis of 53 trials. *Lancet, 358*, 1668–1675.

Freilinger, T., Anttila, V., de Vries, B., Malik, R., Kallela, M., Terwindt, G. M., Pozo-Rosich, P., Winsvold, B., Nyholt, D. R., van Oosterhout, W. P., Artto, V., Todt, U., Hämäläinen, E., Fernández-Morales, J., Louter, M. A., Kaunisto, M. A., Schoenen, J., Raitakari, O., Lehtimäki, T., Vila-Pueyo, M., Göbel, H., Wichmann, E., Sintas, C., Uitterlinden, A. G., Hofman, A., Rivadeneira, F., Heinze, A., Tronvik, E., van Duijn, C. M., Kaprio, J., Cormand, B., Wessman, M., Frants, R. R., Meitinger, T., Müller-Myhsok, B., Zwart, J. A., Färkkilä, M., Macaya, A., Ferrari, M. D., Kubisch, C., Palotie, A., Dichgans, M., van den Maagdenberg, A. M., & International Headache Genetics Consor-tium. (2012). Genome-wide association analysis iden-tifies susceptibility loci for migraine without aura. *Nature Genetics, 44*, 777–782.

Goadsby, P. J., Lipton, R. B., & Ferrarri, M. D. (2002). Migraine-current understanding and treatment. *The New England Journal of Medicine, 24*, 257–270.

Olesen, J., Diener, H. C., Husstedt, I. W., et al. (2004). BIBN 4096 BS clinical proof of concept study group. Calcito-nin gene-related peptide receptor antagonist BIBN 4096 BS for the acute treatment of migraine. *The New England Journal of Medicine, 350*, 1104–1110.

Patel, N. V., Bigal, M. E., Kolodner, K. B., et al. (2004). Prevalence and impact of migraine and probable migraine in a health plan. *Neurology, 63*, 1432–1438.

Reyngoudt, H., Paemeleire, K., Descamps, B., De Deene, Y., & Achten, E. (2011). 31P-MRS demonstrates a reduction in high-energy phosphates in the occipital lobe of migraine without aura patients. *Cephalalgia, 31*, 1243–1253.

Sandor, P. S., Mascia, A., Seidel, L., et al. (2001). Sub-clinical cerebellar impairment in the common types of migraine: A 3-dimensional analysis of reaching move-ments. *Annals of Neurology, 49*, 668–672.

Sandor, P. S., Di Clemente, L., Coppola, G., et al. (2005). Effectiveness of coenzyme Q10 in migraine prophylaxis: A randomized controlled trial. *Neurology, 64*, 713–715.

Schoenen, J. (1994). Pathogenesis of migraine: The biobehavioural and hypoxia theories reconciled. *Acta Neurologica Belgica, 94*, 79–86.

Schoenen, J., Ambrosini, A., Sandor, P. S., et al. (2003). Evoked potentials and transcranial magnetic stimula-tion in migraine: Published data and viewpoint on their pathophysiologic significance. *Clinical Neurophysiol-ogy, 114*, 955–972.

Steiner, T. J., Scher, A. I., Stewart, W. F., et al. (2003). The prevalence and disability burden of adult migraine in England and their relationships to age, gender and ethnicity. *Cephalalgia, 23*, 519–527.

Stewart, W. F., Staffa, J., Lipton, R. B., et al. (1997). Familial risk of migraine: A population-based study. *Annals of Neurology, 41*, 166–172.

The International Classification of Headache Disorders (ICHD-II) (2004). *Cephalalgia 24*, 1–160. (2nd Edn).

Weiller, C., May, A., Limmroth, V., et al. (1995). Brain stem activation in spontaneous human migraine attacks. *Nature Medicine, 1*, 658–660.

Welch, K. M., Levine, S. R., D'Andrea, G., et al. (1989). Preliminary observations on brain energy metabolism in migraine studied by in vivo 3^1-phosphorus NMR spectroscopy. *Neurology, 39*, 538–541.

Migraine, Childhood Syndromes

Eric M. Pearlman
Pediatrics, Mercer University School of
Medicine, The Children's Hospital at Memorial
University Medical Center, Savannah, USA

Synonyms

Childhood migraine; Migraine variants of childhood; Pediatric migraine; Periodic disorders of
childhood that are precursors to migraine

Definition

There are many paroxysmal disorders of
childhood that have been associated with
migraine. The link to migraine is very strong for
some of these childhood syndromes and more
tenuous for others. The International Headache
Society (International Headache Society Classification Subcommittee 2004) considers the following to be linked to migraine:

- Abdominal migraine
- Cyclic vomiting syndrome
- Benign paroxysmal vertigo
- Alternating hemiplegia of childhood
- Familial hemiplegic migraine
- Benign paroxysmal torticollis is probably
associated with migraine, based on genetic
links. ► Ophthalmoplegic migraine, originally
considered to be linked to migraine, is probably
not actually a migraine variant. Other
disorders probably represent unusual auras of
migraine rather than separate entities. These
include acute ► confusional migraine, ► basilar-
type migraine, and ► Alice-in-Wonderland
syndrome.

Characteristics

Migraine with or without aura may differ slightly
between children and adults. The diagnostic
criteria proposed by the International
Classification of Headache (IHS) Disorders II
for children less than 15 years requires headaches
of 1–48 h in duration instead of the 4–72 h in
individuals greater than 15 years of age. The
remainder of the criteria is similar to the adult
diagnostic criteria, including at least five attacks
with either photophobia or phonophobia, nausea
or vomiting, and two symptoms out of unilateral
pain, throbbing or pulsatile pain, moderate or
severe pain intensity, or exacerbation by routine
activity (Table 1).

There are features of headache in children that
are not specifically recognized by the IHS classification, but are commonly noted in adolescents
and children (Table 2). The quality of the pain
may be described as constant or squeezing
instead of throbbing. Children also are more
likely to report bilateral, bifrontal, or
a nondescript location rather than unilateral

Migraine, Childhood Syndromes, Table 1 Classification of adolescent and pediatric migraine with and without
aura (The International Classification of Headache Disorders 2004)

Pediatric migraine without aura	Pediatric migraine with aura
A. At least five distinct attacks	E. Fulfills criteria for migraine without aura
B. Headache attack lasting 1–48 h	F. At least three of the following:
C. Headache has at least two of the following:	1. One or more fully reversible aura symptoms indicating focal cortical and/or brainstem dysfunction
1. Bilateral location (frontal/temporal) or unilateral location	2. At least one aura developing gradually over more than 4 min or two or more symptoms occurring in succession
2. Pulsating quality	3. No aura lasting more than 60 min
3. Moderate to severe intensity	4. Headache follows in less than 60 min
4. Aggravation by routine physical activity	
D. During headache, at least one of the following:	
1. Nausea and/or vomiting	
2. Photophobia and/or phonophobia	

Migraine, Childhood Syndromes, Table 2 Differentiating features in childhood vs. adult migraine

Duration	1–48 h	4–72 h
Quality	May or not be described as throbbing	Throbbing/pulsatile
Location	Unilateral, bilateral, whole head	Unilateral; bilateral less likely
Aura	May overlap with head pain	Resolves prior to head pain
Resolution	Resolves with rest or brief sleep	Resolves with longer sleep
Associated features	Nausea/vomiting, photophobia or phonophobia – may be less prominent	Nausea/vomiting, photophobia, phonophobia – necessary for diagnosis

pain. Adults who have migraine with aura have onset and resolution of their aura before the onset of head pain. When aura occurs in children, it may also overlap with the actual headache in onset. Aura should resolve before the headache phase ends. The onset of the head pain may be quite dramatic in children, with maximal pain achieved within 15 min. Pain also may resolve fairly quickly, that is, in 2–4 h, and may require just a short period of sleep to achieve resolution.

Abdominal Migraine

Abdominal migraine is an idiopathic disorder characterized by discreet episodes of abdominal pain, is poorly localized, is moderate to severe in intensity, and lasts from 1 to 72 h. These episodes are often associated with anorexia, nausea, vomiting, and pallor. Sometimes, headache, photophobia, and phonophobia accompany migraine episodes. Eight percent of school children between 5 and 15 years reported recurrent abdominal pain, and 4.1 % reported episodes that fulfilled the criteria for the diagnosis of abdominal migraine. Abdominal migraine affects both boys and girls equally with a peak age of onset at 10 years (Abu-Arafeh and Russell 1995). Affected children are otherwise normal between attacks. Often, children undergo extensive gastrointestinal evaluations before the diagnosis of abdominal migraine is considered. Individual

attacks will ultimately resolve spontaneously and the syndrome can last from months to years before abating. In these children, there is a strong family history of migraine, and individuals who suffer from abdominal migraine are more likely to develop migraine as adults than the general population.

Cyclic Vomiting Syndrome

Cyclic vomiting syndrome consists of stereotypical attacks of intense nausea and vomiting lasting from 1 h to 5 days. Vomiting can be quite dramatic, occurring at least four times in 1 h. It can lead to significant dehydration and children often require intravenous rehydration. Associated features include lethargy, pallor, headache, photophobia, phonophobia, vertigo, diarrhea, excess salivation, and fever. It is equally distributed between males and females and has a mean age of onset of 5.2 years (Li and Balint 2000). The duration of attacks is usually 2–3 years and is associated with significant morbidity. Children with cyclic vomiting syndrome miss on average 20 days of school each year, and more than 50 % will require hospitalization for rehydration. Almost half of the children with a diagnosis of cyclical vomiting will go on to develop migraine later in life.

Both cyclic vomiting syndrome and abdominal migraine have a periodicity to attacks, so that many families can predict when the next attack might occur. Attacks can be triggered by many of the same triggers as migraine, for example, infection, certain foods, menstruation, sleep changes, physical exertion, and psychological stress.

Benign Paroxysmal Vertigo

Benign paroxysmal vertigo is another paroxysmal disorder linked to migraine. It is characterized by attacks of vertigo, dizziness, or unsteadiness lasting minutes to hours. These attacks occur in toddlers or preschoolers and are associated with nystagmus, irritability, pallor, nausea, and vomiting. Children will describe an unreal sense of movement. Younger children will often refuse to walk, sit down, or cling to a parent until the episode subsides. Attacks are usually brief and do not require any treatment. The key in distinguishing these attacks from seizures or metabolic disturbances is the sparing of consciousness. Attacks often resolve with sleep and can occur in clusters. As with the other paroxysmal disorders, individuals often go on to develop migraine.

Alternating Hemiplegia of Childhood

Alternating hemiplegia of childhood is another rare, perplexing paroxysmal disorder of childhood with a recently identified genetic link to migraine. This syndrome consists of repeated, frequent attacks of hemiparesis, monoparesis, or diparesis usually beginning before 18 months of age. Episodes can last from hours to days and may include dystonia or athetosis along with the weakness, which can be on either side of the body or shift from side to side during an attack. There also can be bulbar dysfunction leading to swallowing and respiratory difficulties, which may result in respiratory failure. All these features disappear during sleep. Attacks may increase in frequency, plateau, and eventually decrease in frequency, often resolving after 5–7 years. The syndrome may occur in the setting of mild developmental delays and can rarely be associated with developmental regression. The exact etiology is unknown but inborn errors of metabolism, mitochondrial dysfunction, and channelopathy have all been suggested. Recently, a genetic link with familial hemiplegic migraine has been identified. In a family with alternating hemiplegia of childhood, affected individuals had a mutation in the ATP1A2 gene, which is an ▶ ATP-dependent Na^+/K^+ pump (Bassi et al. 2004). Interestingly, mutations in this gene are also associated with familial hemiplegic migraine.

Familial Hemiplegic Migraine

Familial hemiplegic migraine is a rare autosomal dominant disorder characterized by an aura that has more stroke-like qualities than typical aura. The aura consists of motor weakness, usually unilateral, along with other neurological changes. These include positive visual changes (e.g., scintillating lights, colors, or lines) or vision loss, sensory changes (e.g., pins and needles sensation

or numbness), or speech difficulties. The aura develops gradually, lasts between 5 min and 24 h, and may or may not be associated with headache, fulfilling the criteria for migraine. Identification of a first- or second-degree relative with similar attacks completes the diagnostic criteria (Thomsen et al. 2002). While there is no diagnostic testing for familial hemiplegic migraine, specific causative gene mutations have been identified. A missense mutation in ▶ P/Q-type calcium channel gene (CACNA1A) on chromosome 19p13 is responsible for the disorder in about 50 % of the families (Kors et al. 2003). Mutations in this gene are also responsible for acetazolamide-responsive episodic ataxia and benign paroxysmal torticollis (Ducros et al. 2001). Other families with familial hemiplegic migraine have been identified with mutations in the ATP1A2 gene, which is an ATP-dependent Na/K pump found on chromosome 1q31. This suggests that there may be multiple mechanisms for the neuronal instability that leads to these unusual auras.

Benign Paroxysmal Torticollis

Benign paroxysmal torticollis is an uncommon disorder occurring in infants. This syndrome consists of recurrent attacks of head tilt, truncal ataxia, and occasionally vomiting. The attacks last from several hours to days. Attacks may be associated with some irritability but no alteration in consciousness. The differential diagnosis for this condition includes congenital torticollis, disorders of the craniocervical junction, intracranial abnormalities, especially in the posterior fossa, and metabolic disorders. A completely normal interictal examination and repeated pattern of attacks is reassuring, but further diagnostic evaluations are certainly appropriate. A strong family history of migraine and tendency to develop migraine later in life suggested a link between benign paroxysmal torticollis and migraine. Recently, a genetic link has been suggested. Giffin and colleagues have found mutations in the CACNA1A gene in four cases of benign paroxysmal torticollis. Mutations in this P/Q-type calcium channel gene have been linked to familial hemiplegic migraine, a rare migraine phenotype (Giffin et al. 2002).

References

Abu-Arafeh, I., & Russell, G. (1995). Prevalence and clinical features of abdominal migraine compared with those of migraine headache. *Archives of Disease in Childhood, 72,* 413–417.

Bassi, M. T., Bresolin, N., Tonelli, A., et al. (2004). A novel mutation in the ATP1A2 gene causes alternating hemiplegia of childhood. *Journal of Medical Genetics, 41,* 621–628.

Ducros, A., Denier, C., Joutel, A., et al. (2001). The clinical spectrum of familial hemiplegic migraine associated with mutations in a neuronal calcium channel. *The New England Journal of Medicine, 345,* 17–24.

Giffin, N. J., Benton, S., & Goadsby, P. J. (2002). Benign paroxysmal torticollis of infancy: Four new cases and linkage to CACNA1A mutation. *Developmental Medicine and Child Neurology, 44,* 490–493.

International Headache Society Classification Subcommittee. (2004). International classification of headache disorders, 2nd edn. *Cephalalgia, 24*(suppl 1):1–160.

Kors, E. E., Haan, J., Giffen, N. J., et al. (2003). Expanding the phenotypic spectrum of the CACNA1A gene T666M mutation: A description of 5 families with hemiplegic migraine. *Archives of Neurology, 60,* 684–688.

Li, B. U., & Balint, J. P. (2000). Cyclic vomiting syndrome: Evolution in our understanding of a brain-gut disorder. *Advances in Pediatrics, 47,* 117–60.

Thomsen, L., Eriksen, M., Roemer, S., et al. (2002). A population-based study of familial hemiplegic migraine suggests revised diagnostic criteria. *Brain, 125,* 1379–1391.

Migraine, Genetics

Anne Ducros
Headache Emergency Department, INSERM, Paris, France

Synonyms

Inherited factors implicated in the mechanisms of migraine

Definition

Migraine is a primary headache disorder. It is a complex disease in which environmental factors interact with genetic factors. The major goal of genetic studies is the identification of susceptibility genes that may give clues to the mechanisms underlying migraine and may help to identify new therapeutic targets.

Characteristics

Migraine is a primary headache disorder responsible for recurrent attacks of disabling pain lasting a few hours to a few days. This condition is characterized by the repetition of such attacks in the absence of any disorder causing secondary headaches. The International Headache Society (IHS) classification distinguishes different varieties of migraine attacks, the most frequent being migraine without aura (MO) and migraine with typical aura (MA). In MO attacks, the headache is the most prominent and disabling symptom. In MA attacks, the headache phase is preceded or accompanied by transient neurological signs including visual, sensory, and language troubles. Some patients have only MO, others have only MA, and some have both types of attacks. In addition, hemiplegic migraine is a rare variety of MA characterized by the presence of a motor deficit during the aura, in addition to one or more of the typical aura symptoms. In familial hemiplegic migraine (FHM), the index case has at least one first- or second-degree relative who also has migraine with aura including motor weakness. In sporadic hemiplegic migraine, there is no such familial history. Migraine is a highly frequent condition affecting about 15 % of the western population. Familial aggregation has long been known and was even classically used as a diagnosis criterion. Genetic epidemiological surveys have demonstrated that this familial aggregation was not observed purely by chance, but was due to the existence of genetic factors.

Twin studies have shown that the "pairwise" concordance rates were significantly higher among monozygous than dizygous twin pairs for MO (Gervil et al. 1999) and for MA (Ulrich et al. 1999), demonstrating the existence of genetic factors in migraine that interact with environmental factors to produce the phenotype. Family studies provided further evidence in favor of the existence of genetic factors in MO and MA, those factors being more important in MA than in MO (Russell and Olesen 1995). In addition, segregation analysis showed that both MO and MA have a non-Mendelian polygenic mode of inheritance (Russell et al. 1995). The possible number of genetic susceptibility loci is still unknown.

FHM is the only variety of migraine in which a monogenic Mendelian mode of inheritance has been clearly established. Two clinical forms of FHM have been described: pure FHM (80 % of the families) and FHM with permanent cerebellar symptoms in which some affected subjects have nystagmus and/or ataxia. In addition to usual attacks, patients may have severe attacks with confusion, coma, fever, and prolonged hemiplegia. Genetic tools are more powerful to identify the genes responsible for monogenic conditions than genes responsible for polygenic conditions. Thus, the only known migraine genes have been identified in FHM. FHM is genetically heterogeneous, with at least three responsible genes of which two have been identified. The first gene, CACNA1A, located on chromosome 19p13, was identified in 1996 (Ophoff et al. 1996). CACNA1A encodes the pore-forming α1A-subunit of voltage-gated neuronal $Ca_v2.1$ (P/Q type) Ca^{2+} channels. CACNA1A is implicated in about 50 % of unselected FHM families and in the vast majority of families with permanent cerebellar symptoms. The second gene, ATP1A2, located on chromosome 1q21-q23, was identified in 2003 (De Fusco et al. 2003). ATP1A2 is implicated in about 20–30 % of unselected FHM families and encodes the catalytic α2-subunit of a transmembrane NA^+/K^+ ATPase. Both known FHM genes encode proteins regulating ion translocation; FHM is thus a channelopathy.

$Ca_v2.1$ channels are found exclusively in neurons including central and peripheral neurons at the neuromuscular junction. They play a major role in neurotransmitter release (mainly glutamate) and in the control of neuronal excitability. CACNA1A is a large gene containing 47 exons.

Thus far, a total of 17 CACNA1A mutations have been identified in 41 FHM1 families (33 affected by FHM and cerebellar symptoms and 8 by pure FHM) and in 4 sporadic cases affected by HM and cerebellar symptoms. Five of these mutations are recurrent, i.e., they have been detected in two or more unrelated families. The most frequent mutation T666M has been detected in 19 families and 2 sporadic cases. All mutations are missense mutations, changing only one of the 2,550 amino acids of the predicted protein. They are located in important functional domains of the subunit, near the ionic pore or within the voltage sensor. Mutations causing FHM1 with cerebellar symptoms are distinct from those causing pure FHM1, and all recurrent mutations are causing FHM1 with cerebellar symptoms. Mutations have been detected in sporadic cases, including a de novo mutation, demonstrating that sporadic HM is due to CACNA1A mutations in at least a part of the cases.

Different kinds of CACNA1A mutations have been identified in two other autosomal dominant neurological disorders. Episodic ataxia type 2 is a paroxysmal neurological condition producing recurrent attacks of major cerebellar ataxia, which is due to CACNA1A mutations (Ophoff et al. 1996). Spinocerebellar ataxia type 6 (SCA6) is a progressive cerebellar disease characterized by an adult onset ataxia, which is due to small expansions of the CAG repeat contained within exon 47 of CACNA1A.

Electrophysiological studies of CACNA1A mutations causing FHM1 have shown that all mutations analyzed so far modify the density and the gating properties of $Ca_v2.1$ currents (Pietrobon and Striessnig 2003). Mutated single $Ca_v2.1$ channels display lower activation thresholds and increased opening probabilities. FHM1 mutations are thus gain-of-function mutations. On the contrary, EA2 mutations have been shown to be loss-of-function mutations, the mutated allele generating no current when expressed in heterologous cells.

Knockout mice that do not have the CACNA1A gene are born with a severe ataxia and die within a few days. Mice carrying different spontaneous mutations within CACNA1A display distinct phenotypes called tottering, leaner, or rocker, characterized by various paroxysmal manifestations (absence epilepsy, motor attacks) always associated with a permanent cerebellar ataxia of variable severity. Leaner mice have lower levels of voltage-dependant glutamate release measurable by cerebral microdialysis. They also have an elevated threshold for initiating cortical spreading depression (CSD) and a slower velocity of CSD propagation. Knockin mice carrying the human pure FHM1 mutation R192Q display multiple gain-of-function phenotypes: increased $Ca_v2.1$ current density in cerebellar neurons, enhanced neurotransmission at the neuromuscular junction, and, in the intact animal, a reduced threshold and increased velocity of cortical spreading depression (van den Maagdenberg et al. 2004). This mouse is the first animal model for FHM. These various abnormalities strongly suggest that FHM1 mutations in the $Ca_v2.1$ channel modify the cortical excitability and increase the susceptibility to CSD, which is the mechanism underlying the aura.

ATP1A2 encodes the catalytic α2-subunit of a transmembrane NA^+/K^+ ATPase. This protein hydrolyzes ATP to extrude Na^+ ions and pump K^+ ions into the cell. This active pumping maintains the Na^+ transmembrane gradient that is essential for the transport of amino acids, neurotransmitters (including glutamate), and Ca^{2+}. In neonates, the α2-subunit is mainly expressed in neurons. In adults, it is mainly expressed in astrocytes. ATP1A2 contains 23 exons. So far, 14 mutations have been identified in 14 families and one sporadic case. These mutations are missense mutations responsible for the substitution of one of the 1,020 amino acids of the protein. Only one mutation (R763H) was found in two unrelated families. In the family with the R689Q mutation, FHM cosegregates with benign familial infantile convulsions. In another family with the T378N mutation, the phenotypic spectrum includes features characteristic of FHM and of alternating hemiplegia. Electrophysiological studies initially suggested that ATP1A2 mutations are loss-of-function mutations leading to haploinsufficiency. Further studies suggest that

the mutations are gain-of-function mutations leading to an abnormal function of the pump.

FHM is characterized by an important clinical variability. The age of onset, the frequency and duration of attacks, the aura features, and the headache characteristics may vary from one patient to another, even among affected members from a given family who are carrying the same mutation in the same gene. This variability suggests complex interactions between the consequences of the FHM causing mutation and environmental factors or modifying genetic factors. However, several studies have shown that the various genotypes play a role in producing this clinical variability. FHM1, due to CACNA1A mutations, is characterized by a higher penetrance of FHM and of permanent cerebellar symptoms, the vast majority of families with permanent cerebellar symptoms being linked to CACNA1A (Ducros et al. 2001). FHM2, due to ATP1A2 mutations, is characterized by a lower penetrance of FHM. Cerebellar symptoms are rarely part of the clinical spectrum of FHM2 (2 families published). In addition, striking correlations between genotype and phenotype have been shown in patients with CACNA1A mutations, including the fact that the most frequent FHM1 causing mutation, T666M, had the highest penetrance of hemiplegic migraine, severe attacks with coma, and nystagmus (Ducros et al. 2001). The existence of different CACNA1A mutations partly accounts for the clinical variability.

The implication of CACNA1A in the more frequent varieties of migraine has been analyzed by the mean of linkage and association studies with contradictory results, three studies concluding in favor and five against. None of the positive studies provided a direct analysis of the CACNA1A gene in order to detect pathogenic mutations. CACNA1A is thus probably not a susceptibility gene for the more common varieties of migraine. Moreover, a subsequent study showed that the 19p13 region does indeed contain a susceptibility locus for MA that is distinct from CACNA1A. The same group showed an association between migraine and five polymorphisms within the insulin receptor (INSR) gene located in 19p13.3/2 (McCarthy et al. 2001). These five polymorphisms had no effect on INSR transcription, translation, and protein expression and INSR-mediated functions. The nature of the implication of the INSR gene in migraine remains to be understood.

With regard to the FHM2 gene, a linkage analysis conducted in a single family suggested the presence of a possible migraine susceptibility locus in 1q31, distinct from the ATP1A2 locus in 1q21-q23. An ATP1A2 mutation (R548H) was found in a small family in which the two affected subjects had migraine with basilar, non-hemiplegic, aura. This mutation was not found in 100 healthy controls or in 77 migrainous controls. Actual data do not permit a conclusion regarding the implication of ATP1A2 in MO/MA.

Three linkage analyses, each conducted in a large panel of MA families, have permitted the mapping of three susceptibility loci for MA: the first on 4q24 (in 50 Finnish families) (Wessman et al. 2002), the second on 11q24 (in 43 Canadian families) (Cader et al. 2003), and the last on 15q11-q13 (in 10 Italian families) (Russo et al. 2004). An important linkage analysis in a group of 289 Icelandic patients with MO identified a susceptibility locus on 4q21 (Bjornsson et al. 2003). This genetic interval overlapped the genetic region identified in the Finnish families on 4q24, suggesting an implication of this chromosome 4q region in both MO and MA. Finally, several linkage analyses, each conducted in a single family, suggested the existence of other migraine susceptibility loci on 1q21-q23, 6p12.2-p21.1, 14q21.2-q22.3, and Xq24-q28 (Estevez and Gardner 2004). None of the results have been replicated.

Finally, numerous association studies have been performed in migraine. A wide range of candidate polymorphisms have been tested in migraine. Association studies compare the frequency of alleles of a polymorphic genetic marker between cases and controls. In the absence of methodological shortcomings, a significant difference means that the polymorphism is located within the susceptibility gene or is in linkage disequilibrium with the susceptibility gene. Association studies do not provide any demonstration of the implication of the tested gene or of an abnormal function of the tested gene in the pathophysiology of the studied disease. Positive associations have been found between migraine and polymorphisms within the dopamine receptor D2 (DRD2) gene, the human serotonin transporter gene, the catechol-o-methyltransferase gene, the endothelin type A gene, the dopamine beta-hydroxylase (DBH) gene, and the 5,10-methylenetetrahydrofolate reductase (MTHFR). These various candidate polymorphisms have been chosen based on the hypothesis that the genes containing them encode proteins that are suspected to be involved in migraine. However, additional studies are needed to determine if the alleles associated with migraine have any biological effect.

References

Bjornsson, A., Gudmundsson, G., Gudfinnsson, E., et al. (2003). Localization of a gene for migraine without aura to chromosome 4q21. *American Journal of Human Genetics, 73*, 986–993.

Cader, Z. M., Noble-Topham, S., Dyment, D. A., et al. (2003). Significant linkage to migraine with aura on chromosome 11q24. *Human Molecular Genetics, 12*, 2511–2517.

De Fusco, M., Marconi, R., Silvestri, L., et al. (2003). Haploinsufficiency of ATP1A2 encoding the Na^+/K^+ pump alpha2 subunit associated with familial hemiplegic migraine type 2. *Nature Genetics, 33*, 192–196.

Ducros, A., Denier, C., Joutel, A., et al. (2001). The clinical spectrum of familial hemiplegic migraine associated with mutations in a neuronal calcium channel. *The New England Journal of Medicine, 345*, 17–24.

Estevez, M., & Gardner, K. L. (2004). Update on the genetics of migraine. *Human Genetics, 114*, 225–235.

Gervil, M., Ulrich, V., Kaprio, J., Olesen, J., & Russell, M. B. (1999). The relative role of genetic and environmental factors in migraine without aura. *Neurology, 53*, 995–999.

McCarthy, L. C., Hosford, D. A., Riley, J. H., et al. (2001). Single-nucleotide polymorphism alleles in the insulin receptor gene are associated with typical migraine. *Genomics, 78*, 135–149.

Ophoff, R. A., Terwindt, G. M., Vergouwe, M. N., et al. (1996). Familial hemiplegic migraine and episodic ataxia type-2 are caused by mutations in the Ca^{2+} channel gene CACNL1A4. *Cell, 87*, 543–552.

Pietrobon, D., & Striessnig, J. (2003). Neurobiology of migraine. *Nature Reviews Neuroscience, 4*, 386–398.

Russell, M. B., & Olesen, J. (1995). Increased familial risk and evidence of genetic factor in migraine. *BMJ, 311*, 541–544.

Russell, M. B., Iselius, L., & Olesen, J. (1995). Inheritance of migraine investigated by complex segregation analysis. *Human Genetics, 96*, 726–730.

Russo, L., Mariotti, P., Sangiorgi, E., et al. (2004). A new susceptibility locus for migraine with aura in the 15q11-q13 genomic region containing three GABA-A receptor genes. *American Journal of Human Genetics, 76*, 2.

Ulrich, V., Gervil, M., Kyvik, K. O., et al. (1999). Evidence of a genetic factor in migraine with aura: A population-based Danish Twin study. *Annals of Neurology, 45*, 242–246.

van den Maagdenberg, A. M., Pietrobon, D., Pizzorusso, T., et al. (2004). A cacna1a knockin migraine mouse model with increased susceptibility to cortical spreading depression. *Neuron, 41*, 701–710.

Wessman, M., Kallela, M., Kaunisto, M. A., et al. (2002). A susceptibility locus for migraine with aura, on chromosome 4q24. *American Journal of Human Genetics, 70*, 652–662.

Migraine, Pathophysiology

Peter Goadsby
Headache Center, Department of Neurology,
University of California, San Francisco,
CA, USA

Introduction

An understanding of the pathophysiology of migraine should be based upon the anatomy and physiology of the pain-producing structures of the cranium integrated with knowledge of central nervous system modulation of these pathways. Headache in general, and in particular migraine (Goadsby et al. 2002) and cluster headache (Goadsby 2002b), is better understood now than has been the case for the last four millennia (Lance and Goadsby 2005). This entry will set out the current understanding of migraine.

Characteristics

Migraine: Explaining the Clinical Features

Migraine is in essence a familial episodic disorder whose key marker is headache with certain associated features (Table 1). It is these features that give clues to its migraine pathophysiology and will ultimately provide insights leading to new treatments.

The essential elements to be considered are:

- Genetics of migraine
- Physiological basis for the ▶ aura
- Anatomy of head pain, particularly that of the ▶ trigeminovascular system
- Physiology and pharmacology of activation of the peripheral branches of ophthalmic branch of the trigeminal nerve
- Physiology and pharmacology of the trigeminal nucleus, in particular its caudal-most part, the ▶ trigeminocervical complex
- ▶ Brainstem and diencephalic modulatory systems that influence trigeminal pain transmission and other sensory modality processing

Migraine involves a form of sensory processing disturbance with wide ramifications within the central nervous system; while pain pathways will be used as an example, it is useful to remember that migraine is not simply a pain problem.

Genetics of Migraine

One of the most important aspects of the pathophysiology of migraine is the inherited nature of the disorder. It is clear from clinical practice that many patients have first-degree relatives who also suffer from migraine (Lance and Goadsby 2005; Silberstein et al. 2002). Transmission of migraine from parents to children has been reported as early as the seventeenth century (Willis 1682), and numerous published studies have reported a positive family history (Russell 1997).

Genetic Epidemiology

Studies of twin pairs are the classical method to investigate the relative importance of genetic and environmental factors. A Danish study included 1,013 monozygotic and 1,667 dizygotic twin pairs of the same gender, obtained from a population-based twin register (Ulrich et al. 1999). The pairwise concordance rate was significantly higher among monozygotic than

Migraine, Pathophysiology, Table 1 International Headache Society defined features of migraine (Headache Classification Committee of The International Headache Society 2004). Repeated episodic headache (4–72 h) with the following features:

Any two of	Any one of
Unilateral	Nausea/vomiting
Throbbing	Photophobia and phonophobia
Worsened by movement	
Moderate or severe	

dizygotic twin pairs ($P < 0.05$). Several studies have attempted to analyze the possible mode of inheritance in migraine families, and conflicting results have been obtained (Lalouel and Morton 1981; Mochi et al. 1993; Russell et al. 1995). Both twin studies and population-based epidemiological surveys strongly suggest that migraine without aura is a multifactorial disorder, caused by a combination of genetic and environmental factors.

Familial Hemiplegic Migraine (FHM)
In approximately 50 % of the reported families, FHM has been assigned to chromosome 19p13 (Joutel et al. 1994; Ophoff et al. 1994). Few clinical differences have been found between chromosome 19 linked and unlinked FHM families. Indeed, the clinical phenotype does not associate particularly with the known mutations (Ducros et al. 2001). The most striking exception is cerebellar ataxia, which occurs in approximately 50 % of the chromosome 19 linked, but in none of the unlinked families (Haan et al. 1994; Joutel et al. 1993, 1994; Ophoff et al. 1994; Teh et al. 1995). Another less striking difference includes the fact that patients from chromosome 19 linked families are more likely to have attacks that can be triggered by minor head trauma or that are associated with coma (Terwindt et al. 1996).

The biological basis for the linkage to chromosome 19 is mutations (Ophoff et al. 1996) involving the $Ca_v2.1$ (P/Q) type ▶ voltage-gated calcium channel (Ertel et al. 2000) *CACNA1A* gene. Now known as FHM-I, this mutation is responsible for about 50 % of identified families.

Mutations in the *ATP1A2* gene (De Fusco et al. 2003; Marconi et al. 2003) have been identified to be responsible for about 20 % of FHM families. Interestingly, the phenotype of some FHM-II families involves epilepsy (Jurkat-Rott et al. 2004; Vanmolkot et al. 2003), while it has also been suggested that alternating hemiplegia of childhood can be due to *ATP1A2* mutations (Swoboda et al. 2004). The latter cases are most unconvincing for migraine.

Taken together, the known mutations suggest that migraine, or at least the neurological manifestations currently called the aura, is caused by a ▶ channelopathy (Goadsby and Ferrari 2001). Linking the channel disturbance for the first time to the aura process has demonstrated that human mutations expressed in a knock-in mouse produce a reduced threshold for ▶ cortical spreading depression (van den Maagdenberg et al. 2004), which has some profound implications for understanding that process (Goadsby 2004).

Migraine Aura
Migraine aura is defined as a focal neurological disturbance manifest as visual, sensory, or motor symptoms (Headache Classification Committee of The International Headache Society 2004). It is seen in about 30 % of patients (Rasmussen and Olesen 1992) and it is clearly neurally driven (Cutrer et al. 1998; Olesen et al. 1990). The case for the aura being the human equivalent of the cortical spreading depression (CSD) of Leão (1944a, b) has been well made (Lauritzen 1994).

In humans, visual aura has been described as affecting the visual field, suggesting the visual cortex, and it starts at the center of the visual field and propagates to the periphery at a speed of 3 mm/min (Lashley 1941). This is very similar to spreading depression described in rabbits (Leao 1944b). Blood flow studies in patients have also shown that a focal hyperemia tends to precede the spreading oligemia (Olesen et al. 1981), and again this is similar to what would be expected with spreading depression. After this passage of oligemia, the cerebrovascular response to hypercapnia in patients is blunted, while autoregulation remains intact (Harer and Kummer 1991; Lauritzen et al. 1983; Sakai and

Meyer 1979). Again this pattern is repeated with experimental spreading depression (Kaube and Goadsby 1994; Kaube et al. 1999; Lambert et al. 1999). Human observations have rendered the arguments reasonably sound that human aura has cortical spreading depression as its equivalent in animals (Hadjikhani et al. 2001). An area of controversy surrounds whether aura in fact triggers the rest of the attack and is indeed painful (Moskowitz et al. 2004). Based on the available experimental and clinical data, this author is not at all convinced that aura is painful (Goadsby 2001), but this does not diminish its interest or the importance of understanding it. Indeed therapeutic developments may shed even further light on these relationships.

Tonabersat is a CSD inhibitor that has entered clinical trials in migraine. Tonabersat (SB-220453) inhibits CSD; CSD induced nitric oxide (NO) release and cerebral vasodilation (Read et al. 1999; Smith et al. 2000). Tonabersat does not constrict isolated human blood vessels (MaassenVanDenBrink et al. 2000), but does inhibit trigeminally induced craniovascular effects (Parsons et al. 2001). Remarkably, topiramate, a proven preventive agent in migraine (Brandes et al. 2004; Diener et al. 2004; Silberstein et al. 2004), also inhibits CSD in cat and rat (Akerman and Goadsby 2004). Tonabersat is inactive in the human NO model of migraine (Tvedskov et al. 2004a) as is propranolol (Tvedskov et al. 2004c), although valproate showed some activity in that model (Tvedskov et al. 2004b). Topiramate inhibits trigeminal neurons activated by nociceptive intracranial afferents (Storer and Goadsby 2004), but not by a mechanism local to the trigeminocervical complex (Storer and Goadsby 2013), and thus CSD inhibition may be a model system to contribute to the development of preventive medicines.

Headache Anatomy
The Trigeminal Innervation of Pain-Producing Intracranial Structures
Surrounding the large cerebral vessels, pial vessels, large venous sinuses, and dura mater is a plexus of largely unmyelinated fibers that arise from the ophthalmic division of the trigeminal ganglion (Liu-Chen et al. 1984) and in the posterior fossa from the upper cervical dorsal roots (Arbab et al. 1986; Fig. 1). Trigeminal fibers innervating cerebral vessels arise from neurons in the trigeminal ganglion that contains substance P and ► calcitonin gene-related peptide (CGRP) (Uddman et al. 1985), both of which can be released when the trigeminal ganglion is stimulated in either humans or the cat (Goadsby et al. 1988). Stimulation of the cranial vessels, such as the superior sagittal sinus (SSS), is certainly painful in humans (Feindel et al. 1960; Wolff 1948). Human dural nerves that innervate the cranial vessels largely consist of small diameter myelinated and unmyelinated fibers (Penfield and McNaughton 1940) that almost certainly subserve a nociceptive function.

Headache Physiology: Peripheral Connections
Plasma Protein Extravasation
Moskowitz (1990) has provided a series of experiments to suggest that the pain of migraine may be a form of sterile neurogenic inflammation. Although this seems clinically implausible, the model system has been helpful in understanding some aspects of trigeminovascular physiology. Neurogenic plasma extravasation can be seen during electrical stimulation of the trigeminal ganglion in the rat (Markowitz et al. 1987). Plasma extravasation can be blocked by ergot alkaloids, indomethacin, acetylsalicylic acid, and the serotonin-5HT$_{1B/1D}$ receptor agonist, sumatriptan (Moskowitz and Cutrer 1993). The pharmacology of abortive anti-migraine drugs has been reviewed in detail (Cutrer et al. 1997). In addition there are structural changes in the dura mater that are observed after trigeminal ganglion stimulation. These include mast cell degranulation and changes in post-capillary venules including platelet aggregation (Dimitriadou et al. 1991, 1992). While it is generally accepted that such changes and particularly the initiation of a sterile inflammatory response would cause pain (Burstein et al. 1998; Strassman et al. 1996), it is not clear whether this is sufficient of itself or requires other stimulators

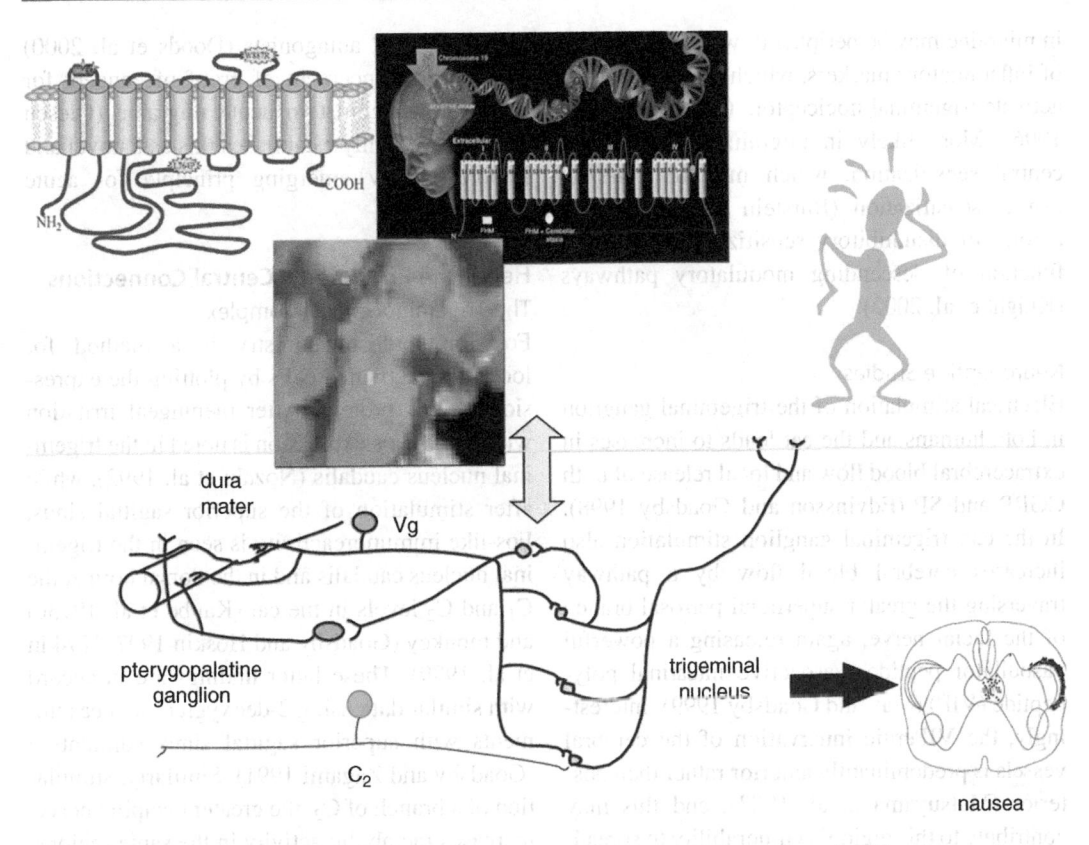

Migraine, Pathophysiology, Fig. 1 Migraine is best considered as an inherited, sub-cortical disturbance of sensory regulation. The elements of this are shown. Vg, trigeminal ganglion

or promoters. Preclinical studies suggest that cortical spreading depression may be a sufficient stimulus to activate trigeminal neurons (Bolay et al. 2002), although this has been a controversial area (Ebersberger et al. 2001; Goadsby 2001; Ingvardsen et al. 1997, 1998; Moskowitz et al. 1993).

Although plasma extravasation in the retina, which is blocked by sumatriptan, can be seen after trigeminal ganglion stimulation in experimental animals, no changes are seen with retinal angiography during acute attacks of migraine or cluster headache (May et al. 1998b). Clearly, blockade of neurogenic plasma protein extravasation is not completely predictive of anti-migraine efficacy in humans, as evidenced by the failure in clinical trials of substance P, neurokinin-1 antagonists (Connor et al. 1998; Diener and The RPR100893 Study Group 2003; Goldstein et al. 1997; Norman et al. 1998),

specific PPE blockers, CP122,288 (Roon et al. 1997) and 4991w93 (Earl et al. 1999), an endothelin antagonist (May et al. 1996), and a neurosteroid (Data et al. 1998).

Sensitization and Migraine

While it is highly doubtful that there is a significant sterile inflammatory response in the dura mater during migraine, it is clear that some form of sensitization takes place during migraine, since allodynia is common. About two thirds of patients complain of pain from non-noxious stimuli, allodynia (Burstein et al. 2000a, b; Selby and Lance 1960). A particularly interesting aspect is the demonstration of allodynia in the upper limbs ipsilateral and contralateral to the pain. This finding is consistent with at least third-order neuronal sensitization, such as sensitization of thalamic neurons, and firmly places the pathophysiology within the central nervous system. Sensitization in migraine may be peripheral, with local release of inflammatory markers, which would certainly activate trigeminal nociceptors (Strassman et al. 1996). More likely in migraine is a form of central sensitization, which may be classical central sensitization (Burstein et al. 1998) or a form of disinhibitory sensitization with dysfunction of descending modulatory pathways (Knight et al. 2002).

Neuropeptide Studies

Electrical stimulation of the trigeminal ganglion in both humans and the cat leads to increases in extracerebral blood flow and local release of both CGRP and SP (Edvinsson and Goadsby 1998). In the cat, trigeminal ganglion stimulation also increases cerebral blood flow by a pathway traversing the greater superficial petrosal branch of the facial nerve, again releasing a powerful vasodilator peptide, vasoactive intestinal polypeptide (VIP) (May and Goadsby 1999). Interestingly, the VIPergic innervation of the cerebral vessels is predominantly anterior rather than posterior (Matsuyama et al. 1983), and this may contribute to this region's vulnerability to spreading depression, explaining why the aura is so very often seen to commence posteriorly. Stimulation of the more specifically vascular pain-producing superior sagittal sinus increases cerebral blood flow and jugular vein CGRP levels. Human evidence that CGRP is elevated in the headache phase of migraine (Gallai et al. 1995; Goadsby et al. 1990), cluster headache (Fanciullacci et al. 1995; Goadsby and Edvinsson 1994), and chronic paroxysmal hemicrania (Goadsby and Edvinsson 1996) supports the view that the trigeminovascular system may be activated in a protective role in these conditions. It is of interest in this regard that compounds which have not shown activity in human migraine, notably the conformationally restricted analogue of sumatriptan, CP122,288 (Knight et al. 1999), and the conformationally restricted analogue of zolmitriptan, 4991w93 (Knight et al. 2001), were both ineffective inhibitors of CGRP release after superior sagittal sinus stimulation in the cat. The recent development of non-peptide, highly

specific CGRP antagonists (Doods et al. 2000) and the announcement of proof of concept for a CGRP antagonist in acute migraine (Olesen et al. 2004) firmly establishes this as a novel and important new emerging principle for acute migraine.

Headache Physiology: Central Connections

The Trigeminocervical Complex

Fos immunohistochemistry is a method for looking at activated cells by plotting the expression of Fos protein. After meningeal irritation with blood, Fos expression is noted in the trigeminal nucleus caudalis (Nozaki et al. 1992), while after stimulation of the superior sagittal sinus, Fos-like immunoreactivity is seen in the trigeminal nucleus caudalis and in the dorsal horn at the C_1 and C_2 levels in the cat (Kaube et al. 1993c) and monkey (Goadsby and Hoskin 1997; Hoskin et al. 1999). These latter findings are in accord with similar data using 2-deoxyglucose measurements with superior sagittal sinus stimulation (Goadsby and Zagami 1991). Similarly, stimulation of a branch of C_2, the greater occipital nerve, increases metabolic activity in the same regions, i.e., trigeminal nucleus caudalis and $C_{1/2}$ dorsal horn (Goadsby et al. 1997). In experimental animals, one can record directly from trigeminal neurons with both supratentorial trigeminal input and input from the greater occipital nerve, a branch of the C_2 dorsal root (Bartsch and Goadsby 2002). Stimulation of the greater occipital nerve for 5 min results in substantial increases in responses to supratentorial dural stimulation, which can last for over an hour (Bartsch and Goadsby 2002). Conversely, stimulation of the middle meningeal artery dura mater with the C-fiber irritant mustard oil sensitizes responses to occipital muscle stimulation (Bartsch and Goadsby 2003). Taken together these data suggest convergence of cervical and ophthalmic inputs at the level of the second-order neuron. Moreover, stimulation of a lateralized structure, the middle meningeal artery, produces Fos expression bilaterally in both cat and monkey brains (Hoskin et al. 1999), a finding that is consistent with the fact that up to one third of patients complain of bilateral pain. This group of neurons from the superficial laminae of trigeminal nucleus caudalis and $C_{1/2}$ dorsal horns should be regarded functionally as the trigeminocervical complex.

These data demonstrate that trigemino-vascular nociceptive information comes by way of the most caudal cells. This concept provides an anatomical explanation for the referral of pain to the back of the head in migraine. Moreover, experimental pharmacological evidence suggests that abortive anti-migraine drugs, such as ergots (Hoskin et al. 1996), acetylsalicylic acid (Kaube et al. 1993b), sumatriptan after blood-brain barrier disruption (Kaube et al. 1993a), eletriptan (Goadsby and Hoskin 1999; Lambert et al. 2002), naratriptan (Cumberbatch et al. 1998; Goadsby and Knight 1997), rizatriptan (Cumberbatch et al. 1997), and zolmitriptan (Goadsby and Hoskin 1996), can have actions at these second-order neurons that reduce cell activity and suggest a further possible site for therapeutic intervention in migraine. This action can be dissected out to involve each of the 5-HT_{1B}, 5-HT_{1D}, and 5-HT_{1F} receptor subtypes (Goadsby and Classey 2003). Interestingly, ▶ triptans also influence the CGRP promoter (Durham et al. 1997) and regulate CGRP secretion from neurons in culture (Durham and Russo 1999). Furthermore, the demonstration that some part of this action is postsynaptic with either 5-HT_{1B} or 5-HT_{1D} receptors located non-presynaptically (Goadsby et al. 2001; Maneesi et al. 2004) offers a prospect of highly anatomically localized treatment options.

Higher Order Processing

Following transmission in the caudal brainstem and high cervical spinal cord information is relayed rostrally.

Thalamus

Processing of vascular nociceptive signals in the thalamus occurs in the ventroposteromedial (VPM) thalamus, medial nucleus of the posterior complex, and the intralaminar thalamus (Zagami and Goadsby 1991). Zagami and Lambert (1991)

have shown by application of capsaicin to the superior sagittal sinus that trigeminal projections with a high degree of nociceptive input are processed in neurons particularly in the ventroposteromedial thalamus and in its ventral periphery. These neurons in the VPM can be modulated by activation of GABA_A inhibitory receptors (Shields et al. 2003) and, perhaps of more direct clinical relevance, by propranolol through a β_1-adrenoceptor mechanism (Shields and Goadsby 2005). Remarkably, triptans can also inhibit VPM neurons locally through $5\text{-HT}_{1B/1D}$ mechanisms, as demonstrated by microiontophoretic application (Shields and Goadsby 2004), suggesting a hitherto unconsidered locus of action for triptans in acute migraine. Human imaging studies have confirmed activation of the thalamus contralateral to pain in acute migraine (Afridi et al. 2005a; Bahra et al. 2001), cluster headache (May et al. 1998a), and SUNCT (short-lasting unilateral neuralgiform headache with conjunctival injection and tearing) (May et al. 1999).

Activation of Modulatory Regions

Stimulation of nociceptive afferents by stimulation of the superior sagittal sinus in the cat activates neurons in the ventrolateral periaqueductal gray matter (PAG) (Hoskin et al. 2001). PAG activation in turn feeds back to the trigemino-cervical complex with an inhibitory influence (Knight and Goadsby 2001). PAG is clearly included in the area of activation seen in PET studies in migraineurs (Weiller et al. 1995). This typical negative feedback system will be further considered below as a possible mechanism for the symptomatic manifestations of migraine.

Another potential modulatory region activated by stimulation of nociceptive trigeminovascular input is the posterior hypothalamic gray (Benjamin et al. 2004). This area is crucially involved in several primary headaches, notably cluster headache (Goadsby 2002b), short-lasting unilateral neuralgiform headache attacks with conjunctival injection and tearing (SUNCT) (May et al. 1999), and hemicrania continua (Matharu et al. 2004b).

Moreover, the clinical features of the premonitory phase (Giffin et al. 2003) and other features of the disorder (Bes et al. 1982; Peroutka 1997) suggest dopamine neuron involvement. Orexinergic neurons in the posterior hypothalamus can be both pro- and antinociceptive (Bartsch et al. 2004), offering a further possible region whose dysfunction might involve the perception of head pain.

Acknowledgment The work of the author has been supported by the Wellcome Trust.

References

Afridi, S., Giffin, N.J., Kaube, H., et al. (2005a). A PET study in spontaneous migraine. *Archives Neurology, 62*, 1270–1275.

Afridi, S., Matharu, M. S., Lee, L., et al. (2005b). A PET study exploring the laterality of brainstem activation in migraine using glyceryl trinitrate. *Brain, 128*, 932–939.

Akerman, S., & Goadsby, P. J. (2004). Topiramate inhibits cortical spreading depression in rat and cat: A possible contribution to its preventive effect in migraine. *Cephalalgia, 24*, 783–784.

Arbab, M. A.-R., Wiklund, L., & Svendgaard, N. A. (1986). Origin and distribution of cerebral vascular innervation from superior cervical, trigeminal and spinal ganglia investigated with retrograde and anterograde WGA-HRP tracing in the rat. *Neuroscience, 19*, 695–708.

Bahra, A., Matharu, M. S., Buchel, C., et al. (2001). Brainstem activation specific to migraine headache. *Lancet, 357*, 1016–1017.

Bartsch, T., & Goadsby, P. J. (2002). Stimulation of the greater occipital nerve induces increased central excitability of dural afferent input. *Brain, 125*, 1496–1509.

Bartsch, T., & Goadsby, P. J. (2003). Increased responses in trigeminocervical nociceptive neurones to cervical input after stimulation of the dura mater. *Brain, 126*, 1801–1813.

Bartsch, T., Levy, M. J., Knight, Y. E., et al. (2004). Differential modulation of nociceptive dural input to [hypocretin] Orexin A and B receptor activation in the posterior hypothalamic area. *Pain, 109*, 367–378.

Benjamin, L., Levy, M. J., Lasalandra, M. P., et al. (2004). Hypothalamic activation after stimulation of the superior sagittal sinus in the cat: A Fos study. *Neurobiology of Disease, 16*, 500–505.

Bes, A., Geraud, A., Guell, A., et al. (1982). Dopaminergic hypersensitivity in migraine: A diagnostic test? *La Nouvelle Presse Medicale, 11*, 1475–1478.

Bolay, H., Reuter, U., Dunn, A. K., et al. (2002). Intrinsic brain activity triggers trigeminal meningeal afferents in a migraine model. *Nature Medicine, 8*, 136–142.

Brandes, J. L., Saper, J. R., Diamond, M., et al. (2004). Topiramate for migraine prevention: A randomized controlled trial. *JAMA: The Journal of the American Medical Association, 291*, 965–973.

Burstein, R., Yamamura, H., Malick, A., et al. (1998). Chemical stimulation of the intracranial dura induces enhanced responses to facial stimulation in brain stem trigeminal neurons. *Journal of Neurophysiology, 79*, 964–982.

Burstein, R., Cutrer, M. F., & Yarnitsky, D. (2000a). The development of cutaneous allodynia during a migraine attack. *Brain, 123*, 1703–1709.

Burstein, R., Yarnitsky, D., Goor-Aryeh, I., et al. (2000b). An association between migraine and cutaneous allodynia. *Annals of Neurology, 47*, 614–624.

Connor, H. E., Bertin, L., Gillies, S., et al. (1998). The GR205171 Clinical Study Group. Clinical evaluation of a novel, potent, CNS penetrating NK_1 receptor antagonist in the acute treatment of migraine. *Cephalalgia, 18*, 392.

Cumberbatch, M. J., Hill, R. G., & Hargreaves, R. J. (1997). Rizatriptan has central antinociceptive effects against durally evoked responses. *European Journal of Pharmacology, 328*, 37–40.

Cumberbatch, M. J., Hill, R. G., & Hargreaves, R. J. (1998). Differential effects of the $5HT_{1B/1D}$ receptor agonist naratriptan on trigeminal versus spinal nociceptive responses. *Cephalalgia, 18*, 659–664.

Cutrer, F. M., Limmroth, V., Waeber, C., et al. (1997). New targets for antimigraine drug development. In P. J. Goadsby & S. D. Silberstein (Eds.), *Headache* (pp. 59–72). Philadelphia: Butterworth-Heinemann.

Cutrer, F. M., Sorensen, A. G., Weisskoff, R. M., et al. (1998). Perfusion-weighted imaging defects during spontaneous migrainous aura. *Annals of Neurology, 43*, 25–31.

Data, J., Britch, K., Westergaard, N., et al. (1998). A double-blind study of ganaxolone in the acute treatment of migraine headaches with or without an aura in premenopausal females. *Headache, 38*, 380.

De Fusco, M., Marconi, R., Silvestri, L., et al. (2003). Haploinsufficiency of ATP1A2 encoding the Na^+/K^+ pump Î ± 2 subunit associated with familial hemiplegic migraine type 2. *Nature Genetics, 33*, 192–196.

Diener, H.-C., & The RPR100893 Study Group. (2003). RPR100893, a substance-P antagonist, is not effective in the treatment of migraine attacks. *Cephalalgia, 23*, 183–185.

Diener, H. C., Tfelt-Hansen, P., Dahlof, C., et al. (2004). Topiramate in migraine prophylaxis -results from a placebo-controlled trial with propranolol as an active control. *Journal of Neurology, 251*, 943–950.

Dimitriadou, V., Buzzi, M. G., Moskowitz, M. A., et al. (1991). Trigeminal sensory fiber stimulation induces morphological changes reflecting secretion in rat dura mater mast cells. *Neuroscience, 44*, 97–112.

Dimitriadou, V., Buzzi, M. G., Theoharides, T. C., et al. (1992). Ultrastructural evidence for neurogenically mediated changes in blood vessels of the rat dura mater and tongue following antidromic trigeminal stimulation. *Neuroscience, 48,* 187–203.

Doods, H., Hallermayer, G., Wu, D., et al. (2000). Pharmacological profile of BIBN4096BS, the first selective small molecule CGRP antagonist. *British Journal of Pharmacology, 129,* 420–423.

Ducros, A., Denier, C., Joutel, A., et al. (2001). The clinical spectrum of familial hemiplegic migraine associated with mutations in a neuronal calcium channel. *The New England Journal of Medicine, 345,* 17–24.

Durham, P. L., & Russo, A. F. (1999). Regulation of calcitonin gene-related peptide secretion by a serotonergic antimigraine drug. *Journal of Neuroscience, 19,* 3423–3429.

Durham, P. L., Sharma, R. V., & Russo, A. F. (1997). Repression of the calcitonin gene-related peptide promoter by 5-HT1 receptor activation. *Journal of Neuroscience, 17,* 9545–9553.

Earl, N. L., McDonald, S. A., Lowy, M. T., & 4991 W93 Investigator Group. (1999). Efficacy and tolerability of the neurogenic inflammation inhibitor, 4991 W93, in the acute treatment of migraine. *Cephalalgia, 19,* 357.

Ebersberger, A., Schaible, H.-G., Averbeck, B., et al. (2001). Is there a correlation between spreading depression, neurogenic inflammation, and nociception that might cause migraine headache? *Annals of Neurology, 41,* 7–13.

Edvinsson, L., & Goadsby, P. J. (1998). Neuropeptides in headache. *European Journal of Neurology, 5,* 329–341.

Ertel, E. A., Campbell, K. P., Harpold, M. M., et al. (2000). Nomenclature of voltage-gated calcium channels. *Neuron, 25,* 533–535.

Fanciullacci, M., Alessandri, M., Figini, M., et al. (1995). Increase in plasma calcitonin gene-related peptide from extracerebral circulation during nitroglycerin-induced cluster headache attack. *Pain, 60,* 119–123.

Feindel, W., Penfield, W., & McNaughton, F. (1960). The tentorial nerves and localization of intracranial pain in man. *Neurology, 10,* 555–563.

Gallai, V., Sarchielli, P., Floridi, A., et al. (1995). Vasoactive peptides levels in the plasma of young migraine patients with and without aura assessed both interictally and ictally. *Cephalalgia, 15,* 384–390.

Giffin, N. J., Ruggiero, L., Lipton, R. B., et al. (2003). Premonitory symptoms in migraine: An electronic diary study. *Neurology, 60,* 935–940.

Goadsby, P. J. (2001). Migraine, aura and cortical spreading depression: Why are we still talking about it? *Annals of Neurology, 49,* 4–6.

Goadsby, P. J. (2002a). Neurovascular headache and a midbrain vascular malformation -evidence for a role of the brainstem in chronic migraine. *Cephalalgia, 22,* 107–111.

Goadsby, P. J. (2002b). Pathophysiology of cluster headache: A trigeminal autonomic cephalgia. *Lancet Neurology, 1,* 37–43.

Goadsby, P. J. (2004). Migraine aura: A knock-in mouse with a knock-out message. *Neuron, 41,* 679–680.

Goadsby, P. J., & Classey, J. D. (2003). Evidence for 5-HT$_{1B}$, 5-HT$_{1D}$ and 5-HT$_{1F}$ receptor inhibitory effects on trigeminal neurons with craniovascular input. *Neuroscience, 122,* 491–498.

Goadsby, P. J., & Edvinsson, L. (1994). Human in vivo evidence for trigeminovascular activation in cluster headache. *Brain, 117,* 427–434.

Goadsby, P. J., & Edvinsson, L. (1996). Neuropeptide changes in a case of chronic paroxysmal hemicrania - evidence for trigemino-parasympathetic activation. *Cephalalgia, 16,* 448–450.

Goadsby, P. J., & Ferrari, M. D. (2001). Migraine: A multifactorial, episodic neurovascular channelopathy? In M. R. Rose & R. C. Griggs (Eds.), *Channelopathies of the nervous system* (pp. 274–292). Oxford: Butterworth Heinemann.

Goadsby, P. J., & Hoskin, K. L. (1996). Inhibition of trigeminal neurons by intravenous administration of the serotonin (5HT)$_{1B/D}$ receptor agonist zolmitriptan (311C90): Are brain stem sites a therapeutic target in migraine? *Pain, 67,* 355–359.

Goadsby, P. J., & Hoskin, K. L. (1997). The distribution of trigeminovascular afferents in the nonhuman primate brain *Macaca nemestrina*: A c-fos immunocytochemical study. *Journal of Anatomy, 190,* 367–375.

Goadsby, P. J., & Hoskin, K. L. (1999). Differential effects of low dose CP122,288 and eletriptan on fos expression due to stimulation of the superior sagittal sinus in cat. *Pain, 82,* 15–22.

Goadsby, P. J., & Knight, Y. E. (1997). Inhibition of trigeminal neurons after intravenous administration of naratriptan through an action at the serotonin (5HT$_{1B / 1D}$) receptors. *British Journal of Pharmacology, 122,* 918–922.

Goadsby, P. J., Edvinsson, L., & Ekman, R. (1988). Release of vasoactive peptides in the extracerebral circulation of man and the cat during activation of the trigeminovascular system. *Annals of Neurology, 23,* 193–196.

Goadsby, P. J., Edvinsson, L., & Ekman, R. (1990). Vasoactive peptide release in the extracerebral circulation of humans during migraine headache. *Annals of Neurology, 28,* 183–187.

Goadsby, P. J., Zagami, A. S., & Lambert, G. A. (1991). Neural processing of craniovascular pain: A synthesis of the central structures involved in migraine. *Headache, 31,* 365–371.

Goadsby, P. J., Hoskin, K. L., & Knight, Y. E. (1997). Stimulation of the greater occipital nerve increases metabolic activity in the trigeminal nucleus caudalis and cervical dorsal horn of the cat. *Pain, 73,* 23–28.

Goadsby, P. J., Akerman, S., & Storer, R. J. (2001). Evidence for postjunctional serotonin (5-HT$_1$) receptors in

M

the trigeminocervical complex. *Annals of Neurology,* *50,* 804–807.

Goadsby, P. J., Lipton, R. B., & Ferrari, M. D. (2002). Migraine -current understanding and treatment. *The New England Journal of Medicine, 346,* 257–270.

Goldstein, D. J., Wang, O., Saper, J. R., et al. (1997). Ineffectiveness of neurokinin-1 antagonist in acute migraine: A crossover study. *Cephalalgia, 17,* 785–790.

Haan, J., Terwindt, G. M., Bos, P. L., et al. (1994). Familial hemiplegic migraine in The Netherlands. *Clinical Neurology and Neurosurgery, 96,* 244–249.

Hadjikhani, N., Sanchez del Rio, M., Wu, O., et al. (2001). Mechanisms of migraine aura revealed by functional MRI in human visual cortex. *Proceedings of the National Academy of Sciences of the United States of America, 98,* 4687–4692.

Harer, C., & Kummer, R. (1991). Cerebrovascular CO_2 reactivity in migraine: Assessment by transcranial Doppler ultrasound. *Journal of Neurology, 238,* 23–26.

Headache Classification Committee of The International Headache Society. (2004). The international classification of headache disorders (2nd edn.). *Cephalalgia* 24, 1–160.

Hoskin, K. L., Kaube, H., & Goadsby, P. J. (1996). Central activation of the trigeminovascular pathway in the cat is inhibited by dihydroergotamine. A c-Fos and electrophysiology study. *Brain, 119,* 249–256.

Hoskin, K. L., Zagami, A., & Goadsby, P. J. (1999). Stimulation of the middle meningeal artery leads to Fos expression in the trigeminocervical nucleus: A comparative study of monkey and cat. *Journal of Anatomy, 194,* 579–588.

Hoskin, K. L., Bulmer, D. C. E., Lasalandra, M., et al. (2001). Fos expression in the midbrain periaqueductal grey after trigeminovascular stimulation. *Journal of Anatomy, 197,* 29–35.

Ingvardsen, B. K., Laursen, H., Olsen, U. B., et al. (1997). Possible mechanism of c-fos expression in trigeminal nucleus caudalis following spreading depression. *Pain, 72,* 407–415.

Ingvardsen, B.K, Laursen, H., Olsen, U.B., et al. (1998). Comment on Ingvardsen et al. (1997). *Pain, 72,* 407–415; Reply to Moskowitz et al. (1998). *Pain, 76,* 266–267.

Joutel, A., Bousser, M. G., Biousse, V. A., et al. (1993). A gene for familial hemiplegic migraine maps to chromosome 19. *Nature Genetics, 5,* 40–45.

Joutel, A., Ducros, A., Vahedi, K., et al. (1994). Genetic heterogeneity of familial hemiplegic migraine. *American Journal of Human Genetics, 55,* 1166–1172.

Jurkat-Rott, K., Freilinger, T., Dreier, J. P., et al. (2004). Variability of familial hemiplegic migraine with novel A1A2 Na$^+$/K$^+$-ATPase variants. *Neurology, 62,* 1857–1861.

Kaube, H., & Goadsby, P. J. (1994). Anti-migraine compounds fail to modulate the propagation of cortical spreading depression in the cat. *European Neurology,* *34,* 30–35.

Kaube, H., Hoskin, K. L., & Goadsby, P. J. (1993a). Inhibition by sumatriptan of central trigeminal neurones only after blood-brain barrier disruption. *British Journal of Pharmacology, 109,* 788–792.

Kaube, H., Hoskin, K. L., & Goadsby, P. J. (1993b). Intravenous acetylsalicylic acid inhibits central trigeminal neurons in the dorsal horn of the upper cervical spinal cord in the cat. *Headache, 33,* 541–550.

Kaube, H., Keay, K. A., Hoskin, K. L., et al. (1993c). Expression of c-*Fos*-like immunoreactivity in the caudal medulla and upper cervical cord following stimulation of the superior sagittal sinus in the cat. *Brain Research, 629,* 95–102.

Kaube, H., Knight, Y. E., Storer, R. J., et al. (1999). Vasodilator agents and supracollicular transection fail to inhibit cortical spreading depression in the cat. *Cephalalgia, 19,* 592–597.

Knight, Y. E., & Goadsby, P. J. (2001). The periaqueductal gray matter modulates trigeminovascular input: A role in migraine? *Neuroscience, 106,* 793–800.

Knight, Y. E., Edvinsson, L., & Goadsby, P. J. (1999). Blockade of CGRP release after superior sagittal sinus stimulation in cat: A comparison of avitriptan and CP122,288. *Neuropeptides, 33,* 41–46.

Knight, Y. E., Edvinsson, L., & Goadsby, P. J. (2001). 4991 W93 inhibits release of calcitonin gene-related peptide in the cat but only at doses with 5HT$_{1B / 1D}$ receptor agonist activity. *Neuropharmacology, 40,* 520–525.

Knight, Y. E., Bartsch, T., Kaube, H., et al. (2002). P / Q-type calcium channel blockade in the PAG facilitates trigeminal nociception: A functional genetic link for migraine? *Journal of Neuroscience, 22,* 1–6.

Lalouel, J. M., & Morton, N. E. (1981). Complex segregation analysis with pointers. *Human Heredity, 31,* 312–321.

Lambert, G. A., Michalicek, J., Storer, R. J., et al. (1999). Effect of cortical spreading depression on activity of trigeminovascular sensory neurons. *Cephalalgia, 19,* 631–638.

Lambert, G. A., Boers, P. M., Hoskin, K. L., et al. (2002). Suppression by eletriptan of the activation of trigeminovascular sensory neurons by glyceryl trinitrate. *Brain Research, 953,* 181–188.

Lance, J. W., & Goadsby, P. J. (2005). *Mechanism and management of headache.* New York: Elsevier.

Lashley, K. S. (1941). Patterns of cerebral integration indicated by the scotomas of migraine. *Archives of Neurology and Psychiatry, 46,* 331–339.

Lauritzen, M. (1994). Pathophysiology of the migraine aura. The spreading depression theory. *Brain, 117,* 199–210.

Lauritzen, M., Skyhoj-Olsen, T., Lassen, N. A., et al. (1983). The changes of regional cerebral blood flow during the course of classical migraine attacks. *Annals of Neurology, 13,* 633–641.

Leao, A. A. P. (1944a). Pial circulation and spreading activity in the cerebral cortex. *Journal of Neurophysiology, 7*, 391–396.

Leao, A. A. P. (1944b). Spreading depression of activity in cerebral cortex. *Journal of Neurophysiology, 7*, 359–390.

Liu-Chen, L.-Y., Gillespie, S. A., Norregaard, T. V., et al. (1984). Co-localization of retrogradely transported wheat germ agglutinin and the putative neurotransmitter substance P within trigeminal ganglion cells projecting to cat middle cerebral. *The Journal of Comparative Neurology, 225*, 187–192.

MaassenVanDenBrink, A., van den Broek, R. W., de Vries, R., et al. (2000). The potential anti-migraine compound SB-220453 does not contract human isolated blood vessels or myocardium; a comparison with sumatriptan. *Cephalalgia, 20*, 538–545.

Maneesi, S., Akerman, S., Lasalandra, M. P., et al. (2004). Electron microscopic demonstration of pre- and postsynaptic 5-HT$_{1D}$ and 5-HT$_{1F}$ receptor immunoreactivity (IR) in the rat trigeminocervical complex (TCC) new therapeutic possibilities for the triptans. *Cephalalgia, 24*, 148.

Marconi, R., De Fusco, M., Aridon, P., et al. (2003). Familial hemiplegic migraine type 2 is linked to 0.9 Mb region on chromosome 1q23. *Annals of Neurology, 53*, 376–381.

Markowitz, S., Saito, K., & Moskowitz, M. A. (1987). Neurogenically mediated leakage of plasma proteins occurs from blood vessels in dura mater but not brain. *Journal of Neuroscience, 7*, 4129–4136.

Matharu, M. S., Bartsch, T., Ward, N., et al. (2004a). Central neuromodulation in chronic migraine patients with suboccipital stimulators: A PET study. *Brain, 127*, 220–230.

Matharu, M. S., Cohen, A. S., McGonigle, D. J., et al. (2004b). Posterior hypothalamic and brainstem activation in hemicrania continua. *Headache, 44*, 462–463.

Matsuyama, T., Shiosaka, S., Matsumoto, M., et al. (1983). Overall distribution of vasoactive intestinal polypeptide-containing nerves on the wall of the cerebral arteries: An immunohistochemical study using whole-mounts. *Neuroscience, 10*, 89–96.

May, A., & Goadsby, P. J. (1999). The trigeminovascular system in humans: Pathophysiological implications for primary headache syndromes of the neural influences on the cerebral circulation. *Journal of Cerebral Blood Flow and Metabolism, 19*, 115–127.

May, A., Gijsman, H. J., Wallnoefer, A., et al. (1996). Endothelin antagonist bosentan blocks neurogenic inflammation, but is not effective in aborting migraine attacks. *Pain, 67*, 375–378.

May, A., Bahra, A., Buchel, C., et al. (1998a). Hypothalamic activation in cluster headache attacks. *Lancet, 352*, 275–278.

May, A., Shepheard, S., Wessing, A., et al. (1998b). Retinal plasma extravasation can be evoked by trigeminal stimulation in rat but does not occur during migraine attacks. *Brain, 121*, 1231–1237.

May, A., Bahra, A., Buchel, C., et al. (1999). Functional MRI in spontaneous attacks of SUNCT: Short-lasting neuralgiform headache with conjunctival injection and tearing. *Annals of Neurology, 46*, 791–793.

Mochi, M., Sangiorgi, S., Cortelli, P., et al. (1993). Testing models for genetic determination in migraine. *Cephalalgia, 13*, 389–394.

Moskowitz, M. A. (1990). Basic mechanisms in vascular headache. *Neurologic Clinics, 8*, 801–815.

Moskowitz, M. A., & Cutrer, F. M. (1993). Sumatriptan: A receptor-targeted treatment for migraine. *Annual Review of Medicine, 44*, 145–154.

Moskowitz, M. A., Nozaki, K., & Kraig, R. P. (1993). Neocortical spreading depression provokes the expression of C-fos protein-like immunoreactivity within the trigeminal nucleus caudalis via trigeminovascular mechanisms. *Journal of Neuroscience, 13*, 1167–1177.

Moskowitz, M. A., Bolay, H., & Dalkara, T. (2004). Deciphering migraine mechanisms: Clues from familial hemiplegic migraine genotypes. *Annals of Neurology, 55*, 276–280.

Norman, B., Panebianco, D., & Block, G. A. (1998). A placebo-controlled, in-clinic study to explore the preliminary safety and efficacy of intravenous L-758,298 (a prodrug of the NK1 receptor antagonist L-754,030) in the acute treatment of migraine. *Cephalalgia, 18*, 407.

Nozaki, K., Boccalini, P., & Moskowitz, M. A. (1992). Expression of c-fos-like immunoreactivity in brainstem after meningeal irritation by blood in the subarachnoid space. *Neuroscience, 49*, 669–680.

Olesen, J., Larsen, B., & Lauritzen, M. (1981). Focal hyperemia followed by spreading oligemia and impaired activation of rCBF in classic migraine. *Annals of Neurology, 9*, 344–352.

Olesen, J., Friberg, L., Skyhoj-Olsen, T., et al. (1990). Timing and topography of cerebral blood flow, aura, and headache during migraine attacks. *Annals of Neurology, 28*, 791–798.

Olesen, J., Diener, H.-C., Husstedt, I.-W., et al. (2004). Calcitonin gene-related peptide (CGRP) receptor antagonist BIBN4096BS is effective in the treatment of migraine attacks. *The New England Journal of Medicine, 350*, 1104–1110.

Ophoff, R. A., Rv, E., Sandkuijl, L. A., et al. (1994). Genetic heterogeneity of familial hemiplegic migraine. *Genomics, 22*, 21–26.

Ophoff, R. A., Terwindt, G. M., Vergouwe, M. N., et al. (1996). Familial hemiplegic migraine and episodic ataxia type-2 are caused by mutations in the Ca^{2+} channel gene CACNL1A4. *Cell, 87*, 543–552.

Parsons, A. A., Bingham, S., Raval, P., et al. (2001). Tonabersat (SB-220453) a novel benzopyran with anticonvulsant properties attenuates trigeminal nerve-induced neurovascular reflexes. *British Journal of Pharmacology, 132*, 1549–1557.

Penfield, W., & McNaughton, F. L. (1940). Dural headache and the innervation of the dura mater. *Archives of Neurology and Psychiatry, 44*, 43–75.

Peroutka, S. J. (1997). Dopamine and migraine. *Neurology, 49*, 650–656.

Rasmussen, B. K., & Olesen, J. (1992). Migraine with aura and migraine without aura: An epidemiological study. *Cephalalgia, 12*, 221–228.

Read, S. J., Smith, M. I., Hunter, A. J., et al. (1999). SB-220453, a potential novel antimigraine compound, inhibits nitric oxide release following induction of cortical spreading depression in the anaesthetized cat. *Cephalalgia, 20*, 92–99.

Roon, K., Diener, H. C., Ellis, P., et al. (1997). CP-122,288 blocks neurogenic inflammation, but is not effective in aborting migraine attacks: Results of two controlled clinical studies. *Cephalalgia, 17*, 245.

Russell, M. B. (1997). Genetic epidemiology of migraine and cluster headache. *Cephalalgia, 17*, 683–701.

Russell, M. B., Iselius, L., & Olesen, J. (1995). Investigation of the inheritance of migraine by complex segregation analysis. *Human Genetics, 96*, 726–730.

Sakai, F., & Meyer, J. S. (1979). Abnormal cerebrovascular reactivity in patients with migraine and cluster headache. *Headache, 19*, 257–266.

Selby, G., & Lance, J. W. (1960). Observations on 500 cases of migraine and allied vascular headache. *Journal of Neurology, Neurosurgery, and Psychiatry, 23*, 23–32.

Shields, K. G., & Goadsby, P. J. (2004). Naratriptan modulates trigeminovascular nociceptive transmission in the ventroposteromedial (VPM) thalamic nucleus of the rat. *Cephalalgia, 24*, 1098.

Shields, K. G., & Goadsby, P. J. (2005). Propranolol modulates trigeminovascular responses in thalamic ventroposteromedial nucleus: A role in migraine? *Brain, 128*, 86–97.

Shields, K. G., Kaube, H., & Goadsby, P. J. (2003). GABA receptors modulate trigeminovascular nociceptive transmission in the ventroposteromedial (VPM) thalamic nucleus of the rat. *Cephalalgia, 23*, 728.

Silberstein, S. D., Lipton, R. B., & Goadsby, P. J. (2002). *Headache in clinical practice.* London: Martin Dunitz.

Silberstein, S. D., Neto, W., Schmitt, J., et al. (2004). Topiramate in migraine prevention: Results of a large controlled trial. *Archives of Neurology, 61*, 490–495.

Smith, M. I., Read, S. J., Chan, W. N., et al. (2000). Repetitive cortical spreading depression in a gyrencephalic feline brain: Inhibition by the novel benzoylamino-benzopyran SB-220453. *Cephalalgia, 20*, 546–553.

Storer, R.J. & Goadsby, P.J. (2013). Topiramate has a locus of action outside of the trigeminocervical complex. *Cephalalgia, 33*, 291–300.

Storer, R. J., & Goadsby, P. J. (2004). Topiramate inhibits trigeminovascular neurons in the cat. *Cephalalgia, 24*, 1049–1056.

Strassman, A. M., Raymond, S. A., & Burstein, R. (1996). Sensitization of meningeal sensory neurons and the origin of headaches. *Nature, 384*, 560–563.

Swoboda, K. J., Kanavakis, E., Xaidara, A., et al. (2004). Alternating hemiplegia of childhood or familial hemiplegic migraine? A novel ATP1A2 mutation. *Annals of Neurology, 55*, 884–887.

Teh, B. T., Silburn, P., Lindblad, K., et al. (1995). Familial cerebellar periodic ataxia without myokymia maps to a 19-cM region on 19p13. *American Journal of Human Genetics, 56*, 1443–1449.

Terwindt, G. M., Ophoff, R. A., Haan, J., et al. (1996). The Dutch Migraine Genetics Research Group. Familial hemiplegic migraine: A clinical comparison of families linked and unlinked to chromosome 19. *Cephalalgia, 16*, 153–155.

Tvedskov, J. F., Iversen, H. K., & Olesen, J. (2004a). A double-blind study of SB-220453 (Tonerbasat) in the glyceryltrinitrate (GTN) model of migraine. *Cephalalgia, 24*, 875–882.

Tvedskov, J. F., Thomsen, L. L., Iversen, H. K., et al. (2004b). The prophylactic effect of valproate on glyceryltrinitrate induced migraine. *Cephalalgia, 24*, 576–585.

Tvedskov, J. F., Thomsen, L. L., Iversen, H. K., et al. (2004c). The effect of propranolol on glyceryl-trinitrate-induced headache and arterial response. *Cephalalgia, 24*, 1076–1087.

Uddman, R., Edvinsson, L., Ekman, R., et al. (1985). Innervation of the feline cerebral vasculature by nerve fibers containing calcitonin gene-related peptide: Trigeminal origin and co-existence with substance P. *Neuroscience Letters, 62*, 131–136.

Ulrich, V., Gervil, M., Kyvik, K. O., et al. (1999). Evidence of a genetic factor in migraine with aura: A population based Danish twin study. *Annals of Neurology, 45*, 242–246.

van den Maagdenberg, A. M. J. M., Pietrobon, D., Pizzorusso, T., et al. (2004). A Cacna1a knock-in migraine mouse model with increased susceptibility to cortical spreading depression. *Neuron, 41*, 701–710.

Vanmolkot, K. R. J., Kors, E. E., Hottenga, J. J., et al. (2003). Novel mutations in the Na^+/K^+-ATPase pump gene ATP1A2 associated with familial hemiplegic migraine and benign familial infantile convulsions. *Annals of Neurology, 54*, 360–366.

Veloso, F., Kumar, K., & Toth, C. (1998). Headache secondary to deep brain implantation. *Headache, 38*, 507–515.

Weiller, C., May, A., Limmroth, V., et al. (1995). Brain stem activation in spontaneous human migraine attacks. *Nature Medicine, 1*, 658–660.

Willis, T. (1682). *Opera omnia.* Amstelaedami: Henricum Wetstenium.

Wolff, H. G. (1948). *Headache and other head pain.* New York: Oxford University Press.

Zagami, A. S., & Goadsby, P. J. (1991). Stimulation of the superior sagittal sinus increases metabolic activity in cat thalamus. In F. C. Rose (Ed.), *New advances in headache research* (Vol. 2, pp. 169–171). London: Smith-Gordon.

Zagami, A. S., & Lambert, G. A. (1991). Craniovascular application of capsaicin activates nociceptive thalamic neurons in the cat. *Neuroscience Letters, 121*, 187–190.

Migraine-Type Headache

Definition

Headache in a patient with migraine (see ▶ Migraine).

Cross-References

▶ Headache Due to Arteritis

Minimal Alveolar Concentration

Synonyms

MAC

Definition

Alveolar concentration of a volatile anesthetic at 1 atm pressure, which abolishes motor response to a painful surgical stimulus in 50 % of patients.

Cross-References

▶ Alpha(α) 2-Adrenergic Agonists in Pain Treatment

Minimal Clinical Important Difference

Synonyms

MCID

Definition

The smallest difference in score in the domain of interest which patients perceive as beneficial.

Cross-References

▶ Oswestry Disability Index

Minimal Sedation

Synonyms

Anxiolysis

Definition

A drug-induced state in which the patient responds normally to verbal commands. Ventilatory and cardiovascular functions are unaffected, although cognitive function and coordination may be impaired.

Cross-References

▶ Pain and Sedation of Children in the Emergency Setting

Minimum Effective Analgesic Serum Concentrations of Fentanyl

Definition

Minimum effective analgesic serum concentrations of fentanyl range from 0.3 to 1.5 ng/ml.

Cross-References

▶ Postoperative Pain, Fentanyl

Minimum Erythema Dose

▶ MED

Minocycline

Definition

Minocycline is used to pharmacologically block microglial activation without directly affecting neurons or astrocytes. As a result, minocycline blocks the production of microglial-derived proinflammatory cytokines. In neuropathic pain, administration of minocycline disrupts exaggerated pain by inhibiting proinflammatory cytokines. Importantly, minocycline has revealed that microglia may be more important in the initiation, rather than the maintenance, of exaggerated pain states.

Cross-References

▶ Cord Glial Activation

Miss Rate

Definition

Miss rate is the probability of response "B" when event A has occurred.

Mitogen-Activated Protein Kinase

Synonyms

MAPK

Definition

MAPK mediate intracellular signal transduction in response to a variety of stimuli, including several hormones and growth factors. Among them, extracellular signal-regulated protein kinase (ERK) is involved in cellular growth and differentiation, whereas p38 MAPK and c-*Jun* N-terminal kinase (JNK) function as mediators of cellular stresses such as inflammation and apoptosis.

Cross-References

▶ ERK Regulation in Sensory Neurons During Inflammation

Mitogen-activated Protein Kinase (MAPK)

▶ Phosphatases and Glia

Mitogen-activated Protein Kinase Phosphatases (MKP)

▶ Phosphatases and Glia

Mittelschmerz

▶ Gynecological Pain, Neural Mechanisms

Mixed Forms of Diabetic Neuropathy

▶ Diabetic Neuropathies

Mixed-Acting Analgesics

▶ Drugs with Mixed Action and Combinations: Emphasis on Tramadol

MKP-1

▶ Phosphatases and Glia

MKP3

▶ Phosphatases and Glia

MMP

▶ Matrix Metalloproteinases

Mnemonic Function

Definition

The hippocampus plays a role in learning and memory, especially declarative learning and memory that involves, at least partly, forming associations between different stimuli.

Cross-References

▶ Nociceptive Processing in the Hippocampus and Entorhinal Cortex, Neurophysiology and Pharmacology

Mobilisation/Mobilization

Definition

Mobilization is the restoration of motion in a joint.

Cross-References

▶ Chronic Pain in Children, Physical Medicine and Rehabilitation
▶ Passive Spinal Mobilization

Mobilization

▶ Passive Spinal Mobilization

Mobilization Following Surgery

▶ Postoperative Pain, Importance of Mobilization

Mobilization with Impulse

▶ Spinal Manipulation, Pain Management

Mobilization with Thrust

▶ Spinal Manipulation, Pain Management

Modalities

Alison Thomas
Lismore, NSW, Australia

Synonyms

Electrotherapy; Interferential therapy; Pulsed electromagnetic therapy; Shortwave diathermy; Therapeutic ultrasound

Definition

The term ▶ modalities is used, in practice, as a generic term to encompass a variety of treatments that share the characteristic of requiring an electrical device to deliver some form of energy to the tissues of the body. That energy may be an electrical current, intended to stimulate or inhibit nerve conduction, or electromagnetic or acoustic energy, intended to heat tissues by inducing agitation of molecules within the tissue.

Characteristics

Shortwave Diathermy
Mechanism
Application of a high-frequency (27.12 MHz) electromagnetic current induces rapid movement of charged ions in the target tissue. The current can be successfully delivered to heat deep tissues (Mercer and Bogduk 2004). Heating tissues

results in the physiological reactions of increased blood flow, stimulation of peripheral nerves, and increased localized metabolic rate (Baxter and Barlos 2002). In vitro studies have identified effects on fibroblast and chondrocyte proliferation in human tissue (Hill et al. 2002).

Applications

Common purported therapeutic effects are the following: acceleration of the resolution of chronic inflammation, acceleration of healing, relief of pain, reduction of muscle spasm, and increase in extensibility of fibrous tissue. Common applications for which shortwave diathermy is used are acute and chronic osteoarthritis, polyarthritis, rheumatoid arthritis, acute and chronic low back pain, and acute and chronic soft tissue injuries including hematomas and sinusitis.

Efficacy

Experimental evidence has shown that the increase in heat in deep tissues by a shortwave diathermy does raise the pain threshold in that region, but only for a period of 30 min (Mercer and Bogduk 2004). It has also been shown that the increase in heat is much less than that generated by moderate or gentle exercise (Mercer and Bogduk 2004) and that the effect on healing rate of wounds is equivocal. Clinical trials have generally been of relatively poor quality.

In essence, there is little experimental or high-quality clinical data at this time to support the purported therapeutic effects of shortwave diathermy. Temporary change in pain threshold due to heating of tissues may be effected more safely and expeditiously through gentle exercise.

Side Effects

Shortwave diathermy has the potential to cause burns (particularly in an area surrounding a metal implant or vascular insufficiency), interfere with cardiac pacemakers, and stimulate further changes in carcinoma or infection. It should not be used over a pregnant uterus (Baxter and Barlos 2002).

Indications

Based on current experimental evidence and efficacy data, there are no indications for the therapeutic use of shortwave diathermy.

Therapeutic Ultrasound

Mechanism

High-frequency sound waves (0.75–1.0 MHz) are generated by vibration of a piezoelectric crystal in the sound head of the machine. These are delivered to tissues by application of the sound head to the skin via a coupling medium (a gel or water). The ultrasound head is moved rhythmically over the area to be treated. There is reasonable evidence that ultrasound heats deep tissues and produces a range of nonthermal effects such as cavitation, acoustic streaming, and standing waves (Mercer and Bogduk 2004).

Applications

Ultrasound therapy is commonly used to treat acute and chronic musculoskeletal symptoms, including joint inflammation or restriction, pain, and related disability (Gam and Johannsen 1995; van der Windt et al. 1999). It is proposed that the underlying mechanisms include increased blood flow, increased extensibility of collagenous tissues, decreased pain, and decreased muscle spasm (Mercer and Bogduk 2004). Ultrasound therapy is also used for the promotion of wound healing and tissue-repair processes in both acute (soft tissue injury) and chronic (ulcerative) conditions (Watson 2000).

Efficacy

There is evidence, from in vitro and cutaneous lesion studies, that ultrasound has proinflammatory effects on damaged tissue. This may justify the use of ultrasound therapy for superficial wound healing and ulcers (Mercer and Bogduk 2004).

Early evidence that ultrasound therapy increases blood flow, increases extensibility of collagen tissue, increases muscle strength, and improves the strength and quality of repair of injured tendons has not been confirmed with subsequent studies (Mercer and Bogduk 2004). Experimental evidence of the physiological effect of ultrasound on peripheral nerve conduction (Baxter and Barlos 2002) has yet to be demonstrated to have an analgesic effect.

The results of three, recent, systematic reviews on the efficacy of ultrasound therapy for musculoskeletal disorders concluded that there was little or no evidence for the clinical effectiveness of ultrasound therapy (Gam and Johannsen 1995; van der Windt et al. 1999; Robertson and Baker 2001).

Side Effects

Ultrasound has the potential to produce unstable cavitation and standing wave formation within the tissues. Unstable cavitation occurs when the gas bubble collapses and reforms. Standing waves are created when the sound waves are reflected from a dense surface, and interfere with incoming waves to produce a high concentration of energy in the tissue. Both may result in tissue damage (Baxter and Barlos 2002).

Ultrasound therapy should not be applied over areas of acute infection, artificial implants, fractures, air filled cavities, rapidly dividing tissues, bony prominences, eyes, testicles and areas of impaired circulation or sensation (Baxter and Barlos 2002).

Indications

Ultrasound may have a role in the healing of superficial tissues. There is no evidence to support the role of ultrasound in the healing of deep tissues. To date, there is no conclusive evidence that ultrasound is efficacious in the treatment of pain associated with musculoskeletal injuries.

Interferential Therapy

Mechanism

Interferential therapy is produced by two different, medium-frequency currents delivered to the tissues in such a way as to produce a low-frequency current in the target tissue. It is generally applied using two pairs of electrodes, either plate (carbon-rubber) or circular suction electrodes. The suction electrodes provide a "massage-like" effect in addition to the current. The rationale for this application has been that the medium-frequency currents overcome the high skin impedance encountered by low-frequency currents. Low-frequency currents can then be delivered to tissues at a greater depth.

Applications

Common purported therapeutic effects include pain relief, stimulation of muscle fibers to produce contraction, increase in blood flow, and promotion of tissue healing (Baxter and Barlos 2002). Interferential therapy has been widely used for the treatment of pain and swelling in soft tissue injuries and to enhance the healing of bone fractures. It is also one of the most commonly used modalities for the management of back pain (Baxter and Barlos 2002).

Efficacy

To date, there has been little experimental data to substantiate the alleged physiological effects of interferential therapy (Watson 2000).

The pivotal rationale in the use of interferential therapy is that it is the effects of the low-frequency current, generated by different swing patterns of the medium-frequency currents, which are therapeutic. It has been shown that the amplitude-modulated frequency was immaterial to the effects of interferential current on sensory or motor nerves (Palmer et al. 1999) and pain-relieving effects are not affected by different swing patterns (Johnson and Tabasam 2003a). This would indicate that the medium-frequency component is the main parameter in stimulation. It appears that interferential offers no advantage over transcutaneous nerve stimulation for pain relief (Johnson and Tabasam 2003b).

Side Effects

It is not recommended that interferential be used over areas of compromised skin sensation, the carotid sinus, pregnant uterus, cardiac pacemakers, eyes, or gonads or for the treatment of undiagnosed pain.

Chemical burns may occur if an excessively high intensity of current is delivered.

When suction cap electrodes are used to deliver the current, injury or bruising may result if the pressure is too high.

Indications
On current available evidence, it appears that interferential therapy is no more efficacious in the treatment of pain than transcutaneous electrical nerve stimulation.

References

Baxter, G. D., & Barlos, P. (2002). Electrophysical agents in pain management. In Strong, Unrah, Wright, & Baxter (Eds.), *Pain: A textbook for therapists* (pp. 207–226). Edinburgh: Harcourt Publishers.

Gam, A. N., & Johannsen, F. (1995). Ultrasound therapy in musculoskeletal disorders: A meta-analysis. *Pain, 63*, 85–91.

Hill, J., Lewis, M., Mills, P., & Kielty, C. (2002). Pulsed short-wave diathermy effects on human fibroblast proliferation. *Archives of Physical Medicine Rehabilitation, 83*, 832–836.

Johnson, M. I., & Tabasam, G. (2003a). A single-blind investigation into the hypoalgesic effects of different swing patterns of interferential currents on cold-induced pain in healthy volunteers. *Archives of Physical Medicine and Rehabilitation, 84*, 350–357.

Johnson, M. I., & Tabasam, G. (2003b). An investigation into the analgesic effects of interferential currents and transcutaneous electrical nerve stimulation on experimentally induced ischemic pain in otherwise pain-free volunteers. *Physical Therapy, 83*, 208–223.

Mercer, S. & Bogduk, N. (2004). Selection and application of treatment. In: Efshaugee, R. K. & Gass, E. (eds.), *Musculoskeletal physiotherapy. Clinical science and practice*, 2nd edn. Butterworth Heinman (in press).

Palmer, S. T., Martin, D. J., Steedman, W. M., & Ravey, J. (1999). Alteration of interferential current and transcutaneous nerve stimulation frequency: Effects on nerve excitation. *Archives of Physical Medicine and Rehabilitation, 80*, 1065–1071.

Robertson, V. J., & Baker, K. G. (2001). A review of therapeutic ultrasound: Effectiveness studies. *Physical Therapy, 81*, 1339–1358.

van der Windt, D. A. W. M., van der Heijden, G. J. M. G., van den Berg, S. G. M., ter Riet, G., de Winter, A. F., & Bouter, L. M. (1999). Ultrasound therapy for musculoskeletal disorders: A systematic review. *Pain, 81*, 257–271.

Watson, T. (2000). The role of electrotherapy in contemporary physiotherapy practice. *Manual Therapy, 5*, 132–141.

Modality

▶ Therapeutic Heat, Microwaves, and Cold

Model of Neuropathic Pain

Definition

Several models of neuropathic pain have been developed in both rats and mice. Neuropathic pain can be induced by injury to a nerve and, depending on the model, results in either or both mechanical and heat hyperalgesia. Frequently used models include tying loose chromic gut ligatures around the sciatic nerve; ligating spinal nerves L5, L6, or both L5 and L6 (and sometimes also transecting the nerve distal to the ligation); partial nerve ligation; and spared nerve preparation (see Boyce-Rustay and Jarvis 2009).

Cross-References

▶ TENS, Mechanisms of Action

References

Boyce-Rustay, J. M., & Jarvis, M. F. (2009). Neuropathic pain: Models and mechanisms. *Current Pharmaceutical Design, 15*, 1711–1716.

Modeling

▶ Modeling, Social Learning in Pain

Modeling, Social Learning in Pain

Christiane Hermann
Department of Clinical Psychology, Justus-Liebig University Giessen, Giessen, Germany

Synonyms

Imitation learning; Modeling; Observational learning; Vicarious learning

Definition

Social learning or learning by observation of others' reaction to pain has been implicated as an important mechanism by which beliefs about pain, pain responses, and complex pain behavior patterns are acquired (Craig and Pillai Riddell 2003; Goubert et al. 2011). By observing their parents and other significant persons, children acquire attitudes about health and health care and learn to perceive and interpret physical symptoms and how to respond to illness and injury (Baranowski and Nader 1985). Hence, forming beliefs about pain and its threat value may not necessarily require direct pain experience. From this perspective, a member in the immediate family environment who suffers from chronic pain may act as a pain model. Exposure to such a pain model at an early age may shape the child's future pain behavior and pain experience. Social learning of maladaptive pain beliefs, pain responses, and pain behavior can presumably act as a risk factor in the development or maintenance of chronic pain.

Characteristics

Learning by observing others, or modeling, enables the observer to acquire new patterns of behaviors and rules for regulating behavior without having to rely on one's own actions or trial-and-error experiences (Bandura 1986). Much of human behavior is presumed to be learned by modeling. This includes cognitive and motor skills, knowledge, attitudes, beliefs, values and judgmental standards, and emotional responses. Social learning involves classical and operant learning principles. Similar to direct ► classical conditioning, fears can be acquired by ► vicarious instigation of emotional arousal when observing a model's fear response to an object (Olsson and Phelps 2007). Social learning can direct attention to stimuli that otherwise would be ignored. A child may increasingly become aware of bodily symptoms and interpret them as threatening by repeated observations of a parent focusing and being preoccupied about somatic symptoms. Similar to direct operant conditioning, modeling can also exert an inhibiting or facilitating influence on previously learned behavior since it entails learning about the functional value of behaviors. For example, by observing the mother receiving much attention and care from her spouse when she is verbally and nonverbally expressing her pain, a child may as an adult be more likely to show high levels of pain behavior. Depending on the model's behavior, social learning can also be protective by immunizing the observer against learning from adverse experiences. For example, young children acquired less fear of fantasy animals when they had previously observed their mothers enjoying to play with these animals (Egliston and Rapee 2007). Modeling is to be inferred primarily by its outcome rather than being directly observable. Also, few learning trials are necessary for social learning to occur. The effects of modeling are not necessarily tangible in close temporal relationship to the learning phase. Whether or not behaviors previously acquired by modeling are performed depends on situational and motivational cues. Thus, beliefs and responses learned by modeling constitute one facet of the social context in which pain occurs.

Two lines of research have provided evidence that social learning may shape pain beliefs and responses to pain. Correlational studies have focused on the co-occurrence of pain problems within families or, under naturalistic conditions, have examined the modulating impact of a positive family history of pain on responses to acute pain and injury. Experimental manipulations of modeling behavior have been used to delineate its direct influence on the observer's response to painful stimuli.

Correlational Studies. Since the late 1970s, several reports have noted that pain problems tend to run in families. Studies have looked at the prevalence of pain models within the families of adult and child pain patients but also at the incidence of pain and illness in the offspring of pain patients. For example, adult chronic pain patients as compared to pain-free chronic patients indicated a greater number of family members also suffering from pain. Familial aggregation of pain problems and the implied role of modeling have also received support in nonclinical samples. Among students, correlations between the reported frequency of current pain episodes and the number of familial pain models have been observed in several studies. The specific influence of a parental pain model on the frequency of pain in children has not been addressed systematically. Also, rather little is known about whether a positive family history of pain determines an individual's response to naturally occurring or injury-related pain. Schrader et al. (1996) assessed in a cohort of individuals the occurrence of neck and other pain 1–3 years after they had experienced a rear-end collision. A positive family history of pain emerged as an important risk factor for the evolution of pain problems subsequent to the injury.

While being consistent with the proposed influence of modeling, mere observation of the co-occurrence of pain problems in families tells little about the involved learning processes. From a social learning perspective, one would expect at least some relationship between the model's and the observer's pain response including pain-related beliefs and coping behavior. Consistent with this assumption, in a group of mothers with chronic pain, significant correlations were observed between the frequency of pain episodes in the child and the mother's perceived pain intensity, subjective pain-related interference in everyday life, and the level of emotional distress. Growing up with a familial pain model may increase the risk for later experiencing pain problems by learning maladaptive pain and illness behaviors and exacerbating responses to pain. However, the social transmission of specific (maladaptive) pain behaviors including coping responses is yet to be clarified. In children with idiopathic juvenile arthritis, children's and parents' pain-related catastrophizing was significantly correlated (Thastum et al. 2001). Moreover, the negative impact of a parental pain history on the health status of children with rheumatic diseases has been shown to be mediated by children's pain-related catastrophizing (Schanberg et al. 2001). Investigating the role of modeling is complicated by the fact that familial

pain models may exert their influence in a sex-dependent manner. Early on, pain models were found to have a greater impact on females than on males. A higher incidence of familial pain models in a female than a male student population has been noted. Moreover, in females, a positive family history of pain was associated with more frequent pain and an enhanced sensitivity to experimental pain stimuli (Fillingim et al. 2000). One explanation has been that females may be more aware of pain in others. Yet, it may also reflect sex-specific effects of modeling. It is well established that social learning depends on the model's attributes and the perceived similarity between observer and model. For example, sex-role information is conveyed by observational learning in a sex-selective manner (Bandura 1986). Since more women are affected by chronic pain, girls may be more likely than boys to grow up with a relevant pain model. The possibly sex-dependent impact of social learning on pain behavior and its interaction with sex-related biologically determined differences in pain susceptibility certainly need to be further explored.

Experimental studies. The pioneering experimental work by Kenneth Craig has been most important in accruing empirical support for the hypothesis that social learning can account for individual differences in pain responding (for a review see Craig 1986). Starting in the late 1960s, Craig has conducted a series of experimental studies using variations of the following experimental paradigm. Students were exposed to series of increasingly painful shocks starting with a noticeable but not painful stimulus. After each shock, the subject rated pain intensity and then observed a confederate rating the same stimulus as either less (i.e., tolerant model) or more (i.e., intolerant model) painful. This procedure continued for the whole series of shocks. Across studies, subjects exposed to a tolerant model as compared to those exposed to an intolerant model had a higher ▶ pain ▶ threshold, were less autonomically aroused, and indicated less subjective stress. Moreover, ▶ signal detection analyses revealed that subjects showed greater discrimination between painful and non-painful stimuli when they had observed an intolerant model. Prior exposure to a pain-tolerant or intolerant model also yields a change of the subjects' *pain tolerance* in the expected direction. Since Craig's seminal work, there have been surprisingly few studies on modeling and its impact on an individual's pain response. Maternal modeling affects children's pain responses during a ▶ cold pressor task (Goodman and McGrath 2003). When mothers had been instructed to exaggerate their pain display during a cold pressor task as their children were watching, their children later had a significantly lower pain threshold as compared to the control children. Following instructions minimized maternal pain behavior did not increase the children's pain threshold. Maternal modeling had also no impact on the children's subjective pain intensity ratings during the task. In a recent study (Helsen et al. 2011), the effects of observational learning were tested using a differential conditioning paradigm. Participants saw videos of a model undergoing two cold pressor tasks with colored water serving as a cue for painfulness. One color indicated that the task was very painful as shown by the model's painful facial expression and served as threat stimulus. The second color indicated that the task was not painful as shown by the model's neutral facial expression. Observers reported significantly greater fear of pain and expected more pain for the cold pressor task with the colored water signaling high pain intensity as compared to the cold pressor task with the colored water signaling less pain. During actual immersion of their hand, participants also reported significantly greater fear of pain in the high pain as compared to the low pain cold pressor task, although the objective water temperature was the same. Yet, participants did not endorse higher pain intensity and unpleasantness (Helsen et al. 2011).

Hence, experimental studies have consistently demonstrated that modeling can modulate the observer's detection of a stimulus as painful and increase fear of pain and expected pain intensity. Whether social learning can also alter the perceived intensity and aversiveness of a painful stimulus is less clear. It should be noted that in all experimental studies, the observers could rely

either on verbal responses or other overt behavioral responses to infer that the model was experiencing pain. Under natural conditions, however, the level of pain experienced by a model is less easily decoded by the observer, and its estimate may not always be accurate (Kappesser and Williams 2010). Clearly, the behavior shown by the model and the situational context will impact on what is learned by observation. Observer characteristics are also crucial determinants. Learning from observing others requires empathizing with the observed person (Goubert et al. 2011). Empathy has been defined as mental perspective taking which is accompanied by affective and behavioral responses. Accordingly, social learning of a fear response involves the activation of brain regions that are also activated by direct conditioning but also of brain regions known to be active when mentalizing about other's thinking and feeling (Olsson and Phelps 2007). Experimental studies further suggest that individual differences in pain sensitivity may impose constraints on the direction and magnitude of altered pain responding due to modeling. Moreover, the impact of a tolerant versus an intolerant model may not be symmetrical (e.g., Goodman and McGrath 2003). In one study (Prkachin and Craig 1986), female students were first selected based on whether they had a high or low pain threshold and were then tested alone or in the presence of another participant with either a high or low pain threshold. Results showed that only women with high pain thresholds lowered their pain threshold when exposed to a low pain threshold model. Modeling had no effect on the pain threshold in subjects with a low pain threshold.

Summary

Taken together, there is converging evidence from correlational and experimental studies that social learning is important in shaping an individual's pain-related beliefs and pain response, even though the exact mediating mechanisms by which this occurs are not fully understood. Data suggest that the effects of social modeling are modulated by observer characteristics such as age, sex, pain sensitivity, or empathic concerns, the model's characteristics and behavior, and the situational context in which observational learning occurs.

Cross-References

▶ Classical Conditioning
▶ Motivational Aspects of Pain
▶ Operant Conditioning
▶ Psychology of Pain and Psychological Treatment

References

Bandura, A. (1986). *Social foundations of thought and action: A social cognitive theory.* Englewood Cliffs: Prentice Hall.
Baranowski, T., & Nader, P. R. (1985). Family health behavior. In D. C. Turk & R. D. Kerns (Eds.), *Health, illness, and families: A life-span perspective* (pp. 51–80). New York: Wiley.
Craig, K. D. (1986). Social modeling influences: Pain in context. In R. A. Sternbach (Ed.), *The psychology of pain* (2nd ed., pp. 67–95). New York: Raven.
Craig, K. D., & Pillai Riddell, R. R. (2003). Social influences, culture, and ethnicity. In P. J. McGrath & G. A. Finley (Eds.), *Pediatric pain: Biological and social context* (pp. 159–182). Seattle: IASP Press.
Egliston, K. A., & Rapee, R. M. (2007). Inhibition of fear acquisition in toddlers following positive modelling by their mothers. *Behaviour Research and Therapy, 45,* 1871–1882.
Fillingim, R. B., Edwards, R. R., & Powell, T. (2000). Sex-dependent effects of reported familial pain history on recent pain complaints and experimental pain responses. *Pain, 86,* 87–94.
Goodman, J. E., & McGrath, P. J. (2003). Mothers' modeling influences children's pain during a cold pressor task. *Pain, 104,* 559–565.
Goubert, L., Vlaeyen, J. W. S., Crombez, G., & Craig, K. D. (2011). Learning about pain from others: An observational learning account. *The Journal of Pain, 12,* 167–174.
Helsen, K., Goubert, L., Peters, M. L., & Vlaeyen, J. W. (2011). Observational learning and pain-related fear: An experimental study with colored cold pressor tasks. *The Journal of Pain, 12,* 1230–1239.
Kappesser, J., & Williams, A. C. (2010). Pain estimation: Asking the right questions. *Pain, 148,* 184–187.
Olsson, A., & Phelps, E. A. (2007). Social learning of fear. *Nature Neuroscience, 10,* 1095–1102.
Prkachin, K. M., & Craig, K. D. (1986). Social transmission of natural variations in pain behaviour. *Behaviour Research and Therapy, 24,* 581–585.
Schanberg, L. E., Anthony, K. K., Gil, K. M., Lefebvre, J. C., Kredich, D. W., & Macharoni, L. M. (2001). Family pain history predicts child health status in children with chronic rheumatic disease. *Pediatrics, 108,* E47.
Schrader, H., Obelieniene, D., Bovim, G., Surkiene, D., Mickeviciene, D., Miseviciene, I., et al. (1996). Natural evolution of late whiplash syndrome outside the medicolegal context. *Lancet, 347,* 1207–1211.
Thastum, M., Zachariae, R., & Herlin, T. (2001). Pain experience and pain coping strategies in children with juvenile idiopathic arthritis. *Journal of Rheumatology, 28,* 1091–1098.

Moderate Sedation

Definition

A drug-induced state of depressed consciousness during which the patient responds purposefully to verbal or light tactile stimulation, while maintaining protective airway reflexes and airway patency. Cardiovascular function is usually maintained.

Cross-References

▶ Pain and Sedation of Children in the Emergency Setting

Modifying Factors

Definition

Pain-related factors that may affect the assessment of pain in the newborn include gestational age (number of days since conception), postnatal age (number of days of life since birth), behavioral state (sleep/wake state), severity of illness, and gender. Pain expression in infants may also be affected by developmental ability, consolability, chronic pain, environmental stimulation and repeated exposure to pain and stress events, medication, and caregiving including direct parent involvement.

Cross-References

▶ Pain Assessment in Neonates

Modulation by P2Y Receptors

▶ TRPV1 Modulation by P2Y Receptors

Molecular Contributions

▶ Molecular Contributions to the Mechanism of Central Pain

Molecular Contributions to the Mechanism of Central Pain

Bryan C. Hains
Department of Neurology, Yale University
School of Medicine, West Haven, CT, USA

Synonyms

Central pain mechanisms; Molecular contributions

Definition

Changes in ion channels, neurotransmitters, receptors, signaling pathways, and gene expression represent important consequences of disease or ischemic and traumatic injury of the central nervous system that ultimately impact the functional state of neurons at different levels of the neuraxis, leading to the development of central pain.

The molecular organization of the structural and functional components of a neuron endows it with certain electrophysiological properties. After injury to the central nervous system, molecular reconfiguration of ion channels, neurotransmitters, and receptors takes place in nociceptive neurons, leading to dysfunctional firing properties and the development and maintenance of central pain syndromes.

Characteristics

The spinal cord dorsal horn contains the primary synapse through which afferent somatosensory information related to touch, pressure, brush, temperature, and noxious stimuli is received from the periphery. Dorsal horn sensory neurons receive this information primarily from the skin, perform a degree of processing, and transmit signals through distinct tracts within the spinal cord to supraspinal structures where pain is perceived. Within peripheral nerves and spinal cord, pain is thought to exist simply as a signal and is subject to a degree of modulation by circuitry that it passes through.

Experimental spinal cord injury (SCI) induces electrophysiological changes in dorsal horn sensory neurons that contribute to pain-like behaviors in animals (Hains et al. 2003b). These include shifts in proportions of cells responding to evoked noxious stimulation, increases and irregularity in spontaneous ▶ background activity/firing, increased ▶ evoked activity to (formerly) innocuous and noxious stimuli, and increases in ▶ afterdischarge activity following stimulation. Since the nature of the applied stimuli does not change, other mechanisms must account for the alterations in stimulus processing that lead to central pain. Central changes in expression of molecules, such as ion channels, neurotransmitters, and their receptors, contribute to altered sensory processing by changing the electrophysiological excitability and therefore output of sensory neurons. Specific molecular changes shown to contribute to the development of central pain in animal models and specifically to the development of pain following spinal cord injury (SCI) are described below.

Ion Channels

At the most fundamental level, ▶ action potential generation and propagation by sensory neurons relies on multiple isoforms of ▶ voltage-gated sodium channels (termed Na_v). The selective expression of ensembles of sodium channels tunes the biophysical properties of each neuron. Within the normal nervous system, properties fundamental to neuronal function such as activation threshold, inactivation and ▶ refractory period, rates of ▶ repriming, and the ability to generate and conduct high-frequency trains of action potentials all depend on the type(s) of sodium channels expressed within a given neuron (Waxman 2000). Similarly, dysregulation of channel expression can abnormally reconfigure neuronal function in disease states. Ten genes encode molecularly distinct voltage-gated sodium channels, at least seven of which are expressed in the rat nervous system. In the adult spinal cord dorsal horn, $Na_v1.1$, $Na_v1.2$, and $Na_v1.6$ are strongly detectable, whereas $Na_v1.3$ expression decreases with development and is downregulated in adults. In the adult spinal cord, $Na_v1.3$ is expressed at very low levels (Fig. 1a) (Felts et al. 1997). Following SCI, expression of $Na_v1.3$ is upregulated in the dorsal horn (Fig. 1b) and colocalizes with NK^1R, a marker of nociceptive neurons. $Na_v1.3$ plays a role in maintaining hyperresponsiveness to peripheral stimulation as well as pain-related behaviors, as evidenced through selective knockdown of $Na_v1.3$ expression via antisense oligodeoxynucleotide administration (Fig. 1c) (Hains et al. 2003a). The $Na_v1.3$ sodium channel produces a rapidly repriming tetrodotoxin-sensitive sodium current that permits neuronal firing at higher than normal frequencies (Cummins et al. 2001). Since stimulus intensity is encoded in the dorsal horn by the rate of firing, this change in how neurons process incoming sensory information serves to amplify incoming signals so that perceived pain thresholds are lowered.

Neurotransmitters and Receptors

Neurotransmitter molecules and their receptors produce signals that cause depolarization of dorsal horn sensory neurons. After SCI, neurotransmitter levels are altered in a way that is facilitative to central pain. ▶ Serotonin (5-HT) is an important molecule that participates in the functional modulation of dorsal horn nociceptive neurons. Released by terminals of fibers descending from cell bodies located within the midline ▶ raphe nuclei of the brain stem, 5-HT can act directly on ion channels or trigger intracellular cascades that cause neuronal depolarization and/or regulate intrinsic excitability. SCI results in a loss of the supply of 5-HT within the spinal cord through interruption of descending sources (Hains et al. 2001). After SCI, replacement of interrupted 5-HT reduces abnormal activity of dorsal horn neurons, as well as pain-related behaviors.

Following SCI, 5-HT receptors also undergo changes in expression levels, a change thought to play a role in the development of central pain. The major class of 5-HT receptor found in the dorsal horn is the $5\text{-}HT_1$ family (Zemlan and Schwab 1991). Of this subset of receptors, $5\text{-}HT_{1A}$ represents a high percentage of all high-affinity 5-HT binding sites, with highest receptor densities in laminae I and II. $5\text{-}HT_{1A}$ couples an intracellular G-protein cascade and is upregulated after SCI (Giroux et al. 1999). Also highly represented in the spinal cord, the $5\text{-}HT_3$ receptor, which gates a nonselective monovalent cation channel, has been found in the substantia gelatinosa at all levels of the spinal cord (Hamon et al. 1989). Following SCI, direct activation of both the $5\text{-}HT_{1A}$ and $5\text{-}HT_3$ (Hains et al. 2002) receptors reduces neuronal hyperresponsiveness and/or pain-related behaviors caudal to the injury site. $5\text{-}HT_3$ receptors are thought to be facilitatory to pain after SCI at levels rostral to the injury site (Oatway et al. 2004). The 5-HT transporter protein, which regulates the activity level of 5-HT, undergoes upregulated expression following SCI (Hains et al. 2001).

Reductions in spinal levels of the inhibitory neurotransmitter gamma-aminobutyric acid (▶ GABA) also support central pain after SCI by decreasing the inhibitory influences within spinal circuitry (Drew et al. 2004). Synthesized locally, GABA is the major inhibitory

Molecular Contributions to the Mechanism of Central Pain, Fig. 1 Na$_v$1.3 mRNA is expressed at low levels in naïve animals, (**a**) but is upregulated in lumbar dorsal horn neurons after spinal cord injury (SCI). (**b**) Treatment with antisense oligodeoxynucleotides against Na$_v$1.3 reduces Na$_v$1.3 transcripts after SCI. (**c**) Corresponding unit recordings show evoked activity to peripheral stimulation (*BR* brush, *PR* press, *PI* pinch, increasing strength von Frey filament stimulation and noxious thermal heating (47 °C)), after SCI (**b**) compared to controls (**a**). After Na$_v$1.3 antisense delivery, evoked activity of dorsal horn neurons resembles that found in naïve levels (**c**)

neurotransmitter and acts via ligand-gated ion channels (GABA$_A$ receptors) and G-protein-coupled (GABA$_B$) receptors. After SCI, reduced GABAergic inhibition results in abnormally exaggerated evoked and spontaneous neuronal firing through the impairment of GABA production or release and/or loss of GABA-releasing cells.

Another mechanism for maintained hyperexcitability of dorsal horn neurons leading to central pain involves changes in the expression of metabotropic glutamate receptors (▶ mGluRs).

mGluRs are G-protein-coupled receptors that have been subdivided into three groups, based on sequence similarity, pharmacology, and intracellular signaling mechanisms. Group I is made up of mGluR1 and 5, Group II is made up of mGluR2 and 3, and Group III is made up of mGluR4, 6, 7, and 8. mGluR1 is found primarily in deeper laminae of the dorsal horn, and mGluR5 is found at highest levels in lamina II. mGluR2/3 is also expressed in high levels within lamina II. The differential distribution of mGluRs within laminae associated with sensory processing implicates them in central pain after SCI. In the spinal cord, mGluRs directly modulate neuronal excitability; for example, group I mGluRs decrease thresholds of activation, inhibit afterhyperpolarization, and induce transient membrane depolarizations. Following SCI, mGluR1 expression increases at the level of injury, and mGluR2/3 expression levels are chronically decreased in and around the lesion site. There is also a chronic increase in mGluR1 in all laminae measured and a decrease in mGluR2/3 in laminae Iii, III, IV, and V (Mills et al. 2001). SCI produces an increase in mGluR1 expression on spinothalamic tract neurons in both the cervical enlargement and the spinal segment just rostral to the injury site (Mills et al. 2002). Treatment with agonists to groups II and III mGluRs affects pain responses following SCI.

Genetic changes that may play a role in pain pathogenesis after SCI have been brought to light by microarray analysis. SCI can lead to altered expression of genes whose transcripts help to determine membrane excitability, including increases in mRNAs for ionotropic glutamate receptors and sodium and calcium channels and decreases in mRNAs for GABA receptors and potassium channels (Nesic et al. 2002).

In summary, molecular events involved in the processing of afferent signals in spinal neurons significantly change following SCI. These changes endow affected neurons with maladaptive functional properties and provide the substrate to both generate and amplify incorrect signals that ultimately contribute to central pain after SCI. Although the majority of studies related to molecular changes following central injury have focused on changes occurring at the level of the spinal cord (following spinal injury), similar changes are likely to be found at supraspinal levels.

References

Cummins, T. R., Aglieco, F., Renganathan, M., et al. (2001). Nav1.3 sodium channels: Rapid repriming and slow closed-state inactivation display quantitative differences after expression in a mammalian cell line and in spinal sensory neurons. *Journal of Neuroscience, 21*, 5952–5961.

Drew, G. M., Siddall, P. J., & Duggan, A. W. (2004). Mechanical allodynia following contusion injury of the rat spinal cord is associated with loss of GABAergic inhibition in the dorsal horn. *Pain, 109*, 379–388.

Felts, P. A., Yokoyama, S., Dib-Hajj, S., et al. (1997). Sodium channel alpha-subunit mRNAs I, II, III, NaG, Na6 and hNE (PN1): Different expression patterns in developing rat nervous system. *Brain Research. Molecular Brain Research, 45*, 71–82.

Giroux, N., Rossignol, S., & Reader, T. A. (1999). Autoradiographic study of alpha[1-] and alpha2-noradrenergic and serotonin1A receptors in the spinal cord of normal and chronically transected cats. *The Journal of Comparative Neurology, 406*, 402–414.

Hains, B. C., Fullwood, S. D., Eaton, M. J., et al. (2001). Subdural engraftment of serotonergic neurons following spinal hemisection restores spinal serotonin, downregulates serotonin transporter, and increases BDNF tissue content in rat. *Brain Research, 913*, 35–46.

Hains, B. C., Willis, W. D., & Hulsebosch, C. E. (2002). Serotonin receptors 5-HT1A and 5-HT3 reduce hyperexcitability of dorsal horn neurons after chronic spinal cord hemisection injury in rat. *Experimental Brain Research, 149*, 174–186.

Hains, B. C., Johnson, K. M., Eaton, M. J., et al. (2003a). Serotonergic neural precursor cell grafts attenuate bilateral hyperexcitability of dorsal horn neurons after spinal hemisection in rat. *Neuroscience, 116*, 1097–1110.

Hains, B. C., Klein, J. P., Saab, C. Y., et al. (2003b). Upregulation of sodium channel Nav1.3 and functional involvement in neuronal hyperexcitability associated with central neuropathic pain after spinal cord injury. *Journal of Neuroscience, 23*, 8881–8892.

Hamon, M., Gallissot, M. C., Menard, F., et al. (1989). 5-HT3 receptor binding sites are on capsaicin-sensitive fibres in the rat spinal cord. *European Journal of Pharmacology, 164*, 315–322.

Mills, C. D., & Hulsebosch, C. E. (2002). Increased expression of metabotropic glutamate receptor subtype 1 on spinothalamic tract neurons following spinal cord injury in the rat. *Neuroscience Letters, 319*, 59–62.

Mills, C. D., Fullwood, S. D., & Hulsebosch, C. E. (2001). Changes in metabotropic glutamate receptor expression following spinal cord injury. *Experimental Neurology, 70*, 244–247.

Nesic, O., Svrakic, N. M., Xu, G. Y., et al. (2002). DNA microarray analysis of the contused spinal cord: Effect of NMDA receptor inhibition. *Journal of Neuroscience Research, 68*, 406–423.

Oatway, M. A., Chen, Y., & Weaver, L. C. (2004). The 5-HT3 receptor facilitates at-level mechanical allodynia following spinal cord injury. *Pain, 110*, 259–268.

Waxman, S. G. (2000). The neuron as a dynamic electrogenic machine: Modulation of sodium-channel expression as a basis for functional plasticity in neurons. *Philosophical Transactions of the Royal Society of London. Series B, Biological Sciences, 355*, 199–213.

Zemlan, F. P., & Schwab, E. F. (1991). Characterization of a novel serotonin receptor subtype (5-HT1S) in rat CNS: Interaction with a GTP binding protein. *Journal of Neurochemistry, 57*, 2092–2099.

Monoamine Oxidase Inhibitors

▶ Antidepressant Analgesics in Pain Management

Monoamines in Pain Pathways

Definition

Classification of neurochemicals based on their chemical structure (usually used to refer to serotonin, norepinephrine, and dopamine).

Cross-References

▶ Stimulation-Produced Analgesia

Monocyte/Macrophage

Definition

Monocytes are circulating immune/inflammatory cells that originate in the bone marrow and differentiate into macrophages after migrating through blood vessel walls into tissues.

Cross-References

▶ Wallerian Degeneration

Monosynaptic

Definition

Pertaining to a single synapse.

Cross-References

▶ Amygdala, Pain Processing and Behavior in Animals

Mood

Definition

DSM4 APA includes mood disorders associated with depression: depressive disorders, bipolar disorders, mood disorders due to a general medical condition, and substance-induced mood disorder, as well as anxiety, especially post-traumatic stress disorder, acute stress disorder, generalized anxiety disorder, anxiety disorder due to a general medical condition, substance- induced anxiety disorder, and anxiety disorder not otherwise specified. Other mood states may include anger, frustration, fear, and disappointment.

Cross-References

▶ Pain Assessment in the Elderly
▶ Psychological Treatment in Acute Pain

MOP

▶ Mu(μ)-Opioid Receptor

MOP Receptor

Definition

The term μ-opioid (μ from *m*orphine) peptide receptor represents the G-protein-coupled receptor that responds selectively to the majority of clinically useful opioid drugs. It is expressed in areas of the nervous system that mediate therapeutic and adverse effects of most opioid drugs. The MOP receptor protein is produced by a single gene. Several mRNA splice variants are known to exist and produce receptor proteins that display different properties when expressed in cells. When activated, the MOP receptor predominantly transduces actions via inhibitory G proteins. The electrophysiological consequences of MOP receptor activation are usually inhibitory.

Cross-References

▶ Mu(μ)-Opioid Receptor
▶ Opioid Electrophysiology in PAG

MOR-1

Definition

A clone encoding a mu opioid receptor.

Cross-References

▶ Opioid Receptors

Moral Hazard

Definition

When the presence of insurance changes a disability claimant's or health-care provider's behavior.

Morbus Sudeck

▶ Complex Regional Pain Syndromes, Clinical Aspects
▶ Sympathetically Maintained Pain in CRPS I, Human Experimentation

Morning Stiffness

Definition

In arthritis, morning stiffness of joints and muscles may last for several hours.

Cross-References

▶ Muscle Pain in Systemic Inflammation (Polymyalgia Rheumatica, Giant Cell Arteritis, Rheumatoid Arthritis)

Morphine

Definition

Morphine is a strong (potent) naturally occurring opiate (a phenanthrene derivative) produced from extracts of the poppy plant. Morphine undergoes extensive hepatic biotransformation by phase II reactions to morphine-3-glucuronide (70 %) and morphine-6-glucuronide (5–10 %), and the remainder undergoes sulfation.

Cross-References

▶ Forebrain Modulation of the Periaqueductal Gray and Its Role in Pain
▶ Postoperative Pain, Morphine

Morphine and Muscle Pain

▶ Opioids and Muscle Pain

Morphine Sulfate

▶ Postoperative Pain, Morphine

Morphine Tolerance

▶ Glutamate Homeostasis and Opioid Tolerance

Morphology, Intraspinal Organization of Visceral Afferents

Yasuo Sugiura
School of Human Care Study, Nagoya University
of Arts and Sciences, Nissin City, Aichi, Japan

Synonyms

Synaptic organization of afferent fibers from
viscera in the spinal cord

Definition

▶ Visceral afferent Aδ- and ▶ C-fibers convey
sensory information from the viscera via vagal
and spinal nerves. Vagal afferents, with cell bodies in the nodose and jugular ganglia, terminate
principally in the brainstem nucleus tractus
solitarius; spinal afferents, with cell bodies in
dorsal root ganglia, terminate in the spinal cord.

Introduction

Visceral primary afferent fibers terminate in the
brain stem and spinal cord, transferring information to second-order neurons. Most visceral afferent input is not consciously perceived, but rather
is associated with what are often termed vegetative functions, including normal baroreceptor
activation, respiration, digestion, and motility.
Visceral afferent input that is consciously perceived is typically restricted to discomfort and
pain (e.g., dyspnea, nausea, pain). Visceral pain
is typically diffuse in character and difficult
to localize, in part because of the intraspinal
organization of visceral afferent terminals in the
spinal cord.

Characteristics

Terminal Distribution in the Spinal Cord

Visceral afferent fibers terminate in the superficial dorsal horn (spinal laminae I and II; see
▶ Spinal Cord Laminae) and deeper ▶ spinal
laminae (IV–V, X) (de Groat et al. 1978).
A single visceral afferent C-fiber can project to
the dorsal, superficial spinal cord and also in
laminae V and X of the contralateral spinal cord
(Sugiura et al. 1989). Visceral afferent C-fibers
have collateral branches with terminal swellings
in the adjacent white mater of the ▶ lateral and
dorsal funiculi, in addition to terminal branches
in the spinal dorsal horn, revealing a distinctly
different pattern of termination from non-visceral
afferent C-fiber terminals (Fig. 1). Along the longitudinal axis of the spinal cord, visceral afferent
C-fibers terminate with about 20 collateral
branches, which are given off at 3–400-μm
spans in several ▶ spinal segments from rostral
and caudal branches. The rostral branch of the
C-fiber shown in Fig. 2 runs in the surface of the
dorsal funiculus, which may project in the dorsal
column to the nucleus gracilis, part of a new tract
for visceral pain (Fig. 2) (Al-Chaer et al. 1998).
Each collateral branch ramifies in a relative narrow sheet consisting of only one or two daughter
branches, which do not form a concentrated nest-like termination commonly seen with non-visceral C-fibers. The collateral branches have
250–300 terminal swellings, forming an array
arranged in the orientation of the neuropil of the
spinal laminae (Sugiura et al. 1993). The number
of terminal swellings is summarized in Table 1.
About 5,000–6,000 enlargements (boutons) were
identified in visceral afferent C-fibers. Over 60 %
of visceral terminal swellings are found in the
superficial dorsal horn, lamina I, and adjacent
area, which seems to be the main region of termination. About 10 % of terminal swellings are
found in deeper dorsal horn (laminae IV and V).
A few visceral fiber boutons are in other laminae.

Comparative Pattern of Unmyelinated Terminals: Cutaneous, Muscle, and Visceral Afferents

Intraspinal terminals of unmyelinated afferent
fibers revealed different patterns depending on

Morphology, Intraspinal Organization of Visceral Afferents, Fig. 1 A transverse reconstruction of the central projection of a visceral afferent C-fiber. One of the collaterals could be traced to synaptic enlargements in lamina II. *Broken lines* indicate borders of laminae I, II, and III; *LF* lateral funiculus, *DF* dorsal funiculus, *CC* central canal. (Sugiura et al. 1989)

tissues of origin – skin, muscle, and viscera. The terminal areas in the spinal cord are schematically drawn in Fig. 3, in which each terminal showed characteristic terminal patterns. Afferents from skin had a distinct, delimited small area of termination whereas afferents from viscera exhibited a large, wide-spread distribution in the spinal cord. Afferents from muscle had areas of termination larger than those from skin, but significantly less than those from viscera.

Synaptic Organization
Light microscopic examination reveals information about intraspinal terminal swellings (boutons). The terminal swellings of visceral afferent C-fibers are small, ranging between 1.6 and 1.7 μm in diameter and 4.3 and 5.7 μm² in area each, relative to non-visceral afferent terminal swellings. The distribution of visceral terminal swellings occupies a smaller fraction (one third to one half) of non-visceral terminal swellings.

In electron microscopy, primary afferent terminals in the spinal cord generally show a central synaptic profile composed of axons and some dendritic spines. In non-visceral nociceptors,

over 75 % of terminals show central synaptic profiles whereas visceral afferent terminals show a ► simple synapse and less than 30 % of terminals show central synaptic profiles (Table 2) (reviewed in Sugiura and Tonosaki 1995).

A great number of visceral afferent cell bodies in dorsal root ganglia contain neuropeptides, especially substance P, calcitonin-gene-related peptide, somatostatin, and vasoactive intestinal peptide, and some are IB-4 positive.

Functional Extrapolations from Morphological Organization
Visceral afferent C-fibers terminate in spinal laminae I and V, similar to terminals of thinly myelinated fibers from non-visceral nociceptors, which are different from C-fiber terminals of skin and muscle (Ling et al. 2003). The laminar arrangement of visceral afferent C-fiber terminals suggests that ► somatovisceral convergence onto second-order neurons in laminae I and V arises principally from non-visceral myelinated and visceral unmyelinated fibers. Somatovisceral convergence relates to referred pain, common in visceral pain conditions. In contrast, unmyelinated afferents of non-visceral origin may thus

10588
Rest.

Morphology, Intraspinal Organization of Visceral Afferents, Fig. 2 Sagittal view of reconstructions of the arborization of a visceral afferent C-fiber. On entering the spinal cord (*arrow head*), the axon bifurcated into rostral and caudal main branches. The main branches ran on the dorsal surface of the dorsal funiculus or in Lissauer's tract, as far rostrally as the 10th thoracic segment (*top left*) and as far caudally as the 2nd lumbar segment (*bottom right*). Many collaterals left these parent branches to terminate in laminae I (I) and II (II). Some collaterals also terminated in the lateral funiculus (LF) and the ipsilateral or contralateral laminae V (V, CV) and X (X, CX). The *asterisk* in a circle (*bottom*) shows the terminal profiles in lamina II, which is indicated by the *circle* in the upper view

Morphology, Intraspinal Organization of Visceral Afferents, Table 1 Number of terminal swellings

	Visceral fibers[a]		Non-visceral fibers	
Case number	10,588	30,488	70,887	72,987
Number of swellings	6,099(100)	5,370(100)	1,482	1,450
Number of terminal areas	22	18	1	2
Mean number of areas	277	298	–	725
Dorsal funiculus	359(5.9)	1,089(20.3)		
Lamina I	3,735(61.2)	2,466(45.9)		
Lamina II	240(3.9)	151(2.8)		
Lamina IV and V	391(6.4)	553(10.3)		
Lateral funiculus	276(4.5)	21(0.4)		
Lamina X	171(2.8)	153(2.8)		
Contralateral X	121(2.0)	61(1.1)		
Contralateral VI and V	108(1.8)	142(2.6)		
Total	5,401(88.6)	4,636(86.3)		

[a]Number and (percentage)

Morphology, Intraspinal Organization of Visceral Afferents, Fig. 3 Schematic representation of terminal patterns of cutaneous, muscle and visceral afferent c-fibers. Cutaneous afferents densely terminated in laminae I and II, and occupy small areas longitudinally and in the medio-lateral dimension in the spinal cord (*top*). Muscle afferents terminated in 2–3 segments in the spinal cord in laminae I, II, and III with two different terminal patterns (Ling et al. 2003) (*middle*). Viscera afferents terminated in a large, wide-spread distribution over multiple segments in the spinal cord (*bottom*). These different patterns of intraspinal termination suggest functional differences between afferents arising from different tissues

principally subserve the transmission of nociceptive information onto second-order neurons in the dorsal horn.

Visceral afferent C-fibers terminate widely in thin sheets of termination characterized by one or two terminal branches with several enlargements at each terminal focus. In contrast, intraspinal terminations of cutaneous afferents are delimited. These features of intraspinal terminations provide the morphological foundation for the poor localization of visceral pain relative to cutaneous pain.

Morphology, Intraspinal Organization of Visceral Afferents, Table 2 Incidence of central terminals of primary C-afferents

	Fraction	Percentage
Non-visceral afferents:		
High-threshold mechanoreceptor	24/32	75.0
Polymodal nociceptor	13/16	81.03
Mechanical cold nociceptor	53/63	84.1
Warming receptor	12/38	31.5
Visceral afferents:		
Superficial layer	3/29	10.3
Deeper layer	10/38	26.3

The intraspinal patterns of distribution of muscle afferents suggest two functions. Similar to both cutaneous and viscera afferents, muscle pain is better localized than is visceral pain, but not as well as cutaneous pain.

The size of boutons on visceral afferent C-fiber terminals is smaller than the boutons found on many kinds of myelinated fiber terminals (Brown 1981) and those of non-visceral afferent C-fibers (Sugiura and Tonosaki 1995). This difference in size may have relevance to the receptor or to the nature of synaptic transmission in the dorsal horn. The size of terminal swellings of visceral afferent fibers seems to reflect the morphological variety of synaptic ultrastructural profiles (Ribeiro da Silva and Coimbra 1982; Alvarez et al. 1993).

Non-visceral C-fiber nociceptor central terminal profiles have various shapes according to the adequate noxious stimulus: high-threshold nociceptor, cold nociceptor, or polymodal nociceptor. These ultrastructural profiles may imply to modulate pain at the synaptic site of central terminals (Wall 1989). Contrary to the non-visceral nociceptors, most visceral afferent C-fiber terminals show simple synaptic profiles (Fig. 4). This

Morphology, Intraspinal Organization of Visceral Afferents, Fig. 4 An ultrastructural profile of a visceral afferent C-fiber. A visceral afferent terminal has simple synaptic contacts with postsynaptic elements. To easily appreciate the synaptic relationship, a schematic illustration is drawn (*right bottom*). Bar: 1 μm (Modified from Sugiura et al. 1992; Sugiura and Tonosaki 1995)

means that these terminals transmit peripheral inputs to second-order neurons or elements without modulating that information at the synaptic site. In this interpretation, non-visceral central terminals may be essential to transmit pain from peripheral non-visceral tissues to the central nervous system, but visceral pain may be transmitted by other mechanisms or mediated by small numbers of central terminals. If we presume that these central terminals mainly mediate visceral pain, visceral pain would be a vague, strange feeling and not definitive for nociception.

References

Al-Chaer, E. D., Feng, Y., & Willis, W. D. (1998). A role for the dorsal column in nociceptive visceral input into the thalamus of primates. *Journal of Neurophysiology, 70,* 3143–3150.

Alvarez, F. J., Kavookjian, A. M., & Light, A. R. (1993). Ultrastructural morphology, synaptic relationships, and CGRP immunoreactivity of physiologically identified C-fiber terminals in the monkey spinal cord. *The Journal of Comparative Neurology, 329,* 472–490.

Brown, A. G. (1981). *Organization in the spinal cord: The anatomy and physiology of identified neurons* (pp. 1–138). New York: Springer.

de Groat, W. C., Nadelhaft, I., Morgan, C., & Schauble, T. (1978). Horseradish peroxidase tracing of visceral afferent and primary afferent pathways in the cat's sacral spinal cord using benzidine processing. *Neuroscience Letters, 10,* 103–108.

Ling, L.-J., Honda, T., Shimada, Y., Ozaki, N., Shiraishi, Y., & Sugiura, Y. (2003). Central projection of unmyelinated (C) primary afferent fibers from gastrocunemus muscle in the guinea pig. *The Journal of Comparative Neurology, 461,* 140–150.

Ribeiro da Silva, A., & Coimbra, A. (1982). Two types of synaptic glomeruli and their distribution in laminae I-III of the rat spinal cord. *The Journal of Comparative Neurology, 209,* 176–186.

Sugiura, Y., & Tonosaki, Y. (1995). Spinal organization of unmyelinated visceral afferent fibers in comparison with somatic afferent fibers. In G. F. Gebhart (Ed.), *Visceral pain, progress in pain research and management* (Vol. 5, pp. 41–59). Seattle: IASP Press.

Sugiura, Y., Terui, N., & Hosoya, Y. (1989). Difference in distribution of central terminals between visceral and somatic unmyelinated (C) primary afferent fibers. *Journal of Neurophysiology, 62,* 834–840.

Sugiura, Y., Tonosaki, Y., Nishiyama, K., Honda, T., & Oda, S. (1992). Organization of the central projections of unmyelinated primary afferent fibers. In R. Inoki, Y. Shigenaga, & M. Tohyama (Eds.), *Processing and inhibition of nociceptive information* (pp. 9–13). Amsterdam: Excerpta Medica.

Sugiura, Y., Terui, N., Hosoya, Y., Tonosaki, Y., Nishiyama, K., & Honda, T. (1993). Quantitative analysis of central terminal projections of visceral and somatic unmyelinated (C) primary afferent fibers in the guinea pig. *The Journal of Comparative Neurology, 332,* 315–325.

Wall, P. D. (1989). Introduction. In P. D. Wall & R. Merzack (Eds.), *Textbook of pain* (2nd ed., pp. 1–18). Edinburgh: Churchill Livingstone.

Motif

Definition

The motif is the region of a protein that is responsible for binding or functional action, which is homologous to other proteins of like function.

Cross-References

▶ Trafficking and Localization of Ion Channels

Motivation

Definition

Motivation is that which contributes or leads to the initiation, direction, persistence, intensity, or termination of behavior.

Cross-References

▶ Motivational Aspects of Pain
▶ Psychology of Pain and Psychological Treatment

Motivational Aspects of Pain

Mark P. Jensen
Department of Rehabilitation Medicine,
University of Washington, Seattle, WA, USA

Synonyms

Drive; Incentive

Definition

▶ Motivation refers to the initiation, direction, persistence, intensity, and termination of behavior (Landy and Becker 1987). According to most theories of motivation, the primary factors that lead to behavior change can be classified as falling into two categories: (1) the perceived importance of behavior change and (2) the belief that behavior change is possible (i.e., ▶ self-efficacy; Bandura 1986).

Characteristics

As the management of chronic pain depends more on what patients do than on what is done to them, patient motivation for pain self-management represents an essential, yet understudied aspect of clinical care (Jensen et al. 2003). The more that clinicians understand motivation and the factors that contribute to motivation for pain self-management, the better they will be at helping patients learn, practice, and use adaptive pain self-management coping strategies.

Theories of Health Behavior and Health Behavior Change

A number of theories and models have been developed that address the motivational issues that impact on health behavior and health behavior change: the operant model, cognitive-behavioral theory, expectancy-value models, the transtheoretical model, and client-centered approaches. The operant learning theory of chronic pain emphasizes the role that environmental ▶ contingencies have on pain behavior (Fordyce 1976). According to this model, patients continue to use maladaptive pain management strategies (such as pain-contingent rest and guarding) because such strategies are followed by reinforcement. Motivation for change therefore results when the contingencies are changed: for example, by encouraging and praising adaptive coping efforts and ignoring maladaptive ones. Cognitive-behavioral theory views behavior as a complex interaction of cognitive structures, processes, and their consequences. Cognitive-behavioral therapy for chronic pain uses various strategies for helping patients change maladaptive cognitions that contribute to both pain and suffering, and also integrates a number of other techniques for modifying, teaching, and encouraging adaptive coping skills (▶ positive reinforcement, ▶ biofeedback, self-control techniques) (Bradley 1996). Expectancy-value models of motivation and health behavior change posit that motivation for behavior and behavior change is determined by patient's beliefs (expectancies) and values (incentives) (Bandura 1986; Janz and Becker 1984; Rogers 1983). According to these models, motivation for behavior change increases as people believe that the benefits of change outweigh the costs, and as they increase their beliefs that change is possible. The transtheoretical model of behavior change (Prochaska and DiClemente 1984) emphasizes the fact that people vary in their readiness (or motivation) to make change, and classify readiness into five distinct stages: ▶ precontemplation, ▶ contemplation, ▶ preparation, action, and maintenance. This model argues that different therapeutic responses may be used to help patients move from one stage of readiness to the next, until the patient is able to maintain the change for good. Although there has been debate concerning whether readiness to change is best thought of as a stage or continuous dimension, few researchers question the importance of the readiness construct, especially when determining the approach best taken with a particular patient (see Jensen et al. 2003). Finally, client-centered approaches to patient motivation, such as motivational interviewing and the patient-centered counseling model, have identified specific therapeutic responses that increase patient motivation for adaptive behavior change, including responses such as developing a collaborative (rather than confrontive) relationship, eliciting and reinforcing patient's self-statements consistent with adaptive change, and increasing patient's awareness of the risks of maladaptive responses, while at the same time encouraging patient's beliefs that adaptive change is possible (Miller and Rollnick 2002; Ockene et al. 1988).

Clinical Implications of the Motivation Models for Enhancing Patient's Self-Management

Clinicians may use the motivational models of health behavior change as guides when helping motivate patients with chronic pain make adaptive changes. For example, based on the transtheoretical model, clinicians would be wise to assess a patient's readiness to participate in whatever specific pain treatment(s) is/are being recommended (e.g., by stating "I think that this intervention is worth trying." and then asking, "What do you think?"). This is particularly important if the treatment involves active patient participation to be effective (such as keeping to a specific medication regimen, relaxation or self-hypnosis training, and active physical or occupational therapy, etc.). If the patient expresses a clear disinterest in the planned treatment approach (i.e., is in the "precontemplation stage" concerning this approach), then minimal participation can be expected, leading to few benefits. In this case, specific motivational interventions should be considered to help increase patient readiness to engage in treatment. However, motivational interventions can be useful at all readiness levels, to maximize patient commitment to active participation throughout treatment.

A variety of clinician behaviors and responses will increase patient motivation for engagement in active pain management. Most of these can be classified as interventions that impact either the importance the patient places on participating in treatment, and/or the beliefs that participation is possible (see Fig. 1).

To increase the importance for engaging in regular exercise, for example, clinicians could start with providing patients with information concerning the long-term (negative) effects of inactivity and the benefits of regular exercise. However, information alone is unlikely to produce a change motivation for exercise, or any other adaptive coping behavior, unless patients incorporate this information into their belief system. Merely lecturing patients can, in fact, have a paradoxical effect of pushing patients into the position of defending their current maladaptive (e.g., inactivity, rest) coping responses, causing them to argue against and therefore resist making adaptive changes (Miller and Rollnick 2002). A more effective strategy is to first provide information, and then encourage the patient to discuss and think through the implications of this information for his or her own pain problem, that is, to elicit from the patient his or her own thoughts concerning the information provided, and especially to pay attention to and reinforce patient

Motivational Aspects of Pain, Fig. 1 Motivational factors that contribute to pain self-management behaviors

Perceived Importance
Beliefs regarding cost/benefit ratio
Learning history
Current contingencies

Self-Efficacy
Personal experience
Modeling
Verbal Persuasion
Perceived barriers

Readiness to change (or maintain) self-management behaviors

Self-Management Behaviors (Coping)
Exercise
Pacing
Relaxation
Assertiveness
Task persistence
Body mechanics
Positive selfstatements
Ignoring pain
Avoid asking for Assistance
Avoid Guarding
Avoid catastrophizing
Avoid pain contingent rest
Avoid pain contingent analgesics

statements indicating incorporation of the information. Patients are much more likely to believe what they themselves say than what is told to them (Miller and Rollnick 2002).

The importance that patients place on any particular pain coping response can also be changed by altering the contingencies surrounding that response (Fordyce 1976). Clinicians can, and should, praise and reinforce approximations toward adaptive coping, and seek to ignore coping responses associated with greater dysfunction. Patients' attention on their efforts toward adaptive coping can be emphasized by asking questions about these efforts, and helping the patient come up with solutions to problems associated with the use of new adaptive coping strategies. Such discussion can help focus patient's attention on adaptive coping efforts. Patients, and the people who are close to the patient, can also be taught operant principles, and be encouraged to integrate regular reinforcement for adaptive coping (e.g., exercise contingent rest, praise for regular exercise) (Fordyce 1976).

Self-efficacy for adaptive coping can also be increased in a number of ways. One of these is to simply encourage patients to engage (slowly at first) in new adaptive coping efforts. As the patient sees himself/herself engage in adaptive coping, he/she learns that such coping is possible. Such direct experience is perhaps the most powerful strategy for increasing self-efficacy (Bandura 1986). Self-efficacy can also be increased by observing others engage in the behavior (► modeling), such as occurs when pain treatment is provided in a group setting. Eliciting from patients their past successes at changing behavior (Miller and Rollnick 2002), and directly addressing perceived barriers, can also increase self-efficacy for adaptive pain self-management.

By understanding that patient motivation is a state that can vary as a function of clinician responses, and by engaging in responses that increase the importance that patients place on adaptive coping and patient's beliefs that adaptive coping is possible, clinicians can help patients increase their motivation for engagement in treatment and adherence to treatment recommendations. To the

extent that the treatment is effective, motivational interventions can therefore lead to improved outcomes for many patients.

References

Bandura, A. (1986). *Social foundations of thought and action: A social cognitive theory*. Englewood Cliffs: Prentice Hall.

Bradley, L. (1996). Cognitive-behavioral therapy for chronic pain. In R. Gatchel & D. Turk (Eds.), *Psychological approaches to pain management: A Practitioner's handbook* (pp. 131–147). New York: Guilford Press.

Fordyce, W. (1976). *Behavioral methods for chronic pain and illness*. Saint Louis: Mosby.

Janz, N., & Becker, M. (1984). The health belief model: A decade later. *Health Education Quarterly, 11*, 1–47.

Jensen, M. P., Nielson, W. R., & Kerns, R. D. (2003). Toward the development of a motivational model of pain self-management. *The Journal of Pain, 4*, 477–492.

Landy, F., & Becker, W. (1987). Motivation theory reconsidered. *Research in Organizational Behavior, 9*, 1–38.

Miller, W., & Rollnick, S. (2002). *Motivational interviewing: Preparing people to change* (2nd ed.). New York: Guilford Press.

Ockene, J., Quirk, M., Goldberg, R., Kristeller, J., Donnelly, G., Kalan, K., Gould, B., Greene, H., Harrison-Atlas, R., Pease, J., Pickens, S., & Williams, J. (1988). A Residents' training program for the development of smoking intervention skills. *Archives of Internal Medicine, 148*, 1039–1045.

Prochaska, J., & DiClemente, C. (1984). *The transtheoretical approach: Crossing traditional boundaries of therapy*. Homewood: Dow Jones Irwin.

Rogers, R. (1983). Cognitive and physiological processes in fear-based attitude change: A revised theory of protection motivation. In J. Cacioppo & R. Petty (Eds.), *Social psychophysiology: A sourcebook* (pp. 153–176). New York: Guilford Press.

Motivational Components

Definition

Motivational components are a group of basic behaviors induced by a homeostatic survival necessity. Awakening, falling asleep, aggressive/defensive reaction, vocalization, flight, freezing, and hypotonic immobility.

Cross-References

▶ Hypothalamus and Nociceptive Pathways

Motivational Reactions

Definition

Motivational reactions are basic behaviors induced by a homeostatic necessity. Awakening, aggression, flight, freezing, or hypotonic immobility in response to a noxious event.

Cross-References

▶ Parabrachial Hypothalamic and Amygdaloid Projections

Motivational-Affective

Definition

Response to nociceptive stimuli involving the emotional (limbic) and visceral components, including response by homeostatic systems responsible for regulation of blood pressure, respiration rate, and responses involving the hypothalamic-pituitary-adrenal axis.

Cross-References

▶ Spinomesencephalic Tract
▶ Spinothalamic Tract Neurons, Descending Control by Brain Stem Neurons

Motivational-Affective Aspects/ Components of Pain

Definition

The pain experience includes motivational and affective aspects/components, such as negative emotions and arousal.

Cross-References

▶ Motivational Aspects of Pain
▶ Parafascicular Nucleus, Pain Modulation
▶ Spinothalamic Input: Cells of Origin (Monkey)
▶ Thalamic Nuclei Involved in Pain: Cat and Rat

Motor Cortex

Definition

Primate precentral cortex receiving inputs from cerebellar-receiving zones of the thalamus and projecting to the spinal cord, striatum, and pontine nuclei. Histologically identified by the presence of large pyramidal cells in deep layers of cortex.

Cross-References

▶ Pain Treatment, Motor Cortex Stimulation

Motor Cortex Stimulation

▶ Pain Treatment, Motor Cortex Stimulation

Motor Cortex, Effect on Pain-Related Behavior

Timothy K. Y. Kaan, Peter T. Ohara and
Luc Jasmin
Department of Anatomy, University of California
San Francisco, San Francisco, CA, USA

Definition

Electrical stimulation of the ▶ primary motor cortex (M1) is occasionally used to relieve intractable pain. Stimulating electrodes are placed over the primary motor cortex to deliver an electrical current, which activates the underlying neurons. For reasons that are not yet well understood, this leads to a reduction in certain types of chronic pain without affecting thoughts, mood, motor function, or other sensations. It can affect, however, acute pain perception such as the heat pain from a laser. The electrodes are left in placed and can be effective for years.

Characteristics

Electrical stimulation of the motor cortex in humans can result in the relief of chronic, intractable pain that is unresponsive to other therapies. Stimulating the motor cortex for pain relief seems counterintuitive as one would assume that the primary somatosensory cortex (S1) is a more logical target since functional imaging studies show that the sensory, not the motor cortex, is activated by nociceptive stimuli (Bushnell et al. 1999). Large lesions of the primary somatosensory cortex in humans, however, rarely disrupt pain sensation and direct electrical stimulation of the sensory cortex in hundreds of cases of conscious surgery that rarely produced pain sensations.

Historical Background
If activation of the cerebral cortex is necessary for the perception of pain, changing the cortical activity should alter the perception of pain. The proof of this emerged in the early 1990s, largely through the pioneering work of Tsubokawa and colleagues (Hirayama et al. 1990; Tsubokawa et al. 1991, 1993) who reported that stimulating the motor cortex is often successful in relieving central pain (i.e., pain due to lesion of the CNS). The effectiveness of motor cortex stimulation might seem puzzling, particularly if one adopts a textbook view of the nervous system and sees the motor and sensory systems as separate entities. However, it has long been known that stimulation of the motor gyrus can produce sensory responses and anatomical studies have shown connections between somatosensory and motor cortices providing support for interaction between the motor and sensory systems.

In a series of articles beginning in 1990, Tsubokawa and colleagues (Hirayama et al. 1990; Tsubokawa et al. 1991, 1993) reported that stimulating the motor cortex resulted in 70 % of chronic central pain patient, who were not responsive to other treatments, reporting pain relief. The procedure has since gained wide acceptance, and inasmuch as any invasive surgical procedure can be called routine, this form of treatment for chronic central or, occasionally, neuropathic pain has become a standard procedure. In most cases, the motor cortex on the side ▶ contralateral to the area of the body that hurts is stimulated. Ipsilateral stimulation is usually reserved for patients in whom the contralateral motor cortex is damaged, most often following an ischemic event.

Surgical Protocol
Motor cortex-stimulating electrodes are implanted using standard surgical techniques. The precise details vary according to the particular preferences of the surgeon, but there is little evidence that any difference here affects the efficacy of the procedure as long as the localization of the electrodes is adequate. The surgery is carried out under general or local anesthesia, and the cranium is opened exposing the dura mater (the thick opaque ▶ meningeal layer covering the brain). The bony opening consists of one or more small burr holes or a small cranial flap. The dura mater can be left intact or opened to allow direct visualization of the cerebral cortex. The cortical area to be stimulated is then localized. With the existing technology, one cannot stimulate the entire motor cortex, which means that for patients with pain that extends to an entire hemibody, the neurosurgeon and patient have to first agree on where the pain is maximal and if relieving that painful area leaving the others untouched would be worthwhile. For instance, in many patients the most bothersome pain is present in the face and hand. Because those two areas are immediately adjacent in the motor cortex, it is possible to concomitantly stimulate them and bring adequate relief. Once the appropriate location of the electrode array is determined (see below), it is fixed to the outer

Motor Cortex, Effect on Pain-Related Behavior,
Fig. 1 Postoperative lateral cranial X-rays of a patient
with stimulating electrodes (**a**) located over the primary
motor cortex. The leads from the electrodes are tunneled
subcutaneously via the neck to the extension wires (**b**),
which connect it to the internal current generator (not
seen)

surface of the dura matter (most neurosurgeons
prefer there to be no direct contact between the
electrode and the cortex because of risk of scar
formation). The electrode array usually consists
of 4–20 individual electrodes. The electrode
wires are tunneled subcutaneously to the anterior
chest where they can be connected to a small
current generator (similar to a pacemaker) and
the entire system is internalized under the skin
(Fig. 1).

Identifying the Motor Cortex

Although the actual implantation of the electrodes is a relatively straightforward procedure,
identifying the motor cortex is not so easily
achieved. Even if the motor cortex can be recognized on strictly anatomical grounds by its location anterior to the central (Rolandic) sulcus, such
identification is not straightforward. As the dura
mater is not always opened and even when it is
there is only a very limited exposure of the

cortex, it is often difficult to identify landmarks
that would ensure proper location. Modern imaging techniques such as ▶ CT scans or preferably
MRI can give high-quality images in which it is
possible to identify the central sulcus and
precentral gyrus with good precision before any
surgery is initiated. In addition to correctly locating the motor cortex, one has to also determine
the ▶ somatotopical location. Functional MRI
allows   one   to   selectively   locate   the
somatotopically relevant areas of the motor cortex by having the subject perform small movements of the hand or foot while in the magnet.
The use of the images obtained preoperatively
combined with an intraoperative computer-
assisted 3D navigation and guidance system
(▶ neuronavigation) and direct electrical stimulation or recording of the cortex allows the surgeon to locate the appropriate parts of the motor
cortex in most cases. Also, the transition from
motor to sensory cortex can be recorded in the
form of a "phase reversal" when using somatosensory evoked potentials. Additional methods to
localize the somatotopically relevant areas are to
record the cortical potentials evoked in response
to stimulating a peripheral nerve or to stimulate
the motor cortex and locate the region of the body
where muscle twitches or movements occur using
standard electromyography. These latter techniques suffer some technical issues that reduce
their reliability.

As soon as the electrodes are implanted, it is
usually preferable to test the electrodes for placement and function. Although the anesthesia often
precludes testing for pain relief, it is good confirmation of proper operation of the electrodes if
some muscle twitch of the targeted area
is obtained on low-frequency stimulation
(<5 cycle/s). Once in place and functioning correctly, the stimulation parameters required for
pain relief are below those that produce any muscle activation or seizures. A wide range of stimulation parameters that have been reported as
effective (Brown and Barbaro 2003; Canavero
and Bonicalzi 2001, 2002) as might be imagined
from the large number of variables involved in
each placement. The stimulation parameters that
have been used and reported to provide relief

range from 0.5 to 1.5 V with frequencies from 5 to 130 Hz and pulse widths from 60 to 460 μs. The times used for stimulation also vary widely from continuously on to 10 min a day. The most usual is several hours at a time spread evenly throughout the day or activated by the patient as required. As with other treatments for central or neuropathic pain, it is sometimes the case that short-term treatment with the cortical stimulators results in long-lasting and occasionally permanent pain relief.

Indications and Reported Efficacy

The reported success rate of this invasive procedure runs the gamut, with some patients reporting no relief while others experiencing complete (i.e., 100 %) analgesia. Fifty to eighty percentage pain relief seems to be the more usual result (Brown and Barbaro 2003; Canavero and Bonicalzi 2001). In terms of numbers of patients helped, Canavero and Bonicalzi (Canavero and Bonicalzi 2001) conclude that approximately 50 % of patients with central pain had satisfactory long-term results. A more recent review of studies reporting outcomes of invasive motor cortex stimulation concluded that overall, approximately 50 % of chronic neuropathic pain patients experienced pain relief greater than 40 % following surgery and for up to a year for those who had postoperative follow-up (Fontaine et al. 2009). Another systematic review reported a positive responder rate of greater than 70 % following motor cortex stimulation (Lima and Fregni 2008). In comparison, the same study reported analgesia from noninvasive brain stimulation, including transcranial magnetic stimulation and transcranial direct-current stimulation, in around 45 % of treated patients. There is a report that used double-blinded testing to determine whether a chronic stimulator should be implanted following electrode placement on the motor cortex and suggested a placebo effect in 35 % of cases (6 out of 17 patients) (Rasche et al. 2006). Recent randomized crossover trials that were carried out by switching the motor cortex stimulator "on" and "off" have provided overall positive results. Despite the small sample sizes of less than 20 patients, two individual studies reported

significant reduction in various clinical pain scales during the double-blinded crossover phase, and the effect persisted for at least 1 year (Nguyen et al. 2008; Velasco et al. 2008). Although another study reported the results of randomized crossover phase to be negative (attributed to carryover and ceiling effects left from the long-lasting efficacy of motor cortex stimulation during and immediately following surgery), efficacy was found during the open long-term trial that lasted 1-year postoperation (Lefaucheur et al. 2009).

Motor cortex stimulation is clearly most applicable in central or neuropathic pain. These conditions include poststroke pain, trigeminal neuropathic pain, brachial plexus and root avulsions, spinal cord injury pain, phantom limb and stump pain, postherpetic neuralgia, and complex regional pain syndrome. The fact that the technique involves surgical intervention and cannot be considered without risks and complications, and that success is not ensured, argues that other less invasive therapies might be considered first. In this context, it would be useful if there were some ways to predict the success of the procedure. Although there have been attempts to find predictive factors, there are not yet any reliable markers. Barbiturate sensitivity and opioid insensitivity have been suggested as possible predictors of response (Tsubokawa et al. 1993; Yamamoto et al. 1997) as has ▶ transcranial magnetic stimulation and motor functions (Migita et al. 1995; Katayama et al. 1998). None of these indicators has been found to be infallible and exceptions to these predictors have been reported.

Overall, use of motor cortex stimulation to relieve refractory neuropathic pain seems to offer encouraging results. However, these results are based on a small number of patients and there are differences in patient selection, evaluation of pain scores, stimulation parameters, and the origin and types of pain treated. These variables can all have a large impact on the effectiveness of the therapy. Therefore, a large double-blinded, placebo-controlled randomized study is still needed to validate the efficacy of motor cortex stimulation to relief pain.

Possible Mechanisms of Action

The mechanism by which cortical stimulation leads to pain reduction is not clearly understood. In their original description, Tsubokawa and colleagues (Tsubokawa et al. 1991) considered that motor cortex stimulation reduced somatosensory cortex and/or thalamic activity through inhibition evoked by non-nociceptive neuron activity. Although this might be a reasonable general explanation, there is little detail on how, or through exactly what pathways, the inhibition might occur. Cortical stimulation might operate at the cortical level through some form of inhibition or gating of somatosensory cortex or other cortical areas known to be associated with pain (Ohara et al. 2005). Alternatively, cortical stimulation could activate subcortical sites directly or indirectly and disrupt nociceptive signals. Currently, there is no consensus on this question, although one can find individual studies and evidence in the literature to support almost any particular hypothesis. Although we often speak of the motor and sensory systems as though they were separate entities, in actuality, we know they are intimately linked functionally and anatomically. It has been thought for some time (Teuber 1966) that activity in the motor cortex directly influences the somatosensory cortex and evidence has been steadily accumulating to support the idea that activity in the motor cortex alters somatosensory cortex activation by direct connections (Nelson 1996). Thus, it is reasonable to postulate that stimulation of the motor cortex gates or damps down activity in the somatosensory cortex. It must be reiterated that there is still a question over how much role the primary sensory cortex plays in nociceptive processing, so even if motor cortex stimulation does affect S1 directly, it is not clear what, if any, the effect on pain would be.

Brain imaging studies have suggested that motor cortex stimulation can affect regional cerebral blood flow in various cortical and subcortical structures that are involved in pain modulation. Therefore, motor cortex stimulation may act as a focal point to activate descending inhibition of pain signaling and modulate the emotional aspects of pain (Saitoh and Yoshimine 2007;

Lima and Fregni 2008; Xie et al. 2009). In addition, enhanced secretion of endogenous opioids in multiple brain regions that include the anterior middle cingulate cortex, periaqueductal gray, prefrontal cortex, and cerebellum has been demonstrated in patients with refractory neuropathic pain following motor cortex stimulation (Maarrawi et al. 2007).

There is considerable anatomical and electrophysiological evidence from animal studies that the motor cortex projects to, and influences the responses of, both the dorsal column nuclei and the spinal cord dorsal horn (Giuffrida et al. 1985; Zhang et al. 1991), both of which are parts of the pathways transmitting sensory information to the thalamus. These, and other subcortical sites such as the periaqueductal gray matter and raphe nuclei, are known to be involved in pain modulation circuits (Millan 2002) and are obvious sites where the motor cortex could exert sensory effects. Here, it should be emphasized that we are concerned specifically with stimulation of the motor cortex. There is substantial experimental evidence that lesion or stimulation of a number of cortical regions such as the cingulate, medial prefrontal, and insular cortices in animals (a majority of this work has been done on rats) can change the behavioral responses to noxious stimuli as well as the activity of spinal nociceptive neurons. In this case, most evidence suggests that descending projections to the regions including the hypothalamus, amygdala, and brain stem mediate the sensory changes.

Animal studies have reported similar efficacy in reducing pain behavior following motor cortex stimulation as in humans. Motor cortex stimulation increased pain thresholds in control and neuropathic rats, induced by deafferentation, chronic constriction of the sciatic nerve, or spinal nerve ligation, using multiple mechanical and thermal nociceptive behavior tests (Vaculin et al. 2008; Fonoff et al. 2009; Pagano et al. 2010; Viisanen and Pertovaara 2010a, b). Pharmacological evidence further indicates the involvement of opioid receptors and the descending serotonergic pathway acting on 5-HT1A receptor (Fonoff et al. 2009; Viisanen and Pertovaara 2010b). Reduced spinal dorsal horn neuronal activity has also been reported following motor cortex stimulation (Senapati et al. 2005; Pagano et al. 2010). On the other hand, stimulation of the posterior parietal or somatosensory cortices did not result in any changes in pain behavior, indicating the effect is specific to the motor cortex (Fonoff et al. 2009). All together, these animal studies seem to correlate with the suggested activation of the limbic and descending pain inhibitory systems following motor cortex stimulation in human studies to result in pain reduction. Further studies will likely demonstrate other underlying mechanisms to fully explain why motor cortex stimulation works.

References

Brown, J. A., & Barbaro, N. M. (2003). Motor cortex stimulation for central and neuropathic pain: Current status. *Pain, 104*, 431–435.

Bushnell, M. C., Duncan, G. H., Hofbauer, R. K., Ha, B., Chen, J. I., & Carrier, B. (1999). Pain perception: Is there a role for primary somatosensory cortex? *Proceedings of the National Academy of Sciences of the United States of America, 96*, 7705–7709.

Canavero, S., & Bonicalzi, V. (2001). Motor cortex stimulation. *Journal of Neurosurgery, 94*, 688–689.

Canavero, S., & Bonicalzi, V. (2002). Therapeutic extradural cortical stimulation for central and neuropathic pain: A review. *The Clinical Journal of Pain, 18*, 48–55.

Fonoff, E. T., Dale, C. S., Pagano, R. L., Paccola, C. C., Ballester, G., Teixeira, M. J., & Giorgi, R. (2009). Antinociception induced by epidural motor cortex stimulation in naive conscious rats is mediated by the opioid system. *Behavioural Brain Research, 196*, 63–70.

Fontaine, D., Hamani, C., & Lozano, A. (2009). Efficacy and safety of motor cortex stimulation for chronic neuropathic pain: Critical review of the literature. *Journal of Neurosurgery, 110*, 251–256.

Giuffrida, R., Sanderson, P., & Sapienza, S. (1985). Effect of microstimulation of movement-evoking cortical foci on the activity of neurons on the dorsal column nuclei. *Somatosensory Research, 2*, 237–247.

Hirayama, T., Tsubokawa, T., Katayama, Y., Yamamoto, Y., & Koyama, S. (1990). Chronic changes in activity of thalamic relay neurons following spinothalamic tractotomy in cat. Effects of motor cortex stimulation. *Pain, 5*, 273.

Katayama, Y., Fukaya, C., & Yamamoto, T. (1998). Poststroke pain control by chronic motor cortex stimulation: Neurological characteristics predicting a favorable response. *Journal of Neurosurgery, 89*, 585–591.

Lefaucheur, J. P., Drouot, X., Cunin, P., Bruckert, R., Lepetit, H., Creange, A., Wolkenstein, P., Maison, P., Keravel, Y., & Nguyen, J. P. (2009). Motor cortex stimulation for the treatment of refractory peripheral neuropathic pain. *Brain, 132*, 1463–1471.

Lima, M. C., & Fregni, F. (2008). Motor cortex stimulation for chronic pain: Systematic review and meta-analysis of the literature. *Neurology, 70*, 2329–2337.

Maarrawi, J., Peyron, R., Mertens, P., Costes, N., Magnin, M., Sindou, M., Laurent, B., & Garcia-Larrea, L. (2007). Motor cortex stimulation for pain control induces changes in the endogenous opioid system. *Neurology, 69*, 827–834.

Migita, K., Uozumi, T., Arita, K., & Monden, S. (1995). Transcranial magnetic coil stimulation of motor cortex in patients with central pain. *Neurosurgery, 36*, 1037–1039 discussion 1039–1040.

Millan, M. J. (2002). Descending control of pain. *Progress in Neurobiology, 66*, 355–474.

Nelson, R. J. (1996). Interactions between motor commands and somatic perception in sensorimotor cortex. *Current Opinion in Neurobiology, 6*, 801–810.

Nguyen, J. P., Velasco, F., Brugieres, P., Velasco, M., Keravel, Y., Boleaga, B., Brito, F., & Lefaucheur, J. P. (2008). Treatment of chronic neuropathic pain by motor cortex stimulation: Results of a bicentric controlled crossover trial. *Brain Stimulation, 1*, 89–96.

Ohara, P. T., Vit, J. P., & Jasmin, L. (2005). Cortical modulation of pain. *Cellular and Molecular Life Sciences, 62*, 44–52.

Pagano, R. L., Assis, D. V., Clara, J. A., Alves, A. S., Dale, C. S., Teixeira, M. J., Fonoff, E. T., & Britto, L. R. (2011). Transdural motor cortex stimulation reverses neuropathic pain in rats: A profile of neuronal activation. *European Journal of Pain, 15*, 268.e1–268.e14. http://onlinelibrary.wiley.com/doi/10.1016/j.ejpain.2010.08.003/abstract;jsessionid=1A9083987229DED3EBAC6A7BFA40C499.d01t04

Rasche, D., Ruppolt, M., Stippich, C., Unterberg, A., & Tronnier, V. M. (2006). Motor cortex stimulation for long-term relief of chronic neuropathic pain: A 10 year experience. *Pain, 121*, 43–52.

Saitoh, Y., & Yoshimine, T. (2007). Stimulation of primary motor cortex for intractable deafferentation pain. *Acta Neurochirurgica. Supplement, 97*, 51–56.

Senapati, A. K., Huntington, P. J., & Peng, Y. B. (2005). Spinal dorsal horn neuron response to mechanical stimuli is decreased by electrical stimulation of the primary motor cortex. *Brain Research, 1036*, 173–179.

Teuber, H.-L. (1966). Alterations of perception after brain injury. In J. C. Eccles (Ed.), *Brain and conscious experience* (pp. 182–216). New York: Springer.

Tsubokawa, T., Katayama, Y., Yamamoto, T., Hirayama, T., & Koyama, S. (1991). Chronic motor cortex stimulation for the treatment of central pain. *Acta Neurochirurgica Supplement (Wien), 52*, 137–139.

Tsubokawa, T., Katayama, Y., Yamamoto, T., Hirayama, T., & Koyama, S. (1993). Chronic motor cortex stimulation in patients with thalamic pain. *Journal of Neurosurgery, 78*, 393–401.

Vaculin, S., Franek, M., Yamamotova, A., & Rokyta, R. (2008). Motor cortex stimulation in rats with chronic constriction injury. *Experimental Brain Research, 185*, 331–335.

Velasco, F., Arguelles, C., Carrillo-Ruiz, J. D., Castro, G., Velasco, A. L., Jimenez, F., & Velasco, M. (2008). Efficacy of motor cortex stimulation in the treatment of neuropathic pain: A randomized double-blind trial. *Journal of Neurosurgery, 108*, 698–706.

Viisanen, H., & Pertovaara, A. (2010a). Antinociception by motor cortex stimulation in the neuropathic rat: Does the locus coeruleus play a role? *Experimental Brain Research, 201*, 283–296.

Viisanen, H., & Pertovaara, A. (2010b). Roles of the rostroventromedial medulla and the spinal 5-HT(1A) receptor in descending antinociception induced by motor cortex stimulation in the neuropathic rat. *Neuroscience Letters, 476*, 133–137.

Xie, Y. F., Huo, F. Q., & Tang, J. S. (2009). Cerebral cortex modulation of pain. *Acta Pharmacologica Sinica, 30*, 31–41.

Yamamoto, T., Katayama, Y., Hirayama, T., & Tsubokawa, T. (1997). Pharmacological classification of central post-stroke pain: Comparison with the results of chronic motor cortex stimulation therapy. *Pain, 72*, 5–12.

Zhang, D. X., Owens, C. M., & Willis, W. D. (1991). Short-latency excitatory postsynaptic potentials are evoked in primate spinothalamic tract neurons by corticospinal tract volleys. *Pain, 45*, 197–201.

Movement Cycles

Definition

Units of behavior counted to program reinforcement. A movement cycle is said to have occurred when the person is in a position to repeat it. Moving the left foot forward is not a movement cycle. Moving the left and then the right puts the person in a position to repeat; hence, left step/ right step is a movement cycle. Movement cycles can often be combined into more easily managed units. For example, left step/right step is a movement cycle. A 50-m lap contains many steps. Those can usually be combined into one lap, and "lap" becomes the unit of behavior.

Cross-References

▶ Training by Quotas

Movement-Related Pain

Definition

Movement-related pain is pain that occurs with movement. This term refers to both breakthrough pain, in which background pain exists in addition to pain during activity, and incident pain, in which background pain is absent or controlled.

Cross-References

▶ Evoked and Movement-Related Neuropathic Pain

MPQ

▶ McGill Pain Questionnaire

MPS

▶ Myofascial Pain Syndrome

MRI

▶ Magnetic Resonance Imaging

MRS

Synonyms

Hydrogen magnetic resonance spectroscopy

Definition

1H-MRS (hydrogen magnetic resonance spectroscopy) is an NMR-based technique for assessing function in the brain. MRS takes advantage of the fact that protons (H) possess slightly different resonant properties depending upon the compound. For a given volume of the brain (typically >1 cm^3), the distribution of these H resonances can be displayed as a spectrum. The area under the peak for each resonance provides a quantitative measure of the relative abundance of that compound.

Cross-References

▶ Thalamus, Clinical Pain, Human Imaging

MS Contin®

Definition

MS Contin® is a controlled (slow)-release tablet formulation of morphine that produces a peak morphine concentration 4–5 h post dose, and therapeutic concentrations persist for about 12 h.

Cross-References

▶ Postoperative Pain, Morphine

MS Contin® Controlled (slow)-release Morphine Sulfate

▶ Postoperative Pain, Morphine

MSDs/MSPs

▶ Disability, Upper Extremity

MSPs

▶ Disability, Upper Extremity

MTrP

▶ Myofascial Trigger Points

MTrP Circuit

Definition

An MTrP circuit comprises of the interneuronal connections in the spinal cord. The circuit cannot only transfer the nociceptive impulses to the brain but may also control referred pain patterns via connections to other MTrP circuits. The occurrence of LTRs is also mediated via the MTrP circuit.

Cross-References

▶ Dry Needling

MTrP Locus

Definition

There are multiple MTrP loci in an MTrP region. An MTrP locus contains two components: the sensitive locus, also known as the local-twitch-response locus (LTR locus), and the active locus, also known as the endplate-noisy locus (EPN locus).

Cross-References

▶ Dry Needling

Mu(μ)- and Kappa(κ)-Opioids

Definition

Opioids that preferentially bind to and act at mu- and kappa- receptors.

Cross-References

▶ Endogenous Opioid Receptors
▶ MOP Receptor
▶ Nitrous Oxide Antinociception and Opioid Receptors
▶ Psychological Aspects of Pain in Women

Mu(μ)-Opioid Receptor

Synonyms

MOP

Definition

Opioid receptors that preferentially bind morphine-like drugs. The mu receptor was originally named for the effects of morphine and is the main site of action of most clinically effective opioid drugs. The main clinical effect of these drugs is analgesia, but they are also used as antidiarrheal agents. Agonists produce nausea and vomiting, respiratory depression, and constipation is an important side effect when used therapeutically for pain. Agonists also produce miosis, an important sign of overdose as well as bradycardia. Mu opioid receptor agonists can also elicit euphoria that is associated with liability for drug abuse, addiction, and dependence.

Cross-References

▶ Opiates During Development
▶ Opioid Receptors
▶ Opioid Rotation in Cancer Pain Management
▶ Postoperative Pain, Appropriate Management
▶ Postoperative Pain, Transition from Parenteral to Oral Drugs

Mucosa

Definition

The smooth inner lining of the large intestine that secretes mucus to lubricate the waste materials.

Cross-References

▶ Animal Models of Inflammatory Bowel Disease

Mucosa of Sexual Organs, Nociception

▶ Nociception in Mucosa of Sexual Organs

Mucosal Receptors

▶ Visceral Mechanoreceptors

Mucositis

Definition

Mucous membrane toxicity often occurs shortly after administration of chemotherapy or radiotherapy. Inflammation, ulceration, and infection can then occur on the oral and other mucus membranes.

Mud Therapy

Definition

Mud therapy involves the application of heated mudpacks to parts of the body.

Cross-References

▶ Spa Treatment

Multiaxial Assessment of Pain

Dennis C. Turk
Department of Anesthesiology & Pain Research,
University of Washington, Seattle, WA, USA

Synonyms

Comprehensive assessment; Interdisciplinary
assessment; Multidisciplinary assessment

Definition

Multiaxial (interdisciplinary) assessment refers
to a comprehensive assessment of individuals
with chronic pain that addresses biomedical, psychosocial, and behavioral domains, as each of
these contribute to the chronic pain and disability
(Turk and Rudy 1987).

Characteristics

The ▶ biopsychosocial model of pain proposes
that dynamic and reciprocal interactions among
biological, psychological, and sociocultural variables shape the pain experience (Flor and Turk
2011). According to this model, the pain experience usually begins when peripheral nociceptive
barrage produces physiological changes, although
there may be central mechanisms involved in the
initiation of pain, and the experience is modulated
by the unique genetic endowment, learning history, beliefs, affective state behavior, and contextual factors.

Reports of pain severity and impact will vary
depending on a range of contributions and are
not solely the result of physical pathology or
perturbations within the nervous system. The
biopsychosocial perspective encourages an

evaluator not only to consider the nature, cause,
and characteristics of the noxious stimulation, but
the presence of the sensations reflected against
a history that preceded symptom onset.

The biopsychosocial model incorporates
cognitive-behavioral (CB) concepts (see
▶ Cognitive-Behavioral Perspective of Pain) in
understanding chronic pain. For example, proponents of this model propose that both the person
and the environment reciprocally determine
behavior. People not only respond to their environment but elicit environmental responses by
their behavior. The person who becomes aware
of a physical event and decides the symptom
requires attention from a health-care provider,
initiates a set of circumstances different from
the individual with the same symptom who
chooses to self-manage symptoms (Turk et al.
1983). Another assumption of the CB perspective
is that people are active agents and capable of
change. The passive role many patients have in
traditional physician-patient relationships often
reinforces their beliefs that they have minimal
ability to impact their own recovery (Flor and
Turk 2011).

In order to understand and appropriately treat
a patient whose primary symptom is pain, the
assessment should begin with a comprehensive
history and physical examination. Patients are
usually asked to describe the characteristics
(e.g., stabbing, burning), location, severity and
pattern (e.g., when worse, what makes it better)
of their pain. Physical examination procedures
and sophisticated laboratory and imaging techniques are readily available for use in detecting
organic pathology (Flor and Turk 2011; Turk and
Robinson 2011). Physical and laboratory abnormalities, however, correlate poorly with patients'
pain reports, and it is often not possible to make
any precise pathological diagnosis or even to
identify an adequate anatomical origin for the
pain. Thus, an adequate pain assessment also
requires clinical interviews and utilization of
assessment tools to assist in the evaluation of
the myriad ▶ psychosocial factors and behavioral
factors that influence the subjective report.

As there is no "pain thermometer" that can provide an objective quantification of the amount or

severity of pain experienced by a patient, it can only be assessed indirectly based on a patient's overt communication, both verbal and behavioral. However, even a patient's communications make pain assessment difficult, as pain is a complex, subjective phenomenon, comprised of a range of factors, and is uniquely experienced by each individual. Wide variability in pain severity, quality, and impact may be noted in reports of patients attempting to describe what appear to be objectively identical phenomena. Patient's descriptions of pain are also colored by cultural and sociological influences.

A patient's beliefs about the cause of symptoms, their trajectory, and beneficial treatments will have important influences on emotional adjustment and adherence to therapeutic interventions (Flor and Turk 2011). Clinical attention should focus on the patient's specific thoughts, behaviors, emotions, and physiological responses that precede, accompany, and follow pain episodes or exacerbations, including environmental and temporal conditions, and consequences associated with the patient's responses (cognitive, emotional, and behavioral, including frequency and specificity/generality across situations) in these situations. It is valuable to note any patterns of ▶ maladaptive thoughts, as they may contribute to a sense of hopelessness, dysphoria, and unwillingness to engage in activity. Topics that can be covered in an assessment interview might include: asking what the patient thinks is wrong with him/her, what the patient thinks about the pain (e.g., cause, impact), what worries the patient has (e.g., exacerbation of symptoms, impairment), problems the patient has experienced due to pain (e.g., marital, financial, occupational), how the patient lets others know when pain is present, what effect the patient believes the pain is having on others, in what situations the patient experiences increased pain, what others think about their pain, how others respond to their pain, medications they are taking, strategies they have used to alleviate pain, what treatments they have endured for pain management, and prior and current stressful life events (Turk et al. 2008).

Central Assessment Questions

Turk and Robinson (2011) have suggested that three central questions should guide assessment of people who report pain: (1) What is the extent of the patient's disease or injury (physical impairment)? (2) What is the magnitude of the illness? That is, to what extent is the patient suffering, disabled, and unable to enjoy usual activities? and (3) Does the individual's behavior seem appropriate to the disease or injury, or is there any evidence of amplification of symptoms for any of a variety of psychological or social reasons or purposes? These three general questions can be conceptualized and address three general domains or axes – biomedical, psychosocial, and behavioral (Turk and Rudy 1987).

A thorough clinical assessment should also involve gathering a detailed history from the patient, including past medical history, drug/alcohol use, history of mental illness, and a list of current medications. It is important for the clinician treating chronic pain patients to understand the use of pain medications, in addition to the side effects associated with them, so as to avoid misinterpreting symptoms and potential misdiagnosis. Patients may also make use of alcohol and illicit drugs to palliate their symptoms. Patients with histories of substance abuse may be at particular risk of becoming dependent on and abusing pain medications. Reviewing the chart and conducting a detailed history of previous and current prescription and substance use may help ascertain whether this area warrants further inquiry.

Self-report Measures

In addition to interviews, a number of assessment instruments designed to evaluate patients' attitudes, beliefs, and expectancies about themselves, their symptoms, and the health-care system have been developed (DeGood and Cook 2011). Standardized assessment instruments have advantages over semi-structured and unstructured interviews. They are easy to administer, can suggest issues to be addressed in more depth during an interview, require less time, and most importantly, they can be submitted to analyses that permit determination of their ▶ reliability and ▶ validity.

The poor reliability and questionable validity of physical examination measures has led to the development of self-report functional status measures that seek to quantify symptoms, function, and behavior directly, rather than inferring them (O'Connor and Dworkin 2011). Another advantage of self-report instruments is that they enable the assessment of a wide range of behaviors that are relevant to the patient, some of which may be private (sexual relations) or unobservable (thoughts, emotional arousal). Even traditional psychological measures have been used to identify psychosocial characteristics of the pain experience. These measures, however, must be used with caution, as they were usually not developed for or standardized on samples of medical patients. As a result, it is always best to corroborate information gathered from the instruments with other sources, such as interviews with the patient and significant others, and chart review.

In addition to self-report instruments, more valid information may be obtained by asking about current levels of pain or pain over the last 24 h, and by having patients maintain regular ▶ diaries of pain intensity with ratings recorded several times each day for several days or weeks. Electronic devices can be used to prompt patients to provide ratings in real time and thus avoiding any retrospective recall. In addition to pain intensity, through such diaries, patients can also record their thoughts, mood, and activities at predetermined intervals.

An additional use of patient diaries is to record medication use over a specified interval. Diaries not only provide information about the frequency and quantity of medication, but they may also permit identification of the antecedent and consequent events of medication use. For example, a patient might note that he took medication after an argument with his wife, and that when she saw him taking the medication she expressed sympathy. Antecedent events might include stress, boredom, or activity. Examination of antecedents is useful in identifying patterns of medication use that may be associated with factors other than pain per se. Similarly, patterns

of response to the use of analgesic may be identified. Does the patient receive attention and sympathy whenever he or she is observed by significant others taking medication? That is, do significant others provide positive reinforcement for the taking of analgesic medication, and thereby unwittingly increase medication use?

The multiaxial approach to assessment is designed to assess and integrate physical, psychosocial, and behavior contributors to pain and disability. There are a large number of measures and procedures that can be used to evaluate each axis. The multiaxial approach to assessment does not specify any particular methods for assessment. The evaluator will determine the appropriate methods based on his or her experience with patients with chronic pain, specific questions that may be of particular importance for a given patient, and presenting symptoms.

Components of Multiaxial Assessment

Components of the multiaxial assessment can be summarized as follows:

Biomedical
- Medical history
- Comorbid medical conditions
- Pain (intensity, quality, duration, location, exacerbating and alleviating factors)
- Physical examination
- Laboratory diagnostic tests
- Imagining procedures
- Response to previous treatment (both positive and adverse)

Psychosocial
- General history (family composition, education)
- Living status (e.g., married, divorced, romantic relationship)
- Quality of relationships with significant others
- Socioeconomic status
- Work history
- Substance abuse history and current use of drugs and alcohol to control pain as well as recreational
- Impact of pain on physical, emotional, and social functioning

- Beliefs, attitudes, expectations, and fears about pain, treatment, their plight, and the future
- Prior and current psychiatric status, degree of emotional distress (fear, depression, anger), treatment of patient and family
- Use of coping methods (e.g., problem solving, distraction)
- Satisfaction with previous and current treatment(s)

Behavioral

- Use of behavioral methods of coping with pain (e.g., medication, alcohol, exercise, withdrawal)
- Observation of ▶ pain behaviors (overt communications of pain, distress, and suffering)
- Response to pain behaviors by significant others
- Use of electronic recording devices to gather activity data (e.g., actigraphic recording)
- Physical, social, and recreational activities
- Vocational status
- Adherence to treatment recommendations (e.g., medication, exercise)

References

Degood, D. E., & Cook, A. J. (2011). Psychosocial assessment: Comprehensive measures and measures specific to pain beliefs and coping. In D. C. Turk & R. Melzack (Eds.), *Handbook of pain assessment* (3rd ed., pp. 67–97). New York: Guilford.

Flor, H., & Turk, D. C. (2011). *Chronic pain: An integrated biobehavioral approach.* Seattle: IASP Press.

O'Connor, A. B., & Dworkin, R. H. (2011). Assessment of pain and health-related quality of life in chronic pain clinical trials. In D. C. Turk & R. Melzack (Eds.), *Handbook of pain assessment* (3rd ed., pp. 474–486). New Yori: Guilford.

Turk, D. C., & Robinson, J. P. (2011). Assessment of patients with chronic pain – a comprehensive approach. In D. C. Turk & R. Melzack (Eds.), *Handbook of pain assessment* (3rd ed., pp. 88–212). New York: Guilford.

Turk, D. C., & Rudy, T. E. (1987). Toward a comprehensive assessment of chronic pain patients. *Behaviour Research and Therapy, 25,* 237–249.

Turk, D. C., Meichenbaum, D., & Genest, M. (1983). *Pain and behavioral medicine: A cognitive-behavioral perspective.* New York: Guilford.

Turk, D. C., Okifuji, A., & Slinner, M. (2008). Asssessment of the person with persistent pain. In K. Hunsley & E. Mash (Eds.), *A guide to assessments that work* (pp. 576–592). New York: Oxford Press.

Multichannel CT

▶ CT Scanning

Multidimensional

Definition

A pain assessment tool that includes measurement of two or more indicators of newborn pain (behavioral, physiological, or biochemical (hormonal)). Multidimensional approaches to pain assessment incorporate the use of pain assessment tools that include measurement of two or more indicators of newborn pain and concurrent consideration of a variety of contextual and modifying factors.

Cross-References

▶ Pain Assessment in Neonates

Multidimensional Model

▶ Chronic Pain, Diathesis-Stress Model of

Multidimensional Scaling

▶ Pain Measurement by Questionnaires, Psychophysical Procedures, and Multivariate Analysis

Multidisciplinary

Definition

Medical disciplines working with the same patient.

Cross-References

▶ Physical Medicine and Rehabilitation, Team-Oriented Approach

Multidisciplinary Assessment

▶ Multiaxial Assessment of Pain

Multidisciplinary Pain Centers, Rehabilitation

Dennis C. Turk
Department of Anesthesiology & Pain Research,
University of Washington, Seattle, WA, USA

Synonyms

Functional restoration program; Interdisciplinary pain rehabilitation programs; Multidisciplinary pain clinics; Pain clinics

Definition

An organization of healthcare professionals and basic and applied scientists that includes research, teaching, and patient-care related to acute and chronic pain. It includes a wide array of health care professionals including physicians, psychologists, nurses, physical therapists, occupational therapist, and other specialty healthcare providers. Multiple therapeutic modalities are available. These centers provide evaluation and treatment and are usually affiliated with major health science institutions. Multidisciplinary pain clinics are similar to multidisciplinary pain centers (MPC), with the exception being that they do not include basic scientists and may not be involved in conducting research.

Characteristics

Although there is no single format for multidisciplinary pain management or the operations of

an MPC, almost every treatment facility of this type has a generic concept and plan.

Concepts of Treatment at Multidisciplinary Pain Clinics

• Reconceptualization of the patient's pain and associated problems from uncontrollable to manageable.
• Overt or covert efforts are made to foster optimism and combat demoralization.
• Flexibility is the norm with attempts to individualize some aspects of treatment to patient needs and unique physical and psychological characteristics.
• Emphasize active patient participation and responsibility.
• Provide educational and training in the use of specific skills such as exercise, relaxation, and problem solving.
• Encourage patient feelings of success, self-control, and self-efficacy.
• Encourages patients to attribute success to their own efforts.

These treatment programs also share general features.

General Features of Multidisciplinary Pain Treatment Teams

• Share a common conceptualization of chronic pain patients.
• Synthesize the diverse sets of information based on their own evaluations, as well as those of outside consultants, into a differential diagnosis and treatment plan customized to meet the specific need of each patients.
• Work together to formulate and implement a comprehensive rehabilitation plan based on available data.
• Share a common philosophy of disability management.
• Act as a functional unit whose members are willing to learn from each other and modify, when appropriate, their own opinions based on the combined observations and expertise of the entire group.

In MPCs, patients are usually treated in groups. Patients work on at least four issues simultaneously: physical, pharmacological, psychological, and vocational. Programs usually emphasize helping patients to gain knowledge about pain and how the body functions, physical conditioning, medication management, acquisition of coping, and vocational skills. Individual and group counseling address patient's needs. In contrast to traditional Western health care, the emphasis is upon what the patient accomplishes, not on what they say. The providers serve as teachers, coaches, and sources of information and support (Loeser and Turk 2001).

Multidisciplinary pain management requires the collaborative efforts of many healthcare providers. The healthcare providers act as a team, with extensive interaction among the team members.

The following list itemizes the goals of multidisciplinary pain management. Every patient will have a different mixture of functional limitations, pain behaviors, affective disturbance, physical disability, and vocational dysfunction. Successful treatment will address each of these general areas.

Goals of Multidisciplinary Pain Management
- Identify and treat unresolved medical issues
- Symptomatic improvement
- Eliminate inappropriate medications, institute desirable medications
- Improve aerobic conditioning, endurance, strength, and flexibility (restoration of physical functioning)
- Eliminate excessive guarding behaviors that interfere with normal activities
- Improve coping skills and psychological well-being
- Alleviate depression
- Foster independence
- Assess patient resources and identify vocational and recreational opportunities
- Educate the patient about pain, anatomy, physiology, and psychology, discriminating hurt from harm
- Educate the patient about prudent healthcare consumption
- Assist the patient to establish realistic goals and to maintain treatment gains

- Restoration of social and occupational functioning – social reintegration, return to productive employment
- Reduction in use of healthcare system

The original MPCs were inpatient based. It is now apparent that outpatient programs can be equally successful if they have adequate intensity and duration (for a review see Turk et al. 1993). There are no controlled studies to determine the optimal duration of treatment and hours per day; nor does the literature reveal which aspects of the various components are most important for a treatment program. It is clear, however, that the effects of an MPC are greater than the sum of its parts. Common features of all programs include physical therapy, medication management, education about how the body functions, psychological treatments (e.g., learning coping skills, problem solving, communication skills training), assessment, and therapies aimed at improving function and the likelihood of return to work. Programs usually have a standard daily and weekly format that providers can tailor to individual patient needs. The overall length of a program depends, in part, upon unique patient requirements. Typical programs operate 8 h per day, 5 days per week, and last 3–4 weeks, although some programs meet less frequently and are of longer duration.

Roles of the Physician
The physician is responsible for the initial history and physical examination; review of outside records and determination of the need for any further diagnostic tests. Detailed assessment of the patient's medication history is also a key physician contribution. The implementation of medication management, including drug tapering by means of a ▶ pain cocktail technique (described below), is also a physician role. Another important task for the physician is to review with the patient the medical issues and the findings of diagnostic tests and imaging studies. The physician also plays an essential role in the education of the patient, and in legitimating all of the other components of the treatment program.

Roles of the Psychologist

The psychologist typically conducts the initial psychological evaluation, monitors and implements the cognitive and behavioral treatment strategies, teaches the patient coping skills, and educates patients about the relationship between thoughts, feelings, behavior, and physiology. The psychologist usually leads both individual and group educational and counseling sessions for the patients. In addition, the psychologist plays a critical role in helping other members of the treatment team to employ sound behavioral principles in designing patient treatment activities (Turk and Gatchel 2002).

Roles of the Nurse

The nurse is a key part of the treatment program, playing a major role in patient education regarding such topics as medication, diet, sleep hygiene, and sexual activity. Another nursing function is assisting patients in the practice of newly learned skills, assessing medication responses, and acting as the focal point of the communication that keep such a program operational. The roles of nurses vary with their skills and their interaction with other providers. Since the nurses tend to be with the patients throughout their entire treatment course, they play a central role in maintaining continuity during the treatment program.

Roles of Physical and Occupational Therapists

Physical and occupational therapists provide assessment and active physical therapies for patients to improve their strength, endurance, and flexibility. They do not provide passive modalities of treatment. They assist the patient in developing proper body mechanics and strategies for coping with the physical demands of a job and everyday life, and function mainly as teachers and coaches.

The occupational therapists review the patient's work history, disabilities, and factors that may play a role in determining who goes back to work and who does not. They help in the establishment of training activities. Some programs heavily emphasize ergonomic issues and utilize high technology in physical therapies; however, the need for this type of treatment is unclear.

Role of the Vocational Counselor

The vocational counselor plays a critical role in the treatment of an individual for whom return to work is a treatment goal. Initial assessment occurs as part of the screening process, but in-depth evaluation of interests, education, aptitude, physical capacities, learning capabilities, work experience, transferable skills, and vocational goals occurs upon entry into the treatment program. The goals are to identify vocational opportunities and barriers to effective return to work. In addition to occupational counseling, counselors provide job seeking skills training, placement counseling, work hardening, information about educational options, and liaison services. Information obtained by the vocational counselor is critical for other team members to establish realistic goals for the patient. In some organizations, rehabilitation nurses provide this service.

Treatment Principles

The goals of multidisciplinary pain management are normally specific, definable, operationalizable, and realistic in nature (see list "Goals of Multidisciplinary Pain Management"). As they have evolved, MPC treatments have become performance-based, goal-directed, and outcome driven. Integration of medical evidence related to patient's pain and physical impairment with information concerning what patients are doing or failing to do because of their pain, how these behaviors influence patient's physical capacity, how others respond to the patient, the influence of psychosocial factors that contribute directly and indirectly to patient's physical and emotional status, and the potential for rehabilitation (i.e., disability) are essential. The treatment team must build an alliance with patients to instill a willingness to accept the need for self-management.

Physical therapy employs behavioral medicine principles (Turk et al. 2000) and, as noted, engages few, if any, passive modalities. The emphasis is upon improving strength, endurance, and flexibility through the patient's physical activities. The therapists provide instruction, guidance, safety, and encouragement. Accomplishments, rather than pain behaviors, receive rewards. Patients maintain graphs of their daily activity and track progress. As patients progress, they enroll in more complex activities that simulate the workplace requirement.

Medications are given on a time contingent basis, so as to uncouple the reinforcement of pain behaviors by medications. In general, patients in an MPC program do not derive adequate pain relief from analgesic medications, and this is why they are usually tapered by means of the pain cocktail technique in some programs. This technique is simply a method of converting all opioids to an equivalent dose of methadone, delivered with a masking vehicle. The dose is then tapered over the period of treatment, always with the full knowledge of the patient. Most medications are discontinued; the common exceptions are antidepressants, which often help chronic pain patients. MPCs discourage long-term use of other medications, both because of their potential side effects and because their use undermines the philosophical concept that the patient must learn to control his or her pain, and not depend upon healthcare providers or their prescriptions.

Psychological strategies generally target altering behavior rather than changing the patient's personality. Patients learn coping skills because this is frequently a deficiency that has led to the patient's many difficulties (see following list). Couples therapy is sometimes appropriate. Issues that patients bring up receive attention in either the group format or in individual therapy, as needed. As depression is so often a component of the chronic pain problem, it warrants both psychological as well as pharmacological interventions. Psychologists provide relaxation and consolidation sessions that allow the patients to work on newly acquired skills, and explore educational topics and new psychological skills (see Table 1).

Coping Skills Taught
- Relaxation
- Distraction (attention diversion) methods

Multidisciplinary Pain Centers, Rehabilitation, Table 1 Issues addressed by psychologists

Impact of pain on life, significant others	Resourcefulness versus helplessness
Communication	Problem solving
Reinforcement and pain behaviors	Relationship between thoughts, feelings, behavior, and physiology
Goal setting, homework, adherence	Coping skills (relaxation, distraction, positive thoughts)
Self-reinforcement	Fear of activity
Stress and pain and homework	Generalization, maintenance, flare-ups, relapse

- Cognitive restructuring – identification and challenging maladaptive thoughts and feelings
- Problem solving
- Anger management
- Desensitization
- Rehearsal and home practice
- Communication skills including assertiveness
- Goal setting
- Pleasant activity planning
- Self-monitoring and self-reinforcement

An important aspect of MPCs is education. This is an activity that is shared by all member of the clinical team. Topics cover a wide array of the problems facing those who experience chronic pain. Topic selection and content is to some degree a function of the needs of each group of patients, but content always includes a core set of issues (see Table 2).

An important issue is the maintaining of gains that have occurred during the treatment program. Surrounded by a team of supportive healthcare providers and other patients, most patient see some gains by the end of treatment. However, many are unable to maintain their gains when they return to their normal family and occupational activities. Each patient must learn strategies for maintaining his or her gains in a less supportive environment. Most programs have established brief follow-up interactions to try to assist patients to keep up their physical and psychological skills and to prevent relapses.

Outcomes
A substantial body of literature supports the assertion that multidisciplinary pain treatment is

Multidisciplinary Pain Centers, Rehabilitation, Table 2 General educational topics covered

Body mechanics, modification of movement patterns	Role of medication
Posture	Sleep hygiene
Energy conservation	Diet
Performance of activities of daily living	Sexuality
Hurt versus harm	Anatomy and physiology
Leisure activities	Gate control theory
Vocational activities	Home practice
Active versus passive exercises and modalities	Progressive exercises
Adherence to recommendations	Management of flare-ups

effective in reducing pain, the use of opioid medication, the use of healthcare services; it increases activity, returns people to work, and aids in the closing of disability claims (e.g., Cutler et al. 1994; Guzman et al. 2001). Moreover, treatment at MPCs targets patients with the most recalcitrant problems, yet the benefits appear to exceed those for conventional treatments including surgery and, in contrast to surgery, there are no known iatrogenic complications of treatment at MPCs (Turk 2002). Not only do MPCs appear to be clinically effective, they also appear to be cost effective, with the potential to provide substantial savings in healthcare and disability payments (Turk 2002).

Conclusions

The team approach to complex chronic pain patients, as found in a multidisciplinary pain treatment facility, has evolved with an underlying set of principles. These include the recognition that Cartesian mind-body dualism is an impediment to effective healthcare. Second, a biopsychosocial model is required to capture all of the relevant factors (Flor & Turk 2011). Third, the treatment must address the pain itself, and not just be a search for hidden causes and specific remedies for these causes. Fourth, the treatment must address the restoration of

well-being and not just aim at the alleviation of symptoms. Finally, the illness is not just chronic pain but is also the failure to work, often ascribed erroneously to the pain instead of the patient or the patient's circumstances.

References

Cutler, R. B., Fishbain, D. A., Rosomoff, H. L., et al. (1994). Does nonsurgical pain center treatment of chronic pain return patients to work? A review and meta-analysis of the literature. *Spine, 19,* 643–652.
Flor, H., & Turk, D. C. (2011). Chronic pain: An integrated biobehavioral approach. Seattle WA: IASP Press.
Guzman, J., Esmail, R., Karjalinen, K., et al. (2001). Multidisciplinary rehabilitation for chronic Low back pain: Systematic review. *BMJ, 322,* 1511–1516.
Loeser, J. D., & Turk, D. C. (2001). Multidisciplinary pain management. In J. D. Loeser, S. D. Butler, C. R. Chapman, & D. C. Turk (Eds.), *Bonica's Management of pain* (3rd ed., pp. 2069–2079). Philadelphia PA: Lippincott Williams and Wilkins.
Turk, D. C. (2002). Clinical effectiveness and cost effectiveness of treatments for chronic pain patients. *Clinical Journal of Pain, 18,* 355–365.
Turk, D. C., & Gatchel, R. J. (2002). *Psychological treatment of chronic pain: Clinical handbook* (2nd ed.). New York: Guilford Press.
Turk, D. C., Rudy, T., & Sorkin, B. (1993). Neglected topics in chronic pain treatment outcome studies: Determination of success. *Pain, 53,* 3–16.
Turk, D. C, Okifuji, A., & Sherman, J. (2000). Behavioral aspects of low back pain. In: J. R. Taylor, L. Twomey (Eds.), *Physical therapy of the low back* (3rd ed., pp. 351–383) Australia: Churchill Livingstone.

Multidisciplinary Pain Clinics

▶ Multidisciplinary Pain Centers, Rehabilitation

Multidisciplinary Pain Treatment for the Elderly

Definition

The elderly are underrepresented in multidisciplinary programs, despite the fact that multimodal treatment is a key to successful practice in geriatric medicine.

Cross-References

▶ Psychological Treatment of Pain in Older Populations

Multidisciplinary Treatment

▶ Complex Chronic Pain in Children, Interdisciplinary Treatment
▶ Multimodal Rehabilitation Treatment and Psychiatric Aspects of Multimodal Treatment for Pain

Multidisciplinary Treatment Program

Definition

Physicians, nurses, physical therapists, and psychologists in one pain-management center. Treatment staff meet weekly to review patient progress in the program, which emphasizes pain reduction and functional outcomes.

Cross-References

▶ Complex Chronic Pain in Children, Interdisciplinary Treatment
▶ Multimodal Rehabilitation Treatment and Psychiatric Aspects of Multimodal Treatment for Pain
▶ Psychological Treatment of Chronic Pain, Prediction of Outcome

Multimodal

Definition

Refers to treatment approaches that use more than one type of therapy (e.g., a combination of drug, psychological, and/or physical therapies).

Cross-References

▶ Complex Chronic Pain in Children, Interdisciplinary Treatment

Multimodal Analgesia

Synonyms

Balanced analgesia

Definition

Multimodal analgesia (also known as balanced analgesia) occurs from a combination of analgesics (multimodal mixture). The rationale is that each drug exerts its analgesic effect via a different mechanism. Thus, a low-dose combination might offer the best therapeutic effect, while minimizing the risk of the unwanted adverse effects seen with high doses of each of these drugs.

Cross-References

▶ Balanced Analgesic Regime
▶ Epidural Infusions in Acute Pain
▶ Multimodal Analgesia in Postoperative Pain
▶ Postoperative Pain, Importance of Mobilization
▶ Postoperative Pain, Transition from Parenteral to Oral Drugs

Multimodal Analgesia in Postoperative Pain

Edward A. Shipton
Department of Anaesthesia, Christchurch School of Medicine and Health Sciences, University of Otago, Christchurch, New Zealand

Synonyms

Balanced analgesia

Definition

▶ Multimodal (or "balanced") analgesia represents an approach to preventing postoperative pain where the patient is administered a combination of opioid and non-opioid analgesic drugs that act at different sites within the central and peripheral nervous systems in an effort to minimize opioid use and, therefore, to decrease opioid-related side effects (White et al. 2007a). Adjuvant analgesics are drugs with non-pain primary indication but which are shown to have analgesic efficacy in painful conditions.

Introduction

Despite its central role in acute pain management, the exclusive use of opioids has been challenged recently in view of its immediate and long-term side effects. Apart from common adverse effects of nausea, vomiting, itching, somnolence and serious respiratory depression, urinary retention, spasm of sphincter of Oddi, and potential delay in bowel movements, long-term consequences of perioperative opioids (development of chronic postsurgical pain syndromes, hyperalgesia, and immunomodulation) are increasingly being recognized (Lui and Ng 2011).

The ultimate goals in acute pain management are evidence-based analgesic efficacy with good safety and tolerability profiles, reduction in opioid-related short- and long-term side effects, potential to reduce chronic postsurgical pain, and applicable in diverse clinical groups of patients (Lui and Ng 2011). Multimodal analgesia is needed for acute postoperative pain management due to adverse effects of opioid analgesics, which can impede recovery; this is a problem that is of increasing concern with the rapid increase in the number of ambulatory surgeries (Buvanendran and Kroin 2009).

Characteristics

The concept of multimodal analgesia first proposed about 15 years ago is now quite well established in clinical practice. Single analgesics, for example, opioids or nonsteroidal anti-inflammatory drugs (▶ NSAIDs) are not able to provide effective pain relief without side effects such as nausea, vomiting, sedation, or bleeding (Jin and Chung 2001). However, NSAIDs combined with intravenous patient-controlled morphine administration may decrease nausea and sedation in patients when compared with those using patient-controlled morphine alone (Elia et al. 2005). Different classes of analgesics using different routes of administration such as intravenous and epidural are used to produce fewer side effects of sedation, nausea, vomiting, pruritus, constipation, and improved pain relief (Vadivelu et al. 2010).

Multimodal Mixture

Increasingly a combination of non-opioid analgesic techniques is being used as first-line therapy for the prevention of pain (White 2008). Many clinical studies have reported an opioid-sparing effect using a wide variety of non-opioid adjuvants (i.e., local anesthetics, paracetamol, ketamine, clonidine, dexmedetomidine, adenosine, gabapentin, pregabalin, glucocorticoids, esmolol, neostigmine, and magnesium).

Multimodal analgesia is recommended, combining different drug classes, for example, an opioid (morphine, pethidine, fentanyl, tramadol, codeine) with a non-opioid (▶ NSAID; ▶ COX-2 inhibitor), delivered through various routes, and including neuraxial use of local anesthetics (bupivacaine, ropivacaine) alone or in combination with other drugs, regional anesthesia (with local anesthetics), antihyperalgesics (ketamine, dextromethorphan), and techniques such as patient-controlled analgesia (PCA) and preventive analgesia (Costantini et al. 2011).

Paracetamol

Commonly, acetaminophen (paracetamol) and NSAIDs are in routine use as components of multimodal analgesia, in combination with opioids or local anesthetic techniques to modulate this (Power and Barratt 1999). The optimal use of acetaminophen (paracetamol) by mouth, rectally, or intravenously (as the prodrug proparacetamol) can improve pain control. The predictable bioavailability and onset of action via the intravenous route is in contrast to the variability in absorption, bioavailability, and onset of effect when administered via the oral and particularly via the rectal routes.

Nonsteroidal Anti-inflammatory Drugs (NSAIDs) (Including the COX-2 Inhibitors)

Recent meta-analyses have confirmed the opioid dose-sparing effect of NSAIDs (including the COX-2 inhibitors) and the associated decreases in the opioid-related side effects and/or improvements in clinically meaningful outcome measures (Ong et al. 2007; White et al. 2007b). Unfortunately, NSAIDs have undesirable effects (related to COX-1 inhibition) that include platelet inhibition (and bleeding), renal impairment, and peptic ulceration. The risk and severity of NSAID-associated side effects are increased in the elderly population (Acute Pain Management: Scientific evidence 2010).

Selective COX-2 inhibitors have been introduced into perioperative clinical practice; they have similar analgesic efficacy to that of NSAIDs, but a lower bleeding risk (no platelet inhibition), and are presumed to have less gastrointestinal toxicity (gastropathy is approximately half of the NSAIDS) (Gajraj and Joshi 2005). Coxibs have similar nephrotoxicity to NSAIDs. The long-term use of coxibs can be associated with an increased risk for cardiovascular adverse events (Shi and Klotz 2008). An increased incidence of thrombotic complications might preclude their use in many patients (Nussmeier et al. 2005). For this reason, two coxibs, rofecoxib and valdecoxib were withdrawn from the market. Currently celecoxib, etoricoxib, and lumiracoxib (in low doses) are used. Parecoxib is the only parenterally administered coxib available to date.

Opioid Sparing

Multimodal analgesia can produce opioid sparing. Yet, the literature on multimodal analgesia often shows variable degrees of success. Other studies have shown, however, that multimodal analgesia may not improve postoperative outcome significantly.

A systematic review of paracetamol and selective and nonselective nonsteroidal anti-inflammatory drugs for the reduction of morphine-related side effects after major surgery did not show a strong case for recommending routine addition of any of the three non-opioids to PCA morphine in the 24 h immediately after surgery, or for favoring one drug class above the others (McDaid et al. 2010).

Gabapentanoids

The value of gabapentin in postoperative pain has been established by several meta-analyses (Ho et al. 2006; Hurley et al. 2006; Seib and Paul 2006) with pooled results from each showing a favorable reduction in postoperative pain scores and opioid consumption when administered at a single preoperative dose of 300–1,200 mg/day, at 1–2.5 h before operation (Lui and Ng 2011). There is dose-dependent dizziness and sedation.

The use of pregabalin in acute postoperative pain was recently reviewed in the Cochrane database (Moore et al. 2009). Three more studies have been published since. The most recent is the effect of pregabalin in reducing chronic postsurgical pain in 228 patients undergoing total knee arthroplasty under combined spinal-epidural technique (Della Valle et al. 2010). The incidence of neuropathic pain was significantly reduced in the pregabalin group at 3 and 6 months after surgery (Della Valle et al. 2010). Pregabalin is a safe and effective medication that may decrease perioperative opioid use in patients with more acute neuropathic pain than acute inflammatory pain. When surgery involves more neuropathic-type acute pain, there is growing evidence that pregabalin may decrease the incidence of chronic pain (Durkin et al. 2010). However, more studies are necessary.

Alpha-2 Adrenoreceptor Agonists

The alpha-2-adrenoceptor agonists, clonidine (imidazole receptor agonist) and dexmedetomidine have antinociceptive activity via peripheral, supraspinal, and primarily spinal cord mechanisms including activation of post-synaptic α 2-adreoceptors of descending noradrenergic pathways (Pöpping et al. 2009). They have synergistic interaction with opioids as well as opioid-sparing effects. Clonidine is opioid sparing but associated with pronounced bradycardia and hypotension in many studies.

Ketamine

Among the ▶ N-Methyl-D-Aspartate (NMDA) antagonists, perioperative low-dose ketamine is both efficacious and preventive as an adjuvant analgesic, with good safety and tolerability profiles. Ketamine is widely used as an adjunct (co-analgesic) to analgesics during the perioperative period. Ketamine in sub-anesthetic doses is effective in reducing morphine requirements in the first 24 h after surgery. Low-dose perioperative ketamine is beneficial as a preventive analgesic with its treatment effect dictated by the type of surgical operation involved (Lui and Ng 2011). At low sub-anesthetic doses (0.15–0.25 mg/kg), ketamine exerts a specific NMDA blockade and, hence, modulates central sensitization induced both by the incision and tissue damage and by perioperative analgesics such as opioids (Vadivelu et al. 2010). Intravenous PCA dose is ketamine 1 mg/ml bolus with morphine 1 mg/ml. Extended use as a PCA remains unresolved as psychotomimetic side effects are dose dependent. Continuous subcutaneous infusions of ketamine 0.1 mg/kg/h provided sufficient analgesia with minimal adverse events in trauma patients (Roelofse 2010).

Ketamine can be very useful in pediatric burn patients, as it is unique in its ability to produce profound analgesia without depressing respiration or causing airway obstruction. The trend toward more ambulatory surgery outside the operating room has created an interest in ketamine because of its acceptably short duration of action, safety profile, administration through almost any route, and sedative/analgesic effects. Ketamine may attenuate tolerance to opioids. It has a definite role in preventing opioid-induced hyperalgesia in patients receiving high doses of opioid for their postoperative pain relief (Vadivelu et al. 2010).

Corticosteroids

Glucocorticoids are well known for both their analgesic and antiemetic effects. They have demonstrated opioid-sparing effects in a number of clinical studies of postoperative pain (Thangaswamy et al. 2010). Dexamethasone and methylprednisolone offer analgesia with anti-emesis. Their doses are limited by steroid-related side effects especially in risk-prone groups such as geriatric patients (Lui and Ng 2011). No convincing documentation of an effective analgesic dose of dexamethasone is available in the literature. Practically, a dose of dexamethasone 0.1 mg/kg is given.

Magnesium

In contrast to earlier evidence, magnesium is beneficial especially in abdominal surgery and toxicity is rare when administered as pre-incision loading followed by infusion under vigilant monitoring (Lui and Ng 2011).

Neostigmine

Neostigmine is a parasympathomimetic agent that has been recently investigated for use as an adjunct analgesic agent in the perioperative and peripartum period (Habib and Gan 2006). While the intrathecal administration of neostigmine produced useful analgesic effects in the postoperative period in some studies, the high incidence of side events, mainly nausea and vomiting, limit the clinical usefulness of this route of administration (Habib and Gan 2006).

Others

Adenosine A(1) and A(2) receptors are widely distributed in the brain and spinal cord and are a novel, non-opioid target for pain management. Adenosine improves postoperative recovery, as indicated by lower pain scores and less opioid consumption (Gan and Habib 2007). Tapentadol is a centrally acting analgesic with a unique dual mode of action as an agonist at the μ-opioid receptor and as a norepinephrine (noradrenaline) reuptake inhibitor (Vadivelu et al. 2010). Further research is necessary to establish its clinical efficacy.

Regional Anesthesia

▶ Regional anesthesia attenuates the endocrine-metabolic effects (the rise in cortisol, catecholamines, glucagon, hyperglycemia, insulin resistance, and negative nitrogen balance) on stress-induced organ dysfunction, and extends analgesia into the postoperative period (Kehlet and Wilmore 2002). The inhibitory effects on catabolic responses are most pronounced when regional anesthesia is provided for up to 24–48 h. The use of local anesthetic techniques, ranging from simple subcutaneous wound infiltration by the surgeon to complex peripheral (plexus) and central (spinal and epidural) blocks by the anesthetist, can markedly improve postoperative pain relief. Local anesthetic infiltration (in tissue, joints, peritoneal cavity) as part of a multimodal regimen offers a simple, safe, and inexpensive alternative to epidural pain control (Schumann et al. 2003). Catheters can be used to provide constant infusions of local anesthetics around peripheral nerves and plexuses (with or without a multimodal mixture). Ultrasonography guidance blocks will probably become the reference technique for local anesthetic injections and regional anesthesia catheter placement (Verghese and Hannallah 2010). Regional block techniques reduce the need for systemic opioid therapy and may be associated with a reduction in opioid-related side effects (Acute Pain Management: Scientific evidence 2010). After major thoracic or abdominal surgery, patients at risk for postoperative respiratory and cardiac complications are offered optimal analgesia with thoracic (or thoracolumbar) epidural infusion of low doses of local anesthetic (e.g., levobupivacaine, ropivacaine), a lipophilic opioid (e.g., fentanyl, sufentanil), and epinephrine (adrenaline) (Breivik 2002).

Perineural patient-controlled regional analgesia (PCRA) allows patients to self-titrate local anesthetic peripheral nerve blocks to achieve comfort. Patient-controlled regional analgesia encompasses a variety of techniques that provide effective postoperative pain relief without systemic exposure to opioids (Vadivelu et al. 2010). The use of a local anesthetic wound infusion is an emerging area with perineural catheters and with wound catheters. The dose should be limited if elderly and with comorbidities. The dose is site specific. It is clear that local anesthetic techniques, particularly peripheral nerve blockade will be one of the cornerstones of postoperative pain management (Vadivelu et al. 2010). Systemic lignocaine improves analgesia, reduces postsurgical inflammation, and accelerates bowel functional recovery in selected studies focused on abdominal surgery (Lui and Ng 2011). Liposome or polymer (microsphere) encapsulation of local anesthetics is being formulated. Long-acting preparations of local anesthetics constitute another area of ongoing research.

Enhanced Recovery Programs

Perioperative care has been improved with newer anesthetic and analgesic techniques and multimodal analgesia, the development of minimally invasive surgery, and drugs to reduce surgical stress (Kehlet 2008). Fast-track surgery or enhanced postoperative recovery programs have been developed by combining these techniques with evidence-based adjustments to the use of nasogastric tubes, drains, and urinary catheters, preoperative bowel preparation, and early initiation of oral feeding and mobilization (Kehlet 2008). Faster recovery, reduced hospital stay, and decreased length of convalescence can occur if multimodal analgesia is combined with a rehabilitation program that is multidisciplinary and multimodal (Gajraj and Joshi 2005).

Future

The application of a multimodal approach, administration of preventive analgesia, and a paradigm shift in surgical techniques all mandate a revisit of evidence-based perioperative pain management (Lui and Ng 2011). The use of perioperative multimodal techniques may provide long-term benefits to patients. The development of newer agents available for postoperative pain control opens up possibilities for newer combinations in multimodal analgesia.

References

Acute pain management: Scientific evidence (3rd ed.) (2010). http://www.anzca.edu.au/resources/books-and-publications/

Breivik, H. (2002). Postoperative pain: Toward optimal pharmacological and epidural analgesia. In M. A. Giamberardino (Ed.), *Pain 2002 – an updated review* (pp. 337–349). Seatlle: IASP Press.

Buvanendran, A., & Kroin, J. S. (2009). Multimodal analgesia for controlling acute postoperative pain. *Current Opinion in Anaesthesiology, 22*(5), 588–593.

Costantini, R., Affaitati, G., Fabrizio, A., & Giamberardino, M. A. (2011). Controlling pain in the post-operative setting. *International Journal of Clinical Pharmacology, 49*(2), 116–127.

Della Valle, C. J., et al. (2010). Perioperative oral pregabalin reduces chronic pain after total knee arthroplasty: A prospective, randomized, controlled trial. *Anesthesia and Analgesia, 110*, 199–207.

Durkin, B., Page, C., & Glass, P. (2010). Pregabalin for the treatment of postsurgical pain. *Expert Opinion on Pharmacotherapy, 11*(16), 2751–2758.

Elia, N., Lysakowski, C., & Tramèr, M. (2005). Does multimodal analgesia with acetaminophen, nonsteroidal antiinflammatory drugs, or selective cyclooxygenase-2 inhibitors and patient-controlled analgesia morphine offer advantages over morphine alone? Meta-analyses of randomized trials. *Anesthesiology, 103*(6), 1296–1304.

Gajraj, N., & Joshi, G. (2005). Role of cyclooxygenase-2 inhibitors in postoperative pain management. *Anesthesiology Clinics of North America, 23*(1), 49–72.

Gan, T. J., & Habib, A. S. (2007). Adenosine as a non-opioid analgesic in the perioperative setting. *Anesthesia and Analgesia, 105*(2), 487–494.

Habib, A. S., & Gan, T. J. (2006). Use of neostigmine in the management of acute postoperative pain and labour pain. *CNS Drugs, 20*(10), 821–839.

Ho, K. Y., Gan, T. J., & Habib, A. S. (2006). Gabapentin and postoperative pain – a systematic review of randomized controlled trials. *Pain, 126*, 91–101.

Hurley, R. W., Cohen, S. P., Williams, K. A., et al. (2006). The analgesic effects of perioperative gabapentin on postoperative pain: A meta-analysis. *Regional Anesthesia and Pain Medicine, 31*, 237–247.

Jin, F., & Chung, F. (2001). Multimodal analgesia for postoperative pain control. *Journal of Clinical Anesthesia, 13*(7), 524–539.

Kehlet, H., & Wilmore, D. W. (2002). Mutlimodal strategies to improve surgical outcome. *American Journal of Surgery, 183*(6), 630–641.

Kehlet, H. (2008). Fast track colorectal surgery. *Lancet, 371*, 791–793.

Lui, F., & Ng, K. F. (2011). Adjuvant analgesics in acute pain. *Expert Opinion on Pharmacotherapy, 12*(3), 363–385.

McDaid, C., Maund, E., Rice, S., Wright, K., Jenkins, B., & Woolacott, N. (2010). Paracetamol and selective and non-selective non-steroidal anti-inflammatory drugs (NSAIDs) for the reduction of morphine-related side effects after major surgery: A systematic review. *Health Technology Assessment, 14*(17), 1–153.

Moore, R. A., Straube, S., & Wiffen, P. J. (2009). Pregabalin for acute and chronic pain in adults. *Cochrane Database of Systematic Reviews, 3*, CD007076.

Nussmeier, N. A., Whelton, A. A., Brown, M. T., et al. (2005). Complications of the COX-2 inhibitors parecoxib and valdecoxib after cardiac surgery. *The New England Journal of Medicine, 352*, 1081–1091.

Ong, C. K., Lirk, P., Tan, C. H., & Seymour, R. A. (2007). An evidence-based update on nonsteroidal anti-inflammatory drugs. *Clinical Medicine & Research, 5*(1), 19–34.

Pöpping, D. M., Elia, N., Marret, E., et al. (2009). Clonidine as an adjuvant to local anesthetics for peripheral nerve and plexus blocks: A meta-analysis of randomized trials. *Anesthesiology, 111*(2), 406–415.

Power, I., & Barratt, S. (1999). Analgesic agents for the postoperative period. Nonopioids. *The Surgical Clinics of North America, 79*(2), 275–295.

Roelofse, J. A. (2010). The evolution of ketamine applications in children. *Paediatric Anaesthesia, 20*(3), 240–245.

Schumann, R., Shikora, W., Weis, J. M., Wurm, H., Strassels, S., & Carr, D. B. (2003). A comparison of multimodal perioperative analgesia to epidural pain management after gastric bypass surgery. *Anesthesia and Analgesia, 96*(2), 469–474.

Seib, R. K., & Paul, J. E. (2006). Preoperative gabapentin for postoperative analgesia: A meta-analysis. *Canadian Journal of Anaesthesia, 53*, 461–469.

Shi, S., & Klotz, U. (2008). Clinical use and pharmacological properties of selective COX-2 inhibitors. *European Journal of Clinical Pharmacology, 64*(3), 233–352.

Thangaswamy, C. R., Rewari, V., Trikha, A., et al. (2010). Dexamethasone before total laparoscopic hysterectomy: A randomized controlled dose–response study. *Journal of Anesthesia, 24*(1), 24–30.

Vadivelu, N., Mitra, S., & Narayan, D. (2010). Recent advances in postoperative pain management. *The Yale Journal of Biology and Medicine, 83*(1), 11–25.

Verghese, S. T., & Hannallah, R. S. (2010). Acute pain management in children. *Journal of Pain Research, 3*, 105–123.

White, P. F., Kehlet, H., Neal, J. M., et al. (2007a). Role of the anesthesiologist in fast-track surgery: From multimodal analgesia to perioperative medical care. *Anesthesia and Analgesia, 104*(6), 1380–1396.

White, P. F., Sacan, O., Tufanogullari, E. M., et al. (2007b). Effect of short-term postoperative celecoxib administration on patient outcome after outpatient laparoscopic surgery. *Canadian Journal of Anesthesia*, 54(5), 342–348.

White, P. F. (2008). Multimodal analgesia: Its role in preventing postoperative pain. *Current Opinion in Investigational Drugs, 9*(1), 1–7.

Multimodal Analgesics

▶ Drugs with Mixed Action and Combinations: Emphasis on Tramadol

Multimodal Rehabilitation Treatment and Psychiatric Aspects of Multimodal Treatment for Pain

David A. Fishbain
Department of Psychiatry, University of Miami, Miami, FL, USA

Synonyms

Multidisciplinary treatment; Multimodal treatment; Psychiatric aspects of multimodal treatment for pain

Definition

1. Multimodal treatment is defined as applying several treatments simultaneously/concurrently or sequentially (one after another) in order to improve pain and facilitate rehabilitation.
2. Multidisciplinary treatment is defined as treatment involving more than one treatment discipline, and as different disciplines usually utilize different treatments, multidisciplinary treatment is by definition multimodal.

Characteristics

Pain Treatment Facilities

Pain treatment facilities developed for a number of reasons. First, people in general put a very high value on living a pain-free life. Second, numerous epidemiological studies determined that a large percentage (2–40 %) of the general population suffered from chronic intractable benign pain. And third, the experience of treating continuous severe pain in battle injured World War II soldiers determined that a coordinated team was required to manage the different types of pain (Rosomoff and Steele-Rosomoff 1991).

Multidisciplinary pain clinics or centers evolved from these concepts in the early 1970s. It was observed that a significant percentage of low-back pain and neck pain patients did not improve with traditional medical treatment but remained disabled (Rosomoff and Steele-Rosomoff 1991). These patients demonstrated a host of behavioral and psychosocial problems in association with their chronic pain. These problems required the intervention of disciplines besides those of neurosurgery, orthopedic surgery, and anesthesiology (Rosomoff and Steele-Rosomoff 1991). In addition, these patients required a concurrent highly integrated multidisciplinary treatment approach that would address all the patient's problems simultaneously, i.e., multimodal treatment (Rosomoff and Steele-Rosomoff 1991).

There were approximately 1,500–2,000 pain treatment facilities in the USA (Rosomoff and Steele-Rosomoff 1991). They differed in their staff composition, size, philosophy, and, most importantly, treatment approach. Because of this problem, the International Association for the Study of Pain (IASP) developed definitions for four types of pain treatment facilities (see below) (Loeser 1991).

IASP Classification of Pain Facilities (Adapted from Loeser 1991)

Modality-Oriented Clinic
- Provides specific type of treatment, e.g., nerve blocks, transcutaneous nerve stimulation, acupuncture, biofeedback
- May have one or more health care disciplines
- Does not provide an integrated, comprehensive approach

Pain Clinic
- Focuses on the diagnosis and management of patients with chronic pain or may specialize in

Multimodal Rehabilitation Treatment and Psychiatric Aspects of Multimodal Treatment for Pain, Table 1 MPC chronic pain treatment meta-analyses

Author/ year	Treatment type and clinical condition	Outcome measure	Author clinical interpretation of meta-analysis data
Malone and Strube (1988)	Pain facility, chronic pain	Activity level. Frequency pain. Index pain. Intensity pain. Medication use. Mood	Treatment effective, good effect sizes
Curtis (1992)	Multidisciplinary pain facility, chronic low-back pain	Physical fitness. Subjective distress. Daily activity increase. Medication decrease	Multidisciplinary treatment programs effective
Flor et al. (1992)	Multidisciplinary pain facilities, chronic low-back pain	Activity level. Pain behavior. Medication use. Return to work	Multidisciplinary pain clinics effective at returning chronic back pain patients to work
Cutler et al. (1994)	Multidisciplinary pain facility, chronic low-back pain	Return to work	Multidisciplinary pain facilities return chronic low-back pain patients to work

specific diagnoses or pain related to a specific region of the body

- Does not provide comprehensive assessment or treatment, an institution offering appropriate consultative and therapeutic services would qualify but never an isolated solo practitioner

Multidisciplinary Pain Clinic

- Specializes in the multidisciplinary diagnosis and management of patients with chronic pain or may specialize in specific diagnoses or pain related to a specific region of the body
- Staffed by physicians of different specialties and other health care providers
- Differs from a multidisciplinary pain center only because it does not include research and teaching

Multidisciplinary Pain Center

- Organization of health care professionals and basic scientists that includes research, teaching, and patient care in acute and chronic pain
- Typically a component of a medical school or a teaching hospital
- Clinical programs supervised by an appropriately trained and licensed director
- Staffed by a minimum of physician, psychologist, occupational therapist, physical therapist, and registered nurse

- Services provided integrated and based on interdisciplinary assessment and management
- Offers both inpatient and outpatient program

Inspection of the list above indicates that there is a clear distinction between modality-oriented clinics, pain clinics, and multidisciplinary facilities. Differences between multidisciplinary pain clinics and multidisciplinary pain centers (MPCs) include research and teaching in the MPCs. These definitions also indicate that the MPCs may be more likely to have inpatient and outpatient treatment and to have larger and more diversified multidisciplinary staffs, including more than one physician specialty. As a consequence, MPCs are likely to offer a wider range of treatments than multidisciplinary pain clinics and thus involve more multimodal treatment. Most of the pain facility treatment outcome studies involve MPCs.

MPC Treatment Outcome Studies

Because MPC developed in the 1970s and because they had to demonstrate treatment efficacy, there are now close to 300 treatment outcome studies in the literature (Fishbain et al. 1993). As such, this literature has been subjected to a number of meta-analyses (Table 1). It is interesting to note that these meta-analyses (four) were consistent in indicating that MPC treatment or multimodal treatment is effective. This meta-analysis literature has recently been reviewed for meta-analytic procedure quality and found to be of acceptable quality (Fishbain et al. 2000). It is also important to note that this literature now indicates that multidisciplinary (multimodal) treatment is superior to single treatment such as medical therapy or physical therapy (Flor et al. 1992).

Which Treatments or Combination Makes MPC Treatment Effective?

For an exhaustive list of treatment by medical specialty available at MPCs, please refer to Fishbain et al. (1997). It is to be noted that some of these treatments, e.g., physical therapy, have been studied in a placebo fashion, resulting in a significant number of studies in the literature. There are meta-analysis results available for some of these (Fishbain et al. 2000). Of the non-pharmacological treatments having meta-analyses, physical therapy, cognitive and behavior therapy, and educational therapy appear to be effective. Of the psychopharmacological treatments, antidepressants, topical NSAIDs, capsaicin, and anticonvulsants are effective (Fishbain et al. 1997, 2000).

Yet, in reality, MPCs may or may not provide these treatments or may provide a greater range of treatments (Fishbain et al. 1997). Pain facilities generally use multiple treatments simultaneously which are integrated into a treatment package (Fishbain et al. 1997). Thus, it is impossible to say whether the positive MA results for MPCs related to the fact that the specific treatments shown to be effective were provided or that other hitherto untested treatments were utilized. This question awaits further research.

Another issue in relation to which treatments make MPC treatment effective is what outcome variables are examined (Fishbain et al. 1993). As an example, a specific treatment such as capsaicin may decrease pain, but if used alone, it may not be enough to return the patient to work. In other words, for the attainment of some outcome variables, such as return to work, a combination of specific treatments may be required. The evidence indicates that treatments under one category, such as behavior, may have equal efficacy. For example, operant conditioning has been found to be as effective as cognitive-behavior treatment (Turner and Clancy 1988) and outpatient group cognitive therapy, relaxation training, and cognitive therapy have all been demonstrated to be equally efficacious (Turner and Jensen 1993).

In addition, different treatments within one category appear not be additive. When a cognitive component was added to operant pain treatment, there was no decrease in patient medical utilization costs or improvement in patient quality of life scores (Goossens et al. 1998). However, treatments from different groups may have an additive effect. The combined package of cognitive-behavioral group pain treatment with physical therapy has been found to be superior to physical therapy alone (Nicholas et al. 1992). This speaks to the apparent advantage of multimodal treatment.

At this time, it is unclear what combination of MPC treatments delivered in a treatment package is effective. It is therefore unclear what combination of treatments is necessary for an effective package. It is also possible that the effectiveness of MPCs rests in their ability to deliver this treatment package and an ability to integrate treatments into a package (Rosomoff and Steele-Rosomoff 1991). This last issue may influence an overall effectiveness that has not been explored in the literature.

Psychiatric Multimodal Treatment within MPCs

Early in the course of development of MPCs, it became clear that the vast majority of chronic pain patients suffered from associated psychiatric comorbidity (Fishbain 1999). In addition, it has recently become clear that drugs commonly utilized by psychiatrists, e.g., antidepressants, anticonvulsants, have strong analgesic properties (Fishbain et al. 2000). Finally, it has become recognized that many chronic pain patients have neuropathic pain for which psychopharmacological treatment is becoming increasingly available (Fishbain et al. 2000). This confluence of factors and the fact that psychiatrists by the nature of their specialty have expertise in psychopharmacological treatment and detoxification (Fishbain 2002) has increased the potential impact of this medical specialty on pain treatment outcome.

The list below presents the functions of the psychiatrist within MPCs and the potential treatments that he/she can initiate in a multimodal fashion. With reference to treatments, it is to be noted that these now reflect the major issue of directing psychopharmacological treatment both at pain relief and at the same time psychiatric comorbidity. Thus, the psychiatrist working at an MPC should be very familiar with the literature addressing the psychopharmacological treatment of pain in order to be able to choose an agent for pain, which will also have impact on the psychiatric comorbidity. This concept is very different from 10 to 15 years ago, when psychiatrists treated psychiatric comorbidity exclusively. In addition, it is to be noted that as there is now a subspecialty psychiatry board in pain management, some psychiatrists are developing expertise in the procedure treatments of chronic pain.

Psychiatric Functions and Multimodal Treatments

Potential Functions
1. Diagnosis of psychiatric comorbidity as per DSM-system.
2. Determination of presence of psychological difficulties, problems, etc., e.g., suicidal ideation, homicidal ideation, marital/system conflicts, etc.
3. Determination presence of somatic comorbidities, e.g., sleep problems, headaches, dizziness, etc.
4. Develop a psychopharmacological and non-psychopharmacological treatment plan for #1–#3 above.
5. Administer behavioral rating scale testing if necessary.
6. According to the history, physical examination and required laboratory workup determine whether the pain is nociceptive, neuropathic, or both.
7. Participate in multidisciplinary case staffing in order to develop a multidisciplinary treatment plan that would encompass the psychiatric treatment plan.
8. Act as consultant to other multidisciplinary staff.
9. Educate members of multidisciplinary staff on psychiatric aspects of chronic pain.
10. Perform clinical research on all aspects of pain.
11. Act as director of the multidisciplinary team.

Multimodal Treatments
1. Psychopharmacological treatment directed at nociceptive pain.
2. Psychopharmacological treatment directed at neuropathic pain.
3. Psychopharmacological treatment directed at psychiatric comorbidity.
4. Detoxify patient if necessary.
5. Provide supportive counseling.
6. Lead behavior rehabilitation groups.
7. Provide program patient education for his/her area of expertise.
8. Participate in the behavior modification (reinforcement) aspects of the multidisciplinary program.
9. If within his/her area of expertise, perform procedures such as trigger point injections, acupuncture, blocks, epidurals, etc.

M

References

Curtis, J. E. (1992). The efficacy of multidisciplinary treatment programs for chronic low back pain: A meta-analysis. *Dissertation Abstracts International, 53*, 4948.

Cutler, R. B., Fishbain, D. A., Rosomoff, H. L., et al. (1994). Does nonsurgical pain center treatment of chronic pain return patients to work? A review and meta-analysis of the literature. *Spine, 19*, 643–652.

Fishbain, D. A. (1999). Approaches to treatment decisions for psychiatric comorbidity in the management of the chronic pain patient. *The Medical Clinics of North America, 83*, 737–759.

Fishbain, D. A. (2002). Opiate, hynopsedative, alcohol, and nicotine detoxification protocols. In C. D. Tollison, J. R. Satterthwaite, & J. W. Tollison (Eds.), *Practical pain management* (pp. 314–329). Philadelphia: Lippincott Williams and Wilkins.

Fishbain, D. A., Rosomoff, H. L., Goldberg, M., et al. (1993). The prediction of return to the workplace after multidisciplinary pain center treatment. *The Clinical Journal of Pain, 9*, 3–15.

Fishbain, D. A., et al. (1997). Pain facilities: A review of their effectiveness and referral selection criteria. *Current Review of Pain, 1*, 107–115.

Fishbain, D., Cutler, R. B., Rosomoff, H. L., et al. (2000).
What is the quality of the implemented meta-analytic
procedures in chronic pain treatment meta-analyses?
The Clinical Journal of Pain, 16, 73–85.

Flor, H., Fydrich, T., & Turk, D. C. (1992). Efficacy of
multidisciplinary pain treatment centers: A
meta-analytic review. *Pain, 49,* 221–230.

Goossens, M. E., Rutten-Van Molken, M. P.,
Kole-Snijders, A. M., et al. (1998). Health economic
assessment of behavioural rehabilitation in chronic
low back pain: A randomised clinical trial. *Health
Economics, 7,* 39–51.

Loeser, J. D. (1991). Desirable characteristics for pain
treatment facilities: Reports of the IASP taskforce. In
M. R. Bond, J. E. Charlton, & C. J. Woolf (Eds.),
Proceedings of the VIth world congress on pain
(pp. 411–415). Amsterdam: Elsevier.

Malone, M. D., & Strube, M. J. (1988). Meta-analysis of
non-medical treatments for chronic pain. *Pain,
34,* 231–244.

Nicholas, M. K., Wilson, P. H., & Goyen, J. (1992).
Comparison of cognitive-behavioral group treatment
and an alternative nonpsychological treatment for
chronic low back pain. *Pain, 48,* 339–347.

Rosomoff, H. L., & Steele-Rosomoff, R. (1991).
Comprehensive multidisciplinary pain center
approach to the treatment of low back pain. *Neurosurgical Clinics of North America, 2,* 877–890.

Turner, J. A., & Clancy, S. (1988). Comparison of operant
behavioral and cognitive-behavioral group treatment
for chronic low back pain. *Clinical Psychologist,
56,* 261–266.

Turner, J. A., & Jensen, M. P. (1993). Efficacy of
cognitive therapy for chronic low back pain. *Pain,
52,* 169–177.

Multimodal Treatment

▶ Multimodal Rehabilitation Treatment and Psy-
chiatric Aspects of Multimodal Treatment for
Pain
▶ Psychological Treatment of Pain in Children

Multiple Sclerosis (MS)

Definition

An idiopathic immune system disease of the
central nervous system in which gradual destruction of myelin occurs in patches throughout the
brain or spinal cord (or both), interfering with
the nerve pathways and causing muscular
weakness, loss of coordination, and speech and
visual disturbances. Sometimes manifests as
radiating pain in a similar distribution as the
sciatic nerve.

Cross-References

▶ Central Pain, Outcome Measures in Clinical
Trials
▶ Sciatica
▶ Trigeminal, Glossopharyngeal, and Geniculate
Neuralgias

Multipolar Cells

Definition

Neuron type whose cell body (soma) issues an
axon and several primary dendrites forming the
dendritic tree.

Cross-References

▶ Trigeminal Brainstem Nuclear Complex,
Anatomy

Multiprofessional

Definition

Healthcare professionals working with the same
patient.

Cross-References

▶ Physical Medicine and Rehabilitation, Team-
Oriented Approach

Multireceptive Neuron (MR)

Definition

A neuron that is activated by a variety of innocuous and noxious mechanical and thermal stimuli applied to its receptive field, which is typically widespread and comprises of cutaneous and deep tissue.

Cross-References

▶ Arthritis Model, Kaolin-Carrageenan-Induced Arthritis (Knee)
▶ Chronic Pain
▶ Thalamic Bursting Activity
▶ Thalamus, Nociceptive Cells in VPI, Cat and Rat

Multireceptive Neurons (MR)

▶ Thalamus, Nociceptive Cells in VPI, Cat and Rat

Multisensory Perceptions

Definition

Perceptions of external objects based on more than one form of sensory input (e.g., sight and smell).

Cross-References

▶ Amygdala, Pain Processing and Behavior in Animals

Munchausen's by Proxy

Definition

A severe type of factitious disorder where the parent consciously causes the child to assume the sick role to satisfy the parent's psychological need.

Muscle Contraction Headache

▶ Headache, Episodic Tension-Type

Muscle Cramp

Definition

Involuntary sudden painful muscle contraction. During the cramp, the muscle is visibly and palpably taut and painful, often with abnormal posture of the affected joint, condition which can be relieved by stretching or massage.

Cross-References

▶ Muscular Cramps

Muscle Cramps

▶ Muscular Cramps

Muscle Discrimination

Definition

This refers to a person's ability to accurately sense the current level of tension within the muscles being monitored.

Cross-References

▶ Psychophysiological Assessment of Pain

Muscle Hyperactivity

▶ Orofacial Pain, Movement Disorders

Muscle Nociceptor

Definition

A free nerve ending with high thresholds to mainly mechanical and chemical stimuli (nociceptive); also high-intensity thermal stimuli may excite muscle nociceptors. Their afferents are small-diameter myelinated (Group III) or unmyelinated (Group IV) muscle nerve fibers. Group IV corresponds to cutaneous C-fibers and group III to Aδ-fibers. Conduction velocities for cat muscle afferent fibers are below 2.5 m/s for group IV and 2.5–30 m/s for group III fibers.

Cross-References

▶ Exogenous Muscle Pain

Muscle Nociceptors, Neurochemistry

Siegfried Mense
Neuroanatomy, Heidelberg University,
Medical Faculty Mannheim, CBTM,
Mannheim, Germany

Synonyms

Nociceptor; Nocisensor; Potentially damaging stimulus

Definition

A nociceptor is a receptive ending that specializes in informing the central nervous system (CNS) about the presence of a tissue-threatening stimulus. A muscle nociceptor has an elevated stimulation threshold just below the noxious level. (The receptor responds to stimulus intensities below the level that causes tissue damage as it also fulfills the function of an alarm system to prevent tissue damage.) In addition to an elevated stimulation threshold, a nociceptor encodes the intensity of a stimulus within the noxious range (i.e., it must not saturate when a stimulus reaches noxious intensities).

A receptor molecule is a protein located within the axonal membrane of a receptive ending. It binds sensitizing and stimulating substances in a highly specific manner and is either coupled to an ion channel (which opens after the binding has occurred) or a G-protein, which starts an intracellular cascade of events that – among other effects – leads to the activation of enzymes such as kinases.

Occasionally, "nocisensor" is used in place of nociceptor. The term "pain receptor" should be avoided because receptive nerve endings are named after the stimulus energy to which they respond (e.g., mechano-receptor, thermo-receptor, or poly/multi-modal receptor if they respond to multiple modalities of stimulation). A nociceptor responds to an actually or potentially damaging stimulus.

Characteristics

Morphology
A muscle nociceptor is an unencapsulated ("free") nerve ending that is connected to the CNS by thin myelinated (group III) or unmyelinated (group IV) nerve fibers (Stacey 1969). These fibers have a slow conduction velocity, with ▶ group III fibers conducting between 2.5 and approximately 20 m/s and ▶ group IV fibers between 0.5 and 2 m/s. Histologically, the nerve ending is not free in the strict sense but is surrounded by a single layer of ▶ Schwann cells. The Schwann cells leave small patches of the axonal membrane uncovered, the so-called exposed axon areas. These areas are assumed to be the sites of action for the chemical stimuli that are present in a pathologically altered muscle.

Neuropeptide Content

There is no one ▶ neuropeptide that can be regarded as specific for sensory fibers from muscle or for muscle nociceptors, although a greater proportion of small-fiber muscle afferents appear to contain *substance P (SP)* and *calcitonin-gene-related peptide (CGRP)* than do skin afferents (O'Brien et al. 1989). Afferent endings in skeletal muscle of the rat have been reported to contain SP, ▶ CGRP, ▶ somatostatin (SOM), ▶ vasoactive intestinal polypeptide (VIP) as well as the neurotrophin ▶ nerve growth factor (NGF). In experimentally inflamed muscle, afferents expressing neuropeptides – particularly SP-positive ones – increased in number (Reinert et al. 1998). With respect to nociception, SP is of particular interest because, in experiments on skin afferents, SP was shown to be present predominantly in nociceptive afferents (Lawson et al. 1997). Neuropeptides are released locally during excitation of the afferent ending and influence the chemical milieu of the receptive ending. SP and CGRP have a strong dilating and permeability-increasing actions on small blood vessels. By these effects, neuropeptides cause an increase in local microcirculation and edema formation at the site of insult. In addition, SP – as well as NGF – has a sensitizing action on nociceptors.

Receptor Molecules in the Membrane of a Nociceptive Ending

Systematic studies on the presence of receptor molecules in muscle nociceptors are not available. One recent article reported that the capsules of muscle spindles contain putative nociceptive fibers that express P2X3, TRPV1, and the metabotropic glutamate receptor MGluR5 (Lund et al. 2010). Based on the responsiveness of muscle nociceptors to intramuscular and intra-arterial injections of an ▶ Algesic Agent/Algesic Chemical, and existing parallels with cutaneous nociceptors, the following types of receptor molecules are likely to be present in skeletal muscle (Caterina and David 1999; Mense and Meyer 1985; McCleskey and Gold 1999):

1. *Receptors for inflammatory substances.* Among these are receptors for ▶ bradykinin (BK; B1 and B2 receptors), 5-hydroxytryptamine (5-HT, ▶ serotonin; e.g., 5-HT3 receptor), and prostaglandins (e.g., PGE2; EP2 receptor). In naive tissue, BK is known to influence the receptive ending via the B2 receptor; in inflamed tissue, the B1 receptor is synthesized in the soma of the nociceptor, transported to the receptive ending, and inserted into its membrane. This is an example of a neuroplastic change in the nociceptive ending. In contrast to the ionotropic 5-HT3 receptor that controls an ion channel, BK activates B1 and B2 receptors, both of which are G-protein-coupled receptors.

2. *Receptors for protons.* Besides immunoreactivity (IR) for acid-sensing ion channels (e.g., ASIC1 and particularly ASIC3; Sluka et al. 2003), IR for the transient receptor potential vanilloid receptor (▶ TRPV1) (Caterina and David 1999) is present in somata of the dorsal root ganglion (DRG) that supply receptive endings in skeletal muscle (J.Reinöhl, U. Hoheisel and S. Mense, unpublished; Fig. 1). The receptor is sensitized by and responds to an increase in H^+-concentration and to heat. The sensitivity of this receptor to protons is important under conditions in which the pH of the tissue is lowered (e.g., exhaustive muscle work, ischemia, inflammation, tonic contraction). A specific stimulant for the TRPV1 receptor molecule is capsaicin, the active ingredient of chili peppers. In inflamed tissue, TRPV1 receptors have been reported to be more frequent than under normal conditions (Carlton and Coggeshall 2001).

3. *Purinergic receptors.* These receptor molecules bind adenosine triphosphate (ATP) and the products of ATP degradation. The P2X3 receptor (Burnstock 2000; Cook and McCleskey 2002) has been demonstrated to be present in cutaneous nociceptors; it has also been shown to exist in ▶ DRG neuron somata supplying the gastrocnemius-soleus muscle of the rat (U. Hoheisel and S. Mense, unpublished).

4. *Receptors for growth factors.* In our group, nerve growth factor (NGF; ▶ TrkA receptor) proved to excite muscle nociceptors, but only those having afferents with tetrodotoxin-resistant (TTX-r) sodium channels. The

Proportion of immunoreactive neurones (n = 139)

Muscle Nociceptors, Neurochemistry, Fig. 1 Neurons in the rat L5 DRG exhibiting immunoreactivity (IR) for the acid-sensing ion channel 1 (ASIC 1) and the transient receptor potential family V member 1 receptor 1 (TRPV1, formerly called vanilloid receptor 1), respectively. All evaluated neurons were retrogradely labeled from the gastrocnemius-soleus (GS) muscle with the fluorescent dye True blue. (a) The same neuron labeled with three different stains: Aa retrograde labeling with True blue from the GS muscle; Ab fluorescent staining with antibodies to TRPV1; Ac fluorescent staining with antibodies to ASIC 1. (b) Proportion of neurons that were IR for ASIC 1 and/or TRPV1. The filled bar shows neurons exhibiting IR for both ASIC 1 and TRPV1 (ASIC 1-IR + TRPV1-IR). Without IR, neurons exhibiting neither ASIC 1-IR nor TRPV1-IR (J. Reinöhl, U. Hoheisel and S. Mense, unpublished)

intramuscular injection of NGF did not elicit immediate pain in humans and no pain-related behavior in non-anesthetized rats (Svensson et al. 2003; Hoheisel et al. 2005). ▶ Brain-derived neurotrophic factor (BDNF; TrkB receptor) had no excitatory effect on muscle nociceptors, but desensitized the endings to mechanical stimuli (i.e., after intramuscular injection of BDNF, the response magnitude of the ending to noxious pressure stimuli was reduced for more than 15 min).

5. *Receptors for excitatory amino acids*. Reports in the literature indicate that nociceptors in the deep tissues around the temporomandibular joint are activated by glutamate (Cairns et al. 1998). Apparently, glutamate receptors – including the NMDA receptor (Ro et al. 2005) – are present in nociceptors of muscle and other deep somatic tissues.

6. *Opioid receptors*. These receptors were found on Aδ- and C-fibers in cutaneous nerves (Stein et al. 1990). They are upregulated during tissue inflammation.

Putative nociceptive afferent fibers from the rat masseter muscle differed from proprioceptive muscle spindle afferents in that the former expressed TTX-r sodium channels (Connor et al. 2005).

Other algesic agents may activate muscle nociceptors without binding to specific receptor molecules. An example of this is the potassium ion which might depolarize and excite nociceptors following muscle trauma if the relatively high intracellular concentration of K^+ is released from muscle cells. High concentrations of Na^+ may likewise have an unspecific mechanism of action. A marked increase in extracellular Na^+ does not occur under pathophysiologic conditions, but is induced in clinical studies of muscle pain mechanisms when hypertonic saline injections or infusions are injected intramuscularly (Graven-Nielsen et al. 1997). In these studies, the high Na^+ concentration – and not the hypertonicity of the solution – appears to be the effective stimulus (see entry "▶ Muscle Pain Model, Ischemia-Induced and Hypertonic Saline-Induced").

Response Properties

Mechanical Stimulation

Upon mechanical stimulation (e.g., pressure), a muscle nociceptor has a high stimulation threshold and requires noxious (tissue-threatening, subjectively painful) intensities of stimulation for excitation. In intact muscle, a nociceptor does not respond to everyday stimuli such as physiologic movements or muscle stretch (Mense 1997). Recordings of the activity of single muscle afferent fibers in cats and rats have shown that nociceptors as described above are present in skeletal muscle (Mense and Meyer 1985). Microneurographic recordings in humans also demonstrated the existence of muscle nociceptors (Marchettini et al. 1996). It is important to note that not all unencapsulated nerve endings in skeletal muscle have nociceptive properties. Many of them can be excited by weak innocuous pressure stimuli and probably mediate subjective pressure sensations.

Chemical Stimulation

BK, 5-HT, and prostaglandin E2 have long been known to excite muscle nociceptors in concentrations that are likely to be present in an inflamed or ischemic muscle (Kumazawa and Mizumura 1977; Mense and Meyer 1985). 5-HT and PGE2 are ubiquitously present in the body and are released under pathologic conditions. BK is cleaved by enzymatic action from kallidin, a plasma protein. Tissue ischemia is a potent stimulus for inducing this cleavage.

In experiments by the author's group, acidic solutions were effective stimulants for muscle receptors with group IV afferent fibers in the rat. Approximately 60 % of the tested receptors were excited by intramuscular injections of an acidic phosphate buffer (pH 6; Fig. 2). Such a lowering in pH is known to occur in inflamed or ischemic tissue. The proton-sensitive nociceptors may be of particular importance for the induction of chronic muscle pain. There is evidence indicating that repeated intramuscular administration of acidic solutions results in long-lasting hyperalgesia and spread of pain by activating ASIC3 receptors (Sluka et al. 2001, 2003).

Muscle   Nociceptors,   Neurochemistry,
Fig. 2 Proportion of rat muscle nociceptors responding to adenosine triphosphate (*ATP*), capsaicin (*Caps.*), acidic phosphate solution (pH 6), and hypertonic saline (NaCl, 5 %). The data were obtained in electrophysiological experiments in which the impulses of single group IV afferent fibers from rat muscle were recorded. The stimulating solutions (injection volume 25 µl) were injected intramuscularly into the mechanosensitive receptive field of the ending. Solutions had a pH of 5.5. ATP*, ATP dissolved in Tyrode's solution with the pH adjusted to neutral (7.4). NaCl 5 % was the most effective stimulus and activated all of the units tested; the other agents excited at least 40 % of the nociceptors (J. Reinöhl, U. Hoheisel, and S. Mense, unpublished). Please note that these substances were not specific stimulants for nociceptors; non-nociceptive (low-threshold mechanosensitive) group IV receptors were also excited

In patients, the pain during ► bruxism, chronic ► dystonia, and some cases of ► tension-type headache may be mediated by the TRPV1 receptor or other acid-sensing ion channels, because these conditions are likely to be associated with low tissue pH and ischemia. Due to the ischemia, BK acting on B2 or B1 receptors could also contribute to this pain. In microneurographic recordings from muscle nerves in humans, muscle nociceptors with moderate to high mechanical thresholds were found that could be activated by intramuscular injections of capsaicin (Marchettini et al. 1996). The capsaicin injections were associated with strong muscle pain. As capsaicin is assumed to be a specific stimulant for the TRPV1 receptor, these data show that TRPV1 is present in human muscle nociceptors.

Muscle Nociceptors, Neurochemistry, Fig. 3 Receptor molecules in the membrane of a nociceptor. The scheme shows three branches of a free nerve ending. Not all of the receptor molecules included in the figure have been proven to exist in muscle nociceptors, some (e.g., the opioid receptor) are known from studies of cutaneous nociceptors. The following processes are of practical importance. Branch (**a**), sensitization. Sensitization is caused by the binding of sensitizing substances (e.g., prostaglandin E2, PG E_2; serotonin, 5-HT) to receptor molecules, which induce intracellular cascades of events that – among other effects – increase the sensitivity of receptor channel proteins by phosphorylation (coupling of an organic phosphate residue to the channel protein) through activated protein kinases (PK) such as PKA. Branch (**b**), neuroplastic changes of the bradykinin (BK) receptor. In naive tissue, BK excites nociceptors by binding to the B2 receptor; in inflamed tissue, BK does so by binding to the B1 receptor. Branch (**c**), other more recently detected receptors. Purinergic receptors (e.g., P2X3) bind adenosine triphosphate (ATP) and its degradation products, and vanilloid receptors (e.g., TRPV1) are sensitive to protons (H+), capsaicin, and heat. ASICs are acid-sensing ion channels (e.g., ASIC 1) that can be opened by low pH. TrkA is the high-affinity receptor for nerve growth factor (NGF), TrkB for brain-derived neurotrophic factor (BDNF)

Nociceptors of the gastrocnemius muscle in the rat have been shown to respond to ATP in concentrations that are present in muscle cells (Reinöhl et al. 2003). This means that every time a muscle cell is damaged, it releases ATP in amounts that can excite muscle nociceptors. Human muscle nociceptors also appear to be equipped with purinergic receptors, since ATP causes pain when injected intramuscularly (Mörk et al. 2003). In patients, ATP may not only be involved in the pain of muscle trauma, but also in cases of sympathetically maintained pain.

The underlying mechanism is that postganglionic sympathetic fibers release ATP as a co-transmitter with norepinephrine.

Among the growth factors studied so far, NGF is of particular interest because data obtained in the author's group indicate that NGF excites muscle nociceptive endings exclusively (i.e., low-threshold mechanosensitive, presumably non-nociceptive endings were not affected). NGF is the only substance known so far that has such an exclusive action on muscle nociceptive endings. Surprisingly, even though NGF evoked a purely nociceptive input to the spinal cord, an intramuscular NGF injection does not elicit immediate pain. An overview of most of the known receptor molecules in the membrane of a nociceptive ending is given in Fig. 3.

Interactions Between Stimulants at the Receptive Nerve Ending

Under pathological conditions, many of the above-mentioned algesic or sensitizing agents are released together and act on several of the membrane receptors simultaneously. A particularly effective combination of algesic substance for exciting muscle nociceptors in animal experiments consists of protons, ATP, and lactate acting on ASIC3, P2X4, and TRPV1 receptors (Light et al. 2008).

Interactions between algesic agents have been mainly studied for BK on the one hand, and PGE_2 or 5-HT on the other. PGE_2 and 5-HT enhance the excitatory action of BK on slowly conducting muscle afferents (Mense 1981). The pain elicited in volunteers by intramuscular injection of a combination of BK and 5-HT is likewise stronger than that caused by either stimulant alone (Babenko et al. 1999; Mörk et al. 2003). These interactions are of clinical significance, because the substances are released together in damaged or pathologically altered tissue.

The concentration of PGE_2 and 5-HT required for enhancing the action of BK is lower than that for exciting nociceptors. Therefore, in the course of tissue inflammation, the receptive endings will first be sensitized and then excited. Clinical observations point in the same direction; in the course of a pathologic alteration, the patient experiences tenderness first (because of nociceptor ► sensitization) and then ongoing pain (because of nociceptor excitation).

Polymodal Response Behavior

Most of the muscle nociceptors studied in single fiber recordings responded to noxious local pressure and injections of BK (and some even additionally to thermal stimuli), but receptors were also found that were activated by one type of noxious stimulation only (mechanical or chemical). This finding indicates that different types of nociceptors are present in skeletal muscle, similar to the skin where mechano-, mechano-heat, and polymodal nociceptors are known to exist (Millan 1999).

References

Babenko, V., Graven-Nielsen, T., Svensson, P., et al. (1999). Experimental human muscle pain and muscular hyperalgesia induced by combinations of serotonin and bradykinin. *Pain, 82*, 1–8.

Burnstock, G. (2000). P2X receptors in sensory neurones. *British Journal of Anaesthesia, 84*, 476–488.

Cairns, B. E., Sessle, B. J., & Hu, J. W. (1998). Evidence that excitatory amino acid receptors within the temporomandibular joint region are involved in the reflex activation of the jaw muscles. *The Journal of Neuroscience, 18*, 8056–8064.

Carlton, S. M., & Coggeshall, R. E. (2001). Peripheral capsaicin receptors increase in the inflamed rat hindpaw: A possible mechanism for peripheral sensitization. *Neuroscience Letters, 310*, 53–56.

Caterina, M. J., & David, J. (1999). Sense and specificity: A molecular identity for nociceptors. *Current Opinion in Neurobiology, 9*, 525–530.

Connor, M., Naves, L. A., & McCleskey, E. W. (2005). Contrasting phenotypes of putative proprioceptive and nociceptive trigeminal neurons innervating jaw muscle in rat. *Molecular Pain, 1*, 31.

Cook, S. P., & McCleskey, E. W. (2002). Cell damage excites nociceptors through release of cytosolic ATP. *Pain, 95*, 41–47.

Graven-Nielsen, T., Arendt-Nielsen, L., Svensson, P., et al. (1997). Experimental muscle pain: A quantitative study of local and referred pain in humans following injection of hypertonic saline. *Journal of Musculoskeletal Pain, 5*, 49–69.

Hoheisel, U., Unger, T., & Mense, S. (2005). Excitatory and modulatory effects of inflammatory cytokines and neurotrophins on mechanosensitive group IV muscle afferents in the rat. *Pain, 114*, 168–176.

Kumazawa, T., & Mizumura, K. (1977). Thin-Fiber receptors responding to mechanical, chemical, and thermal stimulation in the skeletal muscle of the dog. *Journal de Physiologie, 273*, 179–194.

Lawson, S. N., Crepps, B. A., & Perl, E. R. (1997). Relationship of substance p to afferent characteristics of dorsal root ganglion neurons in guinea-pig. *Journal de Physiologie, 505*, 177–191.

Light, A. R., Hughen, R. W., Zhang, J., Rainier, J., Liu, Z., & Lee, J. (2008). Dorsal root ganglion neurons innervating skeletal muscle respond to physiological combinations of protons, ATP, and lactate mediated by ASIC, P2X, and TRPV1. *Journal of Neurophysiology, 100*, 1184–1201.

Lund, J. P., Sadeghi, S., & Athanassiadis, T. (2010). Assessment of the potential role of muscle spindle mechanoreceptor afferents in chronic muscle pain in the rat masseter muscle. *PLoS One, 15*, e11131.

Marchettini, P., Simone, D. A., Caputi, G., & Ochoa, J. L. (1996). Pain from excitation of identified muscle nociceptors in humans. *Brain Research, 40*, 109–116.

McCleskey, E. W., & Gold, M. S. (1999). Ion channels of nociception. *Annual Review Physiology, 61*, 835–856.

Mense, S. (1981). Sensitization of group IV muscle receptors to bradykinin by 5-hydroxytryptamine and prostaglandin E2. *Brain Research, 225*, 95–105.

Mense, S. (1997). Pathophysiologic basis of muscle pain syndromes. *Physical Medicine and Rehabilitation Clinics of North America, 8*, 23–53.

Mense, S., & Meyer, H. (1985). Different types of slowly conducting afferent units in cat skeletal muscle and tendon. *Journal de Physiologie, 363*, 403–417.

Millan, M. J. (1999). The Induction of pain: An integrative review. *Progress in Neurobiology, 57*, 1–164.

Mörk, H., Ashina, M., Bendtsen, L., et al. (2003). Experimental muscle pain and tenderness following infusion of endogenous substances in humans. *European Journal of Pain, 7*, 145–153.

O'Brien, C., Woolf, C. J., Fitzgerald, M., et al. (1989). Differences in the chemical expression of rat primary afferent neurons which innervate skin, muscle or joint. *Neuroscience, 32*, 493–502.

Reinert, A., Kaske, A., & Mense, S. (1998). Inflammation-induced increase in the density of neuropeptide-immunoreactive nerve endings in rat skeletal muscle. *Experimental Brain Research, 121*, 174–180.

Reinöhl, J., Hoheisel, U., Unger, T., et al. (2003). Adenosine triphosphate as a stimulant for nociceptive and non-nociceptive muscle group IV receptors in the rat. *Neuroscience Letters, 338*, 2528.

Ro, J. Y., Nies, M., & Zhang, Y. (2005). The role of peripheral N-methyl-D-aspartate receptors in muscle hyperalgesia. *NeuroReport, 16*, 485–489.

Sluka, K. A., Kalra, A., & Moore, S. A. (2001). Unilateral intramuscular injections of acidic saline produce a bilateral long-lasting hyperalgesia. *Muscle & Nerve, 24*, 37–46.

Sluka, K. A., Price, M. P., Breese, N. M., et al. (2003). Chronic hyperalgesia induced by repeated acid injections in muscle is abolished by the loss of ASIC3, but not ASIC1. *Pain, 106*, 229–239.

Stacey, M. J. (1969). Free nerve endings in skeletal muscle of the cat. *Journal of Anatomy, 105*, 231–254.

Stein, C., Hassan, A. H. S., Przewlocki, R., et al. (1990). Opioids from immunocytes interact with receptors on sensory nerves to inhibit nociception in inflammation. *Proceedings of the National Academy of Sciences of the United States of America, 87*, 5935–5939.

Svensson, P., Cairns, B. E., Wang, K., et al. (2003). Injection of nerve growth factor into human masseter muscle evokes long-lasting mechanical allodynia and hyperalgesia. *Pain, 104*, 241–247.

Muscle Pain

Definition

Pain originating from muscle and usually experienced as dull, deep, diffuse, and aching.

Cross-References

► Myalgia
► Spinothalamic Tract Neurons, Visceral Input

Muscle Pain in Systemic Inflammation (Polymyalgia Rheumatica, Giant Cell Arteritis, Rheumatoid Arthritis)

Henning Bliddal
The Parker Institute, Frederiksberg Hospital, Copenhagen, Denmark

Synonyms

Arthritogenic pain; Giant cell arthritis; Immunoinflammatory muscle pain; Polymyalgia rheumatica; Rheumatoid arthritis

Definition

With the exception of some cases of ► myositis, the inflammatory conditions of joints and muscles have muscle pain as a very prominent symptom. The pain may be caused by changes in the muscles per se, including possible referred pain, or be secondary to the inflammatory changes in the joints. Inflammation in both tendon sheaths and tendon insertions may contribute to the condition, which may be quite complex with regard to pain analysis. Due to its frequent involvement, ► giant cell arteritis is often called ► temporal arteritis.

Characteristics

The inflammatory process involves an increased production of ► cytokines, e.g., the important pro-inflammatory cytokines TNF-α and Il-1. These cytokines are involved in the pathogenesis of pain (Gotoh et al. 2002), and pain in these conditions is accompanied by tiredness and anemia due to systemic actions of the cytokines as in rheumatoid cachexia (Walsmith and Roubenoff 2002).

In the elderly, it may be very difficult to distinguish between ► rheumatoid arthritis (RA) and ► polymyalgia rheumatica/giant cell arteritis (PMR/GCA) with a possible overlap between the diseases (Lange et al. 2000; Salvarani and Hunder 1999).

PMR/GCA is associated with pain and muscle soreness in the shoulder and head regions. The two diseases have similar symptoms from the muscles but may be distinguished by the finding of vasculitis in a biopsy of the temporal artery in the latter. PMR occurs in the elderly population with an incidence rate of about 0.5 per 10,000 population aged more than 50, while GCA is considerably less frequent, although the incidence may vary over the world (Fig. 1).

The muscle pain in PMR/GCA may resemble other conditions in the area, and there are no pathognomonic features of the head and ► neck pain, with the possible exception of jaw claudication present in about half of the cases (Hall et al. 1983). The diagnosis must be suspected in any person above 50 years of age with unexplained changes in pain patterns in the proximal parts of the upper extremity and to some

Muscle Pain in Systemic Inflammation (Polymyalgia Rheumatica, Giant Cell Arteritis, Rheumatoid Arthritis), Fig. 1 Arteritis temporalis. The artery is abnormally swollen and sore on palpation due to immuno-inflammatory changes in the arterial wall. This may eventually lead to obstruction of the lumen and as the process may involve other arteries in the area, including the arteria to the optic nerve, a feared complication is blindness

extent the lower extremity as well. In more than 2/3 of the patients, the pain involves the shoulder region as well as headache. The muscles of the shoulder girdle are sore and exhibit exercise intolerance and fatigue. There is often pain at night and decreased range of motion of the joints in the affected area. The finding of an elevated erythrocyte sedimentation rate substantiates the diagnosis. The signs of systemic inflammatory action, e.g., light fever, malaise, and weight loss, may dominate pain as a secondary phenomenon.

Rheumatoid Arthritis (RA)

RA has a prevalence of about 1 %, with maximum incidence among the 30–50-year-olds and a female to male ratio of 2.5:3. The etiology is unknown, with autoimmune mechanisms involved in the pathogenesis. The disease is most often chronic with fluctuations of disease activity leading to a gradual loss of function. Cardinal symptoms are ► morning stiffness of joints, swollen and tender finger joints, symmetrical distribution of arthritis, and, depending on severity, joint erosions. The symptoms and signs may differ with the age group. All patients suffer from inflammatory pain in synovial tissues including joints, tendon sheaths, and bursae.

In the elderly, the muscle pain may be very prominent, while all patients have various degrees of muscle fatigue. Some of the medications used for treatment of RA may induce myopathy (Le Quintrec and Le Quintrec 1991). The weakness and the associated activity induced pain in the muscles may not be a primary feature as indicated by the lack of correlation with the muscle strength (Schiøttz-Christensen et al. 2001). A specific muscle weakness induced by treatment with ► steroids is well described; it is probable, but not fully established, that this may be accompanied by pain (Danneskiold-Samsøe and Grimby 1986). It is possible for patients with RA to train in spite of their joint disease (Bearne et al. 2002; Lyngberg et al. 1994), and

Muscle Pain in Systemic Inflammation (Polymyalgia Rheumatica, Giant Cell Arteritis, Rheumatoid Arthritis), Fig. 2 Muscle wasting in rheumatoid arthritis. The inflammation of the joints and the arthritogenic pain affect the whole movement segment with pain-induced muscle wasting. The process is irreversible in cases of joint destruction as in this case where subluxations and ulnar deviation are evident in both wrist joint and fingers

exercises to increase muscle strength should be encouraged to maintain general functional abilities Fig. 2.

RA, in 20 % of cases, coexists with ► Sjögren's syndrome, which, apart from the ► sicca syndrome, is characterized by general muscle pain and fatigue. Myositis may occur in Sjögren's as well and is unrelated to the pain, which is of unknown origin (Lindvall et al. 2002). Finally, RA may – as do other immune-inflammatory diseases – induce a secondary fibromyalgia, and the pain quality in the two diseases may be indistinguishable (Burckhardt et al. 1992). In the elderly, an overlap syndrome exists, with muscle pain in the shoulder region as a common sign between PMR/GCA and RA. In younger women, a similar diagnostic problem arises in the lighter cases of RA, which may resemble fibromyalgia for long periods before a certain diagnosis can be given.

There are aspects of the pain in inflammatory rheumatic diseases, which make the arthritogenic pain differ from other chronic pain condition.

1. The rheumatic patient is typically well diagnosed with a clear explanation of the pain. Most often the pain correlates with either obvious clinical changes in the joints, e.g., swelling, readily understandable blood tests, e.g., erythrocyte sedimentation rate (ESR), or imaging of the inflammatory changes in synovial tissue, e.g., by ultrasound Doppler or MRI (Terslev et al. 2003). The patients will often have frequent contacts with a specialist, who can reassure them about the nature of the pain. In consequence, unlike other chronic pain conditions, there is less uncertainty associated with the origin or extent of the rheumatic pain.

Muscle Pain in Systemic Inflammation (Polymyalgia Rheumatica, Giant Cell Arteritis, Rheumatoid Arthritis), Table 1 Medical diseases which may give rise to both joint and muscle pain

Suspected disease	Pain region	Examination
Heart disease	Neck/shoulder/arm	Stethoscopy, EKG
Disease close to diaphragm (e.g., pneumonia, gallbladder problems)	Neck/shoulder/arm	Stethoscopy, palpation of the abdomen
Thyroid diseases	Diffuse pain in muscles and joints	Palpation of the thyroid – other signs of thyroid disease incl. eye changes
Calcium metabolism	Diffuse pain in muscles and joints	In severe cases, cramps and involuntary muscle contractions

2. The rheumatic patient has unique possibilities of pain relief by anti-inflammatory agents. The most pronounced effect is the well-documented, almost immediate effect of glucocorticoids in polymyalgia rheumatica (PMR/GCA), which may even be used to substantiate the diagnosis.

By such procedures, the pain may be explained and even documented to both patients and social contacts, which may save the patients from a deterioration due to an uncertainty of diagnosis, which is otherwise a definite psychological problem in muscle pain of the shoulders (Dyrehag et al. 1998). Nevertheless, the patients with inflammatory diseases may benefit from general therapeutic measures against their pain, including non-medicinal pain management (Evers et al. 2003; Keefe et al. 2001).

Combined muscle and joint pain in a patient involves a number of considerations in internal medicine and in unsettled diagnosis; it must be suggested that both clinical examination and blood tests including calcium and thyroid hormones are performed (Table 1).

References

Bearne, L. M., Scott, D. L., & Hurley, M. V. (2002). Exercise can reverse quadriceps sensorimotor dysfunction that is associated with rheumatoid arthritis without exacerbating disease activity. *Rheumatology, 41*, 157–166.

Burckhardt, C. S., Clark, S. R., & Bennett, R. M. (1992). A comparison of pain perceptions in women with fibromyalgia and rheumatoid arthritis: Relationship to depression and pain extent. *Arthritis Care and Research, 5*, 216–222.

Danneskiold-Samsoe, B., & Grimby, G. (1986). The relationship between the leg muscle strength and physical capacity in patients with rheumatoid arthritis, with reference to the influence of corticosteroids. *Clinical Rheumatology, 5*, 468–474.

Dyrehag, L. E., Widerstrom-Noga, E. G., Carlsson, S. G., et al. (1998). Relations between self-rated musculoskeletal symptoms and signs and psychological distress in chronic neck and shoulder pain. *Scandinavian Journal of Rehabilitation Medicine, 30*, 235–242.

Evers, A. W., Kraaimaat, F. W., Geenen, R., et al. (2003). Stress-vulnerability factors as long-term predictors of disease activity in early rheumatoid arthritis. *Journal of Psychosomatic Research, 55*, 293–302.

Gotoh, M., Hamada, K., Yamakawa, H., et al. (2002). Interleukin-1-induced glenohumeral synovitis and shoulder pain in rotator cuff diseases. *Journal of Orthopaedic Research, 20*, 1365–1371.

Hall, S., Persellin, S., Lie, J. T., et al. (1983). The therapeutic impact of temporal artery biopsy. *Lancet, 2*, 1217–1220.

Keefe, F. J., Affleck, G., Lefebvre, J., et al. (2001). Living with rheumatoid arthritis: The role of daily spirituality and daily religious and spiritual coping. *The Journal of Pain, 2*, 101–110.

Lange, U., Piegsa, M., Teichmann, J., et al. (2000). Ultrasonography of the glenohumeral joints -a helpful instrument in differentiation in elderly onset rheumatoid arthritis and polymyalgia rheumatica. *Rheumatology International, 19*, 185–189.

Le Quintrec, J. S., & Le Quintrec, J. L. (1991). Drug-induced myopathies. *Baillière's Clinical Rheumatology, 5*, 21–38.

Lindvall, B., Bengtsson, A., Ernerudh, J., et al. (2002). Subclinical myositis is common in primary Sjogren's syndrome and is not related to muscle pain. *Journal of Rheumatology, 29*, 717–725.

Lyngberg, K. K., Ramsing, B. U., Nawrocki, A., et al. (1994). Safe and effective isokinetic knee extension training in rheumatoid arthritis. *Arthritis and Rheumatism, 37*, 623–628.

Salvarani, C., & Hunder, G. G. (1999). Musculoskeletal manifestations in a population-based cohort of patients with giant cell arteritis. *Arthritis and Rheumatism, 42*, 1259–1266.

Schiottz-Christensen, B., Lyngberg, K., Keiding, N., et al. (2001). Use of isokinetic muscle strength as a measure of severity of rheumatoid arthritis: A comparison of this assessment method for RA with other assessment methods for the disease. *Clinical Rheumatology, 20*, 423–427.

Terslev, L., Torp-Pedersen, S., Savnik, A., et al. (2003). Doppler ultrasound and magnetic resonance imaging of synovial inflammation of the hand in rheumatoid arthritis: A comparative study. *Arthritis and Rheumatism, 48*, 2434–2441.

Walsmith, J., & Roubenoff, R. (2002). Cachexia in rheumatoid arthritis. *International Journal of Cardiology, 85*, 89–99.

Muscle Pain Model, Inflammatory Agents-Induced

Donald A. Simone and Darryl T. Hamamoto
Department of Diagnostic and Biological Sciences, University of Minnesota, Minneapolis, MN, USA

Synonyms

Animal models of inflammatory muscle pain; Animal models of inflammatory myalgia; Nocifensive behaviors evoked by myositis

Definition

Animal models have been recently developed that may allow further elucidation of the underlying mechanisms of pain associated with inflammation of muscle (i.e., ▶ Myositis). A better understanding of these mechanisms may lead to the development of novel approaches to treating inflammatory ▶ myalgia (i.e., muscle pain). These animal models also provide methods by which the efficacy and potency of novel pharmaceutical agents can be examined.

Animal models of inflammatory muscle pain consist of two parts: a method of inducing inflammation in muscle, and a method of evaluating changes in ▶ nocifensive behaviors. Methods of inducing inflammation in muscle include intramuscular injection of substances, such as carrageenan (Kehl et al. 2000), or components of the inflammatory milieu, such as protons (i.e., acidity) (Sluka et al. 2001). Changes in nocifensive behaviors examined by these models are usually described as ▶ hyperalgesia, because they increase the response to stimuli that are normally considered nociceptive. One method of evaluating changes in nocifensive behaviors involves measuring the amount of force that an animal can produce with the inflamed muscle (e.g., ▶ grip force). Another method examines changes in withdrawal responses to controlled mechanical (▶ Von Frey Hair) stimuli applied to areas of skin, often the hind paw, that are remote from the inflamed muscle.

Characteristics

Protons
Inflammation can result in decreased tissue pH (i.e., acidosis), which activates cutaneous ▶ nociceptors (Steen et al. 1992) and increases withdrawal responses to mechanical stimuli in rats (Hamamoto et al. 2001). In humans, constant infusion of acidic buffer into skin or muscle produces flow-dependent pain that has been shown to correlate with decreased tissue pH, at least in the skin (Steen et al. 1995; Isseberner et al. 1996).

Thus, decreased tissue pH may contribute to pain associated with muscle inflammation. Two injections of acidic saline (pH 4.0–6.0) administered 2 days apart into the gastrocnemius muscle of one hind limb of a rat produced cutaneous mechanical hyperalgesia that persisted for up to 4 weeks (Fig. 1) (Sluka et al. 2001). In this model, mechanical hyperalgesia was examined by applying von Frey monofilaments to the plantar surface of the hind paw, a site remote from the site where the acidic saline was injected. The contralateral hind paw also exhibited mechanical hyperalgesia. After mechanical hyperalgesia had developed, neither pharmacological (i.e., intramuscular injection of lidocaine into the site of acidic saline injection) nor physical (i.e., ▶ dorsal rhizotomy) interruption of input from primary afferent innervating the gastrocnemius muscle abolished the contralateral mechanical hyperalgesia. These results suggest that ▶ central sensitization following nociceptive input from muscle may underlie the ▶ secondary hyperalgesia in this model. Interestingly, injection of acidic saline produced histologic evidence of only mild injury and inflammation of the muscle fibers in some rats, which was attributed to the needle penetration and not to the acidic solution. Thus, this model appears to isolate the effects of one component of inflammation, which is tissue acidity. Of interest is the long-lasting (i.e., 4 weeks) hyperalgesia produced by two injections of acidic saline. Also, the mechanical hyperalgesia was exhibited in a different tissue (i.e., skin) and at a site remote from the acidified muscle tissue. This pattern of referral from deep structures to remote cutaneous sites has been shown in humans (Marchettini et al. 1990). Thus, this model of acid-induced muscle pain may help to determine the mechanisms by which patients with chronic generalized musculoskeletal pain, such as fibromyalgia, experience areas of tenderness called tender points (Sluka et al. 2001). This model has subsequently been used to examine the role of excitatory amino acid receptors in the development and maintenance of mechanical hyperalgesia induced by intramuscular injection of acidic saline (Skyba et al. 2002).

Mechanical Threshold

Muscle Pain Model, Inflammatory Agents-Induced, Fig. 1 Decreased mechanical threshold following two injections of acidic saline into the gastrocnemius muscle of one hind limb of a rat. The injections were spaced 2 or 5 days apart. Decrease in the threshold force is expressed as area under the curve for the 6 weeks after the second injection. A significant decrease in mechanical withdrawal threshold occurred bilaterally for rats injected with pH 4 or pH 5 saline, but not for those injected with pH 6 or 7.2. Data are expressed as mean ± SEM, $N = 8$ rats/group, $*P < 0.05$ (Figure modified from Sluka et al. (2001))

Carrageenan

Injection of carrageenan into muscle evokes a localized myositis, as demonstrated by accumulation of leukocytes around the site of injection (Diehl et al. 1988). In electrophysiological studies, injection of carrageenan into muscle sensitizes nociceptors, both increasing their spontaneous activity and lowering their threshold to activation by mechanical stimuli (Berberich et al. 1988; Diehl et al. 1988). Thus, when inflamed muscle is contracted, nociceptors in the muscle would be activated by lower than normal magnitudes of mechanical stimuli. The animal might then react to this nociceptive input by decreasing contraction of the muscle, in order to decrease mechanical activation of muscle nociceptors. This reduction in effort would reduce the force produced by the muscle. Thus, Kehl and colleagues have developed a model of inflammatory muscle pain that involves bilateral injections of carrageenan into the triceps muscles, and examines the grip force produced by the inflamed muscle in rats and mice (Kehl et al. 2000; Wacnik et al. 2003). This appears to be a clinically relevant model, because patients with muscle pain exhibit reduced grip force (Norsdenskiold and Grimby 1993). In this model, grip force is measured using a strain gauge attached to a wire mesh upon which the rodent is allowed to grab with its fore paws (Fig. 2). The rodent is held by the tail and gently pulled until it releases the wire mesh and the maximum force produced by the rodent is recorded. Grip force is subject to factors, such as hyperalgesia, that influence the behavioral performance of the rodent. Kehl and colleagues have shown that intramuscular injection of carrageenan into the triceps muscle in rodents produces a dose- and time-dependent reduction in grip force (Fig. 3). This reduction in grip force was specific to the inflamed triceps muscles, because the force produced by the hind limbs was not affected. Furthermore, injection of carrageenan into one triceps muscle did not affect the grip force in the contralateral fore limb. Additional support for the clinical relevance of this model of inflammatory muscle pain is the fact that carrageenan-induced reduction in grip force is attenuated by the opioid levorphanol, the nonsteroidal anti-inflammatory drug indomethacin, and the steroid dexamethasone. These drugs represent three classes of analgesic agents used clinically to treat muscle pain. Additionally, experimental analgesic agents such as the noncompetitive NMDA antagonist MK801 and the cannabinoid agonist WIN 55,212-2 have been

shown to have attenuated inflammatory muscle hyperalgesia using this model.

The carrageenan-induced model of muscle pain has several advantages over the model of acidic saline-induced hyperalgesia. First, the carrageenan-induced model of muscle pain is a model of pain produced by inflammation in muscle, whereas injection of individual or multiple inflammatory mediators may not be. Second, the carrageenan model is a model of ► primary hyperalgesia, because it examines the function of the inflamed muscle. In contrast, models in which the site of hyperalgesia is remote from the site of inflammation or injury are models of secondary hyperalgesia. Third, the grip force assay avoids the potential of misinterpreting drug-induced motor dysfunction or sedation with ► antihyperalgesia. Many behavioral assays, such as paw withdrawal from mechanical or thermal stimuli, are based on the assumption that an animal that withdraws less or withdraws at a higher threshold is exhibiting antihyperalgesia. However, drug-induced motor dysfunction and sedation may produce the same behavioral response. That is, it may take a more intense mechanical or thermal stimulus to elicit

Muscle Pain Model, Inflammatory Agents-Induced, Fig. 2 Measurement of grip force in mice. Mice are positioned and allowed to grasp a wire mesh with their fore paws. Mice are held by the tail and gently moved in a rostral-caudal direction. The wire mesh is attached to a strain gauge and the peak grip force produced by the mouse is determined

Muscle Pain Model, Inflammatory Agents-Induced, Fig. 3 Time-response curve for the reduction in forelimb grip force in rats following injection of carrageenan (4 mg) or an equal volume (75 µl) of PBS into the triceps muscles of both forelimbs. Intramuscular carrageenan (Carra) produced a significant treatment- and time-dependent reduction in grip force. Data are expressed as mean ± SEM, N = 15 rats/group, **$P < 0.01$ (Figure modified from Kehl et al. (2000))

a withdrawal response. In contrast, if a drug produces motor dysfunction or sedates an animal, then it will produce less grip force, which will not be misinterpreted as antihyperalgesia. Thus, the carrageenan model appears to be useful for further elucidation of the mechanisms associated with inflammatory muscle pain. Furthermore, the grip force assay provides a method by which the analgesic efficacy and potency of novel pharmaceutical agents can be examined in models of inflammatory muscle pain.

References

Berberich, P., Hoheisel, U., & Mense, S. (1988). Effects of a carrageenan-induced myositis on the discharge properties of group III and IV muscle receptors in the cat. *Journal of Neurophysiology, 59*, 1395–1409.

Diehl, B., Hoheisel, U., & Mense, S. (1988). Histological and neurophysiological changes induced by carrageenan in skeletal muscle of Cat and Rat. *Agents and Actions, 25*, 210–213.

Hamamoto, D. T., Ortiz-Gonzalez, X. R., Honda, J. M., & Kajander, K. C. (2001). Intraplantar injection of hyaluronic acid at low pH into the rat hindpaw produces tissue acidosis and enhances withdrawal responses to mechanical stimuli. *Pain, 74*, 225–234.

Issberner, U., Reeh, P. W., & Steen, K. H. (1996). Pain due to tissue acidosis: A mechanism for inflammatory and ischemic myalgia? *Neuroscience Letters, 208*, 191–194.

Kehl, L. J., Trempe, T. M., & Hargreaves, K. M. (2000). A new animal model for assessing mechanisms and management of muscle hyperalgesia. *Pain, 85*, 333–343.

Marchettini, P., Cline, M., & Ochoa, J. (1990). Innervation territories for touch and pain afferents of single fascicles of the human ulnar nerve. *Brain, 113*, 1491–1500.

Norsdenskiold, U. M., & Grimby, G. (1993). Grip force in patients with rheumatoid arthritis and fibromyalgia and in healthy subjects: A study with the grippit instrument. *Scandinavian Journal of Rheumatology, 22*, 14–19.

Skyba, D. A., King, E. W., & Sluka, K. A. (2002). Effects of NMDA and Non-NMDA ionotropic glutamate receptor antagonists on the development and maintenance of hyperalgesia induced by repeated intramuscular injection of acidic saline. *Pain, 98*, 69–78.

Sluka, K. A., Kalra, A., & Moore, S. A. (2001). Unilateral intramuscular injections of acidic saline produce a bilateral, long-lasting hyperalgesia. *Muscle & Nerve, 24*, 37–46.

Steen, K. H., Issberner, U., & Reeh, P. W. (1995). Pain due to experimental acidosis in human skin: Evidence for

Non-adapting nociceptor excitation. *Neuroscience Letters, 199*, 29–32.

Steen, K. H., Reeh, P. W., Anton, F., & Handwerker, H. O. (1992). Protons selectively induce lasting excitation and sensitization to mechanical stimulation of nociceptors in Rat skin, in vitro. *Journal of Neuroscience, 12*, 86–95.

Wacnik, P. W., Kehl, L. J., Trempe, T. M., Ramnaraine, M. L., Beitz, A. J., & Wilcox, G. L. (2003). Tumor implantation in mouse humerus evokes movement-related hyperalgesia exceeding that evoked by intramuscular carrageenan. *Pain, 101*, 175–186.

Muscle Pain Model, Ischemia-Induced and Hypertonic Saline-Induced

Siegfried Mense
Neuroanatomy, Heidelberg University, Medical Faculty Mannheim, CBTM, Mannheim, Germany

Synonyms

Ischemia-induced muscle pain; Ischemic muscle pain

Definition

Ischemia-induced muscle pain: muscle pain during and following muscle exercise under ischemic conditions. In clinical studies, ischemic conditions were induced by occlusion of blood vessels of an entire limb (e.g., by a tourniquet or by administration of adrenalin to study pain mechanisms in the tooth pulp). In animal experiments, ischemia was induced by several methods:

1. Occlusion of arteries (Lee et al. 2012)
2. Tourniquet around the tail
3. Circulatory arrest
4. Thrombus induction in an artery (Lee 2012)

Hypertonic saline-induced pain: muscle pain induced by intramuscular or intra-arterial (into the muscle artery) injection of hypertonic saline (NaCl) at a concentration of 5–10 % (5.6–11.1 times hypertonic).

Characteristics

Ischemia-Induced Pain

Data from clinical studies and animal experiments indicate that ischemia alone (without muscle contractions) is not an effective stimulus for muscle nociceptors. In clinical studies where a tourniquet was used to induce ischemia in a limb, the pain that occurred after approximately half an hour appeared to be mostly due to the pressure exerted on the nerve by the tourniquet, and not by the ischemia.

Clinical Findings and Results from Studies on Human Subjects

One of the clinically important types of ischemic muscle pain is the pain of ▶ intermittent claudication. It occurs during walking in patients with sclerotic narrowing of the arteries of the lower leg and foot. In earlier clinical studies, a multitude of factors were discussed as being the cause of the pain, e.g., chemical metabolites, accumulation of potassium ions (K^+), and reduced pH due to release of lactic acid. Lactic acid was soon ruled out, because patients with McArdle's disease – who are unable to produce lactic acid due to a gene defect – often present with pain of intermittent claudication or angina pectoris. Another candidate substance was the nonapeptide bradykinin (BK) that is cleaved by kallikrein – a proteinase – from plasma proteins. The importance of BK for the pain of intermittent claudication was supported by the finding that administration of a proteinase inhibitor in claudication patients extended the distance the patients were able to walk without pain (Digiesi et al. 1975). Electron microscopic investigations of bioptic material of claudication patients showed that the muscle exhibited signs of inflammation with infiltration by inflammatory cells and muscle fiber necrosis. The latter finding adds adenosine triphosphate (ATP) to the list of candidate substances, because muscle cells contain high concentrations of ATP, which are released into the interstitial space when a muscle fiber is damaged (see below).

Another clinically relevant ischemic muscle disorder is myalgia often induced by light, but stressful, muscle work. The main reason for the pain appears to be a reduced muscle blood flow and/or oxygenation (Cagnie et al. 2012; Sögaard et al. 2012). In cases of myocardial infarction, platelet activation occurs which leads to excitation of cardiac nerve endings by the release of serotonin and histamine. Furthermore, infarction is associated with increased concentrations of protons from lactate and BK (Fu and Longhurst 2009). All these agents are known to stimulate nociceptive nerve endings.

Data from Animal Experiments

The results obtained from animal studies indicate that the following agents or factors may be involved in ischemic pain:

1. Potassium ions. The potassium concentration is increased in muscles of experimental animals following contractions with and without occlusion of the blood supply.
2. Proton concentration (pH). The interstitial pH is lowered to approximately 6.5 during prolonged ischemia plus exercise. The source of the protons is often the lactate that is produced under ischemic conditions.
3. Kinins such as BK and kallidin. The kinin-forming activity of muscle tissue is increased during prolonged ischemia.
4. Degeneration and necrosis of muscle fibers. If necrosis is present in a chronically ischemic muscle, the ATP concentrations released from the muscle cells are sufficient to activate muscle nociceptors (Reinöhl et al. 2003).

Electrophysiological recordings from single muscle afferent fibers have shown that in skeletal muscle of cat and rat, receptive endings are present that respond specifically to ischemic contractions, in that they are not excited by short-lasting ischemia alone but require ischemia plus contractions for activation (Bessou and Laporte 1958; Mense and Stahnke 1983; Fig. 1). This behavior was interpreted as indicating that the units were nociceptors and may mediate the pain of intermittent claudication, but other authors emphasized a possible function in the adjustment of respiration and circulation to the requirements during muscle work (Kaufman et al. 1984).

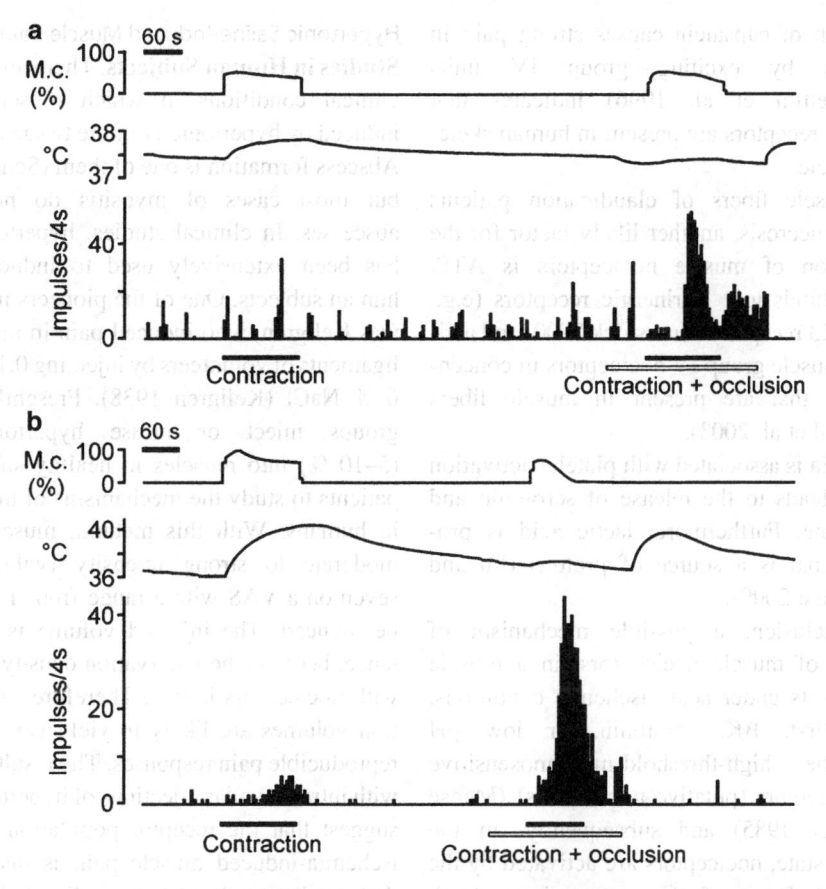

Muscle Pain Model, Ischemia-Induced and Hypertonic Saline-Induced, Fig. 1 Impulse activity of single cat muscle receptors with unmyelinated afferent fibers showing exclusive or predominant activation during ischemic contractions. In both panels, the upper trace indicates the force of the contraction of the gastrocnemius-soleus muscle in percent of maximal contraction (M.c.) induced by electrical stimulation of the muscle nerve, the middle trace shows the temperature inside the muscle, and the lower trace is a histogram of the fiber's activity (bin width 4 s). Periods of contractions are marked by filled bars underneath the abscissa; occlusion of the muscle artery is marked by open bars. In (**a**), the force of contraction was adjusted to approximately 50 % M.c., and in (**b**) to 100 %. Under the latter condition, the muscle could not maintain the contraction force during occlusion of the muscle artery. Note that in both panels, the discharge frequency drops markedly at the end of the contraction period. This indicates that in addition to chemical factors, a mechanical component is involved in the activation of the receptors during ischemic contractions (From Mense and Stahnke 1983)

What Factors Cause the Pain of Intermittent Claudication?

The available data suggest that it is not a single factor that is responsible for ischemic muscle pain but that several stimuli act in combination. Recent findings regarding receptor molecules that are present in the membrane of nociceptors suggest that the following agents could be involved in the induction of ischemic pain:

1. BK acting on the B2 ► bradykinin receptors (unless the nociceptor is sensitized, then BK stimulates the ending by binding to the B1 receptor).

2. The low pH in an ischemic muscle is likely to sensitize the vanilloid receptor 1 (now called ► TRPV1) and to activate other acid-sensing ion channels (ASICs) (Immke and McCleskey 2003; for a review on ASICs as channels mediating the pain of angina pectoris, see Sutherland et al. 2000). ► Capsaicin is assumed to be a specific ligand for the TRPV1 receptor. The fact that intramuscular

injection of capsaicin causes strong pain in humans by exciting group IV units (Marchettini et al. 1996) indicates that TRPV1 receptors are present in human skeletal muscle.

3. As muscle fibers of claudication patients exhibit necrosis, another likely factor for the activation of muscle nociceptors is ATP, which binds to ▶ purinergic receptors (e.g., the P2X3 receptor) (Burnstock 2000) and activates muscle group IV nociceptors in concentrations that are present in muscle fibers (Reinöhl et al. 2003).

4. Ischemia is associated with platelet activation which leads to the release of serotonin and histamine. Furthermore, lactic acid is produced that is a source of protons (Fu and Longhurst 2009).

In conclusion, a possible mechanism of activation of muscle nociceptors, in a muscle that contracts under acute ischemic conditions, is that first, BK, serotonin, or low pH sensitize the ▶ high-threshold mechanosensitive muscle receptors (putative nociceptors) (Mense and Meyer 1985) and subsequently, in the sensitized state, nociceptors are activated by the mechanical force of the contractions. Such a mechanism is suggested by the time course of activation of the single group IV unit shown in Fig. 1a: the unit was not activated by ischemia or contractions alone but gave a strong response during ischemic contractions. Moreover, the activity dropped markedly when the contractions were discontinued but the ischemia maintained. The unit in Fig. 1b was weakly activated at the end of the contraction period without ischemia (in this case the force of contraction was higher: close to 100 % maximal voluntary contraction) but likewise showed a much stronger excitation when the contractions were repeated with the muscle artery occluded.

The release of ATP from damaged muscle fibers appears to occur more under chronic ischemic conditions, but ATP might also be involved in cases of acute ischemic pain, because ATP is a product of any ▶ anaerobic glycolysis (Sutherland et al. 2000).

Hypertonic Saline-Induced Muscle Pain

Studies in Human Subjects. There are just a few clinical conditions in which muscle pain is induced by hypertonicity of the tissue in patients. Abscess formation is one of them (Schade 1924), but most cases of myositis do not exhibit abscesses. In clinical studies, hypertonic saline has been extensively used to induce pain in human subjects. One of the pioneers in this field was Kellgren, who induced pain in muscles and ligaments of volunteers by injecting 0.1–0.3 ml of 6 % NaCl (Kellgren 1938). Presently, several groups inject or infuse hypertonic NaCl (5–10 %) into muscles in healthy subjects and patients to study the mechanisms of muscle pain in humans. With this method, muscle pain of moderate to strong intensity (values around seven on a VAS with a range from 1 to 10) can be induced. The injected volume is of importance, because the innervation density of muscle with nociceptors is low. Therefore, small injection volumes are likely to yield variable or ill-reproducible pain responses. The results obtained with intramuscular injections of hypertonic saline suggest that the receptor population mediating ischemia-induced muscle pain is distinct from that mediating hypertonic saline-induced pain (Graven-Nielsen et al. 2003). In human subjects, hypertonic saline appears to excite predominantly group IV muscle afferent units (as opposed to group III units).

Data from Animal Experiments. Many authors who recorded the activity of single muscle nociceptors used injections of hypertonic saline for receptor activation. Interestingly, 5 % NaCl proved to be the only chemical stimulus that excited all of the nociceptive and non-nociceptive (low-threshold mechanosensitive) group IV receptors that were tested in a study (Hoheisel et al. 2004) (Fig. 2) in which the stimulants were injected into the mechanosensitive receptive field of rat group IV receptors. The other agents excited only a fraction of the receptors. Therefore, in experiments using intramuscular injections, hypertonic saline is a useful stimulus to test if a given receptor is reached by the injected stimulant. In such experiments, a lack of response to an injection could mean

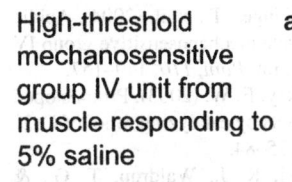

Muscle Pain Model, Ischemia-Induced and Hypertonic Saline-Induced, Fig. 2 Effects of hypertonic saline on a single high-threshold mechanosensitive (presumably nociceptive) group IV afferent fiber from rat gastrocnemius-soleus (GS) muscle. (**a**) Experimental setup. The impulse activity of a single unmyelinated afferent fiber was recorded from a thin nerve filament dissected from the sciatic nerve. (**b**) Response of the unit to 5 % NaCl injected into that region of the muscle where the receptive ending was located with mechanical stimuli (receptive field, RF). The same NaCl solution that had evoked the strong response when injected i.m. was later put on the nerve filament to show that the axonal membrane was not the site of action of the hypertonic saline. Inset: original recording showing the high mechanical threshold of the receptive ending. Mod.p., moderate, innocuous pressure; Nox.p., noxious, painful pressure

that (1) the chemical stimulus did not reach the ending or (2) the stimulus reached the ending but the receptor was not sensitive to the stimulant.

Mechanisms of Nociceptor Activation by Hypertonic Saline

The mechanisms by which muscle nociceptors are excited by hypertonic saline are obscure. Several possibilities exist (for an overview, see Kress and Reeh 1996):

1. Activation by increased tonicity in the interstitial space. The receptive ending might shrink in the hypertonic environment, and stretch-sensitive Na^+ channels could be opened. However, nothing is known about the water permeability of nociceptive endings, and the high mechanical stimulation threshold of muscle nociceptors speaks against this mechanism.

2. Activation by ionic changes. The high Na^+ concentration in the interstitial fluid should have only little influence on the membrane potential of the nociceptive ending, because the Na^+ conductance of an axon is normally low. However, there is evidence indicating that the Na^+ conductance of receptive endings is higher, and therefore, the high extracellular Na^+ concentration could cause an effective depolarization with ensuing excitation of the ending. Thus, the high Na^+ concentration could be the decisive factor for hypertonic saline-induced muscle pain. The high Cl^- concentration appears not to be a stimulus, because in peripheral nerve fibers, Cl^- channels are rare. An additional unknown factor is that – if shrinking of the ending occurs – the intra-axonal concentrations of all ions increase to an unknown extent.

3. Indirect activation of the nociceptor by other algesic agents released from muscle tissue or the nociceptive ending itself. The injection of hypertonic saline has been reported to release glutamate from muscle tissue (Tegeder et al. 2002). Glutamate excites muscle nociceptors and causes muscle pain in humans (Svensson et al. 2003). Hypertonic solutions have also been shown to release substance P from airway sensory cells in vitro (Garland et al. 1995).

4. Other indirect mechanisms that may be involved in muscle pain after injection of hypertonic saline are the opening of the perineurial barrier thus facilitating the access to nociceptive nerve endings for algesic agents (Rittner et al. 2012) and toxic effects on erythrocytes and myocytes at high saline concentrations (exceeding 7 % and 15 %, respectively; Svendsen et al. 2005).

References

Bessou, P., & Laporte, Y. (1958). Activation des fibres afférentes amyéliniques d'Origine musculaire. *Comptes Rendus Society for Biology (Paris), 152*, 1587–1590.

Burnstock, G. (2000). P2X Receptors in sensory neurones. *British Journal of Anaesthesia, 84*, 476–488.

Cagnie, B., Dhooge, F., Van Akeleyen, J., et al. (2012). Changes in microcirculation of the trapezius muscle during a prolonged computer task. *European Journal of Applied Physiology, 112*, 305–312.

Digiesi, V., Bartoli, V., & Dorigo, B. (1975). Effect of a proteinase inhibitor on intermittent claudication or on pain at rest in patients with peripheral arterial disease. *Pain, 1*, 385–389.

Fu, L. W., & Longhurst, J. C. (2009). Regulation of cardiac afferent excitability in ischemia. *Handbook of Experimental Pharmacology, 194*, 185–225.

Garland, A., Jordan, J. E., Necheles, J., Alger, L. E., Scully, M. M., Miller, R. J., Ray, D. W., White, S. R., & Solway, J. (1995). Hypertonicity, but not hyperthermia, elicits substance P release from Rat C-fiber neurons in primary culture. *The Journal of Clinical Investigation, 95*, 2359–2366.

Graven-Nielsen, T., Jansson, Y., Segerdahl, M., Kristensen, J. D., Mense, S., Arendt-Nielsen, L., & Sollevi, A. (2003). Experimental pain by ischaemic contractions compared with pain by intramuscular infusions of adenosine and hypertonic saline. *European Journal of Pain, 7*, 93–102.

Hoheisel, U., Reinöhl, J., Unger, T., et al. (2004). Acidic pH and capsaicin activate mechanosensitive group IV muscle receptors in the rat. *Pain, 110*, 149–157.

Immke, D. C., & McCleskey, E. W. (2003). Protons open acid-sensing Ion channels by catalyzing relief of Ca^{2+} blockade. *Neuron, 37*, 75–84.

Kaufman, M. P., Rybicki, K. J., Waldrop, T. G., & Ordway, G. A. (1984). Effect of ischemia on responses of group III and IV afferents to contraction. *Journal of Applied Physiology, 57*, 644–650.

Kellgren, J. H. (1938). Observations on referred pain arising from muscle. *Clinical Science, 3*, 175–190.

Kress, M., & Reeh, P. (1996). Chemical excitation and sensitization in nociceptors. In C. Belmonte & F. Cervero (Eds.), *Neurobiology of nociceptors* (pp. 259–297). Oxford: Oxford University Press.

Lee, J. H. (2012). A new rat pain model of thrombus-induced ischemia. *Methods in Molecular Biology, 851*, 213–222.

Lee, J. E., Wang, K. C., Chiang, H. Y., et al. (2012). Patterns of nerve injury and neuropathic pain in ischemic neuropathy after ligation-reperfusion of femoral artery in mice. *Journal of the Peripheral Nervous System, 17*, 301–311.

Marchettini, P., Simone, D. A., Caputi, G., & Ochoa, J. L. (1996). Pain from excitation of identified muscle nociceptors in humans. *Brain Research, 40*, 109–116.

Mense, S., & Meyer, H. (1985). Different types of slowly conducting afferent units in cat skeletal muscle and tendon. *Journal de Physiologie, 363*, 403–417.

Mense, S., & Stahnke, M. (1983). Responses in muscle afferent fibres of slow conduction velocity to contractions and ischaemia in the cat. *Journal de Physiologie, 342*, 383–397.

Reinöhl, J., Hoheisel, U., Unger, T., & Mense, S. (2003). Adenosine triphosphate as a stimulant for nociceptive and non-nociceptive muscle group IV receptors in the rat. *Neuroscience Letters, 338*, 25–28.

Rittner, H. L., Amasheh, S., Moshourab, R., et al. (2012). Modulation of tight junction proteins in the perineurium to facilitate peripheral opioid analgesia. *Anesthesiology, 116*, 1323–1334.

Schade, H. (1924). Die molekularpathologie in ihrem verhältnis zur zellularpathologie und zum klinischen krankheitsbild am beispiel der entzündung. *Munchener Medizinische Wochenschrift, 71*, 1–4.

Sögaard, K., Blangsted, A. K., Nielsen, P. K., et al. (2012). Changed activation, oxygenation, and pain response of chronically painful muscles to repetitive work after training interventions: A randomized controlled trial. *European Journal of Applied Physiology, 112*, 173–181.

Sutherland, S. P., Cook, S. P., & McCleskey, E. W. (2000). Chemical mediators of pain due to tissue damage and ischemia. *Progress in Brain Research, 129*, 21–38.

Svendsen, O., Edwards, C. N., Lauritzen, B., et al. (2005). Intramuscular injection of hypertonic saline: In vitro and in vivo muscle tissue toxicity and spinal neurone c-fos expression. *Basic and Clinical Pharmacology and Toxicology, 97*, 52–57.

Svensson, P., Cairns, B. E., Wang, K., Hu, J. W., Graven-Nielsen, T., Arendt-Nielsen, L., & Sessle, B. J. (2003). Glutamate-evoked pain and mechanical allodynia in the human masseter muscle. *Pain, 101*, 221–227.

Tegeder, L., Zimmermann, J., Meller, S. T., & Geisslinger, G. (2002). Release of algesic substances in human experimental muscle pain. *Inflammation Research, 51*, 393–402.

Muscle Pain, Fear-Avoidance Model

Liesbet Goubert[1], Geert Crombez[1] and Johan W. S. Vlaeyen[2,3]
[1]Department of Experimental Clinical and Health Psychology, Ghent University, Ghent, Belgium
[2]Research Group Health Psychology, Catholic University of Leuven, Leuven, Belgium
[3]Department of Clinical Psychological Science, Maastricht University, Maastricht, The Netherlands

Synonyms

Fear of pain; Fear-avoidance model; Kinesiophobia; Pain-related fear

Definition

Fear-avoidance models explain how and why patients experiencing acute pain may become chronic sufferers and become trapped into a vicious circle of more pain and disability. Common in these models is their focus upon the role of ▶ pain-related fear (fear of pain, fear of movement/(re)injury) and dynamic learning processes (avoidance learning).

Characteristics

Pain is more than an unpleasant perceptual and emotional experience. It elicits innate responses and action tendencies that prepare and facilitate

escape from pain (Eccleston and Crombez 1999). Pain is an experience that drives learning. The relevance of learning in chronic pain has been recognized early on in the field of pain (Fordyce 1976) and continues to play a role in more recent biopsychosocial accounts of chronic pain (Asmundson et al. 2004; Gatchel et al. 2007; Vlaeyen and Linton 2000).

Fordyce (1976) was the first to apply the principles of ▶ operant conditioning to the problems of chronic pain patients. The central idea was that "pain behavior" (e.g., verbal complaints, medication use, bed rest, avoidance of home, and work responsibilities) should be conceived of as behaviors that are learned and maintained by their positive consequences (empathy from solicitous spouses, access to pain medications, financial compensation, avoidance of pain-worsening activities, or escape from distressing events). In the context of fear-avoidance models, the principles of ▶ classical conditioning and operant conditioning are of paramount importance. Much behavior is learned because it permits the person to avoid or postpone an aversive experience. What exactly is avoided may vary considerably between persons. An obvious example of ▶ avoidance behavior is that patients learn to avoid pain. Some patients do not avoid activities because of anticipated pain but because they fear that these activities may lead to (re)injury (Kori et al. 1990). It may also be true that pain behaviors are maintained by the avoidance of aversive experiences that are not at all related to pain. A popular example is that pain behavior and, in particular, sick leave may be reinforced and maintained by the avoidance of aversive and unsatisfying work conditions. Whenever these extra benefits outweigh the costs of pain behavior, one talks about secondary gain. Although clinical practice suggests that secondary gain does occur, its incidence is probably overestimated. It is often overlooked that the presence of pain also creates distress and frustration at work. Patients, whose pain interferes with their valued professional activities, may not resume work because they avoid these distressing and frustrating experiences.

Muscle Pain, Fear-Avoidance Model,
Fig. 1 Confrontation to pain

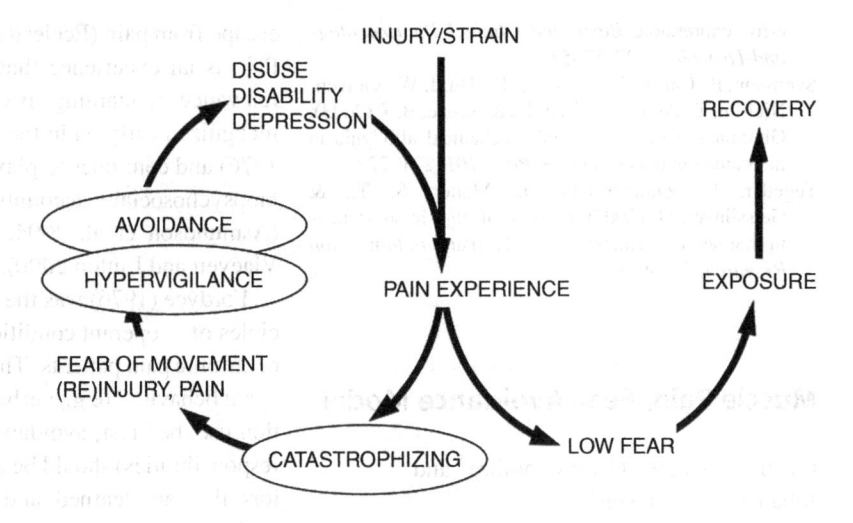

Building upon the model of Lethem et al. (1983), the ▶ cognitive-behavioral model of fear of movement/(re)injury was developed to explain the development of chronic suffering in nonspecific low back pain. Vlaeyen and Linton (2000, 2012) postulated two extreme responses to pain, namely, confrontation and avoidance. As depicted in Fig. 1, a gradual confrontation and resumption of daily activities despite pain is considered as an adaptive response that eventually leads to the reduction of fear, the encouragement of physical recovery, and functional rehabilitation. In contrast, a catastrophic interpretation of pain is considered as a maladaptive response that initiates a vicious circle in which fear of movement/(re)injury and the subsequent avoidance of activities augment functional disability and the pain experience by means of hypervigilance, depression, and disuse (Vlaeyen and Linton 2000; Leeuw et al. 2007). The latter response is also known as ▶ kinesiophobia.

Several empirical studies have provided support for these central ideas. First, ▶ pain catastrophizing, an exaggerated negative orientation toward actual or anticipated pain experiences, has been found to be strongly related to pain-related fear (e.g., Goubert et al. 2004b; Vlaeyen and Linton 2000). Second, fear of movement/(re)injury has been found to be characterized by escape and avoidance behaviors and to lower the ability to accomplish daily living tasks. Patients who are afraid of pain or of (re)injuring

themselves during physical activities tend to avoid physical activities or are reluctant to perform vigorously during standard physical tests (Crombez et al. 1999, see video extract). Because avoidance behaviors occur in anticipation of pain rather than as a response to pain, opportunities are limited to correct (erroneous) beliefs about their pain. As a consequence, pain-related fear and avoidance may become resistant and persistent (e.g., Crombez et al. 1999; Vlaeyen and Crombez 1999). Prospective studies have further substantiated the important and unique role of pain-related fear and avoidance as predictors of chronic pain problems (Jensen et al. 2009; Klenerman et al. 1995; Linton et al. 2000). Third, excessive avoidance of physical activities may have detrimental physical and psychological consequences. A decrease in mobility, decreased muscle strength, and loss of fitness can occur, possibly resulting in a "▶ disuse syndrome" (Verbunt et al. 2003). Avoidance can also result in loss of self-esteem, deprivation of reinforcers, depression, and worrying. Fourth, pain-related fear has been found to be related to a ▶ hypervigilance for pain (Goubert et al. 2004b) and a difficulty in switching attention away from pain (Eccleston and Crombez 1999). This pattern of hypervigilance in fearful chronic pain patients may further diminish the ability to perform everyday activities and hamper the recruitment of pain-coping strategies (Eccleston and Crombez 1999; McCracken and

Gross 1993). Recently, the fear-avoidance model has been applied to other populations, e.g., elderly individuals (Kempen et al. 2009). Also, the model has been argued to be valid to explain the development of persistent complaints after an acute whiplash injury (Vangronsveld et al. 2007).

Fear-avoidance models are dynamic in that both pain-related fear and avoidance are not static or stable characteristics of the individual (Goubert et al. 2004b) but are the result of complex interactions that are typical for a biopsychosocial perspective. Catastrophic thinking about pain and pain-related fear does not emerge within a social and cultural vacuum. In Western cultures, many – even pain-free – persons hold a biomedical view of back pain. As a consequence, many believe that back pain is caused by an injury, that a wrong movement can lead to serious problems if one has back pain, that X-ray and imaging tests can always identify the cause of back pain, and that bed rest is an important part of treatment (Goubert et al. 2004a). Health care providers may further reinforce these beliefs, inadvertently, fuelling pain-related fear and avoidance. Indeed, studies have demonstrated that many health care providers still advise patients to avoid painful movements and believe that pain reduction is a necessary requirement for return to work or that sick leave is an adequate treatment for back pain (Linton et al. 2002; Coudeyre et al. 2006).

One of the clinical implications of the fear-avoidance models is that treatment should directly target pain-related fear and avoidance. Therefore, in analogy with phobia and anxiety disorders, ▶ exposure in vivo has been proposed as a potentially effective treatment of pain-related fear (Vlaeyen et al. 2002). During exposure, patients are gradually exposed to physical activities that are feared or believed to be harmful, in order to correct erroneous cognitions and to extinguish fear and avoidance. At present, several studies have demonstrated the effectiveness of exposure in vivo for reducing pain-related fear, pain catastrophizing, and disability (Leeuw et al. 2008; Vlaeyen et al. 2002).

References

Asmundson, G. J. G., Vlaeyen, J. W. S., & Crombez, G. (2004). *Understanding and treating fear of pain* (p. 367). Oxford: Oxford University Press.

Coudeyre, E., Rannou, F., Tubach, F., Baron, G., Coriat, F., Brin, S., Revel, M., & Poiraudeau, S. (2006). General practitioners' fear-avoidance beliefs influence their management of patients with low back pain. *Pain, 124*, 330–337.

Crombez, G., Vlaeyen, J. W. S., Heuts, P. H. T. G., et al. (1999). Pain-related fear is more disabling than pain itself: Evidence on the role of pain-related fear in chronic back pain disability. *Pain, 80*, 329–339.

Eccleston, C., & Crombez, G. (1999). Pain demands attention: A cognitive-affective model of interruptive function of pain. *Psychological Bulletin, 3*, 356–366.

Fordyce, W. E. (1976). *Behavioral methods for chronic pain and illness*. Saint Louis: The CV Mosby Company.

Gatchel, R. E., Peng, Y. B., Peters, M. L., Fuchs, P. N., & Turk, D. C. (2007). The biopsychosocial approach to chronic pain: Scientific advances and future directions. *Psychological Bulletin, 133*, 581–624.

Goubert, L., Crombez, G., & De Bourdeaudhuij, I. (2004a). Low back pain, disability and back pain myths in a community sample: Prevalence and inter-relationships. *European Journal of Pain, 8*, 385–394.

Goubert, L., Crombez, G., & Van Damme, S. (2004b). The role of neuroticism, pain catastrophizing and pain-related fear in vigilance to pain: A structural equations approach. *Pain, 107*, 234–241.

Jensen, J. N., Albertsen, K., Borg, V., & Nabe-Nielsen, K. (2009). The predictive effect of fear-avoidance beliefs on low back pain among newly qualified health care workers with and without previous low back pain: A prospective cohort study. *BMC Musculoskeletal Disorders, 10*, 117.

Kempen, G. I. J. M., van Haastregt, J. C. M., Mckee, K. J., Delbaere, K., & Zijlstra, G. A. R. (2009). Socio-demographic health-related and psychosocial correlates of fear of falling and avoidance of activity in community-living older persons who avoid activity due to fear of falling. *BMC Public Health, 9*, 170.

Klenerman, L., Slade, P. D., Stanley, M., et al. (1995). The prediction of chronicity in patients with an acute attack of low back pain in a general practice setting. *Spine, 20*, 478–484.

Kori, S. H, Miller, R. P., & Todd, D. D. (1990). Kinesiophobia: A new view of chronic pain behavior. *Pain Manage, 3*, 35–43.

Leeuw, M., Goossens, M. E. J. B., Linton, S. J., Crombez, G., Boersma, K., & Vlaeyen, J. W. S. (2007). The fear-avoidance model of musculoskeletal pain: Current state of scientific evidence. *Journal of Behavioral Medicine, 30*, 77–94.

Leeuw, M., Goossens, M. E. J. B., van Breukelen, G. J. P., de Jong, J. R., Heuts, P. H. T. G., Smeets, R. J. E. M., Koke, A. J. A., & Vlaeyen, J. W. S. (2008). Exposure in vivo versus operant graded activity in chronic low back pain patients: Results of a randomized controlled trial. *Pain, 138*, 192–207.

Lethem, J., Slade, P. D., Troup, J. D., et al. (1983). Outline of a fear-avoidance model of exaggerated pain perception: I. *Behaviour Research and Therapy, 21*, 401–408.

Linton, S. J., Buer, N., Vlaeyen, J. W. S., et al. (2000). Are fear-avoidance beliefs related to the inception of an episode of back pain? A prospective study. *Psychology & Health, 14*, 1051–1059.

Linton, S. J., Vlaeyen, J. W. S., & Ostelo, R. W. (2002). The back pain beliefs of health care providers: Are we fear-avoidant? *Journal of Occupational Rehabilitation, 12*, 223–232.

McCracken, L. M., & Gross, R. T. (1993). Does anxiety affect coping with chronic pain? *The Clinical Journal of Pain, 9*, 253–259.

Vangronsveld, K., Peters, M., Goossens, M., et al. (2007). Applying the fear-avoidance model to the chronic whiplash syndrome. *Pain, 131*, 258–261.

Verbunt, J. A., Seelen, H. A., Vlaeyen, J. W. S., et al. (2003). Disuse and deconditioning in chronic low back pain: Concepts and hypotheses on contributing mechanisms. *European Journal of Pain, 7*, 9–21.

Vlaeyen, J. W. S., & Crombez, G. (1999). Fear of movement/(re)injury, avoidance and pain disability in chronic low back pain patients. *Manual Therapy, 4*, 187–195.

Vlaeyen, J. W. S., & Linton, S. J. (2000). Fear-avoidance and its consequences in chronic musculoskeletal pain: A state of the art. *Pain, 85*, 317–332.

Vlaeyen, J. W. S., & Linton, S. J. (2012). Fear-avoidance model of chronic musculoskeletal pain: 12 years on. *Pain, 153*(6), 1144–1147.

Vlaeyen, J. W. S., de Jong, J. R., Sieben, J. M., et al. (2002). Graded exposure in vivo for pain-related fear. In D. C. Turk & R. J. Gatchel (Eds.), *Psychological approaches to pain management: A practitioner's handbook* (pp. 210–233). New York: The Guilford Press.

Muscle Pain, Fibromyalgia Syndrome (Primary, Secondary)

Irwin Jon Russell
Division of Clinical Immunology and Rheumatology, Department of Medicine, The University of Texas Health Science Center, San Antonio, TX, USA

Synonyms

Chronic widespread allodynia; Fibromyalgia syndrome; Fibrositis syndrome; FMS

Definition

The ► Fibromyalgia Syndrome (FMS) is a common, female gender–predominant, idiopathic clinical syndrome characterized by widespread musculoskeletal pain and reproducible tenderness to deep palpation pressure at anatomically defined soft tissue structures, collectively called ► tender points (TePs, see Table 1) (Wolfe et al. 1990). Newer criteria have been developed (Wolfe et al. 2010) for the diagnosis of fibromyalgia and have been validated for use in community practice. These criteria do not require a formal tender point examination. Data collection for application to these criteria can be accomplished simply by administering a patient self-report questionnaire instrument (see Table 2, in which a diagnostic score is 13 or greater is consistent with a diagnosis of fibromyalgia). The classification of soft tissue pain (STP) disorders, including the FMS, is shown below (Russell 1993).

Characteristics

Taxonomically, FMS is classified among over 100 other clinical conditions that cause pain in soft tissue structures (including nerve, muscle, ligament, tendon, and/or bursae) of the musculoskeletal system. The term "► soft tissue pain" (STP) has been proposed (Russell 1993) as the generic heading for this long list of clinical disorders. As such, the term STP replaces the antiquated term "non-articular rheumatism." The proposed classification of STP as listed below is based on the typical body distributions of the included conditions, rather than upon their pathogenesis. For example:

Localized STP is exemplified by ► tendinitis and ► bursitis; Regionalized STP is exemplified by ► complex regional pain syndrome and ► myofascial pain syndrome with ► trigger points (TrPs); and Generalized STP is exemplified by FMS with tender points (TePs) and by chronic fatigue syndrome, when that condition presents with musculoskeletal pain. As a member of the generalized STP category,

Muscle Pain, Fibromyalgia Syndrome (Primary, Secondary), Table 1 American College of Rheumatology (ACR) Criteria for Classification (Wolfe et al. 1990) *History* – Chronic, widespread (four quadrants) soft tissue pain for 3 months. *Examination* – Pain (1+ or greater severity) induced by 4 kg of palpation pressure at 11 of 18 anatomically defined tender points[*]

Number	Official ACR bilateral tender point sites
1, 2	Occiput: suboccipital muscle insertions
3, 4	Low cervical: anterior aspects of C5-7 intertransverse spaces
5, 6	Trapezius: midpoint of upper border
7, 8	Supraspinatus: origins, above the scapula spine, near medial border
9,10	2nd rib: upper lateral surface of second costochondral junction
11,12	Lateral epicondyle: two cm distal to the epicondyles
13,14	Gluteal: upper outer buttock, anterior fold of muscle
15,16	Greater trochanter: posterior to trochanteric prominence
17,18	Knees: medial fat pad, just proximal to medial condyle

Sensitivity and specificity are >80 %
[*]Anatomically Defined Tender Points in Fibromyalgia Syndrome (FMS). Note that named sites are bilateral, even number refers to right side

FMS is now viewed as the human model for chronic, widespread, ▶ allodynia (see Allodynia (Clinical, Experimental)).

Localized
- Bursitis (subacromial, olecranon, trochanteric, prepatellar, anserine)
- Tenosynovitis (biceps, supraspinatus, infrapatellar, Achilles)
- Enthesopathies (lateral epicondylitis, medial epicondylitis)
- Entrapment syndrome (carpal tunnel, tarsal tunnel, cubital tunnel)

Regionalized
- Myofascial pain syndrome (trapezius, piriformis, iliopsoas)
- Myofascial pain dysfunction syndrome (MPDS, replaces TMD)
- Referred pain (angina or subphrenic abscess to shoulder, hip to thigh)
- Complex regional pain syndrome (CRPS)

Muscle Pain, Fibromyalgia Syndrome (Primary, Secondary), Table 2 Self-administered 2010 American College of Rheumatology Fibromyalgia Diagnostic Criteria (2010 ACR FDC) questionnaire designed to collect information required for a diagnosis of fibromyalgia syndrome (Wolfe et al. 2010). The score for subscale A is the total number of painful sites. The score for subscale B is the sum of the four indicated values. The total score is the simple sum of subscale A and subscale B. A total score of 13 or greater is supportive of a fibromyalgia syndrome diagnosis

A: Widespread Pain Index [WPI, Score 0–19] Time period = past 2 weeks:

Jaw, Rt.	Upper arm, Rt.	Upper back	Upper leg, Rt.
Jaw, Lt.	Upper arm, Lt.	Lower back	Upper leg, Lt.
Neck	Lower arm, Rt.	Hip (buttock, trochanter), Rt.	Lower leg, Rt.
Shoulder girdle, Rt.	Lower arm, Lt.	Hip (buttock, trochanter), Lt.	Lower leg, Lt.
Shoulder girdle, Lt.	Chest	Abdomen	N/A

B: Symptoms Severity Score [SSS, range 0–12] Time period = past 2 weeks:

Tiredness in the morning (0–3 severity, 0 = none, 1 = slight/mild, 2 = moderate, 3 = severe)
Fatigue through the day (0–3 severity, 0 = none, 1 = slight/mild, 2 = moderate, 3 = severe)
Dyscognition (0–3 severity, 0 = None, 1 = slight/mild, 2 = moderate, 3 = severe)
Somatic *symptoms* (0–3 scale, # of other symptoms: 0 = none (0), 1 = few (1–9), 2 = mod (10–29), 3 = many (30+)

- Type 1 – non-nerve injury (replaces reflex sympathetic dystrophy)
- Type 2 – nerve injury (replaces causalgia)

Generalized
- Fibromyalgia syndrome
- Chronic fatigue syndrome
- Hypermobility syndrome

The term "primary FMS," has come into common usage to identify a condition meeting the American College of Rheumatology (ACR) classification criteria based upon widespread pain and TeP tenderness (Wolfe et al. 1990), but lacking evidence for any associated painful condition (Table 1).

It is apparent, however, that patients can exhibit the same ACR classification criteria

Muscle Pain, Fibromyalgia Syndrome (Primary, Secondary), Table 3 Illnesses concomitant with secondary fibromyalgia syndrome, with tests profile to characterize the diagnosis (Russell 1993)[a]

Illness	Tests
Rheumatic disease	
a. Systemic lupus erythematosus	ANA + ESR or CRP[‡]
b. Rheumatoid arthritis	RF + ESR or CRP
c. Polymyositis	Creatine phosphokinase
d. Sjögren's syndrome	Labial salivary gland biopsy
Infection/inflammation	
a. Tuberculosis	PPD, ESR, Chest X-ray
b. Chronic syphilis	VDRL, FTA, CSF
c. Subacute bacterial endocarditis	Culture, ESR
d. Lyme disease	Serology
e. Parvovirus	Serology, change
e. Acquired immunodeficiency syndrome	Serology
f. Breast implant	Serology
g. Inflammatory bowel syndromes	Colonoscopy
Endocrine disorders	
a. Hypothyroidism	Thyroxine, TSH
b. Hypopituitary	Prolactin
Obstructive myelopathy	
a. Whiplash C1,2 or subaxial subluxation	CT or MRI of CSpine
b. Chiari malformation	MRI foramen magnum
c. Syringomyelia	MRI of brainstem/CSpine
d. Spinal stenosis	MRI of CSpine

[a]Adapted with consent
Abbreviations: *ANA* antinuclear antibody, *ESR* Westergren erythrocyte sedimentation rate, *CRP* C-reactive protein, *RF* rheumatoid factor, *PPD* purified protein derivative delayed cutaneous test for exposure to tuberculosis bacilli, *VDRL* serologic screening test for syphilis, *FTA* fluorescent treponemal antibody test for syphilis, *CSF* cerebrospinal fluid tests, *TSH* thyroid-stimulating hormone, *C1,2* cervical spine level 1 and level 2 atlantoaxial, *CT* computerized tomography, *CSpine* cervical spine, *MRI* magnetic resonance tomography

(Wolfe et al. 1990) in association with other painful conditions (see Table 3). That seems to be the case in about 30 % of patients with rheumatoid arthritis, 40 % of patients with systemic lupus erythematosus, and 50 % with Sjögren's syndrome. The FMS has been recognized in association with tuberculosis, syphilis, and Lyme disease. There also seems to be a high prevalence of FMS among women having body pain in association with silicone breast implants. For want of better terminology, the FMS in such settings has been collectively referred to as "secondary FMS."

The ACR classification criteria study (Wolfe et al. 1990) failed to disclose any clinical features in the secondary FMS patients that would clearly distinguish them from primary FMS, so it was recommended that the term "secondary FMS" be abandoned. Despite that, essentially all of the current and past FMS clinical trials, sponsored by industry or by government agencies, have specified enrollment of only primary FMS patients. They accomplish that distinction by excluding all subjects with a concomitant rheumatic disease or organ-related abnormalities.

In each of the secondary FMS situations, it has been unclear what pathogenic role the "other" condition might have had in the development of the FMS. The current belief is that the comorbid inflammatory conditions have not developed from the FMS, nor has the FMS necessarily resulted from them, so the implication of causality in the term "secondary FMS" is likely to be a misnomer. Despite that, the terminology serves a need, and has become entrenched, so it will probably endure, at least for the short term.

Biochemical studies help to support this distinction of a single syndrome developing in two or more different settings. Most people with FMS (both primary and secondary) have elevated spinal fluid (CSF) substance P as an amplifier of afferent pain signals. The distinction seems to be that only primary FMS patients exhibit elevated concentrations of nerve growth factor as a substance P–inducing agent (Giovengo et al. 1999). In secondary FMS, associated with an inflammatory condition, the nerve growth factor levels were found to be normal. The inflammation itself may be responsible for initiating the high CSF substance P levels in those conditions. This concept is important because it provides evidence that there may be a final common pathway (related to the elevated CSF substance P) by which primary FMS and secondary FMS could exhibit the same clinical syndrome (i.e., FMS).

Epidemiology
The FMS exhibits a worldwide distribution. It is viewed as being common, because it affects 2–5 % of the general population. It is female predominant, with 8 of 10 affected individuals being female. It increases in prevalence with age, such that nearly 10 % of women in the 50–60 years of age decade are affected. About one third of FMS patients are so severely affected by their symptoms that they are unable to maintain their usual occupational activities (Wolfe et al. 1997c). The average annual direct cost of this condition in the United States was estimated to be about $2,250 (Wolfe et al. 1997a). The natural history of FMS is to develop a pattern of severity over a relatively short period of time, and then remain relatively unchanged over a period of many years (Wolfe et al. 1997b). It does not usually become another medical condition like a rheumatic disease.

Comorbidities
In addition to the defining widespread musculoskeletal pain and TePs, people with FMS typically exhibit a constellation of painful symptoms or syndromes that can include insomnia (more correctly termed sleep dysfunction), which occurs in more than 70 % (Smythe and Moldofsky 1977; Moldofsky et al. 2010), nocturnal myoclonus, cognitive dysfunction, headache, depression in about 40 % (Ahles et al. 1991), anxiety, autonomic neuropathy (Martinez-Lavin 2003), nocturnal bruxism, myofascial pain dysfunction syndrome (the so-called TMJ syndrome), myofascial pain syndrome (particularly involving the piriformis muscles), hypermobility syndrome in about 30 % (Acasuso-Diaz and Collantes-Estevez 1998), biceps tendinitis, pes anserine bursitis, irritable bowel syndrome in about 40 % (Aaron and Buchwald 2001), and irritable bladder syndrome (interstitial cystitis) in about 10 % (Clauw et al. 1997). Despite overlapping features and prevalence, FMS can be distinguished from rheumatic diseases, the complex regional pain syndromes, the chronic

fatigue syndrome, and the myofascial pain syndrome affecting a variety of muscles, on the basis of its clinical presentation, its consistent epidemiologic pattern, and its predictable prognosis.

Heterogeneity
How should this heterogeneity of associated manifestations be viewed? Should clinicians caring for FMS patients be lumpers or splitters? It is worthy, in this regard, to consider the established situation with systemic lupus erythematosus (SLE). The ACR supports the concept that 11 diagnostic criteria are important to the clinical diagnosis of SLE, but only four of these criteria are required to diagnose SLE in a given individual (Tan et al. 1982). Lupus subgroups are characterized by their patterns of major organ involvement, critical to the prognosis and management of the disorder (Hughes 1978). For example, in various combinations, about 50 % of SLE patients have renal involvement, 50 % exhibit lung involvement, cardiac involvement is seen in about 40 %, neuropsychiatric manifestations are present in about 60 %, autoimmune hemolysis occurs in about 5 %, and about 30 % have concomitant FMS.

The findings from prior psychological studies of FMS, in which two or three distinct clusters (subgroups?) of FMS patients had been identified, were reviewed by Walen and colleagues (Walen et al. 2002). They then conducted extensive evaluations on 600 FMS patients who were members of a health maintenance organization. Cluster analysis on the data from those subjects confirmed the previously described unique clusters of FMS patients that differed from each other with respect to mood disturbance, pain, physical function, and social support. While these subgroup differences were statistically definable, all three subgroups were still much worse than healthy normal controls in the general population.

Electrophysiological assessment methods, such as combinations of quantitative electroencephalography (qEEG) and electromyography (sEMG), have also helped to identify distinct FMS subgroups (Donaldson et al. 2002). Forty patients with FMS, off their usual FMS medications for five half lives, were stratified into three subgroups on the basis of their Symptom Checklist-90-Revised (SCL-90-R) Global Severity Index (GSI) scores (Derogatis 1994). The subgroups were then examined by qEEG activity patterns with the subjects resting, eyes closed. By-subgroup differences in the quantity of the typical qEEG wave forms were found. Across the GSI-defined subgroups, from least to most distressed, alpha brainwave activity (7.5–13 Hz) progressively decreased, while theta activity (3.5–7.5 Hz) progressively increased. Beta activity (13–22 Hz) was highest in the middle group; while delta activity (0.5–3.5 Hz) was consistently low across all three subgroups. The addition of sEMG data in this study demonstrated widely dispersed skeletal muscle co-contraction in response to a distant volitional stimulus, but the magnitude of this abnormality did not differ among the three GSI-defined subgroups.

Muscle

The FMS is probably is a central neurological disorder rather than a disorder of skeletal muscle, but there was a time in the history of the disorder when skeletal muscle was the focus of investigation into the pathogenesis of FMS. Indeed, the clinical complaint most consistently reported by FMS patients is deep, aching, body pain. Whether spontaneously, or as learned from their health care providers, FMS patients have tended to interpret these symptoms as muscle pains, muscle fatigue, and muscle stiffness. That may have been what prompted early FMS researchers to seek some specific pathology in affected muscles. Controlled histological examination of FMS muscle tissue sections by light microscopy and electron microscopy have disclosed minor mitochondrial abnormalities, atrophy of type 2 muscle fibers, ragged red fibers, or moth-eaten fibers (Yunus et al. 1989; Lindman et al. 1995), and lower than normal capillary density, but the histological evidence did not support FMS specificity. In addition, lower levels of ATP and phosphoryl creatine were found in the trapezius and the tibialis anterior muscles of FMS patients compared with HNC (Bengtsson et al. 1986). Clearly, skeletal muscle that is at risk could

serve as a peripheral pain generator in FMS patients who fail to maintain physical fitness of the very muscle groups needed for usual function, and for critical situations of physiological stress that cause the falling injuries observed to be common among people with FMS.

Biochemistry

Biochemical analysis of fluid samples from patients with FMS has also identified apparent subgroups. For example, 84 % of FMS patients had elevated cerebrospinal fluid levels of substance P (Russell et al. 1994). As noted earlier, patients with "primary" FMS had elevated levels of nerve growth factor not seen in normal controls, or in FMS patients with concomitant rheumatic diseases ("secondary" FMS), or in rheumatic disease patients who do not have FMS (Giovengo et al. 1999). Similarly, blood samples from a subgroup of about 50 % of FMS patients exhibit an antibody to an environmental polymer (Wilson et al. 1999), so the antipolymer antibody assay is being proposed as a way to distinguish these two unique subgroups. A genetic predisposition appears to be a factor in a subgroup of FMS (Arnold et al. 2004). The current hope is that the findings from objective tests, which are now in active development, can be leveraged to identify pathogenic mechanisms.

Medications

Finally, attempts at diagnostic treatment regimens have disclosed differences in the responses of FMS patients to ketamine, suggesting that the N-methyl-D-aspartic receptor (which ketamine inhibits) in the spinal cord is important to the perceived pain in FMS (Graven-Nielsen et al. 2000). Patient volunteers in one key study (Sorensen et al. 1997) were treated (randomized serial crossover) with brief intravenous infusions of ketamine, morphine sulfate, lidocaine, or saline placebo. Of the 18 FMS patients studied, two patients were placebo responders (responded with improvement to all three agents and the placebo) and three were nonresponders (no improvement with any of the administered agents). The majority (N = 13) responded to one or more of the active medications but not to placebo. Four patients responded to a single drug (morphine-one, ketamine-three), six responded to two drugs (lidocaine and morphine-four, lidocaine and ketamine-two), and three responded to all three active drugs. Since all of the patients were clinically diagnosed as having FMS, this study implies substantial heterogeneity in the response of FMS patients to treatment. It may further help to explain why some patients fail to respond to a clinician's favorite regimen.

From the preceding discussion, it seems likely that the composite of FMS subjects, identified by the 1990 ACR criteria for the classification of FMS (Wolfe et al. 1990), is heterogeneous in several important respects. Walen and colleagues (Walen et al. 2002) concluded their assessment by saying that: "People with FMS may fall into distinct subgroups; ... (but) the utility of dividing participants into these (sub) groups in planning interventions remains unclear." On the other hand, they also suggest that "the most helpful direction for future research would involve comparing the effects of interventions designed especially for each cluster to a 'nontaylored' intervention." Finally, it always comes back to the hopeful prediction that the better we understand the mechanisms of chronic FMS pain, the more likely we are to find specific and effective therapies.

Pathogenesis
The pathogenesis of FMS is becoming increasingly evident from studies of physiology, pharmacology, neurochemistry, and brain imaging. Available evidence supports the hypothesis that the underlying dysfunction in FMS is within the central nervous system. It is no longer considered to be merely a psychological disorder, a diagnosis to be made by exclusion, or a condition devoid of objective laboratory findings. Patients with FMS exhibit objective abnormalities in ▶ nociception and in neuroendocrine functions, which undoubtedly contribute substantially to their generalized symptoms.

Some of the biological participants in this process include: unmyelinated dorsal horn neurons (like A-delta and C-fibers), excitatory amino acids, neuropeptides, zinc, biogenic amines,

nitric oxide, wide-dynamic range spinal neurons, the limbic system of the midbrain, and several regions of the cerebral cortex. A lowered pain threshold (i.e., allodynia) characterizes the examination findings in FMS. Allodynia can be caused in animal systems by strategic manipulation of nociceptive neurochemicals. Studies of the nociceptive neurochemicals in FMS spinal fluid find them abnormal in concentration and/or correlated with the symptoms. Those observations change the way FMS is viewed, and identify it as a remarkably interesting human syndrome of chronic central neurochemical pain amplification.

Management
Dealing with FMS has been a complicated process for twentieth century medicine, leading to widely conflicting opinions about it. The reasons for the resultant role modeling are buried deep in the fabric of belief system anchoring. Effective contemporary management of FMS requires each of the following (Russell 1996): The unequivocal recognition that pain is always subjective, a willingness to accept FMS as a medical syndrome, the effort to integrate a logical but increasingly complex pathogenesis, and empathetic individualization of increasingly evidence-based therapy. The components of a practical regimen include: accurate diagnosis of the FMS and associated conditions, education about FMS, physical modalities such as exercise, medications for the presenting symptoms, and follow-up assessment to monitor therapeutic progress.

References

Aaron, L. A., & Buchwald, D. (2001). A review of the evidence for overlap among unexplained clinical conditions. *Annals of Internal Medicine, 134,* 868–881.

Acasuso-Diaz, M., & Collantes-Estevez, E. (1998). Joint hypermobility in patients with fibromyalgia syndrome. *Arthritis Care and Research, 11,* 39–42.

Ahles, T. A., Khan, S. A., Yunus, M. B., et al. (1991). Psychiatric status of patients with primary fibromyalgia, patients with rheumatoid arthritis, and subjects without pain: A blind comparison of DSM-III diagnoses. *The American Journal of Psychiatry, 148,* 1721–1726.

Arnold, L. M., Hudson, J. I., Hess, E. V., Ware, A. E., Fritz, D. A., Auchenbach, M. B., Starck, L. O., &

Keck, P. E., Jr. (2004). Family study of fibromyalgia. *Arthritis & Rheumatism*, *50*, 944–952.

Bengtsson, A., Henriksson, K. G., & Larsson, J. (1986). Reduced high energy phosphate levels in the painful muscles of patients with primary fibromyalgia. *Arthritis and Rheumatism, 29*, 817–821.

Clauw, D. J., Schmidt, M., Radulovic, D., et al. (1997). The relationship between fibromyalgia and interstitial cystitis. *Journal of Psychiatric Research, 31*, 125–131.

Derogatis, L. (1994). *SCL-90-R: Administration, scoring, and procedures manual.* (Third Edition) Baltimore: Clinical Psychometric Research.

Donaldson, M., Donaldsson, C. C., Mueller, H. H., et al. (2002). QEEG patterns, psychological status, and pain reports of fibromyalgia sufferers. *American Journal of Pain Management, 13*, 60–73.

Giovengo, S. L., Russell, I. J., & Larson, A. A. (1999). Increased concentrations of nerve growth factor in cerebrospinal fluid of patients with fibromyalgia. *Journal of Rheumatology, 26*, 1564–1569.

Graven-Nielsen, T., Aspegren, K. S., Henriksson, K. G., et al. (2000). Ketamine reduces muscle pain, temporal summation, and referred pain in fibromyalgia patients. *Pain, 85*, 483–491.

Hughes, G. R. V. (1978). Systemic lupus erythematosus. In J. T. Scott (Ed.), *Copeman's textbook of the rheumatic diseases* (pp. 901–922). Edinburgh: T & A Constable.

Lindman, R., Hagberg, M., Bengtsson, A., et al. (1995). Capillary structure and mitochondrial volume density in the trapezius muscle of chronic trapezius myalgia, fibromyalgia and healthy subjects. *Journal of Musculoskeletal Pain, 3*, 5–22.

Martinez-Lavin, M. (2003). Use of the Leeds assessment of neuropathic symptoms and signs questionnaire in patients with fibromyalgia. *Seminars in Arthritis and Rheumatism, 32*, 407–411.

Moldofsky, H., Inhaber, N. H., Guinta, D. R., & varez-Horine, S. B. (2010) Effects of sodium oxybate on sleep physiology and sleep/wake-related symptoms in patients with fibromyalgia syndrome: a double-blind, randomized, placebo-controlled study. *Journal of Rheumatology, 37*, 2156–2166.

Russell, I. J. (1993). A new journal. *Journal of Musculoskeletal Pain, 1*, 1–7.

Russell, I. J. (1996). Fibromyalgia syndrome: Approaches to management. *Bulletin on the Rheumatic Diseases, 45*(3), 1–4.

Russell, I. J., Orr, M. D., Littman, B., et al. (1994). Elevated cerebrospinal levels of substance P patients fibromyalgia syndrome. *Arthritis and Rheumatism, 37*, 1593–1601.

Smythe, H. A., & Moldofsky, H. (1977). Two contributions to understanding of the "Fibrositis" syndrome. *Bulletin on the Rheumatic Diseases, 28*, 928–931.

Sorensen, J., Bengtsson, A., Ahlner, J., et al. (1997). Fibromyalgia – are there different mechanisms in the processing of pain? A double-blind crossover comparison of analgesic drugs. *Journal of Rheumatology, 24*, 1615–1621.

Tan, E. M., Cohen, A. S., Fries, J. F., et al. (1982). The 1992 revised criteria for the classification of systemic lupus erythematosus. *Arthritis and Rheumatism, 25*, 1271–1277.

Walen, H. R., Cronan, T. A., Serber, E. R., et al. (2002). Subgroups of fibromyalgia patients: Evidence for heterogeneity and an examination of differential effects following a community-based intervention. *Journal of Musculoskeletal Pain, 10*, 9–32.

Wilson, R. B., Gluck, O. S., Tesser, J. R., et al. (1999). Antipolymer antibody reactivity in a subset of patients with fibromyalgia correlates with severity. *Journal of Rheumatology, 26*, 402–407.

Wolfe, F., Anderson, J., Harkness, D., et al. (1997a). A prospective, longitudinal, multicenter study of service utilization and costs in fibromyalgia. *Arthritis and Rheumatism, 40*, 1560–1570.

Wolfe, F., Anderson, J., Harkness, D., et al. (1997b). Health status and disease severity in fibromyalgia: Results of a six center longitudinal study. *Arthritis and Rheumatism, 40*, 1571–1579.

Wolfe, F., Anderson, J., Harkness, D., et al. (1997c). Work and disability status of persons with fibromyalgia. *Journal of Rheumatology, 24*, 1171–1178.

Wolfe, F., Smythe, H. A., Yunus, M. B., et al. (1990). The American College of Rheumatology 1990 criteria for the classification of fibromyalgia. *Arthritis and Rheumatism, 33*, 160–172.

Wolfe, F., et al. (2010). Arthritis Care Res 62, 600–610.

Yunus, M. B., Kalyan-Raman, U. P., Masi, A. T., et al. (1989). Electron microscopic studies of muscle biopsy in primary fibromyalgia syndrome: A controlled and blinded study. *Journal of Rheumatology, 16*, 97–101.

Muscle Pain, Referred Pain

Maria Adele Giamberardino
Pathophysiology of Pain Laboratory and Centre for Fibromyalgia and Musculoskeletal Pain, Department of Medicine and Science of Aging, University of Chieti, Chieti, Italy

Synonyms

Musculoskeletal transferred pain; Referred muscle pain

Definition

Pain perceived at muscle level due to a primary algogenic process located at a distance, either in a visceral organ or in a deep somatic structure.

Muscle Pain, Referred Pain, Fig. 1 Classification of referred muscle pain

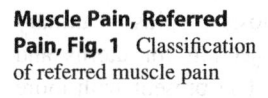

Characteristics

Referred muscle pain can result from an algogenic process in internal organs or in deep somatic structures, e.g., another muscle or a joint (referred muscle pain from viscera or from deep somatic structures) (Fig. 1). In both cases, it may or may not be accompanied by secondary hyperalgesia, i.e., increased local sensitivity to pain/decreased pain threshold (referred muscle pain with or without hyperalgesia) (Vecchiet et al. 1999). When present, hyperalgesia is proportional in extent to the degree of activity of the primary algogenic focus and is often accompanied by local trophic changes, i.e., decreased thickness/sectional area of the muscle (Galletti et al. 1990; Giamberardino et al. 2010; Simons and Mense 2003; Vecchiet et al. 1989). Both central (neuronal sensitization) and peripheral (reflex arc activation) mechanisms are likely to be involved in the pathophysiology of referred phenomena at muscle level (Arendt-Nielsen and Svensson 2001; Cervero and Laird 2004; Mense 1994; Procacci et al. 1986).

Referred Muscle Pain from Viscera

Pain referral occurs constantly in visceral nociception (Cervero and Laird 2004). After the transitory phase of "true visceral pain" (midline, vague, and poorly defined), the sensation is "transferred" to somatic areas neuromerically connected to the specific viscus (Procacci et al. 1986). Secondary hyperalgesia most often arises at this level, especially if pain episodes are recurrent and/or prolonged; this may involve

all three body wall tissues – skin, subcutis, and muscle – but is most frequently confined to muscle, often accompanied by sustained contraction (referred muscle pain from viscera without and with hyperalgesia) (Vecchiet et al. 1989). Trophic changes may also occur in the same sites, i.e., increased thickness of subcutis but mostly decreased thickness of muscle (Giamberardino et al. 2005; Procacci et al. 1986; Vecchiet et al. 1989). In patients with algogenic conditions from a number of internal organs, muscle hyperalgesia in referred zones has been detected clinically (hypersensitivity to digital compression) and quantified instrumentally (decrease in pain thresholds to different stimuli); in a minor percentage of visceral pain patients, muscle trophic changes have also been measured using ultrasounds (Giamberardino et al. 1997, 2005; Vecchiet et al. 1989). Hyperalgesia is accentuated by the repetition of the visceral episodes; although decreasing as they stop, it usually remains significant after cessation of the spontaneous pain and sometimes even after removal of the primary visceral focus. Trophic changes are even more persistent than the hyperalgesia, often remaining unaltered for a long time after extinction of the visceral trigger (Giamberardino et al. 2005).

Clinical Examples
In myocardial infarction, pain is referred to the thoracic region, anteriorly or posteriorly, often extending to the left arm. Hyperalgesia almost always affects the pectoralis major and muscles of the interscapular region and forearm.

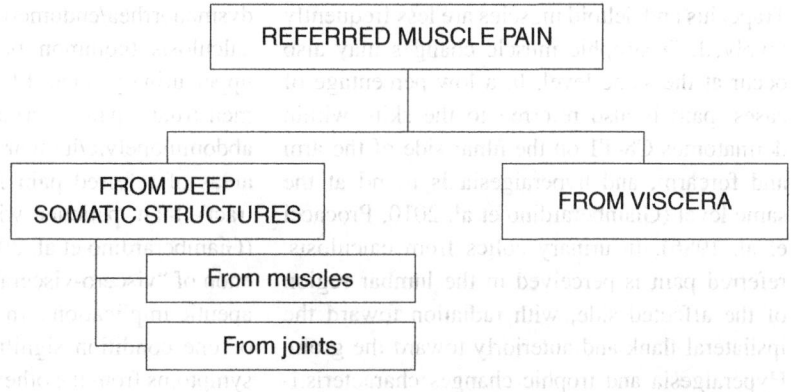

Trapezius and deltoid muscles are less frequently involved. Dystrophic muscle changes may also occur at the same level. In a low percentage of cases, pain is also referred to the skin, within dermatomes C8-T1 on the ulnar side of the arm and forearm, and hyperalgesia is found at the same level (Giamberardino et al. 2010; Procacci et al. 1986). In urinary colics from calculosis, referred pain is perceived in the lumbar region of the affected side, with radiation toward the ipsilateral flank and anteriorly toward the groin. Hyperalgesia and trophic changes characteristically affect muscles of the lumbar and flank area (quadratus lumborum, oblique muscles) (Vecchiet et al. 1989). In biliary calculosis, pain is referred to the upper right quadrant of the abdomen with radiation toward the back. Hyperalgesia and reduced trophism of the rectus abdominis at the cystic point (junction of 10th rib with outer margin of the same muscle) are typical findings (Giamberardino et al. 2005). In dysmenorrhea, pain is referred to the lower abdomen, perineum, and sacral region, with radiation toward the groin and upper part of the thighs. Hyperalgesia and trophic changes can be detected in the lowest part of the rectus abdominis and muscles of the pelvic region (Giamberardino et al. 1997, 2001).

Referred muscle pain and hyperalgesia are typically enhanced in the case of "viscero-visceral hyperalgesia," when concurrent algogenic conditions affect two viscera which share part of their central sensory projection (Giamberardino et al. 2010). Patients affected with coronary artery disease (CAD) plus gallbladder calculosis (common projection for heart and gallbladder, T5) present more angina episodes, biliary colics, and referred muscle hyperalgesia in the left anterior chest area and upper right abdominal quadrant than patients with CAD or gallbladder calculosis only. Women suffering from both dysmenorrhea and irritable bowel syndrome (IBS) (common projection for uterus and colon, T10-L1) report more menstrual pain, intestinal pain, and abdominal/pelvic muscle hyperalgesia (in areas of referral from uterus and intestine) than women with dysmenorrhea or IBS only. Patients with dysmenorrhea/endometriosis plus urinary calculosis (common projection for uterus and upper urinary tract, T10-L1) present with more menstrual pain, urinary colic pain, and abdominopelvic/lumbar muscle hyperalgesia (in areas of referred pain from uterus and urinary tract) than patients with one condition only (Giamberardino et al. 2001, 2010). The phenomenon of "viscero-visceral hyperalgesia" has therapeutic implications. In fact, effective treatment of one condition significantly improves typical symptoms from the other, e.g., decrease in angina pain and muscle hyperalgesia at chest level after cholecystectomy, biliary pain and muscle hyperalgesia in upper right abdominal quadrant after cardiac revascularization, urinary pain and referred lumbar hyperalgesia after hormonal treatment of dysmenorrhea, and menstrual pain and referred abdominopelvic hyperalgesia after dietary treatment of IBS or urinary stone elimination following lithotripsy (Giamberardino et al. 2010).

Pathophysiology

"Referred muscle pain from viscera without hyperalgesia" is explained on the basis of the convergence of visceral and somatic afferent fibers upon the same central neurons (convergence-projection theory) (Cervero and Laird 2004; Procacci et al. 1986). Messages from the viscus are "interpreted" by higher brain centers as coming from muscles due to mnemonic traces of previous experiences of somatic pain, more numerous than those of visceral pain in life. "Referred muscle pain from viscera with hyperalgesia" is contributed to by central mechanisms, i.e., a process of sensitization in the central nervous system (CNS), triggered by the massive afferent visceral barrage involving hyperactivity and hyperexcitability of viscero-somatic convergent neurons, as shown by the results of electrophysiological studies on animal models of referred muscle hyperalgesia from viscera (Cervero and Laird 2004; Refs. in Giamberardino et al. 2010). This process would facilitate the central effect of the normal input coming from the muscle (convergence-facilitation theory). NMDA receptors have been shown to play an important role in the generation of these central hyperexcitability changes (Cervero and Laird 2004).

In addition to central changes, the afferent visceral barrage probably activates a number of viscero-somatic reflex arcs, whose afferent branch is represented by sensory fibers from the internal organ, and whose efferent branch is represented by sympathetic efferences to the skin/subcutis and somatic efferences to the muscle of the referred area (Procacci et al. 1986). Activation of this arc would produce sustained contraction in the skeletal muscle, this, in turn, being responsible for sensitization of nociceptors locally. Studies in a rat model of referred muscle hyperalgesia from artificial ureteric calculosis provide experimental support for this mechanism. Positivity was found for a number of ultrastructural indices of contraction in the hyperalgesic muscle ipsilateral to the affected ureter at lumbar level but not in the contralateral, non-hyperalgesic muscle, and the extent of these indices was proportional to the degree of visceral pain behavior and referred hyperalgesia recorded in the animals. In the same model, c-Fos activation was found in the spinal cord not only in sensory neurons but also motoneurons, significantly more on the affected side (Refs. in Giamberardino et al. 2005).

Mechanisms underlying enhancement of referred phenomena in "viscero-visceral hyperalgesia" are still hypothetical, but probably involve sensitization of viscero-viscero-somatic convergent neurons. Along with the well-documented viscero-somatic convergence onto the same sensory neurons, extensive viscero-visceral convergence has also been found in the CNS, for instance, between the colon/rectum, bladder, vagina, and uterine cervix (Refs. in Vecchiet et al. 1999). The increased afferent barrage from one visceral organ would thus enhance the afferent signal from the second organ projecting to the same neuron and also from the referred area. This hypothesis needs to be verified in electrophysiological studies on animal models of the condition, such as the model of endometriosis plus ureteral calculosis in the female rat in which an exponential enhancement is observed of

both spontaneous pain behavior and referred lumbar muscle hyperalgesia with respect to rats with calculosis only (Refs. in Giamberardino et al. 2010).

Referred Muscle Pain from Deep Somatic Structures
Clinical Examples
Myofascial pain syndromes (MPSs) are classical examples of "referred muscle pain from muscles." An MPS is characterized by regional muscle pain and dysfunction due to the presence, in muscles or their fascia, of active trigger points (TrPs). A trigger point is a site of tissue hyperirritability included in a taut, palpable band of muscle fibers. When stimulated, it gives rise to pain not only locally but also at a distance in an area called "target" because it is often remote from its location (Simons and Mense 2003). The target zone (area of referred pain) is typical and characteristic for each muscle and is the site not only of spontaneous pain but also of sensory changes. These consist of hyperalgesia, which is a function of the degree of hyperirritability of the trigger point, i.e., the more irritable the TrP, the greater the degree of hyperalgesia in the target and the greater the extension of the tissues involved. Latent trigger points (preclinical phase of MPSs) only give rise to muscle hyperalgesia in the target, while active TrPs give rise to hyperalgesia, which extends from the muscle to the overlying subcutaneous/cutaneous tissues also. The dependence of these changes on the presence of the TrPs is testified by their regression after the TrP is extinguished through local infiltration (Giamberardino et al. 2011).

A typical example of "referred muscle pain from joints" is represented by the painful symptomatology in patients with osteoarthritis of the knee (Vecchiet et al. 1999). Pain from this condition is deep, fairly well localized, and of varying intensity, sometimes making walking difficult. It spreads upward to the lower part of the thigh and downward as far as the middle of the calf. It begins when walking, increases as walking continues, and decreases at rest. The skin appears pale and hypothermic to touch in an area covering the anterior surface of the knee. The underlying subcutis is tender and thickened at pincer palpation; the skeletal muscles connected to the joint are tender and tense. Pain thresholds to electrical stimulation of all three tissues of the parietal wall (but mostly the muscle) in the painful area are decreased (hyperalgesia). The sectional area of periarticular muscles (mostly the vastus medialis), measured via ultrasounds, is reduced. The sensory changes in the referred area are reversible when the intra-articular focus is extinguished, testifying their referred nature (Galletti et al. 1990).

Patients with osteoarthritis of the knee have also been shown to present increased pain intensity and larger referred and radiating pain areas after infusion of hypertonic saline into the leg muscles (tibialis anterior) as compared to controls, a finding suggesting the setting up of central sensitization by painful osteoarthritis (Bajaj et al. 2001).

Pathophysiology

As deducible from the two clinical examples provided (MPS and osteoarthritis), referred muscle pain from somatic structures is almost exclusively of the type "referred with hyperalgesia." Similar to what has been described for visceral nociception, referred pain from somatic structures has been attributed mainly to phenomena of central hyperexcitability triggered by the primary algogenic focus (Arendt-Nielsen and Svensson 2001). Animal studies have indeed provided good evidence that dorsal horn neurons become hyperexcitable in response to noxious stimulation of deep tissues (and that NMDA receptors and neurokinin receptors are most probably involved in this mechanism) (Hoheisel et al. 1993; Mense 1994). To account for the phenomenon of referral, however, central hyperexcitability should involve neurons receiving convergent input from the site of injury and the referred zone, whereas it is known that in dorsal horn neurons, there is little convergence from deep tissues (Mense 1994). Thus, referred pain from somatic structures is not easily explained

by the "convergence-facilitation" theory in its original form. A modified theory has therefore been suggested, especially for referred pain from one muscle to another, based on animal studies. Recordings from dorsal horn neurons reveal that noxious stimuli to a specific receptive field in a muscle generate within minutes new muscle receptive fields at a distance from the original one (Hoheisel et al. 1993). On this basis, the explanation proposed is that convergent connections from deep tissues to dorsal horn neurons are not present from the beginning but are opened by nociceptive input from skeletal muscle. Referral to myotomes outside the lesion is due to spread of central sensitization to adjacent spinal segments (Mense 1994). Central mechanisms alone, however, are not adequate to account for all referred phenomena, especially the trophic changes. Thus, like the hypothesis for visceral nociception, it has been suggested that the afferent barrage from the deep focus (in muscle or joint) triggers the activation of reflex arcs toward the periphery (area of referral) via both somatic (toward the muscle) and sympathetic (toward subcutis and skin) efferent fibers, responsible for the local referred changes (Vecchiet et al. 1999).

Diagnosing Referred Muscle Pain

Detection and interpretation of simple referred pain (without hyperalgesia) are relatively easy, as the absence of tenderness at the site of the symptom immediately points out a site of origin of the algogenic impulses at a distance. Correct diagnosis of referred muscle pain with hyperalgesia is, however, much more difficult and represents a major clinical problem, since this form is far more frequent than that of simple referred pain. The problem concerns differentiation from primary muscle pain, i.e., that arising in relation to an algogenic pathology primarily affecting the tissue. Only careful study of the clinical history, accurate physical examination, and complete sensory evaluation of the painful areas can help toward diagnostic orientation, an indispensable step in the institution of a therapeutic strategy that is not merely symptomatic (Vecchiet et al. 1999).

References

Arendt-Nielsen, L., & Svensson, P. (2001). Referred muscle pain: Basic and clinical findings. *The Clinical Journal of Pain, 17*, 11–19.

Bajaj, P., Bajaj, P., Graven-Nielsen, T., et al. (2001). Osteoarthritis and its association with muscle hyperalgesia: An experimental controlled study. *Pain, 93*, 107–114.

Cervero, F., & Laird, J. M. (2004). Understanding the signaling and transmission of visceral nociceptive events. *Journal of Neurobiology, 61*, 45–54.

Galletti, R., Obletter, M., Giamberardino, M. A., et al. (1990). Pain from osteoarthritis of the knee. *Advances in Pain Research Therapy, 13*, 183–191.

Giamberardino, M. A., Berkley, K. J., Iezzi, S., et al. (1997). Pain threshold variations in somatic wall tissues as a function of menstrual cycle, segmental site and tissue depth in non-dysmenorrheic women, dysmenorrheic women and men. *Pain, 71*, 187–197.

Giamberardino, M. A., De Laurentis, S., Affaitati, G., et al. (2001). Modulation of pain and hyperalgesia from the urinary tract by algogenic conditions of the reproductive organs in women. *Neuroscience Letters, 304*, 61–64.

Giamberardino, M. A., Affaitati, G., Lerza, R., et al. (2005). Relationship between pain symptoms and referred sensory and trophic changes in patients with gallbladder pathology. *Pain, 114*, 239–249.

Giamberardino, M. A., Costantini, R., Affaitati, G., et al. (2010). Viscero-visceral hyperalgesia: Characterization in different clinical models. *Pain, 151*, 307–322.

Giamberardino, M. A., Affaitati, G., Fabrizio, A., et al. (2011). Myofascial pain syndromes and their evaluation. *Best Practice & Research. Clinical Rheumatology, 25*, 185–198.

Hoheisel, U., Mense, S., Simons, D. G., et al. (1993). Appearance of new receptive fields in rat dorsal horn neurons following noxious stimulation of skeletal muscle: A model for referral of muscle pain? *Neuroscience Letters, 153*, 9–12.

Mense, S. (1994). Referral of muscle pain. *Journals of the American Physical Society, 3*, 1–9.

Procacci, P., Zoppi, M., & Maresca, M. (1986). Clinical approach to visceral sensation. In F. Cervero & J. F. B. Morrison (Eds.), *Visceral sensation, progress in brain research* (pp. 21–28). Amsterdam: Elsevier.

Simons, D. G., & Mense, S. (2003). Diagnosis and therapy of myofascial trigger points. *Schmerz, 17*, 419–424.

Vecchiet, L., Giamberardino, M. A., Dragani, L., et al. (1989). Pain from renal/ureteral calculosis: Evaluation of sensory thresholds in the lumbar area. *Pain, 36*, 289–295.

Vecchiet, L., Vecchiet, J., & Giamberardino, M. A. (1999). Referred muscle pain: Clinical and pathophysiologic aspects. *Current Review of Pain, 3*, 489–498.

Muscle Relaxants

Nikolai Bogduk
University of Newcastle, Newcastle Bone & Joint Institute, Royal Newcastle Centre, Newcastle, NSW, Australia

Definition

Muscle relaxants are drugs ostensibly designed to relieve pain by reducing painful muscle spasm.

Characteristics

Muscle relaxants have been largely used in the treatment of spinal pain, for it is believed that spasm of the spinal muscles occurs in response to pain, but is itself painful. Most of the literature describes their use for low back pain, but they have also been used and tested for neck pain.

Rationale

The rationale for the use of muscle relaxants is that if they can relieve the spasm, they serve as analgesics by relieving at least part of the pain caused by the muscle spasm. This rationale, however, is not vindicated by the evidence.

There is no evidence that spasm of the back muscles is painful or contributes to the patient's pain. There is no evidence that so-called muscle spasm can be reliably diagnosed. There is no correlation between clinical muscle spasm and any biological parameter such as EMG. Indeed, eminent authorities have decried the wisdom of belief in muscle spasm (Johnson 1989) or lamented its lack of validity (Andersson et al. 1989).

Efficacy

The most recent, systematic review (Van Tulder et al. 2003) concluded that there was strong evidence that muscle relaxants are more effective than placebo for short-term relief of acute low back pain. Inspection of the review revealed that studies included in the tables were not included in the narrative, where the conclusions were drawn. Inspection of the original studies suggests that this conclusion may have been overly generous (Bogduk 2004).

The data are conflicting for orphenadrine (Bogduk 2004). One study found that only 9 out of 20 patients treated with orphenadrine had reduced pain at 48 h after treatment, compared to 4 out of 20 patients treated with placebo. No other, or better, measure of outcome was reported. In contrast, another study found no superiority over placebo.

A low-quality study found diazepam to be more effective than placebo at day 5, but a high-quality study found it to be no more effective than placebo (Bogduk 2004). Dantrolene was considered more effective than placebo on the grounds that it reduced pain during maximum voluntary movement to a greater extent, but not at other times (Bogduk 2004).

Carisoprodol was considered effective because it reduced pain to a greater extent than placebo at 4 days (Bogduk 2004). Methocarbamol was as effective as chlormezanone, but chlormezanone was as effective as an NSAID (Bogduk 2004).

Conflicting data concerning tizanidine have been reported by the same investigators in two separate studies. One found no differences in outcome at 3 and at 7 days between patients treated with tizanidine and those treated with placebo (Bogduk 2004). The other study compared tizanidine plus ibuprofen with tizanidine plus placebo (Bogduk 2004). It found no differences in pain scores between the two groups at 3 days and at 7 days. The success attributed to tizanidine was based on a larger proportion of patients having no pain or only mild pain at night, at 3 days but not at 7 days, and a larger proportion of patients having no pain or only mild pain at rest, both at 3 days and at 7 days. The respective proportions in the latter instance were 90 % versus 72 % at 3 days and 93 % versus 77 % at 7 days. Two other studies found no differences from placebo (Bogduk 2004).

Baclofen is significantly more effective than placebo in reducing pain, but the magnitude of the difference is about 10 % in absolute terms, and 20 % in relative terms, but applies only to assessment at 10 days after treatment (Bogduk 2004).

Cyclobenzaprine is more effective than placebo in so far as it achieved a reduction in pain, 9 days after treatment, of 5.5 points on a 10-point scale, compared with 4 points for placebo (Bogduk 2004). However, cyclobenzaprine is not more effective than NSAIDs (Bogduk 2004). For relieving pain and improving daily activities, it was slightly more effective than placebo at day 7 but not at day 14 (Bogduk 2004). A meta-analysis of studies that compared cyclobenzaprine with placebo (Browning et al. 2001) found that cyclobenzaprine substantially improved local pain and global symptoms when used for 4 days, but the effect declined considerably after the first week and was associated with a 25 % increase in side effects such as drowsiness, dry mouth, and dizziness compared to placebo. That meta-analysis, however, included studies that were rejected by a systematic review on the grounds that they did not explicitly address low back pain (Van Tulder et al. 2003).

These data indicate that for orphenadrine, diazepam, and tizanidine, the evidence is conflicting as to whether the drugs are more effective than placebo, with the balance favoring no superiority over placebo. Methocarbamol has not been shown to be more effective than placebo. Carisoprodol is more effective than placebo but only at 4 days. Baclofen and cyclobenzaprine are each only slightly more effective than placebo but only for a few days, but not thereafter. Meanwhile, the evidence is strong that muscle relaxants have consistently been associated with a greater incidence of central nervous system side effects (Van Tulder et al. 2003). This factor, balanced against their limited attributable effect, gives cause to question the propriety of their use for acute low back pain, especially when other interventions are no less effective, or even more effective, with far less risk of troublesome side effects.

For the treatment of chronic low back pain, the data on muscle relaxants are limited to only a few studies of a few agents. Agents used for acute low back pain have not been studied for chronic low back pain. Those that have been studied show limited or no superiority of placebo (Bogduk 2004).

Two studies each showed that tetrazepam was significantly more effective than placebo in reducing pain at 7 days, but not subsequently (Bogduk 2004). Flupirtine was perceived by physicians as achieving better overall improvement than placebo at 7 days, but improvements in pain or muscle spasm were not significantly better (Bogduk 2004). In terms of global impression of efficacy, tolperisone was not significantly better than placebo (Bogduk 2004).

For the treatment of neck pain, diazepam and phenobarbital have been tested and found to have no greater effect than placebo for the treatment of acute neck pain (Basmajian 1983).

References

Andersson, G., Bogduk, N., Deluca, C., et al. (1989). Muscle: Clinical perspectives. In J. W. Frymoyer & S. L. Gordon (Eds.), *New perspectives on low back pain* (pp. 293–334). Park Ridge: American Academy Of Orthopaedic Surgeons.

Basmajian, J. V. (1983). Reflex cervical muscle spasm: Treatment by diazepam, phenobarbital or placebo. *Archives of Physical Medicine and Rehabilitation, 64,* 121–124.

Bogduk, N. (2004). Pharmacological alternatives for the alleviation of back pain. *Expert Opin Phamacother, 5,* 2091–2098.

Browning, R., Jackson, J. L., & O'Malley, P. G. (2001). Cyclobenzaprine and back pain: A meta-analysis. *Archives of Internal Medicine, 161,* 1613–1620.

Johnson, E. W. (1989). The myth of skeletal muscle spasm (editorial). *American Journal of Physical Medicine, 68,* 1.

Van Tulder, M. W., Touray, T., Furlan, A. D., et al. (2003). Muscle relaxants for nonspecific low back pain: A systematic review within the framework of the cochrane collaboration. *Spine, 28,* 1978–1992.

Muscle Scanning

Definition

A form of psychophysiological assessment that involves making multiple bilateral recordings of muscle tension levels from varied locations.

Cross-References

▶ Biofeedback in the Treatment of Pain
▶ Psychophysiological Assessment of Pain

Muscle Spasm

▶ Chronic Back Pain and Spinal Instability

Muscle Tension

Definition

Muscle tension refers to the level of tension in a muscle that is usually measured by electromyography and is related to the amount of contraction in muscle fibers.

Cross-References

▶ Muscle Cramps
▶ Respondent Conditioning of Chronic Pain

Muscular Cramps

Mariano Serrao
Department of Neurology and Otorinolaringoiatry, University of Rome La Sapienza, Rome, Italy

Synonyms

Muscle cramps

Definition

▶ Muscle cramps are "spasmodic, painful, involuntary, contractions of the skeletal muscle" (Layzer and Rowland 1971). They represent one of the most common medical complaints. Cramps sometimes occur spontaneously at rest, especially when the muscle is slack (i.e., relaxed and shortened). More often, cramps are triggered by a brief contraction or during exercise of a susceptible muscle (Layzer and Rowland 1971). They can occur in any muscle of the body, though the foot and calf muscles are more frequently involved (Jansen et al. 1991). During the cramp, the muscle is visibly and palpably taut and painful, often with abnormal posture of the affected joint, a condition which can be relieved by stretching or massage (Rowland 1985).

Characteristics

The mechanism of muscle cramps is still speculative, because both their expression and their association with numerous diseases vary. Thus, their pathophysiology awaits clarification. A clear etiological distinction should be made between cramps generically named "▶ contractures" present in metabolic myopathies (e.g., McArdle's disease) and cramps associated with most other pathologies, usually named "▶ ordinary cramps." The former type of cramp is characterized by an absence of activity ("electrical silence") on EMG recordings, suggesting a muscular origin (Layzer 1994); in most cramps, however, the EMG shows brief periodic bursts of high-frequency, high-voltage action potentials (Norris et al. 1957), suggesting a neural origin.

The origin of the pain associated with muscle cramps is not known, and there are no consistent studies that have investigated this aspect. The pain present during the muscle cramp seems to be strikingly associated to the muscular contraction, which probably produces a mechanical stimulation of the group III or IV intramuscular fibers (Layzer 1985). Other factors such as an ischemia produced during the muscle contraction, an increase in muscular metabolites, or a direct lesion of the muscular fibers have been hypothesized as well.

Muscle cramps have been classified according to their anatomic site of origin, cause, or clinical features (Layzer and Rowland 1971; McGee 1990; Rowland 1985). Recently, we have proposed a new classification (Parisi et al. 2003) based primarily on their site of pathogenesis and on Layzer's definition of muscle cramp (see below). In accordance with such a definition, the muscle contraction of the cramp must be painful, which therefore excludes involuntary muscle contractions of various origins which are not painful, such as myotonia, stiffness, dystonia, and various movement disorders, e.g., myoclonias, chorea, tremors, tics. Furthermore, the muscle contraction must be sudden and sustained for a period of time ranging from a few second to a few minutes. Any pathological conditions such as antalgic contractures, which develop after a muscle injury and last for several days or weeks, should be excluded from this classification because they, unlike cramps, are neither sudden nor transient. In addition, muscular, articular, or nervous disorders that induce strong, localized pain but are not associated with a real muscle contraction (e.g., neuralgia, contusion, myalgia) should be distinguished from cramps.

Classification of Muscle Cramps
Paraphysiological Cramps
- Occasional cramps
- Cramps during sporting activity
- Cramps during pregnancy

Idiopathic Cramps
- Familial
  - Autosomal dominant cramping disease
  - Familial nocturnal cramps
  - Continuous muscle fibers activity syndrome
- Sporadic
  - Continuous muscle fiber activity syndrome (Isaac's syndrome, stiff-man syndrome, cramp-fasciculation syndrome, myokymia-cramp syndrome)
  - Syndrome of progressive muscle ▶ spasm, alopecia, and diarrhea (Sathoyoshi's syndrome)
  - Idiopathic nocturnal cramps
  - Idiopathic generalized myokymia
  - Myokymia-hyperhidrosis syndrome

- Others (familial insulin resistance with acanthocytosis nigricans and acral hypertrophy, muscle cramps in cancer patients)

Symptomatic Cramps
- Central and peripheral nervous system diseases
  - Motor neuron disease
  - Occupational dystonias
  - Parkinson's disease
  - Tetanus
  - Multiple sclerosis
  - Radiculopathies
  - Plexopathies
  - Peripheral neuropathies (inherited, endocrine-metabolic, infectious, toxic, inflammatory, demyelinating)
  - Other rare diseases (neurolathyrism, familial paroxysmal dystonic choreoathetosis)
- Muscular diseases
  - Metabolic myopathy (deficiency of myophosphorylase, phosphofructokinase, phosphoglyceromutase, phosphoglycerokinase, lactate dehydrogenase (LDH), adenylate deaminase, G6PDH, phosphorylase b-kinase)
  - Mitochondrial myopathy (carnitine deficiency, CPT1 e 2 deficiency)
  - Endocrine myopathy (Hoffman's syndrome)
  - Dystrophinopathies (Duchenne, Becker, others)
  - Myotonia (Thomsen, Becker, rippling syndrome)
  - Inflammatory myopathies (myositis, myopathy with tubular aggregates, rheumatic polymyalgia)
  - Other rare diseases (Lambert-Brody's diseases, Schwartz-Jampel syndrome, eosinophilia-myalgia syndrome, type two muscle fiber myopathy)
- Cardiovascular diseases
  - Venous diseases
  - Arterial diseases
  - Heart diseases
  - Hypertension
- Endocrine-metabolic disease
  - Hypo-hyperthyroidism
  - Hypo-hyperparathyroidism
  - Cirrhosis
  - Isolated deficiency of ACTH accompanied by generalized painful muscle cramp
  - Bartter's syndrome
  - Gitelman's syndrome
  - Conn's disease
  - Addison's disease
  - Uremia and dialysis
- Hydro-electrolyte disorders
  - Dehydration with or without electrolytes imbalance (diarrhea, vomiting, etc.)
  - Hypo-hypernatremia
  - Hypo-hypercalcemia
  - Hypo-hyperkalemia
  - Hypomagnesemia
  - Heat cramps
- Toxic and pharmacological causes
  - Drugs
  - Pesticides
  - Black widow bite
  - Toxic oil syndrome
  - Malignant hyperthermia

Psychiatric Disorders By contrast, all the true "sudden, involuntary, and painful muscular contractions" are included in the definition of muscle cramps. Thus, many diseases in which the cramp is not well recognized or in which the cramp is not the principal symptom, such as multiple sclerosis, Parkinson's disease, and myotonia with painful spasms, have been recognized as causes of muscle cramps.

▶ Paraphysiological cramps are characterized by the occurrence of cramps in healthy subjects during particular conditions such as pregnancy or exercising. We have named occasional cramps (Parisi et al. 2003) those cramps that occasionally occur in healthy subjects in the absence of exercising or pregnancy. In this case, a strong muscular contraction or a sustained abnormal posture may lead to muscle cramps.

▶ Idiopathic cramp is the main symptom of a disease about which little is known, the cause of the cramp being obscure or speculative. Idiopathic cramps can be either sporadic or inherited and are not usually associated with any cognitive, pyramidal, cerebellar, or sensory abnormalities.

Muscular Cramps, Table 1 Drugs inducing muscle cramps

Hypocholesterolemics	Clofibrate
	Fenofibrate
	Bezafibrate
Antiblastics	Vincristine
Antihypertensive	Diuretics
	Beta-blockers
	Angiotensin-converting-enzyme inhibitors, calcium channel blockers
Hormones	Medroxyprogesterone acetate
	Testosterone
	Estrogens and progesterone
Beta agonist	Terbutaline
	Salbutamol
	Pindolol
Others	Penicillamine
	Insulin

In the case of the familial forms, several authors have described an autosomal dominant cramping disease (Jacobsen et al. 1986; Lazaro et al. 1981; Ricker and Moxley 1990; Van den Bergh et al. 1980). Among the idiopathic forms, continuous muscle fiber activity syndrome and other rare diseases show frequent muscle cramps (Arimura et al. 1997; Solimena et al. 1988).

▶ Symptomatic cramps include cramps due to several types of diseases. Both central and peripheral nervous system pathologies as well as numerous muscular diseases are associated with muscle cramps (Parisi et al. 2003). Moreover, muscular cramps are frequent in patients affected by vascular diseases, endocrinopathies, electrolyte imbalance, and psychiatric disorders. Furthermore, toxic and pharmacological causes (Table 1) can induce muscle cramps (Parisi et al. 2003).

References

Arimura, K., Watanabe, O., Kitajima, I., et al. (1997). Antibodies to potassium channels of PC12 in serum of Isaacs' syndrome: Western blot and immunohisto-chemical studies. *Muscle & Nerve, 20,* 299–305.

Jacobsen, J. H., Rosenberg, R. S., Huttenlocher, P. R., et al. (1986). Familial nocturnal cramping. *Sleep, 9,* 54–60.

Jansen, P. H. P., Joosten, E. M., Van Dijck, J., et al. (1991). The incidence of muscle cramps. *Journal of Neurology, Neurosurgery and Psychiatry, 54,* 1124–1125.

Layzer, R. B. (1985). Diagnosis of neuromuscular disorders. In R. B. Layzer (Ed.), *Neuromuscular manifestation of systemic disease* (pp. 19–22). Philadelphia: Davis.

Layzer, R. B. (1994). Muscle pains, cramps and fatigue. In A. G. Engel & A. C. Franzini (Eds.), *Myology* (pp. 3462–3497). New York: McGraw-Hill.

Layzer, R. B., & Rowland, L. P. (1971). Cramps. *The New England Journal of Medicine, 285,* 31–40.

Lazaro, R. P., Rollinson, R. D., & Fenichel, G. M. (1981). Familial cramps and muscle pain. *Archives of Neurology, 38,* 22–24.

McGee, S. R. (1990). Muscle cramps. *Archives of Internal Medicine, 150,* 511–518.

Norris, F. H., Gasteiger, E. L., & Chatfield, P. O. (1957). An electromyographic study of induced and spontaneous muscle cramps. *Electroencephalography and Clinical Neurophysiology, 9,* 139–147.

Parisi, L., Amabile, G., Valente, G., et al. (2003). Muscular cramps: A new proposal of classification. *Acta Neurologica Scandinavica, 107,* 176–186.

Ricker, K., & Moxley, R. T. (1990). Autosomal dominant cramping disease. *Archives of Neurology, 47,* 810–812.

Rowland, L. P. (1985). Cramps, spasms and muscle stiffness. *Revue Neurologique, 141,* 261–273.

Solimena, M., Folli, F., Denis-Donini, S., et al. (1988). Autoantibodies to glutamic acid decarboxylase in a patient with stiff-man syndrome, epilepsy, and type I diabetes mellitus. *The New England Journal of Medicine, 318,* 1012–1020.

Van den Bergh, P., Bulcke, J. A., & Dom, R. (1980). Familial muscle cramps with autosomal dominant transmission. *European Neurology, 19,* 207–212.

Muscular Rheumatism

▶ Fibromyalgia
▶ Myalgia

Musculoskeletal Disorders

Definition

Musculoskeletal disorders are a descriptive term used to label pain associated with repetitive activities.

Cross-References

▶ Disability, Upper Extremity

Musculoskeletal Examination

Wade King
Department of Clinical Research, Royal
Newcastle Centre University of Newcastle,
Newcastle, NSW, Australia

Definition

Musculoskeletal examination means physical examination of the musculoskeletal system.

Characteristics

By tradition, physical examination complements the medical history in clinical assessment. Its main purpose is to elicit objective information about the index condition. It can also foster rapport and enhance the doctor-patient relationship.

The three elements of physical examination are described in an aphorism by Apley (Apley and Solomon 1993) as "look, feel, move," or:

- Inspection
- Palpation
- Movement testing

These elements may be tested in any order. The order suggested by Apley is appropriate for fractures. The sequence "look, move, feel" is more suitable for many mechanical musculoskeletal conditions. The order of testing is at the examiner's discretion and does not matter as long as all three elements are assessed.

Inspection
Observations made by visual inspection of the patient are general and specific. General observations include dynamic posture (gait and other gross movements), static posture (standing and sitting), and bodily deformity. Specific observations of the affected region include skin discoloration, scars, swelling, and other local abnormalities of form or shape.

Palpation
The main finding on palpation is tenderness: It may be focal or diffuse. Other palpatory signs include alterations of skin sensitivity, apparent alteration of soft tissue conformation, and abnormality of bony landmarks.

Movement Testing
Movements are assessed by testing active, passive, and accessory ranges, and challenging restraints to movement.

Ranges of Movement
Active ranges are assessed by the patient assuming a neutral position and then moving the body part in question in each direction as far as they can. For the spine, the standard movements are extension, flexion, side-bending to left and right, and rotation to left and right. For peripheral joints, they are extension, flexion, abduction, adduction, external rotation, and internal rotation. Conventions have been set (Greene and Heckman 1994; Russe et al. 1976) for these tests. Ranges are estimated visually or measured with a goniometer (an instrument like a protractor with arms) or an inclinometer (a device with gravitational reference and a dial showing degrees). The examiner records any limitation of range and any association with pain.

Passive movements are tested through the same ranges with the patient remaining relaxed and the examiner supplying the effort to move the patient's body parts passively.

Accessory movements are movements that typically cannot be performed actively in isolation by a patient but which can be elicited passively by an examiner. They can be translations or rotations, and their ranges are usually small. They are tested by fixing one body part and moving the other relative to it. The accessory movements of a joint all contribute to "joint play."

Challenging Restraints
Restraints to movement include bones, joint capsules, ligaments, tendons, and muscles. They are tested actively by the patient moving a body segment as far as possible. They are tested passively by the examiner moving a joint through its ranges and accessory movements. During testing, the examiner makes a judgment as to what may be limiting the movement. This may be the onset of

pain, an obstruction by bone or soft tissue, tethering by tight ligaments or fascia, or tension in antagonist muscles.

Movements may be limited or impeded before terminal restraints are engaged, if structures adjacent to the moving one produce pain or interfere with movement. A "painful arc" is part of a range through which movement is associated with pain (Kessell and Watson 1977). The "impingement sign" relates to the shoulder. The sign is considered positive if, during flexion of the shoulder, movement is limited by pain as the humeral head and acromion move closer and impinge on tissues between them, but the pain and limitation is abolished after subacromial injection of lignocaine (Neer 1972). The "apprehension sign" occurs when the patient exhibits apprehension as the examiner starts to test restraints to a particular movement. It is said to signify instability, in particular, of a glenohumeral joint (Blazina and Satzman 1969) or a patella (Hughston 1972).

Many other tests of restraints have been developed for spinal segments and peripheral joints. They are described in articles and textbooks dealing with disorders of the various regions of the musculoskeletal system.

Application
Musculoskeletal examination is often overlooked. A study of patients in a teaching hospital (Ahern et al. 1991) showed 55 % had musculoskeletal symptoms and 18 % significant musculoskeletal disorders; yet, musculoskeletal examination was performed on only 15 % compared to examination of the respiratory system on 100 % and cardiovascular examination on 99 %.

Although musculoskeletal examination is often not done, or is done but in a cursory manner, many clinicians perform a **neurological examination** to assess musculoskeletal problems, probably because of confusion about mechanisms of pain. Low back pain and neck pain in particular are widely considered as commonly due to impingement on spinal nerves, even though such beliefs are at variance with the evidence base. There is no good reason for neurological

examination of a patient with musculoskeletal pain unless they also have neurological symptoms. If so, the neurological condition should be addressed first and the pain dealt with when the neurological issues have been clarified.

Clinical Utility
The diagnostic utility of physical tests depends on data from well-designed studies, of which there have been many in recent years. In general, despite the traditional use of musculoskeletal examination, formal studies have shown that few tests have **reliability** and **validity** (Australian Acute Musculoskeletal Pain Guidelines Group 2003). Space permits only a few examples to be outlined here.

Evidence of Reliability
Inspection
The reliability of inspection of any musculoskeletal region is unknown as there are no relevant data.

Palpation
Studies have shown high **reliability** (kappa 0.80–1.0) of eliciting tenderness at nonspecific sites around a painful shoulder or lower back, but the **validity** of such nonspecific tenderness is unknown (Australian Acute Musculoskeletal Pain Guidelines Group 2003). Much lower **reliability** (kappa 0.11–0.38) has been shown for finding focal tenderness in the lumbar region (Australian Acute Musculoskeletal Pain Guidelines Group 2003) and its significance is also uncertain. There are no data on the reliability of focal tenderness in other regions (Australian Acute Musculoskeletal Pain Guidelines Group 2003).

Movement Testing
Ranges of Movement Visual estimations of ranges of shoulder movement are of varying reliability (Australian Acute Musculoskeletal Pain Guidelines Group 2003). Goniometry confers no advantage (Australian Acute Musculoskeletal Pain Guidelines Group 2003). Inclinometry seems better if done by trained practitioners but

is not uniformly reliable (Australian Acute Musculoskeletal Pain Guidelines Group 2003).

Challenging Restraints Some tests of restraints to movement have reasonable reliability for various joints and regions of the body, but for others, reliability is poor or lacking (Australian Acute Musculoskeletal Pain Guidelines Group 2003). This means clinicians can be trained to perform physical tests in similar ways and sometimes achieve consistent results, but it does not mean that the findings have diagnostic significance. That depends on validity data.

Evidence of Validity
Inspection
There are no data on the **validity** of inspection of any part of the musculoskeletal system (Australian Acute Musculoskeletal Pain Guidelines Group 2003).

Palpation
There are no data showing the **validity** of tenderness in any musculoskeletal region (Australian Acute Musculoskeletal Pain Guidelines Group 2003).

Movement Testing
Ranges of Movement There are no data on the validity of testing gross ranges of movement of musculoskeletal structures, so the utility of many such tests is unknown (Australian Acute Musculoskeletal Pain Guidelines Group 2003).

There are data on the testing of passive and accessory intersegmental movements of the cervical zygapophysial joints for the diagnosis of neck pain. The results of one study (Jull et al. 1988) published in 1988 seemed to show that manual examination of the neck based on testing passive and accessory intersegmental movements had 100 % sensitivity and 100 % specificity for identifying painful zygapophysial joints. Those results have been quoted as the rationale for much that has been done in manual therapy and similar fields since. However, the study was small and its authors suggested its results should be reproduced before they could be generalized. The

results of a larger study (King et al. 2007) published in 2007 showed manual examination had a high sensitivity for cervical zygapophysial joint pain, at the segmental levels commonly symptomatic, but its specificity was poor and likelihood ratios barely greater than 1.0 indicated that manual examination lacks validity. Although these results were contrary to those of the previous study, there was a paradoxical lack of statistical difference between the two studies, accounted for by the small sample size of the earlier study.

Challenging Restraints The studies of examination of the cervical spine quoted above included challenging the restraints to movement of the zygapophysial joints and showed the tests were not valid for the diagnosis of neck pain. There are no data on the validity of such tests applied in other regions of the spine.

Evidence on the validity of tests that challenge restraints to movement of peripheral joints is inconsistent. It is exemplified by data relating to the shoulder, on which there have been many studies. One study (Çaliş et al. 2000) of physical examination of the shoulder found "the highly sensitive tests seem to have low specificity values and the highly specific ones to have low sensitivity." This is reflected in the low positive likelihood ratios of all individual tests. Other investigators (MacDonald et al. 2000; Naredo et al. 2002) have reported similarly, even for combinations of tests.

Cross-References

▶ Lumbar Medial Branch Blocks

References

Ahern, M. J., Soden, M., Schulz, D., & Clark, M. (1991). The musculo-skeletal examination: A neglected skill. *Australian and New Zealand Journal of Medicine, 21*, 303–306.

Apley, A. G., & Solomon, L. (Eds.). (1993). *Apley's system of orthopaedics and fractures* (7th ed., pp. 8–12). Oxford: Butterworth-Heinemann, 261–263.

Australian Acute Musculoskeletal Pain Guidelines Group. (2003). *Evidence-based management of acute musculoskeletal pain*. Brisbane: Australian Academic Press. http://www.nhmrc.gov.au

Blazina, M. E., & Satzman, J. S. (1969). Recurrent anterior subluxation of the shoulder in athletics: A distinct entity. *Journal of Bone and Joint Surgery, 51A*, 1037–1038.

Çaliş, M., Akgün, K., Birtane, M., Karacan, I., Çaliş, H., & Tüzün, F. (2000). Diagnostic values of clinical diagnostic tests in subacromial impingement syndrome. *Annals of the Rheumatic Diseases, 59*, 44–47.

Greene, W. B., & Heckman, J. D. (Eds.). (1994). *The clinical measurement of joint motion*. Rosemont: American Academy of Orthopaedic Surgeons.

Hughston, J. C. (1972). Subluxation of the patella in athletes. In *Symposium on sports medicine* (p. 162). St. Louis: CV Mosby.

Jull, G., Bogduk, N., & Marsland, A. (1988). The accuracy of manual diagnosis for cervical zygapophysial joint pain syndromes. *The Medical Journal of Australia, 148*, 233–236.

Kessell, L., & Watson, M. (1977). The painful arc syndrome: clinical classification as a guide to management. *Journal of Bone and Joint Surgery, 59B*, 166.

King, W., Lau, P., Lees, R., & Bogduk, N. (2007). The validity of manual examination in assessing patients with neck pain. *The Spine Journal, 7*, 22–26.

MacDonald, P. B., Clark, P., & Sutherland, K. (2000). An analysis of the diagnostic accuracy of the Hawkins and Neer subacromial impingement signs. *Journal of Shoulder and Elbow Surgery, 9*, 299–301.

Naredo, E., Aguado, P., De Miguel, E., Uson, J., Mayordomo, L., Gijon-Baños, J., & Martin-Mola, E. (2002). Painful shoulder: comparison of physical examination and ultrasonographic findings. *Annals of the Rheumatic Diseases, 61*, 132–136.

Neer, C. S. (1972). Anterior acromioplasty for the chronic impingement syndrome of the shoulder. *Journal of Bone and Joint Surgery, 54A*, 41.

Russe, O., Gerhardt, J. J., & King, P. S. (1976). *An atlas of examination, standard measurements and diagnosis in orthopaedics and traumatology* (2nd ed.). Berne: Hans Huber.

Musculoskeletal Pain

Lars Arendt-Nielsen
Center for Sensory-Motor Interaction,
Department of Health Science and Technology,
School of Medicine, Aalborg University,
Aalborg, Denmark

Characteristics

Manifestations of Musculoskeletal Pain

Musculoskeletal pain is manifested by four main components: (1) the effect on the motor system, (2) local and referred pain, (3) increased or

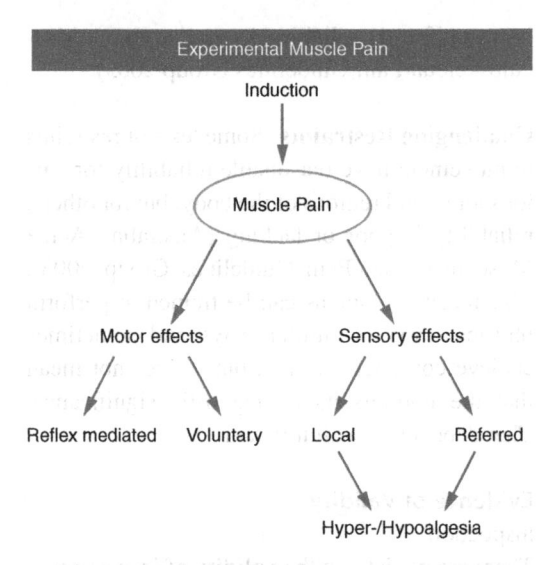

Musculoskeletal Pain, Fig. 1 A sketch of how musculoskeletal pain can be manifested. (1) Muscle pain has effects on the motor system. This involves effects on voluntary contractions, endurance, and muscle coordination and on reflex mediated pathways (e.g., stretch reflexes, proprioception, and motor unit firing characteristics). (2) Muscle pain may result in local as well as referred pain areas. (3) In the local muscle pain area, the muscle may have changed sensibility (e.g., hypersensitive to pressure). (4) In the referred area, various somatosensory changes can take place

decreased muscle sensitivity, and (4) somatosensory changes in the referred pain area (Fig. 1).

In contrast to the localized sharp, burning characteristics of cutaneous pain, muscle pain is described as diffuse aching and cramping. Furthermore, musculoskeletal pain is more complicated to investigate than cutaneous and visceral pain, as muscle pain can consist of both localized pain and referred pain (Fig. 2).

Pain localization is poor in skeletal muscles and patients may be unable to differentiate it from pain arising from tendons, ligaments, and bones as well as from joints and their capsules. The characteristically referred pain pattern was initially observed by Kellgren (1938), who injected hypertonic saline into many skeletal muscles and ligaments and characterized the referred pain patterns (Fig. 3). Similar characterization has been performed clinically when activating ▶ trigger points in various muscles (Travell and Simons 1992).

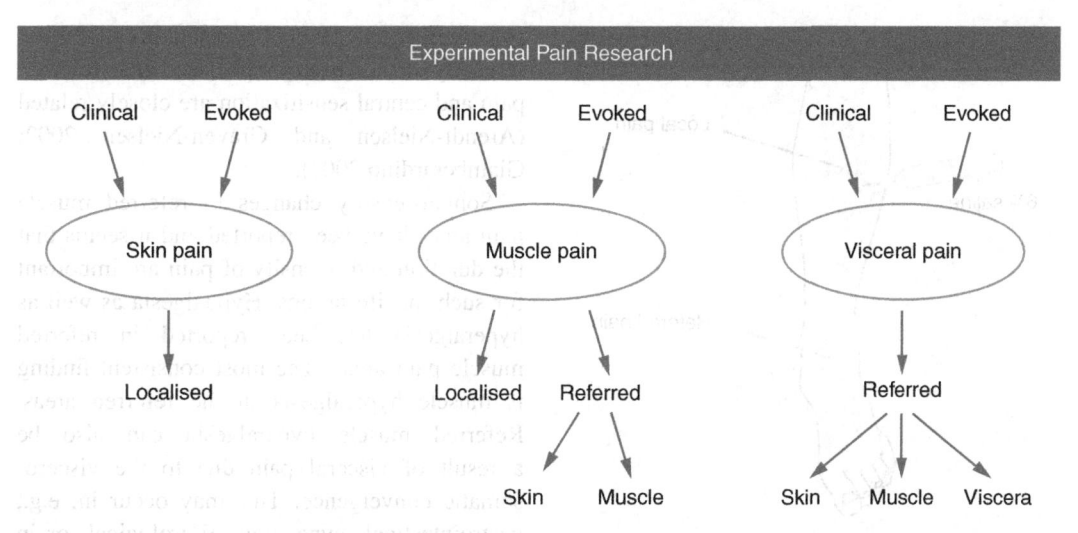

Musculoskeletal Pain, Fig. 2 A simple sketch to illustrate the different characteristics and manifestations of
cutaneous, muscle, and visceral pain. Cutaneous pain is
sharp and localized, whereas pain from deep structures is
more diffuse. Pain from muscles can be localized to

a specific muscle, but can also be referred to another
somatic structure. In this referred structure, changes in
somatosensibility can occur. Visceral pain is always
referred either to somatic structures or via viscero-visceral
convergent neurons

It is obviously important to distinguish the
painful tissue, but it may be very difficult,
due to poor localization and ▶ referred pain.
Examples can be pain from an arthritic hip,
which may refer to the thigh muscles or knee
joint; a carpal tunnel syndrome, which may
refer to forearm muscles; and cervical
spondylosis, which may refer to arm muscles.
Pain from joints and their capsules tends to be
more localized than ▶ myalgia and arthralgia is
often worsened by passive joint movements.
Capsular pain may be present only in specific
joint positions. Bone pain also tends to be poorly
localized but, unlike myalgia, usually has a deep,
boring quality. Furthermore, bone pain is usually
worse at night and tends to be unaffected by
either movement or muscle activity (Newham
et al. 1994). The manifestations related to
musculoskeletal pain can be projected pain,
spread of pain and referred pain, somatosensory
changes in referred pain areas (Vecchiet et al.
1999), and interaction with the motor system
(e.g., muscle coordination and activation,
postural stability, movement initiation, and reflex
pathways). Projected muscle pain is normally
used as terminology in relationship to nerve

fiber damage (e.g., compression) and felt in the
innervation territory/myotome of the damaged
nerve.

Spread of muscle pain is not clearly defined,
but best described by phenomena related to
experimental studies. If the muscle nociceptive
afferent is repetitively, electrically stimulated by
an intramuscular electrode, the area where the
muscle pain is experienced expands or spreads
as the stimulation progresses (Laursen et al.
1997). A similar manifestation is seen after
repetitive visceral electrical stimulation
(Arendt-Nielsen et al. 1997).

Referred muscle pain has been known and
described for more than a century and is used
extensively as a diagnostic tool. Substantial
clinical knowledge exists concerning the patterns
of referral from various skeletal muscles and after
activation of trigger points (Travell and Simons
1992). The referred pain pattern follows the distribution of sclerotomes (muscle, fascia, and bone)
more frequently than the classical dermatomes.
A clear distinction between spread of pain and
referred pain is not possible and these
phenomena may also share common pathophysiological mechanisms. Firm neurophysiologically

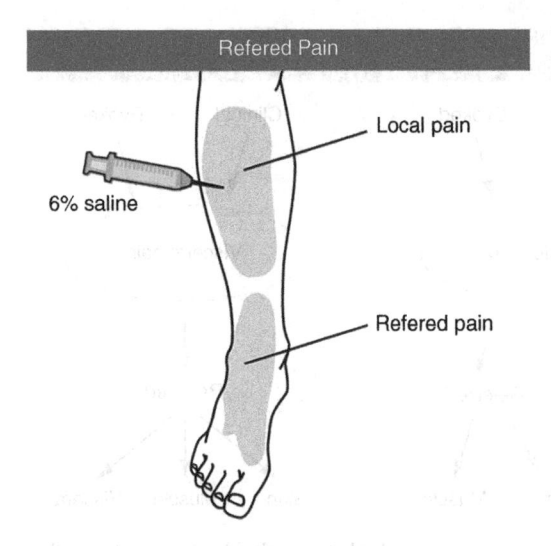

Musculoskeletal Pain, Fig. 3 Examples of local and referred pain patterns in volunteers following intramuscular injections of hypertonic saline (6 %, 0.5 ml) into the anterior tibialis muscle. If the area of referred pain is completely blocked by anesthetic procedures (e.g., a regional block), the referred pain can still be elicited, indicating that referred pain is a central phenomenon. The size of the referred pain area depends on the intensity and duration of the actual muscle stimulus

based explanations for referred pain do not exist, but it has been shown that wide dynamic range neurons and nociceptive specific neurons in the spinal cord and in the brain stem of animals receive convergent afferent input from the mucosa, skin, muscles, joints, and viscera. This may cause a misinterpretation of the afferent information coming from muscle afferents and reaching high levels in the central nervous system and hence be one reason for the diffuse and referred characteristics.

Referred pain is mainly a central phenomenon as it is possible to induce referred pain to limbs with complete sensory loss due to spinal injury or anesthetic blockade. The size of the referred area to experimental muscle pain stimulation depends on the intensity and duration of the nociceptive musculoskeletal input and the evoked referred area is significantly enlarged in patients with painful musculoskeletal disorders (fibromyalgia, whiplash, low back pain and osteoarthritis) (Arendt-Nielsen and Svensson 2001; Koelbaek

Johansen et al. 1999; Sorensen et al. 1998). Recently, it has been emphasized that referred pain and central sensitization are closely related (Arendt-Nielsen and Graven-Nielsen 2002; Giamberardino 2003).

Somatosensory changes in referred muscle pain areas have been reported and it seems that the duration and intensity of pain are important for such manifestations. Hypoalgesia as well as hyperalgesia has been reported in referred muscle pain areas. The most consistent finding is muscle hyperalgesia in the referred areas. Referred muscle hyperalgesia can also be a result of visceral pain due to the viscero-somatic convergence. This may occur in, e.g., gastrointestinal, gynecological/urological, or in chronic visceral painful conditions without known etiology (e.g., irritable bowel diseases and endometriosis). It has been shown that the degree of referred muscle hyperalgesia is related to the severity of the visceral pathology and hence the degree of visceral pain. Persistent referred muscle hyperalgesia can be manifested not only by chronic conditions, but also after recurrent painful visceral attacks, such as in dysmenorrheic women where lumbar muscles are hyperalgesic to pressure or after colic attacks following calculosis of the upper urinary tract, where hyperalgesia to pressure is found in muscles in the left lumbar region (Giamberardino 1999).

Exercise-Related Painful Musculoskeletal Conditions

Muscle pain in relation to exercise may increase in intensity until the exercise/contractions stop and the blood flow is restored. This is termed ischemic pain and is the cause of well-known clinical presentations such as intermittent claudication and angina pectoris. Hypoxia was thought to be the cause of ischemic pain, but it now seems more likely that accumulation of metabolites is responsible, at least in part. Lactic acid was assumed to be the prime algesic substance, but since patients with myophosphorylase deficiency (patients with McArdle's disease do not produce lactic acid) also experience ischemic muscle pain during ► exercise, other substances such as histamine, acetylcholine, serotonin, bradykinin, potassium, and adenosine are most likely involved.

Another manifestation of exercise may be ▶ muscle cramps. Muscle cramps are involuntary, painful, sudden contractions of the skeletal muscles and may also appear after overuse of a muscle. Cramps can occur in normal subjects under a variety of conditions (during a strong voluntary contraction, sleep, sports, and pregnancy) and in several pathologies such as myopathies, neuropathies, motoneurone diseases, in tetanus, metabolic disorders, hydroelectrolyte imbalances, or endocrine pathologies (e.g., glycolytic disorders), or as a side effect after certain drugs (salbutamol, phenothiazine, vincristine, lithium, cimetidine, and bumetanide). Considerable uncertainty is found in the literature regarding the classification and nomenclature of muscle cramps, both because the term "cramp" is used to indicate a variety of clinical features of muscles leading to its use as an imprecise "umbrella" term that includes stiffness, contractures, and local pain and because the spectrum of the diseases in which it appears is wide.

If a muscle is exposed to heavy and unaccustomed and especially eccentric exercise, delayed onset muscle pain may peak 1–2 days later. Delayed onset muscle pain/soreness (postexercise soreness) is associated with tenderness to pressure and pain during movement/contraction. The mechanisms underlying postexercise muscle pain seem different from those of ischemic muscle pain, as considerable damage such as structural, biochemical, and radioisotopic changes are found (Friden 1984). Post-exercise pain can be inhibited and treated by NSAIDs, suggesting that inflammatory mediators belonging to the arachidonic cycle may be involved in this kind of muscle pain.

Selected Painful Musculoskeletal Disorders

Myofascial pain syndromes are regional muscle pain disorders characterized by localized tenderness and pain and are common causes of persistent regional pain such as back pain, neck pain, shoulder pain, headaches, and orofacial pain. The affected muscles often display an increased fatigability, stiffness, subjective weakness, pain in movement, and slightly restricted range of motion that is unrelated to joint restrictions. The exact etiology of myofascial pain syndromes is unclear. The major characteristics of myofascial pain syndromes include tenderness in muscles (trigger points) and local and referred pain. A trigger point is an up to 0.5-cm diameter point of hypersensitivity to pressure in a palpable taut band of skeletal muscle, tendon, and ligament. Active trigger points are hypersensitive and display continuous pain in the zone of reference that can be altered with specific palpation. Latent trigger points display only hypersensitivity to pressure with no continuous pain. ▶ Tender points differ from trigger points in the sense that they are tender to pressure, do not cause referred pain by activation, and can be identified as 1 of 18 designated soft tissue body sites.

Widespread pain is defined as pain lasting for longer than 3 months, presenting in both sides of the body, above and below the waist. In addition, axial skeletal pain (cervical spine, anterior chest, thoracic spine, or low back pain) must be present. Widespread pain includes classes of syndromes such as fibromyalgia, chronic fatigue syndrome, and exposure syndromes (e.g., Gulf War Illnesses). Fibromyalgia is a chronic painful musculoskeletal disorder of unknown etiology and is defined by chronic widespread pain, involving three or more segments of the body plus the finding of at least 11 out of 18 designated tender points. Physical or emotional trauma, infection, and surgery have been reported anecdotally to be precipitating factors in fibromyalgia and it is not uncommon for the patients to report the onset of the syndrome in relation to an accident or an injury. Fibromyalgia and the ▶ whiplash syndrome therefore share some common features.

Studies of the endocrine profile of fibromyalgia patients have indicated elevated activity of corticotropin-releasing hormone (CRH) neurons, which could not only explain some symptoms of fibromyalgia, but may also explain alterations observed in the hormonal axes. Hypothalamic CRH neurons may play a role not only in resetting the various endocrine loops, but possibly also nociceptive and psychological mechanisms as well.

The whiplash syndrome is normally associated with car accidents where the car has been hit from the rear or the side and the persistent chronic pain is localized in the neck, shoulder, and back. Patients with fibromyalgia and whiplash show accentuated reactions to a variety of stimulus modalities, and in general, they are hyperalgesic to experimental muscle stimulation (pressure and infusion of algogenic substances). A feature in many of the chronic musculoskeletal disorders is that the referred pain areas to experimentally induced muscle pain are significantly larger than in pain-free volunteers. Both fibromyalgia (Carli et al. 2002) and whiplash (Curatolo et al. 2001) patients show differentiated hyperalgesia to different sensory stimuli, indicating that only specific parts of the sensory and nociceptive systems are influenced and that the sensory disturbances are accentuated as the syndrome progresses (Carli et al. 2002). In recent years, these generalized chronic musculoskeletal disorders together with work-related muscle-related disorders have become an increasing challenge for the health care system with an increased socioeconomic burden on societies.

Myalgia can be related to a variety of medical conditions (see list below) and common terms for the symptom are stiffness, soreness, aching, spasms, or cramps. The associated pain is often described as having a dull, aching quality and can be exacerbated by muscle contractions. The manifestation of pain is, however, in some of the myalgias not the most prominent problem, as in ▶ myositis (e.g., ▶ polymyositis and ▶ dermatomyositis) where muscle weakness is often the prominent feature.

Clinical Conditions Associated with Myalgia
- Trauma and sports injuries
- Primary infective myositis
- Inflammatory myopathies
  - Polymyositis with and without connective tissue disease
  - Dermatomyositis with and without connective tissue disease
  - Viral myositis
  - Polymyalgia rheumatica
- Myalgia of neurogenic origin

- Muscle cramp
- Impaired muscle energy metabolism
- Drug induced myalgia
- Myalgic encephalomyelitis/chronic fatigue syndrome
- Muscle pain of uncertain cause (repetitive strain injury, fibromyalgia)

Myalgia of neurogenic origin can be difficult to dissociate from other manifestations of neuropathic pain (Newham et al. 1994). Examples can be cervical radiculopathy with pain radiating into the myotomal distribution of the roots or nerve compressions (e.g., carpal tunnel syndrome) where the pain radiates into the muscles in the region. Little is known specifically related to muscle sensitization in relation to neuropathic pain, as normally only the cutaneous manifestations (allodynia, hyperalgesia) are investigated.

Physical Therapies for the Management of Musculoskeletal Pain

A wide range of physical therapies are traditionally used in the repertoire of management regimes offered to patients with chronic musculoskeletal disorders. Such therapies include, e.g., ice, heat, massage, acupuncture, and ▶ transcutaneous electrical nerve stimulation. The efficacy of these therapies is not known and has rarely been compared in controlled studies. Acupuncture, electroacupuncture, and transcutaneous electrical nerve stimulation have been applied in order to inactivate, e.g., trigger points, but again the efficacy is not known. The advantage of, e.g., transcutaneous nerve stimulation is that the patients feel that they have some control when they can take the device home and switch it on when needed.

A variety of physiotherapeutic management regimes are developed for different musculoskeletal disorders and, e.g., the Quebec task force reviewed available literature on the treatment employed by physical therapists in the management of whiplash injuries (Spitzer et al. 1995). Only two of the commonly employed physiotherapy modalities (exercise, mobilization) showed an effect. Unfortunately, only a few of the physical therapies commonly used in the clinic for the management of musculoskeletal pain are scientifically validated. However, exercise with instruction from physical therapists seems to be the modality most often resulting in good outcome.

Gender Differences in Musculoskeletal Pain
Many musculoskeletal disorders and pains are more prevalent in females than males. Sex (referring to biologically determined aspects of femaleness and maleness) and/or gender (referring to modifiable, socioculturally shaped behavior and traits such as femininity and masculinity) differences are present, with female predominance in painful musculoskeletal syndromes such as widespread pain, temporomandibular disorders, neck pain, shoulder pain, back pain, joint pain, fibromyalgia, whiplash, and headache. Psychological factors operative in chronic painful musculoskeletal disorders cannot be understood within any single frame of reference. It is appropriate to examine emotional, interpersonal pain behavior and interpersonal relationships in parallel, not as dichotomous concepts. High rates of comorbidity have been reported between, e.g., temporomandibular dysfunctions and other clinical musculoskeletal disorders (e.g., fibromyalgia) and are again more common in women than men.

Fluctuations in hormonal levels have been implicated in symptom severity in women with rheumatoid arthritis, temporomandibular disorders, and fibromyalgia. One mechanism by which hormones may affect muscle nociceptors sensitization could be related to nerve growth factor (NGF) and one of its high-affinity receptors (trkA). The trkA receptor expression is influenced by gonadal hormones (Liuzzi et al. 1999). Injection of NGF into the muscle tissue causes muscle tenderness to pressure which lasts for weeks (Svensson et al. 2003). There has been some speculation that hormone replacement therapy may increase a woman's risk of developing musculoskeletal pain (LeResche et al. 1997). Numerous studies indicate that women have greater sensitivity to experimental muscle pain stimuli than men and the pressure pain threshold is generally lower in females. There are indications in the literature that males and

females have different results from some drugs. For musculoskeletal pain, the nonsteroidal anti-inflammatory drugs (NSAIDs) are commonly used and some studies have indicated that females experience less pain alleviating effect than males.

Techniques to Assess Sensitivity of Musculoskeletal Structures
Several methods exist to assess the sensitivity of musculoskeletal structures. The methods are based on standardized application or induction of standardized pain to musculoskeletal structures to evaluate how sensitive the structure is to that specific stimulus modality. Such procedures can be applied to healthy volunteers in the laboratory for basic experimental studies or to patients for clinical examinations (Arendt-Nielsen 1997; Arendt-Nielsen et al. 1997). The standardized pain stimulus can be classified as endogenous pain models (the pain arises from muscles without the involvement of external stimuli) or exogenous models (the pain is evoked by external stimuli) (see list below).

Classification of Experimental Stimulus Modalities That Can Be Used to Induce Muscle Pain Experimentally

Endogenous
• Ischemic
• Exercise induced
  – Dynamic concentric contraction
  – Isometric contraction
  – Dynamic eccentric contraction

Exogenous
• Electrical
• Mechanical
• Chemical
• Focused ultrasound

Endogenous Models Ischemic muscle pain is a classically experimental pain model and has been used as an unspecified pain stimulus. A tourniquet is applied and after a period of voluntary muscle contractions, a very unpleasant tonic pain sensation develops. The number of contractions, the level of force, and the duration are important determinants for the resulting pain. This is a very efficient model to induce pain in muscles, but skin, periosteum, and other tissues will contribute. The model is applicable in experimental studies requiring a general tonic pain stimulus for, e.g., PET and fMRI experimental studies. ► Delayed Muscle Soreness is another model used experimentally to investigate endogenous muscle pain and this model is useful for, e.g., evaluating drug effect.

Exogenous Models Intraneural microstimulation of muscle nociceptive afferents can be performed in laboratory studies and can selectively elicit muscle pain accompanied by referred pain, which is dependent on the stimulation time (temporal summation) and the number of stimulated afferents (spatial summation). Repetitive intramuscular electrical stimulation can evoke localized and referred muscle pain. The local and referred pain areas develop during stimulation with the referred area slightly delayed as compared with the local pain area. The areas are dependent on stimulus intensity and duration (Laursen et al. 1997).

Intramuscular infusion of algogenic substances (e.g., hypertonic saline, capsaicin, bradykinin) causes local and referred pain. In most experimental human studies, manual bolus infusions of hypertonic saline are used, but more advanced models such as standardized infusion of small volumes by computer-controlled infusion pumps have been used. When this advanced model is used, the infusion rate can be controlled by the pain intensity rating feedback so that constant pain intensity can be achieved by continuous adjustment of the infusion rate. Muscle pain induced experimentally by intramuscular infusion of algogenic substances causes cutaneous and muscular sensibility changes in referred pain areas. It is not entirely conclusive which somatosensory changes occur as both hypo- and hyperesthesia have been reported. Furthermore, these changes are stimulus modality specific (Graven-Nielsen et al. 1997).

Pressure algometry is the most commonly used technique to induce muscle pain and hence assess tenderness in myofascial tissues and joints such as tender points, fibromyalgia, work-related myalgia, myofascial pain, strain injuries, myositis, chronic fatigue syndrome, arthritis/orthoses, and other muscle/tendon/joint inflammatory conditions (Fischer 1987a, b). The technique is adequate to quantify and follow the development of given diseases, but has also proven to be instrumental for the documentation of treatment outcome such as local/systemic administration of drugs. Pressure algometry is also suitable for the assessment of joint tenderness.

Assessing Sensory Aspects of Musculoskeletal Pain

Assessing the sensory aspects of musculoskeletal pain involves both evaluation of the muscle pain and of the somatic structures related to the referred pain area (Fig. 1). Clinical and experimentally induced muscle pain intensity and the implications on, e.g., physical performance can be assessed quantitatively by various measures (see list below).

Techniques for Assessment of Experimentally Induced Musculoskeletal Pain and for Assessing Clinical Musculoskeletal Pain
Sensory Characterization

Muscle Sensitivity
• Verbal assessment
  – Visual analogue scales
  – Verbal descriptor scales
  – McGill Pain Questionnaire
• Psychophysical tests
  – Pain thresholds
• Referred pain
  – Distribution of pain
  – Area (size)
  – Somatosensory changes

Motor Effects
• Electromyography
  – Surface electrodes
  – Needle electrodes
  – Indwelling wire electrodes

Musculoskeletal Pain,
Fig. 4 A schematic figure
indicating how the
psychophysical thresholds
related to pressure pain
(pressure algometry) are
measured. Using a pressure
algometer, it is possible to
control the rate of force
increase (e.g., 30 kPa/s to
a 1 cm^2 probe) and
determine when the
volunteer/patient experience
the pressure as just painful
(pain detection threshold)
and when he/she will not
tolerate any further increase
in stimulus intensity
(pain tolerance threshold)

• Kinetics and kinematics
  – Sirognathograph (jaw movements)
  – Optoelectronic devices
• Force
  – Dynamometers
  – Force platforms
  – Bite force meters

Musculoskeletal pain has implications in
many aspects of daily life and questionnaires for
assessment of different dimensions have been
developed for general and regional (e.g., back
pain and neck pain) pain problems (general function score, Roland and Morris disability scale,
▶ Oswestry disability index, West Haven-Yale
multidimensional pain inventory, Bournemouth
questionnaire, ▶ fear avoidance beliefs, life
satisfaction) (Turk and Melzack 1992).
A simple and useful technique is pain drawings
where, e.g., tender points, trigger points, localized and referred pain areas, hyper- and
hypoalgesia are marked. Different colors can
eventually be used to characterize different
manifestations.

The psychophysical parameters used for the
quantification of musculoskeletal pain sensitivity
to experimentally applied pressure (pressure
algometry) are pain detection, pain tolerance
thresholds (Fig. 4), and stimulus-response functions. Stimulus-response functions can provide

more information than just a threshold, as sensitization to low as well as high intensities can be
assessed and a shift in parallel of the curve toward
the left together with an increased slope has been
found in patients with myofascial pain.

For monitoring purposes, quantitative parameters are advantageous as compared with manual
palpation. The essence of pressure algometry is
that standardized increase in pressure is applied to
the part of the body that is being investigated and
the outcome is the patient's or volunteer's reaction
to the pressure (Fischer 1987a, b). Pressure rate and
pressure area have been shown to be important
factors for reliable results. The pressure pain
thresholds vary substantially between regions. In
some of the commercially available pressure
algometers, the pressure application rate can be
monitored, which is important for the reliability
of the results. For research purposes, advanced
computer-controlled devices are available where
the pressure rate can be predefined.

Thresholds from various clinical pressure pain
studies are difficult to compare, as different
instrumentations have been used with different
probe diameters and shapes and different force
increase rates. The probe diameter is of utmost
importance as there is not necessarily a simple
relation between diameter and threshold since
spatial summation plays an important role.

The shape and contour of the probe are important, as sharp edges may excite more cutaneous receptors due to high shear forces compared with blunt probes.

Verbal assessments of the musculoskeletal pain intensity and other subjective characteristics of the pain are obviously needed in any clinical and experimental muscle pain studies. Visual analogue scales (VAS), verbal descriptor scales (VDS), McGill pain questionnaire (MPQ), and similar scales and questionnaires may be very helpful for the assessment of perceived pain intensity and quality (Turk and Melzack 1992). Musculoskeletal pain is most frequently characterized by descriptors as "drilling," "aching," "boring," and "taut."

The intensity of musculoskeletal pain is easily measured using visual analogue scales (VAS). However, this is only a one-dimensional aspect of the experienced pain and additional VAS should be applied to monitor, e.g., unpleasantness and soreness. Word descriptors on the VAS are important, as muscle tenderness and muscle pain may not reflect the same mechanisms. In addition to verbal assessments, psychophysical tests are valuable adjuncts for the examination of musculoskeletal pain (Turk and Melzack 1992).

Interaction Between Musculoskeletal Pain and Motor Performance

Musculoskeletal pain has implications for motor performance and different electrophysiological and biomechanical methods and disability questionnaires (e.g., Oswestry pain disability index) have been developed to assess this interaction.

After reviewing articles describing motor function in five chronic musculoskeletal pain conditions (temporomandibular disorders, muscle tension headache, fibromyalgia, chronic lower back pain, and post-exercise muscle soreness), Lund et al. (1991) concluded that the activity of agonist muscles is often reduced and the activity of the antagonist is slightly increased by musculoskeletal pain. As a consequence of these changes, force production and the range and velocity of movement of the affected body part are often reduced. To explain such changes

in motor control, Lund et al. (1991) proposed a neurophysiological model based on the phasic modulation of excitatory and inhibitory interneurons supplied by high threshold sensory afferents. They suggested that the "dysfunction," which is characteristic of several types of chronic musculoskeletal pain, is a normal protective adaptation and is not a cause of pain. This pain adaptation model has also been verified in experimental studies, where motor strategies have been investigated before and after experimentally induced muscle pain. It seems however that the adaptation pattern mainly occurs in relation to gross, high force movements and to a lesser extent for high precision, low force movements.

Johansson and Sojka (1991) suggested another pathophysiological model for chronic muscle pain and tension disorders. The model suggests that, under circumstances when chemosensitive group III-IV muscle afferents, joint afferents, or certain descending pathways are activated, the activity of both primary and secondary muscle spindle afferents would be increased via fusimotor reflexes. This would have several consequences. Initially, an increased activity in γ-motoneurons would reduce the information transmission capacity of muscle spindles. Secondly, an altered activity in the primary muscle spindle afferents would influence proprioception and thereby sensorimotor control at higher levels. This influence could possibly lead to unfavorable work techniques, such as increased co-contractions and insufficient rest periods, which might further increase the interstitial concentration of substances exciting group III–IV afferents. Thirdly, increased levels of these substances might also excite, via chemosensitive group III–IV afferents, the sympathetic outflow to skeletal muscles, which may change the balance between sympathetic vasoconstriction and metabolic vasodilation and thus entail a deficiency in oxygen and nutrients with the consequences.

There is so far little clinical evidence to support this model. Experimental human studies have to some degree verified the interaction between γ-motoneuron excitability and the activation of group III–IV nociceptive afferents.

Experimentally induced muscle pain causes excitation of the human stretch reflex, which may indicate such facilitatory interaction (Matre et al. 1998). In chronic muscle pain conditions where group III–IV muscle nociceptors are assumed to be active, muscle spindle responses and hence proprioception should be affected. Indeed, patients suffering from work-related myalgia often exhibit reduced proprioception, disturbed motor control, and impaired balance.

Cross-References

▶ Disability, Upper Extremity
▶ Opioids and Muscle Pain

References

Arendt-Nielsen, L. (1997). Induction and assessment of experimental pain from human skin, muscle and viscera. In T. S. Jensen, J. A. Turner, & Z. Wiesenfeld-Hallin (Eds.), *Proceedings of the 8th world congress on pain* (pp. 393–425). Seattle: IASP Press.

Arendt-Nielsen, L., & Graven-Nielsen, T. (2002). Deep tissue hyperalgesia. *Journal of Musculoskeletal Pain, 10*, 97–119.

Arendt-Nielsen, L., & Svensson, P. (2001). Referred muscle pain: Basic and clinical findings. *The Clinical Journal of Pain, 17*, 11–19.

Arendt-Nielsen, L., Svensson, P., & Graven-Nielsen, T. (1997). How to assess muscle pain experimentally and clinically. *European Journal of Pain, 1*, 64–65.

Carli, G., Suman, A. L., Biasi, G., et al. (2002). Reactivity to superficial and deep stimuli in patients with chronic musculoskeletal pain. *Pain, 100*, 259–269.

Curatolo, M., Petersen-Felix, S., Arendt-Nielsen, L., et al. (2001). Central hypersensitivity in chronic pain after whiplash injury. *The Clinical Journal of Pain, 17*, 306–315.

Fischer, A. A. (1987a). Pressure algometry over normal muscles. Standard values, validity and reproducibility of pressure threshold. *Pain, 30*, 115–126.

Fischer, A. A. (1987b). Reliability of the pressure algometer as a measure of myofascial trigger point sensitivity. *Pain, 28*, 411–414.

Friden, J. (1984). Muscle soreness after exercise: Implications of morphological changes. *International Journal of Sports Medicine, 5*, 57–66.

Giamberardino, M. A. (1999). Recent and forgotten aspects of visceral pain. *European Journal of Pain, 3*, 77–92.

Giamberardino, M. A. (2003). Referred muscle pain/hyperalgesia and central sensitisation. *Journal of Rehabilitation Medicine, 41*, 85–88.

Graven-Nielsen, T., Arendt-Nielsen, L., Svensson, P., et al. (1997). Stimulus-response functions in areas with experimentally induced referred muscle pain – a psychophysical study. *Brain Research, 2*, 121–128.

Johansson, H., & Sojka, P. (1991). Pathophysiological mechanisms involved in genesis and spread of muscular tension in occupational muscle pain and in chronic musculoskeletal pain syndromes: A hypothesis. *Medical Hypotheses, 35*, 196–203.

Kellgren, J. H. (1938). Observations on referred pain arising from muscle. *Clinical Science, 3*, 175–190.

Koelbaek Johansen, M., Graven-Nielsen, T., Schou Olesen, A., et al. (1999). Generalised muscular hyperalgesia in chronic whiplash syndrome. *Pain, 83*, 229–234.

Laursen, R. J., Graven-Nielsen, T., Jensen, T. S., et al. (1997). Quantification of local and referred pain in humans induced by intramuscular electrical stimulation. *European Journal of Pain, 1*, 105–113.

LeResche, L., Saunders, K., Von Korff, M. R., et al. (1997). Use of exogenous hormones and risk of temporomandibular disorder pain. *Pain, 69*, 153–160.

Liuzzi, F. J., Scoville, S. A., & Bufton, S. M. (1999). Long term estrogen replacement co-ordinately decreases trkA and beta-PPT mRNA levels in dorsal root ganglion neurons. *Experimental Neurology, 155*, 260–267.

Lund, J. P., Donga, R., Widmer, C. G., et al. (1991). The pain-adaptation model: A discussion of the relationship between chronic musculoskeletal pain and motor activity. *Canadian Journal of Physiology and Pharmacology, 69*, 683–694.

Matre, D., Sinkjær, T., Svensson, P., et al. (1998). Experimental muscle pain increases the human stretch reflex. *Pain, 75*, 331–339.

Newham, D. J., Edwards, R. H. T., & Mills, K. R. (1994). Skeletal muscle pain. In P. D. Wall & R. Melzack (Eds.), *Textbook of pain* (3rd ed., pp. 423–440). Edinburgh: Churchill Livingstone.

Sorensen, J., Graven-Nielsen, T., Henriksson, K. G., et al. (1998). Hyperexcitability in fibromyalgia. *Journal of Rheumatology, 25*, 152–1555.

Spitzer, W. O., Skovron, M. L., Salmi, L. R., et al. (1995). Scientific monograph of the Quebec Task Force on Whiplash-Associated Disorders: Redefining "whiplash" and its management. *Spine, 20*, 1–73.

Svensson, P., Cairns, B. E., Wang, K., et al. (2003). Injection of nerve growth factor into human masseter muscle evokes long-lasting mechanical allodynia and hyperalgesia. *Pain, 104*, 241–247.

Travell, J. G., & Simons, D. G. (1992). *Myofascial pain and dysfunction: The trigger point manual* (Vol. 1, 2). Baltimore: Lippincott Williams & Wilkins.

Turk, D. C., & Melzack, R. (Eds.). (1992). *Handbook of pain assessment*. New York: Guilford.

Vecchiet, L., Vecchiet, J., & Giamberardino, M. A. (1999). Referred muscle pain: Clinical and pathophysiologic aspects. *Current Review of Pain, 3*, 489–498.

Musculoskeletal Transferred Pain

▶ Muscle Pain, Referred Pain

Musculoskeletal Type of Pain

Definition

This refers to the types of pain that involve the musculature.

Cross-References

▶ Respondent Conditioning of Chronic Pain

MVD

▶ Microvascular Decompression

Myalgia

Robert D. Gerwin
Department of Neurology, Johns Hopkins
University, Bethesda, MD, USA

Synonyms

Fibromyalgia; Fibrositis; Fibrositis syndrome; Idiopathic myalgia; Muscle pain; Muscular rheumatism; Myofascial pain syndrome; Myofibrositis; Myogelosis

Definition

Myalgia

Myalgia is muscle pain or pain of muscular origin, without regard to cause. ▶ Fibromyalgia (FMS) is a syndrome in which there is chronic,

widespread muscle tenderness related to central hypersensitization. This is now considered a disorder of central pain modulation where descending pain inhibition is impaired. However, Ge et al. (2010) have shown that the tender points that are at the 18 predetermined sites for examination selected by the American College of Rheumatology are actually trigger points and that they reproduce much of the pain of patients with fibromyalgia. The implication is that fibromyalgia has a major component of peripheral nociceptive input that contributes to central sensitization in fibromyalgia. It does not explain the more widespread aspects of fibromyalgia like sleep disorders, irritable bladder and irritable bowel, and unusual fatigue.

Myofascial Pain Syndrome (MPS)

Myofascial pain syndrome (MPS) is a condition in which there are discrete taut bands of hardened muscle that contain zones of exquisite muscle tenderness and which generate pain referred to another, usually distal, site.

Characteristics

Introduction

Myalgia can be a primary chronic pain state in which there are no laboratory abnormalities. Pain can be disabling and is often associated with debilitating fatigue as in FMS. Myalgia can be comorbid with other conditions as in secondary FMS. Myalgia is considered chronic when it has been present for at least 3 months. The most common myalgic syndromes are FMS and MPS. These two conditions both give rise to muscle tenderness, but otherwise, they form two distinct entities. FMS is a syndrome of central sensitization and augmentation that results in widespread musculoskeletal tenderness and pain. MPS is the result of local muscle metabolic stress that is thought to produce an energy crisis. MPS is associated with discrete linear band-like hardness or tautness within one or more muscles, leading to the release of nociceptive substances such as substance P, potassium, and histamine that activate peripheral nociceptive receptors and dorsal horn nociceptive neurons. The region of a taut band is exquisitely tender and can refer pain to another, usually distal, region. Non-restorative sleep in addition to chronic myalgia is characteristic of FMS. Exercise intolerance can be seen with either FMS or MPS. However, muscle pain associated with exercise intolerance and fatigue can be due to an entirely different problem, such as a mutation in the cytochrome *b* gene of mtDNA (Andreu et al. 1999), Lyme disease, or hypothyroidism. Consideration of the differential diagnosis in myalgia is essential. Having made this distinction between myofascial pain and fibromyalgia, recent studies tend to blur the distinction since myofascial pain clearly has a major element of central sensitization that results in referred pain and in allodynia. Moreover, there is now good evidence that peripheral pain input from muscle affects pain and central pain processing in fibromyalgia (Staud et al. 2006, 2009, 2010; Ge et al. 2009). Thus, the distinction between these two conditions has become blurred.

Fibromyalgia: Characteristics

FMS is a chronic, widespread myalgia that involves 3 or 4 quadrants of the body (the American College of Rheumatology criteria (ACR)) (Wolfe et al. 1990). Using the ACR criteria, 3.5 % of women and 0.5 % of men in the United States are estimated to have FMS. The ACR criteria, intended to provide a uniform definition of fibromyalgia for research studies, require that (1) symptoms have been present for at least 3 months and (2) that 11 of 18 specified sites be tender. The clinical diagnosis of FMS in clinical practice was never intended to be as strict. The ACR criteria do not distinguish FMS from chronic, widespread MPS or any other chronic condition where there is widespread muscle tenderness. A new diagnostic tool has been proposed for the diagnosis of FMS, one that relies on symptom severity rather than tender point counts (Wolfe et al. 2010). MPS is the most common condition that must be considered in the differential diagnosis (Gerwin). Consequently, the physical examination performed for the evaluation of myalgia must include palpation for the taut bands

of MPS trigger points as well as for the tender points of FMS. A comprehensive medical evaluation is necessary in order to identify conditions in which diffuse muscle pain occurs secondarily. The diagnosis of FMS is based on the history and physical examination. Laboratory tests and imaging procedures are not useful for making a positive diagnosis but are required to identify comorbid conditions or other causes of chronic myalgia.

Associated Symptoms of FMS

FMS is associated with sleep disturbance, fatigue, headache, morning stiffness, irritable bowel syndrome (IBS), interstitial cystitis (IC), dyspareunia, and mood disturbance. Some of these symptoms are manifestations of referred muscle pain from ▶ myofascial trigger points (headache, ▶ dyspareunia, morning stiffness), and others such as IBS and IC are ▶ viscerosomatic pain syndromes (Gerwin) that occur in up to 70 % of FMS patients. They are by no means unique to FMS and are usually associated with pelvic floor MPS syndromes. Depression may occur in as many as 30 % of FMS patients but is no more common in FMS than in the general population.

Fibromyalgia: Etiology

Tenderness in FMS is related to central sensitization with amplification of nociception, resulting in a broad array of stimuli perceived as being more painful among FMS patients than they are in control subjects (Russell 2001). Alterations in cardiovascular autonomic nervous system function cause orthostatic hypotension or neurally mediated orthostatic tachypnea (Martinez-Lavin 1997). Neuroendocrine abnormalities in the hypothalamic-pituitary-adrenal system and growth hormone deficiency are hormonal deficiency states that may tie together the symptoms of fatigue, pain, and sleep and mood disturbances (Dessein et al. 2000).

Fibromyalgia: Treatment

Treatment of fibromyalgia includes a wide variety of pharmacologic, nutritional, hormonal, behavioral, cognitive, exercise, and physical modalities (Goldenberg et al. 2004). The long-term prognosis for FMS is more favorable than initially thought. Symptoms may persist for years, but patients either learn to cope with the chronic pain or the pain does not progress.

Myofascial Pain Syndrome: Characteristics

MPS is a muscular pain syndrome that arises from a primary dysfunction in muscle. It is associated with central sensitization and a segmental spread within the spinal cord to give rise to the phenomenon of referred pain or pain that is felt at a distance (Mense 2001). The trigger point is acidic, hypoxic, and has elevated neurotransmitters and cytokines at its center (Shah et al. 2005). The clinical picture of MPS is one of musculoskeletal pain, limited mobility, weakness, and referred pain (Simons et al. 1999). The trigger point has both a motor abnormality in which an abnormal hardness in muscle (the taut band) is felt on palpation and a sensory abnormality of exquisite tenderness in the taut band. Identification of the trigger point by physical examination has good interrater reliability (Gerwin et al. 1997). The specifics of MPS depend on which muscles are involved. For example, TrPs in the muscles of the head, neck and shoulders can cause headache. Local or regional myofascial syndrome can spread through the body and become a widespread myofascial syndrome, but does not become fibromyalgia.

Treatment

Treatment of myofascial pain requires the inactivation of MTrPs by manual trigger point compression or by needling, the restoration of normal muscle length and the elimination or correction of the factors that created or perpetuated the trigger points in the first place. Awareness of ergonomic and postural factors is important in developing a treatment plan. Trigger point needling is done with or without the injection of local anesthetic (Cummings 2001). Injection of other materials such as steroids or ketorolac is inappropriate. Trigger point inactivation must be accompanied by correction of mechanical or structural stresses and of psychological and medical contributing factors. Dry needling and injection of lidocaine are equally effective in treating trigger points (Ay et al. 2010).

Mechanical Causes

Mechanical causes of myalgia include ergonomic, structural, and postural pain syndromes.

Hypermobility Syndromes

Hypermobility syndromes produce multiple mechanical stresses. Structural stress occurs as ligamentous laxity results in poor joint stabilization, muscles then being recruited to maintain joint integrity. This can result in a seemingly disproportionate number of hypermobile persons, mostly women, who have focal or generalized myalgia. The affected muscles always have myofascial trigger points. The mechanism of injury appears to be muscular stress or overload that arises from the effort required to maintain joint integrity. The most effective treatment is strengthening.

Forward Head Posture

Forward head posture places stress on the extensor muscles of the neck and shoulder (longissimus cervicis, semispinalis capitis and cervicis, splenius capitis and cervicis, the suboccipital muscles at the base of the skull, and the trapezius and levator scapulae muscles). Posterior cervical muscle and shoulder muscle myalgic syndromes are thus frequently associated with head pain and headache.

Pelvic Torsion-Related Pain

Pelvic torsion-related pain is associated with pseudo-leg-length inequality or with trigger point muscular shortening (pseudoscoliosis). It can be caused by, and in turn either cause or aggravate, myalgia in lumbar muscles and in the pelvic floor muscles.

SI Joint Dysfunction

Sacroiliac joint dysfunction, or sacroiliac joint hypomobility, can cause pelvic and spine dysfunction that results in painful, widespread axial muscle trigger points. Pain may be felt in the sacroiliac joint region (on either the hypomobile or the normal side), referred to the low back or occur because of the secondary development of paraspinal trigger point up the axial spine to the shoulders and neck.

Static Overload

Static overload occurs when mechanically stressful positions are held for prolonged periods of time, causing fatigue and pain in the persistently activated muscles.

Nerve Root Compression

Nerve root compression can present with acute or chronic myofascial trigger points. Trigger point pain syndromes can develop acutely when there is an acute disc herniation. Muscle pain is in the distribution of the affected nerve root. It can precede any neurological impairment such as weakness, sensory loss, paresthesia, or reflex loss.

Delayed Onset Muscle Soreness (DOMS)

Delayed onset muscle soreness (DOMS), as is well known, occurs after eccentric exercise but has been shown to occur after ischemic exercise as well (Barlas et al. 2000). Exercise under these conditions causes injury to the muscle fiber and consequent pain or soreness.

The second major category is systemic medical illness. The relationship of some of these conditions to myalgia has been difficult to confirm. Yet when such an illness is identified and treated and muscle pain improves or resolves, it is tempting to equate the treatment with a successful outcome. Nonetheless, one must be cautious about assuming a causal relationship. The conditions of interest include autoimmune disorders, infectious diseases, allergies, hormonal and nutritional deficiencies, viscerosomatic pain syndromes, and iatrogenic drug-induced myalgic pain syndromes.

Autoimmune Disorders

Muscle pain is a common accompanying symptom of many autoimmune disorders, particularly connective tissue diseases like lupus and Sjögren's syndrome. Polymyalgia rheumatica (PMR) must certainly be considered in any

head, neck, and shoulder regional muscular pain syndrome in an older (>50 years of age) individual. Chewing-induced pain is an important component of both PMR and MPS, the latter when the temporomandibular joint is involved. Muscle pain may precede other signs of Sjögren's syndrome by several years.

Degenerative Diseases

Inclusion body myositis is the most common progressive muscle disease of the older population. Focal or regional muscle pain is a common complaint. The pathology is characterized by mononuclear cell inflammation, vacuolar degeneration of muscle fibers, and congophilic amyloid β deposits and paired helical filaments. The muscle changes are similar to those in Alzheimer Disease brains. No treatment has been found to be effective, although methotrexate has been used. Intravenous IgG may be of some benefit.

Infectious Diseases

Lyme disease is perhaps the most prevalent of the infectious diseases associated with myalgia and arthralgia. Post-Lyme disease syndrome is characterized by diffuse arthralgia, myalgia, fatigue, and subjective cognitive difficulty (Weinstein and Britchkov 2002). Patients diagnosed with this condition do not show evidence of chronic Borrelia infection and they do not respond to a 3-month course of antibiotics any better than a control group treated with placebo. Chronic infections that look like Lyme disease and that may coinfect with Lyme disease are babesiosis, ehrlichiosis, and *Bartonella*. Other infectious diseases thought to be related to myalgia are *mycoplasma* pneumonia and *chlamydia* pneumonia. Interest in these two diseases arises because of a putative association with arthralgia or synovitis and with chronic fatigue. A number of patients with widespread myalgia have parasitic disease, most commonly amebiasis and *giardiasis*.

Metabolic Myopathies

Fatty acid oxidation defects can present with exercise intolerance, myalgias, proximal weakness, and intermittent rhabdomyolysis.

Treatment is dietary supplementation with carnitine (Pitceathly et al. 2010).

Allergies

Cases of widespread myalgia (myofascial pain syndrome) have been associated with persons who have had untreated allergies. When the myalgic syndrome is limited to the head, neck, and shoulders, forward head posture as described above may be related to allergies, and obstruction of the nasal passages may play a role.

Viscerosomatic Pain Syndromes

Internal organs are associated with somatic segmental referred pain syndromes. Endometriosis, for example, is associated with abdominal myofascial pain (Jarrell and Robert 2003). Interstitial cystitis and irritable bowel syndrome are associated with chronic pelvic pain syndromes (Hetrick et al. 2003; Weiss 2001). Liver disease can cause local abdominal and referred shoulder myofascial pain syndromes that present as a regional pain syndrome responsive to treatment of the trigger points by needling or by manual therapy.

Brain Tumor and Base of Skull Pain

Posterior fossa mass lesions (primary and metastatic tumors) can present as focal base of skull or upper cervical pain with identifiable myofascial trigger points that transiently respond to trigger point inactivation.

Nutritional Deficiencies

Vitamin D deficiency was found in 89 % of the persons in a group with chronic musculoskeletal pain (Plotnikof 2003). This is an astounding figure that indicates that vitamin D deficiency is extremely common among those with myalgia.

Iron Deficiency

This causes a metabolic stress that produces fatigue and muscle pain. Muscle is depleted of iron available for energy-producing enzymatic reactions when serum ferritin levels are 15 ng/ml or less. Treatment with iron supplements in women whose serum ferritin level is 20 ng/ml or less results in less fatigue, less coldness, and less muscle pain. Iron insufficiency is also associated with restless leg syndrome (RLS), a cause of sleep disturbance. Sleep deprivation also causes myalgia. Thus, iron deficiency can aggravate muscle pain secondarily by causing RLS.

Drug-Induced Myalgia

This is widespread and diffuse, rather than regional. Drug-induced myalgia may or may not be associated with trigger points. Elevation of CK is a marker of muscle tissue breakdown. However, CK is not necessarily elevated when there is drug-induced myalgia. Drugs known to produce myalgia, regardless of whether or not CK is elevated, include propoxyphene and the statin family of cholesterol-lowering drugs (Thompson et al. 2003). The risk of acquiring myalgia is increased when a statin drug is taken concomitantly with fibric acid derivatives like gemfibrozil, niacin, cyclosporin, azole B antifungal and macrolide antibiotics, protease inhibitors, nefazodone, verapamil, diltiazem, amiodarone, or grapefruit juice (> 1 qt/day). Aromatase inhibitors commonly used in treatment of early hormone-positive breast cancer precipitate a rapid drop in estrogen levels and are associated with an increase in arthralgia and myalgia (Khan et al. 2010).

Conclusion

Muscle pain can be the result of a wide array of clinical conditions. The inflammatory diseases of muscle, such as polymyositis or the inherited myopathies, can also present with pain. Myoadenylate deaminase deficiency is the most common inherited muscle enzyme disorder, but the role that it plays in muscle pain is still no clearer today than it was 10 years ago. In order to treat patients well, the conditions that produce muscle pain must be identified in order to be specifically addressed.

Cross-References

▶ Guillain-Barré Syndrome
▶ Muscle Pain Model, Inflammatory Agents-Induced

References

Andreu, A. L., Hanna, M. G., Reichman, H., et al. (1999). Exercise intolerance due to mutations in the cytochrome *b* gene of mitochondrial DNA. *The New England Journal of Medicine, 341*, 1037–1044.

Ay, S., Evcik, D., & Tur, B. S. (2010). Comparison of injection methods in myofascial pain syndrome: A randomized controlled trial. *Clinical Rheumatology, 29*, 19–23.

Barlas, P., Walsh, D. M., Baxter, G. D., et al. (2000). Delayed onset muscle soreness: Effect of an ischemic block upon mechanical allodynia in humans. *Pain, 87*, 221–225.

Cummings, T., & White, A. (2001). Needling therapies in the management of myofascial trigger point pain: A systematic review. *Archives of Physical Medicine and Rehabilitation, 82*, 986–992.

Dessein, P. H., Shipton, E. A., Joffe, B. I., et al. (2000). Neuroendocrine deficiency-mediated development and persistence of pain in fibromyalgia: A promising paradigm? *Pain, 86*, 213–215.

Ge, H. Y., Nie, H., Madeleine, P., Danneskiod-Samsoe, B., Graven-Nielsen, T., & Arendt-Nielsen, L. (2009). Contribution of the local and referred pain from active myofascial trigger points in fibromyalgia syndrome. *Pain, 147*, 233–240.

Ge, H. Y., Wang, Y., Danneskiood-Samsøe, B., Graven-Nielsen, T., & Arendt-Nielsen, L. (2010). The Predetermined sites of examination for tender points in fibromyalgia are frequently associated with myofascial trigger points. *The Journal of Pain, 11*, 644–651.

Gerwin, R. D., Shannon, S., Hong, C. Z., et al. (1997). Interrater reliability in myofascial trigger point examination. *Pain, 69*, 65–73.

Goldenberg, D. L., Burckhardt, C., & Crofford, L. (2004). Management of fibromyalgia syndrome. *Journal of the American Medical Association, 292*, 2388–2395.

Hetrick, D. C., Ciol, M. A., Rothman, I., et al. (2003). Musculoskeletal dysfunction in men with chronic pelvic pain syndrome type III: A case–control study. *Journal of Urology, 170*, 828–831.

Jarrell, J., & Robert, M. (2003). Myofascial dysfunction and pelvic pain. Canadian Journal of Continuing Medical Education, 107–116.

Khan, Q. J., O'Dea, A. P., & Sharma, P. (2010). Musculoskeletal adverse events associated with adjuvant aromatase inhibitors. *Journal of clinical Oncology, v2010* (n9453), 1–8.

Martinez-Lavin, M. A., Hermosilla, A. G., & Mendoza, C. (1997). Orthostatic sympathetic derangement in subjects with fibromyalgia. *Journal of Rheumatology, 24*, 714–718.

Mense, S., & Simons, D. G. (2001). *Muscle pain* (pp. 205–288). Baltimore: Lippincott Williams & Wilkins.

Pitceathly, R., Rahman, S., Maritz, C., Hanna, M. G., Lachmann, R., & Murphy, E. (2010). PONM12 fatty acid oxidation disorders in adults: A potential treatable cause of muscle disease. *Journal of Neurology, Neurosurgery, and Psychiatry, 81*, e63.

Plotnikoff, G. A., & Quigley, J. M. (2003). Prevalence of severe hypovitaminosis D in patients with persistent, nonspecific musculoskeletal pain. *Mayo Clinic Proceedings, 78*, 1463–1470.

Russell, I. J. (2001). Fibromyalgia syndrome. In S. Mense & D. G. Simons (Eds.), *Muscle pain* (pp. 289–337). Baltimore: Lippincott Williams & Wilkins.

Shah, J. P., Phillips, T. M., Danoff, J. V., & Gerber, L. H. (2005). An in vitro microanalytical technique for measuring the local biochemical milieu of human skeletal muscle. *Journal of Applied Physiology, 99*, 1977–1984.

Simons, D. G., Travell, J. G., & Simons, L. S. (1999). *Myofascial pain and dysfunction: The trigger point manual*. Baltimore: Williams and Wilkins.

Staud, R., Vierck, C., Robinson, M., & Price, D. (2006). Overall fibromyalgia pain is predicted by ratings of local pain and pain-related negative affect – possible role of peripheral tissues. *Rheumatolgy, 45*, 1409.

Staud, R., Nagel, S., Robinson, M. E., & Price, D. D. (2009). Enhanced central pain processing of fibromyalgia patients is maintained by muscle afferent in put: A randomized, double-blind, placebo-controlled study. *Pain, 145*, 96–104.

Staud, R., Robinson, M. E., Weyl, E. E., & Price, D. D. (2010). Pain variability in fibromyalgia is related to activity and rest: Role of peripheral tissue impulse input. *The Journal of Pain, v11 n12*(201012), 376–1383.

Thompson, P. D., Clarkson, P., & Karas, R. H. (2003). Statin-associated myopathy. *Journal of the American Medical Association, 289*, 1681–1690.

Weinstein, A., & Britchkov, M. (2002). Lyme arthritis and post-Lyme disease syndrome. *Current Opinion in Rheumatology, 14*, 383–387.

Weiss, J. M. (2001). Pelvic floor myofascial trigger points: Manual therapy for interstitial cystitis and the urgency-frequency syndrome. *Journal of Urology, 166*, 2226–2231.

Wolfe, F., Smythe, H., Yunus, M. B., et al. (1990). The American college of rheumatology criteria for the classification of fibromyalgia. *Arthritis and Rheumatism, 33*, 160–172.

Wolfe, F., Clauw, D. J., Fitzcharles, M. A., Golderberg, D. L., Katz, R. S., Mease, P., Russell, A. S., Russell, I. J., Winfield, J. B., & Yunus, M. B. (2010). The American college of rheumatology preliminary diagnostic criteria for fibromyalgia and measurement of symptom severity. *Arthritis Care and Research, 62*, 600–610.

Mycobacterium Leprae

Definition

Mycobacterium leprae is the causative agent of Hansen's disease. The bacterium was discovered

in 1873 by a Norwegian physician named
Gerhard Armauer Hansen.

Cross-References

▶ Hansen's Disease

Mycobacterium Species

Definition

Mycobacteria are a large genus of class of bacteria
that are responsible for a multitude of diseases of
many species. The most common mycobacterium
used in Freund's complete adjuvant is Mycobacterium tuberculosis, the organism responsible for
tuberculosis in humans. Mycobacterium
butyricum has also been used in FCA.

Cross-References

▶ Arthritis Model, Adjuvant-induced Arthritis

Myelin

Definition

Myelin is a specialized cell membrane composed
of lipids and proteins that ensheathes axons and
fosters rapid electrical conduction.

Cross-References

▶ Demyelination
▶ Hereditary Neuropathies
▶ Postsynaptic Dorsal Column Projection,
  Anatomical Organization
▶ Spinothalamic Tract Neurons, Morphology
▶ Toxic Neuropathies

Myelinated

Definition

Medium and large nerve fibers can be wrapped
in a variable number of concentric layers of
glial membrane that provide for rapid
neurotransmission.

Cross-References

▶ Postsynaptic Dorsal Column Projection,
  Anatomical Organization

Myelination

Definition

Myelination is a process by which oligodendroglial cells wrap axons with multiple layers of glial
cell membrane forming myelin, which
electrically insulates the axon, increasing axonal
conduction velocity.

Cross-References

▶ Mechanonociceptors

Myelopathy

Definition

Myelopathy refers to any pathological change in
the spinal cord, including spinal cord inflammation (myelitis).

Cross-References

▶ Headache Due to Arteritis

Myelotomy

Definition

Myelotomy is a neurosurgical procedure in which white-matter pathways in the spinal cord are severed, usually in an attempt to relieve severe pain. The cut is sometimes made on the midline (midline myelotomy). Second order spinothalamic tract axons, which carry nociceptive information, cross the spinal cord from one side to the other before ascending to the brain. Midline myelotomy is intended to cut these fibers as they cross.

Cross-References

▶ Midline Myelotomy

Myenteric Plexus

Definition

The myenteric plexus consists of unmyelinated nerve fibers that are spread out in the muscular part of the intestinal wall of the esophagus, stomach, and intestines. The plexus is involved in the regulation of gut motility.

Cross-References

▶ Opioid Therapy in Cancer Pain Management, Route of Administration

Myocardial Ischemia

▶ Visceral Pain Model, Angina Pain

Myoclonus

Definition

Myoclonus refers to irregular, involuntary contraction of a muscle.

Cross-References

▶ Opioids and Reflexes

Myofascial

Definition

The muscle fibers and the fascia (the sheaths) within and around a muscle.

Cross-References

▶ Lower Back Pain, Physical Examination
▶ Myofascial Pain

Myofascial Manipulation

Definition

Myofascial manipulation is the forceful, passive movement of the musculofascial components through their restrictive direction. Treatment is initiated in the superficial layers and moves into the deeper layers while considering the relationship with the joints involved.

Cross-References

▶ Chronic Pelvic Pain, Physical Therapy Approaches, and Myofascial Abnormalities

Myofascial Pain

Robert Bennett
Department of Medicine, Oregon Health and
Science University, Portland, USA

Synonyms

Localized muscle pain; Myotomal pain; Soft tissue rheumatism; Trigger point pain

Definition

Myofascial pain: Pain arising from muscles or related fascia.

Active trigger point: This is a trigger point that results in pain at rest with increased pain on contraction or stretching of the involved muscle.

Latent trigger point: A latent trigger point is a focal area of tenderness and tightness in a muscle that does not result in spontaneous pain. However, a latent trigger point may restrict range of movement and cause weakness of the involved muscle.

Characteristics

Prevalence

Myofascial pain is a universal occurrence commonly developing as a result of muscle injuries, overuse, or repetitive strain. In most instances, the problem resolves within a few days without any need for medical intervention. When pain persists or worsens, necessitating a medical consultation, it is referred to as a myofascial pain syndrome (Travell and Simons 1983). There have not been any epidemiological studies of myofascial pain problems in the general population. Rather, physicians have reported prevalence figures in specialized situations. For instance, in one internal medicine practice, 30 % of patients presenting with pain complaints were diagnosed as having a myofascial pain origin for their symptoms. A report from a clinic specializing in head and neck pain reported a myofascial etiology in 55 % of cases. A report from one pain management center attributed a myofascial origin to the symptomatology in 85 % of patients. On the other hand, latent trigger points have been found in the shoulder girdle muscles of 54 % of female and 45 % of male subjects who were completely asymptomatic (Simons 2001).

Diagnosis

The clinical diagnosis of myofascial pain is critically dependent on the physician being aware of this diagnosis as a possible cause for the patient's pain complaint (Travell and Simons 1983). Myofascial pain syndromes may mimic a large number of other disorders; thus there is a necessity to perform a thorough physical examination, with appropriate investigations. Myofascial pain characteristically presents as a dull deep aching sensation, which is aggravated by use of the involved muscles as well as psychological stressors that cause increased muscle tension (Alvarez and Rockwell 2002). The defining clinical characteristic of myofascial pain is the finding of a trigger point. This is a well-defined point of focal tenderness within a muscle. Sometimes, firm palpation of this focus elicits pain in a referred distribution that reproduces the patient's symptoms. Importantly, referred pain from a trigger point does not follow a nerve root distribution (i.e., it is not dermatomal). Palpation usually reveals a ropelike induration of the associated muscle fibers, often referred to as the "taut band." Sometimes, snapping this band or needling the trigger point produces a localized twitch response of the involved muscle. This twitch response can only be reproducibly elicited in fairly superficial muscles. Importantly, trigger points produce functional consequences in terms of a restriction of range of movement and weakness (probably a reflex inhibition secondary to pain), which is usually associated with easy fatigability of the involved muscle.

Clinical Syndromes

A myofascial pain syndrome may be due to just one trigger point, but more commonly there are several trigger points responsible for any given regional pain problem. It is not uncommon for the problem to be initiated with a single trigger point, with the subsequent development of satellite trigger points, which evolve over time due to the mechanical imbalance resulting from reduced range of movement and pseudo-weakness. The persistence of a trigger point may lead to neuroplastic changes at the level of the dorsal horn, which results in amplification of the pain sensation (i.e., central sensitization) with a tendency to spread beyond its original boundaries (i.e., expansion of receptive fields) (Graven-Nielsen and Arendt-Nielsen 2002). In some instances, segmental central sensitization leads to the phenomena of mirror image pain (i.e., pain on the opposite side of the body in the same segmental distribution), and in other instances a progressive spread of segmental central sensitization gives rise to the widespread pain that characterizes fibromyalgia (Arendt-Nielsen and Graven-Nielsen 2003).

Low Back Pain

Acute low back pain has many causes. Some are potentially serious, such as cancer metastases, osteomyelitis, massive disk herniations (e.g., cauda equina syndrome), vertebral fractures, pancreatic cancer, and aortic aneurysms. However, the commonest cause of acute back pain is so-called lumbosacral strain. In 95 % of cases, this resolves within 3 months. In those cases that do not resolve, the development of a chronic low back pain syndrome is usually accompanied by the finding of active myofascial trigger points. Simons describes 15 torso and pelvic muscles that may be involved in low back pain (Simons 2001). The most commonly involved muscle group is the quadratus lumborum; pain emanating from trigger points in these muscles is felt in the low back with occasional radiation in a sciatic distribution or into the testicles. Trigger points involving the iliopsoas are also a common cause of chronic low back pain. The typical distribution of iliopsoas pain is a vertical band in the low back region and the upper portion of the anterior thigh. Trigger points at the origin of the gluteus medius from the iliac crest are a common cause for low back pain in the sacral and buttock, with a referral pattern to the outer hip region.

Neck and Shoulder Pain

Latent trigger points are universal findings in many of the muscles of the posterior neck and upper back. Active trigger points commonly involve the upper portion of the trapezius and levator scapula. Upper trapezius trigger points refer pain to the back of the neck and not uncommonly to the angle of jaw. Levator scapula trigger points cause pain at the angle of the neck and shoulder; this pain is often described as lancinating, especially on active use of this muscle. As many of the muscles in this area have an important postural function, they are commonly activated in office workers and developmental problems causing spinal malalignment (e.g., short leg syndrome, hemipelvis, and scoliosis). As the upper trapezius and levator scapulae act synergistically with several other muscles in elevation and fixation of the scapula, it is common for a single trigger point in this region to initiate a spread of satellite trigger points through adjacent muscles that are part of the same functional unit.

Hip Pain

Pain arising from disorders of the hip joint is felt in the groin and the lower medial aspect of the anterior thigh. This distribution is uncommon in myofascial pain syndromes except for iliopsoas pain. The great majority of patients complain of hip pain and in fact localize their pain to the outer aspect of the hip. In some patients this is due to a trochanteric bursitis, but in the majority of cases it is related to myofascial trigger points in the adjacent muscles. By far the commonest trigger points giving rise to outer hip pain are those in the attachments of the gluteus medius and minimus muscles into the greater trochanter.

Pelvic Pain

The pelvic floor musculature is a common sight for myofascial trigger points. There is increasing recognition by gynecologists and urologists that pain syndromes described in terms of prostatitis, coccydnia, vulvodynia, and endometriosis are often accompanied by active myofascial trigger points. One of the most commonly involved intrapelvic muscles is the levator ani; its pain distribution is central low buttock.

Headaches

Active myofascial trigger points in the muscles of the shoulder, neck, and face are a common source of headaches (Borg-Stein 2002). In many instances, the headache has the features of so-called tension headache, but there is increasing acceptance that myofascial trigger points may initiate classical migraine headaches or be part of a mixed tension/migraine headache complex. For instance, sternocleidomastoid trigger points refer pain to the anterior face and supraorbital area. Upper trapezius trigger points refer pain to the vertex forehead and temple. Trigger points in the deep cervical muscles of the neck may cause post-occipital and retro-orbital pain.

Jaw Pain

There is a complex interrelationship between temporomandibular joint dysfunction and myofascial trigger points (Fricton et al. 1985) Common trigger points involved in jaw pain syndromes are the massetters, pterygoids, upper trapezius and upper sternocleidomastoid.

Upper Limb Pain

The muscles attached to the scapula are common sites for trigger points that can cause upper limb pain (Gerwin 1997). These include the subscapularis, infraspinatus, teres major, and serratus anterior. It is not uncommon for trigger points in these locations to refer pain to the wrist, hand, and fingers. Extension flexion injuries to the neck often activate a trigger point in the pectoralis minor with a radiating pain, or down the ulnar side of the arm and into the little finger. Myofascial pain syndromes of the upper limbs are often misdiagnosed as frozen shoulder, cervical radiculopathy, or thoracic outlet syndrome (Simons 2001).

Lower Limb Pain

Trigger points in the tensor fascia lata and iliotibial band may be responsible for lateral thigh pain and lateral knee pain, respectively.

Anterior knee pain may result from trigger points in various components of the quadriceps musculature. Posterior knee pain can result from trigger points in the hamstring muscles and popliteus. Trigger points in the anterior tibialis and the peroneus longus muscles may cause pain in the anterior leg and lateral ankle, respectively. Myofascial pain syndromes involving these muscles are often associated with ankle injuries or an excessively pronated foot. Sciatica pain may be mimicked by a trigger point in the posterior portion of the gluteus minimus muscle.

Chest and Abdominal Pain

Disorders affecting intrathoracic and intra-abdominal organs are some of the commonest problems encountered in internal medicine. For instance, anterior chest pain is a frequent cause for emergency room admissions, but in the majority of patients a myocardial infarction is not found. In some cases, the chest pain is caused by trigger points in the anterior chest wall muscles (Travell and Simons 1992). Pectoralis major trigger points cause ipsilateral anterior chest pain with radiation down the ulnar side of the arm – thus mimicking cardiac ischemic pain. A trigger point in the sternalis muscle typically causes a deep substernal aching sensation. Trigger points at the upper and lower insertions of the rectus abdominis muscles may mimic the discomfort of gall bladder and bladder infections, respectively. It is important to note that myofascial trigger points may accompany disorders of intrathoracic and intra-abdominal viscera, and thus a diagnosis of an isolated myofascial cause for symptoms should never be made without an appropriate workup.

Pathogenesis

The precise pathophysiological basis for the trigger point phenomena is still not fully understood. There is a general agreement that electromyographic recordings from trigger points show low-voltage spontaneous activity resembling end plate spike potentials (Rivner 2001). Simons envisions a myofascial trigger point to be "a cluster of numerous microscopic loci of intense abnormality that are scattered throughout the tender nodule" (Simons 2001). It is thought that these loci result from a focal energy crisis (from injury or repetitive use), which results in contraction of focal sarcomeric units due to calcium release from the sarcoplasmic reticulum. A more detailed coverage of this topic is provided in the entry ▶ Myofascial Trigger Point. Factors commonly cited as predisposing to trigger point formation include deconditioning, poor posture, repetitive mechanical stress, mechanical imbalance (e.g., leg length inequality), joint disorders, and non-restorative sleep and vitamin deficiencies.

Prognosis

Uncomplicated myofascial pain syndromes usually resolve with appropriate correction of predisposing factors and myofascial treatment (Alvarez and Rockwell 2002). If the symptoms are persistent, due to ineffective management, the development of segmental central sensitization may lead to a stubbornly recalcitrant pain disorder. In some such cases, the spread of central sensitization leads to the widespread pain syndrome of fibromyalgia.

Treatment

For effective management of myofascial pain, syndromes require attention to the following issues (Alvarez and Rockwell 2002; Rudin 2003):

Postural and Ergonomic

The most critical element in the effective management of myofascial pain syndromes is the correction of predisposing factors (see above). These interfere with the ability of the muscle to fully recover and are the commonest reason for treatment failures.

Stretching

The muscles involved in myofascial pain syndromes are shortened due to the aforementioned focal contractions of sarcomeric units. It is thought that these focal contractions result in a prolonged ATP consumption and that the restoration of a muscle to its full stretch length breaks the link between the energy crisis and contraction of sarcomeric units. Effective stretching is most commonly achieved through the technique of spray and stretch (Rudin 2003). This involves the cutaneous application, along the axis of the muscle, of ethyl chloride spray, while at the same time passively stretching the involved muscle. Other techniques to enhance effective stretching include trigger point pressure release, post-isometric relaxation, reciprocal inhibition, and deep-stroking massage (Simons 2001).

Strengthening

Muscles harboring trigger points usually become weak due to the inhibitory effects of pain. A program of slowly progressive strengthening is essential to restore full function and minimize the risk of recurrence and perpetuation of satellite trigger points.

Trigger Point Injections

Injection of trigger points is generally considered to be the most effective means of direct inactivation. A peppering technique using a fine needle to inactivate all the foci within a trigger point locus is the critical element of successful trigger point therapy (Hong 1994a). Accurate localization of the trigger point is confirmed if a local twitch response is obtained; however, this may not be obvious when needling deeply lying muscles. Successful elimination of the trigger point usually results in a relaxation of the taut band. Although dry needling is effective, the use of a local anesthetic (1 % lidocaine or 1 % procaine) helps confirm the accuracy of the injection and provides instant gratification for patients (Hong 1994b). There is no evidence that the injection of corticosteroids provides any enhanced effect. A beneficial role for botulinum toxin in trigger point injections has not so far been conclusively demonstrated.

Medications

There is currently no evidence that any form of drug treatment eliminates myofascial trigger points (Rudin 2003). NSAIDs and other analgesics usually provide moderate symptomatic relief. Tricyclic antidepressant drugs, which modulate pain at the central level, are often of benefit, especially in those patients with an associated sleep disturbance. In the author's experience, tizanidine (a muscle relaxant that also ameliorates pain by activating alpha-2 adrenergic receptors) is often a useful adjunct in difficult to treat myofascial pain syndromes.

Psychological Techniques

In severe myofascial pain syndromes that are not responding to treatment, patients often become anxious and depressed. These mood disorders need to be recognized and appropriately treated. Persistent muscle tension exacerbates the pain of myofascial trigger points and can often be effectively managed with EMG biofeedback, cognitive behavioral therapy, and hypnotic/meditation relaxation techniques.

References

Alvarez, D. J., & Rockwell, P. G. (2002). Trigger points: Diagnosis and management. *American Family Physician, 65*, 653–660.

Arendt-Nielsen, L., & Graven-Nielsen, T. (2003). Central sensitization in fibromyalgia and other musculoskeletal disorders. *Current Pain and Headache Reports, 7*, 355–361.

Borg-Stein, J. (2002). Cervical myofascial pain and headache. *Current Pain and Headache Reports, 6*, 324–330.

Fricton, J. R., Kroening, R., Haley, D., et al. (1985). Myofascial pain syndrome of the head and neck: A review of clinical characteristics of 164 patients. *Oral Surgery, Oral Medicine, and Oral Pathology, 60*, 615–623.

Gerwin, R. D. (1997). Myofascial pain syndromes in the upper extremity. *Journal of Hand Therapy, 10*, 130–136.

Graven-Nielsen, T., & Arendt-Nielsen, L. (2002). Peripheral and central sensitization in musculoskeletal pain disorders: An experimental approach. *Current Rheumatology Reports, 4*, 313–321.

Hong, C.-Z. (1994a). Considerations and recommendations regarding myofascial trigger point injection. *Journal of Musculoskeletal Pain, 2*, 29–59.

Hong, C.-Z. (1994b). Lidocaine injection versus dry needling to myofascial trigger point. The importance of the local twitch response. *American Journal of Physical Medicine & Rehabilitation, 73*, 256–263.

Rivner, M. H. (2001). The neurophysiology of myofascial pain syndrome. *Current Pain and Headache Reports, 5*, 432–440.

Rudin, N. J. (2003). Evaluation of treatments for myofascial pain syndrome and fibromyalgia. *Current Pain and Headache Reports, 7*, 433–442.

Simons, D. G. (2001). Myofascial pain caused by trigger points. In S. Mense, D. G. Simons, & I. J. Russel (Eds.), *Muscle pain: Understanding its nature, diagnosis, and treatment* (pp. 205–288). Philadelphia: Lippincott Williams and Wilkins.

Travell, J. G., & Simons, D. G. (1983). *Myofascial pain and dysfunction: The trigger point manual*. Baltimore: Williams and Wilkins.

Travell, J., & Simons, D. (1992). *Myofascial pain and dysfunction: The trigger point manual* (Vol. 2). Baltimore: Williams and Wilkins.

Myofascial Pain Syndrome

Synonyms

MPS

Definition

Myofascial pain syndrome is a muscle disorder characterized by the presence of trigger points (TrPs) within the muscle. There is also pain, muscle spasm, tenderness, stiffness, limited range of motion, weakness, and/or autonomic dysfunction. Pressure on the trigger point refers pain to an area distant from the trigger point. However, the trigger point itself could be and usually is painful.

Cross-References

▶ Chronic Back Pain and Spinal Instability
▶ Chronic Pelvic Pain, Musculoskeletal Syndromes
▶ Chronic Pelvic Pain, Physical Therapy Approaches, and Myofascial Abnormalities
▶ Muscle Pain, Fibromyalgia Syndrome (Primary, Secondary)
▶ Myalgia
▶ Nocifensive Behaviors, Muscle and Joint
▶ Opioids and Muscle Pain
▶ Sacroiliac Joint Pain
▶ Stretching
▶ Trigger Point

Myofascial Pain Syndromes

► Chronic Pelvic Pain, Musculoskeletal Syndromes

Myofascial Release

Definition

Myofascial release can be described as a manual soft tissue technique that can be either direct or indirect and is frequently combined. As a treatment technique, it utilizes the principles of biomechanical loading of soft tissue, and the neural reflex changes by stimulation of mechano-receptors in the fascia.

Cross-References

► Chronic Pelvic Pain, Physical Therapy Approaches, and Myofascial Abnormalities

Myofascial Syndrome

► Lower Back Pain, Physical Examination

Myofascial Trigger Point

► Chronic Pelvic Pain, Physical Therapy Approaches, and Myofascial Abnormalities

Myofascial Trigger Point Pain Syndrome

► Chronic Pelvic Pain, Physical Therapy Approaches, and Myofascial Abnormalities

Myofascial Trigger Points

David G. Simons
Department of Rehabilitation Medicine, Emory University, Atlanta, GA, USA

Synonyms

MTrP

Definition

Clinically, a central ► myofascial trigger point (MTrP) is characteristically a very tender, circumscribed, nodule-like spot that is located in the midportion of a ► taut band of skeletal muscle fibers and can cause referred pain. Application of digital pressure to the spot typically induces referred pain that is characteristic of that muscle and is familiar to the subject if the MTrP is active. The indurated tender attachments of taut-band muscle fibers are identified as ► attachment trigger points (see Fig. 1).

Myofascial Trigger Points, Fig. 1 This schematic longitudinal view of muscle illustrates key clinical features of a myofascial trigger point. The central trigger point (CTrP), which is located near the mid-muscle fiber, exhibits outstanding spot tenderness in a palpable taut band and refers pain to a distance. The taut band extends the length of the muscle fibers to its attachments, where the sustained tension of the taut band induces an attachment trigger point (ATrP) enthesopathy (that is sensitive to applied pressure and increased muscle tension) (Adapted from Simons et al. (1999))

Etiologically, the central MTrP is associated with ▶ endplate noise, which originates from a motor end plate that is releasing abnormal amounts of ▶ acetylcholine. Shortened ▶ sarcomeres, in the region of end plates, are associated with local release of sensitizing substances (Shah et al. 2004) that cause pain and tenderness.

Characteristics

Myofascial trigger points (MTrPs) are only one part of the complex neuromusculoskeletal pain picture; however, that part is probably as important as the nervous system and is certainly the most neglected and overlooked part. The term myofascial pain is often used with two different meanings. Myofascial pain caused by MTrPs is a specific etiological diagnosis, whereas myofascial pain, used in the general sense of regional pain of unidentified etiology in myofascial structures, is only a symptom, not a diagnosis. Musculoskeletal pain is one of the major causes of common human aches and pains such as low back pain and tension type headache.

In one study, of the 32.5 % of the unselected population who were experiencing chronic pain, 94.5 % identified it as musculoskeletal (Wolfe et al. 1998). Myofascial trigger points (MTrPs) are often the major cause of the pain, or a major component in association with well-recognized sources of pain. Any of the approximately 500 skeletal muscles can develop MTrPs. For more detailed information on this subject, consult the two latest volumes of The Trigger Point Manual (Simons et al. 1999; Travell and Simons 1992), or the German or Spanish editions, and the Muscle Pain book (Mense et al. 2001).

Active MTrPs cause a clinical pain complaint but latent MTrPs do not – an important clinical distinction. One well-designed article (Hsieh et al. 2000) described 520 muscle examinations of muscles commonly causing low back pain (LBP), both in subjects with LBP and in normal pain-free controls. The authors found MTrPs (some active) in 90 % of low back pain subjects' muscles and found latent MTrPs (not producing

Myofascial Trigger Points, Table 1 Distribution of latent myofascial trigger points (MTrPs) in eight muscles of 13 pain-free, healthy normal control subjects. Subjects were numbered to correspond to their relative ages, which ranged between 23 and 59 years. Number 1 was the youngest. Age does not appear to be an important factor among these adults and showed no linear regression

Subject number	Number of muscles with MTrPs in each subject
2	7
6	7
10	6
8	5
3	5
12	5
5	4
11	3
13	3
4	2
9	2
7	1
1	0

a pain complaint) in 70 % of the control subjects' muscles. Latent MTrPs were common in control subjects.

The natural history of MTrPs is unknown; however, a recent study provided a valuable lead. Among 13 healthy, normal, pain-free hospital personnel, only one subject had no latent MTrPs in the eight muscles examined. Two subjects had MTrPs in seven of the eight muscles (see Table 1). Age was not a significant factor. Note subjects 1 and 2. This small sample suggests that most adults have at least some latent MTrPs; a few adults have many and a few have almost none. This needs verification. If this study is representative, the natural history of MTrPs could be that we are born with the tendency to develop latent MTrPs upon reading adolescence, and, during adulthood, life stresses can convert them into active MTrPs. A simple study, by reasonably skilled investigators, could resolve this issue.

The diagnosis of MTrPs can firstly be approached by answering three questions, which require limited manual skill to determine if MTrPs are a viable differential diagnosis. Could MTrPs of muscle be causing the pain? If so, which muscle(s) should be examined? Does an initial screening exam identify active MTrPs? If so, the patient needs further examination and testing by a skilled practitioner who can also treat the MTrPs as listed below.

Diagnosis of myofascial trigger points (MTrPs)

- Could MTrPs of muscle be causing the pain?
  - Other possibilities: nerves (radiculopathy), joints (somatic dysfunction), central nervous system (fibromyalgia), viscera (renal calculi), etc.
  - Trigger point pain is initiated by sudden, unexpected overload (acute onset from fall or motor vehicle accident) and chronic overload (insidious onset from poor posture, poor ergonomic arrangement, repetitive strain).
- Which muscle(s) should be examined?
  - The muscles that were overloaded (history above) – based on knowledge of muscle functions.
  - Distorted posture or movements.
  - Pain pattern that fits known patterns of suspected muscle(s).
  - Painful restriction in stretch range of motion of involved muscle(s). Involved muscle shows increased stiffness and tension.
- What is a simple initial muscle examination (differential diagnosis screening)?
  - Localized very tender spot in skeletal muscle+
  - Pressure-evoked referred pain that is characteristic of that muscle = a likely latent myofascial trigger point+
  - Patients' recognition of evoked pain as part or all of the primary pain complaint = a likely active myofascial trigger point
- Confirmatory myofascial trigger point examination (training and skill required)
  - Tender spot feels nodular (if located sufficiently superficial).
  - Tender spot located in a palpable taut band, see Fig. 1.

- Tender attachment TrP (enthesopathy) at attachment of taut band.
- Favorable response to specific MTrP treatment.
- Occurrence of a local twitch response evoked incidentally by palpation of, or purposely by needle insertion into, the TrP.

The diagnostic process is complicated by the complex neuromusculoskeletal nature of musculoskeletal pain and the lack of a diagnostic gold standard for MTrPs, which are not identifiable by available laboratory or imaging testing. In addition, the appropriate examination depends on the amount and texture of overlying tissue and the skills of the examiner. Different muscles may require quite different examinations.

Initially, the diagnosis depends on medical history, knowledge of MTrP characteristics, and knowledge of muscle anatomy and function. Effective confirmatory examination and treatment depend on clinical skill. A growing body of research studies and clinical experience increasingly substantiates the guidelines presented here.

Could MTrPs of Muscle Be Causing the Pain?

Commonly, in patients with chronic myofascial pain, the likely diagnoses, except MTrPs, have already been ruled out, often by numerous expensive tests. Chronic progressive MTrP symptoms are perpetuated by mechanical or systemic factors, which need identification and correction – factors that alone, without the activated MTrPs, often do not cause enough symptoms to demand attention. Without a perpetuating factor, acute onset active MTrPs tend to subside to asymptomatic latent MTrPs with gentle, normal, muscle-stretching daily activities.

Which Muscle(s) Should Be Examined?

Patient posture and modified movements is often a key. Special attention should be paid to problems in the feet that commonly reflect dysfunctions, which extend up the body to the head. Active MTrPs of each muscle project a characteristic referred pain pattern that is variable in time and among individuals. Critically important referred pain patterns from active MTrPs throughout the body are described in books (Travell and Simons 1992; Simons et al. 1999; Dejung et al. 2003), also on wall charts and flip charts, and have been updated for masticatory muscles in a journal article (Wright 2000). The pain patterns help greatly to identify the MTrP cause of the pain. The pain that limits stretch is usually located in the involved muscle, rather than in the referred pain zone.

Simple Initial Examination

If palpation with about 3 kg of calibrated consistent pressure on a spot of localized tenderness in a suspected muscle elicits referred pain that is familiar to the patient, that spot is likely an active MTrP. MTrPs are then ruled in as a differential diagnosis that requires a confirmatory examination and, if indicated, treatment by a skilled practitioner. Referral may be necessary. Latent MTrPs can cause muscle inhibition, imbalance, and other motor problems.

Confirmatory Myofascial Trigger Point Examination

This examination demands skills that usually require considerable training and clinical practice and are often critical for effective treatment of MTrPs. Clinically, the taut band is considered by skilled and experienced clinicians to be the most distinctive feature of MTrPs but was one of the most difficult exams in credible interrater reliability studies because of the skill required (Gerwin et al. 1997; Sciotti et al. 2001). Fortunately, confusingly, similar fascial structures are rarely tender. Taut bands are usually indistinguishable in muscle that lies beneath thick layers of fat, firm subcutaneous tissue, and/or overlying muscle tissue. Figure 1 schematically illustrates the relationship between the taut band and its central and attachment ▶ trigger points.

The ▶ local twitch response is a sensory-motor spinal reflex that induces a series of motor unit activations only in taut band fibers. It has high specificity but relatively poor reliability in most hands and can be extremely painful to the patient.

Myofascial Trigger Points, Fig. 2 The integrated hypothesis postulates a positive feedback cycle, which characteristically involves excessive spontaneous release of acetylcholine in a number of motor end plates associated with a myofascial trigger point. Mechanical stress (muscle overload) and increased autonomic activity can cause or increase abnormal spontaneous acetylcholine release. The strong spontaneous local contractile activity increases fiber tension producing the palpable taut band, increased energy demand, and local hypoxia. Together, they deplete the supply of adenosine triphosphate (ATP) and produce tissue distress, which releases substances that sensitize nociceptors causing MTrP pain. Shortage of ATP could reduce recovery of calcium into the sarcoplasmic reticulum sustaining contractile activity and increased fiber tension. See text for alternate pathway (*broken arrows*)

The pathophysiology of MTrPs is not totally unknown, but it is not firmly established and has been controversial. An integrated hypothesis (Simons et al. 1999; Mense et al. 2001) incorporates the available research studies and explains the clinical features of MTrPs. The hypothesis helps to identify what we do know with confidence and what remains to be clarified. Figure 2 outlines and summarizes the hypothesis.

Clinically, compression (radiculopathy) of the motor nerve can facilitate conversion of latent MTrPs to active ones, which are associated with electromyographic evidence of increased acetylcholine (ACh) release. Excessive spontaneous release can be identified by end plate noise (Simons 2001), which is mistakenly considered a normal finding by many electromyographers – a source of controversy. End plate noise is significantly associated with MTrPs – a fact not generally recognized by electromyographers – but it is not uniquely diagnostic of MTrPs (Couppé et al. 2001; Simons et al. 2002).

Microscopic changes in muscle fibers in the region of the endplate include localized regions of severe sarcomere shortening (contraction disks and contraction knots), with evidence of increased tension of some fibers. The microscopic picture also suggests the possibility of abnormal calcium permeability of the muscle fiber membrane (sarcolemma). Increased tension of many fibers can account for the taut band. The pathological changes and natural history of MTrPs suggest the possibility of a genetic aberration of calcium channels of variable penetrance in the sarcolemma and/or neurolemma of the nerve terminal.

Mechanisms by which the sensitizing substances produce local tenderness and referred pain have been described in detail (Mense et al. 2001) (see Fig. 2). Many substances have been demonstrated histochemically (Shah et al. 2004).

An alternate route to complete the feedback cycle (broken arrows) is presented, because several kinds of studies indicate that increased sympathetic nervous system activity increases the end plate noise (or end plate spikes) of a trigger point. Specific mediators have not been adequately identified for locally modulating autonomic effects, which include release of ACh packets from the nerve terminal. Both routes may occur.

Treatment

Three manual treatments of MTrPs – contract-release, trigger point pressure release, and vapocoolant spray and release – are usually effective (Simons 2002). In addition, injection or dry needling provides another valuable approach (Simons et al. 1999). A quite new and remarkably effective modality that operates on a poorly explored mechanism is frequency specific microcurrent therapy (www.frequentcyspecific. com). This has been demonstrated to be effective for MTrPs (McMakin 2004) and in patients who developed fibromyalgia following a whiplash injury (McMakin 2005) as listed below. The patient benefits when treatment begins with manual therapy rather than dry needling or injection.

When patients are shown how to perform the manual therapy on themselves, the experience of immediate pain relief helps to convince them that muscles are causing the pain, gives them a sense of control, and improves compliance for the important home stretching exercises. When the patient receives only dry needling or injection, treatment becomes the clinician's responsibility, not the patient's. A close physician-therapist relationship permits the physician to concentrate on injections when indicated, while the therapist teaches the patient home stretch exercises specific to the involved muscles. Treatments may cause slight discomfort but should always be pain-free (not over about 2 or 3 on a 10 point visual analog scale) to avoid plasticity changes in the central nervous system, which enhances and prolongs pain perception.

Treatment of myofascial trigger points (MTrPs)
- Manual techniques
  - Trigger point pressure release – finger pressure applied to the MTrP
  - Contract-release – alternate voluntary contraction and passive or active stretching of the muscle with the MTrPs
  - Vapocoolant spray and stretch – application of a stream of vapocoolant, or use of a covered edge of ice, to facilitate release
- Needling techniques (require diagnostic skill)
  - Dry needling (as effective as injection, but leaves more post injection soreness)
  - Injection of 1 % lidocaine (reduces postinjection soreness)
  - Injection of botulinum toxin (specifically for muscles with spasticity and MTrPs using electromyographic guidance for end plate noise)
- New possibilities
  - Frequency specific microcurrent therapy
  - Shockwave therapy

Figure 3 shows how ▶ trigger point pressure release can be effective. Shortened sarcomeres are a key part of the cause of MTrP symptoms. The constant-volume balloon is analogous to a sarcomere, showing how gentle but firm

Trigger Point
Pressure Release

Balloon   Sarcomeres

Finger Pressure

Muscle Fiber

Myofascial Trigger Points, Fig. 3 This schematic illustrates the mechanism by which trigger point pressure release can relieve the tension of the taut band by elongating the shortened sarcomeres, which has the effect of localized stretching of the muscle fibers. Like the balloon, a sarcomere has a nearly constant volume, so flattening elongates it. The contracted fibers in the nodular swelling of muscle fibers represent either a contraction knot or contraction disc. The effect of trigger point pressure release is enhanced by including a gentle voluntary contraction of that muscle while holding it at positions of progressive gentle passive stretch

pressure on contraction knots (or contraction discs) will flatten and lengthen (normalize) the shortened sarcomeres. As their tension releases, digital pressure is increased. This local stretch reduces actin and myosin overlap, which reduces contractile activity, energy consumption, and ischemia – all of which tend to break the TrP feedback cycle.

▶ Contract-release (also post-isometric relaxation) requires only recognition of the painfully restricted stretch range of motion to which this treatment is then applied (see Fig. 4). This can be learned and used by patients. If necessary, additional release can often be obtained with vapocoolant spray and stretch, dry needling, injection of the MTrP with 1 % lidocaine (to reduce postinjection soreness), or with reciprocal inhibition. Following treatment, three full cycles of active range of motion help to normalize function of the treated muscle. If relief is still temporary, perpetuating factors must be investigated and resolved.

Finger Pressure

Voluntary
Contraction

Voluntary
Contraction

Tension
Produced

Length
1.5μ   3.0μ
Sarcomere Length

Myofascial Trigger Points, Fig. 4 Schematic showing effect of combining trigger point pressure release (finger pressure) and contract-release, which takes advantage of the reduction in muscle tightness following a gentle (about 10 % of maximum effort) voluntary contraction. Treatment slowly alternates between gentle voluntary contractions held for several seconds, immediately followed by passive or active stretching of the muscle. This takes up slack that developed following contraction. The contraction-release cycles can be repeated rhythmically, as long as each cycle brings progress. The graph of a characteristic length-tension curve for one sarcomere shows that the strongly shortened sarcomeres produce much less tension than the mid-length sarcomeres throughout the rest of the muscle fiber. The mid-length sarcomeres are both stronger and far outnumber the shortened sarcomeres, so a gentle contraction temporarily increases the length of the shortened sarcomeres. Several cycles of full range of motion help to consolidate treatment gains. The two techniques are combined for increased therapeutic effectiveness

Effective injection or dry needling requires precise location of the MTrP to enter it with the needle. Elicited pain and local twitch responses assure a more effective treatment. Concentration on just injection encourages neglect of home exercises and perpetuating factors, and the patient usually does not understand the cause of the pain as well.

Since botulinum toxin inactivates ACh release at the motor end plate, it should be a specific therapy for MTrPs and has been effective. However, it is very expensive, lasts about 3 months, can induce an immune reaction, and is unlikely to be more effective than the other techniques described when they are skillfully applied. It is specifically indicated in muscles that are spastic and also have painful MTrPs. Botulinum toxin is

Myofascial Trigger Points, Table 2 Number of MEDLINE citations (mainstream medical literature) that were indexed in the past 7 years under *low back pain* compared to the number of trigger point citations retrieved by combining *low back pain and trigger points* with a *low back pain and myofascial trigger point* search

Year of publication	Low back pain citations	Trigger point citations	Percent trigger point (%)
1996	552	2	0.4
1997	312	2	0.5
1998	459	2	0.4
1999	561	0	0
2000	527	3	0.6
2001	537	3	0.6
2002	553	5	0.9
1996–2002	3,601	17	0.5

most effectively injected under electromyographic guidance.

Application of shockwave technology for the localization and treatment of MTrPs looks promising (Bauermeister 2005; Müller-Ehrenberg and Licht 2005).

One reason for the neglect of MTrPs as a muscular source of pain is that no medical specialty takes responsibility for research and training in all medical aspects of muscle as an organ. As a result, clinical and basic research on MTrPs is conspicuous for its scarcity, and MTrP training of medical practitioners is rarely adequately covered in schools. It can be very difficult for patients to find practitioners adequately skilled in the diagnosis and treatment of MTrPs.

Unfortunately, many practitioners were not trained to include MTrPs as a differential diagnosis for musculoskeletal pain. Low back pain (LBP) causes much human suffering and health care costs. However, of the 3,501 LBP citations retrieved from MEDLINE covering the past years, less than 1 % were also indexed as including MTrPs (see Table 2). Moreover, much LBP is either caused by MTrPs or has a significant MTrP component. A number of papers that do relate MTrPs and LBP describe MTrPs but do not identify them in the title or abstract. A few articles that identified MTrPs in

LBP were published in journals that are not listed by MEDLINE. Most authors writing about LBP do not include MTrPs as part of their routine differential diagnosis.

A recent series of papers by César Fernández of Spain and colleagues is now appearing in the scientific literature that indicate a strong relationship between MTrPs and several kinds of headache (Fernández et al. 2005, 2006).

Cross-References

► Central Trigger Point
► Chronic Pelvic Pain, Musculoskeletal Syndromes
► Chronic Pelvic Pain, Physical Therapy Approaches, and Myofascial Abnormalities
► Dry Needling
► Myalgia

References

Bauermeister, W. (2005). Diagnose und Therapie des Myofaszialen Triggerpunk Syndroms durch Lokalisierung und Stimulation Sensibilisierter Nozizeptoren mit Focussierten Elektrohydraulischen Stosswellen. *MOT, 5,* 65–74.
Couppé, C., Midttun, A., Hilden, J., et al. (2001). Spontaneous eedle electromyographic activity in myofascial trigger points in the infraspinatus muscle: A blinded assessment. *Journal of Musculoskeletal Pain, 9,* 7–16.
Dejung, B., Gröbli, C., Colla, F., et al. (2003). *Triggerpunkt-therapie (trigger point therapy).* Bern: Hans Huber Verlag.
Fernández de las Peñas, C., Campo, M., Carnero, J., & Page, J., (2005). Manual therapies in myofascial trigger point treatment: A systematic review. *Journal of Bodywork and Movement Therapies, 9,* 27e34.
Fernández-de-las-Peñas, C., Alonso-Blanco, C., Cuadrado, M. L., Gerwin, R. D., Pareja, J. A. (2006). Trigger points in the suboccipital muscles and forward head posture in tension type headache. *Headache, 46,* 454–460.
Gerwin, R. D., Shannon, S., Hong, C.-Z., et al. (1997). Interrater reliability in myofascial trigger point examination. *Pain, 69,* 65–73.
Hsieh, C.-Y., Hong, C.-Z., Adams, A. H., et al. (2000). Interexaminer reliability of the palpation of trigger points in the trunk and lower limb muscles. *Archives of Physical Medicine and Rehabilitation, 81,* 258–264.

McMakin, C. R. (2004). Microcurrent therapy: A novel treatment method for chronic low back myofascial pain. *Journal of Bodywork and Movement Therapies*, *8*, 143–153.

McMakin, C. R., Gregory, W. M., & Phillips, T. M. (2005). Cytokine changes with microcurrent treatment of fibromyalgia associated with cervical spine trauma. *Journal of Bodywork and Movement Therapies*, *9*, 169–176.

Mense, S., Simons, D. G., & Russell, I. J. (2001). *Muscle pain: Its nature, diagnosis, and treatment*. Philadelphia: Lippincott, Williams & Wilkins.

Müller-Ehrenberg, H., & Licht, G. (2005). Diagnosis and therapy of myofascial pain syndrome with focused shock waves (ESWT). *Annals of Rehabilitation Medicine*, *5*, 1–6.

Sciotti, V. M., Mittak, V. L., DiMarco, L., et al. (2001). Clinical precision of myofascial trigger point location in the trapezius muscle. *Pain*, *93*, 259–266.

Shah, J. P., Phillips, T., Danoff, J., et al. (2004). Novel microanalytical technique distinguishes three clinically distinct groups: 1) Subjects without pain and without a myofascial trigger point; 2) Subjects without pain with a myofascial trigger point; 3) Subjects with pain and a myofascial trigger point. *American Journal of Physical Medicine & Rehabilitation*, *83*, 231.

Simons, D. G. (2001). Do endplate noise and spikes arise from normal motor endplates? *American Journal of Physical Medicine & Rehabilitation*, *80*, 134–140.

Simons, D. G. (2002). Understanding effective treatments of myofascial trigger points. *Journal of Bodywork and Movement Therapies*, *6*, 81–88.

Simons, D. G., Travell, J. G., & Simons, L. S. (1999). *Travell & Simons' myofascial pain and dysfunction: The trigger point manual* (2nd ed., Vol. 1). Baltimore: Williams & Wilkins.

Simons, D. G., Hong, C.-Z., & Simons, L. S. (2002). Endplate potentials are common to midfiber myofascial trigger points. *American Journal of Physical Medicine & Rehabilitation*, *81*, 212–222.

Travell, J. G., & Simons, D. G. (1992). *Myofascial pain and dysfunction: The trigger point manual* (The lower extremities, Vol. 2). Baltimore: Williams & Wilkins.

Wolfe, F., Ross, K., Anderson, J., et al. (1998). The prevalence and characteristics of fibromyalgia in the general population. *Arthritis and Rheumatism*, *38*, 19–28.

Wright, E. F. (2000). Referred craniofacial pain patterns in patients with temporomandibular disorder. *Journal of the American Dental Association*, *131*, 1307–1315.

Myofibrositis

▶ Myalgia

Myogelosis

▶ Myalgia

Myositis

Dieter Pongratz
Friedrich Baur Institute, Ludwig Maximilians University, Munich, Germany

Synonyms

Inflammatory myopathies

Definition

Inflammatory myopathies represent a small but, from a therapeutical point of view, important group of acquired muscular disorders. With the exception of infectious causes (viral myositis, bacterial myositis, and parasitic myositis), immunogenic forms have to be considered. There are four different forms:

1. Dermatomyositis
2. Overlap syndromes
3. ▶ Polymyositis
4. ▶ Inclusion body myositis

Characteristics

The most prominent clinical symptoms of all immunogenic inflammatory myopathies are muscle weakness and muscle atrophy. Muscle pain is common in ▶ dermatomyositis and overlap syndromes but is less prominent or even missing in chronic polymyositis and especially inclusion body myositis. From a clinical point of view, there is a predominant involvement of the proximal muscles of the limbs and the arms. Only in inclusion body myositis is the presence of distal muscle weakness, especially of the foot extensors and finger flexors, a diagnostic clue from the beginning. The pharyngeal and neck flexor muscles are often affected in all forms of myositis causing dysphagia and, at times, difficulties in holding up the head. Muscular wasting develops during the course of the illness and is pronounced in chronic polymyositis and inclusion body myositis.

Special Signs of Dermatomyositis

Dermatomyositis is characterized by skin lesions accompanying or preceding muscle symptoms. There is the typical heliotrope erythema (lilac disease), involving especially the eyelids, face, and upper trunk. More chronic skin lesions show depigmentation and hyperpigmentations. As the disease progresses, subcutaneous calcifications occur.

Laboratory Diagnosis and Differential Diagnoses

Laboratory diagnosis is based on:
1. Measurement of muscle enzymes.
2. Electromyography.
3. Muscle biopsy.
4. In some cases, clinical or laboratory signs of an associated connective tissue disease may be an additional aid (▶ overlap syndromes).

Muscle MRI with STIR sequence is helpful in chronic forms to establish the best position for performing a muscle biopsy (muscle edema).

The level of muscle enzymes, especially creatine kinase (CK), usually parallels disease activity and can be elevated in acute stages by as much as 50 × above normal. In rare cases of active polymyositis and dermatomyositis, however, CK can also be within normal range. In inclusion body myositis, the level of CK is not usually elevated more than tenfold, and in some cases, it may be normal.

Needle electromyography shows myopathic potentials characterized by short duration, low amplitude, and polyphasic configuration on voluntary activation in combination with increased spontaneous activity (fibrillations, positive sharp waves, complex repetitive discharges). This pattern also occurs in a variety

of active myopathic processes of other origins and should therefore not be considered diagnostic for the inflammatory myopathies. Mixed myopathic and neurogenic potentials may be present in some cases as a consequence of inclusion body myositis.

Muscle biopsy is the definitive test, not only for establishing the diagnosis of dermatomyositis, polymyositis, or inclusion body myositis, but also for excluding other neuromuscular disorders.

In dermatomyositis, inflammatory infiltrates are found predominantly in perivascular and perifascicular regions, producing the characteristic picture of a myositis of the perifascicular type. There are striking lesions of the small intramuscular blood vessels, with endothelial proliferation and so-called tubulovesicular inclusions seen by electron microscopy. Perifascicular atrophy and fiber damage are diagnostic of dermatomyositis even in the absence of infiltration.

Immunohistological methods show that the cellular infiltrates in this disease consist of B lymphocytes, CD4$^+$ lymphocytes, and macrophages. C5b9 complement deposits within small blood vessels are very characteristic, causing destruction of the capillaries and muscle ischemia.

In polymyositis, infiltrates are predominantly endomysial, producing the picture of a diffuse myositis. There is no evidence of microangiopathy. Immunohistologically, cytotoxic CD8$^+$ lymphocytes are the predominant cells that invade nonnecrotic muscle fibers. The muscle fibers themselves aberrantly express major histocompatibility complex class I (MHC I) antigen, which is absent in normal muscle.

Inclusion body myositis is characterized by endomysial inflammation with CD8$^+$ lymphocytes in particular, similar to those seen in polymyositis. In addition, rimmed vacuoles with eosinophilic cytoplasmatic inclusions can be found, as a rule, within many muscle fibers. On electron microscopy, the vacuoles correspond to the so-called autophagic vacuoles. In addition, filamentous inclusions in the cytoplasm and nuclei are prominent.

High levels of myositis-associated autoantibodies are found in overlap syndromes.

Natural Course

The incidence of polymyositis, dermatomyositis, and inclusion body myositis is approximately 1/100,000. The natural course of polymyositis and dermatomyositis is relatively unknown because patients are almost always treated with steroids. Inclusion body myositis is relatively resistant to all therapies and, as a rule, slowly progressive.

Principles of Therapy

As corticosteroids and immunosuppressive drugs seem to be of benefit, polymyositis and dermatomyositis belong to the treatable group of myopathies. In contrast, inclusion body myositis is resistant to most therapies, although treatment with intravenous immunoglobulins may be of some benefit.

Cross-References

▶ Muscle Pain in Systemic Inflammation (Polymyalgia Rheumatica, Giant Cell Arteritis, Rheumatoid Arthritis)
▶ Muscle Pain Model, Inflammatory Agents-Induced
▶ Nocifensive Behaviors, Muscle and Joint
▶ Opioids and Muscle Pain

Myotomal Pain

▶ Myofascial Pain

Encyclopedia of Pain

Gerald F. Gebhart • Robert F. Schmidt

Editors

Encyclopedia of Pain

Second Edition

Volume 4

N–O

With 795 Figures and 217 Tables

Springer Reference

Editors

Gerald F. Gebhart
Department of Anesthesiology
Center for Pain Research
University of Pittsburgh
Pittsburgh, PA, USA

Robert F. Schmidt
Physiologisches Institut der
Universität Würzburg
Würzburg, Germany

ISBN 978-3-642-28752-7   ISBN 978-3-642-28753-4 (eBook)
ISBN 978-3-642-28754-1 (print and electronic bundle)
DOI 10.1007/978-3-642-28753-4
Springer Heidelberg New York Dordrecht London

Library of Congress Control Number: 2013951551

Printed on acid-free paper

Springer is part of Springer Science+Business Media (www.springer.com)

Preface to the Second Edition

As was pointed out in the Preface to the first edition, pain is the principal reason individuals seek medical and dental attention. Fortunately, acute tissue insults associated with pain – and typically inflammation – readily resolve with appropriate treatment and care. Chronic pain states, in contrast, are notoriously difficult to manage and are commonly associated with comorbidities (e.g., depression, cross-organ sensitization) and reduced quality of life. Indeed, chronic pain has come to be considered by some as a disease in its own right. This edition of the *Encyclopedia of Pain* updates the content of the 1st edition and introduces new topics consistent with advances in our knowledge of underlying mechanisms of pain. Thus, new content on channelopathies, expanded and updated material on the role of glia and immune-competent cell interactions with nociceptive neurons, and advances in imaging and changes in brain structure in chronic pain states are included in this edition.

The cardinal objective of research on pain is the translation of knowledge between the laboratory and the clinic. The present prevailing refrain is "bench to bedside," but in fact many ideas for research in pain derive from clinical observations and the absence of knowledge to explain clinical pain states. In the recent past, basic and clinical science research has seen the application of imaging techniques to visualize areas of the brain that are involved in both the sensory and affective aspects of pain, cloning of pain-related channels and receptors, and the application of elegant genetic manipulations that selectively "label" functionally distinct dorsal root ganglion sensory and spinal neurons, two-photon imaging, and optogenetic approaches.

Accordingly, much is new in this second edition of the *Encyclopedia of Pain*. The number of authors who have contributed to this edition is close to 800. In addition to a print edition, the publisher, Springer-Verlag, will produce an electronic version that will make comprehensive searches easy to manage. The electronic version, to be made available on the online publishing platform SpringerReference, can be updated regularly to ensure that the content remains current.

October 2013
Gerald F. Gebhart   Robert F. Schmidt
Pittsburgh, PA, USA   Würzburg, Germany

Preface to the First Edition

As all medical students know, pain is the most common reason for a person to consult a physician. Under ordinary circumstances, acute pain has a useful, protective function. It discourages the individual from activities that aggravate the pain, allowing faster recovery from tissue damage. The physician can often tell from the nature of the pain what its source is. In most cases, treatment of the underlying condition resolves the pain. By contrast, children born with congenital insensitivity to pain suffer repeated physical damage and die young (see Sweet WH (1981) Pain 10:275).

Pain resulting from difficult to treat or untreatable conditions can become persistent. Chronic pain "never has a biologic function but is a malefic force that often imposes severe emotional, physical, economic, and social stresses on the patient and on the family..." (Bonica JJ (1990) The Management of Pain, vol 1, 2nd edn. Lea & Febiger, Philadelphia, p 19). Chronic pain can be considered a disease in its own right.

Pain is a complex phenomenon. It has been defined by the Taxonomy Committee of the International Association for the Study of Pain as "An unpleasant emotional and sensory experience associated with actual or potential tissue damage, or described in terms of such damage" (Merskey H and Bogduk N (1994) Classification of Chronic Pain, 2nd edn. IASP Press, Seattle). It is often ongoing, but in some cases it may be evoked by stimuli. Hyperalgesia occurs when there is an increase in pain intensity in response to stimuli that are normally painful. Allodynia is pain that is evoked by stimuli that are normally non-painful.

Acute pain is generally attributable to the activation of primary efferent neurons called nociceptors (Sherrington CS (1906) The Integrative Action of the Nervous System. Yale University Press, New Haven; 2nd edn, 1947). These sensory nerve fibers have high thresholds and respond to strong stimuli that threaten or cause injury to tissues of the body. Chronic pain may result from continuous or repeated activation of nociceptors, as in some forms of cancer or in chronic inflammatory states, such as arthritis.

However, chronic pain can also be produced by damage to nervous tissue. If peripheral nerves are injured, peripheral neuropathic pain may develop. Damage to certain parts of the central nervous system may result in central neuropathic pain. Examples of conditions that can cause central neuropathic pain include spinal cord injury, cerebrovascular accidents, and multiple sclerosis.

Research on pain in humans has been an important clinical topic for many years. Basic science studies were relatively few in number until experimental work on pain accelerated following detailed descriptions of peripheral nociceptors and central nociceptive neurons that were made in the 1960s and 1970s, by the discovery of the endogenous opioid compounds and the descending pain control systems in the 1970s and the application of modern imaging techniques to visualize areas of the brain that are affected by pain in the 1990s. Accompanying these advances has been the development of a number of animal models of human pain states, with the goal of using these to examine pain mechanisms and also to test analgesic drugs or non-pharmacologic interventions that might prove useful for the treatment of pain in humans. Basic research on pain now emphasizes multidisciplinary approaches, including behavioral testing, electrophysiology, and the application of many of the techniques of modern cell and molecular biology, including the use of transgenic animals.

The *Encyclopedia of Pain* is meant to provide a source of information that spans contemporary basic and clinical research on pain and pain therapy. It should be useful not only to researchers in these fields but also to practicing physicians and other health care professionals and to health care educators and administrators. The work is subdivided into 35 fields, and the Field Editor of each of these describes the areas covered in each in a brief review entry. The topics included in a field are the subject of a series of short essays, accompanied by key words, definitions, illustrations, and a list of significant references. The number of authors who have contributed to the encyclopedia exceeds 550. The plan of the publisher, Springer-Verlag, is to produce both print and electronic versions of this encyclopedia. Numerous links within the electronic version should make comprehensive searches easy to manage. The electronic version will be updated at sufficiently short intervals to ensure that the content remains current.

The editors thank the staff at Springer-Verlag who have provided oversight for this project, including Rolf Lange, Thomas Mager, Claudia Lange, Natasja Sheriff, and Michaela Bilic. Working with these outstanding individuals has been a pleasure.

July 2006
ROBERT F. SCHMIDT   WILLIAM D. WILLIS
Würzburg, Germany   Galveston, Texas, USA

About the Editors

Gerald F. Gebhart received his PhD from The University of Iowa at Iowa City, Iowa, USA, after which he completed a 2-year fellowship at the Université de Montréal, Montréal, Canada, under the direction of Herbert H. Jasper. He began his academic career in the Department of Pharmacology at The University of Iowa in 1973, where he developed a productive research program and led a Pain Interest Group. His career includes a sabbatical year in Heidelberg, Germany (1981–1982), and leadership as head of the Department of Pharmacology from 1996 to 2006. In 2006, Dr. Gebhart was named director of the Center for Pain Research, Department of Anesthesiology, the University of Pittsburgh at Pittsburgh, Pennsylvania, USA. Dr. Gebhart has served the pain community throughout his career. He was elected to the Board of Directors of the American Pain Society (1991–1994) and subsequently elected president of the APS (1997). In 2000, he became the founding editor of *The Journal of Pain*, the official journal of the APS, serving as editor-in-chief until 2010. Dr. Gebhart also served on the Council and later as president-elect, president, and past president of the International Association for the Study of Pain (2005–2012).

Dr. Gebhart is best known for his research contributions in descending modulation of pain and mechanisms of visceral pain, and was designated in 2004 by Thomson ISI as a Highly Cited Researcher (Neuroscience). His scientific contributions have been recognized with several awards, including a 5-year Bristol Myers-Squibb Award for Excellence in Pain Research (1989–1994), a 10-year MERIT Award from the National Institutes of Health (1993–2003),

the Frederick W.L. Kerr Award from the American Pain Society (1994), the Purdue Pharma Prize for Pain Research (2004), the Janssen Award in Gastroenterology (2005), the Founders Award from the American Academy of Pain Medicine (2006), and the Patrick D. Wall Award from the British Pain Society (2012).

Robert F. Schmidt studied medicine at Heidelberg University (Germany). He graduated in 1959 and was promoted to Dr.med. in the same year. After residencies in clinical disciplines, Dr. Schmidt went to the John Curtin School of Medical Research of the Australian National University in Canberra, ACT, where he worked under the head of the Department of Neurophysiology, Sir John C. Eccles. He received his Ph.D. in 1963, the same year Sir John was awarded the Nobel Prize for Physiology and Medicine.

Returning to Heidelberg Medical School, Dr. Schmidt continued his academic career. He completed his habilitation in 1964, becoming a tenured lecturer (Wissenschaftlicher Rat) in 1966 and an associate professor in 1970. After an 8-month sabbatical with Sir John at the State University of New York, Dr. Schmidt became professor and director of the Department of Physiology at the University of Kiel, Germany. In 1982, he moved to the University of Würzburg to take the same position in the Department of Physiology. In 2000, Dr. Schmidt became professor emeritus at this department.

The research work of Dr. Schmidt and his associates in both Kiel and Würzburg focused on the neurophysiology of fine sensory afferents and their connections and functions in the spinal cord. In Würzburg, he concentrated on the processes leading to the changes in response characteristics of nociceptive afferents (i.e., peripheral sensitization) during inflammation of joints, discovering in the process silent nociceptors, which have subsequently been described in all vertebrates studied, including humans. In addition, Dr. Schmidt has edited, coedited, authored, and coauthored numerous textbooks and monographs, most of them having several editions and translations.

In recognition of his research contributions, Dr. Schmidt received many prizes, awards, and honors, including a D.Sc. honoris causa from the University of New South Wales, Sydney, Australia, and an Honorary Professorship from the University of Tübingen, Germany. He was elected to membership in the Akademie der Wissenschaften und der Literatur, Mainz, Germany. Dr. Schmidt is an Honorary Member of the Colombian Association for the Study of Pain, the Japan Physiological Society, the German Society for the Study of Pain, the German Pain Association, the International Association for the Study of Pain (IASP), and the German Physiological Society (in which he served as president in 1979). In 2000, he was awarded the Bundesverdienstkreuz 1st Class (Order of Merit of the German Federal Republic).

In recognition of his research contributions, Dr. Schmidt received many prizes, awards, and honors, including a D.Sc. honoris causa from the University of New South Wales, Sydney, Australia, and an Honorary Professorship from the University of Tübingen, Germany. He was elected to membership in the Akademie der Wissenschaften und der Literatur, Mainz, Germany. Dr. Schmidt is an Honorary Member of the Colombian Association for the Study of Pain, the Japan Physiological Society, the German Society for the Study of Pain, the German Pain Association, the International Association for the Study of Pain (IASP), and the German Physiological Society, of which he served as President in 1979. In 2003, he was awarded the Bundesverdienstkreuz, 1st Class (Order of Merit of the German Federal Republic).

Section Editors

Ebtesam Ahmed College of Pharmacy and Health Sciences, St. John's University, Department of Pain Medicine and Palliative Care, Beth Israel Medical Center, New York, NY, USA

A. Vania Apkarian Department of Physiology, Northwestern University Feinberg School of Medicine, Chicago, IL, USA

Lars Arendt-Nielsen Center for Sensory-Motor Interaction, Department of Health Science and Technology, School of Medicine, Aalborg University, Aalborg, Denmark

Carlos Belmonte Instituto de Neurociencias, Universidad Miguel Hernandez-CSIC, San Juan de Alicante, Spain

Niels Birbaumer Institute of Medical Psychology and Behavioral Neurobiology, University of Tübingen, Tübingen, Germany

Nikolai Bogduk University of Newcastle, Newcastle Bone & Joint Institute, Royal Newcastle Centre, Newcastle, NSW, Australia

Jin Mo Chung Department of Neuroscience and Cell Biology, University Chair in Neuroscience, University of Texas Medical Branch, Galveston, TX, USA

Joyce DeLeo Dartmouth Hitchcock Medical Center, Dartmouth Medical School, Hanover, NH, USA

Marshall Devor Department of Cell & Developmental Biology, Institute of Life Sciences and Center for Research on Pain, The Hebrew University of Jerusalem, Jerusalem, Givat Ram, Israel

Hans-Christoph Diener Department of Neurology, University Hospital Essen, University Duisburg-Essen, Essen, Germany

Jonathan O. Dostrovsky Department of Physiology, University of Toronto, Toronto, ON, Canada

Ronald Dubner Department of Neural and Pain Sciences, University of Maryland Dental School, Baltimore, MD, USA

Michel Y. Dubois NYU Medical School, NYU Langone Medical Centers, Center for Research & Treatment of Pain, NYU Ambulatory Care Center, New York, NY, USA

Nanna Brix Finnerup Aarhus University Hospital, Danish Pain Research Center, Aarhus, Denmark

Herta Flor Department of Cognitive and Clinical Neuroscience, Central Institute of Mental Health, Medical Faculty Mannheim, Heidelberg University, Mannheim, Germany

Gerald F. Gebhart Department of Anesthesiology, Center for Pain Research, University of Pittsburgh, Pittsburgh, PA, USA

Maria Adele Giamberardino Pathophysiology of Pain Laboratory and
Centre for Fibromyalgia and Musculoskeletal Pain, Department of Medicine
and Science of Aging, University of Chieti, Chieti, Italy

Glenn J. Giesler Department of Neuroscience, University of Minnesota,
Minneapolis, MN, USA

Peter Goadsby Headache Center, Department of Neurology, University of California, San Francisco, San Francisco, CA, USA

Michael S. Gold Department of Anesthesiology, Center for Pain Research, University of Pittsburgh, Pittsburgh, PA, USA

Hermann O. Handwerker Department of Physiology and Pathophysiology, University of Erlangen/Nürnberg, Erlangen, Germany

Burkhard Hinz Institute of Toxicology and Pharmacology, University of Rostock, Rostock, Germany

Wilfrid Jaenig Physiologisches Institut, Christian-Albrechts-Universität zu
Kiel, Kiel, Germany

Michaela Kress Department für Physiologie und Medizinische Physik,
Medizinische Universität Innsbruck, Innsbruck, Austria

Frederick A. Lenz Department of Neurosurgery, Johns Hopkins Hospital, Baltimore, MD, USA

Tonya M. Palermo Seattle Children's Hospital Research Institute, CW8-6 - Child Health, Behavior and Development, University of Washington, Seattle, WA, USA

Frank Porreca Department of Pharmacology, University of Arizona Health
Sciences Center, Tucson, AZ, USA

Russell K. Portenoy Department of Pain Medicine and Palliative Care, Beth
Israel Medical Center, New York, NY, USA

James P. Robinson Department of Anesthesiology & Pain Medicine, University of Washington, Seattle, WA, USA

Hans-Georg Schaible Department of Physiology - Neurophysiology, Friedrich-Schiller-University of Jena, Jena, Germany

Robert F. Schmidt Physiologisches Institut der Universität Würzburg, Würzburg, Germany

Stephan A. Schug School of Medicine and Pharmacology, Royal Perth Hospital, Faculty of Medicine, Dentistry & Health Sciences, The University of Western Australia, UWA Anaesthesiology, Perth, Australia

Barry J. Sessle Faculty of Dentistry, University of Toronto, Toronto, ON, Canada

Bengt H. Sjölund Department of Community Medicine and Rehabilitation Medicine, Umea University, Umea, Sweden

Mark D. Sullivan Psychiatry and Behavioral Sciences, University of Washington, Seattle, WA, USA

Irene Tracey Pain Imaging Neuroscience (PaIN) Group, Department of Physiology, Anatomy and Genetics, Oxford University, Oxford, UK

Dennis C. Turk Department of Anesthesiology & Pain Research, University of Washington, Seattle, WA, USA

Félix Viana Instituto de Neurociencias, Universidad Miguel Hernandez-CSIC, San Juan de Alicante, Alicante, Spain

Katy Vincent Nuffield Department of Obstetrics and Gynaecology, The Women's Centre, John Radcliffe Hospital, University of Oxford, Oxford, UK

Gary A. Walco Department of Anesthesiology and Pain Medicine, University of Washington School of Medicine, Seattle Children's Hospital, Seattle, WA, USA

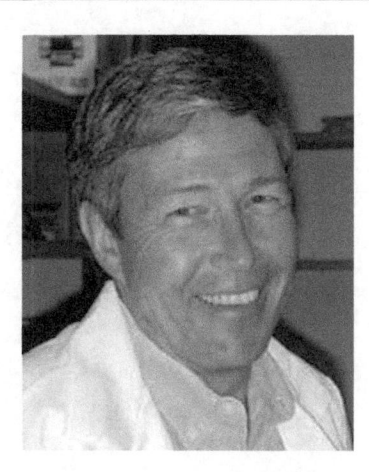

George L. Wilcox Department of Neuroscience, Pharmacology and Dermatology, University of Minnesota Medical School, Minneapolis, MN, USA

Katja Wiech Nuffield Department of Clinical Neurosciences, John Radcliffe Hospital, Oxford, UK

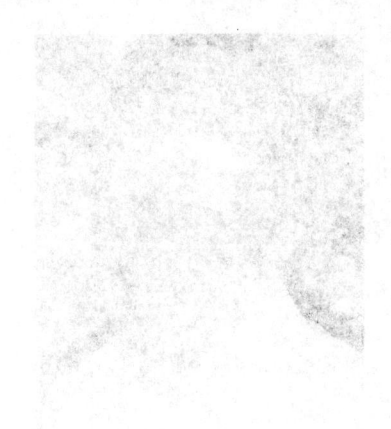

George L. Wilcox Department of Neuroscience, Pharmacology and Dermatology, University of Minnesota Medical School, Minneapolis, MN, USA

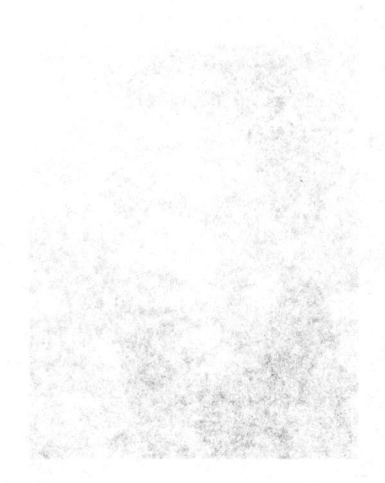

Katja Wiech Nuffield Department of Clinical Neuroscience, John Radcliffe Hospital, Oxford, UK

Contributors

Tipu Aamir The Auckland Regional Pain Service, Auckland Hospital, Auckland, New Zealand

Catherine Abbadie Department of Pharmacology, Merck Research Laboratories, Rahway, NJ, USA

Frances V. Abbott Departments of Psychiatry and Psychology, McGill University, Montreal, QC, Canada

Heather Adams Department of Surgery, McGill University Health Centre, Montreal, QC, Canada

Mary O. Adedoyin Department of Anatomy, University of California San Francisco, San Francisco, CA, USA

Rolf H. Adler University of Berne Medical School, Kehrsatz, Switzerland

Claire D. Advokat Department of Psychology, Louisiana State University, Baton Rouge, LA, USA

Reto M. Agosti Headache Center Hirlsanden, Zurich, Switzerland

Ebtesam Ahmed College of Pharmacy and Health Sciences, St. John's University, Department of Pain Medicine and Palliative Care, Beth Israel Medical Center, New York, NY, USA

Marie-Claire Albanese Department of Psychology, McGill University, Montreal, QC, Canada

Kathryn M. Albers Department of Neurobiology, University of Pittsburgh School of Medicine, Pittsburgh, PA, USA

Phillip J. Albrecht Center for Neuropharmacology and Neuroscience, Albany Medical College, Albany, NY, USA

Elie D. Al-Chaer Center for Pain Research, University of Arkansas for Medical Sciences, Little Rock, AR, USA

Terry Altilio Department of Pain Medicine and Palliative Care, Beth Israel Medical Center, New York, USA

Rainer Amann Medical University Graz, Graz, Austria

Ron Amir Institute of Life Sciences, The Hebrew University of Jerusalem, Jerusalem, Israel

Praveen Anand Peripheral Neuropathy Unit, Imperial College London, London, UK

Karen O. Anderson The University of Texas MD Anderson Cancer Center, Houston, TX, USA

Ole K. Andersen Department of Health Science & Technology, Aalborg University, Aalborg, Denmark

Stanley W. Anderson Department of Neurosurgery, Johns Hopkins Hospital, Johns Hopkins Medical Institutions, Johns Hopkins University, Baltimore, MD, USA

Trevor Anderson Fairview Pain & Palliative Care, University of Minnesota Medical Center, Minneapolis, MN, USA

Karl-Erik Andersson Wake Forest Institute for Regenerative Medicine, Wake Forest University School of Medicine, Winston Salem, USA

Frank Andrasik Department of Psychology, Applied Psychophysiology and Biofeedback, The University of Memphis, Memphis, TN, USA

A. Vania Apkarian Department of Physiology, Northwestern University Feinberg School of Medicine, Chicago, IL, USA

Lars Arendt-Nielsen Center for Sensory-Motor Interaction, Department of Health Science and Technology, School of Medicine, Aalborg University, Aalborg, Denmark

Kirsten Arndt CNS Pharmacology, Boehringer Ingelheim Pharma GmbH & Co. KG, Biberach, Germany

Charles-Daniel Arreto Université Paris Descartes, INSERM UMR 894, France

Erik Askenasy Neurobiology and Anatomy, Health Science Center, University of Texas, Houston, TX, USA

J. H. Atkinson Psychiatry Service, VA San Diego Healthcare System and Department of Psychiatry, University of California, San Diego, CA, USA

Nadine Attal Centre d'Evaluation et de Traitement de la Douleur, Hôpital Ambroise Paré, INSERM U 987 APHP, Boulogne-Billancourt, France

Christoph Aufenberg Functional Neurosurgery, Neurosurgical Clinic University Hospital, Zürich, Switzerland

Beate Averbeck Sanofi-Aventis Deutschland GmbH, Frankfurt, Germany

Qasim Aziz Section of GI Sciences, Hope Hospital University of Manchester, Salford, UK

Cathrine Baastrup Danish Pain Research Center, Aarhus University Aarhus University Hospital, Aarhus, Denmark

F. W. Bach Danish Pain Research Center, Department of Neurology, Aarhus University Hospital, Aarhus, Denmark

S. K. Back Neuroscience Research Institute and Department of Physiology, Korea University, College of Medicine, Seoul, Republic of Korea

Misha-Miroslav Backonja Department of Neurology, University of Wisconsin-Madison, Madison, WI, USA

Banani Banerjee Division of Gastroenterology and Hepatology and Pediatric Gastroenterology, Medical College of Wisconsin, Milwaukee, WI, USA

John D. Banja Department of Rehabilitation Medicine, Center for Ethics Emory University, Atlanta, GA, USA

Carsten Bantel Department of Surgery & Cancer? Anaesthetics Section, Imperial College of Science and Medicine Chelsea and Westminster Campus, London, UK

Andrew Paul Baranowski The Pain Management Centre, National Hospital for Neurology and Neurosurgery, UCL Hospitals NHS Foundation Trust, London, UK

Alex Barling Department of Neurology, Birmingham Muscle and Nerve Centre Queen Elizabeth Hospital, Birmingham, UK

Ralf Baron Department of Neurological Pain Research and Therapy, Christian Albrechts University Kiel, Kiel, Germany

Gordon A. Barr Department of Anesthesiology and Critical Care Medicine, Children's Hospital of Philadelphia University of Pennsylvania, Philadelphia, PA, USA

Emily J. Bartley Department of Psychology, The University of Tulsa, Tulsa, OK, USA

Jeffrey R. Basford Department of Physical Medicine and Rehabilitation, Mayo Clinic and Mayo Foundation, Rochester, NY, USA

Michele C. Battie Department of Physical Therapy, University of Alberta, Edmonton, AB, Canada

Ulf Baumgartner Chair of Neurophysiology, Center for Biomedicine and Medical Technology Mannheim, Heidelberg University, Mannheim, Germany

Nicola Beale Nuffield Department of Anaesthetics, Oxford University Hospitals NHS Trust, UK

Alain Beaudet Montreal Neurological Institute, McGill University, Montreal, QC, Canada

Lino Becerra P.A.I.N. Group, Department of Anesthesiology, Perioperative and Pain Medicine, Boston Children's Hospital Department of Radiology Massachusetts General Hospital Center for Pain and the Brain, Harvard Medicals, Boston, MA, USA

Yaakov Beilin Department of Anesthesiology, and Obstetrics, Mount Sinai School of Medicine, New York, NY, USA

Alvin J. Beitz Department of Veterinary and Biomedical Sciences, University of Minnesota, St. Paul, MN, USA

Miles Belgrade Fairview Pain Management Center, University of Minnesota Medical Center, Minneapolis, MN, USA

Carlo Valerio Bellieni Neonatal Intensive Care Unit, University Hospital, Siena, Italy

Carlos Belmonte Instituto de Neurociencias, Universidad Miguel Hernandez-CSIC, San Juan de Alicante, Spain

Allan J. Belzberg Department of Neurosurgery, Johns Hopkins School of Medicine, Baltimore, MD, USA

Matt Bender Department of Neurosurgery, Johns Hopkins Hospital, Baltimore, MA, USA

Fabrizio Benedetti Department of Neuroscience, Clinical and Applied Physiology Program, University of Turin Medical School, Turin, Italy

David L. H. Bennett Wolfson CARD, Kings College London, London, UK

Nuffield Department of Clinical Neurosciences, John Radclife Hospital, The University of Oxford, Oxford, UK

Robert Bennett Department of Medicine, Oregon Health and Science University, Portland, USA

Edward C. Benzel Department of Neurosurgery, Center for Spine Health Neurological Institute Cleveland Clinic, Cleveland, OH, USA

David A. Bereiter Department of Diagnostic and Biological Sciences, University of Minnesota School of Dentistry, Minneapolis, MN, USA

Elmar Berendes Klinik für Anästhesiologie, Operative Intensivmedizin und Schmerztherapie, Klinikum Krefeld, Krefeld, Germany

Karen J. Berkley Program in Neuroscience, Florida State University, Tallahassee, FL, USA

Peter Berlit Department of Neurology, Alfried Krupp Hospital, Essen, Germany

Jean-François Bernard INSERM UMR-894, UPMC Site Pitié-Salpêtrière, Paris, France

Anne K. Bertelsen Neurology, Aleris Hospital, Bergen, Norway

Jennifer Bickel Integrative Pain Management, Children's Mercy Hospitals and Clinics, University of Missouri-Kansas City School of Medicine, Kansas City, MO, USA

Marcelo E. Bigal Department of Neurology and Montefiore Headache Unit, Albert Einstein College of Medicine, Bronx, NY, USA

Yitzchak M. Binik McGill University, Montreal, QC, Canada

Niels Birbaumer Institute of Medical Psychology and Behavioral Neurobiology, University of Tübingen, Tübingen, Germany

Frank Birklein Department of Neurology, University of Mainz, Mainz, Germany

Kathryn A. Birnie Department of Psychology, Dalhousie University, Halifax, NS, Canada

Thomas Bishop Centre for Neuroscience, King's College London, London, UK

Henning Bliddal The Parker Institute, Frederiksberg Hospital, Copenhagen, Denmark

Nikolai Bogduk University of Newcastle, Newcastle Bone & Joint Institute, Royal Newcastle Centre, Newcastle, NSW, Australia

Jorgen Boivie Department of Neurology, University Hospital, Linköping, Sweden

Michael R. Bond University of Glasgow, Glasgow, UK

Richard L. Boortz-Marx Division of Pain Medicine, Department of Anesthesiology, University of North Carolina, Chapel Hill, North Carolina, USA

James M. Borowczyk Department of Orthopaedics and Musculoskeletal Medicine, University of Otago, Christchurch, Christchurch, New Zealand

David Borsook P.A.I.N. Group, Department of Anesthesiology, Perioperative and Pain Medicine, Boston Children's Hospital Department of Radiology Massachusetts General Hospital Center for Pain and the Brain, Harvard Medicals, Boston, MA, USA

George S. Borszcz Department of Psychology, Wayne State University, Detroit, USA

Jasenka Borzan Department of Anesthesiology and Critical Care Medicine, Johns Hopkins University, Baltimore, USA

Regina M. Botting London, UK

Didier Bouhassira Centre d'Evaluation et de Traitement de la Douleur, Hôpital Ambroise Paré, INSERM U 987 APHP, Boulogne-Billancourt, France

Laurence Bourgeais INSERM UMR 894, Centre de Psychiatrie et Neurosciences, Paris, France

David Bowsher Pain Research Institute, University Hospital Aintree, Liverpool, UK

Emma L. Brandon Anaesthesiology and Pain Medicine, University of Western Australia Royal Perth Hospital, Perth, WA, Australia

Jennifer Brawn P.A.I.N. Group, Center for Pain and the Brain, Boston Children's Hospital, Harvard Medical School, Waltham, MA, USA

Harald Breivik Department of Pain Management and Research, Oslo University Hospital, Rikshospitalet and University of Oslo, Oslo, Norway

Kori L. Brewer Department of Emergency Medicine, Brody School of Medicine East Carolina University, Greenville, NC, USA

Christopher R. Brigham Brigham and Associates, Inc., Portland, USA

Abigail Broderick Nuffield Department of Obstetrics and Gynaecology, University of Oxford, Oxford, UK

Jeffrey A. Brown Department of Neurological Surgery, Wayne State University School of Medicine, Detroit, MI, USA

Stephen C. Brown Department of Anesthesia, The Hospital for Sick Children, Toronto, ON, Canada

Eduardo Bruera Department of Palliative Care and Rehabilitation Medicine, The University of Texas M. D. Anderson Cancer Center, Houston, TX, USA

Pablo Brumovsky Faculty of Biomedical Sciences, Austral University, Buenos Aires, Argentina

Consejo Nacional de Investigaciones Científicas y Técnicas (CONICET), Buenos Aires, Argentina

Kay Brune Department of Experimental and Clinical Pharmacology and Toxicology, Friedrich Alexander University Erlangen-Nürnberg, Erlangen, Germany

Maria Burian Department of General, Visceral and Transplantation Surgery, University of Heidelberg, Heidelberg, Germany

Dan Buskila Rheumatic Disease Unit, Soroka Medical Center, Beer Sheva, Israel

Catherine M. Cahill Anesthesiology & Perioperative Care, University of California, Irvine, Irvine, CA, USA

Brian E. Cairns Faculty of Pharmaceutical Sciences, University of British Columbia, Vancouver, BC, Canada

Nigel A. Calcutt Department of Pathology, University of California San Diego, La Jolla, CA, USA

Margarita Calvo Wolfson CARD, Kings College London, London, UK

James N. Campbell Johns Hopkins University School of Medicine, Baltimore, MD, USA

Lauren Campbell Department of Psychology, York University, Toronto, ON, Canada

Ling Cao Department of Biomedical Sciences, College of Osteopathic Medicine, University of New England, ME, USA

Adam Carinci Department of Anesthesia, Critical Care and Pain Medicine, Massachusetts General Hospital Harvard Medical School, Boston, MA, USA

Giancarlo Carli Department of Physiology, University of Siena, Siena, Italy

Christer P. O. Carlsson Rehabilitation Clinic, Lunds University Hospital, Lund, Sweden

Susan M. Carlton Neuroscience and Cell Biology, University of Texas Medical Branch, Galveston, TX, USA

Richard Carr Clinic for Anaesthesiology, Heidelberg University, Mannheim, Baden-Wuerttemberg, Germany

Benjamin S. Carson Department of Neurosurgery, Johns Hopkins Hospital, Baltimore, MA, USA

Brian Carter University of Missouri at Kansas City Children's Mercy Hospital and Bioethics Center, Kansas City, MO, USA

Kenneth L. Casey Department of Neurology, University of Michigan Medical Center, Ann Arbor, MI, USA

Christine T. Chambers Departments of Pediatrics and Psychology, Dalhousie University, Halifax, NS, Canada

Sandra Chaplan Janssen Research and Development, LLC, San Diego, CA, USA

C. Richard Chapman Department of Anesthesiology, University of Utah School of Medicine, Salt Lake City, UT, USA

Santosh K. Chaturvedi Psychiatry, National Institute of Mental Health & Neurosciences, Bangalore, India

Andrew C. N. Chen Human Brain Mapping and Cortical Imaging Laboratory, Aalborg University, Aalborg, Denmark

Hsinlin Thomas Cheng Department of Neurology, University of Michigan, Ann Arbor, MI, USA

Pavel Cherkas Faculty of Dentistry, University of Toronto, Toronto, ON, Canada

Andrea Cheville Department of Physical Medicine and Rehabilitation, Mayo Clinic, Rochester, MN, USA

Chen Yu Chiang Faculty of Dentistry, University of Toronto, Toronto, ON, Canada

N. L. Chillingworth School of Physiology and Pharmacology, University of Bristol, Bristol, UK

Edward Chow Department of Radiation Oncology, Sunnybrook Health Sciences Centre Odette Cancer Centre (T2-182), Toronto, ON, Canada

Julie A. Christianson Department of Anatomy and Cell Biology, University of Kansas Medical Center, Kansas City, KS, USA

MacDonald J. Christie Discipline of Pharmacology, University of Sydney, Sydney, Australia

Jin Mo Chung Department of Neuroscience and Cell Biology, University Chair in Neuroscience, University of Texas Medical Branch, Galveston, TX, USA

Kyungsoon Chung Department of Neuroscience and Cell Biology, University of Texas Medical Branch, Galveston, TX, USA

Anna K. Clark Wolfson Centre for Age Related Diseases, King's College London, London, UK

W. Crawford Clark College of Physicians and Surgeons, Columbia University, New York, USA

James F. Cleary Pain & Policy Studies Group, UW Carbone Cancer Center, Madison, WI, USA

John De Clercq Department of Rehabilitation Sciences and Physiotherapy, Ghent University Hospital, Ghent, Belgium

Jacqui Clinch Royal National Hospital for Rheumatic Diseases NHS Foundation Trust, Bath, UK

Terence J. Coderre Departments of Anesthesia, Neurology & Neurosurgery and Psychology, McGill University, Montreal, QC, Canada

Robert Coghill Department of Neurobiology and Anatomy, Wake Forest University School of Medicine, Winston-Salem, NC, USA

Lindsey Cohen Psychology, Georgia State University, Atlanta, GA, USA

Alan L. Colledge Utah Labor Commission, International Association of Industrial Accident Boards and Commissions Central Utah Medical Clinic, South Salt Lake City, UT, USA

Beverly J. Collett University Hospitals of Leicester, Leicester, UK

Lesley A. Colvin Department of Anaesthesia, Critical Care and Pain Medicine Western General Hospital, Edinburgh, UK

Kevin C. Conlon Trinity College Dublin, University of Dublin, Dublin, Ireland

Mark Connelly Integrative Pain Management, Children's Mercy Hospitals and Clinics, University of Missouri-Kansas City School of Medicine, Kansas City, MO, USA

Brian Y. Cooper Department of Oral and Maxillofacial Surgery, University of Florida, Gainesville, FL, USA

Christine Cordle University Hospitals of Leicester, Leicester, UK

Laura A. Cousins Psychology, Georgia State University, Atlanta, GA, USA

Michael J. Cousins Department of Anesthesia and Pain Management, Royal North Shore Hospital University of Sydney, Sydney, NSW, Australia

Verne C. Cox Department of Psychology, University of Texas at Arlington, Arlington, TX, USA

George A. Crabill Department of Neurosurgery, Johns Hopkins Hospital, Baltimore, MA, USA

Rebecca M. Craft Psychology, Washington State University, Pullman, WA, USA

Kenneth D. Craig Department of Psychology, University of British Columbia, Vancouver, BC, Canada

Richard Crevenna Medizinische Universität Wien, Wien, Austria

Geert Crombez Department of Experimental Clinical and Health Psychology, Ghent University, Ghent, Belgium

Joan Crook School of Nursing, Faculty of Health Sciences, McMaster University, Hamilton, ON, Canada

Giorgio Cruccu Department of Neurological Sciences, La Sapienza University, Rome, Italy

Michele Curatolo Department of Anesthesiology, Division of Pain Therapy, University Hospital of Bern Inselspital, Bern, Switzerland

Nachum Dafny Neurobiology and Anatomy, Health Science Center, University of Texas, Houston, TX, USA

Jorgen B. Dahl Department of Anaesthesiology, Glostrup University Hospital, Herlev, Denmark

Yi Dai Department of Anatomy and Neuroscience, Hyogo College of Medicine, Hyogo, Japan

Stefaan Van Damme Department of Experimental Clinical and Health Psychology, Ghent University, Ghent, Belgium

Alexandre F. M. DaSilva Biologic & Materials Sciences, University of Michigan Dental School, MI, USA

Steve Davidson Pain Center & Department of Anesthesiology, Washington University School of Medicine, MO, USA

Brian Davis Department of Neurobiology, University of Pittsburgh School of Medicine, Pittsburgh, PA, USA

Karen D. Davis Brain, Imaging and Behaviour – Systems Neuroscience, Toronto Western Research Institute, Toronto, ON, Canada

Department of Surgery, University of Toronto, Toronto, ON, Canada

Richard O. Day Department of Clinical Pharmacology and Toxicology, St Vincent's Hospital, Sydney, NSW, Australia

Isabelle Decosterd Pain Center, University Hospital Center and University of Lausanne, Lausanne, Switzerland

Ruth Defrin Physical Therapy, Faculty of Medicine, Tel-Aviv University, Tel-Aviv, Israel

Antoon F. C. De Laat Department of Oral/Maxillofacial Surgery, Catholic University of Leuven, Leuven, Belgium

Joyce DeLeo Dartmouth Hitchcock Medical Center, Dartmouth Medical School, Hanover, NH, USA

A. Lee Dellon Department of Plastic Surgery and Neurosurgery, Johns Hopkins University, Baltimore, USA

Dellon Institutes for Peripheral Nerve Surgery, Baltimore, MD, USA

Marshall Devor Department of Cell & Developmental Biology, Institute of Life Sciences and Center for Research on Pain, The Hebrew University of Jerusalem, Jerusalem, Givat Ram, Israel

Perry Dhaliwal Department of Neurosurgery, Cleveland Clinic Spine Institute, Cleveland, OH, USA

David Diamant Neurological and Spinal Surgery, Lincoln, NE, USA

Anthony H. Dickenson Department of Pharmacology, University College London, London, UK

Hans-Christoph Diener Department of Neurology, University Hospital Essen, University Duisburg-Essen, Essen, Germany

Natalia Dmitrieva Program in Neuroscience, Florida State University, Tallahassee, FL, USA

L. F. Donaldson School of Life Sciences, University of Nottingham, Nottingham, UK

Henri Doods CNS Pharmacology, Boehringer Ingelheim Pharma GmbH & Co. KG, Biberach, Germany

Victoria Dorf Social Security Administration, Baltimore, MD, USA

Michael J. Dorsi Department of Neurosurgery, Johns Hopkins School of Medicine, Baltimore, MD, USA

Jonathan O. Dostrovsky Department of Physiology, University of Toronto, Toronto, ON, Canada
Faculty of Dentistry, University of Toronto, Toronto, ON, Canada

Robyn Drach Fairview Pain Management Center, Minneapolis, MN, USA

Ron Dressler Utah Labor Commission, International Association of Industrial Accident Boards and Commissions Central Utah Medical Clinic, South Salt Lake City, UT, USA

Paul Dreyfuss Department of Rehabilitation Medicine, University of Washington, Seattle, USA

Peter D. Drummond School of Psychology, Murdoch University, Perth, Australia

Ronald Dubner Department of Neural and Pain Sciences, University of Maryland Dental School, Baltimore, MD, USA

Anne Ducros Headache Emergency Department, INSERM, Paris, France

Gary H. Duncan Département de stomatologie, Faculté de médecine dentaire, Université de Montréal, Montreal, QC, Canada
Department of Neurology and Neurosurgery, McGill University, Montreal, QC, Canada

Mary J. Eaton Research Service (151), Miami VA Health Care System, Miami VAMC, Miami, FL, USA

Andrea Ebersberger Department of Physiology, Friedrich Schiller University of Jena, Jena, Germany

Christopher Eccleston Pain Management Unit, University of Bath, Bath, UK

John Edmeads Sunnybrook and Woman's College Health Sciences Centre, University of Toronto, Toronto, Canada

Edzard Ernst Complementary Medicine, Peninsula Medical School Universities of Exeter and Plymouth, Exeter, UK

Reuben Escorpizo Swiss Paraplegic Research, Nottwil, Switzerland
Department of Health Sciences and Health Policy, University of Lucerne, Lucerne, Switzerland
ICF Research Branch in cooperation with the World Health Organization (WHO) Collaborating Centre for the Family of International Classifications in Germany, Nottwil, Switzerland

Schonstein Eva RehabCo, Fyshwick, ACT, Australia

Joshua C. Eyer Department of Psychology, University of Alabama, Tuscaloosa, AL, USA

Marie Faaborg-Andersen McGill University, Montreal, QC, Canada

Keri L. Fakata College of Pharmacy and Pain Management Center, University of Utah, Salt Lake City, USA

Marie Fallon Department of Oncology, Edinburgh Cancer Center University of Edinburgh, Edinburgh, UK

Melissa Farmer Department of Physiology / Feinberg School of Medicine, Northwestern University, Chicago, IL, USA

Samantha R. Fashler Department of Psychology, University of British Columbia, Vancouver, BC, Canada

Jean-Baptiste Fassier Department of Occupational Health and Medicine, UMRESTTE-Université Claude Bernard Lyon 1, Lyon, France

Charlotte Feinmann Eastman Dental Institute London, University College London, London, UK

Elizabeth Roy Felix The Miami Project to Cure Paralysis, University of Miami Miller School of Medicine, Miami, FL, USA

Bin Feng Center for Pain Research, Department of Anesthesiology, School of Medicine, University of Pittsburgh, Pittsburgh, Pennsylvania, USA

Yi Feng Department of Anesthesiology, Peking University People's Hospital, Peking, China

Andres Fernandez Department of Neurosurgery, Johns Hopkins Hospital, Baltimore, MD, USA

Antonio Vicente Ferrer-Montiel Institute of Molecular and Cell Biology, University Miguel Hernández, Elche, Alicante, Spain

Perry G. Fine Pain Research Center, School of Medicine, University of Utah, Salt Lake City, UT, USA

David J. Fink Department of Neurology, University of Michigan and VA Ann Arbor Healthcare System, Ann Arbor, MI, USA

Nanna Brix Finnerup Aarhus University Hospital, Danish Pain Research Center, Aarhus, Denmark

David A. Fishbain Department of Psychiatry, University of Miami, Miami, FL, USA

Maria Fitzgerald Neuroscience, Physiology & Pharmacology, University College London, London, UK

Per Flisberg Department of Anaesthesia and Intensive Care, Lund University Hospital, Lund, Sweden

Herta Flor Department of Cognitive and Clinical Neuroscience, Central Institute of Mental Health, Medical Faculty Mannheim, Heidelberg University, Mannheim, Germany

Christopher M. Flores Drug Discovery, Johnson and Johnson Pharmaceutical Research and Development, Spring House, PA, USA

Wilbert E. Fordyce Department of Rehabilitation, University of Washington School of Medicine, Seattle, WA, USA

Robert D. Foreman Department of Physiology, College of Medicine, University of Oklahoma Health Sciences Center, Oklahoma City, OK, USA

Gary M. Franklin Department of Environmental and Occupational Health Sciences, University of Washington, Seattle, WA, USA

Department of Health Services, University of Washington, Seattle, WA, USA

Jan Fridén Department of Hand Surgery, Sahlgrenska University Hospital, Gothenborg, Sweden

Justin Friedlander The Arthur Smith Institute for Urology, North Shore-LIJ Healthcare System, Hofstra North Shore LIJ School of Medicine, New Hyde Park, NY, USA

Allan H. Friedman Division of Neurosurgery, Duke University Medical Center, Durham, NC, USA

Perry N. Fuchs Department of Psychology, University of Texas at Arlington, Arlington, TX, USA

Deborah Fulton-Kehoe Department of Environmental and Occupational Health Sciences, University of Washington, Seattle, WA, USA

Andrea D. Furlan Institute for Work & Health, Toronto, ON, Canada

Vasco Galhardo Departamento de Biologia Experimental, Faculdade de Medicina, Universidade do Porto, Portugal

Andreas R. Gantenbein RehaClinic Bad Zurzach, Bad Zurzach, Switzerland

I. Garonzik Department of Neurosurgery, Johns Hopkins Medical Institutions, Baltimore, MD, USA

Robert Gassin Musculoskeletal Medicine Clinic, Frankston, VIC, Australia

Robert J. Gatchel Department of Psychology, The University of Texas, Arlington, TX, USA

Gerald F. Gebhart Department of Anesthesiology, Center for Pain Research, University of Pittsburgh, Pittsburgh, PA, USA

Gerd Geisslinger Department of Clinical Pharmacology, University Frankfurt am Main, Frankfurt, Germany

Robert D. Gerwin Department of Neurology, Johns Hopkins University, Bethesda, MD, USA

Maria Adele Giamberardino Pathophysiology of Pain Laboratory and Centre for Fibromyalgia and Musculoskeletal Pain, Department of Medicine and Science of Aging, University of Chieti, Chieti, Italy

Stephen Gibson National Ageing Research Institute Department of Medicine, University of Melbourne, Parkville, VIC, Australia

Glenn J. Giesler Department of Neuroscience, University of Minnesota, Minneapolis, MN, USA

Philip L. Gildenberg Baylor Medical College, Houston, TX, USA

Myra Glajchen Department of Pain Medicine and Palliative Care, Beth Israel Medical Center Albert Einstein College of Medicine, New York, NY, USA

Hartmut Göbel Kiel Migraine and Headache Center, Kiel Pain Center, Kiel, Germany

Peter Goadsby Headache Center, Department of Neurology, University of California, San Francisco, San Francisco, CA, USA

Philippe Goffaux Faculty of Medicine, University of Sherbrooke, Sherbrooke, QC, Canada

Ana Gomis Instituto de Neurociencias de Alicante, Universidad Miguel Hernández-CSIC, San Juan de Alicante, Spain

Gilbert R. Gonzales Department of Neurology, Memorial Sloan-Kettering Cancer Center, New York, USA

Allan Gordon Wasser Pain Management Centre, Mount Sinai Hospital University of Toronto, Toronto, ON, Canada

Liesbet Goubert Department of Experimental Clinical and Health Psychology, Ghent University, Ghent, Belgium

Jayantilal Govind Pain Clinic, Department of Anaesthesia, University of New South Wales, Sydney, NSW, Australia

Richard H. Gracely Departments of Medicine-Rheumatology and Neurology, University of Michigan Health System, Ann Arbor, MI, USA

Garry G. Graham Department of Pharmacology, School of Medical Sciences, University of New South Wales, Sydney, NSW, Australia

David K. Grandy Department of Physiology and Pharmacology, Oregon Health and Science University, Portland, OR, USA

Thomas Graven-Nielsen Laboratory for Musculoskeletal Pain and Motor Control, Center for Sensory-Motor Interaction, Aalborg University, Aalborg, Denmark

Barry Green Pierce Laboratory, Yale School of Medicine, New Haven, CT, USA

Paul G. Green Oral & Maxillofacial Surgery, University of California San Francisco, San Francisco, CA, USA

Joel D. Greenspan Department of Neural and Pain Sciences, University of Maryland School of Dentistry, Baltimore, MD, USA

John W. Griffin Department of Neurology, Johns Hopkins University School of Medicine, Baltimore, MD, USA

Cornelius Groenwald Department of Anesthesiology and Pain Medicine, Seattle Children's Hospital, University of Washington School of Medicine, Seattle, WA, USA

Douglas P. Gross Department of Physical Therapy, University of Alberta WCB-Alberta Millard Health, Edmonton, AB, Canada

Sabine Grösch Institute of Clinical Pharmacology, Clinical Center of the Goethe University, Frankfurt, Germany

Blair D. Grubb Department of Cell Physiology and Pharmacology, University of Leicester, Leicester, UK

Ruth Eckstein Grunau Pediatrics Dept, University of British Columbia Developmental Neurosciences & Child Health, Child & Family Research Institute, Vancouver, BC, Canada

Christoph Gutenbrunner Department of Physical Medicine and Rehabilitation, Hanover Medical School, Hanover, Germany

Maija Haanpää Department of Neurosurgery, Helsinki University Central Hospital, Helsinki, Finland

Sherri Haas Fairview Pain Management Center, Minneapolis, MN, USA

Ute Habel Department of Psychiatry and Psychotherapy, RWTH Aachen University, Aachen, Germany

Thomas Hachenberg Clinic for Anaesthesiology and Intensive Therapy, Otto-von-Guericke University, Magdeburg, Germany

Nouchine Hadjikhani Martinos Center for Biomedical Imaging, Massachusetts General Hospital Harvard Medical School, Charlestown, MA, USA

Ole Haegg Department of Orthopedic Surgery, Sahlgren University Hospital, Goteborg, Sweden

Bryan C. Hains Department of Neurology, Yale University School of Medicine, West Haven, CT, USA

Else K. B. Hals University of Oslo, Faculty of Dentistry, Oslo, Norway

Darryl T. Hamamoto Diagnostic and Biological Sciences, University of Minnesota, Minneapolis, MN, USA

Donna L. Hammond Department of Anesthesia, University of Iowa, Iowa City, IA, USA

Menachem Hanani Hadassah-Hebrew University Medical Center Hebrew University Medical School, Jerusalem, Israel

Hermann O. Handwerker Department of Physiology and Pathophysiology, University of Erlangen/Nürnberg, Erlangen, Germany

George M. Hanna Department of Anesthesia, Critical Care and Pain Medicine, Massachusetts General Hospital Harvard Medical School, Boston, MA, USA

Jing-Xia Hao Department of Physiology and Pharmacology, Karolinska Institutet, Stockholm, Sweden

Eleni G. Hapidou Department of Psychiatry and Behavioural Neurosciences, McMaster University and Hamilton Health Sciences Chronic Pain Management Unit, Chedoke Hospital, Hamilton, ON, Canada

Geoffrey Harding Sandgate Spinal Medicine Clinic, Sandgate, QLD, Australia

Louise Harding Hospitals NHS Foundation Trust, University College London, London, UK

Kenneth M. Hargreaves Department of Endodontics, University of Texas Health Science Center at San Antonio, San Antonio, USA

D. Paul Harries Pain Treatment Center, Samaritan Hospital Lexington, Lexington, KY, USA

Denise Harrison Nursing, Children's Hospital of Eastern Ontario (CHEO), University of Ottawa, Ottawa, ON, Canada

Steven Harte Anesthesiology and Internal Medicine/Rheumatology, University of Michigan VA Ann Arbor Healthcare System, Ann Arbor, MI, USA

Barbara A. Hastie Community Dentistry & Behavioral Science, University of Florida College of Dentistry, Gainesville, FL, USA

Jennifer A. Haythornthwaite Psychiatry & Behavioral Sciences, Johns Hopkins University, Baltimore, MD, USA

Heinz-Joachim Häbler FH Bonn-Rhein-Sieg, Rheinbach, Germany

John H. Healey Orthopaedic Service, Department of Surgery, Memorial Sloan-Kettering Cancer Center Weill Medical College of Cornell University, New York, NY, USA

Alicia A. Heapy VA Connecticut Healthcare System, Westhaven, CT, USA

Anne M. Heffernan The Adelaide and Meath Hospital Incorporating The National Children's Hospital, Dublin, Ireland

Geert Van der Heijden Julius Center, Department of Epidemiology, University Medical Center Utrecht, Utrecht, The Netherlands

Mary M. Heinricher Departments of Neurological Surgery and Behavioral Neuroscience, Oregon Health & Science University, Portland, USA

Robert D. Helme Department of Medicine, Royal Melbourne Hospital, University of Melbourne, Melbourne, Australia

Kim Helsen Research Group on Health Psychology, KU Leuven Katholieke Universiteit Leuven, Leuven, Belgium

Department of Experimental–Clinical and Health Psychology, Ghent University, Ghent, Belgium

Amin S. Herati The Arthur Smith Institute for Urology, North Shore-LIJ Healthcare System, Hofstra North Shore LIJ School of Medicine, New Hyde Park, NY, USA

Robert D. Herbert Neuroscience Research Australia, Sydney, Australia

Christiane Hermann Department of Clinical Psychology, Justus-Liebig University Giessen, Giessen, Germany

Juan F. Herrero Department of Physiology, Faculty of Medicine University of Alcala, Alcalá de Henares Madrid, Spain

Aki J. Hietaharju Department of Neurology and Rehabilitation, Pain Clinic Tampere University Hospital, Tampere, Finland

Karin Westlund High Department of Physiology, University of Kentucky, Lexington, KY, USA

Marita Hilliges Dalarna University, Halmstad, Sweden

Max J. Hilz University of Erlangen-Nuremberg, Erlangen, Germany

Burkhard Hinz Institute of Toxicology and Pharmacology, University of Rostock, Rostock, Germany

Anthony R. Hobson Section of GI Sciences, Hope Hospital University of Manchester, Salford, UK

Edward Högestätt Department of Laboratory Medicine, Lund University, Division of Clinical Chemistry and Pharmacology, Lund, Sweden

Tomas Hökfelt Department of Neuroscience, Karolinska Institute, Stockholm, Sweden

Sarah V. Holdridge Department of Pharmacology and Toxicology, Queen's University, Kinston, Canada

Kjell Hole University of Bergen, Bergen, Norway

Graham R. Holland Department of Cariology, University of Michigan, Ann Arbor, MI, USA

Chang-Zern Hong Department of Physical Medicine and Rehabilitation, University of California, Irvine, CA, USA

Minoru Hoshiyama Department of Integrative Physiology, National Institute for Physiological Sciences, Okazaki, Japan

Fred M. Howard University of Rochester, Rochester, NY, USA

Sung-Tsang Hsieh Department of Anatomy and Cell Biology, Department of Neurology, National Taiwan University Hospital, National Taiwan University, Taipei, Taiwan

James W. Hu Faculty of Dentistry, University of Toronto, Toronto, Canada

Claire E. Hulsebosch Spinal Cord Injury Research, University of Texas Medical Branch, Galveston, TX, USA

Thomas Hummel Department of Otorhinolaryngology, Technical University of Dresden Medical School, Dresden, Germany

Mark R. Hutchinson School of Medical Sciences, University of Adelaide, Adelaide, SA, Australia

Charles E. Inturrisi Department of Pharmacology, Weill Cornell Medical College, New York, USA

Koji Inui Department of Integrative Physiology, National Institute for Physiological Sciences, Okazaki, Japan

Koichi Iwata Department of Physiology, Nihon University School of Dentistry, Chiyoda-ku, Japan

Suhayl J. Jabbur Anatomy, Cell Biology and Physiology, Faculty of Medicine, American University of Beirut, Beirut, Lebanon

Neuroscience Program, Faculty of Medicine, American University of Beirut, Beirut, Lebanon

Nora Janjan University of Texas M. D. Anderson Cancer Center, Houston, TX, USA

Luc Jasmin Department of Anatomy, University of California San Francisco, San Francisco, CA, USA

Anne Jaumees Royal North Shore Hospital, St Leonards, NSW, Australia

Wilfrid Jaenig Physiologisches Institut, Christian-Albrechts-Universität zu Kiel, Kiel, Germany

Daniel Jeanmonod Functional Neurosurgery, The Center of Ultrasound Functional Neurosurgery, Solothurn, Switzerland

Andrew Jeffreys Director Department of Anaesthesia & Pain Medicine, Western Health, Melbourne, VIC, Australia

Mark P. Jensen Department of Rehabilitation Medicine, University of Washington, Seattle, WA, USA

Troels Staehelin Jensen Danish Pain Research Center, Department of Neurology, Aarhus University Hospital, Aarhus, Denmark

Kurt L. Johnson Department of Rehabilitation Medicine, University of Washington, Seattle, WA, USA

Mark I. Johnson Faculty of Health and Social Sciences, Leeds Pallium Research Group, Leeds Metropolitan University, Leeds, UK

Mark Johnston Hibiscus Coast Highway, Orewa, New Zealand

Edward Jones Center for Neuroscience, University of California, Davies, CA, USA

Jeroen R. De Jong Department of Rehabilitation, University Hospital Maastricht, Maastricht, The Netherlands

Sven-Eric Jordt Department of Pharmacology, Yale University School of Medicine, New Haven, CT, USA

Ellen Jørum The Laboratory of Clinical Neurophysiology, Rikshospitalet University, Oslo, Norway

Stefan Just CNS Pharmacology, Boehringer Ingelheim Pharma GmbH & Co. KG, Biberach, Germany

Timothy K. Y. Kaan Department of Anatomy, University of California San Francisco, San Francisco, CA, USA

Noufissa Kabli Department of Pharmacology and Toxicology, Queen's University, Kinston, Canada

Ryusuke Kakigi Department of Integrative Physiology, National Institute for Physiological Sciences, Okazaki, Japan

Peter C. A. Kam Anaesthesia, Central Clinical School, The University of Sydney, Sydney, Australia

Giresh Kanji The Sports and Pain Clinic, Wellington, New Zealand

Joel Katz Department of Psychology, McGill University, Montreal, QC, Canada

Valerie Kayser NSERM U677 91Bd de l'Hôpital, Paris, France

Francis J. Keefe Pain Prevention and Treatment Research, Department of Psychiatry and Behavioral Sciences, Duke University Medical Center, Durham, NC, USA

Lois J. Kehl Department of Anesthesiology, University of Minnesota, Minneapolis, MN, USA

Maureen Kelley Department of Pediatrics, Bioethics Division, University of Washington School of Medicine, WA, USA

Seattle Children's Hospital, Seattle, WA, USA

Stephen Kennedy Nuffield Department of Obstetrics and Gynaecology, University of Oxford, Oxford, UK

Edmund Keogh Department of Psychology, University of Bath, Bath, UK

Robert D. Kerns VA Connecticut Healthcare System, Westhaven, CT, USA

Sanjay Khanna Department of Physiology, National University of Singapore, Singapore

Kok Eng Khor Department of Pain Management, Prince of Wales Hospital University of New South Wales, Randwick, NSW, Australia

Mina Khoshnejad Department of Neuroscience, University of Montreal, Montreal, QC, Canada

H. J. Kim Department of Life Science, Yonsei University, Wonju, Republic of Korea

Sara King Faculty of Education (School Psychology), Mount Saint Vincent University, Halifax, NS, Canada

Wade King Department of Clinical Research, Royal Newcastle Centre University of Newcastle, Newcastle, NSW, Australia

K. L. Kirsh Behavioral Medicine, The Pain Treatment Center of the Blue grass, Lexington, KY, USA

Henriette Klit Danish Pain Research Center, Aarhus University Hospital, Aarhus C, Denmark

Ricardo J. Komotar Department of Neurological Surgery, Neurological Institute Columbia University, New York, NY, USA

Sungtae Koo Division of Meridian and Structural Medicine, School of Korean Medicine, Pusan National University, Yangsan, Korea

Rhonda K. Kotarinos Oakbrook Terrace, IL, USA

Katalin J. Kovacs Department of Veterinary and Biomedical Sciences, University of Minnesota, St. Paul, MN, USA

Malgorzata Krajnik Chair of Palliative Care, Nicolaus Copernicus University in Torun, Collegium Medicum in Bydgoszcz, Bydgoszcz, Poland

Elliot Krane Anesthesia and Pediatrics, Stanford University School of Medicine Lucile Packard Children's Hospital at Stanford, Stanford, CA, USA

Ajit Krishnaney Department of Neurosurgery, Cleveland Clinic, Cleveland, OH, USA

Greg Krohm International Association of Industrial Accident Boards and Commissions (IAIABC), Utah Valley University, USA

Lawrence Kruger UCLA Geffen School of Medicine, Los Angeles, CA, USA

M. M. ter Kuile Department of Gynecology, Leiden University Medical Center, Leiden, The Netherlands

Promil Kukreja Department of Anesthesiology, University of Alabama at Birmingham, Birmingham, AL, USA

Alice Kvåle Department of Public Health and Primary Health Care, Faculty of Medicine and Dentistry University of Bergen, Bergen, Norway

Gunnvald Kvarstein Department of Pain Management and Pain Research, Oslo University Hospital, Rikshospitalet, Oslo, Norway

Jean-Jacques Labat Neurology and Rehabilitation, Clinique Urologique, Nantes, France

Gaston Labrecque Faculty of Pharmacy, University of Laval, Quebec City, QC, Canada

Marco Lacerenza Pain Medicine Center, Casa di Cura S. Pio X, Opera San Camillo Foundation, Milan, Italy

Uri Ladabaum Gastroenterology & Hepatology, Stanford Cancer Institute, Redwood City, CA, USA

Marie Andree Lahaie McGill University, Montreal, QC, Canada

Miguel J. A. Lainez-Andres Department of Neurology, University of Valencia, Valencia, Spain

Robert H. LaMotte Department of Anesthesiology, Yale School of Medicine, New Haven, CT, USA

James W. Lance Institute of Neurological Sciences, University of New South Wales, Sydney, Australia

Robert G. Large The Auckland Regional Pain Service, Auckland Hospital, Auckland, New Zealand

Alice A. Larson Department of Veterinary and Biomedical Sciences, University of Minnesota, St. Paul, MN, USA

Roberta Lattanzi Department of Physiology and Pharmacology, Sapienza University of Rome, Rome, Italy

Peter Lau Department of Clinical Research, Royal Newcastle Hospital, University of Newcastle, Newcastle, NSW, Australia

Stefan A. Laufer Institute of Pharmacy, Eberhard-Karls University of Tuebingen, Tuebingen, Germany

Stefan Lautenbacher Physiologische Psychologie, Otto-Friedrich-Universität, Bamberg, Germany

Gilles J. Lavigne Facultés de Médecine Dentaire et de Médecine, Université de Montréal, Hopital Sacre Coeur de Montreal. Surgery-trauma dept, Montréal, QC, Canada

Emily Law Center for Child Health, Behavior, and Development, Seattle Children's Research Institute, Seattle, WA, USA

Peter G. Lawlor Our Lady's Hospice, Medical Department, Dublin, Ireland

Michel Lazdunski Institut de Pharmacologie Moléculaire et Cellulaire (IPMC) CNRS/UNS UMR7275, Valbonne-Sophia Antipolis, France

Vincenzo Di Lazzaro Institute of Neurology, Campus Biomedico University, Rome, Italy

Alyssa Lebel Boston Children's Hospital, Boston, MA, USA

Frank Lee Johns Hopkins University, Baltimore, MD, USA

Maaike Leeuw Department of Medical, Clinical and Experimental Psychology, Maastricht University, Maastricht, The Netherlands

Siri Graff Leknes Research Fellow, Psykologisk institutt, Oslo, Norway

Frederick A. Lenz Department of Neurosurgery, Johns Hopkins Hospital, Baltimore, MD, USA

Guillaume Léonard Faculty of Medicine, University of Sherbrooke, Sherbrooke, QC, Canada

Pauline Lesage Department of Pain Medicine and Palliative Care, Beth Israel Medical Center, New York, USA

Isobel Lever Neural Plasticity Unit, Institute of Child Health University College London, London, UK

Khan W. Li Department of Neurosurgery, Johns Hopkins Hospital, Baltimore, MD, USA

Alan R. Light Department of Anesthesiology, University of Utah, Salt Lake City, UT, USA

Michael Lim Department of Neurosurgery and Department of Neurology, Johns Hopkins Hospital, Baltimore, MA, USA

Deolinda Lima Department of Experimental Biology, Faculdade de Medicina da Universidade do Porto, Porto, Portugal

Volker Limmroth Department of Neurology, University of Essen, Essen, Germany

Eric Lingueglia Institut de Pharmacologie Moléculaire et Cellulaire (IPMC) CNRS/UNS UMR7275, Valbonne-Sophia Antipolis, France

Steven J. Linton Department of Occupational and Environmental Medicine, Orebro Medical Center, Orebro, Sweden

Arthur G. Lipman Departments of Pharmscotherpay and Anesthesiology and Pain Management Center, University of Utah, Salt Lake City, UT, USA

Richard B. Lipton Departments of Neurology, Epidemiology and Population Health, Albert Einstein College of Medicine, Bronx, NY, USA

Diana Lisi Department of Psychology, York University, Toronto, ON, Canada

Kenneth M. Little Division of Neurosurgery, Duke University Medical Center, Durham, NC, USA

Edward Littleton Queen Elizabeth Hospital, Birmingham, UK

Spencer S. Liu Virginia Mason Medical Center, Seattle, WA, USA

Anne Elisabeth Ljunggren Section for Physiotherapy Science, Department of Public Health and Primary Health, Faculty of Medicine, University of Bergen, Bergen, Norway

John D. Loeser Department of Neurological Surgery, University of Washington, Seattle, WA, USA

Joern Loetsch Institute for Clinical Pharmacology, Pharmaceutical Center Frankfurt, Johann Wolfgang Goethe University, Frankfurt, Germany

Patrick Loisel Disability Prevention Research and Training Centre, Université de Sherbrooke, Longueuil, Canada

Elisa Lopez-Dolado Servicio de Rehabilitacion, National Hospital of Paraplèjicos SESCAM, Toledo, Spain

Jürgen Lorenz Faculty of Life Science, Competence Center Gesundheit, University of Applied Science Hamburg, Hamburg, Germany

Dayna R. Loyd US Army Institute of Surgical Research, Fort Sam Houston, TX, USA

Daniel Lubelski Cleveland Clinic Lerner College of Medicine, Cleveland, OH, USA

Stephen Lutz Blanchard Valley Regional Cancer Center, Findlay, OH, USA

Bruce Lynn Department of Physiology, University College London, London, UK

Ross MacPherson Department of Anesthesia and Pain Management, Royal North Shore Hospital University of Sydney, Sydney, NSW, Australia

Christophe Maes Manager Innovatie AZ Alma, Belgium

Michel Magnin INSERM, Neurological Hospital, Lyon, France

Bryan Mahony Department of Anesthesiology, Ohio State University Medical Center, Mount Sinai Hospital, New York, NY, USA

Steven F. Maier Department of Psychology and Neuroscience, and The Center for Neuroscience, University of Colorado at Boulder, Boulder, CO, USA

Thorsten J. Maier Institute of Pharmaceutical Chemistry, Goethe University, Frankfurt, Germany

Chris J. Main Keele University, Keele, Staffordshire, UK

Marzia Malcangio Wolfson Centre for Age Related Diseases, King's College London, London, UK

Paolo L. Manfredi Department of Neurology, Memorial Sloan-Kettering Cancer Center, New York, USA

Kaisa Mannerkorpi Department of Rheumatology and Inflammation Research, Institute of Medicine, University of Gothenburg, Göteborg, Sweden

Barton H. Manning Amgen Inc., Thousand Oaks, CA, USA

Patrick W. Mantyh Department of Pharmacology, The University of Arizona, AZ, USA

Jianren Mao Department of Anesthesia and Critical Care, Harvard Medical School, Boston, MA, USA

Serge Marchand Faculty of Medicine, University of Sherbrooke, Sherbrooke, QC, Canada

Paolo Marchettini Pain Medicine Center, Scientific Institute San Raffaele, Milan, Italy

Sarah R. Martin Psychology, Georgia State University, Atlanta, GA, USA

Peggy Mason Department of Neurobiology, Pharmacology and Physiology, University of Chicago, Chicago, IL, USA

Scott Masters Caloundra Spinal and Sports Medicine Centre, Caloundra, QLD, Australia

Laurence E. Mather Department of Anaesthesia and Pain Management, University of Sydney at Royal North Shore Hospital, St. Leonards, Sydney, NSW, Australia

Bill McCarberg Pain Services, Kaiser Permanente, San Diego, USA

Lance McCracken Health Psychology/Psychology Department, King's College London, London, UK

Kathleen McGinn Anesthesiology & Pain Medicine, Seattle Children's Hospital, Seattle, WA, USA

Patricia A. McGrath The Hospital for Sick Children Pain Research Center, Toronto, ON, Canada

Patrick McGrath Departments of Psychology, Pediatrics and Psychiatry, Dalhousie University, Canada

Greg McIntosh Canadian Back Institute, Toronto, ON, Canada

Peter McIntyre Department of Pharmacology, University of Melbourne, Melbourne, Australia

Elspeth M. McLachlan Neuroscience Research Australia, University of New South Wales, Randwick, NSW, Australia

University of Glasgow, Glasgow, UK

Peter A. McNaughton Department of Pharmacology, University of Cambridge, Cambridge, UK

J. Mark Melhorn Section of Orthopaedics, Department of Surgery, University of Kansas School of Medicine, Wichita, KS, USA

Ronald Melzack Department of Psychology, McGill University, Montreal, QC, Canada

Lorne M. Mendell Department of Neurobiology and Behavior, Stony Brook University, Stony Brook, NY, USA

George Mendelson Department of Psychological Medicine, Monash University, Caulfield, VIC, Australia

Siegfried Mense Neuroanatomy, Heidelberg University, Medical Faculty Mannheim, CBTM, Mannheim, Germany

Daniel Menétrey Biomédicale, Paris, France

Sebastiano Mercadante Pain Relief & Palliative Care, La Maddalena Cancer Center University of Palermo, Palermo, Italy

Susan Mercer School of Biomedical Sciences, The University of Queensland, Brisbane, Australia

Karl Messlinger University of Erlangen-Nürnberg, Erlangen, Germany

Richard A. Meyer Department of Neurosurgergy, School of Medicine Johns Hopkins University, Baltimore, MD, USA

Walter J. Meyer III Shriners Hospitals for Children, Shriners Burns Hospital, Galveston, TX, USA

Department of Psychiatry, University of Texas Medical Branch, Galveston, TX, USA

Björn A. Meyerson Department of Clinical Neuroscience, Section of Neurosurgery, Karolinska Institute/University Hospital, Stockholm, Sweden

Martin Michaelis Sanofi-Aventis Deutschland GmbH, Frankurt am Main, Germany

Joseph Mignogna Michael E. DeBakey VA Medical Center, Houston VA HSR&D Center of Excellence (152), Houston, TX, USA

Judith A. Mikacich Department of Obstetrics and Gynecology, University of California Los Angeles Medical Center, Los Angeles, CA, USA

Kensaku Miki Department of Integrative Physiology, National Institute for Physiological Sciences, Okazaki, Japan

Vjekoslav Miletic School of Medicine and Public Health, University of Wisconsin, Madison, WI, USA

Scott R. Miller Memorial Sloan-Kettering Cancer Center, New York, USA

Erin D. Milligan Department of Neurosciences, University of New Mexico, Albuquerque, NM, USA

Michael S. Minett Molecular Nociception Group, Wolfson Institute for Biomedical Research, University College London, London, WC, UK

J. Mocco Department of Neurological Surgery, Neurological Institute Columbia University, New York, NY, USA

Jeffrey S. Mogil Department of Psychology, McGill University, Montreal, QC, Canada

Harvey Moldofsky Sleep Disorders Clinic of the Centre for Sleep and Chronobiology, Toronto, ON, Canada

Robert M. Moldwin The Arthur Smith Institute for Urology, North Shore-LIJ Healthcare System, Hofstra North Shore LIJ School of Medicine, New Hyde Park, NY, USA

Derek C. Molliver Department of Neurobiology, University of Pittsburgh School of Medicine, Pittsburgh, PA, USA

Robert D. Mootz Washington State Department of Labor and Industries, Olympia, WA, USA

Laura Mordillo-Mateos FENNSI Group, Hospital Nacional de Parapléjicos, Toledo, Spain

Anne Morel Laboratory for Functional Neurosurgery, Neurosurgical Clinic University Hospital Zürich, Zürich, Switzerland

Anne Morinville Montreal Neurological Institute, McGill University, Montreal, QC, Canada

Stephen Morley Academic Unit of Psychiatry and Behavioral Sciences, University of Leeds, Leeds, UK

David B. Morris University of Virginia, Charlottesville, VA, USA

G. Lorimer Moseley Sansom Institute for Health Research, University of South Australia Neuroscience Research Australia, Adelaide, Australia

Joachim Mossner Medical Clinic and Policlinic II, University Clinical Center Leipzig AöR, Leipzig, Germany

Eric A. Moulton P.A.I.N. Group, Anesthesiology, Boston Children's Hospital, Harvard Medical School, Waltham, MA, USA

Francis Githae Muriithi Department of Obstetrics and Gynaecology, Aga Khan University, Nairobi, Kenya

Anne Z. Murphy Neuroscience Institute, Georgia State University, Atlanta, GA, USA

Paul M. Murphy Department of Anaesthesia and Pain Management, Royal North Shore Hospital, St. Leonards, NSW, Australia

Alison Murray Foothills Medical Center, Calgary, AB, Canada

H. S. Na Neuroscience Research Institute and Department of Physiology, Korea University, College of Medicine, Seoul, Republic of Korea

Daisuke Naka Department of Integrative Physiology, National Institute for Physiological Sciences, Okazaki, Japan

Yoshio Nakamura Pain Research Center, Department of Anesthesiology University of Utah School of Medicine, Salt Lake City, UT, USA

Barbara Namer Department of Physiology & Pathophysiology, FAU, Erlangen/Nuremberg, Bavaria, Germany

Matti V. O. Närhi Department of Biomedicine/Physiology, Department of Dentistry, University of Eastern Finland, Institute of Medicine, Kuopio, Finland

H. J. W. Nauta University of Texas Medical Branch, Galveston, TX, USA

Lucia Negri Department of Physiology and Pharmacology, Sapienza University of Rome, Rome, Italy

Gary Nesbit Department of Radiology, Oregon Health and Science University, Portland, OR, USA

Timothy Ness Department of Anesthesiology, University of Alabama at Birmingham, Birmingham, AL, USA

Volker Neugebauer Department of Neuroscience & Cell Biology, University of Texas Medical Branch, Galveston, TX, USA

Lawrence C. Newman Albert Einstein College of Medicine, New York, NY, USA

Michael K. Nicholas Royal North Shore Hospital, University of Sydney, St. Leonards, NSW, Australia

Ellen Niederberger Pharmazentrum frankfurt, Institute of Clinical Pharmacology, Goethe-University Hospital, Frankfurt, Germany

Lone Nikolajsen Department of Anaesthesiology, Danish Pain Research Center Aarhus University Hospital, Aarhus, Denmark

Katie Nixdorf Comprehensive Pain Center, Oregon Health and Science University, Portland, OR, USA

Donald R. Nixdorf Department of Diagnostic and Biological Sciences, School of Dentistry University of Minnesota, Minneapolis, MN, USA

Carl E. Noe University of Texas Southwestern Medical Center, Lubbock, TX, USA

Koichi Noguchi Department of Anatomy and Neuroscience, Hyogo College of Medicine, Hyogo, Japan

Richard B. North The Johns Hopkins University, Baltimore, MD, USA

Turo J. Nurmikko Department of Neurological Science, University of Liverpool, and Pain Research Institute, The Walton Centre NHS Foundation Trust, Liverpool, UK

Anne Louise Oaklander Nerve Injury Unit, Departments of Neurology and Pathology, Massachusetts General Hospital Harvard Medical School, Boston, MA, USA

Koichi Obata Department of Anatomy and Neuroscience, Hyogo College of Medicine, Hyogo, Japan

Tim F. Oberlander Division of Developmental Pediatrics, University of British Columbia, Vancouver, BC, Canada

Bruno Oertel Institute for Clinical Pharmacology, Pharmaceutical Center Frankfurt, Johann Wolfgang Goethe University, Frankfurt, Germany

Andrea C. Ogbonna Wolfson Centre for Age Related Diseases, King's College London, London, UK

Alfred T. Ogden Department of Neurological Surgery, Neurological Institute Columbia University, New York, NY, USA

S. Ohara Department of Neurosurgery, Johns Hopkins Hospital, Johns Hopkins Medical Institutions, Johns Hopkins University, Baltimore, MD, USA

Peter T. Ohara Department of Anatomy, University of California San Francisco, San Francisco, CA, USA

Akiko Okifuji Department of Anesthesiology, University of Utah, Salt Lake City, UT, USA

Jean-Louis Oliveras Inserm, Paris, France

Antonio Oliviero FENNSI Group, National Hospital of Parapléjicos, Toledo, Spain

Sean O'Mahony Palliative medicine, Rush University Medical Center, Chicago, IL, USA

Peregrine B. Osborne Dept Anatomy & Neuroscience, University of Melbourne, Melbourne, VIC, Australia

Michael H. Ossipov Department of Pharmacology, University of Arizona College of Medicine, Tucson, AZ, USA

Tonya M. Palermo Seattle Children's Hospital Research Institute, CW8-6 - Child Health, Behavior and Development, University of Washington, Seattle, WA, USA

Shreela Palit Department of Psychology, The University of Tulsa, Tulsa, OK, USA

Juan A. Pareja Department of Neurology, Fundación Hospital Alcorcón, Madrid, Spain

Steven D. Passik Psychiatry and Anesthesiology, Vanderbilt University, Nashville, TN, USA

Gavril W. Pasternak Laboratory of Molecular Neuropharmacology, Memorial Sloan-Kettering Cancer Center, New York, USA

S. H. Patel Department of Neurosurgery, Johns Hopkins Hospital, Baltimore, MD, USA

Gavin Pattullo Department of Anesthesia and Pain Management, Royal North Shore Hospital, University of Sydney, St. Leonards, NSW, Australia

Kevin Pauza Tyler Spine and Joint Hospital, Tyler, TX, USA

Eric M. Pearlman Pediatrics, Mercer University School of Medicine, The Children's Hospital at Memorial University Medical Center, Savannah, USA

Elvira De la Pena Instituto de Neurociencias, Universidad Miguel Hernández-CSIC, Alicante, Spain

Yuan Bo Peng Department of Psychology, University of Texas at Arlington, Arlington, TX, USA

Jose Pereira Foothills Medical Center, Calgary, AB, Canada

Edward Perl Department of Cell and Molecular Physiology, School of Medicine University of North Carolina, Chapel Hill, NC, USA

Tiffany Grace Perry Center for Spine Health, The Cleveland Clinic Foundation, Cleveland, OH, USA

Marie Pertin Pain Center, University Hospital Center and University of Lausanne, Lausanne, Switzerland

Antti Pertovaara Department of Physiology, Institute of Biomedicine University of Helsinki, Helsinki, Finland

Peter Peter Department of Neurobiology and Anatomy, University of Rochester Medical Center, Rochester, NY, USA

Lisa M. Peters Department of Anesthesiology and Pain Medicine, Seattle Children's Hospital, Seattle, WA, USA

Pramit Phal Department of Radiology, Oregon Health and Science University, Portland, OR, USA

Rebecca Pillai Riddell Department of Psychology, York University, Toronto, ON, Canada

Department of Psychiatry, Hospital for Sick Children, Toronto, ON, Canada

Issy Pilowsky University of Adelaide, Adelaide, Australia

Leon Plaghki Institute of Neuroscience (IoNS), Université Catholique de Louvain, Brussels, Belgium

Markus Ploner Department of Neurology, Technische Universität München, Munich, Germany

Esther M. Pogatzki-Zahn Department of Anesthesiology and Intensive Care Medicine, University Muenster, Muenster, Germany

Michael Polydefkis Department of Neurology, Johns Hopkins University School of Medicine, Baltimore, MD, USA

James D. Pomonis Algos Preclinical Services, Roseville, MN, USA

Dieter Pongratz Friedrich Baur Institute, Ludwig Maximilians University, Munich, Germany

Hayley Poole Hadley House, Leicester General Hospital, Leicester, UK

Frank Porreca Department of Pharmacology, University of Arizona Health Sciences Center, Tucson, AZ, USA

Russell K. Portenoy Department of Pain Medicine and Palliative Care, Beth Israel Medical Center, New York, NY, USA

Ian Power Critical Care and Pain Medicine, The University of Edinburgh, Edinburgh, UK

Donald D. Price Oral Surgery and Neuroscience, McKnight Brain Institute University of Florida, Gainesville, FL, USA

Daryl Pullman Medical Ethics, Memorial University of Newfoundland, St. John's, NL, Canada

Yunhai Qiu Department of Integrative Physiology, National Institute for Physiological Sciences, Okazaki, Japan

Michael Quittan Department of Physical Medicine and Rehabilitation, Kaiser-Franz-Joseph Hospital, Vienna, Austria

Raymond M. Quock Department of Pharmaceutical Sciences, Washington State University College of Pharmacy, Pullman, WA, USA

Jennifer Rabbitts Department of Anesthesiology and Pain Medicine, Seattle Children's Hospital, University of Washington School of Medicine, Seattle, WA, USA

Nicole Racine Psychology, York University, Toronto, ON, Canada

Gabor B. Racz Texas Tech University Health Sciences Center, Lubbock, TX, USA

Lukas Radbruch Department of Palliative Medicine, University of Bonn, Bonn, Germany

Robert B. Raffa Department of Pharmaceutical Sciences, Temple University School of Pharmacy, Philadelphia, PA, USA

Vasudeva Raghavendra Algos Therapeutics Inc., Saint Paul, MN, USA

Pierre Rainville Department of Stomatology, Faculty of Dental Medicine, University of Montreal, Montreal, QC, Canada

Srinivasa N. Raja Department of Anesthesiology and Critical Care Medicine, Johns Hopkins University, Baltimore, USA

Alan Randich University of Alabama, Birmingham, AL, USA

Andrea J. Rapkin Department of Obstetrics and Gynecology, University of California Los Angeles Medical Center, Los Angeles, CA, USA

Z. Harry Rappaport Department of Neurosurgery, Rabin Medical Center Tel-Aviv University, Petah Tiqva, Israel

Matthew N. Rasband Department of Molecular and Cellular Biology, Baylor College of Medicine, Farmington, Houston TX, USA

University of Connecticut Health Center, Farmington, CT, USA

Douglas Rasmusson Department of Physiology and Biophysics, Dalhousie University, Halifax, Canada

James P. Rathmell Department of Anesthesiology, University of Vermont College of Medicine, Burlington, USA

K. Ravishankar The Headache and Migraine Clinic, Jaslok Hospital and Research Centre Lilavati Hospital and Research Centre, Mumbai, India

Ke Ren Department of Neural and Pain Sciences, University of Maryland Dental School, Baltimore, MD, USA

Seth Resnick Silver Hill Hospital, New York University School of Medicine, USA

Cielito Reyes-Gibby The University of Texas MD Anderson Cancer Center, Houston, TX, USA

Jamie L. Rhudy Department of Psychology, The University of Tulsa, Tulsa, OK, USA

Alfredo Ribeiro-da-Silva Department of Pharmacology and Therapeutics, McGill University, Montreal, QC, Canada

Frank L. Rice Center for Neuropharmacology and Neuroscience, Albany Medical College, Albany, NY, USA

Matthias Ringkamp Department of Neurosurgergy, School of Medicine Johns Hopkins University, Baltimore, MD, USA

Pierre J. M. Rivière Ferring Research Institute, San Diego, CA, USA

Claude Robert Université Paris Descartes, INSERM UMR 894, France

Roger Robert Neurosurgery, Hotel Dieu, Nantes, CHU, France

James P. Robinson Department of Anesthesiology & Pain Medicine, University of Washington, Seattle, WA, USA

John Robinson Pain Management Centre, Burwood Hospital, Christchurch, New Zealand

Alfonso Romero-Sandoval Department of Pharmaceuticals and Administrative Sciences, Presbyterian College School of Pharmacy, Clinton, SC, USA

David Roselt Aberdovy Clinic, Bundaberg, QLD, Australia

Jackie Ross Early Pregnancy Unit, King's College Hospital Early Pregnancy Unit, London, UK

Shlomo Rotshenker Department of Medical Neurobiology, IMRIC, Faculty of Medicine Hebrew University of Jerusalem, Jerusalem, Israel

Paul C. Rousseau Medical University of South Carolina, Charleston, SC, USA

Ranjan Roy Department of Clinical Health Psychology, Faculty of Medicine, University of Manitoba, Winnipeg, MB, Canada

Carolina Roza Biología de Sistemas (Unidad Fisiología), Universidad de Alcalá, Alcalá de Henares, Madrid, Spain

Todd D. Rozen Department of Neurology, Geisinger Health System, Wilkes-Barre, PA, USA

Roman Rukwied Department of Anaesthesiology and Intensive Care Medicine, Faculty of Clinical Medicine Mannheim University of Heidelberg, Heidelberg, Germany

Irwin Jon Russell Division of Clinical Immunology and Rheumatology, Department of Medicine, The University of Texas Health Science Center, San Antonio, TX, USA

Nayef E. Saadé Anatomy, Cell Biology and Physiology, Faculty of Medicine, American University of Beirut, Beirut, Lebanon

Mary Ann C. Sabino Oral and Maxillofacial Surgery, Hennepin County Medical Center University of Minnesota, Minneapolis, MN, USA

Thomas E. Salt Institute of Ophthalmology, University College London, London, UK

A. Samdani Department of Neurosurgery, Johns Hopkins Medical Institutions, Baltimore, MD, USA

Yasmin Sana King's College Hospital, London, UK

Margarita Sanchez-del-Rio Neurology Department, Hospital Ruber Internacional, Madrid, Spain

Jürgen Sandkühler Department of Neurophysiology, Center for Brain Research Medical University of Vienna, Vienna, Austria

Peter S. Sándor RehaClinic, Cantonal Hospital Baden, Baden, Switzerland

Johannes Sarnthein Functional Neurosurgery, Neurosurgical Clinic University Hospital, Zürich, Switzerland

Susanne K. Sauer Department of Physiology and Pathophysiology, University of Erlangen, Erlangen, Germany

Steven Savvas Biomedical Division, National Ageing Research Institute, Melbourne, VIC, Australia

Jana Sawynok Department of Pharmacology, Dalhousie University, Halifax, Canada

John W. Scadding The National Hospital for Neurology and Neurosurgery, London, UK

Frederieke Geraldine Schaafsma Department of Public and Occupational Health, EMGO+ Institute/VU Medical Center, Amsterdam, MB, The Netherlands

Hans-Georg Schaible Department of Physiology - Neurophysiology, Friedrich-Schiller-University of Jena, Jena, Germany

Steven S. Scherer Department of Neurology, The Perelman School of Medicine at the University of Pennsylvania, Philadelphia, PA, USA

Michael Schäfer Department of Anesthesiology and Intensive Care Medicine, Charité University, Berlin, Germany

Martin Schmelz Department of Anesthesiology in Mannheim, University of Heidelberg, Mannheim, Germany

Robert F. Schmidt Physiologisches Institut der Universität Würzburg, Würzburg, Germany

Suzan Schneeweiss The Hospital for Sick Children, Toronto, ON, Canada

Frank Schneider Department of Psychiatry and Psychotherapy, RWTH Aachen University, Aachen, Germany

Alfons Schnitzler Department of Neurology, Heinrich Heine University, Düsseldorf, Germany

Jean Schoenen Department of Neurology, University of Liège, CHR Citadelle, Belgium

Department of Neuroanatomy, University of Liège, CHR Citadelle, Belgium

Jens Schouenborg Neuronano Research Center, Lund University, Lund, Sweden

Stephan A. Schug School of Medicine and Pharmacology, Royal Perth Hospital, Faculty of Medicine, Dentistry & Health Sciences, The University of Western Australia, UWA Anaesthesiology, Perth, Australia

Patrick Schuss Department of Neurosurgery, University-Hospital Bonn, Bonn, Germany

Anthony C. Schwarzer Faculty of Medicine and Health Sciences, The University of Newcastle, Newcastle, Australia

Whitney Scott Department of Psychology, McGill University, Montreal, QC, Canada

Laura C. Seidman Pediatric Pain Program, Department of Pediatrics David Geffen School of Medicine at UCLA, Los Angeles, CA, USA

Ze'ev Seltzer Faculty of Dentistry, University of Toronto Centre for the Study of Pain, Toronto, ON, Canada

Peter Selwyn Albert Einstein College of Medicine, Bronx, NY, USA

Patrick B. Senatus Department of Neurological Surgery, Neurological Institute Columbia University, New York, NY, USA

Jyoti N. Sengupta Division of Gastroenterology and Hepatology and Pediatric Gastroenterology, Medical College of Wisconsin, Milwaukee, WI, USA

Mariano Serrao Department of Neurology and Otorinolaringoiatry, University of Rome La Sapienza, Rome, Italy

Barry J. Sessle Faculty of Dentistry, University of Toronto, Toronto, ON, Canada

Faculty of Dentistry, University of Toronto, Toronto, ON, Canada

Virginia S. Seybold Department of Neuroscience, University of Minnesota, Minneapolis, MN, USA

T. Shah Royal Perth Hospital and University of Western Australia, Crawley, WA, Australia

Yair Sharav Department of Oral Medicine, School of Dental Medicine, Hebrew University-Hadassah, Jerusalem, Israel

S. Murray Sherman Department of Neurobiology, University of Chicago, Chicago, IL, USA

Yoshio Shigenaga Department of Oral Anatomy and Neurobiology, Osaka University, Osaka, Japan

Susan Shin Neurology, Mount Sinai School of Medicine, New York, NY, USA

Edward A. Shipton Department of Anaesthesia, Christchurch School of Medicine and Health Sciences, University of Otago, Christchurch, New Zealand

Yoram Shir The Alan Edwards Pain Management Unit, McGill University Health Centre, Montreal, QC, Canada

Peter Shrager Department of Neurobiology and Anatomy, University of Rochester Medical Center, Rochester, NY, USA

Marc J. Shulman VA Connecticut Healthcare System, Yale University, New Haven, CT, USA

David M. Sibell Comprehensive Pain Center/Spine Center, Oregon Health and Science University, Portland, OR, USA

Philip J. Siddall Department of Pain Management, Greenwich Hospital, University of Sydney, Sydney, NSW, Australia

Stephen D. Silberstein Jefferson Medical College, Thomas Jefferson University and Jefferson Headache Center, Thomas Jefferson University Hospital, Philadelphia, PA, USA

Donald A. Simone Diagnostic and Biological Sciences, University of Minnesota, Minneapolis, MN, USA

David G. Simons Department of Rehabilitation Medicine, Emory University, Atlanta, GA, USA

David Simpson Clinical Neurophysiology Laboratories, Mount Sinai Medical Center, New York, NY, USA

Marc Sindou Department of Neurosurgery, Hopital Neurologique Pierre Wertheimer, University of Lyon 1, Lyon, France

Soeren H. Sindrup Department of Neurology, Odense University Hospital, Odense, Denmark

Bengt H. Sjölund Department of Community Medicine and Rehabilitation Medicine, Umea University, Umea, Sweden

Carsten Skarke Institute for Translational Medicine and Therapeutics (ITMAT), University of Pennsylvania Perelman School of Medicine, Philadelphia, PA, USA

Paul A. Sloan Department of Anesthesiology, University of Kentucky, Lexington, KY, USA

Kathleen A. Sluka Physical Therapy and Rehabilitation Science Graduate Program, University of Iowa, Iowa City, IA, USA

Seymour Solomon Montefiore Medical Center, Albert Einstein College of Medicine, Bronx, NY, USA

Claudia Sommer Department of Neurology, University of Würzburg, Würzburg, Germany

Linda S. Sorkin Occupational Therapy and Rehabilitation, Anesthesia Research Laboratories, University of California, San Diego, CA, USA

Ingrid Söderback Uppsala University, Uppsala, Sweden

Kari Sørensen Department of Pain Management and Research, Oslo University Hospital, Rikshospitalet, Oslo, Norway

Michael Spaeth Friedrich-Baur-Institute, University of Munich, Munich, Germany

Peregrin O. Spielholz Washington State Department of Labor and Industries, SHARP Program, Olympia, WA, USA

Peter S. Staats Division of Pain Medicine, The Johns Hopkins University, Baltimore, MD, USA

Brett R. Stacey Comprehensive Pain Center, Oregon Health and Science University, Portland, OR, USA

Wendy M. Stein San Diego Hospice and Palliative Care, UCSD, San Diego, CA, USA

Christoph Stein Department of Anesthesiology and Intensive Care Medicine, Freie Universitaet Berlin Charité Campus Benjamin Franklin, Berlin, Germany

Michael P. Steinmetz Department of Neurosurgery, MetroHealth Medical Center, Case Western Reserve University School of Medicine, Cleveland, OH, USA

Jair Stern Functional Neurosurgery, Neurosurgical Clinic University Hospital, Zürich, Switzerland

Bonnie J. Stevens University of Toronto and The Hospital for Sick Children, Toronto, ON, Canada

Carl-Olav Stiller Division of Clinical Pharmacology and Department of Medicine, Karolinska University Hospital, Stockholm, Sweden

Jennifer N. Stinson The Hospital for Sick Children, Toronto, ON, Canada

Henry Stockbridge University of Washington School of Public Health and Community Medicine, Seattle, WA, USA

Christian S. Stohler Baltimore College of Dental Surgery, University of Maryland, Baltimore, MD, USA

Laura Stone Faculty of Dentistry, Alan Edwards Centre for Research on Pain, Montreal, QC, Canada

William Stones Department of Obstetrics and Gynaecology, Aga Khan University, Nairobi, Kenya

Andreas Straube Department of Neurology, Ludwig Maximilians University, Munich, Germany

Jenny Strong University of Queensland, Brisbane, QLD, Australia

Audun Stubhaug Department of Pain Management and Research, Oslo University Hospital, Rikshospitalet and University of Oslo, Oslo, Norway

Gerold Stucki Swiss Paraplegic Research, Nottwil, Switzerland

Department of Health Sciences and Health Policy, University of Lucerne, Lucerne, Switzerland

ICF Research Branch in cooperation with the World Health Organization (WHO) Collaborating Centre for the Family of International Classifications in Germany, Nottwil, Switzerland

Cheryl L. Stucky Department of Cell Biology Neurobiology and Anatomy, Medical College of Wisconsin, Milwaukee, WI, USA

Mathias Sturzenegger Department of Neurology, University Hospital, Inselspital, University of Bern, Bern, Switzerland

Yasuo Sugiura School of Human Care Study, Nagoya University of Arts and Sciences, Nissin City, Aichi, Japan

Mark D. Sullivan Psychiatry and Behavioral Sciences, University of Washington, Seattle, WA, USA

Michael J. L. Sullivan Department of Psychology, McGill University, Montreal, QC, Canada

Rie Suzuki Department of Pharmacology, University College London, London, UK

Kristina B. Svendsen Danish Pain Research Center, Aarhus University Hospital, Aarhus, Denmark

Peter Svensson Clinical Oral Physiology, Aarhus University, Aarhus, Denmark

Peter Szucs Spinal Neuronal Networks, Instituto de Biologia Molecular e Celular - IBMC, Porto, Portugal

A. Taghva Department of Neurosurgery, Johns Hopkins Hospital, Baltimore, MD, USA

Kimberson Cochien Tanco Department of Palliative Care and Rehabilitation Medicine, The University of Texas M. D. Anderson Cancer Center, Houston, TX, USA

Kersi Taraporewalla Royal Brisbane and Womens' Hospital, University of Queensland, Brisbane, QLD, Australia

Irmgard Tegeder Pharmazentrum frankfurt, Institute of Clinical Pharmacology, Goethe-University Hospital, Frankfurt, Germany

Astrid Juhl Terkelsen Danish Pain Research Center, Department of Neurology, Aarhus University Hospital, Aarhus, Denmark

Gregory W. Terman Department of Anesthesiology and the Graduate Program in Neurobiology and Behavior, University of Washington, Seattle, WA, USA

Alison Thomas Lismore, NSW, Australia

Benjamin Thomas Chelsea & Westminster Hospital Foundation Trust, London, UK

Jacob Thomas School of Medical Sciences, University of Adelaide, Adelaide, SA, Australia

Miles Thompson Goldsmiths, University of London, London, UK

Beverly E. Thorn Department of Psychology, University of Alabama, Tuscaloosa, AL, USA

Cindy L. Thurston-Stanfield Department of Biomedical Sciences, University of South Alabama, Mobile, AL, USA

Arne Tjølsen Department of Bomedicine, University of Bergen, Haukeland University Hospital, Bergen, Norway

Andrew J. Todd Institute of Neuroscience and Psychology, College of Medical, Veterinary and Life Sciences, University of Glasgow, Glasgow, UK

Makoto Tominaga Department of Cellular and Molecular Physiology, Mie University School of Medicine, Tsu Mie, Japan

Victor Tortorici Instituto Venezolano de Investigaciones Cientificas (IVIC), Caracas, Venezuela

Irene Tracey Pain Imaging Neuroscience (PaIN) Group, Department of Physiology, Anatomy and Genetics, Oxford University, Oxford, UK

Diep Tuan Tran Department of Integrative Physiology, National Institute for Physiological Sciences, Okazaki, Japan

Rolf-Detlef Treede Institute of Physiology and Pathophysiology, Johannes Gutenberg University, Mainz, Germany

Jennie C. I. Tsao Pediatric Pain Program, Department of Pediatrics David Geffen School of Medicine at UCLA, Los Angeles, CA, USA

Lawrence Tsen Department of Anesthesiology, Perioperative and Pain Medicine, Brigham and Women's Hospital Harvard Medical School, Boston, MA, USA

Jeanna Tsenter Department of Physical Medicine and Rehabilitation, Hadassah University Hospital, Jerusalem, Israel

Maurits Van Tulder Institute for Research in Extramural Medicine, Free University Amsterdam, Amsterdam, The Netherlands

Eldon Tunks Hamilton Health Sciences Corporation, Hamilton, ON, Canada

Susan Tupper Pediatrics, University of Saskatchewan, Saskatoon, SK, Canada

Dennis C. Turk Department of Anesthesiology & Pain Research, University of Washington, Seattle, WA, USA

Judith A. Turner Department of Psychiatry and Behavioral Sciences, University of Washington, Seattle, WA, USA

Department of Rehabilitation Medicine, University of Washington, Seattle, WA, USA

Stephen P. Tyrer Royal Victoria Infirmary, University of Newcastle upon Tyne, Newcastle-on-Tyne, UK

Alexander Tzabazis Department of Anesthesia, Stanford University, Stanford, CA, USA

Kene Ugokwe Department of Neurosurgery, St Elizabeth Health Center, Youngstown, OH, USA

Lindsay S. Uman Department of Psychology, IWK Health Centre, Halifax, NS, Canada

Christiane Vahle-Hinz Department of Neurophysiology and Pathophysiology, University Medical Center Hamburg-Eppendorf, Hamburg, Germany

Muthukumar Vaidyaraman Division of Pain Medicine, The Johns Hopkins University, Baltimore, MD, USA

Todd W. Vanderah Department of Pharmacology, College of Medicine University of Arizona, Tucson, AZ, USA

Guy Vanderstraeten Department of Rehabilitation Sciences and Physiotherapy, Ghent University Hospital, Ghent, Belgium

Horacio Vanegas Instituto Venezolano de Investigaciones Cientificas (IVIC), Caracas, Venezuela

Jean-Jacques Vatine Outpatient and Research Division, Reuth Medical Center, Sackler Faculty of Medicine, Tel Aviv University, Tel Aviv, Israel

Leigh M. Vaughan Medical University of South Carolina, Charleston, SC, USA

Linda K. Vaughn Department of Biomedical Sciences, Marquette University, Milwaukee, WI, USA

Enrique Vazquez Instituto Venezolano de Investigaciones Cientificas (IVIC), Caracas, Venezuela

Louis Vera-Portocarrero Department of Pharmacology, College of Medicine University of Arizona, Tucson, AZ, USA

Paul Verrills Metropolitan Spinal Clinic, Prahran, VIC, Australia

Tine Vervoort Department of Experimental Clinical and Health Psychology, Ghent University, Ghent, Belgium

Félix Viana Instituto de Neurociencias, Universidad Miguel Hernández-CSIC, San Juan de Alicante, Alicante, Spain

Charles J. Vierck Jr. Department of Neuroscience and McKnight Brain Institute, University of Florida College of Medicine, Gainesville, FL, USA

Luis Villanueva INSERM UMR 894, Centre de Psychiatrie et Neurosciences, Paris, France

Marcelo Villar Faculty of Biomedical Sciences, Austral University, Buenos Aires, Argentina

Katy Vincent Nuffield Department of Obstetrics and Gynaecology, The Women's Centre, John Radcliffe Hospital, University of Oxford, Oxford, UK

David Vivian Metro Spinal Clinic, South Caulfield, VIC, Australia

Johan W. S. Vlaeyen Research Group Health Psychology, Catholic University of Leuven, Leuven, Belgium

Department of Clinical Psychological Science, Maastricht University, Maastricht, The Netherlands

Stéphanie Volders Research Group on Health Psychology, Leuven, Belgium

Ernest Volinn Pain Research Center, University of Utah, Salt Lake City, UT, USA

Lucy Vulchanova Department of Veterinary and Biomedical Sciences, University of Minnesota, St. Paul, MN, USA

Paul W. Wacnik Neuromodulation Research, Medtronic Inc., Minneapolis, MN, USA

Gordon Waddell University of Glasgow, Glasgow, UK

Gary A. Walco Department of Anesthesiology and Pain Medicine, University of Washington School of Medicine, Seattle Children's Hospital, Seattle, WA, USA

Suellen M. Walker University of Sydney Pain Management and Research Centre, Royal North Shore Hospital, NSW, Australia

Victoria C. J. Wallace Department of Anaesthetics, Pain Medicine and Intensive Care Chelsea and Westminster Hospital Campus, London, UK

Xiaohong Wang Department of Integrative Physiology, National Institute for Physiological Sciences, Okazaki, Japan

Gunnar Wasner Department of Neurological Pain Research and Therapy, Neurological Clinic, Kiel, Germany

Shoko Watanabe Department of Integrative Physiology, National Institute for Physiological Sciences, Okazaki, Japan

Linda R. Watkins Department of Psychology and Neuroscience, and The Center for Neuroscience, University of Colorado at Boulder, Boulder, CO, USA

Peter N. Watson Department of Medicine, University of Toronto, Toronto, ON, Canada

James Watt North Shore Hospital, Auckland, New Zealand

Stephen G. Waxman Department of Neurology, Yale School of Medicine, New Haven, CT, USA

Inge Wegner Julius Center, Department of Epidemiology, University Medical Center Utrecht, Utrecht, The Netherlands

Feng Wei Department of Neural and Pain Sciences, University of Maryland Dental School, Baltimore, MD, USA

Christian Weidner Department of Physiology and Experimental Pathophysiology, University Erlangen, Erlangen, Germany

P. T. M. Weijenborg Department of Gynecology, Leiden University Medical Center, Leiden, The Netherlands

Robin Weir School of Nursing, Faculty of Health Sciences, McMaster University, Hamilton, ON, Canada

Nirit Weiss Department of Neurosurgery, Mount Sinai Medical Center, New York, NY, USA

Ursula Wesselmann Departments of Neurology and Anesthesiology / Division of Pain Management, University of Alabama at Birmingham, Birmingham, AL, USA

Dagmar Westerling Department of Anesthesiology, Central Hospital of Kristianstad, Kristianstad, Sweden

Karin N. Westlund Department of Physiology, University of Kentucky, Lexington, KY, USA

Thomas M. Wickizer Health Services Management and Policy, Ohio State University, Colombus, OH, USA

Eva Widerström-Noga The Miami Project to Cure Paralysis, Department of Neurological Surgery, University of Miami Miller, School of Medicine, Miami, FL, USA

Julie Wieseler Department of Psychology and Neuroscience, and The Center for Neuroscience, University of Colorado at Boulder, Boulder, CO, USA

Zsuzsanna Wiesenfeld-Hallin Karolinska Institute, Stockholm, Sweden

Ann Wigham McCoull Clinic, Prudhoe Hospital, Northumberland, UK

George L. Wilcox Department of Neuroscience, Pharmacology and Dermatology, University of Minnesota Medical School, Minneapolis, MN, USA

Oliver H. G. Wilder-Smith Pain Centre, Department of Anaesthesiology, Radboud University Nijmegen Medical Centre, Nijmegen, The Netherlands

Kjersti Wilhelmsen Section for Physiotherapy Science, Department of Public Health and Primary Health, Faculty of Medicine, University of Bergen, Bergen, Norway

Victor Wilk Brighton Spinal Group, Brighton, VIC, Australia

W. A. Williams Royal Perth Hospital and University of Western Australia, Perth, WA, Australia

Sara E. Williams Division of Pediatric Gastroenterology, Medical College of Wisconsin, Milwaukee, WI, USA

William D. Willis Department of Neuroscience and Cell Biology, University of Texas Medical Branch, Galveston, TX, USA

John B. Winer Department of Neurology, Birmingham Muscle and Nerve Centre Queen Elizabeth Hospital, Birmingham, UK

Christopher J. Winfree Department of Neurological Surgery, Neurological Institute Columbia University, New York, NY, USA

Janet Winter Novartis Institute for Medical Science, London, UK

Alain Woda Faculté de Chirurgie Dentaire, University Clermont-Ferrand, Clermont-Ferrand, France

John N. Wood Molecular Nociception Group, Wolfson Institute for Biomedical Research, University College London, London, WC, UK

Lee Woodson Shriners Hospitals for Children, Shriners Burns Hospital, Galveston, TX, USA

Department of Anesthesia, University of Texas Medical Branch, Galveston, TX, USA

Anava A. Wren Pain Prevention and Treatment Research, Department of Psychiatry and Behavioral Sciences, Duke University Medical Center, Durham, NC, USA

Xiao-Jun Xu Department of Physiology and Pharmacology, Section of Integrative Pain Rsearch Karolinska Institutet, Stockholm, Sweden

Hiroshi Yamasaki Department of Integrative Physiology, National Institute for Physiological Sciences, Okazaki, Japan

Michael Yelland School of Medicine, Logan campus, Griffith University, Logan, Australia

Sriram Yennurajalingam Department of Palliative Care and Rehabilitation Medicine, The University of Texas M. D. Anderson Cancer Center, Houston, TX, USA

David C. Yeomans Department of Anesthesia, Stanford University, Stanford, CA, USA

Robert P. Yezierski Comprehensive Center for Pain Research and Department of Neuroscience, The McKnight Brain Institute University of Florida, Gainesville, FL, USA

Way Yin Department of Anesthesiology, University of Washington, Seattle, WA, USA

Atsushi Yoshida Department of Oral Anatomy and Neurobiology, Osaka University, Osaka, Japan

Joanna M. Zakrzewska Eastman Dental Hospital UCLH NHS Foundation Trust, London, UK

Jenna E. Zauk St. George's University, School of Medicine, Grenada, West Indies

Hanns Ulrich Zeilhofer Institute for Pharmacology and Toxicology, University of Zurich, and Swiss Federal Institute of Technology (ETH) Zurich, Zürich, Switzerland

Lonnie K. Zeltzer Pediatric Pain Program, Department of Pediatrics David Geffen School of Medicine at UCLA, Los Angeles, CA, USA

Giovambattista Zeppetella St. Clare Hospice, Hastingwood, UK

Xijing Zhang Department of Neuroscience, University of Minnesota, Minneapolis, MN, USA

Min Zhuo Department of Physiology, University of Toronto, Toronto, ON, Canada

C. Zöllner Department of Anesthesiology and Intensive Care Medicine, Charité University, Berlin, Germany

Krina Zondervan Genetic and Genomic Epidemiology Unit, Nuffield Dept of Obstetrics and Gynaecology, Wellcome Trust Centre for Human Genetics, Oxford, UK

Peter Zygmunt Department of Laboratory Medicine, Lund University, Division of Clinical Chemistry and Pharmacology, Lund, Sweden

Zbigniew Zylicz Hildegard Hospiz, Basel Switzerland, Basel, Switzerland

Davis C. Yeomans Department of Anesthesia, Stanford University, Stanford, CA, USA

Robert L. Yezierski Comprehensive Center for Pain Research and Department of Neuroscience, The McKnight Brain Institute, University of Florida, Gainesville, FL, USA

Wei Yin Department of Anesthesiology, University of Washington, Seattle, WA, USA

Atsushi Yoshida Department of Oral Anatomy and Neurobiology, Osaka University, Osaka, Japan

Joanna M. Zakrzewska Eastman Dental Hospital UCLH NHS Foundation Trust, London, UK

Ioana E. Zaric St. George's University School of Medicine, Grenada, West Indies

Hanns Ulrich Zeilhofer Institute for Pharmacology and Toxicology, University of Zurich and Swiss Federal Institute of Technology (ETH) Zurich, Zurich, Switzerland

Lonnie K. Zeltzer Pediatric Pain Program Department of Pediatrics David Geffen School of Medicine at UCLA, Los Angeles, CA, USA

Giovanna Maria Zucchelli St. Clare Hospice, Hastingwood, UK

Yijing Zhang Department of Neuroscience, University of Minnesota, Minneapolis, MN, USA

Min Zhuo Department of Physiology, University of Toronto, Toronto, ON, Canada

C. Zöllner Department of Anesthesiology and Intensive Care Medicine, Charité University, Berlin, Germany

Krina Zondervan Genetic and Genomic Epidemiology Unit, Nuffield Department of Obstetrics and Gynaecology, Wellcome Trust Centre for Human Genetics, Oxford, UK

Peter Zygmunt Department of Laboratory Medicine, Lund University Division of Clinical Chemistry and Pharmacology, Lund, Sweden

Walpurga Zylka Hoffmann-La Roche Ltd., Basel, Switzerland, Basel, Switzerland

N

N/OFQ

▶ Orphanin FQ

N₂O

▶ Nitrous Oxide Antinociception and Opioid Receptors

Na⁺ Channel

Definition

Na⁺ channel is a voltage-dependent permeation pathway for sodium ions.

Cross-References

▶ Trafficking and Localization of Ion Channels

Na⁺K⁺Cl⁻Cotransporter

Definition

A transporter protein whose primary function is to transport Cl^- into the cell. The $Na^+K^+Cl^-$ cotransporter (NKCC) does not directly consume energy (ATP) in the transport of Cl^-. Rather, it takes advantage of the electrochemical driving force on Na^+ to facilitate the movement of Cl^- and K^+ into the cell. It is therefore referred to as a secondary transport mechanism. NKCC1 is the isoform of this transporter present in the nervous system. NKCC1 is developmentally regulated in the central nervous system, expressed at high levels during the extension of neurite processes, but downregulated to very low levels after synapse formation. In contrast, NKCC1 expression is maintained at elevated levels in primary afferents into adulthood, and is therefore thought to contribute to the elevated intracellular Cl^- levels found in primary afferent neurons. As a result, $GABA_A$ receptor activation drives Cl^- efflux and membrane depolarization a phenomenon referred to as primary afferent depolarization.

Cross-References

▶ GABA and Glycine in Spinal Nociceptive Processing

NAcc

▶ Nociceptive Processing in the Nucleus Accumbens, Neurophysiology and Behavioral Studies

G.F. Gebhart, R.F. Schmidt (eds.), *Encyclopedia of Pain*, DOI 10.1007/978-3-642-28753-4,
© Springer-Verlag Berlin Heidelberg 2013

nAChr

Synonyms

Nicotinic receptors

Definition

Acetylcholine (ACh) is released from endothelial cells, keratinocytes, and other cells following trauma and can activate nociceptors. In nociceptors, ACh can interact with either muscanic (mAChr) or nicotinic (nAChr) receptors. The nicotinic type can directly induce action potentials by the gating of ion channels. The muscarinic type acts via G-protein-coupled receptors to either upregulate or downregulate nociceptor excitability.

Cross-References

▶ Nociceptors in the Orofacial Region (Skin/ Mucosa)

Nadolol

Definition

Nadolol is a beta-blocker.

Cross-References

▶ Preventive Migraine Therapy

Naloxone

Definition

Naloxone is an antagonist at the mu-opioid receptor with a short duration of action (30–45 min). It can be administered for rapid reversal of opioid-induced sedation or respiratory depression, but its administration will also reverse opioid-induced analgesia and may precipitate withdrawal symptoms.

Cross-References

▶ Alternative Medicine in Neuropathic Pain

Narcolepsy

Definition

Narcolepsy is a condition marked by an uncontrollable desire for sleep or by sudden attacks of sleep occurring at intervals.

Cross-References

▶ Hypothalamus and Nociceptive Pathways

Narcotic

▶ Opioid Peptide Co-localization and Release

Narcotics

▶ Oral Opioids

Natural History

Definition

Natural course of a disease or a symptom.

Cross-References

▶ Placebo Analgesia and Descending Opioid Modulation

Na$_v$1.3: (NAC3, KIAA1356)

▶ Nociception: Role of Voltage-Gated Sodium Channels

Na$_v$1.7: (PN1, NaS, hNE-Na)

▶ Nociception: Role of Voltage-Gated Sodium Channels

Na$_v$1.8: (SNS, PN3, NaNG)

▶ Nociception: Role of Voltage-Gated Sodium Channels

Na$_v$1.9: (NaN, SNS2, PN5)

▶ Nociception: Role of Voltage-Gated Sodium Channels

NCCP

Synonyms

Non-cardiac chest pain

Definition

NCCP is a functional chest pain of esophageal origin without obvious manifestation of pathology. The pain is very similar to cardiac angina.

Cross-References

▶ Visceral Pain Model; Esophageal Pain

Near-Infrared Spectroscopy

Synonyms

NIRS

Definition

This is a relatively novel technique for measuring functional activation in infants, which is both noninvasive and localized. In order to measure a localized hemodynamic response within the brain in response to peripheral stimulation, paired near-infrared (NIR) light emitters and detectors ("optodes") are placed symmetrically on each side of the head over the somatosensory cortex during stimulation. NIR light at four wavelengths is conveyed through the head, and a controlling computer calculates the changes in optic absorption at each wavelength and converts these into changes in oxyhemoglobin, deoxyhemoglobin, and total hemoglobin using known extinction coefficients. In studies on newborn infants, the presence and amplitude of the hemodynamic response is used to elucidate the maturation of cortical processing of stimuli. NIRS is ideally suited to the study of infants, because the infant skull is thinner than that of the adult so that the optical signal is cleaner and easier to detect.

Cross-References

▶ Infant Pain Mechanisms

Neck Pain

Definition

Localized neck or back pain is the symptom of a protruding disk frequently heralding a cervical or lumbosacral radiculopathy in spondyloarthritis. Nonspecific neck pain is a prominent feature of polymyalgia rheumatica.

Cross-References

▶ Muscle Pain in Systemic Inflammation
  (Polymyalgia Rheumatica, Giant Cell
  Arteritis, Rheumatoid Arthritis)
▶ Radiculopathies

Necrosis

Definition

Necrosis is the localized death of living cells.

Cross-References

▶ Sacroiliac Joint Pain

Negative Affectivity

Definition

The stable disposition to experience negative
affect and low mood (neuroticism).

Cross-References

▶ Hypervigilance and Attention to Pain

Negative Mucosal Potential

Definition

Electrical potential recorded from the respiratory
epithelium that reflects the excitation of nociceptive nerve terminals.

Cross-References

▶ Nociception in Nose and Oral Mucosa

Negative or Punishing Responses

Definition

Negative or punishing responses (e.g., expressions of irritation, ignoring) represent a third category of responses (in addition to solicitous and
distracting responses) to the expression of pain.

Cross-References

▶ Spouse, Role in Chronic Pain

Negative Reinforcement

Definition

Negative reinforcement is the removal of an aversive stimulus (tangible or non-tangible) following a behavior, with the goal of increasing future
incidents of that behavior.

Cross-References

▶ Impact of Familial Factors on Children's
  Chronic Pain
▶ Operant Perspective of Pain

Negative Responding

▶ Spouse, Role in Chronic Pain

Negative Sensory Phenomenon

Definition

Negative sensory phenomenon is a clinical sign
that is interpreted by the patient as less than
when compared to normal bodily function and
experiences.

Cross-References

► Hypoalgesia, Assessment
► Hypoesthesia, Assessment

Nematode

Definition

Nematode is a roundworm, a non-segmented worm phylum.

Cross-References

► Species Differences in Skin Nociception

Neocortical

Definition

Belonging to the top, approximately 2-mm-thick layer of the two hemispheres of the brain.

Cross-References

► Prefrontal Cortex, Effects on Pain-Related Behavior

Neonatal Inflammation

► Visceral Pain Model, Irritable Bowel Syndrome Model

Neonatal Pain

► Visceral Pain Model, Irritable Bowel Syndrome Model

Neonate

► Newborn

Neospinothalamic Tract

Definition

Lateral and phylogenetically younger component of the spinothalamic tract, also known as the lateral spinothalamic tract. It is comprised of the axons of nociceptive specific and wide dynamic range neurons. It projects to the ventral posterolateral nucleus of the thalamus and is responsible for the discriminative aspects of pain (location, intensity, duration).

Cross-References

► Acute Pain Mechanisms
► Parafascicular Nucleus, Pain Modulation
► Somatic Pain

Nerve Block

► Cancer Pain Management, Anesthesiologic Interventions
► Epidural Steroid Injections for Chronic Back Pain

Nerve Blocks by Local Anesthetic Drugs

Definition

Nerve blocks by local anesthetic drugs stop nerve impulse conduction in nerve cells, inhibiting pain impulses from reaching the central nervous system (CNS). They will often also make the pain-free body part numb, with weak or paralyzed muscles.

Cross-References

- ▶ Cancer Pain Management, Anesthesiologic Interventions
- ▶ Epidural Steroid Injections for Chronic Back Pain
- ▶ Postoperative Pain, Acute Pain Management Principles

Nerve Compression

Definition

Nerve compression is caused by mechanical obstruction along the course of a nerve. This may be due to entrapment of the nerve as it passes through a narrow anatomical passage that becomes further narrowed, e.g., as a result of local inflammation. An example is carpal tunnel syndrome. Other causes may be tramatic compression or a tumor. Nerve compression often results in a compression (or entrapment) neuropathy accompanied by negative symptoms such as numbness and muscle weakness. It may also result in dysesthesias and pain.

Cross-References

- ▶ Cancer Pain

Nerve Conduction

Definition

Nerve conduction refers to the ability of nerve fibers (axons) to propagate electrical impulses along the length of the nerve. Complete or partial disruption of nerve conduction due to injury or neuropathy can often be diagnosed clinically using a "nerve conduction study" in which electrical stimuli are applied at one location along the length of the nerve and the propagated compound action potential is recorded at another location. Current protocols can detect abnormalities in the conduction of myelinated nerve fibers, but not unmyelinated fibers.

Cross-References

- ▶ Electrodiagnosis and EMG
- ▶ Hereditary Neuropathies

Nerve Conduction Studies

- ▶ Electrodiagnosis and EMG

Nerve Growth Factor

Synonyms

NGF

Definition

Nerve growth factor (NGF) belongs to a family of polypeptide growth factors. It consists of alpha, beta, and gamma subunits. NGF is a target-derived factor and is essential for survival, differentiation, and maintenance of sympathetic and afferent neurons. In inflamed tissue, NGF biosynthesis is rapidly increased, leading to elevated concentrations of NGF in inflamed tissues. It has been shown that NGF is a mediator of inflammatory hyperalgesia and also a modulator of immune cell function. An enhanced retrograde transport of NGF to the DRG leads to an increase in the production of brain-derived neurotrophic factor (BDNF) at the level of gene expression, mainly in trkA-expressing small- and medium-sized neurons. During embryonic and early postnatal stages, sensory neurons are dependent on NGF for survival. Although adult sensory neurons do not depend on NGF for survival, the functional properties of some nociceptive sensory neurons, such as responsiveness to capsaicin or noxious heat, are modulated by NGF. NGF can exert its actions through either the high-affinity trkA receptor or the low-affinity p75 neurotrophin receptor.

Cross-References

- ▶ Congenital Insensitivity to Pain with Anhidrosis
- ▶ ERK Regulation in Sensory Neurons during Inflammation
- ▶ IB4-Positive Neurons, Role in Inflammatory Pain
- ▶ Immunocytochemistry of Nociceptors
- ▶ Nerve Growth Factor, Sensitizing Action on Nociceptors
- ▶ Neutrophils in Inflammatory Pain
- ▶ Spinal Cord Nociception, Neurotrophins
- ▶ TRPV1, Regulation by Nerve Growth Factor
- ▶ TRPV1, Regulation by Protons
- ▶ Wallerian Degeneration

Nerve Growth Factor (NGF)

- ▶ NGF, Regulation During Inflammation

Nerve Growth Factor Overexpressing Mice as Models of Inflammatory Pain

Kathryn M. Albers and Derek C. Molliver
Department of Neurobiology, University of
Pittsburgh School of Medicine, Pittsburgh,
PA, USA

Synonyms

NGF-OE mice; Transgenic mice

Definition

NGF and Inflammatory Pain

In peripheral tissues, the level of nerve growth factor (NGF) expression is often elevated following inflammation or injury (Heumann et al. 1987; Weskamp and Otten 1987). Studies using rodents have shown that injection of NGF causes behavioral thermal and mechanical ▶ hyperalgesia (Lewin et al. 1993; Lewin et al. 1994). Increased NGF expression is also accompanied by elevation of other inflammatory mediators such as bradykinin, prostaglandins, serotonin, ATP, and protons (Bennett 2001). These changes in the periphery are thought to collectively contribute to sensitization of sensory afferents and central pain-processing pathways.

The link between NGF and inflammatory pain signaling can be examined using a transgenic mouse model (see ▶ Nerve Growth Factor Overexpressing Mice as Models of Inflammatory Pain) in which NGF is overexpressed in the skin, a major target of sensory afferents. In these mice (NGF-OE mice), NGF is overexpressed in basal keratinocytes of stratified, keratinizing tissues such as the skin and oral epithelium, using the human keratin K14 promoter and enhancer region to drive expression of the mouse NGF cDNA (Albers et al. 1994). As described below, the increase in NGF expression causes an increase in the developmental survival of neurons that project to K14-expressing epithelium, altering their physiological properties and the expression of genes related to nociceptive signaling.

Characteristics

Anatomical Characteristics of NGF-OE Transgenic Mice

Mice that overexpress NGF in the skin exhibit hypertrophy of both sensory and sympathetic neurons (Albers et al. 1994; Davis et al. 1994, 1997). NGF-OE mice have an approximate twofold increase in the number of trigeminal and dorsal root ganglion (DRG) sensory neurons and a 2.5-fold increase in the number of sympathetic neurons in the superior cervical ganglia. In addition, preferential increases of unmyelinated and thinly myelinated fibers that project to the skin occur (Davis et al. 1997; Stucky et al. 1999), a finding consistent with the types of axons lost in ▶ ngf −/− mice (Crowley et al. 1994). Immunolabeling of skin and DRG has shown a preferential increase of peptidergic sensory neuron subtypes. For example, the percent of TrkA neurons is doubled, as is the percent of calcitonin gene-related peptide-positive neurons (Goodness et al. 1997). The population of sensory neurons that bind the plant lectin IB4 is not increased, consistent with the finding that glial cell line-derived growth factor (GDNF) is a major contributor to the trophic support of these neurons (Molliver et al. 1997).

Electrophysiologic Properties of NGF-OE Cutaneous Afferents

Electrophysiologic properties of cutaneous sensory afferents in the saphenous nerve of NGF-OE mice were analyzed using a skin-nerve preparation (Stucky et al. 1999). Large myelinated, low-threshold Aβ fibers showed no change in the proportion of slowly adapting (SA) or rapidly adapting (RA) fibers relative to wild-type animals. In addition, no significant difference in the mechanical stimulus-response properties, or conduction velocity, of SA or RA fibers of NGF-OE mice was found.

In contrast to Aβ fibers, both Aδ and C fiber nociceptors of NGF-OE mice had altered properties. The percent of Aδ mechanosensitive (AM) nociceptors was significantly increased from control values of 65 % of all Aδ fibers analyzed to 97 % in NGF-OE mice. Individual AM fibers also showed increased mechanical responsiveness, which was particularly evident at suprathreshold stimuli. A 100–300 mN sustained force evoked discharge rates in NGF-OE AM fibers double those of wild types. Though mechanically sensitized, AM fibers were unchanged with heat sensitivity. No significant difference was measured in the percent of AM fibers that respond to heat, the threshold for a response, or in the mean spikes per heat stimulus.

C fiber afferents showed a 50 % increase in total number in the saphenous nerve of NGF-OE mice (Stucky et al. 1999). Nearly all C fibers (96 %) responded to heat and showed a fourfold increase in the number of heat-evoked action potentials per C fiber. In addition, C fibers in NGF-OE mice exhibited spontaneous activity that was much higher than C fibers of wild-type mice (60 % vs. 6.5 %, respectively). This increase in sensitivity was not global however, since the response of C fibers to mechanical stimulation was half relative to control fibers. Thus, increased NGF in the skin regulates the receptive properties of cutaneous C and Aδ fibers in differential manners.

Behavioral Phenotype of Naïve NGF-OE Mice

To evaluate the response of NGF-OE mice to inflammatory stimuli, the behavioral response of NGF-OE mice to a focused heat source applied to the foot was measured. Two other types of animals were used for comparison in this analysis: littermate control mice (Blk6/C3H strain) and transgenic mice that overexpress GDNF in the skin (GDNF-OE mice). GDNF-OE mice have an enhancement of GDNF-dependent nociceptor neurons (Zwick et al. 2002). GDNF-dependent neurons are peptide-poor neurons, which primarily project to lamina II of the spinal cord (with some overlap in lamina I) and bind the plant lectin IB4 (Vulchanova et al. 2001). During postnatal development, GDNF-dependent neurons switch dependence from NGF to GDNF and express the tyrosine kinase receptor Ret and its coreceptor GFRα1 (Molliver et al. 1997). GDNF-dependent neurons have been proposed to primarily modulate responses to ▶ neuropathic pain as opposed to inflammatory pain (Snider and McMahon 1998). To compare NGF- and GDNF-dependent nociceptor populations, the behavioral response of NGF-OE, GDNF-OE, and wild-type (WT) mice to heat was measured (Fig. 1). This analysis showed that NGF-OE mice had slightly longer latencies (they were hypoalgesic) relative to WT and GDNF-OE mice, which had equivalent baselines (Zwick et al. 2003).

Nerve Growth Factor Overexpressing Mice as Models of Inflammatory Pain, Fig. 1 CFA injections did not cause increased hyperalgesia in NGF-OE and GDNF-OE mice. (**a**) Comparison of all three genotypes. (**b**) Wild-type, (**c**) NGF-OE, and (**d**) GDNF-OE mice were injected with CFA and tested for behavioral heat hyperalgesia over a month time period. Each mouse line exhibited significant hyperalgesia within 3 days of being injected. NGF-OE mice recovered first (by day 5), followed by GDNF-OE mice on day 7. Wild-type mice did not fully recover until day 9. NGF-OE mice also exhibited hypoalgesia on days 15 and 22 (relative to their pre-CFA baseline). v wild-type mice, μ NGF-OE mice, υ GDNF-OE mice, BL baseline value

Response of NGF-OE and GDNF-OE Mice to Inflammatory Stimuli

The response of NGF-OE, GDNF-OE, and WT mice to inflammatory pain was tested by injecting an emulsion of ▶ Complete Freund's Adjuvant (CFA) subcutaneously into the plantar skin of the hind paw (Zwick et al. 2003). Sets of ten animals were tested for heat and mechanical hyperalgesia at various time-points following CFA injection (Fig. 1). WT and NGF-OE mice showed decreased response times 1-day post-injection (Fig. 1b–d). On day 3, all three genotypes displayed hyperalgesic behavior compared to their respective baselines. All groups of animals showed recovery following the 3-day time-point, with WT mice recovering to normal by day 9, GDNF-OE mice recovering by day 7, and

NGF-OE mice recovering by day 5. NGF-OE mice not only recovered faster than wild-type and GDNF-OE mice, but they became hypoalgesic between days 15 and 22 relative to their starting baseline. Thus, the increased number of nociceptors in NGF-OE and GDNF-OE transgenic mice did not cause a hyperalgesic phenotype in the naïve or inflamed state.

The lack of enhanced behavioral hyperalgesia in NGF-OE and GDNF-OE mice suggested compensatory changes developed in each transgenic mouse line in response to the trophin-induced anatomical and physiological changes. To examine how these analgesic effects could be elicited, mRNA expression for selected genes thought to be involved in nociceptive signaling was analyzed in the L4/L5 dorsal horn and DRG of naïve mice (Tables 1 and 2) (Zwick et al. 2003). ▶ Real-time PCR analysis of reverse transcribed total RNA isolated from the dorsal horn of the spinal cord and lumbar DRG was done. No significant change for any of the genes examined was found in dorsal horn mRNA samples (Table 1).

However, in L4/L5 DRG, significant changes were measured for most of the gene products examined (Table 2).

In NGF-OE DRG, changes were found for the opioid receptors MOR1, DOR1, KOR1 and NR1, NR2B, and mGluR1 and the sodium channel Nav1.3. In GDNF-OE DRG, mRNAs encoding DOR1, KOR1, and mGluR1 were changed. Thus, opioid and glutamate signaling in the primary afferent may contribute to the compensatory changes evoked in transgenic OE animals in the naïve state.

How these selected genes changed on the transcriptional level, following CFA injection into the hind paw, was then examined (Molliver et al. 2005). Genes expressed in lumbar DRG of NGF-OE, GDNF-OE, and WT animals were assayed using real-time PCR (Fig. 2). Measures were done at 0 (baseline), 1-day, 4-day, and 15-day time-point, post-CFA treatment. These times coincide with the development, maximal expression, and resolution of thermal hyperalgesia, as indicated by the behavioral measures (Fig. 1). This analysis showed that opioid receptor mRNA abundance is changed in DRG following CFA injection (Fig. 2). Following CFA injection in WT mice, MOR mRNA levels were slightly elevated, in contrast to GDNF-OE mice, where a spike at 4 days occurred followed by a decline. Although the abundance of MOR mRNA was increased (3.2-fold) in NGF-OE mice at baseline, a steady decline in MOR levels was also measured in NGF-OE DRG. DOR mRNA was downregulated in WT mice following CFA injection, though an overall greater decline occurred in both OE lines. In all genotypes, KOR showed a peak rise at day 1, followed by a decline back to near baseline levels.

For the NMDA receptor subunit NR1, a decrease for all genotypes occurred by the 4-day time-point and continued to the 15-day

Nerve Growth Factor Overexpressing Mice as Models of Inflammatory Pain, Table 1 Change in mRNA abundance in mouse dorsal horn

mRNA	WT vs. NGF-OE (fold change)	WT vs. GDNF-OE (fold change)
MOR1	1.0	+1.2
DOR1	1.0	+1.2
KOR1	+1.3	+1.1
NR1	−1.1	1.0
NR2B	+1.3	+1.3
mGluR1	+1.1	+1.1
DREAM	+1.1	+1.1

All values are reported as fold change relative to wild-type (*WT*) measurements. A value of "1" indicates no change. Negative values indicate a decrease. None of the observed changes were statistically significant

Nerve Growth Factor Overexpressing Mice as Models of Inflammatory Pain, Table 2 Change in mRNA abundance in mouse L4-L5 dorsal root ganglia

mRNA	WT vs. NGF-OE (fold change)	WT vs. GDNF-OE (fold change)
MOR1	+3.2*	−1.1
DOR1	−1.6*	−1.5*
KOR1	+1.5*	+2.1*
NR1	−1.8*	1.0
NR2B	−1.8*	1.0
mGluR1	+3.1*	+3.7*
Nav 1.8	+2.6	+2.1
Nav 1.3	−1.4*	+1.1

All values are reported as fold change relative to wild-type (*WT*) measurements. Fold change equal to "1" indicates no change. Negative values indicate a decrease. Asterisk indicates $p < 0.05$

time-point. The decrease in NR1 in the transgenics is particularly profound, given the increased number of nociceptive neurons in these mice. Notably, both lines of transgenic mice recover from CFA-evoked hyperalgesia early: NGF-OEs by day 5 and GDNF-OEs by day 7 (Fig. 1). The NR2B subunit showed no significant change in NGF-OE or GDNF-OE ganglia and was only modestly elevated in WT ganglia at the 15-day time-point. As NR2B may mediate central sensitization, mRNA abundance was assessed in the dorsal horn (DH) of the

Nerve Growth Factor Overexpressing Mice as Models of Inflammatory Pain, Fig. 2 Comparison of the temporal change in mRNA levels of various genes related to nociception in DRG and dorsal horn of wild-type (*green line*), NGF-OE (*red line*), and GDNF-OE (*blue line*) mice following CFA injection in the hind lumbar spinal cord. In WT mice, NR2B was slightly decreased in the dorsal horn at 4 days, with a return to baseline by 15 days. In the GDNF-OE and NGF-OE samples, this pattern of regulation was exaggerated, particularly for NGF-OE samples, which showed a near seven-fold decrease at 4 days. Similar to WT animals,

paw. CFA injection was done at day 0 and the relative abundance of mRNAs for each gene determined using real-time PCR assays. A significant change from the baseline value determined for each animal type is indicated by *an asterisk*

both transgenic samples had a return to baseline by day 15. This suggests that NGF-OE mice compensate for the increased nociceptor input by downregulation of NR2B, which presumably restricts second-messenger potentiation of NMDA currents and inhibits sensitization of spinal synapses contributing to hyperalgesia.

The regulation of the sodium channel (Nav1.8 and Nav1.3) mRNA level in DRG of NGF-OE and GDNF-OE lines was very similar to the pattern of change in WT mice, i.e., a sharp rise in Nav1.8 is seen at 2-day post-CFA, followed by a decline to near normal level by day 4. In contrast to the response in Nav1.8, Nav1.3 mRNA abundance showed a steady decline, which became significantly lower than baseline levels by day 15.

NGF overexpression in skin results in nociceptive primary sensory neurons that are hyperexcitable and present in substantially increased numbers. However, when tested behaviorally, these mice are resistant to inflammatory hyperalgesia and actually become hypoalgesic. Evidence suggests the resistance in each OE line to inflammatory pain is due to compensatory changes in nociceptive signaling, which act to reduce the impact of the increased nociceptive input. Understanding of how these compensatory changes develop and are regulated following injury will provide insight into the role of NGF in inflammatory pain processes. In particular, this model system has allowed identification of genes that are more susceptible to compensatory regulation. For instance, Nav1.8 is essentially the same in all genotypes after CFA, whereas in transgenic mice, the opioid receptors and NMDA receptors exhibit striking alteration, relative to WT mice, in their patterns of transcriptional regulation following an inflammatory challenge. This suggests that specific elements in the transcriptional response to injury are particularly amenable to modulation and that their expression may determine the severity of the injury response, whereas other elements show a more fixed transcriptional response to injury. The NGF-OE mice provide a model system in which to examine this hypothesis. In addition, the OE system provides a means in which to determine how different subpopulations of nociceptive neurons respond to inflammatory stimuli. In this manner, a more global visualization of the role of growth factor expression in primary afferent sensitization following injury and their use as therapeutic targets can be constructed.

References

Albers, K. M., Wright, D. E., & Davis, B. M. (1994). Overexpression of nerve growth factor in epidermis of transgenic mice causes hypertrophy of the peripheral nervous system. *Journal of Neuroscience, 14*, 1422–1432.

Bennett, D. L. (2001). Neurotrophic factors: Important regulators of nociceptive function. *The Neuroscientist, 7*, 13–17.

Crowley, C., Spencer, S. D., Nishimura, M. C., Chen, K. S., Pitts-Meek, S., Armanini, M. P., Ling, L. H., MacMahon, S. B., Shelton, D. L., & Levinson, A. D. (1994). Mice lacking nerve growth factor display perinatal loss of sensory and sympathetic neurons yet develop basal forebrain cholinergic neurons. *Cell, 76*, 1001–1011.

Davis, B. M., Albers, K. M., Seroogy, K. B., & Katz, D. M. (1994). Overexpression of nerve growth factor in transgenic mice induces novel sympathetic projections to primary sensory neurons. *The Journal of Comparative Neurology, 349*, 464–474.

Davis, B. M., Fundin, B. T., Albers, K. M., Goodness, T. P., Cronk, K. M., & Rice, F. L. (1997). Overexpression of nerve growth factor in skin causes preferential increases among innervation to specific sensory targets. *The Journal of Comparative Neurology, 387*, 489–506.

Goodness, T. P., Albers, K. M., Davis, F. E., & Davis, B. M. (1997). Overexpression of nerve growth factor in skin increases sensory neuron size and modulates Trk receptor expression. *European Journal of Neuroscience, 9*, 1574–1585.

Heumann, R., Korsching, S., Bandtlow, C., & Thoenen, H. (1987). Changes of nerve growth factor synthesis in nonneuronal cells in response to sciatic nerve transection. *The Journal of Cell Biology, 104*, 1623–1631.

Lewin, G. R., Ritter, A. M., & Mendell, L. M. (1993). Nerve growth factor-induced hyperalgesia in the neonatal and adult rat. *Journal of Neuroscience, 13*, 2136–2148.

Lewin, G. R., Rueff, A., & Mendell, L. M. (1994). Peripheral and central mechanisms of NGF-induced hyperalgesia. *European Journal of Neuroscience, 6*, 1903–1912.

Molliver, D. C., Wright, D. E., Leitner, M. L., Parsadanian, A. S., Doster, K., Wen, D., Yan, Q., & Snider, W. D. (1997). IB4-Binding DRG neurons switch from NGF to GDNF dependence in early postnatal life. *Neuron, 19*, 849–861.

Molliver, D. C., Lindsay, J., Albers, K. M., & Davis, B. M. (2005). Overexpression of NGF or GDNF alters transcriptional plasticity evoked by inflammation. *Pain, 113*, 277–284.

Snider, W. D., & McMahon, S. B. (1998). Tackling pain at the source: New ideas about nociceptors. *Neuron, 20*, 629–632.

Stucky, C. L., Koltzenburg, M., Schneider, M., Engle, M. G., Albers, K. M., & Davis, B. M. (1999).

Overexpression of nerve growth factor in skin selectively affects the survival and functional properties of nociceptors. *Journal of Neuroscience, 19*, 8509–8516.

Vulchanova, L., Olson, T. H., Stone, L. S., Riedl, M. S., Elde, R., & Honda, C. N. (2001). Cytotoxic targeting of isolectin IB4-binding sensory neurons. *Neuroscience, 108*, 143–155.

Weskamp, G., & Otten, U. (1987). An enzyme-linked immunoassay for Nerve Growth Factor (NGF): A tool for studying regulatory mechanisms involved in NGF production in brain and in peripheral tissues. *Journal of Neurochemistry, 48*, 1779–1786.

Zwick, M., Davis, B. M., Woodbury, C. J., Burkett, J. N., Koerber, H. R., Simpson, J. F., & Albers, K. M. (2002). Glial cell line-derived neurotrophic factor is a survival factor for isolectin B4-positive, but not vanilloid receptor [1]-positive, neurons in the mouse. *Journal of Neuroscience, 22*, 4057–4065.

Zwick, M., Molliver, D. C., Lindsay, J., Fairbanks, C. A., Sengoku, T., Albers, K. M., & Davis, B. M. (2003). Transgenic mice possessing increased numbers of nociceptors do not exhibit increased behavioral sensitivity in models of inflammatory and neuropathic pain. *Pain, 106*, 491–500.

Nerve Growth Factor, Sensitizing Action on Nociceptors

Lorne M. Mendell
Department of Neurobiology and Behavior, Stony Brook University, Stony Brook, NY, USA

Definition

The response of the nociceptive system can be sensitized by exposure to a ▶ neurotrophin molecule called ▶ nerve growth factor (NGF). This sensitization has two components, one peripheral due to an enhanced response to nociceptive stimuli and the other central due to increased action of nociceptive impulses in the dorsal horn.

Characteristics

Nerve Growth Factor is a member of a family of molecules called neurotrophins. Neurotrophins are best known for their function during development, specifically in promoting axonal growth

and in assuring cell survival. Cells affected selectively by NGF express a specific receptor tyrosine kinase called trkA to which NGF binds. Nociceptors express trkA, which makes them sensitive to NGF during development (reviewed in Lewin and Mendell 1993; Mendell et al. 1999). A different postnatal role for NGF has now been established. Administration of NGF to an adult animal results in enhanced responsiveness to noxious heat stimulation (hyperalgesia), which is due partly to direct sensitization of nociceptive afferents, i.e., peripheral sensitization. This sensitizing role for NGF begins gradually during the immediate postnatal period (Zhu et al. 2004) and is associated with upregulation of PI3K (Zhu and Oxford 2011), the signaling molecule mediating the initial acute sensitization as well as ERK1/2 which is associated with the later components of sensitization (Zhuang et al. 2004). Notably no change in trkA expression was reported during this switchover.

Several findings have established the involvement of endogenous NGF in sensitizing the subsequent response to nociceptive inputs after injury (reviewed in Lewin and Mendell 1993; Mendell et al. 1999). First, is the upregulation of NGF in skin and other peripheral tissues after inflammatory injury. A second is the demonstration that administration of exogenous NGF can elicit hyperalgesia. The third is the finding that inflammatory pain can be significantly reduced by interfering with endogenous NGF action, using either an antibody to NGF or an immunoadhesin (trkA-IgG) which sequesters endogenous NGF.

The time course of behavioral hyperalgesia elicited by systemically administered NGF (1 μg/g) has revealed two phases of the response, an initial thermal component beginning just a few minutes after NGF administration and a later one beginning several hours after NGF administration (Lewin et al. 1994). The early response can also be elicited by local injections of NGF into the periphery (Fig. 1), suggesting that exogenous NGF directly sensitizes thermal nociceptive afferents (Shu et al. 1999). These results confirm the results of previous recordings from individual nociceptors using a skin-nerve preparation where

Nerve Growth Factor, Sensitizing Action on Nociceptors, Fig. 1 Administration of NGF to the foot of the rat makes the affected paw hyperalgesic to noxious heat as measured by a reduced latency to withdrawal from a fixed thermal stimulus. The ordinate represents the mean difference in the latency of response of the affected limb compared to the contralateral limb (*negative value* implies that it took less time for the thermal stimulus to reach noxious threshold on the treated foot than on the untreated foot). NGF treatment gave a rapid and consistent thermal hyperalgesia lasting at least 1 day. NT-3 produced no change in response to noxious heat (Adapted from Shu et al. (1999))

it was found that the response to noxious heat is sensitized, measured as a virtually immediate decrease in threshold. In addition to the acute actions of NGF on nociceptors, there is also a later upregulation of numerous channels (Na, Ca, and peptides (BDNF, SP, CGRP) that enhances the effectiveness of NGF as a sensitizing agent, especially in prolonging its action (reviewed in Mantyh et al. 2011). In contrast, there is no acute change in the threshold to mechanical stimulation (Rueff and Mendell 1996) although a reduction in threshold to noxious mechanical stimuli has been observed at a latency similar to the late response to noxious heat (Lewin et al. 1994). This suggests that mechanical hyperalgesia is of central origin (see below) although the possibility of a peripheral contribution by increased discharge of high threshold mechanoreceptors is not ruled out by currently available data.

NGF has also been shown to operate as a sensitizing agent in visceral structures such as the bladder or the gut. As in skin, there is upregulation of NGF message and protein in painful inflammatory conditions, brought on by diseases such as interstitial cystitis or in an experimental model of ulcers (e.g., Lamb et al. 2004). Administration of NGF to the visceral periphery results in enhanced afferent activity. Experimental models of arthritis are also characterized by release of NGF into the synovial fluid, indicating a role in joint hyperalgesia (Manni et al. 2003).

A difficulty in determining the mechanism of acute NGF action from these experiments arises from the multiplicity of cell types in the peripheral target tissues that express trkA (the high affinity receptor for NGF) or that release NGF. Many of these cells are nonneural and are believed to interact closely in the inflammatory cascade. For example, ► mast cells are known to express trkA and to release NGF after injury, and ► keratinocytes have been shown to release NGF in response to histamine produced by mast cells (reviewed in Mendell et al. 1999). Degranulation of mast cells can diminish the sensitizing action of exogenous NGF and other inflammatory mediators (Lewin et al. 1994). In order to investigate the effect of NGF directly on nociceptors, small diameter neurons acutely dissociated from DRG have been studied in culture. The assumption in carrying out such experiments is that the cell body in culture expresses the same receptors as the peripheral terminals in situ. A problem with this approach is that the original target (skin, muscle, viscera) can only be identified if it is prelabeled with a dye transported to the ganglion from the target tissue. However, this still leaves the identity of the receptor type (e.g., for skin: polymodal nociceptor, mechanical nociceptor, and D-Hair) to be determined, since unique molecular identifiers are not yet available at this level of resolution.

NGF is now recognized as an inflammatory mediator whose sensitizing action is similar to other inflammatory mediators such as prostaglandin and bradykinin. The sensitizing effect of NGF has been examined most extensively on the response to capsaicin, which is now known to signal via the recently cloned TRPV1 receptor (previously known as VR1). This receptor can also be activated by physiological stimuli, specifically noxious heat and low pH (rev. in Caterina and Julius 2001). Normally, theTRPV1-mediated response studied in isolated cells is smaller to the second of two capsaicin or noxious heat stimuli (i.e., exhibits tachyphylaxis) that are separated by as much as 10 or 15 min (Galoyan et al. 2003; Shu and Mendell 1999). However, in the presence of NGF (100 ng/ml), tachyphylaxis does not occur in most cells; rather, the second response is larger than the first, i.e., it is sensitized (Shu and Mendell 1999) (Fig. 2). These same studies have revealed that the initial response to noxious heat or capsaicin is larger on the average in the presence of NGF than in its absence. Sensitization by NGF is not accompanied by any systematic change in threshold temperature for the increase in inward current (Galoyan et al. 2003). This indicates that the sensitization at the level of individual nociceptors does not involve a change in the threshold of individual temperature-sensitive channels but rather a recruitment of additional channels from the cytoplasm to the membrane of the cell (Zhang et al. 2005; Stein et al. 2006) to produce a larger inward current. Interestingly NGF- induced sensitization measured in the

skin-nerve preparation does reduce the temperature threshold (Rueff and Mendell 1996) suggesting that NGF-induced sensitization at the organ level is not a property of nociceptors alone; other cells in the skin (keratinocytes, mast cells, etc.) are likely to contribute significantly. It is important to note that administration of NGF alone does not elicit any response from the cell; it merely sensitizes the response evoked by the adequate stimulus, e.g., noxious heat. Immunohistochemical analysis of these cells reveals that the ability of NGF to sensitize these responses is strongly correlated with expression of trkA (Galoyan et al. 2003), indicating that sensitization to noxious heat by NGF involves an interaction between the trkA receptor and the TRPV1 receptor.

NGF also sensitizes the response of nociceptors by increasing their membrane gain, as determined by an enhanced action potential firing in response to an imposed current (Zhang et al. 2002; Nicol and Vasko 2007). This occurs as a result of augmentation of a TTX-resistant Na^+ current known to be expressed in nociceptors. An additional factor underlying this enhanced response to depolarization is inhibition of a K^+ current. NGF mediates these actions on membrane gain by activating the ▶ p75 receptor, rather than trkA which is responsible for enhancing the inward current through TRPV1. The p75 receptor is coupled to the sphingomyelin signaling pathway and exposure to ▶ ceramide, an independent intermediate of this signaling pathway, mimics the effect of NGF on membrane gain. Experiments with independent expression of p75 and TRPV1 in heterologous cells suggest that the p75 receptor is unlikely to be crucial for sensitization of the response of TRPV1 to capsaicin (Chuang et al. 2001). However, some modulatory effect of p75 on the response of trkA is not ruled out by these experiments.

Thus, NGF can sensitize the response of nociceptors to noxious heat both by enhancing the response of the noxious heat sensitive receptor via trkA and by amplifying the gain of the membrane via the p75 receptor, in effect sensitizing the response of the receptor as well as enhancing the gain of the impulse encoder. Longer-term exposure to NGF also induces

Time (S)

Nerve Growth Factor, Sensitizing Action on Nociceptors, Fig. 2 Response of small diameter DRG cell in acute cell culture to noxious heat stimulation. Note that the response to the second pulse of heat (*bottom* traces) measured 10 min after the initial response in the continuous presence of NGF (100 ng/ml) during the 10 min interval was a larger inward current (*top* traces) measured in perforated patch clamp mode. This sensitization is never observed under control conditions (Adapted from Galoyan et al. (2003))

Nerve Growth Factor, Sensitizing Action on Nociceptors, Fig. 3 Schematic diagram illustrating the effect of NGF in causing peripheral sensitization by direct action on nociceptive terminals and indirect central sensitization by upregulating peptides such as brain-derived neurotrophic factor (*BDNF*), substance P (*SP*), and calcitonin gene-related peptide (*CGRP*)

changes in the ▶ P2X3 receptor composition of sensory neurons. By this mechanism, NGF can influence the response of these neurons to ATP which is released by nonneural cells after damage or noxious stimuli (Scholz and Woolf 2002).

An important translational extension of these findings has been the development of humanized antibodies to NGF (reviewed in Mantyh et al. 2011) that have been found to be effective in attenuating pain in certain conditions, particularly the pain associated with bone cancer. This may be because the nociceptive afferents innervating bone are particularly rich in the trkA receptor and thus would be expected to be highly susceptible to being sensitized by NGF whose expression is elevated in these conditions.

The central action of nociceptors is also sensitized by inflammatory stimuli including NGF (Scholz and Woolf 2002). NGF has not been shown to have any direct effect on spinal neurons in the superficial dorsal horn that are involved in transmitting nociceptive signals (Kerr et al. 1999). Rather, exposure of the peripheral terminals to NGF results in internalization of the NGF-trkA complex and transport to the cell body, where it stimulates upregulation of several peptides including substance P, CGRP, and another neurotrophin, ▶ brain-derived neurotrophic factor (BDNF). These peptides are released into the dorsal horn (e.g., Lever et al. 2001) where they can rapidly sensitize the response of dorsal horn neurons in lamina II to subsequent inputs (Garraway et al. 2003). They can also elicit changes in gene expression that are pronociceptive (long-term central sensitization) (Scholz and Woolf 2002).

Together, these studies indicate that the role of NGF in eliciting sensitization of nociceptors is complex with both direct peripheral and indirect central components (Fig. 3).

References

Caterina, M. J., & Julius, D. (2001). The vanilloid receptor: A molecular gateway to the pain pathway. *Annual Review Neuroscience, 24*, 487–517.

Chuang, H. H., Prescott, E. D., Kong, H., et al. (2001). Bradykinin and nerve growth factor release the capsaicin receptor from PtdIns(4,5)P2-mediated inhibition. *Nature, 411*, 957–962.

Galoyan, S. M., Petruska, J., & Mendell, L. M. (2003). Mechanisms of sensitization of the response of single DRG cells from adult rat to noxious heat. *European Journal of Neuroscience, 18*, 535–541.

Garraway, S. M., Petruska, J. C., & Mendell, L. M. (2003). BDNF sensitizes the response of lamina II neurons to high threshold primary afferent inputs. *European Journal of Neuroscience, 18*, 2467–2476.

Kerr, B. J., Bradbury, E. J., Bennett, D. L., et al. (1999). Brain-derived neurotrophic factor modulates nociceptive sensory inputs and NMDA-evoked responses in the rat spinal cord. *Journal of Neuroscience, 19*, 5138–5148.

Lamb, K., Gebhart, G. F., & Bielefeldt, K. (2004). Increased nerve growth factor expression triggers bladder overactivity. *The Journal of Pain, 5*, 150–156.

Lever, I. J., Bradbury, E. J., Cunningham, J. R., et al. (2001). Brain- derived neurotrophic factor is released

in the dorsal horn by distinctive patterns of afferent fiber stimulation. *Journal of Neuroscience, 21*, 4469–4477.

Lewin, G. R., & Mendell, L. M. (1993). Nerve growth factor and nociception. *Trends in Neurosciences, 16*, 353–359.

Lewin, G. R., Rueff, A., & Mendell, L. M. (1994). Peripheral and central mechanisms of NGF-induced hyperalgesia. *European Journal of Neuroscience, 6*, 1903–1912.

Manni, L., Lundeberg, T., Fiorito, S., et al. (2003). Nerve growth factor release by human synovial fibroblasts prior to and following exposure to tumor necrosis factor-alpha, interleukin-1 beta and cholecystokinin-8: The possible role of NGF in the inflammatory response. *Clinical and Experimental Rheumatology, 21*, 617–624.

Mantyh, P. W., Koltzenburg, M., Mendell, L. M., Tive, L., & Shelton, D. L. (2011). Antagonism of nerve growth factor-TrkA signaling and the relief of pain. *Anesthesiology, 115*, 189–204.

Mendell, L. M., Albers, K. M., & Davis, B. M. (1999). Neurotrophins, nociceptors and pain. *Microscopy Research and Technique, 45*, 252–261.

Nicol, G. D., & Vasko, M. R. (2007). Unraveling the story of NGF-mediated sensitization of nociceptive sensory neurons: ON or OFF the Trks? *Molecular Interventions, 7*, 26–41.

Rueff, A., & Mendell, L. M. (1996). Nerve growth factor and NT-5 induce increased thermal sensitivity of cutaneous nociceptors in vitro. *Journal of Neurophysiology, 76*, 3593–3596.

Scholz, J., & Woolf, C. J. (2002). Can we conquer pain? *Nature Neuroscience, 5*, 1062–1067.

Shu, X., & Mendell, L. M. (1999). Nerve growth factor acutely sensitizes the response of adult rat sensory neurons to capsaicin. *Neuroscience Letters, 274*, 159–162.

Shu, X., Llinas, A., & Mendell, L. M. (1999). Effects of trkB and trkC neurotrophin receptor agonists on thermal nociception: A behavioural and electrophysiological study. *Pain, 80*, 463–470.

Stein, A. T., Ufret-Vincenty, C. A., Hua, L., Santana, L. F., & Gordon, S. E. (2006). Phosphoinositide 3-kinase binds to TRPV1 and mediates NGF-stimulated TRPV1 trafficking to the plasma membrane. 1. *Journal of General Physiology, 128*, 509–522.

Zhang, Y. H., Vasko, M. R., & Nicol, G. D. (2002). Ceramide, a putative second messenger for nerve growth factor, modulates the TTX resistant Na(+) current and delayed rectifier K(+) current in rat sensory neurons. *The Journal of Physiology, 544*, 385–402.

Zhang, X., Huang, J., & McNaughton, P. A. (2005). NGF rapidly increases membrane expression of TRPV1 heat-gated ion channels. *The EMBO Journal, 24*, 4211–4223.

Zhu, W., & Oxford, G. S. (2011). Differential gene expression of neonatal and adult DRG neurons correlates with the differential sensitization of TRPV1 responses to nerve growth factor. *Neuroscience Letters, 500*, 192–196.

Zhu, W., Galoyan, S. M., Petruska, J. C., Oxford, G. S., & Mendell, L. M. (2004). A developmental switch in acute sensitization of small dorsal root ganglion (DRG) neurons to capsaicin or noxious heating by NGF *J. Neurophysiology, 92*, 3148–3152.

Zhuang, Z. Y., Xu, H., Clapham, D. E., & Ji, R. R. (2004). Phosphatidylinositol 3-kinase activates ERK in primary sensory neurons and mediates inflammatory heat hyperalgesia through TRPV1 sensitization. *The Journal of Neuroscience, 24*, 8300–8309.

Nerve Inflammation

▶ Inflammatory Neuritis

Nerve Injury

▶ Retrograde Cellular Changes After Nerve Injury

Nerve Ligation

▶ Retrograde Cellular Changes After Nerve Injury

Nerve Pain

▶ Peripheral Neuropathic Pain

Nerve Pain of Joint and Muscle Origin

▶ Neuropathic Pain, Joint and Muscle Origin

Nerve Stump Pain

▶ Neuroma Pain

Nerve Terminals

Definition

These axon endings are found in the dermis around the base of hair follicles and close to the surface of the skin (epidermis) where the hair emerges. These free endings contain specialized receptors that respond to changes in temperature and other events (pH) associated with tissue damage.

Cross-References

▶ Opioid Receptor Localization

Nerve Viral Infection

▶ Viral Neuropathies

Nervus Intermedius Neuralgia

▶ Geniculate Neuralgia

Neural Blockade

▶ Cancer Pain Management, Anesthesiologic Interventions, and Neural Blockade

Neural Compressive Syndrome

▶ Lower Back Pain, Physical Examination

Neural Foramen

Definition

Neural foramen is a foramen in the spinal canal which is bounded by the intervertebral disc, the pedicles and facet joints of the vertebrae above and below, and the posterior aspect of the vertebral bodies above and below. The nerve root exits through this foramen and the dorsal root ganglion is situated in the foramen.

Cross-References

▶ Dorsal Root Ganglionectomy and Dorsal Rhizotomy

Neural Plasticity

Definition

The ability of the brain and/or certain parts of the nervous system to change in order to adapt to new conditions such as an injury and can include changes in synaptic connectivity and strength between cells.

Cross-References

▶ Cytokines, Effects on Nociceptors

Neuralgia

▶ Neuropathic Pain in Children

Neuralgia of Cranial Nerve V: Fifth Nerve Neuralgia

▶ Trigeminal and Cranial Nerve Neuralgias

Neuralgia of Cranial Nerve VII

▶ Trigeminal and Cranial Nerve Neuralgias

Neuralgia, Assessment

Ralf Baron
Department of Neurological Pain Research and Therapy, Christian Albrechts University Kiel, Kiel, Germany

Definition

Neuralgia is defined as a pain in the distribution of a nerve or nerves (Merskey and Bogduk 1994). It is mostly associated with neuropathic pain states that occur after nerve lesion. It is a pure descriptive term that does not imply the etiology of the pain generation, nor the underlying pathophysiological mechanism, nor the characteristic of the pain. According to this definition, neuralgia pain may be located superficially in the skin or also in deep somatic structures; it may be of constant spontaneous type, shooting type, or of evoked type (▶ hyperalgesia, ▶ allodynia).

Note: Common usage, especially in Europe, often implies a paroxysmal quality, but neuralgia should not be reserved for paroxysmal pains.

Although the definition clearly states the fact that the pain of neuralgia occurs within the innervation territories, the symptoms may in individual cases spread to some degree beyond the innervation territories. This is particular true for allodynic pain in, for example, ▶ postherpetic neuralgia or posttraumatic neuralgia that sometime occurs in formerly unaffected dermatomes or peripheral nerve territories. Thus the symptoms, signs, and their distribution can lead to confusion with regard to the diagnosis.

According to the underlying etiology, several different terms for neuralgias are commonly used in pain medicine. The following comprises examples of some syndromes without being complete.

Characteristics

Posttraumatic Neuralgia (PTN)

Traumatic mechanical partial injury to a peripheral nerve may lead to PTN. The cardinal symptoms are spontaneous burning pain, shooting pain, and hyperalgesia and mechanical and, in some cases severe, cold allodynia. These sensory symptoms are confined to the territory of the affected peripheral nerve, although allodynia may extend beyond the border of nerve territories to a certain degree (Wahren and Torebjörk 1992; Wahren et al. 1991, 1995).

Special forms of ▶ posttraumatic neuralgias are chronic compression injuries to peripheral nerves, e.g., spermatic neuralgia and ▶ meralgia paresthetica.

Postherpetic Neuralgia (PHN)

PHN is one of the most common types of neuropathic pain. Its etiology is well known; the recrudescence of the varicella zoster virus (VZV) with inflammation and damage to dorsal root ganglion cells. If the pain lasts more than 3–6 months after the acute shingles, the criteria for PHN are fulfilled. It typically occurs in elderly but otherwise healthy individuals with no previous history of chronic pain. The diagnosis is straightforward, based on the history of a dermatomal rash and the dermatomal distribution of the pain. The incidence of PHN in zoster-affected patients of all age groups is about 15 %.

The pain of PHN appears as the acute viral infection subsides and persists, often indefinitely. The severity is frequently sufficient to completely disrupt the lives of otherwise healthy individuals. Patients with PHN report one or more of the following: a steady, deep aching pain that often has an abnormal quality, a lancinating pain that is brief, intense, and often described in terms reminiscent of ▶ trigeminal neuralgia and finally, dynamic mechanical allodynia, which is the induction of a sharp pain by light, moving, cutaneous stimuli. In individual patients, the most unpleasant aspect of their pain may be either a continuous deep aching pain, lancinating pain or allodynia (Dworkin and Portenoy 1994; Fields et al. 1998).

Cranial Nerve Neuralgias

Trigeminal Neuralgia (TN)

Trigeminal neuralgia (tic douloureux) is a disorder of the fifth cranial (trigeminal) nerve that causes episodes of intense, stabbing, electric shock-like pain in the areas of the face where the branches of the nerve are distributed – lips, eyes, nose, scalp, forehead, upper jaw, and lower jaw (Kapur et al. 2003). The disorder is more common in women than in men and rarely affects anyone younger than 50. The attacks of pain, which generally last several seconds and may be repeated one after the other, may be triggered by talking, brushing teeth, touching the face, chewing, or swallowing. The attacks may come and go throughout the day and last for days, weeks, or months at a time, and then disappear for months or years. Trigeminal neuralgia is not fatal, but it is universally considered to be the most painful affliction known to medical practice (Fields 1996).

Glossopharyngeus Neuralgia (GN)

▶ Glossopharyngeal neuralgia is described as sharp, jabbing, electric, or shock-like pain located deep in the throat on one side. It is generally located near the tonsil, although the pain may extend deep into the ear. It is usually triggered by swallowing or chewing.

Facial (Geniculate) Ganglion Neuralgia (FN)

Pain paroxysms are felt in the depth of the ear, lasting for seconds or minutes or intermittently. A trigger zone is present in the posterior wall of the auditory canal. Disorders of lacrimation, salivation, and taste sometimes accompany the pain. There is a common association with herpes zoster.

Post-sympathectomy Neuralgia (PSN)

Post-sympathectomy neuralgia is a pain syndrome associated with a lesion at the sympathetic nervous system. One to two weeks after lumbar or cervicothoracic sympathectomy, up to 35 % of the patients develop a deep, boring pain. The pain of PSN characteristically has a proximal location within the innervation territory of the sympathectomized nerves. PSN patients describe a variable degree of deep somatic tenderness in the area of pain, which typically responds to oral cyclooxygenase inhibitors. PSN is often nocturnal and typically remits in a few weeks without specific treatment (Baron et al. 1999).

Meralgia Paresthetica
Meralgia paresthetica, a painful mononeuropathy of the lateral femoral cutaneous nerve, is commonly due to focal entrapment of this nerve as it passes through the inguinal ligament. It is a purely sensory nerve and has no motor component. Pain associated with paresthesias and numbness in the area of the anterolateral thigh are common symptoms. Rarely, it has other etiologies such as direct trauma, stretch injury, or ischemia. It typically occurs in isolation. The clinical history and examination is usually sufficient for making the diagnosis. However, the diagnosis can be confirmed by nerve conduction studies. Treatment is usually supportive.

Differential Diagnoses
Atypical Facial Neuralgia and Posttraumatic Facial Pain
The typical symptom is a continuous, unilateral, deep, aching pain, sometimes with a burning component within the face, most commonly in the area of the second trigeminal branch. More than half the patients with nondescript facial pain report its onset after trauma to the face, often surgical trauma. Orbital enucleations, sinus procedures, and complicated dental extractions are the most common procedures that antedate the appearance of pain. Fortunately, for the large majority of the patients, their pain problem is self-limited; within 1–5 years it subsides whether symptomatic treatment is effective or not. The mechanism underlying this disorder presumably involves activation or central pain transmission pathways; how and why this occurs remains to be elucidated (Burchiel 2003; Kapur et al. 2003).

Complex Regional Pain Syndrome Type II (CRPS II, Causalgia)
Injury of a peripheral nerve may lead to CRPS II. In contrast to patients with posttraumatic neuralgias, CRPS II patients exhibit a more complex clinical picture. They show marked swelling and a tendency for progressive spread of symptoms in the entire distal extremity. Spontaneous and evoked pains are felt superficially as well as deep inside the extremity and the intensity of both is dependent on the position of the extremity (Baron et al. 2002; Janig and Baron 2003; Wasner et al. 1998).

Assessment of Neuralgia
Since neuralgia is a pure descriptive term defined as a pain that occurs within the innervation territory of a peripheral nerve or a nerve root, there are no objective diagnostic procedures. However, in addition to the pain history and the clinical symptoms clinical signs that are characteristic for neuropathic pain states are also helpful and should be analyzed with quantitative sensory testing to aid the diagnosis of neuropathy (e.g., in postherpetic neuralgia, posttraumatic neuralgia, meralgia paresthetica) (see below) (Baron 2000). However, it should be recognized that in several forms of neuralgia (e.g., idiopathic trigeminal neuralgia) sensory testing does not reveal any abnormalities.

Symptom-Based Classification of Neuropathic Pain. I. General Definitions

Negative sensory symptoms
- Loss of sensory quality
- Due to system involved: hypoesthesia, hypoalgesia, thermhypesthesia, pallhypoesthesia, etc.
- Bothering, but not painful

Positive sensory symptoms
- Paresthesias
- Dysesthesias
- Spontaneous pain (burning ongoing pain, shock-like pain)
- Evoked pain (see below)
  - Allodynia: a normally non-painful stimulus evokes pain
  - Hyperalgesia: a painful stimulus evokes pain of higher intensity

Symptom-Based Classification of Neuropathic Pain. II. Definition of Different Evoked Pains

- Static mechanical allodynia
  - Gentle static pressure stimuli at the skin evoke pain
  - Present in the area of affected (damaged or sensitized) primary afferent nerve endings (primary zone)
- Punctate mechanical allodynia
  - Normally stinging but not painful stimuli (stiff von Frey hair) evoke pain
  - Present in the primary affected zone and spread beyond into unaffected skin areas (secondary zone)
- Dynamic mechanical allodynia
  - Gentle moving stimuli at the skin (brush) evoke pain
  - Present in the primary affected zone and spread beyond into unaffected skin areas (secondary zone)
- Warm allodynia, heat hyperalgesia
  - Warm or heat stimuli at the skin evoke pain
  - Present in the area of affected (damaged or sensitized) primary afferent nerve endings (primary zone)
- Cold allodynia
  - Cold stimuli at the skin evoke pain
  - Characteristic of posttraumatic neuralgia and some polyneuropathies
- Temporal summation
  - Repetitive application of identical single noxious stimuli (interval <3 s) is perceived as increasing pain sensation

Quantitative Sensory Testing (QST) in Neuralgia

A bedside testing should be part of the physical examination to confirm, e.g., loss of afferent function, as well as evoked pain symptoms (e.g., allodynia and hyperalgesia), i.e., dynamic mechanical allodynia (cotton swab). Additionally, standardized psychophysical tests (von Frey hairs, thermotest) should be used to detect impairment and changes in warm and cold sensation as well as heat and cold pain thresholds. By these means the function of small myelinated and unmyelinated afferent fibers is assessed.

So far, no characteristic sensoric pattern of patients with neuralgia has been identified. However, the analysis is useful to determine and quantify the individual signs of each patient and to document successful response to treatment.

References

Baron, R. (2000). Peripheral neuropathic pain: From mechanisms to symptoms. *The Clinical Journal of Pain, 16*, 12–20.

Baron, R., Levine, J. D., & Fields, H. L. (1999). Causalgia and reflex sympathetic dystrophy: Does the sympathetic nervous system contribute to the generation of pain? *Muscle & Nerve, 22*, 678–695.

Baron, R., Fields, H. L., Janig, W., et al. (2002). National Institutes of Health Workshop: Reflex sympathetic dystrophy/complex regional pain syndromes -state-of-the-science. *Anesthesia and Analgesia, 95*, 812–816.

Burchiel, K. J. (2003). A new classification for facial pain. *Neurosurgery, 53*, 1164–1166; discussion 1166–1167.

Dworkin, R. H., & Portenoy, R. K. (1994). Proposed classification of herpes zoster pain. *Lancet, 343*, 1648.

Fields, H. L. (1996). Treatment of trigeminal neuralgia. *The New England Journal of Medicine, 334*, 1125–1126.

Fields, H. L., Rowbotham, M., & Baron, R. (1998). Postherpetic neuralgia: Irritable nociceptors and deafferentation. *Neurobiology of Disease, 5*, 209–227.

Janig, W., & Baron, R. (2003). Complex regional pain syndrome: Mystery explained? *Lancet Neurology, 2*, 687–697.

Kapur, N., Kamel, I. R., & Herlich, A. (2003). Oral and craniofacial pain: Diagnosis, pathophysiology, and treatment. *International Anesthesiology Clinics, 41*, 115–150.

Merskey, H., & Bogduk, N. (1994). *Classification of chronic pain: Descriptions of chronic pain syndromes and definition of terms* (2nd ed.). Seattle: IASP Press.

Wahren, L. K., & Torebjörk, E. (1992). Quantitative sensory tests in patients with neuralgia 11 to 25 years after injury. *Pain, 48*, 237–244.

Wahren, L. K., Torebjörk, E., & Nystrom, B. (1991). Quantitative sensory testing before and after regional guanethidine block in patients with neuralgia in the hand. *Pain, 46*, 23–30.

Wahren, L. K., Gordh, T., Jr., & Torebjork, E. (1995). Effects of regional intravenous guanethidine in patients with neuralgia in the hand; a follow-up study over a decade. *Pain, 62*, 379–385.

Wasner, G., Backonja, M. M., & Baron, R. (1998). Traumatic neuralgias: Complex regional pain syndromes (reflex sympathetic dystrophy and causalgia): Clinical characteristics, pathophysiological mechanisms and therapy. *Neurologic Clinics, 16*, 851–868.

Neuralgia, Diagnosis

Jayantilal Govind
Pain Clinic, Department of Anaesthesia,
University of New South Wales, Sydney,
NSW, Australia

Synonyms

Neurodynia

Definition

Pain in the distribution of a nerve, ostensibly due to an intrinsic disorder of that nerve (Merskey and Bogduk 1994).

Characteristics

The taxonomical distinction between ▶ neuralgia and ▶ neuropathic pain is contentious. The distinction is largely historical, in that the classical neuralgias were named before the entity of neuropathic pain was popularized. Nevertheless, certain semantic, anatomic, and pathologic distinctions apply.

The term – neuralgia – explicitly means pain along a nerve. It identifies neither the etiology pathophysiology, nor any specific feature, e.g., pain quality. Neuropathic pain implies that the affected nerve has a disease, and typically it is associated with features of abnormal nerve function, such as numbness, hyperesthesia, or allodynia. These latter features are characteristically absent in neuralgias or are subtle and minor.

Neuralgia should also be distinguished from ▶ radicular pain. Although similar to neuralgia in some respects clinically, radicular pain has certain distinguishing features clinically and with respect to etiology and mechanisms (see ▶ Radicular Pain).

Archetypical Conditions

There are two archetypical conditions that are different from one another in many respects,

and each is representative of other, less common conditions. These are ▶ trigeminal neuralgia and its relatives glossopharyngeal and vagal neuralgia and ▶ postherpetic neuralgia, which is probably a unique condition and which may exemplify other dorsal root ganglionopathies, such as tabes dorsalis and possibly Guillain-Barre syndrome.

Site of Lesion

In trigeminal neuralgia, the lesion is in the sensory root. It is not in the ganglion or in the peripheral nerve.

In postherpetic neuralgia, the lesion is largely in the dorsal root ganglion but may also extend into the peripheral and central nervous systems.

Histopathology

Trigeminal neuralgia is a focal disorder of the cell membrane of the sensory root of the nerve. Focal demyelination is the primary pathology, located between the ganglion and the dorsal root entry zone of the nerve into the brainstem (Kerr 1967). The remainder of the root is normal (Rappaport et al. 1997). In the majority of cases, demyelination is due to irritation of the sensory root by an aberrant blood vessel, typically, the superior cerebellar artery (Loeser 2001). Other causes include aberrant veins, angiomas, and tumors of the posterior cranial fossa. Intrinsic demyelination may occur in patients with multiple sclerosis.

In postherpetic neuralgia, the fundamental pathology is intrinsic inflammation of the affected dorsal root ganglion. In time, inflammation is replaced by axonal degeneration. Demyelination, degeneration, and fibrosis occur in the dorsal root ganglion, associated with atrophy of the dorsal horn and the dorsal root, which may extend distally into the peripheral nerve, with loss of axons and a lymphocytic response (Watson et al. 1991).

Pathophysiology

In trigeminal neuralgia, areas of demyelination act as a site for generation of ectopic impulses ("ectogenesis") and/or abnormal impulse traffic. The ignition hypothesis (Devor et al. 2002) calls for ectopic generation of action potentials. The Calvin model only requires abnormal refractory periods and reflection of normally generated impulses (Calvin et al. 1977). In both models, the location of the lesion proximal to the ganglion seems critical. This allows impulses to reflect between the lesion and ganglion, which becomes the basis for the characteristic reverberation of the pain.

In postherpetic neuralgia, inflammation of the affected nerve may cause pain on the basis of neuritis, but progressive necrosis of peripheral axons and the cell bodies in the dorsal root ganglion may result in deafferentation and disinhibition of dorsal horn neurons. Thereby, postherpetic neuralgia converts from a peripheral neuropathic pain to a ▶ central pain.

Epidemiology

The incidence of trigeminal neuralgia is estimated to be 4 per 100,000 persons per year (Katusic et al. 1990). Risk factors include multiple sclerosis and hypertension (Katusic et al. 1990). There is a weak association with multiple sclerosis, but multiple sclerosis could be an incidental finding given that vascular compression of the trigeminal root has been reported in individuals as young as 17 years (Katusic et al. 1991).

While few children develop postherpetic neuralgia, the risk of contracting this disorder and the intractability of pain increases with age. This susceptibility is attributed to a selective decline in cellular immunity to varicella virus, and recurrences are strongly related to immunosuppressive conditions such as HIV and SLE (Head and Campbell 1900).

Clinical Features

Trigeminal neuralgia is characterized by electric shock-like brief stabbing pains, with pain-free intervals between attacks, during which the patient is completely asymptomatic. Onset is usually abrupt. Pain is restricted to the trigeminal nerve distribution, and idiopathic trigeminal neuralgia is not associated with sensory loss. Non-nociceptive triggering of the pain (light touch, hair movement, chewing, speech, wind puffs) is almost ipsilateral to the pain and usually from the perioral region.

Intraoral trigger zones may be accompanied by a decline in general health.

Patients with PHN present with a constellation of painful sensations including burning, dysesthesia, aching, itching, or severe paroxysmal of stabbing pain. Allodynia or hyperpathia may occur, and the sensitivity to touch is the most distressing feature to patients. These various features are consistent with loss of nerve fibers, of all types, and ultimately with central disinhibition.

Treatment

Although trigeminal neuralgia and postherpetic neuralgia are both called neuralgias, they differ in pathology and pathophysiology. Consequently, they respond differently to treatment. There is no treatment universally applicable to all neuralgias. To be effective, treatment should target the known pathomechanisms.

Since trigeminal neuralgia is a membrane disease, membrane stabilizers (anticonvulsants) are the treatment of first choice. Classical agents include phenytoin, clonazepam, and carbamazepine. Of these, only carbamazepine has been vindicated in placebo-controlled trials. Baclofen may be used an adjunct, if required. For resistant cases, gabapentin and lamotrigine are reputedly effective, as is intravenous lignocaine (Sindrup and Jensen 2002).

When the condition is refractory to pharmaceuticals, various surgical interventions are known to be effective. The choice lies with the preference of the operator. They include ganglionolysis, radiofrequency neurotomy, balloon compression, or injection of glycerine or alcohol and microvascular decompression (see ▶ Dorsal Root Ganglionectomy and Dorsal Rhizotomy).

For postherpetic neuralgia, amitriptyline is the drug of first choice (Watson and Evans 1985). It is the only agent for which there is consistent and strong evidence of efficacy, from controlled trials. However, only about 60 % of patients obtain reasonable benefit. For resistant cases, a large variety of interventions have been recommended but few with evidence of efficacy (Kingery 1997). Gabapentin and opioids appear to be effective for resistant cases (Watson 2000). In contrast to their efficacy for trigeminal

Neuralgia, Diagnosis, Table 1 Similarities and differences between trigeminal neuralgia, postherpetic neuralgia, and radicular pain

Domain	Trigeminal neuralgia	Postherpetic neuralgia	Radicular pain
Anatomical site	Sensory root	**Dorsal root ganglion**	**Dorsal root ganglion**
Key pathology	Demyelination	**Inflammation**	**Inflammation**
Etiology	**Extrinsic**	Intrinsic	**Extrinsic**
Mechanism	Reverberating impulses	Disinhibition	Ectopic discharge
Site	Precentral	Central	Peripheral
Pain	Paroxysmal	Constant, burning	Intermittent, lancinating
Associated	Trigger point	Allodynia	Nil
Neurological	Normal	**Sensory loss**	**Sensory loss**
Treatment	**Decompression**		**Decompression**
	Neuro-ablation		
	Anticonvulsants	Tricyclics	Steroids

Bold text indicates features shared by two of the conditions; otherwise, features are unique and distinctive

neuralgia, baclofen, phenytoin, and carbamazepine are usually not helpful in postherpetic neuralgia (Kingery 1997).

Topical applications of local anesthetic or capsaicin can be used to palliate the cutaneous sensory symptoms, but these interventions do not target the fundamental mechanism of postherpetic neuralgia. Topical capsaicin has been vindicated in a placebo-controlled trial (Watson et al. 1993). In extreme cases, surgical interventions can be undertaken (see ▶ Dorsal Root Ganglionectomy and Dorsal Rhizotomy).

Comparison

While the ▶ neuralgias share certain features, they also share some features with radicular pain but are distinct in others (Table 1). The similarities invite some practitioners to assume that the conditions belong to the same class and should respond to the same treatments. However, the differences in pathology and mechanisms predicate distinctly different responses to treatment. Appreciating these differences is pivotal to successful management. Treatments that work for one type of neuralgia will not work for another nor do treatments commonly used for neuralgias work for radicular pain.

Acknowledgement Dr. Jayantilal Govind passed away on June 20, 2009, at age 86. We thank Prof. Nik Bogduk for the update of his entry.

References

Calvin, W. J., Loeser, J. D., & Howe, J. F. (1977). A neurophysiological theory for the pain mechanism of Tic douloureux. *Pain, 3,* 147–154.

Devor, M., Amir, R., & Rappaport, Z. H. (2002). Pathophysiology of trigeminal neuralgia: The ignition hypothesis. *The Clinical Journal of Pain, 18,* 4–13.

Head, H., & Campbell, A. W. (1900). The pathology of herpes zoster and its bearing on sensory localisation. *Brain, 23,* 323–353.

Katusic, S., Beard, M., Bergstralh, M. S., & Kurland, L. T. (1990). The incidence and clinical features of trigeminal neuralgia. Rochester, Minnesota, 1945–1984. *Annals of Neurology, 27,* 89–95.

Katusic, S., Williams, D. B., Beard, M., Bergstralh, E. J., & Kurkland, L. T. (1991). Epidemiology and clinical features of idiopathic trigeminal neuralgia and glossopharyngeal neuralgia; similarities and differences, Rochester, Minnesota, 1945–1984. *Neuroepidemiology, 10,* 276–281.

Kerr, S. W. L. (1967). Pathology of trigeminal neuralgia: Light and electron microscopic observation. *Journal of Neurosurgery, 26*(suppl 6), 151–156.

Kingery, W. (1997). A critical review of controlled trials for peripheral neuropathic pain and complex regional pain syndromes. *Pain, 73,* 123–139.

Loeser, J. D. (2001). Cranial neuralgias. In J. D. Loeser (Ed.), *Bonica's management of pain* (pp. 855–866). Philadelphia: Lippincott, Williams & Wilkins.

Merskey, H., & Bogduk, N. (1994). *Classification of chronic pain. Descriptions of chronic pain syndromes and definitions of pain terms. International association for the study of pain.* Seattle: IASP Press.

Rappaport, Z. H., Govrin-Lippmann, R., & Devor, M. (1997). An electron microscopic analysis of biopsied samples of the trigeminal root taken during microvascular decompressive surgery. *Stereotactic and Functional Neurosurgery, 68*(1/4 Pt 1), 182–186.

Sindrup, S. H., & Jensen, T. S. (2002). Pharmacotherapy of trigeminal neuralgia. *The Clinical Journal of Pain, 18*, 22–27.

Watson, C. P. N. (2000). The treatment of neuropathic pain: Antidepressants and opioids. *The Clinical Journal of Pain, 16*, S49–S55.

Watson, C. P. N., & Evans, R. J. (1985). A comparative trial of amitriptyline and zimelidine in post-herpetic neuralgia. *Pain, 25*, 387–394.

Watson, C. P. N., Deck, J. S., Morshead, C., Van der Koog, D., & Evans, R. J. (1991). Post-herpetic neuralgia: Further post-mortem studies of cases with and without pain. *Pain, 44*, 105–117.

Watson, C. P. N., Tyler, K. L., Bicers, D. R., Millikan, L. E., Smith, S., & Coleman, E. (1993). A randomized vehicle-controlled trial capsaicin in the treatment of post-herpetic neuralgia. *Clinical Therapeutics, 15*, 510–526.

Neuraxial Blocks

Definition

Neuraxial block is a regional anesthetic technique aimed at the spinal cord (spinal block). Such blocks are accomplished by injecting a local anesthetic agent epidurally (epidural block) or intrathecally (intrathecal block, often referred to as "spinal block").

Cross-References

► Multimodal Analgesia in Postoperative Pain

Neuraxial Infusion

Definition

Neuraxial infusion is the delivery of medications via a catheter inserted into the epidural or intrathecal space.

Cross-References

► Cancer Pain Management, Anesthesiologic Interventions, Spinal Cord Stimulation, and Neuraxial Infusion

Neuraxial Morphine

Definition

Morphine administered into the cerebrospinal fluid (or epidurally) to reach the spinal cord directly, causing profound analgesia.

Cross-References

► Postoperative Pain, Acute Pain Team

Neuraxis

Definition

Neuraxis is the term that refers to the entire nervous system, from receptors in periphery to spinal cord and to the subcortical structures and cortex of the brain.

Cross-References

► Hypoalgesia, Assessment
► Hypoesthesia, Assessment

Neurectomy

Definition

Neurectomy is the removal of part of a nerve, implying surgery. This usually refers to the distal or nerve portion farthest from the brain.

Cross-References

► Cancer Pain Management, Neurosurgical Interventions
► Trigeminal Neuralgia: Diagnosis and Treatment

Neuritis

▶ Inflammatory Neuritis

Neuroablation

Definition

Neuroablation is an irreversible surgical technique that permanently blocks nerve pathways to the brain by destroying nerves and tissues at the source of the pain. This may be caused by various means, such as thermal or chemical, and occur in various places, such as in peripheral nerve or the brain.

Cross-References

▶ Dorsal Root Ganglionectomy and Dorsal Rhizotomy
▶ Trigeminal Neuralgia: Diagnosis and Treatment

Neuroactive Substance

Definition

A substance that can activate cells of the nervous system.

Cross-References

▶ Cytokines, Regulation in Inflammation

Neuroaxial

Definition

The epidural and intrathecal (spinal) spaces. Neuroaxial analgesic/anesthetic techniques involve the administration of agents into these spaces.

Cross-References

▶ Postoperative Pain, Appropriate Management

Neuro-Behçet

Definition

Behçet's disease with neurologic manifestations.

Cross-References

▶ Headache Due to Arteritis

Neurobiological Basis of Developing Pain Pathways

▶ Infant Pain Mechanisms

Neurochemical Markers

▶ Immunocytochemistry of Nociceptors

Neurochemistry

▶ Immunocytochemistry of Nociceptors

Neurodegeneration

Definition

Neurodegeneration refers to degenerative changes in the nervous system. The term is used mostly with reference to diseases affecting the central nervous system such as Parkinson's disease and Alzheimer's disease. Degenerative changes typically involve progressive dying of neurons, but may be selective to axons or even to myelin.

Cross-References

▶ Viral Neuropathies

Neurodynia

▶ Neuralgia, Diagnosis

Neurofilament Protein NF200

Definition

Neurofilaments are a class of intermediate filaments that are found in neurons. They form the structure of the cytoskeleton and are particularly abundant in axons. DRG neurons express low (68 kD), medium (155 kD), and high (200 kD) molecular weight neurofilament proteins. Large light DRG neurons are rich in expression of neurofilaments, especially the high molecular weight (200 kD) protein, whereas the small dark neurons are poor in expression of neurofilaments. The phosphorylated form, the 200 kD neurofilament, is localized specifically to the large light cell population, and therefore, the presence of immunoreactivity for this protein can be used to distinguish large light cells from small dark cells. Since a low level of non-phosphorylated 200 kD neurofilament is found in small and large neurons, an antibody against the non-phosphorylated form of the 200 kD subunit does not distinguish well between the large light and small dark neurons.

Cross-References

▶ Immunocytochemistry of Nociceptors

Neurogenic Claudication

▶ Lower Back Pain, Physical Examination

Neurogenic Inflammation

▶ Neurogenic Inflammation and Sympathetic Nervous System
▶ Nociceptor, Axonal Branching

Neurogenic Inflammation and Sympathetic Nervous System

Heinz-Joachim Häbler
FH Bonn-Rhein-Sieg, Rheinbach, Germany

Synonyms

Neurogenic inflammation; Sympathetic nervous system

Definition

There is evidence that ▶ neurogenic inflammation may be influenced by the sympathetic nervous system. This evidence is based on experiments, mainly on animals, in which neurogenic inflammation was elicited by chemical activation of nociceptors, and the extent of either ▶ plasma extravasation (neurogenic edema) or vasodilation (flare) was measured before and after interventions on the sympathetic nervous system. Pharmacological experiments indicate that the interaction between sympathetic neurons and primary afferent neurons, which may be responsible for sympathetic modulation of neurogenic inflammation, can take place proximal and/or distal to the neurovascular junction, i.e., prejunctionally on the nociceptor terminal, and/or postjunctionally at the level of the blood vessels. Only a prejunctional interaction can specifically modulate ▶ neuropeptide release from small-diameter afferent terminals, whereas at the postjunctional level, the interaction would occur nonspecifically and indirectly via changes in blood flow. However, both modes of interaction may be difficult to distinguish in experiments. Here only sympathetic modulation of neurogenic inflammation is considered likely to occur at the prejunctional level.

Characteristics

Capsaicin-Induced Neurogenic Inflammation in Skin

In healthy humans, after topical application of ▶ capsaicin to forearm skin, the area of flare was significantly decreased during whole-body cooling, which enhances sympathetic vasoconstrictor activity to the skin (Baron et al. 1999). However, capsaicin-induced pain and ▶ mechanical hyperalgesia were not changed, indicating an inhibitory effect of sympathetic activity on the signs of neurogenic inflammation, but not on the corresponding pain symptoms.

In contrast, in animal experiments, the flare response after intradermal capsaicin was found to be partially dependent on the sympathetic nervous system. It was reported that capsaicin injected intradermally into the hind paw of rats elicits dorsal root reflexes in peptidergic afferents, which antidromically evoke vasodilation outside the axon reflex area, as far as 20 mm remote from the capsaicin injection site. This part of the flare response was almost abolished by surgical ▶ sympathectomy, unaffected by decentralization of the postganglionic sympathetic neurons supplying the hind paw, and depended on an α_1-adrenoceptor-mediated mechanism (Lin et al. 2003). The sympathetic co-transmitter ▶ neuropeptide Y (NPY) via Y2 receptors also seemed to contribute to the flare (Lin et al. 2004). Vasopressin had no effects on the flare response, ruling out the possibility that sympathetic effects on the flare were indirectly due to the evoked cutaneous vasoconstriction. These findings suggest that sympathetic neurons contribute to neurogenic inflammation and that

the integrity of the sympathetic terminals, but not the ongoing activity of postganglionic neurons, may be the crucial factor. Interestingly, an almost identical sympathetic dependence was reported for mechanical hyperalgesia induced by intradermal capsaicin in the same rat model. Intradermal capsaicin injection led to mechanical hyperalgesia at the injection site (▶ Primary Hyperalgesia) and in areas remote from the injection site (▶ Secondary Hyperalgesia). Capsaicin-induced secondary hyperalgesia was blocked by the α_1-adrenoceptor antagonist prazosin but not by the α_2-adrenoceptor antagonist yohimbine. Surgical sympathectomy before capsaicin injection prevented secondary hyperalgesia (Kinnman and Levine 1995), but decentralization of sympathetic postganglionic neurons did not affect mechanical hyperalgesia.

Neurogenic Inflammation of the Rat Knee Joint

Capsaicin injection into the knee joint of rats produces intra-articular plasma extravasation lasting for 30–40 min, which was found to be reduced after chemical or surgical sympathectomy (Coderre et al. 1989). This indicates that small-diameter afferent-evoked plasma extravasation in the synovia is in part dependent on the sympathetic nervous system. However, release of norepinephrine from postganglionic sympathetic terminals by joint perfusion with 6-hydroxy-dopamine led to a much larger and prolonged plasma extravasation, which was almost abolished after surgical sympathectomy and after pretreatment with indomethacin, restored in the presence of prostaglandin E_2 but unchanged in rats treated with capsaicin neonatally. Similar results were obtained when brady-kinin, instead of 6-hydroxy-dopamine, was injected into the knee joint. Decentralization of postganglionic sympathetic neurons supplying the hind limb left bradykinin-induced plasma extravasation unchanged (Miao et al. 1996). Indirect effects due to changes in blood flow resulting from interventions on the sympathetic nervous system were excluded. These observations led to the concept that neurogenic inflammation in the rat knee joint has two components:

a relatively small component depending on primary afferent terminals and in part dependent on sympathetic neurons, and a second larger component depending on the presence of post-ganglionic sympathetic terminals but not on ongoing sympathetic activity nor on capsaicin-sensitive afferents. Inflammatory mediators such as ► bradykinin are thought to release prostaglandins, and possibly other mediators from postganglionic terminals, to elicit this so-called sympathetically dependent neurogenic inflammation (Green et al. 1998).

In contrast, knee joint inflammation in rats induced by kaolin and carrageenan was found to depend, in part, on primary afferents but was unaffected by sympathectomy (Sluka et al. 1994).

Prejunctional Control of Neuropeptide Release from Nociceptive Afferents by Sympathetic Transmitters

It is well established that in the superficial dorsal horn, synaptic release of glutamate from small-diameter afferents is inhibited by α_2-adrenoceptor agonists (e.g., Pan et al. 2002), indicating the presence of presynaptic α_2-adrenoceptors. Presynaptic inhibition of nociceptors via these receptors is thought to be one mechanism by which descending monoaminergic pathways control input from nociceptors at the level of the spinal cord. The evidence that the release of neuropeptides, such as substance P and CGRP, is inhibited in parallel with that of glutamate at spinal synapses is scarce. However, in vitro pharmacological studies on animals indicate that α_2-adrenoceptors are also present on the peripheral terminals of capsaicin-sensitive afferents (prejunctional receptors) and inhibit stimulation-induced release of neuropeptides.

In the guinea-pig lower airways, capsaicin-sensitive afferents elicit neurogenic inflammation by liberating ► substance P, ► neurokinin A, and ► CGRP. Both CGRP release and neurokinin-evoked bronchoconstriction after a low dose of capsaicin or low-frequency (1 Hz) antidromic stimulation of the vagus nerve were attenuated by α_2-adrenoceptor agonists (Lou et al. 1992). These inhibitory effects were small when high

doses of capsaicin or high-frequency (10 Hz) electrical stimulation of afferents was used. Similar experiments provided evidence for the presence of inhibitory prejunctional NPY (Y2) receptors on small-diameter afferents (see Lundberg 1996). The effects of both prejunctional α_2-adrenoceptors and Y2 receptors are likely to be mediated by the opening of large conductance Ca^{++}-activated K^+ channels (Stretton et al. 1992). As these observations were made on vagal rather than spinal afferents, it may be questioned whether these results can be generalized to all peptidergic small-diameter afferents.

However, results obtained in other animal models are similar. In the perfused mesenteric vascular bed of the rat, in vitro perivascular nerve stimulation elicits vasodilation, which depends on CGRP released from capsaicin-sensitive afferents (Kawasaki et al. 1988) that are mainly of spinal origin. Pharmacological experiments in this model indicate that norepinephrine, released from sympathetic postganglionic terminals, can suppress CGRP release from perivascular afferents by activation of prejunctional α_2-adrenoceptors and that NPY also inhibits CGRP release via prejunctional Y receptors (Kawasaki 2002). Thus, results identical to those obtained in the lower airways of the guinea pig were found in the rat mesentery, and these were confirmed in a number of in vitro studies on the control of isolated autonomic targets by neuropeptides originating from capsaicin-sensitive afferents.

Taken together, evidence indicates that the release of neuropeptides, which can elicit neurogenic inflammation from the peripheral terminals of capsaicin-sensitive afferents, is inhibited via prejunctional receptors for sympathetic transmitters. However, it remains to be shown that these prejunctional receptors play any functional role in the intact organism under the conditions of physiological or pathophysiological regulation. Rates of ongoing activity in sympathetic neurons are normally low, and in particular, the release of NPY from sympathetic terminals requires a high rate of ongoing activity that may rarely occur in vivo.

In conclusion, there is evidence that neurogenic inflammation, induced by the liberation of neurokinins and CGRP from peripheral terminals of capsaicin-sensitive afferents, may be influenced by sympathetic postganglionic neurons by a direct action on peripheral afferent terminals. However, the evidence is not yet conclusive. While in some studies capsaicin-induced neurogenic inflammation in the skin and knee joint of the rat was found to depend, in part, on sympathetic neurons, other studies indicate that there are prejunctional α_2-adrenoceptors and NPY receptors on afferent terminals that may inhibit neurogenic inflammation.

References

Baron, R., Wasner, G., Borgstedt, R., et al. (1999). Effect of sympathetic activity on capsaicin-evoked pain, hyperalgesia, and vasodilatation. *Neurology, 52,* 923–932.

Coderre, T. J., Basbaum, A. I., & Levine, J. D. (1989). Neural control of vascular permeability: Interactions between primary afferents, mast cells, and sympathetic efferents. *Journal of Neurophysiology, 62,* 48–58.

Green, P. G., Miao, F. J., Strausbaugh, H., et al. (1998). Endocrine and vagal controls of sympathetically dependent neurogenic inflammation. *Annals of the New York Academy of Sciences, 840,* 282–288.

Kawasaki, H. (2002). Regulation of vascular function by perivascular calcitonin gene-related peptide-containing nerves. *Japanese Journal of Pharmacology, 88,* 39–43.

Kawasaki, H., Takasaki, K., Saito, A., et al. (1988). Calcitonin gene-related peptide acts as a novel vasodilator transmitter in mesenteric resistance vessels of the rat. *Nature, 335,* 164–167.

Kinnman, E., & Levine, J. D. (1995). Involvement of the sympathetic postganglionic neuron in capsaicin-induced secondary hyperalgesia in the rat. *Neuroscience, 65,* 283–291.

Lin, Q., Zou, X., Fang, L., & Willis, W. D. (2003). Sympathetic modulation of acute cutaneous flare induced by intradermal injection of capsaicin in anesthetized rats. *Journal of Neurophysiology, 89,* 853–861.

Lin, Q., Zou, X., Ren, Y., Wang, J., et al. (2004). Involvement of peripheral neuropeptide Y receptors in sympathetic modulation of acute cutaneous flare induced by intradermal capsaicin. *Neuroscience, 123,* 337–347.

Lou, Y. P., Franco-Cereceda, & Lundberg, J. M. (1992). Variable $\hat{I}\pm_2$-adrenoceptor-mediated inhibition of bronchoconstriction and peptide release upon activation of pulmonary afferents. *European Journal of Pharmacology, 210,* 173–181.

Lundberg, J. M. (1996). Pharmacology of cotransmission in the autonomic nervous system: Integrative aspects on amines, neuropeptide, adenosine triphosphate, amino acids and nitric oxide. *Pharmacological Reviews, 48,* 113–178.

Miao, F. J., Jänig, W., & Levine, J. D. (1996). Role of sympathetic postganglionic neurons in synovial plasma extravasation induced by bradykinin. *Journal of Neurophysiology, 75,* 715–724.

Pan, Y. Z., Li, D. P., & Pan, H. L. (2002). Inhibition of glutamatergic synaptic input to spinal lamina lio neurons by presynaptic $\hat{I}\pm_2$-adrenergic receptors. *Journal of Neurophysiology, 87,* 1938–1947.

Sluka, K. A., Lawand, N. B., & Westlund, K. N. (1994). Joint inflammation is reduced by dorsal rhizotomy and not by sympathectomy or spinal cord transection. *Annals of the Rheumatic Diseases, 53,* 309–314.

Stretton, D., Miura, M., Belvisi, M. G., et al. (1992). Calcium-activated potassium channels mediate prejunctional inhibition of peripheral sensory nerves. *Proceedings of the National Academy of Science of the United States of America, 89,* 1325–1329.

Neurogenic Inflammation, Basis   N

Definition

See also "▶ Axon Reflex." Some C-fibers release vasoactive substances from their stimulated nerve terminals and also from non-stimulated axon collaterals depolarized by the axon reflex. The released neurokinins CGRP (calcitonin-gene-related peptide) and substance P bind to G-protein-coupled membrane receptors in arterioles (mainly CGRP) and venules (mainly SP), and induce vasodilatation and plasma-extravasation. Since these reactions are also part of inflammatory reactions and may trigger other elements of an inflammatory cascade, this phenomenon has been called "neurogenic inflammation." It is still debated to what extent neurogenic inflammation contributes to clinically important inflammatory states and other diseases, for example, migraine or allergic reactions.

Neurogenic Inflammation, Mechanism

A subset of nociceptive Aδ and C afferents contains pro-inflammatory neuropeptides (mainly "sleeping" nociceptors). If a peptidergic C fiber is activated, it releases neuropeptides from all the nerve terminals, which belong to this C fiber (axonal tree). Since always more than one C fiber will be activated by noxious stimulation, a homogenous area of neurogenic inflammation in the vicinity of the painful stimulus occurs. Release of these neuropeptides upon nociceptive activation causes inflammatory responses consisting of protein ► plasma extravasation (edema, mainly induced by substance P and neurokinin A) and vasodilatation (mainly mediated by CGRP). Endothelial activation and secretion, degranulation of perivascular mast cells, and the attraction of leucocytes have additionally been observed in some tissues like the dura mater. This reaction was first described as the axon reflex by Thomas Lewis and underlies the flare-and-wheal response often seen surrounding local tissue damage.

Cross-References

- ► Arthritis Model, Adjuvant-Induced Arthritis
- ► Cytokines, Effects on Nociceptors
- ► Formalin Test
- ► Freezing Model of Cutaneous Hyperalgesia
- ► Functional Imaging of Cutaneous Pain
- ► Inflammation, Modulation by Peripheral Cannabinoid Receptors
- ► Mechano-Insensitive C-Fibers, Biophysics
- ► Neurogenic Inflammation and Sympathetic Nervous System
- ► Nociceptor, Axonal Branching
- ► Nociceptors in the Orofacial Region (Meningeal/Cerebrovascular)
- ► Quantitative Thermal Sensory Testing of Inflamed Skin
- ► Substance P Regulation in Inflammation
- ► Sympathetically Maintained Pain and Inflammation, Human Experimentation

Neurogenic Inflammation, Vascular Regulation

Heinz-Joachim Häbler
FH Bonn-Rhein-Sieg, Rheinbach, Germany

Definition

► Neuropeptides, such as ► neurokinins (► substance P, neurokinin A) and CGRP, are the mediators of ► neurogenic inflammation. Their primary targets are blood vessels within the microvasculature where they elicit vasodilation and, by increasing the leakiness of the blood-tissue-barrier, ► plasma extravasation. Neurokinin effects may, in part, be mediated indirectly by the activation of mast cells and endothelial cells. At the neurovascular junction, neurokinins and ► CGRP interact with sympathetic neurotransmitters, vasoactive hormones, and autacoids produced by the endothelium that are involved in the ongoing regulation of the vasculature.

Characteristics

Vascular Effects of Neurokinins and CGRP

Both neurokinins and CGRP act primarily on blood vessels, but their efficacy in evoking vasodilation and plasma extravasation is different. The neurokinins substance P and neurokinin A are the main mediators of neurogenic plasma extravasation. They act on neurokinin 1 (NK1) receptors on postcapillary venues and within seconds lead to the opening of circular gaps of about 1.5 μm diameter between endothelial cells, exposing the basement membrane and permitting the leakage of plasma proteins into the interstitial space (McDonald 1998). In addition, substance P also degranulates mast cells, which enhances plasma extravasation by an indirect mechanism involving histamine. CGRP alone does not elicit plasma extravasation. However, it seems to cooperate with neurokinins, since it potentiates neurokinin-induced plasma extravasation, possibly resulting from its vasodilator action or from its inhibitory effect on substance P degradation (Gamse and Saria 1985, Escott and Brain 1993). Generally, neurogenic edema requires longer-lasting and higher-frequency stimulation of small-diameter afferents than vasodilation, probably because neurokinin effects are short lived and a higher amount of neuropeptides may be necessary.

The main mediator of vasodilation in neurogenic inflammation is CGRP, acting via CGRP 1 receptors. CGRP is one of the most potent vasodilators known and upon brief stimulation of small-diameter afferents, elicits long-lasting vasodilation by relaxing small arteries, arterioles, and precapillary sphincters. Most of the vasodilation, in particular during the later phase, can be blocked by the CGRP 1 receptor antagonist $CGRP_{8-37}$, but part of the vasodilation remains, indicating that CGRP is not the only vasodilator involved. Substance P and neurokinin A applied exogenously also evoke strong but short-lasting vasodilation. Their role in the vasodilation elicited by adequate or electrical stimulation of small-diameter afferents has been disputed, because in animal models, NK1 receptor antagonists can block vasodilation evoked by exogenous substance P and neurokinin A, but no clear effects of NK1 receptor antagonists on the vasodilation evoked by stimulation of small-diameter afferents were seen (Rinder and Lundberg 1996; Delay-Goyet et al. 1992). However, in the hairy and hairless skin of the rat, it has been possible to demonstrate that an NK1 receptor antagonist can delay antidromic vasodilation elicited by electrical stimulation of small-diameter afferents by several seconds. Furthermore, the NK1 receptor antagonist potentiated the reduction of the amplitude of antidromic vasodilation by $CGRP_{8-37}$, implying that substance P and/or neurokinin A plays a role in the early phase of antidromic vasodilation (Häbler et al. 1999).

Endothelium Dependence of Vascular Neurokinin and CGRP Effects

The effects of substance P and neurokinin A, when applied exogenously, on vasodilation and plasma extravasation, are probably mediated indirectly, at least in part, via NK1 receptors located on vascular endothelium leading to the production of ▶ nitric oxide (NO), because these effects can be reduced by blocking NO synthesis (e.g., Whittle et al. 1989). It is, however, unclear whether the effects of neurokinins released upon adequate or antidromic electrical stimulation of small-diameter afferents are endothelium-dependent and involve NO. As perivascular nerves contact vascular smooth muscle on the adventitial side, it is an open question whether neuropeptides released at the neurovascular junction can penetrate the vascular wall to act on receptors located on the endothelium. Experimental studies addressing this issue are scarce. A study on neurogenic edema elicited by antidromic nerve stimulation in the rat found an involvement of NO in the response, but NO was generated by the neuronal isoform rather than the endothelial isoform of NO synthase (Kajekar et al. 1995). In another study on rats, the NK1 receptor-dependent component of antidromic vasodilation was unaffected by blocking NO synthesis (Häbler et al., unpublished).

In contrast to the neurokinins, CGRP, applied exogenously or released from small-diameter afferents, exerts its vascular effects in a manner independent of the endothelium in most vascular beds, including that of the skin. Exceptions are the rat aorta and the gastric microcirculation of the rat, where the vasodilator effects of CGRP are partially inhibited by blocking NO synthesis (Holzer et al. 1995).

Interaction of Sympathetic Efferents and Small-Diameter Afferents in Vascular Regulation

As the vascular bed of most organs, including skin, is regulated under physiological conditions by low-frequency ongoing activity in sympathetic vasoconstrictor fibers, the question arises as to how this activity interferes with small-diameter afferent-induced vasodilation. In human skin, antidromic vasodilation evoked by transcutaneous electrical stimulation was decreased under the conditions of body cooling, which raises sympathetic vasoconstrictor activity to skin. This effect was abolished by an anesthetic block of the proximal nerves supplying the skin territory. Other stimuli that are known to increase sympathetic vasoconstrictor activity to skin, such as deep breaths or emotional stress, also transiently reduced antidromic vasodilation (Hornyak et al. 1990). In rat hairless skin, antidromic vasodilation elicited by brief stimulation of the corresponding dorsal root was able to override the vasoconstriction evoked by electrical stimulation of the sympathetic chain, up to a frequency of 3 Hz. Higher sympathetic frequencies suppressed antidromic vasodilation, but this suppression could be overcome by longer-lasting stimulation of the afferents at high frequency (Häbler et al. 1997). These studies show that the vasodilation elicited by thin afferents is likely to dominate over sympathetic vasoconstriction under almost all conditions of normal regulation and may be reduced only when sympathetic vasoconstrictor activity is exceptionally high. Pharmacological experiments indicate that the interaction of both neural vasomotor systems occurs mainly at the postjunctional level, but inhibitory prejunctional α_2-adrenoceptors on peripheral terminals of small-diameter afferents may also be involved.

Implication of Neuropeptides Derived from Small-Diameter Afferents in Systemic Vascular Regulation

Neuropeptides are released from small-diameter afferents, not only in the context of noxious stimulation and local neurogenic inflammation but they also appear in the systemic circulation, where they may be of importance for vascular regulation under physiological and pathophysiological conditions. Without any overt noxious stimulation, CGRP is present in the plasma of humans, and CGRP levels increase during exercise (Lind et al. 1996) and in patients with sepsis (Shimizu et al. 2003) and severe hypertension (Edvinsson et al. 1992). Studies on spontaneously hypertensive rats suggest that CGRP release from perivascular nerves may be impaired, implying a role for peptidergic small-diameter afferents in the long-term control of systemic vascular resistance and blood pressure (Kawasaki 2002).

References

Delay-Goyet, P., Satoh, H., & Lundberg, J. M. (1992). Relative involvement of substance P and CGRP mechanisms in antidromic vasodilation in the rat skin. *Acta Physiologica Scandinavica, 146*, 537–538.

Edvinsson, L., Erlinge, D., Ekman, R., et al. (1992). Sensory nerve terminal activity in severe hypertension as reflected by circulating calcitonin gene-related peptide and substance P. *Blood Pressure, 1*, 223–229.

Escott, K. J., & Brain, S. D. (1993). Effect of a calcitonin gene-related peptide antagonist (CGRP8-37) on skin vasodilatation and oedema induced by stimulation of the rat saphenous nerve. *British Journal of Pharmacology, 110*, 772–776.

Gamse, R., & Saria, A. (1985). Potentiation of tachykinin-induced plasma protein extravasation by calcitonin gene-related peptide. *European Journal of Pharmacology, 114*, 61–66.

Häbler, H. J., Wasner, G., & Jänig, W. (1997). Interaction of sympathetic vasoconstriction and antidromic vasodilatation in the control of skin blood flow. *Experimental Brain Research, 113*, 402–410.

Häbler, H. J., Timmermann, L., Stegmann, J. U., et al. (1999). Involvement of neurokinins in antidromic vasodilatation in hairy and hairless skin of the rat hindlimb. *Neuroscience, 89*, 1259–1268.

Holzer, P., Wachter, C., Heinemann, A., et al. (1995). Sensory nerves, nitric oxide and NANC vasodilatation. *Archives Internationales de Pharmacodynamie et de Therapie, 329*, 67–79.

Hornyak, M. E., Naver, H. K., Rydenhag, B., et al. (1990). Sympathetic activity influences the vascular axon reflex. *Acta Physiologica Scandinavica, 139*, 77–84.

Kajekar, R., Moore, P. K., & Brain, S. D. (1995). Essential role for nitric oxide in neurogenic inflammation in rat cutaneous microcirculation. Evidence for an endothelium-independent mechanism. *Circulation Research, 76*, 441–447.

Kawasaki. (2002). Regulation of vascular function by perivascular calcitonin gene-related peptide-containing nerves. *Japanese Journal of Pharmacology, 88*, 39–43.

Lind, H., Brudin, L., Lindholm, L., et al. (1996). Different levels of sensory neuropeptides (calcitonin gene-related peptide and substance P) during and after exercise in man. *Clinical Physiology, 16*, 73–82.

McDonald, D. M. (1998). Endothelial gaps: Plasma leakage during inflammation. *News in Physiological Sciences, 13*, 104–105.

Rinder, J., & Lundberg, J. M. (1996). Effects of hCGRP 8-37 and the NK[1]-receptor antagonist SR 140.333 on capsaicin-evoked vasodilation in the pig nasal mucosa in vivo. *Acta Physiologica Scandinavica, 156*, 115–122.

Shimizu, T., Hanasawa, K., Tani, T., et al. (2003). Changes in circulating levels of calcitonin gene-related peptide and nitric oxide metabolites in septic patients during direct hemoperfusion with polymyxin B-immobilized fiber. *Blood Purification, 21,* 237–243.

Whittle, B. J., Lopez-Belmonte, J., & Rees, D. D. (1989). Modulation of the vasodepressor actions of acetylcholine, bradykinin, substance P and endothelin in the rat by a specific inhibitor of nitric oxide formation. *British Journal of Pharmacology, 98,* 646–652.

Neurogenic Pain

Definition

"Neurogenic pain" is an obsolete term, not much used any more. It has been replaced by "neuropathic pain." "Neurogenic" implies that the pain is generated in the peripheral nerve. Mechanisms of neuropathic pain are now known to be more complicated than that.

Cross-References

▶ Neuropathic Pain, Diagnosis, Pathology, and Management
▶ Thalamotomy for Human Pain Relief

Neurogenic Pain of Joint and Muscle Origin

▶ Neuropathic Pain, Joint and Muscle Origin

Neurogenic Vasodilation

▶ Nociceptor, Axonal Branching

Neuroimaging

Definition

Neuroimaging is the production of images of the brain and/or spinal cord. It can include computerized tomography (CT) scanning, magnetic resonance imaging (MRI), photon emission computerized tomography (SPECT), and positron emission tomography (PET).

Cross-References

▶ Amygdala, Pain Processing and Behavior in Animals
▶ Hippocampus and Entorhinal Complex: Functional Imaging

Neuroimmune Activation

Definition

Neuroimmune activation is the adaptive, specific activation of endothelial cells, microglia, and astrocytes leading to the production of cytokines, chemokines, and the expression of surface antigens.

Cross-References

▶ Viral Neuropathies

Neuroimmune Interaction

Definition

Interactions between the immune system and the nervous system.

Cross-References

▶ Cytokines, Regulation in Inflammation

Neuroinflammation

Definition

Following an immune challenge or an injury in the nervous system, immune cells invade from the vascular system. T-lymphocytes enter the central nervous system where the microglia are activated and express major histocompatibility complexes, particularly class II (MHC II). Blood-derived macrophages also become activated and encroach the perivascular space. In the dorsal root ganglia and sympathetic ganglia, where the vasculature is leakier and microglia are absent, the inflammatory response involves activation of endogenous macrophages and invasion of hematogenous ones and T-cells. The role of the immune cells is controversial. The macrophages/activated microglia are phagocytic if neuronal death occurs, but T-cells may be neuroprotective.

Cross-References

▶ Viral Neuropathies

Neurokinin

Definition

The tachykinins are a family of small biologically active peptides whose principal mammalian members are substance P (11 amino acids) and neurokinin (NK) A and B (10 amino acids). These peptides are derived from precursor proteins, the preprotachykinins, which are encoded by two different genes. Three receptors for tachykinins, the so-called neurokinin receptors NK1, NK2, and NK3, have been cloned and characterized to have seven transmembrane spanning segments, to be coupled to G proteins and to be linked to the phosphoinositide signaling pathway. Although NK1 receptors are considered to be substance P-preferring, NK2 receptors NKA-preferring, and NK3 receptors NKB-preferring, substance P, NKA, and B are full agonists at all three tachykinin receptors.

Cross-References

▶ Neuropeptide Release in Inflammation
▶ Substance P Regulation in Inflammation

Neurological Deficit

Definition

Neurological deficit refers to loss of function related to the nervous system.

Cross-References

▶ Hypoalgesia, Assessment
▶ Hypoesthesia, Assessment

Neurolytic Drugs

Definition

Neurolytic drugs (usually 50–96 % ethanol or 5–7 % phenol) destroy nerve cells and stop pain-impulse transmission for days to months.

Cross-References

▶ Anesthesiologic Interventions
▶ Cancer Pain Management
▶ Neural Blockade
▶ Trigeminal Neuralgia

Neuroma

Definition

When a peripheral nerve is cut across, as by a knife cut or limb amputation, the fibers in the segment distal to the cut degenerate. However, the fibers in the proximal segment, the segment closer to the spinal column, survive. This is because they remain in continuity with their cell body which lies paraspinally (in a sensory ganglion for afferents, in the spinal cord ventral horn or medulla for somatic efferents, or in an autonomic ganglion or nucleus for sympathetic or parasympathetic efferents). Some of the cut axons form a terminal swelling or "endbulb." Others emit regenerating sprouts, many of which form a tangled skein at the cut nerve end. The endbulbs, tangled sprouts and masses of scar tissue at the nerve end, form a swollen bulb called a nerve-end neuroma. If the initial nerve injury was only partial, the neuroma may appear as a bulge on the side of the nerve. This is a neuroma-in-continuity. In some patients, neuromas are an important source of spontaneous ectopic impulse discharge that contributes to spontaneous neuropathic paresthesias and pain. In other individuals, the neuroma is silent. Whether spontaneously active or silent, tapping on a neuroma or otherwise applying mechanical force to it (e.g., from a poorly fitting prosthetic limb) usually evokes impulse discharge. The typical result is an intense stabbing or electric shock-like sensation called the Tinel sign. Surgical excision of a painful nerve-end neuroma usually provides transient relief at best as a new neuroma quickly forms at the freshly cut nerve end. However, surgically moving a neuroma into a location where it is less likely to be subjected to mechanical pressure can provide long-lasting relief from neuroma pain.

Cross-References

▶ Ectopia, Spontaneous
▶ Painful Scars
▶ Sympathetically Maintained Pain in CRPS II, Human Experimentation

Neuroma Endbulb

Definition

Severed axons form swollen terminal endbulbs. This usually occurs as a prelude to sprouting, but endbulbs may persist in the absence of sprouting. See also "neuroma."

Cross-References

▶ Neuroma
▶ Neuroma Pain

Neuroma Model of Neuropathic Pain

▶ Anesthesia Dolorosa Model, Autotomy

Neuroma Pain

James N. Campbell
Johns Hopkins University School of Medicine, Baltimore, MD, USA

Synonyms

Nerve stump pain; Pain associated with traumatic nerve injury; Traumatic nerve endbulb pain

Definition

The peripheral nerve consists of nerve fibers, supporting ▶ Schwann cells, and associated elements such as blood vessels, extracellular matrix molecules, and the nerve sheaths. The cell body of motor fibers (somatic and sympathetic) is in the anterior or intermediolateral part of the gray matter of the spinal cord, whereas the cell body of sensory fibers is contained in the paraspinal, dorsal root ganglion. When the peripheral nerve is cut, the injured fibers form terminal endbulbs and outgrowing sprouts (Fawcett and Keynes 1990; Fried et al. 1991). The nerve fibers distal to the cut are not supported by the cell body and as a result they undergo (Wallerian) degeneration. The Schwann cells survive and begin to divide, a process apparently triggered by denervation. When the denervated Schwann cells are encountered by the sprouting fibers, after nerve repair, for example, regeneration proceeds in an orderly fashion. When Schwann cells guides cannot be accessed by the outgrowing sprouts, as, for example, happens in the event of amputation or if there is a large gap between the two ends of the severed nerve, the fibers entangle into an often bulbous mass at the proximal cut end of the nerve. This nerve end structure is known as a neuroma (Fawcett and Keynes 1990; Sunderland 1978).

Characteristics

Neuromas go through a life cycle as the nerve fibers continue to grow out into the adjoining tissue. The ultimate appearance of the neuroma depends on milieu, how the nerve was injured, and the amount of time that has passed (Campbell 2001). When a nerve is caught by a suture, as may happen inadvertently in surgery, a swollen bulb tends to form at the site, as the outgrowing sprouts are contained by the epineurium and scar tissue. When the nerve is cut and otherwise left alone in a healthy bed of tissue, the neuroma tends to be less discrete, as the outgrowing nerve fibers may advance and spread widely through the host tissue. Variations between full regeneration and a swollen ▶ neuroma endbulb abound, however. These variations may reflect the age of the patient, the nature and point of injury, the nature of the surrounding tissue, vascularization, and genetic factors. All of these parameters can affect the fraction of severed nerve fibers that emit sprouts, the number that regenerates successfully, the number that fans out in local tissues at the nerve end, and the number that becomes trapped within the swollen endbulb (Sunderland 1978). They *may* also affect whether the neuroma will be a source of pain. It is likely that most neuromas are not a source of pain.

Most neuromas, painful and non-painful, are due to nerve trauma caused by penetrating injuries, amputations, burns, bone fractures, and surgery. Tumors and vascular insufficiency are also common causes. The location of some nerves, and perhaps their intrinsic biological properties, makes them particularly prone to generating a (painful) neuroma. Entrapment may cause a neuroma if severe enough to actually sever axons. In these cases, the clinician observes a significant swelling of the nerve just proximal to the entrapment. After the entrapment is relieved, the nerve fibers may regenerate. Morton's "neuroma" actually in most cases represents an entrapment neuropathy and may or may involve neuroma formation. Pain is associated with neuromas of cutaneous nerves and of nerves that serve muscles. It is uncertain how frequently pain originating in other deep tissues and viscera is related to neuroma formation.

Causes of Neuroma Pain

Surgeons in the past have followed the logic:

- Nerve injuries lead to neuromas.
- Nerve injuries are painful.
- *Ergo*, neuromas cause pain.

The exploration of this hypothesis has led to a richly complicated understanding of the basis of neuropathic pain. In short what we can say as of now is that yes, the neuroma is an important element in the genesis and perpetuation of pain, but that the biology of this pain goes well beyond the neuroma. A considerable amount of information has been obtained from observations on humans and on animal models, but these studies still leave unanswered questions. Most informative are electrophysiological recordings from injured nerves, from which we can say the following:

- Ectopic impulse activity in C-fibers, presumably nociceptive afferents, arises from the neuroma. Neuroma A-fibers also become abnormally active (Devor 2005).
- Abnormal spontaneous activity also arises in the dorsal root ganglion, though the prevailing data indicate that this activity is primarily in large cells that presumably give rise to A-beta (non-nociceptive) fibers (Devor 2005).
- The nociceptors that share the innervation territory of the injured nerve also become spontaneously active (Ali et al. 1999; Wu et al. 2001).
- Central abnormalities develop (e.g., in the dorsal horn) following nerve injury. It remains unclear to what extent discharges in nociceptive transmission pathways in the spinal cord and brain are dependent on peripheral inputs (Ji et al. 2003).

These considerations leave open the question of how important the neuroma itself is in generating pain. A very simple human experiment can go a long way towards resolving this issue. One may merely anesthetize the nerve injury site and determine whether the pain goes away for the duration of the block. If the pain goes away, then the neuroma generated the relevant neural signals that led to pain. If the pain does not go away, then the pain generator is elsewhere. Amazingly, there remains no comprehensive study of this question. For sure, there are cases where simply anesthetizing the neuroma gives rise to partial if not complete relief of pain. In other cases, however, pain persists (Campbell 2001).

In the instance where the neuroma does indeed directly play a role in generating pain, we still have a question of major clinical significance to consider. The neuroma may induce pain simply by virtue of (Burchiel et al. 1993) inherent spontaneous activity that arises in the nociceptors and/or it may induce pain when activated by virtue of its location. Neuromas have *mechanosensitivity* (Devor 2005; Koschorke et al. 1991; Koschorke et al. 1994). This means that mechanical stimuli applied to the neuroma produce neural activity and pain (the ► Tinel Sign). Mechanical stimuli can arise in different ways. For example, the neuroma at the end of a stump of an amputated limb may be stimulated by the prosthesis. The neuroma infiltrating muscles or tendons in the hand or adhered to these by scar tissue may be stimulated every time the patient opens and closes his/her hand.

Treatment Approaches

In the clinical situation, where applying a temporary local anesthetic to the neuroma convincingly removes pain, what should be done? One possibility is to simply remove the neuroma. However, neuroma resection is a misguided surgical mission. As long as the sensory nerve fibers in the proximal nerve end are connected to the dorsal root ganglion, the neuroma reforms. Thus, neuroma resection is in fact *neuroma relocation* to the position of the new nerve cut. Will neuroma relocation work? This depends. If the dominant mechanism of pain production from the neuroma is inherent spontaneous activity, relocating the neuroma would not be expected to be effective. If however, the dominant mechanism is ► ectopic mechanosensitivity, then neuroma relocation does indeed make sense, if it is moved to a new location less prone to mechanical stimulation. There are numerous results in the literature about neuroma relocation

surgery. Reports of efficacy range from 30 % to 100 % (Burchiel et al. 1993; Dellon et al. 1995).

If a nerve is only partially severed, intact conducting nerve fibers and severed neuroma sprouts may intermingle forming what may be termed a "► neuroma-in-continuity." Resection in this case may relieve the neuroma pain, but at the cost of residual nerve function. Nerve grafting may be feasible.

Part of the variability of neuroma pain may be due to unexpected peculiarities of the anatomy associated with the nerve's attempts to regenerate. For example, neuromas may form on the wrong (distal) side of an injured nerve (Belzberg and Campbell 1998). Nerves routinely have nerve branches going back and forth to one another. For example, in the forearm the superficial radial nerve has nerve branches to the lateral antebrachial cutaneous nerve. A nerve branch that makes its way to the distal end of another nerve that has been severed upstream encounters denervated Schwann cells. These denervated Schwann cells appear to attract outgrowing sprouts from the intact nerve. Over time the growing fibers could make their way back upstream along the distal portion of the severed nerve to reach the distal end of the cut nerve. The surgeon would do well to consider the possibility of such scenarios in planning neuroma relocation surgery.

Neuromas may be surgically treated in other ways as well. When a major nerve is involved, the primary approach is to repair the nerve. Clinical experience and some data from experimental animals (Lancelotta et al. 2003) suggest that pain is less when the nerve successfully regenerates. This raises an interesting clinical issue. In the instance where motor recovery is not feasible, for example, in the event of proximal lesions involving the lower brachial plexus or sciatic nerve, should nerve repair be considered as a means to relieve pain? The answer is a qualified yes: "qualified" because little data are available to answer this question; "yes" because the rationale is compelling and the other options are less attractive. Painful cutaneous neuromas should be treated with proximal resection (*neuroma relocation surgery*), particularly when diagnostic block indicates that the origin of pain is in the neuroma. This is because the morbidity of the surgical procedure is low regardless of the fact that efficacy may be as low as 30 % (Burchiel et al. 1993; Sunderland 1978).

Other surgical options exist. Surgical sympathectomy may be considered in cases of sympathetically maintained pain. In the case of spinal nerve injury, it might make sense to consider dorsal root rhizotomy or dorsal root ganglionectomy, though these options are notorious for late recurrence of pain. Spinal cord or direct nerve stimulation may have striking palliative benefits. Pain might subside with time. Finally standard pharmacological approaches to neuropathic pain may be useful.

References

Ali, Z., Ringkamp, M., Hartke, T. V., et al. (1999). Uninjured cutaneous C-fiber nociceptors develop spontaneous activity and alpha adrenergic sensitivity following L6 spinal nerve ligation in the monkey. *Journal of Neurophysiology, 81*, 455–466.

Belzberg, A. J., & Campbell, J. N. (1998). Evidence for end-to-side sensory nerve regeneration in a human. Case reports. *Journal of Neurosurgery, 89*, 1055–1057.

Burchiel, K. J., Johans, T. J., & Ochoa, J. (1993). The surgical treatment of painful traumatic neuromas. *Journal of Neurosurgery, 78*, 714–719.

Campbell, J. N. (2001). Nerve lesions and the generation of pain. *Muscle & Nerve, 24*, 1261–1273.

Dellon, A. L., Mont, M. A., Krackow, K. A., et al. (1995). Partial denervation for persistent neuroma pain after total knee arthroplasty. *Clinical Orthopaedics and Related Research, 316*, 145–150.

Devor, M. (2005). Response of nerves to injury in relation to neuropathic pain. In S. L. McMahon & M. Koltzenburg (Eds.), *Wall and Melzack's textbook of pain* (5th ed.). London: Churchill Livingstone. in press.

Fawcett, Y., & Keynes, R. (1990). Peripheral nerve regeneration. *Annual Review of Neuroscience, 13*, 43–60.

Fried, K., Govrin-Lippmann, R., Rosenthal, F., et al. (1991). Ultra-structure of afferent axon endings in a neuroma. *Journal of Neurocytology, 20*, 682–701.

Ji, R. R., Kohno, T., Moore, K. A., et al. (2003). Central sensitization and LTP: Do pain and memory share similar mechanisms? *Trends in Neurosciences, 26*, 696–705.

Koschorke, G., Meyer, R., Tillman, D., et al. (1991). Ectopic excitability of injured nerves in monkey: Entrained responses to vibratory stimuli. *Journal of Neurophysiology, 65*, 693–701.

Koschorke, G. M., Meyer, R. A., & Campbell, J. N. (1994). Cellular components necessary for mechanoelectrical transduction are conveyed to primary afferent terminals by fast axonal transport. *Brain Research, 641*, 99–104.

Lancelotta, M. P., Sheth, R. N., Meyer, R. A., et al. (2003). Severity and duration of hyperalgesia in rat varies with type of nerve lesion. *Neurosurgery, 53*, 1200–1209.

Sunderland, S. (1978). *Nerves and nerve injuries* (2nd ed.). London: Churchill Livingstone.

Wu, G., Ringkamp, M., Hartke, T. V., et al. (2001). Early onset of spontaneous activity in uninjured C-fiber nociceptors after injury to neighboring nerve fibers. *Journal of Neuroscience, 21*, RC140.

Neuroma-in-Continuity

Definition

A bulbous swelling in the nerve formed by sprouting axons that are intermixed with nerve fibers that are in continuity. The nerve appears intact to gross inspection but is swollen at the point of pathology. See also "▶ Neuroma."

Cross-References

▶ Neuroma
▶ Neuroma Pain
▶ Neuropathic Pain Model, Chronic Constriction Injury

Neuromatrix

Definition

Neuromatrix is the term used by Ronald Melzack to describe a proposed cortical/subcortical network of interconnections, including the limbic system, responsible for pain. See Melzack (1996).

Cross-References

▶ Ethics of Pain, Culture, and Ethnicity

References

Melzack, R. (1996). Gate control theory: On the evolution of pain concepts. *Pain Forum, 5*, 128–138.

Neuromodulation

Definition

The delivery of an electric current (neurostimulation) or drugs (intrathecal drug delivery systems) directly to targeted nerve fibers to treat pain.

Cross-References

▶ Cancer Pain Management, Anesthesiologic Interventions, Spinal Cord Stimulation, and Neuraxial Infusion
▶ Dorsal Root Ganglionectomy and Dorsal Rhizotomy
▶ Stimulation Treatments of Central Pain

Neuromodulator

▶ Thalamic Neurotransmitters and Neuromodulators

Neuromodulator(s)

Definition

Neuromodulators are signaling molecules that play a role in the alteration of baseline neural activity. These neural effector molecules can increase or decrease baseline membrane activation. Examples are substance P, dynorphin, enkephalin, galanin, cholecystokinin, and bombesin.

Cross-References

▶ Placebo Analgesia and Descending Opioid Modulation
▶ Spinothalamic Tract Neurons, Peptidergic Input
▶ Thalamic Neurotransmitters and Neuromodulators
▶ Thalamus, Nociceptive Neurotransmission

Neuromodulators

▶ Spinothalamic Tract Neurons, Peptidergic Input

Neuron

Definition

A neuron is a nerve cell.

Cross-References

▶ Allodynia (Clinical, Experimental)
▶ Drugs Targeting Voltage-Gated Sodium and Calcium Channels
▶ Hypoalgesia, Assessment
▶ Lateral Thalamic Pain-Related Cells in Humans
▶ Postherpetic Neuralgia, Pharmacological and Nonpharmacological Treatment Options

Neuronal Architecture

▶ Spinothalamic Tract Neurons, Morphology

Neuronal Dysfunction

Definition

Neuronal dysfunction is a state in which neurons display abnormal properties compared to those observed in normal development. For example, neuronal injuries or various neurodegenerative diseases can change normal neuronal conduction velocity or activation threshold by altering receptor distribution and properties, protein synthesis, activation of secondary messengers, etc.

Cross-References

▶ Dietary Variables in Neuropathic Pain

Neuronal Hyperexcitability

Synonyms

Central hyperexcitabiltiy

Definition

Increased responsiveness of central neurons and may include increased activity in response to stimulation, reduced threshold, increased afterdischarge, and expansion of receptive field size.

Cross-References

▶ Central Pain, Pharmacological Treatments
▶ Post-Seizure Headache

Neuronal Release

▶ Opioid-Induced Release of CCK

Neuronal Structure

▶ Spinoparabrachial Tract Neurons, Anatomical Features
▶ Spinothalamic Tract Neurons, Morphology

Neuronavigation

Definition

A stereotaxic system used for locating internal structures in 3D space without the need for fixing the patient in a stereotaxic frame.

Cross-References

▶ Motor Cortex
▶ Secondary Somatosensory Cortex (S2) and Insula, Effect on Pain-Related Behavior in Animals and Humans

Neuron-Glial Interactions in Descending Circuitry in Persistent Pain

Ke Ren and Feng Wei
Department of Neural and Pain Sciences,
University of Maryland Dental School,
Baltimore, MD, USA

Definition

The brain is immune privileged since it is protected by the blood-brain barrier. Glia, consisting of microglia, astrocytes, and oligodendrocytes, are the major nonneuronal cells in the CNS. Glial cells, microglia and astroglia in particular, release inflammatory cytokines and function as immune cells in the brain. Multiple brain sites and pathways are involved in descending pain modulation. One of the key pathways consists of the midbrain periaqueductal gray (PAG), rostral ventromedial medulla (RVM), and spinal cord. In response to peripheral injury, glia exhibit increased activity and interact with neurons within the descending pain modulatory circuitry. Such interactions facilitate neuronal hyperexcitability and contribute to the development of chronic or persistent pain.

Introduction

Studies suggest that brain stem descending input to the spinal cord plays an important role in the maintenance of persistent pain (Burgess et al. 2002; Ren and Dubner 2008a). Under chronic pain conditions, the net descending modulation becomes facilitatory and involves immune system activity. There is evidence that the signal of peripheral immune activation reaches the brain through afferent nerve input, circulating cytokines and immune cell trafficking (see Ren and Dubner 2010). Meningeal cells may serve as an interface by secreting proinflammatory cytokines into the brain (Wieseler-Frank et al. 2007). The injury-related immune signal derived from the periphery impact on cellular activity in the pain modulatory circuitry. Systemic administration of lipopolysaccharide (LPS), an endotoxin that induces immune responses from the host organism, produces behavioral hyperalgesia that is dependent upon the integrity of the descending circuitry, as indicated by the attenuation of nocifensive behavior after lesions of the nucleus raphe magnus, the major structure of the RVM (Wiertelak et al. 1997).

Characteristics

PAG-RVM Glial Responses to Peripheral Injury

A few studies have demonstrated injury-induced glial hyperactivity in the PAG-RVM circuitry, phenomena similar to that extensively reported at the spinal and trigeminal levels. At 6 days following chronic constriction injury (CCI) of the sciatic nerve in the rat, there is a select upregulation of glial fibrillary acidic protein (GFAP), a marker of astroglia, in the lateral and caudal ventrolateral columns of the PAG (Mor et al. 2010). CCI of the infraorbital nerve induces upregulation of CD11b (cluster of differentiation molecule 11b) and Iba1 (ionized calcium-binding adapter molecule), markers of microglia and GFAP in the RVM (Wei et al. 2008). Raghavendra et al. (2004) show that complete Freund's adjuvant-induced hind paw inflammation upregulates marker proteins of microglia and astroglia in the brain stem samples including pons, medulla, and midbrain. Roberts et al. (2009) further show that upregulation of microglial and astroglial markers occurs in the RVM after carrageenan-induced hind paw inflammation. However, in CX3CR1$^{GFP/+}$ mice with common peroneal nerve ligation, no microglia activation was identified in supraspinal brain regions including PAG and RVM (Zhang et al. 2008). Associated with upregulated glial markers, inflammatory cytokines such as interleukin-1beta (IL-1β) and tumor necrosis factor-alpha (TNFα) were also increased (Raghavendra et al. 2004; Wei et al. 2008).

The injury-related glial hyperactivity in the descending circuitry is correlated with behavioral hyperalgesia. Administration of glial inhibitors into the RVM attenuates or reverses behavioral hyperalgesia after nerve or tissue injury (Roberts et al. 2009; Wei et al. 2008). Interestingly, microglial and astroglial inhibitors attenuated hyperalgesia at 3 days and 14 days after CCI of the infraorbital nerve, respectively, which correlates with an early and transient activation of microglia and prolonged activation of astroglia (Wei et al. 2008). Studies at the spinal level also suggest a differential time-dependent coordinated involvement of microglia and astroglia in persistent pain (See Ren and Dubner 2008b). Further studies are necessary to elucidate differential contribution of different types of glial cells to neuronal hyperexcitability and pain.

It is noted that the upregulation of glial activity in regions involved in pain modulation may not necessarily be tied to pain conditions per se. In the PAG, region-specific upregulation of GFAP was only seen in a subgroup of CCI rats that show persistent behavioral disability in social interactions and sleep-wake cycle, but not in rats with transient or no disability in these behaviors, although all rats have similar levels of pain behaviors after nerve injury (Mor et al. 2010). These findings are consistent with roles of PAG in integrating behavioral responses to debilitating conditions and point to a selective role of astroglia in assisting neuronal activity to cope with injury and stress.

Glia-Cytokine-Neuronal Interactions in Descending Pain Facilitation
Glial cells and glia-derived inflammatory cytokines in descending pathways impact on neuronal activities. For example, upregulated TNFα and IL-1β are found in astroglia after CCI of the infraorbital nerve and receptors for TNFα and IL-1β colocalize with the NR1 subunit of the N-methyl-D-aspartate (NMDA) receptor in RVM neurons (Wei et al. 2008). Nerve injury increased the expression in TNF and IL-1 receptor levels, and NR1 phosphorylation in the RVM. The neutralization of endogenous TNF-α and IL-1β in the RVM significantly attenuated NR1 phosphorylation and reduced CCI-induced behavioral hypersensitivity. Further, intro-RVM administration of recombinant TNF-α or IL-1β upregulated NR1 phosphorylation and produced NMDA receptor-dependent allodynia in normal rats. These findings indicate contribution of supraspinal astrocytes and associated cytokines as well as glia-neuronal interactions to descending facilitation of neuropathic pain.

Glutamate is the major excitatory amino acid neurotransmitter in the nervous system. Astroglia play a major role in maintaining glutamate homeostasis in the brain and actively prevent neuronal hyperexcitability and excitotoxic cascades. The glutamate transporter GLT1 (EAAT2), mainly expressed in astroglia, is crucial for glutamate removal from the synaptic cleft. Recent studies show that there were time-dependent biphasic changes in GLT1 expression in the RVM after nerve or nerve injury: an early upregulation (30 min after hind paw inflammation and 3 days after CCI of the infraorbital nerve) and a prolonged late downregulation (Ren et al. 2011). Kappa B-motif binding phosphoprotein (KBBP), a transcriptional activator of GLT1 (Yang et al. 2009), exhibited similar biphasic changes in RVM. Overexpression of GLT1 in

RVM attenuated neuropathic pain behavior and RNAi of GLT1 and KBBP by GLT1-shRNA and KBBP-shRNA in RVM produced pain hypersensitivity (Ren et al. 2011). The increased GLT1 activity immediately after injury may serve to enhance glutamate uptake, which counteracts increased synaptic activity and dampens the development of hyperalgesia (also see Hayashida et al. 2010). However, a reduction in GLT1 activity may lead to a buildup of glutamate concentration in the synaptic cleft, leading to neuronal hyperexcitability. Consequently, the net descending modulation develops into a facilitatory state as injury persists, which correlates with decreased KBBP and GLT1 expression and the development of hyperalgesia.

Nonconventional Opioid Actions and Glia
Endogenous opioid peptides and their receptors are distributed extensively in descending pathways and normally produce pain inhibition. However, a series of studies reveal that opioids may also act on glial cells through nonclassic sites (see Review by Hutchinson et al. 2011). Instead of producing analgesia, nonconventional action of opioids seems to be related to opioid tolerance and hyperalgesia. In mice with deletion of all three subtypes of opioid receptors, opioid-produced hyperalgesia remains, suggesting nonconventional opioid signaling (see Hutchinson et al. 2011). In general, the nonconventional effects of opioids are mediated by their dextrorotatory isomers. The (+)-Isomer of morphine is anti-analgesia and induces behavioral hyperalgesia and (+)-naloxone and (+)-naltrexone potentiate morphine analgesia and protect against hyperalgesia development. Consistent with an effect of opioids on glia, morphine upregulated GFAP and CD11b in PAG and RVM, which was reversed by AV411 (ibudilast; 3-isobutyryl-2-isopropylpyrazolo-[1,5-a]pyridine), a glial activation inhibitor (Hutchinson et al. 2009).

Summary
Neuron-glial interactions have emerged as critical mechanisms underlying chronic pain

conditions (McMahon and Malcangio 2009; Ren and Dubner 2010). These interactions also occur in the brain pain-modulatory circuitry. Peripheral tissue or nerve injury generates enhanced afferent inputs that lead to long-term changes in the pain-modulatory circuitry including hyperactivity of glial cells and secretion of inflammatory cytokines. Glial cells interact with neurons to tip net descending modulation into facilitatory status and presumably mediate the transition from acute to persistent or chronic pain.

References

Burgess, S. E., Gardell, L. R., Ossipov, M. H., Malan, T. P., Jr., Vanderah, T. W., Lai, J., & Porreca, F. (2002). Time-dependent descending facilitation from the rostral ventromedial medulla maintains, but does not initiate, neuropathic pain. *Journal of Neuroscience, 22*, 5129–5136.

Hayashida, K., Parker, R. A., & Eisenach, J. C. (2010). Activation of glutamate transporters in the locus coeruleus paradoxically activates descending inhibition in rats. *Brain Research, 1317*, 80–86.

Hutchinson, M. R., Lewis, S. S., Coats, B. D., Skyba, D. A., Crysdale, N. Y., Berkelhammer, D. L., Brzeski, A., Northcutt, A., Vietz, C. M., Judd, C. M., Maier, S. F., Watkins, L. R., & Johnson, K. W. (2009). Reduction of opioid withdrawal and potentiation of acute opioid analgesia by systemic AV411 (ibudilast). *Brain, Behavior, and Immunity, 23*, 240–250.

Hutchinson, M. R., Shavit, Y., Grace, P. M., Rice, K. C., Maier, S. F., & Watkins, L. R. (2011). Exploring the neuroimmunopharmacology of opioids: An integrative review of mechanisms of central immune signaling and their implications for opioid analgesia. *Pharmacological Reviews, 63*, 772–810.

McMahon, S. B., & Malcangio, M. (2009). Current challenges in glia-pain biology. *Neuron, 64*, 46–54.

Mor, D., Bembrick, A. L., Austin, P. J., Wyllie, P. M., Creber, N. J., Denyer, G. S., & Keay, K. A. (2010). Anatomically specific patterns of glial activation in the periaqueductal gray of the sub-population of rats showing pain and disability following chronic constriction injury of the sciatic nerve. *Neuroscience, 166*, 1167–1184.

Raghavendra, V., Tanga, F. Y., & DeLeo, J. A. (2004). Complete Freunds adjuvant-induced peripheral inflammation evokes glial activation and proinflammatory cytokine expression in the CNS. *European Journal of Neuroscience, 20*, 467–473.

Ren, K., & Dubner, R. (2008a) Descending control mechanisms. In: A.I. Basbaum, A. Kaneko, G.M. Shepherd, G. Westheimer, & M. C. Bushnell (Eds.), *The senses: A comprehensive reference, Volume 5: Pain* (pp. 723–762). San Diego: Academic Press.

Ren, K., & Dubner, R. (2008b). Neuron-glia crosstalk gets serious: Role in pain hypersensitivity. *Current Opinion in Anaesthesiology, 21*, 570–579.

Ren, K., & Dubner, R. (2010). Interactions between the immune and nervous systems in pain. *Nature Medicine, 16*, 1267–1276.

Ren, K., Guo, W., Gu, M., Zou, S.-P., & Wie, F. (2011). Dubner, R. Kappa B-motif binding phosphoprotein (KBBP), astroglial glutamate transporter GLT1 and behavioral hyperalgesia. *Glia, 59*(Suppl. 1), S113–S113.

Roberts, J., Ossipov, M. H., & Porreca, F. (2009). Glial activation in the rostroventromedial medulla promotes descending facilitation to mediate inflammatory hypersensitivity. *European Journal of Neuroscience, 30*, 229–241.

Wei, F., Guo, W., Zou, S., Ren, K., & Dubner, R. (2008). Supraspinal glial-neuronal interactions contribute to descending pain facilitation. *Journal of Neuroscience, 28*, 10482–10495.

Wiertelak, E. P., Roemer, B., Maier, S. F., & Watkins, L. R. (1997). Comparison of the effects of nucleus tractus solitarius and ventral medial medulla lesions on illness-induced and subcutaneous formalin-induced hyperalgesias. *Brain Research, 748*, 143–150.

Wieseler-Frank, J., Jekich, B. M., Mahoney, J. H., Bland, S. T., Maier, S. F., & Watkins, L. R. (2007). A novel immune-to-CNS communication pathway: Cells of the meninges surrounding the spinal cord CSF space produce proinflammatory cytokines in response to an inflammatory stimulus. *Brain, Behavior, and Immunity, 21*, 711–718.

Yang, Y., Gozen, O., Watkins, A., Lorenzini, I., Lepore, A., Gao, Y., Vidensky, S., Brennan, J., Poulsen, D., Won Park, J., Li Jeon, N., Robinson, M. B., & Rothstein, J. D. (2009). Presynaptic regulation of astroglial excitatory neurotransmitter transporter GLT1. *Neuron, 61*, 880–894.

Zhang, F., Vadakkan, K. I., Kim, S. S., Wu, L. J., Shang, Y., & Zhuo, M. (2008). Selective activation of microglia in spinal cord but not higher cortical regions following nerve injury in adult mouse. *Molecular Pain, 4*, 15.

Neuron-Restrictive Silencer Factor

Synonyms

NRSF

Definition

Is a repressor that is predominantly expressed in non-neuronal cells. NRSF silences neuronal genes in non-neuronal cells by binding to the NRSE (neuron-restrictive silencer element) motif.

Cross-References

▶ Substance P Regulation in Inflammation

Neuropathic Pain

"Neuropathic pain" is defined by the International Association for the Study of Pain (IASP, 2011 update) as pain caused by a lesion or disease of the somatosensory nervous system. The lesion or disease may affect the peripheral nervous system (peripheral neuropathic pain) or the central nervous system (central neuropathic pain, or simply "central pain"). In the event of a sudden injury, pain is felt acutely. In general, however, neuropathic pain is a chronic pain state.

Cross-References

▶ Cathepsin and Microglia
▶ Chronic Neuropathic Pain

Neuropathic Pain (NP)

Definition

Neuropathic pain is pain initiated or caused by a primary lesion or dysfunction in the nervous system. Although neuropathic pain includes a number of different conditions, some characteristics are shared: pain is often described as stabbing or burning, sensory abnormalities are common, and treatment is difficult. Neuropathic (neurogenic pain) has been described in about 1 % of the population. It is caused by functional abnormalities or structural lesions in the peripheral or central nervous system and occurs without peripheral nociceptor stimulation. It is caused by heterogeneous conditions unexplained by a single etiology or anatomic lesion. There are many different causes of neuropathic pain. Neuropathic pain may arise from infection/inflammation (postherpetic neuralgia, HIV-associated neuralgia, postpoliomyelitis, leprosy, interstitial cystitis, spinal arachnoiditis, acute inflammatory polyradiculopathy), noninfectious illness (multiple sclerosis, diabetic neuropathy, thalamic pain syndrome, essential vulvodynia), pain associated with pressure/entrapment (neoplasia, trigeminal and glossopharyngeal neuralgia, carpal tunnel), and injury/trauma (surgery, complex regional pain syndrome, spinal cord injury). The etiology may be classified as localized (ischemic neuropathy, complex regional pain syndrome, phantom limb) or diffuse (toxins, AIDS, alcohol). Damage can affect the peripheral nerves, the cranial nerves, the posterior nerve roots, the spinal cord, and certain regions within the brain. A variety of pain-related phenomena (mechanisms) may be operative in an individual patient necessitating mechanistic-based treatment. Patients with chronic NP are overrepresented amongst those who are refractory to classic analgesic including opioid therapy. Although NP is not always opioid insensitive, the treatments of choice are tricyclic antidepressants (e.g., amitriptyline) and antiepileptic drugs (e.g., gabapentin).

Cross-References

▶ Allodynia (Clinical, Experimental)
▶ Alpha(α) 2-Adrenergic Agonists in Pain Treatment
▶ Amygdala, Pain Processing and Behavior in Animals
▶ Cancer Pain
▶ Cancer Pain Management, Treatment of Neuropathic Components
▶ Complex Chronic Pain in Children, Interdisciplinary Treatment
▶ COX-1 and COX-2 in Pain

Neuropathic Pain in Children

Alyssa Lebel
Boston Children's Hospital, Boston, MA, USA

Synonyms

Central nervous system pain disorder; Neuralgia; Neuropathy; Peripheral and central neuronal sensitization

Definition

The International Association for the Study of Pain presented the most recent definition of neuropathic pain in 2008 as "pain arising as a direct consequence of a lesion or disease affecting the somatosensory system" (Treede et al. 2008). As there is no specific diagnostic tool for neuropathic pain (NP), NP is a clinical description rather than a diagnosis, requiring demonstration of clinical, neurophysiologic, and/or, most recently, neuroimaging equivalents of a lesion or disease within the somatosensory neural network.

Introduction

Pediatric neuropathic pain differs from pain in adults in terms of presentation, epidemiology (King et al. 2011), evidence for treatment efficacy, and prognosis. These differences are predominantly defined by our increasing knowledge of the developmental neurobiology of pain processing, age-appropriate pain assessment, and increasing but still limited data from pediatric clinical trials and practice guidelines (Walker 2008). Adult neuropathic pain disorders, such as diabetic peripheral neuropathy, post-herpetic neuralgia, phantom pain, brachial plexus neuralgia, and, possibly, low back pain, are uncommon in pediatric practice. Other disorders, such as small fiber neuralgia, complex regional pain syndrome, and post-infectious and post-traumatic neuropathic pain are increasingly recognized in children and are often more clinically responsive to therapy than in adults. Rare neuropathic pain syndromes, such as metabolic, neurodegenerative, and toxic neuropathies, are primarily present during childhood. These conditions include Fabry Disease, hereditary sensorimotor neuropathies, lead toxicity, mitochondrial disorders, and primary erythromelagia (Walco et al. 2010). Such rare conditions may provide an investigative window into the pathophysiology of early somatosensory changes in neuropathic pain, as patients often have a defined pathology unassociated with multiple complex pain problems. Pediatric oncology patients frequently present with peripheral and central neuropathic pain related to treatment with chemotherapeutic and biologic agents rather than solid tumor burden. Current experience with neuropathic pain pharmacology is particularly informed by these patients (Fig. 1).

Characteristics

Research
Current data include case reports, clinical series, and some clinical trials. There are no current randomized controlled clinical trials to guide the assessment and treatment of neuropathic pain in children. Therefore, treatment regimens may be derived from adult trials, although the developing nervous system may be both more vulnerable to interventions as well as more plastic and resilient (Stanton-Hicks 2010). Research is limited by the lack of reliable epidemiologic reports and the putative low incidence of neuropathic pain disorders in the pediatric population (Walco et al. 2010). Cancer-related chronic pain has an estimated prevalence rate of 60 %, but is often the result of mixed neuropathic and somatic etiologies, severe and compelling, and thus often reported as an anecdote or small series. The prevalence of chronic limb pain ranges between 4 % and 40 % (King 2011). Despite a paucity of comprehensive data, reports from the past several years show some intriguing findings.

For example, in patients with complex regional pain syndrome (CRPS), lower limb involvement is most common, and most patients improve with a rehabilitative program of physical therapy, desensitization, cognitive reframing, family education, and normalization of functional activity. Therapies often used for adults with CRPS, such as lumbar sympathetic nerve blocks, tricyclic antidepressants, and gabapentin, may also show efficacy in children, but the data remains more limited (Wilder 2006). There may be age-related differences regarding the clinical response and prognosis after nerve injury. Infants who sustain a brachial plexus trauma at birth rarely report features of neuropathic pain in early childhood, although adults and adolescents with traumatic brachial plexus injury often report neuropathic symptoms (Anand and Birch 2002). However, some infants who required surgical exploration of the plexus after birth do demonstrate self-mutilation behavior, considered as a possible human manifestation of autotomy, seen in animal studies of peripheral nerve injury (McCann et al. 2004). One British survey of pediatric records noted that the incidence of post-herpetic neuralgia, trigeminal neuralgia, and phantom limb pain was lowest for patients aged 0–14 years and then increased with age, with no patients with diabetic neuropathy under 14 years (Hall et al. 2006). However, more specific examination, such as using quantitative sensory testing (QST), may demonstrate subclinical neuropathy in these patients (Blankenburg et al. 2012). QST has even demonstrated somatosensory processing abnormalities in children with "growing pains" (Pathirana et al. 2011). Evidence supporting pharmacologic therapies for neuropathic pain is minimal, with limited data for the use of amitriptyline for patients with central neuropathic dysfunction such as chronic headache and abdominal pain, case reports for the use of gabapentin in children with limb pain, and no specific reports using anticonvulsant therapy for neuropathic disorders except primary erythromelagia (Damen et al. 2005; Golden et al. 2006; Rusy et al. 2001). Cognitive and biobehavioral therapies are often more effective than pharmacology (Lee et al. 2002).

Neuropathic Pain in Children, Fig. 1 Adult versus children's brain

Assessment

In addition to the neurologic examination, specifically focusing on sensory testing for allodynia, hyperesthesia, hyperpathia, and summation in cooperative patients, quantitative sensory testing (QST) has been particularly useful in the delineation of cutaneous sensory abnormalities in pediatric patients with neuropathic pain. QST has been quantified in healthy children, providing a reliable baseline for testing in patients (Meier et al. 2001). This measure is now routinely used to better define neuropathic pain in subjects and patients (Sethna et al. 2007). However, neonates and nonverbal patients remain a diagnostic challenge. Noninvasive neuroimaging has provided visualization of developing central neural networks potentially altered by injury and disease processes, as well as allowing the investigator to longitudinally follow allosteric changes within the neural network. Magnetic resonance spectroscopy measures changes in neural chemistry associated with pain. The use of this technology in pediatric pain is nascent but rapidly expanding (Sava et al. 2009). Some examples of imaging techniques are shown below in (Fig. 2).

Specific Examples and Lessons from Pediatric Pain Disorders

▶ Fabry disease is an X-linked recessive glycosphingolipid storage disorder due to deficiency of a lysosomal enzyme, α-galactosidase A, which generally presents during childhood and adolescence with recurrent episodes of debilitating pain in the extremities. This neuropathic pain syndrome is a small fiber peripheral neuropathy, selectively affecting C- and A-δ fibers (Maag et al. 2008), that results from destruction of dorsal root ganglion cells by progressive deposition of abnormal lipid. Morbitity occurs with increasing age as glycosphingolipid is accumulated in endothelial and smooth muscle cells of the vascular system, resulting renal, cardiac, and central vascular dysfunction. Primary treatment of Fabry's disease is modulatory and includes repetitive administration of IV α-galactosidase A (Schiffman et al. 2001). Pharmacologic treatment of neuropathic pain with neuronal channel blocking agents, such as gabapentin (Ries et al. 2003) and lidocaine, is minimally effective, despite a well-defined peripheral neuronal target and early presentation of disease (Walco et al. 2010), suggesting modulation by a more complex somatosensory pain system.

Therefore, it is not surprising that a neuropathic pain disorder that involves less well-defined changes in both the peripheral and central pain transmission systems, ▶ Complex regional pain syndrome (CRPS), requires multimodality therapy (Wilder 2006), including pharmacology, physical therapy, cognitive biobehavioral therapy, and functional rehabilitation (Lee et al. 2002;

Neuropathic Pain in Children, Fig. 2 Examples of imaging techniques

Sherry et al. 1999). There is also some evidence for the persistence of pain relief following lumbar sympathetic neuronal blockade and systemic antinflammatory therapy (Walco 2008). Nascent research in neuroimaging of pediatric pain corroborates changes is a complex central neural network, including sensory, motor, and emotional regions of interest. Using functional MRI, symptomatic and recovered patients with CRPS received mechanical and thermal stimulation to affected and unaffected extremities, and functional activation of endogenous pain modulatory systems was visualized and analyzed. This study highlighted the complexity of central changes in young patients with CRPS and showed persistent changes in central inhibitory and excitatory neuronal circuits despite clinical recovery (LeBel et al. 2008). Subsequently, changes in cortical volume have been shown per fMRI in this patient population as well as alterations in central connectivity (Fig. 3).

Therapy

As previously mentioned, recommended pharmacologic treatment for pediatric patients is currently based on adult data (Bril et al. 2011), anecdote, and small case series (Walco et al. 2010). Suggested dosing is listed below:

Neuropathic pain medication dosing

Medication	Dosing	Max dose	Side effects
Amitriptyline	0.2–0.4 mg/kg/day	100 mg/day	Sedation, weight gain, constipation, dizziness

(continued)

Neuropathic pain medication dosing

Medication	Dosing	Max dose	Side effects
Nortriptyline	0.2–0.4 mg/kg/day	75 mg/day	Tachycardia, weight gain
Gabapentin	5–10 mg/kg/day	2,400–3,000 mg/day	Fatigue, nausea, confusion
Pregabalin	50 mg bid/tid	300 mg/day	Fatigue, confusion
Topiramate	1–3 mg/kg/day	400 mg/day	Confusion, weight loss, acidosis
Zonisamide	2–4 mg/kg/day	400 mg/day	Weight loss, kidney stones, rash
Clonidine	2–4 µg/kg/day	0.4 mg/day	Sedation, bradycardia, hypotension
Mexiletine	2–3 mg/kg/day	600 mg/day	Dizziness, tremor Gi distress

In addition, intravenous agents, such as the NMDA receptor antagonist ketamine and local anesthetic lidocaine, may be administered at time of crises and as loading doses for subsequent oral dosing (Kajiume et al. 2012; Berde et al. 2003). Representative dosing for ketamine is 6–10 mg/kg/dose po, 3–7 mg/kg/dose IM, and 0.5–2 mg/kg/dose IV. Short-term infusions are preferable as adverse effects include sedation, memory loss, increased intracranial pressure, cardiac dysrhythmia, respiratory depression, and hypersalivation. Lidocaine may be administered in bolus IV dosing or by brief continuous infusion, maintaining a serum level less than 5 µg/ml to prevent cardiac and central nervous system toxicity. Acute cardiac toxicity may be reversed by

Neuropathic Pain in Children, Fig. 3 Sava (2010), Abstract at Society for Neuroscience

IV administration of a lipid emulsion, such as Intralipid 20 % at 0.25 ml/kg/min. Recent research regarding local anesthetic agents for chronic pain is focused on developing sensory selective blocking agents derived from biologic nerve toxins, such as tetrodotoxin and saxitoxin (Waxman et al. 1999). Of note, some neuropathic medications have additional local anesthetic effects, such as tricyclic antidepressants, ketamine, and clonidine.

Neuropathic Pain in Children, Fig. 4 LeBel (2011), Abstract from American Headache Society

Genetics

The identification of mutations in the voltage-gated sodium channel Na v1.7, encoded by the SCN9A gene, provides a genetic basis for previously clinically defined neuropathic disorders, such as primary erythromelagia, paroxsysmal extreme pain disorder (PEPD), and idiopathic small fiber neuropathy (I-SFN) (Faber 2011; Faber et al. 2012). These rare pain problems frequently present in infancy or early childhood. Similar sodium "channelopathies" may also be found in nonpainful pediatric disorders, such as benign familial neonatal seizures and severe myoclonic epilepsy of infancy. Ligand-gated channels involving the human glycine receptor have been implicated in hyperexplexia or stiff baby syndrome, but also in neuropathic pain (Gardiner 2006; Benarroch 2011). These rare pain disorders demonstrate the molecular genetics of nervous system channelopathies and suggest possible pharmacologic interventions. As well, the identification of neuropathic-specific genes via high density oligonucleotide microarray technology may provide gene candidates for gene therapy (Fink et al. 2011).

Central Network Disorders

Chronic headache and, possibly, functional abdominal pain are disorders that are generated by central nervous system dysfunction, including dynamic changes in neurochemistry and neurocircuitry within the brain as well as ascending and descending modulatory pathways. Neuroimaging of pediatric patients with these conditions is beginning to demonstrate a "pain fingerprint" neural connectivity that may provide additional targets for pharmacologic and cognitive therapies. Below is data from connectivity studies in pediatric headache (Fig. 4).

Summary

In conclusion, investigating and treating pediatric neuropathic pain is challenging due to a relatively low prevalence of disorders in children and the limitations of ethical research design. However, the prognosis for pediatric patients with neuropathic pain is often more favorable than for adults, and the pathophysiology of this clinical recovery, likely influenced by developmental neurobiology, is of interest regarding the successful treatment of all patients. Using appropriate methods, such as collaborative crossover trials for pharmacologic therapy, and noninvasive techniques, such as functional neuroimaging, we may gain insight into the early changes within the somatosensory system networks which will positively respond to pharmacologic and nonpharmacologic therapies for neuropathic pain.

Acknowledgement I wish to gratefully acknowledge the technical assistance of Sharon Segal, Research Coordinator in the Headache Program, Childrens Hospital.

References

Anand, P., & Birch, R. (2002). Restoration of sensory function and lack of long-term chronic pain syndromes after brachial plexus injury in human neonates. *Brain, 125,* 113–122.

Benarroch, E. E. (2011). Glycine and its synaptic interactions: Functional and clinical implications. *Neurology, 77,* 677–683.

Berde, C. B., Lebel, A. A., & Olsson, G. (2003). Neuropathic pain in children. In N. L. Schechter, C. B. Berde, & M. Yaster (Eds.), *Pain in infants, children, and adolescents* (2nd ed., pp. 620–641). Philadelphia: LIppincott, Williams, and Wilkins.

Blankenburg, M., Kraemer, N., Hirschfeld, G., Krumova, E. K., Maier, C., Hechler, T., Aksu, F., Magerl, W., Reinehr, T., Wiesel, T., & Zernikow, B. (2012). Childhood diabetic neuropathy: Functional impairment and non-invasive screening assessment. *Diabetic Medicine, 29*(11), 1425–1432.

Bril, V., England, J., Franklin, G. M., Backonja, M., Cohen, J., Del Toro, D., Feldman, E., Iverson, D. J., Perkins, B., Russell, J. W., & Zochodne, D. (2011). Evidence-based guideline: Treatment of painful diabetic neuropathy: Report of the American Academy of Neurology, the American Association of Neuromuscular and Electrodiagnostic Medicine, and the American Academy of Physical Medicine and Rehabilitation. *Neurology, 76,* 1758–1765.

Damen, L., Bruijn, J. K., Verhagen, A. P., Berger, M. Y., Passchier, J., & Koes, B. W. (2005). Symptomatic treatment of migraine in children: A systematic review of medication trials. *Pediatrics, 116,* e295–e302.

Faber, C. G., Hoeijmakers, J. G., Ahn, H. S., Cheng, X., Han, C., Choi, J. S., Estacion, M., Lauria, G., Vanhoutte, E. K., Gerrits, M. M., Dibb-Hajj, S., Drenth, J. P., Waxman, S. G., & Merkies, I. S. (2012). Gain of function Nav1.7 mutations in idiopathic small fiber neuropathy. *Annals of Neurology, 71,* 26–39.

Fink, D. J., Wechuck, J., Marina, M., Glorioso, J. C., Goss, J., Krisky, D., & Wolfe, D. (2011). Gene therapy for pain: Results of a phase I clinical trial. *Annals of Neurology, 70,* 207–212.

Gardiner, M. (2006). Molecular genetics of infantile nervous system channelopathies. *Early Human Development, 82,* 775–779.

Golden, A. S., Haut, S. R., & Moshe, S. L. (2006). Nonepileptic uses of antiepileptic drugs in children and adolescents. *Pediatric Neurology, 34,* 421–432.

Hall, G. C., Carroll, D., Parry, D., & McQuay, H. J. (2006). Epidemiology and treatment of neuropathic pain: The UK primary care perspective. *Pain, 122,* 156–162.

Kajiume, T., Sera, Y., Nakanuno, R., Ogura, T., Karakawa, S., Kobayakawa, M., Taguchi, S., Oshita, K., Kawaguchi, H., Sato, T., & Kobayashi, M. (2012). Continuous intravenous infusion of ketamine and lidocaine as adjuvant analgesics in a 5-year-old patient with neuropathic cancer pain. *Journal of Palliative Medicine, 15*(6), 719–722. doi:10.1089/jpm.2011.0097. Epub 2012 Mar 8.

King, S., Chambers, C. T., Huguet, A., MacNevin, R. C., McGrath, P. J., Parker, L., MacDonald, A. J. (2011). The epidemiology of chronic pain in children and adolescents revisited: A systematic review. *Pain, 152,* 2729–2738.

LeBel, A., Becerra, L., Wallin, D., Moulton, E. A., Morris, S., Pendse, G., Jasciewicz, J., Stein, M., Aiello-Lammens, M., Grant, E., Berde, C., & Borsook, D. (2008). fMRI reveals distinct CNS processing during symptomatic and recovered complex regional pain syndrome in children. *Brain, 131,* 1854–1879.

Lee, B. H., Scharff, L., Sethna, N. F., McCarthy, C. F., Scott-Sutherland, J., Shea, A. M., Sullivan, P., Meier, P., Zurakowski, D., Masek, B. J., & Berde, C. B. (2002). Physical therapy and cognitive-behavioral treatment for complex regional pain syndromes. *Journal of Pediatrics, 141,* 135–140.

Maag, R., Binder, A., Maier, C., Scherens, A., Toelle, T., Treede, R. D., & Baron, R. (2008). Detection of a characteristic painful neuropathy in Fabry disease: A pilot study. *Pain Medicine, 9,* 1217–1223.

McCann, M. E., Waters, P., Goumnerova, L. C., & Berde, C. (2004). Self-mutilation in young children following brachial plexus birth injury. *Pain, 110,* 123–139.

Meier, P. M., Berde, C. B., DiCanzio, J., Zurakowski, D., & Sethna, N. F. (2001). Quantitative assessment of cutaneous thermal and vibration sensation and thermal pain detection thresholds in healthy children and adolescents. *Muscle Nerve, 24,* 1339–1345.

Pathirana, S., Champion, D., Jaaniste, T., Yee, A., & Chapman, C. (2011). Somatosensory test responses in children with growing pains. *Journal of Pain Research, 4,* 393–400.

Ries, M., Mengel, E., Kutschke, G., Kim, K. S., Berklein, F., Krummenauer, F., & Beck, M. (2003). Use of Gabapentin to reduce chronic neuropathic pain in Fabry disease. *Journal of Inherited Metabolic Disease, 26,* 413–414.

Rusy, L. M., Troshynski, T. J., & Weisman, S. J. (2001). Gabapentin in phantom limb pain management in children and young adults: Report of seven cases. *Journal of Pain Symptom Management, 21,* 78–82.

Sava, S., LeBel, A. A., Leslie, D. S., Drosos, A., Berde, C., Becerra, L., & Borsook, D. (2009). Challenges of functional imaging research of pain in children. *Molecular Pain, 16,* 5–30.

Schiffman, R., Kopp, J. B., Austin, H. A., Sabnis, S., Moore, D. F., Weibel, T., Balow, J. E., & Brady, R. O. (2001). Enzyme replacement therapy in Fabry disease: A randomized controlled trial. *Journal of the American Medical Association, 285,* 2743–2749.

Sethna, N. F., Meier, P. M., Zurakowski, D., & Berde, C. B. (2007). Cutaneous sensory abnormalities in children and adolescents with complex regional pain syndromes. *Pain, 131,* 153–161.

Sherry, D. D., Wallace, C. A., Kelley, C., Kidder, M., & Sapp, L. (1999). Short- and long-term outcomes of children with complex regional pain syndrome type I treated with exercise therapy. *Clinical Journal of Pain, 15,* 218–223.

Stanton-Hicks, M. (2010). Plasticity of complex regional pain syndrome (CRPS) in children. *Pain Medicine, 11,* 1216–1223.

Treede, R. D., Jensen, T. S., Campbell, J. N., Cruccu, G., Dostrovsky, J. O., Griffin, J. W., Hansson, P., Hughes, R., Nurmikko, T., & Serra, J. (2008). Neuropathic pain: redefinition and a grading system for clinical and research purposes. *Neurology, 70,* 1630–1635.

Walco, G. A. (2008). Needle pain in children: Contextual factors. *Pediatrics, 122*(Suppl 3), S125–S129.

Walco, G. A., Dworkin, R. H., Krane, E. J., LeBel, A. A., & Treede, R. D. (2010). Neuropathic pain in children: Special considerations. *Mayo Clinical Proceedings, 85,* S33–S41.

Walker, S. M. (2008). Pain in children: Recent advances and ongoing challenges. *British Journal of Anaesthesia, 101,* 101–110.

Waxman, S. G., Cummins, T. R., Dib-Hajj, S., Fjell, J., & Black, J. A. (1999). Sodium channels, excitability of primary sensory neurons, and the molecular basis of pain. *Muscle Nerve, 22,* 1177–1187.

Wilder, R. T. (2006). Management of pediatric patients with chronic regional pain syndrome. *Clinical Journal of Pain, 22,* 443–448.

Neuropathic Pain Model, Chronic Constriction Injury, Fig. 1 (a) Rat sciatic nerve. Four chromic gut sutures are tied around the nerve to produce a CCI. (b) Example of CCI in mouse sciatic nerve using three prolene 8.0 ligatures

Neuropathic Pain Model, Chronic Constriction Injury

Claudia Sommer
Department of Neurology, University of Würzburg, Würzburg, Germany

Synonyms

Bennett model; CCI; Chronic constrictive injury

Definition

▶ Chronic constriction injury (CCI) is a partial nerve injury, mostly used in rodents, which is produced by tying several ligatures around a nerve, such that these ligatures slightly constrict the nerve. This induces an incomplete nerve injury, which entails behavioral signs of ▶ hyperalgesia in the animals.

Characteristics

Animals, Technique, and Behavioral Data

Originally, the model was established in rats (Bennett and Xie 1988), but it has since been successfully used in mice (Sommer et al. 1997). Most often, the sciatic nerve is constricted; other nerves that have been used are the infraorbital nerve (Vos et al. 1994) and the median nerve (Day et al. 2001). The nerve is exposed by blunt dissection of the overlying muscles and several ligatures are tied around it such that they slightly constrict the nerve. Originally, four 4.0 chromic gut ligatures were used to constrict the sciatic nerve, which were tied around the nerve above the trifurcation with a spacing of 1 mm (Fig. 1a).

Many investigators have since used modifications of the model, either varying the number of ligatures (between two and four) or the type of ligatures (e.g., silk or vicryl). In mice, prolene (7.0–10.0) has been used (Fig. 1b).

In spite of minor differences, with most variations it has been possible to produce a reliable model of neuropathic pain. The degree of constriction may be more important than the type of suture in determining the degree and duration of behavioral signs, concomitantly with the degree of nerve injury and the time needed for regeneration (Myers et al. 1993; Lindenlaub and Sommer 2000). The measure of constriction is defined as to "just barely constrict" the nerve and to retard, but not arrest circulation through the epineurial vessels (Bennett and Xie 1988). Typically, this type of constriction is associated with a short twitch of the respective limb of the animal. In the original description, paw withdrawal latencies to heat returned to normal between 60 and 90 days after the injury (Bennett and Xie 1988). In another early study, guarding behavior, increased responses to heat, cold, and pressure returned to normal between 6 and 8 weeks after the injury (Attal et al. 1990), which is very much what most investigators report. Some investigators report mechanical allodynia tested with von Frey hairs to occur less reliably in CCI compared to other models (Kim et al. 1997), however, this is in contrast to our own experience. Recently, efforts have been made to establish further readouts more relevant to continuous pain, like automatically recording spontaneous limb movement (Kawasaki-Yatsugi et al. 2012) or, in a related model, the spinal nerve ligation model, using a place-preference paradigm (King et al. 2009).

Neuropathology

Obviously, the amount of nerve damage depends upon the degree of nerve constriction by the ligatures, which makes the model quite variable between laboratories. Even if the initial constriction is only slight, the nerve develops edema, self-strangulates between the ligatures and thus develops additional damage. The temporal course of neuropathology of rat sciatic nerve with CCI has been extensively investigated (Munger et al. 1992; Coggeshall et al. 1993; Guilbaud et al. 1993; Sommer et al. 1993, 1995). In summary, CCI induces the typical process of ► Wallerian degeneration. Most myelinated fibers and about two thirds of the unmyelinated fibers degenerate, with maximum nerve damage during the first 2 weeks and a slow regeneration which is still incomplete 3 months after the injury. Remaining long-term changes include the formation of minifascicles and ► neuroma-in-continuity. Whereas endoneurial macrophages reflect axonal degeneration and phagocytosis, epineurial macrophages are the result of the

foreign body reaction induced by the ligatures, which is one aspect in which CCI differs from models not using foreign material, like crush or partial sciatic nerve transection (Lindenlaub and Sommer 2000). Whether this epineurial inflammation plays a major role in the development of hyperalgesia is still a matter of debate.

CCI induces a large number of anatomical and neurochemical changes in the ► DRG, in the spinal cord, and possibly in the CNS. These include sprouting of sympathetic fibers in the DRG, changes in gene expression in DRG neurons, activation of spinal cord glia, changes in cerebral blood flow, and many others. Expounding upon these is beyond the scope of this entry. Correlations between any of these changes and hyperalgesia have to be considered with caution, since these may be general reactions to a nerve injury, irrespective of whether neuropathic pain develops or not. Important recent findings derived from the CCI model include the relevance of tetrahydrobiopterin (Tegeder et al. 2006), of the chemokine CCL2 (Van Steenwinckel et al. 2011), and of SOCS3 and the JAK/STAT3 signaling pathway in neuropathic pain (Dominguez et al. 2010).

Neurophysiology

After CCI, spontaneous activity develops in all fiber types, but more frequently in myelinated fibers, and can originate from the DRG or directly from the injury site (Xie and Xiao 1990; Kajander and Bennett 1992; Tal and Eliav 1996). Furthermore, spontaneous activity can be recorded in the L5/6 dorsal horn of the spinal cord (Sotgiu and Biella 2000). In addition, DRG neurons from nerves with CCI are more excited by inflammatory mediators (Song et al. 2003). Thus, CCI, like other incomplete nerve injuries, induces a hyperexcitable state in the nerve fibers, which may be underlying the pain-related behavior observed in the animals.

The qualities most often investigated in experimental animals with CCI are the withdrawal latencies to heat (► Thermal Hyperalgesia) and the withdrawal thresholds to von Frey hairs (► Mechanical Allodynia). Several investigators have tried to determine which qualities are transmitted by which fiber type (A or C fibers), and whether peripheral or central changes are responsible for the respective quality in CCI. At present, the questions are not entirely solved. In summary, exaggerated responses to heat and cold are believed to be mediated by C-fibers, mechanical allodynia, and hyperalgesia by $A\beta$ and $A\delta$-fibers (Field et al. 1999). Double and triple spikes have been recorded from C-fibers in CCI rats by microneurography (Serra et al. 2011).

Human Correlates

CCI has been claimed to be a model for different human pathological conditions that are associated with neuropathic pain. At present, no single disease is exactly modeled by CCI. One condition sharing some symptoms and many pathological features is ▶ carpal tunnel syndrome. Here, the median nerve is constricted, which leads to edema of the nerve, initially associated with hyperexcitability and pain only, later with overt nerve damage and neurologic deficits. Recovery is possible if decompression is performed in time. The difference to CCI is the abrupt onset in the animal model as opposed to the chronic course in the human disease, and, of course, the presence of foreign material in CCI.

Some parallels have been found with ▶ CRPS type II (Daemen et al. 1998; Suyama et al. 2002), specifically the presence of ▶ neurogenic inflammation and of osteoporosis. Importantly, CCI entails an incomplete injury of a (usually) mixed peripheral nerve, such that injured and uninjured fibers run in close vicinity in the nerve and that uninjured fibers are exposed to the altered endoneurial microenvironment induced by Wallerian degeneration. This situation is also present in incomplete nerve injuries in humans. One has to keep in mind that, in spite of the ligatures hindering regeneration, peripheral nerve regeneration in a rodent is considerably faster than in humans and that this difference in time course may account for some CCI-specific findings that may not be reflected by human disease.

References

Attal, N., Jazat, F., Kayser, V., & Guilbaud, G. (1990). Further evidence for 'pain-related' behaviours in a model of unilateral peripheral mononeuropathy. *Pain, 41*, 235–251.

Bennett, G. J., & Xie, Y. K. (1988). A peripheral mononeuropathy in rat that produces disorders of pain sensation like those seen in man. *Pain, 33*, 87–107.

Coggeshall, R. E., Dougherty, P. M., Pover, C. M., & Carlton, S. M. (1993). Is large myelinated fiber loss associated with hyperalgesia in a model of experimental peripheral neuropathy in the rat? *Pain, 52*, 233–242.

Daemen, M., Kurvers, H., Bullens, P., Barendse, G., Van Kleef, M., & Van den Wildenberg, F. (1998). Neurogenic inflammation and reflex sympathetic dystrophy (in vivo and in vitro assessment in an experimental model). *Acta Orthopaedica Belgica, 64*, 441–447.

Day, A., Lue, J., Sun, W., Shieh, J., & Wen, C. (2001). A beta-fiber intensity stimulation of chronically constricted median nerve induces c-fos expression in thalamic projection neurons of the cuneate nucleus in rats with behavioral signs of neuropathic pain. *Brain Research, 895*, 194–203.

Dominguez, E., Mauborgne, A., Mallet, J., Desclaux, M., & Pohl, M. (2010). SOCS3-Mediated blockade of JAK/STAT3 signaling pathway reveals its major contribution to spinal cord neuroinflammation and mechanical allodynia after peripheral nerve injury. *Journal of Neuroscience, 30*, 5754–5766. doi:10.1523/jneurosci.5007-09.2010.

Field, M. J., Bramwell, S., Hughes, J., & Singh, L. (1999). Detection of static and dynamic components of mechanical allodynia in rat models of neuropathic pain: Are they signalled by distinct primary sensory neurones? *Pain, 83*, 303–311.

Guilbaud, G., Gautron, M., Jazat, F., Ratinahirana, H., Hassig, R., & Hauw, J. J. (1993). Time course of degeneration and regeneration of myelinated nerve fibers following chronic loose ligatures of the rat sciatic nerve: Can nerve lesions be linked to the abnormal pain-related behaviors? *Pain, 53*, 147–158.

Kajander, K. C., & Bennett, G. J. (1992). Onset of a painful peripheral neuropathy in rat: A partial and differential deafferentation and spontaneous discharge in a beta and a delta primary afferent neurons. *Journal of Neurophysiology, 68*, 734–744.

Kawasaki-Yatsugi, S., Nagakura, Y., Ogino, S., Sekizawa, T., Kiso, T., Takahashi, M., Ishikawa, G., Ito, H., & Shimizu, Y. (2012). Automated measurement of spontaneous pain-associated limb movement and drug efficacy evaluation in a rat model of neuropathic pain. *European Journal of Pain, 16*, 1426–1436.

Kim, K. J., Yoon, Y. W., & Chung, J. M. (1997). Comparison of three rodent neuropathic pain models. *Experimental Brain Research, 113*, 200–206.

King, T., Vera-Portocarrero, L., Gutierrez, T., Vanderah, T. W., Dussor, G., Lai, J., Fields, H. L., & Porreca, F. (2009). Unmasking the tonic-aversive state in neuropathic pain. *Nature Neuroscience, 12*, 1364–1366.

Lindenlaub, T., & Sommer, C. (2000). Partial sciatic nerve transection as a model of neuropathic pain: A qualitative and quantitative neuropathological study. *Pain, 89*, 97–106.

Munger, B., Bennett, G., & Kajander, K. (1992). An experimental painful peripheral neuropathy due to nerve constriction. *Experimental Neurology, 118*, 204–214.

Myers, R. R., Yamamoto, T. Y., Yaksh, T. L., & Powell, H. C. (1993). The role of focal nerve ischemia and Wallerian degeneration in peripheral nerve injury producing hyperesthesia. *Anesthesiology, 78*, 308–316.

Serra, J., Sola, R., Aleu, J., Quiles, C., Navarro, X., & Bostock, H. (2011). Double and triple spikes in C-nociceptors in neuropathic pain states: An additional peripheral mechanism of hyperalgesia. *Pain, 152*, 343–353. doi:10.1016/j.pain.2010.10.039.

Sommer, C., Galbraith, J. A., Heckman, H. M., & Myers, R. R. (1993). Pathology of experimental compression neuropathy producing hyperesthesia. *Journal of Neuropathology and Experimental Neurology, 52*, 223–233.

Sommer, C., Lalonde, A., Heckman, H. M., Rodriguez, M., & Myers, R. R. (1995). Quantitative neuropathology of a focal nerve injury causing hyperalgesia. *Journal of Neuropathology and Experimental Neurology, 54*, 635–643.

Sommer, C., Schmidt, C., George, A., & Toyka, K. V. (1997). A metalloprotease-inhibitor reduces pain associated behavior in mice with experimental neuropathy. *Neuroscience Letters, 237*, 45–48.

Song, X. J., Zhang, J. M., Hu, S. J., & LaMotte, R. H. (2003). Somata of nerve-injured sensory neurons exhibit enhanced responses to inflammatory mediators. *Pain, 104*, 701–709.

Sotgiu, M. L., & Biella, G. (2000). Contribution of central sensitization to the pain-related abnormal activity in neuropathic rats. *Somatosensory & Motor Research, 17*, 32–38.

Suyama, H., Moriwaki, K., Niida, S., Maehara, Y., Kawamoto, M., & Yuge, O. (2002). Osteoporosis following chronic constriction injury of sciatic nerve in rats. *Journal of Bone and Mineral Metabolism, 20*, 91–97.

Tal, M., & Eliav, E. (1996). Abnormal discharge originates at the site of nerve injury in experimental constriction neuropathy (CCI) in the rat. *Pain, 64*, 511–518.

Tegeder, I., Costigan, M., Griffin, R. S., et al. (2006). GTP cyclohydrolase and tetrahydrobiopterin regulate pain sensitivity and persistence. *Nature Medicine, 12*, 1269–1277. doi:10.1038/nm1490.

Van Steenwinckel, J., Reaux-Le Goazigo, A., Pommier, B., et al. (2011). CCL2 Released from neuronal synaptic vesicles in the spinal cord is a major mediator of local inflammation and pain after peripheral nerve injury. *Journal of Neuroscience, 31*, 5865–5875. doi:10.1523/jneurosci.5986-10.2011.

Vos, B. P., Strassman, A. M., & Maciewicz, R. J. (1994). Behavioral evidence of trigeminal neuropathic pain following chronic constriction injury to the rat's infraorbital nerve. *Journal of Neuroscience, 14*, 2708–2723.

Xie, Y., & Xiao, W. (1990). Electrophysiological evidence for hyperalgesia in the peripheral neuropathy. *Science in China. Series B, 33*, 663–672.

Neuropathic Pain Model, Diabetic Neuropathy Model

Nigel A. Calcutt[1] and Sandra Chaplan[2]
[1]Department of Pathology, University of California San Diego, La Jolla, CA, USA
[2]Janssen Research and Development, LLC, San Diego, CA, USA

Synonyms

Diabetic neuropathy model; Experimental diabetic neuropathy

Definition

Diabetes mellitus is characterized by elevated blood glucose levels. This may be a consequence of either insulin deficiency (Insulin-Dependent Diabetes Mellitus or type 1 diabetes) or impaired action (Non-insulin-Dependent Diabetes Mellitus or type 2 diabetes). The term "neuropathy" implies physical degeneration of the nervous system, although it has also been applied more broadly to encompass any neurochemical, physiologic, or structural nerve disorders.

Diabetes can affect any combination of the sensory, motor, or autonomic components of the peripheral nervous systems, and there is accumulating evidence that the central nervous system may also be vulnerable. Classifications of diabetic neuropathy usually reflect the distribution of nerves that show clinical evidence of dysfunction and the presumed etiology of the disorder. The most common form of diabetic neuropathy is a distal symmetrical polyneuropathy, which presents as sensory loss and/or pain in the distal extremities and progresses proximally toward the trunk.

Introduction

Human Diabetic Neuropathy

Peripheral nerves obtained by biopsy or autopsy show pathologic changes in most cell types. Schwann cells exhibit reactive and degenerative changes that are presumed to precede overt pathologic features, such as widening of the nodes of Ranvier and segmental demyelination. Axons undergo ▶ Wallerian degeneration which is followed by inefficient regeneration and subsequent net fiber loss. The ▶ vasa nervorum display endothelial cell hyperplasia and hypertrophy that can reduce lumen size and also reduplication of the basal lamina. In final stages of diabetic neuropathy, there can be almost complete loss of axons, Schwann cells, and blood vessels, with the vacated endoneurial space filled by collagen. Unmyelinated fibers, when viewed by immunocytochemistry in skin biopsies or by confocal microscopy in the cornea, show simplification and retraction of their distal projections. The physiologic consequences of this progressive degenerative neuropathy include nerve ischemic hypoxia, conduction slowing, progressive loss of sensation, muscle weakness, and impaired regulation of organ functions. Some 10–20 % of diabetic patients also describe the occurrence of inappropriate sensations, such as pain or paresthesias, usually starting in the feet. Where painful diabetic neuropathy is present, the pain may be spontaneous or evoked by light touch, persistent or sporadic, and is frequently described as having a burning or lancinating quality.

Characteristics

Animal Models of Diabetes

Diabetes occurs spontaneously in a range of species including domestic companion animals and rodents. The latter have been inbred to provide models of type 1 diabetes, such as the BB Wistar rat and NOD and Akita mice, and models of type 2 diabetes that include the ZDF rat and the db/db mouse. Insulin deficiency may also be induced by treating animals with the β cell toxins streptozotocin (STZ) or alloxan to produce models of severe type 1 diabetes, while hyperglycemia in the presence of normal insulin levels can be induced by dietary overfeeding of sugars such as galactose or fructose. High blood sugar is a feature common to all of these models, and hyperglycemia has been widely assumed to be a primary pathogenic mechanism underlying diabetic neuropathy. Recent studies using animals fed with high fat diets to induce mild insulin resistance also implicate impaired insulin signaling or dyslipidemia in peripheral nerve disorders (Vincent et al. 2009).

The STZ-diabetic rat and mouse remain the most commonly studied animal models of diabetic neuropathy, having the advantages of being relatively cheap and allowing initiation of hyperglycemia at a defined stage of maturity. Disadvantages of the model include the extreme hyperglycemia (20–50 mmol/l) relative to diabetic patients (8–12 mmol/l) that can produce an unrepresentative catabolic dominance and ▶ cachexia. These animals, if untreated, also have a life span of only a few months but minimal insulin replacement therapy can be used to maintain muscle mass and general health of the animals while retaining hyperglycemia for a year or more (see Calcutt 2004). Direct STZ-induced neurotoxicity has been discounted as a contributor to the commonly studied features of neuropathy (Davidson et al. 2009), but it remains advisable to confirm that newly identified nerve disorders in

STZ-diabetic rodents can be prevented by insulin therapy that prevents sustained hyperglycemia.

The relatively short life span of diabetic rodents may help explain why features of overt degenerative neuropathy are infrequently reported. There is no significant demyelination or loss of axons in the major nerve trunks of STZ-diabetic rodents within the time frame commonly studied (1–4 months), and reduced axonal caliber is the most notable pathologic feature of diabetic rats. Longer durations of diabetes may induce degenerative pathology in the most distal nerve trunks of many of the rodent models of diabetes but, in general, diabetic rodents are not faithful models of the myelinated fiber pathology seen in human diabetic neuropathy. In contrast, diabetic rodents do show physiologic disorders similar to those seen in diabetic patients. Most prominently, large myelinated motor and sensory fibers exhibit slowing of nerve conduction velocities (NCV) within weeks of the onset of hyperglycemia, and it has been argued that this disorder reflects early stages of human diabetic neuropathy. In the absence of the obvious axonal loss, nodal widening or segmental demyelination seen in human diabetic neuropathy, the early NCV slowing of diabetic rodents has been attributed to initial neurochemical disorders with a later contribution from developing axonal atrophy and other ultrastructural changes. While NCV slowing in diabetic humans and rodents may not necessarily have the same etiology, the disorder has been studied extensively to identify pathogenic mechanisms and screen potential therapeutic agents. Other physiologic and neurochemical disorders of peripheral nerve that develop within weeks of the onset of hyperglycemia in diabetic rodents include increased glucose metabolism by enzymes of the ▶ polyol pathway impaired mitochondrial function, increased ▶ oxidative stress, reduced nerve blood flow and oxygen tensions, resistance to ischemic conduction block, and decreased slow ▶ axonal transport. Discovery of these disorders has inspired many hypotheses regarding the pathogenesis of diabetic neuropathy and prompted the development of a number of therapeutic strategies, although none as yet have translated successfully to clinical use (reviewed in Calcutt et al. 2009).

While diabetic rodents are generally not good models of the myelinated fiber pathology of diabetic neuropathy, skin biopsies from diabetic rats and mice have revealed simplification and retraction of epidermal C fiber terminal processes that parallel the human condition. This structural pathology of nociceptive fibers is accompanied by altered behavioral responses to sensory stimuli, so that diabetic rodents have been widely used over the past decade to model states of sensory loss and pain. This approach carries all of the caveats associated with animal models of ▶ neuropathic pain with the additional concern that behavioral responses might also be modified by parallel effects of diabetes on motor nerve function or on non-neural systems. For example, diabetic rodents exhibit reduced locomotor activity and, when allowed to progress to a state of extreme untreated hyperglycemia, develop a general behavioral depression. Nevertheless, there is accumulating evidence to suggest that diabetic rodents display some indices of sensory dysfunction that model both the sensory loss and allodynia or hyperalgesia seen in diabetic patients.

Diabetic rats react to light touch of the plantar surface of the hind paws, indicating tactile allodynia of a magnitude similar to other models of neuropathic pain (see Calcutt 2004). This allodynia is frequently used in screening tests for agents designed to alleviate established neuropathic pain. In diabetic mice, the literature is more varied, with reports of allodynia, no change, or hypoalgesia, and it remains to be fully established whether duration of diabetes, mouse strain, or other factors contribute to inconsistencies of neuropathy phenotype between diabetic mouse models (Sullivan et al. 2008). Limb withdrawal from a more substantial mechanical stimulus is also altered in diabetic rodents, with both the tail pinch and paw pressure tests indicating an increased sensitivity, either by a faster withdrawal from a fixed pressure or a response to lower amounts of pressure. As with tactile allodynia, mechanical hyperalgesia in diabetic rodents has been widely used to evaluate the potential of drugs designed to alleviate pain, although whether the disorder has a direct correlate in the human condition is not clear.

Thermal hyperalgesia has been described in some diabetic patients, although thermal hypoalgesia associated with loss of epidermal nociceptors is a more common feature. Studies in diabetic rodents have reported both thermal hypoalgesia and hyperalgesia, and this discord may reflect variations in experimental protocols, such as differences in the species, sex, strain, limb tested, heating method, and degree or duration of diabetes. We have found that when exposing the hind paw plantar surface to a temperature ramp, rising from 30° C at a rate of 1° C per second over 20 s, STZ-diabetic rats exhibit a transient thermal hyperalgesia after 4 weeks of diabetes that progresses to thermal hypoalgesia in 2–3 months (Calcutt et al. 2004). We have also recorded progression to thermal hypoalgesia in STZ-diabetic mice, Akita mice, db/db mice, and ZDF rats (unpublished data). Concerns that the progression to hypoalgesia reflects general behavioral depression of ailing animals may be assuaged by the observation that loss of thermal sensation can be prevented by therapies that target the nervous system and do not alter the systemic indices of insulin-deficient diabetes (Calcutt et al. 2004).

Diabetes also modifies other behavioral tests developed to study mechanisms of hyperalgesia, particularly those that highlight the role of spinal nociceptive processing. The transient thermal hyperalgesia induced by direct application of substance P to the spinal cord of normal rats is prolonged in diabetic rats (Calcutt et al. 2000), while paw flinching after local formalin injection is increased during the periods associated with spinal sensitization of nociceptive processing (see Calcutt 2004). Interestingly, diabetic mice quickly develop a diminished behavioral response to formalin that may correspond to the more rapid onset of epidermal fiber loss (Johnson et al. 2007).

The description of a range of behavioral disorders that are associated with sensory dysfunction in diabetic rodents has stimulated a search for underlying pathogenic mechanisms. In both humans and rodents, the dominant phenotype of diabetic neuropathy is degenerative, and the hypothesis that the progression to thermal hypoalgesia in diabetic rats and mice and loss of formalin-evoked responses in diabetic mice reflects loss of epidermal thermal nociceptors is attractive. Impaired synthesis, transport and spinal release of sensory neuropeptides such as substance P and CGRP that is secondary to impaired ▶ neurotrophic support may also play a role in early stages of hypoalgesia. With regard to the allodynia and hyperalgesia that is frequently shown to be concurrent with thermal hypoalgesia, there is accumulating evidence that disruption of both peripheral and spinal sites of nociceptive processing may be involved. In the periphery, changes in threshold or firing response to stimuli have been reported, along with alterations to assorted voltage-gated sodium and calcium channels that could lead to membrane instability, although the literature remains divided regarding the occurrence of spontaneous activity in sensory fibers of diabetic rats. Moreover, the stimulus-evoked release of spinal excitatory neurotransmitters and spinal activity measured by fMRI are paradoxically reduced, not increased, in diabetic rats (Malmberg et al. 2006; Malisza et al. 2009). This has prompted consideration of the role of spinal cord hypersensitivity in behavioral allodynia and hyperalgesia. Increased spontaneous activity of postsynaptic neurons has been recorded in the spinal cord (Pertovaara et al. 2001) along with activation of spinal inflammatory systems (Daulhac et al. 2006; Ramos et al. 2007) and loss of spinal inhibitory functions (Jolivalt et al. 2008). The effects of diabetes on higher processing centers have been only infrequently studied to date but may include impairment of descending inhibitory systems (Paulson et al. 2007). Taken together, much of the current evidence suggests that diabetic rodents develop an early hyperalgesic state with involvement of peripheral, spinal, and higher CNS systems and that progressive damage to distal terminals can promote either progression to hypoalgesia in some modalities, such as thermal nociception or continued hyperalgesia, where central inhibitory systems are disrupted.

Summary

Diabetic rodents are probably best viewed as modeling early biochemical and physiologic disorders associated with diabetic neuropathy, along with the early retraction of epidermal C fibers. Behavioral studies have shown that they exhibit tactile allodynia, mechanical and chemical hyperalgesia, and changes in thermal discrimination that range from hyperalgesia to hypoalgesia. It is not unusual to read reports of animals that show tactile and chemical hyperalgesia with concurrent thermal hypoalgesia. Both of the drugs currently approved by the FDA to treat painful diabetic neuropathy, gabapentin and duloxetine, showed efficacy in rat models of painful diabetic neuropathy, but it remains to be established whether any of the dozens of other agents reported to alleviate tactile allodynia and thermal hyperalgesia in diabetic rodents will translate to clinical use.

Acknowledgment Supported by DK057629 (NAC).

Cross-References

▶ Wallerian Degeneration

References

Calcutt, N. A. (2004). Modeling diabetic sensory neuropathy in rats. *Methods in Molecular Medicine, 99*, 55–65.

Calcutt, N. A., Freshwater, J. D., & O'Brien, J. S. (2000). Protection of sensory function and antihyperalgesic properties of a prosaposin-derived peptide in diabetic rats. *Anesthesiology, 93*, 1271–1278.

Calcutt, N. A., Freshwater, J. D., & Mizisin, A. P. (2004). Prevention of sensory disorders in diabetic sprague-dawley rats by aldose reductase inhibition or treatment with ciliary neurotrophic factor. *Diabetologia, 47*, 718–724.

Calcutt, N. A., Cooper, M. E., Kern, T. S., & Schmidt, A. M. (2009). Therapies for hyperglycaemia-induced diabetic complications: From animal models to clinical trials. *Nature Reviews Drug Discovery, 8*, 417–429.

Daulhac, L., Mallet, C., Courteix, C., Etienne, M., Duroux, E., Privat, A.-M., Eschalier, A., & Fialip, J. (2006). Diabetes-induced mechanical hyperalgesia involves spinal mitogen-activated protein kinase activation in neurons and microglia via N-methyl-D-aspartate-dependent mechanisms. *Molecular Pharmacology, 70*, 1246–1254.

Davidson, E., Coppey, L., Lu, B., Arballo, V., Calcutt, N. A., Gerard, C., & Yorek, M. (2009). The roles of streptozotocin neurotoxicity and neutral endopeptidase in murine experimental diabetic neuropathy. *Experimental Diabetes Research, 2009*, 431980.

Johnson, M. S., Ryals, J. M., & Wright, D. E. (2007). Diabetes-induced chemogenic hypoalgesia is paralleled by attenuated stimulus-induced fos expression in the spinal cord of diabetic mice. *The Journal of Pain, 8*, 637–649.

Jolivalt, C. G., Lee, C. A., Ramos, K. M., & Calcutt, N. A. (2008). Allodynia and hyperalgesia in diabetic rats are mediated by GABA and depletion of spinal potassium-chloride co-transporters. *Pain, 140*, 48–57.

Malisza, K. L., Jones, C., Gruwel, M. L. H., Foreman, D., Fernyhough, P., & Calcutt, N. A. (2009). Functional magnetic resonance imaging of the spinal cord during sensory stimulation in diabetic rats. *Journal of Magnetic Resonance Imaging, 30*, 271–276.

Malmberg, A. B., O'Connor, W. T., Glennon, J. C., Cesena, R., & Calcutt, N. A. (2006). Impaired formalin-evoked changes of spinal amino acid levels in diabetic rats. *Brain Research, 1115*, 48–53.

Paulson, P. E., Wiley, J. W., & Morrow, T. J. (2007). Concurrent activation of the somatosensory forebrain and deactivation of periaqueductal gray associated with diabetes-induced neuropathic pain. *Experimental Neurology, 208*, 305–313.

Pertovaara, A., Wei, H., Kalmari, J., & Ruotsalainen, M. (2001). Pain behavior and response properties of spinal dorsal horn neurons following experimental diabetic neuropathy in the rat: Modulation by nitecapone, a COMT inhibitor with antioxidant properties. *Experimental Neurology, 167*, 425–434.

Ramos, K. M., Jiang, Y., Svensson, C. I., & Calcutt, N. A. (2007). Pathogenesis of spinally mediated hyperalgesia in diabetes. *Diabetes, 56*, 1569–1576.

Sullivan, K. A., Lentz, S. I., Roberts, J. L., & Feldman, E. L. (2008). Criteria for creating and assessing mouse models of diabetic neuropathy. *Current Drug Targets, 9*, 3–13.

Vincent, A. M., Hinder, L. M., Pop-Busui, R., & Feldman, E. L. (2009). Hyperlipidemia: A new therapeutic target for diabetic neuropathy. *Journal of the Peripheral Nervous System, 14*, 257–267.

Neuropathic Pain Model, Neuritis/Inflammatory Neuropathy

Erin D. Milligan[1], Steven F. Maier[2] and
Linda R. Watkins[2]
[1]Department of Neurosciences, University of
New Mexico, Albuquerque, NM, USA
[2]Department of Psychology and Neuroscience,
and The Center for Neuroscience, University of
Colorado at Boulder, Boulder, CO, USA

Synonyms

Sciatic inflammatory neuritis; Sciatic inflammatory neuropathy; SIN

Definition

Neuropathic pain can arise from frank nerve trauma and/or inflammation of peripheral nerves (Bourque et al. 1985; Said and Hontebeyrie-Joskowicsz 1992). In addition, pathological pain can arise from (a) tissues innervated by neighboring healthy nerves, called extraterritorial pain, and (b) healthy areas contralateral to the site of tissue damage, referred to as mirror pain. Animal models have focused on examining mechanisms underlying neuropathic pain associated with nerve damage (Bennett and Xie 1988; Seltzer et al. 1990; Kim and Chung 1992; DeLeo et al. 1994). The sciatic inflammatory neuropathy (SIN) model was recently developed to examine how inflammatory, extraterritorial, and mirror image pain are created without frank nerve trauma (Chacur et al. 2001; Gazda et al. 2001; Milligan et al. 2003; Spataro et al. 2003).

Prior procedures that create inflammation of the sciatic nerve were used as a foundation to develop the SIN model (Eliav et al. 1999; Eliav et al. 2001). Modifications were made of these prior procedures in order to allow the full-time course of ipsilateral and mirror image allodynias to be productively studied, even at their earliest onset, after induction of peri-sciatic immune activation. Further, these modifications allow peri-sciatic immune activation to be induced in freely moving rats without the use of anesthetics, as these drugs can grossly alter immune responses (Lockwood et al. 1993; Sato et al. 1995). Lastly, SIN can be used to examine the responses of peri-sciatic immune cells and spinal cord neurochemical function underlying acute (less than 24 h) and extended (up to 14 days) allodynias. This entry will describe detailed surgical methods for chronic peri-sciatic catheterization (Milligan et al. 2004) and a brief protocol for induction of inflammation to create SIN.

Characteristics

Tools and Supplies for Preparing Peri-sciatic Catheters, Surgery, Catheter Injectors, and Catheter Cleaners

Peri-sciatic catheters are constructed from:

1. Sterile Gelfoam sheets (approximately 2 × 6 cm; NDC#0009-0315-03; Upjohn, Kalamazoo, MI). Each sheet will produce 4 Gelfoam pieces.
2. Silastic tubing (1.57-mm inner diameter, 2.41-mm outer diameter; Helix Medical Inc., Carpinteria, CA).
3. Sterile silk suture (4–0 and 3–0) with attached needle (cutting FS-2 and FS-1, respectively; Ethicon, Somerville, NJ).
4. Sterile polyethylene tubing (PE-50, Becton Dickinson, Sparks, MD).
5. #11 scalpel blade.
6. Sterile metal metric ruler (15 cm).

Tools used for chronic peri-sciatic catheter surgery are:

1. 2 pairs of micro forceps
2. 1 pair toothed forceps
3. 1 pair blunt dissection scissors
4. Suture hemostats with scissors
5. #11 or #10 scalpel blades
6. 1 scalpel handle
7. 2 small towel clips
8. 4–8 glass Pasteur pipettes with the tips previously melted into small hooks
9. 1 hot glass-bead sterilizer (World Precision Instruments, Sarasota, FL) to sterilize tools between use

10. 1 shaver
11. 70 % alcohol
12. Exidine-2 surgical scrub solution (undiluted, Baxter Healthcare, Deerfield, IL)
13. Sterile gauze (4 × 4 in.) to create a drape around the surgical site
14. ~800 ml of both an antibacterial water mixture and water for cleansing hands between surgeries
15. Sterile autoclave paper (12 × 12 cm) to provide a clean surface for placing sterilized tools
16. 1 sleeve/rat that protects the exteriorized portion of the peri-sciatic catheter (cc-sleeve) described in detail previously (Milligan et al. 1999)

The catheter injector is made from:

1. A sterile 23-gauge, 1-in. hypodermic needle (Becton Dickinson, Franklin Lakes, NJ)
2. PE-50 tubing (Becton Dickinson, Sparks, MD)
3. An autoclaved 100-ul Hamilton glass syringe (Fisher Scientific, Houston, TX)
4. A black permanent fine-tip marker
5. Sterile metal metric ruler (15 cm)
6. Dental probe with a 45° angle (microprobe; Fisher Scientific, Houston, TX)

The catheter cleaners are made from:

1. A sterile 3-cc syringe
2. A 23-gauge needle
3. PE-50 tubing (Becton Dickinson, Sparks, MD)

Chronic Peri-sciatic Catheter Construction, Surgery, Injection, and Cleaning Procedures

The procedures detailed here are for:

- The construction of supplies required for chronic peri-sciatic catheters
- Chronic peri-sciatic catheter surgeries
- Cleaning and injecting drugs around the sciatic nerve

Importantly, all procedures are conducted aseptically. All instruments (forceps, scissors, and scalpels) used to handle or make the supplies are sterile. It is imperative that the catheters for these surgeries are endotoxin-free as well as sterile since (a) endotoxin is not destroyed by autoclaving or gas sterilization and (b) endotoxin and bacterial contamination activates immune cells.

All instruments are sterilized prior to conducting surgery on each animal using a glass-bead mini sterilizer. Hands are washed with antibacterial soap between animals. The skin surrounding the open wound is draped with sterilized gauze or autoclave paper.

Constructing Chronic Peri-sciatic Catheters

All of the following steps should be performed under a sterile cell-culture grade hood:

Step 1: A 7-cm sterile silastic tube is threaded with 10-cm PE-50 tubing (Fig. 1a). One end of the PE-50 tubing remains flush with one end of the silastic tubing. This 10-cm PE-50 tubing serves as a placeholder, to ensure that the silastic tube does not become blocked either by silk sutures used for making the silastic tube + Gelfoam assembly or by the surgical anchoring step (described below).

Step 2: Sterile 4–0 silk suture with needle attached is hooked through a small portion of the silastic tube 3–4 mm from the end, where the PE-50 placeholder is flush with the silastic tube (Fig. 1a).

Use a sterile metal ruler for accuracy. It is critical that the 4–0 silk suture does not pierce the indwelling PE-50 placeholder tubing. The final length of the silk suture is cut to approximately 6 cm, which is needed to easily tie Gelfoam (described below) to the silastic + PE-50 assembly, as well as to tie the Gelfoam ends together after enwrapping the sciatic nerve (described below). Store this unit in a sterile 50-ml conical tube (Fisher Scientific, Houston, TX) until needed.

Step 3: The silastic + PE-50 assemblies are stored in a sterile, dry 50-ml conical tube that is tightly capped until the prepared Gelfoam is ready to be attached. The peri-sciatic Gelfoam is constructed from sterile Gelfoam sheets (6 (L) × 0.5 (W) × 2 (H) cm) cut, using a sterile #11 scalpel blade on a sterile metal metric ruler, into 4 equal parts with approximate dimensions of 3 (L) × 0.5 (W) × 1 (H) cm strips. Use a sterile metal ruler for accuracy. One end of the Gelfoam strip is bisected (0.25 cm W) to a depth of 1.5 cm, creating 2 Gelfoam flaps.

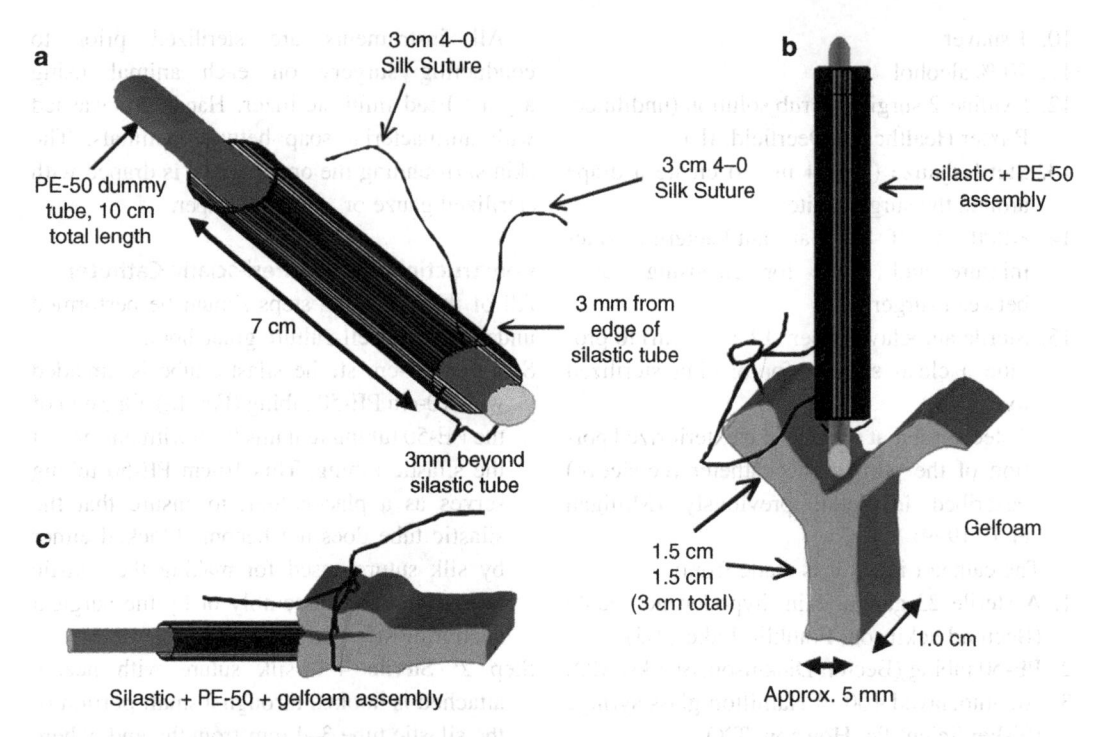

Neuropathic Pain Model, Neuritis/Inflammatory Neuropathy, Fig. 1 (a) The silastic + Pe-50 Gelfoam assembly. This is the assembly immediately prior to attaching it to the Gelfoam. (b) Attachment of the silastic + PE-50 Gelfoam assembly to the Gelfoam. Note that this is the size of the Gelfoam after it has been cut from the Gelfoam sheet. (c) Attachment of the silastic + PE-50 Gelfoam assembly to the Gelfoam. Note that this is the size of the Gelfoam after it has been cut from the Gelfoam sheet. Reprinted with permission (Milligan et al. in press)

Step 4: The 4–0 silk suture end of the silastic + PE-50 assembly is inserted between the Gelfoam flaps to a depth of approximately 1.5 cm (Fig. 1b).

The Gelfoam flaps are tightly closed together around the silastic + PE-50 assembly by tying the 4–0 silk suture around one Gelfoam flap and again around the opposing flap (Fig. 1c).

The silastic + PE-50 + Gelfoam assembly is stored in a sterile, dry 50-ml conical tube (which can store up to 6 silastic + PE-50 + Gelfoam assemblies) until the time of surgery.

Chronic Peri-sciatic Catheter Surgery

Step 1: Surgery is conducted under isoflurane anesthesia (Halocarbon Laboratories, River Edge, NJ), 2.5 vol.% in oxygen, which is chosen because it has minimal effects on immune cell function compared to other commonly used anesthetics (Lockwood et al. 1993; Sato et al. 1995).

Step 2: The dorsal aspect of the rat faces up, with the nose facing away from the surgeon; the left and right hind legs are positioned laterally to the left of the surgeon or splayed laterally to the left and right of the surgeon. The fur is shaved from the left leg and lower back.

Step 3: The exposed skin is cleaned with 70 % alcohol-soaked gauze. A small amount of concentrated Exidine-2 surgical scrub solution applied to fresh gauze is then used to further clean the surgical area.

Step 4: A midline incision at the lower back area is then made with either a #10 or #11 surgical scalpel blade, and the skin around the cut is separated from its underlying loose connective tissue using blunt dissection scissors. This is done to provide space for subcutaneous implantation of the cc-sleeve (described below).

Step 5: A second incision is made along the lateral aspect of the left thigh and separation of the skin from the underlying connective tissue extends beyond the incision site to the midline incision (in Step 4), where the exterior portion of the catheter will be encased by the cc-sleeve.

Step 6: The shaved and cleaned skin surrounding the wound is lightly retracted using small-toothed towel clips and then draped with sterile gauze.

Step 7: Exposure of the sciatic nerve is achieved by blunt dissection, and the connective tissue is gently teased apart using glass Pasteur pipette hooks, sterilized in the glass-bead mini sterilizer before each use.

Step 8: At approximately mid-thigh level, a portion of the sciatic nerve is slightly lifted using a sterile glass Pasteur pipette hook, followed by adding a couple of drops of sterile isotonic, endotoxin-free saline to keep the nerve moist.

Step 9: The Gelfoam of the silastic + PE-50 + Gelfoam assembly is then gently threaded around the sciatic nerve starting from the quadriceps side to maintain a clear view of the implant site (Fig. 2a). The surrounding muscle walls are maneuvered to support the silastic + PE-50 + Gelfoam assembly upright.

Step 10: The 4–0 silk suture that is part of the silastic + PE-50 + Gelfoam assembly (as previously described in Chronic Peri-Sciatic Catheter Construction) is used to tie together the proximal and distal ends of the Gelfoam once it forms a U shape around the sciatic nerve. Precise control is best when using 2 pairs of micro forceps. The surrounding muscle walls are then closed around the Gelfoam-enwrapped sciatic nerve, leaving the silastic + PE-50 portion exteriorized (Fig. 2b).

Step 11: The silastic tube is anchored to the muscle by threading a sterile 4–0 silk suture with attached suture needle through the muscle at the most proximal portion of the dissection site, followed by threading the suture through the silastic tube, avoiding the internal PE-50 placeholder tube, and then through the opposing muscle. The remaining overlying muscle is closed with one or two more sutures through the muscle.

Step 12: The exposed portion of the silastic catheter is tunneled subcutaneously to exit through the lower back incision (Fig. 2c).

Step 13: The skin overlaying the sutured muscle is closed with wound clips.

Step 14: The PE-50 dummy catheter is carefully removed while holding the silastic catheter in place, to ensure that the Gelfoam does not become torn or displaced.

Step 15: The exposed portion of the silastic catheter is threaded through the cc-sleeve. The reader is referred to the methods paper that describes in detail the construction and use of this cc-sleeve (Milligan et al. 1999). The cc-sleeve is then anchored to the muscle overlaying the lumbosacral area, by threading one or two 3–0 silk sutures with attached suture needle through each flange on the cc-sleeve. The overlying skin is then sutured closed with 3–0 silk.

Step 16: The remaining exteriorized portion of the silastic tube is folded into the cc-sleeve and an air-dried concave plug (part of the cc-sleeve) (Milligan et al. 1999) is inserted inside the tip of the sleeve. A small amount of silastic silicon sealant is coated over the end of the plug and cc-sleeve with a moistened Q-tip.

Step 17: The wound areas around the hind leg and lower back are lightly cleaned with 0.9 % saline. Total surgical time is typically 15–20 min.

Step 18: Beginning 4–5 days following surgery, when catheters are used to induce chronic allodynia for extended periods of time, the wound areas of the hind leg and lower back are cleaned with 0.9 % saline every 2 days for as long as 2 weeks. This decreases the amount of inflammation, such as redness, slight bleeding, and scabbing of the skin around the surgical sites. The cc-sleeve and the indwelling peri-sciatic silastic catheter are also

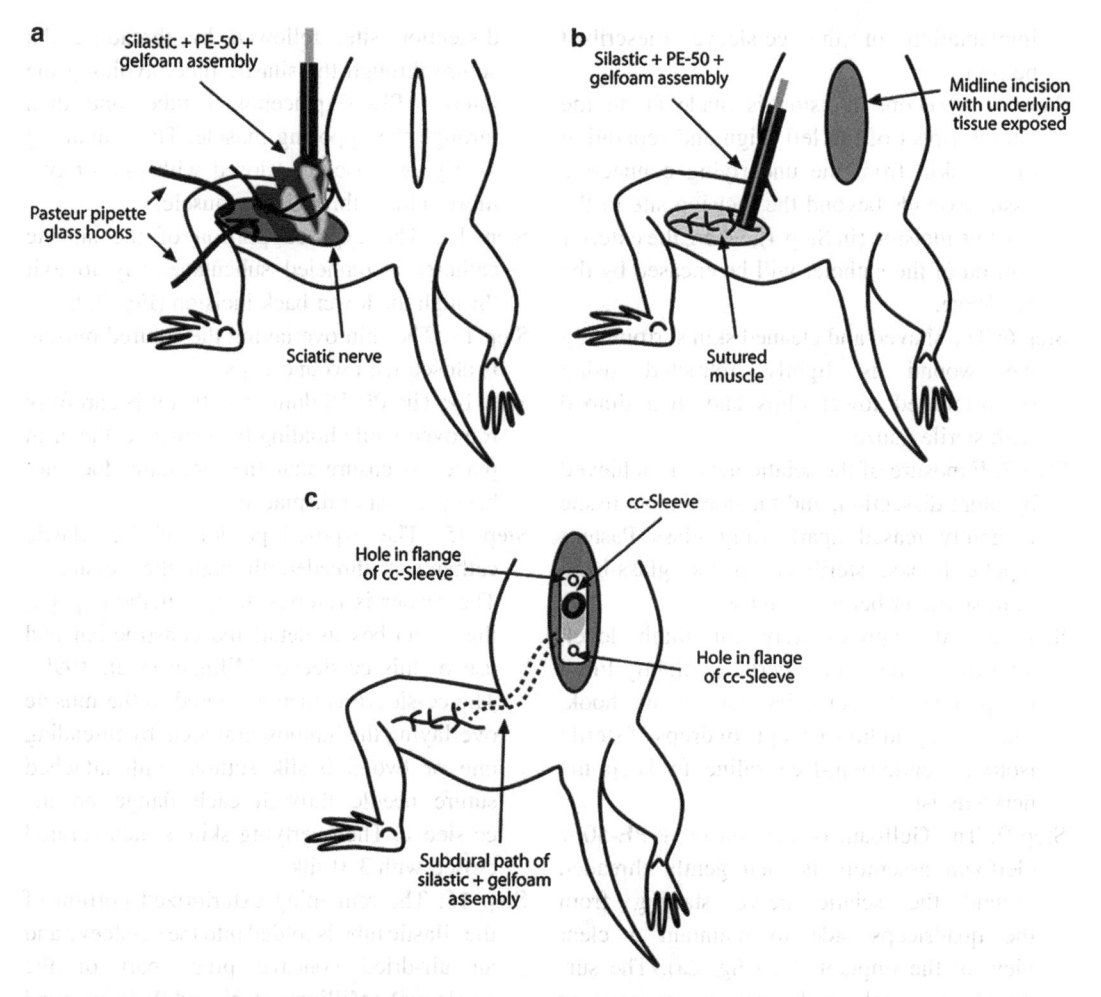

Neuropathic Pain Model, Neuritis/Inflammatory Neuropathy, Fig. 2 (a) The process of enwrapping the Gelfoam portion of silastic + PE-50 + Gelfoam assembly around the sciatic nerve. (b) Illustrating the exteriorized portion of the silastic + PE-50 catheter after the Gelfoam has been implanted and muscle walls sutured closed. (c) cc-Sleeve attachment to the lower back area after the silastic + PE-50 + Gelfoam assembly has been implanted. Reprinted with permission (Milligan et al. in press)

cleaned with separate single-use peri-sciatic catheter cleaners.

Peri-sciatic Catheter Injectors

Peri-sciatic catheter injectors are made using the following steps:

Step 1: The beveled end of a sterile 23-gauge, 1-in. hypodermic needle is inserted into one end of a 30-cm PE-50 tube.

Step 2: A mark is made 7.3 cm from the opposite end, using a black permanent fine-tip marker. The mark must line up with the exterior end of the peri-sciatic silastic catheter upon PE-50 tubing insertion. This alignment assures that the interior end of the PE-50 tubing is 3 mm beyond the indwelling silastic catheter within the Gelfoam.

Step 3: Prepared peri-sciatic catheter injectors are stored in a sterile, dry place (typically, an autoclavable box) until the time of injections.

Peri-sciatic Catheter Injection Procedures

Peri-sciatic catheter injections are conducted using the following steps:

Step 1: The 23-gauge needle is attached to a sterile Hamilton 100-ul microsyringe.

Step 2: The sterile glass Hamilton microsyringe and the peri-sciatic catheter injector are flushed with sterile, endotoxin-free water and tightly connected, making the syringe and injector airtight.

Step 3: An air bubble is then created in the 30-cm PE-50 tubing of the peri-sciatic catheter injector by drawing up 1-μl of air followed by the drug. The length of the injection catheter will vary depending on the volume of drug injection. A 1.0-um volume occupies approximately 0.41 cm of PE-50 tubing.

Step 4: Animals are gently placed in crumpled soft cotton towels and allowed to move freely underneath the towels.

Step 5: The cc-sleeve area is exposed, the indwelling concave rubber plug is removed with a dental probe, and the folded portion of the silastic catheter is exteriorized. The reader is referred to the methods paper that describes in detail the construction and use of this cc-sleeve (Milligan et al. 1999).

Step 6: Fluid that had accumulated in the indwelling silastic catheter is suctioned off with the peri-sciatic catheter cleaner (described below) and discarded.

Step 7: Drug injection is completed using the prepared PE-50 injectors (described above). The PE-50 tubing from the peri-sciatic catheter injector is inserted into the silastic catheter until the 7.3-cm mark on the PE-50 tubing of the peri-sciatic catheter injector is flush with the edge of the silastic catheter. Chronic allodynia is maintained for over 2 weeks by repeated injections every 2 days, which are conducted in steps identical to that described immediately above (steps 1–7).

Peri-sciatic Catheter Cleaning Procedures

In chronic allodynia, an additional step of cleaning the inside of the cc-sleeve with a separate peri-sciatic catheter cleaner is done, to decrease local inflammation/infection around this foreign body. Peri-sciatic catheter and cc-sleeve cleaning followed by drug injections are done every 2 days to maintain chronic allodynia. Using this paradigm, unilateral and bilateral allodynia remain stable during the entire testing period, in terms of both pattern (i.e., unilateral does not change to bilateral nor the reverse) and magnitude. Peri-sciatic catheter cleaners are used to suction out fluid accumulation within the indwelling silastic catheter starting 4–5 days after surgery and prior to drug injections.

Step 1: The catheter cleaners are made from the same supplies as the injectors, except that the Hamilton 100-μl microsyringe is replaced with a sterile 3-cc syringe.

Step 2: The cleaners are constructed in the same way as the injectors (described above) except that the 3-cc syringe is attached to the 23-gauge needle. Prepared peri-sciatic catheter cleaners are stored in a sterile, dry place (typically, an autoclavable box) until the time of injections.

References

Bennett, G. J., & Xie, Y. K. (1988). A Peripheral Mononeuropathy in Rat that Produces Disorders of Pain Sensation like those Seen in Man. *Pain, 33*, 87–107.

Bourque, C. N., Anderson, B. A., Martin del Campo, C., & Sima, A. A. (1985). Sensorimotor Perineuritis - An Autoimmune Disease? *Canadian Journal of Neurological Sciences, 12*, 129–133.

Chacur, M., Milligan, E. D., Gazda, L. S., Armstrong, C. A., Wang, H., Tracey, K. J., Maier, S. F., & Watkins, L. R. (2001). A New Model of Sciatic Inflammatory Neuritis (SIN): Induction of Unilateral and Bilateral Mechanical Allodynia Following Acute Unilateral Peri-Sciatic Immune Activation in Rats. *Pain, 94*, 231–244.

DeLeo, J. A., Coombs, D. W., Willenbring, S., Colburn, R. W., Fromm, C., Wagner, R., & Twitchell, B. B. (1994). Characterization of a Neuropathic Pain Model: Sciatic Cryoneurolysis in the Rat. *Pain, 56*, 9–16.

Eliav, E., Herzberg, U., Ruda, M. A., & Bennett, G. (1999). Neuropathic Pain from an Experimental Neuritis of the Rat Sciatic Nerve. *Pain, 83*, 169–182.

Eliav, E., Benoliel, R., & Tal, M. (2001). Inflammation with no Axonal Damage of the Rat Saphenous Nerve Trunk Induces Ectopic Discharge and Mechanosensitivity in Myelinated Axons. *Neuroscience Letters, 311*, 49–52.

Gazda, L. S., Milligan, E. D., Hansen, M. K., Twining, C. M., Poulos, N. M., Chacur, M., O'Connor, K. A., Armstrong, C. A., Maier, S. F., Watkins, L. R., & Myers, R. R. (2001). Sciatic Inflammatory Neuritis (SIN): Behavioral Allodynia is Paralleled by Peri-Sciatic Proinflammatory Cytokine and Superoxide

Production. *Journal of the Peripheral Nervous System,* 6, 111–129.

Kim, S. H., & Chung, J. M. (1992). An Experimental Model for Peripheral Neuropathy Produced by Segmental Spinal Nerve Ligation in the Rat. *Pain, 50,* 355–363.

Lockwood, L. L., Silbert, L. H., Laudenslager, M. L., Watkins, L. R., & Maier, S. F. (1993). Anesthesia-Induced Modulation of In Vivo Antibody Levels. *Anesthesia and Analgesia, 77,* 769–775.

Milligan, E. D., Hinde, J. L., Mehmert, K. K., Maier, S. F., & Watkins, L. R. (1999). A Method for Increasing the Viability of the External Portion of Lumbar Catheters Placed in the Spinal Subarachnoid Space of Rats. *Journal of Neuroscience Methods, 90,* 81–86.

Milligan, E. M., Twining, C. M., Chacur, M., Biedenkap, J., O'Connor, K. A., Poole, S., Tracey, K. J., Martin, D., Maier, S. F., & Watkins, L. R. (2003). Spinal Glia and Proinflammatory Cytokines Mediate Mirror-Image Neuropathic Pain. *Journal of Neuroscience, 23,* 1026–1040.

Milligan, E. D., Maier, S. F., & Watkins, L. R. (2004). Sciatic Inflammatory Neuropathy in the Rat. In D. Luo (Ed.), *Pain Research: Methods and Protocols.* Totowa, NJ: Humana Press (in press).

Said, G., & Hontebeyrie-Joskowicsz, M. (1992). Nerve Lesions Induced by Macrophage Activation. *Research in Immunology, 143,* 589–599.

Sato, W., Enzan, K., Masaki, Y., Kayaba, M., & Suzuki, M. (1995). The Effect of Isoflurane on the Secretion of TNF-Alpha and IL-1 Beta from LPS-Stimulated Human Peripheral Blood Monocytes. *Masui, 44,* 971–975.

Seltzer, Z., Dubner, G., & Shir, Y. (1990). A Novel Behavioral Model of Neuropathic Pain Disorders Produced in Rats by Partial Sciatic Nerve Injury. *Pain, 43,* 205–218.

Spataro, L., Sloane, E. M., Milligan, E. D., Maier, S. F., Watkins, L. R., (2003). Gap junctions mediate neuropathic pain produced by Sciatic Inflammatory Neuropathy (SIN) and chronic constriction injury. In: *Journal of Pain, 4* (suppl 1). Philadelphia: Churchill Livingstone, p. 52.

Neuropathic Pain Model, Partial Sciatic Nerve Ligation Model

Ze'ev Seltzer
Faculty of Dentistry, University of Toronto
Centre for the Study of Pain, Toronto,
ON, Canada

Synonyms

Null

Definition

The ▶ partial sciatic nerve ligation (PSL) model of neuropathic pain refers to a rodent ▶ neuropathic pain model that is produced by tightly ligating the dorsal third to half of the common sciatic nerve at the upper-thigh level (Seltzer et al. 1990).

Characteristics

Methods for Producing the PSL Model

Animals
Young adult male rats of various strains are commonly used. Like most other behavioral tests, it is helpful for rats to be acclimated for about a week in the laboratory holding facility with free access to food and water before performing any experiments. It is also convenient to keep rats in a room with a reversed light-dark cycle, allowing behavioral tests to be conducted during their active period.

Surgical Operation
Rats are deeply anesthetized with general anesthetics and placed in the prone position. Under a dissection microscope, the dorsum of the sciatic nerve is freed from surrounding connective tissues at a proximal site just distal to the point where the posterior biceps semitendinosus nerve branches off. An 8-0 silk suture is inserted into the nerve with a three eighths curved needle, trapping the dorsal third or half of the nerve. The trapped portion of the nerve dorsal to the suture is tightly ligated. The wound is sutured closed. The original developers of the model (Seltzer et al. 1990) emphasized the importance of the site of ligation, which needs to be proximal enough so that the sciatic nerve is not yet fasciculated into individual nerves. Therefore, sciatic injury is made at a site between the branch point of the posterior biceps semitendinosus and a little fat pad (which is located just distally) since the sciatic nerve becomes fasciculated into branches just distal to the fat pad (Schmalbruch 1986; Seltzer et al. 1990).

Extent of Injury and Behavioral Outcome

The number of fibers injured by this procedure varies between animals (Seltzer et al. 1990). It is possible that some fibers in the unligated portion of the nerve may undergo degeneration due to secondary events such as:

(1) disruption of the perineurium, (2) interference of local blood flow, (3) focal edema, and (4) reaction to the ligature. Therefore, variability in the number of injured fibers after partial sciatic nerve ligation comes from the fact that a variable number of fibers are being ligated and that presumably variable numbers of unligated fibers are being injured by the above secondary causes.

Operated animals normally do not show severe motor deficits, except for the two lateral toes, which are flexed (Seltzer et al. 1990). At rest, the operated rats guard their operated limb somewhat when placing the limb on the floor, yet they show excessive abnormal grooming behaviors on the operated limb, such as repeated licking of the paw. However, none of operated rats show self-mutilating behavior due to the partially deafferented hind paw. During walking, they do not show obvious limping or any other severe locomotive abnormalities.

Practically all operated rats show various behavioral signs of neuropathic pain such as ► ongoing pain, ► heat hyperalgesia, and mechanical as well as ► cold allodynia (Seltzer et al. 1990; Kim et al. 1997). Behaviors believed to represent ► mechanical allodynia can be quantified either by measuring foot withdrawal frequency to mechanical stimuli applied to the paw with von Frey filaments (Kim et al. 1997) or by determining the mechanical threshold (Seltzer et al. 1990). Heat hyperalgesia can also be measured by determining the heat threshold of the paw for foot withdrawals or duration of responses to suprathreshold heat stimuli (Seltzer et al. 1990). In addition, behaviors reflecting spontaneous pain have been shown by measuring the duration of foot withdrawals (guarding behaviors) in the absence of any obvious stimuli. Behavior thought to represent cold allodynia was assessed by measuring frequency of foot withdrawals to cold stimuli

(e.g., acetone droplet application) applied to the paw (Kim et al. 1997).

Kim et al. (1997) made a comparison between the chronic constriction injury (CCI), spinal nerve ligation (SNL), and PSL models. When magnitudes of behaviors were compared with the CCI and SNL models, the PSL model fell in between the two for mechanical allodynia and spontaneous pain, but the magnitude of cold allodynic behavior was similar in all three models. Behaviors in PSL are partially reversed by sympathectomy (Shir and Seltzer 1991; Seltzer and Shir 1991) suggesting that sympathetic abnormality is involved in producing pain behaviors.

Factors Influencing Variability

Many factors influence neuropathic pain behavior in the PSL model as in other models. As shown in the SNL model (Mogil et al. 1999a, b; Yoon et al. 1999), genetic factors influence pain behaviors in the PSL model (Shir et al. 2001). Another important factor is diet. The effect of diet on neuropathic pain behavior has been studied in detail using the PSL model (Shir et al. 1998, 2002). Although the mechanisms are not clear, a diet with high soybean content reduces the expression of neuropathic pain behaviors. Finding such factors is important not only in terms of future studies on underlying mechanisms but also for potential therapeutic implications.

Advantages and Disadvantages of the PSL Model Compared to Others

The PSL model has several advantages over other neuropathic pain models. Tight ligation models such as the PSL provide information about the timing of injury better than loose ligation models. The most attractive feature of the PSL model is that it most closely resembles the original description of causalgia patients with injuries produced by high-velocity missile impact (Mitchell 1872).

Acknowledgement This work was supported by NIH Grants NS 31860 and NS 11255.

References

Kim, K. J., Yoon, Y. W., & Chung, J. M. (1997). Comparison of three rodent neuropathic pain models. *Experimental Brain Research, 113*, 200–206.

Mitchell, S. W. (1872). *Injuries of nerves and their consequences*. Philadelphia: JB Lippincott.

Mogil, J. S., Wilson, S. G., Bon, K., et al. (1999a). Heritability of nociception I: Responses of 11 inbred mouse strains on 12 measures of nociception. *Pain, 80*, 67–82.

Mogil, J. S., Wilson, S. G., Bon, K., et al. (1999b). Heritability of nociception II. "Types" of nociception revealed by genetic correlation analysis. *Pain, 80*, 83–93.

Schmalbruch, H. (1986). Fiber composition of the rat sciatic nerve. *The Anatomical Record, 215*, 71–81.

Seltzer, Z., & Shir, Y. (1991). Sympathetically-maintained causalgiform disorders in a model for neuropathic pain: A review. *Journal of Basic and Clinical Physiology and Pharmacology, 2*, 17–61.

Seltzer, Z., Dubner, R., & Shir, Y. (1990). A novel behavioral model of neuropathic pain disorders produced in rats by partial sciatic nerve injury. *Pain, 43*, 205–218.

Shir, Y., & Seltzer, Z. (1991). Effects of sympathectomy in a model of causalgiform pain produced by partial sciatic nerve injury in rats. *Pain, 45*, 309–320.

Shir, Y., Ratner, A., Raja, S. N., et al. (1998). Neuropathic pain following partial nerve injury in rats is suppressed by dietary soy. *Neuroscience Letters, 240*, 73–76.

Shir, Y., Zeltser, R., Vatine, J. J., et al. (2001). Correlation of intact sensibility and neuropathic pain-related behaviors in eight inbred and outbred rat strains and selection lines. *Pain, 90*, 75–82.

Shir, Y., Campbell, J. N., Raja, S. N., et al. (2002). The correlation between dietary soy phytoestrogens and neuropathic pain behavior in rats after partial denervation. *Anesthesia and Analgesia, 94*, 421–426.

Yoon, Y. W., Lee, D. H., Lee, B. H., et al. (1999). Different strains and substrains of rats show different levels of neuropathic pain behaviors. *Experimental Brain Research, 129*, 167–171.

Neuropathic Pain Model, Spared Nerve Injury

Isabelle Decosterd and Marie Pertin
Pain Center, University Hospital Center and University of Lausanne, Lausanne, Switzerland

Synonyms

SNI model; Spared nerve injury model; sSNI model; Sural spared nerve injury model

Definition

The rodent spared nerve injury model (SNI) consists of the selective injury of two of the three terminal branches of the sciatic nerve (the tibial and common peroneal nerves), leaving the third branch, the sural nerve, intact (Decosterd and Woolf 2000; Bourquin et al. 2006). Rapid and robust pain hypersensitivity to mechanical and thermal external stimuli is produced in the sural nerve skin territory, similar to stimulus-evoked pain observed in clinical ▶ neuropathic pain syndromes.

Characteristics

Description of the Model

Since the original description of ▶ anesthesia dolorosa by Wall et al. (1979), several models of transection-/ligation-/constriction-related injury to peripheral nerves have been described, allowing evaluation of the response to an applied external innocuous or nociceptive stimulus (stimulus-evoked pain) (Mosconi and Kruger 1996; Seltzer et al. 1990; Kim and Chung 1992; Bennett and Xie 1988; Vos et al. 1994). Unlike complete denervation of the paw after a sciatic nerve transection, these models enable assessment of pain sensitivity of the spared intact nerve fibers.

However, in the partial sciatic nerve injury and in the chronic constriction injuries, the degree of nerve damage might be delicate to reproduce, leading to variability within and between laboratories. Developed by Kim and Chung, the tight ligature of L5 and L6 spinal nerves ostensibly leaves all L4 afferents intact, but the surgery itself may damage or produce an inflammatory reaction to the intact L4 spinal nerve. The SNI model has the advantage of a simple surgical procedure and a high degree of reproducibility within and between laboratories, for rats and mice (Decosterd and Woolf 2000; Bourquin et al. 2006). The original model consists of a complete transection of two of the three terminal branches of the sciatic nerve, i.e., the tibial and common peroneal nerves. The third

a

Mechanical allodynia

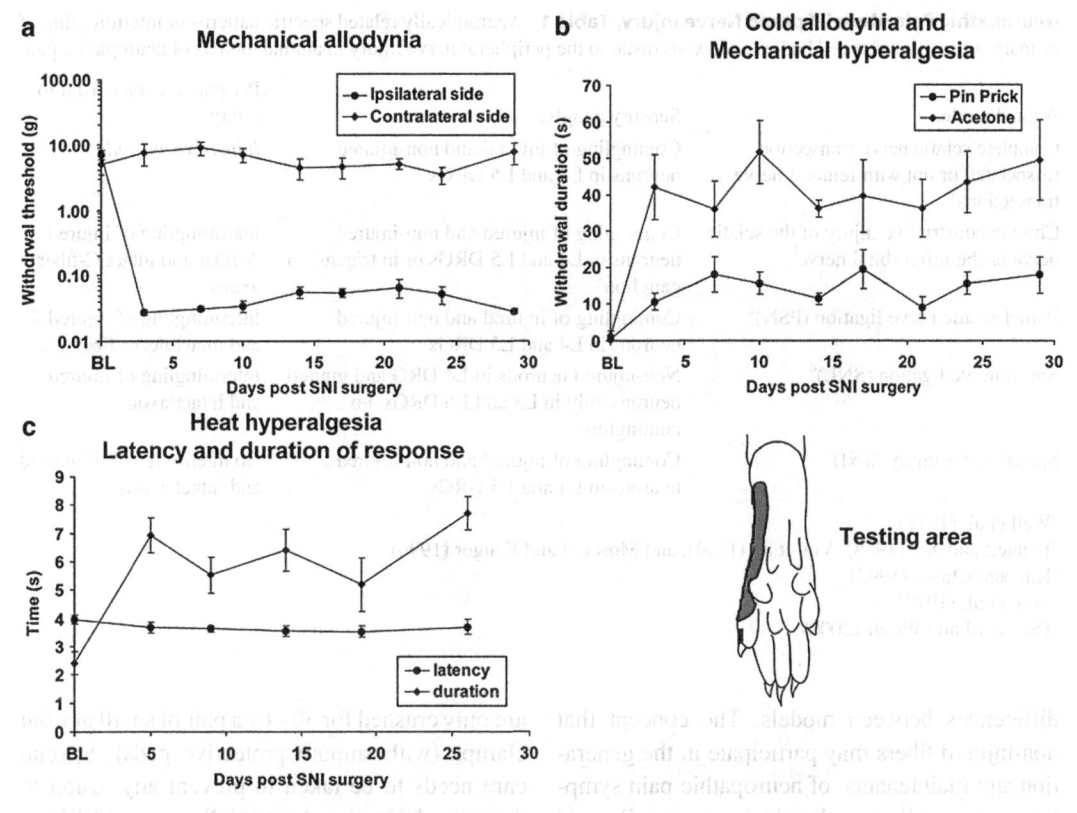

b

Cold allodynia and Mechanical hyperalgesia

c

Heat hyperalgesia Latency and duration of response

Testing area

Neuropathic Pain Model, Spared Nerve Injury, Fig. 1 SNI in rats: mechanical and thermal hypersensitivity recorded in the spared sural nerve territory. (a) Withdrawal threshold in g determined after application of ascending series of von Frey monofilaments. (b) Withdrawal duration in s after application of acetone or pin prick stimulation. (c) Heat sensitivity (withdrawal latency and duration in s). *BL* baseline. Representative series of 12 animals tested before (*BL*) and after SNI surgery. Results are displayed as the mean value+/−SEM

branch, the spared one, is the sural nerve. If the surgical procedure is well executed, tactile hypersensitivity develops within the first days after the injury. Indeed, the onset of allodynia- and hyperalgesia-like behavior is rapid (within 3 days post injury), and sensitivity changes last for months (up to 6 months). Interestingly, allodynia-like behavior is always present in various mouse strains (18 tested in the laboratories of J. Mogil and C. J. Woolf); only the severity of hypersensitivity was varying between strains (Sorge et al. 2012).

Alternatively, sparing the tibial nerve produces less robust allodynia-like behavior that resolves after 4 weeks (Shields et al. 2003; Solway et al. 2011). Indeed, SNI variant in mice has been successfully used in order to circumvent

the floor effect encountered in mechanical allodynia-like behavior with the original model.

Pain hypersensitivity is recorded in the skin territory of the spared sural nerve, preferentially on the plantar surface of the paw (Fig. 1). Withdrawal threshold modifications are also present in the dorsal hairy sural nerve skin territory and to a less extent in the saphenous nerve skin territory. In our hands, no change in the contralateral paw is observable when compared to same age sham-injured animal recordings. A crush injury to the tibial and common peroneal nerves also induces pain hypersensitivity, but within 9 weeks withdrawal thresholds return to baseline value.

Despite technical advantages or disadvantages of each model, a distinct pattern is distinguished in order to better understand mechanisms and

Neuropathic Pain Model, Spared Nerve Injury, Table 1 Anatomically related specific patterns of intermingling of primary sensory neuron cell bodies and axons distal to the peripheral nerve injury in animal models of neuropathic pain

Animal model	Sensory ganglia	Peripheral axons distal to injury
Complete sciatic nerve transection (associated or not with femoral nerve transection)[a]	Comingling of injured and non-injured neurons in L4 and L5 DRGs	*Injured axons only*
Chronic constrictive injury of the sciatic nerve or the infraorbital nerve[b]	Comingling of injured and non-injured neurons in L4 and L5 DRGs or in trigeminal ganglion	Intermingling of injured A-fiber and intact C-fiber axons
Partial sciatic nerve ligation (PSN)[c]	Comingling of injured and non-injured neurons in L4 and L5 DRGs	Intermingling of injured and intact nerve fibers
Spinal nerve ligation (SNL)[d]	Non-injured neurons in L4 DRG and injured neurons only in L5 and L6 DRGs, no comingling	Intermingling of injured and intact axons
Spared nerve injury (SNI)[e]	Comingling of injured and non-injured neurons in L4 and L5 DRGs	No intermingling of injured and intact axons

[a]Wall et al. (1979)
[b]Bennett and Xie (1988), Vos et al. (1994), and Mosconi and Kruger (1996)
[c]Kim and Chung (1992)
[d]Wall et al. (1979)
[e]Decosterd and Woolf (2000)

differences between models. The concept that non-injured fibers may participate in the generation and maintenance of neuropathic pain symptoms, as well as chemical cross talk and nonneuronal cell signaling, makes it important to differentiate models in relation to the comingling of injured primary sensory neurons/afferent fibers and uninjured neurons/fibers (see Table 1).

Surgery

Under general anesthesia, the sciatic nerve portion at thigh level is exposed using a longitudinal section through the biceps femoris muscle. The three terminal branches are easily located. Common peroneal and tibial nerves are delicately dissected in order to separate them from the surrounding tissue. The tip of a fine curved forceps is placed under the common peroneal nerve, avoiding any lift up. In rats, 3 cm of five silk suture is placed into the forceps' tip and slipped under the nerve. Two tight knots are made and the nerve is distally transected, including the removal of a portion of 2–4 mm. The same procedure is repeated for the tibial nerve (further details are described in Pertin et al. 2012). Crush injury is performed as above, except that injured nerves

are only crushed for 30 s by a pair of small arterial clamps (with smooth protective pads). Special care needs to be taken to prevent any lesion to the spared nerve and especially to avoid lifting up, touching, or stretching the spared sural nerve during the surgical procedure.

In mice, the surgical procedure is very similar (for detailed procedure, see Pertin et al. 2012) except:

– The procedure is performed under a binocular microscope.
– The suture is 6.0 silk.
– The tibial and common peroneal nerves are ligated and cut together, i.e., the thread is placed around the two nerves.
– A smaller portion of the nerve (1 mm) is discarded.

Experimental Conditions

In addition to the genetic and sex differences between animals, investigator and environmental factors influence the course of pain behavioral studies (Chesler et al. 2002; Bourquin et al. 2006; Mogil 2012). Like other tests performed in laboratory animals and based on behavioral assessment, efforts are needed to standardize study conditions when using the SNI model.

We try in our laboratory as much as possible to minimize the variability, and here are our standard precautions:

- Studies are performed with animals of the same sex and age. Young adult animals are more docile and easier to habituate. The same vendor and vendor's strain is recommended for each study.
- The same investigator conducts a specific study, and he or she must be blind to the treatment/genotype applied.
- Recordings are performed in a room devoted only to behavioral testing. No other investigator or other person than the investigator enters the behavioral room during the whole period of experiments.
- The animals are housed and tested in a room of constant temperature and light cycle, and they have free access to food (same diet) and water. Animal transportation is avoided immediately before the testing session.
- The behavioral testing is always performed at the same time of the day; the same order of testing is respected in between testing sessions.
- Behavioral assessment of treated and control animals is performed during the same testing session in order to maintain the same experimental environment in both groups.

Assessment of Tactile and Thermal Sensitivity
Several modalities are tested, all in the lateral area of the hind paw (Fig. 1). The animals are daily habituated to the environment, testing material, and investigator 2 weeks before the first recordings. For plantar application, animals are enclosed in a transparent Plexiglas observation chamber ($22 \times 13 \times 13$ cm for rats; $7 \times 7 \times 5$ cm for mice) atop a wire mesh floor (mesh of 0.36 cm^2 for rats, 0.03 cm^2 for mice). For dorsal application, rats are placed on a plane neutral floor. The investigator gently holds the animal and has direct access to the dorsal side of the hind paw.

During a study, we perform on the first day mechanical and cold sensitivities consecutively, with a 30-min time interval in between modalities. Heat assay is performed the second day.

We always keep a day off after heat assays in order to preserve the integrity of the skin.

▶ Mechanical allodynia-like behavior: The classical method uses von Frey monofilaments applied perpendicularly to the skin (Fig. 1a and Decosterd and Woolf 2000 for rats; Fig. 2 and Suter et al. 2012 for mice). A series of monofilaments are applied in ascending force order until a brisk withdrawal response or up and down depending on the response. Several outcomes are used, including:

- Frequency of response to five (rats) or ten (mice) stimuli for each monofilament
- Response threshold to one out of five (rat) or one out of ten (mice) stimulations
- Fifty percent withdrawal threshold (Chaplan et al. 1994) using the up and down method

New electronic or automated tools have been developed and have proven good accuracy (see, for instance, Liu et al. 2012 and Casals-Días et al. 2009 for rats and Sorge et al. 2012 for mice).

▶ Cold allodynia-/hyperalgesia-like behavior: The acetone test is the standard method for cold allodynia. A drop of acetone delicately applied on the skin evaporates and produces a sensation of cold. In the affected side, in rats, acetone induces long-lasting paw withdrawal as well as paw shaking and licking (Fig. 1b and Decosterd and Woolf 2000). Cold plate or thermal place preference systems allow thermal discrimination between innocuous and nociceptive temperatures and permit interesting experimental paradigms (Allchorne et al. 2005; Walczak and Beaulieu 2006; Moqrich et al. 2005; Mishra et al. 2011).

▶ Mechanical hyperalgesia-like behavior: In rats, the simplest method includes stimulation with a safety pin (pinprick test) at a force such that the skin dimpled but is not penetrated. The duration of paw withdrawal is recorded. Response duration is increased for the injured paw, but to a lesser extent than after acetone stimulation (Fig. 1b and Decosterd and Woolf 2000). Two other semiautomated analgesimeters are described: the digital paw pressure Randall-Selitto instrument and the rodent pincher (Luis-Delgado et al. 2006).

▶ Heat hyperalgesia-like behavior: A movable radiant infrared heat source enables

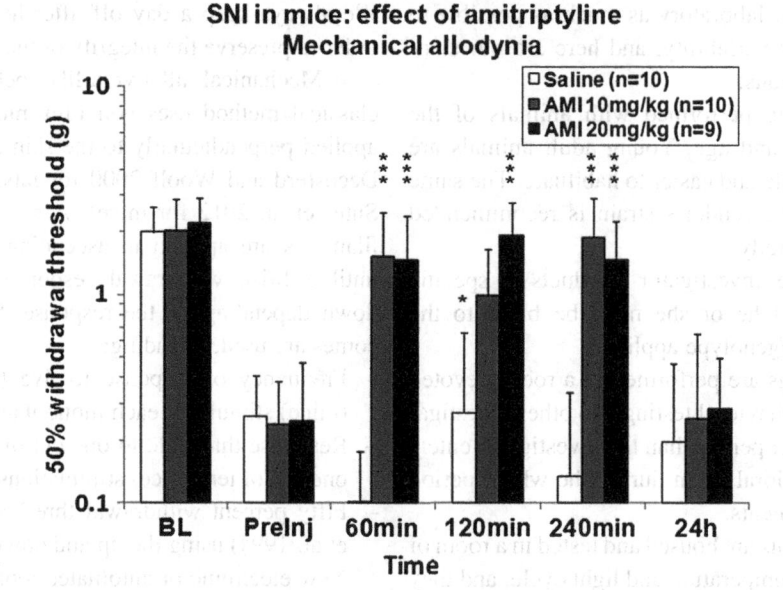

Neuropathic Pain Model, Spared Nerve Injury, Fig. 2 SNI in mice: effect of amitriptyline. SNI induces a large reduction of withdrawal threshold in response to the application of a series of von Frey monofilaments (up and down method). Here, amitriptyline alleviates neuropathic behavior following SNI from 60 to 240 min after injection and lost its effect at 24 h (*$P < 0.05$, **$P < 0.01$, ***$P < 0.001$ vs. pre-injection). *AMI* amitriptyline, *BL* baseline, *DMSO* dimethyl sulfoxide, *Preinj* pre-injection 2 weeks after SNI, *SNI* spared nerve injury. Values are presented as mean ± SD (From Suter et al. 2012)

stimulation of the lateral part of the hind paw (Hargreaves et al. 1988). Withdrawal reflex latency and duration due to heat stimulation are recorded (Fig. 1c and Decosterd and Woolf 2000). Precision of plantar heat stimulation through the transparent floor is difficult after SNI because the animals refrained from weight bearing on the affected paw, increasing the variability in the transmission of the thermal stimulus. Unilateral positioning on a hot plate, a thermal place preference system, and more recently a temperature gradient assay are other alternatives (Moqrich et al. 2005; Mishra et al. 2011).

In summary, the spared sural nerve SNI model is an easily reproducible model of neuropathic pain. Responses to mechanical and thermal stimuli are reasonably and accurately assessable in standard routine procedures for the investigation on mechanisms of neuropathic pain responsible for stimulus-evoked pain hypersensitivity.

References

Allchorne, A. J., Broom, D. C., Woolf, C. J., & Woolf, C. J. (2005). Detection of cold pain, cold allodynia and cold hyperalgesia in freely behaving rats. *Molecular Pain, 1*(1), 36.

Bennett, G. J., & Xie, Y. K. (1988). A peripheral mononeuropathy in rat that produces disorders of pain sensation like those seen in man. *Pain, 33*(1), 87–107.

Bourquin, A. F., Suveges, M., Pertin, M., Gilliard, N., Sardy, S., Davison, A. C., & Decosterd, I. (2006). Assessment and analysis of mechanical allodynia-like behavior induced by spared nerve injury (SNI) in the mouse. *Pain, 122*(1–2), 14.e11–14.e14.

Casals-Diaz, L., Vivo, M., & Navarro, X. (2009). Nociceptive responses and spinal plastic changes of afferent C-fibers in three neuropathic pain models induced by sciatic nerve injury in the rat. *Experimental Neurology, 217*(1), 84–95. Research Support, Non-U.S. Gov't.

Chaplan, S. R., Bach, F. W., Pogrel, J. W., Chung, J. M., & Yaksh, T. L. (1994). Quantitative assessment of tactile allodynia in the rat paw. *Journal of Neuroscience Methods, 53*(1), 55–63.

Chesler, E. J., Wilson, S. G., Lariviere, W. R., Rodriguez-Zas, S. L., & Mogil, J. S. (2002). Influences of laboratory environment on behavior. *Nature Neuroscience, 5*(11), 1101–1102.

Decosterd, I., & Woolf, C. J. (2000). Spared nerve injury: An animal model of persistent peripheral neuropathic pain. *Pain, 87*(2), 149–158.

Hargreaves, K., Dubner, R., Brown, F., Flores, C., & Joris, J. (1988). A new and sensitive method for measuring thermal nociception in cutaneous hyperalgesia. *Pain, 32*(1), 77–88.

Kim, S. H., & Chung, J. M. (1992). An experimental model for peripheral neuropathy produced by segmental spinal nerve ligation in the rat. *Pain, 50*(3), 355–363.

Liu, M., Zhou, L., Chen, Z., & Hu, C. (2012). Analgesic effect of iridoid glycosides from Paederia scandens (LOUR.) MERRILL (Rubiaceae) on spared nerve injury rat model of neuropathic pain. *Pharmacology Biochemistry and Behavior, 102*(3), 465–470. Research Support, Non-U.S. Gov't.

Luis-Delgado, O. E., Barrot, M., Rodeau, J. L., Schott, G., Benbouzid, M., Poisbeau, P., & Lasbennes, F. (2006). Calibrated forceps: A sensitive and reliable tool for pain and analgesia studies. *The Journal of Pain, 7*(1), 32–39.

Mishra, S. K., Tisel, S. M., Orestes, P., Bhangoo, S. K., & Hoon, M. A. (2011). TRPV1-lineage neurons are required for thermal sensation. *EMBO Journal, 30*(3), 582–593. Research Support, N.I.H., Intramural.

Mogil, J. S. (2012). Sex differences in pain and pain inhibition: Multiple explanations of a controversial phenomenon. *Nature Reviews Neuroscience, 13*(12), 859–866.

Moqrich, A., Hwang, S. W., Earley, T. J., Petrus, M. J., Murray, A. N., Spencer, K. S., & Patapoutian, A. (2005). Impaired thermosensation in mice lacking TRPV3, a heat and camphor sensor in the skin. *Science, 307*(5714), 1468–1472.

Mosconi, T., & Kruger, L. (1996). Fixed-diameter polyethylene cuffs applied to the rat sciatic nerve induce a painful neuropathy – ultrastructural morphometric analysis of axonal alterations. *Pain, 64*(1), 37–57.

Pertin, M., Gosselin, R. D., & Decosterd, I. (2012). The spared nerve injury model of neuropathic pain. *Methods in Molecular Biology, 851*, 205–212.

Seltzer, Z., Dubner, R., & Shir, Y. (1990). A novel behavioral model of neuropathic pain disorders produced in rats by partial sciatic nerve injury. *Pain, 43*(2), 205–218.

Shields, S. D., Eckert, W. A., III, & Basbaum, A. I. (2003). Spared nerve injury model of neuropathic pain in the mouse: A behavioral and anatomic analysis. *Journal of Pain, 4*(8), 465–470.

Solway, B., Bose, S. C., Corder, G., Donahue, R. R., & Taylor, B. K. (2011). Tonic inhibition of chronic pain by neuropeptide Y. *Proceedings of the National Academy of Sciences of the United States of America, 108*(17), 7224–7229. Research Support, N.I.H., Extramural.

Sorge, R. E., Trang, T., Dorfman, R., Smith, S. B., Beggs, S., Ritchie, J., & Mogil, J. S. (2012). Genetically determined P2X7 receptor pore formation regulates variability in chronic pain sensitivity. *Nature Medicine, 18*(4), 595–599. doi:10.1038/nm.2710.

Suter, M., Kirschmann, G., Laedermann, C. J., Abriel, H., & Decosterd, I. (2012). Rufinamide attenuates mechanical allodynia in a model of neuropathic pain in the mouse and stabilizes voltage-gated sodium channel inactivated state. *Anesthesiology* (2013).

Vos, B. P., Strassman, A. M., & Maciewicz, R. J. (1994). Behavioral evidence of trigeminal neuropathic pain following chronic constriction injury to the rat's infraorbital nerve. *Journal of Neuroscience, 14*(5 Pt 1), 2708–2723.

Walczak, J. S., & Beaulieu, P. (2006). Comparison of three models of neuropathic pain in mice using a new method to assess cold allodynia: The double plate technique. *Neuroscience Letters, 399*(3), 240–244. Comparative Study.

Wall, P. D., Devor, M., Inbal, R., Scadding, J. W., Schonfeld, D., Seltzer, Z., & Tomkiewicz, M. M. (1979). Autotomy following peripheral nerve lesions: Experimental anaesthesia dolorosa. *Pain, 7*(2), 103–111.

Neuropathic Pain Model, Spinal Nerve Ligation Model

Kyungsoon Chung[1] and Jin Mo Chung[2]
[1]Department of Neuroscience and Cell Biology, University of Texas Medical Branch, Galveston, TX, USA
[2]Department of Neuroscience and Cell Biology, University Chair in Neuroscience, University of Texas Medical Branch, Galveston, TX, USA

Definition

The spinal nerve ligation (SNL) model of ► neuropathic pain refers to a rodent neuropathic pain model that is produced by tightly ligating the lumbar segmental spinal nerve (L5 alone or both L5 and L6) (Kim and Chung 1992). The lumbar segmental spinal nerve refers to a short segment of the peripheral nerve distal to the dorsal root ganglion before it joins with other segmental nerves to form the lumbar plexus. The lumbar segmental spinal nerve divides into a small dorsal ramus and a large ventral ramus, and the SNL model usually ligates the ventral ramus only since the dorsal ramus is removed with paraspinal muscles during the surgery.

Characteristics

Methods to Produce the SNL Model
Animals
Most commonly, young adult male rats of various strains are used (see Factors influencing variability). After purchase, rats are normally acclimated for about a week in the Institutional Animal Care Center, with free access to food and water in a room with a reversed light-dark cycle (dark: 8 A.M.–8 P.M.; light: 8 P.M.–8 A.M.) before experimental manipulation.

Surgical Operation
Figure 1 shows the anatomy of the lumbosacral paraspinal region. Rats are anesthetized with either inhalation gas or intraperitoneal injection of sodium pentobarbital and placed in the prone position. Under sterile conditions, a longitudinal incision is made at the lower lumbar/upper sacral level to expose the paraspinal muscles on the left. Using a pair of small scissors with blunt tips, the paraspinal muscles are isolated and removed from the level of the L5 spinous process to the S1. This opens up the space dorsolateral to the articular processes, dorsal to the L6 transverse process, and medial to the ileum. Connective tissues and remaining muscles are removed with a small scraper. Under a dissecting microscope, the L6 transverse process, which covers the ventral rami of the L4 and L5 spinal nerves, is removed using a small rongeur. Access to the L5 spinal nerve is easier when the transverse process is removed very close to the body of the vertebrae. One can normally visualize the ventral rami of the L4 and L5 spinal nerves (a thin sheet of connective tissue may cover them in some animals) once the L6 transverse process is carefully removed. The L4 spinal nerve usually runs more laterally (or ventrally in some animals) than the L5, and these two nerves join distally in a common epineurial sheet; however, there is a great deal of individual variability where these two nerves join. Thus, the L4 and L5 spinal

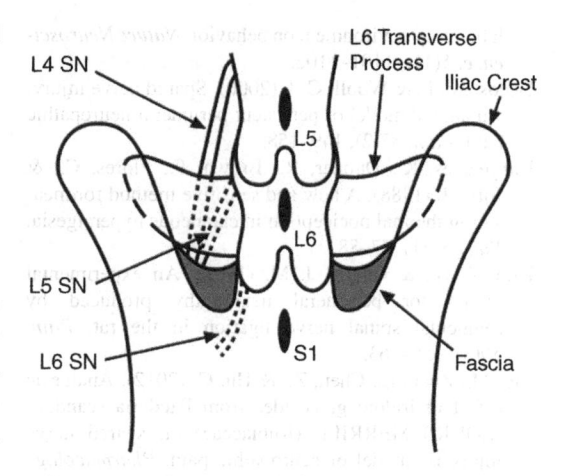

Neuropathic Pain Model, Spinal Nerve Ligation Model, Fig. 1 Schematic diagram showing the dorsal view of the bony structures and spinal nerves at the lumbosacral level after removal of paraspinal muscles

nerves need to be separated in some animals to make the L5 spinal nerve accessible for ligation. It is very important not to damage the L4 nerve during this process, because even a slight damage to the L4 spinal nerve invariably results in a greatly reduced mechanical sensitivity of the foot. Damage to the L4 spinal nerve can occur with a seemingly mild mechanical trauma (excessive touch, gentle stretch, or slight entrapment within the epineurial sheet). When sufficient length of the L5 spinal nerve is freed from the adjacent structure, a piece of 6-0 silk thread is placed around the L5 spinal nerve and the nerve is tightly ligated to interrupt all axons in the nerve. Another option would be to cut the spinal nerve just distal to the ligation to make sure all fibers are interrupted.

The L6 spinal nerve can also be ligated if so desired. The L6 spinal nerve runs underneath the sacrum and is not visible without chipping away a part of the sacrum. Since the sacrum bleeds a lot when chipped, it would be better to approach the L6 spinal nerve blindly without chipping the sacrum. After carefully removing the fascia joining the sacrum to the ileum, one can place a small glass hook underneath the sacrum and gently pull the L6 spinal nerve out into the paravertebral space and ligate it tightly with 6-0 silk thread.

Upon completion of the operation, which normally takes about 10–15 min (after some practice), hemostasis is confirmed and the muscles are sutured in layers using silk thread and the skin is closed with metal clips, anesthesia is then discontinued. Animals are then kept in a cage with warm bedding until they completely recover from anesthesia.

Behavioral outcome of surgery: Successfully operated animals normally do not show any motor deficits beyond a mild inversion of the left hind foot with slightly ventroflexed toes. The most common and obvious motor deficit of unsuccessfully operated animals is dragging the hind limb of the operated side, a sign of paralyzed proximal muscles. This invariably indicates damage to the L4 spinal nerve, since this nerve innervates many proximal muscles of the hind limb.

A successfully operated rat shows various behavioral signs of neuropathic pain such as ongoing pain, heat ▶ hyperalgesia, and mechanical as well as cold ▶ allodynia. Since the SNL model shows a particularly robust sign of mechanical allodynia, one can use the degree of mechanical hypersensitivity to gauge the success of the operation. Mechanical sensitivity is quantified either by measuring response frequency to mechanical stimuli applied with ▶ von Frey filaments (Kim and Chung 1991, 1992) or by determining the mechanical threshold (Chaplan et al. 1994). A successful surgical operation will result in a clear sign of mechanical allodynia demonstrated by either: (1) lowering the foot withdrawal threshold below the normal nociceptor activation threshold [below 1.4 g (Leem et al. 1993)], or (2) frequent foot withdrawals to obviously innocuous stimulations. On the other hand, a sham-operation should not produce any significant changes in mechanical sensitivity, except a mild transient effect lasting 1 or 2 days. A significant and long-lasting hyperalgesia following sham-operation indicates that the surgery induced damage and/or inflammation to the nerve and is thus an unsuccessful operation.

Factors Influencing Variability
Multiple factors seem to influence the behavioral outcome after SNL and hence contribute to variability of data. The strain of rats is an important variable. Not only do different strains of rats show different levels of neuropathic pain behaviors, but also different levels of pain behaviors can also be seen in different substrains of Sprague-Dawley rats obtained from different suppliers (Yoon et al. 1999). Another important factor that influences the sign of mechanical allodynia is the exact testing spot on the paw where the mechanical stimulation is applied. To represent the most intensely painful area of a human patient, one must measure the threshold at the most sensitive area of the rat. The most sensitive area of the hind paw after ligation of the L5 or both the L5 and L6 spinal nerves is the base of the third or fourth toe (Xie et al. 1995). The most sensitive spot of the paw after spinal nerve ligation is confined to a small area and does not vary much between rats, presumably due to stereotyped denervation of the foot by the surgical procedure. When measuring in the most sensitive area, the threshold is usually well below the 1 g range, whereas the threshold ranges from 2 to 3 g if one measures it by stimulating the mid-plantar area (Chung et al. 2004).

Advantages and Disadvantages of the SNL Model Compared to Others
The SNL model has several advantages over other models. These include: (1) the injury is stereotyped and does not allow regeneration, (2) it has successfully adapted to multiple species of animals, and (3) the injured and uninjured afferents are segregated to different spinal segments. Tight ligation is advantageous over loose ligation in terms of knowing the timing of injury, the population of fibers being injured, and the lack of regeneration of injured fibers. In addition, a tight ligation of a specific set of nerves in every animal will reduce the variability among animals. Many neuropathic pain models, including the SNL model, which was originally developed using the rat, are now successfully adapted to the mouse (Mogil et al. 1999). The SNL model has also been successfully applied to the monkey (Carlton et al. 1994). The uniqueness of the SNL model, however, is that spinal inputs of injured and uninjured afferents are segregated at separate

spinal segments. This feature allows the investigation of the contribution of injured and uninjured afferent fibers to neuropathic pain.

There are also some disadvantages to the SNL model. These include: (1) the surgical procedure is invasive, and (2) the model is highly artificial. Because the spinal nerves are located deep, the surgery to expose them requires some level of skill and, hence, may be a source of variability. A particularly technically difficult aspect is to preserve the L4 spinal nerve completely undamaged while ligating the L5 spinal nerve since they are located in close proximity. Since it is rare to find a patient with discrete spinal nerve injury, this model is highly artificial.

Different models tend to produce different behavioral outcomes. A previous study compared the behaviors of three different models [SNL, chronic constriction injury (CCI), and partial sciatic nerve ligation (PSL)] (Kim et al. 1997). The CCI model produced the most robust ongoing pain behaviors, whereas mechanical allodynic behaviors were most prominent in the SNL model.

References

Carlton, S. M., Lekan, H. A., Kim, S. H., & Chung, J. M. (1994). Behavioral manifestations of an experimental model for peripheral neuropathy produced by spinal nerve ligation in the primate. *Pain, 56,* 155–166.

Chaplan, S. R., Bach, F. W., Pogrel, J. W., Chung, J. M., & Yaksh, T. L. (1994). Quantitative assessment of tactile allodynia in the rat paw. *Journal of Neuroscience Methods, 53,* 55–63.

Chung, J. M., Kim, H. K., & Chung, K. (2004). Segmental spinal nerve ligation model of neuropathic pain in pain research: Methods and protocols. In Z. D. Luo (Ed.), *Methods in molecular medicine series* (Serial Ed. J. M. Walker). Totowa, NJ: The Humana Press Inc (in press).

Kim, S. H., & Chung, J. M. (1991). Sympathectomy alleviates mechanical allodynia in an experimental animal model for neuropathy in the rat. *Neuroscience Letters, 134,* 131–134.

Kim, S. H., & Chung, J. M. (1992). An experimental model for peripheral neuropathy produced by segmental spinal nerve ligation in the rat. *Pain, 50,* 355–363.

Kim, K. J., Yoon, Y. W., & Chung, J. M. (1997). Comparison of three rodent neuropathic pain models. *Experimental Brain Research, 113,* 200–206.

Leem, J. W., Willis, W. D., & Chung, J. M. (1993). Cutaneous sensory receptors in the rat foot. *Journal of Neurophysiology, 69,* 1684–1699.

Mogil, J. S., Wilson, S. G., Bon, K., Lee, S. E., Chung, K., Raber, P., et al. (1999). Heritability of nociception I: Responses of 11 inbred mouse strains on 12 measures of nociception. *Pain, 80,* 67–82.

Xie, J., Yoon, Y. W., Yom, S. S., & Chung, J. M. (1995). Norepinephrine rekindles mechanical allodynia in sympathectomized neuropathic rat. *Analgesia, 1,* 107–113.

Yoon, Y. W., Lee, D. H., Lee, B. H., Chung, K., & Chung, J. M. (1999). Different strains and substrains of rats show different levels of neuropathic pain behaviors. *Experimental Brain Research, 129,* 167–171.

Neuropathic Pain Model, Tail Nerve Transection Model

S. K. Back[1], H. J. Kim[2] and H. S. Na[1]
[1]Neuroscience Research Institute and Department of Physiology, Korea University, College of Medicine, Seoul, Republic of Korea
[2]Department of Life Science, Yonsei University, Wonju, Republic of Korea

Synonyms

Dynamic allodynia; Mechanical allodynia; Peripheral neuropathic pain; Rat tail model; Static allodynia; Thermal allodynia

Definition

The tail nerve transection model is produced by partial injury of the nerve innervating the rodent tail. The model displays chronic neuropathic signs like ▶ mechanical allodynia, cold allodynia, and ▶ warm allodynia in the tail.

Characteristics

The tail nerve transection model is one of the ▶ peripheral neuropathic pain animal models. Peripheral nerve injury sometimes results in neuropathic pain. This type of pain is characterized by spontaneous burning pain accompanied by ► hyperalgesia and ► allodynia lasting variable times. Several experimental animal models for neuropathic pain, produced by a partial injury of the nerves supplying the rat hind paw, were developed by Bennett and Xie (1988), Seltzer et al. (1990), and Kim and Chung (1992), respectively. Although these models display clear signs of neuropathic pain, there are some inherent problems in performing behavioral tests due to foot deformity. To avoid these problems, the tail nerve transection model, produced first by transection of the interior caudal trunk at the level between the S3 and S4 spinal nerves, was developed (Na et al. 1994; Kim et al. 1995). This model shows neuropathic signs without tail deformity.

The tail is innervated by the inferior and superior caudal trunks, which are located in the ventral and dorsal parts of the pelvic bone, respectively. Each trunk is composed of the four sacral and the first two caudal spinal nerves. To induce neuropathic pain in the tail, the inferior caudal trunk is exposed carefully from the surrounding tissues and transected at the level between the S3 and S4 spinal nerves. This surgery eliminates the S1–S3 spinal nerve innervation of the tail via the trunk. Figure 1 illustrates schematically how the caudal trunks are composed and the level of the transection of the inferior caudal trunk.

The signs indicative of mechanical allodynia can be sought by applying normally innocuous mechanical stimuli to the tail using ► von Frey hairs. For convenient application of the stimuli, the animal is restrained in a transparent plastic tube and the tail is laid on a plate. The mechanically sensitive area of the tail is first determined by rubbing various areas of the tail with the shank of the von Frey hair, and then, this area is challenged systematically by applying static pressure with the von Frey hair to locate the most sensitive spot. An abrupt tail movement of about 0.5–20 cm in response to the von Frey hair stimulation is considered to be an abnormal response, indicative of mechanical allodynia.

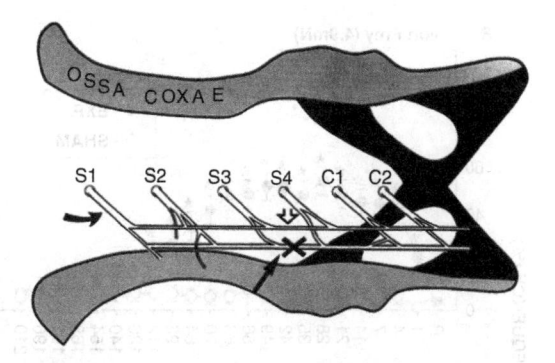

Neuropathic Pain Model, Tail Nerve Transection Model, Fig. 1 A schematic diagram (dorsal view) illustrating how the inferior (*black arrow*) and superior (*open arrow*) caudal trunks are composed and the level of transection (×) of the inferior caudal trunk. The curved arrow indicates the S1 spinal nerve

During repeated trials, the test stimuli can be easily delivered to the same spot without difficulty, since the tail is usually stationary.

Figure 2a and b shows the data obtained with the von Frey hairs (4.9 and 19.6 mN). Prior to the neuropathic surgery, animals rarely respond to von Frey hair stimulation. So, the tail withdrawal frequency is near 0 %. However, after the neuropathic surgery, the frequency increases dramatically from 1 day postoperatively (PO) and lasts for at least 4 months, unlike the frequency in sham-operated animals. These results suggest that the partial injury of the nerves innervating the tail leads to mechanical allodynia in the tail.

The signs indicative of cold and warm allodynia can be sought by immersing the tail into 4 °C and 40 °C water bath, respectively. The rat is restrained in a plastic tube, and the tail is drooped for convenient application of the thermal stimuli. Following the tail immersion, the investigator measures the latency of the tail withdrawal response with a cutoff time of 15 s. A tail withdrawal response within the cutoff time is considered to be an abnormal tail response indicative of thermal allodynia. The tail immersion test is repeated 5 times at 5 min intervals to obtain the average tail response latency.

Figure 2c and d shows the data obtained with the cold (4 °C) and warm (40 °C) water, respectively. Prior to the neuropathic surgery, most rats

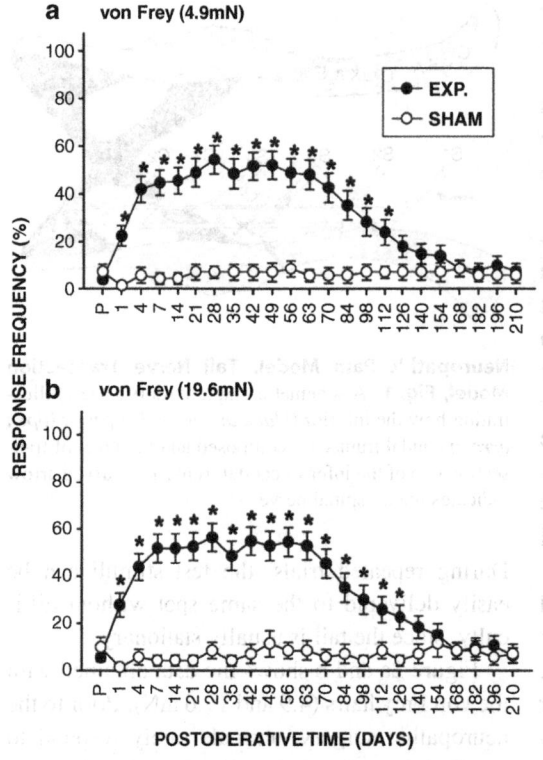

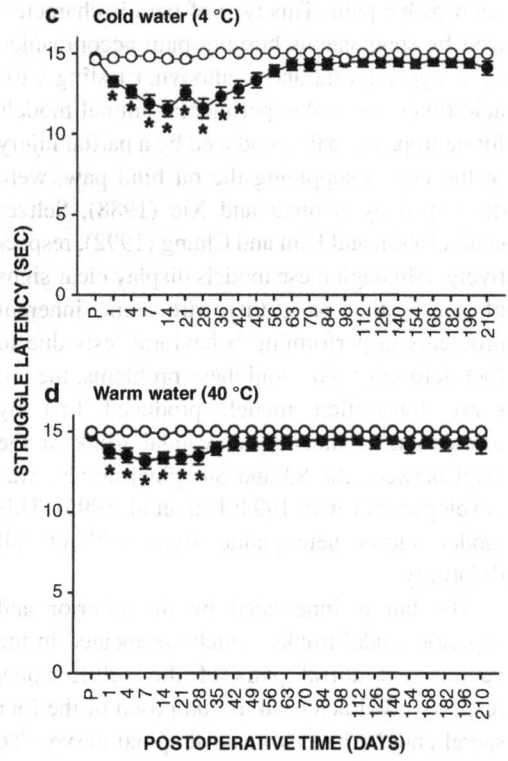

Neuropathic Pain Model, Tail Nerve Transection Model, Fig. 2 Tail responses to mechanical (**a**, **b**), cold (**c**), and warm (**d**) stimuli. The mean (± SEM) response frequency in the case of mechanical stimulation with von Frey hairs (4.9 mN, 19.6 mN) and the mean (± SEM) response latency in the case of cold (4 °C) and warm (40 °C) stimulation of experimental (Exp, n = 44) and sham (n = 14) groups are plotted against the experimental days (P: 1 day before nerve injury). *Asterisks* indicate the scores significantly different from the preoperative value (P < 0.05 by the Friedman test followed by a pairwise post-hoc test)

do not show abnormal tail responses to the cold or warm water stimuli. However, after the neuropathic surgery the tail response latency significantly decreases from 1 day PO and lasts for 5–7 weeks. These results suggest that the partial injury of the nerves innervating the tail leads to cold and warm allodynia in the tail. The possibility that the abnormal tail responses to the 4 °C or 40 °C water immersion are due to the mechanical contact of the tail with the water instead of thermal stimulation is essentially ruled out, since (1) in a vast majority of the cases, 30 °C water does not induce any abnormal tail responses and (2) the 4 °C or 40 °C water-induced responses have latencies greater than at least a few seconds, unlike the von Frey

hair–evoked responses which have virtually no latencies.

The tail nerve transection model offers several advantages in surgical approach and performing the behavioral tests. First, surgical procedures for the model are so simple that even mice (Back et al. 2002) or neonatal rats (Back et al. 2008) can be used. Although other rat models for neuropathic pain have been applied to the mouse (Malmberg and Basbaum 1998; Mansikka et al. 2000; Mogil et al. 1999), there are some problems from the invasiveness of these approaches. Second, since each of the inferior and superior caudal trunks is composed of the four sacral and the two caudal spinal nerves, the number of injured fibers or the spinal level of injury is easily controllable by changing the transection site. In fact, modified nerve injury procedures, such as the transection of superior caudal trunk (Back et al. 2002) or both trunks (Kim et al. 2001), also induced neuropathic changes. This advantage was helpful to elucidate the fact that the extent of sympathetic fiber sprouting in the ▶ dorsal root ganglion (DRG) was related to the number of injured nerve fibers (Kim et al. 2001) and the distance between the DRG and injury site (Kim et al. 1996). Third, application of both mechanical and thermal stimuli to the tail is straightforward. Based on the type of stimulation, such as light static pressure and light stroking, mechanical allodynia can be recognized as static or dynamic (Pappagallo et al. 2000). Unlike the hindpaw models in which animals are placed on a metal mesh floor to stimulate the plantar surface, the tail is laid on a plate; therefore, stroking the tail skin with a brush is not at all obstructed. In addition, tail immersions into cold or warm water make all thermal receptors in the tail receive the same thermal stimulation simultaneously. This advantage has been allowing us to study the distinct mechanisms of static, dynamic, and thermal allodynia (Sung et al. 1998; Back et al. 2003; Li et al. 2010).

In addition, since there is no deformity in the tail after the nerve injury, the mechanically sensitive spot is easily located and blind behavioral tests are available. It is also worth that tails provide the hairy skin. Because almost all the skin of human beings, except for the palms, soles, and lips which are glabrous, is hairy, this tail model may closely reflect much more features of human neuropathic pain.

References

Back, S. K., Sung, B., Hong, S. K., et al. (2002). A mouse model for peripheral neuropathy produced by a partial injury of the nerve supplying the tail. *Neuroscience Letters, 322*, 153–156.

Back, S. K., Kim, J. S., Hong, S. K., et al. (2003). Ascending pathways for mechanical allodynia in a rat model of neuropathic pain. *Neuro Report, 14*, 1623–1626.

Back, S. K., Kim, M. A., Kim, H. J., et al. (2008). Developmental characteristics of neuropathic pain induced by peripheral nerve injury of rats during neonatal period. *Neuroscience Research, 61*, 412–419.

Bennett, G. J., & Xie, Y.-K. (1988). A peripheral mononeuropathy in rat that produces disorders of pain sensation like those seen in man. *Pain, 33*, 87–107.

Kim, S. H., & Chung, J. M. (1992). An experimental model for peripheral neuropathy produced by segmental spinal nerve ligation in the rat. *Pain, 50*, 355–363.

Kim, Y. I., Na, H. S., Han, J. S., et al. (1995). Critical role of the capsaicin-sensitive nerve fibers in the development of the causalgic symptoms produced by transecting some but not all of the nerves innervating the rat tail. *Journal of Neuroscience, 15*, 4133–4139.

Kim, H. J., Na, H. S., Nam, H. J., et al. (1996). Sprouting of sympathetic nerve fibers into the dorsal root ganglion following peripheral nerve injury depends on the injury site. *Neuroscience Letters, 212*, 191–194.

Kim, H. J., Na, H. S., Back, S. K., et al. (2001). Sympathetic sprouting in sensory ganglia depends on the number of injured neurons. *Neuro Report, 12*(16), 3529–3532.

Li, C., Back, SK., Lee, J., et al. (2010). Gastrin-releasing peptide receptor in the spinal cord mediates mechanical allodynia following nerve injury. *Society for Neuroscience Abstracts*, 176.20

Malmberg, A. B., & Basbaum, A. I. (1998). Partial sciatic nerve injury in the mouse as a model of neuropathic pain: Behavioral and neuroanatomical correlates. *Pain, 76*, 215–222.

Mansikka, H., Sheth, R. N., DeVries, C., et al. (2000). Nerve injury-induced mechanical but not thermal hyperalgesia is attenuated in neurokinin-1 receptor knockout mice. *Experimental Neurology, 162*, 343–349.

Mogil, J. W., Wilson, S. G., Bon, K., et al. (1999). Heritability of nociception I: Responses of 11 inbred mouse strains on 12 measures of nociception. *Pain, 80*, 67–82.

Na, H. S., Han, J. S., Ko, K. H., et al. (1994). A behavioral peripheral neuropathy produced in rat's tail by inferior caudal trunk injury. *Neuroscience Letters, 177*, 50–52.

Pappagallo, M., Oaklander, A. L., Quatrano-Piacentini, A. L., et al. (2000). Heterogenous patterns of sensory dysfunction in postherpetic neuralgia suggest multiple pathophysiologic mechanisms. *Anesthesiology, 92*, 691–698.

Seltzer, Z., Dubner, R., & Shir, Y. (1990). A novel behavioral model of neuropathic pain disorders produced in rats by partial sciatic nerve injury. *Pain, 43*, 205–218.

Sung, B., Na, H. S., Kim, Y. I., et al. (1998). Supraspinal involvement in the production of mechanical allodynia by spinal nerve injury in rats. *Neuroscience Letters, 246*, 117–119.

Neuropathic Pain Models, CRPS-I Neuropathy Model

Jean-Jacques Vatine[1], Ze'ev Seltzer[2] and
Jeanna Tsenter[3]
[1]Outpatient and Research Division, Reuth
Medical Center, Sackler Faculty of Medicine,
Tel Aviv University, Tel Aviv, Israel
[2]Faculty of Dentistry, University of Toronto
Centre for the Study of Pain, Toronto,
ON, Canada
[3]Department of Physical Medicine and
Rehabilitation, Hadassah University Hospital,
Jerusalem, Israel

Synonyms

Algodystrophy; CRPS-I neuropathy model;
Reflex sympathetic dystrophy; Sudeck's atrophy

Definition

Complex regional pain syndrome (CRPS) is
a neuropathic pain disorder that usually develops
after a noxious event. Pain is frequently described
as burning and continuous and exacerbated by
movement, continuous stimulation, or stress.
The syndrome includes spontaneous pain and/or
stimulus   evoked   pain   (▶ allodynia   and
▶ hyperalgesia), exceeding in both magnitude
and duration the clinical course expected to follow the inciting event. Regardless of the site of
injury, the symptoms begin and remain most
intense in the distal extremity and are not limited
to the distribution of a single peripheral nerve. At
some point in time, pain may be associated with
edema, changes in skin blood flow, and abnormal
sudomotor activity in the same area, often
resulting in significant impairment of motor function and showing variable progression over time.
Two forms of CRPS have been identified. CRPS-
I usually develops following trauma with only
minor nerve damage or without any demonstrable
nerve lesion whereas CRPS-II is associated with
a clear nerve injury that can be characterized by

abnormal clinical and/or electrodiagnostic findings (Stanton-Hicks et al. 1995). The tetanized
sciatic neuropathy (TSN) model is a preparation
in rodents that results in allodynia, hyperalgesia,
and vasomotor disturbances that mimic CRPS-I.
This model is produced by activating unmyelinated   afferents   (C-fibers)   at   "▶ wind   up"
(Mendell 1996) parameters, using a 10-min electrical stimulation (i.e., tetanization) of an intact
sciatic nerve.

Characteristics

The diagnostic algorithm proposed by Harden and
Bruehl (2005) to diagnose CRPS-I in human was
recently adopted as diagnostic criteria by the IASP
(International Association for the Study of Pain).
This algorithm is based on presentation of at least
one sign in two out of the following four categories: (a) sensory abnormalities: allodynia and/or
hyperalgesia, (b) vasomotor abnormalities: temperature asymmetry and/or skin color changes
and/or asymmetry, (c) sudomotor abnormalities
or edema: swelling and/or altered sweating
and/or sweating asymmetry, (d) motor abnormalities or dystrophy: decreased range of motion
and/or motor dysfunction (weakness and/or
tremor and/or dystonia) and/or trophic changes in
hair and/or skin and/or nails. Edema and autonomic dysregulation are usually seen in the early
stage of the disease, while movement disorders
and trophic changes are more apparent in the
later stage. For many decades, sympathetic blockade was the method of choice for the diagnosis and
treatment of CRPS-I. Pain relief in response to this
procedure has been used as a criterion for diagnosis, hence the term reflex sympathetic dystrophy
(RSD). However, it now appears that not all
patients may have sympathetically maintained
pain in CRPS (Stanton-Hicks et al. 1995).

Several animal models have been developed
to study the mechanisms underlying neuropathic
pain mimicking CRPS-II, including the chronic
constriction injury (CCI), partial sciatic ligation
(PSL), and spinal nerve ligation (SNL) models.
Some of these are described elsewhere in this
section. The common denominator of these

models is that abnormal sensory responses to stimuli and spontaneous behavior, indicative of neuropathic pain, are produced by partial denervation of a paw, tail, or the face. The typical spontaneous pain behaviors include guarding behavior, repeated flicking of the partially denervated paw, excessive licking and holding the paw in the mouth, claw pulling, elevated paw position and antalgic gait, as well as vocalization, reduced appetite, weight loss, and self-mutilation (autotomy). The abnormal sensory responses to stimuli include a reduced withdrawal threshold to a stimulus that is normally non-noxious (allodynia), and exaggerated responses to a stimulus that is normally noxious (hyperalgesia). The latter is manifested as increased duration and robustness of nocifensive responses. A wealth of data on the mechanisms that trigger and underlie the pain in CRPS-II has been gathered from these animal models, since they are all produced by some type of nerve injury that partially denervates a limb. However, since in CRPS-I there is no evidence for such a nerve injury, these models may not be relevant to the study of its mechanisms (Jänig and Baron 2001).

The signals that trigger CRPS-I as well as the mechanisms maintaining the abnormalities that characterize this syndrome are still enigmatic. Since the clinical signs of CRPS-I and CRPS-II are similar, they may be produced by the same triggering input. When sensory fibers are injured, 25–33 % of transected axons emit a barrage of impulses termed ▶ injury discharge (ID). This is the first neural message to notify the CNS that an injury has occurred (Wall et al. 1974). This message comprises a burst of high-frequency discharge in A-fibers and of low frequency firing in C-fibers. This burst decays in most A- and C-fibers within minutes after the injury. However, about 10 % of injured fibers continue to fire for many hours and may not stop for days (Baik-Han et al. 1990; Blenk et al. 1996; Sackstein et al. 1996). Muscular afferents emit a clearly more robust ID than cutaneous afferents. ID is a distinct signal in the "alphabet" of these fibers, unlike their normal response to natural stimuli, since the peak discharge of ID is

twofold to sixfold higher than the response to maximal normal stimulation (Sackstein et al. 1996). ID is an important signal in triggering neuropathic pain disorders in some animal models and possibly in humans as well (reviewed by Kissin 2000).

The TSN model has been developed on the basis of the working hypothesis that CRPS I may be caused by a bodily injury that does not result in frank nerve injury but produces a massive nociceptive input that is similar to ID. The following description provides methodological details of the TSN model.

Animals

The original model was developed in the rat but it can easily be adapted to the mouse.

Preparation of the TSN Model

Surgery

Under inhalation anesthesia and aseptic conditions, the sciatic nerve on both sides is exposed at midthigh level and the surgical field is kept widely open with separators, taking care not to pull the posterior biceps semitendinosus nerve or other thigh nerves. The sciatic nerve is then carefully separated from neighboring tissues (Fig. 1) and a sheet of parafilm is placed under the nerve. A pair of stainless steel stimulating electrode hooks is inserted under the nerve on both sides. The exposed nerves are covered with mineral oil (37 °C) to prevent damage by drying. The side that receives the tetanic stimulation should alternate between individual animals, to minimize the bias of the experimenter when testing.

Tetanization

Electrical stimulation of intact nerves activates leg jerking that may pull the nerves by the hook electrodes. To prevent the injury, both hind paws, the pelvis, and the tail are taped to the surgical board. The tetanization includes a train supramaximally activating C-fibers at a wind-up frequency, in addition to A-fibers (shock duration = 0.5 ms, frequency = 0.5 Hz, intensity = 5 mA, train duration = 10 min, n = 300 shocks). It is noteworthy that no sensory disorders were detected when activating A-fibers

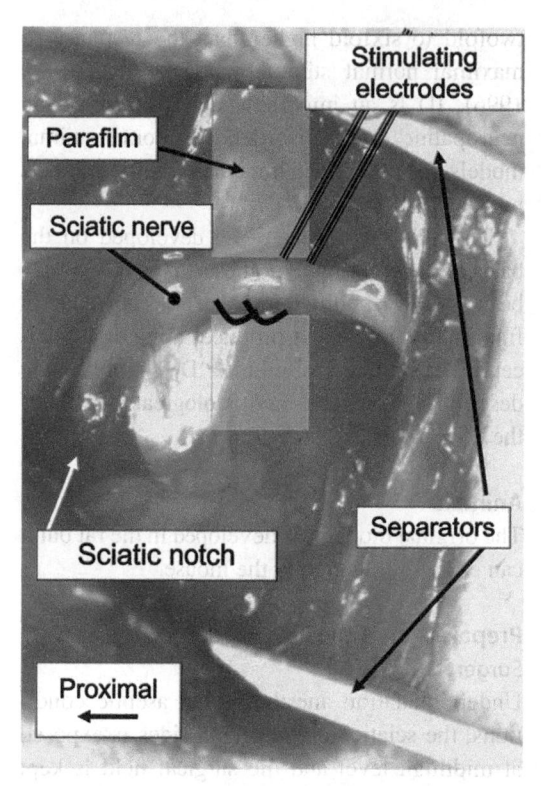

Neuropathic Pain Models, CRPS-I Neuropathy Model, Fig. 1 The sciatic nerve is exposed at midthigh level and the surgical field is kept widely open with separators. A sheet of parafilm separates the nerve from neighboring tissues. A pair of stainless steel stimulating electrode hooks is inserted under the nerve

only (Vatine et al. 1998). Directly observing the surgical field with a dissecting microscope during the tetanization, the experimenter should verify that the nerves are not pulled by the electrodes. The contralateral sciatic nerve receives a sham stimulation at the same time as the ipsilateral nerve is tetanized. A separate control group of bilaterally sham-tetanized animals should optimally be included in every experiment.

Determination of Sensory Disorders

Tactile Allodynia

The animal is placed on top of a metal mesh floor, and covered with an opaque plastic cage. This enables the experimenter to introduce the filaments from underneath, preventing the animal from observing the stimulus approaching. Allodynia is assessed with a set of von Frey hairs.

These hairs are nylon monofilaments of different diameters that exert defined levels of force when pressed against the plantar skin with sufficient force to cause the hair to bend. Each hair is indented five times at a frequency of about 2 Hz. The testing process begins by using the lowest hair in the set, ascending in the series until the animal responds at threshold by lifting the paw, withdrawing from the filament. The set typically ranges from 0.05 to 25 g and needs to be calibrated weekly using a top load balance. Other methods can be used (Bennett et al. 2002). Figure 2a shows that the average group withdrawal threshold (in g) of TSN rats is significantly decreased for a period of about 40 days, compared to sham-tetanized rats.

Heat Hyperalgesia

Several methods can be used, including the Hargreaves instruments or a laser. For the latter method, a painful pulse of infrared energy is beamed from a CO_2 laser (120 ms, 5 W, 150 mCal and 1.5 mm in diameter) to the midplantar area of the hind paw from underneath, targeted by the visual aid of a He/Ne laser beam. This intensity causes sharp stinging pain to humans. Sham-tetanized rats respond by a momentary paw flick or paw lift lasting less than a second. When stimulated at the TSN side, rats typically respond by immediate withdrawal followed by prolonged licking, paw lifting and claw pulling lasting on average up to 10 s, depending on the posttetanization day and genetic and environmental variables. Figure 2b shows that the average group response duration (in sec) to stimulation on the TSN side significantly increased for a period of about 40 days, compared to the sham-tetanized side.

Cold Hyperalgesia

A drop of acetone from a syringe is smeared on the plantar surface of the paw through the mesh floor of the testing chamber. As a control stimulus, a drop of tap water at room temperature is likewise applied, alternating between the acetone and water. The response time to each stimulus is recorded, subtracting the water from the acetone and the net result averaged for the group for each

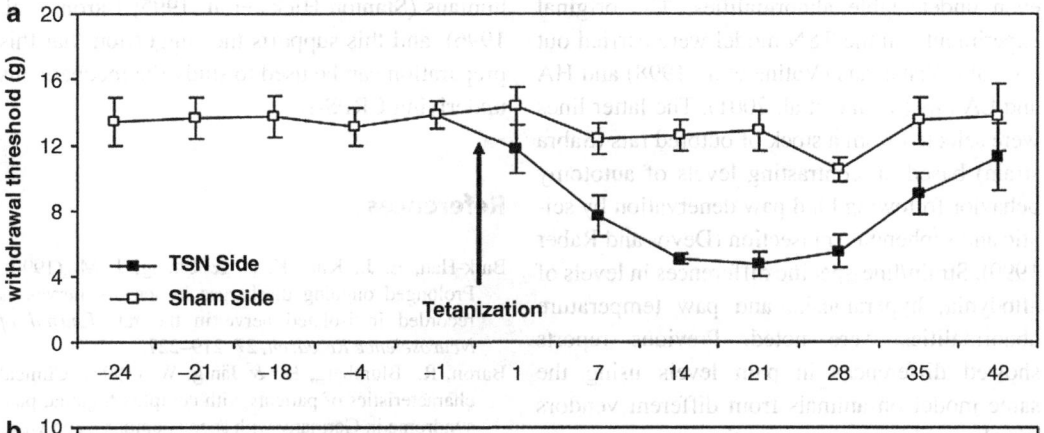

Neuropathic Pain Models, CRPS-I Neuropathy Model, Fig. 2 (a) *Tactile allodynia.* Baseline tactile sensitivity was tested with a set of von Frey hairs. The withdrawal thresholds (g) to repetitive touch on the plantar side of both hind paws were tested on five sessions prior to nerve stimulation (days −24, −21, −18, −4 and −1). On day 0, the rats underwent unilateral tetanic stimulation of the sciatic nerve for 10 min. Tactile allodynia developed ipsilaterally, but not on the sham stimulated side. Increased response duration indicates cold hyperalgesia.

side. (b) *Heat hyperalgesia.* Baseline sensitivity to a noxious heat pulse from a CO_2 laser was tested on days −24, −18, −4 and −1 prior to nerve tetanization in intact rats. On day 0, the rats underwent unilateral tetanic stimulation of the sciatic nerve for 10 min. The significantly increased response duration (in seconds) to the noxious stimulus denotes heat hyperalgesia developed ipsilaterally on the tetanized sciatic side but not on the sham side

Mechanical Hyperalgesia

An increased response duration to pinprick applied from underneath to the midplantar area of the hind paw reflects hyperalgesia.

Determination of Vasomotor Disorders

Since the temperature of the plantar skin area of conscious animals fluctuates, the paw temperature is measured 2 min after the rat is lightly anesthetized with an inhalation gas and 30 s after the temperature stabilizes. Plantar hind paw temperature is recorded bilaterally using a remote infrared sensing thermometer, while the rat is lying on its ventral side in a room with an ambient temperature maintained at $21.0 \pm 0.5°C$. Unilateral tetanic stimulation of the sciatic nerve causes a relative cooling of the tetanized hind paw.

Variables Potentially Affecting the TSN Model

Genetic Considerations

Work done in other animal models of chronic pain showed that the choice of animals might have a critical effect on the outcome, since genetic variation plays an important role. Some lines may develop robust neuropathic pain, while others may develop very weak, short lasting or

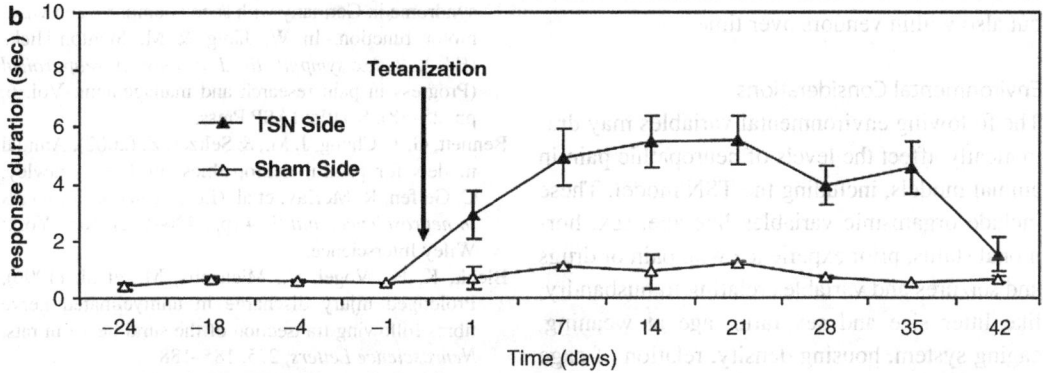

even undetectable abnormalities. The original experiments on the TSN model were carried out on male Wistar rats (Vatine et al. 1998) and HA and LA rats (Vatine et al. 2001). The latter lines were selected from a stock of outbred rats (Sabra strain) based on contrasting levels of autotomy behavior following hind paw denervation by sciatic and saphenous transection (Devor and Raber 1990). Strain/line-specific differences in levels of allodynia, hyperalgesia, and paw temperature abnormalities were noted. Previous reports showed differences in pain levels using the same model on animals from different vendors but also within vendors over time.

Environmental Considerations

The following environmental variables may dramatically affect the levels of neuropathic pain in animal models, including the TSN model. These include organismic variables like age, sex, hormonal status, prior experience with pain or drugs and seizures and variables relating to husbandry, like litter size and sex ratio, age at weaning, caging system, housing density, relation of cage mates, male-male fighting, handling frequency, bedding material, colony health status, prenatal (maternal) stress, maternal deprivation, ambient noise, ambient temperature, and illumination. Also important are the circadian phase, circannual phase, meteorological factors, temperature, humidity, barometric pressure, experimenter, testing apparatus, restraint, and drug injection method. Variables related to the stimulus can be no less important, including type of noxious stimulus, intensity, location, testing apparatus particularities, dependent measure, repeated testing, data transformation, and experimenter's proneness to bias. Investigators should exercise extreme care to control these as much as possible.

The TSN model shows similar types and durations of sensory disorders to those produced by intended nerve injury as in the CCI, PSL, SNL, and PNI models. Lack of an overt nerve injury in the TSN model, combined with the appearance of mechanical allodynia, mechanical and thermal hyperalgesia, and some vasomotor disturbances, resembles the abnormalities of CRPS-I in

humans (Stanton-Hicks et al. 1995; Baron et al. 1996), and this supports the suggestion that this preparation can be used to study the mechanisms underlying CRPS-I.

References

Baik-Han, E. J., Kim, K. J., & Chung, J. M. (1990). Prolonged ongoing discharges in sensory nerves as recorded in isolated nerve in the rat. *Journal of Neuroscience Research, 27*, 219–227.

Baron, R., Blumberg, H., & Jänig, W. (1996). Clinical characteristics of patients with complex regional pain syndrome in Germany with special emphasis on vasomotor function. In W. Jänig & M. Stanton-Hicks (Eds.), *Reflex sympathetic dystrophy: A reappraisal* (Progress in pain research and management, Vol. 6, pp. 25–48). Seattle: IASP Press.

Bennett, G. J., Chung, J. M., & Seltzer, Z. (2002). Animal models for painful neuropathies. In J. N. Crawley, C. Gerfen, R. McKay, et al. (Eds.), *Current protocols in neuroscience, unit 9.14* (pp. 426–442). New York: Wiley Interscience.

Blenk, K. H., Vogel, C., Michaelis, M., et al. (1996). Prolonged injury discharge in unmyelinated nerve fibres following transection of the sural nerve in rats. *Neuroscience Letters, 215*, 185–188.

Devor, M., & Raber, P. (1990). Heritability of symptoms in an experimental model of neuropathic pain. *Pain, 42*, 51–68.

Harden, R., & Bruehl, S. (2005). Diagnostic criteria: The statistical derivation of the four criterion factors. In P. R. Wilson, M. Stanton-Hicks, & R. N. Harden (Eds.), *CRPS: Current diagnosis and therapy* (pp. 45–58). Seattle: IASP Press.

Harden, R. N., Bruehl, S., Perez, R. S., Birklein, F., Marinus, J., Maihofner, C., Lubenow, T., Buvanendran, A., Mackey, S., Graciosa, J., Mogilevski, M., Ramsden, C., Chont, M., & Vatine, J. J. (2010). Validation of proposed diagnostic criteria (the "Budapest Criteria") for Complex Regional Pain Syndrome. *Pain, 150*(2), 268–74.

Jänig, W., & Baron, R. (2001). The value of animal models in research on CRPS. In R. N. Harden, R. Baron, & W. Jänig (Eds.), *Complex regional pain syndrome* (Progress in pain research and management, pp.75–85). Seattle: IASP Press.

Kissin, I. (2000). Preemptive analgesia: How can we make it work? In M. Devor, M. C. Rowbotham, & Z. Wiesenfeld-Hallin (Eds.), *Proceedings of 9th World Congress of Pain. Progress in pain research and management* (pp. 973–985). Seattle: IASP Press.

Mendell, L. M. (1996). Physiological properties of unmyelinated fiber projection to the spinal cord. *Experimental Neurology, 16*, 316–332.

Sackstein, M. J., Ratner, A., & Seltzer, Z. (1996). *Specific patterns of injury discharge are associated with* receptor types in freshly injured sensory myelinated (A-) and unmyelinated (C-) fibers in rat. *Abstracts of the 8th World Congress of the International Association for the Study of Pain* (p. 11). Vancouver: IASP Press.

Stanton-Hicks, M., Janig, W., Hassenbusch, S., et al. (1995). Reflex sympathetic dystrophy: Changing concepts and taxonomy. *Pain, 63*, 127–133.

Vatine, J. J., Argov, R., & Seltzer, Z. (1998). Short electrical stimulation of c-fibers in rats produces thermal hyperalgesia lasting weeks. *Neuroscience Letters, 246*, 125–128.

Vatine, J. J., Tsenter, J., Raber, P., et al. (2001). A model of CRPS I produced by tetanic electrical stimulation of an intact sciatic nerve in the rat: Genetic and dietary effects. In R. N. Harden, R. Baron, & W. Jänig (Eds.), *Complex regional pain syndrome* (Progress in pain research and management, Vol. 22, pp. 53–74). Seattle: IASP Press.

Wall, P. D., Waxman, S., & Basbaum, A. I. (1974). Ongoing activity in peripheral nerve: Injury discharge. *Experimental Neurology, 45*, 576–589.

Neuropathic Pain of Central Origin

Robert P. Yezierski[1] and Nanna Brix Finnerup[2]
[1]Comprehensive Center for Pain Research and Department of Neuroscience, The McKnight Brain Institute University of Florida, Gainesville, FL, USA
[2]Aarhus University Hospital, Danish Pain Research Center, Aarhus, Denmark

Definition

The term central pain was initially considered synonymous with thalamic pain, and for this reason, most descriptions have placed both in the same category. Although thalamic lesions are considered to be a common cause of central pain, it is also recognized that ▶ central neuropathic pain can result from lesions anywhere along the neuraxis from the spinal cord to the cerebral cortex. In 2011 the International Association for the Study of Pain defined central pain as pain caused by a lesion or disease of the central somatosensory nervous system (Jensen et al. 2011).

Characteristics

Epidemiology

Poststroke Pain

The incidence of ▶ central poststroke pain (CPSP) was estimated in 1991 to be approximately 750,000–1,000,000 patients worldwide (Bonica 1991). This calculation was based on a figure for poststroke pain of 1–2 %, but recent reports describe this condition as more frequent than previously thought. Andersen et al. (1995) reported that central pain affected 8 % of 207 central poststroke patients with 5 % having moderate to severe pain, and Klit et al. (2011) reported a prevalence of 7–9 % in a population-based study of 964 stroke patients. The prevalence of CPSP is particularly high following thalamic and brainstem lesions (MacGowan et al. 1997).

Spinal Cord Injury Pain

The prevalence of pain following ▶ spinal cord injury ranged from 34 % to 94 % (mean 69 %) in ten studies published between 1947 and 1988 (Bonica 1991) and from 36 % to 96 % (mean 66 %) in studies from 1975 to 1991 (Yezierski 1996). In recent years, the use of more comprehensive pain assessment strategies has raised the overall prevalence of all categories of SCI pain to 70–80 % (Rintala et al. 1998; Widerström-Noga et al. 1999). The prevalence of neuropathic pain following SCI is estimated to be between 38 % and 52 % (Siddall et al. 2003; Budh et al. 2003; Werhagen et al. 2004) (see ▶ Central Neuropathic Pain from Spinal Cord Injury).

Surgical Lesions

Surgical lesions involving spinal or supraspinal levels of the neuraxis intended to relieve pain can result in the onset of pain. Although Cassinari and Pagni (1969) reported that the incidence of ▶ dysesthesia and central pain following functional neurosurgery was 10–60 %, more conservative numbers have been reported following cordotomy (3–5 %) (Lipton 1989; White and Sweet 1969). The incidence of pain after medullary tractotomies was described by Bonica (1991) as 30 %, with 70–100 % of patients experiencing pain after open mesencephalic tractotomy and 5–10 % after stereotaxic mesencephalic tractotomy (Bonica 1991).

Traumatic Brain Injury

The association of chronic pain with traumatic brain injury (TBI) is a condition that is receiving a lot of attention due in part to the increasing number of casualties in war zones around the world. Traumatic and blast injuries involving the brain have become the signature injuries of these conflicts. For human TBI victims, chronic pain is a common consequence, with an overall prevalence between 22 % and 95 % (Nampiaparampil 2008; Ofek and Defrin 2007; Beetar et al. 1996). Most often, TBI patients complain of headache, followed by neck/shoulder, back, upper limb, and lower limb pain (see also ▶ Central Pain Following Traumatic Brain Injury). Surprisingly, it appears that chronic head pain in mild TBI cases is more prevalent than in patients with more severe injuries. This may be due to cognitive impairment and lack of self-monitoring with severe injury (Nampiaparampil 2008; Ofek and Defrin 2007). However, it appears that the prevalence of chronic pain in other areas of the body is similar between mild and moderate to severe TBI cases (Lahz and Bryant 1996).

Other Neurological Disorders

Other central neurological conditions are known to be associated with chronic pain. For example, multiple sclerosis (see ▶ Central Pain in Multiple Sclerosis) causes central pain in 29–75 % of patients (Bonica 1991; Österberg et al. 2005). Although systematic epidemiological studies for pain associated with other types of neurological disorders have not been carried out, other conditions associated with central pain include ▶ epilepsy, Parkinson's disease, Huntington's disease, and cancer (see also ▶ Central Pain and Cancer).

Clinical Characteristics of Central Pain

Spontaneous, persistent, diffuse and/or intermittent, burning, shooting, icelike, squeezing, lancinating sensations as well as hyperesthesia, hyperalgesia, ▶ allodynia, and hyperpathia are characteristics commonly associated with central pain. Dysesthesias, enhancement of pain by emotion and temperature changes, as well as the radiation of sensation, summation of repeated stimuli, and prolonged aftersensations have also been described as components of central pain (Pagni 1998).

Evoked pain can develop explosively and usually continues long after stimulation. Pain intensity can vary during the day, often due to external (e.g., touch, vibration, cold) and frequently emotional factors. Spontaneous sensations occur in a large number of cases, varying from ▶ paresthesia to uncomfortable aching, shooting, burning ▶ dysesthesia or pain of great intensity. As a rule, spontaneous pains frequently vary in position and may change in character and are aggravated by somatic or visceral stimulation as well as by stress and emotion, especially anxiety. Auditory, visual, olfactory, and visceral stimuli can provoke or exacerbate spontaneous pain. In the assessment of pain, it is important to evaluate both pain intensity as well as sensory symptoms and signs along with the physical and emotional functioning and quality of life (see also ▶ Diagnosis and Assessment of Clinical Characteristics of Central Pain and ▶ Central Pain, ▶ Central Pain, Outcome Measures in Clinical Trials).

Temporal Profile

Central pain can start at any time after a neurological insult, although it usually begins within the first 3 months. The time of onset does not appear to depend on the location of the lesion, and there are no definite correlations between the time of onset and associated pathology. In general, central pain following stroke develops gradually as sensory impairment and weakness improve. In cases involving ▶ ischemia or hematomyelia, pain has been reported to appear suddenly after insult. Shieff and Nashold (1987) described patients with pain from the time of initial insult as well as intervals varying up to 2 years. Andersen et al. (1995) described cases with pain 1 month after stroke, at 1–6 months and more than 6 months. In Tasker's series of patients (1991), central pain of brain origin was found to have a delayed onset in two thirds of the cases, being less than 1 year in 50 % of patients.

Leijon described patients where pain began within 1 day after stroke, during the first month, after 3–12 months, or after 2–3 years (Leijon et al. 1989). Following spinal injury, Siddall and colleagues (2003) reported an early onset ▶ of at-level neuropathic pain, while ▶ below-level neuropathic pain had an onset months after the injury.

Location of Pain

The distribution of central pain is nearly always related to the somatotopic organization of the brain structure damaged by trauma, disease, or vascular insult. Because of this, it is possible to identify the location of the lesion in cases of dorsal horn and bulbar lesions, whereas it is difficult to distinguish between cortical, subcortical, and thalamic lesions. In the majority of cases, central pain coincides with all or part of the territory in which sensory loss is clinically observed or revealed by quantitative sensory testing. Central pain is generally described as having a diffuse distribution; however, it can involve only one extremity or a portion of an extremity, e.g., hand or side of the face, and is therefore more accurately described as extensive rather than diffuse.

Bilateral girdle pain is found in cases of intramedullary tumors or ▶ syringomyelia (for detail see ▶ Pain in Syringomyelia). Spinal injury to the anterolateral quadrant is often referred to the opposite side of the body below the lesion. Dysesthesias from injury to the posterior column or dorsal column nuclei are typically located on the same side, below the lesion, and may be unilateral or bilateral. Pain and dysesthesia due to vascular pontomedullary lesions usually have an alternating distribution, i.e., face on the lesion side and limbs and trunk contralateral to the lesion. This distribution is largely due to the fact that bulbar pain syndromes commonly result from the involvement of the posterior inferior communicating artery. Bulbar lesions can give rise to bilateral facial pain when the lesion impinges on the descending root of the trigeminal nerve on one side and on the crossed trigeminothalamic fibers coming from the other side. With pontine lesions, pain in the face is most often on the side opposite the lesion as is pain experienced in the limbs and trunk. Following mesencephalopontine lesions, pain occurs on the side of the body contralateral to the lesion, typically with a hemiplegic distribution. Pain and dysesthesia due to thalamic lesions also have a hemiplegic distribution and affect the side of the body contralateral to the injured thalamus. Finally, cortical or subcortical lesions result in pain referred to the contralateral distal parts of the body (regions with the most extensive cortical representation).

Central Pain Syndromes

Thalamic Syndrome

The major features of thalamic lesions include severe, often intolerable, persistent, or paroxysmal pain on the side opposite the lesion (Dejerine and Roussy 1906). This syndrome is characterized by slight hemiparesis, persistent superficial ▶ hemianesthesia, mild hemiataxia, and astereognosis. While spontaneous pain may be absent in thalamic syndrome, excessive reaction to stimulation of affected body parts is consistent. Pain or discomfort can be evoked by almost any stimulus capable of arousing a sensation and is commonly characterized as intensely disagreeable and unbearable. Aside from spontaneous variations in pain intensity, fluctuations in pain are often exacerbated by environmental changes (especially cold), emotional stress (sudden fear, joy, anxiety), strong taste or smell, loud noises, bright lights, movements, light touch, smoking, and intellectual concentration. Pain is typically prolonged after stimulation, and stimuli that normally have no obvious affective qualities may elicit a reaction in patients with thalamic syndrome. Patients with thalamic pain often have signs of autonomic impairment, e.g., vasoconstriction, abnormal sweating, and edema (Bowsher et al. 1989).

Poststroke Pain

The most common cause of ▶ central poststroke pain (CPSP) is vascular abnormalities, e.g., ischemic lesions, with an etiology usually including supratentorial thrombotic stroke. Infarcts are not the only vascular disorders causing pain however, as subarachnoid hemorrhage is also associated with the onset of chronic pain (Bowsher et al. 1989; Tasker et al. 1991). CPSP is characterized by sensory deficits involving cold and warm stimuli, pinprick, and to a lesser extent vibration, touch, and 2-point discrimination (Vestergaard et al. 1995). There may be spontaneous or evoked sensory disturbances such as paresthesia, dysesthesias, hyperpathia, and allodynia to cold.

The majority of patients with poststroke pain also have more than one kind of pain which can be described as aching, pricking, shooting, stabbing, throbbing, squeezing, stinging, or lacerating. The pain may be superficial or deep and is typically constant, although it is not uncommon for patients to have intermittent pain and/or pain-free periods lasting a few hours. While there is evidence that a spinothalamic deficit is a necessary condition for poststroke pain (Andersen et al. 1995; Dejerine and Roussy 1906), it is not a sufficient condition, since spinothalamic deficits are seen in more that 50 % of stroke patients who show no signs of pain. However, there is evidence that the development of sensory loss and hyperalgesia in a body part deafferented by stroke is a necessary and sufficient condition for the development of central pain.

Pain Following Spinal Lesions

Painful sensations are a frequent and troublesome sequela of paraplegia and quadriplegia following partial or complete lesions of the spinal cord (further described in ▶ Central Neuropathic Pain from Spinal Cord Injury). The International Spinal Cord Injury Data Set includes a newly updated and validated specific pain data set questionnaire (Widerström-Noga et al. 2008; Jensen et al. 2010), which may assist the physician in collecting the relevant data in a standardized way. For a more detailed pain description, e.g., in clinical trials, various outcome measures for pain after SCI have been evaluated (Bryce et al. 2007) and may serve as a guideline when designing trials. Recently, a new consensus classification, the International Spinal Cord Injury Pain (ISCIP) Classification, has been put forward after review by major SCI

and pain organizations and after validation using vignettes (Bryce et al. 2012). Pain types following SCI are classified as either nociceptive (musculoskeletal, visceral, or others) or neuropathic (pain caused by a lesion or disease of the somatosensory nervous system). Neuropathic pain (NP) following SCI is classified as either NP directly related to the SCI (at-level or below-level pain, where level refers to the neurological level) or other neuropathic pain, which is pain indirectly or not related to the SCI. Musculoskeletal pain has the highest prevalence (58 % of patients 5 years after injury), but 12–42 % report at-level and 23–34 % report below-level NP, with below-level pain being described as the most severe or excruciating (Siddall et al. 2003; Werhagen et al. 2007). Although there is no question concerning the diversity of different pain states associated with spinal injury, of greater importance is the practical impact of SCI pain on a patient's quality of life. Widerstrom-Noga et al. (1999) described 37 % of patients rating SCI pain as very hard to deal with (rating of 7–10 on a scale of 0–10). In this study, a cluster analysis of different consequences of injury showed a strong interrelationship among ratings for pain, spasticity, abnormal sensations, and sadness further supporting the negative impact of pain on quality of life following injury.

Pain Following Traumatic Brain Injury

Central pain following traumatic brain injury (TBI) usually develops within the first 12 months after injury. It may be located uni- or bilateral and is often associated with allodynia (Ofek and Defrin 2007). Chronic pain in TBI patients appears to be localized to one side and resembles the type of central pain seen after spinal cord injury, stroke, and multiple sclerosis. Associations of TBI with stress (Schneiderman et al. 2008), inflammation (Israelsson et al. 2008), and pain (Nampiaparampil 2008) provide a rationale for the hypothesis that sympathetic activation and pain sensitivity are increased after TBI. A recent study showed that 42 % of subjects with a mild TBI history also had posttraumatic stress disorder symptoms (Lew et al. 2009) with other complications including frequent pain and auditory and visual dysfunction. It is acknowledged that multiple transmitter systems are potentially involved in the functional consequences of TBI (e.g., dopaminergic, serotonergic, cholinergic, substance P) and dysfunction in any or all of these systems may contribute to the development of chronic pain following TBI.

Other Central Pain Conditions

Parkinson's disease may be associated with dystonia and musculoskeletal pain but also sometimes with a burning, aching pain in the more affected side (Schestatsky et al. 2007). While there is no clear lesion of the somatosensory system in Parkinson's disease, hypersensitivity to painful stimuli in the "off" condition with attenuation by L-dopa has been reported, and it is possible that alterations in the dopaminergic system cause changes in pain modulation (Schestatsky et al. 2007).

Imaging Central Pain

Central pain patients can be studied with neurometabolic techniques such as ▶ single-photon emission computed tomography (SPECT), ▶ positron emission tomography (PET), ▶ functional magnetic resonance imaging (fMRI), and magnetic resonance spectroscopy (MRS), which together with pharmacologic dissection can be helpful in classifying patients according to the pathophysiological mechanism(s) responsible for producing central pain. In PET studies, patients with chronic pain show a decrease in thalamic activity (Di Piero et al. 1994). These findings may be compatible with a decrease in thalamic neuronal activity between bursts observed in patients with central pain secondary to spinal injury. Cesaro et al. (1991) in a SPECT study using an amphetamine tracer found hyperactivity in the thalamus contralateral to the pain. In another SPECT study, Canavero et al. (1993) observed hypoactivity in the parietal cortex of a patient with central pain, suggesting that under normal conditions the cortex exerts an inhibitory control over thalamic structures. Consistent with this, four patients with central poststroke pain, two with hyperpathia, showed

hyperactivity in the thalamus contralateral to the hyperpathic side. Pattany and colleagues (2002) used proton magnetic spectroscopy to study alterations in metabolites resulting from injury-induced functional changes in thalamic nuclei following SCI. In patients with pain, the concentrations of N-acetyl- and myoinositol were different compared to those without pain, suggesting anatomical and functional changes in the region of thalamus. Using spectroscopy measurements from the brain and processing using wavelet-based feature extraction, biochemical changes have been described that distinguish control subjects from subjects with SCI as well as subdivide the SCI group into those with and without chronic pain. Thalamic changes appeared to be linked more strongly to SCI, while the anterior cingulate cortex and prefrontal cortex changes appeared to be specifically linked to the presence of pain (Stanwell et al. 2010). Cortical reorganization related to central pain following SCI was observed using fMRI (Wrigley et al. 2009) consistent with the changes in regional brain anatomy and connectivity observed using diffusion tensor imaging (Gustin et al. 2010b). "Mental allodynia" has also been reported, where imaging ankle movements caused increased neuropathic pain (Gustin et al. 2010a). In this study, activation of perigenual anterior cingulate cortex and right dorsolateral prefrontal cortex correlated with an increase in pain intensity (Gustin et al. 2010a). In a PET and fMRI study of a patient who developed central pain after an infarct involving the right parietal cortex and anterior cingulate cortex, allodynia was associated with increased activity anteriorly to the infarct on the right insula/SII cortex (Peyron et al. 2000). The authors suggested that this might reflect an abnormal transfer function between a nonnoxious input and allodynia.

Lesions Causing Central Pain

Central pain can be caused by any lesion of the nervous system that affects either completely, incompletely, or subclinically the spinothalamocortical pathway. Based on an extensive review, it was concluded that central pain can be due to lesions localized anywhere along this afferent sensory projection system, irrespective of whether cells or fibers are destroyed (Cassinari and Pagni 1969). Lesions leading to central pain are generally slow developing, and the highest prevalence of central pain is reported in cases of lesions in the spinal cord, lower brainstem, and ventroposterior part of the thalamus (Boivie 1992; Bonica 1991; Tasker 1991; Loh et al. 1981).

The most severe injury to the spinal cord is a complete spinal transection following which patients can experience pain felt below the injury and complain of uncomfortable sensations such as tightness or burning sensations. Severe pain may follow hemisection, but remote pains are rare, usually transient, lasting only a few days and are generally referred to the paralyzed, nonanalgesic side of the body but may be bilateral. Holmes (1919) attributed the spontaneous pain in these patients to local irritative effects of the lesion. Other lesions of the spinal cord causing central pain include (a) ▶ anterolateral cordotomy, (b) dorsal root entry zone coagulation (see also ▶ Junctional DREZ Coagulation), and (c) cordectomy (Pagni 1998). Spinal contusion (see also ▶ Spinal Cord Injury Pain Model, Contusion Injury Model) is the most common cause of spinal injury pain. Spinal tumors can also lead to local pain in the case of extramedullary neoplasms. Local segmental pain with intramedullary tumors is infrequent but does occur in some cases especially when the tumor arises in the posterior gray matter. One of the most pathologically destructive conditions giving rise to central pain is syringomyelia (Madsen et al. 1994, see also ▶ Pain in Syringomyelia). More than half of patients with delayed onset of central pain following SCI have syringomyelia, and it appears that the syrinx rather than the original injury is responsible for the pain (Tasker et al. 1991). The potential involvement of supraspinal mechanisms in pain associated with syringomyelia was evaluated in a combined psychophysical and fMRI study (Ducreux et al. 2006). Recently, a model of below-level mechanical allodynia was proposed involving avulsion of T13 and L1 dorsal roots (Wieseler et al. 2010). This procedure resulted in tissue damage confined to Lissauer's tract, dorsal horn, and dorsal

columns at the site of avulsion with no gross physical changes at other spinal levels. This model suggests that dorsal root avulsion is a model of central neuropathic pain.

The most common brainstem site for the development of central pain is the medulla. Central pain follows thrombosis of the posterior inferior cerebellar artery (PICA), described as Wallenberg's syndrome, and includes analgesia in the trigeminal area on the side of the lesion, which results from damage to the descending nucleus of the fifth nerve and the crossed ascending fibers in the anterolateral system. Bulbar spinothalamic tractotomy and bulbar trigeminal tractotomy (Sjogvist's operation) are also associated with central pain. In general, pain from pontobulbar lesions whether spontaneous or evoked has the same general characteristics as pain of thalamic origin. Pontobulbar pain is aggravated by emotional disturbances and whether facial or remote is often chronic and resistant to pharmacotherapy. The striking fact that central pain of mesencephalic origin is uncommon may well be due to the absence of sensory nuclei in this region. Except for cases of pontomesencephalic tumors, central pain associated with pure midbrain lesions has not been reported, although surgical lesions following spinothalamic tractotomy at mesencephalic levels have been associated with central pain (Pagni 1998).

Within the thalamus, three regions have been implicated in the onset of central pain: (a) the ventroposterior part including the posterior and interior nuclei bordering this region, (b) the medial-intralaminar region, and (c) the reticular nucleus (see ▶ Central Pain). Lesions of the posterior insula and inner parietal operculum may cause central pain with predominant thermoalgesic sensory loss (Garcia-Larrea et al. 2010).

Cortical lesions causing central pain are located primarily in the parietal cortex and perhaps the second sensory cortex where the ▶ spinothalamocortical projections are known to terminate. In general, pain is rare after brain tumors, craniotomies, or thalamotomies for movement disorders (Marshall 1951). As a rule, pain and hyperpathia occur when both sensory cortex and subcortical white matter are damaged, possibly due to the destruction of inhibitory corticothalamic fibers. Several reports have described patients with combined subcortical and cortical lesions leading to central pain, particularly with lesions in the insular region (Schmahmann and Leifer 1992). These lesions include those caused by infarcts, hematomas, meningiomas, and trauma.

Pathophysiology of Central Pain

A number of theories have been proposed to explain central pain: (a) irritation of spinothalamic and lemniscal pathways (Dejerine and Roussy 1906), (b) loss of inhibitory mechanisms controlling pain pathways (Head and Holmes 1911; Jeanmonod et al. 1994), (c) switching of importance from primary to secondary pain pathways (Cassinari and Pagni 1969; Tasker et al. 1980), and (d) the emergence of abnormal spontaneous and hyperexcitable cells (secondary to ▶ Deafferentation) at spinal and/ or supraspinal levels of the neuraxis (Pagni 1989). The potential contribution of altered gene expression following brain and spinal cord injury should also not be ignored (Nesic et al. 2005).

Since most central pain patients have abnormal temperature and pain sensibility but near-normal thresholds to touch, vibration, and joint movement (Boivie 1994), it was concluded that central pain occurs only after lesions of projections to the ventroposterior thalamic region (Pagni 1998). The fact that thalamic involvement is believed to be at the center of the mechanism responsible for the emergence of pain is underscored by the fact that anatomical and functional abnormalities are found at the termination site of pathways in this region of the brain. A critical question regarding the mechanism of central pain concerns the location of neurons responsible for this condition. Neurons within the ventroposterior nuclei have been shown to have increased spontaneous activity characterized by bursts of action potentials in the region of the nucleus representing the painful area of the body (Lenz et al. 1989). Bursting is believed to be

a fundamental characteristic of central pain and is found in both lateral and medial thalamus. Whether this abnormal burst activity is due to loss of excitatory afferent drive on postsynaptic receptors, increased activity at ▶ NMDA receptors, changes in sodium channel expression, or activation of microglia is not known. In patients with thalamic pain, spontaneous neuronal hyperactivity is also found in the mediodorsal, central lateral, central median, and parafascicular nuclei (Rinaldi et al. 1991). In patients with central pain secondary to spinal transection, thalamic cells without receptive fields due to loss of sensory input show increased bursting but decreased firing rates between bursts. These findings support the hypothesis that loss of STT input leads to hyperpolarization of these cells with resulting increased burst firing. Although the hypothesis that abnormal activity in the ventroposterior thalamic region is important for the onset of central pain, one must reconcile the fact that in some patients this region is completely silent due to existing pathology. In fact some authors contend that this region is precisely where a thalamic lesion must be located in order to precipitate central pain (Leijon et al. 1989). Some thalamic lesions are thought to remove the inhibitory influence exerted by the reticular thalamic nucleus on medial and intralaminar nuclei, thereby releasing abnormal activity leading to pain and hypersensitivity (Cesaro et al. 1991).

The pathophysiology of central pain states may also involve the irritation of cells and fibers of sensory pathways and nuclei that develops at the lesion site. A recent prospective study found that dynamic mechanical allodynia, temporal summation of pain, and hyperpathia to skin heating preceded the development of below-level pain in patients with incomplete SCI, suggesting some relation between early signs of neuronal hyperexcitability and later development of central pain (Zeilig et al. 2011). The resulting disruption of normal function is thought to lead to the development of an irritant focus (Dejerine and Roussy 1906; Livingston 1943). This hypothesis however does not explain pain onset following complete destruction of sensory pathways and nuclei or pain due to section of fiber tracts.

One explanation for the sudden disappearance of central pain after focal strokes in the subparietal white matter suggests that central pain is generated by a disturbance in the normal oscillatory mechanisms between cortex and thalamus (Canavero 1994).

Another proposed mechanism of central pain is based on the ► disinhibition hypothesis of Head and Holmes (Craig 2002). This thermosensory disinhibition theory proposes that central pain results from the disinhibition of pain resulting from imbalanced sensory integration caused by the loss of temperature sensation. This theory suggests that central pain is a thermoregulatory disorder that produces a thermal distress signal that is modulated by homoeostatic processing. At the heart of this proposal is the fact that loss of input from the lateral spinothalamic tract unmasks a homeostatic spinobulbothalamic pathway to the medial thalamus that is responsible for the development of central pain. This theory proposes that loss of activity in the thermosensory cortex in the dorsolateral mid/posterior insula disinhibits polymodal activation of the medial dorsal nucleus and anterior cingulate cortex, which produces burning pain.

Several chronic pain models with pathological and behavioral characteristics consistent with the clinical profile of SCI pain have been developed (Christensen et al. 1996; Siddall et al. 1995; Vierck and Light 1999; Xu et al. 1992; Yezierski et al. 1998, ► Central Pain: Outcome Measures in Preclinical Research). The hypersensitivity of dorsal horn ► wide dynamic range neurons (WDR) described after ischemic and excitotoxic injury of the spinal cord reflects changes similar to those described following peripheral injury (see ► Functional Changes in Sensory Neurons Following Spinal Cord Injury). The abnormal bursting patterns and evoked responses of thalamic neurons in patients with SCI support the hypothesis that the functional changes after spinal injury are not limited to the spinal cord but, as with peripheral injury, can also be found at supraspinal sites. Neuronal hyperexcitability, with increased responsiveness of neurons to their normal or subthreshold afferent input, is

thought to be a main substrate for the increase in evoked behavior to sensory stimuli observed in animal models of SCI (Gwak and Hulsebosch 2011) (see also ► Molecular Contributions to the Mechanism of Central Pain). Structural plasticity, functional changes of receptors (including N-methyl-D-aspartate (► NMDA) and other glutamate receptors) and ion channels (including sodium and calcium channels), glia activation, and hypofunction of endogenous inhibitory tone may underline the neuronal hyperexcitability seen following SCI (Berger et al. 2011). Experimental models of ► cell therapy in the treatment of central pain support the role of serotonin, noradrenaline, peptides, and gamma-aminobutyric acid (GABA) in central SCI pain. Attempts to develop models of experimental thalamic pain have included placing electrolytic lesions in different thalamic nuclei (LaBuda et al. 2000; Saade et al. 1999), excitotoxic lesions in the lateral thalamus (LaBuda et al. 2000), giving cortical injections of picrotoxin (Oliveras and Montagne-Clavel 1996), and hemorrhagic lesions (Wasserman and Koeberle 2009). All of these models produce heightened responses to peripheral stimuli, thereby providing support for their use in the study of central pain states.

Treatment of Central Pain

At present a long-term effective treatment for central pain is not available, and for this reason, the strategy for treatment is to try all available treatment modalities in order to systematically determine the best approach for an individual patient. The realistic goal of central pain treatment is to reduce the intensity of pain intensity to a tolerable level. The pharmacological treatment of central pain is discussed elsewhere (► Central Pain, ► Central Pain, Pharmacological Treatments). In brief, tricyclic antidepressants, gabapentin, and pregabalin are considered first-line drugs. Mixed serotonin-noradrenaline reuptake inhibitors may be considered when tricyclic antidepressants are not tolerated. Opioids including tramadol are considered second-line treatments. In refractory cases, ► stimulation treatments of central pain may be considered. Alternative strategies that may complement existing treatments include comprehensive educational, cognitive, and behavioral approaches (Budh et al. 2006) as well as cell and molecular approaches (Eaton 2006).

Conclusion

Chronic pain associated with injury or disease of the central nervous system represents a longstanding enigma that presents a significant challenge to the scientific and health-care communities. As a condition that seems to depend on damage to the very substrate required for pain perception, it is one that has defied effective therapeutic intervention and continues to baffle those searching for an underlying mechanism. In spite of efforts to understand the pathophysiology and underlying etiology, there remain many unanswered questions. Parallels between chronic pain states resulting from injury or disease of peripheral and central substrates offer hope for the future. The fact that there are spontaneous and evoked components of central pain suggests multiple mechanisms underlying these and other divergent clinical findings as well as the varied temporal profile and location of pain in patients with different central lesions. The role of disinhibition, sensitization, denervation, and other plastic changes in central sensory pathways remains to be addressed. Continued efforts to develop experimental models and strategies to study the human condition will hopefully lead to new insights into the progression of anatomical, chemical, and functional changes from the site of injury to higher levels of the neuraxis along with the development of novel treatments.

References

Andersen, G., Vestergaard, K., Ingeman-Nielsen, M., et al. (1995). Incidence of central post-stroke pain. *Pain, 61*, 187–193.

Beetar, J. T., Guilmette, T. J., & Sparadeo, F. R. (1996). Sleep and pain complaints in symptomatic traumatic brain injury and neurologic populations. *Archives of Physical Medicine and Rehabilitation, 77*, 1298–1302.

Berger, J. V., Knaepen, L., Janssen, S. P., et al. (2011). Cellular and molecular insights into neuropathy-induced pain hypersensitivity for mechanism-based treatment approaches. *Brain Research Reviews, 67*, 282–310.

Boivie, J. (1992). Hyperalgesia and allodynia in patients with CNS lesions. In W. D. Willis (Ed.), *Hyperalgesia and allodynia* (pp. 363–373). New York: Raven.

Boivie, J. (1994). Central pain. In P. D. Wall & R. Melzack (Eds.), *Textbook of pain* (pp. 871–902). New York: Churchill Livingston.

Bonica, J. J. (1991). Semantic, epidemiologic and educational issues of central pain. In K. Casey (Ed.), *Pain and central nervous system disease: The central pain syndromes* (pp. 13–29). New York: Raven.

Bowsher, D. R., Foy, P. M., & Shaw, M. D. (1989). Central pain complicating infarction following subarachnoid haemorrhage. *British Journal of Neurosurgery, 3*, 435–441.

Bryce, T. N., Biering-Sorensen, F., Finnerup, N. B., et al. (2012). International spinal cord injury pain (ISCIP) classification: Part I. Background and description. *Spinal Cord, 50*, 413–417.

Bryce, T. N., Budh, C. N., Cardenas, D. D., et al. (2007). Pain after spinal cord injury: An evidence-based review for clinical practice and research. Report of the national institute on disability and rehabilitation research spinal cord injury measures meeting. *The Journal of Spinal Cord Medicine, 30*, 421–440.

Budh, C. N., Kowalski, J., & Lundeberg, T. (2006). A comprehensive pain management programme comprising educational, cognitive and behavioral interventions for neuropathic pain flowing spinal cord injury. *Journal of Rehabilitation Medicine, 38*, 172–180.

Budh, C. N., Lund, I., Ertzgaard, P., Holtz, A., Hultling, C., Levi, R., Werhagen, L., & Lundeberg, T. (2003). Pain in a Swedish spinal cord injury population. *Clinical Rehabilitation, 17*, 685–690.

Canavero, S. (1994). Dynamic reverberation. A unified mechanism for central and phantom pain. *Medical Hypotheses, 42*, 203–207.

Canavero, S., Pagni, C. A., Castellano, G., et al. (1993). The role of cortex in central pain syndromes preliminary results of a long-term technetium-99 hexamethylpropyleneamineoxime single photon emission computed tomography study. *Neurosurgery, 32*, 185–191.

Cassinari, V., & Pagni, C. A. (1969). *Central pain.* Cambridge: Harvard University Press.

Cesaro, P., Mann, M. W., Moretti, J. L., et al. (1991). Central pain and thalamic hyperactivity. A single photon emission computerized tomographic study. *Pain, 47*, 329–336.

Christensen, M. D., Everhart, A. W., Pickeman, J., et al. (1996). Mechanical and thermal allodynia in chronic central pain following spinal cord injury. *Pain, 68*, 97–107.

Craig, A. D. (2002). New and old thoughts on the mechanisms of spinal cord injury pain. In R. P. Yezierski & K. J. Burchiel (Eds.), *Spinal cord injury pain: Assessment, mechanisms, management, progress in pain research and management* (Vol. 23, pp. 237–261). Seattle: IASP Press.

Dejerine, J., & Roussy, G. (1906). Le syndrome thalamique. *Revue Neurologique, 14*, 521–532.

Di Piero, V., Ferracuti, S., Sabatini, U., et al. (1994). A cerebral blood flow study on tonic pain activation in man. *Pain, 56*, 167–173.

Ducreux, D., Attal, N., Parker, F., & Bouhassira, D. (2006). Mechanisms of central neuropathic pain: a combined psychophysical and fMRI study in syringomyelia. *Brain, 129*, 963–976.

Eaton, M. J. (2006). Cell and molecular approaches to the attenuation of pain after spinal cord injury. *Journal of Neurotrauma, 23*, 549–559.

Garcia-Larrea, Perchet C., Creach, C., et al. (2010). Operculo-insular pain (parasylvian pain): A distinct central pain syndrome. *Brain, 133*, 2528–2539.

Gustin, S. M., Wrigley, P. J., Henderson, L. A., & Siddall, P. J. (2010a). Brain circuitry underlying pain in response to imagined movement in people with spinal cord injury. *Pain, 148*, 438–445.

Gustin, S. M., Wrigley, P. J., Siddall, P. J., & Henderson, L. A. (2010b). Brain anatomy changes associated with persistent neuropathic pain following spinal cord injury. *Cerebral Cortex, 20*, 1409–1419.

Gwak, Y. S., & Hulsebosch, C. E. (2011). Neuronal hyperexcitability: A substrate for central neuropathic pain after spinal cord injury. *Current Pain and Headache Reports, 15*, 215–222.

Head, H., & Holmes, G. (1911). Sensory disturbances from cerebral lesions. *Brain, 34*, 102–254.

Holmes, G. (1919). Pain of central origin. In W. Osler (Ed.), *Contributions to medical and biological research* (pp. 235–246). New York: Hoeber.

Israelsson, C., et al. (2008). Distinct cellular patterns of upregulated chemokine expression supporting a prominent inflammatory role in traumatic brain injury. *Journal of Neurotrauma, 25*, 959–974.

Jeanmonod, D., Magnin, M., & Morel, A. (1994). A thalamic concept of neurogenic pain. In G. F. Gebhart, D. L. Hammond, T. S. Jensen (Eds.), *Proceedings 7th world congress of pain*, Vol. 2. IASP Press, Seattle, pp. 767–787.

Jensen, M. P., Widerstrom-Noga, E., Richards, J. S., et al. (2010). Reliability and validity of the international spinal cord injury basic pain data set items as self-report measures. *Spinal Cord, 48*, 230–238.

Jensen, T. S., Baron, R., Haanpää, M., Kalso, E., Loeser, J. D., Rice, A. S., & Treede, R. D. (2011). A new definition of neuropathic pain. *Pain, 152*, 2204–2205.

Klit, H., Finnerup, N. B., Andersen, G., & Jensen, T. S. (2011). Central poststroke pain: A population-based study. *Pain, 152*, 818–824.

LaBuda, C. J., Cutler, T. D., Dougherty, P. M., et al. (2000). Mechanical and thermal hypersensitivity develops following kainite lesion of the ventral posterior lateral thalamus in rats. *Neuroscience Letters, 290*, 79–81.

Lahz, S., & Bryant, R. A. (1996). Incidence of chronic pain following traumatic brain injury. *Archives of Physical Medicine and Rehabilitation, 77*, 889–891.

Leijon, G., Boivie, J., & Johansson, L. (1989). Central post-stroke pain: Neurological symptoms and pain characteristics. *Pain, 36*, 13–25.

Lenz, F. A., Kwan, H. C., Dostrovsky, J. O., et al. (1989). Characteristics of the bursting pattern of action potential that occurs in the thalamus of patients with central pain. *Brain Research, 496*, 357–360.

Lew, H. L., Otis, J. D., Tun, C., Kerns, R. D., Clark, M. E., & Cifu, D. X. (2009). Prevalence of chronic pain, posttraumatic stress disorder, and persistent postconcussive symptoms in OIF/OEF veterans: Polytrauma clinical triad. *Journal of Rehabilitation Research and Development, 46*, 697–702.

Lipton, S. (1989). Percutaneous cordotomy. In P. D. Wall & R. Melzack (Eds.), *Textbook of pain* (2nd ed., pp. 832–839). Edinburgh: Churchill Livingstone.

Livingston, W. K. (1943). *Pain mechanisms*. New York: Macmillan.

Loh, L., Nathan, P. W., & Schott, G. D. (1981). Pain due to lesions of central nervous system removed by sympathetic block. *BMJ, 282*, 1026–1028.

MacGowan, D. J. L., Janal, M. N., Clark, W. C., et al. (1997). Central post-stroke pain in Wallenberg's lateral medullary infarction: Frequency, character and determinants in 63 patients. *Neurology, 49*, 120–125.

Madsen, P. W., Yezierski, R. P., & Holets, V. R. (1994). Syringomyelia: Clinical observations and experimental studies. *Journal of Neurotrauma, 11*, 241–254.

Marshall, J. (1951). Sensory disturbances in cortical wounds with special reference to pain. *The Journal of Neurology, Neurosurgery, and Psychiatry, 14*, 187–204.

Nampiaparampil, D. E. (2008). Prevalence of chronic pain after traumatic brain injury: A systematic review. *Journal of the American Medical Association, 300*, 711–719.

Nesic, O., Lee, J., Johnson, K. M., Ye, Z., Xu, G.-Y., Unabia, G. C., Wood, T. G., McAdoo, D. J., et al. (2005). Transcriptional profiling of spinal cord injury-induced central neuropathic pain. *Journal of Neurochemistry, 95*, 998–1014.

Ofek, H., & Defrin, R. (2007). The characteristics of chronic central pain after traumatic brain injury. *Pain, 131*, 330–340.

Oliveras, J. L., & Montange-Clavel, J. (1996). Picrotoxin produces a "central" pain like syndrome when microinjected into the somato-motor cortex of the rat. *Physiology and Behavior, 60*, 1425–1434.

Österberg, A., Boivie, J., & Thuomas, K. A. (2005). Central pain in multiple sclerosis–prevalence and clinical characteristics. *European Journal of Pain, 9*, 531–542.

Pagni, C. A. (1989). Central pain due to spinal cord and brain stem damage. In P. D. Wall & R. Melzack (Eds.), *Textbook of pain* (2nd ed., pp. 634–655). Edinburgh: Churchill Livingstone.

Pagni, C. A. (1998). *Central pain: A neurosurgical challenge*. Torino: Edizioni Minerva Medica.

Pattany, P. M., Yezierski, R. P., Widerstrom-Noga, E. G., et al. (2002). Proton magnetic resonance spectroscopy of the thalamus in patients with chronic neuropathic pain after spinal cord injury. *American Journal of Neuroradiology, 23*, 901–905.

Peyron, R., Garcia-Larrea, L., Grégoire, M. C., et al. (2000). Parietal and cingulate processes in central pain. A combined positron emission tomography (PET) and functional magnetic resonance imaging (fMRI) study of an unusual case. *Pain, 84*, 77–87.

Rinaldi, P. C., Young, R. F., Albe-Fessard, D., et al. (1991). Spontaneous neuronal hyperactivity in the medial and intralaminar thalamic nuclei of patients with deafferentation pain. *Journal of Neurosurgery, 74*, 415–421.

Rintala, D. H., Loubser, P. G., Castro, J., et al. (1998). Chronic pain in a community-based sample of men with spinal cord injury: Prevalence, severity, and relationship with impairment, disability, handicap, and subjective well-being. *Archives of Physical Medicine and Rehabilitation, 79*, 604–614.

Saade, N. E., Katroum, A. L., Saab, C. Y., et al. (1999). Chronic thalamotomy increases pain-related behavior in rats. *Pain, 83*, 401–409.

Schestatsky, P., Kumru, H., Valls-Sole, J., et al. (2007). Neurophysiologic study of central pain in patients with Parkinson disease. *Neurology, 69*, 2162–2169.

Schmahmann, J. D., & Leifer, D. (1992). Parietal pseudothalamic pain syndrome: Clinical features and anatomic correlates. *Archives of Neurology, 49*, 1032–1037.

Schneiderman, Al, Braver, Er, & Kang, H. K. (2008). Understanding sequelae of injury mechanisms and mild traumatic brain injury incurred during conflicts in Iraq and Afghanistan: Persistent postconcussive symptoms and posttraumatic stress disorder. *American Journal of Epidemiology, 167*, 1446–1452.

Shieff, C., & Nashold, B. S. (1987). Stereotactic mesencephalic tractotomy for thalamic pain. *Neurological Research, 9*, 101–104.

Siddall, P., Xu, C. L., & Cousins, M. (1995). Allodynia following traumatic spinal cord injury in the rat. *Neuroreport, 6*, 1241–1244.

Siddall, P. J., McClelland, J. M., Rutkowski, S. B., & Cousins, M. J. (2003). A longitudinal study of the prevalence and characteristics of pain in the first 5 years following spinal cord injury. *Pain, 103*, 249–257.

Stanwell, P., Siddall, P., Keshava, N., et al. (2010). Neuro magnetic resonance spectroscopy using wavelet decomposition and statistical testing identifies biochemical changes in people with spinal cord injury and pain. *NeuroImage, 53*, 544–552.

Tasker, R. R. (1991). Deafferentation pain syndromes. In B. S. Nashold & J. Ovelmen-Levitt (Eds.), *Deafferentation pain syndromes. Pathophysiology and treatment* (pp. 241–258). New York: Raven.

Tasker, R. R., Organ, L. W., & Hawrylyshyn, P. (1980). Deafferentation and causalgia. In J. J. Bonica (Ed.), *Pain* (pp. 305–329). New York: Raven.

Tasker, R. R., DeCarvalho, G., & Dostrovsky, J. O. (1991). The history of central pain syndromes, with observations concerning pathophysiology and treatment. In K. L. Casey (Ed.), *Pain and central nervous system disease: The central pain syndromes* (pp. 31–58). New York: Raven.

Vestergaard, K., Nielsen, J., Andersen, G., et al. (1995). Sensory abnormalities in consecutive, unselected patients with central post-stroke pain. *Pain, 61*, 177–186.

Vierck, C. J., & Light, A. R. (1999). Effects of combined hemotoxic and anterolateral spinal lesions on nociceptive sensitivity. *Pain, 83*, 447–457.

Wasserman, J. K., & Koeberle, P. D. (2009). Development and characterization of a hemorrhagic rat model of central post-stroke pain. *Neuroscience, 161*, 173–183.

Werhagen, L., Hulting, C., & Molander, C. (2007). The prevalence of neuropathic pain after non-traumatic spinal cord lesion. *Spinal cord, 45*, 609–615.

Werhagen, L., Budh, C. N., Hultling, C., & Molander, C. (2004). Neuropathic pain after traumatic spinal cord injury – Relations to gender, spinal level, completeness, and age at the time of injury. *Spinal Cord, 42*, 665–673.

White, J. C., & Sweet, W. H. (1969). *Pain and the neurosurgeon: A 40-year experience.* Springfield: Charles C Thomas.

Widerström-Noga, E. G., Felipe-Cuervo, E., Broton, J. G., et al. (1999). Perceived difficulty in dealing with consequences of spinal cord injury. *Archives of Physical Medicine and Rehabilitation, 80*, 580–586.

Widerström-Noga, E., Biering-Sørensen, F., Bryce, T., Cardenas, D. D., Finnerup, N. B., Jensen, M. P., Richards, J. S., & Siddall, P. J. (2008). The international spinal cord injury pain basic data set. *Spinal Cord, 46*, 818–823.

Wieseler, J., Ellis, A. L., McFadden, A., Brown, K., Starnes, C., Maire, S. F., Watkins, L. R., & Falci, S. (2010). Below level central pain induced by discrete dorsal spinal cord injury. *Journal of Neurotrauma, 27*, 1697–1707.

Wrigley, P. J., Press, S. R., Gustin, S. M., et al. (2009). Neuropathic pain and primary somatosensory cortex reorganization following spinal cord injury. *Pain, 141*, 52–59.

Xu, X.-J., Hao, J.-X., Aldoskogius, H., et al. (1992). Chronic pain-related syndrome in rats after ischemic spinal cord lesion: A possible animal model for pain in patients with spinal cord injury. *Pain, 48*, 279–290.

Yezierski, R. P. (1996). Pain following spinal cord injury: The clinical problem and experimental studies. *Pain, 68*, 185–194.

Yezierski, R. P., Liu, S., Ruenes, G. L., et al. (1998). Excitotoxic spinal cord injury: Behavioral and morphological characteristics of a central pain model. *Pain, 75*, 141–155.

Zeilig, G., Enosh, S., Rubin-Asher, D., et al. (2011). The nature and course of sensory changes following spinal cord injury: Predictive properties and implications on the mechanism of central pain. *Brain, 135*, 418–430.

Neuropathic Pain, Diagnosis, Pathology, and Management

Robert Gassin
Musculoskeletal Medicine Clinic, Frankston, VIC, Australia

Synonyms

Neurogenic pain; Painful neuropathy

Definition

▶ Neuropathic pain is pain initiated or caused by a primary lesion or dysfunction in the nervous system (Merskey and Bogduk 1994).

This definition allows neuropathic pain to encompass disorders of either the peripheral nervous system or the central nervous system. In the present context, neuropathic pain refers to pain associated with diseases or injuries of peripheral nerves. Pain associated with disorders of the central nervous system is more specifically referred to as ▶ central pain and covered separately (see ▶ Central Pain).

Characteristics

Peripheral neuropathic pain occurs in a variety of diseases of peripheral nerves. These include painful peripheral neuropathies, such as diabetic neuropathy, alcoholic neuropathy, and postherpetic neuralgia. It can also occur after injuries to peripheral nerves, such as avulsions, stretching or crush injury, or nerve transection.

Diagnosis

The diagnosis of neuropathic pain is suggested by the presence of certain features revealed by the ▶ medical history and neurological examination (see ▶ Diagnosis of Pain, Neurological Examination). The history reveals certain features about the nature of the pain and may provide a cause. The examination reveals features of loss of neurological function or exaggerated neurological function.

Neuropathic pain is commonly worse at night. It is often described as shooting, stabbing, lancinating, burning, or searing. It can be continuous but often presents as paroxysms of pain in the absence of any identifiable stimulus. In some conditions, such as nerve entrapment syndromes, the pain follows a nerve distribution, whereas in others such as neuropathies and poststroke pain, the distribution is more diffuse and may affect more than one body region. Pain experienced in a numb or insensate site (anesthesia dolorosa) is highly suggestive of the diagnosis.

Cardinal to the diagnosis are features of disturbed neurological function. These may be in the form of loss of function, such as numbness, which indicates nerve damage directly, or they may be in the form of exaggerated function, which suggest loss of inhibition and imply nerve damage. The latter features include hyperalgesia (increased sensitivity to noxious stimuli), hyperpathia (increase response to minimal noxious stimulation), hyperesthesia (increased sensitivity to touch), and allodynia (touch or brush is perceived as painful). Sometimes, neuropathic pain can be accompanied by the features of ▶ complex regional pain syndrome, such as temperature changes in the skin, swelling, skin color changes, and increased or decreased sweating.

Patients suffering neuropathic pain commonly suffer insomnia, anxiety, depression, and a significant deterioration in their quality of life.

The diagnosis is essentially made on the basis of these clinical features. The diagnosis is consolidated if the patient has an obvious reason to have pain, such as a nerve injury or a metabolic disorder known to cause peripheral neuropathy. Investigations, such as ▶ electrodiagnostic studies, are not required but may help to establish the exact etiology in cases of peripheral neuropathy.

Pathology

Nociceptive pain is caused by noxious stimulation of Aδ fiber mechanothermal and C fiber polymodal nociceptors by algogenic substances. In contrast, neuropathic pain results from damage to, or pathological changes in, the axons of peripheral nerves.

The nature of the pathological changes leading to pain is poorly understood. Much of our understanding is based on animal models of neuropathic pain. However, the relevance of such models to the human experience of neuropathic pain has been questioned.

The physiological and structural changes following nerve injuries have been summarized in several review articles (Siddall and Cousins 1998; Woolf and Mannion 1999). Following peripheral nerve injury, changes include:

1. Hyperexcitability and spontaneous action potential discharges from the damaged primary afferent fibers proximal to the injury, and near the dorsal root ganglion, due to an accumulation of voltage-sensitive sodium channels at the injury site and along the damaged axons
2. Formation of neuromas where a nerve has been cut or microneuromas where a nerve has been partially injured, which generates spontaneous action potentials in nociceptive afferents
3. Sympathetic neo-innervation of the dorsal root ganglia of the damaged primary afferents which may lead to sympathetic activity initiating activity in sensory nerve fibers
4. Changes in the morphology and physiological properties of dorsal horn neurones
5. Decreased inhibitory mechanisms within the spinal cord due to several mechanisms, including a reduction of the inhibitory transmitter GABA and upregulation of the expression of CCK, an inhibitor of opioid receptors, and downregulation of GABA and opioid receptors
6. Sprouting of the central terminals of damaged Aβ touch fibers from Lamina IV of the dorsal horn to lamina II, where normally only nociceptors terminate, and formation of contacts with pain transmission neurones projecting to the brain
7. Spontaneous firing of pain transmission neurones that project to the brain but have lost their normal peripheral afferent input

8. Death of interneurons in lamina II of the dorsal horn

Some combination of these changes is thought to account for at least some of the clinical features of neuropathic pain conditions.

Prevalence

Neuropathic pain has been estimated to affect at least 0.6 % of the United States population (Bennett 1998). However, this is most likely to be an underestimation, as it does not include sufferers of chronic musculoskeletal conditions such as neck and back pain, including postsurgical patients, who often present with symptoms and signs highly suggestive of neuropathic pain. The true prevalence is thought to be closer to 2 %.

Management

The management of neuropathic pain conditions is dependent on the diagnosis. For certain conditions such as nerve entrapment syndromes, management of the underlying cause of pain, whether it be malignant or nonmalignant, often results in rapid improvement in symptoms. With varying degrees of success, neuromas can be resected, buried, or ligated, using a variety of neurosurgical techniques (see ▶ Dorsal Root Ganglionectomy and Dorsal Rhizotomy).

For most other neuropathic pain conditions, management is not as rewarding, and a multimodal approach is required.

Psychological approaches are geared at challenging and improving patterns of negative thoughts and dysfunctional attitudes, promoting healthy thoughts and emotions and encouraging reactivation, as well as educating the patient regarding their illness and in nonmedical pain management strategies such as ▶ relaxation, imagery, and hypnotic techniques.

Physical therapists have a role in maximizing function in deactivated patients and in assessing the potential benefit of splints and other aids.

Medical approaches often form the mainstay of treatment and include the use of pharmacological agents, ▶ nerve blocks, ▶ intravenous infusions, ▶ TENS, ▶ acupuncture, and ▶ spinal cord stimulation.

Pharmacological agents can be administered topically or systemically. Several of the agents presently available for the management of neuropathic conditions are believed to act as blockers of neuronal sodium channels or by their action on noradrenaline and serotonin reuptake.

For most neuropathic pain conditions, ▶ tricyclic antidepressants including amitriptyline and imipramine, and the anticonvulsant gabapentin, are considered first-line agents. Their effectiveness has been confirmed for postherpetic neuralgia and for painful diabetic neuropathy (Backonja et al. 1998; Graff-Radford et al. 2000; Max et al. 1987; Rowbotham et al. 1998; Sindrup et al. 1989).

There is increasing evidence for the effectiveness of topical lignocaine in relieving the pain of postherpetic neuralgia and other neuropathic pain conditions (Galer 2001; Sawynok 2003). In view of its good safety profile, this preparation is being increasingly used and gaining acceptance as a first-line agent.

Other pharmacological agents commonly used in the management of neuropathic pain include topical capsaicin, mexiletine, carbamazepine, sodium valproate, tramadol, slow release opioids, intravenous and intranasal ketamine, and subcutaneous lignocaine.

For patients with intractable pain, a programmable intrathecal drug pump allowing continuous delivery of active agents, or spinal cord stimulation are options.

References

Backonja, M., Beydoun, A., Edwards, K. R., Schwartz, S. L., Fonseca, V., Hes, M., et al. (1998). Gabapentin for the symptomatic treatment of painful neuropathy in patients with diabetes mellitus. A randomized controlled trial. *Journal of the American Medical Association, 280*(21), 1831–1836.

Bennett, G. J. (1998). Neuropathic pain: New insights, new interventions. *Hospital Practice (Office Edition), 33*(10), 95–98. passim; online: http://www.hosppract.com/issues/1998/10/bennett.htm.

Galer, B. (2001). Topical medications. In J. D. Loeser (Ed.), *Bonica's management of pain* (3rd ed., pp. 1736–1742). Philadelphia: Lippincott Williams & Wilkins.

Graff-Radford, S. B., Shaw, L. R., & Naliboff, B. N. (2000). Amitriptyline and fluphenazine in the treatment of postherpetic neuralgia. *The Clinical Journal of Pain, 16*, 188–192.

Max, M. B., Culnane, M., Schafer, S., et al. (1987). Amitriptyline relieves diabetic neuropathy pain in patients with normal or depressed mood. *Neurology, 37*, 589–596.

Merskey, H., & Bogduk, N. (1994). *Classification of chronic pain: Description of chronic pain syndromes and definitions of pain terms* (2nd ed., pp. 209–214). Seattle: IASP Press.

Rowbotham, M., Harden, N., Stacey, B., Bernstein, P., & Magnus Miller, L. (1998). Gabapentin for the treatment of postherpetic neuralgia. *Journal of the American Medical Association, 280*, 1837–1842.

Sawynok, J. (2003). Topical and peripherally acting analgesics. *Pharmacological Reviews, 55*, 1–20.

Siddall, P. J., & Cousins, M. J. (1998). Clinical aspects of present models of neuropathic pain. *Pain Reviews, 5*, 101–123.

Sindrup, S. H., Ejlertsen, B., Frøland, A., Sindrup, E. H., Brøsen, K., & Gram, L. F. (1989). Imipramine treatment in diabetic neuropathy: Relief of subjective symptoms without changes in peripheral and autonomic nerve function. *European Journal of Clinical Pharmacology, 37*, 151–153.

Woolf, C. J., & Mannion, R. J. (1999). Neuropathic pain: Aetiology, symptoms, mechanisms, and management. *Lancet, 353*, 1959–1964.

Neuropathic Pain, Joint and Muscle Origin

Martin Michaelis
Sanofi-Aventis Deutschland GmbH, Frankurt am Main, Germany

Synonyms

Nerve pain of joint and muscle origin; Neurogenic pain of joint and muscle origin

Definition

Pain caused by a lesion or dysfunction of a dorsal root, spinal nerve, or peripheral nerve trunk that is reported by the patient to be originated in muscle or joint tissue (projected neuropathic pain).

Pain caused by a lesion or dysfunction of nerve fibers in consequence of disease processes affecting primarily nonneuronal joint or muscle tissues.

Characteristics

A typical example of a proximal nerve lesion associated with pain felt in deep somatic tissues is lumbar radiculopathy (sciatica) evoked by herniation of an intervertebral disc. The lesion results from mechanical nerve compression in concert with localized inflammation. Depending on the lumbar root affected, the pain is predominantly felt in the low back, anterior or posterior thigh, knee, or lower leg. Many patients describe their pain as aching and cramping, which are characteristics of pain arising from muscle. Others describe their pain as sharp and shooting (Dubuisson 1999). Other injuries affect nerves that supply solely muscle tissue; here the patient's pain is definitely projected neuropathic muscle pain. Examples are iatrogenic injury of a spinal accessory nerve, which occurs as a complication of radical neck dissection or cervical lymph node biopsy (London et al. 1996), and entrapment of a part of the brachial plexus, the subscapular nerve, which is a known cause of the thoracic outlet syndrome (Huang and Zager 2004).

Sciatica is painful because the nerve lesion generates extra action potentials in sensory neurons. The magnitude and pattern of the evoked impulse barrage, along with the innervation territory of the affected nerve fibers, determine the quality of the pain sensation and the tissue domain where the patient feels the pain. Neuronal activity may originate in sensory neurons that are directly affected by the nerve lesion. For example, following axotomy, impulse generation occurs in the neuroma and ▶ dorsal root ganglia (DRGs) containing axotomized neurons. Mechanisms of ▶ ectopic activity generation in lesioned neurons are outlined elsewhere in this volume.

Preclinical work indicates a specific form of projected neuropathic pain of muscle origin. Following surgically induced nerve lesions in rats, muscle afferents proved to be particularly prone to developing ongoing ectopic discharges originating in DRGs, whereas cutaneous afferents did not (Michaelis et al. 2000). Muscle afferents started to produce this abnormal activity even when they were not directly injured. Lesion of neighboring nerves was sufficient, and the more neighboring afferents were lesioned, the higher were the frequency of the ectopic discharges. Some firing muscle afferents probably had nociceptive function. Ongoing activity in these afferents may therefore generate ongoing muscle pain. Ongoing activity in muscle afferents may in addition contribute to the development of central sensitization. Intriguingly, input through small diameter muscle afferents appears to produce central sensitization of longer duration than a similar input through cutaneous C fibers, which may be of clinical relevance beyond neuropathic pain (Wall and Woolf 1984). Furthermore, nerve lesion induces increase in brain-derived neurotrophic factor (BDNF) expression in large-diameter DRG neurons (Michael et al. 1999), which may be another key factor for development of central sensitization after nerve lesion (Thompson et al. 1999).

Neuropathic pain that is exclusively projected to a joint is very unlikely, because relatively few sensory neurons supply joints compared to muscle and the likelihood that a process like disc herniation predominantly affects joint afferents is accordingly low. However, projected neuropathic pain often includes joints.

Traumatic nerve lesions are well-known triggers for the generation of neuropathic pain. The final treatment option for patients with knee or hip osteoarthritis is surgical joint replacement, which inevitably destroys the distal parts of many nerve fibers, particularly those that had supplied subchondral bone of the removed joint. Intriguingly, the vast majority of patients is completely pain-free or report substantial pain reduction soon after replacement surgery (McLain and Weinstein 1999). Thus, joint replacement surgery very probably does not induce neuropathic pain.

The question as to whether or not in the course of chronic painful diseases that affect primarily nonneuronal joint or muscle tissues a neuropathic component is developing is hard to answer with certainty. This is mainly because the clinical diagnosis of neuropathic pain is difficult when an overt neurological disease (e.g., postherpetic neuralgia or traumatic lesion of a major peripheral nerve) has been ruled out. In a recent study, Rasmussen et al. (2004) asked whether signs and symptoms cluster differentially in groups of patients with increasing evidence of neuropathic pain. Patients were first categorized to have definite, possible, or unlikely neuropathic pain. Subsequently, patients were examined using pain descriptors, intensity of five categories of pain, questionnaires, and sensory tests. This resulted in a considerable overlap concerning signs and symptoms; single pain descriptors could not distinguish between the three categories of patients (Rasmussen et al. 2004). In addition, disease markers for neuropathic pain (e.g., biochemical or functional imaging) are not available yet. Regardless, patients with painful osteoarthritis, by far the most common joint disease, or rheumatoid arthritis are traditionally said to have nociceptive or inflammatory pain, not neuropathic pain (Backonja 2003; Rasmussen et al. 2004).

What can we learn from animal models? Recently, two rat models of osteoarthritis with well-characterized and human OA-like joint histopathology have been analyzed for pain-related behavior (Combe et al. 2004; Fernihough et al. 2004). Surprisingly, at the same time that osteoarthritic joint pathology developed, a marked tactile allodynia was detected. Pharmacological investigations in one of the models, OA induced by intra-articular injection of iodoacetate, revealed that morphine and gabapentin, which are effective in the treatment of neuropathic pain, but not diclofenac, were able to diminish the tactile allodynia (Combe et al. 2004; Fernihough et al. 2004). Since tactile allodynia is not a hallmark of OA symptomatology, the relevance of this preclinical result for human OA pain is unclear.

In conclusion, it is not known yet whether neuropathic pain can be caused by a lesion or dysfunction of nerve fibers that innervate joint or muscle tissues.

References

Backonja, M. M. (2003). Defining neuropathic pain. *Anesthesia and Analgesia, 97*, 785–790.

Combe, R., Bramwell, S., & Field, M. J. (2004). The monosodium iodoacetate model of osteoarthritis: A model of chronic nociceptive pain in rats? *Neuroscience Letters, 370*, 236–240.

Dubuisson, D. (1999). Nerve root disorders and arachnoiditis. In P. D. Wall & R. Melzack (Eds.), *Textbook of pain* (4th ed., pp. 851–878). New York: Churchill Livingstone.

Fernihough, J., Gentry, C., Malcangio, M., et al. (2004). Pain related behaviour in two models of osteoarthritis in the rat knee. *Pain, 112*, 83–93.

Huang, J., & Zager, E. L. (2004). Thoracic outlet syndrome. *Neurosurgery, 55*, 897–902.

London, J., London, N. J., & Kay, S. P. (1996). Iatrogenic accessory nerve injury. *Annals of the Royal College of Surgeons of England, 78*, 146–150.

McLain, R. F., & Weinstein, J. N. (1999). Orthopedic surgery. In P. D. Wall & R. Melzack (Eds.), *Textbook of pain* (4th ed., pp. 1289–1306). New York: Churchill Livingstone.

Michael, G. J., Averill, S., Shortland, P. J., et al. (1999). Axotomy results in major changes in BDNF expression by dorsal root ganglion cells: BDNF expression in large trkB and trkC cells, in pericellular baskets, and in projections to deep dorsal horn and dorsal column nuclei. *European Journal of Neuroscience, 11*, 3539–3551.

Michaelis, M., Liu, X., & Jänig, W. (2000). Axotomized and intact muscle afferents but no skin afferents develop ongoing discharges of dorsal root ganglion origin after peripheral nerve lesion. *Journal of Neuroscience, 20*, 2742–2748.

Rasmussen, P. V., Sindrup, S. H., Jensen, T. S., et al. (2004). Symptoms and signs in patients with suspected neuropathic pain. *Pain, 110*, 461–469.

Thompson, S. W., Bennett, D. L., Kerr, B. J., et al. (1999). Brain-derived neurotrophic factor is an endogenous modulator of nociceptive responses in the spinal cord. *Proceedings of the National Academy of Sciences of the United States of America, 96*, 7714–7718.

Wall, P. D., & Woolf, C. J. (1984). Muscle but not cutaneous C-afferent input produces prolonged increases in the excitability of the flexion reflex in the rat. *The Journal of Physiology, 356*, 443–458.

Neuropathic Syndrome

▶ Lower Back Pain, Physical Examination

Neuropathy

Definition

Neuropathy is a disturbance of function or pathological change in a nerve (Merskey and Bogduk 1994). Mononeuropathy refers to affection of a single nerve, mononeuritis multiplex to several nerves, and polyneuropathy to diffuse involvement of the peripheral nerves.

Cross-References

▶ Descending Modulation and Persistent Pain
▶ Guillain-Barré Syndrome
▶ Hereditary Neuropathies
▶ Neuropathic Pain in Children
▶ Toxic Neuropathies

References

Merskey, H., & Bogduk, N. (1994). Task force on taxonomy. Classification of chronic pain. Descriptions of chronic pain syndromes and definitions of pain terms (2nd ed.). International Association for the Study of Pain. Seattle: IASP Press.

Neuropathy Due to HAART

▶ Pain in Human Immunodeficiency Virus Infection and Acquired Immune Deficiency Syndrome

Neuropeptide Nociceptin 1

▶ Orphanin FQ

Neuropeptide Release

▶ Neuropeptide Release in Inflammation

Neuropeptide Release in Inflammation

Beate Averbeck
Sanofi-Aventis Deutschland GmbH, Frankfurt, Germany

Synonyms

Inflammation, Neuropeptide release

Definition

▶ Neuropeptides are a group of small peptides with 4 to more than 40 amino acids found in the central and peripheral nervous system. This entry focuses on neuropeptides that are released into peripheral tissues upon neuronal activation; that is the case, for example, during an ▶ inflammation. The neuropeptides are synthesized in the cell body of primary afferent (sensory) neurons located in the ▶ dorsal root ganglia and are transported through the axons into preferentially the peripheral nerve endings. The best-known neuropeptides in primary afferent neurons are ▶ substance P and ▶ calcitonin-gene-related peptide.

Characteristics

▶ Nociceptors are primary afferent neurons, many of which are characterized by their ability to release neuropeptides such as substance P (SP, 11 amino acids), calcitonin gene-related peptide (CGRP, 37 amino acids), and ▶ neurokinins (10 amino acids) from their peripheral terminals upon noxious stimulation (Maggi 1995). In isolated tissue preparations (e.g., skin and dura), the release of SP and CGRP was shown to be induced by noxious stimulation with ▶ inflammatory mediators (▶ bradykinin, ▶ serotonin, ▶ histamine, ▶ prostaglandins, and protons) (Averbeck and Reeh 2001; Ebersberger et al. 1999). In addition, electrical stimulation of cutaneous nociceptors was demonstrated to cause neuropeptide release from the skin (Kress et al. 1999). The release of SP and CGRP leads to arteriolar vasodilatation, which becomes visible as a flare surrounding a site of injury, and to plasma extravasation from post-capillary venules, which may become apparent as a wheal at the site of injury. Since these inflammatory signs depend on the function and integrity of the peripheral sensory nervous system, the response has been termed ▶ neurogenic inflammation. By means of this mechanism, neuropeptides are able to induce the release of secondary mediators like prostaglandins, cytokines, and histamine from endothelial and inflammatory cells (monocytes, mast cells) thereby maintaining inflammation. In addition, neuropeptides show immunomodulatory and trophic effects and thus play a role in tissue repair. A direct exciting effect on nociceptors could not be found for neuropeptides, whereas sensitizing effects have been described in the literature. SP was shown to sensitize nociceptors to mechanical stimulation in the knee joint of anesthetized cats (Herbert and Schmidt 2001) and preliminary data point to CGRP sensitizing nociceptors to heat stimulation in the isolated rat skin. The biological actions of neuropeptides are limited by degradation caused by neutral endopeptidases located in many tissues surrounding the primary afferent nerve fibers Fig. 1.

Inflammation is a critical protective reaction to irritation, injury, or infection, characterized by redness (*rubor*), heat (*calor*), swelling (*tumor*), loss of function (*funtio lasea*), and pain (*dolor*). During inflammation, inflammatory mediators reach the nerve terminals, causing excitation and sensitization of nociceptors, which results in ▶ hyperalgesia and pain. Due to the fact that inflammatory mediators induce neuropeptide release by acting directly on nociceptors, inflammation is linked to an increased neuropeptide release from inflamed tissue. In addition, inflammation may result in an upregulation of neuropeptide levels in primary afferent nerves, due to an enhanced neuropeptide biosynthesis in the dorsal root ganglion cells. In rats with ▶ adjuvant-induced arthritis, the levels of CGRP were found to be increased in the dorsal root ganglia and in the sciatic nerves, particularly in those fibers that innervate the inflamed area (Kuraishi et al. 1989). In contrast, the synovia of the arthritic joints of these animals showed less immunostaining for SP and CGRP, pointing to an increased neuropeptide release from the nerve terminals to the synovial fluid (Konttinen et al. 1990). The phenomenon of low neuropeptide immunostaining in the synovia and high neuropeptide levels in the synovial fluid was also found in patients with rheumatoid arthritis (Menkes et al. 1993). In experimental ▶ colitis in rats, a significant reduction of CGRP- and SP-immunoreactive nerve fibers was observed in the mucosa, again pointing to an enhanced neuronal neuropeptide release in inflammation (Miampamba and Sharkey 1998). The observation that the intensity and density of neuropeptide-containing nerve fibers increased in the circular muscle 7 days after the induction of colitis suggests their possible involvement in tissue repair (Miampamba and Sharkey 1998). In inflammatory skin diseases, certain neuropeptides have been found to be enhanced, for example, SP and CGRP in ▶ urticaria and SP and pituitary adenylate cyclase-activating polypeptide (PACAP) in psoriasis (Steinhoff et al. 2003). Thus, the increased number of mast cells and their characteristic degranulation observed in early psoriasis might be due to pathological neuropeptide release.

Blocking the neuropeptide response by blocking the neuropeptide binding sites may have anti-inflammatory and analgesic effects. A CGRP antagonist being clinically tested in ▶ migraine headache revealed an analgesic effect (Olesen et al. 2004). Thus, blocking CGRP effects postjunctionally (e.g., on blood vessels) seems to be an effective antinociceptive mechanism, at least in migraine. Another analgesic principle would be the use of anti-inflammatory

Neuropeptide Release in Inflammation, Fig. 1 Neuropeptides in inflammation. Neurogenic inflammation. Painful stimuli on peripheral tissues are detected by primary afferent neurons (nociceptors), the cell bodies of which lie in the dorsal root ganglion (DRG). The painful signal is then transmitted to neurons in the spinal cord and further on to higher centers of the brain. Nociceptor activation results in the release of neuropeptides, which are synthesized in the nociceptors' cell bodies and transported to the central and peripheral nerve endings. The release of neuropeptides, such as substance P (SP) and calcitonin gene-related peptide (CGRP), from the peripheral nerve endings causes arteriolar vasodilatation and plasma extravasation from post-capillary venules, seen as typical signs of neurogenic inflammation: local edema, hyperemia, and an erythema which extends beyond the site of stimulation (so-called flare response). The neuropeptides released maintain inflammation by releasing secondary mediators, for example, histamine by mast cell degranulation (Modified from Mantyh et al. 2002; reprinted by permission from Nat Rev Cancer 2: 201–209 copyright 2002 Macmillan Magazines Ltd)

neuropeptides. Somatostatin (SOM, 14 amino acids) which is synthesized in primary afferent neurons, but also in most major peripheral organs, revealed anti-inflammatory effects in inflammatory pain models (Heppelmann and Pawlak 1997). SOM, like SP and CGRP, is released upon nerve fiber activation and in inflammation. In rats, SOM was demonstrated to be released into the blood circulation upon electrical stimulation of nociceptors, revealing a systemic anti-inflammatory action (Szolcsànyi et al. 1998). As the clinical use of SOM is limited by its rapid degradation after systemic injection and its neurotoxicity when applied intrathecally, a new analgesic strategy for inflammatory pain involving the modulation of localized release of SOM is under discussion.

References

Averbeck, B., & Reeh, P. W. (2001). Interactions of inflammatory mediators stimulating release of calcitonin gene-related peptide, substance P and prostaglandin E2 from isolated rat skin. *Neuropharmacology, 40*, 416–423.

Ebersberger, A., Averbeck, B., Messlinger, K., et al. (1999). Release of substance P, calcitonin gene-related peptide and prostaglandin E2 from rat dura mater encephali following electrical and chemical stimulation in vitro. *Neuroscience, 89*, 901–907.

Heppelmann, B., & Pawlak, M. (1997). Inhibitory effect of somatostatin on the mechanosensitivity of articular afferents in normal and inflamed knee joints of the rat. *Pain, 73*, 377–382.

Herbert, M. K., & Schmidt, R. F. (2001). Sensitization of group III articular afferents to mechanical stimuli by substance P. *Inflammation Research, 50*, 257–282.

Konttinen, Y. T., Rees, R., Hukkanen, M., et al. (1990). Nerves in inflammatory synovium:

Immunohistochemical observations on the adjuvant arthritis rat model. *Journal of Rheumatology, 17,* 1586–1591.

Kress, M., Guthmann, C., Averbeck, B., et al. (1999). Calcitonin gene-related peptide, substance P and prostaglandin E2 release induced by antidromic nerve stimulation from rat skin, in vitro. *Neuroscience, 89,* 303–310.

Kuraishi, Y., Nanayama, T., Ohno, H., et al. (1989). Calcitonin gene-related peptide increases in the dorsal root ganglia of adjuvant arthritic rat. *Peptides, 10,* 447–452.

Maggi, C. A. (1995). Tachykinins and calcitonin gene-related peptide (CGRP) as cotransmitters released from peripheral endings of sensory nerves. *Progress in Neurobiology, 45,* 1–98.

Mantyh, P. W., Clohisy, D. R., Koltzenburg, M., et al. (2002). Molecular mechanisms of cancer pain. *Nature Reviews Cancer, 2,* 201–209.

Menkes, C. J., Renoux, M., Laoussadi, S., et al. (1993). Substance P levels in the synovium and synovial fluid from patients with rheumatoid arthritis and osteoarthritis. *Journal of Rheumatology, 20,* 714–717.

Miampamba, M., & Sharkey, K. A. (1998). Distribution of calcitonin gene-related peptide, Somatostatin, substance P and vasoactive intestinal polypeptide in experimental colitis in rats. *Neurogastroenterology and Motility, 10,* 315–329.

Olesen, J., Diener, H. C., Husstedt, I. W., et al. (2004). Calcitonin gene-related peptide receptor antagonist BIBN 4096 BS for the acute treatment of migraine. *New England Journal of Medicine, 350,* 1104–1110.

Steinhoff, M., Ständer, S., Seeliger, S., et al. (2003). Modern aspects of cutaneous neurogenic inflammation. *Archives of Dermatology, 139,* 1478–1488.

Szolcsànyi, J., Helyes, Z., Oroszi, G., et al. (1998). Release of somatostatin and its role in the mediation of the anti-inflammatory effect induced by antidromic stimulation of sensory fibres of rat sciatic nerve. *British Journal of Pharmacology, 123,* 936–942.

Neuropeptide Y

Synonyms

NPY

Definition

Neuropeptide Y (NPY) is a 36-amino acid peptide containing many tyrosine (Y) residues. NPY is a cotransmitter of noradrenaline in sympathetic neurons, with long-lasting vasoconstrictor effects and potent influences on the immune system. It is not contained in normal sensory afferents. It is the most abundant neuropeptide known in the brain. NPY modulates pain sensation as well as other neuronal processes and is involved in the regulation of food intake and energy expenditure, in the central control of reproductive hormones, and in anxiolytic and antiepileptic control circuits.

Cross-References

▶ Peptides in Neuropathic Pain States

Neuropeptides

Definition

Neuropeptides are bioactive molecules consisting of a varying number of amino acids (2 to >40). Neuropeptides are synthesized (mostly, but not exclusively) in and released from neurons. The final, pharmacologically active peptides are cleaved from large precursor peptides. Over the past two decades, more than 40 biologically active polypeptides, termed neuropeptides, have been found in the central and peripheral nervous system. Neuropeptides such as substance P (SP), calcitonin gene-related peptide (CGRP), neurokinin (NK) A and B, somatostatin (SOM), vasoactive intestinal peptide (VIP), neuropeptide Y, and cholecystokinin (CCK) are synthesized in the cell body of autonomic and sensory neurons located in the autonomic or sensory (dorsal root) ganglia. As they are released upon neuronal activation, neuropeptides act centrally as neurotransmitters or neuromodulators. In the periphery, neuropeptides have various functions such as vasodilatation, plasma extravasation, sweat gland stimulation, mast cell degranulation, immunomodulation, and trophic effects, thereby playing an important role in inflammation. Neuropeptides are metabolically more stable than amine transmitters; therefore, they have a longer duration of action and can diffuse over longer distances.

Neurons may contain several neuropeptides coexisting with small molecule transmitters.

Cross-References

▶ Cytokines, Effects on Nociceptors
▶ Descending Circuitry, Transmitters and Receptors
▶ Neuropeptide Release in Inflammation
▶ Nociceptor Generator Potential
▶ Opioid Modulation of Nociceptive Afferents In Vivo
▶ Retrograde Cellular Changes after Nerve Injury
▶ Sensitization of Muscular and Articular Nociceptors
▶ Spinothalamic Tract Neurons, Peptidergic Input
▶ Thalamic Neurotransmitters and Neuromodulators
▶ Thalamus, Nociceptive Neurotransmission

Neuropeptides in Neuropathic Pain States

▶ Peptides in Neuropathic Pain States

Neuropile

Definition

Neuropile is the complex network of axonal, dendritic, and glial arborizations in nuclei and laminae of the CNS containing the cell bodies.

Cross-References

▶ Deafferentation Pain

Neuroplasticity

Definition

Neuroplasticity is a general term referring to persistent changes in neural activity or function.

A neuroplastic change can be caused by frequent usage of a neuron or a neuronal connection, which stays for a longer period of time after the end of the neuronal activity that started the change. Memory processes and the transition from acute to chronic pain are thought to be examples of neuroplastic changes.

Cross-References

▶ Amygdala, Pain Processing and Behavior in Animals
▶ Arthritis Model, Kaolin-Carrageenan-Induced Arthritis (Knee)
▶ Metabotropic Glutamate Receptors in Spinal Nociceptive Processing
▶ Nociceptive Processing in the Amygdala: Neurophysiology and Neuropharmacology
▶ Sensitization of Muscular and Articular Nociceptors
▶ Transition from Acute to Chronic Pain
▶ Trigeminal Brain Stem Nuclear Complex, Immunohistochemistry, and Neurochemistry

Neurosecretion

Definition

Neurosecretion is the active process of releasing signaling molecules from nerve terminals by exocytosis.

Cross-References

▶ Inflammation, Modulation by Peripheral Cannabinoid Receptors

Neurosensory Testing

Synonyms

NST

Definition

This is a form of Quantitative Sensory Testing that refers to the use of the Pressure-Specified Sensory Device to measure the cutaneous pressure threshold of the large myelinated fibers for both moving (quickly adapting fibers) and static (slowly adapting fibers) touch. The pressure required to discriminate one from two static touch stimuli is the first to become abnormal in the patient with a chronic nerve compression, like carpal or tarsal tunnel syndrome. All QST, by definition, is subjective, requiring cognitive input from the person being tested to identify the end point of the test.

Cross-References

▶ Carpal Tunnel Syndrome

Neurosteroids

Definition

Steroid hormones with neurotransmitter-like actions produced in the central nervous system via metabolism of parent steroids or by local synthesis from cholesterol.

Cross-References

▶ Premenstrual Syndrome

Neurostimulation

▶ Stimulation Treatments of Central Pain

Neurosurgery for Pain

Definition

Neurosurgery for pain is a part of so-called Functional Neurosurgery. Functional Neurosurgery was

defined in 1956 by P. Wertheimer from LYON as follows: "Functional Neurosurgery aims at correcting the functional disorders which cannot be normalized by direct cure of the causative lesion. Operations are based on neurophysiological information. Procedures consist of removing irritative foci or interrupting excitatory pathways to compensate failing inhibitory systems" (Pierre Wertheimer 1956).

Cross-References

▶ Brachial Plexus Avulsion and Surgery in the Dorsal Root Entry Zone (DREZ)

Neurosurgery for Pain in the DREZ

▶ Brachial Plexus Avulsion and Surgery in the Dorsal Root Entry Zone (DREZ)

Neurosurgical Treatment of Pain

Frederick A. Lenz
Department of Neurosurgery, Johns Hopkins Hospital, Baltimore, MD, USA

Synonyms

Complex regional pain syndrome (CRPS); Cordotomy; Sympathectomy; Sympathetically maintained pain (SMP)

Introduction

Since the 1970s there have been significant changes in neurosurgical approaches to the treatment of pain. Specifically, there has been a substantial decrease in the numbers of ablative procedures and an increase in the numbers of stimulation or augmentative procedures for the treatment of chronic pain. The decrease in ablative procedures is in part due to the recognition of the occurrence of ▶ neuropathic pain following surgical injuries to the nervous system (Boivie 1999; Tasker 1984). Ablative procedures, which have survived, have often been associated with a clearer understanding of the rationale and indications for the surgery. This is so in the case of sympathetically maintained pain treated by ▶ sympathectomy. Other current ablative procedures that are associated with a high degree of efficacy are ▶ cordotomy for cancer pain or surgical treatments for tic douloureux. This review is derived, in part, from the topical essays included in this section.

Characteristics

Ablative Neurosurgical Procedures with Improved Rationales
The practice of surgery on the sympathetic nervous system has been altered by advances in the understanding of the ▶ sympathetically maintained pain (SMP), a subset of patients with ▶ complex regional pain syndrome (CRPS). CRPS typically occurs after trauma without (CRPS type 1) or with concomitant nerve injury (CRPS type 2). In SMP the distal extremities, areas rich in sympathetic innervation, can be affected with edema, hyperalgesia, and cool and sweaty skin. The dramatic relief of pain that occurs with selective blockage of the sympathetic nervous system defines SMP.

The mechanism of SMP depends upon α_1-adrenergic receptors as demonstrated by the relief of hyperalgesia produced by application of an α_2 agonist to the painful skin in patients with SMP (Davis et al. 1991). Binding at the α_2-adrenergic receptors, located on sympathetic terminals, blocks norepinephrine release. When phenylephrine, a selective α_1-adrenergic agonist, was applied to the clonidine treated area, pain was rekindled. The density of α_1-adrenoceptors in the epidermis of the hyperalgesic skin of patients with CRPS is increased (Drummond et al. 1996). Thus, the data suggest that the α_1-adrenergic receptor plays a pivotal role in SMP. Sympathetic

mechanisms may also play a role in other types of pain. For example, injection of noradrenaline around stump neuromas or skin in patients with postherpetic neuralgia induces an increase in spontaneous pain (Chabal et al. 1992; Choi and Rowbotham 1997; Raja et al. 1998).

The gold standard for diagnosing SMP has been the determination of the response to blockade of the appropriate level of the sympathetic chain (▶ Sympathetic Blocks) with local anesthetic. An intravenous infusion of phentolamine, an α-adrenergic antagonist (▶ phentolamine test) given systemically, has proven to be a safe, specific test for SMP. Skin temperature is monitored. If the skin temperature does not rise with phentolamine, then a higher dose of phentolamine may have to be given. A positive result is the finding that phentolamine relieves pain, which is the result that identifies patients with SMP (Arner 1991; Raja et al. 1991).

By definition, SMP is relieved by performance of a sympathetic block. The pain relief of sympathetic block often outlives the pharmacologic action of the block. Similar, long-lasting pain relief has also been reported following systemic phentolamine infusion (Galer et al. 1992). In cases where sympathetic blockade provides only transient pain relief, surgical ▶ sympathectomy may offer lasting pain relief (Singh et al. 2003). Thus, recent research findings have advanced our understanding of the basis of SMP and rationalized surgical therapy, which has led to a resurgence of this surgical treatment.

Ablative Neurosurgical Procedures with High Efficacy
Percutaneous cordotomy produces relief of pain by interrupting the transmission of signals in the spinothalamic tractz (STT) from below the level of intervention and caused by cancer. It is not helpful for the steady burning element of neuropathic pain. It may, however, be useful for the relief of allodynia, hyperpathia, and neuralgic pain associated with ▶ neuropathic pain syndromes (Tasker and Dostrovsky 1989; Tasker et al. 1992). It is usually unsuccessful if done in

the lower cervical area in relieving pain above the C5 dermatome.

Published data suggests a 63–77 % range of complete, 68–96 % significant, contralateral pain relief (Tasker 1988). Complete pain relief was found in 90 % of patients immediately postoperatively, 84 % after 3 months, 61 % at 1 year, 43 % between 1 and 5 years, and 37 % between 5 and 10 years (Tasker 1988). If the expected rate of immediate significant pain relief with unilateral cordotomy is 80 %, then that after bilateral surgery is 80 % × 80 % or 64 %. If the first procedure interfered with automatic respiration ipsilaterally, it is unwise to proceed on the other side. The efficacy of this technique in the face of intense, refractory pain due to cancer makes it a viable procedure in patients with a limited survival.

The surgery of ▶ tic douloureux has been revolutionized by a number of "minimally destructive procedures" such as stereotactic radiosurgery and percutaneous ganglion-level procedures including RF thermocoagulation, glycerolysis, and balloon compression. The surgery of tic douloureux also includes ▶ microvascular decompression (MVD) procedures. Some of these latter procedures, such as MVD, for the treatment of intracranial neuralgias were reinvented in the 1960s and found to be effective and safe procedures, although they are not without significant risks (McLaughlin et al. 1999; Patel et al. 2002).

These procedures are strikingly effective for the treatment of typical pain of tic douloureux. Ganglion-level procedures had high initial levels of initial pain relief (91–99 %) with subsequent recurrence rates of 10–25 % over the various study times (25 % at 14 years) (Peters and Nurmikko 2002). Stereotactic radiosurgery has recently been promoted and found to produce complete pain relief without medication in 57 % at 1 year and 55 % at 3 years in a group of whom 60 % had previously been operated on. The long-term effects of radiation therapy close to the brainstem are unknown. The first line of treatment for young healthy patients is MVD and glycerol rhizotomy procedures (Peters and Nurmikko 2002; Pollock et al. 2002). The striking effectiveness of these procedures in

the treatment of the medically intractable pain of typical tic douloureux has supported the present high rates of performance of these procedures.

Ablative Neurosurgical Procedures with Tighter Indications

Spinal and ▶ nucleus caudalis DREZ lesioning procedures can be an effective means of treating deafferentation pain syndromes in carefully selected patients. Traumatic ▶ brachial plexus avulsions frequently result in a characteristic pain syndrome, which must be confirmed clinically, as peripheral nerve injury pain does not respond to DREZ lesions. This is the best indication for the ▶ DREZ procedure. In spinal cord injury, radicular or segmental pain syndromes occurring in the partially deafferented levels adjacent to the level of injury also respond well to DREZ procedures. Diffuse pain occurring below the level of injury, especially constant burning pain in the sacral dermatomes, does not respond to DREZ procedures. In the case of injuries to the conus medullaris and cauda equina at the T12 to L1 levels, the best results are observed in patients with incomplete neurological deficits and those with "electric" pain. Finally, patients suffering from phantom pain after limb amputation respond well to DREZ procedures, significantly better than those with stump pain.

Overall, larger series report long-term pain relief in over 60–90 % (Dreval 1993; Rath et al. 1997; Sindou et al. 2001; Thomas and Kitchen 1994). Variations in results can be attributed to differences between criteria for patient selection, outcome measures, times of follow-up, and techniques (Dreval 1993; Friedman et al. 1988; Iskandar and Nashold 1998; Nashold and El-Naggar 1992; Sampson et al. 1995; Sindou 2002; Spaic et al. 2002; Thomas and Kitchen 1994). Careful selection of patients is critical since complication rates can be significant, especially following DREZ lesions for brachial plexus avulsion pain. Of this group, 41 % experienced objective sensory deficits and 41 % objective motor deficits (Friedman et al. 1988). Overall, the better defined indications for DREZ procedures have led to better outcomes and a resurgence of interest in these procedures.

Selection of the proper patients for intracranial ablative procedures is focused on patients with pain involving the head and/or neck or upper extremities or pain that is widespread throughout the body. The etiology of the pain must be clearly defined and the severity of the pain must be consistent with the etiology (Gildenberg 1973; Gildenberg and DeVaul 1985). Ordinarily, such surgery is reserved for persistent pain accompanying serious conditions, such as head or neck cancer. Patients must be free of significant psychological issues.

Intracranial procedures include lesions in the spinothalamic tract by stereotactic ▶ mesencephalotomy (Gildenberg 1974) and lesions of the mesencephalic central gray (Nashold et al. 1969). The spinoreticulothalamic pathways may be lesioned in the intralaminar and centromedian nuclei and posteriorly adjacent nuclei (Gybels and Sweet 1989; Jeanmonod et al. 1993). A common target for psychosurgery of the limbic system is the anterior portion of the cingulate bundle as it wraps around the anterior end of the corpus callosum. Ablation of that same area has been successful in alleviating severe persistent pain, particularly that of cancer and particularly in patients with severe emotional distress (Hassenbusch 1998). Thus the improved indications for surgery have led to an improved rationale for these uncommon procedures.

Stimulation Procedures: Improved Indications and Demonstrated Efficacy

Since the 1970s stimulation of the nervous system has largely supplanted lesioning of the nervous system for control of pain (Gybels and Sweet 1989; White and Sweet 1969). The increase in the numbers of augmentative or stimulation procedures for the treatment of chronic pain arises from their effectiveness and reversibility. The term augmentative refers to the characteristics of procedures that are not destructive, as in the case of implanted nervous system stimulators and drug pumps. Some of these procedures are characterized by the use of clinical or pharmacologic criteria to identify patients who are candidates for these surgeries. For example, successful pain relief by ▶ motor cortex stimulation may be predicted if patients have significant pain relief in response to infusions of intravenous thiamylal (Yamamoto et al. 1997). Additionally, an intact motor system, but not an intact somatosensory system, is required.

The most commonly used stimulation modality is ▶ spinal cord stimulation (SCS), which is indicated only when pharmacologic or surgical treatment options for chronic pain have been exhausted. In general, good results, defined as >50 % reduction in chronic pain, are reported by 60–70 % of the patients (Meyerson and Linderoth 2000; Simpson 1994). Large numbers of retrospective analyses have demonstrated reduction in chronic pain as demonstrated by reduction in analgesic medication and by patient satisfaction.

The most common indication is lumbosacral rhizopathy, a diagnosis that often represents a mix of nociceptive, neuropathic, and inflammatory pain located in the lumbar region. This "low back pain" is less likely to respond to SCS than is the "radiating leg pain" that is amenable to SCS (North et al. 1993). The second common indication is pain following peripheral nerve injury or disease. Of the many forms of neuropathy due to metabolic disease, ▶ diabetic polyneuropathy is the most common and it is likely to respond to SCS. Pain due to peripheral nerve injury, which sometimes presents as ▶ complex regional pain syndrome (CPRS), is also considered to be a good indication (Kumar et al. 1997). A recent trend is the treatment of the pain of ▶ peripheral vascular disease or ▶ angina pectoris by SCS; this is performed at a small number of centers, mostly in Europe.

Stimulation of the thalamus or midbrain for the treatment of chronic pain has a 50-year history. Patient selection for placement of DBS is an important part of current treatment using this modality (Hosobuchi 1986; Levy et al. 1987; Young and Rinaldi 1997). In many published studies, patients have been assessed by intravenous morphine infusion tests, based on the hypothesis that nociceptive but not neuropathic pain responds to opioids. Then ▶ nociceptive pain is treated by ▶ periaqueductal gray (PAG) stimulation, and ▶ neuropathic pain is treated with thalamic stimulation.

There are a number of large studies demonstrating that DBS can be effective for both neuropathic and nociceptive pain (Hosobuchi 1986; Levy et al. 1987; Young and Rinaldi 1997). A meta-analysis of 13 studies (1,114 patients) evaluating DBS for the treatment of chronic pain reported that 50 % of all patients experienced long-term pain relief. Patients with nociceptive pain experienced a 60 % long-term relief from pain with PAG stimulation, while those with neuropathic pain experienced a 56 % long-term success rate with Vc stimulation.

Stimulation of motor cortex for relief of neuropathic pain of the head, neck, or upper extremity has recently emerged as an option for patients with chronic pain. The first series, based on studies in animals, was carried out by Tsubokawa who reviewed a series of 11 patients with central pain after putaminal or thalamic hemorrhage treated with ► motor cortex stimulation for 2 years with significant (>80 %) pain relief, sustained in 45 % of patients (Tsubokawa et al. 1993). As in the stimulation modalities described above, there are well-described protocols for selection of patients to be treated by motor cortical stimulation. For example, Yamamoto and coworkers (Yamamoto et al. 1997) noted that successful pain relief by motor cortex stimulation could be predicted if patients responded by at least 40 % pain relief to incremental infusions of intravenous thiamylal to a maximum dose of 250 mg, but not to morphine in doses of up to 18 gm given over 5 h (Yamamoto et al. 1997). Additionally, motor cortex stimulation requires an intact motor system to be effective, but not an intact somatosensory system. Thus current stimulation procedures are based upon improved indications and demonstrated efficacy, as in the case of current ablative procedures.

It seems likely that the future will reflect the fact that the conditions we are now treating surgically are all ultimately dependent upon the chemical mechanisms. Surgical treatment of these conditions will be elaborations of the currently available drug pump technology. These therapies will involve selective intrathecal administration to of a drug or drugs (Rainov et al. 2001) specific to the condition being treated (Penn 2003;

Weiss et al. 2003). Examples of such tailored drug administration are found in the case of patients with pain due to spasticity (Middleton et al. 1996) or with pain following spinal cord injury or of patients in opiate withdrawal (Lorenz et al. 2002). The possibility of anatomic as well as chemical approaches to surgical targets within the forebrain will shortly be a possibility. Intraaxial administration is becoming practical for delivery of drugs to anatomically or physiologically defined structures. The feasibility of this approach for selectively lesioning neurons but not axons by convection delivery through an intracerebral catheter has been demonstrated in primate models of Parkinson's disease (Lieberman et al. 1999). The intracerebral delivery of neurotransmitters or proteins, such as growth factors or neurotransmitters, into defined structures can also be accomplished by stereotactically placed catheters or by implantation of other novel drug delivery systems (Gouhier et al. 2002; Pappas et al. 1997). These technologies promise to revolutionize the neurosurgical treatment of pain in the future.

Acknowledgement Supported by grants to FAL from the NIH: NS39498 and NS40059

References

Arner, S. (1991). Intravenous phentolamine test: Diagnostic and prognostic use in reflex sympathetic dystrophy. *Pain, 46,* 17–22.

Boivie, J. (1999). Central pain. In P. D. Wall & R. Melzack (Eds.), *Textbook of pain* (pp. 879–914). Edinburgh: Churchill Livingstone.

Chabal, C., Jacobson, L., Russell, L. C., et al. (1992). Pain response to perineuromal injection of normal saline, epinephrine, and lidocaine in humans. *Pain, 49,* 9–12.

Choi, B., & Rowbotham, M. C. (1997). Effect of adrenergic receptor activation on post-herpetic neuralgia pain and sensory disturbances. *Pain, 69,* 55–63.

Davis, K. D., Treede, R. D., Raja, S. N., et al. (1991). Topical application of clonidine relieves hyperalgesia in patients with sympathetically maintained pain. *Pain, 47,* 309–317.

Dreval, O. N. (1993). Ultrasonic DREZ-operations for treatment of pain due to brachial plexus avulsion. *Acta Neurochirurgica, 122,* 76–81.

Drummond, P. D., Skipworth, S., & Finch, P. M. (1996). Alpha 1-adrenoceptors in normal and hyperalgesic human skin. *Clinical Science, 91,* 73–77.

Friedman, A. H., Nashold, B. S., Jr., & Bronec, P. R. (1988). Dorsal root entry zone lesions for the treatment of brachial plexus avulsion injuries: A follow-up study. *Neurosurgery, 22*, 369–373.

Galer, B. S., Rowbotham, M. C., Von Miller, K., et al. (1992). Treatment of inflammatory, neuropathic and sympathetically maintained pain in a patient with SjÃ¶gren's syndrome. *Pain, 50*, 205–208.

Gildenberg, P. L. (1973). General and psychological assessment of the pain patient. In G. T. Tindall, P. R. Cooper, & D. L. Barrow (Eds.), *The practice of neurosurgery* (pp. 2987–2996). Baltimore: Williams and Wilkins.

Gildenberg, P. L. (1974). Percutaneous cervical cordotomy. *Clinical Neurosurgery, 21*, 246–256.

Gildenberg, P. L., & DeVaul, R. A. (1985). *The chronic pain patient. Evaluation and management.* Basel: Karger.

Gouhier, C., Chalon, S., Aubert-Pouessel, A., et al. (2002). Protection of dopaminergic nigrostriatal afferents by GDNF delivered by microspheres in a rodent model of Parkinson's disease. *Synapse, 44*, 124–131.

Gybels, J. M., & Sweet, W. H. (1989). *Neurosurgical treatment of persistent pain. Physiological and pathological mechanisms of human pain.* Basel: Karger.

Hassenbusch, S. J. (1998). Cingulotomy for cancer pain. In P. L. Gildenberg & R. R. Tasker (Eds.), *Stereotactic and functional neurosurgery* (pp. 1447–1451). New York: McGraw-Hill.

Hosobuchi, Y. (1986). Subcortical electrical stimulation for control of intractable pain in humans. Report of 122 cases (1970–1984). *Journal of Neurosurgery, 64*, 543–553.

Iskandar, B. J., & Nashold, B. S. (1998). Spinal and trigeminal DREZ lesions. In P. L. Gildenberg & R. R. Tasker (Eds.), *Textbook of stereotactic and functional neurosurgery* (pp. 1573–1583). New York: McGraw-Hill, Health professional division.

Jeanmonod, D., Magnin, M., & Morel, A. (1993). Thalamus and neurogenic pain: Physiological, anatomical and clinical data. *NeuroReport, 4*, 475–478.

Kumar, K., Nath, R. K., & Toth, C. (1997). Spinal cord stimulation is effective in the management of reflex sympathetic dystrophy. *Neurosurgery, 40*, 503–508.

Levy, R. M., Lamb, S., & Adams, J. E. (1987). Treatment of chronic pain by deep brain stimulation: Long term follow-up and review of the literature. *Neurosurgery, 21*, 885–893.

Lieberman, D. M., Corthesy, M. E., Cummins, A., et al. (1999). Reversal of experimental parkinsonism by using selective chemical ablation of the medial globus pallidus. *Journal of Neurosurgery, 90*, 928–934.

Lorenz, M., Hussein, S., & Verner, L. (2002). Continuous intraventricular clonidine infusion in controlled morphine withdrawal – Case report. *Pain, 98*, 335–338.

McLaughlin, M. R., Jannetta, P. J., Clyde, B. L., et al. (1999). Microvascular decompression of cranial nerves: Lessons learned after 4400 operations. *Journal of Neurosurgery, 90*, 1–8.

Meyerson, B. A., & Linderoth, B. (2000). Spinal cord stimulation. In J. D. Loeser (Ed.), *Bonica's management of pain* (pp. 1857–1987). Philadelphia: Lippincott Williams & Wilkins.

Middleton, J. W., Siddall, P. J., Walker, S., et al. (1996). Intrathecal clonidine and baclofen in the management of spasticity and neuropathic pain following spinal cord injury: A case study. *Archives of Physical Medicine and Rehabilitation, 77*, 824–826.

Nashold, B. S., & El-Naggar, A. O. (1992). Dorsal root entry zone (DREZ) lesioning. In S. S. Rengachary & R. H. Wilkins (Eds.), *Neurosurgical operative atlas* (pp. 9–24). Baltimore: Williams & Wilkins.

Nashold, B. S., Jr., Wilson, W. P., & Slaughter, D. G. (1969). Stereotaxic midbrain lesions for central dysesthesia and phantom pain. Preliminary report. *Journal of Neurosurgery, 30*, 116–126.

North, R. B., Kidd, D. H., Zahurak, M., et al. (1993). Spinal cord stimulation for chronic intractable pain: Experience over two decades. *Journal of Neurosurgery, 32*, 384–395.

Pappas, G. D., Lazorthes, Y., Bes, J. C., et al. (1997). Relief of intractable cancer pain by human chromaffin cell transplants: Experience at two medical centers. *Neurological Research, 19*, 71–77.

Patel, A., Kassam, A., Horowitz, M., et al. (2002). Microvascular decompression in the management of glossopharyngeal neuralgia: Analysis of 217 cases. *Neurosurgery, 50*, 705–710.

Penn, R. D. (2003). Intrathecal medication delivery. *Neurosurgery Clinics of North America, 14*, 381–387.

Peters, G., & Nurmikko, T. J. (2002). Peripheral and gasserian ganglion-level procedures for the treatment of trigeminal neuralgia. *The Clinical Journal of Pain, 18*, 28–34.

Pollock, B. E., Phuong, L. K., Gorman, D. A., et al. (2002). Stereotactic radiosurgery for idiopathic trigeminal neuralgia. *Journal of Neurosurgery, 97*, 347–353.

Rainov, N. G., Heidecke, V., & Burkert, W. (2001). Long-term intrathecal infusion of drug combinations for chronic back and leg pain. *Journal of Pain and Symptom Management, 22*, 862–871.

Raja, S. N., Treede, R. D., Davis, K. D., et al. (1991). Systemic alpha-adrenergic blockade with phentolamine: A diagnostic test for sympathetically maintained pain. *Anesthesiology, 74*, 691–698.

Raja, S.N., Abatzis, V., & Frank, S. (1998). Role of Î± -adrenoceptors in neuroma pain in amputees. *American Society of Anesthesiologists*, Abstracts.

Rath, S. A., Seitz, K., Soliman, N., et al. (1997). DREZ coagulations for deafferentation pain related to spinal and peripheral nerve lesions: Indication and results of 79 consecutive procedures. *Stereotactic and Functional Neurosurgery, 68*, 161–167.

Sampson, J. H., Cashman, R. E., Nashold, B. S., Jr., et al. (1995). Dorsal root entry zone lesions for intractable pain after trauma to the conus medullaris and cauda equina. *Journal of Neurosurgery, 82*, 28–34.

Simpson, B. A. (1994). Spinal cord stimulation. *Pain Reviews, 1*, 199–230.

Sindou, M. P. (2002). Dorsal root entry zone lesions. In K. J. Burchiel (Ed.), *Surgical management of pain* (pp. 701–713). New York: Thieme Medical Publishers.

Sindou, M., Mertens, P., & Wael, M. (2001). Microsurgical DREZotomy for pain due to spinal cord and/or cauda equina injuries: Long-term results in a series of 44 patients. *Pain, 92*, 159–171.

Singh, B., Moodley, J., Shaik, A. S., et al. (2003). Sympathectomy for complex regional pain syndrome. *Journal of Vascular Surgery, 37*, 508–511.

Spaic, M., Markovic, N., & Tadic, R. (2002). Microsurgical DREZotomy for pain of spinal cord and Cauda equina injury origin: Clinical characteristics of pain and implications for surgery in a series of 26 patients. *Acta Neurochirurgica, 144*, 453–462.

Tasker, R. R. (1984). Deafferentation. In P. D. Wall & R. Melzack (Eds.), *Textbook of pain* (pp. 119–132). Edinburgh: Churchill Livingstone.

Tasker, R. R. (1988). Percutaneous cordotomy: The lateral high cervical technique. In H. H. Schmidek & W. H. Sweet (Eds.), *Operative neurosurgical techniques indications, methods, and results* (pp. 1191–1205). Philadelphia: Saunders WB.

Tasker, R. R., & Dostrovsky, J. O. (1989). Deafferentation and central pain. In P. D. Wall & R. Melzack (Eds.), *Textbook of pain* (pp. 154–180). Edinburgh: Churchill Livingstone.

Tasker, R. R., DeCarvalho, G. T., & Dolan, E. J. (1992). Intractable pain of spinal cord origin: Clinical features and implications for surgery. *Journal of Neurosurgery, 77*, 373–378.

Thomas, D. G., & Kitchen, N. D. (1994). Long-term follow up of dorsal root entry zone lesions in brachial plexus avulsion. *Journal of Neurology, Neurosurgery, and Psychiatry, 57*, 737–738.

Tsubokawa, T., Katayama, Y., Yamamoto, T., et al. (1993). Chronic motor cortex stimulation in patients with thalamic pain. *Journal of Neurosurgery, 78*, 393–401.

Weiss, N., North, R. B., Ohara, S., et al. (2003). Attenuation of cerebellar tremor with implantation of an intrathecal baclofen pump: The role of gamma-aminobutyric acidergic pathways. Case report. *Journal of Neurosurgery, 99*, 768–771.

White, J. C., & Sweet, W. H. (1969). *Pain and the neurosurgeon: A forty year experience*. Springfield: Charles C Thomas.

Yamamoto, T., Katayama, Y., Hirayama, T., et al. (1997). Pharmacological classification of central post-stroke pain: Comparison with the results of chronic motor cortex stimulation therapy. *Pain, 72*, 5–12.

Young, R. F., & Rinaldi, P. C. (1997). Brain stimulation. In R. B. North & R. M. Levy (Eds.), *Neurosurgical management of pain* (pp. 283–301). New York: Springer-Verlag.

Neurotomy

Definition

The dissection, or anatomy, of the nervous system.

Cross-References

► Facet Joint Pain

Neurotransmitter

Definition

A neurotransmitter is a chemical released from a presynaptic neuron into the synaptic cleft, which can trigger a response in the adjacent postsynaptic neuron on the opposite side of the cleft. Neurotransmitters may excite, inhibit, or otherwise influence the activity of cells.

Cross-References

► Cell Therapy in the Treatment of Central Pain
► Pain Treatment, Implantable Pumps for Drug Delivery
► Somatic Pain

Neurotransmitter Receptors

Definition

Neurotransmitter receptors are membrane proteins to which synaptic transmitters bind, leading to a physiological response in the postsynaptic cell. Neurotransmitter receptors can be ionotropic and cause a change in membrane conductance by an action on membrane channels or they can be metabotropic, causing activation of intracellular second messenger systems which then may influence ion channels. Metabotropic receptors are often coupled to metabolic pathways through G proteins.

Cross-References

▶ Spinothalamic Tract Neurons, Role of Nitric Oxide

Neurotransmitters Used by the Axons that Form Synapses Onto These Neurons

▶ Spinothalamic Tract Neurons: Neurotransmitter Inputs and Receptors

Neurotrophic Factors

Definition

Molecules by which tissues or cells affect nerve cell survival and/or phenotype.

Cross-References

▶ Wallerian Degeneration

Neurotrophic Support

Definition

Both developing and mature neurons require the ongoing delivery of a range of factors such as cytokines and neurotrophic factors to maintain survival and the phenotype of the neuron. This neurotrophic support may be supplied directly by cells of the target organ for the neuron, other cells adjacent to the axon or cell body or via the blood supply. Changes in neurotrophic support can induce changes in phenotype such as altered

patterns of neurotransmitter synthesis or potential death of the neuron. Loss of neurotrophic support has been implicated in the etiology of diabetic neuropathy.

Cross-References

▶ Neuropathic Pain Model, Diabetic Neuropathy Model

Neurotrophin

Definition

Neurotrophins are dimeric growth factors that regulate the development and maintenance of central and peripheral neurons. Members of this protein family include nerve growth factor (NGF), neurotrophin-3 (NT-3), brain-derived neurotrophic factor (BDNF), and neurotrophin-4/5 (NT-4/5). They regulate growth, survival, differentiation of neurons, and many other neuroectoderm tissues. All bind with low affinity to the p75 receptor and with high affinity to three structurally related receptors, named Trk receptors. NGF specifically activates TrkA, whereas BDNF and NT-4/5 specifically activate TrkB. NT-3 primarily activates TrkC, but recognizes both TrkA and TrkB to a lesser extent. Trk receptors consist of an extracellular ligand-binding domain, a single transmembrane region, and an intracellular tyrosine kinase (TK) domain. Ligand binding to Trk receptors results in dimerization of receptor molecules followed by autophosphorylation of their cytoplasmic tyrosine residues.

Cross-References

▶ Cell Therapy in the Treatment of Central Pain
▶ Congenital Insensitivity to Pain with Anhidrosis
▶ Nerve Growth Factor, Sensitizing Action on Nociceptors

▶ NGF, Regulation During Inflammation
▶ NGF, Sensitization of Nociceptors
▶ Spinal Cord Nociception, Neurotrophins
▶ Trigeminal Brain Stem Nuclear Complex, Immunohistochemistry, and Neurochemistry

Neurotrophin Receptors

Definition

Two types of neurotrophin receptors are involved in retrograde neurotrophin signaling: the low affinity p75 (NTR) receptor, also referred to as Nerve Growth Factor Receptor (NGFR), which binds with all the above neurotrophins, and the high affinity tyrosine kinase (Trk) receptors. The receptor tyrosine kinase TrkA is the signaling receptor for NGF, TrkB is the signaling receptor for BDNF and NT-4/-5, and TrkC is the primary receptor for NT-3, although NT-3 also binds to TrkA and TrkB, yet with lower affinity.

Neurotrophins

▶ NGF, Regulation During Inflammation
▶ Spinal Cord Nociception, Neurotrophins

Neurotrophins in Spinal Cord Nociception

▶ Spinal Cord Nociception, Neurotrophins

Neutral Cell (RVM)

Definition

Third class of RVM neurons, defined by the lack of reflex-related activity. Neutral cells do not respond to mu-opioid agonists. RVM serotonergic neurons behave as neutral cells, although some authors prefer to consider serotonergic cells as a fourth class of RVM neurons, distinct from neutral cells. Neutral cells as defined in vivo are likely to be a subset of primary cells defined in vitro.

Cross-References

▶ Opiates, Rostral Ventromedial Medulla, and Descending Control

Neutral Medical Examination

▶ Independent Medical Examinations

Neutrophils

▶ Neutrophils in Inflammatory Pain

Neutrophils in Inflammatory Pain

Rainer Amann
Medical University Graz, Graz, Austria

Synonyms

Inflammatory pain; Neutrophils; Polymorphonuclear neutrophil granulocytes (PMN)

Definition

The initial stages of the inflammatory response are characterized by the vascular reaction and the local biosynthesis of mediators that sensitize primary afferent neurons. Thereafter, immune cells are attracted to the site of injury, initiating processes that lead to either tissue repair or destruction. Immigration of polymorphonuclear neutrophil granulocytes (PMN), already in the early cellular phase of inflammation, serves to destroy infectious agents and/or cellular debris. In addition, PMNs are sources of various mediators that can affect, directly or indirectly, the sensitivity of primary afferent neurons.

Characteristics

Tissue injury triggers the release of a large number of mediators from neuronal and nonneuronal resident cells, resulting in microvascular changes and increased afferent neuronal sensitivity, effects that provide the basis of key symptoms of inflammation, redness, edema, and pain. Arachidonic acid metabolites are central in this initial response. Conversion of arachidonic acid through the cyclooxygenase pathway leads to enhanced biosynthesis of ► prostanoids, which increase tissue perfusion and sensitize primary afferent neurons. Pharmacological inhibition of cyclooxygenase activity by ► NSAIDs survey (NSAIDs) is, therefore, effectively used to alleviate inflammatory pain and edema.

In addition to mediators that serve the initial vascular and neuronal defense reaction, compounds with chemotactic activity are produced by injured tissue. This leads to immigration of PMN to the site of injury, a process that involves a cascade of events, primarily consisting of leukocyte rolling along the endothelial layer, adherence to, and transmigration through the endothelium and vascular wall. This process is under control of various ► cytokines, ► chemoattractants, and ► chemokines (Wagner and Roth 2000).

PMNs can produce a wide range of mediators (e.g., arachidonic acid metabolites, various cytokines, proteases, reactive oxygen intermediaries, and nitric oxide) (Sampson 2000) that serve to initiate the immune response and destroy infectious agents and/or cellular debris. It is evident that, in chronic inflammatory disease, the severity of the cellular inflammatory response and accompanying tissue destruction correlates with inflammatory pain. This provides a rationale for the use of ► disease-modifying antirheumatic drugs (DMARDs), which

primarily target immune cell function, in diseases such as rheumatoid arthritis. However, there are reasons to assume that PMNs are involved in the development of inflammatory pain at earlier stages of inflammation, a concept that is primarily based on a number of studies by Levine and associates (Levine et al. 1984, 1986).

With regard to the role of PMN in inflammatory pain, one chemoattractant is of particular interest, ► leukotriene (LT) B_4, which has been shown to be present in inflammatory exudates in human inflammatory diseases such as ► rheumatoid arthritis (Davidson et al. 1983). LTB_4 is a product of the 5-lipoxygenase pathway of arachidonic acid metabolism; it is a potent chemoattractant for neutrophils, which, in turn, are also major sources of LTB_4.

In experimental inflammation, it has been shown that LTB_4 produces ► hyperalgesia, which is not prevented by ► cyclooxygenase inhibitors (Levine et al. 1984). In the presence of LTB_4, PMNs release a 15-lipoxygenase product of arachidonic acid, (8R, 15S)-dihydroxyeicosa-(5E-9,11,13Z)-tetraenoic acid [(8R,15S)-diHETE], which directly sensitizes primary afferent neurons (Levine et al. 1986). Since it has been shown that the hyperalgesic activity of exogenous LTB_4 is dependent on the presence of PMN (Levine et al. 1984), it seems likely that the final mediator of LTB_4-induced hyperalgesia is a product of PMN 15-lipoxygenase. According to this concept, PMNs would promote inflammatory hyperalgesia in a dual fashion, as sources and as targets of LTB_4.

In view of these experimental data, inhibition of LTB_4 biosynthesis would be expected to show analgesic effects. In fact, there are studies showing analgesic effects of 5-lipoxygenase inhibition in some models of experimental inflammation. Thus, it has been shown in rodents that ► nerve growth factor (NGF) elicits ► thermal hyperalgesia, which is dependent on PMNs and is blocked by inhibition of the 5-lipoxygenase (Amann et al. 1996, Bennett et al. 1998, Schuligoi 1998). The demonstration of PMN-dependent hyperalgesia induced by NGF may be important, because NGF has been shown to contribute to inflammatory hyperalgesia and bronchial hyperreflexia (Renz 2001). However, the studies mentioned above were conducted using exogenous NGF and, therefore, did not provide direct evidence for the presence of a PMN-dependent, 5-lipoxygenase-mediated mechanism in inflammatory pain.

In a more recent study, Cunha et al. (2003) have shown that mechanical hypersensitivity, induced by antigen challenge in rats immunized with ovalbumin, is suppressed by inhibition of 5-lipoxygenase and a LTB_4 receptor antagonist. Although in this study the participation of PMNs as sources of LTB_4 was not addressed, the observation that, at later stages of the inflammatory response, LTB_4 was the principal mediator of hyperalgesia points to the involvement of PMNs.

In contrast to these experimental studies in rodents, which are suggestive of involvement of PMN 5-lipoxygenase in inflammatory hyperalgesia, there is no firm evidence that 5-lipoxygenase inhibition can attenuate inflammatory pain in patients, although selective inhibitors of 5-lipoxygenase have been available for a number of years. Therefore, the pathophysiological relevance of PMN lipoxygenases as sources of hyperalgesia-inducing factors in inflammatory pain remains doubtful.

In recent years a novel concept has emerged, suggesting that PMNs play an important role in the generation of lipoxins, lipid mediators that promote the resolution of the inflammatory process (Levy et al. 2001). According to this concept, 15-lipoxygenase products of arachidonic acid [15S-hydroxyperoxyeicosatetraenoic acid (15S-H(p)ETE), or the reduced alcohol form 15S-hydroxyeicosatetraenoic acid (15S-HETE)], derived from epithelial cells, eosinophils, or monocytes serve as substrates for PMN 5-lipoxygenase, which results in the generation of lipoxin $(LX)A_4$ and LXB_4. During the synthesis of lipoxins, leukotriene synthesis is blocked at the 5-lipoxygenase level in PMN, resulting in an inverse relationship of lipoxin and LT biosynthesis. An alternative pathway involves an interaction between PMN and platelets, which convert PMN 5-lipoxygenase-derived LTA_2

via 12-lipoxygenase to LXA_4 and LXB_4 (Serhan 1997).

Lipoxins contribute to the resolution of inflammation by inhibiting neutrophil chemotaxis and transmigration (Takano et al. 1997), stimulation of macrophage clearance of apoptotic PMN from an inflammatory focus (Godson et al. 2000), inhibition of cell proliferation, and modulation of metalloproteinase activity (Sodin-Semrl et al. 2000). Although there are no studies on the possible effects of lipoxins on the nociceptive threshold of primary afferent neurons, it can be expected that, by attenuating the inflammatory response, they can indirectly reduce inflammatory pain.

Since PMN 5-lipoxygenase is a key enzyme in the biosynthesis of lipoxins, pharmacological inhibition of 5-lipoxygenase in PMNs may in fact counteract processes that are involved in the resolution of inflammation. Theoretically, this may also be one possible explanation for the absence of obvious analgesic effects of 5-lipoxygenase inhibition in most types of inflammation.

References

Amann, R., Schuligoi, R., Lanz, I., & Peskar, B. A. (1996). Effect of a 5-lipoxygenase inhibitor on nerve growth factor-induced thermal hyperalgesia in the rat. *European Journal of Pharmacology, 306*, 89–91.

Bennett, G., al-Rashed, S., Hoult, J. R., & Brain, S. D. (1998). Nerve growth factor induced hyperalgesia in the rat hind paw is dependent on circulating neutrophils. *Pain, 77*, 315–322.

Cunha, J. M., Sachs, D., Canetti, C. A., Poole, S., Ferreira, S. H., & Cunha, F. Q. (2003). The critical role of leukotriene B4 in antigen-induced mechanical hyperalgesia in immunised rats. *British Journal of Pharmacology, 139*, 1135–1145.

Davidson, E. M., Rae, S. A., & Smith, M. J. (1983). Leukotriene B4, a mediator of inflammation present in synovial fluid in rheumatoid arthritis. *Annals of the Rheumatic Diseases, 42*, 677–679.

Godson, C., Mitchell, S., Harvey, K., Petasis, N. A., Hogg, N., & Brady, H. R. (2000). Cutting edge: Lipoxins rapidly stimulate nonphlogistic phagocytosis of apoptotic neutrophils by monocyte-derived macrophages. *Journal of Immunology, 164*, 1663–1667.

Levine, J. D., Lau, W., Kwiat, G., & Goetzl, E. J. (1984). Leukotriene B4 produces hyperalgesia that is dependent on polymorphonuclear leukocytes. *Science, 225*, 743–745.

Levine, J. D., Lam, D., Taiwo, Y. O., Donatoni, P., & Goetzl, E. J. (1986). Hyperalgesic properties of 15-lipoxygenase products of arachidonic acid. *Proceedings of the National Academy of Sciences of the United States of America, 83*, 5331–5334.

Levy, B. D., Clish, C. B., Schmidt, B., Gronert, K., & Serhan, C. N. (2001). Lipid mediator class switching during acute inflammation: Signals in resolution. *Nature Immunology, 2*, 612–619.

Renz, H. (2001). The role of neurotrophins in bronchial asthma. *European Journal of Pharmacology, 429*, 231–237.

Sampson, A. P. (2000). The role of eosinophils and neutrophils in inflammation. *Clinical and Experimental Allergy: Journal of the British Society for Allergy and Clinical Immunology, 30*(Suppl 1), 22–27.

Schuligoi, R. (1998). Effect of colchicine on nerve growth factor-induced leukocyte accumulation and thermal hyperalgesia in the rat. *Naunyn-Schmiedeberg's Archives of Pharmacology, 358*, 264–269.

Serhan, C. N. (1997). Lipoxins and novel aspirin-triggered 15-epi-lipoxins (ATL): A jungle of cell-cell interactions or a therapeutic opportunity? *Prostaglandins, 53*, 107–137.

Sodin-Semrl, S., Taddeo, B., Tseng, D., Varga, J., & Fiore, S. (2000). Lipoxin A4 inhibits il-1 beta-induced il-6, il-8, and matrix metalloproteinase-3 production in human synovial fibroblasts and enhances synthesis of tissue inhibitors of metalloproteinases. *Journal of Immunology, 164*, 2660–2666.

Takano, T., Fiore, S., Maddox, J. F., Brady, H. R., Petasis, N. A., & Serhan, C. N. (1997). Aspirin-triggered 15-epi-lipoxin a4 (lxa4) and lxa4 stable analogues are potent inhibitors of acute inflammation: Evidence for anti-inflammatory receptors. *The Journal of Experimental Medicine, 185*, 1693–1704.

Wagner, J. G., & Roth, R. A. (2000). Neutrophil migration mechanisms, with an emphasis on the pulmonary vasculature. *Pharmacological Reviews, 52*, 349–374.

New Daily Persistent Headache

Todd D. Rozen
Department of Neurology, Geisinger Health
System, Wilkes-Barre, PA, USA

Synonyms

Chronic daily headache; Daily persistent headache

Definition

New daily persistent headache (NDPH) is a form of chronic daily headache along with ► chronic migraine, ► chronic tension-type headache, and ► hemicrania continua. The defining symptom of NDPH is a daily headache from onset, typically in an individual with minimal or no prior headache history. The headache will start one day and in most instances continue as daily unremitting pain.

Characteristics

New daily persistent headache (NDPH) was first described by Vanast in 1986 as a benign form of chronic daily headache that improved without therapy (Vanast 1986). Only recently has NDPH been recognized as a distinct entity by headache specialists. The syndrome is unique in that the headache begins daily from onset, oftentimes in a patient with no prior headache history (thus this can be their first ever headache), and can continue for years, without any sign of alleviation despite aggressive treatment (Rozen 2011). Overall NDPH has a female predominance (gender ratio range 1.4–2.5:1). The age of onset is younger in women than men and many will develop the syndrome in their second to third decade of life. Pain location is bilateral in most sufferers although the pain can occur anywhere on the head. Pain intensity is typically moderate to severe in most patients. Pain duration is constant typically without any pain-free time. Recognized triggers for NDPH include infection, surgical procedures, and stressful life events although almost 50 % will be unable to document a triggering event. Associated symptoms include migrainous features which are common in this population. This is an area of contention as current ICHD-2 criteria for NDPH exclude migrainous symptoms (Headache Classification Subcommittee of the International Headache Society 2004). There appears to be two main subtypes of NDPH, a self-limited form, which typically goes away within several months up to 1 year without any treatment and rarely presents to a physician's office, and a refractory form, which is basically resistant to aggressive outpatient and inpatient treatment schemes. There also maybe a recurrent form of NDPH that relapses and remits.

Epidemiology

Even though NDPH has probably been around for centuries, it has only recently been diagnosed as an entity separate from chronic tension-type headache, hemicrania continua, and chronic migraine. Several studies have documented the prevalence of NDPH; Castillo et al. (1999) looked at the prevalence of CDH in 2,252 subjects in Spain and found that 4.7 % of the population has CDH, of which 0.1 % had NDPH. Bigal et al. (2002) noted that 10.8 % of 638 patients with CDH in a headache specialty clinic had NDPH, while Koenig et al. (2002) found that 13 % of a pediatric CDH population, surveyed from selected pediatric headache specialty clinics, had NDPH.

Etiology of NDPH

As at least a third of NDPH patients have a cold or flu-like illness when their headaches begin, thus an infectious etiology has been hypothesized. Some authors have linked Epstein-Barr virus (EBV) infection with NDPH. Diaz-Mitoma et al. (1987) identified oropharyngeal secretions of EBV in 20 of 32 patients with NDPH compared with 4 of 32 age- and gender-matched controls. A history of mononucleosis was identified in 12 of the patients with NDPH. Almost 85 % of the NDPH patients were found to have an active EBV infection as opposed to 8 in the control group. The authors hypothesized that activation of a latent EBV infection may have been the trigger for the development of a chronic daily headache from onset. Santoni and Santoni-Williams (1993) demonstrated evidence of systemic infection in 108 patients with NDPH including *Salmonella*, adenovirus, toxoplasmosis, herpes zoster, EBV, and *E. coli* urinary tract infections.

There are two new studies which have provided some further insight into the pathogenesis of NDPH.

(a) Elevation of CSF Tumor Necrosis Factor Alpha Levels in NDPH Patients

As a certain percentage of NDPH patients have their headaches start after an infection, the possibility of a persistent state of systemic or CNS inflammation comes into question. Tumor necrosis factor alpha (TNF alpha) is a proinflammatory cytokine involved in brain immune and inflammatory activities, as well as in pain initiation. Rozen and Swidan (Rozen 2006) looked at TNF alpha levels in the CSF of NDPH patients. Twenty patients were studied and TNF alpha levels were elevated in 19 of the 20 samples. The authors suggested a role for TNF alpha in the pathogenesis of NDPH. Glial cells are known manufacturers of TNF alpha in the CNS. Interestingly in laboratory animals, recognized triggers of glial cell activation and cytokine production include infection, stress, and surgical procedures (O'Conner et al. 2003). These are the recognized triggering events of NDPH in humans. There is recent evidence that TNF alpha will induce calcitonin gene-related peptide production, which is a known factor in the pathogenesis of other primary headaches including migraine and cluster headache (Durham 2006) and thus may play a role on the pain of NDPH.

(b) Cervical Spine Joint Hypermobility as a Predisposing Factor for the Development of NDPH

Rozen et al. (2005) noted a similar body habitus in NDPH patients at their headache specialty clinic of tall height, thin weight, and a long neck reminiscent of the physical characteristics seen in individuals with hereditary connective tissue disorders. Thus the authors hypothesized that joint hypermobility, especially of the cervical spine, may be a predisposing factor for the development of NDPH. Eleven of 12 NDPH patients were studied and found to have cervical spine joint hypermobility on evaluation. Ten of the 12 NDPH patients also had evidence of widespread joint hypermobility utilizing the Beighton score. The authors concluded that cervical spine joint hypermobility may be a predisposing factor for the development of NDPH. Evidence exists that there is a convergence of trigeminal and cervical afferents in the trigeminal nucleus caudalis (TNC) (Piovesan et al. 2003). Thus, cervical spine pathology can present as head pain typically in a trigeminal nerve V1 distribution. Cervical spine joint hypermobility in some manner may influence cervical afferent input into the TNC with the subsequent development of head pain. It may also set up upper cervical facet and/or atlantoaxial joint inflammation leading to head and neck pain.

Differential Diagnosis of NDPH

A diagnosis of primary NDPH is made only after secondary causes have been ruled out. Two disorders in particular can mimic the presentation of NDPH, ▶ spontaneous cerebrospinal fluid leak (CSF) and cerebral venous sinus thrombosis. Spontaneous CSF leaks typically present as a daily headache with a positional component (headache improved in a supine position, worsens in a sitting or standing position). However, the longer a patient suffers with a CSF leak-induced headache, the less pronounced the positional component becomes. Thus if a patient is seen in a physician's office months to years after onset of a CSF leak, that patient may not even divulge a history of positional headaches, as that trigger may not have been evident to the patient for a very long time. In this setting, the CSF leak headache may mimic a primary NDPH picture.

In the patient who presents with new daily headache and is subsequently found to have cerebral vein thrombosis, in many instances none of the typical features recognized for cerebral venous clots are present, including no history of new onset seizures, focal neurological deficits, change of consciousness, cranial nerve palsies, or bilateral cortical signs and no evidence of papilledema on fundi examination. The patient

will just have a new headache that is daily from onset.

The evaluation of an NDPH patient should include neuroimaging, specifically brain MRI without and with gadolinium and MR venography (MRV). Gadolinium must be given to look for the diffuse pachymeningeal enhancement associated with spontaneous CSF leaks, while MRV will help make the diagnosis of cerebral venous thrombosis. If a new daily headache begins after the age of 50 years, then giant cell arteritis must be ruled out. Headache is the most common reported symptom of the disorder occurring in up to 90 % of individuals.

Treatment

Even with aggressive treatment, many NDPH patients do not improve. Patients with NDPH can fail every possible class of abortive and preventive medications without any sign of pain relief (The International Classification of Headache Disorders 2004). At present no specific treatment strategy can be suggested for primary NDPH based on clinical evidence. Most headache specialists will treat NDPH with the same acute and preventive medications that they use to treat chronic migraine although based on nonresponse to most of these medications, NDPH and chronic migraine are two disparate syndromes.

Very few therapies for NDPH have been documented in the literature:
1. Antiepileptic medications

   Rozen (2002) presented five patients in which successful treatment of NDPH was obtained with gabapentin or topiramate but these agents do not work in the majority of cases.
2. Tetracycline derivatives

   Rozen (2008) in abstract form reported on the use of daily oral doxycycline which is a TNF alpha inhibitor (see above). Four patients with treatment-resistant NDPH and elevated CSF TNF α levels were treated with doxycycline 100 mg 2× per day in an open label fashion for 3 months. Headache frequency and pain intensity levels were assessed. All patients had failed at least five preventive agents and thus were deemed treatment refractory. Three of four patients failed inpatient headache treatment while another failed outpatient infusion therapy. Duration of NDPH prior to doxycycline therapy ranged from 8 months to 3 years. An infection precipitated NDPH in three of four patients. Four out of four patients had a positive response to doxycycline treatment. Two patients became pain-free. One patient had an 80 % improvement in daily pain intensity, but did not achieve any pain-free time. One patient had a slight improvement in daily pain intensity, but had a >50 % reduction in frequency of severe pain episodes. Average time to see improvement on doxycycline was after 2 months of therapy; however, one patient responded within 2 weeks of starting treatment. The two patients with the highest CSF TNF α levels became pain-free with doxycycline. This study has yet to be repeated.

3. Mexiletine

Marmura et al. (2008) looked at mexiletine in three patients with NDPH who had been deemed treatment refractory. All three patients showed a reduced severity of pain on mexiletine but only one showed a reduced frequency of headaches. Significant side effects were noted on the medication.

4. IV Corticosteroids

Prakash and Shah (2010) reported on nine patients with post-infectious NDPH. All patients were given high-dose intravenous methylprednisolone for 5 days, and six patients were given oral steroids for 2–3 weeks. All patients improved with seven patients experiencing almost complete pain relief within 2 weeks, while two patients needed between 6 and 8 weeks. The results of this study are promising but there are several drawbacks. First off none of the patients fit the ICHD-2 criteria for NDPH which requires headache duration of at least 3 months. As the patients were treated only several weeks after headaches onset, we do not know if these patients would have improved

on their own (self-limited form of NDPH) even without treatment.

References

Bigal, M. E., Sheftell, F. D., Rapoport, A. M., et al. (2002). Chronic daily headache in a tertiary care population: Correlation between the international headache society diagnostic criteria and proposed revisions of criteria for chronic daily headache. *Cephalalgia, 22,* 432–438.

Castillo, J., Munoz, P., Guitera, V., et al. (1999). Epidemiology of chronic daily headache in the general population. *Headache, 38,* 497–506.

Diaz-Mitoma, F., Vanast, W. J., & Tyrell, D. L. (1987). Increased frequency of Epstein-Barr virus excretion in patients with new daily persistent headaches. *The Lancet, 1,* 411–415.

Durham, P. L. (2006). Calcitonin gene-related peptide (CGRP) and migraine. *Headache, 46*(suppl 1), 3–8.

Headache Classification Subcommittee of the International Headache Society (2004). International Classification of Headache Disorders (ICHD-II) (2nd ed.). *Cephalalgia, 24*(suppl 1), 1–160.

Koenig, M. A., Gladstein, J., McCarter, R. J., et al. (2002). The pediatric committee of the American Headache Society chronic daily headache in children and adolescents presenting to tertiary headache clinics. *Headache, 42,* 491–500.

Marmura, M. J., Passero, F. C., & Young, W. B. (2008). Mexiletine for refractory chronic daily headache: A report of nine cases. *Headache, 48,* 1506–1510.

O'Conner, K. A., Johnson, J. D., Hansen, M. K., et al. (2003). Peripheral and central proinflammatory cytokine response to a severe acute stressor. *Brain Research, 991,* 123–132.

Piovesan, E. J., Kowacs, P. A., & Oshinsky, M. L. (2003). Convergence of cervical and trigeminal sensory afferents. *Current Headache Reports, 2,* 155–161.

Prakash, S., & Shah, N. D. (2010). Post-infectious new daily persistent headache may respond to intravenous methylprednisolone. *The Journal of Headache and Pain, 1,* 59–66.

Rozen, T. D. (2002). Successful treatment of new daily persistent headache with gabapentin and topiramate. *Headache, 42,* 433.

Rozen, T. D. (2008). Doxycycline for treatment resistant new daily persistent headache. *Neurology, 70*(suppl 1), A348.

Rozen, T. D. (2011). New daily persistent headache. *Current Opinion in Neurology, 24,* 211–216.

Rozen, T. D., Roth, J. M., & Denenberg, N. (2005). Joint hypermobility as a predisposing factor for the development of new daily persistent headache. *Headache, 45,* 828–829.

Rozen, T. D., Swidan Rozen, T. D., & Swidan, S. (2006). Elevation of CSF tumor necrosis factor alpha levels in

NDPH and treatment refractory chronic daily migraine. *Headache, 46,* 897.

Santoni, J. R., & Santoni-Williams, C. J. (1993). Headache and painful lymphadenopathy in extracranial or systemic infection: Etiology of new daily persistent headaches. *Internal Medicine, 32,* 530–533.

Vanast, W. J. (1986). New daily persistent headaches: Definition of a benign syndrome. *Headache, 26,* 317.

Newborn

Synonyms

Neonate

Definition

Newborn infant who is less than 1 month postnatal age.

Cross-References

▶ Pain Assessment in Neonates

NGF

▶ Nerve Growth Factor

NGF −/− Mice

Definition

Mice that lack a functional gene encoding nerve growth factor, i.e., NGF knockout mice.

Cross-References

▶ Nerve Growth Factor Overexpressing Mice as Models of Inflammatory Pain

NGF, Regulation During Inflammation

Rainer Amann
Medical University Graz, Graz, Austria

Synonyms

Inflammation; Neurotrophins; Nerve Growth Factor (NGF)

Definition

In an inflamed tissue, there is an increased expression of ▶ nerve growth factor (NGF), which affects afferent neuron function and contributes to the development and resolution of the inflammatory process. Pharmacological tools that can modify inflammation-induced NGF biosynthesis are, therefore, of potential therapeutic value.

Characteristics

Nerve growth factor (NGF) belongs to a family of structurally related ▶ neurotrophins. During inflammation, there is a rapid increase in local NGF biosynthesis. The local increase in NGF in inflamed tissues leads to changes in the phenotype of a subset of primary afferent neurons, with consequences for the transmission of noxious afferent input (Mendell et al. 1999).

In addition to its neurotrophic properties, NGF has been shown to affect immune cell function (Aloe et al. 1999). Direct neuronal effects (Mendell et al. 1999), as well as effects on immune cells (Bennett et al. 1998; Schuligoi 1998), seem to contribute to NGF-induced sensitization of primary afferent neurons, which manifests itself as inflammatory ▶ hyperalgesia in the skin or bronchial hyperreactivity in the respiratory system (Renz 2001). Furthermore, it has been suggested that, by promoting keratinocyte proliferation and vascular neoangiogenesis, NGF contributes to cutaneous morphogenesis, wound healing, and tissue response to inflammation (Aloe 2004).

Inhibition of NGF Biosynthesis

Attenuation of the inflammation-induced rise in tissue NGF may be effective in preventing longer-lasting hyperalgesic effects of NGF. Therefore, a number of studies were conducted in order to investigate the effects of anti-inflammatory drugs on NGF biosynthesis in inflamed tissue.

Glucocorticoids have been shown to suppress inflammatory edema as well as the inflammation-induced increase in NGF biosynthesis. This is in contrast to non-inflamed tissues, where acute glucocorticoid treatment seems to have no inhibitory effect on NGF biosynthesis (Safieh-Garabedian et al. 1995).

Nonsteroidal anti-inflammatory drugs (▶ NSAIDs), belonging to the group of ▶ cyclooxygenase (COX) inhibitors, suppress prostanoid biosynthesis, thereby reducing edema and pain. Therefore, it seems interesting that inhibition of prostanoid biosynthesis, although suppressing inflammatory edema, has no significant effect on NGF expression in inflamed paw tissue of the rat (Amann et al. 1996). This suggests that the potency of anti-inflammatory drugs to inhibit inflammatory edema is not necessarily predictive of their ability to reduce the inflammation-induced increase in NGF. In contrast to the absence of significant influence on inflammation-induced NGF expression by COX inhibition, high-dose NSAID treatment of rats has been shown to be effective in decreasing NGF in inflamed tissue (Safieh-Garabedian et al. 1995), an effect that may be ascribed to drug actions not related to COX inhibition (Tegeder et al. 2001).

Drugs Acting at Adrenergic Receptors

It has been shown that treatment of rats with adrenergic antagonists inhibits the increase in NGF observed in endotoxin-induced paw inflammation (Safieh-Garabedian et al. 2002), suggesting that endogenous adrenergic tone can

be sufficient to stimulate local NGF biosynthesis (see "Stimulation of NGF Biosynthesis").

However, a participation of endogenous adrenergic mechanisms in stimulating NGF biosynthesis in inflamed tissues does not seem to be a general feature of the inflammatory response. In contrast, there are even models of inflammation where adrenergic agonists can attenuate the inflammation-induced rise in NGF:

1. In allergic inflammation of the rat paw or respiratory system, the beta adrenoceptor agonist terbutaline attenuates edema as well as the increase in local NGF formation. In contrast to its inhibitory effect in allergic inflammation, terbutaline does not significantly affect NGF in carrageenan-induced inflammation of the rat paw, suggesting no general inhibitory effects of terbutaline on NGF biosynthesis in inflammation (Amann et al. 2001).

2. Neurogenic inflammation is caused by experimental stimulation of capsaicin-sensitive primary afferent neurons, leading to the peripheral release of neuropeptides (e.g., calcitonin gene-related peptide and ▶ tachykinins) that cause vasodilation, plasma protein extravasation, and also an increase in the NGF content of the innervated skin (see "Stimulation of NGF Biosynthesis"). It can be shown that terbutaline inhibits the increase in NGF of rat skin in capsaicin-induced neurogenic inflammation and the NGF increase following local injection of the tachykinin NK1 receptor agonist substance P (Amann and Schuligoi 2004).

At the transcriptional level, beta adrenergic agonists can stimulate rather than decrease the expression of NGF (see "Stimulation of NGF Biosynthesis"). The observed inhibitory effects on NGF biosynthesis of beta adrenergic stimulation in the abovementioned studies are, therefore, not caused by specifically inhibiting NGF transcription, but by interference with the inflammatory stimulus itself.

Tachykinin NK1 receptor antagonists inhibit the increase in NGF caused by experimental neurogenic inflammation, but do not detectably affect the NGF response to carrageenan or allergic inflammation. In non-inflamed skin, treatment with tachykinin NK1 receptor antagonists has no detectable effect on NGF biosynthesis (Amann et al. 2000).

Stimulation of NGF Biosynthesis

Beta Adrenergic Agonists

A number of studies have shown that, depending on cell type, beta adrenergic agonists stimulate the expression of NGF mRNA and protein. These effects seem to be caused by an agonist-induced rise in intracellular cAMP and activation of ▶ protein kinase A, which causes an increase in early inducible genes such as ▶ c-fos and the associated AP-1 binding. The AP-1 element is present within intron 1 of the NGF gene. In addition, the transcription factor C/EBPδ is activated by elevated cAMP levels and is one of the transcription factors for induced NGF expression (Riaz and Tomlinson 2000).

Vitamin D Receptor Agonists

There are several studies showing that vitamin D analogs, acting at vitamin D receptors, stimulate the expression of NGF mRNA, probably due to increased AP-1 binding. Vitamin D analogs may indirectly modulate NGF mRNA levels through increasing intracellular Ca^{2+} and promoting c-fos and c-jun induction; these ▶ immediate early genes are postulated to enhance NGF expression (Riaz and Tomlinson 2000).

Tachykinin NK1 Receptor Agonists

It has been shown in rodents that an intraplantar injection of the endogenous NK1 receptor agonist substance P stimulates NGF biosynthesis in the paw skin. The effect of substance P is blocked by a selective tachykinin NK1 receptor antagonist, indicating that NK1 receptor activation stimulates the NGF increase. Further experiments have provided evidence of the involvement of NK1 receptors in the NGF increase caused by neurogenic inflammation, suggesting that endogenous substance P, released from primary afferent nerve terminals during stimulation, acts on NK1 receptors to increase tissue NGF (Amann et al. 2000).

It seems to be of particular interest that primary afferent neuron activation, through local release of neurotransmitters, augments NGF biosynthesis. This may not only constitute a regulatory feed back by which enhanced peripheral release of neuropeptides is signaled to dorsal root ganglia where NGF stimulates neuropeptide biosynthesis (Donnerer et al. 1993) but may be of importance in inflammation and/or the local response to tissue injury. However, the ▶ cellular targets of substance P have not been determined so far.

The vanilloid VR1 receptor agonist ▶ capsaicin is a potent stimulant to induce excitation and local neuropeptide release from small-diameter primary afferent neurons and thus induces neurogenic inflammation. When applied either by intraplantar injection (Saade et al. 2002 or by topical application to the paw skin (Amann and Schuligoi 2004 capsaicin stimulates NGF biosynthesis in the skin. NK1 receptor antagonists prevent this effect of capsaicin, indicating that substance P, released from primary afferent neurons, is the mediator of the capsaicin-induced NGF response. Capsaicin would seem, therefore, to be a suitable pharmacological tool to increase skin NGF biosynthesis, especially since it is a constituent of various drug formulations made for topical administration to the skin. However, a single topical application of capsaicin to the skin results in reduced responsiveness to subsequent applications of capsaicin (Amann and Schuligoi 2004). Desensitization may, therefore, limit the use of vanilloid receptor agonists as tools for producing longer-lasting stimulation of skin NGF biosynthesis.

References

Aloe, L. (2004). Nerve growth factor, human skin ulcers and vascularization. Our experience. *Progress in Brain Research, 146*, 515–522.

Aloe, L., Simone, M. D., & Properzi, F. (1999). Nerve growth factor: A neurotrophin with activity on cells of the immune system. *Microscopy Research and Technique, 45*, 285–291.

Amann, R., & Schuligoi, R. (2004). Beta adrenergic inhibition of capsaicin-induced, NK1 receptor-mediated nerve growth factor biosynthesis in rat skin. *Pain, 112*, 76–82.

Amann, R., Schuligoi, R., Herzeg, G., & Donnerer, J. (1996). Intraplantar injection of nerve growth factor into the rat hind paw: Local edema and effects on thermal nociceptive threshold. *Pain, 64*, 323–329.

Amann, R., Egger, T., & Schuligoi, R. (2000). The tachykinin NK(1) receptor antagonist SR140333 prevents the increase of nerve growth factor in rat paw skin induced by substance P or neurogenic inflammation. *Neuroscience, 100*, 611–615.

Amann, R., Peskar, B. A., & Schuligoi, R. (2001). Effects of terbutaline on NGF formation in allergic inflammation of the rat. *British Journal of Pharmacology, 133*, 186–192.

Bennett, G., al-Rashed, S., Hoult, J. R., & Brain, S. D. (1998). Nerve growth factor induced hyperalgesia in the rat hind paw is dependent on circulating neutrophils. *Pain, 77*, 315–322.

Donnerer, J., Schuligoi, R., Stein, C., & Amann, R. (1993). Upregulation, release and axonal transport of substance P and calcitonin gene-related peptide in adjuvant inflammation and regulatory function of nerve growth factor. *Regulatory Peptides, 46*, 150–154.

Mendell, L. M., Albers, K. M., & Davis, B. M. (1999). Neurotrophins, nociceptors, and pain. *Microscopy Research and Technique, 45*, 252–261.

Renz, H. (2001). The role of neurotrophins in bronchial asthma. *European Journal of Pharmacology, 429*, 231–237.

Riaz, S. S., & Tomlinson, D. R. (2000). Pharmacological modulation of nerve growth factor synthesis: A mechanistic comparison of vitamin D receptor and beta(2)-adrenoceptor agonists. *Molecular Brain Research, 85*, 179–188.

Saade, N. E., Massaad, C. A., Ochoa-Chaar, C. I., Jabbur, S. J., Safieh-Garabedian, B., & Atweh, S. F. (2002). Upregulation of proinflammatory cytokines and nerve growth factor by intraplantar injection of capsaicin in rats. *The Journal of Physiology, 545*, 241–253.

Safieh-Garabedian, B., Poole, S., Allchorne, A., Winter, J., & Woolf, C. J. (1995). Contribution of interleukin-1 beta to the inflammation-induced increase in nerve growth factor levels and inflammatory hyperalgesia. *British Journal of Pharmacology, 115*, 1265–1275.

Safieh-Garabedian, B., Poole, S., Haddad, J. J., Massaad, C. A., Jabbur, S. J., & Saade, N. E. (2002). The role of the sympathetic efferents in endotoxin-induced localized inflammatory hyperalgesia and cytokine upregulation. *Neuropharmacology, 42*, 864–872.

Schuligoi, R. (1998). Effect of colchicine on nerve growth factor-induced leukocyte accumulation and thermal hyperalgesia in the rat. *Naunyn-Schmiedeberg's Archives of Pharmacology, 358*, 264–269.

Tegeder, I., Pfeilschifter, J., & Geisslinger, G. (2001). Cyclooxygenase-independent actions of cyclooxygenase inhibitors. *The FASEB Journal, 15*, 2057–2072.

NGF, Sensitization of Nociceptors

David L. H. Bennett
Wolfson CARD, Kings College London, London, UK
Nuffield Department of Clinical Neurosciences, John Radclife Hospital, The University of Oxford, Oxford, UK

Synonyms

Inflammatory Pain and NGF; Neurotrophin

Definition

NGF is a secreted protein of molecular mass of 13 kD which exists as a homodimer. It is a member of the neurotrophin family, which also includes brain-derived neurotrophic factor (BDNF), neurotrophin-3 (NT3), and neurotrophin-4/neurotrophin-5 (NT4/5). NGF binds to both a high-affinity tyrosine kinase receptor trkA and a low-affinity receptor p75. Nociceptors are primary afferent neurons that respond to potentially tissue-damaging stimuli. NGF can sensitize these neurons so that they show an increased response to thermal and chemical stimuli.

Characteristics

It has now been established that NGF is a key mediator involved in the generation of inflammatory pain; for a full discussion of this, please see the entry "Inflammatory Pain and NGF." Administration of NGF either locally or systemically results in both thermal and mechanical hyperalgesia. NGF administration produces both a peripheral sensitization (increased response of primary afferent neurons to noxious stimuli) and central sensitization (a facilitation of the spinal processing of noxious stimuli). This entry will deal with the mechanisms by which NGF sensitizes nociceptors.

An in vitro skin-nerve preparation has been used to study the effects of NGF on cutaneous nociceptors. Direct application of NGF to the receptive fields of these afferents increased their sensitivity to noxious thermal stimuli but had no effect on their mechanical sensitivity (Rueff and Mendell 1996). In particular some fibers, which were thermally insensitive, developed a novel heat sensitivity after NGF treatment. NGF administration did not induce ongoing activity. Application of NGF to dissociated DRG cells results in a sensitization of these neurons to noxious heat and capsaicin that occurs within minutes (Shu and Mendell 2001; Galoyan et al. 2003). NGF potentiates the response to a given heat stimulus and prevents the tachyphylaxis normally seen when paired stimuli are given. Visceral afferents innervating the bladder have been shown to display increased mechanical sensitivity following NGF administration (Dmitrieva et al. 1997), and high-threshold group IV muscle afferents develop ongoing activity following intramuscular NGF injection (Hoheisel et al. 2005).

We have studied the role of NGF in producing primary afferent sensitization following intraplantar carrageenan, which evokes an acute inflammatory reaction (Koltzenburg et al. 1999). A trkA-IgG fusion protein was used to sequester endogenous NGF. Three hours after carrageenan administration, there was a marked increase in spontaneous activity (50 % of nociceptors innervating inflamed skin vs. 4 % in control). This could be almost completely prevented by administration of trkA-IgG (Fig. 1). Interestingly, as described above, NGF administration alone does not induce spontaneous activity in nociceptors, and therefore, in the context of inflammation, it must be acting cooperatively with other chemical mediators. Inflammation also produced an increased response of nociceptors to a standard heat ramp, and the proportion of fibers responding to the inflammatory mediator bradykinin increased. Both of these changes could be prevented by administration of trkA-IgG (Fig. 1). We did not find any significant mechanical sensitization of cutaneous nociceptors following inflammation. Other groups have found varying degrees of mechanical

NGF, Sensitization of Nociceptors, Fig. 1 The effects of carrageenan inflammation on the properties of primary afferent nociceptors. Recordings were made from an isolated skin-nerve preparation. Carrageenan inflammation led to increased spontaneous activity in primary afferents (**a**), which could be blocked by trkA-IgG treatment. Afferents were also tested for their response to a ramp increase in skin temperature (**b**). Afferents innervating inflamed skin showed an enhanced response to this heat ramp. This thermal sensitization was prevented by administration of a trkA-IgG, which sequesters endogenous NGF (Figure adapted from Koltzenburg et al. 1999)

sensitization following inflammation. It is likely that the mechanical hyperalgesia seen after NGF administration is due to central mechanisms.

Cell Types Involved in the Generation of NGF-Mediated Hyperalgesia

As described in entry "Inflammatory Pain and NGF," there are two receptors for NGF, a high-affinity tyrosine kinase receptor trkA and a low-affinity receptor p75. It is likely that the hyperalgesic effects of NGF are principally mediated by binding to trkA. Animals that lack p75 continue to develop thermal and mechanical hyperalgesia in response to NGF. A substantial proportion of peripheral nociceptor terminals express the trkA receptor, and I will discuss the direct sensitizing action of NGF on these neurons in the following section. A number of other cellular elements within peripheral tissues also express trkA, and some of the sensitizing actions of NGF may therefore be indirect (for a schematic representation of the sensitizing effects of NGF on nociceptors, see Fig. 2).

Mast cells expressing trkA and NGF can induce mast cell degranulation (resulting in the release of histamine and serotonin) and the expression of a number of cytokines including Il-3, Il-4, Il-10, and TNFα. Mast cell degranulators and serotonin antagonists have been shown to reduce the thermal, but not the mechanical, hyperalgesia that occurs following NGF administration (Lewin et al. 1994; Woolf et al. 1996). There may also be an interaction between NGF and sympathetic efferents, which also express the trkA receptor. Surgical or chemical sympathectomy can reduce the short-latency thermal and mechanical

hyperalgesia evoked by NGF (Andreev et al. 1995; Woolf et al. 1996). The production of eicosanoids by sympathetic efferents has been suggested to contribute to hyperalgesia under some inflammatory conditions. This does not appear to be the case for NGF-induced hyperalgesia as this is unaffected by treatment with nonsteroidal anti-inflammatory agents.

A further mechanism by which NGF produces peripheral sensitization is through activation of the 5-lipoxygenase pathway. The enzyme 5-lipoxygenase converts arachidonic acid into leukotrienes. Leukotriene B_4 (LTB_4) is a powerful chemotactic factor for neutrophils and has been shown to sensitize nociceptive afferents to thermal and mechanical stimuli. NGF increases the production of (LTB_4) in the skin, and inhibitors of the 5-lipoxygenase enzyme prevent the development of thermal

NGF, Sensitization of Nociceptors, Fig. 2 Schematic representation of the role of NGF in nociceptor sensitization. Following tissue injury there is a rapid increase in NGF expression, which is secondary to increased levels of Il-1 and TNFα. Some of the effects of NGF on nociceptor terminals are mediated indirectly via mast cells, sympathetic efferents, and neutrophil chemotaxis. NGF can also act directly on nociceptors following binding to the trkA

receptor. This leads to activation of multiple second messenger pathways resulting in the potentiation of a number of ion channels including TRPV1, Nav 1.8, and ASIC3. An activated trkA-NGF complex is also transported to the cell nucleus and modulates gene expression causing more long-term effects on nociceptor excitability and function. See text for details

hyperalgesia following intraplantar NGF injection (Amann et al. 1996).

Direct Actions of NGF on Nociceptor Function

Forty percent of adult DRG cells express trkA and it is selectively expressed in a population of small-diameter DRG cells which express neuropeptides such as CGRP and which are predominantly nociceptors. The recent rapid advances in characterizing both the second messenger pathways and the molecular identities of signal transducing elements within nociceptors have led to a much greater understanding of the sensitizing actions of NGF on these neurons.

NGF binding to trkA results in the cross-linking of two adjacent receptors. This triggers each trkA molecule to phosphorylate tyrosine residues on the cytoplasmic domain of its cross-linked neighbor. Phosphorylation causes conformational changes, which expose binding sites for proteins with SH2 domains. Multiple second messenger pathways are subsequently activated including phospholipase C-γ (PLC-γ), protein kinase C, and members of the mitogen-activated protein kinase (MAPK) pathway. NGF can have acute effects on ion channel function via posttranslational modification such as phosphorylation and long-term effects based on transcriptional regulation and transport of ion channels to the nerve terminal.

One of the most intensively studied areas which demonstrate these multiple levels of regulation by NGF is the potentiation of TRPV1 function. TRPV1 is expressed in nociceptors and is a capsaicin- and proton-sensitive cation-selective channel that transduces noxious heat stimuli. Although its role in thermal nociception in the normal state is controversial, it has been shown to be important in the generation of thermal hyperalgesia following inflammation and following NGF administration (Chuang et al. 2001). NGF rapidly (within minutes) sensitizes TRPV1 responses to low pH, capsaicin, and noxious heat. This sensitizing effect of NGF is clearly established; however, there is some confusion in the literature as to the exact second messenger pathways mediating this effect. One proposed pathway is via activation of phospholipase C (PLC). PLC-γ activation results in reduced levels of phosphatidylinositol-4,5-bisphosphate (PtdIns(4,5)P2) in the plasma membrane. As PtdIns(4,5)P2 exerts a tonic inhibitory action on TRPV1, this can rapidly enhance the sensitivity of TRPV1 and this effect has been demonstrated both in heterologous cells expressing TRPV1 and DRG cells (Chuang et al. 2001; Galoyan et al. 2003). Another second messenger pathway activated by NGF is phosphatidyl 3-kinase (PI3K). This is a lipid kinase that phosphorylates the D-3 position of phosphatidylinositol lipids to produce PI (3,4,5)P3. One of the downstream effects of PI3K activation is activation of extracellular signal-regulated protein kinase (ERK), a member of the MAPK family. NGF activates PI3K and ERK in DRG cells. These pathways have been shown to have a role in mediating both the potentiation of TRPV1 current and the thermal hyperalgesia induced by NGF (Zhuang et al. 2004). Another group studying the sensitization of capsaicin-induced $[Ca^{2+}]_i$ by NGF in neonatal DRG cells also reported an important role for PI3K and also for PKC, but did not find any effect of blocking PLC (Bonnington and McNaughton 2003).

Following binding of NGF to trkA at the nerve terminal, the complex is internalized and retrogradely transported in a signaling endosome to the cell body, where it regulates further second messenger pathways. Signals may also be activated by NGF at the cell surface and then transported back to the cell body in the absence of NGF. Once this signal has reached the cell body, it can modulate gene transcription and the targeting of signal transducing molecules back to nociceptor terminals. Twenty-four hours after induction of inflammation, there is an activation of p38 MAP kinase within DRG cell bodies that has been shown to be NGF dependent. This activation of p38 by NGF results in increased levels of TRPV1 protein (but not mRNA), which is transported to the peripheral terminals of C-fibers. This increased TRPV1 expression makes a contribution to the late phase of thermal hyperalgesia during inflammation (Ji et al. 2002).

A second group of molecular targets within sensory neurons that are regulated by NGF are the tetrodotoxin resistant (TTXR) sodium channels (Nav 1.8 and 1.9). These are exclusively expressed within nociceptive neurons. Inflammatory mediators have been shown to increase the TTXR current in DRG cells. Posttranslational modulation of TTXR is via phosphorylation of the channel mediated by both PKA and PKC (Gold et al. 1998). NGF may also regulate the expression of Nav 1.8 in DRG cells. The importance of this mechanism in mediating the sensitizing effects of NGF on nociceptive neurons is demonstrated by the fact that animals that carry a null mutation of the Nav 1.8 gene show reduced thermal hyperalgesia following NGF administration (Kerr et al. 2001). NGF can also regulate the expression of other ion channels such as the acid-sensing ion channel ASIC 3 within nociceptors. This and the potentiation in TRPV1 function (which is also activated by protons) are likely to increase the sensitivity of nociceptors to the low pH seen in inflammatory tissue.

In summary, NGF has been shown convincingly to sensitize nociceptive afferents to thermal stimuli and in the context of inflammation also contributes to the development of ongoing activity. There are likely to be a number of direct and indirect mechanisms that lead to such sensitization. A greater understanding of this process, and in particular the means by which NGF potentiates the function of TRPV1 and Nav 1.8 in nociceptors, may ultimately lead to novel analgesic therapies for inflammatory pain.

References

Amann, R., Schuligoi, R., Lanz, I., et al. (1996). Effect of a 5-lipoxygenase inhibitor on nerve growth factor-induced thermal hyperalgesia in the rat. *European Journal of Pharmacology, 306*, 89–91.

Andreev, N. Y., Dimitrieva, N., Koltzenburg, M., et al. (1995). Peripheral administration of nerve growth factor in the adult rat produces a thermal hyperalgesia that requires the presence of sympathetic postganglionic neurones. *Pain, 63*, 109–115.

Bonnington, J. K., & McNaughton, P. A. (2003). Signalling pathways involved in the sensitisation of mouse nociceptive neurones by nerve growth factor. *The Journal of Physiology, 551*, 433–446.

Chuang, H. H., Prescott, E. D., Kong, H., et al. (2001). Bradykinin and nerve growth factor release the capsaicin receptor from PtdIns(4,5)P2-mediated inhibition. *Nature, 411*, 957–962.

Dmitrieva, N., Shelton, D., Rice, A. S., et al. (1997). The role of nerve growth factor in a model of visceral inflammation. *Neuroscience, 78*, 449–459.

Galoyan, S. M., Petruska, J. C., & Mendell, L. M. (2003). Mechanisms of sensitization of the response of single dorsal root ganglion cells from adult rat to noxious heat. *The European Journal of Neuroscience, 18*, 535–541.

Gold, M. S., Levine, J. D., & Correa, A. M. (1998). Modulation of TTX-R INa by PKC and PKA and their role in PGE2-induced sensitization of rat sensory neurons in vitro. *The Journal of Neuroscience, 18*, 10345–10355.

Hoheisel, U., Unger, T., & Mense, S. (2005). Excitatory and modulatory effects of inflammatory cytokines and neurotrophins on mechanosensitive group IV muscle afferents in the rat. *Pain, 114*, 168–176.

Ji, R. R., Samad, T. A., Jin, S. X., et al. (2002). p38 MAPK activation by NGF in primary sensory neurons after inflammation increases TRPV1 levels and maintains heat hyperalgesia. *Neuron, 36*, 57–68.

Kerr, B. J., Souslova, V., McMahon, S. B., et al. (2001). A role for the TTX-resistant sodium channel Nav 1.8 in NGF-induced hyperalgesia, but not neuropathic pain. *Neuroreport, 12*, 3077–3080.

Koltzenburg, M., Bennett, D. L., Shelton, D. L., et al. (1999). Neutralization of endogenous NGF prevents the sensitization of nociceptors supplying inflamed skin. *The European Journal of Neuroscience, 11*, 1698–1704.

Lewin, G. R., Rueff, A., & Mendell, L. M. (1994). Peripheral and central mechanisms of NGF-induced hyperalgesia. *The European Journal of Neuroscience, 6*, 1903–1912.

Rueff, A., & Mendell, L. M. (1996). Nerve growth factor NT-5 induce increased thermal sensitivity of cutaneous nociceptors in vitro. *Journal of Neurophysiology, 76*, 3593–3596.

Shu, X., & Mendell, L. M. (2001). Acute sensitization by NGF of the response of small-diameter sensory neurons to capsaicin. *Journal of Neurophysiology, 86*, 2931–2938.

Woolf, C. J., Ma, Q. P., Allchorne, A., et al. (1996). Peripheral cell types contributing to the hyperalgesic action of nerve growth factor in inflammation. *The Journal of Neuroscience, 16*, 2716–2723.

Zhuang, Z. Y., Xu, H., Clapham, D. E., et al. (2004). Phosphatidylinositol 3-kinase activates ERK in primary sensory neurons and mediates inflammatory heat hyperalgesia through TRPV1 sensitization. *The Journal of Neuroscience, 24*, 8300–8309.

NGF-OE Mice

► Nerve Growth Factor Overexpressing Mice as Models of Inflammatory Pain

Nick Model of Cutaneous Pain and Hyperalgesia

Esther M. Pogatzki-Zahn
Department of Anesthesiology and Intensive
Care Medicine, University Muenster, Muenster,
Germany

Synonyms

Brennan pain model; Incision model for postoperative pain; Plantar incision model; Postoperative pain model

Definition

Postoperative pain is a unique and common form of acute pain. There is ample evidence that pain caused by inflammation, nerve injury, or incision is based on different pathophysiological mechanisms. This explains why many treatment strategies are efficacious only against specific types of persistent pain (Hunt and Mantyh 2001). Recognizing this gap between preclinical models of persistent pain and postsurgical pain, new rodent skin incision models have been developed (Brennan et al. 1996; Martin et al. 2004; Pogatzki et al. 2002b). In the plantar incision model developed by Brennan and co-workers, a simple incision is made in the plantar aspect of the hind paw of a rat (Brennan et al. 1996). The plantar hind paw incision is performed under halogenated volatile anesthesia using sterile technique. After induction of anesthesia, a 1 cm longitudinal incision is made with a number 11 blade, through skin and fascia of the paw, starting 0.5 cm from the proximal edge of the heel and extending toward the digits (Fig. 1a). Subsequently, the underlying flexor muscle is elevated with the forceps and incised longitudinally with the scalpel blade. The muscle origin and insertion remain intact. After hemostasis with gentle pressure, the skin is apposed with two mattress sutures of 5-0 nylon and the wound site is covered with antibiotic ointment. The rats recover for 30 min^{-1} h from anesthesia in clean bedding. Nylon is used for closure to minimize the inflammatory response to the suture material. The sutures are removed on postoperative day 2 to prevent any persistent responses caused by the sutures.

Mouse models of postoperative pain have been performed (Pogatzki and Raja 2003). The behaviors caused by plantar incision in the mouse roughly parallel those in the rat.

Characteristics

Primary and Secondary Hyperalgesia After Incision Injury

An incision made in the plantar aspect of the rat hind paw causes persistent, reduced withdrawal thresholds to mechanical stimuli suggesting hyperalgesia (decreased pain threshold to suprathreshold stimuli). Primary mechanical and heat hyperalgesia, enhanced pain to mechanical and thermal stimuli in the area of the incision, is present after a surgical incision in this model (Fig. 1b, c). Whereas primary mechanical hyperalgesia after plantar incision in rodents lasts for 2–3 days, primary heat hyperalgesia lasts up to 7 days after incision (Zahn and Brennan 1999b).

► Secondary hyperalgesia, enhanced pain to stimuli applied adjacent to the area of incision, was observed to punctate mechanical stimuli but not to a blunt mechanical probe or to a thermal stimulus (Pogatzki et al. 2002b; Zahn and Brennan 1999b).

Increased mechanical sensitivity after incision in rodents has striking similarities to an experimental incision in humans (Kawamata et al. 2002) and to patients' pain reports after surgery (Stubhaug et al. 1997). Hyperalgesia from the area of incision (► Primary Hyperalgesia)

Nick Model of Cutaneous Pain and Hyperalgesia, Fig. 1 Mechanical and thermal hyperalgesia after incision. (**a**) Schematic of the incision in the plantar aspect of the rat foot. (**b**) Punctate mechanical hyperalgesia adjacent to the incision. The results are expressed as median (*horizontal line*) with first and third quartiles (*boxes*) and 10th and 90th percentiles (*vertical lines*). Primary punctate hyperalgesia after an incision of skin, fascia, and muscle. The site of incision and of testing is illustrated in a schematic accompanying the graph. *P < 0.05 vs. pre; †P < 0.05 vs. sham. (**c**) Heat hyperalgesia after incision. Primary heat hyperalgesia after an incision of skin, fascia, and muscle. The site of testing and incision is illustrated to the right of the graph. * P < 0.05 vs. pre; †P < 0.05 vs. sham. The symbols represent the mean ± standard deviation (SD)

probably contributes to mechanically evoked pain in postoperative patients. A reduction of primary hyperalgesia with a specific treatment should therefore affect pain induced by mechanical stimuli (e.g., coughing, movement, or ambulation after surgery).

Peripheral and Central Sensitization After Incision Injury

A number of neurophysiological studies using the rodent incision models have revealed some of the underlying mechanisms inherent in incision-induced pain. From these studies mechanical hyperalgesia after incision appears to involve activation and sensitization of primary afferent ► nociceptors (► Peripheral Sensitization) (Pogatzki et al. 2002a) and neurons in the spinal dorsal horn (central sensitization) (Vandermeulen and Brennan 2000; Zahn and Brennan 1999a).

Peripheral Sensitization After Incision

In general, the characteristic features of experimentally induced peripheral sensitization of primary afferent fibers are a lowering of response threshold, an increase in response magnitude to suprathreshold stimuli, an increase in spontaneous activity, or an increase in RF size (Fig. 2a–c). Recording mechanosensitive afferent fibers innervating the plantar aspect of the rat hind paw using standard teased-fiber techniques revealed an increase in the ► receptive field size and the occurrence of spontaneous activity of A-delta and C-fibers after incision; activation thresholds of A-delta fibers, but not of C-fiber nociceptors, decreased after incision (Pogatzki et al. 2002a). Importantly, the conversion of mechanical insensitive ► silent nociceptors to mechanically activated fibers was indicated 24 h after incision (Pogatzki et al. 2002a). A-beta fibers did not sensitize after an incision.

In conclusion, spatial summation of modestly increased response magnitude may contribute to the reduced withdrawal threshold after incision. Spontaneous activity in A-delta and C-fibers may not only account for spontaneous pain behavior but may also contribute to mechanical hyperalgesia by amplifying responses centrally. Furthermore, the conversion of mechanical insensitive silent nociceptors to mechanically activated fibers probably has a role in the maintenance of hyperalgesia after an incision.

Spinal Sensitization After Incision

Recording action potentials from dorsal horn neurons (DHN) after incision revealed unique characteristics of ► spinal sensitization defined by decreased withdrawal thresholds, increased neuronal activity, enlarged receptive field (RF) size, and increased background activity (Fig. 2d). DHNs receiving input from the plantar aspect of

the hind paw using natural stimuli in anesthetized rats were characterized as ► wide-dynamic-range (WDR) and ► high-threshold (HT) neurons based on their responses to brush and pinch. WDR neurons respond to both brush and pinch, whereas HT neurons respond only to pinch. The results from DHN recording demonstrate that a plantar incision caused dorsal horn cell activation and central sensitization (Vandermeulen and Brennan 2000; Zahn and Brennan 1999a). Both WDR and HT neurons have increased background activity that is driven by activated primary afferent fibers from the incision area (Pogatzki et al. 2002c) and probably transmits evidence for nonevoked ongoing pain like guarding behaviors. Because the threshold of HT neurons did not decrease to the range of the withdrawal responses in behavioral experiments, particular WDR dorsal horn neurons probably contribute to the reduced withdrawal threshold observed in behavioral experiments. In contrast to other tissue injuries, there is no evidence that after an incision HT neurons convert to WDR cells or low-threshold neurons are sensitized.

Pharmacology of Incisional Pain

Excitatory amino acids (EAA), such as glutamate and aspartate, are important in the processing of nociceptive inputs in the dorsal horn of the spinal cord. These EAA, contained in primary afferent fibers and spinal interneurons, activate spinal N-methyl-D-aspartate (NMDA), non-NMDA, and metabotropic EAA receptors to facilitate transmission of sensory inputs and contribute to the enhanced excitability of nociceptive pathways in the dorsal horn of the spinal cord in persistent pain states. These neurotransmitters are also candidates to facilitate dorsal horn neuron activity and contribute to the hyperalgesia following surgical injury.

Pharmacological studies (Zahn et al. 2002) demonstrated that intrathecally administered non-NMDA receptor antagonists blocked nonevoked pain behaviors and mechanical hyperalgesia after incision (Fig. 3b). In contrast to other tissue injuries, both the ongoing pain behavior and primary mechanical hyperalgesia that occurs after incision were found not to be

a Spontaneous activity: C-fiber

b Aδ - and C-fibers

c Response threshold: Aδ -fibers

d Dorsal horn neuron activity

Nick Model of Cutaneous Pain and Hyperalgesia, Fig. 2 (a–d) Sensitization of peripheral afferent fibers and dorsal horn neurons after incision. (a, b) Spontaneous activity of afferent fibers 1 day after plantar incision. A total of 67 fibers have been recorded. None of 39 fibers in the sham group had spontaneous activity, whereas 11 of 28 fibers (39 %) in the incision group were spontaneously active. Both A-delta and C-fibers in incised rats exhibited spontaneous activity. An example of a spontaneously

active C-fiber with a high rate of activity is shown in (a); mean firing frequency in this example was 33.2 imp/s. The mechanical receptive field (RF) of this fiber (*black area*) includes the incision. (b) The mean activity (imp/s averaged over 5 min) of 11 spontaneously active fibers in the incision group is shown. Two of five spontaneously active A-delta fibers and four of six spontaneously active C-fibers had firing rates greater than 15 imp/s. (c) Mechanical response thresholds of 39 A-delta fibers. Median

dependent on activation of spinal NMDA receptors or spinal metabotropic EAA receptors (Fig. 3c). Furthermore, studying the role of different glutamate receptors for secondary hyperalgesia after incision indicates that non-NMDA receptors, but not NMDA receptors, are important for enhanced mechanical responses outside the area of an incision. Recently it has been demonstrated that spinally administered Joro spider toxin (JSTX), an antagonist for calcium-permeable non-NMDA receptors, reduced secondary but not primary hyperalgesia after incision. This, again, differs from other types of tissue injuries. Furthermore, these data indicate that different spinal receptors are involved in maintaining primary and secondary hyperalgesia after an incision (Pogatzki et al. 2003).

Thus, the mechanism(s) for maintaining pain behaviors following an incision is different from mechanisms described for inflammatory or neuropathic models of hyperalgesia; only the non-NMDA EAA receptors are required for the

maintenance of pain behaviors after plantar incision. These data also suggest that models that rely on different receptor systems for the development and maintenance of pain behavior may not predict analgesia for patients with postoperative pain.

Other spinally administered drugs producing analgesia in this model include L-type calcium channel receptor antagonists (Wang et al. 2000), alpha-2 adrenoceptor antagonists (Onttonen and Pertovaara 2000), and prostaglandin receptor antagonists (Omote et al. 2002).

Preemptive Analgesia

Because central sensitization may be important for postoperative pain, many have proposed that a blockade of noxious input to the spinal cord before the tissue injury will reduce postoperative pain more than blockade after injury (▶ Preemptive Analgesia).

The usefulness of preemptive analgesia has been reported in several animal models of

Nick Model of Cutaneous Pain and Hyperalgesia, Fig. 3 Effect of IT vehicle, NBQX (non-NMDA receptor antagonist), and MK-801 (NMDA receptor antagonist) on punctate mechanical hyperalgesia caused by incision. The results are expressed as median (*horizontal line*) with first and third quartiles (boxes) and 10th and 90th percentiles

(*vertical lines*). Withdrawal threshold after incision in rats treated with vehicle (**a**), 5 nmol of NBQX (**b**), and 40 nmol MK-801 (**c**) on the day of surgery. *P < 0.05 vs. 0 min by Friedman and Dunnett's test. †P < 0.05 vs. saline by Kruskal-Wallis and Dunnett's test. *POD1* postoperative day 1

Nick Model of Cutaneous Pain and Hyperalgesia, Fig. 2 (continued) response thresholds of A-delta fibers after incision (n = 13) were significantly decreased compared to A-delta fiber response thresholds after sham procedures (n = 26). The results are expressed as median

(*horizontal line*) with first and third quartiles (boxes) and 10th and 90th percentiles (*vertical lines*). †P < 0.05 vs. sham. (**d**) Recording of dorsal horn neurons. Increased activity to mechanical stimuli of one WDR dorsal horn neurons and 1 h after incision

inflammation, chemical irritation, or neuropathic pain, but results from clinical studies on postoperative pain have been disappointing (Moiniche et al. 2002). In the incision model, spinal administration of morphine, bupivacaine, and EAA receptor antagonists did not reduce pain behaviors beyond the expected duration of the analgesic effect (Zahn et al. 2002). These data indicate that when the early effect of a pharmacological treatment diminishes, the surgical wound appears capable of generating pain behaviors equivalent to the untreated group. In fact, ongoing input from the periphery after incision, not the afferent barrage during the injury, is critical for the expression of behaviors and sensitization of DHN after incision (Pogatzki et al. 2002c). This may explain why clinical studies using treatments attempting to prevent the development of postoperative pain have yielded mostly negative results. Duration of treatment, rather than time initiated, may be important.

Conclusion

In conclusion, it is important to recognize that pain caused by different tissue injuries is probably a result of distinct neurochemical and electrophysiological mechanisms. Basic scientific studies on mechanisms of incisional pain have called attention to postoperative pain as an important clinical problem, distinguish postoperative pain from other experimental models by mechanism(s), and have facilitated pharmaceutical research aimed specifically toward treating incisional pain. Pain and hyperalgesia even from a simple cutaneous incision are only beginning to be understood; therefore, it is not surprising that despite the simple nature of incisional pain, postoperative pain remains a costly, poorly understood problem.

References

Brennan, T. J., Vandermeulen, E. P., & Gebhart, G. F. (1996). Characterization of a rat model of incisional pain. *Pain, 64*, 493–501.

Hunt, S. P., & Mantyh, P. W. (2001). The molecular dynamics of pain control. *Nature Reviews. Neuroscience, 2*, 83–91.

Kawamata, M., Watanabe, H., Nishikawa, K., et al. (2002). Different mechanisms of development and maintenance of experimental incision-induced hyperalgesia in human skin. *Anesthesiology, 97*, 550–559.

Martin, T. J., Buechler, N. L., Kahn, W., et al. (2004). Effects of laparotomy on spontaneous exploratory activity and conditioned operant responding in the rat: A model for postoperative pain. *Anesthesiology, 101*, 191–203.

Moiniche, S., Kehlet, H., & Dahl, J. B. (2002). A qualitative and quantitative systemic review of preemptive analgesia for postoperative pain relief: The role of timing of analgesia. *Anesthesiology, 96*, 725–741.

Omote, K., Yamamoto, H., Kawamata, T., et al. (2002). The effects of intrathecal administration of an antagonist for prostaglandin E receptor subtype EP(1) on mechanical and thermal hyperalgesia in a rat model of postoperative pain. *Anesthesia and Analgesia, 95*, 1708–1712.

Onttonen, T., & Pertovaara, A. (2000). The mechanical antihyperalgesic effect of intrathecally administered MPV-2426, a novel alpha2 -adrenoceptor agonist, in a rat model of postoperative pain. *Anesthesiology, 92*, 1740–1745.

Pogatzki, E. M., & Raja, S. N. (2003). A mouse model of incisional pain. *Anesthesiology, 99*, 1023–1027.

Pogatzki, E. M., Gebhart, G. F., & Brennan, T. J. (2002a). Characterization of A-δ and C-fibers innervating the plantar rat hindpaw one day after an incision. *Journal of Neurophysiology, 87*, 721–731.

Pogatzki, E. M., Niemeier, J. S., & Brennan, T. J. (2002b). Persistent secondary hyperalgesia after gastrocnemius incision in the rat. *European Journal of Pain, 6*, 295–305.

Pogatzki, E. M., Vandermeulen, E. P., & Brennan, T. J. (2002c). Effect of plantar local anesthetic injection on dorsal horn neuron activity and pain behaviors caused by incision. *Pain, 97*, 151–161.

Pogatzki, E. M., Niemeier, J. S., Sorkin, L. S., et al. (2003). Spinal glutamate receptor antagonists differentiate primary and secondary mechanical hyperalgesia caused by incision. *Pain, 105*, 97–107.

Stubhaug, A., Breivik, H., Eide, P. K., et al. (1997). Mapping of punctate hyperalgesia around a surgical incision demonstrates that ketamine is a powerful suppressor of central sensitization to pain following surgery. *Acta Anaesthesiologica Scandinavica, 41*, 1124–1132.

Vandermeulen, E. P., & Brennan, T. J. (2000). Alterations in ascending dorsal horn neurons by a surgical incision in the rat foot. *Anesthesiology, 93*, 1294–1302.

Wang, Y. X., Pettus, M., Gao, D., et al. (2000). Effects of intrathecal administration of ziconotide, a selective neuronal N-type calcium channel blocker, on mechanical allodynia and heat hyperalgesia in a rat model of postoperative pain. *Pain, 84*, 151–158.

Zahn, P. K., & Brennan, T. J. (1999a). Incision-induced changes in receptive field properties of rat dorsal horn neurons. *Anesthesiology, 91*, 772–785.

Zahn, P. K., & Brennan, T. J. (1999b). Primary and secondary hyperalgesia in a rat model of postoperative pain. *Anesthesiology, 90*, 863–872.

Zahn, P. K., Pogatzki, E. M., & Brennan, T. J. (2002). Mechanisms for pain caused by incisions. *Regional Anesthesia and Pain Medicine, 27*, 514–516.

Nicotinic Receptors

▶ nAChr

NIRS

▶ Near-Infrared Spectroscopy

Nitric Oxide (NO)

Definition

Nitric oxide is a colorless, radical-free gas that reacts rapidly with O_2 to form other nitrogen oxides (e.g., NO_2, N_2O_2, and N_2O_4) and ultimately is converted to nitrite (NO_2) and nitrate (NO_3-). Nitric oxide itself is formed from L-arginine in bone, brain, endothelium, granulocytes, pancreatic beta cells, and peripheral nerves by a constitutive nitric oxide synthase, and in hepatocytes, Kupffer cells, macrophages, and smooth muscle by an inducible nitric oxide synthase (e.g., induced by endotoxin). NO activates soluble guanylate cyclase. Nitric oxide is a gaseous mediator of cell-to-cell communication and acts as a second messenger in some neurons. It can diffuse from one neuron to another, and so participates not only in intracellular signaling but also in intercellular signaling. It may be the first known retrograde neurotransmitter. Nitric oxide is also a potent vasodilator, mediates penile erection, and it can participate as a reactive oxygen species in free radical actions.

Cross-References

▶ Headache Attributed to a Substance or its Withdrawal

▶ Opiates During Development

▶ Spinothalamic Tract Neurons, Role of Nitric Oxide

Nitric Oxide Synthase

Synonyms

NOS

Definition

Nitric oxide synthase is an enzyme that synthesizes nitric oxide. The reaction involves arginine, which is transformed into citrulline accompanied by the release of nitric oxide.

Cross-References

▶ Spinothalamic Tract Neurons, Role of Nitric Oxide

Nitrous Oxide Antinociception and Opioid Receptors

Linda K. Vaughn[1] and Raymond M. Quock[2]
[1]Department of Biomedical Sciences, Marquette University, Milwaukee, WI, USA
[2]Department of Pharmaceutical Sciences, Washington State University College of Pharmacy, Pullman, WA, USA

Synonyms

Laughing gas; N_2O

Definition

An inorganic chemical gas with clinical anesthetic, analgesic, and anxiolytic properties, as well as potential for inhalant abuse and chemical dependency.

Characteristics

N_2O is a pharmacologically active inorganic gas that has been known since the early 1800s. Among its more clinically important properties is the ability to relieve pain. The mechanism of this analgesic action has long been speculated and studied. The modern era of nitrous oxide research began in the mid-1970s when Berkowitz and coworkers reported that N_2O-induced ▶ antinociception in the mouse abdominal constriction test was partly antagonized by the ▶ opioid antagonist naloxone, thus implicating an opioid mechanism (Berkowitz et al. 1976). They also discovered that rodents tolerant to morphine were cross-tolerant to N_2O, but animals tolerant to N_2O were not cross-tolerant to morphine. The explanation for this unilateral cross-tolerance was that tolerance to N_2O resulted from exhaustion of the reservoir of opioid peptide, leaving the response to exogenously applied morphine intact (Berkowitz et al. 1979). They formulated a hypothesis that N_2O caused an indirect activation of central opioid receptors by stimulating the neuronal release of endogenous opioid peptides.

Methionine-Enkephalin (ME)

The first chemical evidence of N_2O-induced release of endogenous opioid peptide was a twofold increase in levels of immunoreactive (i.r.) methionine-enkephalin (ME) in fractions of artificial cerebrospinal fluid (aCSF) perfusate collected from urethane-anesthetized, ventricular-cisternally-perfused rats exposed to 75 % N_2O for 60 min (Quock et al. 1985). At the end of the N_2O exposure, the levels of i.r. ME returned to baseline.

This led to the conclusion that N_2O was capable of inducing the neuronal release of either ME itself or a ME-like peptide in the rat brain. During exposure to N_2O, the rate of release of methionine-enkephalin-like activity into the ventricles rose from a resting state of 3.4–8.3 pg/min (Swartz and Quock 1987). Similarly in dogs with chronically implanted ventricular catheters, exposure to 66–75 % N_2O significantly raised levels of i.r. ME and ME-Arg[6]-Phe[7] but not leucine-enkephalin, β-endorphin (β-EP), or dynorphin $(DYN)_{1-17}$ in samples of CSF, compared to samples collected under room air (Finck et al. 1990).

Changes in CNS tissue levels of methionine-enkephalin in response to N_2O vary with region. In a variety of experimental designs, exposure to N_2O increased tissue concentrations of i.r. ME or preproenkephalin (PPE) in the brainstem (Quock et al. 1986; Kugel et al. 1991), hypothalamus (Quock et al. 1986; Kugel et al. 1991; Agarwal et al. 1996), and corpus striatum (Quock et al. 1986). No appreciable changes were found in the cerebral cortex and diencephalon (Quock et al. 1986) or dorsal raphé, medial thalamus, periaqueductal gray (PAG), raphé magnus, and locus coeruleus (Morris and Livingston 1984). Mixed results were found in the spinal cord, with N_2O causing an increase in i.r. ME (Quock et al. 1986) but no change in PPE (Tanimoto et al. 1998).

While these findings indicate that N_2O is, indeed, capable of stimulating neuronal release of ME in the CNS, its relationship to the antinociceptive effect of N_2O is debatable. Exposure of rats to 70 % N_2O causes a prolongation in the latency time to response in the 52 °C hot plate test; this antinociceptive effect of N_2O was not antagonized by intracerebroventricular (i.c.v.) pretreatment with rabbit antiserum to rat ME in doses as great as 400 μg/rat (Hara et al. 1994). Further, N_2O still retains its antinociceptive effect in $Penk2^{-/-}$ mice with disrupted PPE gene (Kingery et al. 2001). However, intrathecal (i.t.) pretreatment with ME or DYN antisera, but not β-EP antiserum, antagonized N_2O antinociception in the mouse abdominal constriction test (Cahill et al. 2000). This would be consistent with earlier observations that i.c.v.-administered κ opioid agonist-induced antinociception is blocked by i.t. administration of antisera against both DYN and ME (Tseng and Collins 1993), suggesting that activation of supraspinal κ opioid receptors causes antinociception involving release of DYN and ME in the spinal cord.

β-Endorphin (β-EP)

Zuniga and associates (Zuniga et al. 1986) found that a 60-min exposure to 60 % or 80 % N_2O caused a concentration-related increase in levels of i.r. β-EP in the mediobasal region of the hypothalamus, diencephalon, and PAG. The same research team also reported that exposure to N_2O increased the concentration of β-EP as well as the level of i.r. $ACTH_{1-39}$ staining in the arcuate pro-opiomelanocortin (POMC) neuronal system (Zuniga et al. 1987a). Since they also found that a similar exposure of rats to N_2O resulted in an increase in opiocortin immunoreactivity with $ACTH_{1-29}$ antiserum (Zuniga et al. 1986), they proposed that N_2O produces a generalized stimulation of the central pro-opiomelanocortin system in vivo in rats and that the antinociceptive effect of N_2O might be due to stimulated release of β-EP in the brain. This was subsequently confirmed in an in vitro system when they demonstrated that exposure to 80 % N_2O stimulated the release of β-EP from basal hypothalamic cells that were dispersed on cytodex microcarrier beads in an in vitro superfusion model (Zuniga et al. 1987b).

Kugel and coworkers found that exposure of female rats to 30 % N_2O for 8 h/day over an estrous cycle produced an increase in β-EP in the pituitary but not in the brainstem (Kugel et al. 1991). In a more recent investigation, exposure to 70 % N_2O increased levels of β-EP in aCSF perfusate collected from urethane-anesthetized, ventricular-cisternally-perfused rats exposed to 70 % N_2O (Zelinski et al. 2009).

The role of N_2O-induced release of β-EP in antinociception is indirectly supported by our earlier finding that N_2O-induced antinociception in the rat hot plate test was antagonized in

dose-dependent fashion by intracerebroventricular (i.c.v.) pretreatment with a rabbit antiserum against rat β-EP but not by an antiserum against ME (Hara et al. 1994). Furthermore, the antinociceptive effect of N_2O in the rat hot plate test was also attenuated in dose-related manner by i.c.v. pretreatment with lower doses of $β-EP_{1-27}$ (Hodges et al. 1994) – a truncated form of $β-EP_{1-31}$ that is a partial antagonist of β-EP (Takemori and Portoghese 1993) – and also by intra-PAG microinjection with the μ-opioid antagonist CTOP (D-Phe-Cys-Tyr-D-Trp-Orn-Thr-Pen-Thr-NH₂) (Hawkins et al. 1989). These findings implicate a N_2O-induced neuronal release of β-EP in the antinociceptive response to N_2O.

The evidence for a N_2O-induced release of β-EP has been most widely studied and is strongest in rats. In dogs, N_2O produced no change in the concentration of β-EP in CSF (Finck et al. 1990). In mice, β-EP antiserum was ineffective at reducing the antinociceptive effect of N_2O in the abdominal constriction test (Cahill et al. 2000; Branda et al. 2000).

Whether exposure to N_2O induces release of β-EP in human subjects remains uncertain. CSF levels of i.r. β-EP or leucine-enkephalin were unaltered 5 min after induction of anesthesia by thiopental and 70 % N_2O, 10 min after further addition of halothane and throughout the period of surgery to emergence from anesthesia (Way et al. 1984). Treatment with 30 % N_2O failed to alter plasma concentrations of i.r. β-EP or ME in pain-free, unstressed volunteers (Evans et al. 1985).

Dynorphin (DYN)

There is considerable evidence that N_2O produces its antinociceptive action in mice in the abdominal constriction test through κ-opioid mechanisms. Intraperitoneal injection of a dilute solution of acetic acid induces abdominal constrictions – lengthwise stretches of the torso with concave arching of the back. N_2O causes a concentration-dependent reduction in the number of abdominal constrictions. The κ-selective opioid antagonist norbinaltorphimine (nor-BNI) was able to reduce the magnitude of N_2O-induced antinociception, implicating κ-opioid receptors in the N_2O antinociception (Quock et al. 1990a; Koyama and Fukuda 2010). This antinociceptive effect of N_2O was sensitive to antagonism by naloxone but not by the μ-selective antagonist β-funaltrexamine (Quock et al. 1990a) or the δ-selective antagonist naltrindole (Koyama and Fukuda 2010; Quock and Vaughn 1995). N_2O also continued to evoke an antinociceptive effect in μ-opioid-receptor-deficient CXBK/By transgenic mice, suggesting that the antinociceptive effect was independent of μ-opioid receptors (Quock et al. 1993).

If N_2O works indirectly on κ-opioid receptors, then it is likely that N_2O stimulates the neuronal release of the endogenous κ-opioid ligand, DYN (Chavkin et al. 1982). This was confirmed by studies in i.c.v.-administered DYN antisera but not antisera against β-EP or ME antagonized N_2O-induced antinociception (Branda et al. 2000). Antagonism was noted with both an antiserum against DYN_{1-8} and another against DYN_{1-13}. Since DYN_{1-8} antiserum does not react with longer DYN fragments, and the DYN_{1-13} antiserum does not react with shorter DYN fragments, it was suggested that N_2O may be capable of releasing both long and short forms of DYN to evoke the antinociceptive response. Spinal as well as supraspinal release of DYN may be involved since the antinociceptive effect of N_2O was also antagonized by i.t. administration of dynorphin antisera (Cahill et al. 2000).

Experiments on the effect of N_2O on DYN release and subsequent activation of κ-opioid receptors have predominantly been performed in mice in the abdominal constriction model. In the rat hot plate test, N_2O-induced antinociception involves β-endorphin and μ-opioid receptors. In the rat warm water tail withdrawal test, N_2O-induced antinociception is seemingly mediated by κ- and δ-opioid receptors (Quock et al. 1990b). Furthermore, in rats, N_2O suppressed intraplantar formalin injection-induced increase of preprodynorphin mRNA (Tanimoto et al. 1998). These differences may reflect a species and/or test specificity for which opioid systems are recruited into play by N_2O (Tyers 1980).

Other Actions of N_2O

Direct Effects on Receptors

Electrophysiological experiments have demonstrated that N_2O has effects on ion channel receptors such as the $GABA_A$, NMDA, $5HT_3$, glycine, and nicotinic cholinergic receptors (Mennerick et al. 1998; Suzuki et al. 2002, 2003; Hapfelmeier et al. 2001; Yamakura and Harris 2000). The subsequent changes by N_2O in inhibition and excitation of neurons within antinociceptive pathways could be responsible for the release of endogenous opioid peptides. There is strong evidence that N_2O acts, in part, by disinhibiting a descending noradrenergic pathway projecting from the PAG and pons to the spinal cord that modulates spinal nociceptive processing [for review see (Fujinaga and Maze 2002)]. N_2O-induced antinociception is attenuated by lesioning of the PAG (Zuniga et al. 1987c) or intra-PAG microinjections of opioid antagonists (Hodges et al. 1994; Fang et al. 1997). Other brain areas that project to the PAG, such as the hypothalamus (Zuniga et al. 1987b) and cortex (Gyulai et al. 1997), or to which the PAG projects, such as the pons (Sawamura et al. 2000) and the spinal cord (Fujinaga and Maze 2002), may also be targets of N_2O.

N_2O Induction of Another Messenger: NO and CRF

Substantial evidence has also been amassed indicating that N_2O-induced antinociception is mediated by the release of nitric oxide (NO). Recently, it was reported that exposure to 70 % N_2O increased levels of β-EP as well as amounts of the NO metabolites nitrite (NO_2^-) and nitrate (NO_3^-) in aCSF perfusate collected from urethane-anesthetized, ventricular-cisternally-perfused rats exposed to 70 % N_2O (Zelinski et al. 2009). When the rats were pretreated systemically with an inhibitor of nitric oxide synthase (NOS), L-N^G-nitro arginine methyl ester (L-NAME), both the increases in β-EP and NO metabolites were reversed. One interpretation of these results is that the N_2O-stimulated neuronal release of β-EP is NO dependent. This had previously been suggested following a reduction in ME release by i.c.v. β-EP in i.t.-perfused rats (Hara et al. 1995). Preliminary studies using microdialysis have also demonstrated that exposure of rats to 70 % N$_2$O increases the levels of nitrite, nitrate, and β-EP in the arcuate nucleus and the PAG; these increases were simultaneously blocked by L-NAME (Ohgami et al. 2010). The feasibility of this concept is strong as NO has been implicated as a modulator in the neuronal release of neurotransmitters and endocrine hormones (Kodama and Koyama 2006; Uretsky et al. 2003; Rubinek et al. 2005). A role for NO in modulating excitatory and inhibitory synaptic transmission in antinociceptive pathways is also possible and such effects have been demonstrated elsewhere (Wang et al. 2007).

Another potential intermediary between N$_2$O and the release of endogenous opioid peptides is corticotrophin-releasing factor (CRF). It was found that exposure to N$_2$O induced a dose-dependent increase in expression of c-Fos in CRF-positive neurons in the paraventricular nucleus of the hypothalamus (Sawamura et al. 2003). I.c.v. pretreatment with α-helical CRF$_{9-14}$, a CRF antagonist, antagonized the N$_2$O-induced c-Fos expression in the locus coeruleus as well as its antinociceptive effect in the tail flick test (Sawamura et al. 2003). It was suggested that N$_2$O-induced neuronal release of CSF precedes the neuronal release of opioid peptides, meaning that the stimulated release of opioid peptides is actually a secondary or indirect effect of N$_2$O (Sanders et al. 2008).

Summary

In summary, there is evidence that N$_2$O can stimulate the neuronal release of endogenous opioid peptides. It is also evident that the recruitment of endogenous opioid mechanisms may be species or test dependent. N$_2$O seems likely to produce antinociception in the rat hot plate test through stimulating the neuronal release of β-EP to activate μ-opioid receptors, whereas N$_2$O may produce antinociception in the mouse abdominal constriction test by stimulating the neuronal release of DYN to activate κ-opioid receptors. The net effect of N$_2$O activation of opioid receptors appears to be disinhibition of a descending antinociceptive noradrenergic pathway, which, in turn, modulates pain processing through α$_2$-adrenergic and GABA mechanisms in the spinal cord. Roles for NO and CRF as intermediaries between N$_2$O and opioid peptide release have been suggested.

References

Agarwal, R. K., Kugel, G., Karuri, A., Gwosdow, A. R., & Kumar, M. S. (1996). Effect of low and high doses of nitrous oxide on preproenkephalin mRNA and its peptide methionine enkephalin levels in the hypothalamus. *Brain Research, 730*, 47–51.

Berkowitz, B. A., Ngai, S. H., & Finck, A. D. (1976). Nitrous oxide "analgesia": Resemblance to opiate action. *Science, 194*, 967–968.

Berkowitz, B. A., Finck, A. D., Hynes, M. D., & Ngai, S. H. (1979). Tolerance to nitrous oxide analgesia in rats and mice. *Anesthesiology, 51*, 309–312.

Branda, E. M., Ramza, J. T., Cahill, F. J., Tseng, L. F., & Quock, R. M. (2000). Role of brain dynorphin in nitrous oxide antinociception in mice. *Pharmacology, Biochemistry, and Behavior, 65*, 217–222.

Cahill, F. J., Ellenberger, E. A., Mueller, J. L., Tseng, L. F., & Quock, R. M. (2000). Antagonism of nitrous oxide antinociception in mice by intrathecally administered opioid peptide antisera. *Journal of Biomedical Science, 7*, 299–303.

Chavkin, C., James, I. F., & Goldstein, A. (1982). Dynorphin is a specific endogenous ligand of the k-opiate receptor. *Science, 215*, 413–415.

Evans, S. F., Stringer, M., Bukht, M. D., Thomas, W. A., & Tomlin, S. J. (1985). Nitrous oxide inhalation does not influence plasma concentrations of beta-endorphin or Met-enkephalin-like immunoreactivity. *British Journal of Anaesthesia, 57*, 624–628.

Fang, F., Guo, T. Z., Davies, M. F., & Maze, M. (1997). Opiate receptors in the periaqueductal gray mediate analgesic effect of nitrous oxide in rats. *European Journal of Pharmacology, 336*, 137–141.

Finck, A. D., Samaniego, E., & Ngai, S. H. (1990). Nitrous oxide selectively releases Met^5-enkephalin and Met^5-enkephalin-Arg^6-Phe^7 into canine third ventricular cerebrospinal fluid. *Anesthesia and Analgesia, 80*, 664–670.

Fujinaga, M., & Maze, M. (2002). Neurobiology of nitrous oxide-induced antinociceptive effects. *Molecular Neurobiology, 25*, 167–189.

Gyulai, F. E., Firestone, L. L., Mintun, M. A., & Winter, P. M. (1997). In vivo imaging of nitrous oxide-induced changes in cerebral activation during noxious heat stimuli. *Anesthesiology, 86*, 538–548.

Hapfelmeier, G., Haseneder, R., Kochs, E., Beyerle, M., & Zieglgansberger, W. (2001). Coadministered nitrous oxide enhances the effect of isoflurane on GABAergic transmission by an increase in open-channel block. *Journal of Pharmacology and Experimental Therapeutics, 298,* 201–208.

Hara, S., Gagnon, M. J., Quock, R. M., & Shibuya, T. (1994). Effect of opioid peptide antisera on nitrous oxide antinociception in rats. *Pharmacology, Biochemistry, and Behavior, 48,* 699–702.

Hara, S., Kuhns, E. R., Ellenberger, E. A., Mueller, J. L., Shibuya, T., Endo, T., & Quock, R. M. (1995). Involvement of nitric oxide in intracerebroventricular β-endorphin-induced neuronal release of methionine-enkephalin. *Brain Research, 675,* 190–194.

Hawkins, K. N., Knapp, R. J., Lui, G. K., Gulya, K., Kazmierski, W., Wan, Y. P., Pelton, J. T., Hruby, V. J., & Yamamura, H. I. (1989). [^{3}H]-[H-D-Phe-Cys-Tyr-D-Trp-Orn-Thr-Pen-Thr-NH$_2$] ([^{3}H]CTOP), a potent and highly selective peptide for mu opioid receptors in rat brain. *Journal of Pharmacology and Experimental Therapeutics, 248,* 73–80.

Hodges, B. L., Gagnon, M. J., Gillespie, T. R., Breneisen, J. R., O'Leary, D. F., Hara, S., & Quock, R. M. (1994). Antagonism of nitrous oxide antinociception in the rat hot plate test by site-specific mu and epsilon opioid receptor blockade. *Journal of Pharmacology and Experimental Therapeutics, 269,* 596–600.

Kingery, W. S., Sawamura, S., Agashe, G. S., Davies, M. F., Clark, J. D., & Zimmer, A. (2001). Enkephalin release and opioid receptor activation does not mediate the antinociceptive or sedative/hypnotic effects of nitrous oxide. *European Journal of Pharmacology, 427,* 27–35.

Kodama, T., & Koyama, Y. (2006). Nitric oxide from the laterodorsal tegmental neurons: Its possible retrograde modulation on norepinephrine release from the axon terminal of the locus coeruleus neurons. *Neuroscience, 138,* 245–256.

Koyama, T., & Fukuda, K. (2010). Involvement of the k-opioid receptor in nitrous oxide-induced analgesia in mice. *Journal of Anesthesia, 24,* 297–299.

Kugel, G., Zive, M., Agarwal, R. K., Beumer, J. R., & Kumar, A. M. (1991). Effect of nitrous oxide on the concentrations of opioid peptides, substance P, and LHRH in the brain and beta-endorphin in the pituitary. *Anesthesia Progress, 38,* 206–211.

Mennerick, S., Jevtovi-Todorovi, V., Todorovi, S. M., Shen, W., Olney, J. W., & Zorumski, C. F. (1998). Effect of nitrous oxide on excitatory and inhibitory synaptic transmission in hippocampal cultures. *The Journal of Neuroscience, 18,* 9716–9726.

Morris, B., & Livingston, A. (1984). Effects of nitrous oxide exposure on Met-enkephalin levels in discrete areas of rat brain. *Neuroscience Letters, 45,* 11–14.

Ohgami, Y., Chung, E., & Quock, R. M. (2010). Nitrous oxide-induced NO-dependent neuronal release of β-endorphin form the rat arcuate nucleus and periaqueductal gray. *Brain Research, 1366,* 38–43.

Quock, R. M., & Vaughn, L. K. (1995). Nitrous oxide: Mechanism of its analgesic action. *Analgesia, 1,* 151–159.

Quock, R. M., Kouchich, F. J., & Tseng, L. F. (1985). Does nitrous oxide induce release of brain opioid peptides? *Pharmacology, 30,* 95–99.

Quock, R. M., Kouchich, F. J., & Tseng, L. F. (1986). Influence of nitrous oxide upon regional brain levels of methionine-enkephalin-like immunoreactivity in rats. *Brain Research Bulletin, 16,* 321–323.

Quock, R. M., Best, J. A., Chen, D. C., Vaughn, L. K., Portoghese, P. S., & Takemori, A. E. (1990a). Mediation of nitrous oxide analgesia in mice by spinal and supraspinal κ-opioid receptors. *European Journal of Pharmacology, 175,* 97–100 corrigendum 187: 564.

Quock, R. M., Walczak, C. K., Henry, R. J., & Chen, D. C. (1990b). Effect of subtype-selective opioid receptor blockers on nitrous oxide antinociception in rats. *Pharmacological Research, 22,* 351–357.

Quock, R. M., Mueller, J. L., & Vaughn, L. K. (1993). Strain-dependent differences in responsiveness of mice to nitrous oxide (N$_2$O) antinociception. *Brain Research, 614,* 52–56.

Rubinek, T., Rubinfeld, H., Hadani, M., Barkai, G., & Shimon, I. (2005). Nitric oxide stimulates growth hormone secretion from human fetal pituitaries and cultured pituitary adenomas. *Endocrine, 28,* 209–216.

Sanders, R. D., Weimann, J., & Maze, M. (2008). Biologic effects of nitrous oxide. A mechanistic and toxicologic review. *Anesthesiology, 109,* 707–722.

Sawamura, S., Kingery, W. S., Davies, M. F., Agashe, G. S., Clark, J. D., Kobilka, B. K., Hashimoto, T., & Maze, M. (2000). Antinociceptive action of nitrous oxide is mediated by stimulation of noradrenergic neurons in the brainstem and activation of α_{2B} adrenoceptors. *Journal of Neuroscience, 20,* 9242–9251.

Sawamura, S., Obara, M., Takeda, K., Maze, M., & Hanaoka, K. (2003). Corticotropin-releasing factor mediates the antinociceptive action of nitrous oxide in rats. *Anesthesiology, 99,* 708–715.

Suzuki, T., Koyama, H., Sugimoto, M., Uchida, I., & Mashimo, T. (2002). The diverse actions of volatile and gaseous anesthetics on human-cloned 5-hydroxytryptamine$_3$ receptors expressed in Xenopus oocytes. *Anesthesiology, 96,* 699–704.

Suzuki, T., Ueta, K., Sugimoto, M., Uchida, I., & Mashimo, T. (2003). Nitrous oxide and xenon inhibit the human $(\alpha_7)_5$ nicotinic acetylcholine receptor expressed in Xenopus oocyte. *Anesthesia and Analgesia, 96,* 443–448.

Swartz, C. M., & Quock, R. M. (1987). Kinetic modeling of hormone release: Application to time-averaged met-enkephalin release. *Psychoneuroendocrinology, 12,* 377–384.

Takemori, A. E., & Portoghese, P. S. (1993). The mixed antinociceptive agonist–antagonist activity of beta-endorphin$_{1-27}$ in mice. *Life Sciences, 53,* 1049–1052.

Tanimoto, M., Fukuoka, T., Miki, K., Tokunaga, A., Tashiro, C., & Noguchi, K. (1998). Effects of halothane, ketamine and nitrous oxide on dynorphin mRNA expression in dorsal horn neurons after peripheral tissue injury. *Brain Research, 811,* 88–95.

Tseng, L. F., & Collins, K. A. (1993). Spinal involvement of both dynorphin A and metenkephalin in the antinociception induced by intracerebroventricularly administered bremazocine but not morphine in the mouse. *Journal of Pharmacology and Experimental Therapeutics, 266*, 1430–1438.

Tyers, M. B. (1980). A classification of opiate receptors that mediate antinociception in animals. *British Journal of Pharmacology, 69*, 503–512.

Uretsky, A. D., Weiss, B. L., Yunker, W. K., & Chang, J. P. (2003). Nitric oxide produced by a novel nitric oxide synthase isoform is necessary for gonadotropin-releasing hormone-induced growth hormone secretion via a cGMP-dependent mechanism. *Journal of Neuroendocrinology, 15*, 667–676.

Wang, W., Paton, J. F., & Kasparov, S. (2007). Differential sensitivity of excitatory and inhibitory synaptic transmission to modulation by nitric oxide in rat nucleus tractus solitarii. *Experimental Physiology, 92*, 371–382.

Way, W. L., Hosobuchi, Y., Johnson, B. H., Eger, E. I., 2nd, & Bloom, F. E. (1984). Anesthesia does not increase opioid peptides in cerebrospinal fluid of humans. *Anesthesiology, 60*, 43–45.

Yamakura, T., & Harris, R. A. (2000). Effects of gaseous anesthetics nitrous oxide and xenon on ligand-gated ion channels. Comparison with isoflurane and ethanol. *Anesthesiology, 93*, 1095–1101.

Zelinski, L. M., Ohgami, Y., & Quock, R. M. (2009). Exposure to nitrous oxide stimulates a nitric oxide-dependent neuronal release of β-endorphin in ventricular cisternally-perfused rats. *Brain Research, 1300*, 37–40.

Zuniga, J. R., Knigge, K. K., & Joseph, S. A. (1986). Central beta-endorphin release and recovery after exposure to nitrous oxide in rats. *Journal of Oral and Maxillofacial Surgery, 44*(9), 714–718.

Zuniga, J., Joseph, S., & Knigge, K. (1987a). Nitrous oxide analgesia: Partial antagonism by naloxone and total reversal after periaqueductal gray lesions in the rat. *European Journal of Pharmacology, 142*, 51–60.

Zuniga, J. R., Joseph, S. A., & Knigge, K. M. (1987b). The effects of nitrous oxide on the central endogenous pro-opiomelanocortin system in the rat. *Brain Research, 420*, 57–65.

Zuniga, J. R., Joseph, S. A., & Knigge, K. M. (1987c). The effects of nitrous oxide on the secretory activity of pro-opiomelanocortin peptides from basal hypothalamic cells attached to cytodex beads in a superfusion in vitro system. *Brain Research, 420*, 66–72.

NK

▶ Substance P Regulation in Inflammation

NK-1 Blockers

Definition

Neurokinin-1 (NK-1) receptor inhibitors lessen emesis after cisplatin, ipecac, copper sulfate, apomorphine, and radiation therapy, e.g., aprepitant.

Cross-References

▶ Cancer Pain Management, Chemotherapy

NK-1 Receptor

Definition

One of the three types of G-protein-coupled receptors where tachykinins act. Substance P is the preferred ligand, although neurokinin A (NKA) can also activate the NK-1 receptor.

Cross-References

▶ Windup of Spinal Cord Neurons

NMDA

Definition

N-methyl-D-aspartate (NMDA), a chemical analogue of glutamate that gives its name to the receptor. This receptor is formed of at least four subtypes, one of which is the NR2B.

Cross-References

▶ GABA and Glycine in Spinal Nociceptive Processing
▶ N-methyl-D-aspartate
▶ Opioids in the Spinal Cord and Central Sensitization

NMDA Glutamate Receptor(s)

Synonyms

N-methyl-D-aspartate receptor; NMDA receptor

Definition

One of the four general classes of glutamate receptors. Activated by N-methyl-D-aspartate with high affinity. It is involved in inducing long-term changes in the function of the neurons due to an influx of Ca^{2+} through it. This receptor requires a release of glutamate (ligand-gated) and a change in the membrane voltage (voltage-gated) concomitantly to be activated. The NMDA receptor is found in high concentrations in the anterior horn of the spinal cord, where it is associated with the process of central sensitization, one of the precursors of neuropathic pain. Agonists include glutamate and aspartate, and antagonists include ketamine and dextromethorphan and to a much lesser extent methadone. The receptors may be the site of action of dissociative anesthetics such as ketamine. The NMDA ion channel has a relatively high Ca^{2+} permeability, and this may be important in the initiation of plastic changes necessary for several types of learning and memory.

Cross-References

▶ Acute Pain Mechanisms
▶ Amygdala, Pain Processing and Behavior in Animals
▶ Descending Modulation and Persistent Pain
▶ GABA and Glycine in Spinal Nociceptive Processing
▶ Metabotropic Glutamate Receptors in the Thalamus
▶ NSAIDs, Adverse Effects
▶ Opiates During Development
▶ Opioids in the Spinal Cord and Central Sensitization
▶ Pain Control in Children with Burns

▶ Pain-Modulatory Systems, History of Discovery
▶ Postoperative Pain, Transition from Parenteral to Oral Drugs
▶ Somatic Pain
▶ Thalamus, Nociceptive Neurotransmission
▶ Windup of Spinal Cord Neurons

NMDA Receptor

▶ NMDA Glutamate Receptor(s)

NMDA Receptors in Spinal Nociceptive Processing

Horacio Vanegas[1] and Hans-Georg Schaible[2]
[1]Instituto Venezolano de Investigaciones Cientificas (IVIC), Caracas, Venezuela
[2]Department of Physiology - Neurophysiology, Friedrich-Schiller-University of Jena, Jena, Germany

Synonyms

Calcium channel; Glutamate receptors; Ionotropic receptors; Spinal dorsal horn

Definition

The NMDA receptor (NMDAR) is a tetrameric molecule provided with a channel which, upon glutamate binding, allows the outflow of potassium ions as well as the inflow of sodium and, characteristically, calcium ions across the neuronal membrane. The NMDAR is responsible for excitatory synaptic transmission in nociceptive neurons, circuits and pathways.

Characteristics

Glutamate is the main excitatory transmitter in the central nervous system. When released by presynaptic terminals, glutamate reaches the postsynaptic membrane, where various types of receptor molecules may be found. Glutamate receptor molecules can be divided in two broad types: the ionotropic type, which possesses an ion channel that opens upon glutamate binding; and the metabotropic type, which has no channel but is coupled to G-proteins that transduce glutamate binding into modifications of intracellular messengers and enzymes.

Molecular Biology and Biophysics of NMDARs

The ionotropic glutamate receptors (Kandel and Siegelbaum 2000) are named after their main pharmacological ligands, that is, AMPA (α-amino-3-hydroxy-5-methylisoxazole-4-propionic acid), kainate, and NMDA (N-methyl-D-aspartate). In all three types, the ion channel permits the flow of sodium and potassium ions. Although a minority of AMPA receptors also permit the flow of calcium ions, a large calcium inflow is characteristic of all NMDARs and constitutes the key to their physiological role. NMDARs are also characterized by the fact that, at resting membrane potential (ca. -65 mV), the entrance to the ion channel is blocked by a magnesium ion. At resting membrane potential, glutamate binding does not lead to any ion current through the NMDAR; first the magnesium ion must be dislodged from this channel. The ionotropic glutamate receptors then function in the following manner:

The glutamate released from presynaptic terminals first activates AMPA and kainate receptors (usually called non-NMDARs), which leads to a fast flow of sodium and potassium ions and thus to depolarization of the postsynaptic neuron. If this depolarization is large enough, the magnesium ion will no longer be attracted into the NMDA channel entrance, and the channel will finally become free for sodium and potassium to flow through; this causes a prolonged depolarization. Most importantly, the unblocked NMDA channel now lets large amounts of calcium ions flow into the neuron.

Ionotropic glutamate receptor molecules are made of two pairs of subunits, each subunit in turn with four intramembranous segments.

Subunits in NMDARs are called NR1, NR2 and NR3. The predominant NR2 subunit is the NR2B. These subunits can be sensitized through phosphorylation by protein kinases (see below). NR1 binds glycine, and NR2 binds glutamate (Furukawa et al. 2005). Glycine is essential for the NMDAR to function. Normally there is enough glycine in the extracellular fluid, and the function of the NMDAR then depends only on the glutamate released presynaptically and on the membrane potential of the postsynaptic neuron (Kandel and Siegelbaum 2000).

Involvement of NMDARs in Normal Nociception

Some authors have found that NMDAR antagonists, whether given systemically or intrathecally onto the spinal cord, do not modify behavioral baseline responses to acutely painful stimuli (Yaksh 1999). Also, conditional deletion of the NR1 subunit in the spinal cord had no influence upon behavioral responses to tactile stimulation or to acute, high-intensity non-damaging thermal stimuli applied to the skin (South et al. 2003). However, recordings from spinal cord nociceptive neurons have shown (Dougherty et al. 1992; Neugebauer et al. 1993a, b) that NMDARs participate in responses to various types of cutaneous and deep somatic noxious stimuli. It therefore seems that NMDARs will contribute to normal nociception whenever the postsynaptic neurons become sufficiently depolarized for the magnesium ion to be dislodged from the receptor. This will of course be more intense in cases of persistent nociception.

Involvement of NMDARs in Persistent Nociception

In cases of persistent damage to a peripheral tissue, such as during inflammation, trauma or nerve lesions, the continuous and often large barrage of action potentials coming into the spinal cord causes a considerable release of glutamate and this in turn, via non-NMDARs, causes a sufficiently large neuronal depolarization for the magnesium ion to be evicted from the NMDA channel. Calcium then flows into the postsynaptic neurons and a whole series of events is triggered which result in a considerable and long-lasting increase in spinal neuronal excitability (central sensitization). The hyperalgesia (increased pain upon noxious stimulation) and allodynia (pain elicited by normally non-painful stimuli) of chronic pain conditions are thus thought to arise not only from an increased nociceptive input into the spinal cord, but also from an increased synaptic relay of nociceptive messages towards supraspinal structures.

It must be made clear that activation of NMDARs is not the only mechanism whereby central sensitization comes about. Concomitant mechanisms are, for example: (1) activation of receptors for pronociceptive neuropeptides such as substance P (NK1), neurokinin A (NK2), calcitonin gene-related peptide (CGRP receptor), cholecystokinin (mainly CCK_B), and vasoactive intestinal peptide (VIP receptor); and (2) induction of cyclooxygenase-2 (Vanegas and Schaible 2001), with the resulting enhancement of the pronociceptive effects of prostaglandins and thromboxanes. It must also be made clear that, in addition to the NMDAR, there are voltage-activated calcium channels (Vanegas and Schaible 2000) through which considerable amounts of calcium can flow into the depolarized neuron. Compounds that antagonize NMDARs or the other mechanisms just mentioned have been shown to attenuate pain messages and are of course potential analgesic agents.

Electrically Induced Spinal Neuronal Hyperexcitability

Two types of electrophysiological manipulations have shown that the NMDAR plays a key role in spinal neuronal hyperexcitability. Both of these involve stimulating the primary afferents with electrical pulses whose intensities are generally large enough to fire all fiber types, from $A\beta$ to C. Each stimulus pulse elicits in spinal cord nociceptive neurons a depolarizing synaptic potential that lasts up to 20 s and is due to the activation of non-NMDA, NMDA and substance P receptors (Woolf 1996). If the pulse is applied about once per second, these potentials summate one on top of the other thus giving rise to increasingly larger neuronal discharges. This is known as windup,

and shows that the excitability of neurons may increase as a result of previous activity (thus a form of "learning"), that this involves NMDARs, and that low frequency but persistent discharges in pain afferent fibers may cause increasing pain. Windup is not a product of neuronal plasticity; it happens by virtue of mechanisms that are normally present in the spinal cord. The enhanced excitability quickly returns to baseline if the electrical stimulation is terminated.

On the other hand, an increase in synaptic strength that may last for hours can be obtained if the conditioning pulse is applied to the primary afferents about 100 times per second during a few seconds. This is known as long-term potentiation (LTP), and is the most widely and deeply studied form of activity-dependent synaptic enhancement (the basis of learning) (Ikeda et al. 2003; Sandkühler 2000). On the other hand, in nociceptive dorsal spinal neurons that project to the midbrain periaqueductal gray matter, LTP is induced only when the conditioning pulse is applied twice per second, which is more similar than 100/s to the firing rate of nociceptive C-fiber afferents (Ikeda et al. 2006). In either case the presence of LTP is shown after the inducing pulse train by applying a single test pulse to the presynaptic axons every few minutes – each test pulse now elicits a larger excitatory postsynaptic response in the recorded neuron. As mentioned above, in spinal nociceptive neurons NMDARs, NK1 receptors and T-type voltage dependent calcium channels, as well as calcium-permeable AMPA receptors and calcium release from intracellular stores due to activation of metabotropic glutamate receptors, all contribute to the elevation of intracellular calcium concentration that results in LTP.

Induction of Hyperexcitability by Persistent Tissue Damage

In the spinal cord, LTP of responses to test electrical pulses can be induced by previously applied strong "natural" noxious stimuli (burning or crushing of paws, crushing of nerves) in anesthetized rats (Sandkühler 2000), provided that the nociceptive inhibition that normally descends from the brain stem is blocked. Another way of investigating the role of NMDARs in central sensitization and hyperalgesia is to use natural noxious stimuli, instead of electrical pulses, to test the responses of dorsal horn nociceptive neurons in anesthetized animals or the behavioral nociceptive responses in awake animals. As mentioned above, inflammation or trauma of peripheral tissues, or lesions to peripheral nerves, induce an NMDAR-dependent enhancement of spinal nociceptive responses in animals that is akin to the hyperalgesia and allodynia of clinical chronic pain conditions. The paramount role of NMDAR activation has been demonstrated in numerous studies where application of pharmacological antagonists or deletion of the NR1 subunit of NMDARs in the spinal cord completely prevented induction of LTP or central sensitization by persistent tissue damage (Sandkühler 2000; South et al. 2003; Woolf and Salter 2000).

Plasticity of the NMDAR

Noxious stimulation or peripheral inflammation enhance the synaptic expression, and induce phosphorylation, of the NR1 subunit of NMDARs in superficial dorsal horn neurons (Yang et al. 2009). This requires an initial activation of the NMDAR, followed by feed-forward steps where protein kinase A (PKA) and protein kinase C (PKC) are involved in the phosphorylation mechanisms. During persistent damage or LTP the key to an increase in neuronal excitability is an increase in intracellular calcium concentration. As mentioned above, this is brought about not only by the inflow of calcium through the NMDAR channel, but also by inflow through voltage-dependent calcium channels or calcium-permeable AMPA receptors, and by activation of G-protein-coupled receptors (to glutamate, to prostaglandins, to substance P, etc.), which results in calcium inflow and/or release from intracellular stores towards the cytosol (Ikeda et al. 2003; Woolf and Salter 2000). The increased intracellular calcium leads to activation of calcium-calmodulin-dependent protein kinase II (CaMKII), PKA and PKC, as well as nitric oxide synthase and the cyclooxygenases

(Sandkühler 2000; Woolf 1996; Woolf and Salter 2000). This in turn leads to several events, including sensitization of NMDARs and non-NMDARs.

In spinal cord neurons, one important point of convergence for various intracellular pathways is PKC. CaMKII, another serine/threonine kinase that plays a key role in hippocampal and neocortical plasticity, is less important in the spinal dorsal horn (Woolf and Salter 2000). Protein kinases phosphorylate the NMDAR, and thereby sensitize it and partially dislodge the magnesium ion from the channel (Woolf 1996); as a result, binding of glutamate to the NMDAR will more easily cause calcium inflow. It must be noted that increases in intracellular calcium as well as other forms of protein kinase activation do not necessarily result from an activation of the NMDAR, and yet they feed forward and sensitize it. Indeed, activation of NK1, glutamate metabotropic and tyrosine kinase receptors (e.g., the trkB receptor, see below) may lead to sensitization of nociceptive neurons (Ikeda et al. 2003), and this has been shown to be associated with phosphorylation of the NMDAR (Guo et al. 2004a, b).

Particularly interesting is the association of group I metabotropic glutamate receptors (mGluRs) and the NMDARs (Guo et al. 2004b; Yang et al. 2011). In cases of peripheral inflammation, dorsal spinal group I mGluRs contribute to the increased presynaptic glutamate release plus the phosphorylation of the NR2B subunit and the augmented sensitivity of postsynaptic NMDARs. Again this occurs via an increase in intracellular calcium concentration and an activation of the Src tyrosine kinase, and it contributes to inflammatory hyperalgesia.

Spinal neuroglia are important factors in central sensitization and NMDAR plasticity (Scholz and Woolf 2007). During peripheral inflammation or nerve damage, dorsal horn microglia and astrocytes are activated and release several mediators such as ATP, interleukin (IL)-1β, IL-6, tumor necrosis factor-α (TNFα), brain derived neurotrophic factor (BDNF) and the chemokine CCL2 (MCP-1). CCL2 induces IL-1β release by microglia and this in turn sensitizes

NMDARs and contributes to hyperalgesia (Baamonde et al. 2011; Kawasaki et al. 2008). IL-1β can be released by astrocytes, and this leads to increased presynaptic glutamate release, to phosphorylation of the NR1 subunit and to sensitization of postsynaptic NMDARs (Kawasaki et al. 2008; Zhang et al. 2008). Also TNFα induces an increase in NMDAR sensitivity (Kawasaki et al. 2008). In response to a peripheral nerve injury, expression of P2X$_4$ receptors (P2X$_4$Rs) to ATP is enhanced in activated microglia (Tsuda et al. 2003; Ulmann et al. 2008). Activation of P2X$_4$Rs leads to microglial release of BDNF. Also peripheral inflammation induces expression of BDNF in primary afferents and spinal neuroglia. BDNF in turn activates trkB receptors in neurons and induces phosphorylation of the NR1 and NR2B subunits plus behavioral hyperalgesia and allodynia (Geng et al. 2010; Guo et al. 2004a).

Activation of spinal neuronal NMDA, glutamate metabotropic, NK1 and trkB receptors as a result of peripheral noxious stimulation, inflammation and nerve damage eventually leads to activation of ERK (extracellular signal regulated protein kinase, a mitogen-activated protein, or MAP, kinase). ERK activation, by way of the cAMP responsive element binding (CREB) protein in the cell nucleus, results in gene expression mediated by the cAMP responsive element (CRE). This may in turn lead to medium- and long-term persistence of central sensitization to painful stimulation (Kawasaki et al. 2004) and to transition to chronicity in clinical pain.

References

Baamonde, A., Hidalgo, A., & Menendez, L. (2011). Involvement of glutamate NMDA and AMPA receptors, glial cells and IL-1β in the spinal hyperalgesia evoked by the chemokine CCL2 in mice. *Neuroscience Letters, 502*, 178–181.

Dougherty, P. M., Palecek, J., Sorkin, L. S., & Willis, W. D. (1992). The role of NMDA and non-NMDA excitatory amino acid receptors in the excitation of primate spinothalamic tract neurons by mechanical, chemical, thermal, and electrical stimuli. *Journal of Neuroscience, 12*, 3025–3041.

Furukawa, H., Singh, S. K., Mancusso, R., & Gouaux, E. (2005). Subunit arrangement and function in NMDA receptors. *Nature (Lond.), 438*, 185–192.

Geng, S. J., Liao, F. F., Dang, W. H., Liu, X. D., Cai, J., Han, J. S., Wan, Y., & Xing, G. G. (2010). Contribution of the spinal cord BDNF to the development of neuropathic pain by activation of the NR2B-containing NMDA receptors in rats with spinal nerve ligation. *Experimental Neurology, 222*, 256–266.

Guo, W., Wei, F., Zou, S.-P., Dubner, R., & Ren, K. (2004a). Effect of brain-derived neurotrophic factor on N-methyl-D-aspartate receptor subunit NR2B tyrosine phosphorylation in the rat spinal dorsal horn. *Abstr. View./Itin. Plann., Soc. Neurosci.*, Progr. No. 864.811.

Guo, W., Wei, F., Zou, S.-P., Robbins, M. T., Sugiyo, S., Ikeda, T., Tu, J.-C., Worley, P. F., Dubner, R., & Ren, K. (2004b). Group I metabotropic glutamate receptor NMDA receptor coupling and signaling cascade mediate spinal dorsal horn NMDA receptor 2B tyrosine phosphorylation associated with inflammatory hyperalgesia. *Journal of Neuroscience, 24*, 9161–9173.

Ikeda, H., Heinke, B., Ruscheweyh, R., & Sandkühler, J. (2003). Synaptic plasticity in spinal lamina I projection neurons that mediate hyperalgesia. *Science, 299*, 1237–1240.

Ikeda, H., Stark, J., Fischer, H., Wagner, M., Drdla, R., Jäger, T., & Sandkühler, J. (2006). Synaptic amplifier of inflammatory pain in the spinal dorsal horn. *Science, 312*, 1659–1662.

Kandel, E. R., & Siegelbaum, S. A. (2000). Synaptic integration. In E. R. Kandel, J. H. Schwartz, & T. M. Jessell (Eds.), *Principles of neural science* (pp. 207–228). New York: McGraw-Hill.

Kawasaki, Y., Kohno, T., Zhuang, Z.-Y., Brenner, G. J., Wang, H., Van Der Meer, C., Befort, K., Woolf, C. J., & Ji, R.-R. (2004). Ionotropic and metabotropic receptors, protein kinase a, protein kinase C, and Src contribute to C-fiber-induced ERK activation and cAMP response element-binding protein phosphorylation in dorsal horn neurons, leading to central sensitization. *Journal of Neuroscience, 24*, 8310–8321.

Kawasaki, Y., Zhang, L., Cheng, J.-K., & Ji, R.-R. (2008). Cytokine mechanisms of central sensitization: Distinct and overlapping role of interleukin-1β, interleukin-6, and tumor necrosis factor-α in regulating synaptic and neuronal activity in the superficial spinal cord. *Journal of Neuroscience, 28*, 5189–5194.

Neugebauer, V., Lücke, T., & Schaible, H.-G. (1993a). N-methyl-D-aspartate (NMDA) and non-NMDA receptor antagonists block the hyperexcitability of dorsal horn neurons during development of acute arthritis in rat's knee joint. *Journal of Neurophysiology, 70*, 1365–1377.

Neugebauer, V., Lücke, T., & Schaible, H.-G. (1993b). Differential effects of N-methyl-D-aspartate (NMDA) and non-NMDA receptor antagonists on the responses of rat spinal neurons with joint input. *Neuroscience Letters, 155*, 29–32.

Sandkühler, J. (2000). Learning and memory in pain pathways. *Pain, 88*, 113–118.

Scholz, J., & Woolf, C. J. (2007). The neuropathic pain triad: Neurons, immune cells and glia. *Nature Neuroscience, 10*, 1361–1368.

South, S. M., Kohno, T., Kaspar, B. K., Hegarty, D., Vissel, B., Drake, C. T., Ohata, M., Jenab, S., Sailer, A. W., Malkmus, S., Masuyama, T., Horner, P., Bogulavsky, J., Gage, F. H., Yaksh, T. L., Woolf, C. J., Heinemann, S. F., & Inturrisi, C. E. (2003). A conditional deletion of the NR1 subunit of the NMDA receptor in adult spinal cord dorsal horn reduces NMDA currents and injury-induced pain. *Journal of Neuroscience, 23*, 5031–5040.

Tsuda, M., Shigemoto-Mogami, Y., Koizumi, S., Mizokoshi, A., Kohsaka, S., Salter, M. W., & Inoue, K. (2003). P2X$_4$ receptors induced in spinal microglia gate tactile allodynia after nerve injury. *Nature, 424*, 763–778.

Ulmann, L., Hatcher, J. P., Hughes, J. P., Chaumont, S., Green, P. J., Conquet, F., Buell, G. N., Reeve, A. J., Chessell, I. P., & Rassendren, F. (2008). Up-regulation of P2X$_4$ receptors in spinal microglia after peripheral nerve injury mediates BDNF release and neuropathic pain. *Journal of Neuroscience, 28*, 11263–11268.

Vanegas, H., & Schaible, H.-G. (2000). Effects of antagonists to high-threshold calcium channels upon spinal mechanisms of pain, hyperalgesia and allodynia. *Pain, 85*, 9–18.

Vanegas, H., & Schaible, H.-G. (2001). Prostaglandins and cyclooxygenases in the spinal cord. *Progress in Neurobiology, 64*, 327–363.

Woolf, C. J. (1996). Windup and central sensitization are not equivalent. *Pain, 66*, 105–108.

Woolf, C. J., & Salter, M. W. (2000). Neuronal plasticity: Increasing the gain in pain. *Science, 288*, 1765–1768.

Yaksh, T. L. (1999). Spinal systems and pain processing: Development of novel analgesic drugs with mechanistically defined models. *Trends in Neurosciences, 20*, 329–337.

Yang, X., Yang, H. B., Xie, O. J., & Hu, X. D. (2009). Peripheral inflammation increased the synaptic expresssion of NMDA receptors in spinal dorsal horn. *Pain, 144*, 162–169.

Yang, K., Takeuchi, K., Wei, F., Dubner, R., & Ren, K. (2011). Activation of group I mGlu receptors contributes to facilitation of NMDA receptor membrane current in spinal dorsal horn neurons after hind paw inflammation in rats. *European Journal of Pharmacology, 670*, 509–518.

Zhang, R. X., Li, A., Liu, B., Wang, L., Ren, K., Zhang, H., Berman, B. M., & Lao, L. (2008). IL-1ra alleviates inflammatory hyperalgesia through preventing phosphorylation of NMDA receptor NR-1 subunit in rats. *Pain, 135*, 232–239.

N-methyl-D-aspartate

Synonyms

NMDA

Definition

N-methyl-D-aspartate (NMDA), a chemical analogue of glutamate that selectively binds to and activates a subpopulation of ionotropic glutamate receptors. Because of this selectivity, these receptors are referred to as NMDA receptors. NMDA receptors have unique biophysical properties relative to the two other classes for ionotropic glutamate receptor that are also defined by their selective ligands: AMPA and kainate receptors. Because of its unique properties, the NMDA receptor is able to play a particularly import role in driving synaptic plasticity responsible for such phenomenon as learning and memory. This receptor also plays an important role in the plasticity observed in the spinal cord dorsal horn following tissue injury, a process referred to as central sensitization, thought to be responsible for much of the increase in pain and sensitivity associated with tissue injury.

Cross-References

▶ GABA and Glycine in Spinal Nociceptive Processing
▶ Opioids in the Spinal Cord and Central Sensitization
▶ Spinal Cord Nociception, Neurotrophins

N-Methyl-D-aspartate (NMDA) Antagonist

Definition

There is accumulating evidence to implicate the importance of N-methyl-D-aspartate (NMDA) receptors to the induction and maintenance of central sensitization during pain states. However, NMDA receptors may also mediate peripheral sensitization and visceral pain. NMDA receptors are composed of NR1, NR2 (A, B, C, and D), and NR3 (A and B) subunits, which determine the functional properties of native NMDA receptors. Antagonists acting at the N-methyl-D-aspartate (NMDA) receptor can block the development of tolerance to the analgesic effects of [mu] opioid receptor (MOR) ligands, such as morphine, and can also enhance the analgesic efficacy of opioids. The last decade has seen significant progress in our understanding of the NMDA receptor complex and the site(s) of action of various uncompetitive antagonists. This has led to the development of a family of low-affinity, uncompetitive, cation channel antagonists that seem to offer many of the benefits of the older channel blockers but with a more acceptable adverse effect profile. Drugs such as memantine have shown beneficial effects in clinical trials for Alzheimer's disease and ischemia, with few adverse effects. Likewise, the NMDA receptor NR2B subunit antagonists derived from drugs such as ifenprodil have proven beneficial in the treatment of neuropathic pain and are also associated with few adverse effects.

Cross-References

▶ Multimodal Analgesia in Postoperative Pain

N-Methyl-D-aspartate (NMDA) Receptor

Definition

Activation of the NMDA receptor sets in motion a series of events that increase the responsiveness of the nociceptive system (central sensitization). NMDA receptors are composed of NR1, NR2 (A, B, C, and D), and NR3 (A and B) subunits, which determine the functional properties of native NMDA receptors. Central sensitization lowers the activation thresholds of spinal neurons and is characterized by windup, whereby repeated C-fiber volleys result in a progressive increase in discharge of secondary dorsal horn nociceptive neurons. This contributes to the hyperalgesia. The N-methyl-D-aspartate (NMDA) receptor, at physiological Mg^{++} levels, is initially unresponsive to glutamate, but following depolarization at the amino methylpropionic acid (AMPA) receptor by glutamate or the neurokinin-1 receptor by substance P and the trkB receptor by brain-derived neurotrophic factor (BDNF), it becomes responsive to glutamate, allowing Ca^{++} influx. The action of glutamate on the metabotropic receptor (modulated by glycine) stimulates G-protein-mediated activation of phospholipase C (PLC), which catalyzes the hydrolysis of phosphatidylinositol 4,5-biphosphonate (PIP2) to produce inositol triphosphate (IP3) and diacylglycerol (DAG). DAG stimulates production of protein kinase C (PKC), which is activated in the presence of high levels of intracellular Ca^{++}. IP3 stimulates the release of Ca^{++} from intracellular stores within the endoplasmic reticulum. Increased PKC induces a sustained increase in membrane permeability, with Ca^{++} leading to the expression of proto-oncogenes (c-fos, c-jun). The proteins produced encode a number of peptides (enkephalins, dynorphins, tachykinins). Increased intracellular Ca^{++} leads to the activation of calcium-/calmodulin-dependent protein kinase (which briefly increases membrane permeability), to the activation of phospholipase A-2 (PLA-2), as well as to the activation of nitric oxide synthase (NOS) (via a calcium/calmodulin mechanism). The conversion of phosphatidylcholine to prostaglandins and thromboxane is catalyzed by PLA2. The lipoxygenase pathway also produces leukotrienes. NOS catalyzes the production of nitric oxide (NO) and L-citrulline from L-arginine. NO activates soluble guanylate cyclase, increasing the intracellular content of cyclic GMP, leading to the production of protein kinases and alterations in gene expression. NO diffuses out of the cell to the primary afferent terminal, where, via a guanylate cyclase/cyclic-GMP mechanism, it increases glutamate release. NO is responsible for the cell death demonstrated after prolonged activation of nociceptor afferents. Among NMDA receptor subtypes, the NR2B subunit-containing receptors appear particularly important for nociception, leading to the possibility that NR2B-selective antagonists may be useful in the treatment of chronic pain.

Cross-References

▶ Opioid Responsiveness in Cancer Pain Management
▶ Postoperative Pain, Persistent Acute Pain
▶ Postoperative Pain, Postamputation Pain, Treatment, and Prevention
▶ Spinothalamic Tract Neurons, Glutamatergic Input

N-Methyl-D-Aspartate Receptor

▶ NMDA Glutamate Receptor(s)

NMR

▶ Magnetic Resonance Imaging

NNT

Definition

The NNT of oxycodone 15 mg is 2.3.

Cross-References

▶ Number Needed to Treat
▶ Postoperative Pain, Oxycodone

NOC

▶ Orphanin FQ

Nocebo

Definition

Nocebo (Latin for "I shall harm") can be thought of as an agent or intervention that results in harm, either in the form of adverse outcomes or adverse side effects. A nocebo effect is the effect that such an agent or intervention ostensibly exerts.

The term "nocebo effect" is most commonly used to describe side effects that occur in response to interventions for which there is no physiological mechanism, such as a placebo. These side effects are the same as, and may be more prevalent than, those encountered with the active medication. Interestingly, the side-effect profile of placebos mimic the active drug that it is being compared with and also appear to be disease specific (e.g., dizziness in psychiatric disorders, headache in angina, gastrointestinal disturbance in peptic disease (Barsky et al. 2002). This is thought to be due to patients being informed, and therefore cued, to expect potential adverse effects of the medication that they might be taking in the research project. Suggestion, expectation, and conditioning are the likely reasons. It is commonly seen in clinical practice when some patients assiduously study their consumer product information in their medicine packets and read about all the possible side effects of the medicine they have been prescribed. Upon taking the medication, they believe that they experience the expected side effect.

The nocebo effect can be reduced by some simple steps (Weihrauch and Gauler 1999). Firstly, if patients with negative expectations can be identified, especially in conditions such as anxiety, depression, and somatization, it is worthwhile attempting to shift their cognition to more positive expectations. For example, with serotonin uptake inhibitors, telling patients that the nausea they are likely to experience is a good sign (since it shows serotonin is increasing and, therefore, the drug is working) makes it more likely they will continue taking the medication. Secondly, commencing with low doses and increasing them slowly reduces the risk of side effects. Thirdly, using other health-care professionals who will reinforce the messages given to the patient lessens the risk of nocebo.

The term nocebo response is used in another fashion to describe the response of patients who know that they have been given or allocated to an inferior treatment. Under those conditions, they report failure to improve in order to indicate indirectly their disaffection with the way that they have been treated.

References

Barsky, A. J., Saintfort, R., Rogers, M. P., & Borus, J. F. (2002). Nonspecific medication side effects and the nocebo phenomenon. *JAMA, 287*(5), 622-627.
Weihrauch, T. R., & Gauler, T. C. (1999). Placebo-efficacy and adverse effects in controlled clinical trials. *Arzneimittelforschung, 49*(5), 385-393.

Nocebo Effect

Definition

Harmful effects occurring from nonspecific interventions as opposed to positive effects (placebo).

Cross-References

▶ Placebo

Nociceptin

▶ Orphanin FQ

Nociception

Definition

Nociception (from the Latin word *nocere*, to injure) is the transduction, encoding, and

transmission of neural information about tissue damage or impending tissue damage, which would occur if a stimulus was maintained over time. Cognitive central processing identifies the location of the stimulus, its general character, and the severity of the associated tissue injury. Some of the biological participants in this process include unmyelinated dorsal horn neurons (like ▶ A delta and C fibers), excitatory amino acids, neuropeptides, zinc, biogenic amines, nitric oxide, wide- dynamic range spinal neurons, the limbic system of the brain, and several regions of the cerebral neocortex. Normal nociception depends on a delicate balance between pronociceptive and antinociceptive forces. Chronic noxious stimulation causes ▶ central sensitization in which there is an increase in sensitivity of nociceptive ▶ neurons, and new, semi-permanent neural connections develop (neuroplasticity).

Cross-References

▶ Allodynia and Allokinesis
▶ Amygdala, Pain Processing and Behavior in Animals
▶ Consciousness and Pain
▶ Ethics of Pain, Culture, and Ethnicity
▶ Forebrain Modulation of the Periaqueductal Gray and Its Role in Pain
▶ Muscle Pain, Fibromyalgia Syndrome (Primary, Secondary)
▶ Nociceptive Withdrawal Reflex
▶ Pain in Humans, Psychophysical Law
▶ Psychological Aspects of Pain in Women
▶ Secondary Somatosensory Cortex (S2) and Insula, Effect on Pain-Related Behavior in Animals and Humans
▶ Spinothalamic Tract Neurons, in Deep Dorsal Horn
▶ Spinothalamocortical Projections from SM
▶ Transition from Acute to Chronic Pain

Nociception and Sleep

▶ Orofacial Pain, Sleep Disturbance

Nociception in Genital Mucosa

▶ Nociception in Mucosa of Sexual Organs

Nociception in Mucosa of Sexual Organs

Marita Hilliges
Dalarna University, Halmstad, Sweden

Synonyms

Genital mucosa, nociception; Mucosa of sexual organs, nociception; Nociception in genital mucosa

Definition

Pain perception and nociceptors in mucosal lining, including epithelium and underlying connective tissue, of human external genital organs. The external genital organs covered by mucosa comprise vulvar vestibule, clitoris, and glans penis. The vagina belongs to the internal female genital organs but is included since it has partly somatic innervation.

Characteristics

Pain Perception in Sexual Mucosa

The vestibule is by origin visceral tissue but is considered to have a somatic innervation (Cervero 1994). Sensations for touch, temperature, and pain are therefore similar to sensations evoked in skin. Heat pain threshold is approximately 43°C, which is slightly below heat pain thresholds in skin (45–46°C). There are limited data on cold pain threshold. Mechanical pain assessment has been performed using various methods and devices. Normal values with von Frey monofilaments are defined in two independent studies using slightly different experimental

setups. The results show a wide range in punctuate mechanical pain thresholds (185–430 mN) in healthy fertile women (Bohm-Starke et al. 2001; Pukall et al. 2002; Lowenstein et al. 2004; Giesecke et al. 2004). Pain and perception thresholds in the urethral meatus have been compared using a neurometer CPT (current perception threshold) device (Kenton et al. 2003). Pain thresholds were three to six times higher than perception thresholds.

Experimental data on pain thresholds in the vagina are sparse. However, in clinical practice, it is obvious that thermal pain perception exists, although spatial discrimination is poor. Furthermore, while it is possible to perform unanesthetized surgical interventions such as biopsies without eliciting major pain in the proximal vagina, pain is easily evoked in the distal part close to the introitus. Mean mechanical pain threshold in the lateral vaginal wall using a 12-mm diameter probe is 10.7 N (range 4.3–25.3) in healthy premenopausal women (Baguley et al. 2003). In healthy pre- and postmenopausal women (mean age 37.2), the pressure pain threshold in the anterior and posterior wall using a 1 cm^2 algometer is 1.65 kg/cm^2 ($\pm$0.64 SD) (Tu et al. 2008). The vagina, although considered a visceral organ, is not sensitive to distension (Bohm-Starke et al. 2001).

In the literature, no reports on pain sensitivity of the clitoris are found, whereas tactile thresholds have been studied (Connell et al. 2005).

On the glans penis, mechanically as well as temperature-induced pain can be elicited. Mechanical detection thresholds seem to coincide with pain thresholds (Halata and Munger 1986).

Nociceptors in Sexual Mucosa

The external genital area is supplied both with myelinated and unmyelinated nerves. Free nerve endings are most frequently found at the labia minora lateral to Hart's line (Krantz 1958). Biopsies from young healthy women taken from the posterior vestibule revealed that the intraepithelial free nerve endings rarely branch and penetrate two-thirds of the epithelium. These intraepithelial nerve fibers are not evenly distributed. Some areas are almost devoid of nerves, whereas other parts of the epithelium are densely innervated. In addition, one population of free nerve endings terminates within the basal layer of the epithelium. The intraepithelial nerves are immunoreactive to CGRP. Nerve fibers are also encountered in the subepithelial connective tissue. In this area, it is, however, more difficult to evaluate the existence of free nerve endings. Most nerve fibers are found in connection to vascular or glandular elements (Bohm-Starke et al. 1998; Bohm-Starke et al. 1999; Tympanidis et al. 2003). The clitoris and the female urethral meatus are richly innervated by free nerve endings. They are located in or just below the basal layer of the epithelium (Krantz 1958).

Free nerve endings in the vagina are rare and found in the distal part, the anterior wall being more densely innervated than the posterior. Most of these nerve terminals end after penetrating two-thirds of the epithelium; however, some terminate just a few cell layers from the surface (Hilliges et al. 1995). The hymenal ring is richly supplied with free nerve endings (Krantz 1958).

The human glans penis is richly innervated by free nerve endings. These endings are found in almost every connective tissue papilla (Halata and Munger 1986).

References

Baguley, S. D., Curnow, J. S., Morrison, G. D., et al. (2003). Vaginal algometer: Development and application of a device to monitor vaginal wall pressure pain threshold. *Physiological Measurement, 24*, 833–836.

Bohm-Starke, N., Hilliges, M., Falconer, C., et al. (1998). Increased intraepithelial innervation in women with vulvar vestibulitis syndrome. *Gynecologic and Obstetric Investigation, 46*, 256–260.

Bohm-Starke, N., Hilliges, M., Falconer, C., et al. (1999). Neurochemical characterization of the vestibular nerves in women with vulvar vestibulitis syndrome. *Gynecologic and Obstetric Investigation, 48*, 270–275.

Bohm-Starke, N., Hilliges, M., Brodda-Jansen, G., et al. (2001). Psychophysical evidence of nociceptor sensitization in vulvar vestibulitis syndrome. *Pain, 94*, 177–183.

Cervero, F. (1994). Sensory innervation of the viscera: Peripheral basis of visceral pain. *Physiological Reviews, 74*, 95–138.

Connell, K., Guess, M. K., La Combe, J., et al. (2005). Evaluation of the role of pudendal nerve integrity in female sexual function using noninvasive techniques. *Obstetrics and Gynecology, 192*, 1712–1717.

Giesecke, J., Reed, B. D., Haefner, H. K., et al. (2004). Quantitative sensory testing in vulvodynia patients and increased peripheral pressure pain sensitivity. *Obstetrics and Gynecology, 104*, 126–133.

Halata, Z., & Munger, B. L. (1986). The neuroanatomical basis for the protopathic sensibility of the human glans penis. *Brain Research, 371*, 205–230.

Hilliges, M., Falconer, C., Ekman-Ordeberg, G., et al. (1995). Innervation of the human vaginal mucosa as revealed by PGP 9.5 immunohistochemistry. *Acta Anatomica, 153*, 119–126.

Kenton, K., Fuller, E., & Benson, J. T. (2003). Current perception threshold evaluation of the female urethra. *International Urogynecology Journal, 14*, 133–135.

Krantz, K. E. (1958). Innervation of the human vulva and vagina. *Obstetrics and Gynecology, 12*, 382–396.

Lowenstein, L., Vardi, Y., Deutsch, M., et al. (2004). Vulvar vestibulitis severity – assessment by sensory and pain testing modalities. *Pain, 107*, 47–53.

Pukall, C. F., Binik, Y. M., Khalifé, S., et al. (2002). Vestibular tactile and pain thresholds in women with vulvar vestibulitis syndrome. *Pain, 96*, 163–175.

Tu, F. F., Fitzgerald, C. M., Kuiken, T., et al. (2008). Vaginal pressure-pain thresholds: Initial validation and reliability assessment in healthy women. *The Clinical Journal of Pain, 24*, 45–50.

Tympanidis, P., Terenghi, G., & Dowd, P. (2003). Increased innervation of the vulval vestibule in patients with vulvodynia. *British Journal of Dermatology, 148*, 1021–1027.

Nociception in Nose and Oral Mucosa

Barry Green[1] and Thomas Hummel[2]
[1]Pierce Laboratory, Yale School of Medicine, New Haven, CT, USA
[2]Department of Otorhinolaryngology, Technical University of Dresden Medical School, Dresden, Germany

Definition

Nociception in the oral and nasal cavities is predominantly mediated through fibers of the ▶ trigeminal nerve. Different from other areas of the body, nociceptive afferents are easily accessible as they are not covered, for example, by a corneal layer of epidermis. Trigeminal input is intimately involved in the processing of both olfactory and gustatory information. In turn, trigeminal sensitivity appears to depend on olfactory/gustatory stimulation that may play a role in, for example, ▶ Burning Mouth Syndrome.

Characteristics

Nasal Cavity

Anatomy of Nasal Trigeminal Afferents
The nasal cavity is innervated by the ophthalmic (first branch of the fifth cranial nerve: V_1) and maxillary (V_2) branches of the trigeminal nerve. V_1 (anterior ethmoidal and infraorbital nerves) innervates the anterior portion of the nasal cavity; the posterior part of the nasal cavity is innervated by V_2 (posterior superior medial nasal and nasopalatine nerves). Cell bodies of trigeminal afferents lie in the Gasserian ganglion. Axons project to the trigeminal sensory nucleus extending from the rostral spinal cord to the midbrain. Chemosensory fibers from the nasal cavity have been shown to project to the subnucleus caudalis and subnucleus interpolaris (Anton et al. 1991). Trigeminal information is relayed to the amygdala via the lateral parabrachial complex. Neurons of the spinal nucleus project to the ventral posterior medial, intralaminar, and mediodorsal nuclei of the thalamus. While most ascending fibers cross to the contralateral side, some fibers also ascend ipsilaterally (Barnett et al. 1995) – similar to the olfactory system. Apart from projections to the ▶ primary somatosensory cortex (SI), trigeminal stimulation also produces activation of SII. Trigeminal stimulation leads to activity in the insular cortex and the ventral orbital cortex, with stronger right-sided activity, and to activation of "olfactory" structures like the piriform cortex.

Intranasal Pain-Related Fibers
The nasal mucosa is highly sensitive to painful stimuli. This seems to be partly due to the fact that, other than in the skin, nociceptors innervating the mucosa are not covered by squamous epithelium, giving nociceptive stimuli almost direct access to the nerve endings (Finger et al. 1990). Compared to noxious thermal or mechanical stimuli, this is of particular importance with regard to chemical irritants. Trigeminal chemoreceptors act as a sentinel of the airways, where they prevent inhalation of potentially life-threatening substances, with intranasal trigeminal activation producing an inspiratory stop. This correlates with the finding of an area of increased chemosensitivity in the anterior third of the nasal cavity, whereas the posterior nasal mucosa is more sensitive to mechanical stimuli.

The physicochemical properties of most chemicals (e.g., molecular size, lipophilicity) determine the degree to which they activate the intranasal trigeminal system (Abraham et al. 1998). Nevertheless, chemical stimulation may also activate specific receptor types. Stinging sensations are likely to be mediated by A_{delta}-fibers, whereas burning sensations are largely mediated by ▶ C fiber. In addition, a variety of receptors are involved in the coding of different qualities of trigeminal-mediated sensations, for example, tingling, stinging, or burning. These receptive structures include the ▶ TRPV1 receptor, the nicotinic receptor (which can be activated in a stereoselective manner), the ASIC receptors, the M2 receptor, and the ▶ P2X receptor. Most excitingly, solitary chemosensory receptor cells have been found to be attached to trigeminal afferents (Finger et al. 2003). Thus, it seems that the trigeminal system allows discrimination between numerous different chemical stimuli – although the number of discriminable stimulus qualities is an order of magnitude below that of the olfactory system.

Measures of Nasal Nociceptive Sensation
Techniques to study intranasal trigeminal function in humans include the psychophysical lateralization paradigm, electrophysiological recordings of the ▶ negative mucosal potential, or recordings of the ▶ EEG-based ▶ event-related potential (Hummel 2000). However, especially in a clinical context, to date there is no rapid and reliable standardized test of trigeminal function in humans. A major limitation of

many studies, however, is that only CO_2 has been used as the chemical pain stimulant, although this gas has the advantages of being virtually odorless, inexpensive, and nontoxic. Other stimuli of the intranasal trigeminal system include ► capsaicin, ► menthol, or nicotine.

Recent research on the genetic variability of trigeminal receptors (e.g., TRPV1 polymorphisms), or opioid receptor polymorphisms, helps to explain the heterogeneity of individual responses to trigeminal stimulation.

Intranasal Trigeminal Sensations and Smell

Intranasal trigeminal activation has been shown to influence the perception of odors, although suppression between the olfactory and the trigeminal systems can be mutual. Importantly, odors typically produce trigeminally mediated sensations when presented at higher stimulus concentrations. In a clinical context, it has been shown that olfactory activation increases sensitivity to trigeminally mediated stimuli. In turn, ► loss of olfactory function may result in a decrease of trigeminal responsiveness. Several possible mechanisms have been identified by which trigeminal activity may influence olfactory processing (Hummel and Livermore 2002). The systems may interact centrally (e.g., mediodorsal thalamus), at the level of the olfactory bulb, at the level of the olfactory epithelium through the release of substance P/other peptides from trigeminal fibers, or indirectly via reflexes designed to minimize potentially damaging exposure of the olfactory epithelium to noxious substances (e.g., alteration of nasal patency, change of the constitution of the olfactory mucus).

Oral Cavity

Anatomy of Oral Pain

Three cranial nerves serve oral pain. The two most important are the trigeminal (V) and the glossopharyngeal (IX) nerves. The trigeminal nerve, which also innervates the nasal cavity, the face, and much of the scalp, provides sensitivity to temperature, touch, and pain from the base of the tongue forward. The trigeminal nerve projects to the tongue via the lingual nerve, which it shares with the chorda tympani nerve (VII), a taste nerve that innervates the front of the

tongue. The density of trigeminal endings is greatest in the fungiform taste ► papillae at the tip and sides of the tongue, where they greatly outnumber chorda tympani nerve endings. Two other branches of V, the maxillary (V2) and the mandibular (V3) nerves, innervate the upper and lower surfaces of the anterior oral cavity (e.g., the gingiva, buccal mucosa, and lips) and thus mediate all non-lingual oral pain, including dental pain. The glossopharyngeal nerve innervates the posterior 1/3 of the tongue as well as the soft palate, palatal arch, and anterior oral pharynx. Glossopharyngeal innervation is particularly dense in circumvallate taste papillae on the back of the tongue. Classified as a visceral nerve, the glossopharyngeal differs from the trigeminal in that it contains gustatory as well as somatosensory fibers. The receptive field of the vagus nerve (X) overlaps with the glossopharyngeal in the oral pharynx and posterior soft palate. The vagus is also sensitive to touch, temperature, and pain, but appears not to contribute to taste perception in humans. The vagus, nevertheless, contributes to the perception of the somatosensory qualities of all foods and beverages as they are swallowed. Accordingly, IX and X function as the final gatekeepers of the upper alimentary canal and trigger the protective reflexes of gagging and coughing.

Intraoral Pain-Related Fibers

Although concentrated in the fungiform and circumvallate papillae, lingual pain fibers are present throughout the oral mucosa and gingiva. Evidence of TRPV1 and peptidergic C-fiber nociceptors has been found in both types of papillae (Ishida et al. 2002). In addition to signaling acute pain, neurons that express TRPV1 also contribute to hyperalgesia to temperature caused by inflammation. Evidence of other types of nociceptors, including some that respond to cold, menthol, carbonation, nicotine, and high concentrations of salts (e.g., NaCl, KCl) and acids (e.g., citric acid), has also been found in various lingual nerve preparations (Wang et al. 1993).

Consistent with the variety of nociceptors that have been identified in the lingual nerve, salts and acids as well as menthol, carbonation, cinnamic acid, and nicotine can produce burning, stinging, or tingling when applied to the anterior tongue. Collectively, the sensitivity to all chemicals that produce sensations, other than (or in addition to) taste or smell, is referred to as ▶ chemesthesis (Green 2002).

Measures of Oral Nociception

By far the best studied oral pain stimulus is the vanilloid capsaicin, which has been used as a model for oral pain. As it occurs naturally in red ("hot") peppers, capsaicin is also of interest as a flavor stimulus. Not surprisingly, capsaicin, which acts via TRPV1, is perceived most acutely in the areas of the mouth most heavily innervated by the trigeminal nerve, such as the tongue tip. In addition, as TRPV1 is expressed on C-polymodal nociceptors, capsaicin also serves as an indicator of sensitivity to nonchemical pain stimuli such as intense heat. Thus, heat sensitivity is also greatest in the front of the mouth, particularly on the tongue tip and lips. Pain sensitivity is lowest in the superficial buccal mucosa. The sensitivity to chemical irritants and heat is also less acute on the back of the tongue, where most glossopharyngeal pain fibers are located in the basement membranes of the circumvallate papillae. Interestingly, however, swallowing capsaicin or piperine (black pepper) produces burning sensations in the throat, which are at least as strong as those produced on the front of the tongue (Rentmeister-Bryant and Green 1997), indicating that the glossopharyngeal and vagus nerves contribute significantly to chemical pain during consumption.

As capsaicin has the capacity to desensitize the neurons it stimulates (Szolcsanyi 1993), desensitization has been used to determine the importance of capsaicin-sensitive neurons for perception of chemical irritants. The results have shown that much, but not all, of oral chemesthesis is mediated by such fibers.

Oral Chemesthesis and Taste

Like olfactory stimuli, most taste stimuli evoke chemesthetic sensations as well as taste, particularly at high concentrations. Notable in this regard are acids, salts, and alcohols, which can produce burning, stinging, or tingling even at moderate concentrations. On the other hand, some taste stimuli (particularly sucrose) have been shown to partially suppress the burn of capsaicin, suggesting that taste stimulation has the potential to inhibit some types of oral pain.

Oral Pain Syndromes

Injuries to cranial nerves V, IX, or X can result in oral neuropathologies that range in severity from annoying to debilitating. Trigeminal and glossopharyngeal neuralgias can occur following injuries or insults to the deep tissue of the mouth or mandible (e.g., dental extractions) or inflammatory disorders such as herpes zoster. However, trigeminal neuralgias most often affect perioral and facial skin rather than oral structures. Glossopharyngeal neuralgias, which are relatively rare, are usually secondary to pathologies such as abscesses and tumors that affect IX and are characterized by pain at the base of the tongue, the pharynx, or soft palate. Perhaps the most common oral pain disorder is Burning Mouth Syndrome (BMS), which manifests as consistent or episodic burning or tingling localized in the front of the mouth, particularly the tongue and/or lips (Kapur et al. 2003). BMS generally occurs in older individuals (>50 years old) and is much more common in women than in men. Although it is believed to have more than one etiology, recent research has pointed to an association between BMS and a loss of taste function, particularly in postmenopausal women (Grushka et al. 2003). Such a link is consistent with other evidence that the taste system may normally exert an inhibitory effect on the oral pain system.

References

Abraham, M. H., Kumarsingh, R., Cometto-Muniz, J. E., & Cain, W. S. (1998). An algorithm for nasal pungency tin man. *Archives of Toxicology, 72*, 227–232.

Anton, F., Peppel, P., Euchner, I., et al. (1991). Controlled noxious chemical stimulation: Responses of rat trigeminal brainstem neurones to CO_2 pulses applied to the nasal mucosa. *Neuroscience Letters, 123*, 208–211.

Barnett, E. M., Evans, G. D., Sun, N., et al. (1995). Anterograde tracing of trigeminal afferent pathways from the murine tooth pulp to cortex using herpes simplex virus type I. *Journal of Neuroscience, 15,* 2972–2984.

Finger, T. E., Getchell, M. L., Getchell, T. V., et al. (1990). Affector and effector functions of peptidergic innervation of the nasal cavity. In B. G. Green, J. R. Mason, & M. R. Kare (Eds.), *Chemical senses: Irritation* (pp. 1–20). New York: Marcel Dekker.

Finger, T. E., Bottger, B., Hansen, A., et al. (2003). Solitary chemoreceptor cells in the nasal cavity serve as sentinels of respiration. *Proceedings of the National Academy of Sciences, 100,* 8981–8986.

Green, B. G. (2002). Psychophysical assessment of oral chemesthesis. In S. A. Simon & M. A. Nicolelis (Eds.), *Methods in chemosensory research* (pp. 3–20). New York: CRC Press.

Grushka, M., Epstein, J. B., & Gorsky, M. (2003). Burning mouth syndrome and other oral sensory disorders: A unifying hypothesis. *Pain Research & Management, 8,* 133–135.

Hummel, T. (2000). Assessment of intranasal trigeminal function. *International Journal of Psychophysiology, 36,* 147–155.

Hummel, T., & Livermore, A. (2002). Intranasal chemosensory function of the trigeminal nerve and aspects of its relation to olfaction. *International Archives of Occupational and Environmental Health, 75,* 305–313.

Ishida, Y., Ugawa, S., Ueda, T., et al. (2002). Vanilloid receptor subtype-1 (VR1) is specifically localized to taste papillae. *Brain Research. Molecular Brain Research, 107,* 17–22.

Kapur, N., Kamel, I. R., & Herlich, A. (2003). Oral and craniofacial pain: Diagnosis, pathophysiology, and treatment. *International Anesthesiology Clinics, 41,* 115–150.

Rentmeister-Bryant, H., & Green, B. G. (1997). Perceived irritation during ingestion of capsaicin or piperine: Comparison of trigeminal and non-trigeminal areas. *Chemical Senses, 22,* 257–266.

Szolcsanyi, J. (1993). Actions of capsaicin on sensory receptors. In J. N. Wood (Ed.), *Capsaicin in the study of pain* (pp. 1–26). New York: Academic.

Wang, Y., Erickson, R. P., & Simon, S. A. (1993). Selectivity of lingual nerve fibers to chemical stimuli. *Journal of General Physiology, 101,* 843–866.

Nociception Induced by Injection of Dilute Formaldehyde

▶ Formalin Test

Nociception: Role of Voltage-Gated Sodium Channels

Michael S. Minett and John N. Wood
Molecular Nociception Group, Wolfson Institute for Biomedical Research, University College London, London, WC, UK

Synonyms

$Na_v1.3$: (NAC3, KIAA1356); $Na_v1.7$: (PN1, NaS, hNE-Na); $Na_v1.8$: (SNS, PN3, NaNG); $Na_v1.9$: (NaN, SNS2, PN5); VGSC

Definition

The voltage-gated sodium channel (VGSC) gene family comprises 9 homologous members (*SCN1A-SCN11A*), which encode the sodium-selective ion channels $Na_v1.1$-$Na_v1.9$. One α subunit alone is sufficient to form a functional channel, however, α-subunits can also associate with β-subunits (*SCN1B-SCN4B*), which modulate channel biophysics and trafficking. The nine different voltage-gated sodium channel α-subunits are differentially expressed. Subunits $Na_v1.3$, 1.7, 1.8, and 1.9 are either highly expressed or confined to primary afferent neurons of the dorsal root ganglia (DRG) and have all been shown to play important roles in nociception and pain.

Characteristics

$Na_v1.3$ (*SCN3A*)

In contrast to many of the other VGSC genes, there are as-yet no monogenic disease associated with an *SCN3A* mutation. Animal studies have focused on a possible role of $Na_v1.3$ in neuropathic pain. Following axotomy and inflammation in mice, $Na_v1.3$ transcript levels increase in sensory neurons (Waxman et al. 1994). Anti-sense knockdown of $Na_v1.3$ expression attenuates pain-related behavior associated with spinal cord injury and chronic constriction injury

(Hains 2004) but not allodynia in the spared nerve injury model (Lindia et al. 2005). However, $Na_V1.3$ knockout mice show normal development of pain in the Chung model of neuropathic pain (Nassar et al. 2006). Although several lines of investigation have implicated $Na_V1.3$ as a candidate drug target to treat neuropathic pain, its precise role in neuropathic pain is still unclear.

$Na_V1.7$ (SCN9A)

$Na_V1.7$ is expressed in peripheral sensory neurons innervating the skin, viscera, and orofacial region, as well as sympathetic neurons (Toledo-Aral et al. 1997). A number of human heritable pain disorders map to mutations in SCN9A, the gene encoding $Na_V1.7$. Dominant gain-of-function mutations lead to inherited primary erythromelalgia (IEM) that is characterized by symmetrical burning ▶ pain of the feet/lower legs and hands, elevated skin temperature of affected areas, and reddened extremities (Yang et al. 2004). Additionally, dominant gain-of-function mutations can cause paroxysmal extreme pain disorder (PEPD) that is characterized by episodic burning pain of the rectum, ocular, and mandibular regions (Fertleman et al. 2006). Rarer recessive loss-of function conditions can cause an inability to experience pain (Cox et al. 2006).

Biophysical characterization of the $Na_V1.7$ mutations present in erythromelalgia patients shows a significant hyperpolarizing shift in voltage-dependence of activation (Cummins et al. 2007), resulting in gain-of-function. The $Na_V1.7$ mutations underlying PEPD, where mechanical stimulation evokes excruciating pain (Fertleman et al. 2006), attenuate the fast inactivation of $Na_V1.7$, resulting in persistent sodium currents. Such a deficit in inactivation is predicted to produce prolonged bursts of action potentials, leading to increased nociceptive signaling. In mouse studies, selective knockout of $Na_V1.7$ expression in $Na_V1.8$-positive nociceptors leads to a loss of acute noxious mechanosensation and inflammatory pain (Nassar et al. 2004), while deletion of $Na_V1.7$ in all sensory neurons leads to additional loss of noxious thermosensation (Minett et al. 2012). This suggests that $Na_V1.7$ expressed

within $Na_V1.8$-positive DRG neurons is critical for acute noxious mechanosensation, while $Na_V1.7$ expressed within $Na_V1.8$-negative DRG neurons is critical for acute noxious thermosensation. This highlights a novel way of identifying modality specific subpopulations. Furthermore, no effect on neuropathic pain behavior was observed in mice without expression of $Na_V1.7$ in $Na_V1.8$-positive sensory neurons (Nassar et al. 2005). This is also true for mice in which $Na_V1.7$ has been deleted from all DRG neurons. In contrast, mice in which $Na_V1.7$ is deleted from all DRG neurons as well as sympathetic neurons show a dramatic reduction of mechanical hypersensitivity following a surgical model of neuropathic pain. This demonstrates an essential role for $Na_V1.7$ within sympathetic neuron in the development of neuropathic pain (Minett et al. 2012). Overall, the role of $Na_V1.7$ in human as well as animal pain perception highlights $Na_V1.7$ as an important analgesic drug target.

$Na_V1.8$ (SCN10A)

$Na_V1.8$ is a tetrodotoxin-resistant (TTXr) sodium channel subtype that is expressed in nociceptive sensory neurons (Akopian et al. 1999) and acts as a major contributor to the upstroke of action potentials (Renganathan et al. 2001). $Na_V1.8$ is essential in maintaining the excitability of nociceptors at low temperatures (Zimmermann et al. 2007), becoming the sole electrical impulse generator at temperatures below 10 °C. This is caused by enhanced slow inactivation of TTX-sensitive (TTXs) channels in response to cooling, whereas inactivation of $Na_V1.8$ is cold resistant. Behavioral studies of mice in which $Na_V1.8$-expressing sensory neurons are ablated show loss of response to noxious cold and mechanical stimuli, as well as inflammatory pain (Abrahamsen et al. 2008). Antisense oligonucleotides attenuate the development and maintenance of neuropathic pain in rats (Joshi et al. 2006). However, $Na_V1.8$ knockout mice as well as $Na_V1.7/1.8$ double knockout mice show normal neuropathic pain behavior (Nassar et al. 2005). However, selective blockers of $Na_V1.8$ such as A-803467 (Jarvis et al. 2007) successfully suppress various pain symptoms and neuropathic pain in rats.

Na$_v$1.9 (SCN11A)

Na$_v$1.9 is expressed within primary nociceptors (Fang et al. 2002) and in the enteric nervous system (Rugiero et al. 2003). The biophysical characteristics of Na$_v$1.9 are unique among the sodium channels since it generates TTX-resistant currents that have very slow gating kinetics (Dib-Hajj et al. 2002). The current generated by Na$_v$1.9 is "persistent" and can be activated at potentials close to resting membrane potential (~ -60mV). Although the activation kinetics are too slow to contribute to the upstroke of an action potential, the channel acts as a modulator of membrane excitability by contributing regenerative inward currents over a strategic membrane potential range both negative to and overlapping with the voltage-threshold for other transient sodium channels.

Na$_v$1.9 knockout mice exhibit a clear analgesic phenotype in inflammatory pain states (Amaya et al. 2006). This appears to be explicable by changes in the properties of distal primary afferents. The response to inflammatory mediators is suppressed in Nav1.9 knockout mice. This is consistent with the immunocytochemical localization of Nav1.9 at unmyelinated nerve endings (Padilla et al. 2007), and the remarkable functional plasticity of the current, known to be under G-protein pathway control via protein kinase C (Baker, 2005).

References

Abrahamsen, B., Zhao, J., Asante, C. O., Cendan, C. M., Marsh, S., Martinez-Barbera, J. P., et al. (2008). The cell and molecular basis of mechanical, cold, and inflammatory pain. *Science, 321*(5889), 702–705.

Akopian, A. N., Souslova, V., England, S., Okuse, K., Ogata, N., Ure, J., et al. (1999). The tetrodotoxin-resistant sodium channel SNS has a specialized function in pain pathways. *Nature Neuroscience, 2*(6), 541–548.

Amaya, F., Wang, H., Costigan, M., Allchorne, A. J., Hatcher, J. P., Egerton, J., et al. (2006). The voltage-gated sodium channel Na(v)1.9 is an effector of peripheral inflammatory pain hypersensitivity. *Journal of Neuroscience, 26*(50), 12852–12860.

Baker, M. (2005). Protein kinase C mediates up regulation of tetrodotoxin resistant, persistent Na+ current in rat and mouse sensory neurones. *Journal of Physiology, 567*(3):851–867.

Cox, J. J., Reimann, F., Nicholas, A. K., Thornton, G., Roberts, E., Springell, K., et al. (2006). An SCN9A channelopathy causes congenital inability to experience pain. *Nature, 444*(7121), 894–898.

Cummins, T. R., Sheets, P. L., & Waxman, S. G. (2007). The roles of sodium channels in nociception: Implications for mechanisms of pain. *Pain, 131*(3), 243–257.

Dib-Hajj, S., Black, J. A., Cummins, T. R., & Waxman, S. G. (2002). NaN/Nav1.9: A sodium channel with unique properties. *Trends in Neurosciences, 25*(5), 253–259.

Fang, X., Djouhri, L., Black, J., Dib-Hajj, S., Waxman, S. G., & Lawson, S. N. (2002). The presence and role of the tetrodotoxin-resistant sodium channel Nav1. 9 (NaN) in nociceptive primary afferent neurons. *The Journal of Neuroscience, 22*(17), 7425–7433.

Fertleman, C. R., Baker, M. D., Parker, K. A., Moffatt, S., Elmslie, F. V., Abrahamsen, B., et al. (2006). SCN9A Mutations in paroxysmal extreme pain disorder: Allelic variants underlie distinct channel defects and phenotypes. *Neuron, 52*(5), 767–774.

Hains, B. C. (2004). Altered sodium channel expression in second-order spinal sensory neurons contributes to pain after peripheral nerve injury. *Journal of Neuroscience, 24*(20), 4832–4839.

Jarvis, M. F., Honore, P., Shieh, C.-C., Chapman, M., Joshi, S., Zhang, X.-F., et al. (2007). A-803467, a potent and selective Nav1.8 sodium channel blocker, attenuates neuropathic and inflammatory pain in the rat. *Proceedings of National Academic Sciences USA, 104*(20), 8520–8525.

Joshi, S., Mikusa, J., Hernandez, G., Baker, S. (2006). Involvement of the TTX-resistant sodium channel Nav 1.8 in inflammatory and neuropathic, but not post-operative, pain states. *Pain, 123*(1–2), 75–82.

Lindia, J. A., Kohler, M. G., Martin, W. J., & Abbadie, C. (2005). Relationship between sodium channel NaV1.3 Expression and neuropathic pain behavior in rats. *Pain, 117*(1–2), 145–153.

Minett, M. S., Nassar, M. A., Clark, A. K., Passmore, G., Dickenson, A. H., Wang, F., et al. (2012). Distinct Nav1.7-dependent pain sensations require different sets of sensory and sympathetic neurons. *Nature Communications, 3*, 791. Nature Publishing Group.

Nassar, M. A., Stirling, L. C., Forlani, G., Baker, M. D., Matthews, E. A., Dickenson, A. H., et al. (2004). Nociceptor-specific gene deletion reveals a major role for Nav1.7 (PN1) in acute and inflammatory pain. *Proceedings of National Academic Sciences USA, 101*(34), 12706–12711.

Nassar, M. A., Levato, A., Stirling, L. C., & Wood, J. N. (2005). Neuropathic pain develops normally in mice lacking both Na(v)1.7 and Na(v)1.8. *Molecular Pain, 1*, 24.

Nassar, M. A., Baker, M. D., Levato, A., Ingram, R., Mallucci, G., McMahon, S. B., et al. (2006). Nerve injury induces robust allodynia and ectopic discharges in Nav1.3 null mutant mice. *Molecular Pain, 2*, 33.

Padilla, F., Couble, M.-L., Coste, B., Maingret, F., Clerc, N., Crest, M., et al. (2007). Expression and localization of the Nav1.9 sodium channel in enteric neurons and in trigeminal sensory endings: Implication for intestinal reflex function and orofacial pain. *Molecular and Cellular Neurosciences, 35*(1), 138–152.

Renganathan, M., Cummins, T., & Waxman, S. G. (2001). Contribution of Nav1.8 sodium channels to action potential electrogenesis in DRG neurons. *Journal of Neurophysiology, 86*, 629–640.

Rugiero, F., Mistry, M., Sage, D., Black, J. A., Waxman, S. G., Crest, M., et al. (2003). Selective expression of a persistent tetrodotoxin-resistant Na+ current and NaV19 subunit in myenteric sensory neurons. *Journal of Neuroscience, 23*(7), 2715–2725.

Toledo-Aral, J. J., Moss, B. L., He, Z. J., Koszowski, A. G., Whisenand, T., Levinson, S. R., et al. (1997). Identification of PN1, a predominant voltage-dependent sodium channel expressed principally in peripheral neurons. *Proceedings of National Academic Sciences USA, 94*(4), 1527–1532.

Waxman, S. G., Kocsis, J. D., Black, J. A. (1994). Type III sodium channel mRNA is expressed in embryonic but not adult spinal sensory neurons, and is reexpressed following axotomy. *Journal of Neurophysiology, 72*(1), 466–470.

Yang, Y., Wang, Y., Li, S., Xu, Z., Li, H., Ma, L., et al. (2004). Mutations in SCN9A, encoding a sodium channel alpha subunit, in patients with primary erythermalgia. *Journal of Medical Genetics, 41*(3), 171–174.

Zimmermann, K., Leffler, A., Babes, A., Cendan, C. M., Carr, R. W., Kobayashi, J.-I., et al. (2007). Sensory neuron sodium channel Nav1.8 is essential for pain at low temperatures. *Nature, 447*(7146), 855–858.

Nociceptive Circuitry in the Spinal Cord

Andrew J. Todd
Institute of Neuroscience and Psychology, College of Medical, Veterinary and Life Sciences, University of Glasgow, Glasgow, UK

Definition

Pain is generally perceived as a result of stimuli that either damage or threaten to damage peripheral tissues. Sensory information resulting from tissue damage in the limbs and trunk is transmitted through nociceptive ▶ primary afferent axons to the dorsal horn of the spinal cord and subsequently relayed to various sites in the brain, including the thalamus. The term nociceptive circuitry is used to describe the arrangement and functional interconnections of neurons that are responsible for conveying this information. Nociceptive circuits, including those in the dorsal horn, also involve inhibitory control mechanisms.

Characteristics

Background

Nociceptive primary afferent axons terminate in the dorsal horn of the spinal cord, and this region therefore contains the first synapse in pathways that convey nociceptive information to the brain, as well as those involved in reflexes. The dorsal horn is also an important site for modulation of sensory input. This modulation involves both local neurons and axons that descend from the brainstem and is thought to contribute to various types of analgesia. Changes affecting the central terminals of primary afferents and dorsal horn neurons occur in certain pathological states and are likely to contribute to the abnormal pain that follows peripheral inflammation and certain types of nerve injury. A knowledge of the nociceptive circuits in the dorsal horn would therefore be of fundamental importance for our understanding of the mechanisms underlying pain and analgesia and also for the development of new treatments for pain. However, despite extensive research on this subject, we still know relatively little about the neuronal organization and circuitry of the spinal dorsal horn.

Neuronal Components in the Dorsal Horn

Rexed (1952) divided the gray matter of the dorsal horn into six parallel laminae, and this scheme is generally used in anatomical studies (Fig. 1). The dorsal horn contains four different neuronal components: (1) central terminals of primary afferent axons, (2) ▶ projection neurons,

Nociceptive Circuitry in the Spinal Cord, Fig. 1 A diagrammatic representation of the laminar distribution of primary afferent axons in the spinal cord. The dorsal horn is divided into six parallel laminae, and each type of afferent has a characteristic area of termination in one or more of these laminae. Note that the mediolateral and rostrocaudal distribution of afferents is related to the region of the body that they innervate. The body is thus "mapped" onto the spinal cord in these two dimensions. C-LTMR, C-low-threshold mechanoreceptor

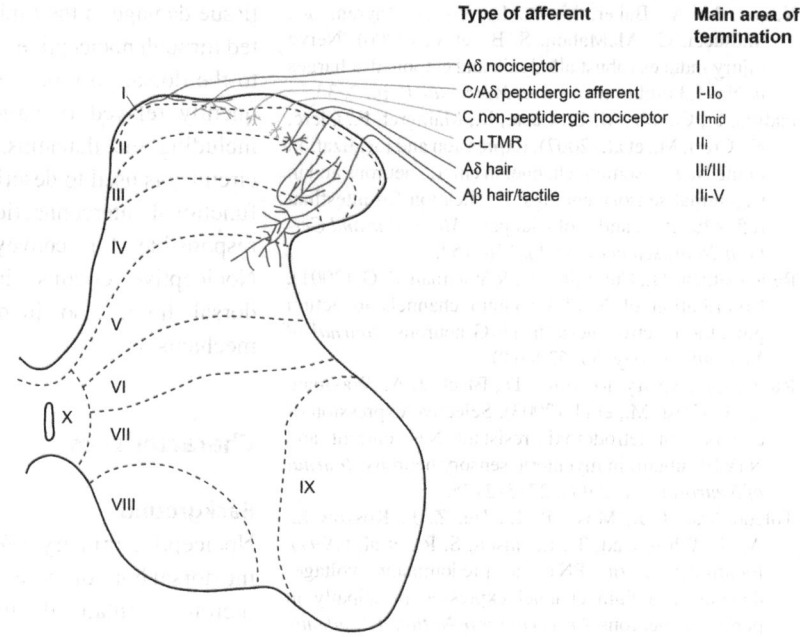

Type of afferent	Main area of termination
Aδ nociceptor	I
C/Aδ peptidergic afferent	I-IIo
C non-peptidergic nociceptor	IImid
C-LTMR	IIi
Aδ hair	IIi/III
Aβ hair/tactile	IIi-V

(3) ▶ interneurons, and (4) axons that descend from various brain regions.

All primary afferent axons that innervate tissues in the trunk and limbs form synapses in the spinal dorsal horn. Many different classes of primary afferent can be recognized based on the peripheral tissue that they innervate, the types of stimulus that activate them, their axonal diameter, and the chemical messengers that they use (Todd 2010). Each type of primary afferent has a characteristic distribution pattern within the dorsal horn (Fig. 1). The majority of nociceptive afferents have unmyelinated or fine myelinated axons, and these are referred to as C and Aδ fibers, respectively. Intra-axonal injection of Aδ nociceptive afferents has shown that these terminate mainly in lamina I, although they also give rise to terminals in lamina V (Light and Perl 1979). Because of their small size, intracellular labeling of C fibers has proved extremely difficult, and most anatomical studies of these afferents have used an indirect "neurochemical" approach. Many fine afferents, including approximately half of those that give rise to C fibers, express neuropeptides, such as ▶ substance P (SP) and ▶ calcitonin gene-related peptide (CGRP). In the rat, all CGRP-containing axons

in the dorsal horn are primary afferents, and these can be identified by immunocytochemistry. Most of the C fibers that do not express peptides can be labeled with a lectin (IB4) derived from *Bandeiraea simplicifolia*, and in the mouse these axons have been shown to express Mas-related G protein-coupled receptor member D (Mrgprd), which allows their peripheral and central terminals to be visualized (Zylka et al. 2005). Both peptidergic and Mrgprd-expressing C fibers are thought to function as nociceptors, and these differ in their peripheral distribution. The Mrgprd afferents are restricted to the epidermal layer of the skin, while peptidergic nociceptors are distributed to many different tissues, including deeper regions of the skin, muscles, joints, and viscera. Not all C fibers are nociceptors, and a population of C-low-threshold mechanoreceptors (C-LTMRs) has been identified, based on the expression of the vesicular glutamate transporter 3 (VGLUT3) (Seal et al. 2009). Within the dorsal horn, the peptidergic afferents (many of which are C fibers) terminate mainly in lamina I and the outer part of lamina II (lamina IIo), non-peptidergic C nociceptors in the central part of lamina II, and C-LTMRs in the inner part of lamina II (lamina IIi). Tactile and hair afferents convey innocuous (low-threshold) information from the skin and most have myelinated axons. In some pathological conditions (e.g., ► neuropathic pain), activation of these afferents can be perceived as painful. The majority of low-threshold cutaneous afferents have large myelinated (Aβ) fibers and these terminate in a zone extending from lamina IIi to lamina V. Fine myelinated afferents that innervate down hairs have a more restricted central distribution in laminae IIi and III.

Projection neurons have axons that ascend to the brain and form the major output from the dorsal horn. An important group consists of cells with axons that cross the midline and travel rostrally in the contralateral ventrolateral white matter. This collection of neurons is often referred to as the ascending anterolateral tract (ALT). The cell bodies of ALT neurons are concentrated in lamina I and scattered throughout the deeper parts of the gray matter (laminae III–VII) and the lateral spinal nucleus. Their axons travel to several brain regions, including the thalamus, the periaqueductal gray matter of the midbrain, the lateral parabrachial area in the pons, and various parts of the brainstem reticular formation (Willis and Coggeshall 2004; Villanueva and Bernard 1999). Those cells with axons that project directly to the thalamus are referred to as the spinothalamic tract. Many ALT neurons innervate more than one of these brain regions by means of axon collaterals (Todd 2010). In all parts of the spinal cord projection neurons are greatly outnumbered by interneurons. For example, it has been estimated that only around 5 % of neurons in lamina I are projection cells. ALT neurons in lamina I have received considerable attention, as they are relatively numerous and most respond to noxious stimulation. The majority of these cells, as well as some of those in laminae III and IV, express the neurokinin 1 (NK1) receptor, on which substance P acts. Substance P is present in many nociceptive primary afferents and is released into the dorsal horn following noxious stimulation. Intrathecal administration of substance P conjugated to the cytotoxin saporin leads to death of NK1 receptor-expressing neurons in the dorsal horn, as well as a dramatic reduction of ► hyperalgesia in both inflammatory and neuropathic conditions (Mantyh et al. 1997). This suggests that projection neurons with the NK1 receptor play an important role in the generation of hyperalgesia. Not all ALT projection neurons in lamina I express the NK1 receptor, for example, there is a class of giant cells that lack the receptor, but which are also activated by noxious stimuli (Polgár et al. 2008).

Interneurons make up the great majority of the neurons in each lamina of the dorsal horn. Although some of these have relatively long axons, it is likely that most (or all) have axonal arbors close to the cell body. These cells are therefore involved in local processing of information. Interneurons can be divided into two main classes: inhibitory cells, which use GABA and/or glycine as their neurotransmitter(s), and excitatory (glutamatergic) cells. Laminae I–III of the dorsal horn contain a particularly high density of small interneurons, and the majority of these (60–75 %, depending on the lamina) are excitatory. The axons of inhibitory interneurons form two types of synaptic connection: (1) axoaxonic synapses on primary afferent axon terminals, which mediate presynaptic inhibition, and (2) axodendritic or axosomatic synapses onto other dorsal horn neurons, which are responsible for postsynaptic inhibition. Blocking the actions of GABA or glycine with intrathecal antagonists leads to tactile ► allodynia (Yaksh 1989), and this suggests that one of the roles of inhibitory interneurons is to prevent activity transmitted by tactile and hair afferents from being perceived as painful. Inhibitory interneurons are also important for limiting the extent and severity of pain following noxious stimuli (Sandkuhler 2009) and for inhibiting the sensation of itch (Ross et al. 2010). Much less is known about the function of excitatory interneurons, although it is known that they convey information from primary afferents to other dorsal horn neurons through polysynaptic pathways. For example, they are likely to be responsible for transmitting nociceptive information from C fibers (which terminate mainly in laminae I–II) to neurons in deeper laminae and also for conveying tactile inputs to lamina I projection neurons in pathological pain states (Torsney and MacDermott 2006). Dorsal horn interneurons are heterogeneous and have proved difficult to assign to distinct functional populations (Graham et al. 2007; Todd 2010). For example, Grudt and Perl (2002) recorded from lamina II neurons in spinal cord slices and described several different populations, based on the shapes of their cell bodies and dendritic trees, and their firing pattern in response to injected depolarizing current. However, a recent study that combined patch-clamp recording with post hoc identification of neurotransmitter type reported that while some distinct morphological classes of inhibitory and excitatory interneuron could be recognized in lamina II, many cells of each type did not belong to any recognized morphological class (Yasaka et al. 2010). It is also possible to identify interneuron populations using neurochemical characteristics, for example, patterns of neuropeptide or receptor expression. It has recently been shown that four non-overlapping populations of inhibitory interneurons in laminae I–III can be identified based on expression of galanin, neuropeptide Y (NPY), neuronal nitric oxide synthase (nNOS), and the calcium-binding protein parvalbumin (Tiong et al. 2011). It is therefore likely that electrophysiological, morphological, and neurochemical criteria will all need to be considered in order to produce a comprehensive classification scheme for these cells.

Among the various populations of descending axons, those that contain the monoamine transmitter serotonin or norepinephrine have attracted particular attention, because of their role in stimulation-produced analgesia. Serotoninergic axons arise from cells in the raphe nuclei of the medulla, while those with norepinephrine are derived from cells in and around the locus coeruleus of the pons. Both types of axon are distributed throughout the dorsal horn, but are concentrated in the superficial laminae (I–II). The monoamine transmitters are likely to act through ▶ volume transmission, and therefore, knowing the distribution of monoamine receptors on dorsal horn neurons will be important for understanding their roles in modulating sensory transmission.

What We Know About Neuronal Circuitry in the Spinal Cord

As stated above, our knowledge of specific neuronal circuits within the dorsal horn is still rather limited. It has been shown that primary afferents that contain substance P form numerous synapses with ALT projection neurons in laminae I and III that express the NK1 receptor, accounting for around half of the excitatory (glutamatergic) synapses on these cells (Todd 2010). Since all substance P-containing afferents respond to noxious stimulation (Lawson et al. 1997), this provides a powerful direct route through which nociceptors can activate brain regions involved in pain. The nociceptive afferents release both glutamate and substance P and these act through different mechanisms: glutamate will be released across the synaptic cleft and act on receptors in the postsynaptic membrane, whereas substance P will diffuse to nearby NK1 receptors (▶ volume transmission). The organization of synaptic inputs to projection neurons is highly specific, as giant lamina I cells (which lack the NK1 receptor) appear to receive few if any synapses from primary afferents. Nociceptive primary afferents also form synapses with both excitatory and inhibitory interneurons, although much less is known about these connections.

There is evidence that different populations of inhibitory interneurons have specific postsynaptic targets. NK1 receptor-expressing ALT neurons in lamina III are densely innervated by axons that contain NPY and GABA, which are thought to originate from local NPY-expressing inhibitory interneurons, and these axons do not form synapses with another population of projection cells that occupy the same laminae, those belonging to the ▶ postsynaptic dorsal column pathway (Polgár et al. 1999) (Fig. 2). The lamina I giant cells are selectively innervated by nNOS-containing GABAergic axons, and again these are thought to be derived from local interneurons (Todd 2010). The NPY- and nNOS-containing interneurons that give rise to these inputs therefore presumably suppress the transmission of nociceptive inputs through the dorsal horn. The parvalbumin-containing inhibitory interneurons, which belong to a morphological class known as

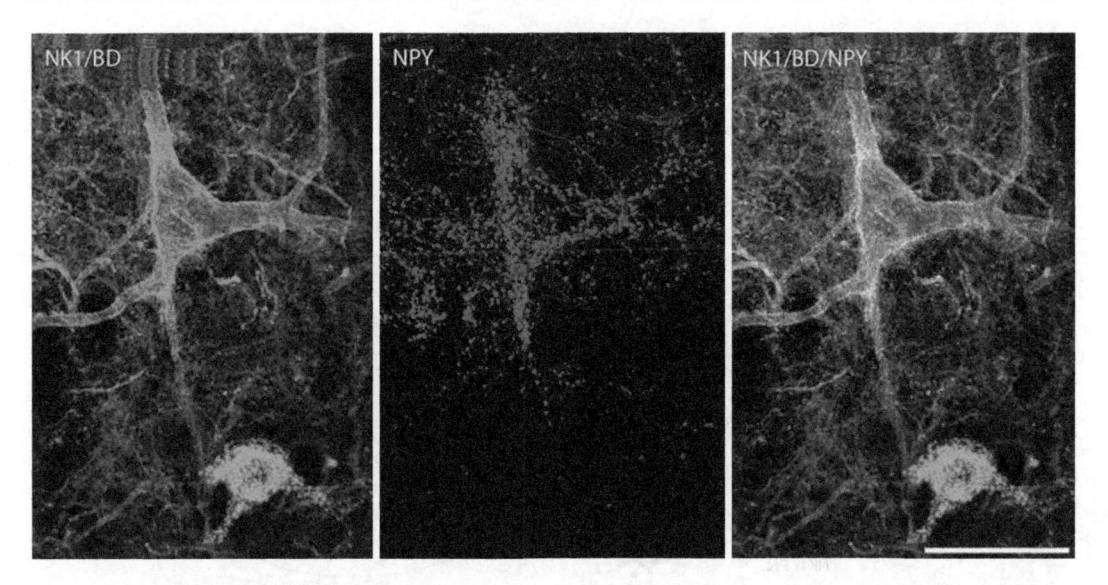

Nociceptive Circuitry in the Spinal Cord, Fig. 2 This figure shows selective innervation of one of two different types of projection neuron in lamina III of the dorsal horn by axons belonging to a population of inhibitory interneurons. Two cells can be seen in the left image: the large cell in the upper part of the field is stained with an antibody against the NK1 receptor (*green*). All cells of this type are projection neurons. The lower cell was retrogradely labeled with biotin dextran (BD; *blue*) injected into the gracile nucleus and therefore belongs to the postsynaptic dorsal column (PSDC) pathway. The middle image shows the same field scanned to reveal axons that contain neuropeptide Y (NPY), and the image on the right is a merge of the three colors. NPY-containing axons in the dorsal horn are derived from a population of GABAergic interneurons in laminae I and II, and these axons can be seen to form numerous contacts with the NK1 receptor-immunoreactive cell, but not with the PSDC neuron. Scale bar = 50 μm (Modified from Polgár et al. 1999. Copyright 1999 by the Society for Neuroscience)

islet cells, have recently been shown to give rise to axoaxonic synapses on myelinated low-threshold mechanoreceptive afferents, which also provide input to these cells, and this indicates that they have a role in regulating tactile inputs (Hughes et al. 2012). The inhibitory interneurons that presynaptically inhibit non-peptidergic C nociceptors have not yet been identified, but they are known to be different from those that innervate myelinated afferents (Todd 2010). There is limited information concerning the synaptic connections between excitatory interneurons and projection neurons. The lamina III ALT neurons that express the NK1 receptor are selectively innervated by a population of excitatory interneurons that contain the peptide dynorphin (Baseer et al. 2012), while a morphologically defined class of excitatory lamina II interneurons known as vertical cells provides synaptic input to ALT neurons in lamina I (Lu and Perl 2005). The giant lamina I ALT neurons receive a very high density of synapses from glutamatergic axons that are thought to originate

from local excitatory interneurons, although their source has not yet been identified (Polgár et al. 2008). We know even less about the organization of synaptic circuits that link different populations of interneurons. However, these are also likely to be organized in an orderly way, and certain "modular" patterns have already been identified (Lu and Perl 2005).

Some of the neuronal circuits that have been identified within the dorsal horn are illustrated in Fig. 3. Although the picture is still far from complete, what we know so far indicates that the connections between different dorsal horn neurons are not random, but are organized in an orderly (although complex) way. A major limitation to unraveling the circuitry has been the lack of a comprehensive classification scheme for both the interneurons and the projection cells. Without this, it is very difficult to obtain the quantitative information about synaptic connections that is needed to produce a circuit diagram for the dorsal horn.

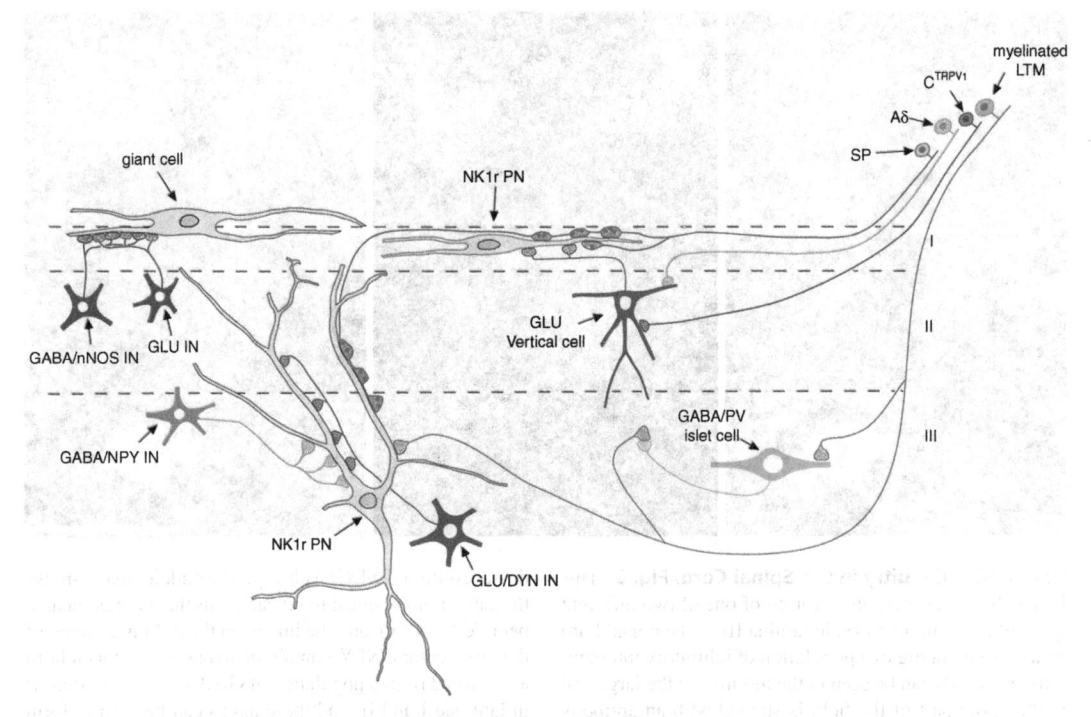

Nociceptive Circuitry in the Spinal Cord, Fig. 3 Some of the neuronal circuits that have been identified in the dorsal horn. Three anterolateral tract (ALT) projection neurons are shown: two of these express the neurokinin 1 receptor (NK1r PN), while the other one is a giant cell in lamina I, which lacks the NK1r. Substance P-containing nociceptive primary afferents form numerous synapses on the NK1r-expressing ALT neurons in laminae I and III (SP), while the lamina III cells also receive a weaker input from myelinated low-threshold mechanoreceptors (LTM). The myelinated LTM afferents also synapse with GABAergic islet cells that contain parvalbumin (PV), and the axons of these cells form axoaxonic synapses on the same type of afferent. Lamina I giant cells are densely innervated by GABAergic inhibitory interneurons, many of which contain neuronal nitric oxide synthase (GABA/ nNOS IN), and also by glutamatergic excitatory interneurons (GLU IN). The lamina III NK1r-expressing ALT neurons are selectively innervated by inhibitory interneurons that contain NPY (GABA/NPY IN) and also by excitatory interneurons containing dynorphin (GLU/ DYN IN). Vertical cells, a class of glutamatergic (GLU) interneuron, receive monosynaptic input from Aδ nociceptors and from C fibers that possess TRPV1 receptors. Their axons form synapses with NK1r-expressing lamina I ALT neurons

Cross-References

▶ Nociceptive Processing in the Spinal Cord

References

Baseer, N., Polgár, E., Watanabe, M., Furuta, T., Kaneko, T., & Todd, A. J. (2012). Projection neurons in lamina III of the rat spinal cord are selectively innervated by local dynorphin-containing excitatory neurons. *The Journal of Neuroscience, 32,* 11854–11863.

Graham, B. A., Brichta, A. M., & Callister, R. J. (2007). Moving from an averaged to specific view of spinal cord pain processing circuits. *Journal of Neurophysiology, 98,* 1057–1063.

Grudt, T. J., & Perl, E. R. (2002). Correlations between neuronal morphology and electrophysiological features in the rodent superficial dorsal horn. *The Journal of Physiology, 540,* 189–207.

Hughes, D. I., Sikander, S., Kinnon, C. M., Boyle, K. A., Watanabe, M., Callister, R. J., & Graham, B. A. (2012). Morphological, neurochemical and electrophysiological features of parvalbumin-expressing cells: a likely source of axo-axonic inputs in the mouse spinal dorsal horn. *Journal of Physiology, 590,* 3927–3951.

Lawson, S. N., Crepps, B. A., & Perl, E. R. (1997). Relationship of substance P to afferent characteristics of dorsal root ganglion neurones in guinea-pig. *The Journal of Physiology, 505,* 177–191.

Light, A. R., & Perl, E. R. (1979). Spinal termination of functionally identified primary afferent neurons with slowly conducting myelinated fibers. *The Journal of Comparative Neurology, 186*, 133–150.

Lu, Y., & Perl, E. R. (2005). Modular organization of excitatory circuits between neurons of the spinal superficial dorsal horn (laminae I and II). *The Journal of Neuroscience, 35*, 3900–3907.

Mantyh, P. W., Rogers, S. D., Honore, P., et al. (1997). Ablation of lamina I spinal neurons expressing the substance P receptor profoundly inhibits hyperalgesia. *Science, 278*, 275–279.

Polgár, E., Shehab, S. A. S., Watt, C., et al. (1999). GABAergic neurons that contain neuropeptide Y selectively target cells with the neurokinin 1 receptor in laminae III and IV of the rat spinal cord. *The Journal of Neuroscience, 19*, 2637–2646.

Polgár, E., Al-Khater, K. M., Shehab, S., Watanabe, M., & Todd, A. J. (2008). Large projection neurons in lamina I of the rat spinal cord that lack the neurokinin 1 receptor are densely innervated by VGLUT2-containing axons and possess GluR4-containing AMPA receptors. *The Journal of Neuroscience, 28*, 13150–13160.

Rexed, B. (1952). The cytoarchitectonic organization of the spinal cord in the cat. *The Journal of Comparative Neurology, 96*, 415–495.

Ross, S. E., Mardinly, A. R., McCord, A. E., Zurawski, J., Cohen, S., Jung, C., Hu, L., Mok, S. I., Shah, A., Savner, E. M., Tolias, C., Corfas, R., Chen, S., Inquimbert, P., Xu, Y., McInnes, R. R., Rice, F. L., Corfas, G., Ma, Q., Woolf, C. J., & Greenberg, M. E. (2010). Loss of inhibitory interneurons in the dorsal spinal cord and elevated itch in Bhlhb5 mutant mice. *Neuron, 65*, 886–898.

Sandkuhler, J. (2009). Models and mechanisms of hyperalgesia and allodynia. *Physiological Reviews, 89*, 707–758.

Seal, R. P., Wang, X., Guan, Y., Raja, S. N., Woodbury, C. J., Basbaum, A. I., & Edwards, R. H. (2009). Injury-induced mechanical hypersensitivity requires C-low threshold mechanoreceptors. *Nature, 462*, 651–655.

Tiong, S. Y. X., Polgár, E., van Kralingen, J. C., Watanabe, M., & Todd, A. J. (2011). Galanin-immunoreactivity identifies a distinct population of inhibitory interneurons in laminae I-III of the rat spinal cord. *Molecular Pain, 7*, 36.

Todd, A. J. (2010). Neuronal circuitry for pain processing in the dorsal horn. *Nature Reviews. Neuroscience, 11*, 823–836.

Torsney, C., & MacDermott, A. B. (2006). Disinhibition opens the gate to pathological pain signaling in superficial neurokinin 1 receptor-expressing neurons in rat spinal cord. *The Journal of Neuroscience, 26*, 1833–1843.

Villanueva, L., Bernard, J.-F., Lydic, R., & Baghdoyan, H. A. (1999). The multiplicity of ascending pain pathways. In R. Lydic & H. A. Baghdoyan (Eds.), *Handbook of behavioral state control: Cellular and molecular mechanisms* (pp. 569–585). Boca Raton: CRC Press.

Willis, W. D., & Coggeshall, R. E. (2004). *Sensory mechanisms of the spinal cord* (3rd ed.). New York: Kluwer Academic/Plenum Publishers.

Yaksh, T. L. (1989). Behavioral and autonomic correlates of the tactile evoked allodynia produced by spinal glycine inhibition: Effects of modulatory receptor systems and excitatory amino acid antagonists. *Pain, 37*, 111–123.

Yasaka, T., Tiong, S. Y. X., Hughes, D. I., Riddell, J. S., & Todd, A. J. (2010). Populations of excitatory and inhibitory interneurons in lamina II of the adult rat spinal dorsal horn revealed by a combined electrophysiological and anatomical approach. *Pain, 151*, 475–488.

Zylka, M. J., Rice, F. L., & Anderson, D. J. (2005). Topographically distinct epidermal nociceptive circuits revealed by axonal tracers targeted to *Mrgprd*. *Neuron, 45*, 17–25.

Nociceptive Coding in Lateral Thalamus

▶ Thalamic Nuclei Involved in Pain, Human and Monkey

Nociceptive Masseter Muscle Afferents

▶ Nociceptors in the Orofacial Region (Temporomandibular Joint and Masseter Muscle)

Nociceptive Nerve Endings

Definition

The terminal branches of the peripheral axon of nociceptive neurons located in sensory ganglia.

Cross-References

▶ Nociceptor Generator Potential

Nociceptive Neuroplasticity

Definition

At its most general level, nociceptive neuroplasticity denotes the changes in nervous system processing resulting from nociceptive inputs. Used in this way, the term includes both functional and structural and reversible and irreversible changes. Other groups would use this term in a narrower sense and only include alterations in nervous system function that are due to structural change.

Cross-References

▶ Central Sensitization
▶ Quantitative Sensory Testing

Nociceptive Pain

Definition

Pain caused by ongoing activation of Aδ and C nociceptors in response to a noxious stimulus of somatic or visceral structures such as inflammation, trauma, or disease.

Cross-References

▶ Analgesic Guidelines for Infants and Children
▶ Cancer Pain Management, Treatment of Neuropathic Components
▶ Cancer Pain, Goals of a Comprehensive Assessment
▶ Complex Chronic Pain in Children, Interdisciplinary Treatment
▶ Guillain-Barré Syndrome
▶ Opioid Responsiveness in Cancer Pain Management
▶ Opioids in Geriatric Application

Nociceptive Pathways

Definition

Neural circuits, including long sensory tracts, which convey information related to noxious stimuli, are called nociceptive pathways. The consequences of nociceptive processing can include pain sensation, motivational-affective responses, reflex behavior, endocrine changes, and learning and memory of painful events.

Cross-References

▶ Nociceptive Circuitry in the Spinal Cord
▶ Spinothalamic Tract Neurons, Descending Control by Brain Stem Neurons

Nociceptive Peripheral Afferent Fibers

▶ Nociceptor, Categorization

Nociceptive Primary Afferents

Definition

A distinct subset of peripheral sensory fibers that respond to tissue damaging stimuli, including thermal and mechanical events or chemical agents.

Cross-References

▶ Opioid Receptors at Postsynaptic Sites

Nociceptive Processing

▶ Encoding of Noxious Information in the Spinal Cord

Nociceptive Processing in the Amygdala, Behavioral and Pharmacological Studies

▶ Amygdala, Pain Processing and Behavior in Animals

Nociceptive Processing in the Amygdala: Neurophysiology and Neuropharmacology

Volker Neugebauer
Department of Neuroscience & Cell Biology,
University of Texas Medical Branch, Galveston,
TX, USA

Synonyms

Amygdala, Nociceptive processing

Definition

The amygdala is an almond-shaped structure in the medial temporal lobe and consists of several functionally and pharmacologically distinct nuclei. The laterocapsular division of the central nucleus of the amygdala (CeA), which has been designated as the "nociceptive amygdala," plays an important role in pain processing and pain modulation.

Characteristics

As part of the limbic system, the amygdala plays a key role in attaching emotional significance to sensory stimuli, emotional learning and memory, and affective states and disorders. The amygdala receives information from all sensory modalities; it also processes nociceptive information and projects to pain modulatory systems through forebrain and brainstem connections. Accumulating evidence suggests that the amygdala integrates nociceptive information with affective content, contributes to the emotional response to pain, and serves as a neuronal interface for the reciprocal relationship between pain and affective states and disorders.

Anatomy and Circuitry

The amygdala includes at least 12 different nuclei. The lateral, basolateral, and central nuclei of the amygdala (LA, BLA, and CeA, respectively) are of particular importance for the processing and evaluation of sensory information (Fig. 1). The LA-BLA network represents the input region, receiving sensory information from the thalamus, particularly the posterior areas, and from the cortex, including insular, anterior cingulate, and association cortical areas (LeDoux 2000; Orsini and Maren 2012; Pape and Pare 2010; Price 2003). The LA-BLA network represents the initial site of sensory convergence, processing and associative learning, and plasticity in the amygdala (LeDoux 2000; Pape and Pare 2010; Shi and Davis 1999). This highly processed information, which is a key element of the fear- and anxiety-related circuitry, is then transmitted to other amygdaloid nuclei, including the CeA (LeDoux 2000; Orsini and Maren 2012; Pape and Pare 2010).

The CeA serves as the output nucleus of major amygdala functions. The CeA integrates inputs from other amygdala nuclei without forming reciprocal intra-amygdaloid connections. Sensory information reaches the CeA from the LA-BLA network as well as from the brainstem (parabrachial area, PB) and spinal cord (Bernard et al. 1996; Burstein and Potrebic 1993; Gauriau and Bernard 2002; Neugebauer et al. 2004, 2009). Hippocampal input to the BLA provides contextual representations (LeDoux 2000). The CeA forms widespread connections with various forebrain and brainstem areas that are involved in emotional behavior and emotional experience and regulate autonomic and somatomotor functions. Targets of CeA projections include the cholinergic basal forebrain nuclei and the ▶ bed nucleus of the stria terminalis, midline and mediodorsal thalamic nuclei and paraventricular hypothalamus via the

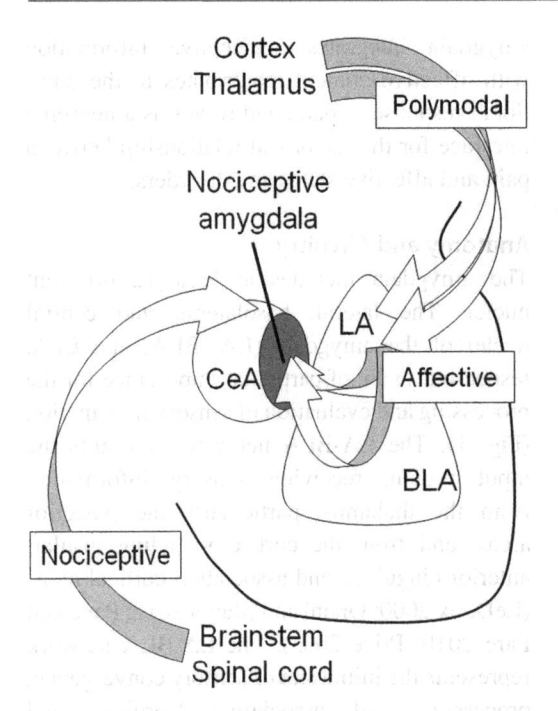

Nociceptive Processing in the Amygdala: Neurophysiology and Neuropharmacology, Fig. 1 Circuitry of information processing in the principal sensory nuclei of the amygdala. The lateral nucleus of the amygdala (LA) receives and integrates polymodal information from thalamic and cortical areas. This highly processed information with affective content is then distributed to other amygdaloid nuclei, including the central nucleus (CeA), either directly or through the basolateral (BLA) nuclei. The CeA is the major output nucleus of the amygdala and forms widespread connections with forebrain and brainstem areas. The laterocapsular division of the CeA represents the "nociceptive amygdala"

▶ stria terminalis, and lateral hypothalamus and brainstem areas such as periaqueductal gray (PAG) and parabrachial area (PB) via the ▶ ventral amygdaloid pathway (LeDoux 2000; Neugebauer et al. 2004, 2009; Price 2003).

Nociception and Nociceptive Plasticity

Within the CeA, the laterocapsular division is defined as the "nociceptive amygdala" because of the high content of neurons that respond exclusively or predominantly to noxious stimuli (Bernard et al. 1996; Neugebauer et al. 2004, 2009). The laterocapsular CeA receives nociceptive-specific information from the spinal cord and brainstem

through the spino-parabrachio-amygdaloid pain pathway (Bernard et al. 1996; Gauriau and Bernard 2002) as well as through direct projections from the spinal cord (Burstein and Potrebic 1993).

Electrophysiological single-unit analysis in anesthetized rats has shown several characteristics of neurons in the nociceptive amygdala (Bernard et al. 1996; Gauriau and Bernard 2002; Neugebauer et al. 2004; Neugebauer and Li 2002). The majority of these neurons (80 %) respond either exclusively ("nociceptive-specific" [NS] neurons) or predominantly ("multireceptive" [MR] neurons) to noxious stimuli. More neurons are excited than inhibited by noxious stimuli. A significant number of "nonresponsive" (NR) neurons (up to 20 %) also exist in the laterocapsular CeA; they do not respond to somatic stimuli. NS and MR CeA neurons have large, mostly symmetrical bilateral receptive fields in the superficial and deep tissue; they respond to mechanical and thermal stimuli. Their stimulus–response functions are not monotonically increasing linear but sigmoidal. These properties argue against a sensory-discriminative function of CeA neurons. Among neurons with predominantly cutaneous input, there appear to be more NS than MR neurons, whereas a larger percentage of MR neurons can be found among neurons with receptive fields mainly in the deep tissue. It is believed that NS neurons receive input from the spino-parabrachio-amygdaloid pathway, whereas MR neurons integrate nociceptive information with affective content from the polymodal LA-BLA circuitry (Fig. 1). MR neurons have also been recorded in the BLA (Ji et al. 2010).

Neurons in the nociceptive amygdala develop plasticity in models of arthritic pain (Bird et al. 2005; Fu and Neugebauer 2008; Han et al. 2005; Neugebauer and Li 2003; Neugebauer et al. 2003), formalin-induced pain (Adedoyin et al. 2010), visceral pain (Han and Neugebauer 2004), and chronic neuropathic pain (Ikeda et al. 2007). Extracellular single-unit recordings in anesthetized rats showed that MR neurons and NR neurons, but not NS neurons, become sensitized to afferent inputs in the ▶ kaolin-carrageenan-induced arthritis pain model. Characteristics of the pain-related sensitization of MR neurons are as follows: the processing of mechanical, but not thermal, pain-related information is increased (upward shift of the stimulus–response functions), responses to stimulation of the arthritic knee as well as of non-injured tissue in other parts of the body are enhanced, the total size of the receptive field expands, a constant input evoked by orthodromic electrical stimulation in the PB produces greater activation, and background activity is increased. Unlike changes in the peripheral nervous system and spinal cord in this arthritis model, changes in MR amygdala neurons develop with a biphasic time course; the first phase (1–3 h) reflects changes at the spinal cord and brainstem levels, whereas the persistent plateau phase (>5h) involves intra-amygdala plasticity. MR neurons serve to integrate and evaluate sensory-affective information in the context of pain. NS neurons would continue to distinguish between noxious and Innocuous Input/Stimulus at the stage of plasticity.

Evidence that the sensitization of amygdala neurons involves plastic changes within the amygdala comes from electrophysiological studies in brain slices in vitro. Whole-cell patch-clamp recordings in coronal brain slices from rats with a ▶ kaolin-carrageenan-induced arthritis (Bird et al. 2005; Fu and Neugebauer 2008; Han et al. 2005; Neugebauer et al. 2003), colitis (Han and Neugebauer 2004), formalin-induced inflammation (Adedoyin et al. 2010), or nerve injury (spinal nerve ligation; Ikeda et al. 2007) showed enhanced synaptic transmission and increased neuronal excitability of CeA neurons with input from the PB and from the BLA (resembling the MR neurons in vivo). Similar changes were also observed in BLA neurons in the ▶ kaolin-carrageenan-induced arthritis pain model (Ji et al. 2010). Synaptic plasticity in the reduced preparation is thus maintained at least in part independently of continuous input from the site of injury.

Pharmacology of Nociception and Plasticity
The roles of ionotropic and metabotropic glutamate receptors and neuropeptide receptors in

brief nociceptive processing and persistent pain have been studied in CeA neurons.

Ionotropic Glutamate Receptors
Extracellular single-unit recordings of CeA neurons in anesthetized animals showed that non-NMDA receptors are involved in the responses to innocuous and brief (15 s) noxious stimuli, whereas NMDA receptors contribute only to the processing of nociceptive information (Li and Neugebauer 2004b). In the ▶ kaolin-carrageenan-induced arthritis pain model (6-h post induction), activation of NMDA and non-NMDA receptors is required for the pain-related sensitization of CeA neurons. In these studies, antagonists at NMDA receptors (AP5) and non-NMDA (NBQX) receptors were administered into the CeA by microdialysis.

Pain-related synaptic plasticity recorded (patch-clamp) in the CeA in brain slices from arthritic rats involves enhanced function of postsynaptic NMDA receptors through PKA-dependent phosphorylation of the NR1 subunit (Bird et al. 2005). In contrast, there is no evidence for a contribution of NMDA receptors to synaptic plasticity of CeA neurons in a neuropathic pain model (Ikeda et al. 2007).

Metabotropic Glutamate Receptors
Electrophysiological studies of amygdala neurons in vivo and in vitro have shown an important role of group I metabotropic glutamate receptors (mGluRs), which include the mGluR1 and mGluR5 subtypes and couple to G proteins to activate phospholipase C, PKC, and MAP kinases such as ERK (Kolber et al. 2010; Li and Neugebauer 2004a; Li et al. 2011; Neugebauer et al. 2003). Extracellular single-unit recordings of CeA neurons in anesthetized rats suggest a change of mGluR1 function in the amygdala in pain-related sensitization, whereas mGluR5 contributes to both brief and prolonged nociception. Activation of mGluR1 and mGluR5 by the agonist DHPG enhances the responses of CeA neurons to brief (15 s) innocuous and noxious stimuli under normal conditions. This effect can be mimicked by an mGluR5 agonist (CHPG). In the kaolin/carrageenan arthritis pain

model, antagonists for mGluR1 (CPCCOEt) or
mGluR5 (MPEP) inhibit the responses of
sensitized CeA neurons. Under normal
conditions, CPCCOEt has no effect, but MPEP
inhibits brief nociceptive responses. Agonists and
antagonists were administered into the CeA by
microdialysis.

The contribution of group I mGluRs to normal
synaptic transmission and pain-related synaptic
plasticity in the amygdala was analyzed in brain
slices in vitro using whole-cell voltage-clamp
recordings of neurons in the nociceptive
amygdala (Neugebauer et al. 2003). Synaptic
transmission was studied at the nociceptive
PB-CeA synapse and the polymodal-affective
BLA-CeA synapse (Fig. 1). An agonist for
mGluR1 and mGluR5 (DHPG) and an mGluR5
agonist (CHPG) potentiate normal synaptic
transmission in CeA neurons in slices from normal rats. Evidence suggests that ERK is an
important signaling mechanism of mGluR5. An
antagonist for mGluR1 (CPCCOEt) has no effect
on synaptic transmission in CeA neurons in slices
from normal rats but inhibits synaptic plasticity
in slices from arthritic rats. An mGluR5
antagonist (MPEP) inhibits basal synaptic transmission in CeA neurons in slices from normal rats
and synaptic plasticity in slices from arthritic
rats. Thus, enhanced receptor activation of
mGluR1 appears to be a key mechanism of
pain-related synaptic plasticity in the CeA. Additional evidence suggests that mGluR1 may act
presynaptically, whereas mGluR5 can increase
neuronal excitability through a postsynaptic site
of action.

Neuropeptides
Calcitonin gene-related peptide (CGRP) and
corticotropin-releasing factor (CRF) have been
shown to contribute to amygdala plasticity in the
kaolin/carrageenan pain model (Fu and
Neugebauer 2007; Han et al. 2005; Ji and
Neugebauer 2007). The CeA contains particularly
high levels of CGPR and CRF. In electrophysiological and pharmacological in vitro (patch-clamp
recordings) and in vivo (extracellular single-unit
recordings) studies, selective CGRP1 receptor
antagonists (CGRP(8–37) and BIBN4096BS)

inhibit pain-related plasticity in the kaolin/
carrageenan arthritis model. The data further
suggest that CGRP1 receptors contribute to
pain-related plasticity through a protein kinase
A (PKA)-dependent postsynaptic mechanism
that involves NMDA receptors. Similarly,
blockade of CRF1, but not CRF2, receptors
with selective antagonists (NBI27914 and
Asterssin-2B, respectively) inhibits enhanced
synaptic transmission and central sensitization
of CeA neurons in the kaolin/carrageenan arthritis pain model.

Pain Modulation by the Amygdala
As part of the pain system and a key player in
affective states and disorders, the amygdala
contributes to the emotional response to pain and
its modulation by affective state (Fig. 2). The
CeA, including the laterocapsular division,
forms direct and indirect connections with
descending pain-modulating systems in the
brainstem (Neugebauer et al. 2004). Descending
pain control can be facilitatory (pro-nociceptive)
and inhibitory (anti-nociceptive) (Gebhart 2004;
Ossipov et al. 2010). Using pharmacological and
optogenetic manipulations of amygdala output,
recent behavioral studies suggest that the
consequence of pain-related plasticity in the
CeA is increased pain behavior. In other words,
amygdala activity correlates positively with pain
behaviors, and manipulations that reduce amygdala hyperactivity also inhibit pain, whereas
increasing amygdala output facilitates pain
responses. Pharmacologic targets used to
correlate amygdala activity with behavior include
mGluR1, mGluR5, CGRP1, and CRF1 receptors
(Fu and Neugebauer 2007; Han and Neugebauer
2005; Han et al. 2005; Kolber et al. 2010).

Activity in the amygdala can be modified by
negative and positive emotions, which in turn are
known to reduce (acute stress, pleasant music) or
enhance (anxiety, depression) pain (Fig. 2). The
dependence of amygdala activity on affective
state and the dual coupling of the amygdala to
inhibitory and facilitatory pain control may
explain some of the differential effects of
amygdala stimulation and/or activation on
pain behavior.

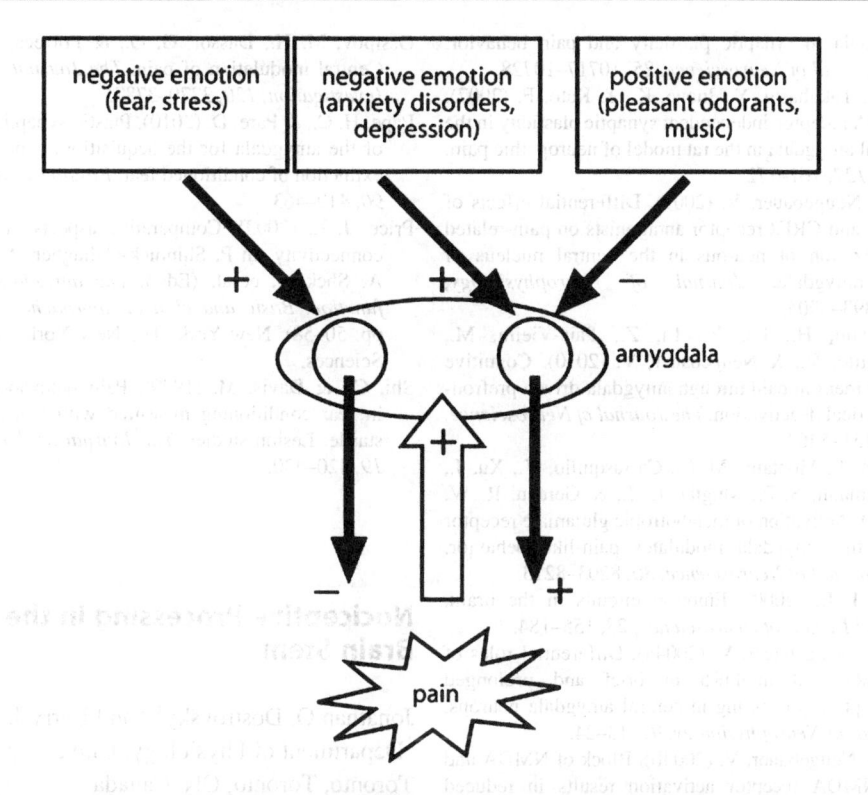

Nociceptive Processing in the Amygdala: Neurophysiology and Neuropharmacology, Fig. 2 Pain, emotions, and the amygdala: a hypothetical model. Pain produces plastic changes in the amygdala. Affective states also modify activity in the amygdala; negative emotions generally increase amygdala activity, whereas positive emotions have been shown to deactivate the amygdala. The amygdala is linked to facilitatory and inhibitory pathways to modulate pain. Negative emotions associated with pain reduction (fear and stress) would activate amygdala-linked inhibitory control systems, whereas negative affective states that correlate with increased pain (depression and anxiety disorders) would activate pain-facilitating pathways. Positive emotions inhibit amygdala coupling to pain facilitation (Reprinted with permission of Sage Publications, Inc., from Neugebauer et al. (2004) The amygdala and persistent pain. Neuroscientist 10, p 232)

References

Bernard, J.-F., Bester, H., & Besson, J. M. (1996). Involvement of the spino-parabrachio-amygdaloid and -hypothalamic pathways in the autonomic and affective emotional aspects of pain. *Progress in Brain Research, 107,* 243–255.

Bird, G. C., Lash, L. L., Han, J. S., Zou, X., Willis, W. D., & Neugebauer, V. (2005). PKA-dependent enhanced NMDA receptor function in pain-related synaptic plasticity in amygdala neurons. *The Journal of Physiology, 564*(3), 907–921.

Burstein, R., & Potrebic, S. (1993). Retrograde labeling of neurons in the spinal cord that project directly to the amygdala or the orbital cortex in the rat. *The Journal of Comparative Neurology, 335,* 469–485.

Fu, Y., & Neugebauer, V. (2008). Differential mechanisms of CRF1 and CRF2 receptor functions in the amygdala in pain-related synaptic facilitation and behavior. *The Journal of Neuroscience, 28,* 3861–3876.

Gauriau, C., & Bernard, J.-F. (2002). Pain pathways and parabrachial circuits in the rat. *Experimental Physiology, 87,* 251–258.

Gebhart, G. F. (2004). Descending modulation of pain. *Neuroscience and Biobehavioral Reviews, 27,* 729–737.

Han, J. S., & Neugebauer, V. (2004). Synaptic plasticity in the amygdala in a visceral pain model in rats. *Neuroscience Letters, 361,* 254–257.

Han, J. S., & Neugebauer, V. (2005). mGluR1 and mGluR5 antagonists in the amygdala inhibit different components of audible and ultrasonic vocalizations in a model of arthritic pain. *Pain, 113,* 211–222.

Han, J. S., Li, W., & Neugebauer, V. (2005). Critical role of calcitonin gene-related peptide 1 receptors in the amygdala in synaptic plasticity and pain behavior. *The Journal of Neuroscience, 25,* 10717–10728.

Ikeda, R., Takahashi, Y., Inoue, K., & Kato, F. (2007). NMDA receptor-independent synaptic plasticity in the central amygdala in the rat model of neuropathic pain. *Pain, 127,* 161–172.

Ji, G., & Neugebauer, V. (2007). Differential effects of CRF1 and CRF2 receptor antagonists on pain-related sensitization of neurons in the central nucleus of the amygdala. *Journal of Neurophysiology, 97,* 3893–3904.

Ji, G., Sun, H., Fu, Y., Li, Z., Pais-Vieira, M., Galhardo, V., & Neugebauer, V. (2010). Cognitive impairment in pain through amygdala-driven prefrontal cortical deactivation. *The Journal of Neuroscience, 30,* 5451–5464.

Kolber, B. J., Montana, M. C., Carrasquillo, Y., Xu, J., Heinemann, S. F., Muglia, L. J., & Gereau, R. W. (2010). Activation of metabotropic glutamate receptor 5 in the amygdala modulates pain-like behavior. *The Journal of Neuroscience, 30,* 8203–8213.

LeDoux, J. E. (2000). Emotion circuits in the brain. *Annual Review of Neuroscience, 23,* 155–184.

Li, W., & Neugebauer, V. (2004a). Differential roles of mGluR1 and mGluR5 in brief and prolonged nociceptive processing in central amygdala neurons. *Journal of Neurophysiology, 91,* 13–24.

Li, W., & Neugebauer, V. (2004b). Block of NMDA and non-NMDA receptor activation results in reduced background and evoked activity of central amygdala neurons in a model of arthritic pain. *Pain, 110,* 112–122.

Li, Z., Ji, G., & Neugebauer, V. (2011). Mitochondrial reactive oxygen species are activated by mGluR5 through IP3 and activate ERK and PKA to increase excitability of amygdala neurons and pain behavior. *The Journal of Neuroscience, 31,* 1114–1127.

Neugebauer, V., & Li, W. (2002). Processing of nociceptive mechanical and thermal information in central amygdala neurons with knee-joint input. *Journal of Neurophysiology, 87,* 103–112.

Neugebauer, V., & Li, W. (2003). Differential sensitization of amygdala neurons to afferent inputs in a model of arthritic pain. *Journal of Neurophysiology, 89,* 716–727.

Neugebauer, V., Li, W., Bird, G. C., et al. (2003). Synaptic plasticity in the amygdala in a model of arthritic pain: Differential roles of metabotropic glutamate receptors 1 and 5. *The Journal of Neuroscience, 23,* 52–63.

Neugebauer, V., Li, W., Bird, G. C., et al. (2004). The amygdala and persistent pain. *The Neuroscientist, 10,* 221–234.

Neugebauer, V., Galhardo, V., Maione, S., & Mackey, S. C. (2009). Forebrain pain mechanisms. *Brain Research Reviews, 60,* 226–242.

Orsini, C. A., & Maren, S. (2012). Neural and cellular mechanisms of fear and extinction memory formation. *Neuroscience and Biobehavioral Reviews, 36,* 1773–1802.

Ossipov, M. H., Dussor, G. O., & Porreca, F. (2010). Central modulation of pain. *The Journal of Clinical Investigation, 120,* 3779–3787.

Pape, H. C., & Pare, D. (2010). Plastic synaptic networks of the amygdala for the acquisition, expression, and extinction of conditioned fear. *Physiological Reviews, 90,* 419–463.

Price, J. L. (2003). Comparative aspects of amygdala connectivity. In P. Shinnick-Gallagher, A. Pitkanen, A. Shekhar, et al. (Eds.), *The amygdala in brain function. Basic and clinical approaches* (Vol. 985, pp. 50–58). New York: The New York Academy of Sciences.

Shi, C., & Davis, M. (1999). Pain pathways involved in fear conditioning measured with fear-potentiated startle: Lesion studies. *The Journal of Neuroscience, 19,* 420–430.

Nociceptive Processing in the Brain Stem

Jonathan O. Dostrovsky[1,2] and Barry J. Sessle[2]
[1]Department of Physiology, University of Toronto, Toronto, ON, Canada
[2]Faculty of Dentistry, University of Toronto, Toronto, ON, Canada

Introduction

The craniofacial region is principally innervated by the ► trigeminal (V) nerve, which terminates in the ► trigeminal brain stem nuclear complex (TBNC). The caudal portion of the TBNC is largely homologous to the spinal cord ► dorsal horn in terms of anatomy, neurochemistry, and physiology and is termed the ► subnucleus caudalis (Vc) or medullary dorsal horn. It is the major region involved in the processing of pain and temperature sensations from the head. However, the more rostrally located subnuclei ► interpolaris (Vi) and oralis (Vo) also receive some nociceptive afferents and contain neurons responding to noxious stimuli. There is evidence that these subnuclei contribute to the perception of pain and especially pain of intraoral origin. There is no obvious homologous region in the spinal cord.

The craniofacial region contains several unique structures which include the tooth pulp (▶ nociceptors in the dental pulp) and the ▶ cornea (▶ ocular nociceptors), as well as several other deep and intraoral tissues such as the ▶ temporomandibular joint (TMJ) (▶ Nociceptors in the Orofacial Region (Temporomandibular Joint and Masseter Muscle)), the intracranial meninges and vessels (▶ Nociceptors in the Orofacial Region (Meningeal/Cerebrovascular)), and the intraoral mucosa (▶ Nociceptors in the Orofacial Region (Skin/Mucosa); ▶ Nociception in Nose and Oral Mucosa). Pathological changes in these structures or their central representations can result in several pain conditions unique to the trigeminal region, and these are described in the entry on ▶ Orofacial Pain. There is mounting evidence for sex differences in pain and analgesia, e.g., ▶ temporomandibular disorders (TMD) and ▶ migraine headaches are much more prevalent in females. Experimental studies in both animals and humans are revealing clear sex differences in peripheral and central neural processes underlying craniofacial nociception (e.g., Cairns et al. 2003; Okamoto et al. 2003).

Characteristics

Role of the Vc in Orofacial Nociceptive Processing

Several of the entries point out that most of the small-diameter trigeminal (V) primary afferents terminate in the Vc, which is a laminated structure with many morphological and functional similarities with the spinal dorsal horn (although see below) (▶ Trigeminal Brainstem Nuclear Complex, Anatomy; ▶ Trigeminal Brainstem Nuclear Complex, Physiology, ▶ Trigeminal Brain Stem Nuclear Complex, Immunohistochemistry, and Neurochemistry). There are seven lines of evidence that the Vc is critical in TBNC nociceptive processing. These include a direct projection to higher brain centers involved in pain perception and other aspects of pain behavior. In addition, by virtue of these ascending projections, the Vc has

recently been documented to be essential for the expression of central sensitization in rostral elements of the V brain stem complex (e.g., subnucleus oralis Vo) and ventrobasal thalamus (Chiang et al. 2002; Chiang et al. 2003). The extensive range of convergent afferent inputs to the Vc contributes to the development of the central sensitization that can be induced by inflammation or injury of peripheral tissues or nerves (Iwata et al. 2011; Sessle 2011). Moreover, most Vc nociceptive neurons have in addition to a cutaneous receptive field (RF) also a deep RF (e.g., in the TMJ, muscle, tooth pulp, dura). The particular efficacy of deep nociceptive afferent inputs in inducing central sensitization, including an expansion of both their cutaneous and deep RFs, represents neuronal properties that may explain the poor localization of deep pain, as well as contribute to the spread and referral of pains that are typical of deep pain conditions such as TMD, toothaches, and headaches (see ▶ Orofacial Pain). In recent years, there has been great interest in and progress made in understanding the role that glia in the TBNC play in orofacial inflammatory and neuropathic pain conditions (see ▶ Glial Cells in Orofacial Pathological Pain Mechanisms), as well as emerging interest in the brain stem processes underlying the influences of genetic and environmental factors in orofacial acute and chronic pain states (see Chiang et al. 2011; Seltzer and Mogil 2008; Sessle 2011; Woda and Salter 2008).

The neurons in the TBNC and particularly those in the Vc are subject to modulatory influences originating locally within the Vc or in more rostral parts of the TBNC as well as descending modulatory influences from brain stem, in particular the rostral ventromedial medulla (RVM) (Dubner and Ren 2004). These descending influences are very similar to those directed at the spinal dorsal horn and are described in descending modulation of nociceptive processing.

As mentioned above, a unique feature of the V system is the processing of nociceptive inputs from afferents supplying structures not found elsewhere in the body; these include the tooth pulp, periodontal tissues, cornea, and nasal mucosa. Another unique feature is the dual organization of the Vc in the representation of some orofacial tissues. Some noxious stimuli (e.g., to cornea or TMJ) evoke a bimodal distribution of c-Fos-labeled neurons in the Vc that includes c-fos expression in the rostral Vc/caudal Vi transitional zone (Ren and Dubner 2011) as well as in the transitional region between the caudal part of the Vc and the upper cervical spinal cord (see Bereiter et al. 2000; Dubner and Ren 2004). These findings are consistent with electrophysiological evidence of neurons responsive to corneal or TMJ inputs in one or both of these regions. The caudal Vc merges without clear boundaries with the cervical dorsal horn, while the rostral Vc forms a distinctive transition region with the Vi. This transition region has a ventral Vi/Vc transition area, which is especially clear in rodents, and a dorsomedial transition area. It is becoming apparent that these two areas of the rostral Vc and the caudal Vc each have their own unique morphological and functional features that may be differentially involved in contributing to perceptual, autonomic, endocrine, and muscle reflex responses to noxious stimulation of different orofacial tissues, in particular involving ophthalmic structures (Bereiter et al. 2000; Dubner and Ren 2004; Ren and Dubner 2011).

Further investigation of these different areas within the Vc, as well as caudal to the Vc, is needed to determine their specific functional roles in orofacial pain processes. It has been over 50 years since the trigeminal spinal nucleus was divided into Vo, Vi, and Vc subdivisions and over 25 years since the structural homologies between the Vc and the spinal dorsal horn were emphasized. These studies have shaped discussions of the special relationship between the Vc and craniofacial pain (see Sessle 2000, 2005), although this useful homology may need some revision as recent evidence cited above indicates that select portions of the Vc are organized differently from the spinal system.

The region ventral to the Vc, which is part of the medullary reticular formation and is outside the main projection area of trigeminal primary afferents, also contains neurons that respond to noxious facial stimulation. Their receptive fields are usually larger than those in Vc, often bilateral and sometimes include the entire body (e.g., see Fujino et al. 1996; Yokota et al. 1991). In the rat, a population of neurons with large nociceptive receptive fields including the face is located in the subnucleus reticularis dorsalis. It has been suggested that this region is involved in circuits mediating diffuse noxious inhibitory controls (Villanueva et al. 1996).

Role of Rostral Components of the TBNC
Morphological, physiological, neurochemical, and behavioral evidence support the involvement of more rostral components of the TBNC in orofacial nociceptive processing, especially in the case of the Vo. For example, lesions of rostral components may disrupt some pain behaviors, and substantial numbers of NS and WDR neurons exist in the rostral components. These neurons have cutaneous RFs that are usually localized to intraoral or perioral areas, and many can be activated by tooth pulp stimulation. These findings have raised the possibility that the rostral nociceptive neurons, particularly those in the Vo, are principally involved in intraoral and perioral nociceptive processing. Moreover, it is probable that their nociceptive afferent inputs are predominantly relayed through more caudal regions, such as the Vc, which exerts a considerable modulatory influence over Vo nociceptive processes.

Although some rostral neurons may directly receive primary afferent inputs from the tooth pulp, it is not clear that these inputs are nociceptive. How then can nociceptive phenomena occur in the rostral components when, especially in the case of the Vo, they lack an anatomical and in many instances a neurochemical substrate generally considered necessary for nociceptive processing? One possible explanation is that these substrates, although typical of the Vc and the spinal dorsal horn, are not essential for all types of nociceptive phenomena. In support of this possibility are the observations that nociceptive neurons contributing to nociceptive behavior occur in some spinal cord and brain stem areas devoid of some of these substrates (e.g., lateral cervical nucleus, reticular formation) (Dubner and Bennett 1983).

Another explanation lies in the anatomical and neurochemical framework for nociceptive processing that typifies the Vc and in its ascending projections to the rostral components of the TBNC. Parada et al. (1997) have argued that the cutaneous C fiber-related nociceptive responses of Vo cutaneous nociceptive neurons that can be blocked by systemic administration of the NMDA antagonist MK-801 may depend on the well-documented ascending projection from the Vc that exerts a net facilitatory modulatory influence on Vo neurons (for a review, see Sessle 2000), since the Vc does have the features (NMDA receptors, C fiber afferent terminals, substantia gelatinosa) considered necessary for these nociceptive phenomena. Furthermore, local application of MK-801 or morphine to the Vc can block the C fiber-related activity of some Vo nociceptive neurons. An analogous argument has also been used as an explanation for the neuroplastic changes that have been documented in the Vo and in the main sensory nucleus subsequent to C fiber depletion induced by the neonatal application of capsaicin. Nonetheless, MK-801 applied directly to the Vo itself can antagonize Vo nociceptive neuronal changes induced by afferent inputs evoked from the tooth pulp (Park et al. 2001), which suggests that NMDA receptor mechanisms do operate within the Vo (see Sessle 2000, 2011 for references and further details).

Collectively, these various findings raise the possibility that, on the one hand, the cutaneous RF and response properties of rostral nociceptive neurons, particularly in the Vo, may be dependent on caudal regions such as the Vc for the relay of nociceptive signals from primary afferents supplying superficial craniofacial tissues. On the other hand, some of the extensive pulp afferent inputs to and effects upon neurons in the rostral TBNC may be dependent on relays in both rostral and caudal components. Such a view, nonetheless, is still largely speculative, and the relative roles of the rostral and caudal TBNC components, not only in cutaneous nociceptive mechanisms but also in nociceptive responses to noxious stimulation of deep craniofacial tissues

and tooth pulp, represent an important issue requiring further research.

Ascending Projections and Higher Level Processing of Trigeminal Nociceptive Inputs

The nociceptive signals from the TBNC are relayed on to higher levels and in particular to the thalamus and from there to cerebral cortex. The thalamus receives direct contralateral input from each of the TBNC subnuclei (Kemplay and Webster 1989; Mantle-St John and Tracey 1987). The majority of the TBNC neurons projecting to the contralateral thalamus are found in the trigeminal principal nucleus, terminate in the ventroposterior medial nucleus (VPM), and are primarily involved in tactile sensation. The remaining neurons are located in the Vi, the Vo, and the Vc. However, the major thalamic projection related to pain and temperature perception arises from neurons in the contralateral Vc and is equivalent to the spinothalamic tract (for references, see Discussion in Craig 2004; Dostrovsky and Craig 2012).

The trigeminothalamic neurons in the Vc are located primarily in laminae I and V. Those in lamina V have a major projection to the VPM (or its borders in the cat), but those in lamina I terminate in several other regions, which are species dependent. For example, in the monkey, lamina I neurons project primarily to the posterior ventromedial nucleus (VMpo), as well as to the ventrocaudal medialis dorsalis (MDvc), but have only a sparse projection to the VPM (Craig 2004). In the cat, the main projections of lamina I neurons are to the ventral border of the VPM and adjacent nuclei, the dorsomedial VPM, and the nucleus submedius (Dostrovsky and Craig 2012). In the rat, lamina I neurons project largely to the VPM, the nucleus submedius, the posterior nucleus, and the posterior triangular nucleus (Dostrovsky and Craig 2012; Iwata et al. 1992; Jasmin et al. 2004; Yoshida et al. 1991).

The main cortical targets of thalamic nuclei receiving nociceptive inputs are the insula, the primary and secondary somatosensory cortices, and the cingulate (Dostrovsky and Craig 2012). There are extensive connections between cortex and thalamus, which play an important although poorly understood role in processing nociceptive inputs. Most of the electrophysiological studies on responses of thalamic and cortical neurons to noxious stimuli have focused on inputs from the limbs and have revealed neurons with both NS and WDR type responses. These responses are generally similar to those of neurons in the spinal dorsal horn and the Vc but tend to have increased and fluctuating spontaneous activity (see the review by Kenshalo and Willis 1991). The encyclopedia entries "▶ Nociceptive Processing in the Thalamus" and "▶ Cortical and Limbic Mechanisms Mediating Pain and Pain-Related Behavior" provide further details of the roles of thalamus and cortex in pain.

In addition to the thalamus, TBNC neurons also project to several diencephalic and brain stem areas that are involved in regulation of autonomic, endocrine, affective, and motor functions. In the rat, all TBNC subnuclei contain neurons that project directly to the hypothalamus (Malick and Burstein 1998). Most of these hypothalamic tract neurons in the Vc and C1-2 respond preferentially or exclusively to noxious mechanical and thermal stimulation to the facial skin and to electrical, mechanical, and chemical stimulation of the dura mater (Burstein et al. 1998). There also are projections from the TBNC to the parabrachial and Kölliker-Fuse nuclei (Bernard et al. 1989; Feil and Herbert 1995; Hayashi and Tabata 1990). In particular, neurons in the Vc, including those in the superficial laminae, project to the external portion of the lateral parabrachial area, where many respond exclusively to noxious stimuli (Hayashi and Tabata 1990). It has been proposed that this projection is part of a trigeminopontoamygdaloid pathway involved in the affective, behavioral, and autonomic reactions to noxious stimuli (Bernard et al. 1989). Anatomical studies also have shown that there are projections from the TBNC to the adjacent reticular formation, the RVM, the periaqueductal gray, various brain stem autonomic nuclei, the superior colliculus, the ipsilateral cerebellum, the contralateral inferior olive, and the nucleus of the solitary tract (Bruce et al. 1987; Dostrovsky and Craig 2012; Jacquin et al. 1989; Mantle-St John and

Tracey 1987; Marfurt and Rajchert 1991; Keay and Bandler 1998; Veinante et al. 2000). Several of these projections are likely to be important in mediating nonperceptual (e.g., autonomic) effects elicited by nociceptive stimuli.

For further details, see the encyclopedia entries: ▶ Trigeminal Brain Stem Nuclear Complex, Anatomy, ▶ Trigeminal Brain Stem Nuclear Complex, Physiology, ▶ Trigeminal Brain Stem Nuclear Complex, Immunohistochemistry, and Neurochemistry, ▶ Glial Cells in Orofacial Pathological Pain Mechanisms, ▶ Nociceptors in the Dental Pulp, ▶ Ocular Nociceptors, ▶ Nociceptors in the Orofacial Region (Temporomandibular Joint and Masseter Muscle), ▶ Nociceptors in the Orofacial Region (Meningeal/Cerebrovascular), ▶ Nociceptors in the Orofacial Region (Skin/Mucosa), and ▶ Orofacial Pain.

Summary

A great deal of information regarding the representation and processing of craniofacial nociceptive inputs in the TBNC and higher levels has been gained in recent years. Although there are many similarities between the spinal and trigeminal systems, there are also some important differences, and some of these relate to unique structures and likely contribute to the pain conditions that are unique to this region. Nevertheless, there are still important gaps in our knowledge, and further experimental studies are necessary to fully understand the mechanisms underlying the normal and pathological processing in the TBNC of nociceptive inputs from the craniofacial region.

References

Bereiter, D. A., Hirata, H., & Hu, J. W. (2000). Trigeminal subnucleus caudalis: Beyond homologies with the spinal dorsal horn. *Pain, 88,* 221–224.

Bernard, J. F., Peschanski, M., & Besson, J. M. (1989). A possible spino (trigemino)-ponto-amygdaloid pathway for pain. *Neuroscience Letters, 100,* 83–88.

Bruce, L. L., McHaffie, J. G., & Stein, B. E. (1987). The organization of trigeminotectal and trigeminothalamic neurons in rodents: A double-labeling study with fluorescent dyes. *The Journal of Comparative Neurology, 262*, 315–330.

Burstein, R., Yamamura, H., Malick, A., et al. (1998). Chemical stimulation of the intracranial dura induces enhanced responses to facial stimulation in brain stem trigeminal neurons. *Journal of Neurophysiology, 79*, 964–982.

Cairns, B. E., Wang, K., Hu, J. W., et al. (2003). The effect of glutamate-evoked masseter muscle pain on the human jaw-stretch reflex differs in men and women. *Journal of Orofacial Pain, 17*, 317–325.

Chiang, C. Y., Hu, B., Hu, J. W., et al. (2002). Central sensitization of nociceptive neurons in trigeminal subnucleus oralis depends on integrity of subnucleus caudalis. *Journal of Neurophysiology, 88*, 256–264.

Chiang, C. Y., Hu, B., Park, S. J. et al. (2003). Purinergic and NMDA-receptor mechanisms underlying tooth pulp stimulation-induced central sensitization in trigeminal nociceptive neurons. In *Proceedings of the 10th World Congress on pain*. Seattle: IASP Press, pp 345–354.

Chiang, C. Y., Dostrovsky, J. O., Iwata, K., & Sessle, B. J. (2011). Role of glia in orofacial pain. *The Neuroscientist, 17*, 303–320.

Craig, A. D. (2004). Distribution of trigeminothalamic and spinothalamic lamina I terminations in the macaque monkey. *The Journal of Comparative Neurology, 477*, 119–148.

Dostrovsky, J. O., Craig, A. D. (2012). Ascending projection systems. In S. B. McMahon, & M. Koltzenburg (Eds.), *Wall and Mezack's Textbook of Pain*. Elsevier, Churchill Livingstone, Edinburgh, Ch 11.

Dubner, R., & Bennett, G. J. (1983). Spinal and trigeminal mechanisms of nociception. *Annual Review of Neuroscience, 6*, 381–418.

Dubner, R., & Ren, K. (2004). Brainstem mechanisms of persistent pain following injury. *Journal of Orofacial Pain, 18*, 299–305.

Feil, K., & Herbert, H. (1995). Topographic organization of spinal and trigeminal somatosensory pathways to the rat parabrachial and Kolliker-Fuse nuclei. *The Journal of Comparative Neurology, 353*, 506–528.

Fujino, Y., Koyama, N., & Yokota, T. (1996). Differential distribution of three types of nociceptive neurons within the caudal bulbar reticular formation in the cat. *Brain Research, 715*, 225–229.

Hayashi, H., & Tabata, T. (1990). Pulpal and cutaneous inputs to somatosensory neurons in the parabrachial area of the cat. *Brain Research, 511*, 177–179.

Iwata, K., Kenshalo, D. R., Jr., Dubner, R., et al. (1992). Diencephalic projections from the superficial and deep laminae of the medullary dorsal horn in the rat. *The Journal of Comparative Neurology, 321*, 404–420.

Iwata, K., Imamura, Y., Honda, K., et al. (2011). Physiological mechanisms of neuropathic pain: The orofacial region. *International Review of Neurobiology, 97*, 227–250.

Jacquin, M. F., Barcia, M., & Rhoades, R. W. (1989). Structure-function relationships in rat brainstem subnucleus interpolaris: IV. Projection neurons. *The Journal of Comparative Neurology, 282*, 45–62.

Jasmin, L., Granato, A., & Ohara, P. T. (2004). Rostral agranular insular cortex and pain areas of the central nervous system: A tract-tracing study in the rat. *The Journal of Comparative Neurology, 468*, 425–440.

Keay, K. A., & Bandler, R. (1998). Vascular head pain selectively activates ventrolateral periaqueductal gray in the cat. *Neuroscience Letters, 245*, 58–60.

Kemplay, S., & Webster, K. E. (1989). A quantitative study of the projections of the gracile, cuneate and trigeminal nuclei and of the medullary reticular formation to the thalamus in the rat. *Neuroscience, 32*, 153–167.

Kenshalo, D. R., Jr., & Willis, W. D., Jr. (1991). The role of the cerebral cortex in pain sensation. In E. G. Jones & A. Peters (Eds.), *Cerebral cortex* (pp. 151–212). New York: Plenum.

Malick, A., & Burstein, R. (1998). Cells of origin of the trigeminohypothalamic tract in the rat. *The Journal of Comparative Neurology, 400*, 125–144.

Mantle-St John, L. A., & Tracey, D. J. (1987). Somatosensory nuclei in the brainstem of the rat: Independent projections to the thalamus and cerebellum. *The Journal of Comparative Neurology, 255*, 259–271.

Marfurt, C. F., & Rajchert, D. M. (1991). Trigeminal primary afferent projections to "non-trigeminal" areas of the rat central nervous system. *The Journal of Comparative Neurology, 303*, 489–511.

Okamoto, K., Hirata, H., Takeshita, S., et al. (2003). Response properties of TMJ units in superficial laminae at the spinomedullary junction of female rats vary over the estrous cycle. *Journal of Neurophysiology, 89*, 1467–1477.

Parada, C. A., Luccarini, P., & Woda, A. (1997). Effect of an NMDA receptor antagonist on the wind-up of neurons in the trigeminal oralis subnucleus. *Brain Research, 761*, 313–320.

Park, S. J., Chiang, C. Y., Hu, J. W., et al. (2001). Neuroplasticity induced by tooth pulp stimulation in trigeminal subnucleus oralis involves NMDA receptor mechanisms. *Journal of Neurophysiology, 85*, 1836–1846.

Ren, K., & Dubner, R. (2011). The role of trigeminal interpolaris-caudalis transition zone in persistent orofacial pain. *International Review of Neurobiology, 97*, 207–225.

Seltzer, Z., & Mogil, J. S. (2008). Pain and genetics. In B. J. Sessle, G. J. Lavigne, R. Dubner, & J. P. Lund (Eds.), *Orofacial pain* (2nd ed., pp. 69–78). Chicago: Quintessence.

Sessle, B. J. (2000). Acute and chronic craniofacial pain: Brainstem mechanisms of nociceptive transmission and neuroplasticity, and their clinical correlates. *Critical Reviews in Oral Biology and Medicine, 11*, 57–91.

Sessle, B. J. (2005). Orofacial pain. In H. Merskey (Ed.), *Pathways of pain*. Seattle: IASP Press.

Sessle, B. J. (2011). Peripheral and central mechanisms of orofacial inflammatory pain. *International Review of Neurobiology, 97*, 179–206.

Veinante, P., Jacquin, M. F., & Deschênes, M. (2000). Thalamic projections from the whisker-sensitive regions of the spinal trigeminal complex in the rat. *The Journal of Comparative Neurology, 420*, 233–243.

Villanueva, L., Bouhassira, D., & Le, B. D. (1996). The medullary subnucleus reticularis dorsalis (SRD) as a key link in both the transmission and modulation of pain signals. *Pain, 67*, 231–240.

Woda, A., & Salter, M. W. (2008). Mechanisms of neuropathic pain. In B. J. Sessle, G. J. Lavigne, R. Dubner, & J. P. Lund (Eds.), *Orofacial pain* (2nd ed., pp. 53–60). Chicago: Quintessence.

Yokota, T., Koyama, N., Nishikawa, Y., et al. (1991). Trigeminal nociceptive neurons in the subnucleus reticularis ventralis. I. Response properties and afferent connections. *Neuroscience Research, 11*, 117.

Yoshida, A., Dostrovsky, J. O., Sessle, B. J., et al. (1991). Trigeminal projections to the nucleus submedius of the thalamus in the rat. *The Journal of Comparative Neurology, 307*, 609–625.

Nociceptive Processing in the Cingulate Cortex

▶ Cingulate Cortex, Nociceptive Processing, Behavioral Studies in Animals

Nociceptive Processing in the Hippocampus and Entorhinal Cortex, Neurophysiology and Pharmacology

Sanjay Khanna
Department of Physiology, National University of Singapore, Singapore

Introduction

The hippocampus is the simplest part of the cortex, the ▶ allocortex, which, in humans, is arched around the mesencephalon, while in rodents it arches over the thalamus (Amaral and Witter 1995). The various fields of the hippocampus and its layers are illustrated in Fig. 1. The principal neurons in the hippocampus are the

▶ pyramidal cells that are localized in the stratum pyramidale (Fig. 1).

Both anatomical and physiological studies suggest that the entorhinal cortex is a major source of afferent input to the hippocampus, either directly or indirectly via the dentate gyrus (Amaral and Witter 1995). Indeed, stimulation of perforant path fibers from the entorhinal cortex evokes both short- and long-latency excitatory responses in CA1, which are suggested to involve direct entorhinal to CA1 projection and a ▶ sequential relay from the entorhinal cortex to CA1 via the dentate gyrus and CA3, respectively.

Besides the entorhinal cortex, the other major source of afferent input to the hippocampus is the medial septum-diagonal band of Broca region (MS; Risold 2004). The region also provides afferent input to the entorhinal cortex (Risold 2004). The septo-hippocampal projection includes neurons of different phenotype, including cholinergic, GABAergic, and glutamatergic. Indeed, the region is the major source of ▶ cholinergic input to the hippocampus. Physiologically, the MS facilitates the generation of the ▶ theta rhythm in the septo-hippocampal-entorhinal network, theta being a sinusoidal rhythmic extracellular oscillations that reflect rhythmic oscillations of neurons in processing of information.

Characteristics

Facilitation of Pain

Based largely on stimulation and lesion studies in animals, Melzack and Casey (1968) proposed that the limbic forebrain structures, including the hippocampus, are involved in "aversive drive and affect that comprise the motivational dimension of pain." Indeed, lesion or drug-induced manipulation of the hippocampal network in animals affect a range of pain-related behaviors in a fashion that is consistent with the idea that the network facilitates pain, especially persistent and neuropathic pain. The lines of evidence include:

(i) Lesions of the hippocampus-dentate gyrus region reduced aversive foot shock-induced

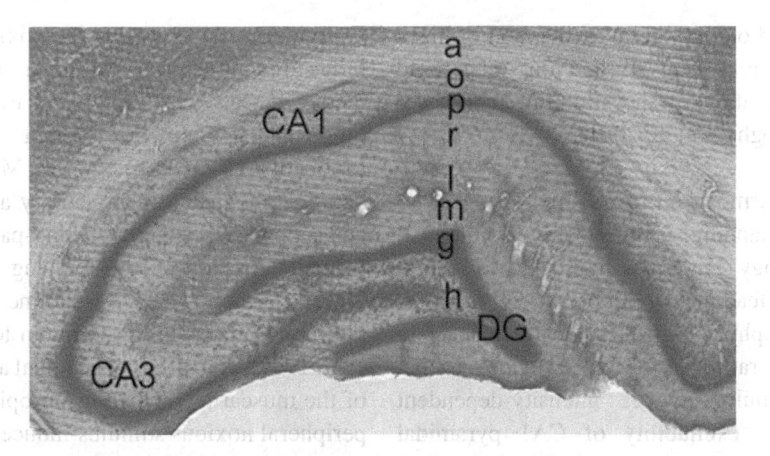

Nociceptive Processing in the Hippocampus and Entorhinal Cortex, Neurophysiology and Pharmacology, Fig. 1 Digitized image of Nissl-stained transverse section taken through dorsal hippocampus and dentate gyrus (*DG*) of rat. The hippocampus is subdivided into various fields, of which the prominent ones are *CA1* and *CA3*. In a dorsal to ventral order in the transverse section, the layers of the hippocampus and *DG* are *a* alveus (which is as a fiber bundle marking the outer boundary of the hippocampus), *o* stratum oriens, *p* stratum pyramidale, *r* stratum radiatum, *l* stratum lacunosum-moleculare, *m* stratum moleculare, *g* stratum granulosum, and *h* hilus

▶ conditioned place avoidance (Selden et al. 1991).

(ii) Intrahippocampal administration of an NMDA receptor antagonist attenuated spontaneous nociceptive behaviors to hind-paw injection of the algogen formalin (▶ formalin test), a model of persistent clinical inflammatory pain (McKenna and Melzack 2001).

(iii) Inactivating the MS region upstream to hippocampus also attenuated formalin-induced pain behaviors (Lee et al. 2011). The effects of MS inactivation with the GABA$_A$ receptor agonist, muscimol, were marked on the second phase of the formalin behaviors that are partly driven by central plasticity. Furthermore, the antinociceptive effects of muscimol were accompanied by a decrease in spinal nociceptive processing (Lee et al. 2011).

(iv) Changes in hippocampal neurochemistry that follow nerve damage in models of ▶ neuropathic pain facilitate peripheral hypersensitivity. For example, the hippocampal levels of ▶ tumor necrosis factor-alpha (TNFα), a cytokine, and indoleamine 2,3-dioxygenase

1 (IDO1), an enzyme involved in tryptophan metabolism, are increased in models of neuropathic pain (Ignatowski et al. 1999; Ren et al. 2011; Kim et al. 2012). In context of TNFα, increasing the levels of TNFα per se in the hippocampus with bilateral microinjection of TNF nanoplasmidexes induces peripheral hypersensitivity (Martuscello et al. 2012; but see Ren et al. 2011). Conversely, inhibiting the synthesis of TNFα with intraperitoneal thalidomide attenuated mechanical allodynia in a chronic pain model (Ren et al. 2011). In context of IDO1, intrahippocampal microinjection of 1-methyl-tryptophan (1-MT), a pharmacological inhibitor of IDO1, decreased thermal hyperalgesia and depressive behavior in a model of hind-paw arthritis (Kim et al. 2012).

Imaging studies in humans also bear out the idea that the hippocampal network is involved in pain. Thus, anxiety-induced hyperalgesia in man is associated with bilateral activation in the entorhinal area, which correlated with activity in anterior cingulate cortex and insular cortex (Ploghaus et al. 2001). Furthermore, the hippocampus is also activated on peripheral application of high-intensity noxious heat stimulation, relative to similar application of the stimulus at a lower intensity (Ploghaus et al. 2001).

Noxious Information Processing

Functional mapping, using electrophysiological and cell biology techniques, has identified several nociceptive features of hippocampal neurons. Thus, electrophysiological studies conducted in anesthetized rat have provided evidence that noxious stimulus evokes intensity-dependent changes in ▶ excitability of CA1 pyramidal cells (Khanna and Sinclair 1989; Wei et al. 2000) that is ▶ non-topographic (Khanna and Sinclair 1992). Consistently, ▶ c-Fos mapping techniques in anesthetized and behaving rats also indicated that neural changes in hippocampus, especially in field CA1 and medial entorhinal cortex, are noxious intensity dependent (Funahashi et al. 1999; Khanna et al. 2004). The changes in CA1 were bilateral and were observed along the length of the region. However, the noxious stimuli-induced effect in entorhinal cortex was significant only ipsilateral to the stimuli, though a trend was also observed in the contralateral entorhinal cortex (Funahashi et al. 1999). Collectively, the foregoing findings are consistent with a role of the hippocampal network as a central intensity monitor.

Indeed, the hippocampal neurons are particularly affected by more intense and damaging noxious stimuli. For example, Khanna (1997) reported that hind-paw injection of formalin, but not brief peripheral noxious heat stimulus in anesthetized rat, evoked a "signal-to-noise" processing by the hippocampal dorsal field CA1. The "signal-to-noise" processing was marked by a selective excitation of a population of dorsal CA1 putative pyramidal cells against the background of widespread pyramidal cell suppression which enhanced the contrast of the excitatory signal. Interestingly, the "signal-to-noise" processing was observed in parallel with hippocampal theta activation (Khanna 1997; Tai et al. 2006).

Evidence suggests that the noxious stimulus-induced suppression and theta activation are mediated, at least in part, by the septo-hippocampal cholinergic afferents. In this regard, lesion of MS or destruction of MS cholinergic neurons attenuated theta activity and pyramidal cell suppression induced on hind-paw injection of formalin (Khanna 1997; Zheng and Khanna 2001). Furthermore, acetylcholine is released in the hippocampus in the formalin test (Ceccareli et al. 1999), and intrahippocampal administration of the muscarinic antagonist atropine attenuated peripheral noxious stimulus-induced suppression of CA1 pyramidal cell synaptic excitability (Khanna and Sinclair 1992).

From a broad perspective, the cholinergic-mediated CA1 pyramidal cell suppression is proposed to facilitate the performance of CA1 neural network, especially in encoding of information in relation to memory functions of the network, by suppressing irrelevant processing (Hasselmo and McGaughy 2004). Indeed, changes in electrophysiological and cellular markers of learning and memory are observed in models of neuropathic pain. However, the patterns of changes are time dependent. For example, amputation of a paw digit or the tip of the tail induced an increase in levels of the transcription protein Egr1 in hippocampal field CA1 between 15 min and 2 h after injury (Wei et al. 2000). Likewise, peripheral injury facilitated the generation of afferent stimulation-induced ▶ long-term potentiation (LTP) of excitatory synaptic transmission in the region (Wei et al. 2000), LTP being a cellular model of learning and memory. The enhanced level of Egr1 in CA1 was linked to facilitation of long-term potentiation (LTP) such that the injury-induced facilitation of LTP was attenuated in mutant mice lacking Egr1.

However, a facilitation of LTP is not observed in the hippocampus at 15 h or more after nerve injury (Ren et al. 2011). Rather, both the late phase of LTP and frequency facilitation are attenuated. The electrophysiological changes are accompanied by decreased density of presynaptic boutons and impaired performance of animals in tests of working memory. Further, both the electrophysiological effects and behavioral effects were mimicked by intracerebroventricular and intrahippocampal administration of TNFα, while intraperitoneal administration of thalidomide, an inhibitor of synthesis of TNFα, reversed the various effects of nerve damage (Ren et al. 2011).

Summary

In summary, the evidence so far suggests that nociceptive processing in hippocampus and the entorhinal cortex is, at least in part, distributed, non-topographic, and noxious stimulus intensity dependent, which is in line with the postulated role of these regions in affect motivation to pain. Moreover, intense and damaging noxious stimuli lead to longer-term changes in the hippocampal network that might be a basis for comorbid deficits in cognition and mood disorders.

References

Amaral, D. G., & Witter, M. P. (1995). Hippocampal formation. In G. Paxinos (Ed.), *The rat nervous system* (pp. 443–493). San Diego: Academic.

Ceccareli, I., Casamenti, F., Massafra, C., et al. (1999). Effects of novelty and pain on behavior and hippocampal extracellular ach levels in male and female rats. *Brain Research, 815*, 169–176.

Funahashi, M., He, Y.-F., Sugimoto, T., et al. (1999). Noxious tooth pulp stimulation suppresses C-Fos expression in the rat hippocampal formation. *Brain Research, 827*, 215–220.

Hasselmo, M. E., & McGaughy, J. (2004). High acetylcholine levels set circuit dynamics for attention and encoding and low acetylcholine levels set dynamics for consolidation. *Progress in Brain Research, 145*, 207–231.

Ignatowski, T. A., Covey, W. C., Knight, P. R., et al. (1999). Brain-derived TNFÎ ± mediates neuropathic pain. *Brain Research, 841*, 70–77.

Khanna, S. (1997). Dorsal hippocampus field CA1 pyramidal cell responses to a persistent versus an acute nociceptive stimulus and their septal modulation. *Neuroscience, 77*, 713–721.

Khanna, S., & Sinclair, J. G. (1989). Noxious stimuli produce prolonged changes in the CA1 region of the rat hippocampus. *Pain, 39*, 337–343.

Khanna, S., & Sinclair, J. G. (1992). Responses in the CA1 region of the rat hippocampus to a noxious stimulus. *Experimental Neurology, 117*, 28–35.

Khanna, S., Chang, S., Jiang, F., et al. (2004). Nociception-driven decreased induction of Fos protein in ventral hippocampus field CA1 of the rat. *Brain Research, 1004*, 167–176.

Kim, H., Chen, L., Grewo, L., et al. (2012). Brain indoleamine 2,3-dioxygenase contributes to the comorbidity of pain and depression. *The Journal of Clinical Investigation, 122*, 2940–2954.

Lee, A. T.-H., Ariffin, M. Z., Zhou, M., et al. (2011). Forebrain medial septum region facilitates nociception in a rat formalin model of inflammatory pain. *Pain, 152*, 2528–2542.

Martuscello, R. T., Spengler, R. N., Bonoiu, A. C., et al. (2012). Increasing TNF levels solely in the rat hippocampus produces persistent pain-like symptoms. *Pain, 153*, 1871–1882.

McKenna, J. E., & Melzack, R. (2001). Blocking NMDA receptors in the hippocampal dentate gyrus with AP5 produces analgesia in the formalin pain test. *Experimental Neurology, 172*, 92–99.

Melzack, R., & Casey, K. L. (1968). Sensory, motivational and central control determinants of pain. In D. R. Kenshalo (Ed.), *The skin senses* (pp. 423–443). Springfield: Thomas.

Ploghaus, A., Narain, C., Beckmann, C. F., et al. (2001). Exacerbation of pain by anxiety is associated with activity in a hippocampal network. *The Journal of Neuroscience, 21*, 9896–9903.

Ren, W.-J., Liu, Y., Zhou, L.-J., et al. (2011). Peripheral nerve injury leads to working memory deficits and dysfunction of the hippocampus by upregulation of TNFa in rodents. *Neuropsychopharmacology, 36*, 979–992.

Risold, P. Y. (2004). The septal region. In G. Paxinos (Ed.), *The rat nervous system* (pp. 605–632). Amsterdam/Boston: Elsevier Academic.

Selden, N. R. W., Everitt, B. J., Jarrard, L. E., et al. (1991). Complementary roles for the amygdala and hippocampus in aversive conditioning to explicit and contextual cues. *Neuroscience, 42*, 335–350.

Tai, S. K., Huang, F. D., Moochhala, S., et al. (2006). Hippocampal theta state in relation to formalin nociception. *Pain, 121*, 29–42.

Wei, F., Xu, Z. C., Qu, Z., et al. (2000). Role of EGR1 in hippocampal synaptic enhancement induced by tetanic stimulation and amputation. *The Journal of Cell Biology, 149*, 1325–1333.

Zheng, F., & Khanna, S. (2001). Selective destruction of medial septal cholinergic neurons attenuates pyramidal cell suppression, but not excitation in dorsal hippocampus field CA1 induced by subcutaneous injection of formalin. *Neuroscience, 103*, 985–998.

Nociceptive Processing in the Nucleus Accumbens, Neurophysiology and Behavioral Studies

Lino Becerra and David Borsook
P.A.I.N. Group, Department of Anesthesiology,
Perioperative and Pain Medicine, Boston
Children's Hospital Department of Radiology
Massachusetts General Hospital Center for Pain
and the Brain, Harvard Medicals, Boston,
MA, USA

Synonyms

NAcc; Nucleus accumbens

Definition

The nucleus accumbens (NAcc) is a brain structure that forms part of the ventral striatum. It has been established that the NAcc is a critical structure for the rewarding and reinforcing properties of addictive drugs. It also appears to play an important role in modulating pain sensations.

Characteristics

The human ventral striatum encompasses a series of structures of which the nucleus accumbens (NAcc) is the primary one (Nolte 2002). This definition is functionally based on its connectivity. In rodents, the NAcc has been clearly defined by cytoarchitecture, neuroactive transmitters, and receptor distributions (Otake and Nakamura 2000). Two main substructures have been identified in the rodent NAcc: a core and a shell. The core is located in the dorsolateral portion of the ventral striatum and the shell is in the medioventral portion (Otake and Nakamura 2000). The core projects directly to the lateral ▶ ventral tegmental area (VTA), while the shell projects to the ventromedial ventral pallidum, which in turn projects to the VTA. In primates,

however, the core region appears contiguous with the rest of the striatum and cannot be easily distinguished. The shell has several histochemical features that make it different from the rest of the striatum (Prensa et al. 2003). The NAcc in humans is located at the base of the ▶ caudate nucleus and the ▶ putamen (Fig. 1) (see Buchsbaum et al. 1998 for a systematic approach to identify the NAcc in MRI images). In humans, several histological and chemical studies have given mixed results and it has been concluded that separation of core and shell substructures is difficult, probably due to the complexity and heterogeneity of the structure and its innervation (Prensa et al. 2003).

The NAcc is involved in evaluating probability assessments and the emotional valence of information. Disturbances at this level have been implicated in a number of affective disorders including drug abuse (Altier and Stewart 1999). Recent functional imaging investigations in humans have shown that the NAcc is activated in situations related to drug-associated reward and to monetary and other rewarding stimuli (Breiter et al. 1997; Zink et al. 2004).

Opioids such as morphine and heroin and psychostimulants such as amphetamine are known drugs of potential abuse. However, it is also clinically known that these drugs are effective in the treatment of pain (Altier and Stewart 1999). These results suggest that reward and aversion share a common neural substrate. Opioids are known to inhibit neural systems that transmit nociceptive information (Fields 2004); the descending pain pathway relays information from the ▶ periaqueductal gray (PAG) to the ▶ rostroventral medulla (RVM) and through the spinal cord to the dorsal horn, where peripheral nociceptive information is inhibited from reaching supraspinal structures (Fields 2004). Microinjections of morphine into the PAG produce deep analgesia allowing the performance of major surgery on subjects. Interestingly, opioids as well as psychostimulant drugs increase transmission in mesocorticolimbic dopamine (DA) neurons that are known to be activated by rewards such as food or sex (Altier and Stewart 1999). These neurons arise from cell bodies in the

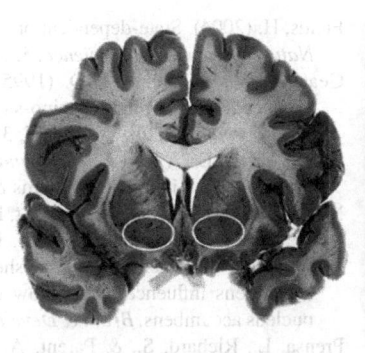

Nociceptive Processing in the Nucleus Accumbens, Neurophysiology and Behavioral Studies, Fig. 1 Histological and MRI slices of the NAcc. *Left panel* depicts a brain slice through the anterior part of the caudate nucleus and putamen with the NAcc bridging both structures (Adapted from Haines (2004), Neuroanatomy). *Right panel* shows a corresponding MRI slice of the NAcc

ventral tegmental area (VTA) and project to various sites, among them the NAcc, amygdala, and frontal cortex.

A number of studies have suggested a significant role for the NAcc in the processing of pain and analgesia. Experiments in rodents subjected to the ▶ formalin test (subcutaneous injection of diluted formalin into an animal's hind paw to produce persistent nociceptive pain) have been performed to elucidate the role of NAcc in analgesia. Injection of morphine, amphetamine, or substance P agonist into the VTA produces analgesia, i.e., the animal displays fewer effects of pain due to the formalin injection. However, lesions of the NAcc or injection of DA receptor antagonists in the NAcc diminishes the analgesia produced following intra-VTA injections (Altier and Stewart 1999).

The role of the NAcc in aversive stimuli such as pain may be as a result of both dopaminergic projections from the VTA region and inputs from spinal levels. Dopamine release in the NAcc modulates information about tonic noxious stimulation coming from the periphery. The broader neuroanatomical circuitry underlying this effect may be explained by looking at the NAcc afferent and efferent connections. One possibility arises from direct projections of spinal cord neurons into the NAcc (Cliffer et al. 1991). Several studies have implicated an ascending pain control pathway that produces analgesia if stimulated by intense noxious stimuli or by

peripheral injection of local anesthetic or opioids (Gear and Levine 1995). Another possibility arises from NAcc projections into the medial thalamus known to receive input from dorsal horn neurons sensitive to noxious stimuli. The NAcc has projections to the amygdala, which has connection to the brainstem and from there to the spinal cord (Altier and Stewart 1999).

Other mechanisms, besides dopaminergic ones, have been found to be involved in processing pain by the NAcc. In a series of experiments, animals were subjected to intense noxious stimuli (injection of high concentrations of capsaicin subcutaneously or immersion of hind paws in hot water) and tested for pain in a different body part (Gear and Levine 1995). Under intense stimuli, animals displayed less pain in the stimulated body part than without the stimuli. The mechanism is based on an ascending pain pathway projecting to the NAcc that disinhibits opioid neurons in the NAcc and produces antinociception. There is no direct connection between the NAcc and the RVM; however, the NAcc projects to other structures such as the hypothalamus, which in turn sends projections to the brainstem (Gear and Levine 1995).

In animal studies, it has been established that the NAcc plays an important role in processing pain and several mechanisms have been elucidated and are continuously studied. Human studies, however, present a much larger challenge, given that none of the manipulations

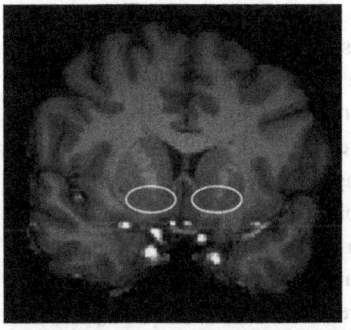

done in animals are possible. Yet new neuroimaging techniques have made possible the study of responses in the CNS, including the NAcc, non-invasively in conscious human subjects (Becerra et al. 2001; Breiter et al. 1997). Most of the studies reporting NAcc activation have been devoted to the role of NAcc in processing rewarding stimuli such as drugs of abuse (Breiter et al. 1997) or monetary reward (Zink et al. 2004). For positive valence hedonic stimuli (e.g., food, money), the NAcc was found to produce a positive signal change in ▶ functional MRI (fMRI) studies (Zink et al. 2004).

Recently, fMRI studies of noxious stimuli in normal subjects have indicated that the NAcc displays negative signal changes (Becerra et al. 2001), opposite to those observed for rewarding stimuli. Further studies have also allowed the dissection of the role of the NAcc in chronic pain (Baliki et al. 2010); in chronic back pain patients, the response in the NAcc had distinctive responses in patients compared to healthy controls. NAcc activity might reflect maladaptive processing of motivational and emotional information in chronic pain such as depression and anxiety.

References

Altier, N., & Stewart, J. (1999). The role of dopamine in the nucleus accumbens in analgesia. Life Sciences, 65, 2269–2287.

Baliki, M. N., Geha, P. Y., Fields, H. L., et al. (2010). Predicting value of pain and analgesia: Nucleus accumbens response to noxious stimuli changes in the presence of chronic pain. Neuron, 66, 149–160.

Becerra, L., Breiter, H. C., Wise, R., et al. (2001). Reward circuitry activation by noxious thermal stimuli. Neuron, 32, 927–946.

Breiter, H. C., Gollub, R. L., Weisskoff, R. M., et al. (1997). Acute effects of cocaine on human brain activity and emotion. Neuron, 19, 591–611.

Buchsbaum, M. S., Fallon, J. H., Wei, T. C., et al. (1998). A method of basal forebrain anatomical standardization for functional image analysis. Psychiatry Research, 84, 113–125.

Cliffer, K. D., Burstein, R., & Giesler, G. J., Jr. (1991). Distributions of spinothalamic, spinohypothalamic, and spinotelencephalic fibers revealed by anterograde transport of PHA-L in rats. The Journal of Neuroscience, 11, 852–868.

Fields, H. (2004). State-dependent opioid control of pain. Nature Reviews. Neuroscience, 5, 565–575.

Gear, R. W., & Levine, J. D. (1995). Antinociception produced by an ascending spino-supraspinal pathway. The Journal of Neuroscience, 15, 3154–3161.

Haines, D. E. (2004). Neuroanatomy (6th ed.). Philadelphia: Lippincott Williams & Wilkins.

Nolte, J. (2002). The human brain. St. Louis: Mosby, Inc.

Otake, K., & Nakamura, Y. (2000). Possible pathways through which neurons of the shell of the nucleus accumbens influence the outflow of the core of the nucleus accumbens. Brain & Development, 22, 17–26.

Prensa, L., Richard, S., & Parent, A. (2003). Chemical anatomy of the human ventral striatum and adjacent basal forebrain structures. The Journal of Comparative Neurology, 460, 345–367.

Zink, C. F., Pagnoni, G., Martin-Skurski, M. E., et al. (2004). Human striatal responses to monetary reward depend on saliency. Neuron, 42, 509–517.

Nociceptive Processing in the Secondary Somatosensory Cortex

Frederick A. Lenz[1], Rolf-Detlef Treede[2] and Ulf Baumgartner[3]
[1]Department of Neurosurgery, Johns Hopkins Hospital, Baltimore, MD, USA
[2]Institute of Physiology and Pathophysiology, Johannes Gutenberg University, Mainz, Germany
[3]Chair of Neurophysiology, Center for Biomedicine and Medical Technology Mannheim, Heidelberg University, Mannheim, Germany

Synonyms

S2; Second somatosensory cortex; Second somatic area; Secondary somatosensory cortex; SII; SII/PV

Definition

The term "secondary somatosensory cortex" refers to a cortical representation of the body outside the primary somatosensory cortex (SI). Such a secondary representation was initially found in two separate locations (for detail see Burton 1986; Caselli 1993), in the human parietal lobe (supplementary sensory area) and within the parietal ▶ operculum in the superior bank of the Sylvian or lateral fissure (secondary somatosensory area SII); additional somatosensory areas are situated in the superior parietal lobule and in the ▶ insula. The somatosensory area in the parietal operculum (Fig. 1) has retained the designation as SII, because historically it was the first example of a second representation of any of the senses in the brain (Adrian 1941).

Characteristics

SII is located on the lateral curvature of parietal cortex in rodents and rabbits, in the anterior ectosylvian gyrus in cats and dogs, and along the superior bank of the ▶ Sylvian fissure in primates (Burton 1986). Human SII does not usually reach the convexity of the brain, but was explored by Wilder Penfield during procedures in which it did reach the convexity or in which the fissure was split (Penfield and Jasper 1954). Functional localization of SII in humans with imaging techniques such as PET and fMRI confirmed its location predominantly in the upper bank of the Sylvian fissure (Burton et al. 1993), opposite the auditory cortex in Heschl's convolution (Özcan et al. 2005).

The SII region probably contains more than one somatosensory area as suggested by the presence of two separate face areas in the lateral parietal operculum and a common foot area in medial parietal operculum. SII proper is in the caudal (posterior) part of this region, the parietal ventral area PV may be in its rostral (anterior) portion (Disbrow et al. 2000; Fitzgerald et al. 2004). Cytoarchitectonic classification of the parietal operculum has been undertaken only recently (Young et al. 2004), yielding four subregions OP1–OP4, of which SII is probably situated in OP/2 and PV in OP3/4. Brodmann areas are not defined in this region. In the stereotactic coordinates of the atlas of Talairach and Tournoux, SII extends 40–60 mm lateral of the midline, 15–30 mm behind the anterior

commissure (AC), and 5–25 mm above the AC-PC line. The Sylvian fissure runs obliquely through this coordinate system and its location varies considerably across individuals (Özcan et al. 2005). Therefore, either the auditory cortex or the Sylvian fissure and the central sulcus should be used as anatomical landmarks when referring to the location of SII.

The secondary somatosensory cortex receives direct thalamocortical input from the ventral posterior nuclear group, in particular the ventro-postero-inferior nucleus VPI (Apkarian and Shi 1994; Gauriau and Bernard 2004; Jones and Burton 1976). In addition, there is a prominent projection from SI to SII, particularly in primates (Friedman et al. 1986; Jones et al. 1978).

Single neuron recordings in SII have largely focused on the tactile representation (Fitzgerald et al. 2004; Robinson and Burton 1980). Although most neurons in SII have contralateral ▶ receptive fields, a sizable proportion of bilateral receptive fields has been observed. Most imaging and electrophysiological studies in humans have found a bilateral response to unilateral stimulation, with a contralateral preponderance. Functionally, SII is considered to play a role in tactile object recognition and memory (Caselli 1993), as well as the perception of vibrotactile stimuli (Burton et al. 1993; Ferrington and Rowe 1980). The cortico-cortical output connections of SII into the insula and parahippocampal gyrus are similar to those of the inferior temporal cortex, which is implicated in visual object recognition (Friedman et al. 1986).

In human imaging studies, enhanced perfusion as a sign of activation by phasic nociceptive stimuli was found more regularly in SII and adjacent parts of the insula than in SI (Peyron et al. 2002; Treede et al. 2000). Evoked potential recordings showed that SII was activated simultaneously with or even earlier than SI (Schlereth et al. 2003). Hence, SII appears as a major site for the early cortical encoding of pain in the human brain. SII is considered to be important for the recognition of the noxious nature of a painful stimulus, for intensity coding and other sensory-discriminative aspects of pain, and as

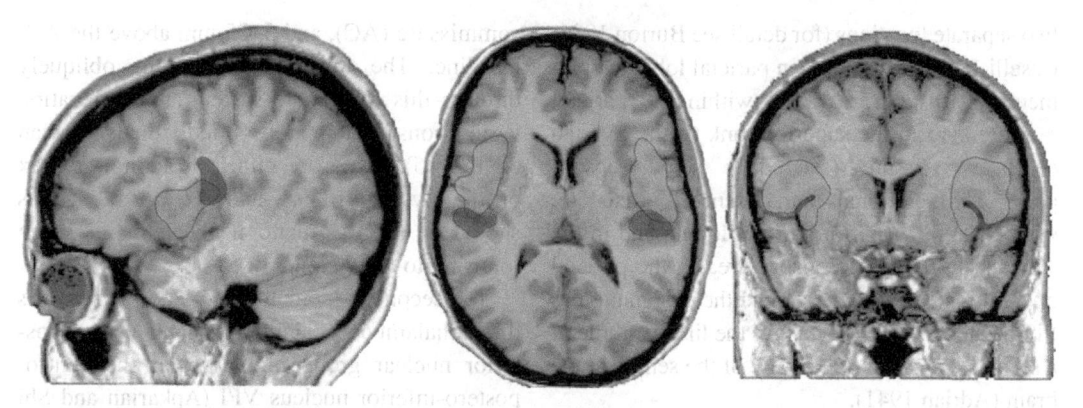

Nociceptive Processing in the Secondary Somatosensory Cortex, Fig. 1 Location of the secondary somatosensory cortex in the human brain. This figure shows schematic drawings of the classical secondary somatosensory area SII as identified using tactile stimulation (*blue*) in comparison with those areas in the operculo-insular cortex that respond to nociceptive stimulation (*orange*). Note that the nociceptive cortex extends further anterior than the tactile areas and includes both opercular and insular parts. *Left*: sagittal section passing slightly medial of SII in order to show the insula also. *Center*: horizontal section through SII. *Right*: coronal section through nociceptive regions in the anterior insula and in the inner vertical face of the frontal operculum

part of a sensory-limbic pathway for pain memory and affective motivational aspects of pain (Berthier et al. 1988; Lenz et al. 1997; Treede et al. 2000; Ohara et al. 2004).

In contrast to the abundance of evidence for nociceptive activation of the SII region from human evoked potential and imaging studies, there are few single neuron recordings in this region showing specific nociceptive responses (Dong et al. 1994). In monkey, some cells in the SII region respond to the approach of a sharp object to the face, suggesting that this region may represent the position of a painful or threatening stimulus in extrapersonal space. These neurons, however, were located in area 7b, which is adjacent to SII in monkey but not in humans, in whom Brodmann areas 39 and 40 separate SII and area 7b. Since search stimuli in all these animal studies were tactile, an intriguing alternative possibility is that tactile and nociceptive inputs are represented in different areas within the SII region (Treede et al. 2000).

The insula has been suggested to contain such a separate representation of nociception at its dorsal junction with the parietal and frontal operculum (Craig 2002). The insula subserves sensory integrative functions for pain, taste, and other visceral sensations, as well as visceral motor and other autonomic functions (Treede et al. 2000). Direct comparisons of vibrotactile and painful heat stimulation in humans showed activation of SII by both stimuli and a more pronounced activation in anterior insula by painful stimuli (Coghill et al. 1994). It has been suggested that the dorsal and anterior insula may be part of a multisensory interoceptive pathway signalling the internal state of the body. In this view, pain is an emotion or interoceptive sensation resulting from disequilibrium of this internal state (Craig 2002). This hypothesis is called into question by recent studies suggesting that interoceptive sensations and emotions are infrequently evoked by stimulation of this system (Lenz et al. 2004).

Subdural and depth electrophysiological recordings in patients undergoing epilepsy surgery have identified a region in the inner vertical face of the frontal operculum that receives nociceptive input with a latency of about 150 ms, corresponding to the earliest portion of the whole cerebral response (Frot and Mauguière 2003; Vogel et al. 2003). The center of this region is located about 15–20 mm anterior of the tactile SII area and lateral of the anterior insula which lies on the other side of the circular sulcus of the insula (Fig. 1). Because of this proximity, many neuroimaging studies may mislabel the inner vertical face of the frontal operculum as either SII or anterior insula. In current meta-analyses of neuroimaging studies (PET, fMRI, MEG, and EEG source analysis), it was found that the resolution of the techniques was not sufficient to separate insular from opercular activity (Peyron et al. 2002). More studies are needed to clarify how many functionally and cytoarchitectonically separate areas are contained within the parasylvian cortex. It is evident, however, that the parasylvian cortex plays an important role in the cortical representation of pain. This region includes SII but extends further anterior into the frontal operculum and medially into the insula.

Acknowledgment Drs. Treede, Baumgärtner, and Lenz and some of the studies reviewed here were supported by DFG Tr 236/13-3 and by NIH-NINDS 38493.

References

Adrian, E. D. (1941). Afferent discharges to the cerebral cortex from peripheral sense organs. *The Journal of Physiology, 100*, 159–191.

Apkarian, A. V., & Shi, T. (1994). Squirrel monkey lateral thalamus. I. Somatic nociresponsive neurons and their relation to spinothalamic terminals. *The Journal of Neuroscience, 14*, 6779–6795.

Berthier, M., Starkstein, S., & Leiguarda, R. (1988). Asymbolia for pain: A sensory-limbic disconnection syndrome. *Annals of Neurology, 24*, 41–49.

Burton, H. (1986). Second somatosensory cortex and related areas. In E. G. Jones & A. Peters (Eds.), *Sensory-motor areas and aspects of cortical connectivity* (pp. 31–98). New York: Plenum Press. Cerebral Cortex, vol 5.

Burton, H., Videen, T. O., & Raichle, M. E. (1993). Tactile-vibration-activated foci in insular and parietal-opercular cortex studied with positron emission tomography – mapping the 2nd somatosensory area in humans. *Somatosensory & Motor Research, 10*, 297–308.

Caselli, R. J. (1993). Ventrolateral and dorsomedial somatosensory association cortex damage produces distinct somesthetic syndromes in humans. *Neurology, 43*, 762–771.

Coghill, R. C., Talbot, J. D., Evans, A. C., et al. (1994). Distributed processing of pain and vibration by the human brain. *The Journal of Neuroscience, 14*, 4095–4108.

Craig, A. D. (2002). How do you feel? Interoception: The sense of the physiological condition of the body. *Nature Reviews, 3*, 655–666.

Disbrow, E., Roberts, T., & Krubitzer, L. (2000). Somatotopic organization of cortical fields in the lateral sulcus of homo sapiens: Evidence for SII and PV. *The Journal of Comparative Neurology, 418*, 1–21.

Dong, W. K., Chudler, E. H., Sugiyama, K., et al. (1994). Somatosensory, multisensory, and task-related neurons in cortical area 7b (PF) of unanesthetized monkeys. *Journal of Neurophysiology, 72*, 542–564.

Ferrington, D. G., & Rowe, M. (1980). Differential contributions to coding of cutaneous vibratory information by cortical somatosensory areas I and II. *Journal of Neurophysiology, 43*, 310–331.

Fitzgerald, P. J., Lane, J. W., Thakur, P. H., et al. (2004). Receptive field properties of the macaque second somatosensory cortex: Evidence for multiple functional areas. *The Journal of Neuroscience, 24*, 11193–11204.

Friedman, D. P., Murray, E. A., O'Neill, J. B., et al. (1986). Cortical connections of the somatosensory fields of the lateral sulcus of macaques: Evidence for a corticolimbic pathway for touch. *The Journal of Comparative Neurology, 252*, 323–347.

Frot, M., & Mauguière, F. (2003). Dual representation of pain in the operculo-insular cortex in humans. *Brain, 126*, 438–450.

Gauriau, C., & Bernard, J. F. (2004). Posterior triangular thalamic neurons convey nociceptive messages to the secondary somatosensory and insular cortices in the rat. *The Journal of Neuroscience, 24*, 752–761.

Jones, E. G., & Burton, H. (1976). Areal differences in the laminar distribution of thalamic afferents in cortical fields of the insular, parietal and temporal regions of primates. *The Journal of Comparative Neurology, 168*, 197–248.

Jones, E. G., Coulter, J. D., & Hendry, S. H. C. (1978). Intracortical connectivity of architectonic fields in the somatic sensory, motor and parietal cortex of monkeys. *The Journal of Comparative Neurology, 181*, 291–348.

Lenz, F. A., Gracely, R. H., Zirh, A. T., et al. (1997). The sensory-limbic model of pain memory. *Pain Forum, 6*, 22–31.

Lenz, F. A., Ohara, S., Gracely, R. H., et al. (2004). Pain encoding in the human forebrain: Binary and analog exteroceptive channels. *The Journal of Neuroscience, 24*, 6540–6544.

Ohara, S., Crone, N. E., Weiss, N., Treede, R.-D., & Lenz, F. A. (2004). Amplitudes of laser evoked potential recorded from primary somatosensory, parasylvian and medial frontal cortex are graded with stimulus intensity. *Pain, 110*, 318–328.

Özcan, M., Baumgärtner, U., Vucurevic, G., et al. (2005). Spatial resolution of fMRI in the human parasylvian cortex: Comparison of somatosensory and auditory stimulation. *NeuroImage, 25*, 877–887.

Penfield, W., & Jasper, H. (1954). *Epilepsy and the functional anatomy of the human brain*. Boston: Little Brown.

Peyron, R., Frot, M., Schneider, F., et al. (2002). Role of operculoinsular cortices in human pain processing: Converging evidence from PET, fMRI, dipole modeling, and intracerebral recordings of evoked potentials. *NeuroImage, 17*, 1336–1346.

Robinson, C. J., & Burton, H. (1980). Organization of somatosensory receptive fields in cortical areas 7b, retroinsula, postauditory and granular insula of M. fascicularis. *The Journal of Comparative Neurology, 192*, 69–92.

Schlereth, T., Baumgärtner, U., Magerl, W., et al. (2003). Left-hemisphere dominance in early nociceptive processing in the human parasylvian cortex. *NeuroImage, 20*, 441–454.

Treede, R. D., Apkarian, A. V., Bromm, B., et al. (2000). Cortical representation of pain: Functional characterization of nociceptive areas near the lateral sulcus. *Pain, 87*, 113–119.

Vogel, H., Port, J. D., Lenz, F. A., et al. (2003). Dipole source analysis of laser-evoked subdural potentials recorded from parasylvian cortex in humans. *Journal of Neurophysiology, 89*, 3051–3060.

Young, J. P., Herath, P., Eickhoff, S., et al. (2004). Somatotopy and attentional modulation of the human parietal and opercular regions. *The Journal of Neuroscience, 24*, 5391–5399.

Nociceptive Processing in the Spinal Cord

Hans-Georg Schaible
Department of Physiology - Neurophysiology, Friedrich-Schiller-University of Jena, Jena, Germany

Characteristics

Nociceptive Functions of the Spinal Cord

Nociceptive primary afferent fibers project to the spinal cord, where they form synapses with second-order neurons. The spinal cord has several functions in nociceptive processing. Firstly, ascending axons of nociceptive neurons in the gray matter project to supraspinal sites. These projections activate neuronal circuits in the brain stem that are involved in descending inhibition and facilitation, and they activate the thalamocortical system that produces the complex conscious pain response.

Secondly, the spinal cord generates motor responses to noxious stimuli. Motor responses are nociceptive reflexes and also more complex motor behavior such as avoidance of movements. Thirdly, the spinal cord is involved in the generation of autonomic reflexes that are elicited by noxious stimuli.

Nociceptive sensory spinal cord neurons are located in the superficial and deep dorsal horn. Neurons with nociceptive properties are also found in the ventral horn. The majority of nociceptive neurons are interneurons that do not project to supraspinal sites (Willis and Coggeshall 2004).

Structure of the Dorsal Horn and Spinal Projection of Nociceptive Afferents

The spinal cord gray matter contains neuronal cell bodies and fibers, whereas the white matter consists of fibers including axons of ascending and descending tracts. The gray matter has been divided into laminae I–X by ▶ Rexed's laminae (1952, 1954). The dorsal horn consists of laminae I–VI. For a detailed description, see ▶ nociceptive circuitry in the spinal cord.

Laminae I–VI

Lamina I (▶ laminae I and V neurons), also called the marginal zone of the dorsal horn, is characterized by large horizontal neurons (the Waldeyer cells) and a plexus of numerous horizontally arranged fine axons and dendrites. However, many cells appear much smaller. Lamina II (▶ Lamina II Outer and Lamina II Inner), also called ▶ substantia gelatinosa, consists of numerous small neurons and their processes; myelinated axons are almost absent. Stalked cells and islet cells have been identified, but many interneurons have other morphologies. ▶ Lamina III forms a broad band across the dorsal horn, and it has slightly larger and more widely spaced cells. Lamina III contains many myelinated fibers. ▶ Lamina IV is relatively thick and contains many small and numerous large neurons. Lamina V (▶ Laminae I and V Neurons) extends as a thick band across the narrowest part of the dorsal horn. Its lateral boundary is

indistinct because of the many bundles of myelinated fibers coursing longitudinally through this
area, giving the lateral part of this lamina
a reticulated appearance. Cells are more varied
than in lamina IV. Lamina VI cells are arranged
much like those in lamina V (Rexed 1952, 1954;
Willis and Coggeshall 2004).

Spinal Termination of Non-nociceptive and Nociceptive Primary Afferent Fibers

In general, non-nociceptive and nociceptive
primary afferent fibers have different termination
sites in the dorsal horn. Nociceptive Aδ-fibers
project mainly to lamina I (and II). Some
Aδ-fibers have further projections into lamina
V. C fibers (▶ C afferent axons/fibers) project
mainly to lamina II. By contrast, non-nociceptive
primary afferents with Aβ-fibers (▶ A Afferent
Fibers (Neurons)) project to lamina III and deeper
(Willis and Coggeshall 2004).

It should be noted that the patterns of inputs
from different tissues are not identical. The main
inputs into lamina II are primary afferents from
skin, but visceral and muscular unmyelinated
afferents project to lamina II as well (Sugiura
et al. 1989). However, visceral and muscular
afferents also terminate in laminae I and deeper
laminae. It is thought that visceral afferents
distribute to a wider area of the cord, but that
numbers of terminals for each fiber are much
sparser for visceral than for cutaneous fibers
(Sugiura et al. 1989).

Inputs of Neurons in Laminae I–VI

Lamina I neurons get inputs from primary
afferent fibers, interneurons, and descending
fibers. Approximately 50 % of the synapses are
lost following dorsal rhizotomy and thus arise
presumably from primary afferents (Chung
et al. 1989). All of the primary afferents entering
lamina I arise from Aδ- and C fibers. The input to
lamina II consists of unmyelinated fine primary
afferent fibers arising from the tract of Lissauer
and the marginal plexus, from propriospinal
neurons and from descending fibers. About
50 % of the synapses in lamina II survive dorsal
rhizotomy (Chung et al. 1989). The primary

afferent input into lamina III is largely from collaterals of large sensory axons (Scheibel and
Scheibel 1968). These are from different types
of non-nociceptive primary afferents such as
hair follicle afferents, Pacinian corpuscle afferents, rapidly and slowly adapting afferents,
etc. However, some projection of fine primary
fiber input into lamina III has also been noted
(Sugiura et al. 1986). Many of the large neurons
in lamina IV and lamina V have long spiny
dendrites that pass dorsally, laterally, and
ventrally. The dorsal dendrites penetrate the
substantia gelatinosa to be contacted by axons
from interneurons and from fine primary
afferents. In addition, they receive inputs from
terminal ramifications of large myelinated
primary afferent fibers that make synaptic
contacts to lamina IV cells. There is not much
fine primary afferent input directly into lamina
IV. Neurons in lamina VI often send dendrites
across the width of the dorsal horn, but the
dendrites do not penetrate laminae I and II.
The primary afferent input is from collaterals
from primary afferent axons destined to reach
ventral horn cells (for references, see Willis and
Coggeshall 2004).

Output of Neurons in Laminae I–VI

Most neurons are interneurons with axons ending
in the same or adjacent laminae. For example,
some of the laminae III–V cells are antenna-type
neurons that send dendrites to lamina II and are
thus output neurons from lamina II. However,
lamina I includes the dendrites to lamina III–VI
neurons that form long axons projecting to
supraspinal sites (Willis and Coggeshall 2004).
Lamina I neurons project to various sites in
the brain stem. Within laminae III–VI are
spinocervical tract neurons and ▶ postsynaptic
dorsal column cells. Neurons in laminae IV–VI
send their axons into the lateral white column or
across the midline, presumably to the opposite
▶ spinothalamic tract through the anterior white
commissure. They may bifurcate before they go
to the white matter, and collaterals of these axons
ramify in laminae III–V and deeper laminae,
including the contralateral side and lamina X.

Response Properties of Nociceptive Spinal Cord Neurons

Recordings have shown responses of spinal cord neurons to electrical stimulation of nerve fibers and to natural innocuous and noxious stimulation (for references, see Willis and Coggeshall 2004). Upon electrical nerve stimulation, lamina I neurons were excited by cutaneous Aδ-fibers and sometimes by volleys in C fibers, in line with the morphological studies. However, some neurons were excited (polysynaptically?) by Aβ-fiber stimulation. Neurons in the substantia gelatinosa were activated primarily by C fibers; however, Aβ- and Aδ-fiber stimulation also activated some of them. Neurons in laminae III to VI respond either to A and C fiber or only to A fiber stimulation. Many neurons show convergence of Aβ-, Aδ-, and C fiber inputs.

It is noteworthy that neurons in laminae I and V in the thoracic and sacral spinal cord also show responses to stimulation of visceral nerves, and neurons in the superficial and deep lumbosacral enlargement are activated by cutaneous fibers, muscular group III fibers, and joint afferents.

Neuronal responses were also recorded during natural stimulation (for references, see Willis and Coggeshall 2004). It appeared that neurons in the different laminae are heterogeneous in their response properties. This is not surprising on the one hand, because each of the Aβ-, Aδ-, and C fiber classes has different modalities (mechanoreception, nociception, thermoreception). On the other hand, the response properties do not seem simply to reflect the input into the laminae. This has to be expected because only about 50 % of the synapses are from primary afferent fibers.

Lamina I contains neurons that are only activated by intense mechanical stimulation of the skin, neurons that are activated by intense mechanical stimulation of skin and noxious heat applied to the skin, and non-nociceptive thermoreceptive neurons that respond to innocuous warming or cooling. Other neurons are wide dynamic range of neurons responding weakly to innocuous and strongly to noxious stimulation. Furthermore, neurons in lamina I show convergent inputs from cutaneous and deep tissue or convergence from cutaneous and visceral inputs.

Information on identified lamina II neurons is sparse. Cells were excited by innocuous or noxious mechanical stimuli. However, several authors reported that spontaneously active neurons in this lamina were inhibited by natural stimulation or they were inhibited and then excited by weak mechanical stimuli. Furthermore, neurons showed phenomena such as habituation, long afterdischarges following brief stimuli and variable receptive fields.

Neurons in laminae III–VI are of different types. Concerning thresholds and encoding ranges, some neurons are low threshold, responding only to innocuous stimulation. Many of these neurons are located in laminae III and IV. Most neurons are wide dynamic range, responding weakly to innocuous stimuli and strongly to noxious stimuli, and a further group of neurons are high threshold responding only to noxious intensities. Thus, at least the wide dynamic range neurons seem to receive inputs from non-nociceptive as well as from nociceptive afferents, in line with morphological studies showing projection of non-nociceptive afferents into deep laminae, extension of dendrites of deep cells up to the superficial laminae, and interneurons transmitting information from superficial into deeper laminae. Many neurons in laminae III–VI receive inputs from deep tissue. These cells are solely excited by deep tissue stimulation or show convergent inputs from deep tissue and skin, or they are excited from skin and viscera. All neurons with visceral input seem to receive input from the skin as well. Lamina X neurons are also either low threshold, wide dynamic range, or high threshold. Many of these neurons respond to visceral stimulation such as colorectal distension, in addition to cutaneous stimuli. Some cells in lamina X have bilateral receptive fields.

Encoding of Noxious Stimuli in the Spinal Cord

The understanding of the encoding of nociceptive information in the spinal cord is not very advanced. On the basis of single neuron recordings, both ► wide dynamic range neurons and ► nociceptive specific neurons seem to be suitable to encode the intensity of a noxious stimulus to a specific site. However, wide dynamic range neurons in particular often have large receptive fields, and a stimulus of a defined intensity may elicit differently strong responses when applied to different sites in the receptive field. It is questionable whether the activity of a single neuron reflects more the magnitude of a stimulus or the site in the receptive area where a stimulus of a defined intensity is applied. Furthermore, the situation becomes more complicated when inputs from different tissues to a neuron are studied. It seems impossible that, e.g., a single wide dynamic range neuron with convergent cutaneous and visceral inputs can unequivocally signal that the skin or a visceral organ has been challenged with a noxious stimulus. This uncertainty in the message of a neuron could in fact be the reason why pain in viscera and to some extent in the deep tissue is often referred to a cutaneous area, namely into a so-called heat zone. However, it is quite clear that the precise location of a noxious stimulus, its intensity, and character cannot be encoded by a single nociceptive neuron. It is assumed, therefore, that encoding of a noxious stimulus is only achieved by a population of nociceptive neurons (for further discussion, see Price et al. 2003). This topic is addressed in detail in ▶ Encoding of Noxious Information in the Spinal Cord.

Samples of neurons were studied by the recording of field potentials in the dorsal horn, but these data have not contributed to the understanding of encoding. Furthermore, changes in metabolic activity and the expression of immediate early genes such as ▶ C-Fos have been used to map regions in the spinal cord that are activated by a noxious stimulus. The expression of FOS protein (▶ c-fos) has been used extensively because individual neurons can be visualized. The expression of C-FOS in a neuron is thought to show its activation (Willis and Coggeshall 2004). For example, noxious heat stimulation evokes expression of C-FOS in the superficial dorsal horn within a few minutes and staining shifts to deeper laminae of the dorsal horn thereafter (Menétrey et al. 1989; Williams et al. 1990). Noxious visceral stimulation evokes C-FOS expression in laminae I, V, and X, thus

resembling the projection area of visceral afferent fibers and injection of mustard oil into muscle elicits C-FOS expression in laminae I and IV to VI (Hunt et al. 1987; Menétrey et al. 1989). These data show, therefore, in which spinal laminae and segments neurons were activated by noxious stimulation. It should be noted, however, that excitatory as well as inhibitory neurons may express C-FOS.

Plasticity in the Nociceptive Processing in the Spinal Cord

Importantly, spinal cord neurons show changes in their response properties, including the size of their receptive fields, when the peripheral tissue is sufficiently activated by noxious stimuli, or when thin fibers in a nerve are electrically stimulated. In general, it is thought that plasticity in the spinal cord contributes significantly to clinically relevant pain states.

Wind-Up

▶ Wind-up is the increase in the response of a spinal cord neuron when electrical stimulation of C fibers is repeated at intervals of about 1 s (Mendell 1966; Mendell and Wall 1965). The basis of wind-up is a prolonged EPSP that builds up as a result of a repetitive C fiber volley, and thus it rests on temporal summation of synaptic potentials within the cord (Sivilotti et al. 1993). Other neurons show wind-down (Alarcon and Cervero 1990; Fitzgerald and Wall 1980; Woolf and King 1987). Wind-up disappears quickly when repetitive stimulation is stopped. Wind-up is likely to contribute to short-lasting increases in response to painful stimulation (see ▶ Windup of Spinal Cord Neurons).

Long-term Potentiation (LTP) and Long-term Depression (LTD)

These are long-lasting changes in synaptic activity after peripheral nerve stimulation (Randic et al. 1993; Rygh et al. 1999; Sandkühler and Liu 1998; Svendsen et al. 1997). They can be observed as increases in field potentials in the superficial dorsal horn. The most pronounced ▶ LTP with a short latency can be elicited after application of a high-frequency train of electrical stimuli that are suprathreshold for C fibers when the spinal cord has been transected in order to interrupt descending inhibitory influences from the brain stem. However, LTP can also be elicited with natural noxious stimulation, although the time course is much slower (Rygh et al. 1999). By contrast, LTD in the superficial dorsal horn is elicited by electrical stimulation of Aδ-fibers. This latter form of plasticity may be a basis of inhibitory mechanisms that counteract responses to noxious stimulation (Sandkühler et al. 1997). LTP and LTD will be addressed in detail in the entry ► Long-Term Potentiation and Long-Term Depression in the Spinal Cord.

Central Sensitization in the Course of Inflammation and Nerve Damage

Changes in responses of spinal cord neurons have been studied in models of inflammation and neuropathy. Pronounced changes in response properties of neurons in the superficial dorsal horn, the deep dorsal horn, and the ventral cord have been described. ► Central sensitization, originally described by Woolf (1983), has been observed in neurons with cutaneous input during cutaneous inflammation and other forms of cutaneous irritation such as capsaicin application. Pronounced central sensitization of spinal cord neurons with deep input has been shown during inflammation in joints, muscle, and viscera. Typical changes in responses of individual neurons are (a) increased responses to noxious stimulation of inflamed tissue, (b) lowering of the threshold of spinal cord neurons with an initially high threshold (initially nociceptive spinal neurons will change into wide dynamic range neurons), (c) increased responses to stimuli applied to noninflamed tissue surrounding the inflamed site, and (d) expansion of the receptive field. In particular, the enhanced responses to stimuli applied to noninflamed tissue in the vicinity of the inflamed zone indicate that the sensitivity of the spinal cord neurons is enhanced so that previously subthreshold input is sufficient to activate the neuron under inflammatory conditions. The sensitization of individual spinal cord neurons will lead to an increased percentage of neurons in a segment that respond to stimulation of an inflamed tissue.

Thus the population of responsive neurons increases. Central sensitization can persist for weeks, judging from the recording of neurons at different stages of acute and chronic inflammation (for review, see Coderre et al. 1993; Dubner and Ruda 1992; Mense 1993; Schaible and Grubb 1993).

In ► neuropathic pain states, findings in the spinal cord are dependent on the neuropathic model used. Evidence for central sensitization has been observed in neuropathic pain states in which conduction in the nerve remains present, and thus a receptive field of neurons can be identified. In these models, more neurons show ongoing discharges, and on average higher responses can be elicited by innocuous stimulation of receptive fields (Laird and Bennett 1993; Palacek et al. 1992a, b). In some models of neuropathy, neurons with abnormal discharge properties can be observed. For more information, see entries on Neuropathic Pain.

During inflammation and neuropathy, more neurons that express C-FOS are observed in the spinal cord, supporting the finding that a large number of neurons are activated. At least at some time points, enhanced ► metabolism can be seen in the spinal cord during inflammation and neuropathy. Both of these findings underscore the plasticity that occurs in the spinal cord under these conditions (Price et al. 1991; Schadrack et al. 1999).

Transmitters and Receptors Involved in the Spinal Nociceptive Processing

Numerous transmitters and receptors are involved in spinal nociceptive processing. They mediate processing of noxious information arising from noxious stimulation of normal tissue, and they are involved in plastic changes in spinal cord neuronal responses when the peripheral tissue is inflamed or when a nerve is damaged in a neuropathic fashion. Other transmitters are inhibitory and control spinal processing. In general, transmitter actions have either fast kinetics (action of glutamate on ionotropic AMPA and kainate receptors, action of ATP at ionotropic purinergic receptors, action of GABA at ionotropic GABA receptors) or slower kinetics (in particular neuropeptides that act through G-protein-coupled metabotropic receptors). Actions with fast kinetics evoke immediate and short-term effects on neurons, thus encoding the input to the neuron, whereas actions with slow kinetics rather modulate synaptic processing (Millan 1999; Willis and Coggeshall 2004). The following paragraphs summarize the main findings on synaptic transmission of nociceptive information (for references, see Willis and Coggeshall 2004).

Glutamate

This excitatory amino acid is a principal transmitter in the spinal cord that produces fast synaptic transmission. ▶ Glutamate is a transmitter of primary afferent neurons and of dorsal horn interneurons. Many spinal cord interneurons are excited by glutamate and by agonists at glutamate receptors. Glutamate activates ionotropic AMPA/kainate (non-NMDA) and ▶ NMDA receptors as well as ▶ metabotropic glutamate receptors. They are expressed all over the spinal cord gray matter, although differences in regional densities are found. Glutamate receptors are found in both excitatory and inhibitory interneurons.

Glutamate receptors are involved in the excitation of substantia gelatinosa neurons by Aδ- and C fibers in slice preparations. Usually these actions are mainly blocked by CNQX, an antagonist at non-NMDA receptors, whereas ▶ NMDA receptor antagonists usually cause a small reduction in EPSPs and reduce later components of the EPSP. In vivo recordings showed that both non-NMDA and NMDA receptors are involved in the synaptic activation of neurons by noxious stimuli. In addition, both of these receptors are involved in forms of functional plasticity. For example, both wind-up and central sensitization by inflammation are blocked by spinal application of NMDA antagonists (and non-NMDA antagonists). Metabotropic glutamate receptors can potentiate the action of ionotropic glutamate receptors (for review, see Fundytus 2001; Millan 1999; Willis and Coggeshall 2004). A detailed discussion of glutamate receptors is provided by the entries "▶ NMDA Receptors in Spinal Nociceptive

Processing" and "▶ Metabotropic Glutamate Receptors in Spinal Nociceptive Processing."

Adenosine Triphosphate

▶ ATP depolarizes some dorsal horn neurons in the superficial dorsal horn. ATP has been implicated in the fast synaptic transmission of innocuous mechanoreceptive input, but evidence has also been provided for an involvement in nociceptive synaptic transmission. Some spinal cord neurons seem to express purinergic receptors for actions of ATP, but other reports rather described presynaptic actions of ATP that caused an enhanced release of glutamate. The latter finding is consistent with the localization of purinergic receptors in dorsal root ganglion cells. ▶ P2X3 receptor immunoreactivity in the inner part of lamina II is reduced following dorsal rhizotomy (for review, see Willis and Coggeshall 2004).

GABA and Glycine

▶ GABA is probably the most important fast inhibitory transmitter in the spinal cord. Application of GABA to neurons causes IPSPs and inhibition of the activity of spinal cord neurons. GABA occurs in inhibitory interneurons throughout the spinal cord. ▶ GABAergic inhibitory interneurons can be synaptically activated by primary afferent fibers, and this explains why strong nociceptive inputs can induce, in addition to excitation, inhibition of neurons (usually following initial excitation) that are under the control of these inhibitory interneurons. Noxious stimuli can cause FOS expression in GABAergic interneurons. Some of the GABAergic interneurons also contain other mediators such as glycine, acetylcholine, enkephalin, galanin, neuropeptide Y, or nitric oxide synthase (NOS).

Both the ionotropic ▶ $GABA_A$ receptor and the metabotropic ▶ $GABA_B$ receptor are located presynaptically on primary afferent neurons or postsynaptically on dorsal horn neurons. Responses to both innocuous mechanical and noxious stimuli can be reduced by GABA receptor agonists (for review, see Willis and Coggeshall 2004). It is under discussion whether reduced inhibition may be a mechanism of neuropathic pain (Polgár et al. 2004).

Some of the inhibitory effects are due to glycine, and indeed, the ventral and the dorsal horn contain numerous glycinergic neurons. Glycine may be colocalized with GABA in synaptic terminals. The roles of GABA and glycine are addressed in more detail in ► GABA and Glycine in Spinal Nociceptive Processing.

Acetylcholine

Many small DRG neurons, some large DRG ones, and some neurons in the dorsal horn are cholinergic. Vice versa, many DRG neurons and neurons in the dorsal horn express nicotinergic and muscarinergic receptors. Application of ► acetylcholine to the skin is pronociceptive (via nicotinic and muscarinergic receptors), whereas spinal acetylcholine produces pro- or antinociception (for review, see Willis and Coggeshall 2004).

Excitatory Neuropeptides

A number of peptides are colocalized with excitatory transmitters, in particular with glutamate. Excitatory neuropeptides evoke EPSPs, but these differ from EPSPs evoked by glutamate in several respects. Usually they occur after a latency of seconds, but they last longer. They may not be sufficient to evoke action potential generation. Because glutamate and excitatory peptides are coreleased from synaptic endings, they are thought to act in a synergistic way (Urban et al. 1994).

Substance P

This excitatory peptide is colocalized with glutamate in a proportion of thin-diameter primary afferents and in a proportion of spinal cord interneurons. SP-containing endings are concentrated in laminae I and II and in lamina X. They terminate on cell bodies and dendrites of dorsal horn neurons. SP is released mainly in the superficial dorsal horn following electrical stimulation of unmyelinated fibers and during noxious mechanical, thermal, or chemical stimulation of the skin and deep tissues such as the joints. In part, release of SP is dependent on NMDA receptors on primary afferent endings (for review, see Willis and Coggeshall 2004).

SP acts on ► neurokinin-1 (NK-1) receptors that are located on dendrites and cell bodies of dorsal horn neurons in laminae I, IV–VI, and X. Fewer neurons in laminae II and III have NK-1 receptors. The vast majority of neurons with NK-1 receptors are excitatory, while a few contain GABA and glycine and are thus inhibitory. Some of the neurons with NK-1 receptors are projection neurons including spinothalamic, spinoreticular, and spinobrachial neurons. Upon strong activation by SP, NK-1 receptors can be internalized. Such an internalization is blocked by NK-1 receptor antagonists and by NMDA receptor antagonists.

NK-1 receptors are G-protein-coupled receptors, i.e., the action of SP on ion channels is indirect. Application of SP evokes a prolonged excitation of nociceptive dorsal horn neurons. These depolarizations are presumably caused by Ca^{2+} inward currents and inhibition of K^+ currents and possibly by other currents. Several second messengers such as PKA and PKC are involved.

There is general agreement that SP and NK-1 receptors are involved in the plasticity of nociceptive processing, whereas the involvement of this system in normal nociception is controversial. Responses of dorsal horn neurons to C fiber volleys and to noxious stimulation of skin and deep tissue are enhanced by SP, and receptive fields can show an expansion. Antagonists at NK-1 receptors reduce responses to C fiber volleys and to noxious stimulation of skin and deep tissue, and they attenuate central sensitization. Mice with a deletion of preprotachykinin A have intact responses to mildly noxious stimuli but reduced responses to moderate and intensely noxious stimuli. Mice with a deletion of the gene responsible for the production of NK-1 receptors respond to acutely painful stimuli but lack intensity coding for pain and wind-up (for review, see Willis and Coggeshall 2004).

Neurokinin A

In addition to substance P, neurokinin A (NKA) is found in small DRG cells and in the spinal dorsal horn. NKA is released in the spinal cord upon noxious stimulation. Due to its resistance to enzymatic degradation, NKA spreads throughout the gray matter. Interestingly, it has not so far been possible to identify NK-2 receptors immunohistochemically in the dorsal horn, leading to the unresolved question as to where NKA acts. It was proposed that NKA activates NK-1 receptors, because these are internalized after application of both SP and NKA. However, specific NK-2 receptor antagonists suggest that specific binding sites for NKA should be present.

The literature on NKA effects is controversial because some authors found an involvement of NKA in nociceptive processing, whereas others did not. Intrathecal application of NKA produces nocifensive behavior, and iontophoretic application of NKA activates nociceptive and non-nociceptive dorsal horn neurons. NKA facilitates behavioral nociceptive responses to heat stimulation, which are blocked by NK-2 receptor antagonists. While some authors could not antagonize responses of dorsal horn neurons to noxious mechanical stimulation, others were able to show such an effect. Antagonists at NK-2 receptors were able to attenuate central sensitization during knee inflammation and colon inflammation (for review, see Willis and Coggeshall 2004).

Neurokinin B
This peptide is found in synaptic terminals and dendrites in laminae I–III, independent of SP-containing elements and is not contained in dorsal root ganglion cells. NK-3 receptors are found in the most superficial part of the dorsal horn. Their function is unclear.

Calcitonin Gene-Related Peptide (CGRP)
This peptide is found in many small DRG neurons and is often colocalized with substance P. Probably the primary afferent neurons are the only source of CGRP in the dorsal horn. CGRP-containing afferents project mainly to laminae I, II, and V. However, CGRP is also contained in motoneurons. CGRP is released in the spinal cord by electrical stimulation of thin fibers and by noxious mechanical and thermal stimulation. During joint inflammation, the pattern of CGRP release changes in that innocuous stimuli to the joint are sufficient to elicit CGRP release.

CGRP-binding sites are concentrated in lamina I and in the deep dorsal horn. CGRP enhances actions of substance P. It inhibits enzymatic degradation of SP, and it seems to potentiate release of SP. Enhanced Ca^{2+} influx may be important in this respect. CGRP activates nociceptive dorsal horn neurons with a slow time course. Blockade of CGRP effects reduces nociceptive responses and attenuates inflammation-evoked central sensitization. The effects of CGRP are discussed in more detail in the entry ▶ CGRP and Spinal Cord Nociception.

Vasoactive Intestinal Polypeptide (VIP)
▶ Vasoactive intestinal polypeptide is found in small-diameter afferent fibers especially in the sacral spinal cord. Terminals with VIP are concentrated in laminae I, II, V, VII, and X. In addition, neurons with VIP are located in laminae II–IV and X. VIP binding sites are concentrated in laminae I and II. VIP excites nociceptive dorsal horn neurons.

Neurotensin
Neurotensin is located in interneurons in lamina I and the substantia gelatinosa, and neurotensin binding sites are present in the dorsal horn. The peptide excites neurons in the superficial dorsal horn.

Cholecystokinin (CCK)
▶ Cholecystokinin (CCK) is located in DRG neurons and in neurons in several laminae of the dorsal horn. Binding sites reach their highest concentration in laminae I and II. CCK can excite neurons in laminae I–VII, and an antagonist at CCK-B receptors is antinociceptive. Antinociceptive effects of CCK have also been described.

Thyrotropin-Releasing Hormone (TRH)
Thyrotropin-releasing hormone (TRH) is located in the ventral and dorsal horn. Many TRH-containing neurons also contain GABA. TRH facilitates responses of nociceptive neurons to glutamate at NMDA receptors, wind-up, and responses to noxious stimuli.

Corticotropin-Releasing Hormone (CRH)

CRH-immunoreactive fibers are present in the sacral spinal cord (laminae I, V–VII, X, intermediolateral column), and CRH immunostaining is abolished after dorsal rhizotomy. CRH binding sites are found in the superficial dorsal horn.

Pituitary Adenylate Cyclase-Activating Polypeptide (PACAP)

▶ Pituitary adenylate cyclase-activating polypeptide (PACAP) is localized in small DRG cells and many axons in the superficial dorsal horn. PACAP is released after intrathecal capsaicin, and intrathecal PACAP causes nocifensive behavior. Nociceptive dorsal horn neurons are excited by PACAP (for review, see Willis and Coggeshall 2004).

Inhibitory Neuropeptides

Numerous neuropeptides are inhibitory. They may reduce release of transmitters by presynaptic actions or inhibit postsynaptic neurons.

Opioid Peptides

The dorsal horn contains leu-enkephalin and met-enkephalin, ▶ dynorphin, and ▶ endomorphins 1 and 2. Enkephalin-containing neurons are particularly located in laminae I and II and dynorphin-containing neurons in laminae I, II, and V. Endomorphin II has been visualized in terminals of primary afferent neurons in the superficial dorsal horn and in dorsal root ganglia but also in postsynaptic neurons.

▶ Opiate receptors (μ, δ, κ) are concentrated in the superficial dorsal horn, and in particular μ- and δ-receptors are not only located in interneurons but also on primary afferent fibers. The activation of these opiate receptors reduces release of mediators from primary afferents (presynaptic effect). This effect is mediated by inhibition of Ca^{2+} channels. Other opiate receptors are located on intrinsic spinal cord neurons and mediate postsynaptic effects. The activation of a K^+ conductance could be the relevant mechanism. In general, enkephalins are ligands at δ-receptors, endomorphins are ligands at μ-receptors, and dynorphin is a ligand at κ-receptors. However, dynorphin may also activate NMDA receptors. Actions of all opiates are antagonized by naloxone. Specific antagonists at different receptors are available (Waldhoer et al. 2004).

Application of opioids into the dorsal horn reduces responses to (innocuous) and noxious stimulation and responses of neurons to iontophoretic application of excitatory amino acids, showing postsynaptic effects of opioids. Depending on the site of application (superficial or deep laminae), μ-, κ-, or δ-receptor ligands are more or less effective in producing neuronal effects. In addition, many dorsal horn neurons are hyperpolarized by opiates. While agonists at μ- and δ-receptors usually evoke inhibitory effects, dynorphin may produce either inhibitory or excitatory effects.

In addition to these "classical" opiate receptors, nociceptin/orphanin FQ receptors (see ▶ Orphanin FQ) have recently been discovered. These proteins share greater than 90 % sequence identity and about 60 % homology with the classical opiate receptors (Waldhoer et al. 2004). An endogenous ligand at these receptors is ▶ nociceptin. This peptide has similar cellular actions to classical opioid peptides. It causes presynaptic inhibition of glutamate release in the spinal cord and reduces FOS expression in the superficial dorsal horn. However, pronociceptive effects have also been described. A related peptide is nocistatin. At present, it is unknown at which receptor nocistatin acts (for review, see Willis and Coggeshall 2004).

Somatostatin

This peptide is expressed in primary afferent neurons, in dorsal horn interneurons, and in axons that descend from the medulla. ▶ Somatostatin is released mainly in the substantia gelatinosa by heat stimulation. Actions of somatostatin on nociceptive neurons in the dorsal horn are inhibitory. It is an intriguing question as to whether inhibitory somatostatin is released in the spinal cord from primary afferent fibers or from interneurons.

Galanin

This peptide is expressed in a subpopulation of small DRG neurons, and galanin binding sites are also expressed on DRG neurons. Both facilitatory and inhibitory effects of galanin have been described in inflammatory and neuropathic pain states.

Neuropeptide Y

This is normally only expressed at very low levels in DRG neurons, but DRG neurons express Y1 and Y2 receptors. It was proposed that Y1 and Y2 receptors contribute to presynaptic inhibition.

Other Mediators

A number of other mediators influence synaptic transmission in the spinal dorsal horn. Most attention has given to ► NO, ► prostaglandins, and ► neurotrophins. These mediators have actions in non-neuronal tissues as well as in neuronal tissues in the peripheral and central nervous systems including the spinal cord. They play significant roles in pathophysiological pain states. The role of prostaglandins is addressed in another section, and a recent review (Vanegas and Schaible 2001) provides a comprehensive summary. The role of neurotrophins is addressed in the entry ► Spinal Cord Nociception, Neurotrophins. Furthermore, glia activation provides cytokines (see below).

Involvement of Calcium Channels in Release of Transmitter and Postsynaptic Excitability

The release of transmitters is dependent on the influx of Ca^{2+} into the presynaptic ending through ► voltage-dependent calcium channels. In addition, Ca^{2+} also regulates excitability of postsynaptic neurons. High voltage-activated N-type channels, which are mainly located presynaptically but also on the postsynaptic side, and P-/Q-type channels that are located on the presynaptic side are important for nociceptive processing. In particular, blockers of N-type channels reduce responses of spinal cord neurons, and behavioral responses to noxious stimulation of normal and inflamed tissue and blockade of N-type channels can also reduce neuropathic pain. There is some evidence that P-/Q-type channels are mainly involved in the generation of pathophysiological pain states. A role for high voltage-activated L-type channels and low

voltage-activated T-type channels has also been discussed. The role of calcium channels is addressed in detail in the entry ► Calcium Channels in the Spinal Processing of Nociceptive Input.

Activation of Glia

Glial cells can contribute to pain by interacting with neurons. Both astrocytes and microglia undergo changes during pain states which are thought to contribute to the features of chronic pain. These changes are very prominent in neuropathic pain states. Whether, and if, to what extent, they occur in inflammatory pain states is a matter of research. Activation is defined by morphological changes, and this parameter may not be sensitive enough for subtle activation changes (Cao and Zhang 2008; Gosselin et al. 2010; Marchand et al. 2005; McMahon and Malcangio 2009; Milligan and Watkins 2009).

Astrocytes have a housekeeping function under resting conditions. They control ion homeostasis, neurotransmitter reuptake and release, control of blood brain barrier, etc. When they are activated they show increased expression of GFAP, enlarged astrocytic processes, reduced glutamate uptake, and a release of various neuromodulatory molecules (Gosselin et al. 2010). Astrocytic reactions start several days postinjury and are longlasting (Gosselin et al. 2010).

Microglia are resident macrophage-related cells in the central nervous system which generate innate immune responses (Cao and Zhang 2008). In their resting state, they perform a constant immune surveillance. Upon activation they show reduced ramification, a release of proinflammatory factors, and an enhanced expression of proteins such as major histocompatibility complex (MHC) proteins (Gosselin et al. 2010; Marchand et al. 2005). Microglial activation may occur transiently during the early phase of nerve damage but may also be long lasting in some models. It was suggested that microglial activation may be important in the initiation of chronic pain whereas astrocyte activation may be responsible for its long-lasting maintenance (Gosselin et al. 2010).

It is thought that glial cells are activated by a sustained release of neurotransmitters, ATP, neuropeptides, and chemokines, e.g., fractalkine which is cleaved from sensory neurons. Once activated, glial cells release cytokines such as TNF-α, IL-1ß, IL-18, IL-6; prostaglandins; neurotrophins such as BDNF; and chemokines such as CCL2/MCP-1. These mediators can either reduce inhibitory GABAergic currents or sensitize excitatory neurons, and BDNF can reduce the expression of the potassium-chloride exporter KCC2 and, consequently, reverse the polarity of GABA-induced currents leading to excitatory currents (Gosselin et al. 2010; McMahon and Malcangio 2009).

Efferent Functions of the Spinal Cord

The nervous system not only produces joint pain, it also influences the inflammation in the joint. Effects of the nervous system on joints are mediated by (a) primary afferent neurons which release neuropeptides in peripheral tissue (Brain et al. 1992; Schaible et al. 2005), (b) postganglionic sympathetic efferent axons in the joint nerve (Schaible and Grubb 1993; Straub and Cutolo 2001), and (c) neuroendocrine systems, e.g., the hypothalamo-pituitary-adrenal (HPA) axis (Straub and Cutolo 2001; Waldburger and Firestein 2010). Additionally, the parasympathetic nervous system was reported to modify joint inflammation.

The spinal cord is the first interface between afferent and efferent mechanisms. Spinal cord activity can produce neurogenic inflammation (Willis 1999) and regulate neutrophil recruitment (Waldburger and Firestein 2010), and thus evoke acute as well as protracted inflammatory events. Here dorsal root reflexes can be generated (Willis 1999), and the cell bodies of preganglionic sympathetic neurons are located, which form the efferent branch of sympathetic reflexes (Jänig 2005).

Final Conclusions

This review concentrated on spinal cord neurons that are excited by noxious stimuli applied to the peripheral tissue. The significance of particular types of neurons in the transmission of sensory information to supraspinal sites through ascending tracts will be addressed in another section. The modulation of responses of spinal cord cells by descending inhibition and facilitation is described in another section. In addition, the particular aspects concerning deep somatic and visceral pain are also covered elsewhere. As briefly outlined here, the spinal cord is subject to considerable changes during pathophysiological pain states. These changes will also be addressed in other sections. Moreover, the spinal cord is the major site at which antinociceptive compounds work. Again this will be addressed in other sections.

References

Alarcon, G., & Cervero, F. (1990). The effects of electrical stimulation of A and C visceral afferent fibres on the excitability of viscerosomatic neurones in the thoracic spinal cord of the cat. Brain Research, 509, 24–30.
Brain, S. D., Cambridge, H., Hughes, S. R., et al. (1992) Evidence that CGRP contributes to inflammation in the skin and in joints. Annals of the New York Academy of Sciences, 165, 412–419.
Cao, H., & Zhang, Y.-Q. (2008). Spinal glia activation contributes to pathological pain states. Neuroscience and Biobehavioral Reviews, 32, 972–983.
Chung, K., McNeill, D. L., Hulsebosch, C. R., et al. (1989). Changes in dorsal horn synaptic disc numbers following unilateral dorsal rhizotomy. The Journal of Comparative Neurology, 283, 568–577.
Coderre, T. J., Katz, J., Vaccarino, A. L., et al. (1993). Contribution of central neuroplasticity to pathological pain: Review of clinical and experimental evidence. Pain, 52, 259–285.
Dubner, R., & Ruda, M. A. (1992). Activity-dependent neuronal plasticity following tissue injury and inflammation. Trends in Neurosciences, 15, 96–103.
Fitzgerald, M., & Wall, P. D. (1980). The laminar organization of dorsal horn cells responding to peripheral C fibre stimulation. Experimental Brain Research, 41, 36–44.
Fundytus, M. E. (2001). Glutamate receptors and nociception. CNS Drugs, 15, 29–58.
Gosselin, R.-D., Suter, M. R., Ji, R.-R., & Decosterd, I. (2010). Glial cells and chronic pain. The Neuroscientist, 16, 519–531.
Hunt, S. P., Pini, A., & Evan, G. (1987). Induction of c-fos-like protein in spinal cord neurons following sensory stimulation. Nature, 328, 632–634.
Jänig, W. (2005). The integrative action of the autonomic nervous system. Cambridge: Cambridge University Press.

Laird, J. M. A., & Bennett, G. J. (1993). An electrophysiological study of dorsal horn neurons in the spinal cord of rats with an experimental peripheral neuropathy. *Journal of Neurophysiology, 69*, 2072–2085.

Marchand, F., Perretti, M., & McMahon, S. B. (2005). Role of the immune system in chronic pain. *Nature Reviews Neuroscience, 6*, 521–532.

McMahon, S. B., & Malcangio, M. (2009). Current challenges in glia-pain biology. *Neuron, 64*, 46–54.

Mendell, L. M. (1966). Physiologial properties of unmyelinated fiber projection to the spinal cord. *Experimental Neurology, 16*, 316–332.

Mendell, L. M., & Wall, P. D. (1965). Responses of single dorsal cord cells to peripheral cutaneous unmyelinated fibers. *Nature, 206*, 97–99.

Menétrey, D., Gannon, J. D., Levine, J. D., et al. (1989). Expression of c-fos protein in interneurons and projection neurons of the rat spinal cord in response to noxious somatic, articular, and visceral stimulation. *The Journal of Comparative Neurology, 285*, 177–195.

Mense, S. (1993). Nociception from skeletal muscle in relation to clinical muscle pain. *Pain, 54*, 241–289.

Millan, M. J. (1999). The induction of pain: An integrative review. *Progress in Neurobiology, 57*, 1–164.

Milligan, E. D., & Watkins, L. R. (2009). Pathological and protective roles of glia in chronic pain. *Nature Review Neuroscience, 10*, 23–36.

Palacek, J., Paleckova, V., Dougherty, P. M., et al. (1992a). Responses of spinothalamic tract cells to mechanical and thermal stimulation of skin in rats with experimental peripheral neuropathy. *Journal of Neurophysiology, 67*, 1562–1573.

Palacek, J., Dougherty, P. M., Kim, S. H., et al. (1992b). Responses of spinothalamic tract neurons to mechanical and thermal stimuli in an experimental model of peripheral neuropathy in primates. *Journal of Neurophysiology, 68*, 1951–1966.

Polgár, E., Gray, S., Riddell, J. S., et al. (2004). Lack of evidence for significant neuronal loss in laminae I-III of the spinal dorsal horn of the rat in the chronic constriction injury model. *Pain, 111*, 144–150.

Price, D. D., Mao, J., Coghill, R. C., et al. (1991). Regional changes in spinal cord glucose metabolism in a rat model of painful neuropathy. *Brain Research, 564*, 314–318.

Price, D. D., Greenspan, J. D., & Dubner, R. (2003). Neurons involved in the exteroceptive function of pain. *Pain, 106*, 215–219.

Randic, M., Jiang, M. C., & Cerne, R. (1993). Long-term potentiation and long-term depression of primary afferent neurotransmission in the rat spinal cord. *Journal of Neuroscience, 13*, 5228–5241.

Rexed, B. (1952). The cytoarchitectonic organization of the spinal cord in the rat. *The Journal of Comparative Neurology, 96*, 415–466.

Rexed, B. (1954). A cytoarchitectonic atlas of the spinal cord in the cat. *The Journal of Comparative Neurology, 100*, 297–380.

Rygh, L. J., Svendson, F., Hole, K., et al. (1999). Natural noxious stimulation can induce long-term increase of spinal nociceptive responses. *Pain, 82*, 305–310.

Sandkühler, J., & Liu, X. (1998). Induction of long-term potentiation at spinal synapses by noxious stimulation or nerve injury. *European Journal of Neuroscience, 10*, 2476–2480.

Sandkühler, J., Chen, J. G., Cheng, G., et al. (1997). Low-frequency stimulation of afferent AÎ′-fibers induces long-term depression at primary afferent synapses with substantia gelatinosa neurons in the rat. *Journal of Neuroscience, 17*, 6483–6491.

Schadrack, J., Neto, F. L., Ableitner, A., et al. (1999). Metabolic activity changes in the rat spinal cord during adjuvant monoarthritis. *Neuroscience, 94*, 595–605.

Schaible, H.-G., Del Rosso, A., & Matucci-Cerinic, M. (2005). Neurogenic aspects of inflammation. *Rheumatic Diseases Clinics North America, 31*, 77–101.

Schaible, H.-G., & Grubb, B. D. (1993). Afferent and spinal mechanisms of joint pain. *Pain, 55*, 5–54.

Scheibel, M. E., & Scheibel, A. B. (1968). Terminal axon patterns in cat spinal cord: II. The dorsal horn. *Brain Research, 9*, 32–58.

Sivilotti, L. G., Thompson, S. W. N., & Woolf, C. J. (1993). The rate of rise of the cumulative depolarization evoked by repetitive stimulation of small-calibre afferents is a predictor of action potential windup in rat spinal neurons in vitro. *Journal of Neurophysiology, 69*, 1621–1631.

Straub, R. H., & Cutolo, M. (2001). Involvement of the hypothalamic-pituitary-adrenal/gonadal axis and the peripheral nervous system in rheumatoid arthritis. *Arthritis Rheumatism, 44*, 493–507.

Sugiura, Y., Lee, C. L., & Perl, E. R. (1986). Central projections of identified, unmyelinated (C) afferent fibres innervating mammalian skin. *Science, 234*, 358–361.

Sugiura, Y., Terui, N., & Hosoya, Y. (1989). Difference in the distribution of central terminals between visceral and somatic unmyelinated (C) primary afferent fibres. *Journal of Neurophysiology, 62*, 834–840.

Svendsen, F., Tjolsen, A., & Hole, K. (1997). LTP of spinal A δ and C-fibre evoked responses after electrical sciatic nerve stimulation. *Neuroreport, 8*, 3427–3430.

Urban, L., Thompson, S. W. N., & Dray, A. (1994). Modulation of spinal excitability: Cooperation between neurokinin and excitatory amino acid transmitters. *Trends in Neurosciences, 17*, 432–438.

Vanegas, H., & Schaible, H.-G. (2001). Prostaglandins and cyclooxygenases in the spinal cord. *Progress in Neurobiology, 64*, 327–363.

Waldburger, J.-M., & Firestein. G. S. (2010). Regulation of peripheral inflammation by the central nervous system. *Current Rheumatology Reports, 12*, 370–378.

Waldhoer, M., Bartlett, S. E., & Whistler, J. L. (2004). Opioid receptors. *Annual Review of Biochemistry, 73*, 953–990.

Williams, S., Ean, G. L., & Hunt, S. P. (1990). Changing pattern of c-fos induction following thermal cutaneous stimulation in the rat. *Neuroscience, 36*, 73–81.

Willis, W. D., Jr. (1999). Dorsal root potentials and dorsal root reflexes: A double-edged sword. *Experimental Brain Research, 124*, 395–421.

Willis, W. D., & Coggeshall, R. E. (2004). *Sensory mechanisms of the spinal cord* (3rd ed., Vol. 1). New York: Kluwer/Plenum.

Woolf, C. J. (1983). Evidence for a central component of post-injury pain hypersensitivity. *Nature, 306*, 686–688.

Woolf, C. J., & King, A. E. (1987). Physiology and morphology of multireceptive neurons with C-afferent inputs in the deep dorsal horn of the rat lumbar spinal cord. *Journal of Neurophysiology, 58*, 460–479.

Nociceptive Processing in the Thalamus

A. Vania Apkarian
Department of Physiology, Northwestern
University Feinberg School of Medicine,
Chicago, IL, USA

Characteristics

Spinothalamic Targets

The differential projections for innocuous and nociceptive inputs to the thalamus were first noted when the functional distinctions between the dorsal columns and the anterolateral tract were observed in humans following cordotomies, over 100 years ago. The terminations of the spinothalamic pathway still remain controversial. There is disagreement as to the terminations of the pathway in the lateral and medial thalamus. In 1980s, the ▶ nucleus submedius (SM) in the medial thalamus was claimed to be the nociceptive-specific nucleus of the thalamus (Craig and Burton 1981), since it was thought to receive inputs only from spinal cord lamina I nociceptive neurons. This idea has been mostly discounted, at least in the rat and monkey. However, the primary medial thalamic termination site, which has traditionally been reported to be in ▶ MD, has been put into doubt in recent reports as well; instead it is claimed that the main spinothalamic input is to ▶ CL. Lateral thalamic terminations have classically been reported to target VP, VPI, and PO and there is ample

evidence for this idea. However, this was recently challenged by the claimed existence of another nociception, thermoception, and itch-specific nucleus ▶ VMpo within the lateral thalamus, which seems to receive spinal cord and trigeminal lamina I inputs somatotopically and is most evident in the monkey and man (Craig et al. 1994). This notion has recently been questioned as well, by demonstrating that VMpo may simply be part of the periphery of ▶ VPM. At one level, these disputes seem simply a consequence of disagreements over how one delineates various borders of thalamic nuclei, which are invariably ambiguous and poorly defined, especially the transition zones between nuclei. On the other hand, they reflect philosophical differences as to the rules of the organization of the central nervous system regarding pain perception. The claim of the existence of nociceptive-specific nuclei in the thalamus has the consequence of implying that these regions together with their inputs and output targets constitute the pain-specific network of the brain. This claim in turn denies the contribution of other thalamic nuclei and their inputs and outputs to pain perception (see ▶ Spinothalamocortical Projections from SM).

Periphery of VP, Rod, and Matrix of VP and VMpo

Staining thalamic tissue for ▶ cytochrome oxidase (CO) showed that lateral thalamic somatosensory regions, VP and its surrounding nuclei, can be subdivided into two compartments, a region densely labeled with CO and a surrounding periphery (VPp) labeled weakly with CO. This observation was first made in Kniffki's lab for the cat thalamus, in the late 1980s. Independently, E.G. Jones noted the same CO staining-based distinction in the monkey lateral thalamus and extended the parcellation by combining CO staining with two calcium binding proteins (▶ Calbindin and ▶ parvalbumin), which preferentially label the CO dense and CO sparse domains of ▶ VP and its periphery. Jones advanced the hypothesis that the CO dense region, dubbed the rod region, was involved in sensory information transmission to the cortex by terminating in layer IV, while the

CO weakly labeled region, dubbed the matrix region, projected mainly to layer I of the cortex and was involved in modulating cortical responses. A subsequent study did not substantiate this idea (Shi et al. 1993). Kniffki and colleagues hypothesized, mainly by relating terminations of the spinothalamic tract to the CO sparse region of the cat VP, that neurons in the VP periphery (VPp) are mainly nociceptive, while the CO dense region receives inputs from the medial lemniscus and signals innocuous somatosensory information to the cortex. A series of studies indicated that the cat VP does not have nociceptive cells. However, the opposite claim has been harder to prove since the number of nociceptive cells characterized in the VPp of the cat has remained small (Horn et al. 1999; Martin et al. 1990). To further explore the segregation of VP and its periphery in the monkey, electrophysiological studies contrasted nociceptive neurons in the squirrel monkey VP and VPI (equivalent to VPp in the cat) and observed that most VP nociceptive cells were WDR type, while VPI nociceptive cells were both NS and WDR types. Given the distinct white matter localization of spinal cord lamina I vs. deeper spinothalamic cells and the evidence that most NS-type cells in the spinal cord are localized to lamina I, it was proposed that the monkey periphery of VP preferentially or exclusively receives inputs from spinal cord lamina I cells, while the VP proper receives inputs from deeper laminae spinothalamic cells. This claim remains unproven and does not match with earlier reports (in the macaque) that VP nociceptive cells are of both NS and WDR types; it also does not match with the prevalences of NS- and WDR-type spinothalamic cells in the monkey, which do not seem to differ between spinal cord laminae. Craig and colleagues extended these ideas to the extreme and proposed the existence of VMpo, a unique lateral thalamic nucleus that provides pain-specific inputs to the insular cortex. The corollary to this idea is that nociceptive inputs to VP proper are modulatory in nature and do not signal nociceptive sensory information. This idea has been criticized by various groups and remains controversial. Human imaging shows robust

activation of SII and posterior insula and more variable responses in SI for experimental pain in healthy humans. A recent meta-analysis, however, indicates no significant difference in activation incidence between these structures (Apkarian et al. 2005). Even if one assumes that posterior insular activity is due to inputs from VMpo, nociceptive inputs to SI and SII are undoubtedly from VP and VPI and hence nociceptive inputs to the latter nuclei participate in the cortical activity associated with pain perception (see essays ▶ Corticothalamic and Thalamocortical Interactions; ▶ Spinothalamic Terminations, Core and Matrix; ▶ Thalamic Nuclei Involved in Pain, Human and Monkey; ▶ Spinothalamic Projections in Rat; ▶ Spinothalamic Input: Cells of Origin (Monkey); ▶ Thalamic Nuclei Involved in Pain: Cat and Rat; ▶ Spinothalamocortical Projections to Ventromedial and Parafascicular Nuclei; ▶ Thalamus, Nociceptive Cells in VPI, Cat and Rat).

Generally, this debate is the most modern version of specificity vs. pattern theories of coding for pain in the nervous system, a debate that goes back to Helmholz, Von Frey, and Goldsheider, who were battling pain representation models based on the discovery of punctate receptive fields on the human skin. In the 1960s, the debate moved from psychophysics to the properties of spinal cord neurons and the contribution of nociceptive specific vs. wide dynamic range-type cells to pain perception. One group of scientists staked the claim that the wide dynamic range neurons were necessary and sufficient for pain perception, while another group claimed that nociceptive-specific cells were all that was needed for pain perception. There was probably always a silent majority of scientists who simply accepted that both types of neurons are involved in pain perception and that these cell types complement each other in the range of stimuli that could evoke pain perception. Thus, VMpo and its connectivity represent the latest effort in pinpointing a specific group of cells from the skin to the cortex that uniquely signal pain from neurons that are specifically involved in coding noxious, thermal, and itch stimuli. The opponents of this idea question whether VMpo exists as a unique thalamic nucleus and argue that there is ample evidence that other spinal cord neurons, thalamic nuclei, and cortical regions have repeatedly been demonstrated to have all the necessary characteristics to encode nociceptive information.

Brainstem Inputs

Spinoreticulothalamic projections, which are pathways conveying spinal cord inputs to different brainstem targets and then in turn projecting to the thalamus, have been studied best in the rat. Some of these pathways seem specifically involved in conveying whole-body nociceptive information to the thalamus. In the medullary brainstem, the subnucleus reticularis dorsalis (SRD) seems to be one of the main nociceptive relays to the thalamus. Neurons from the SRD project to VMl and PF in the ▶ medial thalamus and neurons in VMl in turn project to the layer I of the ventrolateral cortex, while projections from PF project to the ▶ basal ganglia, the subthalamic nucleus, and parts of the motor and parietal cortex (see ▶ Brainstem Subnucleus Reticularis Dorsalis Neuron).

Another brainstem-thalamic projection, described in the rat, is through the internal lateral parabrachial nucleus, which receives nociceptive inputs from deep laminae of the spinal cord and projects to medial thalamic nuclei PC, CM, and PF. Traditionally, CM and PF have been grouped together and implicated in affective modulation of pain. A large proportion of PF cells respond to nociceptive stimuli and stimulation in the region evokes pain-like reports in humans and pain-like behavior in animals. In humans, medial thalamic stimulation or lesioning has been used for pain relief and has targeted the CM-PF region. Such procedures report a fair incidence of success. PF and amygdala receive serotonergic inputs from ventral PAG and the three structures seem to interact in modulating the affective component of pain. Suppression of rats' affective reaction to noxious stimuli by injection of morphine into the ventral PAG is reversed by serotonin antagonists applied either to PF or amygdala (see ▶ Parafascicular Nucleus, Pain Modulation, ▶ Thalamus, Nociceptive Inputs in the Rat (Spinal), ▶ Thalamo-Amygdala Interactions and Pain).

Spinothalamocortical Connectivity

Although the suggested role of the nucleus submedius (SM) in nociception has diminished over the years, its projection targets show that SM neurons (in the rat) terminate in the ventrolateral orbital cortex (VLO), a region where nociceptive neurons have been described in the rat. The VLO nucleus also receives inputs from the ventral ▶ periaqueductal gray (PAG) and dorsal raphe. A similar but smaller brainstem projection seems to exist for the SM as well. The rat SM does not receive spinal cord lamina I inputs as originally claimed; instead its inputs are from deeper laminae. Although a nucleus equivalent to SM was originally described in the monkey, this claim has been repeatedly refuted (see ▶ Thalamus, Nociceptive Inputs in the Rat (Spinal); ▶ Spinothalamocortical Projections from SM).

Potential connectivity between thalamocortical projecting neurons and spinothalamic terminations has been studied probabilistically at the light microscopy level (Gingold et al. 1991). Hand primary somatosensory cortex projecting cells are labeled retrogradely with a given marker and spinothalamic terminations from the upper cervical enlargement labeled anterogradely with a different marker. Given the known dendritic branching pattern of thalamocortical cells, one can then calculate how many of these cells can potentially receive spinothalamic inputs on their dendritic perimeter. Although such an analysis cannot establish the presence or absence of synapses, it provides quantitative bounds as to the influence of nociceptive inputs through the spinothalamic projections to the cortical target. The analysis shows that 87 % of cervical enlargement spinothalamic terminations are localized to VPL, VPI, and CL and 24 % of the hand region of the primary somatosensory cortex is putatively contacted by these spinothalamic terminations. A more recent study examined connectivity between thalamic cells and spinothalamic afferents by intracellularly labeling individual thalamic neurons and examining the proximity of the labeled dendrites to spinothalamic terminals

(Shi and Apkarian 1995). A similar study has also been done electrophysiologically and indicates that the probability of encountering spinothalamic terminations in the vicinity of nociceptive cells in VP is 33 %, while in VPI it is 73 % (Apkarian and Shi 1994).

Synaptic Connectivity

Synaptic morphology, using electron microscopy, has been studied for spinothalamic inputs and contrasted to dorsal column inputs in VP (Ralston 2003). The study examined synapses for spinothalamic projections onto VP proper cells and contrasted them to medial lemniscal synapses. Spinothalamic synapses were found to be mainly on dendrites of projecting cells, in contrast to medial lemniscal synapses that formed triads between terminals and projecting and local GABAergic cells. This distinction between afferent input types and synapses is congruent with physiological connectivity differences observed for VP cell groups with and without nociceptive inputs (see ► Thalamus, Dynamics of Nociception). More recently, similar electron microscopic studies were also done for spinal cord lamina I inputs to VMpo (Beggs et al. 2003). The synaptic profiles in VMpo were almost always triadic. Thus, the lamina I spinothalamic inputs to VMpo are different from spinothalamic inputs to VP. The difference is partly attributed to the targets and partly to inputs. Most likely, most of the inputs examined in VP reflected terminations from deeper laminae than lamina I. These differences are consistent with the light microscopic observation regarding proximity of spinothalamic terminations to nociceptive cells in VP vs. VPI.

Species Differences

There are important species differences regarding the spinothalamic pathway, its terminations in the thalamus and response properties of thalamic nociceptive cells. The rat spinothalamic tract (see ► Spinothalamic Projections in Rat) is composed of a third fewer cells than that of the monkey. Rat lateral thalamic nuclei are devoid of local interneurons. Thus, one can assert that terminations in these nuclei are synapses on cortex projecting cells. Also, the rat VP was not traditionally subdivided into a core and a periphery; as a result, there is no doubt that some spinothalamic terminals are on VP cells that project to the primary or secondary somatosensory cortex. Rat spinothalamic terminations in the medial thalamus generally target the same nuclei as in the cat or monkey, perhaps with the exception of SM, which does not seem to receive spinal cord lamina I inputs. In the spinal cord, the lateral cervical nucleus (LCN) seems more prominent in the rat and these cells do project to the thalamus, although their functional role has remained unclear.

More recent studies have unraveled differences in terminations of superficial laminar spinal cord cells and deeper ones as to their targets in the rat thalamus. Lamina I inputs seem to be limited to lateral thalamic targets, where the region is subdivided into VP, VPpc, Po, and PoT, where VPpc and PoT probably correspond to different portions of VP periphery. Direct deeper laminae projections in the rat seem to be limited to PoT and CL (see ► Thalamus, Nociceptive Inputs in the Rat (Spinal) ► Spinothalamic Projections in Rat).

Neurotransmitters and Neuromodulators

Like other sensory inputs to the thalamus, glutamatergic neurotransmission is assumed to transmit nociceptive information. Although such transmission has been demonstrated for other sensory modalities, as in somatosensory transmission, it is not proven in the case of nociception. The thalamocortical efferent pathway is made of neurons that are glutamatergic. Cortical inputs to the thalamus seem to be mediated through ► AMPA and mGLU receptors. ► GABAergic inhibitory interneurons and GABAergic thalamic reticular nucleus innervation provide inhibition on projecting neurons through ► GABAA and ► GABAA receptors. Most current knowledge regarding nociceptive neurotransmission is based on studies of VB neuronal properties and indicates that ► NMDA receptors signal acute thermal and mechanical responses. There is also evidence that thalamic mGLU receptor mechanisms are important in inflammation-induced hyperalgesia and in the expression of such behavior (see ► Metabotropic Glutamate Receptors in the Thalamus).

Eight metabotropic glutamate receptor subtypes (mGLU1–mGLU8) have been characterized, are divided into three groups (I, II, III), and are all found in the thalamus. Group I receptors mediate mainly postsynaptic actions, while Groups II and III regulate presynaptic transmitter release. Different subtypes are important in either corticothalamic inputs or thalamic reticular neuronal modulation of GABAergic transmission. Thus, mGLU receptors have a minimal role in ascending sensory transmission but are more important in modulating this transmission (see ► Thalamus, Nociceptive Neurotransmission).

Spinothalamic neurons contain glutamate and various ► neuropeptides. Of the large list of neuropeptides seen in spinothalamic cells, ► substance P (SP) is found abundantly in the medial thalamus, some of which may be due to spinothalamic inputs. Cholinergic, ► serotonergic, and ► noradrenergic inputs from the brainstem are also found in the thalamus, all of which are probably part of the arousal modulatory system. Intrinsic SP neurons are described in thalamic regions receiving spinothalamic inputs. ► CGRP neurons are found in the periphery of VP and described as projecting to the amygdala or insula (see ► Thalamus, Visceral Representation ► Parafascicular Nucleus, Pain Modulation ► Thalamic Neurotransmitters and Neurochemical Effector Molecules).

Somatic Representation

Nociceptive representation in the thalamus has been described in multiple species, primarily in rat, cat, and monkey and mainly in anesthetized preparations. The overall number of nociceptive cells described remains relatively small and the properties of cells located in medial in contrast to lateral thalamic nuclei seem distinct. Nociceptive cells found in the medial thalamic nuclei tend to have more nociceptive-specific responses, with response patterns that are modulated with the sleep-wakefulness cycle, level of anesthesia, and attentional manipulations. In contrast,

nociceptive cells in the lateral thalamus have more divergent inputs; they can be nociceptive specific or wide dynamic range types, with response properties that seem more reproducible and less dependent on attentional manipulations. The receptive field size, location, and properties seem labile in medial thalamic cells, while in lateral thalamic cells they seem more constant and correspond to the properties of similar cells described in the spinal cord or trigeminal nuclei. These response differences are generally consistent with the notion that medial thalamic nociceptive information may be providing cortical signals regarding the affective properties of pain and also providing a more general modulatory signal that may be important in biasing the cortex and in modulating the attentional circuitry of the cortex. In contrast, lateral thalamic nociceptive signals are consistent with the general idea that these are the sensory-discriminative information being transmitted to cortical regions specifically involved in pain perception. Even though both notions may generally be true, there are important deviations regarding input-output properties and response properties of nociceptive cells in specific thalamic nuclei, implying that the medial vs. lateral thalamic functional differentiation is most probably too simplistic (see ► Parafascicular Nucleus, Pain Modulation ► Thalamus, Nociceptive Inputs in the Rat (Spinal) ► Spinothalamic Input: Cells of Origin (Monkey)).

References

Apkarian, A. V., & Shi, T. (1994). Squirrel monkey lateral thalamus I. Somatic nociresponsive neurons and their relation to spinothalamic terminals. *The Journal of Neuroscience, 14*, 6779–6795.

Apkarian, A. V., Bushnell, M. C., Treede, R. D., et al. (2005). Human brain mechanisms of pain perception and regulation in health and disease. *European Journal of Pain, 9*, 463–484.

Beggs, J., Jordan, S., Ericson, A. C., et al. (2003). Synaptology of trigemino- and spinothalamic lamina I terminations in the posterior ventral medial nucleus of the macaque. *The Journal of Comparative Neurology, 459*, 334–354.

Craig, A. D., Jr., & Burton, H. (1981). Spinal and medullary lamina I projection to nucleus submedius in medial thalamus: A possible pain center. *Journal of Neurophysiology, 45*, 443–466.

Craig, A. D., Bushnell, M. C., Zhang, E. T., et al. (1994). A thalamic nucleus specific for pain and temperature sensation. *Nature, 372*, 770–773.

Gingold, S. I., Greenspan, J. D., & Apkarian, A. V. (1991). Anatomic evidence of nociceptive inputs to primary somatosensory cortex: Relationship between spinothalamic terminals and thalamocortical cells in squirrel monkeys. *The Journal of Comparative Neurology, 308*, 467–490.

Horn, A. C., Vahle-Hinz, C., Bruggemann, J., et al. (1999). Responses of neurons in the lateral thalamus of the cat to stimulation of urinary bladder, colon, esophagus, and skin. *Brain Research, 851*, 164–174.

Martin, R. J., Apkarian, A. V., & Hodge, C. J., Jr. (1990). Ventrolateral and dorsolateral ascending spinal cord pathway influence on thalamic nociception in cat. *Journal of Neurophysiology, 64*, 1400–1412.

Ralston, H. J., III. (2003). Pain, the brain, and the (calbindin) stain. *The Journal of Comparative Neurology, 459*, 329–333.

Shi, T., & Apkarian, A. V. (1995). Morphology of thalamocortical neurons projecting to the primary somatosensory cortex and their relationship to spinothalamic terminals in the squirrel monkey. *The Journal of Comparative Neurology, 361*, 1–24.

Shi, T., Stevens, R. T., Tessier, J., et al. (1993). Spinothalamocortical inputs nonpreferentially innervate the superficial and deep cortical layers of SI. *Neuroscience Letters, 160*, 209–213.

Nociceptive Projecting Neurons

▶ Spinothalamic Tract Neurons, Visceral Input

Nociceptive Reflex

Definition

A reflex that is elicited by noxious stimuli. It is mediated by nociceptive (Aδ, C) afferents and exerts a defense reaction.

Cross-References

▶ Jaw-Muscle Silent Periods (Exteroceptive Suppression)
▶ Nociceptive Withdrawal Reflex

Nociceptive Selective Stimulation

Definition

Stimulation which selectively activates nociceptive (Aδ, C) fibers without simultaneously activating tactile-mediated large myelinated Aβ fibers, for example, laser stimulation.

Cross-References

▶ Insular Cortex, Neurophysiology, and Functional Imaging of Nociceptive Processing

Nociceptive Sensory Neurons

Definition

Nociceptive sensory neurons are specialized sensory nerve fibers innervating peripheral tissues that are normally only activated by noxious stimuli (i.e., the stimuli capable of causing tissue damage) and not innocuous stimuli.

Cross-References

▶ Inflammation, Modulation by Peripheral Cannabinoid Receptors
▶ Nociceptors
▶ Nociceptors, Action Potentials, and Post-Firing Excitability Changes

Nociceptive System

Definition

Peripheral, spinal, and cerebral neurons and neuronal systems involved in the processing of noxious stimuli.

Cross-References

▶ Noxious Stimulus
▶ Primary Somatosensory Cortex (S1), Effect on Pain-Related Behavior in Humans

Nociceptive Temporomandibular Joint Afferents

▶ Nociceptors in the Orofacial Region (Temporomandibular Joint and Masseter Muscle)

Nociceptive Threshold

Definition

Nociceptive thresholds are usually defined as the temperature or mechanical force at which a withdrawal response is evoked, measured either by active withdrawal of a limb or the tail. In an anesthetized subject, measurement of the electrical activity of a muscle can be used.

Cross-References

▶ Arthritis Model, Adjuvant-Induced Arthritis

Nociceptive Transduction

Definition

Generation of the nociceptive information in nociceptors in response to noxious stimuli by generation of depolarizing currents.

Cross-References

▶ Nociceptor Generator Potential
▶ NSAIDs, Mode of Action

Nociceptive Transmission

▶ Metabotropic Glutamate Receptors in Spinal Nociceptive Processing

Nociceptive Withdrawal Reflex

Definition

Nociceptive withdrawal reflexes denote an integrated reflex to avoid potential tissue injury. The reflex response is dependent on stimulus site, stimulus intensity, and functional context. During standardized experimental conditions, the reflex is correlated to the pain intensity.

Cross-References

▶ Pain in Humans, Electrical Stimulation (Skin, Muscle, and Viscera)

Nociceptive-Specific Neurons

Synonyms

NS neuron

Definition

Nociceptive-specific sensory neurons are a type of nociceptive neuron, which are excited by stimulus intensities that are sufficiently intense to cause injury to the body if sustained for a sufficiently long period of time. In the absence of injury, this type of neuron will not normally respond to innocuous thermal or innocuous mechanical stimulation of its receptive field. This type of neuron is defined physiologically by its response properties and can be found at spinal, thalamic, and cortical sites important in the processing of noxious information.

Cross-References

Nociceptive-Specific Neurons (NS)

Nociceptor Accommodation

Nociceptor Desensitization

Definition

Decreased sensitivity to noxious stimuli elicited by application of capsaicin. Short-term desensitization is due to the inactivation of TRPV1, preventing the generation of action potentials. On the other hand, long-term desensitization evoked by large doses of capsaicin onto polymodal nociceptor endings is due to the destruction of a subset of small-diameter primary afferent fibers and their cell bodies.

Cross-References

Nociceptor Generator Potential

Carlos Belmonte
Instituto de Neurociencias, Universidad Miguel Hernandez-CSIC, San Juan de Alicante, Spain

Synonyms

Generator current; Receptor potential

Definition

The local change in ▶ membrane potential caused by the opening of ion channels in the peripheral terminals of nociceptor neurons when natural stimuli (mechanical, thermal, chemical) activate their transduction mechanisms.

Characteristics

Transduction of natural stimuli by the specialized membrane of sensory receptor cells leads ultimately to the opening and closing (gating) of ion channels and to the generation of local electrical signals. These were already described in early studies using invertebrate stretch receptor cells (Eyzaguirre and Kufler 1955) as well as in the receptor cells of specialized sensory organs in mammals, such as the cochlear hair cells, the olfactory neurons, or the retinal photoreceptors, taking advantage of their accessibility to direct electrophysiological recording (for review, see Gardner and Martin 2000). In all cases, the gating of transduction channels triggered by the stimulus caused charge transfer across the membrane and a gradual depolarization (or hyperpolarization) of an amplitude proportional to the intensity of the stimulus called the "receptor or generator potential."

The receptor endings of primary sensory neurons in mammals are not easily accessible to the conventional biophysical methods that were successfully applied to the cell soma. An indirect approach aimed at recording the small current flows associated with the opening of transduction channels in sensory endings was made in the middle of the twentieth century using the Pacinian corpuscle, a specialized mechanoreceptor that responds to very low mechanical forces and can be easily visualized and isolated from the mesentery of the cat. The pacinian corpuscle is formed by the nerve terminal of a large myelinated sensory axon surrounded by a number of concentric lamellae, which loses its myelin sheath and Schwann cells upon entering the corpuscle, running subsequently as a straight, bare nerve ending. In a series of classical studies (Gray and Sato 1953; Loewenstein 1959), the pacinian corpuscle isolated "in vitro" was employed to record extracellularly the ▶ membrane potential changes associated with mechanical stimulation of the corpuscle surface, establishing that in the most distal part of the nerve terminal, the stimulus evoked a flow of generator current of an amplitude proportional to the magnitude of indentation (Fig. 1). ▶ Generator currents could summate temporally and propagate electrotonically (i.e., declining exponentially with distance) along the length of the axon. When the amplitude of the depolarization reaching the first node of Ranvier (located inside the corpuscle near to the point of entrance of the nerve) attained a critical level, an ▶ action potential was generated (Fig. 1). The conclusion of these and subsequent studies was that the generator process in sensory receptor fibers takes place in a terminal portion of the nerve membrane that is not electrically excitable, i.e., cannot support a regenerative change in sodium conductance, and that the process of transduction that leads to the generator potential is spatially separated from the point where propagated action potentials are produced. The firing frequency and the duration of the impulse discharge are proportional to the amplitude and duration of the generator potential (for details, see Patton 1966; Gardner and Martin 2000).

It has been speculated that the sequence of phenomena observed in the transduction process of mechanoreceptor fibers is general to all mammalian sensory receptor endings, including nociceptive terminals, where the transduction

channels opened by mechanical, thermal, or chemical stimuli are thought to generate a local receptor or generator potential that will ultimately lead to impulse firing in the parent axon (Belmonte 1996). However, direct evidence for this extrapolation is still lacking, due to the difficulty of applying the intracellular or extracellular recording techniques used in other receptor classes to nociceptive nerve fibers intimately embedded in their surrounding tissues and with a diameter below 1 μ.

In recent years, new technical approaches have offered indirect evidence of the presence of generator currents in nociceptive terminals. Extracellular activity of single ▶ nociceptive nerve endings was successfully recorded in the cornea of the eye (Fig. 2) using a large tip microelectrode tightly applied against the corneal surface (Brock et al. 1998). Terminal sensory branches run between corneal epithelium cells ending close to the most superficial epithelium layers; the high resistance seal formed by the electrode allowed the recording of the spontaneous and stimulus-evoked propagated impulse activity of the ending located immediately below the electrode tip. In these experiments, a depolarization preceding the propagated nerve terminal impulse, suggestive of a local generator current, was occasionally observed in polymodal nociceptor endings (Brock et al. 1998). Likewise, in single nociceptive fibers of the rat skin, Sauer et al. (2005) using the "threshold tracking technique" reported that heat and ▶ bradykinin stimulation of polymodal nociceptor endings was preceded by a reduction of threshold suggestive of a local depolarization presumably corresponding to the generator potential.

Nevertheless, the location in nociceptors of the transformation site where generator currents give rise to propagated impulses when the depolarization exceeds threshold remains unknown. It has been suggested, based on morphological evidence, that the patches of axonal membrane devoid of Schwann cell coating observed in the terminal portion of knee joint nociceptors correspond to the

Nociceptor Generator Potential, Fig. 1 (**a**) The pacinian corpuscle. (**b**) Generator potentials recorded extracellularly in a pacinian corpuscle, "in vitro" *a*, *b*, and *c*, generator potentials (*e*) elicited by mechanical compressions of increasing magnitude shown in *m*. *d*. Short mechanical pulses elicit both "on" and "off" responses which sum (**a**, From Schmidt and Thews 1990; **b**, After Gray 1953)

▶ transduction sites where receptor potentials would be generated (Heppelmann et al. 1990) while action potentials occur at a more central point. However, in corneal polymodal nociceptor endings, it has been shown that the most distant portion of the terminal possesses sufficient density of ▶ tetrodotoxin-resistant

sodium channels to sustain propagated action potentials (Brock et al. 1998), a characteristic that is not shared by cold-sensitive nerve fibers, whose peripheral terminals seem to lack regenerative properties (Brock et al. 2001). Thus, the possibility exists that in nociceptor terminals, the ion channels responsible for

Nociceptor Generator Potential, Fig. 2 Recording of nerve terminal impulses in the guinea pig cornea. Schematic diagram of recording setup (**a**) and photomicrograph (**b**) showing the location of the recording electrode. (**c**) Confocal micrograph of nerve terminals in the cornea. (**d**) Averages of spontaneously occurring nerve terminal impulses recorded from a mechano-nociceptor (*upper trace*) and a polymodal nociceptor (*lower trace*)

generator currents and those sustaining the production of propagated action potentials are not spatially segregated. This may have a functional significance in the profusely ramified nociceptor fibers. Action potentials originated at the peripheral endings by a direct action of the stimulus also propagate antidromically and invade terminals of the same parent axon that were not directly excited by the stimulus. A large proportion of nociceptor terminals contain vesicles filled with ▶ neuropeptides (CGRP, SP) that are released by the entrance of calcium ions driven by the invading antidromic action potential, thereby contributing to neurogenic inflammation.

References

Belmonte, C. (1996). Signal transduction in nociceptors: General principles. In C. Belmonte & F. Cervero (Eds.), *Neurobiology of nociceptors* (pp. 243–257). Oxford: Oxford University Press.

Brock, J., McLachlan, E. M., & Belmonte, C. (1998). Tetrodotoxin-resistant impulses in single nerve terminals signalling pain. *The Journal of Physiology, 512*, 211–217.

Brock, J. A., Pianova, S., & Belmonte, C. (2001). Differences between nerve terminal impulses of polymodal nociceptors and cold sensory receptors of the guinea-pig cornea. *The Journal of Physiology, 533*, 493–501.

Eyzaguirre, C., & Kuffler, S. W. (1955). Processes of excitation in the dendrites and in the soma of single isolated sensory nerve cells of the lobster and crayfish. *Journal of General Physiology, 39*, 87–119.

Gardner, E. P., & Martin, J. H. (2000). Coding of sensory information. In E. Kandel, J. H. Schwartz, & T. M. Jessell (Eds.), *Principles of neural sciences* (4th ed., pp. 411–625). New York: McGraw-Hill.

Gray, J. A. B., & Sato, M. (1953). Properties of the receptor potential in Pacinian corpuscles. *The Journal of Physiology, 122*, 610–636.

Heppelmann, B., Messlinger, K., Neiss, W. F., & Schmidt, R. F. (1990). Ultrastructural three-dimensional reconstruction of group III and group IV sensory nerve endings ("free nerve endings") in the knee joint capsule of the cat: Evidence for multiple receptive sites. *The Journal of Comparative Neurology, 292*, 103–116.

Loewenstein, W. R. (1959). The generation of electric activity in a nerve ending. *Annals of the New York Academy of Sciences, 81*, 367–387.

Patton, D. H. (1966). Receptor mechanisms. In T. C. Ruch & H. D. Patton (Eds.), *Physiology and biophysics* (pp. 95–112). Philadelphia/London: WB Saunders.

Sauer, S. K., Weidner, C., Carr, R. W., Averbeck, B., Nesnidal, U., Reeh, P. W., & Handwerker, H. O. (2005). Can receptor potentials be detected with threshold tracking in rat cutaneous nociceptive terminals? *Journal of Neurophysiology, 94*, 219–25.

Schmidt, R. F., & Thews, G. (1990). *Physiologie des Menschen, 24. Auflage*. Berlin: Springer.

Nociceptor Inactivation

▶ Nociceptor, Adaptation

Nociceptor Sensitization

Definition

Process by which there is a decrease in nociceptor threshold and enhanced responses to suprathreshold stimuli. This phenomenon is an increment of the excitability of the nociceptor, due to a metabolic change induced by sensitizing agents such as proinflammatory mediators.

Cross-References

▶ Capsaicin Receptor
▶ Polymodal Nociceptors, Heat Transduction
▶ Thalamus, Clinical Pain, Human Imaging

Nociceptor(s)

Definition

Harmful stimuli activate the peripheral endings of primary afferent neurons, also called nociceptors. Their cell bodies lie in the dorsal root ganglia (DRG) or the trigeminal ganglia. Distinct classes of nociceptors encode discrete intensities and modalities of noxious stimuli. Receptor molecules that lend these specific properties to diverse classes of nociceptors and mediate transduction have been cloned. One important molecule is the vanilloid receptor TRPV1, which serves as a transducer of noxious thermal and chemical (e.g., protons) stimuli and can be activated by capsaicin, the active ingredient of hot chili peppers. Conduction of nociceptive signals in nociceptors is mediated via activation of voltage-gated sodium channels. A family of nociceptor-specific tetrodotoxin (TTX)-resistant sodium channels modulates the excitability of primary afferents and likely mediates pathophysiological alterations thereof.

Cross-References

▶ Acute Pain Mechanisms
▶ Cancer Pain Management, Nonopioid Analgesics
▶ Cancer Pain, Animal Models
▶ COX-1 and COX-2 in Pain
▶ Cytokines, Effects on Nociceptors
▶ Descending Modulation and Persistent Pain
▶ Dorsal Root Ganglionectomy and Dorsal Rhizotomy

Nociceptor, Adaptation

Matthias Ringkamp and Richard A. Meyer
Department of Neurosurgergy, School of
Medicine Johns Hopkins University, Baltimore,
MD, USA

Synonyms

Nociceptor accommodation; Nociceptor inactivation

Definition

The gradual decrease over time in the response of a nociceptor to a maintained noxious stimulus of fixed intensity.

Characteristics

The response of nociceptors to a constant-temperature heat stimulus adapts with time. Unmyelinated nociceptors innervating the hairy skin of monkey can be separated into two classes based on the rate of adaptation to heat stimuli (Fig. 1). In response to a 53 °C stimulus, the discharge rate of quickly adapting C fibers is 20 % of the peak response within 4 s, whereas slowly adapting C fibers take more than 15 s to reach this level (Meyer and Campbell 1981;

Nociceptor, Adaptation, Fig. 1 Adaptation of response to heat (53 °C) in C-fiber nociceptor afferents in the monkey (From Meyer and Campbell 1981)

Johanek et al. 2008). Myelinated fibers can also be separated into two classes based on their heat response: type II fibers adapt quickly in a manner similar to quickly adapting C fibers, whereas type I fibers actually exhibit an increase in their response with time (Treede et al. 1998).

Nociceptors also exhibit a slowly adapting response to mechanical stimuli applied to their receptive field (Slugg et al. 2000). An exception to this rule exists for mechanically insensitive nociceptors, although they can develop a response to tonic pressure (Schmidt et al. 2000). The mechanisms underlying adaptation in nociceptors are not well understood, but calcium-dependent and calcium-independent mechanisms appear to be involved.

References

Johanek, L. M., Meyer, R. A., Friedman, R. M., Greenquist, K. W., Shim, B., Borzan, J., LaMotte, R. H., & Ringkamp, M. (2008). A role for polymodal C-fiber afferents in non-histaminergic itch. *Journal of Neuroscience, 28*, 7659–7669.

Meyer, R. A., & Campbell, J. N. (1981). Evidence for two distinct classes of unmyelinated nociceptive afferents in monkey. *Brain Research, 224*, 149–152.

Schmidt, R., Schmelz, M., Torebjörk, H. E., et al. (2000). Mechano-insensitive nociceptors encode pain evoked by tonic pressure to human skin. *Neuroscience, 98*, 793–800.

Slugg, R. M., Meyer, R. A., & Campbell, J. N. (2000). Response of cutaneous A- and C-fiber nociceptors in the monkey to controlled-force stimuli. *Journal of Neurophysiology, 83*, 2179–2191.

Treede, R.-D., Meyer, R. A., & Campbell, J. N. (1998). Myelinated mechanically insensitive afferents from monkey hairy skin: Heat-response properties. *Journal of Neurophysiology, 80*, 1082–1093.

Nociceptor, Axonal Branching

Martin Schmelz
Department of Anesthesiology in Mannheim, University of Heidelberg, Mannheim, Germany

Synonyms

Axon reflex; Flare; Neurogenic inflammation; Neurogenic vasodilation; Protein extravasation

Definition

Single nociceptive nerve fibers branch extensively in the periphery to form their receptive fields: branches can measure up to 9 cm in human skin. In addition, their terminal endings can also inhibit extensive branching in the micrometer range. Axonal branching is the structural basis for antidromic action potential propagation (axon reflex), leading to neurogenic inflammation.

Characteristics

Innervation Territories of Nociceptors

The receptive field of primary afferent nociceptors has been found to be very small in rodents. However, in humans, skin innervation territories measuring up to 9 cm in diameter have been found (Schmelz et al. 1997). Most of the available data on the structure of innervation territories and branching derive from skin, as they can more easily be analyzed as compared to deep somatic or visceral nociceptors (Graven-Nielsen and Mense 2001; Marchettini et al. 1996). Extensive branching of skin nociceptors has also been found in monkey skin, with branching points being rather proximal from the actual receptive field; interestingly, frequently unmyelinated branches of A delta fibers were found, which had a length of about 5 cm (Peng et al. 1999).

Axonal Properties of Different Nociceptor Classes

In human skin, unmyelinated nociceptors fall into two basic classes: the majority of the fibers are mechano-heat-sensitive polymodal nociceptors; however, about 20 % are unresponsive to mechanical stimulation (Schmidt et al. 1995). These "silent" or "sleeping" nociceptors differ from conventional polymodal nociceptors in their receptive properties, their biophysical characteristics, and their function. They have higher activation thresholds for heat and are not activated even by intense mechanical stimuli (Schmidt et al. 1995). Their innervation

Nociceptor, Axonal Branching, Fig. 1 Response latency of action potentials between electrical stimulation of the receptive endings in the skin and recording in the peripheral nerve (n. peroneus) is shown. *Left panel*: Electrical stimulation at increasing frequency provokes increased response latency for different classes of afferent C-fibers (activity-dependent slowing). The activity-dependent slowing is most pronounced in mechano-insensitive C-nociceptors which slow down in conduction velocity even at low stimulatory frequencies of 0.125 or 0.25 Hz. The slowing of traditional mechano-responsive C-nociceptors ("polymodal") is far less pronounced and clearly separates between the two nociceptor classes (Modified from Weidner et al. 1999). *Right panel*: At higher stimulation frequencies of 2 Hz, polymodal nociceptors show a pronounced activity-dependent slowing which clearly separates them from cold-sensitive C-fibers (C-cold receptor), which only slightly increase their response latency when stimulated at 2 Hz (Modified from Serra et al. 1999)

territories in the leg are larger (Schmidt et al. 2002; Schmelz et al. 1997) (6 vs. 2 cm^2) and conduction velocity is lower (0.8 vs. 1 m/s) than in polymodal fibers (Schmidt et al. 1995). Most interestingly, their high transcutaneous electrical thresholds and their activity-dependent hyperpolarization by far exceed the values observed in polymodal nociceptors. Although based on the axonal properties, analysis of activity-dependent hyperpolarization allows classification of these two nociceptor classes (Weidner et al. 1999; Bostock et al. 2003) and thus predicts sensory properties of their endings. Moreover, activity-dependent hyperpolarization has been shown to also separate C-cold thermoreceptors from C-polymodal nociceptors (Fig. 1). It should be pointed out that this unexpected correlation between specific *axonal* properties and characteristics of *sensory endings* has a variety of implications as axonal excitability is also modulating sensory sensitivity as shown for mechanical responsiveness in the rat (Taguchi et al. 2010; Decol et al. 2012).

Neurogenic Inflammation

Decades ago, Thomas Lewis described the erythema arising in human skin in the surroundings of trauma as part of the "triple response" to noxious stimuli (Lewis et al. 1927). This "flare response" is dependent on the integrity of primary afferent nerves but not on their central nervous connections. From his own findings and earlier work, Lewis developed the concept of "axon reflex flare," i.e., the notion that "nocifensive" nerve fibers excited by a trauma send impulses not only into the central nervous system, but also via axon branches into the surrounding skin, where they trigger the release of a vasodilating substance from the nerve endings. Neuropeptides are now held responsible for the vasodilatation, which are produced by small dorsal root ganglion cells and transported in

Nociceptor, Axonal Branching, Fig. 2 Schematical drawing of mechanisms involved in dermal neurogenic inflammation. Noxious stimulation of the skin (*hatched area*) results in generation of action potentials in nociceptors. The action potentials reach the arborizations of the axonal tree via the axon reflex (*black arrows*). By depolarization of the terminals, neuropeptides (*black circles*) are released. Key effects of neuropeptides are given in *black squares*. The involvement of mast cell mediators (*MC, open circles*) in vasodilation and protein extravasation in neurogenic inflammation is controversial

their thin axons to the central and peripheral nerve terminals. The main vasodilatory agent is probably CGRP which induces vasodilatation but not plasma extravasation (Brain 1996). The edema, caused by increased permeability of the endothelia for plasma proteins (neurogenic protein extravasation), can be attributed to the release of substance P (Holzer 1998). However, a variety of other neuropeptides like neurokinin A, neurokinin B, somatostatin, galanin, and recently, endomorphins have been also be found in primary afferent neurons. It should also be noted that in addition to the acute vascular effects, neuropeptides have important trophic functions and modulate the activity of local immune cells (Schmelz and Petersen 2001) (Fig. 2).

Analysis of Neurogenic Inflammation

Chemical, thermal, and electrical stimulations have been widely used to elicit neurogenic

inflammation, and direct evidence of neuropeptide release has been obtained using capsaicin as well as antidromic electrical nerve stimulation (Kilo et al. 1997; Kress et al. 1999). In rodents, activation of polymodal nociceptors was sufficient to cause neurogenic inflammation (Gee et al. 1997). In contrast, mechano-insensitive but heat- and chemosensitive C-nociceptors have been found responsible for the neurogenic vasodilation in pig skin (Gee et al. 1997; Lynn et al. 1996) and human skin (Sauerstein et al. 2000; Schmelz et al. 2000). The extent of the neurogenic erythema nicely matches the large cutaneous receptive fields of mechano-insensitive nociceptors, and their high electrical thresholds also match the strong currents required to provoke neurogenic flare electrically. Thus, neurogenic inflammation in humans differs from rodents, in which the neurogenic inflammation can be elicited by polymodal C nociceptors and consists of a combination of vasodilation and

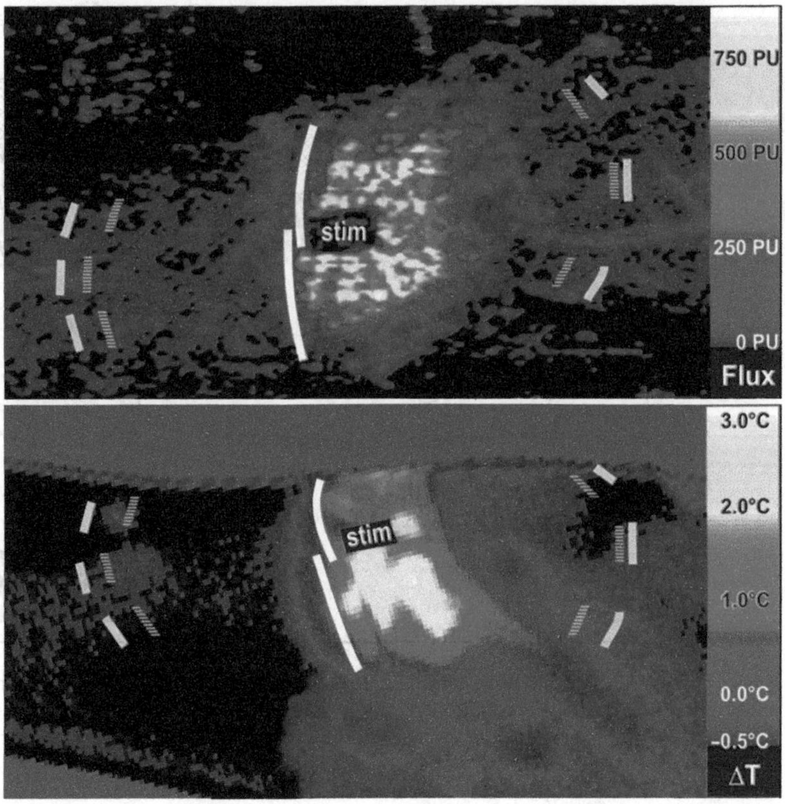

Nociceptor, Axonal Branching,
Fig. 3 Specimen of transcutaneous electrical stimulation (1 Hz, 50 mA, 0.5 ms; stimulation site ("stim") is marked by a *rectangle*) provoking an area of increased superficial blood flow as assessed with a laser Doppler scanner (*upper panel*) and with an infrared thermocamera (*lower panel*). An anesthetic strip was induced by perfusing two intradermal microdialysis membranes (*vertical white lines*) with 2 % lidocaine. The borders of hyperalgesia to punctate stimuli (*short solid lines*) and to light stroking (*short dotted lines*) are shown in the laser Doppler scan and the thermogram (Modified from Klede et al. 2003)

protein extravasation (Sauerstein et al. 2000). In healthy volunteers, no neurogenic protein extravasation could be induced (Sauerstein et al. 2000; Weidner et al. 2000), but they may develop under pathophysiological conditions (Weber et al. 2001).

Neurogenic Inflammation and Secondary Hyperalgesia

The areas of vasodilation and warming around a noxious stimulation site have been found to be similar to the areas of secondary mechanical hyperalgesia (punctate hyperalgesia) (Serra et al. 1998). However, recent results would suggest, that by blocking axonal action potential propagation by a local anesthetic, only the spread of axon reflex vasodilation and warming can be blocked. In contrast, areas of punctate hyperalgesia developed symmetrically, even beyond a peripherally located "anesthetic strip" (Fig. 3).

Epidermal Axonal Branching

Under physiological conditions, unmyelinated human nerve fibers entering the epidermis are found to be oriented straight up and reach the outermost layers of viable skin without pronounced branching (Hilliges et al. 1995). Interestingly, in some human skin diseases, extensive branching of epidermal nerve fibers has been described. Increased intradermal nerve fiber density has been found in patients with chronic pruritus (Urashima and Mihara 1998). In addition, increased epidermal levels of neurotrophin 4 (NT4) have been found in patients with atopic dermatitis (Grewe et al. 2000) and massively increased serum levels of NGF and SP have been found to correlate with the severity of the disease in such patients (Toyoda et al. 2002). Increased fiber density and higher local NGF concentrations were also found in patients with contact dermatitis (Kinkelin et al. 2000). Most interestingly, intraepidermal sprouting has also been found as a physiological response to circular skin incision (Rajan et al. 2003): "collateral sprouting" from axons at the incision margins lead to centripedal reconstituting of skin innervation probably due to higher local NGF concentrations in the denervated skin (Rajan et al. 2003). A similar mechanism might also explain findings in patients with diabetic neuropathy characterized by reduced epidermal innervation density, but higher degree of epidermal branching (Sumner et al. 2003; Polydefkis et al. 2003). It will be of major interest in the future to assess the effects of the local sprouting on sensory function of the nociceptors. There is already evidence for increased epidermal nerve fiber sprouting in vulvodynia (Bohm-Starke et al. 1998; Tympanidis et al. 2004;Tympanidis et al. 2003) and moreover, signs of nociceptor sensitization (Bohm-Starke et al. 2001). Taken together, these data would suggest that local inflammatory processes could initiate nociceptor sprouting and sensitization by increased NGF production. NGF-induced sensitization has recently been established as human pain model (Obreja et al. 2011; Rukwied et al. 2010) and the clinical analgesic effect of anti-NGF strategies (Lane et al. 2010) suggest a major clinical effect.

References

Bohm-Starke, N., Hilliges, M., Falconer, C., & Rylander, E. (1998). Increased intraepithelial innervation in women with vulvar vestibulitis syndrome. *Gynecologic and Obstetric Investigation, 46*, 256–260.

Bohm-Starke, N., Hilliges, M., Brodda-Jansen, G., Rylander, E., & Torebjörk, H. E. (2001). Psychophysical evidence of nociceptor sensitization in vulvar vestibulitis syndrome. *Pain, 94*, 177–183.

Bostock, H., Campero, M., Serra, J., & Ochoa, J. (2003). Velocity recovery cycles of C fibres innervating human skin. *The Journal of Physiology, 553*, 649–663.

Brain, S. D. (1996). Sensory neuropeptides in the skin. In P. Geppetti & P. Holzer (Eds.), *Neurogenic inflammation* (pp. 229–244). Boca Raton: CRC Press.

Decol, R., Messlinger, K., & Carr, R. W. (2012). Repetitive activity slows axonal conduction velocity and concomitantly increases mechanical activation threshold in single axons of the rat cranial dura. *The Journal of Physiology, 590*, 725–736.

Gee, M. D., Lynn, B., & Cotsell, B. (1997). The relationship between cutaneous C fibre type and antidromic vasodilatation in the rabbit and the rat. *Journal of Physiology (London), 503*, 31–44.

Graven-Nielsen, T., & Mense, S. (2001). The peripheral apparatus of muscle pain: Evidence from animal and human studies. *The Clinical Journal of Pain, 17*, 2–10.

Grewe, M., Vogelsang, K., Ruzicka, T., Stege, H., & Krutmann, J. (2000). Neurotrophin-4 production by human epidermal keratinocytes: Increased expression in atopic dermatitis. *The Journal of Investigative Dermatology, 114*, 1108–1112.

Hilliges, M., Wang, L., & Johansson, O. (1995). Ultrastructural evidence for nerve fibers within all vital layers of the human epidermis. *The Journal of Investigative Dermatology, 104*, 134–137.

Holzer, P. (1998). Neurogenic vasodilatation and plasma leakage in the skin. *General Pharmacology, 30*, 5–11.

Kilo, S., Harding-Rose, C., Hargreaves, K. M., & Flores, C. M. (1997). Peripheral CGRP release as a marker for neurogenic inflammation: A model system for the study of neuropeptide secretion in rat paw skin. *Pain, 73*, 201–207.

Kinkelin, I., Motzing, S., Koltenzenburg, M., & Brocker, E. B. (2000). Increase in NGF content and nerve fiber sprouting in human allergic contact eczema. *Cell and Tissue Research, 302*, 31–37.

Klede, M., Handwerker, H. O., & Schmelz, M. (2003). Central origin of secondary mechanical hyperalgesia. *Journal of Neurophysiology, 90*, 353–359.

Kress, M., Guthmann, C., Averbeck, B., & Reeh, P. W. (1999). Calcitonin Gene-related peptide and prostaglandin E2 but not substance P release induced by antidromic nerve stimulation from rat skin, in vitro. *Neuroscience, 89*, 303–310.

Lane, N. E., Schnitzer, T. J., Birbara, C. A., et al. (2010). Tanezumab for the treatment of pain from osteoarthritis of the knee. *The New England Journal of Medicine, 363*, 1521–1531.

Lewis, T., Harris, K. E., & Grant, R. T. (1927). Observations relating to the influence of the cutaneous nerves on various reactions of the cutaneous vessels. *Heart, 14*, 1–17.

Lynn, B., Schutterle, S., & Pierau, F. K. (1996). The vasodilator component of neurogenic inflammation is caused by a special subclass of heat-sensitive nociceptors in the skin of the pig. *Journal of Physiology (London), 494*, 587–593.

Marchettini, P., Simone, D. A., Caputi, G., & Ochoa, J. L. (1996). Pain from excitation of identified muscle nociceptors in humans. *Brain Research, 740*, 109–116.

Obreja, O., Kluschina, O., Mayer, A., et al. (2011). NGF enhances electrically induced pain, but not axon reflex sweating. *Pain, 152*, 1856–1863.

Peng, Y. B., Ringkamp, M., Campbell, J. N., & Meyer, R. A. (1999). Electrophysiological assessment of the cutaneous arborization of Adelta- fiber nociceptors. *Journal of Neurophysiology, 82*, 1164–1177.

Polydefkis, M., Griffin, J. W., & McArthur, J. (2003). New insights into diabetic polyneuropathy. *Journal of the American Medical Association, 290*, 1371–1376.

Rajan, B., Polydefkis, M., Hauer, P., Griffin, J. W., & McArthur, J. C. (2003). Epidermal reinnervation after intracutaneous axotomy in man. *The Journal of Comparative Neurology, 457*, 24–36.

Rukwied, R., Mayer, A., Kluschina, O., Obreja, O., Schley, M., & Schmelz, M. (2010). NGF induces non-inflammatory localized and lasting mechanical and thermal hypersensitivity in human skin. *Pain, 148*, 407–413.

Sauerstein, K., Klede, M., Hilliges, M., & Schmelz, M. (2000). Electrically evoked neuropeptide release and neurogenic inflammation differ between rat and human skin. *The Journal of Physiology, 529*, 803–810.

Schmelz, M., & Petersen, L. J. (2001). Neurogenic inflammation in human and rodent skin. *News in Physiological Sciences, 16*, 33–37.

Schmelz, M., Schmidt, R., Bickel, A., Handwerker, H. O., & Torebjörk, H. E. (1997). Specific C-receptors for itch in human skin. *Journal of Neuroscience, 17*, 8003–8008.

Schmelz, M., Michael, K., Weidner, C., Schmidt, R., Torebjörk, H. E., & Handwerker, H. O. (2000). Which nerve fibers mediate the axon reflex flare in human skin? *NeuroReport, 11*, 645–648.

Schmidt, R., Schmelz, M., Forster, C., Ringkamp, M., Torebjörk, H. E., & Handwerker, H. O. (1995). Novel classes of responsive and unresponsive C nociceptors in human skin. *Journal of Neuroscience, 15*, 333–341.

Schmidt, R., Schmelz, M., Weidner, C., Handwerker, H. O., & Torebjörk, H. E. (2002). Innervation territories of mechano-insensitive C nociceptors in human skin. *Journal of Neurophysiology, 88*, 1859–1866.

Serra, J., Campero, M., & Ochoa, J. (1998). Flare and hyperalgesia after intradermal capsaicin injection in human skin. *Journal of Neurophysiology, 80*, 2801–2810.

Serra, J., Campero, M., Ochoa, J., & Bostock, H. (1999). Activity-dependent slowing of conduction differentiates functional subtypes of C fibres innervating human skin. *The Journal of Physiology, 515*, 799–811.

Sumner, C. J., Sheth, S., Griffin, J. W., Cornblath, D. R., & Polydefkis, M. (2003). The spectrum of neuropathy in diabetes and impaired glucose tolerance. *Neurology, 60*, 108–111.

Taguchi, T., Ota, H., Matsuda, T., Murase, S., & Mizumura, K. (2010). Cutaneous C-fiber nociceptor responses and nociceptive behaviors in aged Sprague–Dawley rats. *Pain, 151*, 771–782.

Toyoda, M., Nakamura, M., Makino, T., Hino, T., Kagoura, M., & Morohashi, M. (2002). Nerve growth factor and substance P are useful plasma markers of disease activity in atopic dermatitis. *British Journal of Dermatology, 147*, 71–79.

Tympanidis, P., Terenghi, G., & Dowd, P. (2003). Increased innervation of the vulval vestibule in patients with vulvodynia. *British Journal of Dermatology, 148*, 1021–1027.

Tympanidis, P., Casula, M. A., Yiangou, Y., Terenghi, G., Dowd, P., & Anand, P. (2004). Increased vanilloid receptor VR1 innervation in vulvodynia. *European Journal of Pain, 8*, 129–133.

Urashima, R., & Mihara, M. (1998). Cutaneous nerves in atopic dermatitis. A histological, immunohistochemical and electron microscopic study. *Virchows Arch. 432*, 363–370.

Weber, M., Birklein, F., Neundorfer, B., & Schmelz, M. (2001). Facilitated neurogenic inflammation in complex regional pain syndrome. *Pain, 91*, 251–257.

Weidner, C., Schmelz, M., Schmidt, R., Hansson, B., Handwerker, H. O., & Torebjörk, H. E. (1999). Functional attributes discriminating mechano-insensitive and mechano-responsive C nociceptors in human skin. *Journal of Neuroscience, 19*, 10184–10190.

Weidner, C., Klede, M., Rukwied, R., et al. (2000). Acute effects of substance P and calcitonin gene-related peptide in human skin - a microdialysis study. *The Journal of Investigative Dermatology, 115*, 1015–1020.

Nociceptor, Categorization

Edward Perl
Department of Cell and Molecular Physiology, School of Medicine University of North Carolina, Chapel Hill, NC, USA

Synonyms

Nociceptive peripheral afferent fibers; Nociceptors: high-threshold afferents; Pain afferents; Pain sense organs

Introduction

Noxious and Nociceptor

Definition of certain classes of stimuli as noxious and creation of the term nociceptor (noci-receptor) were outgrowths of the dispute in the late nineteenth century about the sensory nature of pain. Physicians and physiologists of those days generally accepted pain to be a sensation. On the other hand, philosophical critics of this idea argued that in contrast to accepted sensations, pain does not have a well-defined physical or chemical stimulus (Perl 1996).

Mechanical events, heat, cold, and chemical agents can all produce it. Charles Sherrington (1906), an eminent physiologist, proposed an answer to this criticism with the logic that pain ordinarily results from tissue injury. Consequently, tissue damage represents a common denominator for natural stimuli evoking pain. He suggested that events producing disruption of tissue or representing a physical threat to its integrity could be labeled noxious regardless of their nature, thereby providing an encompassing definition for stimuli producing pain. In this concept, sense organs signaling the presence of noxious events are called noci-receptors (now shortened to nociceptors).

Designation of stimuli as noxious creates its own problems. Tissues of the mammalian body are diverse with widely differing mechanical and thermal characteristics. This means that quantitatively the intensity of environmental or circumstantial events necessary to cause damage of given tissues varies over a substantial range. For example, compare mechanical characteristics of the cornea of the human eye to the skin on the sole of a human foot. Furthermore, subcutaneous tissues and organs are protected from many environmental changes and potential insults; their exposure to some conditions that are innocuous for the skin surface would lead to tissue injury and thereby should be considered noxious. Thus, the nature of noxious events and their signaling by sense organs must be considered in the context of tissue type and location.

Classification of Sense Organs as Nociceptors

A primary afferent neuron is appropriately considered to be a nociceptor if the intensity of the most effective "natural" stimulus that evokes conducted action potentials approaches or exceeds the noxious (damaging) level for the innervated tissue. This criterion implies that such sensory units respond weakly or not all to innocuous stimuli of any type. Since the nature and intensity of noxious stimuli will vary for different tissues, the responsive characteristics of nociceptors will differ from one tissue to another.

Observations demonstrating nociceptors to be distinctive categories of somatic sense organs provide evidence that more than one type innervate many tissues. How are these different types distinguished? Actually, nociceptor classification has evolved as information about them expands and the changing criteria sometimes have led to ambiguity. Given the view that the function of nociceptors is to transmit to the central nervous system information about events dangerous to the physical integrity of the tissue they innervate, a first order in their classification, and one commonly used, is the nature of effective stimuli or the events signaled.

Definition

Rationale

Why separate primary afferent neurons into groups? As Lynn (1996) points out, classifying a neuronal population into categories on the basis of shared characteristics does more than one service. It facilitates communication by providing a shorthand to designate subsets with certain features. It also provides a way to deal with the large number of neurons in the mammalian nervous system that serves common functions. Moreover, appropriate classification of neurons can help facilitate concepts on development, the functional organization of nervous mechanisms, in pathology, the mechanisms of disease, and therapy.

Characteristics

Classification of Nociceptors by Effective Stimuli

Much information about nociceptors has come from study of the innervation of epithelial tissue, particularly the skin. Early in the documentation of cutaneous nociceptors as distinctive types of sense organs, it became evident that the skin is innervated by more than one type (Perl 1996). As already suggested, nociceptors can be distinguished and classified according to the nature of stimuli activating them. On this basis, several categories are demonstrable in mammalian skin (Perl 1996; Campbell and Meyer 1996). In terms of effective stimuli, the cornea, another epithelial tissue, is innervated by closely similar varieties (Belmonte and Gallar 1996). One kind of cutaneous nociceptor responds vigorously to strong mechanical stimulation, positively grading the frequency and number of impulses in proportion to stimulus intensity. Extreme temperatures (e.g., noxious heat, freezing) excite such mechanical nociceptors (high-threshold mechanoreceptors) only after a delay. A second category of cutaneous nociceptor responds more globally, being promptly activated by heat, mechanical distortion, and irritant chemicals including protons. The latter response pattern led to the designation of this class as polymodal nociceptors. A third kind, prominent in the innervation of glabrous skin, is excited by mechanical distortion and elevated skin temperature (mechanical-heat), but does not promptly respond to surface application of irritant chemicals or acid. A fourth type responds both to low skin temperatures and to noxious mechanical stimuli (mechanical-cold). In addition to these four categories, evidence exists for a class of primary afferents (labeled "silent" nociceptors) that are only excited by mechanical stimuli when sensitized by local inflammation and for another category selectively excited by ► histamine (pruritus receptors).

Characterization by Conduction Velocity
Whereas the nature of effective stimuli represents an approach to classifying nociceptors that relates to function, it is not the only important criterion. Indications that peripheral stimuli evoke pain after distinctly different delays existed prior to documentation of nociceptors as a special set of peripheral sense organs. Transient application of a noxious mechanical or heat stimulus to distally located skin anecdotally and experimentally was noted to provoke a double pain response, one of short latency and a second delayed; these differences in latency can be attributed to differences in ► conduction velocity of peripheral nerve fibers

responsible for the afferent messages. In addition to these distinctions of delay, "first" and "second" pain is reported to differ in quality of the sensation. This circumstantial evidence for conduction differences is consistent with findings that some categories of nociceptors have myelinated (A) and others unmyelinated (C) afferent fibers. Those with ► A-fibers conduct much more rapidly (10–50 m/s) than the ► C-fiber groups, and even though most A-nociceptors have thinly myelinated fibers (Aδ), a number from distal limb regions are in the medium myelinated range with Aβ (35–50 m/s) conduction velocities. A-fiber and C-fiber nociceptors also differ in other ways. For instance, several sets of C-fiber nociceptors express peptide mediators (e.g., substance-P, CGRP) that are apparently absent in myelinated nociceptors. Furthermore, the central projections of A- and C-fiber nociceptors differ. These distinctions are consistent with certain differences in functions initiated by the A- and C-fiber categories.

Characterization by Tissue of Origin
Primary afferent neurons with responsive features of nociceptors innervate many mammalian tissues or organs, both somatic and visceral. In addition to skin and cornea, these include teeth, skeletal muscle, tendon, joints, bone, urethra, ureter, urinary bladder, blood vessels, bronchi, heart, pleura and peritoneum, segments of the alimentary tract, meninges, and testis (Cervero 1996). These tissues differ substantially in physical and chemical attributes, differences that are reflected in part in the responsive and signaling features of innervating nociceptive fibers. The testis is innervated by nociceptors mimicking the broad activation of cutaneous polymodal nociceptors by being responsive to noxious mechanical, heat, and chemical stimuli (Kumazawa 1996). In contractile tissues, unusually, high tension is an effective excitant for part of the nociceptive innervation. Similarly, lowered pH (protons), by itself or in combination with anoxemia, activate or enhance the responsiveness of certain nociceptive afferents of skeletal muscle (Mense 1996). Circumstantial evidence

suggests that subcutaneous tissues such as joints and muscle contain a number of primary afferent fibers that are unresponsive to mechanical or thermal stimuli until injury has induced inflammation and its chemical environment (Schaible and Schmidt 1996). The latter can be considered types of chemoreceptor. Thus, the classification of nociceptors must take into account the tissue innervated in addition to effective stimuli, and the diameter (conduction velocity) of the afferent fiber.

Characterization by Molecular Features

A presumption underlying hypotheses about differentiation and specialization of biological cells, in our case neurons, is that these processes are guided and controlled by the presence and expression of particular molecular entities. Relating factors associated with molecular expression to functionally important features of nociceptors is an ongoing effort and at present represents at most an emerging story with promise for future insights.

In one example, the heat responsiveness of polymodal nociceptors is attributed to a membrane receptor, ▶ TRPV1 (Caterina and Julius 2001). TRPV1 donates such reactivity when expressed in heterologous cells. TRPV1 is the endogenous receptor that is selectively activated by capsaicin, the substance that gives the sensation of heat upon ingestion of "hot" pepper. A structurally related membrane receptor, ▶ TRPV2, is predominantly expressed in different primary afferent neurons than TRPV1, and is proposed to provide a higher-threshold heat response for a set of nociceptors different from the polymodal type. TRPV2 is neither excited by capsaicin nor acid (Caterina and Julius 2001).

Other relationships between molecular features and categories of nociceptors include the immunocytochemical labeling of a subset of small diameter dorsal root ganglia (DRG) neurons and their processes for the peptides, ▶ substance-P and ▶ CGRP. Both circumstantial and direct correlations indicate that at least some peptide-labeled elements are polymodal or mechanical-heat nociceptors (Lawson 1996;

Lawson et al. 1997). The small substance-P-containing neurons are mostly distinct from an isolectin IB4–binding population of presumed nociceptors (Lawson 1996). Thus, evidence for the common presence of TRPV1, TRPV2, or any other unique cellular constituent links the nociceptors in question by a shared feature, and represents a basis for categorization.

Characterization by Central Projection

Nociceptors, as primary afferent neurons, project to the spinal cord or the brainstem with a heavy concentration of synaptic terminations in the superficial dorsal horn or the trigeminal equivalent. The bulk of nociceptors enter the spinal cord through the ▶ lateral division (of the spinal dorsal root). There is a distinct layering of terminations from different nociceptor subsets as defined by other criteria. Substance-P (and CGRP)-labeled endings concentrate in the marginal zone ▶ (lamina-I) and outer ▶ substantia gelatinosa (lamina-II), as do the non-peptide terminations of cutaneous myelinated-fiber, mechanical nociceptors (Light 1992; Perl 1984). IB4-labeled fibers generally terminate more deeply in lamina-II than the concentration of substance-P terminals (Woodbury et al. 2000). In addition to these superficial dorsal horn terminations, nociceptive primary afferents contribute to other dorsal horn regions. Whereas these terminal synaptic regions are not definitive markers of receptive category, they have been used as guides to relate participation by particular types of nociceptors in experimental analyses.

Summary

Categorization of nociceptors is a multifactorial task. To be of value in facilitating understanding of their biology and place in function, categorization of nociceptive afferents should consider the nature of events signaled, the structural and functional details of the neuron (including its afferent fiber), the presence of unique molecules, and features of the peripheral and central terminations.

References

Belmonte, C., & Gallar, J. (1996). Corneal nociceptors. In C. Belmonte & F. Cervero (Eds.), *Neurobiology of nociceptors* (pp. 146–183). Oxford: Oxford University Press.

Campbell, J. N., & Meyer, R. A. (1996). Cutaneous nociceptors. In C. Belmonte & F. Cervero (Eds.), *Neurobiology of nociceptors* (pp. 117–145). Oxford: Oxford University Press.

Caterina, M. J., & Julius, D. (2001). The vanilloid receptor: A molecular gateway to the pain pathway. *Annual Review of Neuroscience, 24*, 487–517.

Cervero, F. (1996). Visceral nociceptors. In C. Belmonte & F. Cervero (Eds.), *Neurobiology of nociceptors* (pp. 220–242). Oxford: Oxford University Press.

Kumazawa, T. (1996). Sensitization of polymodal receptors. In C. Belmonte & F. Cervero (Eds.), *Neurobiology of nociceptors* (pp. 325–345). Oxford: Oxford University Press.

Lawson, S. N. (1996). Neurochemistry of cutaneous nociceptors. In C. Belmonte & F. Cervero (Eds.), *Neurobiology of nociceptors* (pp. 72–91). Oxford: Oxford University Press.

Lawson, S. N., Crepps, B. A., & Perl, E. R. (1997). Relationship of substance P to afferent characteristics of dorsal root ganglion neurones in guinea pig. *The Journal of Physiology, 505*, 177–191.

Light, A. R. (1992). *The initial processing of pain and its descending control: Spinal and trigeminal systems.* Basel: Karger.

Lynn, B. (1996). Principles of classification and nomenclature relevant to studies of nociceptive neurones. In C. Belmonte & F. Cervero (Eds.), *Neurobiology of nociceptors* (pp. 1–2). Oxford: Oxford University Press.

Mense, S. (1996). Nociceptors in skeletal muscle and their reaction to pathological tissue changes. In C. Belmonte & F. Cervero (Eds.), *Neurobiology of nociceptors* (pp. 184–201). Oxford: Oxford University Press.

Perl, E. R. (1984). Pain and nociception. In I. Darian-Smith (Ed.), *Handbook of physiology. The nervous system* (Vol. 3, pp. 915–975). Bethesda, MD: American Physiological Society.

Perl, E. (1996). Pain and discovery of nociceptors. In C. Belmonte & F. Cervero (Eds.), *Neurobiology of nociceptors* (pp. 5–36). Oxford: Oxford University Press.

Schaible, H.-G., & Schmidt, R. F. (1996). Neurobiology of articular nociceptors. In C. Belmonte & F. Cervero (Eds.), *Neurobiology of nociceptors* (pp. 202–219). Oxford: Oxford University Press.

Sherrington, C. S. (1906). *The integrative action of the nervous system.* New York: Scribner.

Woodbury, C. J., Ritter, A. M., & Koerber, H. R. (2000). On the problem of lamination in the superficial dorsal horn of mammals: A reappraisal of the substantia gelatinosa in postnatal life. *The Journal of Comparative Neurology, 417*, 88–102.

Nociceptor, Fatigue

Matthias Ringkamp and Richard A. Meyer
Department of Neurosurgery, School of
Medicine Johns Hopkins University, Baltimore,
MD, USA

Synonyms

Deactivation; Desensitization; Suppression; Tachyphylaxis

Definition

The decrement in response of nociceptors to natural stimuli that occurs upon repeated stimulation.

Characteristics

The response of a primary afferent nociceptor to a given stimulus is greatly reduced when the stimulus is applied a second time. For example, the response of a nociceptor to a heat stimulus applied to its receptive field is reduced by about 40 % when presented a second time within 60 s of the first presentation (Fig. 1). Nociceptors exhibit fatigue to all stimuli that activate them, including heat, mechanical and chemical stimuli (Liang et al. 2001; Slugg et al. 2000; Tillman 1992). The magnitude of fatigue is dependent on the number of action potentials evoked by the conditioning stimulus, but is independent of the frequency of the evoked action potentials. Consequently, the magnitude of fatigue increases with the intensity of the applied stimulus (LaMotte and Campbell 1978; Peng et al. 2003). The time course for recovery from fatigue is slow, with full recovery taking more than 5 min (Fig. 2). Fatigue is also seen for repeated stimuli that have been applied to isolated dorsal root ganglion cells (Schwarz et al. 2000).

The psychophysical correlate of fatigue is suppression. Suppression corresponds to the

Nociceptor, Fatigue, Fig. 1 Normalized response of C-fiber mechano-heat sensitive nociceptors to five presentations of a heat stimulus with a repetition period of 60s. The response decreased between trials 1 and 2 and stabilized to about 60 % of the response to the first heat stimulus (Adapted from Peng et al. (2003))

Nociceptor, Fatigue, Fig. 2 Slow recovery from fatigue. The magnitude of fatigue is dependent on the time between the conditioning and the test stimulus. The response to the test stimulus approaches the response to the initial identical conditioning stimulus as the interstimulus interval increases. The fatigue to heat stimuli is greater than the fatigue to mechanical stimuli (Adapted from Slugg et al. (2000), Tillman (1992))

decrement in pain rating to a noxious stimulus that occurs upon repeated stimulation at the same location. For example, the pain rating to a heat stimulus applied to the hand is significantly lower than the pain rating to the same stimulus applied 30 s earlier (LaMotte and Campbell 1978). The time for full recovery is greater than 5 min.

Most nociceptors are polymodal and respond to more than one stimulus modality. Cross-modal fatigue occurs such that stimulation with one stimulus modality (e.g., mechanical stimuli) will lead to a decrement in response to another stimulus modality (e.g., heat). In addition, stimulation to one part of the receptive field can lead to fatigue in another part of the receptive field, presumably due to ▶ antidromic invasion of the adjacent terminals. The magnitude of fatigue is less, and recovery time is quicker, for cross-modal fatigue compared to unimodal fatigue (Peng et al. 2003).

Fatigue may occur at the stimulus transduction and/or at the spike initiation site and may involve calcium-dependent and calcium-independent mechanisms.

References

LaMotte, R. H., & Campbell, J. N. (1978). Comparison of responses of warm and nociceptive C-fiber afferents in monkey with human judgements of thermal pain. *Journal of Neurophysiology, 41*, 509–528.

Liang, Y. F., Haake, B., & Reeh, P. W. (2001). Sustained sensitization and recruitment of rat cutaneous nociceptors by bradykinin and a novel theory of its excitatory action. *The Journal of Physiology, 532*, 229–239.

Peng, Y. B., Ringkamp, M., Meyer, R. A., et al. (2003). Fatigue and paradoxical enhancement of heat response in C-fiber nociceptors from cross-modal excitation. *The Journal of Neuroscience, 23*, 4766–4774.

Schwarz, S., Greffrath, W., Busselberg, D., et al. (2000). Inactivation and tachyphylaxis of heat-evoked inward currents in nociceptive primary sensory neurones of rats. *The Journal of Physiology, 528*, 539–549.

Slugg, R. M., Meyer, R. A., & Campbell, J. N. (2000). Response of cutaneous A- and C-fiber nociceptors in the monkey to controlled-force stimuli. *Journal of Neurophysiology, 83*, 2179–2191.

Tillman, D. B. (1992). Heat response properties of unmyelinated nociceptors. Ph.D. dissertation, The Johns Hopkins University, pp. 1–187.

Nociceptors and Activity-Dependent Changes in Axonal Conduction Velocity

Richard Carr
Clinic for Anaesthesiology, Heidelberg
University, Mannheim, Baden-Wuerttemberg,
Germany

Synonyms

Activity-dependent slowing; Recovery cycles;
Spike initiation and conduction in nociceptors

Introduction

> "By tacit general consent so far, the axon does not
> think. It only ax." George Bishop (1965)

Voltage-gated sodium channels (NaVs) generate
and conduct axonal action potentials and it is the
number and subtype of NaVs together with
the passive electrical properties of the axon that
determine the speed of action potential conduction. For C-fibers, axonal conduction velocities
are typically quite slow at around 1 m/s.
However, the axonal conduction velocity of
C-fibers can change dynamically according to
the immediate firing history of the axon. Changes
in axonal conduction velocity modulate both
the frequency of action potentials arriving at the
spinal cord as well as the sensitivity of nerve
endings to sensory stimuli. Remarkably, within
the population of C-fibers, the magnitude and
time course of post-firing changes in excitability
can be used to predict the receptive properties of
individual axons.

Definition

Nociceptive sensory neurons (nociceptors) use
action potential frequency to encode potentially
or overtly damaging thermal, mechanical, and
chemical stimuli. In restoring the electrical
and ionic perturbations associated with each
action potential, axonal C-fiber excitability
cycles through a series of elevated and depressed
phases of up to several 10 s of seconds. These
excitability changes modulate the frequency at
which action potentials arrive at central terminals
in the spinal dorsal horn and also regulate the
sensitivity of the receptive nerve terminal to
further stimuli.

Characteristics

Action Potential Initiation

Extracellular recordings from individual sensory
nerve terminals in the cornea have shown that
both TTX-s and TTX-r voltage-gated sodium
channels (NaVs) contribute to action potential
initiation (Brock et al. 1998). The site at which
NaVs generate action potentials in unmyelinated
endings is not fixed. Instead, there appear to be
multiple discrete sites at which action potentials
can initiate and that the site of action potential
initiation can move dependent upon the strength
of the prevailing stimulus. For example, in
cold-sensitive nerve terminals cooling depolarizes the nerve ending and the point of action
potential initiation moves centripetally to more
proximal sites (Carr et al. 2009).

Axonal Action Potentials

Extracellular spikes from non-nociceptive
afferents (such as cold-sensitive C-fibers) are
narrower than those from polymodal nociceptors,
while spikes from mechanically insensitive
nociceptors are very broad (Gee et al. 1999;
Serra et al. 1999). Whether these differences
reflect differences in NaV isoform expression is
unclear. However, direct intracellular recordings
from cultured rat DRG neurites show that both
TTX-s and TTX-r sodium currents contribute
during an action potential and that TTX-r channels alone can support conduction (Vasylyev and
Waxman 2012). In mature neurons, conduction in

peripheral C-fibers is predominantly dependent upon TTX-s NaVs with a small percentage of axons potentially being able to support pure TTX-r action potentials, particularly within the dorsal root (Brock et al. 1998; Pinto et al. 2008; Quasthoff et al. 1995).

Somal Action Potentials

Intracellular recordings from dorsal root ganglion neurons indicate that action potentials in nociceptive neurons have a slower time course than in mechanoreceptor neurons with similar axonal conduction velocity (Lawson 2002). Compared to non-nociceptive neurons, the somata of nociceptive neurons contain much higher levels of the TTX-s sodium channel NaV1.7 as well as the TTX-r NaV 1.8 and NaV 1.9. Since TTX-r NaVs have slower activation kinetics than TTX-s channels, their expression is very likely to contribute to the longer duration action potentials in nociceptive neurons. Somal action potentials of nociceptive neurons also often have a distinct inflection (or "shoulder") on the falling phase which is further indicative of a TTX-r current contribution (Blair and Bean 2002). Using the action potential recorded from each neuron as a command voltage, TXX-r current was found to comprise up to 60 % of the total sodium current during the action potential. One interesting observation is that following inflammation in the cutaneous innervation territory of the neuron, the somal spike of nociceptive neurons became shorter in duration (Djouhri et al. 2001).

Activity-Dependent Changes in Axonal Conduction Velocity

In the wake of each action potential, three sequential phases of altered axonal excitability are usually observed (Fig. 1). Immediately after an action potential is a period of subnormality (3–20 ms) including an absolute refractory period in which a second action potential cannot be generated at all. A transient and for some axons facultative period of supernormality (ca. 10–300 ms) typically follows, in which the

second of two action potentials conducts more rapidly than the first. The recovery cycle ends with a period of subnormality that may persist for several tens of seconds before the baseline level of excitability is reestablished. A readily observed effect of these excitability changes is that subsequent action potentials propagate with a slowed conduction velocity during periods of subnormality and with a faster conduction velocity during the supernormal period (Fig. 1).

In general, supernormality in C-fiber axons increases with increasing stimulus rate and this increase is greatest for those axons showing the most activity-dependent slowing (see below). While the majority of C-fibers always show a period of supernormality, the incidence of a supernormal period among mechanoreceptive C-fibers is inversely related to their mechanical threshold. This means that while most high-threshold (mechanically insensitive) axons exhibit supernormality at all stimulus rates, supernormality typically only emerges in mechanically sensitive nociceptors above 1 Hz (Bostock et al. 2003; Weidner et al. 2000) and in low-threshold mechanoreceptors at around 20 Hz (Gee et al. 1996). Perhaps the absence of a supernormal phase increases fidelity by reducing the likelihood of doublets. This feature may be important in low-threshold receptors that by virtue of their sensitivity to mechanical stimuli presumably fire relatively often.

Mechanisms Contributing to Activity-Dependent Changes in Axonal Conduction Velocity

Immediately after an action potential, axons are absolutely refractory because insufficient NaVs have recovered from fast inactivation to generate a second action potential. As repolarization proceeds, inactivation of NaVs is relieved and the magnitude of subnormality decreases.

In the majority of C-fibers, a transient supernormal period follows the initial subnormal period and this results from a residual depolarizing charge on the capacitance of the

Nociceptors and Activity-Dependent Changes in Axonal Conduction Velocity, Fig. 1 Characteristic changes in axonal excitability as a function of time in unmyelinated axons. Three phases of altered excitability manifesting as change in threshold (excitability) or conduction velocity (normality) are observed after an action potential before the initial state of the axon (*broken red line*) is recovered. During the initial subexcitable/subnormal phase, absolute refractoriness (*orange*) gives way to a relative refractory period (*yellow*). An action potential initiating in this period will slow relative to its predecessor during conduction and if the conduction distance is sufficient will arrive centrally with a delay corresponding to the end of the relative refractory period (first zero crossing). A superexcitable/supernormal phase (*blue*) follows. An action potential initiating in this period will speed relative to its predecessor and if the

membrane (Bostock et al. 2003). The proposal is that if, during the action potential, the amount of Na^+ entering the axon is greater than the K^+ leaving the axon, a depolarizing afterpotential gives rise to a supernormal phase. Similarly, a hyperpolarizing afterpotential is generated and subnormal conduction results if K^+ egress exceeds Na^+ entry. An action potential conducting along a hyperpolarized axon is more likely to result in excess intraaxonal sodium and thus a supernormal phase, while a depolarized axon would be expected to show subnormality in this period. The magnitude and sign of subnormality/supernormality therefore serves as an indirect index of axonal membrane potential, while the time constant of decay of subnormality/supernormality is representative of the membrane time constant. In general, the magnitude of supernormality increases with hyperpolarization and decreases with depolarization and this property has been used for assessing C-fiber excitability in isolated segments of peripheral nerve.

The prolonged period of late subnormality is additive and in response to modest rates of stimulation (ca. 2 Hz) gives rise to activity-dependent slowing (ADS). The primary mechanisms contributing to ADS are a reduction in sodium channel availability and hyperpolarization of the membrane potential. The sodium action current is reduced with repetitive activity in subsets of DRG neurons (Snape et al. 2010) because TTX-s and TTX-r NaVs enter slow inactivated states reducing the number of available NaVs and slowing axonal conduction (De Col et al. 2008). The time constants of recovery from slow inactivation are consistent with the extended time course of late supernormality in C-fibers. In addition, changes in membrane potential during repetitive activity in C-fibers

are determined by the hyperpolarizing effect of the sodium pump (Na^+/K^+-ATPase) and the depolarizing effect of hyperpolarization-activated inward currents (Ih), which when blocked enhances slowing in C-fibers (Takigawa et al. 1998). In the special case of sympathetic C-fibers, activation of Ih may even marginally reverse conduction velocity slowing (Campero et al. 2004).

Another putative factor which may contribute to conduction slowing during and after bursts of action potentials is an increase in intracellular sodium concentration. Although this possibility has not yet been examined experimentally, it is not unlikely that an increase in the intraaxonal sodium concentration reduces the sodium current during an action potential. This factor is not relevant for neuronal somata, but could be important for thin axons especially in the nerve terminal region where the small intraaxonal volumes may result in substantial changes in ion concentration.

Analysis of Activity-Dependent Changes in Axonal Conduction Velocity

Dynamic changes in axonal conduction velocity have the intriguing effect of causing subsequent action potentials to either catch each other up or drift apart. This means that action potentials generated in the terminal at, say, 50 ms intervals (20 Hz) may arrive, after traversing the axon in the supernormal period, only 10 ms (100 Hz) apart at their central presynaptic terminal. This increase in frequency can potentiate the effectiveness of presynaptic release. On the other hand, action potentials initiating at lower frequencies, say less than 5 Hz, will encounter the late subnormal period and be subject to a slight decrease in firing rate during axonal propagation which may allow the axon to

Nociceptors and Activity-Dependent Changes in Axonal Conduction Velocity, Fig. 1 (continued) conduction distance is sufficient will also arrive centrally with a delay corresponding to the end of the relative refractory period (first zero crossing). A late subexcitable/subnormal phase (*green*) follows before complete recovery and an action potential initiating in this period will slow relative to its predecessor. This slowing is additive and gives rise to the phenomenon of activity-dependent slowing

enhance contrast, increasing firing frequency for strong stimuli and reducing firing rates evoked by weaker stimuli (Weidner et al. 2002).

The pattern of ADS is also indicative of the receptive properties of individual axons. In people, low-threshold tactile afferents show very little latency walking (~1 % at 2 Hz; Campero et al. 2011), cold-sensitive afferents slow a little more (~5 % at 2 Hz; Campero et al. 2001), mechanically sensitive, often also heat sensitive, polymodal nociceptors slow even more (~23 % at 2 Hz, C-MH) and mechanically insensitive nociceptors slow considerably (~35 % at 2 Hz, C-Mi, Namer et al. 2009). Sympathetic efferent axons also show modest slowing (~5 % at 2 Hz) but can be identified by their characteristic rapid initial slowing followed by a plateau or even a slight reversal of slowing (Campero et al. 2004).

Beyond its utility in the classification of individual C-fibers, the process responsible for axonal conduction velocity gives rise to a parallel increase in sensory threshold (De Col et al. 2012). Activity-dependent slowing serves thus as an index of sensory adaptation with the result that axons exhibiting modest ADS remain receptive to and can encode more or less the same stimulus intensities irrespective of their firing history. In contrast, axons with prominent slowing are less likely to sustain firing preferring to reduce their sensitivity and encode strong stimuli in bursts.

A limited number of studies have examined C-fiber function in patients with small fiber neuropathy. Using indices based on activity-dependent slowing, ongoing activity has been detected in cold nociceptors in patients with cold allodynia (Ochoa et al. 2005) and similarly activity was detected in C-Mi fibers in patients with inflammatory symptoms (Ørstavik and Jorum 2010). Abnormal C-Mi fibers are also apparently more abundant and exhibit less slowing in patients with diabetic neuropathy. In contrast, normal aging tends to increase activity-dependent slowing in nociceptors (Namer et al. 2009). As one of many potential factors contributing to changes in C-fiber conduction, nerve growth factor (NGF) can regulate NaV isoform expression in sensory neurons

(Fang et al. 2005; Gould et al. 2000) and CMi-nociceptors in pig show less activity-dependent slowing 1 week after topical NGF (Obreja et al. 2011).

Summary

Action potentials in nociceptors are of longer duration than in other classes of somatosensory afferent neurons and they show more profound conduction velocity slowing during repetitive activation ($\leq$ 2 Hz). Both of these features may reflect specific combinations of NaV expression, in particular isoforms with slow kinetics of activation and inactivation may be more prominent.

The post-spike cycle of altered axonal conduction velocity results from the restoration of electrical, ionic, and metabolic perturbations associated with each action potential. The reasons why post-spike excitability cycles differ for instance between mechanically insensitive and mechanically sensitive nociceptors is not known. However, the differential role of different classes of nociceptors in chronic pain conditions makes these differences worthy of further study. A tantalizing possibility is that differences in electrophysiological properties between nociceptors and non-nociceptors, or even possibly between subclasses of nociceptor, are attributable for instance to specific NaV isoforms and this might offer a basis for the development of novel and specific analgesic drugs.

References

Bishop, G. H. (1965). My life among the axons. *Annual Review of Physiology, 27,* 1–18.
Blair, N. T., & Bean, B. P. (2002). Roles of tetrodotoxin (TTX)-sensitive Na + current, TTX-resistant Na + current, and Ca2+ current in the action potentials of nociceptive sensory neurons. *Journal of Neuroscience, 22,* 10277–10290.
Bostock, H., Campero, M., Serra, J., & Ochoa, J. (2003). Velocity recovery cycles of C fibres innervating human skin. *The Journal of Physiology, 553,* 649–663.
Brock, J. A., McLachlan, E. M., & Belmonte, C. (1998). Tetrodotoxin-resistant impulses in single nociceptor

nerve terminals in guinea-pig cornea. *The Journal of Physiology*, *512*, 211–217.

Campero, M., Serra, J., Bostock, H., & Ochoa, J. L. (2001). Slowly conducting afferents activated by innocuous low temperature in human skin. *The Journal of Physiology*, *535*, 855–865.

Campero, M., Serra, J., Bostock, H., & Ochoa, J. L. (2004). Partial reversal of conduction slowing during repetitive stimulation of single sympathetic efferents in human skin. *Acta Physiologica Scandinavica*, *182*, 305–311.

Campero, M., Bostock, H., Baumann, T. K., & Ochoa, J. L. (2011). Activity-dependent slowing properties of an unmyelinated low threshold mechanoreceptor in human hairy skin. *Neuroscience Letters*, *493*, 92–96.

Carr, R. W., Pianova, S., McKemy, D. D., & Brock, J. A. (2009). Action potential initiation in the peripheral terminals of cold-sensitive neurones innervating the guinea-pig cornea. *The Journal of Physiology*, *587*, 1249–1264.

De Col, R., Messlinger, K., & Carr, R. W. (2008). Conduction velocity is regulated by sodium channel inactivation in unmyelinated axons innervating the rat cranial meninges. *The Journal of Physiology*, *586*, 1089–1103.

De Col, R., Messlinger, K., & Carr, R. W. (2012). Repetitive activity slows axonal conduction velocity and concomitantly increases mechanical activation threshold in single axons of the rat cranial dura. *The Journal of Physiology*, *590*, 725–736.

Djouhri, L., Dawbarn, D., Robertson, A., Newton, R., & Lawson, S. N. (2001). Time course and nerve growth factor dependence of inflammation-induced alterations in electrophysiological membrane properties in nociceptive primary afferent neurons. *Journal of Neuroscience*, *21*, 8722–8733.

Fang, X., Djouhri, L., McMullan, S., Berry, C., Okuse, K., Waxman, S. G., & Lawson, S. N. (2005). trkA is expressed in nociceptive neurons and influences electrophysiological properties via Nav1.8 expression in rapidly conducting nociceptors. *Journal of Neuroscience*, *25*, 4868–4878.

Gee, M. D., Lynn, B., & Cotsell, B. (1996). Activity-dependent slowing of conduction velocity provides a method for identifying different functional classes of C-fibre in the rat saphenous nerve. *Neuroscience*, *73*, 667–675.

Gee, M. D., Lynn, B., Basile, S., Pierau, F. K., & Cotsell, B. (1999). The relationship between axonal spike shape and functional modality in cutaneous C-fibres in the pig and rat. *Neuroscience*, *90*, 509–518.

Gould, H. J., Gould, T. N., England, J. D., Paul, D., Liu, Z. P., & Levinson, S. R. (2000). A possible role for nerve growth factor in the augmentation of sodium channels in models of chronic pain. *Brain Research*, *854*, 19–29.

Lawson, S. N. (2002). Phenotype and function of somatic primary afferent nociceptive neurones with C-, Adelta- or Aalpha/beta-fibres. *Experimental Physiology*, *87*, 239–244.

Namer, B., Barta, B., Orstavik, K., Schmidt, R., Carr, R., Schmelz, M., & Handwerker, H. O. (2009). Microneurographic assessment of C-fibre function in aged healthy subjects. *The Journal of Physiology*, *587*, 419–428.

Obreja, O., Ringkamp, M., Turnquist, B., Hirth, M., Forsch, E., Rukwied, R., Petersen, M., & Schmelz, M. (2011). Nerve growth factor selectively decreases activity-dependent conduction slowing in mechano-insensitive C-nociceptors. *Pain*, *152*, 2138–2146.

Ochoa, J. L., Campero, M., Serra, J., & Bostock, H. (2005). Hyperexcitable polymodal and insensitive nociceptors in painful human neuropathy. *Muscle & Nerve*, *32*, 459–472.

Ørstavik, K., & Jorum, E. (2010). Microneurographic findings of relevance to pain in patients with erythromelalgia and patients with diabetic neuropathy. *Neuroscience Letters*, *470*, 180–184.

Pinto, V., Derkach, V. A., & Safronov, B. V. (2008). Role of TTX-sensitive and TTX-resistant sodium channels in Adelta- and C-fiber conduction and synaptic transmission. *Journal of Neurophysiology*, *99*, 617–628.

Quasthoff, S., Grosskreutz, J., Schroder, J. M., Schneider, U., & Grafe, P. (1995). Calcium potentials and tetrodotoxin-resistant sodium potentials in unmyelinated C fibres of biopsied human sural nerve. *Neuroscience*, *69*, 955–965.

Serra, J., Campero, M., Ochoa, J., & Bostock, H. (1999). Activity-dependent slowing of conduction differentiates functional subtypes of C fibres innervating human skin. *The Journal of Physiology*, *515*, 799–811.

Snape, A., Pittaway, J. F., & Baker, M. D. (2010). Excitability parameters and sensitivity to anemone toxin ATX-II in rat small diameter primary sensory neurones discriminated by *Griffonia simplicifolia* isolectin IB4. *The Journal of Physiology*, *588*, 125–137.

Takigawa, T., Alzheimer, C., Quasthoff, S., & Grafe, P. (1998). A special blocker reveals the presence and function of the hyperpolarization-activated cation current IH in peripheral mammalian nerve fibres. *Neuroscience*, *82*, 631–634.

Vasylyev, C. V., & Waxman, S. G. (2012). Membrane properties and electrogenesis in the distal axons of small dorsal root ganglion neurons in vitro. *Journal of Neurophysiology*, *108*, 729–740.

Weidner, C., Schmidt, R., Schmelz, M., Hilliges, M., Handwerker, H. O., & Torebjork, H. E. (2000). Time course of post-excitatory effects separates afferent human C fibre classes. *The Journal of Physiology*, *527*, 185–191.

Weidner, C., Schmelz, M., Schmidt, R., Hammarberg, B., Orstavik, K., Hilliges, M., Torebjork, H. E., & Handwerker, H. O. (2002). Neural signal processing: The underestimated contribution of peripheral human C-fibers. *Journal of Neuroscience*, *22*, 6704–6712.

N

Nociceptors in the Dental Pulp

Matti V. O. Närhi
Department of Biomedicine/Physiology,
Department of Dentistry, University of Eastern
Finland, Institute of Medicine, Kuopio, Finland

Synonyms

Intradental nociceptors; Pulpal nociceptors

Definition

▶ Nociceptors that are located inside the tooth in the ▶ dental pulp and in the ▶ dentinal tubules in the most inner layers of ▶ dentin. The intradental afferent innervation consists of both myelinated (20–30 %) and unmyelinated (70–80 %) nerve fibers, which are mostly, if not exclusively, nociceptive. The nerve fibers originate from the mandibular (lower jaw) and maxillary (upper jaw) branches of the trigeminal nerve, and have their cell bodies in the ▶ trigeminal ganglion.

Characteristics

Pulpal inflammation can be extremely painful. Also, the intensity of the pain responses induced from teeth, for example, from exposed dentin, by external stimulation can reach the maximum level of any pain score. The structure of the intradental innervation gives a basis for such high sensitivity. It should also be noted that pain is the predominant, if not the only, sensation that can be evoked by activation of intradental nerves.

Structure of Intradental Innervation

The innervation of the dental pulp is exceptionally rich (Fig. 1). Close to a thousand nerve fibers enter each tooth (Beasley and Holland 1978; Byers and Närhi 1999). Approximately 20–30 % of them are myelinated, mostly

Nociceptors in the Dental Pulp, Fig. 1 Schematic presentation of the intradental innervation. A few branches from the alveolar nerve enter the pulp through the apical foramen. These bundles extend further through the root pulp and branch extensively, especially in the coronal pulp. The terminal branches of the axons are free nerve endings, and are located either in the pulp (especially the C-fibers) or in the tubules of the predentin and inner layers of dentin (many of the A-fibers). See text for more details

of smaller diameter, A-δ type, although there are also some larger A-β type axons. A great majority of the pulp nerve fibers are unmyelinated (C-fibers). A small proportion, approximately 10 %, of the unmyelinated fibers are sympathetic efferents (see Byers and Närhi 1999). Their activation causes vasoconstriction and consequently reduction in the pulpal blood flow (Olgart 1996).

All intradental axons terminate as free nerve endings, the C-fibers in the pulp proper, and A-fibers both in the pulp, and a great number of them also in the ▶ predentin and inner layers of dentin (Fig. 1). The myelinated nerve fibers branch abundantly, and one axon may have endings in more than a hundred dentinal tubules (Byers 1984). The maximum distance that the fibers penetrate into the tubules is approximately 100–150 μm in the horns or tip of the coronal pulp. The pulp horns are also the most densely innervated areas of the pulp. At the pulp tip, approximately 50 % of the tubules contain nerve terminals, some of them even multiple endings (Byers 1984). As there are 30,000–40,000 tubules/mm^2 of dentin (Brännström 1981), the density of the innervation of the pulp-dentin border at the tip of the coronal pulp is exceptionally dense. However, there are fewer nerve endings at the pulp-dentin border of the cervical region, and yet exposed dentin in these areas can be extremely sensitive, with innervation in the root being especially sparse (Fig. 1) (Brännström 1981).

Function of Intradental Nociceptors

The similarity of the structure of the intradental innervation in human teeth to that in cats, dogs, and monkeys (Byers 1984; Byers and Närhi 1999) gives a morphological basis for studies, where the function of intradental nerves in experimental animals has been compared to the sensations evoked by dental stimulation in man. Electrophysiological recordings have revealed that A- and C-fibers of the pulp are functionally different (Närhi 1985; Byers and Närhi 1999). Comparison of those results to the sensory responses evoked by stimulation of human teeth also indicated that activation of pulpal A- and C-fibers may induce different types of pain sensations, namely, sharp and dull pain, respectively (Ahlquist et al. 1985; Jyväsjärvi and Kniffki 1987). The intradental nerve activity recordings in human teeth indicate that the nerve function resembles that of experimental animals (Edwall and Olgart 1977). The results of the single fiber recordings in cat and dog teeth also indicate that A-fibers are responsible for the sensitivity of dentin (Närhi 1985; Närhi et al. 1992), and that the intradental Aβ- and Aδ-fibers respond in a similar way to noxious dental stimulation (Närhi et al. 1992). Accordingly, they belong to the same functional group (Närhi et al. 1992, 1996). However, recently it has been proposed that part of the pulpal Aβ-fibers may form a separate group of low-threshold "algoneurons" responding to very light stimuli, such as air blasts or gentle mechanical stimuli, directed to exposed dentin (Fried et al. 2011). The existence of those "algoneurons" would be a unique characteristic for the dental pulp compared to any other tissues and give an explanation for the high pain sensitivity of the exposed dentin even to extremely light stimuli directed to its exposed surface.

Pulpal A-fibers can be activated by a number of different stimuli applied at the tooth surface. Their responses are greatly enhanced if the dentin is exposed and the dentinal tubules are open (Närhi et al. 1992). Heat, cold, hypertonic solutions of various chemicals and desiccating air blasts, for example, applied to exposed dentin, evoke nerve responses with quite a similar pattern. The nerve firing starts immediately, or within a couple of seconds after the stimulus is applied, far before the stimuli, heat or cold, have reached the most inner layers of dentin and the pulp, where the nociceptors are located. So, the responses cannot be caused by a direct effect of the applied stimuli on the intradental nerve endings.

The stimuli, which induce pain from human teeth and activate the intradental nerves in experimental animals, are able to induce fluid flow in the dentinal tubules (Brännström 1981). Moreover, the fluid flow and the nerve responses induced by hydrostatic pressure are directly related (Vongsavan and Matthews 1994). Also, both the induction of the fluid flow, and the sensitivity of dentin are, to a great extent, dependent on the openness of the dentinal tubules (Brännström 1981; Närhi et al. 1992). Accordingly, the final stimulus for the nociceptors, which are responsible for the sensitivity of dentin (intradental A-fibers), seems to be the fluid flow in the tubules and, probably, the consequent mechanical deformation of the tissue and nerve endings in the pulp-dentin border (the so-called hydrodynamic mechanism of pulp nerve activation, Brännström 1981) (Fig. 2).

The hydrodynamic fluid flow is based on the strong capillary forces in the thin dentinal tubules (Brännström 1981). If dentinal fluid is extracted from the outer end of an open tubule by any stimulus, it is immediately replaced by a rapid outward fluid shift, for example, air-drying of dentine induces outward fluid flow and intense firing of the pulpal A-fibers (Närhi 1985). It also causes disruption of the tissues in the peripheral

Nociceptors in the Dental Pulp, Fig. 2 The dental hydrodynamic fluid flow

Nociceptors in the Dental Pulp, Fig. 2 The dental hydrodynamic fluid flow

pulp (Brännström 1981). Thus, the capillary and hydrodynamic forces can considerably intensify the effect of the applied stimuli, and even a light stimulus, such as an air blast, can turn out to be noxious to the pulp.

The pulpal C-nociceptors are polymodal, and only activated if the external stimuli reach the pulp proper (Närhi 1985; Närhi et al. 1996). They do not respond to hydrodynamic stimulation. The C-fibers are activated by intense cold and heat applied to the tooth crown (Närhi et al. 1982, 1996; Jyväsjärvi and Kniffki 1987). The response latencies are quite long, because the thermal stimuli have to reach the pulp where the nerve endings are located (Närhi 1985; Jyväsjärvi and Kniffki 1987; Närhi et al. 1996). The responses seem to be induced by a direct effect of heat and cold on the nerve endings. The C-nociceptors also respond to intense mechanical stimulation as well as to bradykinin, histamine, and capsaicin applied to the exposed pulp (Närhi 1984; Närhi et al. 1992, 1996).

The possible role of the odontoblasts as receptor cells has been discussed widely (see Byers and Närhi 1999). It has been proposed that these cells would be activated by various stimuli applied to the tooth and thereafter transmit the responses to the closely located intradental nerve endings in the dentinal tubules and in the odontoblast layer (Magloire et al. 2009, 2010). So far, the results for and against the role of the odontoblasts as sensory receptors have been variable and also

conflicting, but the recent findings (Magloire et al. 2009, 2010) have revealed interesting findings indicating that these cells may share many characteristics common to mechanosensitive receptors.

Inflammation-Induced Changes in Pulpal Nociceptor Function

In healthy teeth, the intradental nociceptors are well protected by the dental hard tissues and difficult to activate. Thus, pain is seldom induced from teeth during everyday activities. However, this is not the case when dentin with open tubules is exposed, because the effects of external stimuli are intensified by the hydrodynamic forces (Brännström 1981; Närhi et al. 1992). As in other tissue injuries, inflammation of the dental pulp can considerably sensitize the nociceptors (Brännström 1981; Närhi et al. 1996; Byers and Närhi 1999). This may lead to extremely intense spontaneous pain (toothache), and exaggerated painful responses to external stimuli, for example, to hot or cold food or drinks. The change in pulpal nociceptor activity and sensitivity is caused by a number of different inflammatory mediators, which are released and/or synthesized in response to the tissue injury (Närhi et al. 1996; Byers and Närhi 1999).

Intradental A- and C-nociceptors are activated and sensitized by inflammatory mediators in the initial stages of inflammation, which include neurogenic reactions, for example, ▶ axon reflexes

(Närhi 1985; Närhi et al. 1992, 1996; Olgart 1996). However, there are other longer-term neurogenic mechanisms which may contribute to the increased sensitivity (Byers and Närhi 1999). These include activation of silent pulpal nociceptors (Närhi et al. 1996; Byers and Närhi 1999), that is, the proportion of pulpal A-fibers responding to dentinal stimulation is significantly higher in inflamed compared to normal teeth, and is especially profound among the slow-conducting Aδ fibers (Närhi et al. 1996). Moreover, the receptive fields of the single nerve fibers in inflamed teeth become expanded (Närhi et al. 1996). Both of these changes may be related to sprouting of the nociceptive nerve fibers (Kimberly and Byers 1988) and/or activation of normally unresponsive nerve terminals. These changes may contribute to the increased sensitivity of inflamed teeth.

In view of the dense innervation of the dental pulp, and the structural neural changes that can occur in inflamed teeth, it is clinically puzzling to find that pulpitis may frequently be almost or even completely asymptomatic. This can be partially due to central regulatory mechanisms, which may inhibit the impulse transmission in the trigeminal pain pathways (Sessle 2000). However, local mechanisms also seem to exist in the dental pulp itself, which may be important not only for the regulation of the inflammatory reactions, but also the sensitivity of the nociceptors (Närhi et al. 1996; Olgart 1996; Byers and Närhi 1999; Dionne et al. 2001; Närhi 2001). Endogenous opioids may inhibit the nerve activation (Närhi 2001) and neuropeptide release from the nociceptive nerve endings. Also, somatostatin may have an inhibitory effect on the nociceptor activation. Furthermore, sympathetic nerve fibers seem to prevent the release of the neuropeptides, by virtue of their preterminal connections to the nociceptive nerve endings (Olgart 1996). Accordingly, a number of different central and peripheral regulatory mechanisms may be operating, and contribute to the great variability of the pain symptoms connected with pulpal inflammation.

References

Ahlquist, M. L., Franzen, O. G., Edwall, L. G. A., Forss, U. G., & Haegerstam, G. A. T. (1985). Quality of pain sensations following local application of algogenic agents on the exposed human tooth pulp: A psycho physiological and electrophysiological study. In H. L. Fields (Ed.), Advances in pain research and therapy (Vol. 9, pp. 351–359). New York: Raven.

Beasley, W. L., & Holland, G. R. (1978). A quantitative analysis of the innervation of the pulp of cat's canine tooth. The Journal of Comparative Neurology, 178, 487–494.

Brännström, M. (1981). Dentin and pulp in restorative dentistry. Nacka, Sweden: Dental Therapeutics AB.

Byers, M. R. (1984). Dental sensory receptors. International Review of Neurobiology, 25, 39–94.

Byers, M. R., & Närhi, M. (1999). Dental injury models: Experimental tools for understanding neuroinflammatory interactions and polymodal nociceptor function. Critical Reviews in Oral Biology and Medicine, 10, 4–39.

Dionne, R. A., Lepinski, A. M., Jaber, L., Gordon, S. M., Brahim, J. S., & Hargreaves, K. M. (2001). Analgesic effects of peripherally administered opioids in clinical models of acute and chronic inflammation. Clinical Pharmacology and Therapeutics, 70, 66–73.

Edwall, L., & Olgart, L. M. (1977). A new technique for recording of intradental sensory nerve activity in man. Pain, 3, 121–126.

Fried, K., Sessle, B., & Devor, M. (2011). The paradox of pain from the tooth pulp: Low-threshold "Algoneurons"? Pain, 152, 2685–2689.

Jyväsjärvi, E., & Kniffki, K.-D. (1987). Cold stimulation of teeth: A comparison between the responses of cat intradental AÎ′ and C fibresand human sensation. The Journal of Physiology, 391, 193–207.

Kimberly, C. L., & Byers, M. R. (1988). Inflammation of rat molar pulp and periodontium causes increased calcitonin gene-related peptide and axonal sprouting. Anatomical Record, 222, 289–300.

Magloire, H., Couble, M.-L., Thivichon-Prince, B., Maurin, J.-C., & Bleicher, F. (2009). Odontoblast: A mechanosensory cell. Journal of Experimental Zoology Part B: Molecular and Developmental Evolution, 312B, 416–424.

Magloire, H., Maurin, J. C., Couble, M. L., Shibukawa, Y., Tsumura, M., Thivichon-Prince, B., & Bleicher, F. (2010). Topical review. Dental pain and odontoblasts: Facts and hypotheses. Journal of Orofacial Pain, 24, 335–349.

Närhi, M. V. O. (1985). The characteristics of intradental sensory units and their responses to stimulation. Journal of Dental Research, 64, 564–571.

Närhi, M. V. O. (2001). Local application of morphine inhibits the intradental nociceptor responses to mustard oil but not to hydrodynamic stimulation of dentin. Society of Oral Physiology, Abstracts.

Närhi, M., Jyväsjärvi, E., Hirvonen, T., & Huopaniemi, T. (1982). Activation of heat sensitive nerve fibres in the dental pulp of the cat. *Pain, 14*, 317–326.

Närhi, M., Kontturi-Närhi, V., Hirvonen, T., & Ngassapa, D. (1992). Neurophysiological mechanisms of dentin hypersensitivity. *Proceedings of the Finnish Dental Society, 88*, 15–22.

Närhi, M., Yamamoto, H., & Ngassapa, D. (1996). Function of intradental nociceptors in normal and inflamed teeth. In M. Shimono, T. Maeda, H. Suda, & K. Takahashi (Eds.), *Dentin/pulp complex* (pp. 136–140). Tokyo: Quintessence.

Olgart, L. (1996). Neurogenic components of pulp inflammation. In M. Shimono, T. Maeda, H. Suda, & K. Takahashi (Eds.), *Dentin/pulp complex* (pp. 169–175). Tokyo: Quintessence.

Sessle, B. J. (2000). Acute and chronic orofacial pain: Brainstem mechanisms of nociceptive transmission and neuroplasticity and their clinical correlates. *Critical Reviews in Oral Biology and Medicine, 11*, 57–91.

Vongsavan, N., & Matthews, B. (1994). The relationship between fluid flow in dentine and the discharge of intradental nerves. *Archives of Oral Biology, 39*, 140S.

Nociceptors in the Mucosa

▶ Nociceptors in the Orofacial Region (Skin/Mucosa)

Nociceptors in the Orofacial Region (Meningeal/Cerebrovascular)

Karl Messlinger
University of Erlangen-Nürnberg, Erlangen, Germany

Synonyms

Dural sensory receptors; Intracranial nociceptors; Meningeal afferents; Meningeal nociceptors

Definition

Trigeminal afferents that respond to noxious (painful) stimulation of intracranial structures (cranial meninges and intracranial blood vessels).

These afferents originate in the trigeminal ganglion and project centrally to the spinal trigeminal nucleus and to the spinal dorsal horn of the first cervical segments. Activation of meningeal afferents produces the sensation of headache.

Characteristics

The afferent innervation of the meninges and intracranial blood vessels has long been associated with the generation of ▶ headaches. Intraoperative exploration of patients undergoing open head surgery has revealed that intracranial structures are differentially sensitive to various stimuli (Ray and Wolff 1940). Noxious mechanical, thermal, or electrical stimulation of dural blood vessels and the main cerebral arteries, but not other intracranial tissues, caused painful sensations. Since headache-like pain was the only sensation evoked from stimulation of these intracranial structures, meningeal afferents are generally attributed a nociceptive function. This concept is in accordance with morphological findings. Afferent innervation of intracranial structures is restricted to Aδ and C fibers, which are known to terminate as ▶ noncorpuscular sensory endings (free nerve endings). These nerve endings do not form distinct corpuscular end structures but disperse as small bundles of sensory fibers, partly encased by Schwann cells, along dural blood vessels and into the dural connective tissue (Meßlinger et al. 1993; Fricke et al. 2001). Immunohistochemical preparations have shown that a considerable proportion of intracranial afferent fibers contain vasoactive neuropeptides, particularly ▶ calcitonin gene-related peptide (CGRP) (Fig. 1). This vasodilatory neuropeptide is released from activated intracranial afferents and has been suggested to be involved in the generation of ▶ migraine and other primary headaches (Edvinsson and Goadsby 1998). Experimentally, CGRP release from cranial dura mater can be quantitatively assessed in vitro and used as a measure for the activation of meningeal afferents by noxious stimuli. Noxious stimuli like inflammatory mediators,

Nociceptors in the Orofacial Region (Meningeal/ Cerebrovascular), Fig. 1 Meningeal afferents. CGRP immunoreactive nerve fibers in the rat dura mater encephali. Bundles and single immunopositive nerve fibers (*arrows*) run along the middle meningeal artery (MMA) and its branches and terminate close to blood vessels (*arrowheads*). Scale bar 100 μm (Modified from Meßlinger et al. (1993))

capsaicin, protons, and heat are able to release CGRP from rodent cranial dura (Ebersberger et al. 1999) and activate subclasses of meningeal afferents (Strassman et al. 1996). CGRP is in the center of interest since it is regarded as a key mediator in migraine and other primary headaches (Arulmani et al. 2004). CGRP has been measured at elevated concentrations in jugular venous blood and saliva samples of patients suffering from primary headaches (Ashina et al. 2000; Bellamy et al. 2006), whereas substance P, which is probably coreleased together with CGRP at smaller amounts, plays probably not an important role here. Peripheral release of substance P is known to cause ▶ neurogenic inflammation, characterized by protein plasma extravasation, as well as other endothelial and perivascular changes. Although neurogenic inflammation has been proposed to be a key factor in migraine pathophysiology (Moskowitz and Macfarlane 1993), substance P was not found elevated during migraine attacks (Edvinsson and Goadsby 1998). Likewise in animal experiments, noxious stimulation of the meninges has failed to release detectable amounts of substance P in the periphery (Ebersberger et al. 1999). However, in the spinal trigeminal nucleus, increased substance P release during stimulation of rat dura mater with acidic solutions has been shown with a sensitive microprobe technique (Schaible et al. 1997). Iontophoretic administration of CGRP into the cervical dorsal horn in rats has been found to increase responses to L-glutamate of second-order neurons with afferent input from the meninges (Storer et al. 2004). Therefore, substance P and CGRP released from the central terminals of activated meningeal afferents may contribute to nociceptive transmission in the central trigeminal system.

Few direct electrophysiological recordings from intracranial afferents have been made in animal experiments. The teased fiber technique has been used to record from intracranial afferents running in the nasociliary nerve, which innervate frontomedial parts of the supratentorial dura mater of the guinea pig (Bove and Moskowitz 1997). Another approach has been made in the rat to record with microelectrodes from the trigeminal ganglion, selecting neurons with dural receptive fields located around the middle meningeal artery or the large sinuses (Dostrovsky et al. 1991; Strassman et al. 1996). A new in vitro preparation of rodent dura mater allows recordings from single meningeal afferent fibers (De Col et al. 2008). These studies have shown afferent activation caused by noxious chemical stimuli (inflammatory mediators, capsaicin, and acidic buffer), hot and cold stimuli applied to the exposed dura mater, as well as mechanical stimulation of dural receptive fields. Thus many, if not most, meningeal afferents seem to be ▶ polymodal nociceptors, which can be sensitized to mechanical stimuli through a cAMP-mediated intracellular mechanism (Levy and Strassman 2002). There is evidence that most meningeal sensory endings express tetrodotoxin-resistant sodium channels, as has been reported frequently for visceral nociceptors (Strassman and Raymond 1999). More insight into the response properties of meningeal afferents to stimuli such as nitric oxide and histamine, which are suggested to be key mediators in the generation of primary headaches, may ameliorate our knowledge about peripheral factors involved in headache generation.

Extracellular recordings from second-order neurons in the cat and the rat spinal trigeminal nucleus have shown that there is high convergence of afferent input from the meninges, deep extracranial structures like periost and cranial muscles, and the orofacial region (Davis and Dostrovsky 1988; Schepelmann et al. 1999). This observation has led investigators to assume that ▶ referred pain and hyperalgesia associated with headache may result from ▶ central sensitization (Burstein et al. 1998; Yamamura et al. 1999; Ellrich et al. 1999). The morphological and electrophysiological data taken from meningeal afferents suggest that intracranial pain (headache) may share more characteristics with visceral, rather than somatic, nociception and pain.

References

Arulmani, U., Maassenvandenbrink, A., Villalon, C. M., & Saxena, P. R. (2004). Calcitonin gene-related peptide and its role in migraine pathophysiology. *European Journal of Pharmacology, 500*, 315–330.

Ashina, M., Bendtsen, L., Jensen, R., Schifter, S., Jansen-Olesen, I., & Olesen, J. (2000). Plasma levels of calcitonin gene-related peptide in chronic tension-type headache. *Neurology, 55*, 1335–1340.

Bellamy, J. L., Cady, R. K., & Durham, P. L. (2006). Salivary levels of CGRP and VIP in rhinosinusitis and migraine patients. *Headache, 46*, 24–33.

Bove, G. M., & Moskowitz, M. A. (1997). Primary afferent neurons innervating guinea pig dura. *Journal of Neurophysiology, 77*, 299–308.

Burstein, R., Yamamura, H., Malick, A., & Strassman, A. M. (1998). Chemical stimulation of the intracranial dura induces enhanced responses to facial stimulation in brain stem trigeminal neurons. *Journal of Neurophysiology, 79*, 964–982.

Davis, K. D., & Dostrovsky, J. O. (1988). Responses of feline trigeminal spinal tract nucleus neurons to stimulation of the middle meningeal artery and sagittal sinus. *Journal of Neurophysiology, 59*, 648–666.

De Col, R., Messlinger, K., & Carr, R. W. (2008). Conduction velocity is regulated by sodium channel inactivation in unmyelinated axons innervating the rat cranial meninges. *The Journal of Physiology, 586*, 1089–1103.

Dostrovsky, J. O., Davis, K. D., & Kawakita, K. (1991). Central mechanisms of vascular headaches. *Canadian Journal of Physiology and Pharmacology, 69*, 652–658.

Ebersberger, A., Averbeck, B., Messlinger, K., & Reeh, P. W. (1999). Release of substance P, calcitonin gene-related peptide and prostaglandin E$_2$ from rat dura mater encephali following electrical and chemical stimulation, in vitro. *Neuroscience, 89*, 901–907.

Edvinsson, L., & Goadsby, P. J. (1998). Neuropeptides in headache. *European Journal of Neurology, 5*, 329–341.

Ellrich, J., Andersen, O. K., Messlinge, R. K., & Arendt-Nielsen, L. A. (1999). Convergence of meningeal and facial afferents onto trigeminal brainstem neurons: An electrophysiological study in rat and man. *Pain, 82*, 229–237.

Fricke, B., Andres, K. H., & von Düring, M. (2001). Nerve fibres innervating the cranial andspinal meninges: Morphology of nerve fiber terminals and their structural integration. *Microscopy Research andTechnique, 53*, 96–105.

Levy, D., & Strassman, A. M. (2002). Distinct sensitizing effects of the cAMP-PKA second messenger cascade on rat dural mechanonociceptors. *The Journal of Physiology, 538*(2), 483–493.

Meßlinger, K., Hanesch, U., Baumgärtel, M., Trost, B., & Schmidt, R. F. (1993). Innervation of the dura mater encephali of cat and rat: Ultrastructure and CGRP/SP-like immunoreactivity. *Anatomy and Embryology, 188*, 219–237.

Moskowitz, M. A., & Macfarlane, R. (1993). Neurovascular and molecular mechanisms in migraine headaches. *Cerebrovascular and Brain Metabolism Reviews, 5*, 159–177.

Ray, B. S., & Wolff, H. G. (1940). Experimental studies on headache: Pain sensitive structures of the head and their significance in headache. *Archives of Surgery, 1*, 813–856.

Schaible, H.-G., Ebersberger, A., Peppel, P., Beck, U., & Meßlinger, K. (1997). Release of immunoreactive substance P in the trigeminal brain stem nuclear complex evoked by chemical stimulation of the nasal mucosa and the dura mater encephali – a study with antibody microprobes. *Neuroscience, 76*, 273–284.

Schepelmann, K., Ebersberger, A., Pawlak, M., Oppmann, M., & Messlinger, K. (1999). Response properties of trigeminal brain stem neurons with input from dura mater encephali in the rat. *Neuroscience, 90*, 543–554.

Storer, R. J., Akerman, S., & Goadsby, P. J. (2004). Calcitonin gene-related peptide (CGRP) modulates nociceptive trigeminovascular transmission in the cat. *British Journal of Pharmacology, 142*, 1171–1181.

Strassman, A. M., & Raymond, S. A. (1999). Electrophysiological evidence for tetrodotoxin-resistant sodium channels in slowly conducting dural sensory fibres. *Journal of Neurophysiology, 81*, 413–424.

Strassman, A. M., Raymond, S. A., & Burstein, R. (1996). Sensitization of meningeal sensory neurons and the origin of headaches. *Nature, 384*, 560–564.

Yamamura, H., Malick, A., Chamberlin, N. L., & Burstein, R. (1999). Cardiovascular and neuronal responses to head stimulation reflect central sensitization and cutaneous allodynia in a rat model of migraine. *Journal of Neurophysiology, 81*, 479–493.

Nociceptors in the Orofacial Region (Skin/Mucosa)

Alfredo Ribeiro-da-Silva[1] and Brian Y. Cooper[2]
[1]Department of Pharmacology and Therapeutics, McGill University, Montreal, QC, Canada
[2]Department of Oral and Maxillofacial Surgery, University of Florida, Gainesville, FL, USA

Synonyms

High-threshold mechanoreceptor; Mechanoheat nociceptor; Mechanonociceptor; Nociceptors in the mucosa; Nociceptors in the skin; Polymodal nociceptor

Definition

Nociceptors detect and encode stimuli with actual or potential tissue damaging properties. Many are also responsive to chemicals released from traumatized tissue, inflammatory, and immune system cells. Nociceptors are notable among sensory afferents for their plasticity. Detection and encoding capacities undergo rapid quantitative and slower qualitative upregulatory adaptations (sensitization). These contribute to both peripheral and central nervous system changes, which mediate increased pain sensitivity following tissue or nervous system injury. These adaptations serve to protect tissue from further damage and promote healing. Orofacial nociceptors supply a diverse set of tissues that include skin, muscle, bone, and joints but also highly specialized structures such as cornea, mucosa (oral and nasal), and teeth. Accordingly, some of the anatomic and physiologic features are unique. The reader should examine the essays on corneal and tooth pulp nociceptors for further details.

Characteristics

Trigeminal nociceptors of skin and mucosa are a diverse population of mechanical, thermal, and chemically responsive afferents that detect and encode intense physical, thermal, and chemical events associated with actual or near tissue damage. The trigeminal root ganglion (TRG, Gasserian ganglion, semilunar ganglion) contains cell bodies of orofacial skin and mucosal nociceptors. Trigeminal nociceptors are anatomically distinct from spinal nociceptors (derived from dorsal root ganglia) in that their central projections terminate in the brainstem rather than in the spinal cord. In other respects, they are very similar (Hu 2000; Sessle 2000). In addition to detection and encoding of noxious stimuli, some nociceptors of the skin and mucosa are able to release the neuropeptides substance P and calcitonin gene-related peptide (CGRP) at both the peripheral and central processes. Peripherally, these peptides are involved in neurogenic inflammation (plasma extravasation, vasodilatation, and mast cell degranulation), while centrally, they are released with primary neurotransmitters (e.g., glutamate) to modify nociceptive transmission at the first synapse (Cooper and Sessle 1993). These peptidergic nociceptors include myelinated and unmyelinated nociceptors (Aδ and C class) and have been described in both facial and oral mucosa tissues – Aδ high-threshold mechanoreceptors (HTM), Aδ mechanoheat nociceptors (MH), Aδ polymodal nociceptors (PMN), and C PMN (Cooper et al. 1991; Bongenhielm and Robinson 1996; Toda et al. 1997; Flores et al. 2001). Many nociceptors do not express neuropeptides and so are termed nonpeptidergic. These represent an heterogeneous population; the most abundant population comprises mostly unmyelinated afferents and can be recognized by their binding of the isolectin B4 (IB4) and immunoreactivity for the purinergic receptor P2X3 (Taylor et al. 2009), whereas a smaller population represent myelinated nociceptors and are not easily identifiable except by electrophysiology. The latter have been identified in the spinal system and represent myelinated nonpeptidergic HTM (Light and Perl 1979; Woodbury et al. 2008) and are thought to occur in the trigeminal system as well.

In facial hairy skin, peptidergic afferents terminate in the dermis and epidermis as free nerve

Nociceptors in the Orofacial Region (Skin/Mucosa), Fig. 1 Nociceptive innervation of the lower lip skin and oral mucosa. (a) and (b) represent confocal images of the skin of rat lower lip and oral mucosa, respectively, showing the innervation pattern by nociceptive peptidergic (CGRP-immunoreactive – in *red*) and nonpeptidergic (P2X3-immunoreactive – in *green*) fibers. In (a) note that these fibers travel in bundles in small cutaneous nerves in dermis (*arrowhead*) and branch extensively. Both populations penetrate the epidermis (*arrows*), but the P2X3-immunoreactive population is clearly more abundant in the epidermis (*white arrow*) than the peptidergic (*yellow arrow*). In the oral mucosa (b), note that the nonpeptidergic innervation of the epithelium is clearly less abundant than in skin (*white arrow*). (c) represents an electron micrograph showing the termination of two substance P-immunoreactive fibers (*arrows*) in the proximity of a mast cell (*M*). In the fibers, note both agranular and dense-core vesicles. Dense precipitate represents substance P immunoreactivity (a and b were adapted and modified from Taylor et al. J. Comp. Neurol. 514 (6):555–566, 2009; c was reproduced with permission from Ribeiro-da-Silva and Hökfelt, Neuropeptides 34 (5):256–271, 2000)

endings (Fig. 1a) (Ruocco et al. 2001). In the dermis, they innervate blood vessels, hair follicles, and glands and are found in close proximity to mast cells (Fig. 1c). At the ultrastructural level, the terminals of nociceptive afferents in facial skin display numerous granular and agranular synaptic vesicles (Fig. 1c) (Ruocco et al. 2001; Ribeiro-da-Silva and Hökfelt 2000), suggesting the peripheral release of both neuropeptides and the classical transmitter glutamate. Nonpeptidergic nociceptive afferents, as recognized by means of immunoreactivity for P2X3, innervate mostly the epidermis and adjacent region of dermis (Fig. 1a) (Vulchanova et al. 1998; Taylor et al. 2009) and have also been detected in the epithelium of the cornea (Vulchanova et al. 1998). In contrast to the peptidergic fibers, P2X3-immunoreactive fibers

Nociceptors in the Orofacial Region (Skin/Mucosa), Fig. 2 Mucosal nociceptors respond differentially to mechanical, chemical, and thermal stimulation. (a) Brisk response of an Aδ HTM of the rat mucosa to mechanical stimulation. The fiber is less responsive to other modalities. (b) An Aδ MH responds to both mechanical and thermal stimuli (Reprinted from Neuroscience Letters 228, K. Toda, N. Ishii and Y. Nakamura, "Characteristics of mucosal nociceptors in the rat oral cavity: an in vitro study" pp. 95–98, 1997 with permission from Elsevier. *BK* bradykinin 0.1 μM)

seldom terminate in vessel walls (Taylor et al. 2009). It is still unknown whether the terminals of nonpeptidergic afferents display synaptic vesicles in the periphery. In the oral mucosa, the innervation by P2X3-immunoreactive fibers is much less abundant than in the skin (Fig. 1b) (Taylor et al. 2009), suggesting that the epithelium of the oral mucosa is mostly innervated by peptidergic afferents (Kruger et al. 1989). The termination in skin and oral mucosa of HTM afferents, which are nonpeptidergic nociceptors, is at present unknown, although it is assumed that like in the rest of the body, they terminate as free nerve endings and also penetrate the epidermis (Kruger et al. 1981). In the trigeminal ganglion, the small- and medium-sized cells represent the main populations of nociceptive neurons. The large neurons of the TRG may also contain

nociceptive populations, although they represent a minority.

Nociceptors are highly specialized to detect chemical agents associated with tissue trauma and may sensitize following exposure to chemical algesics. Both peptidergic and nonpeptidergic nociceptive afferents express cholinergic (both nicotinic and muscarinic; Liu et al. 1993; Carstens et al. 2000; Steen and Reeh 1993; Bernardini et al. 2002) or acid-sensitive receptors and ion channels (ASIC; Ichikawa and Sugimoto 2002), whereas the expression of P2X3 is restricted to nonpeptidergic afferents (Taylor et al. 2009). These channels enable nociceptors to detect the presence of ATP, acetylcholine (ACh), or protons that are associated with tissue damage and inflammation. ATP and ACh are released from damaged cells, while high levels of protons may be

associated with infection or ischemia due to compromise of the vascular supply. Many of these
same populations express the capsaicin-sensitive
protein TRPV1, which transduces noxious heat
stimuli and is critical to the development of heat
sensitization (Ichikawa and Sugimoto 2000;
Stenholm et al. 2002). Responses to cooling, bradykinin, PGE2, and histamine have also been
described (Toda et al. 1997; Viana et al. 2002)
(Fig. 2).

References

Bernardini, N., Roza, C., Sauer, S. K., Gomeza, J.,
Wess, J., & Reeh, P. W. (2002). Muscarinic M2
receptors on peripheral nerve endings: A molecular
target of antinociception. Journal of Neuroscience,
22, RC229.
Bongenhielm, U., & Robinson, P. P. (1996). Spontaneous
and mechanically evoked afferent activity originating
from myelinated fibres in ferret inferior alveolar nerve
neuromas. Pain, 67, 399–406.
Carstens, E., Simons, C. T., Dessirier, J. M., Carstens,
M. I., & Jinks, S. L. (2000). Role of neuronal nicotinic-acetylcholine receptors in the activation of neurons in trigeminal subnucleus caudalis by nicotine
delivered to the oral mucosa. Experimental Brain
Research, 132, 375–383.
Cooper, B. Y., Ahlquist, M., Friedman, R. M., Loughner,
B., & Heft, M. (1991). Properties of high-threshold
mechanoreceptors in the oral mucosa. I. Responses to
dynamic and static pressure. Journal of Neurophysiology, 66, 1272–1279.
Cooper, B. Y., & Sessle, B. J. (1993). Physiology of
nociception in the trigeminal system. In J. Olesen, P.
Tfelt-Hansen, & K. M. A. Welch (Eds.), The headaches (pp. 87–92). New York: Raven.
Flores, C. M., Leong, A. S., Dussor, G. O., Harding-Rose,
C., Hargreaves, K. M., & Kilo, S. (2001). Capsaicin-
evoked CGRP release from rat buccal mucosa: Development of a model system for studying trigeminal
mechanisms of neurogenic inflammation. European
Journal of Neuroscience, 14, 1113–1120.
Hu, J. W. (2000). Neurophysiological mechanism of head,
face and neck pain. In H. Vernon (Ed.), The craniocervical syndrome: Mechanisms, assessment and
treatment (pp. 31–48). Boston: Oxford.
Ichikawa, H., & Sugimoto, T. (2000). Vanilloid receptor
1-like receptor-immunoreactive primary sensory neurons in the rat trigeminal nervous system. Neuroscience, 101, 719–725.
Ichikawa, H., & Sugimoto, T. (2002). The co-expression
of ASIC3 with calcitonin gene-related peptide and
parvalbumin in the rat trigeminal ganglion. Brain
Research, 943, 287–291.

Kruger, L., Perl, E. R., & Sedivec, M. J. (1981). Fine
structure of myelinated mechanical nociceptor endings
in cat hairy skin. The Journal of Comparative Neurology, 198, 137–154.
Kruger, L., Silverman, J. D., Mantyh, P. W., Sternini, C.,
& Brecha, N. C. (1989). Peripheral patterns of calcitonin gene-related peptide general somatic sensory
innervation: Cutaneous and deep terminations. The
Journal of Comparative Neurology, 280, 291–302.
Light, A. R., & Perl, E. R. (1979). Spinal termination of
functionally identified primary afferent neurons with
slowly conducting myelinated fibers. The Journal of
Comparative Neurology, 186, 133–150.
Liu, L., Pugh, W., Ma, H., & Simon, S. A. (1993). Identification of acetylcholine receptors in adult rat trigeminal ganglion neurons. Brain Research, 617, 37–42.
Ribeiro-da-Silva, A., & Hökfelt, T. (2000). Neuroanatomical localisation of Substance P in the CNS and sensory
neurons. Neuropeptides, 34, 256–271.
Ruocco, I., Cuello, A. C., Shigemoto, R., & Ribeiro-da-
Silva, A. (2001). Light and electron microscopic study
of the distribution of substance P-immunoreactive
fibers and neurokinin-1 receptors in the skin of the rat
lower lip. The Journal of Comparative Neurology,
432, 466–480.
Sessle, B. J. (2000). Acute and chronic craniofacial pain:
Brainstem mechanisms of nociceptive transmission
and neuroplasticity, and their clinical correlates. Critical Reviews in Oral Biology and Medicine, 11, 57–91.
Steen, K. H., & Reeh, P. W. (1993). Actions of cholinergic
agonists and antagonists on sensory nerve endings
in rat skin, in vitro. Journal of Neurophysiology,
70, 397–405.
Stenholm, E., Bongenhielm, U., Ahlquist, M., & Fried, K.
(2002). VR1- and VRL-1-like immunoreactivity in normal and injured trigeminal dental primary sensory
neurons of the rat. Acta Odontologica Scandinavica,
60, 72–79.
Taylor, A. M., Peleshok, J. C., & Ribeiro-da-Silva, A.
(2009). Distribution of P2X3-immunoreactive fibers
in hairy and glabrous skin of the rat. The Journal of
Comparative Neurology, 514, 555–566.
Toda, K., Ishii, N., & Nakamura, Y. (1997). Characteristics of mucosal nociceptors in the rat oral cavity: An
in vitro study. Neuroscience Letters, 228, 95–98.
Viana, F., de la Pena, E., & Belmonte, C. (2002). Specificity of cold thermotransduction is determined by
differential ionic channel expression. Nature Neuroscience, 5, 254–260.
Vulchanova, L., Riedl, M. S., Shuster, S. J., Stone, L. S.,
Hargreaves, K. M., Buell, G., Surprenant, A., North,
R. A., & Elde, R. (1998). P2X3 is expressed by DRG
neurons that terminate in inner lamina II. European
Journal of Neuroscience, 10, 3470–3478.
Woodbury, C. J., Kullmann, F. A., McIlwrath, S. L., &
Koerber, H. R. (2008). Identity of myelinated cutaneous sensory neurons projecting to nocireceptive laminae following nerve injury in adult mice. The Journal
of Comparative Neurology, 508, 500–509.

Nociceptors in the Orofacial Region (Temporomandibular Joint and Masseter Muscle)

Brian E. Cairns
Faculty of Pharmaceutical Sciences, University of British Columbia, Vancouver, BC, Canada

Synonyms

Nociceptive masseter muscle afferents; Nociceptive temporomandibular joint afferents

Definition

Primary afferent fibers that innervate the ▶ temporomandibular joint (TMJ) and masticatory muscles, and are activated by noxious mechanical, chemical, or thermal stimuli applied to these tissues. These afferent fibers transduce and convey information about potential or actual tissue injury from the orofacial region to the central nervous system.

Introduction

Nociceptors are the first link in the pain pathway from the TMJ and masticatory muscles to the central nervous system. These nerve fibers have nonspecialized endings which respond to mechanical and/or chemical stimuli that cause pain in humans. Increases in the level of various pain-causing or algogenic compounds can alter the mechanical sensitivity of nociceptors that innervate TMJ and masticatory muscles, leading to the phenomenon of ▶ peripheral sensitization. The following sections describe the characteristics of nociceptors that innervate these craniofacial structures.

Characteristics

TMJ Nociceptors
The TMJ is innervated by thinly myelinated and unmyelinated afferent fibers with nonspecialized

endings that contain clear and dense core vesicles, which suggests some of these afferent fibers release neurotransmitters and neuropeptides, such as ▶ calcitonin gene related peptide (CGRP) and ▶ substance P, from their terminal endings (Kido et al. 1995). These small diameter afferent fibers project, via the gasserian or ▶ trigeminal ganglion, to the trigeminal ▶ subnucleus interpolaris and ▶ caudalis, areas of the caudal brain stem which appear to be most important for the integration of nociceptive input from deep orofacial tissues (Capra 1987; Widenfalk and Wiberg 1990; Casatti et al. 1999). Electrophysiological studies have confirmed the projection of a subpopulation of TMJ afferent fibers with conduction velocities of less than 25 m/s (Ad and C fibers) to the caudal brain stem (Cairns et al. 2001b). These fibers are activated by noxious mechanical and/or chemical stimuli and appear to function as nociceptors (Cairns et al. 2001b).

TMJ afferent fibers identified as mechanical nociceptors by their response to noxious protrusion or lateral movement of the TMJ have been described (Loughner et al. 1997; Cairns et al. 2001b). These fibers are not activated by innocuous jaw opening but begin to discharge as lateral rotation of the jaw exceeds the normal range and exhibit a progressively increased discharge with supranormal rotation of the jaw. Some of these nociceptors also appear to encode rate of jaw rotation (Loughner et al. 1997). The threshold of TMJ nociceptors to noxious mechanical rotation of the jaw is lower in females than males; however, this is apparently due to sex-related differences in the biomechanical properties of the TMJ tissues (Loughner et al. 1997).

TMJ nociceptors respond to injection of algogenic substances, such as ▶ capsaicin, potassium chloride, and glutamate into the TMJ, which also evokes a nociceptive reflex response in the jaw muscles (Cairns et al. 2001b; Lam et al. 2009). The activity of TMJ nociceptors precedes but has a markedly shorter duration than reflex jaw muscle activity evoked by injection of glutamate into the TMJ. This finding has led to the speculation that a brief activation of TMJ nociceptors by algogenic compounds is sufficient to induce central sensitization, a period of prolonged increase in the excitability of trigeminal subnucleus caudalis neurons (Cairns et al. 2001b). Such a phenomenon may explain the diffuse referral pattern of TMJ pain, which may spread to include the masticatory and neck muscles, and why acute joint pain can sometimes significantly outlast the duration of nociceptive stimulation.

Sex-related differences in the chemical response characteristics of TMJ nociceptors have also been noted. The greatest response to algogenic compounds has been observed in small, mechanically sensitive afferent fibers with conduction velocities of less than 10 m/s, which suggests that these particular fibers function as polymodal nociceptors, that is, nociceptors which respond to more than one type of noxious stimulation. Sex-related differences in response to injection of glutamate and adenosine triphosphate into the TMJ have been best characterized. Injection of these compounds into the TMJ has been found to evoke greater responses in females than in males (Cairns 2007). Such sex-related differences in TMJ nociceptor excitability may in part explain the increased prevalence of certain orofacial pain syndromes in women.

Algogenic compounds, such as mustard oil, capsaicin, adenosine triphosphate, and glutamate, excite TMJ nociceptors in part through activation of peripheral NMDA and non-NMDA receptors (Lam et al. 2009; Watanabe et al. 2010). Although this suggests that peripheral glutamate receptor antagonists may be of use in modifying the excitability of TMJ nociceptors under certain pathological conditions, injection of ketamine into the TMJ of patients with ▶ arthralgia was not found to be effective for the reduction of pain (Ayesh et al. 2008).

Masticatory Muscle Nociceptors

Anatomical and electrophysiological studies indicate that the masseter and temporalis (jaw closer) muscles are also innervated by thinly myelinated and unmyelinated trigeminal afferent fibers with nonspecialized endings that project to the trigeminal subnucleus interpolaris and caudalis (Nishimori et al. 1986; Capra and

Wax 1989; Cairns et al. 2001a, 2002, 2003). These fibers are activated by noxious mechanical and/or chemical stimuli and appear to function as nociceptors (Cairns et al. 2001a, 2002, 2003).

About one third of masticatory muscle afferent fibers that project to subnucleus caudalis have mechanical thresholds that exceed the human pressure pain threshold. In uninjured masseter muscle, these nociceptors are predominantly Aδ fibers with conduction velocities of less than 12 m/s (Cairns et al. 2002, 2003). Most of these nociceptors exhibit slowly adapting responses to sustained noxious mechanical stimulation (Fig. 1). A significant sex-related difference in the mechanical threshold of these nociceptors has not been found (Cairns 2007).

Masseter muscle mechanical nociceptors also respond to the injection of many algogenic substances, such as hypertonic saline, capsaicin, adenosine triphosphate, and glutamate, into their mechanoreceptive field (Dong et al. 2009; Lam et al. 2009). The afferent discharge evoked by these algogenic substances is greatest in slowly conducting afferent fibers with moderate-to-high mechanical thresholds, which suggests that these particular fibers function as polymodal nociceptors.

Glutamate consistently evokes significantly greater nociceptor discharges and pain responses in females than in males (Cairns 2007) (see examples, Fig. 1). Glutamate-evoked discharge in masseter nociceptors is predominantly mediated through activation of peripheral NMDA receptors (Cairns et al. 2003). This sex-related difference has been demonstrated to be due to an estrogen-related increased expression of peripheral NMDA receptors by masticatory muscle nociceptors (Dong et al. 2007). Prolonged mechanical sensitization of the masseter muscle and its nociceptors has also been demonstrated to occur after injection of glutamate and a number of other algogenic substances into the masseter muscle, although there do not appear to be sex-related differences in this phenomenon (Dong et al. 2007, 2009). These findings may provide a rationale for the increased prevalence of masticatory muscle pain conditions suffered by women.

Nociceptors in the Orofacial Region (Temporomandibular Joint and Masseter Muscle), Fig. 1 Examples of deep orofacial tissue nociceptor response characteristics. (a) The *line* drawing illustrates antidromic action potentials (*), evoked by electrical stimulation of the caudal brain stem, to confirm the central projection target of these masseter muscle Aδ nociceptors. (b) Sustained noxious mechanical stimulation of the masseter muscle with an electronic Von Frey hair (*lower trace*) resulted in a slowly adapting discharge (*upper trace*). (c) The peristimulus histograms illustrate the effect of injection of the algogenic substance, glutamate, into the masseter muscle. Note that glutamate-evoked nociceptor discharge was markedly greater in the female than in the male

Summary

The role of orofacial nociceptors is to transduce and convey information about the intensity and quality of orofacial pain. The characteristics of TMJ and masseter muscle nociceptors suggest that they may play a role not only in the development but also in the maintenance of certain types of orofacial pain. Certain orofacial pain syndromes, such as temporomandibular disorders, which involve TMJ and masticatory muscle pain upon palpation and/or during chewing, appear to be more prevalent in women than men (Cairns 2007). This sex-related difference in orofacial pain may, in part, also be explained by sex-related differences in the response characteristics of TMJ and masseter nociceptors.

Acknowledgment The author would like to acknowledge research support from the National Institutes of Health and the Canadian Institutes of Health Research.

References

Ayesh, E. E., Jensen, T. S., & Svensson, P. (2008). Effects of intra-articular ketamine on pain and somatosensory function in temporomandibular joint arthralgia patients. *Pain, 137*, 286–294.

Cairns, B. E. (2007). The influence of gender and sex steroids on craniofacial nociception. *Headache, 47*, 319–324.

Cairns, B. E., Hu, J. W., Arendt-Nielsen, L., Sessle, B. J., & Svensson, P. (2001a). Sex-related differences in human pain perception and rat afferent discharge evoked by injection of glutamate into the masseter muscle. *Journal of Neurophysiology, 86*, 782–791.

Cairns, B. E., Sessle, B. J., & Hu, J. W. (2001b). Characteristics of glutamate-evoked temporomandibular joint afferent activity in the rat. *Journal of Neurophysiology, 85*, 2446–2454.

Cairns, B. E., Gambarota, G., Svensson, P., Arendt-Nielsen, L., & Berde, C. B. (2002). Glutamate-induced sensitization of rat masseter muscle fibers. *Neuroscience, 109*, 389–399.

Cairns, B. E., Svensson, P., Wang, K., et al. (2003). Activation of peripheral NMDA receptors contributes to human pain and rat afferent discharges evoked by injection of glutamate into the masseter muscle. *Journal of Neurophysiology, 90*, 2098–2105.

Capra, N. F. (1987). Localization and central projections of primary afferent neurons that innervate the temporomandibular joint in cats. *Somatosensory Research, 4*, 201–213.

Capra, N. F., & Wax, T. D. (1989). Distribution and central projections of primary afferent neurons that innervated the masseter muscle and mandibular periodontium: A double-label study. *The Journal of Comparative Neurology, 279*, 341–352.

Casatti, C. A., Frigo, L., & Bauer, J. A. (1999). Origin of sensory and autonomic innervation of the rat temporomandibular joint: A retrograde axonal tracing study with the fluorescent dye fast blue. *Journal of Dental Research, 78*, 776–783.

Dong, X. D., Mann, M. K., Kumar, U., et al. (2007). Sex-related differences in NMDA-evoked rat masseter muscle afferent discharge result from estrogen-mediated modulation of peripheral NMDA receptor activity. *Neuroscience, 146*, 822–832.

Dong, X.-D., Svensson, P., & Cairns, B. E. (2009). The analgesic action of topical diclofenac may be mediated through peripheral NMDA receptor antagonism. *Pain, 147*, 36–45.

Kido, M. A., Kiyoshima, T., Ibuki, T., Shimizu, S., Kondo, T., Terada, Y., & Tanaka, T. (1995). A topographical and ultrastructural study of sensory trigeminal nerve endings in the rat temporomandibular joint as demonstrated by anterograde transport of wheat germ agglutinin-horseradish peroxidase (WGA-HRP). *Journal of Dental Research, 74*, 1353–1359.

Lam, D. K., Sessle, B. J., & Hu, J. W. (2009). Glutamate and capsaicin effects on trigeminal nociception I: Activation and peripheral sensitization of deep craniofacial nociceptive afferents. *Brain Research, 1251*, 130–139.

Loughner, B., Miller, J., Broumand, V., & Cooper, B. (1997). The development of strains, forces and nociceptor activity in retrodiscal tissues of the temporomandibular joint of male and female goats. *Experimental Brain Research, 113*, 311–326.

Nishimori, T., Sera, M., Suemune, S., et al. (1986). The distribution of muscle primary afferents from the masseter nerve to the trigeminal sensory nuclei. *Brain Research, 372*, 375–381.

Watanabe, T., Tsuboi, Y., Sessle, B. J., Iwata, K., & Hu, J. W. (2010). P2X and NMDA receptor involvement in temporomandibular joint-evoked reflex activity in rat jaw muscles. *Brain Research, 1346*, 83–91.

Widenfalk, B., & Wiberg, M. (1990). Origin of sympathetic and sensory innervation of the temporomandibular joint. A retrograde axonal tracing study in the rat. *Neuroscience Letters, 109*, 30–35.

Nociceptors in the Skin

▶ Nociceptors in the Orofacial Region (Skin/Mucosa)

Nociceptors, Action Potentials, and Post-Firing Excitability Changes

Bruce Lynn
Department of Physiology, University College London, London, UK

Synonyms

Axonal action potentials; Different nociceptor populations; Primary afferent

Definition

The electrophysiological properties of nociceptive neurons.

Characteristics

This entry will consider the electrical properties of mammalian nociceptive neurons, covering both the action potential itself and the changes in excitability that follow it. Much of the work in this area has been concerned with major differences between nociceptors on the one hand, and non-nociceptive afferent neurons on the other. There has also been a relatively small amount of work showing clear differences within nociceptor subclasses. As well as describing differences in such properties as time course or amplitude, an attempt will be made to look at likely differences in underlying membrane properties, especially in the subclasses of ▶ ion channel present. Differences in electrical properties directly affect the ability of nociceptive neurons to encode information about stimuli, for example, by limiting firing rates. In addition, however, the finding of differences in the key proteins controlling excitability between nociceptors and non-nociceptors, and between subclasses of nociceptor, opens up important possibilities for the development of selective analgesic agents.

Action Potentials in Nociceptive Neurons

Axonal Action Potentials

Extracellular recordings from unmyelinated fibers have revealed that there are differences in the action potential time course between subclasses of nociceptor (Fig. 1). The only extensive study has been in the pig (Gee et al. 1999), where two major classes of C-fiber nociceptor have been found in the skin. Polymodal nociceptors respond to all types of nociceptive stimuli, whereas heat nociceptors respond to heat and chemicals, but not to pressure. The heat nociceptors have ▶ spikes of longer duration and with longer after-potentials than do the polymodal nociceptors (Fig. 1). In the rat, there are mainly polymodal nociceptors, and these have wide spikes compared with non-nociceptive afferents (Gee et al. 1999). However, no comparisons have been made with other classes of C-fiber nociceptor in this species. In both pig and rat, there are also A-fiber nociceptors, and these have narrow spikes compared with C-fiber nociceptors, as one would expect from the generally faster excitability changes seen in myelinated compared with unmyelinated ▶ axons. In man, microneurographic recordings have again revealed longer duration action potentials in C-nociceptor axons compared with C-cold afferent axons (Serra et al. 1999). Interestingly, a group of mechanically insensitive axons with small, short duration spikes was also observed (Serra et al. 1999). The functional identity of these fibers has not been established, but they might be a second group of very mechanically insensitive nociceptors with different axonal properties.

Soma Action Potentials

Intracellular recordings from dorsal root ganglion cells reveal that, as described above for the axons, nociceptive neurons have spikes with a slower time course than mechanoreceptor neurons with a similar conduction velocity (Lawson 2002). This is true for the ▶ soma spikes of neurons with myelinated and unmyelinated axons. Often the broad spikes of nociceptive neurons have a distinct inflection (or "shoulder") on the falling phase. After-potentials are also of longer duration. No studies have been made of soma spike shape for subclasses of nociceptor. One interesting observation is that following peripheral inflammation in the innervation territory of the neuron, the soma spike becomes shorter in duration (Djouhri and Lawson 1999).

Mechanism of Differences in Action Potential Duration

The longer action potential durations in nociceptive neurons may be due to a combination of slower sodium channel kinetics and an additional calcium current (Blair and Bean 2002). The somas of nociceptive neurons contain much higher levels of the sodium channels NaV1.7, NaV 1.8, and NaV 1.9 (Djouhri et al. 2003a, b; Fang et al. 2002). Of these, the ▶ TTX-resistant channel NaV 1.8 (SNS) is probably the most important contributor to the action potential (Wood et al. 2004). This channel has slower kinetics than the TTX-sensitive channels that dominate in non-nociceptive neurons, so this may be part of the reason that spikes have a slower rise time and longer durations. There is also evidence that C-fiber, presumed nociceptor, axons have both TTX-resistant sodium current and calcium transients, so similar factors may operate at an axonal level as in the soma (Grosskreutz et al. 1996; Mayer et al. 1999).

Post-Firing Excitability Changes in Nociceptive Neurons

Following an action potential, the excitability of neurons undergoes a series of fluctuations. Typically, a period of subnormal excitability is followed by a supernormal period of slightly increased excitability, and then sometimes a further period of slightly reduced excitability.

Nociceptors, Action Potentials, and Post-Firing Excitability Changes, Fig. 1 Axonal action potentials recorded from the saphenous nerve of the anaesthetized pig. (a) Examples of three C-fiber spikes recorded from the same filament. Two (*thick lines*) were polymodal nociceptor units, and one (*wider spike with thinner line*) was a heat nociceptor. (b) Duration of the main peak of the action potential at half-maximum amplitude for different types of C-fiber (mean ± s.e.). Analysis of variance showed that heat nociceptors had significantly longer duration spikes than polymodal nociceptors, and that both classes of nociceptor had longer duration spikes than the two non-nociceptor classes. (c) Duration of action potential undershoots at half-maximum amplitude. Analysis of variance showed that the two classes of nociceptor had longer duration after-potentials than the non-nociceptors, but no significant difference between heat and polymodal nociceptors. Class labels: inex unit with no afferent field to pressure or heat, mech sensitive mechanoreceptor (non-nociceptive), PMN polymodal nociceptor, heat heat nociceptor. Number of units in brackets below the class labels. From Gee et al. (1999)

The effects of one action potential can be detected in some neurons for several minutes. A readily observed effect of these excitability changes is that subsequent action potentials propagate with a slowed conduction velocity during periods of subnormal excitability, and with a faster conduction velocity during the supernormal period.

Analysis of Conduction Velocity Slowing in C-Fibers

Conduction velocity changes during repetitive firing of both A-fiber and C-fiber afferents are clearly greater in nociceptors than in non-nociceptive axons (Gee et al. 1996). In human nerves, there are also differences between subclasses of nociceptor. Conduction velocity slowing is more marked in mechanically inexcitable C-fiber afferents (CMi) than in those with noxious pressure receptive fields (C-responsive or CMR). Note that such neurons are also usually heat sensitive, so correspond to C-polymodal nociceptors or may be designated CMH (Weidner et al. 1999). Conduction velocity slowing is also more marked in terminal branches of C-fibers than in the axons within the main nerve trunk (Weidner et al. 2003).

A supernormal period is also seen at all stimulus frequencies in CMi afferents, while it was only present at faster stimulation frequencies, with greater preexisting slowing, in CMH fibers. The supernormal period in C-nociceptor axons has the intriguing effect of causing action potentials to catch each other up at high frequencies of stimulation, from approximately 20 Hz upward. This means that two stimuli given at, say, 50-ms intervals evoke action potentials that are only 10 ms apart by the time they reach the central nervous system (they cannot catch up completely as the short-term subnormal period corresponding to the relative refractory period sets an upper limit to firing rate). This substantial change in frequency of firing could well increase the effectiveness of such inputs at synapses. In contrast, at lower frequencies, typically less than 10 Hz, spikes are in the late subnormal period and some slight decrease in final firing rate will occur during propagation of spikes to the central nervous system. Weidner et al. (2002) suggest that this phenomenon may lead to a degree of contrast enhancement, with the apparent firing rate for strong stimuli being enhanced relative to the firing evoked by weaker stimulation.

Mechanisms Underlying Post-Firing Changes in Nociceptor Excitability

The slow changes in excitability after an action potential will reflect: (a) which particular ion channels are operating, (b) any accumulation of ions intracellularly or in the immediate extracellular space, and (c) hyperpolarization due to increased activity of the ▶ sodium pump. Immediate post-firing subnormal excitability is caused by residual sodium channel inactivation plus delay in closing of depolarization-activated potassium channels. The supernormal period appears to be a passive consequence of the long time constant (around 100 ms) of the C-fiber membrane (Bostock et al. 2003; Weidner et al. 2002). In this respect, mammalian C-fibers resemble the internodal membrane of myelinated fibers. The very long-term subnormality appears to reflect the hyperpolarizing action of the sodium pump (Bostock et al. 2003). The studies to date have not examined why CMi neurons have greater subnormality at long interstimulus intervals than CMR neurons, but this may reflect either some difference in sodium pumping or other very slowly changing ion channels. In sympathetic C-fibers, hyperpolarization-activated inward currents (Ih) appear to play a role. However, double pulse experiments with nociceptive C-fibers did not show any changes attributable to Ih (Bostock et al. 2003).

Summary

The action potentials, both in the axons and the soma, are of longer duration in nociceptors than in other classes of somatosensory afferent neuron. Intriguing differences also exist between subclasses of nociceptor in their action potential shape and in their post-firing excitability changes. In general, the mechanically insensitive nociceptors show longer duration axonal spikes and more profound depression of excitability by slow (2 Hz or less) activation. The differences between nociceptors and non-nociceptors may reflect the presence of slower TTX-resistant currents, plus significant calcium influx, during nociceptor action potentials. The reasons why CMi differ from CMR nociceptors are not known. The importance of CMi fibers for secondary hyperalgesia (Serra et al. 2004) and their involvement in chronic pain conditions (Orstavik et al. 2003) makes these differences worthy of further study. A tantalizing possibility is that differences in electrophysiological properties between nociceptors and non-nociceptors, or even possibly between subclasses of nociceptor, might provide a basis for novel analgesic drugs. Some work on sodium channel blockers is already published and this is likely to remain an active research field (Wood et al. 2004).

References

Blair, N. T., & Bean, B. P. (2002). Roles of tetrodotoxin (TTX)-sensitive Na^+ current, TTX-resistant Na^+ current, and Ca^{2+} current in the action potentials of nociceptive sensory neurons. *Journal of Neuroscience, 22*, 10277–10290.

Bostock, H., Campero, M., Serra, J., et al. (2003). Velocity recovery cycles of C fibres innervating human skin. *The Journal of Physiology, 553*, 649–663.

Djouhri, L., & Lawson, S. N. (1999). Changes in somatic action potential shape in guinea-Pig nociceptive primary afferent neurones during inflammation *in vivo. The Journal of Physiology, 520*, 565–576.

Djouhri, L., Fang, X., Okuse, K., et al. (2003a). The TTX-resistant sodium channel Nav1.8 (SNS/PN3): Expression and correlation with membrane properties in rat nociceptive primary afferent neurons. *The Journal of Physiology, 550*, 739–752.

Djouhri, L., Newton, R., Levinson, S. R., et al. (2003b). Sensory and electrophysiological properties of guinea-pig sensory neurones expressing Nav 1.7 (PN1) Na^+ channel alpha subunit protein. *The Journal of Physiology, 546*, 565–576.

Fang, X., Djouhri, L., Black, J. A., et al. (2002). The presence and role of the tetrodotoxin-resistant sodium channel Na(v)1.9 (NaN) in nociceptive primary afferent neurons. *Journal of Neuroscience, 22*, 7425–7433.

Gee, M. D., Lynn, B., & Cotsell, B. (1996). Activity-dependent slowing of conduction velocity provides a method for identifying different functional classes of C-fibre in the rat saphenous nerve. *Neuroscience, 73*, 667–675.

Gee, M. D., Lynn, B., Basile, S., et al. (1999). The relationship between axonal spike shape and functional modality in cutaneous C-fibres in the pig and rat. *Neuroscience, 90*, 509–518.

Grosskreutz, J., Quasthoff, S., Kuhn, M., et al. (1996). Capsaicin blocks tetrodotoxin-resistant sodium potentials and calcium potentials in unmyelinated C fibres of biopsied human sural nerve in vitro. *Neuroscience Letters, 208*, 49–52.

Lawson, S. N. (2002). Phenotype and function of somatic primary afferent nociceptive neurones with C-, adelta-or aalpha/beta-fibres. *Experimental Physiology, 87*, 239–244.

Mayer, C., Quasthoff, S., & Grafe, P. (1999). Confocal imaging reveals activity-dependent intracellular Ca^{2+} transients in nociceptive human C fibres. *Pain, 81*, 317–322.

Orstavik, K., Weidner, C., Schmidt, R., et al. (2003). Pathological C-fibres in patients with a chronic painful condition. *Brain, 126*, 567–578.

Serra, J., Campero, M., Ochoa, J., et al. (1999). Activity-dependent slowing of conduction differentiates functional subtypes of C fibres innervating human skin [see comments]. *The Journal of Physiology, 515*, 799–811.

Serra, J., Campero, M., Bostock, H., et al. (2004). Two types of C nociceptors in human skin and their behavior in areas of capsaicin-induced secondary hyperalgesia. *Journal of Neurophysiology, 91*, 2770–2781.

Weidner, C., Schmelz, M., Schmidt, R., et al. (1999). Functional attributes discriminating mechano-insensitive and mechano-responsive C nociceptors in human skin. *Journal of Neuroscience, 19*, 10184–10190.

Weidner, C., Schmelz, M., Schmidt, R., et al. (2002). Neural signal processing: the underestimated contribution of peripheral human C-fibers. *Journal of Neuroscience, 22*, 6704–6712.

Weidner, C., Schmidt, R., Schmelz, M., et al. (2003). Action potential conduction in the terminal arborisation of nociceptive C-fibre afferents. *The Journal of Physiology, 547*, 931–940.

Wood, J. N., Boorman, J. P., Okuse, K., et al. (2004). Voltage-gated sodium channels and pain pathways. *Journal of Neurobiology, 61*, 55–71.

Nociceptors, Cold Thermotransduction

Elvira De la Pena[1] and Félix Viana[2]
[1]Instituto de Neurociencias, Universidad Miguel Hernández-CSIC, Alicante, Spain
[2]Instituto de Neurociencias, Universidad Miguel Hernandez-CSIC, San Juan de Alicante, Spain

Synonyms

Cold nociception; Cold thermotransduction; Noxious cold receptor

Definition

A large fraction of nociceptors can be excited by application of cold temperatures to their peripheral endings. Most have the functional properties of C-type polymodal nociceptors. The molecular sensors involved in transducing strong cooling stimuli into an electrical signal are still unresolved; ► TRPA1 channels are contested candidates. Noxious and thermal signals are further processed in the brain to establish the intensity and quality of the sensation. Peripheral nerve injury can modify the process to give rise to ► cold allodynia or ► hyperalgesia.

Characteristics

Humans can feel a wide range of ambient temperatures. This capacity is fundamental for tactile recognition of objects and thermoregulation. Within the cold temperature range, the qualities of sensations evoked vary from pleasantly cool to extremely painful. The neutral zone, evoking no sensation upon temperature change of the skin ranges between 35 °C and 31 °C. Cutaneous temperatures of 30–15 °C are generally perceived as cool to cold. Upon further temperature reduction, the perceptual qualities of the sensation change, becoming painful. The sensation of cold pain can have a burning, aching, prickling, or stinging quality, depending on temperature and stimulus duration, possibly reflecting the activation of different classes of afferents. In contrast to the sharp threshold temperature for heat pain, the threshold for cold pain is less clearly defined and influenced by several factors such as rate of temperature change and stimulus area, indicating that mechanisms of temporal and spatial summation participate in encoding these sensations. Generally, the sensation turns painful only after a considerable delay, many seconds after cold application, which further complicates the definition of a threshold. Focal skin temperatures above approximately 43–45 °C also evoke a ► paradoxical cold sensation in many individuals.

While significant advances in the cellular and molecular mechanisms responsible for temperature transduction by nerve terminals have taken place in recent years (reviewed by Jordt et al. 2003; Patapoutian et al. 2003; Reid 2005), many important aspects of the function of ► cold thermoreceptors and nociceptors and how cold pain is encoded remain obscure. Thus, it is unknown which molecular and cellular factors determine the different temperature thresholds of individual sensory terminals. Also, the mechanisms involved in the development of pain by moderate cooling after nerve injury (► cold hyperalgesia) remain mysterious. These are important questions with implications in the treatment of disorders in which cold temperatures evoke pain.

The existence of separate, small, cold, and warm cutaneous sensory spots has been known since the late nineteenth century. These findings lend support to von Frey's specificity theory of somesthesia, according to which sensory nerve fibers of the skin were sensitive to only one form of stimulation and acted as "labeled lines" for the transmission of information encoding a single perceptual quality. In contrast to the strict labeled line hypothesis, many studies also indicated that the perceptual quality of cold-evoked painful sensations is determined by the integrated activity of both nociceptive and non-nociceptive systems.

Psychophysical studies suggest that pure cooling and cold pain sensations result from the activation of different populations of receptors. In support of this view, humans show better detection ability in the innocuous cold compared to the noxious cold temperature range. Compression block of myelinated fiber conduction in cold-sensitive afferents shifts the pain threshold of cold stimulation toward higher temperatures, pointing to a convergent processing of thermal and nociceptive inputs (Yarnitsky and Ochoa 1990). The same occurs in certain diseases, including peripheral nerve lesions and neuropathic pain syndromes, which are often associated with cold hyperalgesia.

The anatomical substrates for these cold-sensitive spots are free nerve endings branching inside the superficial skin layers. Functional studies suggest that receptors sensing noxious cold may be located more deeply within the skin, some located along vein walls (Klement and Arndt 1992). The difference in location has important implications for interpreting experimental findings; the deeper location of nociceptive endings will result in a large discrepancy between the actual temperature of the receptor and the readings of the surface probe used to apply the cold stimulus. This lag in thermal readings will overestimate the apparent low temperature of activation of nociceptors to cold. As a matter of fact, temperature recordings inside the skin indicate that cold pain may be evoked with intracutaneous temperatures as high as 28 °C (Klement and Arndt 1992).

Cold-Sensitive Fibers

Extending the psychophysical studies, single-fiber recordings in various species, including primates, have identified a population of myelinated Aδ fibers excited by moderate decreases in the temperature of their receptive field. These fibers are insensitive to mechanical stimuli. They are the prototypical "low-threshold" cold thermoreceptors (Hensel 1981). At normal skin temperatures (33–34 °C), cold thermoreceptors show a static ongoing discharge. On sudden cooling, they display a transient peak in firing that adapts to a new static level, often characterized by short bursts of impulses separated by silent intervals. Rewarming of the skin leads to silencing of the receptors. A mirror response is observed in warm receptors. Activation of these receptors is the probable mechanism responsible for the sensation of innocuous cold. In humans, microneurographic recordings confirmed the existence of cold thermoreceptors; the principal difference in primates was the lower conduction velocity and the regular pattern of discharge (Campero et al. 2001). ► Menthol, a natural substance found in leaves of certain plants, evokes cold sensations when applied at low doses to skin and mucosae. This effect is due to sensitization of cold thermoreceptors; they shift their threshold of activation to higher temperatures. However, other studies also show that topical application of high concentrations of menthol can sensitize nociceptors and evoke pain (Green 1992; Wasner et al. 2004).

Greater decreases in temperature recruit an additional population of receptors. Many of these high-threshold cold-sensitive fibers are also heat- and mechano-sensitive and conduct slowly, which would classify them as C-type polymodal nociceptors. In humans, they are characterized by low and irregular firing rates (<1 impulse/s) during cooling (Campero et al. 1996). Activation of these nociceptors is thought to underlie the sensation of cold pain. However, other fiber types also augment their discharge during strong cold stimuli, including high-threshold cold receptors and a fraction of slowly adapting low-threshold mechanoreceptors. Thus, it is still an open question as to the contribution of these various fibers to noxious cold sensations. Terminal stumps of damaged sensory fibers can be excited by moderate cooling stimuli. The majority are C-type and many are also menthol sensitive (Roza et al. 2006).

Parallel input from primary sensory neurons carrying innocuous and noxious thermal information to the brain is suggested by the existence of anatomically and functionally distinct second-order neurons in lamina I of the spinal cord responsive to innocuous cooling, pure nociceptive stimuli, or multimodal thermal and mechanical stimuli.

Cold-Sensitive Neurons

To investigate the cellular mechanisms underlying cold sensing, many laboratories have turned to models that use primary sensory neurons maintained in culture (reviewed by Reid 2005). Activity in these cells can be monitored with calcium-sensitive fluorescent dyes (Fig. 1). Only a small fraction (10–15 %) of sensory neurons respond to application of cold stimuli with an elevation in their intracellular calcium concentration. Many of these cold-sensitive neurons are also activated by cooling compounds like menthol (Fig. 1), suggesting that they are indeed cold thermoreceptors (Viana et al. 2002; McKemy et al. 2002; Reid et al. 2002). These studies further showed that cold-sensitive

neurons have distinct electrophysiological properties. A hallmark is their high excitability; they require minute excitatory currents to reach firing threshold. The increased excitability reflects the relative low expression of subthreshold voltage-gated potassium currents in comparison with other sensory neurons. A high percentage of cold- and menthol-sensitive neurons are also excited by the algesic compound capsaicin, which can be interpreted as further evidence for the expression and role of ▶ TRPM8 channels in polymodal nociceptive neurons (Viana et al. 2002; McKemy et al. 2002).

Molecular Sensors for Cold

Work in the 1970s attributed the activation of nerve terminals by cooling to the depolarization produced by inhibition of the Na^+/K^+ pump. However, the realization that individual terminals can be activated by variable low temperatures hinted at the existence of more specific mechanisms, such as membrane ion channels with distinct temperature thresholds of activation, as likely molecular sensors of innocuous and painful cold signals. In principle, excitation of nerve terminals by a cold temperature could involve opening of cation channels or closure of resting potassium channels (Fig. 2). In either case, the net result is a depolarizing generator potential and firing of action potentials by the terminal.

Following on the landmark identification of ▶ TRPV1, a cation channel activated by heat and capsaicin (Caterina et al. 1997), two research groups cloned two additional transient receptor potential (TRP) channels activated by decreases in temperature. These two ion channels, known as ▶ TRPM8 and ▶ TRPA1, are expressed in discrete, nonoverlapping, subpopulations of primary sensory neurons (Story et al. 2003). So far, the best characterized channel in the molecular machinery for neuronal cold sensing is TRPM8. The data argue strongly for a primary role of TRPM8 in non-noxious cold detection. TRPM8 is a calcium permeable, voltage gated, nonselective cation channel that is activated by temperature and natural cooling compounds like menthol and eucalyptol (McKemy et al. 2002;

Nociceptors, Cold Thermotransduction, Fig. 1 Identification and response characteristics of cold-sensitive trigeminal neurons in culture. (**a**) A neuronal culture (1 day in vitro) obtained from the trigeminal ganglion showing the bright-field image (*top*) and (in descending order) the pseudocolor images of the fura-2 ratio intensity at 35 °C, 20 °C, and 35 °C in the presence of 200 μM menthol. Two of the neurons (marked with *arrowheads*) increased their resting calcium level during the cold stimulus, while the remaining cells did not change their calcium levels. (**b**) Intracellular calcium response to a cold ramp and 200 μM menthol application in the two neurons marked in (**a**). Only one of the cells responded to the stimuli. (**c**) Whole cell current clamp recordings of a cold-sensitive neuron showing the potentiation of the cold response and the shift in temperature threshold by menthol

Peier et al. 2002). TRPM8 is expressed selectively in a small population (10–15 %) of primary sensory neurons of small diameter and most cold-sensitive neurons are also excited by menthol. Many of the same neurons also express TRPM8 mRNA transcripts. Moreover, many cold-sensitive neurons manifest a nonselective cation current with many biophysical and pharmacological properties consistent with the properties of TRPM8-dependent currents (reviewed by Reid 2005).

One important difference between native cold-activated currents and cloned ▶ TRPM8 channels is the temperature threshold (de la Pena et al. 2005). Activation threshold of TRPM8 channels is around 25 °C, a surprisingly low value. This is not a trivial issue for two reasons. First, it leaves unexplained the ability of many cold thermoreceptors to sense temperature decreases in the 33–26 °C range. Second, the high threshold observed suggests that TRPM8 may also be a candidate for thermal nociception as well. As already mentioned, this hypothesis is supported by some psychophysical studies in humans (Wasner et al. 2004; Green 1992). In turn, the discrepancy in thresholds may be explained by intrinsic modulation of TRPM8 channels by endogenous factors.

Nociceptors, Cold Thermotransduction, Fig. 2 Cold-sensing molecules in peripheral nerve endings. A schematic representation of a thin sensory fiber innervating the skin. The transduction of cold temperatures into an electrical signal takes place at free nerve endings. A free nerve ending with all the putative cold-sensing channels and their temperature thresholds. In the case of K^+ channels, closed by low temperature, the "threshold" value represents the lowest temperature at which the channels display significant activity. At normal skin temperatures, the terminal is kept hyperpolarized by the background activity of K^+ channels. Activity in the electrogenic Na^+-K^+ pump contributes to the hyperpolarization. Upon cooling, K^+ channels close and thermosensitive TRPs open, with a decrease in Na^+-K^+ pump activity, leading to a depolarizing receptor potential and spike firing

The identification of cold-sensitive neurons that are insensitive to menthol points to additional cellular mechanisms in neuronal cold sensing (reviewed by Reid 2005). As a matter of fact, many high-threshold cold-sensitive neurons lack TRPM8 expression. The second cold-sensitive TRP channel identified, ▶ TRPA1, is activated by much lower temperatures (< 17 °C) than TRPM8 and has therefore been suggested to be important in the transduction of high threshold, painful, cold stimuli (Story et al. 2003). This channel is not activated by menthol but is potently activated by pungent compounds like cinnamaldehyde and isothiocyanates present in cinnamon oil, wasabi, and mustard oil. However, other investigators have not been able to replicate the cold sensitivity of TRPA1 channels (Jordt et al. 2004). Furthermore, the population of high threshold, menthol insensitive, cold-activated neurons is not activated by mustard oil, which would suggest that their activation is not dependent upon TRPA1 activity (Babes et al. 2004). These findings make TRPA1 an uncertain candidate as a molecular sensor for noxious cold. In an interesting twist of events, recent studies suggest that TRPA1 channels are part of the mechanotransduction complex of vertebrate hair cells.

Cooling also activates ENaC channels, a member of the amiloride-sensitive epithelial sodium channel family. However, a cold-sensitive current matching the pharmacological profile of ENaC channels has not been documented in sensory neurons.

Alternatively, neuronal cold sensing may involve the closure of background potassium channels by cold (Fig. 2). TREK-1, TREK-2, and TRAAK, ▶ Two Pore Domain K + Channels activated by fatty acids and mechanical stimuli, expressed in sensory neurons and with high sensitivity to temperature are good molecular candidates for this role (Kang et al. 2005; Maingret et al. 2000). The steep, rapid, and gradual decrease in current flowing through these channels in a broad physiological temperature range (44–24 °C) make them excellent candidates for thermal sensing in those terminals harboring them. In contrast to the direct temperature sensitivity of ► TRP channels, cell integrity is required for temperature sensitivity of these K⁺ channels. Other background K⁺ channels (i.e., TASK) are minimally affected by temperature. Experimental data implicating the closure of background potassium channels in cold transduction have been obtained in trigeminal and DRG neurons (Viana et al. 2002; Reid and Flonta 2001). However, the pharmacology or molecular nature of the conductance closed by cold temperature has not been addressed directly.

It is interesting that blockade of certain types of voltage-gated K⁺ channels can render a population of sensory neurons cold sensitive. Experiments in trigeminal neurons showed that a slowly inactivating potassium current can act as a brake on excitability, reducing cold sensitivity (Viana et al. 2002). Neurons insensitive to cold and ► menthol could be transformed into cold-sensitive neurons in the presence of low concentrations of 4-AP, a blocker for these channels.

It is important to emphasize that the various ionic mechanisms postulated in cold transduction are not mutually exclusive; if present in the same nerve terminal they could act synergistically to expand the dynamic range of temperature detection. Alternatively, activity of thermosensitive K⁺ channels in those terminals with TRP channels opened by heat (i.e., TRPV2, TRPV3, TRPV4, TRPV1) would act as a brake on the excitatory actions of the latter.

Unfortunately, most recent data on thermosensitive ion channels have been obtained during in vitro animal studies, precluding a direct translation of these results to human physiology and pathophysiology. Much remains to be learnt about the differential expression pattern of the different thermosensitive channels and the functional properties of the sensory fibers expressing them. It is likely that other thermosensitive channels will be uncovered in the next few years, expanding the palette of putative molecular candidates for cold sensing.

In summary, psychophysical studies indicate that input from non-noxious thermal systems is essential for the thermal quality and the intensity of the pain sensation evoked by cold. Thermosensory afferent input normally inhibits cold-evoked pain. At the molecular level, the diverse functional characteristics and broad range of temperature thresholds in the different afferent fibers suggests that each class of nerve terminal may operate with a combinatorial code of sensory receptors. The available evidence supports an important role for TRPM8 in sensing of innocuous cold temperatures in peripheral receptors (evidence reviewed by Reid 2005). In addition, the participation of TRPM8 channels in certain forms of cold pain is a distinct possibility. Lacking genetic evidence (i.e., TRPM8-deficient mice) or specific pharmacological tools, this conclusion is not firm. The role of TRPA1 channels in the transduction of noxious cold pain is uncertain. Activity of background K⁺ channels can certainly influence cold sensitivity, but their role as primary transducers has not been addressed directly. Not surprisingly, interest in the pharmacological profile of thermosensitive channels is very high. It is anticipated that modulators of these channels will provide new therapeutic options with which to treat certain forms of pain, including cold hyperalgesia and allodynia.

References

Babes, A., Zorzon, D., & Reid, G. (2004). Two populations of cold-sensitive neurons in rat dorsal root ganglia and their modulation by nerve growth factor. *European Journal of Neuroscience, 20,* 2276–2282.

Campero, M., Serra, J., & Ochoa, J. L. (1996). C-polymodal nociceptors activated by noxious low temperature in human skin. *The Journal of Physiology, 497,* 565–572.

Campero, M., Serra, J., Bostock, H., et al. (2001). Slowly conducting afferents activated by innocuous low temperature in human skin. *The Journal of Physiology*, *535*, 855–865.

Caterina, M. J., Schumacher, M. A., Tominaga, M., et al. (1997). The capsaicin receptor: A heat-activated ion channel in the pain pathway. *Nature, 389*, 816–824.

de la Pena, E., Malkia, A., Cabedo, H., et al. (2005). The contribution of TRPM8 channels to cold sensing in mammalian neurones. *The Journal of Physiology, 567*, 415–426.

Green, B. G. (1992). The sensory effects of l-menthol on human skin. *Somatosensory & Motor Research, 9*, 235–244.

Hensel, H. (1981). Thermoreception and temperature regulation. *Monographs of the Physiological Society, 38*, 1–321.

Jordt, S. E., McKemy, D. D., & Julius, D. (2003). Lessons from peppers and peppermint: The molecular logic of thermosensation. *Current Opinion in Neurobiology, 13*, 487–492.

Jordt, S. E., Bautista, D. M., Chuang, H. H., McKemy, D. D., Zygmunt, P. M., Högestätt, E. D., Meng, I. D., & Julius, D. (2004). Mustard oils and cannabinoids excite sensory nerve fibres through the TRP channel ANKTM1. *Nature, 427*(6971), 260–5.

Kang, D., Choe, C., & Kim, D. (2005). Thermosensitivity of the two-pore domain K + channels TREK-2 and TRAAK. *The Journal of Physiology, 564*, 103–116.

Klement, W., & Arndt, J. O. (1992). The role of nociceptors of cutaneous veins in the mediation of cold pain in man. *The Journal of Physiology, 449*, 73–83.

Maingret, F., Lauritzen, I., Patel, A. J., et al. (2000). TREK-1 is a heat-activated background K(+) channel. *EMBO Journal, 19*, 2483–2491.

McKemy, D. D., Neuhausser, W. M., & Julius, D. (2002). Identification of a cold receptor reveals a general role for TRP channels in thermosensation. *Nature, 416*, 52–58.

Patapoutian, A., Peier, A. M., Story, G. M., et al. (2003). ThermoTRP channels and beyond: Mechanisms of temperature sensation. *Nature Reviews Neuroscience, 4*, 529–539.

Peier, A. M., Moqrich, A., Hergarden, A. C., et al. (2002). A TRP channel that senses cold stimuli and menthol. *Cell, 108*, 705–715.

Reid, G. (2005). ThermoTRP channels and cold sensing: What are they really up to? *Pflügers Archiv, 451*, 250–263.

Reid, G., & Flonta, M.-L. (2001). Cold current in thermoreceptive neurons. *Nature, 413*, 480.

Reid, G., Babes, A., & Pluteanu, F. (2002). A cold- and menthol-activated current in rat dorsal root ganglion neurones: Properties and role in cold transduction. *The Journal of Physiology, 545*, 595–614.

Roza, C., Belmonte, C., & Viana, F. (2006). Cold sensitivity in axotomized fibers of experimental neuromas in mice. *Pain, 120*, 24–35.

Story, G. M., Peier, A. M., Reeve, A. J., et al. (2003). ANKTM1, A TRP-like channel expressed in nociceptive neurons, is activated by cold temperatures. *Cell, 112*, 819–829.

Viana, F., de la Peña, E., & Belmonte, C. (2002). Specificity of cold thermotransduction is determined by differential ionic channel expression. *Nature Neuroscience, 5*, 254–260.

Wasner, G., Schattschneider, J., Binder, A., et al. (2004). Topical menthol – a human model for cold pain by activation and sensitization of C nociceptors. *Brain, 127*, 1159–1171.

Yarnitsky, D., & Ochoa, J. L. (1990). Release of cold-induced burning pain by block of cold-specific afferent input. *Brain, 113*, 893–902.

Nociceptors: High-Threshold Afferents

▶ Nociceptor, Categorization

Nocifensive

Definition

Denoting a process or mechanism that acts to protect the body from injury.

Cross-References

▶ Nocifensive Behaviors of the Urinary Bladder
▶ Secondary Somatosensory Cortex (S2) and Insula, Effect on Pain-Related Behavior in Animals and Humans

Nocifensive Behavior

Definition

Nocifensive behaviors are those that are evoked by stimuli that activate the nociceptive sensory apparatus. They are associated with protection against insult and injury typically in response to a noxious stimulus. Responses to noxious stimuli in animals may include behaviors resembling responses to pain in humans, such as limping, flinching, vocalization, and reflexive withdrawal. Other specific pain-related responses in animals include tail and paw flicks, licking, and scratching. Responses to increased deep muscle and joint pain may include reduced exploration activity. In the viscera, nocifensive behaviors can be produced by hollow organ distension, ischemia, traction on the mesentery or stimulation of inflamed organs.

Cross-References

► Muscle Pain Model, Inflammatory Agents-Induced
► Nocifensive Behaviors, Muscle and Joint
► Sensitization of Visceral Nociceptors

Nocifensive Behaviors

► Nocifensive Behaviors, Gastrointestinal Tract

Nocifensive Behaviors Evoked by Myositis

► Muscle Pain Model, Inflammatory Agents-Induced

Nocifensive Behaviors of the Urinary Bladder

Timothy Ness
Department of Anesthesiology, University of Alabama at Birmingham, Birmingham, AL, USA

Synonyms

Pseudaffective response; Pseudoaffective response; Urinary Bladder Nocifensive Behaviors

Definition

► Nocifensive reflexes are protective reflexes evoked by noxious stimuli typically to escape a potentially tissue-damaging event. Most acute motor responses to noxious cutaneous stimuli are nocifensive. For visceral systems in general, and the urinary bladder-related systems in particular, similar reflexes are not obvious. Micturition reflexes enhanced by local inflammation can expel toxic urinary contents and so may serve a protective function, but may also increase pain (unlike cutaneous nocifensive reflexes). Many responses to urinary bladder stimulation in nonhuman animals correlate with nociception in humans, and in this way are similar to nocifensive reflexes. These include biting/scratching behaviors, changes in heart rate, blood pressure, respiration, and abdominal muscular tone. These behaviors/reflexes are collectively called pseud(o)affective responses.

Characteristics

Clinical Observations and Psychophysical Studies
The urinary bladder is a common site of visceral pain generation in humans. Urinary tract infections, inflammation produced by radiation, the action of irritating constituents of urine, or acute obstruction of outflow from the bladder can all lead to severe pain localized to the lower abdomen, suprapubic region, and perineum. At the same time, there is typically a sense of urgency (a need to void) and increased frequency of urination. When the bladder is inflamed, the act of urinating may become painful and frequently, there is a sensation of incomplete emptying. Autonomic responses such as heart rate changes, sweating, and increases or decreases in blood pressure may precede, accompany, and outlast the discomforting sensations. Psychophysical studies using fluid distension of the bladder via a urethral catheter evoke reports of pain in humans as well as increases in blood pressure (e.g., Fig. 1), increases in abdominal tone, and respiratory changes (Ness et al. 1998).

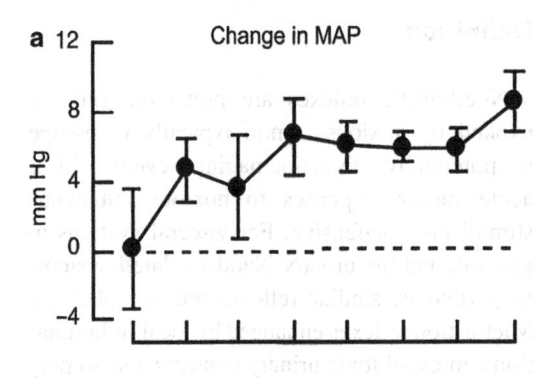

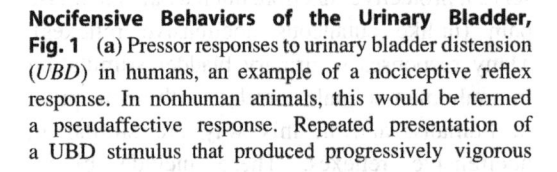

Nocifensive Behaviors of the Urinary Bladder, Fig. 1 (a) Pressor responses to urinary bladder distension (*UBD*) in humans, an example of a nociceptive reflex response. In nonhuman animals, this would be termed a pseudaffective response. Repeated presentation of a UBD stimulus that produced progressively vigorous increases in mean arterial blood pressure (*MAP*). b Reports of pain to the same stimulus paralleled the pressor response with increased vigor of responses with repeated presentation of the UBD stimulus (Data from Ness et al. (1998))

Functional imaging has also demonstrated activation of numerous cortical sites of sensory processing, including sensory and motor cortices, the anterior cingulate cortex, and the insular cortex (Athwal et al. 2001). Chemical stimulation of the bladder using ▶ intravesical capsaicin has similarly produced reports of pain and autonomic responses (e.g., Giannantoni et al. 2002). Similar physiological responses have been noted in nonhuman animals.

Micturition Reflexes

Micturition reflexes are reflexes involved in the emptying of the urinary bladder when full. They consist of a reflex contraction of the bladder itself, coupled with a relaxation of the sphincters and associated pelvic floor musculature which block outflow from the bladder. A lack of coordination of these two components can lead to painful contractions of the bladder and incomplete bladder emptying. When the urinary bladder becomes inflamed by local infection or other irritant action, micturition reflexes become enhanced with lower volume/pressure thresholds for activation. Micturition reflexes enhanced by noxious bladder stimulation could be viewed as "nocifensive" in that they may remove the irritating stimulus causing the pain by emptying the bladder. However, micturition reflexes normally occur in the absence of

pain-producing stimuli, and the presence of irritating factors may paradoxically increase pain by increasing mechanical stimulation of the bladder through enhanced contractile activity. In this way, micturition reflexes may enhance nociception unlike cutaneous nocifensive reflexes.

As an assessment of micturition reflexes, distension of the bladder may be allowed to occur secondary to urine production and measures made of spontaneous voiding frequency and volume. Evoked micturition reflexes can be produced by the introduction of a catheter, which infuses fluid directly into the bladder at a rate higher than that typically produced by spontaneous urine formation. This catheter may be placed in a relatively noninvasive fashion via the urethra, or in nonhuman animals may also be surgically placed via an abdominal incision (Abelli et al. 1989). A cystometrogram can then be performed by slow infusion of fluid into the bladder with simultaneous measurement of intravesical pressure. Micturition reflexes are identified as sharp increases in intravesical pressure coupled with the appearance of urine. Irritant chemicals such as potassium chloride will evoke increased frequency and magnitude of urine expulsion, thereby demonstrating micturition reflexes to be nocifensive in nature (Chuang et al. 2003).

Pseud(o)Affective Reflexes

Highly localized motor reflexes such as a flexion-withdrawal response to a pinprick of the foot are nocifensive. The limb is reflexively pulled away from the damaging stimulus, the threat to the body is removed, and pain generation is minimized. In visceral systems, a similar, direct link between tissue damage and protective behaviors is not obvious. There are typical behaviors in nonhuman animals evoked by visceral stimuli that would be pain-producing in humans, but these responses do not generally remove the threat to the organism. Abdominal licking, contractions of abdominal muscles, and changes in heart rate, blood pressure, and respiratory pattern can all be evoked by stimulation of the urinary bladder of rats and mice. These types of responses were called pseudaffective responses by Sherrington, in that they appeared to be reflex correlates to emotional responses to painful stimuli (Woodworth and Sherrington 1904). Pseudaffective responses are not specific to pain-producing stimuli, but when it can be demonstrated that they are inhibited by pharmacological manipulations known to be analgesic (i.e., morphine), the case for their correlation with accepted nocifensive reflexes becomes stronger.

There have been numerous characterizations of responses to noxious urinary bladder stimulation in nonhuman animals, many of which have been utilized as models of urinary bladder pain. The stimuli for these models can be broadly stratified into those that use mechanical stimuli to activate afferents arising in the bladder (typically by air or fluid distension using a catheter), and those which chemically activate/sensitize afferents arising in the bladder using irritants. These irritants may be administered directly into the bladder using a catheter, or indirectly by the renal excretion of irritating urinary constituents.

Responses to Urinary Bladder Distension (UBD)

Pseudaffective responses to UBD (vigorous cardiovascular and visceromotor reflexes) have been demonstrated in numerous species including rats, mice, guinea pigs, rabbits, cats, dogs, and monkeys. Such studies related to nociception have commonly used a progressively increasing UBD stimulus, similar to that used in studies of micturition but at a much faster rate of filling, or have produced a phasic (on-off) stimulus through the use of a rapid, near instantaneous, infusion of air or saline. In anesthetized animals, vigorous neuronal and reflex responses can be evoked (e.g., Westropp and Buffington 2002). These responses can be inhibited by analgesics. Reliable and reproducible pressor responses and contractions of the abdominal musculature (▶ visceromotor responses measured by electromyogram) evoked by UBDs have been characterized (e.g., Ness and Gebhart 2001) (Fig. 2) with gender differences (see ▶ Gender and Pain) and gonadal hormonal influences apparent (Robbins et al. 2010). Genetic models of painful bladder disorders such as the feline interstitial cystitis model have utilized such responses to bladder stimulation as endpoints (Westropp and Buffington 2002). Sensitizing chemicals associated with inflammation when administered into the urinary bladder lead to more vigorous responses to UBD, particularly at low intensities of stimulation (Dmitrieva and McMahon 1996).

Responses to the Intravesical Instillation of Proinflammatory Compounds

Inflammation of the bladder commonly produces reports of pain and urgency in patients suffering from a urinary tract infection. Enhanced micturition reflexes lead to spasms of pain. Experiments in nonhuman animals have artificially inflamed or sensitized the bladder with the intravesical administration of inflammation-inducing compounds such as lipopolysaccharide/uropathogenic *E. coli* (Stemler et al. 2013), zymosan (Randich et al. 2006), turpentine, mustard oil, croton oil (McMahon and Abel 1987) and neurotrophic agents. Such irritation has been demonstrated to produce alterations in the spontaneous activity of primary afferent neurons encoding for bladder stimuli (e.g., Dmitrieva and McMahon 1996) and spontaneous behavioral responses. Most reflex studies have been performed in rats, although behavioral and neurophysiological

Nocifensive Behaviors of the Urinary Bladder, Fig. 2 Examples of cardiovascular (arterial pressure *AP*; *above*) and distension (UBD) in a halothane-anesthetized rat. Visceromotor responses can be measured by direct visualization or electromyographically (*EMGs*), as in the figure. Responses are graded in response to graded UBD (Data from Ness et al. (2001))

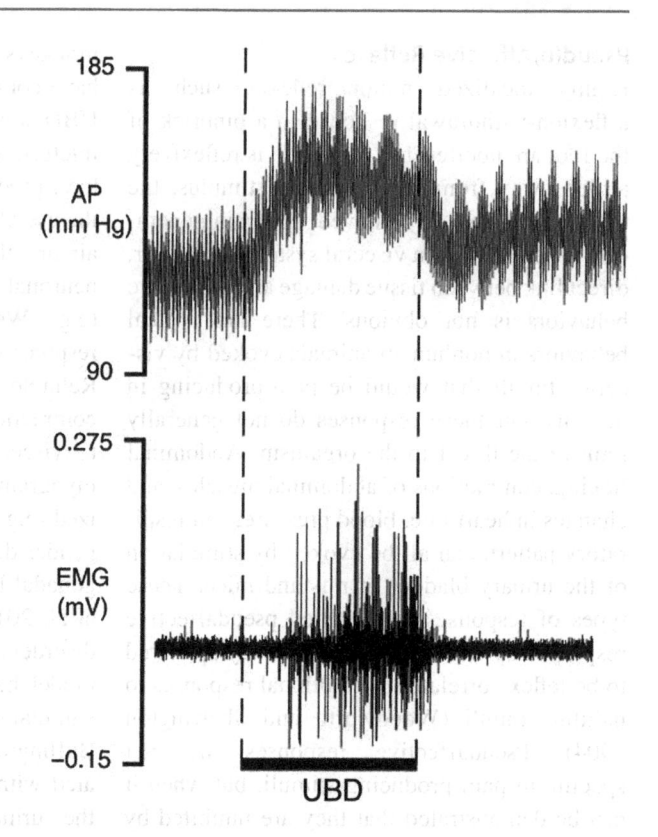

studies have been performed in primates and cats. Effects of neonatal bladder inflammatory treatments have been demonstrated in multiple labs (DeBerry et al. 2010; Sengupta et al. 2013) to produce neurochemical changes, enhanced responses to urinary bladder distension, altered micturition responses, and neuronal alterations in adult subjects.

McMahon and Abel (1987) characterized visceromotor and altered micturition reflexes in chronically decerebrate rats, and demonstrated that following the administration of an irritant, the magnitude of responses to UBD correlated with measures of inflammation such as tissue edema, plasma extravasation, and leukocyte infiltration. While becoming more sensitive to bladder-related stimuli, the rats at the same time became hypersensitive to noxious stimuli applied to the lower abdomen, perineum, and tail, as measured by the number of "kicks" evoked by a given stimulus. The model of McMahon and Abel (1987) has been modified to examine pharmacologically novel mechanisms

related to visceral hyper-reflexia (altered cystometrograms) or secondary hyperalgesia (decreased thresholds to heat stimuli in hindlimbs). Modulatory effects of glutamate-receptor antagonists, nitric oxide synthase inhibitors, neurotrophic antagonists, cannabinoids, and bradykinin receptor antagonists have all been noted (e.g., Jaggar et al. 1999).

Responses to the Direct Application of Irritants

Administration of intravesical irritants via a chronically implanted intravesical catheter was described by Abelli et al. (1989) and modified by Craft et al. (1993). An injection of xylene, capsaicin, or its related compound resiniferatoxin is performed via an intravesical cannula placed a day earlier. Immediate responses include abdominal/perineal licking, headturns, hindlimb hyperextension, head grooming, biting, vocalization, defecation, scratching, and salivation, all of which can be independently measured and graded. The time to onset, incidence, and number of individual behavioral responses are recorded for 15 min or more following intravesical drug administration. Baclofen, mu/kappa/delta opioid receptor agonists, and intravesical tetracaine have all been demonstrated to inhibit these behavioral responses. Bladder denervation abolishes all behavioral responses.

Responses to Cyclophosphamide (Chemotherapy-Induced Cystitis)

Subacute, irritant-induced urinary bladder inflammation occurs secondary to antineoplastic regimens that include the chemotherapeutic agent cyclophosphamide (CP). In humans, pain and increased urgency/frequency are sequelae of this treatment, although use of sodium 2-mercaptoethane sulfonate reduces the metabolite of CP that is the actual irritant, acrolein. In nonhuman animal models, CP is administered intraperitoneally, which evokes behavioral responses such as hypolocomotion and alterations in position (e.g., Lanteri-Minet et al. 1995) or enhanced visceromotor responses to urinary bladder distension (Lai et al. 2011). Chronic (2 week) treatments with CP have also been described with numerous physiological and biochemical alterations (e.g., Vizzard et al. 1996). Olivar and Laird (1999) have characterized similar responses to acutely administered CP in mice.

Conclusions

Numerous responses to urinary bladder stimulation have been noted. Other than enhanced micturition reflexes, none of these responses fit the precise criteria of nocifensive reflexes, in that they are not generally protective reflexes that remove the noxious stimulus. They are nociceptive reflexes in that they do appear to be representative of the sensory processing related to pain from the bladder, and should be expected to be predictive of pharmacological or procedural manipulations intended to treat urinary bladder pain.

References

Abelli, L., Conte, B., Somme, V., Maggi, C. A., Girliani, S., & Meli, A. (1989). A method for studying pain arising from the urinary bladder in conscious, freely-moving rats. *Journal of Urology, 141*, 148–151.

Athwal, B. S., Berkley, K. J., Hussain, I., et al. (2001). Brain responses to changes in bladder volume and urge to void in healthy men. *Brain, 124*, 369–377.

Chuang, Y. C., Chancellor, M. B., Seki, S., Yoshimura, N., Tyagi, P., Huang, L., Lavelle, J. P., DeGroat, W. C., & Fraser, M. O. (2003). Intravesical protamine sulfate and potassium chloride as a model for bladder hyperactivity. *Urology, 61*, 664–670.

Craft, R. M., Carlisi, V. J., Mattia, A., Herman, R. M., & Porreca, F. (1993). Behavioral characterization of the excitatory and desensitizing effects of intravesical capsaicin and resiniferatoxin in the rat. *Pain, 55*, 205–215.

DeBerry, J., Randich, A., Shaffer, A. D., Robbins, M. T., & Ness, T. J. (2010). Neonatal bladder inflammation produces functional changes and alters neuropeptide content in bladders of adult female rats. *The Journal of Pain, 11*, 247–255.

Dmitrieva, N., & McMahon, S. B. (1996). Sensitisation of visceral afferents by nerve growth factor in the adult rat. *Pain, 66*, 87–97.

Giannantoni, A., Di Stasi, S. M., Stephen, R. L., Navarra, P., Scivoletto, G., Mearini, E., & Porena, M. (2002). Intravesical capsaicin versus resiniferatoxin in patients with detrusor hyperreflexia: A prospective randomized study. *Journal of Urology, 167*, 1710–1714.

Jaggar, S. I., Scott, H. C. F., & Rice, A. S. C. (1999). Inflammation of the rat urinary bladder is associated with a referred hyperalgesia which is NGF dependent. *British Journal of Anaesthesia, 83*, 442–448.

Lai, H. H., Qiu, C. S., Crock, L. W., Morales, M. E., Ness, T. J., & Gereau, R. W., 4th. (2011). Activation of spinal extracellualr signal-regulated kinases (ERK) 1/2 is associated with the development of visceral hyperalgesia of the bladder. *Pain, 152*, 2117–2124.

Lanteri-Minet, M., Bon, K., de Pommery, J., Michiels, J. F., & Menetrey, D. (1995). Cyclophosphamide cystitis as a model of visceral pain in rats: Model elaboration and spinal structures involved as revealed by the expression of c-Fos and krox-24 proteins. *Experimental Brain Research, 105*, 220–232.

McMahon, S. B., & Abel, C. (1987). A model for the study of visceral pain states: Chronic inflammation of the chronic decerebrate rat urinary bladder by irritant chemicals. *Pain, 28*, 109–127.

Ness, T. J., & Gebhart, G. F. (2001). Methods in visceral pain research. In L. Kruger (Ed.), *Methods in pain research* (pp. 93–108). New York: CRC Press.

Ness, T. J., Richter, H. E., Varner, R. E., et al. (1998). A psychophysical study of discomfort produced by repeated filling of the urinary bladder. *Pain, 76*, 61–69.

Ness, T. J., Lewis-Sides, A., & Castroman, P. J. (2001). Characterization of pseudaffective responses to urinary bladder distension in the rat: Sources of variability and effect of analgesics. *Journal of Urology, 165*, 968–974.

Olivar, T., & Laird, J. M. (1999). Cyclophosphamide cystitis in mice: Behavioral characterization and correlation with bladder inflammation. *European Journal of Pain, 3*, 141–149.

Randich, A., Uzzell, T., Cannon, R., & Ness, T. J. (2006). Inflammation and enhanced nociceptive responses to bladder distension produced by intravesical zymosan in the rat. *BMC Urology, 6*, 2.

Robbins, M. T., Mebane, H., Ball, C. L., Shaffer, A. D., & Ness, T. J. (2010). Effect of estrogen on bladder nociception in rats. *Journal of Urology, 183*, 1201–1205.

Sengupta, J. N., Pochiraju, S., Kannampalli, P., Bruckert, M., Addya, S., Yadav, P., Miranda, A., Shaker, R., & Banerjee, B. (2013). MicroRNA-mediated GABA (a-alpha-1) receptor subunit down-regulation in adult spinal cord following neonatal cystitis-induced chronic visceral pain in rats. *Pain, 154*, 59–70.

Stemler, K. M., Crock, L. W., Lai, H. H., Mills, J. C., Gereau, R. W., 4th, & Mysorekar, I. U. (2013). Prot-amine sulfate induced bladder injury protects from distention induced bladder pain. *Journal of Urology, 189*, 343–351.

Vizzard, M. A., Erdman, S. L., & de Groat, W. C. (1996). Increased expression of neuronal nitric oxide synthe-tase in bladder afferent pathways following chronic bladder irritation. *The Journal of Comparative Neurology, 370*, 191–202.

Westropp, J. L., & Buffington, C. A. T. (2002). In vivo models of interstitial cystitis. *Journal of Urology, 167*, 694–702.

Woodworth, R. S., & Sherrington, C. S. (1904). A pseudaffective reflex and its spinal path. *Journal of Physiology (London), 31*, 234–243.

Nocifensive Behaviors, Gastrointestinal Tract

Gerald F. Gebhart
Department of Anesthesiology, Center for Pain Research, University of Pittsburgh, Pittsburgh, PA, USA

Synonyms

Gastrointestinal tract; Nocifensive behaviors; Pseudaffective; Pseudoaffective

Definition

Nocifensor, a term introduced by Lewis (1936; see Lewis 1942; LaMotte 1992 for a discussion), describes a system of "nerves" associated with local defense against injury. Nocifensive has since expanded as a term to describe behaviors associated with protection against insult and injury. Nocifensive behaviors are more complex than simple nociceptive flexor withdrawal reflexes, such as the tailflick reflex, and the term is particularly appropriate in the visceral realm, where stimuli considered to be adequate (e.g., hollow organ distension, ischemia, traction on the mesentery) are different from those Sherrington (1906) defined as adequate for activation of cutaneous nociceptors. The nocifensive behaviors produced by visceral stimulation are also considered pseudaffective (Sherrington 1906) (pseudoaffective) because responses to visceral stimulation are organized supraspinally.

Characteristics

Balloon distension of gastrointestinal tract organs has been widely employed in both human and nonhuman animal studies. Significantly, balloon distension of the esophagus, stomach, or large bowel reproduces in humans the distribution of referred sensations, as well as the quality and intensity of discomfort and pain, arising from pathological visceral disorders (see Ness and Gebhart 1990 for a review). In addition, balloon distension of hollow organs is a stimulus that is easy to control in terms of onset, duration, and intensity, as opposed, for example, to a chemical or ischemic visceral stimulus. The nocifensive behaviors produced by balloon dis-tension of hollow organs include changes in blood pressure, heart rate, and respiration and visceromotor reflexes (▶ Visceromotor Reflex). All of these responses are absent in spinally transected animals, but present following midcollicular decerebration, thus revealing that they are responses integrated in the brainstem. All of these response measures are intensity dependent and can be quantified as indices of visceral nociception.

Although balloon distension of hollow organs has been widely used, it was not until Ness parametrically characterized in the rat responses to colorectal distension that this model of visceral pain became widely accepted and widely used (Ness and Gebhart 1988). Similar parametric evaluation of responses to colorectal distension in the mouse (Kamp et al. 2003) and gastric distension in the rat (Ozaki et al. 2002) have since been described.

Colorectal Distension (CRD)

Balloons of different lengths and varying durations of distension have been reported in the literature. As Lewis (1942) noted, distension of hollow organs is most painful in humans when long, continuous segments of the gut are distended, which emphasizes the importance of ▶ spatial summation as an important consideration when using CRD as a stimulus. Results using different length balloons in the rat are qualitatively similar, although greater intensities of stimulation are generally required with smaller balloons to produce quantitatively equivalent responses. The human literature (see Ness and Gebhart 1990 for citations and discussion) also established that constant pressure distension, rather than constant volume, was the appropriate stimulus. It should be appreciated that hollow organs throughout the viscera, and particularly within the gastrointestinal tract, accommodate as they fill. Thus, constant volume distension

produces an inconstant intensity of stimulation as the organ musculature relaxes.

Balloon CRD produces contraction of the peritoneal musculature, representing what has been termed a visceromotor reflex. Among the responses to CRD, the visceromotor reflex recorded from the external oblique peritoneal musculature is perhaps the most reliable and experimentally least complicated measure to quantify. Indwelling arterial catheters to measure changes in blood pressure or heart rate produce equally robust and quantifiable measures of response to distension but are more difficult to maintain over time than electromyographic electrodes sewn into the peritoneal musculature. Figure 1 illustrates graded visceromotor responses to increasing intensities of CRD in the mouse. The response threshold for CRD is typically between 20 mmHg and 25 mmHg in normal rat and mouse colon. In consideration of the response threshold for a subset of pelvic nerve afferent fibers that innervate the colon and spinal dorsal horn neuron responses to CRD, the intensity of CRD in the rat considered to be noxious is ≥30 mmHg. This interpretation is consistent with results from psychophysical studies in humans (e.g., Ness et al. 1990), in which increases in heart rate and blood pressure were produced by colon distension at intensities of distension less than reported as painful. Accordingly,

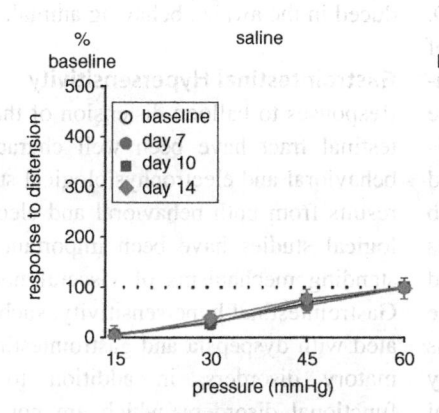

Nocifensive Behaviors, Gastrointestinal Tract, Fig. 1 Visceromotor responses recorded as electromyographic activity in the peritoneal musculature of the mouse to graded intensities of colon distension at times after intracolonic instillation of saline (*left*) or zymosan (*right*), which produces a weeks-long colorectal hypersensitivity (From Shinoda et al. 2009). Distending pressures (15–60 mmHg, 10 s duration) and responses to colorectal distension (normalized to baseline) are illustrated

quantification of pseudaffective responses to visceral stimulation requires interpretation with respect to the quality of the nocifensive behavior produced (e.g., visceromotor response). That is, changes in response measures may be apparent before a behavior is nocifensive, although thresholds for response are similar. Details regarding CRD in the mouse (manufacture of balloons, data acquisition and analysis, etc.) are available in Christianson and Gebhart (2007).

Gastric Distension (GD)

Unlike CRD, in which a lubricated balloon can be inserted via the anus into the descending colon and rectum of a rodent, insertion of balloons for GD requires a surgical procedure in advance of the experiment. Like CRD, responses to balloon GD in the rat are graded with stimulus intensity (Ozaki et al. 2002). The visceromotor response in studies of GD is recorded from the acromiotrapezius muscle; responses to GD recorded in other muscles (e.g., external oblique peritoneal muscles, spinotrapezius muscles, or sternomastoideus muscles) were unreliable and not graded with stimulus intensity. In addition to being able to surgically implant a balloon into the stomach, one can surgically place a small diameter polyethylene tube into the stomach to deliver chemical substances. Thus, one can examine the effect of chemical stimulation of the stomach in the absence or presence of an ulcer and/or before and after GD. For example, intragastric administration of hydrochloric acid produces concentration-dependent (0.05–0.3 mol/l) activity in the acromiotrapezius muscle, which can be quantified and used to assess a chemically evoked nocifensive behavior. Interestingly, we (Lamb et al. 2003) were able to show that responses to balloon gastric distension were conveyed to the central nervous system by the splanchnic innervation of the stomach, whereas chemonociceptive stimulation of the stomach by hydrochloric acid was conveyed to the central nervous system via the vagus nerve. The importance of these observations relates to support for a role of the vagus nerve in chemonociception, an area of growing investigation.

Small Bowel Distension

Balloon distension of the small bowel has also been studied in rodents. Colburn and colleagues (1989) described a model of duodenal distension in which responses to distension were graded from 0 to 4, based on interpretation of the nocifensive behavior produced. A difficulty with models of small bowel distension is that permanent implantation of balloons tends to lead to obstruction. This is avoided in studies of GD by placing the balloon in the antrum, where it does not significantly interfere with ingestion of food or weight gain.

Pseudaffective Reflexes and Anesthesia

The concept of a nocifensive behavior implies evaluation of response to stimulation in the awake animal. However, changes in blood pressure, heart rate, or the visceromotor response can also be interpreted as a behavior, and responses to gastrointestinal distension have been carried out in anesthetized animals. As initially reported by Ness and Gebhart (1988), the presence of anesthesia typically converts pressor and tachycardic responses to CRD to depressor and bradycardic responses and attenuates the visceromotor response to distension. Depressor and bradycardic responses to CRD in anesthetized rats have been quantified and reported, but their relevance to nociception and pain is less clear than vigorous nocifensive behaviors produced in the awake, behaving animal.

Gastrointestinal Hypersensitivity

Responses to balloon distension of the gastrointestinal tract have been well characterized in behavioral and electrophysiological studies (and results from both behavioral and electrophysiological studies have been important to understanding mechanisms of visceral nociception). Gastrointestinal hypersensitivity, such as associated with dyspepsia and gastrointestinal inflammatory disorders, in addition to so-called functional disorders, which are not associated with either biochemical or pathological markers of organ pathology, comprises a significant proportion of patients reporting gastrointestinal discomfort and pain. A variety of models of gastrointestinal organ irritation and insult have been reported accordingly and thus to study the presumed processes underlying visceral hypersensitivity. Experimental gastric ulceration or gastritis produced by ingestion of a dilute solution of iodoacetamide both produced long-lasting gastric hypersensitivity, as assessed by balloon distension and quantification of the visceromotor response. Importantly, the gastric hypersensitivity to balloon distension in these models long outlasts recovery from ulceration or gastritis (Ozaki et al. 2002). In the colon, a variety of chemicals have been instilled into the lumen to produce insult ranging from mild irritation to ulceration. In such models, in which substances such as mustard oil, turpentine, zymosan, and trinitrobenzene sulfonic acid have been placed intracolonically, hypersensitivity to balloon CRD can be shown within hours and can last for days to weeks. For example, in the mouse, intracolonic instillation of zymosan can produce hypersensitivity to balloon CRD that lasts for 5–7 weeks and does so in the absence of histological evidence of damage to the colon (e.g., Feng et al. 2012). In addition to these models of relatively acute to chronic colon hypersensitivity, models have also been developed which are intended to, or closely replicate, contributions to gastrointestinal hypersensitivity in humans. For example, about 30 % of individuals with irritable bowel syndrome report a previous gastrointestinal infectious event, and models of gastrointestinal infection (e.g., *Trichinella spiralis*) have been developed in which both changes in gastrointestinal motility and sensation are altered (Barbara et al. 1997). Patients that suffer from irritable bowel syndrome also often report that symptoms are exacerbated by stressful life events, and a model of maternal separation of rat pups produced visceral hypersensitivity in adult rats as assessed by visceromotor responses to CRD (Coutinho et al. 2002). Al Chaer and colleagues (2000) exposed rat pups to either mechanical balloon distension of the colon or treatment with mustard oil during a critical neonatal period and were able to document hypersensitive responses to CRD in adult rats.

Conclusions

Balloon distension, at constant pressure, produces robust and readily quantifiable pseudaffective responses and nocifensive behaviors. Development and use of these models has improved understanding of visceral nociceptive mechanisms, visceral hypersensitivity, and modulation by pharmacological and other manipulations.

References

Al Chaer, E. D., Kawasaki, M., & Pasricha, P. J. (2000). A New model of chronic visceral hypersensitivity in adult rats induced by colon irritation during postnatal development. *Gastroenterology, 119,* 1276–1285.

Barbara, G., Vallance, B. A., & Collins, S. M. (1997). Persistent intestinal neuromuscular dysfunction after acute nematode infection in mice. *Gastroenterology, 113,* 1224–1232.

Christianson, J. C., & Gebhart, G. F. (2007). Assessment of colon sensitivity by luminal distension in mice. *Nature Protocols, 2,* 2624–2631.

Colburn, R. W., Coombs, D. W., Degnan, C. C., et al. (1989). Mechanical visceral pain model: Chronic intermittent intestinal distension in the Rat. *Physiology & Behavior, 45,* 191.

Coutinho, S. V., Plotsky, P. M., Sablad, M., et al. (2002). Neonatal Maternal separation alters stress-induced responses to viscerosomatic nociceptive stimuli in Rat. *American Journal of Physiology, 282,* G307.

Feng, B., La J. -H., Tanaka, T., McMurray, T. P., Schwartz, E. S., & Gebhart, G. F. (2012). Long-term sensitization of mechanosensitive and insensitive afferents in mice with persistent colorectal hypersensitivity. *Am J Physiol.* April 1; 302(7): G676–G683. Published online 2012 January 19. doi: 10.1152/ajpgi.00490.2011 (http://www.ncbi.nlm.nih.gov/pubmed/22268098)

Kamp, E. H., Jones, R. C. W., 3rd, Tillman, S. R., et al. (2003). Quantitative Assessment and characterization of visceral nociception and hyperalgesia in the mouse. *American Journal of Physiology. Gastrointestinal and Liver Physiology, 284,* G434–G444.

Lamb, K., Kang, Y. M., Gebhart, G. F., et al. (2003). Gastric inflammation triggers hypersensitivity to acid in awake rats. *Gastroenterology, 125,* 1410–1418.

LaMotte, R. H. (1992). Subpopulations of "Nocifensor Neurons" contributing to pain and allodynia, itch and allokinesis. *American Pain Society Journal, 1,* 115–126.

Lewis, T. (1936). Experiments Relating to cutaneous hyperalgesia and its spread through somatic nerves. *Clinical Science, 2,* 373–423.

Lewis, T. (1942). *Pain.* London: Macmillan.

Ness, T. J., & Gebhart, G. F. (1988). Colorectal Distension as a noxious visceral stimulus: Physiologic and pharmacologic characterization of pseudaffective reflexes in the rat. *Brain Research, 450,* 153–169.

Ness, T. J., & Gebhart, G. F. (1990). Visceral pain: A review of experimental studies. *Pain, 41*, 167–234.

Ness, T. J., Metcalf, A. M., & Gebhart, G. F. (1990). A psychophysiological study in humans using phasic colonic distension as a noxious visceral stimulus. *Pain, 43*, 377–386.

Ozaki, N., Bielfeldt, K., Sengupta, J. N., et al. (2002). Models of gastric hyperalgesia in the rat. *American Journal of Physiology. Gastrointestinal and Liver Physiology, 283*, G666–G676.

Sherrington, C. (1906). *The integrative action of the nervous system*. New York: Scribner.

Shinoda, M., Feng, B., & Gebhart, G. F. (2009). Peripheral and central P2X3 receptor contributions to colon mechanosensitivity and hypersensitivity in the mouse. *Gastroenterology, 137*, 2096–2104.

Nocifensive Behaviors, Muscle and Joint

Karin N. Westlund[1] and Kathleen A. Sluka[2]
[1]Department of Physiology, University of Kentucky, Lexington, KY, USA
[2]Physical Therapy and Rehabilitation Science Graduate Program, University of Iowa, Iowa City, IA, USA

Synonyms

Guarding; Hyperalgesia; Spontaneous pain

Definition

Nocifensive behaviors are the response of the animal to noxious or painful stimuli.

Characteristics

There are several animal models of muscle and joint pain utilized to study ► nocifensive behaviors. These include carrageenan (or kaolin/carrageenan) inflammation of the knee joint or the muscle (gastrocnemius, masseter, triceps). Carrageenan inflammation results in an initial acute phase of inflammation that converts to a chronic phase within 1 week after induction and persists 6–8 weeks (Radhakrishan et al. 2003). Alternatively, ► complete Freund's adjuvant (CFA) or capsaicin can be injected into the joint or muscle to produce inflammation and behavioral changes (Yu et al. 2002; Sluka 2002). A noninflammatory model of muscle pain is induced by two injections of acidic saline (Sluka et al. 2001). Comparable findings have been reported in human studies.

Joint Inflammation
Nocifensive responses include spontaneous pain behaviors, ► primary and secondary hyperalgesia and ► allodynia. Spontaneous pain behaviors are observed following knee joint inflammation with carrageenan. These include decreased weight bearing and guarding of the limb, which are most pronounced during the acute phase of inflammation (Sluka and Westlund 1993; Schott et al. 1994; Min et al. 2001; Radhakrishan et al. 2003) (Fig. 1). When the temporomandibular joint (TMJ) is inflamed with CFA, changes in meal pattern occur such that there is an increase in duration of the meal, and a decrease in food intake (Kerine et al. 2003).

Primary Hyperalgesia
Primary hyperalgesia of the knee joint is measured by the response to compression applied to the medial and lateral aspects of the knee joint

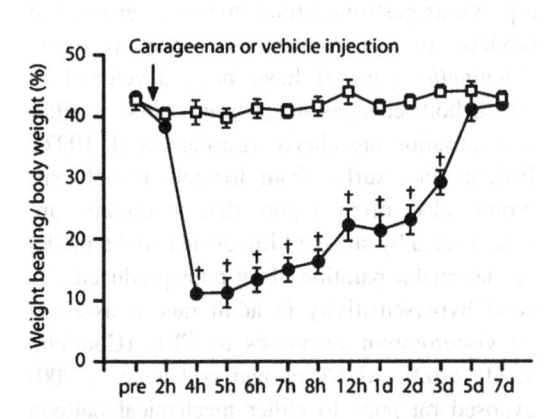

Nocifensive Behaviors, Muscle and Joint, Fig. 1 Time course of the changes in weight bearing of the inflamed hindlimb (*filled circles*) after unilateral injection of carrageenan into the knee joint (From Min et al. (2001). Reprinted with permission of Elsevier Science)

Kaolin/Carrageenan joint inflammation

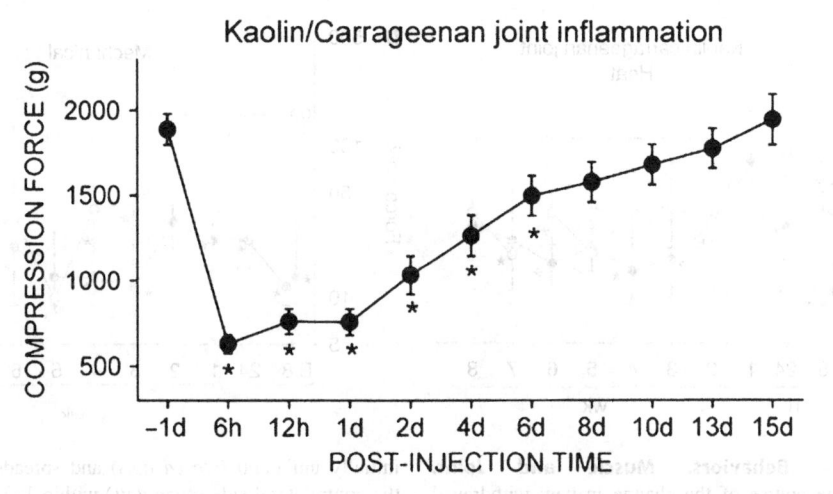

Nocifensive Behaviors, Muscle and Joint, Fig. 2 The time course for the decreased vocalization threshold in response to compression of the knee joint. The compression force was applied to the knee before and after injection of kaolin and carrageenan (*KC*) into the knee joint. There is a decrease in vocalization threshold within 6 h that lasts for approximately 1 week.

Post-injection time is expressed as days (d) after KC injection. Pre-injection control was taken 1 day before the injection (−1 day). Asterisks indicate values significantly different from the pre-injection control value by one-way ANOVA followed by the Dunnett's post-hoc test (n = 10) (From Yu et al. (2002). Reprinted with permission of Elsevier Science)

with a pair of tweezers. This results in a lower threshold to vocalization within 4 h after induction of knee joint inflammation with kaolin and carrageenan, and lasts for several days (Yu et al. 2002) (Fig. 2). Similar responses are observed after injection of CFA into the knee joint (Yu et al. 2002).

Secondary Hyperalgesia

Secondary hyperalgesia to heat stimuli is measured with a radiant heat source applied to the paw following inflammation of the knee joint. Secondary mechanical hyperalgesia is measured by assessing the withdrawal threshold to von Frey filaments applied to the paw. Following carrageenan (3 %) inflammation, there is decreased withdrawal latency to heat and decreased threshold to mechanical stimuli, ipsilaterally, that is long-lasting (weeks). Within 1–2 weeks, a time that corresponds to the conversion to chronic inflammation, the hyperalgesia to both mechanical and heat stimuli spreads to the contralateral limb (Radhakrishan et al. 2003) (Fig. 3). The duration of hyperalgesia and the contralateral spread are dose dependent, such that lower concentrations of carrageenan produce shorter

lasting hyperalgesia that remains unilateral (Radhakrishan et al. 2003). Capsaicin is also utilized to inflame the joint and produces an acute inflammatory response that is associated with a decreased withdrawal threshold to mechanical stimuli, and an increased withdrawal latency to heat stimuli outside the site of injury (Sluka 2002). This secondary hyperalgesia to mechanical stimuli lasts for weeks and spreads to the contralateral limb. The time course and contralateral spread following joint inflammation are distinctly different from the hyperalgesia associated with injection of capsaicin into the skin. Injection of capsaicin into the skin produces a short-lasting (hours) hyperalgesia and remains unilateral (Sluka 2002).

Muscle Pain

Two models of muscle pain, inflammatory and noninflammatory, are utilized to study nocifensive behaviors resulting from muscle insult. Inflammation is induced by injection of carrageenan or capsaicin into muscle tissue. Injection of carrageenan into the muscle results in an initial acute inflammatory response, which converts to a chronic inflammation by 1 week and

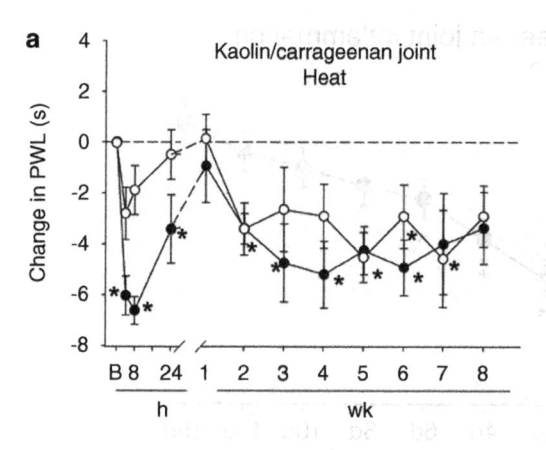

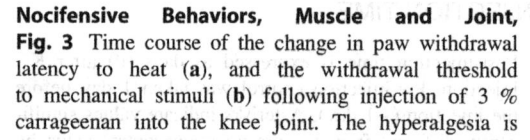

Nocifensive Behaviors, Muscle and Joint, Fig. 3 Time course of the change in paw withdrawal latency to heat (**a**), and the withdrawal threshold to mechanical stimuli (**b**) following injection of 3 % carrageenan into the knee joint. The hyperalgesia is initially unilateral (*closed dots*) and spreads to include the contralateral side (*open dots*) within 1–2 weeks after induction of inflammation (Modified from Radhakrishan et al. (2003). Reprinted with permission of Elsevier Science)

persists for 8 weeks (Radhakrishan et al. 2003). Spontaneous pain behaviors assessed by limb guarding occur during the first 24–48 h, and are less severe than those following the same dose of carrageenan injected into the knee joint.

Primary Hyperalgesia

Primary hyperalgesia of the muscle is measured by examining grip force of the limb. There is a decrease in grip force after carrageenan inflammation of the triceps muscle within 12 h that lasts for 36 h (Fig. 4). Similarly, intramuscular injection of tumor necrosis factor alpha (TNF-α) or formalin reduces grip force for up to 1 day (Schäfers et al. 2003). Alternatively, measurement of withdrawal/vocalization threshold to pressure applied over the belly of the muscle has also been utilized to assess primary hyperalgesia (Schäfers et al. 2003). When TNF-α or formalin is injected into the muscle, decreases in the mechanical threshold applied to the muscle persist for at least 1 day.

Secondary Hyperalgesia

Secondary hyperalgesia is measured in a manner similar to that described above for joint inflammation, but outside the site of injury such as on the paw of the rat. Specifically, after inflammation of the gastrocnemius muscle with

carrageenan, there is a decrease in mechanical withdrawal threshold and decreased latency to radiant heat applied to the paw ipsilateral to the inflamed knee joint within 4 h that lasts for weeks (Fig. 5) (Radhakrishan et al. 2003). After 1–2 weeks, when the inflammation becomes chronic, this hyperalgesia to mechanical and heat stimuli spreads to the contralateral limb (Radhakrishan et al. 2003). A similar long-lasting, bilateral mechanical hyperalgesia occurs when capsaicin is injected into the plantar muscles of the paw (Sluka 2002). The effect of carrageenan muscle injection, and particularly the bilateral spread of hyperalgesia, is dose dependent. Lower doses of intramuscular carrageenan produce only unilateral hyperalgesia that is shorter in duration (Radhakrishan et al. 2003). As a model of chronic noninflammatory muscle pain, two injections of acidic saline injected 2–5 days apart results in a bilateral, long-lasting mechanical, but not heat, hyperalgesia of the paw (Fig. 6) (Sluka et al. 2001). This hyperalgesia is not associated with muscle tissue damage or inflammation, and once developed, the hyperalgesia is independent of continued primary afferent input (Sluka et al. 2001). In contrast, injection of TNF-α or formalin into the muscle does not produce secondary hyperalgesia to mechanical or heat stimuli,

Nocifensive Behaviors, Muscle and Joint,
Fig. 4 Time course of the decrease in grip force that occurs after inflaming the triceps surae muscle with carrageenan. There is a decrease in grip force by 12 h after induction that lasts for 36 h (From Kehl et al. (2000). Reprinted with permission of Elsevier Science)

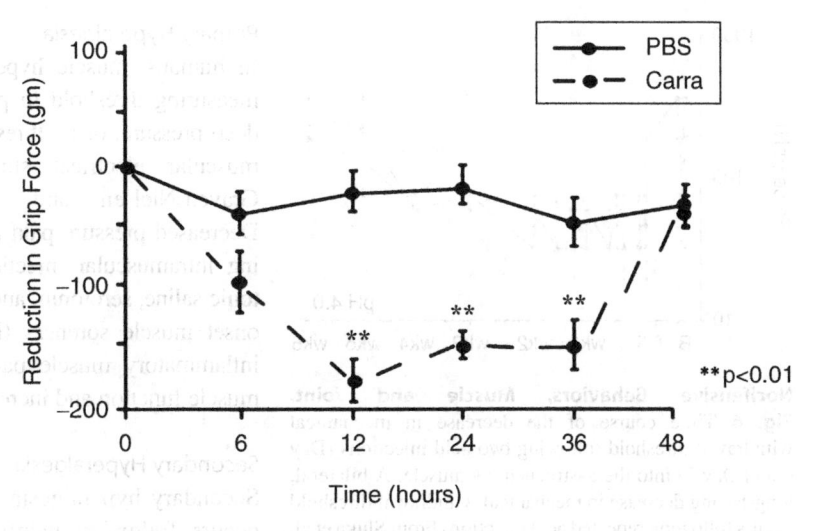

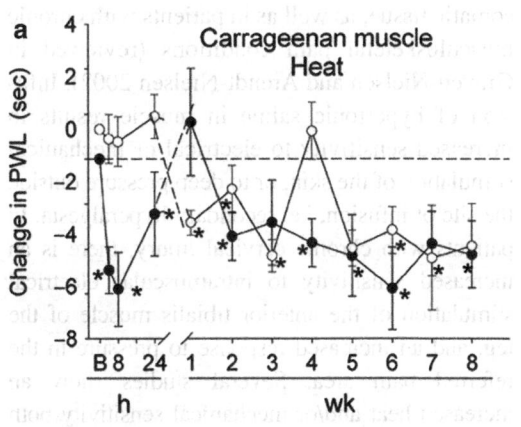

Nocifensive Behaviors, Muscle and Joint, Fig. 5 Time course of the change in paw withdrawal latency to heat (**a**) and the withdrawal threshold to mechanical stimuli (**b**) following injection of 3 % carrageenan into the gastrocnemius muscle. The hyperalgesia

as observed with the other models listed above (i.e., carrageenan, acid) (Schäfers et al. 2003).

In summary, nocifensive behaviors after injury to muscle or joint include spontaneous pain behaviors, primary hyperalgesia, and secondary hyperalgesia. These behaviors are longer lasting than when the same stimulus is applied to the skin, and there is a bilateral spread of the secondary hyperalgesia. The development of hyperalgesia (primary vs. secondary OR mechanical vs. heat) and contralateral hyperalgesia is stimulus dependent and dose dependent.

is initially unilateral (*filled circles*) and spreads to include the contralateral side (*unfilled circles*) within 1–2 weeks after induction of inflammation (Modified from Radhakrishan et al. (2003). Reprinted with permission of Elsevier Science)

Musculoskeletal Pain in Humans

Musculoskeletal pain can arise from a variety of disorders including ▶ myofascial pain, ▶ fibromyalgia, ▶ myositis, or ▶ arthritis. The quality of pain associated with injury to a muscle or joint differs from that associated with injury to the skin. Injury to deep structures results in diffuse, difficult to localize, aching pain (reviewed in Graven-Nielsen and Arendt-Nielsen 2002). In contrast, injury to skin usually produces well-localized, sharp, stabbing, or burning pain. Using microneurography, intraneuronal

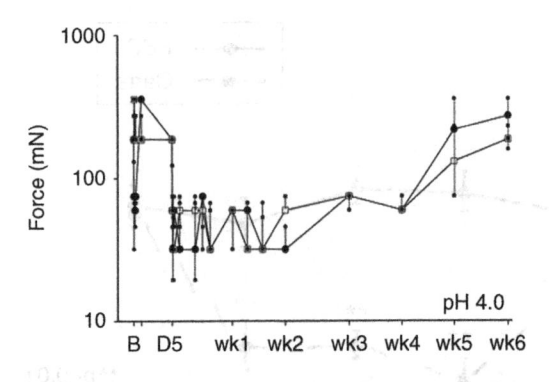

Nocifensive Behaviors, Muscle and Joint, Fig. 6 Time course of the decrease in mechanical withdrawal threshold following two acid injections (Day 0 and Day 5) into the gastrocnemius muscle. A bilateral, long-lasting decrease in mechanical withdrawal threshold occurs following repeated acid injection (From Sluka et al. (2001). Reprinted with permission of Wiley Publishing)

stimulation of muscle nerve fascicles produces only the sensation of deep cramp-like pain (in contrast to stimulation of cutaneous C-fibers [Group IV] which produces burning pain). For muscle pain, the size of the area of ▶ referred pain correlates with the intensity and duration of the primary muscle pain (reviewed in Graven-Nielsen and Arendt-Nielsen 2002). In human subjects, painful intramuscular stimulation is rated as more unpleasant than painful cutaneous stimulation (Svensson et al. 1997), pain is longer lasting, and referred pain is more frequent (see Graven-Nielsen and Arendt-Nielsen 2002). In patients with fibromyalgia or knee osteoarthritis, there is more pain, and a larger and more diffuse area of referred pain following infusion of hypertonic saline into the tibialis anterior muscle when compared to subjects without pain. Similarly, in patients with chronic whiplash pain, there is an extended area of referred pain following infusion of hypertonic saline into the infraspinatus muscle of the shoulder or a distal muscle of the leg (tibialis anterior muscle), suggesting changes in central processing. Thus, pain associated with injury to skin would be expected to differ in quality to pain associated with injury to deeper tissues such as muscle, and results in diffuse widespread increases in sensitivity to noxious stimuli.

Primary Hyperalgesia

In humans, muscle hyperalgesia is assessed by measuring threshold to pain from application of deep pressure, or by threshold to pain from intramuscular electrical stimulation (reviewed in Graven-Nielsen and Arendt-Nielsen 2002). Decreased pressure pain thresholds occur following intramuscular injection of capsaicin, hypertonic saline, serotonin, and bradykinin. ▶ Delayed onset muscle soreness (DOMS) as a model of inflammatory muscle pain, results in decreased muscle function and increased pain to pressure.

Secondary Hyperalgesia

Secondary hyperalgesia in deep tissue and skin occurs following painful stimulation of deep somatic tissue, as well as in patients with chronic musculoskeletal pain conditions (reviewed in Graven-Nielsen and Arendt-Nielsen 2002). Infusion of hypertonic saline in muscle results in increased sensitivity to electrical or mechanical stimulation of the skin, or to deep pressure outside the site of infusion, i.e. secondary hyperalgesia. In patients with chronic cervical injury, there is an increased sensitivity to intramuscular electrical stimulation of the anterior tibialis muscle of the leg, and an increased response to pressure in the referred pain area. Several studies show an increased heat and/or mechanical sensitivity both at the site of an arthritic joint and at a distance from the joint (O'Driscoll and Jayson 1974; Farrell et al. 2000; Kosek and Ordeberg 2000; reviewed in Graven-Nielsen and Arendt-Nielsen 2002; Hendiani et al. 2003). The increased sensitivity to mechanical pressure is found in patients with arthritic pain for greater than 5 years, but not those with arthritic pain for less than 1 year. However, people with arthritis have higher mechanical sensation thresholds (measured with Frey filaments) simultaneously with lower mechanical pain thresholds (measured with von Frey filaments), suggesting that there is an increased activation of descending inhibitory processes in addition to central sensitization.

Thus, secondary hyperalgesia to heat and mechanical stimuli is found in patients with deep tissue pain, similar to that found in animal models of pain.

References

Farrell, M., Gibson, S., McMeeken, J., & Helme, R. (2000). Pain and hyperalgesia in osteoarthritis of the hands. *Journal of Rheumatology, 27*, 441–447.

Graven-Nielsen, T., & Arendt-Nielsen, L. (2002). Peripheral and central sensitization in musculoskeletal pain disorders: An experimental approach. *Current Rheumatology Reports, 4*, 313–321.

Hendiani, J. A., Westlund, K. N., Lawand, N., Goel, N., Lisse, J., & McNearney, T. (2003). Mechanical sensation and pain thresholds in patients with chronic arthropathies. *The Journal of Pain, 4*, 203–211.

Kehl, L. J., Trempe, T. M., & Hargreaves, K. M. (2000). A new animal model for assessing mechanisms and management of muscle hyperalgesia. *Pain, 85*, 333–343.

Kerine, C. A., Carlson, D. S., McIntosh, J. E., & Bellinger, L. L. (2003). Meal pattern changes associated with temperomandibular joint inflammation/pain in rats: Analgesic effects. *Pharmacology Biochemistry and Behavior, 75*, 181–189.

Kosek, E., & Ordeberg, G. (2000). Abnormalities of somatosensory perception in patients with painful osteoarthritis normalize following successful treatment. *European Journal of Pain, 4*, 229–238.

Min, S. S., Han, J. S., Kim, Y. I., Na, H. S., Yoon, Y. W., Hong, S. K., & Han, H. C. (2001). A novel method for convenient assessment of arthritic pain in voluntarily walking rats. *Neuroscience Letters, 308*, 95–98.

O'Driscoll, S. L., & Jayson, M. I. V. (1974). Pain threshold analysis in patients with osteoarthritis of the hip. *British Medical Journal, 3*, 714–715.

Radhakrishan, R., Moore, S. A., & Sluka, K. A. (2003). Unilateral carrageenan injection into muscle or joint induces chronic bilateral hyperalgesia in rats. *Pain, 104*, 567–577.

Schäfers, M., Sorkin, L. S., & Sommer, C. (2003). Intramuscular injection of tumor necrosis factor-alpha induces muscle hyperalgesia in rats. *Pain, 104*, 579–588.

Schott, E., Berge, O. G., Angeby-Moller, K., Hammarstrom, G., Dalsgaard, C. J., & Brodin, E. (1994). Weight bearing as an objective measure of arthritic pain in the rat. *Journal of Pharmacological and Toxicological Methods, 31*, 79–83.

Sluka, K. A. (2002). Stimulation of deep somatic tissue with capsaicin produces long-lasting mechanical allodynia and heat hypoalgesia that depends on early activation of the cAMP pathway. *Journal of Neuroscience, 2*, 5687–5693.

Sluka, K. A., & Westlund, K. N. (1993). Behavioral and immunocytochemical changes in an experimental arthritis model in rats. *Pain, 55*, 367–377.

Sluka, K. A., Kalra, A., & Moore, S. A. (2001). Repeated intramuscular injections of acidic saline produce a bilateral, long-lasting hyperalgesia. *Muscle & Nerve, 4*, 37–6.

Svensson, P., Beydoun, A., Morrow, T. J., & Casey, K. L. (1997). Human intramuscular and cutaneous pain: Psychophysical comparisons. *Experimental Brain Research, 114*, 390–392.

Yu, Y. C., Koo, S. T., Kim, C. H., Lyu, Y., Grady, J. J., & Chung, J. M. (2002). Two variables that can be used as pain indices in experimental animal models of arthritis. *Journal of Neuroscience Methods, 115*, 107–113.

Nocisensor

▶ Muscle Nociceptors, Neurochemistry

Nocistatin

▶ Orphanin FQ

Nodes of Ranvier

Definition

Axonal zones within gaps in myelin in which excitation occurs.

Cross-References

▶ Trafficking and Localization of Ion Channels

Nodose Ganglia

Definition

Inferior ganglia of vagus nerve containing the cell bodies of vagal afferent neurons.

Cross-References

▶ Visceral Pain Model; Esophageal Pain

Nomogenic Symptoms and Signs

▶ Nonorganic Symptoms and Signs

Nonanatomic Symptoms and Signs

▶ Nonorganic Symptoms and Signs

Non-cardiac Chest Pain

▶ NCCP

Noncompetitive Antagonist

Definition

Noncompetitive antagonist is a compound that inhibits receptor function without directly acting at the endogenous ligand-binding site.

Cross-References

▶ Metabotropic Glutamate Receptors in Spinal Nociceptive Processing

Non-corpuscular Sensory Endings

Synonyms

Free nerve endings

Definition

The peripheral end structures of unmyelinated (C or Group IV) and thinly myelinated (Aδ or Group III) sensory fibers that encode noxious, cold, or warm stimuli; that consist of either single fibers or bundles of sensory axons associated with peripheral glia (Schwann cells); and that are characterized by their lack of perineurium and the loss of myelination in Aδ endings. Typically, the sensory axons of nociceptive sensory endings have exposed membrane areas that are not covered by Schwann cell processes and can be regarded as chemoreceptive sites.

Cross-References

▶ Nociceptors in the Orofacial Region (Meningeal/Cerebrovascular)
▶ Toxic Neuropathies

Nondrug Treatment

▶ Psychological Treatment of Headache

Nonexertional Capability

Definition

Capability to perform any of the nonexertional activities, which are activities other than any of the seven strength demands of work.

Cross-References

▶ Disability Evaluation in the Social Security Administration

Nonexertional Limitations and Restrictions

Definition

Limitations or restrictions that affect the capability to perform a work-related function that is not exertional, i.e., a limitation or restriction of mental abilities, vision, hearing, speech, climbing, balancing, stooping, kneeling, crouching, crawling, reaching, handling, fingering, and feeling, or an environmental restriction.

Cross-References

▶ Disability Evaluation in the Social Security Administration

Non-migraine Headaches

Hans-Christoph Diener
Department of Neurology, University Hospital Essen, University Duisburg-Essen, Essen, Germany

Definition

▶ Migraine is the best investigated and understood headache entity. About 50 % of all headache publications deal with migraine. The International Headache Society created a headache classification in 1988 (Headache Classification Committee of the International Headache Society 1988), which was updated in 2004 (Olesen et al. 2004). This classification contains all other headaches in addition to migraine and provides operational criteria for diagnosis. Headaches are grouped into primary and secondary headaches. Primary headaches are migraine, ▶ tension-type headache, and the trigemino-autonomic headaches.

Characteristics

Tension-type headache is the most frequent headache. The new classification distinguishes infrequent from frequent episodic tension-type headache. ▶ Chronic tension-type headache is diagnosed when >180 headache days/year are occurring. The group of trigemino-autonomic headaches is characterized by headaches with autonomic features, e.g., conjunctival injection, lacrimation, nasal congestion, rhinorrhea, forehead and facial sweating, miosis, ptosis, or eyelid edema (Goadsby and Lipton 1997). This group of headaches includes paroxysmal and chronic cluster headaches, episodic and ▶ chronic paroxysmal hemicrania (Sjaastad and Dale 1976), and short-lasting unilateral neuralgiform headache attacks with conjunctival injection and tearing (▶ SUNCT) (Sjaastad et al. 1991). Other primary headaches include primary stabbing headache (Pareja et al. 1996), ▶ primary cough headache (Symonds 1956), ▶ primary exertional headache, primary headache associated with sexual activity (Pascual et al. 1996), and ▶ hypnic headache. In all headaches with provocation by physical activity, a secondary headache such as subarachnoidal hemorrhage has to be excluded. Hypnic headache occurs in elderly women during the nighttime and responds to treatment with lithium (Raskin 1988; Newman et al. 1990). New entities in this group are ▶ hemicrania continua (Sjaastad and Spierings 1984), a chronic headache and ▶ new daily persistent headache (Li and Rozen 2001). New daily persistent headache has similar characteristics to chronic tension headache but an abrupt onset without improvement.

Secondary headaches are related to physical changes to the brain, the head, or the neck. Headache attributed to head and/or neck trauma is divided into acute and chronic headache. This group includes acute and chronic posttraumatic headache (Young and Packard 1997), acute headache attributed to whiplash injury, headache attributed to intracranial hematomas (subdural, epidural), headache attributed to other head and/or neck trauma, and Post-craniectomy headache.

A large group of secondary headaches are headaches attributed to cranial or cervical vascular disorders. Headache in these cases can be due to cerebral ischemia, cerebral bleedings, or an unruptured AV malformation or aneurysm. Other reasons for headache are arteritis (giant cell and intracranial arteritis), cerebral venous thrombosis, and dissections of the carotid or vertebral arteries. Migraine is a risk factor for stroke and may lead to an ischemic stroke during the aura phase (Diener et al. 2004).

The next group in the classification defines headaches attributed to intracranial nonvascular causes. High (Wall 1990) as well as low (Lay et al. 1997) intracranial pressure causes headache. ► Headache attributed to low cerebrospinal fluid pressure can occur spontaneously or be the result of lumbar puncture. Increased intracranial pressure is due in most cases to intracranial neoplasms or hydrocephalus. Headache attributed to noninfectious inflammatory disease can be due to neurosarcoidosis, aseptic meningitis, or lymphocytic hypophysitis. Another interesting association exists between epilepsy and migraine (Leniger et al. 2003). Migraine auras can lead to a seizure and a tonic-clonic seizure may trigger a migraine attack.

A major revision took place in the classification of ► headaches attributed to substance use or withdrawal. This section includes acute headache provoked by drugs or chemical substances such as nitric oxide donors, phosphodiesterase inhibitors, or carbon monoxide, but also food components, e.g., alcohol or monosodium glutamate. Recreational drugs such as heroin, cocaine, and cannabis can cause headache. A new concept is medication overuse headache (Diener and Limmroth 2004). The frequent or regular use of drugs to treat acute headaches can lead to a chronic headache or deterioration of a preexisting headache. Medication overuse headache can be due to analgesics, combination drugs, ergots, triptans, or opioids. Withdrawal of substances, e.g., caffeine, estrogen, or opioids can also result in headache.

Headache attributed to infections is associated with bacterial meningitis, lymphocytic meningitis, encephalitis, brain abscess, or subdural empyema. Headache can also be due to systemic infection with bacteria, viruses (including HIV) or other systemic infections.

Group 10 summarizes headaches associated with disorders of homeostasis. In this group are headaches due to hypoxia and/or hypercapnia, to arterial hypertension, hypothyroidism, and fasting. Headache and facial pain attributed to disorders of cranium, neck, eye, ears nose, sinuses, teeth, mouth, or other facial or cranial structures will be covered in other entries of this encyclopedia.

References

Diener, H. C., & Limmroth, V. (2004). Medication-overuse headache: A worldwide problem. *Lancet Neurology, 3*, 475–483.

Diener, H., Welch, K., & Mohr, J. P. (2004). Migraine and stroke. In J. Mohr, D. Choi, J. Grotta, et al. (Eds.), *Stroke. Pathophysiology, diagnosis and management.* Philadelphia: Churchill Livingstone.

Goadsby, P. J., & Lipton, R. B. (1997). A review of paroxysmal hemicranias, SUNCT syndrome and other short-lasting headaches with autonomic features, including new cases. *Brain, 120*, 193–209.

Headache Classification Committee of the International Headache Society. (1988). Classification and diagnostic criteria for headache disorders, cranial neuralgias and facial pain. *Cephalalgia, 8* (Suppl. 7), 1–93.

Lay, C. L., Campbell, J. K., & Mokri, B. (1997). Low cerebrospinal fluid pressure headache. In S. D. Silberstein & P. Goadsby (Eds.), *Headache* (pp. 359–368). Newton: Butterworth-Heinemann.

Leniger, T., von den Driesch, S., Isbruch, K., et al. (2003). Clinical characteristics of patients with comorbidity of migraine and epilepsy. *Headache, 43*, 672–677.

Li, D., & Rozen, T. D. (2001). The clinical characteristics of new daily persistent headache. *Neurology, 56*(Suppl 3), A452–A453.

Newman, L. C., Lipton, R. B., & Solomon, S. (1990). The hypnic headache syndrome: A benign headache disorder of the elderly. *Neurology, 40*, 1904–1905.

Olesen, J., Bousser, M.-G., Diener, H. et al. for the International Headache Society. (2004). The international classification of headache disorders (2nd ed.). *Cephalalgia, 24* (Suppl. 1), 1–160.

Pareja, J. A., Ruiz, J., Deisla, C., et al. (1996). Idiopathic stabbing headache (jabs and jolts syndrome). *Cephalalgia, 16*, 93–96.

Pascual, J., Iglesias, F., Oterino, A., et al. (1996). Cough, exertional, and sexual headaches: An analysis of 72 benign and symptomatic cases. *Neurol, 46*, 1520–1524.

Raskin, N. H. (1988). The hypnic headache syndrome. *Headache, 28*, 534–536.

Sjaastad, O., & Dale, I (1976) A new (?) clinical headache entity "chronic paroxysmal hemicrania." *Acta Neurologica Scandinavica, 54*, 140–159.

Sjaastad, O., & Spierings, E. L. (1984). Hemicrania continua: Another headache absolutely responsive to indomethacin. *Cephalalgia, 4*, 65–70.

Sjaastad, O., Zhao, J. M., Kruszewski, P., et al. (1991). Short-lasting unilateral neuralgiform headache attacks with conjunctival injection, tearing, etc. (SUNCT): III. Another Norwegian case. *Headache, 31*, 175–177.

Symonds, C. (1956). Cough headache. *Brain, 79*, 557–568.

Wall, M. (1990). The headache profile of idiopathic intracranial hypertension. *Cephalalgia, 10*, 331–335.

Young, W. B., & Packard, R. C. (1997). Posttraumatic headache and posttraumatic syndrome. In P. J. Goadsby & S. D. Silberstein (Eds.), *Headache.* New York: Butterworth-Heinemann.

Nonneurogenic Inflammation

Definition

Inflammatory processes that are not specifically initiated by activity in primary afferent C fibers; typically, this includes inflammatory processes initiated by mediators released from resident or circulating monocytes and leukocytes, including prostaglandins, prostacyclins, leukotrienes, and cytokines, to name a few.

Cross-References

▶ Formalin Test

Nonnutritive Sucking

Definition

Placing an object (e.g., pacifier, non-lactating nipple) into an infant's mouth to stimulate sucking behaviors during a painful event.

Cross-References

▶ Acute Pain Management in Infants

Nonopioid Analgesia

Definition

Type of pain inhibition not meeting any of the criteria for opiate-mediated effects (including blockade by opiate antagonists and cross-tolerance with opiate agonists).

Cross-References

▶ Pain-Modulatory Systems, History of Discovery

Nonopioid Analgesic Drug

Definition

Drugs that relieve pain by other mechanisms than interacting with opioid receptors, e.g., paracetamol and NSAIDs.

Cross-References

▶ Cancer Pain Management
▶ NSAIDs, Survey
▶ Postoperative Pain, Acute Pain Management Principles

Non-opioid Analgesics

▶ Nonsteroidal Anti-Inflammatory Drugs (NSAIDs)

Nonorganic Aspects of Visceral Pain

▶ Psychiatric Aspects of Visceral Pain

Nonorganic Physical Findings (Waddell Signs)

Definition

A group of low back pain physical signs that were identified in the past as being nonorganic and, therefore the presence of which has historically indicated that the patient is either suffering from conversion disorder or malingering.

Nonorganic Symptoms and Signs

Gordon Waddell
University of Glasgow, Glasgow, UK

Synonyms

Behavioral descriptions of symptoms; Behavioral responses to examination; Inappropriate symptoms and signs; Inconsistent symptoms and signs; Medically incongruent symptoms and signs; Nomogenic symptoms and signs; Nonanatomic symptoms and signs; Waddell signs

Definition

Descriptions of symptoms and responses to clinical examination in patients with low back pain, which relate more to cognitive-behavioral processes than to physical pathology.

Characteristics

Clinical assessment and diagnosis are based on the recognition of patterns of symptoms and signs, which in most patients fit to a greater or lesser degree with anatomy, mechanics, and pathology. Occasional patients, however, present symptoms and signs that are not only vague and ill-localized, lack the normal relationship with time and activity, and appear out of proportion to the physical injury or pathology, but positively contradict normal anatomic boundaries and biomechanics.

Nonorganic signs in low back pain have been recognized since the beginning of the twentieth century, initially in the context of compensation assessment, where any clinical findings that were judged excessive or not entirely consistent with physical injury, were interpreted as evidence of malingering. That oversimplification gradually became discredited. The modern description by Waddell et al. (1980) standardized a battery of

Nonorganic Symptoms and Signs, Table 1 Standardized battery of nonorganic signs in low back pain

Tenderness	Superficial
	Nonanatomic
Simulation	Axial loading
	Simulated rotation
Distraction	Straight leg raising
Regional	Weakness
	Sensory disturbance
Overreaction to examination	Subsequently replaced by overt pain behavior (Keefe and Block 1982)
	Grimacing, sighing, guarding, bracing, rubbing

Waddell et al. (1984) subsequently developed a corresponding battery of nonorganic or behavioral descriptions of symptoms in low back pain (see list below), though these have never received as much attention or use as the nonorganic signs

nonorganic signs (Table 1), and reinterpreted them in the light of modern understanding of illness behavior.

Nonorganic or behavioral descriptions of symptoms in low back pain:

- Pain at the tip of the tailbone
- Whole leg pain
- Whole leg numbness
- Whole leg giving way
- Complete absence of any spells with very little pain in the past year
- Intolerance of, or reactions to, many treatments
- Emergency admission(s) to hospital with nonspecific low back pain

These groups of nonorganic symptoms and signs were demonstrated to be reliable, internally consistent, separable from the standard symptoms and signs of physical pathology, and related to psychological distress and other measures of illness behavior. Independent studies (reviewed in detail in Waddell 2004) have since confirmed that they relate to: measures of physical impairment, severity of pain, and poorer physical performance (though there is debate about the extent to which these reflect physical impairment or illness behavior); affective measures of pain and psychological distress; other measures of illness behavior (the pain drawing, overt pain behavior, UAB pain behavior scale, and various scales of the Illness Behavior Questionnaire); disability and incapacity; and the prediction of clinical outcomes (but not occupational outcomes) of conservative and surgical treatment and rehabilitation. Nonorganic symptoms and signs were initially observed in patients with chronic low back pain, and illness behavior was considered to be the consequence of chronicity. It is now clear that they can occur at a much earlier stage, reflecting the involvement of illness behavior in the process of chronification. Illness behavior is learned and a form of communication between patients and health professionals.

Nonorganic symptoms and signs can help to clarify clinical assessment by:

• Distinguishing and hence permitting separate assessment of physical and behavioral elements in the clinical presentation

• Providing a simple clinical screen for psychological and behavioral issues in the clinical presentation, indicating the need for more detailed psychological assessment

• Clarifying clinical decision making – helping to direct physical treatment to treatable pathology, and reducing inappropriate physical treatment; helping to identify patients who may need more psychological support

However, there are a number of important caveats to their use (Main and Waddell 1998; Waddell 2004):

1. Diagnostic triage should be performed first, to exclude serious spinal pathology (such as tumor or infection), widespread neurological disorders (involving multiple nerve roots) or major psychiatric disorders (e.g., psychosis or major depression) before assessing nonorganic symptoms and signs.

2. Clinical observation of illness behavior depends on careful technique and it is important to minimize observer bias.

3. Isolated behavioral symptoms and signs are common in normal patients and of no clinical, psychological, or behavioral significance. Only multiple positive findings (preferably of several different dimensions of illness behavior) are clinically significant.

4. Behavioral symptoms and signs do not provide any information about the original cause of the pain: they do not mean that the patient's pain is "not real," "psychogenic," or "hysteric."

5. It is not a differential diagnosis between physical disease and illness behavior: most patients have both a physical problem in their backs and varying degrees of illness behavior. Inability to demonstrate the physical basis of the pain does not mean that the pain is psychogenic, any more than the presence of illness behavior excludes a treatable physical problem. Diagnosis of psychological dysfunction and illness behavior depends on positive psychological and behavioral findings.

6. Nonorganic symptoms and signs are only a preliminary clinical screening test, indicating the need for more detailed clinical and psychological assessment. Clinical observations of illness behavior do not provide a complete psychological assessment or a psychological or psychiatric diagnosis. They are less sensitive than psychometric measures for detecting distress, so there is a significant false negative rate.

7. Nonorganic symptoms and signs are not lie-detectors, but observations of normal human behavior in illness. They do not necessarily mean that the patient is acting, faking, or malingering. Most illness behavior occurs in pain patients who have no claim for compensation or any question of secondary gain.

This was summarized in the original article (Waddell et al. 1980):

It is safer to assume that all patients complaining of back pain have a physical source of pain in their back. Equally, all patients with pain show some emotional and behavioral reaction. Physical pathology and nonorganic reactions are discrete and yet frequently interacting dimensions; they are not alternative diagnoses but should each be assessed separately.

Nonorganic signs are widely used – and misused – in surgical, compensation, and medicolegal practice, but there is continuing controversy about their value and interpretation.

There are two main criticisms (which are mutually exclusive):

1. Pain specialists have argued that modern neurophysiologic and clinical understanding of chronic pain provides a purely biologic explanation for these findings (Merskey 1988; Margoles 1990; Fishbain et al. 2003). Neurophysiologic mechanisms can produce spread of pain and tenderness, hypersensitivity, altered sensation, and inhibition of motor activity, in the absence of other evidence of illness behavior. That is why nonorganic symptoms and signs must be elicited and interpreted with care, and isolated findings should not be overinterpreted. However, nonorganic symptoms and signs often spread far beyond any likely neurophysiologic mechanism and fit better with body image patterns, they form part of a constellation of other illness behaviors and correlate with other psychological findings.

2. On the contrary, many medicolegal experts claim that nonorganic symptoms and signs demonstrate conscious and deliberate attempts to deceive the examiner, and evidence of faking or malingering. The scientific evidence in this area is weak (Fishbain et al. 1999; Waddell 2004) but this is a legal rather than a clinical matter. Case law leaves no doubt that nonorganic symptoms and signs can occur in patients with other evidence of lack of credibility and represent conscious attempts to exaggerate the severity of pain and disability. However, there is a wealth of clinical and legal evidence that illness behavior can also be produced by unconscious, psychological mechanisms. Thus, nonorganic signs per se do not provide sufficient evidence to prove malingering. As in the clinical setting, nonorganic signs are a screening tool: they may raise the question of credibility, but they do not provide an answer – that depends on thorough assessment and judgment of all the clinical, psychological, and legal evidence.

Nonorganic symptoms and signs have only been standardized in white Anglo-Saxon patients of working age with chronic low back pain. Further research would be required before they can be used in younger or older patients, in nonwhite patients, or in different cultures. In principle, it appears likely that the concept of nonorganic symptoms and signs may be equally applicable to other pain conditions, although there have only been a few preliminary studies in neck pain, other musculoskeletal conditions and cardiac pain and their use without low back pain is not established.

Nonorganic symptoms and signs are a powerful tool to extend the scope of routine clinical assessment of patients with low back pain, provided care is taken to define what they are and what they are not, and to recognize their strengths and their limitations. Like all clinical tools, they must be used with care and compassion.

References

Fishbain, D. A., Cutler, R., Rosomoff, H. L., & Rosomoff, R. S. (1999). Chronic pain disability exaggeration/ malingering and submaximal effort research. *The Clinical Journal of Pain, 15*, 244–274.
Fishbain, D. A., Cole, B., Cutler, R. B., Lewis, J., Rosmoff, H. L., & Rosomoff, R. S. (2003). A structured, evidence-based review of the meaning of non-organic physical signs: Waddell signs. *Pain Medicine, 4*, 141–181.
Keefe, F. J., & Block, A. R. (1982). Development of an observation method for assessing pain behavior in chronic low back pain patients. *Behavior Therapy, 13*, 363–375.
Main, C. J., & Waddell, G. (1998). Behavioral responses to examination. A reappraisal of the interpretation of "non-organic signs". *Spine, 23*, 2367–2371.
Margoles, M. S. (1990). Clinical assessment and interpretation of abnormal illness behaviour in Low back pain. Letter to the editor. *Pain, 42*, 258–259.
Merskey, H. (1988). Regional pain is rarely hysterical. *Archives of Neurology, 45*, 915–918.
Waddell, G. (2004). *The back pain revolution* (2nd ed.). Edinburgh: Churchill Livingstone.
Waddell, G., McCulloch, J. A., Kummel, E., & Venner, R. M. (1980). Non-organic physical signs in low back pain. *Spine, 5*, 117–125.
Waddell, G., Main, C. J., Morris, E. W., Di Paola, M. P., & Gray, I. C. M. (1984). Chronic low back pain, psychological distress, and illness behavior. *Spine, 9*, 209–213.

Non-pharmacologic Pain Management

Definition

Non-pharmacologic pain management includes all approaches to pain control that do not involve drug therapies, such as distraction (e.g., music), psychological interventions (e.g., imaging/visualization, relaxation, meditation), biofeedback, massage, aromatherapy, physical modalities (electrical stimulation, heat, cold), etc.

Cross-References

▶ Opioids in Geriatric Application

Non-Pharmacological Interventions

▶ Psychological Treatment of Pain in Children

Non-pharmacological Treatment

Definition

Nonmedical, psychologically based treatments that teach individuals skills so that they may handle pain in a more optimal manner.

Cross-References

▶ Biofeedback in the Treatment of Pain
▶ Psychological Treatment of Headache

Nonselective COX Inhibitors

▶ NSAIDs, Mode of Action

Nonselective NSAIDs

▶ NSAIDs and Their Indications

Nonspecific Low Back Pain

Definition

Low back pain that is not due to specific pathology such as tumor, fracture, inflammation, osteoporosis, or rheumatoid arthritis.

Cross-References

▶ Back Pain in the Workplace
▶ Lumbar Traction

Nonsteroidal Anti-Inflammatory Drugs

▶ Nonsteroidal Anti-Inflammatory Drugs (NSAIDs)
▶ NSAIDs and Cancer
▶ NSAIDs, Adverse Effects
▶ NSAIDs, Mode of Action

Nonsteroidal Anti-Inflammatory Drugs (NSAIDs)

Gerd Geisslinger
Department of Clinical Pharmacology,
University Frankfurt am Main, Frankfurt,
Germany

Synonyms

Aspirin-like drugs; Non-opioid analgesics; Nonsteroidal anti-inflammatory drugs; Nonsteroidal antirheumatic drugs; NSAIDs; NSAR

Definition

Non-opioid analgesic agents can be divided into two groups. The first group contains substances having anti-inflammatory effects in addition to their analgesic and antipyretic activity and is called nonsteroidal anti-inflammatory drugs (NSAIDs). The members of this group, and one substance (lumiracoxib) within the group of selective inhibitors of cyclooxygenase 2 (COX-2), are acids. Acidic NSAIDs, which include salicylates, derivatives of acetic acid and propionic acid, and oxicams, among others, comprise molecules containing a lipophilic and a hydrophilic region and are more than 99 % bound to plasma proteins. The second group of non-opioid analgesics, which are not classified as NSAIDs, consists of substances that lack anti-inflammatory properties, such as phenazones, metamizole (= dipyrone), and paracetamol. Their molecules are neutral or weakly basic, have no hydrophilic polarity, and are much less strongly bound to plasma proteins than NSAIDs (see ▶ NSAIDs, chemical structure and molecular mode of action).

Characteristics

Mechanism of Action

In the 1970s, NSAIDs were shown to interfere with the biosynthesis of prostaglandins (Vane 1971; Fereira et al. 1971). NSAIDs block cyclooxygenases (COX) that catalyze the formation of cyclic endoperoxides from arachidonic acid. Cyclic endoperoxides are precursors of the prostaglandins, thromboxane A2, and prostacyclin. Prostaglandins have a major role in the pathogenesis of pain, fever, and inflammation. Inhibition of their biosynthesis would therefore be expected to result in analgesic, antipyretic, and anti-inflammatory activity. However, since prostaglandins are synthesized in most tissues and have a variety of physiological functions, inhibition of their biosynthesis also causes unwanted effects. The clinically most important of these are gastrointestinal erosion and ulceration with bleeding and perforation and kidney disorders

with retention of sodium ions and water. About 30 years ago, the principle mode of the antinociceptive action of NSAIDs was related to their anti-inflammatory activity and attributed to the inhibition of the production of prostaglandins at the peripheral site of inflammation. The traditional belief in the exclusively peripheral action of NSAIDs, however, has been challenged by the growing evidence that prostaglandins are key players in spinal nociceptive processing. Thus, NSAIDs have a peripheral and central component of antinociceptive activity (see essays on ▶ History of Analgesics; ▶ COX-1 and COX-2 in Pain; ▶ NSAIDs and their Indications; ▶ Prostaglandins, Spinal Effects; ▶ NSAIDs, Adverse Effects).

Moreover, there are cyclooxygenase-independent effects of NSAIDs potentially contributing to the activity of NSAIDs (for review, see Tegeder et al. 2001) (see also ▶ NSAIDs, COX-Independent Actions; ▶ NSAIDs and Cancer).

The identification of two distinct types of cyclooxygenases in 1990 (Fu et al. 1990) encouraged the search for NSAIDs devoid of the side effects associated with COX-1 inhibition. The cyclooxygenase isoform COX-1 is physiologically expressed in the stomach, platelets, and the kidney and is therefore responsible for the synthesis of prostaglandins needed for normal organ function. Its inhibition by conventional NSAIDs causes side effects, for example, inhibition of prostaglandin synthesis in the gastrointestinal tract results in a loss of protection in the gastrointestinal mucosa and ulcerations. The cyclooxygenase isoform, COX-2, is rapidly induced by various factors including cytokines, and its expression is triggered by inflammation, pain, or tissue damage. It is clear from this division of cyclooxygenases into COX-1 and an inducible COX-2 that the anti-inflammatory, analgesic, and antipyretic effects of the NSAIDs are mainly attributable to inhibition of COX-2, whereas inhibition of COX-1 is associated with most of the unwanted effects of the NSAIDs. It follows that drugs that selectively inhibit COX-2 should cause fewer side effects than those that inhibit both COX-1 and COX-2. At therapeutic doses, all currently available NSAIDs, with the exception of coxibs, are nonselective and inhibit both COX isoforms (see ▶ NSAIDs, Chemical Structure and Molecular Mode of Action).

Newer research has shown that the assignment of physiological activity exclusively to COX-1 and pathophysiological activity exclusively to COX-2 is not strictly valid, since COX-2 is expressed constitutively in organs such as spinal cord, kidney, and uterus. Furthermore, COX-2 is formed during various physiological adaptation processes, such as the healing of wounds and ulcers.

Clinical Use and Side Effects

NSAIDs are indicated in the treatment of:
- Various pain states (e.g., headache, toothache, and migraine), primarily pathophysiological pain involving nociceptors, for example, rheumatic pain and pain caused by bone metastases
- Defects of the ductus arteriosus Botalli (short circuit connection between arteria pulmonalis and aorta, non-closure after birth)
- Fever (see ▶ NSAIDs and their Indications)

Side effects of NSAIDs include:
- Gastrointestinal disorders (e.g., dyspepsia), gastrointestinal erosion with bleeding, ulceration, and perforation.
- Kidney malfunctions with retention of sodium and water, causing hypertension.
- Inhibition of platelet aggregation.
- Central nervous symptoms such as dizziness and headache.
- Disturbance of uterine motility.
- Skin reactions.
- Triggering of asthma attacks in asthmatics. This side effect is a pseudoallergic reaction where COX inhibition increases the availability of substrates for lipoxygenase that are converted to bronchoconstrictive leukotrienes (see ▶ NSAIDs, Adverse Effects and ▶ NSAID-Induced Lesions of the Gastrointestinal Tract).

Nonselective inhibitors of prostaglandin synthesis are contraindicated:
- In gastric and duodenal ulcer
- In asthma
- In bleeding disorders

- During the last few weeks of pregnancy because of the danger of early sealing of the ductus Botalli

Glucocorticoids increase the risk of gastrointestinal complications. Considerable caution is necessary when using NSAIDs in patients with severe liver and kidney damage, and they should not be combined with coumarins. Owing to the limited experience obtained, these precautions and contraindications also apply to COX-2 selective inhibitors.

The following drug interactions are the most important that can occur when conventional NSAIDs are co-administered with other agents:
- The uricosuric effect of probenecid is reduced.
- The diuretic effect of saluretics is weakened.
- The blood glucose lowering effect of oral antidiabetics is increased.
- The elimination of methotrexate is delayed, and its toxicity is increased.
- The elimination of lithium ions is delayed.
- The anticoagulation effect of coumarin derivatives is enhanced.
- The antihypertensive effect of ACE inhibitors is reduced (see ▶ NSAIDs, Pharmacokinetics).

Due to the short period of clinical use, the interaction profile of COX-2 selective inhibitors cannot be described at the present time.

Derivatives of Salicylic Acid

Salicylic acid for systemic use has been replaced by acetylsalicylic acid, amides of salicylic acid (salicylamide, ethenzamide, salacetamide), salsalate, and diflunisal.

Acetylsalicylic Acid (Aspirin)

The esterification of the phenolic hydroxyl group in salicylic acid with acetic acid results in an agent not only with improved local tolerability but also with greater antipyretic and anti-inflammatory activity and, in particular, more marked inhibitory effects on platelet aggregation (inhibition of thromboxane A2 synthesis). Because of these qualities, acetylsalicylic acid is one of the most frequently used non-opioid analgesics and the most important inhibitor of platelet aggregation.

Acetylsalicylic acid irreversibly inhibits both COX-1 and COX-2 by acetylating the enzymes.

Since mature platelets lack a nucleus, they are unable to synthesize new enzyme. The antiplatelet effects of acetylsalicylic acid therefore persist throughout the lifetime of the platelet, and the half-life of this effect is thus much longer than the elimination half-life of acetylsalicylic acid (15 min). Since new platelets are continuously launched into the circulation, the clinically relevant antiplatelet effect of aspirin lasts for up to 5 days. This is the reason why low doses of acetylsalicylic acid (ca. 100 mg per day) are sufficient in the prophylaxis of heart attacks.

After oral administration, acetylsalicylic acid is rapidly and almost completely absorbed, but in the intestinal mucosa, it is partly deacetylated to salicylic acid, which also exhibits analgesic activity. The plasma half-life of acetylsalicylic acid is approximately 15 min, whereas that of salicylic acid at therapeutic dosages of acetylsalicylic acid is 2–3 h. Salicylic acid is eliminated more slowly when acetylsalicylic acid is administered at high dose rates because of the saturation of the liver enzymes. The metabolites are mainly excreted via the kidney. The dosage of acetylsalicylic acid in the treatment of pain and fever is 1.5–3 g daily and in the prophylaxis of heart attacks 30–100 mg daily.

Side effects of acetylsalicylic acid administration include buzzing in the ears, loss of hearing, dizziness, nausea, vomiting, and most importantly gastrointestinal bleeding and gastrointestinal ulcerations including gastric perforation. The administration of acetylsalicylic acid in children with viral infections can, in rare cases, produce Reye's syndrome, involving liver damage, encephalopathy, and a mortality rate exceeding 50 %. Acute salicylate poisoning results in hyperventilation, marked sweating, and irritability followed by respiratory paralysis, unconsciousness, hyperthermia, and dehydration.

Derivatives of Acetic Acid
Indomethacin
Indomethacin is a strong inhibitor of both cyclooxygenase isoforms with a slight stronger effect in the case of COX-1. It is rapidly and almost completely absorbed from the gastrointestinal tract and has high plasma protein binding.

The plasma half-life of indomethacin varies from 3 to 11 h due to intense enterohepatic cycling. Only about 15 % of the substance is eliminated unchanged in the urine, the remainder being eliminated in urine and bile as inactive metabolites (O-demethylation, glucuronidation, N-deacylation). The daily oral dose of indomethacin is 50–150 mg (up to 200 mg).

Indomethacin treatment is associated with a high incidence (30 %) of side effects typical for those seen with other NSAIDs (see above). Gastrointestinal side effects, in particular, are more frequently observed after indomethacin than after administration of other NSAIDs. The market share of indomethacin (approximately 5 %) is therefore low as compared to that for other nonsteroidal antirheumatic agents.

Diclofenac
Diclofenac is an exceedingly potent cyclooxygenase inhibitor slightly more efficacious against COX-2 than COX-1. Its absorption from the gastrointestinal tract varies according to the type of pharmaceutical formulation used. The oral bioavailability is only 30–80 % due to a first-pass effect. Diclofenac is rapidly metabolized (hydroxylation and conjugation) and has a plasma half-life of 1.5 h. The metabolites are excreted renally and via the bile.

Epidemiological studies have demonstrated that diclofenac causes less serious gastrointestinal complications than indomethacin. However, a rise in plasma liver enzymes occurs more frequently with diclofenac than with other NSAIDs. The daily oral dose of diclofenac is 50–150 mg. Diclofenac is also available as eyedrops for the treatment of nonspecific inflammation of the eye and for the local therapy of eye pain (see ► NSAIDs, Pharmacokinetics).

Derivatives of Arylpropionic Acid
2-arylpropionic acid derivatives possess an asymmetrical carbon atom, giving rise to S- and R-enantiomers. The S-enantiomer inhibits cyclooxygenases 2–3 orders of magnitude more potently than the corresponding R-enantiomer. This finding has led to the marketing of pure S-enantiomers (e.g., S-ibuprofen and

S-ketoprofen) in some countries in addition to the racemates where the R-enantiomer is considered as "ballast." However, it is not yet proven whether 2-arylpropionic acids are better tolerated when given as S-enantiomer than as the racemate. Naproxen, for example, which is clinically available only as the S-enantiomer, does not cause less serious gastrointestinal side effects than, for example, ibuprofen racemate.

Ibuprofen is the most thoroughly researched 2-arylpropionic acid. It is a relatively weak, nonselective inhibitor of COX. In epidemiological studies, ibuprofen compared to all other conventional NSAIDs, has the lowest relative risk of causing severe gastrointestinal side effects. Because of this, ibuprofen is the most frequently used OTC ("over-the-counter," sale available without prescription) analgesic. Ibuprofen is highly bound to plasma proteins and has a relatively short elimination half-life (approximately 2 h). It is mainly glucuronidated to inactive metabolites that are eliminated via the kidney. The typical single oral dose of ibuprofen as an OTC analgesic is 200–400 mg and 400–800 mg when used in antirheumatic therapy. The corresponding maximum daily doses are 1,200 or 2,400 mg, respectively, but the dose in antirheumatic therapy in some countries can be as high as 3,200 mg daily.

Other arylpropionic acids include naproxen, ketoprofen and flurbiprofen. They share most of the properties of ibuprofen. The daily oral dose of ketoprofen is 50–150 mg, 150–200 mg for flurbiprofen and 250–1,000 mg for naproxen. Whereas the plasma elimination half-lives of ketoprofen and flurbiprofen are similar to that of ibuprofen (1.5–2.5 h and 2.4–4 h, respectively), naproxen is eliminated much more slowly with a half-life of 13–15 h (see ▶ NSAIDs, Pharmacokinetics).

Oxicams

Oxicams, for example, piroxicam, tenoxicam, meloxicam, and lornoxicam, are nonselective inhibitors of cyclooxygenases. Like diclofenac, meloxicam inhibits COX-2 slightly more potently than COX-1. This property can be exploited clinically with doses up to 7.5 mg per day, but at higher doses, COX-1 inhibition becomes clinically relevant. Since the dose of meloxicam commonly used is 15 mg daily, this agent cannot be regarded as a COX-2 selective NSAID, and considerable caution needs to be exercised when making comparisons between the actions of meloxicam and those of other conventional NSAIDs. The average daily dose in antirheumatic therapy is 20 mg for piroxicam and tenoxicam, 7.5–15 mg for meloxicam, and 12–16 mg for lornoxicam. Some oxicams have long elimination half-lives (lornoxicam 3–5 h, meloxicam approximately 20 h, piroxicam approximately 40 h, and tenoxicam approximately 70 h).

COX-2 Selective NSAIDs (COXIBs)

The development of the COXIBs has been based on the hypothesis that COX-1 is the physiological COX and COX-2 the pathophysiological isoenzyme. Inhibition of the pathophysiological COX-2 only is assumed to result in fewer side effects as compared to nonselective inhibition of both COX isoenzymes. Rofecoxib and celecoxib were the first COX-2 selective substances approved followed by etoricoxib, valdecoxib, parecoxib, and lumiracoxib. Some of these substances were subsequently withdrawn due to cardiovascular (rofecoxib, valdecoxib) or hepatotoxic (lumiracoxib) side effects.

Unlike conventional NSAIDs, with the exception of lumiracoxib, the "COXIBs" have no functional acidic group. The indications for these agents are in principle identical to those of the nonselective NSAIDs, although they have not yet received approval for the whole spectrum of indications of the conventional NSAIDs. Because they lack COX-1-inhibiting properties, COX-2 selective inhibitors show fewer side effects than conventional NSAIDs. However, they are not free of side effects because COX-2 has physiological functions that are blocked by the COX-2 inhibitors. The most frequently observed side effects are infections of the upper respiratory tract, diarrhea, dyspepsia, abdominal discomfort, and headache. Peripheral edema is as frequent as with conventional NSAIDs. The frequency of gastrointestinal complications is approximately half that observed with conventional NSAIDs. The precise side effect profile of the selective COX-2 inhibitors, however, will only be known after several years of clinical use (see ▶ NSAID-Induced Lesions of the Gastrointestinal Tract).

On September 30, 2004, MSD voluntarily withdrew rofecoxib from the market because of a colon cancer prevention study (APPROVe) suggesting that rofecoxib nearly doubled the rate of myocardial infarction and strokes as compared to placebo. Celecoxib, a less potent selective COX-2 inhibitor with a shorter half-life, showed a similar dose-related problem in another colon cancer prevention trial (APC), but did not show an increased risk in an Alzheimer prevention trial (ADAPT) or in another colon cancer study (PreSAP). The reasons for these discrepancies are a matter of scientific debate (Tegeder and Geisslinger 2006). Many scientists favor a group effect of all coxibs, because the balance between two fatty acids, prostacyclin and thromboxane A2, that control blood clotting and vasodilation may be disturbed after intake of selective COX-2 inhibitors. As a result of this possible imbalance, the risk for cardiovascular events may increase. However, there are also data suggesting that nonselective NSAIDs are also not safe with respect to thromboembolic events. In the aforementioned ADAPT trial, the nonselective NSAID naproxen significantly increased the risk of heart attacks and stroke (see ▶ NSAIDs, Chemical Structure and Molecular Mode of Action; ▶ NSAIDs and Their Indications; ▶ NSAIDs, Pharmacokinetics; ▶ NSAIDs, Adverse Effects). Finally, comparable rates of thrombotic cardiovascular events have been reported for the highly selective COX-2 inhibitor etoricoxib and the traditional NSAID diclofenac in the MEDAL study program.

More than 20 years ago, it was shown for the first time that chronic use of NSAIDs reduces the risk of colon cancer. The molecular mechanisms of the anticarcinogenic effects of NSAIDs are still not fully understood. Predominantly, these effects have been suggested to be due to their COX-inhibiting activity. This notion is based on the observation that COX-2 is overexpressed in more than 80 % of colon carcinomas and other cancer types and that an enhanced production of prostaglandins plays a crucial role in cell proliferation and angiogenesis. However, several in vitro and in vivo results cannot be explained by an enhanced COX-2 expression and PG synthesis indicating that COX-2-independent mechanisms must also be involved. Until now five selective COX-2 inhibitors have been developed and partly introduced into clinical practice. Of these, only celecoxib has been approved by the FDA for adjuvant treatment of patients with familial adenomatous polyposis.

It has been shown that the antiproliferative effects of celecoxib are at least in part mediated through induction of a cell-cycle block (in G_1 phase) and apoptosis. These effects occurred in COX-2 expressing as well as in COX-2-deficient colon carcinoma cells (see ▶ NSAIDs, COX-Independent Actions and ▶ NSAIDs and Cancer).

Cross-References

▶ Pharmacology of Cyclooxygenase Inhibitors

References

Fereira, S. H., Moncada, S., & Vane, J. R. (1971). Indomethacin and aspirin abolish prostaglandin release from the spleen. *Nature: New Biology, 231*, 237–239.

Fu, J. Y., Masferrer, J. L., Seibert, K., et al. (1990). The induction and suppression of prostaglandin H2 synthase (cyclooxygenase) in human monocytes. *Journal of Biological Chemistry, 265*, 16737–16740.

Tegeder, I., & Geisslinger, G. (2006). Cardiovascular risk with cyclooxygenase inhibitors: General problem with substance specific differences? *Naunyn-Schmiedeberg's Archives of Pharmacology, 373*, 1019.

Tegeder, I., Pfeilschifter, J., & Geisslinger, G. (2001). Cyclooxygenase-independent actions of cyclooxygenase inhibitors. *The FASEB Journal, 15*, 2057–2072.

Vane, J. R. (1971). Inhibition of prostaglandin synthesis as a mechanism of action for aspirin-like drugs. *Nature: New Biology, 231*, 232–235.

Nonsteroidal Anti-Inflammatory Drugs NSAIDs

▶ NSAIDs, Survey

Nonsteroidal Antirheumatic Drugs

▶ Nonsteroidal   Anti-Inflammatory   Drugs
(NSAIDs)
▶ NSAIDs, Adverse Effects

Non-structural Disorders

▶ Visceral Pain Model, Irritable Bowel Syndrome Model

Nonsynaptic Functional Coupling

▶ Cross Excitation

Nonsynaptic Neuron-to-Neuron Coupling

▶ Cross Excitation

Nonsystemic (Isolated) Vasculitic Neuropathy (NSVN)

Synonyms

NVNP

Definition

A peripheral neuropathy, usually involving several nerves, associated with perivascular inflammatory changes.

Cross-References

▶ Vascular Neuropathies

Nonthermal Effects

Definition

Effects produced by nominally thermal agents that do not directly result from heating or cooling. Examples of nonthermal effects include cavitation (the production of bubbles in liquids and tissues) in ultrasound, as well as the proposed and established benefits of low-intensity pulsed delivery of the diathermies.

Cross-References

▶ Therapeutic Heat, Microwaves, and Cold

Non-topographic

Definition

Non-topographic organization contrasts with topographic/somatotopic organization and favors the notion that the region is involved in affect-motivation. Khanna and Sinclair (1992) reported that a noxious heat stimulus, whether applied to the hind paw or the tail, suppressed CA3 stimulation-elicited CA1 pyramidal cell excitation at a given site in the hippocampus.

Cross-References

▶ Nociceptive Processing in the Hippocampus and Entorhinal Cortex, Neurophysiology and Pharmacology

References

Khanna, S., Sinclair, J. G. (1992) Responses in the CA1 region of the rat hippocampus to a noxious stimulus. *Experimental Neurology, 117*(1), 28–35.

Nontraditional Medicine

▶ Alternative Medicine in Neuropathic Pain

NOP Receptor

Definition

The term NOP or ORL1 receptor (N for ▶ Nociceptin or ▶ orphanin-FQ N/OFQ) represents the G-protein coupled receptor that is closely related to MOP, DOP, and KOP receptors, but responds to the peptide N/OFQ rather than any of the classical opioid drugs or peptides. It is expressed in many areas of the nervous system and has effects that may be analgesic or hyperalgesic depending on the anatomical region. The NOP receptor protein is produced by a single gene. When activated, the NOP receptor predominantly transduces cellular actions via inhibitory G-proteins. The electrophysiological consequences of NOP receptor activation are usually inhibitory.

Cross-References

▶ Opioid Electrophysiology in PAG

Noradrenaline

Definition

Noradrenaline is a catecholamine that acts as a neurotransmitter both centrally and peripherally by binding to adrenergic receptors. It is also known as norepinephrine.

Cross-References

▶ Descending Circuitry, Transmitters and Receptors

Noradrenergic

Definition

Neurons containing norepinephrine.

Cross-References

▶ Stimulation-Produced Analgesia

Noradrenergic and Serotonergic Inhibitory Pathways

Definition

Monoamine neurotransmitters norepinephrine (noradrenalin; NE) and serotonin (5-HT) were implicated in the 1970s and 1980s in spinal mechanisms of descending inhibition of spinal nociceptive processing. Electrical stimulation or morphine microinjection in the brainstem was shown to release NE and/or 5-HT in the lumbar spinal cord, and brainstem inhibition of spinal nociceptive processing was blocked when NE and 5-HT receptor antagonists were administered intrathecally at the lumbar enlargement. Serotonin is derived from axons descending from raphe nuclei in the rostroventral medial medulla, and NE is derived from axons descending from the dorsolateral pons.

Cross-References

▶ Postoperative Pain, Transition from Parenteral to Oral Drugs

Norepinephrine-Dopamine Reuptake Inhibitors

▶ Antidepressant Analgesics in Pain Management

Norepinephrine-Dopamine Transporter Inhibitors

▶ Antidepressant Analgesics in Pain Management

NOS

▶ Nitric Oxide Synthase

Noxious Cold Receptor

▶ Nociceptors, Cold Thermotransduction

Noxious Stimulus

Definition

A stimulus which will or potentially will produce damage to the tissue to which it is applied. Such stimuli are usually high or low temperatures, high-intensity mechanical stimuli, and some types of chemicals and activate nociceptors and elicit nociceptive responses.

Noxious Stimulus Intensity

Definition

The physical magnitude of a potentially injurious stimulus that is applied to the body. This could be the amount of energy deposited from a mechanical or thermal stimulus, or an application of a noxious chemical.

Cross-References

▶ Encoding of Noxious Information in the Spinal Cord

Noxious Stimulus Location

Definition

The body region that is being affected by a potentially injurious stimulus.

Cross-References

▶ Encoding of Noxious Information in the Spinal Cord

NPY

▶ Neuropeptide Y

NRS

▶ Numerical Rating Scale

NRSF

▶ Neuron-Restrictive Silencer Factor

NS Neuron

▶ Nociceptive-Specific Neurons

NSAID

▶ Postoperative Pain, Nonsteroidal Anti-Inflammatory Drugs

NSAID-Induced Lesions of the Gastrointestinal Tract

Joachim Mossner
Medical Clinic and Policlinic II, University
Clinical Center Leipzig AöR, Leipzig, Germany

Synonyms

Anticoagulants; Comorbidities; Lipophilic Substances; Prostaglandins

Definition

Definition of gastrointestinal complications due to application of nonsteroidal anti-inflammatory drugs (▶ NSAIDs): erosions, ulcers, bleeding erosions or bleeding ulcers, and ulcer perforations. These lesions are mostly located in the stomach, less common in the upper duodenum. However, lesions are possible along the entire gastrointestinal tract, i.e., small gut and colon.

Characteristics

Pathogenesis of NSAID-Induced Peptic Ulcers

Inhibition of prostaglandin synthesis in the stomach by NSAIDs seems to be the key pathogenetic factor. Prostaglandins of the E-type stimulate gastric blood circulation, mucus secretion, and cell regeneration and inhibit acid secretion. Furthermore, NSAIDs exert direct side effects at the gastric mucosa besides this systemic effect. NSAIDs are weak acids. Within an acidic environment, NSAIDs are not dissociated. As lipophilic substances, they may penetrate the mucus and exert direct damaging effects.

Gastrointestinal Side Effects Due to Therapy with Nonsteroidal Anti-inflammatory Drugs (NSAIDs)

Treatment with NSAIDs may cause life-threatening complications in the gastrointestinal tract such as bleeding or perforation. In many cases, no symptoms precede. In a study by Sing et al. including 1,921 patients, 81 % had no preceding symptoms (Sing et al. 1996). Very often complications affect rather old patients with comorbidities. Thus, a rather high mortality may be a consequence. In patients on NSAIDs, accessory risk factors increase the odds ratio of risk of peptic ulcer bleeding (Weil et al. 2000). In a case control study on 1,121 patients with bleeding ulcers, therapy with anticoagulants increased the risk by 7.8, history of peptic ulcer by 3.8, heart insufficiency by 5.9, oral glucocorticosteroids by 3.1, and smoking by 1.6. The odds ratio increased multiplicatively when in addition to these risk factors NSAIDs were administered (Weil et al. 2000).

Primary Prophylaxis, Therapy, and Secondary Prophylaxis of NSAID-Induced Gastrointestinal Lesions

In primary prophylaxis of NSAID-induced gastrointestinal lesions, one has to discuss the roles of coxibs, ▶ Helicobacter pylori eradication, and prophylactic therapy with either misoprostol, histamine-2-receptor antagonists (H2 blockers), or proton pump inhibitors (PPIs). In acute treatment of bleeding ulcers, endoscopic therapy is one of the most important mainstays. For treatment of NSAID-induced gastrointestinal lesions, one has to compare the effectiveness of H2 blockers with that of misoprostol or PPIs. In secondary prophylaxis, H2 blockers have again to be compared with misoprostol, PPIs, and *H. pylori* eradication. Fifteen to twenty percent of ulcers rebleed after successful endoscopic therapy. In primary prevention of NSAID-induced gastrointestinal lesions, one has to discuss the role of coxibs, antagonists, and proton pump inhibitors (PPIs).

Intravenous omeprazole reduced this risk of recurrent bleeding after successful endoscopic treatment of bleeding peptic ulcers (Lau et al. 2000).

Two hundred forty patients were randomly assigned to either placebo or omeprazole (80 mg bolus intravenously followed by 8 mg/h for 72 h). Thereafter, both groups received omeprazole 20 mg orally for 4 weeks. In the PPI group, eight patients rebled (6.7 %) within 30 days versus 27 (22.5 %) (p < 0.001) in the placebo group. Five patients died in the PPI group (4.2 %) as compared to 12 (10 %) in the placebo group. This difference did not reach statistical significance (p = 0.13).

In primary prevention of diclofenac-associated ulcers and dyspepsia, omeprazole was compared with triple therapy in H. pylori-positive patients (Labenz et al. 2002). Patients had no history of ulcer disease. Patients were on continuous NSAID therapy (diclofenac 2 × 50 mg/day). They received either French triple therapy for H. pylori eradication (PPI plus clarithromycin plus amoxicillin) followed by placebo or omeprazole or omeprazole alone or placebo alone. Ulcer incidence after 5 weeks was 1.2 % versus 1.2 % versus 0 % versus 5.8 %, respectively. Thus, persistence of H. pylori gastritis and no inhibition of acid secretion were associated with an increased risk of ulcer development due to diclofenac therapy. In another study from China, H. pylori eradication decreased the risk of ulcer development when NSAID-naive patients received long-term NSAID therapy (Chan et al. 2002a).

A key question in prophylaxis is not only prevention of gastric or duodenal ulcers in patients on NSAID therapy but also prevention of serious ulcer complications such as bleeding or perforation. For misoprostol, it has been clearly demonstrated that this substance reduces serious gastrointestinal complications in patients with rheumatoid arthritis receiving non-steroidal anti-inflammatory drugs (Silverstein et al. 1995). Eight thousand eight hundred forty three patients on NSAIDs were randomly assigned to either misoprostol (4 × 200 µg/day) versus placebo. Twenty five out of 4,404 patients on misoprostol developed ulcer complications as compared to 42 out of 4,439 in the placebo group (risk reduction around 40 %; odds ratio 0.598 (95 % CI, 0.364–0.982; P < 0.049)). However, 20 % of patients on misoprostol terminated the therapy due to side effects such as severe diarrhea, as compared to 15 % on placebo.

Therapy of NSAID-Induced Peptic Ulcers

H2 antagonists, misoprostol, and PPIs have been shown to be effective in healing gastric and duodenal ulcers. Omeprazole has been compared with the prostaglandin analogue misoprostol for ulcers associated with nonsteroidal anti-inflammatory drugs (Hawkey et al. 1998). Omeprazole was superior to misoprostol with regard to the percentage of healed ulcers after 4 weeks. A concomitant H. pylori gastritis favors the healing rates with a PPI. This effect could be due to endogenous prostaglandin synthesis induced by the gastritis and/or to the gastric acid buffering by ammonia produced by H. pylori.

When omeprazole was compared with ranitidine, the PPI therapy was significantly superior to the H2-receptor antagonist therapy in healing NSAID-induced gastric or duodenal ulcers (Yeomans et al. 1998).

Secondary Prophylaxis of Gastrointestinal Complications Due to NSAIDs

The most efficient prophylaxis is certainly termination of any NSAID administration. However, this is not possible in many cases due to the underlying diseases. Further options are H. pylori eradication since an underlying H. pylori gastritis may increase the risk of NSAID-induced lesions. Other options are administration of inhibitors of acid secretion such as PPIs or H2 blockers, treatment with misoprostol, or switching from nonselective NSAIDs to coxibs. The choice of the most efficient prophylaxis can be based on the results of several controlled prospective studies, but the reduction of risk by a switch to coxibs and additional acid inhibition by PPIs has not been studied.

H. pylori Eradication Versus PPI

Four hundred patients with a history of a bleeding ulcer due to NSAIDs or aspirin and concomitant H. pylori gastritis were randomly assigned to either H. pylori eradication or acid inhibition by omeprazole (20 mg per day). NSAID therapy (250 patients, naproxen 2 × 500 mg per day) or aspirin (150 patients, aspirin 80 mg per day) was continued for 6 months (Chan et al. 2001). The bleeding probability within 6 months in the aspirin group was 0.9 % when omeprazole was used and 1.9 % after *H. pylori* eradication. The difference was not statistically significant. Thus, *H. pylori* eradication is as efficient as inhibition of acid secretion when low-dose aspirin is used. However, in the naproxen group, 18.8 % rebled when *H. pylori* was eradicated as compared to 4.4 % when omeprazole was used as prophylaxis. Thus, *H. pylori* eradication does not protect against ulcer recurrences when NSAID therapy is continued. In another study, patients with *H. pylori* gastritis who developed an ulcer complication under long-term therapy with aspirin (<325 mg per day) were enrolled (Lai et al. 2002). *H. pylori* was eradicated and after healing of the ulcers aspirin (100 mg per day) was continued. The PPI lansoprazole (30 mg per day) was compared to placebo for up to 12 months or until the occurrence of the next complication. Nine out of 61 (14.8 %) patients developed an ulcer complication in the placebo group versus 1 out of 62 (1.2 %) in the PPI group. Four out of 10 patients had a reinfection by *H. pylori*, and 2 out of 10 used further NSAIDs. Thus, in countries with a rather high risk of *H. pylori* reinfection or in cases of uncontrolled concomitant NSAID use, acid inhibition by PPIs seems to be more effective than *H. pylori* eradication alone when aspirin is continued. In a recent study from Hong Kong, patients who experienced ulcer bleeding after aspirin were switched to either clopidogrel (75 mg per day) or aspirin (80 mg per day) together with the pure enantiomer of omeprazole esomeprazole (2 × 20 mg per day). Both groups were *H. pylori* negative. Thirteen out of 161 rebled in the clopidogrel group versus 1 out of 159 in the aspirin plus esomeprazole group. Thus, a switch to clopidogrel does not offer any protection in contrast to efficient acid inhibition by esomeprazole (Chan et al. 2005).

Misoprostol Versus PPI

When omeprazole was compared with misoprostol for ulcers associated with nonsteroidal anti-inflammatory drugs, misoprostol was slightly less effective than omeprazole in healing gastric and duodenal ulcers. In secondary prevention, omeprazole was superior. Furthermore, patients given omeprazole experienced fewer side effects (Hawkey et al. 1998).

Selective Cyclooxygenase-2 (COX-2) Inhibitors, So-Called Coxibs, Solution of the Problem?

Coxibs do not inhibit aggregation of platelets. The gastrointestinal bleeding risk seems to be decreased. However, these drugs may increase the risk of cardiovascular diseases such as myocardial infarction and stroke. Prostaglandins of the E type are gastroprotective. They stimulate gastrointestinal blood flow, cell regeneration, and mucus secretion and are weak inhibitors of gastric acid secretion. Selective COX-2 inhibitors, so-called coxibs, are supposed mainly to inhibit prostaglandin synthesis in inflammation but not constitutive prostaglandin synthesis by cyclooxygenase-1. Thus, coxibs should offer less gastrointestinal toxicity.

Celecoxib was compared with diclofenac in long-term management of rheumatoid arthritis (Emery et al. 1999). Six hundred fifty-five patients with rheumatoid arthritis received either celecoxib (2 × 200 mg per day) or diclofenac (slow release 2 × 75 mg) for 24 weeks. Gastroduodenal ulcers developed in 4 % in the celecoxib group and in 15 % given diclofenac. The CLASS study evaluated gastrointestinal toxicity with celecoxib versus nonsteroidal anti-inflammatory drugs in patients suffering from osteoarthritis and rheumatoid arthritis. 4,573 patients received either celecoxib (2 × 400 mg per day) or ibuprofen (3 × 800 mg per day) or diclofenac (2 × 75 mg per day) for 6 months. Aspirin (<375 mg per day) was permitted. Ulcer complications were seen in the celecoxib group in 0.76 % versus 1.45 % in the NSAID group. The difference was not significant (p < 0.09). Symptomatic ulcers were fewer in the celecoxib group, 2.08 % versus 3.54 % (p < 0.02). Less gastrointestinal toxicity was only seen in patients who did not take aspirin in addition (complicated ulcers, no aspirin 0.44 % vs. 1.27 % (p < 0.04); symptomatic ulcers, 1.40 % vs. 2.91 % (p < 0.02); with aspirin

2.01 % vs. 2.12 % (p < 0.92) and 4.70 % vs. 6.00 % (p < 0.49)) (Silverstein et al. 2000). The data after 1 year of treatment (not published but available via the Internet) are less impressive with regard to less gastrointestinal toxicity with celecoxib. Moreover, celecoxib was evaluated in a meta-analysis (Goldstein et al. 2000). Fourteen multicenter studies on 11,008 patients who had been treated for 2–24 weeks and 5,155 patients on therapy for up to 2 years demonstrated ulcer complications such as bleeding, perforation, and obstruction in the placebo group in none out of 1,864 patients, in the celecoxib group in 2 out of 6,376, and in the NSAID group in 9 out of 2,768 patients. Thus, the incidence of ulcer complications is reduced by a factor of 8 by celecoxib. Rofecoxib (25 and 50 mg per day) was compared with ibuprofen (3 × 800 mg per day) and placebo in 775 patients with osteoarthritis (Hawkey et al. 2000). The end point of this prospective study was gastric or duodenal ulcers. In the ibuprofen group, 29.2 % of patients developed ulcers after 12 weeks and 46.8 % after 24 weeks versus 5.3 % and 8.8 % in the low-dose rofecoxib group and 9.9 % and 12.4 % in the 50 mg group. Thus, rofecoxib is clearly much safer as regards gastroduodenal ulcers than ibuprofen. However, not all endoscopically demonstrated ulcers may be clinically relevant.

In another study, rofecoxib (50 mg per day) was compared with naproxen (2 × 500 mg per day) with special emphasis on ulcer complications such as bleeding, stenosis, perforation, or symptomatic ulcers in patients with rheumatoid arthritis (Bombardier et al. 2000). Eight thousand seventy six patients older than 50 years, or 40 years when on steroids in addition, were enrolled. In the rofecoxib group, 2.1 ulcers (0.6 life threatening) developed within 9 months in 100 patient years and 4.1 (1.4 threatening) with naproxen. However, myocardial infarction was observed in 0.4 % versus 0.1 % with naproxen. When valdecoxib (2 × 40 mg per day), another coxib, was compared with naproxen (2 × 500 mg per day) and placebo in a multicenter study with 62 patients per group, gastroduodenal ulcerations were seen in 3 % with placebo, 0 % with valdecoxib, and 18 % with naproxen (Goldstein et al. 2003). Thus, the risk of developing gastroduodenal ulcerations is clearly lower during treatment with coxibs as compared to nonselective NSAIDs. However, a larger prospective trial is still warranted to clarify the key question as to whether the rate of life-threatening ulcer complications is really markedly decreased when treatment is with coxibs versus nonselective NSAIDs. The risk reduction for gastroduodenal ulcers has to be compared with the risk elevation of cardiovascular side effects attributed to treatment with coxibs. This risk has led to the withdrawal of rofecoxib from the market. The elevated risk of cardiovascular side effects is possibly a class effect of coxibs and not solely seen under rofecoxib. A trial to reduce the incidence of new colon adenomas by celecoxib after endoscopic polypectomy had to be stopped. In this trial (ca. 2,000 patients), with placebo, 6 cardiovascular events were seen versus 15 in the low-dose celecoxib group (2 × 200 mg per day) and 20 in the high-dose celecoxib group (2 × 400 mg per day). In another trial in which celecoxib was used to treat Alzheimer disease, no elevated cardiovascular risk was seen. In a population-based observation, 1,005 patients used COX-2 inhibitors and 5,245 patients used a non-naproxen NSAID. Of the 6,250 patients, 70 % were female, 50 % were African American, and 30 % were older than 50 years. Overall, 12 % of the patients had at least one cardiovascular thrombotic event after treatment within the follow-up period. The propensity adjusted odds ratio showed no significant effect of COX-2 inhibitor use on this percentage of patients (odds ratio, 1.09; 95 % confidence interval, 0.90–1.33). The authors conclude that coxibs do not increase cardiovascular risk over non-naproxen NSAIDs in a high-risk Medicaid population (Shaya et al. 2005).

Several questions are still not solved. NSAIDs may lead to lesions in the jejunum and ileum that cannot be prevented by the addition of proton pump inhibitors. In a small study on 21 chronic NSAID users and 20 controls, a small video capsule was used to look for lesions in the small intestine. Small-bowel injury was seen in 71 % of NSAID users compared with 10 % of controls (P < 001). The injury was mild (few or no erosions, absence of large erosions/ulcers) in ten NSAID users compared with two controls. Five NSAID users had major (>4 erosions or large ulcers) damage compared with none in the control group. The authors conclude that endoscopically evident small intestinal mucosal injury is very common among chronic NSAID users (Graham et al. 2005). However, the clinical significance and quantitative risk of lesions in the small intestine due to nonselective NSAIDs is still not known.

Co-treatment with aspirin to overcome the coxibs' possible cardiovascular risk does not solve the problem since the gastroprotective effect is lost (Laine et al. 2004). Low-dose aspirin plus rofecoxib resulted in ulcers within 12 weeks in 16 %, similar to aspirin plus ibuprofen. Furthermore, it is not clear whether the elevated cardiovascular risk with rofecoxib can be decreased by aspirin.

Addition of proton pump inhibitors to nonselective NSAIDs causes a similar risk reduction to that seen under treatment with coxibs. This has been demonstrated in a double-blinded randomized comparison of celecoxib versus omeprazole and diclofenac for secondary prevention of ulcer bleeding in chronic NSAID users (Chan et al. 2002b). Rebleeding occurred in the celecoxib group in 4.9 % (7/144 patients) versus 6.4 % (9/143 patients) in the omeprazole plus diclofenac group. The difference was not significant. Finally, one has to be aware that prostaglandins are needed for ulcer healing. Thus, coxibs may delay gastric ulcer healing when one switches from non selective NSAIDs to coxibs due to abdominal pain, not knowing whether an ulcer is present or not.

In summary, coxibs decrease the risk of gastroduodenal ulcers when compared with nonselective NSAIDs. However, this risk reduction has to be balanced against the possible elevated risk of severe cardiovascular side effects. Addition of proton pump inhibitors to NSAIDs causes a similar risk reduction to that of coxibs. However, lesions in the small intestine due to

NSAIDs cannot be prevented by PPIs. The overall risk and clinical relevance of lesions in the small intestine has still to be clarified in larger trials.

References

Bombardier, C., Laine, L., Reicin, A., VIGOR Study Group, et al. (2000). Comparison of upper gastrointestinal toxicity of rofecoxib and naproxen in patients with rheumatoid arthritis. *The New England Journal of Medicine, 343*, 1520–1528. VIGOR Study Group.

Chan, F. K., Chung, S. C., Suen, B. Y., et al. (2001). Preventing recurrent upper gastrointestinal bleeding in patients with *Helicobacter pylori* infection who are taking low-dose aspirin or naproxen. *The New England Journal of Medicine, 344*, 967–973.

Chan, F. K., To, K. F., Wu, J. C., et al. (2002a). Eradication of *Helicobacter pylori* and risk of peptic ulcers in patients starting long-term treatment with nonsteroidal anti-inflammatory drugs: A randomised trial. *Lancet, 359*, 9–13.

Chan, F. K., Hung, L. C., Suen, B. Y., et al. (2002b). Celecoxib versus diclofenac and omeprazole in reducing the risk of recurrent ulcer bleeding in patients with arthritis. *The New England Journal of Medicine, 347*, 2104–2110.

Chan, F. K., Ching, J. Y., Hung, L. C., et al. (2005). Clopidogrel versus aspirin and esomeprazole to prevent recurrent ulcer bleeding. *The New England Journal of Medicine, 352*, 238–244.

Emery, P., Zeidler, H., Kvien, T. K., et al. (1999). Celecoxib versus diclofenac in long-term management of rheumatoid arthritis: Randomised double-blind comparison. *Lancet, 354*, 2106–2111.

Goldstein, J. L., Silverstein, F. E., Agrawal, N. M., et al. (2000). Reduced risk of upper gastrointestinal ulcer complications with celecoxib, a novel COX-2 inhibitor. *The American Journal of Gastroenterology, 95*, 1681–1690.

Goldstein, J. L., Kivitz, A. J., Verburg, K. M., et al. (2003). A comparison of the upper gastrointestinal mucosal effects of valdecoxib, naproxen and placebo in healthy elderly subjects. *Alimentary Pharmacology & Therapeutics, 18*, 125–132.

Graham, D. Y., Opekun, A. R., Willingham, F. F., et al. (2005). Visible small-intestinal mucosal injury in chronic NSAID users. *Clinical Gastroenterology and Hepatology, 3*, 55–59.

Hawkey, C. J., Karrasch, J. A., Szczepanski, L., et al. (1998). Omeprazole compared with misoprostol for ulcers associated with nonsteroidal antiinflammatory drugs. Omeprazole versus Misoprostol for NSAID-induced Ulcer Management (OMNIUM) Study Group. *The New England Journal of Medicine, 338*, 727–734.

Hawkey, C., Laine, L., Simon, T., et al. (2000). Comparison of the effect of rofecoxib (a cyclooxygenase

2 inhibitor), ibuprofen, and placebo on the gastroduodenal mucosa of patients with osteoarthritis: A randomized, double-blind, placebo-controlled trial. The Rofecoxib Osteoarthritis Endoscopy Multinational Study Group. *Arthritis and Rheumatism, 43,* 370–377.

Labenz, J., Blum, A. L., Bolten, W. W., et al. (2002). Primary prevention of diclofenac associated ulcers and dyspepsia by omeprazole or triple therapy in *Helicobacter pylori* positive patients: A randomised, double blind, placebo controlled, clinical trial. *Gut, 51,* 329–335.

Lai, K. C., Lam, S. K., Chu, K. M., et al. (2002). Lansoprazole for the prevention of recurrences of ulcer complications from long-term low-dose aspirin use. *The New England Journal of Medicine, 346,* 2033–2038.

Laine, L., Maller, E. S., Yu, C., et al. (2004). Ulcer formation with low-dose enteric-coated aspirin and the effect of COX-2 selective inhibition: A double-blind trial. *Gastroenterology, 127,* 395–402.

Lau, J. Y., Sung, J. J., Lee, K. K., et al. (2000). Effect of intravenous omeprazole on recurrent bleeding after endoscopic treatment of bleeding peptic ulcers. *The New England Journal of Medicine, 343,* 310–316.

Shaya, F. T., Blume, S. W., Blanchette, C. M., et al. (2005). Selective cyclooxygenase-2 inhibition and cardiovascular effects: An observational study of a Medicaid population. *Archives of Internal Medicine, 24,* 181–186.

Silverstein, F. E., Graham, D. Y., Senior, J. R., et al. (1995). Misoprostol reduces serious gastrointestinal complications in patients with rheumatoid arthritis receiving nonsteroidal anti-inflammatory drugs. A randomized, double-blind, placebo-controlled trial. *Annals of Internal Medicine, 123,* 241–249.

Silverstein, F. E., Faich, G., Goldstein, J. L., et al. (2000). Gastrointestinal toxicity with celecoxib vs nonsteroidal anti-inflammatory drugs for osteoarthritis and rheumatoid arthritis: The CLASS study: A randomized controlled trial. Celecoxib long-term arthritis safety study. *JAMA: The Journal of the American Medical Association, 284,* 1247–1255.

Sing, G., Ramey, D. R., Morfeld, D., et al. (1996). Gastrointestinal tract complications of nonsteroidal anti-inflammatory drug treatment in rheumatoid arthritis. A prospective observational cohort study. *Archives of Internal Medicine, 156,* 1530–1536.

Weil, J., Langman, M. J., Wainwright, P., et al. (2000). Peptic ulcer bleeding: Accessory risk factors and interactions with non-steroidal anti-inflammatory drugs. *Gut, 46,* 27–31.

Yeomans, N. D., Tulassay, Z., Juhasz, L., et al. (1998). A comparison of omeprazole with ranitidine for ulcers associated with nonsteroidal antiinflammatory drugs. Acid Suppression Trial: Ranitidine versus Omeprazole for NSAID-associated Ulcer Treatment (ASTRONAUT) Study Group. *The New England Journal of Medicine, 338,* 719–726.

NSAIDs

▶ COX-2 Inhibitor
▶ Nonsteroidal Anti-Inflammatory Drugs (NSAIDs)
▶ NSAIDs and Cancer
▶ NSAIDs, Adverse Effects
▶ NSAIDs, Mode of Action

NSAIDs and Cancer

Sabine Grösch[1] and Thorsten J. Maier[2]
[1]Institute of Clinical Pharmacology, Clinical Center of the Goethe University, Frankfurt, Germany
[2]Institute of Pharmaceutical Chemistry, Goethe University, Frankfurt, Germany

Synonyms

Cyclooxygenase inhibitors; Nonsteroidal anti-inflammatory drugs; NSAIDs

Definition

NSAIDs are inhibitors of the ▶ cyclooxygenases-1 (COX-1) and/or ▶ cyclooxygenases-2 (COX-2). While COX-1 is constitutively expressed in a wide range of tissues, COX-2 is predominantly cytokine and stress inducible. Interestingly, in numerous cancer types, the regulatory mechanisms of COX-2 expression are abrogated, leading to an overexpression of the COX-2 protein and enhanced prostaglandin production. Especially in the case of colon cancer, an upregulation of COX-2 has been shown to occur already in the early adenoma stage and is therefore one of the first deregulation mechanisms in tumor development (Shiff and Rigas 1999). Enhanced COX-2 expression has been attributed a key role in the development of cancer by promoting cell division, inhibiting ▶ apoptosis, altering ▶ cell adhesion, and enhancing ▶ metastasis and

NSAIDs and Cancer, Fig. 1 The COX-2 pathway. The cyclooxygenase-2 catalyzes the conversion of arachidonic acid to prostaglandin H_2 (PGH$_2$) which is further converted by different prostaglandin synthases to PGE$_2$, PGD$_2$, PGF$_{2\alpha}$, and TXA$_2$. These prostaglandins promote cell division, metastasis, and angiogenesis and inhibit apoptosis leading to an enhanced tumor growth

neovascularization (Fig. 1) (Trifan and Hla 2003). Therefore, inhibition of COX-2 activity by NSAIDs thwarts all these effects and accounts for the anticarcinogenic effect of these drugs.

Introduction

Publications in recent years, investigating the molecular pathway of the anticarcinogenic effects of NSAIDs, revealed that next to COX-2 inhibition, COX-2-independent mechanisms are also responsible for the antineoplastic effects of these substances. The main arguments for this hypothesis are:

1. NSAIDs inhibit growth of tumor cells which do not express COX-2 in vitro and in vivo (Grösch et al. 2001).
2. NSAIDs which exhibit no COX-2 inhibitory efficacy nevertheless cause anticarcinogenic effects (Grösch et al. 2003).

In the literature, various molecular mechanisms were described to be involved in the anticarcinogenic effect of different NSAIDs, whereas each NSAID seems to have its specific COX-independent mode of action. An overview is given in a separate chapter of this issue (Tegeder and Niederberger) and also in Tegeder et al. 2001.

Characteristics

For cancer therapy, it is important to bear in mind that anticarcinogenic treatment regimens are usually administered for a long period of time (months to years). Therefore, it is of vital importance that the therapies used should exhibit a reasonable benefit-risk ratio. Unfortunately, long-term use of classical nonselective NSAIDs, which inhibit both COX-1 and COX-2, is associated with serious gastrointestinal side effects such as *ulceration* of the gastric mucosa. These side effects are thought to arise from the inhibition of the constitutive cyclooxygenase isoform (COX-1), which is mainly responsible for the maintenance of the physiologically important prostaglandin synthesis in the gastrointestinal mucosa. To circumvent this problem, selective COX-2 inhibitors such as celecoxib, rofecoxib, etoricoxib, valdecoxib, and lumiracoxib have been developed and introduced into clinical practice. However, several toxic side effects of coxibs have led to the withdrawal of rofecoxib (2004) and valdecoxib (2005) from the pharmaceutical market and in 2007 stopped approval of lumiracoxib. Interestingly, each coxib had a different serious side effect. Rofecoxib was withdrawn due to cardiovascular side effect, valdecoxib was withdrawn due to serious skin toxicities, and lumiracoxib was stopped due to liver toxicity. In recent years, considerable efforts have been made to investigate whether these toxicities are specific for coxibs or whether other NSAIDs also show such side effects. Very recently, data from a nationwide cohort study of patients with first-time myocardial infarction (MI) in Denmark demonstrated that the use of any NSAID (especially celecoxib, rofecoxib, ibuprofen, diclofenac) in the years following MI was persistently associated with an increased risk of death, regardless of the time elapsed after first-time MI (Schjerning Olsen et al. 2011). So, not only must the group of coxibs be regarded as posing a potential risk for cardiovascular side effects, but also other NSAIDs increase cardiovascular toxicity. Moreover, all other main side effects of coxibs have been shown also to occur after long-term treatment with NSAIDs, such as ibuprofen, diclofenac, and acetaminophen/paracetamol (Laine and White 2008). This prompted the European Medicines Agency (EMEA), as well as the Food and Drug Administration (FDA), to oblige pharmaceutical companies to label their NSAIDs with contraindications and warnings. Nevertheless, celecoxib is the only NSAID to have been approved (in December 1999) by the FDA for ▶ adjuvant treatment of patients with ▶ familial adenomatous polyposis *(FAP)* and was, therefore, the only NSAID to be approved for cancer therapy (Koehne and Dubois 2004). Due to the lack of new confirmatory data, subsequent to the first investigations, Pfizer withdrew the FAP indication for *Celebrex* in the USA in February 2011 and pulled the claim for *Onsenal* in Europe in March.

The anticarcinogenic effects of celecoxib appear to be unique and are based on its ability to inhibit cell cycle progression and ▶ angiogenesis and to induce apoptosis. Its COX-2-inhibiting effects, as mentioned above, undoubtedly make a contribution, but on the other hand, COX-2-independent mechanisms also seem to play a crucial role.

Inhibition of Cell Cycle Progression

The cell cycle transition is controlled by different regulatory proteins such as cyclins and cyclin-dependent kinases as well as by cell cycle inhibitors (Fig. 2). Treatment of different tumor cell lines with celecoxib caused downregulation of cyclins and cyclin-dependent kinases and an increase in the expression of various cell cycle inhibitors (Table 1), resulting in a cell cycle block in the G_1-phase. However, the precise target(s) responsible for these effects is still unknown.

Induction of Apoptosis

Apoptosis, or programmed cell death, can be induced either by the extrinsic pathway via activation of death receptors or by the intrinsic pathway via releasing of cytochrome c from the mitochondria (Green and Evan 2002). Both pathways finally lead to the activation of various ▶ caspases which, as the executors of apoptosis,

cleave different cellular substrates and cause DNA-fragmentation (Fig. 3). After treatment of cancer cells with celecoxib, it has been shown that the expression of various antiapoptotic proteins such as ▶ Bcl-2, ▶ Mcl-1, and ▶ survivin decreases, whereas the expression of the proapoptotic protein Bad is upregulated. This was further accompanied by a rapid release of cytochrome c from mitochondria and subsequent activation of caspase 3, caspase 8, and caspase 9 (see also Table 1). These data provide evidence that after celecoxib treatment apoptosis is induced by activation of the intrinsic pathway. Currently, several cellular pathways have been identified which play a role for the apoptosis-inducing potency of celecoxib (Grösch et al. 2006). It has been shown that celecoxib treatment leads to inhibition of the 3-phosphoinositide-dependent-kinase 1 (PDK1) or its downstream substrate protein kinase B (PKB/AKT). Due to the antiapoptotic potency of these enzymes, inhibition of these kinases promotes apoptosis. Another recently identified target of celecoxib is the endoplasmatic reticulum Ca^{2+}-ATPase, which is strongly inhibited by celecoxib. As a consequence, Ca^{2+}-reuptake in the endoplasmatic reticulum is prevented, leading to an elevated cytoplasmatic Ca^{2+}-concentration. Ca^{2+} may then trigger the induction of apoptosis by activating Ca^{2+}-sensitive proteases, endonucleases, and caspases and by opening Ca^{2+}-sensitive mitochondrial permeability transition pores, resulting in cytochrome c release.

Furthermore, nanomolar concentrations of celecoxib and valdecoxib have been shown to inhibit members of the carbonic anhydrase (CA) family such as CA I, II, IV, and IX. This effect is closely associated with the presence of an arylsulfonamide group in both drugs, which seems to be responsible for this effect (Weber et al. 2004). CA II and CA IX have been described to play a pivotal role in tumor growth and development and represent potential biomarkers for various tumor types. Carbonic anhydrases catalyze the reversible hydration of carbon dioxide, thereby buffering protons that protect the cell from undergoing intracellular

NSAIDs and Cancer, Fig. 2 A simplified diagram of cell cycle control. The cell cycle is divided in four phases: the G_1-phase, S-phase, G_2-phase, and M-phase. The cell cycle transition is controlled by different cyclins, cyclin-dependent kinases (CDKs), and cell cycle inhibitors such as p21 and p27

NSAIDs and Cancer, Table 1 Effect of celecoxib on proteins involved in apoptosis, cell cycle regulation, and angiogenesis

Apoptosis	Cell cycle	Angiogenesis/metastasis	Kinases	Transcription factors
Caspase 3 ↑	Cyclin D1 ↓	VEGF↓	PDK1 activity ↓	NF-κB ↓
Caspase 8 ↑	Cyclin A ↓	Phospho-EGFR ↓	AKT/PKB activity ↓	PPARγ ↑
Caspase 9 ↑	Cyclin B ↓	MMP-1 ↓	Phospho-SAPK ↓	SP-1 ↓
Bcl-2 ↓	p21 ↑	MMP-2 ↓		Egr-1 ↓
Bcl-xL ↓	p27 ↑	MMP-3 ↓		c-Fos ↓
Survivin ↓	Rb ↑	MMP-9 ↓		
Mcl-1 ↓	Rb ↑			
Bad ↑	Ornithine decarboxylase ↓			
Ceramide ↑				
Apaf-1 ↑				

acidification. Since the growth of tumor cells requires a high bicarbonate flux, inhibition of carbonic anhydrases by celecoxib could lead to an acidification and subsequently to growth regression.

Inhibition of Angiogenesis

Early tumor growth is divided into several stages. In stage one, the malignant cells form small tumors of limited size due to an inadequate supply of oxygen (▶ hypoxia). In stage two, hypoxia triggers a drastic change in gene expression leading to the formation of new blood vessels. Overexpression of COX-2 in tumor cells has a decisive impact on angiogenesis: (a) The eicosanoid TXA_2 stimulates endothelial cell migration, and (b) prostaglandin E_2 (PGE_2) causes a release of the vascular endothelial growth factor (VEGF), thereby promoting endothelial cell proliferation. Both mechanisms play a pivotal role in

NSAIDs and Cancer, Fig. 3 A simplified diagram of apoptosis. Apoptosis (programmed cell death) is a succession of controlled molecular events leading to the suicide of the cell. Apoptosis can be initiated from the cell surface membrane (extrinsic pathway) via activation of death receptors and subsequent activation of caspase 8 (and others) or by intracellular stress (e.g., DNA damage) leading to the release of cytochrome c from the mitochondria and activation of the apoptosom (intrinsic pathway). Both pathways finally lead to caspase-mediated cleavage of different cellular substrates (proteins, DNA) and subsequent cell death

the formation of new blood vessels. Therefore, inhibition of COX-2 activity by celecoxib represses the effects mentioned above and leads to inhibition of angiogenesis and diminished tumor growth (Gately and Li 2004). Nevertheless, COX-2-independent mechanisms also contribute to the antiangiogenic effect of celecoxib. It has been shown that celecoxib inhibits the activation of the early growth response factor (Egr-1) (Table 1). Egr-1 is a ▶ transcription factor which is rapidly activated by multiple extracellular agonists (such as growth factors and cytokines) and environmental stresses (hypoxia, vascular injury). As a major activator of proangiogenic genes, such as fibroblast growth factor (FGF), cytokines, and receptors, Egr-1 is essentially involved in angiogenesis and promotion of tumor development. Furthermore, it has been shown that inhibition of angiogenesis after celecoxib treatment is associated with suppression of VEGF expression due to the inhibition of the transcription factor Sp1. Celecoxib treatment reduced both Sp1 DNA-binding activity and transactivating activity, which correlated with reduced Sp1 protein expression and its phosphorylation. The ▶ promoter region of VEGF contains a $Sp1$-binding site that seems to be crucial for VEGF expression. Taken together, COX-2-independent mechanisms, such as the inhibition of Sp1 and Erg-1 transcriptional activity, may explain the antiangiogenic effects of celecoxib.

Implications for Clinical Management

In the light of the cardiovascular risks associated with COX-2-selective inhibitors, attitudes towards their use in clinical practice have changed dramatically. In recent years, many studies have been performed to investigate the anticarcinogenic effects of coxibs and other NSAIDs in different tumor types. In December 2004, the National Cancer Institute cancelled the "Adenoma Prevention with Celebrex" (APC) Study due to the occurrence of a statistically significant increase in cardiovascular side effects under celecoxib treatment. Also, in the Prevention of Colorectal Sporadic Adenomatous Polyps (PreSAP) trial, celecoxib significantly increased the risk of renal/hypertension events and cardiac disorders. Even in this study, though, the 5-year cumulative measures of incidence of new and advanced adenomas were significantly lower in the celecoxib group than in the placebo group. The data demonstrated also that over the 2 years after treatment, patients in the celecoxib group were more likely to have new adenomas than in the placebo group, possibly to release from COX-2 inhibition (Aber et al. 2011). A recent analysis of randomized trials of daily aspirin for prevention of vascular events found a substantial reduction in overall cancer mortality during follow-up occurring after 5 years on aspirin (Jacobs et al. 2012). So the story of NSAIDS as cancer preventing drugs has taken a roller-coaster ride, and the current discussion is tending towards a more personalized approach. Patients with an increased risk of developing ulcerations or cardiovascular problems should not be treated with NSAIDs or coxibs. Instead, patients without such risks, but with an increased risk of cancer probably may benefit from the chemopreventive efficacy of NSAIDs and coxibs.

Acknowledgment We thank Prof. Mike Parnham for critical reading this chapter.

References

Aber, N., Spicak, J., et al. (2011). Five-year analysis of the prevention of colorectal sporadic adenomatous polyps trial. *The American Journal of Gastroenterology, 106*, 1135–1146.

Gately, S., & Li, W. W. (2004). Multiple roles of COX-2 in tumor angiogenesis: A target for antiangiogenic therapy. *Seminars in Oncology, 31*, 2–11.

Green, D. R., & Evan, G. I. (2002). A matter of life and death. *Cancer Cell, 1*, 19–30.

Grösch, S., Tegeder, I., Niederberger, E., et al. (2001). COX-2 independent induction of cell cycle arrest and apoptosis in colon cancer cells by the selective COX-2 inhibitor celecoxib. *The FASEB Journal, 15*, 2742–2744.

Grösch, S., Tegeder, I., Schilling, K., et al. (2003). Activation of c-Jun-N-terminal-kinase is crucial for the induction of a cell cycle arrest in human colon carcinoma cells caused by flurbiprofen enantiomers. *The FASEB Journal, 17*, 1316–1318.

Grösch, S., Maier, T. J., et al. (2006). Cyclooxygenase-2 (COX-2)-independent anticarcinogenic effects of selective COX-2 inhibitors. *Journal of the National Cancer Institute, 98*, 736–747.

Jacobs, E. J., Newton, C. C., et al. (2012). Daily aspirin use and cancer mortality in a large US cohort. *Journal of the National Cancer Institute, 104*, 1208–1217.

Koehne, C. H., & Dubois, R. N. (2004). COX-2 inhibition and colorectal cancer. *Seminars in Oncology, 31*, 12–21.

Laine, L., & White, W. B. (2008). COX-2 selective inhibitors in the treatment of osteoarthritis. *Seminars in Arthritis and Rheumatism, 28*, 165–187.

Schjerning Olsen, A. M., Fosbøl, E. L., et al. (2011). Duration of treatment with nonsteroidal anti-inflammatory drugs and impact on risk of death and recurrent myocardial infarction in patients with prior myocardial infarction: A nationwide cohort study. *Circulation, 123*, 2226–2235.

Shiff, S. J., & Rigas, B. (1999). The role of cyclooxygenase inhibition in the antineoplastic effects of nonsteroidal anti-inflammatory drugs (NSAIDs). *The Journal of Experimental Medicine, 190*, 445–450.

Tegeder, I., Pfeilschifter, J., & Geisslinger, G. (2001). Cyclooxygenase-independent actions of cyclooxygenase inhibitors. *The FASEB Journal, 15*, 2057–2072.

Trifan, O. C., & Hla, T. (2003). Cyclooxygenase-2 modulates cellular growth and promotes tumorigenesis. *Journal of Cellular and Molecular Medicine, 7*, 207–222.

Weber, A., Casini, A., Heine, A., et al. (2004). Unexpected nanomolar inhibition of carbonic anhydrase by COX-2-selective celecoxib: New pharmacological opportunities due to related binding site recognition. *Journal of Medicinal Chemistry, 47*, 550–557.

NSAIDs and Coxibs

▶ Coxibs and Novel Compounds, Chemistry

NSAIDs and Their Indications

Garry G. Graham[1] and Richard O. Day[2]
[1]Department of Pharmacology, School of
Medical Sciences, University of New South
Wales, Sydney, NSW, Australia
[2]Department of Clinical Pharmacology and
Toxicology, St Vincent's Hospital, Sydney,
NSW, Australia

Synonyms

Antipyretic analgesics; Aspirin-like drugs;
Nonselective NSAIDs; Selective COX-2
inhibitors; Simple analgesics

Definition

The term nonsteroidal anti-inflammatory drugs
(NSAIDs) refers to a group of drugs whose
major therapeutic activities are the suppression
of pain (analgesia), reduction of body temperature in fever (antipyresis), and a decrease in signs
of inflammation (anti-inflammatory activity).

Characteristics

Classes of NSAIDs

The NSAIDs can be separated into three major
groups by their effects on the synthesis of prostaglandins and related compounds (Fig. 1):
- Nonselective NSAIDs, such as aspirin,
  diclofenac, and indomethacin. These drugs
  have anti-inflammatory, analgesic, and antipyretic activities. They are also well known
  to produce gastrointestinal damage, inhibit
  platelet aggregation, decrease kidney function
  in some patients, and precipitate aspirin-
  induced asthma. The activity of these drugs
  is due to inhibition of two central enzymes
  (cyclooxygenase-1 (COX-1) and COX-2)
  involved in the synthesis of ► prostaglandins
  and related compounds. The prostaglandins
  are mediators of pain, fever, and

inflammation, inhibition of platelet aggregation, as well as gastroprotection. Consequently, inhibition of their synthesis leads to
the characteristic actions and adverse actions
of the NSAIDs.
- COX-2 selective inhibitors (coxibs or COX-1-
  sparing agents, CSIs) have similar therapeutic
  activities to the nonselective NSAIDs but have
  improved gastrointestinal tolerance, no significant antiplatelet effect, and little or no tendency to produce asthma in aspirin-sensitive
  patients.
- Drugs which are selective inhibitors of the
  oxidation of endocannabinoids by COX-2 but
  do not inhibit the oxidation of arachidonic acid
  by COX-2 (Fig. 1). This group includes the
  R enantiomers of the profens (ibuprofen,
  naproxen, ketoprofen, and flurbiprofen). The
  S enantiomers of these drugs are typical
  nonselective NSAIDs, but the R enantiomers
  have contrasting pharmacological and clinical
  effects. Their inhibition of COX-2 may
  maintain endocannabinoid tone while simultaneously decreasing the production of prostaglandin-cannabinoid intermediates (Fig. 1).
  These effects may be responsible for their
  effects in reducing neuropathic pain (Duggan
  et al. 2011).

There are two other NSAIDs which require
further discussion:
- Aspirin is a nonselective NSAID by irreversible inhibition of COX-1 and COX-2 (Fig. 1),
  but its metabolism to salicylate provides
  a second mechanism of action.
- Paracetamol (acetaminophen) is an analgesic
  and antipyretic agent which has weak anti-
  inflammatory activity. Paracetamol is often
  used as an alternative to both classes of
  NSAIDs, and for this reason, this general
  chapter on NSAIDs contains some comparisons of the effects of NSAIDs with those of
  paracetamol.

Clinical Uses

Inflammation

Inflammatory rheumatic disorders, such as rheumatoid arthritis, are important indications for the
nonselective NSAIDs and COX-2 selective

2-ARACHIDONOYLGLYCEROL   PHOSPHOLIPID

Monoacylglycerol lipase   cPLA$_2$

ARACHIDONIC ACID   ENDOCANNABINOIDS
(2-ARACHIDONOYLGLYCEROL,
ANANDAMIDE)

Cyclooxygenase

COX-1   PGG$_2$ COX-2   PGG$_2$-C

Peroxidase

PGH$_2$ PGH$_2$-C

PROSTAGLANDINS   PROSTAGLANDIN-Cs
THROMBOXANE A$_2$ THROMBOXANE A$_2$-C
PROSTACYCLIN   PROSTACYCLIN-C

NSAIDs and Their Indications, Fig. 1 Synthesis of prostaglandins and related compounds. Cyclooxygenase-1 (*COX*-1) catalyzes the synthesis of prostaglandin (*PG*) intermediates, PGG$_2$ and PGH$_2$. COX-2 not only catalyzes the synthesis of PG intermediates from arachidonic acid but also converts endogenous cannabinoids to prostaglandin-cannabinoid intermediates (PGG$_2$-C and PGH$_2$-C). COX-1 and COX-2 are bifunctional enzymes with cyclooxygenase and peroxide functions. The nonselective NSAIDs inhibit the formation of PG intermediates by both COX-1 and COX-2, whereas the COX-2 selective inhibitors block the synthesis of PG and PG cannabinoid intermediates by COX-2. R enantiomers of the profens selectively inhibit the synthesis of PGH$_2$-C by COX-2. The final production of PGs and PG-Cs is synthesized by specific enzymes

inhibitors. The symptoms of rheumatoid arthritis are relieved by both classes of NSAIDs, but neither class has any clear effect on the progression of the disease (Barraclough 2002; McCormack 2011). Joint damage appears to continue. Consequently, the NSAIDs are very commonly used with the disease-modifying antirheumatic drugs which may slow the degeneration of joints. Both the nonselective NSAIDs and the selective COX-2 inhibitors are valuable in the treatment of other inflammatory, painful, arthritic states, including acute gout (Rubin et al. 2004; Schumacher et al. 2002), ankylosing spondylitis, and psoriatic arthritis.

Treatment of Pain

Acute Pain Both major classes of NSAIDs (nonselective and selective COX-2 inhibitors) have good activity in the relief of pain and inflammation, particularly after oral surgery to remove third molar (wisdom) teeth (Moore et al. 2011). There is no significant difference between the mean various nonselective NSAIDs or between the nonselective and selective COX-2 inhibitors

(Moore et al. 2011). However, patients may prefer individual NSAIDs to other NSAIDs (Brooks and Day 1991).

NSAIDs have been used as part of the analgesic regimen in more substantial surgery, such as knee replacement. In these cases, the selective COX-2 inhibitors are particularly useful because COX-2 selective inhibitors, unlike the nonselective NSAIDs, do not inhibit platelet aggregation so that blood clots normally (Buvanendran et al. 2003). This property is particularly useful with the major shift to "day-only" surgery. Renal function and hypertension may, however, be worsened by NSAIDs after surgery (see below), and the NSAIDs should be used carefully in this situation.

Alone, the NSAIDs are not useful for severe, acute pain, for example, the pain of bone fractures, very painful surgical procedures, or myocardial infarction. In these cases, the opioids, such as morphine, are more effective analgesics. The opioids are very useful for the treatment of the pain of cancer although they may be used with the NSAIDs in this indication. Although very effective analgesics, the opioids have well-known addictive properties, actions which are not shown by the NSAIDs. In the elderly, the use of the opiates is associated with increased falls which can produce long-term difficulties (Solomon et al. 2010).

Good pain relief has been achieved with NSAIDs in conditions of painful often obstructed contraction of smooth muscle. Examples are renal and biliary colic. This is because the smooth muscle contractions are dependent on prostaglandin synthesis. NSAIDs even outperform opioids which are widely used for these indications.

Osteoarthritis and Musculoskeletal Pain These are the major indications for NSAIDs. Surveys have shown that 15–20 % of individuals over the age of 65 years take NSAIDs regularly and this is largely for osteoarthritis and low back pain. Despite their wide use, orally administered NSAIDs are only modestly effective at relieving pain and the muscle stiffness associated with these disorders (Bjordal et al. 2004; Roelofs et al. 2008). As outlined below, paracetamol is often the preferred initial treatment because of its lesser adverse effects. Non-pharmacological therapies, such as exercise, muscle strengthening, and weight loss, are also important in the overall management of patients. The combination of pharmacological and non-pharmacological measures makes up the modern treatment of osteoarthritis and low back pain.

The so-called "soft tissue" rheumatic problems are also highly prevalent and include muscle strains and aches, tennis elbow, and many others. The first choice for all these conditions is normally paracetamol. The NSAIDs are, in general, more effective, but, again, paracetamol is often tried first because of its better tolerance.

Migraine The nonselective NSAIDs and the COX-2 selective inhibitors have an important place in the treatment of migraine (Goadsby et al. 2002). Their use soon after onset of an acute attack of migraine is most effective. Combination with an antiemetic is often required in order to suppress vomiting.

Dysmenorrhea Dysmenorrhea is dependent on prostaglandin synthesis and, consequently, amenable to treatment with both classes of NSAIDs. The COX-2 selective drugs have also been useful as they have no significant antiplatelet effect and therefore do not increase menstrual bleeding (Daniels et al. 2002; Marjoribanks et al. 2010).

Topical Preparations Topical preparations of NSAIDs are widely used in the short term for acute muscle pain (sprains, minor sports injuries etc.) and in the long term for osteoarthritis. For acute pain from sprains and minor injuries, the topical preparations are fairly effective (Massey et al. 2010), but the available preparations (gels) are less effective for the treatment of osteoarthritis (Derry et al. 2012). Nevertheless, these products are recommended by some groups because of their excellent tolerability which results from the low systemic exposure (NICE clinical guideline 59. www.nice.org.uk/CG059). Topical solutions containing diclofenac and high concentrations of dimethyl sulfoxide are more effective than gels of diclofenac in the treatment of osteoarthritis of the knee (Derry et al. 2012), but these solutions are not available in many countries.

Other Clinical Uses
The nonselective NSAIDs have antithrombotic effects due to their inhibition of the formation of thromboxane A_2 by platelets. This mediator is synthesized by a COX-1-dependent pathway in platelets and induces platelet aggregation which is a first step in blood coagulation (thrombosis). All the nonselective NSAIDs have antithrombotic activity. Aspirin is, however, the preferred antithrombotic agent because of its long duration of action. Consistent with their lack of effect on COX-1 at therapeutic doses, the COX-2 selective inhibitors do not have significant antithrombotic activity. This lack of antithrombotic activity may be beneficial after surgery but, conversely, may be a cause of a major adverse reaction of these new NSAIDs, as is outlined below.

Both the nonselective NSAIDs and the COX-2 selective inhibitors appear useful in decreasing the progression of several tumors, particularly cancer of the colon. Celecoxib, a COX-2 selective inhibitor, is now used as part of the treatment of patients with ► familial polyposis coli.

Comparisons with Paracetamol

It is well known that paracetamol has analgesic and antipyretic activities. It is regarded as a weak anti-inflammatory agent. It does, however, reduce swelling after oral surgery but not in rheumatoid arthritis. Although the NSAIDs are, on average, somewhat more efficacious than paracetamol (Pincus et al. 2004), paracetamol is still recommended as first-line treatment for osteoarthritis, not only because it is effective in many patients, but also because it is better tolerated than the NSAIDs. Although little work has been done to evaluate the utility of combining NSAIDs with paracetamol, this practice has no obvious disadvantages and may control pain better than either drug alone.

Paracetamol has been regarded as having a distinct mechanism of action from the nonselective NSAIDs and the selective COX-2 inhibitors. The reason is that, in broken cell preparations, paracetamol only inhibits prostaglandin synthesis at supratherapeutic concentrations. In many intact cells, however, paracetamol is a potent inhibitor of prostaglandin synthesis when the levels of hydroperoxides are low (Graham and Scott 2005). This selectivity may be the reason for the weak anti-inflammatory activity of paracetamol. Apart from its weak anti-inflammatory activity, the clinical and adverse effects of paracetamol resemble those of the selective COX-2 inhibitors.

Adverse Effects

Gastrointestinal

Prostaglandins are cytoprotective in the stomach and small intestine. Consequently, the nonselective NSAIDs commonly cause a variety of serious adverse reactions in the gastrointestinal tract, most importantly, perforations, ulceration, and bleeding. Serious cases of gastrointestinal damage affect nearly 1 % of chronic users of the nonselective NSAIDs per year. Older, sicker patients, particularly those with previous peptic ulceration, are most at risk.

Gastrointestinal tolerance of the nonselective NSAIDs is improved with enteric coating, co-prescription of antacids, ingestion with food, or rectal or parental routes of administration, but there is still a considerable risk of serious upper gastrointestinal bleeding. Another approach is to use the nonselective NSAIDs with prostaglandin analogues, such as misprostol which are cytoprotective, or the NSAIDs are administered with proton pump inhibitors which decrease acid secretion by the stomach (Latimer et al. 2009). A further approach is to use the COX-2 selective inhibitors or paracetamol. The COX-2 selective inhibitors were developed in order to decrease the gastrointestinal toxicity of the NSAIDs. This was successful, to a degree, and the use of the selective inhibitors, in preference to the nonselective NSAIDs, reduces the incidence of serious gastrointestinal damage (Bombardier et al. 2000). However, the selective COX-2 inhibitors still produce gastrointestinal damage in some patients (Latimer et al. 2009), and it is now recommended that selective COX-2 inhibitors should be taken with proton pump inhibitors (NICE clinical guideline 59 www.nice.org.uk/CG059).

A particular problem arises for patients who require long-term dosage with both low-dose aspirin as an antithrombotic agent and an NSAID. It now appears that low doses of aspirin block the gastrointestinal-sparing effects of the COX-2 selective inhibitors. These patients are now often prescribed a selective COX-2 inhibitor together with a cytoprotective drug or an inhibitor of acid secretion as well as aspirin.

Thrombosis

As discussed above, thromboxane A_2 is synthesized by a COX-1-dependent pathway in platelets, and platelet aggregation is therefore not affected by the COX-2 selective inhibitors. However, the selective COX-2 inhibitors block the synthesis of prostacyclin, a vasodilator and antithrombotic factor that is largely synthesized through COX-2. Thus, there has been considerable concern that the selective COX-2 inhibitors may increase the incidence of thrombosis, causing myocardial infarction, for example. This concern has been confirmed by the withdrawal of one COX-2 selective inhibitor, rofecoxib, because of the increased occurrence of stroke and myocardial infarction during long-term therapy (Fitzgerald 2004). However, the major currently used COX-2 selective inhibitor, celecoxib, has only a small tendency to lead to thrombosis.

Non-selective NSAIDs, apart from diclofenac, block the anti-thrombotic effect of aspirin. COX-2 selective inhibitors and paracetamol do not and can be taken with aspirin. A proton pump inhibitor should be added if the patient is at risk of gastrointestinal damage.

Renal Impairment and Hypertension

Both the nonselective NSAIDs and the selective COX-2 inhibitors may precipitate renal failure.

Risk factors include:

- Age over 60
- Preexisting renal impairment
- Dehydration
- Cirrhosis
- Congestive cardiac failure
- Salt-restricted diets
- Concomitant treatment with diuretics or inhibitors of angiotensin formation or action
- General anesthesia

The renal function of patients in these situations is considered to be more dependent on the function of prostaglandins than normal subjects, and consequently, inhibition of prostaglandin synthesis by NSAIDs may markedly reduce renal function. As far as possible, NSAIDs should be avoided in patients with substantial risk factors. If NSAID is used, greater care is required with increasing risk factors for renal impairment.

Blood pressure may rise, in some cases, quite substantially during treatment with either the nonselective NSAIDs or the COX-2 selective agents (Barraclough 2002; Whelton et al. 2002). Consequently, it is now recommended that blood pressure should be monitored if dosage with the nonselective NSAIDs or the COX-2 selective drugs is commenced in patients taking antihypertensives and the dosage of antihypertensives adjusted accordingly, particularly if NSAIDs are taken in the long term. NSAIDs should be avoided in patients with unstable blood pressure.

Asthma

Asthma is precipitated in up to 20 % of asthmatics by aspirin and other nonselective NSAIDs (Jenkins et al. 2004). This reaction appears to be produced by inhibition of COX-1 because asthma is not induced by the selective COX-2 inhibitors (West and Fernandez 2003). Paracetamol is also safer in aspirin-sensitive asthmatics but does produce mild asthma in occasional patients (Jenkins et al. 2004).

Summary

The NSAIDs are widely used drugs. They decrease inflammation of rheumatoid arthritis and gout and are useful in the treatment of fever and mild to moderate pain, but their adverse effects are well-known, and their use should be restricted. Aspirin is a nonselective NSAID but is largely used in present-day medicine for its antiplatelet effects.

References

Barraclough, D. R. (2002). Considerations for the safe prescribing and use of COX-2-specific inhibitors. *Medical Journal of Australia, 176*, 328–331.

Bjordal, J., Ljunggren, A., Klovning, A., & Slørdal, L. (2004). Non-steroidal anti-inflammatory drugs, including cyclo-oxygenase-2 inhibitors, in osteoarthritic knee pain: Meta-analysis of randomised placebo controlled trials. *British Medical Journal, 329*, 1317–1320.

Bombardier, C., Laine, L., Reicin, A., Shapiro, D., Burgos-Vargas, R., Davis, B., et al. (2000). Comparison of upper gastrointestinal toxicity of rofecoxib and naproxen in patients with rheumatoid arthritis. VIGOR Study Group. *New England Journal of Medicine, 343*(21), 1520–1530.

Brooks, P. M., & Day, R. O. (1991). Nonsteroidal anti-inflammatory drugs – differences and similarities. *New England Journal of Medicine, 324*, 1716–1725.

Buvanendran, A., Kroin, J. S., Tuman, K. J., Lubenow, T. R., Elmofty, D., Moric, M., et al. (2003). Effects of perioperative administration of a selective cyclooxygenase 2 inhibitor on pain management and recovery of function after knee replacement: A randomized controlled trial. *JAMA, 290*(18), 2411–2418.

Daniels, S. E., Talwalker, S., Torri, S., Snabes, M. C., Recker, D. P., & Verburg, K. M. (2002). Valdecoxib, a cyclooxygenase-2-specific inhibitor, is effective in treating primary dysmenorrhea. *Obstetrics & Gynecology, 100*(2), 350–358.

Derry, S., Moore, R. A., & Rabbie, R. (2012). Topical NSAIDs for chronic musculoskeletal pain in adults. *Cochrane Database of Systematic Reviews, 9,* CD007400.

Duggan, K. C., Hermanson, D. J., Musee, J., Prusakiewicz, J. J., Scheib, J. L., Carter, B. D., et al. (2011). (R)-Profens are substrate-selective inhibitors of endocannabinoid oxygenation by COX-2. *Nature Chemical Biology, 7*(11), 803–809.

Fitzgerald, G. A. (2004). Coxibs and cardiovascular disease. *New England Journal of Medicine, 351*(17), 1709–1711.

Goadsby, P. J., Lipton, R. B., & Ferrari, M. D. (2002). Migraine - current understanding and treatment. *New England Journal of Medicine, 346,* 257–270.

Graham, G. G., & Scott, K. F. (2005). Mechanism of action of paracetamol. *American Journal of Therapeutics, 12,* 46–55.

Jenkins, C., Costello, J., & Hodge, L. (2004). Systematic review of prevalence of aspirin induced asthma and its implications for clinical practice. *British Medical Journal, 328,* 434.

Latimer, N., Lord, J., Grant, R., O'Mahony, R., Dickson, J., & Conaghan, P. (2009). Cost effectiveness of COX 2 selective inhibitors and traditional NSAIDs alone or in combination with a proton pump inhibitor for people with osteoarthritis. *British Medical Journal, 339,* b2538.

Marjoribanks, J., Proctor, M., Farquhar, C., & Derks, R. S. (2010). Nonsteroidal anti-inflammatory drugs for dysmenorrhoea. *Cochrane Database of Systematic Reviews, 1,* CD001751.

Massey, T., Derry, S., Moore, R. A., & McQuay, H. J. (2010). Topical NSAIDs for acute pain in adults. *Cochrane Database of Systematic Reviews, 6,* CD007402.

McCormack, P. L. (2011). Celecoxib: A review of its use for symptomatic relief in the treatment of osteoarthritis, rheumatoid arthritis and ankylosing spondylitis. *Drugs, 71*(18), 2457–2489.

Moore, R. A., Derry, S., McQuay, H. J., & Wiffen, P. J. (2011). Single dose oral analgesics for acute postoperative pain in adults. *Cochrane Database of Systematic Reviews, 9,* CD008659.

(2013). NICE clinical guideline 59 Osteoarthritis: The care and management of osteoarthritis in adults. www.nice.org.uk/CG059

Pincus, T., Koch, G., Lei, H., Mangal, B., Sokka, T., Moskowitz, R., et al. (2004). Patient Preference for Placebo, Acetaminophen (paracetamol) or Celecoxib Efficacy Studies (PACES): Two randomised, double blind, placebo controlled, crossover clinical trials in patients with knee or hip osteoarthritis. *Annals of the Rheumatic Diseases, 63*(8), 931–939.

Roelofs, P., Deyo, R., Koes, B., Scholten, R., & van Tulder, M. (2008). Non-steroidal anti-inflammatory drugs for low back pain. *Cochrane Database of Systematic Reviews, 1,* CD000396.

Rubin, B. R., Burton, R., Navarra, S., Antigua, J., Londono, J., Pryhuber, K. G., et al. (2004). Efficacy and safety profile of treatment with etoricoxib 120 mg once daily compared with indomethacin 50 mg three times daily in acute gout: A randomized controlled trial. *Arthritis and Rheumatism, 50*(2), 598–606.

Schumacher, H. R., Boice, J. A., Daikh, D. I., Mukhopadhyay, S., Malmstrom, K., Ng, J., et al. (2002). Randomised double blind trial of etoricoxib and indometacin in treatment of acute gouty arthritis. *British Medical Journal, 324,* 1488–1492.

Solomon, D. H., Rassen, J. A., Glynn, R. J., Garneau, K., Levin, R., Lee, J., et al. (2010). The comparative safety of opioids for nonmalignant pain in older adults. *Archives of Internal Medicine, 170*(22), 1979–1986.

West, P. M., & Fernandez, C. (2003). Safety of COX-2 inhibitors in asthma patients with aspirin hypersensitivity. *Annals of Pharmacotherapy, 37,* 1497–1501.

Whelton, A., White, W. B., Bello, A. E., Puma, J. A., & Fort, J. G. (2002). Effects of celecoxib and rofecoxib on blood pressure and edema in patients > or =65 years of age with systemic hypertension and osteoarthritis. *American Journal of Cardiology, 90*(9), 959–963.

NSAIDs, Adverse Effects

Irmgard Tegeder
Pharmazentrum frankfurt, Institute of Clinical Pharmacology, Goethe-University Hospital, Frankfurt, Germany

Synonyms

Cyclooxygenase inhibitors; Nonsteroidal anti-inflammatory drugs; Nonsteroidal antirheumatic drugs; NSAIDs; NSAIDs, Side effects

Definition

NSAIDs constitute a large group of chemically diverse substances that inhibit ▶ cyclooxygenases activity and thereby prostaglandin synthesis. Traditional NSAIDs inhibit both COX-isoenzymes (COX-1 and COX-2). Novel NSAIDs ("coxibs") inhibit only COX-2. NSAIDs are mainly used as analgesics.

Characteristics

Overview

▶ Adverse effects of NSAIDs arise from the difficulty to inhibit exclusively the synthesis of ▶ prostaglandins that cause pain and inflammation. The inhibition of cyclooxygenases will always also affect the synthesis of prostaglandins and thromboxanes that are needed for physiological processes. In addition, COX inhibition may shift arachidonic acid metabolism to ▶ leukotriene synthesis because of the excess supply of substrate. This particularly applies to traditional NSAIDs. In addition, COX-isoenzymes are involved in the production of anti-inflammatory lipid mediators which are required for the resolution of inflammation so that permanent use of NSAIDs may interfere with endogenous healing processes.

COX-1 and COX-2 perform different tasks due to the different localization and regulation. COX-1 is expressed in all tissues and produces prostaglandins and thromboxanes which are needed for the maintenance of physiological functions. COX-2 is not expressed in most healthy tissue but is upregulated after stimulation, which may be any kind of tissue damage. Hence, exclusively targeting COX-2 will not affect COX-1-derived physiological prostaglandins and will therefore avoid some of the adverse effects that are typical for traditional NSAIDs. However, COX-2 is also constitutively expressed (see ▶ Constitutive Gene) in some tissues, so that COX-2 selective NSAIDs do not completely spare physiological prostaglandin production and may shift the balance towards COX-1-derived thromboxane and consequent cardiovascular side effects. In addition to common adverse effects, some substance-specific side effects may occur, so the individual tolerability of NSAIDs may vary among patients.

Gastrointestinal Toxicity

Physiological prostaglandin E2 (PGE2) and prostacyclin (PGI2) in the stomach play an important role in the defense mechanisms that protect the gastric epithelium from the acidic environment. PGE2 increases the production of gastric mucus, which builds a protective layer on the epithelium, while PGI2 maintains gastric blood flow. Inhibition of PG synthesis in the stomach causes serious adverse effects such as gastric erosions, bleeding, ulceration, and perforation. A single dose of aspirin is sufficient to cause small erosions. The risk for serious GI toxicity increases considerably with long-term use of traditional NSAIDs and concomitant use of glucocorticoids.

Multiple clinical trials have demonstrated that COX-2 selective inhibitors are associated with a lower frequency of serious gastrointestinal side effects than traditional NSAIDs. Serious toxicity is reduced to the placebo level, suggesting that the physiological PG production in the stomach is mediated primarily by COX-1. However, experimental studies have revealed that gastric damage only occurs if both COX-enzymes are inhibited collectively, but not with either COX-1 or COX-2 inhibition alone. This indicates that COX-2 also is important for protection of the gastric mucosa (Halter et al. 2001). This idea is further supported by studies showing that COX-2 is ▶ upregulated in the stomach in the case of ulceration (Schmassmann et al. 1998) or other epithelial damage such as ▶ Helicobacter pylori infection (Seo et al. 2002) and contributes to the production of prostaglandins that are involved in healing processes such as PGJ2 (Gilroy et al. 1999). Hence, although COX-2 selective inhibitors are relatively safe for the healthy stomach, they may impair ulcer healing (Jones et al. 1999).

Renal Toxicity

COX-2 is constitutively expressed in the kidney and is highly regulated in response to alterations in intravascular volume (Harris et al. 1994). COX-2-derived PGs elicit the release of ▶ renin from the renal ▶ juxtaglomerular apparatus, especially during volume depletion. PGs also maintain renal blood flow and regulate salt and water excretion.

COX-2 inhibition, both with traditional NSAIDs or selective COX-2 inhibitors, may transiently decrease urine sodium excretion in some subjects and cause mild to moderate elevations of systolic and diastolic blood pressure and thereby interferes with antihypertensive drugs, increase the risk for stroke and heart failure in patients with diseases of the heart. Substance-specific differences have been suggested. For example, rofecoxib users were at a significantly increased relative risk of new-onset hypertension compared with patients taking celecoxib, nonspecific NSAIDs, or no NSAID (Solomon et al. 2004). The risk for renal side effects is increased in patients with preexisting renal or heart disease.

Platelet Aggregation and Cardiovascular System

Thrombocyte aggregation depends on thromboxane A2 (TXA2), which is produced by the COX-1 pathway in platelets. Hence, traditional NSAIDs inhibit platelet aggregation, and particularly aspirin, the irreversible unselective COX-inhibitor, causes a long-lasting prolongation of the bleeding time, which may increase the risk of bleeding, e.g., during or after surgery. On the other hand, inhibition of thrombocyte aggregation may be the desired therapeutic effect of aspirin for patients with increased risk of thrombosis such as coronary heart disease, or it may be a welcome side effect of traditional NSAIDs. In addition, specifically aspirin triggers the production of lipoxin A4 which is a potent anti-inflammatory mediator (Schwab et al. 2007).

The activation of platelets by TXA2 is counterbalanced by vascular prostacyclin (PGI2) under physiological conditions. Systemic PGI2 is mainly produced through the COX-2 pathway (McAdam et al. 1999). Specific COX-2 inhibitors may therefore shift the balance between TXA2 and PGI2 towards TXA2-mediated thrombocyte aggregation. This may result in an increased risk of thrombotic events in predisposed patients. Clinical trials have revealed that the risk for thrombotic cardiovascular effects is increased with rofecoxib and other COX-2 inhibitors leading to a restrictive use of COX-2 selective agents in patients with cardiovascular diseases or risk factors such as diabetes or hypertension.

Bone Healing

PGs participate in inflammatory responses in the bone, increased ▶ osteoclast activity and subsequent bone resorption, and increased ▶ osteoblast activity and new bone formation.

Data from animal studies suggest that both nonspecific and specific inhibitors of cyclooxygenases impair fracture healing. This is due to the inhibition of COX-2 (Zhang et al. 2002). Although these data raise concerns about the use of traditional NSAIDs and COX-2-specific inhibitors as anti-inflammatory or anti-analgesic drugs in patients undergoing bone repair, clinical reports have been inconclusive.

Aspirin Asthma

Inhibition of cyclooxygenase activity reduces arachidonic acid metabolism for prostaglandin synthesis and hence yields more substrate for leukotriene and endocannabinoid synthesis. While ▶ endocannabinoids may enhance the analgesic effects of certain NSAIDs (Ates et al. 2003), overproduction of cysteinyl leukotrienes may cause bronchial constriction and trigger asthma attacks in "aspirin-sensitive" patients (Szczeklik et al. 2001). Aspirin asthma has also been observed with other nonselective NSAIDs but not with COX-2 selective drugs. Similar mechanisms also account for the NSAID-evoked ▶ urticaria (Mastalerz et al. 2004).

Ototoxicity

A recent study suggested that salicylate-induced tinnitus is mediated through an indirect activation of ▶ NMDA receptors in the cochlea, causing an increase of spontaneous auditory nerve activity. NMDA receptor activation is probably mediated by inhibition of prostaglandin synthesis (Guitton et al. 2003) and hence is not specific for salicylate. ▶ Tinnitus occurs at high plasma concentrations.

Pregnancy

COX-2-derived prostaglandins play a prominent role at all stages of female reproduction, from ovulation to implantation, ▶ decidualization, and delivery. The regulation of prostaglandin release is mediated by transcriptional control of COX-2 and microsomal prostaglandin E synthase. Elevated uterine PGs in combination with heightened sensitivity of the myometrium to PGs lead to contractions and labor. Hence, traditional NSAIDs as well as COX-2 inhibitors may prolong parturition or may be used in the treatment of preterm labor (McWhorter et al. 2004).

In addition, prostaglandins regulate the transition to pulmonary respiration following birth that requires closure and remodeling of the ▶ ductus arteriosus. The maintenance of the ductus arteriosus in the open, or patent, state is dependent on prostaglandin synthesis, and the neonatal drop in prostaglandin E2 that triggers ductal closure is sensed through the EP4 receptor (Nguyen et al. 1997). Hence, NSAIDs may cause a premature DA closure if taken during late pregnancy or may be used to induce DA closure in preterm infants.

References

Ates, M., Hamza, M., Seidel, K., et al. (2003). Intrathecally applied flurbiprofen produces an endocannabinoid-dependent antinociception in the rat formalin test. *The European Journal of Neuroscience, 17*, 597–604.

Gilroy, D. W., Colville-Nash, P. R., Willis, D., et al. (1999). Inducible cyclooxygenase may have anti-inflammatory properties. *Nature Medicine, 5*, 698–701.

Guitton, M. J., Caston, J., Ruel, J., et al. (2003). Salicylate induces tinnitus through activation of cochlear NMDA receptors. *The Journal of Neuroscience, 23*, 3944–3952.

Halter, F., Tarnawski, A. S., Schmassmann, A., et al. (2001). Cyclooxygenase-2 implications on maintenance of gastric mucosal integrity and ulcer healing: Controversial issues and perspectives. *Gut, 49*, 443–453.

Harris, R. C., McKanna, J. A., Akai, Y., et al. (1994). Cyclooxygenase-2 is associated with the macula densa of rat kidney and increases with salt restriction. *The Journal of Clinical Investigation, 94*, 2504–2510.

Jones, M. K., Wang, H., Peskar, B. M., et al. (1999). Inhibition of angiogenesis by nonsteroidal anti-inflammatory drugs: Insight into mechanisms and implications for cancer growth and ulcer healing. *Nature Medicine, 5*, 1418–1423.

Mastalerz, L., Setkowicz, M., Sanak, M., et al. (2004). Hypersensitivity to aspirin: Common eicosanoid alterations in urticaria and asthma. *The Journal of Allergy and Clinical Immunology, 113*, 771–775.

McAdam, B. F., Catella-Lawson, F., Mardini, I. A., et al. (1999). Systemic biosynthesis of prostacyclin by cyclooxygenase (COX)-2: The human pharmacology of a selective inhibitor of COX-2. *Proceedings of the National Academy of Sciences of the United States of America, 96*, 272–277.

McWhorter, J., Carlan, S. J., Richichi, K., et al. (2004). Rofecoxib versus magnesium sulfate to arrest preterm labor: A randomized trial. *Obstetrics and Gynecology, 103*, 923–930.

Nguyen, M., Camenisch, T., Snouwaert, J. N., et al. (1997). The prostaglandin receptor EP4 triggers remodelling of the cardiovascular system at birth. *Nature, 390*, 78–81.

Schmassmann, A., Peskar, B. M., Stettler, C., et al. (1998). Effects of inhibition of prostaglandin endoperoxide synthase-2 in chronic gastro-intestinal ulcer models in rats. *British Journal of Pharmacology, 123*, 795–804.

Schwab, J. M., Chiang, N., Arita, M., Serhan, C. N., et al. (2007). Resolvin E1 and protectin D1 activate inflammation-resolution programmes. *Nature, 447*(7146), 869–874.

Seo, J. H., Kim, H., & Kim, K. H. (2002). Cyclooxygenase-2 expression by transcription factors in helicobacter pylori-infected gastric epithelial cells: Comparison between HP 99 and NCTC 11637. *Annals of the New York Academy of Sciences, 973*, 477–480.

Solomon, D. H., Schneeweiss, S., Levin, R., et al. (2004). Relationship between COX-2 specific inhibitors and hypertension. *Hypertension, 44*, 140–145.

Szczeklik, A., Nizankowska, E., Sanak, M., et al. (2001). Aspirin-induced rhinitis and asthma. *Current Opinion in Allergy and Clinical Immunology, 1*, 27–33.

Zhang, X., Schwarz, E. M., Young, D. A., et al. (2002). Cyclooxygenase-2 regulates mesenchymal cell differentiation into the osteoblast lineage and is critically involved in bone repair. *The Journal of Clinical Investigation, 109*, 1405–1415.

NSAIDs, Chemical Structure and Molecular Mode of Action

Stefan A. Laufer
Institute of Pharmacy, Eberhard-Karls University of Tuebingen, Tuebingen, Germany

Definition

Structure and Metabolic Function of COX-1

COX-1 is a 70-kD enzyme catalyzing the reaction of ▶ arachidonic acid to PGG_2 (cyclooxygenase

NSAIDs, Chemical Structure and Molecular Mode of Action, Fig. 1 Reaction catalyzed by COX enzymes

NSAIDs, Chemical Structure and Molecular Mode of Action, Fig. 2 A ribbon representation of the Co^{3+}-oPGHS-1 monomer with AA bound in the COX channel. The EGF domain, MBD, and catalytic domain are shown in green, orange, and blue, respectively; Co^{3+}-protoporphyrin IX is depicted in *red*, disulfide bonds (Cys36–Cys47, Cys37–Cys159, Cys41–Cys57, Cys59–Cys69, and Cys569–Cys575) in *dark blue*, and side chain atoms for COX channel residues Arg120, Tyr355, and Tyr385 in *magenta* (From Malkowski et al. (2000))

reaction) and consecutively PGG_2 to PGH_2 (peroxidase reaction) as outlined in Fig. 1.

There are distinct active sites for the ► cyclooxygenases (COX) and the peroxidase reactions (Fig. 2).

Characteristics

Inhibitors of Cyclooxygenases

Different chemical classes can provide the structural features necessary to mimic arachidonic acid at the active site. The substrate, arachidonic acid, is a C_{20} carboxylic acid with 4 isolated double bonds at positions 5, 8, 11, and 14. For the enzyme reaction, arachidonic acid must adapt to a "folded" conformation, allowing the oxygen to insert between C_9 and C_{11} and the ring closure between C_8 and C_{12} (Fig. 3).

To fix arachidonic acid in such a conformation, several interactions with the active site of the enzyme are necessary, e.g., ionic interaction

(a salt bridge) between the carboxylic group and arginine 120, π-π interactions between the double bonds of arachidonic acid and aromatic amino acids, and numerous hydrophobic interactions (Fig. 4).

All these structural features can be identified in many ► NSAIDs. Most acidic NSAIDs are therefore believed to mimic arachidonic acid in its folded conformation at the active site of COX. Structure activity relationships follow these structural constrictions closely. The activity against COX-1 clearly correlates with torsion angles around the π-electron systems and the overall lipophilicity of the molecule (Moser et al. 1990).

Two classes of compounds however have a distinctly different molecular mode of action:
- ASS acetylates serine 530 irreversibly at the active site of the enzyme.
- The oxicames are believed to interfere with the peroxidase active site, which also explains their structural difference.

NSAIDs, Chemical Structure and Molecular Mode of Action, Fig. 3 Mechanistic sequence for converting AA to PGG$_2$. Abstraction of the 13-proS hydrogen by the tyrosyl radical leads to the migration of the radical to C-11 on AA. The attack of molecular oxygen, coming from the base of the COX channel, occurs on the side interfacial to hydrogen abstraction. As the 11R-peroxyl radical swings over C-8 for an R-side attack on C-9 to form the endoperoxide bridge, C-12 is brought closer to C-8 via rotation about the C-10/C-11 bond, allowing the formation of the cyclopentane ring. The movement of C-12 also positions C-15 optimally for addition of a second molecule of oxygen, formation of PGG$_2$, and the migration of the radical back to Tyr385 (From Malkowski et al. (2000))

Most of the currently used NSAIDs, including diclofenac, ibuprofen, naproxen, piroxicam, and indomethacin, for instance, may produce full inhibition of both COX-1 and COX-2 with relatively poor selectivity under therapeutic conditions (Warner et al. 1999).

Acidic NSAIDs like diclofenac accumulate particularly in blood, liver, milt, and bone marrow, but also in tissue with acidic extracellular pH values. Such tissues are mainly inflamed tissues such as gastric tissue and the manifolds of the kidney. In inflamed tissue, NSAIDs inhibit the pathological overproduction of prostaglandins. In contrast, neutral NSAIDs (paracetamol) and weakly acidic NSAIDs (metamizol) distribute themselves quickly and homogeneously in the organism. They also penetrate the blood–brain barrier.

Fenamate Group

The core structure is 2-aminobenzoic acid (anthranilic acid). The 2-amino group is substituted with aromatic residues:
- Flufenamic acid
- Mefenamic acid
- Meclofenamic acid
- Niflumic acid (core structure: 2-amino-pyridyl-3-carboxylic acid)

For topical application, the carboxylic acid group is esterified with diethyleneglycol Fig. 5:
- Etofenamate

NSAIDs, Chemical Structure and Molecular Mode of Action, Fig. 4 A schematic of interactions between AA and COX channel residues. Carbon atoms of AA are *yellow*, oxygen atoms *red*, and the 13proS hydrogen *blue*. All *dashed lines* represent interactions within 4.0 Å between the specific side chain atom of the protein and AA (From Malkowski et al. (2000))

NSAIDs, Chemical Structure and Molecular Mode of Action, Fig. 5 Chemical structures of fenamate group NSAIDs

NSAIDs, Chemical Structure and Molecular Mode of Action, Fig. 6 Chemical structures of fenac group NSAIDs

Fenac structure   Diclofenac   Felbinac

Heteroaryl Acetic Acid core structure

Indomethacin   Tolmetin   Acemetacine

Lonazolac   Proglumetacine

NSAIDs, Chemical Structure and Molecular Mode of Action, Fig. 7 Chemical structures of heteroaryl acetic acid group NSAIDs

Profen group

Ibuprofen   Ketoprofen   Thiaprofen

Naproxen   Ketorolac

NSAIDs, Chemical Structure and Molecular Mode of Action, Fig. 8 Chemical structures of profen group NSAIDs

1,2-Benzothiazine derivatives

Piroxicam

Meloxicam

Droxicam

Sudoxicam

Cinnoxicam

1,2-Thienothiazine derivatives

Tenoxicam

Lornoxicam

NSAIDs, Chemical Structure and Molecular Mode of Action, Fig. 9   Chemical structures of oxicam group NSAIDs

NSAIDs, Chemical Structure and Molecular Mode of Action, Fig. 10 Chemical structure of pyrazolone group NSAIDs

Pyrazolone core structure

Propyphenazone

Metamizole-Na

Phenazone

Fenac Group

The core structure is 2-aminophenylacetic acid. The 2-amino group is substituted with aromatic residues Fig. 6:

- Diclofenac
- Felbinac (only used topically)

Heteroaryl Acetic Acid Group

- Indomethacin (Fig. 7)
- Acemetacin
- Proglumetacin
- Tolmetin (and its ring-closed analog ketorolac)
- Lonazolac

Profen Group

The core structure is 2-arylpropionic acid Fig. 8:

- Ibuprofen
- Ketoprofen
- Thiaprofen
- Naproxen
- Ketorolac (can be seen formally as a ring-closed profen)

Oxicam Group

The core structure is 1,2-benzothiazine or 1,2 thienothiazine Fig. 9:

- Piroxicam
- Tenoxicam

NSAIDs, Chemical Structure and Molecular Mode of Action, Fig. 11 Chemical structures of pyrazolidindione group NSAIDs

Pyrazolidindione core structure

Phenylbutazone

Mofebutazone

a PGHS-1

Tyr 385

Ser 530

Ile 523

His 513

Arg 120

b PGHS-2

Tyr 371

Ser 516

Val 509

Arg 499

Arg 106

NSAIDs, Chemical Structure and Molecular Mode of Action, Fig. 12 Comparison of the active site of COX-1 (PGHS-1) and COX-2 (PGHS-2) (From Wong et al. 1997)

- Lornoxicam
- Droxicam
- Cinnoxicam
- Sudoxicam
- Meloxicam

Pyrazolone Group

The mode of action of the pyrazolones remains unclear. It is thought that they may not be involved in the inhibition of COX-1 or COX-2. The compounds of the pyrazol-3-on series at least

Sulfonamide structures

Celecoxib

Valdecoxib

Methylsufone structures

Rofecoxib

Etoricoxib

Aryl Acetic Acid

Lumiracoxib   **others**

Parecoxib

NSAIDs, Chemical Structure and Molecular Mode of Action, Fig. 13 Chemical structures of coxibs

are neutral molecules with no acidity. A central mode of action is suggested. They also act antispasmodically and are effective in visceral pain. In the past, pyrazolones were very frequently used nonsteroidal anti-inflammatory drugs. They show a high plasma protein binding and therefore have a high rate of interaction with other pharmaceuticals. Agranulocytosis is a rare but severe side effect.

The core structure is 3H-pyrazol-3-one Fig. 10:
- Propyphenazone
- Metamizole-Na
- Phenazone

Pyrazolidindione Group
The core structure is pyrazolidin-3,5-dione Fig. 11:
- Phenylbutazone
- Mofebutazone

COX-2 Selective Inhibitors
The isoform 2 of the COX enzyme catalyzes the identical reaction AA to PGG_2; the active site however is slightly different from COX-1 (Fig. 12).

Isoleucine 523 is replaced by valine 509, making the active site of COX-2 more "spacious." This difference can be used to generate COX-2 selective inhibitors, as this active site tolerates more bulky molecules. Celecoxib is capable of producing full inhibition of COX-1 and COX-2. However, it shows a preferential selectivity toward COX-2 (>5-fold). The newer coxibs like rofecoxib strongly inhibit COX-2 with only weak activity against COX-1 (Warner et al. 1999).

A common pharmacophore cannot be identified; however, vicinal diaryl systems (celecoxib, rofecoxib, valdecoxib) and sulfone or sulfonamide groups seem to be advantageous (Dannhardt and Laufer 2000). Lumiracoxib however is an excellent example of the fact that spatial demanding substituents (bulky groups) alone are sufficient to generate selectivity, even with a diclofenac-like pharmacophore.

Structural Features of Selective COX-2 Inhibitors

Sulfonamide structure:
- Celecoxib
- Valdecoxib

Methylsulfone structure:
- Rofecoxib
- Etoricoxib

Aryl Acetic acid:
- Lumiracoxib

Others Fig. 13:
- Parecoxib (water-soluble prodrug for parenteral application, rapidly metabolized to valdecoxib)

References

Dannhardt, G., & Laufer, S. (2000). Structural approaches to explain the selectivity of COX-2 inhibitors: Is there a common pharmacophore? *Current Medicinal Chemistry, 7*, 1101–1112.

Malkowski, M., Ginell, S. L., Smith, W. L., & Garavito, R. M. (2000). The productive conformation of arachidonic acid bound to prostaglandin synthase. *Science, 289*, 1933–1937.

Moser, P., Sallmann, A., & Wiesenberg, L. (1990). Synthesis and quantitative structure-activity relationships of diclofenac analogues. *Journal of Medicinal Chemistry, 33*, 2358–2368.

Warner, T., Giuliano, F., Vojnovic, I., et al. (1999). Nonsteroid drug selectivities for cyclo-oxygenase-1 rather than cyclo-oxygenase 2 are associated with human gastrointestinal toxicity: A full in vitro analysis. *Proceedings of the National Academy of Sciences of the United States of America, 96*, 7563–7568.

Wong, E., Bayly, E., Waterman, H. L., et al. (1997). Conversion of prostaglandin G/H synthase-1 into an enzyme sensitive to PGHS-2-selective inhibitors by a double His^{513}â†'Arg and Ile^{523}â†'Val mutation. *The Journal of Biological Chemistry, 272*, 9280–9286.

NSAIDs, COX-Independent Actions

Irmgard Tegeder and Ellen Niederberger
Pharmazentrum frankfurt, Institute of Clinical Pharmacology, Goethe-University Hospital, Frankfurt, Germany

Synonyms

COX; Cyclooxygenase; PGH synthase; Prostaglandin H synthase

Definition

Non-steroidal anti-inflammatory drugs, NSAIDs are first choice drugs for the therapy of inflammation and pain. Their main mechanism of action is based on inhibition of cyclooxygenase (COX) 1 and 2 enzymes which result in reduced prostaglandin synthesis. In addition to traditional NSAIDs (tNSAIDs) which unselectively inhibit both COX-1 and COX-2, selective COX-2 inhibitors (coxibs) have been developed to prevent typical gastrointestinal side effects of tNSAIDs. Since aspirin was reported to inhibit nuclear factor kappa B (NF-kB) activation (Kopp and Ghosh 1994), there is increasing evidence that certain NSAIDs have various biological effects that are independent of cyclooxygenase activity and prostaglandin synthesis and may account at least in part for their analgesic, anti-inflammatory, and antiproliferative effects. These effects occur mostly at drug concentrations beyond the IC_{50} for COX-inhibition and therefore probably occur primarily at high concentrations. Various mechanisms have been shown to be involved (Tegeder et al. 2001; Grösch et al. 2006) and are summarized in this entry.

A schematic overview of these mechanisms is shown in Fig. 1.

Characteristics

Effects on Transcription Factors
Nuclear Factor Kappa B (NF-κB)
NF-κB is an important mediator of the cellular response to a variety of extracellular stress stimuli.

NSAIDs, COX-Independent Actions, Fig. 1 Systematic overview of COX-independent NSAIDs effects → activation or induction; —| inhibition ● inhibited by NSAIDs; ○ activated by NSAIDs Abbreviations: Akt/*PKB*, protein kinase B; *APC* adenomatous polyposis coli tumor suppressor gene; *AP*-1 activator protein-1; *Cdk*, cyclin-dependent kinase; *CREB*, cAMP response element binding protein; *Erk*, extracellular signal-regulated kinase; *GSK3β*, glycogen synthase kinase 3 beta; *Hsp*, heat shock protein; *IKK*, I-kappa kinase; *JNK*, Jun N-terminal kinase; *MAPK*, mitogen-activated kinase; *MEK/MKK*, mitogen-activated protein kinase; *NIK*, nuclear factor-kappaB-inducing kinase; *NF-κB*, nuclear factor kappa B; *PDK*, phosphoinositol-dependent kinase; PI-3 K, phosphatidylinositol-3-kinase; *PKC*, protein kinase C; *PPAR*, peroxisome proliferator-activated receptor; *pRb*, retinoblastoma protein; p90*RSK*, ribosomal S6 kinase; *Tcf*, T cell factor

As homodimers and heterodimers, Rel/NF-κB proteins bind to DNA target sites and regulate gene transcription of proinflammatory mediators and proteins that are involved in cell death or survival. Various NSAIDs including salicylates, sulindac, ibuprofen, and R- and S-flurbiprofen inhibit NF-κB activation. While aspirin, ibuprofen, and sulindac also inhibit COX activity, R-flurbiprofen and salicylic acid are inactive in this regard and therefore do not cause typical NSAID-evoked side effects that are due to COX-inhibition. Indomethacin, ketoprofen, and ketorolac do not inhibit NF-κB. Hence, in contrast to COX-inhibition, NF-κB inhibition is not a "class effect." The COX-2 selective inhibitors rofecoxib and celecoxib have different effects on NF-κB. While rofecoxib inhibits its activation in RAW 264.7 macrophages (Niederberger et al. 2003), celecoxib further increases LPS-induced NF-κB activation. The latter effect of celecoxib results in a loss of its anti-inflammatory efficacy at high doses in an experimental inflammatory model in rats (Niederberger et al. 2003), suggesting that effects on NF-κB are important for the anti-inflammatory efficacy of some COX-inhibitors. In contrast, it has been shown, e.g., in chondrocytes, that celecoxib inhibits NO production by inactivation of JNK and NF-κB (Tsutsumi et al. 2008).

NF-κB is also involved in the regulation of cell death and survival, either as being essential for the induction of ▶ apoptosis or more commonly as an inhibitor of apoptosis. Whether NF-κB promotes or inhibits apoptosis appears to depend on the cell type and the type of inducer. NF-κB is persistently active in numerous human cancer cells. This is suggested to contribute to increased resistance towards chemo- or radiotherapy. NSAIDs that inhibit NF-κB may eliminate this resistance and thereby re-increase cancer cell sensitivity toward apoptosis inducing treatments. Hence, inhibition of NF-κB in tumor cells may contribute to the observed antitumor activity of various NSAIDs including S- and R-flurbiprofen (Wechter et al. 1997), celecoxib (Grosch et al. 2001), sulindac, and aspirin (Thun et al. 1991).

AP-1

The transcription factor AP-1 is a homo- or heterodimer of Jun, Fos, and Fra oncogenes. AP-1 is activated by various stimuli including UV irradiation, growth factors, and inflammatory cytokines. Some of the AP-1 target genes are involved in the immune and inflammatory responses or tumor formation and progression. AP-1-regulated genes partially overlap with genes that are regulated by NF-κB. Inhibition of AP-1 has been shown for aspirin, sodium salicylate, piroxicam, R-flurbiprofen, and the selective COX-2 inhibitors celecoxib and NS398 in various cell types. However, it is not known to what extent effects on AP-1 contribute to the anti-inflammatory or antiproliferative effects of NSAIDs, because different AP-1 homo- and heterodimers may have either stimulating or inhibiting effects on gene transcription (Grosch et al. 2003), and the number of genes that are regulated by AP-1 is high. Therefore, the result of NSAID-induced AP-1 inhibition may vary between tissues and cell types.

Effects on Cellular Kinases

Inhibitor Kappa B Kinase Complex (IKK)

In most cells NF-κB exists in the cytoplasm in an inactive complex bound to inhibitor IκB proteins. NF-κB is activated upon phosphorylation and subsequent proteasome-mediated proteolysis of IκB. The key regulatory step in this pathway is the activation of an IκB kinase (IKK) complex. IKK consists of two catalytic subunits, IKK-α and IKK-β, and a regulatory subunit (IKK-γ) that regulates binding of activators. Liberated NF-κB translocates from the cytoplasm to the nucleus where it binds to the κB-sites of target genes and regulates their transcription. Various NSAIDs including aspirin, sodium salicylate, and sulindac inhibit the ATP binding to IKKβ and thereby its catalytic activity (Yin et al. 1998). However, the IKK inhibitory potency of these NSAIDs is low, and the specificity of aspirin-induced IKK, and thereby NF-κB inhibition, has been doubted.

Mitogen-Activated Protein (MAP)-kinase cascade

MAP-kinases (MAPK) play a central role in the differentiation and proliferation of several cell types and can be activated by various extracellular stimuli. AP-1 is one downstream target of the MAP-kinase family members including extracellular signal-regulated kinases (Erk-1 and −2; p42/p44 MAPK), c-Jun N-terminal kinases (JNK1-3), and p38 MAPK. Erk activation has been implicated in growth promotion and cell survival, whereas JNK and p38 MAPK activation are associated with stress responses and apoptosis. Recent results have shown that the effect of NSAIDs on MAP-kinases largely depends on the cellular context. In cancer cells, the ability of NSAIDs to modulate MAPK activities may play an important role in the cytotoxicity and induction of apoptosis.

Aspirin and sodium salicylate were shown to inhibit activation of Erk-1 and Erk-2 under certain circumstances. Inhibition of angiogenesis by COX-2 selective and unselective NSAIDs has been shown to be mediated through inhibition of Erk-2 activity and interference with its nuclear translocation in vascular endothelial cells

(Tsujii et al. 1998). p38 MAPK was reported to be activated by sodium salicylate in human fibroblasts, leading to induction of apoptosis. TNF-α-induced JNK activation was inhibited by salicylate in human fibroblasts. Oppositely, in HT-29 colon cancer and COS-1 cells, salicylate treatment resulted in activation of JNK.

Protein Kinase B (PKB/Akt)
The protein kinase B (PKB/Akt) promotes cell proliferation and survival, thereby contributing to cancer progression. Activation of Akt occurs by translocation of the kinase to the cell membrane and phosphorylation through phosphoinositide-dependent kinase 1 (PDK1). The COX-2 selective inhibitor celecoxib has been shown to induce apoptosis in prostate carcinoma (Hsu et al. 2000), hepatocarcinoma, and colon carcinoma cells by inhibiting the phosphorylation of PKB/Akt, thereby blocking its antiapoptotic activity. The effects of celecoxib on Akt depended in part on the inhibition of PDK1. Inhibition of PKB/Akt by celecoxib has also been observed in vascular smooth muscle cells leading to inhibition of neointima formation after balloon injury. Similarly, sulindac has been reported to inhibit Akt phosphorylation in lung adenocarcinoma cells. Aspirin has been shown to either activate or inhibit Akt activation, dependent on the cell type.

AMP-Activated Kinase (AMPK)
The AMP-activated kinase is a metabolic sensor involved in cellular energy homeostasis, cell growth, and metabolism. Its activation has been associated with beneficial effects in inflammation, diabetes, or obesity and is induced by increased AMP levels or pharmacologically active drugs such as AICAR or metformin. A recent study showed that aspirin at high doses is also able to activate AMPK (Hawley et al. 2012).

Effects on Cell Cycle Proteins
The progression through the various phases of the cell cycle is regulated by cyclins and cyclin-dependent kinases (Cdks). The function of cyclins is primarily controlled by changes in cyclin transcription, while Cdks are regulated by phosphorylation. The activity of the Cdk/cyclin complex is inhibited by p21 and p27. Sodium salicylate inhibits the proliferation of vascular smooth muscle cells through upregulation of these cell cycle inhibitors. This is probably caused by salicylate-induced upregulation of p53, which is the primary regulator of p21 transcription. Similar to salicylates, sulindac, sulindac sulfide, and celecoxib inhibited the proliferation of colon carcinoma cells and caused them to accumulate in the G_0/G_1 phase. This effect was attributed to inhibition of cyclin-dependent kinases and/or upregulation of p27 and p21.

Modulation of the Activity of Nuclear Receptor Family Members
Activation of Peroxisome Proliferator-Activated Receptor (PPAR)
PPARs-α, PPARs-δ, and PPARs-γ are nuclear hormone receptors that control the transcription of genes involved in energy metabolism, cell differentiation, apoptosis, and inflammation. PPARs bind to sequence-specific DNA response elements as a heterodimer with the retinoic acid receptor (RXR). PPAR-γ is highly expressed in adipose tissue and plays an important role in the regulation of genes involved in lipid utilization and storage, adipocyte differentiation, and insulin action and inflammation. Indomethacin binds to PPAR-γ and induces the differentiation of mesenchymal stem cells into adipocytes ► in vitro. Some other NSAIDs including ibuprofen, fenoprofen, flurbiprofen and flufenamic acid also bind and activate PPAR-γ.

In addition to the role in adipogenesis and inflammation, PPAR-γ is highly expressed in normal large intestine and in breast, colon, and prostate cancer. PPAR-γ agonists such as troglitazone and 15deoxy-PGJ$_2$ were able to induce differentiation and apoptosis in tumor cells, suggesting that PPAR-γ suppresses tumor cell proliferation. Indomethacin was shown to reduce the clonogenic activity of prostate cancer cells and increased the antiproliferative effect of 5-fluorouracil in colon cancer cells. This effect was supposed to be mediated through activation of PPAR-γ.

PPAR-δ is a nuclear transcription factor that is activated by prostacyclin. Inhibition of COX activity with aspirin or other NSAIDs causes inhibition of PPAR-δ, which has been identified as one of the downstream targets of the WNT-β-catenin pathway (He et al. 1999). This pathway plays a crucial role in embryonic development and carcinogenesis. PPAR-δ expression is normally controlled by the APC tumor suppressor. However, during colon carcinogenesis, APC function is almost always lost, leading to a dysfunction of β-catenin and uncontrolled PPAR-δ expression. This is considered as a crucial initiating step in tumor transformation. The suppression of PPAR-δ activity by various NSAIDs, including aspirin and sulindac, can compensate for the loss of APC or β-catenin dysfunction and thereby reduce colon carcinogenesis. Hence, the inhibition of PPAR-δ activity may contribute to the chemopreventive effects of some NSAIDs.

Other Targets

Intracellular Carbonic Anhydrase

Carbonic anhydrases play an important role in the extracellular acidification. Several studies suggest a possible involvement of these enzymes in tumor progression, resulting from the acidic extracellular pH. Celecoxib and valdecoxib inhibit carbonic anhydrases. This effect is supposed to depend on their sulfonamide structure and is therefore not shared by rofecoxib or etoricoxib.

Acid Sensing Ion Channels (ASICs)

H^+-gated currents mediated by acid sensing ion channels (ASICs) are involved in acidosis which occurs under inflammatory conditions and in tumors. Various NSAIDs including aspirin, salicylic acid, flurbiprofen, ibuprofen, and diclofenac are inhibitors of these H^+-gated channels, thereby inhibiting acid-induced pain reaction and inflammatory responses.

Ca^{2+} Release

Treatment with some COX-2 inhibitors increased intracellular calcium levels in osteoblasts, PC-12, and HUVEC cells. This effect might be mediated

by a block of endoplasmic reticulum Ca^{2+}-ATPases and may increase the risk of cardiovascular events in predisposed patients.

References

Grosch, S., Tegeder, I., Niederberger, E., et al. (2001). COX-2 independent induction of cell cycle arrest and apoptosis in colon cancer cells by the selective COX-2 inhibitor celecoxib. *The FASEB Journal, 15*, 2742–2744.

Grosch, S., Tegeder, I., Schilling, K., et al. (2003). Activation of c-Jun-N-terminal-Kinase is crucial for the induction of a cell cycle arrest in human colon carcinoma cells caused by flurbiprofen enantiomers. *The FASEB Journal, 17*, 1316–1318.

Grösch, S., Maier, T. J., Schiffmann, S., & Geisslinger, G. (2006). Cyclooxygenase-2 (COX-2)-independent anticarcinogenic effects of selective COX-2 inhibitors. *Journal of the National Cancer Institute, 98*, 736–747.

Hawley, S. A., Fullerton, M. D., Ross, F. A., et al. (2012). The ancient drug salicylate directly activates AMP-activated protein kinase. *Science, 336*, 918–922.

He, T. C., Chan, T. A., Vogelstein, B., et al. (1999). PPARdelta is an APC-regulated target of nonsteroidal anti-inflammatory drugs. *Cell, 99*, 335–345.

Hsu, A. L., Ching, T. T., Wang, D. S., et al. (2000). The cyclooxygenase-2 inhibitor celecoxib induces apoptosis by blocking Akt activation in human prostate cancer cells independently of Bcl-2. *The Journal of Biological Chemistry, 275*, 11397–11403.

Kopp, E., & Ghosh, S. (1994). Inhibition of NF-Kappa B by sodium salicylate and aspirin. *Science, 265*, 956–959.

Niederberger, E., Tegeder, I., Schafer, C., et al. (2003). Opposite effects of rofecoxib on nuclear factor-kappa B and activating protein-1 activation. *The Journal of Pharmacology and Experimental Therapeutics, 304*, 1153–1160.

Tegeder, I., Pfeilschifter, J., & Geisslinger, G. (2001). Cyclooxygenase-independent actions of cyclooxygenase inhibitors. *The FASEB Journal, 15*, 2057–2072.

Thun, M. J., Namboodiri, M. M., & Heath, C. W., Jr. (1991). Aspirin use and reduced risk of fatal colon cancer. *The New England Journal of Medicine, 325*, 1593–1596.

Tsujii, M., Kawano, S., Tsuji, S., et al. (1998). Cyclooxygenase regulates angiogenesis induced by colon cancer cells. *Cell, 93*, 705–716.

Tsutsumi, R., Ito, H., Hiramitsu, T., Nishitani, K., Akiyoshi, M., Kitaori, T., Yasuda, T., & Nakamura, T. (2008). Celecoxib inhibits production of MMP and NO via down-regulation of NF-kappaB and JNK in a PGE2 independent manner in human articular chondrocytes. *Rheumatology International, 28*, 727–736.

Wechter, W. J., Kantoci, D., Murray, E. D., Jr., et al. (1997). R-flurbiprofen chemoprevention and treatment of intestinal adenomas in the APC(Min)/+ mouse model: Implications for prophylaxis and treatment of colon cancer. *Cancer Research, 57,* 4316–4324.

Yin, M. J., Yamamoto, Y., & Gaynor, R. B. (1998). The anti-inflammatory agents aspirin and salicylate inhibit the activity of I(kappa)B kinase-beta. *Nature, 396,* 77–80.

NSAIDs, Mode of Action

Maria Burian
Department of General, Visceral and
Transplantation Surgery, University of
Heidelberg, Heidelberg, Germany

Synonyms

Aspirin-like drugs; COX; Cyclooxygenase; Nonselective COX inhibitors; Nonsteroidal anti-inflammatory drugs; NSAIDs; PGHS; Prostaglandin H synthase

Definition

Nonsteroidal anti-inflammatory drugs (NSAIDs) are among the most widely prescribed and used drugs for the management of pain, especially of pain in inflammatory conditions. Despite the wide use of NSAIDs over the last century, little was known on the mode of actions of these drugs for a long time. Initially, the principle mode of the antinociceptive action of NSAIDs was related to their anti-inflammatory activity and attributed to the inhibition of the production of prostaglandins at the peripheral site of inflammation (peripheral mode of action) (Vane 1971). The traditional belief in the exclusively peripheral action of NSAIDs, however, has been challenged by the growing evidence showing the dissociation between the anti-inflammatory and antinociceptive effects of NSAIDs (McCormack and Brune 1991). This is the basis for the hypothesis of additional antinociceptive mechanisms existing with NSAIDs, where the inhibition of prostaglandin synthesis in CNS appears to

be universally applicable for all NSAIDs (central mode of action).

While inhibiting prostanoid synthesis, NSAIDs do not typically elevate the normal pain threshold, but mainly normalize the increased pain threshold observed in ► inflammatory pain (► hyperalgesia), so that their antinociceptive effect should rather be defined as antihyperalgesic instead of analgesic.

Characteristics

Prostanoid Synthesis

Despite the diverse chemical structure of aspirin-like drugs, all NSAIDs bear a common pharmacological property of inhibiting the formation of prostanoids. Prostanoids are formed by most cells and act as lipid mediators. They are synthesized from membrane-released arachidonic acid mobilized by phospholipases (PLA_2) when cells are activated by mechanical trauma, cytokines, growth factors, etc. (Fig. 1). Conversion of arachidonic acid to prostanoid end products occurs by cyclooxygenases (COX), an enzyme also known as prostaglandin H synthase (PGHS), at two different sites of the enzyme. It is initially cyclized and oxidized to the endoperoxide PGG_2 at the cyclooxygenase site of the COX. This product is then reduced to a second endoperoxide, PGH_2, at the peroxidase site of the COX enzyme. Subsequent formation of prostaglandin end products from PGH_2 depends on the presence of the specific synthase enzymes that produce functionally important prostanoids PGD_2, PGE_2, $PGF_{2\alpha}$, PGI_2 (prostacyclin), and TXA_2 (thromboxane), which mediate their effects through the specific receptors: PGD_2 (DP_1, DP_2 receptors), PGE_2 (EP_1, EP_2, EP_3, EP_4 receptors), $PGF_{2\alpha}$ (FP receptor), PGI_2 (IP receptors), and TXA_2 (TP_α and TP_β receptors).

Inhibition of Cyclooxygenases by NSAIDs

NSAIDs inhibit the formation of prostanoids by several different effects on COX, including irreversible inactivation of COX (e.g., aspirin) or reversible competitive inhibition

NSAIDs, Mode of Action, Fig. 1 Prostanoids are synthesized from membrane-released arachidonic acid mobilized by phospholipases (PLA_2). Conversion of arachidonic acid to prostanoid end products occurs by COX-1 and COX-2 at two different sites of the enzyme. It is initially cyclized and oxidized to the endoperoxide PGG_2 at the cyclooxygenase site of the COX and then reduced to a second endoperoxide, PGH_2, at the peroxidase site of the COX enzyme. Tissue-specific isomerases catalyze subsequent formation of prostaglandin end products including PGD_2, PGE_2, $PGF_{2\alpha}$, PGI_2 (prostacyclin), and TXA_2 (thromboxane). These prostanoids exert their effects by acting through the specific receptors: PGD_2 (DP_1, DP_2 receptors), PGE_2 (EP_1, EP_2, EP_3, EP_4 receptors), $PGF_{2\alpha}$ (FP receptor), PGI_2 (IP receptors), and TXA_2 (TP_α and TP_β receptors). The "relaxant" receptors IP, DP_1, EP_2, and EP_4 signal through Gs-mediated increases in intracellular cyclic adenosine monophosphate (cAMP). The "contractile" receptors EP_1, FP, and TP signal through Gq-mediated increases in intracellular calcium. The EP_3 receptor is regarded as an "inhibitory" receptor that couples to Gi and decrease cAMP formation

(e.g., ibuprofen). COX is represented by two isoforms, COX-1 and COX-2, which are membrane-associated enzymes with a 63 % amino acid sequence similarity. The identification of two isoforms of COX in the early 1990s offered a simple and attractive hypothesis: COX-1, being found in almost all cells, is the constitutive "house-keeping" enzyme responsible for production of basal "beneficial" PGs, which are vital for protecting the stomach through mucus production or maintenance of renal blood flow. In contrast, COX-2, in which expression is low or undetectable in most cells but increased dramatically in a variety of pathological conditions, is the inducible enzyme responsible for "pathological" production of prostanoids in different conditions ranging from inflammation to cancer.

NSAIDs are nonselective inhibitors of COX (both COX-1 and COX-2). Initially, the favorable anti-inflammatory, antinociceptive, and antipyretic effects of NSAIDs were solely attributed to the inhibition of COX-2, while the concomitant inhibition of COX-1 was supposed to lead to adverse reactions of the drugs (gastrointestinal and renal toxicities). With time, especially along with the introduction/gaining experience of selective COX-2 inhibitors (e.g., celecoxib, rofecoxib), this concept turned out to be more complicated than initially thought, indicating that both COX-1 and COX-2 have physiological and pathological roles, so that the inhibition of both isoforms may be responsible for favorable and unfavorable pharmacological effects of NSAIDs.

Cyclooxygenases and Prostanoids at the Peripheral Site

In injured tissue, COX-2 is the predominant iso-form expressed and a main source of prostanoids during inflammation. The significant induction of COX-2 is found in activated polymorphonuclear leucocytes, phagocytosing mononuclear cells, and fibroblasts, which are abundantly present at the site of inflammation. However, COX-1 is also involved in the modulation of the inflammatory response and is mainly increased in circulating monocytes and stimulated mast cells at the early inflammatory phase. Thus, both COX isoforms are involved in the inflammatory reaction in the periphery and may contribute to the genera-tion and maintenance of inflammatory pain. The earliest prostanoid response is dependent on COX-1, but COX-2 becomes a major source of prostanoids along with the progression of inflam-mation. NSAIDs-mediated inhibition of COX-1 and COX-2 at the site of inflammation results in the attenuation of peripheral ▶ sensitization asso-ciated with inflammatory pain (Fig. 2).

Cyclooxygenase and Prostanoids in CNS

In contrast to the periphery, in the CNS both COX-1 and COX-2 are expressed constitutively (Beiche et al. 1996). These isoforms are present in neurons and nonneuronal elements of the spinal cord and brain (Maihofner et al. 2000). The peripheral inflammatory reactions associated with tissue injury result in the release of proinflammatory cytokines, such as IL-1β, which may enhance the upregulation of COX-2 in the CNS (Samad et al. 2001). This upregulation is paralleled to the substantial elevation of prostanoids in the cerebrospinal fluid and typical nociceptive behavior of the animals in experimental pain models. Therefore, COX-2 appears to be mainly responsible for the central processing of pain after peripheral inflammation. COX-1, however, can also become the source of spinal prostaglandins in response to peripheral inflammation under specific conditions, as has been particularly demonstrated in COX-2-deficient knockout mice. NSAIDs-mediated inhibition of COX-1 and COX-2 in CNS results in the attenuation

NSAIDs, Mode of Action, Fig. 2 The inhibition of prostaglandin synthesis by NSAIDs takes place at both the site of peripheral inflammation (*1*) and at the spinal level (*2*), indicating that peripheral and central mecha-nisms may be responsible for their antinociceptive action. NSAIDs-mediated inhibition of COX-1 and COX-2 at the peripheral site results in the attenuation of peripheral sensitization, whereas inhibition of the upregulated COX-2 in CNS leads to the attenuation of central hyperalgesia. Peripheral and central hyperalgesia are hallmarks of inflammatory pain

of central sensitization associated with inflamma-tory pain (Fig. 2).

Mechanisms of Prostaglandin-Mediated Hyperalgesia

Prostaglandins are potent sensitizing agents, which are able to modulate multiple sites along the nociceptive pathway enhancing transduction (peripheral sensitization), as well as transmission (central sensitization), of the nociceptive infor-mation (Woolf 1983).

Peripheral Sensitization

Direct and indirect mechanisms of the peripheral sensitizing action of prostanoids have been suggested. The direct effects are mediated by their action upon prostaglandin receptors and modulation of ion channels in ▶ nociceptors. The indirect effects are produced through enhancing the sensitivity of nociceptors to nox-ious agents, including heat and bradykinin. Both direct and indirect sensitizing effects of prostanoids lead to the enhanced transduction of the nociceptive information and manifest as peripheral hyperalgesia. NSAIDs-mediated inhibition of prostanoid synthesis in the periphery results in the attenuation of peripheral hyperalgesia associated with inflammatory pain (peripheral ▶ antihyperalgesic effect).

Central Sensitization

The sensitizing effects of prostanoids in CNS are mediated by their action on presynaptic and postsynaptic membranes of the primary afferent synapse in the dorsal horn of the spinal cord. Acting via prostanoid receptors located on the presynaptic membrane, prostanoids may cause the enhancement of nociception via facilitation of the spinal release of the excitatory neurotransmitter ▶ glutamate and neuropeptides (▶ substance P and/or ▶ calcitonin-gene-related peptide). At the postsynaptic level, prostanoids can directly activate deep dorsal horn neurons and/or block the inhibitory glycinergic neurotransmission onto dorsal horn neurons. All this leads to the enhanced transmission of the nociceptive activity to the brain and manifests as central hyperalgesia. NSAIDs-mediated inhibition of prostanoid synthesis in CNS results in the attenuation of central hyperalgesia associated with inflammatory pain (central antihyperalgesic effect).

Central and Peripheral Antihyperalgesic Effects of NSAIDs

From the pharmacological point of view, the contribution of the central versus peripheral mechanisms to the overall antihyperalgesic effects of NSAIDs depends on the following:

- The site of drug delivery (e.g., systemic, local-peripheral, epidural, spinal intracerebroventricular)
- Uptake and distribution from the site of drug delivery, as determined by factors including the drug's physical and chemical properties, specific transport mechanisms, local and systemic blood flow, and tissue barriers to drug permeation, such as the blood–brain barrier

One of the accepted approaches to prove that NSAIDs act upon CNS to alleviate pain is an assessment of their antinociceptive activity,

following a direct application of NSAIDs to spinal or supraspinal structures. This approach has gained its particular importance in behavioral animal studies. The intrathecal delivery of various NSAIDs has been shown to be effective in the reduction of behavioral hyperalgesia in several animal models of acute, short-term, and long-term inflammatory pain (Brune et al. 1991). The antihyperalgesic effects have been observed with doses that are significantly lower than those needed to produce the similar degree of antinociception by systemic administration. Clinical relevance of the central antinociceptive mechanisms of the antinociceptive action of NSAIDs has been demonstrated in patients with intractable pain due to various types of cancer conditions. In these patients, intrathecal administration of small doses of lysine acetylsalicylate (equivalent to acetylsalicylic acid 500 µg/kg) has been shown to bring rapid and prolonged pain relief (Pellerin et al. 1987).

The evidence for a clinically relevant peripheral antinociceptive action has been obtained with locally applied NSAIDs, where the effective antinociceptive effect of NSAIDs versus placebo has been established with intra-articular and topical applications of NSAIDs (Romsing et al. 2000).

The contribution of central versus peripheral mechanisms to the total antihyperalgesic effect of NSAIDs has been studied in the human experimental pain model of "freeze lesion." The experimental hyperalgesia in this model was produced by short-lasting freezing of volunteer's skin. The relative contribution of the central component of orally administered diclofenac has accounted for approximately 40 % of the total antihyperalgesic efficacy of the drug (Burian et al. 2003).

Conclusion

NSAIDs are potent antinociceptive agents, whose efficacy in reducing pain is widely recognized in various pain conditions, including postsurgical pain and persistent pain states, such as arthritis and cancer. Although NSAIDs have long been used in clinical practice, the mechanism of their antihyperalgesic action remains controversial. It appears that the inhibition of prostaglandin synthesis by NSAIDs takes place both at the site of peripheral inflammation and at the spinal level, indicating that peripheral and central mechanisms are responsible for their antinociceptive action. The contribution of peripheral and central COX-dependent mechanisms to the overall antinociceptive action of NSAIDs is individual for each drug and is dependent on the site of drug delivery, as well as pharmacokinetic characteristics of the drug determining its penetration to the sites of action (e.g., peripheral and spinal COX).

References

Beiche, F., Scheuerer, S., Brune, K., et al. (1996). Upregulation of cyclooxygenase-2 mRNA in the rat spinal cord following peripheral inflammation. *FEBS Letters, 390*, 165–169.

Brune, K., Beck, W. S., Geisslinger, G., et al. (1991). Aspirin-like drugs may block pain independently of prostaglandin synthesis inhibition. *Experientia, 47*, 257–261.

Burian, M., Tegeder, I., Seegel, M., et al. (2003). Peripheral and central antihyperalgesic effects of diclofenac in a model of human inflammatory pain. *Clinical Pharmacology and Therapeutics, 74*, 113–120.

Maihofner, C., Tegeder, I., Euchenhofer, C., et al. (2000). Localization and regulation of cyclooxygenase-1 and -2 and neuronal nitric oxide synthase in mouse spinal cord. *Neuroscience, 101*, 1093–1108.

McCormack, K., & Brune, K. (1991). Dissociation between the antinociceptive and anti-inflammatory effects of the non-steroidal anti-inflammatory drugs. A survey of their analgesic efficacy. *Drugs, 41*, 533–547.

Pellerin, M., Hardy, F., Abergel, A., et al. (1987). Chronic refractory pain in cancer patients. Value of the spinal injection of lysine acetylsalicylate. 60 cases. *Presse Médicale, 16*, 1465–1468.

Romsing, J., Moiniche, S., Ostergaard, D., et al. (2000). Local infiltration with NSAIDs for postoperative analgesia: Evidence for a peripheral analgesic action. *Acta Anaesthesiologica Scandinavica, 44*, 672–683.

Samad, T. A., Moore, K. A., Sapirstein, A., et al. (2001). Interleukin-1beta-mediated induction of cox-2 in the CNS contributes to inflammatory pain hypersensitivity. *Nature, 410*, 471–475.

Vane, J. R. (1971). Inhibition of prostaglandin synthesis as a mechanism of action for aspirin-like drugs. *Nature: New Biology, 231*, 232–235.

Woolf, C. J. (1983). Evidence for a central component of post-injury pain hypersensitivity. *Nature, 306*, 686–688.

NSAIDs, Pharmacogenetics

Carsten Skarke[1] and Patrick Schuss[2]
[1]Institute for Translational Medicine and Therapeutics (ITMAT), University of Pennsylvania Perelman School of Medicine, Philadelphia, PA, USA
[2]Department of Neurosurgery, University-Hospital Bonn, Bonn, Germany

Synonyms

Inherited variability of drug response

Definition

Pharmacogenetics seeks to explore how genetic variants influence the pharmacokinetic and pharmacodynamic properties of a given drug by determining how mutations in the genes which encode drug-metabolizing enzymes, drug targets, and drug transporters influence drug response.

Characteristics

Nonsteroidal anti-inflammatory drugs (NSAID) block the formation of prostaglandins by inhibiting the rate-limiting cyclooxygenase (COX) enzymes, COX-1 and COX-2, also known as prostaglandin H_2 synthases (PGHS1 and PGHS2). Since prostaglandins participate in mediating the inflammatory response, the pharmacological activity of NSAIDs consists mainly of antinociceptive, anti-inflammatory, and antipyretic properties.

Variation of this pharmacological activity can arise as a basic principle from mutations in proteins, which (i) influence the bioavailability of a drug, (ii) vary the binding affinity to the drug target, or (iii) modify drug elimination. Since NSAIDs possess a high solubility and high permeability (Yazdanian et al. 2004) and were not found to be substrate of drug efflux transporters, it is unlikely that the disposition varies on a genetic basis. Far more expected are mutations in the cyclooxygenase enzymes as drug targets and the metabolizing enzymes as determinant of drug elimination. The majority of such mutations are single-nucleotide polymorphisms (SNP), which consist of an exchange of a nucleotide. If this SNP is located within the coding region of the gene, the exon, the change in nucleotide sequence is mostly but not always translated to an altered amino acid sequence of the resulting protein.

The field of pharmacogenetics started off with the promise to tailor an analgesic therapy to the genetic makeup of the individual patient, both with regard to harvesting the intended drug effects while limiting potential adverse effects. This remains a challenge for pain and its analgesic treatment with NSAIDs. Pain itself varies substantially between patients, as do the pharmacodynamic effects of NSAIDs. The latter is impressively demonstrated in a mechanistic study by Fries et al. in healthy volunteers (Fries et al. 2006). The observed variability in the ex vivo indices of COX-1 and COX-2 suppression, thromboxane A_2 and prostaglandin E_2, respectively, in response to single therapeutic doses of celecoxib or rofecoxib, was attributed in about 30 % to between-subject differences. Arguably, this is a strong indicator for genetic differences, and indeed a mutation in the COX-1 and CYP2C9*2 genes, a major metabolizing enzyme of celecoxib, was associated with a change in drug response. On the other hand, this study brought up a crucial aspect of clinical trial design: Choosing the sample size adequately powered to detect pharmacogenetics differences. In a subset, the authors exposed healthy subjects five replicate times to celecoxib, rofecoxib, or placebo and noted a within-subject variability in thromboxane A_2 and prostaglandin E_2 measurements of the same magnitude as measured cross-sectional between subjects for the three interventions (Fries et al. 2006).

Note: The reference SNP ID numbers, "rs," are provided where retrievable.

Polymorphisms in the Cyclooxygenase-1 Gene

The COX-1 gene is located on chromosome 9 and consists of 11 exons. Although nearly 950 genomic variants are listed in the NCBI dbSNP database as of January 2013, only a limited number have been associated so far with a potential functional consequence (Halushka et al. 2003; Lee et al. 2007).

Using an oxygen consumption assay as in vitro indicator of metabolic activity, several variants in the COX-1 gene, R53H, R78W, K185T, G230S, and L237M (rs5789), demonstrated about 50–70 % less activity for the altered enzyme compared to control. Enzymes formed from COX-1 carrying either one of the two variants P17L (rs3842787) or G230S but not from control facilitated inhibition by indomethacin, evident as a decrease in IC_{50} values. Further studies identified a disruption of the active conformation of COX-1 as a potential mechanism for one of the variants, the G230S (Lee et al. 2007).

An earlier computerized approach evaluated the likelihood that a resulting amino acid exchange leads to a phenotypic alteration (Ulrich et al. 2002). Qualifying candidates were in exon 2, the SNPs 22C>T (arginine to tryptophan at position 8, R8W) and 50C>T (proline to leucine at position 17, P17L), and in exon 7 the SNPs 688G>A (glycine to serine at position 230, G230S) and 709C>A (leucine to methionine at position 237, L237M). Using human platelets as a system to study COX-1 activity, heterozygous carriers of the SNP 50C>T showed greater inhibition of prostaglandin H_2 production after aspirin exposure than carriers of the CC50 genotype. The possible mechanism for the increased sensitivity to aspirin was seen in a decrease of COX-1 enzyme levels. Since the 50C>T was found in complete linkage disequilibrium with −842A>G, a mutation in the COX-1 promoter, this SNP may also account for the functional impact, possibly because gene transcription is repressed (Halushka et al. 2003). The high frequency for the minor allele of 50C>T in Europeans, up to 18 %, raises the expectation that genotype-phenotype associations should easily become evident in even small cohorts under study. Indeed, healthy carriers of the 50C>T variant of the COX-1 enzyme showed less of an inhibitory thromboxane A_2 response to celecoxib and rofecoxib, supporting the functional role for this polymorphism (Fries et al. 2006).

Japanese researchers recently associated the SNP $-1676T>C$ (rs1330344) in the COX-1 gene promoter with a higher incidence of gastroduodenal ulcers among NSAID users (Arisawa et al. 2007). In a later study, the $-1676T$ allele of this SNP was linked to an increased risk to develop functional dyspepsia among Japanese women suffering from an epigastric pain syndrome (EPS) (Arisawa et al. 2008). Both studies suggest a functional relationship between the SNP $-1676T>C$ and NSAID use, pain and NSAID-related side effects, but since pharmacodynamic indices of cyclooxygenase enzyme activity were not assessed, this remains speculative.

In male Swedish septuagenarians, two SNPs in the COX-1 gene were linked to affect enzyme activity. Carriers of the T allele for the $4466A>T$ variant (rs10306135) as well as carriers of the C allele for variant $-8202C>T$ (rs883484) showed a gene-dose-dependent decrease in prostaglandin F_{2a} formation. Both SNPs were associated with a lower risk for cardiovascular disease; however, pain or NSAID use was not assessed (Helmersson et al. 2009).

In conclusion, some work has been accomplished to point at variants in the COX-1 gene that may alter analgesic drug response to NSAIDs, but overall this field seems still far away to include pharmacogenetics into the decision-making process of clinical pain management.

Over the past years, COX-1 gene variants have been associated with risk modulation in a number of diseases. Among these are ischemic stroke $(-1006G>A)$ (Lee et al. 2008) and colorectal tumors $(639C>A,$ rs5788) (Kury et al. 2008; Frank et al. 2010).

Polymorphisms in the Cyclooxygenase-2 Gene

The COX-2 gene (prostaglandin-endoperoxide synthase 2 (PTGS2) as common alias) on chromosome 1 displays a substantial number of genomic variants in the promoter and the ten exons. To date (January 2013), almost 500 different entries are listed in the NCBI dbSNP database (www.ncbi.nlm.nih.gov/SNP/). Allele frequencies remain unknown for the majority of these SNPs. For another large fraction, the frequency of the minor allele is so low, generally below 1 %, that an impact on drug action is likely to stay at an anecdotal level. A handful of SNPs have been consistently reported with higher allele frequency. Among those, several polymorphisms have been associated with modulating disease risk or outcome by altering the gene product (Skarke et al. 2007). In patients with lung cancer, carriers of the minor allele of the variant $837T>C$ (rs5275) were associated with a lower risk for severe pain (Reyes-Gibby et al. 2009), suggesting that pain intensity might be modulated by this genetic variant. Whether this polymorphism interferes with NSAIDs cannot be determined from this study as patients were managed with opioids.

One polymorphism in particular, the 765C (rs20417) in the COX-2 promoter, is drawing substantial interest initiated by early reports that the $G>C$ nucleotide exchange at position -765 causes variation in the transcription factor binding site of stimulatory protein 1 which results in lower promoter activity for carriers of the $-765C$ allele (Hernandez et al. 2004; Papafili et al. 2002). Although a study in patients with myocardial infarction has shown decreases COX-2 protein expression (Szczeklik et al. 2004), consistent with the expectation based on the proposed mechanism, other studies have seen either an increase (Cipollone et al. 2004) or no changes (Skarke et al. 2006) in gene transcription, protein expression, and lipopolysaccharide-stimulated prostaglandin E_2 (PGE_2) production in ex vivo blood monocytes. Notably, evidence is accumulating that the rs20417 polymorphism is associated with various disease risks, although some studies challenge this (Montali et al. 2010; Hegener et al. 2006). A brief overview is provided in Table 1.

A single study assessed whether the rs20417 polymorphism has any impact on the analgesic effects of NSAIDs. Lee et al. used removal of an impacted molar tooth as an acute pain model in otherwise healthy dental patients and found that the pain relief response varies by genotype.

NSAIDs, Pharmacogenetics, Table 1 Accumulating evidence associating the rs20417 polymorphism with disease risks

Disease	First author	Year	Reference
Asthma	Szczeklik	2004	(Szczeklik et al. 2004)
Alzheimer's disease	Abdullah	2006	(Abdullah et al. 2006)
	Fehér	2010	(Feher et al. 2010)
	Listì	2010	(Listi et al. 2010)
Ischemic stroke	Maguire	2011	(Maguire et al. 2011)
	Lee	2008	(Lee et al. 2008)
	Kohsaka	2008	(Kohsaka et al. 2008)
	Colaizzo	2006	(Colaizzo et al. 2006)
	Cipollone	2004	(Cipollone et al. 2004)
	Sharma V	2013	(Sharma et al. 2013)
Coronary artery disease	Ol	2011	(Ol et al. 2011)
	Li	2009	(Li et al. 2009)
	Kohsaka	2008	(Kohsaka et al. 2008)
Myocardial infarction	Cipollone	2004	(Cipollone et al. 2004)
Diabetes mellitus II	Konheim	2003	(Konheim and Wolford 2003)
Glioma (WHO grades II–IV)	Amirian	2009	(Kim et al. 2009)
Colorectal tumors	Daraei	2012	(Daraei et al. 2012)
	Ueda	2008	(Ueda et al. 2008)
	Ulrich	2005	(Ulrich et al. 2005)
Esophageal tumors	Liang	2011	(Liang et al. 2011)
	Zhang	2005	(Zhang et al. 2005)
Osteoarthritis	Schneider	2011	(Schneider et al. 2011)

In detail, rofecoxib was more effective than ibuprofen 48 h after surgery in carriers of the −765G allele, suggesting that the observed higher expression of COX-2 forms the basis of improved efficacy for the selective COX-2 inhibitor (Lee et al. 2006). Publication of this study was accompanied with an optimistic commentary about individualizing analgesic therapy based on cyclooxygenase gene variants (Halushka and Halushka 2006). Further breakthroughs, however, have not been achieved so far, so that the personalization of NSAID analgesia based on pharmacogenetics information still lacks momentum.

Genome-Wide Association Studies (GWAS)

Over the past years, genome-wide association studies have become a powerful tool to scale up the search for associations between genetic loci and disease phenotypes. To date, the online repository for GWAS at the National Human Genome Research Institute (NHGRI), NIH, lists a total of two GWAS for the search string "NSAID and/or pain" (Kim et al. 2009; Peters et al. 2012). Only one of these was designed to measure pain relief in response to a nonselective COX inhibitor (Kim et al. 2009). The authors used molar tooth extraction as an acute postsurgical pain model in young dental patients, 17–35 years of age. Onset of NSAID analgesia with i.v. ketorolac tromethamine was assessed by visual analog scale in European Americans stratified for gender. Candidate SNPs in any gene of the cyclooxygenase pathway were not found. Only a single SNP (rs2562456) in a yet unannotated gene (*LOC400680*) withstood correction for multiple comparisons. This SNP existed in linkage disequilibrium with a gene (*ZNF429*) encoding a zinc finger protein which has a relevance in DNA binding. The authors note that the biological function of *ZNF429* may play a role in forming an individual's response to analgesic drugs. As the sample size was small in this

GWAS, a little over 100, and ethnic diversity was constrained, it remains to be seen if future endeavors in this field will produce candidate mutations within the cyclooxygenase pathway.

After all, the field might benefit from efforts in other domains. Designed primarily to explore associations between SNPs in candidate genes and metabolic, inflammatory, and cardiovascular phenotypes, the IBC array (ITMAT-Broad-CARe array) identifies among its 45,000 SNPs also mutations located in genes of the cyclooxygenase pathway (Kim et al. 2009). Implementation in large cohorts well characterized for pain and analgesic intake is now the challenging next step to adapt this methodological platform to pain and NSAID use.

Polymorphisms in Metabolizing Enzyme Genes

Many NSAIDs are hepatically metabolized by the cytochrome P450 (CYP) system. Substrates for one of the most important isoforms, CYP2C9, are diclofenac, ibuprofen, flurbiprofen, naproxen, piroxicam, tenoxicam, meloxicam, mefenamic acid, and celecoxib. Polymorphisms in the CYP2C9 gene are recognized to account for variable NSAID pharmacokinetics. Among numerous mutations, the alleles CYP2C9*2 (Cys144, Ile359, Asp360) and CYP2C9*3 (Arg144, Leu359, Asp360) are of particular importance in the Caucasian population due to the reduced intrinsic metabolic activity combined with a high allele frequency of 8–14 % and 4–16 %, respectively. For flurbiprofen, which is exclusively metabolized by CYP2C9, most of the pharmacokinetic variability could be explained by the CYP2C9 genotype with most pronounced effects in carriers of the CYP2C9*3 allele (Lee et al. 2003). That the CYP2C9*3 allele is mostly responsible for the interindividual variability was also seen when the oral clearance of celecoxib was more than twofold reduced in homozygous carriers of the CYP2C9*3 allele as compared to the nonmutated volunteers, while no significant influence was determined for the CYP2C9*2 allele (Kirchheiner et al. 2003). In the presence of nonfunctional CYP2C9 alleles, other

cytochrome isoforms (CYP3A4) might compensate by increasing their contribution. This might be the reason why no evidence was seen that CYP2C9 mutations were a determinant for the diclofenac-induced hepatotoxicity (Aithal et al. 2000). Significant pharmacokinetic differences were also seen for racemic and S-(+)-ibuprofen between carriers of one or two CYP2C9*3 alleles and nonmutated alleles (Kirchheiner et al. 2002), while the CYP2C9*2 variant only displayed a compromised metabolic activity when found in combination with the CYP2C8*3 (K139, R399) mutation (Garcia-Martin et al. 2004). In fact, the two alleles CYP2C9*2 and CYP2C8*3 were shown to be in linkage disequilibrium (Yasar et al. 2002).

For celecoxib, higher plasma concentrations associated with carriers of CYP2C9*2; however, a significant impact on drug response was not observed (Fries et al. 2006). Important to note is that this study in healthy volunteers was not designed to primarily assess drug-SNP interactions.

A cross-sectional study supports the concept that loss of NSAID metabolizing function in carriers of CYP2C8*3 and CYP2C9*2 is associated with a higher incidence of acute gastrointestinal bleeding. Consistent with this, these CYP mutations were not overrepresented in patients with gastrointestinal bleeding but receiving NSAIDs which were not substrates of CYP2C8 and CYP2C9 (Blanco et al. 2008). However, a systematic review on the association between increase NSAID-induced gastrointestinal bleeds and variants in the metabolizing cytochrome enzymes concluded that the evidence to date is too scarce to prove a causal relationship (Estany-Gestal et al. 2011).

The metabolism of NSAIDS further involves glucuronidation by uridine 5′-diphosphoglucose glucuronosyl transferase (UGT) enzymes. Since most NSAIDs first undergo CYP450-mediated transformation to inactive metabolites, it is rather unlikely that UGT alleles with a compromised catalytic activity cause a change in drug response due to drug accumulation. However, this perception may be challenged by rofecoxib, which after

UGT2B15-mediated glucuronidation with minor contribution of UGT2B7 and UGT1A9 (Zhang et al. 2003) and deglucuronidation in the lower gastrointestinal tract may cycle enterohepatically and reappear in the plasma again as rofecoxib. This explained the observed second maximum concentration peak in the concentration-time curves (Baillie et al. 2001) and serves as an example that mutations rendering the UGT less metabolic active could gain clinical importance in special circumstances. In support of a potential clinical role for variants in the UGT2B7 gene is a study in patients with diclofenac-induced hepatotoxicity after medication with this NSAID over 4 years on average. In these patients, the UGT2B7*2 allele occurred with a higher frequency than in controls. The authors note that UGT2B7 is part of other variant alleles, IL-4, IL-10, and ABCC2, commonly associated with increased risk for liver toxicity (Daly et al. 2007). Future work is likely to address the contribution of these individual variants to the genetic susceptibility for diclofenac-related liver injury.

Substrates of the N-acetyltransferase 2 (NAT2) show three different acetylation phenotypes depending on the possession of two nonmutated alleles (fast), two mutated alleles (slow), or one nonmutated allele combined with a mutated allele (intermediate). NAT2 plays an important role in the detoxification of sulfasalazine metabolites; hence accumulation in slow acetylator genotypes was associated with the onset of adverse reactions such as infectious mononucleosis-like syndrome (Ohtani et al. 2003), acute pancreatitis (Tanigawara et al. 2002), discoid (Sabbagh et al. 1997), systemic lupus erythematosus (Gunnarsson et al. 1997), or allergic reactions (Tanaka et al. 2002). Metamizole (dipyrone) also qualifies as a NAT2 substrate, but differences in drug response according to the acetylation phenotype have not yet been reported. The distribution of the polymorphic alleles in the NAT2 gene is an example for the relevance of the ethnic background, since frequency of genotypes associated with fast or intermediate acetylation ranges from approx.

40 % in the European population to approx. 90 % in Japanese.

Impact of Nonfunctional Alleles on Drug-Drug Interaction

A patient genotyped as a compound carrier of nonfunctional CYP2C9*2/*3 alleles presented with normal INR values after therapeutic warfarin dosing. But when a concomitant analgesic therapy was introduced with celecoxib, the INR rapidly increased (>10) with extensive ecchymosis (Malhi et al. 2004). The impaired warfarin-metabolizing capacity of the CYP2C9*2/*3 alleles had no clinically significant effect on the INR, but when these CYP2C9*2/*3 alleles were challenged by the second substrate celecoxib, which in addition exhibits a high affinity for CYP2C9, the metabolizing rate of (S)-warfarin via CYP2C9 rapidly decreased, while the metabolism of the two to five times less potent (R)-warfarin by CYP 1A2 and 2C19 was not affected. This observation should represent a possible mechanism, how a drug interaction may be elicited in the drug-metabolizing system. However, in this particular observation, this has to be further elucidated, since the warfarin-celecoxib interaction was not seen in healthy volunteers (Karim et al. 2000).

Celecoxib was shown to inhibit the CYP2D6-mediated metabolism of metoprolol in healthy volunteers with a more pronounced rise in the plasma concentration-time profile in carriers of two functional alleles as compared to one allele. Such an effect is not anticipated with two mutated alleles leading to a minimum of CYP2D6 catalytic function (poor metabolizer) (Werner et al. 2003). Although increasing information is available of such induced drug interactions, further research may elucidate more, since a considerable number of drugs are listed as CYP2D6 substrates. More information is available on the P450 Drug Interaction Table, Division of Clinical Pharmacology, Department of Medicine, Indiana University School of Medicine (http://medicine.iupui.edu/clinpharm/ddis/table.aspx).

Summary

Little is known about the impact of pharmacogenetics on the therapeutic effect and adverse event profile of NSAIDs. Evidence suggests that polymorphisms in the major drug-metabolizing enzyme cytochrome P450 2C9 modify the pharmacokinetic profile of its substrates of which diclofenac, ibuprofen, naproxen, and celecoxib represent the most important ones. An increasing number of studies relate polymorphisms in the drug targets COX-1 and COX-2 to an altered drug response or safety profile, but due to the preliminary character of such data, it is yet too early to estimate the clinical relevance. Continued research efforts will have to show on what terms information on the genotype-phenotype relationship will enter the clinical domain to personalize pain management decisions.

References

Abdullah, L., et al. (2006). The cyclooxygenase 2-765 C promoter allele is a protective factor for Alzheimer's disease. *Neuroscience Letters, 395*, 240–243.

Aithal, G. P., Day, C. P., Leathart, J. B., & Daly, A. K. (2000). Relationship of polymorphism in CYP2C9 to genetic susceptibility to diclofenac-induced hepatitis. *Pharmacogenetics, 10*, 511–518.

Arisawa, T., et al. (2007). Association between genetic polymorphisms in the cyclooxygenase-1 gene promoter and peptic ulcers in Japan. *International Journal of Molecular Medicine, 20*, 373–378.

Arisawa, T., et al. (2008). Genetic polymorphisms of cyclooxygenase-1 (COX-1) are associated with functional dyspepsia in Japanese women. *Journal of Women's Health (Larchmt), 17*, 1039–1043.

Baillie, T. A., et al. (2001). Mechanistic studies on the reversible metabolism of rofecoxib to 5-hydroxyrofecoxib in the rat: Evidence for transient ring opening of a substituted 2-furanone derivative using stable isotope-labeling techniques. *Drug Metabolism and Disposition, 29*, 1614–1628.

Blanco, G., et al. (2008). Interaction of CYP2C8 and CYP2C9 genotypes modifies the risk for nonsteroidal anti-inflammatory drugs-related acute gastrointestinal bleeding. *Pharmacogenetics and Genomics, 18*, 37–43.

Cipollone, F., et al. (2004). A polymorphism in the cyclooxygenase 2 gene as an inherited protective factor against myocardial infarction and stroke. *JAMA, 291*, 2221–2228.

Colaizzo, D., et al. (2006). The COX-2 G/C-765 polymorphism may modulate the occurrence of cerebrovascular ischemia. *Blood Coagulation & Fibrinolysis, 17*, 93–96.

Daly, A. K., Aithal, G. P., Leathart, J. B., Swainsbury, R. A., Dang, T. S., & Day, C. P. (2007). Genetic susceptibility to diclofenac-induced hepatotoxicity: Contribution of UGT2B7, CYP2C8, and ABCC2 genotypes. *Gastroenterology, 132*, 272–281.

Daraei, A., Salehi, R., & Mohamadhashem, F. (2012). PTGS2 (COX2)-765 G>C gene polymorphism and risk of sporadic colorectal cancer in Iranian population. *Molecular Biology Reports, 39*, 5219–5224.

Estany-Gestal, A., Salgado-Barreira, A., Sanchez-Diz, P., & Figueiras, A. (2011). Influence of CYP2C9 genetic variants on gastrointestinal bleeding associated with nonsteroidal anti-inflammatory drugs: A systematic critical review. *Pharmacogenetics and Genomics, 21*, 357–364.

Feher, A., Juhasz, A., Rimanoczy, A., Kalman, J., & Janka, Z. (2010). Association study of interferon-gamma, cytosolic phospholipase A2, and cyclooxygenase-2 gene polymorphisms in Alzheimer disease. *The American Journal of Geriatric Psychiatry, 18*, 983–987.

Frank, B., Hoffmeister, M., Klopp, N., Illig, T., Chang-Claude, J., & Brenner, H. (2010). Polymorphisms in inflammatory pathway genes and their association with colorectal cancer risk. *International Journal of Cancer, 127*, 2822–2830.

Fries, S., et al. (2006). Marked interindividual variability in the response to selective inhibitors of cyclooxygenase-2. *Gastroenterology, 130*, 55–64.

Garcia-Martin, E., Martinez, C., Tabares, B., Frias, J., & Agundez, J. A. (2004). Interindividual variability in ibuprofen pharmacokinetics is related to interaction of cytochrome P450 2C8 and 2C9 amino acid polymorphisms. *Clinical Pharmacology and Therapeutics, 76*, 119–127.

Gunnarsson, I., Kanerud, L., Pettersson, E., Lundberg, I., Lindblad, S., & Ringertz, B. (1997). Predisposing factors in sulphasalazine-induced systemic lupus erythematosus. *British Journal of Rheumatology, 36*, 1089–1094.

Halushka, M. K., & Halushka, P. V. (2006). Toward individualized analgesic therapy: Functional cyclooxygenase 1 and 2 haplotypes. *Clinical Pharmacology and Therapeutics, 79*, 404–406.

Halushka, M. K., Walker, L. P., & Halushka, P. V. (2003). Genetic variation in cyclooxygenase 1: Effects on response to aspirin. *Clinical Pharmacology and Therapeutics, 73*, 122–130.

Hegener, H. H., Diehl, K. A., Kurth, T., Gaziano, J. M., Ridker, P. M., & Zee, R. Y. (2006). Polymorphisms of prostaglandin-endoperoxide synthase 2 gene, and prostaglandin-E receptor 2 gene, C-reactive protein concentrations and risk of atherothrombosis: A nested case-control approach. *Journal of Thrombosis and Haemostasis, 4*, 1718–1722.

Helmersson, J., Arnlov, J., Axelsson, T., & Basu, S. (2009). A polymorphism in the cyclooxygenase 1 gene is associated with decreased inflammatory prostaglandin F2alpha formation and lower risk of cardiovascular disease. *Prostaglandins, Leukotrienes, and Essential Fatty Acids, 80*, 51–56.

Hernandez, E. M., Chan, C. H., Xu, B., Notario, V., & Richert, J. R. (2004). Role of an internal ribosome entry site in the translational control of the human transcription factor Sp3. *International Journal of Oncology, 24*, 719–724.

Karim, A., et al. (2000). Celecoxib does not significantly alter the pharmacokinetics or hypoprothrombinemic effect of warfarin in healthy subjects. *Journal of Clinical Pharmacology, 40*, 655–663.

Kim, H., Ramsay, E., Lee, H., Wahl, S., & Dionne, R. A. (2009). Genome-wide association study of acute postsurgical pain in humans. *Pharmacogenomics, 10*, 171–179.

Kirchheiner, J., Meineke, I., Freytag, G., Meisel, C., Roots, I., & Brockmoller, J. (2002). Enantiospecific effects of cytochrome P450 2C9 amino acid variants on ibuprofen pharmacokinetics and on the inhibition of cyclooxygenases 1 and 2. *Clinical Pharmacology and Therapeutics, 72*, 62–75.

Kirchheiner, J., Stormer, E., Meisel, C., Steinbach, N., Roots, I., & Brockmoller, J. (2003). Influence of CYP2C9 genetic polymorphisms on pharmacokinetics of celecoxib and its metabolites. *Pharmacogenetics, 13*, 473–480.

Kohsaka, S., et al. (2008). Increased risk of incident stroke associated with the cyclooxygenase 2 (COX-2) G-765C polymorphism in African-Americans: The Atherosclerosis risk in communities study. *Atherosclerosis, 196*, 926–930.

Konheim, Y. L., & Wolford, J. K. (2003). Association of a promoter variant in the inducible cyclooxygenase-2 gene (PTGS2) with type 2 diabetes mellitus in Pima Indians. *Human Genetics, 113*, 377–381.

Kury, S., et al. (2008). Low-penetrance alleles predisposing to sporadic colorectal cancers: A French case-controlled genetic association study. *BMC Cancer, 8*, 326–341.

Lee, C. R., Pieper, J. A., Frye, R. F., Hinderliter, A. L., Blaisdell, J. A., & Goldstein, J. A. (2003). Differences in flurbiprofen pharmacokinetics between CYP2C9*1/*1, *1/*2, and *1/*3 genotypes. *European Journal of Clinical Pharmacology, 58*, 791–794.

Lee, Y. S., Kim, H., Wu, T. X., Wang, X. M., & Dionne, R. A. (2006). Genetically mediated interindividual variation in analgesic responses to cyclooxygenase inhibitory drugs. *Clinical Pharmacology and Therapeutics, 79*, 407–418.

Lee, C. R., et al. (2007). Identification and functional characterization of polymorphisms in human cyclooxygenase-1 (PTGS1). *Pharmacogenetics and Genomics, 17*, 145–160.

Lee, C. R., North, K. E., Bray, M. S., Couper, D. J., Heiss, G., & Zeldin, D. C. (2008). Cyclooxygenase polymorphisms and risk of cardiovascular events: The Atherosclerosis Risk in Communities (ARIC) study. *Clinical Pharmacology and Therapeutics, 83*, 52–60.

Li, W., et al. (2009). Cyclooxygenase-2 (COX-2) G-765C is a protective factor for coronary artery disease but not for ischemic stroke: A meta-analysis. *Atherosclerosis, 207*, 492–495.

Liang, Y., Liu, J. L., Wu, Y., Zhang, Z. Y., & Wu, R. (2011). Cyclooxygenase-2 polymorphisms and susceptibility to esophageal cancer: A meta-analysis. *The Tohoku Journal of Experimental Medicine, 223*, 137–144.

Listi, F., et al. (2010). Role of cyclooxygenase-2 and 5-lipoxygenase polymorphisms in Alzheimer's disease in a population from northern Italy: Implication for pharmacogenomics. *Journal of Alzheimer's Disease, 19*, 551–557.

Maguire, J., et al. (2011). Impact of COX-2 rs5275 and rs20417 and GPIIIa rs5918 polymorphisms on 90-day ischemic stroke functional outcome: A novel finding. *Journal of Stroke and Cerebrovascular Diseases, 20*, 134–144.

Malhi, H., Atac, B., Daly, A. K., & Gupta, S. (2004). Warfarin and celecoxib interaction in the setting of cytochrome P450 (CYP2C9) polymorphism with bleeding complication. *Postgraduate Medical Journal, 80*, 107–109.

Montali, A., et al. (2010). Functional rs20417 SNP (-765G>C) of cyclooxygenase-2 gene does not predict the risk of recurrence of ischemic events in coronary patients: Results of a 7-year prospective study. *Cardiology, 115*, 236–242.

Ohtani, T., Hiroi, A., Sakurane, M., & Furukawa, F. (2003). Slow acetylator genotypes as a possible risk factor for infectious mononucleosis-like syndrome induced by salazosulfapyridine. *British Journal of Dermatology, 148*, 1035–1039.

Ol, K. K., Agachan, B., Gormus, U., Toptas, B., & Isbir, T. (2011). Cox-2 gene polymorphism and IL-6 levels in coronary artery disease. *Genetics and Molecular Research, 10*, 810–816.

Papafili, A., et al. (2002). Common promoter variant in cyclooxygenase-2 represses gene expression: Evidence of role in acute-phase inflammatory response. *Arteriosclerosis, Thrombosis, and Vascular Biology, 22*, 1631–1636.

Peters, M. J., et al. (2012). Genome-wide association study meta-analysis of chronic widespread pain: Evidence for involvement of the 5p15.2 region. *Annals of the Rheumatic Diseases, 72(3)*, 427–3.

Reyes-Gibby, C. C., et al. (2009). Role of inflammation gene polymorphisms on pain severity in lung cancer patients. Cancer epidemiology, biomarkers & prevention: A publication of the American Association for Cancer Research. *American Society of Preventive Oncology, 18*, 2636–2642.

Sabbagh, N., Delaporte, E., Marez, D., Lo-Guidice, J. M., Piette, F., & Broly, F. (1997). NAT2 genotyping and efficacy of sulfasalazine in patients with chronic discoid lupus erythematosus. *Pharmacogenetics, 7,* 131–135.

Schneider, E. M., et al. (2011). The (-765 G–>C) promoter variant of the COX-2/PTGS2 gene is associated with a lower risk for end-stage hip and knee osteoarthritis. *Annals of the Rheumatic Diseases, 70,* 1458–1460.

Sharma, V., Kaul, S., Al-Hazzani, A., Alshatwi, A. A., Jyothy, A., & Munshi, A. (2013). Association of COX-2 rs20417 with aspirin resistance. *Journal of Thrombosis and Thrombolysis, 35,* 95–99.

Skarke, C., et al. (2006). The cyclooxygenase 2 genetic variant -765G>C does not modulate the effects of celecoxib on prostaglandin E2 production. *Clinical Pharmacology and Therapeutics, 80,* 621–632.

Skarke, C., Schuss, P., Kirchhof, A., Doehring, A., Geisslinger, G., & Lotsch, J. (2007). Pyrosequencing of polymorphisms in the COX-2 gene (PTGS2) with reported clinical relevance. *Pharmacogenomics, 8,* 1643–1660.

Szczeklik, W., Sanak, M., & Szczeklik, A. (2004). Functional effects and gender association of COX-2 gene polymorphism G-765C in bronchial asthma. *The Journal of Allergy and Clinical Immunology, 114,* 248–253.

Tanaka, E., et al. (2002). Adverse effects of sulfasalazine in patients with rheumatoid arthritis are associated with diplotype configuration at the N-acetyltransferase 2 gene. *Journal of Rheumatology, 29,* 2492–2499.

Tanigawara, Y., et al. (2002). N-acetyltransferase 2 genotype-related sulfapyridine acetylation and its adverse events. *Biological and Pharmaceutical Bulletin, 25,* 1058–1062.

Ueda, N., Maehara, Y., Tajima, O., Tabata, S., Wakabayashi, K., & Kono, S. (2008). Genetic polymorphisms of cyclooxygenase-2 and colorectal adenoma risk: The self defense forces health study. *Cancer Science, 99,* 576–581.

Ulrich, C. M., et al. (2002). Cyclooxygenase 1 (COX1) polymorphisms in African-American and Caucasian populations. *Human Mutation, 20,* 409–410.

Ulrich, C. M., et al. (2005). PTGS2 (COX-2) -765G > C promoter variant reduces risk of colorectal adenoma among nonusers of nonsteroidal anti-inflammatory drugs. *Cancer Epidemiology, Biomarkers & Prevention, 14,* 616–619.

Werner, U., Werner, D., Rau, T., Fromm, M. F., Hinz, B., & Brune, K. (2003). Celecoxib inhibits metabolism of cytochrome P450 2D6 substrate metoprolol in humans. *Clinical Pharmacology and Therapeutics, 74,* 130–137.

Yasar, U., et al. (2002). Linkage between the CYP2C8 and CYP2C9 genetic polymorphisms. *Biochemical and Biophysical Research Communications, 299,* 25–28.

Yazdanian, M., Briggs, K., Jankovsky, C., & Hawi, A. (2004). The "high solubility" definition of the current FDA guidance on biopharmaceutical classification system may be too strict for acidic drugs. *Pharmaceutical Research, 21,* 293–299.

Zhang, J. Y., Zhan, J., Cook, C. S., Ings, R. M., & Breau, A. P. (2003). Involvement of human UGT2B7 and 2B15 in rofecoxib metabolism. *Drug Metabolism and Disposition, 31,* 652–658.

Zhang, X., et al. (2005). Identification of functional genetic variants in cyclooxygenase-2 and their association with risk of esophageal cancer. *Gastroenterology, 129,* 565–576.

NSAIDs, Pharmacokinetics

Joern Loetsch and Bruno Oertel
Institute for Clinical Pharmacology,
Pharmaceutical Center Frankfurt, Johann
Wolfgang Goethe University, Frankfurt,
Germany

Synonyms

Pharmacokinetics of the NSAIDs; Plasma concentration versus time profiles

Definition

The pharmacokinetics describes the journey of a drug molecule through the body. The journey includes its release from the drug product, its ► absorption into the body system, for some substances its bio-activation via ► metabolism, the ► distribution to its site of action and back into the blood, and its ► elimination from the body either via transformation into inactive metabolites, which are then excreted, or direct excretion of the active entity. The goal of pharmacokinetics is to describe the time course of the drug concentrations in the organism in order to derive dosing regimens that provide most effective clinical drug actions with least side effects.

Characteristics

NSAIDs are mainly administered orally. Formulations for intravenous, intramuscular, topical, rectal, or intraocular administration are also available for some NSAIDs. The pharmacokinetics of the NSAIDs is best described by the

LADME model, which describes the Liberation, Absorption, Distribution, Metabolism, and Elimination of a drug. The concentration versus time profiles of the drug depends on these five processes.

The ▶ liberation of the active ingredient from a pharmaceutical product is mainly the result of the galenic engineering (tablet coating, tablet disintegration, particle size, etc.). Enteric- and sustained release coatings are often used with NSAIDs to reduce their gastrointestinal toxicity. While the effectiveness of the coating for reduction of the toxicity is doubted, its influence on the absorption of the drug due to delayed release is clear. Once freed from the pharmaceutical formulation, the absorption of a drug is mainly defined by its physicochemical properties. Most of the NSAIDs are carbonic acids, or at least have an acidic function in their molecular structure. They pass the gastrointestinal wall by a passive ▶ diffusion process and are rapidly and extensively absorbed from the stomach and proximal small intestine, with peak plasma concentrations generally occurring within 2–3 h post-administration, or within 30 min in the case of fast release formulations. However, although absorption is extensive, some NSAIDs (e.g., diclofenac) have a low ▶ bioavailability because they are subject to a considerable ▶ first-pass metabolism that takes place in the intestinal wall and in the liver. Most NSAIDs are metabolized by ▶ phase-1 metabolic reactions such as oxidation, hydroxylation, demethylation, deacetylation, and hepatic conjunctions (▶ phase-2 metabolic reactions such as glucuronidation and sulphation), or both, with subsequent excretion into urine or bile. In addition, acetyl salicylic acid is deacetylated directly in the blood. The lower the bioavailability is, the fewer molecules reach the circulation and are available for transport to their site of action. Once the molecules of the NSAIDs have reached the blood, they are extensively bound to plasma proteins, especially to albumin. Their ▶ volume of distribution (Vd) is usually small, mainly between 0.1 and 0.3 l/kg body weight, which approximates plasma volume. After excretion into the bile, several NSAIDs undergo an ▶ enterohepatic recirculation. That is,

they are reabsorbed from the intestine after cleavage of the phase-2-conjugates (i.e., glucuronides or sulphates) by intestinal human or bacterial glucuronidases or sulfatases. Depending on the elimination process (metabolism or excretion via the kidney or the bile), NSAIDs differ in their speed of elimination, which is usually numerically characterized by the ▶ elimination half-life values ($t_{1/2}$). The half-life of the NSAIDs can vary within and between individuals due to organ damage (i.e., kidney or liver failure) or due to genetic polymorphisms of the enzymes involved in the metabolism of the NSAIDs. For example, some NSAIDs are metabolized via cytochrome P450 2 C9 (CYP2C9), for which mutations resulting in a less-functional or nonfunctional enzyme have been described. Furthermore, pharmacokinetic drug-drug interactions can occur that result in inhibition of enzymes or of transporters involved in the elimination of the NSAIDs, with the consequence of altered, often decreased, rates of elimination, and thus increased half-lives of the NSAIDs (Davies and Skjodt 2000).

In the following, the pharmacokinetic properties will be described in more detail for the particular class of NSAIDs.

Salicylic Acid Derivatives

The salicylic acid derivatives (Table 1) are rapidly and completely absorbed after oral administration. The reduced bioavailability and very short elimination half-life of acetyl salicylic acid is the result of an extensive first-pass metabolism by hydrolysis of acetyl salicylic acid to salicylic acid and acetic acid (Fig. 1), which takes place while passing the gastrointestinal mucosa, in the liver and in the blood. Due to the irreversible acetylation of the amino acid serine at position 530 of the COX-1 protein in platelets, the half-life of the antithrombotic effect of acetyl salicylic acid is much longer than the half-life of acetyl salicylic acid in plasma. Thus, a clinically relevant aggregation of platelets will only be reestablished after enough new platelets have been produced, which is several days after administration of acetyl salicylic acid, at a time when the drug has already completely been eliminated from plasma for some days.

NSAIDs, Pharmacokinetics, Table 1 Pharmacokinetics of the salicylic acid derivates

Drug	F (%)	PB (%)	t_{Max} (h)	Vd (L/kg)	CL (L/h/kg)	$t_{1/2}$ (h)	Active metabolites	Remarks
Salicylic acid (SA)	~100	80–90	0.5–2	0.17	4.2	2–3 (−30)	No	Dose-dependent half-life due to saturation of liver enzymes involved in the metabolism
Aspirin (ASA)	~68	85–95	0.4–0.5	0.15	0.6–3.6	14–20 min	Salicylic acid (SA)	Hydrolysis through nonspecific esterases while passing the gut wall in plasma and liver
Diflunisal	~100	99.8	2–3	0.10	0.0066	5–20	No	Dose-dependent half-life due to saturation of liver enzymes involved in the metabolism
Sulfasalazine (Prodrug)	<15	>99.3	3–12	NR	~1	4–11	5-Aminosalicylic acid	Bacteria-mediated splitting in anti-inflammatory active 5-Aminosalicylic acid and sulfapyridine
							(F = 10–30 %, t_{Max} ~ 10 h)	Half-life dependent on slow or fast acetylation by polymorphic enzyme
Salsalate (Prodrug)	~100	NR	~1.5	NR	NR	1.1(−16)	Salicylic acid (SA)	Rapid esterase hydrolysis in two molecules of salicylic acid in the small intestine and plasma

F oral bioavailability, *PB* binding to plasma proteins, t_{Max} time from administration until maximum plasma concentrations are reached, *Vd/F* volume of distribution divided by bioavailability (because data from systemic administration that are needed to calculate the true Vd are usually not available, and the volumes are derived from data after oral drug administration, and have therefore been corrected for bioavailability), *CL* drug clearance, describing the speed of its elimination from the body, $t_{1/2}$ half-life in plasma

NSAIDs, Pharmacokinetics, Fig. 1 Hydrolysis of acetyl salicylic acid

Aspirin (ASA)   Hydrolysis →   Salicylic acid   +   Acetic acid

Salicylic acid is an active metabolite of acetyl salicylic acid. It has a longer elimination half-life than acetyl salicylic acid. During treatment with high or repetitive doses of salicylic acid or diflunisal, the elimination half-life can increase due to saturation of liver enzymes involved in the metabolism (nonlinear kinetic). Salicylic acid and its metabolites are excreted via the kidney. The renal elimination rate of salicylic acid is influenced by the urinary pH. Therefore, acidifying agents (e.g., ammonium chloride) decrease its excretion, while alkalinizing agents, such as sodium bicarbonate, increase the urinary excretion. Salicylates can displace the anticonvulsants phenytoin and valproic acid from their plasma protein binding sites and prevent the elimination of valproic acid due to inhibition of its main metabolic pathway, the β-oxidation. To prevent intoxication due to increased free plasma levels, combination of the drugs should be avoided (Brouwers and de Smet 1994; Needs and Brooks 1985).

In contrast to other salicylic acid derivatives, sulfasalazine and its active metabolite 5-aminosalicylic acid are only poorly absorbed and, therefore, remain mainly within the gastrointestinal tract. Sulfasalazine and salsalate are prodrugs and are mainly used for the treatment of ulcerative colitis and Crohn's disease. While salsalate is an ester of two salicylic acid molecules, which is rapidly absorbed and then hydrolyzed, the anti-inflammatory effect of sulfasalazine is probably attributed to its metabolite 5-aminosalicylic acid.

Arylpropionic Acids

Arylpropionic acids are chiral compounds. Chirality results when three-dimensional repositioning produces different forms (enantiomers) of the same molecule. A chiral drug exists as a pair of molecules that are each other's mirror image, called S-enantiomer and R-enantiomer. The most common example of chirality is a sp^3-hybridized tetrahedral carbon atom, to which four different atoms are attached (Fig. 2). Such a sp^3-hybridized tetrahedral chiral carbon atom is the common structural feature of the arylpropionic acids and is located within their propionic acid side chain. Most of the arylpropionic acids are marketed as racemats, i.e., as mixtures of both enantiomers. In addition, pure S-enantiomers are available for ibuprofen and ketoprofen. Naproxen is available and clinically used only as S-enantiomer. This is because the S-enantiomers have been shown to possess almost the whole pharmacologic activity. However, more recent studies have also demonstrated some pharmacologic effects of the R-enantiomers.

NSAIDs, Pharmacokinetics, Fig. 2 Chirality of the arylpropionic acids

Enantiomers of arylpropionic acids have different physical properties such as water solubility and differ in their pharmacokinetics. For example, a stereoselectivity, i.e., a different pharmacokinetic behavior of the S- and R-enantiomers of ibuprofen and flurbiprofen has been reported (Davies and Skjodt 2000). Some arylpropionic acids undergo a unidirectional inversion of the inactive R-enantiomer to the active S- enantiomer (Fig. 3). The extent of the inversion is variable from drug to drug and is most substantial for R-(−)-ibuprofen, while it is negligible for R-(−)-flurbiprofen (Table 2) (Geisslinger et al. 1994).

The arylpropionic acids get rapidly and nearly completely absorbed. They have no extensive first-pass metabolism and their elimination is, with the exception of naproxen, quite fast. Ibuprofen and naproxen bind in a concentration-dependent manner to plasma proteins. There is an increase in the unbound fraction of the drug at doses greater than 600 mg and 500 mg, respectively, resulting in an increased ▶ clearance and reduced area under curve (AUC) of the total drug. A decreased clearance of S-(+)-ibuprofen is further reported in carriers of certain CYP2C9 genetic polymorphisms resulting in a decreased or absent functionality of the enzyme (see ▶ NSAIDs, Pharmacogenetics).

Heteroaryl Acetic Acids

Diclofenac, aceclofenac, ketorolac, and lumiracoxib are heteroaryl acetic acids (Table 3). They are rapidly and nearly completely absorbed. Due to an extensive first-pass metabolism, diclofenac has a decreased systemic availability. The primary metabolite of diclofenac, 4'-hydroxy-diclofenac (Fig. 4), is produced by the genetically polymorphic CYP2C9. The amount of 4'-hydroxy-diclofenac excreted in urine and bile accounts for 30 % and 10–20 %, respectively, of an oral dose of diclofenac (van der Marel et al. 2004). Data from experiments in laboratory animals suggest that 4'-hydroxy-diclofenac has 30 % of the anti-inflammatory and antipyretic activity of

Diclofenac   Lumiracoxib

NSAIDs, Pharmacokinetics, Fig. 3 Unidirectional inversion of the arylpropionic acids

diclofenac. Aceclofenac, an ester of diclofenac, appears to inhibit both COX isoforms through conversion into diclofenac and its metabolite 4'-hydroxy-diclofenac (Hinz et al. 2003). Ketorolac is a chiral NSAID and is marketed as a racemic mixture of the S-(−)- and the R-(+)-enantiomeric isoforms. The S-(−)-form possesses the analgesic and ulcerogenic activity. There is no evidence for an inversion of R-(+)-ketorolac to S-(−)-ketorolac in man, but the pharmacokinetics of ketorolac shows enantioselectivity. The S-(−)-form has a two times shorter plasma half-life and greater clearance in adults than the R-(+)-Form, and the elimination half-life of S-(−)-ketorolac seems to be further increased in children (Kauffman et al. 1999; Mroszczak et al. 1996). While the other heteroaryl acetic acids are classical NSAIDs, lumiracoxib is a selective inhibitor of COX-2. Its molecular structure is very similar to that of diclofenac (Fig. 5) and very different from the other COX-2 selective agents

NSAIDs, Pharmacokinetics, Table 2 Pharmacokinetics of the arylpropionic acid derivates

Drug	F (%)	PB (%)	t_{Max} (h)	Vd (L/kg)	CL (L/h/kg)	$t_{1/2}$ (h)	Active metabolites	Remarks
Ibuprofen (Rac/S)[a]	~100	98–99	1–2	0.15	0.045	1.5–3	No	Unidirectional inversion to S-(+)-enantiomer with an inversion-rate R:S = 50–80 %
								Metabolizing enzymes (Phase 1): CYP2C9
								Dose-dependent binding to plasma proteins
Flurbiprofen (Rac)[a]	~100	>99	~1.5	0.1	0.018	3–6	No	Unidirectional inversion to S-(+)-enantiomer
								Inversion-rate R:S = 0–5 %
Ketoprofen (Rac/S)[a]	81–84	98.7	0.5–3	0.11	0.072	2–4	No	Unidirectional inversion to S-(+)-enantiomer
								Inversion-rate R:S ~10 %
Naproxen (S)[a]	~100	>99	2–4	0.10–0.16	0.0042	12–15	No	Dose-dependent binding to plasma proteins
Tiaprofenic acid (Rac)[a]	~100	98	NR	0.4–1	0.036–0.084	3–6	No	Negligible R to S conversion upon oral administration
Fenoprofen (Rac)[a]	NR	>99	~2	0.08–0.11	NR	1.5–3	NR	

F oral bioavailability, *PB* binding to plasma proteins, t_{Max} time from administration until maximum plasma concentrations are reached, *Vd/F* volume of distribution divided by bioavailability (because data from systemic administration that are needed to calculate the true Vd are usually not available, and the volumes are derived from data after oral drug administration, and have therefore been corrected for bioavailability), *CL* drug clearance, describing the speed of its elimination from the body, $t_{1/2}$ half-life in plasma

[a]Available as racemate and/or single S-(+)-enantiomer

NSAIDs, Pharmacokinetics, Table 3 Pharmacokinetics of the heteroaryl acetic acids

Drug	F (%)	PB (%)	t_{Max} (h)	Vd (L/kg)	CL (L/h/kg)	$t_{1/2}$ (h)	Active metabolites	Remarks
Diclofenac	30–80	>99	0.37–89	0.1–0.2	0.26–0.45	1.1–1.7	4'-hydroxy-diclofenac (30 % activity of diclofenac in animal models)	Extensive first-pass metabolism
								Metabolizing enzymes (Phase 1): CYP2C9, CYP3A4, CYP3A5
								Prodrug of diclofenac: aceclofenac
Tolmetin	~100	~99	0.5–1	~0.1	NR	~2 (~5)	No	Biphasic elimination with a rapid phase ($t_{1/2} = 2$ h), followed by a slow phase ($t_{1/2} = 5$ h)
Ketorolac (Rac)	80–100	>99	0.3–1	0.1–0.3	~0.03	2.12.9(S) 3.3–6.7®	No	S-(−)-ketorolac is the active enantiomer
								Stereoselective metabolism with increased clearance for the active S-(−)-enantiomer
Lumiracoxib	66–80.8	>98	1–4	7.3–10.7 (L)	NR	3–6	4'-hydroxy-lumiracoxib (Potency and selectivity is similar to lumiracoxib)	Metabolizing enzymes (Phase 1): CYP2C9

F oral bioavailability, PB binding to plasma proteins, t_{Max} time from administration until maximum plasma concentrations are reached, Vd/F volume of distribution divided by bioavailability (because data from systemic administration that are needed to calculate the true Vd are usually not available, and the volumes are derived from data after oral drug administration, and have therefore been corrected for bioavailability), CL drug clearance, describing the speed of its elimination from the body, $t_{1/2}$ half-life in plasma

(diarylheterocycles). This difference is reflected in the pharmacokinetic properties of lumiracoxib, which are more similar to those of the classical acidic NSAIDs than to the diarylheterocycles. In 2007, lumiracoxib was withdrawn from markets in the wake of a handful of serious liver adverse events associated with drug. Novartis recently seems to have found a genetic marker that allows identifying patients at risk of certain liver-related side effects from lumiracoxib and making them ineligible for therapy. Novartis is currently trying to resubmit lumiracoxib, in combination with the genetic marker, to European regulatory authorities for the treatment of osteoarthritis symptoms.

Diarylheterocycles

The diarylheterocycles (celecoxib, etoricoxib, and oxaprozin) (Table 4) have a higher selectivity to the COX-2 isoform compared to the classical NSAIDs. With the exception of oxaprozin, the diarylheterocycles do not possess an acidic function in their molecule. They are very lipophilic and poorly water soluble. Since absorption of the drug molecules requires their dissolution in fluids, special formulations have had to be developed to enhance their water solubility, and thus to ensure their absorption. Compared to the classical NSAIDs, diarylheterocycles show very different pharmacokinetics, particularly the volume of distribution, the plasma clearance, and

NSAIDs, Pharmacokinetics, Fig. 4 Primary metabolite of diclofenac

NSAIDs, Pharmacokinetics, Fig. 5 Structural similarity of lumiracoxib to classical NSAIDs

NSAIDs, Pharmacokinetics, Table 4 Pharmacokinetics of the diarylheterocycles

Drug	F (%)	PB (%)	t_{Max} (h)	Vd (L)	CL (L/h)	$t_{1/2}$ (h)	Active metabolites	Remarks
Celecoxib	NR (22–40 in dogs)	>97	2–3	339–571	23.7–27.8	8–12	No	Metabolizing enzymes (Phase 1): CYP2C9, CYP3A4
Etoricoxib	~100	NR	0.5–1.5	82–156	4.92–8.04	18.9–30.9	NR	–
Oxaprozin	~95	99	2.4–3.1	0.16–0.24 (L/kg)	0.15–0.3 (L/h/kg)	41.4–54.9	No	–

F oral bioavailability, *PB* binding to plasma proteins, t_{Max} time from administration until maximum plasma concentrations are reached, *Vd/F* volume of distribution divided by bioavailability (because data from systemic administration that are needed to calculate the true Vd are usually not available, and the volumes are derived from data after oral drug administration, and have therefore been corrected for bioavailability), *CL* drug clearance, describing the speed of its elimination from the body, $t_{1/2}$ half-life in plasma

NSAIDs, Pharmacokinetics, Table 5 Pharmacokinetics of the enolic acids

Drug	F (%)	PB (%)	t_{Max} (h)	Vd (L/kg)	CL (L/h/kg)	$t_{1/2}$ (h)	Active metabolites	Remarks
Piroxicam	NR (~100)	~99	2–3	0.1–0.2	0.002–0.003	30–70	No	Prodrugs of piroxicam: ampiroxicam, droxicam, pivoxicam
Meloxicam	~100	>99.5	3–9	0.1–0.2	0.0066	~20	No	Metabolizing enzymes (Phase 1): CYP2C9
Tenoxicam	~100	>98.5	~2	~0.15	0.001–0.002	49–81	No	Metabolizing enzymes (Phase 1): CYP2C9
Isoxicam	~100	~98	~10	0.1–0.2	~0.3 (L/h)	30–50	No	Increased $t_{1/2}$ for about 10 % of a population, eventually due to polymorphic enzyme
Lornoxicam	NR	99.7	0.5–2.5	0.1–0.2	1.5–3.4 (L/h)	3–5	NR	Metabolizing enzymes (Phase 1): CYP2C9
Azapropazone	60–100	>99.5	3–6	8.4–15.4 (L)	0.48–0.73 (L/h)	11.5–17.1	NR	–
Phenylbutazone	90	>98	NR	0.02–0.15	0.09 (L/h)	29–175	Oxyphenbutazone γ-hydroxyphenylbutazone	Dose-dependent half-life
Oxyphenbutazone	NR	97–98	NR	~0.15	NR	27–64	NR	–

F oral bioavailability, PB binding to plasma proteins, t_{Max} time from administration until maximum plasma concentrations are reached, Vd/F volume of distribution divided by bioavailability (because data from systemic administration that are needed to calculated the true Vd are usually not available and the volumes are derived from data after oral drug administration and have therefore be corrected for bioavailability), CL drug clearance, describing the speed of its elimination from the body, $t_{1/2}$ half-life in plasma the elimination half-life (Ahuja et al. 2003; Alsalameh et al. 2003).

Enolic Acids

Members of the enolic acids family (piroxicam, meloxicam, tenoxicam, lornoxicam) (Table 5) are weakly acidic by virtue of the enolic 4-hydroxy substituent (Fig. 6). They are well absorbed and extensively bound to plasma proteins. Due to this plasma protein binding, their apparent volumes of distribution are small. They are mainly eliminated by hepatic metabolism (Olkkola et al. 1994). The polymorphic CYP2C9 provides the major catabolic pathway for tenoxicam and meloxicam and are therefore

candidates for an altered pharmacokinetic due to genetic polymorphisms (see ▶ NSAIDs, Pharmacogenetics). The elimination is usually slow. The elimination half-lives of the oxicams are long, with the exception of lornoxicam. Therefore, the oxicams have a tendency to accumulate in patients.

Phenylbutazone and oxyphenbutazone tend to accumulate due to the slow metabolism and renal elimination. In addition, they have a high potential to interact with other drugs, particularly with oral anticoagulants, anticonvulsants, and oral antihyperglycemic agents, by either inhibiting metabolic pathways or by displacement from plasma protein–binding sites (Brouwers and de Smet 1994).

Indole and Indene Acetic Acids

Indomethacin and its prodrugs acemetacin, sulindac, and etodolac are all accounted to the indole and indene acetic acids (Table 6). Their bioavailability is high and their binding to plasma proteins after absorption is extensive. They undergo an extensive enterohepatic circulation

NSAIDs, Pharmacokinetics, Fig. 6 Acidic nature of piroxicam

NSAIDs, Pharmacokinetics, Table 6 Pharmacokinetics of the indole and indene acetic acids

Drug	F (%)	PB (%)	t_{Max} (h)	Vd (L/kg)	CL (L/h/kg)	$t_{1/2}$ (h)	Active metabolites	Remarks
Indomethacin	~100	>90	1–2	0.12	0.044–0.11	3–11	No	Biphasic elimination due to enterohepatic cycling
								Prodrug of indomethacin: acemetacin
Sulindac (Prodrug/ Rac)	>90	93.1	~1	NR	NR	1.7–7	Sulindac sulfide (PB = 95.4, t_{Max} = 2–4, $t_{1/2}$ = 16–18)	Enterohepatic cycling
								Slower elimination in liver failure
Etodolac (Rac)	(~100)	~99	1–2	~1.6(S) ~0.21[®]	~0.3(S) ~0.02[®]	~4.3(S) ~6.6[®]	No	S-(+)-etodolac is the active enantiomer
								Less binding to plasma proteins, higher volume of distribution, and clearance for S-(+)-etodolac

F oral bioavailability, *PB* binding to plasma proteins, t_{Max} time from administration until maximum plasma concentrations are reached, *Vd/F* volume of distribution divided by bioavailability (because data from systemic administration that are needed to calculate the true Vd are usually not available, and the volumes are derived from data after oral drug administration, and have therefore been corrected for bioavailability), *CL* drug clearance, describing the speed of its elimination from the body, $t_{1/2}$ half-life in plasma

Nabumeton   6-Methoxy-2-naphtylacetic acid (6-MNA)

NSAIDs, Pharmacokinetics, Fig. 7 Bioactivation of sulindac

NSAIDs, Pharmacokinetics, Fig. 8 Hepatic activation of nabumetome

Piroxicam

NSAIDs, Pharmacokinetics, Table 7 Pharmacokinetics of the alkanones

Drug	F (%)	PB (%)	t_{Max} (h)	Vd/F (L/kg)	CL (L/h/kg)	$t_{1/2}$ (h)	Active metabolites	Remarks
Nabumetone[a] (Prodrug)	80 (35)[a]	>99	2.5–4	~0.83	NR	20–30	6-methoxy-2-naphthylacetic acid (6-MNA)	Nabumetone is inactive but is rapidly converted to the active 6-MNA
								80 % of a radiolabeled nabumetone dose can be found as metabolites in urine; 35 % of nabumetone gets converted in the active 6-MNA

F oral bioavailability, *PB* binding to plasma proteins, t_{Max} time from administration until maximum plasma concentrations are reached, *Vd/F* volume of distribution divided by bioavailability (because data from systemic administration that are needed to calculate the true Vd are usually not available, and the volumes are derived from data after oral drug administration, and have therefore been corrected for bioavailability), *CL* drug clearance, describing the speed of its elimination from the body, $t_{1/2}$: half-life in plasma
[a]Kinetic Parameters related to 6-MNA

resulting in a prolonged elimination half-life compared to other NSAIDs because the already eliminated drug is reabsorbed. About 60 % of an oral dose of indomethacin is excreted in the urine, while about 40 % is excreted in the feces after biliary secretion (Helleberg 1981).

Sulindac has a chiral sulfur atom and is marketed as racemat. It is a prodrug and is bioactivated in the kidney and liver to its active sulfide metabolite (Fig. 7), which has an increased elimination half-life compared to sulindac. It tends to accumulate after repeated administration, with an even greater tendency in elderly patients (Davies and Watson 1997). Etodolac is a chiral NSAID due to a chiral carbon atom and is marketed as the racemate. Similar to other chiral NSAIDs, it possesses some unique disposition features due to its stereoselective pharmacokinetics. The volume of distribution for the racemic etodolac is greater than

NSAIDs, Pharmacokinetics, Table 8 Pharmacokinetics of the sulfanilides

Drug	F (%)	PB (%)	t_Max (h)	Vd/F (L/kg)	CL (L/h/kg)	t_1/2 (h)	Active metabolites	Remarks
Nimesulide	NR	99	1.2–2.7	0.18–0.39	0.03–0.11	1.8–4.7	Yes, NR	Dose reduction (4–5 ×) in patients with hepatic impairment is needed, due to increased elimination half-life

F oral bioavailability, PB: binding to plasma proteins, t_{Max} time from administration until maximum plasma concentrations are reached, Vd/F volume of distribution divided by bioavailability (because data from systemic administration that are needed to calculate the true Vd are usually not available, and the volumes are derived from data after oral drug administration, and have therefore been corrected for bioavailability), CL drug clearance, describing the speed of its elimination from the body, $t_{1/2}$ half-life in plasma

NSAIDs, Pharmacokinetics, Table 9 Pharmacokinetics of the anthranilic acids

Drug	F (%)	PB (%)	t_Max (h)	Vd/F (L/kg)	CL (L/h)	t_1/2 (h)	Active metabolites	Remarks
Mefenamic acid	NR	~99	2–4	~1.06	~21.23	2–4	NR	Metabolizing enzymes (Phase 1): CYP2C9
Meclofenamate sodium	NR	>99	0.5–2	0.13–0.62	7.6–20.5	0.8–5.3	3-hydroxymethyl-MA (20 % activity of MA, $t_{1/2}$ = 15 h)	Extensively metabolized to the still active metabolite 3-hydroxymethyl-MA
Flufenamic acid	NR	NR	~1.5	NR	4.8–9	5–22	NR	Large interindividual variations in the pharmacokinetic parameters

F oral bioavailability, PB binding to plasma proteins, t_{Max} time from administration until maximum plasma concentrations are reached, Vd/F volume of distribution divided by bioavailability (because data from systemic administration that are needed to calculate the true Vd are usually not available, and the volumes are derived from data after oral drug administration, and have therefore been corrected for bioavailability), CL drug clearance, describing the speed of its elimination from the body, $t_{1/2}$ half-life in plasma

that of other NSAIDs, mainly because of the extensive distribution of the active S-(+)-etodolac. Therefore, the concentrations of the inactive R-(−)-enantiomer are about tenfold higher than the active S-(+)-enantiomer in plasma (Brocks and Jamali 1994).

Alkanones

After absorption from the gastrointestinal tract, the prodrug nabumetone undergoes an extensive hepatic metabolism to its active metabolite 6-methoxy-2-naphthylacetic acid (6-MNA) (Fig. 8) (Table 7), which is structurally related to naproxen. Approximately 35 % of a single dose is converted into 6-MNA. Coadministration of food seems to increase the bioavailability of nabumetone and subsequent the appearance of 6-MNA. 6-MNA is highly bound to plasma proteins.

Sulfanilides

Nimesulide inhibits preferably the COX-2 enzyme (Table 8). Its selectivity for the COX-2 enzyme seems to be as high as that of celecoxib. It is largely eliminated via the metabolism into several minor active metabolites with anti-inflammatory and analgesic actions. Hepatic insufficiency decreases the elimination of nimesulide pharmacokinetics, and therefore a dose-reduction (4–5 ×) for patients with hepatic impairment is required (Bernareggi 1998; Warner and Mitchell 2004).

Anthranilic Acids

The anthranilic acids (Table 9) are structurally related to the salicylic acids and heteroaryl acetic acids. Their physicochemical properties and their analgesic activities are very similar to the other NSAIDs. Since they do not show comparable anti-inflammatory activity, but have a similar occurrence of side effects, their application in the clinic is restricted.

References

Ahuja, N., Singh, A., & Singh, B. (2003). Rofecoxib: An update on physicochemical, pharmaceutical, pharmacodynamic and pharmacokinetic aspects. *Journal of Pharmacy and Pharmacology, 55,* 859–894.

Alsalameh, S., Burian, M., Mahr, G., et al. (2003). Review Article: The pharmacological properties and clinical use of valdecoxib, a New cyclo-oxygenase-2-selective inhibitor. *Alimentary Pharmacology & Therapeutics, 17,* 489–501.

Bernareggi, A. (1998). Clinical pharmacokinetics of nimesulide. *Clinical Pharmacokinetics, 35,* 247–274.

Brocks, D. R., & Jamali, F. (1994). Etodolac clinical pharmacokinetics. *Clinical Pharmacokinetics, 26,* 259–274.

Brouwers, J. R., & de Smet, P. A. (1994). Pharmacokinetic-pharmacodynamic drug interactions with nonsteroidal anti-inflammatory drugs. *Clinical Pharmacokinetics, 27,* 462–485.

Davies, N. M., & Skjodt, N. M. (2000). Choosing the right nonsteroidal anti-inflammatory drug for the right patient: A pharmacokinetic approach. *Clinical Pharmacokinetics, 38,* 377–392.

Davies, N. M., & Watson, M. S. (1997). Clinical pharmacokinetics of sulindac. A dynamic old drug. *Clinical Pharmacokinetics, 32,* 437–459.

Geisslinger, G., Lotsch, J., Menzel, S., et al. (1994). Stereoselective disposition of flurbiprofen in healthy subjects following administration of the single enantiomers. *British Journal of Clinical Pharmacology, 37,* 392–394.

Helleberg, L. (1981). Clinical pharmacokinetics of indomethacin. *Clinical Pharmacokinetics, 6,* 245–258.

Hinz, B., Rau, T., Auge, D., et al. (2003). Aceclofenac spares cyclooxygenase-1 as a result of limited but sustained biotransformation to diclofenac. *Clinical Pharmacology and Therapeutics, 74,* 222–235.

Kauffman, R. E., Lieh-Lai, M. W., Uy, H. G., et al. (1999). Enantiomer-selective pharmacokinetics and metabolism of ketorolac in children. *Clinical Pharmacology and Therapeutics, 65,* 382–388.

Mroszczak, E., Combs, D., Chaplin, M., et al. (1996). Chiral kinetics and dynamics of ketorolac. *The Journal of Clinical Pharmacology, 36,* 521–539.

Needs, C. J., & Brooks, P. M. (1985). Clinical pharmacokinetics of the salicylates. *Clinical Pharmacokinetics, 10,* 164–177.

Olkkola, K. T., Brunetto, A. V., & Mattila, M. J. (1994). Pharmacokinetics of oxicam nonsteroidal anti-inflammatory agents. *Clinical Pharmacokinetics, 26,* 107–120.

van der Marel, C. D., Anderson, B. J., Romsing, J., et al. (2004). Diclofenac and metabolite pharmacokinetics in children. *Paediatric Anaesthesia, 14,* 443–451.

Warner, T. D., & Mitchell, J. A. (2004). Cyclooxygenases: New forms, new inhibitors, and lessons from the clinic. *The FASEB Journal, 18,* 790–804.

NSAIDs, Side Effects

▶ NSAIDs, Adverse Effects

NSAIDs, Survey

Giresh Kanji
The Sports and Pain Clinic, Wellington, New Zealand

Synonyms

Anti-inflammatories; Nonsteroidal anti-inflammatory drugs NSAIDs

Definition

Nonsteroidal anti-inflammatory drugs (▶ NSAIDs) are a group of drugs derived from salicylates, which occur naturally in the bark of the willow tree. They encompass salicylic acid (aspirin) and a variety of more recent, synthesized agents.

Characteristics

NSAIDs have analgesic, anti-inflammatory, antipyretic properties and inhibit the aggregation of thrombocytes. They have traditionally been used to treat pain and inflammation. They do not alter the disease process that is giving rise to pain, but interfere with the mechanisms that produce pain and inflammation.

The original NSAID was salicylic acid, marketed as aspirin. It has been synthesized by

Bayer for over 100 years. As the mechanism of action of NSAIDs has been progressively elucidated, new agents have been developed, giving rise to different families of NSAIDs.

Mechanism of Action

The therapeutic and adverse effects of NSAIDs result from decreased production of prostaglandins from arachidonic acid, due to the inhibition of the cyclooxygenase (COX). The COX enzyme has two isotypes: COX 1 and COX 2. COX 1 is constitutively expressed by various tissues, including the gastrointestinal tract, the kidney, and the platelet, where its functions include preserving the gastric mucosa, diminishing renal vascular resistance, and maintaining homeostasis, respectively. COX 2 is an enzyme induced by injury and produces large amounts of prostanoids involved in the pain and inflammation pathways (Lipsky 1999).

Nonselective, earlier NSAIDs (e.g., diclofenac, ibuprofen, indomethacin, and naproxen) inhibit both COX 1 and COX 2 isotypes. Newer agents (e.g., celecoxib, rofecoxib) specifically inhibit the COX 2 isotype.

Central analgesic effects, independent of COX inhibition, have also been described for NSAIDs, but their mechanisms are not well understood. Analgesic effects are a combination of central and peripheral actions.

Routes of Administration

Oral, rectal, intramuscular, intravenous, and topical routes of administration are available. The oral is the preferred route.

Applications

NSAIDs are commonly used for a variety of pain states, including somatic and visceral pain. They are used temporarily for immediate relief of acute pain, such as occurs after muscle sprains, gout, and dysmenorrhea; for headache; and for postoperative pain. They are used long term for persisting pain, such as occurs in osteoarthritis, rheumatoid arthritis, and other arthritides. Among individuals over the age of 65, the use of NSAIDs is in the order of 20 %, due

to the common occurrence of osteoarthritis (Day et al. 1999).

Side Effects

Side effects of NSAIDs are related to the inhibition of the constitutively expressed COX 1 enzyme, which in turn suppresses the synthesis of prostanoids that have several important homeostatic functions. Gastric ulceration, inhibition of platelet aggregation, and renal dysfunction are the most important side effects.

Prostaglandin E is known to protect the gastrointestinal mucosa and limit gastric acid output (Raskin 1999).Gastroduodenal ulceration is the most common side effect of nonspecific COX 1- and COX 2-inhibiting NSAIDs, with up to 2 % of patients taking NSAIDs for 12 months developing a significant gastrointestinal bleed (Hawkey et al. 2000). The prevalence of gastric and duodenal ulcers varies from 9 % to 22 %, with one in ten being complicated by obstruction, perforation, or hemorrhage (Raskin 1999). In studies assessing gastroduodenal ulceration using endoscopy, approximately 45 % of patients taking a nonspecific NSAID developed an ulcer greater than 3 mm in length, compared to approximately 9 % taking a specific COX 2 NSAID (Hawkey et al. 2000).

In Australia, with a population of 18 million, it is estimated that there are 4,500 hospital admissions per year for serious upper gastrointestinal side effects, and it is estimated that 10 % of these may die with between 200 and 400 deaths per annum (Day et al. 1999). Estimates in the United States of NSAID-induced gastroduodenal injury are 107,000 hospitalizations and approximately 16,000 deaths per year (Singh and Rosen-Ramey 1998).

Prevention of gastroduodenal side effects center on the use of proton pump inhibitors such as omeprazole, H2-receptor agonists such as ranitidine, and the PGE1 analogue misoprostol. Omeprazole seems the most effective in prophylaxis and treatment of NSAID-induced gastroduodenal ulceration (Raskin 1999). Prophylaxis is, however, not cost-effective for all patients, with the cost-effectiveness of misoprostol estimated to be £27,300 (sterling) per gastrointestinal event avoided (Freemantle 2000).

NSAIDs can inhibit normal platelet function. Thromboxane A2 is a platelet activator that is suppressed by COX 1 inhibition. Aspirin irreversibly inactivates the COX enzyme and its action lasts for the lifespan of the platelet, 7–10 days. Other NSAIDs are reversible inhibitors of COX, and durations of action depend on clearance of drug from the circulation. Bleeding is not a significant side effect, as thromboxane A2 is only one of several mediators of platelet activation (Schafer 1999). The antithrombotic effect is important in the presence of coexisting bleeding disorders and the simultaneous use of alcohol or anticoagulants.

The role of renal prostaglandin production for maintenance of stable renal hemodynamic function is limited. The prevalence of nephrotoxicity from NSAIDs is relatively low, but the risk is greater when renal perfusion is reduced, as in the aged population, in cardiovascular disease, and during dehydration. In these circumstances, a variety of clinical syndromes can include fluid and electrolyte imbalance, acute renal dysfunction, nephrotic syndrome, interstitial nephritis, and renal papillary necrosis (Whelton 1999).

NSAIDs adversely affect blood pressure control for those taking angiotensin-converting enzyme inhibitors, diuretics, and beta-blockers. The risk of developing congestive cardiac failure also increases with NSAIDs in patients receiving diuretics.

Efficacy

Systematic reviews have found no important differences in analgesic effect between different NSAIDS (Gotzsche 2000), and the newer specific COX 2 inhibitors are of similar efficacy to older NSAIDs (Cannon et al. 2000). However, for specific conditions, the data on efficacy varies.

A Cochrane review for back pain (Van Tulder et al. 2003) collected evidence from 51 trials. It found that NSAIDs are more effective than placebo for short-term symptomatic relief in patients with acute low back pain. Sufficient evidence on chronic low back pain is still lacking.

Qualitative analysis showed there is conflicting evidence that NSAIDs are more effective than paracetamol for acute low back pain.

For osteoarthritis of the knee, a review of randomized controlled trials (Towheed and Hochberg 1997) concluded that NSAIDs were superior to placebo in all short-term trials. Acetaminophen was also found to be superior to placebo and comparably efficacious to low-dose naproxen and ibuprofen. Few studies have shown superiority of NSAIDs when compared to other analgesics (Watson et al. 2003).

A Cochrane review of lateral elbow pain concluded there is insufficient evidence to recommend or discourage the use of oral NSAIDs to relieve lateral elbow pain (Green et al. 2003). They found there was some evidence for the use of topical NSAIDs and that a trial comparing oral administration with topical administration has not been performed.

Cost

The high comparative cost of newer COX 2 selective NSAIDs compared to nonselective older NSAIDs would significantly increase healthcare costs. The number needed to treat has been calculated as 133 for 6 months regular therapy with a COX 2 rather than a traditional NSAID to avoid a serious gastrointestinal event and 1,333 to avoid a death. It appears that it is not cost-effective to change all patients from older nonselective medications to newer COX 1 selective agents.

Indications

In most conditions such as osteoarthritis, back pain, muscle strains, and tendon strains, it is well accepted that simple analgesics should be tried before trying NSAIDs. The American College Of Rheumatology guidelines (ACR subcommittee on Osteoarthritis Guidelines 2000) recommend acetaminophen as first-line therapy for the treatment of symptomatic osteoarthritis due to the significant side effects of NSAIDs and the lack of data confirming the superior efficacy of NSAIDs over simple analgesics.

References

ACR subcommittee on Osteoarthritis Guidelines. (2000). Recommendations for the medical management of osteoarthritis of the hip and knee. *Arthritis and Rheumatism, 43*, 1905–1915.

Cannon, G. W., Caldwell, J. R., Holt, P., Mclean, B., Seidenberg, B., Bolognese, J., et al. (2000). Rofecoxib, a specific inhibitor of cyclooxygenase 2, with clinical efficacy comparable with that of diclofenac sodium. *Arthritis and Rheumatism, 43*, 978–987.

Day, R., Rowett, D., & Roughead, E. E. (1999). Towards the safer use of non-steroidal inflammatory drugs. *Journal of Quality in Clinical Practice, 19*, 51–53.

Freemantle, P. (2000). Cost-effectiveness of non-steroidal anti-inflammatory drugs (NSAIDs) – what makes a NSAID good value for money. *Rheumatology, 39*, 232–234.

Gotzsche, P. C. (2000). Extracts from clinical evidence: Non-steroidal anti-inflammatory drugs. *British Medical Journal, 320*, 1058–1061.

Green, S., Buchbinder, R., Barnsley, L., Hall, S., White, M., Smidt, N., et al. (2003). Non-Steroidal Anti-Inflammatory Drugs (NSAIDs) for treating lateral elbow pain in adults (Cochrane Review). In: Cochrane Library, Issue 3. Update Software, Oxford.

Hawkey, C., Laine, L., Simon, T., Beaulieu, A., Maldonado-Cocco, J., Acevedo, E., et al. (2000). Comparison of the effect of rofecoxib (a cyclooxygenase 2 inhibitor), ibuprofen, and placebo on the gastroduodenal mucosa of patients with osteoarthritis: A randomized, double-blind, placebo-controlled trial. *Arthritis and Rheumatism, 2*, 370–377.

Lipsky, P. E. (1999). The clinical potential of cyclooxygenase-2-specific inhibitors. *The American Journal of Medicine, 106*(5B), 51S–57S.

Raskin, J. B. (1999). Gastrointestinal effects of non-steroidal anti-inflammatory therapy. *The American Journal of Medicine, 106*(5B), 3S–12S.

Schafer, A. I. (1999). Effects of non-steroidal anti-inflammatory therapy on platelets. *The American Journal of Medicine, 106*(5B), 25S–36S.

Singh, G., & Rosen-Ramey, D. (1998). NSAID induced gastrointestinal complications: The ARAMIS perspective 1997. *The Journal of Rheumatology. Supplement, 51*, 8–16.

Towheed, T. E., & Hochberg, M. C. (1997). A systematic review of randomised controlled trials of pharmacological therapy in osteoarthritis of the knee, with an emphasis on trial methodology. *Seminars in Arthritis and Rheumatism, 2*, 755–770.

Van Tulder, M. W., Scholten, R. J. P. M., Koes, B. W., & Deyo. R. A. (2003). Non-steroidal anti-inflammatory drugs for low back pain (Cochrane Review). In: Cochrane Library, Issue 3. Update Software, Oxford.

Watson, M. C., Brookes, S. T., Kirwan, J. R., & Faulkner, A. (2003). Non aspirin non-steroidal anti-inflammatory drugs for treating osteoarthritis of the knee (Cochrane Review). In: The Cochrane Library, Issue 3. Update Software, Oxford.

Whelton, A. (1999). Nephrotoxicity of non-steroidal anti-inflammatory drugs: Physiologic foundations and clinical implications. *The American Journal of Medicine, 106*(5B), 13S–24S.

NSAR

▶ Nonsteroidal Anti-Inflammatory Drugs (NSAIDs)

NST

▶ Neurosensory Testing

NT-3 Neurotrophin 3

Definition

Neurotrophic factor that belongs to the neurotrophin family.

Cross-References

▶ Spinal Cord Nociception, Neurotrophins

NTT

▶ Attributable Effect and Number Needed to Treat

Nuclear Magnetic Resonance

▶ Magnetic Resonance Imaging

Nucleoplasty

Definition

A non-heat-driven process using Coblation technology, where a bipolar radiofrequency device is used to dissolve the disc nucleus, thus decreasing the disc volume and decompressing the disc.

Cross-References

▶ Discogenic Back Pain

Nucleotide

Definition

Originally a combination of a nucleic acid (purine or pyrimidine), one sugar (usually ribose or deoxyribose), and a phosphoric group; by extension, any compound containing a heterocyclic compound bound by an N-glycosol line (e.g., adenosine monophosphate, NAD+).

Cross-References

▶ Headache Attributed to a Substance or its Withdrawal

Nucleotide Receptors

▶ Purine Receptor Targets in the Treatment of Neuropathic Pain

Nucleus Accumbens

▶ Nociceptive Processing in the Nucleus Accumbens, Neurophysiology and Behavioral Studies

Nucleus Caudalis DREZ

▶ DREZ Procedures

Nucleus Gelatinosus

▶ Spinothalamocortical Projections from SM

Nucleus Gracilis

Definition

An area of cell bodies within the medulla that receives inputs from large afferent fibers, relaying touch and vibration sensation. In addition, this area of cell bodies receives a direct projection from neurons throughout the length of the ipsilateral dorsal horn. These projections can carry nociceptive input to the region.

Cross-References

▶ Peptides in Neuropathic Pain States

Nucleus Pulposus

Definition

The nucleus pulposus is a centrally located gelatinous mass that comprises of the central portion of the intervertebral disk. It serves to resist compressive forces and allows the spine to have increased mobility. The nucleus pulposus is encircled by layers of collagen referred to as the annulus fibrosis. Together, the nucleus pulposus and the annulus fibrosis comprise the intervertebral disk.

Cross-References

▶ Cytokines as Targets in the Treatment of Neuropathic Pain
▶ Evoked and Movement-Related Neuropathic Pain

Nucleus Raphe Magnus

Synonyms

RVM, rostral ventromedial medulla

Definition

The nucleus raphe magnus is one of several midline structures in the brain stem that contain serotonergic neurons, among others. Neurons in this region are hypothesized to provide descending modulation of pain.

Cross-References

▶ On-Cells (RVM)
▶ Opiates, Rostral Ventromedial Medulla, and Descending Control
▶ Pain-Modulatory Systems, History of Discovery
▶ Stimulation-Produced Analgesia
▶ Vagal Input and Descending Modulation

Nucleus Submedius (SM)

Definition

A small oblong nucleus in the medial thalamus of rats and cats postulated to play a role in nociception, because trigeminal and spinal nociceptive neurons project to this thalamic region.

Cross-References

▶ Spinothalamocortical Projections from SM

Nucleus Tractus Solitarius

Definition

Bilateral sensory nuclei of the caudal medulla that receive input from several cranial nerves including the facial nerve (CN VII), glossopharyngeal nerve (CN IX), and the vagus nerve (CN X). It is especially involved in the relay and processing of taste, but also has roles in reflexes of the respiratory and alimentary tracts (e.g., coughing, swallowing).

Cross-References

▶ Vagal Input and Descending Modulation

Null

▶ Neuropathic Pain Model, Partial Sciatic Nerve Ligation Model

Number Needed to Harm

Definition

The number of patients needed to treat with a certain drug before one patient will experience a defined degree of side effects, e.g., drop out of the drug trial due to side effects. It is calculated as the reciprocal value of the difference in drop-out rate on active treatment and placebo.

Cross-References

▶ Antidepressants in Neuropathic Pain

Number Needed to Treat

Synonyms

NNT

Definition

This is a measure of clinical meaningfulness in clinical trials, and is defined as the number of patients needed to be treated to obtain one patient with moderate or better improvement (50 % or greater improvement) over and above placebo. This is a useful concept that allows the comparison, efficacy of analgesics that have different modes of action, as well as those from similar groups. It describes the magnitude of the difference between the active drug and the control. It can be calculated from:

$$NNT = 1/((IMP_{act}/TOT_{act}) - (IMP_{con}/TOT_{con})),$$

Where:

IMP_{act} = number of patients given active treatment achieving target (e.g., 50 % pain relief);

TOT_{act} = total number of patients given active treatment;

IMP_{con} = number of patients given control treatment achieving target;

TOT_{con} = total number of patients given control treatment.

The NNT of oxycodone 15 mg is 2.3.

Cross-References

▶ Antidepressants in Neuropathic Pain
▶ Attributable Effect and Number Needed to Treat
▶ Central Pain, Pharmacological Treatments
▶ Postherpetic Neuralgia, Etiology, Pathogenesis, and Management
▶ Postoperative Pain, COX-2 Inhibitors
▶ Postoperative Pain, Oxycodone

Numerical Rating Scale

Synonyms

NRS

Definition

An NRS allows a person to describe the intensity of his/her pain as a number usually ranging from

0 to 10, where "0" means "no pain" and "10" means pain as "bad as it could be." An NRS is a pain rating scale. Another commonly used scale is the visual analog scale (VAS).

Cross-References

▶ Central Pain, Outcome Measures in Clinical Trials

Nursing Home Residents

Definition

Chronic pain is more frequent in nursing home residents than in a community sample. Uncontrolled pain is a frequent cause of admission to nursing homes.

Cross-References

▶ Psychological Treatment of Pain in Older Populations

Nutraceuticals

Scott Masters
Caloundra Spinal and Sports Medicine Centre, Caloundra, QLD, Australia

Synonyms

Avocado-soybean unsaponifiables; Chondroitin; Glucosamine

Definition

▶ Nutraceuticals are loosely defined as foods with a health benefit. They are naturally occurring substances that can be used as drugs in order to treat specific symptoms or to modify disease processes. In the field of osteoarthritis, four nutraceuticals have been recognized and studied: ▶ glucosamine, ▶ chondroitin, glycosaminoglycan polysulfuric acid, and ▶ avocadosoybean unsaponifiables (ASU).

Glucosamine is a hexosamine sugar that is a component of almost all human tissues. It is one of the two molecules that form the repeating units of certain glycosaminoglycans, which in turn form the matrix of all connective tissues (Deal and Moskowitz 1999). Glycosaminoglycans, and therefore glucosamine, form a large component of articular cartilage, which is the tissue that is primarily damaged in osteoarthritis.

Chondroitin sulfate is the principal glycosaminoglycan found in articular cartilage. It is composed of a long unbranched polysaccharide chain, with a repeating disaccharide structure of N-acetylgalactosamine and glucuronic acid (Deal and Moskowitz 1999). Chondroitin sulfate is a strongly charged polyanion, which endows cartilage with water-binding properties. Functionally, this allows the cartilage matrix to absorb compression forces and thereby protect the underlying bone from damage.

Glycosaminoglycan polysulfuric acid is an extract from bovine cartilage and bone marrow, which contains a variety of glycosaminoglycans, including chondroitin and chondroitin sulfate (Pavelka et al. 1995, 2000).

ASUs are unsaponifiable fractions of one third avocado oil and two thirds soybean oil (Maheu et al. 1998; Blotman et al. 1997; Appleboom et al. 2001).

Characteristics

Mechanism
Glucosamine has been shown to reach the articular cartilage after oral, intramuscular, and intravenous administration (Deal and Moskowitz 1999; McAlindon et al. 2000; Pavelka et al. 2003). It is preferentially incorporated by human chondrocytes into GAGS (Deal and Moskowitz 1999; Pavelka et al. 2003)

and stimulates the synthesis of proteoglycans (Deal and Moskowitz 1999; McAlindon et al. 2000; Pavelka et al. 2003; Reginster et al. 2001).

Glucosamine and chondroitin sulfate have been shown to have anti-inflammatory effects (Deal and Moskowitz 1999; Pavelka et al. 2003; Reginster et al. 2001), positive effects on cartilage metabolism in vitro, and antiarthritic effects in animal models (Deal and Moskowitz 1999; McAlindon et al. 2000; Towheed et al. 2002). These results suggest possible structure-modifying and disease-modifying roles for glucosamine and chondroitin in osteoarthritis.

In laboratory studies, glycosaminoglycan polysulfuric acid has been found to stimulate cartilage metabolism and to inhibit the catabolic effects of interleukin-1 (Pavelka et al. 1995, 2000).

ASUs have also shown some anti-osteoarthritis properties both in vitro and in vivo (Maheu et al. 1998; Blotman et al. 1997; Appleboom et al. 2001).

Application
Nutraceuticals have been promoted for the treatment of osteoarthritis, on the grounds that they are natural substances that might promote healing or impede further deterioration of damaged cartilage.

Nutraceuticals are mainly taken by the oral route, as a tablet capsule or powder, and are also available as a cream. Although the majority of trials have focused on osteoarthritis of the knee, nutraceuticals are marketed for relief in a wide variety of conditions, e.g., arthritis pain, fibromyalgia, and joint swelling.

Efficacy
One pragmatic (Deal and Moskowitz 1999) and four systematic reviews (McAlindon et al. 2000; Towheed et al. 2002; Richy et al. 2003; Leeb et al. 2000) in the last 5 years on the use of glucosamine and chondroitin in osteoarthritis reached similar conclusions. A number of randomized controlled trials showed benefits from these agents greater than that of placebo, in terms of reduction of pain and increased function. Others showed equivalent or better efficacy than treatment with ▶ nonsteroidal anti-inflammatory drugs (NSAIDs).

However, the reviews offer caution regarding the methodological problems associated with many of the studies and advise that the treatment effects are probably exaggerated.

Important points raised by these reviews include:

- The ▶ effect sizes in the larger and better quality studies are smaller than those of other studies.
- Many trials suffer from inadequate blinding and absence of intention-to-treat analysis of results.
- Publication bias may apply, in that studies with statistically significant and positive results are more likely to be published than studies with negative results.
- Manufacturer support was prevalent in many of the early trials.

These problems are not unique to trials of nutraceuticals, for they have also been noted in trials of drugs for osteoarthritis. Nevertheless, they constitute grounds for caution when interpreting or stating the results of trials.

A further factor is that all of the early trials with positive results were of European origin. Later studies, conducted in the USA and in England, have found glucosamine to be no more effective than placebo for the relief of pain (Rindone et al. 2000; Hughes and Carr 2002).

Less contentious is the effect of glucosamine on the prevention of disease progression. Two randomized controlled trials have assessed the long-term effects of glucosamine sulfate on knee osteoarthritis (Pavelka et al. 2003; Reginster et al. 2001). Both studies compared the effects of placebo with that of 1,500 mg glucosamine sulfate daily for 3 years. In both studies, patients treated with glucosamine showed greater reductions in pain, and greater improvements in function, as measured by the Western Ontario and McMaster Universities osteoarthritis index (WOMAC). Both studies also demonstrated significantly less loss of joint space width in those patients treated with glucosamine. The second study (Pavelka et al. 2003) showed that the ▶ NNT for preventing clinically substantial loss of joint space was 11.

For glycosaminoglycan polysulfuric acid, the evidence has not been favorable. For the relief of pain, it is no more effective than placebo (Pavelka et al. 1995). It does not protect against loss of joint space (Pavelka et al. 2000).

Three studies of ASUs found greater reductions in pain and greater improvements in function than those following treatment with placebo ((Maheu et al. 1998; Blotman et al. 1997; Appleboom et al. 2001). These agents also had significant effects in reducing the need of patients to use NSAIDs.

Safety

No major toxicity problems have emerged with the use of these nutraceuticals. In particular, glucosamine has been shown to be safe in the two trials with 3-year follow-up (Pavelka et al. 2003; Reginster et al. 2001). Side effects were similar to those of placebo, and no specifically adverse effects were uncovered.

Some chondroitin preparations are derived from bovine cartilage. Due to the recent European epidemic of bovine spongiform encephalopathy (BSE) and its transmission to humans (resulting in variant Creutzfeldt-Jakob disease), all bovine-derived products are being reevaluated for potential transmission risks. On the List of Tissues with Suspected Infectivity (World Health Organization), bovine cartilage is listed as Category IV (no detectable infectivity). Most other commercially available chondroitin products are derived from shark cartilage.

Conclusions

The evidence for the use of nutraceuticals in the treatment of OA of the knee is growing. However, doubts remain about their effect size when compared with placebo, and the efficacy of these agents has not been compared with other regimens of long-term treatment of osteoarthritis. It is still not evident if they are a cost-effective substitute for treatment with NSAIDs or exercise, or if they are a worthwhile addition to such treatment. Nor has their efficacy been determined for osteoarthritis of joints other than those studied to date or for other painful conditions.

References

Appleboom, T., Scheuermans, J., Verbrugger, G., Henroitin, Y., & Reginster, J. Y. (2001). Symptoms modifying effect of avocado/soybean unsaponifiables (ASU) in knee osteoarthritis. A double blind, prospective, placebo-controlled study. *Scandinavian Journal of Rheumatology, 30*, 242–247.

Blotman, F., Maheu, E., Wulwik, A., Caspard, H., & Lopez, A. (1997). Efficacy and safety of avocado/soybean unsaponifiables in the treatment of symptomatic osteoarthritis of the knee and hip. A prospective, multicenter, three-month, randomized, double-blind, placebo-controlled trial. *Revue du Rhumatisme, 64*, 825–834.

Deal, C. L., & Moskowitz, R. W. (1999). Nutraceuticals as therapeutic agents in osteoarthritis. The role of glucosamine, chondroitin sulfate, and collagen hydrolysate. *Rheumatic Diseases Clinics of North America, 25*, 379–395.

Hughes, R., & Carr, A. (2002). A randomised, double-blind, placebo-controlled trial of glucosamine sulphate as an analgesic in osteoarthritis of the knee. *Rheumatology, 41*, 279–284.

Leeb, B., Schweitzer, H., & Mantag, S. J. S. (2000). A metaanalysis of chondroitin sulfate in the treatment of osteoarthritis. *The Journal of Rheumatology, 27*, 205–211.

Maheu, E., Mazieres, B., Valat, J. P., Loyau, G., Le Loet, X., Bourgeois, P., et al. (1998). Symptomatic efficacy of ASU in the treatment of osteoarthritis of the knee and hip. *Arthritis and Rheumatism, 41*, 81–91.

McAlindon, D. M., LaValley, M. P., Gulin, J. P., & Felson, D. T. (2000). Glucosamine and chondroitin for treatment of osteoarthritis. *Journal of the American Medical Association, 283*, 1469–1475.

Pavelka, K., Sedlackova, M., Gatterova, J., Becvar, R., & Pavelka, K. (1995). Glycosaminoglycan polysulfuric acid (GAGPS) in osteoarthritis of the knee. *Osteoarthritis and Cartilage, 3*, 15–23.

Pavelka, K., Gatterova, J., Gollarova, V., Urbanova, Z., Sedlackova, M., & Altman, R. D. (2000). A 5-year randomised controlled, double-blind study of glycosaminoglycan polysulfuric acid complex (Rumalon®) as a structure modifying therapy in osteoarthritis of the hip and knee. *Osteoarthritis and Cartilage, 8*, 335–342.

Pavelka, K., Gatterova, J., Olejarova, M., Machacek, S., Giacovelli, G., & Rovati, L. (2003). Glucosamine sulphate use and delay of progression of knee osteoarthritis. *Archives of Internal Medicine, 162*, 2113–2123.

Reginster, J. Y., Deroisy, R., Rovati, L. C., Lee, R. L., Lejeune, E., Bruyere, O., et al. (2001). Long-term effects of glucosamine sulphate on osteoarthritis progression: A randomised, placebo-controlled clinical trial. *Lancet, 357*, 251–256.

Richy, F., Bruyere, O., Ethgen, O., Cucherat, M., Henrotin, Y., & Reginster, J. Y. (2003). Structural and symptomatic efficacy of glucosamine and chondroitin in knee osteoarthritis: A comprehensive metaanalysis. *Archives of Internal Medicine, 163*, 1514–1522.

Rindone, J. P., Hiller, D., Collacott, E., Nordhaugen, N., & Ariola, G. (2000). Randomized, controlled trial of glucosamine for treating osteoarthritis of the knee. *The Western Journal of Medicine, 172*, 91–94.

Towheed, T. E., Anastassiades, T. P., Shea, B., Houpt, J., Welch, V., & Hochberg, M. C. (2002). Glucosamine therapy for treating osteoarthritis (Cochrane Review). In: The Cochrane Library, Issue 1, Update Software, Oxford.

Nutriceutical

Definition

The use of supplements and other substances like herbs and foods that are ingested or otherwise absorbed into the tissues to promote health.

Cross-References

▶ Alternative Medicine in Neuropathic Pain

Nutritional Neuropathies

▶ Metabolic and Nutritional Neuropathies

NVNP

▶ Nonsystemic (Isolated) Vasculitic Neuropathy (NSVN)

O

κ-opioid Receptors

▶ Opioids, Kappa Receptors, and Visceral Pain

40 Hz Oscillations

▶ Corticothalamic and Thalamocortical Interactions

O'Leary-Sant Interstitial Cystitis Symptom Index

▶ ICSI

Objective Factors

Definition

Objective factors are those that contribute to disability and can be assessed objectively by an examiner. They include laboratory findings, imaging findings, and some physical examination findings.

Cross-References

▶ Impairment, Pain-Related

Objective Judgment

Definition

Objective judgment is the process of observing and transmitting impersonal knowledge that is free from bias in judgment and that is readily quantifiable.

Cross-References

▶ Pain Assessment in Neonates

Objective Medical Evidence

Definition

Medical signs and laboratory findings.

Cross-References

▶ Disability Evaluation in the Social Security Administration

OBP

▶ Plexus Injuries and Deafferentation Pain

G.F. Gebhart, R.F. Schmidt (eds.), *Encyclopedia of Pain*, DOI 10.1007/978-3-642-28753-4,
© Springer-Verlag Berlin Heidelberg 2013

Observation of Pain Behavior

▶ Assessment of Pain Behaviors

Observational Learning

▶ Modeling, Social Learning in Pain

Obsolete: Hyperesthesia

▶ Allodynia (Clinical, Experimental)

Obstetric and Gynecological Pain

Katy Vincent
Nuffield Department of Obstetrics and
Gynaecology, The Women's Centre,
John Radcliffe Hospital, University of Oxford,
Oxford, UK

Introduction

Despite being the specialist interest of very few obstetricians and gynecologists, pain is an important issue in all areas of the specialty and for women through all stages of their reproductive lives and beyond into the menopause. In many cases, obstetric or gynecological pain is acute and associated with a specific experience or procedure such as labor and delivery, surgery, early pregnancy complications, and assisted reproduction techniques (ART). For pains such as these, the recent advances have mainly been in management as little opportunity exists for prevention. However, in the context of obstetrics and gynecology, such acute pain is often associated with major life events, and therefore the impact of adequate pain management during these experiences on the woman, and often her partner, should not be underestimated: enhancing the positive experiences such as childbirth and

minimizing one negative component of a traumatic experience such as pregnancy loss. Unfortunately, gynecology is not without its chronic pain conditions: chronic pelvic pain syndrome and vulval pain syndromes where no pathology is identified, and pain associated with specific pathologies, most notably endometriosis. Many advances have been made in this area in recent years, such that we now have a better understanding of how pain is generated and also of the psychological, behavioral, and brain structural and functional alterations associated with these conditions. Sadly, however, few advances in treatment have been made.

The aim of this section is to give an overview of the entries in this section and to bring together common themes. It is hoped that in the future, obstetricians and gynecologists will think more about the importance of pain within their specialty, while basic and clinical scientists will realize the potential for research in this field.

Characteristics

Acute Pain in Obstetrics and Gynecology
Analgesia during labor and delivery has long been an important specialist area of anesthesia, and while some women have a strong desire to experience the pain of labor, for the many who request labor analgesia, effective, safe options now exist. Epidural analgesic techniques have developed over recent years such that in many cases complete relief of pain can be achieved while minimizing motor block. Furthermore, it is now accepted that epidural analgesia does not cause an increase in caesarean section rate (ACOG 2006). Although there may be an association between epidural use and operative vaginal deliveries (Chestnut 1997; Cutura et al. 2011), this relationship is complex, with many of the obstetric factors that are associated with an increased instrumental rate also being associated with more painful or prolonged labors, e.g., persistent occipito-posterior position and macrosomia.

The increasing frequency of assisted reproduction worldwide (de Mouzon et al. 2009), on the other hand, has opened up a relatively novel area of gynecological anesthesia. Women undergoing a cycle of ART experience a number of procedures (including transvaginal scans, daily injections, oocyte retrieval, and embryo transfer) and complications (such as ovarian hyperstimulation syndrome) that could be expected to cause pain. Additionally, anxiety and expectation may also play a role in amplifying the pain experience (Chen et al. 2004). Furthermore, these patients are exposed to rapid alterations in hormonal milieu during an ART cycle, and the influence of these changes on pain perception and modulation remains unknown.

For women in the early stages of pregnancy, pain plays an important role in signaling potential problems. The maxim "abdominal pain in a woman of reproductive age is an ectopic pregnancy until proven otherwise" should not be forgotten: ectopic pregnancy is still a significant cause of maternal morbidity and mortality (Cantwell et al. 2011), and its incidence is increasing with the increase in ART. However, pain can be a symptom of many other pathologies in early pregnancy, as well as a common experience in healthy, ongoing pregnancies. Dedicated early pregnancy assessment units and education of Emergency Department clinicians have markedly improved the diagnosis and management of early pregnancy loss; nonetheless it is important to remember that other conditions causing pain can still occur (and may even be more common) in pregnancy, including acute appendicitis, ovarian torsion, and fibroid degeneration. Pain during pregnancy presents a particular diagnostic challenge, given that normal levels of many commonly ordered laboratory measures are altered (Bacq et al. 1996; Lurie et al. 2008), and many imaging techniques are relatively or absolutely contraindicated (Patel et al. 2007).

Drug prescribing in pregnancy is also an important consideration. In early pregnancy it is mandatory to avoid known teratogens, and because ethical considerations restrict human studies, there are many drugs whose safety in pregnancy remains unknown (BNF 2010). Even in the third trimester, many drugs should still be avoided because of their effects on the fetus, including drugs that do not cross the placenta as

the fetus can still be harmed by alterations in maternal physiology. For example, fetal hypoxia can rapidly ensue if maternal blood pressure falls precipitously. During the postnatal period, normal prescribing practice may still not be appropriate if the woman is breast-feeding and there is significant transfer of the drug into breast milk (BNF 2010).

Chronic Obstetric and Gynecological Pain

Women are overrepresented in almost all chronic pain conditions (LeResche 2000); however, they also experience female-specific conditions such as dysmenorrhea and vulvodynia. Although it is interesting how often it is forgotten quite how prevalent chronic pelvic pain (CPP) is in men (Baranowski 2012). While it is clear that CPP in women is common, with estimates suggesting a community prevalence similar to that of back pain and asthma (Zondervan et al. 1999), studies of the epidemiology of pelvic pain have been hampered by a lack of consensus on a definition. Over recent years the taxonomy of pelvic pain has been reassessed, moving away from an end-organ classification and considering both conditions with a known pathology and pain "syndromes" where no obvious causative pathology is identified (Fall et al. 2010). This work has culminated in a recent revision of the IASP classification of chronic pain http://www.iasp-pain.org/Content/NavigationMenu/Publications/FreeBooks/Classification_of_Chronic_Pain/default.htm.

Given the evidence supporting a conceptual shift towards chronic pain being considered as a disease in its own right rather than as a symptom (Tracey and Bushnell 2009), it is probably unsurprising that over recent years it has become apparent that gynecological pain conditions share many features with other chronic pain disorders. Thus vulvodynia, endometriosis, and chronic pelvic pain without identifiable pathology have been associated with reductions in quality of life (QoL) (Mathias et al. 1996; Breivik et al. 2006; Petrelluzzi et al. 2008) and altered psychology (Low and Edelmann 1990; Jelovsek et al. 2008; Romao et al. 2009), brain volume (Schweinhardt et al. 2008;

As-Sanie et al. 2012), and behavioral (Pukall et al. 2005; He et al. 2010) and central responses to noxious stimuli (Pukall et al. 2005). Perhaps more concerning is the recent evidence that dysmenorrhea, a symptom affecting up to 90 % of adolescents (French 2005), with one in eight adult women reporting severe pain (Donaldson 2009), is similarly associated with long-lasting changes. Moreover, these alterations are present throughout the month, not just at the time of menstruation. Thus recent studies have shown women with dysmenorrhea (when compared to pain-free controls) to have reduced QoL (Unsal et al. 2010; Vincent et al. 2011); increased sensitivity to noxious stimuli, both somatic (Bajaj et al. 2002; Vincent et al. 2011) and visceral (Brinkert et al. 2007); altered central processing of noxious stimuli (Vincent et al. 2011); and altered brain volume (Tu et al. 2010). Additionally, all the conditions mentioned above are also associated with reduced activity of the hypothalamic-pituitary-adrenal (HPA) axis (Heim et al. 1997; Petrelluzzi et al. 2008; Ehrstrom et al. 2009; Vincent et al. 2011), potentially exacerbating pain and increasing the likelihood of other somatic symptoms (Straub and Cutolo 2001).

Obstetric pain is usually considered to be relatively short lived. For some women, pain becomes an increasing problem as pregnancy progresses (e.g., symphysis pubis dysfunction, lower back pain), and it is almost universally associated with labor and delivery. However, the entry on postpartum pain highlights how common pain is in the immediate postnatal period and that this can persist beyond the first year in a not insignificant proportion of women (Eisenach et al. 2013).

Mechanisms of Pain Generation and Modulation in Women

Traditionally gynecological pain was thought mainly to originate in the pelvic viscera, explaining many of the associated features including poor localization, referred pain, and comorbid symptoms (Berkley 2005). However, a large proportion of women with CPP will have a musculoskeletal component to their pain

(Prendergast and Weiss 2003), as do those with postpartum pain (Kanakaris et al. 2011), dyspareunia (Rosenbaum and Owens 2008), and interstitial cystitis (bladder pain syndrome) (Giamberardino 2003). Additionally, somatic (e.g., vulvodynia (Cervero 1994)) and neuropathic sensations (e.g., pudendal neuralgia (Labat et al. 2008)) can be important in generating pain.

Of course in many chronic pain conditions, no peripheral nociceptive input is identified, or the pain experienced may be disproportionate to the extent of any pathology that is found. The central changes described earlier may contribute to these clinical observations, and the entry on pain mechanisms in endometriosis elegantly demonstrates how this pathology in particular can engage the nervous system via a variety of mechanisms (Stratton and Berkley 2011).

No matter how pain is initially generated, it is well known that a huge number of factors can influence the pain experience (Tracey and Mantyh 2007). Psychological factors may be of particular importance in obstetric and gynecological pain because of the association with life events (menarche, pregnancy/childbirth, menopause) and very personal experiences such as sexual relationships and fertility. However, it is important not to forget the influence of sex steroid hormones on factors associated with pain (including psychology/mood (Rubinow and Schmidt 2006), cognition (Craig et al. 2008), and inflammation (Bird et al. 2008)) and on the pain experience itself (Martin 2009). While behavioral studies investigating the relationship between hormones and pain have been difficult to interpret and frequently contradictory (Sherman and LeResche 2006), imaging studies, especially those assessing absolute levels of hormones rather than assigning arbitrary menstrual cycle phases, are beginning to reveal significant interactions between sex steroid hormones and endogenous pain inhibitory mechanisms in healthy women. Thus, estradiol appears to interact with the endogenous opioid system, with low estradiol levels being associated with reduced endogenous opioid tone (Smith et al. 2006), while both progesterone and estradiol influence GABAergic inhibition (Epperson et al. 2002). Interestingly, testosterone is related to activity in descending inhibitory pathways, but only in women in a low estradiol state (Vincent et al. 2013). To date, these relationships have not, however, been investigated in women with chronic pain conditions.

Management of Chronic Pain in Obstetrics and Gynecology

Unfortunately, although our understanding of chronic pain conditions has increased, there have been few advances in therapeutic options for these women. The strength of the evidence supporting some of the regularly used treatments has increased however (Yap et al. 2004; Davis et al. 2007; Wong et al. 2009; Marjoribanks et al. 2010). A number of entries in this section (e.g., clitoral pain, dyspareunia and vaginismus, vulvodynia) draw attention to the intimate and embarrassing nature of these symptoms and the fact that without allowing the woman to reveal all her pain symptoms and their associations in detail, adequate management cannot be achieved. Given how many women with CPP describe not being believed or made to feel that the pain is "all in their head" (Grace 1995; Price et al. 2006), the importance of the doctor-patient relationship cannot be over emphasized (Selfe et al. 1998). However, doctors should not expect to manage chronic pain on their own. The pain conditions seen in obstetrics and gynecology are no different from those presenting to other specialties, in that their management requires multidisciplinary input. For example, the prevalence of musculo-skeletal factors almost always warrants a detailed assessment by a specially trained women's health physiotherapist (Prendergast and Weiss 2003), while psychologists and psychosexual counselors are also likely to be key members of the team (Fall et al. 2010).

Summary

It can be seen that pain is an important problem within obstetrics and gynecology, which is only recently beginning to be given the attention it deserves, both clinically and scientifically. As many patients are reluctant to reveal the details of their pain, either through fear of not being believed or because of the intimate nature of their symptoms, an increased awareness among obstetricians and gynecologists is required. However, other specialists who may also see pain in women need to be aware of the influence of sex steroid hormones on the pain experience and of the prescribing issues relevant to a fertile, pregnant, or nursing population. For further advances to be made, however, it is crucial that clinicians and basic scientists collaborate to integrate research findings into clinical practice.

References

ACOG. (2006). ACOG committee opinion. No. 339: Analgesia and cesarean delivery rates. *Obstetrics and Gynecology, 107*(6), 1487–1488.

As-Sanie, S., Harris, R. E., et al. (2012). Changes in regional gray matter volume in women with chronic pelvic pain: A voxel-based morphometry study. *Pain, 153*(5), 1006–1014.

Bacq, Y., Zarka, O., et al. (1996). Liver function tests in normal pregnancy: A prospective study of 103 pregnant women and 103 matched controls. *Hepatology, 23*(5), 1030–1034.

Bajaj, P., Bajaj, P., et al. (2002). A comparison of modality-specific somatosensory changes during menstruation in dysmenorrheic and nondysmenorrheic women. *The Clinical Journal of Pain, 18*(3), 180–190.

Baranowski, A. P. (2012). Urogenital/pelvic pain in men. *Current Opinion in Supportive and Palliative Care, 6*(2), 213–219.

Berkley, K. J. (2005). A life of pelvic pain. *Physiology and Behavior, 86*(3), 272–280.

Bird, M. D., Karavitis, J., et al. (2008). Sex differences and estrogen modulation of the cellular immune response after injury. *Cellular Immunology, 252*(1–2), 57–67.

BNF. (2010). *British national formulary 60*. London: BMJ Group and RPS.

Breivik, H., Collett, B., et al. (2006). Survey of chronic pain in Europe: Prevalence, impact on daily life, and treatment. *European Journal of Pain, 10*(4), 287–333.

Brinkert, W., Dimcevski, G., et al. (2007). Dysmenorrhoea is associated with hypersensitivity in the sigmoid colon and rectum. *Pain, 132*(1), S46–S51.

Cantwell, R., Clutton-Brock, T., et al. (2011). Saving mothers' lives: Reviewing maternal deaths to make motherhood safer: 2006–2008. The eighth report of the confidential enquiries into maternal deaths in the United Kingdom. *BJOG: An International Journal of Obstetrics and Gynaecology, 118*(1), 1–203.

Cervero, F. (1994). Sensory innervation of the viscera: Peripheral basis of visceral pain. *Physiological Reviews, 74*(1), 95–138.

Chen, T. H., Chang, S. P., et al. (2004). Prevalence of depressive and anxiety disorders in an assisted reproductive technique clinic. *Human Reproduction, 19*(10), 2313–2318.

Chestnut, D. H. (1997). Does epidural analgesia during labor affect the incidence of cesarean delivery? *Regional Anesthesia, 22*(6), 495–499.

Craig, M. C., Fletcher, P. C., et al. (2008). Physiological variation in estradiol and brain function: A functional magnetic resonance imaging study of verbal memory across the follicular phase of the menstrual cycle. *Hormones and Behavior, 53*(4), 503–508.

Cutura, N., Soldo, V., et al. (2011). Optimal dose of an anesthetic in epidural anesthesia and its effect on labor duration and administration of vacuum extractor and forceps. *Clinical and Experimental Obstetrics and Gynecology, 38*(3), 247–250.

Davis, L., Kennedy, S., et al. (2007). Modern combined oral contraceptives for pain associated with endometriosis. *Cochrane Database of Systematic Reviews, 3*, CD001019 (Online).

de Mouzon, J., Lancaster, P., et al. (2009). World collaborative report on assisted reproductive technology, 2002. *Human Reproduction, 24*(9), 2310–2320.

Donaldson, L. (2009). *150 Years of the annual report of the chief medical officer: On the state of public health 2008*. London: DoH.

Ehrstrom, S., Kornfeld, D., et al. (2009). Chronic stress in women with localised provoked vulvodynia. *Journal of Psychosomatic Obstetrics and Gynaecology, 30*(1), 73–79.

Eisenach, J. C., Pan, P., et al. (2013). Resolution of pain after childbirth. *Anesthesiology, 118*(1), 143–151.

Epperson, C. N., Haga, K., et al. (2002). Cortical gamma-aminobutyric acid levels across the menstrual cycle in healthy women and those with premenstrual dysphoric disorder: A proton magnetic resonance spectroscopy study. *Archives of General Psychiatry, 59*(9), 851–858.

Fall, M., Baranowski, A. P., et al. (2010). EAU guidelines on chronic pelvic pain. *European Urology, 57*(1), 35–48.

French, L. (2005). Dysmenorrhea. *American Family Physician, 71*(2), 285–291.

Giamberardino, M. A. (2003). Referred muscle pain/hyperalgesia and central sensitisation. *Journal of Rehabilitation Medicine, 41*, 85–88.

Grace, V. M. (1995). Problems women patients experience in the medical encounter for chronic pelvic pain: A New Zealand study. *Health Care for Women International, 16*(6), 509–519.

He, W., Liu, X., et al. (2010). Generalized hyperalgesia in women with endometriosis and its resolution following a successful surgery. *Reproductive Sciences, 17*(12), 1099–1111.

Heim, C., Ehlert, U., et al. (1997). Psychoendocrinological observations in women with chronic pelvic pain. *Annals of the New York Academy of Sciences, 821*, 456–458.

http://www.iasp-pain.org/Content/NavigationMenu/Publications/FreeBooks/Classification_of_Chronic_Pain/default.htm

Jelovsek, J. E., Walters, M. D., et al. (2008). Psychosocial impact of chronic vulvovagina conditions. *The Journal of Reproductive Medicine, 53*(2), 75–82.

Kanakaris, N. K., Roberts, C. S., et al. (2011). Pregnancy-related pelvic girdle pain: An update. *BMC Medicine, 9*, 15.

Labat, J. J., Riant, T., et al. (2008). Diagnostic criteria for pudendal neuralgia by pudendal nerve entrapment (Nantes criteria). *Neurourology and Urodynamics, 27*(4), 306–310.

LeResche, L. (2000). Epidemiologic perspectives on sex differences in pain. In R. B. Fillingim (Ed.), *Sex, gender and pain* (pp. 233–249). Seattle: IASP Press.

Low, W. Y., & Edelmann, R. J. (1990). Psychosocial aspects of endometriosis: A review. *Journal of Psychosomatic Obstetrics and Gynaecology, 12*, 3–12.

Lurie, S., Rahamim, E., et al. (2008). Total and differential leukocyte counts percentiles in normal pregnancy. *European Journal of Obstetrics, Gynecology, and Reproductive Biology, 136*(1), 16–19.

Marjoribanks, J., Proctor, M. L., et al. (2010). Nonsteroidal anti-inflammatory drugs for dysmenorrhoea. *Cochrane Database of Systematic Reviews, 20*(1), CD001751.

Martin, V. T. (2009). Ovarian hormones and pain response: A review of clinical and basic science studies. *Gender Medicine, 6*(2), 168–192.

Mathias, S. D., Kuppermann, M., et al. (1996). Chronic pelvic pain: Prevalence, health-related quality of life, and economic correlates. *Obstetrics and Gynaecology, 87*(3), 321–327.

Patel, S. J., Reede, D. L., et al. (2007). Imaging the pregnant patient for nonobstetric conditions: Algorithms and radiation dose considerations. *Radiographics, 27*(6), 1705–1722.

Petrelluzzi, K. F. S., Garcia, M. C., et al. (2008). Salivary cortisol concentrations, stress and quality of life in women with endometriosis and chronic pelvic pain. *Stress, 11*(5), 390–397.

Prendergast, S. A., & Weiss, J. (2003). Screening for musculoskeletal causes of pelvic pain. *Clinical Obstetrics and Gynecology, 46*(4), 773–782.

Price, J., Farmer, G., et al. (2006). Attitudes of women with chronic pelvic pain to the gynaecological consultation: A qualitative study. *British Journal of Obstetrics and Gynaecology, 113*(4), 446–452.

Pukall, C. F., Strigo, I. A., et al. (2005). Neural correlates of painful genital touch in women with vulvar vestibulitis syndrome. *Pain, 115*(1–2), 118–127.

Romao, A. P., Gorayeb, R., et al. (2009). High levels of anxiety and depression have a negative effect on quality of life of women with chronic pelvic pain. *International Journal of Clinical Practice, 63*(5), 707–711.

Rosenbaum, T. Y., & Owens, A. (2008). The role of pelvic floor physical therapy in the treatment of pelvic and genital pain-related sexual dysfunction (CME). *The Journal of Sexual Medicine, 5*(3), 513–523. quiz 524–515.

Rubinow, D. R., & Schmidt, P. J. (2006). Gonadal steroid regulation of mood: The lessons of premenstrual syndrome. *Frontiers in Neuroendocrinology, 27*(2), 210–216.

Schweinhardt, P., Kuchinad, A., et al. (2008). Increased gray matter density in young women with chronic vulvar pain. *Pain, 140*(3), 411–419.

Selfe, S. A., Matthews, Z., et al. (1998). Factors influencing outcome in consultations for chronic pelvic pain. *Journal of Women's Health, 7*(8), 1041–1048.

Sherman, J. J., & LeResche, L. (2006). Does experimental pain response vary across the menstrual cycle? A methodological review. *American Journal of Physiology – Regulatory, Integrative and Comparative Physiology, 291*(2), R245–R256.

Smith, Y. R., Stohler, C. S., et al. (2006). Pronociceptive and antinociceptive effects of estradiol through endogenous opioid neurotransmission in women. *Journal of Neuroscience, 26*(21), 5777–5785.

Stratton, P., & Berkley, K. J. (2011). Chronic pelvic pain and endometriosis: Translational evidence of the relationship and implications. *Human Reproduction Update, 17*(3), 327–346.

Straub, R. H., & Cutolo, M. (2001). Involvement of the hypothalamic-pituitary-adrenal/gonadal axis and the peripheral nervous system in rheumatoid arthritis: Viewpoint based on a systemic pathogenetic role. *Arthritis and Rheumatism, 44*(3), 493–507.

Tracey, I., & Bushnell, M. C. (2009). How neuroimaging studies have challenged us to rethink: Is chronic pain a disease? *The Journal of Pain, 10*(11), 1113–1120.

Tracey, I., & Mantyh, P. W. (2007). The cerebral signature for pain perception and its modulation. *Neuron, 55*(3), 377–391.

Tu, C.-H., Niddam, D. M., et al. (2010). Brain morphological changes associated with cyclic menstrual pain. *Pain, 150*(3), 462–468.

Unsal, A., Ayranci, U., et al. (2010). Prevalence of dysmenorrhea and its effect on quality of life among a group of female university students. *Upsala Journal of Medical Sciences, 115*(2), 138–145.

Vincent, K., Warnaby, C., et al. (2011). Dysmenorrhoea is associated with central changes in otherwise healthy women. *Pain, 152*(9), 1966–1975.

Vincent, K., Warnaby, C., et al. (2013). Brain imaging reveals that engagement of descending inhibitory pain pathways in healthy women in a low endogenous estradiol state varies with testosterone. *Pain, 154*(4), 515–524.

Wong, C. L., Farquhar, C., et al. (2009). Oral contraceptive pill for primary dysmenorrhoea. *Cochrane Database of Systematic Reviews, 7*(4), CD002120.

Yap, C., Furness, S., et al. (2004). Pre and post operative medical therapy for endometriosis surgery. *Cochrane Database of Systematic Reviews, 3*, CD003678 (online).

Zondervan, K. T., Yudkin, P. L., et al. (1999). Prevalence and incidence of chronic pelvic pain in primary care: Evidence from a national general practice database. *British Journal of Obstetrics and Gynaecology, 106*(11), 1149–1155.

Obstetric Brachial Palsy

▶ Plexus Injuries and Deafferentation Pain

Obstipation

Definition

Obstipation refers to severe constipation or reduction in bowel motility possibly due to bowel obstruction.

Cross-References

▶ Opioids in Geriatric Application

Occlusal Force

Definition

Force is generated between the teeth during voluntary, usually maximal, activation of the jaw-closing muscles. Measurement requires insertion of a transducer between the upper and lower teeth.

Cross-References

▶ Orofacial Pain, Movement Disorders

Occupational Behavior

Definition

Occupational behavior is defined as the clients' amount and types of activities and the meaningfulness of their activities (i.e., enjoyability, autonomy, and competency) and is often related to their life satisfaction. Included are emotional reactions, e.g., emotional, stress, and physical reactions to occupational demands, work organization, and psychosocial work environment. Important occupational behavior components for performing job tasks are motivation, initiative, concentration, attendance, attention span, decision-making, responsibility for production, physical stamina, and ability to accept supervision and criticism, as well as interpersonal relationships with workmates and manager.

Cross-References

▶ Vocational Counseling

Occupational Capacity Evaluation

▶ Disability, Functional Capacity Evaluations

Occupational Pain

▶ Pain in the Workplace, Risk Factors for Chronicity, Job Demands

Occupational Therapist

Definition

An occupational therapist is a health-care professional that aims to enable individuals to perform meaningful and purposeful activities across the lifespan. He/she thus works primarily in teams with the activity domain in rehabilitation, not only with activities of daily living but also with activity analyses, with trying out technical aids and with occupational training. The therapist focuses on adapting the environment, modifying the task, teaching the skill, and educating the client/family in order to increase participation in and performance of activities. The profession stems from the 1920s, has developed during wartime, usually needs a state license to practice, and has an academic training in about half of western countries.

Cross-References

▶ Physical Medicine and Rehabilitation, Team-Oriented Approach

Occupational Therapy in Children

▶ Chronic Pain in Children, Physical Medicine and Rehabilitation

Ocular Nociceptors

Carlos Belmonte
Instituto de Neurociencias, Universidad Miguel Hernandez-CSIC, San Juan de Alicante, Spain

Synonyms

Eye nociceptors; Eye pain receptors; Ocular pain receptors

Definition

Sensory afferent fibers that innervate the various tissues of the eye and are activated by physical and chemical stimuli of intensity near or within the noxious range, giving rise to sensations of discomfort or pain referred to the eye and/or the periocular region.

Characteristics

Anatomy of Ocular Innervation

The cell bodies of the ocular sensory afferents, most of which are of small or medium size, are located in the ipsilateral ▶ trigeminal ganglion chiefly clustered in the ophthalmic (medial) region of the ganglion. The axons of these pseudomonopolar neurons divide into a peripheral branch that projects to the peripheral target tissues and a central branch that enters the brainstem to reach the trigeminal brainstem sensory nuclear complex. Sensory nerve fibers directed to the ocular region run with the ophthalmic nerve, which branches into nasociliary, frontal, and lacrimal nerves. The ciliary nerves originate in the nasociliary nerve and travel to the eyeball, where they innervate connective tissue, epithelia, and blood vessels of the uvea, ciliary body, choroid, scleral spur, lids, sclera, cornea, and bulbar conjunctiva. The richest nerve density is found in the cornea. The lens and the retina do not receive direct trigeminal sensory innervation. The remaining ophthalmic nerve branches carry sensory nerve fibers that originate in orbital tissues, extraocular muscles, palpebral conjunctiva, lids, lacrimal gland, periocular skin and mucosae of the nose, frontal sinus, and the lacrimal sac.

A small proportion of ocular sensory fibers are large-diameter, myelinated axons that possess morphological specializations at their terminal portion typical of mechanoreceptors. Most of them however are thin myelinated and unmyelinated fibers.

In the cornea, sensory axons are grouped in a variable number of radially oriented nerve bundles that branch extensively and form the subbasal plexus below the basal epithelial cells. From this plexus, naked single fibers ascend vertically between the epithelial cells ending at variable depths (for review see Müller et al. 2003).

Immunocytochemical staining reveals the presence of different ▶ neuropeptides in the cell soma and axons of the ocular sensory neurons. About 50 % of corneal neurons are immunoreactive to calcitonin gene-related peptide (▶ CGRP), of which 20 % also contain ▶ substance P (SP) (de Felipe et al. 1999). Small amounts of other neuropeptides such as cholecystokinin, vasopressin, neurotensin, and ▶ β-endorphin have been also reported in ocular sensory neurons. Neurotrophins such as ▶ nerve growth factor (NGF), glial-cell-line-derived ▶ neurotrophic factor (GDNF, artemin), ▶ brain-derived neurotrophic factor (BDNF), ▶ neurotrophin-3 (NT-3), opioid growth factor (OGF), ciliary neurotrophic factor (CNTF), and pigment-epithelium-derived factor (PEDF) are found in the cornea (for review, see Müller et al. 2003). They may participate, as target-tissue-derived growth factors, in the development, maintenance, and regeneration of nerves (Bonini et al. 2003). Also, an accumulating evidence suggests that neuropeptides, in particular SP and CGRP, and neurotrophins (especially NGF) promote corneal healing and/or maintenance of epithelial integrity either alone or in combination with other growth factors (like insulin-like growth factor) (Gallar et al. 1990; Garcia-Hirschfeld et al. 1994; Lambiase et al. 2000; Bonini et al. 2003; Müller et al. 2003).

Functional Properties of Ocular Sensory Receptors

Functional studies of ocular sensory fibers have been performed mainly in the cornea. The majority of corneal sensory nerve fibers, about 70 %, are ▶ polymodal nociceptors. They are activated by near-noxious mechanical energy, heat, and chemical irritants and by a large variety of endogenous chemical mediators released by damaged corneal tissue, resident inflammatory cells or which leak from vessels in the periphery of the cornea (the limbus). Some of the polymodal nociceptor fibers belong to the group of thin myelinated (A-delta) nerve fibers, but most of them are of the C type. Polymodal nociceptors respond to their natural stimuli with a continuous, irregular discharge of nerve impulses that persist as long as the stimulus is maintained. They have a firing frequency roughly proportional to the intensity of the stimulating force. Therefore, the impulse discharge of polymodal nociceptors does not only signal the presence of a noxious stimulus, but it also encodes to a certain degree its intensity and duration. Polymodal nociceptors have a mechanical threshold slightly lower than ▶ mechanonociceptors (see below), and when stimulated with heat, they begin to fire at temperatures over 39–40°C. A fraction of polymodal fibers (around 50 %) also increase their firing rate when the corneal temperature is reduced below 29°C. Many chemical agents known to excite ▶ polymodal nociceptors of other territories also activate ocular nociceptors. Acidic solutions (of pH 5.0–6.5), or gas jets containing increasing concentrations of CO_2 (carbonic acid formation at the corneal surface drops the local pH), evoke an impulse discharge in corneal polymodal nociceptors (e.g., Belmonte et al. 1991; Gallar et al. 1993; MacIver and Tanelian 1993; Chen et al. 1995; Acosta et al. 2001; for review, see Belmonte et al. 1997, 2004; Belmonte and Gallar 2011).

About 15–20 % of the axons innervating the cornea (all thin myelinated) respond only to mechanical forces in the order of magnitude close to that required to damage corneal epithelial cells. Accordingly, they belong to the mechanonociceptor type. The fibers of this class of receptor fire one or a few nerve impulses either in response to brief or sustained indentations of the corneal surface and often also when the stimulus is removed. Thus, they are "phasic sensory receptors" that signal the presence of the stimulus and, in a very limited degree, its intensity and duration (Belmonte et al. 1997; MacIver and Tanelian 1993). The threshold force required to activate mechanonociceptors is apparently low (about 0.6 mN), far below the force that activates mechanonociceptor fibers in the skin. However, this intensity might be sufficient to damage unkeratinised corneal epithelium. Mechanonociceptors in the cornea are probably responsible for the acute, sharp

sensation of pain produced by touching the corneal surface. The aftersensations of pain elicited by noxious stimuli are probably explained by the more sustained activity exhibited by polymodal nociceptors.

Another category of corneal nerve fibers that represents 10–15 % of the total population are cold-sensitive thermal receptors. These are ▶ A-delta and ▶ C fibers that discharge spontaneously at rest and increase their firing rate when the normal temperature of the corneal surface (around 33°C) is reduced, while they are transiently silenced upon warming (Tanelian and Beuerman 1984; Gallar et al. 1993; Brock et al. 1998). They also increase their firing rate as soon as the temperature of the cornea drops because of evaporation at the corneal surface, application of cold solutions, or blowing of cold air on the cornea. Cold receptor fibers are able to detect and encode as a change in impulse frequency small temperature variations of 0.1°C or less (Gallar et al. 1993, 2003), thus allowing for the perception of decreases in corneal temperature as a conscious sensation of innocuous cooling. (Acosta et al. 2001).

Finally, it has been suggested that the cornea possesses ▶ mechanically insensitive, "silent" nociceptors, i.e., nerve terminals that are not activated by mechanical or thermal stimuli when the tissue is intact, but in the case of local inflammation, they become responsive to these exogenous stimuli as well as to a variety of endogenous chemicals. Although the experimental evidence for their presence in the cornea is only indirect, they have been identified in virtually all other somatic tissues. Thus, it seems likely that such nociceptors also exist in the cornea.

Variable activation of the different populations of corneal sensory afferents by stimuli of different modalities evokes sensations that also differ qualitatively. Sensations resulting from the excitation of mechano- and polymodal nociceptors always have an irritative component. Only when cold fibers are stimulated separately with small temperature decreases are nonnoxious sensations of cooling evoked. Larger temperature

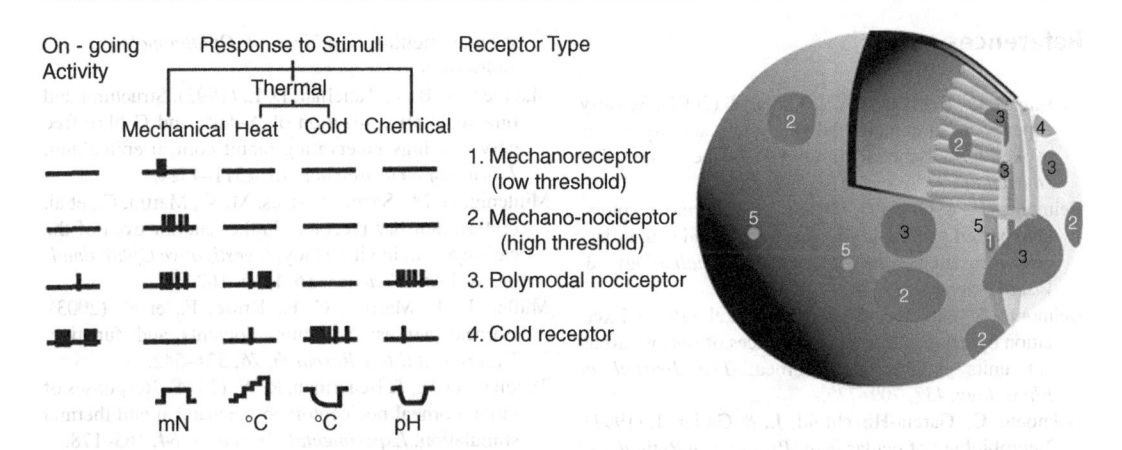

Ocular Nociceptors, Fig. 1 Functional types of sensory fibers innervating the eye. The *left-hand side* of the figure represents the pattern of impulse discharge, either spontaneous or evoked by the different modalities of stimuli, produced by each functional receptor type. The *right-hand side* shows schematically the location and relative size of the receptive fields of the various classes of ocular sensory receptors (Modified from Belmonte et al. 1997)

decreases also recruit polymodal nociceptors, resulting in an irritating cooling sensation (Acosta et al. 2001).

Electrophysiological studies dedicated to identify sensory receptor types in ocular structures have shown that the same main functional classes of sensory afferents identified in the cornea and the episclera, i.e., mechanonociceptors, polymodal nociceptors, and cold receptors are also present in the bulbar conjunctiva, the scleral surface, the iris, and the ciliary body (e.g., Gallar 1991; Mintenig et al. 1995; for review, see Belmonte and Gallar 2011). Cold fibers with response properties similar to the thermal receptors present in the cornea and limbus are also found in unexposed areas of the iris and posterior sclera. It has been hypothesized that such scleral cold receptors contribute to the detection of choroidal and retinal blood flow changes for reflex blood flow regulation rather than to the production of conscious thermal sensations (Gallar et al. 2003). Likewise, endings in the chamber angle with the appearance of mechanosensory terminals have been associated with neural regulation of intraocular pressure (Belmonte et al. 1973). Figure 1 represents schematically the functional types of sensory endings identified electrophysiologically in the various structures of the cat's eye.

Polymodal nociceptors of the cornea and the uvea not only encode the intensity of noxious stimuli but also become sensitized following injurious stimulation (e.g., Gallar et al. 1993; Mintenig et al. 1995). These characteristics explain the acute pain sensations and the development of primary hyperalgesia following injuries of the ocular surface, as well as the pain reported when the iris is touched accidentally during ocular surgery or often when argon laser pulses are applied to the posterior uvea (for review, see Belmonte et al. 2004; Belmonte and Gallar 2011).

Intense ocular pain is also observed in parallel with the rise in intraocular pressure that takes place in congestive glaucoma (see below). However, artificial increases of intraocular pressure up to 100 mmHg in the intact cat resulted in transient impulse discharges in most corneal, scleral, and iridal sensory fibers with only a small proportion of them giving a sustained firing (Zuazo et al. 1986). Possibly, in the noninflamed eye very high intraocular pressures are required to effectively stimulate the mechanosensory and polymodal fibers that innervate the eye coats. When uveal inflammation develops as during congestive glaucoma, ▶ sensitization of nociceptors increases their excitability to mechanical stimulation, making them more responsive to pressure.

References

Acosta, M. C., Belmonte, C., & Gallar, J. (2001). Sensory experiences in humans and single-unit activity in cats evoked by polymodal stimulation of the cornea. *The Journal of Physiology, 534,* 511–525.

Belmonte, C., & Gallar, J. (2011). Cold thermoreceptors, unexpected players in tear production and ocular dryness sensations. *Investigative Ophthalmology & Visual Science, 52,* 3888–3892.

Belmonte, C., Gallar, J., Pozo, M. A., et al. (1991). Excitation by irritant chemical substances of sensory afferent units in the cat's cornea. *The Journal of Physiology, 437,* 709–725.

Belmonte, C., Garcia-Hirschfeld, J., & Gallar, J. (1997). Neurobiology of ocular pain. *Progress in Retinal and Eye Research, 16,* 117–156.

Belmonte, C., Acosta, M. C., & Gallar, J. (2004). Neural basis of sensation in intact and injured corneas. *Experimental Eye Research, 78,* 513–525. doi: 10.1016/j.exer.2003.09.023

Bonini, S., Rama, P., Olzi, D., et al. (2003). Neurotrophic keratitis. *Eye, 17,* 989–995.

Brock, J. A., McLachlan, E. M., & Belmonte, C. (1998). Tetrodotoxin-resistant impulses in single nociceptor nerve terminals in guinea-pig cornea. *The Journal of Physiology, 512,* 211–217.

Chen, X., Gallar, J., Pozo, M. A., et al. (1995). CO_2 stimulation of the cornea: A comparison between human sensation and nerve activity in polymodal nociceptive afferents of the cat. *European Journal of Neuroscience, 7,* 1154–1163.

De Felipe, C., González, G. G., Gallar, J., et al. (1999). Quantification and immunocytochemical characteristics of trigeminal ganglion neurons projecting to the cornea: Effect of corneal wounding. *European Journal of Pain, 3,* 31–39.

Gallar, J., Pozo, M. A., Rebollo, I., & Belmonte, C. (1990). Effects of capsaicin on corneal wound healing. *Investigative Ophthalmology & Visual Science, 31,* 1968–1974.

Gallar, J. (1991). Caracterización electrofisiológica de los receptores sensoriales del ojo del gato. PhD Thesis, Universidad de Alicante, Spain.

Gallar, J., Pozo, M. A., Tuckett, R. P., et al. (1993). Response of sensory units with unmyelinated fibres to mechanical, thermal and chemical stimulation of the cat's cornea. *The Journal of Physiology, 468,* 609–622.

Gallar, J., Acosta, M. C., & Belmonte, C. (2003). Activation of scleral cold thermoreceptors by temperature and blood flow changes. *Investigative Ophthalmology & Visual Science, 44,* 697–705.

Garcia-Hirschfeld, J., Lopez-Briones, L.G., & Belmonte, C. (1994). Neurotrophic influences on corneal epithelial cells. *Experimental Eye Research, 59,* 597–605.

Lambiase, A., Bonini, S., Aloe, L., & Rama, P. (2000a). Anti-inflammatory and healing properties of nerve growth factor in immune corneal ulcers with stromal melting. *Archives of Ophthalmology, 118,* 1446–1449.

MacIver, M. B., & Tanelian, D. L. (1993). Structural and functional specialization of A-delta and C fibre free nerve endings innervating rabbit corneal epithelium. *Journal of Neuroscience, 13,* 4511–4524.

Mintenig, G. M., Sanchez-Vives, M. V., Martin, C., et al. (1995). Sensory receptors in the anterior uvea of the cat's eye. An in vitro study. *Investigative Ophthalmology & Visual Science, 36,* 1615–1624.

Müller, L. J., Marfurt, C. F., Kruse, F., et al. (2003). Corneal nerves: Structure, contents and function. *Experimental Eye Research, 76,* 521–542.

Tanelian, D. L., & Beuerman, R. W. (1984). Responses of rabbit corneal nociceptors to mechanical and thermal stimulation. *Experimental Neurology, 84,* 165–178.

Zuazo, A., Ibañez, J., & Belmonte, C. (1986). Sensory nerve responses elicited by experimental ocular hypertension. *Experimental Eye Research, 43,* 759–769.

Ocular Pain Receptors

▶ Ocular Nociceptors

Odds Ratio

Definition

This is a calculation of the degree to which the probability of a given outcome in a sample exceeds the probability of the outcome in the control group.

Cross-References

▶ Psychiatric Aspects of the Epidemiology of Pain

ODI

▶ Oswestry Disability Index

Odorant

Definition

Molecule capable of eliciting responses from receptors in the olfactory mucosa.

Cross-References

▶ Perireceptor Elements

Off-Cells (RVM)

Definition

RVM neurons are located in the rostroventral medulla and are defined by a sudden pause in firing that begins just prior to execution of the tail flick, paw withdrawal, or other nocifensive reflex. Spontaneous activity is variable with time, and an off-cell can only be identified if it is firing at the time at which the reflex is evoked. Off-cells are not directly responsive to mu-opioids, but are activated indirectly by mu-opioids given systemically or by microinjection into the PAG, basolateral amygdala, or in the RVM itself. RVM off-cells, as defined in vivo, are presumed to be a subset of primary cells defined in vitro.

Cross-References

▶ Opiates: Pharmacology of Pain
▶ Opiates, Rostral Ventromedial Medulla, and Descending Control
▶ Pain-Modulatory Systems, History of Discovery

OFQ

▶ Orphanin FQ

OFQ/N

▶ Orphanin FQ

Older People

▶ Pain Assessment in the Elderly

Oligonucleotide Array

Definition

Also called "gene chip," this is a technology that permits simultaneous semiquantitative evaluation of the degree of expression of large numbers of genes in a tissue sample of choice.

Cross-References

▶ Central Changes after Peripheral Nerve Injury

On-Cells (RVM)

Definition

On-Cells are neurons in the rostral ventromedial medulla (RVM) defined by a sudden burst of activity that begins just prior to execution of the tail flick, paw withdrawal, or other nocifensive reflex. Spontaneous activity is variable with time, so that the reflex-related burst may not be discernible if the reflex is evoked at a time when the neuron is already firing spontaneously. On-cells are directly sensitive to mu-opioid agonists and are inhibited by morphine or mu-agonists given systemically or by microinjection into the periaqueductal gray, basolateral amygdala, or in the RVM itself.

Cross-References

▶ Opiates, Rostral Ventromedial Medulla, and Descending Control
▶ Opiates: Pharmacology of Pain
▶ Pain-Modulatory Systems, History of Discovery

Oncological Pain

▶ Cancer Pain

Ondansetron

Definition

Ondansetron, a $5HT_3$ receptor antagonist and an antiemetic agent commonly used for postanesthetic nausea and vomiting, reduces the overall analgesic effect of tramadol by approximately 25 %, probably by blocking spinal $5HT_3$ receptors.

Cross-References

▶ Postoperative Pain, Tramadol

On-demand Analgesia

▶ Postoperative Pain, Patient-Controlled Analgesia Devices, Parenteral

On-demand Analgesia Computer

▶ Postoperative Pain, Patient-Controlled Analgesia Devices, Parenteral

On-Demand Epidural Analgesia

▶ Postoperative Pain, Patient-Controlled Analgesia Devices, Epidural

Ondine's Curse

Definition

Ondine's curse is the loss of involuntary respiration by an injury to the reticulospinal tract that may be produced by a lesion of the spinothalamic tract that is located nearby. Radio-frequency current lesion making: localized heating of neural or other tissue by delivery of a radio-frequency lesion through an electrode temporarily placed in that tissue for this purpose.

Ongoing Pain

Definition

Baseline pain without any obvious external stimuli, which is constant in nature, described as dull, aching, and throbbing in character, and with progressive severity in accordance with disease.

Cross-References

▶ Cancer Pain, Animal Models
▶ Neuropathic Pain Model, Partial Sciatic Nerve Ligation Model

On-the-Job Site Evaluation

Definition

On-the-job evaluation is performed in a variety of job-related realistic work environments, ideally at the person's employer in his/her real workplace or alternatively in a (artificial) workplace or a training kitchen at a rehabilitation clinic. The client is systematically observed during performance of simulated work tasks or work samples. Job analysis and reports of the client's health-related occupational behavior are recorded. It should be underlined that there are now scientifically valid methods for observation and documentation of behaviors, by (1) identifying and operationalizing a target behavior, (2) identifying a measurement strategy, (3) establishing interrater reliability, and (4) reporting scores and graph data.

Cross-References

▶ Vocational Counseling

Oocyte

Definition

An oocyte is a female gametocyte that develops into an ovum after two meiotic divisions.

Cross-References

▶ Perireceptor Elements

OP1 Receptors

▶ Delta(δ) Opioid Receptor(s)

Operant

Definition

Operant theory (also called behavior analysis) focuses on the relationship between behavior and the contingent consequences (reinforcers)

associated with it. The theory distinguishes four broad classes of reinforcement: positive, negative, punishment, and extinction. Positive and negative reinforcement contingencies increase the likelihood of a behavior occurring; punishment and extinction contingencies decrease the likelihood of the preceding behavior. Operant theory also analyses the events that signal that contingencies between the behavior and the consequences are active (discriminative stimuli and setting events).

Cross-References

▶ Operant Perspective of Pain
▶ Psychology of Pain, Efficacy of Cognitive-Behavioral Treatment

Operant Conditioning

Definition

Operant conditioning is a psychological theory in which learning occurs when a response to a stimulus is reinforced. If a reward or reinforcement follows the response to a stimulus, then the response becomes more probable in the future. If a punishment occurs, or a reward is removed, the response becomes less probable.

Cross-References

▶ Impact of Familial Factors on Children's Chronic Pain
▶ Muscle Pain, Fear-Avoidance Model

Operant Conditioning Approaches to Chronic Pain

▶ Behavioral Therapies to Reduce Disability

Operant Conditioning Model of Chronic Pain

Definition

The operant conditioning model of chronic pain hypothesizes that pain and pain-related disability may be maintained by environmental contingencies for overt, observable demonstrations of pain.

Cross-References

▶ Spouse, Role in Chronic Pain

Operant Conditioning or Learning Models of Pain

▶ Operant Perspective of Pain

Operant Escape

Definition

An intentional response that terminates aversive stimulation. Operant conditioning involves learning to manipulate the environment in order to achieve a desired (motivated) result.

Cross-References

▶ Opioids, Effects of Systemic Morphine on Evoked Pain

Operant Perspective of Pain

Michael K. Nicholas
Royal North Shore Hospital, University of Sydney, St. Leonards, NSW, Australia

Synonyms

Operant conditioning or learning models of pain

Definition

The behaviors displayed by people in pain may be modified and maintained by their consequences, to some extent independent of any nociceptive or neuropathic stimuli.

Characteristics

Theory

The ▶ operant model of behavior holds that behaviors can be learned and maintained by their consequences. Two main types of consequences are ▶ positive reinforcement, where the consequence involves the addition of a stimulus or event, and ▶ negative reinforcement, where the consequence involves a stimulus or event being withdrawn. Thus, in positive reinforcement, the consequence might be attention from others or food. In negative reinforcement, the consequence might be cessation of a noxious stimulus. Both types of reinforcement act to strengthen or maintain the behavior they follow.

The learning process of behaviors being acquired through reinforcing consequences is known as operant ▶ conditioning. This is contrasted with ▶ respondent conditioning, in which a behavior or response is elicited by a previously neutral stimulus that has been paired with a stimulus that normally elicits the behavior, as described initially by Pavlov (1927). In Pavlov's study, dogs learned to salivate to the sound of a bell after the bell had been repeatedly rung shortly before the dogs' food was presented. While more recent studies of operant and respondent conditioning have led to further elaboration of these learning processes, the fundamental relationship between responses and subsequent stimuli (not necessarily immediate) remains a key feature of operant conditioning.

Fordyce (1976) argued that while pain was a subjective experience, patients in pain often display behaviors that communicate the presence of pain to others. He termed such behaviors as "▶ pain behaviors." Fordyce proposed that, like other overt behaviors, pain behaviors could be subject to the influence of operant conditioning. As a result, he argued that pain behaviors could become maintained by reinforcement and, to some extent, independent of nociception.

Fordyce proposed that, unlike the ▶ subjective pain experience, pain behaviors were amenable to observation and thus recording and measurement. Pain behaviors in Fordyce's view included verbal complaints of pain (this would include pain intensity ratings), paraverbal sounds (e.g., moans), body posturing and gestures (e.g., limping, grimacing), display of functional limitations (e.g., lying down during the day), and behaviors concerned with pain reduction (e.g., use of analgesic medication, treatment seeking). Actual delineation of pain behaviors varies from study to study, with some focussing on specific behaviors, such as counting the number of grimaces or moans displayed in a given period (e.g., Keefe and Block 1982), while others address larger categories of behavior, such as general activity avoidance or withdrawal from social activities. It should be recognized, however, that caution should be used in defining a pain behavior a priori (or independent of context) as other explanations for the same behavior are often possible (e.g., a hip injury can cause limping).

Fordyce identified attention from others (positive reinforcement), relief of pain (negative reinforcement), and financial gain (positive reinforcement) as potential reinforcers for

▶ chronic pain behavior, but the list of possible reinforcers is long and varies between individuals.

Treatment Implications

In treatments based on the operant model, the focus of treatment is not the original pathological cause of nociception, or ongoing neuropathic processes, but rather the identification and reversal of the reinforcement contingencies (the relationship between the behaviors and their reinforcing consequences) that are thought to be maintaining the persisting pain behaviors. At the same time, well behaviors (activities normally incompatible with pain behaviors, such as exercise and work-related activities) are encouraged instead (e.g., Gil et al. 1988). Well behaviors may be reinforced directly by praise from treatment providers, but they may also be intrinsically reinforcing for pain patients, especially when they represent improvement or the achievement of personally relevant goals. Treatments for chronic pain utilizing operant methods have reported considerable success in increasing the activity levels of chronic pain patients, as well as decreasing distress and pain behaviors, such as medication use, lying down during the day, and reports of pain severity (Morley et al. 1999).

One reinforcer of pain behavior that has been successfully targeted in operant treatments is the verbal response of others (Latimer 1982). White and Sanders (1986), for example, found that the pain reports of four chronic pain patients were significantly reduced by systematically reinforcing "well" talk with positive verbal responses by staff. Similarly, Cairns and Pasino (1977), using a sample of chronic pain patients, found that significant increases in daily walking and exercise tolerance were obtained by verbal reinforcement alone and in combination with visual feedback using graphs.

Experimental Studies

Despite the effectiveness of the operant treatments in modifying pain behaviors, treatment studies cannot demonstrate that reinforcement contingencies played a role in the development and maintenance of pain behaviors nor can they control any noxious stimuli that may be contributing to the patients' pain. In contrast, laboratory studies allow closer control of the noxious stimulus and extraneous variables. However, laboratory studies are also limited by ethical constraints and the degree to which they can simulate the environment of the chronic pain sufferer.

Acknowledging the limitations of laboratory studies, Linton and Gotestam (1985) reported inducing learned pain responses in healthy individuals with no existing learned pain responses. Over fifteen trials, they verbally reinforced (with praise) the pain reports of subjects in the presence of a noxious stimulus (a blood pressure cuff which was inflated to a painful level). In their first experiment, they found that pain reports could be operantly conditioned to increase and decrease, relative to a control condition, even though the intensity of the cuff pressure was kept stable across trials. In their second experiment, the experimenters found that, with verbal reinforcement for reports of increased pain, the pain reports of subjects were maintained and increased even when the intensity of the noxious stimulus decreased. In contrast, for those in the control condition who were not given verbal reinforcement for reports of increased pain, subjects' pain reports decreased as the intensity of the painful stimulus decreased.

Schmidt et al. (1989) argued that instead of providing support for the reinforcement hypothesis, Linton and Gotestam's findings could have been interpreted as evidence of augmentation and amplification of sensations. However, this is unlikely, since Linton and Gotestam (1985) found that the pain reports of subjects could be positively reinforced up and down in the presence of a stable noxious stimulus.

Lousberg et al. (1996) reported a replication study based on Linton and Gotestam's (1985) paradigm but using brief electric shocks as the noxious stimulus with a sample of healthy volunteers. In addition to pain ratings, Lousberg et al. also used psychophysiological measures of skin conductance response and skin conductance level. The results replicated those found previously by Linton and Gotestam, with the

experimental group reporting significantly increased pain levels, relative to the control group. Interestingly, the experimenters also found a similar effect in skin conductance responses, although not on skin conductance levels. Thus, Lousberg et al.'s results revealed an effect of verbal conditioning not only on pain ratings but on a physiological measure as well.

Interestingly, Lousberg (1994) reported failure at an attempted replication of their results. Lousberg et al. (1996) also reported failure in an attempt to "down condition" (train reductions in pain reporting). They considered that these findings may have been due to modified "punishing responses" (expressions of surprise that pain was lower) for lower pain ratings in the experimental group during "up-conditioning." However, the variable findings clearly called for further investigation.

Flor et al. (2002) replicated Linton and Gotestam's earlier findings with a sample of chronic pain patients and normal, matched control subjects. Like Lousberg et al. (1996), they used an aversive electrical stimulus but also recorded EEG, EOG, heart rate, skin conductance, and muscle tension levels. Flor et al. used a similar reinforcement paradigm to that described by Linton and Gotestam, but the reinforcer in this case was a "smiley" face on a computer screen whose mouth could be modified (up for positive feedback and down for negative feedback). Flor et al. also provided (or withdrew) tokens that could be exchanged for small amounts of money as additional sources of reinforcement. Flor et al. also examined reinforcement for increased and decreased pain reports. Their results showed that both patient and normal groups responded according to the reinforcement contingencies, in both directions (increased and decreased pain). Interestingly, Flor et al. also found that extinction of both the (previously reinforced) increased verbal pain reports and associated cortical responses (measured N150 component of somatosensory evoked potentials) took longer in the patient group. The patient group also displayed prolonged elevated electromyogram levels in this task, relative to the normal controls. Flor et al. concluded that their data provided support for the operant conditioning of pain responses and indicated that chronic back pain patients are more easily influenced by reinforcement contingencies than healthy controls.

Consistent with these earlier studies, Jolliffe and Nicholas (2004) also replicated Linton and Gotestam's original findings. They used a similar paradigm but a larger sample size and tests for the influence of other possible confounding factors, such as awareness of the response-reinforcement contingencies, heightened somatic awareness, anxiety, gender, and locus of control. None of these variables were found to influence the results that supported those reported by the earlier researchers. More recently, a series of studies by Becker et al. (2011) confirmed that operant learning processes with pain can work when implicit, that is, without awareness by the subjects.

At the social level, Chambers et al. (2002) demonstrated that the interactions of mothers with their daughters, but not sons, could also modulate (up and down) the experience of pain in children subjected to cold pressor pain in the laboratory. Importantly, Chambers et al. demonstrated this effect with observed facial activity as a nonverbal pain measure and not just self-reported pain ratings.

Taken together, these studies provide general support for the thesis that pain reports can be operantly conditioned and this effect can be reflected in measures of skin conductance responses, facial activity, and cortical responses as well. These studies also support Fordyce's original contention that pain responses (or behaviors) could be, to some extent, independent of any noxious stimulus.

The treatment implications are that those interacting with chronic pain patients should be alert to the possibility that their responses to reports of pain by patients could well act to reinforce and maintain such behaviors and may complicate the clinical presentation and assessment/management of such cases. Interventions based on these ideas have been effective in modifying pain behaviors, mood, and disability, but it is also true to say that other factors need to be taken into account as well. These include the roles of classical conditioning and cognitions.

References

Becker, S., Kleinböhl, D., Baus, D., & Hölzl, R. (2011). Operant learning of perceptual sensitization and habituation is impaired in fibromyalgia patients with and without irritable bowel syndrome. *Pain, 152*, 1408–1417.

Cairns, D., & Pasino, J. A. (1977). Comparison of verbal reinforcement and feedback in the operant treatment of disability due to chronic low back pain. *Behavior Therapy, 8*, 621–630.

Chambers, C. T., Craig, K. D., & Bennett, S. M. (2002). The impact of maternal behavior on children's pain experiences: An experimental analysis. *Journal of Pediatric Psychology, 27*, 293–301.

Flor, H., Knost, B., & Birbaumer, N. (2002). The role of operant conditioning in chronic pain: An experimental investigation. *Pain, 95*, 111–118.

Fordyce, W. E. (1976). *Behavioural methods for chronic pain and illness*. St. Louis: Mosby.

Gil, K. M., Ross, S. L., & Keefe, F. J. (1988). Behavioral treatment of chronic pain: Four pain management protocols. In R. D. France & K. R. R. Krishnan (Eds.), *Chronic pain* (pp. 376–413). Washington: American Psychiatric Press.

Jolliffe, C. D., & Nicholas, M. K. (2004). Verbally reinforcing pain reports: An experimental test of the operant model of chronic pain. *Pain, 107*, 167–175.

Keefe, F. J., & Block, A. R. (1982). Development of an observation method for assessing pain behavior in chronic low back pain patients. *Behavior Therapy, 13*, 363–375.

Latimer, P. R. (1982). External contingency management for chronic pain: Critical review of evidence. *The American Journal of Psychiatry, 139*, 1308–1312.

Linton, S. J., & Gotestam, K. G. (1985). Controlling pain reports through operant conditioning: A laboratory demonstration. *Perceptual and Motor Skills, 60*, 427–437.

Lousberg, R. (1994). *Chronic pain: Multiaxial diagnostics and behavioral mechanisms*. PhD thesis. Universitaire Pers Maastricht, Maastricht.

Lousberg, R., Groenman, N. H., Schmidt, A. J. M., et al. (1996). Operant conditioning of the pain experience. *Perceptual and Motor Skills, 83*, 883–900.

Morley, S., Eccleston, C., & Williams, A. (1999). A Systematic Review and Meta Analysis of Randomised Control Trials of Cognitive Behavioural Therapy for Chronic Pain in Adults, Excluding Headache. *Pain 80*:1–13.

Pavlov, I. P. (1927). *Conditioned Reflexes*: An Investigation of the Physiological Activity of the Cerebral Cortex. Translated and Edited by G. V. Anrep. London: Oxford University Press.

Schmidt, A. J. M., Gierlings, R. E. H., & Peters, M. L. (1989). Environmental and interoceptive influences on chronic low back pain behaviour. *Pain, 38*, 137–143.

White, B., & Sanders, S. H. (1986). The influence of patient's pain intensity ratings of antecedent reinforcement of pain talk or well talk. *Journal of Behavior Therapy and Experimental Psychiatry, 17*, 155–159.

Operant Treatment

▶ Psychological Treatment of Pain in Children

Operant Treatment of Chronic Pain

Herta Flor
Department of Cognitive and Clinical
Neuroscience, Central Institute of Mental Health,
Medical Faculty Mannheim,
Heidelberg University, Mannheim, Germany

Synonyms

Behavioral treatment; Operant-behavioral
treatment

Definition

The operant treatment of pain is based on
Fordyce's operant conditioning model of pain
and involves behavioral exercises to reduce pain
behaviors and to increase healthy behaviors in
many areas of life, including medication
reduction.

Characteristics

The operant conditioning formulation proposed
by Fordyce (1976, 1988) has substantially
contributed to our understanding of chronic pain
and has had a significant impact on treatment and
rehabilitation. The operant model of chronic pain
is presented in the essay ▶ Operant Perspective of
Pain perspective on pain. This article will focus
on operant treatment. Based on the assumption
that pain behaviors have been positively
reinforced and health behaviors have not been
adequately reinforced, extinguished, or punished,
the operant approach focuses on the restructuring
of pain-incompatible behaviors and the reduction

of pain behaviors such as lack of physical activity, guarding and bracing, and excessive use of
medication and emphasizes the responses of the
patients' significant others to pain. Three areas of
intervention are primarily addressed: (1) modification of the response of significant others to the
patient's pain-related and health-related behaviors; (2) decrease of observable pain behaviors
and increase of activity levels of the patient in
areas such as activities at home, work-related
behaviors, social and leisure time behaviors
(including social skills training), and interaction
with significant others and health-care providers;
and (3) modification of medication-intake and
other illness-related behaviors.

In the original studies, the operant treatment
was usually conducted in an inpatient setting
because of greater control of the environment
and the contingencies of reinforcement. More
recently, operant treatments have also been used
in outpatient settings. They have the advantage
that the home is a more natural environment than
the hospital, the patient can practice in his or her
own home environment, and it is easier to involve
significant others in treatment. Thus, outpatient
treatment gains should be more readily
maintained. It is, moreover, more cost-effective,
especially if it is delivered in a group setting.
Group settings have the additional advantage
that patients can be models to each other and
that the group members can be involved in the
treatment as additional reinforcing agents.

Fordyce (1976) suggested that patients should
only be accepted into operant treatment if they
had a spouse who was willing to become involved
in the treatment, if they were willing to reduce
pain-related medication, and if no compensation
payments were involved. The latter point is not as
strictly adhered to by everyone, since patients can
improve their personal pain coping skills without
having to give up pain-related compensation, if
this status is preferable to them. Although
primarily designed to alter observable pain
behaviors, successful operant treatment should
also change the patients' attitude towards pain
and the psychophysiological response to pain.
Just like biofeedback and cognitive-behavioral

treatments, the operant treatment involves altering the patients' responses from passive recipients of health care into individuals who actively engage in altering their own behavior.

A first important task in an operant group treatment program is the identification and reduction of ▶ pain behaviors and the identification and increase of well behaviors. The best identification of pain behaviors comes from viewing one's own behavior on a videotape. The patients need to understand that many exercises are necessary because pain behaviors have had a long time to become established and have become so automated that neither the patient nor the spouse can easily detect and modify them. Significant others are involved in the treatment because they can give important assistance in modifying pain behaviors in the patients' everyday lives. A brief list of topics that are addressed and practiced in a standard outpatient operant treatment program is listed below (see Flor and Turk 2011 for more details):

- How to deal with medication
- Increase in physical activity, correct body posture
- Decrease of interference in the family
- Decrease of interference in work/housework
- Decrease of interference in leisure time
- Decrease of interference in everyday activities/social activities
- How to deal with the health-care system

Group treatment is preferred to individual therapy sessions because patients can assist each other and there is an opportunity for modeling of appropriate behavior to occur. It is important to convey to the patients that they are not in a pain but a "health" group. All behaviors that are good for the patients' health are to be rewarded; all behaviors that increase pain are ignored. To make this reinforcement process easier and to keep everybody thinking about those goals, red and green cards can be introduced that signify "pain behavior!" (red card) or "well behavior!" (green card). Therapists and patients may use these cards and give them to the patients when they observe the respective behavior. The group can obtain larger reinforcers by completely eliminating red cards.

Increasing physical activity levels is a core goal of the operant approach but is often met with considerable resistance by the patients who are often fearful that their pain will become more severe or that they will injure themselves. The most important principle in the increase of activity is that pain and activity have to be separated in time. Consider the behavior of a typical pain patient: With the advice of the physician to continue as long as they can, they often pursue activities well beyond the pain limit and subsequently stop the activity, which in turn reduces the pain. This is the classical case for negative reinforcement of pain behaviors: a negative state – pain – is reduced by the cessation of activity; hence, the inactivity and rest increases in frequency. The solution to this negative reinforcement problem is the temporal separation of pain and activity. The patients and the therapists define activity goals that do not excessively increase the patient's pain, and the patient rests not when the pain increases but at certain fixed time intervals. An additional complicating factor related to activity is that over time patients develop fear of movement and subsequently avoid all movement or any activity that has been found to be pain-eliciting or is thought to be pain-eliciting (Vlaeyen et al. 1999). Fear of movement and of movement-related pain is a main predictor of levels of patient disability (Vlaeyen and Linton 2000, 2012).

Physical therapists have been routinely included in operant treatment programs. It is, however, important that they understand the principles of operant conditioning to the extent that they implement physical exercises in an operant-compatible way. Most important here is that they also adhere to the quota system and do not let the patient rest contingent on the expression of pain. The increase of tolerance levels regarding physical activity and pain is a treatment goal to which the physical therapists can make significant contributions.

The significant other is usually the most important reinforcing agent for the chronic pain patient. Altering ▶ significant others' responses to pain is therefore one of the most important goals in operant pain treatment. For patients who do not have spouses, a close friend or a child may take the role of the significant other in the operant group treatment. In the initial stage, the therapists should let all dyads display typical pain behaviors and typical responses to the pain. The next step is to have the significant others ignore pain behaviors and subsequently have the patients stop displaying pain behaviors and let the spouses positively reinforce the absence of pain and the presence of well behaviors. The therapists can then provide some examples of responses to pain behaviors taken from the dyads' ▶ role-plays that had been videotaped for analysis and list: (1) what were visible pain behaviors and (2) how were the responses (reinforcing, punishing, ignoring, distracting). Therapists and patients then think of alternate responses to the ones that have been observed, responses that would not reinforce pain behaviors, and the significant others and patients practice these in role-plays. It is important to emphasize throughout that the non-response to pain behaviors serves the sole purpose of reducing pain and pain behaviors and is therefore a positive response by the significant other. It is important to then engage in the next step of teaching the patient the cessation of pain behaviors and the display of well behaviors and have the significant other positively reinforce this step. The therapists and couples should construct lists of alternative behaviors to pain behaviors and also construct lists of reinforcing the significant other's responses. Then these alternate behaviors should be practiced in role-plays.

In addition to the modification of specific significant others' behaviors, another important goal of the operant group training is to enlist significant others' support for the behavioral changes that have been planned for the patients. The significant others need to be informed in detail about the treatment goals that have been defined with the patient, and they need to be able to voice concerns about certain areas of change they view as problematic. In the later sessions, the focus should shift from simple pain behaviors and their reinforcement to family and marital interaction around the pain and potential problems in sexual interaction. It is important to check if the reinforcement the spouse is giving is really genuine or if it is a "mixed message." Spouses need to learn to genuinely reinforce well behaviors and to express concerns or criticisms directly, not by means of pain-related responses. Another area of joint goal formulation and practice are leisure activities. Often the pain has been very detrimental for all leisure and social activities and they may have been reduced to a minimum.

The intake of medication may often be a behavior that has been negatively reinforced. If a patient waits until the pain increases to a certain level (often prompted by a well-meaning physician who may instruct the patients to take the medication only when he or she really needs it), then the intake of medication reduces the pain, an aversive state is terminated, and the behavior that led to the end of the unpleasant state, the taking of medication, will increase in frequency and will over time occur earlier and earlier in order to prevent pain onset and/or exacerbation. From a pharmacological point of view, the prescription of analgesic medication as needed (prn) is just as problematic as from a behavioral learning point of view. If the medication is only taken when the pain level is very high, the maximum effectiveness of the medication may not be achieved. Thus, switching the patients from a ▶ prn to a ▶ time-locked medication intake is the first goal. As suggested by Fordyce (1976), a prn baseline is taken by having the patients keep a medication diary or, if they are inpatients, by carefully noting the medication that is dispensed to them by the staff who have previously been instructed to give medication as needed.

To what extent all medication should be eliminated, or to what extent medication should be used as an adjunct therapy, is a matter of much debate. Analgesic medication should be continued when the underlying disease process prevents the patients from relying on alternative methods or when the patient is unable to obtain sufficient pain relief. In inpatient treatment, the

▶ "pain cocktail" as described by Fordyce (1976), is usually implemented to reduce medication intake. The pain cocktail consists of a colorless and taste-masking vehicle such as raspberry syrup to which the medication the patient is to receive is added. The amount of active medication in the cocktail is successively reduced on a predetermined basis, which has been decided upon in a joint agreement with the attending physician. The most important characteristic of the pain cocktail is that it is given not on a prn but at consistent time intervals around the clock. The amount of active ingredient initially consists of about the same or slightly exceeds the amount recorded in the baseline phase, with the only difference that time-contingent delivery is used. In outpatient treatment, a pain cocktail is usually not employed, rather a prescribed time-contingent medication usage and subsequent medication reduction on a fixed schedule, as determined by the therapist in cooperation with the patient and his or her physician, are used. Prerequisite for this reduction regiment is a very exact medication baseline the patient notes in his or her pain diary which should cover a period of at least 3 weeks. A simplified explanation of both the psychological and pharmacological reasons for giving medication on a time-contingent basis should be given.

Each patient receives an individualized plan for medication reduction that has been previously determined from the patient's diary and discussions with his or her physician. The intake of medication must be noted in the pain diary. One aspect that may be considered in the new medication reduction regimen is that the intake of medication may itself have become a conditioned activity, and many intake-associated aspects such as the sight or feel of the medication, the carton, the place where it is usually taken may have become conditioned stimuli and should be changed.

An extension of the increase in overall activity levels applies to activity at the workplace. Before the generalization of increased activity levels to the workplace can be accomplished, a very careful analysis of the activities the patient typically engages in at the workplace is necessary. Most important here is the assessment of repetitive work, of the number and distribution of breaks, of the patient's activity during the breaks, of the responses of colleagues and supervisors to the patient's behavior, and of the overall level of stress at the workplace.

Based on this, increases in work-related activity can be planned. Very often, however, it is less an increase of activity but in general a better timing of activity and rest phases is needed. In certain cases where the patient has problems with the supervisor, the therapist may offer to talk to him or her. In general, however, the patients are well able to defend the activity plans that have been devised for them, especially when assertiveness training has been included in the treatment.

For patients who are unemployed, on compensation, or involved in litigation, it will have to be determined to what extent return to work can be a desirable goal. This decision depends on numerous factors such as the patient's willingness to work, his or her age and level of qualification, the current state of the economy, or the cooperation of employers. If the patients are close to retirement age and engaged in physically or psychologically stressful employment, and the patients do not show a major desire to continue work, early retirement may be an option while at the same time teaching patients how to better cope with the remaining pain problem. In younger and work-motivated patients, high emphasis should be placed on the acquisition of skills that will allow them to return to work.

A major problem of chronic pain patients is the over-reliance on the health-care system and constant doctor shopping. An important treatment goal is therefore to reduce health-care-seeking behaviors with respect to the pain problem. A first step is to exclude any additional treatments from being delivered during the course of the pain-treatment program, except for active physical therapy or active occupational therapy. This will give the patient an opportunity to discover that he or she can do without these additional interventions. Moreover, problems of attribution of the treatment success can thus be avoided. Specific exercises involving dealing with the health-care system are used to teach the patients more reliance on their own means and more independence from passive treatments. For example, role-plays are used to help the patient express the wish to discontinue additional treatment or to refuse the treatment with additional analgesic medication.

It is important for the therapist to note that the pain problems of the participants are chronic, that there is usually no danger of injury, nor is there a high probability of complete recovery (that is freedom from pain). The patients must, however, be careful to note new types of pain or unusual exacerbations and report those to their physician.

An important aspect of the ▶ transfer and generalization of operant behavior change is that the new behaviors are sufficiently reinforced. In the group sessions, reinforcement is provided by the therapist and the group members. An additional possibility for reinforcement is ▶ self-reinforcement by the patients. The patients can practice self-reinforcement skills in role-plays.

Several studies have shown that operant-behavioral treatment is effective in reducing pain behaviors and excess disability. Several controlled studies demonstrated the efficacy of operant-behavioral treatments for ▶ low back pain (e.g., Kole-Snijders et al. 1999; Nicholas et al. 1991; Turner and Clancy 1988; Vlaeyen et al. 1995), although operant treatment may not be superior to other types of interventions (see Ostelo et al. 2005). In their ▶ meta-analysis Flor et al. (1992) found operant-behavioral treatments to be the most effective treatment for back pain, and a Cochrane review (van Tulder et al. 2001) also reported the efficacy of behavioral treatment. In ▶ fibromyalgia patients, Thieme et al. (2003) and Thieme et al. (2007) reported very high effect sizes for an operant behavioral compared to a standard medical treatment. New developments of the operant approach emphasize the exposure ▶ Fear Reduction through Exposure in Vivo to feared movements (de Jong et al. 2012) and behaviors, and there are also attempts to use behavioral treatments to modify pain memories (Flor 2000, 2009).

References

de Jong, J. R., Vlaeyen, J. W., van Eijsden, M., Loo, C., & Onghena, P. (2012). Reduction of pain-related fear and increased function and participation in work-related upper extremity pain (WRUEP): effects of exposure in vivo. *Pain, 153*(10), 2109–2118.

Flor, H. (2000). The functional organization of the brain in pain. *Progress in Brain Research, 129*, 313–322.

Flor, H. (2009). Extinction of pain memories: importance for the treatment of chronic pain. In J. Castro-Lopes (Ed.), *Current topics in pain: 12th world congress on pain* (pp. 221–244). Seattle: IASP Press.

Flor, H., Fydric, T., & Turk, D. C. (1992). Efficacy of multi-disciplinary pain treatment centers: A meta-analytic review. *Pain, 49*, 221–230.

Flor, H., & Turk, D. C. (2011). *Chronic pain: An integrated biobehavioral approach*. Seattle, WA: IASP Press.

Fordyce, W. E. (1976). *Behavioral methods in chronic pain and illness*. St. Louis: Mosby.

Fordyce, W. E. (1988). Pain and suffering: A reappraisal. *American Psychologist, 43*, 276–283.

Kole-Snijders, A. M., Vlaeyen, J. W., Goossens, M. E., et al. (1999). Chronic low-back pain: What does cognitive coping skills training add to operant behavioral treatment? Results of a randomized clinical trial. *Journal of Consulting and Clinical Psychology, 67*, 931–944.

Nicholas, M. K., Wilson, P. H., & Goyen, J. (1991). Operant-behavioural and cognitive-behavioural treatment for chronic low back pain. *Behaviour Research and Therapy, 29*, 225–238.

Ostelo, R. W., van Tulder, M. W., Vlaeyen, J. W., et al. (2005). Behavioural treatment for chronic low-back pain. *Cochrane Database of Systematic Reviews, 1*, CD002014.

Thieme, K., Gromnica-Ihle, E., & Flor, H. (2003). Operant behavioral treatment of fibromyalgia: A controlled study. *Arthritis and Rheumatism, 49*, 314–320.

Thieme, K., Turk, D. C., & Flor, H. (2007). Responder criteria for operant and cognitive-behavioral treatment of fibromyalgia syndrome. *Arthritis Rheum-Arthritis Care Res., 57*(5), 830–836.

Turner, J. A., & Clancy, S. (1988). Comparison of operant behavioral and cognitive-behavioral treatment for chronic low-back pain. *Journal of Consulting and Clinical Psychology, 56*, 261–266.

van Tulder, M. W., Ostelo, R., Vlaeyen, J. W., et al. (2001). Behavioral treatment for chronic low back pain: A systematic review within the framework of the Cochrane back review group. *Spine, 26*, 270–281.

Vlaeyen, J. W. S., Haazen, I. W. C. J., Kole-Snijders, A. M. J., et al. (1995). Behavioural rehabilitation of chronic low back pain: Comparison of an operant treatment, an operant-respondent treatment. *British Journal of Clinical Psychology, 34*, 95–118.

Vlaeyen, J. W., & Linton, S. J. (2000). Fear-avoidance and its consequences in chronic musculoskeletal pain: A state of the Art. *Pain, 85*, 317–332.

Vlaeyen, J. W., & Linton, S. J. (2012). Fear-avoidance model of chronic musculoskeletal pain: 12 years on. *Pain, 153*(6), 1144–1147.

Vlaeyen, J. W., Seelen, H. A. M., Peters, M., de Jong, P., Aretz, E., Beisiegel, E., et al. (1999). Fear of movement/(re)injury and muscular reactivity in chronic low back pain patients: an experimental investigation. *Pain, 82*(3), 297–304.

Operant-Behavioral Treatment

▶ Operant Treatment of Chronic Pain

Operculum

Definition

Part of the brain that covers the insula, consisting of frontal, parietal, and temporal operculum as part of the corresponding lobes of the cerebral cortex.

Cross-References

▶ Insular Cortex, Neurophysiology, and Functional Imaging of Nociceptive Processing
▶ Nociceptive Processing in the Secondary Somatosensory Cortex

Operidine

▶ Postoperative Pain, Pethidine

Ophthalmic Migraine

▶ Clinical Migraine with Aura

Ophthalmodynia Periodica

▶ Primary Stabbing Headache

Ophthalmoplegic Migraine

Definition

This condition consists of painful ophthalmoplegia with or without headache. Usually the ophthalmoplegia is transient; however, it can become permanent, especially after repeated attacks. Since headache was often associated with attacks, it was felt to be a variant of migraine. This is no longer felt to be true at least in Western Europe and North America, because many reports have shown abnormalities of the ocular motor nerve using magnetic resonance imaging (MRI) in children with recurrent painful ophthalmoplegia fulfilling the previous criteria for ophthalmoplegic "migraine."

Cross-References

▶ Migraine, Childhood Syndromes

Opiate

▶ Opioid Electrophysiology in PAG
▶ Opioid Peptide Co-localization and Release
▶ Postoperative Pain, Opioids

Opiate Receptors

▶ Opioid Receptors

Opiates During Development

Gordon A. Barr
Department of Anesthesiology and Critical
Care Medicine, Children's Hospital of
Philadelphia University of Pennsylvania,
Philadelphia, PA, USA

Definition

Opiates and Development

Opioids are a class of peptides that bind to specific membrane-bound receptors. There are three major receptor types for opioids, μ (MOP), ∂- (DOP), and κ (KOP), and a number of subtypes for each of the major classes. The peptides that interact with each receptor do so preferentially but not exclusively. These are the endomorphins (μ), enkephalins (∂-), and dynorphins (κ). β-Endorphin also interacts preferentially with the MOP receptor. A historic review of the opioid field has been published (Snyder and Pasternak 2003).

Historically, infant patients have been under-medicated for pain. In part, this has been because it was assumed that infants, due to the immaturity of the nervous system, were incapable of experiencing pain. Moreover, the untoward side effects of opiates, respiratory depression, chest wall rigidity, and constipation, complicated treatment of severely ill infants. Indeed, there is still ongoing debate as to whether or not neonates and prematurely born infants can experience a complex perception such as pain (Anand and Craig 1996; Derbyshire 2001; van Lingen et al. 2002). What is not debatable any longer is that adequate and even aggressive opiate treatment for invasive procedures reduces mortality and morbidity (Anand and Hickey 1987; Anand et al. 1987). Nonetheless, even with convincing evidence that adequate analgesia is medically necessary for appropriate clinical care, the use of opiates is complicated because of the immaturity of the infant.

Characteristics

Development of Opioid Receptors and Peptides

The development of opioid peptides and their receptors is precocious. The exact age at which they are first detectable, and the pattern in which they develop, is dependent on the specific receptor or peptide and on the region of the central nervous system under consideration. In general, MOP and KOP receptors develop early in rodents and humans and the ∂- receptor appears later. In the human fetus, μ and κ opioid receptors appear at the start of the second trimester and mature through prenatal life. Delta opioid receptors are generally absent prenatally in the human, although they can be transiently expressed. In the rat and other non-primate species, results are comparable: early appearance of MOP and KOP opioid receptors that peak between 1 and 2 weeks after birth, at which time the numbers decrease until adult levels are reached shortly thereafter. DOP receptors are first detected at the end of the first week after birth and are not likely to be fully functional until even later. In all cases, receptor affinity remains unchanged with age.

Peptide development in general is also early and regionally specific, although the endomorphins mature relatively late. Met-enkephalin-, dynorphin-, and β-endorphin-like proteins can be detected by day 11.5 in the rat embryo, preceding the maturation of the receptors, although in some regions opioids appear at the end of the first postnatal week.

Pharmacokinetics of Morphine

Morphine kinetics differ with both age and body weight. In the adult, the major active metabolites of morphine are morphine-3-glucuronide (M3G) and morphine-6-glucuronide (M6G). M3G may have anti-opioid properties whereas m6G has analgesic effects (Christrup 1997). In the infant, M3G is the major metabolite. Although neither M3G nor M6G readily cross the blood-brain barrier in the adult, whether they do or not in the infant is not known. Steady-state morphine levels require 24–48 h of treatment for morphine and at least 5 days of treatment for the metabolites (Saarenmaa et al. 2000).

At 6 months of age, the full-body clearance rate for morphine is about 80 % of that of the adult (Bouwmeester et al. 2004). Metabolite clearance likewise increases with age. Thus, morphine doses in the infant must, out of necessity, be lower for the premature infant or neonate than the older infant. Like morphine, fentanyl clearance is slower in the infant than in the older child and adult.

Development of Analgesia

In animal models (mostly rodent), the infant is generally more responsive to noxious stimuli (e.g., more likely to experience "pain") than the older animal and is differentially responsive to opiate-induced analgesia. Although it is not known for certain why the infant is more sensitive to injury, the organization of the nervous system is clearly different in the infant than the adult (Fitzgerald and Walker 2009). Receptive fields of the infant, human or rodent, are larger and more diffuse. The afferent input of noxious stimulation in the infant is by A fibers rather than C fibers, and the A fibers terminate in lamina II of the spinal cord dorsal horn. In the rat, these immaturely targeted fibers retract and are replaced by C fibers over the first 3 weeks of life. Further, inhibitory processes are not mature in the infant. Descending facilitatory processes predominate early in the infant rat, before the end of the second week of life, although opiates injected into several brain regions are analgesic for mild pain in rostral body (Hathway et al. 2009; Barr and Wang 2013). Diffuse noxious inhibitory processes within the spinal cord, where injury to one limb reduces nociception in a heterotopic region, also do not appear until sometime in the third week of life. Thus, in the absence of multiple inhibitory processes, the infant is unable to dampen the effects of injury (Fitzgerald and Walker 2009).

Opiates, despite the immaturity of the opioid peptides and receptors, are quite effective analgesics in the infant. This has been demonstrated numerous times in many paradigms.

Morphine and fentanyl are the most effective treatments for pain in human neonates undergoing invasive medical treatment and postoperative care. Infants treated aggressively with opiates during and after surgery show decreased morbidity and mortality (Anand and Hickey 1987; Anand et al. 1987). Nonetheless, little detail is known of the best opiate, dose, dosing regimen, and match of opiate to type of pain. Issues of gestational age, body weight, and criticalness of illness only add to these complications (van Lingen et al. 2002).

The immaturity of the blood-brain barrier may also contribute to the differential effectiveness of the opiate in the infant, by allowing more opiate to enter the central nervous system. There may also be yet undefined pharmacodynamic factors that account for this increased sensitivity. However, in the infant even more than the adult, the effectiveness of particular opiates is dependent on the type of nociceptive test and the intensity of the noxious stimulus. Intense thermal stimuli are less amenable to relief by morphine in the newborn rat pup, whereas analgesia to lower the intensity of mechanical noxious stimuli occurs quite early. The complexity is well illustrated by data showing that the relative analgesic effectiveness of buprenorphine, fentanyl, and morphine in the infant compared to the adult is dependent on the type of analgesic test. Morphine is most effective in the infant in thermal tests, but buprenorphine is most effective in tests of pain that are models for inflammatory insult (McLaughlin and Dewey 1994).

Moreover, there is regional specificity to the actions of the opiates. For example, due to the lack of descending inhibition, brain administration is less effective than spinal administration in the infant rat, although that is dependent on the type of analgesic test. In rats, administration of opiates in a manner that targets the spinal cord is quite effective as an analgesic (Barr et al. 1992; Marsh et al. 1999). In human children, these procedures are not common, but the clinical literature suggests that they are effective (Walker and Yaksh 2012).

In both the adult and the infant, there are untoward side effects that affect treatment.

One of these is respiratory depression, a well-known effect of opiates. Initially, part of the reluctance to treat infants was out of concern for these actions. Although still a concern, infants are not likely to be more sensitive to the respiratory effects of opiates than are adults. Two other concerns are dependence and tolerance that occur with repeated exposure.

Dependence

As in the adult, opiates such as morphine produce physical dependence. The human infant can become dependent due to medical treatment with opiates (Arnold et al. 1990; French and Nocera 1994; Anand et al. 2010) or by passive exposure through placental transfer, if the mother uses heroin, methadone, or other opiates (Franck and Vilardi 1995; Jones et al. 2012). When therapeutic opiate treatment is discontinued or when the opiate-dependent mother gives birth, the infant shows a clear withdrawal syndrome that consists of increased irritability, disruption of sleep/wake states, incessant crying, and other physical signs (Franck and Vilardi 1995). This is termed the "narcotic abstinence syndrome." As might be expected, higher doses and longer exposure increase risk for this syndrome.

In the rodent, withdrawal exists quite early, even in utero. In the infant rat, withdrawal is clearly different from withdrawal in the adult rat and is characterized by increased separation-induced ultrasonic vocalizations (crying), increased activity, and a unique set of behaviors not normally seen in pups not in withdrawal. The dysphoric state induced by opiate withdrawal occurs later, around the second week of life in the infant rat. Autonomic signs are minimal. In the adult, NMDA glutamate receptor antagonists, when given concomitantly with morphine, reduce withdrawal; but in the infant, that treatment is ineffective. In contrast, AMPA blockers or nitric oxide inhibitors reduce dependence in the infant as they do in the adult. Thus, it is likely that the fundamental mechanisms mediating the chronic effects of opiates differ in the neonate (Zhu and Barr 2001; Barr et al. 2011). Treatment with opiates is the preferred initial therapy for narcotic abstinence syndrome and the only treatment for

which there is evidence of efficacy. Morphine and methadone are currently the most commonly prescribed opioids to treat infant withdrawal, but which is more effective has not been determined. Buprenorphine given to pregnant women is more effective than methadone in reducing the infant abstinence syndrome (Jones et al. 2012). Clinically, there are no accepted or proven non-opioid treatments for the withdrawal syndrome. Barbiturate, benzodiazepines, and other sedatives are ineffective. The preclinical data suggest that NMDA antagonists, effective in the adult, are not likely to work in the human infant. There is preliminary evidence that clonidine may reduce abstinence symptoms and shorten hospital stays for methadone-exposed newborns (Esmaeili et al. 2010).

Tolerance

Tolerance also occurs to chronic opiate use in both human and animal infants. The tolerance is substantially less than that seen in the adult for reasons that are not understood. There are differences in intracellular signaling mechanisms that are age dependent that might account for this; as for dependence, tolerance to morphine in the infant is not NMDA receptor mediated (Barr et al. 2011).

Summary

Opioid receptors and peptides mature early during development but in a complex manner. Opiates are effective well-characterized drugs for the treatment of pain in the premature and full-term infant. Regardless of the philosophical debate as to whether or not the infant can experience "pain," opiate treatment reduces morbidity and mortality in the infant patient. Opiates tend to be cleared more slowly in the premature or newborn full-term infant, and issues of the unique developmental effects of metabolites remain unaddressed. Whether or not different opiates are differentially effective for different painful procedures is not known. As in the adult, there are problems with iatrogenically induced tolerance and dependence and with dependence in infants born of opiate-using mothers. The most effective treatments remain opiate-based.

References

Anand, K. J., & Craig, K. D. (1996). New perspectives on the definition of pain [editorial] [see comments]. *Pain, 67*, 3–6. discussion 209-211.

Anand, K. J., & Hickey, P. R. (1987). Pain and its effects in the human neonate and fetus. *The New England Journal of Medicine, 317*, 1321–1329.

Anand, K. J., Sippell, W. G., & Aynsley-Green, A. (1987). Randomised trial of fentanyl anaesthesia in preterm babies undergoing surgery: Effects on the stress response. *Lancet, 1*, 62–66.

Anand, K., Willson, D., Berger, J., Harrison, R., Meert, K., Zimmerman, J., et al. (2010). Tolerance and withdrawal from prolonged opioid use in critically ill children. *Pediatrics, 125*, e1208–e1225.

Arnold, J., Truog, R., Orav, E., Scavone, J., & Hershenson, M. (1990). Tolerance and dependence in neonates sedated with fentanyl during extracorporeal membrane oxygenation. *Anesthesiology, 73*, 1136–1140.

Barr, G. A., & Wang, S. (2013). Analgesia induced by localized injection of opiate peptides into the brain of infant rats. *European Journal of Pain, 17*(5):676–691.

Barr, G. A., Miya, D. Y., & Paredes, W. (1992). Analgesic effects of intraventricular and intrathecal injection of morphine and ketocyclazocine in the infant rat. *Brain Research, 584*, 83–91.

Barr, G. A., McPhie-Lalmansingh, A., Perez, J., & Riley, M. A. (2011). Changing mechanisms of opiate tolerance and withdrawal during early development: Animal models of the human experience. *ILAR Journal, 52*, 329–341.

Bouwmeester, N. J., Anderson, B. J., Tibboel, D., & Holford, N. H. (2004). Developmental pharmacokinetics of morphine and its metabolites in neonates, infants and young children. *British Journal of Anaesthesia, 92*, 208–217.

Christrup, L. L. (1997). Morphine metabolites. *Acta Anaesthesiologica Scandinavica, 41*, 116–122.

Derbyshire, S. W. (2001). Fetal pain: An infantile debate. *Bioethics, 15*, 77–84.

Esmaeili, A., Keinhorst, A. K., Schuster, T., Beske, F., Schlosser, R., & Bastanier, C. (2010). Treatment of neonatal abstinence syndrome with clonidine and chloral hydrate. *Acta Paediatrica, 99*, 209–214.

Fitzgerald, M., & Walker, S. M. (2009). Infant pain management: A developmental neurobiological approach. *Nature Clinical Practice. Neurology, 5*, 35–50.

Franck, L., & Vilardi, J. (1995). Assessment and management of opioid withdrawal in ill neonates. *Neonatal Network, 14*, 39–48.

French, J. P., & Nocera, M. (1994). Drug withdrawal symptoms in children after continuous infusions of fentanyl. *Journal of Pediatric Nursing, 9*, 107–113.

Hathway, G. J., Koch, S., Low, L., & Fitzgerald, M. (2009). The changing balance of brainstem-spinal cord modulation of pain processing over the first weeks of rat postnatal life. *The Journal of Physiology, 587*, 2927–2935.

Jones, H. E., Finnegan, L. P., & Kaltenbach, K. (2012). Methadone and buprenorphine for the management of opioid dependence in pregnancy. *Drugs, 72*, 747–757.

Marsh, D., Dickenson, A., Hatch, D., & Fitzgerald, M. (1999). Epidural opioid analgesia in infant rats I: Mechanical and heat responses. *Pain, 82*, 23–32.

McLaughlin, C. R., & Dewey, W. L. (1994). A comparison of the antinociceptive effects of opioid agonists in neonatal and adult rats in phasic and tonic nociceptive tests. *Pharmacology, Biochemistry, and Behavior, 49*, 1017–1023.

Saarenmaa, E., Neuvonen, P. J., Rosenberg, P., & Fellman, V. (2000). Morphine clearance and effects in newborn infants in relation to gestational age. *Clinical Pharmacology and Therapeutics, 68*, 160–166.

Snyder, S. H., & Pasternak, G. W. (2003). Historical review: Opioid receptors. *Trends in Pharmacological Sciences, 24*, 198–205.

van Lingen, R. A., Simons, S. H., Anderson, B. J., & Tibboel, D. (2002). The effects of analgesia in the vulnerable infant during the perinatal period. *Clinics in Perinatology, 29*, 511–534.

Walker, S. M., & Yaksh, T. L. (2012). Neuraxial analgesia in neonates and infants: A review of clinical and preclinical strategies for the development of safety and efficacy data. *Anesthesia and Analgesia, 115*, 638–662.

Zhu, H., & Barr, G. (2001). Opiate withdrawal during development: Are NMDA receptors indispensable? *Trends in Pharmacological Sciences, 22*, 404–408.

Opiates, Rostral Ventromedial Medulla, and Descending Control

Mary M. Heinricher
Departments of Neurological Surgery and
Behavioral Neuroscience, Oregon Health &
Science University, Portland, USA

Introduction

The ▶ Rostral Ventromedial Medulla (RVM) and Descending Control

The recognition that supraspinal structures can exert control over sensory processing at the level of the spinal cord has a long history. However, the realization that descending controls have a specific role in the regulation of spinal pain processing achieved prominence in 1969 with the demonstration of analgesia during electrical stimulation of the midbrain periaqueductal gray (PAG) (Reynolds 1969). This phenomenon came to be called "stimulation-produced analgesia" (SPA). Subsequent intensive investigations of SPA led to the definition of a central pain modulatory network, now known to span the neuraxis, with critical links in the PAG and RVM (Fields et al. 2006). Nonselective activation of the RVM was shown to inhibit behavioral responses to noxious stimuli, and this antinociception is in large part due to interference with nociceptive processing at the level of the spinal cord. Moreover, further investigation has now demonstrated that the RVM exerts bidirectional control over nociception, and as described below, the best current evidence is that two different physiologically defined populations of RVM neurons, "off-cells" and "on-cells," mediate antinociceptive and pronociceptive effects of RVM manipulation.

The anatomical substrate for spinal modulation of nociception from the RVM is a large, diffuse projection from this region to the dorsal horn. The projection travels through the dorsolateral funiculus and terminates at all levels of the spinal cord in laminae known to be involved in nociception. The specifics of how the RVM projection interfaces with nociceptive circuitry at the level of the dorsal horn to alter nociception remain unresolved. A major input to the RVM is the PAG, which channels input from higher structures, including hypothalamus and amygdala, to the RVM.

The delineation of this brainstem pain-modulating system has thus been a major breakthrough in defining pain modulation as a separable function of the nervous system and in showing that the influence of psychological variables on pain could have an understandable neural basis.

Definition

Rostral Ventromedial Medulla (RVM)
Defined in terms of function and connectivity, the rostral ventromedial medulla includes the nucleus raphe magnus and adjacent reticular formation at the level of the facial nucleus.

On-Cell, RVM
RVM neurons defined by a sudden burst of activity beginning just prior to execution of the tail flick, paw withdrawal, or other nocifensive reflex. Spontaneous activity is variable with time, so that the reflex-related burst may not be discernible if the reflex is evoked at a time when the neuron is already firing spontaneously. On-cells are directly sensitive μ-opioid agonists and are inhibited by morphine or μ-agonists given systemically or by microinjection into the PAG, in the basolateral amygdala, or in the RVM itself. RVM on-cells as defined in vivo are assumed to map to secondary cells defined in vitro.

Off-Cell, RVM
RVM neurons defined by a sudden pause in firing beginning just prior to execution of the tail flick, paw withdrawal, or other nocifensive reflex. Spontaneous activity is variable with time, and an Off-cell can only be identified if it is firing at the time at which the reflex is evoked. Off-cells are not directly responsive to μ-opioids, but are activated indirectly by μ-opioids given systemically or by microinjection into the PAG, in the basolateral amygdala, or in the RVM itself. RVM Off-cells as defined in vivo are presumed to be a subset of primary cells defined in vitro.

Neutral Cell, RVM
Third class of RVM neurons, defined by the lack of reflex-related activity. Neutral cells do not respond to μ-opioid agonists. RVM serotonergic neurons behave as neutral-cells, although some authors prefer to consider serotonergic cells as a fourth class of RVM neurons, distinct from neutral cells. Neutral cells as defined in vivo are likely be a subset of primary cells defined in vitro.

Primary Cell, RVM
Primary cells are a class of RVM neurons defined in vitro by lack of response to μ-opioid agonists. Primary cells presumably correspond to off-cells and neutral cells defined in vivo.

Secondary Cell, RVM
Secondary cells are a class of RVM neurons defined in vitro by response to μ-opioid

agonists. These neurons were originally proposed to be unresponsive to κ agonists, but this characteristic has since been questioned. Secondary cells presumably correspond to on-cells recorded in vivo.

Characteristics

RVM and Mu-Opioid Analgesia

Interest in the PAG/RVM descending modulatory network was heightened when it became apparent that this system was also an important substrate for opioid analgesia (Heinricher and Ingram 2008). The RVM, and especially the PAG, is rich in opioid receptors and opioid peptides. The RVM is required for the full analgesic action of systemically administered morphine, and focal application of morphine or μ-opioid agonists within the RVM is sufficient to produce potent behavioral analgesia (Yaksh et al. 1988).

The neural basis of μ-opioid analgesia within the RVM has been studied in depth. There is now clear evidence that the antinociceptive effects of μ-opioid agonists are mediated by the class of RVM neurons referred to as "▶ off-cells." These neurons are the inhibitory output neurons of the RVM and are defined physiologically, by a pause in firing when an organism is responding to a noxious stimulus (Fig. 1). This period of inactivity is due to GABAergic inhibition of the off-cells and effectively releases dorsal horn nociceptive transmission circuits from inhibition, thus permitting a behavioral response. Mu-opioid agonists and morphine block the GABA release responsible for the off-cell pause. This presynaptic disinhibition of the off-cells in turn leads to a positive feedback process resulting in further off-cell activation through excitation. However, it is the opioid-induced elimination of the characteristic pause in firing that is critical for behavioral analgesia (Fig. 2, Heinricher et al. 1994, 2010).

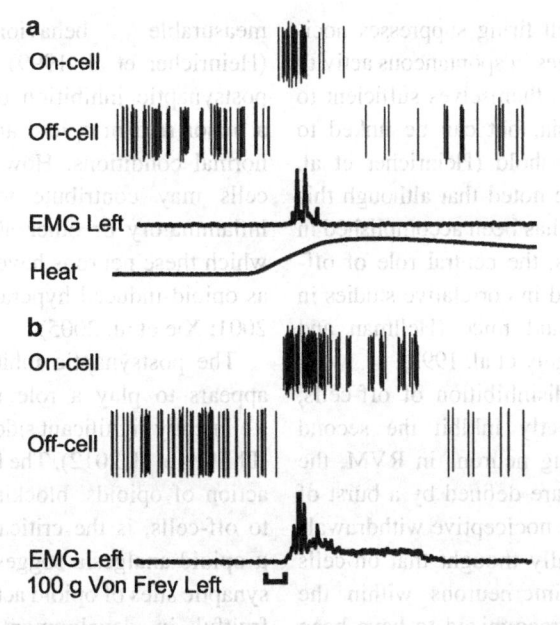

Opiates, Rostral Ventromedial Medulla, and Descending Control, Fig. 1 Off-cells and on-cells in the RVM are defined by sudden cessation in firing ("pause") or activation ("burst") associated with withdrawal reflexes evoked by noxious somatic stimuli. (a) Simultaneously recorded on-cell and off-cell during heat-evoked withdrawal. Contact heat stimulus applied using a Peltier device applied to the plantar hindpaw.

Temperature trace shown below the cell activity shows increase at 1.8 °C from 35 °C holding temperature. Maximum temperature was 53 °C. Withdrawal was measured from EMG. (b) The same neuron pair also responded during withdrawal evoked by application of a von Frey fiber (100 g) to the left hindpaw (Adapted from Carlson et al. 2007)

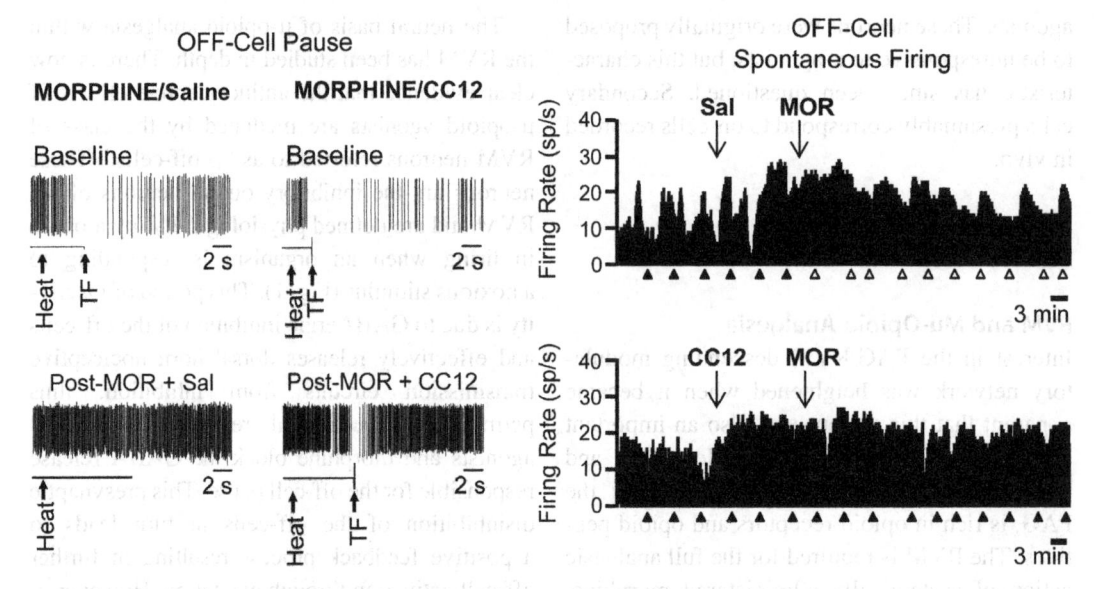

Opiates, Rostral Ventromedial Medulla, and Descending Control, Fig. 2 Protection of the off-cell pause prevents opioid antinociception. In this experiment, the P450 epoxygenase inhibitor CC12 or saline was given prior to morphine. CC12 interfered with morphine effects on the off-cell pause (*left*) and prevented behavioral antinociception (not shown) but failed to block morphine effect on spontaneous activity (*right*) (Adapted from Heinricher et al. 2010)

Thus, continuing off-cell firing suppresses nociceptive behavior. Changes in spontaneous activity (Fig. 2, right) are not in themselves sufficient to produce potent analgesia, but can be linked to modest changes in threshold (Heinricher et al. 1989). It should also be noted that although this detailed circuit analysis has been accomplished in lightly anesthetized rats, the central role of off-cells has been confirmed in correlative studies in awake behaving rats and mice (Hellman and Mason 2012; McGaraughty et al. 1993).

In addition to the disinhibition of off-cells, μ-opioid agonists directly inhibit the second class of pain-modulating neurons in RVM, the "▶ on-cells." On-cells are defined by a burst of activity associated with nociceptive withdrawals (Fig. 1). It was originally thought that on-cells served as inhibitory interneurons within the RVM, but this is now recognized to have been incorrect (Cleary et al. 2008; Heinricher and McGaraughty 1998). On-cells are now instead recognized as a separate, parallel output from the RVM to the dorsal horn which exerts a net pro-nociceptive effect. Given that suppression of on-cell discharge does not by itself produce measurable behavioral antinociception (Heinricher et al. 1999), it is unlikely that the postsynaptic inhibition of these neurons plays a major role in opioid analgesia, at least under normal conditions. However, inhibition of on-cells may contribute to anti-hyperalgesia in inflammatory or other abnormal pain states in which these neurons have been implicated, such as opioid-induced hyperalgesia (Vanderah et al. 2001; Xie et al. 2005).

The postsynaptic inhibition of on-cells also appears to play a role respiratory depression, a clinically significant side effect of opioid agents (Phillips et al. 2012). The fact that the presynaptic action of opioids, blocking the inhibitory input to off-cells, is the critical RVM mechanism of μ-opioid analgesia suggests that a focus on presynaptic sites of opioid action is likely to be more fruitful in development of analgesic agents (Fig. 3, Heinricher et al. 2010). Evidence that pre- and postsynaptic opioid effects are mediated by different signal transduction mechanisms gives hope that the two actions can be separated pharmacologically (Heinricher and Ingram 2008).

Dorsal Horn

Opiates, Rostral Ventromedial Medulla, and Descending Control, Fig. 3 Off-cell activation is triggered by presynaptic inhibition of a GABAergic input originating from outside of the RVM ("GABAergic neuron"). Parallel postsynaptic inhibition of on-cells by the opioid likely contributes to antinociception, although under normal conditions on-cell inhibition is not sufficient to produce antinociception. Mu-opioid receptors are thus located both presynaptically (on inhibitory inputs to off-cells) and postsynaptically (on on-cells)

A third class of RVM neurons, "neutral ► cells," do not respond to noxious stimulation or to opioid application. Their role in pain modulation is unknown. It is likely that neutral cells comprise a heterogeneous group, physiologically, neurochemically, and functionally. It has been proposed that all serotonergic neurons in the RVM fall within the neutral cell class (Mason 1997). This has presented something of a conundrum, since pharmacological studies point to descending serotonergic projections as central to descending pain modulation, including opioid analgesia. One hypothesis reconciling the behavioral and physiological observations is that the serotonergic outflow enables or "gates" other descending influences that are directly inhibitory or facilitatory (Heinricher 1998). However, the dogma that neutral cells do not respond to noxious inputs was recently challenged by Gau and colleagues (2013), who showed that some serotonergic neurons have on-cell or off-cell

properties, although with on-like cells encountered more frequently. The existence of serotonergic on-cells and off-cells would be easily reconciled with the well-documented contribution of serotonin to both descending facilitation and inhibition. In addition, and possibly more interesting, this new finding argues that neurotransmitter content does not map to physiological function in the RVM.

Delta-Opioid Actions in the RVM
Under basal conditions, δ-opioid agonists in RVM have only modest effects on nociceptive behavior or activity of local neurons (Heinricher and Fields 2003). However, the contribution of the δ receptor becomes significant during persistent inflammation, with both direct effects and interactions with μ-mediated antinociception (Hurley and Hammond 2000, 2001). First, the effects of exogenous ligands are enhanced during chronic inflammation, presumably due to increased trafficking of δ receptors to the cell membrane. In addition to these presumed changes in δ *receptors*, levels of δ-preferring endogenous *peptides* are increased in the RVM during persistent inflammation. Intriguingly, this novel contribution of δ receptors during inflammation underlies the enhanced potency of μ-selective agents in this condition, since the interaction of δ- and μ-receptor activation is supra-additive (Hurley and Hammond 2001).

Kappa-Opioid Actions in the RVM
The role of the κ receptor within the RVM is somewhat controversial. Focal application of κ agonists within the RVM has been reported to produce potent analgesia by one group (Ackley et al. 2001), while other investigators find that κ compounds have no effect by themselves on nociception, but interfere with the analgesic actions of μ-opioids (Pan 1998). While the sex of the subjects may account for some of the discrepant results, it does not appear to be a complete explanation. As seen with the δ receptor, there is evidence that the anti-hyperalgesic effect of κ receptor activation in the RVM is enhanced in inflammatory states (Schepers et al. 2008).

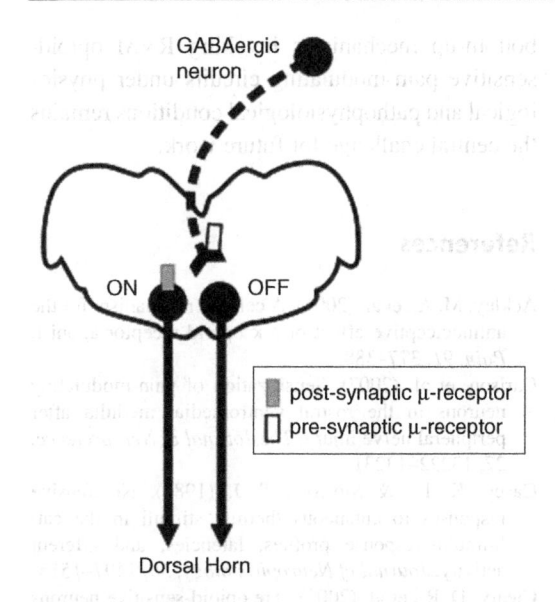

Like the behavioral analyses, electrophysiological studies of κ effects in the RVM provide a complex picture. Pan and colleagues found that μ and κ receptors are expressed by separate populations of RVM neurons (referred to as "► secondary" and "primary" cells, respectively). They therefore proposed that the behavioral antianalgesic effect of κ agonists given by microinjection in the RVM could be explained by κ inhibition of off-cells, the inhibitory output neurons of the RVM (Pan 1998). This is consistent with the inhibition of RVM off-cells by κ agonists reported in vivo (Meng et al. 2005). However, κ agonists suppressed the firing of on-cells and even some neutral cells as well as off-cells, which accords with work in the RVM slice by Marinelli et al. (2002). These authors showed that μ- and κ-opioid receptor types are commonly co-expressed by spinally projecting RVM neurons, with only a small subset demonstrating κ responses in the absence of μ responses. Still other workers have shown that κ agonists act presynaptically within the RVM (Ackley et al. 2001).

Physiological Role of Opioids Within the RVM

What is the physiological role of opioids within the RVM? Under basal conditions, opioid receptor antagonists microinjected into the RVM do not alter nociceptive responses, indicating that endogenous opioids do not maintain a significant ongoing tone. Endogenous opioid-mediated effects are likely to be important in stress-induced analgesia, although relatively few studies have focused specifically on the RVM. As already mentioned, plasticity in δ and κ function appears to be important in inflammatory pain states (Hurley and Hammond 2000; Schepers et al. 2008). More subtle variations in nociceptive responding appear to be associated with fluctuations in the ongoing activity of on-cells and off-cells (Heinricher et al. 1989), which is presumed to underlie behavioral weighting of pain relative to other motivations. For example, interactions between pain and hunger have been demonstrated, with opioid-dependent suppression of pain behaviors when highly palatable food is available (Casey and Morrow 1983; Dum and Herz 1984, recently replicated by Foo and Mason 2005). Understanding top-down as well as

bottom-up mechanisms invoking RVM opioid-sensitive pain-modulating circuits under physiological and pathophysiological conditions remains the central challenge for future work.

References

Ackley, M. A., et al. (2001). A cellular mechanism for the antinociceptive effect of a κ opioid receptor agonist. *Pain, 91*, 377–388.

Carlson, et al. (2007). Sensitization of pain-modulating neurons in the rostral ventromedial medulla after peripheral nerve injury. *The Journal of Neuroscience, 27*, 13222–13231.

Casey, K. L., & Morrow, T. J. (1983). Nocifensive responses to cutaneous thermal stimuli in the cat: Stimulus-response profiles, latencies, and afferent activity. *Journal of Neurophysiology, 50*, 1497–1515.

Cleary, D. R., et al. (2008). Are opioid-sensitive neurons in the rostral ventromedial medulla inhibitory interneurons? *Neuroscience, 151*, 564–571.

Dum, J., & Herz, A. (1984). Endorphinergic modulation of neural reward systems indicated by behavioral changes. *Pharmacology, Biochemistry, and Behavior, 21*, 259–266.

Fields, H. L., et al. (2006). Central nervous system mechanisms of pain modulation. In S. McMahon & M. Koltzenburg (Eds.), *Wall and Melzack's textbook of pain* (5th ed., pp. 125–142). London: Elsevier.

Foo, H., & Mason, P. (2005). Sensory suppression during feeding. *Proceedings of the National Academy of Sciences of the United States of America, 102*, 16865–16869.

Gau, R., et al. (2013). Noxious stimulation excites serotonergic neurons: A comparison between the lateral paragigantocellular reticular and the raphe magnus nuclei. *Pain, 154*, 647–659.

Heinricher, M. M. (1998). Organizational characteristics of supraspinally mediated responses to nociceptive input. In T. L. Yaksh (Ed.), *Anesthesia: Biologic foundations* (pp. 643–651). Philadelphia: Lippincott-Raven.

Heinricher, M. M., & Fields, H. L. (2003). The δ opioid receptor and brain pain-modulating circuits. In K. J. Chang, F. Porreca, & J. Woods (Eds.), *The δ receptor: Molecular and effect of δ opioid compounds* (pp. 467–480). New York: Marcel Dekker.

Heinricher, M. M., & Ingram, S. L. (2008). The brainstem and nociceptive modulation. In M. C. Bushnell & A. I. Basbaum (Eds.), *The science of pain* (pp. 593–626). San Diego, CA: Academic.

Heinricher, M. M., & McGaraughty, S. (1998). Analysis of excitatory amino acid transmission within the rostral ventromedial medulla: Implications for circuitry. *Pain, 75*, 247–255.

Heinricher, M. M., et al. (1989). Putative nociceptive modulating neurons in the rostral ventromedial medulla of the rat: Firing of on- and off-cells is related to nociceptive responsiveness. *Somatosensory & Motor Research, 6*, 427–439.

Heinricher, M. M., et al. (1994). Disinhibition of off-cells and antinociception produced by an opioid action within the rostral ventromedial medulla. *Neuroscience, 63*, 279–288.

Heinricher, M. M., et al. (1999). The role of excitatory amino acid transmission within the rostral ventromedial medulla in the antinociceptive actions of systemically administered morphine. *Pain, 81*, 57–65.

Heinricher, M. M., et al. (2010). Physiological basis for inhibition of morphine and improgan antinociception by CC12, a P450 epoxygenase inhibitor. *Journal of Neurophysiology, 104*, 3222–3230.

Hellman, K. M., & Mason, P. (2012). Opioids disrupt pro-nociceptive modulation mediated by raphe magnus. *The Journal of Neuroscience, 32*, 13668–13678.

Hurley, R. W., & Hammond, D. L. (2000). The analgesic effects of supraspinal m and d opioid receptor agonists are potentiated during persistent inflammation. *The Journal of Neuroscience, 20*, 1249–1259.

Hurley, R. W., & Hammond, D. L. (2001). Contribution of endogenous enkephalins to the enhanced analgesic effects of supraspinal μ opioid receptor agonists after inflammatory injury. *The Journal of Neuroscience, 21*, 2536–2545.

Marinelli, S., et al. (2002). Rostral ventromedial medulla neurons that project to the spinal cord express multiple opioid receptor phenotypes. *The Journal of Neuroscience, 22*, 10847–10855.

Mason, P. (1997). Physiological identification of pontomedullary serotonergic neurons in the rat. *Journal of Neurophysiology, 77*, 1087–1098.

McGaraughty, S., et al. (1993). Two distinct unit activity responses to morphine in the rostral ventromedial medulla of awake rats. *Brain Research, 604*, 331–333.

Meng, I. D., et al. (2005). Kappa opioids inhibit physiologically identified medullary pain modulating neurons and reduce morphine antinociception. *Journal of Neurophysiology, 93*, 1138–1144.

Pan, Z. Z. (1998). m-Opposing actions of the κ-opioid receptor. *Trends in Pharmacological Sciences, 19*, 94–98.

Phillips, R. S., et al. (2012). Pain-facilitating medullary neurons contribute to opioid-induced respiratory depression. *Journal of Neurophysiology, 108*, 2393–2404.

Reynolds, D. V. (1969). Surgery in the rat during electrical analgesia induced by focal brain stimulation. *Science, 154*, 444–445.

Schepers, R. J., et al. (2008). Inflammation-induced changes in rostral ventromedial medulla μ and κ opioid receptor mediated antinociception. *Pain, 136*, 320–330.

Vanderah, T. W., et al. (2001). Tonic descending facilitation from the rostral ventromedial medulla mediates opioid-induced abnormal pain and antinociceptive tolerance. *The Journal of Neuroscience, 21*, 279–286.

Xie, J. Y., et al. (2005). Cholecystokinin in the rostral ventromedial medulla mediates opioid-induced hyperalgesia and antinociceptive tolerance. *The Journal of Neuroscience, 25*, 409–416.

Yaksh, T. L., et al. (1988). Sites of action of opiates in production of analgesia. *Progress in Brain Research, 77*, 371–394.

Opiates: Pharmacology of Pain

Frank Porreca
Department of Pharmacology, University of Arizona Health Sciences Center, Tucson, AZ, USA

Introduction

The efficacy of the opioid alkaloids, morphine, and codeine, as well as that of their derivatives, as analgesics in conditions of acute pain is well established in clinical practice. Morphine, the major opiate alkaloid, is the prototypic analgesic and serves as the reference standard against which the analgesic activity of pain treatments is measured in clinical and preclinical paradigms. Opiate analgesics are routinely used in postoperative pain and in the treatment of many painful conditions of short- and medium-term duration and are gaining acceptance in the treatment of some chronic pain states as well. A complicating factor in the use of opioids to manage chronic pain conditions is the decrease in analgesic activity observed over time, at least in some patients, indicative of the phenomenon of analgesic tolerance. Opioids exert their antinociceptive activity through peripheral, spinal, and supraspinal sites, and adaptations induced by opioids at these multiple levels of the neuraxis may be contributing factors in the expression of tolerance. There is evidence to indicate that long-term opioid exposure may result in activation of pain facilitatory systems from medullary sites that oppose the analgesic actions of these compounds. The engagement of descending pain facilitatory mechanisms can paradoxically result in the expression of a hyperalgesic state, which can act as a "physiological antagonist" of analgesia, in essence manifesting as "antinociceptive tolerance." The entries included in this section discuss many different aspects of opioid-mediated action and the neurophysiological adaptations that can contribute to the effects of these compounds over time.

Characteristics

Opioid Receptors and Endogenous Opioids

The discovery of a receptor selective for opioids was reported almost simultaneously by different groups in 1973 (Pert and Snyder 1973a, b; Simon 1973) and precipitated an increased interest in the field of pain research. This discovery was rapidly followed by the identification and characterization of endogenous ligands for this receptor. Almost immediately thereafter, the pentapeptide ► enkephalins (Hughes et al. 1975; Kosterlitz and Hughes 1975), the ► endorphins (Cox et al. 1975), and ► dynorphin (Goldstein et al. 1979) were described. Concurrently, behavioral studies and experiments performed with isolated tissue preparations revealed the existence of three subtypes of the ► opioid receptor, identified as mu, kappa, and delta (Lord et al. 1977; Martin et al. 1976; Waterfield et al. 1978); this was later confirmed through cloning techniques (Zaki et al. 1996). Antibodies to these receptors allowed for anatomical location (Arvidsson et al. 1995; Kalyuzhny et al. 1996; Mansour et al. 1994, 1995) and localization of these receptors in pain pathways including the central and peripheral terminations of sensory fibers and in ascending pain transmission pathways as well as in descending pain modulatory systems. Collectively, these findings suggested that opioids may exert their antinociceptive activity through several mechanisms, providing their overall analgesic actions through anatomical synergy. These discoveries also provided strong evidence for the existence of an endogenous pain control system, with implications for the understanding of tolerance, dependence, and withdrawal.

The Distributions of the Opioid Receptors are Consistent with Pain Pathways

Spinal Sites of Action

The distribution and anatomical localization of the opioid receptors have been extensively explored through immunohistochemistry, autoradiography, radioligand binding, and in situ hybridization of message for the receptors (Bzdega et al. 1993; Mansour et al. 1995). The distribution of opioid receptors is consistent with known pathways related to the processing of nociceptive signals. The opioid receptors are predominant in the outer laminae of the spinal dorsal horn, with a less extensive distribution in lamina V and around the central canal, and are sparse within the intermediate laminae (Besse et al. 1991; Quirion et al. 1983). These sites are consistent with the distribution of the terminals of nociceptors from the periphery and viscera. Studies employing a variety of techniques established that a substantial proportion of the opioid receptors found in the outer laminae reside on the terminals of primary afferent nociceptive C-fibers and that message for the opioid receptors is found principally in nociceptive C- and A-delta dorsal root ganglia (DRG) neurons (Arvidsson et al. 1995; Besse et al. 1991; Ji et al. 1995). This distribution is consistent with a role in the regulation of nociceptive inputs without altering innocuous sensory signals. Accordingly, it was found that activation of mu-opioid receptors predictably inhibited Ca^{++} channels of small diameter nociceptors and not of large diameter myelinated A-beta cells and suggested that mu-receptor activation selectively inhibits the activity of C-fibers (Taddese et al. 1995). Approximately 20–30 % of spinal opioid receptors reside either on interneurons or on cell bodies of second-order neurons that transmit nociceptive inputs to supraspinal sites that process nociceptive signals (Besse et al. 1991).

The presence of opioid receptors in the spinal cord suggests a direct spinal antinociceptive action of opioids. The direct spinal application of morphine or of enkephalins into the spinal space produced dose-dependent, naloxone-reversible antinociception to noxious thermal stimuli (Yaksh et al. 1977a). Mid-thoracic spinalization reduced the potency of systemic, but not of spinal, morphine against a noxious evoked spinal reflex (Advokat 1989). The iontophoretic application of morphine into the outer laminae of the dorsal horns of the spinal cord also attenuated the responses of dorsal horn units to noxious stimuli (Duggan et al. 1976). These observations supported the hypothesis that opioids may act directly at spinal sites to modulate nociceptive spinal reflexes and nociceptive inputs, and the spinal administration of opiates has now become routine medical practice in the treatment of pain.

Supraspinal Sites of Action

The microinjection of opioids into the cerebral ventricles has been shown to produce dose-dependent antinociception in several species, including mice and rats, suggesting a supraspinal modulation of nociceptive inputs (Erspamer et al. 1989; Jiang et al. 1990; Miaskowski et al. 1991; Porreca et al. 1984). Autoradiographic studies have demonstrated that there are significant levels of opioid-receptor mRNA in many cortical, diencephalic, and brainstem regions in addition to spinal loci (Mansour et al. 1994; Mansour et al. 1995). In particular, significant expression of message for opioid receptors was found in the periaqueductal gray (PAG) and the rostral ventromedial medulla (RVM), regions that are critical to expression of supraspinal antinociceptive manipulations (Basbaum et al. 1978; Basbaum and Fields 1978; Fields and Anderson 1978). The microinjection of morphine into the PAG produced dose-dependent, naloxone-reversible antinociception to peripheral noxious stimuli (Lewis and Gebhart 1977; Yaksh et al. 1976). Moreover, PAG sites that were responsive to morphine also produced antinociception in response to electrical stimulation (Lewis and Gebhart 1977; Yeung et al. 1977). The microinjection of morphine into the PAG attenuated the activity of projection neurons in the dorsal horn in response to peripheral nociceptive stimuli (Bennett and Mayer 1979).

The RVM is recognized as a critical region with respect to nociceptive processing and modulation, receiving inputs from the spinal dorsal horn and from rostral sites as well (Fields and Basbaum 1999; Fields and Heinricher 1985; Fields et al. 1983). Electrophysiological studies on the responses of RVM neurons to noxious thermal stimulation have identified the existence of ▶ on-cells and ▶ off-cells (Fields 1992; Fields and Basbaum 1999; Heinricher et al. 2003). The off-cells are tonically active and pause in firing immediately before the animal withdraws from the noxious thermal stimulus, whereas the on-cells accelerate firing immediately before the nociceptive reflex occurs. An additional class, the "neutral" cells were initially characterized by the absence of response to noxious thermal stimulation. It is now generally understood that the activity of the off-cells correlates with inhibition of nociceptive input and nocifensive responses and these neurons may be the source of descending inhibition of nociceptive inputs (Fields 1992; Fields and Basbaum 1999; Heinricher et al. 2003). In contrast, the response characteristics of the on-cells suggest that these neurons are the source of descending facilitation of nociception (Fields 1992; Fields and Basbaum 1999; Heinricher et al. 2003; McNally 1999). Accordingly, manipulations that facilitate responses to nociceptive stimuli also increase on-cell activity (Fields and Basbaum 1999; Fields 2000; Heinricher and Roychowdhury 1997; Heinricher et al. 2003). For example, prolonged delivery of a noxious thermal stimulus produced increased on-cell activity along with a facilitation of nociceptive reflexes (Morgan and Fields 1994). Moreover, inactivation of RVM neuronal activity with lidocaine blocked the facilitated withdrawal response (Morgan and Fields 1994). It is now generally accepted that a spino-bulbo-spinal loop may be important to the development and maintenance of exaggerated pain behaviors produced by persistent noxious stimuli (Heinricher et al. 2003; Ossipov et al. 2001; Suzuki et al. 2002; Suzuki et al. 2004). Morphine microinjection or electrical stimulation in the RVM has produced naloxone-sensitive antinociception by activating spinopetal mechanisms (Kiefel et al. 1993; McGowan and Hammond 1993; Rossi et al. 1993). Studies employing retrograde tracing methods demonstrated the existence of opioid expressing neurons that project from the RVM to the spinal cord to provide a descending inhibition of nociceptive inputs (Kalyuzhny et al. 1996). It was also proposed that opioids exert indirect effects on RVM projection neurons and have both direct and indirect effects on bulbospinal neurons (Kalyuzhny et al. 1996).

Synergistic Actions at Spinal/Supraspinal Sites

An important aspect of morphine-mediated antinociception is the existence of a synergistic interaction between morphine given spinally and supraspinally. The administration of a 1:1 fixed ratio of morphine administered intrathecally and into the cerebral ventricles produced an approximate 30-fold increase in potency when evaluated in the tail-flick test and a 45-fold increase in the hot-plate test compared to supraspinal morphine alone (Yeung and Rudy 1980). Isobolographic analyses employing several dose ratios demonstrated a hyperbolic function with a strong degree of curvature, and it was concluded that potentiation would occur at all possible combinations of spinal and supraspinal levels of morphine after systemic injection (Miyamoto et al. 1991; Roerig and Fujimoto 1988; Yeung and Rudy 1980). Situations where the antinociceptive effect of morphine was diminished corresponded with a loss of spinal/supraspinal synergy. For example, repeated exposure to systemic morphine abolished the synergistic antinociceptive effect of morphine given into the PAG and spinally (Siuciak and Advokat 1989). Similarly, rats with peripheral nerve injury demonstrate a loss of spinal morphine potency along with a loss of the spinal/supraspinal synergy (Bian et al. 1999). Restoration of the spinal site of morphine activity with an NMDA antagonist restored supraspinal-spinal synergy and increased the potency of systemically given morphine in animals with nerve injury (Bian et al. 1999).

Paradoxical Opioid-Induced Hyperalgesia

A number of anecdotal clinical reports exist that show that opioid administration can paradoxically elicit abnormally heightened pain sensations in response to a stimulus. Long-term spinal morphine provoked hyperesthesias and allodynia that were unrelated to the original pain complaint in cancer patients (Ali 1986; Arner et al. 1988; De Conno et al. 1991). The spinal infusion of sufentanil in a patient with neuropathic pain secondary to arachnoiditis and laminectomy evoked hyperesthesias in the lower extremities (Devulder 1997). This abnormal pain state was described as being qualitatively different from the original complaint and included the back, abdomen, and both legs. Cancer patients who received high doses of intrathecal morphine by bolus injections also reported paradoxical intense pain within 0.5 h of the injections (Stillman et al. 1987).

Animal studies have clearly demonstrated that opioid administration may produce an abnormal, paradoxical hyperalgesic state (Gardell et al. 2002b; Mao et al. 1994; Mayer et al. 1995a; Vanderah et al. 2000, 2001b). The repeated daily injection of spinal or systemic morphine produced enhanced responses of the tail or hind paw to noxious thermal stimuli within 8 days (Mayer et al. 1995b; Trujillo and Akil 1994). The continuous exposure to opioids produced behavioral signs of exaggerated pain, and importantly, such pain occurred while the opioid was continuously present in the system (Gardell et al. 2002b; Vanderah et al. 2000, 2001b). The continuous spinal infusion of [D-Ala^2,N-Me-Phe^4, Gly-ol^5]enkephalin (DAMGO) delivered through an osmotic minipump or of morphine administered by subcutaneous pellets produced antinociceptive tolerance to spinal DAMGO or morphine (Vanderah et al. 2000). These animals expressed tactile and thermal hypersensitivities (Vanderah et al. 2000, 2001b) while the opioids were still being administered (Vanderah et al. 2000, 2001b), and changes did not result from changes in metabolism or metabolites since blood levels of opiates were constant over the period of infusion (Ossipov, Stiller and Porreca, unpublished observations).

This paradoxical hyperalgesia and the neurobiological adaptations which underlie this state may play a role in the requirement for increased levels of opioids to maintain a constant degree of antinociception, perhaps reflecting the expression of antinociceptive tolerance (Gardell et al. 2002b; Vanderah et al. 2000, 2001a, b). Clinically, the need for increasing doses of opioids in cases of chronic pain, at least in some patients, is well documented and can present a major obstacle to providing adequate pain relief over a long period of time (Cherney and Portenoy 1999; Foley 1993; Foley 1995). In spite of much intensive research, however, the mechanisms that underlie the development of tolerance to the analgesic effects of opioids remain largely unknown. Many studies have focused on changes occurring at the cellular level in order to gain an appreciation of the mechanisms that drive the development of antinociceptive tolerance (e.g., Childers 1991; Sabbe and Yaksh 1990). Although alterations in subcellular processes appear potentially to contribute to the phenomenon of tolerance, the present level of understanding of such processes is insufficient to allow for the direct correlation of intracellular changes with those occurring at the level of the neuronal circuits mediating antinociception or analgesia. Nevertheless, opioid-induced upregulation of peptidic neuromodulators such as cholecystokinin (CCK) or dynorphin may supply endogenous physiological antagonists to endogenous or exogenous opioid activity. These changes may be related to the activation of descending facilitation of nociception. Consequently, hyperalgesia may be considered as a physiological antagonist of analgesia, and increased states of pain require increased levels of pain-relieving opiate, resulting in "opiate tolerance" (Vanderah et al. 2001a).

Opioid-Induced Paradoxical Pain is Promoted by Descending Facilitation from the RVM

As noted above, the on-cells of the RVM are an important source of descending pain facilitatory projections. The microinjection of CCK into the RVM has enhanced nociceptive input and attenuated the morphine-induced reduction of on-cell responses to nociception, suggesting that these neurons may be activated by CCK (Heinricher and McGaraughty 1996). The microinjection of lidocaine into the RVM of rats with persistent exposure to morphine produced a reversible

block of both tactile and thermal hyperesthesias and abolished antinociceptive tolerance to morphine (Vanderah et al. 2001b). Prolonged morphine exposure elicited a five-fold increase in basal CCK in the RVM (Xie et al. 2005). The microinjection of CCK into the RVM produced tactile and thermal hyperesthesias whereas a CCK_2 antagonist in the RVM abolished opioid-induced hyperesthesias (Xie et al. 2005). Moreover, the microinjection of a CCK_2 antagonist into RVM or lesions of the dorsolateral funiculus abolished both behavioral signs of enhanced pain and antinociceptive tolerance in response to persistent morphine exposure (Xie et al. 2005).

Spinal Dynorphin and Opioid-Induced Paradoxical Pain

Considerable evidence has demonstrated that enhanced expression of spinal dynorphin is pronociceptive and promotes facilitated pain states. A single spinal injection of dynorphin has produced long-lasting tactile allodynia in rats and mice (Laughlin et al. 1997; Vanderah et al. 1996). Elevations in spinal dynorphin content are also seen in animals with prolonged, constant exposure to opioids either systemically or spinally (Gardell et al. 2002b; Vanderah et al. 2001a, b). Spinal infusion of DAMGO elicited enhanced responses to tactile and noxious thermal stimuli and also caused an elevation in dynorphin content in the lumbar cord (Vanderah et al. 2000). The spinal injection of antiserum to dynorphin blocked enhanced responses in the DAMGO-treated rats and unmasked the antinociceptive action of the DAMGO that was still infused (Vanderah et al. 2000). Furthermore, dynorphin antiserum blocked the rightward displacement of the dose-effect curve for spinal morphine in DAMGO-infused rats, indicating a blockade of antinociceptive tolerance (Vanderah et al. 2000). Bilateral lesions of the DLF, which were shown to block abnormal pain and tolerance to the antinociceptive effect of morphine, also prevented the upregulation of spinal dynorphin (Gardell et al. 2003). Thus, manipulations that block opioid-induced hyperalgesia, in this case due to spinal infusion of opioid, also block the behavioral manifestation of antinociceptive tolerance. The data show that sustained opioid administration leads to elevated spinal dynorphin content, which in turn promotes an abnormal hyperalgesic state and increases the requirement for opioid dose in order to produce a comparable antinociceptive effect to that in animals without increased nociception, resulting in an apparent manifestation of antinociceptive tolerance.

The precise mechanisms through which increased spinal dynorphin expression promotes hyperalgesia and consequently the manifestation of opioid tolerance remain to be elucidated but appear to be related, in part, to an enhanced release of transmitters from primary afferent terminals. Dynorphin produced a dose-dependent release of glutamate and aspartate elicited by exogenous dynorphin in the hippocampus and spinal cord (Faden 1992; Skilling et al. 1992). More recently, the capsaicin-stimulated release of calcitonin gene-related peptide (CGRP) was potentiated by dynorphin $A_{(2-13)}$, a non-opioid fragment, in spinal cord sections in vitro (Claude et al. 1999; Gardell et al. 2002a, 2003). In addition, dynorphin also facilitated capsaicin-evoked substance P release from trigeminal nuclear slices, and this effect was abolished by MK-801, but not by opioid antagonists (Arcaya et al. 1999). Most recently, it was demonstrated that persistent exposure to morphine pellets implanted subcutaneously produced enhanced capsaicin-evoked release of CGRP from spinal tissue (Gardell et al. 2002b, 2003). This enhanced, evoked release was blocked by the addition of antiserum to dynorphin in the perfusion medium. Moreover, the disruption of descending facilitation from supraspinal sites by selective ablation of RVM neurons that express the mu-opioid receptor or by surgical lesions of the DLF prevented opioid-induced abnormal pain, spinal dynorphin upregulation, and enhanced capsaicin-evoked release of CGRP (Gardell et al. 2002b, 2003). Finally, enhanced capsaicin-evoked release of CGRP was also blocked by the NMDA antagonist MK-801 (Gardell et al. 2002b, 2003). The introduction of NMDA, dynorphin $A_{(1-17)}$, or dynorphin $A_{(2-17)}$ into the lumbar spinal cord through a microdialysis catheter elicited a prolonged release of prostaglandin E2 and of

excitatory amino acids (Koetzner et al. 2004). These observations provide indirect links to a mechanism through which pathologically elevated levels of spinal dynorphin may promote enhanced pain (Koetzner et al. 2004).

Summary

The opioid analgesics have been employed throughout our history for the treatment and control of pain. The opioids, exemplified by the prototype morphine, represent the most efficacious means of controlling pain at the present time. The analgesic activity of opioids is mediated through activity at spinal and supraspinal sites. Opioids may act directly at the spinal level to block incoming nociceptive inputs. Alternatively, opioids may activate supraspinal sites that activate descending pain inhibitory systems to block the transmission of nociceptive signals to supraspinal structures. Paradoxically, the very same substances that are so efficacious against pain may actually cause an abnormal pain phenomenon themselves by eliciting the activation of endogenous pronociceptive systems. Persistent exposure to the opioids causes an increase in the presence of the pronociceptive neurotransmitter CCK, which acts as an endogenous modulator of antinociceptive activity. One means through which this is achieved is through the activation of a descending facilitation from the RVM. These findings reveal the complex neurobiological adaptations that result from opioid administration and raise questions as to the optimal use of opioids for the management of pain in humans, as well as suggesting alternative approaches to improve the use of this class of compounds for the management of chronic pain states.

References

Advokat, C. (1989). Tolerance to the antinociceptive effect of morphine in spinally transected rats. *Behavioral Neuroscience, 103*, 1091–1098.

Ali, N. M. (1986). Hyperalgesic response in a patient receiving high concentrations of spinal morphine. *Anesthesiology, 65*, 449.

Arcaya, J. L., Cano, G., Gomez, G., et al. (1999). Dynorphin A increases substance P release from trigeminal primary afferent C-fibers. *European Journal of Pharmacology, 366,* 27–34.

Arner, S., Rawal, N., & Gustafsson, L. L. (1988). Clinical experience of long-term treatment with epidural and intrathecal opioids -a nationwide survey. *Acta Anaesthesiologica Scandinavica, 32,* 253–259.

Arvidsson, U., Dado, R. J., Riedl, M., et al. (1995). delta-opioid receptor immunoreactivity: Distribution in brainstem and spinal cord, and relationship to biogenic amines and enkephalin. *Journal of Neuroscience, 15,* 1215–1235.

Basbaum, A. I., & Fields, H. L. (1978). Endogenous pain control mechanisms: Review and hypothesis. *Annals of Neurology, 4,* 451–462.

Basbaum, A. I., Clanton, C. H., & Fields, H. L. (1978). Three bulbospinal pathways from the rostral medulla of the cat: An autoradiographic study of pain modulating systems. *The Journal of Comparative Neurology, 178,* 209–224.

Bennett, G. J., & Mayer, D. J. (1979). Inhibition of spinal cord interneurons by narcotic microinjection and focal electrical stimulation in the periaqueductal central gray matter. *Brain Research, 172,* 243–257.

Besse, D., Lombard, M. C., & Besson, J. M. (1991). Autoradiographic distribution of mu, delta and kappa opioid binding sites in the superficial dorsal horn, over the rostrocaudal axis of the rat spinal cord. *Brain Research, 548,* 287–291.

Bian, D., Ossipov, M. H., Ibrahim, M., et al. (1999). Loss of antiallodynic and antinociceptive spinal/supraspinal morphine synergy in nerve-injured rats: Restoration by MK-801 or dynorphin antiserum. *Brain Research, 831,* 55–63.

Bzdega, T., Chin, H., Kim, H., et al. (1993). Regional expression and chromosomal localization of the delta opiate receptor gene. *Proceedings of the National Academy of Sciences USA, 90,* 9305–9309.

Cherney, N. I., & Portenoy, R. K. (1999). Practical issues in the management of cancer pain. In P. D. Wall & R. Melzack (Eds.), *Textbook of pain* (pp. 1479–1522). Edinburgh: Churchill Livingstone.

Childers, S. R. (1991). Opioid receptor-coupled second messenger systems. *Life Sciences, 48,* 1991–2003.

Claude, P., Gracia, N., Wagner, L., et al. (1999). Effect of dynorphin on ICGRP release from capsaicin-sensitive fibers. *Abstracts of the 9th World Congress on Pain, 9,* 262.

Cox, B. M., Opheim, K. E., Teschemacher, H., et al. (1975). A peptide-like substance from pituitary that acts like morphine. 2. Purification and properties. *Life Sciences, 16,* 1777–1782.

De Conno, F., Caraceni, A., Martini, C., et al. (1991). Hyperalgesia and myoclonus with intrathecal infusion of high-dose morphine. *Pain, 47,* 337–339.

Devulder, J. (1997). Hyperalgesia induced by high-dose intrathecal sufentanil in neuropathic pain. *Journal of Neurosurgical Anesthesiology, 9,* 146–148.

Duggan, A. W., Davies, J., & Hall, J. G. (1976). Effects of opiate agonists and antagonists on central neurons of the cat. *Journal of Pharmacology and Experimental Therapeutics, 196,* 107–120.

Erspamer, V., Melchiorri, P., Falconieri-Erspamer, G., et al. (1989). Deltorphins: A family of naturally occurring peptides with high affinity and selectivity for delta opioid binding sites. *Proceedings of the National Academy of Sciences USA, 86,* 5188–5192.

Faden, A. I. (1992). Dynorphin increases extracellular levels of excitatory amino acids in the brain through a non-opioid mechanism. *Journal of Neuroscience, 12,* 425–429.

Fields, H. L. (1992). Is there a facilitating component to central pain modulation? *Journal of American Physical Society, 1,* 71–78.

Fields, H. L. (2000). Pain modulation: Expectation, opioid analgesia and virtual pain. *Progress in Brain Research, 122,* 245–253.

Fields, H. L., & Anderson, S. D. (1978). Evidence that raphe-spinal neurons mediate opiate and midbrain stimulation-produced analgesias. *Pain, 5,* 333–349.

Fields, H. L., & Basbaum, A. I. (1999). Central nervous system mechanisms of pain modulation. In P. D. Wall & R. Melzack (Eds.), *Textbook of pain* (pp. 309–329). Edinburgh: Churchill Livingstone.

Fields, H. L., & Heinricher, M. M. (1985). Anatomy and physiology of a nociceptive modulatory system. *Philosophical Transactions of the Royal Society of London. Series B, Biological Sciences, 308,* 361–374.

Fields, H. L., Bry, J., Hentall, I., et al. (1983). The activity of neurons in the rostral medulla of the rat during withdrawal from noxious heat. *Journal of Neuroscience, 3,* 2545–2552.

Foley, K. M. (1993). Opioids. *Neurologic Clinics, 11,* 503–522.

Foley, K. M. (1995). Misconceptions and controversies regarding the use of opioids in cancer pain. *Anti-Cancer Drugs, 6,* 4–13.

Gardell, L. R., Burgess, S. E., Dogrul, A., et al. (2002a). Pronociceptive effects of spinal dynorphin promote cannabinoid-induced pain and antinociceptive tolerance. *Pain, 98,* 79–88.

Gardell, L. R., Wang, R., Burgess, S. E., et al. (2002b). Sustained morphine exposure induces a spinal dynorphin-dependent enhancement of excitatory transmitter release from primary afferent fibers. *Journal of Neuroscience, 22,* 6747–6755.

Gardell, L. R., Vanderah, T. W., Gardell, S. E., et al. (2003). Enhanced evoked excitatory transmitter release in experimental neuropathy requires descending facilitation. *Journal of Neuroscience, 23,* 8370–8379.

Goldstein, A., Tachibana, S., Lowney, L. I., et al. (1979). Dynorphin-(1–13), an extraordinarily potent opioid peptide. *Proceedings of the National Academy of Sciences USA, 76,* 6666–6670.

Heinricher, M. M., & McGaraughty, S., (1996). CCK modulates the antinociceptive actions of opioids by an action within the rostral ventromedial medulla: A combined electrophysiological and behavioral study. In *Abstracts of the 8th World Congress on Pain* (p. 472), Vancouver. Seattle: IASP Press.

Heinricher, M. M., & Roychowdhury, S. M. (1997). Reflex-related activation of putative pain facilitating neurons in rostral ventromedial medulla requires excitatory amino acid transmission. *Neuroscience, 78,* 1159–1165.

Heinricher, M. M., Pertovaara, A., & Ossipov, M. H. (2003). Descending modulation after injury. In D. O. Dostrovsky, D. B. Carr, & M. Koltzenburg (Eds.), *Proceedings of the 10th world congress on pain* (pp. 251–260). Seattle: IASP Press.

Hughes, J., Smith, T., Morgan, B., et al. (1975). Purification and properties of enkephalin -the possible endogenous ligand for the morphine receptor. *Life Sciences, 16,* 1753–1758.

Ji, R. R., Zhang, Q., Law, P. Y., et al. (1995). Expression of mu-, delta-, and kappa-opioid receptor-like immunoreactivities in rat dorsal root ganglia after carrageenan-induced inflammation. *Journal of Neuroscience, 15,* 8156–8166.

Jiang, Q., Mosberg, H. I., & Porreca, F. (1990). Antinociceptive effects of [D-Ala2]deltorphin II, a highly selective delta agonist in vivo. *Life Sciences, 47,* PL43–PL47.

Kalyuzhny, A. E., Arvidsson, U., Wu, W., et al. (1996). Mu-opioid and delta-opioid receptors are expressed in brainstem antinociceptive circuits: Studies using immunocytochemistry and retrograde tract-tracing. *Journal of Neuroscience, 16,* 6490–6503.

Kiefel, J. M., Rossi, G. C., & Bodnar, R. J. (1993). Medullary mu and delta opioid receptors modulate mesencephalic morphine analgesia in rats. *Brain Research, 624,* 151–161.

Koetzner, L., Hua, X. Y., Lai, J., et al. (2004). Nonopioid actions of intrathecal dynorphin evoke spinal excitatory amino acid and prostaglandin E2 release mediated by cyclooxygenase-1 and −2. *Journal of Neuroscience, 24,* 1451–1458.

Kosterlitz, H. W., & Hughes, J. (1975). Some thoughts on the significance of enkephalin, the endogenous ligand. *Life Sciences, 17,* 91–96.

Laughlin, T. M., Vanderah, T. W., Lashbrook, J., et al. (1997). Spinally administered dynorphin A produces long-lasting allodynia: Involvement of NMDA but not opioid receptors. *Pain, 72,* 253–260.

Lewis, V. A., & Gebhart, G. F. (1977). Morphine-induced and stimulation-produced analgesias at coincident periaqueductal central gray loci: Evaluation of analgesic congruence, tolerance, and cross-tolerance. *Experimental Neurology, 57,* 934–955.

Lord, J. A., Waterfield, A. A., Hughes, J., et al. (1977). Endogenous opioid peptides: Multiple agonists and receptors. *Nature, 267,* 495–499.

Mansour, A., Fox, C. A., Burke, S., et al. (1994). Mu, delta, and kappa opioid receptor mRNA expression in the rat CNS: an in situ hybridization study. *The Journal of Comparative Neurology, 350,* 412–438.

Mansour, A., Fox, C. A., Akil, H., et al. (1995). Opioid-receptor mRNA expression in the rat CNS: Anatomical and functional implications. *Trends in Neurosciences, 18,* 22–29.

Mao, J., Price, D. D., & Mayer, D. J. (1994). Thermal hyperalgesia in association with the development of morphine tolerance in rats: Roles of excitatory amino acid receptors and protein kinase C. *Journal of Neuroscience, 14,* 2301–2312.

Martin, W. R., Eades, C. G., Thompson, J. A., et al. (1976). The effects of morphine- and nalorphine- like drugs in the nondependent and morphine-dependent chronic spinal dog. *Journal of Pharmacology and Experimental Therapeutics, 197,* 517–532.

Mayer, D. J., Mao, J., & Price, D. D. (1995). The development of morphine tolerance and dependence is associated with translocation of protein kinase C. *Pain, 61,* 365–374.

McGowan, M. K., & Hammond, D. L. (1993). Antinociception produced by microinjection of L-glutamate into the ventromedial medulla of the rat: Mediation by spinal GABAA receptors. *Brain Research, 620,* 86–96.

McNally, G. P. (1999). Pain facilitatory circuits in the mammalian central nervous system: Their behavioral significance and role in morphine analgesic tolerance. *Neuroscience and Biobehavioral Reviews, 23,* 1059–1078.

Miaskowski, C., Taiwo, Y. O., & Levine, J. D. (1991). Contribution of supraspinal mu- and delta-opioid receptors to antinociception in the rat. *European Journal of Pharmacology, 205,* 247–252.

Miyamoto, Y., Morita, N., Kitabata, Y., et al. (1991). Antinociceptive synergism between supraspinal and spinal sites after subcutaneous morphine evidenced by CNS morphine content. *Brain Research, 552,* 136–140.

Morgan, M. M., & Fields, H. L. (1994). Pronounced changes in the activity of nociceptive modulatory neurons in the rostral ventromedial medulla in response to prolonged thermal noxious stimuli. *Journal of Neurophysiology, 72,* 1161–1170.

Ossipov, M. H., Lai, J., Malan, T. P., Jr., et al. (2001). Tonic descending facilitation as a mechanism of neuropathic pain. In P. T. Hansson, H. L. Fields, R. G. Hill, et al. (Eds.), *Neuropathic pain: Pathophysiology and treatment* (pp. 107–124). Seattle: IASP Press.

Pert, C. B., & Snyder, S. H. (1973a). Opiate receptor: Demonstration in nervous tissue. *Science, 179,* 1011–1014.

Pert, C. B., & Snyder, S. H. (1973b). Properties of opiate-receptor binding in rat brain. *Proceedings of the National Academy of Sciences USA, 70,* 2243–2247.

Porreca, F., Mosberg, H. I., Hurst, R., et al. (1984). Roles of mu, delta and kappa opioid receptors in spinal and supraspinal mediation of gastrointestinal transit effects and hot-plate analgesia in the mouse. *Journal of Pharmacology and Experimental Therapeutics, 230,* 341–348.

Quirion, R., Zajac, J. M., Morgat, J. L., et al. (1983). Autoradiographic distribution of mu and delta opiate receptors in rat brain using highly selective ligands. *Life Sciences, 33*(Suppl. 1), 227–230.

Roerig, S. C., & Fujimoto, J. M. (1988). Morphine antinociception in different strains of mice: Relationship of supraspinal-spinal multiplicative interaction to tolerance. *Journal of Pharmacology and Experimental Therapeutics, 247*, 603–608.

Rossi, G. C., Pasternak, G. W., & Bodnar, R. J. (1993). Synergistic brainstem interactions for morphine analgesia. *Brain Research, 624*, 171–180.

Sabbe, M. B., & Yaksh, T. L. (1990). Pharmacology of spinal opioids. *Journal of Pain and Symptom Management, 5*, 191–203.

Simon, E. J. (1973). In search of the opiate receptor. *The American Journal of the Medical Sciences, 266*, 160–168.

Siuciak, J. A., & Advokat, C. (1989). The synergistic effect of concurrent spinal and supraspinal opiate agonisms is reduced by both nociceptive and morphine pretreatment. *Pharmacology Biochemistry and Behavior, 34*, 265–273.

Skilling, S. R., Sun, X., Kurtz, H. J., et al. (1992). Selective potentiation of NMDA-induced activity and release of excitatory amino acids by dynorphin: Possible roles in paralysis and neurotoxicity. *Brain Research, 575*, 272–278.

Stillman, M. J., Moulin, D. E., & Foley, K. M., (1987). Paradoxical pain following high-dose spinal morphine. *Pain, 4*:S389

Suzuki, R., Morcuende, S., Webber, M., et al. (2002). Superficial NK^1-expressing neurons control spinal excitability through activation of descending pathways. *Nature Neuroscience, 5*, 1319–1326.

Suzuki, R., Rahman, W., Hunt, S. P., et al. (2004). Descending facilitatory control of mechanically evoked responses is enhanced in deep dorsal horn neurones following peripheral nerve injury. *Brain Research, 1019*, 68–76.

Taddese, A., Nah, S. Y., & McCleskey, E. W. (1995). Selective opioid inhibition of small nociceptive neurons. *Science, 270*, 1366–1369.

Trujillo, K. A., & Akil, H. (1994). Inhibition of opiate tolerance by non-competitive N-methyl-D-aspartate receptor antagonists. *Brain Research, 633*, 178–188.

Vanderah, T. W., Laughlin, T., Lashbrook, J. M., et al. (1996). Single intrathecal injections of dynorphin A or des-Tyr-dynorphins produce long-lasting allodynia in rats: Blockade by MK-801 but not naloxone. *Pain, 68*, 275–281.

Vanderah, T. W., Gardell, L. R., Burgess, S. E., et al. (2000). Dynorphin promotes abnormal pain and spinal opioid antinociceptive tolerance. *Journal of Neuroscience, 20*, 7074–7079.

Vanderah, T. W., Ossipov, M. H., Lai, J., et al. (2001a). Mechanisms of opioid-induced pain and antinociceptive tolerance: Descending facilitation and spinal dynorphin. *Pain, 92*, 5–9.

Vanderah, T. W., Suenaga, N. M., Ossipov, M. H., et al. (2001b). Tonic descending facilitation from the rostral ventromedial medulla mediates opioid-induced abnormal pain and antinociceptive tolerance. *Journal of Neuroscience, 21*, 279–286.

Waterfield, A. A., Lord, J. A., Hughes, J., et al. (1978). Differences in the inhibitory effects of normorphine and opioid peptides on the responses of the vasa deferentia of two strains of mice. *European Journal of Pharmacology, 47*, 249–250.

Xie, J. Y., Herman, D. S., Stiller, C. O., et al. (2005). Cholecystokinin in the rostral ventromedial medulla mediates opioid-induced hyperalgesia and antinociceptive tolerance. *Journal of Neuroscience, 25*, 409–416.

Yaksh, T. L., Yeung, J. C., & Rudy, T. A. (1976). Systematic examination in the rat of brain sites sensitive to the direct application of morphine: Observation of differential effects within the periaqueductal gray. *Brain Research, 114*, 83–103.

Yaksh, T. L., Huang, S. P., & Rudy, T. A. (1977a). The direct and specific opiate-like effect of met5-enkephalin and analogues on the spinal cord. *Neuroscience, 2*, 593–596.

Yaksh, T. L., Kohl, R. L., & Rudy, T. A. (1977b). Induction of tolerance and withdrawal in rats receiving morphine in the spinal subarachnoid space. *European Journal of Pharmacology, 42*, 275–284.

Yeung, J. C., & Rudy, T. A. (1980). Multiplicative interaction between narcotic agonisms expressed at spinal and supraspinal sites of antinociceptive action as revealed by concurrent intrathecal and intracerebroventricular injections of morphine. *Journal of Pharmacology and Experimental Therapeutics, 215*, 633–642.

Yeung, J. C., Yaksh, T. L., & Rudy, T. A. (1977). Concurrent mapping of brain sites for sensitivity to the direct application of morphine and focal electrical stimulation in the production of antinociception in the rat. *Pain, 4*, 23–40.

Zaki, P. A., Bilsky, E. J., Vanderah, T. W., et al. (1996). Opioid receptor types and subtypes: The delta receptor as a model. *Annual Review of Pharmacology and Toxicology, 36*, 379–401.

Opioid

Definition

A class of drugs with a molecular structure related to opium. Opiates have analgesic effects and other effects, some of them adverse. Examples of opiates include morphine, codeine, and fentanyl. Opioid is a generic term that refers to all molecules, either natural or synthetic, either small molecule or peptide, which exert morphine-like pharmacological actions. The predominant pharmacologic property of therapeutic interest is analgesia, the selective relief of pain. Other pharmacologic actions include sedation, respiratory depression, decreased gastrointestinal motility, nausea, and vomiting. Opioids exert their actions through a family of seven-transmembrane, G_i/G_o-coupled receptors, traditionally identified as μ, κ, and δ (a.k.a. MOR, KOR, and DOR). Morphine is the prototypical μ opioid agonist.

Cross-References

▶ Analgesia During Labor and Delivery
▶ Cancer Pain Management, Cancer-Related Breakthrough Pain, Therapy
▶ Cancer Pain Management, Principles of Opioid Therapy, Dosing Guidelines
▶ Cancer Pain Management, Principles of Opioid Therapy, Drug Selection
▶ Cytokine Modulation of Opioid Action
▶ Deep Brain Stimulation
▶ Forebrain Modulation of the Periaqueductal Gray and Its Role in Pain
▶ Hot Plate Test (Assay)
▶ Opioid Peptide Co-Localization and Release
▶ Opioid Receptor Localization
▶ Opioids, Clinical Opioid Tolerance
▶ Pain Treatment, Implantable Pumps for Drug Delivery
▶ Postoperative Pain, Acute Pain Management Principles
▶ Postoperative Pain, Opioids
▶ Postoperative Pain, Postamputation Pain, Treatment, and Prevention

Opioid Addiction

▶ Psychiatric Aspects of the Epidemiology of Pain

Opioid Adverse Effects

Definition

Unwanted and unpleasant effects of opioid drugs when used in pain management.

Cross-References

▶ Cancer Pain Management, Principles of Opioid Therapy, Drug Selection

Opioid Analgesia, Genetic Factors Contributing to

▶ Opioid Analgesia, Strain Differences

Opioid Analgesia, Genotypic Influences on

▶ Opioid Analgesia, Strain Differences

Opioid Analgesia, Strain Differences

Jeffrey S. Mogil
Department of Psychology, McGill University, Montreal, QC, Canada

Synonyms

Opioid analgesia, genetic factors contributing to; Opioid analgesia, genotypic influences on

Definition

The efficacy of the pain-inhibitory neural circuitry activated by opioid compounds differs among individuals. The contribution of inherited genetic factors to such interindividual variability in responses to drugs (i.e., ▶ Pharmacogenetics) is most easily studied by comparing subpopulations of a species, especially ▶ inbred strains. Robust strain differences in the magnitude and neurochemical mediation of opioid analgesia have been observed in both mice and rats. The genes underlying these strain differences are starting to be identified and appear to play similar roles in analgesic variability in humans.

Characteristics

Morphine at standard doses exhibits a wide range of clinical efficacies against postoperative and chronic pain. A human twin study revealed that alfentanil inhibition of experimental pain is ▶ heritable (Angst et al. 2012), and successful selective breeding has demonstrated the heritability of opioid analgesia in mice (see Mogil et al. 1996b). The identification of genes (and their common DNA sequence variants, or ▶ Alleles) underlying the variable pharmacodynamic and pharmacokinetic properties of opioid analgesics would allow individualized treatment of pain, to maximize efficacy and minimize side effects in each patient.

Differential sensitivity of laboratory mouse and rat populations to opioid drugs has long been noted (see Mogil 1999). The genetic contribution to such variability can be assessed using inbred strains, in which repeated brother x sister matings have eliminated genetic ▶ heterozygosity, rendering each individual virtually isogenic to (i.e., a clone of) all others of that strain. By comparing multiple strains tested simultaneously, genetic components (differences between strain means) and environmental components (within-strain error) of analgesic variability can be partitioned. Large "strain surveys" of opioid analgesia have been performed in the mouse (Wilson et al. 2003) and rat (Morgan et al. 1999). Robust differences in half-maximal analgesic doses (AD50s) were noted, in one study of morphine ranging almost fourfold among 12 inbred strains (see Fig. 1). Strain differences can be observed regardless of the route of opioid administration. The magnitude of strain differences has been shown to depend on opioid efficacy and stimulus intensity (Morgan et al. 1999). Genotype dependence of analgesia is not limited to exogenous opioids, with strain differences also being documented for stress-induced analgesia and non-opioid drug analgesia (see Mogil 1999). In the few studies that have ever looked, pharmacokinetic considerations do not appear to explain these strain differences (e.g., Mas et al. 2000).

In addition to analgesic magnitude, the physiological processing of analgesia also appears to be strain dependent. For example, forced swims in 15 °C water produce opioid-mediated (naloxone-sensitive) analgesia in male DBA/2J mice but non-opioid-mediated analgesia in male C57BL/6J mice (Mogil and Belknap 1997). Similarly, analgesia from the κ-opioid receptor agonist, U50,488, is reversed by the N-methyl-D-aspartate receptor antagonist, MK-801, in male mice of a number of strains but not in C3HeB/FeJ mice (unpublished data). Elegant pharmacological studies have shown that heroin analgesia is predominantly mediated by mu-opioid receptors in three inbred strains, by κ-opioid receptors in two others, and by δ-opioid receptors in yet another (Rady et al. 1999). Such qualitative differences among strains suggest differential neural circuit organization, as has been demonstrated for coeruleospinal projections (Clark and Proudfit 1992).

The strongest predictor of strain-dependent morphine AD50 is the baseline sensitivity to the noxious stimulus being inhibited by morphine. Generally speaking, strains sensitive to nociception are resistant to analgesia, and strains resistant to nociception are sensitive to analgesia (Mogil et al. 1996a). This might imply that the underlying genetic contribution is ▶ opioid "tone" and/or related to fractional receptor occupancy of opioid receptors (Elmer et al. 1998). The phenomenon is not limited to morphine analgesia, however, having been demonstrated also for clonidine, epibatidine, U50,488, and the cannabinoid receptor agonist, WIN55,212-2 (Wilson et al. 2003). In fact, these five analgesics show a high degree of genetic correlation with each

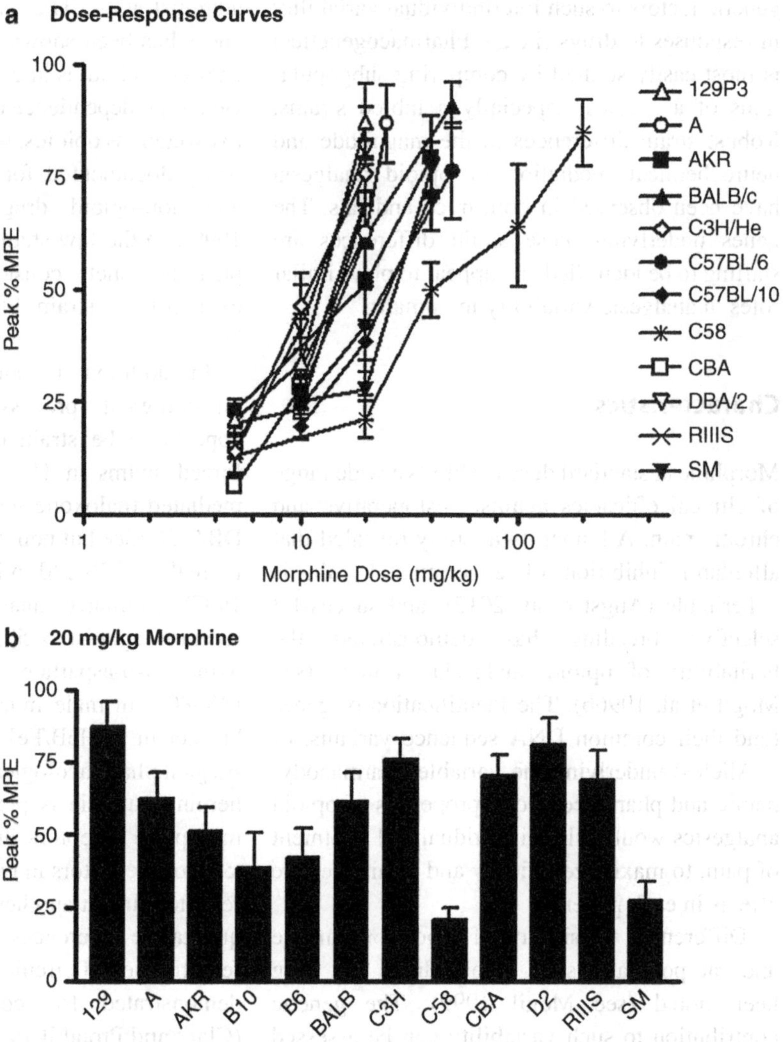

Opioid Analgesia, Strain Differences,

Fig. 1 Genotype-dependent sensitivity to systemic morphine analgesia in 12 inbred mouse strains (all "J" substrains; from Wilson et al. 2003). All mice were tested for baseline sensitivity on the 49 °C hot-water tail-withdrawal assay, administered 5–200 mg/kg (i.p.) morphine sulfate, and retested 15, 30, and 60 min postinjection. Symbols in **a** represent mean ± SEM peak (over the 60 min time course) percentage of the maximum possible effect (%MPE) at each dose. Associated AD50s range from 10.5 mg/kg (129P3) to 38.5 mg/kg (C58). The 12 strains appear to group into three clusters: a sensitive group (*unfilled symbols*), a resistant group (*filled symbols* and *cross*), and the very resistant C58 strain (*asterisk*). The peak %MPE at one dose common to all strains, 20 mg/kg, is shown in **b**

other, suggestive of the existence of "master" genes responsible for variability in both nociception and analgesia across a wide range of compounds (Wilson et al. 2003). One such master gene has been found: the *Kcnj9* gene coding for the Kir 3.3 inwardly rectifying potassium channel subunit (GIRK3) (Smith et al. 2008).

When male and female subjects have been tested simultaneously for opioid analgesia, a strong interaction between genotype and sex has been noted (Mogil 2003). Sex differences can be demonstrated in some genotypes but not others, and even the direction of the sex difference may depend on the subject's genotype. The converse is, of course, also true: some

strain differences are much larger in one sex than the other.

Different physiological effects of a drug may have independent genetic bases. For example, in a study comparing the sensitivity of three inbred strains on six behavioral effects of morphine (analgesia, hyperlocomotion, hypothermia, muscular rigidity, antidiuresis, and constipation), almost every possible rank ordering of strain was observed (Belknap et al. 1989). With respect to analgesia specifically, strains sensitive to morphine inhibition of thermal nociception on the tail-withdrawal or hot-plate tests are *not* necessarily the same strains sensitive to morphine inhibition of chemical nociception on the

Opioid Analgesia, Strain Differences, Table 1 QTLs for morphine inhibition of hot-plate nociception in DBA/2 and C57BL/6 mice

Chrom.	Loc.[a]	LOD[b]	Candidate gene/protein[c]	Reference
1	10 cM	4.7[d]	*Oprk1*, κ-opioid receptor (6 cM)	–
9[e]	20 cM	5.2[d]	–	–
9[e]	42 cM	4.5	*Htr1b*, serotonin-1B receptor (46 cM)	Hain et al. (1999) Journal of Pharmacological Experimental Therapy 291, 444–449
10[e]	9 cM	7.5	*Oprm*, μ-opioid receptor (8 cM)	Belknap et al. (1995) Life Science 57, PL117–PL124

[a]Location in centiMorgans (cM; approximately one million base pairs) from the proximal end of the chromosome of the peak statistical evidence for genetic linkage. The 95 % confidence intervals in this type of study are generally very large, however

[b]Logarithm of the odds (LOD) score for linkage of morphine analgesia to the defined genomic region. The conservative threshold for "significant" linkage is LOD = 4.3

[c]Gene and the protein it codes that may underlie the QTL, supported by evidence presented in the cited reference. The gene's chromosomal location in cM is given in parentheses

[d]Significant linkage in female mice only; no evidence for linkage whatsoever was obtained in male mice

[e]The existence of these QTLs was independently replicated in a different laboratory using different progenitor strains (A/J and SM/J) (*Mizuo et al., International Narcotics Research Conference, Seattle, 2000*)

formalin or abdominal constriction tests (Mogil et al. 1996a; Elmer et al. 1998). The dependence of analgesia pharmacogenetics on the type of pain being inhibited is a generalizable phenomenon, also applicable to clonidine (Wilson et al. 2003) and pregabalin (Chesler et al. 2003).

Of opioid analgesics, only morphine and U50,488 have been subjected to ▶ quantitative trait locus (QTL) mapping, a technique allowing the broad localization of trait variability genes to chromosomal regions as the first step towards their identification. Once the genes underlying the QTLs are identified, one can search for the DNA sequence variants, (e.g., ▶ Single-Nucleotide Polymorphisms) responsible for the strain difference. Although it is not necessarily true that murine polymorphisms will be preserved in humans, there is some reason to expect that similar genes will be trait relevant and genetically variable in both rodents and humans. For morphine inhibition of thermal (hot plate) nociception, using the sensitive DBA/2 and resistant C57BL/6 strains (see Fig. 1) as parental genotypes, four significant QTLs have been uncovered that together account for two-thirds of the additive genetic variance and one-quarter of the overall trait variance (Bergeson et al. 2001). These QTLs, and the genes likely

underlying each QTL, are shown in Table 1. As can be seen, in two cases the QTLs are sex specific. A female-specific QTL was also uncovered for U50,488 inhibition of hot-water tail-withdrawal nociception (Mogil et al. 2003). This QTL, on distal mouse chromosome 8, was determined by convergent mouse mutant and pharmacological data to be *Mc1r*, encoding the melanocortin-1 receptor (Mogil et al. 2003).

Success in identifying genes responsible for mouse or rat strain differences will be of purely academic interest unless those same genes are shown to be relevant to human individual differences. Thus, more attention is being paid to "translational" genetic strategies. If a particular gene is postulated to be important for a trait and that gene has known sequence variants in humans, an ▶ association or allele dosage study can be performed. A number of studies have observed associations with the *OPRM* (μ-opioid receptor) and *OPRD* (δ-opioid receptor) genes, although these associations remain controversial (Mogil 2012). Identification of robust genetic associations will allow more effective usage of existing analgesics, identify new molecular targets for drug development, and facilitate the regulatory approval of new drugs based on the ability to identify likely "responders."

References

Angst, M. S., Phillips, N. G., Drover, D. R., et al. (2012). Pain sensitivity and opioid analgesia: A pharmacogenomic twin study. *Pain, 153*, 1397–1409.

Belknap, J. K., Mogil, J. S., Helms, M. L., Richards, S. P., O'Toole, L. A., Bergeson, S. E., & Buck, K. J. (1995). Localization to chromosome 10 of a locus influencing morphine analgesia in crosses derived from C57BL/6 and DBA/2 mouse strains. *Life Science, 57*, PL117–124.

Belknap, J. K., Noordewier, B., & Lamé, M. (1989). Genetic dissociation of multiple morphine effects among C57BL/6J, DBA/2J and C3H/HeJ inbred mouse strains. *Physiology and Behavior, 46*, 69–74.

Bergeson, S. E., Helms, M. L., O'Toole, L. A., Jarvis, M. W., Hain, H. S., Mogil, J. S., et al. (2001). Quantitative trait loci influencing morphine antinociception in four mapping populations. *Mammalian Genome, 12*, 546–553.

Chesler, E. J., Ritchie, J., Kokayeff, A., Lariviere, W. R., Wilson, S. G., & Mogil, J. S. (2003). Genotype-dependence of gabapentin and pregabalin sensitivity: The pharmacogenetic mediation of analgesia is specific to the type of pain being inhibited. *Pain, 106*, 325–335.

Clark, F. M., & Proudfit, H. K. (1992). Anatomical evidence for genetic differences in the innervation of the rat spinal cord by noradrenergic locus coeruleus neurons. *Brain Research, 591*, 44–53.

Elmer, G. I., Pieper, J. O., Negus, S. S., & Woods, J. H. (1998). Genetic variance in nociception and its relationship to the potency of morphine-induced analgesia in thermal and chemical tests. *Pain, 75*, 129–140.

Hain, H. S., Belknap, J. K., & Mogil, J. S. (1999). Pharmacogenetic evidence for the involvement of 5-hydroxytryptamine (serotonin)-1B receptors in the mediation of morphine antinociceptive sensitivity. *Journal of Pharmacological Experimental Therapy, 291*, 444–449.

Mas, M., Sabater, E., Olaso, M. J., Horga, J. F., & Faura, C. C. (2000). Genetic variability in morphine sensitivity and tolerance between different strains of rats. *Brain Research, 866*, 109–115.

Mogil, J. S. (1999). The genetic mediation of individual differences in sensitivity to pain and its inhibition. *Proceedings of the National Academy of Sciences of the United States of America, 96*, 7744–7751.

Mogil, J. S. (2003). Interaction between sex and genotype in the mediation of pain and pain inhibition. *Seminars in Pain and Medicine, 1*, 197–205.

Mogil, J. S. (2012). Pain genetics: Past, present and future. *Trends in Genetics, 28*, 258–266.

Mogil, J. S., & Belknap, J. K. (1997). Sex and genotype determine the selective activation of neurochemically-distinct mechanisms of swim stress-induced analgesia. *Pharmacology, Biochemistry, and Behavior, 56*, 61–66.

Mogil, J. S., Kest, B., Sadowski, B., & Belknap, J. K. (1996a). Differential genetic mediation of sensitivity to morphine in genetic models of opiate antinociception: Influence of nociceptive assay. *The Journal of Pharmacology and Experimental Therapeutics, 276*, 532–544.

Mogil, J. S., Sternberg, W. F., Marek, P., Sadowski, B., Belknap, J. K., & Liebeskind, J. C. (1996b). The genetics of pain and pain inhibition. *Proceedings of the National Academy of Sciences of the United States of America, 93*, 3048–3055.

Mogil, J. S., Wilson, S. G., Chesler, E. J., et al. (2003). The melanocortin-1 receptor gene mediates female-specific mechanisms of analgesia in mice and humans. *Proceedings of the National Academy of Sciences of the United States of America, 100*, 4867–4872.

Morgan, D., Cook, C. D., & Picker, M. J. (1999). Sensitivity to the discriminative stimulus and antinociceptive effects of m opioids: Role of strain of rat, stimulus intensity, and intrinsic efficacy at the m opioid receptor. *The Journal of Pharmacology and Experimental Therapeutics, 289*, 965–975.

Rady, J. J., Elmer, G. I., & Fujimoto, J. M. (1999). Opioid receptor selectivity of heroin given intracerebroventricularly differs in six strains of inbred mice. *The Journal of Pharmacology and Experimental Therapeutics, 288*, 438–445.

Smith, S. B., Marker, C. L., Perry, C., et al. (2008). Quantitative trait locus and computational mapping identifies Kcnj9 (GIRK3) as a candidate gene affecting analgesia from multiple drug classes. *Pharmacogenetics and Genomics, 18*, 231–241.

Wilson, S. G., Smith, S. B., Chesler, E. J., et al. (2003). The heritability of antinociception: Common pharmacogenetic mediation of five neurochemically distinct analgesics. *The Journal of Pharmacology and Experimental Therapeutics, 304*, 547–559.

Opioid Analgesic

Definition

Opioid analgesics are pharmacologic agents whose mechanism of action involves interaction with receptors for endogenous opioids. The analgesic actions of these agonists are antagonized by naloxone, an opioid receptor antagonist. Opioid receptors located in the central nervous system, with their endogenous ligands, including the enkephalins, β-endorphin, and dynorphin, form part of an endogenous pain modulating system.

Opioid Analgesic Actions Outside the Central Nervous System

▶ Opioids in the Periphery and Analgesia

Opioid Antagonists

Definition

Naloxone is the classical opioid antagonist, a drug that possesses affinity but lacks efficacy for opioid receptors. At sufficient doses, naloxone will block mu, delta, and kappa opioid receptors; low doses are more selective for mu opioid receptors. Naltrexone is another opioid receptor antagonist drug of clinical importance. Experimental molecules with different degrees of selectivity are available for opioid receptor subtypes.

Cross-References

▶ Nitrous Oxide Antinociception and Opioid Receptors

Opioid Dependency

▶ Psychiatric Aspects of the Epidemiology of Pain

Opioid Dose Titration

Definition

The clinical practice of increasing opioid dosage to achieve analgesia while diminishing unwanted adverse effects.

Cross-References

▶ Cancer Pain Management, Principles of Opioid Therapy, Drug Selection

Opioid Electrophysiology in PAG

MacDonald J. Christie[1] and
Peregrine B. Osborne[2]
[1]Discipline of Pharmacology, University of Sydney, Sydney, Australia
[2]Dept Anatomy & Neuroscience, University of Melbourne, Melbourne, VIC, Australia

Synonyms

Central gray; Central gray/central grey; Opiate; Periaqueductal gray; Periaqueductal grey

Definition

The midbrain periaqueductal gray (PAG) is one of the major brain targets for the analgesic actions of opioid drugs and endogenously released opioids. The PAG contributes to a descending inhibitory neural network. When PAG output neurons are activated, nociceptive neurotransmission at the level of the dorsal horn of the spinal cord is inhibited. Opioids are thought to produce analgesia in the PAG by a disinhibitory mechanism via direct inhibition of GABAergic neurotransmission impinging on descending output neurons.

Characteristics

The PAG, as the name suggests, is a cell-dense region surrounding the cerebral aqueduct extending from the third ventricle to the pontine division of the fourth ventricle. Anatomical, physiological, and behavioral studies all indicate that the PAG is organized into distinct functional columns that extend along the rostrocaudal axis. The different functional columns are considered to be involved in the expression of distinct patterns of autonomic and behavioral responses (Bandler and Shipley 1994).

All of the major types of opioid receptor, μ- (▶ MOP), δ- (▶ DOP), and κ- (KOP) and N/OFQ (▶ NOP) receptors, are expressed abundantly in PAG (Mansour et al. 1995). The first three of these are considered to mediate antinociceptive actions in PAG, with the μ-receptor mediating the analgesic actions of most clinically useful opioids. The N/OFQ receptor is closely related to μ-, δ-, and κ-receptors, but is not activated by classical opioid agonists. Activation of the N/OFQ receptor in PAG functionally antagonizes the effects of morphine. Endogenous opioids derived from each of the major prohormone precursors are also richly expressed within the PAG (Fallon and Leslie 1986). These include pro-opiomelanocortin (POMC), which produces ▶ β-endorphin from nerve terminals of hypothalamic neurons, proenkephalin from intrinsic neurons and projections from several brain regions, and pro-dynorphin also from intrinsic neurons and projections from several brain regions.

Direct microinjection of opioid receptor agonists into the PAG produces analgesia in experimental animals. Agonists for each receptor type have antinociceptive actions when microinjected into PAG (Ossipov et al. 1995). However, μ-receptor agonists are more efficacious than δ- or κ-agonists. Microinjected opioids produce antinociception more effectively in the ventrolateral PAG, particularly in the most caudal area, than in lateral or dorsal columns (Jensen and Yaksh 1986). Electrical stimulation of the PAG in experimental animals and humans also produces analgesia that can be reversed by opioid receptor antagonists such as naloxone but only when stimulation is located in the ventrolateral area. This implies that electrical stimulation of ventrolateral PAG releases endogenous opioids. Stimulation of other areas of the PAG can also produce antinociception, but the effects are generally not reversed by opioid antagonists (see Bandler and Shipley 1994).

The mechanisms by which opioids modulate electrical excitability in PAG are summarized in Fig. 1. These mechanisms provide a general explanation for the analgesic actions of μ- and κ-opioid agonists in PAG, as well as the anti-opioid actions of N/OFQ.

Postsynaptic Actions of Opioids in PAG

Opioid receptors expressed on the somatic membrane inhibit electrical excitability of PAG neurons (see Williams et al. 2001, for review). Inhibition occurs via actions that are typical of the subfamily of inhibitory G-protein-coupled receptors, to which all of the opioid receptors belong. Opioid agonists ctivate inhibitory G proteins (predominantly Gi) to simultaneously reduce currents through voltage-gated calcium channels and enhance currents through inwardly rectifying potassium channels in PAG neurons. Both actions inhibit neuronal excitability. The principal calcium channels inhibited by opioids are the N and ▶ P/Q channels, which play a major role in regulation of membrane excitability as well as neurotransmitter release. The potassium channel is probably a ▶ GIRK subtype that hyperpolarizes the membrane to reduce excitability. Opioid agonists also inhibit adenylyl cyclase in PAG. While this has little effect on excitability under basal conditions, it reduces currents through a protein kinase A-dependent, nonselective cation channel that is activated when adenylyl cyclase, and hence cAMP levels are elevated in the neuron. The latter mechanism of opioid action is also inhibitory.

Opioids display the actions described above only in a subpopulation of neurons in PAG. Actions mediated by μ-receptors have been observed in brain slices and isolated neurons from rat PAG. Although μ-agonists act on approximately 30 % of all lateral and ventrolateral PAG neurons, only 14 % of those in the ventrolateral PAG that project to the ▶ rostral ventromedial medulla (RVM) are sensitive to μ-agonists (Osborne et al. 1996). At least one subpopulation of GABAergic neurons (GIN neurons) in PAG is directly inhibited by opioids (Park et al. 2010), but whether this extends to all GABAergic neurons is not known. This

Opioid Electrophysiology in PAG, Fig. 1 Cellular actions of opioids in PAG

observation supports the proposal (see Williams et al. 2001) that μ-opioids act predominantly on GABAergic interneurons in PAG and not on neurons that form part of the descending projection to the rostral ventromedial medulla. However, it should be noted that in the same study, 56 % of descending projection neurons in the lateral PAG were sensitive to μ-agonists, so the interpretation may only be true for the ventrolateral PAG which mediates opioid antinociception.

Almost all PAG neurons of the rat are inhibited via the same mechanisms described above by the ORL1 agonist, N/OFQ. By contrast, no electrophysiological effects of κ- or δ-receptor agonists were observed in rat PAG, although both receptors are expressed in the region. Failure of δ-receptors to affect membrane currents is probably due to their largely intracellular localization (Commons 2003). Density of κ-receptors in rat PAG might not be sufficiently

high to produce robust cellular physiological effects in rats. In mouse PAG, opioids act on μ-, δ-, and κ-receptors in about 80 %, 20 %, and 20 % of PAG neurons, respectively (Vaughan et al. 2003). It should be noted that κ- and δ-receptor agonists produce physiological effects including antinociception when microinjected into rat PAG, although these are generally weaker than μ-receptor actions. Failure to detect actions on membrane currents or synaptic transmission (see below) may, therefore, indicate that these cellular measures are not sufficiently sensitive to detect small effects on large populations of cells or during specific physiological conditions such as stress, which induces translocation of δ-receptors to the surface membrane in PAG (Commons 2003).

Presynaptic Actions of Opioids in PAG

Opioid receptors are also abundantly expressed on nerve terminals in PAG (Commons 2003).

Opioids acting on μ-receptors presynaptically inhibit both GABAergic and glutamatergic synapses impinging on all PAG neurons, and the maximum inhibition is similar for both types of synapse (Vaughan et al. 1997b). Inhibition is due to a reduction in probability of transmitter release that reduces both electrically evoked and spontaneous synaptic transmission in both terminal types. However, the intracellular mechanisms of opioid action differ between GABAergic and glutamatergic synapses. In glutamatergic synapses, the mechanisms coupling opioid receptors to reduced probability of quantal release of glutamate have not been resolved, with the exception of some dependence on inhibition of voltage-gated calcium channels under conditions of exaggerated calcium entry into the nerve terminal. In GABAergic nerve terminals, presynaptic inhibition is mediated by activation of a voltage-gated potassium channel (Kv) and that inhibition can be prevented by the Kv blockers, 4-aminopyridine and α-dendrotoxin (Vaughan et al. 1997b). The mechanism that couples opioid receptors to the Kv channel involves activation of phospholipase A_2 that mobilizes arachidonic acid, followed by formation of its metabolites via 12-lipoxygenase. Inhibitors of 5-lipoxygenase and cyclooxygenase potentiate presynaptic inhibition by opioids in PAG, presumably by shunting metabolism of arachidonic acid through the 5-lipoxygenase pathway. This process could provide one of the mechanisms of nonsteroidal analgesia in the central nervous system. In addition to this mechanism, μ-opioids inhibit GABAergic synapses in PAG via inhibition of voltage-gated calcium channels, under conditions of exaggerated calcium entry into the nerve terminal (CW Vaughan, unpublished observations).

Opioids acting on κ- and δ-receptors do not produce presynaptic inhibition in rat PAG, and ORL1 agonists inhibit synaptic transmission in about 50 % of recordings (Vaughan et al. 1997a). In mouse PAG, μ-opioids inhibit GABAergic synapses in all neurons, while κ-receptor agonists do so in about 50 % of neurons, and δ-agonists are without effect. Local circuit neurons may not be the only source of inputs forming the GABAergic

synapses targeted by opioids in PAG. As is common in brainstem nuclei, projection neurons can also be GABAergic and could give rise to local recurrent axon collaterals. GABAergic neurons also provide most of the projections from the central nucleus of the amygdala (Day et al. 1999; Oka et al. 2008), which function to modulate descending analgesia in the context of fear, anxiety, stress, and placebo analgesia. PAG also receives GABAergic projections from the ventral tegmental area (Kirouac et al. 2004).

Functional Consequences of Cellular Actions of Opioids in PAG

Although opioids inhibit both inhibitory (GABAergic) and excitatory (glutamatergic) synapses, the dominant action is probably on GABAergic synapses, because there is a large GABAergic tone in PAG (Chieng and Christie 1994). This would explain observations (e.g., Behbehani et al. 1990) that μ-opioids produce excitatory actions on PAG neurons under recording conditions that enhance GABAergic tone. Opioids also inhibit the soma of very few neurons in ventrolateral PAG that descend to the rostral ventromedial medulla, suggesting that the major effect of opioids is to disinhibit these output neurons. Disinhibition of descending output from the PAG in vivo is known to produce antinociception, so this is presumably the mechanism of action of μ-opioids to produce analgesia in PAG. However, the actions of μ-opioids on glutamatergic synapses, descending projection neurons in the lateral PAG, and observations that at least some projections from PAG to both the ▶ rostral ventromedial medulla (Morgan et al. 2008) and midbrain ▶ ventral tegmental area (VTA: Omelchenko and Sesack 2010) are GABAergic still require explanation.

Stimulation of ORL1 receptors inhibits GABAergic and glutamatergic synapses in some PAG neurons and might be expected to produce antinociception (Vaughan et al. 1997a). However, these receptors are also located on the cell bodies of all PAG neurons, including output neurons. Thus, any disinhibition produced presynaptically on output neurons would be expected to be blocked by the direct somatic inhibitory actions of N/OFQ on output neurons. This organization of receptors would predict that ORL receptor stimulation in PAG would be expected to block the antinociceptive actions of opioids, which has been observed in microinjection studies (Morgan et al. 1997).

In summary, all of the major opioid receptor types are expressed in PAG neurons. The membrane actions are summarized above. Opioids couple to inhibition of voltage-gated calcium channels and activation of potassium channels via Gi-protein βγ-subunits in the membrane of PAG neurons and nerve terminals, although different potassium channels are involved in the soma (GIRK) and nerve terminal (Kv). Opioids inhibit both soma and nerve terminal excitability of intrinsic (presumably GABAergic) neurons, but not neurons that project the rostral ventromedial medulla. This disinhibits these output neurons to produce antinociception. The exception is the ORL1 receptor, which directly inhibits the output neurons to block opioid antinociception.

References

Bandler, R., & Shipley, M. T. (1994). Columnar organization in the midbrain periaqueductal gray: Modules for emotional expression. *Trends in Neurosciences, 17,* 379–389.

Behbehani, M. M., Jiang, M., & Chandler, S. D. (1990). The effect of [Met]enkephalin on the periaqueductal gray neurons of the rat: An in vitro study. *Neuroscience, 38,* 373–380.

Chieng, B., & Christie, M. J. (1994). Inhibition by opioids acting on mu-receptors of GABAergic and glutamatergic postsynaptic potentials in single rat periaqueductal gray neurones in vitro. *British Journal of Pharmacology, 113,* 303–309.

Commons, K. G. (2003). Translocation of presynaptic delta opioid receptors in the ventrolateral periaqueductal gray after swim stress. *The Journal of Comparative Neurology, 464,* 197–203.

Day, H. E., Curran, E. J., & Watson, S. J. (1999). Distinct neurochemical populations in the rat central nucleus of the amygdala and bed nucleus of the stria terminalis: Evidence for their selective activation by interleukin-1β. *The Journal of Comparative Neurology, 413,* 113–128.

Fallon, J. H., & Leslie, F. M. (1986). Distribution of dynorphin and enkephalin peptides in the rat brain. *The Journal of Comparative Neurology, 249,* 293–336.

Jensen, T. S., & Yaksh, T. L. (1986). Comparison of antinociceptive action of morphine in the periaqueductal gray, medial and paramedial medulla in rat. *Brain Research, 36,* 99–113.

Kirouac, G. J., Li, S., & Mabrouk, G. (2004). GABAergic projection from the ventral tegmental area and substantia nigra to the periaqueductal gray region and the dorsal raphe nucleus. *The Journal of Comparative Neurology, 469,* 170–184.

Mansour, A., Fox, C. A., Akil, H., & Watson, S. J. (1995). Opioid-receptor mRNA expression in the rat CNS: Anatomical and functional implications. *Trends in Neurosciences, 18,* 22–29.

Morgan, M. M., Grisel, J. E., Robbins, C. S., & Grandy, D. K. (1997). Antinociception mediated by the periaqueductal gray is attenuated by orphanin FQ. *Neuroreport, 8,* 3431–3434.

Morgan, M. M., Whittier, K. L., Hegarty, D. M., & Aicher, S. A. (2008). Periaqueductal gray neurons project to spinally projecting GABAergic neurons in the rostral ventromedial medulla. *Pain, 140,* 376–386.

Oka, T., Tsumori, T., Yokota, S., & Yasui, Y. (2008). Neuroanatomical and neurochemical organization of projections from the central amygdaloid nucleus to the nucleus retroambiguus via the periaqueductal gray in the rat. *Neuroscience Research, 62,* 286–298.

Omelchenko, N., & Sesack, S. R. (2010). Periaqueductal gray afferents synapse onto dopamine and GABA neurons in the rat ventral tegmental area. *Journal of Neuroscience Research, 88,* 981–991.

Osborne, P. B., Vaughan, C. W., Wilson, H. I., & Christie, M. J. (1996). Opioid inhibition of rat periaqueductal gray neurones with identified projections to rostral ventromedial medulla in vitro. *Journal of Physiology (London), 490,* 383–389.

Ossipov, M. H., Kovelowski, C. J., Nichols, M. L., Hruby, V. J., & Porreca, F. (1995). Characterization of supraspinal antinociceptive actions of opioid delta agonists in the rat. *Pain, 62,* 287–293.

Park, C., Kim, J. H., Yoon, B. E., Choi, E. J., Lee, C. J., & Shin, H. S. (2010). T-type channels control the opioidergic descending analgesia at the low threshold-spiking GABAergic neurons in the periaqueductal gray. *Proceedings of the National Academy of Sciences of the United States of America, 107,* 14857–14862.

Vaughan, C. W., Ingram, S. L., & Christie, M. J. (1997a). Actions of the ORL1 receptor ligand nociception on membrane properties of rat periaqueductal gray neurons in vitro. *The Journal of Neuroscience, 17,* 996–1003.

Vaughan, C. W., Ingram, S. L., Connor, M. A., & Christie, M. J. (1997b). How opioids inhibit GABA-mediated neurotransmission. *Nature, 390,* 611–614.

Vaughan, C. W., Bagley, E. E., Drew, G. M., Schuller, A., Pintar, J. E., Hack, S. P., & Christie, M. J. (2003). Cellular actions of opioids on periaqueductal gray neurons from C57B16/J mice and mutant mice lacking MOR-1. *British Journal of Pharmacology, 139,* 362–367.

Williams, J. T., Christie, M. J., & Manzoni, O. (2001). Cellular and synaptic adaptations mediating opioid dependence. *Physiological Reviews, 81,* 299–343.

Opioid Hyperexcitability

Definition

Opioid hyperexcitability is a syndrome that includes whole-body hyperalgesia and allodynia, abdominal muscle spasms, and symmetrical jerking of legs.

Cross-References

▶ Opioid

Opioid Hypoalgesia

▶ Opioids, Effects of Systemic Morphine on Evoked Pain

Opioid Modulation by Cytokines

▶ Cytokine Modulation of Opioid Action

Opioid Modulation of Nociceptive Afferents In Vivo

David C. Yeomans and Alexander Tzabazis
Department of Anesthesia, Stanford University, Stanford, CA, USA

Synonyms

First-order sensory neurons; Opioid therapy for primary afferents

Definition

Modulation of primary afferents by endogenous or exogenous opioids. Primary afferents are sensory neurons with one peripheral process innervating a peripheral tissue area (▶ receptive field (RF)) and one central process that enters the dorsal horn of the spinal cord via a dorsal root. The cell bodies of primary afferents are located in the ▶ dorsal root ganglia (DRG) or trigeminal ganglia (TG). Opioid targets are either receptors on pre- or postsynaptic membranes of central terminals in the dorsal horn or peripheral opioid receptors on nociceptive endings in the innervated tissue. Modulation can be achieved by systemic or local application of drugs (agonists or antagonists) or gene therapy (de novo or altered expression of an "antinociceptive" gene or knockdown of a "pronociceptive" gene).

Characteristics

Physiology of the Spinal and Peripheral Opioid System

Yaksh and Rudy discovered in 1976 that spinally administered opioids can produce analgesia. Leading to reduced side effects, this new administration route of opioids became the gold standard for severe pain relief. Antinociceptive effects of spinally applied opioids are mediated predominantly by pre- and/or postsynaptic μ-, δ-, and κ-opioid receptors in laminae 1 and 2 (Rexed) of the dorsal horn of the spinal cord. Methionine (met)- and leucine (leu)-enkephalin, β-endorphin, dynorphin, α-neoendorphin, and the endomorphins I and II are all endogenous opioids with different affinities to the opioid receptors. The preferential endogenous ligand for μ-opioid receptors are β-endorphin and endomorphins I and II. Enkephalins favor δ-opioid receptors, and dynorphins have the highest affinity for κ-opioid receptors. However, one should keep in mind that receptor affinity depends on the type of fragment that is released by the enzymatic cleavage from the precursor, which again depends on the tissue where this process takes place (Zaki et al. 1996; Wall and Melzack 1999).

Opioid receptors can also be found on peripheral nociceptive endings (Fields et al. 1980). These receptors are upregulated by peripheral inflammation which enhances the potency of endogenous and exogenous opioids. The appearance of opioid receptors within minutes to hours after onset of inflammation suggests their preexistence although they cannot be readily detected in uninjured tissue. In cultured sensory neurons, bradykinin – a mediator of inflammation – enhanced the responsiveness of μ-opioid receptors to DAMGO, an opioid with high μ-opioid receptor specificity (Berg et al. 2007). A possible explanation for this sudden availability of opioid receptors on nociceptive terminals might be an unblocking of the perineurium at a very early stage of inflammation (within 12 h (Antonijevic et al. 1995)) and/or at a later time point by sprouting of the terminals. This theory is supported by a recently published study showing that opening tight junctions in the perineurial barrier using hypertonic saline leads to enhanced effects of peripherally administered opioid receptor agonists on primary afferents (Rittner et al. 2012). After nerve injury, expression of opioid receptors on primary afferents undergoes dynamic changes as well (Lee et al. 2011), and peripherally administered opioid receptor agonist have been shown to attenuate neuropathic pain in rats (Guan et al. 2008).

Studies in knockout mice reveal different contributions of opioid receptors and endogenous opioids in mediating nociceptive behavior (Table 1) (Kieffer and Gaveriaux-Ruff 2002). Activation of the μ-opioid receptor induces both desirable analgesic effects and undesirable effects of exogenous opioids like respiratory depression and constipation, making this receptor a difficult but widely used therapeutic target for systemic drug treatment.

Opioid receptors are ▶ type II receptors, i.e., their action is carried forward by a second-messenger G protein. Stimulation of presynaptic μ- and δ2-receptors is associated with a reduction in calcium influx (Ca^{2+}) leading to a hyperpolarization of the terminal and a reduced release of neurotransmitters such as ▶ glutamate and ▶ neuropeptides such as

▶ substance P. Activated postsynaptic opioid receptors stabilize membrane potentials by enhancing the outward flow of potassium (K^+), decreasing the amplitude of excitatory postsynaptic potentials.

Intrathecally administered pertussis toxin (PTX) produces thermal hyperalgesia and allodynia in mice (Womer et al. 1997) by inactivation of the inhibitory G_i and G_0 proteins that are involved in the intracellular signalling process. This disinhibition leads to predominance of excitation, ▶ windup, and reduces the analgesic potency of many μ-agonists, like morphine and DAMGO. Antinociception mediated by mixed μ-agonists/μ-antagonists like buprenorphine, however, is less affected. The effectiveness of buprenorphine as opposed to other exogenous μ-agonists even after intrathecal PTX indicates that the mechanism of action of buprenorphine does not entirely depend on PTX-sensitive, intracellular signalling pathways, which might explain its special role in treatment of ▶ neuropathic pain (McCormack et al. 1998).

Modulation of Central Projections

As with endogenous opioids, intrathecally applied exogenous opioids have different affinities for different opioid receptors as well (Table 2). Whereas κ-agonists do not attenuate either Aδ- or C-thermonociceptor-mediated responses, μ-, δ1-, and δ2-opioid agonists are effective on both. DPDPE (δ1-selective agonist) showed parallel dose–response curves for Aδ- and C-mediated antinociception, indicating similar mechanisms of action. On the other hand, μ-agonists (DAMGO and morphine) and a δ2-selective agonist (DSLET) produced nonparallel dose–response curves indicating that the mechanisms by which these drugs act are different for the two types of nociception. Specifically, μ- and δ2-agonists seem to act presynaptically on C-fiber afferent terminals, but they attenuate responses to Aδ-mediated nociception on dorsal horn neurons, i.e., postsynaptically (Lu et al. 1997). It has been shown recently (Jones et al. 2003) that knockdown of μ-opioid receptors in primary afferents by way of a recombinant herpes vector attenuates the

Opioid Modulation of Nociceptive Afferents In Vivo, Table 1 Changes in nociception as observed in different knockout mice (Kieffer and Gaveriaux-Ruff 2002)

Pain/hyperalgesia	MOR−/−	DOR−/−	KOR−/−	βend-	Penk-	Pdyn-
Thermal	↑	↔	↔	↔	↑	↑
Mechanical	↔	↔	↔	n/a	n/a	↔
Chemical	↓	n/a	↑	n/a	↓	↔
Inflammatory	Shorter duration of hyperalgesia	n/a	n/a	n/a	n/a	n/a
Neuropathic	n/a	n/a	↑[a]	n/a	n/a	Shorter duration of hyperalgesia

MOR μ-opioid receptor, *DOR* δ-opioid receptor, *KOR* κ-opioid receptor, *βend* β-endorphin, *Penk* preproenkephalin, *Pdyn* predynorphin

[a]Based on Xu et al. (2004)

Opioid Modulation of Nociceptive Afferents In Vivo, Table 2 Antinociceptive potency of exogenous opioids on Aδ- and C-fiber-mediated behavior in the paw withdrawal test. Parallel dose–response curves indicate similar mechanisms of action, which allows comparison of potencies on Aδ- and C-mediated analgesia

Substance	Aδ	C	Dose–response curves
Morphine (μ-agonist)	++[a]	++[a]	Nonparallel
DAMGO (μ-agonist)	+++[a]	+++[a]	Nonparallel
DPDPE (δ1-agonist)	+[a]	++[a]	Parallel
DSLET (δ2-agonist)	+[a]	+[a]	Nonparallel
U50488 (κ-agonist)	0	0	–

[a]Indicates reversibility by μ-, δ1-, and δ2-antagonists, respectively

potency of μ-receptor-selective agonists for C, but not for Aδ-fibers. These findings support the observations of Lu et al., as herpes viruses usually do not jump synapses, i.e., they act on primary afferents only. Consistent with these results are observations of Taddese et al. that activation of the μ-opioid receptor inhibited presynaptic calcium channels preferentially on small nociceptors but had minimal effects on larger nociceptors, whereas ▶ somatostatin had the opposite effects (Taddese et al. 1995).

Several groups (Wilson et al. 1999; Bras et al. 2001; Wilson and Yeomans 2002; Glorioso et al. 2003) have used herpes virus-based vectors for delivery of transgenes to the nervous system. The natural property of the herpes simplex virus to enter dorsal root ganglia via peripheral infection and to become latent there allows a phenotypic alteration of the first cell in the pathway of nociception. Pain therapy, based on this technology, should minimize adverse side effects of opioids like constipation or respiratory depression. Delivery of a transgene-encoding opioid peptide precursor (human preproenkephalin) leads to synthesis of enkephalin peptides in sensory neurons in mice (Wilson et al. 1999). Baseline withdrawal latencies to noxious radiant heat were similar in animals infected with proenkephalin-encoding and control viruses. However, after sensitization of C fibers with ▶ capsaicin and Aδ-fibers with ▶ dimethyl sulfoxide (DMSO), thermal hyperalgesia was reduced or eliminated in mice infected with proenkephalin-encoding virus for at least 7 weeks (Fig. 1).

Modulation of Nociceptive Terminals

Peripheral localized administered endogenous and exogenous opioids have been shown to mediate antinociception in models of inflammatory and neuropathic pain (Stein et al. 2003). For example, intra-articular application of morphine during knee surgery in a dose that does not lead to systemic effects produced a significant, naloxone-reversible, and long-lasting analgesic effect (Stein 1995).

Similarly, after infection of primary afferents with a viral vector encoding for

Opioid Modulation of Nociceptive Afferents In Vivo, Fig. 1 Human proenkephalin immunofluorescence in primary afferent fibers of mouse dorsal root 6 weeks after application of a human preproenkephalin-encoding herpes vector to the skin

preproenkephalin A (HSVLatEnk1), the majority of met-enkephalin-like material accumulated at the proximal side of the ligatured sciatic nerve, while much less was seen in ligatured L4–L5 roots (Bras et al. 2001). In HSVLatEnk1-infected animals without sciatic nerve ligation, subcutaneous microdialysis showed significantly higher basal levels of met-enkephalin-like material in the interstitial fluid of the hindpaw indicating that overexpressed peptides have reached a releasable compartment. Electrical stimulation of the sciatic nerve revealed an approximately threefold higher overflow of met-enkephalin-like material in HSVLatEnk1-infected rats as compared to control rats. In a rodent model of visceral pain, applying a herpes-based virus encoding for enkephalin on the surface of the pancreas resulted in an increase in exploratory behavior, increased enkephalin immunoreactivity in both the pancreas and spinal cord, as well as reduced c-fos staining in the dorsal horn of the spinal cord (Lu et al. 2007). Taken together, these data suggest that overexpression of antinociceptive genes in primary afferents

leads to a combined analgesic effect, i.e., effects on nociceptive terminals and central projections.

Expression and release of opioid peptides can very likely be induced by hormones (e.g., corticotropin-releasing hormone, CRH) and cytokines (interleukin 1, IL-1) in ▶ immune cells (B and T lymphocytes, macrophages, and monocytes) as well (Stein et al. 2003). The released endogenous opioids from immune cells can interact with peripheral receptors on nociceptive terminals. As many painful conditions are accompanied or sustained by inflammatory processes, inducing overexpression of endogenous opioids in immune cells could provide a promising therapeutic approach to the treatment of pain.

In a phase I clinical study (Fink et al. 2011), the overexpression of human preproenkephalin after intradermal injections of a herpes-based viral vector for the treatment of intractable cancer pain has been shown to be well tolerated, and there seemed to be a dose-dependent analgesic effect that is currently further investigated (ClinicalTrials.gov, NCT01291901).

References

Antonijevic, I., Mousa, S. A., Schäfer, M., & Stein, C. (1995). Perineurial defect and peripheral opioid analgesia in inflammation. *The Journal of Neuroscience, 15,* 165–172.

Berg, K. A., Patwardhan, A. M., Sanchez, T. A., Silva, Y. M., Hargreaves, K. M., & Clarke, W. P. (2007). Rapid modulation of micro-opioid receptor signaling in primary sensory neurons. *The Journal of Pharmacology and Experimental Therapeutics, 321*(3), 839–847.

Bras, J.-M. A., Becker, C., Bourgoin, S., Lombard, M.-C., Cesselin, F., Hamon, N., et al. (2001). Met-enkephalin is preferentially transported into the peripheral processes of primary afferent fibres in both control and HSV1-driven proenkephalin A overexpressing rats. *Neuroscience, 103,* 1073–1083.

Fields, H. L., Emson, P. C., Leigh, B. K., Gilbert, R. F. T., & Iversen, L. L. (1980). Multiple opiate receptor sites on primary afferent fibers. *Nature, 284,* 351–353.

Fink, D. J., Wechuck, J., Mata, M., Glorioso, J. C., Goss, J., Krisky, D., et al. (2011). Gene therapy for pain: Results of a phase I clinical trial. *Annals of Neurology, 70*(2), 207–212.

Glorioso, J. C., Mata, M., & Fink, D. J. (2003). Exploiting the neurotherapeutic potential of peptides: Targeted delivery using HSV vectors. *Expert Opinion on Biological Therapy, 3*, 1233–1239.

Guan, Y., Johanek, L. M., Hartke, T. V., Shim, B., Tao, Y. X., Ringkamp, M., et al. (2008). Peripherally acting mu-opioid receptor agonist attenuates neuropathic pain in rats after L5 spinal nerve injury. *Pain, 138*(2), 318–329.

Jones, T. L., Sweitzer, S. M., Wilson, S. P., & Yeomans, D. C. (2003). Afferent fiber-selective shift in opiate potency following targeted opioid receptor knockdown. *Pain, 106*, 365–371.

Kieffer, B. L., & Gaveriaux-Ruff, C. (2002). Exploring the opioid system by gene knockout. *Progress in Neurobiology, 66*, 285–306.

Lee, C. Y., Perez, F. M., Wang, W., Guan, X., Zhao, X., Fisher, J. L., et al. (2011). Dynamic temporal and spatial regulation of mu opioid receptor expression in primary afferent neurons following spinal nerve injury. *European Journal of Pain, 15*(7), 669–675.

Lu, Y., Pirec, V., & Yeomans, D. C. (1997). Differential antinociceptive effects of spinal opioids on foot withdrawal responses evoked by C fibre or Adelta nociceptor activation. *British Journal of Pharmacology, 121*, 1210–1216.

Lu, Y., McNearney, T. A., Lin, W., Wilson, S. P., Yeomans, D. C., & Westlund, K. N. (2007). Treatment of inflamed pancreas with enkephalin encoding HSV-1 recombinant vector reduces inflammatory damage and behavioral sequelae. *Molecular Therapy, 15*(10), 1812–1819.

McCormack, K., Prather, P., & Chapleo, C. (1998). Some new insights into the effects of opioids in phasic and tonic nociceptive tests. *Pain, 78*, 79–98.

Rittner, H. L., Amasheh, S., Moshourab, R., Hackel, D., Yamdeu, R. S., Mousa, S. A., et al. (2012). Modulation of tight junction proteins in the perineurium to facilitate peripheral opioid analgesia. *Anesthesiology, 116*(6), 1323–1334.

Stein, C. (1995). The control of pain in peripheral tissue by opioids. *The New England Journal of Medicine, 332*, 1685–1690.

Stein, C., Schäfer, M., & Machelska, H. (2003). Attacking pain at its source: New perspectives on opioids. *Nature Medicine, 9*, 1003–1008.

Taddese, A., Nah, S.-Y., & McCleskey, E. W. (1995). Selective opioid inhibition of small nociceptive neurons. *Science, 270*, 1366–1369.

Wall, P. D., & Melzack, R. (1999). *Textbook of pain.* London: Churchill Livingstone.

Wilson, S. P., & Yeomans, D. C. (2002). Virally mediated delivery of enkephalin and other neuropeptide transgenes in experimental pain models. *Annals of the New York Academy of Sciences, 917*, 515–521.

Wilson, S. P., Yeomans, D. C., Bender, M. A., Lu, Y., Goins, W. F., & Glorioso, J. C. (1999). Antihyperalgesic effects of infection with a preproenkephalin-encoding herpes virus. *Proceedings of the National Academy of Sciences of the United States of America, 96*, 3211–3216.

Womer, D. E., DeLapp, N. W., & Shannon, H. E. (1997). Intrathecal pertussis toxin produces hyperalgesia and allodynia in mice. *Pain, 70*, 223–228.

Xu, M., Petraschka, M., McLaughlin, J. P., Westenbroek, R. E., Caron, M. G., Lefkowitz, R. J., et al. (2004). Neuropathic pain activates the endogenous kappa opioid system in mouse spinal cord and induces opioid receptor tolerance. *The Journal of Neuroscience, 24*(19), 4576–4584.

Zaki, P. A., Bilsky, E. J., Vanderah, T. W., Lai, J., Evans, C. J., & Porecca, F. (1996). Opioid receptor types and subtypes: The delta receptor as a model. *Annual Review of Pharmacology and Toxicology, 36*, 379–401.

Opioid Peptide

Definition

The endogenous opioid ligands consist of four peptide families. Three are derived from the known precursors proopiomelanocortin (POMC; encoding β-endorphin), proenkephalin (encoding Met-enkephalin and Leu-enkephalin), and prodynorphin (encoding dynorphins). These peptides contain the common Tyr-Gly-Gly-Phe-[Met/Leu] sequence at their amino terminals, known as the opioid motif. β-Endorphin and the enkephalins are potent antinociceptive agents acting at μ and δ receptors. Dynorphins can elicit both antinociceptive effects via κ-opioid receptors and pronociceptive effects through unknown mechanisms. Opioid peptides are expressed throughout the central and peripheral nervous system, in neuroendocrine tissues, and in immune cells.

Cross-References

▶ Opioid Peptide Co-Localization and Release
▶ Opioid Peptides from the Amphibian Skin
▶ Opioid Receptors at Postsynaptic Sites
▶ Opioids in the Periphery and Analgesia
▶ Stimulation-Produced Analgesia

Opioid Peptide Co-localization and Release

Catherine M. Cahill[1], Sarah V. Holdridge[2] and Noufissa Kabli[2]
[1]Anesthesiology & Perioperative Care, University of California, Irvine, Irvine, CA, USA
[2]Department of Pharmacology and Toxicology, Queen's University, Kinston, Canada

Synonyms

Narcotic; Opiate; Opioid; Opioid peptide release

Definition

Opioid peptides are a family of neuropeptides processed from three large precursor proteins known as proopiomelanocortin, preproenkephalin, and preprodynorphin. The gene products ► beta-endorphin (END), ► enkephalin (ENK), and ► dynorphin (DYN) share a common N-terminal sequence (Tyr-Gly-Gly-Phe-Leu/Met), which interacts with ► mu (μ)-opioid receptor, ► delta-opioid receptors, and ► kappa-opioid receptors (OR) with varying, but relatively nonselective, binding affinities for the different receptor types. The subsequent discovery of a novel group of peptides, the ► endomorphins (endomorphin-1 and endomorphin-2), has expanded the opioid peptide family. The endomorphins are unique within the opioid peptide family as they possess an atypical peptide sequence and a very high selectivity for the mu-OR. A further opioid-like 17-amino-acid peptide that resembles DYN, named "► orphanin FQ" or "► nociceptin," was also described in recent years and acts at opioid receptor-like (ORL-1) receptors.

Characteristics

The discovery of various endogenous opioid peptides, their anatomical distribution, and the characteristics of the multiple receptors with

which they interact has enhanced our understanding of the role of opioid peptide systems, including their integration in neural circuits that modulate the cognitive and affective aspects of the pain experience (for review see Przewlocki and Przewlocka (2001) and citations therein).

Localization and Co-expression

Opioid peptides have distinct but overlapping distribution in the central nervous system (CNS) (Mansour et al. 1988). Nociceptin/orphanin FQ and the opioid peptides show overlapping distributions, but not co-localization, in pain-modulatory brain regions (Schulz et al. 1996). Immunohistochemical labeling revealed a similar pattern of nerve fiber and terminal labeling within the spinal cord dorsal horn, sensory trigeminal complex, raphe nuclei, locus coeruleus, periaqueductal gray, amygdala, and hypothalamic region reflecting their potential role in many major processes including perception of pain; autonomic, neuroendocrine, and homeostatic functions; and affective states. However, while the distributions are similar, critical evaluation of immunopositive labeling via confocal microscopy showed no instances of co-localization of nociceptin and opioid peptides in any of the spinal or brain regions examined (Schulz et al. 1996). Similarly, Martin-Schild and colleagues reported that although often present in the same brain nuclei, endomorphin-1- and endomorphin-2-immunopositive neuronal elements were not identically distributed in the CNS (Martin-Schild et al. 1999).

While reports from rodent and primate studies have not identified co-localization of opioid peptides, despite their similarity in regional distribution particularly in nociceptive pathways, co-localization was found to be induced under certain conditions. For example, experimental arthritis or other chronic inflammatory conditions resulted in detection of opioid peptide or peptide mRNA co-localization. Hence, co-localization of pro-enkephalin and pro-dynorphin opioid peptides was identified in deep dorsal horn neurons of polyarthritic, but not normal, rats (Weihe et al. 1988). Various studies have subsequently demonstrated that chronic

O

pain states, including nerve section, produce dramatic changes in neuronal activity in response to the injury, leading to changes in gene expression of neuropeptides (Calza et al. 1998). Changes in opioid peptide expression are also dramatically modulated by various injuries. Peripheral inflammation, chronic arthritis, peripheral nerve injury, or damage to the spinal cord elicit profound alterations in dynorphin biosynthesis. While the functional significance of changes in opioid peptide expression is yet to be fully elucidated, it will undoubtedly provide further insight into their functional role in pain modulation.

Release

Hökfelt proposed that peptide release requires bursting or high-frequency activities, whereas individual action potentials firing at low frequencies will not induce the release of peptides (Hökfelt 1991). The facilitation of peptide release by high-frequency stimulation was considered to be due to the lengthening of the action potential duration, together with the increase in frequency, leading to an increase in calcium accumulation within the terminal and therefore dictating the amount of secretion per unit of action potential. This rule also applies to most, if not all, opioid peptides released from afferent terminals in various regions of the CNS. However, recent evidence suggests that frequency requirements for opioid peptide release can vary for different peptides. For example, analgesia was induced by low-frequency (4 Hz), but not by high-frequency (200 Hz), stimulation, and was inhibited by low doses of the opioid antagonist naloxone, suggesting that low-frequency stimulation can increase the release of opioid peptides in the CNS (Han 2003).

In recent years, a novel approach to the study of opioid peptide release was undertaken. Initially, the discovery of opioid peptides and their receptors made it possible, in principle, to trace causal links between stimuli, opioid release, OR activation, and analgesic responses. However, traditional approaches for studying opioid release have been limited by their inability to predict, (i) OR activation, (ii) precise concentration of the peptide at the receptor, and (iii) the individual actions of the opioid peptide, since more than one opioid peptide may be released. Using an approach effectively used to study the activation of neurokinin-1 receptors by neurokinins following noxious stimulation, OR internalization has been used successfully to assess the release of endogenous opioids by various stimuli. The inherent limitation of this technique lies in the promiscuity of opioid peptides to activate multiple OR types, with the exception of the highly selective endomorphins.

Mu-OR internalization attributable to opioid release was demonstrated in hypothalamic neurons after estrogen treatment (Eckersell et al. 1998) and in myenteric plexus neurons following electrical stimulation (Sternini et al. 2000), but was not evident in dorsal spinal cord neurons following primary afferent stimulation or noxious mechanical stimulation (Trafton et al. 2000), although both of these stimuli were shown to evoke the spinal release of ENK. In a more recent study, mu-OR internalization was noted following dorsal horn stimulation of rat spinal cord slices, but only in the presence of peptidase inhibitors (Song and Marvizon 2003). The internalization was attributed to opioid release, because it was abolished by the selective MOR antagonist CTAP (D-Phe-Cys-Tyr-D-Trp-Arg-Thr-Pen-Thr-NH_2) and was shown to be dependent on extracellular calcium. These findings support the notion that peptidases normally prevent the activation of extra-synaptic mu-OR in dorsal horn neurons, causing degradation prior to their activation of receptors to elicit an effect. This latter report suggests that there is no significant difference in the potency at which an opioid peptide activates its receptor to elicit a pharmacological effect and the ability to cause ligand-dependent receptor internalization.

Opioid peptide release has also been extensively studied outside the CNS. Localized inflammation of a rat's hindpaw elicits an accumulation of END-containing immune cells, which release END following stimulation via proinflammatory cytokines (Cabot et al. 1997). In this study, END release was receptor-specific, calcium-dependent, and mimicked by potassium, consistent with vesicular release and thus similar to release in the CNS. It has been proposed that the activation of opioid peptide biosynthesis and release from immune cells may validate the development of peripherally acting opioid analgesics (Machelska and Stein 2000). Moreover, as such drugs would target events in peripheral injured tissue and not cross the blood-brain barrier, these novel analgesics should lack unwanted central side effects typically associated with most clinically available opiates to date.

Other Considerations

In addition to the localization of opioid peptides in nociceptive pathways, they are also present in many CNS nuclei including the hippocampus, where they have been proposed to play a role in functional plasticity (Morris and Johnson 1995). High-frequency stimulation of presynaptic fibers in various regions of the hippocampus produces a long-lasting increase in the response of the postsynaptic neurons to subsequent presynaptic activity (▶ long-term potentiation, LTP). The mossy fibers of the hippocampus, in addition to containing the excitatory amino acid glutamate, also release opioid peptides. It has been documented that all granule cells express the pro-dynorphin gene, and a high level of DYN is observed in the mossy fiber terminals. In addition, a subset of granule cells in the dentate gyrus expresses the pro-enkephalin gene. Activation of OR in the hippocampus via ENKs acting at mu-OR produces an excitatory action via disinhibition of GABAergic interneurons, resulting in a reduction in the number of high-frequency stimuli required to induce LTP. However, activation of kappa-OR following release of DYN has been shown to produce the opposite effect on excitability, and hence, the endogenous opioid peptides have opposite effects on modulating synaptic strength.

Opposing effects of opioid peptides have also been noted in nociceptive pathways, whereby intracerebroventricular administration of nociceptin produces hyperalgesia, but ENKs produce analgesia. Nociceptin and ENKs are released from distinct terminals yet localized in similar brain nuclei, suggesting that nociceptin is likely to modulate pain signals opposite to other endogenous opioids. Due to the inherent ability of opioid peptides to produce opposing actions in the CNS and their potential to modulate synaptic efficacy, changes in opioid peptide translation and release may contribute to the genesis of chronic pain states, as synaptic modulation and modification have been identified in such pathological conditions.

References

Cabot, P. J., Carter, L., Gaiddon, C., Zhang, Q., Schafer, M., Loeffler, J. P., et al. (1997). Immune cell-derived beta-endorphin production, release, and control of inflammatory pain in rats. *Journal of Clinical Investigation, 100*, 142–148.

Calza, L., Pozza, M., Zanni, M., Manzini, C. U., Manzini, E., & Hokfelt, T. (1998). Peptide plasticity in primary sensory neurons and spinal cord during adjuvant-induced arthritis in the rat: An immunocytochemical and in situ hybridization study. *Neuroscience, 82*, 575–589.

Eckersell, C. B., Popper, P., & Micevych, P. E. (1998). Estrogen-induced alteration of mu-opioid receptor immunoreactivity in the medial preoptic nucleus and medial amygdala. *Journal of Neuroscience, 18*, 3967–3976.

Han, J.-S. (2003). Acupuncture: Neuropeptide release produced by electrical stimulation of different frequencies. *Trends in Neurosciences, 26*, 17–22.

Hökfelt, T. (1991). Neuropeptides in perspective: The last ten years. *Neuron, 7*, 867–879.

Machelska, H., & Stein, C. (2000). Pain control by immune-derived opioids. *Clinical and Experimental Pharmacology and Physiology, 27*, 533–536.

Mansour, A., Khachaturian, H., Lewis, M. E., Akil, H., & Watson, S. J. (1988). Anatomy of CNS opioid receptors. *Trends in Neurosciences, 11*, 308–314.

Martin-Schild, S., Gerall, A. A., Kastin, A. J., & Zadina, J. E. (1999). Differential distribution of endomorphin 1- and endomorphin 2-like immunoreactivities in the CNS of the rodent. *The Journal of Comparative Neurology, 405*, 450–471.

Morris, B. J., & Johnson, H. M. (1995). A role for hippocampal opioids in long-term functional plasticity. *Trends in Neurosciences, 18*, 350–355.

Przewlocki, R., & Przewlocka, B. (2001). Opioids in chronic pain. *European Journal of Pharmacology, 429*, 79–91.

Schulz, S., Schreff, M., Nuss, D., Gramsch, C., & Höllt, V. (1996). Nociceptin/orphanin FQ and opioid peptides show overlapping distribution but not Co-localization in pain-modulatory brain regions. *NeuroReport, 7*, 3021–3025.

Song, B., & Marvizon, J. C. (2003). Peptidases prevent mu-opioid receptor internalization in dorsal horn neurons by endogenously released opioids. *Journal of Neuroscience, 23,* 1847–1858.

Sternini, C., Brecha, N. C., Minnis, J., D'Agostino, G., Balestra, B., Fiori, E., & Tonini, M. (2000). Role of agonist-dependent receptor internalization in the regulation of mu-opioid receptors. *Neuroscience, 98,* 233–241.

Trafton, J. A., Abbadie, C., Marek, K., & Basbaum, A. I. (2000). Postsynaptic signaling via the [mu]-opioid receptor: Responses of dorsal horn neurons to exogenous opioids and noxious stimulation. *Journal of Neuroscience, 20,* 8578–8584.

Weihe, E., Millan, M. J., Leibold, A., Nohr, D., & Herz, A. (1988). Co-localization of proenkephalin- and prodynorphin-derived opioid peptides in laminae IV/V spinal neurons revealed in arthritic rats. *Neuroscience Letters, 85,* 187–192.

Opioid Peptide Release

▶ Opioid Peptide Co-localization and Release

Opioid Peptides from the Amphibian Skin

Lucia Negri and Roberta Lattanzi
Department of Physiology and Pharmacology,
Sapienza University of Rome, Rome, Italy

Definition

Amphibian opioids are small peptides (▶ Amphibian Peptides) found in the skin of South American hylid frogs belonging to the subfamily *Phyllomedusinae* (*Phyllomedusa, Agalychnis,* and *Pachymedusa* species). They are codified by different genes and secreted by the syncytial cells forming the luminal wall of serous holocrine glands in the frog integument. Two subfamilies of opioid peptides have been distinguished according to their selectivity for the mu and delta opioid receptors, the mu-selective dermorphins and the delta-selective deltorphins.

All amphibian opioid peptides contain the N-terminal sequence Tyr-D-Xaa-Phe, where the amino acids Tyr^1 and Phe^3 are of L-configuration and D-Xaa^2 is a D-amino acid. Because the D-enantiomer is encoded by the codon for the L-isomer in the precursor cDNA, the translated L-Xaa^2 must be converted to D-Xaa^2 by an unusual posttranslational reaction that takes place in the precursor itself. C-terminal amino acids in deltorphin and dermorphin sequences are usually amidated. The presence of a D-amino acid in the N-terminal sequence restricts peptide conformation and together with C-terminal amidation confers resistance against enzyme degradation.

Characteristics

Although Vittorio Erspamer and his colleagues discovered these opiates (Erspamer 1994) in Amazonian frogs only in the 1980s, the Matses of the upper Amazonian basin unveiled the pharmacological properties of amphibian skin opiates long ago. For centuries they had habitually applied the dried skin secretions of *Phyllomedusa bicolor,* called "sapo" (the Spanish word for "toad"), to cuts in their skin during shamanistic hunting rituals. The abundance of deltorphins and dermorphins in "sapo" probably caused the hunters analgesia and behavioral excitation.

The first to be discovered were two dermorphins (Table 1; 1 and 2) identified in the skin of *Ph. sauvagei, Ph. rohdei,* and *Ph. burmeisteri* (Broccardo et al. 1981; Montecucchi et al. 1981). The very high dermorphin content in the skin of these *Phyllomedusinae,* about 50–80 μg/g of fresh skin, made it easy to purify and sequence the peptides.

Eight years later, deltorphins, a family of highly selective delta opioid peptides, were identified either by cloning the cDNA from the skin of *Ph. sauvagei* (Table 1; 8)

Opioid Peptides from the Amphibian Skin, Table 1 Naturally occurring opioid peptides and their origin

Dermorphins		
1 Tyr-D-Ala-Phe-Gly-Tyr-Pro-Ser-NH_2	Dermorphin	*Ph. sauvagei, burmeisteri, rohdei,*
2 Tyr-D-Ala-Phe-Gly-Tyr-Hyp-Ser-NH_2	[Hyp^6]derm	*hypochondrialis, tarsius, Agalychnis* (Montecucchi et al. 1981; Broccardo et al. 1981)
3 Tyr-D-Ala-Phe-Gly-Tyr-Pro-Lys-OH	[Lys^7-OH]-derm	*Ph. bicolor* (Negri et al. 1992)
4 Tyr-D-Ala-Phe-Trp-Tyr-Pro-Asn-OH	[Trp^4,Asn^7-OH]-derm	
5 Tyr-D-Ala-Phe-Trp-Asn-OH	[Trp^4,Asn^5-OH]-derm	
6 Dermorphin-Gly-Glu-Ala-OH	Y10A	*Ph. sauvagei* (Erspamer 1994; Kovolevski
7 Dermorphin-Gly-Glu-Ala-Lys-Lys-Ile-OH	Y131	et al. 1999)
Deltorphins		
8 Tyr-D-Met-Phe-His-Leu-Met-Asp-NH_2	D-Met-deltorphin	*Ph. sauvagei, burmeisteri, rohdei, tarsius* (Kreil et al. 1989)
9 Tyr-D-Ala-Phe-Asp-Val-Val-Gly-NH_2	D-Ala-deltorphin I	*Ph. bicolor* (Erspamer et al. 1989)
10 Tyr-D-Ala-Phe-Glu-Val-Val-Gly-NH_2	D-Ala-deltorphin II	
11 Tyr-D-Leu-Phe-Ala-Asp-Val-Ala-Ser-Thr-Ile-Gly-Asp-Phe-Phe-His-Ser-Ile-NH_2	D-Leu-deltorphin	*Ph. burmeisteri* (Barra et al. 1994)
12 Tyr-D-Ile-Phe-His-Leu-Met-Asp-NH_2	D-Ile-deltorphin	*Pachymedusa dacnicolor, Agalychnis annae* (Wechselberger et al. 1998)

(Kreil et al. 1989) or by extracting the mature peptides from the skin of *Ph. bicolor* (Table 1; 9 and 10) (Erspamer et al. 1989). Chromatographic separation of skin extracts and screening of cDNA libraries from the skin of various species of *Phyllomedusinae* led to the identification of other opioid peptides belonging to the dermorphin and deltorphin families (see Table 1) (Negri et al. 1992; Barra et al. 1994; Wechselberger et al. 1998).

Despite the common N-terminal tripeptide (Tyr-D-Xaa-Phe), the two families of amphibian opioids differ widely in ▶ receptor selectivity but bind to their own receptors with similar affinities (▶ Receptor Affinity). The N-terminal domain contains the minimum sequence essential for opioid receptor binding, whereas the C-terminal domain contains the address requisites for receptor selectivity. The presence of negatively charged amino acid residues (Asp 114, Asp 147) that contribute to ligand binding, within the putative transmembrane domains II and III of the mu receptor protein, may explain why positively charged dermorphins have high mu opioid receptor selectivity and why amidation of

their terminal carboxyl group increases their affinity and potency. Similarly, the negatively charged C-terminal tetrapeptide of deltorphins enhances deltorphin selectivity for the delta opioid receptor by electrostatic attraction to the positively charged binding site of the delta receptor (Arg 292) and electrostatic repulsion from the negatively charged mu receptor site. Electrostatic attraction within the peptide molecule may explain why deltorphins have a folded structure, whereas dermorphins have a more distended and flexible structure. Unlike the positively charged C terminus of dermorphins, the negatively charged C-terminal tail of deltorphins comes into close contact with the positively charged N-terminal tripeptide of the molecule and folds the backbone, thus placing the Tyr^1 and Phe^3 aromatic rings in definite orientations best suited for delta receptor docking (Negri and Giannini 2003).

Dermorphin affinity for mu opioid receptors and its opioid potency in guinea pig ileum preparations (GPI) are 20 and 100 times higher, respectively, than those of morphine. Among other dermorphins, [Lys7]dermorphin shows an affinity and selectivity for mu opiate receptors six times higher than dermorphin and 100 times higher than morphine (Table 2). Like mu opiate agonists, dermorphins produce antinociception but also catalepsy, respiratory depression, constipation, tolerance, and dependence, albeit to a lower degree than morphine. Dermorphin-induced antinociception takes place at supraspinal and spinal levels. Owing to its low CNS permeability and bioavailability, the analgesic potency of dermorphin is about 250 times higher than that of morphine after icv injection but comparable to that of morphine after sc injection. [Lys7]dermorphin and some synthetic analogs bearing a hydrophilic group (a basic amino acid or a glycosyl residue) at the C-terminal end of the molecule enter the CNS in seven to ten times higher amounts than dermorphin, suggesting facilitated transport across the blood-brain barrier by a carrier or endocytosis. The amphibian opioid with the highest analgesic potency and efficacy is [Lys7]dermorphin (AD_{50} 30 pmol/rat by the intracerebroventricular route (icv), 0.22 μmol/kg by the subcutaneous route (sc)). The ratio of antinociceptive to respiratory depressant ED_{50} doses was 17 times lower for [Lys7]dermorphin than for morphine. Indeed, [Lys7]dermorphin, in the range of analgesic doses (36–120 pmol, icv, 0.12–4.7 μmol/kg, sc), significantly increases respiratory frequency and minute volume of rats breathing air or hypoxic inspirates. The early-onset ventilatory stimulation produced by [Lys7]dermorphin and ascribed to the intact peptide is mediated by serotonergic descending excitatory pathways that stimulate neurons of the brainstem respiratory network (Negri et al. 1998). In rats and mice, central or peripheral administration of the dermorphin-like peptides induces a significantly slower development of tolerance to the antinociceptive effect than with morphine. Withdrawal symptoms precipitated by naloxone are less intense in peptide-dependent than in morphine-dependent rats (Negri et al. 2000).

The discovery of deltorphins provided the tools for the functional characterization of the delta opiate system. While D-Ala^2-deltorphins have delta-binding affinity similar to D-Met^2-deltorphin, they consistently have the highest delta opiate selectivity (Table 2). The rank order of selectivity ($K_i\delta$ / $K_i\mu$) is D-Ala^2-deltorphin I = D-Ala^2-deltorphin II > D-Met^2-deltorphin >>D-Ile^2-deltorphin >>D-Leu^2-deltorphin heptadecapeptide or its N-terminal decapeptide fragment. Both D-Met^2-deltorphin and D-Ala^2-deltorphins are highly resistant to enzyme degradation. D-Ala^2-deltorphins I and II cross the blood-brain barrier in vivo and in vitro (Fiori et al. 1997). Recently, D-Ala^2-deltorphin II was identified as a transport substrate of organic anion-transporting polypeptides (Oatp/OATP), a family of polyspecific membrane transporters, strongly expressed in the rat and human blood-brain barrier (Negri and Giannini 2003).

In mice, D-Ala^2-deltorphin II by the icv route (EC_{50} = 2.1 nmol/mouse) is half as potent as morphine, and the analgesic effect is antagonized by the delta-selective antagonist naltrindole. Repeated injection of D-Ala2-deltorphin II induces tolerance to the antinociceptive effect. Isobolographic analysis shows that the supraspinal antinociception induced by the delta opioid agonist DPDPE and the spinal antinociception induced by D-Ala^2-deltorphin II are synergistic in many nociceptive tests, suggesting that the compounds act on distinct delta opiate receptors (Kovolevski et al. 1999). In rats, intrathecal injections of D-Ala^2-deltorphin II (from 0.6 nmol/rat) produce a naltrindole-reversible antinociception, by inhibiting the nociceptive neurons in the superficial and deeper dorsal horn of the medulla. Conversely, when injected icv in rats, D-Ala2-deltorphin II was a weak partial agonist (>30 nmol/rat) and induces fleeting naloxone-sensitive antinociception. Both D-Ala2-deltorphin I and D-Met2-deltorphin at doses between 6.5 nmol and 52 nmol/rat induce a naloxone-sensitive analgesia. Data suggest that the delta agonists play a predominantly modulatory role in antinociception rather than a primary role. In mice and in rats, the intensity of delta opiate analgesia depends on coactivation of mu opiate receptors by endogenous or exogenous opiates. Deltorphin-induced analgesia is weaker in homozygote mice with a disrupted mu opiate receptor gene than in wild-type mice.

Opioid Peptides from the Amphibian Skin, Table 2 Affinity and selectivity for mu and delta opioid receptors, biological activities on guinea pig ileum (GPI) and mouse vas deferens (MVD) preparations, and analgesic potency of dermorphins and deltorphins

Compounds	K_i, nM (mean ± S.E.)			IC_{50}, nM (mean ± S.E.)		Analgesia AD_{50} nmol/rat
	μ-System	δ-System	μ/δ	GPI	MVD	
Morphine	11 ± 1.5	500 ± 48	22×10^{-3}	150 ± 18.5	1,215 ± 115	8.7 ± 1.1
Dermorphin						
Dermorphin	0.6 ± 0.02	929 ± 41	6.1×10^{-4}	1.29 ± 0.11	16.5 ± 1.3	0.035 ± 0.01
[Hyp⁶]derm	0.7 ± 0.03	1,200 ± 131	5.8×10^{-4}	1.6 ± 0.12	18.1 ± 2.9	
[Lys⁷]derm	0.09 ± 0.008	1,105 ± 185	8.1×10^{-5}	1.15 ± 0.13	13.6 ± 1.5	0.026 ± 0.009
[Lys⁷-OH]derm	5.7 ± 0.51	1,150 ± 172	4.9×10^{-3}	3.82 ± 0.45	56.3 ± 7.8	
[Trp⁴,Asn⁵]derm	0.9 ± 0.052	480 ± 45	1.9×10^{-3}	5.00 ± 0.52	73.7 ± 9.1	0.43 ± 0.04
[Trp⁴,Asn⁷]derm	0.32 ± 0.026	690 ± 57	4.6×10^{-4}	0.58 ± 0.06	6.6 ± 0.9	2.086 ± 0.31
Deltorphin						
D-Ala²-delt I	1,985 ± 224	0.78 ± 0.08	2,545	1,239 ± 203	0.18 ± 0.02	15 ± 3.0
D-Ala²-delt II	2,222 ± 233	1.03 ± 1.09	2,157	2,500 ± 170	0.37 ± 0.03	54 ± 6.0
D-Ala²,Gly⁴-delt	13.5 ± 2.1	3.26 ± 0.7	4.14	22 ± 3	2.62 ± 0.32	2.5 ± 0.3
D-Met²-delt	693 ± 37	1.18 ± 0.21	587	1,476 ± 185	0.97 ± 0.05	20 ± 5.0
D-Leu²-delt	>10,000	>10,000	–	>5,000	2,480 ± 378	
D-Leu²-delt(1–10)	–	–	–	1,684 ± 403	37.0 ± 3.9	
D-Ile²-delt	1,021 ± 57	24 ± 3	42.6	4,200 ± 475	7.0 ± 0.9	6.7 ± 1.0
D-alle²-delt	452 ± 68	54 ± 65	8.4	3,200 ± 307	70 ± 8.2	9.8 ± 1.2

K_i, equilibrium inhibition constant of the competing ligand; mu opioid receptors have been labeled with [³H]DAMGO, 0.5 nM; delta opioid receptors have been labeled with [³H] deltorphin I, 0.3 nM; μ/δ represents selectivity for the mu receptors versus delta receptors

Injections of the deltorphins into the rat brain ventricles, ventral tegmental area, and nucleus accumbens at doses 10–100 times lower than those inducing analgesia (0.063.8 nmol/rat) invariably increase locomotor activity and induce stereotyped behavior. Deltorphin-induced motor activity is antagonized by the delta-selective antagonist naltrindole. Repeated icv injections of D-Ala^2-deltorphin II in naive rats induce tolerance to the stimulant effects. High doses (10–50 nmol/rat) of all the D-Met^2-deltorphin analogs and His^4-substituted D-Ala^2-deltorphins induce non-opioid motor dysfunction that is completely blocked by the noncompetitive NMDA antagonist dextrorphan. Because the deltorphin peptides do not induce dependence and because they stimulate respiratory activity and leave the rate of transit through the small intestine unchanged (Negri et al. 2000; Negri and Giannini 2003), they are excellent candidates as drugs to relieve acute or chronic pain in humans.

References

Barra, D., Mignogna, G., Simmaco, M., et al. (1994). [D-Leu^2]deltorphin, a 17 amino acid opioid peptide from the skin of the Brazilian hylid frog, *Phyllomedusa burmeisteri*. *Peptides, 15*, 199–202.

Broccardo, M., Erspamer, V., Falconieri-Erspamer, G., et al. (1981). Pharmacological data on dermorphins, a new class of potent opioid peptides from amphibian skin. *British Journal of Pharmacology, 73*, 625–631.

Erspamer, V. (1994). Bioactive secretions of the amphibian integument. In H. Heatwole (Ed.), *Amphibian biology* (pp. 178–350). Chipping Norton: Surrey Beatty & Son.

Erspamer, V., Melchiorri, P., Falconieri-Erspamer, G., et al. (1989). Deltorphins: A family of naturally occurring peptide with high affinity and selectivity for delta-opioid binding sites. *Proceedings of the National Academy of Sciences of the United States of America, 86*, 5188–5192.

Fiori, A., Cardelli, P., Negri, L., et al. (1997). Deltorphin transport across the blood-brain barrier. *Proceedings of the National Academy of Sciences of the United States of America, 94*, 9469–9474.

Kovolevski, C. J., Bian, D., Ruby, H., et al. (1999). Selective opioid delta agonists elicit antinociceptive supraspinal/spinal synergy in the rat. *Brain Research, 843*, 12–17.

Kreil, G., Barra, D., Simmaco, M., et al. (1989). Deltorphin, a novel amphibian skin peptide with high selectivity and affinity for delta-opioid receptors. *European Journal of Pharmacology, 162*, 123–128.

Montecucchi, P. C., De Castiglione, R., & Erspamer, V. (1981). Identification of dermorphin and hypdermorphin in skin extract of the Brazilian frog Phyllomedusa rohdei. *International Journal of Peptide and Protein Research, 17*, 316–321.

Negri, L., & Giannini, E. (2003). Deltorphins. In K. J. Chang, F. Porreca, & J. H. Woods (Eds.), *The delta receptors, molecule and effects of delta opioid compounds* (pp. 175–189). New York: Marcel Dekker Book Chapters.

Negri, L., Falconieri-Erspamer, G., Severini, C., et al. (1992). Dermorphin related peptides from the skin of phyllomedusa bicolor and their amidated analogs activate two mu-opioid receptor subtypes which modulate antinociception and catalepsy, in the rat. *Proceedings of the National Academy of Sciences of the United States of America, 89*, 7203–7207.

Negri, L., Lattanzi, R., Tabacco, F., et al. (1998). Respiratory and cardiovascular effects of the mu-opioid receptor agonist [Lys^7]dermorphin in awake rats. *British Journal of Pharmacology, 124*, 345–355.

Negri, L., Melchiorri, P., & Lattanzi, R. (2000). Pharmacology of amphibian opiate peptides. *Peptides, 21*, 1639–1647.

Wechselberger, C., Severini, C., Kreil, G., et al. (1998). A new opioid peptide predicted from clone cDNAs from skin of pachymedusa dacnicolor and agalychnis annae. *FEBS Letters, 429*, 41–43.

Opioid Pseudo-pharmacological Ceiling Dose

▶ Cancer Pain Management, Principles of Opioid Therapy, Drug Selection

Opioid Receptor Localization

Gavril W. Pasternak[1] and Catherine Abbadie[2]
[1]Laboratory of Molecular Neuropharmacology, Memorial Sloan-Kettering Cancer Center, New York, USA
[2]Department of Pharmacology, Merck Research Laboratories, Rahway, NJ, USA

Definition

Opioids have long played a major role in pharmacology, representing one of the oldest classes of clinically important pharmaceuticals (Pasternak 1993; Reisine and Pasternak 1996). Opioid receptor binding to brain membranes was first demonstrated in 1973, and represents one of the earliest neurotransmitter receptors localized within the central nervous system using autoradiographic techniques. Presently, three distinct classes of opioid receptors have been identified. All three classes of opioid receptors are members of the G-protein-coupled receptor family, and have the typical seven transmembrane domains seen with this large family of membrane receptors. Opioids bind to the extracellular face of the receptor and activate intracellular G-proteins. The opioid receptors are almost exclusively inhibitory, activating primarily G_o and G_i. The three opioid receptor classes are structurally homologous, but each is encoded by a separate gene.

Early autoradiographic studies of the brain, defining the regional distribution of opioid receptors, used a variety of radioligands that we now know are not very selective. Although these initial studies firmly established the presence of opioid binding sites in brain regions known to be important in mediating opioid actions, it is almost certain that they were labeling more than one class of binding site. With the cloning of the receptors (for review see Pasternak 2001; Kieffer and Gaveriaux-Ruff 2002; Wei and Loh 2002), it is now possible to selectively localize them at the mRNA level using in situ hybridization, and at the protein level immunohistochemically.

Mu Receptors

Morphine and most clinical opioid analgesics act through the mu-▶ opiate receptors, making these receptors particularly important. However, this class of receptor is complex, with over a dozen different splice variants of the mu receptor having been cloned. While these may provide potential insights into early pharmacological studies implying different mu receptor subtypes for morphine analgesia versus respiratory depression or constipation, it complicates the interpretation their distributions.

Delta Receptors

Delta receptors were first proposed shortly after the discovery of the enkephalins, their endogenous ligand. The development of highly selective, stable delta compounds has greatly facilitated studies of the pharmacology of this receptor, and the cloning of the DOR-1 has enabled studies of this receptor at the molecular level. Delta opioids produce analgesia, but with far less respiratory depression and constipation. There is strong pharmacological evidence for delta receptor subtypes, but functional splice variants have not yet been reported.

Kappa₁ Receptors

Kappa receptors were initially suggested from classical in vivo pharmacological studies long before the identification of its endogenous ligand, dynorphin A. The synthesis of highly selective ligands has helped to define the $kappa_1$ receptor, both biochemically and pharmacologically. $Kappa_1$ receptors can produce analgesia, but through mechanisms distinct from those of the other receptor classes. However, the clinical use of kappa receptor ligands has been limited by their psychotomimetic side effects.

A kappa receptor with the appropriate pharmacological profile has been cloned, KOR-1, and it has high homology with the other opiate receptors. In addition to the nervous system, $kappa_1$ receptors are also present in other tissues, particularly immune cells.

Other Kappa Receptors

Several other kappa receptor classes have been proposed. U50,488H is a potent, and highly selective, $kappa_1$ receptor agonist. $Kappa_2$ receptors, initially defined by their insensitivity to traditional mu, delta, and $kappa_1$ opioids, display a unique binding profile and may correspond to dimers of KOR-1 and DOR-1 receptors. $Kappa_3$ receptors represent another class of U50,488H-insensitive receptors totally distinct from any other class.

Opioid-Receptor-Like (ORL-1) Receptor

A fourth member of the opiate receptor family has been cloned that is highly selective for the peptide orphanin FQ/nociceptin (▶ OFQ/N). The pharmacology of OFQ/N is complex, with actions depending upon both dose and route. Despite the similarity of the cloned receptor with other members of the opioid receptor family, OFQ/N shows no appreciable affinity for traditional opioid receptors, and opioids bind very poorly to the ORL-1 receptor. However, the ORL-1 receptor will dimerize with MOR-1, yielding a complex with a profoundly different binding profile, in which traditional mu opioids potently compete with OFQ/N and vice versa.

Conclusion

The opioid system is comprised of three major classes of receptors, with each being encoded by a different gene, as well as the ORL-1 receptor. Furthermore, the gene encoding MOR-1 generates a large number of variants, many of which bind morphine and related mu opioids with high affinity. Their close structural and ligand-binding characteristics make it difficult to localize a single variant.

Characteristics

The first attempt to look at opioid distribution in the brain used autoradiographic approaches. These early studies defining their distributions used various opioid ligands, and quickly established the presence of opioid binding sites in brain regions presumed to be important in mediating opioid actions. However, as our understanding of opioid receptors has expanded, it has become apparent that opioids act through a family of receptors, as described earlier. Equally important, many of the ligands initially thought to be "selective" are now known not to be, complicating the interpretation of these earlier studies.

Following the cloning of the opioid receptors, their distributions within the brain were evaluated using in situ hybridization and immunohistochemistry. Unlike traditional receptor autoradiography, the selectivity of both techniques is excellent. In situ hybridization labels mRNA and identifies the cell bodies responsible for synthesizing the receptors. However, the receptors themselves are often transported long distances along axons. Thus, it is also important to localize with subcellular resolution the receptors at protein level. The epitopes identified from the cloned receptors have provided unique, and very selective, tools for immunohistochemistry. However, even some of antisera do not distinguish among some of the splice variants. Thus, even now, localization studies must be interpreted cautiously, taking into account the possibility that a specific epitope might be contained within more than one variant.

A complete description of the distribution of opioid receptors is beyond the scope of this entry. Detailed descriptions of the mRNA of MOR-1 have been published (Mansour et al. 1994), as have the distributions of the peptides (Ding et al. 1996), DOR-1 (Arvidsson et al. 1995a), and KOR-1 (Arvidsson et al. 1995b). The descriptions below refer primarily to opioid receptor localization in relationship to pain pathways and are based primarily upon studies with rodents, although the distributions in human tissues are quite similar (Fig. 1).

Peripheral Opioid Receptors

Mu, delta, and kappa opioid receptors have been identified in rat and human ▶ nerve terminals in the skin (for review see Stein et al. 2001). These fibers are the peripheral terminals of neuronal cell bodies that are localized in the ▶ dorsal root ganglia (DRG). Within the DRG, approximately 50 % of all neurons express MOR 1, whereas only 20 % express DOR-1 or KOR-1. MOR-1 and KOR-1 are found almost exclusively in small diameter myelinated (Aδ fibers) and unmyelinated (C fibers) axons, whereas DOR-1 is also found in large diameter fibers (Aαβ fibers). Opioid receptors are synthesized in the DRG and then transported to the nerve endings, both in the periphery and centrally in the ▶ spinal cord, via axonal transport. In inflammatory states, all three types of receptors have been shown to be upregulated, facilitating peripheral-mediated opioid analgesia.

It has been suggested that opioid receptors are also located on sympathetic postganglionic terminals, and that they contribute to opioid

Opioid Receptor Localization, Fig. 1 Opioid receptor distribution in relationship with pain pathways. These schematics represent the ascending (*arrows* on the *left side*) and descending (*arrows* on the *right side*) pain pathways. Distributions of MOR-1, DOR-1, and KOR-1, based on in situ hybridization and immunohistochemistry studies are indicated by the blue shading. Nociceptive information is transmitted from the periphery to the spinal cord. Pain perception mainly depends upon the activation of the spinothalamic tract. Neurons in the spinal cord project to thalamic nuclei, which in turn project to subcortical and cortical areas, particularly those within the limbic system. Nociceptive information is also transmitted by the spino-parabrachial tract (not represented in this schematic) which projects to the amygdala and hypothalamus. There are also spino-limbic pathways that contribute to different aspects of pain (emotional, autonomic) by influencing different cortical areas. These ascending pain sensory pathways are subject to modulatory control by descending modulatory circuits. These circuits originate in the limbic forebrain, and via brain stem connections, control the transmission of pain signals at their first central relay in the spinal cord. These modulatory circuits mediate the action of morphine and their action is mediated by the release of endogenous opioid peptides. Abbreviations, I–II, superficial laminae of the spinal cord; CL, centrolateral thalamic nucleus; DR, dorsal raphe nucleus; DRG, dorsal root ganglia; Fr, frontal cortex; G, gelatinosus nucleus; Gi, gigantocellular reticular nucleus; GiA, gigantocellular reticular nucleus part alpha; MnR, median raphe nucleus; Pa, parietal cortex; PAG, periaqueductal gray; Pf, prefrontal cortex; PnO, pontine reticular nucleus; Po, posterior thalamic nuclear group; py, pyramidal tract; RMg, raphe magnus nucleus; Sp5, spinal trigeminal nucleus (Schematics are modified from Paxinos and Watson (1997))

antinociception. However, direct attempts to localize opioid receptors in sympathetic ganglia have failed. Additionally, chemical sympathectomy with 6-hydroxydopamine does not modify opioid receptor expression in the dorsal root ganglia, nor does it change the analgesic effect of either mu, delta, or kappa agonists in an inflammatory model of pain. Immune cells have also been shown to express opioid receptors; however, their role in mediating analgesia remains unclear.

Spinal Cord

In the spinal dorsal horn, expression of mu, delta, and kappa opioid receptors is intense in the superficial laminae (laminae I–II), where nociceptive Aδ and C fibers terminate. Opioid receptors are located presynaptically on axons from the DRG neurons, as well as postsynaptically on cell bodies that can be visualized by in situ hybridization, and immunohistochemically. In deeper laminae of the dorsal horn, and in the ventral horn, mu and kappa opioid receptor expression is weaker, whereas delta receptors are found throughout the dorsal and ventral horns including motoneurons. Two splice variants of the mu opioid receptor, MOR-1C and MOR-1D, are expressed almost exclusively in unmyelinated axons in the superficial laminae of the spinal cord (Abbadie et al. 2001). Furthermore, confocal studies indicate that neurons in the dorsal horn express either MOR-1 or MOR-1C, but rarely both.

In pain pathways, ascending projections originating from the deep laminae terminate in different areas than projections originating from lamina I (Gauriau and Bernard 2002). Deep laminae neurons project to caudal reticular areas, including the lateral reticular nucleus, the subnucleus reticularis dorsalis, and the gigantocellular lateral paragigantocellular reticular nuclei. Lamina I neurons project to the thalamus, including the ventral posterolateral nucleus, ventral posteromedial, posterior nuclear group, and triangular posterior nuclei (Willis and Westlund 1997). Lamina I neurons also project to the lateral parabrachial area.

The distribution of opioid receptors in the spinal trigeminal nucleus is similar to the distribution in the spinal cord.

Brain Stem

Moderate expression of MOR-1 is observed in regions of the pons and medulla, particularly those involved in mediating analgesia, such as the raphe magnus, intermediate reticular, gigantocellular reticular, and lateral paragigantocellular nuclei. Opioids also act on regions of the brain stem to activate descending pain inhibitory systems, which consist of descending serotoninergic and noradrenergic neurons (Basbaum and Fields 1984). Indeed microinjections of morphine into the lateral paragigantocellular nucleus, raphe magnus nucleus, and periaqueductal gray induce profound analgesic effects. The periaqueductal gray, medial and dorsal raphe nuclei also contain moderate MOR-1, but intense KOR-1 labeling.

Intense expression of MOR-1 is also seen in the nucleus of the solitary tract, nucleus ambiguous, and parabrachial nucleus. These brain stem structures mediate the respiratory and cardiovascular side effects of opioids. Neurons in the lateral parabrachial area project to the extended amygdala, the hypothalamus, the periaqueductal gray matter, and the ventrolateral medulla and have been implicated in emotional and autonomic aspects of pain (Willis and Westlund 1997; Gauriau and Bernard 2002). DOR-1 expression is low or absent in all hindbrain and midbrain structures, such as the periaqueductal gray, raphe nucleus, or parabrachial nucleus.

Midbrain and Forebrain

MOR-1 expression has been demonstrated in the thalamic nuclei involved in the transmission of nociceptive information, including the centrolateral, ventrolateral, and ventromedial thalamic nuclei and posterior thalamic nuclear group. Only the centrolateral and geniculate nuclei express KOR. Little or no expression of DOR-1 is observed in the rodent thalamus. In the hypothalamus, one splice variant of MOR-1 is found in various nuclei, whereas MOR-1 distribution is limited (Abbadie et al. 2000).

In the ► cerebral cortex, MOR-1 is weakly expressed in layers II–IV, while DOR-1 expression is moderate to intense in layers II, III, V, and VI, and KOR-1 is only found in layer VI.

Conclusion

Overall, there is an excellent correlation between the localization of opioid receptors and pain pathways. Opioid receptors are distributed at various sites from nerve endings in the skin, to the cerebral cortex and various nuclei located along nociceptive pathways, such as the dorsal raphe or the lateral parabrachial nucleus.

References

Abbadie, C., Pan, Y. X., & Pasternak, G. W. (2000). Differential distribution in rat brain of mu-opioid receptor carboxy terminal splice variants MOR-1C-like and MOR-1-like immunoreactivity: Evidence for region-specific processing. *The Journal of Comparative Neurology, 419*, 244–256.

Abbadie, C., Pasternak, G. W., & Aicher, S. A. (2001). Presynaptic localization of the carboxy-terminus epitopes of the mu-opioid receptor splice variants MOR-1C and MOR-1D in the superficial laminae of the rat spinal cord. *Neuroscience, 106*, 833–842.

Arvidsson, U., Dado, R. J., Riedl, M., Lee, J. H., Law, P. Y., Loh, H. H., Elde, R., & Wessendorf, M. W. (1995a). delta-Opioid receptor immunoreactivity: Distribution in brainstem and spinal cord, and relationship to biogenic amines and enkephalin. *Journal of Neuroscience, 15*, 1215–1235.

Arvidsson, U., Riedl, M., Chakrabarti, S., Vulchanova, L., Lee, J. H., Nakano, A. H., Lin, X., Loh, H. H., Law, P. Y., Wessendorf, M. W., et al. (1995b). The kappa-opioid receptor is primarily postsynaptic: Combined immunohistochemical localization of the receptor and endogenous opioids. *Proceedings of the National Academy of Sciences of the United States of America, 92*, 5062–5066.

Basbaum, A. I., & Fields, H. L. (1984). Endogenous pain control systems: Brainstem spinal pathways and endorphin circuitry. *Annual Review of Neuroscience, 7*, 309–338.

Ding, Y. Q., Kaneko, T., Nomura, S., & Mizuno, N. (1996). Immunohistochemical localization of mu-opioid receptors in the central nervous system of the rat. *The Journal of Comparative Neurology, 367*, 375–402.

Gauriau, C., & Bernard, J. F. (2002). Pain pathways and parabrachial circuits in the rat. *Experimental Physiology, 87*, 251–258.

Kieffer, B. L., & Gaveriaux-Ruff, C. (2002). Exploring the opioid system by gene knockout. *Progress in Neurobiology, 66*, 285–306.

Mansour, A., Fox, C. A., Burke, S., Meng, F., Thompson, R. C., Akil, H., & Watson, S. J. (1994). Mu, delta, and kappa opioid receptor mRNA expression in the rat CNS: An in situ hybridization study. *The Journal of Comparative Neurology, 350*, 412–438.

Pasternak, G. W. (1993). Pharmacological mechanisms of opioid analgesics. *Clinical Neuropharmacology, 16*, 1–18.

Pasternak, G. W. (2001). Insights into mu-opioid pharmacology: The role of mu-opioid receptor subtypes. *Life Sciences, 68*, 2213–2219.

Paxinos, G., & Watson, C. (1997). *The rat brain in stereotaxic coordinates*. San Diego: Academic.

Reisine, T., & Pasternak, G. W. (1996). Opioid analgesics and antagonists. In J. G. Hardman & L. E. Limbird (Eds.), *Goodman and Gilman's: The pharmacological basis of therapeutics* (pp. 521–556). New York: McGraw-Hill.

Stein, C., Machelska, H., & Schafer, M. (2001). Peripheral analgesic and anti-inflammatory effects of opioids. *Zeitschrift für Rheumatologie, 60*, 416–424.

Wei, L. N., & Loh, H. H. (2002). Regulation of opioid receptor expression. *Current Opinion in Pharmacology, 2*, 69–75.

Willis, W. D., & Westlund, K. N. (1997). Neuroanatomy of the pain system and of the pathways that modulate pain. *Journal of Clinical Neurophysiology, 14*, 2–31.

Opioid Receptor Recruitment

▶ Opioid Receptor Trafficking in Pain States

Opioid Receptor Redistribution

▶ Opioid Receptor Trafficking in Pain States

Opioid Receptor Sorting

▶ Opioid Receptor Trafficking in Pain States

Opioid Receptor Targeting

▶ Opioid Receptor Trafficking in Pain States

Opioid Receptor Trafficking in Pain States

Alain Beaudet[1], Catherine M. Cahill[2] and Anne Morinville[1]
[1]Montreal Neurological Institute, McGill University, Montreal, QC, Canada
[2]Anesthesiology & Perioperative Care, University of California, Irvine, Irvine, CA, USA

Synonyms

Opioid receptor recruitment; Opioid receptor redistribution; Opioid receptor sorting; Opioid receptor targeting; Trafficking

Definition

Receptor ▶ trafficking is the constitutive or regulated movement of a receptor protein within the cell, whether between subcellular compartments or towards or away from plasma membranes. The presence of opioid receptors (OR) at the cell surface is essential for regulating opioid signal transduction and subsequent cellular functions. Following agonist activation, receptor ▶ internalization plays an important role in cellular responsiveness, by depleting the cell surface of receptors and contributing to the processes of receptor ▶ desensitization and re-sensitization. Likewise, receptor insertion in plasma membranes through recycling of internalized receptors and/or targeting of reserve receptors from intracellular stores is critical for controlling the number of plasma membrane receptors accessible for stimulation and thereby for regulating various neuronal responses including pain transmission.

Characteristics

Background

Three genes have been identified to encode for three opioid receptors (OR): ▶ mu (μ), ▶ delta (δ), and ▶ kappa (κ). All OR inhibit the transmission of pain and are located on peripheral and central branches of somatic and visceral sensory afferents, as well as at various locations within the central nervous system including the spinal cord, midbrain, and cerebral cortex. Within the last few decades, great advances have been made in our understanding of the physiological and pharmacological properties of OR, as well as of their implication in mechanisms of acute and chronic pain. By comparison, much less is known about the mechanisms of OR intraneuronal trafficking, specifically, the physiological function or maladaptive processes that are linked to these trafficking events (for reviews see Roth et al. 1998; von Zastrow et al. 2003).

The first documentation of OR trafficking in models of persistent and chronic pain was generated by studies on axonal transport in peripheral sensory afferents. Thus, autoradiographic and immunohistochemical studies have demonstrated that induction of inflammatory pain via local injection of chemical irritants, antigens, or cytokines resulted in an increase in the synthesis and expression of OR in the cell bodies of sensory afferent neurons (▶ Dorsal Root Ganglia, DRG). This increased expression led, in turn, to enhanced axonal transport of OR in the sciatic nerve and hence to an increase in their density in the inflamed tissue (Hassan et al. 1993; Ji et al. 1995). Ligation of the nerve following induction of tissue inflammation resulted in an accumulation of all three OR in the sciatic nerve, both proximal and distal to site of ligation, indicating both antero- and ▶ retrograde transport of multiple OR. This enhanced transport was postulated to underlie the augmented antinociceptive potency of OR agonists following either spinal or local administration of opiates in models of inflammation (Antonijevic et al. 1995; Cahill et al. 2003). Similarly, μOR were found to be upregulated in DRG ▶ ipsilateral to the site of nerve injury in a model of neuropathic pain (Truong et al. 2003). In this latter study, the OR protein was trafficked to axonal end bulbs of Cajal just proximal to the site of nerve injury, within aberrantly regenerating small axons in the epineurial sheath, and in residual small axons distal to the nerve constriction. These changes resulted in an increase in anti-allodynic and anti-hyperalgesic effects of locally or peripherally applied opioid agonists, suggesting that the accumulation of μOR to sites of neuromas and concomitant disruption of the blood-nerve barrier provided a means for enhanced agonist access to the receptors (Antonijevic et al. 1995).

Cell Trafficking of δOR in Chronic Pain

We postulated that, likewise, changes in receptor trafficking might underlie the enhanced antinociceptive potency of δOR agonists administered intrathecally in rodents with unilateral hind paw inflammation (Hylden et al. 1991; Qiu et al. 2000). To test this hypothesis, we used electron microscopic immunocytochemistry to quantify the ultrastructural distribution of δOR in neurons of

Opioid Receptor Trafficking in Pain States, Fig. 1 Trafficking of δOR to neuronal plasma membranes of postsynaptic neurons in the dorsal spinal cord following induction of peripheral inflammation. Electron microscopic distribution of δOR in dendrites from the ipsilateral dorsal spinal cord of sham-injected (**a**) versus CFA-injected (**b**) rats. Silver-intensified immunogold labeling of δOR demonstrates a predominantly intracellular localization of the receptor in both conditions. However, the number of gold particles is higher over the plasma membrane of dendrites in CFA-treated than in sham-injected animals (*arrows*). (**c**) Quantitative analysis of the subcellular distribution of immunoreactive δOR in dendrites from the lumbar spinal cord of untreated wild-type (WT), CFA-treated WT, and μOR knockout (KO) mice. Data are expressed as the percentage of dendrite-associated gold particles present on plasma membranes. Note that the percentage of membrane-associated δOR is significantly higher in CFA-injected than in sham-treated WT mice and that this effect is absent in CFA-treated μOR KO mice. *D* dendrite, *At* axon terminal. Scale bar = 0.4 μm

the dorsal spinal cord of rats and mice, subjected or not to peripheral inflammation via intraplantar injection of complete Freund's adjuvant (CFA). In non-treated animals, the bulk of δOR immunoreactivity was associated with neuronal cell bodies and ▶ dendrites, in conformity with in situ hybridization and radioligand binding data (Cahill et al. 2001a). Within these structures, only a small proportion of immunoreactive receptors were found on the plasma membrane, the majority being associated with intracellular vesicular stores (Cahill et al. 2001a; Cheng et al. 1997). By contrast, 72 h after CFA injection, the proportion of δOR associated with plasma membranes was significantly increased in both species within dendrites from laminae III–V of the dorsal horn ipsilateral to the side of inflammation, compared to both untreated controls and to the ▶ contralateral spinal cord (Cahill et al. 2003; Morinville et al. 2004a). This change in subcellular compartmentalization was positively correlated with an augmented antinociceptive effect of δOR agonists following spinal administration (Cahill et al. 2003). The increase in δOR plasma membrane density was partly due to recruitment of reserve receptors from intracellular reserve stores, since it was accompanied by a statistical decrease in the mean distance separating intracellular receptors from the closest plasma membrane (Fig. 1).

A similar increase in the targeting of δOR to neuronal plasma membranes had previously been documented by us in rats and mice subjected to sustained treatment with morphine for 48 h (Cahill et al. 2001b; Morinville et al. 2003). Similar to the results obtained in CFA-treated mice, this increase in δOR targeting was not accompanied by corresponding augmentations in either δOR mRNA, protein expression, or $[^{125}I]$-deltorphin binding levels, suggesting that it was not due to enhanced receptor production but solely to increased recruitment to the surface (Morinville et al. 2004b). Most importantly, this targeting effect of morphine was the result of a selective stimulation of μOR, as it was no longer observed in μOR KO mice (Morinville et al. 2003).

To determine whether the CFA-induced targeting of δOR to dendritic membranes was also due to stimulation of μOR, we repeated the CFA experiments in transgenic μOR knockout animals. The CFA-induced recruitment of δOR to neuronal plasma membranes was totally abolished in these animals, suggesting that it was dependent on the activation of μOR through pain-induced release of endogenous μOR-acting peptides (Morinville et al. 2004a).

Functional Significance

In summary, alterations in the subcellular distribution of OR, as the result either of acute or chronic stimulation or of adaptive changes in response to injury, can have dramatic pathophysiological consequences on pain transmission. In the case of δOR, the increase in cell surface receptor density induced by the chronic inflammatory pain process has major pharmacological implications in allowing for increased therapeutic efficacy of analgesic drugs selective for the δOR. In other pain model systems, OR have been demonstrated to be involved in activity-dependent synaptic plasticity, a process known to be fundamental to the development of chronic pain states (Woolf and Salter 2000). Studying phenomena of receptor trafficking and internalization will hopefully provide another venue that allows either better diagnostic or alternative novel, strategies for the treatment of chronic pain syndromes.

References

Antonijevic, I., Mousa, S. A., Schafer, M., & Stein, C. (1995). Perineurial defect and peripheral opioid analgesia in inflammation. *Journal of Neuroscience, 15*, 165–172.

Cahill, C. M., McClellan, K. A., Morinville, A., Hoffert, C., Hubatsch, D., O'Donnell, D., O'Donnell, D., & Beaudet, A. (2001a). Immunocytochemical distribution of delta opioid receptors in rat brain: Antigen-specific sub-cellular compartmentalization. *The Journal of Comparative Neurology, 440*, 65–84.

Cahill, C. M., Morinville, A., Lee, M.-C., Vincent, J. P., & Beaudet, A. (2001b). Targeting of delta opioid receptor to the plasma membrane following chronic morphine treatment. *Journal of Neuroscience, 21*, 7598–7607.

Cahill, C. M., Morinville, A., Hoffert, C., Hubatsch, D., O'Donnell, D., & Beaudet, A. (2003). Up-regulation and trafficking of δOR in a model of chronic inflammatory pain; Positive correlation with enhanced D-[Ala2]deltorphin-induced antinociception. *Pain, 101*, 199–208.

Cheng, P. Y., Liu-Chen, L. Y., & Pickel, V. M. (1997). Dual ultrastructural immunocytochemical labeling of mu and delta opioid receptors in the superficial layers of the rat cervical spinal cord. *Brain Research, 778*, 367–380.

Hassan, A. H., Ableitner, A., Stein, C., & Herz, A. (1993). Inflammation of the rat paw enhances axonal transport of opioid receptors in the sciatic nerve and increases their density in the inflamed tissue. *Neuroscience, 55*, 185–195.

Hylden, J. L., Thomas, D. A., Iadarola, M. J., Nahin, R. L., & Dubner, R. (1991). Spinal opioid analgesic effects are enhanced in a model of unilateral inflammation/hyperalgesia: Possible involvement of noradrenergic mechanisms. *European Journal of Pharmacology, 194*, 135–143.

Ji, R. R., Zhang, Q., Law, P. Y., Loh, H. H., Elde, R., & HÅkfelt, T. (1995). Expression of mu-, delta-, and kappa-opioid receptor-like immunoreactivities in rat dorsal root ganglia after carrageenan-induced inflammation. *Journal of Neuroscience, 15*, 8156–8166.

Morinville, A., Cahill, C. M., Esdaile, M. J., Aibak, H., Collier, B., Kieffer, B. L., & Beaudet, A. (2003). Regulation of delta opioid receptor trafficking through mu-opioid receptor stimulation: Evidence from mu-opioid receptor knock-out mice. *Journal of Neuroscience, 23*, 4888–4898.

Morinville, A., Cahill, C. M., Kieffer, B., Collier, B., & Beaudet, A. (2004a). Mu opioid receptor knockout prevents changes in delta-opioid receptor trafficking induced by chronic inflammatory pain. *Pain, 109*, 266–273.

Morinville, A., Cahill, C. M., Aibak, H., et al. (2004b). Morphine-induced changes in delta opioid receptor trafficking are linked to somatosensory processing in the rat spinal cord. *Journal of Neuroscience, 24*, 5549–5559.

Qiu, C., Sora, I., Ren, K., Uhl, G., & Dubner, R. (2000). Enhanced delta-opioid receptor-mediated antinociception in mu-opioid receptor-deficient mice. *European Journal of Pharmacology, 387*, 163–169.

Roth, B. L., Willins, D. L., & Kroeze, W. K. (1998). G Protein-Coupled Receptor (GPCR) trafficking in the central nervous system: Relevance for drugs of abuse. *Drug and Alcohol Dependence, 51*, 73–85.

Truong, W., Cheng, C., Xu, Q. G., Li, X. Q., & Zochodne, D. W. (2003). Mu opioid receptors and analgesia at the site of a peripheral nerve injury. *Annals of Neurology, 53*, 366–375.

von Zastrow, M., Svingos, A., Haberstock-Debic, H., & Evans, C. (2003). Regulated endocytosis of opioid receptors: Cellular mechanisms and proposed roles in physiological adaptation to opiate drugs. *Current Opinion in Neurobiology, 13*, 348–353.

Woolf, C. J., & Salter, M. W. (2000). Neuronal plasticity: Increasing the gain in pain. *Science, 288*, 1765–1769.

Opioid Receptors

Gavril W. Pasternak
Laboratory of Molecular Neuropharmacology,
Memorial Sloan-Kettering Cancer Center,
New York, USA

Synonyms

DOR-1; KOR-1; MOR-1; Opiate receptors

Definition

Opioid receptors are proteins located on the surface of cells that bind opiate and opioid peptide drugs selectively and with high affinity.

Characteristics

Opioids act by interacting with specific recognition sites on the surface of the cell termed receptors. The concept of opiate receptors goes back over 50 years. The first opioids, morphine and codeine, were isolated from opium over a hundred years ago, which was followed by the synthesis of thousands of derivatives in an effort to dissociate analgesia from side effects. The rigid structure-activity relationships of these compounds led to the suggestion of very specific binding sites. Indeed, these structure-activity relationships led to models of the binding pocket decades before the receptors were cloned. The classification of these receptors became more complex with the observations of the interactions between nalorphine and morphine. Differing only by the substitution of the N-methyl group by an N-allyl group, the two drugs were quite distinct pharmacologically. Whereas morphine is an effective analgesic, nalorphine was one of the first antagonists. However, if given at sufficiently high doses, nalorphine was also able to elicit analgesia, leading Martin to propose two subtypes of opioid receptors. He suggested that nalorphine was an antagonist at the M, or morphine receptor, and an agonist at the N, or nalorphine receptor. Martin extended this classification based upon detailed pharmacological observations in vivo, to propose specific receptors for morphine (mu) and for ketocyclazocine (kappa) (Martin et al. 1976). The subsequent discovery of the endogenous opioid peptides (Table 1) led Kosterlitz and co-workers to propose an ▶ enkephalin-selective receptor, termed delta (Lord et al. 1977) (▶ Delta(δ) Opioid Receptor(s)). Goldstein and colleagues then identified dynorphin A as the endogenous ligand for the kappa receptor

Opioid Receptors, Table 1 Selected opioid peptides

Opioid peptide	Amino acid sequence
[Leu^5]enkephalin	Tyr-Gly-Gly-Phe-Leu
[Met^5] enkephalin	**Tyr-Gly-Gly-Phe-Met**
Dynorphin A	**Tyr-Gly-Gly-Phe-Leu**-Arg-Arg-Ile-Arg–Pro-Lys-Leu-Lys-Trp-Asp-Asn-Gln
Dynorphin B	**Tyr-Gly-Gly-Phe-Leu**-Arg-Arg-Gln-Phe-Lys-Val-Val-Thr
α-Neoendorphin	**Tyr-Gly-Gly-Phe-Leu**-Arg-Lys-Tyr-Pro-Lys
β$_h$-Endorphin	**Tyr-Gly-Gly-Phe-Met**-Thr-Ser-Glu-Lys-Ser-Gln-Thr-Pro-Leu-Val-Thr-Leu-Phe-Lys-Asn-Ala-Ile-Ile-Lys-Asn-Ala-Tyr-Lys-Lys-Gly-Glu
Endomorphin-1	Tyr-Pro-Trp-Phe-NH$_2$
Endomorphin-2	Tyr-Pro-Phe-Phe-NH$_2$

(▶ Kappa-Opioid Receptors). Pharmacological studies have suggested these three major families of receptors, as well as subtypes (Table 2).

Opiate-Receptor Binding Sites

The initial concept of opioid receptors and the suggestion of receptor classes had been inferred by classical pharmacological studies. Receptor binding studies provided the verification of these sites at the biochemical level. Opioid binding sites within the nervous system were first reported in 1973 using mu ligands (reviewed in Snyder and Pasternak 2003). In these initial studies, the receptors were labeled by incubating brain membranes with radiolabeled opioids of high specific activity and washing away the unbound drug. Specificity of the labeled sites was established by demonstrating that only active opioids completed the binding. A number of studies then explored the nature of the binding, demonstrating its protein nature and differences in the binding of agonists and antagonists. As the kappa and delta classes of opioid receptors were described and suitable radioligands became available, they, too, were verified in binding studies.

During the course of these studies exploring the binding of a variety of opioids, evidence surfaced for subpopulations of sites, starting with mu receptors (Wolozin and Pasternak 1981;

Opioid Receptors, Table 2 Pharmacological classification of opioid receptors and their actions

Receptor	Clone	Actions
Mu	MOR-1	Sedation
Mu$_1$		Supraspinal and peripheral analgesia
		Prolactin release; feeding
Mu$_2$		Spinal analgesia; respiratory depression
		Inhibition of gastrointestinal transit
		Guinea pig ileum bioassay; feeding
Kappa		
Kappa$_1$	KOR-1	Analgesia; dysphoria; diuresis; feeding
Kappa$_2$	(KOR-1/DOR-1 dimer)	Unknown
Kappa$_3$		Analgesia
Delta	DOR-1	Mouse vas deferens bioassay, feeding
Delta$_1$		Supraspinal analgesia
Delta$_2$		Spinal and supraspinal analgesia

Some actions assigned to general families have not yet been correlated with a specific subtype. Some pharmacologically defined subtypes have not been correlated with specific clones

Pasternak and Snyder 1975). Early binding studies identified a very high-affinity binding component, which was subsequently found to be uniquely sensitive to the antagonists naloxonazine and naloxazone. Under conditions in vivo, in which this high-affinity binding site was blocked, naloxonazine was able to antagonize selected mu actions, such as supraspinal analgesia, but not others, such as respiratory depression or the inhibition of gastrointestinal transit. The mu binding and mu actions sensitive to naloxonazine were termed mu$_1$, and those mu sites and functions that were insensitive were termed mu$_2$ (Wolozin and Pasternak 1981). The complexity of mu subtypes increased with studies of the morphine metabolite morphine-6β-glucuronide, a highly selective mu opioid with a pharmacology distinct from that of morphine (Pasternak 2001a, b).

The suggestion of receptor subtypes was not limited to mu receptors. Using selective delta ligands, investigators also demonstrated differences suggesting the existence of delta receptor subtypes (Mattia et al. 1991). Kappa receptor subtypes were proposed by several groups. These distinctions were based in a large part upon the kappa-selective drug U50,488H. First, the kappa receptors were separated into U50,488H-sensitive and U50,488H-insensitive sites (Zukin et al. 1988) The U50,488H-sensitive sites were designated as kappa$_1$. These kappa$_1$ receptors were further subdivided by their sensitivity towards several endogenous opioids, α-neoendorphin and dynorphin B (Clark et al. 1989). Dynorphin A is the endogenous ligand for kappa$_1$ and competes at all kappa$_1$ sites with high affinity. In contrast, there are subpopulations of kappa$_1$ binding sites that are sensitive to dynorphin B and α-neoendorphin and those that are not. Two subdivisions of the U50,488H-insensitive sites have also been suggested. The kappa$_2$ receptors were defined by the benzomorphan ethyl ketocyclazocine and appear to consist of a heterodimer of the delta and kappa$_1$ receptors DOR-1 and KOR-1, while the kappa$_3$ receptor was defined in both binding studies and pharmacologically using the novel opioid naloxone benzoylhydrazone. Recent work indicates that kappa$_3$ sites contain novel, truncated six transmembrane domain splice variants of the mu receptor (MOR-1) gene (Majumdar et al. 2012). Similar conclusions regarding four subdivisions of kappa receptors were proposed detailed computer modeling (Rothman et al. 1992).

Molecular Biology of Opioid Receptors

The various opioid receptor classes were initially defined pharmacologically and in receptor binding assays. It took almost 20 years for the receptors to be cloned (Uhl et al. 1994). The delta receptor (▶ DOR-1) was the first one to be cloned (Evans et al. 1992; Kieffer et al. 1992), followed quickly by mu (▶ MOR-1) and kappa$_1$ (▶ KOR-1) receptors. All the opioid receptors are traditional G-protein-coupled receptors, with seven transmembrane domains, an extracellular amino (N) terminus, and an intracellular carboxy (C) terminus (Fig. 1). The three clones predicted very similar proteins, particularly within the

Opioid Receptors, Fig. 1 Schematic of the opioid receptors. Like traditional G-protein-coupled receptors, the opioid receptors have seven transmembrane (*TM*) domains. DOR-1 and KOR-1 both have three exons that encode the receptor protein. The first TM (*green*) is encoded by the first translated exon, the next three TM (*orange*) by the second, and the last three TM (*blue*) by the third translated exon. Although MOR-1 has this same exon structure from the extracellular N terminus to the seven TM regions, MOR-1 has an additional exon that encodes the terminal 12 amino acids in the C terminus or alternative splicing at this position (*red*). The structure of MOR-1 is complex with extensive 3′ and 5′ splicing

transmembrane domains. DOR-1 and KOR-1 are comprised of three exons that encode the full protein, with the first encoding the NH$_2$ terminus through the first transmembrane domain, the second encoding the next three transmembrane domains, and the third encoding the last three transmembrane domains and the COOH terminus. MOR-1 is similar in that there are three exons that encode the NH$_2$ terminus, all seven transmembrane domains, and most of the intracellular COOH terminus. However, it differs from the others in that there are 12 additional amino acids at the tip of the COOH terminus encoded by exon 4. The binding pocket is within the membrane and is comprised of the seven transmembrane domains of the receptor. When expressed, each receptor showed the anticipated selectivity in receptor binding assays and sensitivity in functional assays when looking at their ability to activate G proteins.

Pharmacological studies strongly implied the existence of subtypes of the receptors. With

Opioid Receptors,
Fig. 2 Schematic of
selected human MOR-1
variants. MOR-1 undergoes
alternative splicing at the
C terminus located inside
the cell and at the
N terminus. The exons
responsible for encoding
the terminal amino acids
and its sequence are
presented

Mouse MOR-1 3'-Splice variants

	Exons	C-Terminus Amino Acid Sequence
MOR-1	4	LENLEAETAPLP
MOR-1A		VRSL
MOR-1B	5	KIDLF
MOR-1C	7,8,9	PTLAVSVAQIFTGYPSPTHV EKPCKSCMDRGMRNLLPD DGPRQESGEGOLGR
MOR-1D	8,9	RNEEPSS
MOR-1E	6,7,8,9	KKKLDSQRGCVQHPV
MOR-1F	10,6,7,8,9	APCACVPGANRGQTKASDL LDLELETVGSHQADAETNP GPYEGSKCAEPLAISLVPLY

regard to mu receptors, this concept is supported
with the isolation of a number of different splice
variants of MOR-1 in mice, rats, and humans
(Pasternak and Pan, 2013). There is both 5' and
3' splicing of MOR-1 that generates three general
classes of splice variants. The predominant splice
variants are full-length ones that contain exon 1
and all seven transmembrane domains, with
differing C termini due to 3' splicing (Fig. 2).
A second set of variants produced through
a different promoter lack exon 1 due to 5' splicing
and thus generate only a truncated version of
MOR-1 that lacks the first transmembrane
domain and contains only the remaining six.
The importance of these variants has been

established through knockout mouse models
(see below). All the full-length variants contain
the same first three exons that encode transmembrane domains that define the binding pocket.
Thus, it is not surprising that they all show high
affinity and selectivity for mu opioids. However,
functional studies have shown differences in the
sensitivity of these variants to different opioid
ligands. Thus, the concept of multiple ▶ mu (μ)
opioid receptor has been confirmed at the
molecular level, although the relationship
between these variants and the pharmacologically defined subtypes is still not certain.

The situation is less clear with the delta
receptor subtypes. Although DOR-1 undergoes

splicing, no functional splice variants have been uncovered. KOR-1 corresponds to the kappa$_1$ subtype. Again, despite the presence of alternative splicing, there are no additional functional KOR-1 variants. There is evidence to suggest that kappa$_2$ receptors are generated by a dimer of KOR-1/DOR-1. The kappa$_3$ receptors were originally proposed based upon receptor binding studies, but more recent work indicates that they are comprised, in part, by the truncated six transmembrane domain variants.

Functional Studies

The overall pharmacology of these agents is quite complex (Reisine and Pasternak 1996; Pasternak and Pan, 2013). Highly selective agonists and antagonists are available for each receptor family, permitting their evaluation pharmacologically. All the opioid receptor classes have been implicated in analgesia, with many involved with other actions as well. With mu opioids, these actions include respiratory depression, sedation, and the inhibition of gastrointestinal transit. The actions of the kappa and delta drugs are less well studied clinically.

Analgesia results from the activation of opioid receptors in the brain, spinal cord, or periphery. However, the simultaneous activation of more than one site, such as when the drugs are given systemically, leads to more profound responses due to synergistic interactions among the sites. This was first demonstrated between the brain and the spinal cord and has been extended to peripheral sites as well. Chronic administration of opioids leads to a progressive decline in activity or ▸ tolerance. Drugs within the same class show cross-tolerance, but agents from different classes do not. ▸ Dependence is also seen with chronic administration. Unlike ▸ addiction, which implies drug-seeking behaviors, dependence is a physiological response to the chronic administration of the agents and is observed with all subjects.

Correlating these opioid actions with the cloned receptors has been a major goal. Shortly after the initial cloning of MOR-1, antisense approaches demonstrated the importance of the MOR-1 gene in morphine analgesia. This was further confirmed using knockout models (▸ Knockout Mice). Several different animal models were generated that were deficient in MOR-1. All the animal models targeting exon 1 or exon 2 revealed the absence of morphine analgesia as well as other morphine effects. However, one of these animal models was quite unusual. Despite the disruption of exon 1, there was a persistent expression of the exon 11-associated variants that included the 6-TM truncated variants. While morphine was inactive in this animal model, some mu opioids, including morphine-6β-glucuronide, fentanyl, and heroin, retained their analgesic activity. More recently, a new knockout with a deletion of exon 11 was reported (Pan et al. 2009; Majumdar et al. 2012). In this animal, the full-length MOR-1 variants were still expressed, explaining why morphine analgesia was unaffected. However, the analgesic activity of a number of other mu agents, including heroin, was diminished, as were drugs acting through kappa3 sites.

Knockout models have also been evaluated for delta (DOR-1) and kappa$_1$ (KOR-1) receptors. The loss of the receptors in both cases led to the loss of activity of their respective classes of drugs. Together, these experiments confirm the importance of each of these receptors in the analgesic actions of the drugs involved.

Conclusion

The concept of opiate receptors goes back many years. Initial studies, based upon traditional pharmacological approaches, implied the existence of several families of receptors and subtypes which have now been confirmed with the cloning of these receptors. The concept of subtypes of mu receptors is supported by the identification of splice variants of MOR-1.

Acknowledgement This work is supported by grants (DA02615, DA07242 and DA06241) from the National Institute of Drug Abuse of the National Institutes of Health.

Cross-References

▶ Opioids in the Periphery and Analgesia
▶ Opioids in the Spinal Cord and Modulation of Ascending Pathways (N. gracilis)
▶ Postoperative Pain, Opioids

References

Clark, J. A., Liu, L., Price, M., Hersh, B., Edelson, M., & Pasternak, G. W. (1989). Kappa opiate receptor multiplicity: Evidence for two U50,488-sensitive kappa$_1$ subtypes and a novel kappa$_3$ subtype. *Journal of Pharmacology and Experimental Therapeutics, 251*, 461–468.

Evans, C. J., Keith, D. E., Jr., Morrison, H., Magendzo, K., & Edwards, R. H. (1992). Cloning of a delta opioid receptor by functional expression. *Science, 258*, 1952–1955.

Kieffer, B. L., Befort, K., Gaveriaux-Ruff, C., & Hirth, C. G. (1992). The Î´-opioid receptor: Isolation of a cDNA by expression cloning and pharmacological characterization. *Proceedings of the National Academy of Sciences USA, 89*, 12048–12052.

Lord, J. A. H., Waterfield, A. A., Hughes, J., & Kosterlitz, H. W. (1977). Endogenous opioid peptides: Multiple agonists and receptors. *Nature, 267*, 495499.

Majumdar, S., Grinnell, S., LeRouzic, V., Burgman, M., Polikar, L., Ansonoff, M., Pintar, J., Pan, Y. X., & Pasternak, G. W. (2012). Truncated G- protein-coupled mu opioid receptor MOR-1 splice variants are targets for highly potent opioid analgesics lacking side effects. *Proceedings of the National Academy of Sciences USA, 108*, 19776–19783.

Martin, W. R., Eades, C. G., Thompson, J. A., Huppler, R. E., & Gilbert, P. E. (1976). The effects of morphine and nalorphine-like drugs in the nondependent and morphine-dependent chronic spinal dog. *Journal of Pharmacology and Experimental Therapeutics, 197*, 517–532.

Mattia, A., Vanderah, T., Mosberg, H. I., & Porreca, F. (1991). Lack of antinociceptive cross tolerance between [D-Pen2, D-Pen5]enkephalin and [D-Ala2] deltorphin II in mice: Evidence for delta receptor subtypes. *Journal of Pharmacology and Experimental Therapeutics, 258*, 583–587.

Pan, Y. X., Xu, J., Xu, M., Rossi, G. C., Matulonis, J. E., & Pasternak, G. W. (2009). Involvement of exon 11-associated variants of the mu opioid receptor MOR-1 in heroin, but not morphine actions. *Proceedings of the National Academy of Sciences USA, 106*, 4917–4922.

Pasternak, G. W. (2001a). Incomplete cross tolerance and multiple mu opioid peptide receptors. *Trends in Pharmacological Sciences, 22*, 67–70.

Pasternak, G. W. (2001b). Insights into mu opioid pharmacology – The role of mu opioid receptor subtypes. *Life Sciences, 68*, 2213–2219.

Pasternak, G.W., & Pan, Y.X. (2013). Mu opioids and their receptors. *Pharmacological Reviews*. submitted.

Pasternak, G. W., & Snyder, S. H. (1975). Identification of a novel high affinity opiate receptor binding in rat brain. *Nature, 253*, 563–565.

Reisine, T., & Pasternak, G. W. (1996). Opioid analgesics and antagonists. In J. G. Hardman & L. E. Limbird (Eds.), *Goodman & Gilman's the pharmacological basis of therapeutics* (pp. 521–556). New York: McGraw-Hill.

Rothman, R. B., Bykov, V., Xue, B. G., Xu, H., De Costa, B. R., Jacobson, A. E., Rice, K. C., Kleinman, J. E., & Brady, L. S. (1992). Interaction of opioid peptides and other drugs with multiple kappa receptors in rat and human brain. Evidence for species differences. *Peptides, 13*, 977–987.

Snyder, S. H., & Pasternak, G. W. (2003). Historical review: Opioid receptors. *Trends in Pharmacological Sciences, 24*, 198–205.

Uhl, G. R., Childers, S., & Pasternak, G. W. (1994). An opiate-receptor gene family reunion. *Trends in Neurosciences, 17*, 89–93.

Wolozin, B. L., & Pasternak, G. W. (1981). Classification of multiple morphine and enkephalin binding sites in the central nervous system. *Proceedings of the National Academy of Sciences USA, 78*, 6181–6185.

Zukin, R. S., Eghbali, M., Olive, D., Unterwald, E., & Tempel, A. (1988). Characterization and visualization of rat and guinea pig brain kappa opioid receptors: Evidence for kappa$_1$ and kappa$_2$ opioid receptors. *Proceedings of the National Academy of Sciences USA, 85*, 4061–4065.

Opioid Receptors at Postsynaptic Sites

Andrew J. Todd
Institute of Neuroscience and Psychology, College of Medical, Veterinary and Life Sciences, University of Glasgow, Glasgow, UK

Synonyms

Postsynaptic opioid receptors

Definition

Postsynaptic opioid receptors are those that are located on the cell bodies and dendrites of neurons, rather than on their axon terminals.

Opioids administered to the spinal cord can produce analgesia, and part of this effect is mediated by opioid receptors on the terminals of ▶ nociceptive primary afferents. Activation of these receptors reduces the release of excitatory neurotransmitter from the nociceptive afferents, and this is referred to as a presynaptic action. However, opioid receptors are also present on certain neurons in the spinal cord (postsynaptic opioid receptors), and these are thought to contribute to opioid analgesia.

Characteristics

Anatomical Distribution of Opioid Receptors in the Spinal Cord

The distribution of opioid receptors in the spinal cord was initially investigated by using ▶ radioligand binding. Numerous studies were carried out with this method, and these showed that the highest concentration of receptors was in the superficial part of the dorsal horn (▶ Rexed's Laminae I and II). Transection of dorsal roots led to a substantial reduction of this opioid binding, presumably due to loss of presynaptic receptors on primary afferent terminals. However, significant binding was still seen in the superficial dorsal horn after rhizotomy, and this was thought to result from the presence of opioid receptors on spinal cord neurons.

The cloning of μ, δ, and κ opioid receptors meant that antibodies could be raised against them and used to investigate their distribution in the spinal cord by means of ▶ immunocytochemistry (Arvidsson et al. 1995a, b, c; Kemp et al. 1996; Spike et al. 2002; Chen and Marvizon 2009). With this approach, all 3 receptors have been found to be concentrated in laminae I and II of the dorsal horn, which matches the results obtained with radioligand binding. Immunocytochemical studies have shown that μ and κ receptors are present on both dorsal horn neurons (postsynaptic opioid receptors) and primary afferents (presynaptic receptors) (Arvidsson et al. 1995b, c; Kemp et al. 1996; Spike et al. 2002; Chen and Marvizon 2009), while δ receptors in the dorsal horn appear to be restricted to the terminals of primary afferents (Arvidsson et al. 1995b). Relatively little is known about postsynaptic κ opioid receptors except that they have been seen on a few neuronal cell bodies in laminae I and II and that these cells are associated with axons that express either enkephalin or dynorphin (Arvidsson et al., 1995c). In contrast, several studies have demonstrated μ opioid receptors on dorsal horn neurons. Since μ opioid receptors play a fundamental role in opioid analgesia and more is known about their distribution, they will be the main focus of this entry. Most of the studies on μ opioid receptors have been carried out with an antibody raised against the first identified form of the protein (MOR1), and these have shown that numerous immunoreactive cells are present in the superficial part of the dorsal horn (Arvidsson et al. 1995b; Kemp et al. 1996; Spike et al. 2002; Chen and Marvizon 2009) (Fig. 1). MOR1 immunoreactivity is present on small neurons that are generally located in lamina II, where they make up approximately 10 % of the neuronal population (Spike et al. 2002). As very few ▶ projection neurons are seen in this lamina (Todd 2010), it is likely that the cells with MOR1 are interneurons, with axons that terminate locally in the spinal cord. Lamina II contains both excitatory (▶ Glutamatergic) and inhibitory (▶ GABAergic) interneurons (Todd 2010; Yasaka et al. 2010). The great majority of cells with MOR1 immunoreactivity do not contain GABA (Kemp et al. 1996), and it is therefore likely that most of them are excitatory interneurons. The responses of MOR1-expressing neurons in lamina II have been investigated by examining activation of the immediate early gene ▶ c-fos after different forms of noxious stimulation (Spike et al. 2002). It was found that although approximately 15 % of these cells upregulated *c-fos* in response to noxious thermal stimulation, very few did so following mechanical or chemical noxious stimuli. This suggests that at least some of the MOR1-expressing neurons in lamina II of the spinal dorsal horn may be selectively activated by noxious thermal stimuli.

Although most immunocytochemical studies have used antibodies raised against MOR1, it is

Opioid Receptors at Postsynaptic Sites, Fig. 1 The appearance of MOR1 in the rat spinal cord revealed with immunocytochemistry. (**a**) In a transverse section of the spinal cord, MOR1 immunoreactivity is seen to be highly concentrated in the upper part of the dorsal horn (laminae I and II). (**b**) At higher magnification it is possible to see several MOR1-immunoreactive cell bodies (some of which are marked with *arrows*). (**c**) This image shows part of a longitudinal (*parasagittal*) section through the spinal cord. The cell body (*arrowhead*) and dendrites (*arrows*) of a MOR1-immunoreactive neuron can be seen. Scale bars: a = 500 μm; b, c = 20 μm (Reprinted from Neuroscience, 75, Kemp et al. The μ-opioid receptor (MOR-1) is mainly restricted to neurons that do not contain GABA or glycine in the superficial dorsal horn of the rat spinal cord, pp 1,231–1,238, Copyright (1996), with permission from Elsevier)

known that there are other ▶ splice variants of the μ opioid receptor, and two of these, known as MOR1C and MOR1D, have been identified in spinal cord of both rat and mouse (Abbadie et al. 2000). Like MOR1 itself, these splice variants were both present at highest concentrations in the superficial laminae, but significant amounts of MOR1C were also observed near the central canal. MOR1C (like MOR1) was found on cell bodies in lamina II and was also present on primary afferents. In contrast, MOR1D immunoreactivity was seen on neurons in lamina I, but not on primary afferents (Abbadie et al. 2000). Lamina I contains many projection neurons, and this therefore raises the possibility that some of these might express MOR1D, which would allow a direct inhibitory action of μ opioid agonists on the output cells of the dorsal horn.

However, it has since been reported that the MOR1C and MOR1D splice variants do not occur in the rat, raising the possibility that the immunostaining represents a cross-reaction with unrelated proteins (Schnell and Wessendorf 2009).

In situ hybridization histochemistry has also been used to investigate the locations of cells that express the 3 opioid receptors (Maekawa et al. 1994; Mansour et al. 1994; Schafer et al. 1994); however, the results with this approach have been inconsistent. For example, one study reported that neurons with mRNA for the μ receptor are present throughout the dorsal horn, with the highest levels in the superficial laminae (Maekawa et al. 1994), while another found that they were present in the deeper laminae of the dorsal horn with few in the superficial part

(Mansour et al. 1994). Neurons with mRNA for the κ receptor have been seen concentrated in the superficial laminae (Schafer et al. 1994), throughout the spinal gray matter (Maekawa et al. 1994), and in the deeper parts of the dorsal horn but not the superficial part (Mansour et al. 1994). The reasons for these discrepancies are unknown.

Responses of Dorsal Horn Neurons to Drugs that Act on the μ Opioid Receptor

The actions of opioid agonists on neurons in the superficial dorsal horn have been studied in vitro in slices of spinal cord and brainstem. Yoshimura and North (1983) demonstrated that ▶ opioid peptides could hyperpolarize many neurons in lamina II of the spinal cord and that this effect was mediated by an increase in K$^+$ conductance. Subsequent studies have shown that this involves μ (rather than δ) opioid receptors (Jeftinija 1988; Grudt and Williams 1994; Schneider et al. 1998), which is consistent with anatomical observations that μ, but not δ, receptors are present on dorsal horn neurons. Hyperpolarization resulting from activation of μ opioid receptors will have an inhibitory effect on these neurons. This, taken together with the fact that many of the lamina II cells with MOR1 are excitatory (Kemp et al. 1996) and that at least some respond to noxious stimuli (Spike et al. 2002), has led to the suggestion that excitatory interneurons with μ opioid receptors in lamina II normally play a role in conveying nociceptive information to projection cells in other laminae of the dorsal horn. Inhibition of these interneurons by μ opioid agonists would therefore reduce the input to projection neurons following noxious stimuli, and this would result in analgesia (reduced perception of pain).

Although the hyperpolarizing effect of μ opioid agonists on lamina II neurons appears at first sight to be compatible with the results of anatomical studies, there are certain discrepancies. Firstly, while only 10 % of lamina II neurons are MOR1 immunoreactive (Spike et al. 2002), effects of μ agonists are seen on a much higher proportion of neurons. The proportion of lamina II neurons hyperpolarized by μ agonists has been variously estimated as 39 % (Schneider et al. 1998), 50 % (Yoshimura and North 1983), or 75 % (Jeftinija 1988) for the rat spinal cord, 60 % for hamster spinal cord (Schneider et al. 1998), and 86–90 % for the trigeminal nucleus of rat and guinea pig (Grudt and Williams 1994). While sampling bias in recording experiments may contribute to this effect, it is unlikely to explain such a dramatic difference. It is possible that MOR1C is present on some neurons that are not MOR1 immunoreactive, and this could increase the proportion of cells with functional μ opioid receptors (but see above). Secondly, a "postsynaptic" action of μ agonists has been reported on many cells in deeper laminae of the dorsal horn, in areas where radioligand binding and immunocytochemistry both suggest that there are only low levels of opioid receptors.

Internalization of MOR1 as a Means of Studying Its Activation

Activation of receptors on the surface of neurons sometimes causes them to leave the plasma membrane and enter the cytoplasm on small membrane-bound vesicles (endosomes). Since internalized receptors can be identified with immunocytochemistry, this phenomenon can be used to investigate various aspects of receptor activation, for example, dose-response relationships and time course. It has been found that various μ-selective agonists cause internalization of MOR1 on lamina II neurons, although morphine did not produce this effect (Trafton et al. 2000). The doses of μ agonists that were needed to cause MOR1 internalization were similar to those that have been shown to produce hyperpolarization of lamina II neurons (Grudt and Williams 1994) and also to doses that produced analgesia in behavioral tests (Trafton et al. 2000). Although this does not prove that postsynaptic actions of μ opioid drugs contribute to spinal opioid analgesia, they are compatible with this suggestion. Interestingly, noxious stimulation did not evoke MOR1 internalization, either in normal animals or in those with a chronic inflammation (Trafton et al. 2000), and this was taken as evidence that endogenous opioid peptides that act on μ receptors are not released in sufficient amounts to activate this system following noxious stimulation. However, it is thought that the failure to see internalization of opioid receptors may have resulted from rapid breakdown of the opioid peptides, and MOR1 internalization has been observed following noxious mechanical stimulation and acute (but not chronic) inflammation of the hindlimb when peptidase inhibitors were administered into the spinal subarachnoid space (Lao et al 2008; Chen and Marvizon 2009). It has also been shown that direct electrical stimulation of the dorsal horn in vitro can cause internalization of MOR1 in lamina II neurons, presumably due to release of opioid peptides by local neurons (Song and Marvizon 2003).

Opioid Receptors and Itch

Opioid receptors have complex roles in itch, since μ opioid agonists can induce itching, while κ agonists can inhibit itch. The itch-inducing action of μ agonists is thought to involve receptors in the spinal dorsal horn (Cevikbas et al 2010), although it is not known whether these are located at pre- and/or postsynaptic sites. The mechanism (and site) of the antipruritic action of κ agonists is not yet known. It has recently been suggested that the itch-inducing action of μ opioid agonists is mediated through the MOR1D splice variant in mice and that this results from the formation of heterodimers of MOR1D with the gastrin-releasing peptide receptor (GRPR) on neurons in lamina I of the spinal dorsal horn (Liu et al 2011).

References

Abbadie, C., Pan, Y., Drake, C. T., et al. (2000). Comparative immunohistochemical distributions of carboxy terminus epitopes from the Î¼-opioid receptor splice variants MOR-1D, MOR-1 and MOR-1C in the mouse and Rat CNS. *Neuroscience, 100*, 141–153.

Arvidsson, U., Dado, R. J., Riedl, M., et al. (1995a). δ-opioid receptor immunoreactivity: Distribution in brainstem and spinal cord, and relationship to biogenic amines and enkephalin. *Journal of Neuroscience, 15*, 1215–1235.

Arvidsson, U., Riedl, M., Chakrabarti, S., et al. (1995b). Distribution and targeting of a μ-opioid receptor (MOR1) in brain and spinal cord. *Journal of Neuroscience, 15*, 3328–3341.

Arvidsson, U., Riedl, M., Chakrabarti, S., Vulchanova, L., Lee, J. H., Nakano, A. H., Lin, X., Loh, H. H., Law, P. Y., Wessendorf, M. W., & Elde, R. (1995c). The κ-opioid receptor is primarily postsynaptic: Combined immunohistochemical localization of the receptor and endogenous opioids. *Proceedings of the National Academy of Sciences of the United States of America, 92*, 5062–5066.

Cevikbas, F., Steinhoff, M., & Ikoma, A. (2010). Role of spinal neurotransmitter receptors in itch: New insights into therapies and drug development. *CNS Neuroscience and Therapeutics, 17*, 742–749.

Chen, W., & Marvizon, J. C. G. (2009). Acute inflammation induces segmental, bilateral and supraspinally mediated opioid release in the rat spinal cord, as measured by μ-opioid receptor internalization. *Neuroscience, 161*, 157–172.

Grudt, T. J., & Williams, J. T. (1994). μ-opioid agonists inhibit spinal trigeminal substantia gelatinosa neurons in guinea pig and rat. *Journal of Neuroscience, 14*, 1646–1654.

Jeftinija, S. (1988). Enkephalins modulate excitatory synaptic transmission in the superficial dorsal horn by acting at Î¼-opioid receptor sites. *Brain Research, 460*, 260–268.

Kemp, T., Spike, R. C., Watt, C., et al. (1996). The Î¼-opioid receptor (MOR1) is mainly restricted to neurons that do not contain GABA or Glycine in the superficial dorsal horn of the rat spinal cord. *Neuroscience, 75*, 1231–1238.

Lao, L., Song, B., Chen, W., & Marvizon, J. C. (2008). Noxious mechanical stimulation evokes the segmental release of opioid peptides that induce mu-opioid receptor internalization in the presence of peptidase inhibitors. *Brain Research, 1197*, 85–93.

Liu, X. Y., Liu, Z.-C., Sun, Y.-G., Ross, M., Kim, S., Tsai, F.-F., et al. (2011). Unidirectional cross-activation of GRPR by MOR1D uncouples itch and analgesia induced by opioids. *Cell, 147*, 447–458.

Maekawa, K., Minami, M., Yabuuchi, K., et al. (1994). In situ hybridization study of Î¼- and Î°-opioid receptor mRNAs in the rat spinal cord and dorsal root ganglia. *Neuroscience Letters, 168*, 97–100.

Mansour, A., Fox, C. A., Burke, S., et al. (1994). Mu, delta, and kappa opioid receptor mRNA expression in the rat CNS: An in situ hybridization study. *The Journal of Comparative Neurology, 350*, 412–438.

Schafer, M. K., Bette, M., Romeo, H., et al. (1994). Localization of Î°-opioid receptor mRNA in neuronal subpopulations of rat sensory ganglia and spinal cord. *Neuroscience Letters, 167*, 137–140.

Schneider, S. P., Eckert, W. A., III, & Light, A. R. (1998). Opioid-activated postsynaptic, inward rectifying potassium currents in whole cell recordings in substantia gelatinosa neurons. *Journal of Neurophysiology, 80*, 2954–2962.

Schnell, S. A., & Wessendorf, M. W. (2009). Lack of evidence for the μ-opioid receptor splice variant MOR1C in rats. *The Journal of Comparative Neurology, 517*, 452–458.

Song, B., & Marvizon, J. C. (2003). Dorsal horn neurons firing at high frequency, but not primary afferents, release opioid peptides that produce μ-opioid receptor internalization in the rat spinal cord. *Journal of Neuroscience, 23*, 9171–9184.

Spike, R. C., Puskar, Z., Sakamoto, H., et al. (2002). MOR-1-immunoreactive neurons in the dorsal horn of the rat spinal cord: Evidence for nonsynaptic innervation by substance P-containing primary afferents and for selective activation by noxious thermal stimuli. *European Journal of Neuroscience, 15*, 1306–1316.

Todd, A. J. (2010). Neuronal circuitry for pain processing in the dorsal horn. *Nature Reviews Neuroscience, 11*, 823–836.

Trafton, J. A., Abbadie, C., Marek, K., et al. (2000). Postsynaptic signaling via the μ-opioid receptor: Responses of dorsal horn neurons to exogenous opioids and noxious stimulation. *Journal of Neuroscience, 20*, 8578–8584.

Yasaka, T., Tiong, S. Y. X., Hughes, D. I., Riddell, J. S., & Todd, A. J. (2010). Populations of inhibitory and excitatory interneurons in lamina II of the adult rat spinal dorsal horn revealed by a combined electrophysiological and anatomical approach. *Pain, 151*, 475–488.

Yoshimura, M., & North, R. A. (1983). Substantia gelatinosa neurones hyperpolarized in vitro by enkephalin. *Nature, 305*, 529–530.

Opioid Responsiveness

Definition

Opioid responsiveness is the probability that adequate analgesia without intolerable and unmanageable side effects can be obtained during opioid dose titration.

Cross-References

▶ Cancer Pain Management, Principles of Opioid Therapy, Drug Selection
▶ Opioid Rotation
▶ Opioid Rotation in Cancer Pain Management

Opioid Responsiveness in Cancer Pain Management

Sebastiano Mercadante
Pain Relief & Palliative Care, La Maddalena
Cancer Center University of Palermo,
Palermo, Italy

Definition

Cancer pain is complex and multidimensional and often requires the use of opioids for pain of moderate-to-severe intensity. Opioid doses are commonly increased to obtain satisfactory analgesia. The response to opioids is determined by several factors and could be defined as the degree of analgesia obtained following dose escalation to an endpoint determined by either analgesia or intolerable adverse effects. It is a continuum of response rather than a quantal, yes or no, phenomenon, consistent with the interindividual variability that characterizes opioid analgesia, and the occurrence of dose-dependent effects (Portenoy et al. 1990).

Characteristics

Chronic opioid therapy has been recognized as standard management for cancer pain. Analgesia can be achieved with different opioid dosages in individuals, as there is effectively no ceiling to their analgesic effect. However, the proportion of adverse effects that the patient can tolerate may limit this process. Several factors can interfere with an appropriate analgesic opioid response. The unpredictable individual response in terms of analgesia and toxicity depends on patient-related factors, drug-selective effects, and pain-related factors (Mercadante and Portenoy 2001a). The need for opioid escalation may indicate an abrupt change in the underlying disease, or reveal a previously unknown complication, indicating a change in the relation between dose and response.

Primary tumors may have different evolutions in terms of distant metastases and then pain mechanisms, and can also modify the opioid response through the intervention of some metabolic factors, like cytokines. Increases in opioid doses during a chronic morphine treatment are commonly related to a progression of disease, generating an increase in pain. On the other hand, tolerance to analgesic and non-analgesic effects commonly occur, although other factors may be operant that allow the patient to tolerate higher doses of opioids (Mercadante and Portenoy 2001c). Opioid response may also be drug-selective, a possibility suggested by the remarkable variability of individual patients and their response to different drugs.

The most frequently used outcome measurements have included prognostic factors, as well as monitoring of pain relief, pain intensity, adverse effects, satisfaction, and quality of life. Evaluation instruments and definitions vary from study to study and the lack of universal definition and accepted pain outcomes has created difficulties in evaluating care. Recently, initial pain intensity, initial pain relief, incident pain, localization of pain, cancer diagnosis, and age were predictors of difficult pain conditions (Knudsen et al. 2012).

Initial pain intensity has been found as a contributing factor in requiring a longer time to achieve stable pain control, high final opioid doses, and more complicated analgesic regimes (Fainsinger et al. 2009; Knudsen et al. 2012). However, this may simply mean that patients were less intensively treated, and could be related to the timing. Indeed, other surveys suggest that patients with highest pain scores may paradoxically perceive a better pain relief with an appropriate opioid dose titration (Mercadante 2011; Dalai et al. 2012).

The pain mechanism may influence the opioid response. ▶ Neuropathic cancer pain (NcP) is considered to be less responsive to opioid treatment in comparison to ▶ nociceptive pain (Bruera et al. 1995). The opioid response in NcP has been a controversial topic in the last decades. NcP has been described as a possible negative predictive prognostic factor in cancer pain

therapy, which is associated with a relatively decreased responsiveness to systemic opioids, although neuropathic mechanism does not result in an inherent resistance to opioids but may decrease the likelihood of a favorable outcome (Mercadante et al. 1992). Indeed, patients with NcP did not show a particular disadvantage compared to those exhibiting nociceptive pain, unless associated with neurological impairment (Mercadante et al. 2009). Clinical observation suggests that the change in opioid responsiveness depends on the context and probably on the specific kind of nerve injury. Moreover, the response attained with one drug cannot be generalized to other opioids.

Thus, NcP can still be responsive to analgesic treatment and does not result in an inherent resistance to opioids (Portenoy et al. 1990; Mercadante and Portenoy 2001a). On the other hand, diagnosis of NcP is often disputable, frequently no data about assessment have been reported, and no universally accepted validated clinical diagnostic criteria for diagnosis exist. NcP is not a single entity, but a heterogeneous group of conditions that differ in etiology and location, and symptoms do neither respect cause nor anatomical site (Bennett 2012). The concept that chronic pain may be more or less neuropathic is relatively recent and fits well with basic science opinion regarding chronic pain mechanisms, and the continuum of responses obtainable with different opioids. This is even true for cancer pain, due to the multifactorial mechanisms underlying cancer pain syndromes. A grading system of definite, probable, and possible neuropathic pain has been recently proposed by a group of experts for clinical and research purposes. It was developed in recognition that there is no gold standard and is based on pain distribution, causality, clinical examination, and diagnostic tests.

The temporal pattern of pain may also limit the opioid response. ▶ Breakthrough pain, episodic pain that interrupts basal analgesia, is commonly associated with a poor prognosis (Portenoy and Hagen 1990). The difficulty in the treatment consists in the temporal pattern of this kind of pain, involving almost the same site of basal pain. ▶ Breakthrough pain may be spontaneous, occurring without an identifiable precipitating event, and precipitant factors, volitional and non-volitional, have been identified in more than 50 % of patients (Davies et al. 2009). The most well-known form of breakthrough pain is incident pain, due to movement and commonly caused by bone metastases. The difficulty with incident pain is not a lack of response to systemic opioids, but rather that the doses required to control the incident pain produce unacceptable adverse effects when the patient is at rest or when pain spontaneously stops (Mercadante et al. 1992).

Absorption, metabolism, and elimination may alter the opioid responsiveness, as well as receptor disposition. All these factors depend on individual genetic profile and polymorphisms, which are obvious clinical implications. Unfortunately, studies have never determined the level of influence of such differences (Klepstad et al. 2011). Metabolic changes may obviously modify the response. For example, the accumulation of toxic hydrophilic metabolites during opioid therapy with morphine or hydromorphone can reduce the opioid response, leading to severe and intolerable adverse effects, even in patients with apparently normal renal function (Mercadante and Arcuri 2004).

Strategies to Improve the Opioid Response

Considering the importance of opioids in the management of chronic pain, countermeasures should be taken to limit the interference with opioid responsiveness. The countermeasures begin with a comprehensive evaluation of the causes that can contribute to worsening pain or altered pain perception, including recurrent disease, psychological distress, and the history of previous opioid use. These approaches may produce either a leftward shift of the analgesia curve or a rightward shift of the toxicity curve (Mercadante and Portenoy 2001b).

One approach to the patient with pain that is poorly responsive to opioids is the co-administration of adjuvants, such as antidepressants and anticonvulsants. These groups of drugs may be helpful in improving the analgesic effects when added to opioids, although their use can be associated with more adverse effects (Bennett 2012). Anti-inflammatory drugs have some opioid sparing effect, and may allow the lowering of opioid doses, reducing the risk of toxicity (Mercadante et al., 2002). However, due to their organ toxicity their use should be monitored and individualized for specific patients who show a relevant analgesic effect when added to opioids.

Another strategy to address poor opioid response is the aggressive management of adverse effects. This approach may open the therapeutic window and allow higher and more effective opioid doses. Different symptomatic drugs have been used, although few studies could demonstrate the validity of this approach (Cherny et al. 2001).

Alternately, the poor opioid response could be managed targeting analgesic tolerance and attempt to reduce or prevent it. Agents that block the activity of NMDA-receptors may provide new tools for the treatment of poorly responsive pain syndromes, particularly neuropathic ones. There are clinical reports about the use of ketamine, although this drug is hard to manage due to its excitatory adverse effects and should be used by skilled people (Mercadante and Portenoy 2001b, c).

The treatment of breakthrough pain is challenging for physicians. A rescue dose of an opioid can provide a means to treat breakthrough pain in patients already stabilized on a baseline opioid regimen. Most opioids administered by oral route, including morphine, oxycodone, and hydromorphone, have a relatively slow onset of effect (about 30–45 min), and the pain onset may be so rapid that an oral dose may not provide sufficiently prompt relief, even if taken as needed (Mercadante 2012). Subcutaneous or intravenous administration is timely and an effective route of administration. Oral transmucosal dosing is a noninvasive approach to the rapid onset of analgesia. Fentanyl, incorporated in a hard matrix on a handle, is rapidly absorbed and has been shown to have an onset of pain relief similar to intravenous morphine, that is, within 10 min. Titration of the rescue dose should be attempted in an individual way to identify the most appropriate dose, as no relationship between basal dose and rescue dose has been found in studies with transmucosal fentanyl (Davies et al. 2009). However, there are many new data suggesting that a predictable dose proportional to the doses used for background analgesia is safe, effective, and more practical (Mercadante 2011).

Most patients receive morphine as the first drug for their pain conditions. The ratio of metabolites to the parent compound is less during parenteral administration than oral administration. In some situations, this difference in the production of metabolites could account for the occurrence of fewer adverse effects during parenteral morphine administration. Thus, ▶ switching from oral administration to a continuous parenteral administration, both subcutaneously or intravenously may reduce the metabolite-morphine relationship and decrease the toxicity due to accumulation of the metabolites. Adverse effects may also be reduced in intensity with the use of the *spinal route*, although the administration of local anesthetics with an opioid may provide additional analgesia and permit the recapture of patients unresponsive to spinal morphine administered alone (Mercadante et al. 2007).

Different receptorial attitudes may explain these differences among opioids, in terms of ▶ efficacy and tendency to produce tolerance. Asymmetric tolerance among opioids exists and has been reported as the principal rational for opioid switching. As opioids have differential effects on selective subsets of opioid receptors in the central nervous system, and cross-tolerance between opioids is incomplete, a shift from one opioid to another is a useful option when the side effect-analgesic relationship is inconvenient. This pharmacological approach is named opioid rotation or switching. To restore a more advantageous analgesia-toxicity relationship, sequential trials of different opioids have been suggested. The degree of cross-tolerance may change as opioid doses are escalated, and care must be taken in applying an equianalgesic dose table to patients on high doses of opioids (Mercadante and Bruera 2006).

Finally, cancer pain is a complex experience, which involves personality, learning, and situational components. Although difficult to assess, psychological status plays an important role in the experience of cancer pain and may influence the need for opioid escalation. The degree of psychological distress has been reported as a major negative prognostic factor for opioid responsiveness (Knudsen et al. 2012). It is possible to attain better analgesia following the introduction of psychological interventions or psychotropic medication.

References

Bennett, M. I. (2012). Effectiveness of antiepileptic or antidepressant drugs when added to opioids for cancer pain: Systematic review. *Palliative Medicine, 153*, 359–365.

Bruera, E., Schoeller, T., Wenk, R., MacEachern, T., Marcellino, S., Hanson, J., & Suarez-Almazor, M. (1995). A prospective multicenter assessment of the Edmonton staging system for cancer pain. *Journal of Pain and Symptom Management, 10*, 348–355.

Cherny, N., Ripamonti, C., Pereira, J., et al. (2001). Strategies to manage the adverse effects of oral morphine: An evidence-based report. *Journal of Clinical Oncology, 19*, 2542–2554.

Dalai, S., Hui, D., Nguyen, L., Chacko, R., et al. (2012). Achievement of personalized pain goal in cancer patients referred to a supportive care clinic at a comprehensive cancer center. *Cancer, 118*(15), 3869–3877, in press.

Davies, A. D., Dickman, A., Reid, C., Stevens, A., & Zeppetella, G. (2009). The management of cancer-related breakthrough pain: Recommendations of a task group of the science committee of the association for palliative medicine of Great Britain and Ireland. *European Journal of Pain, 13*, 331–338.

Fainsinger, R., Fairchild, A., Nekolaichuk, C., Lawlor, P., Lowe, S., & Hanson, J. (2009). Is pain intensity a predictor of the complexity of cancer pain management? *Journal of Clinical Oncology, 27*, 585–590.

Klepstad, P., Fladvad, T., Skorpen, F., et al. (2011). Influence from genetic variability on opioid use for cancer pain: A European genetic association study of 2,294 cancer pain patients. *Pain, 152*, 1139–1145.

Knudsen, A. K., Brunelli, C., Klepstad, P., Aass, N., Apolone, G., Corli, O., Montanari, M., Caraceni, A., & Kaasa, S. (2012). Which domains should be included in a cancer pain classification system? Analyses of longitudinal data. *Pain, 153*, 696–703.

Mercadante, S. (2011). The use of rapid onset opioids for breakthrough cancer pain: The challenge of its dosing. *Critical Reviews in Oncology/Hematology, 80*, 460–465.

Mercadante, S. (2012). Pharmacotherapy for breakthrough cancer pain. *Drugs, 72*, 181–190.

Mercadante, S., & Arcuri, E. (2004). Opioids and renal function. *The Journal of Pain, 5*, 2–19.

Mercadante, S., & Bruera, E. (2006). Opioid switching: A systematic and critical review. *Cancer Treatment Reviews, 32*, 304–315.

Mercadante, S., & Portenoy, R. K. (2001a). Opioid poorly responsive cancer pain. Part 1. Clinical considerations. *Journal of Pain and Symptom Management, 21*, 144–150.

Mercadante, S., & Portenoy, R. K. (2001b). Opioid poorly responsive cancer pain. Part 3. Clinical strategies to improve opioid responsiveness. *Journal of Pain and Symptom Management, 21*, 338–354.

Mercadante, S., & Portenoy, R. K. (2001c). Opioid poorly responsive cancer pain. Part 2. Basic mechanisms that could shift dose–response for analgesia. *Journal of Pain and Symptom Management, 21*, 255–264.

Mercadante, S., Maddaloni, S., Roccella, S., & Salvaggio, L. (1992). Predictive factors in advanced cancer pain treated only by analgesics. *Pain, 50*, 51–155.

Mercadante, S., Fulfaro, F., & Casuccio, A. (2002). A randomized controlled study on the use of anti-inflammatory drugs in patients with cancer pain on morphine therapy: Effects on dose-escalation and a pharmacoeconomic analysis. *European Journal of Cancer, 38*, 1358–1363.

Mercadante, S., Intravaia, G., Villari, P., Ferrera, P., Riina, S., David, F., & Mangione, S. (2007). Intrathecal treatment in cancer patients unresponsive to multiple trials of systemic opioids. *The Clinical Journal of Pain, 23*, 793–798.

Mercadante, S., Gebbia, V., David, F., Aielli, F., Verna, L., Casuccio, A., Porzio, G., Mangione, S., & Ferrera, P. (2009). Tools for identifying cancer pain of predominantly neuropathic origin and opioid responsiveness in cancer patients. *The Journal of Pain, 10*, 594–600.

Portenoy, R. K., & Hagen, N. A. (1990). Breakthrough pain: Definition, prevalence, and characteristics. *Pain, 41*, 273–281.

Portenoy, R. K., Foley, K. M., & Inturrisi, C. E. (1990). The nature of opioid responsiveness and its implications for neuropathic pain: New hypotheses derived from studies of opioid infusions. *Pain, 43*, 273–286.

Opioid Rotation

Charles E. Inturrisi
Department of Pharmacology, Weill Cornell Medical College, New York, USA

Definition

▶ Opioid rotation refers to the substitution or switching from one opioid to another to achieve a more favorable therapeutic outcome.

Characteristics

The individualization of the dose of an opioid is essential for the appropriate and successful management of pain with these drugs (Inturrisi 2002). When gradual ▶ dose titration is limited by adverse effects, opioid rotation is one of the alternative therapeutic strategies that should be employed (Indelicato and Portenoy 2002). This approach is based on clinical observations that the interindividual response to opioids varies widely and that the switch to an alternative opioid often results in an "opening in the therapeutic window" by reducing limiting adverse effects (Pereira et al. 2001; Indelicato and Portenoy 2002; Inturrisi 2002). Unfortunately, while there is general agreement about the utility of opioid rotation, the factors underlying interindividual variation cannot yet be readily identified in most patients, and therefore, no consensus has been reached on the steps that should be used in converting the dose from one opioid to the alternative (Ripamonti et al. 1998; Mercadante 1999; Pereira et al. 2001; Indelicato and Portenoy 2002). However, a rational pharmacological approach begins with the recognition that opioid rotation requires knowledge of the ▶ relative potency relationship between the two opioids involved in the therapeutic situation. Relative potency refers to the ratio of doses of two drugs required to produce the same level of effect (▶ Equianalgesic Dose) (Houde et al. 1965). As defined, it can only be obtained from dose-response data. The most reliable data is obtained from randomized, double-blind, cross-over studies, where the dose-response curves are generated and comparisons are made in the equianalgesic effect range to reduce errors associated with extrapolation (Houde et al. 1965). An equianalgesic dose table provides evidence-based values for estimates of the relative potency of different opioids administered by commonly used routes of administration (e.g., see Table 1). Importantly, the table simplifies comparisons, by relating the potency of each opioid to that of a 10 mg intramuscular dose of morphine. Nevertheless, an equianalgesic dose table does not provide all of the essential information

Opioid Rotation, Table 1 Opioid analgesics used for severe pain

Name	Equianalgesic im dose[a]	Starting oral dose range (mg)[a]	Comments	Precautions
Morphine	10	30	Standard of comparison for opioid analgesics. Several sustained-release dosage forms	Lower doses for aged patients; impaired ventilation; bronchial asthma; increased intracranial pressure; liver failure
Hydromorphone	1.5	4–8	Slightly shorter acting. HP im dosage form for tolerant patients	Like morphine
Oxycodone	10	15–30	Immediate-release and sustained-release dosage forms	Like morphine
Methadone	10	2.5[b]	Good oral potency; long plasma half-life	Like morphine; may accumulate with repetitive dosing causing excessive sedation
Levorphanol	2	2–4	Like methadone	Like methadone
Fentanyl	0.1	–	Transdermal preparation. See package insert for equianalgesic dosing	Transdermal creates skin reservoir of drug – 12-h delay in onset and offset
				Fever increases absorption
Meperidine	75	Not recommended	Slightly shorter acting; used orally for less severe pain	Normeperidine (toxic metabolite) accumulates with repetitive dosing causing CNS excitation; not for patients with impaired renal function or receiving monoamine oxidase inhibitors

For these equianalgesic im doses (also see comments), the time of peak analgesia in nontolerant patients ranges from 1/2 h to 1 h and the duration from 4 h to 6 h. The peak analgesic effect is delayed and the duration prolonged after oral administration

Im intramuscular, *po* oral

Adapted from, Table 1, Inturrisi (2002)

[a]These doses are recommended starting doses from which the optimal dose for each patient is determined by titration and the maximal dose limited by adverse effects. For single IV bolus doses, use half the im dose

[b]See Table 3 of Ripamonti et al. (1998) for the approximate rotation ratios for switching from morphine to methadone

needed for dose conversion in individual patients. The table (Table 1) presents the mean estimate of relative potency, but not the confidence limits which define the range of values where the estimate will fall 95 % of the time. If the confidence limits are wide, as they are for the estimates used to construct the table (Houde et al. 1965), then any ratio number derived from the table and used as the sole means of calculating the equianalgesic dose for opioid rotation will be subject to this variation. The table (Table 1) was originally intended as the starting point for the estimation of the conversion dose ratio, with the consideration of individual patient factors

and clinical experience to guide the final dose selection. Some of the factors affecting opioid response, and therefore the equianalgesic dose, are intrinsic to the patient, and some are intrinsic to the drug. For example, with repetitive dosing, the opioid will accumulate until steady state is reached. Opioids with a short ▶ elimination half-life will reach steady state sooner than those with a longer elimination half-life (Inturrisi 2002). Even at an apparent steady state, the analgesic responses vary widely among cancer patients. Estimates of the concentration of methadone required to produce 50 % pain relief at steady state vary nearly ten fold

(Inturrisi et al. 1990). For some opioids, including morphine, active metabolites can accumulate with repetitive dosing, and their contribution to the therapeutic and adverse effects of these opioids remains controversial (Penson et al. 2000; Skarke et al. 2003). Some opioids have nonopioid effects that can influence opioid responses. Methadone has NMDA receptor antagonist activity in animal studies that are not seen with morphine or hydromorphone. NMDA receptor antagonists prevent or reverse opioid tolerance and are antihyperalgesic in models of injury-induced pain (Davis and Inturrisi 1999). Factors intrinsic to the patient that can influence ▶ opioid responsiveness include the mechanism and intensity of the patient's pain. It is generally recognized that pain generated by injury to the nervous system, i.e., neuropathic pain, is often less responsive to opioids (Portenoy et al. 1990). Advanced age and decreased renal function can increase opioid responsiveness, while the development of opioid tolerance decreases the response to opioids (Inturrisi 2002). Current information is not sufficient to assure us that any one of these factors equally affects the response to each opioid in the table and therefore can be ignored while calculating a ratio. Thus, it is generally accepted that tolerance does not appear to develop at the same rate to all opioids (Pasternak 2001). Indeed, this concept of incomplete cross-tolerance among opioids provides one of the rationales for opioid rotation but also needs to be accounted for in the calculation of the rotation dose. Many of the pharmacokinetic and pharmacodynamic factors outlined above can be expected to make a significant contribution to the response seen after multiple dosing with the alternate drug rather than after a single dose. A systemic review of estimates of the ▶ equianalgesic dose ratios reported in studies that utilized chronic opioid administration was conducted by Pereira et al. (2001). This review identified many of the issues discussed above and also noted that the equianalgesic dose ratio may change according to the direction of the opioid switch. This complication is neither well appreciated nor has it been systematically evaluated.

Given these caveats, two principles that clearly emerge are that the rotation dose should be some fraction of the equianalgesic dose calculated from Table 1 and that some form of ▶ dose titration is required to provide a margin of safety (Ripamonti et al. 1998; Mercadante 1999; Pereira et al. 2001; Indelicato and Portenoy 2002). The overarching consideration in selecting a dosing schedule for opioid rotation is first to limit the risk of overdose as one opioid is discontinued and the substitute is introduced. The dose of the new opioid is calculated from the equianalgesic dose table (Table 1). This estimated dose is then reduced by 25–50 % for opioids other than methadone (Indelicato and Portenoy 2002; Inturrisi 2002). For methadone, the estimated dose is reduced by 75–90 % (Indelicato and Portenoy 2002). This estimated dose may also be adjusted based on the medical condition of the patient and the severity of the pain (Indelicato and Portenoy 2002). Ripamonti (Ripamonti et al. 1998) observed that in cancer patients the estimated equianalgesic dose ratio of morphine to methadone increased as a function of the prior morphine dose so that no single dose ratio was appropriate for both opioid-naïve patients and patients who were receiving short- or longer-duration morphine therapy at the time they were switched to methadone. The data indicate that those patients who were receiving the highest doses of morphine were relatively more sensitive to the analgesic effects of methadone than predicted by the dose ratio derived from the table. Therefore, patients with prior morphine experience require a greater reduction in the estimated methadone dose than relatively morphine-naïve patients. It is not known whether this variability in the estimated dose ratio between morphine and methadone is unidirectional or if it should also be considered when switching from methadone to morphine. Recognition of the complexity of estimating the equianalgesic dose has led several investigators to conduct both prospective and retrospective studies aimed at better defining the equianalgesic dose ratios. Thus, from retrospective studies in palliative care patients, the ratio of morphine to hydromorphone was estimated to be approximately five to one (Pereira et al. 2001). However, this ratio (morphine to hydromorphone) was only approximately 3.5 to one when patients were shifted from hydromorphone to morphine (Pereira et al. 2001). This led to the suggestion that the values in Table 1 should be modified to reflect these estimates (Pereira et al. 2001). This change in the relative potency estimates could decrease or eliminate the need to reduce the dose calculated from the table in the manner described above. Whether this provides a safer and more convenient approach remains to be verified. Nevertheless, anyone planning to use opioid rotation should carefully consider these approaches (Ripamonti et al. 1998; Mercadante 1999; Pereira et al. 2001).

The ▶ dosing regimen to be employed should also reflect the understanding that the estimated dose is just that a starting dose from which the optimal dose for that patient will be achieved by careful dose titration up or down. Alternatively, the dose can be fixed and the ▶ dosing interval adjusted to optimize the response. Usually some combination of these two approaches provides the necessary flexibility. Where both opioids have relatively short ▶ elimination half-lives (e.g., morphine and hydromorphone), the final estimated dose of the new opioid can be substituted for current opioid, provision made for a rescue dose (usually 5–15 % of the total daily dose) and titration with the new opioid begun (Indelicato and Portenoy 2002). Often the new opioid is administered on a fixed interval schedule as was the previous opioid (Indelicato and Portenoy 2002). An important exception to this approach are the opioids with relatively long elimination half-lives (see Inturrisi 2002). The most important of these is methadone, whose half-life is much longer than that of morphine and subject to significant interindividual variation (Inturrisi 2002). Early studies employed a patient-controlled dosage regimen of oral methadone, which fixed the dose of methadone but allowed the dosing interval to be determined by the duration of the patient's pain relief after

methadone (Sawe et al. 1981). This "▶ as needed dosage regimen during the titration phase of rotation to methadone is still advocated" (Foley and Houde 1998). More recent versions of the fixed dose with an "as-needed" interval approach for methadone rotation have been described (Morley et al. 1993; Mercadante 1999). Others have proposed that the switch from morphine to methadone be more gradual, over 3–4 days, and that a portion of the morphine dose (one third) be substituted with methadone each day (Bruera et al. 1995; Ripamonti et al. 1998). These approaches also incorporate our current understanding of the need to further reduce the dose of methadone from the estimate obtained using the unmodified table (Table 1) (Morley et al. 1993; Bruera et al. 1995; Mercadante 1999), and they take into account the complex relationship discussed above between prior morphine experience and the response to methadone (Ripamonti et al. 1998).

This discussion has focused on the rotation of opioids by the oral route of administration because most of the studies have employed this route.

However, some data are available on opioid rotation that involves the transdermal and the parenteral routes (Mercadante 1999; Pereira et al. 2001; Indelicato and Portenoy 2002).

Given the current limitations in our knowledge of dose conversion strategies, additional controlled studies of opioid rotation are essential so that more specific guidelines for opioid rotation can be developed, and this useful method of optimizing pain relief can be both improved and simplified.

References

Bruera, E., Watanabe, S., Fainsinger, R. L., Spachynski, K., Suarez-Almazor, M., & Inturrisi, C. (1995). Custom-made capsules and suppositories of methadone for patients on high-dose opioids for cancer pain. *Pain, 62*, 141–146.

Davis, A. M., & Inturrisi, C. E. (1999). d-methadone blocks morphine tolerance and N-methyl-D-aspartate-induced hyperalgesia. *Journal of Pharmacology and Experimental Therapeutics, 289*, 1048–1053.

Foley, K. M., & Houde, R. W. (1998). Methadone in cancer pain management: Individualize dose and titrate to effect. *Journal of Clinical Oncology, 16*, 3213–3215.

Houde, R. W., Wallenstein, S. L., & Beaver, W. T. (1965). Clinical measurement of pain. In G. de Stevens (Ed.), *Analgesics* (pp. 75–122). New York: Academic Press.

Indelicato, R. A., & Portenoy, R. K. (2002). Opioid rotation in the management of refractory cancer pain. *Journal of Clinical Oncology, 20*, 348–352.

Inturrisi, C. E. (2002). Clinical pharmacology of opioids for pain. *The Clinical Journal of Pain, 18*, S3–S13.

Inturrisi, C. E., Portenoy, R. K., Max, M. B., Colburn, W. A., & Foley, K. M. (1990). Pharmacokinetic-pharmacodynamic relationships of methadone infusions in patients with cancer pain. *Clinical Pharmacology and Therapeutics, 47*, 565–577.

Mercadante, S. (1999). Opioid rotation for cancer pain: Rationale and clinical aspects. *Cancer, 86*, 1856–1866.

Morley, J. S., Watt, J. W., Wells, J. C., Miles, J. B., Finnegan, M. J., & Leng, G. (1993). Methadone in pain uncontrolled by morphine. *Lancet, 342*, 1243.

Pasternak, G. W. (2001). Incomplete cross tolerance and multiple mu opioid peptide receptors. *Trends in Pharmacological Sciences, 22*, 67–70.

Penson, R. T., Joel, S. P., Bakhshi, K., Clark, S. J., Langford, R. M., & Slevin, M. L. (2000). Randomized placebo-controlled trial of the activity of the morphine glucuronides. *Clinical Pharmacology and Therapeutics, 68*, 667–676.

Pereira, J., Lawlor, P., Vigano, A., Dorgan, M., & Bruera, E. (2001). Equianalgesic dose ratios for opioids. A critical review and proposals for long-term dosing. *Journal of Pain and Symptom Management, 22*, 672–687.

Portenoy, R. K., Foley, K. M., & Inturrisi, C. E. (1990). The nature of opioid responsiveness and its implications for neuropathic pain: New hypotheses derived from studies of opioid infusions. *Pain, 43*, 273–286.

Ripamonti, C., Groff, L., Brunelli, C., Polastri, D., Stavrakis, A., & De Conno, F. (1998). Switching from morphine to oral methadone in treating cancer pain: What is the equianalgesic dose ratio? *Journal of Clinical Oncology, 16*, 3216–3221.

Sawe, J., Hansen, J., Ginman, C., Hartvig, P., Jakobsson, P. A., Nilsson, M. I., Rane, A., & Anggard, E. (1981). Patient-controlled dose regimen of methadone for chronic cancer pain. *British Medical Journal (Clinical Research Ed.), 282*, 771–773.

Skarke, C., Darimont, J., Schmidt, H., Geisslinger, G., & Lotsch, J. (2003). Analgesic effects of morphine and morphine-6-glucuronide in a transcutaneous electrical pain model in healthy volunteers. *Clinical Pharmacology and Therapeutics, 73*, 107–121.

Opioid Rotation in Cancer Pain Management

Russell K. Portenoy[1] and Ebtesam Ahmed[2]
[1]Department of Pain Medicine and Palliative Care, Beth Israel Medical Center, New York, NY, USA
[2]College of Pharmacy and Health Sciences, St. John's University, Department of Pain Medicine and Palliative Care, Beth Israel Medical Center, New York, NY, USA

Definition

Converting from one opioid to an alternative opioid with the goal of achieving a more favorable balance between analgesia and side effects

Characteristics

Chronic opioid therapy remains the cornerstone of treatment for moderate to severe cancer pain. Optimization of therapy relies on individualization of opioid dosing. This process involves gradual dose titration until adequate analgesia or the development of unmanageable side effects has been reached. ▶ Opioid responsiveness refers to the likelihood that a favorable balance between pain relief and side effects can be achieved during dose titration (Portenoy 2000). Poor responsiveness can be due to a number of factors, including comorbid medical conditions that predispose to toxicity, pain pathophysiology associated with limited analgesic response, and pharmacological effects such as active metabolite accumulation associated with dehydration or renal insufficiency (Portenoy 1999; Mercadante and Portenoy 2001a, b).

The management of a patient who experiences a poor response to an opioid drug begins with a comprehensive pain assessment, including physical examination. Based on this assessment, the clinician may implement one of the following strategies (Indelicato and Portenoy 2002):

- More aggressive therapy for side effects
- Administration of a coanalgesic (nonopioid or ► Adjuvant Analgesic) with the goal of reducing the systemic opioid requirement
- Intraspinal therapy to reduce the systemic opioid requirement
- Nonpharmacological interventions, such as transcutaneous nerve stimulation (TENS), neural blockade, cognitive approaches, or complementary therapies to reduce systemic opioid requirement
- Opioid rotation to a different opioid with the hope of a more favorable balance between analgesia and side effects

Opioid rotation is based on the variability of response to different opioids. As a result of this variation, a switch to an alternative drug may produce a better balance between analgesia and side effects (Galer et al. 1992; Cherny et al. 1995; Bruera et al. 1996).

To reduce the risk of overdosing or underdosing as one opioid is discontinued and another is started, the clinician must have a working knowledge of the ► equianalgesic dose table. This dosing table provides evidence-based values for the relative potencies among different opioid drugs and routes of administration. These values were established from well-controlled single-dose assays conducted in cancer populations with limited exposure to opioid analgesics (Houde et al. 1966). The standard to which all relative potencies are compared is defined as morphine 10 mg parenterally.

The equianalgesic table offers a broad guide for dose selection when switching from one opioid to another. The clinician must first calculate the total daily dosage of the current opioid, including the fixed schedule dosing and the supplemental doses taken as needed for breakthrough pain.

In many cases, the calculated dose of the new opioid must be reduced to minimize the risk of overdosage (Derby et al. 1998; Anderson et al. 2001). The usual starting point for dose reduction from the calculated equianalgesic dose is 25–50 %. The following guidelines should be followed when converting from one opioid analgesic to another (Indelicato and Portenoy 2002):

1. Calculate the equianalgesic dose of the new opioid based on the equianalgesic table.
2. If switching to any opioid other than methadone or fentanyl, decrease the equianalgesic dose by 25–50 %.
3. If switching to methadone, reduce the dose by 75–90 %.
4. If switching to transdermal fentanyl (Duragesic®), do not reduce the equianalgesic dose.
5. Consider further changes in the adjusted equianalgesic dose based on medical condition and pain.

If the patient is elderly or has significant cardiopulmonary, hepatic, or renal disease, consider further dose reduction.

If the patient has severe pain, consider a lesser dosage reduction or forego any dose reduction.

1. Calculate a rescue dose as 5–15 % of the total daily opioid dose, and administer at an appropriate interval.
2. Reassess and titrate the new opioid as needed. This reduction is justified by:

- The potential for incomplete cross-tolerance between opioid drugs (potentially leading to effects, including side effects that would be greater than expected when a switch to a new analgesic is implemented)
- The interindividual variability in the relative potencies among opioids (ratios listed on the equianalgesic table may be more or less than the ratio that would be found if a single-dose study was done in the individual patient)
- The need for dosage adjustment for conditions that increase opioid risk (including advanced age and medical comorbidities)
- The possible difference in relative potencies in single-dose assays compared to repeated dose assays

There are exceptions to the dosage reductions previously mentioned. The first is a conversion to a transdermal fentanyl system (TFS). When this formulation was developed, conversion guidelines were created that incorporated a safety factor, precluding the need for additional dose reduction in most patients.

The second exception occurs with a conversion to methadone. A larger dosage reduction in the calculated equianalgesic dose (75–90 %) is supported by data that demonstrate a much higher potency than expected when switching to methadone from another pure mu agonist, such as morphine (Ripamonti et al. 1998; Bruera et al. 1996; Shimoyama et al. 1997). These data suggest that the potency of methadone following a switch from another mu agonist is dependent on the dose of the prior drug (Bruera et al. 1996; Shimoyama et al. 1997); a dose of 500 mg or more of morphine requires a larger dose reduction (90 %) than a smaller dose. This greater-than-expected potency is thought to be related to the d-isomer which represents 50 % of the commercially available racemic mixture in the United States. This isomer blocks the N-methyl-D-aspartate (NMDA) receptor and in turn may produce analgesic effects and partially reverse opioid tolerance (Elliot et al. 1994; Hagen and Wasylenko 1999).

Conversion Example

Mr. J was a 64-year-old man with a history of prostate cancer and metastases to the thoracic spine. His back pain had been well controlled with modified-release oxycodone 80 mg every 12 h and oxycodone 5 mg, one to two tablets every 4 h as needed for breakthrough pain. He usually took four tablets per day. He developed an increase in the intensity of his back pain. Physical exam and MRI ruled out spinal cord compression. He had no other medical comorbidities. His dosage of oxycodone was initially increased to 80 mg every 12 h, and his oxycodone dose was increased to four to five tablets per dose, which he took twice daily. This change improved pain relief, but he developed sedation and dizziness. The plan was to switch to oral morphine.

1. Calculate the 24-h total of oxycodone:
   (Modified-release oxycodone 80 mg × 2) + (oxycodone 20 mg × 2)
   160 mg + 40 mg = 200 mg of Oxycodone/day

2. Convert to the oral morphine equivalent using the equianalgesic table:
   Oxycodone 20 mg = morphine 30 mg
   Oxycodone 200 mg = morphine 300 mg

3. Decrease the morphine dosage by 25 %:
   Morphine 300 mg × 25 % = 75 mg
   Morphine 300 mg − 75 mg = 225 mg

4. Convert the 24-h total to a fixed schedule:
   Morphine 225 mg/24 h = modified-release morphine 120 mg every 12 h. The dosage has been rounded up given the tablet dosage size available.

5. Calculate a rescue dose. The rescue dose may be calculated as a dose that is 5–15 % of the total daily dose:
   Morphine 240 mg/day × 5 % = 12 mg; morphine 240 mg/day × 10 % = 24 mg; and morphine 240 mg/day × 15 % = 36 mg
   A rescue dose of 30 mg every 4 h as needed is reasonable.

References

Anderson, R., Saiers, J. H., Abram, S., et al. (2001). Accuracy in equianalgesic dosing: Conversion dilemmas. *Journal of Pain and Symptom Management, 21*, 397–406.

Bruera, E. B., Pereira, J., Watanabe, S., et al. (1996). Systemic opioid therapy for chronic cancer pain: Practical guidelines for converting drugs and routes. *Cancer, 78*, 852–857.

Cherny, N. I., Chang, V., Frager, G., et al. (1995). Opioid pharmacotherapy in the management of cancer pain: A survey of strategies used by pain physicians for the selection of analgesic drugs and routes of administration. *Cancer, 76*, 1288–1293.

Derby, S., Chin, J., & Portenoy, R. K. (1998). Systemic opioid therapy for chronic cancer pain: Practical guidelines for converting drugs and routes of administration. *CNS Drugs, 9*, 99–109.

Elliot, K., Hynanski, A., & Inturrisi, C. E. (1994). Dextromethorphan attenuates and reverses analgesic tolerance to morphine. *Pain, 59*, 361–368.

Galer, B. S., Coyle, N., Pasternak, G. W., et al. (1992). Individual variability in the response to different opioids: Report of five cases. *Pain, 49*, 87–91.

Hagen, N. A., & Wasylenko, E. (1999). Methadone: Outpatient titration and monitoring strategies in cancer patients. *Journal of Pain and Symptom Management, 18*, 369–375.

Houde, R. W., Wallenstein, S. L., & Beaver, W. T. (1966). Evaluation of analgesics in patients with cancer pain. In L. Lasagna (Ed.), *International encyclopedia of pharmacology and therapeutics* (pp. 59–98). Oxford: Pergamon.

Indelicato, R. A., & Portenoy, R. K. (2002). Opioid rotation in the management of refractory cancer pain. *Journal of Clinical Oncology, 20*, 348–352.

Mercadante, S., & Portenoy, R. K. (2001a). Opioid poorly responsive cancer pain. Part 3: Clinical strategies to improve opioid responsiveness. *Journal of Pain and Symptom Management, 21*, 338–354.

Mercadante, S., & Portenoy, R. K. (2001b). Opioid poorly-responsive cancer pain. Part 1: Clinical considerations. *Journal of Pain and Symptom Management, 21*, 144–150.

Portenoy, R. K. (1999). Managing cancer pain poorly responsive to systemic opioid therapy. *Oncology, 13*, 25–29.

Portenoy, R. K. (2000). *Contemporary diagnosis and management of pain in oncologic and AIDS patients* (3rd ed.). Newtown: Handbooks in Health Care Company.

Ripamonti, C., Groff, L., Brunelli, C., et al. (1998). Switching from morphine to oral methadone in treating cancer pain: What is the equianalgesic ratio? *Journal of Clinical Oncology, 16*, 3216–3221.

Shimoyama, N., Shimoyama, M., Megumi, S., et al. (1997). D-Methadone is antinociceptive in rat formalin test. *Journal of Pharmacology and Experimental Therapeutics, 283*, 648–652.

Opioid Switch

Definition

Clinical practice of changing from one opioid to another (see also ► Opioid Rotation).

Cross-References

► Cancer Pain Management, Principles of Opioid Therapy, Drug Selection

Opioid Therapy for Primary Afferents

► Opioid Modulation of Nociceptive Afferents In Vivo

Opioid Therapy in Cancer Pain Management, Route of Administration

Lukas Radbruch
Department of Palliative Medicine, University of Bonn, Bonn, Germany

Synonyms

Application route; Delivery route

Definition

Before they can affect their analgesic activity, opioids must be transferred to their sites of action, usually the opioid receptors in the dorsal horn of the spinal cord or brain. The route of administration is the method by which the opioid enters the body. By altering the pharmacokinetics of the opioid, the route also influences the effects (or pharmacodynamics). The course of analgesia is different for oral, subcutaneous, or intrathecal application (Table 1). The side effect profile may be different for different application routes. Preference for a given application route in an individual patient may influence the choice of the opioid.

Characteristics

Oral Administration

The administration of opioids "by the mouth" is one of the major recommendations of the World Health Organization (1996). The oral route as the first-line approach has been confirmed by other cancer pain guidelines (Agency for Health Care Policy and Research, US Department of Health and Human Services 1994; Hanks et al. 2001). The approach is easy and comfortable and finds high acceptance with the patients. No technical equipment or special skills are needed. For opioid therapy, where dosage has to be titrated individually and sometimes up to high ranges, the

Opioid Therapy in Cancer Pain Management, Route of Administration, Table 1 Potency of different opioids used in cancer pain management in relationship to the application route

Opioid	Route					
	Oral	Transdermal	Subcutaneously	Intravenously	Epidurally	Intrathecal
Morphine	1		2–3	2–3	30	100
Oxycodone	2					
Hydromorphone	5–7.5		7.5	7.5		
Fentanyl	Transmucosal: 50	100		100	100	1000
Buprenorphine	Sublingual: 50	50		50		
Methadone	4–12					
Levomethadone	8–20					

inherent safety of the oral route provides an additional advantage. Sedation and nausea are the predominant side effects with an overdose, and with increasing somnolence, patients will sleep through subsequent doses, and with nausea and vomiting, opioid uptake will be reduced from the gastrointestinal tract. The oral route is ideally suited for patients treated at home, provided that they can swallow the medication and have no impediment of the enteral uptake of the drug.

Opioids given orally are absorbed predominantly in the upper ileum. They may bind to opioid receptors in the ▶ myenteric plexus before they are taken up by the circulation of the portal venous system. For some opioids, such as morphine or fentanyl, the first-pass effect in the liver reduces bioavailability considerably.

In most countries, morphine is the opioid used predominantly for the oral route of application. With a normal-release form, the time to the onset of the analgesic action is usually approximately 30 min, and the time to peak concentration is about 1 h. The analgesic effect lasts only 3–4 h. Modified-release formulations have an onset time of 1 h, time to peak concentration of up to 4 h, and duration of up to 24 h. Oxycodone is also available in normal-release and modified-release forms. The modified-release form has a biphasic profile of action, providing an initial rapid onset in less than 1 h and a prolonged second phase providing analgesia to 12 h.

Patients with cancer of the head and neck region or the esophagus and patients with neurological diseases such as amyotrophic lateral sclerosis may have difficulties swallowing medication due to progression of the disease. Oral opioids can be continued as long as the intake of fluids is still possible. Sometimes, the oral formulation must be changed, for example, from tablets to liquids. For morphine, a modified-release solution is available in some countries, which can provide analgesia for up to 24 h.

For patients receiving nutrition and fluids via gastroenteral tubes or catheters, opioids can be administered through these lines. As long as the catheter tip lies in the stomach or duodenum, pharmacokinetics with the enteral administration will not be different from the oral. Tablets must be crushed to pass through the line, and modified-release kinetics will be lost with crushing. However, the modified-release solution of morphine, as well as capsules that can be opened without changing the modified-release properties of the small beads inside, can be used via feeding tube without losing the sustained effects.

Transdermal Administration

The introduction of a transdermal delivery system with fentanyl 20 years ago offered an alternative to the oral route. A transdermal delivery system with buprenorphine has been introduced recently. The transdermal systems are noninvasive, easy to apply, safe, and comfortable. They may provide stable analgesia for 3–7 days.

The transdermal therapeutic system is applied on the skin, usually on the chest or back, upper arms, or upper legs. The fentanyl patch is available either as a matrix or a depot system, buprenorphine as a matrix system. From the reservoir in the patch system, fentanyl diffuses through the adhesive layer into the skin. A rate control membrane in the patch system prevents excessive fluctuations in the delivery rate. The matrix system has the opioid dissolved in the adhesive layer.

The main barrier in transdermal opioid delivery is the penetration of the intact skin. The pharmacological properties of fentanyl and buprenorphine allow sufficient diffusion rates to provide analgesia, as these opioids are small, lipophilic molecules with high analgesic potency. However, diffusion rates are low and the opioid pools in the uppermost skin layer, the ▶ stratum corneum. From this depot, the opioid slowly diffuses to lower skin regions, where it is taken up via capillary blood vessels into the systemic circulation. The depot in the stratum corneum is responsible for the slow pharmacokinetics of the transdermal application. It takes 2–4 h after application of the patch until measurable opioid concentrations are found in the blood, and up to 24 h to reach stable serum concentrations. The elimination of the opioid after removal of the patch is also slow, with a half-life of 16 h for fentanyl, as the intracutaneous depot will only be drained slowly. The slow pharmacokinetic profile may provide stable analgesia but also has to be considered during dose titration, as increases and decreases of the dosage will be effective only after 24 h.

The transdermal systems with fentanyl and buprenorphine are available in different sizes. Several patches may be combined reaching dosages of more than 10 mg/day fentanyl. The highest dosage used in our pain clinic for a cancer pain patient was 28.8 mg/day fentanyl. Patches must be applied on intact skin. If the skin at the application site is damaged, uptake rates may increase considerably. Heating of the application site may also increase the diffusion rate, and case reports have described an overdose with excessive heating.

The available transdermal systems are only indicated for patients with continuous stable pain syndromes. In addition to the transdermal opioids, an oral or parenteral opioid with a fast onset should be available for the treatment of breakthrough pain. More than half of the patients with continuous cancer pain will need additional medication for ▶ breakthrough pain every day or nearly every day. As examples, oral transmucosal fentanyl citrate or intranasal fentanyl spray has often been used with transdermal fentanyl, and immediate-release morphine or sublingual buprenorphine has been used with transdermal buprenorphine.

Transmucosal Administration
Opioid application is also possible through the mucosa of the oral cavity or the nose. However, the pharmacokinetics seems to be more variable than with oral application. Most patients will swallow part of the drug with the saliva, meaning that only a fraction of the opioid will be taken up through the oral mucosa, whereas the swallowed fraction will undergo the same pharmacokinetics seen with oral application.

Opioids have also been applied successfully via the rectal route. Opioid uptake is possible through the rectal mucosa as easily as through other intestinal sections. However, the rectal mucosa is drained by several venous systems, with the upper section processed through the enterohepatic circulation, and the lower sections drained directly into the inferior vena cava. Therefore, the amount of first-pass effect in the liver varies considerably after rectal administration, resulting in even larger variability of bioavailability compared to oral application. For chronic cancer pain, repeated rectal application is uncomfortable and will be accepted by only a few patients (Ripamonti and Bruera 1991).

Sublingual and Buccal
Buprenorphine has a very low bioavailability via the oral route. However, sublingual application results in fast uptake of approximately 60 % of the opioid. The tablets will take nearly 20 min to dissolve, and patients may find it cumbersome to have the tablets beneath the tongue for longer times. Sublingual forms for other opioids have to be prepared by the pharmacist, and sublingual

morphine has been used with good effect in postoperative pain in children. However, uptake through the mucosa of oral cavity or nose is low for hydrophilic opioids such as morphine, making the use of lipophilic opioids such as buprenorphine or fentanyl advantageous for these application routes.

A transmucosal system uses fentanyl citrate that is pressed around a plastic stick, with dosages ranging from 200 to 1,600 μg per unit. After being placed in the oral cavity, about one-third of the fentanyl is absorbed through the oral mucosa. The rest is swallowed with the saliva, and approximately half of that amount is absorbed from the intestinal tract. The onset of analgesia is fast, and the transmucosal system has been recommended for treatment of breakthrough pain. Clinical trials have found little correlation between the dosage of the continuous opioid drug and transmucosal fentanyl (Coluzzi et al. 2001). This means that individual titration of transmucosal breakthrough medication is required.

A buccal tablet and a sublingual tablet with fentanyl have also been introduced. These tablets are rapidly dissolving and have a faster onset than oral morphine. Similar to the oral transmucosal fentanyl citrate, individual dose titration is necessary, as the adequate dosage of the breakthrough medication does not correlate to the continuous opioid dosage (Davis 2010).

Intranasal

More recently, therapeutic systems with fentanyl for intranasal application have been developed (Panagiotou and Mystakidou 2010). The uptake via the nasal mucosa is rapid, with onset of analgesia in 5–10 min and time to peak effect in less than 15 min. One of the available systems uses pectin to form a gel with the fentanyl solution on the mucosa. This ensures that the solution does not flow down the back of the throat and also provides a lockout time for subsequent boli, as the pectin gel will slow down the resorption of additional solution. The intranasal spray has a duration of effect of only 1 h, making it the most suitable system for treatment of breakthrough pain.

Subcutaneous Administration

In a large series of cancer pain patients, oral administration was no longer possible in half of the patients at the last follow-up before death (Zech et al. 1995). Patients suffering from nausea or vomiting in spite of adequate antiemetic therapy may not be able to retain oral drugs in the gastrointestinal tract long enough to obtain adequate analgesia. Gastrointestinal obstruction may also impede the uptake of drugs from the gastrointestinal tract. Patients with ▶ dysphagia due to cancer progression in the head and neck area may not be able to swallow tablets or even liquids.

For these patients, the subcutaneous route offers an easy and effective alternative. For conversion from oral to subcutaneous morphine, an average ratio between 1:2 and 1:3 has been recommended (Hanks et al. 2001). An oral daily dosage of 60 mg morphine would be equianalgesic to 20–30 mg subcutaneously. Conversion ratios between the oral and subcutaneous routes depend on oral bioavailability, and opioids with higher oral bioavailability, such as hydromorphone or methadone, will have lower conversion ratios compared to morphine. Morphine is used most frequently for subcutaneous application, but other opioids, such as buprenorphine, fentanyl, hydromorphone, methadone, or oxycodone, have also been used with good effect. Subcutaneous administration of analgesics is usually only indicated for the last days of life, but long-term treatment for several months has also been reported (Drexel et al. 1989; Vermeire et al. 1998).

For long-term analgesic therapy, opioids can be administered by continuous subcutaneous infusion or with repeated bolus injections. Indwelling subcutaneous needles can be used for up to 7 days and may be used not only for continuous infusions but also for repeated bolus injections. With continuous infusions, other drugs such as dipyrone or haloperidol can be combined with opioids. Portable pumps with patient-controlled boli, with continuous infusion, or with a combination of both allow for flexible adaptation of analgesic treatment in the home care setting. For patients requiring parenteral fluid substitution, 500–1,000 ml saline or dextrose infusion can be administered daily via the subcutaneous route. Addition of hyaluronidase may be necessary to prevent induration in the subcutaneous tissue with larger amounts of infusion.

Intravenous Administration

The intravenous route offers another alternative if oral administration is no longer possible. Compared to subcutaneous administration, the intravenous route offers little advantage and requires higher technical standards. Dislocation of an intravenous line will burden the patient more than dislocation of a subcutaneous needle. The faster onset of analgesia after intravenous application provides no additional benefit for patients with continuous pain who require continuous opioid medication (Moulin et al. 1991).

However, in some patients, subcutaneous application may be impaired by generalized edema, poor peripheral circulation, coagulation disorders, or skin conditions. In these patients, intravenous administration can be advantageous. Patients may already have an indwelling intravenous line, for example, for parenteral nutrition. For these patients, opioids may be implemented in the intravenous infusion regimen. As with the subcutaneous route, the average relative potency ratio of oral to intravenous morphine is between 1:2 and 1:3 (Hanks et al. 2001). For hydromorphone, the average relative potency ratio of oral to intravenous application is the same as for oral to subcutaneous.

The short time to peak concentration after intravenous bolus injection may be advantageous for patients with severe pain exacerbations. Intravenous titration with repeated bolus injection will provide pain relief more rapidly than titration via other application routes. Experienced physicians can perform intravenous titration at the bedside or even at home.

Spinal Administration

Opioid receptors are found throughout the central nervous system. An area with high density of opioid receptors is the ▶ substantia gelatinosa of the dorsal horn, where the receptors are located on short interneurons, inhibiting pain transmission. Systemic administration delivers opioids into the bloodstream, which reach these receptors in the spinal cord from the circulation. Delivering the opioid directly into the intrathecal or epidural space will result in higher concentrations at the receptors than with systemic application. This means that less of the drug is required to provide analgesia, and lower opioid dosages may be correlated with fewer side effects. However, spinal administration is not devoid of side effects, and some side effects such as pruritus may be reported even more frequently with spinal application than with systemic opioid therapy. Technical and staff resources are required for the initiation of spinal therapy, and special training is required for maintenance. Costs will be higher in the initial phase, as the implantation of catheter and pump systems is expensive. With long-term treatment over several months, lower drug costs will make spinal treatment cheaper than systemic opioid therapy. However, patients with intractable cancer pain and an indication for spinal opioid therapy often have poor life expectancies.

The place of spinal opioid therapy in cancer pain management is still under discussion. In a large series of 2,118 cancer patients treated in a tertiary pain management unit, only 74 patients (3 %) required epidural or intrathecal opioids for a mean duration of 41 ± 48 days (Zech et al. 1995).

Intrathecal

The pharmacokinetics of spinal opioid application varies in relation to lipophilicity and site of application. The intrathecal space is filled with the cerebrospinal fluid and is separated by the dura mater, a fibrinous membrane, from the epidural space filled with fatty tissue and from blood vessels.

Lipophilic opioids such as fentanyl will penetrate easily into dorsal horn to the receptors in the substantia gelatinosa following injection into the subarachnoidal space. Analgesic activity may be restricted to the spinal cord levels around the site of injection. The duration of analgesia is approximately 4–6 h after a fentanyl bolus. Intrathecal fentanyl may be advantageous for patients with regional pain problems such as pain from lumbar plexus infiltration.

Hydrophilic opioids such as morphine are slower to penetrate the dorsal horn. A large fraction of morphine is transported rostrally with the cerebrospinal fluid circulation. This fraction eventually reaches supraspinal opioid receptors in the brainstem and the cortex. Rostral transport and supraspinal analgesic activity will provide analgesia that is not restricted to the level of injection. However, rostral transport is also responsible for side effects and complications from brainstem receptor sites. Late respiratory depression has been reported 24 h after intrathecal injection of morphine. Nausea and vomiting as well as pruritus are also late side effects related to higher morphine concentrations reaching receptors in the brainstem. Distribution throughout the central nervous system is also responsible for the long duration of action, as a single injection of morphine will provide analgesia for up to 24 h.

In many countries, morphine is the only opioid released for intrathecal treatment of chronic pain. However, many other opioids have been used for intrathecal application, including hydromorphone (hydrophilic), meperidine and methadone (moderately lipophilic), and buprenorphine, fentanyl, and sufentanil (very lipophilic). Diacetylmorphine (heroin) is frequently used in the United Kingdom for intrathecal opioid therapy. Diacetylmorphine has moderate lipophilicity, making it a compromise between morphine and fentanyl.

Long-term intrathecal application usually requires continuous administration via an intrathecal catheter and implanted pump system. With these systems, the intrathecal catheter is connected with a reservoir pump implanted subcutaneously in the abdominal or thoracic wall. As the whole system is concealed under the skin, and the skin barrier is only broken when the reservoir of the pump is filled, infections such as meningoencephalitis as well as loss of cerebrospinal fluid from catheter dislocation are rare. Dislocation and obstruction are possible complications,

requiring replacement of the catheter. Modern pumps use ▶ telemetric programming, so that the delivery rate can be changed without an invasive procedure. However, implanted pumps will deliver only continuous rates, and patients have to use systemic opioids for breakthrough pain. Patients will also be dependent on specialist care, as refilling and programming of the pumps have to be done in specialized centers.

Intrathecal opioid therapy is indicated for patients with severe and intractable stable cancer pain, whose pain is not relieved by systemic opioid therapy with adequate doses and administration times, or when systemic opioid dosage cannot be increased into an effective dose range because of intolerable side effects.

Epidural
Following epidural injection, a hydrophilic opioid such as morphine penetrates the dura mater more easily than a lipophilic opioid such as fentanyl. However, even though a larger amount of morphine will pass to the subarachnoidal space, only a small fraction will proceed into the spinal dorsal horn, while the rest is transported rostrally with the cerebrospinal fluid. On the other hand, lipophilic opioids such as fentanyl or sufentanil will, to a large extent, be taken up by the blood vessels in the epidural space. Some studies have found comparable serum concentrations after epidural and intravenous injection of fentanyl (Baxter et al. 1994) and sufentanil (Camann et al. 1992; Miguel et al. 1994). As with intrathecal application, hydrophilic opioids will have a slow onset (morphine: 30–90 min) and a long duration of action (morphine: 12–24 h) compared to lipophilic opioids (fentanyl onset: 5–10 min, duration 2–4 h) (Grass 1992). Side effects, such as nausea, pruritus, urinary retention, or respiratory depression, are more frequent with hydrophilic opioids. Nausea may be reported by as much as 60 % of the patients, and prevalence of pruritus may be even higher (Grass 1992).

As with intrathecal application, epidural application of opioids can be combined with other drugs, such as local anesthetics or clonidine. For example, patients with neuropathic pain from tumor infiltration of the lumbar plexus may benefit from these combinations.

Dosages for epidural opioid therapy are about tenfold higher compared to intrathecal application, and accordingly larger infusion volumes are required. Implanted pumps are not able to deliver these volumes or would require refilling too often. Epidural application is, therefore, usually delivered through an external pump connected either via a subcutaneous port or directly to the epidural catheter. Catheters should be tunneled subcutaneously to prevent infections along the catheter line. External pumps allow continuous infusion rates and also bolus injections for treatment of breakthrough pain and for dose titration with pain exacerbations. Long-term epidural therapy has been reported with treatment durations of several months. However, complications such as catheter dislocation, obstruction, or infection are not rare with long-term therapy.

Epidural opioid therapy is indicated for intractable chronic cancer pain that is not relieved with systemic opioid therapy with adequate dosage and administration times or with intolerable side effects preventing further dose increases. Epidural therapy with an external pump has the advantage that the dosage can easily be adapted with changing requirements in patients with progressive disease and that even patient-controlled epidural anesthesia is possible. As the insertion of an epidural catheter is a standard skill for anesthesiologists, technical skills for epidural opioid application should be available in every hospital. Transient epidural opioid treatment may be useful to provide pain relief with pain exacerbations. Once adequate pain relief has been accomplished, changing back to oral or subcutaneous application is often possible.

Intraventricular

Intracerebroventricular infusion of opioids has also been used for cancer pain management. In a review of case series (Ballantyne et al. 1996), analgesic efficacy was comparable to intrathecal and better than epidural application. Transient side effects such as nausea or respiratory depression were more frequent with intracerebroventricular application, but persistent side effects such as nausea, urinary retention, or pruritus were less frequent. Complications such as infections or catheter obstruction were reported only rarely. However, no controlled trial has compared this application route directly with other routes, and the technical skills for intracerebroventricular opioid therapy are often not available.

Topical Administration

As opioid receptors are located in the spinal and supraspinal nervous system, topical application seems to have little use in pain management. However, recent studies have demonstrated that opioid receptors are expressed in peripheral tissues with ongoing inflammation. Intra-articular injections of morphine or buprenorphine have been used for postoperative pain after arthroscopy and for osteoarthritis pain with good effect (reviewed in Gupta et al. 2001). No reports have been published on intra-articular opioid application for cancer pain patients.

A mouthwash with morphine was used successfully for the treatment of mucositis following chemotherapy in patients with head and neck cancer (Cerchietti et al. 2002).

Inhalation

Nebulized application of opioids such as morphine has been reported for the treatment of breathlessness (reviewed in Chandler 1999), but not for pain management. Opioid receptors are located throughout the respiratory tract, predominantly in the alveolar walls. The effect of opioids at these receptors is not known. The results of the clinical trials are not clear, and the use of nebulized opioids cannot be recommended at this stage, even if this application route can relieve breathlessness in selected patients.

Change of Route

For patients with inadequate pain relief with opioid therapy in adequate doses and administration times, a switch to another opioid has been recommended. After switching, most patients report improved pain relief and fewer side effects, even if the dosage of the new opioid is lower than expected from accepted equianalgesic potency ratios (Pereira et al. 2001). A possible explanation for the effect is incomplete cross-tolerance between opioids (the preference of different opioid receptor populations by different opioids and accordingly the development of tolerance only to specific subpopulations with each opioid). Opioid switching has become a standard in cancer pain management in recent years.

Switching of the route of administration has also been recommended to improve pain relief and reduce side effects (Cherny et al. 2001). Changing from oral morphine to transdermal fentanyl or buprenorphine combines opioid switching with a change of the application route. However, only a few small trials have reported the effect of route switching, and larger trials are needed to establish the value of this method.

References

Agency for Health Care Policy and Research, US Department of Health and Human Services. (1994). *Clinical practice guideline no. 9: Management of cancer pain.* Washington, DC: U.S. Dept. of Health and Human Services, Public Health Service, Agency for Health Care Policy and Research.

Ballantyne, J. C., Carr, D. B., Berkey, C. S., et al. (1996). Comparative efficacy of epidural, subarachnoid, and intracerebroventricular opioids in patients with pain due to cancer. *Regional Anesthesia, 21,* 542–556.

Baxter, A. D., Laganiere, S., Samson, B., et al. (1994). A comparison of lumbar epidural and intravenous fentanyl infusions for post-thoracotomy analgesia. *Canadian Journal of Anaesthesia, 41,* 184–191.

Camann, W. R., Denney, R. A., Holby, E. D., et al. (1992). A comparison of intrathecal, epidural, and intravenous sufentanil for labor analgesia. *Anesthesiology, 77,* 884–887.

Cerchietti, L. C., Navigante, A. H., Bonomi, M. R., et al. (2002). Effect of topical morphine for mucositis-associated pain following concomitant chemoradiotherapy for head and neck carcinoma. *Cancer, 95,* 2230–2236.

Chandler, S. (1999). Nebulized opioids to treat dyspnea. *American Journal of Hospice and Palliative Care, 16,* 418–422.

Cherny, N., Ripamonti, C., Pereira, J., Davis, C., Fallon, M., McQuay, H., Mercadante, S., Pasternak, G., & Ventafridda, V. (2001). Strategies to manage the adverse effects of oral morphine: An evidence-based report. *Journal of Clinical Oncology, 19*(9), 2542–2554.

Coluzzi, P. H., Schwartzberg, L., Conroy, J. D., et al. (2001). Breakthrough cancer pain: A randomized trial comparing oral transmucosal fentanyl citrate (OTFC) and morphine sulfate immediate release (MSIR). *Pain, 91,* 123–130.

Davis, M. P. (2010). Recent development in therapeutics for breakthrough pain. *Expert Review of Neurotherapeutics, 10,* 757–773.

Drexel, H., Dzien, A., Spiegel, R. W., et al. (1989). Treatment of severe cancer pain by low-dose continuous subcutaneous morphine. *Pain, 36,* 169–176.

Grass, J. A. (1992). Fentanyl: Clinical use as postoperative analgesic-epidural/intrathecal route. *Journal of Pain and Symptom Management, 7,* 419–430.

Gupta, A., Bodin, L., Holmstrom, B., et al. (2001). A systematic review of the peripheral analgesic effects of intra-articular morphine. *Anesthesia and Analgesia, 93,* 761–770.

Hanks, G. W., deConno, F., Cherny, N., et al. (2001). Morphine and alternative opioids in cancer pain: The EAPC recommendations. *British Journal of Cancer, 84,* 587–593.

Miguel, R., Barlow, I., Morrell, M. E., et al. (1994). A prospective, randomized, double-blind comparison of epidural and intravenous sufentanil infusions. *Anesthesiology, 81,* 346–352. Discussion 25A–26A.

Moulin, D. E., Kreeft, J. H., Murray-Parsons, N., et al. (1991). Comparison of continuous subcutaneous and intravenous hydromorphone infusions for management of cancer pain. *Lancet, 337,* 465–468.

Panagiotou, I., & Mystakidou, K. (2010). Intranasal fentanyl: From pharmacokinetics and bioavailability to current treatment applications. *Expert Review of Anticancer Therapy, 10,* 1009–1021.

Pereira, J., Lawlor, P., Vigano, A., et al. (2001). Equianalgesic dose ratios for opioids. A critical review and proposals for long-term dosing. *Journal of Pain and Symptom Management, 22,* 672–687.

Ripamonti, C., & Bruera, E. (1991). Rectal, buccal, and sublingual narcotics for the management of cancer pain. *Journal of Palliative Care, 7,* 30–35.

Vermeire, A., Remon, J. P., Rosseel, M. T., et al. (1998). Variability of morphine disposition during long-term subcutaneous infusion in terminally Ill cancer patients. *European Journal of Clinical Pharmacology, 53,* 325–330.

World Health Organization. (1996). *Cancer pain relief: With a guide to opioid availability.* Geneva: World Health Organization.

Zech, D. F., Grond, S., Lynch, J., et al. (1995). Validation of world health organization guidelines for cancer pain relief: A 10-year prospective study. *Pain, 63,* 65–76.

Opioid Therapy in Cancer Patients with Substance Abuse Disorders, Management

Steven D. Passik[1] and K. L. Kirsh[2]
[1]Psychiatry and Anesthesiology, Vanderbilt University, Nashville, TN, USA
[2]Behavioral Medicine, The Pain Treatment Center of the Blue grass, Lexington, KY, USA

Definition

Substance Abuse

The intentional misuse of a medication, either overuse or taken for a purpose not prescribed (i.e., mood alteration).

Addiction

The development of a maladaptive pattern of medication use that leads to clinically significant impairment or distress in personal or occupational roles. This syndrome also includes a great deal of time used to obtain the medication, use the medication, or recover from its effects; loss of control over medication use; and continuation of medication use after medical or psychological adverse effects have occurred.

Chemical Coping

The occasional use of a patient's medications in nonprescribed ways to cope with stress. The hallmark of chemical coping is the overly central place occupied by the procurement of drugs for pain and inflexibility about nondrug components of care. Medication use becomes central to life, while other interests become less important. As a result, chemical copers in treatment often fail to move forward toward stated psychosocial goals.

Pseudoaddiction

The term was originally coined in a paper that appeared in *Pain* in 1989. Spanning four pages, it is a single case study of a 17-year-old man presenting with leukemia and complaining of chest wall pain. The case is made that undertreated pain created behavioral changes which looked much like iatrogenic addiction in the patient and resulted in a crisis regarding the clinical interactions between the health-care team and patient. Later definitions expanded to include an explanation for when a pain patient engages in worrisome behaviors with opioids (i.e., increasing the dose of medication on their own, doctor shopping, mixing opioids with alcohol) in a desperate effort to obtain the needed analgesia while not rising to meet the criteria for addiction.

Tolerance

The development of a need to take increasing doses of a medication to obtain the same effect; tachyphylaxis is the term used when this process happens quickly.

Physical Dependence

A state of adaptation to a drug class marked by the development of substance-specific symptoms of withdrawal after the abrupt stopping of a medication; these symptoms are physiological in nature (i.e., diarrhea, sweating, restlessness, etc., after opioid cessation).

Least Divertible Units (LDUs)

While there is a notion that long-acting opioids may be a better choice than short-acting opioids in the treatment of cancer pain in general and in those with abuse or addiction histories, there is little empirical data to actually support this. The LDU concept bypasses the need for empirical evidence to argue that keeping the total number of prescribed opioids low (i.e., 10 transdermal fentanyl patches or 60 tapentadol ER doses versus 240 hydrocodones for a 30-day period) is much preferred in the case that any abuse or diversion subsequently does occur.

Characteristics

The abuse of prescription medications, particularly opioids that are commonly used to treat cancer pain, has increased over the last decade to a level that some have described as "epidemic" (NCASA 2006). While ► substance abuse and ► addiction are commonly believed to have lower base rates in cancer patients than in society at large, these occurrences are highly problematic and worthy of attention. Also, the relatively low prevalence of substance abuse among cancer patients treated in tertiary care hospitals may reflect institutional biases or a tendency for patient under-reporting in these settings.

Consideration of substance abuse in the cancer population also highlights the need to diagnose and understand the less-obvious aberrant drug-taking behaviors sometimes in evidence in the treatment of patients without formal psychiatric histories of substance use disorders. Such behaviors can be manifest in the form of chemical coping or ► pseudoaddiction (Passik et al. 2011; Kirsh et al. 2007). Once these aberrant behaviors are identified, the clinician must decide on a course of action that is fair, appropriate, consistent with laws and regulations, and in the best interests of the patient.

With the pressure of regulatory scrutiny and the duty to treat pain but contain abuse or diversion, clinicians often feel that they must avoid being duped by those abusing prescription pain medications. However, the clinician attempting to diagnose the meaning of aberrant drug-related behaviors during pain management need not be correct or be forced into playing the role of a detective. The clinician has an obligation to be thorough, thoughtful, logically consistent, and careful (not to mention humane and caring), but not necessarily correct. Indeed, there are multiple possibilities in the differential diagnosis of aberrant drug-taking behaviors, with criminal intent and diversion being only one of the more remote possibilities.

Clinical management can be tailored for the multiple possibilities that might give rise to the behaviors noted in the assessment, and asserting control over prescriptions can be accomplished without necessarily terminating the prescribing of controlled substances entirely. In the treatment of cancer pain, clinicians do not have the same latitude to not prescribe simply because of abuse concerns. The clinical and moral imperative to treat pain in these patients can create dilemmas if the patient, or someone in their social circle, is abusing medications. These situations defy simple solutions.

Concerns over Current Definitions

► Tolerance and ► physical dependence are not necessarily signs of addiction or abuse in cancer patients with pain. In fact, physical dependence is to be expected with chronic dosing of opioids, and the need for dose escalation to maintain effect is usually a sign of disease progression and worsening pain (and cannot be ascribed to tolerance) (Schneider and Kirsh 2010).

The ability to categorize questionable behaviors (e.g., consuming a few extra doses of a prescribed opioid, or using an opioid prescribed for pain as a nighttime hypnotic) as outside the social or cultural norm also presupposes that there is certainty about the parameters of normative behavior. Less aberrant behaviors (such as aggressively complaining about the need for medications) are more likely to reflect untreated distress of some type, rather than addiction-related concerns. Conversely, more aberrant behaviors (such as injection of an oral formulation) are more likely to reflect true addiction. Although empirical studies are needed to validate this conceptualization, it may be a useful model when evaluating aberrant behaviors.

A more appropriate model definition of addiction notes that it is a chronic disorder characterized by compulsive use of a substance despite harm (which can be psychological, physical, or social) with continued use over time despite that harm (Savage et al. 2008). Although this definition was originally developed from addicted populations without medical illness, it appropriately emphasizes that addiction is a psychological and behavioral syndrome. Any appropriate definition of addiction must include the concepts of loss of control over drug use, compulsive use, and continued use despite harm.

Clinical Management

Out-of-control aberrant drug-taking among patients with cancer pain represents a serious and complex clinical occurrence. The principles outlined herein help the clinician establish structure, control, and monitoring so that they can prescribe freely and without prejudice.

First, a multidisciplinary team approach is recommended for the management of substance abuse in the palliative care setting. Mental health professionals with specialization in the addictions can be instrumental in helping palliative care team members develop strategies for management and treatment compliance. Second, the first member of the medical team to suspect problematic drug-taking or a history of drug abuse should alert the patient's palliative care team, thus beginning the assessment and management process, which includes use of empathic and truthful communication (Savage et al. 2008). This approach entails starting the assessment interview with broad questions about the role of drugs (e.g., nicotine, caffeine) in the patient's life and gradually becoming more specific in focus to include illicit drugs. Third, the development of clear treatment goals is essential for the management of drug abuse. The distress of coping with a life-threatening illness and the availability of prescription drugs for symptom control can make complete abstinence an unrealistic goal (Savage et al. 2008). Rather, a harm reduction approach should be employed. A written agreement between the team and patient helps to provide structure to the treatment plan, establishes clear expectations, and outlines the consequences of aberrant drug-taking, and the inclusion of spot urine toxicology screens and pill counts in the agreement can be useful in maximizing compliance. Fourth, the team should consider using longer acting drugs (e.g., transdermal fentanyl patch or modified-release opioids). The longer duration and slow onset may help reduce aberrant drug-taking behaviors when compared to the rapid onset and increased frequency of dosage associated with short-acting drugs. In addition, this type of choice is helpful in that it keeps in mind the notion of least divertible units. Fifth, the team should make plans to frequently reassess the adequacy of pain and symptom control. Finally, the team should

involve family members and friends in the treatment to help bolster social support and functioning. Becoming familiar with the family may help the team identify family members who are themselves drug abusers and who may potentially divert the patient's medications and contribute to the patient's noncompliance.

Managing addiction problems in patients with cancer is labor intensive and time consuming. Clinicians must recognize that virtually any centrally acting drug, and any route of administration, can potentially be abused. The problem does not lie in the drugs themselves. The effective management of patients with pain who engage in aberrant drug-taking behavior necessitates a comprehensive approach and provides practical means to manage risk, treat pain effectively, and assure patient safety.

References

Kirsh, K. L., Jass, C., Bennett, D. S., Hagen, J. E., & Passik, S. D. (2007). Initial development of a survey tool to detect issues of chemical coping in chronic pain patients. *Palliative & Supportive Care, 5*, 219–226.

National Center on Addiction and Substance Abuse. (2006). Controlled prescription drug abuse at epidemic level. *Journal of Pain & Palliative Care Pharmacotherapy, 20*, 61–64.

Passik, S. D., Webster, L., & Kirsh, K. L. (2011). Pseudoaddiction revisited: A commentary on clinical and historical considerations. *Pain Management, 1*(3), 239–248.

Savage, S. R., Kirsh, K. L., & Passik, S. D. (2008). Challenges in using opioids to treat pain in persons with substance use disorders. *Addiction Science & Clinical Practice, 4*(2), 4–25. Available at: http://www.drugabuse.gov/PDF/ascp/vol4no2/Challenges.pdf.

Schneider, J., & Kirsh, K. L. (2010). Defining clinical issues around tolerance, hyperalgesia and addiction: A quantitative and qualitative outcome study of long-term opioid dosing in a chronic pain practice. *Journal of Opioid Management, 6*(6), 385–395.

Opioid Tolerance

▶ Opioids, Clinical Opioid Tolerance

Opioid Tolerance and Glutamate Homeostasis

▶ Glutamate Homeostasis and Opioid Tolerance

Opioid Tone

Definition

Tonic physiological or pathological release of endogenous opioid peptides demonstrated by blocking opioid receptors with naloxone. The presumed effect of opioid tone would be to produce a constant state of analgesia, as the opioids bind to and activate opioid receptors. The sometimes observed hyperalgesic effect of the opioid antagonist, naloxone, can reveal opioid tone in some settings.

Cross-References

▶ Opioid Analgesia, Strain Differences

Opioid-Induced Analgesia

Definition

Analgesia produced by administration of opioids.

Cross-References

▶ Forebrain Modulation of the Periaqueductal Gray and Its Role in Pain

Opioid-Induced Neurotoxicity

Definition

A syndrome of neuropsychiatric consequences of opioid administration. The features of opioid-induced neurotoxicity include cognitive impairment, severe sedation, hallucinosis, delirium, myoclonus, seizures, hyperalgesia, and allodynia.

Cross-References

▶ Cancer Pain Management, Opioid Side Effects, Cognitive Dysfunction

Opioid-Induced Release of CCK

Carl-Olav Stiller
Division of Clinical Pharmacology and Department of Medicine, Karolinska University Hospital, Stockholm, Sweden

Synonyms

Increased extracellular levels; Neuronal release; Release of CCK

Definition

Cholecystokinin (CCK) was first identified as a gastrointestinal hormone, contracting the gallbladder and stimulating the secretion of the exocrine pancreas (Jorpes and Mutt 1966). CCK (or gastrin-like immunoreactivity) was demonstrated in the CNS (Vanderhaeghen et al. 1975). The predominant form of CCK in gastrointestinal tissue is CCK-33, whereas the predominant form in the CNS is the sulfated CCK-8 (Rehfeld 1978).

"Release of CCK" refers to neuronal release. However, a direct measurement of synaptic release of CCK in nervous tissue has so far not been possible. Instead, information on release has to be obtained by indirect measures. This essay is based on in vitro release (▶ Release of CCK), in vivo release (▶ Release of CCK), and receptor ▶ binding studies. Studies of the behavioral effects of CCK antagonists in combination with opioids are not addressed here.

Characteristics

In Vitro Release Studies

The effect of opioids on CCK-like immunoreactivity (CCK-LI) release in vitro may seem confusing, since not only inhibition (Table 1) but also potentiation (Table 2) or a biphasic effect (Table 3) on potassium-induced release has been reported.

Even though an opioid-induced inhibition of potassium-induced release of CCK in tissue sections may seem to be a good indicator for the drug effect in vivo, it has to be kept in mind that potassium stimulation induces a depolarization of both inhibitory and excitatory neurons. Thus, the interpretation of inhibited potassium-induced release by opioids is difficult.

In Vivo Release Studies

With regard to clinical relevance, an in vivo situation, preferably using awake animals, may seem more appropriate. Furthermore, studies on the effect of opioids on CCK release should focus on the basal non-stimulated release.

Systemic administration of morphine, at doses ranging from 2.5 to 10 mg/kg, has been demonstrated to induce an increased outflow of CCK-LI in the frontal cortex (Becker et al. 1999), the nucleus accumbens (Hamilton et al. 2000), and the dorsal horn of the spinal cord in the rat in vivo (de Araujo Lucas et al. 1998; Gustafsson et al. 1999). However, increasing intraperitoneal doses of morphine ranging from 7 to 30 mg/kg over a period of 6 days did not induce any significant change of the extracellular level of CCK-LI in the PAG of awake rats compared to saline-treated animals (Nieto et al. 2001). In the same publication, a single dose of naloxone (1 mg/kg s.c.), which precipitated withdrawal did not induce statistically significant changes of the CCK-LI level.

Opioid Receptors Mediating CCK: LI Release

In order to determine which opioid receptor mediates CCK-LI release in vivo, the effect of different opioid antagonists on morphine-induced release has been studied (Table 4).

δ-opioid antagonists blocked the CCK-LI release of morphine, and δ-opioid agonists mimicked the CCK-LI releasing effect of morphine. A δ-opioid receptor mediated inhibition of CCK-LI release is also supported by some in vitro experiments (Table 2).

With regard to the effect of μ-opioid receptor activation on CCK-LI release, this opioid receptor does not seem to induce CCK-LI release. The in vitro data indicate an inhibitory effect on the potassium-evoked release (Table 1). We could not demonstrate an increased non-stimulated CCK-LI level in the dorsal horn in vivo during local (1 μM in the microdialysis perfusion medium) or systemic (1 mg/kg s.c.) administration of the selective μ-opioid receptor agonist DAMGO (Gustafsson et al. 2001). However,

Opioid-Induced Release of CCK, Table 1 Opioid-induced inhibition of potassium-induced release of CCK-LI in vitro

Region (species)	Opioid agonist	Agonist concentration in perfusate (M)	Effect reversed by antagonist	Reference
Hypothalamus (cat/rat)	Nonselective: morphine	$10^{-11} - 10^{-8} + 10^{-5}$	Nonselective: naloxone	(Micevych et al. 1982, 1984, 1985)
	δ-opioid agonist: DADL	10^{-12}–10^{-10}	Nonselective: naloxone	
Substantia nigra (rat)	μ-Opioid agonist:			(Benoliel et al. 1992)
	DAGO	10^{-5}	Nonselective: naloxone	
	Morphiceptin (PL017)	10^{-5}	Nonselective: naloxone	
Dorsal lumbar spinal cord (rat)	μ-Opioid agonist: DAGO	10^{-8}–10^{-5}	Nonselective: naloxone	(Benoliel et al. 1994)

Opioid-Induced Release of CCK, Table 2 Opioid-induced potentiation of potassium-induced release of CCK-LI in vitro

Region (species)	Opioid agonist	Agonist concentration in perfusate (M)	Effect reversed by antagonist	Reference
Substantia nigra (rat)	δ-agonist:			(Benoliel et al. 1992)
	D-Pen2, D-Pen5-enkephalin	5×10^{-5}		
	Tyr-D-Thr-Gly-Phe-Leu-Thr (DTLET)	3×10^{-6}	δ-Antagonist: ICI-154129	
Dorsal lumbar spinal cord (rat)	Nonselective: morphine	10^{-5}		(Benoliel et al. 1991)

Opioid-Induced Release of CCK, Table 3 Biphasic effect of opioids on potassium-induced release of CCK-LI in vitro

Region (species)	Opioid agonist	Agonist concentration for inhibition (M)	Agonist concentration for excitation (M)	Antagonist for reversal of effect	Reference
Dorsal lumbar spinal cord (rat)	δ-agonist: DTLET	10^{-8}–3×10^{-6}	10^{-5}	δ-antagonist: naltrindole ICI 154129	(Benoliel et al. 1991, 1994)
	Nonselective: morphine	10^{-8}–10^{-7}	10^{-5}	Inhibition: naloxone	
				Excitation: naltrindole ICI 154129	

Opioid-Induced Release of CCK, Table 4 Reversal of opioid-induced in vivo release of CCK-LI by opioid antagonists

Region (species)	Opioid agonist (dose)	Effect reversed by antagonist	Effect not affected by	Reference
Frontal cortex (rat)	Nonselective: morphine (10 mg/kg i.p.)	Nonselective: naloxone (1.5 mg/kg i.p.) (10 μM)	μ-antagonist: CTOP (10 μM)	(Becker et al. 1999)
		δ-antagonist: naltrindole:10 μM	κ-antagonist: nor-BNI:10 μM	
		δ2-antagonist: naltriben:10 μM	δ1-antagonist: BNTX:10 μM	
	δ2-agonist: [D-Ala²] deltorphin II (1 μM)	δ2-antagonist: naltriben:10 μM		
Dorsal lumbar spinal cord (rat)	Nonselective: morphine (5 mg/kg s.c.)	Nonselective: naloxone: 2 mg/kg s.c.; 10 μM	μ-antagonist: CTOP:10 μM	(Gustafsson et al. 1999, 2001)
		δ-Antagonist: naltrindole:10 μM	κ-antagonist: nor-BNI:10 μM	
	δ-agonist: BW373U86 (1 mg/kg s.c.)	δ-antagonist: naltrindole; 10 μM		
	δ2-agonist: [D-Ala²] deltorphin II (1 μM)			

since the basal CCK-LI level in the microdialysate was close to the detection limit of the RIA, the possibility of μ-opioid receptor mediated CCK release should not be excluded.

An indirect indication of δ-opioid receptor-mediated release of CCK is also provided by a receptor-binding study with a selective CCK-B radioligand (Ruiz-Gayo et al. 1992). Following an intracerebroventricular (ICV) injection of the δ-opioid receptor agonist BUBUC, but not the μ-opioid receptor agonist DAMGO, the specific binding of the CCK-B radioligand decreased. According to the author's interpretation, an endogenous release of CCK competes with the radioligand for the receptor and decreases the specific binding.

The combined experimental evidence indicates that agonist activation of δ-opioid receptors, probably of the δ_2-opioid receptor subtype, induces release of CCK. A decrease of CCK release by opioids is only reported in conditions of potassium stimulation.

Mechanisms of Opioid-Induced Release of CCK-LI

The exact mechanisms responsible for the opioid-induced release of CCK-LI have not been determined so far. The morphine-induced release of CCK-LI is calcium dependent, indicating a neuronal origin. Furthermore, we have demonstrated that an influx of calcium at presynaptic terminals, through voltage-dependent calcium channels (VDCCs) of the L and N type, is essential for morphine-induced release of CCK-LI in the dorsal horn (Gustafsson et al. 1999).

Since opioid receptor activation normally induces inhibition of neuronal activity and reduces transmitter release, a disinhibition of local inhibitory neurons (i.e., GABA neurons), a mechanism which has been demonstrated in the periaqueductal gray in vivo (Stiller et al. 1996), could be an alternative. However, no evidence for this mechanism in the dorsal horn has been provided so far. Furthermore, we could not demonstrate that perfusion of the dialysis probe with the GABA-A antagonist bicuculline (100 μM in the perfusion medium) increased the

extracellular levels of CCK-LI in the spinal cord (Gustafsson 2001). An opioid-mediated disinhibition of other inhibitory neurotransmitters (e.g., glycine) could still be an option.

However, opioid receptor activation may not only mediate an inhibition but also excitation (for review, see Crain and Shen 1990). If δ-opioid receptors are present in CCK-containing nerve cells in the dorsal horn, a direct activation of δ-opioid receptors should release CCK-LI in this region. Such a mechanism should be insensitive to the sodium channel-blocking agent tetrodotoxin (TTX). However, TTX inhibited the morphine-induced CCK-LI release (Gustafsson et al. 1999), suggesting that a local propagation of action potentials in the dorsal horn seems to be a prerequisite for the opioid-induced release of CCK. Thus, it is possible that the opioid-induced CCK-LI release involves several neurons.

An opioid-induced release of excitatory amino acids, which in turn stimulate the release of CCK, is indicated by the observation that administration of NMDA or AMPA antagonists prevents the morphine-induced release of CCK-LI in the dorsal horn (Gustafsson 2001). In addition, spinal morphine (100 μM in the perfusion fluid) induced a several-fold increase of the extracellular glutamate concentration in the dorsal horn in vivo (Gustafsson 2001).

References

Becker, C., Hamon, M., Cesselin, F., & Benoliel, J. J. (1999). Delta(2)-opioid receptor mediation of morphine-induced CCK release in the frontal cortex of the freely moving rat. *Synapse, 34*, 47–54.

Benoliel, J. J., Bourgoin, S., Mauborgne, A., Legrand, J. C., Hamon, M., & Cesselin, F. (1991). Differential inhibitory/stimulatory modulation of spinal CCK release by mu and delta opioid agonists, and selective blockade of mu-dependent inhibition by kappa receptor stimulation. *Neuroscience Letters, 124*, 204–207.

Benoliel, J. J., Mauborgne, A., Bourgoin, S., Legrand, J. C., Hamon, M., & Cesselin, F. (1992). Opioid control of the in vitro release of cholecystokinin-like material from the rat substantia nigra. *Journal of Neurochemistry, 58*, 916–922.

Benoliel, J. J., Collin, E., Mauborgne, A., Bourgoin, S., Legrand, J. C., Hamon, M., et al. (1994). Mu and delta opioid receptors mediate opposite modulations by morphine of the spinal release of cholecystokinin-like material. *Brain Research, 653*, 81–91.

Crain, S. M., & Shen, K. F. (1990). Opioids can evoke direct receptor-mediated excitatory effects on sensory neurons. *Trends in Pharmacological Sciences, 11*, 77–81.

de Araujo Lucas, G., Alster, P., Brodin, E., & Wiesenfeld-Hallin, Z. (1998). Differential release of cholecystokinin by morphine in rat spinal cord. *Neuroscience Letters, 245*, 13–16.

Gustafsson, H. (2001). *Opioid-induced cholecystokinin release in the CNS-neurochemical mechanisms and effects of sciatic nerve lesion.* Ph.D. thesis, Karolinska Institutet, ISBN 91-628-4440-7.

Gustafsson, H., Afrah, A., Brodin, E., & Stiller, C. O. (1999). Pharmacological characterization of morphine-induced in vivo release of cholecystokinin in rat dorsal horn: Effects of ion channel blockers. *Journal of Neurochemistry, 73*, 1145–1154.

Gustafsson, H., Afrah, A. W., & Stiller, C. O. (2001). Morphine-induced in vivo release of spinal cholecystokinin is mediated by delta-opioid receptors-effect of peripheral axotomy. *Journal of Neurochemistry, 78*, 55–63.

Hamilton, M. E., Redondo, J. L., & Freeman, A. S. (2000). Overflow of dopamine and cholecystokinin in rat nucleus accumbens in response to acute drug administration. *Synapse, 38*, 238–242.

Jorpes, E., & Mutt, V. (1966). Cholecystokinin and pancreozymin, one single hormone? *Acta Physiologica Scandinavica, 66*, 196–202.

Micevych, P. E., Yaksh, T. L., & Go, V. L. (1982). Opiate-mediated inhibition of the release of cholecystokinin and substance P, but not neurotensin from cat hypothalamic slices. *Brain Research, 250*, 283–289.

Micevych, P. E., Yaksh, T. L., & Go, V. L. (1984). Studies on the opiate receptor-mediated inhibition of K+-stimulated cholecystokinin and substance P release from cat hypothalamus in vitro. *Brain Research, 290*, 87–94.

Micevych, P. E., Yaksh, T. L., Go, V. W., & Finkelstein, J. A. (1985). Effect of opiates on the release of cholecystokinin from In vitro hypothalamus and frontal cortex of Zucker lean (Fa/–) and obese (fa/fa) rats. *Brain Research, 337*, 382–385.

Nieto, M. M., Wilson, J., Cupo, A., Roques, B. P., & Noble, F. (2001). Chronic morphine treatment modulates the extracellular levels of endogenous enkephalins in rat brain structures involved in opiate dependence: A microdialysis study. *Journal of Neuroscience, 22*(3), 1034–1041.

Rehfeld, J. F. (1978). Localisation of gastrins to neuro- and adenohypophysis. *Nature, 271*, 771–773.

Ruiz-Gayo, M., Durieux, C., Fournié-Zaluski, M. C., & Roques, B. P. (1992). Stimulation of delta-opioid receptors reduces the in vivo binding of the cholecystokinin (CCK)-B-selective agonist [3H]pBC 264:

Evidence for a physiological regulation of CCKergic systems by endogenous enkephalins. *Journal of Neurochemistry, 59*, 1805–1811.

Stiller, C. O., Bergquist, J., Beck, O., Ekman, R., & Brodin, E. (1996). Local administration of morphine decreases the extracellular level of GABA in the periaqueductal gray matter of freely moving rats. *Neuroscience Letters, 209*, 165–168.

Vanderhaeghen, J. J., Signeau, J. C., & Gepts, W. (1975). New peptide in the vertebrate CNS reacting with antigastrin antibodies. *Nature, 257*, 604–605.

Opioids and Bladder Pain/Function

Karl-Erik Andersson[1] and Dagmar Westerling[2]
[1]Wake Forest Institute for Regenerative Medicine, Wake Forest University School of Medicine, Winston Salem, USA
[2]Department of Anesthesiology, Central Hospital of Kristianstad, Kristianstad, Sweden

Definition

Opioids are ligands (agonists and antagonists) at opioid receptors (κ, δ, and μ or OR 1–3). Opioid receptors are mainly located on primary afferents in the dorsal horn and in several supraspinal centers of importance for micturition control, such as the periaqueductal gray. There are also opioid receptors on primary afferents in peripheral tissues, including the bladder.

Bladder pain may be evoked from distension, inflammation, and/or physical damage (surgery, bladder stone, tumor growth, etc.). Bladder pain is a type of ▶ visceral nociception and pain. Opioid receptor agonists may be effective in the treatment of some types of bladder pain.

Bladder function comprises collection, storage, and emission of urine. The bladder and urethra constitute a functional unit, which is under spinal and supraspinal nervous control.

Administration of analgesic doses of opioid agonists may cause pain relief but also urinary retention due to a dose-dependent depression of micturition reflexes, by actions at local, spinal, and/or supraspinal sites.

Characteristics

Bladder Function

Anatomy

The lower urinary tract consists of the ▶ urinary bladder and the urethra. The bladder can be divided into two main components: the bladder body, which is located above the ureteral orifices, and the base, consisting of the trigone, urethrovesical junction, deep detrusor, and the anterior bladder wall (Fig. 1). The bladder is a hollow smooth muscle organ, lined by a mucous membrane, and covered on its outer aspect partly by peritoneal serosa and partly by fascia. Its muscular wall is formed of smooth muscle cells, which comprise the detrusor muscle. The detrusor is structurally and functionally different from, for example, trigonal and urethral smooth muscle. The urethra contains both smooth and striated muscles, the latter forming the external urethral sphincter (rhabdosphincter).

Innervation of the Urinary Bladder

The urinary bladder receives sympathetic, parasympathetic, sensory, and somato-motor innervation via the hypogastric, the pelvic, and the pudendal nerves (Morrison et al. 2002) (Fig. 2). The smooth muscle layers of the bladder wall receive autonomic (mainly sympathetic but also parasympathetic fibers) and afferent innervation. The striated muscle of the external sphincter is under voluntary somato-motor control. There is a considerable overlap in the source of the innervation. Primary afferent nerves from the bladder have their cell bodies in the dorsal root ganglia of the thoracolumbar and sacral spinal cord. From the dorsal root ganglia, sensory nerve cell bodies project both to the bladder, where information is received, and to the spinal cord, where the efferent pathways originate and where the ▶ primary afferents of the urinary bladder connect to the spinothalamic tracts.

The primary afferents are myelinated (Aδ) and thin unmyelinated (C) fibers, with nerve endings in the lamina propria and in the smooth muscle layers of the bladder and urethra. Retrograde tracing studies have shown that most of the sensory innervation of the bladder and urethra

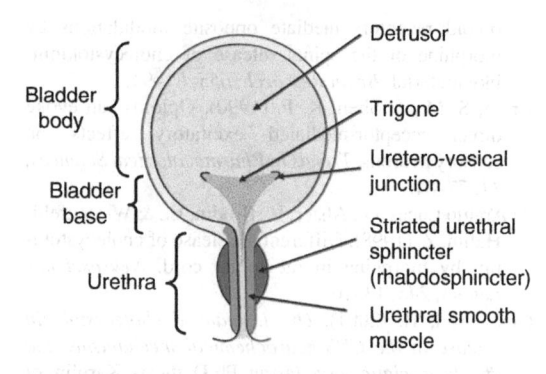

Opioids and Bladder Pain/Function, Fig. 1 Anatomy of the urinary bladder. The bladder can be divided into two main components: the bladder body, which is located above the ureteral orifices, and the base, consisting of the trigone, urethrovesical junction, and the detrusor

originates in the thoracolumbar region and travels via the pelvic nerve. In addition, some afferents originating in ganglia at the thoracolumbar level of the sympathetic outflow project via the hypogastric nerve. The sensory nerves of the striated muscle in the rhabdosphincter travel in the pudendal nerve to the sacral region of the spinal cord. Sacral sensory nerve terminals are uniformly distributed to all areas of the detrusor and urethra, whereas lumbar sensory nerve endings are most frequent in the trigone and scarce in the bladder body. In the urinary bladder of both humans and animals, sensory nerves have been identified suburothelially as well as in the detrusor muscle. Suburothelially, they form a nerve plexus, which lies immediately beneath the epithelial lining. Some terminals may even be located within the basal parts of the urothelium. This suburothelial plexus is relatively sparse in the dome of the bladder but becomes progressively denser near the bladder neck, and it is particularly prominent in the trigone.

The hypogastric and pelvic pathways are implicated, in the sensations associated not only with normal bladder filling but also with bladder pain. The pelvic and pudendal pathways are also concerned with the sensation that micturition is imminent and with thermal sensations from the urethra. The incoming information controls the activity in the parasympathetic, sympathetic, and somatic efferent nerves to the lower urinary tract. Under normal conditions, there is little ongoing

Opioids and Bladder Pain/Function,
Fig. 2 Innervation of the
urinary bladder. The
urinary bladder receives
efferent sympathetic
(hypogastric nerve),
parasympathetic (pelvic
nerve), and somato-motor
(pudendal nerve)
innervation. Afferent
(sensory) information is
conveyed via the
hypogastric, pelvic, and
pudendal nerves. In the
figure, only afferent
pathways in the pudendal
nerve have been indicated
(*dotted line*)

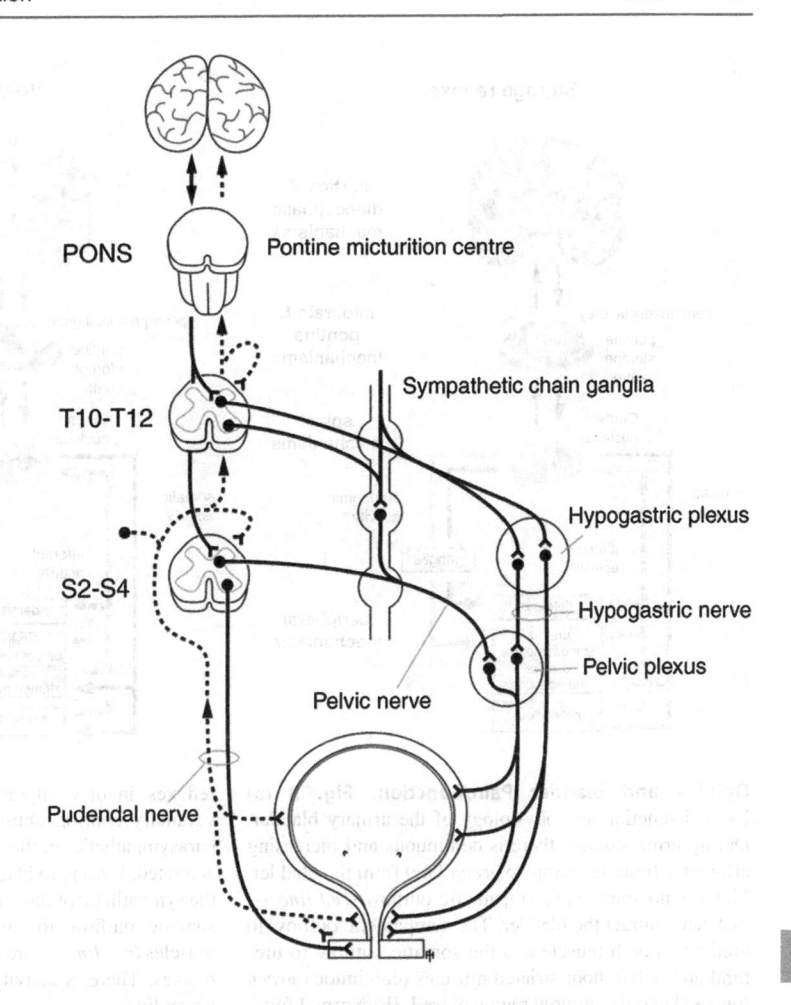

activity in the primary afferent nerves corresponding to little or no conscious sensations from the empty and slowly filling urinary bladder.

Physiology: Initiation of Micturition

Bladder pressure increases very modestly during normal collection and storage of urine. When bladder pressure reaches an individual level of around 25–35 cm H_2O, there is an urge to void. Under normal conditions, this is a non-painful sensation. The smooth muscle of the detrusor contracts and urine is passed if the bladder neck and internal and external sphincters are relaxed at the same time. All of this requires a complex combination of nervous stimulation and inhibition (Fig. 3).

Myogenic activity, distension of the detrusor, and signals from the urothelium may initiate voiding (Andersson 2002; Birder and de

Groat 2007). The normal stimulus for initiating micturition is distension of the bladder, activating mechanoreceptors in the bladder wall. A high level of activity can be recorded in small myelinated afferent nerves (Aδ), which reach the lumbosacral spinal cord via the dorsal root ganglia. These Aδ afferents connect to a spinobulbospinal reflex consisting of an ascending limb from the lumbosacral spinal cord; an integration center in the rostral brain stem, which is known as the pontine micturition center (PMC); and a descending limb from the PMC back to the parasympathetic nucleus in the lumbosacral spinal cord (Kanai and Andersson 2010). In the fetus and neonate, afferent information is conveyed by small unmyelinated (C-fiber) vesical afferents, which have a high mechanical threshold. Activity in these fibers is suppressed during development ("silent C fibers") but may be activated when the

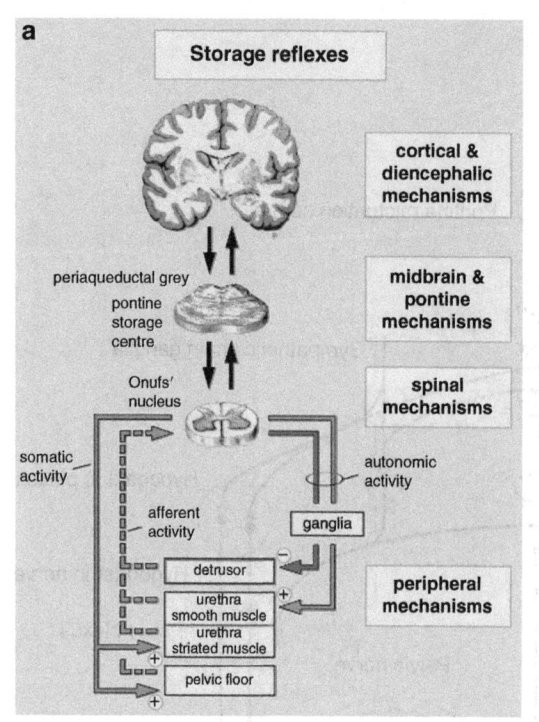

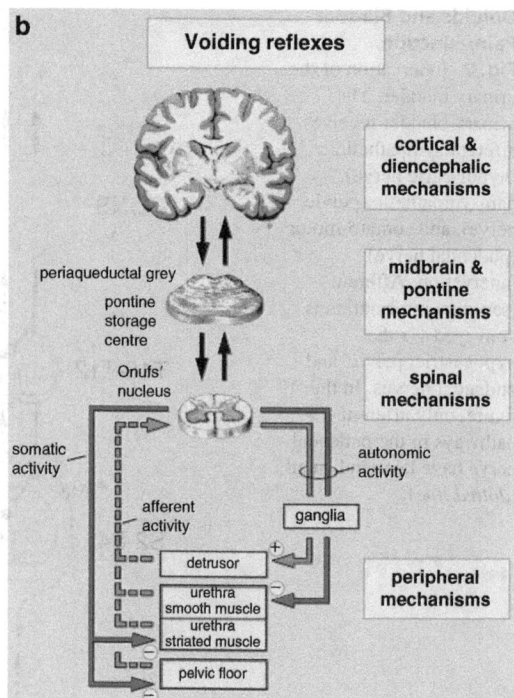

Opioids and Bladder Pain/Function, Fig. 3 (a) Normal function and physiology of the urinary bladder. During urine storage, there is continuous and increasing afferent activity (interrupted *green line*) from the bladder. There is no spinal parasympathetic outflow (*red line* −) that can contract the bladder. The sympathetic outflow to urethral smooth muscle and the somatic outflow to urethral and pelvic floor striated muscles (continuous *green lines* +) keep the outflow region closed. (**b**) Normal function and physiology of the urinary bladder. Voiding reflexes involve supraspinal pathways and are under voluntary control. During bladder emptying, the spinal parasympathetic outflow (continuous *green line* +) is activated, leading to bladder contraction. Simultaneously, the sympathetic outflow to urethral smooth muscle and the somatic outflow to urethral and pelvic floor striated muscles (*red line* −) are turned off, and the outflow region relaxes. There is activity in afferent nerves (interrupted *green line*)

long reflex is damaged, as in spinal cord injuries, or by inflammation of the bladder mucosa, for example.

Extracellular adenosine triphosphate (ATP) has been found to mediate excitation of small-diameter sensory neurons via P2X$_3$ receptors, and it has been shown that bladder distension causes release of ATP from the urothelium. In turn, ATP can activate P2X$_3$ receptors on suburothelial afferent nerve terminals to evoke a neural discharge. However, it is not only ATP but also a cascade of inhibitory and stimulatory transmitters/mediators that are most probably involved in the transduction mechanisms underlying the activation of afferent fibers during bladder filling. The urothelium may serve as a mechanosensor which, by producing NO, ATP, and other mediators, can control the activity in afferent nerves and thereby the initiation of the micturition reflex. The ▶ firing of suburothelial afferent nerves, and the threshold for bladder activation, may be modified by both inhibitory (e.g., NO) and stimulatory (e.g., ATP, tachykinins, prostanoids) mediators. These mechanisms can be involved in the generation of bladder overactivity, causing urge, frequency, and incontinence, but also in bladder pain.

Pathophysiology: Effect of Inflammation and Distension

If micturition is prevented and bladder pressure increases above 25–35 cm H$_2$O, the sensation of bladder fullness becomes more and more unpleasant, and most subjects describe this as painful (Ness et al. 1998). The pressure in the bladder, although painful, is not even near pressure levels that potentially might be harmful to the tissues. Pain arising from the bladder is usually reported from the lower abdomen above the symphysis but may also be referred to the inner thigh, groin, and knees.

Inflammation of the bladder may be caused by infections (acute bacterial or viral cystitis), or there may be no identifiable causing agent (e.g., ▶ Interstitial Cystitis). Exposure to chemicals (poisoning, cytostatics, etc.) and radiation may also lead to inflammation. Bladder inflammation causes distress, severe pain, and micturition disturbances with an increase in voiding frequency and a decrease in micturition volume. The bladder lamina propria is invaded by inflammatory cells, and there is edema and extravasation of plasma proteins.

Inflammation of the bladder causes increased firing in the afferent nerve fibers "silent" under normal conditions, which leads to increased sensitivity in the dorsal horn. This is due to activation of several receptors/systems, including N-methyl-D-aspartate (NMDA) receptors, nitric oxide (NO), nerve growth factor (NGF), and substance P (SP). The increased sensitivity may be persistent, and the central nervous system (CNS) may continue to signal pain, even in the absence of ongoing stimuli from the urinary bladder.

As the neurons in the spinothalamic tract receive painful stimuli from bladder afferents, there are also reports of pain from somatic structures at the same spinal level as the bladder (L4–S2). This phenomenon is defined as referred pain and is due to the convergence of visceral primary afferents upon the neurons of the spinothalamic tract, where somatic afferent signalling is also projecting. Referred hyperalgesia with a lowering of pain thresholds in the areas of referred pain may also be detected (McMahon et al. 1995), as well as referral of hyperalgesia to other visceral organs (viscero-visceral hyperalgesia) (Giamberardino 1999).

Bladder Pain: Animal Models
Instillation of, for example, capsaicin, prostaglandin E_2, turpentine, or mustard oil into the bladder causes pain behavior in rats. A widely used pain model is intraperitoneal injection of cyclophosphamide into rats and mice. Cyclophosphamide is metabolized to acrolein, which is excreted in urine and produces severe inflammation of the bladder (hemorrhagic cystitis).

Cat models simulating interstitial cystitis in humans have been developed.

Bladder Pain: Clinical States
Pain from the bladder is often diffuse and poorly located. The pain may be felt in the midline above the symphysis. It may also be referred to the inner thighs, the groin, or even to the knees. Pain may not be severe but is rather described as unpleasant and dull (Ness et al. 1998). Bladder pain may be accompanied by signs of activation of the autonomous nervous system, like sweating, nausea, or dizziness. Pain may be continuous, with or without exacerbations, or appear in intervals, which may not be possible to anticipate. Typically, the contractions of an inflamed bladder cause severe, sudden, repeated cramping pain diffusely located to the midline in the lower abdomen. Bladder pain may be acute (acute infectious cystitis) or long lasting (chronic interstitial cystitis, malignancies, bladder stones, etc.). According to the International Association for the Study of Pain (IASP) nomenclature, a painful process lasting longer than 3 months or more, is defined as chronic.

Bladder Function: Effects of Opioids
Morphine, given by various routes of administration to animals and humans, can increase bladder capacity or block bladder contractions. Furthermore, given intrathecally (i.t.) to anesthetized rats and intravenously (i.v.) to humans, the μ-opioid receptor antagonist naloxone has been shown to stimulate micturition, suggesting that a tonic activation of μ-opioid receptors has a depressant effect on the micturition reflex.

Morphine given i.t. was effective in patients with detrusor overactivity due to spinal cord lesions (Herman et al. 1988) but was associated with side effects such as nausea and pruritus. Further side effects of opioid receptor agonists include respiratory depression, constipation, and abuse. Attempts have been made to reduce these side effects by increasing selectivity towards one of the different opioid receptor types. Theoretically, selective receptor actions, or modifications of effects mediated by specific opioid receptors, may have useful therapeutic effects for micturition control.

Tramadol, a well-known analgesic, is a very weak μ-receptor agonist. Tramadol is metabolized to several different compounds, some of them almost as effective as morphine at the μ receptor. However, the drug also inhibits serotonin (5-HT) and noradrenaline reuptake (Raffa and Friderichs 1996). This profile is of particular interest, since both μ-receptor agonism and amine reuptake inhibition may be useful principles for the treatment of detrusor overactivity (Pandita et al. 2003).

Central stimulation of δ-opioid receptors in anesthetized cats and rats inhibited micturition and parasympathetic neurotransmission in cat bladder ganglia. In humans, nalbuphine, a μ-receptor antagonist and κ-receptor agonist, increased bladder capacity (Malinovsky et al. 1998). Buprenorphine (a partial μ-receptor agonist and κ-receptor antagonist) decreased micturition pressure and increased bladder capacity more than morphine (Malinovsky et al. 1998). Animal studies have shown that nociceptin or orphanin FQ, the endogenous ligand of opioid-like receptors-1 (ORL1), may have an inhibitory effect on the micturition reflex in the rat (Lecci et al. 2000). In patients with neurogenic detrusor overactivity, nociceptin/orphanin FQ given intravenously significantly increased bladder capacity and volume threshold for the appearance of detrusor overactivity (Lazzeri et al. 2003). Further exploration of these non-μ-opioid-receptor-mediated actions on micturition seems motivated.

Symptoms from the bladder and urinary tract are rarely included in questionnaires for assessment of pain and quality of life and may therefore be overlooked. Dysuria, defined as bladder dysfunction, disturbances in micturition, incomplete elimination of urine, or full bladder with inability to void (Bremnor and Sadovsky 2002), was reported in 15 % of advanced cancer patients treated with opioids (Mercadante et al. 2011) and is thus a common problem in this large group of patients.

Bladder Pain: Effect of Opioids

Clinically used opioid analgesics produce pain relief by activating opioid receptors. Most opioid analgesics are μ agonists. κ and δ receptors and other receptor systems (NMDA, monoaminergic transmission) may also be affected by some opioid analgesics and may thus be of special interest in the treatment of bladder pain. By binding to opioid receptors, opioid analgesics provide pain relief through actions on GABA interneurons. This causes activation of descending bulbospinal tracts. There is an increase in intracellular potassium, inhibition of some types of voltage-gated calcium channels and of adenylyl cyclase, thereby decreasing the production of cAmp. Transmitter release is inhibited, neuronal firing rate is decreased, and gene expression changes, all of which modulates the transmission of afferent pain signals from the bladder, as well as from other visceral and somatic structures.

Nociceptive pain is generally more sensitive to opioids than neuropathic pain. However, distension of the bladder produces nociceptive pain, although the low-pressure range of the bladder is not noxious. Painful bladder contractions are a type of incident pain, which may be difficult to treat due to its fast onset, as even the most fast-acting opioid analgesics have an onset time of at least 1–2 min after injection.

Clinical Considerations: Routes of Administration of Opioids

Opioid analgesics are indicated in opioid-sensitive bladder pain caused by, for example, operations, trauma, or cancer (Guidelines on Pain Management 2009). Dull continuous pain from the urinary bladder may be reduced by opioid analgesics. In patients with bladder pain dominated by painful, frequent bladder contractions, opioid analgesics may provide dramatic pain relief.

Opioid treatment of chronic painful conditions like bladder pain in interstitial cystitis may provide pain relief and improve quality of life.

However, there are few, long-term studies of conventional analgesics and interstitial cystitis (Fall et al. 2010; Hanno et al. 2011). Before long-term opioid analgesic treatment of chronic benign conditions is initiated, careful consideration must be taken to inherent problems of tolerance, side effects, and the goals of the treatment. Drug (legal or illegal) dependence, in the past or present, constitutes a relative contraindication to long-term opioid treatment and complicates any analgesic therapy.

Opioid-sensitive pain from the bladder should be treated promptly with adequate doses using a suitable route of administration. The oral route of administration is the preferred route for short- and long-term therapy. Transdermal fentanyl or buprenorphine patches may be convenient alternatives for special groups of patients.

Injections of opioids or patient-controlled analgesia (PCA) may be more effective in patients with malignant disease and bladder pain who have difficulties in eating and are nauseated, or in patients who suffer from severe, sudden incident pain caused by severe bladder contractions.

Spinal (epidural or i.t.) administration of opioids may be extremely effective, combined with local anesthetics or clonidine for superior effect for postoperative pain relief and for long-term use in patients with malignant disease. However, urinary retention is common after spinal administration of opioids. Bladder catheters may be necessary but carry an increased risk of infection.

Intravesical administration of morphine has been recommended in children undergoing vesicoureteral surgery (Duckett et al. 1997). The rationale for intravesical administration of comparatively low doses of morphine in opioid-naïve patients would be to activate peripheral opioid receptors with little or no side effects. In a controlled randomized study of postoperative pain, however, no advantages of a continuous bladder infusion of morphine compared to placebo could be demonstrated (El-Ghoneimi et al. 2002).

Pain relief in a suffering subject is dependent on the qualities of the pain itself, qualities within the person, and of the particular opioid at hand depending on the route of administration. Acute and subacute side effects and long-term effects of the analgesic treatment may ultimately determine the success or failure of therapy aimed at relief of bladder pain.

References

Andersson, K.-E. (2002). Bladder activation: Afferent mechanisms. *Urology, 59*, 43–50.

Birder, L. A., & de Groat, W. (2007). Mechanisms of disease: Involvement of the urothelium in bladder dysfunction. *Nature Clinical Practice Urology, 4*(1), 46–54.

Bremnor, J. D., & Sadovsky, R. (2002). Evaluation of dysuria in adults. *American Family Physician, 65*(8), 1589–1596.

Duckett, J. W., Cangiano, T., Cubina, M., et al. (1997). Intravesical morphine analgesia after bladder surgery. *Journal of Urology, 157*, 1407–1409.

El-Ghoneimi, A., Deffarges, C., Hankard, R., et al. (2002). Intravesical morphine analgesia is not effective after bladder surgery in children: Results of a randomised double-blind study. *Journal of Urology, 168*, 694–697.

Fall, M., Baranowski, A. P., Elnei, S., et al. (2010). EAU guidelines on chronic pelvic pain. *European Urology, 57*, 35–48.

Giamberardino, M. A. (1999). Recent and forgotten aspects of visceral pain. *European Journal of Pain, 3*, 77–92.

Guidelines on Pain Management. (2009). *European Association of Urology*, 2009.

Hanno, P. M., Burks, D. A., Clemens, J. Q., et al. (2011). AUA guideline for the diagnosis and treatment of interstitial cystitis/bladder pain syndrome. *Journal of Urology, 185*, 2162–2170.

Herman, R. M., Wainberg, M. C., delGiudice, P. F., et al. (1988). The effect of a low dose of intrathecal morphine on impaired micturition reflexes in human subjects with spinal cord lesions. *Anesthesiology, 69*, 313–318.

Kanai, T., & Andersson, K. E. (2010). Bladder afferent signaling: Recent findings. *Journal of Urology, 183*(4), 1288–1295.

Lazzeri, M., Calo, G., Spinelli, M., et al. (2003). Urodynamic effects of intravesical nociceptin/orphanin FQ in neurogenic detrusor overactivity: A randomized, placebo-controlled, double-blind study. *Urology, 61*, 946–950.

Lecci, A., Giuliani, S., Meini, S., et al. (2000). Nociceptin and the micturition reflex. *Peptides, 21*, 1007–1021.

Malinovsky, J. M., Le Normand, L., Lepage, J. Y., et al. (1998). The urodynamic effects of intravenous opioids and ketoprofen in humans. *Anesthesia and Analgesia, 87*, 456–461.

McMahon, S. B., Dmitrieva, N., & Koltzenburg, M. (1995). Visceral pain. *British Journal of Anaesthesia, 75*, 132–144.

Mercadante, S., Ferrera, P., & Casuccio, A. (2011). Prevalence of opioid-related dysuria in patients with advanced cancer having pain. *The American Journal of Hospice & Palliative Care, 28*(1), 27–30.

Morrison, J., Steers, W. D., Brading, A., et al. (2002). Neurophysiology and neuropharmacology. In P. Abrams, S. Khoury, & A. Wein (Eds.), *Incontinence, 2nd International consultation on incontinence* (pp. 85–161). Plymouth: Plymbridge Distributors Ltd.

Ness, T. J., Richter, H. E., Varner, R. E., et al. (1998). A psychophysical study of discomfort produced by repeated filling of the urinary bladder. *Pain, 76*, 61–69.

Pandita, R. K., Pehrson, R., Christoph, T., et al. (2003). Actions of tramadol on micturition in awake, freely moving rats. *British Journal of Pharmacology, 139*, 741–748.

Raffa, R. B., & Friderichs. (1996). The basic science aspect on tramadol hydrochloride. *Pain Reviews, 3*, 249–271.

Opioids and Gene Therapy

David J. Fink
Department of Neurology, University of Michigan and VA Ann Arbor Healthcare System, Ann Arbor, MI, USA

Synonyms

Gene therapy and opioids; Therapeutic gene transfer; Viral vectors

Definition

Gene therapy refers to the use of ▶ vectors to deliver genetic material in order to express a foreign gene ("▶ Transgene") in vivo. Gene delivery vectors are commonly constructed from viruses that have been modified to reduce their pathogenic potential but that retain the ability of the native virus to deliver RNA or DNA (Table 1). Vector-mediated gene transfer of the genes coding for opioid peptides and of genes coding for opiate receptors has been used to achieve or enhance regional analgesic effects in animal models of chronic pain.

Characteristics

Drugs acting at opioid receptors remain the most potent analgesic agents available, but their effective use in severe chronic pain is limited by three phenomena: (1) the widespread distribution of receptors within and outside the neuraxis results in side effects unrelated to analgesia that limit the maximum dose that may be administered, (2) continued use of opiate drugs results in tolerance, and (3) addiction and abuse are problematic. Three different gene therapy approaches to the treatment of chronic pain have been described (Fig. 1).

The first strategy uses injection of a gene transfer vector into CSF to transduce cells of the meninges in order to provide the continuous release of natural opioid peptides into the CSF. A replication-deficient ▶ adenoviral (Ad) vector containing the coding sequence for human beta-endorphin injected into the subarachnoid space transduces cells of the pia mater, from which beta-endorphin is released into the cerebrospinal fluid. An analgesic effect could be demonstrated by attenuation of inflammatory hyperalgesia (Finegold et al. 1999). Indirect evidence suggests that the transgene product was acting a non-μ, non-δ, and non-κ1 site in the spinal cord.

The second approach involves direct injection of a gene transfer vector into the dorsal root ganglia (DRG) to transduce DRG neurons in vivo. A recombinant ▶ adeno-associated viral (AAV) vector, containing the cDNA coding for the μ-opioid receptor (MOR), inoculated directly into the DRG, produces a persistent increase in MOR expression in DRG to enhance the efficacy of morphine, measured by a 50 % reduction in ED_{50} of morphine in both normal rats and rats with inflammatory pain (Xu et al. 2003). The major disadvantage of this strategy is "that direct injection of rAAV into the DRG could incur tissue damage and thus limits its application in gene therapy."

Vectors constructed from recombinant ▶ herpes simplex virus (HSV) are particularly suitable to target transgene expression to DRG in an anatomically predictable manner from peripheral inoculation. HSV is a neurotropic virus that

Opioids and Gene Therapy, Table 1 Summary of the principal viral vectors used for gene transfer to the nervous system

Vector	Genome size (kb)	Genome type	Insert capacity (kb)	Integration	Characteristics
Retrovirus	8	ss-RNA	8	Yes	Requires cell division for gene transfer. Can be used to transduce cells that are then transplanted back into host
Adenovirus	36	ds-DNA	8	No	Highly efficient gene transfer following direct injection into tissue or inoculation into CSF; immunogenicity limits applications
Adeno-associated virus	4.7	ss-DNA	4.7	Yes	Non-immunogenic; prolonged transgene expression; limited insert capacity
Herpes simplex	152	ds-DNA	50	No	Effective transduction of DRG neurons from subcutaneous application; vector does not integrate into host genome
Lentivirus	9.4	ss-RNA	8	Yes	Efficient gene transfer following direct inoculation; prolonged high-level expression of transgenes

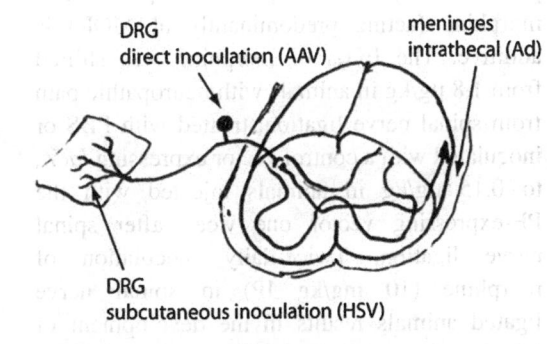

Opioids and Gene Therapy, Fig. 1 Three different approaches to gene transfer of opioid peptides or receptors have been described. Intrathecal inoculation of an Ad vector can be used to transduce cells of the meninges to release peptide into the CSF. Direct injection of an AAV vector into DRG can be used to overexpress opioid receptors in primary sensory neurons to increase opiate sensitivity. Subcutaneous inoculation of HSV vectors can be used to transduce primary sensory neurons in DRG; release of peptide from terminals in the spinal cord provides a regional analgesic effect

naturally infects skin and mucous membranes. Following the initial epithelial infection, wild-type HSV is taken up by nerve terminals in the skin and carried by retrograde axonal transport to the DRG. Uptake and transport of the virion from the skin is an efficient process, mediated first by interactions between specific viral envelope glycoproteins and high-affinity receptors in the sensory nerve terminals in the skin, followed by specific interactions of capsid and tegument proteins with dynein molecules in the axon, to mediate the retrograde transport along microtubules to the cell body. Deletion of a single gene (e.g., HSV thymidine kinase, tk) from the viral genome impairs the ability of the recombinant to replicate in neurons. Deletion of one or more essential immediate early genes allows the construction of HSV-based vectors that are incapable of replicating and therefore appropriate for application to human gene therapy.

A tk-deleted recombinant HSV vector containing the human proenkephalin (PE) gene under the control of the HSV latency associated promoter expresses PE RNA in DRG following subcutaneous inoculation in the skin and increased levels of enkephalin in DRG (Antunes Bras et al. 1998). A similar tk-deleted recombinant HSV vector, delivered by subcutaneous inoculation into the dorsum of the foot, has been shown to reduce thermal hyperalgesia after sensitization with capsaicin or dimethyl sulfoxide but does not alter baseline foot withdrawal responses to noxious radiant heat mediated by Aδ and C fibers (Wilson et al. 1999), a phenomenon that is reversed by naloxone. In a mouse model of polyarthritis, subcutaneous inoculation of the PE-expressing tk-deleted HSV vector into the foot reduces thermal hyperalgesia and improves locomotion in the treated animals (Braz et al. 2001). In the CFA polyarthritis model, vector-mediated enkephalin production not only reduces pain-related behaviors but also significantly reduces the arthritic destruction of joint space and bone measured radiographically (Braz et al. 2001).

A replication-incompetent recombinant HSV vector, deleted for the essential immediate early HSV gene ICP4, has also been demonstrated to provide a pain-relieving effect in rodent models. Despite being unable to replicate in epithelial cells, the PE-expressing replication-incompetent HSV vector injected subcutaneously establishes a "latent" state in DRG (Goss et al. 2001) and reduces spontaneous pain behavior during the delayed phase in the formalin test of inflammatory pain (10 min to 1 h after injection of formalin) without affecting the acute pain score. This effect is reversed by intrathecal administration of naltrexone (Goss et al. 2001). The analgesic effect was limited to the injected limb; formalin testing of the limb contralateral to the injection reveals no analgesic effect on that side (unpublished observation).

The vector-mediated analgesic effect of the PE-expressing vector persists for several weeks before waning. Animals tested 4 weeks after vector inoculation have shown no significant reduction in pain-related behavior during the delayed phase of the formalin test (Goss et al. 2001), but animals reinoculated with the vector 4 weeks after the initial inoculation, and then tested with formalin injection at 5 weeks, show a substantial and significant return of the antinociceptive effect (Goss et al. 2001).

In the spinal nerve ligation model of neuropathic pain, subcutaneous inoculation of the PE-expressing HSV vector into the foot 1 week after spinal nerve ligation produces an antiallodynic effect that lasts for several weeks before waning (Hao et al. 2003). Reinoculation of the vector 6 weeks after the initial inoculation reestablishes the antiallodynic effect; the magnitude of the effect produced by the reinoculation is at least as great as that produced by the initial injection and persists for a longer time after the reinoculation than the effect produced by the initial inoculation. Intraperitoneal naloxone reverses the antiallodynic effect.

The effect of the PE-expressing HSV vector-mediated antiallodynic effect in neuropathic pain is continuous throughout the day. Animals tested repeatedly at different times through the day show a similar elevation in threshold at all times tested (Hao et al. 2003). Intraperitoneal morphine produces a greater antiallodynic effect than the vector alone, but inoculation of the maximum dose of morphine produces an antiallodynic effect that persists for only 1–2 h before waning. The effect of vector-mediated enkephalin (acting predominantly at δ opioid receptors) and morphine (acting predominantly at MOR) is additive. The ED_{50} of morphine was shifted from 1.8 μg/kg in animals with neuropathic pain from spinal nerve ligation, treated with PBS or inoculated with a control vector expressing *lacZ*, to 0.15 μg/kg in animals injected with the PE-expressing vector one week after spinal nerve ligation. Twice-daily inoculation of morphine (10 mg/kg IP) in spinal nerve ligated animals results in the development of tolerance by 1 week; beyond that time, continued twice-daily administration of morphine has no antiallodynic effect. Animals inoculated with the PE-expressing vector 1 week after spinal nerve ligation continue to demonstrate the antiallodynic effect of the vector, despite the induction of tolerance to morphine (Hao et al. 2003).

In a rodent model of pain resulting from cancer in bone, created by implantation of NTCT 2472 cells into the distal femur, subcutaneous inoculation of the PE-expressing replication-incompetent HSV vector into the plantar surface of the foot, 1 week after tumor injection, produces a significant reduction in ambulatory pain score compared to control vector-inoculated tumor-bearing animals. This effect is reversed by intrathecal naltrexone (Goss et al. 2002).

The experimental evidence demonstrates that replication-incompetent HSV vectors can be used to target delivery of genes with analgesic potential to neurons of the DRG, resulting in the local spinal release of analgesic substances to produce focal analgesic effects. The efficacy of this intervention in different models, which recapitulate essential features of the several subtypes of chronic pain, sets the stage for further development in two directions. The same approach may be used to express other molecules that might be anticipated to have significant analgesic effects at the spinal level, such as other inhibitory neurotransmitters, anti-inflammatory cytokines, or neurotrophic factors that produce analgesic effects in specific types of chronic pain. The second direction will be to determine whether HSV-mediated gene transfer will prove effective in the treatment of pain in patients. Replication-incompetent HSV vectors are appropriate for human use, and a proposal for a phase I human trial to examine the safety and tolerability of subcutaneous inoculation of an HSV vector deleted for the IE genes ICP4, ICP22, ICP27, and ICP41, and expressing human PE under the control of the HCMV IEp, was presented to the Recombinant Advisory Committee at the NIH in June 2002 and is now under review by the US Food and Drug Administration.

References

Antunes Bras, J. M., Epstein, A. L., Bourgoin, S., Hamon, M., Cesselin, F., & Pohl, M. (1998). Herpes simplex virus 1-mediated transfer of preproenkephalin a in Rat dorsal root ganglia. *Journal of Neurochemistry, 70,* 1299–1303.

Braz, J., Beaufour, C., Coutaux, A., Epstein, A. L., Cesselin, F., Hamon, M., & Pohl, M. (2001). Therapeutic efficacy in experimental polyarthritis of viral-driven enkephalin overproduction in sensory neurons. *Journal of Neuroscience, 21,* 7881–7888.

DeLuca, N. A., McCarthy, A. M., & Schaffer, P. A. (1985). Isolation and characterization of deletion

mutants of herpes simplex virus type 1 in the gene encoding immediate-early regulatory protein ICP4. *Journal of Virology, 56,* 558–570.

Finegold, A. A., Mannes, A. J., & Iadarola, M. J. (1999). A paracrine paradigm for in vivo gene therapy in the central nervous system: Treatment of chronic pain. *Human Gene Therapy, 10,* 1251–1257.

Fink, D. J., & Glorioso, J. C. (1997). Engineering herpes simplex virus vectors for gene transfer to neurons. *Nature Medicine, 3,* 357–359.

Goss, J. R., Mata, M., Goins, W. F., Wu, H. H., Glorioso, J. C., & Fink, D. J. (2001). Antinociceptive effect of a genomic herpes simplex virus-based vector expressing human proenkephalin in rat dorsal root ganglion. *Gene Therapy, 8,* 551–556.

Goss, J. R., Harley, C. F., Mata, M., O'Malley, M. E., Goins, W. F., Hu, X.-P., Glorioso, J. C., & Fink, D. J. (2002). Herpes vector-mediated expression of proenkephalin reduces pain-related behavior in a model of bone cancer pain. *Annals of Neurology, 52,* 662–665.

Hao, S., Mata, M., Goins, W., Glorioso, J. C., & Fink, D. J. (2003). Transgene-mediated enkephalin release enhances the effect of morphine and evades tolerance to produce a sustained antiallodynic effect. *Pain, 102,* 135–142.

Kim, S. H., & Chung, J. M. (1992). An experimental model for peripheral neuropathy produced by segmental spinal nerve ligation in the rat. *Pain, 50,* 355–363.

Krisky, D. M., Wolfe, D., Goins, W. F., Marconi, P. C., Ramakrishnan, R., Mata, M., Rouse, R. J., Fink, D. J., & Glorioso, J. C. (1998). Deletion of multiple immediate-early genes from herpes simplex virus reduces cytotoxicity and permits long-term gene expression in neurons. *Gene Therapy, 5,* 1593–1603.

Sah, D. W., Ossipo, M. H., & Porreca, F. (2003). Neurotrophic factors as novel therapeutics for neuropathic pain. *Nature Reviews Drug Discovery, 2,* 460–472.

Wilson, S. P., Yeomans, D. C., Bender, M. A., Lu, Y., Goins, W. F., & Glorioso, J. C. (1999). Antihyperalgesic effects of infection with a preproenkephalin-encoding herpes virus. *Proceedings of the National Academy of Sciences of the United States of America, 96,* 3211–3216.

Wu, N., Watkins, S. C., Schaffer, P. A., & DeLuca, N. A. (1996). Prolonged gene expression and cell survival after infection by a herpes simplex virus mutant defective in the immediate-early genes encoding ICP4, ICP27 and ICP22. *Journal of Virology, 70,* 6358–6368.

Xu, Y., Gu, Y., Xu, G. Y., Wu, P., Li, G. W., & Huang, L. Y. (2003). Adeno-associated viral transfer of opioid receptor gene to primary sensory neurons: A strategy to increase opioid Antinociception. *Proceedings of the National Academy of Sciences of the United States of America, 100,* 6204–6209.

Opioids and Inflammatory Pain

Michael Schäfer and C. Zöllner
Department of Anesthesiology and Intensive
Care Medicine, Charité University, Berlin,
Germany

Synonyms

Inflammatory pain and opioids; Opioids and
opioid receptor function in inflammation

Definition

In addition to the traditional belief that opioids
elicit antinociception via opioid receptors within
the central nervous system, recent research
has shown that they also elicit antinociception
via opioid receptors located on peripheral
nerve terminals of sensory neurons. Such
antinociceptive effects in the periphery are
particularly prominent under painful inflammatory conditions. In addition, opioids exert direct
anti-inflammatory effects preventing local
edema, immigration of immune cells, and
cytokine release.

Characteristics

Peripheral Opioid Receptors and Inflammation

Pain can be relieved by systemic administration
of opioids acting on specific opioid receptors
(OR) within the central nervous system.
However, substantial side effects such as sedation, nausea, vomiting, respiratory depression, or
respiratory arrest can occur following i.v. opioid
administration. In animal experiments, it was
shown that local application of μ-opioid receptor
(MOR) agonists into inflamed tissue elicited
a pronounced antinociceptive effect (Mousa
et al. 2007). These results suggest that MOR

agonists have a peripheral site of action within
inflamed tissue. All three opioid receptor types
(μ, δ, and κ) are present on peripheral sensory
neurons and are functionally involved in analgesia (Stein et al. 2003). As already known from the
central side of action, peripheral OR increase
potassium currents and decrease calcium currents
in the cell bodies of sensory neurons. These
effects can diminish the neuronal firing of
peripheral sensory neurons (Woolf and Salter
2000). In addition, opioids can activate inhibitory
G proteins ($G_{i/o}$) which lead to a decrease in the
second messenger adenylate cyclase (cAMP)
content (Zöllner et al. 2008). Saturation and
competition experiments indicate that the
pharmacological characteristics of these peripheral OR are very similar to those in the brain.

In primary sensory neurons, the neuronal cell
body is located in the ▶ dorsal root ganglion
(DRG) (Fig. 1). OR have been found on
small-to-medium-diameter neuronal cell bodies
within the DRG and on central as well as peripheral nerve terminals of primary afferent neurons
(Mousa et al. 2013). Peripheral analgesic effects
of opioids are greatly enhanced within inflamed
tissue. Following the induction of inflammation,
MOR mRNA increases within the lumbar spinal
cord (Maekawa et al. 1996). In addition, the
▶ axonal transport of MOR from DRG to the
peripheral nerve terminals is increased (Mousa
et al. 2007). In addition, the number of peripheral
sensory nerve terminals is increased in inflamed
tissue, a phenomenon known as sprouting
(Stein et al. 2003). It might be that inflammation
can also disrupt the perineurium, which is
normally a rather impermeable barrier. This
allows opioid agonists easier access to peripheral
MOR (Hackel et al. 2012). In conclusion, OR are
present in the periphery and mediate potent
analgesic effects after peripheral opioid application. An inflammatory process can result in
a profound increase in the density of MOR in
the DRG, an increase in the axonal transport of
MOR to the periphery, and in the enhanced
sprouting of MOR$^+$ peripheral sensory nerve
terminals. This is in contrast to other painful

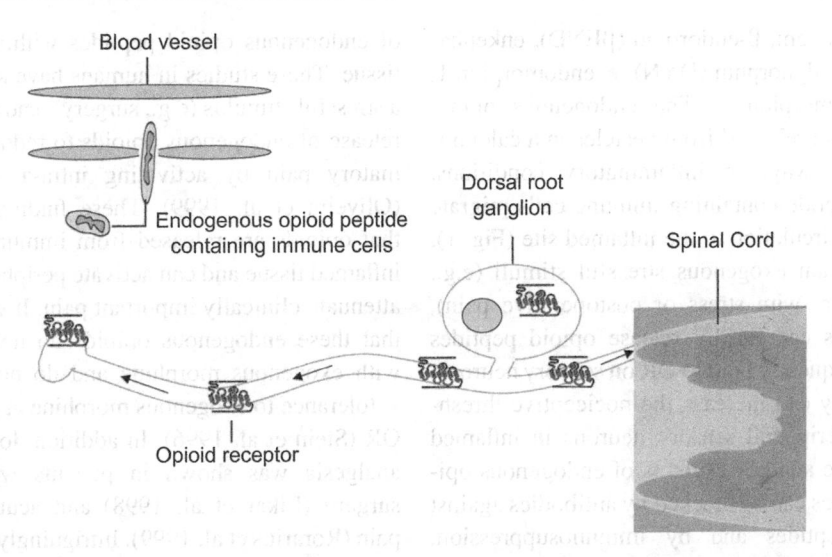

Blood vessel

Endogenous opioid peptide
containing immune cells

Opioid receptor

Dorsal root
ganglion

Spinal Cord

Opioids and Inflammatory Pain, Fig. 1 Primary afferent sensory neuron with its cell body (DRG). Opioid receptors are synthesized in the cell body of the DRG and transported towards its central (*right*) and peripheral (*left*) terminals. Under inflammatory conditions, immune cells migrate into the tissue. Exogenous (e.g., stressful stimuli) or endogenous (e.g., corticotropin-releasing hormone, interleukin 1) stimuli release opioid peptides from monocytic cells or lymphocytes. The endogenous ligands for opioid receptors reduce the excitability of the primary afferent neuron

diseases such as diabetic neuropathy in which sensory neuron MOR are downregulated because of lysosomal degradation (Mousa et al. 2013).

Inflammation and Exogenous Opioids

Quantification of MOR binding sites by ligand-binding experiments confirmed an increase in MOR within the DRG during inflammation. Together with immunohistochemistry experiments, it was shown that this increase in MOR binding sites of DRG was due to an increase in both the number of MOR⁺ neurons and the density of MOR⁺ staining per neuron (Mousa et al. 2007). The affinity of certain opioid agonists (e.g., ▶ DAMGO, ▶ buprenorphine) to MOR remained unchanged during inflammation. However, in a G protein coupling assay, it was shown that the alteration in the density of MOR in the DRG and the periphery during inflammation could lead to an increase in the number of ▶ G proteins activated (Zöllner et al. 2008). Interestingly, these changes did not occur either at spinal, supraspinal (e.g., hypothalamus), or at

contralateral DRG regions. Subsequent to the MOR upregulation in DRG, more OR are axonally transported towards the peripheral nerve endings. With a certain delay, an upregulation in MOR can also be seen within inflamed subcutaneous tissue. In addition, neuronal OR in the inflamed paw might undergo changes owing to the specific milieu of inflamed tissue (e.g., low pH), which could contribute to an increase in opioid efficacy (Selley et al. 1993). Opioid receptors are also located on various immune cells and are potential targets for immune-modulating effects. Indeed, it has been shown that opioids exert direct anti-inflammatory effects preventing local edema, immigration of immune cells, and cytokine release (Sacerdote et al. 2012).

Inflammation and Endogenous Opioid Peptides

The natural ligands for OR are opioid peptides. So far, five different peptides have been described in the neuroendocrine and central nervous system: β-endorphin (βEND), enkephalin (ENK), dynorphin (DYN), ▸ endomorphin 1, and endomorphin 2. The endogenous opioid peptides are released from vesicles in a calcium-dependent way. In inflammatory conditions, opioid peptide-containing immune cells migrate from the circulation to the inflamed site (Fig. 1). Upon certain exogenous stressful stimuli (e.g., cold water swim stress or postoperative pain), these cells can locally release opioid peptides that subsequently bind to OR on sensory neurons, where they can increase the nociceptive thresholds of peripheral sensory neurons in inflamed tissue. The analgesic effects of endogenous opioid peptides can be blocked by antibodies against opioid peptides and by immunosuppression, suggesting involvement of the immune system in inflammatory pain control (Machelska et al. 2002). It has been shown that opioid-containing cells are predominantly granulocytes during early and monocytes or macrophages during later stages of inflammation (Rittner et al. 2001). The endogenous opioid-mediated analgesia increases in parallel with the ▸ immune cell recruitment and the degree of inflammation. Other endogenous stimuli have been identified to trigger opioid secretion from immune cells. For instance, corticotropin-releasing hormone (CRH) stimulates the release of opioid peptides from immune cells that subsequently activate OR on sensory nerve endings (Schäfer et al. 1997). These results indicate that the activation of the endogenous opioid production and release from immune cells might be a new approach to the development of peripherally acting analgesics.

Peripheral Opioids and Clinical Implications

Since the first demonstration of the analgesic efficacy of intra-articular morphine in patients (Stein et al. 1991), many clinical studies have confirmed that local application of low, systemically inactive doses of opioids in patients with acute and chronic arthritic pain resulted in significant pain reduction. These effects were shown to be dose-dependent and reversible by naloxone, indicating a specific effect at OR. In addition, the intra-articular application of naloxone after knee arthroscopy increased pain in the presence

of endogenous opioid peptides within synovial tissue. These studies in humans have shown that a stressful stimulus (e.g., surgery) leads to a tonic release of endogenous opioids to reduce inflammatory pain by activating intra-articular OR (Oliveira et al. 1999). These findings suggest that opioids are released from immune cells in inflamed tissue and can activate peripheral OR to attenuate clinically important pain. It was shown that these endogenous opioids do not interfere with exogenous morphine and do not produce ▸ tolerance to exogenous morphine at peripheral OR (Stein et al. 1996). In addition, local opioid analgesia was shown in patients with dental surgery (Likar et al. 1998) and acute visceral pain (Rorarius et al. 1999). Intriguingly, systemic opioids in patients with chronic non-cancer pain such as arthritis produced statistically significant but clinically insufficient degrees of pain relief and did not improve the overall quality of life (Stein et al. 2010) which might be due to a possible preponderance of the supraspinal and spinal over the peripheral opioid effects (Khalefa et al. 2012).

Summary

In summary, local application of opioids significantly reduces pain in patients with acute and chronic inflammatory diseases. In comparison to central-acting opioids, the peripheral administration of morphine did not show any side effects like respiratory depression, sedation, or nausea.

References

Hackel, D., Krug, S. M., Sauer, R. S., et al. (2012). Transient opening of the perineural barrier for analgesic drug delivery. *Proceedings of the National Academy of Sciences of the United States of America, 109*, E2018–27.

Khalefa, B. I., Shaqura, M., Al-Khrasani, M., et al. (2012). Relative contributions of peripheral versus supraspinal and spinal opioid receptors to the Antinociception of systemic opioids. *European Journal of Pain, 16*, 690–705.

Likar, R., Sittl, R., Gragger, K., et al. (1998). Peripheral morphine analgesia in dental surgery. *Pain, 76*, 145–150.

Machelska, H., Mousa, S. A., Brack, A., et al. (2002). Opioid control of inflammatory pain regulated by intercellular adhesion molecule-1. *Journal of Neuroscience, 22*, 5588–5596.

Maekawa, K., Minami, M., Masuda, T., et al. (1996). Expression of mu- and kappa-, but not delta-, opioid receptor mRNA is enhanced in the spinal dorsal horn of the arthritic rats. *Pain, 64*, 365–371.

Mousa, S. A., Cheppudira, B. P., Shaqura, M., et al. (2007). Nerve growth factor governs the enhanced ability of opioids to suppress inflammatory pain. *Brain, 130*, 502–513.

Mousa S.A., Shaqura, M., Khalefa, B.I., Zöllner, C. et al. (2013). Rab7 silencing prevents μ-opioid receptor lysosomal targeting and rescues opioid responsiveness to strengthen diabetic neuropathic pain therapy. *Diabetes*, in press.

Oliveira, L., Paiva, A. C., & Vriend, G. (1999). A Low resolution model for the interaction of G proteins with G protein-coupled receptors. *Protein Engineering, 12*, 1087–1095.

Rittner, H., Brack, A., Machelska, H., et al. (2001). Opioid peptide-expressing leukocytes: Identification, recruitment, and simultaneously increasing inhibition of inflammatory pain. *Anesthesiology, 95*, 505–508.

Rorarius, M., Suominen, P., Baer, G., et al. (1999). Peripherally administered sufentanil inhibits pain perception after postpartum tubal ligation. *Pain, 79*, 83–88.

Sacerdote, P., Franchi, S., Panerai, A.E., (2012). Nonanalgesic effects of opioids: Mechanisms and potential clinical relevance of opioid-induced immunosuppression. *Curr Pharm Des*, in print.

Schäfer, M., Mousa, S. A., & Stein, C. (1997). Corticotropin-releasing factor in Antinociception and inflammation. *European Journal of Pharmacology, 323*, 1–10.

Selley, D. E., Breivogel, C. S., & Childers, S. R. (1993). Modification of G protein-coupled functions by Low-pH pretreatment of membranes from NG108-15 cells: Increase in opioid agonist efficacy by decreased inactivation of G proteins. *Molecular Pharmacology, 44*, 731–741.

Stein, C., Comisel, K., Haimerl, E., et al. (1991). Analgesic effect of intraarticular morphine after arthroscopic knee surgery. *The New England Journal of Medicine, 325*, 1123–1126.

Stein, C., Pfluger, M., Yassouridis, A., et al. (1996). No tolerance to peripheral morphine analgesia in presence of opioid expression in inflamed synovia. *The Journal of Clinical Investigation, 98*, 793–799.

Stein, C., Schäfer, M., & Machelska, H. (2003). Attacking pain at its source. *Nature Medicine, 9*, 1003–1008.

Woolf, C. J., & Salter, M. W. (2000). Neuronal plasticity: Increasing the gain in pain. *Science, 288*, 1765–1769.

Zöllner, C., Mousa, S. A., Fischer, O., et al. (2008). Chronic morphine use does not induce peripheral tolerance in a rat model of inflammatory pain. *The Journal of Clinical Investigation, 118*, 1056–1073.

Opioids and Muscle Pain

Kathleen A. Sluka
Physical Therapy and Rehabilitation Science
Graduate Program, University of Iowa, Iowa
City, IA, USA

Synonyms

Morphine and muscle pain; Musculoskeletal pain

Definition

Muscle pain is pain that originates in muscle tissue and results in an increased sensitivity of the muscle to noxious stimuli. This pain can be a result of tissue injury, inflammation, exercise, or ischemia. Opioids are a class of compounds utilized to treat pain that activate opioid receptors. There are three types of opioid receptors, μ, δ, and κ, which are located peripherally and throughout the central nervous system.

Characteristics

Muscle pain is characterized as a deep aching pain and is associated with increased pain to deep pressure applied over the muscle, as well as ▶ referred pain to areas remote from the site of injury. Clinically, ▶ fibromyalgia, ▶ myofascial pain, ▶ myositis, and ▶ strain are common forms of pain associated with tissue injury and/or pain in the muscle.

Animal models of muscle pain include inflammatory, induced by injection of carrageenan into the muscle tissue, and noninflammatory, induced by repeated injections of acid into the muscle or by combinations of fatiguing stimuli with muscle insult (Kehl et al. 2000; Sluka et al. 2001). Models of muscle pain result in ▶ secondary hyperalgesia to mechanical (von Frey filaments) and heat stimuli. Further, ▶ primary hyperalgesia, as measured by a decrease in grip force or by withdrawal to muscle pressure, is present following carrageenan inflammation of the muscle (Kehl et al. 2000; Radhakrishnan et al., 2004). The primary and secondary hyperalgesia associated with carrageenan muscle inflammation is reversed by systemic administration of the opioid agonists morphine and levorphanol (Kehl et al. 2000; Radhakrishnan et al. 2004). In the noninflammatory model of muscle pain, produced by repeated intramuscular acid injections, systemic morphine and intrathecal μ- and δ-opioid agonists reverse the secondary mechanical hyperalgesia (Sluka et al. 2002). Thus, these data show that animal models of muscle pain are responsive to μ- and δ-opioid activation at receptors that could be located peripherally, spinally, or supraspinally.

In humans, muscle pain can be produced by infusion of hypertonic saline into the muscle, direct electrical stimulation of the muscle, or eccentric contraction of the muscle (▶ DOMS). The pain associated with intramuscular hypertonic saline infusion is reversed by epidural injection of the opioid agonist fentanyl (50 or 100 μg) (Eichenberger et al. 2003), suggesting a role for spinal opioid receptors in pain reduction. The pain associated with DOMS is reduced by activation of peripheral opioid receptors with morphine-6-β-glucuronide (M6Gl; a major metabolite of morphine that does not cross blood–brain barrier) (Tegeder et al. 2003), but not by systemic administration of several opioid agonists (Barlas et al. 2000), suggesting a role for peripheral opioid receptors. The pain from electrical stimulation of the muscle is inhibited by the systemic administration of the opioid agonist remifentanil. Remifentanil produces a greater inhibition of pain induced by electrical stimulation of the muscle when compared to the inhibition of pain from electrical stimulation of the skin, suggesting that muscle pain is more sensitive to opioids than cutaneous pain (Curatolo et al. 2000). Thus, there is evidence that in human subjects muscle pain is responsive to opioid agonists, and these effects could be as a result of activation of receptors located peripherally, spinally, or supraspinally.

The use of opioids in the treatment of clinical muscle pain, either acute or chronic, is not common. However, there are a few studies, done predominately with patients with fibromyalgia or acute muscle sprain, which suggest treatment of muscle pain with opioids is effective. Mild to moderate pain resulting from acute muscle strain of the low back or soft tissue injury is effectively treated with a mild opioid (Tylenol with codeine), reducing pain by approximately 50 % (Brown et al. 1986; Muncie et al. 1986). However, when compared to treatment with an anti-inflammatory, there was an equivalent reduction and more side effects with the opioid (Brown et al. 1986). In patients with fibromyalgia, a more chronic pain condition, systemic administration of opioid agonists reduced spontaneous pain, as measured by a visual analogue scale (Biasi et al. 1998) and ▶ temporal summation (Price et al. 2002). Spinal administration (epidural) of opioid agonists also reduces spontaneous pain, tenderness, and lower limb fatigue (Bengtsson et al. 1989). Tramadol, a weak opioid agonist and combined with a norepinephrine reuptake inhibitor, is recommended in clinical guidelines for fibromyalgia (Buckhardt et al. 2005). Thus, clinically, the use of opioid treatment for muscle pain conditions is supported by results from basic science, experimental muscle pain models, and clinical studies using people with muscle sprain and fibromyalgia. Short-term use of opioids is common in those with acute muscle pain. However, long-term use of opioids is generally not recommended due to significant side effects, tolerance, and addiction. Long-term studies on usage of opioids in muscle pain conditions, or effectiveness in a number of acute muscle pain conditions, such as myositis and myofascial pain, have not been investigated.

References

Barlas, P., Craig, J. A., Robinson, J., Walsh, D. M., Baxter, G. D., & Allen, J. M. (2000). Managing delayed-onset muscle soreness: Lack of effect of selected oral systemic analgesics. *Archives of Physical Medicine and Rehabilitation, 81*, 966–972.

Bengtsson, M., Bengtsson, A., & Jorfeldt, L. (1989). Diagnostic epidural opioid blockade in primary fibromyalgia at rest and during exercise. *Pain, 39*, 171–180.

Biasi, G., Manca, S., Manganelli, S., & Marcolongo, R. (1998). Tramadol in the fibromyalgia syndrome: A controlled clinical trial versus placebo. *International Journal of Clinical Pharmacology Research, 18*, 13–19.

Brown, F. L., Jr., Bodison, S., Dixon, J., Davis, W., & Nowoslawski, J. (1986). Comparison of diflunisal and acetaminophen with codeine in the treatment of initial or recurrent acute low back strain. *Clinical Therapeutics, 9*, C 52–C 58.

Buckhardt, C. S., Goldenberg, D., Crofford, L., Gerwin, R., Gowens, S., Jackson, K., Kugel, P., McCarberg, W., Rudin, N., Schanberg, L., Taylor, A. G., Taylor, J., & Turk, D. (2005). *Guideline for the management of fibromyalgia syndrome pain in adults and children.* Glenview: American Pain Society (APS). 109 p. (Clinical practice guideline? no. 4).

Curatolo, M., Petersen-Felix, S., Gerber, A., & Arendt-Nielsen, L. (2000). Remifentanil inhibits muscular more than cutaneous pain in humans. *British Journal of Anaesthesia, 85*, 529–532.

Eichenberger, U., Giani, C., Petersen-Felix, S., Graven-Nielsen, T., Arendt-Nielsen, L., & Curatolo, M. (2003). Lumbar epidural fentanyl: Segmental spread and effect on temporal summation and muscle pain. *British Journal of Anaesthesia, 90*, 467–473.

Kehl, L. J., Trempe, T. M., & Hargreaves, K. M. (2000). A new animal model for assessing mechanisms and management of muscle hyperalgesia. *Pain, 85*, 333–343.

Muncie, H. L., Jr., King, D. E., & DeForge, B. (1986). Treatment of mild to moderate pain of acute soft tissue injury: Diflunisal vs. acetaminophen with codeine. *Journal of Family Practice, 23*, 125–127.

Price, D. D., Staud, R., Robinson, M. E., Mauderli, A. P., Cannon, R., & Vierck, C. J. (2002). Enhanced temporal summation of second pain and its central modulation in fibromyalgia patients. *Pain, 99*, 49–59.

Radhakrishnan, R., Bement, M., Skyba, D., Kehl, L., & Sluka, K. (2004). Models of muscle pain: Carrageenan model and acidic saline model. In S. J. Enna, M. Williams, J. Ferkany, T. Kenakin, R. Porsolt, & J. Sullivan (Eds.), *Current protocols in pharmacology* (pp. 5.35.1–5.35.28). Hoboken: Wiley.

Sluka, K. A., & Rasmussen, L. A. (2010). Fatiguing exercise enhances hyperalgesia to muscle inflammation. *Pain, 148*, 188–197.

Sluka, K. A., Kalra, A., & Moore, S. A. (2001). Unilateral intramuscular injections of acidic saline produce a bilateral, long-lasting hyperalgesia. *Muscle & Nerve, 24*, 37–46.

Sluka, K. A., Rohlwing, J. J., Bussey, R. A., Eikenberry, S. A., & Wilken, J. M. (2002). Chronic muscle pain induced by repeated acid injection is reversed by spinally administered μ and δ - but not κ, opioid receptor agonists. *Journal of Pharmacology and Experimental Therapeutics, 302*, 1146–1150.

Tegeder, I., Meier, S., Burian, M., Schmidt, H., Geisslinger, G., & Lotsch, J. (2003). Peripheral opioid analgesia in experimental human pain models. *Brain, 126*, 1092–1102.

Opioids and Nucleus Gracilis

▶ Opioids in the Spinal Cord and Modulation of Ascending Pathways (N. gracilis)

Opioids and Opioid Receptor Function in Inflammation

▶ Opioids and Inflammatory Pain

Opioids and Reflexes

Claire D. Advokat
Department of Psychology, Louisiana State University, Baton Rouge, LA, USA

Synonyms

Hindlimb flexor reflex; Reflexes and opioids; Spinal nociceptive tail flick withdrawal reflex

Definition

The tail flick (TF) withdrawal reflex is a spinally mediated standard measure of pain sensitivity. In the most common procedure, a beam of high-intensity light is focused on the tail, and the response time is automatically measured and defined as the interval between the onset of the thermal stimulus and the abrupt flick of the tail. Typically, several determinations are made, and the mean score is taken as the response latency. Increases in latency are interpreted to indicate an antinociceptive, i.e., analgesic, response. To reduce tissue damage, animals not responding after a predetermined cutoff score are assigned the maximum score.

The hindlimb flexor reflex is a spinally mediated withdrawal response of the hindlimb. Typically, flexor reflexes are elicited by electrical stimulation of the hindpaw. The reflex has two components: a short latency component that appears within 20–100 ms and is elicited by low-threshold, non-nociceptive stimulation and a longer latency component that appears at about 150–450 ms, in response to high threshold, nociceptive, C-fiber stimulation ($>6.8 + 0.2$ mA).

Characteristics

Modern views of opiate action in the central nervous system (CNS) originated with the classic experiments of Wikler and colleagues, who demonstrated that systemically administered opiates could suppress polysynaptic spinal nociceptive responses, such as the hindlimb flexor reflex and the tail withdrawal reflex, after ▶ spinal transection (Wikler 1950; Advokat and Burton 1987). While this showed that opiates could act directly at the level of the spinal cord, the doses required in the spinal animal were larger than the doses needed in the intact animal, to produce the same effect. This observation led to the hypothesis that opiate analgesia was not only due to a direct action of the drug in the brain and the spinal cord but was also mediated by an additional, indirect, effect of opiates in the brain, which increased descending supraspinal inhibition of spinal nociceptive reflexes (Advokat 1988).

This view was supported by studies of the analgesic effect of morphine, administered concomitantly into the ▶ intrathecal space and the third cerebral ventricle (Yeung and Rudy 1980) or the ▶ periaqueductal gray (Siuciak and Advokat 1989a) of rats. Concurrent morphine administration to these sites had a ▶ synergistic effect on spinal antinociception, measured with the tail flick or ▶ hot plate assay. The total amount of morphine concurrently administered to the brain and spinal cord, necessary to produce an ED_{50} in these studies, was substantially less than the amount required separately at each site to produce the same effect. Furthermore, the total amount administered concomitantly to both sites was similar to the total concentration at these sites, which produced the same effect after systemic morphine administration (Advokat 1988).

This confirmed that morphine-induced analgesia depended on a combined action at both spinal and supraspinal sites, but it did not indicate the nature of that interaction.

Some insight into this process was provided by studies in which morphine was administered intrathecally to the spinal cord of acute (1 day) spinally transected rats (Siuciak and Advokat 1989b). In this situation, the antinociceptive effect on spinal reflexes was considerably more potent than in the intact animal (Fig. 1a). These data demonstrated that the analgesic effect of spinal morphine at the spinal level was increased by either spinal transection or by supraspinally administered morphine. The interpretation of this phenomenon was that spinal transection and intracerebral morphine administration produced the same effect on spinal opiate analgesia, because they eliminated (spinal transection) or suppressed (supraspinal morphine administration) an inhibitory action exerted by descending pathways on opiate-induced spinal antinociception.

The fact that systemically administered morphine was *less* potent in the acute spinal, compared to the intact animal, appeared to conflict with this hypothesis. However, it was subsequently shown that the amount of morphine reaching the spinal cord after acute spinalization was less than the concentration obtained at the same doses in intact rats (Advokat and Gulati 1991), demonstrating that the apparent decrease in potency of systemically administered morphine might have been due to a disruption in the distribution of the drug to the CNS.

Several implications follow from these findings. First, they suggest that mechanisms responsible for descending control of spinal reflexes, per se, may be different from those which mediate descending control of the effects of opiates on those same reflexes. A corollary of this proposition is that supraspinal mechanisms may also mediate the development of tolerance to opiate inhibition of spinal reflexes. That is, repeated administration of morphine may not only produce tolerance locally, at supraspinal and spinal sites, chronic opiate exposure may also reduce the inhibitory effect of morphine acting through descending pathways. This hypothesis

Opioids and Reflexes, Fig. 1 (a) Dose–response functions to intrathecal (*IT*) morphine in intact (*open circles*) and acute spinal rats (*filled circles*) on the tail flick test. The data show that 24 h after spinal transection, the antinociceptive effect of spinally administered morphine is profoundly increased (Siuciak and Advokat 1989b). (b) Dose–response functions to subcutaneous morphine in intact rats and in rats spinally transected for either 3, 10, 20, or 30 days. The data show a gradual decline in the antinociceptive effect of systemic morphine within a month after spinal transection (Advokat and Burton 1987). (c) Dose–response functions to intrathecal morphine in rats that were spinally transected either 1 day, 20, or 30 days previously. The data show a gradual decline in the antinociceptive effect of intrathecal morphine within a month after spinal transection (Siuciak and Advokat 1989b)

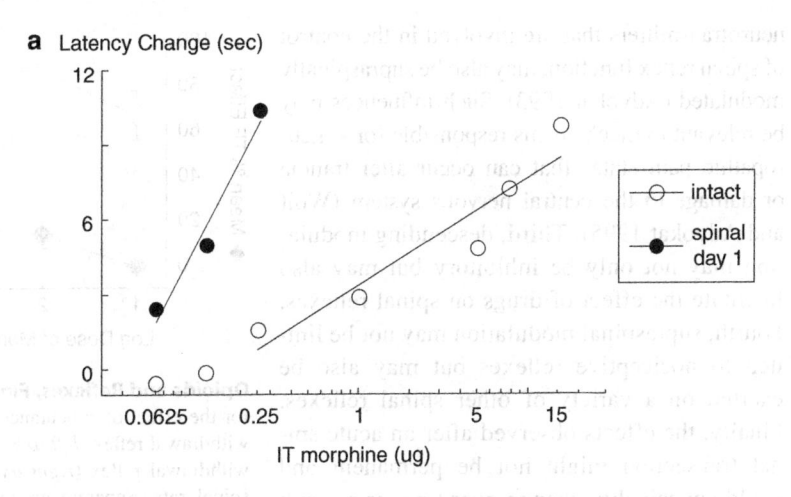

a Latency Change (sec)

IT morphine (ug)

— intact
— spinal day 1

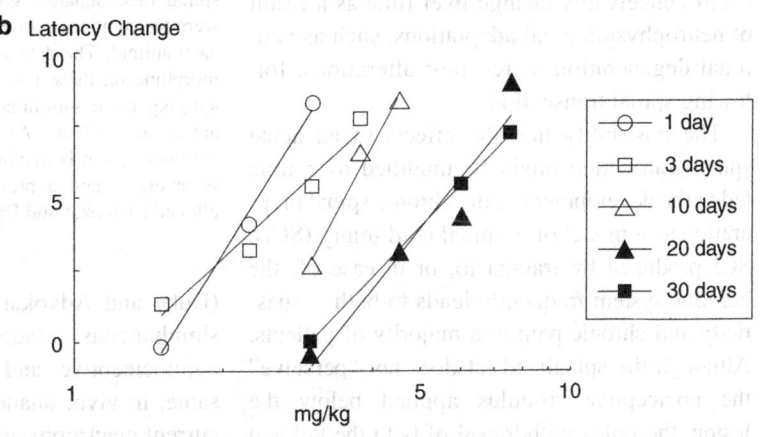

b Latency Change

mg/kg

— 1 day
— 3 days
— 10 days
— 20 days
— 30 days

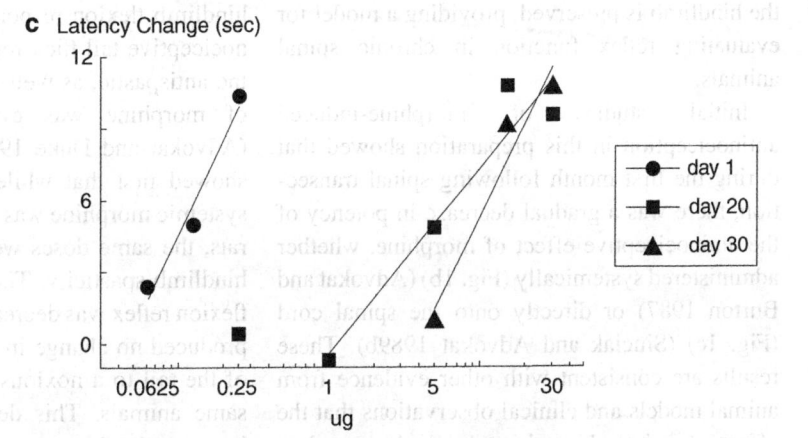

c Latency Change (sec)

ug

— day 1
— day 20
— day 30

would be consistent with the observation that tolerance to systemic morphine administration does not necessarily confer tolerance to the effect of morphine at the level of the spinal cord (Siuciak and Advokat 1989a, b; Advokat et al. 1987; Advokat 1989). Second, there is no reason to assume that opiates would be unique in producing effects at the spinal cord by a combination of direct and indirect mechanisms. The spinal action of other pharmacological agents, or endogenous neurotransmitters that are involved in the control of spinal reflex function, may also be supraspinally modulated (Advokat 1993). Such influences may be relevant to mechanisms responsible for ▶ neuropathic pain states that can occur after trauma or damage to the central nervous system (Wolf and Advokat 1995). Third, descending modulation may not only be inhibitory but may also facilitate the effect of drugs on spinal reflexes. Fourth, supraspinal modulation may not be limited to nociceptive reflexes but may also be exerted on a variety of other spinal reflexes. Finally, the effects observed after an acute spinal transection might not be permanent and could conceivably change over time as a result of neurophysiological adaptations, such as neuronal degeneration or receptor alterations, following spinal transection.

The possibility that the effects of an acute spinal transection might be modified over time led to the development of the chronic spinal preparation as a model of ▶ spinal cord injury (SCI). SCI produced by trauma to, or disease of, the nervous system frequently leads to both ▶ spasticity and chronic pain in a majority of patients. Although the spinalized rat does not "perceive" the nociceptive stimulus applied below the lesion, the reflex withdrawal of both the tail and the hindlimb is preserved, providing a model for evaluating reflex function in chronic spinal animals.

Initial studies of morphine-induced antinociception in this preparation showed that during the first month following spinal transection, there was a gradual decrease in potency of the antinociceptive effect of morphine, whether administered systemically (Fig. 1b) (Advokat and Burton 1987) or directly onto the spinal cord (Fig. 1c) (Siuciak and Advokat 1989b). These results are consistent with other evidence from animal models and clinical observations that the effect of opiate drugs is at least attenuated in neuropathic pain states produced by damage to the nervous system.

As with SCI in humans, a permanent and stable spastic response of the hindlimbs was also observed, developing in chronic spinal rats during the first few weeks following transection

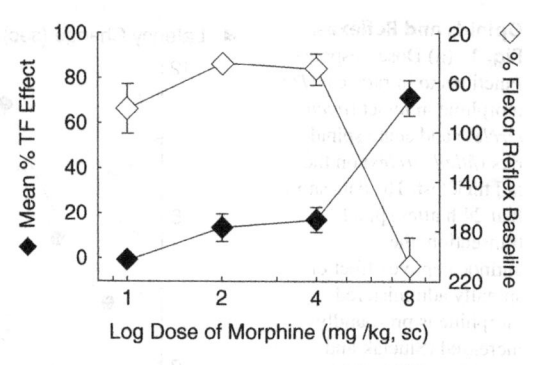

Opioids and Reflexes, Fig. 2 Dose–response functions for the effect of subcutaneous morphine on the tail flick withdrawal reflex (*left axis, filled symbols*) and the flexor withdrawal reflex (*right axis, open symbols*) of chronic spinal rats. Separate groups of spinally transected rats were tested at each dose; both tests were administered to each animal. The data show a separation in the effect of morphine on these two measures. At low doses of 1 to 4 mg/kg, there is minimal antinociception and a maximum antispastic effect. At the highest, 8 mg/kg, dose, antinociception is maximum while the antispastic action is reversed, and a profound hyperreactive response is elicited (Advokat and Duke 1999)

(Duke and Advokat 2000). This permitted the simultaneous assessment of the effect of antinociceptive and antispastic agents in the same, in vivo, unanesthetized animals. By concurrent electrophysiological measurement of the hindlimb flexion response, and recording of the nociceptive tail flick reflex in the same animals, the antispastic, as well as antinociceptive, effect of morphine was evaluated in this model (Advokat and Duke 1999). The results (Fig. 2) showed first that while the analgesic effect of systemic morphine was reduced in chronic spinal rats, the same doses were still effective against hindlimb spasticity. The hyperreactive hindlimb flexion reflex was decreased by 50 % at doses that produced no change in the nociceptive response of the tail to a noxious thermal stimulus, in the same animals. This demonstrated a separation between the dose–response functions of the two behaviors, indicating that the antispastic action occurs at a dose that does not produce an analgesic effect.

Second, the data in Fig. 2 also show that at the highest dose (8 mg/kg), there was a significant reversal of the effect of morphine on both

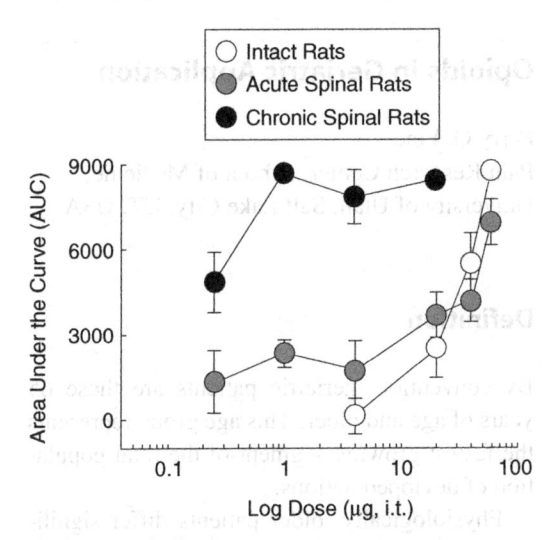

Opioids and Reflexes, Fig. 3 Dose–response functions to intrathecal clonidine on the tail flick test, in intact (*open circles*), acute spinal (*shaded circles*), and chronic spinal rats (*solid circles*). The data represent the mean ± S.E.M. of the area under the time-effect curve (at 30, 60, and 90 min) for separate groups of rats (n = 3–6) tested after administration of the indicated doses of clonidine. There was a significant dose-dependent effect of clonidine within each of the three experimental conditions. Clonidine-induced antinociception in intact rats was the same as that in acute spinal rats but was greatly increased in chronic spinal rats (Advokat 2002)

Opioids and Reflexes, Fig. 4 Dose–response to intrathecal clonidine on the flexor reflex of acute spinal (*shaded circles*) and chronic spinal rats (*solid circles*). The data were obtained from the same animals whose tail flick test results are summarized in Fig. 3. There was no difference among the doses in acute spinal rats and no group differed from the saline condition. However, the same doses produced a significant dose-dependent effect in chronic spinal rats (Advokat 2002)

behaviors. While this dose was large enough to produce an antinociceptive effect, it also *increased* the flexion response to more than 200 % of baseline. That is, morphine simultaneously decreased the reaction to a normally painful stimulus and increased the spastic response to a non-painful stimulus. This may be an example of the clinical phenomenon of "opioid-related" ▶ myoclonus that has been reported in cancer patients, which was "highly" associated with nerve dysfunction due to spinal cord lesions.

The observation that a single drug can modulate nociception and spasticity in the chronic spinal rat is not limited to morphine but is relevant to the action of other pharmacological agents used in treating pain and spasticity in SCI. Studies with the antispastic agent baclofen have also shown a separation between the antinociceptive action of low doses and an antispastic, muscle-relaxant effect of higher doses in chronic spinal rats (Advokat et al. 1999). The usefulness of the

chronic spinal rat has also been demonstrated in recent studies of the alpha(α) 2-adrenergic agonist clonidine on spinal reflex function (Advokat 2002). Acute spinalization did not change the antinociceptive effect of this drug on spinal reflexes (Fig. 3), nor did clonidine have any effect on the non-nociceptive flexor response in this situation (Fig. 4). However, 1 month after spinalization, there was a significant increase in the antinociceptive effect of clonidine on the thermally elicited tail withdrawal reflex, as well as a dose-dependent response on the spastic hindlimb flexion reflex. These data provide behavioral evidence of ▶ supersensitivity to alpha2-adrenoceptor agonists in chronic spinal rats. Such results illustrate the insights provided by careful consideration of differential pharmacological modulation of spinal reflex function.

References

Advokat, C. (1988). The role of descending inhibition in morphine-induced analgesia. *Trends in Pharmacological Sciences, 9*, 330–334.
Advokat, C. (1989). Tolerance to the antinociceptive effect of systemic morphine in spinally transected rats. *Behavioral Neuroscience, 103*, 1091–1098.

Advokat, C. (1993). Intrathecal co-administration of serotonin (5HT) and morphine differentially modulates the tail flick reflex of intact and spinal rats. *Pharmacology, Biochemistry, and Behavior, 45*, 871–879.

Advokat, C. (2002). Spinal transection increases the potency of clonidine on the tail flick and hindlimb flexion reflexes. *European Journal of Pharmacology, 437*, 63–67.

Advokat, C., & Burton, P. (1987). Antinociceptive effect of systemic and intrathecal morphine in spinally transected rats. *European Journal of Pharmacology, 139*, 335–343.

Advokat, C., & Duke, M. (1999). Comparison of morphine-induced effects on thermal nociception, mechanoreception and hindlimb flexion in chronic spinal rats. *Experimental and Clinical Psychopharmacology, 7*, 219–225.

Advokat, C., & Gulati, A. (1991). Spinal transection reduces both spinal antinociception and CNS concentration of systemically administered morphine in rats. *Brain Research, 555*, 251–258.

Advokat, C., Burton, P., & Tyler, C. B. (1987). Investigation of tolerance to chronic intrathecal morphine infusion in the rat. *Physiology & Behavior, 39*, 161–168.

Advokat, C., Duke, M., & Zeringue, R. (1999). Dissociation of (−)baclofen-induced effects on the tail withdrawal and hindlimb flexor reflexes of chronic spinal rats. *Pharmacology, Biochemistry, and Behavior, 63*, 527–534.

Duke, M., & Advokat, C. (2000). Pentobarbital-induced modulation of flexor and H-reflexes in spinal rats. *Brain Research, 881*, 217–221.

Siuciak, J., & Advokat, C. (1989a). The synergistic effect of concurrent spinal and supraspinal opiate agonisms is reduced by both nociceptive and opiate pretreatment. *Pharmacology, Biochemistry, and Behavior, 34*, 265–273.

Siuciak, J., & Advokat, C. (1989b). Antinociceptive effect of intrathecal morphine injections in tolerant and nontolerant spinal rats. *Pharmacology, Biochemistry, and Behavior, 34*, 445–452.

Wikler, A. (1950). Sites and mechanisms of action of morphine and related drugs in the central nervous system. *Pharmacological Reviews, 2*, 435–506.

Wolf, E. A., III, & Advokat, C. (1995). The N-methyl-D-aspartate antagonist dextrorphan acts like ketamine by selectively increasing tail flick latencies in spinally transected, but not intact rats. *Analgesia, 1*, 161–170.

Yeung, J. C., & Rudy, T. A. (1980). Multiplicative interaction between narcotic agonisms expressed at spinal and supraspinal sites of antinociceptive action as revealed by concurrent intrathecal and intracerebroventricular injections of morphine. *The Journal of Pharmacology and Experimental Therapeutics, 215*, 633–642.

Opioids in Geriatric Application

Perry G. Fine
Pain Research Center, School of Medicine, University of Utah, Salt Lake City, UT, USA

Definition

By convention, geriatric patients are those 65 years of age and older. This age group represents the fastest growing segment of the total population of devloped nations.

Physiologically, older patients differ significantly from those in younger age groups in a variety of ways. The physiologic changes of aging, along with acute and chronic pathologic conditions and altered drug disposition, frequently lead to increased pain-producing problems; this is in addition to changing sensory perception and the ability to communicate or cope with distress, and these normal and pathologic changes influence drug disposition. The rate of aging is influenced by genetic and environmental factors, so the trajectory of changes is different from person to person. Nevertheless, with the passage of time, senescence is inevitable, and aging is marked by several common traits:

Rates of gene transcription, lipofuscin and extracellular matrix cross-linking, and protein oxidation are altered, among other biochemical and tissue changes.

Physiologic capacity diminishes.

Adaptive processes to physiologic stress (e.g., increases in cardiac output, thermoregulation, GI transit rate) are blunted.

Susceptibility and vulnerability to diseases are increased.

There is an accelerating rise in mortality with advancing age after maturation occurs.

These factors must be taken into account when assessing older patients with significant pain. Decisions about goals of therapy, when and how to prescribe opioid ▶ analgesics, and what effects to anticipate will be strongly influenced by these circumstances. They are oftentimes considerably different than the types of concerns relevant to younger patients.

Characteristics

Pain Prevalence, Etiology, and Features

Persistent and recurrent episodic pains are highly prevalent in older patients. About 5 % of older Americans reportedly take prescription analgesics for most of a given year (Cooner and Amorosi 1997). Causes of pain in older patients are often from multiple etiologies. Most commonly, these include skeletal diseases such as osteoarthritis and degenerative spinal disorders and ▶ neuropathic pain syndromes such as ▶ postherpetic neuralgia and painful ▶ diabetic polyneuropathy. Also, pain-producing cardiovascular ailments and malignancies gain prevalence, adding to the overall burden of aging.

Surveys over the last several years have shown that serious pain is a frequent finding among both community dwelling and institutionalized older individuals, and it is largely undertreated in most settings except hospice, where comfort is a focus of care (Helme and Gibson 1999). Between 45 % and 80 % of nursing home residents are reported to have moderate to severe pain on a regular basis that is not adequately treated.

Although the adverse consequences of pain in older persons are not necessarily unique, they are oftentimes amplified, due to coexisting problems, including cognitive impairment, balance disturbances, deconditioning, sensory impairments, and other comorbidities. As a result, the ill effects of pain, such as mood disorders (especially depression), social isolation, poor sleep, gait disturbance, and inability to perform self-care and routine activities of daily living (ADLs), are compounded. The use of multiple medications for concurrent disease and symptom management creates an additional confound in the management of pain in geriatrics. Drug-drug and drug-disease interactions are, as a result, a far

more significant concern in this group of patients. The consequences of drug-related adverse effects, such as falls, confusion, or ▶ obstipation, can have much graver consequences than in younger patients. Overall, however, as a pharmacologic class, the risks associated with opioid analgesics are considerably less compared to other commonly prescribed drugs in older patients (Doucet et al. 1996).

Aging and Opioid Pharmacology

Consistent with current theories of aging, the great individual variability observed in older persons is more likely to be the result of "tightly controlled but individually different cell and tissue-specific patterns of gene expression," rather than age-specific genetic instability (Arking 2001). Each patient's history, environmental exposure, and genetic makeup contributes to a unique pattern of organ function (or dysfunction) and subjective response to therapy (Barja 1998). Evidence suggests that sensitivity to drugs that act in the central nervous system, best demonstrated in studies of opioid analgesics, increases with age (Rooke et al. 2002). Age has long been known to be an independent predictor of response to opioids (Belleville et al. 1971). In decile groupings, after the age of 40, there appears to be a linear age-related increase in responsiveness to fixed morphine doses with regard to pain intensity differences in postoperative patients. This correlates with changes in drug absorption, distribution, metabolism, and clearance, all of which are influenced by body composition changes (e.g., muscle fat), reduced serum protein levels, cardiac output, and organ perfusion (Fine 2004). Notwithstanding the multiple pharmacokinetic (▶ pharmacokinetics) alterations that influence opioid blood levels over time, the cause of increased sensitivity to opioids appears to be mostly a function of reduced central nervous system resilience (Guay et al. 2002). In addition, it has been postulated that increased sensitivity to the effects of opioids, both therapeutic and toxic, are related to the high prevalence of subclinical malnutrition in community and institutional dwelling geriatric patients. In the few studies that have specifically evaluated pharmacodynamic (▶ pharmacodynamics) effects of analgesic drugs in older patients, the rate of drug delivery, rather than the absolute dose of drug over time, influences both analgesia and adverse effects, including the most feared risk of life-threatening respiratory depression (Aburun et al. 2002). Opioids have been shown to affect immune system responses and neuroendocrine function. A systematic review of the literature did not reveal any conclusive age-specific, or age-related, effects of pain or opioids on human immune functioning (Page 2003). However, this is an area that requires further study, since patients are being treated for more protracted periods of time with opioids, and overall life expectancy is increasing. The potential for chronic opioid therapy to reduce testosterone levels in males requires ongoing assessment, in addition to a risk-benefit analysis, in terms of pain treatment, hormone depletion, and hormone replacement.

Indications and Guidelines

Opioid analgesics are most effectively used to achieve positive therapeutic goals when they are incorporated into a comprehensive non-pharmacologic, pharmacologic, behavioral, and functional improvement (rehabilitative) pain management plan of care (Fine 2004) (see also ▶ non-pharmacologic pain management). The American Geriatrics Society has recently produced a clinical guideline based upon a systematic review of the literature, which outlines the principles for analgesic use in older patients (Fine 2001). Key points with regard to opioids include (AGSPanel on Persistent Pain in Older Patients 2002):

- All older patients with functional impairment or diminished quality of life as a result of persistent pain are candidates for pharmacologic therapy.
- There is no role for ▶ placebos in the management of pain. Their use is unethical.
- The least toxic means of achieving pain relief should be used. When systemic

medications are indicated, noninvasive routes should be considered first.

- Opioid analgesic drugs may help relieve moderate to severe pain, especially ▶ nociceptive pain (As an update, though, it should be noted that recent studies have demonstrated the efficacy of opioids in the management of various neuropathic pain conditions (Rowbotham et al. 2003)).
- Opioids for episodic (noncontinuous) pain should be prescribed as needed, rather than around the clock.
- Long-acting or sustained-release analgesic preparations should be used for continuous pain.
- ▶ Breakthrough pain should be identified and treated by the use of fast-onset, short-acting preparations.
- Constipation and opioid-related gastrointestinal symptoms should be prevented. Assessment of bowel function should be an initial and ongoing process during every follow-up visit for patients receiving analgesics.
- Non-opioid pain-modulating medications may be appropriate for some patients with neuropathic pain and some other chronic pain conditions.
- Patients taking analgesic medications should be monitored closely.
- Patients should be reevaluated frequently for drug efficacy and side effects during initiation, titration, or any change in dose of analgesic medications.
- Patients should be reevaluated regularly for drug efficacy and side effects throughout long-term analgesic drug maintenance.
- Patients on long-term opioid therapy should be evaluated periodically for inappropriate or dangerous drug-use patterns.
- Clinical endpoints should be decreased pain, increased function, and improvements in mood and sleep, not decreased drug dose.

Non-Self-Reporting Patients

One of the more significant challenges in geriatric care is the management of symptoms in patients who are unable to provide an adequate history or narrative of their complaints. Most typically, these are patients suffering from severe cognitive impairments with associated verbal loss from advanced dementing illnesses (e.g., Alzheimer's disease, multi-infarct dementia). In these cases, it is incumbent upon practitioners to anticipate and appreciate the myriad coexisting conditions that typically cause pain in this population and to develop the skills necessary to recognize symptoms, assess them, and sufficiently treat and monitor outcomes of therapy in patients who cannot self-report. Behavioral disturbances can be, and oftentimes are, mistaken as part and parcel of the dementing illness, rather than a manifestation of pain. Changes in usual behaviors, alterations in eating/sleeping/ interpersonal response patterns, vocalizations, and various forms of agitation have been shown to be associated with pain perception that is modifiable by both pharmacologic and non-pharmacologic means. The decision to initiate and titrate opioid therapy must be based upon a high index of suspicion, combined with a failure to provide comfort through other means. An "N of 1" trial of opioid analgesics may be the best means of testing the hypothesis that a non-self-reporting patient is experiencing pain. This approach has been shown to be effective at distinguishing analgesic- from non-analgesic-responsive behaviors, while significantly decreasing inappropriate use of psychotropic drugs that may only mask symptoms (Kovach et al. 1999). By sorting out causality in this way and actively treating their pain, an important contribution to the maintenance of these vulnerable patients' dignity and basic humanistic care can be made.

References

Aburun, F., Monsel, S., Langeron, O., Coriat, P., & Riou, B. (2002). Postoperative titration of intravenous morphine in the elderly patient. *Anesthesiology, 96,* 17–23.

AGS Panel on Persistent Pain in Older Persons. (2002). The management of persistent pain in older persons. *Journal of the American Geriatrics Society, 50,* S205–S224.

Arking, R. (2001). The biology of aging: What is it and when it will become useful. *Infertility and Reproductive Medicine Clinics of North America, 12,* 469–487.

Barja, G. (1998). Mitochondrial free radical production and aging in mammals and birds. *Annals of the New York Academy of Sciences, 854,* 224–238.

Belleville, J. W., Forrest, W. H., Miller, E., et al. (1971). Influence of age on pain relief from analgesics: A study of postoperative patients. *Journal of the American Medical Association, 217,* 1835–1841.

Cooner, E., & Amorosi, S. (1997). *The study of pain in older Americans.* New York: Louis Harris and Associates.

Doucet, J., Chassagne, P., Trivalle, C., et al. (1996). Drug-drug interactions related to hospital admissions in older adults: A prospective study of 1000 patients. *Journal of the American Geriatrics Society, 44,* 944–948.

Fine, P. G. (2001). Opioid analgesics in older people. In B. A. Ferrell (Ed.), *Clinics in geriatric medicine: Pain management in the elderly* (pp. 479–485). Philadelphia: WB Saunders.

Fine, P. G. (2004). Pharmacological management of persistent pain in older patients. *The Clinical Journal of Pain, 20,* 220–226.

Guay, D. R. P., Lackner, T. E., & Hanlon, J. T. (2002). Pharmacologic management. In D. K. Weiner, K. Herr, & T. E. Rudy (Eds.), *Persistent pain in older adults* (pp. 160–187). New York: Springer.

Helme, R. D., & Gibson, S. J. (1999). Pain in older people. In I. K. Crombie, P. R. Croft, S. J. Linton, et al. (Eds.), *Epidemiology of pain* (pp. 103–312). Seattle: IASP Press.

Kovach, C. R., Weissman, D. E., Griffie, J., et al. (1999). Assessment and treatment of discomfort for people with late-stage dementia. *Journal of Pain and Symptom Management, 18,* 412–419.

Page, G. G. (2003). The immune-suppressive effects of pain. *Advances in Experimental Medicine and Biology, 521,* 117–125.

Rooke, G. A., Reeves, J. G., & Rosow, C. (2002). Anesthesiology and geriatric medicine: Mutual needs and opportunities. *Anesthesiology, 96,* 2–4.

Rowbotham, M. C., Twilling, L., Davies, M. S., Reisner, L., Taylor, K., & Mohr, D. (2003). Oral opioid therapy for chronic peripheral and central neuropathic pain. *The New England Journal of Medicine, 348,* 1223–1232.

Opioids in Sympathetically Maintained Pain

▶ Opioids in the Management of Complex Regional Pain Syndrome

Opioids in the Management of Complex Regional Pain Syndrome

Srinivasa N. Raja[1] and Frank Lee[2]
[1]Department of Anesthesiology and Critical Care Medicine, Johns Hopkins University, Baltimore, USA
[2]Johns Hopkins University, Baltimore, MD, USA

Synonyms

Causalgia and reflex sympathetic dystrophy; Opioids in sympathetically maintained pain

Definition

Our understanding of the mechanisms of pain in complex regional pain syndrome (CRPS) remains inadequate. The role of chronic use of opioids for the treatment of non-cancer pain conditions, like CRPS, continues to be debated. Here, we aim to focus our efforts on highlighting the body of evidence around the use of opioids in treating CRPS and other neuropathic conditions.

CRPS is a chronic pain state that usually follows a traumatic event and is characterized by ongoing pain and/or hypersensitivity to exogenous stimuli along with symptoms and signs of autonomic dysregulation over a region of the body that extends beyond the expected course of recovery from the injury. Motor and trophic changes may also be part of the presentation of CRPS. CRPS was initially defined in 1994 by the International Association for the Study of Pain (IASP), and then a more stringent diagnostic criterion was developed in 2003. The new diagnostic criteria differentiates between clinical and research diagnosis of CRPS and adds an extra category of diagnosis CRPS-NOS (Not Otherwise Specified) (Table 1). Because of the limited studies that focus specifically on the chronic use of opioids in CRPS, we have extrapolated from studies that involve other pain conditions of neuropathic origin.

Characteristics

The perception that neuropathic pain conditions like CRPS are resistant to opioid therapy compared to nociceptive pain has been a commonly accepted notion and is the basis for general avoidance of opioids for treatment of neuropathic pain by many physicians today. This concept was supported, in part by an early study that examined the comparative efficacy of opioids in neuropathic and nociceptive pains (Arner and Meyerson 1988). The authors concluded, based on the effectiveness of infusions of opioid and placebo in a mixed group of 48 patients with neuropathic, nociceptive, and idiopathic pain, that while nociceptive pain was opioid-responsive, neuropathic pain appeared to be resistant. Twelve patients in this group had neuropathic pain; 4 of these 12 patients had combined neuropathic and nociceptive pain. Only 1 of the 8 patients with neuropathic pain alone, and 1 in the combined neuropathic/nociceptive group, responded positively to the opioid infusion test. These observations were in contrast with the reduction in pain in all of the 15 subjects with nociceptive pain after opioid infusions. Other small subsequent studies appeared to support this finding. Many of these studies were criticized for problems of selection bias, poor study design, and small sample size (Dellemijn 1999). In general, studying neuropathic conditions is a more challenging task due to their diverse etiology and poorly understood mechanisms for pain, in contrast to nociceptive pain that is easier to define.

Some have argued that although resistant, opioid therapy could be effective in treating neuropathic pain with optimal titration and that the drug doses required to attenuate neuropathic pain may be higher than that required to relieve nociceptive pain. Several controlled trials have shown that neuropathic pain states, such as postherpetic neuralgia, diabetic neuropathy, and post-amputation pain states, respond to opioids during a 4–12-week study period. The evidence supporting either side-potential benefit or the lack of efficacy has been less than definitive, and there are considerable gaps in the literature concerning the long-term effectiveness of opioids for

Opioids in the Management of Complex Regional Pain Syndrome, Table 1   New CRPS diagnostic criteria

General definition of the syndrome
CRPS describes an array of painful conditions that are characterized by a continuing (spontaneous and/or evoked) regional pain that is seemingly disproportionate in time or degree to the usual course of any known trauma or other lesion. The pain is regional (not in a specific nerve territory or dermatome) and usually has a distal predominance of abnormal sensory, motor, sudomotor, vasomotor, and/or trophic findings. The syndrome shows variable progression over time
To make the *clinical* diagnosis, the following criteria must be met:
1. Continuing pain, which is disproportionate to any inciting event
2. Must report at least one symptom in *three of the four* following categories:
Sensory: Reports of hyperesthesia and/or allodynia
Vasomotor: Reports of temperature asymmetry and/or skin color changes and/or skin color asymmetry
Sudomotor/edema: Reports of edema and/or sweating changes and/or sweating asymmetry
Motor/trophic: Reports of decreased range of motion and/or motor dysfunction (weakness, tremor, dystonia) and/or trophic changes (hair, nail, skin)
3. Must display at least one sign **at time of evaluation** in *two or more* of the following categories:
Sensory: Evidence of hyperalgesia (to pinprick) and/or allodynia (to light touch and/or temperature sensation and/or deep somatic pressure and/or joint movement)
Vasomotor: Evidence of temperature asymmetry (>1 °C) and/or skin color changes and/or asymmetry
Sudomotor/edema: Evidence of edema and/or sweating changes and/or sweating asymmetry
Motor/trophic: Evidence of decreased range of motion and/or motor dysfunction (weakness, tremor, dystonia) and/or trophic changes (hair, nail, skin)
4. There is no other diagnosis that better explains the signs and symptoms

non-cancer patients with chronic pain. Long-term opioid therapy may also be associated with a myriad of potential side effects that include nausea, vomiting, constipation, cognitive impairment, respiratory depression, tolerance, dependence, hyperalgesia, immune suppression, and hormonal effects. Hence, it is prudent that much individual consideration should be given when weighing the risks and benefits of chronic opioid therapy for neuropathic pain like CRPS.

Effects of Opioids in CRPS

There is a paucity of studies examining the chronic use of opioids specifically in CRPS. Although the published literature is limited, growing clinical experience along with clinical reevaluation of issues related to safety, efficacy, and addiction or abuse shows that some patients can achieve analgesia and improved function and quality of life without the occurrence of intolerable side effects. Because the pathophysiology of CRPS is still being investigated, the overall goals of the treatment strategy are symptomatic relief of pain and functional rehabilitation. In practice, it is not uncommon to find opioids being

prescribed to alleviate ongoing pain when more conservative therapies fail and to help the patient participate in rehabilitative treatments.

An area of promise is the use of opioids along with adjuvant drugs. In a recent double-blind randomized study in patients with CRPS affecting the upper extremity, therapy with morphine alone was compared to that with morphine and the NMDA antagonist memantine. The combination therapy, but not morphine alone, reduced pain at rest and during movement, and disability (Gustin et al. 2010). In addition, fMRI studies in these patients showed that activation in the contralateral primary somatosensory and anterior cingulate cortex was significantly reduced when the affected hand was moved. These data suggest that the combination of morphine and an NMDA antagonist affects cerebral processing of nociceptive signals in CRPS.

Since there are no controlled trials of chronic opioid use in patients with CRPS, we will attempt to extrapolate from results of studies in patients with neuropathic pain. It is important to note that in general, the vast majority of research subjects in studies for neuropathic pain are patients with diabetic peripheral neuropathy or postherpetic neuralgia. Therefore, we caution against direction application of results from such studies on patients with CRPS.

Effects of Opioids in Other Neuropathic Pain States

Initial proof of the principle of the effectiveness of opioids on neuropathic pain was achieved using brief intravenous (IV) infusion studies (Rowbotham et al. 1991), comparing the effects of IV infusions of morphine, lidocaine, and placebo in 19 patients with postherpetic neuralgia, using a double-blind, crossover design. A 33 % decline in pain intensity was observed after morphine (0.3 mg/kg) infusion, compared to a 13 % reduction with placebo. Pain relief ratings were also higher during the morphine sessions than lidocaine or placebo sessions.

Several subsequent studies have confirmed the beneficial role of opioids in the treatment of various neuropathic pain states. In a double-blind crossover trial, the infusion of the opioid fentanyl was more effective in reducing the intensity of neuropathic pain, compared to the active placebo, diazepam, and the inert placebo, saline (Dellemijn and Vanneste 1997). Two recent additional randomized, controlled, crossover studies have reported beneficial effects of intravenous opioids on central pain and post-amputation pains (Attal et al. 2002; Wu et al. 2002). All four trials described above report a mean pain relief of 30–55 % with opioids in the group of patients studied, but significant individual variations were observed. In contrast, the placebo response varied from an increase in pain of 5 % to a 25 % decrease in pain.

Infusion studies help demonstrate that opioids are likely to provide analgesia in patients with neuropathic pain. However, they may not predict whether therapy with oral opioids is likely to be similarly beneficial. A number of controlled trials published during the last two decades have provided evidence for a beneficial effect of oral opioids in chronic neuropathic pain. In a crossover trial, the effects on pain of twice-daily controlled-release oxycodone treatment were studied in 50 patients with postherpetic

neuralgia (Watson and Babul 1998). Subjects began with either placebo or 10 mg oxycodone twice a day and titrated to 60 mg per day. A significant decrease in overall pain intensity and pain relief was observed in the oxycodone treatment period as compared to the placebo period. Fifty-eight percent of patients experienced at least moderate pain relief with oxycodone as compared to 18 % with placebo. In addition, disability scores were lower during treatment with oxycodone. A similar decrease in pain and greater pain relief compared to placebo was observed in a multicenter, randomized, placebo-controlled trial in patients with distal symmetric diabetic neuropathy, treated with the weak opioid agonist tramadol (Harati et al. 1998), compared with the change in pain intensity and pain relief with opioids and tricyclic antidepressants in patients with postherpetic neuralgia (Raja et al. 2002). They observed similar reductions in pain intensity with both drugs, but patients reported greater satisfaction with the opioid therapy as compared to the therapy with tricyclic antidepressants (50 % vs. 34 %). Rowbotham et al. (2003) compared treatment with low (0.15 mg)- and high (0.75 mg)-strength capsules of levorphanol in a heterogenous group of patients with peripheral or central neuropathic pain states. They reported a greater reduction in intensity with higher doses of opioids than with lower doses. The mean dose of levorphanol in the high-strength group was 8.9 mg/day, approximately equivalent to 135–270 mg of oral morphine and 90–135 mg of oxycodone. A double-blind placebo-controlled trial in neuropathic patients who were being treated with spinal cord stimulators, however, failed to demonstrate a significant effect of morphine on pain at lower doses, between 60 and 90 mg/day (Harke et al. 2001).

Another case for use of opioids for chronic neuropathic pain was presented by a meta-analysis (Eisenberg et al. 2005 JAMA) that claimed a 20–30 % greater reduction in neuropathic pain with opioids than with placebo in eight intermediate term RCT (less than 8 weeks). Some of the studies already mentioned were included in this meta-analysis. However, as is the case for most meta-analysis, there was a wide range of types of neuropathic pain, categories of opioids and dosages, and study end points. In the same study, 14 short-term RCT (less than 4 weeks) showed equivocal evidence for efficacy of opioids for reducing pain.

Summary

In summary, several studies have demonstrated that oral therapy with opioids can result in a reduction in neuropathic pain intensity. The category and quantity of opioids to achieve this pain reduction, however, is still being discovered. Studies also indicate that therapy with opioids can be associated with side effects, although mild, but the risk-benefit ratio needs to be evaluated carefully. More high-quality studies need to be conducted to examine if patients with CRPS are likely to benefit from therapy with opioids similar to that observed in patients with other neuropathic pain states.

References

Arner, S., & Meyerson, B. A. (1988). Lack of analgesic effect of opioids on neuropathic and idiopathic forms of pain. *Pain, 33*, 11–23.

Attal, N., Guirimand, F., Brasseur, L., Gaude, V., Chauvin, M., & Bouhassira, D. (2002). Effects of IV morphine in central pain – A randomized placebo-controlled study. *Neurology, 58*, 554–563.

Dellemijn, P. (1999). Are opioids effective in relieving neuropathic pain? *Pain, 80*, 453–462.

Dellemijn, P. L., & Vanneste, J. A. (1997). Randomised double-blind active-placebo-controlled crossover trial of intravenous fentanyl in neuropathic pain. *Lancet, 349*, 753–758.

Eisenberg, E., McNicol, E. D., & Carr, D. B. (2005). Efficacy and safety of opioid agonists in the treatment of neuropathic pain of nonmalignant origin: Systematic review and meta-analysis of randomized controlled trials. *Journal of the American Medical Association, 293*, 3043–3052.

Gustin, S. M., et al. (2010). NMDA-receptor antagonist and morphine decrease CRPS-pain and cerebral pain representation. *Pain, 151*, 69–76.

Harati, Y., Gooch, C., Swenson, M., Edelman, S., Greene, D., Raskin, P., Donofrio, P., Cornblath, D., Sachdeo, R., Siu, C. O., & Kamin, M. (1998). Double-blind randomized trial of tramadol for the treatment of the pain of diabetic neuropathy. *Neurology, 50*, 1842–1846.

Harke, H., Gretenkort, P., Ladleif, H. U., Rahman, S., & Harke, O. (2001). The response of neuropathic pain and pain in complex regional pain syndrome I to carbamazepine and sustained-release morphine in patients pretreated with spinal cord stimulation: A double-blinded randomized study. *Anesthesia and Analgesia, 92*, 488–495.

Raja, S. N., Haythornthwaite, J. A., Pappagallo, M., Clark, M. R., Travison, T. G., Sabeen, S., Royall, R. M., & Max, M. B. (2002). Opioids versus antidepressants in postherpetic neuralgia: A randomized, placebo-controlled trial. *Neurology, 59*, 1015–1021.

Rowbotham, M. C., Reisner-Keller, L. A., & Fields, H. L. (1991). Both intravenous lidocaine and morphine reduce the pain of postherpetic neuralgia. *Neurology, 41*, 1024–1028.

Rowbotham, M. C., Twilling, L., Davies, P. S., Reisner, L., Taylor, K., & Mohr, D. (2003). Oral opioid therapy for chronic peripheral and central neuropathic pain. *The New England Journal of Medicine, 348*, 1223–1232.

Watson, C. P., & Babul, N. (1998). Efficacy of oxycodone in neuropathic pain: A randomized trial in postherpetic neuralgia. *Neurology, 50*, 1837–1841.

Wu, C. L., Tella, P., Staats, P. S., Vaslav, R., Kazim, D. A., Wesselmann, U., & Raja, S. N. (2002). Analgesic effects of intravenous lidocaine and morphine on post-amputation pain: A randomized double-blind, active-placebo-controlled, crossover trial. *Anesthesiology, 96*, 841–848.

Opioids in the Modulation of Ascending Pathways

▶ Opioids in the Spinal Cord and Modulation of Ascending Pathways (N. gracilis)

Opioids in the Periphery and Analgesia

Christoph Stein
Department of Anesthesiology and Intensive Care Medicine, Freie Universitaet Berlin Charité Campus Benjamin Franklin, Berlin, Germany

Synonyms

Opioid analgesic actions outside the central nervous system; Peripheral mechanisms of opioid analgesia; Peripheral opioid analgesia

Definition

Opioid analgesia produced outside the central nervous system by interaction of endogenous or exogenous opioids with opioid receptors on peripheral sensory neurons.

Characteristics

Opioids can produce potent analgesia by activating opioid receptors outside the central nervous system, thus avoiding centrally mediated unwanted effects such as nausea, addiction, or respiratory depression. Peripheral opioid receptors are localized on primary afferent neurons. The cell bodies of these neurons in dorsal root ganglia express mu-, delta-, and kappa-opioid receptor mRNAs and proteins. ▶ Opioid receptors are intra-axonally transported into the neuronal processes and are expressed on peripheral sensory nerve terminals both in animals and in humans. Co-localization and electrophysiological studies have confirmed the presence of opioid receptors on ▶ C and A Fibers, on transient receptor potential vanilloid subtype-1 (TRPV-1) carrying fibers, and on neurons expressing isolectin B4, ▶ substance P, and/or ▶ calcitonin gene-related peptide, consistent with the phenotype of ▶ nociceptors. Sympathetic neurons and immune cells can also express opioid receptors, but their functional role in pain control is unclear. The binding characteristics (affinity) of peripheral and central opioid receptors are similar (Stein et al. 2003; Stein and Machelska 2011).

Opioid Receptor Signaling in Primary Afferent Neurons

All three types of opioid receptors mediate the inhibition of high-voltage activated calcium currents in cultured primary afferent neurons. These effects are transduced by G-proteins (G_i and/or G_o). In addition, opioids – via inhibition of adenylyl cyclase – suppress tetrodotoxin-resistant Na^+-, TRPV1-, and other nonselective cation currents stimulated by inflammatory agents, which may account for the notable efficacy of peripheral opioids in inflammatory and

▶ neuropathic pain. Consistent with their effects on ion channels, opioids attenuate the excitability of peripheral nociceptor terminals, the propagation of action potentials, the release of excitatory proinflammatory neuropeptides (substance P, calcitonin gene-related peptide) from peripheral sensory nerve endings, and vasodilatation evoked by stimulation of C fibers. These mechanisms result in analgesia and/or anti-inflammatory actions (Stein and Machelska 2011; Stein and Küchler 2012).

Peripheral Opioid Receptors and Tissue Injury

Peripheral opioid analgesic effects are particularly prominent in inflamed tissue (Kalso et al. 2002; Stein et al. 2003). Under such conditions, the synthesis and expression of opioid receptors in dorsal root ganglia is elevated. Subsequently, the axonal transport and membrane-directed trafficking of opioid receptors increases, leading to their upregulation on peripheral neuron terminals. These events are dependent on neuronal electrical activity, cytokines, and nerve growth factor from the damaged tissue. In mechanical nerve injury leading to neuropathic pain, opioid receptors accumulate proximal and distal to the lesion, indicating anterograde and retrograde transport (Labuz et al. 2009). Inflammatory milieu (low pH, prostanoid release) augments opioid receptor function by more efficient G-protein coupling and inhibition of elevated neuronal cyclic adenosine monophosphate production. Inflammation also leads to sprouting of sensory nerve terminals and disruption of the perineurial barrier, thus facilitating the access of opioid agonists to their receptors. Endogenous opioid ligands derived from inflammatory cells stimulate recycling of opioid receptors to the membrane of sensory neurons, which can prevent the development of ▶ tolerance to peripherally active opioid agonists. Consistently, clinical studies have indicated a lack of cross-tolerance between peripheral exogenous and endogenous opioids in synovial inflammation. All of these mechanisms can contribute to enhanced antinociceptive efficacy of opioid agonists in injured tissue (Zöllner et al. 2008; Stein and Machelska 2011).

Endogenous Ligands of Peripheral Opioid Receptors

Concurrent with the development of inflammation, opioid peptide-producing immune cells are recruited to the site of injury. The most thoroughly characterized peptides are β-endorphin and enkephalins deriving from the respective precursors proopiomelanocortin (POMC) and proenkephalin. Transcripts and peptides derived from POMC and proenkephalin, as well as the prohormone convertases PC1/3 and PC2, necessary for their posttranslational processing were detected in such cells. The expression of immune-derived opioids is stimulated by viruses, endotoxins, cytokines, corticotropin-releasing hormone (CRH), and adrenergic agonists. In painful tissue inflammation and neuropathy, POMC mRNA, β-endorphin, met-enkephalin, and dynorphin are detectable in circulating cells and lymph nodes and are upregulated in resident lymphocytes, monocytes/macrophages, and granulocytes. Circulating opioid-containing leukocytes migrate to injured tissue attracted by ▶ adhesion molecules, chemokines, and neurokinins. In inflamed tissue, opioid-containing leukocytes, vascular P-selectin, ICAM-1, and PECAM-1 are simultaneously upregulated. Blocking chemokines, selectins, or ICAM-1 reduces the extravasation of opioid-containing cells and increases inflammatory and neuropathic pain. Immunosuppression similarly exacerbates pain (Labuz et al. 2009; Stein and Machelska 2011).

Stimuli such as environmental stress, noradrenaline, CRH, interleukin-1β, chemokines, or mycobacteria can elicit opioid peptide release from immune cells via specific receptors and a regulated secretory pathway. Depending on the cell type and agent, intracellular Ca^{++} release from endoplasmic reticulum or extracellular Ca^{++} is required. In vivo, the secreted opioid peptides bind to opioid receptors on sensory neurons and elicit analgesia in injured tissue and neuropathy (Labuz et al. 2009; Rittner et al. 2009). Not only stimulated but also tonic release of opioids from immune cells decreases pain in animals (Rittner et al. 2009) and in humans (Stein et al. 1993). Thus, the development of inflammatory and neuropathic pain is counteracted by immune cells producing and secreting opioid peptides. Gene therapeutic approaches are aiming to increase the production of opioid peptides and receptors in inflammatory cells and peripheral sensory neurons, respectively (Stein and Machelska 2011; Raja 2012). Preventing the extracellular degradation of endogenous opioid peptides by peptidase inhibitors has been shown to diminish inflammatory pain (Roques et al. 2012; Schreiter et al. 2012).

Preclinical Studies on Peripheral Opioid Analgesics

This basic research has stimulated the development of novel opioid ligands acting exclusively in the periphery without central side effects. A common approach is the use of hydrophilic compounds with minimal capability to cross the blood-brain barrier. Among the first compounds were the mu-agonist loperamide (known as an antidiarrheal drug) and the kappa-agonist asimadoline. Peripheral restriction was also achieved with glucuronidation, arylacetamide (ADL 10-0101), morphinan-based (TRK-820, HS-731), triazaspiro (DiPOA), and peptidic compounds (DALDA, FE200665, CR845). While earlier attempts to demonstrate peripheral opioid analgesia in normal tissue failed, they were much more successful in models of pathological pain (Stein 1993). For example, in subcutaneous inflammation, the local injection of low, systemically inactive doses of mu-, delta-, and kappa-agonists produces dose-dependent and opioid receptor-specific antinociception and, in some conditions, anti-inflammatory effects. Possible underlying mechanisms of the latter include a reduced release of proinflammatory neuropeptides or cytokines and a diminished expression of adhesion molecules (Stein and Küchler 2012). Potent antinociception was also shown in models of nerve damage, visceral, thermal, cancer, and bone pain (Stein and Machelska 2011).

Clinical Studies on Peripheral Opioid Analgesics

The most extensively examined clinical application is the intra-articular injection of

Opioids in the Periphery and Analgesia, Fig. 1 Opioid peptide (*OP*)-containing leukocytes migrate into inflamed tissue. OP release from polymorphonuclear granulocytes is triggered by, e.g., the chemokine CXCL2/3 (insert). This requires elevation of intracellular Ca^{2+} through efflux from the endoplasmic reticulum (ER) and (in part) phosphoinositol-3-kinase (PI3K). OP can also be released by noradrenaline (*NA*; derived from sympathetic neurons) activating adrenergic receptors (*AR*). Opioid receptors (*OR*) and neuropeptides (substance P; *sP*) are synthesized in the dorsal root ganglion and transported along intra-axonal microtubules into central and peripheral processes of the primary afferent neuron. The permeability of the perineurium is increased in inflamed tissue. Upon activation by endogenous (OP) or exogenous opioids (*EO*), OR couple to inhibitory G-proteins (*Gi/o*). This leads to decreased cAMP and inhibition (−) of Ca^{2+}, TRPV1, and/or Na^+ currents, resulting in reduced excitability of the neuron and in attenuated sP release. Activation of OR stimulates receptor phosphorylation, arrestin (*Arr*) binding, and internalization (insert). Subsequently, OR can either be degraded in lysosomes or dephosphorylated and recycled to the cell membrane. Recycling counteracts the development of tolerance to opioid agonists. *cAMP* cyclic adenosine monophosphate, *IP3* inositol 1,4,5-triphosphate, *PLC* phospholipase C (Adapted from Stein and Machelska 2011)

morphine. Both in human and veterinary medicine, numerous controlled clinical studies have demonstrated dose-dependent and peripherally mediated reduction of pain and/or supplemental analgesic consumption without significant side effects (Kalso et al. 2002). Intra-articular morphine is effective in acute (postoperative) and chronic (arthritic) pain, its effect is similar to intra-articular local anesthetics and steroids, and it is long lasting, possibly due to anti-inflammatory activity. Locally applied opioids were also effective in dental pain, skin ulcers, corneal abrasions, and visceral pain (Sawynok 2003; Vadivelu et al. 2011; Farley 2011). Some studies found no peripheral effects of opioids, e.g., after injection into the noninflamed environment along nerve trunks. This suggests that intra-axonal opioid receptors are "in transit" and not available as functional receptors at the membrane. Peripherally restricted opioids (morphine-6-glucuronide, CR845) have been shown to reduce visceral and postoperative pain with limited central side effects and similar efficacy as morphine (van Dorp et al. 2008; Stein and Machelska 2011). Such compounds may provide additional advantages such as anti-inflammatory actions and absence of cardiovascular and gastrointestinal side effects typical for ▶ NSAIDs (Fig. 1).

Acknowledgement Supported by Deutsche Forschungsgemeinschaft, Bundesministerium für Bildung und Forschung, International Anesthesia Research Society, and European Society of Anaesthesiology.

References

Farley, P. (2011). Should topical opioid analgesics be regarded as effective and safe when applied to chronic cutaneous lesions? *The Journal of Pharmacy and Pharmacology, 63*, 747–756.

Kalso, E., Smith, L., McQuay, H. J., & Moore, R. A. (2002). No pain, no gain: Clinical excellence and scientific rigour – lessons learned from IA morphine. *Pain, 98*, 269–275.

Labuz, D., Schmidt, Y., Schreiter, A., Rittner, H. L., Mousa, S. A., & Machelska, H. (2009). Immune cell-derived opioids protect against neuropathic pain in mice. *Journal of Clinical Investigation, 119*, 278–286.

Raja, S. N. (2012). Modulating pain in the periphery: Gene-based therapies to enhance peripheral opioid analgesia. *Regional Anesthesia and Pain Medicine, 37*(2), 210–214.

Rittner, H. L., Hackel, D., Voigt, P., Mousa, S. A., Stolz, A., Labuz, D., Schäfer, M., Schaefer, M., Stein, C., & Brack, A. (2009). Mycobacteria attenuate nociceptive responses by formyl peptide receptor triggered opioid peptide release from neutrophils. *PLoS Pathogens, 5*, e1000362.

Roques, B. P., Fournie-Zaluski, M. C., & Wurm, M. (2012). Inhibiting the breakdown of endogenous opioids and cannabinoids to alleviate pain. *Nature Reviews Drug Discovery, 11*, 292–310.

Sawynok, J. (2003). Topical and peripherally acting analgesics. *Pharmacological Reviews, 55*, 1–20.

Schreiter, A., Gore, C., Labuz, D., Fournie-Zaluski, M. C., Roques, B. P., Stein, C., & Machelska, H. (2012). Pain inhibition by blocking leukocytic and neuronal opioid peptidases in peripheral inflamed tissue. *The FASEB Journal, 26*(12), 5161–5171.

Stein, C. (1993). Peripheral mechanisms of opioid analgesia. *Anesthesia and Analgesia, 76*, 182–191.

Stein, C., & Küchler, S. (2012). Non-analgesic effects of opioids: Peripheral opioid effects on inflammation and wound healing. *Current Pharmaceutical Design, 18*(37), 6053–6069.

Stein, C., & Machelska, H. (2011). Modulation of peripheral sensory neurons by the immune system: Implications for pain therapy. *Pharmacological Reviews, 63*, 860–881.

Stein, C., Hassan, A. H. S., Lehrberger, K., Giefing, J., & Yassouridis, A. (1993). Local analgesic effect of endogenous opioid peptides. *Lancet, 342*, 321–324.

Stein, C., Schäfer, M., & Machelska, H. (2003). Attacking pain at its source: New perspectives on opioids. *Nature Medicine, 9*, 1003–1008.

Vadivelu, N., Mitra, S., & Hines, R. L. (2011). Peripheral opioid receptor agonists for analgesia: A comprehensive review. *Journal of Opioid Management, 7*, 55–68.

van Dorp, E. L., Morariu, A., & Dahan, A. (2008). Morphine-6-glucuronide: Potency and safety compared with morphine. *Expert Opinion on Pharmacotherapy, 9*, 1955–1961.

Zöllner, C., Mousa, S. A., Fischer, O., Rittner, H. L., Shaqura, M., Brack, A., Shakibaei, M., Binder, W., Urban, F., Stein, C., & Schäfer, M. (2008). Chronic morphine use does not induce peripheral tolerance in a rat model of inflammatory pain. *Journal of Clinical Investigation, 118*, 1065–1073.

Opioids in the Spinal Cord and Central Sensitization

Anthony H. Dickenson
Department of Pharmacology, University College London, London, UK

Definition

Activation of the N-methyl-D-aspartate (▶ NMDA) receptor causes a major excitatory drive in the dorsal horn of the spinal cord where ▶ wind-up and associated ▶ central sensitization is induced. Once the NMDA receptor is activated, inhibitory controls may be compromised, simply since greater levels of excitability will require greater inhibitory controls. Combinations of NMDA antagonists plus opioids therefore predictably synergize to produce marked inhibitory effects.

Characteristics

Some of the very first studies on glutamate as a transmitter involved the spinal cord. The NMDA receptor has become an increasingly important target site for potential analgesics, as evidence accumulates for a role of the receptor in the enhancement of spinal processing of painful messages and in many long-term events in the brain, including ▶ long-term potentiation and excitotoxic cell death. However, this important role of the receptor in a number of CNS functions can lead to problems in terms of side effects.

There is now little doubt that glutamate seen in primary afferent terminals in laminae I, III, and IV of the dorsal horn is a releasable transmitter. In the case of C fibers, glutamate may coexist with peptides such as substance P and CGRP, which would make it highly likely that a noxious stimulus releases both peptides and excitatory amino acids from the afferent nociceptive fibers (Battaglia and Rustioni 1988). Large A fibers, terminating in deeper laminae, have glutamate, yet do not normally contain peptides. Thus, glutamate is involved in the transmission of both high and low threshold information from afferents into the spinal cord. In addition to a key role in transmission from afferent fibers, there is further evidence that transmission, at least from trigemino- and spinothalamic tract cells, also involves glutamate (Dickenson 1997).

Although a number of studies have demonstrated that glutamate is released from both low and high threshold afferents, electrophysiological studies would indicate that, in general, only noxious events, at least under normal conditions, activate the NMDA receptor. It could be suggested that peptides may allow the differentiation between large A fiber and C fiber inputs, since only C fiber stimulation will elicit peptide release. Consequently, C fiber induced release of excitatory peptides may provide the required

depolarization to remove the Mg^{++} block of the receptor and allow NMDA receptor activation (Dickenson 1997).

Wind-Up

Wind-up is the term given to a short-term increase in excitability of spinal neurones. When a constant C fiber stimulus is delivered, the response of deep dorsal horn neurone increases rapidly after the first few stimuli and then decays slowly after cessation of the stimulus. If a train of stimuli are given, at frequencies of about 0.5 Hz and above, the responses of the neurone to the first few stimuli remain constant. However, as the stimulus continues, there is a rapid incremental increase in firing of the neurone, and cumulative long slow depolarization such as those mediated by peptides including substance P seen with intracellular recordings, associated with increased action potentials. The neurone firing can increase by up to 20-fold the initial rate. Although wind-up is only induced by C fiber stimulation, once the process has occurred, all responses of the neurones are enhanced and a post-discharge, firing following the C fiber latency band, is evoked. Thus, noxious stimuli, applied at sufficient intensities, can enhance spinal excitatory events by mechanisms that are restricted to the spinal cord. Wind-up, in normal animals, is only produced by stimulation at C fiber, and possibly also A-delta, stimulation, but not by low-intensity stimuli (Dickenson 1997). With regard to the functional significance of wind-up, it has been shown that high-frequency C fiber stimulation results in a marked and prolonged increase in the ▶ flexion withdrawal reflex, recorded from motoneurones in spinal rats. It is evident that a number of physiological inputs, causing repeated C fiber activity, should activate the NMDA receptor, if the intensity and area of stimulus is sufficient. This has turned out to be the case, and there is evidence for involvement of the NMDA receptor in inflammatory pain, neuropathic pain, allodynia and ischaemic pain, and all processes in which the receptor alters the normal relationship between stimulus and response (Dickenson 1997; Price et al. 1994). In these persistent pain states, the NMDA receptor is vital both in establishing the augmented pain state and in maintaining this state. Higher frequency stimulation of afferents can lead to NMDA-dependent ▶ long-term potentiation in the spinal cord, where the enhanced responses now last for hours (Ikeda et al. 2003; Rygh et al. 2000). This pivotal role of the NMDA receptor in the plasticity of the pain signaling system makes it an attractive target for development of new analgesics.

The NMDA receptor is of course not restricted to spinal pain pathways, and it is not surprising that NMDA receptor antagonists, such as the channel blocker ketamine and competitive antagonists, are associated with a range of adverse effects. In volunteers and patients, key roles of the receptor have been shown in states of capsaicin-induced central hypersensitivity and after nerve injury response (Dickenson 1997; Price et al. 1994). The increased excitability leads not simply to increased pain ratings but may well also increase the area of pain as has been seen in animal studies. However, the clinical use of NMDA receptor antagonists has been fraught by the simple fact that as therapeutic levels are attained, side effects are on the verge of unacceptable. One possible approach to avoid the side effects associated with global block of NMDA receptors is to target a particular receptor type by its subunit makeup. This was justified by animal studies that demonstrated that NR2B-selective antagonists may have clinical utility for the treatment of pain conditions with a reduced side effect profile compared to existing NMDA receptor antagonists (Boyce et al. 1999).

NMDA Receptor Interactions with Opioids

Accumulating evidence indicates plasticity in opioid controls. The degree of effectiveness of morphine analgesia is subject to modulation by other transmitter systems in the spinal cord and by pathological changes induced by peripheral nerve injury. In neuropathic pain states, a number of explanations can be given for this reduction in opioid actions. A potential marked loss of presynaptic opioid receptors as a result of nerve section can occur, although this appears not to be a factor in less severe nerve injuries. Possibly, more importantly, an upregulation of the "anti-opioid peptide" ▶ cholecystokinin, after damage to peripheral nerve, could also contribute to the reduced morphine analgesia. In reality, a combination of factors is likely to be the cause of the particular problems that arise in the control of neuropathic pains with opioid drugs (Dickenson and Suzuki 1999). The CNS operates by a balance between excitation and inhibition. Thus, the profound excitations produced by NMDA receptor involvement in pain-related responses of spinal cord neurones can further shift the balance in favor of excitation. The enhanced levels of firing will then lead to a reduced effectiveness of a given dose of an opioid. Reductions in opioid controls can be overcome by dose-escalation if excessive side effects do not intervene, and one approach could be spinal delivery where high local levels can be achieved.

An approach that could be considered would be to use a combination of an NMDA antagonist with an opioid such as morphine. The dual actions of these pharmacological agents could be beneficial. The fact that NMDA receptor antagonists are effectively anti-hyperalgesic agents, reducing the hyperexcitability, means that the addition of an analgesic such as an opioid would enable the baseline response to also be inhibited. Interestingly, some opioids such as methadone, pethidine, and ketobemidone appear to have both mu opioid receptor actions and weaker, but potentially functionally relevant, NMDA receptor blocking actions (Ebert et al. 1995). This additional action of these opioids, noncompetitive NMDA antagonism, has been revealed by both binding and electrophysiological approaches, but it remains unclear as to whether the non-opioid component of their actions contributes at all to their in vivo profile (Carpenter et al. 2000).

The spinal actions of opioids and their mechanisms of analgesia involve (1) reduced transmitter release (e.g., glutamate and peptides) from nociceptive C fibers (75 % of spinal opioid receptors are presynaptic) so that spinal neurones are less excited by incoming painful messages and

(2) postsynaptic inhibitions of neurones conveying information from the spinal cord to the brain (Dickenson and Suzuki 1999). From electrophysiological studies, a moderate dose of morphine, or indeed any opioid, will initially profoundly inhibit neuronal activity evoked by C fiber stimulation, delaying or abolishing activity indicative of presynaptic inhibition of transmitter release. However, as the stimulation continues, as wind-up is induced, the neuronal excitations produced by postsynaptic NMDA receptor activation break through the opioid inhibitions so that at moderate doses, opioids can only delay the onset of wind-up without inhibiting the process itself (Chapman and Dickenson 1992). Higher doses of opioid, presumably activating postsynaptic receptors as well, will abolish responses.

At supraspinal sites, opiate actions are becoming increasingly well understood, but exact mechanisms are still elusive. Morphine can act to alter descending pathways from the brain to the cord which involve noradrenaline and serotonin, and these pathways then act to reduce spinal nociceptive activity. Actions of opioids on descending systems may be of particular relevance to pain states where supraspinal facilitatory pathways are superimposed on spinal hypersensitivity (Porreca et al. 2002; Suzuki et al. 2002).

Marked inhibitions can be produced through synergism between the combination of low threshold doses of morphine with low doses of NMDA receptor antagonists (Chapman and Dickenson 1992) using neuronal measures. In a model of neuropathic pain using behavior against which morphine fails to be antinociceptive, the combined application of an NMDA antagonist, in this case MK-801, with morphine, restored the ability of morphine to inhibit the response (Yamamoto and Yaksh 1992). A human study on the flexion reflex in volunteers confirmed this idea but indicated that the combination of ketamine and morphine had actions that rely on the type of afferent input (Bossard et al. 2002). These studies suggest that in the absence of NMDA receptor antagonists devoid of side effects, the coadministration of a low dose of an NMDA antagonist with a low dose

of an opioid may deliver a good pain control with minor side effects. However, there are some other important issues. Neurones in lamina I of the spinal cord have projections to ▶ parabrachial/PAG areas, whereas many deep cells project in the ▶ spinothalamic tract (Todd 2002). NMDA-dependent wind-up is clear and obvious in deep cells and almost absent in lamina I cells, although both neuronal types support LTP when high-frequency stimuli are given (Dickenson 1997; Ikeda et al. 2003; Rygh et al. 2000). This suggests that in pain states other than those activated by very high-frequency stimuli, spinothalamic inputs will be potentiated through wind-up like mechanisms, whereas inputs to emotional areas will not. This may then lead to dissociation between the emotional and sensory-discriminative aspects of pain. However, opioids have identical dose-response curves for ▶ laminae I and V neurones which would agree well with their ability to modulate both components of the sensation. As a consequence, the opioid-NMDA interaction will only be on lamina V activity, so that the sensory responses to pain could be expected to be altered more than the emotional aspects by a combination. However, an added complexity regarding lamina V is that once LTP is induced, morphine, at doses that completely suppress LTP, does not alter the underlying process. Thus, reversal of the opioid inhibition is only to the post-LTP response level (Rygh et al. 2000). Therefore, complex interactions of opioids with spinal mechanisms of hyperexcitability are likely, but further exploration of this area could lead to useful clinical advances.

References

Battaglia, G., & Rustioni, A. (1988). Coexistence of glutamate and substance P in dorsal root ganglion cells of the rat and monkey. *The Journal of Comparative Neurology, 277,* 302–312.

Bossard, A. E., Guirimand, F., Fletcher, D., et al. (2002). Interaction of a combination of morphine and ketamine on the nociceptive flexion reflex in human volunteers. *Pain, 98,* 47–57.

Boyce, S., Wyatt, A., Webb, J., et al. (1999). Selective NMDA NR2B antagonists induce antinociception without motor dysfunction: Correlation with restricted localisation of NR2B subunit in dorsal horn. *Neuropharmacology, 38*, 611–623.

Carpenter, K., Chapman, V., & Dickenson, A. H. (2000). Neuronal inhibitory effects of methadone are predominantly opioid receptor mediated in the rat spinal cord in vivo. *European Journal of Pain, 4*, 19–26.

Chapman, V., & Dickenson, A. H. (1992). The combination of NMDA antagonism and morphine produces profound antinociception in the rat dorsal horn. *Brain Research, 573*, 321–323.

Dickenson, A. H. (1997). Mechanisms of central hypersensitivity: Excitatory amino acid mechanisms and their control. In A. H. Dickenson & J. M. Besson (Eds.), *The pharmacology of pain* (pp. 167–196). Berlin/Heidelberg/New York: Springer.

Dickenson, A. H., & Suzuki, R. (1999). Function and dysfunction of opioid receptors in the spinal cord. In E. Kalso, H. J. McQuay, & Z. Wiesenfeld-Hallen (Eds.), *Opioid sensitivity of chronic noncancer pain, progress in pain research and management* (Vol. 14, pp. 1–28). Seattle: IASP Press.

Ebert, B., Andersen, S., & Krogsgaard-Larsen, P. (1995). Ketobemidone, methadone and pethidine are noncompetitive N-methyl-d-aspartate antagonists in the rat cortex and spinal cord. *Neuroscience Letters, 187*, 165–168.

Ikeda, H., Heinke, B., Ruscheweyh, R., et al. (2003). Synaptic plasticity in spinal lamina I projection neurons that mediate hyperalgesia. *Science, 299*, 1237–1240.

Porreca, F., Ossipov, M., & Gebhart, G. (2002). Chronic pain and medullary descending facilitation. *Trends in Neurosciences, 25*, 319–325.

Price, D. D., Mao, J., & Mayer, D. J. (1994). Central neural mechanisms of normal and abnormal pain states. In H. Fields et al. (Eds.), *Pharmacological approaches to the treatment of chronic pain: New concepts and critical issues* (Progress in pain research and management, Vol. 1, pp. 61–84). Seattle: IASP Press.

Rygh, L., Green, M., Athauda, N., et al. (2000). The effect of spinal morphine following long-term potentiation of wide dynamic range neurones in the rat. *Anesthesiology, 92*, 140–146.

Suzuki, R., Morcuende, S., Webber, S., et al. (2002). Superficial NK1 expressing neurons control spinal excitability through activation of descending pathways. *Nature Neuroscience, 5*, 1319–1326.

Todd, A. J. (2002). Anatomy of primary afferents and projection neurones in the rat spinal dorsal horn with particular emphasis on substance P and the neurokinin 1 receptor. *Experimental Physiology, 87*, 245–249.

Yamamoto, T., & Yaksh, T. L. (1992). Studies on the spinal interaction of morphine and the NMDA antagonist MK-801 on the hyperesthesia observed in a rat model of sciatic mononeuropathy. *Neuroscience Letters, 135*, 67–70.

Opioids in the Spinal Cord and Modulation of Ascending Pathways (N. gracilis)

Anthony H. Dickenson and Rie Suzuki
Department of Pharmacology, University
College London, London, UK

Synonyms

Opioids and nucleus gracilis; Opioids in the modulation of ascending pathways

Definition

Spinally applied opioids have been shown to produce powerful ▶ antinociception behaviorally, both in acute pain models (tail flick, hot plate, von Frey) and in more chronic pain states (for review, see Dickenson and Suzuki 1999). These actions are mediated through the activation of ▶ opioid receptors, which are found in abundance in the ▶ superficial dorsal horn of the spinal cord, the periphery, and at various supraspinal sites, such as the nucleus gracilis (NG), parabrachial area (PB), periaqueductal gray (PAG), rostroventral medulla (RVM), and thalamus.

Characteristics

The NG is an area in the brainstem that receives afferent input from direct projections of myelinated and unmyelinated primary afferent fibers, as well as second-order projections from ▶ postsynaptic dorsal column (PSDC) neurones of the spinal cord (laminae III–V) (Willis and Coggeshall 2004). A large proportion of NG neurones project to the contralateral thalamus via the medial lemniscus, although a smaller population has axons terminating within the ▶ dorsal column nuclei, and hence appear to be interneurones. Substantial evidence exists for a role of dorsal column nuclei in the processing of visceral nociceptive input (Willis and Coggeshall 2004). This structure is of particular interest since it has been involved in various pathological conditions (Miki et al. 2000, 2002; Schwark and Ilyinsky 2001). The observation that ipsilateral lesions of the dorsal column or microinjection of lidocaine into the NG suppresses ▸ tactile allodynia (but not ▸ thermal hyperalgesia) (Sun et al. 2001) led studies to speculate that the NG may be implicated in mediating tactile hypersensitivity following peripheral nerve injury. In parallel with these findings, plasticity in NG neurones has been reported after tissue and nerve injury, which include electrophysiological and neurochemical alterations (Al-Chaer et al. 1997; Ma and Bisby 1998; Miki et al. 1998; Suzuki and Dickenson 2002). Clinically, a limited midline myelotomy at T8-10 level has been shown to relieve cancer pain in patients (for refs, see Willis and Coggeshall 2004).

To study how activity in the NG can be modulated by spinal opioids, in vivo electrophysiological approaches were employed to record the responses of deep dorsal horn neurones after peripheral nerve injury. Spinal application of morphine was shown to selectively modulate noxious cutaneous input to the NG (Suzuki and Dickenson 2002). Morphine attenuated the noxious Aδ- and C-fiber-evoked responses of NG neurones, as well as responses evoked by mechanical punctate and heat stimuli. In contrast, the innocuous brush-evoked responses and Aβ-fiber-evoked activity remained relatively unaltered, in keeping with the proposed specific action of morphine on noxious inputs. Remarkably, responses evoked by mechanical punctate stimuli (von Freys of 2 and 9 g) were dramatically attenuated by morphine administration in nerve injured animals, an effect that was not observed in sham control animals. These von Frey filaments are normally innocuous when applied to the rat hindpaw; however, after nerve injury, rats show aversive behaviors to the stimulus, a sign indicative of mechanical allodynia. The finding that neuronal responses evoked by normally innocuous stimuli are robustly inhibited by morphine after neuropathy is interesting, since this suggests an alteration in spinal opioid control,

particularly on low threshold signaling systems. While the efficacy of opioids in acute pain conditions is well established, the issue of opioid sensitivity in chronic neuropathic pain remains a much debated subject (Rowbotham 2001). Although previous behavioral studies have demonstrated a general lack of effect of intrathecally administered morphine after neuropathy, morphine administered via the systemic route was shown to reverse tactile hypersensitivity (Bian et al. 1995; Lee et al. 1995), suggesting that these actions may be mediated through the supraspinal activation of opioid receptors and subsequent activation of descending modulatory systems (Bian et al. 1995). Interestingly, the neuronal counterpart of this activity, as assessed by innocuous von Frey-evoked activity recorded in NG neurones, is sensitive to intrathecal morphine treatment and shows robust inhibition after nerve injury (Suzuki and Dickenson 2002).

The exact mechanism by which spinal morphine inhibits NG neuronal activity remains unclear; however, several possibilities exist. One possibility is that spinal morphine, directly or indirectly, inhibits the PSDC pathway that originates from laminae III–V of the dorsal horn. The PSDC pathway projects via the dorsal funiculus and dorsolateral funiculus to the dorsal column nuclei, and the conduction velocities of their axons indicate they are small- to medium-sized myelinated fibers (Willis and Coggeshall 2004). Dendrites of PSDC neurones have been shown to extend to laminae I and II, allowing them to receive monosynaptic input from small diameter fibers. Morphine applied spinally can, therefore, modulate the activity of PSDC neurones, through presynaptic inhibition of neurotransmitter release from nociceptive afferents. Alternatively, C-fibers can send collaterals through the dorsal column from the lumbar segment. In this case, morphine may dampen excitability in these collateral fibers running to the GN via hyperpolarization of the nerve terminals of small diameter fibers. The finding that low-intensity mechanically evoked responses of NG neurones are selectively attenuated by spinal morphine after nerve injury possibly suggests an acquired de novo opioid sensitivity of Aβ-fibers (through a phenotypic switch).

Another electrophysiological study investigating the role of the PSDC pathway in visceral nociceptive processing demonstrated that morphine administration into the sacral cord robustly suppressed the responses of NG neurones to colorectal distension (CRD), but not to noxious cutaneous stimuli (Al-Chaer et al. 1996), suggesting that the latter responses are largely conveyed by the ascending collaterals of primary afferent fibers. Recordings were also made in PSDC neurones, and morphine was similarly effective in reducing the responses of PSDC neurones to CRD and also to noxious cutaneous stimuli. A recent study on one of the hallmarks of neuropathic pain, the upregulation of the alpha-2 delta subunit of calcium channels, showed changes in the lumbar spinal cord where the nerve injury was located. However, the upregulation could be traced up the dorsal columns to the NG (Bauer et al. 2009). Since this subunit is key to transmitter release which is modulated by opioids, the study provides evidence for a link between nerve injury and the observed changes in NG neuronal properties.

Thus, the NG represents an important area of the brainstem that is involved in the relay of nociceptive visceral and somatic inputs to the VPL nucleus of the thalamus. Input to the NG is mediated through direct projections of primary afferent fibers and indirectly through the PSDC system which originates from laminae III–V of the spinal cord. Furthermore, these inputs are sensitive to spinally administered opioids and can be modulated by agents such as morphine. Since the NG has been implicated in tactile hypersensitivity that accompanies pathological conditions such as neuropathy, the fact that spinal morphine can modulate the activity of these neurones may have important clinical implications, particularly in the treatment of static allodynia.

References

Al-Chaer, E., Lawand, N., Westlund, K., et al. (1996). Pelvic visceral input into the nucleus gracilis is largely mediated by the postsynaptic dorsal column pathway. *Journal of Neurophysiology, 76,* 2675–2690.

Al-Chaer, E., Westlund, K., & Willis, W. (1997). Sensitization of postsynaptic dorsal column neuronal responses by colon inflammation. *Neuroreport, 8,* 3267–3273.

Bauer, C. S., Nieto-Rostro, M., Rahman, W., et al. (2009). The increased trafficking of the calcium channel subunit alpha2delta-1 to presynaptic terminals in neuropathic pain is inhibited by the alpha2delta ligand pregabalin. *The Journal of Neuroscience: The Official Journal of the Society for Neuroscience, 29,* 4076–4088.

Bian, D., Nichols, M. L., Ossipov, M. H., et al. (1995). Characterization of the antiallodynic efficacy of morphine in a model of neuropathic pain in rats. *Neuroreport, 6,* 1981–1984.

Dickenson, A., & Suzuki, R. (1999). Function and dysfunction of opioid receptors in the spinal cord. In E. Kalso, H. McQuay, & Z. Wiesenfeld-Hallin (Eds.), *Opioid sensitivity of chronic noncancer pain. Progress in pain research and management* (pp. 17–44). Seattle: IASP Press.

Lee, Y. W., Chaplan, S. R., & Yaksh, T. L. (1995). Systemic and supraspinal, but not spinal, opiates suppress allodynia in a rat neuropathic pain model. *Neuroscience Letters, 199,* 111–114.

Ma, W., & Bisby, M. A. (1998). Partial and complete sciatic nerve injuries induce similar increases of neuropeptide Y and vasoactive intestinal peptide immunoreactivities in primary sensory neurons and their central projections. *Neuroscience, 86,* 1217–1234.

Miki, K., Iwata, K., Tsuboi, Y., et al. (1998). Responses of dorsal column nuclei neurons in rats with experimental mononeuropathy. *Pain, 76,* 407–415.

Miki, K., Iwata, K., Tsuboi, Y., et al. (2000). Dorsal column thalamic pathway is involved in thalamic hyperexcitability following peripheral nerve injury: A lesion study in rats with experimental mononeuropathy. *Pain, 85,* 263–271.

Miki, K., Zhou, Q.-Q., Guo, W., et al. (2002). Changes in gene expression and neuronal phenotype in brain stem pain modulatory circuitry after inflammation. *Journal of Neurophysiology, 87,* 750–760.

Rowbotham, M. (2001). Efficacy of opioids in neuropathic pain. In P. Hansson, H. Fields, R. Hill, et al. (Eds.), *Neuropathic pain: Pathophysiology and treatment* (pp. 203–213). Seattle: IASP Press.

Schwark, H., & Ilyinsky, O. (2001). Inflammatory pain reduces correlated activity in the dorsal column nuclei. *Brain Research, 889,* 295–302.

Sun, H., Ren, K., Zhong, C., et al. (2001). Nerve injury-induced tactile allodynia is mediated via ascending spinal dorsal column projections. *Pain, 90,* 105–111.

Suzuki, R., & Dickenson, A. (2002). Nerve injury-induced changes in opioid modulation of wide dynamic range dorsal column nuclei neurones. *Neuroscience, 111,* 215–228.

Willis, W., & Coggeshall, R. (2004). Sensory pathways in the dorsal funiculus. In W. Willis & R. Coggeshall (Eds.), *Sensory mechanisms of the spinal cord* (pp. 597–664). New York: Kluwer Academics/Plenum Press.

Opioids, Clinical Opioid Tolerance

Misha-Miroslav Backonja
Department of Neurology, University of
Wisconsin-Madison, Madison, WI, USA

Synonyms

Analgesic tolerance; Opioid tolerance

Definition

Tolerance is the term used to define the phenomenon in which an organism is less susceptible to
the effect of a drug as a consequence of its prior
administration (Foley 1993; Management of
Cancer Pain Guideline Panel 1994).

Characteristics

Though the definition of tolerance is relatively
straightforward, tolerance to ▶ opioids is the
need to increase dose requirements over time to
maintain pain relief (Management of Cancer
Pain Guideline Panel 1994). There are many
clinical circumstances when opioid tolerance is
suspected, and defining opioid analgesic tolerance in a clinical setting is much more subtle
and certainly more challenging. This entry is
a review of the issues and questions rather than
a source of answers, and this is largely because
the evidence-based publications addressing this
topic are lacking, and consensus statements
related to opioids by experts and professional
societies are thus far focused on regulatory issues
of prescribing opioids. Unfortunately, basic
science related to opioid analgesia, including
tolerance, in spite of remarkable advances, has
contributed very little to the opioid tolerance
problems in the clinical setting. The primary
reason for this shortcoming results from the
fact that the majority of studies about opioid
tolerance utilize models of acute tolerance,
which is not a main issue in the clinical setting.

Opioid analgesic tolerance during long-term therapy for chronic pain is the problem that first needs
to be defined in its full complexity before it can be
understood.

In most general terms, even in the clinic, there
is a clear distinction between acute and chronic
opioid tolerance.

Acute tolerance is the term used to describe
tolerance that develops very rapidly following
either a single dose or a few doses given over
a short period of time. Acute opioid tolerance is
an everyday event for patients who are receiving
opioids for perioperative and postoperative
analgesia; it is always anticipated, and with the
introduction of postoperative patient-controlled
analgesia (PCA), pain relief is individualized
and in most cases successful in providing analgesia. In addition, postoperative acute pain as a rule
dramatically improves as healing takes place
over a short period of time and consequently
tolerance is not an issue. For patients who take
opioids on a chronic basis (the commonly used
term in this type of scenario is "opioid tolerant")
prior to operative procedures, the preoperative
dose has to be taken into account, and there are
efforts to assist clinicians in this task (de Leon-
Casasola 2002; Gordon and Ward 1995). In spite
of these practical successes, incidence of acute
opioid analgesic tolerance and mechanisms of
acute tolerance in clinical settings, perioperative
or otherwise, is not known.

Issues related to clinical opioid tolerance
could be categorized as pharmacological, pathophysiological, psychological, social, and societal,
spanning from mechanisms to perceptions. Long-
term treatment with opioids (a.k.a. chronic opioid
therapy) for treatment of chronic pain presents
a significant clinical problem. Incidence of opioid
analgesic tolerance is not known and systematic
studies on this issue are severely lacking.

Two ways of defining opioid analgesic tolerance in the case of the chronic opioid setting have
emerged: first as a need to increase the dose of
opioid analgesic to maintain the same level of
pain relief and the second as increasing intensity
of pain at the same dose of medication. In either
case, opioid analgesic tolerance is not only the
result of pharmacological properties and effects

of opioids, but it is also under the direct influence of underlying pain mechanisms.

The fact that underlying pain mechanisms are an important factor can be demonstrated in a case of neuropathic pain, which represents an example when pain mechanisms are the basis of labeling this type of pain as "opioid resistant" (Arner and Meyerson 1988), although this view was proven to be wrong when a number of clinical trials demonstrated evidence of analgesia for neuropathic pain (Dworkin et al. 2003). In practical terms, even neuropathic pain can be relieved with titration of the opioid analgesic until the therapeutic effect.

Another example related to underlying pain mechanisms is the natural course of the disease, which can make pain symptoms worsen as the disease progresses, such as in the case of progressive HIV neuropathy. This should be simply termed as worsening of pain due to progression of the disease. Another frequently cited clinical phenomenon is breakthrough pain, which refers to the underlying metastatic spread of neoplastic disease and associated mechanisms in cancer pain practice that require an increase in the dose to maintain the previous degree of pain relief. The most common scenario in the clinical setting of chronic noncancer pain treatment is flare-up of symptoms, which present natural fluctuations of chronic noncancer pain symptoms without any evidence of disease progression. Unfortunately, this spontaneous or activity precipitated fluctuation of pain symptoms is frequently erroneously termed as breakthrough pain, while in reality it is a fluctuation and flare-up of symptoms requiring flare-up management and not an increase in the dose of opioid ▶ analgesics. It is difficult to determine which of the factors are at play in each particular patient to cause more pain at the same dose of the opioid analgesics and it requires ongoing and very skillful clinical assessment.

There are numerous clinical scenarios when interpreting the possibilities, which are not biological but rather due to psychosocial factors. A common scenario is undertreated pain, where a patient is asking for higher and higher doses, and many clinicians interpret this behavior as

▶ addiction, although this is now a well-recognized phenomenon of pseudo-addiction. Further challenges stem from the assessment of severity of pain, which in many instances has a lot to do with the hidden meaning of pain rather than severity of any particular sensation (Williams et al. 2000), which then introduces issues of psychosocial influences.

There are a couple of terms that confuse the lay public as well as uninformed clinicians when it comes to the use of opioids in clinical practice, and they are withdrawal symptoms and addiction and for many it is equated as tolerance. Only some patients who take opioids regularly at any dose and for any length of time and stop abruptly experience withdrawal symptoms, which include diarrhea, sweating, insomnia, and goose bumps. Withdrawal symptoms are never the result of tolerance. Neither opioid tolerance nor withdrawal symptoms are inevitable but should be anticipated and managed appropriately if they appear in patients who take opioids for pain control. The most bothersome ▶ withdrawal symptom in most patients with chronic pain who take opioids for pain relief is severe exacerbation of pain and associated anxiety.

Opioid tolerance is one of the defining characteristics of opioid addiction, which is confusing and a serious concern to the lay public as much as to uninformed clinicians. The fundamental difference between chronic pain and addiction related to opioids is that patients who have pain obtain and maintain clinically significant pain relief that allows them to maintain and improve their function, while patients with opioid addiction that take opioids as part of their addiction experience and demonstrate a decline in function and well-being. Additional and unfortunate confusion arises from the impression of language when opioids are termed as narcotics. It should be noted that narcotics is a regulatory term referring to substances of illegal and criminal trade, while opioid analgesics are pharmacological agents (drugs) used for pain relief, which when illegally used become narcotics.

Psychology studies of opioid use for pain relief led to the identification of associative and nonassociative tolerance. Associative tolerance is best expressed with low doses of drugs at long interdose intervals and is readily modified by behavioral or environmental interventions. Nonassociative tolerance is best expressed with high doses of drugs at short interdose intervals and is not modified by behavioral or environmental interventions. It is thought that associative tolerance results from the learning of drug-environment associations, whereas nonassociative tolerance and dependence can be viewed simply as adaptive changes resulting from direct drug actions.

Better understanding of clinical opioid tolerance has to address the complex interaction between the underlying mechanisms and psychosocial factors that ultimately lead to the prescribing and administering of opioids for the treatment of chronic pain. Unfortunately, this issue has not been addressed in any publication or expert forum. In addition to this complexity, myths and confusion, not only in the lay public but among many clinicians, are further mired by frequent imprecision of language equating the use of opioids for pain relief with unavoidable development of tolerance and addiction, which then result in an unnecessary fear of opioids.

Opioid tolerance includes an additional range of clinical phenomena. In most general terms, these phenomena can be viewed as tolerance to adverse events, and this type of tolerance should be viewed as desirable, which is a contrast to tolerance to the analgesic effect, which is not desirable. Most significant tolerance to adverse events is tolerance to the respiratory depression effect, which develops relatively quickly after the initial exposure, especially in the presence of severe pain. In contrast, tolerance to gut effects, which leads to constipation, rarely develops.

Once other factors that can mimic opioid analgesic tolerance are excluded and managed, then clinical strategy for management, i.e., treatment of opioid analgesic tolerance, is ▶ opioid rotation. This is possible because opioid cross-tolerance is incomplete. Opioid rotation is done on the basis of equianalgesic dose calculation.

In summary, although clinical opioid analgesic tolerance is much less frequent than

is feared by the lay public and inexperienced clinicians, opioid tolerance is a very complex and dynamic process, under influence of many biological and psychosocial factors. In spite of its complexity, it is manageable by means of opioid rotation.

References

Arner, S., & Meyerson, B. A. (1988). Lack of analgesic effect of opioids on neuropathic and idiopathic forms of pain. *Pain, 33*, 11–23.

de Leon-Casasola, O. A. (2002). Cellular mechanisms of opioid tolerance and the clinical approach to the opioid tolerant patient in the post-operative period. *Best Practice & Research. Clinical Anaesthesiology, 16*, 521–525.

Dworkin, R. H., Backonja, M., Rowbotham, M. C., et al. (2003). Advances in neuropathic pain: Diagnosis, mechanisms, and treatment recommendations. *Archives of Neurology, 60*, 1524–1534.

Foley, K. M. (1993). Changing concepts of tolerance to opioids: What the cancer patient has taught us. In C. R. Chapman & K. M. Foley (Eds.), *Current and emerging issues in cancer pain: Research and practice*. New York: Lippincott-Raven.

Gordon, D. B., & Ward, S. E. (1995). Correcting patient misconceptions about pain. *American Journal of Nursing, 95*, 43–45.

Management of Cancer Pain Guideline Panel. (1994). *AHCPR clinical practice guideline: Management of cancer pain*. Rockville: U.S. Department of Health and Human Services.

Williams, A. C. D., Davies, H. T., & Chadury, Y. (2000). Simple pain rating scales hide complex idiosyncratic meanings. *Pain, 85*, 457–463.

Opioids, Effects of Systemic Morphine on Evoked Pain

Charles J. Vierck Jr.
Department of Neuroscience and McKnight Brain Institute, University of Florida College of Medicine, Gainesville, FL, USA

Synonyms

Evoked pain and morphine; Opioid hypoalgesia; Systemic morphine effects on evoked pain

Definition

Systemic morphine activates opiate receptors throughout the nervous system, attenuating a clinically important component of pain sensations and modulating a variety of pain reactions. Due to these effects, opiate agonists are the prototypical pharmacological treatment for many chronic pain conditions. Without addressing actions or relative potencies of opiate agonists acting at different receptors, the effects of systemic administration of agonists will be outlined. These actions depend upon dosage, which is an important experimental and clinical consideration.

Characteristics

The effects of systemic morphine on pain were first discovered in humans and were utilized clinically before experimental attempts to understand mechanisms of opioid hypoalgesia were initiated. A discrepancy between the substantial effects of morphine on clinical pain and meager attenuation of experimental pain in humans became apparent early (Beecher 1957). Potential explanations for results such as these were the following: (1) that there is a critical difference between chronic (clinical) pain and phasic (experimental) pain or (2) that morphine reduces psychological suffering produced by pain with a lesser effect on the sensory component of pain. Subsequent research has revealed attenuation of a component of pain not assessed by early experimental methods and also suppression of emotional and motivational reactions and associated physiological responses to pain.

Clinicians have noted that pain is rarely eliminated by systemic morphine. Chronic pain can be well controlled for a patient who is resting comfortably, but a procedure such as changing a dressing can produce intense pain. Observations such as this went unexplained until experimental studies asked subjects to separately rate early and late pain sensations that result from phasic activation of myelinated (A delta) and unmyelinated (C) ► nociceptors. For example,

Opioids, Effects of Systemic Morphine on Evoked Pain, Fig. 1 Average tracings of 6 subjects who followed the time course and rated the magnitude of first and second pain elicited by 50 ms of electrocutaneous stimulation (1.28 mA/mm^2; 0.5 ms pulses at 200 Hz) delivered to one lateral calf. Time course estimates utilized a finger-span device attached to a potentiometer, and the magnitudes of first and second pain were rated verbally, using ► free magnitude estimation. Administration of 10-mg morphine sulfate 1 h before testing did not alter the magnitude of first pain. In contrast, second pain (*dashed line*) was delayed in onset and substantially reduced in magnitude and duration by morphine, relative to control measurements (*solid line*) (from Cooper et al. 1986)

when a preheated thermode briefly taps the skin, distinct early (► first pain) and late (► second pain) sensations are produced (Cooper et al. 1986). Prior i.m. injection of a therapeutic dose of morphine (approximately 0.1 mg/kg) dramatically reduces second pain but has no detectable effect on first pain (Fig. 1). This differential result is also seen for brief mechanical stimulation of the skin that produces first and second pain. Similarly, chemical stimulation of C nociceptors by capsaicin produces a sensation of heat pain that is substantially attenuated by systemic morphine. These effects are specific to nociceptor input and do not apply to sensations of warmth that are subserved by activation of unmyelinated afferents (Cooper et al. 1986).

Neurophysiological studies have shown that input from C nociceptors to spinal neurons is selectively reduced by systemic morphine (Jurna and Heinz 1979). This effect involves both presynaptic and postsynaptic inhibition

(Zieglgänsberger and Bayerl 1976). Thus, pain reduction by opiate agonists (presumably including endogenous opiate agonists) ingeniously attenuates tonic pain from activation of C nociceptors by tissue injury, providing relief and facilitating sleep, but inappropriate activities (e.g., walking on a broken leg) are discouraged by preservation of phasic pain from activation of fast conducting A-delta nociceptors. Several implications of the inhibitory effects of morphine at the entry level of nociceptive processing are that (1) spinal application of opiate agonists might be comparable to systemic administration and (2) that the effects of morphine on pain can be modeled behaviorally by observing spinal reflexes. Neither of these suppositions is correct.

Reflex response latencies to heat or electrical stimulation and thresholds for mechanical stimulation can be increased by administration of opiate agonists. However, ▶ intrathecal application does not entirely account for systemic effects (Dickenson and Sullivan 1986), in part because opiate binding within the brain stem initiates descending modulation of spinal reflexes (Gebhart 1982). Despite these actions at brain stem and spinal levels, high doses of systemic morphine (minimally 3 mg/kg) are required to attenuate nociceptive reflexes. An important determinant of this relative insensitivity to systemic morphine is that nociceptive reflexes ordinarily function to withdraw a limb or tail from a tissue-threatening stimulus in response to myelinated afferent input, well before input from C nociceptors reaches the spinal cord. However, cutaneous stimulation with a low level of nociceptive heat (e.g., 43–45 °C) preferentially excites C nociceptors (Yeomans and Proudfit 1996), particularly for non-hairy skin. Prolonged stimulation at such temperatures can be used to evaluate whether reflex responses can appropriately reveal low-dose modulation by morphine.

Stimulation of rodents' feet by nociceptive heat produces licking and guarding reflexes that are mediated by spinal-brain stem-spinal circuits and do not require cerebral processing of nociception. These, and other responses (e.g., simple withdrawal, jumping, vocalizing), are preserved after decerebration (Woolfe and MacDonald 1944).

However, ▶ operant escape from the same stimulus requires learning of a complex adaptive strategy within a specific environmental context, which involves multiple cerebral circuits for recognition of stimulus significance and for planning an appropriate motor sequence. After i.m. injection of 0.5 mg/kg morphine in rats, escape from 44 °C stimulation is significantly decreased, as expected, but lick/guard responding is increased (Vierck et al. 2002). Similar to many experiments that have observed reflex responses to stimulation of myelinated afferents, lick/guard responding to low-level nociceptive heat is not attenuated below a systemic dose of 3 mg/kg. A similar disparity between the sensitivity of reflex and operant escape behaviors to morphine has been reported for monkeys (Yeomans et al. 1995). Thus, in order to evaluate the effects of systemic morphine on nociceptive processing, it is important to utilize a behavioral response that requires cerebral processing and a stimulus that activates C nociceptors.

Diverse supraspinal actions of systemic morphine can be inferred from a widespread distribution of opiate receptors. In addition to effects on cerebral projection neurons at the spinal level by peripheral (Stein et al. 1989), spinal, and descending modulation, opiate influences involve numerous systems of control over physiological and psychological reactions to nociceptive input. Considering first the brain stem, opiate agonists activate a variety of nuclei with descending projections not only to the spinal dorsal horns but also to the ventral and intermediolateral gray matter. Thus, opiate administration has direct sensory, motor, and autonomic effects. For example, modulation of sympathetic outflow affects thermoregulation, which can influence nociceptive reactions to thermal stimulation (Tjolsen and Hole 1992). Parasympathetic modulation includes opiate actions within the vagal projection system of the brain stem (e.g., nucleus solitarius), and vagal activation can exert inhibitory influences on nociception (Ammons et al. 1983). Other brain stem effects involve modulation of defensive motor actions associated with nociception (e.g., fight or flight behaviors associated with rage or fear) (Braez 1994). Similarly, vocalizations related or unrelated to pain are suppressed via influences within the periaqueductal gray and associated structures (Cooper and Vierck 1986). In general, opiate influences on the brain stem are inhibitory for behavioral output and physiological reactions, but excitation can result, depending on the dose and site of pharmacological application.

In addition to specific effects on nociceptive reactions, regulation of arousal by brain stem projection systems is susceptible to opiate modulation. Patterns of cortical EEG activity are modified by opiate actions within the brain stem and rostrally (Matejcek et al. 1988), and a generalized behavioral suppression results from therapeutic doses of morphine, particularly for primates (Vierck et al. 1984). Low doses of systemic morphine are sedative, promoting sleep and aiding recuperation from painful injuries. These effects on arousal can appear to represent analgesia by suppression of attention, particularly when high doses of morphine are administered.

Cerebral projection of nociception is susceptible to modulation via opiate agonist application to the thalamus and somatosensory cortex, and these effects are diverse (Wamsley 1983). For example, opiate receptor distribution in the medial thalamus is particularly high, and opiate receptors are present in cingulate, insular, and prefrontal cortical regions. Opiate modulation at these sites is consistent with observations that systemic morphine reduces affective (emotional) and motivational responses to pain (Beecher 1957). Opiate actions within limbic projection systems (e.g., the septal area, amygdaloid nuclei, and hypothalamus) are likely to be responsible for rewarding effects of opiates such as euphoria (Koob 1992). Also, nociceptive activation of the limbic system accesses brain stem systems of descending modulation, to influence autonomic and hormonal responses integral to diverse ▶ homeostatic adaptations to nociceptive input. Systemic morphine actions on limbic system processing include influences on the hypothalamic-pituitary-adrenal axis and modulation of hypothalamic output of numerous hormonal

releasing factors and of pituitary hormonal secretion (Wood and Iyengar 1988). Relevance of these actions to a comprehensive regulation of behavioral and physiological reactions to pain is indicated by demonstrations of extensive nociceptive projections to the limbic system.

The multiple actions of opiate agonists make apparent some important considerations concerning experimental evaluation of systemic morphine's effects relevant to control over clinical pain. Reflex tests are relatively insensitive and do not reveal numerous cerebral actions of morphine on pain sensations and reactions. Also, observation of responses to phasic input from myelinated nociceptors is not useful. Evaluation of responses to C nociceptor input appears to be necessary and is especially informative if ▶ temporal summation (windup) of pain is produced. Attenuation of central sensitization produced by prolonged C nociceptor input is likely to be an important component of the clinical utility of systemic agonists (Price et al. 1985). Also, tests of responses to stimulation of inflamed tissue have revealed low-dose effects of systemic morphine on behavioral tests that are otherwise insensitive (Stein et al. 1989), indicating that peripheral actions constitute an important component of the effectiveness of systemic morphine for certain forms of clinical pain. In contrast, systemic morphine is relatively ineffective for control of chronic pain that can result from lesions within the central nervous system that interrupt nociceptive input to sites that contain opiate receptors (Arner and Meyerson 1988). Thus, morphine modulates nociceptive transmission to and within the central nervous system but is less effective for attenuation of abnormal "spontaneous" activity that can result from deafferentation of pain transmission systems.

References

Ammons, W., Blair, R., & Foremen, R. (1983). Vagal afferent inhibition of primate thoracic spinothalamic neurons. *Journal of Neurophysiology, 50*, 926–940.
Arner, S., & Meyerson, B. A. (1988). Lack of analgesic effect of opioids on neuropathic and idiopathic forms of pain. *Pain, 33*, 11–23.

Beecher, H. K. (1957). Measurement of pain; Prototype for the quantitative study of subjective sensations. *Pharmacological Reviews, 9*, 59–209.

Braez, J. (1994). Neuroanatomy and neurotransmitter regulation of defensive behaviors and related emotions in mammals. *Brazilian Journal of Medical and Biological Research, 27*, 811–829.

Cooper, B. Y., & Vierck, C. J., Jr. (1986). Vocalizations as measures of pain in monkeys. *Pain, 26*, 393–408.

Cooper, B. Y., Vierck, C. J., Jr., & Yeomans, D. C. (1986). Selective reduction of second pain sensations by systemic morphine in humans. *Pain, 24*, 93–116.

Dickenson, A. H., & Sullivan, A. F. (1986). Electrophysiological studies on the effects of intrathecal morphine on nociceptive neurones in the rat dorsal horn. *Pain, 24*, 211–222.

Gebhart, G. (1982). Opiate and opioid peptide effects on brain stem neurons: Relevance to nociception and antinociceptive mechanisms. *Pain, 12*, 93–140.

Jurna, I., & Heinz, G. (1979). Differential effects of morphine and opioid analgesics on A and C fibre-evoked activity in ascending axons of the rat spinal cord. *Brain Research, 171*, 573–576.

Koob, G. (1992). Drugs of abuse: Anatomy, pharmacology and function of reward pathways. *Trends in Pharmacological Sciences, 13*, 177–184.

Matejcek, M., Pokorny, R., Ferber, G., et al. (1988). Effect of morphine on the electroencephalogram and other physiological and behavioral parameters. *Neuropsychobiology, 19*, 202–211.

Price, D. D., Von der Gruen, A., Miller, J., et al. (1985). A psychophysical analysis of morphine analgesia. *Pain, 22*, 261–269.

Stein, C., Millan, M. J., Shippenberg, T. S., et al. (1989). Peripheral opioid receptors mediating antinociception in inflammation. Evidence for involvement of *Mu, Delta* and *Kappa* receptors. *The Journal of Pharmacology and Experimental Therapeutics, 248*, 1269–1275.

Tjolsen, A., & Hole, K. (1992). The effect of morphine on core and skin temperature in rats. *Neuroreport, 3*, 512–514.

Vierck, C. J., Jr., Cooper, B. Y., Cohen, R. H., et al. (1984). Effects of systemic morphine on monkeys and man: Generalized suppression of behavior and preferential inhibition of pain elicited by unmyelinated nociceptors. In C. von Euler, O. Franzen, U. Lindblom, et al. (Eds.), *Somatosensory mechanisms* (pp. 309–323). London: MacMillan Press.

Vierck, C. J., Acosta-Rua, A., Nelligan, R., et al. (2002). Low dose systemic morphine attenuates operant escape but facilitates innate reflex responses to thermal stimulation. *The Journal of Pain, 3*, 309–319.

Wamsley, J. (1983). Opioid receptors: Autoradiography. *Pharmacological Reviews, 35*, 69–83.

Wood, P. L., & Iyengar, S. (1988). Central actions of opiates and opioid peptides. In G. W. Pasternak (Ed.), *The Opiate Receptors* (pp. 307–356). Totowa, NJ: The Humana Press.

Woolfe, G., & MacDonald, A. D. (1944). The evaluation of the analgesic action of pethidine hydrochloride (Demerol). *The Journal of Pharmacology and Experimental Therapeutics, 80*, 300–307.

Yeomans, D. C., & Proudfit, H. K. (1996). Nociceptive responses to high and low rates of noxious cutaneous heating are mediated by different nociceptors in the rat: Electrophysiological. *Pain, 68*, 141–150.

Yeomans, D. C., Cooper, B. Y., & Vierck, C. J., Jr. (1995). Comparisons of dose-dependent effects of systemic morphine on flexion reflex components and operant avoidance responses of awake non-human primates. *Brain Research, 670*, 297–302.

Zieglgänsberger, W., & Bayerl, H. (1976). The mechanism of inhibition of neuronal activity by opiates in the spinal cord of cat. *Brain Research, 115*, 111–128.

Opioids, Kappa Receptors, and Visceral Pain

Pierre J. M. Rivière
Ferring Research Institute, San Diego, CA, USA

Synonyms

Kappa receptors and visceral pain; Kappa-opioid receptors; KOR; κ-opioid receptors

Definition

κ (kappa)-opioid receptors are a subtype of opioid receptors. They are differentiated from other opioid receptor subtypes (mu and delta) by distinct gene, protein, tissue expression pattern, functional properties, and side effect profile. Kappa-opioid receptor agonists are particularly effective analgesics in visceral pain models.

Characteristics

Kappa-Opioid Receptors, Pharmacological Subtypes, and Cloned Receptor

Pharmacological studies have long established the existence of κ (kappa)-opioid receptors that are functionally differentiated from μ (mu)- and δ (delta)-opioid receptor subtypes

(Martin et al. 1976). Radioligand studies have also provided evidence of heterogeneity of kappa binding sites in brain membrane preparations, characterizing two main binding sites termed κ_1 and κ_2, each of them being further subdivided into high (κ_{1a}-, κ_{2a}-) and low affinity (κ_{1b}-, κ_{2b}-) binding sites (Rothman et al. 1990). However, only one kappa-opioid receptor (KOR1) has so far been cloned in human and rodents (Simonin et al. 1995). Its pharmacology is virtually identical to the previously characterized κ_1-receptor. KOR1 is a seven-transmembrane domain (7TM) receptor, coupled to G-proteins (G-protein-coupled receptors, GPCR) and negatively coupled to adenylate cyclase.

Kappa-Agonists

Most of the kappa-agonists available so far have been optimized at the κ_1 binding site or the cloned KOR1. Thus, they are all κ_1 selective, with agonist potencies in the nanomolar range (typically 0.1–10 nM). Prototypic representatives of this class of compounds include both organic (U50,488, enadoline/CI-977, asimadoline/EMD61753, ADL 10–010, ADL 10–0116) and peptidic molecules (FE 200665 and FE 200666) (Binder et al. 2001; Rivière et al. 1999). U50,488 and enadoline are brain-penetrating compounds (Rivière et al. 1999), while the others have various degrees of peripheral selectivity, ranging from moderate (asimadoline) (Rivière et al. 1999) (ADL 10–0101, ADL 10–0116) (Murphy et al. 2000) to high (FE 200665, FE 200666) (Rivière et al. 1999).

Opioid Receptors and Pain Pathways

Opioid receptors are expressed on nerves involved in pain transmission (ascending sensory pathways) and modulation (descending inhibitory pathways) in the periphery, the spinal cord, and the brain. Opioid receptors are mainly present on peptide-rich C fibers of primary sensory afferents, where they prevent the activation and sensitization of these fibers and inhibit the release of pain transmitters. Upon activation by selective agonists, each opioid receptor subtype elicits different degrees of analgesia that varies greatly

relative to the experimental conditions including receptor subtype selectivity of the pharmacological agent, somatic versus visceral pain, supraspinal versus spinal or peripheral site of action, and acute versus inflammatory or chronic pain.

Analgesic Effects of Kappa-Agonists in Visceral Pain Models

Kappa-agonists are particularly potent analgesics after systemic administration in a wide variety of visceral pain models (Burton and Gebhart 1998; Su et al. 2002). The antinociceptive effects of kappa-agonists in visceral pain are consistent across a multitude of experimental conditions, irrespective of species (rats or mice), targeted visceral organs (duodenum, colon, bladder, vagina, uterus, or peritoneum), nature of noxious stimuli (distension or chemical irritant), nature of measured endpoint (cardiovascular, visceromotor, or electrophysiological responses, anesthetized or conscious animals, basal or inflammatory pain), and chemical nature of kappa-agonists (organic molecules or peptides). Experimental visceral inflammation decreases pain thresholds, increases pain response, and enhances the analgesic potency of kappa-agonists (Burton and Gebhart 1998; Langlois et al. 1994, 1997; Sengupta et al. 1999). Peritoneal irritation-induced pain is also associated with gastrointestinal transit inhibition (Friese et al. 1997a; Riviere et al. 1993, 1994). In these conditions, blockade by kappa-agonists of peritoneal irritation-induced pain results in a reversal of intestinal ileus. The ability of kappa-agonists to reverse peritoneal irritation-induced ileus is correlated with their antinociceptive potency in this model (Friese et al. 1997a). These data suggest that kappa-agonists might be appropriate to treat postoperative pain associated with ileus.

Involvement of Peripheral κ_1-Receptors in Kappa-Agonists Induced Visceral Analgesia in Intact Animals

In visceral pain models using intact animals, kappa-agonists are active at relatively low doses (typically 0.010–0.5 mg/kg, systemic route), and the rank order for analgesic potencies

(Rivière et al. 1999; Burton and Gebhart 1998; Friese et al. 1997b) is consistent with that for agonist activity at the cloned KOR1. In these conditions, the analgesic effects of kappa-agonists are generally blocked by opioid antagonists (Burton and Gebhart 1998; Diop et al. 1994a, b; Friese et al. 1997a; Langlois et al. 1994, 1997), validating the involvement of opioid receptors, likely κ_1, considering the selectivity of the agonists for κ_1-receptors. Furthermore, in these models, the effects of kappa-agonists are generally blocked by peripherally restricted opioid antagonists, and/or peripherally restricted kappa-agonists are also equally effective compared to brain-penetrating compounds in these models (Rivière et al. 1999; Burton and Gebhart 1998; Friese et al. 1997b; Sandner-Kiesling et al. 2002), suggesting that the κ_1-receptors involved in mediating kappa-agonist responses are located in the periphery.

Involvement of Non-opioid Blockade of Sodium Currents in Kappa-Agonists Direct Effects on Visceral Sensory Input

Kappa-agonists have a direct inhibitory effect on visceral sensory input in the periphery. They inhibit the firing of decentralized pelvic afferents activated by colorectal distension (CRD) (Joshi et al. 2000; Sengupta et al. 1996, 1999; Su et al. 1997, 2002). These effects require higher doses than those needed in visceral pain models using intact animals (about 10-fold higher doses, typically 5–10 mg/kg, arterial administration) (Joshi et al. 2000; Sengupta et al. 1996, 1999; Su et al. 1997, 2002). The inhibitory activity on pelvic afferents is not correlated with the rank order of potency at the cloned KOR1. The response to kappa-agonists cannot be completely antagonized by high doses of nonselective opioid antagonists or selective kappa antagonists (Sengupta et al. 1996; Su et al. 1997). KOR1 antisense oligodeoxynucleotide pretreatment does not alter the response to kappa-agonists (Joshi et al. 2000) and different enantiomers of U50,488, despite marked differences in agonist activity at κ_1-receptors; they are all equally potent and effective in inhibiting the firing of pelvic afferents (Su et al. 2002). Taken together,

these data indicate the involvement of a non-opioid mechanism in the response to kappa-agonists in decentralized pelvic afferents. Whole cell patch-clamp experiments performed on rat colonic sensory neurons have established that organic, but not peptidic, kappa-agonists, when used at micromolar concentrations (i.e., about a 1,000-fold higher concentration than those for kappa-agonists activity), have non-opioid sodium channel blocking properties (Joshi et al. 2003).

Clinical Data on Kappa-Agonists in Visceral Pain

Very limited clinical information is available regarding kappa-agonists in visceral pain. Fedotozine was the first compound with kappa-agonist activity to be evaluated for visceral pain in a clinical setting. The compound was superior to placebo in reducing abdominal pain and bloating in non-ulcer dyspepsia (Fraitag et al. 1994) and irritable bowel syndrome (IBS) (Dapoigny et al. 1995). It also increased pain perception thresholds to colonic distension in IBS patients (Delvaux et al. 1999). Fedotozine is an atypical kappa-agonist, being a mixed kappa/mu ligand (Allescher et al. 1991), with high affinity for the κ_{1a} binding site (Lai et al. 1994), and relatively low affinity for the cloned KOR1 (Lai et al. 1994).

More recently, a pilot study reported that the kappa-agonist, ADL 10–0101, was effective in reducing pain in patients suffering from chronic pancreatitis (Eisenach et al. 2003). ADL 10–0101 is a classical κ_1/KOR1-selective agonist combined with peripheral selectivity. The analgesic response appeared to be robust and was not associated with the CNS side effects of brain-penetrating kappa-agonists, suggesting that the compound did not reach the brain, and therefore supporting the hypothesis that its effects were mediated in the periphery. The onset of analgesia was immediate, reaching a plateau within 60 min, and remaining at maximal levels for the duration of the monitoring period (4 h). During this time interval, the plasma levels of the compound are estimated to have ranged from high nanomolar (first hour, during and after administration) to intermediate/low nanomolar levels (~100 nM or less, 3rd and 4th hours). The plasma levels, at least during the last 2 h of the study, when visceral analgesia was still maximal, were probably too low to elicit a nonspecific sodium channel blockade, but were perfectly suitable to activate κ_1-receptors. Despite these preliminary and encouraging results, more definitive clinical proof of the concept for the therapeutic relevance of peripheral κ_1-receptors in visceral pain is still needed.

Side Effect Profile of Kappa-Agonists

The development and use of kappa-agonists has been limited by their side effects, which include primarily sedation and dysphoria (Rivière and Junien 2000). Mu-agonists unlike kappa agonists do not inhibit intestinal transit and induce euphoria, addiction, or respiratory depression (Riviere and Junien, 2000).

Kappa-Agonists, Targeted Clinical Profile for Visceral Pain

Since the unwanted side effects of kappa-agonists (sedation, dysphoria) are restricted to kappa receptors located in the CNS, i.e., beyond the blood–brain barrier, and because the targeted kappa receptors in visceral pain are located in the periphery, kappa-agonists with high peripheral selectivity are expected to be efficacious and safe for the treatment of visceral pain. Therapeutic indications for peripherally restricted kappa-agonists may include abdominal surgery associated with postoperative pain and ileus, pancreatitis pain, dysmenorrhea, labor pain, and irritable bowel syndrome.

References

Allescher, H. D., Ahmad, S., Classen, M., & Daniel, E. E. (1991). Interaction of trimebutine and Jo-1196 (fedotozine) with opioid receptors in the canine ileum. *The Journal of Pharmacology and Experimental Therapeutics, 257*, 836–842.

Binder, W., Machelska, H., Mousa, S., Schmitt, T., Riviere, P. J., Junien, J. L., Stein, C., & Schafer, M. (2001). Analgesic and anti-inflammatory effects of two novel kappa-opioid peptides. *Anesthesiology, 94*, 1034–1044.

Burton, M. B., & Gebhart, G. F. (1998). Effects of kappa-opioid receptor agonists on responses to colorectal distension in rats with and without acute colonic inflammation. *The Journal of Pharmacology and Experimental Therapeutics, 285*, 707–715.

Dapoigny, M., Abitbol, J. L., & Fraitag, B. (1995). Efficacy of peripheral kappa agonist fedotozine versus placebo in treatment of irritable bowel syndrome. A multicenter dose–response study. *Digestive Diseases and Sciences, 40*, 2244–2249.

Delvaux, M., Louvel, D., Lagier, E., Scherrer, B., Abitbol, J. L., & Frexinos, J. (1999). The kappa agonist fedotozine relieves hypersensitivity to colonic distention in patients with irritable bowel syndrome. *Gastroenterology, 116*, 38–45.

Diop, L., Riviere, P. J., Pascaud, X., Dassaud, M., & Junien, J. L. (1994a). Role of vagal afferents in the antinociception produced by morphine and U-50,488H in the colonic pain reflex in rats. *European Journal of Pharmacology, 257*, 181–187.

Diop, L., Riviere, P. J., Pascaud, X., & Junien, J. L. (1994b). Peripheral kappa-opioid receptors mediate the antinociceptive effect of fedotozine on the duodenal pain reflex in rat. *European Journal of Pharmacology, 271*, 65–71.

Eisenach, J. C., Carpenter, R., & Curry, R. (2003). Analgesia from a peripherally active kappa-opioid receptor agonist in patients with chronic pancreatitis. *Pain, 101*, 89–95.

Fraitag, B., Homerin, M., & Hecketsweiler, P. (1994). Double-blind dose–response multicenter comparison of fedotozine and placebo in treatment of nonulcer dyspepsia. *Digestive Diseases and Sciences, 39*, 1072–1077.

Friese, N., Diop, L., Lambert, C., Riviere, P. J., & Dahl, S. G. (1997a). Antinociceptive effects of morphine and U-50,488H on vaginal distension in the anesthetized rat. *Life Sciences, 61*, 1559–1570.

Friese, N., Chevalier, E., Angel, F., Pascaud, X., Junien, J. L., Dahl, S. G., & Riviere, P. J. (1997b). Reversal by kappa-agonists of peritoneal irritation-induced ileus and visceral pain in rats. *Life Sciences, 60*, 625–634.

Joshi, S. K., Su, X., Porreca, F., & Gebhart, G. F. (2000). Kappa-opioid receptor agonists modulate visceral nociception at a novel, peripheral site of action. *The Journal of Neuroscience, 20*, 5874–5879.

Joshi, S. K., Lamb, K., Bielefeldt, K., & Gebhart, G. F. (2003). Arylacetamide kappa-opioid receptor agonists produce a tonic- and use-dependent block of tetrodotoxin-sensitive and -resistant sodium currents in colon sensory neurons. *The Journal of Pharmacology and Experimental Therapeutics, 307*, 367–372.

Lai, J., Ma, S. W., Zhu, R. H., Rothman, R. B., Lentes, K. U., & Porreca, F. (1994). Pharmacological characterization of the cloned kappa-opioid receptor as a kappa 1b subtype. *Neuroreport, 5*, 2161–2164.

Langlois, A., Diop, L., Riviere, P. J., Pascaud, X., & Junien, J. L. (1994). Effect of fedotozine on the cardiovascular pain reflex induced by distension of the

O

irritated colon in the anesthetized rat. *European Journal of Pharmacology, 271*, 245–251.

Langlois, A., Diop, L., Friese, N., Pascaud, X., Junien, J. L., Dahl, S. G., & Riviere, P. J. (1997). Fedotozine blocks hypersensitive visceral pain in conscious rats: Action at peripheral kappa-opioid receptors. *European Journal of Pharmacology, 324*, 211–217.

Martin, W. R., Eades, C. G., Thompson, J. A., Huppler, R. E., & Gilbert, P. E. (1976). The effects of morphine- and nalorphine-like drugs in the nondependent and morphine-dependent chronic spinal dog. *The Journal of Pharmacology and Experimental Therapeutics, 197*, 517–532.

Murphy, D. M., Koblisch, M., Gauntner, E. K., Little, P. J., Gottshall, S. L., Garver, D. D., & DeHaven-Hudkins, D. L. (2000). A screening process for in vivo assessment of blood–brain barrier penetration for kappa (k) opioid receptor agonists. *Drug Metabolism Reviews, 32*(S2), 179, Abstract.

Rivière, P. J. M., & Junien, J. L. (2000). Opioid receptors, targets for new gastrointestinal drug development. In T. S. Gaginella & A. Guglietta (Eds.), *Drug development. Molecular targets for GI diseases* (pp. 203–238). Totowa: Humana Press.

Riviere, P. J., Pascaud, X., Chevalier, E., Le Gallou, B., & Junien, J. L. (1993). Fedotozine reverses ileus induced by surgery or peritonitis: Action at peripheral kappa-opioid receptors. *Gastroenterology, 104*, 724–731.

Riviere, P. J., Pascaud, X., Chevalier, E., & Junien, J. L. (1994). Fedotozine reversal of peritoneal-irritation-induced ileus in rats: Possible peripheral action on sensory afferents. *The Journal of Pharmacology and Experimental Therapeutics, 270*, 846–850.

Rivière, P. J. M., Vanderah, T. W., Porreca, F., Houghton, R., Schteingart, C., Trojnar, J., & Junien, J. L. (1999). Novel peripheral peptidic kappa-agonists. *Acta Neurobiologiae Experimentalis, 59*(3), 186, Abstract.

Rothman, R. B., Bykov, V., De Costa, B. R., Jacobson, A. E., Rice, K. C., & Brady, L. S. (1990). Interaction of endogenous opioid peptides and other drugs with four kappa-opioid binding sites in guinea pig brain. *Peptides, 11*, 311–331.

Sandner-Kiesling, A., Pan, H. L., Chen, S. R., James, R. L., DeHaven-Hudkins, D. L., Dewan, D. M., & Eisenach, J. C. (2002). Effect of kappa-opioid agonists on visceral nociception induced by uterine cervical distension in rats. *Pain, 96*, 13–22.

Sengupta, J. N., Su, X., & Gebhart, G. F. (1996). Kappa, but not mu or delta, opioids attenuate responses to distention of afferent fibers innervating the rat colon. *Gastroenterology, 111*, 968–980.

Sengupta, J. N., Snider, A., Su, X., & Gebhart, G. F. (1999). Effects of kappa-opioids in the inflamed rat colon. *Pain, 79*, 175–185.

Simonin, F., Gaveriaux-Ruff, C., Befort, K., Matthes, H., Lannes, B., Micheletti, G., Mattei, M. G., Charron, G., Bloch, B., & Kieffer, B. (1995). Kappa-opioid receptor in humans: CDNA and genomic cloning, chromosomal assignment, functional expression, pharmacology, and expression pattern in the central nervous system. *Proceedings of the National Academy of Sciences of the United States of America, 92*, 7006–7010.

Su, X., Sengupta, J. N., & Gebhart, G. F. (1997). Effects of opioids on mechanosensitive pelvic nerve afferent fibers innervating the urinary bladder of the rat. *Journal of Neurophysiology, 77*, 1566–1580.

Su, X., Joshi, S. K., Kardos, S., & Gebhart, G. F. (2002). Sodium channel blocking actions of the kappa-opioid receptor agonist U50,488 contribute to its visceral antinociceptive effects. *Journal of Neurophysiology, 87*, 1271–1279.

Oral Dysesthesia

▶ Burning Mouth Syndrome

Oral Dystonia, Dyskinesia

▶ Orofacial Pain, Movement Disorders

Oral Opioids

Kok Eng Khor
Department of Pain Management, Prince of Wales Hospital University of New South Wales, Randwick, NSW, Australia

Synonyms

Major analgesics; Narcotics

Definition

Opioids are drugs chemically related to morphine. They are strong analgesics, used to relieve moderate to severe chronic pain when ▶ paracetamol and nonsteroidal anti-inflammatory drugs (NSAIDs) are not effective or limited by adverse reactions.

Characteristics

Historical Background

In the form of opium, opioids have been used for centuries to provide relief from pain. In the modern era, the active agents of opium have been isolated, and related compounds have been synthesized in an effort to produce drugs with different potencies and duration of action (Table 1).

Application

Opioids are routinely used to provide relief of pain during and immediately after surgical procedures. They are used to relieve persistent pain in patients with cancer. They are used to provide relief of severe pain following acute injury. In these applications, there is little or no controversy. Indeed, the use of opioids for these conditions could be regarded as a hallmark of conventional medical practice.

More contentious is the use of opioids for persistent noncancer pain, and for which there appears to be no other effective treatment. The vast majority of patients in this state suffer from musculoskeletal and neuropathic pain. In the 1950s, medical opinion considered that the risk of addiction far outweighed the benefits of long-term therapy with opioids, but in 1986, Portenoy and Foley (Portenoy and Foley 1986) reported that opioids could be used effectively and safely for noncancer pain. Nevertheless, their use has remained controversial. Concerns persist about efficacy, adverse effects, development of tolerance, addiction, drug diversion, and scrutiny of provider prescription by governmental health agencies (Portenoy 1996; Fanciullo and Cobb 2001). The use of opioid analgesics should be in the context of a multimodal treatment plan including nonpharmacological, physical therapies, psychological strategies, and interventional approaches, taking into account not only biophysical but also psychosocial contributors to the persistent pain condition (Passik 2009).

Routes of Administration

Opioids can be administered by a variety of routes. Continuous or intermittent administration,

Oral Opioids, Table 1 Opioids commonly used in pain medicine

Natural opium derivatives	Semisynthetic derivatives	Synthetic compounds
Morphine	Hydromorphone	Methadone
Codeine	Oxycodone	Meperidine (pethidine)
	Buprenorphine	Fentanyl

by subcutaneous or intravenous injections are typically used for postoperative pain, acute pain (from medical diseases or major injuries), and in some cases of unstable cancer pain. Intrathecal or epidural routes can be used for postoperative pain and for the more complex cancer or other chronic noncancer pain. Some preparations are available for transdermal, sublingual, and intranasal administration. Preparations for oral use are the most common and available for patients who are able to swallow and who can obtain adequate analgesia by this route of administration. It is the preferred route for long-term administration.

Adverse Effects

Opioid side effects are well known and include respiratory depression, nausea/vomiting, sedation, constipation, pruritus, urinary retention, and euphoria/dysphoria. However, most of these side effects, with the exception of constipation, subside with chronic use.

Common adverse effects leading to cessation of therapy include constipation, somnolence, confusion, nausea, and dizziness (Roth et al. 2000; Caldwell et al. 1999; Peloso et al. 2000). Stable dosing of opioids without concurrent use of other centrally acting drugs generally does not lead to cognitive and psychomotor impairment. Good management of pain often leads to improvement in cognitive function (Fanciullo and Cobb 2001). Long-term use also results in analgesic tolerance resulting in reduced efficacy and necessitating an increase in dose and at times leading to refractoriness to the opioid (Ballantyne and Shin 2008).

However, long-term opioid use may also lead to suppression of hypothalamic-pituitary-adrenal and gonadal axis (resulting in amenorrhea, erectile dysfunction, osteoporosis), suppression of immune function, and development of opioid-induced hyperalgesia (Ballantyne 2007).

In some patients, use of opioids leads to the problem of misuse, abuse, addiction, and diversion. These patients exhibit problematic and aberrant drug-taking behaviors. These include multiple unsanctioned dose escalations, prescription forgery, seeking prescriptions from multiple sources, stealing or borrowing drugs, injecting oral formulations, and repeated resistance to changes in therapy despite drug-related side effects. Although it was originally thought to be rare for patients to develop opioid abuse, studies now show aberrant drug-taking behaviors can be common and often undetected by treating physicians. Physicians therefore need to identify likelihood that a patient will exhibit aberrant behaviors preferably before opioid therapy commenced so that an appropriate structured program to minimize risk and guide monitoring can be instituted (Passik 2009).

Efficacy

For postoperative pain, cancer pain, and the pain of acute injury, the efficacy of opioids is not questioned. Their obvious and virtually universal effectiveness obviates the need for randomized controlled trials. The only issues in these arenas are securing an optimum dose, which may be very high in some cases, and combating side effects that may arise.

In the context of chronic noncancer (predominantly musculoskeletal and neuropathic) pain, opioids have not been uniformly successful in relieving pain and improving function in the long term. Most randomized controlled trials show short-term (up to 12 weeks) analgesic efficacy and a small improvement in functional status with moderate doses of opioids (Ballantyne and Shin 2008; Furlan et al. 2006). It is clear that a subpopulation of patients is able to achieve sustained pain relief without substantial toxicity, functional deterioration, or aberrant behaviors (Portenoy 1996; Fanciullo and Cobb 2001; Graziotti and Goucke 1997). The opioid doses required are generally low, and stabilize during long-term administration, suggesting that tolerance is rarely the "driving force" for dose escalation or diminishing efficacy (Roth et al. 2000; Arkinstall et al. 1995; Jamison et al. 1998). A significant number of patients (30–56 %) will cease therapy early because of intolerance to adverse events or ineffectiveness (Roth et al. 2000; Caldwell et al. 1999; Moulin et al. 1996; Ballantyne et al. 2008). Some will achieve relief when different opioids are used.

Osteoarthritis

Codeine and oxycodone are significantly more effective than placebo (Roth et al. 2000; Caldwell et al. 1999; Peloso et al. 2000) when used for osteoarthritis. Analgesia is accompanied by an improvement in physical function and sleep (Roth et al. 2000; Caldwell et al. 1999; Peloso et al. 2000; Fleischman et al. 1999). After an initial titration period, analgesia is maintained long term (up to 18 months) (Roth et al. 2000). However, typical opioid related side effects complicate treatment and account for a number of patients ceasing therapy (Roth et al. 2000; Caldwell et al. 1999). Since they act by a different mechanism, opioids are synergistic with NSAIDs, and can be used to spare increases in NSAID dose, and thereby avoiding serious adverse effects of NSAIDs (Caldwell et al. 1999).

Low Back Pain

Two trials, although not a third, showed codeine, morphine, and methadone to be more effective than placebo for the relief of pain (Arkinstall et al. 1995; Moulin et al. 1996; Maier et al. 2002). Others have shown that opioids achieve significantly better relief of pain than usual care (Haythornthwaite et al. 1998) or NSAIDs (Jamison et al. 1998).

The degree of relief, however, was only modest. Opioids do not abolish pain. Furthermore, although some have found improvement in function and mood (Arkinstall et al. 1995), others found no improvement in these outcome measures (Moulin et al. 1996). ▶ Multidisciplinary treatment may be required to accomplish reductions in disability and suffering related to chronic pain (Haythornthwaite et al. 1998).

Neuropathic Pain

Contrary to traditional belief, most types of neuropathic pain do respond to opioids, although a higher dose is often required. A recent meta-analysis considered nine good quality randomized control trials of up to 12 weeks duration utilizing morphine, oxycodone, methadone, and levophanol. All trials showed opioid efficacy compared to placebo with a clear analgesic dose-dependent response. Opioids were able to reduce both spontaneous as well as evoked pain such as allodynia. However, no consistent impact was found on quality of life measures and disability with positive improvements being balanced by studies showing no difference (Eisenberg et al. 2006). When compared to gabapentin or the tricyclic antidepressants, opioids seemed to exhibit better efficacy although this did not reach statistical significance (Raja et al. 2002; Gilron et al. 2005). The study by Gilron et al. (2005) may lend support to the concept of combination therapy being superior to unimodal therapies in the treatment of chronic neuropathic pain.

Management Strategies

Numerous guidelines (Portenoy 1996; Fanciullo and Cobb 2001; Graziotti and Goucke 1997; Furlan et al. 2010) are available for physicians to identify those patients who might benefit from the long-term use of opioids, to achieve optimal outcomes.

A comprehensive assessment should indicate that the patient have objective evidence of organic pathology causing pain; and be psychologically stable, with appropriate mood and behavior; be reliable; be in a stable social environment; and have health-support systems and no prior history of substance abuse (Graziotti and Goucke 1997). A screen for risk of aberrant drug use or addiction using a screening tool (e.g., Opioid Risk Tool) should be conducted (Furlan et al. 2010). An "opioid contract" (which functions more as an opioid agreement) can be used to inform the patient of risks and benefits of opioid therapy, explains what is expected of the patient, drug storage, acceptable and unacceptable drug-taking behaviors, setting limits of opioid use, and reasons for discontinuation of therapy (Portenoy 1996; Fanciullo and Cobb 2001; Passik 2009). Its efficacy is still uncertain.

Only one doctor who has an established therapeutic relationship with the patient should prescribe the opioid. Realistic goals should be set with the patient and follow-up conducted after a monitored trial to assess the effectiveness of the opioid therapy. Failure to reach predetermined goals is an indication to cease therapy (Graziotti and Goucke 1997; Furlan et al. 2010).

Sustained-release (morphine, oxycodone, hydromorphone) or long-acting (methadone) preparations are the drugs of choice because of single or twice-daily dosing and stable blood concentration. Immediate release preparations may be used for dose finding before commencement of sustained-release agents and are useful for breakthrough pain. Codeine or parenteral (e.g., intramuscular) administration of opioids are not suitable because of short half-lives (Graziotti and Goucke 1997).

A trial of oral opioids can be conducted over 4–6 weeks, with clearly defined goals and endpoints. Patient should be advised to avoid driving until a stable dose is achieved and cautioned about alcohol and other sedative medications. A low dose of a sustained-release morphine or oxycodone is started twice daily or sustained-release hydromorphone started once daily and the dose titrated gradually to analgesic effect (or side effects). The outcome is assessed after 1–2 weeks. Patients with fluctuating pain conditions require a variable dosing regimen, with immediate release morphine, oxycodone, or hydromorphone every 4 h. Improvement in analgesia should be the minimal requirement, and patients should show improved function with no significant side effects. Persistence with opioid therapy is contraindicated if the patient fails to achieve at least partial analgesia at a moderate dose, or if the dose rapidly escalates in opioid-naïve patients within a month of starting treatment. At the end of the trial period, if expected outcomes have not been achieved, the drug dose is tapered over a few days and ceased.

Patients who are prescribed opioids on an ongoing basis should be reviewed monthly by the prescribing physician, with a detailed review by a pain management center undertaken annually. During monitoring of long-term opioid therapy, analgesic efficacy, level of function, adverse effects, patient's affect, and aberrant behavior should be assessed and documented (Graziotti and Goucke 1997; Furlan et al. 2010).

References

Arkinstall, W., Sandler, A., Goughnour, B., et al. (1995). Efficacy of controlled-release codeine in chronic non-malignant pain: A randomized, placebo-controlled clinical trial. *Pain, 62*, 169–178.

Ballantyne, J. C., (2007). Opioid Analgesia: Perspectives on Right Use and Utility. Pain Physician 2007; 10:479–491.

Ballantyne, J. C., & Shin, N. S. (2008). Efficacy of opiids for chronic pain. *The Clinical Journal of Pain, 24*(6), 469–478.

Caldwell, J. R., Hale, M. E., Boyd, R. E., et al. (1999). Treatment of osteoarthritis pain with controlled release oxycodone or fixed combination oxycodone plus acetaminophen added to non-steroidal anti-inflammatory drugs: A double-blind, randomized, multicenter, placebo controlled trial. *Journal of Rheumatology, 26*, 862–869.

Eisenberg, E., McNicol, E. D., & Carr, D. B. (2006). Opioids for neuropathic pain. *Cochrane Database of Systematic Reviews* (3), Art.No.: CD006146. (2009).

Fanciullo, G., & Cobb, J. (2001). The use of opioids for chronic non-cancer pain. *International Journal of Pain Medicine and Palliative Care, 1*, 49–55.

Fleischman, R. M., Kamin, M., Olson, W. H., et al. (1999). Safety and efficacy of tramadol for the signs and symptoms of osteoarthritis. *Arthritis & Rheumatism, 42*, S144.

Furlan, A. D., Reardon, R., & Weppler, C. (2010). Opioids for chronic noncancer pain: A new Canadian practice guideline. *CMAJ: Canadian Medical Association Journal, 182*, 923–930.

Furlan, A. D., Sandoval, J. A., Mailis-Gagnon, A., Tunks E (2006). Opioids for chronic non-cancer pain: A meta-analysis of effectiveness and side effects. *The Canadian Medical Association Journal, 174*, 1589–1594.

Gilron, I., Bailey, J. M., Tu, D., et al. (2005). Morphine, gabapentin, or their combination for neuropathic pain. *The New England Journal of Medicine, 352*, 1324–1334.

Graziotti, P. J., & Goucke, C. R. (1997). The use of oral opioids in patients with chronic non-cancer pain. *MJA: Medical Journal of Australia, 167*, 30–34.

Haythornthwaite, J. A., Menefee, L. A., Quatrano-Piacentini, B. A., et al. (1998). Outcome of chronic opioid therapy for non-cancer pain. *Journal of Pain and Symptom Management, 15*, 185–194.

Jamison, R. N., Raymond, S. A., Slawsby, E. A., et al. (1998). Opioid therapy for chronic non-cancer back pain. *Spine, 23*, 2591–2600.

Maier, C., Hildebrandt, J., Klinger, R., et al. (2002). Morphine responsiveness, efficacy and tolerability in patients with chronic non-tumor associated pain – results of a double-blind placebo-controlled trial (MONTAS). *Pain, 97*, 223–233.

Moulin, D. E., Lezzi, A., Amireh, R., et al. (1996). Randomised trial of oral morphine for chronic non-cancer pain. *Lancet, 347*, 143–147.

Passik, S. D. (2009). Issues in long-term opioid therapy: Unmet needs, risks and solutions. *Mayo Clinic Proceedings, 84*, 593–601.

Peloso, P. M., Bellamy, N., Bensen, W., et al. (2000). Double-blind randomized placebo control trial of controlled release codeine in the treatment of osteoarthritis of the hip or knee. *Journal of Rheumatology, 27*, 764–771.

Portenoy, R. K. (1996). Opioid therapy for chronic non-malignant pain: A review of the critical issues. *Journal of Pain and Symptom Management, 11*, 203–217.

Portenoy, R., & Foley, K. (1986). Chronic use of opioid analgesics in non-malignant pain: Report of 38 cases. *Pain, 25*, 171–186.

Raja, S. N., Haythornthwaite, J. A., Pappagallo, M., et al. (2002). Opioids versus antidepressants in postherpetic neuralgia: A randomized, placebo-controlled trial. *Neurology, 59*, 1015–1021.

Roth, S. H., Fleishmann, R. M., Burch, F. X., et al. (2000). Around-the-clock, controlled-release oxycodone therapy for osteoarthritis-related pain. *Archives of Internal Medicine, 160*, 853–860.

Ordinary Cramp

Definition

The term "ordinary" usually includes conditions most frequently referred to as muscle cramps in normal or pathologic conditions, characterized by high-frequency, high-amplitude discharge of potentials on the EMG recordings.

Cross-References

▶ Muscular Cramps

Ordinary Headache

► Headache, Episodic Tension-Type

Ordine®-Morphine-Hydrochloride

► Postoperative Pain, Morphine

Organ or Body Part Dysfunction

► Impairment, Pain-Related

Orofacial Pain

Barry J. Sessle[1] and Jonathan O. Dostrovsky[1,2]
[1]Faculty of Dentistry, University of Toronto, Toronto, ON, Canada
[2]Department of Physiology, University of Toronto, Toronto, ON, Canada

Introduction

Although pain arising from the craniofacial region shares many features with the pain originating in other body regions, there are several pain conditions that are unique to this region. One of the best known and most prevalent is toothache, which results primarily from activation of nociceptors in the tooth pulp or dentin. A description of the clinical features, pathophysiology, and management of dental pain is provided in the entry ► Dental Pain, Etiology, Pathogenesis, and Management. Two other well-known conditions unique to the trigeminal system are trigeminal neuralgia and temporomandibular disorders (TMD), and these are described in the entries ► Postherpetic Neuralgia, Etiology, Pathogenesis, and Management; ► Temporomandibular Joint Disorders; and ► Orofacial Pain, Movement Disorders.

Although postherpetic neuralgia can occur outside the orofacial region, facial tissues are often affected; its features are discussed in ► Postherpetic Neuralgia, Etiology, Pathogenesis, and Management. In addition to these well-described pain states, there are several other conditions that are more difficult to classify, diagnose, and manage. The best examples of these poorly defined conditions are atypical facial pain and burning mouth syndrome (BMS), which are described in the entry ► Atypical Facial Pain: Etiology, Pathogenesis, and Management. The problem of classification and taxonomy of orofacial pain is discussed in the entry ► Orofacial Pain: Taxonomy/Classification. The relationships and interactions of pain with sleep and with oral movement disorders (e.g., tremor, dystonia, and dyskinesia) and parafunctions such as bruxism are discussed in the entries ► Orofacial Pain, Sleep Disturbance and ► Orofacial Pain, Movement Disorders. Headaches are of course also very common in the orofacial region and originate primarily as a result of activity in nociceptive afferents innervating the intracerebral vessels and meninges. The clinical features and underlying mechanisms of headaches are covered by many entries (e.g., ► Headache; ► Migraine; ► Migraine Epidemiology; ► Migraine Without Aura; ► Migraine, Genetics; ► Migraine, Pathophysiology; ► Preventive Migraine Therapy).

Characteristics

The Scope of Orofacial Pain

In addition to the wide range of pain conditions manifested in the orofacial area, epidemiological studies (e.g., LeResche and Drangsholt 2008; Lipton et al. 1993) have revealed the very high prevalence of orofacial pain. Fortunately, some of the chronic conditions that are extremely painful or bothersome (e.g., trigeminal neuralgia, postherpetic neuralgia) are relatively uncommon (see ► Trigeminal Neuralgia: Diagnosis and Treatment, ► Postherpetic Neuralgia, Pharmacological and Nonpharmacological Treatment Options). However, conditions such as TMD,

BMS, headaches, and toothaches are quite common and collectively have a prevalence of 10–15 % or more in different populations (see ▶ Temporomandibular Joint Disorders; ▶ Atypical Facial Pain: Etiology, Pathogenesis, and Management; ▶ Dental Pain, Etiology, Pathogenesis, and Management; ▶ Migraine, Pathophysiology and ▶ Non-Migraine Headaches). Translated into worldwide figures, this means that TMD, for example, affects half a billion people around the world at any onetime!

Studies in humans have drawn attention to the female predominance and to the importance of psychosocial influences as predisposing or modulating influences in many of these conditions, especially those that are chronic in nature, and to the enormous economic as well as social costs of orofacial pain (Feinmann and Newton-John 2004; LeResche and Drangsholt 2008; Sessle 2000; 2011a). For example, changes in mental state (e.g., pain affect, depression) and sleep are frequent accompaniments of pain, as noted in the entries ▶ Orofacial Pain, Sleep Disturbance, and ▶ Temporomandibular Joint Disorders, and there is evidence that in some situations pain may be the cause of these changes, in other cases the consequence. Also, the annual *direct* health-care costs plus *indirect costs* (e.g., lost productivity, compensation) for pain are *estimated to be more than $500B* in the USA alone (Institute of Medicine 2011; and see ▶ Pain in the Workplace, Compensation, and Disability Management). By conservative estimates, orofacial pain including headache would represent a third of these costs, and so the cost to the US economy of orofacial pain each year is approximately $150B! Added to this economic burden is the personal suffering and disrupted quality of life of the orofacial pain patient and their interactions with family and friends. As pointed out in the various entries cited above, there have been some important advances in our knowledge of the clinical features and possible underlying mechanisms, although much remains to be learned and adequate therapy is still lacking in many cases.

Etiology and Pathogenesis of Orofacial Pain Conditions

The etiology and underlying processes of most acute pain conditions occurring in the face and mouth are now reasonably well understood, although there are still several points requiring clarification, as outlined in ▶ Nociceptive Processing in the Brain Stem. We now know, for example, that trauma, dental caries, and infection are the most common causes of acute pain from the tooth, as pointed out in ▶ Dental Pain, Etiology, Pathogenesis, and Management, although the precise mechanisms and factors involved in the activation or sensitization of pulpal or dentinal afferents are not fully resolved (see ▶ Nociceptors in the Dental Pulp). However, the biological processes giving rise to chronic orofacial pain are still poorly understood, and it is clear from several entries on this topic that the etiology of most of the chronic pain conditions expressed in the orofacial region is unclear and that their pathogenesis is still unresolved. This has impacted upon their diagnosis and management and even upon the classification schemes used to describe these conditions (see below and also ▶ Orofacial Pain: Taxonomy/Classification). The predisposing or risk factors related to each of these conditions are also poorly understood, and this is exemplified by our limited ability to explain the female predominance in most of these chronic pain conditions. Nonetheless, with the recent emphasis on the need for a scientific underpinning and for an evidence basis for diagnostic and management approaches advocated for orofacial pain, some long-held beliefs are being laid to rest. For example, the past emphasis on occlusal factors, including sleep bruxism, as being of prime etiological importance in TMD and related conditions has not held up to *rigorous* scientific scrutiny, as the entries ▶ Temporomandibular Joint Disorders and ▶ Orofacial Pain, Sleep Disturbance and recent systematic reviews (e.g., Forssell and Kalso 2004; Stohler 2008) have pointed out. But we clearly have a long way to go to identify the etiological mechanisms and risk factors predisposing to most of the chronic orofacial pain states. There is in particular a need for longitudinal population-based studies to help define the factors that have been implicated for most conditions (e.g., gender, genetic, age, comorbidity, trauma).

While the past 10 years or so has seen the development of animal models of chronic orofacial inflammatory and neuropathic pain (see below), there is still a need for a greater emphasis on the development of animal models to help improve our understanding of the etiology and pathogenesis of these conditions. Many of the current concepts on orofacial pain mechanisms draw largely upon findings from spinal models of chronic pain. Given the uniqueness of some of the orofacial pain conditions (e.g., trigeminal neuralgia, BMS), there is a need to apply and test these concepts within the framework of the trigeminal system by developing chronic orofacial pain models *that mimic these specific conditions*. Recent approaches that show much promise for improving our understanding of inflammatory or neuropathic orofacial pain conditions include the use of inflammatory irritants or complete Freund's adjuvant in inducing musculoskeletal pain (e.g., Dubner and Ren 2004; Dubner et al. 2012; Henry 2004; Sessle 2011b; Woda and Pionchon 2000) and of trigeminal nerve damage in evoking nociceptive pain (Iwata et al. 2004, 2011; Woda and Pionchon 2000; Woda and Salter 2008). From the limited trigeminal studies to date (see below) and from analogous studies in animal models of pain in the spinal system (see ▶ Animal Models and Experimental Tests to Study Nociception and Pain and ▶ Peripheral Neuropathic Pain), the mechanisms that appear to be involved in the pathogenesis of these chronic orofacial pain states include ectopic impulses generated in damaged afferent fibers, peripheral sensitization of afferent fibers, central sensitization of central nociceptive neurons, changes in segmental and descending inhibitory and facilitatory influences on nociceptive transmission and phenotypic changes in afferent fibers, and central nociceptive neurons, including enhanced sympathetic modulation of afferent

fibers and central sprouting of afferent fibers and the involvement of nonneural (glial) cells (see ▶ Nociceptive Processing in the Brain Stem). It has been suggested for instance that abnormal firing of trigeminal primary afferents leading to changes in central nociceptive pathways may be important in trigeminal neuralgia and neuropathic orofacial pain conditions (e.g., Devor et al. 2002; Iwata et al. 2004, 2011; Woda and Salter 2008). Also, findings from inflammatory trigeminal models have led to suggestions (Khan et al. 2008; Sessle 2000, 2011) that both peripheral sensitization and central sensitization may explain the allodynia, hyperalgesia, diffuse and often referred pain, and limitations of jaw movements that are characteristic of TMD (see ▶ Orofacial Pain, Movement Disorders and ▶ Temporomandibular Joint Disorders) or several types of toothache including the "hot" tooth (see ▶ Dental Pain, Etiology, Pathogenesis, and Management). The animal models and correlated human studies also draw attention to the close interplay between pain and other CNS functions, such as sleep, anxiety, and depression (see ▶ Orofacial Pain, Sleep Disturbance), and between sensory and motor systems in the expression of acute or chronic orofacial pain (see ▶ Orofacial Pain, Movement Disorders). While Svensson (see ▶ Orofacial Pain, Movement Disorders) also notes that a pain adaptation may apply in TMD, future research of this model needs to explain the cause of the pain and address the long-term motor consequences of chronic pain.

Classification and Diagnostic Approaches and Issues

There is clearly a need for a practical, comprehensive, and unified classification system for the variety of pain conditions that occur in the orofacial region. Zakrzewska (2004) has noted that classification schemes need to be valid, reliable, comprehensive, flexible, and generalizable, but this is certainly not the case for orofacial pain. As revealed in ▶ Orofacial Pain, Taxonomy/Classification, several classification schemes for orofacial pain have been developed, but there is disagreement or inconsistency between them on a number of points. There has recently been a move toward a mechanism-based classification, but application of such a scheme to orofacial pain or indeed pain elsewhere in the body is stymied by a lack of clear definition of the specific mechanisms applicable to the development and maintenance of each of these conditions, especially many of the neuropathic pain states (see ▶ Peripheral Neuropathic Pain and ▶ Neuropathic Pain of Central Origin).

Also related to the classification issues are the difficulties associated with differential diagnosis of orofacial pain conditions. While some orofacial pain states are reasonably well defined and readily discernable (e.g., trigeminal neuralgia, postherpetic neuralgia see ▶ Postherpetic Neuralgia, Etiology, Pathogenesis, and Management), many are complicated by vague or varying symptomatology, referral of pain, depression, and concomitant pain elsewhere in the body (e.g., some toothaches, TMD; see ▶ Dental Pain, Etiology, Pathogenesis, and Management and ▶ Temporomandibular Joint Disorders) or are diagnosed by exclusion (e.g., atypical facial pain, see ▶ Atypical Facial Pain: Etiology, Pathogenesis, and Management). As a consequence of these features, it is not uncommon for a condition to be misdiagnosed and inappropriate therapy instituted (Truelove 2004; Vickers et al. 1998; also see below).

There are also available a multitude of diagnostic and assessment approaches that have been developed for orofacial pain (see ▶ Dental Pain, Etiology, Pathogenesis, and Management; ▶ Atypical Facial Pain: Etiology, Pathogenesis, and Management; ▶ Temporomandibular Joint Disorders; ▶ Orofacial Pain, Movement Disorders). Clinical examination procedures, radiographic scans, and common sensory assessment techniques (e.g., tooth pulp vitality tests) have long been mainstays of clinical diagnostic approaches. Over the past three decades, there have been substantive improvements in a number of these approaches that have been specifically developed for chronic as well as acute orofacial pains. While many of these

particular approaches are not part of standard clinical practice, they have found extensive application in experimental or clinical studies of these conditions and include pressure pain or tolerance thresholds, visual analogue scales, McGill pain questionnaire, quantitative sensory testing (QST), and magnetic resonance imaging (e.g., Eliav et al. 2004; Essick 2004; Jaaskelainen 2004; Matos et al. 2011; Schweinhardt et al. 2008; Svensson et al. 2004). Although some of the orofacial pain states have reliable and validated diagnostic approaches, for example, the research diagnostic criteria for TMD (see ▶ Temporomandibular Joint Disorders), there are not yet any "gold standards" for diagnosis of most of the conditions. Clearly there is room for further development and standardization and validation of assessment and diagnostic techniques for orofacial pain.

Management Approaches and Issues

Some orofacial pain conditions, especially those that are chronic in nature, are often dealt with by medical specialists, but the dentist is the frontline clinician for most orofacial pain conditions. Indeed, dentists have been at the forefront in developing therapeutic procedures to control pain, a prime example being the introduction of gas anesthesia 160 years ago by dentists. Through these and subsequent developments, nearly all acute orofacial pain and procedural pain can be readily controlled and managed by general anesthetics, local anesthetics, analgesic drugs, etc. A case in point is the variety of management approaches available to the clinician for acute toothache (see ▶ Dental Pain, Etiology, Pathogenesis, and Management). Similarly, for the chronic orofacial pain conditions, a multitude of approaches are available, and several so-called alternative medicine approaches (e.g., acupuncture, herbal remedies) have additionally become widely available over the past 25 years. The entries on ▶ Temporomandibular Joint Disorders; ▶ Atypical Facial Pain: Etiology, Pathogenesis, and Management; and ▶ Postherpetic Neuralgia, Etiology, Pathogenesis, and Management as well as recent reviews (e.g., Forssell and Kalso 2004; Stohler 2008; Truelove 2004; Watson 2004; Widmer 2008) have noted that while some of these techniques have proven efficacy for certain orofacial pain conditions (e.g., trigeminal neuralgia, postherpetic neuralgia), the efficacy of most management approaches currently in use has not been validated or fully tested for their specificity and sensitivity. Indeed, several common management strategies have a limited scientific underpinning, and so in a sense it could be argued that some conditions are often overtreated or have inappropriate treatment rendered. One example is the case of occlusal adjustments for TMD. Another is the not uncommon misdiagnosis of cases of atypical facial pain or trigeminal neuralgia where as a consequence the therapy instituted may be inappropriate and may even be associated with an exacerbation of the pain and suffering of the patient (e.g., Truelove 2004; Vickers et al. 1998) (also see ▶ Atypical Facial Pain: Etiology, Pathogenesis, and Management).

Such considerations emphasize the need for well-designed cohort studies, randomized controlled trials (RCTs), etc., to test the efficacy, validity, specificity, sensitivity, and reliability of a number of management and diagnostic strategies currently in use for pain in the face and mouth, as well as the need for an increased focus on enhancing pain educational programs for health professionals and students who will be called upon to care for patients with orofacial pain (Attanasio 2002; Sessle 2011a; Widmer 2008).

Concluding Remarks

The challenges ahead in clarifying the mechanisms underlying the etiology and pathogenesis of orofacial pain offer the reward of better diagnostic and management approaches that currently exist. Management approaches for many of these conditions are relatively ineffective or unproven or are varied in view of the so-called multifactorial nature of the condition (e.g., TMD). Nonetheless, emerging technologies in imaging, sensory testing (e.g., QST), biological markers, and genetic and molecular biology are starting to be applied to the orofacial area and hold out promise of improved therapeutic approaches for these conditions.

See also ▶ Psychiatric Aspects of Pain and Dentistry

Cross-References

▶ Orofacial Pain: Taxonomy/Classification
▶ Psychiatric Aspects of Pain and Dentistry

References

Attanasio, R. (2002). The study of temporomandibular disorders and orofacial pain from the perspective of the predoctoral dental curriculum. *Journal of Orofacial Pain, 16*, 176–180.

Devor, M., Amir, R., & Rappaport, Z. H. (2002). Pathophysiology of trigeminal neuralgia: The ignition hypothesis. *The Clinical Journal of Pain, 18*, 4–13.

Dubner, R., & Ren, K. (2004). Brainstem mechanisms of persistent pain following injury. *Journal of Orofacial Pain, 18*, 299–305.

Dubner, R., Ren, K., & Sessle, B. J. (2012). Sensory mechanisms of orofacial pain and their clinical implications. In D. M. Laskin, & C. S. Greene (Eds.), *Treatment of temporomandibular disorders: Bridging the gap between. Advances in basic research and clinical patient management.* Chicago: Quintessence (in press).

Eliav, E., Gracely, R. H., Nahlieli, O., & Benoliel, R. (2004). Quantitative sensory testing in trigeminal nerve damage assessment. *Journal of Orofacial Pain, 18*, 339–344.

Essick, G. K. (2004). Psychophysical assessment of patients with posttraumatic neuropathic trigeminal pain. *Journal of Orofacial Pain, 18*, 345–354.

Feinmann, C., & Newton-John, T. (2004). Psychiatric and psychological management consideration associated with nerve damage and neuropathic trigeminal pain. *Journal of Orofacial Pain, 18*, 360–365.

Forssell, H., & Kalso, E. (2004). Application of principles of evidence-based medicine to occlusal treatment for temporomandibular disorders: Are there lessons to be learned? *Journal of Orofacial Pain, 18*, 9–22.

Henry, J. L. (2004). Future basic science directions into mechanisms of neuropathic pain. *Journal of Orofacial Pain, 18*, 306–310.

Institute of Medicine. (2011). *Relieving pain in America: A blueprint for transforming prevention, care, education, and research* (pp. 1–364). Washington, DC: The National Academies Press.

Iwata, K., Tsuboi, Y., Shima, A., et al. (2004). Central neuronal changes after nerve injury: Neuroplastic influences of injury and aging. *Journal of Orofacial Pain, 18,* 293–298.

Iwata, K., Imamura, Y., Honda, K., et al. (2011). Physiological mechanisms of neuropathic pain: The orofacial region. *International Review of Neurobiology, 97,* 227–250.

Jaaskelainen, S. K. (2004). The utility of clinical neurophysiological and quantitative sensory testing for trigeminal neuropathy. *Journal of Orofacial Pain, 18,* 355–359.

Khan, A. A., Ren, K., & Hargreaves, K. M. (2008). Neurochemical factors in injury and inflammation of orofacial tissues. In B. J. Sessle, G. J. Lavigne, R. Dubner, & J. P. Lund (Eds.), *Orofacial Pain* (2nd ed., pp. 45–52). Chicago: Quintessence.

LeResche, L., & Drangsholt, M. (2008). Epidemiology of orofacial pain: Prevalence, incidence, and risk factors. In B. J. Sessle, G. J. Lavigne, R. Dubner, & J. P. Lund (Eds.), *Orofacial Pain* (2nd ed., pp. 13–18). Chicago: Quintessence.

Lipton, J. A., Ship, J. A., & Larach-Robinson, D. (1993). Estimated prevalence and distribution of reported orofacial pain in the United States. *The Journal of the American Dental Association, 124,* 115–121.

Matos, R., Wang, K., Jensen, J. D., Jensen, T., Neuman, B., Svensson, P., & Arendt-Nielsen, L. (2011). Quantitative sensory testing in the trigeminal region: Site and gender differences. *Journal of Orofacial Pain, 25,* 161–169.

Schweinhardt, P., Loggia, M. L., Villemure, C., et al. (2008). Psychologic state and pain perception. In B. J. Sessle, G. J. Lavigne, R. Dubner, & J. P. Lund (Eds.), *Orofacial Pain* (2nd ed., pp. 87–92). Chicago: Quintessence.

Sessle, B. J. (1999). The neural basis of temporomandibular joint and masticatory muscle pain. *Journal of Orofacial Pain, 13,* 238–245.

Sessle, B. J. (2000). Acute and chronic craniofacial pain: Brainstem mechanisms of nociceptive transmission and neuroplasticity, and their clinical correlates. *Critical Reviews in Oral Biology and Medicine, 11,* 57–91.

Sessle, B. J. (2011a). Unrelieved pain: a crisis. *Pain Research Management, 16,* 416–420.

Sessle, B. J. (2011b). Peripheral and central mechanisms of orofacial inflammatory pain. *International Review of Neurobiology, 97,* 179–206.

Stohler, C. S. (2008). Management of persistent orofacial pain. In B. J. Sessle, G. J. Lavigne, R. Dubner, & J. P. Lund (Eds.), *Orofacial Pain* (2nd ed., pp. 153–160). Chicago: Quintessence.

Svensson, P., Baad-Hansen, L., Thygesen, T., et al. (2004). Overview on tools and methods to assess neuropathic trigeminal pain. *Journal of Orofacial Pain, 18,* 332–338.

Truelove, E. (2004). Management issues of neuropathic trigeminal pain from a dental perspective. *Journal of Orofacial Pain, 18,* 374–380.

Vickers, E. R., Cousins, M. J., Walker, S., et al. (1998). Analysis of 50 patients with atypical odontalgia. A preliminary report on pharmacological procedures for diagnosis and treatment. *Oral Surgery, Oral Medicine, Oral Pathology, Oral Radiology, and Endodontics, 85,* 24–32.

Watson, C. P. N. (2004). Management issues of neuropathic trigeminal pain from a medical perspective. *Journal of Orofacial Pain, 18,* 366–373.

Widmer, C. G. (2008). Current beliefs and educational guidelines. In B. J. Sessle, G. J. Lavigne, R. Dubner, & J. P. Lund (Eds.), *Orofacial Pain* (2nd ed., pp. 19–26). Chicago: Quintessence.

Woda, A., & Pionchon, P. (2000). A unified concept of idiopathic orofacial pain: Pathophysiologic features. *Journal of Orofacial Pain, 14,* 196–212.

Woda, A., & Salter, M. W. (2008). Mechanisms of neuropathic pain. In B. J. Sessle, G. J. Lavigne, R. Dubner, & J. P. Lund (Eds.), *Orofacial pain* (2nd ed., pp. 53–60). Chicago: Quintessence.

Zakrzewska, J. M. (2004). Classification issues related to neuropathic trigeminal pain. *Journal of Orofacial Pain, 18,* 325–331.

Orofacial Pain, Movement Disorders

Peter Svensson
Clinical Oral Physiology, Aarhus University, Aarhus, Denmark

Synonyms

Bruxism; Mandibular dysfunction; Muscle hyperactivity; Oral dystonia, dyskinesia; Temporomandibular disorders; Tremor

Definition

Perturbations of normal jaw functions that are either likely to cause pain in orofacial tissues or are the consequences of pain, i.e., a broad and unspecific term that covers a variety of pathophysiological mechanisms and clinical conditions.

Introduction

Normal jaw-motor functions involve fast, precise, highly coordinated, and pain-free movement of the mandible during chewing, swallowing, speech, yawning, etc. Deviation from normal jaw movements can be recognized as altered range of motion, irregular movements, postural changes, and eventually noises such as clicking or grating sounds from the ▶ temporomandibular joint (TMJ). There has been a longstanding "chicken or egg" debate in the dental community, whether changes in jaw-motor function will cause pain or pain will cause changes in jaw-motor function. However, syntheses of recent experimental and clinical studies have provided a better description of the relationship between different types of orofacial pain complaints and jaw-motor function.

Characteristics

Temporomandibular Disorders (TMD)

TMD pains are considered a cluster of related pain conditions in the masticatory muscles, TMJ, and associated structures, i.e., a form of orofacial musculoskeletal pain. The chewing pattern in patients with TMD pains differs in several characteristic ways compared to control subjects (Stohler 1999). For example, the movements of the mandible are smaller and tend to be more irregular, and the average and maximum opening velocities are slower. In accordance, it has been shown that TMD pain patients have a significantly longer duration of the chewing cycle. Furthermore, the jaw-closing muscles (masseter, temporalis) are significantly less activated during the agonist phase (jaw-closing), but significantly increased during the antagonist phase (jaw-opening) in TMD pain patients. The maximal voluntary ▶ occlusal force and electromyographic (EMG) activity are also reduced, and the endurance time at submaximal contractions is significantly shorter when compared with control subjects (Svensson and Graven-Nielsen 2001). More recent studies have indicated that the variability of the jaw movements seems to be very sensitive to pain and most notably in subjects with high levels of pain ▶ catastrophizing.

With the mandible in its rest or ▶ postural position (normally 2–3-mm distance between the teeth), increased EMG activity, frequently referred to as muscle hyperactivity, has sometimes been reported in the jaw-closing muscles of TMD pain patients when compared to control subjects. However, firm conclusions have not been reached due to confounding factors in the studies, such as EMG cross talk from facial muscles, inadequate control groups, and prevalence of bruxism (see later). Recently, very small (1–2 μV) EMG increases have been found in well-controlled studies, although the diagnostic and pathophysiological significance of this finding remains unclear. It has also been reported that TMD pain patients may have abnormalities in mandibular kinesthesia, which could influence the rest position of the mandible, and possibly underlie the common perception in TMD pain patients that the teeth do not fit together properly.

Traditionally, the interaction between pain and jaw-motor function has been explained by the vicious cycle concept, where muscle hyperactivity leads to pain and pain reinforces the muscle hyperactivity (Travell et al. 1942). The dental version of the vicious cycle was thought to be initiated by misalignment of the teeth (malocclusion), leading to changes in postural activity and chewing patterns. However, this stereotyped relation between pain and jaw-motor function was challenged in a series of critical papers (Lund et al. 1991). Furthermore, it was pointed out that comparable findings with slower movements and less EMG activity in the agonist phase, and more EMG activity in the antagonist phase, could also be observed during other dynamic motor tasks like gait and other repetitive movements in various musculoskeletal pain conditions (Graven-Nielsen et al. 2000). Analyses of the literature led to the formulation of the pain-adaptation model, which strongly contrasts with the vicious cycle model (Lund et al. 1991) (Fig. 1). The pain-adaptation model predicts that the consequences of orofacial nociceptive inputs to premotor neurons in the brainstem during agonist function are a facilitation of inhibitory pathways to the alpha-motor neurons, and during antagonist function a facilitation of excitatory pathways. The essential prerequisite for the

Orofacial Pain, Movement Disorders, Fig. 1 Highly schematic presentation of the pain-adaptation model (Modified after Lund et al. (1991)). Thin nociceptive afferents (*dotted lines*) from orofacial tissues influence both the central motor program and the central pattern generator (CPG) and networks with excitatory and inhibitory interneurons in the brainstem. There is a shift in the normal drive to the alpha-motor neuron pool so that the jaw-closing muscles are less activated (*dotted line*) during pain in their agonist phase and more activated in their antagonist phase. The same is true for the jaw-opening muscles. The consequences of this reorganization of the motor function are slower and smaller jaw movements

model is a collection of premotor neurons, constituting the ▶ central pattern generator in the brainstem and groups of inhibitory and excitatory interneurons. The pain-adaptation model explains many of the jaw-motor consequences of TMD pain, but does not provide any explanation for the origin of the pain. Nevertheless, the pain-adaptation model has made it clear that no causal link between changes in jaw-motor function (mandibular dysfunction) and pain can be derived from cross-sectional clinical studies, because pain in itself has a significant influence on jaw-motor function. Johansson and Sojka (1991) presented an alternative model to explain muscle tension and spread of muscle pain, which

integrates the gamma-motor neuron system and, most recently the sympathetic nervous system in the pathophysiological mechanisms. Further research will be necessary to test the hypotheses and pivotal parts of these different models. Recently, a modified pain-adaptation model was proposed – the integrated pain-adaptation model – which highlights the influence of the brain in shaping the unique motor responses during painful jaw movements (Murray and Peck 2007).

Human experimental pain studies have nevertheless helped to clarify the interaction between orofacial pain and jaw-motor function. It has been shown that painful injections of, e.g., hypertonic saline into the human masseter muscle causes a reduction of the agonist EMG burst during empty open-close jaw movements and during gum-chewing (Lund et al. 1991; Svensson et al. 1996). Experimental masseter pain also reduces the maximum displacement of the mandible in the lateral and vertical axes, and slows down the maximum velocity during jaw-opening and jaw-closing. Thus, experimental muscle pain studies have consistently shown a decrease in agonist EMG activity in the range of 10–15 %, a small increase in the antagonist EMG activity and modest reductions in maximum displacements. These findings are in accordance with other experimental pain studies and recording of muscle activity during gait and repetitive shoulder movements in humans (Graven-Nielsen et al. 2000). In relation to the postural EMG activity, a small, but significant, increase in the EMG activity of the jaw-closing muscles, both during hypertonic saline-evoked pain and during the imagination of pain, has been demonstrated (Stohler 1999). Furthermore, ▶ experimental jaw muscle pain is associated with a short (<30–60 s) increase in EMG activity recorded with either intramuscular wire or surface EMG electrodes (Svensson and Graven-Nielsen 2001). Generally, the magnitude of these increases in human EMG activity is comparatively small in the jaw-closing muscles and relatively short-lived, which does not seem to justify the term muscle hyperactivity.

The experimental findings in humans are furthermore supported by observations in animal preparations of mastication. In decerebrated rabbits with cortically driven mastication, noxious pressure applied to the zygoma or intramuscular injection of hypertonic saline causes a significant reduction in the agonist EMG burst, significant increases in the duration of the masticatory cycle, and significantly smaller amplitudes (Westberg et al. 1997). Thus, dynamic jaw-motor function seems to be reflexly changed by nociceptive activity. The relatively small EMG changes in the human jaw-closing muscles with the mandible at rest contrast with the significant EMG increases (300–400 %) observed in animal studies using intramuscular EMG electrodes to record nocifensive reflex jaw responses following noxious stimulation of TMJ or muscle tissue (e.g., Hu et al. 1993; Cairns et al. 1998). The evidence of facilitation of EMG activity in both jaw-closing and jaw-opening muscles of rats and cats, following injection of various algesic substances into the deep craniofacial tissues, indicates that nociceptive afferents supplying these tissues activate excitatory pathways projecting to the alpha-motor neuron pools of the jaw-opening and jaw-closing muscles (Sessle 2000). The co-contraction of the jaw-opening and jaw-closing muscles may serve as a "splinting" effect and reduce jaw movements (Sessle 2000), in accordance with the pain-adaptation model (Lund et al. 1991), although the finding that noxious stimulation induces co-activation of these muscles at rest may not be entirely consistent with this model.

However, animal and human studies could differ in some aspects. For example, human pain studies are performed in conscious human beings, and the influence of higher-order brain centers cannot be excluded. The finding that chewing sometimes increases the perceived intensity of pain also suggests that higher-order brain centers could contribute to pain-induced changes in chewing. Furthermore, human studies are performed with a food bolus in the mouth, whereas many of the animal studies have usually involved empty open-close jaw movements. The

motor programs related to such different types of jaw movements could also be differentially influenced by pain. Nevertheless, the human experimental and animal studies are in general agreement with each other, and in general, support the pain-adaptation model.

In summary, the sensory-motor integration in patients with TMD pain seems to apply to the pain-adaptation model. Several human experimental pain studies and animal trials substantiate this hypothesis. However, there are still important issues that will need to be explained. For example, the pain-adaptation model cannot explain the origin of pain. Moreover, the long-term consequences of an adapted jaw-motor function to a chronic painful input are unknown, and there might be conditions where the jaw muscles are not "allowed" to adapt, e.g., when the same load or work is required. The integrated pain-adaptation model (Murray and Peck 2007) with the renewed focus on the importance of the brain in regulating the unique motor responses may be useful to guide future research.

Bruxism and Oral Parafunctions
Bruxism is considered an involuntary activity of the jaw muscles, which is characterized in subjects who are awake by jaw clenching and sometimes tooth gnashing and grinding (Lavigne et al. 2003). In sleep bruxism, both jaw clenching and tooth grinding are observed. The jaw muscle activity can be classified as either rhythmic (phasic), sustained (tonic), or a mixture of both types (Lavigne et al. 2003). It is estimated that about 8 % of the adult population are sleep bruxers, but as many as 60 % will demonstrate some form of rhythmic masticatory muscle activity. Thus, sleep bruxism can be viewed as an exaggerated normal behavior, which in some cases can be associated with significant destruction of the teeth. A point of discussion has been if sleep bruxism is related to TMD pain. About 20–30 % of sleep bruxers report jaw muscle pain, which could either be due to a kind of localized post-exercise muscle soreness, or alternatively, a generalized muscle pain condition like fibromyalgia.

However, several studies have now found that patients with the most jaw muscle EMG activity during sleep are actually those with the fewest muscle symptoms. One consideration is that protective mechanisms in painful conditions will also prevent the overuse of muscles during sleep.

Human experimental studies have nevertheless clearly shown that sustained voluntary EMG activity (grinding or clenching) of a certain intensity and duration can lead to symptoms both in jaw muscles and TMJs (e.g., Arima et al. 1999); however, the level of pain is rather low and short-lived in accordance with the notion that jaw muscles are extremely fatigue resistant. The current view is, therefore, that bruxism and other parafunctional jaw activities are only modest risk factors for the development or maintenance of TMD pain (Svensson et al. 2008). Furthermore, it may be important to distinguish between sleep bruxism and awake bruxism as well as to better classify the jaw muscle activity (duration, amplitude, frequency).

Tremor, Orofacial Dystonia, and Dyskinesia

The human mandible trembles at about 3–8 Hz when it is in its rest position with the jaw muscles relaxed, although the amplitude of this movement is too low to be detected visually (Jaberzadeh et al. 2003). There is evidence that the mandible in the rest position is under a "pulsatile control," where fluctuating activity in central neural pulse generators activate both the jaw-opening and jaw-closing alpha-motor neurons. Recently, it was shown that hypertonic saline-evoked pain in the masseter muscle caused a reduction in the power of the resting jaw tremor (Jaberzadeh et al. 2003). This indicates that jaw muscle pain is capable of tonically modulating the amplitude of the outputs from the central "pulsatile control" generators that drive the alternating activation of antagonistic muscles, which produce jaw tremor at rest and during jaw movements.

For completeness, neurological conditions such as oral dyskinesia and dystonia will also be briefly mentioned, since they can be associated with orofacial pain complaints. Mandibular dyskinesia is involuntary, often continuous, repetitive movements of the tongue, lips, cheeks, and jaw (Clark and Takeuchi 1995), and can be linked to withdrawal of neuroleptics (tardive dyskinesia). Orofacial dystonia is also involuntary, intermittent, and momentarily sustained contraction of the orofacial or jaw muscles, which are normally only present during wakefulness. The pathophysiology may involve disorders of the central motor pathways and in particular the basal ganglia. Pain in these conditions is traditionally thought to be due to excessive muscle contractions.

In summary, movement disorders and orofacial pain cover a wide range of clinical conditions with different pathophysiological mechanisms and manifestations.

Summary

This entry has briefly summarized the relationship between orofacial pain and various types of movement disorders. In particular, three clinical situations have been described: (1) TMD pain and the impact on jaw movements, (2) bruxism and pain, and (3) dyskinesia and dystonias and orofacial pain. The clinician needs to distinguish between these different manifestations using validated history and clinical examination methods. Relatively rarely will it be necessary with additional testing, e.g., electromyographic recordings or jaw-trackings.

References

Arima, T., Svensson, P., & Arendt-Nielsen, L. (1999). Experimental grinding in healthy subjects: A model for post-exercise jaw muscle soreness. *Journal of Orofacial Pain, 13*, 104–114.

Cairns, B. E., Sessle, B. J., & Hu, J. W. (1998). Evidence that excitatory amino acid receptors within the temporomandibular joint region are involved in the reflex activation of the jaw muscles. *Journal of Neuroscience, 18*, 8056–8064.

Clark, G. T., & Takeuchi, H. (1995). Temporomandibular dysfunction, chronic orofacial pain and oral motor disorders in the 21st century. *Journal of the California Dental Association, 23*, 41–50.

Graven-Nielsen, T., Svensson, P., & Arendt-Nielsen, L. (2000). Effect of muscle pain on motor control: A human experimental approach. *Advances in Physiotherapy, 2*, 26–38.

Hu, J. W., Yu, X.-M., Vernon, H., & Sessle, B. J. (1993). Excitatory effects on neck and jaw muscle activity of inflammatory irritant applied to cervical paraspinal tissues. *Pain, 55*, 243–250.

Jaberzadeh, S., Svensson, P., Nordstrom, M. A., & Miles, T. S. (2003). Differential modulation of tremor and pulsatile control of human jaw and finger by experimental muscle pain. *Experimental Brain Research, 150*, 520–524.

Johansson, H., & Sojka, P. (1991). Pathophysiological mechanisms involved in genesis and spread of muscular tension in occupational muscle pain. *Medical Hypotheses, 135*, 196–203.

Lavigne, G. J., Kato, T., Kolta, A., & Sessle, B. J. (2003). Neurobiological mechanisms involved in sleep bruxism. *Critical Reviews in Oral Biology and Medicine, 14*, 30–46.

Lund, J. P., Donga, R., Widmer, C. G., & Stohler, C. S. (1991). The pain-adaptation model: A discussion of the relationship between chronic musculoskeletal pain and motor activity. *Canadian Journal of Physiology and Pharmacology, 69*, 683–694.

Murray, G. M., & Peck, C. C. (2007). Orofacial pain and jaw muscle activity: A new model. *Journal of Orofacial Pain, 21*, 263–278.

Sessle, B. J. (2000). Acute and chronic craniofacial pain: Brainstem mechanisms of nociceptive transmission and neuroplasticity, and their clinical correlates. *Critical Reviews in Oral Biology and Medicine, 11*, 57–91.

Stohler, C. S. (1999). Craniofacial pain and motor function: Pathogenesis, clinical correlates, and implications. *Critical Reviews in Oral Biology and Medicine, 10*, 504–518.

Svensson, P., Arendt-Nielsen, L., & Houe, L. (1996). Sensory-motor interactions of human experimental unilateral jaw muscle pain: A quantitative analysis. *Pain, 64*, 241–249.

Svensson, P., & Graven-Nielsen, T. (2001). Craniofacial muscle pain: Review of mechanisms and clinical manifestations. *Journal of Orofacial Pain, 15*, 117–145.

Svensson, P., Jadidi, F., Arima, T., Baad-Hansen, L., & Sessle, B. J. (2008). Relationships between craniofacial pain and bruxism. *Journal of Oral Rehabilitation, 35*, 524–547.

Travell, J., Rinzler, S., & Herman, M. (1942). Pain and disability of the shoulder and arm. Treatment by intramuscular infiltration with procaine hydrochloride. *Journal of the American Medical Association, 120*, 417–422.

Westberg, K.-G., Clavelou, P., Schwartz, G., & Lund, J. P. (1997). Effects of chemical stimulation of masseter muscle nociceptors on trigeminal motoneuron and interneuron activities during fictive mastication in the rabbit. *Pain, 73*, 295–308.

Orofacial Pain, Sleep Disturbance

Gilles J. Lavigne
Facultés de Médecine Dentaire et de Médecine, Université de Montréal, Hopital Sacre Coeur de Montreal. Surgery-trauma dept, Montréal, QC, Canada

Synonyms

Nociception and sleep; Poor sleep quality; Sleep disturbance

Definition

Chronic pain is a major cause of sleep disturbances and complaints. Its major influence is to increase the magnitude and/or the frequency of arousal and ▸ awakening in sleep. A day with intense pain could be followed by sleep of poor quality (unrefreshing sleep), and poor sleep may be followed by more pain on the next day.

Introduction

Sleep is a physiological state usually characterized by a partial isolation from the environment, except when an unpleasant, potentially harmful, or life-threatening event occurs (Lavigne et al. 2007, 2011a). During sleep, sensory perception is attenuated to prevent sleep disruption by nonrelevant input in order to promote sleep continuity (i.e., less perturbation as possible on ultradian non-REM to REM cycles). The perception of pain in sleep should rather be termed nociception, since sleep is associated with an altered state of consciousness. The presence of pain during wakefulness, as well as the intrusion of pain in the sleeping period, is potentially associated with fatigue and lower sleep quality (e.g., complaints of nonrestorative sleep), daytime sleepiness and risk of accidents, low memory performance, etc. The influence of chronic orofacial muscle and/or temporoman-dibular joint pain on sleep does not differ from that of pain in other parts of the body. Both masticatory myofascial pain and ▶ trigeminal neuralgia can induce pain-related awakenings and be associated with insomnia or sleep-related breathing disorders such as apnea-hypopnea (Lavigne et al. 2007; Smith et al. 2009; Benoliel et al. 2009).

Characteristics

Epidemiology and Risk Factors
It is reported that ≈2/3 of chronic ▶ orofacial pain patients may report poor sleep quality described in terms of unrefreshing sensation or that sleep is too short or interrupted but series of awakenings over night (Dao et al. 1994; Morin et al. 1998; Riley et al. 2001; Yatani et al. 2002). The exacerbation of poor sleep in chronic pain patients could be associated with several risk factors (Lavigne et al. 2011a):
1. Anxiety, ▶ catastrophizing, depression, etc.
2. Alcohol, caffeine, nicotine, and medications known to disturb sleep (e.g., cardioactive drugs, opioids)
3. Sleep disorders such as insomnia, sleep apnea, and periodic limb movement in sleep

Chronic pain patients frequently experience insomnia with an odds ratio of 1.5–5.0 depending on age and type of pain (Moldofsky 2001; Ohayon 2005). Risk factors in chronic orofacial muscle pain (e.g., myofascial pain alone or with ▶ temporomandibular arthralgia) include a history of trauma, jaw clenching habit, ▶ somatization, and female gender (Huang et al. 2002; Macfarlane et al. 2003). However, the link between orofacial pain and tooth grinding during sleep, termed sleep bruxism, is much less clear and debated (Macfarlane et al. 2003; Smith et al. 2009; Davis et al. 2010). Using polygraphic sleep measures, we noticed that a subgroup of patients with sleep bruxism, the ones with low frequency of rhythmic jaw muscle contraction, was 3.9 times more at risk of reporting morning masticatory muscle pain (Rompré et al. 2007). Sleep-related headache (namely, tension type) is not

uncommon. Sleep complaints with generalized pain sensitivity are not specific to the orofacial pain population, since they are found in the presence of other symptoms such as bowel complaints and frequent headaches that occur with chronic fatigue syndrome, ▶ fibromyalgia, and ▶ temporomandibular disorders (Aaron et al. 2000; Moldofsky 2001; Lavigne et al. 2011a).

Interaction Between Pain and Sleep
In most cases, a new pain episode will precede complaints of poor sleep (Morin et al. 1998; Riley et al. 2001). By contrast, when chronic pain sets in (e.g., burn pain after a few days, fibromyalgia), a vicious circle is reported: A day with high pain is followed by a night of poor sleep, and sleep of poor quality is followed by reports of higher pain the next day (Raymond et al. 2001; Lavigne et al. 2011a). However, this interaction only explains a low percentage of the variability (less than 30 %) in pain and sleep complaints. Other influences such as insomnia, physical and/or psychological disabilities, fatigue, depression, and the risk factors mentioned above also have to be factored into the equation (Nicassio et al. 2002; McCracken and Iverson 2002; Yatani et al. 2002; Tang et al. 2007).

Sleep Architecture and Pain
A normal sleep period is characterized by alternated ▶ sleep stages (light St 1 and 2 to deep St 3 and 4 to rapid eye movement (REM) sleep) that occur three to five times (ultradian rhythm) in a normal sleep period of 7–9 h. We usually spend approximately 50–65 % of a night in St 1 and 2, 20–25 % in St 3 and 4, and 10–20 % in REM sleep. The roles of sleep are to recover from fatigue, maintain cognitive function (e.g., memory consolidation and performance, concentration), and help overall biological regeneration. Sleep disruption (e.g., fragmentation) that interferes with sleep continuity is reported to induce some complaints of fatigue, mood alteration, spontaneous pain complaints with dysfunctional endogenous pain-inhibitory process, expression of prostaglandins, and poor cognitive performance the next days (Smith et al. 2007; Haack et al. 2009; Edwards et al. 2009). Furthermore, when

Orofacial Pain, Sleep Disturbance, Table 1 Sleep changes that could be produced by pain (low to moderate level of evidence)

Lower sleep efficacy (% time asleep over total time in bed: usually >85 % in normal subjects)
Longer duration of light sleep stage 1 and shorter duration of stage 2
Lower duration of deep sleep (stages 3 and 4)
Lower density or power of slow-wave EEG activity usually associated with unrefreshing sleep
Higher density of K complexes and lower frequency of spindles in EEG activity
More frequent sleep stage shift (transition from deep sleep to light sleep) and unstable deep sleep
Presence of more than 15 microarousals/h (3–10 s EEG changes with heart rate increase and possible rise in muscle tone) in young adults; more than 29/h of sleep in older adults
In the presence of a usual number of microarousals/h of sleep, a rise in sympathetic-cardiac activity, a high CAP rate, and alpha EEG intrusions in deep sleep (stages 3 and 4)
High rate of awakening in sleep (e.g., >4/h sleep)

sleep is restricted to 4 and 6 h per day for 14 days, subjects showed deficits in cognitive performance (Van Dongen et al. 2003). The impact of chronic pain on cognitive function (e.g., memory, concentration, fatigue, driving ability) during wakefulness and the benefit of a nap (e.g., <20 min with mainly light sleep vs. 60–90 min nap that allows the occurrence of deep and REM sleep) or pain medications need to be investigated (Brousseau et al. 2003; Lavigne et al. 2011a).

The duration of sleep stages in most chronic pain patients is normal, although some disruption in the microstructure of sleep has been found, e.g., sleep transition (termed sleep stage shift), frequent and long arousals termed awakenings, shorter stage 2, and lower number of non-REM to REM ultradian cycle (see Table 1) (Lavigne et al. 2011a). The quality of deep sleep (St 3 and 4) of chronic pain patients was initially described to be perturbed, due to the intrusion of so-called fast alpha waves (as estimated on the electroencephalographic (EEG) sleep traces; see also Glossary for definition), but this concept has been revisited. It is currently suggested that the sleep of chronic pain patients (chronic widespread pain – fibromyalgia) is under the influence of an EEG "protective" mechanism termed ► cyclic alternating pattern (CAP) (Moldofsky 2001; Rizzi et al. 2004). CAP is a natural rhythm in sleep in which every 20–60 s there is a brief arousal (named a ► microarousal), which allows the individual to adjust his/her body temperature and heart and respiratory rate to the environment.

This physiological activity is like a "sentinel" that protects and/or prepares the body for a rapid awakening in the presence of a threatening situation. In the presence of sleep disruptive influences (e.g., sound, sleep apnea, periodic limb movement, and pain), arousal or CAP rate per hour of sleep could be increased.

Sleep is also a state normally associated with a reduction in heart rate variability, due to a change in the balance of components of the autonomic nervous system. In light and deep sleep, a parasympathetic dominance "slows down" the cardiac activity, while during the awake or REM sleep state, there is a cardiac-sympathetic dominance, characterized by a higher cardiac activity/variability. The absence of a reduction in cardiac activity during light or deep sleep may cause unrefreshing or nonrestorative sleep. This suggestion was supported by findings showing that fibromyalgia patients maintain a high sympathetic-cardiac activity during sleep but was not reproduced when analysis was done excluding high incidence of body movements and arousals were controlled (Martínez-Lavín et al. 1998; Lavigne et al. 2011b).

Experimental Pain and Sleep Fragmentation Models

During sleep, in comparison to the waking state, most sensory stimulation needs to be more intense to evoke a physiological response such as microarousal, awakening, or a behavioral reaction during awakening. An auditory alarm needs to be over 65 dB to waken a sleeping subject, although sound in the range of 40 dB is known to cause sleep perturbations. In an experimental setting, sleep can be disrupted in two ways: ▶ sleep fragmentation or deprivation (Lavigne et al. 2000, 2004; Smith et al. 2007). Sleep fragmentation is a brief intrusion in sleep that causes a transitional change in the sleep process (sleep stage shift, microarousal, or awakening without a conscious response from the subject) or is induced by forced awakening every hour of sleep. Sleep deprivation is either a total prevention of sleep or is limited to a specific sleep stage (e.g., St 2 or 3 and 4 or REM). To be able to induce sleep deprivation, the subject has to be monitored with polysomnography recordings. Both fragmentation and deprivation have potential consequences for functioning the next day (e.g., fatigue, sleepiness, boredom, irritability, poor memory performance) and change in prostaglandins that could influence pain reports or clinician assessment of pain (Smith et al. 2007; Haack et al. 2009). For research purposes, the effects of three sensory pain modalities on sleep have been investigated: thermal (heat-cold thermode, laser heat), mechanical (finger-joint pressure), and chemical (hypertonic saline or other algesic substance injection) (Drewes et al. 1997; Lavigne et al. 2000, 2004; Bastuji et al. 2008). Electric shock has also been used to test nociceptive-polysynaptic flexion reflex latency and amplitude (Sandrini et al. 2001). It appears that the duration of the stimulation needs to be long enough (over 60 s) to trigger a response that is similar to that observed in a situation associated with awakening in sleep, e.g., "I was suddenly woken up by a toothache." The use of these models has confirmed that pain perception remains active during sleep, similar to other sensory stimulations during sleep, and is lower in deep (stages 3 and 4) and REM sleep in comparison to light sleep (stage 2). However, if the nociceptive stimulation is potentially harmful (duration seems to be more important than only intensity), it will evoke arousal and awakening independently of deepness of sleep or sleep stage (Lavigne et al. 2004).

Management

Sleep and pain interactions can be managed with classical approaches (see Table 2) (Brousseau et al. 2003; Lavigne et al. 2011a). First, educate the patient about sleep "hygiene" to identify bad life habits that could exacerbate the problem (e.g., late day exercise, late meal, alcohol, poor sleep schedule, TV, or computer in bedroom). In the presence of musculoskeletal pain-related problems, physical management (e.g., physical therapy, osteopathy, massage, or fasciatherapy) could be beneficial. Light exercise is also recognized as being beneficial. However, an assessment needs to be made of the advantage and risk of a daytime nap; too long a nap could disrupt the circadian rhythm and induce prolonged sleep inertia/low cognitive functioning in the hours after the nap.

Second, if the presence of a sleep disorder is in doubt, it is strongly recommended that consultation with a sleep medicine physician be arranged in order to exclude sleep disorders that create health risks (e.g., sleep apnea patients have a moderate to high risk of vehicle accident and hypertension with a shorter life expectancy) or exacerbate poor sleep (e.g., periodic limb movement, nightmare, upper airway resistance). The retrognathia, deep palate, macroglossia, and large tonsils or presence of adenoids needs to be assessed since they are risk factors for sleep disorder breathing. Psychological support and CBT may also be considered in some cases to improve relaxation and life style or manage concomitant conditions.

Third, some medication (analgesics, muscle relaxants, sleep facilitators) could be used on a short-term basis to improve sleep and reduce pain. Opioid effects on sleep need to be further characterized, since some evidence suggests a disruption of the so-called restorative deep sleep (also termed slow-wave activity sleep) and breathing (avoid if presence of sleep apnea is suspected). In the presence of a severe sleep and pain problem, antidepressants could improve sleep and assist in the management of concomitant psychological conditions (e.g., anxiety, depression), but caution is suggested, since there is limited evidence bearing on their use for this purpose.

Orofacial Pain, Sleep Disturbance, Table 2 Management guidelines for sleep and pain interaction (most have low level of evidence level)

Sleep hygiene – lifestyle advices up to cognitive-behavioral therapy, sleep diagnosis, and medications
Bedroom should be an oasis, good mattress, etc.
Avoid caffeine, nicotine, and alcoholic products in evening
Relaxation (e.g., abdominal breathing, imagery, hot bath hydrotherapy)
Use a regular sleep schedule as much as possible
Avoid heavy meals in the evening
Daytime nap should be in the afternoon and no more than 20 min
Light exercise program may be beneficial; avoid intense exercise late in the day
Physical therapy and cognitive-behavioral therapy (CBT) adapted for sleep and pain
Sleep disorders diagnosis
Respiratory: sleep apnea and hypopnea, upper airway resistance, snoring
Movement: periodic limb movement, sleep bruxism, abnormal swallowing, Parkinson-related activity, REM behavior disorder, etc.
Circadian – sleep cycle related: insomnia, hypersomnia, sleep-wake schedule shift, drug or alcohol abuse, narcolepsy, etc.
Medications (most off label for pain and sleep interaction)
Analgesics properties: AINS (Anti-inflammatory nonsteroidal analgesics), morphine (risk of sleep disruption/avoid if sleep apnea), gabapentin, pregabalin
Muscle relaxants and sedative properties: cyclobenzaprine (1/2 co) or clonazepam (low dose, not every evening) in early evening to prevent morning sleepiness and risk of vehicle accident
Sleep facilitators: zaleplon (triazolam), temazepam, zopiclone, zolpidem, and gabapentin at low dose (empirical basis so far)
Antidepressants: amitriptyline (low dose), trazodone, nefazodone and mirtazapine, duloxetine, and milnacipran (for fibromyalgia = some benefit in pain management)
Others: pramipexole and sodium oxybate (some evidences but for more severe condition)
If breathing disorders during sleep: CPAP (continuous positive airway pressure), mandibular advancement appliance
CAM approaches: valerian, lavender, kava, Cannabis, melatonin (Additive effect possible; Consult: www.nccam.nih.gov) plus massage, Qigong, etc.

Summary

Pain can interfere with perception of sleep quality and disrupt its continuity. Chronic pain patients, including the ones with orofacial and head pain, may also report poor sleep or sudden awakening due to pain. The clinicians need to revise sleep hygiene and sleep habits; identify the presence of comorbid conditions such as anxiety, depression, or sleep disorders; and finally plan management with cognitive-behavioral therapy, medications, and complementary alternative approaches (CAM) when indicated.

Acknowledgment Research supported by the Canada Research Chair in pain, sleep, and trauma; FRSQ; and CIHR grants plus CFI for equipment.

References

Aaron, L. A., Burke, M. M., & Buchwald, D. (2000). Overlapping conditions among patients with chronic fatigue syndrome, fibromyalgia, and temporomandibular disorder. *Archives of Internal Medicine, 160*, 221–227.

Bastuji, H., Perchet, C., Legrain, V., Montes, C., & Garcia-Larrea, L. (2008). Laser evoked responses to painful stimulation persist during sleep and predict subsequent arousals. *Pain, 31*, 589–599.

Benoliel, R., Eliav, E., & Sharav, Y. (2009). Self-reports of pain-related awakenings in persistent orofacial pain patients. *Journal of Orofacial Pain, 23*(4), 330–338.

Brousseau, M., Manzini, C., Thie, N. M. R., & Lavigne, G. J. (2003). Understanding and managing the interaction between sleep and pain: An update for the dentist. *Journal of the Canadian Dental Association, 69*, 437–442.

Dao, T. T. T., Lavigne, G. J., Feine, J. S., Lund, J. P., & Goulet, J.-P. (1994). Quality of life and pain in myofascial pain patients and bruxers. *Journal of Dental Research, 73,* 2047.

Davis, C. E., Carlson, C. R., Studts, J. L., Curran, S. L., Hoyle, R. H., Sherman, J. J., & Okeson, J. P. (2010). Use of a structural equation model for prediction of pain symptoms in patients with orofacial pain and temporomandibular disorders. *Journal of Orofacial Pain, 24*(1), 89–100.

Drewes, A. M., Nielsen, K. M., Arendt-Nielsen, L., Birket-Smith, L., & Hansen, L. M. (1997). The effect of cutaneous and deep pain on the electroencephalogram during sleep – an experimental study. *Sleep, 20,* 623–640.

Edwards, R. R., Grace, E., Peterson, S., Klick, B., Haythornthwaite, J. A., & Smith, M. T. (2009). Sleep continuity and architecture: Associations with pain-inhibitory processes in patients with temporomandibular joint disorder. *European Journal of Pain, 13*(10), 1043–1047.

Haack, M., Lee, E., Cohen, D. A., & Mullington, J. M. (2009). Activation of the prostaglandin system in response to sleep loss in healthy humans: Potential mediator of increased spontaneous pain. *Pain, 145,* 136–141.

Huang, G. J., Leresche, L., Critchlow, C. W., Martin, M. D., & Drangholt, M. T. (2002). Risk factors for diagnostic subgroups of painful temporomandibular disorders (TMD). *Journal of Dental Research, 81,* 284–288.

Lavigne, G. J., Zucconi, M., Castronovo, C., Manzini, C., Marchettini, P., & Smirne, S. (2000). Sleep arousal response to experimental thermal stimulation during sleep in human subjects free of pain and sleep problems. *Pain, 84,* 283–290.

Lavigne, G. J., Brousseau, M., Kato, T., Mayer, P., Manzini, C., Guitard, F., & Montplaisir, J. Y. (2004). Experimental pain perception remains equally active over all sleep stages. *Pain, 110,* 646–655.

Lavigne, G. J., Sessle, B. J., Choiniere, M., & Soja, P. J. (2007). *Editors, Sleep and pain* (p. 473). Seattle: IASP Press.

Lavigne, G. J., Smith, M. T., Denis, R., & Zucconi, M. (2011a). Pain & sleep. In H. M. Kryger, T. Roth, & W. C. Dement (Eds.), *Principles & practice of sleep medicine* (pp. 1442–1451). Philadelphia: Elsevier Saunders.

Lavigne, G. J., Okura, K., Abe, S., Huynh, N., Montplaisir, J. Y., Marchand, S., Culombo, R., et al. (2011b). Gender specificity of the slow wave sleep lost in chronic widespread musculoskeletal pain. *Sleep Medicine, 12*(2), 179–185.

Macfarlane, T. V., Blinkhorn, A. S., Davies, R. M., & Worthington, H. V. (2003). Association between local mechanical factors and orofacial pain: Survey in the community. *Journal of Dentistry, 31*(8), 535–542.

Martínez-Lavín, M., Hermosillo, A. G., Rosas, M., & Soto, M.-E. (1998). Circadian studies of autonomic nervous balance in patients with fibromyalgia: A heart rate variability analysis. *Arthritis and Rheumatism, 41,* 1966–1971.

McCracken, L. M., & Iverson, G. L. (2002). Disrupted sleep patterns and daily functioning in patients with chronic pain. *Pain Research & Management, 7,* 75–79.

Moldofsky, H. (2001). Sleep and pain: Clinical review. *Sleep Medicine Reviews, 5,* 387–398.

Morin, C. M., Gibson, D., & Wade, J. (1998). Self-reported sleep and mood disturbance in chronic pain patients. *The Clinical Journal of Pain, 14,* 311–314.

Nicassio, P. M., Moxham, E. G., Schuman, C. E., & Gervirtz, R. N. (2002). The contribution of pain, reported sleep quality, and depressive symptoms to fatigue in fibromyalgia. *Pain, 100,* 271–279.

Ohayon, M. M. (2005). Relationship between chronic painful physical condition and insomnia. *Journal of Psychiatric Research, 39,* 151–159.

Raymond, I., Nielsen, T. A., Lavigne, G. J., Manzini, C., & Choinière, M. (2001). Quality of sleep and its daily relationship to pain intensity in hospitalized adult burn patients. *Pain, 92,* 381–388.

Riley, J. L., III, Benson, M. B., Gremillion, H. A., Myers, C. D., Robinson, M. E., Smith, C. L., & Waxenberg, L. B. (2001). Sleep disturbances in orofacial pain patients: Pain-related or emotional distress? *The Journal of Craniomandibular Practice, 19,* 106–113.

Rizzi, M., Sarzi-Puttini, P., Atzeni, F., Capsoni, F., Andreoli, A., Pecis, M., Colombo, S., Carrabba, M., & Sergi, M. (2004). Cyclic alternating pattern: A new marker of sleep alteration in patients with fibromyalgia? *Journal of Rheumatology, 31*(6), 1193–1199.

Rompré, P. H., Daigle-Landry, D., Guitard, F., Montplaisir, J. Y., & Lavigne, G. J. (2007). Identification of a sleep bruxism subgroup with a higher risk of pain. *Journal of Dental Research, 86*(9), 837–842.

Sandrini, G., Milanov, I., Rossi, B., Murri, L., Alfonsi, E., Moglia, A., & Nappi, G. (2001). Effects of sleep on spinal nociceptive reflexes in humans. *Sleep, 24,* 13–17.

Smith, M. T., Edwards, R. R., McCann, U. D., & Haythornthwaite, J. A. (2007). The effects of sleep deprivation on pain inhibition and spontaneous pain in women. *Sleep, 30*(4), 494–505.

Smith, M. T., Wickwire, E. M., Grace, E. G., Edwards, R. R., Buenaver, L. F., Peterson, S., Klick, B., & Haythornthwaite, J. A. (2009). Sleep disorders and their association with laboratory pain sensitivity in temporomandibular joint disorder. *Sleep, 32*(6), 779–790.

Tang, N. Y., Wright, K. J., & Salkovskis, P. M. (2007). Prevalence and correlates of clinical insomnia co-occurring with chronic back pain. *Journal of Sleep Research, 16,* 85–95.

Van Dongen, H. P., Maislin, G., Mullington, J. M., & Dinges, D. F. (2003). The cumulative cost of additional wakefulness: Dose-response effects on neurobehavioral functions and sleep physiology from chronic sleep restriction and total sleep deprivation. *Sleep, 26,* 117–126.

Yatani, H., Studts, J., Cordova, M., et al. (2002). Comparison of sleep quality and clinical and psychologic characteristics in patients with temporomandibular disorders. *Journal of Orofacial Pain, 16,* 221–228.

Orofacial Pain: Taxonomy/Classification

Antoon F. C. De Laat
Department of Oral/Maxillofacial Surgery,
Catholic University of Leuven, Leuven, Belgium

Synonyms

Diagnostic subdivisions; Orofacial pain; Taxonomy

Definition

The development and establishment of widely accepted definitions and a classification system for the different orofacial pain conditions.

Characteristics

Composing and validating a classification is a continuously ongoing process, driven by new scientific data or better insights into pathophysiological processes. Ideally, a classification should be complete, and all the syndromes described should have clear exclusion and inclusion criteria. Unfortunately, this system of "complete truth" does not exist and will probably never be reached. Although this may sound frustrating, it provides, however, the necessary flexibility (also in regard of syndromes or definitions under debate) to convince as many users as possible to share a common language.

One of the first classification schemes for orofacial pain was suggested in 1962, by the Ad Hoc Committee on Classification of Headache of the National Institute of Neurological Diseases and Blindness (Friedman et al. 1962). It was based on clinical symptoms and is not widely used.

An extraordinary effort by the International Headache Society (IHS) resulted in the classical "classification and diagnostic criteria for headache disorders, cranial ▶ neuralgias, and facial pain" (1988). It was the first time that sets of clear inclusion criteria for the different diagnoses were put together (Olesen 1988). The focus, however, was on headache, and many of the common sources for orofacial pain were grouped under its section 11: Headache or facial pain associated with a disorder of ▶ cranium, neck, eyes, ears, nose, sinuses, teeth, mouth, or other facial or cranial structures. The second edition of this IHS classification (Olesen et al. 2004) is more clear regarding temporomandibular joint (TMJ) pain but still relates masticatory muscle pain (MMP) only as a possible initiating factor of tension-type headache (Sharav and Benoliel 2008).

In order to incorporate and clarify the different subgroups of ▶ temporomandibular disorders (TMD) in the IHS classification, the American Academy of Orofacial Pain (AAOP) provided specific inclusion and exclusion criteria (Okeson 1996). Even if clear descriptions of the different subgroups of TMD have been provided, the inclusion and exclusion criteria have been only partially operational. This is in contrast with the meticulous inclusion criteria for muscle sites and the separation of MMP into with or without limitation of opening in the Research diagnostic criteria (RDC). The RDC were developed by an expert committee focusing on a subset of the most common temporomandibular disorders (Dworkin and LeResche 1992). A double-axis system has been used: a somatic assessment based on primary signs and symptoms (pain and tenderness) allows classification into 3 groups (muscle problems, ▶ disc displacement, and joint disorders), while a second axis examines and scores the pain severity, the psychological status of the patient, and the related disability. The RDC-TMD allows for multiple diagnoses, has clear and testable inclusion and exclusion criteria, and has been operationally validated in several countries and languages. Recently, extensive research has been devoted to test its validity, specificity and sensitivity, and several suggestions for improvement were implemented (Ohrbach et al. 2010).

The task force on taxonomy of the IASP composed a second edition of the Classification of

Chronic Pain in 1994 (Merskey and Bogduk 1994). In the section of "Relatively Localized Syndromes of the Head and the Neck," each pain condition has been coded, defined, and specified regarding location, system involved, main clinical and technical features, and diagnostic criteria.

In 1996, the AAOP expanded the original set of subdiagnoses of TMD to all conditions that could be associated with orofacial pain (Merskey and Bogduk 1994). Seven subgroups have been distinguished:

1. Intracranial pain disorders comprise tumors, hemorrhage, abscess, hematoma, or edema. Due to their possible life-threatening character, they should always be ruled out expeditiously in the process of differential diagnosis.

2. The neurovascular disorders (primary headache disorders) include the variety of migraines, cluster headache, chronic paroxysmal hemicrania, the tension-type headaches, carotidynia, and also temporal arteritis.

3. Neurogenic pain disorders find their etiology within the nervous system itself: the neurogenic pain may be paroxysmal (as in the typical neuralgias of the trigeminal, glossopharyngeal, nervus intermedius, or superior laryngeal nerves) or continuous, comparable to deafferentation pain syndromes: peripheral neuritis, postherpetic, or posttraumatic neuralgias. The AAOP also includes in this category sympathetically maintained pain. The latter groups are characterized by their unremitting and mostly burning character.

4. Many tissues may be involved in intraoral pain: the ▶ dental pulp, the periodontium as well as the mucogingival tissues, and the tongue are sources for a variety of pain conditions.

5. The group of temporomandibular disorders has been subdivided in more detail by the AAOP into masticatory muscle disorders (myofascial pain, myositis, myospasm, unclassified local myalgia, myofibrotic contracture, and neoplasia) and articular disorders (congenital and developmental disorders, disc

derangement disorders, dislocation of the temporomandibular joint, inflammatory disorders, noninflammatory osteoarthritis, ankylosis, and fractures). Other musculoskeletal pain may be of cervical origin and referred to the orofacial area.

6. Associated structures such as the ears, eyes, nose and paranasal sinuses, the throat, salivary glands, and other soft tissues may be involved in pain felt in the orofacial region. Differential diagnosis is important here, and – as in many of the other categories – correct referral to the medical discipline mostly involved in the referring area is warranted.

7. In rare cases, mental disorders are the origin of orofacial pain, e.g., somatoform disorders and pain syndromes of psychogenic origin.

As for many other classification systems, a debate, also in the orofacial area, is still going on regarding pain syndromes without a clear definition or whose underlying pathophysiology is less well understood. "Atypical facial pain" was originally used as a "wastebasket" for those orofacial pain syndromes that did not correspond to any of the suggested and defined diagnoses. More recently, this term – like "atypical odontalgia" and others – has been associated with a neuropathy-like pain, with attention drawn to their similarities regarding female predominance, clinical manifestation (but in different orofacial tissues), and lack of clear underlying pathophysiology.

Both the AAOP and the workgroup aiming at expanding the RDC for TMD are currently working on clinical diagnostic criteria (CDC) for the complete orofacial pain field, preferably combining it with the IHS classification. This kind of discussion and "regrouping" would probably take place more often if the principles of a "mechanism-based" classification (Woolf et al. 1998) were ever implemented. In a mechanism-based classification, the subdivisions are not driven by anatomical location but merely by the underlying pathophysiological process. Grouping pain syndromes in this way could allow more general and mechanism-oriented treatment approaches and bridge the boundaries of specific medical disciplines. At present, however, too little scientific data are available in this regard.

References

Dworkin, S. F., & LeResche, L. (1992). Research diagnostic criteria for temporomandibular disorders. *Journal of Craniomandibular Disorders, 6*, 301–355.

Friedman, A. P., Finlay, K. H., Graham, J. R., et al. (1962). Classification of headache. Special report of the Ad Hoc committee. *Archives of Neurology, 6*, 173–176.

Merskey, H., & Bogduk, N. (1994). *Classification of chronic pain. Descriptions of chronic pain syndromes and definitions of pain terms* (pp. 59–92). Seattle: IASP Press.

Ohrbach, R., List, T., Goulet, J. P., & Svensson, P. (2010). Recommendations from the international consensus workshop: Convergence on an orofacial pain taxonomy. *Journal of Oral Rehabilitation, 37*(10), 807–812, Mar 29 (EPub ahead of print).

Okeson, J. P. (1996). *Orofacial pain: Guidelines for assessment, diagnosis and management* (pp. 45–52). Chicago: Quintessence.

Olesen, J. (1988). Classification and diagnostic criteria for headache disorders, cranial neuralgias and facial pain. *Cephalalgia, 8*, 1–96.

Olesen, J., Bousser, M.-G., Diener, H. C., et al. (2004). The International classification of headache disorders, 2nd edition. *Cephalalgia, 24*(Suppl 1), 24–50.

Sharav, Y., & Benoliel, R. (Eds.). (2008). *Orofacial pain and headache*. Edinburgh: Mosby Elsevier.

Woolf, C. J., Bennett, G. J., Doherty, M., et al. (1998). Towards a mechanism-based classification of pain? *Pain, 77*, 227–229.

Orphan G Protein-Coupled Receptor

Definition

Orphan G protein-coupled receptor is deduced from nucleic acid sequence to be a G protein-coupled receptor but of unknown ligand specificity.

Cross-References

▶ Orphanin FQ

Orphanin FQ

David K. Grandy
Department of Physiology and Pharmacology, Oregon Health and Science University, Portland, OR, USA

Synonyms

N/OFQ; Neuropeptide nociceptin 1; NOC; Nociceptin; Nocistatin; OFQ; OFQ/N; PNOC (Homo sapiens)

Definition

Orphanin FQ, discovered in 1995, is a naturally occurring, 17 amino acid long, kappa opioid-like peptide that is evolutionarily conserved among vertebrates. The name "orphanin" was chosen because the peptide, isolated from pig hypothalamic extracts, was found to be a potent agonist of an "orphan" opiate receptor-like G protein-coupled receptor (referred to as ORL1; NOP1; KOR3, LC132, MOR-C, OFQR, Oprl, XOR1; ROR-C). The "FQ" was added to reflect that this peptide's N-terminal amino acid is phenylalanine (written as an "F" in the single letter amino acid code), and its C-terminal amino acid is glutamine (Q). A peptide of identical sequence was simultaneously and independently discovered in rat brain extracts by another group of researchers, who named it "nociceptin" based on its pronociceptive characteristics in stressed mice.

Characteristics

The cloning of the mouse delta (δ) opiate receptor cDNA quickly led to the cloning of cDNAs and genes, from a variety of vertebrate species, that code for the mu (μ) opiate receptor, the kappa (κ) opiate receptor, and an opiate receptor-like ▶ orphan G protein-coupled receptor (Bunzow et al. 1994; Mollereau et al. 1994;

Darland et al. 1998; Mogil and Pasternak 2001) whose official human gene symbol is OPRL1. Based on the extensive amino acid sequence shared by OPRL1 and the 3 classic opiate receptor subtypes, it was predicted this orphan opiate-like receptor would be activated by a peptide ligand resembling the opioid peptides β-endorphin, Met/Leu enkephalin, and dynorphin. Furthermore, it was predicted that when activated this receptor would couple to second messenger systems similar, if not identical, to those modulated by the classic opiate receptors.

The search for an endogenous agonist of OPRL1 culminated with the co-discovery of a dynorphin-like heptadecapeptide (PheGlyGlyPheThrGlyAlaArgLysSerAlaArgLysLeuAlaAsnGln) from pig hypothalamic and rat brain extracts that was named ► orphanin FQ (Reinscheid et al. 1995) and ► nociceptin (Meunier et al. 1995), respectively. At the cellular level, synthetic OFQ/N inhibits cAMP production in vitro, activates inwardly rectifying potassium channels, and inhibits a variety of voltage-dependent calcium channels including the L, P, and N types (Darland et al. 1998) in tissue slices, activities shared by the classic opiate receptors. In the high picomolar to low nanomolar range, OFQ/N is a potent and selective agonist of OPRL1; however, at concentrations in the high micromolar range, the peptide can activate $\kappa > > \mu = \delta$ opiate receptors expressed heterologously in vitro (Zhang et al. 1997).

In spite of the many physical and physiological similarities OFQ/N and its receptor share with the opioid peptides and their cognate receptors, respectively, the pair differ notably in that their interaction is insensitive to naloxone antagonism. Not surprisingly, this remains the principal reason OFQ/N and its receptor are still not generally accepted as bona fide members of the opioid peptide/opiate receptor family. Several relatively selective peptide and small-molecule receptor antagonists have been developed (Mogil and Pasternak 2001; Fioravanti and

Vanderah 2008) and are being used in efforts to clarify the functions of OPRL1-mediated actions of OFQ/N in vivo.

Efforts to determine the biosynthetic origin of OFQ/N led to the identification and characterization of its precursor polypeptide prepro-orphanin FQ/nociceptin (PNOC [Homo sapiens gene symbol]; ppOFQ/N; N23K, N23K/N27K; Npnc1; proorphanin; prepronociceptin; proprononociceptin; Mollereau et al. 1996; Nothacker et al. 1996). In this precursor polypeptide, the OFQ/N sequence is flanked by paired basic amino acids that are recognized and cleaved by trypsin-like endopeptidases. Further processing of OFQ/N by a carboxypeptidase may also be important in vivo, since OFQ/N-derived peptides truncated by as many as five amino acid residues have been found to retain agonist activity, at least at OPRL1 expressed in vitro (Butour et al. 1997; Reinscheid et al. 1996).

Analysis of amino acid sequences deduced from *mus-*, *rattus-*, and *homo-*derived cDNAs reveals ppOFQ/N to be evolutionarily conserved across species, as is its receptor. Furthermore, the presence of numerous putative peptide domains flanked by endopeptidase cleavage recognition sites distributed throughout ppOFQ/N and the organization of its gene recapitulate organizational schemes also present in the three opioid peptide precursors: proopiomelanocortin, preproenkephalin, and preprodynorphin (Darland et al. 1998). Taken together, these features led to the prediction and eventual demonstration that ppOFQ/N codes for additional bioactive peptides, a result that suggests caution when interpreting outcomes of genetic manipulations that affect expression of the entire precursor.

Nucleic acid and antibody probes specific for ppOFQ/N and OPRL1 have been used to demonstrate the widespread, albeit unequal, distribution of the peptide and its receptor throughout the central nervous system and periphery (Darland et al. 1998; Mogil and Pasternak 2001). Of particular note in the context of pain and nociception is the significant presence of both peptide and receptor in the periaqueductal gray (PAG), raphe nuclei, and the spinal cord (Darland et al. 1998; Mogil and Pasternak 2001).

As with the opioid peptides, early attempts to study the actions of OFQ/N in vivo were plagued by its sensitivity to proteolysis, membrane impermeability, and failure to cross the blood–brain barrier. Until recently, interpretation of in vivo studies was complicated due to the lack of selective, small-molecule antagonists of the OPRL1. Despite these limitations, the actions of OFQ/N in rodents and primates have been documented.

When OFQ/N is administered intracerebro-ventricularly (icv), the most common observation is that it dose-dependently interferes with several opioid-mediated, naloxone-sensitive responses including antinociception. The literature also contains reports of the peptide producing hyperalgesia as well as analgesia (extensively reviewed in Mogil and Pasternak 2001). In con-trast to its apparent anti-opioid actions supraspinally, when administered intrathecally OFQ/N potentiates opioid analgesia, although the gender and physiological status of the subject (e.g., pregnancy) appear to influence the peptide's effects on nociceptive sensitivity (Dawson-Basoa and Gintzler 1997). To date, there are few reports describing OFQ/N's effects on nociceptive processing when administered directly into brain tissue. Microinjection of OFQ/N into the midbrain PAG region of rats produces a mild hyperalgesia (Bytner et al. 2001) while blocking the antinociceptive effects of intra-PAG administered morphine (Morgan et al. 1997). In contrast, the focal application of OFQ/N into the rostral ventral medulla produced either no effect (two studies) or analgesia (one study) (Heinricher 2003).

Although the literature contains numerous contradictory reports regarding the effects of exogenous OFQ/N on various nociceptive behav-iors, the consensus emerging from recent efforts to integrate these data is that the dynorphin-like peptide OFQ/N is not "anti-opioid" or "pronociceptive." Rather, OFQ/N is being viewed as more of a modulator of opioid peptide/opiate alkaloid actions, exerting a potent, inhibitory effect on a wide range of cell types involved with nociceptive processing in the brain and spinal cord via the naloxone-insensitive activation of its cognate receptor. Therefore, the behavioral manifestations of exog-enous OFQ/N are a reflection of the neuronal circuits present along the route of administration, as well as the species, strain, gender, and physi-ological status of the experimental subject (Harrison and Grandy 2000). What role, if any, endogenous OFQ/N plays in nociceptive processing remains to be established.

References

Bunzow, J. R., Saez, C., Mortrud, M., Bouvier, C., Williams, J. T., Low, M., & Grandy, D. K. (1994). Molecular cloning and tissue distribution of a putative member of the rat opioid receptor gene family that is not a mu, delta or kappa opioid receptor. *FEBS Letters, 347*, 284–288.

Butour, J. L., Moisand, C., Mazarguil, H., et al. (1997). Recognition and activation of the opioid receptor-like ORL 1 receptor by nociceptin, nociceptin analogs and opioids. *European Journal of Pharmacology, 321*, 97–103.

Bytner, B., Huang, Y. H., Yu, L. C., et al. (2001). Nociceptin/orphanin FQ into the rat periaqueductal gray decreases the withdrawal latency to heat and loading, an effect reversed by (Nphe(1))nociceptin (1–13)NH(2). *Brain Research, 922*, 118–124.

Darland, T., Heinricher, M. M., & Grandy, D. K. (1998). Orphanin FQ/nociceptin: A role in pain and analgesia, but so much more. *Trends in Neurosciences, 21*, 215–221.

Dawson-Basoa, M., & Gintzler, A. R. (1997). Nociceptin (orphanin FQ) abolishes gestational and ovarian sex steroid-induced antinociception and induces hyperalgesia. *Brain Research, 750*, 48–52.

Fioravanti, B., & Vanderah, T. W. (2008). The ORL-1 receptor system: Are there opportunities for antago-nists in pain therapy? *Current Topics in Medicinal Chemistry, 8*, 1442–1451.

Harrison, L. M., & Grandy, D. K. (2000). Opiate modu-lating properties of nociceptin/orphanin FQ. *Peptides, 21*, 151–172.

Heinricher, M. M. (2003). Orphanin FQ/nociceptin: From neural circuitry to behavior. *Life Sciences, 73*, 813–822.

Meunier, J. C., Mollereau, C., Toll, L., et al. (1995). Isolation and structure of the endogenous agonist of opioid receptor-like ORL1 receptor. *Nature, 377,* 532–535.

Mogil, J. S., & Pasternak, G. W. (2001). The molecular and behavioral pharmacology of the orphanin FQ/ nociceptin peptide and receptor family. *Pharmacological Reviews, 53,* 381–415.

Mollereau, C., Parmentier, M., Mailleux, P., Butour, J. L., Moisand, C., Chalon, P., Caput, D., Vassart, G., & Meunier, J. C. (1994). ORL1, A novel member of the opioid receptor family. Cloning, functional expression and localization. *FEBS Letters, 341,* 33–38.

Mollereau, C., Simons, M. J., Soularue, P., et al. (1996). Structure, tissue distribution, and chromosomal localization of the prepronociceptin gene. *Proceedings of the National Academy of Sciences of the United States of America, 93,* 8666–8670.

Morgan, M. M., Grisel, J. E., Robbins, C. S., et al. (1997). Antinociception mediated by the periaqueductal gray is attenuated by orphanin FQ. *Neuroreport, 8,* 3431–3434.

Nothacker, H. P., Reinschei, R. K., Mansour, A., et al. (1996). Primary structure and tissue distribution of the orphanin FQ precursor. *Proceedings of the National Academy of Sciences of the United States of America, 93,* 8677–8682.

Reinscheid, R. K., Nothacker, H. P., Bourson, A., et al. (1995). Orphanin FQ: A neuropeptide that activates an opioid-like G protein-coupled receptor. *Science, 270,* 792–794.

Reinscheid, R. K., Ardati, A., Monsma, F. J., Jr., et al. (1996). Structure-activity relationship studies on the novel neuropeptide orphanin FQ. *The Journal of Biological Chemistry, 271,* 14163–14168.

Zhang, G., Murray, T. F., & Grandy, D. K. (1997). Orphanin FQ has an inhibitory effect on the guinea pig ileum and the mouse vas deferens. *Brain Research, 772,* 102–106.

Orthodromic Activation

Definition

Electrical activity due to propagation of action potentials in the same direction to that observed when the neuron is naturally excited.

Cross-References

▶ Corticothalamic and Thalamocortical Interactions

Orthologues

Definition

A paralogue is a homologous member of a multigene family that may share some functional characteristics with the query sequence, while an orthologue is the closest relative of the query sequence in a different species and is likely to share most or all functional characteristics with the protein encoded by the query sequence.

Cross-References

▶ TRPV1 Receptor, Species Variability

Orthostatic Component of Pain

Definition

Change of posture or "intrabody" pressure (coughing, sneezing, straining) can influence headaches due to low intracranial pressure.

Cross-References

▶ Headache Due to Low Cerebrospinal Fluid Pressure

Orthostatic Hypotension

Definition

Orthostatic hypotension is a drop in blood pressure when standing up from a lying/sitting position.

Cross-References

▶ Cancer Pain Management, Anesthesiologic Interventions, and Neural Blockade

Oscillations (Neuronal)

Definition

Oscillations are spike potentials that occur repeatedly at relatively fixed intervals.

Cross-References

▶ Thalamus, Dynamics of Nociception

Oscillatory Mobilization

▶ Passive Spinal Mobilization

Osmolysis

Definition

Osmolysis is the rupture of a cell membrane due to excessive accumulation of solvent (water), produced by a decrease in concentration of ions inside the cell (hypo-osmolarity).

Cross-References

▶ Poststroke Pain Model, Thalamic Pain (Lesion)

Osmosensitive

Definition

Sensitive to changes in the osmolarity of the surrounding medium.

Cross-References

▶ Mechanonociceptors

Osmotic

Definition

A dissolved chemical that acts by causing an osmotic shock to cells, leading to the release of proinflammatory substances. Osmotics include concentrated solutions of glucose, glycerine, and zinc sulphate.

Cross-References

▶ Pain Management
▶ Prolotherapy

Osteoarthritic Pain

Definition

Joint pain associated with osteoarthritis. Osteoarthritic pain involves peripheral and central sensitization, and a neuropathic compound is being discussed.

Cross-References

▶ Perireceptor Elements

Osteoarthritis

Definition

Osteoarthritis is a usually painful joint disease with the hallmark of breakdown of the articular cartilage (the cartilage on the ends of bones which assists the smooth movement of joints). Joints may become deformed and stiff. In most of cases, joints exhibit synovitis, and sometimes bone marrow edema is observed.

Cross-References

▶ Arthritis Model, Osteoarthritis
▶ NSAIDs and Their Indications

Osteoarthritis Model

▶ Arthritis Model, Osteoarthritis

Osteoblast

Definition

Osteoblasts are specialized cells that lay down osseous matrix for bone formation.

Cross-References

▶ Adjuvant Analgesics in Management of Cancer-Related Bone Pain
▶ NSAIDs, Adverse Effects

Osteoblastoma

Definition

Osteoblastoma is a rare, benign, locally recurrent tumor of the bone with a predilection for the spine.

Cross-References

▶ Chronic Back Pain and Spinal Instability

Osteoclast

Definition

A member of the macrophage-monocyte lineage of cells that is responsible for the absorption and removal of bone. Resorption is mediated by demineralization of bone by protons secreted by the osteoclast. Enhanced osteoclast activity can cause osteoporosis.

Cross-References

▶ Adjuvant Analgesics in Management of Cancer-Related Bone Pain
▶ Cancer Pain, Animal Models
▶ NSAIDs, Adverse Effects

Osteolysis

Definition

The process of bone resorption that occurs during normal bone homeostasis but may become pathologic in the presence of cancer cells.

Cross-References

▶ Cancer Pain, Animal Models

Osteolytic Fibrosarcoma

Definition

Osteolytic fibrosarcoma is a tumor of mesenchymal cell origin that may present as an osteolytic bone tumor. The primary tumor is comprised of malignant fibroblasts. NCTC 2472 sarcoma cells that produce osteolytic fibrosarcoma have been used to model cancer pain in mice. The clinical presentation of osteolytic fibrosarcoma features breakthrough pain as a prominent component.

Cross-References

▶ Evoked and Movement-Related Neuropathic Pain

Osteopathy

▶ Spinal Manipulation, Characteristics

Osteophyte

Definition

A bony excrescence or osseous outgrowth. It is a typical feature of advanced osteoarthritis.

Cross-References

▶ Lumbar Traction

Oswestry Disability Index

Ole Haegg
Department of Orthopedic Surgery, Sahlgren
University Hospital, Goteborg, Sweden

Synonyms

ODI; Oswestry disability questionnaire; Oswestry low back pain disability questionnaire

Definition

The Oswestry Disability Index (ODI) is a self-rating condition-specific outcome measure for evaluation of low back pain ▶ disability. It was introduced in 1980 by Fairbank et al. (1980), and consists of ten sections with six response alternatives describing functional impairment in a series of daily activities. There are two authorized versions, version 1.0 (Fig. 1) from 1980 and version 2.0 (Fig. 2) from 1989 (Pynsent et al. 1993). The main difference between the versions is the construction of the section describing pain intensity.

Scoring

For each section of six statements the total score is 5; if the first statement is marked the score = 0; if the last statement is marked the score = 5. Intervening statements are scored according to rank. If more than one box is marked in each section, take the highest score. If all ten sections are completed the score is calculated as follows: Example: 16 (total score) of 50 (total possible score) × 100 = 32 %. If one section is missed (or not applicable) the score is calculated as follows: Example: 16 (total score)/45 (total possible score) × 100 = 35.6 %. The final score may be summarized as: (total score/ (5 × number of questions answered (×100 %))). It is suggested that the total percentage is rounded to a whole number (Fairbank and Pynsent 2000). A low score = low degree of disability, a high score = high degree of disability.

Versions in Non-English Languages

The ODI is available in five validated translations (Finnish, Greek, Norwegian, Spanish, and Japanese) and another five non-validated translations (Dutch, French, German, Danish, and Swedish).

Characteristics

Validity

The ODI has been criticized for not being properly validated according to modern standards. This predicament is shared with many other outcome measures consisting of more than pure physical function, where external criteria by performance tests are available. There are no external criteria for benchmarking of outcome measures including more complex social/ private activities such as personal care, sex life, or social life. Validation of this type of measure is accomplished by assessing the behavior of the outcome measure in conditions with generally accepted/established degrees of disability.

The ODI score was significantly associated with physical performance tests (Gronblad et al. 1993; Reneman et al. 2002). Radiculopathy patients had significantly greater scores than

SECTION 1 - PAIN INTENSITY
☐ I can tolerate the pain I have without having to use painkillers.
☐ The pain is bad but I manage without taking painkillers.
☐ Painkillers give complete relief from pain.
☐ Painkillers give moderate relief from pain.
☐ Painkillers give very little relief from pain.
☐ Painkillers have no effect on the pain and I do not use them.

SECTION 2 - PERSONAL CARE (washing, dressing etc.)
☐ I can look after myself normally, without causing extra pain.
☐ I can look after myself normally, but it causes extra pain.
☐ It is painful to look after myself and I am slow and careful.
☐ I need some help, but manage most of my personal care.
☐ I need help every day in most aspects of self-care.
☐ I do not get dressed, wash with difficulty and stay in bed.

SECTION 3 - LIFTING
☐ I can lift heavy weights without extra pain.
☐ I can lift heavy weights, but it gives extra pain.
☐ Pain prevents me from lifting heavy weights off the floor, but I can manage if they are conveniently positioned (e.g., on a table).
☐ Pain prevents me from lifting heavy weights but I can manage light to medium weights if they are conveniently positioned.
☐ I can lift only very light weights.
☐ I cannot lift or carry anything at all.

SECTION 4 - WALKING
☐ Pain does not prevent my walking any distance.
☐ Pain prevents me walking more than 1 mile.
☐ Pain prevents me walking more than ½ of mile.
☐ Pain prevents me walking more than ¼ mile.
☐ I can only walk using a stick or crutches.
☐ I am in bed most of the time and have to crawl to the toilet.

SECTION 5 - SITTING
☐ I can sit in any chair as long as I like.
☐ I can sit in my favourite chair as long as I like.
☐ Pain prevents me sitting more than 1 hour.
☐ Pain prevents me from sitting more than ½ an hour.
☐ Pain prevents me from sitting more than 10 minutes.
☐ Pain prevents me from sitting at all.

SECTION 6 - STANDING
☐ I can stand as long as I want without extra pain.
☐ I can stand as long as I want but it gives me extra pain.
☐ Pain prevents me from standing for more than 1 hour.
☐ Pain prevents me from standing for more than 30 minutes.
☐ Pain prevents me from standing for more than 10 minutes.
☐ Pain prevents me from standing at all.

SECTION 7 - SLEEPING
☐ Pain does not prevent me from sleeping well.
☐ I can sleep well only by using tablets.
☐ Even when I take tablets, I have less than 6 hours sleep.
☐ Even when I take tablets, I have less than 4 hours sleep.
☐ Even when I take tablets, I have less than 2 hours sleep.
☐ Pain prevents me from sleeping at all.

SECTION 8 - SEX LIFE (If applicable)
☐ My sex life is normal and causes no extra pain.
☐ My sex life is normal but causes some extra pain.
☐ My sex life is nearly normal but is very painful.
☐ My sex life is severely restricted by pain.
☐ My sex life is nearly absent because of pain.
☐ Pain prevents any sex life at all.

SECTION 9 - SOCIAL LIFE
☐ My social life is normal and gives me no extra pain.
☐ My social life is normal, but increases the degree of pain.
☐ Pain has no significant effect on my social life apart from limiting my more energetic interests, e.g., dancing, etc.
☐ Pain has restricted my social life and I do not go out as often.
☐ Pain has restricted my social life to my home.
☐ I have no social life because of pain.

SECTION 10 - TRAVELLING
☐ I can travel anywhere without extra pain.
☐ I can travel anywhere but it gives extra pain.
☐ Pain is bad but I manage journeys over 2 hours.
☐ Pain restricts me to journeys of less than 1 hour.
☐ Pain restricts me to short necessary journeys under 30 minutes.
☐ Pain prevents travel except to the doctor or hospital.

Oswestry Disability Index, Fig. 1 Oswestry Disability Index Version 1.0; Patient name: File...; Date:... This questionnaire has been designed to give the doctor information as to how your back pain has affected your ability to manage in everyday life. Please answer every section and mark in each section only the ONE box that applies to you. We realize you may consider that two of the statements in any one section relate to you, but please just mark the box, which most closely describes your problem

patients with low back pain (49 vs. 33) (Leclaire et al. 1997). "Normal" individuals scored lower (10) than chronic back pain (43), neurogenic claudication (37), and sciatica (45) (Fairbank and Pynsent 2000).

Reliability
Test-retest ▶ reliability assessed with the Intraclass Correlation Coefficient (ICC) was 0.89 (1 week, n = 32) (Pratt et al. 2002), 0.94 (2 weeks, n = 37) (Holm et al. 2003), 0.84/0.92 (6 weeks, n = 47/16) (Davidson and Keating 2002), 0.83 (1 week, n = 20) (Gronblad et al. 1993).

Responsiveness
Score changes associated with different levels of patient perceived improvement/deterioration are summarized in Table 1.

SECTION 1 - PAIN INTENSITY

☐ I have no pain at the moment.
☐ The pain is very mild at the moment.
☐ The pain is moderate at the moment.
☐ The pain is fairly severe at the moment.
☐ The pain is very severe at the moment.
☐ The pain is the worst imaginable at the moment.

SECTION 2 - PERSONAL CARE (washing, dressing etc.)

☐ I can look after myself normally, without causing extra pain.
☐ I can look after myself normally, but it is very painful.
☐ It is painful to look after myself and I am slow and careful.
☐ I need some help, but manage most of my personal care.
☐ I need help every day in most aspects of self-care.
☐ I do not get dressed, wash with difficulty and stay in bed.

SECTION 3 - LIFTING

☐ I can lift heavy weights without extra pain.
☐ I can lift heavy weights, but it gives extra pain.
☐ Pain prevents me from lifting heavy weights off the floor, but I can manage if they are conveniently positioned (e.g., on a table).
☐ Pain prevents me from lifting heavy weights but I can manage light to medium weights if they are conveniently positioned.
☐ I can lift only very light weights.
☐ I cannot lift or carry anything at all.

SECTION 4 - WALKING

☐ Pain does not prevent me walking any distance.
☐ Pain prevents me walking more than 1 mile.
☐ Pain prevents me walking more than ½ of mile.
☐ Pain prevents me walking more than 100 yards.
☐ I can only walk using a stick or crutches.
☐ I am in bed most of the time and have to crawl to the toilet.

SECTION 5 - SITTING

☐ I can sit in any chair as long as I like.
☐ I can sit in my favourite chair as long as I like.
☐ Pain prevents me from sitting for more than 1 hour.
☐ Pain prevents me from sitting more than ½ an hour.
☐ Pain prevents me from sitting more than 10 minutes.
☐ Pain prevents me from sitting at all.

SECTION 6 - STANDING

☐ I can stand as long as I want without extra pain.
☐ I can stand as long as I want but it gives me extra pain.
☐ Pain prevents me from standing for more than 1 hour.
☐ Pain prevents me from standing for more than ½ an hour.
☐ Pain prevents me from standing for more than 10 minutes.
☐ Pain prevents me from standing at all.

SECTION 7 - SLEEPING

☐ My sleep is never disturbed by pain.
☐ My sleep is occasionally disturbed by pain.
☐ Because of pain, I have less than 6 hours of sleep.
☐ Because of pain, I have less than 4 hours of sleep.
☐ Because of pain, I have less than 2 hours of sleep.
☐ Pain prevents me from sleeping at all.

SECTION 8 - SEX LIFE (if applicable)

☐ My sex life is normal and causes no extra pain.
☐ My sex life is normal but causes some extra pain.
☐ My sex life is nearly normal but is very painful.
☐ My sex life is severely restricted by pain.
☐ My sex life is nearly absent because of pain.
☐ Pain prevents any sex life at all.

SECTION 9 - SOCIAL LIFE

☐ My social life is normal and causes me no extra pain.
☐ My social life is normal, but increases the degree of pain.
☐ Pain has no significant effect on my social life apart from limiting my more energetic interests, e.g., sports, etc.
☐ Pain has restricted my social life and I do not go out as often.
☐ Pain has restricted my social life to my home.
☐ I have no social life because of pain.

SECTION 10-TRAVELLING

☐ I can travel anywhere without pain.
☐ I can travel anywhere but it gives extra pain.
☐ Pain is bad but I manage journeys over 2 hours.
☐ Pain restricts me to journeys of less than 1 hour.
☐ Pain restricts me to short necessary journeys under 30 minutes.
☐ Pain prevents me from travelling except to receive treatment.

Oswestry Disability Index, Fig. 2 Oswestry Disability Index 2.0; Patient name: File...; Date:... Please read instructions: Could you please complete this questionnaire? It is designed to give us information as to how your back (or leg) trouble has affected your ability to manage in everyday life. Please answer every section. Mark one box only in each section that most closely describes you today

Oswestry Disability Index, Table 1 Score changes related to different levels of patient (or combined patient/clinician) perceived treatment effect

Level of perceived effect	Reference study			
	Beurskens et al. (1996) (n = 81)	Davidson and Keating (2002) (n = 101)	Taylor et al. (1999) (n = 225)	Hagg et al. (2002) (n = 270)
Completely gone	n/a	33	n/a	n/a
Much better	n/a	16	n/a	28
Better/improved	12	9	18	10
Unchanged/non-improved	0	0	2	0
Worse	n/a	−6	−9	−6
Much worse	n/a	−4	n/a	n/a

n/a not applicable

Effect Size

Patient (or combined patient/clinician) perceived improvement was associated with effect size of 0.82–0.87 (n = 970) (Walsh et al. 2003), 0.80 (n = 81) (Beurskens et al. 1996), 0.52 (n = 106) (Davidson and Keating 2002), 0.3–0.8 (n = 318) (Taylor et al. 1999), and 0.86/1.87 (n = 75/69) (Hagg et al. 2002).

Receiver Operating Characteristics (ROC)

The area below the ROC curve, as a measure of the ability to discriminate between subjects with and without back pain, was 0.76 (n = 76) (Leclaire et al. 1997), 0.78 (n = 99) (Davidson and Keating 2002), 0.78 (n = 88) (Stratford et al. 1994), and 0.72 (n = 970) (Walsh et al. 2003).

Minimal Clinically Important Difference (MCID)

Based on patient (or combined patient/clinician) perceived effect of treatment. For improvement the MCID was 10 (n = 154) and for deterioration −6 (n = 123) (Hagg et al. 2003).

Error of Measurement

The ► error of measurement at repeated measurements was, with 95 % tolerance interval, 10 (n = 289) (Hagg et al. 2003), and with 90 % tolerance interval 11/15 (n = 16/47) (Davidson and Keating 2002).

References

Beurskens, A. J., de Vet, H. C., & Koke, A. J. (1996). Responsiveness of functional status in low back pain: A comparison of different instruments. *Pain, 65*, 71–76.

Davidson, M., & Keating, J. L. (2002). A comparison of five low back disability questionnaires: Reliability and responsiveness. *Physical Therapy, 82*, 8–24.

Fairbank, J. C., & Pynsent, P. B. (2000). The Oswestry Disability Index. *Spine, 25*, 2940–2952.

Fairbank, J. C., Couper, J., Davies, J. B., et al. (1980). The Oswestry low back pain disability questionnaire. *Physiotherapy, 66*, 271–273.

Gronblad, M., Hupli, M., Wennerstrand, P., et al. (1993). Intercorrelation and test-retest reliability of the Pain Disability Index (PDI) and the Oswestry Disability Questionnaire (ODQ) and their correlation with pain intensity in low back pain patients. *The Clinical Journal of Pain, 9*, 189–195.

Hagg, O., Fritzell, P., Oden, A., et al. (2002). Simplifying outcome measurement: Evaluation of instruments for measuring outcome after fusion surgery for chronic low back pain. *Spine, 27*, 1213–1222.

Hagg, O., Fritzell, P., & Nordwall, A. (2003). The clinical importance of changes in outcome scores after treatment for chronic low back pain. *European Spine Journal, 12*, 12–21.

Holm, I., Friis, A., Storheim, K., et al. (2003). Measuring self-reported functional status and pain in patients with chronic low back pain by postal questionnaires: A reliability study. *Spine, 28*, 828–833.

Leclaire, R., Blier, F., Fortin, L., et al. (1997). A cross-sectional study comparing the Oswestry and Roland-Morris Functional Disability Scales in two populations of patients with low back pain of different levels of severity. *Spine, 22*, 68–71.

Pratt, R. K., Fairbank, J. C., & Virr, A. (2002). The reliability of the shuttle walking test, the Swiss Spinal Stenosis Questionnaire, the Oxford Spinal Stenosis Score, and the Oswestry Disability Index in the assessment of patients with Lumbar Spinal Stenosis. *Spine, 27*, 84–91.

Pynsent, P., Fairbank, J., & Carr, A. (1993). *Outcome measures in orthopaedics*. Oxford: Butterworth-Heinemann.

Reneman, M. F., Jorritsma, W., Schellekens, J. M., et al. (2002). Concurrent validity of questionnaire and performance-based disability measurements in patients with chronic nonspecific low back pain. *Journal of Occupational Rehabilitation, 12*, 119–129.

Stratford, P. W., Binkley, J., Solomon, P., et al. (1994). Assessing change over time in patients with low back pain. *Physical Therapy, 74*, 528–533.

Taylor, S. J., Taylor, A. E., Foy, M. A., et al. (1999). Responsiveness of common outcome measures for patients with low back pain. *Spine, 24*, 1805–1812.

Walsh, T. L., Hanscom, B., Lurie, J. D., et al. (2003). Is a condition-specific instrument for patients with low back pain/leg symptoms really necessary? The responsiveness of the Oswestry Disability Index, MODEMS, and the SF-36. *Spine, 28*, 607–615.

Oswestry Disability Questionnaire

► Oswestry Disability Index

Oswestry Low Back Pain Disability Questionnaire

► Oswestry Disability Index

Other Cranial Nerve Neuralgias: Neuralgia of Cranial Nerve IX With or Without Cranial Nerve X

▶ Trigeminal and Cranial Nerve Neuralgias

Other Types of Nerve Injury

▶ Retrograde Cellular Changes After Nerve Injury

Outcome Expectancies

▶ Psychology of Pain, Self-efficacy

Outcome Measures

Victor Wilk
Brighton Spinal Group, Brighton, VIC, Australia

Synonyms

Measuring tools; Outcome questionnaires; Outcomes assessment tools; Pain questionnaires

Definition

An outcome measure is an aspect of a patient's health status that can be quantified in order to assess the benefits of a treatment. Domains that are typically assessed include physical functioning, strength, range of movement, endurance, aerobic capacity, pain, disability, medication usage, work status, work capacity, and quality of life. The term "outcome measure" is increasingly being used to refer to self-administered questionnaires, which seek to measure the change in status of the patient over time.

Characteristics

Principles

Self-administered questionnaires have three broad purposes. They can discriminate between subjects; predict outcome; or evaluate change over time. The outcome measure should be reliable, valid, responsive to change, simple to administer, quick to fill out (less than 5 min), quick and easy to score (less than 1 min), and low in cost. In addition to the technical and scientific issues regarding the collating of outcome measures, there are also ethical and moral issues involved in the collection of data, interpretation of data, and sharing of data with others. The main domains of assessment commonly measured are the following:

- Physical assessment by clinician
- Functional capacity, aerobic capacity, work capacity
- Pain
- Disability
- Quality of life
- Cognitive impairment
- Psychological measures such as depression, anxiety, fear, coping ability, and various personality traits
- Utilization of services such as medical visits, hospitalization, drugs, allied health visits, and home help
- Employment status

Applications

Traditional physical assessment, including noting the range of motion of the spine or joints, does not often correlate well with the level of pain and function. Pain, disability, and quality-of-life measures assessed by self-administered questionnaires are best at predicting meaningful outcomes such as return to work.

Functional capacity is used increasingly in assessing fitness to return to work, and may involve measurement of aerobic capacity, standing and sitting tolerances, walking distance, walking speed, bending, and lifting.

Measuring Pain
The three dimensions of pain that may be assessed are pain intensity, pain quality (including affect),

and pain location. Pain intensity may be simply
defined as "how much a person hurts." Pain quality
refers to the type of pain (dull vs. sharp, etc.) while
affect is more complex and relates to "the emotional feeling of pain." Pain location simply relates
to the site and spread of the pain.

Pain Intensity

Pain intensity is most commonly measured by
rating scale methods including the verbal rating
scales (Likert scale), numerical rating scales
(NRS), the visual analogue scale (VAS), and the
more recently devised visual analogue thermometer (VAT).

Verbal rating scales use descriptors such as
none, slight, mild, moderate, severe, strong,
intense, and extremely intense.

Numerical rating scales (NRS) involve rating
pain on a scale of 0–10 (or 0–100) with no pain
equal to 0 and the worst pain imaginable equal to
10 (or 100). The 0–10 numerical scale is
recommended for its ease of use and patient
compliance.

The visual analogue scale (VAS) consists of
a 10-cm line drawn on the page or administered
by a slide rule, where the subject is asked to rate
their pain from nil to worst imaginable (Fig. 1).

Visual analogue scales have the advantage of
the subject not being able to recall their previous
response, but are more difficult to score, as the
researcher needs to measure the mark on the line
in millimeters each time; and they cannot be
administered over the telephone like the NRS
(Jensen and Karoly 1992).

Faces scales employ a series of caricatures
(faces), expressing different degrees of happiness
and/or distress and are used mainly with children,
and have acceptable levels of correlation with
other scales.

Pain Affect

Pain affect has been assessed using the ▶ McGill
Pain Questionnaire (MPQ) (Melzack 1975) and
others, where patients are asked to choose words
from a list to describe their pain, e.g. dull, sharp,
and stabbing. The use of emotive terms, such as
agonizing or cruel, signify psychological distress
associated with severe pain.

No Pain |————————————————————| Worst Pain
                                                        Imaginable

Outcome Measures, Fig. 1 An example of a visual
analogue scale for pain

Pain Location

Pain location is best assessed using pain
drawings, where the patient is asked to mark
their areas of discomfort. Recent studies have
demonstrated some discriminating properties
(Ohnmeiss et al. 2000) (Fig. 2).

Measures of Disability

Many disability scales are currently in use. The
most common include the ▶ Oswestry Disability
Questionnaire (ODQ), the Roland Morris Questionnaire (RMQ) for assessing low back pain, and
the Neck Disability Index (NDI) for assessing
neck problems. The ODQ has been the most
studied questionnaire, having also been compared to a number of other measures, including
the Pain Disability Index (PDI), the McGill Pain
Questionnaire, and the Aberdeen Back Disability
Questionnaire.

*The Oswestry Low Back Pain Disability Ques-
tionnaire (ODQ)* measures the extent to which
a person's daily activities are restricted by back
or leg pain (Fairbank et al. 1980). The first section
rates the intensity of pain, and the remaining
sections rate difficulties with personal care,
lifting, walking, sitting, standing, sleeping, sex
life, social life, and traveling.

The Roland Morris Questionnaire (RMQ) was
derived from the Sickness Impact Profile (SIP),
reducing the items down from 136 to a more
practical 24 (Roland and Morris 1983). The
scale consists of yes/no responses to 24 items.

The Neck Disability Index (NDI) was first
published in 1991 and was largely based on the
Oswestry Disability Questionnaire (ODQ) for
low back pain. Likewise, the Northwick Park
Questionnaire is based on the Oswestry questionnaire with nine questions using a 5-point scale, as
well as a transitional question on the patients'
global assessment.

Two questionnaires commonly in use to assess
arthritis are the Lequesne (for osteoarthritis of hip

Outcome Measures,
Fig. 2 An example of
a pain drawing

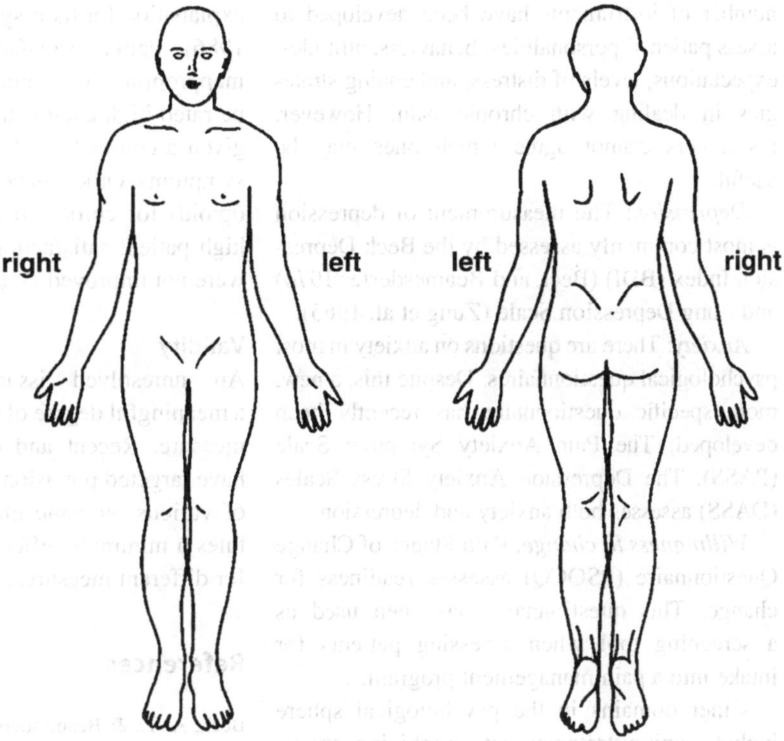

and knee) and the Western Ontario and McMaster
Universities Osteoarthritis Index (WOMAC).

Function-specific questionnaires ask the
patient what particular activities they are limited
in and then to rate that limitation over time. They
are not useful in giving an overall measure of
disability but are very useful in assessing the
usefulness of a treatment over time for
a particular individual. They are reliable and
valid for that subject and that condition, but it is
difficult to compare the results with other subjects
and other populations.

Quality of Life

The best known of these are the Short Form 36
Questionnaire (SF-36) and the Sickness Impact
Profile (SIP).

The SF-36 includes one multi-item scale that
assesses eight health concepts: (1) physical activity limitation, (2) social activity limitation,
(3) usual role activities limitation due to physical
problems, (4) bodily pain. (5) general mental
health, (6) role activities limitation due to emotional problems, (7) vitality, and (8) general

health perceptions (McHornery et al. 1993). The
SF-36 was developed for a normal population and
scored between 0 and 100, the higher the score
the better the health. A cut down version of
12 items has been developed called the SF-12.

The Sickness Impact Profile (SIP) underwent
a long and rigorous development phase and was
one of the first global health measures. However,
as it contains 136 items covering 12 domains of
health, it is not really practical, except in
a clinical trial setting, and thus has not been
used much recently. All of the items utilize binary
scales with yes/no responses which contribute to
its length.

One questionnaire measuring global health
status for a specific condition is the Fibromyalgia
Impact Questionnaire (FIQ). It is a brief 10-item,
self-administered instrument, measuring physical
functioning, work status, depression, anxiety,
sleep, pain, stiffness, fatigue, and well-being.

Psychological Measures

Pain, suffering, and disability each depend upon
a range of cognitive and affective factors. A large

number of instruments have been developed to assess patients' personalities, behaviors, attitudes, expectations, levels of distress, and coping strategies in dealing with chronic pain. However, researchers cannot agree which ones may be useful.

Depression: The measurement of depression is most commonly assessed by the Beck Depression Index (BDI) (Beck and Beamesderfer 1974) and Zung Depression Scale (Zung et al. 1965).

Anxiety: There are questions on anxiety in most psychological questionnaires. Despite this, a new, more specific questionnaire has recently been developed: The Pain Anxiety Symptom Scale (PASS). The Depression Anxiety Stress Scales (DASS) assesses both anxiety and depression.

Willingness to change: Pain Stages of Change Questionnaire (PSOCQ) assesses readiness for change. This questionnaire has been used as a screening tool when assessing patients for intake into a pain management program.

Other domains in the psychological sphere include coping strategies, catastrophizing, abnormal illness behavior/beliefs. A Fear-Avoidance Beliefs Questionnaire (FABQ) was developed by Waddell et al. (Waddell et al. 1993), based on theories of fear and avoidance behavior, and focused specifically on patients' beliefs about how physical activity and work affected their low back pain. Another questionnaire, bridging the gap between physical and psychological impact of chronic pain, is the West Haven-Yale Multidimensional Pain Inventory (WHYMPI).

Service Utilization

This outcome measure requires no specific instruments. It catalogues the incidence and nature of medical consultations, hospitalization, use of drugs, allied health visits, and home help assistance.

Socioeconomic

Questions in this domain cover social, employment, financial, litigation, and compensation issues.

Patient Satisfaction

Deyo et al. found that the most important factor in patient satisfaction was receiving an adequate explanation for their symptoms (Deyo and Diehl 1986). Patient satisfaction may, therefore, be inappropriate as an outcome measure, as it may be rated high even if the patient may have been given a completely absurd explanation for their symptoms. Others have commented on the use of opioids for chronic back pain where there was high patient satisfaction, but pain and disability were not improved (Wallace et al. 2009).

Validity

An unresolved issue is what constitutes a meaningful degree of change with any outcome measure. Recent and ongoing research efforts have targeted precision and sensitivity to change of various outcome measures and what constitutes a minimally effective change. These differ for different measures.

References

Beck, A. T., & Beamesderfer, A. (1974). Assessment of depression: The depression inventory. *Modern Problems of Pharmacopsychiatry, 7*, 151–169.

Deyo, R. A., & Diehl, A. K. (1986). Patient satisfaction with medical care for low-back pain. *Spine, 11*, 28–30.

Fairbank, J. C., Couper, J., Davies, J. B., & O'Brien, J. P. (1980). The Oswestry low back pain disability questionnaire. *Physiotherapy, 66*, 271–273.

Jensen, M. P., & Karoly, P. (1992). Self reporting scales and procedures for assessing pain in adults. In D. C. Turk & R. Melzack (Eds.), *Handbook of pain assessment* (pp. 135–151). New York: The Guilford Press.

McHomery, C. A., Ware, J. E., & Raczek, A. E. (1993). The MOS 36-item short-form health survey (SF-36): II. Psychometric and clinical tests of validity in measuring physical and mental health constructs. *Medical Care, 31*, 247–263.

Melzack, R. (1975). The McGill pain questionnaire: Major properties and scoring methods. *Pain, 1*, 277–279.

Ohnmeiss, D. D., Vanharanta, H., Estlander, A. M., & Jamsen, A. (2000). The relationship of disability (Oswestry) and pain drawings to functional testing. *European Spine Journal, 9*, 208–212.

Roland, M., & Morris, R. (1983). A study of the natural history of back-pain. Part 1: Development of a reliable and sensitive measure of disability in low back pain. *Spine, 8*, 141–144.

Waddell, G., Newton, M., Henderson, I., Somerville, D., & Main, C. J. (1993). A fear-avoidance beliefs questionnaire (FABQ) and the role of fear-avoidance beliefs in chronic low back pain and disability. *Pain, 52*, 157–168.

Wallace, A. S., Freburger, J. K., Darter, J. D., Jackman, A. M., & Carey, T. S. (2009). Comfortably numb? Exploring satisfaction with chronic back pain visits. *The Spine Journal, 9*(9), 721–728.

Zung, W. W., Richards, C. B., & Short, M. J. (1965). Self-rating depression scale in an outpatient clinic. Further validation of the SDS. *Archives of General Psychiatry, 13*, 508–515.

Outcome Questionnaires

▶ Outcome Measures

Outcomes Assessment Tools

▶ Outcome Measures

Ovariectomy

Definition

Surgical removal of both ovaries.

Cross-References

▶ Visceral Pain Models, Female Reproductive Organ Pain

Overalertness

▶ Hypervigilance and Attention to Pain

Overlap Syndrome

Definition

Overlap syndromes are a group of inflammatory myositis in association with other signs of a connective tissue disease. Myositis-associated autoantibodies are a characteristic.

Cross-References

▶ Myositis

Over-the-Counter Analgesics

▶ Simple Analgesics

Oxidative Stress

Definition

Oxidative stress occurs when the production of free radical species such as superoxide, hydrogen peroxide, and nitric oxide exceeds the capacity of cells to remove them via their antioxidant defense mechanisms. The excess free radicals attach to proteins, lipids, and nucleic acids and alter or disrupt the normal function of these cellular constituents. Oxidative stress has been implicated in normal aging and many pathologic processes, including diabetic neuropathy.

Cross-References

▶ Neuropathic Pain Model, Diabetic Neuropathy Model

Oxycodone

Synonyms

Hydroxy-7.8-Dihydrocodeinone

Definition

Oxycodone (14-Hydroxy-7.8-Dihydrocodeinone) is a semisynthetic derivative of the opium alkaloid thebaine, with mu and kappa activity.

Cross-References

▶ Postoperative Pain, Oxycodone

OxyContin

Definition

OxyContin is a controlled-release formulation of oxycodone that provides controlled release/

delivery of oxycodone over 12 h, with an onset of analgesia within 1 h.

Cross-References

▶ Postoperative Pain, Oxycodone

Oxynorm®

▶ Postoperative Pain, Oxycodone

Encyclopedia of Pain

Gerald F. Gebhart • Robert F. Schmidt
Editors

Encyclopedia of Pain

Second Edition

Volume 5

P

With 795 Figures and 217 Tables

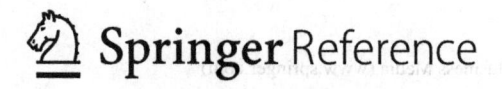 Springer Reference

Editors
Gerald F. Gebhart
Department of Anesthesiology
Center for Pain Research
University of Pittsburgh
Pittsburgh, PA, USA

Robert F. Schmidt
Physiologisches Institut der
Universität Würzburg
Würzburg, Germany

ISBN 978-3-642-28752-7   ISBN 978-3-642-28753-4 (eBook)
ISBN 978-3-642-28754-1 (print and electronic bundle)
DOI 10.1007/978-3-642-28753-4
Springer Heidelberg New York Dordrecht London

Library of Congress Control Number: 2013951551

Preface to the Second Edition

As was pointed out in the Preface to the first edition, pain is the principal reason individuals seek medical and dental attention. Fortunately, acute tissue insults associated with pain – and typically inflammation – readily resolve with appropriate treatment and care. Chronic pain states, in contrast, are notoriously difficult to manage and are commonly associated with comorbidities (e.g., depression, cross-organ sensitization) and reduced quality of life. Indeed, chronic pain has come to be considered by some as a disease in its own right. This edition of the *Encyclopedia of Pain* updates the content of the 1st edition and introduces new topics consistent with advances in our knowledge of underlying mechanisms of pain. Thus, new content on channelopathies, expanded and updated material on the role of glia and immune-competent cell interactions with nociceptive neurons, and advances in imaging and changes in brain structure in chronic pain states are included in this edition.

The cardinal objective of research on pain is the translation of knowledge between the laboratory and the clinic. The present prevailing refrain is "bench to bedside," but in fact many ideas for research in pain derive from clinical observations and the absence of knowledge to explain clinical pain states. In the recent past, basic and clinical science research has seen the application of imaging techniques to visualize areas of the brain that are involved in both the sensory and affective aspects of pain, cloning of pain-related channels and receptors, and the application of elegant genetic manipulations that selectively "label" functionally distinct dorsal root ganglion sensory and spinal neurons, two-photon imaging, and optogenetic approaches.

Accordingly, much is new in this second edition of the *Encyclopedia of Pain*. The number of authors who have contributed to this edition is close to 800. In addition to a print edition, the publisher, Springer-Verlag, will produce an electronic version that will make comprehensive searches easy to manage. The electronic version, to be made available on the online publishing platform SpringerReference, can be updated regularly to ensure that the content remains current.

October 2013
Gerald F. Gebhart   Robert F. Schmidt
Pittsburgh, PA, USA   Würzburg, Germany

Preface to the First Edition

As all medical students know, pain is the most common reason for a person to consult a physician. Under ordinary circumstances, acute pain has a useful, protective function. It discourages the individual from activities that aggravate the pain, allowing faster recovery from tissue damage. The physician can often tell from the nature of the pain what its source is. In most cases, treatment of the underlying condition resolves the pain. By contrast, children born with congenital insensitivity to pain suffer repeated physical damage and die young (see Sweet WH (1981) Pain 10:275).

Pain resulting from difficult to treat or untreatable conditions can become persistent. Chronic pain "never has a biologic function but is a malefic force that often imposes severe emotional, physical, economic, and social stresses on the patient and on the family..." (Bonica JJ (1990) The Management of Pain, vol 1, 2nd edn. Lea & Febiger, Philadelphia, p 19). Chronic pain can be considered a disease in its own right.

Pain is a complex phenomenon. It has been defined by the Taxonomy Committee of the International Association for the Study of Pain as "An unpleasant emotional and sensory experience associated with actual or potential tissue damage, or described in terms of such damage" (Merskey H and Bogduk N (1994) Classification of Chronic Pain, 2nd edn. IASP Press, Seattle). It is often ongoing, but in some cases it may be evoked by stimuli. Hyperalgesia occurs when there is an increase in pain intensity in response to stimuli that are normally painful. Allodynia is pain that is evoked by stimuli that are normally non-painful.

Acute pain is generally attributable to the activation of primary efferent neurons called nociceptors (Sherrington CS (1906) The Integrative Action of the Nervous System. Yale University Press, New Haven; 2nd edn, 1947). These sensory nerve fibers have high thresholds and respond to strong stimuli that threaten or cause injury to tissues of the body. Chronic pain may result from continuous or repeated activation of nociceptors, as in some forms of cancer or in chronic inflammatory states, such as arthritis.

However, chronic pain can also be produced by damage to nervous tissue. If peripheral nerves are injured, peripheral neuropathic pain may develop. Damage to certain parts of the central nervous system may result in central neuropathic pain. Examples of conditions that can cause central neuropathic pain include spinal cord injury, cerebrovascular accidents, and multiple sclerosis.

Research on pain in humans has been an important clinical topic for many years. Basic science studies were relatively few in number until experimental work on pain accelerated following detailed descriptions of peripheral nociceptors and central nociceptive neurons that were made in the 1960s and 1970s, by the discovery of the endogenous opioid compounds and the descending pain control systems in the 1970s and the application of modern imaging techniques to visualize areas of the brain that are affected by pain in the 1990s. Accompanying these advances has been the development of a number of animal models of human pain states, with the goal of using these to examine pain mechanisms and also to test analgesic drugs or non-pharmacologic interventions that might prove useful for the treatment of pain in humans. Basic research on pain now emphasizes multidisciplinary approaches, including behavioral testing, electrophysiology, and the application of many of the techniques of modern cell and molecular biology, including the use of transgenic animals.

The *Encyclopedia of Pain* is meant to provide a source of information that spans contemporary basic and clinical research on pain and pain therapy. It should be useful not only to researchers in these fields but also to practicing physicians and other health care professionals and to health care educators and administrators. The work is subdivided into 35 fields, and the Field Editor of each of these describes the areas covered in each in a brief review entry. The topics included in a field are the subject of a series of short essays, accompanied by key words, definitions, illustrations, and a list of significant references. The number of authors who have contributed to the encyclopedia exceeds 550. The plan of the publisher, Springer-Verlag, is to produce both print and electronic versions of this encyclopedia. Numerous links within the electronic version should make comprehensive searches easy to manage. The electronic version will be updated at sufficiently short intervals to ensure that the content remains current.

The editors thank the staff at Springer-Verlag who have provided oversight for this project, including Rolf Lange, Thomas Mager, Claudia Lange, Natasja Sheriff, and Michaela Bilic. Working with these outstanding individuals has been a pleasure.

July 2006
ROBERT F. SCHMIDT   WILLIAM D. WILLIS
Würzburg, Germany   Galveston, Texas, USA

About the Editors

Gerald F. Gebhart received his PhD from The University of Iowa at Iowa City, Iowa, USA, after which he completed a 2-year fellowship at the Université de Montréal, Montréal, Canada, under the direction of Herbert H. Jasper. He began his academic career in the Department of Pharmacology at The University of Iowa in 1973, where he developed a productive research program and led a Pain Interest Group. His career includes a sabbatical year in Heidelberg, Germany (1981–1982), and leadership as head of the Department of Pharmacology from 1996 to 2006. In 2006, Dr. Gebhart was named director of the Center for Pain Research, Department of Anesthesiology, the University of Pittsburgh at Pittsburgh, Pennsylvania, USA. Dr. Gebhart has served the pain community throughout his career. He was elected to the Board of Directors of the American Pain Society (1991–1994) and subsequently elected president of the APS (1997). In 2000, he became the founding editor of *The Journal of Pain*, the official journal of the APS, serving as editor-in-chief until 2010. Dr. Gebhart also served on the Council and later as president-elect, president, and past president of the International Association for the Study of Pain (2005–2012).

Dr. Gebhart is best known for his research contributions in descending modulation of pain and mechanisms of visceral pain, and was designated in 2004 by Thomson ISI as a Highly Cited Researcher (Neuroscience). His scientific contributions have been recognized with several awards, including a 5-year Bristol Myers-Squibb Award for Excellence in Pain Research (1989–1994), a 10-year MERIT Award from the National Institutes of Health (1993–2003),

the Frederick W.L. Kerr Award from the American Pain Society (1994), the Purdue Pharma Prize for Pain Research (2004), the Janssen Award in Gastroenterology (2005), the Founders Award from the American Academy of Pain Medicine (2006), and the Patrick D. Wall Award from the British Pain Society (2012).

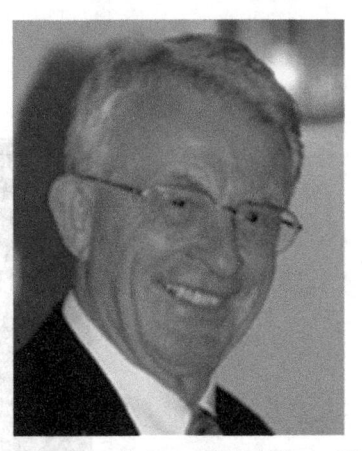

Robert F. Schmidt studied medicine at Heidelberg University (Germany). He graduated in 1959 and was promoted to Dr.med. in the same year. After residencies in clinical disciplines, Dr. Schmidt went to the John Curtin School of Medical Research of the Australian National University in Canberra, ACT, where he worked under the head of the Department of Neurophysiology, Sir John C. Eccles. He received his Ph.D. in 1963, the same year Sir John was awarded the Nobel Prize for Physiology and Medicine.

Returning to Heidelberg Medical School, Dr. Schmidt continued his academic career. He completed his habilitation in 1964, becoming a tenured lecturer (Wissenschaftlicher Rat) in 1966 and an associate professor in 1970. After an 8-month sabbatical with Sir John at the State University of New York, Dr. Schmidt became professor and director of the Department of Physiology at the University of Kiel, Germany. In 1982, he moved to the University of Würzburg to take the same position in the Department of Physiology. In 2000, Dr. Schmidt became professor emeritus at this department.

The research work of Dr. Schmidt and his associates in both Kiel and Würzburg focused on the neurophysiology of fine sensory afferents and their connections and functions in the spinal cord. In Würzburg, he concentrated on the processes leading to the changes in response characteristics of nociceptive afferents (i.e., peripheral sensitization) during inflammation of joints, discovering in the process silent nociceptors, which have subsequently been described in all vertebrates studied, including humans. In addition, Dr. Schmidt has edited, coedited, authored, and coauthored numerous textbooks and monographs, most of them having several editions and translations.

In recognition of his research contributions, Dr. Schmidt received many prizes, awards, and honors, including a D.Sc. honoris causa from the University of New South Wales, Sydney, Australia, and an Honorary Professorship from the University of Tübingen, Germany. He was elected to membership in the Akademie der Wissenschaften und der Literatur, Mainz, Germany. Dr. Schmidt is an Honorary Member of the Colombian Association for the Study of Pain, the Japan Physiological Society, the German Society for the Study of Pain, the German Pain Association, the International Association for the Study of Pain (IASP), and the German Physiological Society (in which he served as president in 1979). In 2000, he was awarded the Bundesverdienstkreuz 1st Class (Order of Merit of the German Federal Republic).

Section Editors

Ebtesam Ahmed College of Pharmacy and Health Sciences, St. John's University, Department of Pain Medicine and Palliative Care, Beth Israel Medical Center, New York, NY, USA

A. Vania Apkarian Department of Physiology, Northwestern University Feinberg School of Medicine, Chicago, IL, USA

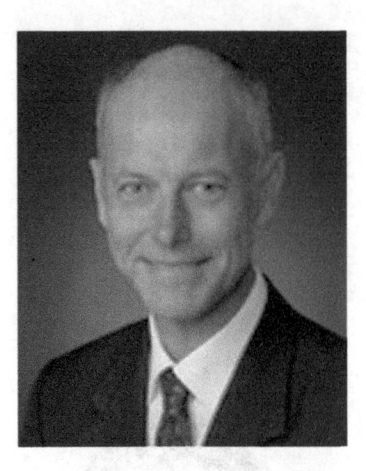

Lars Arendt-Nielsen Center for Sensory-Motor Interaction, Department of Health Science and Technology, School of Medicine, Aalborg University, Aalborg, Denmark

Carlos Belmonte Instituto de Neurociencias, Universidad Miguel Hernandez-CSIC, San Juan de Alicante, Spain

Niels Birbaumer Institute of Medical Psychology and Behavioral Neurobiology, University of Tübingen, Tübingen, Germany

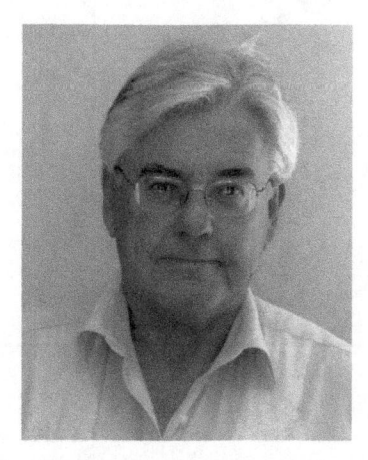

Nikolai Bogduk University of Newcastle, Newcastle Bone & Joint Institute,
Royal Newcastle Centre, Newcastle, NSW, Australia

Jin Mo Chung Department of Neuroscience and Cell Biology, University Chair in Neuroscience, University of Texas Medical Branch, Galveston, TX, USA

Joyce DeLeo Dartmouth Hitchcock Medical Center, Dartmouth Medical School, Hanover, NH, USA

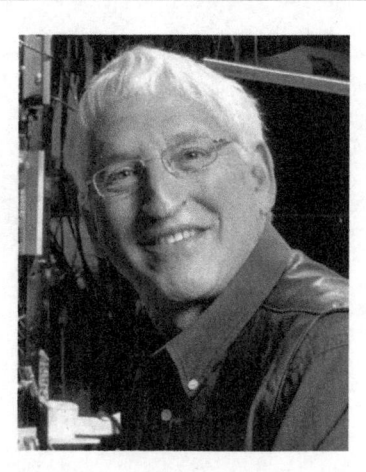

Marshall Devor Department of Cell & Developmental Biology, Institute of Life Sciences and Center for Research on Pain, The Hebrew University of Jerusalem, Jerusalem, Givat Ram, Israel

Hans-Christoph Diener Department of Neurology, University Hospital Essen, University Duisburg-Essen, Essen, Germany

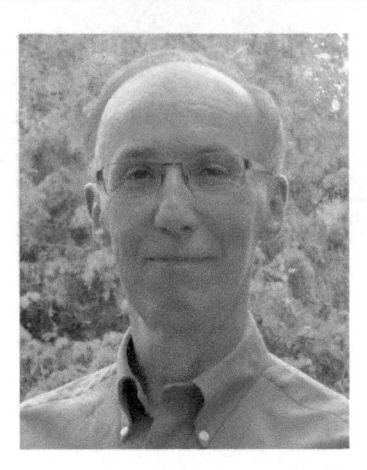

Jonathan O. Dostrovsky Department of Physiology, University of Toronto, Toronto, ON, Canada

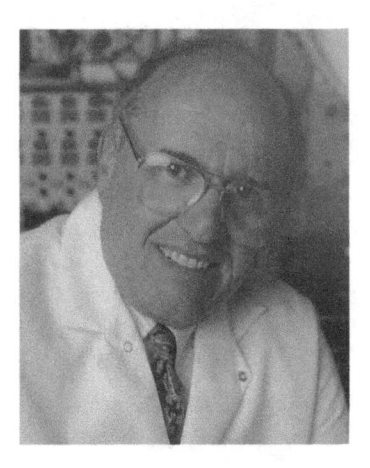

Ronald Dubner Department of Neural and Pain Sciences, University of Maryland Dental School, Baltimore, MD, USA

Michel Y. Dubois NYU Medical School, NYU Langone Medical Centers, Center for Research & Treatment of Pain, NYU Ambulatory Care Center, New York, NY, USA

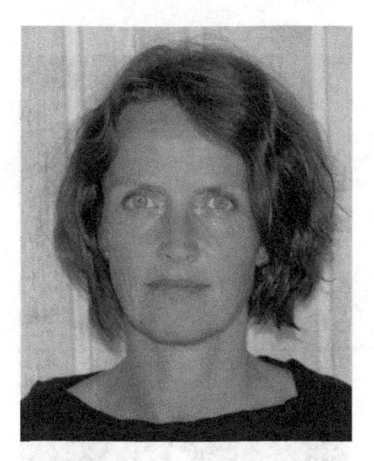

Nanna Brix Finnerup Aarhus University Hospital, Danish Pain Research Center, Aarhus, Denmark

Herta Flor Department of Cognitive and Clinical Neuroscience, Central
Institute of Mental Health, Medical Faculty Mannheim, Heidelberg University,
Mannheim, Germany

Gerald F. Gebhart Department of Anesthesiology, Center for Pain
Research, University of Pittsburgh, Pittsburgh, PA, USA

Maria Adele Giamberardino Pathophysiology of Pain Laboratory and Centre for Fibromyalgia and Musculoskeletal Pain, Department of Medicine and Science of Aging, University of Chieti, Chieti, Italy

Glenn J. Giesler Department of Neuroscience, University of Minnesota, Minneapolis, MN, USA

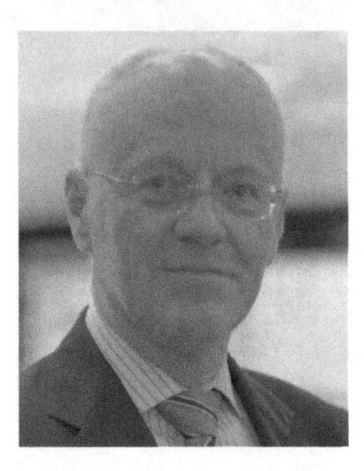

Peter Goadsby Headache Center, Department of Neurology, University of California, San Francisco, San Francisco, CA, USA

Michael S. Gold Department of Anesthesiology, Center for Pain Research, University of Pittsburgh, Pittsburgh, PA, USA

Hermann O. Handwerker Department of Physiology and Pathophysiology, University of Erlangen/Nürnberg, Erlangen, Germany

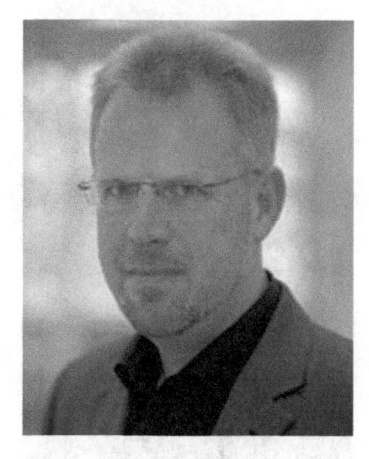

Burkhard Hinz Institute of Toxicology and Pharmacology, University of Rostock, Rostock, Germany

Wilfrid Jaenig Physiologisches Institut, Christian-Albrechts-Universität zu Kiel, Kiel, Germany

Michaela Kress Department für Physiologie und Medizinische Physik, Medizinische Universität Innsbruck, Innsbruck, Austria

Frederick A. Lenz Department of Neurosurgery, Johns Hopkins Hospital, Baltimore, MD, USA

Tonya M. Palermo Seattle Children's Hospital Research Institute, CW8-6 - Child Health, Behavior and Development, University of Washington, Seattle, WA, USA

Frank Porreca Department of Pharmacology, University of Arizona Health Sciences Center, Tucson, AZ, USA

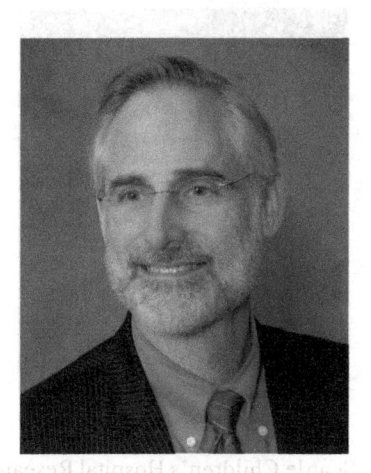

Russell K. Portenoy Department of Pain Medicine and Palliative Care, Beth Israel Medical Center, New York, NY, USA

James P. Robinson Department of Anesthesiology & Pain Medicine, University of Washington, Seattle, WA, USA

Hans-Georg Schaible Department of Physiology - Neurophysiology, Friedrich-Schiller-University of Jena, Jena, Germany

Robert F. Schmidt Physiologisches Institut der Universität Würzburg,
Würzburg, Germany

Stephan A. Schug School of Medicine and Pharmacology, Royal Perth
Hospital, Faculty of Medicine, Dentistry & Health Sciences, The University
of Western Australia, UWA Anaesthesiology, Perth, Australia

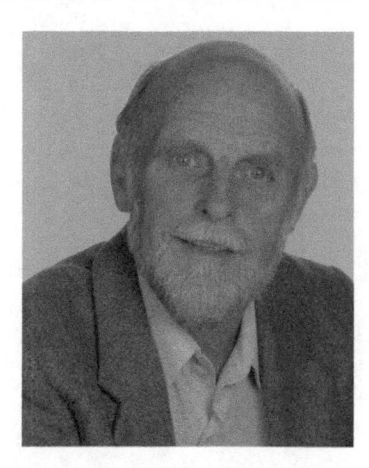

Barry J. Sessle Faculty of Dentistry, University of Toronto, Toronto, ON, Canada

Bengt H. Sjölund Department of Community Medicine and Rehabilitation Medicine, Umea University, Umea, Sweden

Mark D. Sullivan Psychiatry and Behavioral Sciences, University of Washington, Seattle, WA, USA

Irene Tracey Pain Imaging Neuroscience (PaIN) Group, Department of Physiology, Anatomy and Genetics, Oxford University, Oxford, UK

Dennis C. Turk Department of Anesthesiology & Pain Research, University of Washington, Seattle, WA, USA

Félix Viana Instituto de Neurociencias, Universidad Miguel Hernandez-CSIC, San Juan de Alicante, Alicante, Spain

Katy Vincent Nuffield Department of Obstetrics and Gynaecology, The Women's Centre, John Radcliffe Hospital, University of Oxford, Oxford, UK

Gary A. Walco Department of Anesthesiology and Pain Medicine, University of Washington School of Medicine, Seattle Children's Hospital, Seattle, WA, USA

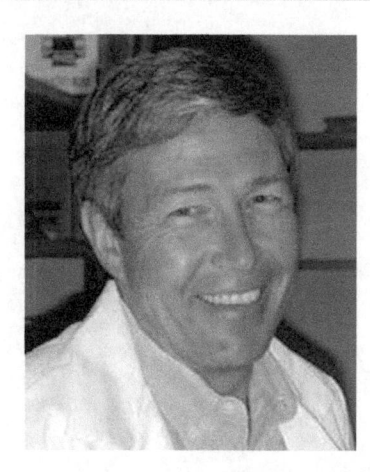

George L. Wilcox Department of Neuroscience, Pharmacology and Dermatology, University of Minnesota Medical School, Minneapolis, MN, USA

Katja Wiech Nuffield Department of Clinical Neurosciences, John Radcliffe Hospital, Oxford, UK

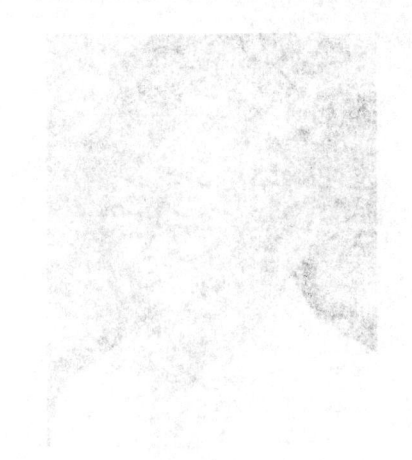

George L. Wilson Department of Neuroscience, Pharmacology and Dermatology, University of Minnesota Medical School, Minneapolis, MN, USA

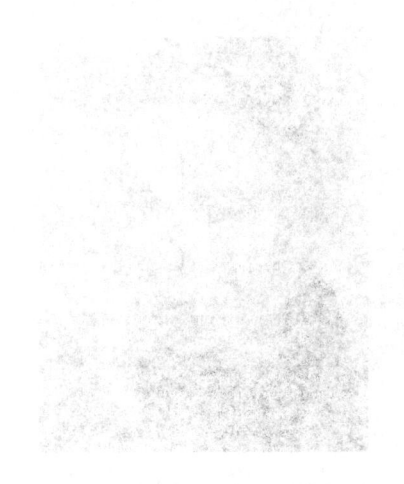

Kaye Wreck Nuffield Department of Clinical Neurosciences, John Radcliffe Hospital, Oxford, UK

Contributors

Tipu Aamir The Auckland Regional Pain Service, Auckland Hospital, Auckland, New Zealand

Catherine Abbadie Department of Pharmacology, Merck Research Laboratories, Rahway, NJ, USA

Frances V. Abbott Departments of Psychiatry and Psychology, McGill University, Montreal, QC, Canada

Heather Adams Department of Surgery, McGill University Health Centre, Montreal, QC, Canada

Mary O. Adedoyin Department of Anatomy, University of California San Francisco, San Francisco, CA, USA

Rolf H. Adler University of Berne Medical School, Kehrsatz, Switzerland

Claire D. Advokat Department of Psychology, Louisiana State University, Baton Rouge, LA, USA

Reto M. Agosti Headache Center Hirlsanden, Zurich, Switzerland

Ebtesam Ahmed College of Pharmacy and Health Sciences, St. John's University, Department of Pain Medicine and Palliative Care, Beth Israel Medical Center, New York, NY, USA

Marie-Claire Albanese Department of Psychology, McGill University, Montreal, QC, Canada

Kathryn M. Albers Department of Neurobiology, University of Pittsburgh School of Medicine, Pittsburgh, PA, USA

Phillip J. Albrecht Center for Neuropharmacology and Neuroscience, Albany Medical College, Albany, NY, USA

Elie D. Al-Chaer Center for Pain Research, University of Arkansas for Medical Sciences, Little Rock, AR, USA

Terry Altilio Department of Pain Medicine and Palliative Care, Beth Israel Medical Center, New York, USA

Rainer Amann Medical University Graz, Graz, Austria

Ron Amir Institute of Life Sciences, The Hebrew University of Jerusalem, Jerusalem, Israel

Praveen Anand Peripheral Neuropathy Unit, Imperial College London, London, UK

Karen O. Anderson The University of Texas MD Anderson Cancer Center, Houston, TX, USA

Ole K. Andersen Department of Health Science & Technology, Aalborg University, Aalborg, Denmark

Stanley W. Anderson Department of Neurosurgery, Johns Hopkins Hospital, Johns Hopkins Medical Institutions, Johns Hopkins University, Baltimore, MD, USA

Trevor Anderson Fairview Pain & Palliative Care, University of Minnesota Medical Center, Minneapolis, MN, USA

Karl-Erik Andersson Wake Forest Institute for Regenerative Medicine, Wake Forest University School of Medicine, Winston Salem, USA

Frank Andrasik Department of Psychology, Applied Psychophysiology and Biofeedback, The University of Memphis, Memphis, TN, USA

A. Vania Apkarian Department of Physiology, Northwestern University Feinberg School of Medicine, Chicago, IL, USA

Lars Arendt-Nielsen Center for Sensory-Motor Interaction, Department of Health Science and Technology, School of Medicine, Aalborg University, Aalborg, Denmark

Kirsten Arndt CNS Pharmacology, Boehringer Ingelheim Pharma GmbH & Co. KG, Biberach, Germany

Charles-Daniel Arreto Université Paris Descartes, INSERM UMR 894, France

Erik Askenasy Neurobiology and Anatomy, Health Science Center, University of Texas, Houston, TX, USA

J. H. Atkinson Psychiatry Service, VA San Diego Healthcare System and Department of Psychiatry, University of California, San Diego, CA, USA

Nadine Attal Centre d'Evaluation et de Traitement de la Douleur, Hôpital Ambroise Paré, INSERM U 987 APHP, Boulogne-Billancourt, France

Christoph Aufenberg Functional Neurosurgery, Neurosurgical Clinic University Hospital, Zürich, Switzerland

Beate Averbeck Sanofi-Aventis Deutschland GmbH, Frankfurt, Germany

Qasim Aziz Section of GI Sciences, Hope Hospital University of Manchester, Salford, UK

Cathrine Baastrup Danish Pain Research Center, Aarhus University Aarhus University Hospital, Aarhus, Denmark

F. W. Bach Danish Pain Research Center, Department of Neurology, Aarhus University Hospital, Aarhus, Denmark

S. K. Back Neuroscience Research Institute and Department of Physiology, Korea University, College of Medicine, Seoul, Republic of Korea

Misha-Miroslav Backonja Department of Neurology, University of Wisconsin-Madison, Madison, WI, USA

Banani Banerjee Division of Gastroenterology and Hepatology and Pediatric Gastroenterology, Medical College of Wisconsin, Milwaukee, WI, USA

John D. Banja Department of Rehabilitation Medicine, Center for Ethics Emory University, Atlanta, GA, USA

Carsten Bantel Department of Surgery & Cancer? Anaesthetics Section, Imperial College of Science and Medicine Chelsea and Westminster Campus, London, UK

Andrew Paul Baranowski The Pain Management Centre, National Hospital for Neurology and Neurosurgery, UCL Hospitals NHS Foundation Trust, London, UK

Alex Barling Department of Neurology, Birmingham Muscle and Nerve Centre Queen Elizabeth Hospital, Birmingham, UK

Ralf Baron Department of Neurological Pain Research and Therapy, Christian Albrechts University Kiel, Kiel, Germany

Gordon A. Barr Department of Anesthesiology and Critical Care Medicine, Children's Hospital of Philadelphia University of Pennsylvania, Philadelphia, PA, USA

Emily J. Bartley Department of Psychology, The University of Tulsa, Tulsa, OK, USA

Jeffrey R. Basford Department of Physical Medicine and Rehabilitation, Mayo Clinic and Mayo Foundation, Rochester, NY, USA

Michele C. Battie Department of Physical Therapy, University of Alberta, Edmonton, AB, Canada

Ulf Baumgartner Chair of Neurophysiology, Center for Biomedicine and Medical Technology Mannheim, Heidelberg University, Mannheim, Germany

Nicola Beale Nuffield Department of Anaesthetics, Oxford University Hospitals NHS Trust, UK

Alain Beaudet Montreal Neurological Institute, McGill University, Montreal, QC, Canada

Lino Becerra P.A.I.N. Group, Department of Anesthesiology, Perioperative and Pain Medicine, Boston Children's Hospital Department of Radiology Massachusetts General Hospital Center for Pain and the Brain, Harvard Medicals, Boston, MA, USA

Yaakov Beilin Department of Anesthesiology, and Obstetrics, Mount Sinai School of Medicine, New York, NY, USA

Alvin J. Beitz Department of Veterinary and Biomedical Sciences, University of Minnesota, St. Paul, MN, USA

Miles Belgrade Fairview Pain Management Center, University of Minnesota Medical Center, Minneapolis, MN, USA

Carlo Valerio Bellieni Neonatal Intensive Care Unit, University Hospital, Siena, Italy

Carlos Belmonte Instituto de Neurociencias, Universidad Miguel Hernandez-CSIC, San Juan de Alicante, Spain

Allan J. Belzberg Department of Neurosurgery, Johns Hopkins School of Medicine, Baltimore, MD, USA

Matt Bender Department of Neurosurgery, Johns Hopkins Hospital, Baltimore, MA, USA

Fabrizio Benedetti Department of Neuroscience, Clinical and Applied Physiology Program, University of Turin Medical School, Turin, Italy

David L. H. Bennett Wolfson CARD, Kings College London, London, UK

Nuffield Department of Clinical Neurosciences, John Radclife Hospital, The University of Oxford, Oxford, UK

Robert Bennett Department of Medicine, Oregon Health and Science University, Portland, USA

Edward C. Benzel Department of Neurosurgery, Center for Spine Health Neurological Institute Cleveland Clinic, Cleveland, OH, USA

David A. Bereiter Department of Diagnostic and Biological Sciences, University of Minnesota School of Dentistry, Minneapolis, MN, USA

Elmar Berendes Klinik für Anästhesiologie, Operative Intensivmedizin und Schmerztherapie, Klinikum Krefeld, Krefeld, Germany

Karen J. Berkley Program in Neuroscience, Florida State University, Tallahassee, FL, USA

Peter Berlit Department of Neurology, Alfried Krupp Hospital, Essen, Germany

Jean-François Bernard INSERM UMR-894, UPMC Site Pitié-Salpêtrière, Paris, France

Anne K. Bertelsen Neurology, Aleris Hospital, Bergen, Norway

Jennifer Bickel Integrative Pain Management, Children's Mercy Hospitals and Clinics, University of Missouri-Kansas City School of Medicine, Kansas City, MO, USA

Marcelo E. Bigal Department of Neurology and Montefiore Headache Unit, Albert Einstein College of Medicine, Bronx, NY, USA

Yitzchak M. Binik McGill University, Montreal, QC, Canada

Niels Birbaumer Institute of Medical Psychology and Behavioral Neurobiology, University of Tübingen, Tübingen, Germany

Frank Birklein Department of Neurology, University of Mainz, Mainz, Germany

Kathryn A. Birnie Department of Psychology, Dalhousie University, Halifax, NS, Canada

Thomas Bishop Centre for Neuroscience, King's College London, London, UK

Henning Bliddal The Parker Institute, Frederiksberg Hospital, Copenhagen, Denmark

Nikolai Bogduk University of Newcastle, Newcastle Bone & Joint Institute, Royal Newcastle Centre, Newcastle, NSW, Australia

Jorgen Boivie Department of Neurology, University Hospital, Linköping, Sweden

Michael R. Bond University of Glasgow, Glasgow, UK

Richard L. Boortz-Marx Division of Pain Medicine, Department of Anesthesiology, University of North Carolina, Chapel Hill, North Carolina, USA

James M. Borowczyk Department of Orthopaedics and Musculoskeletal Medicine, University of Otago, Christchurch, Christchurch, New Zealand

David Borsook P.A.I.N. Group, Department of Anesthesiology, Perioperative and Pain Medicine, Boston Children's Hospital Department of Radiology Massachusetts General Hospital Center for Pain and the Brain, Harvard Medicals, Boston, MA, USA

George S. Borszcz Department of Psychology, Wayne State University, Detroit, USA

Jasenka Borzan Department of Anesthesiology and Critical Care Medicine, Johns Hopkins University, Baltimore, USA

Regina M. Botting London, UK

Didier Bouhassira Centre d'Evaluation et de Traitement de la Douleur, Hôpital Ambroise Paré, INSERM U 987 APHP, Boulogne-Billancourt, France

Laurence Bourgeais INSERM UMR 894, Centre de Psychiatrie et Neurosciences, Paris, France

David Bowsher Pain Research Institute, University Hospital Aintree, Liverpool, UK

Emma L. Brandon Anaesthesiology and Pain Medicine, University of Western Australia Royal Perth Hospital, Perth, WA, Australia

Jennifer Brawn P.A.I.N. Group, Center for Pain and the Brain, Boston Children's Hospital, Harvard Medical School, Waltham, MA, USA

Harald Breivik Department of Pain Management and Research, Oslo University Hospital, Rikshospitalet and University of Oslo, Oslo, Norway

Kori L. Brewer Department of Emergency Medicine, Brody School of Medicine East Carolina University, Greenville, NC, USA

Christopher R. Brigham Brigham and Associates, Inc., Portland, USA

Abigail Broderick Nuffield Department of Obstetrics and Gynaecology, University of Oxford, Oxford, UK

Jeffrey A. Brown Department of Neurological Surgery, Wayne State University School of Medicine, Detroit, MI, USA

Stephen C. Brown Department of Anesthesia, The Hospital for Sick Children, Toronto, ON, Canada

Eduardo Bruera Department of Palliative Care and Rehabilitation Medicine, The University of Texas M. D. Anderson Cancer Center, Houston, TX, USA

Pablo Brumovsky Faculty of Biomedical Sciences, Austral University, Buenos Aires, Argentina

Consejo Nacional de Investigaciones Científicas y Técnicas (CONICET), Buenos Aires, Argentina

Kay Brune Department of Experimental and Clinical Pharmacology and Toxicology, Friedrich Alexander University Erlangen-Nürnberg, Erlangen, Germany

Maria Burian Department of General, Visceral and Transplantation Surgery, University of Heidelberg, Heidelberg, Germany

Dan Buskila Rheumatic Disease Unit, Soroka Medical Center, Beer Sheva, Israel

Catherine M. Cahill Anesthesiology & Perioperative Care, University of California, Irvine, Irvine, CA, USA

Brian E. Cairns Faculty of Pharmaceutical Sciences, University of British Columbia, Vancouver, BC, Canada

Nigel A. Calcutt Department of Pathology, University of California San Diego, La Jolla, CA, USA

Margarita Calvo Wolfson CARD, Kings College London, London, UK

James N. Campbell Johns Hopkins University School of Medicine, Baltimore, MD, USA

Lauren Campbell Department of Psychology, York University, Toronto, ON, Canada

Ling Cao Department of Biomedical Sciences, College of Osteopathic Medicine, University of New England, ME, USA

Adam Carinci Department of Anesthesia, Critical Care and Pain Medicine, Massachusetts General Hospital Harvard Medical School, Boston, MA, USA

Giancarlo Carli Department of Physiology, University of Siena, Siena, Italy

Christer P. O. Carlsson Rehabilitation Clinic, Lunds University Hospital, Lund, Sweden

Susan M. Carlton Neuroscience and Cell Biology, University of Texas Medical Branch, Galveston, TX, USA

Richard Carr Clinic for Anaesthesiology, Heidelberg University, Mannheim, Baden-Wuerttemberg, Germany

Benjamin S. Carson Department of Neurosurgery, Johns Hopkins Hospital, Baltimore, MA, USA

Brian Carter University of Missouri at Kansas City Children's Mercy Hospital and Bioethics Center, Kansas City, MO, USA

Kenneth L. Casey Department of Neurology, University of Michigan Medical Center, Ann Arbor, MI, USA

Christine T. Chambers Departments of Pediatrics and Psychology, Dalhousie University, Halifax, NS, Canada

Sandra Chaplan Janssen Research and Development, LLC, San Diego, CA, USA

C. Richard Chapman Department of Anesthesiology, University of Utah School of Medicine, Salt Lake City, UT, USA

Santosh K. Chaturvedi Psychiatry, National Institute of Mental Health & Neurosciences, Bangalore, India

Andrew C. N. Chen Human Brain Mapping and Cortical Imaging Laboratory, Aalborg University, Aalborg, Denmark

Hsinlin Thomas Cheng Department of Neurology, University of Michigan, Ann Arbor, MI, USA

Pavel Cherkas Faculty of Dentistry, University of Toronto, Toronto, ON, Canada

Andrea Cheville Department of Physical Medicine and Rehabilitation, Mayo Clinic, Rochester, MN, USA

Chen Yu Chiang Faculty of Dentistry, University of Toronto, Toronto, ON, Canada

N. L. Chillingworth School of Physiology and Pharmacology, University of Bristol, Bristol, UK

Edward Chow Department of Radiation Oncology, Sunnybrook Health Sciences Centre Odette Cancer Centre (T2-182), Toronto, ON, Canada

Julie A. Christianson Department of Anatomy and Cell Biology, University of Kansas Medical Center, Kansas City, KS, USA

MacDonald J. Christie Discipline of Pharmacology, University of Sydney, Sydney, Australia

Jin Mo Chung Department of Neuroscience and Cell Biology, University Chair in Neuroscience, University of Texas Medical Branch, Galveston, TX, USA

Kyungsoon Chung Department of Neuroscience and Cell Biology, University of Texas Medical Branch, Galveston, TX, USA

Anna K. Clark Wolfson Centre for Age Related Diseases, King's College London, London, UK

W. Crawford Clark College of Physicians and Surgeons, Columbia University, New York, USA

James F. Cleary Pain & Policy Studies Group, UW Carbone Cancer Center, Madison, WI, USA

John De Clercq Department of Rehabilitation Sciences and Physiotherapy, Ghent University Hospital, Ghent, Belgium

Jacqui Clinch Royal National Hospital for Rheumatic Diseases NHS Foundation Trust, Bath, UK

Terence J. Coderre Departments of Anesthesia, Neurology & Neurosurgery and Psychology, McGill University, Montreal, QC, Canada

Robert Coghill Department of Neurobiology and Anatomy, Wake Forest University School of Medicine, Winston-Salem, NC, USA

Lindsey Cohen Psychology, Georgia State University, Atlanta, GA, USA

Alan L. Colledge Utah Labor Commission, International Association of Industrial Accident Boards and Commissions Central Utah Medical Clinic, South Salt Lake City, UT, USA

Beverly J. Collett University Hospitals of Leicester, Leicester, UK

Lesley A. Colvin Department of Anaesthesia, Critical Care and Pain Medicine Western General Hospital, Edinburgh, UK

Kevin C. Conlon Trinity College Dublin, University of Dublin, Dublin, Ireland

Mark Connelly Integrative Pain Management, Children's Mercy Hospitals and Clinics, University of Missouri-Kansas City School of Medicine, Kansas City, MO, USA

Brian Y. Cooper Department of Oral and Maxillofacial Surgery, University of Florida, Gainesville, FL, USA

Christine Cordle University Hospitals of Leicester, Leicester, UK

Laura A. Cousins Psychology, Georgia State University, Atlanta, GA, USA

Michael J. Cousins Department of Anesthesia and Pain Management, Royal North Shore Hospital University of Sydney, Sydney, NSW, Australia

Verne C. Cox Department of Psychology, University of Texas at Arlington, Arlington, TX, USA

George A. Crabill Department of Neurosurgery, Johns Hopkins Hospital, Baltimore, MA, USA

Rebecca M. Craft Psychology, Washington State University, Pullman, WA, USA

Kenneth D. Craig Department of Psychology, University of British Columbia, Vancouver, BC, Canada

Richard Crevenna Medizinische Universität Wien, Wien, Austria

Geert Crombez Department of Experimental Clinical and Health Psychology, Ghent University, Ghent, Belgium

Joan Crook School of Nursing, Faculty of Health Sciences, McMaster University, Hamilton, ON, Canada

Giorgio Cruccu Department of Neurological Sciences, La Sapienza University, Rome, Italy

Michele Curatolo Department of Anesthesiology, Division of Pain Therapy, University Hospital of Bern Inselspital, Bern, Switzerland

Nachum Dafny Neurobiology and Anatomy, Health Science Center, University of Texas, Houston, TX, USA

Jorgen B. Dahl Department of Anaesthesiology, Glostrup University Hospital, Herlev, Denmark

Yi Dai Department of Anatomy and Neuroscience, Hyogo College of Medicine, Hyogo, Japan

Stefaan Van Damme Department of Experimental Clinical and Health Psychology, Ghent University, Ghent, Belgium

Alexandre F. M. DaSilva Biologic & Materials Sciences, University of Michigan Dental School, MI, USA

Steve Davidson Pain Center & Department of Anesthesiology, Washington University School of Medicine, MO, USA

Brian Davis Department of Neurobiology, University of Pittsburgh School of Medicine, Pittsburgh, PA, USA

Karen D. Davis Brain, Imaging and Behaviour – Systems Neuroscience, Toronto Western Research Institute, Toronto, ON, Canada

Department of Surgery, University of Toronto, Toronto, ON, Canada

Richard O. Day Department of Clinical Pharmacology and Toxicology, St Vincent's Hospital, Sydney, NSW, Australia

Isabelle Decosterd Pain Center, University Hospital Center and University of Lausanne, Lausanne, Switzerland

Ruth Defrin Physical Therapy, Faculty of Medicine, Tel-Aviv University, Tel-Aviv, Israel

Antoon F. C. De Laat Department of Oral/Maxillofacial Surgery, Catholic University of Leuven, Leuven, Belgium

Joyce DeLeo Dartmouth Hitchcock Medical Center, Dartmouth Medical School, Hanover, NH, USA

A. Lee Dellon Department of Plastic Surgery and Neurosurgery, Johns Hopkins University, Baltimore, USA

Dellon Institutes for Peripheral Nerve Surgery, Baltimore, MD, USA

Marshall Devor Department of Cell & Developmental Biology, Institute of Life Sciences and Center for Research on Pain, The Hebrew University of Jerusalem, Jerusalem, Givat Ram, Israel

Perry Dhaliwal Department of Neurosurgery, Cleveland Clinic Spine Institute, Cleveland, OH, USA

David Diamant Neurological and Spinal Surgery, Lincoln, NE, USA

Anthony H. Dickenson Department of Pharmacology, University College London, London, UK

Hans-Christoph Diener Department of Neurology, University Hospital Essen, University Duisburg-Essen, Essen, Germany

Natalia Dmitrieva Program in Neuroscience, Florida State University, Tallahassee, FL, USA

L. F. Donaldson School of Life Sciences, University of Nottingham, Nottingham, UK

Henri Doods CNS Pharmacology, Boehringer Ingelheim Pharma GmbH & Co. KG, Biberach, Germany

Victoria Dorf Social Security Administration, Baltimore, MD, USA

Michael J. Dorsi Department of Neurosurgery, Johns Hopkins School of Medicine, Baltimore, MD, USA

Jonathan O. Dostrovsky Department of Physiology, University of Toronto, Toronto, ON, Canada

Faculty of Dentistry, University of Toronto, Toronto, ON, Canada

Robyn Drach Fairview Pain Management Center, Minneapolis, MN, USA

Ron Dressler Utah Labor Commission, International Association of Industrial Accident Boards and Commissions Central Utah Medical Clinic, South Salt Lake City, UT, USA

Paul Dreyfuss Department of Rehabilitation Medicine, University of Washington, Seattle, USA

Peter D. Drummond School of Psychology, Murdoch University, Perth, Australia

Ronald Dubner Department of Neural and Pain Sciences, University of Maryland Dental School, Baltimore, MD, USA

Anne Ducros Headache Emergency Department, INSERM, Paris, France

Gary H. Duncan Département de stomatologie, Faculté de médecine dentaire, Université de Montréal, Montreal, QC, Canada

Department of Neurology and Neurosurgery, McGill University, Montreal, QC, Canada

Mary J. Eaton Research Service (151), Miami VA Health Care System, Miami VAMC, Miami, FL, USA

Andrea Ebersberger Department of Physiology, Friedrich Schiller University of Jena, Jena, Germany

Christopher Eccleston Pain Management Unit, University of Bath, Bath, UK

John Edmeads Sunnybrook and Woman's College Health Sciences Centre, University of Toronto, Toronto, Canada

Edzard Ernst Complementary Medicine, Peninsula Medical School Universities of Exeter and Plymouth, Exeter, UK

Reuben Escorpizo Swiss Paraplegic Research, Nottwil, Switzerland

Department of Health Sciences and Health Policy, University of Lucerne, Lucerne, Switzerland

ICF Research Branch in cooperation with the World Health Organization (WHO) Collaborating Centre for the Family of International Classifications in Germany, Nottwil, Switzerland

Schonstein Eva RehabCo, Fyshwick, ACT, Australia

Joshua C. Eyer Department of Psychology, University of Alabama, Tuscaloosa, AL, USA

Marie Faaborg-Andersen McGill University, Montreal, QC, Canada

Keri L. Fakata College of Pharmacy and Pain Management Center, University of Utah, Salt Lake City, USA

Marie Fallon Department of Oncology, Edinburgh Cancer Center University of Edinburgh, Edinburgh, UK

Melissa Farmer Department of Physiology / Feinberg School of Medicine, Northwestern University, Chicago, IL, USA

Samantha R. Fashler Department of Psychology, University of British Columbia, Vancouver, BC, Canada

Jean-Baptiste Fassier Department of Occupational Health and Medicine, UMRESTTE-Université Claude Bernard Lyon 1, Lyon, France

Charlotte Feinmann Eastman Dental Institute London, University College London, London, UK

Elizabeth Roy Felix The Miami Project to Cure Paralysis, University of Miami Miller School of Medicine, Miami, FL, USA

Bin Feng Center for Pain Research, Department of Anesthesiology, School of Medicine, University of Pittsburgh, Pittsburgh, Pennsylvania, USA

Yi Feng Department of Anesthesiology, Peking University People's Hospital, Peking, China

Andres Fernandez Department of Neurosurgery, Johns Hopkins Hospital, Baltimore, MD, USA

Antonio Vicente Ferrer-Montiel Institute of Molecular and Cell Biology, University Miguel Hernández, Elche, Alicante, Spain

Perry G. Fine Pain Research Center, School of Medicine, University of Utah, Salt Lake City, UT, USA

David J. Fink Department of Neurology, University of Michigan and VA Ann Arbor Healthcare System, Ann Arbor, MI, USA

Nanna Brix Finnerup Aarhus University Hospital, Danish Pain Research Center, Aarhus, Denmark

David A. Fishbain Department of Psychiatry, University of Miami, Miami, FL, USA

Maria Fitzgerald Neuroscience, Physiology & Pharmacology, University College London, London, UK

Per Flisberg Department of Anaesthesia and Intensive Care, Lund University Hospital, Lund, Sweden

Herta Flor Department of Cognitive and Clinical Neuroscience, Central Institute of Mental Health, Medical Faculty Mannheim, Heidelberg University, Mannheim, Germany

Christopher M. Flores Drug Discovery, Johnson and Johnson Pharmaceutical Research and Development, Spring House, PA, USA

Wilbert E. Fordyce Department of Rehabilitation, University of Washington School of Medicine, Seattle, WA, USA

Robert D. Foreman Department of Physiology, College of Medicine, University of Oklahoma Health Sciences Center, Oklahoma City, OK, USA

Gary M. Franklin Department of Environmental and Occupational Health Sciences, University of Washington, Seattle, WA, USA

Department of Health Services, University of Washington, Seattle, WA, USA

Jan Fridén Department of Hand Surgery, Sahlgrenska University Hospital, Gothenborg, Sweden

Justin Friedlander The Arthur Smith Institute for Urology, North Shore-LIJ Healthcare System, Hofstra North Shore LIJ School of Medicine, New Hyde Park, NY, USA

Allan H. Friedman Division of Neurosurgery, Duke University Medical Center, Durham, NC, USA

Perry N. Fuchs Department of Psychology, University of Texas at Arlington, Arlington, TX, USA

Deborah Fulton-Kehoe Department of Environmental and Occupational Health Sciences, University of Washington, Seattle, WA, USA

Andrea D. Furlan Institute for Work & Health, Toronto, ON, Canada

Vasco Galhardo Departamento de Biologia Experimental, Faculdade de Medicina, Universidade do Porto, Portugal

Andreas R. Gantenbein RehaClinic Bad Zurzach, Bad Zurzach, Switzerland

I. Garonzik Department of Neurosurgery, Johns Hopkins Medical Institutions, Baltimore, MD, USA

Robert Gassin Musculoskeletal Medicine Clinic, Frankston, VIC, Australia

Robert J. Gatchel Department of Psychology, The University of Texas, Arlington, TX, USA

Gerald F. Gebhart Department of Anesthesiology, Center for Pain Research, University of Pittsburgh, Pittsburgh, PA, USA

Gerd Geisslinger Department of Clinical Pharmacology, University Frankfurt am Main, Frankfurt, Germany

Robert D. Gerwin Department of Neurology, Johns Hopkins University, Bethesda, MD, USA

Maria Adele Giamberardino Pathophysiology of Pain Laboratory and Centre for Fibromyalgia and Musculoskeletal Pain, Department of Medicine and Science of Aging, University of Chieti, Chieti, Italy

Stephen Gibson National Ageing Research Institute Department of Medicine, University of Melbourne, Parkville, VIC, Australia

Glenn J. Giesler Department of Neuroscience, University of Minnesota, Minneapolis, MN, USA

Philip L. Gildenberg Baylor Medical College, Houston, TX, USA

Myra Glajchen Department of Pain Medicine and Palliative Care, Beth Israel Medical Center Albert Einstein College of Medicine, New York, NY, USA

Hartmut Göbel Kiel Migraine and Headache Center, Kiel Pain Center, Kiel, Germany

Peter Goadsby Headache Center, Department of Neurology, University of California, San Francisco, San Francisco, CA, USA

Philippe Goffaux Faculty of Medicine, University of Sherbrooke, Sherbrooke, QC, Canada

Ana Gomis Instituto de Neurociencias de Alicante, Universidad Miguel Hernández-CSIC, San Juan de Alicante, Spain

Gilbert R. Gonzales Department of Neurology, Memorial Sloan-Kettering Cancer Center, New York, USA

Allan Gordon Wasser Pain Management Centre, Mount Sinai Hospital University of Toronto, Toronto, ON, Canada

Liesbet Goubert Department of Experimental Clinical and Health Psychology, Ghent University, Ghent, Belgium

Jayantilal Govind Pain Clinic, Department of Anaesthesia, University of New South Wales, Sydney, NSW, Australia

Richard H. Gracely Departments of Medicine-Rheumatology and Neurology, University of Michigan Health System, Ann Arbor, MI, USA

Garry G. Graham Department of Pharmacology, School of Medical Sciences, University of New South Wales, Sydney, NSW, Australia

David K. Grandy Department of Physiology and Pharmacology, Oregon Health and Science University, Portland, OR, USA

Thomas Graven-Nielsen Laboratory for Musculoskeletal Pain and Motor Control, Center for Sensory-Motor Interaction, Aalborg University, Aalborg, Denmark

Barry Green Pierce Laboratory, Yale School of Medicine, New Haven, CT, USA

Paul G. Green Oral & Maxillofacial Surgery, University of California San Francisco, San Francisco, CA, USA

Joel D. Greenspan Department of Neural and Pain Sciences, University of Maryland School of Dentistry, Baltimore, MD, USA

John W. Griffin Department of Neurology, Johns Hopkins University School of Medicine, Baltimore, MD, USA

Cornelius Groenwald Department of Anesthesiology and Pain Medicine, Seattle Children's Hospital, University of Washington School of Medicine, Seattle, WA, USA

Douglas P. Gross Department of Physical Therapy, University of Alberta WCB-Alberta Millard Health, Edmonton, AB, Canada

Sabine Grösch Institute of Clinical Pharmacology, Clinical Center of the Goethe University, Frankfurt, Germany

Blair D. Grubb Department of Cell Physiology and Pharmacology, University of Leicester, Leicester, UK

Ruth Eckstein Grunau Pediatrics Dept, University of British Columbia Developmental Neurosciences & Child Health, Child & Family Research Institute, Vancouver, BC, Canada

Christoph Gutenbrunner Department of Physical Medicine and Rehabilitation, Hanover Medical School, Hanover, Germany

Maija Haanpää Department of Neurosurgery, Helsinki University Central Hospital, Helsinki, Finland

Sherri Haas Fairview Pain Management Center, Minneapolis, MN, USA

Ute Habel Department of Psychiatry and Psychotherapy, RWTH Aachen University, Aachen, Germany

Thomas Hachenberg Clinic for Anaesthesiology and Intensive Therapy, Otto-von-Guericke University, Magdeburg, Germany

Nouchine Hadjikhani Martinos Center for Biomedical Imaging, Massachusetts General Hospital Harvard Medical School, Charlestown, MA, USA

Ole Haegg Department of Orthopedic Surgery, Sahlgren University Hospital, Goteborg, Sweden

Bryan C. Hains Department of Neurology, Yale University School of Medicine, West Haven, CT, USA

Else K. B. Hals University of Oslo, Faculty of Dentistry, Oslo, Norway

Darryl T. Hamamoto Diagnostic and Biological Sciences, University of Minnesota, Minneapolis, MN, USA

Donna L. Hammond Department of Anesthesia, University of Iowa, Iowa City, IA, USA

Menachem Hanani Hadassah-Hebrew University Medical Center Hebrew University Medical School, Jerusalem, Israel

Hermann O. Handwerker Department of Physiology and Pathophysiology, University of Erlangen/Nürnberg, Erlangen, Germany

George M. Hanna Department of Anesthesia, Critical Care and Pain Medicine, Massachusetts General Hospital Harvard Medical School, Boston, MA, USA

Jing-Xia Hao Department of Physiology and Pharmacology, Karolinska Institutet, Stockholm, Sweden

Eleni G. Hapidou Department of Psychiatry and Behavioural Neurosciences, McMaster University and Hamilton Health Sciences Chronic Pain Management Unit, Chedoke Hospital, Hamilton, ON, Canada

Geoffrey Harding Sandgate Spinal Medicine Clinic, Sandgate, QLD, Australia

Louise Harding Hospitals NHS Foundation Trust, University College London, London, UK

Kenneth M. Hargreaves Department of Endodontics, University of Texas Health Science Center at San Antonio, San Antonio, USA

D. Paul Harries Pain Treatment Center, Samaritan Hospital Lexington, Lexington, KY, USA

Denise Harrison Nursing, Children's Hospital of Eastern Ontario (CHEO), University of Ottawa, Ottawa, ON, Canada

Steven Harte Anesthesiology and Internal Medicine/Rheumatology, University of Michigan VA Ann Arbor Healthcare System, Ann Arbor, MI, USA

Barbara A. Hastie Community Dentistry & Behavioral Science, University of Florida College of Dentistry, Gainesville, FL, USA

Jennifer A. Haythornthwaite Psychiatry & Behavioral Sciences, Johns Hopkins University, Baltimore, MD, USA

Heinz-Joachim Häbler FH Bonn-Rhein-Sieg, Rheinbach, Germany

John H. Healey Orthopaedic Service, Department of Surgery, Memorial Sloan-Kettering Cancer Center Weill Medical College of Cornell University, New York, NY, USA

Alicia A. Heapy VA Connecticut Healthcare System, Westhaven, CT, USA

Anne M. Heffernan The Adelaide and Meath Hospital Incorporating The National Children's Hospital, Dublin, Ireland

Geert Van der Heijden Julius Center, Department of Epidemiology, University Medical Center Utrecht, Utrecht, The Netherlands

Mary M. Heinricher Departments of Neurological Surgery and Behavioral Neuroscience, Oregon Health & Science University, Portland, USA

Robert D. Helme Department of Medicine, Royal Melbourne Hospital, University of Melbourne, Melbourne, Australia

Kim Helsen Research Group on Health Psychology, KU Leuven Katholieke Universiteit Leuven, Leuven, Belgium

Department of Experimental–Clinical and Health Psychology, Ghent University, Ghent, Belgium

Amin S. Herati The Arthur Smith Institute for Urology, North Shore-LIJ Healthcare System, Hofstra North Shore LIJ School of Medicine, New Hyde Park, NY, USA

Robert D. Herbert Neuroscience Research Australia, Sydney, Australia

Christiane Hermann Department of Clinical Psychology, Justus-Liebig University Giessen, Giessen, Germany

Juan F. Herrero Department of Physiology, Faculty of Medicine University of Alcala, Alcalá de Henares Madrid, Spain

Aki J. Hietaharju Department of Neurology and Rehabilitation, Pain Clinic Tampere University Hospital, Tampere, Finland

Karin Westlund High Department of Physiology, University of Kentucky, Lexington, KY, USA

Marita Hilliges Dalarna University, Halmstad, Sweden

Max J. Hilz University of Erlangen-Nuremberg, Erlangen, Germany

Burkhard Hinz Institute of Toxicology and Pharmacology, University of Rostock, Rostock, Germany

Anthony R. Hobson Section of GI Sciences, Hope Hospital University of Manchester, Salford, UK

Edward Högestätt Department of Laboratory Medicine, Lund University, Division of Clinical Chemistry and Pharmacology, Lund, Sweden

Tomas Hökfelt Department of Neuroscience, Karolinska Institute, Stockholm, Sweden

Sarah V. Holdridge Department of Pharmacology and Toxicology, Queen's University, Kinston, Canada

Kjell Hole University of Bergen, Bergen, Norway

Graham R. Holland Department of Cariology, University of Michigan, Ann Arbor, MI, USA

Chang-Zern Hong Department of Physical Medicine and Rehabilitation, University of California, Irvine, CA, USA

Minoru Hoshiyama Department of Integrative Physiology, National Institute for Physiological Sciences, Okazaki, Japan

Fred M. Howard University of Rochester, Rochester, NY, USA

Sung-Tsang Hsieh Department of Anatomy and Cell Biology, Department of Neurology, National Taiwan University Hospital, National Taiwan University, Taipei, Taiwan

James W. Hu Faculty of Dentistry, University of Toronto, Toronto, Canada

Claire E. Hulsebosch Spinal Cord Injury Research, University of Texas Medical Branch, Galveston, TX, USA

Thomas Hummel Department of Otorhinolaryngology, Technical University of Dresden Medical School, Dresden, Germany

Mark R. Hutchinson School of Medical Sciences, University of Adelaide, Adelaide, SA, Australia

Charles E. Inturrisi Department of Pharmacology, Weill Cornell Medical College, New York, USA

Koji Inui Department of Integrative Physiology, National Institute for Physiological Sciences, Okazaki, Japan

Koichi Iwata Department of Physiology, Nihon University School of Dentistry, Chiyoda-ku, Japan

Suhayl J. Jabbur Anatomy, Cell Biology and Physiology, Faculty of Medicine, American University of Beirut, Beirut, Lebanon

Neuroscience Program, Faculty of Medicine, American University of Beirut, Beirut, Lebanon

Nora Janjan University of Texas M. D. Anderson Cancer Center, Houston, TX, USA

Luc Jasmin Department of Anatomy, University of California San Francisco, San Francisco, CA, USA

Anne Jaumees Royal North Shore Hospital, St Leonards, NSW, Australia

Wilfrid Jaenig Physiologisches Institut, Christian-Albrechts-Universität zu Kiel, Kiel, Germany

Daniel Jeanmonod Functional Neurosurgery, The Center of Ultrasound Functional Neurosurgery, Solothurn, Switzerland

Andrew Jeffreys Director Department of Anaesthesia & Pain Medicine, Western Health, Melbourne, VIC, Australia

Mark P. Jensen Department of Rehabilitation Medicine, University of Washington, Seattle, WA, USA

Troels Staehelin Jensen Danish Pain Research Center, Department of Neurology, Aarhus University Hospital, Aarhus, Denmark

Kurt L. Johnson Department of Rehabilitation Medicine, University of Washington, Seattle, WA, USA

Mark I. Johnson Faculty of Health and Social Sciences, Leeds Pallium Research Group, Leeds Metropolitan University, Leeds, UK

Mark Johnston Hibiscus Coast Highway, Orewa, New Zealand

Edward Jones Center for Neuroscience, University of California, Davies, CA, USA

Jeroen R. De Jong Department of Rehabilitation, University Hospital Maastricht, Maastricht, The Netherlands

Sven-Eric Jordt Department of Pharmacology, Yale University School of Medicine, New Haven, CT, USA

Ellen Jørum The Laboratory of Clinical Neurophysiology, Rikshospitalet University, Oslo, Norway

Stefan Just CNS Pharmacology, Boehringer Ingelheim Pharma GmbH & Co. KG, Biberach, Germany

Timothy K. Y. Kaan Department of Anatomy, University of California San Francisco, San Francisco, CA, USA

Noufissa Kabli Department of Pharmacology and Toxicology, Queen's University, Kinston, Canada

Ryusuke Kakigi Department of Integrative Physiology, National Institute for Physiological Sciences, Okazaki, Japan

Peter C. A. Kam Anaesthesia, Central Clinical School, The University of Sydney, Sydney, Australia

Giresh Kanji The Sports and Pain Clinic, Wellington, New Zealand

Joel Katz Department of Psychology, McGill University, Montreal, QC, Canada

Valerie Kayser NSERM U677 91Bd de l'Hôpital, Paris, France

Francis J. Keefe Pain Prevention and Treatment Research, Department of Psychiatry and Behavioral Sciences, Duke University Medical Center, Durham, NC, USA

Lois J. Kehl Department of Anesthesiology, University of Minnesota, Minneapolis, MN, USA

Maureen Kelley Department of Pediatrics, Bioethics Division, University of Washington School of Medicine, WA, USA

Seattle Children's Hospital, Seattle, WA, USA

Stephen Kennedy Nuffield Department of Obstetrics and Gynaecology, University of Oxford, Oxford, UK

Edmund Keogh Department of Psychology, University of Bath, Bath, UK

Robert D. Kerns VA Connecticut Healthcare System, Westhaven, CT, USA

Sanjay Khanna Department of Physiology, National University of Singapore, Singapore

Kok Eng Khor Department of Pain Management, Prince of Wales Hospital University of New South Wales, Randwick, NSW, Australia

Mina Khoshnejad Department of Neuroscience, University of Montreal, Montreal, QC, Canada

H. J. Kim Department of Life Science, Yonsei University, Wonju, Republic of Korea

Sara King Faculty of Education (School Psychology), Mount Saint Vincent University, Halifax, NS, Canada

Wade King Department of Clinical Research, Royal Newcastle Centre University of Newcastle, Newcastle, NSW, Australia

K. L. Kirsh Behavioral Medicine, The Pain Treatment Center of the Blue grass, Lexington, KY, USA

Henriette Klit Danish Pain Research Center, Aarhus University Hospital, Aarhus C, Denmark

Ricardo J. Komotar Department of Neurological Surgery, Neurological Institute Columbia University, New York, NY, USA

Sungtae Koo Division of Meridian and Structural Medicine, School of Korean Medicine, Pusan National University, Yangsan, Korea

Rhonda K. Kotarinos Oakbrook Terrace, IL, USA

Katalin J. Kovacs Department of Veterinary and Biomedical Sciences, University of Minnesota, St. Paul, MN, USA

Malgorzata Krajnik Chair of Palliative Care, Nicolaus Copernicus University in Torun, Collegium Medicum in Bydgoszcz, Bydgoszcz, Poland

Elliot Krane Anesthesia and Pediatrics, Stanford University School of Medicine Lucile Packard Children's Hospital at Stanford, Stanford, CA, USA

Ajit Krishnaney Department of Neurosurgery, Cleveland Clinic, Cleveland, OH, USA

Greg Krohm International Association of Industrial Accident Boards and Commissions (IAIABC), Utah Valley University, USA

Lawrence Kruger UCLA Geffen School of Medicine, Los Angeles, CA, USA

M. M. ter Kuile Department of Gynecology, Leiden University Medical Center, Leiden, The Netherlands

Promil Kukreja Department of Anesthesiology, University of Alabama at Birmingham, Birmingham, AL, USA

Alice Kvåle Department of Public Health and Primary Health Care, Faculty of Medicine and Dentistry University of Bergen, Bergen, Norway

Gunnvald Kvarstein Department of Pain Management and Pain Research, Oslo University Hospital, Rikshospitalet, Oslo, Norway

Jean-Jacques Labat Neurology and Rehabilitation, Clinique Urologique, Nantes, France

Gaston Labrecque Faculty of Pharmacy, University of Laval, Quebec City, QC, Canada

Marco Lacerenza Pain Medicine Center, Casa di Cura S. Pio X, Opera San Camillo Foundation, Milan, Italy

Uri Ladabaum Gastroenterology & Hepatology, Stanford Cancer Institute, Redwood City, CA, USA

Marle Andree Lahaie McGill University, Montreal, QC, Canada

Miguel J. A. Lainez-Andres Department of Neurology, University of Valencia, Valencia, Spain

Robert H. LaMotte Department of Anesthesiology, Yale School of Medicine, New Haven, CT, USA

James W. Lance Institute of Neurological Sciences, University of New South Wales, Sydney, Australia

Robert G. Large The Auckland Regional Pain Service, Auckland Hospital, Auckland, New Zealand

Alice A. Larson Department of Veterinary and Biomedical Sciences, University of Minnesota, St. Paul, MN, USA

Roberta Lattanzi Department of Physiology and Pharmacology, Sapienza University of Rome, Rome, Italy

Peter Lau Department of Clinical Research, Royal Newcastle Hospital, University of Newcastle, Newcastle, NSW, Australia

Stefan A. Laufer Institute of Pharmacy, Eberhard-Karls University of Tuebingen, Tuebingen, Germany

Stefan Lautenbacher Physiologische Psychologie, Otto-Friedrich-Universität, Bamberg, Germany

Gilles J. Lavigne Facultés de Médecine Dentaire et de Médecine, Université de Montréal, Hopital Sacre Coeur de Montreal. Surgery-trauma dept, Montréal, QC, Canada

Emily Law Center for Child Health, Behavior, and Development, Seattle Children's Research Institute, Seattle, WA, USA

Peter G. Lawlor Our Lady's Hospice, Medical Department, Dublin, Ireland

Michel Lazdunski Institut de Pharmacologie Moléculaire et Cellulaire (IPMC) CNRS/UNS UMR7275, Valbonne-Sophia Antipolis, France

Vincenzo Di Lazzaro Institute of Neurology, Campus Biomedico University, Rome, Italy

Alyssa Lebel Boston Children's Hospital, Boston, MA, USA

Frank Lee Johns Hopkins University, Baltimore, MD, USA

Maaike Leeuw Department of Medical, Clinical and Experimental Psychology, Maastricht University, Maastricht, The Netherlands

Siri Graff Leknes Research Fellow, Psykologisk institutt, Oslo, Norway

Frederick A. Lenz Department of Neurosurgery, Johns Hopkins Hospital, Baltimore, MD, USA

Guillaume Léonard Faculty of Medicine, University of Sherbrooke, Sherbrooke, QC, Canada

Pauline Lesage Department of Pain Medicine and Palliative Care, Beth Israel Medical Center, New York, USA

Isobel Lever Neural Plasticity Unit, Institute of Child Health University College London, London, UK

Khan W. Li Department of Neurosurgery, Johns Hopkins Hospital, Baltimore, MD, USA

Alan R. Light Department of Anesthesiology, University of Utah, Salt Lake City, UT, USA

Michael Lim Department of Neurosurgery and Department of Neurology, Johns Hopkins Hospital, Baltimore, MA, USA

Deolinda Lima Department of Experimental Biology, Faculdade de Medicina da Universidade do Porto, Porto, Portugal

Volker Limmroth Department of Neurology, University of Essen, Essen, Germany

Eric Lingueglia Institut de Pharmacologie Moléculaire et Cellulaire (IPMC) CNRS/UNS UMR7275, Valbonne-Sophia Antipolis, France

Steven J. Linton Department of Occupational and Environmental Medicine, Orebro Medical Center, Orebro, Sweden

Arthur G. Lipman Departments of Pharmscotherpay and Anesthesiology and Pain Management Center, University of Utah, Salt Lake City, UT, USA

Richard B. Lipton Departments of Neurology, Epidemiology and Population Health, Albert Einstein College of Medicine, Bronx, NY, USA

Diana Lisi Department of Psychology, York University, Toronto, ON, Canada

Kenneth M. Little Division of Neurosurgery, Duke University Medical Center, Durham, NC, USA

Edward Littleton Queen Elizabeth Hospital, Birmingham, UK

Spencer S. Liu Virginia Mason Medical Center, Seattle, WA, USA

Anne Elisabeth Ljunggren Section for Physiotherapy Science, Department of Public Health and Primary Health, Faculty of Medicine, University of Bergen, Bergen, Norway

John D. Loeser Department of Neurological Surgery, University of Washington, Seattle, WA, USA

Joern Loetsch Institute for Clinical Pharmacology, Pharmaceutical Center Frankfurt, Johann Wolfgang Goethe University, Frankfurt, Germany

Patrick Loisel Disability Prevention Research and Training Centre, Université de Sherbrooke, Longueuil, Canada

Elisa Lopez-Dolado Servicio de Rehabilitacion, National Hospital of Paraplèjicos SESCAM, Toledo, Spain

Jürgen Lorenz Faculty of Life Science, Competence Center Gesundheit, University of Applied Science Hamburg, Hamburg, Germany

Dayna R. Loyd US Army Institute of Surgical Research, Fort Sam Houston, TX, USA

Daniel Lubelski Cleveland Clinic Lerner College of Medicine, Cleveland, OH, USA

Stephen Lutz Blanchard Valley Regional Cancer Center, Findlay, OH, USA

Bruce Lynn Department of Physiology, University College London, London, UK

Ross MacPherson Department of Anesthesia and Pain Management, Royal North Shore Hospital University of Sydney, Sydney, NSW, Australia

Christophe Maes Manager Innovatie AZ Alma, Belgium

Michel Magnin INSERM, Neurological Hospital, Lyon, France

Bryan Mahony Department of Anesthesiology, Ohio State University Medical Center, Mount Sinai Hospital, New York, NY, USA

Steven F. Maier Department of Psychology and Neuroscience, and The Center for Neuroscience, University of Colorado at Boulder, Boulder, CO, USA

Thorsten J. Maier Institute of Pharmaceutical Chemistry, Goethe University, Frankfurt, Germany

Chris J. Main Keele University, Keele, Staffordshire, UK

Marzia Malcangio Wolfson Centre for Age Related Diseases, King's College London, London, UK

Paolo L. Manfredi Department of Neurology, Memorial Sloan-Kettering Cancer Center, New York, USA

Kaisa Mannerkorpi Department of Rheumatology and Inflammation Research, Institute of Medicine, University of Gothenburg, Göteborg, Sweden

Barton H. Manning Amgen Inc., Thousand Oaks, CA, USA

Patrick W. Mantyh Department of Pharmacology, The University of Arizona, AZ, USA

Jianren Mao Department of Anesthesia and Critical Care, Harvard Medical School, Boston, MA, USA

Serge Marchand Faculty of Medicine, University of Sherbrooke, Sherbrooke, QC, Canada

Paolo Marchettini Pain Medicine Center, Scientific Institute San Raffaele, Milan, Italy

Sarah R. Martin Psychology, Georgia State University, Atlanta, GA, USA

Peggy Mason Department of Neurobiology, Pharmacology and Physiology, University of Chicago, Chicago, IL, USA

Scott Masters Caloundra Spinal and Sports Medicine Centre, Caloundra, QLD, Australia

Laurence E. Mather Department of Anaesthesia and Pain Management, University of Sydney at Royal North Shore Hospital, St. Leonards, Sydney, NSW, Australia

Bill McCarberg Pain Services, Kaiser Permanente, San Diego, USA

Lance McCracken Health Psychology/Psychology Department, King's College London, London, UK

Kathleen McGinn Anesthesiology & Pain Medicine, Seattle Children's Hospital, Seattle, WA, USA

Patricia A. McGrath The Hospital for Sick Children Pain Research Center, Toronto, ON, Canada

Patrick McGrath Departments of Psychology, Pediatrics and Psychiatry, Dalhousie University, Canada

Greg McIntosh Canadian Back Institute, Toronto, ON, Canada

Peter McIntyre Department of Pharmacology, University of Melbourne, Melbourne, Australia

Elspeth M. McLachlan Neuroscience Research Australia, University of New South Wales, Randwick, NSW, Australia

University of Glasgow, Glasgow, UK

Peter A. McNaughton Department of Pharmacology, University of Cambridge, Cambridge, UK

J. Mark Melhorn Section of Orthopaedics, Department of Surgery, University of Kansas School of Medicine, Wichita, KS, USA

Ronald Melzack Department of Psychology, McGill University, Montreal, QC, Canada

Lorne M. Mendell Department of Neurobiology and Behavior, Stony Brook University, Stony Brook, NY, USA

George Mendelson Department of Psychological Medicine, Monash University, Caulfield, VIC, Australia

Siegfried Mense Neuroanatomy, Heidelberg University, Medical Faculty Mannheim, CBTM, Mannheim, Germany

Daniel Menétrey Biomédicale, Paris, France

Sebastiano Mercadante Pain Relief & Palliative Care, La Maddalena Cancer Center University of Palermo, Palermo, Italy

Susan Mercer School of Biomedical Sciences, The University of Queensland, Brisbane, Australia

Karl Messlinger University of Erlangen-Nürnberg, Erlangen, Germany

Richard A. Meyer Department of Neurosurgergy, School of Medicine Johns Hopkins University, Baltimore, MD, USA

Walter J. Meyer III Shriners Hospitals for Children, Shriners Burns Hospital, Galveston, TX, USA

Department of Psychiatry, University of Texas Medical Branch, Galveston, TX, USA

Björn A. Meyerson Department of Clinical Neuroscience, Section of Neurosurgery, Karolinska Institute/University Hospital, Stockholm, Sweden

Martin Michaelis Sanofi-Aventis Deutschland GmbH, Frankurt am Main, Germany

Joseph Mignogna Michael E. DeBakey VA Medical Center, Houston VA HSR&D Center of Excellence (152), Houston, TX, USA

Judith A. Mikacich Department of Obstetrics and Gynecology, University of California Los Angeles Medical Center, Los Angeles, CA, USA

Kensaku Miki Department of Integrative Physiology, National Institute for Physiological Sciences, Okazaki, Japan

Vjekoslav Miletic School of Medicine and Public Health, University of Wisconsin, Madison, WI, USA

Scott R. Miller Memorial Sloan-Kettering Cancer Center, New York, USA

Erin D. Milligan Department of Neurosciences, University of New Mexico, Albuquerque, NM, USA

Michael S. Minett Molecular Nociception Group, Wolfson Institute for Biomedical Research, University College London, London, WC, UK

J. Mocco Department of Neurological Surgery, Neurological Institute Columbia University, New York, NY, USA

Jeffrey S. Mogil Department of Psychology, McGill University, Montreal, QC, Canada

Harvey Moldofsky Sleep Disorders Clinic of the Centre for Sleep and Chronobiology, Toronto, ON, Canada

Robert M. Moldwin The Arthur Smith Institute for Urology, North Shore-LIJ Healthcare System, Hofstra North Shore LIJ School of Medicine, New Hyde Park, NY, USA

Derek C. Molliver Department of Neurobiology, University of Pittsburgh School of Medicine, Pittsburgh, PA, USA

Robert D. Mootz Washington State Department of Labor and Industries, Olympia, WA, USA

Laura Mordillo-Mateos FENNSI Group, Hospital Nacional de Paraplejicos, Toledo, Spain

Anne Morel Laboratory for Functional Neurosurgery, Neurosurgical Clinic University Hospital Zürich, Zürich, Switzerland

Anne Morinville Montreal Neurological Institute, McGill University, Montreal, QC, Canada

Stephen Morley Academic Unit of Psychiatry and Behavioral Sciences, University of Leeds, Leeds, UK

David B. Morris University of Virginia, Charlottesville, VA, USA

G. Lorimer Moseley Sansom Institute for Health Research, University of South Australia Neuroscience Research Australia, Adelaide, Australia

Joachim Mossner Medical Clinic and Policlinic II, University Clinical Center Leipzig AöR, Leipzig, Germany

Eric A. Moulton P.A.I.N. Group, Anesthesiology, Boston Children's Hospital, Harvard Medical School, Waltham, MA, USA

Francis Githae Muriithi Department of Obstetrics and Gynaecology, Aga Khan University, Nairobi, Kenya

Anne Z. Murphy Neuroscience Institute, Georgia State University, Atlanta, GA, USA

Paul M. Murphy Department of Anaesthesia and Pain Management, Royal North Shore Hospital, St. Leonards, NSW, Australia

Alison Murray Foothills Medical Center, Calgary, AB, Canada

H. S. Na Neuroscience Research Institute and Department of Physiology, Korea University, College of Medicine, Seoul, Republic of Korea

Daisuke Naka Department of Integrative Physiology, National Institute for Physiological Sciences, Okazaki, Japan

Yoshio Nakamura Pain Research Center, Department of Anesthesiology University of Utah School of Medicine, Salt Lake City, UT, USA

Barbara Namer Department of Physiology & Pathophysiology, FAU, Erlangen/Nuremberg, Bavaria, Germany

Matti V. O. Närhi Department of Biomedicine/Physiology, Department of Dentistry, University of Eastern Finland, Institute of Medicine, Kuopio, Finland

H. J. W. Nauta University of Texas Medical Branch, Galveston, TX, USA

Lucia Negri Department of Physiology and Pharmacology, Sapienza University of Rome, Rome, Italy

Gary Nesbit Department of Radiology, Oregon Health and Science University, Portland, OR, USA

Timothy Ness Department of Anesthesiology, University of Alabama at Birmingham, Birmingham, AL, USA

Volker Neugebauer Department of Neuroscience & Cell Biology, University of Texas Medical Branch, Galveston, TX, USA

Lawrence C. Newman Albert Einstein College of Medicine, New York, NY, USA

Michael K. Nicholas Royal North Shore Hospital, University of Sydney, St. Leonards, NSW, Australia

Ellen Niederberger Pharmazentrum frankfurt, Institute of Clinical Pharmacology, Goethe-University Hospital, Frankfurt, Germany

Lone Nikolajsen Department of Anaesthesiology, Danish Pain Research Center Aarhus University Hospital, Aarhus, Denmark

Katie Nixdorf Comprehensive Pain Center, Oregon Health and Science University, Portland, OR, USA

Donald R. Nixdorf Department of Diagnostic and Biological Sciences, School of Dentistry University of Minnesota, Minneapolis, MN, USA

Carl E. Noe University of Texas Southwestern Medical Center, Lubbock, TX, USA

Koichi Noguchi Department of Anatomy and Neuroscience, Hyogo College of Medicine, Hyogo, Japan

Richard B. North The Johns Hopkins University, Baltimore, MD, USA

Turo J. Nurmikko Department of Neurological Science, University of Liverpool, and Pain Research Institute, The Walton Centre NHS Foundation Trust, Liverpool, UK

Anne Louise Oaklander Nerve Injury Unit, Departments of Neurology and Pathology, Massachusetts General Hospital Harvard Medical School, Boston, MA, USA

Koichi Obata Department of Anatomy and Neuroscience, Hyogo College of Medicine, Hyogo, Japan

Tim F. Oberlander Division of Developmental Pediatrics, University of British Columbia, Vancouver, BC, Canada

Bruno Oertel Institute for Clinical Pharmacology, Pharmaceutical Center Frankfurt, Johann Wolfgang Goethe University, Frankfurt, Germany

Andrea C. Ogbonna Wolfson Centre for Age Related Diseases, King's College London, London, UK

Alfred T. Ogden Department of Neurological Surgery, Neurological Institute Columbia University, New York, NY, USA

S. Ohara Department of Neurosurgery, Johns Hopkins Hospital, Johns Hopkins Medical Institutions, Johns Hopkins University, Baltimore, MD, USA

Peter T. Ohara Department of Anatomy, University of California San Francisco, San Francisco, CA, USA

Akiko Okifuji Department of Anesthesiology, University of Utah, Salt Lake City, UT, USA

Jean-Louis Oliveras Inserm, Paris, France

Antonio Oliviero FENNSI Group, National Hospital of Paraplégicos, Toledo, Spain

Sean O'Mahony Palliative medicine, Rush University Medical Center, Chicago, IL, USA

Peregrine B. Osborne Dept Anatomy & Neuroscience, University of Melbourne, Melbourne, VIC, Australia

Michael H. Ossipov Department of Pharmacology, University of Arizona College of Medicine, Tucson, AZ, USA

Tonya M. Palermo Seattle Children's Hospital Research Institute, CW8-6 - Child Health, Behavior and Development, University of Washington, Seattle, WA, USA

Shreela Palit Department of Psychology, The University of Tulsa, Tulsa, OK, USA

Juan A. Pareja Department of Neurology, Fundación Hospital Alcorcón, Madrid, Spain

Steven D. Passik Psychiatry and Anesthesiology, Vanderbilt University, Nashville, TN, USA

Gavril W. Pasternak Laboratory of Molecular Neuropharmacology, Memorial Sloan-Kettering Cancer Center, New York, USA

S. H. Patel Department of Neurosurgery, Johns Hopkins Hospital, Baltimore, MD, USA

Gavin Pattullo Department of Anesthesia and Pain Management, Royal North Shore Hospital, University of Sydney, St. Leonards, NSW, Australia

Kevin Pauza Tyler Spine and Joint Hospital, Tyler, TX, USA

Eric M. Pearlman Pediatrics, Mercer University School of Medicine, The Children's Hospital at Memorial University Medical Center, Savannah, USA

Elvira De la Pena Instituto de Neurociencias, Universidad Miguel Hernández-CSIC, Alicante, Spain

Yuan Bo Peng Department of Psychology, University of Texas at Arlington, Arlington, TX, USA

Jose Pereira Foothills Medical Center, Calgary, AB, Canada

Edward Perl Department of Cell and Molecular Physiology, School of Medicine University of North Carolina, Chapel Hill, NC, USA

Tiffany Grace Perry Center for Spine Health, The Cleveland Clinic Foundation, Cleveland, OH, USA

Marie Pertin Pain Center, University Hospital Center and University of Lausanne, Lausanne, Switzerland

Antti Pertovaara Department of Physiology, Institute of Biomedicine University of Helsinki, Helsinki, Finland

Peter Peter Department of Neurobiology and Anatomy, University of Rochester Medical Center, Rochester, NY, USA

Lisa M. Peters Department of Anesthesiology and Pain Medicine, Seattle Children's Hospital, Seattle, WA, USA

Pramit Phal Department of Radiology, Oregon Health and Science University, Portland, OR, USA

Rebecca Pillai Riddell Department of Psychology, York University, Toronto, ON, Canada

Department of Psychiatry, Hospital for Sick Children, Toronto, ON, Canada

Issy Pilowsky University of Adelaide, Adelaide, Australia

Leon Plaghki Institute of Neuroscience (IoNS), Université Catholique de Louvain, Brussels, Belgium

Markus Ploner Department of Neurology, Technische Universität München, Munich, Germany

Esther M. Pogatzki-Zahn Department of Anesthesiology and Intensive Care Medicine, University Muenster, Muenster, Germany

Michael Polydefkis Department of Neurology, Johns Hopkins University School of Medicine, Baltimore, MD, USA

James D. Pomonis Algos Preclinical Services, Roseville, MN, USA

Dieter Pongratz Friedrich Baur Institute, Ludwig Maximilians University, Munich, Germany

Hayley Poole Hadley House, Leicester General Hospital, Leicester, UK

Frank Porreca Department of Pharmacology, University of Arizona Health Sciences Center, Tucson, AZ, USA

Russell K. Portenoy Department of Pain Medicine and Palliative Care, Beth Israel Medical Center, New York, NY, USA

Ian Power Critical Care and Pain Medicine, The University of Edinburgh, Edinburgh, UK

Donald D. Price Oral Surgery and Neuroscience, McKnight Brain Institute University of Florida, Gainesville, FL, USA

Daryl Pullman Medical Ethics, Memorial University of Newfoundland, St. John's, NL, Canada

Yunhai Qiu Department of Integrative Physiology, National Institute for Physiological Sciences, Okazaki, Japan

Michael Quittan Department of Physical Medicine and Rehabilitation, Kaiser-Franz-Joseph Hospital, Vienna, Austria

Raymond M. Quock Department of Pharmaceutical Sciences, Washington State University College of Pharmacy, Pullman, WA, USA

Jennifer Rabbitts Department of Anesthesiology and Pain Medicine, Seattle Children's Hospital, University of Washington School of Medicine, Seattle, WA, USA

Nicole Racine Psychology, York University, Toronto, ON, Canada

Gabor B. Racz Texas Tech University Health Sciences Center, Lubbock, TX, USA

Lukas Radbruch Department of Palliative Medicine, University of Bonn, Bonn, Germany

Robert B. Raffa Department of Pharmaceutical Sciences, Temple University School of Pharmacy, Philadelphia, PA, USA

Vasudeva Raghavendra Algos Therapeutics Inc., Saint Paul, MN, USA

Pierre Rainville Department of Stomatology, Faculty of Dental Medicine, University of Montreal, Montreal, QC, Canada

Srinivasa N. Raja Department of Anesthesiology and Critical Care Medicine, Johns Hopkins University, Baltimore, USA

Alan Randich University of Alabama, Birmingham, AL, USA

Andrea J. Rapkin Department of Obstetrics and Gynecology, University of California Los Angeles Medical Center, Los Angeles, CA, USA

Z. Harry Rappaport Department of Neurosurgery, Rabin Medical Center Tel-Aviv University, Petah Tiqva, Israel

Matthew N. Rasband Department of Molecular and Cellular Biology, Baylor College of Medicine, Farmington, Houston TX, USA

University of Connecticut Health Center, Farmington, CT, USA

Douglas Rasmusson Department of Physiology and Biophysics, Dalhousie University, Halifax, Canada

James P. Rathmell Department of Anesthesiology, University of Vermont College of Medicine, Burlington, USA

K. Ravishankar The Headache and Migraine Clinic, Jaslok Hospital and Research Centre Lilavati Hospital and Research Centre, Mumbai, India

Ke Ren Department of Neural and Pain Sciences, University of Maryland Dental School, Baltimore, MD, USA

Seth Resnick Silver Hill Hospital, New York University School of Medicine, USA

Cielito Reyes-Gibby The University of Texas MD Anderson Cancer Center, Houston, TX, USA

Jamie L. Rhudy Department of Psychology, The University of Tulsa, Tulsa, OK, USA

Alfredo Ribeiro-da-Silva Department of Pharmacology and Therapeutics, McGill University, Montreal, QC, Canada

Frank L. Rice Center for Neuropharmacology and Neuroscience, Albany Medical College, Albany, NY, USA

Matthias Ringkamp Department of Neurosurgergy, School of Medicine Johns Hopkins University, Baltimore, MD, USA

Pierre J. M. Rivière Ferring Research Institute, San Diego, CA, USA

Claude Robert Université Paris Descartes, INSERM UMR 894, France

Roger Robert Neurosurgery, Hotel Dieu, Nantes, CHU, France

James P. Robinson Department of Anesthesiology & Pain Medicine, University of Washington, Seattle, WA, USA

John Robinson Pain Management Centre, Burwood Hospital, Christchurch, New Zealand

Alfonso Romero-Sandoval Department of Pharmaceuticals and Administrative Sciences, Presbyterian College School of Pharmacy, Clinton, SC, USA

David Roselt Aberdovy Clinic, Bundaberg, QLD, Australia

Jackie Ross Early Pregnancy Unit, King's College Hospital Early Pregnancy Unit, London, UK

Shlomo Rotshenker Department of Medical Neurobiology, IMRIC, Faculty of Medicine Hebrew University of Jerusalem, Jerusalem, Israel

Paul C. Rousseau Medical University of South Carolina, Charleston, SC, USA

Ranjan Roy Department of Clinical Health Psychology, Faculty of Medicine, University of Manitoba, Winnipeg, MB, Canada

Carolina Roza Biología de Sistemas (Unidad Fisiología), Universidad de Alcalá, Alcalá de Henares, Madrid, Spain

Todd D. Rozen Department of Neurology, Geisinger Health System, Wilkes-Barre, PA, USA

Roman Rukwied Department of Anaesthesiology and Intensive Care Medicine, Faculty of Clinical Medicine Mannheim University of Heidelberg, Heidelberg, Germany

Irwin Jon Russell Division of Clinical Immunology and Rheumatology, Department of Medicine, The University of Texas Health Science Center, San Antonio, TX, USA

Nayef E. Saadé Anatomy, Cell Biology and Physiology, Faculty of Medicine, American University of Beirut, Beirut, Lebanon

Mary Ann C. Sabino Oral and Maxillofacial Surgery, Hennepin County Medical Center University of Minnesota, Minneapolis, MN, USA

Thomas E. Salt Institute of Ophthalmology, University College London, London, UK

A. Samdani Department of Neurosurgery, Johns Hopkins Medical Institutions, Baltimore, MD, USA

Yasmin Sana King's College Hospital, London, UK

Margarita Sanchez-del-Rio Neurology Department, Hospital Ruber Internacional, Madrid, Spain

Jürgen Sandkühler Department of Neurophysiology, Center for Brain Research Medical University of Vienna, Vienna, Austria

Peter S. Sándor RehaClinic, Cantonal Hospital Baden, Baden, Switzerland

Johannes Sarnthein Functional Neurosurgery, Neurosurgical Clinic University Hospital, Zürich, Switzerland

Susanne K. Sauer Department of Physiology and Pathophysiology, University of Erlangen, Erlangen, Germany

Steven Savvas Biomedical Division, National Ageing Research Institute, Melbourne, VIC, Australia

Jana Sawynok Department of Pharmacology, Dalhousie University, Halifax, Canada

John W. Scadding The National Hospital for Neurology and Neurosurgery, London, UK

Frederieke Geraldine Schaafsma Department of Public and Occupational Health, EMGO+ Institute/VU Medical Center, Amsterdam, MB, The Netherlands

Hans-Georg Schaible Department of Physiology - Neurophysiology, Friedrich-Schiller-University of Jena, Jena, Germany

Steven S. Scherer Department of Neurology, The Perelman School of Medicine at the University of Pennsylvania, Philadelphia, PA, USA

Michael Schäfer Department of Anesthesiology and Intensive Care Medicine, Charité University, Berlin, Germany

Martin Schmelz Department of Anesthesiology in Mannheim, University of Heidelberg, Mannheim, Germany

Robert F. Schmidt Physiologisches Institut der Universität Würzburg, Würzburg, Germany

Suzan Schneeweiss The Hospital for Sick Children, Toronto, ON, Canada

Frank Schneider Department of Psychiatry and Psychotherapy, RWTH Aachen University, Aachen, Germany

Alfons Schnitzler Department of Neurology, Heinrich Heine University, Düsseldorf, Germany

Jean Schoenen Department of Neurology, University of Liège, CHR Citadelle, Belgium

Department of Neuroanatomy, University of Liège, CHR Citadelle, Belgium

Jens Schouenborg Neuronano Research Center, Lund University, Lund, Sweden

Stephan A. Schug School of Medicine and Pharmacology, Royal Perth Hospital, Faculty of Medicine, Dentistry & Health Sciences, The University of Western Australia, UWA Anaesthesiology, Perth, Australia

Patrick Schuss Department of Neurosurgery, University-Hospital Bonn, Bonn, Germany

Anthony C. Schwarzer Faculty of Medicine and Health Sciences, The University of Newcastle, Newcastle, Australia

Whitney Scott Department of Psychology, McGill University, Montreal, QC, Canada

Laura C. Seidman Pediatric Pain Program, Department of Pediatrics David Geffen School of Medicine at UCLA, Los Angeles, CA, USA

Ze'ev Seltzer Faculty of Dentistry, University of Toronto Centre for the Study of Pain, Toronto, ON, Canada

Peter Selwyn Albert Einstein College of Medicine, Bronx, NY, USA

Patrick B. Senatus Department of Neurological Surgery, Neurological Institute Columbia University, New York, NY, USA

Jyoti N. Sengupta Division of Gastroenterology and Hepatology and Pediatric Gastroenterology, Medical College of Wisconsin, Milwaukee, WI, USA

Mariano Serrao Department of Neurology and Otorinolaringoiatry, University of Rome La Sapienza, Rome, Italy

Barry J. Sessle Faculty of Dentistry, University of Toronto, Toronto, ON, Canada

Faculty of Dentistry, University of Toronto, Toronto, ON, Canada

Virginia S. Seybold Department of Neuroscience, University of Minnesota, Minneapolis, MN, USA

T. Shah Royal Perth Hospital and University of Western Australia, Crawley, WA, Australia

Yair Sharav Department of Oral Medicine, School of Dental Medicine, Hebrew University-Hadassah, Jerusalem, Israel

S. Murray Sherman Department of Neurobiology, University of Chicago, Chicago, IL, USA

Yoshio Shigenaga Department of Oral Anatomy and Neurobiology, Osaka University, Osaka, Japan

Susan Shin Neurology, Mount Sinai School of Medicine, New York, NY, USA

Edward A. Shipton Department of Anaesthesia, Christchurch School of Medicine and Health Sciences, University of Otago, Christchurch, New Zealand

Yoram Shir The Alan Edwards Pain Management Unit, McGill University Health Centre, Montreal, QC, Canada

Peter Shrager Department of Neurobiology and Anatomy, University of Rochester Medical Center, Rochester, NY, USA

Marc J. Shulman VA Connecticut Healthcare System, Yale University, New Haven, CT, USA

David M. Sibell Comprehensive Pain Center/Spine Center, Oregon Health and Science University, Portland, OR, USA

Philip J. Siddall Department of Pain Management, Greenwich Hospital, University of Sydney, Sydney, NSW, Australia

Stephen D. Silberstein Jefferson Medical College, Thomas Jefferson University and Jefferson Headache Center, Thomas Jefferson University Hospital, Philadelphia, PA, USA

Donald A. Simone Diagnostic and Biological Sciences, University of Minnesota, Minneapolis, MN, USA

David G. Simons Department of Rehabilitation Medicine, Emory University, Atlanta, GA, USA

David Simpson Clinical Neurophysiology Laboratories, Mount Sinai Medical Center, New York, NY, USA

Marc Sindou Department of Neurosurgery, Hopital Neurologique Pierre Wertheimer, University of Lyon 1, Lyon, France

Soeren H. Sindrup Department of Neurology, Odense University Hospital, Odense, Denmark

Bengt H. Sjölund Department of Community Medicine and Rehabilitation Medicine, Umea University, Umea, Sweden

Carsten Skarke Institute for Translational Medicine and Therapeutics (ITMAT), University of Pennsylvania Perelman School of Medicine, Philadelphia, PA, USA

Paul A. Sloan Department of Anesthesiology, University of Kentucky, Lexington, KY, USA

Kathleen A. Sluka Physical Therapy and Rehabilitation Science Graduate Program, University of Iowa, Iowa City, IA, USA

Seymour Solomon Montefiore Medical Center, Albert Einstein College of Medicine, Bronx, NY, USA

Claudia Sommer Department of Neurology, University of Würzburg, Würzburg, Germany

Linda S. Sorkin Occupational Therapy and Rehabilitation, Anesthesia Research Laboratories, University of California, San Diego, CA, USA

Ingrid Söderback Uppsala University, Uppsala, Sweden

Kari Sørensen Department of Pain Management and Research, Oslo University Hospital, Rikshospitalet, Oslo, Norway

Michael Spaeth Friedrich-Baur-Institute, University of Munich, Munich, Germany

Peregrin O. Spielholz Washington State Department of Labor and Industries, SHARP Program, Olympia, WA, USA

Peter S. Staats Division of Pain Medicine, The Johns Hopkins University, Baltimore, MD, USA

Brett R. Stacey Comprehensive Pain Center, Oregon Health and Science University, Portland, OR, USA

Wendy M. Stein San Diego Hospice and Palliative Care, UCSD, San Diego, CA, USA

Christoph Stein Department of Anesthesiology and Intensive Care Medicine, Freie Universitaet Berlin Charité Campus Benjamin Franklin, Berlin, Germany

Michael P. Steinmetz Department of Neurosurgery, MetroHealth Medical Center, Case Western Reserve University School of Medicine, Cleveland, OH, USA

Jair Stern Functional Neurosurgery, Neurosurgical Clinic University Hospital, Zürich, Switzerland

Bonnie J. Stevens University of Toronto and The Hospital for Sick Children, Toronto, ON, Canada

Carl-Olav Stiller Division of Clinical Pharmacology and Department of Medicine, Karolinska University Hospital, Stockholm, Sweden

Jennifer N. Stinson The Hospital for Sick Children, Toronto, ON, Canada

Henry Stockbridge University of Washington School of Public Health and Community Medicine, Seattle, WA, USA

Christian S. Stohler Baltimore College of Dental Surgery, University of Maryland, Baltimore, MD, USA

Laura Stone Faculty of Dentistry, Alan Edwards Centre for Research on Pain, Montreal, QC, Canada

William Stones Department of Obstetrics and Gynaecology, Aga Khan University, Nairobi, Kenya

Andreas Straube Department of Neurology, Ludwig Maximilians University, Munich, Germany

Jenny Strong University of Queensland, Brisbane, QLD, Australia

Audun Stubhaug Department of Pain Management and Research, Oslo University Hospital, Rikshospitalet and University of Oslo, Oslo, Norway

Gerold Stucki Swiss Paraplegic Research, Nottwil, Switzerland

Department of Health Sciences and Health Policy, University of Lucerne, Lucerne, Switzerland

ICF Research Branch in cooperation with the World Health Organization (WHO) Collaborating Centre for the Family of International Classifications in Germany, Nottwil, Switzerland

Cheryl L. Stucky Department of Cell Biology Neurobiology and Anatomy, Medical College of Wisconsin, Milwaukee, WI, USA

Mathias Sturzenegger Department of Neurology, University Hospital, Inselspital, University of Bern, Bern, Switzerland

Yasuo Sugiura School of Human Care Study, Nagoya University of Arts and Sciences, Nissin City, Aichi, Japan

Mark D. Sullivan Psychiatry and Behavioral Sciences, University of Washington, Seattle, WA, USA

Michael J. L. Sullivan Department of Psychology, McGill University, Montreal, QC, Canada

Rie Suzuki Department of Pharmacology, University College London, London, UK

Kristina B. Svendsen Danish Pain Research Center, Aarhus University Hospital, Aarhus, Denmark

Peter Svensson Clinical Oral Physiology, Aarhus University, Aarhus, Denmark

Peter Szucs Spinal Neuronal Networks, Instituto de Biologia Molecular e Celular - IBMC, Porto, Portugal

A. Taghva Department of Neurosurgery, Johns Hopkins Hospital, Baltimore, MD, USA

Kimberson Cochien Tanco Department of Palliative Care and Rehabilitation Medicine, The University of Texas M. D. Anderson Cancer Center, Houston, TX, USA

Kersi Taraporewalla Royal Brisbane and Womens' Hospital, University of Queensland, Brisbane, QLD, Australia

Irmgard Tegeder Pharmazentrum frankfurt, Institute of Clinical Pharmacology, Goethe-University Hospital, Frankfurt, Germany

Astrid Juhl Terkelsen Danish Pain Research Center, Department of Neurology, Aarhus University Hospital, Aarhus, Denmark

Gregory W. Terman Department of Anesthesiology and the Graduate Program in Neurobiology and Behavior, University of Washington, Seattle, WA, USA

Alison Thomas Lismore, NSW, Australia

Benjamin Thomas Chelsea & Westminster Hospital Foundation Trust, London, UK

Jacob Thomas School of Medical Sciences, University of Adelaide, Adelaide, SA, Australia

Miles Thompson Goldsmiths, University of London, London, UK

Beverly E. Thorn Department of Psychology, University of Alabama, Tuscaloosa, AL, USA

Cindy L. Thurston-Stanfield Department of Biomedical Sciences, University of South Alabama, Mobile, AL, USA

Arne Tjølsen Department of Bomedicine, University of Bergen, Haukeland University Hospital, Bergen, Norway

Andrew J. Todd Institute of Neuroscience and Psychology, College of Medical, Veterinary and Life Sciences, University of Glasgow, Glasgow, UK

Makoto Tominaga Department of Cellular and Molecular Physiology, Mie University School of Medicine, Tsu Mie, Japan

Victor Tortorici Instituto Venezolano de Investigaciones Cientificas (IVIC), Caracas, Venezuela

Irene Tracey Pain Imaging Neuroscience (PaIN) Group, Department of Physiology, Anatomy and Genetics, Oxford University, Oxford, UK

Diep Tuan Tran Department of Integrative Physiology, National Institute for Physiological Sciences, Okazaki, Japan

Rolf-Detlef Treede Institute of Physiology and Pathophysiology, Johannes Gutenberg University, Mainz, Germany

Jennie C. I. Tsao Pediatric Pain Program, Department of Pediatrics David Geffen School of Medicine at UCLA, Los Angeles, CA, USA

Lawrence Tsen Department of Anesthesiology, Perioperative and Pain Medicine, Brigham and Women's Hospital Harvard Medical School, Boston, MA, USA

Jeanna Tsenter Department of Physical Medicine and Rehabilitation, Hadassah University Hospital, Jerusalem, Israel

Maurits Van Tulder Institute for Research in Extramural Medicine, Free University Amsterdam, Amsterdam, The Netherlands

Eldon Tunks Hamilton Health Sciences Corporation, Hamilton, ON, Canada

Susan Tupper Pediatrics, University of Saskatchewan, Saskatoon, SK, Canada

Dennis C. Turk Department of Anesthesiology & Pain Research, University of Washington, Seattle, WA, USA

Judith A. Turner Department of Psychiatry and Behavioral Sciences, University of Washington, Seattle, WA, USA

Department of Rehabilitation Medicine, University of Washington, Seattle, WA, USA

Stephen P. Tyrer Royal Victoria Infirmary, University of Newcastle upon Tyne, Newcastle-on-Tyne, UK

Alexander Tzabazis Department of Anesthesia, Stanford University, Stanford, CA, USA

Kene Ugokwe Department of Neurosurgery, St Elizabeth Health Center, Youngstown, OH, USA

Lindsay S. Uman Department of Psychology, IWK Health Centre, Halifax, NS, Canada

Christiane Vahle-Hinz Department of Neurophysiology and Pathophysiology, University Medical Center Hamburg-Eppendorf, Hamburg, Germany

Muthukumar Vaidyaraman Division of Pain Medicine, The Johns Hopkins University, Baltimore, MD, USA

Todd W. Vanderah Department of Pharmacology, College of Medicine University of Arizona, Tucson, AZ, USA

Guy Vanderstraeten Department of Rehabilitation Sciences and Physiotherapy, Ghent University Hospital, Ghent, Belgium

Horacio Vanegas Instituto Venezolano de Investigaciones Cientificas (IVIC), Caracas, Venezuela

Jean-Jacques Vatine Outpatient and Research Division, Reuth Medical Center, Sackler Faculty of Medicine, Tel Aviv University, Tel Aviv, Israel

Leigh M. Vaughan Medical University of South Carolina, Charleston, SC, USA

Linda K. Vaughn Department of Biomedical Sciences, Marquette University, Milwaukee, WI, USA

Enrique Vazquez Instituto Venezolano de Investigaciones Cientificas (IVIC), Caracas, Venezuela

Louis Vera-Portocarrero Department of Pharmacology, College of Medicine University of Arizona, Tucson, AZ, USA

Paul Verrills Metropolitan Spinal Clinic, Prahran, VIC, Australia

Tine Vervoort Department of Experimental Clinical and Health Psychology, Ghent University, Ghent, Belgium

Félix Viana Instituto de Neurociencias, Universidad Miguel Hernández-CSIC, San Juan de Alicante, Alicante, Spain

Charles J. Vierck Jr. Department of Neuroscience and McKnight Brain Institute, University of Florida College of Medicine, Gainesville, FL, USA

Luis Villanueva INSERM UMR 894, Centre de Psychiatrie et Neurosciences, Paris, France

Marcelo Villar Faculty of Biomedical Sciences, Austral University, Buenos Aires, Argentina

Katy Vincent Nuffield Department of Obstetrics and Gynaecology, The Women's Centre, John Radcliffe Hospital, University of Oxford, Oxford, UK

David Vivian Metro Spinal Clinic, South Caulfield, VIC, Australia

Johan W. S. Vlaeyen Research Group Health Psychology, Catholic University of Leuven, Leuven, Belgium

Department of Clinical Psychological Science, Maastricht University, Maastricht, The Netherlands

Stéphanie Volders Research Group on Health Psychology, Leuven, Belgium

Ernest Volinn Pain Research Center, University of Utah, Salt Lake City, UT, USA

Lucy Vulchanova Department of Veterinary and Biomedical Sciences, University of Minnesota, St. Paul, MN, USA

Paul W. Wacnik Neuromodulation Research, Medtronic Inc., Minneapolis, MN, USA

Gordon Waddell University of Glasgow, Glasgow, UK

Gary A. Walco Department of Anesthesiology and Pain Medicine, University of Washington School of Medicine, Seattle Children's Hospital, Seattle, WA, USA

Suellen M. Walker University of Sydney Pain Management and Research Centre, Royal North Shore Hospital, NSW, Australia

Victoria C. J. Wallace Department of Anaesthetics, Pain Medicine and Intensive Care Chelsea and Westminster Hospital Campus, London, UK

Xiaohong Wang Department of Integrative Physiology, National Institute for Physiological Sciences, Okazaki, Japan

Gunnar Wasner Department of Neurological Pain Research and Therapy, Neurological Clinic, Kiel, Germany

Shoko Watanabe Department of Integrative Physiology, National Institute for Physiological Sciences, Okazaki, Japan

Linda R. Watkins Department of Psychology and Neuroscience, and The Center for Neuroscience, University of Colorado at Boulder, Boulder, CO, USA

Peter N. Watson Department of Medicine, University of Toronto, Toronto, ON, Canada

James Watt North Shore Hospital, Auckland, New Zealand

Stephen G. Waxman Department of Neurology, Yale School of Medicine, New Haven, CT, USA

Inge Wegner Julius Center, Department of Epidemiology, University Medical Center Utrecht, Utrecht, The Netherlands

Feng Wei Department of Neural and Pain Sciences, University of Maryland Dental School, Baltimore, MD, USA

Christian Weidner Department of Physiology and Experimental Pathophysiology, University Erlangen, Erlangen, Germany

P. T. M. Weijenborg Department of Gynecology, Leiden University Medical Center, Leiden, The Netherlands

Robin Weir School of Nursing, Faculty of Health Sciences, McMaster University, Hamilton, ON, Canada

Nirit Weiss Department of Neurosurgery, Mount Sinai Medical Center, New York, NY, USA

Ursula Wesselmann Departments of Neurology and Anesthesiology / Division of Pain Management, University of Alabama at Birmingham, Birmingham, AL, USA

Dagmar Westerling Department of Anesthesiology, Central Hospital of Kristianstad, Kristianstad, Sweden

Karin N. Westlund Department of Physiology, University of Kentucky, Lexington, KY, USA

Thomas M. Wickizer Health Services Management and Policy, Ohio State University, Colombus, OH, USA

Eva Widerström-Noga The Miami Project to Cure Paralysis, Department of Neurological Surgery, University of Miami Miller, School of Medicine, Miami, FL, USA

Julie Wieseler Department of Psychology and Neuroscience, and The Center for Neuroscience, University of Colorado at Boulder, Boulder, CO, USA

Zsuzsanna Wiesenfeld-Hallin Karolinska Institute, Stockholm, Sweden

Ann Wigham McCoull Clinic, Prudhoe Hospital, Northumberland, UK

George L. Wilcox Department of Neuroscience, Pharmacology and Dermatology, University of Minnesota Medical School, Minneapolis, MN, USA

Oliver H. G. Wilder-Smith Pain Centre, Department of Anaesthesiology, Radboud University Nijmegen Medical Centre, Nijmegen, The Netherlands

Kjersti Wilhelmsen Section for Physiotherapy Science, Department of Public Health and Primary Health, Faculty of Medicine, University of Bergen, Bergen, Norway

Victor Wilk Brighton Spinal Group, Brighton, VIC, Australia

W. A. Williams Royal Perth Hospital and University of Western Australia, Perth, WA, Australia

Sara E. Williams Division of Pediatric Gastroenterology, Medical College of Wisconsin, Milwaukee, WI, USA

William D. Willis Department of Neuroscience and Cell Biology, University of Texas Medical Branch, Galveston, TX, USA

John B. Winer Department of Neurology, Birmingham Muscle and Nerve Centre Queen Elizabeth Hospital, Birmingham, UK

Christopher J. Winfree Department of Neurological Surgery, Neurological Institute Columbia University, New York, NY, USA

Janet Winter Novartis Institute for Medical Science, London, UK

Alain Woda Faculté de Chirurgie Dentaire, University Clermont-Ferrand, Clermont-Ferrand, France

John N. Wood Molecular Nociception Group, Wolfson Institute for Biomedical Research, University College London, London, WC, UK

Lee Woodson Shriners Hospitals for Children, Shriners Burns Hospital, Galveston, TX, USA

Department of Anesthesia, University of Texas Medical Branch, Galveston, TX, USA

Anava A. Wren Pain Prevention and Treatment Research, Department of Psychiatry and Behavioral Sciences, Duke University Medical Center, Durham, NC, USA

Xiao-Jun Xu Department of Physiology and Pharmacology, Section of Integrative Pain Rsearch Karolinska Institutet, Stockholm, Sweden

Hiroshi Yamasaki Department of Integrative Physiology, National Institute for Physiological Sciences, Okazaki, Japan

Michael Yelland School of Medicine, Logan campus, Griffith University, Logan, Australia

Sriram Yennurajalingam Department of Palliative Care and Rehabilitation Medicine, The University of Texas M. D. Anderson Cancer Center, Houston, TX, USA

David C. Yeomans Department of Anesthesia, Stanford University, Stanford, CA, USA

Robert P. Yezierski Comprehensive Center for Pain Research and Department of Neuroscience, The McKnight Brain Institute University of Florida, Gainesville, FL, USA

Way Yin Department of Anesthesiology, University of Washington, Seattle, WA, USA

Atsushi Yoshida Department of Oral Anatomy and Neurobiology, Osaka University, Osaka, Japan

Joanna M. Zakrzewska Eastman Dental Hospital UCLH NHS Foundation Trust, London, UK

Jenna E. Zauk St. George's University, School of Medicine, Grenada, West Indies

Hanns Ulrich Zeilhofer Institute for Pharmacology and Toxicology, University of Zurich, and Swiss Federal Institute of Technology (ETH) Zurich, Zürich, Switzerland

Lonnie K. Zeltzer Pediatric Pain Program, Department of Pediatrics David Geffen School of Medicine at UCLA, Los Angeles, CA, USA

Giovambattista Zeppetella St. Clare Hospice, Hastingwood, UK

Xijing Zhang Department of Neuroscience, University of Minnesota, Minneapolis, MN, USA

Min Zhuo Department of Physiology, University of Toronto, Toronto, ON, Canada

C. Zöllner Department of Anesthesiology and Intensive Care Medicine, Charité University, Berlin, Germany

Krina Zondervan Genetic and Genomic Epidemiology Unit, Nuffield Dept of Obstetrics and Gynaecology, Wellcome Trust Centre for Human Genetics, Oxford, UK

Peter Zygmunt Department of Laboratory Medicine, Lund University, Division of Clinical Chemistry and Pharmacology, Lund, Sweden

Zbigniew Zylicz Hildegard Hospiz, Basel Switzerland, Basel, Switzerland

David C. Yeomans Department of Anesthesia, Stanford University, Stanford, CA, USA

Robert P. Yezierski Comprehensive Center for Pain Research and Department of Neuroscience, The McKnight Brain Institute, University of Florida, Gainesville, FL, USA

Mary Yin Department of Anesthesiology, University of Washington, Seattle, WA, USA

Atsushi Yoshida Department of Oral Anatomy and Neurobiology, Osaka University, Osaka, Japan

Joanna M. Zakrzewska Eastman Dental Hospital UCLH NHS Foundation Trust, London, UK

Jenna E. Zoch St. George's University School of Medicine, Grenada, West Indies

Hanns Ulrich Zeilhofer Institute for Pharmacology and Toxicology, University of Zurich, and Swiss Federal Institute of Technology (ETH) Zurich, Zurich, Switzerland

Frank K. Zeller Pr H Rosenm, Department of Pediatrics David Geffen School of Medicine at UCLA, Los Angeles, CA, USA

Constantin Zeppetella St. Clara Hospice, Hampstead, UK

Sijian Zhang Department of Neuroscience, University of Minnesota, Minneapolis, MN, USA

Min Zhuo Department of Physiology, University of Toronto, Toronto, ON, Canada

C. Zöllner Department of Anesthesiology and Intensive Care Medicine, Charité, Berlin, Germany

Krina Zondervan Genetic and Genomic Epidemiology Unit, ... of Obstetrics and Gynaecology, Wellcome Trust Centre for Human Genetics, Oxford, UK

Peter Zygmunt Department of Laboratory Medicine, Lund University, Division of Clinical Chemistry and Pharmacology, Lund, Sweden

Zbigniew Zylicz Hildegard Hospiz Basel, ... Basel, Switzerland

P

P/Q-Type Calcium Channel

Definition

Voltage-dependent Ca^{2+} channels not only mediate the entry of Ca^{2+} ions into excitable cells but are also involved in a variety of Ca^{2+}-dependent processes, including muscle contraction, hormone or neurotransmitter release, and gene expression. The alpha-1A isoform is abundantly expressed in neuronal tissue and corresponds to the P/Q Ca^{2+} channel type. The P/Q Ca^{2+} channel is responsible for presynaptic neurotransmitter release. Abnormalities in this channel could lead to abnormal neurotransmitter release. The affected CACNA1A gene encodes the pore-forming α_{1A} subunit of the P/Q-type channel ($Ca_V2.1$), a predominant mediator of voltage-gated Ca^{2+} entry and transmitter release at many synapses in the central nervous system. Alterations in synaptic signaling are of likely importance for cross talk between an ascending pain transmission pathway and inhibition by a powerful descending modulatory system, interactions that ultimately govern the generation of headache pain.

Cross-References

▶ Migraine, Childhood Syndromes

P1 Receptors

Definition

P1 receptors are seven transmembrane, G protein-coupled receptors that are activated by adenosine.

Cross-References

▶ Purine Receptor Targets in the Treatment of Neuropathic Pain

P2X Receptor

Definition

Extracellular adenosine 5'-triphosphate (ATP) evokes cation currents in primary afferent neurons. P2X receptors are believed to be ligand-gated cation channels with two membrane-spanning domains that are activated by the binding of extracellular ATP. P2X purinoceptors have been cloned from different tissues and P2X1-7 cDNA have been identified. Among them, P2X3 subunits have been found to be important receptors that sense peripheral information.

G.F. Gebhart, R.F. Schmidt (eds.), *Encyclopedia of Pain*, DOI 10.1007/978-3-642-28753-4,
© Springer-Verlag Berlin Heidelberg 2013

Cross-References

▶ ERK Regulation in Sensory Neurons During Inflammation
▶ Purine Receptor Targets in the Treatment of Neuropathic Pain
▶ TRPV1 Modulation by p2Y Receptors

P2X$_3$ Receptor

Definition

The P2X$_3$ receptor is a ligand-gated channel activated by adenosine triphosphate (ATP) and mediates fast excitatory transmission. The P2X$_3$ channel is selectively expressed by unmyelinated (C-fiber) nociceptors and exists as either P2X$_3$ homomers or P2X$_{2/3}$ heteromers. P2X$_3$ receptor expression overlaps substantially with the IB4-positive population of small-diameter sensory neurons.

Cross-References

▶ IB4-Positive Neurons, Role in Inflammatory Pain
▶ Nerve Growth Factor, Sensitizing Action on Nociceptors

P2X7 Receptor

▶ Cathepsin and Microglia

P2Y Receptors

Definition

P2Y receptors are seven transmembrane, G protein-coupled receptors activated by ATP and other nucleotides.

Cross-References

▶ Purine Receptor Targets in the Treatment of Neuropathic Pain
▶ TRPV1 Modulation by p2Y Receptors

p38 MAP Kinase

Synonyms

P38 MAPK

Definition

p38 mitogen-activated protein kinase is involved in intracellular signaling pathways that control both (a) the production of proinflammatory cytokines and (b) cellular activation caused in response to proinflammatory cytokines binding to their receptors. p38 MAPK is activated by stress signals, such as inflammatory cytokines, heat shock, and ischemia. Although an activity-dependent p38 activation occurs in neurons, the contribution of p38 MAPK to nociception and pain hypersensitivity is still under investigation. Peripheral inflammation induces p38 activation in small DRG neurons and contributes to the maintenance of inflammatory pain hypersensitivity.

Cross-References

▶ ERK Regulation in Sensory Neurons during Inflammation
▶ Proinflammatory Cytokines

P38 MAPK

▶ p38 MAP Kinase

p75 Receptor

Definition

p75 is the low-affinity nerve growth factor (NGF) receptor that binds and internalizes nerve growth factor (NGF), neurotrophin-3 (NT-3), and brain-derived neurotrophic factor (BDNF). It is normally expressed in sensory neurons, mainly of small diameter, and on the terminals of sympathetic and small sensory neurons in peripheral tissues such as skin and vascular muscle. These target tissues produce neurotrophins, which are necessary for the maintenance of many sympathetic and sensory neurons. After peripheral nerve injury, p75 expression is reduced in neurons and upregulated on satellite glia in dorsal root ganglia.

Cross-References

▶ Nerve Growth Factor, Sensitizing Action on Nociceptors
▶ Sympathetic and Sensory Neurons after Nerve Lesions, Structural Basis for Interactions

PACAP

Definition

PACAP (pituitary adenylate cyclase-activating polypeptide), a 38-amino acid peptide, was characterized from hypothalamic tissue as a potent stimulator of adenylyl cyclase. PACAP shows a high sequence homology with VIP. It has a wide distribution in the brain and spinal cord, as well as in sensory and autonomic ganglia and in inhibitory neurons of the enteric nervous system. PACAP in endings that innervate most endocrine glands and lymphatic tissue acts as an anti-inflammatory peptide. Immune cells also contain PACAP.

Pachymeningeal Enhancement

Definition

A distinct MRI picture that, in the setting of an orthostatic headache, is almost diagnostic of a CSF leak. After gadolinium, there is diffuse enhancement limited to the pachymeninges without leptomeningeal involvement.

Cross-References

▶ New Daily Persistent Headache

Pachymeningeal Enhancement in MRI with Gadolinium

Definition

Typically linear, thick, and diffuse enhancement of the dura mater in MR images. Whereas the leptomeninges have blood–brain barriers, the pachymeninges do not and therefore accumulate the contrast medium.

Cross-References

▶ Headache Due to Low Cerebrospinal Fluid Pressure

PAG

▶ Descending Modulation of Nociceptive Transmission During Persistent Damage to Peripheral Tissues
▶ Periaqueductal Gray

Paget's Disease

Definition

It is a localized bone-remodeling disorder that affects widespread, noncontiguous areas of the skeleton. The process is initiated by overactive osteoclastic bone resorption followed by a compensatory increase in osteoblastic new bone formation resulting in structurally disorganized mosaic of woven and lamellar bone. The bone is expanded, less compact, and more vascular.

Cross-References

▶ Chronic Low Back Pain: Definitions and Diagnosis

Pain

Definition

Pain is an unpleasant sensory and emotional experience associated with actual or potential tissue damage or described in terms of such damage (definition of the International Association for the Study of Pain, IASP). The inability to communicate verbally does not negate the possibility that an individual is experiencing pain and is in need of appropriate pain-relieving treatment. Pain is a symptom, not an impairment. A symptom, of itself, is neither exertional nor nonexertional. It is the nature of the limitations or restrictions caused by the symptom that is exertional or nonexertional. In a given individual, pain may cause exertional limitations and restrictions, nonexertional limitations and restrictions, or a combination of both.

Cross-References

▶ Cancer Pain Management Radiotherapy
▶ Disability Evaluation in the Social Security Administration

▶ Disability, Upper Extremity
▶ Ethics of Pain in the Newborn Human
▶ Lateral Thalamic Pain-Related Cells in Humans
▶ Pain Assessment in Neonates
▶ Pain Assessment in the Elderly
▶ Pain Treatment, Intracranial Ablative Procedures
▶ Physical Exercise
▶ Tourniquet Test

Pain Affect

▶ Cingulate Cortex, Nociceptive Processing, Behavioral Studies in Animals

Pain Afferents

▶ Nociceptor, Categorization

Pain and Sedation of Children in the Emergency Setting

Suzan Schneeweiss
The Hospital for Sick Children, Toronto, ON, Canada

Synonyms

Pediatric procedural analgesia; Pediatric sedation

Definition

Procedural sedation is a technique of administering sedatives or dissociative agents with or without analgesics to induce a state that allows the patient to tolerate unpleasant procedures while maintaining cardiorespiratory function. Procedural sedation and analgesia are intended to result in a depressed level of consciousness while allowing the patient to maintain airway control independently and continuously. Specifically, the drug doses and techniques used are intended to maintain protective airway reflexes (American College of Emergency Physicians 1998; American College of Emergency Physicians 2004, 2005).

Characteristics

In recent years, there has been increasing recognition of the need to provide adequate analgesia and anxiolysis to children undergoing procedures in the Emergency Department (Kennedy and Luhmann 1999). Although children often report fear of a procedure to be the most anxiety-provoking aspect of their emergency room visit, they often receive less analgesia than their adult counterparts in the emergency setting (Petrack et al. 1997). Fear of adverse events has often prevented the use of sedative and analgesic agents in children (Cote et al. 2000; Pena and Krauss 1999). Emergency physicians, skilled in the care of critically ill children as well as in the management of the pediatric airway, are increasingly providing children with better pain and anxiety control (Hoffman et al. 2002; Krauss and Green 2000; Maher et al. 2007). The Emergency Department provides a unique situation with regard to the patient population and the environment. Most children presenting with an acute illness or injury requiring an urgent or emergent procedure are healthy.

Although many procedures performed in the Emergency Department are not highly painful, such as an intravenous cannulation, they may provoke a high degree of anxiety in children and not only warrant pain relief, but also require some degree of anxiolysis. Even non-painful diagnostic imaging procedures such as CT scanning in which the child must lie motionless can provoke a high degree of anxiety. There are procedures, however, that are both highly painful and anxiety provoking, requiring both a higher degree of anxiolysis and analgesia (e.g., fracture reduction).

The therapeutic goals of procedural analgesia and anxiolysis are:
1. To ensure the patient's safety and welfare
2. To minimize physical discomfort or pain during a diagnostic or therapeutic procedure
3. To minimize the negative psychosocial responses to treatment
4. To enhance cooperativeness of the child
5. To return the patient to a state whereby safe discharge, as determined by recognized criteria, is possible

Guidelines for Procedural Analgesia

Guidelines are available from various sources regarding the general practice of procedural sedation and analgesia by non-anesthetists in order to provide safe, consistent, and appropriate sedation to children in the emergency setting undergoing invasive and noninvasive procedures (American College of Emergency Physicians 1998, 2004; American Academy of Pediatrics 1992, 2002, 2006; American Society of Anesthesiologists 2002; Innes et al. 1999). Central to this is the notion of the *sedation continuum* whereby any procedural sedative agent can result in a dose-dependent alteration in the level of consciousness from ► *minimal sedation* progressing to ► *moderate sedation* and finally to ► *deep sedation* and ► *general anesthesia*. In the emergency setting, the minimum amount of drug to achieve the desired depth of sedation is achieved by careful intravenous titration of the sedative and analgesic agents.

Agents used for moderate sedation require personnel skilled in management of potentially life-threatening airway complications. These children require a thorough pre-procedural assessment including a focused history and physical examination including a detailed airway assessment for potential complications. Children deemed high risk should be referred to an anesthetist for further assessment and management. Generally, children classified as ASA I or ASA II by the American Society of Anaesthesiologists are suitable to undergo moderate sedation by non-anesthetist physicians skilled in airway management. There must be skilled personnel available to manage potential adverse events. The children should be continuously monitored with frequent vital signs to detect early onset of reversible complications. Practice guidelines must be established in each individual emergency setting ensuring appropriate training of personnel to recognize potential adverse events and skills necessary to intervene and manage all potential complications.

Suitable monitoring and emergency equipment should be available for children of all ages and sizes being treated. Emergency equipment including oxygen, suction, positive pressure bag-valve-mask, and emergency resuscitation drugs should be immediately available and functional during sedation and recovery period. Capnography allows for continuous monitoring of ventilatory status and is useful in detecting early airway or respiratory compromise (Lightdale et al. 2006). An assessment and documentation of food and fluid intake should be done prior to the procedure. Sound clinical judgment should be exercised by the physician in determining the appropriate time interval between the last oral intake and procedural sedation. In urgent, emergent, or other situations where gastric emptying is impaired, the potential for aspiration of gastric contents must be considered in determining the timing of the intervention and the degree of sedation/analgesia. If possible, such patients may benefit from delaying the procedure. When proper fasting cannot be assured, the increased risks of sedation must be carefully weighed against its benefits, and the lightest sedation should be used (American College of Emergency Physicians 2008). Following the procedure, strict discharge criteria are required as increased sedation may occur when the painful stimulus is discontinued or if there is prolonged sedative effects because of the pharmacodynamic or pharmacokinetic properties of the medication.

Classification of the American Society of Anesthesiologists

Class	Description	Suitability for sedation
1	Normally healthy patient	Excellent
2	A patient with mild systemic disease (no functional limitation)	Generally good

(continued)

Class	Description	Suitability for sedation
3	A patient with severe systemic disease (definite functional impairment)	Intermediate to poor; consider benefits relative to risks
4	A patient with severe systemic disease that is a constant threat to life	Poor, benefits rarely outweigh risks
5	A moribund patient who is not expected to survive without the operation	Extremely poor

Non-pharmacological Management of Pain and Anxiety

A child-centered approach to the assessment and management of painful procedures in the emergency setting should be promoted. Parents play a key role both in their response to their child's pain and in their care of the child in the Emergency Department. The child and family should be encouraged to be active participants in the procedure. Allowing parents or family members to act as positive assistants rather than negative restraints helps to reduce stress in both children and parents and minimizes the pain experience (Merritt et al. 1998). Emergency personnel must be sensitive to parental needs while at the same time ensuring parents do not interfere with the procedure. Preparation of the family for the procedure is another key factor in reducing parental anxiety and facilitating the procedure. An explanation of the procedure to be performed as well as an accurate description of the amount of pain to be experienced should be provided. Use of medical jargon should be avoided when discussing the procedure with the family. If more than one attempt at a procedure may be required, this should be communicated to the child.

Age-appropriate distraction techniques are useful adjuncts to reduce procedural pain and anxiety and may also result in reduction in procedure time and number of personnel required. They may be provided by nurses, parents, or child life specialist (Aldock et al. 1985). For younger children, music, bubbles, puppetry, story, and activity books may be very useful. Older children may benefit from music through headphones, directed breathing, and calming coping statements such as "I can do it" or "I am calm."

More sophisticated methods such as imagery require skilled personnel and can be particularly effective for children who require multiple procedures.

Pharmacological Management of Pain and Anxiety

The practitioner must be familiar with the various pharmacological agents available, their side effect profile as well as possible synergistic effects occurring when more than one agent is used (American College of Emergency Physicians 2004). Some agents such as ketamine provide both analgesia and anxiolysis while other agents may provide only analgesia (e.g., morphine) or anxiolysis (e.g., midazolam). Shorter-acting opioid agents such as fentanyl are preferred to the traditional long-acting agents such as morphine or meperidine because of the faster onset of action and shorter duration of action without histamine release. The dissociative agent, ketamine, is commonly used because it is characterized by profound analgesia, sedation, amnesia, and immobilization while preserving upper-airway muscular tone and protective airway reflexes (Green and Krauss 2004; Green et al. 2011). Intravenous propofol has a rapid onset of action and rapid recovery, however, it can quickly result in deep sedation with cardiopulmonary depression and thus requires vigilant monitoring by highly skilled practitioners (Mallory et al. 2011). In the pediatric population, propofol is most commonly used outside the operating room for non-painful diagnostic procedures such as MRI and CT scan. For painful procedures it may be combined with analgesic agents such as fentanyl and ketamine although, the risk of potential adverse events is increased. The addition of ketamine to propofol (ketofol) has the potential advantage of lowered doses of each drug thereby reducing the nauseant and recovery agitation effects of ketamine with slightly shorter recovery time. It has not been shown to reduce the incidence of respiratory depression or produce superior sedation to either drug alone (Shah et al. 2011).

The pharmacologic agent(s) of choice and route of administration are dependent on the type of procedure, non-painful versus moderate to severe pain, and level of sedation required to successfully perform the procedure. The intravenous route of administration allows for careful titration of the agent to the desired effect. Combination of drugs may act synergistically and accentuate the potential side effects of each drug individually. Each agent should be administered individually in small incremental doses to the desired effect. Sufficient time must elapse between doses to allow the effect of each dose to be assessed before subsequent drug administration. The propensity for combinations of sedative and analgesic agents to potentiate respiratory depression emphasizes the need to appropriately reduce the dose of each component, as well as the need to continually monitor respiratory function. Specific benzodiazepine and opioid antagonists (reversal agents) are available to reverse possible respiratory depression from oversedation improving the safety of procedural sedation. Inhalational agents such as nitrous oxide are useful adjunctive agents for painful procedures or may be used alone for less painful procedures such a laceration repair. Nitrous oxide can be given by demand flow system with a handheld mask or by continuous flow with a nose mask, has a strong safety profile, short onset, and duration of action, but does require an appropriate scavenging system. Oral and transmucosal routes (nasal and rectal) are less invasive and useful for procedures with less pain or requiring some motion control. Intranasal route with agents such as fentanyl and midazolam is becoming increasingly utilized as a non-painful method of drug delivery with rapid absorption (Wolfe and Braude 2010).

Local and regional anesthesia may be useful adjuncts to procedural analgesia. They have no systemic effects and do not result in compromises to the airway. Newer topical agents are now available and are particularly useful anesthetics for wound closure and intravenous insertion (Bishai et al. 1999). A mixture of lidocaine, epinephrine, and tetracaine (LET) has been proven to be an effective topical anesthetic agent for wounds, particularly wounds over face and scalp. Topical agents such as EMLA®

and amethocaine and liposomal lidocaine are used on intact skin for procedures requiring needle insertion such as lumbar punctures and venipunctures. Newer agents such as liposomal lidocaine have the advantage over EMLA with shorter application time (30 min vs. 60 min) and no vasoactive properties. Cautious use of EMLA is required in the newborn period due to reduced methemoglobin reductase activity which can lead to methemoglobinemia. Oral sucrose solution has proven effective adjunctive agent in reducing pain with minor procedures such as venipuncture in the newborn period (Stevens et al. 2010).

References

Aldock, D. S., Feldman, W., Goodman, J. T., et al. (1985). Evaluation of child life intervention in emergency department suturing. *Pediatric Emergency Care, 1*, 111–115.

American Academy of Pediatrics. (1992). Guidelines for monitoring and management of pediatric patients during and after sedation for diagnostic and therapeutic procedures. *Pediatrics, 89*, 1110–1115.

American Academy of Pediatrics. (2002). Guidelines for monitoring and management of pediatric patients during and after sedation for diagnostic and therapeutic procedures: Addendum. *Pediatrics, 110*, 836–838.

American Academy of Pediatrics. (2006). Guidelines for monitoring and management of pediatric patients during and after sedation for diagnostic and therapeutic procedures: An update. *Pediatrics, 116*(6), 2587–2602.

American College of Emergency Physicians. (1998). Clinical policy for procedural sedation and analgesia in the emergency department. *Annals of Emergency Medicine, 31*, 663–677.

American College of Emergency Physicians. (2004). Clinical policy: Evidence-based approach to pharmacologic agents used in pediatric sedation and analgesia in the emergency department. *Annals of Emergency Medicine, 44*(4), 342–377.

American College of Emergency Physicians. (2005). Clinical policy. Procedural sedation in the emergency department. *Annals of Emergency Medicine, 45*, 177–196.

American College of Emergency Physicians. (2008). Clinical policy: Critical issues in the sedation of pediatric patients in the emergency department. *Annals of Emergency Medicine, 51*(4), 378–399.e51.

American Society of Anesthesiologists Task Force on Sedation and Analgesia by Non-Anesthesiologists. (2002). Practice guidelines for sedation and analgesia by non-anesthesiologists. *Anesthesiology, 96*, 1004–1017.

Bishai, R., Taddio, A., Bar-Oz, B., et al. (1999). Relative efficacy of lignocaine-prilocaine cream and amethocaine gel for local analgesia before venipuncture in children. *Pediatrics, 104*, e31.

Cote, C. J., Notterman, D. A., Karl, H. W., et al. (2000). Adverse sedation events in pediatrics: A critical incident analysis of contributing factors. *Pediatrics, 105*, 805–814.

Green, S. M., & Krauss, B. (2004). Clinical practice guideline for emergency department ketamine dissociative sedation in children. *Annals of Emergency Medicine, 44*(460), 460–471.

Green, S. M., Roback, M. G., Kennedy, R. M., & Krauss, B. (2011). Clinical practice guideline for emergency department ketamine dissociative sedation: 2011 Update. *Annals of Emergency Medicine, 57*(5), 449–461.

Hoffman, G. M., Nowakowski, R., Troshynski, T. J., et al. (2002). Risk reduction in pediatric procedural sedation by application of an American Academy of Pediatrics/American Society of Anesthesiologists. *Pediatrics, 109*, 236–243.

Innes, G., Murphy, M., Nijssen-Jorden, C., et al. (1999). Procedural sedation and analgesia in the emergency department. Canadian consensus guidelines. *The Journal of Emergency Medicine, 17*, 145–156.

Kennedy, R. M., & Luhmann, J. D. (1999). The ouchless emergency department. Getting closer: Advances in decreasing distress during painful procedures in the emergency department. *Pediatric Clinics of North America, 46*, 1215–1247.

Krauss, B., & Green, S. M. (2000). Sedation and analgesia for procedures in children. *The New England Journal of Medicine, 342*, 938–945.

Lightdale, J. R., Goldmann, D. A., Feldman, H. A., Newburg, A. R., DiNardo, J. A., & Fox, V. L. (2006). Microstream capnography improves patient monitoring during moderate sedation: A randomized, controlled trial. *Pediatrics, 117*, e1170–e1178.

Mallory, M. D., Baxter, A. L., Yanosky, D. J., et al. (2011). Emergency physician – administered propofol sedation: A report on 25,433 sedations from the pediatric sedation research consortium. *Annals of Emergency Medicine, 57*(5), 462–468.

Merritt, K. A., Sargent, J. R., & Osborn, L. M. (1998). Attitudes regarding parental presence during medical procedures. *Am J Dis Child, 144*, 270–271.

Pena, B. M., & Krauss, B. (1999). Adverse events of procedural sedation and analgesia in pediatric emergency department. *Annals of Emergency Medicine, 34*(4), 483–491.

Petrack, E. M., Christopher, N. C., & Kriwinsky, J. (1997). Pain management in the emergency department: Patterns of analgesic utilization. *Pediatrics, 99*, 711–714.

Shah, A., Mosdossy, G., & McLeod, S. et al. (2011). A blinded, randomized controlled trial to evaluate ketamine/propofol versus ketamine alone for procedural sedation in children. *Annals of Emergency Medicine*, 57(5), 425–433.

Stevens, B., Yamada, J., & Ohlsson, A. (2010). Sucrose for analgesia in newborn infants undergoing painful procedures. *Cochrane Database of Systematic Reviews*, (3), CD001069.

Wolfe T. R., & Braude D. A. (2010). Intranasal medication delivery for children: A brief review and update. *Pediatrics*, 126(3), 532–537.

Pain as a Cause of Psychiatric Illness

Tipu Aamir and Robert G. Large
The Auckland Regional Pain Service, Auckland Hospital, Auckland, New Zealand

Synonyms

Pain as a stressor; Psychiatric illness; Psychological consequences of pain; Psychopathology caused by pain

Definition

The title suggests a linear model of causality from pain to psychiatric illness. Careful reading of the studies addressing this issue, however, suggests that it is not that simple.

Pain cannot be fully explained in terms of a pure sensory phenomenon. Melzack's and Wall (1965) gate control theory of pain proposed a balance between sensory and central inputs, with a tonic input to the gating systems in the spinal cord from higher centers. At around the same time, various explanations based on psychoanalytic principles started to dominate the literature. The assumption, underlying the psychoanalytic principles, is that pain is a defense against unconscious psychic conflict. Most of the literature was based on anecdotal accounts (Merskey and Spear 1967).

In the 1970s (Gamsa 1994), behavioral therapies and ▶ cognitive theories were employed to explain the relationship of psychological factors and pain. Fordyce looked at pain as a reflexive response to an antecedent stimulus (tissue damage). He also postulated that pain behaviors depend on contingent reinforcement and can be learned by observing "pain models." In this behavioral model, factors like emotional state and other psychosocial variables are not considered. Cognitive approaches to pain look at the meaning of pain to the patients themselves. This cognitive model also looks at the beliefs, self-efficacy, problem-solving, and coping abilities of the person. In the last 25 years, both cognitive and behavioral approaches have been combined and used extensively in pain management programs.

This brief overview highlights the shift from a linear to a multicausal model. In a biopsychosocial model, psychological and emotional factors play an important role in the continual experience of pain. The fact that psychosocial factors are considered so important in maintaining pain leads to the assumption that psychiatric illnesses will be more common in patients with pain.

Characteristics

Relationship Between Pain and Psychiatric Disorder

Studies of psychiatric diagnoses in pain populations have revealed a high incidence of ▶ psychiatric disorder. In one study, 94 % of chronic pain patients suffered from a psychiatric disorder; 30 % of these were diagnosed as suffering from depression (Large 1980). Polatin and Mayer (1996) showed that, after excluding ▶ somatoform pain disorder, 77 % of patients met lifetime diagnostic criteria and 59 % showed current symptoms for at least one psychiatric diagnosis. The most common diagnoses were of major depressive disorder, substance abuse, and anxiety disorders; 51 % of patients in this study met criteria for at least one ▶ personality disorder. The prevalence rates were significantly higher compared to the rates in the general population. This study also showed that out of the patients with a lifetime diagnosis of a psychiatric disorder, 54 % with major depressive disorder, 94 % with substance abuse disorder, and 95 % with anxiety disorder had experienced the disorders before the onset of their back pain. Hotopf et al. (1998) showed a robust link between psychiatric disorder and presence of physical symptoms. "Psychiatric disorder was associated with 2.9-6.9-fold increase in the odds of each physical symptom. Back pain, abdominal pain and chest pain were all associated with an increased likelihood of new onset psychiatric symptoms aged 43 years." The best estimate of 1-year prevalence of any psychiatric disorder between age group 18 and 54 years, diagnosed using DSM III R, is 21.0 %. The rate of major depressive episode is 6.5 %. Hence, there is a definite association between psychiatric disorder and pain.

Proposed Model of Development of Psychopathology

Gatchel (1996) has proposed a model of the development of psychopathology in low back pain over a period of time. This model looks at pain in three stages. Stage 1 represents the initial psychological distress, which is associated with fear and anxiety. This is a natural emotional reaction. When the pain continues beyond the acute period (2–4 months), it leads into stage 2. This stage is associated with phenomena like depression, distress, anger, somatization, and learned helplessness. Stage 3 is the acceptance of the "► sick role" and consolidation of "► abnormal illness behavior."

Gatchel (1996) hypothesizes that the specific nature of the psychological problems that develop in stage 2 depends on the individual's premorbid personality and psychological characteristics. They suggest that there is no specific "► pain-prone personality." The patient who suffers from persistent emotional problems passes onto stage 3, where the "sick role" is established and the patient is excused from his/her duties. This reinforces the sick role. This situation can be further complicated by compensation issues, which can become critical in the persistence of disability.

Banks and Kerns (1996) proposed a ► Diathesis-Stress framework for the development of depression in patients suffering from chronic pain. Diathesis refers to constitutional, biological, or psychological vulnerability. The stressors are any environmental or life event perceived by the individual as threatening to his or her physical or psychological well-being, and exceeding his or her capacities to cope. This model hypothesizes that a constitutional vulnerability in the presence of ► stress can perpetuate chronic pain and trigger a depressive disorder.

Pain as a Stressor

Pain, unlike some other chronic conditions, e.g., hypertension, is symptomatic (Banks and Kerns 1996). This quality of pain makes it noxious and aversive. The experience of pain is closely associated with affective distress. This may range through anxiety, fear, anger, and depression. Pain becomes an overriding and ever present part of a person's life.

At a psychological level, pain signifies danger. This is an instinctive phenomenon, which is ingrained in animals to propel them into self-preservation behavior. This behavior includes retraction from danger, reduced activity, and rest. The fact that pain represents danger raises the level of anxiety. In the longer term, pain promotes ► impairment and ► disability.

Impairment and disability lead to the restriction of a person's ability to participate in day-to-day social activities. Over a period of time, cumulative effects of ongoing impairment and disability lead to a series of events nominated as secondary losses (Banks and Kerns 1996).

Secondary loss refers to the effect of pain in various domains of daily life. These include loss of work and leisure activities, interference in and loss of intimate relationships, financial strife, unemployment, and loss of self-esteem. Thus, it can be inferred that pain is not an isolated phenomenon; rather it affects a person globally.

A consideration of the models above, together with the fact that pain can act as a stressor in its own right, enables us to postulate a model in which all these factors can interrelate (Fig. 1). An individual can have underlying vulnerabilities. These vulnerabilities can be a biological

Pain as a Cause of Psychiatric Illness, Fig. 1 The conceptualization of pain as a magnifying lens in development in psychopathology

predisposition to various psychiatric disorders or premorbid personality traits, e.g., paranoid, antisocial, borderline, histrionic, narcissistic, dependent, etc. People during their lifetime can pass through various life events, e.g., physical/sexual abuses, deprivation, loss, financial difficulties, etc. In an individual with premorbid vulnerability, any of these life events can become a precipitating factor for psychiatric disorder. When such a vulnerable individual suffers from pain, it can act as a strong stressor as described above, and can precipitate or magnify a mental disorder. Thus, pain acts as a magnifying lens. This relationship works both ways. Conditions like depression, anxiety, and somatization can increase pain, which can lead to an increase in stress and further deterioration of mental state. The underlying mechanisms at work are both cognitive and behavioral.

Relationship of Individual Psychiatric Disorders and Pain

Pain and Substance-Related Disorders
The relationship between substance-related disorders and pain has been widely reported in the literature. The rate of current substance-related disorders ranges between 15 % and 28 %, and the lifetime rate lies between 23 % and 41 %. Lifetime prevalence in the general population is 16.7 % (Dersh et al. 2002). These

disorders are more prevalent in males in both the pain sufferers and the general population.

Various pathways have been suggested to show the underlying interaction between the two conditions. Polatin et al. (1993) showed that 94 % of chronic pain patients with lifetime substance-related disorder had the onset of these disorders before the start of the pain disorder. This has been explained by an increased risk of accidents leading to chronic pain. Brown et al. (1996) have observed an increased risk for new substance-related disorder 5 years after the onset of chronic pain. It can be due to iatrogenic reasons. This risk is greater in individuals with a previous history of substance-related disorder.

Pain and Depressive Disorder
The relationship between pain and depression has generated quite a lot of interest, yet research on this subject is fraught with problems. These problems have arisen due to the different criteria used to define depression. Furthermore, DSM criteria of depression consist of a number of somatic features of depression, e.g., sleep disturbance, sexual dysfunction, etc. These features can also be symptomatic of organic illnesses (Dersh et al. 2002). However, despite these methodological difficulties, various hypotheses regarding the relationship between pain and depression have been studied.

Fishbain et al. (1997) conducted a literature review exploring the relationship between pain and depressive disorder. They considered the following hypotheses:

1. Antecedent: This hypothesis was very weakly supported by the literature.
2. Consequence: This was very strongly supported by the literature.
3. Scar: This hypothesis suggested the episode of depression before the onset of pain predisposing to depression. Literature supporting this hypothesis was not consistent.
4. Cognitive-behavioral mediation model: This hypothesis presupposes the role of cognitive distortions mediating the relationship between depression and chronic pain. This model was supported by the literature.

Based on the currently available evidence, depression can be seen as a consequence of chronic pain in patients.

Pain and Anxiety Disorder
Dersh et al. (2002) have highlighted the high rates of anxiety disorders in chronic pain patients. Rates between 16.5 % and 28.8 % have been observed. Factors leading to anxiety disorders can be premorbid vulnerability and anxiety in stage I (Gatchel 1996). Kinesophobia and operant conditioning (avoidance of feared situation reinforcing the disability) can maintain the disorder. Anxiety disorders can be explained on ▶ diathesis-stress model. Anxiety disorders have high rates in both acute and chronic pain states.

Pain and Somatoform Disorders
Pain and somatoform disorders are closely related. Recently, literature has suggested an increased focus on internal stimuli. There has been no study published recently on this relationship using DSM IV. However, studies in the past suggested a rate of 73 % (Dersh et al. 2002).

Pain and Personality Disorder
The literature shows a high rate of personality disorders in chronic pain patients. The prevalence ranges from 31 % to 81 % (Dersh et al. 2002). However, no specific personality disorder predicts chronicity. Research has shown that

parameters on various personality assessments change from pretreatment to posttreatment in chronic pain subjects, with a reduction in personality disorder diagnoses being made post treatment. It may be that the stress of pain can exacerbate the patient's vulnerability, leading to the expression of psychopathology.

Conclusion
The existing body of research shows that there is a definite association between psychiatric disorders and pain. The relationship between these conditions is complicated and efforts are being made to unravel the complexities of their interaction. It is important for clinicians to be aware of this relationship, because failure to identify psychopathology can adversely affect the outcome of treatment and our patients' overall well-being.

References

Banks, S. M., & Kerns, R. D. (1996). Explaining high rates of depression chronic pain: A diathesis-stress framework. *Psychological Bulletin, 119*, 95–110.

Brown, R. L., Patterson, J. J., Rounds, L. A., & Papasouliotis, O. (1996). Substance use among patients with chronic back pain. *The Journal of Family Practice, 43*, 152–160.

Dersh, J., Polatin, P. B., & Gatchel, R. J. (2002). Chronic pain and psychopathology: Research findings and theoretical considerations. *Psychosomatic Medicine, 64*, 773–786.

Fishbain, D. A., Cutler, R., Rosomoff, H. L., & Rosomoff, R. S. (1997). Chronic pain-associated depression: Antecedent or consequence of chronic pain? A review. *The Clinical Journal of Pain, 13*, 116–137.

Gamsa, A. (1994). The role of psychological factors in chronic pain-I. A half-century of study. *Pain, 57*, 5–15.

Gatchel, R. J. (1996). Psychological Disorders and Chronic Pain: Cause and Effect Relationships. In R. J. Gatchel & D.C. Turk. (Eds.), Psychological Approaches to Pain Management: A Practitioner's Handbook (pp. 33–52). New York: Guilford Publications.

Hotopf, M., Mayou, R., Wadsworth, M., & Wessely, S. (1998). Temporal relationships between physical symptoms and psychiatric disorder. Results from a national birth cohort. *The British Journal of Psychiatry, 173*, 255–261.

Large, R. G. (1980). The psychiatrist and the chronic pain patient: 172 anecdotes. *Pain, 9*, 253–263.

Melzack, R., & Wall, P. D. (1965). Pain mechanisms: A new theory. *Science, 150*, 971–979.

Merskey, H., & Spear, F. G. (1967). The concept of pain. *Journal of Psychosomatic Research, 11*, 59–67.

Polatin, P. B., & Mayer, T. G. (1996). Occupational disorders and the management of chronic pain. *The Orthopedic Clinics of North America, 27*, 881–890.

Polatin, P. B., Kinney, P. K., Gatchel, R. J., Lillo, E., & Mayer, T. G. (1993). Psychiatric illness and chronic low-back pain. The mind and the spine – Which goes first? *Spine, 18*, 66–71.

Pain as a Stressor

▶ Pain as a Cause of Psychiatric Illness

Pain Assay

▶ Central Pain: Outcome Measures in Preclinical Research

Pain Assessment

Definition

Pain assessment is the systematic process of evaluating and making pain treatment decisions, obtained either through the use of self-report or by proxy (a person other than the person experiencing pain, e.g., health-care professional). The process incorporates the consistent use of standardized pain assessment tools to obtain a valid and reliable measure of pain and takes into consideration contextual, situational, and familial factors. As it pertains to newborns, frequent assessment of pain and associated stress is required and especially relevant when newborns are on analgesics or sedatives.

Cross-References

▶ Pain Assessment in Children
▶ Pain Assessment in Neonates
▶ Pain Assessment in the Elderly

Pain Assessment in Children

Patricia A. McGrath
The Hospital for Sick Children Pain Research Center, Toronto, ON, Canada

Synonyms

Pediatric pain measures; Pediatric pain scales

Definition

Pain assessment is an integral component of pain management. At diagnosis, we need a reliable measure of a child's pain and an understanding of the factors that cause or contribute to pain and disability. Subsequent assessments of pain intensity enable us to determine when treatments are effective and to identify those children for whom they are most effective. Thus, pain assessment is a dynamic process that begins with a diagnostic examination and culminates with a clinical decision that a child's pain has improved sufficiently. Although the key aspects of assessment are the same throughout a treatment program, the assessment tools or specific measures differ depending on the treatment phase – diagnosis, routine clinical monitoring of treatment efficacy, and evaluation at discharge to define whether a child has achieved the treatment objectives (McGrath and Koster 2001).

Characteristics

The criteria for an accurate pain measure for children are similar to those required for any measuring instrument. A pain measure must be ▶ valid, in that it unequivocally measures a specific dimension of a child's pain so that changes in pain ratings reflect meaningful differences in a child's pain experience. The measure must be ▶ reliable, in that it provides consistent and trustworthy pain ratings regardless of the time of testing, the clinical setting, or who is administering the measure. The measure must be relatively ► free from bias in that children should be able to use it similarly, regardless of differences in how they may wish to please adults. The pain measure should be practical and versatile for assessing different types of pain (e.g., disease-related, procedural pain) in many different children (according to age, cognitive level, and cultural background) and for use in diverse clinical and home settings.

An extensive array of pain measures have been developed and validated for use with infants, children, and adolescents (Gaffney et al. 2003; McGrath and Gillespie 2001; RCN Institute 1999; Stevens et al. 2000). Children's pain measures are classified as physiological (► Physiological Pain Measure), behavioral (► Behavioral Pain Scale), and psychological (► Psychological Pain Measures) depending on what is monitored: physical parameters, such as heart or respiration rate, and distress behaviors, such as crying or facial expression or children's own descriptions of what they are experiencing. Both physiological and behavioral measures provide indirect estimates of pain because the presence or strength of pain is inferred solely from the type and magnitude of responses to a noxious stimulus. In contrast, psychological measures can provide direct estimates for many different dimensions of pain (i.e., intensity, quality, affect, duration, and frequency) and provide valuable information on the impact of pain.

Behavioral Pain Scales

Most behavioral pain scales include checklists of the different distress behaviors (e.g., crying, grimacing, guarding) that children exhibit when they experience a certain type of pain (McGrath 1998; McGrath and Gillespie 2001; Stevens et al. 2000). Health-care providers complete such pain scales by noting which of the listed behaviors they see and by ranking their intensity. Most behavioral scales were developed to measure acute procedural or postoperative pain in otherwise healthy children. More recently, attention is focusing on validating pain scales for children with developmental disabilities (Breau et al. 2002; Hadden and von Baeyer 2002;

Hunt et al. 2004; Oberlander and Craig 2003; Stallard et al. 2002; Terstegen et al. 2003).

Physiological Pain Scales

Physiological parameters that have been monitored in infants and children as potential pain measures include heart rate, respiration rate, blood pressure, palmar sweating, cortisol and cortisone levels, O_2 levels, vagal tone, and endorphin concentrations (Sweet and McGrath 1998). Although physiological responses can provide valuable information about a child's distress state, more research is required to develop a sensitive system for interpreting how these parameters reflect the quality or intensity of a child's pain. At present, physiological parameters do not constitute valid clinical pain measures for children.

Psychological Pain Scales

Psychological or self-report pain scales directly capture a child's subjective experience of pain. Interviews, questionnaires, adjective checklists, and numerous pain intensity scales are available for children, each with some evidence of validity and reliability (Champion et al. 1998; McGrath and Gillespie 2001). These self-report measures can provide valuable diagnostic information about the causes and contributing factors for a child's pain as well as provide practical tools for regularly assessing pain intensity to ensure that children receive adequate pain control.

When possible, health-care providers should ask children directly about the location, quality, intensity, duration, or frequency of pain using a semistructured format (a few basic questions asked in a consistent manner with different follow-up questions depending on the child's responses). Clinical interviews are ideally suited for learning about the sensory characteristics of pain, the aversive component, and contributing cognitive, behavioral, and emotional factors. Interviews should also include a simple rating scale to document pain strength. Children can use many analogue, facial, and verbal rating scales (samples shown in Fig. 1). Children select a level on the scale (i.e., a number, a mark on a visual analogue scale, a face from a series of faces varying in emotional expression,

Verbal Descriptor Scales

No pain   None
Mild pain   A little bit
Moderate pain A medium amount
Strong pain   A lot
Intense pain   A real lot

Facial Affective Scale (staff side)

Colored Analog Scale Staff side (left) and child side (right)

2nd international
Paediatric
Pain Symposium
McGill, 1991

MOST PAIN

NO PAIN

Pain Assessment in Children, Fig. 1 Three scales for assessing children's pain: word descriptor scale, facial affective scale, and colored analogue scale

or a particular word from an adjective list) to match the strength of their own pain. These scales are easy to administer, requiring only a few seconds once children understand how to use them. Table 1 lists some analogue scales developed for children, along with the type of pain used to initially develop the scale, the scale type, pain feature measured, and the resulting pain score.

Pain Ratings: What Do the Numbers Mean?

Although children can use many pain scales, the resulting pain scores from the different scales are not necessarily equivalent. Thus, it is important to know what the pain scores mean on different scales. Some scales simply use numbers to represent different pain intensity levels (e.g., no pain equals 0, mild pain equals 1, moderate pain

Pain Assessment in Children, Table 1 Analogue pain scales for children

Name	Pain type	Method	Pain characteristic	Scale type/pain score	Comments
Eland color tool	Postoperative	Projective rating scale on body outline	Intensity, location	Quantitative (probably interval)	Children select colors to represent three levels of pain and shade body outline
Eland and Anderson (1977)					
Poker chip tool	Acute pain (immunization)	Object scale comprised of four poker chips	Intensity	Quantitative, interval scale	Children ages 4–7 year
Hester et al. 1990				Pain scores 0–4	
Pain thermometer	Treatment (burn dressing change)	Analogue scale, shaped like thermometer	Intensity	Quantitative, may be interval or ratio scale, pain scores 0–10	Children and adolescents, ages 8–17
Szyfelbein et al. (1985)					
Colored analogue scale (CAS)	Acute trauma, postoperative, recurrent, chronic pain	Analogue scale, varying in length, hue, and area	Intensity	Quantitative, ratio scale properties, pain scores 0–10.0	Psychometric properties demonstrated
McGrath et al. (1996)					Children ages 5 year and older, versatility demonstrated for clinical and home use
Multiple size poker chip tool (MSPCT)	Procedural (immunization)	Object scale comprised of 4 poker chips, varying in size (2–3.8 cm)	Intensity	Quantitative, interval scale	Children ages 4–6 year
St. Laurent-Gagnon et al. (1999)				Pain scores 0–4	

equals 2, and strong pain equals 3). The numbers are often interpreted as if they represent absolute and accurate amounts of pain. But, unless an investigator studies the relationship between a child's pain level and the numbers he or she uses to rate – the psychometric properties – we only know that larger numbers mean stronger pain, and we do not know how much stronger. As a result, when evaluating treatment effects, we cannot assume that a child who reports that a treatment lessened a pain from a score of 2 to 1 (moderate to mild) is equivalent to a reduction from 3 to 2 (strong to moderate) for another child. Similarly, when a child rates a pain as a 3, the pain level may not really be $3\times$ the strength of the pain he/she rates as a 1. Nevertheless, most pain scores are interpreted as if they represent numbers on equal interval or ratio scales.

The four types of measurement scales (nominal, as in the numbers designating players on a sports team; ordinal, as in the rank ordering of children according to height; interval, as in the Fahrenheit temperature scale; and ratio, as in a yardstick) refer to four different relationships between the properties of an event or perception (e.g., pain intensity score) and the number or metric system. Ratio scales have all the properties of the three other scales; they represent a set position or order between numbers, they show the magnitude of the difference between numbers, and the numbers reflect true ratios of magnitude. Each scale has a certain number of permissible mathematical calculations, so it is important to understand which type of scale one is using when measuring a child's pain or evaluating analgesic efficacy.

Pain Assessment in Children,
Table 2 Primary components of pain assessment

Sensory characteristics	*Cognitive factors*
Onset	Understanding of pain source
Location	Understanding of diagnosis, treatment, and prognosis
Intensity	Expectations
Quality	Perceived control
Duration	Relevance of disease or painful treatments
Spread to other sites (consistent with neurological pattern)	Knowledge of pain control
Temporal pattern	
Accompanying symptoms	
Medical/surgical	*Behavioral factors*
Investigations conducted	General coping style
Radiological and laboratory results	Learned pain behaviors
Consult results	Overt distress level
Analgesic and adjuvant medications (type, dose, frequency, route)	Parent's behaviors
	Physical activities and limitations
	Social activities and limitations
Clinical factors	*Emotional factors*
Environmental features	Frustration
Roles of medical and associated health professionals	Anxiety
Nature of interventions	Fear
Documentation of pain	Denial
Criteria for determining analgesic effectiveness	Sadness
	Anger
	Depression

Guidelines for Selecting a Pain Scale for Children

More than 50 pain scales have been developed for children. Yet, no one scale is appropriate for all children and for all situations in which they experience pain. The critical issue is how to select the most appropriate measure for the age and cognitive level of the child and one that satisfies your clinical objectives. Many simple, easy-to-use pain scales provide meaningful values that reflect a child's pain intensity and are ideal measures for evaluating treatment effectiveness throughout a child's treatment. But, they are not adequate for teaching us about the nature of a child's pain or for identifying the primary and secondary situational factors that affect pain.

Health-care providers must first consider the age and cognitive ability of a child when

selecting a pain scale. Behavioral scales and physiological distress indices are required for infants and for children who are unable to communicate verbally. Although caution must always be used when interpreting children's pain solely from their scores on these scales because children's behaviors are not simply passive reflections of their pain intensity, these scales can provide valuable information. Because a child's pain behaviors naturally vary in relation to the type of pain experienced, different behavioral scales are usually required for acute and chronic pain.

Most toddlers (approximately 2 years of age) first express the "hurting" aspect of pain using a few words learned from their parents to describe the sensations they feel when they hurt themselves. Gradually children learn to differentiate and describe three levels of pain intensity – "a little, some or medium, and a lot." By the age of five, most children can differentiate a wide range of pain intensities, and many children at this age can use simple quantitative scales to rate their pain intensity. These pain scales can provide meaningful values that reflect a child's pain intensity and are ideal measures for evaluating treatment effectiveness throughout a child's treatment.

But, pain intensity scales alone cannot teach us about the nature of a child's pain or enable us to identify the primary and secondary situational factors that affect pain (McGrath and Hillier 2003). Health-care providers who treat a wide range of children's pain problems need a flexible pain measurement inventory comprised of a more comprehensive assessment instrument and pain diaries, as well as simple pain intensity scales. Diaries are particularly useful for treating children with recurrent or chronic pain; children record prospectively the frequency, intensity, quality, and duration of pain, as well as what they do to reduce the pain and its effectiveness. Diaries can be flexible instruments to fit the needs of the specific therapy program; health-care providers may monitor school attendance, physical activity, peer activities, medication use, and use of nondrug strategies using simple observation forms.

In summary, evaluating a child's pain requires an integrated approach. Health-care providers should always ask a child directly about his or her pain experience and evaluate the sensory characteristics to facilitate an accurate diagnosis. Pain onset, location, frequency (if recurring), quality, intensity, accompanying physical symptoms, and pain-related disability should be assessed as part of a child's clinical examination. Health-care providers should also assess relevant situational factors in order to modify their pain-exacerbating impact, especially the factors listed in Table 2. Then, health-care providers should assess pain intensity or pain behaviors regularly using brief quantitative scales to monitor the effectiveness of their therapy. Only previously validated pain scales should be used, unless clinicians are willing to conduct the rigorous research necessary to prove that a measure is valid and reliable and to determine its psychometric properties.

References

Breau, L. M., McGrath, P. J., Camfield, C. S., et al. (2002). Psychometric properties of the non-communicating children's pain checklist-revised. *Pain, 99*, 349–357.

Champion, G. D., Goodenough, B., von Baeyer, C. L., et al. (1998). Measurement of pain by self-report. In G. A. Finley & P. J. McGrath (Eds.), *Measurement of pain in infants and children. Progress in pain research and management* (pp. 123–160). Seattle: IASP.

Eland, J. M., & Anderson, J. E. (1977). The experience of pain in children. In A. K. Jacox (Ed.), *Pain: A source book for nurses and other health professionals* (pp. 453–471). Boston: Little, Brown.

Gaffney, A., McGrath, P. J., & Dick, B. (2003). Measuring pain in children: Developmental and instrument issues. In N. L. Schechter, C. B. Berde, & M. Yaster (Eds.), *Pain in infants, children and adolescents* (2nd ed., pp. 128–141). Philadelphia: Lippincott.

Hadden, K. L., & von Baeyer, C. L. (2002). Pain in children with cerebral palsy: Common triggers and expressive behaviors. *Pain, 99*, 281–288.

Hester, N. O., Foster, R., & Kristensen, K. (1990). Measurement of pain in children: Generalizability and validity of the pain ladder and the poker chip tool. In D. C. Tyler & E. J. Krane (Eds.), *Pediatric pain* (pp. 79–84). New York: Raven.

Hunt, A. M., Goldman, A., Seers, K., Crichton, N., Mastroyannopoulou, K., et al. (2004). Clinical validation of the paediatric pain profile. *Developmental Medicine and Child Neurology, 46*, 9–18.

Institute, R. C. N. (1999). *Clinical guideline for the recognition and assessment of acute pain in children: Recommendations.* London: RCN Publishing.

McGrath, P. J. (1998). Behavioral measures of pain. In G. A. Finley & P. J. McGrath (Eds.), *Measurement of pain in infants and children. Progress in pain research and management* (pp. 83–102). Seattle: IASP.

McGrath, P. A., & Gillespie, J. M. (2001). Pain assessment in children and adolescents. In D. C. Turk & R. Melzack (Eds.), *Handbook of pain assessment* (pp. 97–118). New York: Guilford.

McGrath, P. A., & Hillier, L. M. (2003). Modifying the psychological factors that intensify children's pain and prolong disability. In N. L. Schechter, C. B. Berde, & M. Yaster (Eds.), *Pain in infants, children, and adolescents* (pp. 85–104). Baltimore: Lippincott.

McGrath, P. A., & Koster, A. L. (2001). Headache measures for children: A practical approach. In P. A. McGrath & L. M. Hillier (Eds.), *The child with headache: Diagnosis and treatment* (pp. 29–56). Seattle: IASP.

McGrath, P. A., Seifert, C. E., Speechley, K. N., et al. (1996). A new analogue scale for assessing children's pain: An initial validation study. *Pain, 64*, 435–443.

Oberlander, T. F., & Craig, K. D. (2003). Pain and children with developmental disabilities. In N. L. Schechter, C. B. Berde, & M. Yaster (Eds.), *Pain in infants, children and adolescents* (2nd ed., pp. 599–619). Philadelphia: Lippincott.

St. Laurent-Gagnon, T., Bernard-Bonnin, A. C., & Villeneuve, E. (1999). Pain evaluation in preschool children and by their parents. *Acta Paediatrica, 88*, 422–427.

Stallard, P., Williams, L., Velleman, R., et al. (2002). The development and evaluation of the pain indicator for communicatively impaired children (PICIC). *Pain, 98*, 145–149.

Stevens, B. J., Johnston, C. C., & Gibbins, S. (2000). Pain assessment in neonates. In K. J. S. Anand, B. J. Stevens, & P. J. McGrath (Eds.), *Pain in neonates* (2nd ed., pp. 101–134). Amsterdam: Elsevier.

Sweet, S., & McGrath, P. J. (1998). Physiological measures of pain. In G. A. Finley & P. J. McGrath (Eds.), *Measurement of pain in infants and children. Progress in pain research and management* (pp. 59–82). Seattle: IASP.

Szyfelbein, S. K., Osgood, P. F., & Carr, D. B. (1985). The assessment of pain and plasma beta-endorphin immunoactivity in burned children. *Pain, 22*, 173–182.

Terstegen, C., Koot, H. M., de Boer, J. B., et al. (2003). Measuring pain in children with cognitive impairment: Pain response to surgical procedures. *Pain, 103*, 187–198.

Pain Assessment in Neonates

Rebecca Pillai Riddell[1,2], Diana Lisi[1] and Lauren Campbell[1]
[1]Department of Psychology, York University, Toronto, ON, Canada
[2]Department of Psychiatry, Hospital for Sick Children, Toronto, ON, Canada

Synonyms

Babies or infants; Pain judgment; Pain measurement; Pain measures; Pain tools

Definition

What is Pain Assessment in Neonates?

Pain assessment is the evaluation of the existence, intensity, and duration of pain following a nociceptive event (Powell et al. 2010).

The term "neonate" pertains to work focusing predominantly on premature born infants (infants born prior to 37 weeks) and healthy full-term infants in the period directly after birth (approximately one month).

Introduction

Regardless of age, pain experience is fundamentally a subjective phenomenon. Moreover, the transduction of a pain experience into a pain expression is also largely subjective. Two individuals experiencing the same amount of objective pain stimulation may react very differently, while two individuals experiencing different amounts of pain stimulation may act very similarly. In verbal populations the idiosyncratic nuances of one's pain experience and pain expressions can at least be contextualized by self-report. Accordingly, a recommended pain assessment procedure is to ask patients for their pain levels (Turk 1993; American Pain Society 2008). However, clarification of how a particular infant actually perceives a painful stimulation or sensation is not possible due to developing language abilities.

Accurate pain assessment is the first step in adequate pain management (Anand 2012). Thus, pain assessment in the neonate is dependent on caregivers accessing assessment tools that are psychometrically sound and regularly integrating them into routine patient care. A number of more extensive comprehensive reviews on pain assessment during the neonatal and infancy periods have been published (Duhn and Medves 2004; Stevens et al. 2007b; Anand 2012); thus, the focus of this entry will be to offer a brief survey of the extant literature.

Given the suboptimal implications of unmanaged pain during infancy (e.g., Taddio et al. 2002) and its dependence on accurate pain assessment, it is crucial to take into account a number of factors when selecting an appropriate neonatal pain assessment tool. Following is a description of these factors and then, a table providing common neonatal assessment measures that have some evidence of psychometric soundness.

Characteristics

Dimensions to Consider in Tool Selection

Type of Pain Insult

Pain in early infancy can broadly be conceptualized into three main types of pain: acute, acute-prolonged, and chronic-prolonged. Acute pain is the direct physiological and behavioral response from painful stimuli (American Academy of Pediatrics 2001), and is typically the quickest of pain types to subside. It is often the easiest to discern because it is easily attributable to noxious stimuli such as immunization needle or heel lance, and it quickly abates as a function of time (e.g., a needle poke). Measures that are ideal for this type of pain focus on a short burst of physiological or behavioral reaction. Higher scores will be based on the presence or intensity of indicators associated with pain or pain-related distress.

Postoperative pain is a common example of acute-prolonged, which is experienced for a longer period after a more intense noxious event such as a surgical procedure, and is accompanied by the expectation that this pain will subside after some time has passed, most often three to five days (Obrecht and Andreoni 2011). It is more difficult than acute pain to be properly assessed due to the prolonged nature, and thus, it is suggested that observation of physiological and behavioral pain assessment be based on a more prolonged observation period. In addition to those indicators that are often seen in acute pain measures, items that reflect a longer period of distress are also appropriate, such as sleep dysregulation and prolonged hypersensitivity to impacted parts of the body (Emde et al. 1971; Peters et al. 2005).

The assessment of chronic-prolonged or persistent pain is a developing area in the infancy literature, with widely accepted definitions lacking (Anand et al. 2005). Generally, this type of pain may pertain to infants who are born with a chronic health condition that results in either prolonged iatrogenic pain due to repetitive procedures (e.g., infants with cardiac conditions who are subjected to numerous, repetitive painful procedures such as surgeries, venipunctures, and line insertions) or infants with a painful condition that is constant or episodic (such as epidermis bullosa or sickle cell disease). Exploratory work in the area suggests that depending on the temporal spacing of painful insults (regardless of them being within the body or outside the body), infants may become hyporeactive to painful stimulation or hyperreactive to both painful and non-painful stimulation (Pillai Riddell et al. 2009). Chronic pain timelines in older children and adults, such as pain duration greater than one or three months (Task Force on Taxonomy 1994), are not relevant given the brief lifespan of neonates and infants. Until more definitive definitional parameters of this longer type of pain exist, researchers and clinicians are likely best served by using assessment measures validated in the acute-prolonged pain context.

Usage in Research Versus Clinical Settings

Pain assessment during the neonatal period is conducted for both research and clinical purposes. Clinical measures are those that consist of pain indicators that can be readily accessible to parents and/or health professionals. Administration of measures must be feasible in a busy clinical environment.

Research assessment measures are those that while having established psychometrics lack feasibility due to more labor-intensive coding systems that require processing outside of the clinical environment (e.g., microcoding of discrete behaviors from video footage).

An ideal measure during early infancy is one that can be readily applied to both research and clinical settings, as the knowledge transfer between the two is more readily achieved.

Unidimensional Versus Multidimensional Measures Given the lack of specific verbal self-report during infancy, many assessment tools take a multidimensional approach to improve validity and reliability. Multidimensional approaches examine multiple domains of pain reactivity (e.g., facial reactions, physiological reactions, body movements, vocalizations).

Unidimensional measures of pain examine one type of indicator (e.g., cry vocalization), or multiple indicators for one specific domain (e.g., several facial actions within the facial activity domain).

Very often the choice between a multidimensional or unidimensional measure is based on location, age, and resources available. Research with healthy infants can be conducted in primary care settings or in less-equipped hospital settings (e.g., hospital rooms or birthing rooms); thus, access to physiological indicators (e.g., heart rate variability, oxygen saturation) is limited. Moreover, given that healthy neonates have well-established vigorous behavioral indicators of pain (e.g., Grunau and Craig 2010), there is less of a need for multidimensional measures that include physiological indicators. However, with premature and neurologically compromised infants who may have more subtle overt pain behaviors, the need for other sensitive, albeit not specific, indicators is warranted (Johnston et al. 1999; Stevens et al. 2007a). Measurement of physiological indicators are standard care in Neonatal Intensive Care Units, thus, obtention is often not an issue.

Behavioral Versus Physiological Domains Two broad domains of pain indicators (behavioral and physiological) were alluded to above. Behavioral indicators of pain can be defined as directly observable reactions of a pained individual that require minimal inference on the part of the assessor. Examples of such behaviors in the neonatal pain context include crying, body movements, and facial expressions.

Physiological indicators measure variability in bodily processes, either directly or indirectly (Stevens et al. 1996), and are readily quantified. Common ones used in neonatal pain assessment are heart rate variability, oxygen saturation, and cortisol levels.

Owing to the lack of verbal report from the infant patient and demonstrated lack of convergent validity between behavioral and physiological indicators (e.g., Slater et al. 2010) during neonatal pain assessment, it is important to refrain from treating the divergence of indicators in a known painful situation (e.g., the lack of

change in either behavioral or physiological indicators) as sole evidence that the neonate is not experiencing pain.

Commonly Used Neonatal Pain Assessment Tools

Table 1 lists the characteristics of some of the more commonly used pain assessment measures for infants and neonates.

Beyond the Assessment Tool: Challenges to Neonatal and Infant Pain Assessment

The pragmatics of neonatal pain assessment necessitate the use of reliable and valid assessment tools. However, there are a number of broader conceptual issues that must be further understood in order to move the clinical science forward.

Specificity of Pain Responses

A commentary advocating for the central role of patient self-report in an adult context once noted, "it is highly unlikely that we will ever be able to evaluate pain without reliance on the individual's perceptions" (Turk 1993). The implications on the nonverbal infant context are clearly seen when considering the lack of specificity of any indicators, behavioral or physiological.

For example, infant facial expressions are commonly used as behavioral indicators of infant pain in both clinical and research settings because they are considered a valid indicator of pain in lieu of verbal report (Stevens et al. 2007b). But it is often difficult to distinguish an expression of pain from that of hunger, agitation, or other potentially negative states in the neonate. Some researchers argue in favor of a general distress state in the neonatal period rather than attribute distinctive negative emotional experiences such as anger, fear, and pain (Ahola Kohut, in press). Evidence in the field of infant pain supports this debate. In older infants, pain-related distress and non-pain-related distress were not notably distinguishable using a gold standard pain facial coding measure (Ahola and Pillai Riddell 2009). Most assessment tool studies, when discussing validity, refer either to convergent validity with other measures of pain/distress or divergent validity

Pain Assessment in Neonates, Table 1 Characteristics of some of the more commonly used pain assessment measures for infants and neonates

Measures	Age group	Pain context	Primary use	Main parameters
Neonatal Facial Coding System (NFCS; Grunau and Craig 1987, 1990)	Preterm and full-term infants, older infants	Acute, acute-prolonged	Research and clinical (bedside version)	Behavioral
Modified Behavioral Pain Scale (MBPS; Taddio et al. 1995)	Full-term and older infants (0–6 months)	Acute, acute-prolonged	Research	Behavioral
The Neonatal Pain and Discomfort Scale [Echelle Douleur Inconfort Nouveau-Ne] (EDIN; Debillon et al. 2001)	Preterm infants (26–36 weeks GA)	Acute, acute-prolonged, and chronic-prolonged	Clinical	Behavioral
Neonatal Pain, Agitation, and Sedation Scale (N-PASS; Hummel et al. 2008)	Preterm (<28 weeks to 35 weeks GA) and full-term infants	Acute, acute-prolonged, and chronic-prolonged	Clinical	Physiological and behavioral
The COMFORT Scale (Ambuel et al. 1992; van Dijk et al. 2000; Caljouw et al. 2007)	Preterm infants (adapted COMFORT scale; 28–37 weeks GA) and full-term infants (0–18 months)	Acute-prolonged	Research and clinical	Physiological and behavioral
Crying Requires oxygen Increased vital signs Expression Sleep (CRIES; Krechel and Bildner 1995)	Preterm and full-term infants (32–60 weeks GA)	Acute-prolonged	Clinical	Physiological and behavioral
Premature Infant Pain Profile (PIPP; Stevens et al. 1996, 2010)	Preterm and full-term infants (28–40 weeks GA)	Acute, acute-prolonged	Research and clinical	Physiological and behavioral
Neonatal Infant Pain Score (NIPS; Lawrence et al. 1993)	Preterm and full-term infants (24–40 weeks GA)	Acute	Research and clinical	Physiological and behavioral

with an indication of neutral or positive state. To date, no assessment tool to our knowledge has shown clear discriminant validity in different types of distress contexts. To aid in disentangling the ambiguity of these types of assessments, it is critical to assume that if the infant has undergone a painful stimulation that would be painful to an adult, then the distress is pain related (Anand 2012). Moreover, using an efficient method to eliminate other causes of distress (e.g., hunger, overstimulation, proximity-seeking) by process of elimination can help in the assessment of the neonate.

Caregiver Influences
Endemic to an assessment context with no access to self-report, it is important to recognize that one cannot properly assess the infant without understanding the role of the caregiver.

Even when utilizing a valid and reliable tool, a number of factors moderate an observer's experience and actions toward another individual in pain (Goubert et al. 2005). Research with neonates and older infants suggests that factors such as psychological distress and health care profession play significant roles in determining the amount of pain assessed during a painful experience (Pillai Riddell et al. 2007; Elias et al. 2008). Other work has suggested that extensive experience with infants in pain either leads to higher infant pain judgments (Pillai Riddell and Craig 2007) or lower pain judgments (Xavier Balda et al. 2000). Ensuring that caregivers utilize a validated assessment tool at regulated time intervals (when appropriate) can help to reduce biases in pain assessments.

The other side of this clinical confound is that not only have differences been noted in how different types of caregivers make infant pain assessments on the same infant, the same infant may actually have different pain reactions depending on what caregivers are present and how the caregivers are acting in the pain context (Pillai Riddell and Racine 2009). Young infants have been shown to be able to associate caregiving context with painful experiences and thus, demonstrate higher distress within contexts or with caregivers they associate with pain (Taddio et al. 2002). This scenario is most likely in a hospital environment; thus, having different people administer painful procedures (other than the regular nurse caregiver) may help avoid negative associations with any particular nurse or physician. In addition, allowing a neonate to recover fully from one painful stimulus before administering the next one may also aid in preventing anticipatory conditioning.

Summary

Pain assessment is a critical step in the management of infant pain, and of particular importance for neonates. Pain type, setting, and number/types of indicators utilized are all important factors for selecting an assessment tool. A number of psychometrically sound options are available to measure neonatal pain in acute, acute-prolonged, and chronic-prolonged pain contexts. Although there are a number of assessment tools with very good support of their psychometrics, challenges to assessing this age group must be recognized. In a population devoid of self-report and a developing emotional repertoire, we are faced with the fact that all current measures of neonatal pain lack specificity to only pain-related distress. Moreover, the expression of distress in infancy can be highly dependent on which caregivers are present and the context of the painful stimulation. These limitations should not discourage neonatal pain assessment but rather motivate both clinicians and researchers to be more cognizant of the nuances involved when assessing a newborn child.

Acknowledgment This entry was supported by operating funds and/or salary support from the Canadian Institutes of Health Research, the Ontario Ministry of Research and Innovation, and the Lillian Wright Maternal-Child Health Graduate Scholars Program.

References

Ahola Kohut, S., Pillai Riddell, R. R., Flora, D., & Oster, H. (in press). A longitudinal analysis of the development of negative emotional facial expressions. *Pain.*

Ahola, S., & Pillai Riddell, R. R. (2009). Investigating the discriminative validity of the neonatal facial coding system. *The Journal of Pain, 10*(2), 213–220.

Ambuel, B., Hamlett, K. W., Marx, C. M., & Blumer, J. L. (1992). Assessing distress in pediatric intensive care environments: The COMFORT scale. *Journal of Pediatric Psychology, 17*(1), 95–109.

American Academy of Pediatrics. (2001). The assessment and management of acute pain in infants, children, and adolescents. *Pediatrics, 108*(3), 793–797.

American Pain Society. (2008). *Principles of analgesic use in the treatment of acute pain and cancer pain* (6th ed.). Glenview, IL: American Pain Society.

Anand, K. J. S. (2012). Assessment of neonatal pain. In J. A. Garcia-Prats & M. S. Kim (Eds.), *UpToDate.* UpToDate: Waltham, MA.

Anand, K. J., Aranda, J. V., Berde, C. B., Buckman, S., Capparelli, E. V., Carlo, W. A., Hummel, P., Lantos, J., Johnston, C. C., Lehr, V. T., Lynn, A. M., Maxwell, L. G., Oberlander, T. F., Raju, T. N., Soriano, S. G., Taddio, A., & Walco, G. A. (2005). Analgesia and anesthesia for neonates: study design and ethical issues. *Clinical Therapeutics, 27*(6), 814–843.

Caljouw, M. A. A., Kloos, M. A. C., Olivier, M. Y., Heemskerk, I. W., Pison, W. C. R., Stigter, G. D., & Verhoef, A. J. H. (2007). Measurement of pain in premature infants with a gestational age between 28 to 37 weeks: Validation of the adapted COMFORT scale. *Journal of Neonatal Nursing, 13*(1), 13–18.

Debillon, T., Zupan, V., Ravault, N., Magny, J. F., & Dehan, M. (2001). Development and initial validation of the EDIN scale, a new tool for assessing prolonged pain in preterm infants. *Archives of Diseases in Childhood, Fetal and Neonatal Edition, 85*(1), F36–F41.

Duhn, L. J., & Medves, J. M. (2004). A systematic integrative review of infant pain assessment tools. *Advance Neonatal Care, 4*(3), 126–140.

Elias, L. S., Guinsburg, R., Peres, C. A., Balda, R. C., & Santos, A. M. (2008). Disagreement between parents and health professionals regarding pain intensity in critically ill neonates. *Jornal de Pediatria, 84*(1), 35–40.

Emde, R. N., Harmon, R. J., Metcalf, D., Koenig, K. L., & Wagonfeld, S. (1971). Stress and neonatal sleep. *Psychosomatic Medicine, 33*(6), 491–498.

Goubert, L., Craig, K. D., Vervoort, T., Morley, S., Sullivan, M. J., de C Williams, A. C., Cano, A., & Crombez, G. (2005). Facing others in pain: The effects of empathy. *Pain, 118*(3), 285–288.

Grunau, R. V., & Craig, K. D. (1987). Pain expression in neonates: Facial action and cry. *Pain, 28*(3), 395–410.

Grunau, R. V., & Craig, K. D. (1990). Facial activity as a measure of neonatal pain expression. In D. C. Tyler & E. J. Krane (Eds.), *Advances in pain research and therapy* (Pediatric pain, Vol. 15, pp. 147–156). New York: Raven.

Grunau, R. V., & Craig, K. D. (2010). Neonatal facial coding system: Revised training manual. Unpublished monograph and coding manual, University of British Columbia.

Hummel, P., Puchalski, M., Creech, S. D., & Weiss, M. G. (2008). Clinical reliability and validity of the N-PASS: Neonatal pain, agitation and sedation scale with prolonged pain. *Journal of Perinatology, 28*(1), 55–60.

Johnston, C. C., Stevens, B. J., Franck, L. S., Jack, A., Stremler, R., & Platt, R. (1999). Factors explaining lack of response to heel stick in preterm newborns. *Journal of Obstetric, Gynecologic, and Neonatal Nursing, 28*(6), 587–594.

Krechel, S. W., & Bildner, J. (1995). CRIES: A new neonatal postoperative pain measurement score – Initial testing of validity and reliability. *Paediatric Anaesthesia, 5*(1), 53–61.

Lawrence, J., Alcock, D., McGrath, P., Kay, J., MacMurray, B., & Dulberg, C. (1993). The development of a tool to assess neonatal pain. *Neonatal Network, 12*, 59–66.

Obrecht, J., & Andreoni, V. A. (2011). Pain Management. In N. L. Potts & B. L. Mandleco (Eds.), *Pediatric nursing: Caring for children and their families* (3rd ed., pp. 581–605). Clifton Park, NY: Delmar Learning.

Peters, J. W., Schouw, R., Anand, K. J., van Dijk, M., Duivenvoorden, H. J., & Tibboel, D. (2005). Does neonatal surgery lead to increased pain sensitivity in later childhood? *Pain, 114*(3), 444–454.

Pillai Riddell, R. R., & Craig, K. D. (2007). Judgments of infant pain: The impact of caregiver identity and infant age. *Journal of Pediatric Psychology, 32*(5), 501–511.

Pillai Riddell, R., & Racine, N. (2009). Assessing pain in infancy: The caregiver context. *Pain Research & Management, 14*(1), 27–32.

Pillai Riddell, R. R., Stevens, B. J., Cohen, L. L., Flora, D. B., & Greenberg, S. (2007). Predicting maternal and behavioral measures of infant pain: The relative contribution of maternal factors. *Pain, 133*(1–3), 138–149.

Pillai Riddell, R. R., Stevens, B. J., McKeever, P., Gibbins, S., Asztalos, L., Katz, J., Ahola, S., &

Din, L. (2009). Chronic pain in hospitalized infants: Health professionals' perspectives. *The Journal of Pain, 10*(12), 1217–1225.

Powell, R. A., Downing, J., Ddungu, H., & Mwangi-Powell, F. N. (2010). Pain history and pain assessment. In A. Kopf & N. B. Patel (Eds.), *Guide to pain management in low-resource settings*. Seattle: IASP Press.

Slater, R., Cornelissen, L., Fabrizi, L., Patten, D., Yoxen, J., Worley, A., Boyd, S., Meek, J., & Fitzgerald, M. (2010). Oral sucrose as an analgesic drug for procedural pain in newborn infants: A randomised controlled trial. *Lancet, 376*(9748), 1225–1232.

Stevens, B., Johnston, C., Petryshen, P., & Taddio, A. (1996). Premature infant pain profile: Development and initial validation. *The Clinical Journal of Pain, 12*(1), 13–22.

Stevens, B., McGrath, P., Gibbins, S., Beyene, J., Breau, L., Camfield, C., Finley, A., Franck, L., Howlett, A., Johnston, C., McKeever, P., O'Brien, K., Ohlsson, A., & Yamada, J. (2007a). Determining behavioural and physiological responses to pain in infants at risk for neurological impairment. *Pain, 127* (1–2), 94–102.

Stevens, B. J., Pillai Riddell, R. R., Oberlander, T. E., & Gibbins, S. (2007b). Assessment of pain in neonates and infants. In K. J. S. Anand, B. J. Stevens, & P. J. McGrath (Eds.), *Pain in neonates and infants* (3rd ed.). Amsterdam: Elsevier.

Stevens, B., Johnston, C., Taddio, A., Gibbins, S., & Yamada, J. (2010). The premature infant pain profile: Evaluation 13 years after development. *The Clinical Journal of Pain, 26*(9), 813–830.

Taddio, A., Nulman, I., Koren, B. S., Stevens, B., & Koren, G. (1995). A revised measure of acute pain in infants. *Journal of Pain and Symptom Management, 10*(6), 456–463.

Taddio, A., Shah, V., Gilbert-MacLeod, C., & Katz, J. (2002). Conditioning and hyperalgesia in newborns exposed to repeated heel lances. *JAMA: The Journal of the American Medical Association, 288*(7), 857–861.

Task Force on Taxonomy. (1994). *Classification of chronic pain: Descriptions of chronic pain syndromes and definitions of pain terms* (2nd Ed.) (Merskey, H., & Bogduk, N. (Eds.). Seattle: IASP Press.

Turk, D. C. (1993). Assess the person, not just the pain. *Pain: Clinical Updates, I*(3), 1–8.

van Dijk, M., de Boer, J. B., Koot, H. M., Tibboel, D., Passchier, J., & Duivenvoorden, H. J. (2000). The reliability and validity of the COMFORT scale as a postoperative pain instrument in 0 to 3-year-old infants. *Pain, 84*(2–3), 367–377.

Xavier Balda, R., Guinsburg, R., de Almeida, M. F., Peres, C., Miyoshi, M. H., & Kopelman, B. I. (2000). The recognition of facial expression of pain in full-term newborns by parents and health professionals. *Archives in Pediatric and Adolescent Medicine, 154*(10), 1009–1016.

Pain Assessment in the Elderly

Robert D. Helme
Department of Medicine, Royal Melbourne
Hospital, University of Melbourne, Melbourne,
Australia

Synonyms

Measurement; Older people; Pain

Definition

Pain assessment is undertaken in a number of different situations. Here I have made comment on epidemiologic, psychophysical, and clinical studies in older people.

Characteristics

The overriding theme of this essay is the need to carefully select validated measurement tools of pain, mood, activity, and cognitions. Some of these domains have yet to be properly explored.

Epidemiologic Studies

One of the most quoted prevalence studies of pain, in which the effect of age was examined, is that by Crook et al. (1984). This was a telephone survey of a random sample obtained from a group general practitioner list. It demonstrated increased persistent pain complaints with increasing age and highlighted the importance of pain for a large number of older people. There were very few participants over the age of 80 years, a problem common to most community-based studies that seek to explore issues relevant to older people. The questions asked in this study regarding the temporal nature of pain did not follow the usual pattern of descriptions of acute and chronic pain.

However, other studies have not necessarily replicated these results (Gibson 2003).

Collectively, the studies suggest a peak or plateau in the prevalence of chronic pain by the age of 65 years and a decline in reported pain in those over 80 years of age. This is surprising, given an age-related increase in pain-associated disease, which continues into the ninth decade of life.

Many reasons can be put forward as to why reported prevalence figures for pain in older people vary so widely (Gibson 2003). They include methodological variations such as sample bias and choice of psychometric instruments; number of subjects; response rates, especially for the oldest old; and the unreliability of memory for pain. Perhaps the most likely reason for the variation in absolute prevalence figure is the nature of the questions asked: the pain time interval; the time in pain within this interval; the severity or interference of the pain in daily life, often recorded as whether the pain is "troubling" or "bothersome"; and the effect of cueing (summation of pain at different sites).

A number of studies have examined pain at particular body sites, and some consistent trends have emerged. A summary view is that age is associated with an increase in prevalence of chronic pain up to the age of 65 years but not acute pain. Visceral pain is reduced among older people, and joint pain is increased.

Reasons for Age-Related Changes in Pain Prevalence

The overriding factor contributing to increased pain complaints with age is pathological load. Many acute and chronic illnesses increase in prevalence with advancing age, and the interplay of multiple diseases in an individual case is the hallmark of aged care medicine.

Reasons for the relative decline in reported pain among older people are more numerous but also more speculative (Gibson 2003). The report of pain by an older individual may be suppressed by more significant life events such as death of a spouse, concern regarding loss of independence, fear that reporting of pain will lead to unwanted investigations, treatments, or the finding of serious pathology. There is a general perception that older persons are more stoic, cautious in attribution, and more likely to misattribute pain symptoms to the aging process itself (Yong et al. 2001). They may not even use the term pain preferring discomfort, hurting, or aching. The unpleasantness of pain in older people has never been examined.

Pain Physiology

Psychophysical studies using physiologic stimuli have generally shown a progressive decrease in pain sensitivity with advancing age (Gibson 2003), especially if the duration of the stimulus is brief. Using differential nerve fiber blockade, Chakour and Gibson et al. (1996) showed that older persons rely on less well-localized C fiber activation before reporting the presence of pain, whereas younger adults utilize the additional information from A delta nociceptive fibers. Moreover, when A delta fiber input was blocked in young adults, the observed age-associated differences in pain threshold and subjective ratings of pain quality essentially disappeared. Age differences in the temporal summation of nociceptive input also vary as a function of nociceptive fiber type (Harkins et al. 1996) and rate of stimulation. Differential age effects on A delta and C fiber function may help explain some of the disparity in psychophysical findings. On the other hand, diminished physiologic reserve in descending inhibitory pathways may also lead to reduced pain tolerance in older people (Washington et al. 2000).

Studies of clinical pain states have also indicated that older persons may exhibit a relative absence of pain in the presentation of certain visceral disease states, such as ischemic heart pain and abdominal pain associated with acute infection (Gibson 2003). Unfortunately, most of the clinical studies are difficult to interpret, because the severity of pathology is seldom known, although controlled investigations of myocardial pain during exercise-induced ischemia generally support the view of a clinically significant decrease in ischemic pain perception with advancing age (Miller et al. 1990).

In summary, evidence from physiological studies, psychophysical investigations, and age variations in disease presentation all suggests age-related alterations in the function of

peripheral and CNS nociceptive pathways. These changes are likely to influence the sensitivity to painful sensation and would be expected to contribute to a decline in pain reporting in persons of advanced age. However, most of the evidence of age differences in nociceptive function is indirect, and the clinical relevance of reduced pain sensitivity to experimental pain stimuli is still the subject of debate.

Clinical Pain, Suffering, and Disability

Older people can successfully complete validated self-report measures of pain, mood, activity, and pain and disease impact, as long as adequate time is allowed to complete the task and other factors such as visual impairment, hearing loss, and physical disability of the hands are taken into account. Older persons are more likely to spend time deliberating over a question, rather than to respond in an impulsive manner, as often observed in younger individuals. However, visual analogue scales may be less likely to show age-related effects than numerical scales and the McGill pain questionnaire (Gagliese and Katz 2003).

Behavioral indicators of pain include measures of medication use, number of different treatments, and visits to physicians and other healthcare providers. These have rarely been utilized, as specificity is likely to be lower in older people with increased comorbidity. Another category of behavioral measures includes motor activities, which indicate pain complaints such as moaning, limping, rubbing, and facial expressions. These have been validated in older people with severe dementia who have language impairment (Hadjistavropoulos et al. 2000). Other measures include the use of walking aids, collars, and cushions. Functional measures such as sleep duration, uptime, mobility, self-care, and recreational activity are also relevant and can be reliably assessed in older people. There is a preference for behavioral indicators to be performance measures undertaken during direct observation, rather than self-report. Behavioral measures have been shown to be objective, sensitive to treatment effects, and clinically relevant in geriatric medicine. However, such measures are often time consuming and are strongly influenced by social context. Inter- and intraobserver reliability for these measures is strong if observer training has been undertaken.

Due to the high incidence of physical disease in older persons, somatic markers are inappropriate for psychological measures. There is also some evidence to suggest that older adults may underreport negative mood states, thereby emphasizing the need for age-appropriate norms. A more comprehensive psychological assessment should include attention to the cognitive coping style of the older person as well as assessment of pain beliefs and degree of somatic preoccupation. Unfortunately, standardized age-specific instruments for monitoring these domains have yet to be developed.

Management Outcomes

A careful evaluation of the individual is likely to identify the likely pathophysiological basis of chronic pain in the older person, whether nociceptive, neuropathic, psychogenic, or mixed. This in turn leads to the selection of an appropriate management regimen and is helpful in determining prognosis. In clinical practice, most chronic pain in the elderly is related to nociceptive factors, in particular degenerative musculoskeletal disease. However, clinical signs of central sensitization need to be sought in all patients and may well be different in older people. No studies have yet explored this hypothesis.

Multidisciplinary Pain Management

The literature on treatment outcome reveals overwhelming support for multidisciplinary treatment of older chronic pain patients. However, no study has used appropriate control groups or randomization procedures in the evaluation of treatment efficacy. There is a lack of standardized measurement tools for documenting disease and disability. It is not always clear whether the outcome measures have been validated for use with older people and the sample size in evaluation studies is often small. The goals for success, however, may not necessarily include reduced pain. Older people will often prefer to increase ambulation and socialization rather than rest to decrease pain. Others will accept pain if it allows independence

in their own home, rather than opting for less pain in the sedentary environment of an old-age home. These factors need to be accounted for in the measurement tools.

Dementia

The interaction between the pathology causing dementia and the perception of pain in these individuals has been reviewed (Farrell et al. 1996). A major clinical distinction lies between individuals who maintain communicative skills and those who have lost them, particularly in the advanced stages of Alzheimer's disease. Verbal patients, even with severe dementia, can report pain consistently using a variety of psychometric instruments, although simple word descriptors, such as the present pain intensity of the McGill pain questionnaire, appear to have greater clinical utility. Nonverbal patients may come to attention because of changes in behavior, such as agitation, diminished physical activity, or the appearance of distress. Apart from pain, altered behavior may be due to related factors including urinary infection, constipation, decubitus ulcers, depression, and reactions to medications.

Conclusions

Pain is a common complaint of older people. It should be assessed and managed in the same way as for younger people but with close attention to the impact of underlying diseases on the pain experience prior to and following treatment. There is evidence that the pain experience of older people is different, but this pertains more to the experimental laboratory than to the clinic.

References

Chakour, M. C., Gibson, S. J., Bradbeer, M., et al. (1996). Effect of age on A-delta fibre modulation of thermal pain perception. *Pain, 64*, 143–152.

Crook, J., Rideout, E., & Browne, G. (1984). The prevalence of pain complaints in a general population. *Pain, 18*, 299–314.

Farrell, M. J., Katz, B., & Helme, R. D. (1996). The impact of dementia on the pain experience. *Pain, 67*, 7–15.

Gagliese, L., & Katz, J. (2003). Age differences in postoperative pain are scale dependent: A comparison of measures of pain intensity and quality in younger and older surgical patients. *Pain, 103*, 11–20.

Gibson, S. J. (2003). Pain and aging: The pain experience over the adult life span. In Dostrovsky, J. O., Carr, D. B., Koltzenburg, M. (Eds.), *Proceedings of the 10th world congress on pain, progress in pain research and management* (Vol. 24, pp. 767–790). Seattle: IASP Press.

Hadjistavropoulos, T., LaChapelle, D. L., MacLeod, F. K., et al. (2000). Measuring movement exacerbated pain in cognitively impaired frail elders. *The Clinical Journal of Pain, 16*, 54–63.

Harkins, S. W., Davis, M. D., Bush, F. M., et al. (1996). Suppression of first pain and slow temporal summation of second pain in relation to age. *Journal of Gerontology, 51*, M260–M265.

Miller, P. F., Sheps, D. S., Bragdon, E. E., et al. (1990). Age and pain perception in ischemic heart disease. *American Heart Journal, 120*, 22–30.

Washington, L. J., Gibson, S. J., & Helme, R. D. (2000). Age related differences in endogenous analgesia to repeated cold water immersion in human volunteers. *Pain, 89*, 89–96.

Yong, H.-H., Gibson, S. J., Horne, D. Jde. L., et al. (2001). Development of a pain attitudes questionnaire to assess stoicism and cautiousness for possible age differences. *Journal of Gerontology, 56*, P279–P284.

Pain Associated with Assisted Reproduction

Lawrence Tsen
Department of Anesthesiology, Perioperative and Pain Medicine, Brigham and Women's Hospital
Harvard Medical School, Boston, MA, USA

Introduction

Since the birth of the first in vitro fertilized baby in 1978, the use of techniques known collectively as assisted reproductive technologies (ART) has dramatically increased; in 2002, over 600,000 cycles of ART were performed globally (de Mouzon et al. 2009).

Characteristics

Neurobiology of Pain During ART

Follicular hormonal stimulating agents have significantly increased the probability of a live birth and are consequently used for most ART cycles (Tsen 2007). These agents typically include gonadotropin-releasing hormone agonists (GNRH) to induce pituitary and ovarian suppression, followed by follicle-stimulating hormone (FSH) and human menopausal gonadotropin (hMG) to encourage the development and growth of multiple ovarian follicles. Human chorionic gonadotropin (hCG) is added later to induce maturation of the oocyte, and after retrieval, progesterone is often provided to support the growing pregnancy. As a result of these agents, progesterone and particularly estrogen levels are increased acutely, traditionally over a period of 2 weeks (Tsen et al. 2001b).

Hormones have been observed to modulate pain perception. The central and peripheral nervous systems undergo significant changes under the influence of hormones, and animal studies have demonstrated individual nerve fibers to be more sensitive to local anesthetic blockade during pregnancy (Datta et al. 1983) or in oophorectomized rabbits treated chronically, but not acutely, with exogenous progesterone (Bader et al. 1990). Estrogen receptors have been demonstrated in areas (laminar spinal trigeminal nucleus and spinal gray matter) involved in the processing of primary afferent nociceptive information (Amandusson et al. 1995), and decreases in estrogen (through ovariectomy) have been associated with decreases in the density of mu-opioid receptors, when compared to ovariectomized, estrogen-repleted mice or normally cycling mice (Joshi et al. 1993). By contrast, estrogen exposure has been observed to suppress kappa-opioid receptor-mediated functions (Wagner et al. 1994).

The human ovary is innervated by sensory, sympathetic, and a small number of parasympathetic fibers (Gioia 1996). Pain from the ovary is conducted through afferent fibers which enter the spinal cord at T10 and T11 levels (Lamvu and Steege 2006). The vaginal fornix and the cervix/perineum conduct their nociceptive signaling processing through spinal nerves originating at T10-L2, S2-S4, and S2-4, respectively. The integration of sympathetic and parasympathetic output from the pelvis, inclusive of the ovaries, uterus, vagina, and perineum, is through the inferior (particularly uterovaginal plexus) and superior hypogastric plexus, although the ovary also has its own visceral innervation with parasympathetic and sympathetic supply from the vagus nerve and preaortic plexus, respectively (Lamvu and Steege 2006).

Pain Associated with Oocyte Stimulation

Hormonal changes provoked clinically during an ART cycle have been associated with an increased sensitivity to cold thermal, but not pressure, noxious stimuli (Tsen et al. 2001a). Anxiety is not uncommon in individuals undergoing ART (Chen et al. 2004), especially prior to oocyte retrieval (Serafini et al. 2009), and has been observed to be an important predictor for postoperative pain and analgesic consumption in obstetric and gynecologic surgery (Ip et al. 2009). The anxious state has been associated with lowering the pain threshold (Rhudy and Meagher 2000) facilitating an overestimation of pain intensity and activating the entorhinal cortex of the hippocampal formation (Ploghaus et al. 2001).

Preoperative pain has been associated with greater postoperative pain (Ip et al. 2009). During oocyte stimulation, frequent blood sampling and daily injections, mostly subcutaneously but at least one intramuscular, of ART medications create patient discomfort. Hormonal stimulation may exacerbate preexisting conditions associated with abdominal pain, many of which may have contributed to the etiology of infertility, including polycystic ovary syndrome or endometriosis (Zivi et al. 2010). As follicular development occurs, ovarian expansion frequently results in discomfort (Garcia-Velasco et al. 2010) and increases the incidence of ovarian torsion (Kang et al. 2006). Mild ovarian hyperstimulation syndrome (OHSS) is common with stimulation regimens, with symptoms including lower abdominal pain, nausea, vomiting, and diarrhea; ovarian enlargement is a discernible, key indicator of the severity of the condition (Humaidan et al. 2010). OHSS may progress to moderate and severe forms which include hemorrhage from ovarian rupture, thromboembolism, and other life-threatening complications (Humaidan et al. 2010).

Pain Associated with Procedures Related to Oocyte Retrieval and Embryo Transfer

Most commonly performed via transvaginal needle aspiration with ultrasonographic guidance, oocyte retrieval also can be performed transabdominally or with laparoscopy, particularly when immediate tubal transfer is planned (i.e., gamete or zygote intrafallopian transfer) (Tsen 2007). The 17-gauge needle used for oocyte retrieval causes pain as it moves through the wall of the vaginal fornix (or abdomen) and into the ovarian capsule and tissues; on occasion, the body of uterus is penetrated to reach the ovary. When transvaginal oocyte retrieval is performed with light sedation, analgesia, paracervical block, and/or acupuncture, a moderate amount of discomfort (average VAS 20 to 50/100) can still be experienced intra- and postoperatively (Stener-Victorin 2005), leading some practitioners to embrace the use of general or spinal anesthesia for these procedures (Tsen 2007; Tsen et al. 2001b).

Embryos resulting from in vitro fertilization are transferred either directly into the fallopian tubes via laparoscopy or, more commonly, into the uterine cavity. When embryo transfer is performed transcervically into the uterine cavity, it is typically conducted without analgesia or sedation; by contrast, when laparoscopic techniques are employed, a general anesthetic is most often used, although spinal anesthesia can also be used.

Post-procedure recovery for most assisted reproductive procedures typically occurs in the first few days, with minimal postoperative analgesia required (Martin et al. 1999). On rare occasions, pain may be more persistent, particularly when accompanied by complications, such as OHSS.

Summary

Assisted reproductive technologies and procedures are associated with varying degrees of emotional, somatic, and visceral pain, which begin prior to the first use of stimulation medications and in some cases persist beyond the immediate post-procedure period. Clinicians and researchers should be sensitive to this relationship and investigate methods to mitigate the discomfort and pain.

References

Amandusson, A., Hermanson, O., & Blomqvist, A. (1995). Estrogen receptor-like immunoreactivity in the medullary and spinal dorsal horn of the female rat. *Neuroscience Letters, 196*(1–2), 25–28.

Bader, A. M., Datta, S., Moller, R. A., & Covino, B. G. (1990). Acute progesterone treatment has no effect on bupivacaine-induced conduction blockade in the isolated rabbit vagus nerve. *Anesthesia and Analgesia, 71*(5), 545–548.

Chen, T. H., Chang, S. P., Tsai, C. F., & Juang, K. D. (2004). Prevalence of depressive and anxiety disorders in an assisted reproductive technique clinic. *Human Reproduction, 19*(10), 2313–2318. doi:10.1093/humrep/deh414deh414 [pii].

Datta, S., Lambert, D. H., Gregus, J., Gissen, A. J., & Covino, B. G. (1983). Differential sensitivities of mammalian nerve fibers during pregnancy. *Anesthesia and Analgesia, 62*(12), 1070–1072.

de Mouzon, J., Lancaster, P., Nygren, K. G., Sullivan, E., Zegers-Hochschild, F., & Mansour, R. (2009). World collaborative report on assisted reproductive technology, 2002. *Human Reproduction, 24*(9), 2310–2320. doi:dep098 [pii]10.1093/humrep/dep098.

Garcia-Velasco, J. A., Motta, L., Lopez, A., Mayoral, M., Cerrillo, M., & Pacheco, A. (2010). Low-dose human chorionic gonadotropin versus estradiol/progesterone luteal phase support in gonadotropin-releasing hormone agonist-triggered assisted reproductive technique cycles: Understanding a new approach. *Fertility and Sterility, 94*(7), 2820–2823. doi:S0015-0282(10)00988-X [pii]10.1016/j.fertnstert.2010.06.035.

Gioia, L. (1996). Control of ovarian innervation. *Zygote, 4*(4), 295–298.

Humaidan, P., Quartarolo, J., & Papanikolaou, E. G. (2010). Preventing ovarian hyperstimulation syndrome: Guidance for the clinician. *Fertility and Sterility, 94*(2), 389–400. doi:S0015-0282(10)00465-6 [pii] 10.1016/j.fertnstert.2010.03.028.

Ip, H. Y., Abrishami, A., Peng, P. W., Wong, J., & Chung, F. (2009). Predictors of postoperative pain and analgesic consumption: A qualitative systematic review. *Anesthesiology, 111*(3), 657–677. doi:10.1097/ALN.0b013e3181aae87a.

Joshi, D., Billiar, R. B., & Miller, M. M. (1993). Modulation of hypothalamic mu-opioid receptor density by estrogen: A quantitative autoradiographic study of the female C57BL/6 J mouse. *Brain Research Bulletin, 30* (5–6), 629–634.

Kang, H. J., Davis, O. K., & Rosenwaks, Z. (2006). Simultaneous bilateral ovarian torsion in the follicular phase after gonadotropin stimulation. *Fertility and Sterility, 86*(2), 462 e413–464. doi:S0015-0282(06)00678-9 [pii] 10.1016/j.fertnstert.2005.12.063.2005.12.063.

Lamvu, G., & Steege, J. F. (2006). The anatomy and neurophysiology of pelvic pain. *Journal of Minimally Invasive Gynecology, 13*(6), 516–522. doi:S1553-4650(06)00337-2 [pii] 10.1016/j.jmig.2006.06.021.

Martin, R., Tsen, L. C., Tzeng, G., Hornstein, M. D., & Datta, S. (1999). Anesthesia for in vitro fertilization: The addition of fentanyl to 1.5 % lidocaine. *Anesthesia and Analgesia, 88*(3), 523–526.

Ploghaus, A., Narain, C., Beckmann, C. F., Clare, S., Bantick, S., & Wise, R. (2001). Exacerbation of pain by anxiety is associated with activity in a hippocampal network. *Journal of Neuroscience, 21*(24), 9896–9903. doi:21/24/9896 [pii].

Rhudy, J. L., & Meagher, M. W. (2000). Fear and anxiety: Divergent effects on human pain thresholds. *Pain, 84*(1), 65–75. doi:S0304-3959(99)00183-9 [pii].

Serafini, P., Lobo, D. S., Grosman, A., Seibel, D., Rocha, A. M., & Motta, E. L. (2009). Fluoxetine treatment for anxiety in women undergoing in vitro fertilization. *International Journal of Gynaecology and Obstetrics, 105*(2), 136–139. doi:S0020-7292(08)00603-6 [pii] 10.1016/j.ijgo.2008.12.013.

Stener-Victorin, E. (2005). The pain-relieving effect of electro-acupuncture and conventional medical analgesic methods during oocyte retrieval: A systematic review of randomized controlled trials. *Human Reproduction, 20*(2), 339–349. doi:deh595 [pii] 10.1093/humrep/deh595.

Tsen, L. C. (2007). Anesthesia for assisted reproductive technologies. *International Anesthesiology Clinics, 45*(1), 99–113. doi:10.1097/AIA.0b013e31802b8c00 00004311-200704510-00008 [pii].

Tsen, L. C., Natale, M., Datta, S., & Eappen, S. (2001a). Can estrogen influence the response to noxious stimuli? *Journal of Clinical Anesthesia, 13*(2), 118–121. doi:S0952-8180(01)00227-6 [pii].

Tsen, L. C., Schultz, R., Martin, R., Datta, S., & Bader, A. M. (2001b). Intrathecal low-dose bupivacaine versus lidocaine for in vitro fertilization procedures. *Regional Anesthesia and Pain Medicine, 26*(1), 52–56. doi:S1098-7339(01)45173-X [pii] 10.1053/rapm.2001.18185.

Wagner, E. J., Manzanares, J., Moore, K. E., & Lookingland, K. J. (1994). Neurochemical evidence that estrogen-induced suppression of kappa-opioid-receptor-mediated regulation of tuberoinfundibular dopaminergic neurons is prolactin-independent. *Neuroendocrinology, 59*(3), 197–201.

Zivi, E., Simon, A., & Laufer, N. (2010). Ovarian hyperstimulation syndrome: Definition, incidence, and classification. *Seminars in Reproductive Medicine, 28*(6), 441–447. doi:10.1055/s-0030-1265669.

Pain Associated With Traumatic Nerve Injury

▶ Neuroma Pain

Pain Asymbolia

Definition

Pain asymbolia is a neurological disorder characterized by damage to the limbic system in which patients are capable of perceiving the sensory qualities of pain but fail to exhibit appropriate affective and motor reactions to noxious stimuli; "pain without painfulness" is also considered a sensory-limbic disconnection syndrome.

Cross-References

▶ Thalamo-Amygdala Interactions and Pain

Pain Attacks

▶ Pain Paroxysms

Pain Behavior

Definition

Pain behaviors refer to overt, observable, and measurable expressions and manifestations of pain and include distorted ambulation (i.e., use of a prosthesis, limping), facial/audible expressions of pain (e.g., verbal complaints of pain, grimacing), expressions of affective distress (e.g., sighing, crying), and seeking help (e.g., asking for assistance, taking pain medications).

Cross-References

▶ Assessment of Pain Behaviors
▶ Catastrophizing
▶ Multiaxial Assessment of Pain
▶ Multidisciplinary Pain Centers, Rehabilitation
▶ Spouse, Role in Chronic Pain

Pain Behaviors

▶ Spouse, Role in Chronic Pain

Pain Belief

Definition

Pain belief is a technical term for what psychologists see as patients' cognitions about the cause, duration, and likely outcomes of chronic pain. Cognition, for example, catastrophizing, is usually understood to include an affective dimension.

Cross-References

▶ Ethics of Pain, Culture, and Ethnicity

Pain Catastrophizing

Definition

Pain catastrophizing is an exaggerated negative orientation toward actual or anticipated pain experiences, consisting of negative pain-related cognitions.

Cross-References

▶ Catastrophizing
▶ Disability, Fear of Movement
▶ Fear and Pain
▶ Fear Reduction Through Exposure In Vivo
▶ Hypervigilance and Attention to Pain
▶ Muscle Pain, Fear-Avoidance Model
▶ Psychology of Pain, Assessment of Cognitive Variables

Pain Centralization

Definition

The hypothesis that persistent intense pain can leave a durable pain inducing trace in the CNS, which is no longer amenable to relief by blocking afferent input from the periphery is known as pain centralization.

Cross-References

▶ Central Changes After Peripheral Nerve Injury

Pain Chronification

▶ Transition from Acute to Chronic Pain

Pain Clinics

▶ Multidisciplinary Pain Centers, Rehabilitation

Pain Cocktail

Definition

A method of converting all opioids or other analgesic medications to an equivalent dose of methadone and delivering it with a masking vehicle. Pretreatment levels of medicine consumption are first determined and subsequently, medicines are suspended in a flavored liquid that camouflages the type and amount of the analgesic drug. Thus, the patient does not know the exact content of the medicine being consumed. The dose is then systematically reduced over the period of treatment, always with the full knowledge of the patient.

Cross-References

▶ Multidisciplinary Pain Centers, Rehabilitation
▶ Operant Treatment of Chronic Pain

Pain Communication

▶ Empathy and Pain

Pain Control in Children with Burns

Walter J. Meyer III[1,2] and Lee Woodson[1,3]
[1]Shriners Hospitals for Children, Shriners Burns Hospital, Galveston, TX, USA
[2]Department of Psychiatry, University of Texas Medical Branch, Galveston, TX, USA
[3]Department of Anesthesia, University of Texas Medical Branch, Galveston, TX, USA

Synonyms

Burn pain control in children; Pediatric burns; Recognizing issues contributing to pain secondary to burn injury and its treatment; Treating pediatric burns

Definition

Burn pain encompasses wound pain and anxiety related to the cause of the injury as well as pain and anxiety related to the treatment of the burn wound. Late forms of burn pain include the intense itching of the healing skin and phantom limb pain when digits or limbs are lost with the burn injury.

Burn wounds are intensely painful and pain associated with wound care can be unbearable without the use of potent sedatives and analgesics. Appropriate sedation and analgesia for patients with major burn injuries not only reduces suffering but also shortens hospital stay and improves clinical outcome (Ratcliff et al. 2006).

Characteristics

The first 48 h after a burn injury are characterized by a decrease in blood flow to organs and tissues because of shock and extravasation of fluid into the burned tissue. During that period, intravenous pain and anxiety medications often do not reach the burned tissue and elimination of those medications by the kidney and liver is impaired. Non-IV medications are not absorbed well and should not be used for children. There is delayed delivery to the desired site of action, delayed peak concentration, and delayed effect, so short acting rapidly cleared medications are preferable to avoid overdosing children. For example, lorazepam is preferred to diazepam because it has a smaller volume of distribution and its metabolism is not affected by the liver.

By 72 h, a hypermetabolic state exists, characterized by increased oxygen consumption, hyperpyrexia, hypercortisolemia, increased sympathetic tone, tachycardia, and increased blood flow to organs and tissues. Medications not cleared in the initial shock phase are now mobilized and their plasma concentration may rise rapidly into the overdose range. Intravascular proteins are extravasated into the wound with resultant edema and hypoalbuminemia; this leads to an increased free fraction of some drugs, such as benzodiazepines. In contrast, ▶ acute phase proteins like alpha$_1$-acid glycoprotein may double in concentration after large burn injury. These proteins bind basic drugs (e.g., propranolol or imipramine). Since drug-binding proteins may respond in opposite ways to burn injury, changes in drug binding, response, and clearance will depend on which protein binds the drug in question. The most important principle to remember is to monitor response and/or plasma levels, then titrate the drug doses appropriately.

Undertreated pain in burn patients not only results in unnecessary suffering but also enhances these hypermetabolic physiological changes with further rises in cortisol and catecholamines. Significant behavioral consequences are increased agitation and motor restlessness that can result in sheared grafts, dislodged vascular cannulas, or endotracheal tubes.

Pharmacological Treatment for Pediatric Inpatients

When mechanically ventilated children do not receive adequate sedation and analgesia, it is difficult to synchronize the child's respiratory effort with the ventilator. The child "fights" the ventilator and increases the risk of laryngeal injury, increased airway pressures, ▶ barotrauma, and ▶ ARDS. Sedation and analgesia are preferred to muscle relaxation for ventilated children. Pharmacological relaxation makes it difficult to titrate analgesia and sedation since health-care providers have difficulty evaluating children's responses to determine awareness. In addition, prolonged relaxation is associated with myopathy and children with burns already suffer debilitation due to motor weakness from muscle wasting.

In the longer term, inadequate sedation produces delirium, decreased pain tolerance, decreased cooperation with therapy, and contributes to persistent anxiety and psychiatric disability. Treatment of burn pain in children is made more complex by problems with communication that make it difficult to assess pain. Pain in pediatric patients is also difficult to separate from anxiety. When the child can communicate, a variety of assessment tools are available and each hospital should be consistent with whichever tool is used (Meyer et al. 2012). When the patient is unable to communicate, then physiological variables (e.g., hypertension, tachycardia, and agitation) must be used to assess the effectiveness of sedation and analgesia.

In most burn centers, standard therapy includes morphine and a benzodiazepine (Ratcliff et al. 2006; Stoddard et al. 2002). Some centers use acetaminophen successfully to treat pain of small burns and in small children in order to reduce narcotic requirements and avoid the possible respiratory depression and other problems caused by morphine (Meyer et al. 1996). Please see Table 1 for the starting doses used in the Shriners Hospitals for Children, Galveston, TX.

When morphine and a benzodiazepine do not control pain and anxiety adequately because of hyperalgesia associated with morphine tolerance

**Pain Control in Children with Burns,
Table 1** Medications used in pediatric burn survivors
for pain and anxiety

Medication	Dose
Background pain	
Acetaminophen	15 mg kg^{-1} PO q 4 h
Morphine	0.05 mg kg^{-1} IV q 4 h
Morphine	0.3 mg kg^{-1} PO q 4 h
Methadone	0.1 mg kg^{-1} orally q 6–12 h
Hydrocodone/ Acetaminophen	0.2 mg hydrocodone kg^{-1}/dose
Procedural pain	
Morphine	0.05–0.1 mg kg^{-1}/dose IV
Fentanyl	10–20 μg kg^{-1}/dose oral transmucosal
Ketamine	1–2 mg kg^{-1}/dose IV
Anxiety	
Lorazepam	0.05 mg kg^{-1} PO or IV q 4 h
Diazepam	0.1 mg kg^{-1} PO q 8 h
Clonidine	3–5 μg kg^{-1} PO q 8–12 h

or complications such as bowel obstruction,
a change to methadone or to a different class of
drugs may be necessary (Williams et al. 1998).
Methadone has both pharmacokinetic and pharmacodynamic differences from morphine. The
hyperalgesia associated with morphine may
involve central ▶ NMDA receptor pathways and
methadone's ability to antagonize NMDA receptors may contribute to its efficacy in pain no
longer controlled with morphine. In addition,
methadone's longer half-life provides
a smoother and more continuous control of background pain without wide swings in drug concentration and level of analgesia seen with the
shorter acting morphine bolus doses.

Ketamine is another NMDA receptor antagonist that can be very useful in pediatric burn
patients. Ketamine is unique in its ability to
produce profound analgesia without depressing
respiration or causing airway obstruction (Lin
and Durieux 2005). Ketamine has been found to
be safe for sedating children for painful procedures such as staple removal (Dimick et al. 1993)
and is invaluable in the management of acute
burn wounds in children. Since ketamine blocks
NMDA receptors, infusion of low dose ketamine
will often reduce opioid requirements in patients

who have become tolerant (Kissin et al. 2000;
Stoddard et al. 2002). Higher doses of ketamine
alone have been infused for more than 3 weeks
continuously with good effect in patients refractory to other drug regimens.

Alpha(α) 2-Adrenergic Agonist provide
a relatively new class of sedative analgesic
drugs and experience with these drugs in the
ICU is rapidly increasing (Kamibayashi and
Maze 2000). Like ketamine, these drugs possess
the unique ability to provide sedation and analgesia with minimal respiratory depression; in
addition, at lower doses patients can be aroused
to cooperate with therapy. Clonidine, a partial
alpha 2 adrenergic agonist (▶ partial agonist),
has a relatively long duration of action, can be
given orally and intravenously and is inexpensive. Dexmedetomidine, a full agonist, is only
available as an injectable dosage form, has
a shorter duration of action and is much more
selective for alpha-2 adrenergic receptors than
clonidine. Dexmedetomidine can be used as the
sole sedative analgesic in mechanically ventilated patients or, in much higher doses, as
a general anesthetic agent for surgical procedures. Administration of alpha 2 adrenergic
agonists can dramatically reduce opioid requirements and case reports show effective analgesic
action in burn patients tolerant to and poorly controlled with morphine (Lyons et al. 1996).
In pediatric patients scheduled for burn reconstructive surgery, intranasal dexmedetomidine
(2 mcg/kg) is as effective as oral midazolam
(0.5 mg/kg) for preoperative sedation (Talon
et al. 2009).

Reports of metabolic acidosis and increased
mortality have limited the use of propofol infusions in children (Ellett 2010). However,
propofol infusions have been found safe and
helpful for limited periods of time (less than
24 h), while other sedatives are weaned in preparation for extubation (Sheridan et al. 2003).

**Non-pharmacological Treatment for Pediatric
Inpatients**

Psychological interventions include distractions,
relaxation, and active participation in burn dressing changes. These are often very helpful in

reducing pain and anxiety in the patient who can cooperate. Hypnosis is very effective but no longer used because of the excessive personnel time used. A new form of distraction from the pain and anxiety associated with burn wound care is the use of immersive virtual reality. Immersive virtual reality involves the use of sensory input from sight, sound, and touch to give the participant a strong sense of being part of the computer-generated environment. The few case reports published indicate that it is effective in reducing pain ratings (Hoffman et al. 2011).

Surgical approaches to managing pain emphasize covering the wound. For second-degree burns, synthetic dressings (e.g., Biobrane^{RX}, Opsite^{RX}, Tegaderm^{RX}, and Duoderm^{RX}) have been very successful in relieving pain almost immediately while promoting healing. Cultured allogeneic keratinocyte sheets accelerated healing and thereby reduced pain and suffering compared to Opsite^{RX} treatment (Duinslaeger et al. 1997). The donor sites used for grafts can also be quite painful.

Treatment for Pediatric Outpatients

The medications to be used in outpatient pain management should probably be initiated as an inpatient to assure a pain-free transition. As with inpatients, both background and procedural pain should be covered. Usually at this stage, in addition to pain and anxiety, itching has become a major issue. Pain and anxiety also go hand in hand. During rehabilitation, painful stretching of scar-contracted soft tissues such as muscle and tendon causes pain. Diazepam facilitates the muscle psychological relaxation needed for rehabilitation therapy.

Pain management as an outpatient can usually be managed with oral hydrocodone/acetaminophen combinations like Vicodin^{Rx} or Lortab^{Rx} dosed to provide 0.2 mg kg^{-1}/q 4 h of hydrocodone and acetaminophen at no more than 20 mg kg^{-1}/dose or 3,000 mg day^{-1}. These types of medication may require their extra-strength formulation when the amount of acetaminophen is too high. For the same reason they should not be given with Tylenol^{Rx}. However, ibuprofen (10 mg kg^{-1}) PO can be given as an alternate, especially for muscle and bone pain associated with exercise or splints.

There are special outpatient situations that might require extra medication including wound cleaning. ▶ Transmucosal fentanyl (10 µg kg^{-1}) is often effective in reducing the pain (Robert 2003). If the procedure is very painful, such as staple removal, then ketamine will be needed to control pain effectively. Another adjunctive medication is the anesthetic gas mixture of 50 % nitrous oxide and 50 % oxygen, self-administered by patient-held mask during the wound debridement and dressing. Pain is usually effectively controlled but significant side effects have been reported including euphoria, nausea, vomiting, neurological changes, liver function abnormalities, and anemia. Adequate gas scavenging equipment is essential to reduce chronic trace gas exposure to the caregivers.

Phantom Limb Pain

Many children with severe burns have to have amputations. About one third will have significant phantom limb pain. This is usually controlled by amitriptyline (1–3 mg kg^{-1}) (Thomas et al. 2003). In addition, the damage to the skin nerve endings which occurs in burn injury is an important source of pain. Amitriptyline is often effective in treating this as well. Gabapentin (10–35 mg/kg in divided doses) is an important newer medication for treating this form of pain.

Itching

Refractory itching is an unusual form of pain that does not respond to standard pain or anxiety medications. In fact, morphine sometimes makes it worse. A standard assessment instrument is recommended, just as it is for pain and anxiety. The attached figure provides such an instrument (▶ ITCH MAN) based on the drawing of one of our patients (Ratcliff et al. 2006). Itching is usually treated with diphenhydramine HCl 1 mg kg^{-1} and hydroxyzine 0.5 mg kg^{-1} alone or in combination when Itch Man scores are greater than 2 out of 4. Most recently, gabapentin has been reported to be very helpful in controlling itching (Goutos et al. 2010).

Topical medications are also used; they include preparation H, benadryl cream, Eucerin cream, aquaphor, and rarely pyridoxine cream 5 %. In children, pyridoxine cream is hazardous, since too much is absorbed. Some children do not respond well to any regimen and only receive significant relief after the scars mature in several years.

Sleeping Difficulties

Sleep management in the outpatient is often very difficult. Children recovering from burn injury may be having pain at night due to lying in one position a long time or trying to sleep in special splints. Itching and pain causes of the poor sleep must be addressed. The child may be afraid of having nightmares or suffering from other manifestations of ▶ acute stress disorder. Or they may just be itching. Individual assessment must be done. If the child is afraid to go to sleep because of intrusive thoughts or nightmares, imipramine (1 mg kg^{-1} q hs) or fluoxetine (0.3 mg kg^{-1} q am) often provide relief (Tcheung et al. 2005).

Summary

Children who suffer burn injury will have significant pain that requires aggressive treatment. Treating children's burn pain is an important part of their recovery, but is complicated by the major changes in children's metabolism associated with the burn injury. Unfortunately, the medications can also be associated with significant adverse effects including blood pressure changes, respiratory depression, bowel obstruction, or physical dependence. Oversedation can cause immobility that can slow rehabilitation and even lead to pressure sores. A large number of drugs in several drug classes are useful in providing sedation and analgesia for burn patients (Tobias 1999). The goal is to provide individualized drug therapy optimizing sedation and analgesia, while avoiding adverse effects. The treatment plan needs to follow the child throughout the rehabilitation phase of recovery.

References

Dimick, P., Helvig, E., Heimbach, D., et al. (1993). Anesthesia assisted procedures in a burn intensive care unit procedure room: Benefits and complications. *The Journal of Burn Care & Rehabilitation, 14*, 446–449.

Duinslaeger, L. A. Y., Verbeken, G., Vanhalle, S., & Vanderkelen, A. (1997). Cultured allogeneic keratinocyte sheets accelerate healing compared to op site treatment of donor sites in burns. *The Journal of Burn Care & Rehabilitation, 18*, 545–551.

Ellett, M. L. (2010). A literature review of the safety and efficacy of using propofol for sedation in endoscopy. *Gastroenterology Nursing, 33*, 111–117.

Goutos, I., Eldardire, M., Khan, A. A., et al. (2010). Comparative evaluation of antipruritic protocols in acute burns. The emerging value of gabapentin in the treatment of burns pruritus. *Journal of Burn Care & Research, 31*, 57–63.

Hoffman, H. G., Chambers, G. T., Meyer, W. J., et al. (2011). Vitual reality as an adjunctive non-pharmacologic analgesic for acute burn pain during medical procedures. *Annals of Behavioral Medicine, 41*, 183–219.

Kamibayashi, T., & Maze, M. (2000). Clinical uses of Î±2-adrenergic agonists. *Anesthesiology, 93*, 1345–1349.

Kissin, I., Bright, C. A., & Bradley, E. L. (2000). The effect of ketamine on opioid-induced acute tolerance: Can it explain reduction of opioid consumption with ketamine-opioid analgesic combinations? *Anesthesia and Analgesia, 91*, 1483–1488.

Lin, C., & Durieux, M. E. (2005). Ketamine and kids: An update. *Pediatric Anesthesia, 15*, 91–97.

Lyons, B., Casey, W., Doherty, P., et al. (1996). Pain relief with low-dose intravenous clonidine in a child with severe burns. *Intensive Care Medicine, 22*, 249–251.

Meyer, W. J., Wiechman, S., Woodson, L., et al. (2012). Management of pain and other discomforts in burned patients. In D. N. Herndon (Ed.), *Total Burn Care* (4th ed., pp. 715–731). London: Saunders Elsevier.

Meyer, W. J., Nichols, R., Cortiella, J., et al. (1996). Acetaminophen in the management of background pain in children post burn. *Journal of Pain and Symptom Management, 13*, 50–55.

Ratcliff, S. L., Brown, A., Rosenberg, L., et al. (2006). The effectiveness of a pain and anxiety protocol to treat the acute pediatric burn patient. *Burns, 32*, 554–562.

Robert, R., Brack, A., Blakeney, P., et al. (2003). A double-blind study of the analgesic efficacy of oral transmucosal fentanyl citrate and oral morphine in pediatric patients undergoing burn dressing change and tubbing. *The Journal of Burn Care & Rehabilitation, 24*, 351–355.

Sheridan, R. L., Keaney, T., Stoddard, F., et al. (2003). Short-term propofol infusion as an adjunct to extubation in burned children. *The Journal of Burn Care & Rehabilitation, 24*, 356–360.

Stoddard, F. J., Sheridan, R. L., Saxe, G. N., et al. (2002). Treatment of pain in acutely burned children. *The Journal of Burn Care & Rehabilitation, 23,* 135–156.

Talon, M. D., Woodson, L. C., et al. (2009). Intranasal dexmedetomidine premedication is comparable with midazolam in burn children undergoing reconstructive surgery. *Journal of Burn Care & Research, 30,* 599–605.

Tcheung, W. J., Robert, R., Rosenberg, L., et al. (2005). Early treatment of acute stress disorder in children with major burn injury. *Pediatric Critical Care Medicine, 6,* 676–681.

Thomas, C., Brazeal, B., Rosenberg, L., et al. (2003). Phantom limb pain in pediatric burn survivors. *Burns, 29,* 139–142.

Tobias, J. D. (1999). Sedation and analgesia in pediatric intensive care units: A guide to drug selection and use. *Paediatric Drugs, 1,* 109–126.

Williams, P. I., Sarginson, R. E., & Ratcliffe, J. M. (1998). Use of methadone in the morphine-tolerant burned pediatric patient. *British Journal of Anaesthesia, 80,* 92–95.

Pain Control in Children with Cancer

▶ Cancer Pain, Palliative Care in Children

Pain Coping Skills in Children

Definition

Cognitive and behavioral strategies that children may use to assist in coping with pain. Such strategies are often categorized as active, approach-oriented, or problem-focused coping as distinct from passive, avoidance-oriented, or emotion-focused coping. Active coping strategies include information seeking, specific question asking, problem solving, and social support. Avoidant, passive, or emotion-focused strategies include withdrawal, negative thinking, catastrophizing, passive adherence, and self-criticism.

Cross-References

▶ Experimental Pain in Children

Pain Detection and Pain Tolerance Thresholds

Definition

These are the levels at which the applied stimulus first feels painful to the subject (first pain, pricking pain, A-delta fiber pain), or the level at which the pain first becomes intolerable (second pain, burning pain, C-fiber pain).

Cross-References

▶ Quantitative Sensory Testing

Pain Diaries

Definition

Pain diaries are kept by patients to record their mood, activity, sleep, and pain on a daily basis, which provides an actual temporal relationship between psychosocial variables and pain.

Cross-References

▶ Pain Inventories

Pain Disability Index

Definition

The Pain Disability Index was developed to be a brief measure of the degree to which chronic pain interferes with normal role functioning. It includes seven items that assess perceived disability in each of seven areas of normal role functioning.

Cross-References

▶ Pain Inventories

Pain Disorder

▶ Pain, Psychiatry, and Ethics

Pain Drawings

Definition

Pain drawings consist of an outline of a human figure on which patients darken the areas where they are experiencing pain. These drawings can facilitate the assessment of the location and nature of the pain and assist with the decisions for appropriate intervention.

Cross-References

▶ Pain Inventories

Pain Due to Cancer

▶ Cancer Pain

Pain Evaluation

▶ McGill Pain Questionnaire

Pain Evaluation, Psychophysical Methods

Leon Plaghki
Institute of Neuroscience (IoNS), Université
Catholique de Louvain, Brussels, Belgium

Synonyms

Response-dependent   method;   Stimulus-
dependent method

Definition

Psychophysical methods for pain research are methods that allow the study of the relationship between intensity of physical stimuli and intensity of pain sensation.

Characteristics

The fundamental question of classical psychophysics is stated in the following equation (Falmagne 1985):

$$P_{a.b} = F(u(a) - u(b)) \qquad (1)$$

In words, if (a) and (b) are values denoting stimulus intensity on some physical scale (e.g., electrical current, temperature, pressure), the probability (P) of choosing stimulus intensity (a) over stimulus intensity (b) increases as a function (F) of the difference between these stimuli on the sensory scale (u), taken to be a measure of the intensity of sensation evoked by the stimuli. Such a function is traditionally referred to as a ▶ psychometric function.

The value of (b) can represent some fixed value or "background noise," but it may also involve a different sensory modality. The fundamental equation can then be rewritten as follows:

$$P_b(a) = F(u(a) - u(b)) \qquad (2)$$

In words, for a fixed stimulus of intensity (b) (the "standard" or "reference" stimulus), $P_b(a)$ is the probability that stimulus intensity (a) (the "comparison" stimulus) is judged as exceeding stimulus intensity (b) on the sensory scale (u), taken to be a measure of the intensity of sensation evoked by the stimuli. Typically, $P_b(a)$ increases continuously as a function of value (a) in a characteristic ogival way (Fig. 1).

The fundamental question may be restated as follows: On a stimulus continuum (a), what is the smallest increment $\Delta(a)$ such that $a + \Delta(a)$ just noticeably exceeds (a)? This minimum increment of stimulus intensity is usually called the "▶ just-noticeable difference" (JND or "difference

Pain Evaluation, Psychophysical Methods, Fig. 1 Graph of an ideal psychometric function as specified by Eq. 2 (Adapted from Falmagne 1985)

threshold" or "difference limen"). A variant, or particular case, of this question is as follows. What is the minimum value of a stimulus that is "just detectable" by a subject? This value is called the "absolute threshold."

However, as biological systems are intrinsically variable in their reaction, the sequence of stimulus-response pairs must be regarded as a stochastic process (assumed to be stationary during the experiment). Consequently, whatever the stimulus intensity, when a subject is repeatedly presented with the same stimulus, he is likely to respond "yes" on some trials and "no" on other trials. Therefore, a + Δ(a) cannot be defined as the stimulus value below which detection never occurs and above which detection always occurs. An empirical criterion (π) has to be taken such that Δ_π(a) represents a correct determination if a + Δ_π(a) is judged as exceeding (a) on a proportion (π) of trials. Typically, criterion π is defined as the stimulus value which is detectable in 50 % of the trials ($\pi = 0.5$). The experimental procedures used to determine Δ_π(a) are described and discussed in the following sections.

Traditional Psychophysical Methods

It was Fechner (1860) who developed three methods of psychophysical measurement: the

▶ method of limits, the method of constant stimuli, and the ▶ method of adjustment (Gescheider 1985).

Method of Limits
This method is the most frequently used. It is less precise than the other two methods but is also less time consuming and yields satisfactory results if errors characteristic of the method are corrected.

Absolute Threshold To measure the absolute threshold, the experimenter starts with a stimulus well above or well below the threshold. On each successive trial, the threshold is approached by changing the stimulus intensity by a small amount until the boundary of sensation is reached. The stimuli are changed in either an ascending or descending series. Each transition point obtained from a number of ascending and descending series can be considered as reflecting threshold value. The threshold is thus estimated by averaging these values (Fig. 2). Subjects may show a tendency to repeat the same response within series (errors of habituation). This may falsely increase the threshold in ascending series and decrease the threshold in descending series. A technique used to prevent such anticipation is to vary the starting point of each series.

Difference Threshold To measure the difference threshold, standard and comparison stimuli are presented in pairs. On each successive presentation, the comparison stimulus is changed by a small amount in the direction of the standard stimulus. During each series, whether ascending or descending, two transition points are obtained which are called the "upper limen" (Lu) and "lower limen" (Ll). Lu is the point on the physical dimension where "greater" responses change to "equal" responses, and similarly Ll is the point where the "lesser" responses change to "equal" responses. The range on the stimulus dimension on which a subject cannot perceive a difference between the comparison and the standard stimulus is called the interval of uncertainty (IU) and is computed by subtracting the average Ll from the average Lu. The best estimate of the difference

Pain Evaluation, Psychophysical Methods, Fig. 2 Determination of the absolute threshold by the method of limits

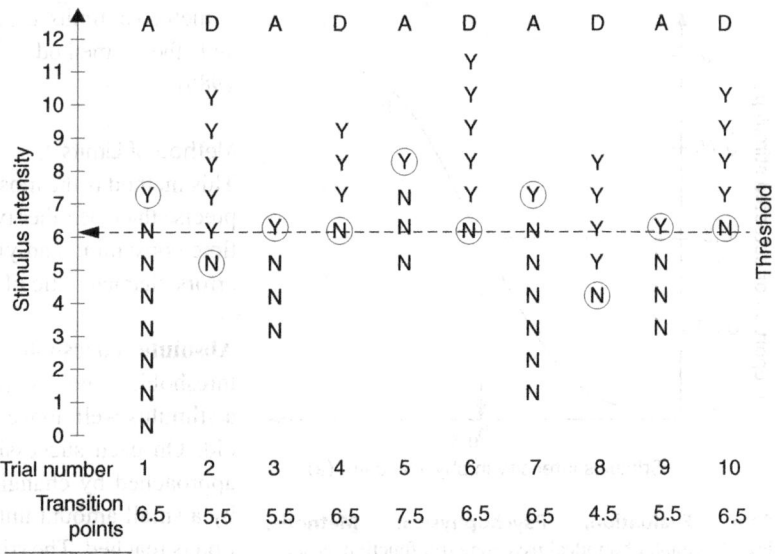

A = ascending series of stimuli and D = descending series of stimuli

limen is taken as half the IU. The point of subjective equality (PSE) is given by (Lu + Ll)/2.

Method of Constant Stimuli

This procedure consists of repeatedly presenting the same set of stimuli (between 5 and 9 different intensities, each presented a large number of times) in a random order throughout the experiment. Preliminary observations are needed for locating the approximate range of stimulus values.

Absolute Threshold During the experiment, the number of "yes" responses for each stimulus intensity level is recorded. The proportion (p) for each stimulus intensity level is then computed, and a graph of the psychometric function is constructed as shown in the upper panel of Fig. 3. The absolute threshold is usually defined as the stimulus intensity for which the proportion of trials resulting in a "yes" response is 0.50. A curve may be fitted to the data points, and the threshold may then be estimated by reading the stimulus intensity corresponding to $p = 0.5$. An ogival function has the property of becoming linear when the proportion of responses for each stimulus value is transformed by the following operation: logit $p = Ln\ (p/(1-p))$ (Ashton 1972). The result of such transformation is illustrated in

the lower panel of Fig. 3. The threshold may then be estimated more precisely by determining the best-fitting straight line. The straight line crosses the abscissa at threshold value. The steepness of the psychometric function depends on the subject's capacity of discrimination and may therefore be used as a measure of sensitivity.

Difference Threshold The method of constant stimuli can also be used to measure difference thresholds. The subject's task is then to examine pairs of stimuli and to judge which stimulus produces a sensation of greater magnitude. One of the stimuli is assigned a fixed value (the "standard" or "reference" stimulus). The value of the other stimulus, called the "comparison" stimulus, is changed from trial to trial. Usually 5–9 values of the comparison stimulus are given in random order. Spatial location or temporal order of stimulus pair presentations should be counterbalanced.

For each stimulus pair, the proportion (p) of "greater" responses is computed, and a graph is constructed in a similar way as shown in Fig. 3. When the subject cannot perceive any difference, we expect the proportions of "greater" and "lesser" responses to be approximately equal. This point on the psychometric function is called the "point of subjective equality" (PSE). As for

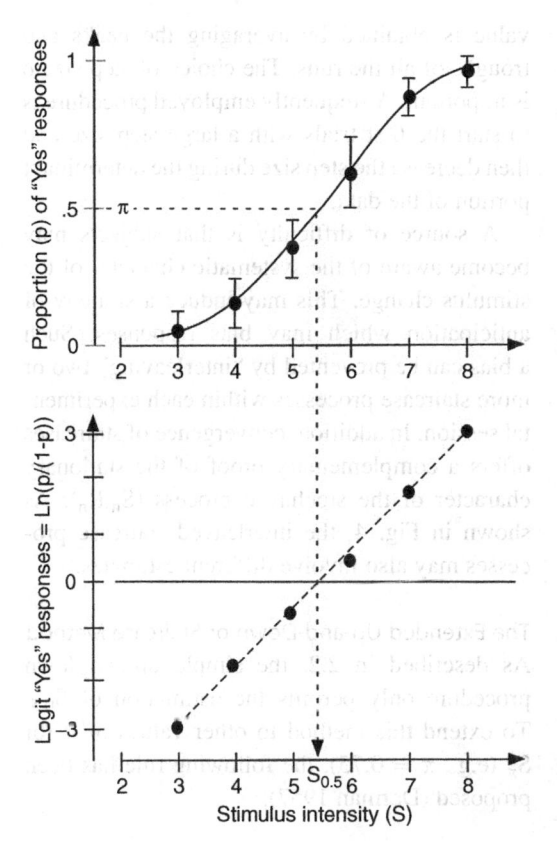

Pain Evaluation, Psychophysical Methods, Fig. 3 Determination of the absolute threshold by the method of constant stimuli. *Upper panel* shows the psychometric function obtained in a fictitious experiment with 6 levels of stimulus intensity repeated n times. The *dashed line* points to the absolute threshold (value $S_{0.5}$) defined as the stimulus intensity for which the proportion of trials resulting in a "yes" response is 0.5. The *lower panel* shows the result of the logit transformation performed to linearize the psychometric function. The *straight line* fitted to the data crosses the abscissa at the threshold value. The steepness of the psychometric function may be used as a measure of the subject's sensitivity

measuring absolute thresholds, the psychometric function for the difference threshold can be expressed in response proportions as well as in logit units, and the PSE can therefore be determined using the same methods of interpolation.

Method of Adjustments
This method resembles the method of limits, with the difference that it is the subject who adjusts the value of the "comparison" stimulus and sets it to apparent equality with the "reference" stimulus.

The adjustment can be made continuously by turning, for instance, a dial. Repeating this procedure allows computing an experimental distribution of values that is used to estimate the threshold or JND. One advantage of this method is the active participation of the subject, which may help prevent boredom and maintain high performance.

As clearly stated by Falmagne (1958), the three traditional methods suffer from one or more of the following defects:

- There is no control over the criterion π (methods 1.1 and 1.3).
- The estimates may be systematically biased in several ways (methods 1.1 and 1.3).
- A large amount of data is often wasted (especially in method 1.2).

Some of these shortcomings can be overcome by using more sophisticated and computerized versions of these methods. Those will be succinctly described and discussed in the next section.

Variations of the Traditional Methods: Adaptive Methods
The originality of adaptive methods is that the course of the experiment is determined by data collected during the ongoing experiment: The stimulus presented on trial n depends on the subject's response to one or more preceding trials.

Let the stimulus presented at trial n be denoted S_n and the subject's response R_n. coded as 1 if the stimulus (a) is not judged as exceeding S_n and 0 if otherwise.

In the following adaptive methods, the succession of stimuli is governed by an equation of the form:

$$S_{n+1} = S_n + f(\pi, n, R_n, R_{n-1}, S_{n-1}, \ldots) \quad (3)$$

In words, the value of the next stimulus is equal to the sum of the value of the present stimulus and another value, which may vary as a function of the probability π assigned to the target value (S_π, i.e., absolute or difference threshold), the trial number n, the subject's response to the present trial (R_n), and eventually stimulus-response pairs collected on earlier trials.

Stochastic Approximation Method

After choosing criterion π, the first stimulus (S_1) is set somewhere in the neighborhood of the threshold (S_π). Accuracy of the initial stimulus intensity is not critical. Stimulus S_1 is then presented to the subject. Intensity of the second stimulus (S_2) is determined by the following rule:

$$\text{if } R_1 = 0 \; S_2 = S_1 + c.\pi$$
$$\text{else } S_2 = S_1 - c(1 - \pi)$$

where c is some constant > 0.

Intensity of the following stimuli is determined by the rule:

$$S_{n+1} = S_n - c.n^{-1}.(R_n - \pi) \qquad (4)$$

where n is the trial number.

By using the value c/n, accuracy of the estimation process is enhanced across successive stimuli and is therefore only limited by the resolution of the stimulator. However, the counterpart to this resolution enhancement is a slowing down of the convergence process. This disadvantage can be overcome by changing the value of (c) as a function of subject's responses on trials preceding trial n.

The Simple Up-and-Down or Staircase Method

In this method, the value of the stimulus on each trial is changed either positively or negatively by a fixed amount. The direction of the change is a function of Eq. 3. The fixed increment by which the stimulus is increased or decreased is referred to as a step, and a sequence of steps in the same direction is called a run.

An educated guess is made when choosing intensity of the initial stimulus (S_1). The successive values (S_{n+1}) are obtained by the following rule:

$$S_{n+1} = S_n + \delta(1 - 2R_n) \qquad (5)$$

In words, δ is the step size, and S_{n+1}, the upcoming stimulus, is increased by δ if $R_n = 0$ and is decreased by δ if $R_n = 1$.

The value of S_π (for $\pi = 0.5$) is estimated by computing the mean of S_n values (n must be sufficiently large). A practical estimate of this value is obtained by averaging the peaks and troughs of all the runs. The choice of step size δ is important. A frequently employed procedure is to start the first trials with a large step size and then decrease the step size during the determinant portion of the data.

A source of difficulty is that subjects may become aware of the systematic character of the stimulus change. This may induce a strategy of anticipation which may bias responses. Such a bias can be prevented by "interleaving" two or more staircase processes within each experimental session. In addition, convergence of staircases offers a complementary proof of the stationary character of the stochastic process (S_n, R_n). As shown in Fig. 4, the interleaved staircase processes may also involve different estimates.

The Extended Up-and-Down or Staircase Method

As described in 2.2, the simple up-and-down procedure only permits the estimation of $S_{0.5}$. To extend this method to other values of π for S_π (e.g., $\pi = 0.75$), the following rule has been proposed (Derman 1957):

$$S_{n+1} = S_n + \delta(1 - 2R_n Y_\pi) \qquad (6)$$

where Y_π is a random variable taking value 1 with probability $(\pi/2)$ and value 0 with probability $(1 - \pi/2)$. By introducing this random variable, a possible strategy of anticipation is rendered even more difficult for the subject.

Conclusions

As compared to conventional psychophysical methods, adaptive methods offer the following advantages:

- There is no need to model the psychometric function and to fit the data to that model.
- The total number of stimuli to deliver is significantly reduced.
- The number of painful and potentially harmful stimuli to deliver is reduced.
- Interleaving several staircases allows a better control of possible psychological bias and gives the possibility of performing different estimates within the same experimental run (see example in Fig. 4).

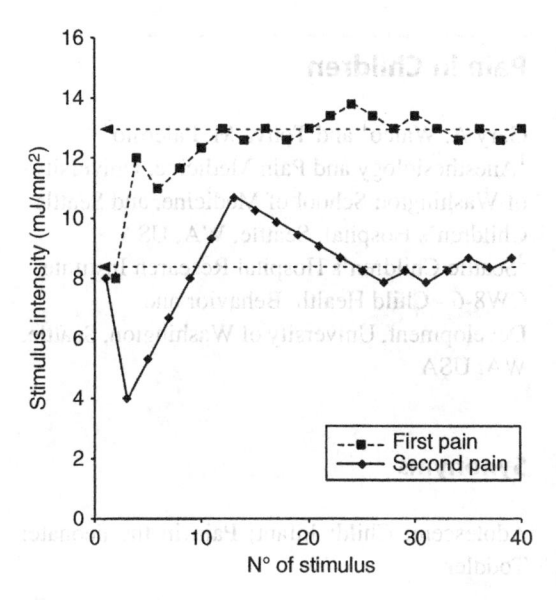

Pain Evaluation, Psychophysical Methods, Fig. 4 Concurrent determination of the heat threshold for "first" pain (i.e., pricking sensation and reaction times < 700 ms) and for "second" pain (i.e., burning sensation and reaction times ≥ 700 ms) by an interleaved adaptive accelerated stochastic procedure. Absence of reaction time and detections with reaction time > 1,800 ms are considered as trials without detection ($R_n = 0$). Laser stimuli of 50-ms duration and 25-mm² surface areas were directed to the hand dorsum with a stimulus interval of 15–20 s. The procedure uses four steps. Step 1: Choose the parameters for the adaptive procedure (e.g., $\pi = 0.75$, $c = 2$, $\delta = 0.4$ mJ/mm², and $S_1 = S_2 = 8$ mJ/mm²). Step 2: The first staircase (composed of the odd trials) starts with stimulus S_1 and reaction times of < 700 ms. The second staircase (composed of the even trials) starts with S_2 and reaction times of ≥ 700 ms. The rule of Eq. 4 is applied for the successive stimuli keeping c/n constant during a run of identical responses (i.e., the accelerated stochastic approximation). This rule is maintained until the maximum resolution of the laser stimulus is achieved. Step 3: Switch to the rule of Eq. 6 until convergence is achieved. Step 4: To estimate the threshold of "first" and "second" pain, compute the average of odd S_n trials and even S_n trials recorded during step 3. For the present subject, the threshold for "first" and "second" pain, defined as the energy density needed to detect the laser stimulus with a probability of 0.75, was 13.0 and 8.2 mJ/mm², respectively

Response-Dependent Methods

In the psychophysical methods described here above, stimulus intensity (or duration) is used as the dependent variable for a specific subjective magnitude of sensation. There is a second class of

psychophysical methods that performs the reverse operation. It uses psychophysical judgments of magnitude of sensation as a dependent variable for specific stimulus intensities. These procedures present a large range of stimulus intensities, repeated several times in random fashion, and record a subjective estimate for each stimulus presented. The sensations produced may be scaled by different response methods, like visual analog scales, discrete numerical scales, cross modality matches (e.g., handgrip force), or choosing among verbal descriptors quantified in preliminary experiments. These scaling methods use the mean response magnitude as a direct measure of perceptual magnitude. However, these measures may be susceptible to response bias, an effect in which a different response is used to describe the same sensation (Coppola and Gracely 1983). To minimize bias effects, the direct scaling methods should rely on appropriate experimental designs.

References

Ashton, W. D. (1972). *The logit transformation*. London: Griffin.

Coppola, R., & Gracely, R. H. (1983). Where is the noise in SDT pain assessment? *Pain, 17*, 257–266.

Derman, C. (1957). Non-parametric up-and-down experimentation. *Annals of Mathematical Statistics, 28*, 795–797.

Falmagne, J. P. (1985). *Elements of psychophysical theory*. New York: Oxford University Press.

Gescheider, G. A. (1985). *Psychophysics method, theory and application* (2nd ed.). Hillsdale: Lawrence Erlbaum Ass.

Pain Facilitatory Pathways

▶ Descending Circuitry, Opioids

Pain Facilitatory Systems

▶ Descending Modulation of Nociceptive Processing

Pain Flares

▶ Breakthrough Pain (BP)
▶ Cancer Pain Management, Cancer-Related Breakthrough Pain, Therapy
▶ Evoked and Movement-Related Neuropathic Pain
▶ Incident Pain

Pain Generator

Definition

The pain generator is an anatomic structure identified at the primary pathological focus of pain production.

Cross-References

▶ Dorsal Root Ganglionectomy and Dorsal Rhizotomy

Pain History

Definition

A pain history is an evaluation of the history of the current pain. This includes a careful description of the pain detailing the sensory characteristics, intensity, quality, location, duration, variability, predictability, exacerbating and alleviating factors, and impact of pain on daily life (e.g., sleeping, eating, school, social and physical activities, family and peer interactions).

Cross-References

▶ Central Pain, Outcome Measures in Clinical Trials

Pain in Children

Gary A. Walco[1] and Tonya M. Palermo[2]
[1]Anesthesiology and Pain Medicine, University of Washington School of Medicine, and Seattle Children's Hospital, Seattle, WA, USA
[2]Seattle Children's Hospital Research Institute, CW8-6 - Child Health, Behavior and Development, University of Washington, Seattle, WA, USA

Synonyms

Adolescent; Child; Infant; Pain in the neonate; Toddler

Definition

In its most basic definition, childhood is the period of life from birth through adulthood. It is rather arbitrary to designate at exactly what chronological age childhood ends; certainly roles and responsibilities that may demark the end of childhood vary a great deal by culture. In 2005, UNICEF estimated that there were 2.2 billion children in the world (UNICEF 2004), about one-third of the estimated 6.4 billion people in the population that year. Given these numbers, it is a bit striking that many textbooks on pain simply list children under "special populations." Often age groups are broken down further into premature and full-term neonates (0–4 weeks), infancy (through 12 months), toddlerhood (preverbal years), childhood (5–12 years), and adolescence (13–18 years).

Introduction

There is an old adage that "children are not little adults." For our purposes, this means that one cannot simply extrapolate knowledge of pain assessment and interventions from adults and adjust them for children. Instead, one must embrace the important elements and nuances of development in order to optimally evaluate and treat children. For example, adult doses of medications may not simply be linearly reduced in line with body mass with expectations of equal safety and efficacy, as metabolic processes may vary significantly over key developmental periods. Similarly, assessment tools must be validated separately in children to ensure that demands of language and cognition are appropriate. Thus, if the goal is to truly understand the nature of pain, its assessment, and remediation in children, one must consider the development of nociceptive systems; the development of various physiological systems, including drug metabolizing mechanisms; and the development of language, cognition, and social relationships. By better understanding these developmental processes and longitudinal sequelae, we will likely come to realize that adults are big children and many lessons learned from pediatrics will be applied to older patients.

Characteristics

Epidemiology

The epidemiology of acute and chronic pain in childhood is not fully known. A recent review on the epidemiology of chronic pain in children indicated very broad prevalence rates with ranges of headache 8–83 %, abdominal pain 4–53 %, back pain 14–24 %, musculoskeletal pain 4–40 %, multiple pains 4–49 %, and other pains 5–88 % (King et al. 2011). Clearly, the manner in which chronic pain is defined and studied has a major impact on estimates of incidence. It is even more difficult to quantify acute pain problems, as one would need to include the bumps and bruises of childhood, needle pain, postoperative pain, as well as an array of pain problems related to illnesses (pharyngitis, otitis media) and treatments (e.g., chemotherapy, physical therapy). Notwithstanding the challenges in understanding the epidemiology of childhood pain, significant progress has been made in characterizing the impact of pain on children and families (Palermo 2000) and in defining longitudinal risk for pain into adulthood (Walker et al. 2010).

Progress in the Field

The evolution of the field of pediatric pain exemplifies the power of truly interdisciplinary approaches to complicated medical issues. There is no discipline that dominates the endeavor, but rather anesthesiologists, pediatricians, psychologists, nurses, rehabilitation specialists, and others have interfaced with basic scientists, notably developmental physiologists, to advance the field substantially over the past few decades. Indeed, it was less than 30 years ago that surgery was performed on young infants without anesthesia, based principally on the notion that nociceptive systems were not sufficiently developed to invoke a pain response, and, if so, surely the risks of treatment outweighed the risks of poorly treated pain. These myths have been refuted as we have come to understand the ontogenesis of nociceptive systems; have documented the deleterious effects of poorly treated pain, especially during sensitive developmental periods; have validated psychometrically sound means of assessing pain, even for the youngest humans; and have begun to implement ethical and feasible means of demonstrating the safety and efficacy of various preventive and remedial interventions for pain in the young (Berde et al. 2012; McGrath 2011; Walco 2008). Progress in pediatric pain management may be demonstrated by the number of publications in the field. A Medline search on the term "pain" limiting to age groups 0–18 years yielded the following:

Years	Number of publications
1951–1970	1,014
1971–1980	3,895
1981–1990	10,002
1991–2000	19,027
2001–2010	33,362

Contemporary Research in Pediatric Pain

The process of nociceptive network development early in life is one of differentiation and integration. Not only do nerve roots continue to branch, but there are also mechanisms to integrate networks so that pain responses may be specific, leading to coordinated and adaptive responses.

The analogy may be applied to the field of pediatric pain as well. There has been an explosion of research efforts examining clinical aspects of pain in infants and children, as well as a proliferation of basic science studies on peripheral and central mechanisms and their developmental trajectories. Several particularly important directions in the field include understanding the transition from acute to chronic pain, measuring central pain processing mechanisms, furthering the evidence base for pharmacological and behavioral interventions for a range of pediatric pain conditions, and developing strategies to improve access to specialized pediatric pain care.

Now is the time for more integration efforts to begin to translate what is learned in the laboratory to clinical applications as well as more feedback from the developmentally sensitive clinicians back to those focused on basic research. Based on history, if there is any field that is likely to see synergy between disciplines, it is the study of pain in our youngest and most vulnerable humans.

Summary

The field of pediatric pain medicine is rapidly advancing. Clinical research representing an array of disciplines continues to advance our insights into factors that may be impacted to better prevent and reduce unnecessary pain in the young. Basic and translational researchers continue to shed light on key peripheral and central mechanisms that belie the pain experience. The first 25 years of study has laid an excellent foundation. We now look forward to more integration of clinical efforts with those conducting basic research on the developmental physiology of pain perception to expand our field as we move forward.

References

Berde, C. B., Walco, G. A., Krane, E. J., Anand, K. J., Aranda, J. V., Craig, K. D., & Zempsky, W. T. (2012). Pediatric analgesic clinical trial designs, measures, and extrapolation: Report of an FDA scientific workshop. *Pediatrics, 129*(2), 354–364. doi:10.1542/peds.2010-3591.

King, S., Chambers, C. T., Huguet, A., MacNevin, R. C., McGrath, P. J., Parker, L., & MacDonald, A. J. (2011). The epidemiology of chronic pain in children and adolescents revisited: A systematic review. *Pain, 152*(12), 2729–2738. doi:10.1016/j.pain.2011.07.016.

McGrath, P. J. (2011). Science is not enough: The modern history of pediatric pain. *Pain, 152*(11), 2457–2459. doi:10.1016/j.pain.2011.07.018.

Palermo, T. M. (2000). Impact of recurrent and chronic pain on child and family daily functioning: A critical review of the literature. *Journal of Developmental and Behavioral Pediatrics, 21*(1), 58–69.

UNICEF. (2004). The state of the world's children 2005. New York, NY: UN Publications.

Walco, G. (2008). Pain and the primary pediatric practitioner. In G. G. Walco & K. Goldschneider (Eds.), *Pain in children: A practical guide for primary care* (pp. 3–8). Totowa, NJ: Humana Press.

Walker, L. S., Dengler-Crish, C. M., Rippel, S., & Bruehl, S. (2010). Functional abdominal pain in childhood and adolescence increases risk for chronic pain in adulthood. *Pain, 150*(3), 568–572. doi:10.1016/j.pain.2010.06.018.

Pain In Children with Cognitive Impairment

▶ Pain in Children with Disabilities

Pain in Children with Disabilities

Kenneth D. Craig[1] and Tim F. Oberlander[2]
[1]Department of Psychology, University of British Columbia, Vancouver, BC, Canada
[2]Division of Developmental Pediatrics, University of British Columbia, Vancouver, BC, Canada

Synonyms

Communication limitations or neurological impairment; Pain in children with cognitive impairment

Definition

The term "children with disabilities" encompasses diverse and overlapping conditions that are developmental (i.e., arising from genetic disorder) or acquired (e.g., ischemia, traumatic brain damage, or infection) and associated with impaired cognitive, motor, social, or communication capabilities. Some conditions are relatively prevalent (e.g., Down syndrome, autism, cerebral palsy, significant neurological impairment from traumatic brain injury) and others are relatively rare (e.g., genetic or metabolic disorders) (see Oberlander and Symons 2006, for review of pain in individuals with neurological impairments).

Introduction

Pain is a frequent experience for children with disabilities, yet its presentation can be ambiguous and recognition by even experienced caregivers is highly variable and dependent upon subjective judgment. Interpretation of the presentation can be confounded by the child's functional limitations and the underlying neurological condition itself. Even when pain-specific behaviors are thought to be present, they may be altered and blunted or confused with other sources of generalized arousal (McGrath et al. 1998; Oberlander et al. 1999). This presents a challenge for their families, clinicians, and researchers. Until recently, pain in children with developmental disabilities received very little scientific attention and they usually were systematically excluded from research. Neglect was in part based on erroneous beliefs. Because expression of pain may not be clear, for example, self-report may not be possible, individuals with disabilities came to be characterized as indifferent, or worse, insensitive to pain (e.g., Baranek and Berkson 1994). However, regardless of the degree of disability, pain is now recognized as a feature of daily life in these populations and altered pain behaviors cannot be represented as pain insensitivity or indifference (Oberlander and Craig 2003), despite some systematic differences in clinical presentation (Dubois et al. 2010; Defrin, Lotan and Pick 2006). For example, Tordjman et al. (2009) demonstrated that children with autism experience more vigorous heart rate response

and higher plasma b-endorphin levels during venepuncture compared to children without autism.

With recognition of the importance of pain in children with disabilities, attention has turned to questions about the nature, extent, and functional impact of pain for these children and their caregivers. The past 15 years has seen exponential growth in research activity in this field, spanning exploration of pain in preterm infants, children with cognitive and communication impairments, and adolescents with developmental disabilities (McGrath et al. 1998; Oberlander and Craig 2003; Vargus-Adams and Martin 2011).

Characteristics

Research on pain in children with disabilities initially was restricted by difficulties with the definition of pain itself. The International Association for the Study of Pain (IASP) definition of pain as "an unpleasant sensory and emotional experience associated with actual or potential tissue damage or described in terms of such damage" (Merskey 1986) should be applicable to anyone regardless of disability. However, an emphasis on self-report and the subjective experience of pain in the definition appeared to necessitate a capacity for effective verbal communication. This led Anand and Craig (1996) to argue that all features of the behavioral response to pain should be accepted as evidence of its presence and nonverbal expression should not be discounted as a "surrogate measure" of pain. IASP recently made a substantial change to its definition by adding a note indicating that "The inability to communicate verbally does not negate the possibility that an individual is experiencing pain and is in need of appropriate pain-relieving treatment" (IASP Task Force on Taxonomy). This broadening of research and clinical approaches to pain should contribute to better understanding of the thoughts, feelings, and sensory features of pain in individuals with language and cognitive disabilities.

Sources and Epidemiology

Reports on the epidemiology of pain indicate that pain is commonplace among these children (Engel et al. 2002; Parkinson et al. 2010). We have less of an understanding of the prevalence of chronic pain typical for all children (e.g., headache, abdominal pain, musculoskeletal pain), given diagnostic and assessment challenges, but children with disabilities are at least as vulnerable to acute pain. They have a substantially greater risk of painful injury compared to children without developmental disabilities (Lee et al. 2008), as they not only suffer the falls, burns, earaches, or toothaches that are a common part of childhood, but cognitive, motor, and communication impairments leave them more susceptible to environmental risk factors. They also suffer from specific conditions that are often painful (e.g., spasticity and contractures, reflux oesophagitis, constipation) and there is increased exposure to painful medical procedures (e.g., surgery for hydrocephalous shunts or motor spasticity). For example, gastroesophageal reflux disease is present in over 50 % of children and adults with developmental disabilities (Bohmer et al. 2000). The likelihood of repeated pain contributes to the possibility that these children will be sensitized to pain and display fear and adverse behavioral reactions; it is possible that degeneration of white matter and the frontal lobe are related to increased sensitivity to pain in adults with Down syndrome (White et al. 2003; Jernigan et al. 1993) through loss of inhibitory controls. Compounding these painful experiences is difficulty in communicating distress, thereby increasing exposure to and duration of pain and reducing the likelihood others will engage in early intervention or provide adequate care (Gilbert-Macleod et al. 2000).

Assessment and Treatment

There has been considerable recent effort to develop multidimensional and empirically based assessment instruments useful in home, clinical, and research settings, given that global proxy judgments tend to underestimate pain. A number of inventories relying upon observation of the behavior of children and adults with significant neurological disabilities or communication and cognitive impairments (Stallard et al. 2002) have demonstrated adequate psychometric features (i.e., reliability and validity) (Breau et al. 2002; Collignon and Giusiano 2001; Hadden and von Baeyer 2005; McGrath et al. 1998; Valkenburg et al. 2010; Voepel-Lewis et al. 2008), although further development and validation often is needed. For example, scales that rely upon coding of the facial expression of pain often yield different scores as they provide dramatically different characterization of the display of pain, leading to variation in observer judgements (Sheu et al. 2011). Research to understand the clinical utility of these tools and the impact of developmental status is needed. Self-report tools can be useful for some children with cognitive impairments (Fanurik et al. 1998), although there are inherent difficulties associated with self-report scales in this population and the evidence is inconclusive (von Baeyer 2006; Breau and Burkitt 2009). Research tools focusing upon sensitive and specific features of nonverbal pain display (e.g., facial expression, Nader et al. 2004) and biobehavioral reactivity (Oberlander et al. 1999), such as heart rate variability (Oberlander and Saul 2002), near-infrared spectroscopy (Slater et al. 2007), and skin conductance (Storm 2008), are also becoming available. With refinement in instruments that document how children with disabilities express pain, we will increasingly better understand the nature of their experience of pain. Irrespective of the instruments used, it is clear that pain assessment should be routinely undertaken, regardless of the disability, particularly when extraordinary behavior or context dictates the possibility that pain is present.Studies of novel treatment strategies that recognize unique features of pain in this setting are emerging. Combined pharmacological and psychosocial interventions that focus on the sensitivity of pain to environmental, situational, and emotional factors need to be explored in a multidisciplinary context. It is becoming more universally accepted that pain management should be an inherent feature of the care of children with disabilities, especially when they are likely to encounter painful conditions (e.g., gastrointestinal pain, orthopedic pain, pain due to respiratory infections) (Oberlander and Craig 2003; Taddio and Oberlander 2006). In 2010, Buie et al. published the conclusions of a multidisciplinary panel affirming that individuals with autism spectrum disorders should receive the same treatment for gastrointestinal symptoms as individuals without autism spectrum disorders. Interventions need to reflect the nature of the underlying condition and associated impairment, as well as individual differences in pain reactivity, development, and life circumstances. It is important to recognize that pharmacological treatment of conditions associated with developmental disabilities (e.g., anticonvulsants, medications used to treat gastrointestinal reflux, etc.) may alter pharmacodynamics and pharmacokinetics of analgesics, thereby making pain management even more challenging.

Functional Impact of Pain

Studies of pain in this setting typically focus on describing symptoms, duration, and intensity of pain without accounting for the functional consequences of the symptom, even though chronic pain is likely to have a major impact on daily activities. If we are to fully understand pain in individuals with disabilities, we need an improved understanding of the impact of pain on the physiological, psychological, behavioral, familial, social, and occupational features of daily life.

Attempts to characterize the functional impairment associated with chronic daily pain in the presence of a disability have received increasing attention as a novel dimension of a pain assessment in this setting (O'Donnell 2006). Application of the World Health Organization's International Classification of Functioning, Disability and Health (ICF; WHO 2001) may offer a useful model to help ensure comprehensiveness in pain assessment and management. ICF offers clinicians a way to understand a health condition (such as pain), along with personal factors (e.g., an individual's resources) and environmental factors (e.g., family, friends), which influence health outcomes such as psychological functioning, activities of daily life and participation in social activities. There have been efforts to characterize functional disability using the dimensions of function and disability: health condition (e.g., disease, disorder), body structure and function, activity, participation, and context (e.g., personal, environment). Using this multidimensional framework for assessing the functional impact of pain in, for example, a youth with cerebral palsy, can be characterized in the following fashion. The pathophysiology of the health condition would be the central lesions (presumably, prenatal ischemic neurological injury) accompanied by painful complaints such as chronic headache, which may or may not be related to factors associated with the disease (e.g., hydrocephalus/shunt malfunction or a concurrent migraine disorder). Here, the disease and pain together receive consideration as potentially interrelated but separate pathophysiological entities. The impairment would include the neuromotor impairment (spasticity) and the frequency and intensity of the symptom (head pain), regardless of the etiology. Compounding the impairment are the cognitive and communication limitations that are a concurrent consequence of the central impairment. The functional limitations to daily activities that follow may be a consequence of motor, cognitive, and communication limitations, as well as the impact of the pain on activities of daily living (e.g., sleep, self-care, relationship with his mother, peer relationships) (Breau et al. 2007). This individual's disability could cause difficulty in carrying on life role such as an adolescent attending school or maintaining age-appropriate social relationships. Finally, the societal limitations would be barriers placed by society, which limit participation even in a modified classroom setting, where teachers may have difficulty understanding, managing, and coping with a student's pain. An individual may also be denied appropriate and timely pain management because caregivers may not accept that the pain is "real" when the ability to communicate experiences with discrete words and motor movements is limited. The social contexts in which children with disabilities live often dictate levels of care and support. Caregiver stress and burden can often be a limiting factor in the quality of care provided. This schema might be an oversimplification, but it begins to illustrate how pain and disability intersect.

Using this approach, one can appreciate how uncontrolled acute pain or chronic pain could play important roles as a determinant of well-being in the context of a disability. There is considerable potential for a cascading spiral of compounding, disabling processes when pain becomes important in the life of a child with a disability. It becomes obvious that one should consider not only the presence and severity of pain, but also its functional consequences and how these relate to any underlying neurological condition. This approach also encourages one to distinguish individuals who experience pain with little or no functional limitations from those that are more severely restricted in function. While the two may be separate in an etiological sense, assessment and management of pain in this setting is invariably influenced by the functional aspects of daily life (i.e., a limited capacity to communicate pain) associated with the neurological impairment. Understanding chronic pain and its impact on daily life is limited and much work is still needed to understand the deleterious impact of pain on child and family functioning in this setting.

Summary

While pain in individuals with disability has received substantially increased attention over the past decade, numerous questions remain. At a very basic level, our understanding of relationships between altered neurological function and the pain system remains incomplete. Similarly, research needs to address questions about how social and cognitive impairments influence the expression of pain and the implications this might have for assessment and intervention strategies to reduce suffering where individuals experience repetitive painful events. Further, work needs to be directed toward understanding the impact of family and social influences, mental illness, and culture on the child's ongoing pain

and daily function. Finally, we need to seek ways to translate research findings into meaningful clinical and public health policies that address pain in individuals with disabilities. Considerable work on recognition, assessment, and management of pain in individuals with disabilities remains to be done.

References

Anand, K. J. S., & Craig, K. D. (1996). Editorial: New perspectives on the definition of pain. *Pain, 67*, 3–6.

Baranek, G. T., & Berkson, G. (1994). Tactile defensiveness in children with developmental disabilities: Responsiveness and habituation. *Journal of Autism and Developmental Disorders, 24*, 457–471.

Bohmer, C. J., Klinkenberg-Knol, E. C., Niezen-de Boer, M. C., et al. (2000). Gastroesophageal reflux disease in intellectually disabled individuals: How often, how serious, how manageable? *The American Journal of Gastroenterology, 95*, 1868–1872.

Breau, L. M., & Burkitt, C. (2009). Assessing pain in children with intellectual disabilities. *Pain Research & Management, 14*, 116–120.

Breau, L. M., McGrath, P. J., Camfield, C. S., et al. (2002). Psychometric properties of the non-communicating children's checklist-revised. *Pain, 99*, 349–357.

Breau, L. M., Camfield, C. S., McGrath, P. J., et al. (2007). Pain's impact on adaptive functioning. *Journal of Intellectual Disability Research, 51*, 125–134.

Buie, T., Fuchs, G. J., III, Furuta, G. T., et al. (2010). Recommendations for evaluation and treatment of common gastrointestinal problems in children with ASDs. *Pediatrics, 125*, S19–S29.

Collignon, P., & Giusiano, B. (2001). Validation of a pain evaluation scale for patients with severe cerebral palsy. *European Journal of Pain, 5*, 433–442.

Defrin, R., Lotan, M., & Pick, C. G. (2006). The evaluation of acute pain in individuals with cognitive impairment: A differential effect of the level of impairment. *Pain, 124*, 312–320.

Dubois, A., Capdevilla, X., Bringuier, S., et al. (2010). Pain expression in children with an intellectual disability. *European Journal of Pain, 14*, 654–660.

Engel, J. M., Kartin, D., & Jensen, M. P. (2002). Pain treatment in persons with cerebral palsy: Frequency and helpfulness. *American Journal of Physical Medicine & Rehabilitation, 81*, 291–296.

Fanurik, D., Koh, J. L., Harrison, R. D., et al. (1998). Pain assessment in children with cognitive impairment: An exploration of self-report skills. *Clinical Nursing Research, 7*, 103–124.

Gilbert-Macleod, C. A., Craig, K. D., Rocha, E. M., et al. (2000). Everyday pain responses in children with and without developmental delays. *Journal of Pediatric Psychology, 25*, 301–308.

Hadden, K. L., & von Baeyer, C. L. (2005). Global and specific behavioural measures of pain in children with cerebral palsy. *The Clinical Journal of Pain, 21*, 140–146.

IASP Task Force on Taxonomy, (n.d.). IASP Taxonomy. Retrieved 6 Jul 2011, from http://www.iasp-pain.org/Content/NavigationMenu/GeneralResourceLinks/Pain Definitions/default.htm.

Jernigan, T. L., Bellugi, U., Sowell, E., et al. (1993). Cerebral morphologic distinctions between Williams and Down syndromes. *Archives of Neurology, 50*, 186–191.

Lee, L. C., Harrington, R. A., Chang, J. J., et al. (2008). Increased risk of injury in children with developmental disabilities. *Research in Developmental Disabilities, 29*, 247–255.

McGrath, P. J., Rosmus, C., Canfield, C., et al. (1998). Behaviours caregivers use to determine pain in non-verbal, cognitively impaired individuals. *Developmental Medicine and Child Neurology, 40*, 340–343.

Merskey, H. (1986). Classification of chronic pain: Descriptions of chronic pain syndromes and definitions of pain terms. *Pain, 3*, 51.

Nader, R., Oberlander, T. F., Chambers, C. T., et al. (2004). The expression of pain in children with autism. *The Clinical Journal of Pain, 20*, 88–97.

O'Donnell, M. (2006). Using the WHO's international classification of functioning, disability and health: A framework in the management of chronic pain in children and youth with developmental disabilities. In T. F. Oberlander & F. Symons (Eds.), *Pain in children and adults with developmental disabilities* (pp. 139–148). Baltimore: Paul H Brookes.

Oberlander, T. F., & Craig, K. D. (2003). Pain and children with developmental disabilities. In N. M. Schechter, C. B. Berde, & M. Yaster (Eds.), *Pain in infants, children and adolescents* (2nd ed., pp. 599–619). Baltimore: Williams & Wilkins.

Oberlander, T., & Saul, J. P. (2002). Methodological considerations for the use of heart rate variability as a measure of pain reactivity in vulnerable infants. *Clinics in Perinatology, 29*, 427–443.

Oberlander, T. F., & Symons, F. (2006). *Pain in individuals with developmental disabilities*. Baltimore: Paul H Brookes.

Oberlander, T. F., Gilbert, C. A., Chambers, C. T., et al. (1999). Biobehavioral responses to acute pain in adolescents with a significant neurological impairment. *The Clinical Journal of Pain, 15*, 201–209.

Parkinson, K. N., Gibson, L., Dickinson, H. O., et al. (2010). Pain in children with cerebral palsy: A cross-sectional multicentre European study. *Acta Paediatrica, 99*, 446–451.

Sheu, S., Versloot, J., Nader, R., et al. (2011). Pain in the elderly: Validity of facial expression components of observational measures. *The Clinical Journal of Pain, 27*, 593–601.

Slater, R., Fitzgerald, M., & Meek, J. (2007). Can cortical responses following noxious stimulation inform us about pain processing in neonates? *Seminars in Perinatology, 31*, 298–302.

Stallard, P., Williams, L., Velleman, R., et al. (2002). The development and evaluation of the pain indicator for communicatively impaired children (PICIC). *Pain, 98*, 145–149.

Storm, H. (2008). Changes in skin conductance as a tool to monitor nociceptive stimulation and pain. *Current Opinion in Anaesthesiology, 21*, 796–804.

Taddio, A., & Oberlander, T. F. (2006). Pharmacological management of pain in children and youth with significant neurological impairments. In T. F. Oberlander & F. Symons (Eds.), *Pain in children and adults with developmental disabilities*. Baltimore: Paul H Brookes.

Tordjman, S., Anderson, G. M., Botbol, M., et al. (2009). Pain reactivity and plasma b-endorphin in children and adolescents with autistic disorder. *PLoS One, 4*, e52–e89.

Valkenburg, A. J., van Dijk, M., de Klein, A., et al. (2010). Pain management in intellectually disabled children: Assessment, treatment, and translational research. *Developmental Disabilities Research Reviews, 16*, 248–257.

Vargus-Adams, J. N., & Martin, L. K. (2011). Domains of importance for parents, medical professionals and youth with cerebral palsy considering treatment outcomes. *Child Care Health and Development, 37*, 276–281.

Voepel-Lewis, T., Malviya, S., Tait, A. R., et al. (2008). A comparison of the clinical utility of pain assessment tools for children with cognitive impairment. *Pediatric Anesthesia, 106*, 72–78.

von Baeyer, C. L. (2006). Children's self-reports of pain intensity: Scale selection, limitations and interpretation. *Pain Research & Management, 11*, 157–162.

White, N. S., Alkire, M. T., & Haire, R. J. (2003). A voxel-based morphometric study of nondemented adults with Down syndrome. *NeuroImage, 20*, 393–403.

WHO. (2001). International classification of functioning, disability and health. Retrieved 6 Sep 2011, from http://www.who.int/classifications/icf/en/

Pain in Early Pregnancy

Jackie Ross[1] and Yasmin Sana[2]
[1]Early Pregnancy Unit, King's College Hospital Early Pregnancy Unit, London, UK
[2]King's College Hospital, London, UK

Synonyms

Ache in first trimester or in early gestation; Discomfort

Definition

Pain in early pregnancy refers to abdominal pain experienced by women in the first trimester of pregnancy up to 14 weeks gestation.

Introduction

Abdominal pain is one of the commonest symptoms in early pregnancy. It is the presenting symptom of a large number of underlying conditions that can either be directly related to pregnancy or that can be unrelated and may or may not be exacerbated by the pregnancy. The various conditions presenting with pain range from minor problems to life-threatening emergencies. However, the majority of patients who present with abdominal pain have a normally developing pregnancy.

A patient presenting with pain in early pregnancy poses a unique diagnostic challenge for the clinicians due to the extensive differential diagnosis, normal physiological changes during pregnancy, and diagnostic tools that are altered limited due to pregnancy. The management of pain in early pregnancy is also affected by the limitation of pharmacological and surgical options. This chapter reviews some of the common pregnancy- and nonpregnancy-related conditions that may present with pain in early pregnancy and how best to investigate and treat.

Characteristics

Physiological Effects of Pregnancy

Assessment of patients during pregnancy is modified by the fact that the gravid uterus displaces the abdominal viscera. Clinical signs and symptoms of various conditions may alter as the pregnancy advances making diagnosis of even common conditions such as appendicitis more challenging. Physiological changes alter the laboratory values from early on in pregnancy and should be taken into consideration when evaluating a pregnant patient. White cell count rises in pregnancy from the first trimester onward (Rahamim et al. 2008).

Plasma C reactive protein (CRP) values vary throughout pregnancy between patients; median levels however remain consistently elevated (Picklesimer et al. 2008). One study showed that CRP rises as early as 4 weeks into pregnancy (Sacks et al. 2004). This study included women who became pregnant as result of IVF treatment. It is therefore not clear whether these results could have been affected by the administration of drugs used during IVF treatment and any preexisting conditions contributing to subfertility.

Liver function also alters, and values of various markers of liver function change during pregnancy. For example, plasma alkaline phosphatase is also a placental enzyme, and thus serum levels rise gradually from the second trimester onward with levels which are most marked in the third trimester. First trimester values are slightly lower than nonpregnant subjects which are attributed to hemodilution. All other LFT values are slightly lower during all trimesters of pregnancy (Giriing et al. 1997; Bacq et al. 1996).

Differential Diagnosis

The list of conditions that may present with pain in early pregnancy is extensive, but the most common are listed below:

- *Causes related to pregnancy*
  - Ectopic pregnancy
  - Miscarriage
- *Gynecological causes potentially exacerbated by pregnancy*
  - Ovarian torsion/ovarian cyst rupture (due to luteal cysts)
  - *Fibroid degeneration*
- *Gynecological causes unrelated to pregnancy*
  - Pelvic infection
  - Rare causes such as fallopian tube torsion, uterine torsion
  - Idiopathic chronic pelvic pain
- *Nongynecological causes potentially exacerbated by pregnancy*
  - Urinary tract infection
  - Reflux esophagitis
  - Thrombosis (iliac vein, mesenteric vein, ovarian vein – extremely rare)
  - Sickle cell crises
  - Porphyria

- *Nongynecological causes unrelated to pregnancy*
  - Appendicitis
  - Intestinal obstruction
  - Abdominal trauma
  - Acute intestinal ischemia
  - Renal colic
  - Abdominal aortic aneurysm
  - Psoas abscess
  - Gastric Volvulus
  - Pheochromocytoma
- *Preexistent conditions*
  - Inflammatory bowel disease (Crohn's disease, ulcerative colitis)
  - Diverticulitis
  - Irritable bowel syndrome
  - Musculoskeletal conditions such as rheumatoid arthritis

For the purpose of this chapter, some of the common conditions are discussed below.

Ectopic Pregnancy

Ectopic pregnancy is a classic gynecological emergency, and although maternal deaths due ectopic pregnancy are at lowest since 1988, it remains a significant cause of maternal mortality and morbidity (Cantwell et al. 2011). In ectopic pregnancy, the implantation site is uterine endometrial cavity. 95 % of ectopic pregnancies are implanted in the fallopian tube and 5 % implant in the uterine interstitium, cervix, and ovaries or in the abdominal cavity (Luciano et al. 2001). Lower abdominal or pelvic pain with or without bleeding is the first presenting symptom in about 70 % patients with ectopic pregnancy (Solima and Luciano 1997). Patients usually present between 5 and 9 weeks of pregnancy. Pain may be poorly localized and vague initially and severe and generalized if ectopic pregnancy has ruptured. Signs of hypovolemic shock may accompany in cases of ruptured ectopic pregnancy. Patients may also present with gastrointestinal symptoms such as diarrhea, vomiting, and dizziness (Cantwell et al. 2011). Up to 9 % patients, who have ectopic pregnancy, have no pain, and 36 % have no adnexal tenderness on examination (Farquhar 2005). History and clinical examination findings are nonspecific and cannot be relied

on to confirm or exclude the diagnosis of ectopic pregnancy (Tay et al. 2000).

Traditionally, laparoscopy was considered the gold standard for the diagnosis of ectopic pregnancy. Ultrasound (TVS) has become the first investigation of choice in clinically stable patients in recent years. Diagnosis on TVS should be based on positive identification of adnexal mass or gestational sac outside the uterine cavity. The sensitivity of ultrasound has been reported to be between 87 % and 93 % for the correct diagnosis of ectopic pregnancy prior to surgery (Cacciatore et al. 1990; Shalev et al. 1998).

Management options for ectopic pregnancy include surgery, expectant, or medical management using methotrexate. Success rate of expectant management varies from 48 % to 100 % (Kirk et al. 2006). This variation is due to different inclusion criteria used in different studies based on initial serum human chorionic gonadotrophin (hCG) level, size of the ectopic mass, and rate of fall of hCG level. Methotrexate, a folate antagonist, is used for medical management of ectopic pregnancy. As with the expectant management, success rates are variable depending on the inclusion criteria used. Surgery remains the most definitive treatment option. In the majority of women who are stable, surgery can be carried out laparoscopically and has several advantages over laparotomy such as shorter recovery time, less need for use of analgesia, and reduced blood loss (Seeber and Barnhart 2006). In patients where ectopic pregnancy is ruptured or the woman is hemodynamically unstable, the surgical procedure that prevents further blood loss most quickly should be performed. Laparoscopic surgery in women who have significant hemoperitoneum due to ruptured ectopic pregnancy is however feasible and has advantages over laparotomy in expert hands (Rizzuto et al. 2008).

Miscarriage

One of the commonest causes of pain in early pregnancy is miscarriage. However, the cardinal symptom of miscarriage is bleeding rather than pain, so isolated pain should not be referred to as a threatened miscarriage. The term is commonly used where failed intrauterine pregnancy is diagnosed on ultrasound scan as well as when products of conception are passed through the cervical os either partially or completely. The incidence of miscarriage is 12–15 % of clinically recognizable pregnancies.

Pain typically associated with miscarriage is crampy in nature, in the lower abdomen or pelvis. Associated back pain is also very common. Analgesia such as paracetamol, opioids, and NSAIDS can be used for the pain associated with miscarriage, once an ongoing pregnancy has been excluded. For women who have an incomplete miscarriage where the cervix is open and products of conception are present at the cervical os, removal of tissue from cervix during examination can relieve the pain.

Ovarian/Adnexal Torsion

Adnexal torsion is a less common but significant cause of acute abdominal pain in early pregnancy. It occurs in about 1 in 1,800 pregnancies (Johnson and Woodruff 1986). Another study by Houry and Abbott (2001) reported an incidence of 3 %. The ovary is enlarged in the majority of cases of torsion, usually due to ovarian cysts or tumors. Most of these cysts are functional. The common nonfunctional cysts are benign cystic teratomas (dermoid cysts), serous or mucinous cystadenomas, or ovarian fibromas. A review of 54 patients who required exploratory surgery during pregnancy for adnexal mass showed that less than 6 % of these adnexal tumors were malignant (Hess et al. 1988). Typical presentation of adnexal torsion is unilateral, colicky pain of sudden onset with associated nausea and vomiting. Right-sided adnexal torsion is more common than the left (McGowan 1983) and be difficult to differentiate from appendicitis. Patients can also have intermittent or incomplete torsion and may therefore present with intermittent symptoms.

Clinical signs include unilateral abdominal tenderness, palpable abdominal mass, and rebound tenderness. Leukocytosis is often present but is nonspecific.

Ultrasound examination is a valuable investigation in combination with clinical symptoms and signs. It cannot however be used in isolation to confirm or exclude the diagnosis even when combined with Doppler assessment of the adnexal blood supply. On gray-scale ultrasound, the most common finding in ovarian torsion is an enlarged ovary. The ovarian stroma may be heterogeneous due to hemorrhage and edema. Free fluid can be seen in the pelvis (Albayram and Hamper 2001). Comparison with morphological features of the contralateral ovary and local tenderness (elicited with transvaginal probe) can aid in diagnosis.

An aggressive approach toward early diagnosis is required in order to achieve ovarian preservation. For years, salpingoophorectomy rather than detorsion has been the treatment of choice. This was due to possible risks of thromboembolism into the systemic circulation after untwisting the ovarian pedicle. However, several recent studies have demonstrated that the conservative approach of untwisting the adnexa can result in preservation of ovarian function despite a necrotic appearance of the twisted ovary at the time of surgery (Breech and Hillard 2005; Oelsner and Shashar 2006).

Rupture of ovarian cyst can also present with similar symptoms. Pain however tends to be constant and more generalized in this situation. Ultrasound can show presence of fluid in the pelvis with or without an ovarian cyst.

Fibroids

Uterine fibroids are the most common benign tumors in women of reproductive age. The incidence varies depending on the age and race of women. The incidence among black women is two to three times higher compared to white women (Marshall et al. 1997). Fibroids can be a cause of significant pain in pregnancy, which is most commonly as a result of fibroid degeneration but can also be caused by torsion of a pedunculated fibroid or fibroid impaction (Exacoustos and Rosati 1993, Katz et al. 1989). The pain typically occurs in the second trimester. Presenting symptoms of fibroid degeneration are abdominal pain, which can be localized over the mass. Nausea, vomiting, and mild fever may also be present. Ultrasound can confirm the presence of fibroids. Red degeneration of fibroid is self-limiting, and management is conservative with analgesia and correction of any dehydration.

Urinary Tract Infections

Urinary tract infection (UTI) is one of the most frequently seen medical problems in pregnancy and can also present with pain in the lower abdomen. The incidence of asymptomatic bacteriuria in pregnancy has been reported to be 6 %. This incidence is similar to age-matched nonpregnant women, however; the risk of complications is higher if it is not treated during pregnancy, the likely cause for which is increased urinary stasis and reduced mucosal immunity during pregnancy (Delzell and Lefevre 2000).

The incidence of acute cystitis has been reported as 1.3 % (Harris and Gilstrap 1981) and that of pyelonephritis as 2 % (Gilstrap et al. 1981). In contrast to asymptomatic bacteriuria, patients with acute cystitis present with dysuria, increased urinary frequency, hematuria, and discomfort in the lower abdomen/suprapubic region. Dysuria is less common in pregnant women, and abdominal pain may be the first symptom of UTI in these women. These patients are usually afebrile and do not have systemic symptoms. The pain in patients with acute pyelonephritis is typically in the flank/renal angle, and there are associated symptoms of systemic illness such as nausea, vomiting, fever and rigors, etc. In pyelonephritis, renal angle tenderness can be elicited on examination.

Urinalysis should be performed, as a first investigation. Blood culture (in patients with systemic symptoms of infection) and MSU should be taken before commencing antimicrobial therapy. *E. coli* is the most common organism found on culture in asymptomatic bacteriuria, acute cystitis, and pyelonephritis (Millar and Cox 1997, Delzell and Lefevre 2000).

Patients with acute pyelonephritis usually require admission and parenteral antimicrobial therapy. Treatment with broad-spectrum antibiotics can be initiated empirically, and this should be reviewed once results of culture and sensitivity reports become available.

Appendicitis

Acute appendicitis is the commonest surgical emergency in pregnancy with an incidence of about 1 in 1,500 pregnancies (Mourad et al. 2000;

Hee and Viktrup 1999). This incidence is not different from nonpregnant population. It can occur in any part of pregnancy, and incidence in the first trimester ranges from 19 % to 36 % (Tracey and Fletcher 2000; Al-Mulhim 1996; Anderson and Nielsen 1999).

The classical presentation of pain in the periumbilical area that migrates to right lower abdomen may not be seen in pregnant women and they can present with poorly localized pain in the right lower abdomen. Pain could also be in the upper abdomen or flank as the appendix may be displaced upward and laterally due to enlargement of the gravid uterus. This is however more relevant in second and third trimester of pregnancy, and pain in right lower quadrant is still by far the most common presentation. (Mourad et al. 2000). Associated symptoms of nausea, vomiting, pyrexia, and anorexia may also be present but are nonspecific. Abdominal tenderness and rebound tenderness are the most common signs on examination. Classical signs of acute appendicitis such as Rovsing's and psoas signs have not been shown to be of any clinical significance in pregnant women (Al-Mulhim 1996). Leukocytosis (up to 16,000 cells/ml) could be physiological in pregnancy and is therefore not a reliable marker (Maslovitz et al. 2003). Diagnosis of this condition in pregnancy can therefore be very difficult and is often missed. Graded compression ultrasound can be used to aid diagnosis and has been shown to have sensitivity of 67–100 % and specificity of 83–96 %. This is however highly dependent on the operator's skill and experience (Williams and Shaw 2007). A positive ultrasound may be useful in the diagnosis, but patients with negative ultrasound should be observed or further investigated until symptoms resolve or an alternate diagnosis is reached. Use of CT scan provides more accurate diagnosis, but there is associated risk of fetal radiation exposure (Castro et al. 2001). MRI is an alternative valuable imaging technique for the evaluation of patients who are clinically suspected to have acute appendicitis. Its use is however complex due to issues such as patient claustrophobia, greater examination time, patient transport, limited availability out of hours, and the cost involved and should therefore be reserved for patients where ultrasound has failed to show the diagnosis and MRI is expected to provide more information.

The definitive treatment for appendicitis is surgery. Both open and laparoscopic approaches are safe and possible in the first trimester of pregnancy. Laparoscopy has additional advantages of more surgical exposure allowing more prompt diagnosis and treatment in addition to reduced recovery time. No adverse effects of pneumoperitoneum (with pressure limited to 10–12 mm of Hg for less than 60 min) have been shown (Moreno-Sanz et al. 2007).

Gastroesophageal Reflux and Peptic Ulcer Disease

Heartburn is very common in pregnancy, and its incidence could be as high as 80 % in certain populations (Richter 2003). The symptoms usually occur toward the end of the first trimester. Incidence of gastroesophageal reflux disease is also equally high and is related to relaxation of lower esophageal sphincter due to the effects of female sex hormones. Serious complications of reflux disease are however rare in pregnancy. Peptic ulcer disease (PUD) is uncommon in pregnancy, and preexisting PUD may improve (Olans and Wolf 1994). The symptoms of the above are water brash, dyspepsia, and upper abdominal/epigastric discomfort. Modifications in lifestyle should be the first-line management of gastroesophageal disease in pregnancy. Antacids are the first-line medical therapy, and H2 receptor antagonists can be used if symptoms persist. Proton pump inhibitors should be used only in women with intractable symptoms or with complicated gastroesophageal disease. Lansoprazole should be the preferred option as it has been reported to be safe in humans in addition to its safety profile in animals (Richter 2003).

Intestinal Obstruction

Acute intestinal obstruction is the second most common surgical emergency in pregnant women and has an incidence of 1 in 1,500 pregnancies (Connolly et al. 1995; Perdue et al. 1992). The classical symptoms of intestinal obstruction are abdominal pain, vomiting, and intractable constipation. A pregnant patient with suspected intestinal obstruction should be assessed and managed in the same way as a nonpregnant patient except that the decisions should be made rather more urgently due to the potential risks to both mother and fetus (Connolly et al. 1995).

Inflammatory Bowel Disease (IBD)

The inflammatory bowel diseases (Crohn's disease and ulcerative colitis) have a peak incidence in the reproductive age group. Pregnancy has little effect on the course of inflammatory bowel disease (Mogadam et al. 1981). Signs and symptoms of inflammatory bowel disease include nausea, vomiting, abdominal discomfort, and constipation. These symptoms are commonly seen in any pregnancy, and it may be difficult to differentiate these from symptoms of IBD. Pyrexia, passage of blood or mucus per rectum, and weight loss in women known to have inflammatory bowel disease suggest acute exacerbations.

Musculoskeletal Pain

In an unpublished retrospective study of 126 patients, virtually every patient was found to have some degree of musculoskeletal pain during the course of pregnancy (Heckman and Sassard 1994). The most common type of pain patients experience during pregnancy is low back pain as a result of lordosis of pregnancy creating mechanical pressure on lower back. Some preexisting conditions like rheumatoid arthritis, ankylosing spondylitis, and scoliosis may be affected by the pregnancy. Rheumatoid arthritis improves in about 70 % of women. The improvement usually begins in early pregnancy and continues until about 6 weeks postpartum (Nelson and Ostensen 1997). On the contrary, the course of spondylarthritis and ankylosing spondylitis remains unchanged during pregnancy.

Idiopathic Pain in Pregnancy

Crampy period like pain in early pregnancy is extremely common complaint, and the majority of women will have an ongoing intrauterine pregnancy, and pain resolves spontaneously. It is most likely due to high-frequency uterine contractions experienced by some women (Tadmor et al. 1995). These patients need careful assessment to evaluate the cause of pain. Munchausen's disease (a factitious disorder with physical manifestations including abdominal pain) can also occur in pregnancy and should be considered in women who have repeated presentations with abdominal pain in pregnancy (Edi-Osagie et al. 1998).

Imaging in Pregnancy

Imaging in pregnant patients requires consideration for safety of the fetus. Pelvic ultrasound remains the first line investigation of choice due to its relatively low cost, safety, and availability. It can provide valuable information about the viability of the pregnancy and its location in the majority of cases and can identify other pelvic pathologies such as fibroids, ovarian cysts, etc. Pain mapping can be also performed with the help of transvaginal probe. While ultrasound is an incredibly useful imaging tool, it is highly dependent on the operator's experience and skill and may be hampered by the factors such as patients' obesity and presence of bowel gas. Further imaging may be required in patients where ultrasound is inconclusive or fails to provide a diagnosis. MRI is considered safe and has been regarded as the imaging technique of choice where further imaging is warranted in the assessment of patient with pain. Most of the studies to evaluate the safety of MRI have not shown any harmful effects. Pregnant women can receive MRI as long as the risk-benefit ratio for the patients warrants that study (Chen et al. 2008). The use of contrast media (gadolinium) with MRI is however controversial. There are no studies assessing its effects on human fetuses, and animal studies have shown harmful effects (Meadows 2001).

A CT scan is an extremely useful investigation in the diagnosis of several non-gynecological causes of pain in early pregnancy such as appendicitis where ultrasound scan or MRI has not been conclusive. There are however concerns with regard to associated risk of teratogenicity and childhood cancers. The developing fetus is most susceptible to teratogenic effects of ionizing radiation between the 2nd and 20th week of pregnancy. These effects include microcephaly, mental retardation, microphthalmia, growth restriction, behavioral defects, and cataracts (De Santis et al. 2005). Risk of teratogenicity is minimal if the dose of radiation does not exceed the minimum threshold. Exact doses below which no teratogenic effects occur are not known, but are estimated to be 0.05–0.15 Gy (5–15 rad) (Berlin 1996). The typical radiation dose from a pelvic CT is variable depending on the gestation and scanning parameters but is likely to be below the threshold level for teratogenicity. The other main concern with the use of CT scanning in pregnancy is the risk of childhood cancers. The relationship between gestational age at the time of radiation exposure and risk of childhood cancer is not clear. According to the Oxford survey of childhood cancers, the risk is higher in the first trimester compared to the second and third trimesters of pregnancy (Gilman et al. 1988). The potential risks may outweigh the benefits in certain situations, and CT can provide life saving diagnostic information (such as in patients where diagnosis of appendicitis cannot be made clinically or by ultrasound and MRI scan). Patients should be counseled about the benefits and risks of radiation exposure. Another concern with the use of CT is the use of iodinated contrast. The contrast crosses the placenta barrier, and thus the fetus is exposed to its effects with maternal administration. There are no studies assessing the risk in the human fetus, and the harmful effects associated with its use, if any, are not known. Intravenous contrast therefore should be used only in exceptional circumstances after counseling of the patient and assessment of the risk-benefit ratio (Patel et al. 2007).

Use of Analgesia in Pregnancy

There is plenty of discussion and evidence regarding the treatment of pain in labor, but there is little addressing the use of analgesics in pregnancy otherwise. Prescribing in pregnancy has to be considered in the context of maternal physiological changes and potential harmful effects on the fetus. Gut motility is reduced in pregnancy which can lead to delayed onset of analgesia but increased absorption of orally administered drugs due to prolonged time in the gut. In early pregnancy, esophageal reflux and emesis can reduce the absorption of drugs. These changes can therefore result in increased or decreased availability of the drugs depending on the dose, form, and route of administration. The risk of harmful effects of drugs is variable depending on the stage of pregnancy. The fetal liver can attain adult enzymatic function between 11 and 18 weeks but during the first trimester is unable to metabolize drugs that involve cytochrome p450 enzymes.

Paracetamol has been used for pain relief in pregnancy for years and is generally considered to be safe. However, there is now evidence to suggest an association between prenatal exposure to paracetamol and asthma in infancy, childhood, and adult life. Further evaluation is required to establish whether this relationship is causal (Eyers et al. 2011).

The use of NSAID (non steroidal anti inflammatory drugs) has a well-documented association with premature closure of ductus arteriosus in late pregnancy. They have been used for the treatment of polyhydramnios in pregnancy but there is a lack of evidence for their effects on a normally developing pregnancy for pain control. There is one study that has shown increased risk of gastroschisis with use of the cyclooxygenase inhibitors aspirin and ibuprofen in pregnancy (Torfs et al. 1996). In exceptional circumstances, NSAIDS can be used in early pregnancy for pain control such as arthritic pain where there are relative contraindications to use of steroids.

For opioid analgesics, there is evidence of safety for use during pregnancy for both acute and chronic pain in pregnancy. Experience of women maintained on methadone buprenorphine and morphine for the treatment of addiction provides additional information to support the safety of this group of drugs (Wunsch et al. 2003).

There is no well-documented evidence for the safety of use of cyclooxygenase-2 inhibitors and neuroleptic membrane stabilizers (gabapentin and topiramate) in pregnancy. A study by Cappon, Cook, and Hurtt (2003) has shown an increased risk of congenital heart anomalies in animals in association with cyclooxygenase-1 inhibitors, but no such effects were seen with exposure to cyclooxygenase-2 inhibitors.

Gabapentin has been shown to be associated with risk of growth restriction and congenital anomalies in animals (Parkash et al. 2008). Limited data of pregnant women who were exposed to gabapentin during pregnancy has not shown any adverse effects. The number of patients in this study was however not sufficient to draw definitive conclusions with regard to the safety of gabapentin during pregnancy (Montouris 2003).

Summary

A large number of conditions can present with pain in early pregnancy, and presentation may be as an acute emergency. A detailed history, clinical examination combined with laboratory investigations, and transvaginal ultrasound provide the diagnosis in majority of cases. Although the differential diagnosis is extremely wide, the old teaching "any pregnant patient presenting with abdominal pain has an ectopic pregnancy unless proved otherwise" is still the key principle while assessing these patients. Laboratory values should be interpreted with caution keeping in mind maternal physiological changes. Similarly, potential effects on the fetus should be taken into consideration before any pharmacotherapy and further radiological imaging. In patients presenting with an acute abdomen, immediate assessment of the patient to reach a diagnosis should be made, and joint assessment by gynecology/surgical team may be needed. In a clinically unwell patient, where diagnosis is unclear, benefits of exploratory surgery may outweigh the risks of delayed diagnosis.

References

Albayram, F., & Hamper, U. M. (2001). Ovarian and adnexal torsion: Spectrum of sonographic findings with pathologic correlation. *Journal of Ultrasound in Medicine, 20*(10), 1083–1089.

Al-Mulhim, A. A. (1996). Acute appendicitis in pregnancy. A review of 52 cases. *International Surgery, 81*(3), 295–297.

Anderson, B., & Nielsen, T. F. (1999). Appendicitis in pregnancy, diagnosis management and complications. *Acta Obstetricia at Gynecologica Scandinavica, 78*(9), 758–762.

Bacq, Y., Zarka, O., Brechot, J., Mariotte, N., Vol, S., Tichet, J., & Weill, J. (1996). Liver function tests in normal pregnancy: A prospective study of 103 patients and 103 matched controls. *Hepatology, 23*(5), 1030–1034.

Berlin, L. (1996). Radiation exposure and the pregnant patient. *American Journal of Roentgenology, 167*(6), 1377–1379.

Breech, L. L., & Hillard, P. J. A. (2005). Adnexal torsion in pediatric and adolescent girls. *Current Opinion in Obstetrics and Gynaecology, 17*(5), 483–489.

Cacciatore, B., Stenman, U. H., & Ylostalo, P. (1990). Diagnosis of ectopic pregnancy by vaginal ultrasonography in combination with a discriminatory serum hCG level of 1000 IU/L (IRP). *British Journal of Obstetrics and Gynaecology, 97*(10), 904–908.

Cantwell, R., Clutton-Brock, T., Cooper, G., Dawson, A., Drife, J., Garrod, D., Harper, A., Hulbert, D., Lucas, S., McClure, J., Millward-Sadler, H., Neilson, J., Nelson-Piercy, C., Norman, J., O'Herlihy, C., Oates, M., Shakespeare, J., de Swiet, M., Williamson, C., Beale, V., Knight, M., Lennox, C., Miller, A., Parmar, D., Rogers, J., & Springett A. (2011). Saving mothers' lives. Reviewing maternal deaths to make motherhood safer: (2006–2008) *British Journal of Obstetrics and Gynaecology 118*(1), 1–203.

Cappon, G. D., Cook, J. C., & Hurtt, M. E. (2003). Relationship between cyclooxygenase 1 and 2 selective inhibitors and fetal development when administered to rats and rabbits during the sensitive periods for heart development and midline closure. *Birth Defects Research. Part B, Developmental and Reproductive Toxicology, 68*(1), 47–56.

Castro, M. A., Shipp, T. D., Castro, E. E., Ouzounian, J., & Rao, P. (2001). The use of helical computed tomography in pregnancy for the diagnosis of acute appendicitis. *American Journal of Obstetric Gynecology, 184*(5), 954–957.

Chen, M. M., Coakley, F. V., Kaimal, A., & Laros, R. K., Jr. (2008). Guidelines for computed tomography and magnetic resonance imaging use during pregnancy and lactation. *Obstetrics and Gynaecology, 112*(2 Part 1), 333–340.

Connolly, M. M., Unit, J. A., & Nora, P. F. (1995). Bowel obstruction in pregnancy. *The Surgical Clinics of North America, 75*(1), 101–113.

De Santis, M., Di Gianantonio, E., Straface, G., Cavaliere, A. F., Caruso, A., Schiavon, F., Berletti, R., & Clementi, M. (2005). Ionizing radiations in pregnancy and teratogenesis: A review of literature. *Reproductive Toxicology, 20*(3), 323–329.

Delzell, J. E., Jr., & Lefevre, M. L. (2000). Urinary tract infections during pregnancy. *American Family Physician, 61*(3), 713–721.

Edi-Osagie, E. C. O., Hopkins, R. E., & Edi-Osagie, N. E. (1998). Munchhausen's syndrome in obstetrics and gynecology: A review. *Obstetrical and Gynecological Survey, 53*(1), 45–49.

Exacoustos, C., & Rosati, P. (1993). Ultrasound diagnosis of uterine myomas and complications in pregnancy. *Obstetrics and Gynecology, 82*(1), 97–101.

Eyers, S., Weatherall, M., Jefferies, S., & Beasley, R. (2011). Paracetamol in pregnancy and the risk of wheezing in offspring: A systematic review and meta-analysis. *Clinical and Experimental Allergy, 41*(4), 482–489.

Farquhar, C. M. (2005). Ectopic pregnancy. *Lancet, 366*(9485), 583–591.

Gilman, E. A., Kneale, G. W., Knox, E. G., & Stewart, A. M. (1988). Pregnancy X-rays and childhood cancers: Effects of exposure age and radiation dose. *Journal of Radiological Protection, 8*, 3–8.

Gilstrap, L. C., 3rd, Cunningham, F. G., & Whalley, P. J. (1981). Acute pyelonephritis in pregnancy, an anterospective study. *Obstetrics and Gynecology, 57*(4), 409–413.

Giriing, J. C., Dow, E., & Smith, J. H. (1997). Liver function tests in pre-eclampsia: Importance of comparison with reference range derived for normal pregnancy. *British Journal of Obstetrics and Gynaecology, 104*(2), 246–250.

Harris, R. E., & Gilstrap, L. C., 3rd. (1981). Cystitis during pregnancy: A distinct clinical entity. *Obstetrics and Gynecology, 57*(5), 578–580.

Heckman, J. D., & Sassard, R. (1994). Musculoskeletal considerations in pregnancy. *Journal of Bone and Joint Surgery, 76*, 1720–1730.

Hee, P., & Viktrup, L. (1999). The diagnosis of appendicitis during pregnancy and maternal and fetal outcomes after appendectomy. *International Journal of Gynecology & Obstetrics, 65*(2), 129–135.

Hess, L. W., Peaceman, A., O'Brien, W. F., Winkel, C. A., Cruikshank, D. P., & Morrison, J. C. (1988). Adnexal mass occurring with intrauterine pregnancy: Report of 54 patients requiring laparotomy for definitive management. *American Journal of Obstetrics and Gynecology, 158*(5), 1029–1034.

Houry, D., & Abbott, J. T. (2001). Ovarian Torsion, a fifteen-year review. *Annals of Emergency Medicine, 38*(2), 156–159.

Johnson, T. R., Jr., & Woodruff, J. D. (1986). Surgical emergencies of uterine adnexae during pregnancy. *International Journal of Gynecology & Obstetrics, 24*(5), 331–335.

Katz, V. L., Dotters, D. J., & Droegemeuller, W. (1989). Complications of uterine leiomyomas in pregnancy. *Obstetrics and Gynecology, 73*(4), 593–596.

Kirk, E., Condous, G., & Bourne, T. (2006). The non-surgical management of ectopic pregnancy. *Ultrasound in Obstetrics & Gynecology, 27*(1), 91–100.

Luciano, A. A., Roy, J., & Solima, E. (2001). Ectopic pregnancy- from surgical emergency to medical management. *Annals of the New York Academy of Sciences, 943*, 235–254.

Marshall, L. M., Spiegelman, D., Barbieri, R. L., Goldman, M. B., Manson, J. E., Colditz, G. A., Willet, W. C., & Hunter, D. J. (1997). Variation in the incidence of uterine leiomyoma among premenopausal women by age and race. *Obstetrics and Gynecology, 90*(6), 967–973.

Maslovitz, S., Gutman, G., Lessing, J. B., Kupferminck, M. J., & Gamzu, R. (2003). The significance of clinical signs and blood indices for the diagnosis of appendicitis during pregnancy. *Gynecologic and Obstetric Investigations, 56*(4), 188–191.

McGowan, L. (1983). Surgical diseases of the ovary in pregnancy. *Clinical Obstetrics and Gynecology, 26*(4), 843–852.

Meadows, M. (2001). Pregnancy and the drug dilemma. *FDA Consumer Magazine, 35*(3), 16–20.

Millar, L. K., & Cox, S. M. (1997). Urinary tract infections complicating pregnancy. *Infectious Disease Clinics of North America, 11*(1), 13–26.

Mogadam, M., Korelitz, B. I., Ahmed, S. W., Dobbins, W. O., 3rd, & Baiocco, P. J. (1981). The course of inflammatory bowel disease during pregnancy and postpartum. *American Journal of Gastroenterology, 75*(4), 265–9.

Montouris, G. (2003). Gabapentin exposure in human pregnancy: Results from the gabapentin pregnancy registry. *Epilepsy & Behavior, 4*(3), 310–317.

Moreno-Sanz, C., Pascual-Pedreno, A., Picazo-Yeste, J. S., & Seoane-Gonzalez, J. B. (2007). Laparoscopic appendectomy during pregnancy: Between personal experience and scientific evidence. *Journal of the American College of Surgeons, 205*(1), 37–42.

Mourad, J., Elliot, J. P., Erickson, L., & Lisboa, L. (2000). Appendicitis in pregnancy: New information that contradicts long- held clinical beliefs. *American Journal of Obstetrics and Gynecology, 82*(5), 1027–1029.

Nelson, J. L., & Ostensen, M. (1997). Pregnancy and rheumatoid arthritis. *Rheumatology Disease Clinics of North America, 23*(1), 195–212.

Oelsner, G., & Shashar, D. (2006). Adnexal torsion. *Clinical Obstetrics and Gynecology, 49*(3), 459–463.

Olans, L. B., & Wolf, J. L. (1994). Gastroesophageal reflux disease in pregnancy. *Gastrointestinal Endoscopic Clinics of North America, 4*(4), 699–712.

Parkash, P. L. V., Rai, R., Pai, M. M., Yadav, S. K., Madhyastha, S., Goel, R. K., Singh, G., & Nasar, M. A. (2008). Teratogenic effects of the anticonvulsant Gabapentin in mice. *Singapore Medical Journal, 49*(1), 47–53.

Patel, S. J., Reede, D. L., Katz, D. S., Subramaniam, R., & Amorosa, J. K. (2007). Imaging the pregnant patient for non obstetric conditions: Algorithms and radiation dose considerations. *RadioGraphics, 27*, 1705–1722.

Perdue, P. W., Johnson, H. W., Jr., & Stafford, P. W. (1992). Intestinal obstruction complicating pregnancy. *The American Journal of Surgery, 164*(4), 384–388.

Picklesimer, A. H., Jared, H. L., Moss, K., Offenbacher, S., Beck, J. D., & Boggess, K. A. (2008). Racial differences in C-reactive protein levels during normal pregnancy. *American Journal of Obstetrics and Gynecology, 199*(5), 523. e.1-6.

Rahamim, E., Piper, I., Golan, A., & Sadan, O. (2008). Total and differential leukocyte count percentiles in normal pregnancy. *European Journal of Obstetrics, Gynecology, and Reproductive Biology, 136*(1), 16–19.

Richter, J. E. (2003). Gastro esophageal reflux disease during pregnancy. *Gastroenterology Clinics of North America, 32*(1), 235–261.

Rizzuto, M. I., Oliver, R., & Odejinmi, F. (2008). Laparoscopic management of ectopic pregnancy in the presence of a significant haemoperitoneum. *Archives of Gynecology and Obstetrics, 277*(5), 433–436.

Sacks, G. P., Seyani, L., Lavery, S., & Trew, G. (2004). Maternal C- reactive protein levels are raised at 4 weeks gestation. *Human Reproduction, 19*(4), 1025–1030.

Seeber, B. E., & Barnhart, K. T. (2006). Suspected ectopic pregnancy. *Obstetrics and Gynecology, 107*(2 Part 1), 399–413.

Shalev, E., Yarom, I., Bustan, M., & Ben-Shlomo, I. (1998). Transvaginal sonography as the ultimate diagnostic tool for the management of ectopic pregnancy: experience with 840 cases. *Fertility and Sterility, 69*(1), 62–65.

Solima, E., & Luciano, A. A. (1997). Ectopic pregnancy. *Annals of the New York Academy of Sciences, 828*, 300–315.

Tadmor, O. P., Rabinowitz, R., & Diamant, Y. Z. (1995). Ultrasonic demonstration of local myometrial thickening in early intrauterine pregnancy. *Ultrasound in Obstetrics & Gynecology, 5*(1), 44–46.

Tay, J. I., Moore, J., & Walker, J. J. (2000). Ectopic pregnancy. *BMJ, 320*(7239), 916–919.

Torfs, C. P., Katz, E. A., Bateson, T. F., Lam, P. K., & Curry, C. J. R. (1996). Maternal medications and environmental exposures as risk factors for gastroschisis. *Teratology, 54*(2), 84–92.

Tracey, M., & Fletcher, H. S. (2000). Appendicitis in pregnancy. *The American Journal of Surgery, 66*(6), 555–559. discussion 559–560.

Williams, R., & Shaw, J. (2007). Ultrasound scanning in the diagnosis of acute appendicitis in pregnancy. *Emergency Medicine Journal, 24*(5), 359–360.

Wunsch, M. J., Standard, V., & Sidney, H. S. (2003). Treatment of pain in pregnancy. *The Clinical Journal of Pain, 19*(3), 148–155.

Pain in Ganglionopathy

▶ Peripheral Neuropathic Pain

Pain in HIV/AIDS

▶ Cancer Pain and Pain in HIV/AIDS

Pain in Human Immunodeficiency Virus Infection and Acquired Immune Deficiency Syndrome

Peter Selwyn
Albert Einstein College of Medicine, Bronx, NY, USA

Synonyms

AIDS and pain; HIV and pain; HIV-associated neuropathy; Neuropathy due to HAART; Zidovudine (AZT)-induced headache

Definition

Neuropathic and nociceptive pain is a common result of Human Immunodeficiency Virus (HIV) infection and its treatment. Despite the fact that pain is associated with psychological and functional morbidity, it remains undertreated in patients with HIV/Acquired Immune Deficiency Syndrome (AIDS) (Breitbart et al. 1996). Both patient and provider barriers contribute to this situation.

Etiologies of Pain

As with cancer, pain in HIV/AIDS may be related to the disease itself, therapy, or diagnostic procedures related to the disease, or may be independent of the disease or disease-related interventions. Also, like cancer, the severity, distribution, and effect on quality of life all increase as the disease progresses (Singer et al. 1999). HIV infection is a progressive illness that results in severe immunodeficiency (declining CD4+ T-lymphocyte count), characterized by opportunistic infections and malignancies. The likelihood of developing particular pain problems is directly related to the CD4+ count of the patient, which reflects the likelihood of susceptibility to infections and complications of the disease (Fig. 1).

Prevalence

Prevalence data on pain in patients with HIV/AIDS from the United States and other developed countries were primarily collected before the era (1996–present) of highly active antiretroviral therapy (HAART). Different populations were studied at varying stages of disease, and different methodologies of assessment were used. In addition, the prevalence data for children and adolescents with HIV/AIDS are sparse at best. As a result, it is difficult to interpret how well the data reflects current reality in either developed countries, where HAART is now widely available, or developing countries, where HAART has been largely inaccessible to date. Data on peripheral neuropathy and encephalopathy illustrate how different factors may be related to risk of pain and/or other HIV-related sequelae, in ways that are not always predictable. The incidence of peripheral neuropathy is increased in people with long-term infection, those with decreased CD4+ counts, and those with increased viral load. In countries where HAART is widely used, there is not only a decreasing incidence of both HIV-associated neurological disease and central nervous system (CNS) opportunistic infections, but also a trend toward increasing incidences of drug-induced peripheral neuropathy (Sacktor 2002). In contrast, a recent study reported that 14.9 % of inpatients in Hospital Kuala Lumpur, who were seropositive for toxoplasmosis, had active toxoplasmic encephalitis (Nissapatorn et al. 2003). Additionally, there is evidence that pain prevalence differs in injection drug users (IDUs) and non-IDUs (Martin et al. 1999). A Swedish survey of outpatients infected with HIV found that in non-IDUs, there was a strong correlation between disease state, CD4+ levels, mortality, and pain. However, there was no significant correlation between these disease parameters and pain in IDUs. In those with symptomatic HIV disease, there was no difference between IDUs and non-IDUs in the prevalence of pain or number of pain sites. However, among those with asymptomatic HIV infection, IDUs reported both a significantly higher prevalence of pain and a significantly higher number of concurrent pain

Pain in Human Immunodeficiency Virus Infection and Acquired Immune Deficiency Syndrome, Fig. 1 The chronology of HIV-induced disease and related painful conditions (Adapted from Stewart et al. 1997)

sites. If these data are borne out in further studies, they will provide clues to which populations are at risk for pain, and at which stage of illness.

Due to all of these factors, estimates of overall pain prevalence in HIV/AIDS have ranged widely, from 30 % (Larue et al. 1997) to 88 % (Frich and Borgbjerg 2000), depending on the population studied and the methodology employed. In one influential study, about 45 % of pain was said to be directly related to the disease, about 15–30 % was due to therapies or diagnostic procedures, and about 25–40 % was attributed to unrelated conditions and treatments (Hewitt et al. 1997). Seventy-one percent of patients had somatic pain, 29 % had visceral pain, 46 % had neuropathic pain, and 46 % experienced headache. These data must be interpreted as cautiously as all other prevalence data. A wide array of pain syndromes has been documented, in every major region and system of the body (O'Neill and Sherrard 1996).

Characteristics

Types of Pain in HIV/AIDS

HIV/AIDS patients most commonly present with four major types of pain: headache, musculoskeletal pain, visceral pain, and neuropathy. Secondary headache is generally due to infection

(e.g., cryptococcal meningitis) or neoplasm (e.g., central nervous system lymphoma), though zidovudine (AZT) is known to induce headache in some patients. Interestingly, in one study, primary migraine decreased in frequency over time, while primary tension-type headache increased in frequency, independent of antiretroviral treatment or CD4+ count (Evers et al. 2000).

Infectious agents, neoplasms, pharmaceuticals, diagnostic procedures, and unrelated conditions may all cause pain in patients with HIV/AIDS. The most common pain-producing infectious agents are the HIV virus itself, cytomegalovirus (CMV), hepatitis B and C, candidiasis, toxoplasmosis, *mycobacterium avium intracellulare* (MAI), cryptococcosis, and cryptosporidiosis. Each of these diseases can cause pain in a variety of organ systems (see Table 1). The most common pain-producing neoplasms are lymphoma and Kaposi's sarcoma (KS). By direct invasion or metastasis, these neoplasms may cause pain almost anywhere in the body (Table 1). Some antivirals and antiretrovirals used to treat the HIV virus, as well as isoniazid (INH) used in the treatment of tuberculosis, may produce painful neuropathy (Table 2). Radiation and surgery used to treat neoplasms may also cause pain. Preexisting conditions such as diabetic neuropathy, osteoarthritis, and degenerative disk disease add to the possible etiologies.

Pain in Human Immunodeficiency Virus Infection and Acquired Immune Deficiency Syndrome, Table 1 Pain-producing infections and neoplasms

Pain site/type	Oropharynx	Esophagus	Abdomen	HA	Chest	Biliary	Cutaneous	Anorectal	Peripheral neuropathy	Arthritis/ arthralgia	Myopathy/ polymyositis
Infectious etiology											
HBV			X							X	
HSV	X	X					X	X			
HCV	X						X			X	
CMV	X	X	X			X			X		
HIV	X		X						X	X	X
EBV	X	X	X								
Toxoplasmosis			X								
Fungi, e.g., Candida	X	X	X								
Microsporidiosis			X			X					X
MAI			X			X					
Cryptococcus		X									
Cryptosporidiosis			X			X					
VZV							X				
Salmonella, Shigella, Campylobacter			X								
Neoplastic etiology											
Kaposi's sarcoma	X	X	X								
Lymphoma	X	X	X								

HBV Hepatitis B virus, *HPV* Human papilloma virus, *HCV* Hepatitis C virus, *HSV* Herpes simplex virus, *CMV* Cytomegalovirus, *HIV* Human immunodeficiency virus, *EBV* Epstein Barr virus, *MAI* Mycobacterium avium intracellulare, *VZV* Varicella zoster virus

Pain in Human Immunodeficiency Virus Infection and Acquired Immune Deficiency Syndrome, Table 2 HIV-related therapeutic agents that may cause pain[a]

Class/agent	Headache	Peripheral neuropathy	Pancreatitis	Abdominal pain	Hepatitis	Nephrolithiasis
NRTIs						
AZT	X			X	X	
3TC			X		X	
DDI		X	X	X		
DDC		X	X			
D4T		X	X			
Abacavir				X		
NTRTI						
Tenofovir				X		
NNRTI						
Nevirapine					X	
Efavirenz					X	
Delavirdine					X	
PI						
Ritonavir				X	X	
Indinavir				X		X
Saquinavir				X		
Atazanavir					X	
ANTIMYCOBACTERIAL						
Isoniazid		X				
ANTI-NEO-PLASTICS						
Vincristine		X				

NRTI nucleoside reverse transcriptase inhibitor, *AZT* zidovudine, *3TC* lamivudine, *DDI* didanosine, *DDC* zalcitabine, *D4T* stavudine, *NTRTI* nucleotide reverse transcriptase inhibitor, *NNRTI* non-nucleoside reverse transcriptase inhibitor, *PI* protease inhibitor

[a] "Xs" indicate most commonly reported pain syndromes in association with different HIV-related therapeutic agents, although this may be seen with lesser frequency with agents not so indicated here

Symptoms may also occur with other protease inhibitors, but are most commonly reported for these PIs

Neuropathic Pain Syndromes

Symptomatic neuropathies are present in about 10 % of patients with HIV/AIDS, although pathologic evidence is found in almost all end-stage patients. The six major types of HIV-associated neuropathy are distal sensory neuropathy (DSP), toxic neuropathy, acute and chronic inflammatory demyelinating polyradiculoneuropathy (AIDP and CIDP), progressive polyradiculopathy (PP), mononeuropathy multiplex (MM), and autonomic neuropathy (AN). Table 3 gives an overview of their clinical features (Verma 2001).

DSP is perhaps the most prevalent of the peripheral neuropathies. Treatment of DSP has been disappointing to date. Gabapentin and opioids may be useful in patients with less advanced disease, but generally fail in patients with end-stage disease, when this condition is most prevalent. In a 48-week open-label study of recombinant nerve growth factor for 200 patients with DSP, there was a "significant trend" toward reduction in subjective report of pain. Clinically significant differences in function or Quantitative Sensory Testing, however, were absent (Schiffito et al. 2001). Antiretrovirals play a paradoxical role in DSP. On the one hand, decreasing viral load by directly treating HIV can alleviate peripheral neuropathic symptoms. On the other hand, zalcitabine (ddC), didanosine (ddI), and stavudine (d4T) all cause a dose-dependent DSP in a significant minority of patients. This appears to be related to the spectrum of mitochondrial

Pain in Human Immunodeficiency Virus Infection and Acquired Immune Deficiency Syndrome, Table 3 HIV-associated neuropathies

Type	Stage of disease	Course/frequency	Symptoms
DSP	Mid to late	Indolent, protracted Common	Distal pain, paresthesias, and numbness especially in feet
Toxic	All	Common	Same
IDP: Acute	Any stage	Rare	Global limb weakness
Chronic	Mid to late	Rare	Global limb weakness
PP	Advanced	Rare, may show rapid progression	Cauda equina syndrome. Saddle anesthesia
MM	Early, late	Infrequent	Self-limited multifocal motor sensory, or mixed somatic or cranial neuropathy
AN	All	Common (subclinical)	Varied

DSP distal sensory polyneuropathy, *IDP* inflammatory demyelinating polyneuropathy, *PP* progressive polyradiculopathy, *MM* mononeuropathy multiplex, *AN* autonomic neuropathy

toxicity induced by these agents. Drug withdrawal does not always lead to symptom resolution, so early dose reduction or substitution of another antiretroviral agent is recommended.

Immunomodulatory therapy is the mainstay of treatment for IDP. PP is usually caused by CMV infection: Irreversible neurological deficits occur in the absence of rapid diagnosis and treatment with ganciclovir, foscarnet, cidofovir, or valganciclovir. A 4-week course is generally sufficient, though maintenance therapy may need to be taken if symptoms recur. MM is usually self-limited, and treatment is supportive. AN in its subclinical form is common. Symptomatic AN usually occurs in late-stage disease. Symptoms depend on which components and levels of the autonomic system are involved. Additional HIV-induced neurological diseases include vacuolar myelopathy, which may involve both sensory and motor loss, and may be associated with painful muscle spasms.

Barriers to Treatment of Pain

In developing countries where resources such as HAART, antibiotics, and other agents are scarce or unavailable, lack of resources is the primary barrier to treatment. In a 1999 study in the United States, clinicians identified lack of knowledge, reluctance to prescribe opioids, inadequate access to pain specialists and issues related to drug addiction and abuse as the major barriers to providing better pain management to patients

with HIV/AIDS (Breitbart et al. 1999). Additionally, the paucity of clinical trials evaluating the efficacy of pain treatment in those with HIV/AIDS negatively impacts on clinician's willingness to prescribe medication.

There are limited studies of patient barriers to pain management. However, a 1998 report identified non-Caucasian racial status, fewer years of education, greater numbers of physical symptoms, and greater psychological distress as independent factors associated with barriers to adequate pain management in AIDS (Breitbart et al. 1998). Racial disparities in access to care, along with comorbid medical and psychiatric disorders, further complicate the management of pain in HIV/AIDS. Of note, people living in predominantly nonwhite neighborhoods in the United States have difficulties filling prescriptions for opioids (Morrison et al. 2000).

Conclusions

Despite the problems of determining actual prevalence, it is clear that pain is a common symptom in AIDS patients. Although there is often an identifiable etiology and treatment strategies exist, pain remains undertreated, and quality of life is adversely affected by lack of attention to this burdensome symptom. Issues of addiction and treating pain in patients with active or remote substance abuse need to be carefully addressed. As the goals of care shift from arresting the course of the disease to chronic disease management and palliation, the primary importance of pain management increases. The better a patient's pain is controlled, the more likely it is that he or she will adhere to other HIV-specific therapy and enjoy the highest quality of life possible.

References

Breitbart, W., Kaim, M., & Rosenfeld, B. (1999). Clinicians' perceptions of barriers to pain management in AIDS. *Journal of Pain and Symptom Management, 18*, 203–212.

Breitbart, W., Rosenfeld, B. D., Passik, S. D., et al. (1996). The undertreatment of pain in ambulatory AIDS patients. *Pain, 65*, 243–249.

Breitbart, W., Passik, S., McDonald, M. V., et al. (1998). Patient-related barriers to pain management in ambulatory AIDS patients. *Pain, 76*, 9–16.

Evers, S., Wibbeke, B., Reichelt, D., et al. (2000). The impact of HIV infection on primary headache. Unexpected findings from retrospective, cross-sectional and prospective analyses. *Pain, 85*, 191–200.

Frich, L. M., & Borgbjerg, F. M. (2000). Pain and pain treatment in AIDS patients: A longitudinal study. *Journal of Pain and Symptom Management, 19*, 339–347.

Hewitt, D., McDonald, M., Portenoy, R., et al. (1997). Pain syndromes and etiologies in ambulatory AIDS patients. *Pain, 70*, 117–123.

Larue, F., Fontaine, A., & Colleau, S. M. (1997). Underestimation and undertreatment of pain in HIV disease: Multicenter study. *BMJ, 4*, 23–28.

Martin, C., Pehrson, P., Osterberg, A., et al. (1999). Pain in ambulatory HIV-infected patients with and without intravenous drug use. *European Journal of Pain, 3*, 157–164.

Morrison, R. S., Wallenstein, S., Natale, D. K., et al. (2000). We don't carry that. Failure of pharmacies in predominantly non-white neighborhoods to stock opioids. *The New England Journal of Medicine, 342*, 1023–1036.

Nissapatorn, V., Lee, C. K., & Khairul, A. A. (2003). Seroprevalence of toxoplasmosis among AIDS patients in Hospital Kuala Lumpur, 2001. *Singapore Medical Journal, 44*, 194–196.

O'Neill, W. M., & Sherrard, J. S. (1996). Pain in human immunodeficiency virus disease. A review. *Pain, 68*, 323–328.

Sacktor, N. (2002). The epidemiology of human immunodeficiency virus-associated neurological disease in the era of highly active antiretroviral therapy. *Journal of Neurovirology, 8S2*, 115–121.

Schiffito, G., Yainnoutsos, C., Simpson, D. M., et al. (2001). Long-term treatment with recombinant nerve growth factor for HIV-associated sensory neuropathy. *Neurology, 57*, 1313–1316.

Singer, E. J., Zorilla, C., Fahy-Chandon, B., et al. (1999). Painful symptoms reported by ambulatory HIV infected men in a longitudinal study. *Pain, 54*, 15–19.

Verma, A. (2001). Epidemiology and clinical features of HIV-1 associated neuropathies. *Journal of the Peripheral Nervous System, 6*, 8–13.

Pain in Humans, EEG Documentation

Andrew C. N. Chen
Human Brain Mapping and Cortical Imaging Laboratory, Aalborg University, Aalborg, Denmark

Synonyms

EEG (electroencephalography); MEG

Definition

EEG (electroencephalography) and its magnetic counter part of MEG (magnetoencephalography) is a web of spontaneous electrical oscillation in the cortex and the subcortical brain. A review is made of how it can be studied for understanding pain in the human brain.

Neurophysiology of EEG

Physiology

EEG has been classically decomposed into broad bands (δ, θ, α, β, γ) and some sub-bands for α or β. The exact physiology for the simultaneous or separate genesis of these fundamental waves is not fully understood. Nevertheless, the physiology of EEG, and its relation to sleep-awake regulation, has been well documented. The cortical biophysics in EEG may encompass both focal transient events with the interaction of global wave oscillations (da Silva 1999).

Functions

In addition to arousal and attention, increased interest is given to the functional aspects of EEG in memory, emotion, and cognitive processing. All of these functions are intimately involved in pain processing of the brain.

Pain and Consciousness

Pain and Awareness in Multifactors of EEG Measurement

Pain is primarily a somatic sensation of aversive quality in conscious awareness, as such; it may elicit, and can in turn be influenced by, emotional reactions (anxiety, fear, anger, depression). Both behaviors (bodily reflex, gesture, facial expression, motor act) and cognitive contingency (expectation, interpretation, copying) can also affect pain processing in the brain. The scientific and clinical interest lies on the neural representation of human pain in its sensory, perceptual, emotional, and cognitive dimensions (if they are really separable), using the advanced EEG/MEG mapping and advent of PET/fMRI neuroimaging (Chen 2001).

EEG Mapping

EEG is sensitive to various states from unconsciousness, sleep, and drug sedation to awareness and wakefulness. For drug-induced EEG changes, the "bispectral index," using only frontal EEG measures, has gained special attention and wide practice in neuro-anesthesia from wakefulness to sedation and from sedation to conscious awareness. Based on these measures, a field theory in neurophysics of consciousness has been proposed (John 2002). To clearly capture the brain states, recent work has been improved from clinical 19-ch EEG to high-resolution EEG of 128-ch, to increase spatial resolution and modeling of its underlying anatomical substrates. EEG, but less so in MEG, is highly dependent on the reference sites recorded as well as being sensitive to distortion by brain, CSF, skull, and scalp resistance. EEG mapping is influenced by the selected reference sites or averaged reference. Large individual differences in EEG activation should be considered, e.g., that some 10 % of healthy subjects have little or no α activity and one subject may exhibit 20× fold the power of the other. Also, the majority of EEG records are often contaminated with various sorts of artifacts. Thus, proper EEG acquisition, signal processing, data transformation, and statistical methods used can greatly influence the accuracy and reliability of the

outcome of studies. Three major parameters in EEG measurements are band power, coherence, and current source density.

Characteristics

EEG in Experimental Pain

Readers are reminded of the complexity of human pain in the brain and the multifactors (technical, physical, physiological, and psychological conditions) determining the EEG/MEG measurements. To control these complexities, experimental human pain has been divided into three major categories of nociceptors: thermal, chemical, and mechanical stimuli. All of these stimuli can induce phasic changes in EEG, as well as tonic effects of EEG.

Phasic Pain

The microstate of EEG changes in ms range can be examined by frequency-domain analysis of averaged evoked potentials, such as the band-power density (Chen and Rappelsberger 1994), coherence of laser evoked potentials, wavelet analysis of galvanic evoked potentials (Chen and Herrmann 2001), and cortical imaging (Babiloni et al. 2002) in somatosensory evoked potentials to median nerve stimulation. The meso-state of event-related synchronization (ERS) and event-related desynchronization (ERD) in 250 ms range can also be calculated by EEG power densities. The results indicated (a) increased contralateral somatosensory and vertex coherence linkage and a web of centro-parietal and frontal instantaneous coherence in response to laser pain and (b) temporospatial dynamics, initiated at the contralateral hand area, then to the posterior parietal area, and terminated in the frontal area, by low to high gamma activation in response to painful median nerve stimulation.

Tonic Pain

To simulate the natural pain experience at the macro-state at a minute by minute level, methods of ► thermal stimulation often by cold-pressor test, chemical stimulation by capsaicin or

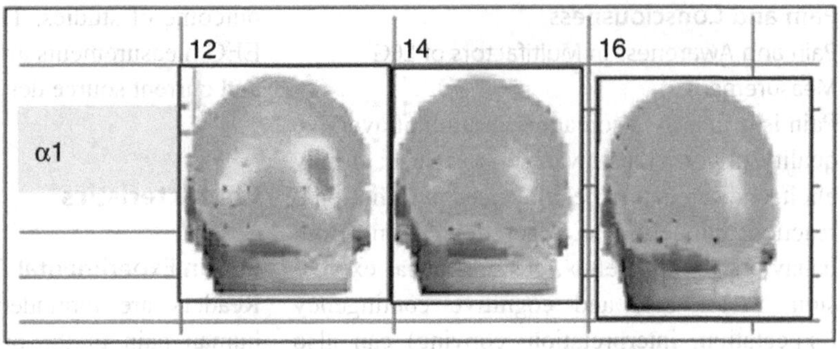

**Pain in Humans, EEG
Documentation,
Fig. 1** Effect of tonic pain
on brain topographic
mapping

hypertonic saline irritants, and ▶ mechanical stimulation by cuff-pressure test have been reported. These works were reviewed in a classic paper (Chen 1993) and have now been extended into spatiotemporal dynamics (Chen 2002; Chang et al. 2004). For tonic pain and EEG, the results showed several general agreements (for review see Chang et al. 2002): (a) increase of δ-power, (b) increase in θ-power, (c) reduction of α-power, and (d) increase of β-power.

Topographic Spatial Effects

The effect of tonic pain (cuff-pressure test) on brain topographic mapping is shown in Fig. 1. The EEG of α-power can be seen to be systematically reduced when between non-pain (VAS intensity-2, I2) and pain (I4, slight pain and I6, moderate pain). The posterior parietal alpha activity has been modeled as sensorimotor activity of hand area and partly reflects reduction of parietal attention processing, largely in the right hemisphere, due to pain.

EEG in Clinical Pain

Adding pathophysiology of the diseases and psychological alterations in patients, measurement of clinical pain by EEG can be very daunting. An early review (Chen 1993) reported (a) focal or generalized bursts of spikes, slow waves, or both in 61 % of the cases and (b) nonspecific changes, diffused delta-theta without bursts in 35 % of the cases, through accumulated literature on headaches. The predominance of EEG abnormality differs with the type of headache (63.5 % in

posttraumatic headache, 46.9 % in common migraine, 47.0 % in mixed headache, 45.5 % in childhood headache, 36.0 % in classic migraine, 17.5 % in cluster migraine, and 12.0 % in muscular headache), as compared to 10–15 % observed in non-headache controls. These EEG changes may be due to pathophysiology rather than pain in the patients per se. EEG changes in pain patients have been reported in cognitive disorders, headache, fibromyalgesia, and chronic pain syndrome.

Cognitive Disorders

This occurs when attention memory mechanisms in the brain are deteriorated, such as in Alzheimer patients in pain stimulation (Benedetti et al. 1999). Patients with cognitive impairment who are insensitive to pain showed high δ peak and low α peak to galvanic and ischemic pain, indicating altered cognitive and affective factors. Likewise, patients of a psychiatric nature, with borderline personality disorders, also exhibited altered EEG activities between self-injury with reported pain experience, from those without reported pain experience in total EEG power. Further, there was a significant correlation between the theta activity, negatively with pain rating and positively with dissociative score (Russ et al. 1999).

Headache

No specific EEG signals in migraine or headache pain have been characterized. One report suggested a general state of altered neuronal excitability, while another showed a significant decrement of alpha rhythm power at the posterior areas of the brain during IPS at 20 c/s. The diagnostic use of EEG was advocated in children suffering from migraine, with aura during the ictal phase, but not the interictal phase. Contradictory reports from a literature review did not support the effectiveness of EEG in headache. The quality of EEG data gathering, signal processing, study design, and data analyses of past research in this field may hamper a proper evaluation of EEG and headache pain.

Fibromyalgesia

EEG has been reported (Moldofsky 2002) to reflect disturbances in circadian sleep-wake rhythms: phasic (alpha-delta)/tonic alpha non-REM sleep disorders, the periodic K alpha cycling alternating pattern disorder in fibromyalgia patients. However, no drug has restored these EEG effects. The EEG characteristics include an increased EEG power density in the higher frequency band and a reduced EEG power density in the lower frequency bands, as well as α K-complex as indicators of fragmented sleep. The fibromyalgia symptoms may relate to a non-restorative sleep disorder associated with the α-EEG sleep anomalies. However, the specific etiology of sleep disorder in EEG, such as "alpha-delta NREM sleep abnormality" in fibromyalgia and chronic fatigue syndrome, has been challenged.

Chronic Pain

It is difficult to characterize EEG in chronic pain, largely due to the lack of specificity in defining chronic pain, as it is often associated with depression, disability, and dysfunctional illness behaviors. Nevertheless, the following reports may be consulted: EEG and injury model of neuropathic pain, EEG and chronic pain, chronic low back pain, and irritable bowel syndrome.

New Perspectives of EEG Measurement for Human Pain

Basic Sciences

EEG inherently has a large variance in genetic basis, contributing to individual characteristics of EEG patterns and may relate to individual sensitivity or tolerance to pain (Chen et al. 1989). The increased availability of the new generation of MEG (Kakigi et al. 2000) can be used more extensively, but there are hardly any reports of spontaneous MEG studies so far. There is also a new focus on how EEG can be related to other PET/fMRI neuroimaging modalities. Current EEG technology and software allow high-resolution scalp mapping (Chen 2002), cortical imaging, and current source imaging (Fernandez-Bouzas et al. 2001) to define structure-function relations. Recently, exciting advances in frequency-domain source analysis of ongoing EEG has contributed to further understanding of brain dynamics in EEG. In principle, pain is a mental experience based on neural events and, as such, should be measurable by neurophysiological activation in the brain.

Clinical Applications

In the near future, it will be questioned if advanced human brain mapping of EEG parameters and dynamics will provide (a) differential diagnosis of pain conditions and (b) predictive values in prognosis of pain changes or modulation efficacies.

Pain Treatment Efficacy and Mechanisms

This is done possibly to gain new understanding of EEG in altered states of consciousness, with focus on EEG patterns of some basic treatment efficacy and mechanisms of pain control in; (a) physical stimulation, such as transcranial magnetic stimulation or deep brain stimulation, transcutaneous electrical stimulation, or acupuncture, (b) psychological pain control, such as self-efficacy, expectation, placebo, relaxation, imagery, hypnosis, religious belief, and (c) various major categories of pharmacological treatments, such as NSAIDs, opiates, or tricyclic antidepressants, etc.

In conclusion, this chapter indicates some, but not enough, reporting on EEG and human pain to date. Advances in EEG analyses will be likely to open a new era of research, and understanding of the natural state of spontaneous EEG dynamics, in acute and chronic pain in humans in the future.

References

Babiloni, C., Babiloni, F., Carducci, F., Cincotti, F., Rosciarelli, F., Arendt-Nielsen, L., Chen, A. C., & Rossini, P. M. (2002). Human brain oscillatory activity phase-locked to painful electrical stimulations: a multi-channel EEG study. *Human Brain Mapping, 15,* 112–123.

Benedetti, F., Vighetti, S., Ricco, C., et al. (1999). Pain threshold and tolerance in Alzheimer's disease. *Pain, 80,* 377–382.

Chang, P., Arendt-Nielsen, L., Graven-Nielsen, T., & Chen, A. C. N. (2002). Differential cerebral responses to aversive auditory arousal versus muscle pain: specific EEG patterns are associated with human pain processing. *Experimental Brain Research, 147,* 387–393. http://link.springer.com/content/pdf/10.1007/s00221–002–1272–9.pdf. doi:10.1007/s00221-002-1272-9.

Chang, P. F., Arendt-Nielsen, L., Graven-Nielsen, T., et al. (2004). Comparative EEG activation to skin pain and muscle pain induced by capsaicin injection. *International Journal of Psychophysiology, 51,* 117–126.

Chen, A. C. (1993). Human brain measures of clinical pain. A review: I. Topographic mappings. *Pain, 54,* 115–132.

Chen, A. C. (2001). New perspectives in EEG/MEG brain mapping and PET/fMRI neuroimaging of human pain. *International Journal of Psychophysiology, 42,* 147–159.

Chen, A. C. (2002). EEG/MEG brain mapping of human pain: Recent advances. *Recent Advances in Human Brain Mapping, 1232,* 5–16.

Chen, A. C., Dworkin, S. F., Hang, J., et al. (1989). Topographic brain measures of human pain and pain responsivity. *Pain, 37,* 129–141.

Chen, A. C., & Herrmann, C. S. (2001). Perception of pain coincides with the spatial expansion of electroencephalographic dynamics in human subjects. *Neuroscience Letters, 297,* 183–186.

Chen, A. C., & Rappelsberger, P. (1994). Brain and human pain: Topographic EEG amplitude and coherence mapping. *Brain Topography, 7,* 129–140.

Da Silva, F. (1999). Biophysical aspects of EEG and magnetoencephalogram generation. In E. Niedermeyer & L. da Silva (Eds.), *Electroencephalography: Principles, clinical applications, and related fields* (4th ed., pp. 76–92). Baltimore: Williams & Wilkins.

Fernandez-Bouzas, A., Harmony, T., Fernandez, T., et al. (2001). Cerebral blood flow and sources of abnormal EEG activity (VARETA) in neurocysticercosis. *Clinical Neurophysiology, 112,* 2281–2287.

John, E. R. (2002). The neurophysics of consciousness. *Brain Research Reviews, 39,* 1–28.

Kakigi, R., Hoshiyama, M., Shimojo, M., et al. (2000). The somatosensory evoked magnetic fields. *Progress in Neurobiology, 61,* 495–523.

Moldofsky, H. (2002). Management of sleep disorders in fibromyalgia. *Rheumatic Diseases Clinics of North America, 28,* 353–365.

Russ, M. J., Campbell, S. S., Kakuma, T., Harrison, K., & Zanine, E. (1999). EEG theta activity and pain insensitivity in self-injurious borderline patients. *Psychiatry Research, 89,* 201–214.

Pain in Humans, Electrical Stimulation (Skin, Muscle, and Viscera)

Ole K. Andersen
Department of Health Science & Technology, Aalborg University, Aalborg, Denmark

Definition

Electrical stimulation initiates activity in nerve fibers directly without activating receptors. The stimulus intensity determines the size of the current field in the tissue and thereby the amount of fibers activated.

Characteristics

Electrical stimulation leads to a depolarization of the membrane of the nerves, and thereby action potentials are generated. Membrane depolarization occurs at the negative electrode (the cathode). Unipolar, constant current pulses with a rectangular shape, are the most commonly used technique in pain research and treatment. The threshold for initiating an action potential is dependent on the diameter of the nerve fiber, with thick fibers having a lower ▶ activation threshold than thin fibers (Blair and Erlanger 1933). Hence, for a rectangular pulse applied to the skin (i.e., to nervous tissue underlying the skin), thick fibers mediating mechanoreceptive input are activated at the lowest stimulus intensities. Increasing the stimulus intensity leads to concurrent activation of thin myelinated fibers (Aδ fibers), and eventually, C fibers will also be activated at high intensities. Selective activation of nociceptive fibers using electrical stimulation is therefore not possible. The same fiber diameter-activation threshold relationship is also present in muscle tissue and in the viscera. The activation threshold is dependent on the duration of the stimulus, with higher currents needed for short pulses. Further, all fibers have a blocking threshold, meaning a current level at which action potentials are blocked due to positive potentials in the tissue at a distance from the cathode. These positive potentials enforce membrane polarization, and thereby, action potential conduction is stopped (anodal blocking). The blocking threshold is dependent on nerve fiber diameter, with thick fibers having the lowest blocking threshold. Further, the distance between the electrode and the nerve is important, with higher currents needed for the blocking of distant fibers. Selective activation of nociceptors using anodal blocking therefore only occurs in fibers close to the electrode and not for fibers that are further away, e.g., other fascicles of the nerve.

Theoretical and experimental studies have indicated that stimulus shapes with a slowly rising waveform may activate thin fibers prior to thick fibers when increasing the stimulus intensity (Grill and Mortimer 1995; Zimmermann 1968). The underlying mechanism is the ability of nerve fibers to accommodate to the stimuli, which is more prevalent in fibers with low excitation threshold (i.e., thick myelinated fibers). This also includes low frequency sinusoidal waveforms. However, these techniques are highly dependent on the distance from the electrode to the nerve and are therefore difficult to use for assessing the integrity of nociceptors selectively.

Various thresholds may be assessed for electrical stimuli. The detection threshold is defined as the lowest current amplitude perceived by the subject, whereas the pain threshold is normally defined as the current amplitude that evokes pain in 50 % of stimuli. Different paradigms have been developed for estimating the thresholds, often involving some kind of ascending and descending staircases with varying intensity steps. The relationship between the detection threshold and the pain threshold depends on

electrode configuration and stimulus site. Using small, invasive microelectrodes, the current density is high near the electrode, and the pain threshold is approximately five times higher than the detection threshold (Burke et al. 1975). The pain threshold is associated with activation of Aδ fibers. Using surface electrodes on the volar forearm with a large area (28 mm^2), the pain threshold is ten times higher than the detection threshold determined as 0.3 mA (Sang et al. 2003). In the latter study, both detection and pain thresholds were stable over both time and electrode position.

Electrical stimulation is used extensively for testing the sensitivity of the pain system in studies activating cutaneous structures (Andersen et al. 1994), muscular structures (Laursen et al. 1999), and in visceral structures (Arendt-Nielsen et al. 1997). In deep structures, repetitive stimulation by rectangular burst above the pain threshold evokes areas of referred pain. In the viscera, it is difficult to determine the pain threshold to a single stimulus, whereas the pain threshold is easily determined if a train of stimuli is used. Further, the referred pain area gradually expands if stimulation is continued for 120 s (Arendt-Nielsen et al. 1997). Repetitive stimulation also allows quantification of central neuronal integration (Arendt-Nielsen et al. 2000). Hence, variation in stimulus burst frequency may be used for testing ▶ temporal summation, whereas stimulus intensity tests ▶ spatial summation as the current field in the tissue is enlarged for more intense stimuli, and thereby, additional nerve fibers are activated concurrently. Electrical stimulation is widely used clinically for quantifying various neuropathies (Krarup 2002) or for assessing detection and pain thresholds (Gracely et al. 1992).

Long-lasting electrical conditioning stimulation has been shown to enhance or reduce the sensitivity of the nociceptive system for several minutes, depending on stimulus parameters and electrode configuration. In the skin, the nerve endings of nociceptors have the most superficial innervation. Using needle like electrodes pressed against the skin, it is possible to activate Aδ and C fibers over several minutes, at tolerable levels, via an array consisting of 16 of these electrodes in

Rectangular pulses   Slowly rising pulse

duration

amplitude

amplitude

Inter-pulse interval   duration

Pain in Humans, Electrical Stimulation (Skin, Muscle, and Viscera), Fig. 1 Rectangular shaped pulses are most often used for electrical stimulation. Other pulse shapes with slower current rise rates have been introduced for stimulation of thinner fibers without activation of thick myelinated fibers

a 6 × 6 cm matrix. Continuous stimulation, by alternating between the electrodes in the array, results in inhibition of itch and reduced heat pain sensitivity (Nilsson et al. 2003). In another study, ▶ long-term potentiation in the sensitivity of the skin was induced experimentally by intense electrical stimulation via an array of 10 circular electrodes with a diameter of 200 μm. The conditioning stimulus consisted of five trains of 100 ▶ rectangular pulses separated by 10 s. The stimulus intensity was up to 20 times the perception threshold. Enhanced pain sensitivity was observed for more than 60 min. after the conditioning (Klein et al. 2003).

Electrical stimulation may also be used for "transcutaneous electrical nerve stimulation" (▶ TENS) for alleviating pain by robust activation of large myelinated fibers innervating mechanoreceptors. The underlying mechanism for the reduction in pain is the gait control theory proposed by Melzack and Wall. In TENS stimulation at frequencies up to 120 Hz, and normally using stimulus intensities below the pain threshold, it has been shown to relieve pain efficiently (Hansson and Lundeberg 1999). Large electrodes (several cm²) are normally used in order to activate thick myelinated fibers in a large skin area, and treatment is continued for at least 30–45 min (Hansson and Lundeberg 1999).

Electrical stimulation may also be used for evoking physiological correlates to pain, evoked potentials, or ▶ nociceptive withdrawal reflexes. In fact, human nociceptive withdrawal reflexes are most efficiently elicited using electrical stimulation in an experimental setup. A train of five electrical stimuli over a period of 20 ms delivered to a sensory nerve, or directly to the skin on the foot, evoke an integrated withdrawal of the limb (Willer 1977; Andersen et al. 1994). For graded stimulus intensity, pain intensity and the size of the withdrawal reflex correlate under standardized experimental conditions (Chan and Dallaire 1989).

Continuous stimulation at the same site may lead to decreased responsiveness of the nerve fibers due to ▶ fatigue (Peng et al. 2003). Repetitive stimulation may also lead to a decrease in the evoked responses due to habituation, a central mechanism. Habituation is commonly seen following regular stimulation at low to medium pain intensities.

In conclusion, electrical stimulation is useful for testing the nociceptive system in experimental and clinical settings. Electrical stimulation is not nociceptive specific, but different techniques (electrode configuration, stimulus waveform, stimulus intensity, etc.) may enhance the selectivity (Fig. 1).

References

Andersen, O. K., Jensen, L. M., Brennum, J., et al. (1994). Evidence for central summation of C and AÎ´ nociceptive activity in man. *Pain, 59*, 273–280.

Arendt-Nielsen, L., Drewes, A. M., Hansen, J. B., et al. (1997). Gut pain reactions in man: An experimental investigation using short and long duration transmucosal electrical stimulation. *Pain, 69*, 255–262.

Arendt-Nielsen, L., Sonnenborg, F. A., & Andersen, O. K. (2000). Facilitation of the withdrawal reflex by repeated transcutaneous electrical stimulation: An experimental study on central integration in humans. *European Journal of Applied Physiology, 81*, 165–173.

Blair, E. A., & Erlanger, J. (1933). A comparison of the characteristics of axons through their individual electrical responses. *American Journal of Physiology, 6*, 524–564.

Burke, D., Mackenzie, R. A., Skuse, N. F., et al. (1975). Cutaneous afferent activity in median and radial nerve fascicles: A microelectrode study. *Journal of Neurology, Neurosurgery, and Psychiatry, 38*, 855–864.

Chan, C. W. Y., & Dallaire, M. (1989). Subjective pain sensation is linearly correlated with the flexion reflex in man. *Brain Research, 479*, 145–150.

Gracely, R. H., Lynch, S. A., & Bennett, G. J. (1992). Painful neuropathy: Altered central processing maintained dynamically by peripheral input. *Pain, 51*, 175–194.

Grill, W. M., & Mortimer, J. T. (1995). Stimulus waveforms for selective neural stimulation. *IEEE Engineering in Medicine and Biology, 14*, 375–385.

Hansson, P., & Lundeberg, T. (1999). Transcutaneous electrical nerve stimulation, vibration and acupuncture as pain-relieving measures. In P. D. Wall & R. Melzack (Eds.), *Textbook of pain* (pp. 1341–1351). Edinburgh: Churchill Livingstone.

Klein, T., Magerl, W., Mantzka, U., et al. (2003). Perceptual correlates of long-term potentiation in the spinal cord. In J. Dostrovsky, D. B. Carr, & M. Koltzenburg (Eds.), *Proceedings of the 10th world congress on pain* (pp. 407–415). Seattle: IASP Press.

Krarup, C. (2002). Nerve conduction studies in selected peripheral nerve disorders. *Current Opinion in Neurology, 15*, 579–593.

Laursen, R. J., Graven-Nielsen, T., Jensen, T. S., et al. (1999). The effect of compression and regional anaesthetic block on referred pain intensity in humans. *Pain, 80*, 257–263.

Nilsson, H. J., Psouni, E., & Schouenborg, J. (2003). Long term depression of human nociceptive skin senses induced by thin fibre stimulation. *European Journal of Pain, 7*, 225–233.

Peng, Y. B., Ringkamp, M., Meyer, R. A., et al. (2003). Fatigue and paradoxical enhancement of heat response in c-fiber nociceptors from cross-modal excitation. *Journal of Neuroscience, 23*, 4766–4774.

Sang, C. N., Max, M. B., & Gracely, R. H. (2003). Stability and reliability of detection thresholds for human A-beta and A-delta sensory afferents determined by cutaneous electrical stimulation. *Journal of Pain and Symptom Management, 25*, 64–73.

Willer, J. C. (1977). Comparative study of perceived pain and nociceptive flexion reflex in man. *Pain, 3*, 69–80.

Zimmermann, M. (1968). Selective activation of c-fibers. *Pflügers Archiv, 301*, 329–333.

Pain in Humans, Psychophysical Law

Donald D. Price
Oral Surgery and Neuroscience, McKnight Brain Institute University of Florida, Gainesville, FL, USA

Synonyms

Psychophysical law

Definition

Similar to other ► sensory modalities, experimentally induced pain has been shown to follow the power law, as formulated in ► psychophysics (Stevens 1975; Marks 1974). The ► power law implies a very simple and parsimonious concept "that equal stimulus ratios produce equal subjective ratios" (Stevens 1975). To the non-psychophysicist, this statement means that proportions of perceived stimulus intensities remain stable. For example, the perceived relations between the lighter or darker parts of a photograph are the same under bright or dim illumination. The power law can be expressed by the following simple equation:

$$\psi = k(S - S_0)^X$$

where ψ = perceived magnitude, k is a constant, S_0 is the stimulus magnitude at threshold (e.g., pain threshold), and X is an exponent for a given sense modality under a standard set of conditions. A power function whose exponent is 1.0 (i.e., $\psi = k(S-S_0)^{1.0}$) predicts that two stimuli, whose ratio of measured intensities is 2:1, will result in two perceived intensities whose ratio is 2:1. Using this same stimulus ratio (2:1) in the case of a power function whose exponent is 2 (i.e., $\psi = k(S-S_0)^{2.0}$), results in a ratio of perceived magnitudes that is 4:1. The relationship of perceived ratios to stimulus ratios would be the same for a given power function, regardless of the units of measurement of stimulus intensity or the stimulus magnitudes used to make up the stimulus intensity ratios. Different sense modalities have characteristic power function exponents. For example, that for visual intensity is about 0.7 (Stevens 1975; Marks 1974). This means that it bends slightly downward in linear coordinates. Perceived warmth has an exponent ranging from 0.5 to about 1.6, depending on the area of the stimulus (Marks 1974).

Characteristics

Contact heat-induced pain has been shown repeatedly to follow a positively accelerating (bending upward) power function (Price 1999; Price et al. 1992, 1996), as illustrated in Figs. 1 and 2. The exponent of the power function is consistently greater than 2.0, regardless of whether pain sensation intensity or unpleasantness (i.e., pain affect) is rated. Thus, for the data presented in Fig. 2, the exponent for pain unpleasantness is significantly higher than that for pain sensation intensity, and both functions have exponents greater than 2.0 (Price et al. 1996). The functions were obtained using a ▶ visual analogue scale (VAS) and were very consistent across different groups of subjects, including both pain patients and pain-free volunteer groups (Fig. 1). The power function exponent for contact heat-induced pain is also independent of area. Thus, the stimulus-response functions of contact heat-induced pain differ in this respect from those of warmth, whose power function exponents decrease systematically as a function of area. They decrease from 0.5 to 1.6 as the stimulus area increases (Marks 1974). However, the exponent for pain produced by radiant heat is close to 1.0 for different areas of stimulation (Adair et al. 1968; Price et al. 1975). Similar to radiant heat, cold pain also has an exponent close to 1.0 (Stevens 1975).

When subjects rate both the unpleasantness and the sensory intensity of contact-induced heat pain, the power function exponent for unpleasantness is consistently higher than that

pain sensation intensity (Price 1999; Price et al. 1992, 1996). A higher exponent for pain affect compared with pain sensation reflects the fact that unpleasantness ratings are systematically lower than sensory ratings, when both dimensions of pain are rated on the same length visual analogue scales (Price et al. 1996). This has the effect of increasing the slope when plotted in logarithmic coordinates. However, unpleasantness ratings are subject to a good deal of influence by psychological factors. The lower ratings in this case are due to the assurances made to the participants that the stimuli would be brief, would not produce tissue damage, and would remain within tolerable limits. Very different ratings would likely occur if subjects were made anxious about the pain stimuli.

These characteristics demonstrate that heat-induced pain is similar to other sensory modalities in that it follows a power function. This attribute helps to characterize heat-induced pain as a ▶ somatosensory submodality. In addition to larger exponents for contact heat-induced pain (>2.0) than for contact-induced warmth (e.g., .5–1.6), other major differences also exist between these sensory modalities. The perception of warmth intensity and the threshold for warmth are dependent on the rate of rise of skin temperature, whereas pain threshold and pain intensity are far less influenced by such factors (Price and Browe 1975; Kenshalo et al. 1967; Stevens and Marks 1971; Hardy et al. 1952). There are also differences in ▶ spatial summation attributes between pain and warmth (see below). These clear differences in psychophysical attributes indicate that heat-induced pain cannot be construed as a simple extension of warmth and must be considered a separate sensory dimension. That warmth and pain are processed separately is also supported by considerable neurophysiological evidence, that separate neuronal populations process warmth and ▶ nociception (Price et al. 1992; Price and Browe 1975).

Spatial Summation
Both the power functions for heat-induced pain and heat-induced warmth are affected by spatial summation (Price 1999; Price et al. 1992;

Pain in Humans, Psychophysical Law, Fig. 1 Stimulus-response functions for noxious temperature stimuli (5-s duration) applied in random order to the ventral surface of the forearm and skin overlying the masseter area in normal volunteers and in pain patients. The *upper* graphs are the functions in linear coordinates, and the *lower* graphs are the same functions expressed in double logarithmic coordinates (Price 1999)

PAIN SENSATION INTENSITY (VAS)

FOREARM
MASSETER

STIMULUS TEMPERATURE (°C)

UNPLEASANTNESS (VAS)

FOREARM
MASSETER

STIMULUS TEMPERATURE (°C)

▲ LOW BACK
○ MPD
● PAIN FREE CONTROLS

Ln PAIN SENSATION INTENSITY (VAS)

FOREARM
MASSETER

Ln STIMULUS TEMPERATURE (T-34°C)

● CONTROLS
○ MPD

Ln UNPLEASANTNESS (VAS)

FOREARM
MASSETER

Ln STIMULUS TEMPERATURE (T-34°C)

PAIN SENSATION INTENSITY (VAS)

STIMULUS TEMPERATURE (°C)

UNPLEASANTNESS (VAS)

STIMULUS TEMPERATURE (°C)

Pain in Humans, Psychophysical Law, Fig. 2 Spatial summation of pain, showing sensory (*left*) and affective (*right*) VAS ratings of noxious temperatures. Functions are plotted in linear (*top*) and double logarithmic coordinates (Price et al. 1989)

Ln PAIN SENSATION INTENSITY (VAS)

Ln STIMULUS TEMPERATURE (T-35°C)

STIMULUS AREA
○ 3 CM²
○ 2 CM²
○ 1 CM²

Ln UNPLEASANTNESS (VAS)

Ln STIMULUS TEMPERATURE (T-35°C)

Kenshalo et al. 1967; Stevens and Marks 1971; Hardy et al. 1952) (Fig. 2). Summation of warmth takes place generously over large areas of the body surface, and as stimulus area increases, the overall perceived intensity of warmth increases. However, the slope of the stimulus response function, obtained when relating stimulus temperature to perceived warmth (in double logarithmic coordinates), decreases in magnitude with increasing stimulus area (Kenshalo et al. 1967; Stevens and Marks 1971). Therefore, the same change in temperature over a small area has a proportionately greater effect than that same change spread over a larger area. Furthermore, the perceived warmth intensity functions converge at a common point near pain threshold (Marks 1974).

In contrast to spatial summation of warmth, which decreases with increasing stimulus temperature, spatial summation of heat-induced pain increases with increasing stimulus temperature (Fig. 2). Thus, for intensely painful stimuli, stimulus area is a particularly critical factor in determining pain level. The pattern of spatial summation of pain occurs for small areas of skin (Douglass et al. 1992; Greene and Hardy 1958; Price et al. 1989), much larger areas that extent across several ▶ dermatomes (Arendt-Nielson et al. 1997), and even for multiple stimuli that are spatially separated by as much as 20 cm (Price et al. 1992). That the extent of spatial summation of pain is a function of both area and nociceptive stimulus intensity also has adaptive significance, because these features are consistent with a mechanism that is sensitive to the total amount of biological threat to the integrity of body tissues. Given its potentially critical role in understanding the neural mechanisms of pain, as well as the obvious medical implications of this property of pain, it is astonishing that it is only within the last 15 years that direct scaling analyses have been applied to testing spatial summation of pain.

In the separate studies that used direct scaling methods, human observers made direct ratings of pain sensation intensity and/or pain unpleasantness in response to contact heat (Price et al. 1992; Douglass et al. 1992) or contact cold (Douglass et al. 1992). Considerable spatial summation occurred in both the intensity and unpleasantness dimensions of pain (Price et al. 1992; Douglass et al. 1992). Although spatial summation was evident throughout a wide range of temperatures, it was far larger at temperatures that were suprathreshold for pain (47–50 °C) than at those near the pain threshold (43–46 °C), and this result was consistent across multiple studies. The psychophysical features of spatial summation are consistent with a pain encoding mechanism that is heavily dependent on central neuronal recruitment.

The spatial summation of heat-induced pain is characterized by upward parallel displacement of the stimulus-response function in double logarithmic coordinates. A similar parallel displacement in double logarithmic coordinates is evident, regardless of whether the various sized stimulus areas are small, large, extend across several dermatomes, or are spatially separated. It also indicates that the power function exponent, the slope of the curve in double logarithmic coordinates, is stable across different areas and different spatial patterns of stimulation. This consistent power function exponent differs markedly from that of warmth, whose slope and power-function exponent decrease with increasing area.

The minimal spatial summation near the pain threshold is consistent with previous studies, which concluded that spatial summation of pain threshold was minimal or nonexistent (Hardy et al. 1952; Greene and Hardy 1958) and helps to explain why the property of spatial summation of pain has long eluded investigators. This spatial summation property of pain is very different from that of warmth, for which spatial summation occurs maximally near the threshold. This difference, in turn, may be related to the fact that the steepest portion of the stimulus response curve for warmth is near the warmth threshold, whereas that for heat-induced pain is well above the pain threshold. There is also a possibility that central neural integrative mechanisms are quite different for pain and warmth. Spatial summation is undoubtedly a clinically important attribute of pain. Spatial summation is not a phenomenon confined to cutaneous pain, for there is evidence that considerable spatial summation takes place for visceral and muscular pain (Arendt-Nielson et al. 1997).

Pain Intensity Discriminability

If one can detect the intensity of a noxious stimulus, one must also be able to detect differences in noxious intensities. What is surprising, however, is the sensitivity of this capability. It has been demonstrated that both humans and monkeys can detect temperature shifts of 0.2–0.3 °C at nociceptive (>47 °C) intensity levels (Dubner et al. 1986). This discriminative ability is even greater than that found for warmth. For example, temperature shifts of 0.3–0.5 °C, from a baseline of 39 °C, are required for trained human observers to detect differences in warmth. The ability to detect very small differences of 0.2–0.3 °C within the nociceptive range (45–51 °C) is generally consistent with the presence of 21 levels of just noticeable differences obtained when studying radiant heat-induced pain (Hardy et al. 1952). Contrary to the notion that discrimination of different magnitudes of pain is poor and unlike other sensory modalities, discrimination of intensity and intensity differences within the nociceptive stimulus intensity range is extremely refined. Both psychophysical studies and studies of neurons at all levels within pain-related neural pathways consistently demonstrate capacities of neurons and human observers to discriminate 0.2–0.3 °C differences (Price et al. 1992; Dubner et al. 1986).

The Relationship of the Power Law to Validation of Pain Scales as Ratio Measurement Scales

The reliable power functions established for pain are interrelated with the validation of a specific type of pain scale as having ▶ ratio scale properties. An example of a pain scale that at least approximates a ratio scale level of measurement is the visual analogue scale (VAS). A critical test of whether VAS has ratio scale properties is to ask subjects to rate different noxious stimuli and then to adjust different stimulus intensities to obtain separate judgments of ratios of intensity. Ratings of noxious stimuli and independent judgments of

ratios should agree if the method represents a ratio scale. Price (1999) asked subjects to rate contact heat-induced temperatures, thereby producing a stimulus response curve like that shown in Fig. 1. Subjects then received a standard test stimulus of 47 °C, after which the temperature increased in intensity to a level judged to be twice as intense as this standard. According to the stimulus-response regression curve, they should have judged 49.6 °C as twice as intense as the standard. The mean temperature chosen was 49.8 °C, a value very close to that predicted. This test was replicated using other standard temperatures within the noxious range. Importantly, when this experiment was repeated using a numerical rating scale of 0–10, the predicted and observed judgments were significantly different, indicating that this scale does not have ratio scale properties. This negative finding is important, considering that the 0–10 numerical rating scale is the most common pain scale used by healthcare professionals.

At least some forms of experimentally induced pain follow the power law. Specific characteristics of their stimulus-response functions distinguish them as a sensory modality that is distinct from other somatosensory submodalities. These characteristics help explain some features of clinical pain, such as spatial summation, and help to establish the ability to measure pain in clinical contexts.

References

Adair, E. E., Stevens, J. C., & Marks, L. E. (1968). Thermally induced pain: The dol scale and the psychophysical power law. *The American Journal of Psychology, 81*, 147–164.

Arendt-Nielson, L., Graven-Nielsen, T., Svennson, P., & Jensen, T. S. (1997). Temporal summation in muscles and referred pain areas: An experimental human study. *Muscle & Nerve, 10*, 1311–1313.

Douglass, D. K., Carstens, E., & Watkins, L. R. (1992). Spatial summation in human pain perception: Comparison within and between dermatomes. *Pain, 50*, 197–202.

Dubner, R., Bushnell, M. C., & Duncan, G. H. (1986). Sensory-discriminative capacities of nociceptive pathways and their modulation by behavior. In T. L. Yaksh (Ed.), *Spinal afferent processing* (pp. 321–331). New York: Plenum.

Greene, L. C., & Hardy, J. D. (1958). Spatial summation of pain. *Journal of Applied Physiology, 13*, 457–464.

Hardy, J. D., Wolff, H. G., & Goodell, H. (1952). *Pain sensations and reactions.* Baltimore: Williams and Wilkins.

Kenshalo, D. R., Decker, T., & Hamilton, A. (1967). Spatial summation on the forehead, forearm, and back produced by radiant and conducted heat. *Journal of Comparative and Physiological Psychology, 63,* 510–515.

Marks, L. W. (1974). *Sensory processes. The new psychophysics.* New York: Academic.

Price, D. D. (1999). *Psychological mechanisms of pain and analgesia.* Seattle: IASP Press.

Price, D. D., & Browe, A. C. (1975). Responses of spinal cord neurons to graded noxious and non-noxious stimuli. *Experimental Neurology, 48,* 201–221.

Price, D. D., Bush, F. M., Long, S., & Harkins, S. W. (1996). A comparison of pain measurement characteristics of mechanical visual analogue and simple numerical rating scales. *Pain, 56,* 217–226.

Price, D. D., McHaffie, J. G., & Larson, M. A. (1989). Spatial summation of heat induced pain: Influence of stimulus area and spatial separation of stimuli on perceived pain sensation intensity and unpleasantness. *Journal of Neurophysiology, 62,* 1270–1279.

Price, D. D., McHaffie, J. G., & Stein, B. E. (1992). The psychophysical attributes of heat-induced pain and their relationships to neural mechanisms. *Journal of Cognitive Neuroscience, 4,* 1–14.

Stevens, S. S. (1975). *Psychophysics. Introduction to its perceptual, neural, and social prospects.* New York: Wiley.

Stevens, J. C., & Marks, L. E. (1971). Spatial summation and the dynamics of warmth sensation. *Perception & Psychophysics, 9,* 291–298.

Pain in Humans, Sensory-Discriminative Aspects

Richard H. Gracely
Departments of Medicine-Rheumatology and Neurology, University of Michigan Health System, Ann Arbor, MI, USA

Synonyms

Sensory-discriminative aspects

Definition

Pain is defined by the International Association for the Study of Pain (IASP) as "an unpleasant sensory and emotional experience associated with actual or potential tissue damage or described in terms of such damage" (Merskey and Bogduk 1994). The sensory aspect is essential, as the IASP definition elaborates: "Unpleasant abnormal experiences (dysesthesias) may also be pain but are not necessarily so because, subjectively, they may not have the usual sensory qualities of pain," pain "is unquestionably a sensation in a part or parts of the body" (Merskey and Bogduk 1994).

Characteristics

Somatosensory sensations such as pain can be described by four major dimensions: intensity, quality, location, and duration. The intensity of pain sensation has been the primary outcome for a majority of basic and clinical pain studies. Pain intensity is intimately associated with pain quality, which can vary over a number of dimensions including mechanical, temporal, spatial, and thermal. A pain sensation may be composed of only a single quality such as burning, with the pain intensity referring to the amount of burning. Alternatively, pain sensation may be multidimensional and described by a number of simultaneous qualities, each described in turn by a specific sensory intensity. In musical terms, pain can resemble a solo instrument with a specific loudness or an ensemble with a separate and varying loudness for each type of instrument. In this case, the ensemble can be described by an overall intensity and also separate intensities can be assigned to individual qualities.

In addition to intensity and quality, pain can vary in the location and extent of the sensation. The pain may be restricted to a small area or occupy a large region of the body. It may be superficial or deep. It may be accurately localized or be diffuse and poorly localized. Finally, pain can be brief or prolonged and can wax and wane. In the acute situation, the duration of an evoked pain may be increased during ▶ central sensitization or during loss of tonic inhibition. In chronic conditions, prolonged pain may be accompanied by neuroplastic changes, resulting in altered and new mechanisms that are not present during acute pain events.

Together, these features of intensity, quality, location, and duration define a rich variety of painful experiences. These sensory-discriminative aspects of pain probably serve a number of important functions. They guide vermin detection (i.e., swat a mosquito) and guide conscious withdrawal behaviors. They provide predictions about the possibility of impending tissue damage (stepping barefoot onto hot rocks, continue or retreat?) and they provide information important for integration with other systems. Examples include modification of gait associated with a sprained ankle, an interaction that may be further modified by other considerations (chasing prey, being chased by a predator).

Discriminating the Presence of Pain: The Pain Threshold

The simplest discrimination of pain is the determination of its presence. In the clinic, the presence of pain is often determined by the response to a standard stimulus. For example, metal objects at room temperature feel cool in the normal situation but evoke painful sensations in clinical conditions of cold allodynia. In the laboratory, the investigator is interested in the exact value of the lowest stimulus intensity needed to evoke pain or simply, the pain threshold.

In sensory psychophysics, the threshold describes a fundamental difference in sensory quality. As the magnitude of a stimulus is increased, the threshold marks the transition from the absence of sensation to the presence of sensation. This transition is not succinct. Subjects fail to report sensations to the same level of stimulus magnitude consistently and will report sensations when no stimulus is presented. Thus, there is a band of stimulus intensities, extending down to zero, in which the presence of a sensation is uncertain. As stimulus intensity is increased, the uncertainty decreases until levels are reached at which the presence of sensation is certain and unmistakable. Mathematically, the probability of sensory detection increases as stimulus intensity

is increased. The resulting function is usually described by an ogive (an accumulated normal Gaussian distribution); the probability of detection is near zero for a range of values, then the slope increases as stimulus intensity increases, reaching a peak slope, and becoming nearly linear through the region of 50 % probability. Further increases in stimulus intensity result in a reduced slope as 100 % probability of detection is approached.

Since there is no single sensory detection threshold, the evaluation of such thresholds is based on determining a stimulus intensity that evokes a report of a sensation a certain percentage of the time, with 50 % and sometimes 75 % used as obvious rational standards.

These properties of the sensory detection threshold apply also to the determination of the pain threshold. However, there is a fundamental difference between determination of sensory thresholds and pain thresholds. In almost all cases, increasing stimulus intensity of a noxious stimulus does not lead to a transition from the absence of sensation to a pain sensation. In between there is a range of non-painful sensation. Heat stimuli become warm and hot before they begin to hurt. These pre-pain sensations present a measurement problem. In the psychophysical evaluation of sensory thresholds, the investigator can deliver a blank stimulus and be confident that the effect is the same as the effect of delivering a subthreshold, unperceived stimulus and also be confident that the blank stimulus will be confused with a real stimulus. The ability to present such blank trials provides the basis for robust methods such as sensory-decision-theory and two-alternative-forced-choice that provide relatively bias-free measures of sensory sensitivity. The pain investigator is hampered in this regard, because a true blank stimulus will rarely be confused with a painful stimulus.

Measurement of the Pain Threshold

Due to concerns about administering excessively painful stimulation, the pain threshold is usually approached from the bottom, by administering continuous stimulation increasing in intensity or an ascending series of discrete stimuli. These methods lack controls for psychophysical errors that are incorporated into more sophisticated psychophysical methods (Engen 1971). These measures are therefore vulnerable to a number of biases. Even more sophisticated methods cannot ensure that subjects follow instructions. The very multi-qualitative nature of pain sensation provides opportunities for irrelevant responses. As stimulus intensity increases, the evoked sensation may change in quality, may spread, or may increase in time. There is nothing to prevent subjects from using any aspect of the stimulus-evoked sensations as their personal criterion for pain, and producing orderly responses by using this point as their pain threshold. In contrast, detection thresholds are immune to this problem, since there is no pre-pain sensation to use as an artifactual criterion.

Subjects may be motivated to provide irrelevant responses. They may wish to minimize pain and discomfort, to be "tough" to an opposite gender experimenter (Levine and DeSimone 1991) or they may strive to give the "correct" answer. In the assessment of an analgesic, they may use perceived side effects as a clue that an active agent has been delivered and change their responses accordingly.

The Fundamental Problem of Pain Measurement

Sophisticated methods can control for many of the biases that influence simple threshold determinations. For example, clinical measures of the pressure pain threshold can be influenced significantly by psychological distress, while suprathreshold measures using random stimulation are independent of distress (Petzke et al. 2003). However, even the most sophisticated methods share a fundamental problem with simple threshold measures. Do differences in reports of pain intensity between individuals or within individuals over time represent altered experience or just altered labels used to describe the same experience? One attempt at solving this problem uses discrimination ability, essentially a behavioral measure of performance, as a correlate of pain sensitivity.

Does Pain Discrimination Solve the Problem of Measuring Pain Intensity?

The psychophysical process can be broken down into two parts. One is discrimination sensitivity, the ability to distinguish between two different sensations. The other is the choice of labels used to describe these differences. All psychophysical judgments involve a "response criterion" which in the case of the pain threshold can be defined as the specific attributes of a stimulus-evoked sensation that result in the label "painful." This criterion can vary from highly expressive to stoical. Two fundamental problems in pain evaluation are the static case, e.g., "do differences in pain reports from different patients represent different sensory magnitudes or just different labels applied to similar sensory experiences?" and the dynamic case, e.g., "does a reduction in pain after a therapeutic treatment represent a reduction in pain sensation or a reduction only in the labels used to describe an essentially unchanged pain sensation?"

The role of the response criterion and the separate assessment of the criterion and sensory performance are explicitly addressed by the use of signal detection and ► sensory decision theory (SDT) (Swets 1964). Basically, SDT analytical procedures consider a threshold judgment in terms of whether the stimulus is present and whether it is reported as present. The two correct responses are a hit (a stimulus is reported present when it is present) and a correct rejection (stimulus reported absent when it is absent). The two errors are called a miss (stimulus reported absent when present) and a false alarm (stimulus reported present when absent). In statistical terms, SDT treats a threshold determination like a t-test. The measure of discrimination is directly analogous to the t statistic; it is the measure of difference in the means of two sensory distributions divided by their variability. It describes the overall sensitivity of the system. The response criterion parameter is directly related to the alpha level set by the experimenter. It is the probability of a Type I error, which is the same (and more meaningfully described) as a false alarm or false positive. As with the t-test, the setting of this criterion to minimize this error increases the other type of error, referred to as Type II or beta error (or as a miss or false negative). In the original application of SDT to wartime radar operators, the response criterion parameter was set by the operator and in this case determined by the demand characteristics. During wartime, it is obviously beneficial to adapt a liberal criterion, to call anything remotely resembling an enemy plane an enemy plane. This vigilant strategy minimizes the misses at the expense of increased false alarms.

SDT has been applied in many studies of pain control interventions. The results of these studies were consistent with an interpretation of d' as pain sensitivity and the response criterion as response bias. Administration of placebo resulted in only a shift in the response criterion without a change in d' while the active interventions reduced d' with variable effects on the response criterion (Chapman 1977; Chapman et al. 1973; Clark and Yang 1974; Yang et al. 1979).

Proponents of SDT for pain assessment claimed that the fundamental problem in pain assessment, separating physiological pain sensitivity from labeling behavior, had been solved. Certainly, the separation of pain responses into these two parameters is a logical, appealing, and good way of describing pain-rating behavior. However, this interpretation has not met with universal acceptance (Rollman 1977). Others note that the SDT measure of discrimination sensitivity is a statistic and using it to measure pain sensitivity is similar to using the t or F statistic or using the statistical p value to assess effect size. These statistics are related to the effects we want to measure, but this effect is divided by a measure of variance. In the case of pain, this variance can be in the physiological sensory system and also in the genitive system of choosing labels to describe the sensory experience (Coppola and Gracely 1983). Pain discrimination and experimental effects on pain discrimination can be measured by a number of robust methods, but the relationship of discrimination to "how much it hurts" is a clear topic for further investigation.

As mentioned above, pain discrimination is not limited to discrimination of intensities.

Discrimination applies also to qualities, duration, and locus. Quality is assessed by well-known instruments such as the McGill Pain Questionnaire (Melzack 1975) or the short form of this questionnaire (Melzack 1987). Lenz et al. (1994) have developed a very rapid method for intraoperative brain stimulation and recording. The temporal and locus aspects of sensation are included in these instruments with qualities such as spreading or flickering. Spatial discrimination may also be assessed by methods such as two-point discrimination. This brief entry is confined to one aspect of psychophysical pain discrimination. From the classic review of Price and Dubner (1977) to the current state of the art, evidence for the mechanisms that mediate the sensory-discriminative aspects of pain processing forms another large and growing story of the multiple systems that discriminate pain sensations and the plasticity of these systems when pain persists.

References

Chapman, C. R. (1977). Sensory decision theory methods in pain research: A reply to Rollman. *Pain, 3*, 295–305.

Chapman, C. R., Murphy, T. M., & Butler, S. H. (1973). Analgesic strength of 33 percent nitrous oxide: A signal detection evaluation. *Science, 179*, 1246–1248.

Clark, W. C., & Yang, J. C. (1974). Acupunctural analgesia? Evaluation by signal detection theory. *Science, 184*, 1096–1098.

Coppola, R., & Gracely, R. H. (1983). Where is the noise in SDT pain assessment? *Pain, 17*, 257–266.

Engen, T. (1971). Psychophysics I. Discrimination and detection. In J. W. Kling & L. A. Riggs (Eds.), *Experimental psychology* (3rd ed., pp. 11–46). New York: Holt.

Lenz, F. A., Gracely, R. H., Hope, E. J., et al. (1994). The sensation of angina can be evoked by stimulation of the human thalamus. *Pain, 59*, 119–125.

Levine, F. M., & DeSimone, L. L. (1991). The effects of experimenter gender on pain report in male and female subjects. *Pain, 44*, 69–72.

Melzack, R. (1975). The McGill pain questionnaire: Major properties and scoring methods. *Pain, 1*, 277–299.

Melzack, R. (1987). The short-form McGill pain questionnaire. *Pain, 30*, 191–197.

Merskey, H., & Bogduk, N. (1994). *Classification of chronic pain, IASP task force on taxonomy* (2nd ed., pp. 209–214). Seattle: IASP Press.

Petzke, F., Gracely, R. H., Park, K. M., et al. (2003). What do tender points measure? Influence of distress on 4 measures of tenderness. *Journal of Rheumatology, 30*, 567–574.

Price, D. D., & Dubner, R. (1977). Neurons that subserve the sensory-discriminative aspects of pain. *Pain, 3*, 307–338.

Rollman, G. B. (1977). Signal detection theory measurement of pain: A review and critique. *Pain, 3*, 187–211.

Swets, J. A. (1964). *Signal detection and recognition by human observers.* New York: Wiley.

Yang, J. C., Clark, W. C., Ngai, S. H., et al. (1979). Analgesic action and pharmaco-kinetics of morphine and diazepam in man: An evaluation by sensory decision theory. *Anesthesiology, 51*, 495–502.

Pain in Humans, Sleep Disturbances

Harvey Moldofsky
Sleep Disorders Clinic of the Centre for Sleep and Chronobiology, Toronto, ON, Canada

Synonyms

Sleep disturbances

Definition

This entry summarizes evidence on how an understanding of sleep and its disorders helps in the assessment and management of common pain problems, that is, headache, migraine, backache, and common rheumatic ailments, such as osteoarthritis, rheumatoid arthritis, and fibromyalgia. The pathophysiology of how pain and insomnia reinforce one another is described. Not only does pain have an adverse effect upon sleep, mood, cognitive functioning and behavior, but disturbances to sleep also promote pain and fatigue as the result of exogenous physical or psychological noxious events or endogenous primary sleep disorders, for example, sleep apnea and ▶ restless legs syndrome (RLS). These sleep- /wake-related issues are especially important in the evolution and management of headache, back pain, and various articular and nonarticular rheumatic illnesses.

Characteristics

The Interaction of Pain and Disordered Sleep

Pain and insomnia are among the most common complaints in our society; so the likelihood that the two conditions coincide in the same person should not be surprising. Our common experience is that any painful condition will disturb sleep, and if prolonged, it could have a negative effect upon mood, thinking, energy, and behavior. In a recent Gallop Poll Survey, 56 million Americans complained that nighttime pain interfered with their falling asleep or promoted awakenings during the night or awakenings in the early morning. A Canadian population survey indicated that 44 % of people with any painful disorder have sleep-related problems. This epidemiological study showed that the greater the severity of pain, the higher the likelihood of having insomnia or unrefreshing sleep (Moldofsky 2001). An experimental study employed to determine whether specific stages in electroencephalographic (EEG) sleep were affected by pain showed that all stages of sleep are disrupted by noxious stimulation of muscles and that, quality of sleep was impaired (Lavigne et al. 2004).

While painful conditions may interfere with sleep, the corollary is also true. That is, in healthy people, who were exposed experimentally to several nights of noise-induced arousals from stages 3 and 4 (slow wave or deep), non-rapid eye movement (non-REM) sleep caused them to experience unrefreshing sleep, nonspecific generalized muscle aching, and fatigue (Moldofsky 2001; Moldofsky and Scarisbrick 1976; Moldofsky et al. 1975). These symptoms receded when they were allowed to sleep undisturbed. Localized effects on muscular function, that is, jaw muscle activity, masseter pain, and tenderness were not induced using a similar experimental design but without any changes in the duration of sleep time and EEG stages 1 and 2 sleep (Arima et al. 2001). Indeed, total sleep deprivation for 40 h as well as partial deprivation of slow wave sleep and rapid eye movement (REM) sleep increases pain sensitivity. Only the increase in slow wave sleep during recovery from sleep deprivation resulted in a decrease in pain sensitivity. Furthermore, sleep deprivation counteracts analgesic effects of drugs that affect opioidergic and serotoninergic neural mechanisms (Kundermann et al. 2004).

The sleep-related problems that occur with chronic pain may also stem from the depression and anxiety that occur as the result of physical, psychosocial, vocational, and economic concerns. Patients' negative beliefs and attitudes about themselves and their chronic pain problems benefit from improvement in their sleep habits, cognitive behavioral therapy, and a suitable aerobic fitness program. While traditional analgesic, anti-anxiety, or antidepressant medications are often used empirically to address pain and mood symptoms, the potential adverse effects of such medications upon sleep and daytime functioning should be considered in the assessment and overall management program (Moldofsky 2002).

Headache and Sleep

People who wake up from sleep with headache often have something wrong with their sleep (Goder et al. 2003). For example, heavy snorers and those with sleep apnea syndrome are more likely to wake up with morning headache than people in the general population. Successful treatment of the sleep apnea is helpful in controlling the frequency of such headaches. Furthermore, complaints of headache at awakening and during the day are three times more likely to occur among sufferers of restless legs syndrome (RLS). Similarly, patients with this sleep disorder experience improvement in their sleep and pain from successful treatment of RLS with gabapentin (Garcia-Borreguero et al. 2002).

The physiology of sleep is intimately linked to migraine. Some people are awakened from rapid eye movement (REM) or the dreaming state of sleep with migraine. Sleep deprivation or excessive sleep may trigger migraine attacks. Restful sleep may be associated with relief of migraine attacks, so that migraine sufferers who improve their sleep habits have reduced incidence and duration of their symptoms.

Back Pain and Sleep

Low back pain is the most common of all musculoskeletal disorders in young adults, but most back strain injuries resolve within 3 months. Delay in falling asleep and higher overall sleep disturbance are found in chronic back pain sufferers who have degenerative spinal disease or post-laminectomy syndrome. These sleep-related problems are more likely to be the result of poor physical functioning, duration of pain and age rather than pain intensity and depressed mood.

Sleep physiological studies show that people with chronic low back pain have an arousal disturbance in EEG sleep that interferes with restful sleep (Moldofsky 2001). This alpha (7.5–11 Hz) non-REM EEG pattern during sleep might be caused by painful stimuli from injured muscles that intrude into sleep. When painful stimuli are applied to muscles, such EEG arousal patterns are seen in slow wave sleep. These differ from EEG sleep frequencies that are experimentally induced in joints and skin (Drewes et al. 1997). These sleep physiological disturbances interfere with the natural restorative properties of sleep, and hence there is an adverse effect upon daytime functioning. Despite their engaging in a rehabilitation program, those workers who had suffered a painful soft tissue strain or sprain injury and have persistent sleep disturbances have a poor prognosis for returning to work (Crook and Moldofsky 1996).

Arthritic Disease and Sleep

About 75 % of people with various painful rheumatic disorders report sleep-related problems. Fatigue, which is almost universal in people with rheumatic disorders, is largely explained by pain, sleep disturbance, and depression.

Osteoarthritis and Sleep

Light, restless sleep is common in patients with osteoarthritis or ankylosing spondylitis. While it can be assumed that proper management of osteoarthritic joint pain with pain remedies or anti-inflammatory drugs would reduce the nocturnal restlessness, sleep disturbances may contribute to unrefreshing sleep, pain, and fatigue symptoms. For example, restless legs and sleep-related involuntary leg movements can influence awakening in the morning with pain and stiffness, which is not the direct result of arthritis. Such people may benefit from treatment of their RLS with special medications for controlling their restlessness and periodic involuntary limb movements during sleep, so that they can wake up with a refreshed feeling and not be troubled by the pain and stiffness.

Rheumatoid Arthritis and Sleep

Sleep disturbance is a major cause in patients with rheumatoid arthritis (Moldofsky 2001). Their fatigue is associated with poor sleep, functional disability, joint pain, and depression. Along with increased weakness and diminished energy, there is an alpha EEG brain wave pattern riding in non-REM sleep that indicates an arousal disturbance during the sleep of acutely ill arthritic patients. Other primary sleep disorders, including periodic involuntary limb movements, RLS, and sleep apnea also contribute to the pain and fatigue symptoms. In depressed rheumatoid patients, impaired sleep is associated with increased pain and increased human lymphocyte antigen DR T-cell monoclonal antibodies that may perpetuate the disease (Cakirbay et al. 2004).

Fibromyalgia and Related Disorders and Sleep

People with fibromyalgia typically complain of generalized aching in their bodies, unrefreshing sleep, fatigue, and cognitive and emotional difficulties that interfere with their day-to-day functioning. Such symptoms along with a hypersensitivity to various food, environmental, and tactile noxious stimuli are also found in patients with other painful complaints, including chronic headache, migraine, chronic fatigue syndrome, irritable bowel syndrome, and temporomandibular joint disorder where unrefreshing sleep is common. The myalgia and multiple tender points in specific anatomic regions that are characteristic of fibromyalgia are related to the poor quality of sleep. The greater the number of tender points, the poorer the sleep in patients with fibromyalgia. This light unrefreshing sleep is associated with an alpha EEG brain wave pattern

that is indicative of an arousal disturbance in sleep (Moldofsky 2003) or periodic arousals in the sleep EEG (Rizzi et al. 2004), which may be accompanied by sleep-related periodic limb movements, or respiratory disturbances. The finding of the alpha EEG sleep disorder in children with fibromyalgia and in their mothers who also have this disorder suggests the possibility of a familial or genetic influence for the disorder (Roizenblatt et al. 1997).

References

Arima, T., Svensson, P., Rasmussen, C., et al. (2001). The relationship between selective sleep deprivation, nocturnal jaw-muscle activity and pain in healthy men. *Journal of Oral Rehabilitation, 28*, 140–148.

Cakirbay, H., Bilici, M., Kavakci, O., et al. (2004). Sleep quality and immune functions in rheumatoid arthritis patients with and without major depression. *International Journal of Neuroscience, 114*, 245–256.

Crook, J., & Moldofsky, H. (1996). The clinical course of musculoskeletal pain in empirically derived groupings of injured workers. *Pain, 67*, 427–433.

Drewes, A. M., Nielsen, K. D., Arendt-Nielsen, L., et al. (1997). The effect of cutaneous and deep pain on the electroencephalogram during sleep: An experimental study. *Sleep, 20*, 632–640.

Garcia-Borreguero, D., Larrosa, O., de la Llave, Y., et al. (2002). Treatment of restless legs syndrome with gabapentin: A double-blind, cross-over study. *Neurology, 59*, 1573–1579.

Goder, R., Friege, L., Fritzer, G., et al. (2003). Morning headaches in patients with sleep disorders: A systematic polysomnographic study. *Sleep Medicine, 4*, 385–391.

Kundermann, B., Krieg, J. C., Schreiber, W., et al. (2004). The effect of sleep deprivation on pain. *Pain Research & Management, 9*, 25–32.

Lavigne, G., Brousseau, M., Kato, T., et al. (2004). Experimental pain perception remains equally active over all sleep stages. *Pain, 110*, 646–655.

Moldofsky, H. (2001). Sleep and pain. *Sleep Medicine Reviews, 5*, 387–398.

Moldofsky, H. (2002). Management of sleep disorders in fibromyalgia. *Rheumatic Diseases Clinics of North America, 28*, 353–365.

Moldofsky, H. (2003). Fibromyalgia and chronic fatigue syndrome: The role of sleep disturbances, Chapter 60. In M. Billiard (Ed.), *Sleep physiology, investigations and medicine*. Kluwer Academic/Plenum.

Moldofsky, H., & Scarisbrick, P. (1976). Induction of neurasthenic musculoskeletal pain syndrome by selective sleep stage deprivation. *Psychosomatic Medicine, 38*, 35–44.

Moldofsky, H., Scarisbrick, P., England, R., et al. (1975). Musculoskeletal symptoms and non-REM sleep disturbance in patients with "fibrositis syndrome" and healthy subjects. *Psychosomatic Medicine, 37,* 341–351.

Rizzi, M., Sarzi-Puttini, P., Atzeni, F., et al. (2004). Cyclic alternating pattern: A new marker of sleep alteration in patients with fibromyalgia? *Journal of Rheumatology, 31,* 1193–1199.

Roizenblatt, S., Tufik, S., Goldenberg, J., et al. (1997). Juvenile fibromyalgia clinical and polysomnographic aspects. *Journal of Rheumatology, 24,* 579–585.

Pain in Humans, Thermal Stimulation (Skin, Muscle, Viscera), Laser, Peltier, Cold (Cold Pressure), Radiant, Contact

Leon Plaghki
Institute of Neuroscience (IoNS), Université
Catholique de Louvain, Brussels, Belgium

Synonyms

Cold pressure; Cold stimulation; Contact; Laser stimulation; Peltier; Radiant stimulation; Thermal stimulation (skin, muscle, viscera)

Definition

▶ Thermal stimulation consists of transferring (adding or subtracting) calorific energy between the skin (mucosa, muscle, viscera) and its surroundings. There are three mechanisms by which calorific energy can be transferred: conduction, convection, and radiation. These mechanisms can occur in isolation or in combination.

Characteristics

Some Fundamentals of Thermal Physics

Two systems are said to be in thermal equilibrium when they are at the same temperature. This equilibrium is dynamic (e.g., the body continuously radiates to the surroundings and continuously absorbs radiation from the surroundings). When a temperature gradient is created between two systems placed in thermal contact (i.e., such as to allow thermal transfer), changes in temperature initially occur in both systems. Eventually changes cease and both systems are then said to have regained thermal equilibrium. The temperature gradient determines the heat flow between systems through the property of thermal conductivity. The calorific energy added to (subtracted from) the system will raise (decrease) its temperature, depending on the system's heat capacity. The transfer of radiation energy between systems is complicated by the properties of reflectivity, transparency, and absorbance, which are, among other factors, a function of wavelength.

Thermal stimulation consists in creating a temperature gradient between the skin (mucosa, muscle, viscera) and its surroundings in such a way as to allow a net flow of calorific energy toward (heating) or from (cooling) the skin. The flow of energy occurs through three mechanisms, which may operate in isolation or in combination: conduction, convection, and radiation. Both conduction and convection require the presence of matter. Conduction requires thermal motions to diffuse through the contact between bodies. Convection involves fluid movements around the skin. In contrast, radiation requires no intervening medium.

Depending on the main mechanism of heat transfer, different kinds of heat sources (or sinks) can be distinguished. Their advantages and disadvantages are based on the following essential features: specificity for the investigated modality, control precision of stimulus parameters such as intensity (power/duration/surface area), reproducibility, and safety (i.e., avoidance of tissue damage). These will be briefly presented.

Stimulators Based on Heat Transfer by Conduction

The most commonly used heat/cold ▶ contact stimulators rely on the Peltier principle (a thermoelectric effect in which passage of an electric current through a junction between two different solids causes heat to be produced or absorbed at

Pain in Humans, Thermal Stimulation (Skin, Muscle, Viscera), Laser, Peltier, Cold (Cold Pressure), Radiant, Contact, Fig. 1 Stimulators based on heat transfer by conduction. A thermode operating on the Peltier principle and positioned on the right thenar region is shown in the *left panel*. Starting at base skin temperature (Tr), the temperature can be either increased or decreased. The subject's task consists of pressing a button when a given sensation (depending on the instructions received) is perceived. The button press reverses the temperature ramp to return to baseline skin temperature. Recordings of thermal perception thresholds for absolute warmth or heat pain thresholds (*upward*) and cold or cold pain thresholds (*downward*) are displayed in the *right panel*

that junction, according to the direction of the current) (Fig. 1). Advantages of such stimulators are numerous; the stimulus is natural, can be of long duration (seconds), and can cover large surface areas (several cm^2). Cooling and heating rise times, as well as the baseline temperature of the device, can be very precisely controlled (Frustorfer et al. 1976).

Disadvantages of contact heat stimulators are: the mechanical activation of the skin, the variability of heat conductivity (which relies on the quality of the interface), and the slow rise time (around 2 °C/s for most commercially available stimulators). The stimulator usually consists of a rigid surface, which does not always allow a good contact with the skin. The pressure applied against the skin may influence heat flow. Several recent and successful attempts have been made to improve some of these limitations such as increasing rise time (for a review, see Arendt-Nielsen and Chen 2003).

Stimulators Based on Heat Transfer by Conduction and Convection

By immersing a limb (or part of a limb) in a stirred and thermostatically controlled water bath, depending on the temperature of the water, one can transfer calorific energy to or from the skin (Fig. 2).

The cold-pressor test (CPT), a technique widely used in laboratory pain studies, consists in immersing a limb in very cold water (usually below 5 °C). It produces a severe pain that quickly increases toward a maximum, which, in most cases, is tolerated until the end of the test (a few minutes). If the test is prolonged, the pain usually declines. As the pain rises progressively, it is possible to rate pain intensity continuously using a device such as an electronic visual analogue scale interfaced to a computer. Collected data allows computing the pain threshold, the peak pain intensity, and the area under the pain intensity-time curve (Jones et al. 1988). However, results show that measurement of pain threshold, withdrawal (or tolerance) threshold, and pain intensity are subject to great variability (Blasco and Bayes 1988).

Stimulators Based on Heat Transfer by Radiation

The first generation of radiant heat stimulators was based on focused light from a light bulb

Pain in Humans, Thermal Stimulation (Skin, Muscle, Viscera), Laser, Peltier, Cold (Cold Pressure), Radiant, Contact, Fig. 2 Stimulators based on heat transfer by convection. Transfer of calorific energy to or from the skin is obtained by immersing part of a limb in a stirred and thermostatically controlled water bath. By using a device such as an electronic visual analogue scale interfaced to a computer, it is possible to rate pain intensity continuously as shown in the *right panel*. The collected data allow the determination of the pain threshold, the peak pain intensity, and the area under the pain intensity time curve

(Hardy et al. 1967). The key advantage of using radiant heat is the absence of contact and therefore the absence of concurrent mechanical activation of the skin. The disadvantage of these conventional sources is their low power. As long (seconds) exposure times are required, these stimulators do not allow the recording of time-locked responses such as reaction times or event-related potentials. An additional disadvantage is that these sources emit in the visible and near-infrared spectral band. At these wavelengths, human skin energy absorption is poor while reflectivity is important and a function, among other factors, of skin pigmentation. This problem can be partly circumvented by blackening the skin (e.g., with India ink). More powerful devices have been constructed (e.g., xenon flash lamp) bypassing some of these limitations (Andersen et al. 1994).

In the mid-1970s, Mor and Carmon (1975) introduced the laser in pain research. A laser (acronym for light amplification by stimulated emission of radiation) is a unique light source. In comparison to classical incandescent light sources, which emit their radiative energy in all spatial directions and in a large spectrum of wavelengths, the laser energy is confined to a narrow beam of nearly parallel monochromatic electromagnetic waves. This results in a high power density (radiation per unit area). The combination of these characteristics makes the laser a light source with a spectral energy density (radiation per unit wavelength) several orders of magnitude higher than any known light source. This is an essential characteristic if fast and high transfer of radiation energy is needed. To achieve a perfectly defined geometrical configuration of the laser beam, the laser is usually built to operate according to the fundamental mode TEM_{00} (transverse electromagnetic mode). This mode exhibits the so-called Gaussian amplitude profile; the distribution of calorific power in a beam section perpendicular to the optical axis follows a Gaussian distribution centered on that axis. The irradiance profile and beam diameter are measured at the target (Fig. 3). Different laser emission sources have been developed for pain research (e.g., CO_2, argon, thulium-YAG, and laser diodes) (see Arendt-Nielsen and Chen 2003). The CO_2 laser is currently the most

Pain in Humans, Thermal Stimulation (Skin, Muscle, Viscera), Laser, Peltier, Cold (Cold Pressure), Radiant, Contact, Fig. 3 Stimulators based on heat transfer by radiation. Block diagram of a laser thermal stimulator. The infrared heat source is a CO_2 laser. A He-Ne laser beam is mixed with the IR beam to provide a visual image of the target skin area. The beam size is determined by an optical collimator and guided to the skin. Online control of the laser power output is obtained by reflecting a small fraction of the laser beam toward a radiometer

commonly used source for both psychophysical and electrophysiological studies, as it is technically simple, provides high power, is relatively cheap, and is easy to control. The light beam is outside the human visible spectrum. At the wavelength of 10.6 μm, the skin acts as a black body (Hardy and Muschenheim 1934): Reflectivity is negligible (<1 %), absorption is nearly complete, and transparency is very low. For these reasons, the calorific energy remains confined to the upper skin layers (where transducer nerve terminals sensitive to thermal variations are located) and intracutaneous temperature profiles may be easily modeled (Haimi-Cohen et al. 1983) in both time and space (Fig. 4). There is therefore no need for intracutaneous temperature measurements using invasive thermocouples, which are unreliable as the probes themselves modify the thermal field. The short- and long-term stability of the output power can be made very constant. The surface area of the stimulus can be easily adjusted (beam diameter 0.2–20 mm). For the purpose of skin stimulation, the radiant heat can be emitted continuously at low power or, more interestingly, as a pulse whose duration (tens of ms) is controlled by instantaneous power (e.g., 0.2–20 W (Meyer et al. 1976)). High power allows the

production of very steep temperature ramps (thousands °C/s). The result is a selective, very synchronized, and quasi-simultaneous (Fig. 4) activation of Aδ- and C-nociceptors (Bromm and Treede 1984), allowing the recording of time-locked neural responses such as laser-evoked brain potentials. The versatility of laser stimulators allows exploitation of the differences between Aδ and C fiber peripheral conduction velocities, nociceptor activation thresholds, and nociceptor distribution density to study the respective effects of both afferents on central targets in isolation or in combination (Plaghki and Mouraux 2003).

Experimental Studies Using Thermal Stimulation

Heat stimulators are used for the determination of warmth and heat-pain detection thresholds. Using supraliminal stimuli, it is possible to characterize the gain of the sensory channel and such cutaneous hypersensitivity phenomena as hyperesthesia, hyperalgesia, and allodynia. However, results are dependent on numerous factors such as anatomic site, age, sex, and physiological condition (Dyck et al. 1993; Hilz and NeundÃrfer 1990; Verdugo and Ochoa 1992).

Pain in Humans, Thermal Stimulation (Skin, Muscle, Viscera), Laser, Peltier, Cold (Cold Pressure), Radiant, Contact, Fig. 4 Simulation of skin temperature after exposing skin to a heat pulse produced by high power CO_2 laser stimulator. The graph displays skin temperature as a function of time for various skin depths. The parameters for the simulation were set as follows: power $= 0.23$ W/mm^2, beam diameter $= 10$ mm, stimulus duration $= 50$ ms, distance from beam axis $r = 0$, initial skin temperature $= 30\,°C$. The horizontal arrows show the approximate arrival time of nociceptive information at the first spinal relay taking a peripheral nerve conduction distance of 1 m (e.g., stimulation of the right hand dorsum) and a nerve conduction velocity of 8 m/s for Aδ-nociceptors and 1 m/s for C-nociceptors

Variability of results may be explained by differences in stimulation devices (e.g., contact vs. radiant heat, rate of thermal change, stimulus surface area), testing environment (e.g., examiner, room temperature, quietness), testing algorithms (e.g., methods of limits vs. methods of levels), biophysical parameters (e.g., thickness of tissue overlying receptors, spatial distribution of receptors, skin base temperature, and for radiative sources on skin reflectance, absorption, and transparence) and psychological factors (e.g., motivation, affect, suggestion, attention, fatigue, learning). It is clear from this list that it is not recommended to extrapolate reference values obtained by one laboratory to another. There is a risk in comparing studies that are not rigidly controlled in methodology, examiner performance, and testing format (Shy et al. 2003).

Clinical Studies Using Thermal Stimulation

Once thermal stimulators are used for the clinical assessments of sensory dysfunction in humans, it is important to remember the recommendations made by the American Academy of Neurology (Shy et al. 2003). Briefly stated, results of sensory testing with heat stimulators should not be used as sole criteria to diagnose peripheral or central structural neuropathies. Abnormalities must be interpreted in the context of a detailed clinical examination and other appropriate neurological testing procedures available to the clinician. Each laboratory engaged in sensory testing with heat stimuli should use standardized instructions to patients and allow only adequately trained personnel to perform the tests in a designated quiet room. They should be able to demonstrate reproducible results on normal and diseased subjects.

References

Andersen, O. K., Jensen, L. M., Brennum, J., & Arendt-Nielsen, L. (1994). Evidence for central summation of C and AÎ´ nociceptive activity in man. *Pain, 59*, 273–280.

Arendt-Nielsen, L., & Chen, A. (2003). Lasers and other thermal stimulators of skin nociceptors in humans. *Neurophysiologie Clinique, 33*, 259–268.

Blasco, T., & Bayes, R. (1988). Unreliability of the cold pressor test method in pain studies. *Methods and Findings in Experimental and Clinical Pharmacology, 10*, 767–772.

Bromm, B., & Treede, R. D. (1984). Nerve fibre discharges, cerebral potentials and sensations induced by CO_2 laser stimulation. *Human Neurobiology, 3*, 33–40.

Dyck, P. J., Zimmerman, I., Gillen, D. A., et al. (1993). Cool, warm and heat-pain detection threshold: Testing methods and interferences about anatomic distribution of receptors. *Neurology, 43*, 1500–1508.

Frustorfer, H., Lindblom, U., & Schmidt, W. G. (1976). A method for quantitative estimation of thermal thresholds in patients. *Journal of Neurology, Neurosurgery & Psychiatry, 39*, 1071–1075.

Haimi-Cohen, R., Cohen, A., & Carmon, A. (1983). A model for the temperature distribution in skin noxiously stimulated by a brief pulse of CO_2 laser radiation. *Journal of Neuroscice Methods, 8*, 127–37.

Hardy, J. D., & Muschenheim, C. (1934). The emission, reflection and transmission of infra-red radiation by human skin. *The Journal of Clinical Investigation, 13*, 817–831.

Hardy, J. D., Wolff, H. G., & Goodell, H. (1967). *Pain sensations and reactions.* New York: Hafner Publishing Company.

Hilz, C. D., & NeundÄrfer, B. (1990). Thermal discrimination thresholds: A comparison of methods. *Acta Neurologica Scandinavica, 81*, 533–540.

Jones, S. F., McQuay, H. J., Moore, R. A., et al. (1988). Morphine and ibuprofen compared using the cold pressor test. *Pain, 34*, 117–122.

Meyer, R. A., Walker, R. E., & Mountcastle, V. B. (1976). A laser stimulator for the study of cutaneous thermal and pain sensations. *IEEE Transactions on Biomedical Engineering, 23*, 54–60.

Mor, J., & Carmon, A. (1975). Laser emitted radiant heat for pain research. *Pain, 1*, 233–237.

Plaghki, L., & Mouraux, A. (2003). How do we selectively activate skin nociceptors with a high power infrared laser? Physiology and biophysics of laser stimulation. *Neurophysiologie Clinique, 33*, 26277.

Shy, M. E., Frohman, E. M., So, Y., et al. (2003). Quantitative sensory testing. Report of the therapeutics and technology assessment subcommittee of the American academy of neurology. *Neurology, 60*, 898–904.

Verdugo, R., & Ochoa, J. L. (1992). Quantitative somatosensory thermotest. A key method for functional evaluation of small caliber afferent channels. *Brain, 115*, 893–913.

Pain in Humans, Thresholds

Joel D. Greenspan
Department of Neural and Pain Sciences, University of Maryland School of Dentistry, Baltimore, MD, USA

Synonyms

Pain thresholds; Tolerance thresholds

Definition

The earliest fundamental concept in ▶ psychophysics was that of threshold (Boring 1942). Simply stated, threshold is the minimal level of energy needed to evoke a sensation. Accordingly, thresholds are reported in terms of stimulus values, such as temperature levels (in °C) or mechanical forces (kg equivalent weight or Newtons). An alternative measure is the time it takes a constant, sustained stimulus to be perceived as painful, thus measuring pain threshold in terms of seconds. This simple idea of threshold served the field of psychophysics well for more than a century, until it was recognized that threshold was not a stationary phenomenon nor solely based on a person's sensory apparatus. More specifically, psychophysically measured thresholds are dependent upon the neurophysiological signals evoked by a stimulus and the person's evaluation of those signals (Gescheider 1997).

At the other end of the continuum from pain threshold is pain tolerance. Simply stated, pain tolerance is the maximum level of stimulation that a person will voluntarily endure. As with threshold, tolerance is expressed in terms of a stimulus value or, alternatively, the time that a person tolerates a constant, sustained stimulus.

Introduction

The assessment of pain threshold is more complicated than other sensory thresholds because of the nature of pain and people's concept of it. In other sensory modalities, threshold is recognized by the transition between no sensation and sensation. Thus, the subject is attempting to distinguish between sensing something versus nothing. For pain threshold, the subject is instead distinguishing between two types of sensation, one he considers painful and one non-painful. Thus, a critical element in pain threshold determination is the particular sensory experience that an individual considers painful.

Characteristics

Pain threshold, as other pain measures, can be influenced by a variety of biological, psychological, and situational factors, including the subject's pain experience history and the instructions given by the experimenter. For instance, thresholds are likely to be different if a subject is instructed to indicate when he perceives "pain" versus when he perceives "a sharp or burning sensation" or "an uncomfortable sensation." It is also possible that a subject's criterion for judging what sensation is painful changes in the course of an evaluation session. Furthermore, the likelihood and extent of such changes can vary, depending upon the range and number of stimuli applied (Gracely and Naliboff 1996). Several other factors can influence pain threshold values, including features of the psychophysical protocol (the specific design, stimulation parameters, the threshold calculation procedure), which makes it questionable to compare threshold values across studies that use different protocols.

Another important fact to recognize is that threshold is a statistical entity. While we are accustomed to representing thresholds as very precise values (i.e., heat pain thresholds expressed as temperature at a 0.1 °C level of precision), it is not the case that weaker stimuli are necessarily painless and stronger ones are always painful. Instead, there is a range of stimuli for which the lower values are less frequently painful and the higher values are more frequently painful. The threshold value, in

principle, is the midpoint of that range (Fig. 1). This concept applies whether one is considering a single person tested repeatedly or a group of people.

Heat pain threshold has been found to be fairly consistent across many body sites (Hardy et al. 1952). However, heat pain thresholds are significantly higher on ▶ glabrous skin than on the hairy skin of the extremities (Taylor et al. 1993). This relative consistency across the body allows one to assess regional pain threshold abnormalities by comparing thresholds between two body sites, when one of them is accepted as a reference. This approach is often used by comparing thresholds between homolateral body sites in the cases of unilateral sensory abnormalities (Kemler et al. 2000).

Pain tolerance is less frequently used than pain threshold in scientific studies, and it has some significant disadvantages:

1. For some forms of stimulation, pain tolerance cannot be reached without risking tissue injury
2. Pain tolerance generally shows greater variability than threshold, both within and across subjects.
3. It is more widely altered by subject bias or past experience than threshold.

However, pain tolerance measures a qualitatively different aspect of the pain experience than pain threshold does. Arguably, pain tolerance is a measure more reflective of the affective and motivational aspects of the pain experience, while threshold is a measure of the discriminative aspect. One common form of this test is the "cold pressor test," involving submersion of a body part in ice water (Hilgard et al. 1974; Chen et al. 1989) or the ischemic technique, using a pressure cuff on the subject's arm (Smith et al. 1972). In both cases, tolerance is measured as the time that the subject will allow exposure to the stimulation.

Summary

Pain threshold, while similar in principle to other sensory thresholds, is unique in that it represents the transition point between two sensory

Idealized Threshold Function

Realistic Threshold Function

Pain in Humans, Thresholds, Fig. 1 Ideally, pain threshold can be envisioned as the stimulus level that divides non-painful from painful intensities of stimulation (*left*). In reality, however, there is a range of stimulus intensities that are sometimes perceived as painful and sometimes as non-painful. Thus, pain threshold is typically regarded as the stimulus intensity that is painful 50 % of the time (*right*)

experiences (non-painful and painful) rather than the detection of a sensory event per se. While pain thresholds are sometimes considered to be more stable and consistent than other pain measures, they too are influenced by the same factors that influence any pain experience: individual constitutive differences, experiential effects, and situational influences. Threshold is not a fixed entity, but probabilistic, and derived statistically. Pain tolerance (occasionally referred to as tolerance threshold) is the upper end of the pain experience continuum.

References

Boring, E. G. (1942). *Sensation and perception in the history of experimental psychology*. New York: Appleton-Century-Crofts.

Chen, A. C. N., Dworkin, S. F., Haug, J., et al. (1989). Human pain responsivity in a tonic pain model: Psychological determinants. *Pain, 37*, 143–160.

Gescheider, G. A. (1997). *Psychophysics: The fundamentals*. Mahwah: Lawrence Erlbaum Associates.

Gracely, R. H., & Naliboff, B. D. (1996). Measurement of pain sensation. In L. Kruger (Ed.), *Pain and touch* (pp. 243–313). San Diego: Academic.

Hardy, J. D., Wolff, H. G., & Goodell, H. (1952). *Pain sensations and reactions*. Baltimore: The Williams & Wilkins.

Hilgard, E. R., Ruch, J. C., Lange, A. F., Lenox, J. R., Morgan, A. H., & Sachs, L. B. (1974). The psychophysics of cold pressor pain and its modification through hypnotic suggestion. *The American Journal of Psychology, 87*, 17–31.

Kemler, M. A., Schouten, H. J. A., & Gracely, R. H. (2000). Diagnosing sensory abnormalities with either normal values or values from contralateral skin – Comparison of two approaches in complex regional pain syndrome I. *Anesthesiology, 93*, 718–727.

Smith, G. M., Lowenstein, E., Hubbard, J. H., et al. (1972). Experimental pain produced by the submaximum effort tourniquet technique: Further evidence of validity. *Journal of Pharmacology and Experimental Therapeutics, 163*, 468–474.

Taylor, D. J., McGillis, S. L. B., & Greenspan, J. D. (1993). Body site variation of heat pain sensitivity. *Somatosensory and Motor Research, 10*, 455–466.

Pain in Radiculopathy

▶ Peripheral Neuropathic Pain

Pain in Superficial Tissues

Hermann O. Handwerker
Department of Physiology and Pathophysiology,
University of Erlangen/Nürnberg, Erlangen,
Germany

Characteristics

Significance of Nociception in Superficial Tissues

Superficial tissues constitute an interface between the body and the outer world. They comprise the cornea and conjunctiva of the eye, the nasal mucosa, the oral mucosa (see ▶ Nociception in Nose and Oral Mucosa), the teeth, and the mucosa of the external sexual organs (see ▶ Nociception in Mucosa of Sexual Organs). Of course, the most important of these organs is the skin, which constitutes by far the largest sensory organ of our body. All these tissues have in common that they guard and protect the body's interior against the environment. Obviously, a protective, i.e., "nociceptive" neuronal system is mandatory for this aim. This becomes very evident in the rare cases of ▶ congenital insensitivity to pain with anhidrosis, the CIPA syndrome, which is due to a mutation of the gene encoding for the trkA receptor, the high affinity receptor for the nerve growth factor (NGF). Children suffering from this syndrome are characterized by multiple unattended bruises, burns, etc., due to pertinent maltreatment.

To fulfill the protective role, the body surface is densely innervated, and indeed, most afferent nerve fibers providing input to the CNS come from it; far fewer nerves come from muscles and joints, and comparatively, little input has its origin in the interior organs of the body. The central nervous projections of the afferent input from the body surface are well organized in a specific thalamocortical system characterized by somatotopy of the neuronal populations receiving input from different sites on the body surface. However, pain processing in the CNS is obviously not confined to somatosensory projection regions. It has been shown by ▶ functional imaging of cutaneous pain that pain is processed in a network of multiple brain locations including those not somatotopically organized.

The largest "superficial tissue" of the body, the skin, fulfils a second task besides protecting the interior organs, namely, the temperature exchange necessary to keep the internal temperature of homoeothermic animals at a more or less constant level. To this end, the skin has a sophisticated vascular system controlled by sympathetic efferent nerves. Many cutaneous nociceptors release vasoactive neuropeptides, e.g., CGRP (calcitonin gene-related peptide) that contribute to the regulation of blood flow, particularly in the superficial skin layers.

Pain Theories

Not surprisingly, the skin was the first organ for which hypotheses were created to explain how nerve excitation leads to pain. In the late nineteenth century, Goldscheider and Blix independently of each other discovered the "cold points" in the skin, defined spots where the touch of a small cold probe induces cold sensations, whereas only touch sensations are perceived from other skin sites. M.v. Frey extended these findings to "touch" and "pain" points and thus inaugurated the "specificity hypothesis" which postulates that separate sensory channels from the skin mediate the sensations of cold, warmth, touch, and pain (von Frey 1896). This hypothesis replaced the older "intensity hypothesis" that assumed that strong excitation of a common, not further specified, cutaneous sensory system leads to pain. Sherrington coined the term "nociceptor" for the sensory endings of the nerve fibers mediating pain (Sherrington 1906). However, the term nociceptor remained a functionally, rather than a structurally, defined entity.

With the advent of refined electrophysiological techniques in the middle of the twentieth century, Erlanger and Gasser established three populations of cutaneous afferent nerve fibers, those with fast conducting and thick myelinated axons, labeled $A\alpha,\beta$; slower conducting, thin myelinated axons of the $A\delta$ class; and the unmyelinated C fibers. These three classes of nerve fibers are clearly distinguished by their different conduction velocities, in the range of 30–70 m/s for Aα,β fibers, 2–30 m/s for Aδ, and less than 2 m/s for the unmyelinated C fibers. Erlanger and Gasser were able to establish that pain is only induced in normal skin when the slower conducting Aδ and/or C fibers are excited (Gasser 1935).

M.v. Frey had tried to correlate sensory structures in the skin with the different sensory modalities, albeit only with partial success. The myelinated nerve fibers connected to Merkel cells were soon identified as part of the "touch system," but an equally convincing association of the nerves at "pain points" with associated characteristic cells failed. The terminal endings in the skin of the thin nerve fibers are not generally endowed with corpuscular structures but were regarded as "naked nerve endings." However, not all "naked nerve endings" are nociceptors. Wedell and coworkers, studying the structure of nerve endings in the cornea, found that in this tissue, only "naked nerve endings" are found (Wedell et al. 1959). On the other hand, from their psychophysical experiments they came to the conclusion that sensations of cooling, touch, and pain can be induced from the cornea. This led them to rebuff the "specificity hypothesis" and to inaugurate a "pattern hypothesis" which flourished until single nerve fiber electrophysiology was established and electrophysiologists were unable to find particular patterns of nerve fiber discharges characteristic for pain input. Nevertheless, this line of thinking survived somehow in form of the "▶ gate control" hypothesis of Melzack and Wall (1965). According to this, pain is dependent on the ratio of input between myelinated "touch" and unmyelinated "pain" afferents through a presynaptic gate mechanism at the entrance to the spinal cord by which myelinated fibers close the door of a pain pathway whereas unmyelinated fibers open it. The gate control hypothesis thus postulated a "spatial" pattern instead of the temporal pattern of Wedell and coworkers.

During the last two decades, the interest in the time-honored debate on pain mechanisms declined. It is now quite clear that pain is initiated by a dedicated class of peripheral nerve fibers, the ▶ nociceptors, at least when the tissues and the nervous system have not been altered before being affected by a painful stimulus. However, nociceptor excitation does not always lead to pain, and not all kinds of pathological pain are mediated by nociceptors (see below). The term "nociceptor" itself has changed. Research is no longer focusing on specific morphological structures and not even much on thresholds to natural stimuli but rather on biochemical and molecular features of neurons.

Properties of Nociceptors

Cutaneous sensory receptors may roughly be subdivided into four categories:

1. Predominantly myelinated afferents that respond to gentle deformations of the skin. Many of the respective sensory nerve endings are connected to corpuscular structures. These afferents are referred to as low threshold mechanoreceptors (LTM) and provide information for touch and gentle pressure. Seen under the aspect of perceived objects, they deal with the shape and texture of objects coming in contact with the skin. The molecular mechanisms of their action are still unknown.

2. Afferents that are selectively sensitive to gentle or moderate cooling stimuli. These "cold fibers" mediate sensations of cold and are thinly myelinated or unmyelinated. Their terminals appear as "naked nerve endings." Recently, membrane structures have been described which might explain the cold sensitivity, i.e., certain cation channels of the TRP family that are cold and menthol sensitive, namely the TRPM8 channels (McKemy et al. 2002) and certain temperature-sensitive potassium channels (Viana et al. 2002).

3. "Warm fibers" which are C fibers that respond to mild heating stimuli and account for the sensation of warmth. They are usually capsaicin sensitive and therefore may be equipped with TRPV1 receptors, which also characterize many heat-sensitive nociceptors.

4. A broad category of units with high threshold receptors for mechanical and thermal stimuli. Because these receptors respond most strongly

to noxious mechanical and thermal stimuli and to chemical agents which either might damage the skin (e.g., acids) or are released as a consequence of skin damage (e.g., ► Inflammatory Mediators), they are labeled "nociceptors" using Sherrington's term. These afferents may be thinly myelinated or unmyelinated and are thought to be responsible for the sensations of pain, itch, and prickle.

Nociceptors were first discovered in the skin, and their properties were described for the body surface. The first to record from nociceptive nerve fibers was Y. Zotterman (1939). Later, A. Iggo achieved single unit recordings and used a spike collision technique to prove the nociceptive properties of individual C fibers (Iggo 1958). Nociceptors are found among thinly myelinated (Aδ) and unmyelinated (C-) afferent nerve fibers, but the unmyelinated units are much more frequent than the faster conducting thinly myelinated units. Though the general properties of nociceptors are similar among mammals, there are species differences. This chapter refers mainly to the nociceptor populations in humans and in other primates.

Different criteria may be used to classify nociceptors:

1. Unmyelinated versus myelinated afferent nerve fibers
2. Adequate stimuli
3. Type of response (e.g., rapidly adapting vs slowly adapting)
4. Immunocytochemistry (e.g., expression and secretion of neuropeptides or expression of receptors for nerve growth factors NGF and BDNF, IB4)

Identification of the adequate stimuli (i.e., the stimuli that most readily activate the respective unit) and determination of the conduction velocity serve most commonly as a basis for characterizing types of nociceptors. Superficial tissues provide an important advantage for studying nociceptor properties since they are readily accessible for psychophysical studies in humans and behavioral studies in animals. Therefore, in the case of cutaneous nociceptors, a relationship can be established between sensations and stimulus magnitude (► Psychophysics) and between

discharge rates of individual fiber types and sensations (psychophysiology). A unique possibility of correlating sensations and nerve input into the CNS is provided by ► microstimulation (Ochoa and Torebjörk 1989; Torebjörk 1993) in the context of microneurography (► Microneurography and Intraneural Microstimulation) experiments.

CMHs and AMHs

One important distinction of nociceptor class is that between units of different conduction velocity. Generally, units conducting faster than 2 m/s are regarded as myelinated, whereas slower conducting units are unmyelinated. Since mechanical and heat stimuli were most often used to discriminate between cutaneous receptor classes, the names "AMH" for units with A fibers which are mechanically and heat sensitive and "CMH" for the respective C fiber units were introduced for the most common types of nociceptors. Since most of these units also respond to inflammatory mediators and other chemical stimuli, i.e., to algogenic substances, they were also labeled "polymodal" nociceptors.

CMHs in different types of skin respond similarly to heat and mechanical stimuli. The AMHs, however, respond in either of two distinct ways to heat. Type I AMHs, found in both hairy and glabrous skin (i.e., palmar and plantar skin of the extremities), have relatively high thresholds to heating stimuli but become sensitized following a burning trauma or topical capsaicin application to the skin. After sensitization, their heat threshold drops by several degrees centigrade. Type II AMHs occur only in hairy skin and have lower heat thresholds (mean, 46 °C) (Meyer et al. 1985). Type I AMHs are also termed high threshold mechanoreceptors in the literature (HTMs (Perl 1968)), since their initial unconditioned threshold to heating often exceeds the limitation of stimuli set to avoid skin damage or to exceed the tolerance of subjects in microneurography experiments. A comparable class of CM units has been described which do not respond to heating stimuli up to 50 °C. It is unclear whether these units are in other respects different from the more common CMH units.

Only rarely were nociceptors characterized according to their sensitivity to noxious cold. Many CM units, but also CMH and AMH units, respond to cooling to temperatures lower than 20 °C and may be essential for the initiation of cold pain. However, the thresholds vary over a wide range and constant cooling to temperatures close to 0 °C blocks axonal conduction.

Mechano-insensitive Nociceptors

An interesting class of cutaneous nociceptor has been described in the monkey (Cohen et al. 1990) and may be characteristic of primate (and perhaps pig) skin. These A or C units do not respond to strong mechanical stimuli and were called MIA (mechano-insensitive units). Similar mechano-insensitive C fibers were characterized in microneurography experiments on human skin nerves and named CM_i units. These units become responsive to mechanical stimulation when the skin is inflamed. Since sensitization "awakens" these units to mechanical responsiveness, they have been poetically called "sleeping nociceptors." One has to note, however, that many of these units are responsive to heating stimuli without preceding sensitization and most of them are excited by ► capsaicin, the specific ligand of the VR1 receptor. Therefore, they can also be regarded as heat or chemo-nociceptors. In the region innervated by the superficial peroneal nerve of man, they comprise about 30 % of the afferent C fibers (Schmidt et al. 1995).

In man, these units generally have larger receptive fields and slower conduction velocities than the more common CMH units. Interestingly, the axons of Cm_i units are characterized by postexcitatory subnormal and supranormal periods that are distinct from those of CMHs and CMs. This may indicate different voltage-dependent membrane channels and/or pump mechanisms (see ► Mechano-Insensitive C-Fibers, Biophysics) (Weidner et al. 1999).

Nociceptors and Pain

A painful stimulus affecting the skin with a fast rise time (e.g., a pinprick) often induces two pain phenomena. A sharp "first" pain is followed by a "second" pain of burning, aching, or throbbing character. The "two pains" phenomenon is characteristic for stimuli applied to skin sites in the distal extremities at large conduction distances from the CNS. The "first pain" is generally attributed to AMHs, in particular AMH II units; the "second pain" may be mainly due to the conduction in C fibers.

In psychophysical studies, the discharges of cutaneous nociceptors to controlled ramp-like or steplike stimuli in monkey or human skin were compared with the pain sensation in human volunteers. There is ample evidence in support of a role of CMHs in heat pain sensation, which includes:

1. Pain increases monotonically with stimulus intensities between 40 °C and 50 °C (LaMotte and Campbell 1978; Meyer and Campbell 1981a). Since also the responses of CMHs increase monotonically over this temperature range, they are likely to be crucial for the transmission of heat pain.

2. Human judgments of pain in response to stimuli over the range 41–49 °C correlate well with the activity of CMH nociceptors over this range.

3. The ability of human subjects to discriminate between heating stimuli of different strengths correlates with the differences in discharge rates of CMHs (Gybels et al. 1979).

4. Selective A-fiber ischemic blocks indicate that C-fiber function is necessary and sufficient for heat pain perception (Torebjörk and Hallin 1973).

5. Stimulus interaction effects observed in psychophysical experiments (Adriaensen et al. 1980) are also observed in recordings from CMHs.

6. The latency to pain sensation on glabrous skin following step temperature changes is long and therefore consistent with conduction in CMHs (Campbell and LaMotte 1983).

7. In patients with congenital insensitivity to pain, microscopic examination of the peripheral nerves indicates the absence of C fibers (see ► Congenital Insensitivity to Pain with Anhidrosis).

8. The heat threshold for activation of CMHs recorded in conscious humans is just below

the pain threshold (Gybels et al. 1979; Van Hees and Gybels 1981).

9. A linear relationship exists between responses of CMHs recorded in conscious humans and of pain over the temperature range 39–50 °C (Torebjörk et al. 1984).

AMH units also contribute to heat pain sensations. This has been demonstrated for long duration heating applied to the glabrous skin of the hand in human subjects, which evoked substantial pain for the duration of the stimulus. CMHs have a prominent discharge during the early phase of the stimulus, but this response adapts within seconds to a low level. In contrast, type I AMHs are initially unresponsive but then discharge vigorously. Therefore, type I AMHs probably contribute to the pain during a sustained high-intensity heat stimulus (Meyer and Campbell 1981b).

On the other hand, the "first pain" of a heat stimulus with fast onset is probably mediated by type II AMHs in the hairy skin, and the absence of a "first pain" sensation to heat stimuli in the glabrous skin of the hand is paralleled by the absence of AMH II in this region (Campbell and LaMotte 1983).

In addition to these types of "conventional" nociceptors, mechano-insensitive CM_i units also contribute to pain sensations:

1. Long-lasting tonic pressure stimuli applied to the skin induce slowly increasing pain. However, the discharges of CMHs adapt quickly after an initial burst of activity. In contrast, CM_i units are unresponsive in the beginning and get slowly activated during the course of such a stimulus, in parallel with the pain sensations (Torebjörk et al. 1996).

2. Similarly, intracutaneous injection of a small bolus (0.1 μl of a 1 % solution) of capsaicin induces an initial high frequency burst of discharges in CMHs and renders them mechano- and heat-insensitive thereafter. In contrast, CM_i units are activated and show activity that waxes and wanes for many minutes in parallel with the burning sensations.

Since different types of nociceptors contribute to different types of pain sensations, it is of interest to know which types of sensation are induced by stimulation of just one receptor type. This aim

can be achieved by intraneural microstimulation (Ochoa and Torebjörk 1983).

Stimuli employing this technique have shown that single impulses induced in CMHs often do not induce any sensation, indicating that a central nervous threshold that requires temporal and/or spatial summation has to be exceeded. Frequent electrical stimulation of presumed single CMHs induces pain, however, often burning in character (Torebjörk and Ochoa 1980). The sensations induced by A-delta fiber stimulation are more stinging in character. Whereas single electrical shocks cannot be resolved when a C-unit is stimulated, this is easily possible when A fibers are stimulated.

Pain and Itch

In the early twentieth century, M.v. Frey who had discovered the "pain points" in the skin (see above) and thus had found evidence for a specific peripheral pain pathway proposed an "intensity" hypothesis for explaining itch sensations. He came to this conclusion from the observation that weak stimulation of the "pain points" in the skin leads with a long latency to itching but stronger stimulation to pain.

The intensity hypothesis was supported by the observation that ▶ anterolateral cordotomy abolished itch and pain likewise. The similarity of the neuronal apparatus for itch and pain was also demonstrated by the contemporary finding that itch and heat pain (but not the sensations of touch and of pinprick) are abolished in a skin patch desensitized with capsaicin. This lead to the conclusion that both sensory modalities are mainly supported by unmyelinated, capsaicin-sensitive primary afferents.

However, an "intensity" hypothesis cannot be brought in line with the well-established phenomenon that itch is suppressed by simultaneously induced pain and with the differential effect of opioids which are known to suppress pain but to enhance itching. Therefore alternative hypotheses had to be considered. An "occlusion" hypothesis postulated that a smaller population of chemosensitive primary afferents supports itch, whereas a larger population supports pain. Itch stimuli like histamine excite only the small population of primary afferents when brought into the superficial layers of human skin and thus induce itching. However, pain stimuli (e.g., heating, strong pressure, or pinpricks) excite this small population of units sensitive to pruritogens and in addition a much larger population of nociceptors which are insensitive to pruritogens, and this would induce pain and occlude itch by appropriate synaptic mechanisms in the CNS.

Some years ago, a class of slowly conducting C fibers was described which were highly sensitive to histamine. Since the time course of the excitation of these units by pruritogenic agents strongly resembles the time course of itching induced by these agents, they were regarded as "itch units" (▶ Itch/Itch Fibers). These units belong to the mechano-insensitive C units and are characterized by extremely slow conduction velocities and very large receptive fields (Schmelz et al. 1997). The notion of a specific "itch channel" in the PNS was supported by the finding of a corresponding specific central pathway in the cat by Craig and Andrew (Andrew and Craig 1999).

It has been shown above that not all sorts of pain can be explained by one single class of nociceptors. Likewise, not all forms of itching can be simply explained by the excitation of histamine-sensitive CM_i units which may, however, play an important role in physiological (like itching after insect bites) or pathological forms of itching (like itching from urticaria), which are sensitive to antihistaminic drugs (H1 blockers). Other forms of itching, e.g., in atopic dermatitis, are insensitive to antihistaminic therapy, and in these patients, itch can be induced by stimuli which do not excite CM_i units. Itching then may be caused by "rechanneling" of impulses from nerve fibers normally mediating pain sensations. Microneurography experiments with cowhage (mucuna pruriens) an agent inducing non-histaminergic itch revealed that this agent excited not the histamine-sensitive itch fibers but rather mechanoresponsive CMH units (Namer et al. 2008).

Even in experimental situations on normal subjects, itch induced by histamine iontophoresis can be rekindled by gently stroking the affected skin site and its environment with a soft brush,

exciting only sensitive mechanoreceptor units (see ▶ Allodynia and Alloknesis). This clearly indicates that impulses from – presumably myelinated – mechanoreceptor units can be "gated" into the itch pathway in analogy to the gating of this kind of input into the pain pathway in allodynia.

Hyperalgesias

Two basic forms of hyperalgesia have to be distinguished: ▶ primary hyperalgesia confined to the place of tissue injury and ▶ secondary hyperalgesia extending into the closer, or even the more distant, surroundings of a trauma. There has been a long-standing debate about the relative importance of peripheral and central neuronal mechanisms in generating hyperalgesia. Here again, we have to distinguish between hyperalgesias due to inflammatory processes in the tissues and those due to alteration of the conductile membranes of peripheral nerves. Initially, the excitability increase of either peripheral or central neural pathways was favored. Recently, it has been suggested that peripheral and central mechanisms may contribute differentially depending on the type of tissue injury and sensory modality being tested (Kilo et al. 1994). This is particularly important for hyperalgesia affecting the skin, with its highly differentiated innervation. Therefore, one has to distinguish again between hyperalgesia to heating, to different forms of mechanical stimulation, and to cold:

1. *Hyperalgesia to Heating.* In several studies on human and monkey skin, it has been shown that topical capsaicin treatment is able to lower the heat thresholds of CMH units by 5 °C or more (Baumann et al. 1991; LaMotte et al. 1992). Similar changes were seen after burning lesions. Such sensitization of nociceptors may well explain primary hyperalgesia to heating. Probably CM and CM_i units, by becoming sensitized to heating, contribute to spatial summation at central synapses. The increased impulse rate in sensitized C units may lead to co-activation of neuropeptide (NK and CGRP) receptors and different types of glutamate receptors, including NMDA receptors, thus enhancing synaptic

efficacy in the pain pathway and inducing ▶ primary heat hyperalgesia. Secondary heat hyperalgesia around a heating trauma or a capsaicin-treated skin patch has not been found in most studies. If it exists, it is not very prominent.

2. *Hyperalgesia to Tonic Blunt Pressure.* Hyperalgesia to tonic blunt pressure was found only in the immediate area of a heating or freezing trauma or of capsaicin application and not in the surroundings (primary hyperalgesia) (Koltzenburg et al. 1992; Kilo et al. 1994). It is known to persist when the A fibers are differentially blocked, and it is diminished by cooling the skin, which reduces the activity in sensitized C fibers. However, the discharges of CMH units to blunt pressure stimulation are not increased after topical capsaicin (LaMotte et al, 1992). As in normal tissue, these units tend to adapt after an initial burst of activity. Therefore, the recruitment of sensitized CH and CM_i units may be most important for this type of hyperalgesia. A-delta HTM units might also contribute to it.

3. *Pinprick Hyperalgesia.* Although hyperalgesia to punctuate stimuli with pins or hypodermic needles has rarely been studied in inflamed tissue (primary form), it clearly extends into the surroundings of a local trauma (secondary hyperalgesia). No clear signs of nociceptor sensitization to this form of stimulation were found in microneurography experiments. Therefore, it probably represents an example of ▶ central sensitization due to changes in responsiveness of centrally projecting units caused by the initial nociceptor impulse barrage, but it can be maintained for some time in the absence of such input (Kilo et al. 1994). It has been shown that the units mediating the secondary hyperalgesia to pinprick are myelinated (A-delta units) and not capsaicin sensitive (Magerl et al. 2002).

4. *Touch-Evoked Hyperalgesia.* This form of hyperalgesia is best evoked by gently stroking the skin, e.g., with a swap of cotton wool. It is often found in inflamed skin, e.g., after sunburn (primary form), but also in the surroundings of a focal inflammation (secondary form).

The latter has been clearly demonstrated in a wide area around a capsaicin injection site, though the area of touch-evoked hyperalgesia is smaller than that of pinprick hyperalgesia. Recordings from nociceptors around capsaicin injection sites in monkeys (Baumann et al. 1991) have not shown any evidence of peripheral sensitization to tactile stimuli. Instead, experiments with selective nerve blocks revealed that the tactile hyperalgesia was only detectable in the presence of intact conduction in large myelinated fibers and could be mimicked by intraneural electrical microstimulation of low threshold mechanoreceptive afferents that normally signal non-painful touch (Torebjörk et al. 1992). The notion that this form of hyperalgesia is mediated by myelinated sensitive mechanoreceptor units has lead to the denomination of touch-evoked hyperalgesia as "▶ allodynia" (i.e., pain induced by another form of sensory impact). This hypothesis is supported by the finding that neurons in the spinal cord of a monkey receiving nociceptor input displayed increased excitability to A-fiber input following capsaicin injections (Simone et al. 1991). It appears that ▶ touch-evoked hyperalgesia or ▶ allodynia is related to a sensitization of central neurons that is critically dependent on ongoing activity in nociceptive afferents and the hyperalgesia quickly disappears when such input is stopped. Thus, in microneurography experiments after the topical application of mustard oil, if the ongoing activity in sensitized nociceptors is stopped by cooling the skin, the touch-evoked hyperalgesia disappears. The linear relationship between background pain and touch-evoked hyperalgesia has also been observed in patients in whom touch-evoked hyperalgesia was a symptom of neuropathy (Koltzenburg et al. 1994). It is therefore independent of its underlying pathophysiology. From these observations, it appears that touch-evoked hyperalgesia is due to an ongoing nociceptor barrage gating the A-fiber impulses into the pain pathway. This nociceptive input does not need to be very strong, since the phenomenon of transient touch-evoked hyperalgesia can be demonstrated (Cervero et al. 1994).

5. *Hyperalgesia to cold* is not apparent at inflamed skin sites. In contrast, cooling relieves the pain, due to suppression of ongoing and evoked C-fiber discharges. Cold pain is, however, often observed in neuropathic pain states (Wahren et al. 1991). The pathophysiology of cold hyperalgesia is still enigmatic. A partial explanation is, however, provided by the following observations. The sensation of cooling the skin is mediated mainly by thin myelinated A-delta cold units. When these units are differentially blocked in a healthy subject leaving conduction in C fibers intact, strong cold stimuli paradoxically induce a burning sensation (Yarnitsky and Ochoa 1990). This has been reported to be typical for patients suffering from cold hyperalgesia. In ► quantitative sensory testing, these patients often show elevated thresholds for cooling, thus combining cold *hypo*esthesia with *hyper*algesia. These observations suggest that cold hyperalgesia may be due to a central nervous disinhibition of the input from cold responsive nociceptive C fibers.

6. *Head Zones.* Pain referred from inner organs to the skin is the classical example of a pain phenomenon caused by central nervous mechanisms. Henry Head made the observation that the skin sites to which the transfer occurs are typical for certain inner organs. Referred pain has been ascribed to convergent projections from inner organs and the respective skin sites to the same spinal and supraspinal projection neurons. Therefore, hyperalgesia within aching Head zones is centrally mediated. One should therefore expect preferentially those forms of hyperalgesia that are caused by central nervous mechanisms in other pathophysiologies. Unfortunately, systematic studies are still lacking.

Conclusion

The various forms and qualities of cutaneous pain cannot be explained by one single form of nociceptor input. One has to define exactly the stimulus conditions for physiological pain forms and the types of pathological changes for pathological pain. This applies even more strikingly to the different forms of hyperalgesia to which peripheral and central nervous processing contribute in different ways.

References

Adriaensen, H., Gybels, J., Handwerker, H. O., et al. (1980). Latencies of chemically evoked discharges in human cutaneous nociceptors and of the concurrent subjective sensations. *Neuroscience Letters, 20*, 55–59.

Andrew, D., & Craig, A. D. (1999). Histamine-selective lamina I spinothalamic tract neurons. *Society of Neuroscience Abstract, 25*, 149.

Baumann, T. K., Simone, D. A., Shain, C. N., et al. (1991). Neurogenic hyperalgesia: the search for the primary cutaneous afferent fibers that contribute to capsaicin-induced pain and hyperalgesia. *Journal of Neurophysiology, 66*, 212–227.

Campbell, J. N., & LaMotte, R. H. (1983). Latency to detection of first pain. *Brain Research, 266*, 203–208.

Cervero, F., Meyer, R. A., & Campbell, J. N. (1994). A psychophysical study of secondary hyperalgesia – Evidence for increased pain to input from nociceptors. *Pain, 58*, 21–28.

Cohen, R. H., Meyer, R. A., Davis, K. D., et al. (1990). Mechanically insensitive afferents (MIAs) in cutaneous nerves of monkey. *Pain. Supplement, 5*, S105–S101.

Gasser, H. S. (1935). Conduction in nerves in relation to fiber types. *Research Publications - Association for Research in Nervous and Mental Disease, 15*, 35–59.

Gybels, J., Handwerker, H. O., & Van Hees, J. (1979). A comparison between the discharges of human nociceptive nerve fibres and the subject's ratings of his sensations. *The Journal of Physiology, 292*, 193–206.

Iggo, A. (1958). The electrophysiological identification of single nerve fibres, with particular reference to the slowest-conducting vagal afferent fibres in the cat. *The Journal of Physiology, 142*, 110–126.

Kilo, S., Schmelz, M., Koltzenburg, M., et al. (1994). Different patterns of hyperalgesia induced by experimental inflammation in human skin. *Brain, 117*, 385–396.

Koltzenburg, M., Lundberg, L. E. R., & Torebjörk, H. E. (1992). Dynamic and static components of mechanical hyperalgesia in human hairy skin. *Pain, 51*, 207–219.

Koltzenburg, M., Kees, S., Budweiser, S., et al. (1994). The properties of unmyelinated nociceptive afferents change in a painful chronic constriction neuropathy. In G. F. Gebhart, D. L. Hammond, & T. S. Jensen (Eds.), *Proceedings of the 7th world congress on pain* (pp. 511–522). Seattle: IASP Press.

LaMotte, R. H., & Campbell, J. N. (1978). Comparison of responses of warm and nociceptive C-fiber afferents in monkey with human judgments of thermal pain. *Journal of Neurophysiology, 41*, 509–528.

LaMotte, R. H., Lundberg, L. E. R., & Torebjörk, H. E. (1992). Pain, hyperalgesia and activity in nociceptive-C units in humans after intradermal injection of capsaicin. *The Journal of Physiology, 448*, 749–764.

Magerl, W., Fuchs, P., Meyer, R. A., et al. (2002). Roles of capsaicin-insensitive nociceptors in cutaneous pain and secondary hyperalgesia. *Brain, 124*, 1754–1764.

McKemy, D. D., Neuhausser, W. M., & Julius, D. (2002). Identification of a cold receptor reveals a general role for TRP channels in thermosensation. *Nature, 416*, 52–58.

Melzack, R., & Wall, P. D. (1965). Pain mechanisms: A new theory. *Science, 150*, 971–978.

Meyer, R. A., & Campbell, J. N. (1981a). Evidence for two distinct classes of unmyelinated nociceptive afferents in monkey. *Brain Research, 224*, 149–152.

Meyer, R. A., & Campbell, J. N. (1981b). Myelinated nociceptive afferents account for the hyperalgesia that follows a burn to the hand. *Science, 213*, 1527–1529.

Meyer, R. A., Campbell, J. N., & Raja, S. N. (1985). Peripheral neural mechanisms of cutaneous hyperalgesia. In H. L. Fields, R. Dubner, & F. Cervero (Eds.), *Advances in pain research and therapy* (Vol. 9, pp. 53–71). New York: Raven.

Namer, B., Carr, J. Johanek, L. M., Schmelz, M., Handwerker, H. O., Ringkamp, M. (2008). Separate peripheral pathways for pruritus in man. *Journal of Neurophysiology, 100*, 2062–2069.

Ochoa, J. L., & Torebjörk, H. E. (1983). Sensations evoked by intraneural microstimulation of single mechanoreceptor units innervating the human hand. *The Journal of Physiology, 342*, 633–654.

Ochoa, J. L., & Torebjörk, H. E. (1989). Sensations evoked by intraneural microstimulation of C nociceptor fibres in human skin nerves. *The Journal of Physiology, 415*, 583–599.

Perl, E. R. (1968). Myelinated afferent fibers innervating the primate skin and their response to noxious stimuli. *The Journal of Physiology, 197*, 593–615.

Schmelz, M., Schmidt, R., Bickel, A., et al. (1997). Specific C-receptors for itch in human skin. *Journal of Neuroscience, 17*, 8003–8008.

Schmidt, R., Schmelz, M., Forster, C., et al. (1995). Novel classes of responsive and unresponsive C nociceptors in human skin. *Journal of Neuroscience, 15*, 333–341.

Sherrington, C. S. (1906). *The integrative action of the nervous system.* New Haven: Yale University Press.

Simone, D. A., Sorkin, L. S., Oh, U., et al. (1991). Neurogenic hyperalgesia: central neural correlates in responses of spinothalamic tract neurons. *Journal of Neurophysiology, 66*, 228–246.

Torebjörk, H. E. (1993). Human microneurography and intraneural microstimulation in the study of neuropathic pain. *Muscle & Nerve, 16*, 1063–1065.

Torebjörk, H. E., & Hallin, R. G. (1973). Perceptual changes accompanying controlled preferential blocking of A and C fibre responses in intact human skin nerves. *Experimental Brain Research, 16*, 321–332.

Torebjörk, H. E., LaMotte, R. H., & Robinson, C. J. (1984). Peripheral neural correlates of magnitude of cutaneous pain and hyperalgesia: simultaneous recordings in humans of sensory judgments of pain and evoked responses in nociceptors with C- fibers. *Journal of Neurophysiology, 51*, 325–339.

Torebjörk, H. E., Lundberg, L. E. R., & LaMotte, R. H. (1992). Central changes in processing of mechanoreceptive input in capsaicin-induced secondary hyperalgesia in humans. *The Journal of Physiology, 448*, 765–780.

Torebjörk, H. E., & Ochoa, J. L. (1980). Specific sensations evoked by activity in single identified sensory units in man. *Acta Physiologica Scandinavica, 110*, 445–447.

Torebjörk, H. E., Schmelz, M., & Handwerker, H. O. (1996). Functional properties of human cutaneous nociceptors and their role in pain and hyperalgesia. In C. Belmonte & F. Cervero (Eds.), *Neurobiology of nociceptors* (pp. 349–369). Oxford: Oxford University Press.

Van Hees, J., & Gybels, J. (1981). C nociceptor activity in human nerve during painful and non painful skin stimulation. *J Neurol Neurosurg Psychiat, 44*, 600–607.

Viana, F., de la Pena, E., & Belmonte, C. (2002). Specificity of cold thermotransduction is determined by differential ionic channel expression. *Nature Neuroscience, 5*, 254–260.

von Frey, M. (1896). Untersuchungen über die Sinnesfunktionen der menschlichen Haut. Erste Abhandlung: Druckempfindung und Schmerz. *Abhandlungen der mathematisch-physischen Klasse der Königl. Sächsischen Gesellschaft der Wissenschaften, 23*(208–217), 239.

Wahren, L. K., Torebjörk, H. E., & Nystrom, B. (1991). Quantitative sensory testing before and after regional guanethidine block in patients with neuralgia in the hand. *Pain, 46*, 23–30.

Wedell, G., Palmer, E., & Taylor, D. (1959). The significance of the peripheral anatomical arrangements of the nerves which subserve pain and itch. In G. E. W. Wolstenholme (Ed.), *Pain and Itch*. London: Churchill.

Weidner, C., Schmelz, M., Schmidt, R., et al. (1999). Functional attributes discriminating mechano-insensitive and mechano-responsive C nociceptors in human skin. *Journal of Neuroscience, 19*, 10184–10190.

Yarnitsky, D., & Ochoa, J. L. (1990). Release of cold-induced burning pain by block of cold-specific afferent input. *Brain, 113*, 893–902.

Zotterman, Y. (1939). Touch, pain and tickling. An electrophysiological investigation on cutaneous sensory nerve. *The Journal of Physiology, 95*, 1–28.

Pain in Syringomyelia

Nadine Attal and Didier Bouhassira
Centre d'Evaluation et de Traitement de la
Douleur, Hôpital Ambroise Paré, INSERM U 987
APHP, Boulogne-Billancourt, France

Synonyms

Syringomyelia and pain

Definition

Pain is frequently associated with syringomyelia
and generally consists of neuropathic (central)
pain. Its main characteristics are similar to those
of other central pain conditions, which suggest
that common mechanisms may be involved in
these pain disorders. However, the nature of
these mechanisms may differ depending on the
clinical phenotypes. Currently, first-line drug
treatments for this condition include tricyclic
antidepressants, SNRI antidepressants,
gabapentin, or pregabalin, while opioids are considered second line. Stimulation techniques may
be proposed in refractory patients.

Introduction

Syringomyelia is a chronic progressive lesion of
the spinal cord. It consists of a cystic cavitation of
the central spinal cord, commonly in cervical or
cervicodorsal region, which sometimes extends
upward into the medulla oblongata and pons
(syringobulbia). In rare cases, the syrinx is confined to the dorsolumbar segments. More than 90
% of cases are associated with developmental
disorders, particularly Chiari I malformation
with or without scoliosis (Fernández et al.
2009). Acquired cases are most commonly due
to traumatic spinal cord injury. In these cases,
the syrinx is situated above the original injury
(0.2–4.5 % of spinal cord trauma) (Carroll and
Brackenridge 2005; Rossier et al. 1985;

Schurch et al. 1996). Other acquired cases are
more rarely due to spinal arachnoiditis,
myelomalacia due to compression of the spinal
cord, and some cases remain idiopathic.

Syringomyelia is clinically characterized
by segmental sensory loss, generally of
a dissociated type (loss of thermal and pain
sensations while retaining tactile and proprioce-
tive sensation), and is associated with segmental
amyotrophy (generally of the hands) in severe
cases. It is frequently associated with pain,
particularly neuropathic central pain (Treede
et al. 2008), which is often extremely difficult to
treat (Attal et al. 2009, 2010; Milhorat et al. 1996;
Finnerup et al. 2010). The main clinical characteristics of such pain are not specific to syringomyelia, but rather similar to those observed in
most other neuropathic pain disorders (Attal
et al. 2008). This suggests that there may be
common mechanisms involved in these various
conditions. However, as in other painful conditions, syringomyelic patients do not represent
a homogeneous group (Hatem et al. 2010).
Thus, these patients may have different combinations of symptoms and signs, which may depend
on distinct mechanisms and possibly respond
differentially to treatments.

This chapter is devoted to pain associated with
syringomyelia/bulbia with a particular emphasis
on neuropathic pain. The clinical characteristics
and theoretical mechanisms of the condition
itself are beyond the scope of this review (Heiss
et al. 1999; Levine 2004; Victor and Ropper
2001).

Characteristics

Clinical Symptomatology: Description and Assessment

Pain is the most common symptom associated
with syringomyelia. It is mainly neuropathic and
is considered a direct consequence of spinal
cord lesion (central pain) (Treede et al. 2008).
Generally, the pain is strictly or predominantly
unilateral and most commonly located in the
hand, shoulder, thorax, or neck in cervicodorsal
cavities and sometimes in the lower limbs in

dorsolumbar syringomyelia (Bouhassira et al. 2000; Boivie 2003, 2006; Williams 1990). Typically, there is a long delay (from months to years) between injury and the onset of pain, particularly in cases of posttraumatic syringomyelia (Rossier et al. 1985; Shannon et al. 1981; Vernon et al. 1982; Williams 1990), although the pain may also be the initial symptom of the condition (Boivie 2006).

Central (Neuropathic) Pain

As is the case for neuropathic pain in general, neuropathic pain associated with syringomyelia includes a variety of symptoms: spontaneous ongoing/paroxysmal pain and allodynia/ hyperalgesia, often associated with paraesthesia and dysaesthesia (Boivie 2006; Baron et al. 2010). Neuropathic pain is generally homolateral to the cavitation in cases of paracentral or eccentric cavities (Bouhassira et al. 2000; Boivie 2006; Milhorat et al. 1995, 1996).

Description of Symptoms Spontaneous pain refers to pain in the absence of any stimulus, and it may be ongoing (superficial or deep) or paroxysmal. Its exact prevalence is unknown in the absence of epidemiological data on the condition, but it has been estimated between 37 % (Milhorat et al. 1996) and 67 % (Ducreux et al. 2006). Superficial burning pain is the most common symptom, but the pain has also been described as nonburning (i.e., squeezing, pressing, aching, cold, etc.) (Milhorat et al. 1996; Boivie 2006; Hatem et al. 2010). There may also be paroxysmal pain, which has been described as shooting, electric shock-like, or stabbing and evoked pain symptoms. In many cases, different types of pain may coexist in the same patient, occurring in the same region or in different parts of the body. The pain is often related to or enhanced by coughing, straining, or Valsalva maneuver, probably because of the dynamic characteristics of the syrinx fluid and flow (Attal et al. 2004; Rossier et al. 1985). Abnormal nonpainful positive symptoms are frequent (paresthesia or dysesthesia) and are often described as "tingling," "pins and needles," and sometimes "numbness." Psychological factors, such as emotion, anxiety, and stress, may worsen the pain, as seen for other pain conditions.

Screening and Assessment of Neuropathic Pain in Syringomyelia In recent years, several screening tools based on verbal pain description with, or without, limited clinical examination have been developed and validated in neuropathic pain including central pain (Haanpää et al. 2011; Bouhassira and Attal 2011). These screening tools share many features despite being developed by different groups in different contexts. These include the *Douleur Neuropathique en 4 questions* (DN4) which has been initially validated in patients with peripheral and central neuropathic pain, the PainDetect questionnaire, the leeds assessment of neuropathic symptoms and Signs (LANSS), the neuropathic pain questionnaire (NPQ), and the ID pain. One study comparing 4 of these tools in patients with spinal cord injury pain found that the diagnostic accuracy for the tools was for DN4 88 %, PainDetect 78 %, NPQ 65 %, and LANSS 55 % (Hallstrom and Norrbrink 2011). Screening tools are helpful to identify patients with possible neuropathic pain, particularly when used by nonspecialists, and this is probably their chief clinical strength. However, screening tools fail to identify about 10–20 % of patients with clinician-diagnosed neuropathic pain indicating that they may offer guidance for further diagnostic evaluation and pain management but cannot replace clinical judgment.

Whereas assessment of neuropathic pain intensity in patients with spinal cord injury pain is generally achieved through VAS, numerical or categorical scales (Haanpää et al. 2011). Moreover several specific tools have been validated specifically for the assessment of neuropathic pain quality and have been used in spinal cord injury patients including syringomyelia patients (Hatem et al. 2010; Ducreux et al. 2006). These include the Neuropathic Pain Scale and the Neuropathic Pain Symptom Inventory (Bouhassira and Attal 2011; Haanpää et al. 2011). Validated neuropathic pain quality measures are probably useful to discriminate between various pain mechanisms associated with distinct dimensions of neuropathic pain experience and are sensitive to change. They are currently recommended to evaluate treatment effects on neuropathic symptoms or their combination (Haanpää et al. 2011).

Is the Neuropathic Pain of Syringomyelia Distinct From or Similar to that of Other Neuropathic Pain Conditions? In a large cohort of patients with various peripheral or central NeP syndromes, we tested whether specific neuropathic combination of symptoms assessed by the NPSI were associated with specific etiology or locations of neurological lesions (Attal et al. 2008). Univariate analyses showed that the five distinct dimensions (i.e., combination of symptoms) of neuropathic pain identified with the NPSI (burning pain, deep squeezing/pressing pain, paroxysmal pain, evoked pain, and paraesthesia/dysaesthesia) were present, in similar proportions, in all the etiologies, with a few exceptions. In particular, a similar proportion of syringomyelic patients to those with other central pain disorders (poststroke pain, spinal cord trauma with below-level pain, multiple sclerosis) had the same proportion of symptoms and dimensions (Table 1). These data were further confirmed by a multiple correspondence analysis which showed no association between the neuropathic symptoms and the etiology or location of the lesion. Thus, the principal differences between the neuropathic pain of syringomyelia and other peripheral or central lesions are the location of the pain and the associated neurological symptoms and signs. These results show that there is more similarity than differences between the neuropathic pain of syringomyelia and other NeP syndromes.

Other Painful Syndromes
Coexisting painful symptoms are frequent in syringomyelia depending on the etiology of the syrinx (50–90 % of cases) (Milhorat et al. 1996, 1999). Suboccipital headache (generally a benign exertional headache, i.e., a headache that is intensified by Valsalva maneuvers such as coughing, sneezing, or bowel movement), migraine, or neck pain (which also usually increases with Valsalva maneuvers) are frequent in patients with Chiari type I malformation (Buzzi et al. 2003; Milhorat et al. 1999; Taylor and Larkins 2002; Fernandez

et al. 2009). Benign exertional headache is thought to be related to the increased intrathecal pressure caused by the free flow of CSF being obstructed in the subarachnoid space and may respond to surgical treatment of the Chiari (Sansur et al. 2003; Taylor and Larkins 2002). However, migraine and benign exertional headache may coexist in the same patient and respond similarly to antimigraine pharmacologic treatment (Buzzi et al. 2003). Trigeminal neuralgia and cluster-like facial pain have also been described and may also respond to surgical treatment of the syrinx (Rosetti et al. 1999). Visceral pain (due to renal calculus, bowel or sphincter dysfunction, etc.) and musculoskeletal pain (due to bone, joint, muscle trauma or inflammation, mechanical instability, muscle spasms, and secondary overuse syndrome) may be observed in syringomyelia related to spinal cord trauma (Siddall et al. 1997; Siddall and Loeser 2001). Finally, peripheral neuropathic pain due to nerve root entrapment or arachnoiditis is common in syringomyelia caused by meningitis.

Sensory Deficits and Evoked Pains
All patients with syringomyelia have some degree of thermoalgesic deficit in their painful area. Such a deficit may be unilateral or bilateral and has a segmental distribution, generally over the neck, shoulder, trunk, and arms, but may also affect the face. It is located or predominates ipsilaterally to the paracentral extension of the syrinx in cases of eccentric or paracentral cavities (Ducreux et al. 2006; Milhorat et al. 1996; Hatem et al. 2010). It may range from mild hypoesthesia or hypoalgesia (diminished pain in response to a normally painful stimulus) to complete anesthesia or analgesia. Sensory deficit can be detected and quantified best by quantitative sensory testing (QST), which indicates an impairment of thermal sensation in nearly 100 % of cases. There are almost always increased warm and cold detection thresholds and to a lesser extent increased pain thresholds (Boivie 2003, 2006; Ducreux et al. 2006; Hatem et al. 2010). This is similar to that observed in other central pain conditions (Boivie 2003, 2006; Bowsher 1996; Bowsher et al. 1998; Defrin et al. 2001; Eide et al. 1996; Finnerup et al. 2003;

Pain in Syringomyelia, Table 1 Prevalence of neuropathic dimensions (e.g., symptom combinations) as assessed by the Neuropathic Pain Symptom Inventory (a specific neuropathic pain questionnaire) across the main etiologies of neuropathic pain (Adapted from Attal et al. 2008). The prevalence of neuropathic dimensions in patients with syringomyelia (in bold) is roughly similar to that observed in other neuropathic pain syndromes. The only exceptions are postherpetic neuralgia (with paucity of deep pain and paresthesia) and trigeminal neuralgia (characterized by pain paroxysms nearly exclusive). Multivariate analyses confirmed the lack of association between etiologies and neuropathic symptoms

	PHN $n = 49$	Diabetic PPN $n = 35$	Nondiabetic PPN $n = 53$	Nerve trauma $n = 110$	Radiculopathy $n = 43$	Trigeminal neuralgia $n = 18$	Spinal trauma $n = 25$	MS $n = 32$	Syrinx $n = 40$	Stroke $n = 31$
Burning	89.8	62.8	58.5	51.1	65.1	16.7	76	56.2	**75**	74.2
Deep pain	28.5	68.6	62.3	58	51.2	22.2	74	62.5	**60**	64.5
Paroxysmal pain	63.2	62.8	62.3	66.3	72	89.9	72	65.6	**65**	58
Evoked pain	91.9	51.5	64.1	76	44.2	61.1	70	75	**62.5**	74
Paresthesia/dysesthesia	30	82.9	84.9	86	81.4	33	80	84.4	**87.5**	83.9

Abbreviations: *PHN* postherpetic neuralgia, *PPN* painful polyneuropathy, *MS* multiple sclerosis

Vestergaard et al. 1995; Osterberg and Boivie 2010). QST is also useful to follow the outcome of syringomyelic patients (Attal et al. 2004). Thus, we have found that the thermoalgesic deficit generally does not improve after surgical decompression of the syrinx, except in patients operated on early (i.e., less than 2 years) after the onset of their symptoms (Attal et al. 2004).

Proprioceptive/tactile sensation is usually considered to be not affected. However, in up to 50 % of patients, loss of proprioceptive/tactile sensation has been observed (Honan and Williams 1993). Increased vibration and/or mechanical thresholds may be observed by quantitative sensory testing (Attal et al. 2004; Ducreux et al. 2006; Hatem et al. 2010). Other signs of dorsal column dysfunction, such as an impairment of graphesthesia and movement direction, may also be observed (Ducreux et al. 2006; Hatem et al. 2010). Proprioceptive deficits, assessed using quantitative sensory tests, respond better than the thermal deficits to surgical treatment of the syrinx (Attal et al. 2004). Loss of vibration and position sense may also be present in the lower limbs, particularly in cases of Chiari malformation.

Patients with painful syringomyelia also present with signs of evoked pain, such as allodynia, which is "pain due to a stimulus which does not normally provoke pain," or hyperalgesia, which is "an increased response to a stimulus which is normally painful" (Merskey and Bogduk 1994) (64 % of painful patients in a psychophysical study of 46 consecutive patients with or without pain) (Ducreux et al. 2006). The pain is most frequently evoked by brushing (dynamic mechanical allodynia), punctate stimuli using a pinprick or von Frey filaments (punctate mechanical allodynia/hyperalgesia), or cold stimuli (cold allodynia/hyperalgesia) (Ducreux et al. 2006; Hatem et al. 2010). Less commonly, it can be evoked by static pressure (static mechanical allodynia/hyperalgesia) or heat stimuli (heat allodynia/hyperalgesia). In most cases, allodynia coexists with spontaneous pain, but "pure" mechanical allodynia has also been described after intraspinal lesion (Attal et al. 1998).

Evoked pain may also persist after stimulation (aftersensation), appear some time after

stimulation (delayed sensation), spread beyond the site of stimulation (radiation), or be increased or provoked by repetitive stimuli (temporal summation). Many patients suffer from hyperpathia, which is "a painful syndrome characterised by an abnormally painful reaction to a stimulus, especially a repetitive stimulus, as well as an increased threshold" (Merskey and Bogduk 1994). Hyperpathia can therefore be considered as a combination of hyperalgesia, temporal summation, and postsensation.

Other Symptoms and Signs

Generally, there is no motor impairment in the early stages of the disease. By contrast, as the cavity enlarges, there may be corticospinal tracts and/or segmental impairment of the anterior horn, leading to pyramidal syndrome and muscular amyotrophy, respectively. Tendon reflexes are generally absent, except in some patients with Chiari malformation. In these patients, the tendon reflexes may be normal or increased. Depending on the etiology of the syrinx, patients may also have variable clinical features, including nystagmus, vertigo, mild neurosensorial hearing low, cerebellar ataxia, and lower cranial nerve impairment such as palatal and vocal cord paresis, particularly in Chiari malformation (with or without syringobulbia) (Fernandez et al. 2009). Kyphoscoliosis is present in many cases, and overt cervico-occipital malformation is present in nearly a quarter of patients (Victor and Ropper 2001). Finally, patients may also have signs of autonomic impairment on the painful side. The painful area may be cooler and vasoconstricted or warmer, and occasional changes in sweating have been reported (Bowsher 1996).

Mechanism of Central Pain in Syringomyelia

Syringomyelia is a typical model of a "pure" intraspinal lesion predominantly affecting the spinothalamic tract. It is therefore of particular interest for studying the mechanisms underlying central pain, particularly with regard to the disputed role of spinothalamic lesion in neuropathic pain. In animal models developed over the last few years to study the mechanisms of central pain, lesions are usually induced by excitotoxic or ischemic injury, leading to damage of the spinal gray matter. These models present pathological characteristics, including neuronal loss, glial response, and syrinx formation, resembling those observed in syringomyelia, particularly posttraumatic cavities (e.g., Christensen and Hulsebosch 1997; Vierck et al. 2000; Yezierski et al. 1998; Yezierski 2001; Hao and Xu 2003). Therefore, they are of particular interest for studying the cellular, molecular, anatomical, and physiological consequences of an intraspinal injury.

Here, we will review the main theories of the mechanisms of central pain and their possible relevance for syringomyelia, the role of spinal versus supraspinal structures in the pain of syringomyelia, and then present evidence suggesting that the clinical symptoms of syringomyelia may be sustained by distinct mechanisms.

The Main Hypotheses of Central Pain Mechanisms and Their Relevance to Syringomyelia

The mechanisms of central pain have been studied for over 100 years but still remain largely unknown. The principal hypotheses fall into two major categories: central imbalance and/or disinhibition and central sensitization. Both phenomena can trigger abnormal neuronal hyperexcitability of spinal and/or supraspinal structures (Refs. in Finnerup et al. 2003; Finnerup and Jensen 2004; Finnerup 2008). Such hyperexcitability has been observed in animal models of spinal cord injury (Vierck et al. 2000; Yezierski 2001) but also in some patients with central pain, in the dorsal horn (Edgar et al. 1993; Falci et al. 2002; Loeser et al. 1968) and in the lateral and medial thalamus (Jeanmonod et al. 1993, 1996; Lenz et al. 1989, 1994, 2000).

Central Imbalance

Patients with central pain nearly always have abnormal temperature and pain sensibility, but may have normal touch and vibration sensation (Boivie 2006). This suggests that the presence of a spinothalamic dysfunction is a necessary condition for the occurrence of central pain, whereas the dorsal columns/medial lemniscus system is less commonly affected. This is further supported

by studies showing abnormalities in laser-evoked potentials (reflecting the function of A delta nociceptive fibers) on the painful side of patients with poststroke pain (Boivie 2006) and syringomyelia (Kakigi et al. 1991; Treede et al. 1991; Hatem et al. 2010), whereas sensory-evoked potentials served by large diameter fibers, correlated to the decrease in touch and vibration sensibility, may be preserved (Beric et al. 1988). It has been proposed that below level pain associated with spinal cord injury is induced by an imbalance of integration between the unaffected dorsal column/medial lemniscus activity and the affected spinothalamic tract (Beric 1993; Beric et al. 1988). However, a number of psychophysical studies of patients with spinal cord injury with or without below-level pain have shown that there is a similar degree of thermoalgesic deficits in painful and painless patients (e.g., Defrin et al. 2001; Finnerup et al. 2003). Similarly, the extent and magnitude of thermal deficits, assessed in patients with syringomyelia, are similar in painful and painless patients (Bouhassira et al. 2000; Hatem et al. 2010; Ducreux et al. 2006). These data suggest that the spinothalamic lesion is necessary but not sufficient to account for the development of central pain.

Disinhibition Theories

Since the original hypothesis of Head and Holmes (1911), the notion of a central disinhibition, particularly in the thalamus, has been one of the most favored pathophysiological theories of central pain (Casey 1991; Cassinari and Pagni 1969; Jeanmonod et al. 1996; Pagni 1989). One hypothesis, the "thermosensory disinhibition" theory of Craig (Craig 1998, 2000, 2003 for Refs.), which is based on the observation that thermosensory loss is the basic feature of nearly all central pain patients, proposes that central pain (particularly burning pain and cold allodynia) may be due to reduction of the physiological inhibition of thermal (cold) systems on nociceptive neurons. In contrast with traditional conceptions, this theory views pain not only as a sensation but also as a "homeostatic emotion" and an aspect of interoception, defined as the physiological condition of the body.

We tested this hypothesis in a first psychophysical study of 46 patients with syringomyelia, of whom 31 suffered from neuropathic pain (Ducreux et al. 2006). We did not observe any correlation between the magnitude of thermal deficits and the amount of pain in the entire group of painful patients. However, patients with spontaneous pain only (without evoked pains) had more severe thermal deficits than those with allodynia, and the intensity of burning pain was correlated to the extent of warm and cold deficits. These patients also had more asymmetrical thermal deficits compared to those with allodynia and painless patients, which was confirmed by a further study (Hatem et al. 2010). This might reflect an imbalance in thermoregulation which involves bilateral brain integration (Craig 1998). Thus, these data are compatible with the disinhibition theory of central pain in a subgroup of syringomyelic patients (i.e., patients with spontaneous pain only). By contrast, patients with allodynia had significantly reduced thermal deficits compared with those having spontaneous pain only (Ducreux et al. 2006).

Another form of central disinhibition has been proposed to account more specifically for the pain of syringomyelia. In a study of 137 patients with syringomyelia, Milhorat et al. (1996) observed that the syrinx extended onto the dorsolateral quadrant of the spinal cord on the same side of the pain in 84 % of patients. It was suggested that pain was due to a disturbance of pain-modulating centers in the dorsolateral quadrant of the spinal cord. However, it was not stated whether this pattern was less common in painless patients.

Central Sensitization

The hyperactivity of nociceptive neurons in central pain may also result from direct modifications of their electrophysiological properties, i.e., central sensitization (for review see Latremoliere and Woolf 2009). This may be caused by damage to central neurons caused by excitatory amino acids related to NMDA receptor activation (Eide 1998; Vierck et al. 2000), loss of tonic inhibition by gamma-aminobutyric acid

interneurons or descending tracts (Hains et al. 2002; Zhang et al. 1994), neuroinflammation (Yu et al. 2003), microglia activation (Hains and Waxman 2006; Gwak and Hulsebosch 2009), and possibly also abnormal expression of sodium channels (Hains et al. 2003; Hains and Waxman 2007). Indirect evidence for the role of central sensitization in spinal cord injury pain is provided by the beneficial effects of NMDA antagonists, GABA-A receptor blockers, and sodium channel blockers in patients with spinal cord injury pain (Attal et al. 2000, 2002; Eide et al. 1995; Finnerup et al. 2005; Kvarnstom et al. 2004; Canavero an Bonicalzi 2004) including syringomyelia (Attal et al. 2000, 2002).

Role of Spinal and Supraspinal Structures in the Pain of Syringomyelia

Several studies on animals and human have suggested that the primary pain generator in at-level and below-level neuropathic pain associated with spinal cord injury pain may be the injury region itself (Refs. in Finnerup and Jensen 2004; Finnerup et al. 2003, 2007; Hari et al. 2009). The observation that pain and sensory deficits associated with syringomyelia predominate ipsilaterally to the extension of the cavity in cases of paracentral or eccentric cavities (Bouhassira et al. 2000) is consistent with this hypothesis. Similarly, spinal drug treatment trials have shown that spinal anesthesia with lidocaine reduced at-level and below-level SCI pain (Loubser and Donovan 1991). More specifically, the ipsilateral dorsal horn or spinothalamic fibers before they cross the midline may be important sites for pain generation (Boivie 2006). Animal models of intraspinal excitotoxic lesion have also shown spontaneous nociceptive behavior and enhancement of nociceptive responses in dermatomes ipsilateral to the lesion (Refs. in Vierck et al. 2000; Yezierski 2001). Electrophysiological recordings in these models have demonstrated abnormal neuronal excitation in the dorsal horn adjacent to the injury (Christensen and Hulsebosch 1997; Vierck et al. 2000; Yezierski 2001; Yezierski and Park 1993; Yezierski et al. 2004). The possible mechanisms of such segmental hyperexcitability are probably not specific and may involve sensitization or disinhibition.

However, both animal and human studies also point to the role of supraspinal structure particularly the thalamus, in SCI pain. Studies in humans with spinal cord injury pain have pointed to thalamocortical dysrhythmia, differences in metabolites in the thalamus region, reorganization, and bursting activity (Refs. in Finnerup et al. 2009). Primary somatosensory cortex reorganization and regional brain anatomical changes have also been demonstrated, at least in patients with complete thoracic SCI and below-level pain, but whether this observation is applicable to pain in syringomyelia has not been confirmed (Henderson et al. 2011; Gustin et al. 2010).

Potential Relationships Between the Clinical Phenotypes of Syringomyelia and Underlying Mechanisms

Several studies have suggested that the various painful symptoms of syringomyelia, particularly with regard to the subtypes of allodynia (mechanical or thermal), may be sustained by distinct mechanisms. Thus, pharmacological studies have shown preferential effects of intravenous administration of the sodium channel blocker lidocaine on mechanical (dynamic and static) but not cold allodynia in patients with central pain, including several syringomyelic patients (Attal et al. 2000). In the same line, a functional imaging study of patients with syringomyelia presenting with cold- or brush-induced allodynia has revealed distinct activation patterns depending on the submodality of allodynia (Ducreux et al. 2006). Thus, cold allodynia activates a "cold pain network" including activation of insular and cingular areas, which is similar to the pattern induced by noxious cold in healthy controls (Craig 2000). In contrast, brush-evoked allodynia did not induce significant activation of several structures of the "pain matrix," particularly the insula and the ACC. The different patterns of activation between the two subtypes of allodynia might reflect distinct pathophysiological mechanism (Fig. 1).

In another study in patients with central pain associated with cervical syringomyelia, we used a systematic multimodal assessment including a clinical examination with quantitative sensory testing and pain questionnaires (NPSI), an assessment of the functionality of the nociceptive pathways with laser-evoked potentials and an assessment of spinal cord damage by diffuse tensor imaging and fiber tracking (DTI-FT) of the cervical spinal cord (Hatem et al. 2009, 2010). We found that patients with both spontaneous and evoked pain clearly differed from patients with spontaneous pain only. Patients with spontaneous pain only had more severe spinal cord damage, as also indicated by individual examples (Fig. 2). Furthermore, we evidenced a direct relationship between the spinal structural damage assessed by diffuse tensor imaging and the intensity of deep pain (measured with the NPSI) in this group of patients only (Hatem et al. 2010). By contrast, patients with both spontaneous and evoked pain had not only less structural spinal cord damage (Fig. 2) but also better preserved spinothalamic and lemniscal tracts on quantitative sensory testing and electrophysiological testing. Although it is difficult to speculate about the cellular and molecular mechanisms, these data strongly suggest that different mechanisms are at play in these two populations defined on the basis of the combination of symptoms and therefore that these patients should respond differently to treatment. In patients with spontaneous pain, only pain, particularly deep pain, might be related to central deafferentation, whereas in patients with spontaneous pain and allodynia, other mechanisms including central sensitization and independent on the severity of the lesion may be at play.

Treatment of Central Pain Associated with Syringomyelia

As with other chronic neuropathic pain syndromes, pain associated with syringomyelia has a major impact on quality of life and affective state (Widerstrom-Noga 2002; Attal et al. 2011b). Therefore, the treatment of such pain should include management of disability and the

Pain in Syringomyelia, Fig. 1 (a) Cerebral activation (in *yellow*) using functional MRI during application of a normally nonpainful cold stimulus in six syringomyelia patients with cold allodynia. (b) Cerebral activation induced by normally nonpainful tactile stimulus in six patients with syringomyelia and brush-evoked allodynia. The anterior cingulate area was only activated during cold stimuli, while the thalamus was activated during brush stimuli. In both cases, a significant activation of the dorsolateral prefrontal cortex was observed. Abbreviations: *S2* secondary somatosensory cortex, *ant insula* anterior insula, *mid/post insula* cortex median/posterior insula, *inf parietal* inferior parietal cortex, *ACC* anterior cingulate cortex, *med frontal* medial frontal cortex, *DLPFC* dorsolateral prefrontal cortex, *thal* thalamus (Adapted from Ducreux et al. 2006)

affective comorbid conditions through the cognitive behavioral therapies and rehabilitation programs, some of which have been found useful in spinal cord injury patients (Sjolund 2002; Norrbrink Budh et al. 2006).

Is the Surgical Treatment of the Syrinx Effective on Neuropathic Pain?

Although surgery (generally foramen magnum decompression in patients with Chiari malformation) is commonly proposed to syringomyelic patients with progressive neurological deterioration, very few prospective studies have evaluated its effects on neuropathic pain (Attal et al. 2004). These studies generally reported disappointing results, in contrast to the positive effects seen for the pain associated with Chiari malformation, such as headache and neck pain. In a prospective 2-year study of 15 patients with syringomyelia, of whom eight presented with neuropathic pain, we found that surgery of the syrinx produced no overall significant effect on ongoing pain intensity at 2 years, whereas pain induced by straining, as well as coexisting dynamic symptoms (i.e., pains induced by coughing, Valsalva maneuver), was relieved (Attal et al. 2004). The latter pains are probably more dependent on the dynamic characteristics of the syrinx fluid and flow and may be particularly affected by surgical decompression which reduces the subarachnoid pressure. Furthermore, no correlation was found between the effects of surgery on the syrinx (foramen stenosis, syrinx diameter, syrinx/canal index) and the outcome of neuropathic pain (Attal et al. 2004).

Pharmacological Treatment

There are no controlled trials specifically devoted to the neuropathic pain of syringomyelia. Therefore, recommended treatments must rely on studies carried out for other neuropathic pain

Pain in Syringomyelia, Fig. 2 T2-weighted MR images (a–d) and diffusion tensor imaging with three-dimensional fiber tracking (DTI-FT) (e–h) in three syringomyelia patients with and in one healthy control. Images are sagittal views of the cervical spinal cord. (a, e) 24-year-old man with a syringeal cavity extending from C0 to L2. He presented with a severe thermonociceptive sensory deficit of both upper limbs. Tactile sensations were preserved. The patient had no neuropathic pain. DTI-FT of the cervical spinal cord showed a severe disorganization of fiber tracts. (b, f) 36-year-old woman with a syringeal cavity extending from C2 to T11. She presented with an asymmetrical deficit of thermal and tactile sensations of the upper limbs, predominating on the right side. She had moderate spontaneous pain of the right forearm and hand (c, g). Woman, 29 years of age, with a syringeal cavity extending from C2 to T2. She presented with a thermonociceptive sensory deficit of the right C5–C8 dermatome which was maximal over the C8 radicular territory. She described severe allodynia to dynamic tactile stimuli and moderate burning and deep pain over the right arm. DTI-FT of the cervical spinal cord showed a moderate disorganization of fiber tracts (d, h). DTI-FT of the cervical spinal cord in a 34-year-old healthy woman (Adapted from Hatem et al. 2010)

conditions (see Central Pain, this volume). On the basis of randomized controlled trials in spinal cord injury pain, only pregabalin to a lesser extent, duloxetine, and amitriptyline have been shown effective in treating SCI neuropathic pain, while lower evidence for efficacy exists for tramadol, gabapentin, and opioids (Attal et al. 2009, 2010; Finnerup et al. 2010; Siddall 2009). IV administration of single-dose opioids, local anesthetics (lidocaine), and NMDA receptor antagonists (ketamine) has also been found useful, but there is no serious evidence for their prolonged efficacy (for reviews see Attal et al. 2010; Finnerup et al. 2010; Siddall 2009). Recommended drugs as first line for spinal cord injury pain including syringomyelia are tricyclic antidepressants (mainly because of their established efficacy in other neuropathic indications), or pregabalin/gabapentin, and SNRIs if TCAs are not tolerated (Attal et al. 2010). Topical lidocaine may also be proposed in patients with posttraumatic syringomyelia or arachnoiditis, who may present with peripheral at level neuropathic pain (Wasner et al. 2007). Intrathecal therapy using opioids combined to clonidine (Siddall et al. 2000) or ziconotide (Schmidtko et al. 2010) may be proposed in refractory cases.

Importantly, the etiology or topography of the central lesion does not seem to play a role in the response to these treatments. In controlled trials of intravenous morphine and lidocaine carried out in patients with central pain, the response to these drugs appeared similar in patients with syringomyelia, spinal cord trauma, and poststroke pain (Attal et al. 2000, 2002). Treatments may also have similar efficacy on at-level and below-level neuropathic pain due to spinal cord injury, as recently shown in a placebo-controlled trial of IV lidocaine (Finnerup et al. 2005). The response to treatments may rather be influenced by the clinical symptoms or their combination, suggesting distinct pain mechanisms (Attal et al. 2008, 2011a). Pharmacological studies in larger cohorts of patients using an in-depth assessment of phenotypic profiles are needed to confirm these data and define possible predictors of the response to treatments (Attal et al. 2011a).

Neurostimulation and Neuroablative Techniques

Transcutaneous electrical nerve stimulation may sometimes be beneficial in spinal cord injury patients with at-level pain (Norrbrink 2009). Repetitive transcranial magnetic stimulation and direct electrical stimulation have been found useful in a few randomized controlled trials of patients with traumatic injury and below-level pain (e.g., Defrin et al. 2007; Soler et al. 2010), but there are no data regarding the maintenance of their analgesic effect over time.

Spinal cord stimulation may be an option in patients suffering from steady at-level neuropathic pain, who have incomplete spinal lesions and preservation of dorsal column function (Sindou and Mertens 2000). Syringomyelic patients may be possible candidates for such techniques.

Extradural stimulation of the motor cortex is increasingly used in patients with neuropathic pain including spinal cord injury pain (Nguyen et al. 2011) and may be particularly useful in patients with syringomyelia, as these patients generally have unilateral pain or predominantly unilateral pain involving one superior limb and trunk. Although no published data are available for patients with syringomyelia, personal observations from our group suggest the benefit of this technique in individual patients with syringomyelia.

Among neuroablative surgical procedures, dorsal root entry zone (DREZ) may be beneficial in at-level neuropathic pain after incomplete spinal cord injury, particularly paroxysmal and evoked pain (Sindou et al. 2001). However, there are very little data concerning the effects of this technique in syringomyelic patients (Prestor 2001). Moreover, DREZ may occasionally aggravate the sensory deficits which are generally incomplete in patients with syringomyelia. Therefore, this technique has a marginal place in the treatment of pain associated with syringomyelia.

Summary

Pain is frequently associated with syringomyelia and generally consists of neuropathic (central) pain. Its main characteristics (burning, deep pain, paroxysmal pain, and allodynia/hyperalgesia) are similar to those of other central pain conditions, which suggest that common mechanisms (either central sensitization or disinhibition) may be involved in these pain disorders. However, the nature of these mechanisms may differ, depending on the clinical presentation, as suggested from psychophysical, electrophysiological, functional MRI, and morphological studies of syringomyelic patients. These data may have important implications with regard to management.

References

Attal, N., Brasseur, L., Chauvin, M., & Bouhassira, D. (1998). A case of "pure" dynamic mechano-allodynia due to a lesion of the spinal cord: pathophysiological considerations. *Pain, 75*, 399–404.

Attal, N., Bouhassira, D., Baron, R., Dostrovsky, J., Dworkin, R. H., Finnerup, N., Gourlay, G., Haanpaa, M., Raja, S., Rice, A. S., Simpson, D., & Treede, R. D. (2011a). Assessing symptom profiles in neuropathic pain clinical trials: Can it improve outcome? *European Journal of Pain, 15*, 441–443.

Attal, N., Cruccu, G., Baron, R., Haanpää, M., Hansson, P., Jensen, T. S., Nurmikko, T., & European Federation of Neurological Societies, (2010). EFNS guidelines on the pharmacological treatment of neuropathic pain: 2010 revision. *European Journal of Neurology, 17*, 1113–e88.

Attal, N., Fermanian, C., Fermanian, J., Lanteri-Minet, M., Alchaar, H., & Bouhassira, D. (2008). Neuropathic pain: Are there distinct subtypes depending on the aetiology or anatomical lesion? *Pain, 138*, 343–353.

Attal, N., Gaude, V., Brasseur, L., Dupuy, M., Guirimand, F., Parker, F., & Bouhassira, D. (2000). Intravenous lidocaine in central pain. A double-blind placebo-controlled psycho-physical study. *Neurology, 544*, 564–574.

Attal, N., Guirimand, F., Brasseur, L., Gaude, V., Chauvin, M., & Bouhassira, D. (2002). Effects of IV morphine in central pain: A randomized placebo-controlled study. *Neurology, 58*, 554–563.

Attal, N., Parker, F., Tadie, M., Aghakani, N., & Bouhassira, D. (2004). Effects of surgery on the sensory deficits of syringomyelia and predictors of outcome: A long-term prospective study. *Journal of Neurology, Neurosurgery, and Psychiatry, 75*, 1025–1030.

Attal, N., Mazaltarine, G., Perrouin-Verbe, B., Albert, T., & SOFMER French Society for Physical Medicine and Rehabilitation. (2009). Chronic neuropathic pain management in spinal cord injury patients. What is the efficacy of pharmacological treatments with a general mode of administration? (oral, transdermal, intravenous). *Annals of Physical and Rehabilitation Medicine, 52*, 124–141.

Attal, N., Lanteri-Minet, M., Laurent, B., Fermanian, J., & Bouhassira, D. (2011b). The specific disease burden of neuropathic pain: Results of a French nationwide survey. *Pain, 152*, 2836–2843.

Baron, R., Binder, A., & Wasner, G. (2010). Neuropathic pain: Diagnosis, pathophysiological mechanisms, and treatment. *Lancet Neurology, 9*, 807–819.

Beric, A. (1993). Central pain: "New" syndromes and their evaluation. *Muscle & Nerve, 16*, 1017–1024.

Beric, A., Dimitrijevic, M. R., & Lindblom, U. (1988). Central dysesthesia syndrome in spinal cord injury patients. *Pain, 34*, 109–116.

Boivie, J. (2003). Central pain and the role of quantitative sensory testing (QST) in research and diagnosis. *European Journal of Pain, 7*, 339–343.

Boivie, J. (2006). Central pain. In S. B. McMahon & M. Koltzenburg (Eds.), *Wall and Melzack's textbook of pain* (5th ed., pp. 1057–1074). Elsevier: Churchill Livingstone.

Bouhassira, D., & Attal, N. (2011). Diagnosis and assessment of neuropathic pain: The saga of clinical tools. *Pain, 152*(Suppl), S74–S83.

Bouhassira, D., Attal, N., Parker, F., & Brasseur, L. (2000). Quantitative sensory evaluation of painful and painless patients with syringomyelia. In M. Devor, M. C. Rowbotham, & Z. Wiesenfeld-Hallin (Eds.), *Proceedings of the IXth world congress on pain* (pp. 401–411). Seattle: IASP Press.

Bowsher, D. (1996). Central pain: Clinical and physiological characteristics. *Journal of Neurology, Neurosurgery, and Psychiatry, 61*, 62–69.

Bowsher, D., Leijon, G., & Thuomas, K. A. (1998). Central poststroke pain: Correlation of MRI with clinical pain characteristics and sensory abnormalities. *Neurology, 51*, 1352–1358.

Buzzi, M. G., Formisano, R., Colonnese, C., & Pierelli, F. (2003). Chiari-associated exertional, cough, and sneeze headache responsive to medical therapy. *Headache, 43*, 404–406.

Canavero, S., & Bonicalzi, V. (2004). Intravenous subhypnotic propofol in central pain: A double-blind, placebo-controlled, crossover study. *Clinical Neuropharmacology, 27*, 182–186.

Carroll, A. M., & Brackenridge, P. (2005). Post-traumatic syringomyelia: A review of the cases presenting in a regional spinal injuries unit in the North East of England over a 5-year period. *Spine, 30*, 1206–1210.

Casey, K. L. (1991). *Pain and central nervous disease: The central pain syndromes*. New York: Raven.

Cassinari, V., & Pagni, C. A. (1969). *Central pain: A neurological survey*. Cambridge, MA: Harvard University.

Christensen, M. D., & Hulsebosch, C. E. (1997). Chronic central pain after spinal cord injury. *Journal of Neurotrauma, 14*, 517–537.

Craig, A. D. (1998). A new version of the thalamic disinhibition hypothesis of central pain. *Pain Forum, 7*, 1–14.

Craig, A. D. (2000). The functional anatomy of lamina I and its role in post-stroke central pain. *Progress in Brain Research, 129*, 137–151.

Craig, A. D. (2003). A new view of pain as a homeostatic emotion. *Trends in Neurosciences, 26*, 303–307.

Defrin, R., Grunhaus, L., Zamir, D., & Zeilig, G. (2007). The effect of a series of repetitive transcranial magnetic stimulations of the motor cortex on central pain after spinal cord injury. *Archives of Physical Medicine and Rehabilitation, 88*, 1574–1580.

Defrin, R., Ohry, A., Blumen, N., & Urca, G. (2001). Characterization of chronic pain and somatosensory function in spinal cord injury subjects. *Pain, 89*, 253–263.

Ducreux, D., Attal, N., Parker, F., & Bouhassira, D. (2006). Mechanisms of central neuropathic pain: A combined psychophysical and fMRI study in syringomyelia. *Brain, 129*, 963–976.

Edgar, R. E., Best, L. G., Quail, P. A., & Obert, A. D. (1993). Computer-assisted DREZ microcoagulation: Posttraumatic spinal deafferentation pain. *Journal of Spinal Disorders, 6*, 48–56.

Eide, P. K. (1998). Pathophysiological mechanisms of central neuropathic pain after spinal cord injury. *Spinal Cord, 36*, 601–612.

Eide, P. K., Stubhaug, A., & Stenehjem, A. E. (1995). Central dysesthesia pain after traumatic spinal cord injury is dependent on N-methyl-D-aspartate receptor activation. *Neurosurgery, 37*, 1080–1087.

Eide, P. K., Jorum, E., & Stenehjem, A. E. (1996). Somatosensory findings in patients with spinal cord injury and central dysesthesia pain. *Journal of Neurology, Neurosurgery, and Psychiatry, 60*, 411–415.

Falci, S., Best, L., Bayles, R., Lammertse, D., & Starnes, C. (2002). Dorsal root entry zone microcoagulation for spinal cord injury-related central pain: Operative intramedullary electrophysiological guidance and clinical outcome. *Journal of Neurosurgery, 97*, 193–200.

Fernández, A. A., Guerrero, A. I., Martínez, M. I., Vázquez, M. E., Fernández, J. B., Chesa i Octavio, E., de Labrado, J. L., Silva, M. E., de Araoz, M. F., García-Ramos, R., Ribes, M. G., Gómez, C., Valdivia, J. I., Valbuena, R. N., & Ramón, J. R. (2009). Malformations of the craniocervical junction (Chiari type I and syringomyelia: Classification, diagnosis and treatment). *BMC Musculoskeletal Disorders, 10* (Suppl 1), S1.

Finnerup, N. B. (2008). A review of central neuropathic pain states. *Current Opinion in Anaesthesiology, 21*(5), 586–589.

Finnerup, N. B., & Jensen, T. S. (2004). Spinal cord injury pain-mechanisms and treatment. *European Journal of Neurology, 11*, 73–82.

Finnerup, N. B., Johannesen, I. L., Fuglsang-Frederiksen, A., Bach, F. W., & Jensen, T. S. (2003). Sensory function in spinal cord injury patients with and without central pain. *Brain, 126*, 57–70.

Finnerup, N. B., Biering-Sorensen, F., Johannesen, I. L., Terkelsen, A. J., Juhl, G. I., Kristensen, A. D., Sindrup, S. H., Bach, F. W., & Jensen, T. S. (2005). Intravenous lidocaine relieves spinal cord injury pain: A randomized controlled trial. *Anesthesiology, 102*, 1023–1030.

Finnerup, N. B., Sørensen, L., Biering-Sørensen, F., Johannesen, I. L., & Jensen, T. S. (2007). Segmental hypersensitivity and spinothalamic function in spinal cord injury pain. *Experimental Neurology, 207*, 139–149.

Finnerup, N. B., Sindrup, S. H., & Jensen, T. S. (2010). The evidence for pharmacological treatment of neuropathic pain. *Pain, 150*, 573–581.

Gustin, S. M., Wrigley, P. J., Siddall, P. J., & Henderson, L. A. (2010). Brain anatomy changes associated with persistent neuropathic pain following spinal cord injury. *Cerebral Cortex, 20*, 1409–1419.

Gwak, Y. S., & Hulsebosch, C. E. (2009). Remote astrocytic and microglial activation modulates neuronal hyperexcitability and below-level neuropathic pain after spinal injury in rat. *Neuroscience, 161*, 895–903.

Haanpää, M., Attal, N., Backonja, M., Baron, R., Bennett, M., Bouhassira, D., Cruccu, G., Hansson, P., Haythornthwaite, J. A., Iannetti, G. D., Jensen, T. S., Kauppila, T., Nurmikko, T. J., Rice, A. S., Rowbotham, M., Serra, J., Sommer, C., Smith, B. H., & Treede, R. D. (2011). NeuPSIG guidelines on neuropathic pain assessment. *Pain, 152*, 14–27.

Hains, B. C., Everhart, A. W., Fullwood, S. D., & Hulsebosch, C. E. (2002). Changes in serotonin, serotonin transporter expression and serotonin denervation supersensitivity: Involvement in chronic central pain after spinal hemisection in the rat. *Experimental Neurology, 175*, 347–362.

Hains, B. C., Klein, J. P., Saab, C. Y., Craner, M. J., Black, J. A., & Waxman, S. G. (2003). Upregulation of sodium channel Nav1.3 and functional involvement in neuronal hyperexcitability associated with central neuropathic pain after spinal cord injury. *Journal of Neuroscience, 23*, 8881–8892.

Hains, B. C., & Waxman, S. G. (2006). Activated microglia contribute to the maintenance of chronic pain after spinal cord injury. *Journal of Neuroscience, 26*, 4308–4317.

Hains, B. C., & Waxman, S. G. (2007). Sodium channel expression and the molecular pathophysiology of pain after SCI. *Progress in Brain Research, 161*, 195–203.

Hallström, H., & Norrbrink, C. (2011). Screening tools for neuropathic pain: Can they be of use in individuals with spinal cord injury? *Pain, 152*, 772–779.

Hao, J. X., & Xu, X. J. (2003). Animal models of spinal cord injury pain and their implications for pharmacological treatments. *Journal of Rehabilitation Medicine, 41*(Suppl), 81–84.

Hari, A. R., Wydenkeller, S., Dokladal, P., & Halder, P. (2009). Enhanced recovery of human spinothalamic function is associated with central neuropathic pain after SCI. *Experimental Neurology, 216*, 428–430.

Hatem, S. M., Attal, N., Ducreux, D., Gautron, M., Parker, F., Plaghki, L., & Bouhassira, D. (2009). Assessment of spinal somatosensory systems with diffusion tensor imaging in syringomyelia. *Journal of Neurology, Neurosurgery, and Psychiatry, 80,* 1350–1356.

Hatem, S. M., Attal, N., Ducreux, D., Gautron, M., Parker, F., Plaghki, L., & Bouhassira, D. (2010). Clinical, functional and structural determinants of central pain in syringomyelia. *Brain, 133,* 3409–3422.

Head, H., & Holmes, G. (1911). Sensory disturbances from cerebral lesions. *Brain, 34,* 102–154.

Heiss, J. D., Patronas, N., DeVroom, H. L., Shawker, T., Ennis, R., Kammerer, W., Eidsath, A., Talbot, T., Morris, J., Eskioglu, E., & Oldfield, E. H. (1999). Elucidating the pathophysiology of syringomyelia. *Journal of Neurosurgery, 91,* 553–562.

Henderson, L. A., Gustin, S. M., Macey, P. M., Wrigley, P. J., & Siddall, P. J. (2011). Functional reorganization of the brain in humans following spinal cord injury: Evidence for underlying changes in cortical anatomy. *Journal of Neuroscience, 31,* 2630–2637.

Honan, W. P., & Williams, B. (1993). Sensory loss in syringomyelia: Not necessarily dissociated. *Journal of the Royal Society of Medicine, 86,* 519–520.

Jeanmonod, D., Magnin, M., & Morel, A. (1993). Thalamus and neurogenic pain: Physiological, anatomical and clinical data. *Neuroreport, 4,* 475–478.

Jeanmonod, D., Magnin, M., & Morel, A. (1996). Low-threshold calcium spike bursts in the human thalamus. Common physiopathology for sensory, motor and limbic positive symptoms. *Brain, 119,* 363–375.

Kakigi, R., Shibasaki, H., Kuroda, Y., Neshige, R., Endo, C., Tabuchi, K., & Kishikawa, T. (1991). Pain-related somatosensory evoked potentials in syringomyelia. *Brain, 114,* 1871–1889.

Kvarnström, A., Karlsten, R., Quiding, H., & Gordh, T. (2004). The analgesic effect of intravenous ketamine and lidocaine on pain after spinal cord injury. *Acta Anaesthesiologica Scandinavica, 48,* 498–506.

Latremoliere, A., & Woolf, C. J. (2009). Central sensitization: A generator of pain hypersensitivity by central neural plasticity. *The Journal of Pain, 10,* 895–926.

Lenz, F., Kwan, H. C., Dostrovsky, J. O., & Tasker, R. R. (1989). Characteristics of the bursting pattern of action potentials that occurs in the thalamus of patients with central pain. *Brain Research, 496,* 357–360.

Lenz, F. A., Kwan, H. C., Martin, R., Tasker, R., Richardson, R. T., & Dostrovsky, J. O. (1994). Characteristics of somatotopic organization and spontaneous neuronal activity in the region of the thalamic principal sensory nucleus in patients with spinal cord transection. *Journal of Neurophysiology, 72,* 1570–1587.

Lenz, F. A., Lee, J. I., Garonzik, I. M., Rowland, L. H., Dougherty, P. M., & Hua, S. E. (2000). Plasticity of pain-related neuronal activity in the human thalamus. *Progress in Brain Research, 129,* 259–273.

Levine, D. N. (2004). The pathogenesis of syringomyelia associated with lesions at the foramen magnum: A critical review of existing theories and proposal of a new hypothesis. *Journal of Neurological Sciences, 220,* 3–21.

Loeser, J. D., Ward, A. A. J. R., & White, L. E., Jr. (1968). Chronic deafferentation of human spinal cord neurons. *Journal of Neurosurgery, 29,* 48–50.

Loubser, P. G., Donovan, W. H. (1991). Diagnostic spinal anaesthesia in chronic spinal cord injury pain. *Paraplegia, 29,* 25–36.

Merskey, H., & Bogduk, N. (Eds.). (1994). *Classification of chronic pain.* Seattle: IASP Press.

Milhorat, T. H., Johnson, R. W., Milhorat, R. H., Capocelli, A. L., & Pevsner, P. H. (1995). Clinicopathological correlations in syringomyelia using axial magnetic resonance imaging. *Neurosurgery, 37,* 206–213.

Milhorat, T. H., Kotzen, R. M., Mu, H. T. M., Capocelli, A. L., & Milhorat, R. H. (1996). Dysesthetic pain in patients with syringomyelia. *Neurosurgery, 38,* 940–947.

Milhorat, T. H., Chou, M. W., Trinidad, E. M., Kula, R. W., Mandell, M., Wolpert, C., & Speer, M. C. (1999). Chiari I malformation redefined: Clinical and radlographic findings for 364 symptomatic patients. *Neurosurgery, 44,* 1005–1017.

Nguyen, J. P., Nizard, J., Keravel, Y., & Lefaucheur, J. P. (2011). Invasive brain stimulation for the treatment of neuropathic pain. *Nature Reviews Neurology, 7,* 699–709.

Norrbrink, C. (2009). Transcutaneous electrical nerve stimulation for treatment of spinal cord injury neuropathic pain. *Journal of Rehabilitation Research and Development, 46,* 85–93.

Norrbrink Budh, C., Kowalski, J., & Lundeberg, T. (2006). A comprehensive pain management programme comprising educational, cognitive and behavioural interventions for neuropathic pain following spinal cord injury. *Journal of Rehabilitation Medicine, 38,* 172–180.

Osterberg, A., & Boivie, J. (2010). Central pain in multiple sclerosis – sensory abnormalities. *European Journal of Pain, 14,* 104–110.

Pagni, C. A. (1989). Central pain due to spinal cord and brainstem damage. In P. D. Wall & R. Melzack (Eds.), *Textbook of pain* (pp. 634–655). Edinburgh: Churchill Livingstone.

Prestor, B. (2001). Microsurgical junctional DREZ coagulation for treatment of deafferentation pain syndromes. *Surgical Neurology, 56,* 259–265.

Rosetti, P., Ben Taib, N. O., Brotchi, J., & De Witte, O. (1999). Arnold Chiari type I malformation presenting as a trigeminal neuralgia: Case report. *Neurosurgery, 44,* 1122–1123.

Rossier, A. B., Foo, D., Shillito, J., & Dyro, F. M. (1985). Posttraumatic cervical syringomyelia. Incidence, clinical presentation, electrophysiological studies, syrinx protein and results of conservative and operative treatment. *Brain, 108,* 439–461.

Sansur, C. A., Heiss, J. D., DeVroom, H. L., Eskioglu, E., Ennis, R., & Oldfield, E. H. (2003). Pathophysiology of headache associated with cough in patients with Chiari I malformation. *Journal of Neurosurgery, 98*, 453–458.

Schmidtko, A., Lötsch, J., Freynhagen, R., & Geisslinger, G. (2010). Ziconotide for treatment of severe chronic pain. *The Lancet, 375*, 1569–1577.

Schurch, B., Wichmann, W., & Rossier, A. B. (1996). Post-traumatic syringomyelia (cystic myelopathy): A prospective study of 449 patients with spinal cord injury. *Journal of Neurology, Neurosurgery, and Psychiatry, 60*, 61–67.

Shannon, N., Symon, L., Logue, V., Cull, D., Kang, J., & Kendall, B. (1981). Clinical features, investigation and treatment of post-traumatic syringomyelia. *Journal of Neurology, Neurosurgery, and Psychiatry, 44*, 35–42.

Siddall, P. J. (2009). Management of neuropathic pain following spinal cord injury: Now and in the future. *Spinal Cord, 47*, 352–359.

Siddall, P. J., & Loeser, J. D. (2001). Pain following spinal cord injury. *Spinal Cord, 39*, 63–73.

Siddall, P. J., Taylor, D. A., & Cousins, M. J. (1997). Classification of pain following spinal cord injury. *Spinal Cord, 35*, 69–75.

Siddall, P. J., Molloy, A. R., Walker, S., Mather, L. E., Rutkowski, S. B., & Cousins, M. J. (2000). The efficacy of intrathecal morphine and clonidine in the treatment of pain after spinal cord injury. *Anesthesia and Analgesia, 91*, 1493–1498.

Sindou, M., & Mertens, P. (2000). Neurosurgical management of neuropathic pain. *Stereotactic and Functional Neurosurgery, 75*, 76–80.

Sindou, M., Mertens, P., & Wael, M. (2001). Microsurgical DREZotomy for pain due to spinal cord and/or cauda equina injuries: Long-term results in a series of 44 patients. *Pain, 92*, 159–171.

Sjolund, B. H. (2002). Pain and rehabilitation after spinal cord injury: The case of sensory spasticity? *Brain Research Reviews, 40*, 250–256.

Soler, M. D., Kumru, H., Pelayo, R., Vidal, J., Tormos, J. M., Fregni, F., Navarro, X., & Pascual-Leone, A. (2010). Effectiveness of transcranial direct current stimulation and visual illusion on neuropathic pain in spinal cord injury. *Brain, 133*, 2565–2577.

Taylor, F. R., & Larkins, M. V. (2002). Headache and Chiari I malformation: Clinical presentation, diagnosis, and controversies in management. *Current Pain and Headache Reports, 6*, 331–337.

Treede, R. D., Lankers, J., Frieling, A., Zangemeister, W. H., Kunze, K., & Bromm, B. (1991). Cerebral potentials evoked by painful, laser stimuli in patients with syringomyelia. *Brain, 114*, 1595–1607.

Treede, R. D., Jensen, T. S., Campbell, J. N., Cruccu, G., Dostrovsky, J. O., Griffin, J. W., Hansson, P., Hughes, R., Nurmikko, T., & Serra, J. (2008). Neuropathic pain: Redefinition and a grading system for clinical and research purposes. *Neurology, 70*, 1630–1635.

Vernon, J. D., Silver, J. R., & Ohry, A. (1982). Post-traumatic syringomyelia. *Paraplegia, 20*, 339–364.

Vestergaard, K., Nielsen, J., Andersen, G., Ingeman-Nielsen, M., Arendt-Nielsen, L., & Jensen, T. S. (1995). Sensory abnormalities in consecutive, unselected patients with central post-stroke pain. *Pain, 61*, 177–186.

Victor, M., & Ropper, A. H. (2001). Diseases of the spinal cord. In *Adams and Victor's principles of neurology* (7th ed., pp. 1293–1345). New York: McGraw Hill.

Vierck, C. J., Jr., Siddall, P., & Yezierski, R. P. (2000). Pain following spinal cord injury: Animal models and mechanistic studies. *Pain, 89*, 1–5.

Wasner, G., Naleschinski, D., & Baron, R. (2007). A role for peripheral afferents in the pathophysiology and treatment of at-level neuropathic pain in spinal cord injury? A case report. *Pain, 131*, 219–225.

Widerstrom-Noga, E. G. (2002). Evaluation of clinical characteristics of pain and psychosocial factors after spinal cord injury. In R. P. Yezierski & K. J. Burchiel (Eds.), *Spinal cord injury pain: Assessment, mechanisms, management* (Progress in pain research and management, Vol. 23, pp. 53–82). Seattle: IASP Press.

Williams, B. (1990). Post-traumatic syringomyelia, an update. *Paraplegia, 28*, 296–313.

Yezierski, R. P. (2001). Pain following spinal cord injury: Pathophysiology and central mechanisms. In R. P. Yezierski & K. J. Burchiel (Eds.), *Progress in brain research* (Vol. 129, pp. 429–448). New York: Elsevier.

Yezierski, R. P., & Park, S. H. (1993). The mechanosensitivity of spinal sensory neurons following intraspinal injections of quisqualic acid in the rat. *Neuroscience Letters, 157*, 115–119.

Yezierski, R. P., Liu, S., Ruenes, G. L., Kajander, K. J., & Brewer, K. L. (1998). Excitotoxic spinal cord injury: Behavioral and morphological characteristics of a central pain model. *Pain, 75*, 141–155.

Yezierski, R. P., Yu, C. G., Mantyh, P. W., Vierck, C. J., & Lappi, D. A. (2004). Spinal neurons involved in the generation of at-level pain following spinal injury in the rat. *Neuroscience Letters, 361*, 232–236.

Yu, C. G., Fairbanks, C. A., Wilcox, G. L., & Yezierski, R. P. (2003). Effects of agmatine, interleukin-10, and cyclosporin on spontaneous pain behavior after excitotoxic spinal cord injury in rats. *The Journal of Pain, 4*, 129–140.

Zhang, A. L., Hao, J. X., Seiger, A., Xu, X. J., Wiesenfeld-Hallin, Z., Grant, G., & Aldskogius, H. (1994). Decreased GABA immunoreactivity in spinal cord dorsal horn neurons after transient spinal cord ischemia in the rat. *Brain Research, 656*, 187–190.

Pain in the Neonate

▶ Pain in Children

Pain in the Workplace, Compensation, and Disability Management

Patrick Loisel
Disability Prevention Research and Training Centre, Université de Sherbrooke, Longueuil, Canada

Introduction

Syringomyelia is a chronic progressive lesion of the spinal cord. It consists of a cystic cavitation of the central spinal cord, commonly in cervical or cervicodorsal region, which sometimes extends upward into the medulla oblongata and pons (syringobulbia). In rare cases, the syrinx is confined to the dorsolumbar segments. More than 90 % of cases are associated with developmental disorders, particularly Chiari I malformation with or without scoliosis (Fernández et al. 2009). Acquired cases are most commonly due to traumatic spinal cord injury. In these cases, the syrinx is situated above the original injury (0.2–4.5 % of spinal cord trauma) (Carroll and Brackenridge 2005; Rossier et al. 1985; Schurch et al. 1996). Other acquired cases are more rarely due to spinal arachnoiditis, myelomalacia due to compression of the spinal cord, and some cases remain idiopathic.

Syringomyelia is clinically characterized by segmental sensory loss, generally of a dissociated type (loss of thermal and pain sensations while retaining tactile and proprioceptive sensation) and is associated with segmental amyotrophy (generally of the hands) in severe cases. It is frequently associated with pain, particularly neuropathic central pain (Treede et al. 2008) which is often extremely difficult to treat (Attal et al. 2009, 2010; Milhorat et al. 1996; Finnerup et al. 2010). The main clinical characteristics of such pain are not specific to syringomyelia, but rather similar to those observed in most other neuropathic pain disorders (Attal et al. 2008). This suggests that there may be

common mechanisms involved in these various conditions. However, as in other painful conditions, syringomyelic patients do not represent a homogeneous group (Hatem et al. 2010). Thus, these patients may have different combinations of symptoms and signs, which may depend on distinct mechanisms and possibly respond differentially to treatments.

This entry is devoted to pain associated with syringomyelia/bulbia with a particular emphasis on neuropathic pain. The clinical characteristics and theoretical mechanisms of the condition itself are beyond the scope of this entry (Heiss et al. 1999; Levine 2004; Victor and Ropper 2001).

Characteristics

Clinical Symptomatology: Description and Assessment

Pain is the most common symptom associated with syringomyelia. It is mainly neuropathic and is considered a direct consequence of spinal cord lesion (central pain) (Treede et al. 2008). Generally, the pain is strictly or predominantly unilateral and most commonly located in the hand, shoulder, thorax, or neck in cervicodorsal cavities, and sometimes in the lower limbs in dorsolumbar syringomyelia (Attal and Bouhassira 2000; Boivie 2003, 2006; Williams 1990). Typically, there is a long delay (from months to years) between injury and the onset of pain, particularly in cases of posttraumatic syringomyelia (Rossier et al. 1985; Shannon et al. 1981; Vernon et al. 1982; Williams 1990), although the pain may also be the initial symptom of the condition (Boivie 2006).

Central (Neuropathic) Pain

As is the case for neuropathic pain in general, neuropathic pain associated with syringomyelia includes a variety of symptoms: spontaneous ongoing/paroxysmal pain, and allodynia/hyperalgesia, often associated with paresthesia and dysesthesia (Boivie 2006; Baron et al. 2010). Neuropathic pain is generally homolateral to the cavitation in cases of paracentral or eccentric cavities (Attal and Bouhassira 2000; Boivie 2006; Milhorat et al. 1995, 1996).

Description of Symptoms Spontaneous pain refers to pain in the absence of any stimulus and it may be ongoing (superficial or deep) or paroxysmal. Its exact prevalence is unknown in the absence of epidemiological data on the condition, but it has been estimated between 37 % (Milhorat et al. 1996) and 67 % (Ducreux et al. 2006). Superficial burning pain is the most common symptom, but the pain has also been described as nonburning (i.e., squeezing, pressing, aching, cold, etc.) (Milhorat et al. 1996; Boivie 2006; Hatem et al. 2010). There may also be paroxysmal pain, which has been described as shooting, electric shock-like or stabbing, and evoked pain symptoms. In many cases, different types of pain may coexist in the same patient, occurring in the same region or in different parts of the body. The pain is often related to or enhanced by coughing, straining, or Valsalva's maneuver, probably because of the dynamic characteristics of the syrinx fluid and flow (Attal et al. 2004; Rossier et al. 1985). Abnormal nonpainful positive symptoms are frequent (paresthesia or dysesthesia) and are often described as "tingling," "pins and needles," and sometimes "numbness." Psychological factors, such as emotion, anxiety, and stress, may worsen the pain, as seen for other pain conditions.

Screening and Assessment of Neuropathic Pain in Syringomyelia In recent years, several screening tools based on verbal pain description with, or without, limited clinical examination have been developed and validated in neuropathic pain including central pain (Haanpää et al. 2011; Bouhassira and Attal 2011). These screening tools share many features despite being developed by different groups in different contexts. These include the *Douleur Neuropathique en 4 questions* (DN4) which has been initially validated in patients with central pain, the PainDetect questionnaire, the Leeds Assessment of Neuropathic Symptoms and Signs (LANSS), the Neuropathic Pain Questionnaire (NPQ), and the ID Pain. One study

comparing 4 of these tools in patients with spinal cord injury pain found that the diagnostic accuracy for the tools was for DN4 88 %, PainDetect 78 %, NPQ 65 %, and LANSS 55 % (Hallström and Norrbrink 2011). Screening tools are helpful to identify patients with possible neuropathic pain, particularly when used by nonspecialists and this is probably their chief clinical strength. However, screening tools fail to identify about 10–20 % of patients with clinician diagnosed neuropathic pain, indicating that they may offer guidance for further diagnostic evaluation and pain management but cannot replace clinical judgment.

Whereas assessment of neuropathic pain intensity in patients with spinal cord injury pain is generally achieved through VAS, numerical, or categorical scales (Haanpää et al. 2011), several specific tools have been validated specifically for the assessment of neuropathic pain quality and have been used in spinal cord injury patients including syringomyelia patients (Hatem et al. 2010; Ducreux et al. 2006). These include the Neuropathic Pain Scale and the Neuropathic Pain Symptom Inventory (Bouhassira and Attal 2011; Haanpää et al. 2011). Validated neuropathic pain quality measures are probably useful to discriminate between various pain mechanisms associated with distinct dimensions of neuropathic pain experience and are sensitive to change. They are currently recommended to evaluate treatment effects on neuropathic symptoms or their combination (Haanpää et al. 2011).

Is the Neuropathic Pain of Syringomyelia Distinct from or Similar to that of Other Neuropathic Pain Conditions? In a large cohort of patients with various peripheral or central NeP syndromes, we tested whether specific neuropathic combination of symptoms assessed by the NPSI were associated with specific etiology or locations of neurological lesions (Attal et al. 2008). Univariate analyses showed that the five distinct dimensions (i.e., combination of symptoms) of neuropathic pain identified with the NPSI (burning pain, deep squeezing/pressing pain, paroxysmal pain, evoked pain, and paresthesia/dysesthesia) were present, in similar proportions, in all the etiologies, with a few exceptions. In particular, a similar proportion of syringomyelic patients to those with other central pain disorders (poststroke pain, spinal cord trauma with below-level pain, multiple sclerosis) had the same proportion of symptoms and dimensions. These data were further confirmed by a Multiple Correspondence Analysis which showed no association between the neuropathic symptoms and the etiology or location of the lesion. Thus, the principal differences between the neuropathic pain of syringomyelia and other peripheral or central lesions are the location of the pain and the associated neurological symptoms and signs. These results show that there is more similarity than differences between the neuropathic pain of syringomyelia and other NeP syndromes.

Other Painful Syndromes

Coexisting painful symptoms are frequent in syringomyelia depending on the etiology of the syrinx (50–90 % of cases) (Milhorat et al. 1996, 1999). Suboccipital headache (generally a benign exertional headache, i.e., a headache that is intensified by Valsalva maneuvers such as coughing, sneezing, or bowel movement), migraine, or neck pain (which also usually increases with Valsalva maneuvers) are frequent in patients with Chiari type I malformation (Buzzi et al. 2003; Milhorat et al. 1999; Taylor and Larkins 2002; Fernández et al. 2009). Benign exertional headache is thought to be related to the increased intrathecal pressure caused by the free flow of CSF being obstructed in the subarachnoid space and may respond to surgical treatment of the Chiari (Sansur et al. 2003; Taylor and Larkins 2002). However, migraine and benign exertional headache may coexist in the same patient and respond similarly to antimigraine pharmacologic treatment (Buzzi et al. 2003). Trigeminal neuralgia and cluster-like facial pain have also been described and may also respond to surgical treatment of the syrinx (Rosetti et al. 1999). Visceral pain (due to renal calculus, bowel, or sphincter dysfunction, etc.) and musculoskeletal pain (due to bone, joint, muscle trauma or inflammation, mechanical instability, muscle spasms, and

secondary overuse syndrome) may be observed in syringomyelia related to spinal cord trauma (Siddall et al. 1997; Siddall and Loeser 2001). Finally, peripheral neuropathic pain due to nerve root entrapment or arachnoiditis is common in syringomyelia caused by meningitis.

Sensory Deficits and Evoked Pains

All patients with syringomyelia have some degree of thermoalgesic deficit in their painful area. Such a deficit may be unilateral or bilateral and have a segmental distribution, generally over the neck, shoulder, trunk, and arms, but may also affect the face. It is located or predominates ipsilaterally to the paracentral extension of the syrinx in cases of eccentric or paracentral cavities (Ducreux et al. 2006; Milhorat et al. 1996; Hatem et al. 2010). It may range from mild hypoesthesia or hypoalgesia (diminished pain in response to a normally painful stimulus) to complete anesthesia or analgesia. Sensory deficit can be detected and quantified best by quantitative sensory testing (QST), which indicates an impairment of thermal sensation in nearly 100 % of cases. There are almost always increased warm and cold detection thresholds and to a lesser extent increased pain thresholds (Boivie 2003, 2006; Boivie et al. 1994; Ducreux et al. 2006; Hatem et al. 2010). This is similar to that observed in other central pain conditions (Boivie 2003, 2006; Bowsher 1996; Bowsher et al. 1998; Defrin et al. 2001; Eide et al. 1996; Finnerup et al. 2003; Vestergaard et al. 1995; Osterberg and Boivie 2010). QST is also useful to follow the outcome of syringomyelic patients (Attal et al. 2004). Thus, we have found that the thermoalgesic deficit generally does not improve after surgical decompression of the syrinx, except in patients operated on early (i.e., less than 2 years) after the onset of their symptoms (Attal et al. 2004).

Proprioceptive/tactile sensation is usually considered to be not affected. However, in up to 50 % of patients, loss of proprioceptive/tactile sensation has been observed (Honan and Williams 1993). Increased vibration and/or mechanical thresholds may be observed by quantitative sensory testing (Attal et al. 2004; Ducreux et al. 2006; Hatem et al. 2010). Other signs of dorsal column dysfunction, such as an impairment of graphesthesia and movement direction, may also be observed (Ducreux et al. 2006; Hatem et al. 2010). Proprioceptive deficits, assessed using quantitative sensory tests, respond better than the thermal deficits to surgical treatment of the syrinx (Attal et al. 2004). Loss of vibration and position sense may also be present in the lower limbs, particularly in cases of Chiari malformation.

Patients with painful syringomyelia also present with signs of evoked pain, such as allodynia, which is "pain due to a stimulus which does not normally provoke pain" or hyperalgesia, which is "an increased response to a stimulus which is normally painful" (Merskey and Bogduk 1994) (64 % of painful patients in a psychophysical study of 46 consecutive patients with or without pain) (Ducreux et al. 2006). The pain is most frequently evoked by brushing (dynamic mechanical allodynia), punctate stimuli using a pinprick or Von Frey filaments (punctate mechanical allodynia/hyperalgesia), or cold stimuli (cold allodynia/hyperalgesia) (Ducreux et al. 2006; Hatem et al. 2010). Less commonly, it can be evoked by static pressure (static mechanical allodynia/hyperalgesia) or heat stimuli (heat allodynia/hyperalgesia). In most cases, allodynia coexists with spontaneous pain, but "pure" mechanical allodynia has also been described after intraspinal lesion (Attal et al. 1998).

Evoked pain may also persist after stimulation (aftersensation), appear some time after stimulation (delayed sensation), spread beyond the site of stimulation (radiation), or be increased or provoked by repetitive stimuli (temporal summation). Many patients suffer from hyperpathia, which is "a painful syndrome characterised by an abnormally painful reaction to a stimulus, especially a repetitive stimulus, as well as an increased threshold" (Merskey and Bogduk 1994). Hyperpathia can therefore be considered as a combination of hyperalgesia, temporal summation, and post-sensation.

Other Symptoms and Signs
Generally, there is no motor impairment in the early stages of the disease. By contrast, as the cavity enlarges, there may be corticospinal tracts and/or segmental impairment of the anterior horn, leading to pyramidal syndrome and muscular amyotrophy, respectively. Tendon reflexes are generally absent, except in some patients with Chiari malformation. In these patients, the tendon reflexes may be normal or increased. Depending on the etiology of the syrinx, patients may also have variable clinical features, including nystagmus, vertigo, mild neurosensorial hearing low, cerebellar ataxia, and lower cranial nerve impairment such as palatal and vocal cord paresis, particularly in Chiari malformation (with or without syringobulbia). Kyphoscoliosis is present in many cases and overt cervico-occipital malformation is present in nearly a quarter of patients (Victor and Ropper 2001). Finally, patients may also have signs of autonomic impairment on the painful side. The painful area may be cooler and vasoconstricted or warmer and occasional changes in sweating have been reported (Bowsher 1996).

Mechanisms of Central Pain in Syringomyelia
Syringomyelia is a typical model of a "pure" intraspinal lesion predominantly affecting the spinothalamic tract. It is therefore of particular interest for studying the mechanisms underlying central pain, particularly with regard to the disputed role of spinothalamic lesion in neuropathic pain. In animal models developed over the last few years to study the mechanisms of central pain, lesions are usually induced by excitotoxic or ischemic injury, leading to damage of the spinal gray matter. These models present pathological characteristics, including neuronal loss, glial response, and syrinx formation, resembling those observed in syringomyelia, particularly posttraumatic cavities (e.g., Christensen and Hulsebosch 1997; Vierck et al. 2000; Yezierski et al. 1998; Yezierski 2001; Hao and Xu 2003). Therefore, they are of particular interest for studying the cellular, molecular, anatomical, and physiological consequences of an intraspinal injury.

Here, we will review the main theories of the mechanisms of central pain and their possible relevance for syringomyelia, the role of spinal versus supraspinal structures in the pain of syringomyelia, and then present evidence suggesting that the clinical symptoms of syringomyelia may be sustained by distinct mechanisms.

The Main Hypotheses of Central Pain Mechanisms and Their Relevance to Syringomyelia

The mechanisms of central pain have been studied for over 100 years but still remain largely unknown. The principal hypotheses fall into two major categories: central imbalance and/or disinhibition and central sensitization. Both phenomena can trigger abnormal neuronal hyperexcitability of spinal and/or supraspinal structures (refs in Finnerup et al. 2003; Finnerup and Jensen 2004; Finnerup 2008). Such hyperexcitability has been observed in animal models of spinal cord injury (Vierck et al. 2000; Yezierski 2001) but also in some patients with central pain, in the dorsal horn (Edgar et al. 1993, Falci et al. 2002; Loeser et al. 1968) and in the lateral and medial thalamus (Jeanmonod et al. 1993, 1996; Lenz et al. 1989, 1994, 2000).

Central Imbalance

Patients with central pain nearly always have abnormal temperature and pain sensibility, but may have normal touch and vibration sensation (Boivie 2006). This suggests that the presence of a spinothalamic dysfunction is a necessary condition for the occurrence of central pain, whereas the dorsal columns/medial lemniscus system is less commonly affected. This is further supported by studies showing abnormalities in laser-evoked potentials (reflecting the function of A-delta nociceptive fibers) on the painful side of patients with poststroke pain (Boivie 2006) and syringomyelia (Kakigi et al. 1991; Treede et al. 1991; Hatem et al. 2010), whereas sensory-evoked potentials served by large diameter fibers, correlated to the decrease in touch and vibration sensibility, may be preserved (Beric et al. 1988). It has been proposed that below-level pain associated with spinal cord injury is induced by an imbalance of integration between the unaffected dorsal column/medial lemniscus activity and the affected spinothalamic tract (Beric 1993; Beric et al. 1988). However, a number of

psychophysical studies of patients with spinal cord injury with or without below-level pain have shown that there is a similar degree of thermoalgesic deficits in painful and painless patients (e.g., Defrin et al. 2001; Finnerup et al. 2003). Similarly, the extent and magnitude of thermal deficits, assessed in patients with syringomyelia, are similar in painful and painless patients (Bouhassira and Attal 2000; Hatem et al. 2010; Ducreux et al. 2006). These data suggest that the spinothalamic lesion is necessary but not sufficient to account for the development of central pain.

Disinhibition Theories

Since the original hypothesis of Head and Holmes (1911), the notion of a central disinhibition, particularly in the thalamus, has been one of the most favored pathophysiological theories of central pain (Casey 1991; Cassinari and Pagni 1969; Jeanmonod et al. 1996; Pagni 1989). One hypothesis, the "thermosensory disinhibition" theory of Craig (1998, 2000, 2003 for refs), which is based on the observation that thermosensory loss is the basic feature of nearly all central pain patients, proposes that central pain (particularly burning pain and cold allodynia) may be due to reduction of the physiological inhibition of thermal (cold) systems on nociceptive neurons. In contrast with traditional conceptions, this theory views pain not only as a sensation but also as a "homeostatic emotion" and an aspect of interoception, defined as the physiological condition of the body.

We tested this hypothesis in a first psychophysical study of 46 patients with syringomyelia, of whom 31 suffered from neuropathic pain (Ducreux et al. 2006). We did not observe any correlation between the magnitude of thermal deficits and the amount of pain in the entire group of painful patients. However, patients with spontaneous pain only (without evoked pains) had more severe thermal deficits than those with allodynia, and the intensity of burning pain was correlated to the extent of warm and cold deficits. These patients also had more asymmetrical thermal deficits compared to those with allodynia and painless patients, which was confirmed by a further study (Hatem et al. 2010). This might reflect an imbalance in thermoregulation which involves bilateral brain integration (Craig 1998). Thus, these data are compatible with the disinhibition theory of central pain in a subgroup of syringomyelic patients (i.e., patients with spontaneous pain only). By contrast, patients with allodynia had significantly reduced thermal deficits compared with those having spontaneous pain only (Ducreux et al. 2006).

Another form of central disinhibition has been proposed to account more specifically for the pain of syringomyelia. In a study of 137 patients with syringomyelia, Milhorat et al. (1996) observed that the syrinx extended onto the dorsolateral quadrant of the spinal cord on the same side of the pain in 84 % of patients. It was suggested that pain was due to a disturbance of pain-modulating centers in the dorsolateral quadrant of the spinal cord. However, it was not stated whether this pattern was less common in painless patients.

Central Sensitization
The hyperactivity of nociceptive neurons in central pain may also result from direct modifications of their electrophysiological properties, i.e., central sensitization (for review see Latremoliere and Woolf 2009). This may be caused by damage to central neurons caused by excitatory amino acids related to NMDA receptor activation (Eide 1998; Vierck et al. 2000), low of tonic inhibition by gamma-aminobutyric acid interneurons or descending tracts (Hains et al. 2002); neuroinflammation (Yu et al. 2003); microglia activation (Hains and Waxman 2006; Gwak and Hulsebosch 2009); and possibly also abnormal expression of sodium channels (Hains et al. 2003; Hains and Waxman 2007). Indirect evidence for the role of central sensitization in spinal cord injury pain is provided by the beneficial effects of NMDA antagonists, GABA-A receptor blockers, and sodium channel blockers in patients with spinal cord injury pain (Attal et al. 2000, 2002; Eide et al. 1995; Finnerup et al. 2005; Kvarnström et al. 2004; Canavero an Bonicalzi 2004) including syringomyelia (Attal et al. 2000, 2002).

Role of Spinal and Supraspinal Structures in the Pain of Syringomyelia
Several studies on animals and human have suggested that the primary pain generator in at-level and below-level neuropathic pain associated with spinal cord injury pain may be the injury region itself (refs in Finnerup and Jensen 2004; Finnerup et al. 2003, 2007; Hari et al. 2009). The observation that pain and sensory deficits associated with syringomyelia predominate ipsilaterally to the extension of the cavity in cases of paracentral or eccentric cavities (Bouhassira and Atta 2000) is consistent with this hypothesis. Similarly, spinal drug treatment trials have shown that spinal anesthesia with lidocaine reduced at-level and below-level SCI pain (Loubser and Donovan 1991). More specifically, the ipsilateral dorsal horn or spinothalamic fibers before they cross the midline may be important sites for pain generation (Boivie 2006). Animal models of intraspinal excitotoxic lesion have also shown spontaneous nociceptive behavior and enhancement of nociceptive responses in dermatomes ipsilateral to the lesion (refs in Vierck et al. 2000; Yezierski 2001). Electrophysiological recordings in these models have demonstrated abnormal neuronal excitation in the dorsal horn adjacent to the injury (Christensen and Hulsebosch 1997; Vierck et al. 2000; Yezierski 2001; Yezierski and Park 1993). The possible mechanisms of such segmental hyperexcitability are probably not specific and may involve sensitization or disinhibition.

However, both animal and human studies also point to the role of supraspinal structure particularly the thalamus, in SCI pain. Studies in humans with spinal cord injury pain have pointed to thalamocortical dysrhythmia, differences in metabolites in the thalamus region, reorganization, and bursting activity (refs in Finnerup et al. 2009). Primary somatosensory cortex reorganization and regional brain anatomical changes have also been demonstrated, at least in patients with complete thoracic SCI and below-level pain, but whether this observation is applicable to pain in syringomyelia has not been confirmed (Henderson et al. 2011; Gustin et al. 2010).

Potential Relationships Between the Clinical Phenotypes of Syringomyelia and Underlying Mechanisms

Several studies have suggested that the various painful symptoms of syringomyelia, particularly with regard to the subtypes of allodynia (mechanical or thermal), may be sustained by distinct mechanisms. Thus, pharmacological studies have shown preferential effects of intravenous administration of the sodium channel blocker lidocaine on mechanical (dynamic and static) but not cold allodynia in patients with central pain, including several syringomyelic patients (Attal et al. 2000). In the same line, a functional imaging study of patients with syringomyelia presenting with cold or brush-induced allodynia has revealed distinct activations patterns depending on the submodality of allodynia (Ducreux et al. 2006). Thus, cold allodynia activates a "cold pain network" including activation of insular and cingular areas, which is similar to the pattern induced by noxious cold in healthy controls (Craig 2000). In contrast, brush-evoked allodynia did not induce significant activation of several structures of the "pain matrix," particularly the insula and the ACC. The different patterns of activation between the two subtypes of allodynia might reflect distinct pathophysiological mechanisms.

In another study in patients with central pain associated with cervical syringomyelia, we used a systematic multimodal assessment including a clinical examination with quantitative sensory testing and pain questionnaires (NPSI), an assessment of the functionality of the nociceptive pathways with laser-evoked potentials, and an assessment of spinal cord damage by diffuse tensor imaging and fiber tracking (DTI-FT) of the cervical spinal cord (Hatem et al. 2009, 2010). We found that patients with both spontaneous and evoked pain clearly differed from patients with spontaneous pain only. Patients with spontaneous pain only had more severe spinal cord damage, as also indicated by individual examples. Furthermore, we evidenced a direct relationship between the spinal structural damage assessed by diffuse tensor imaging and the intensity of deep pain (measured with the NPSI) in this group of patients

only (Hatem et al. 2010). By contrast, patients with both spontaneous and evoked pain had not only less structural spinal cord damage, but also better preserved spinothalamic and lemniscal tracts on quantitative sensory testing and electrophysiological testing. Although it is difficult to speculate about the cellular and molecular mechanisms, these data strongly suggest that different mechanisms are at play in these two populations defined on the basis of the combination of symptoms and therefore that these patients should respond differently to treatment. In patients with spontaneous pain only pain, particularly deep pain, it might be related to central deafferentation, whereas in patients with spontaneous pain and allodynia, other mechanisms including central sensitization and independent on the severity of the lesion may be at play.

Treatment of Central Pain Associated with Syringomyelia

As with other chronic neuropathic pain syndromes, pain associated with syringomyelia has a major impact on quality of life and affective state (Widerstrom-Noga 2002; Attal et al. 2011). Therefore, the treatment of such pain should include management of disability and the affective comorbid conditions through the cognitive behavioral therapies and rehabilitation programs, some of which have been found useful in spinal cord injury patients (Sjolund 2002; Norrbrink Budh et al. 2006).

Is the Surgical Treatment of the Syrinx Effective on Neuropathic Pain?

Although surgery (generally foramen magnum decompression in patients with Chiari malformation) is commonly proposed to syringomyelic patients with progressive neurological deterioration, very few prospective studies have evaluated its effects on neuropathic pain (Attal et al. 2004). These studies generally reported disappointing results, in contrast to the positive effects seen for the pain associated with Chiari malformation, such as headache and neck pain. In a prospective 2 year study of 15 patients with syringomyelia, of whom eight presented with neuropathic pain, we found that surgery of the syrinx produced no overall significant effect on ongoing pain intensity at 2 years, whereas pain induced by straining, as well as coexisting dynamic symptoms (i.e., pains induced by coughing, Valsalva's maneuver) was relieved (Attal et al. 2004). The latter pains are probably more dependent on the dynamic characteristics of the syrinx fluid and flow, and may be particularly affected by surgical decompression which reduces the subarachnoid pressure. Furthermore, no correlation was found between the effects of surgery on the syrinx (foramen stenosis, syrinx diameter, syrinx/canal index) and the outcome of neuropathic pain (Attal et al. 2004).

Pharmacological Treatment

There are no controlled trials specifically devoted to the neuropathic pain of syringomyelia. Therefore, recommended treatments must rely on studies carried out for other neuropathic pain conditions (see central pain, this volume). On the basis of randomized controlled trials in spinal cord injury pain, only pregabalin, to a lesser extent duloxetine and amitriptyline, has been shown to be effective in treating SCI neuropathic pain while lower evidence for efficacy exists for tramadol, gabapentin, and opioids (Attal et al. 2009, 2010; Finnerup et al. 2010; Siddall 2009). Intravenous administration of single-dose opioids, local anesthetics (lidocaine), and NMDA receptor antagonists (ketamine) has also been found useful, but there is no serious evidence for their prolonged efficacy (for reviews, see Attal et al. 2006, 2010; Finnerup et al. 2010; Siddall 2009). Recommended drugs as first line for spinal cord injury pain including syringomyelia are tricyclic antidepressants (mainly because of their established efficacy in other neuropathic indications), or pregabalin/gabapentin, and SNRIs if TCAs are not tolerated (Attal et al. 2010). Topical lidocaine may also be proposed in patients with posttraumatic syringomyelia or arachnoiditis, who may present with peripheral at-level neuropathic pain (Wasner et al. 2007). Intrathecal therapy using opioids combined to clonidine (Siddall et al. 2000) or ziconotide (Schmidtko et al. 2010) may be proposed in refractory cases.

Importantly, the etiology or topography of the central lesion does not seem to play a role in the response of these treatments. In controlled trials of intravenous morphine and lidocaine carried out in patients with central pain, the response to these drugs appeared similar in patients with syringomyelia, spinal cord trauma, and poststroke pain (Attal et al. 2000, 2002). Treatments may also have similar efficacy on at-level and below-level neuropathic pain due to spinal cord injury, as recently shown in a placebo-controlled trial of intravenous lidocaine (Finnerup et al. 2005). The response to treatments may rather be influenced by the clinical symptoms or their combination, suggesting distinct pain mechanisms (Attal et al. 2008, 2011). Pharmacological studies in larger cohorts of patients using an in-depth assessment of phenotypic profiles are needed to confirm these data and define possible predictors of the response to treatments (Attal et al. 2011).

Neurostimulation and Neuroablative Techniques

Transcutaneous electrical nerve stimulation may sometimes be beneficial in spinal cord injury patients with at-level pain (Norrbrink 2009). Repetitive transcranial magnetic stimulation and direct electrical stimulation have been found useful in a few randomized controlled trials of patients with traumatic injury and below-level pain (e.g., Defrin et al. 2007; Soler et al. 2010), but there are no data regarding the maintenance of their analgesic effect over time.

Spinal cord stimulation may be an option in patients suffering from steady at-level neuropathic pain, who have incomplete spinal lesions and preservation of dorsal column function (Sindou and Mertens 2000). Syringomyelic patients may be possible candidates for such techniques.

Extradural stimulation of the motor cortex is increasingly used in patients with neuropathic pain including spinal cord injury pain (Nguyen et al. 2011) and may be particularly useful in patients with syringomyelia, as these patients generally have unilateral pain or predominantly unilateral pain involving one superior limb and trunk. Although no published data are available for patients with syringomyelia, personal observations from our group suggest the benefit of this technique in individual patients with syringomyelia.

Among neuroablative surgical procedures, dorsal root entry zone (DREZ) may be beneficial in at-level neuropathic pain after incomplete spinal cord injury, particularly paroxysmal and evoked pain (Sindou et al. 2001). However, there are very little data concerning the effects of this technique in syringomyelic patients (Prestor 2001). Moreover, DREZ may occasionally aggravate the sensory deficits which are generally incomplete in patients with syringomyelia. Therefore, this technique has a marginal place in the treatment of pain associated with syringomyelia.

Conclusion

Pain is frequently associated with syringomyelia and generally consists of neuropathic (central) pain. Its main characteristics (burning, deep pain, paroxysmal pain, and allodynia/hyperalgesia) are similar to those of other central pain conditions, which suggests that common mechanisms (either central sensitization or disinhibition) may be involved in these pain disorders. However, the nature of these mechanisms may differ, depending on the clinical presentation, as suggested from psychophysical, electrophysiological, functional MRI, and morphological studies of syringomyelic patients. These data may have important implications with regard to management.

Summary

Pain is frequently associated with syringomyelia and generally consists of neuropathic (central) pain. Its main characteristics are similar to those of other central pain conditions, which suggests that common mechanisms may be involved in these pain disorders. However, the nature of these mechanisms may differ depending on the clinical phenotypes. Currently, first-line drug treatments for this condition include tricyclic antidepressants, gabapentin or pregabalin, and SNRI antidepressants, while opioids are considered second line. Neurostimulation techniques may be proposed in refractory patients.

References

Attal, N, & Bouhassira, D. (2000). Recent pharmacologic approaches to pain. Annales Pharmaceutiques Francaises, 58(2):121–34. http://www.ncbi.nlm.nih.gov/pubmed/10790605.

Attal, N., Brasseur, L., Chauvin, M., & Bouhassira, D. (1998). A case of "pure" dynamic mechano-allodynia due to a lesion of the spinal cord: Pathophysiological considerations. Pain, 75, 399–404.

Attal, N., Gaude, V., Brasseur, L., Dupuy, M., Guirimand, F., Parker, F., & Bouhassira, D. (2000). Intravenous lidocaine in central pain: A double-blind placebo-controlled psycho-physical study. Neurology, 544, 564–574.

Attal, N., Guirimand, F., Brasseur, L., Gaude, V., Chauvin, M., & Bouhassira, D. (2002). Effects of IV morphine in central pain: A randomized placebo-controlled study. Neurology, 58, 554–563.

Attal, N., Parker, F., Tadie, M., Aghakani, N., & Bouhassira, D. (2004). Effects of surgery on the sensory deficits of syringomyelia and predictors of outcome: A long term prospective study. Journal of Neurology, Neurosurgery, and Psychiatry, 75, 1025–1030.

Attal, N., Fermanian, C., Fermanian, J., Lanteri-Minet, M., Alchaar, H., & Bouhassira, D. (2008). Neuropathic pain: Are there distinct subtypes depending on the aetiology or anatomical lesion? Pain, 138, 343–353.

Attal, N., Mazaltarine, G., Perrouin-Verbe, B., Albert, T., & SOFMER French Society for Physical Medicine and Rehabilitation. (2009). Chronic neuropathic pain management in spinal cord injury patients. What is the efficacy of pharmacological treatments with a general mode of administration? (Oral, transdermal, intravenous). Annals of Physical and Rehabilitation Medicine, 52, 124–141.

Attal, N., Cruccu, G., Baron, R., Haanpää, M., Hansson, P., Jensen, T. S., Nurmikko, T., & European Federation of Neurological Societies. (2010). EFNS guidelines on the pharmacological treatment of neuropathic pain: 2010 revision. European Journal of Neurology, 17, 1113–e88.

Baron, R., Binder, A., & Wasner, G. (2010). Neuropathic pain: Diagnosis, pathophysiological mechanisms, and treatment. Lancet Neurology, 9, 807–819.

Beric, A. (1993). Central pain: "New" syndromes and their evaluation. Muscle & Nerve, 16, 1017–1024.

Beric, A., Dimitrijevic, M. R., & Lindblom, U. (1988). Central dysesthesia syndrome in spinal cord injury patients. Pain, 34, 109–116.

Boivie, J. (1994). Central pain. In P. D. Wall & R. Melzack (Eds.), Textbook of Pain (pp. 871–892). London: Churchill Livingstone.

Boivie, J. (2003). Central pain and the role of quantitative sensory testing (QST) in research and diagnosis. *European Journal of Pain, 7*, 339–343.

Boivie, J. (2006). Central pain. In S. B. McMahon & M. Koltzenburg (Eds.), *Wall and Melzack's textbook of pain* (5th ed., pp. 1057–1074). Philadelphia: Churchill Livingstone/Elsevier.

Bouhassira, D., & Attal, N. (2000). Recent pharmacologic approaches to pain. *Ann Pharm Fr., 58*(2):121–34.

Bouhassira, D., & Attal, N. (2011). Diagnosis and assessment of neuropathic pain: the saga of clinical tools. *Pain,* 152(3 Suppl):S74–83.

Bowsher, D. (1996). Central pain: Clinical and physiological characteristics. *Journal of Neurology, Neurosurgery, and Psychiatry, 61*, 62–69.

Bowsher, D., Leijon, G., & Thuomas, K. A. (1998). Central poststroke pain: Correlation of MRI with clinical pain characteristics and sensory abnormalities. *Neurology, 51*, 1352–1358.

Buzzi, M. G., Formisano, R., Colonnese, C., & Pierelli, F. (2003). Chiari-associated exertional, cough, and sneeze headache responsive to medical therapy. *Headache, 43*, 404–406.

Canavero, S., & Bonicalzi, V. (2004). Intravenous subhypnotic propofol in central pain: A double-blind, placebo-controlled, crossover study. *Clinical Neuropharmacology, 29*, 182–186.

Carroll, A. M., & Brackenridge, P. (2005). Post-traumatic syringomyelia: A review of the cases presenting in a regional spinal injuries unit in the North East of England over a 5-year period. *Spine, 30*, 1206–1210.

Casey, K. L. (1991). *Pain and central nervous disease: The central pain syndromes.* New York: Raven.

Cassinari, V., & Pagni, C. A. (1969). *Central pain: A neurological survey.* Cambridge, MA: Harvard University Press.

Christensen, M. D., & Hulsebosch, C. E. (1997). Chronic central pain after spinal cord injury. *Journal of Neurotrauma, 14*, 517–537.

Craig, A. D. (1998). A new version of the thalamic disinhibition hypothesis of centrapain. *Pain Forum, 7*, 1–14.

Craig, A. D. (2000). The functional anatomy of lamina I and its role in post-stroke central pain. *Progress in Brain Research, 129*, 137–151.

Craig, A. D. (2003). A new view of pain as a homeostatic emotion. *Trends in Neurosciences, 26*, 303–307.

Defrin, R., Ohry, A., Blumen, N., & Urca, G. (2001). Characterization of chronic pain and somatosensory function in spinal cord injury subjects. *Pain, 89*, 253–263.

Defrin, R., Grunhaus, L., Zamir, D., & Zeilig, G. (2007). The effect of a series of repetitive transcranial magnetic stimulations of the motor cortex on central pain after spinal cord injury. *Archives of Physical Medicine and Rehabilitation, 88*, 1574–1580.

Ducreux, D., Attal, N., Parker, F., & Bouhassira, D. (2006). Mechanisms of central neuropathic pain: A combined psychophysical and fMRI study in syringomyelia. *Brain, 129*, 963–976.

Edgar, R. E., Best, L. G., Quail, P. A., & Obert, A. D. (1993). Computer-assisted DREZ microcoagulation: Posttraumatic spinal deafferentation pain. *Journal of Spinal Disorders, 6*, 48–56.

Eide, P. K. (1998). Pathophysiological mechanisms of central neuropathic pain after spinal cord injury. *Spinal Cord, 36*, 601–612.

Eide, P. K., Stubhaug, A., & Stenehjem, A. E. (1995). Central dysesthesia pain after traumatic spinal cord injury is dependent on N-methyl-D-aspartate receptor activation. *Neurosurgery, 37*, 1080–1087.

Eide, P. K., Jorum, E., & Stenehjem, A. E. (1996). Somatosensory findings in patients with spinal cord injury and central dysesthesia pain. *Journal of Neurology, Neurosurgery, and Psychiatry, 60*, 411–415.

Falci, S., Best, L., Bayles, R., Lammertse, D., & Starnes, C. (2002). Dorsal root entry zone microcoagulation for spinal cord injury-related central pain: Operative intramedullary electrophysiological guidance and clinical outcome. *Journal of Neurosurgery, 97*, 193–200.

Fernández, A. A., Guerrero, A. I., Martínez, M. I., Vázquez, M. E., Fernández, J. B., Chesa i Octavio, E., Labrado Jde, L., Silva, M. E., de Araoz, M. F., García-Ramos, R., Ribes, M. G., Gómez, C., Valdivia, J. I., Valbuena, R. N., & Ramón, J. R. (2009). Malformations of the craniocervical junction (Chiari type I and syringomyelia: Classification, diagnosis and treatment). *BMC Musculoskeletal Disorders, 10*(1), S1.

Finnerup, N. B. (2008). A review of central neuropathic pain states. *Current Opinion in Anaesthesiology, 21*(5), 586–589.

Finnerup, N. B., & Jensen, T. S. (2004). Spinal cord injury pain-mechanisms and treatment. *European Journal of Neurology, 11*, 73–82.

Finnerup, N. B., Johannesen, I. L., Fuglsang-Frederiksen, A., Bach, F. W., & Jensen, T. S. (2003). Sensory function in spinal cord injury patients with and without central pain. *Brain, 126*, 57–70.

Finnerup, N. B., Biering-Sorensen, F., Johannesen, I. L., Terkelsen, A. J., Juhl, G. I., Kristensen, A. D., Sindrup, S. H., Bach, F. W., & Jensen, T. S. (2005). Intravenous lidocaine relieves spinal cord injury pain: A randomized controlled trial. *Anesthesiology, 102*, 1023–1030.

Finnerup, N. B., Sørensen, L., Biering-Sørensen, F., Johannesen, I. L., & Jensen, T. S. (2007). Segmental hypersensitivity and spinothalamic function in spinal cord injury pain. *Experimental Neurology, 207*, 139–149.

Finnerup, N. B., Grydehoj, J., Bing, J., et al. (2009). Levetiracetam in spinal cord injury pain: a randomized controlled trial. Spinal Cord, 47, 861–7.

Finnerup, N. B., Sindrup, S. H., & Jensen, T. S. (2010). The evidence for pharmacological treatment of neuropathic pain. *Pain, 150*, 573–581.

Gustin, S. M., Wrigley, P. J., Siddall, P. J., & Henderson, L. A. (2010). Brain anatomy changes associated with persistent neuropathic pain following spinal cord injury. *Cerebral Cortex, 20*, 1409–1419.

Gwak, Y. S., & Hulsebosch, C. E. (2009). Remote astrocytic and microglial activation modulates neuronal hyperexcitability and below-level neuropathic pain after spinal injury in rat. *Neuroscience, 161*, 895–903.

Haanpää, M., Attal, N., Backonja, M., Baron, R., Bennett, M., Bouhassira, D., Cruccu, G., Hansson, P., Haythornthwaite, J. A., Iannetti, G. D., Jensen, T. S., Kauppila, T., Nurmikko, T. J., Rice, A. S., Rowbotham, M., Serra, J., Sommer, C., Smith, B. H., & Treede, R. D. (2011). NeuPSIG guidelines on neuropathic pain assessment. *Pain, 152*, 14–27.

Hains, B. C., & Waxman, S. G. (2006). Activated microglia contribute to the maintenance of chronic pain after spinal cord injury. *Journal of Neuroscience, 26*, 4308–4317.

Hains, B. C., & Waxman, S. G. (2007). Sodium channel expression and the molecular pathophysiology of pain after SCI. *Progress in Brain Research, 161*, 195–203.

Hains, B. C., Everhart, A. W., Fullwood, S. D., & Hulsebosch, C. E. (2002). Changes in serotonin, serotonin transporter expression and serotonin denervation supersensitivity: Involvement in chronic central pain after spinal hemisection in the rat. *Experimental Neurology, 175*, 347–362.

Hains, B. C., Klein, J. P., Saab, C. Y., Craner, M. J., Black, J. A., & Waxman, S. G. (2003). Upregulation of sodium channel Nav1.3 and functional involvement in neuronal hyperexcitability associated with central neuropathic pain after spinal cord injury. *Journal of Neuroscience, 23*, 8881–8892.

Hallström, H., & Norrbrink, C. (2011). Screening tools for neuropathic pain: Can they be of use in individuals with spinal cord injury? *Pain, 152*, 772–779.

Hao, J. X., & Xu, X. J. (2003). Animal models of spinal cord injury pain and their implications for pharmacological treatments. *Journal of Rehabilitation Medicine, 41*(Suppl.), 81–84.

Hari, A. R., Wydenkeller, S., Dokladal, P., & Halder, P. (2009). Enhanced recovery of human spinothalamic function is associated with central neuropathic pain after SCI. *Experimental Neurology, 216*, 428–430.

Hatem, S. M., Attal, N., Ducreux, D., Gautron, M., Parker, F., Plaghki, L., & Bouhassira, D. (2009). Assessment of spinal somatosensory systems with diffusion tensor imaging in syringomyelia. *Journal of Neurology, Neurosurgery, and Psychiatry, 80*, 1350–1356.

Hatem, S. M., Attal, N., Ducreux, D., Gautron, M., Parker, F., Plaghki, L., & Bouhassira, D. (2010). Clinical, functional and structural determinants of central pain in syringomyelia. *Brain, 133*, 3409–3422.

Head, H., & Holmes, G. (1911). Sensory disturbances from cerebral lesions. *Brain, 34*, 102–154.

Heiss, J. D., Patronas, N., DeVroom, H. L., Shawker, T., Ennis, R., Kammerer, W., Eidsath, A., Talbot, T., Morris, J., Eskioglu, E., & Oldfield, E. H. (1999). Elucidating the pathophysiology of syringomyelia. *Journal of Neurosurgery, 91*, 553–562.

Henderson, L. A., Gustin, S. M., Macey, P. M., Wrigley, P. J., & Siddall, P. J. (2011). Functional reorganization of the brain in humans following spinal cord injury: Evidence for underlying changes in cortical anatomy. *Journal of Neuroscience, 31*, 2630–2637.

Honan, W. P., & Williams, B. (1993). Sensory loss in syringomyelia: Not necessarily dissociated. *Journal of the Royal Society of Medicine, 86*, 519–520.

Jeanmonod, D., Magnin, M., & Morel, A. (1993). Thalamus and neurogenic pain: Physiological, anatomical and clinical data. *Neuroreport, 4*, 475–478.

Jeanmonod, D., Magnin, M., & Morel, A. (1996). Low-threshold calcium spike bursts in the human thalamus. Common physiopathology for sensory, motor and limbic positive symptoms. *Brain, 119*, 363–375.

Kakigi, R., Shibasaki, H., Kuroda, Y., Neshige, R., Endo, C., Tabuchi, K., & Kishikawa, T. (1991). Pain-related somatosensory evoked potentials in syringomyelia. *Brain, 114*, 1871–1889.

Kvarnström, A., Karlsten, R., Quiding, H., & Gordh, T. (2004). The analgesic effect of intravenous ketamine and lidocaine on pain after spinal cord injury. *Acta Anaesthesiologica Scandinavica, 48*, 498–506.

Latremoliere, A., & Woolf, C. J. (2009). Central sensitization: A generator of pain hypersensitivity by central neural plasticity. *The Journal of Pain, 10*, 895–926.

Lenz, F., Kwan, H. C., Dostrovsky, J. O., & Tasker, R. R. (1989). Characteristics of the bursting pattern of action potentials that occurs in the thalamus of patients with central pain. *Brain Research, 496*, 357–360.

Lenz, F. A., Kwan, H. C., Martin, R., Tasker, R., Richardson, R. T., & Dostrovsky, J. O. (1994). Characteristics of somatotopic organization and spontaneous neuronal activity in the region of the thalamic principal sensory nucleus in patients with spinal cord transection. *Journal of Neurophysiology, 72*, 1570–1587.

Lenz, F. A., Lee, J. I., Garonzik, I. M., Rowland, L. H., Dougherty, P. M., & Hua, S. E. (2000). Plasticity of pain-related neuronal activity in the human thalamus. *Progress in Brain Research, 129*, 259–273.

Levine, D. N. (2004). The pathogenesis of syringomyelia associated with lesions at the foramen magnum: A critical review of existing theories and proposal of a new hypothesis. *Journal of Neurological Sciences, 220*, 3–21.

Loeser, J. D., Ward, A. A. J. R., & White, L. E., Jr. (1968). Chronic deafferentation of human spinal cord neurons. *Journal of Neurosurgery, 29*, 48–50.

Loubser, P. G., & Donovan, W. H. Â. (1991). Diagnostic spinal anaesthesia in chronic spinal cord injury pain. *A Paraplegia, 29*:36.

Merskey, H., & Bogduk, N. (Eds.). (1994). *Classification of chronic pain*. Seattle: IASP Press.

Milhorat, T. H., Johnson, R. W., Milhorat, R. H., Capocelli, A. L., & Pevsner, P. H. (1995). Clinicopathological correlations in syringomyelia using axial magnetic resonance imaging. *Neurosurgery, 37*, 206–213.

Milhorat, T. H., Kotzen, R. M., Mu, H. T. M., Capocelli, A. L., & Milhorat, R. H. (1996). Dysesthetic pain in patients with syringomyelia. *Neurosurgery, 38*, 940–947.

Milhorat, T. H., Chou, M. W., Trinidad, E. M., Kula, R. W., Mandell, M., Wolpert, C., & Speer, M. C. (1999). Chiari I malformation redefined: Clinical and radiographic findings for 364 symptomatic patients. *Neurosurgery, 44*, 1005–1017.

Nguyen, J. P., Nizard, J., Keravel, Y., & Lefaucheur, J. P. (2011). Invasive brain stimulation for the treatment of neuropathic pain. *Nature Reviews. Neurology, 7*, 699–709.

Norrbrink, C. (2009). Transcutaneous electrical nerve stimulation for treatment of spinal cord injury neuropathic pain. *Journal of Rehabilitation Research and Development, 46*, 85–93.

Norrbrink Budh, C., Kowalski, J., & Lundeberg, T. (2006). A comprehensive pain management programme comprising educational, cognitive and behavioural interventions for neuropathic pain following spinal cord injury. *Journal of Rehabilitation Medicine, 38*, 172–180.

Osterberg, A., & Boivie, J. (2010). Central pain in multiple sclerosis – Sensory abnormalities. *European Journal of Pain, 14*, 104–110.

Pagni, C. A. (1989). Central pain due to spinal cord and brainstem damage. In P. D. Wall & R. Melzack (Eds.), *Textbook of pain* (pp. 634–655). Edinburgh: Churchill Livingstone.

Prestor, B. (2001). Microsurgical junctional DREZ coagulation for treatment of deafferentation pain syndromes. *Surgical Neurology, 56*, 259–265.

Rosetti, P., Ben Taib, N. O., Brotchi, J., & De Witte, O. (1999). Arnold Chiari Type I malformation presenting as a trigeminal neuralgia: Case report. *Neurosurgery, 44*, 1122–1123.

Rossier, A. B., Foo, D., Shillito, J., & Dyro, F. M. (1985). Posttraumatic cervical syringomyelia. Incidence, clinical presentation, electrophysiological studies, syrinx protein and results of conservative and operative treatment. *Brain, 108*, 439–461.

Sansur, C. A., Heiss, J. D., DeVroom, H. L., Eskioglu, E., Ennis, R., & Oldfield, E. H. (2003). Pathophysiology of headache associated with cough in patients with Chiari I malformation. *Journal of Neurosurgery, 98*, 453–458.

Schmidtko, A., Lötsch, J., Freynhagen, R., & Geisslinger, G. (2010). Ziconotide for treatment of severe chronic pain. *Lancet, 375*, 1569–1577.

Schurch, B., Wichmann, W., & Rossier, A. B. (1996). Post-traumatic syringomyelia (cystic myelopathy): A prospective study of 449 patients with spinal cord injury. *Journal of Neurology, Neurosurgery, and Psychiatry, 60*, 61–67.

Shannon, N., Symon, L., Logue, V., Cull, D., Kang, J., & Kendall, B. (1981). Clinical features, investigation and treatment of post-traumatic syringomyelia. *Journal of Neurology, Neurosurgery, and Psychiatry, 44*, 35–42.

Siddall, P. J. (2009). Management of neuropathic pain following spinal cord injury: Now and in the future. *Spinal Cord, 47*, 352–359.

Siddall, P. J., & Loeser, J. D. (2001). Pain following spinal cord injury. *Spinal Cord, 39*, 63–73.

Siddall, P. J., Taylor, D. A., & Cousins, M. J. (1997). Classification of pain following spinal cord injury. *Spinal Cord, 35*, 69–75.

Siddall, P. J., Molloy, A. R., Walker, S., Mather, L. E., Rutkowski, S. B., & Cousins, M. J. (2000). The efficacy of intrathecal morphine and clonidine in the treatment of pain after spinal cord injury. *Anesthesia and Analgesia, 91*, 1493–1498.

Sindou, M., & Mertens, P. (2000). Neurosurgical management of neuropathic pain. *Stereotactic and Functional Neurosurgery, 75*, 76–80.

Sindou, M., Mertens, P., & Wael, M. (2001). Microsurgical DREZotomy for pain due to spinal cord and/or cauda equina injuries: Long-term results in a series of 44 patients. *Pain, 92*, 159–171.

Sjolund, B. H. (2002). Pain and rehabilitation after spinal cord injury: The case of sensory spasticity ? *Brain Research Reviews, 40*, 250–256.

Soler, M. D., Kumru, H., Pelayo, R., Vidal, J., Tormos, J. M., Fregni, F., Navarro, X., & Pascual-Leone, A. (2010). Effectiveness of transcranial direct current stimulation and visual illusion on neuropathic pain in spinal cord injury. *Brain, 133*, 2565–2577.

Taylor, F. R., & Larkins, M. V. (2002). Headache and Chiari I malformation: Clinical presentation, diagnosis, and controversies in management. *Current Pain and Headache Reports, 6*, 331–337.

Treede, R. D., Lankers, J., Frieling, A., Zangemeister, W. H., Kunze, K., & Bromm, B. (1991). Cerebral potentials evoked by painful, laser stimuli in patients with syringomyelia. *Brain, 114*, 1595–1607.

Treede, R. D., Jensen, T. S., Campbell, J. N., Cruccu, G., Dostrovsky, J. O., Griffin, J. W., Hansson, P., Hughes, R., Nurmikko, T., & Serra, J. (2008). Neuropathic pain: Redefinition and a grading system for clinical and research purposes. *Neurology, 70*, 1630–1635.

Vernon, J. D., Silver, J. R., & Ohry, A. (1982). Posttraumatic syringomyelia. *Paraplegia, 20*, 339–364.

Vestergaard, K., Nielsen, J., Andersen, G., Ingeman-Nielsen, M., Arendt-Nielsen, L., & Jensen, T. S. (1995). Sensory abnormalities in consecutive, unselected patients with central post-stroke pain. *Pain, 61*, 177–186.

Victor, M., & Ropper, A. H. (2001). Diseases of the spinal cord. In *Adams and Victor's principles of neurology* (7th ed., pp. 1293–1345). New York: McGraw Hill.

Vierck, C. J., Jr., Siddall, P., & Yezierski, R. P. (2000). Pain following spinal cord injury: Animal models and mechanistic studies. *Pain, 89*, 1–5.

Wasner, G., Naleschinski, D., & Baron, R. (2007). A role for peripheral afferents in the pathophysiology and treatment of at-level neuropathic pain in spinal cord injury? A case report. *Pain, 131*, 219–225.

Widerstrom-Noga, E. G. (2002). Evaluation of clinical characteristics of pain and psychosocial factors after spinal cord injury. In R. P. Yezierski & K. J. Burchiel (Eds.), *Spinal cord injury pain: Assessment, mechanisms, management* (Progress in pain research and management, Vol. 23, pp. 53–82). Seattle: IASP Press.

Williams, B. (1990). Post-traumatic syringomyelia, an update. *Paraplegia, 28*, 296–313.

Yezierski, R. P. (2001). Pain following spinal cord injury: Pathophysiology and central mechanisms. In R. P. Yezierski & K. J. Burchiel (Eds.), *Progress in brain research* (Vol. 129, pp. 429–448). New York: Elsevier.

Yezierski, R. P., & Park, S. H. (1993). The mechanosensitivity of spinal sensory neurons following intraspinal injections of quisqualic acid in the rat. *Neuroscience Letters, 157*, 115–119.

Yezierski, R. P., Liu, S., Ruenes, G. L., Kajander, K. J., & Brewer, K. L. (1998). Excitotoxic spinal cord injury: Behavioral and morphological characteristics of a central pain model. *Pain, 75*, 141–155.

Yu, C. G., Fairbanks, C. A., Wilcox, G. L., & Yezierski, R. P. (2003). Effects of agmatine, interleukin-10, and cyclosporin on spontaneous pain behavior after excitotoxic spinal cord injury in rats. *The Journal of Pain, 4*, 129–140.

Pain in the Workplace, Risk Factors for Chronicity, and Demographics

Dennis C. Turk
Department of Anesthesiology & Pain Research,
University of Washington, Seattle, WA, USA

Synonyms

Age and chronicity; Education and chronicity; Household income and chronicity; Marital status and chronicity; Sex and chronicity

Definition

Statistical characteristics of populations (such as age, educational attainment, or income) that predispose an individual or population to develop

a problem, in this usage, the ▶ probability of developing chronic pain sufficiently intense to impede the ability to perform work-related activities.

Characteristics

Identification of characteristics of individuals who have an increased ▶ risk of developing chronic pain is important, as it has the potential to (1) guide research on the pathogenesis of chronic pain, (2) guide the design of interventions to prevent the transition from acute to chronic pain, and (3) permit delivery of interventions to those who are most in need of preventive efforts and subsequently to reductions in physical and occupational disability, emotional distress, and health-care utilization.

The transition from acute to chronic pain has generated a great deal of interest, especially in the area of work-related injuries and back pain. The interest in the transition from acute to chronic back pain has been fostered primarily by three factors: (1) the dramatic increases in disability associated with back injuries; (2) high costs of back injury in expenditures for medical treatment, indemnity costs, and lost productivity; and (3) the observation that, whereas up to 90 % of individuals who sustained back injuries with accompanying pain returned to functioning within a brief interval (usually days or weeks), the remaining 10 % developed long-term disabilities, often never returning to pre-injury levels of functioning (Troup et al. 1981). It is the observation that a relatively small percentage of patients with back injuries fail to return to functioning, thus consuming a disproportionate amount of health-care resources and indemnity costs that has served, perhaps, as the greatest impetus toward the question of what factors predict – risk factors – for chronicity.

Although a number of ▶ demographic factors have been examined, the role of age, sex, marital status, and educational attainment has most frequently been considered as potential predictors of chronicity following a back injury.

Age as a Predictor of Chronicity and Work-Related Disability

Individual studies (e.g., Elfing et al. 2009) and reviews (Turk 1997; Turner et al. 2000). However, several more recent studies did not provide confirming evidence supporting the role of workers age in prolonged disability (e.g., Turner et al. 2008). Age may have different roles and at different times in the trajectory from acute symptoms to chronicity. For example, one study found old age to predict poorer outcomes at 6 months (Gatchel et al. 1995a), but not at 1 year (Gatchel et al. 1995b). Only two population-based studies did not find the outcome to be worse in older workers. One found that workers aged 50 and over returned to work sooner and lost fewer days of work; however, age was not associated with work status 3 months after injury (Reid et al. 1997). There are a number of limitations to this study, including a low percentage of eligible people willing to participate and the small number of participants not working at the 3-month follow-up. In the other study, age was not significantly associated with duration of wage replacement benefits among workers on benefits for at least 4 weeks (Hogg-Johnson et al. 1999). This study did not report results for workers on wage replacement for shorter periods of time. Another contradictory result for age was obtained, which found a trend toward earlier return to work in older patients (Lehmann et al. 1993), but again, the sample was small and had limited statistical power and precision.

Specific to the issue of age is the actual age of the sample of participants included in the study, as the age of participants included will define "older" and "younger." For example, Weickgenant et al. (1994) reported that age was not a predictor of chronicity, but the sample of participants in the study consisted of male Navy service men, the majority of whom were under age 40. The definition of older varies substantially across studies, with some studies including participants up to age 65. Participants over age 40 would be defined as older in the Weickgenant et al. (1994) study, but not necessarily in studies with samples of much older individuals.

When observed, several factors may contribute to older workers' poorer outcomes. Older workers may be less able to recover from injuries and may be less likely to find different jobs if they are unable to return to the job they held at the time of injury. It is also possible that workers within a few years of retirement age may lack incentives for returning to work and view workers' compensation benefits as a bridge to social security disability or social security income.

Sex as a Predictor of Chronicity and Work-Related Disability

Even more than the case of age, the results for sex as a predictor of chronicity are quite inconsistent. Studies have reported that males are more likely to develop chronic back problems than females, that females are more likely to become disabled, and that sex did not predict disability at all (Turk 1997). To make things even more confusing, Crook and Moldofsky (1994) found that males were more likely to return to work; however, females had a higher probability of remaining at work once they returned. By contrast, Baldwin et al. (1996) reported that although women experienced shorter work absences, they were less likely to remain employed.

The majority of studies found no statistically significant association between outcome and worker sex (e.g., Turner et al. 2008; DuBois et al. 2009), although several studies observed poorer outcomes for women (Elfing et al. 2009; Heymans et al. 2009). There are a number of factors that may contribute to the inconsistencies observed, and the role of sex may be much more complex than simply whether one is male or female. For example, in one study, sex did not predict work status at 6 months (Gatchel et al. 1995a), but women were less likely to be working at 1 year (Gatchel et al. 1995b). Rossignol et al. (1988) found that men had significantly more days of compensation than did women; however, when a logistic regression analysis that also included occupation was performed, this difference was no longer statistically significant. Shaw et al. (2005) noted that females reported more functional limitation but greater likelihood of return to worth than males. Sex differences may reflect the difficulty of modifying tasks in more physically demanding occupations that are predominantly filled by males (e.g., construction). As a group, these results suggest that interactions or confounding of sex with other variables such as occupation and job physical demands or injury severity may be significant, and underscore the need for studies with large samples and multivariate statistical analyses.

Marital Status as a Predictor of Chronicity and Work-Related Disability

Intuitively, it might be expected that the presence of a spouse would provide social support and thus should serve to buffer against the stresses associated with acute pain. Social support is a vague construct that can be measured in many ways, only one being the presence of a spouse. The simple presence of a spouse, however, does not indicate anything about the quality of the marital relationship. Moreover, a supportive spouse may actually reinforce the sick role and unwittingly contribute to chronicity (Turk 1997).

As was the case for both age and sex, the results on marital status as a predictor of chronicity and work-related disability have been inconsistent and contradictory. Several studies found no association between marital status and outcomes (Turk 1997; Turner et al. 2000); however, other studies found worse outcomes in workers who were single (Hazard et al. 1996; Lehmann et al. 1993), divorced (Cheadle et al. 1994; Hogg-Johnson et al. 1999), or married more than once (Hazard et al. 1996). Still other studies found interactions among marital status, sex, and number of children (Baldwin et al. 1996; Volinn et al. 1991) to be significant in predicting outcomes. It is possible that associations between family status and chronicity are due, at least in part, to differences in wage compensation rates for various family status categories. However, Volinn et al. (1991) found that although divorced and widowed workers with no children had the same wage replacement rates as did workers who had never married, they were over twice as likely to be chronically disabled. This suggests that being divorced or widowed with no children is indeed a risk factor for chronic disability.

Two studies (Baldwin et al. 1996; Crook et al. 1998) demonstrated a potentially important interaction between sex and being single. They reported that being single predicted disability for men but not for women. Conversely, they found that being married was a predictor of disability for women but not for men.

Educational Level as a Predictor of Chronicity and Work-Related Disability

The education level of the individual with acute pain has also been examined to determine whether it predicted the development of chronicity. Yet again, the results are equivocal. Almost an equal number of studies have suggested that educational attainment was either a significant predictor or failed to predict disability (Turk 1997).

Two population-based studies found that less educated workers were at higher risk of chronic disability (Baldwin et al. 1996; Tate 1992), but the four non-population-based study articles reported no association between education and disability (Turner et al. 2000). Educational level likely bares a significant association with type of work and physical load and job demands (e.g., manual labor vs. managerial) (Leclerc et al. 2009). Thus, it is difficult to separate these to disentangle the relative unique contribution of each variable to chronicity.

Household Income as a Predictor of Chronicity and Work-Related Disability

Several studies have considered the role of income in contributing to the transition from acute to chronic pain. Volinn et al. (1991) and Hasenbring et al. (1994) noted that lower income levels were a significant predictor, whereas Murphy and Cornish (1984) and Weickgenant et al. (1994) failed to find that income-predicted chronicity. Thus, no conclusions can be drawn regarding the role of pre-pain income. It is important to acknowledge that an increased level of income is likely to interact with education and job held. It is difficult to tease apart these factors to determine whether any of them make an independent contribution to chronicity.

Race/Ethnicity as a Predictor of Chronicity and Work-Related Disability

Only one study reported the results of analyses of ethnic and racial groups as predictors. Among patients with acute back pain, non-Caucasians were less likely than Caucasians to be working at 6 months (Gatchel et al. 1995a) and 1 year (Gatchel et al. 1995b) later.

Predictors of Recurrent Work Disability

Very little is known about risk factors for recurrences of work disability in workers who do return to work. Only one study (Baldwin et al. 1996) presented data on the question of what happens to workers after they return to work. The documentation in this study of substantial disability recurrence rates in the first few years after an injured worker returns to work points to the need for research to identify predictors not only of length of initial disability but also of whether or not a worker remains at work after returning. Such studies need to examine both recurrences of the original type of injury and new injuries.

Conclusions

Based on the review of demographic predictors, it is difficult to draw any firm conclusions regarding the role of these factors in the evolution of chronic pain following acute back pain. In general, demographic factors have not been shown to be good predictors of return to work or chronicity following an injury or pain onset. The failure to find a consistent role for demographic variables as predictors may be accounted for by differences in a myriad of prognostic factors that vary across studies – time from injury or pain onset (e.g., 7 days, 6 week, 3 months, 1 year); differences in methods of recruiting participants (e.g., population-based, volunteer sample); sample characteristics (e.g., age range 18–50 vs. 21–65; education levels, income range, living/family status, and ethnicity/race); work place factors (size of employer, job tenure, work/injury history, see "▶ Pain in the Workplace, Risk Factors for Chronicity, and Workplace Factors" entry); country where study was conducted (e.g., United States, Sweden, United Kingdom); disability definition (e.g., off work, time to return to work, limitation

of physical function); measures of disability (e.g., questionnaire, employee records); treatment prescribed or received; psychological factors (e.g., mood, expectations for recovery, fear of pain or injury, see "▶ Pain in the Workplace, Risk Factors for Chronicity, and Psychosocial Factors" entry); and so forth, as well as interactions among relevant physical, psychological, social, and environmental variables. Variability in the statistical methods used across studies may also contribute to the discrepant findings reported. In addition to the methodological issues noted above, it is important to realize that demographic factors may interact. For example, age may be associated with income, and educational level may be highly correlated. Typically, demographic variables have been treated as independent predictors. Future research should consider the interaction among these variables when attempting to determine the best demographic set of predictors. An important caveat about the available research is that statistical associations found in analyses of large databases may or may not generalize to individual workers.

References

Baldwin, M. L., Johnson, W., & Butler, R. J. (1996). The error of using returns-to-work to measure the outcomes of health care. *American Journal of Industrial Medicine, 29*, 632–641.

Cheadle, A., Franklin, G., Wolfhagen, C., et al. (1994). Factors influencing the duration of work-related disability: A population-based study of Washington state workers' compensation. *American Journal of Public Health, 84*, 190–196.

Crook, J., & Moldofsky, H. (1994). The probability of recovery and return to work from work disability as a function of time. *Quality of Life Research, 3*, S97–S109.

Crook, J., Moldofsky, H., & Shannon, H. (1998). Determinants of disability after a work related musculoskeletal injury. *Journal of Rheumatology, 25*, 1570–1577.

DuBois, M., Szpalski, M., & Donceel, P. (2009). Patients at risk for long-term sick leave because of low back pain. *Spine Journal, 9*, 350–359.

Elfing, A., Mannion, A. F., Jacobshagen, N., et al. (2009). Beliefs about back pain predict the recovery rate over 52 consecutive weeks. *Scandinavian Journal of Work, Environment and Health, 35*, 437–445.

Gatchel, R., Polatin, P., & Kinney, R. (1995a). Predicting outcome of chronic back pain using clinical predictors of psychopathology: A prospective analysis. *Health Psychology, 14*, 415–420.

Gatchel, R. J., Polatin, P. B., & Mayer, T. G. (1995b). The dominant role of psychosocial risk factors in the development of chronic low back pain disability. *Spine, 20*, 2702–2709.

Hasenbring, M., Marienfeld, G., Kuhlendahl, D., et al. (1994). Risk factors of chronicity in lumbar disc patients. *Spine, 19*, 2759–2765.

Hazard, R. G., Haugh, L. D., Reid, S., et al. (1996). Early prediction of chronic disability after occupational low back injury. *Spine, 21*, 945–951.

Heymans, M. W., Anema, J. R., Van Buuren, S., et al. (2009). Return to work in a cohort of low back pain patients: Development and validation of a clinical prediction rule. *Journal of Occupational Rehabilitation, 19*, 155–165.

Hogg-Johnson, S., Cole, D. C., & Group ECCD. (1999). *Early prognostic factors for duration on benefits among workers with compensated occupational soft tissue injuries*. Toronto: Institute for Work and Health.

Leclerc, A., Gourmelen, J., Pouvier, S., et al. (2009). Level of education and back pain in France: The role of demographic, lifestyle and physical work factors. *International Archives of Occupational and Environmental Health, 82*, 643–652.

Lehmann, T. R., Spratt, K. F., & Lehmann, K. K. (1993). Predicting long-term disability in low back injured workers presenting to a spine consultant. *Spine, 18*, 1103–1112.

Murphy, K. A., & Cornish, R. D. (1984). Prediction of chronicity in acute low back pain. *Archives of Physical Medicine and Rehabilitation, 65*, 334–337.

Reid, S., Haugh, L. D., & Hazard, R. G. (1997). Occupational low back pain: Recovery curves and factors associated with disability. *Journal of Occupational Rehabilitation, 7*, 1–14.

Rossignol, M., Suissa, S., & Abenhaim, L. (1988). Working disability due to occupational back pain: Three-year follow-up of 2,300 compensated workers in Quebec. *Journal of Occupational Medicine, 30*, 502–505.

Shaw, W. S., Pransky, G., Patterson, W., et al. (2005). Early disability risk factors for low back pain assessed at outpatient occupational health clinics. *Spine, 30*, 572–580.

Tate, D. G. (1992). Workers' disability and return to work. *American Journal of Physical Medicine & Rehabilitation, 71*, 92–96.

Troup, J. D. G., Martin, J. W., & Lloyd, D. C. E. F. (1981). Back pain in industry: A prospective study. *Spine, 6*, 61–69.

Turk, D. C. (1997). Transition from acute to chronic pain: Role of demographic and psychosocial factors. In T. S. Jensen, J. A. Turner, & Z. Wiesenfeld-Hallin (Eds.), *Proceedings of the 8th world congress on pain, progress in pain research and management* (pp. 185–214). Seattle: IASP Press.

Turner, J. A., Franklin, G., Fulton-Kehoe, D., et al. (2008). Early predictors of chronic work disability. A prospective, population-based study of workers with back injuries. *Spine, 33*, 2809–2818.

Turner, J. A., Franklin, G., & Turk, D. C. (2000). Predictors of long-term disability in injured workers: A systematic literature synthesis. *American Journal of Industrial Medicine, 38*, 707–722.

Volinn, E., Van Koevering, D., & Loeser, J. D. (1991). Back sprain in industry: The role of socioeconomic factors in chronicity. *Spine, 16*, 542–548.

Weickgenant, A. L., Slater, M. A., & Atkinson, J. H. (1994). A longitudinal analysis of coping and the development of chronic low back pain. *Abstract, 15th Scientific Sessions of the Society of Behavioral Medicine*, Boston.

Pain in the Workplace, Risk Factors for Chronicity, and Psychosocial Factors

Dennis C. Turk
Department of Anesthesiology & Pain Research, University of Washington, Seattle, WA, USA

Synonyms

Emotional factors; Personality; Psychological factors

Definition

Psychosocial factors, including individual differences (i.e., personality), premorbid psychological diagnoses, coping resources and strategies, cognitive (e.g., beliefs, attitudes, perceptions, expectancies), emotional (i.e., depression, ▶ anxiety), and behavioral (i.e., responses by significant others) variables, all have the potential to predispose one to, and to affect, chronicity following a work-related injury.

Characteristics

As is so often the case in healthcare, when physical factors are insufficient to predict symptoms, psychological factors are considered. For example, Burton et al. (1995) demonstrated that, for acute back pain patients presenting in a primary care setting, conventional clinical information predicted outcome disability scores for only 10 % of the cases, compared to psychosocial measures that accounted for 59 % of the cases. The failure of injury severity to account for the majority of the variance in development of chronic disability following work-related injuries (Crook et al. 2002) has led some to postulate that a range of psychological variables may be relevant. A number of reviews have evaluated the role of psychological variables, namely, prior history of emotional problems, current history of psychological distress, personality factors, and alcohol and substance abuse, as well as patients' perceptions of ▶ stress, coping behaviors, attitudes and beliefs, health in general, and their current symptoms (e.g., Van der Windt et al. 2007; Iles et al. 2008; Hartvigsen et al. 2004). The majority of the studies have focused on low back pain, perhaps the most common work-related painful condition. Reviews of these studies have not always arrived at the same conclusions. For example, Linton (2001), Pincus et al. (2002) and de Leeuw et al. (2007) concluded that there was substantial support for the predictive ability of psychological factors on pain and work-related disability; whereas Hartvigsen et al. (2004) and Iles et al. (2008) came to the opposite conclusion. Moreover, there is some inconsistency in the prognostic utility of psychological variable depending on the type of pain, back pain vs. shoulder pain, where van der Windt et al. (2007) found that psychological factors did predict disability for back pain but not for shoulder pain.

One problem with attempting to draw conclusion about the role of psychological factors as a set is that it incorporates a wide range of disparate variables including beliefs, expectations, emotional states, coping styles, personality traits, and current and history of psychiatric disorders. It may not be possible and likely to be inappropriate to collapse all of these factors in an attempt to make general comments about the predictive utility of such a heterogeneous set of variables; rather, they need to be looked at individually as well as how they may interact.

Alcohol and Substance Abuse as Predictors of Chronicity

History of alcohol and substance abuse has consistently been shown to be predictors of chronicity (Turk 1997). Gatchel et al. (1995a, b), however, did not find that the presence of a substance abuse disorder predicted disability following a work-related injury. Overall, it does appear that prior or current use of alcohol or illicit drugs indicates a relatively poor prognosis for those who suffer an acute episode of back pain, especially if the pain is attributed to a work-related injury.

Affective Distress as Predictor of Chronicity

History of anxiety, depression, or psychological treatment has been identified as being involved in the transition from acute to chronic pain in several studies (Turk 1997). A number of studies, but not all (e.g., Van der Windt et al. 2007; Iles et al. 2008), have also demonstrated that levels of anxiety and depression at the time of acute pain predicted persistence of symptoms, time lost from work, and healthcare expenditures following work-related injuries (Crook et al. 2002; Turk 1997).

Pincus et al. (2006) found that emotional distress was a significant predictor of progression from acute to chronic low back pain; however, none of the studies contained in the review included return to work as an outcome. In contrast, reviews by Truchon and Fillion (2000) and Steenstra et al. (2005a), both of which focused on work outcomes, did not find that depression was a useful indicator of longer work absence. It is possible that depression is more likely to be a consequence of work disability rather than a cause, and while depression may have a role in the development of chronic low back pain, depression does not predict work disability and chronicity. There is also limited evidence that anxiety is predictive of poor work outcome (Steenstra et al. 2005b; Iles et al. 2008).

Crook et al. (2002) reported that workers with greater psychological distress returned to work significantly later. Gatchel and colleagues found that higher baseline patient scores on the hysteria scale of the MMPI predicted work disability (Gatchel et al. 1995a, b). Diagnosis of a personality disorder predicted work disability at 6 months, but not at 1 year.

Less pervasive psychological distress described as lassitude, malaise, and loneliness when assessed during acute pain onset have been shown to predict the transition from acute to chronic pain (Turk 1997). These constructs appear to be measuring the emotional state of the individual at the time of assessment, and it is not clear whether they should be viewed as more pervasive characteristics rather than transient states. Diverse measures have been used to assess these states, and thus, one must be cautious in comparing and interpreting the results from different studies.

Fear Avoidance

Psychological beliefs have been identified as important factors in relation to the course of low back pain and accompanying sick leave in a large number of studies (e.g., de Leeuw et al. (2007; Melloh et al. 2009), although the effects albeit statistically significant are relatively modest accounting for small but potentially important amounts of the variance in work disability (Iles et al. 2008). However, the conclusions are somewhat contradictory; several reviews have failed to find the avoidance of activity based on patients fears of increasing pain, injuring themselves further if they engage in various activities, exacerbating existing physical pathology, or reinjury when symptoms have resolved may play a small role in contributing to or predicting chronicity or beyond other predictive factors (e.g., Hartvigsen et al. 2004; Pincus et al. 2006) and mixed results depending on painlocagion – back vs. shoulder (Van der Windt et al. 2007).

Patient Expectations

Summarizing the results of a large number of studies, several systematic reviews

(e.g., Mondloch et al. 2001; Iles et al. 2008) determined that there was "strong evidence" supporting the recovery expectations of back-injured workers on chronicity. In one recent study, examining beliefs about the adverse consequences of back pain, along with work-related fear-avoidance beliefs, was an important predictor of poor recovery of individuals with back pain over a period of 1 year, although the impact of these beliefs on work status was not directly examined (Elfering et al. 2009).

Health Perceptions as Predictors of Chronicity

A number of investigators have attempted to assess patients' general somatic preoccupations and perceptions of their own health, as well as their beliefs about the severity of their current medical condition during periods of acute pain, to determine whether these perceptions and beliefs predicted subsequent chronic pain. A number of studies have supported the role of these cognitive appraisals in predicting chronicity (Turk 1997; Elfering et al. 2009). Two early studies, however, failed to support the importance of the subject's appraisals (Murphy and Cornish 1984; Lehmann et al. 1993). The operationalizations of these cognitive measures make comparisons across studies difficult. Since these psychological variables appear to have some predictive utility, additional research is needed to more carefully examine these factors.

Conclusion

If there is one factor that characterizes the results of the studies seeking to identify psychosocial predictors of the transition from acute to chronic pain and maintenance of disability, it is inconsistency. One explanation for the inconsistency is the predictor measures used, and that despite the fact that the variables and constructs included may appear comparable in different studies, they may be operationalized with different measures that have different characteristics. It cannot be assumed that two scales designed to measure the same thing are actually comparable.

In addition to the factors described above, the design of the predictor studies can influence the outcomes. Cross-sectional and retrospective designs have numerous inherent problems. The inclusion of small samples of nonrepresentative subjects will also influence the nature of the outcome. In addition, most of the studies on the transition from acute to chronic pain have treated the predictors as if they were independent, but many predictors are not independent.

Many studies have demonstrated that psychosocial factors are better predictors of chronicity than are clinical or physical factors (Turner et al. 2002; de Leeuw et al. 2007). Clearly, chronicity is determined by a multitude of factors that interact in a reciprocal way over time. It appears that chronicity, especially accompanying work-related injuries, is a function of a complex interaction among demographic, physical, psychological, social, and economic factors.

References

Burton, A. K., Tillotson, K. M., Main, C. J., et al. (1995). Psychosocial predictors of outcome in acute and subchronic low back trouble. *Spine, 20,* 722–728.

Crook, J., Milner, R., Schulz, I. Z., et al. (2002). Determinants of occupational disability following Low back pain: A critical review of the literature. *Journal of Occupational Rehabilitation, 12,* 277–295.

de Leeuw, M., Goossens, M., Linton, S., et al. (2007). The fear-avoidance model of musculoskeletal pain: Current state of scientific evidence. *Journal of Behavioral Medicine, 30,* 77–94.

Elfering, A., Mannion, A. F., Jacobshagen, N., et al. (2009). Beliefs about back pain predict the recovery rate over 52 consecutive weeks. *Scandinavian Journal of Work, Environment & Health, 35,* 437–445.

Gatchel, R. J., Polatin, P. B., & Kinney, R. K. (1995a). Predicting outcome of chronic back pain using clinical predictors of psychopathology: A prospective analysis. *Health Psychology, 14,* 415–420.

Gatchel, R. J., Polatin, P. B., & Mayer, T. G. (1995b). The dominant role of psychosocial risk factors in the development of chronic low back pain disability. *Spine, 20,* 2702–2709.

Hartvigsen, J., Lings, S., Leboeuf-Yde, C., et al. (2004). Psychosocial factors at work in relation to low back pain and consequences of low back pain; a systematic critical review of prospective cohort studies. *Occupational and Environmental Medicine, 61,* e2.

Iles, R. A., Davidson, M., & Taylor, N. F. (2008). Psychosocial predictors of failure to return to work in non-chronic non-specific low back pain: A systematic review. *Occupational and Environmental Medicine, 65,* 507–517.

Lehmann, T. R., Spratt, K. F., & Lehmann, K. K. (1993). Predicting long-term disability in low back injured workers presenting to a spine consultant. *Spine, 18,* 1103–1112.

Linton, S. J. (2001). Occupational psychological factors increase the risk for back pain: A systematic review. *Journal of Occupational Rehabilitation, 15,* 459–474.

Melloh, M., Eldering, A., Egli, C., et al. (2009). Identification of prognostic factors for chronicity in patients with low back pain: A review of screening instruments. *International Orthopaedics, 33,* 301–313.

Mondloch, M. V., Cole, D. C., & Frank, J. W. (2001). Does how you do depend on how you think you'll do? A systematic review of the evidence for a relation between patients' recovery expectations and health outcomes. *CMAJ: Canadian Medical Association Journal, 165,* 174–179.

Murphy, K. A., & Cornish, R. D. (1984). Prediction of chronicity in acute low back pain. *Archives of Physical Medicine and Rehabilitation, 65,* 334–337.

Pincus, T., Burton, A. K., Vogel, S., et al. (2002). A systematic review of psychological factors as predictors of chronicity/disability in prospective cohorts of low back pain. *Spine, 27,* E109–E120.

Pincus, T., Vogel, S., Burton, A. K., et al. (2006). Fear avoidance and prognosis in back pain: A systematic review and synthesis of current evidence. *Arthritis & Rheumatism, 54,* 3999–4010.

Steenstra, I. A., Verbeek, J. H., Heymans, M. W., et al. (2005a). Prognostic factors for duration of sick leave in patients sick listed with acute low back pain: A systematic review of the literature. *Occupational and Environmental Medicine, 62,* 851–860.

Steenstra, I. A., Koopman, F. S., Knol, D. L., et al. (2005b). Prognostic factors for duration of sick leave due to low-back pain: An inception cohort study in Dutch health care professionals. *Journal of Occupational Rehabilitation, 15,* 591–605.

Truchon, M., & Fillion, L. (2000). Biopsychosocial determinants of chronic disability and low-back pain: A review. *Journal of Occupational Rehabilitation, 10,* 117–142.

Turk, D. C. (1997). Transition from acute to chronic pain: Role of demographic and psychosocial factors. In T. S. Jensen, J. A. Turner, & Z. Wiesenfeld-Hallin (Eds.), *Proceedings of the 8th world congress on pain, progress in pain research and management* (pp. 185–214). Seattle: IASP Press.

Turner, J. A., Franklin, G., & Turk, D. C. (2002). Predictors of long-term disability in injured workers: A systematic literature synthesis. *American Journal of Industrial Medicine, 38,* 707–722.

Van der Windt, D. A. W. M., Kuijpers, T., Jellema, P., et al. (2007). Do psychological factors predict outcome in both low-back pain and shoulder pain? *Annals of the Rheumatic Diseases, 66,* 313–319.

Pain in the Workplace, Risk Factors for Chronicity, and Workplace Factors

Dennis C. Turk
Department of Anesthesiology & Pain Research, University of Washington, Seattle, WA, USA

Synonyms

Risk factors for chronicity; Socioeconomic factors; Work environment; Workplace; Workplace factors

Definition

In the context of a work-related injury, socioeconomic factors specifically related to the workplace (e.g., job satisfaction, control of work pace) and the economic environment (e.g., unemployment rate, extent of ▶ wage replacement) may influence time off work following a work-related injury.

Characteristics

Identification of characteristics of individuals who have an increased risk for developing chronic pain and disability is important as it has the potential to (1) guide research on the pathogenesis of chronic pain, (2) inform the design of interventions to prevent the transition from acute to chronic pain, and (3) permit delivery of interventions to those who are most in need of preventive efforts and subsequently to reductions in physical and occupational disability, emotional distress, and health-care utilization.

The majority of the studies on workplace risk factors have focused on people with work-related back injuries. This emphasis is not surprising since it has been estimated that up to 30 % of all back pain can be attributed to occupational exposure

(Punnett et al. 2005). Occupational low back pain has a number of important characteristics that make it distinct from non-work-related conditions – especially sudden onset, nature of impairment, and involvement with worker compensation disability (e.g., Lehmann et al. 1993). Thus, researchers must be careful in generalizing results derived from this population to back pain more broadly defined or to other pain disorders (e.g., carpal tunnel syndrome).

It might seem reasonable to expect that injury severity, pain severity, perceived disability, as well as ergonomic factors required to perform job responsibilities would be risk factors for both injury and chronicity. The proportion of the variance in health and functional outcomes including work extracted from administrative records and functional tests have been disappointing. A number of job characteristics and workplace factors have been examined in an attempt to supplement other variables to determine their role in the transition from acute to chronic pain and work-related disability (e.g., Post et al. 2005) and to understand the factors that contribute to chronicity and prolonged disability (e.g., Hazard et al. 1996; Infante-Rivard and Lortie 1996; Shaw et al. 2001; Mielenz et al. 2008).

The interest in the transition from acute to chronic pain has been fostered primarily by three factors: (1) the dramatic increase in disability associated with work-related injuries; (2) high costs of work-related injury in expenditures for medical treatment, ▶ indemnity costs, and lost productivity; and (3) the observation that, whereas up to 90 % of workers who sustained work-related injuries with accompanying pain return to functioning within a brief interval (usually days or weeks), the remaining 10 % develop long-term disabilities, often never returning to preinjury levels of functioning. It is the data suggesting that a relatively small percentage of patients with back injuries fail to return to functioning, consuming a disproportionate amount of health-care resources and indemnity costs that has served, perhaps, as the greatest impetus toward the question of what factors are risk factors for chronicity.

Work Load

The potential role of a number of ergonomic factors, biomechanical loads, and actual physical demands (e.g., awkward body posture, manual exertion, hand force) required to perform job responsibilities and the perception o the job environment (e.g., hectic, little autonomy or control) have been investigated as potential risk factors for both injury and chronicity; however, the evidence appears to be inconclusive regarding the predictive ability of any of these factors (e.g., Mielenz et al. 2008; Heyman et al. 2009; Vanderfrift et al. 2012). Even when it has been shown to affect the presence of persistent pain, psychosocial factors have often accounted for a significant additional proportion of the variance (Pincus et al. 2002; Harkness et al. 2004; see ▶ Pain in the Workplace, Risk Factors for Chronicity, and Psychosocial Factors entry).

Psychosocial Work Factors

The role of psychosocial work factors (e.g., job satisfaction, organization of work, coworker and supervisor support) in the presence of pain and an association with chronicity have all been examined in a number of studies and have been reviewed in many papers over the past few decades. The results, however, have been inconsistent. The inconclusive nature of these results may be related to limitation in variability in definitions, criteria, and quality of the methodologies used (Hartvigsen et al. 2004; Iles et al. 2008). At this time, no clear picture of the relationship between any of the work-related psychosocial factors examined and prolonged disability and delayed return to work.

Job Accommodation

The availability of a modified job or workplace accommodations (e.g., light duty, reduced hours) may be an important factor in facilitating return to work (e.g., Fransen et al. 2002; Hogg-Johnson and Cole 2003; Shaw et al. 2005; Turner et al. 2008). In two studies, workers who had a modified job available to them returned to work significantly sooner (Crook et al. 1998, 2002; Hogg-Johnson et al. 1999). Hogg-Johnson et al. (1999) also found that when the employer did not offer accommodations for return to work, workers whose pain did not change, or worsened between baseline and 4 weeks, remained on benefits longer than did workers whose pain improved. When the employer did offer arrangements for return to work, the relationship between change in pain and time on benefits was much weaker. These results were replicated by Turner et al. (2008) who found that adjusting for other predictors, workers who were not offered job accommodations by approximately 3 weeks after submitting a worker compensation claim had almost twice the odds of prolonged work disability. This suggests that job modifications may be particularly critical for workers whose pain is not improving within a relatively brief time following pain onset.

Availability of Wage Replacement

In many countries, people who experience work-related injuries are eligible for disability compensation. The wage replacement benefit is a ratio of the expected disability benefit to the preinjury wage. High benefit replacement rates are expected to provide greater incentives not to return to work. Some have suggested that the availability and high levels of wage replacement would serve as a disincentive for return to work and consequently reinforce disability (e.g., Hadler et al. 1995; Walsh and Dumitru 1988). Several studies have attempted to determine the validity of this assumption. The results are somewhat mixed; whereas some studies have supported the role of wage replacement in fostering chronicity, others have not (Turk 1997; Turner et al. 2000). To confuse matters even further, Ruser (1987) reported that wage replacement predicted chronicity for males, but Baldwin et al. (1996) found that it predicted only for females and not for males. The results on wage replacement illustrate the potential for interactions among predictor variables, in this case sex and levels of wage replacements.

Many authors have suggested that compensation paid to injured workers affects symptoms, reports of disability, time to return to work, and response to treatment. At least regarding response to treatment, the results are inconsistent.

The role of ▶ worker compensation as a predictor of disability is a contentious topic that has generated much debate (e.g., Hadler et al. 1995;

Walsh and Dumitru 1988). For example, Greenough and Fraser (1989) reported that patients who received compensation reported greater pain, disability, psychological disturbance, and unemployment at follow-up than patients who did not receive compensation. The authors concluded that the compensation acts directly and powerfully against the long-term interests of the patients (see also Hadler et al. 1995).

Dworkin (1990) suggested that the studies that have examined the role of compensation neglect to consider the fact that compensation may be a consequence of chronic pain. Dworkin suggests that the cause of compensation includes greater pain, greater disability, greater psychological disturbance, inability to work, and poorer response to treatment – the very same factors that have been associated in some studies but that are almost invariably interpreted as consequences of compensation, rather than causes. Most likely, worker's compensation interacts with other variables described above.

It is important to acknowledge that the generic terms "compensation" and "wage replacement" are quite heterogeneous. Compensation may or may not be time limited. The funding may come from different sources. For example, in the United States, disability compensation may be received from the Veterans Administration, Social Security Disability, or Worker Compensation Company depending upon a number of factors and the potential cause of pain onset. It would not be a surprise if such differences led to different outcomes, and the predictors of chronicity are likely to vary depending upon the type and source of compensation (e.g., Jamison et al. 1988). Studies on the role of compensation in chronicity have rarely made distinctions between the type or nature of compensation that patients received.

Despite limitations in the available studies, it does appear that compensation factors may contribute to delayed recovery and reinforce the sick role. It is important to consider not only whether compensation status affects disability but under what conditions. A detailed review of the issues involved in compensation and litigation as predictors of disability is beyond the scope of this entry (see Turner et al. 2000).

Company Size

Several studies reported that workers in smaller companies had poorer outcomes than workers in larger companies (n > 500) (e.g., Shaw et al. 2005); however, firm size did not predict cumulative compensated work absence in others (Turner et al. 2000). Larger firms may have greater resources for employing specialists in disability management and be better able to offer injured workers different jobs or job modifications to promote return to work.

It is reasonable to expect that biomechanical, workplace (e.g., job satisfaction, support from coworkers and supervisors), and psychosocial factors (e.g., job satisfaction, job demand, control of pace) interact in important ways to contribute to and maintain chronicity and related disability (MacDonald et al. 2001) and that no one factor by itself will account for a substantial proportion of the variance in chronicity and prolonged work-related disability. Prospective research with large samples is necessary to explicate the complex interactions among the disparate but potentially interrelated and synergistic set of factors.

References

Baldwin, M. L., Johnson, W. G., & Butler, R. J. (1996). The error of using returns-to-work to measure the outcome of health care. *American Journal of Industrial Medicine, 29,* 632–641.

Crook, J., Moldofsky, H., & Shannon, H. (1998). Determinants of disability after a work related musculoskeletal injury. *Journal of Rheumatology, 25,* 1570–1577.

Crook, J., Milner, R., Schulz, I. Z., & Stringer, B. (2002). Determinants of occupational disability following low back pain: A critical review of the literature. *Journal of Occupational Rehabilitation, 12,* 277–295.

Dworkin, R. H. (1990). Compensation in chronic pain patients: Cause or consequence? *Pain, 43,* 387–388.

Fransen, M., Woodward, M., & Norton, R. (2002). Risk factors associated with the transition from acute to chronic occupational back pain. *Spine, 27,* 982–989.

Greenough, C. G., & Fraser, R. D. (1989). The effects of compensation on recovery from low-back injury. *Spine, 14,* 947–955.

Hadler, N. M., Carey, T. S., Garrett, J., & The North Carolina Back Project. (1995). The influence of indemnification by workers' compensation insurance on recovery from acute backache. *Spine, 20,* 2710–2715.

Harkness, E. F., Macfarlane, G. J., Nahit, E., et al. (2004). Mechanical injury and psychosocial factors in the work place predict the onset of widespread body pain: A two-year prospective study among cohorts of newly employed workers. *Arthritis and Rheumatism, 50*, 1655–1664.

Hartvigsen, J., Lings, S., Leboeuf-Yde, D., et al. (2004). Psychosocial factors at work in relation to low back pain and consequences of low back pain: A systematic, critical review of prospective cohort studies. *Occupational and Environmental Medicine, 61*, e2.

Hazard, R. G., Haugh, L. D., Reid, S., Preble, J. B., & MacDonald, L. (1996). Early prediction of chronic disability after occupational low back injury. *Spine, 1*, 945–951.

Heyman, M. W., Anema, J. R., van Buuren, S., et al. (2009). Return to work in a cohort of low back pain patients: Development and validation of a critical prediction rule. *Journal of Occupational Rehabilitation, 19*, 155–165.

Hogg-Johnson, S., & Cole, D. C. (2003). Early prognostic factors for duration on temporary total benefits in the first year among workers with compensation occupational soft tissue injuries. *Occupational and Environmental Medicine, 60*, 244–253.

Hogg-Johnson, S., Cole, D. C., Group ECCD, & Workgroup PM. (1999). *Early prognostic factors for duration on benefits among workers with compensated occupational soft tissue injuries*. Toronto: Institute for Work & Health.

Iles, R. A., Davidson, M., & Taylor, N. F. (2008). Psychosocial predictors of failure to return to work in non-chronic non-specific low back pain: A systematic review. *Occupational and Environmental Medicine, 65*, 507–517.

Infante-Rivard, C., & Lortie, M. (1996). Prognostic factors for return to work after a first compensated episode of back pain. *Occupational and Environmental Medicine, 53*, 488–494.

Jamison, R. N., Matt, D. A., & Paris, W. C. V. (1988). Effects of time-limited vs. unlimited compensation on pain behavior and treatment outcome in low back pain patients. *Journal of Psychosomatic Research, 32*, 277–283.

Lehmann, T. R., Spratt, K. F., & Lehmann, K. K. (1993). Predicting long-term disability in low back injured workers presenting to a spine consultant. *Spine, 18*, 1103–1112.

MacDonald, L. A., Karasek, R. A., & Punnett, L. (2001). Covariation between workplace physical and psychosocial stressors: Evidence and implications for occupational health research and prevention. *Ergonomics, 44*, 696–718.

Mielenz, T. J., Garrett, J. M., & Carey, T. S. (2008). Association of psychosocial work characteristics with low back pain outcomes. *Spine, 33*, 1270–1275.

Pincus, T., Burton, A. K., Vogel, A. P., et al. (2002). A systematic review of psychosocial factors as predictors of chronic disability in prospective cohorts of low back pain. *Spine, 27*, E109–E120.

Post, M., Krol, B., & Groothoff, J. W. (2005). Work-related determinants of return to work of employees on long-term sickness absence. *Disability and Rehabilitation, 27*, 481–488.

Punnett, L., Pruss-Ustun, A., & Nelson, D. (2005). Estimated global burden of low back pain attributable to combined occupational exposure. *American Journal of Medicine, 48*, 459–469.

Ruser, J. W. (1987). Workers' compensation and occupational injuries and illnesses. *Journal of Labor Economics, 9*, 325–350.

Shaw, W. S., Pransky, G., Fitzgerald, T. E. (2001). Early prognosis for low back pain disability: Intervention strategies for health care providers. *Disability and Rehabilitation, 23*, 815–828.

Shaw, W. S., Pransky, G., Patterson, W., et al. (2005). Early disability risk factors for low back pain assessed at outpatient occupational health clinics. *Spine, 30*, 572–580.

Turk, D. C. (1997). Transition from acute to chronic pain: Role of demographic and psychosocial factors. In T. S. Jensen, J. A. Turner, & Z. Wiesenfeld-Hallin (Eds.), *Proceedings of the 8th world congress on pain, progress in pain research and management* (pp. 185–214). Seattle: IASP.

Turner, J. A., Franklin, G., & Turk, D. C. (2000). Predictors of long-term disability in injured workers: A systematic literature synthesis. *American Journal of Industrial Medicine, 38*, 707–722.

Turner, J. A., Franklin, G., Fulton-Kehoe, D., et al. (2008). Early predictors of chronic work disability. A prospective, population-based study of workers with back pain. *Spine, 33*, 2809–2818.

Vanderfrift, J. L., Gold, J. E., Hanlon, A., et al. (2012). Physical and psychosocial ergonomic risk factors for low back pain in automobile manufacturing workers. *Occupational and Environmental Medicine, 69*, 39–44.

Walsh, N. E., & Dumitru, D. (1988). The influence of compensation on recovery from low back pain. *Occupational Medicine: State of the Art Reviews, 3*, 109–121.

Pain in the Workplace, Risk Factors for Chronicity, Job Demands

Greg McIntosh

Canadian Back Institute, Toronto, ON, Canada

Synonyms

Ergonomics; Occupational pain; Possible contributors; Risk factors for chronicity; Work duties

Definition

Pain in the workplace refers to pain that occurs at a person's place of gainful employment. The pain may be limited to just the hours spent working or may be experienced at home too, even though the major exacerbating factors occur at work.

Risk factors are conditions, identified in healthy people, which are associated with an increased risk of developing pain; the outcome of interest is the onset of pain. Risk factors and prognostic factors are often regarded as synonymous; however, prognostic factors are conditions, present in people already known to have pain, which are associated with an increased risk of recovery from an injury. The most common outcomes are pain reduction, functional improvement, and/or return to work.

Job demands refer to both the physical and psychosocial components people need to perform daily tasks at work. Supervisors, suppliers, competitors, coworkers, the company culture, and the workers themselves all contribute to required job demands.

Characteristics

Pain in the Workplace

The workplace provides us with financial status and security, defines who we are and our role in society in many cases. Unfortunately, the workplace can be associated with pain. In Canada and the USA, for example, occupational back injuries account for approximately one third of all work-related injuries that lead to protracted work disability (Cheadle et al. 1994). Of the 5 % of American adults who experience a new episode of low back pain each year, 64 % attribute their problem to an occupational injury (Frymoyer et al. 1987). Approximately 1 % of all Americans are permanently disabled by back pain.

Over the last 40 years, low back pain prevalence in the general population has remained stable, but there has been a marked increase in disability rates. From 1960 to 1980, there was a 2,700 % increase in back pain disability awards by the US Social Security Administration (Bigos 1993). The disability rate increased at 14 times the rate of population growth between 1977 and 1981. Recent data indicates that the prevalence of work-related low back pain is no longer rising (Murphy and Volinn 1999), but it has stabilized at a very high level. For example, an estimated 22.4 million back pain cases are responsible for an estimated 149.1 million lost workdays per year (Guo et al. 1995), and an estimated 2.4 million Americans were disabled by back pain in 1990. This leads to a substantial socioeconomic impact in terms of medical costs and compensation; annual costs as of the mid-1980s were estimated at $4.6 billion and $16 billion, respectively, in the USA (Frymoyer et al. 1987).

Despite the staggering numbers, there is a lack of scientific evidence that working is physically harmful to the back. Back pain is work related to the extent that people of working age commonly get back pain and it has an impact on their work. However, ongoing symptoms and disability do not appear to be related to physical stress or occupational exposure. Hall et al. (1998) found that low back pain occurs without a precipitating event in two thirds of cases. They concluded that ▶ spontaneous onset is part of the natural history of back pain.

Risk Factors

The study of risk factors has received a lot of attention. Research in occupational settings has revealed many risk factors for developing low back pain including (NIOSH; Waddell 1998):

Physical
- Heavy manual work
- Lifting and twisting
- Postural stress (sitting)
- Whole body vibration (driving)

Psychosocial
- Monotonous work
- Lack of personal control
- Job dissatisfaction

Physiologic
- Poor fitness
- Inadequate trunk strength

Unfortunately, the contribution of most of these factors is poorly understood in this multi-factorial condition (Deyo 1994). The literature regarding the ▶ correlation between physical demands and back pain is contradictory. While many cross-sectional studies have shown a relationship between heavy physical work load and the occurrence of back pain, other research (including ▶ longitudinal studies) have refuted this conclusion. While physical work may not cause the pain, it is certainly more difficult to continue a heavy job once back pain has occurred. The association between ▶ sedentary work and back pain occurrence is equally inconclusive; multiple studies present evidence for and against this association. Nachemson (1992) has stated that low back pain occurs with about the same frequency among those performing heavy labor and those with sedentary occupations.

There is some evidence, but mostly there is insufficient scientific data from quality studies, to establish strong associations between these risk factors and back pain occurrence. Such a diverse list implies that most workers will get back pain at some time in their life, which is true, but this does not mean that the condition will definitely lead to ▶ chronic pain and/or disability.

Job Demands

Pain involves both physical and emotional components. Addressing one and not the other often leads to treatment failure. There are important job demands and employment conditions to consider when planning for an injured worker to return to the workplace. Recognition of job demands that may become barriers to recovery can enhance recuperation. Addressing these circumstances soon after pain onset may: (a) avoid costly and unnecessary medical investigation and treatment, (b) avoid the frustration of trying to find medical answers for nonmedical problems, and (c) shift the focus toward altering the circumstances in an effort to prevent the development of chronicity and/or disability.

Feuerstein (1993) emphasized that as time away from the workforce due to injury increases, the number of factors contributing to the work disability also increases. Previous research has helped define seven barriers to recovery presented in this entry.

Is There Job Security?

The present economy may leave an injured worker with no job in which to return. This lack of a clear goal or endpoint to the rehabilitative process can negatively affect a person's desire for a timely recovery. Job availability must be determined early on, and monitored constantly throughout the rehabilitative process.

Key questions for the injured worker that help facilitate return to work and maintain employment:

1. When was the last time you spoke to your employer?
2. Is there a job to return to?
3. Does your employer know when you are planning to return?

Is There Worker Motivation?

Even when a job exists, an injured worker's motivation to return to the same employment must be determined. Poor working conditions, poor labor relations or strained relations with coworkers may negatively influence motivation to return to work and/or stay there. The association between job dissatisfaction and occupational back pain is weak; this probably reflects the fact that research on ▶ job satisfaction has been a crude measure that fails to ascertain accurately employees' attitudes about work. Often, the rehabilitation process becomes less effective as motivation to go back to the job decreases; however, it is rarely advisable to tell workers to stay off work, change their job, or give up work completely because of simple back pain (Waddell 1998).

Key questions
1. Do you like your job, fellow workers, boss?
2. Does your company treat you well?

Are There Physical Limitations?

A workplace that requires function beyond the physical capabilities of an individual will produce pain, frustration, and absenteeism. A person who does not have the physical ability to accomplish specific job demands, may not be eager to return to work. There are no quality studies to support the notion that a worker's back pain is job related, or that his/her job demands are bad for the spine. Nevertheless, injured workers must regain a certain functional ability in order to resume their job demands.

Key questions
1. What are the physical requirements of your job?
2. Do you currently have those physical capabilities?
3. When will you be able to meet your job requirements?

Is a Graduated Return to Work Acceptable?

An injured worker usually benefits from any workplace modifications, but these changes should be time limited, with the eventual goal of a return to full duties. The practice of placing ▶ permanent restrictions on an injured worker may have a negative effect on return to work, especially when employers prohibit anything other than a return to full duties. Ergonomic and work-style modification efforts may contribute to an increase in successful return to work, but this finding is not supported in the medical literature. Clinicians often tell workers that their pain is job related and they should stay off work until they are better. Proactive return to work programs is effective because they allow workers gradually to reenter the workplace and improve relationships between workers and employers; thus, good industrial relations make successful rehabilitation and return to work more likely.

Key questions
1. Will your employer accommodate a graduated return to work in terms of progressive hours, progressive loads, and/or rotating positions?
2. What kind of cooperation is required?

Is There Unrecognized Conflict?

Unrecognized or unresolved conflict between the injured worker and the employer (or others outside the work setting) presents a significant barrier to recovery. A worker's unhappiness with the employer or payor regarding the goals

of rehabilitation may prolong recovery. Patients, clinicians, and payors each tend to have different definitions of success. Conflicting goals concerning the expected outcome will hinder return to work and may create disappointment, anxiety, and anger when recovery does not proceed as planned. The goals of everyone involved must be identified and the objectives of the rehabilitation process clear to all.

Litigation can also serve to heighten the conflict between the injured worker and the employer or payor. Unfortunately, our current ▶ disability-litigation system adds greatly to the epidemic of work-related pain; and injured workers are often rewarded for nonfunction.

Key questions
1. Do you have good relations with your family, employer, third party payor, and/or fellow workers?
2. Do you agree with the goals of your rehabilitation?
3. Is there a lawyer involved?
4. Do you need your pain to get your settlement?

Is There Secondary Gain?

Secondary gain is not ▶ malingering. Workers with secondary gain issues have genuine symptoms but are not able to cope well, and may continue in the sick role to varying degrees. Secondary gain may result in a resistance to health that can affect recovery, and is not only related to financial benefits but may include affection, control, and escape.

Financial. People who believe that they are entitled to a financial reward because of an accident have little desire to recover quickly. Financial supplements for an injury may enable people to continue their current lifestyles. Unfortunately, this often removes the economic incentive to return to work (Frank 1995).

Attention. Families often spend more time with a family member who has incurred an injury. They pay more attention to and empathize more with the injured person in an effort to make him/her feel better. An extended duration of this adjusted family situation can eventually serve to reinforce the injured person's sickness behavior. The secondary gain is love from a family that is only trying to help.

Control. The increased attention received by some injured workers can prompt an image of dependency that may give them a sense of control and manipulation over family, friends, and peers. They may become accustomed to a life free of responsibilities, a life only possible when recovery remains unachievable.

Escape. Pain may allow a person to escape from the everyday pressures and responsibilities of life. Should the duration of this escape continue for too long and allow a worker to become too far removed from reality, the family's effort to help only reinforces the negative behavior.

Key questions
1. Is wage replacement sufficient to pay the bills?
2. How long will compensation last?

Is There Physician Interference?
A physician must be part of an injured worker's treatment rather than part of the problem (Amadio 1988). A misinformed or overprotective doctor may prolong the rehabilitation effort unnecessarily. Returning to a normal lifestyle, including work, is a priority.

Key questions
1. Is the person's pain sufficient to cancel returning to work?
2. Has return to work been delayed until the pain disappears?
3. Will returning to work result in increased harm?
4. How many doctors/specialists have been seen?

As this list of questions indicates, the job demands that may contribute to work disability range from obvious physical ones to subtle psychosocial and economic ones.

Finding answers to these pertinent questions helps identify job demands and other factors that may act as barriers to recovery for an injured worker. Also, these answers will help enable clinicians to quantify their instincts, and injured workers to prepare for workplace reintegration. Resolution of possible barriers and rapid recovery allows the injured worker to return to a more normal lifestyle.

References

Amadio, P., Jr., Cummings, D. M., & Amadio, P. B. (1988). A framework for management of chronic pain. *American Family Physician, 38*, 155–160.

Bigos, S. J. (1993). Primary prevention for incumbent workers. Conference: The sports medicine approach to spinal disorders in the workplace.

Cheadle, A., Franklin, G., Wolfhagen, C., Savarino, J., Liu, P. Y., Salley, C., & Weaver, M. (1994). Factors influencing the duration of work-related disability: A population-based study of washington state workers' compensation. *American Journal of Public Health, 84*, 190–196.

Deyo, R. A. (1994). Magnetic resonance imaging of the lumbar spine: Terrific test or tar baby? *The New England Journal of Medicine Editorial, 331*, 115–116.

Feuerstein, M. (1993). Musculoskeletal injuries: Causes and effects. *Rehab Management, 6*, 30–36.

Frank, J. W., Pulcins, I. R., Kerr, M. S., Shannon, H. S., & Stansfeld, S. A. (1995). Occupational back pain – an unhelpful polemic. *Scandinavian Journal of Work, Environment & Health, 21*, 3–14.

Frymoyer, J. W., & Cats-Baril, W. (1987). Predictors of low back pain disability. *Clinical Orthopaedics, 221*, 89–98.

Guo, H. R., Tanaka, S., Cameron, L. L., Seligman, P. J., Behrens, V. J., Ger, J., Wild, D. K., & Putz-Anderson, V. (1995). Back pain among workers in the United States: National estimates and workers at high risk. *American Journal of Industrial Medicine, 28*, 591–602.

Hall, H., McIntosh, G., Wilson, L., & Melles, T. (1998). Spontaneous onset of back pain. *The Clinical Journal of Pain, 14*, 129–133.

Murphy, P. L., & Volinn, E. (1999). Is occupational low back pain on the rise? *Spine, 24*, 691–697.

Nachemson, A. (1992). Newest knowledge of low back pain: A critical look. *Clinical Orthopaedics, 279*, 8–20.

Waddell, G. (1998). *The back pain revolution*. Edinburgh: Churchill Livingstone.

Pain Inhibitory Pathways

▶ Descending Circuitry, Opioids

Pain Inhibitory Systems

▶ Descending Modulation of Nociceptive Processing

Pain Inventories

Robert D. Kerns[1] and Marc J. Shulman[2]
[1]VA Connecticut Healthcare System,
Westhaven, CT, USA
[2]VA Connecticut Healthcare System,
Yale University, New Haven, CT, USA

Synonyms

Pain measurement

Definition

There are a variety of instruments that assess the direct physiological mechanisms of pain as well as affect, beliefs, coping styles, external influences, and psychosocial functioning associated with pain. These instruments are particularly valuable because the measurement of the intensity of pain, and determination of the extent to which pain affects an individual, can be quite difficult. There are no physiological indicators that allow for an objective measurement of the quantity of pain that an individual experiences, because pain is a complex, subjective, and multidimensional phenomenon. Consequently, by providing a holistic assessment of pain, ► pain inventories can serve as vital tools for understanding underlying pain mechanisms and for the development of techniques to control pain.

Characteristics

Self-report Pain Measures

► Self-report pain measures provide the most direct method for assessing aspects of pain such as pain intensity, pain affect, pain quality, and pain location, as well as constructs related to chronic pain. They can be particularly valuable for use in clinical and research settings because they produce standardized, reliable, and valid assessments. Additionally, they are cost- and time-efficient for administration and scoring

and are particularly sensitive to treatment-related changes. Jensen and Karoly (2001) concluded that, although additional research is required to answer vital issues pertaining to the quality and measurement of the pain experience, most of the instruments that are now available have adequately demonstrated their excellent ► reliability and validity. They recommend that it is essential for clinicians and researchers to select instruments with a full understanding of their ► psychometric strengths and weaknesses and to ensure that they are consistent with their conceptualization of pain.

The Medical Outcomes Study 36-Item Short-Form Health Survey (SF-36)

The SF-36 was developed as a general measure of perceived health status (Ware and Sherbourne 1992) and is generally self-administered. The measure contains 36 items that are combined to form eight scales: physical functioning, physical role functioning, bodily pain, general health, vitality, social functioning, emotional role functioning, and mental health. Respondents use yes-no or 5- or 6-point scales to endorse the presence and degree of specific symptoms, problems, and concerns. Scores on the scales range from 0 to 100, with higher scores indicating better health status and functioning. The measure takes about 10–15 min to complete. The SF-36 has been extensively validated with large samples from the general population and across several demographic subgroups, including samples of healthy persons over 65 (Kazis et al. 1998). However, the SF-36 has only recently begun to be studied in ► chronic pain populations, including use as an outcome measure in pain intervention trials (Katz et al. 1992).

West Haven-Yale Multidimensional Pain Inventory (WHYMPI)

The ► West Haven-Yale Multidimensional Pain Inventory (WHYMPI) (Kerns et al. 1985) is designed to measure psychosocial and behavioral aspects of chronic pain and is useful across a variety of clinical populations. It is a multidimensional self-report instrument that consists of 52 questions, each scored on a 7-point

▶ Likert scale (0–6). There are three sections: (1) pain experience (five scales that measure perceived intensity and the impact of pain on several aspects of patients' lives), (2) pain-relevant significant other responses (three scales that measure patient's perceptions of the responses of significant others to their communication of pain), and (3) daily activities (four scales that measure performance of common activities). There is a version of the WHYMPI, where patients' significant others complete items that measure their responses to the patients' pain behaviors (Kerns and Rosenberg 1995). It takes approximately 10–15 min to complete the WHYMPI and is written at a fifth-grade reading level. The WHYMPI has demonstrated good reliability, and scores from the WHYMPI have been found to correlate with several other pain measures.

Pain Disability Index (PDI)

The ▶ Pain Disability Index (Pollard 1984) was developed as a brief measure of the degree to which chronic pain interferes with normal role functioning. While most data come from patients with heterogeneous pain conditions, the PDI has been used to measure function/disability in a number of specific painful conditions. The PDI includes seven items assessing perceived disability in each of seven areas of normal role functioning: family/home responsibilities, recreation, social activity, occupation, sexual behavior, self-care, and life-support activity. Each item is rated on a 10-point scale (0 – no disability to 10 – total disability), and the responses are summed. The PDI shows excellent reliability and ▶ validity as a brief measure of pain-related disability in younger individuals (Tait et al. 1990). There is minimal evidence for the psychometric properties of this scale for elderly individuals.

Chronic Illness Problem Inventory (CIPI)

The ▶ Chronic Illness Problem Inventory (Kames et al. 1984) is a 65-item instrument developed to measure patient functioning in the areas of physical limitations, psychosocial functioning, healthcare behaviors, and marital adjustment. Each CIPI item describes a problem in

functioning, and each patient rates the degree to which each item applies on a 5-point scale. Romano et al. (1992) concluded that its ease of administration and scoring, and its preliminary psychometric properties, make this a useful instrument but that its 18 scales can become unwieldy. Therefore, the CIPI demonstrates potential as an alternative measure of pain dysfunction but should be used with caution and with additional measures of dysfunction, until further validation is achieved.

Pain Drawings

▶ Pain drawings consist of an outline of a human figure on which patients darken the areas where they are experiencing pain. These drawings can facilitate the assessment of the location and nature of the pain and assist with the decisions for appropriate intervention. Additionally, their visual representation enables clinicians to detect patterns of pain that may otherwise be overlooked. However, only very limited correlations have been found between the percentage of the body surface area with pain and the measures of current psychological state and ▶ premorbid functioning. Therefore, firm conclusions about psychological illness cannot be reached based solely on the involvement of multiple areas of pain drawings (Ginzberg et al. 1988).

Pain Diary

▶ Pain diaries can provide additional information beyond traditional assessment tools, which rely on recollections of the past and composite measures of critical variables. Jacob and Kerns (2001) point out that they provide an actual temporal relationship between psychosocial variables and pain, by having patients record their mood, activity, sleep, and pain on a daily basis, rather than relying on patients' global, retrospective self-reports. They also note that the reliability and validity of the pain diaries has been demonstrated across several studies and for a variety of pain-related difficulties.

UAB Pain Behavior Scale

The ▶ UAB Pain Behavior Scale (Richards et al. 1982) consists of 10 target behaviors, each of which contributes equally to a total score. Therefore, there is a range of possible scores from 0 to 10. The 10 target behaviors were selected because they were judged to be the most relevant, frequently observed, and reliably measurable pain behaviors in a chronic pain population. Pain behavior is assessed by a trained clinician, who observes a series of behaviors demonstrated by the patient. Ratings are based on frequency estimates of pain behaviors: none (0), occasional (1/2), and frequent (1). Patients are asked to walk a short distance and stand for a few seconds. They are also asked to move from a sitting to a standing position and vice versa. Frequency of both verbal and nonverbal (i.e., moans, groans, etc.) pain behaviors is scored. The total process typically takes 5 min per patient. Although the UAB scale is a relatively basic instrument that can be administered rapidly by a variety of professionals, it has demonstrated good reliability and validity. However, this tool must be viewed cautiously and used in combination with other measures, since additional research may be necessary to modify some of its pain behavior categories.

Given that these inventories assess different types of pain behaviors and experiences and vary to the extent that they examine how individuals respond to pain, a comprehensive assessment may involve the use of a combination of several instruments, in order to achieve a valid and thorough understanding of the multidimensional nature of an individual's pain experience. For example, when forming an appropriate treatment plan for a patient with chronic pain, it may be useful for a clinician to administer the WHYMPI or CIPI to determine the extent that pain results in disability or interferes with daily functioning across multiple domains while also encouraging the patient to keep a pain diary throughout treatment in order to track and monitor self-reported changes in pain behaviors and coping responses over time. However, in settings that require time-limited, brief screenings of pain, providers may choose instead to administer shorter measures such as the PDI, UAB Pain Behavior Scale, or pain drawings, or a combination of these tools, to inform strategies and goals for treatment. Individual inventories can be useful tools on their own in both clinical and research settings; what is critical is the implementation of ongoing assessment of pain, whether through self-report measures or behavioral tools, in order to evaluate short-term and long-term effectiveness of intervention methods and to improve treatment outcomes for patients with chronic pain.

References

Ginzberg, B. M., Merskey, H., & Lau, C. L. (1988). The relationship between pain drawings and the psychological state. *Pain, 35*, 141–146.

Jacob, M. C., & Kerns, R. D. (2001). Assessment of the psychosocial context of the experience of chronic pain. In D. C. Turk & R. M. Melzack (Eds.), *Handbook of pain assessment* (pp. 15–34). New York: Guilford Press.

Jensen, M. P., & Karoly, P. (2001). Self-report scales and procedures for assessing pain in adults. In D. C. Turk & R. M. Melzack (Eds.), *Handbook of pain assessment* (pp. 15–34). New York: Guilford Press.

Kames, L. D., Naliboff, B. D., Heinrich, R. L., et al. (1984). The chronic illness problem inventory: Problem-oriented psychosocial assessment of patients with chronic illness. *International Journal of Psychiatry in Medicine, 14*, 65–75.

Katz, J. N., Harris, T. M., Larson, M. G., et al. (1992). Predictors of functional outcomes after arthroscopic partial meniscectomy. *Journal of Rheumatology, 19*, 1938–1942.

Kazis, L. E., Miller, D. R., Clark, J., et al. (1998). Health related quality of life in patients served by the department of veterans affairs: Results from the veterans health study. *Archives of Internal Medicine, 158*, 626–632.

Kerns, R. D., & Rosenberg, R. (1995). Pain-relevant responses from significant others: Development of a significant-other version of the WHYMPI scales. *Pain, 61*, 245–249.

Kerns, R. D., Turk, D. C., & Rudy, T. E. (1985). The West Haven-Yale Multidimensional Pain Inventory (WHYMPI). *Pain, 23*, 345–356.

Pollard, C. A. (1984). Preliminary validity study of pain disability index. *Perceptual and Motor Skills, 59*, 974.

Richards, J. S., Nepomuceno, C., Riles, M., et al. (1982). Assessing pain behavior: The UAB pain behavior scale. *Pain, 14*, 93–98.

Romano, J. M., Turner, J. A., & Jensen, M. P. (1992). The chronic illness problem inventory as a measure of dysfunction in chronic pain patients. *Pain, 49*(1), 71–75.

Tait, R. C., Chibnall, J. T., & Krause, S. (1990). The pain disability index: Psychometric properties. *Pain, 40*, 171–182.

Ware, J. E., & Sherbourne, C. D. (1992). The MOS 36-item short-form health survey (SF-36), I: Conceptual framework and item selection. *Medical Care, 30*, 473–478.

Pain Judgment

▶ Pain Assessment in Neonates

Pain Management

Definition

Pain management is an activity in health care whose objective is to prevent, limit, avoid, or eliminate the consequences of pain. It involves the systematic assessment of pain and its association stress and consequences, and the application of pharmacological and nonpharmacological interventions. Interested readers may click of the following website link for an official policy statement of the American Academy of Pediatrics on the prevention and management of and pain and stress in the neonate: http://aappolicy. aappublications.org/cgi/content/fukll/pediatrics; 105/2/454.

Cross-References

▶ Pain Assessment in Neonates

Pain Management in Infants

▶ Acute Pain Management in Infants

Pain Management Programs

▶ Behavioral Therapies to Reduce Disability

Pain Mapping

▶ Chronic Pelvic Pain, Laparoscopic Pain Mapping

Pain Measurement

Definition

A method for rating the intensity and subjective qualities of pain by means of questionnaires or behavioral assessment procedures. As pain is a personal experience, subjective report is the only way to assess it. Conventional pain measurement practice assumes that patients and research subjects can exercise introspection to gauge the magnitude of some features of pain, assign numbers accordingly, and report such numbers without bias.

Cross-References

▶ Consciousness and Pain
▶ McGill Pain Questionnaire
▶ Pain Assessment in Neonates
▶ Pain Inventories
▶ Pain Measurement by Questionnaires, Psychophysical Procedures, and Multivariate Analysis

Pain Measurement by Questionnaires, Psychophysical Procedures, and Multivariate Analysis

W. Crawford Clark
College of Physicians and Surgeons,
Columbia University, New York, USA

Synonyms

Cluster analysis; Multidimensional scaling; Pain measurement; Pain Questionnaire; Pain threshold; Ratio scaling; Sensory decision theory

Definition

If we are to diagnose and treat pain and suffering in a rational manner, we must measure them accurately. Quantification is extremely difficult because what is commonly labeled "pain" is a multidimensional experience that possesses a large number of qualities or dimensions and varies over a wide range of intensities. Measurement is concerned with the process and rationale involved in the construction of a measuring instrument and the properties that can be ascribed to it. Measurement provides the bridge between the empirical phenomena themselves and the laws and theories that interpret them. The increasing sophistication of data collection procedures and measurement models is the hallmark of the steady growth of psychology from speculative philosophy to a scientific discipline. Unfortunately, the application of new measurement models in the field of pain has lagged far behind their successful application elsewhere.

This entry reviews the advantages of recently developed data collection and analytical methods used to quantify three major classes of pain measurement models:

1. Questionnaires and rating scales
2. Psychophysical procedures, traditional thresholds, sensory or statistical decision theory, and ratio scaling
3. Multivariate procedures, including multidimensional scaling and cluster analysis

These quantitative procedures yield numerical values; however, the arithmetical and hence statistical operations that can be performed on them differ according to the level of measurement that they meet, as follows. For nominal scales, the basic empirical operation is the determination of equality, e.g., a player is a member of team A or of team B and the number on the player's jacket is unrelated to any physical quantity. An example of an appropriate statistical test is Chi^2. For ordinal scales, the basic empirical operation is the determination of greater or lesser; the numbers represent order, but the distance between the numbers is unknown, e.g., hardness of minerals and rank order data on Likert scales. Examples of appropriate statistical tests include nonparametric measures such as the Wilcoxon ranked sum test and rank order correlations. For interval scales, the basic empirical operation is the equality of the interval or distance, between

the numbers, e.g., Fahrenheit and centigrade scales and standard deviation scores on intelligence tests. Examples of appropriate statistical tests include the t-test and product-moment correlations. For ► ratio scales, the empirical operation is the determination of equality of ratios, e.g., 2 in. is to 4 in. as 40 in. is to 80 in., that is, one half; one cannot make this statement for the commonly used temperature scales. Ratio scales require both equal interval distances and a true zero origin. Examples are degrees Kelvin, weight, length, watts, and lumens. Appropriate statistical tests include, in addition to tests appropriate for the three levels of measurement already described, coefficient of variation and decibel transformations. All four levels of measurement have been used to study pain.

The clinician must rely on the patient's report because, unlike blood pressure, body temperature, etc., pain is a personal experience that cannot be physically observed. Although the patient's own descriptions of pain are important, questionnaires are needed to ensure that all aspects have been covered and to construct a defined body of data for clinical and research use. The patient's response to questionnaires is an important addition to physical information for a number of reasons, accurate differential diagnosis, e.g., whether a patient's back pain has a neurological or muscular origin, and optimal choice of treatment strategies, including the choice of medications, dosage, radiation, and psychological intervention. The patient's responses are also used to measure treatment progress and its success or failure, as well as the outcome of clinical trials and the evaluation of the health of populations. The collection of quantitative data also serves to increase our understanding of neural mechanisms, as well as psychosocial, gender, and ethnocultural differences. This basic knowledge leads to improved medications and treatment regimens.

Characteristics

Questionnaires and Rating Scales
The number of items on a pain test varies greatly. The long questionnaires may have over 100

questions or items, short questionnaires have 12–30 items, while a unidimensional rating scale has a single item. Each of these procedures has a role in pain assessment, depending upon the information needed and the time available. The advantage of unidimensional rating scales is that they are brief; however, they are typically used to evaluate only one of the many dimensions of pain, the pain intensity dimension. Short questionnaires take less time but provide less information than the longer forms. The advantage of the longer questionnaires, which are simply a large set of rating scales, is that they assess many dimensions of pain, sensory qualities, negative and positive emotions, motivation, etc.; however, they are too lengthy for frequent use.

Unidimensional Scales

There are three types of unidimensional scales. Visual analogue (VAS) and numerical rating scales (NRS) are anchored by descriptors at each end, e.g., from "no pain" to "pain as bad as it could be." The patient responds to the VAS by marking a point on a line between the anchors, and to the NRS by choosing a number from 1 to 10. The particular descriptor chosen for the anchor does not appear to be critical. Seymour et al. (1985) compared a number of anchors, "troublesome," "miserable," "intense," "unbearable," and "worst pain imaginable" and found that the exact wording made little difference to the pain intensity ratings obtained. Furthermore, patients with less education, the aged, and patients in some other cultures find the VAS confusing (Kremer et al. 1981). The accuracy of the VAS has been likened to a thermometer without numbers. The problem with the VAS and NRS scales is that one patient may feel that the proper location for "faint pain" is close to the "no pain" anchor while another may believe that it should be towards the middle of the scale. This problem is effectively resolved by the verbal category scale (VCS). To take an example from a threshold study of calibrated heat intensities, patients select a word from a set of descriptors ordered with respect to intensity, e.g., "no sensation," "warm," "moderate heat," "hot but not painful," "faint pain," "moderate pain," "severe

pain," and "withdraw." The facial expression category scale is mostly used with children and ranges from a smiling face and the statement "no hurt" through a frowning face, "hurts even more" to a crying face, "hurts worse." Scales with an even number of items are superior because the patient is forced to choose between the lower and the upper half of the scale; an indifferent patient cannot avoid a difficult decision by picking the middle item.

A serious problem with all of the unidimensional pain intensity scales is that the patient's emotional state greatly influences the sensory pain intensity rating. Clark et al. (2002) studied postoperative pain in patients recovering from surgery who responded to both the 101-item MAPS, a multidimensional pain questionnaire, and a unidimensional pain numerical rating scale (PNRS). Multiple linear regression analysis revealed that scores on four of the eight subclusters in the MAPS emotional pain supercluster anxiety, depressed mood, fear, and anger significantly predicted the patients' score on the pain rating scale. Even more surprisingly, none of the 17 subclusters in the somatosensory pain supercluster predicted scores on the PNRS. It was concluded that patient scores on unidimensional pain intensity scales reflect the emotional impact of pain much more than the sensory intensity of the pain. Obviously, such pain scales are poor indicators of the intensity of the patient's sensory pain and of the analgesic dosage required. These findings also suggest that the patient's anxiety and depression are poorly measured and hence inadequately treated. A treatment decision between increasing the dosage of an analgesic and instead adding a psychotropic drug cannot be made on the basis of a score on a pain intensity rating scale alone; clearly, as a minimum, the patient also should respond to unidimensional rating scales of negative emotions in addition to a pain scale. This was done by Knotkova et al. (2004), who found, in a study of patients experiencing cancer related pain, that scores on a pain PNRS were significantly correlated with items in the MAPS somatosensory pain supercluster and that scores on the unidimensional anxiety and depression NRS were respectively correlated with the anxiety and depression items in the MAPS emotional pain supercluster. Thus, the MAPS questionnaire yielded much more information; the rating scales were sufficient to determine which of the patients with a high score on the pain intensity rating scale should have an increase in analgesic dosage and which should have an increase in psychotropic medication.

Short Questionnaires

The clinical use of long questionnaires is limited by their length and the burden on incapacitated patients; however, attempts to circumvent this practical problem by resorting to a single pain rating scale fail to yield sufficient information for effective treatment strategies. Accordingly, a short questionnaire fills an important niche in clinical practice and research because it takes less time than the long form and provides much more information than a single rating scale.

Short questionnaires are usually constructed by reducing the number of items in an existing instrument (Coste et al. 1997). Shortening an existing instrument is preferable to constructing a new one because it allows one to bypass the selection of items from a larger untested pool of items that must be defined anew. Typically, the longer form is shortened by a group of experts who select what they believe to be relevant items for inclusion. This is a dubious practice because experts have been found to disagree on two-thirds of the descriptors regarded as relevant to a particular patient population (Sonty et al. 2004). A separate problem is that most short questionnaires are considered to be applicable to any illness. This results in a serious loss of information in the short form because many of the descriptors are irrelevant for a particular disease. This loss is reflected in the small number of factors found with short forms. For example, studies with the long MPQ have yielded from four to as many as seven factors compared with one or two factors found with short forms of the MPQ (Wright et al. 2001). To be effective, the items on a short questionnaire should be tailored to contain items that are relevant to a specific patient population.

Instead of relying on the opinions of experts, item analysis should be used to shorten questionnaires. Item analysis is a superior way to construct a short questionnaire from a long version because an objective procedure determines the most relevant items. Furthermore, the patients themselves, not "experts," determine which items are relevant. Griswold and Clark (2005) used item analysis to examine the responses of 100 outpatients with cancer to the long 101 MAPS questionnaire. The item analysis procedure yields measures of inter-item consistency, the average of the degree to which the score on each item is correlated with the score on every other item, and ▶ discriminability, how well each item contributes to the total score, that is, to the desired construct. In addition, Cronbach's alpha demonstrated that the items within the 101 MAPS subclusters were homogeneous; thus, any item within a cluster is fairly representative of the entire cluster. To construct the 30-item short MAPS (SMAPS) questionnaire, the item within each cluster that best met the item analysis criteria was selected. Representation of each of the 30 subclusters ensured that the structure of the MAPS dendrogram obtained by cluster analysis was maintained in the 30-item SMAPS.

Psychophysical Procedures: Traditional Thresholds, Statistical or Sensory Decision Theory, and Ratio Scaling

Traditional Thresholds

Grossberg and Grant (1978) provide an excellent, detailed examination of the topics presented in this section. The traditional methods of serial exploration and constant stimuli are two methods commonly used to determine the pain threshold or its inverse, sensitivity. In the method of serial exploration or limits, the subject responds "not painful" or "painful" to stimuli that are presented in serial order with respect to intensity. The pain detection threshold is the interpolated stimulus intensity between the last report of "not painful" and the first report of "pain," while the pain tolerance threshold lies between the highest intensity that yields a "severe pain" response and the intensity that causes a withdrawal response. Usually five or more thresholds are averaged to obtain a mean threshold value. The advantage of the method of limits is that it is quick and because the ascending series begins at a relatively low intensity, it avoids the shock to patients who might experience pain from a very low intensity stimulus (e.g., patients with causalgia).

In the method of constant stimuli, the subject responds "not painful" or "painful" to a set of about five different stimulus intensities, selected by the investigator, that are presented randomly with respect to intensity. The method of constant stimuli is clearly superior to the method of limits because it avoids the errors of anticipation (premature report of an expected perceptual change) and habituation (repetition of the previous report). Both the methods of limits and constant stimuli yield thresholds expressed in physical units (degree centigrade, watts, lumens, etc.) at the interval scale level of measurement. Although the methods of limits and constant stimuli may be useful in clinical practice, their use in research studies is indefensible because the traditional threshold obtained by these methods is an unresolvable mixture of neurosensory sensitivity and attitudinal or response bias, the tendency to favor a particular response over another. Thus, as Clark (2003c) emphasizes, it is impossible to determine whether a high threshold (few pain reports) is caused by sensory deficit or by a psychological attitude (e.g., stoicism).

Sensory Decision Theory (SDT)

The statistical decision-making or ▶ sensory decision theory model provides a quantitative solution to the problem of distinguishing sensory sensitivity $P(A)$ or d', from response bias B or Lx. These numerical values are at the interval scale level of measurement. The discrimination parameter, $P(A)$ or d', quantifies the observer's ability to distinguish event A from event B, where the events may be higher and lower stimulus intensities, different physiological states (e.g., present or absent gastric contractions), previously viewed or expressed words, and new words (pain memory). $P(A)$ is a nonparametric measure, while d' assumes normal distributions of noise and signal plus noise. In the instance of laboratory-induced pain, the discriminability index, $P(A)$ or d', provides a pure measure of the functioning of the

neural pathways mediating pain; for example, it has been found to be decreased by analgesics. The other SDT parameter is response bias, B or Lx, which quantifies the decision maker's tendency to favor the reporting of one of the events over the other. It reflects the patient's attitude concerning which stimulus intensities are to be called "painful" or "not painful"; it has been shown to be influenced by suggestion and by placebos described as analgesics. Differences in traditional thresholds between ethnocultural groups, between old and young subjects, and between patients and normals have been demonstrated to be related to report bias, B or Lx. It is obvious that the commonly held belief that traditional thresholds are pure measures of sensory sensitivity uninfluenced by the subject's attitude is patently false; the threshold is an unresolvable conglomerate of physiological and psychological variables (see review by Clark 2003a).

Ratio Scaling

Baird and Noma (1978) provide an excellent review of ratio scaling. The most commonly used ratio scaling technique is magnitude estimation (Stevens 1975). Here the investigator presents a particular stimulus intensity and assigns it a numerical value (the modulus). The subject is then presented with stimuli of various intensities and after each presentation judges the intensity ratio between the stimulus and the modulus, e.g., three times greater, two times less. A plot of the ratio judgments averaged over a number of subjects (ordinate) against an arithmetic scale of the physical intensities (abscissa) yields an exponential function, termed a power function, that describes the increase along the response scale (R) that corresponds to an increase along the stimulus intensity scale (S). In brief, equal stimulus intensity ratios (S) produce equal subjective ratios (R).

The equation expressing this relationship is,

$$R = \lambda \, S^n$$

The constant λ is simply a way to change the unit of measurement; for example, we multiply by 100 to convert meters to centimeters.

When this equation is rewritten logarithmically, it yields the linear equation,

$$\log R = n \log S + \log \lambda$$

A plot of R against S on double logarithmic paper yields a best-fitting straight line with a slope of n, the exponent, and an intercept λ. Some examples of sensitivity (slope) to various stimuli are electrical pain, 3.5; heaviness, 1.1; length of lines, 1.0; and brightness 0.3. The size of the pain exponent has been questioned; studies report slopes of 1.8 and 2.2. Stevens later predicted that, since pain is mediated by a separate neurosensory system, noxious stimulus intensities should yield a different equation and slope. A radically different approach to psychophysical scaling described below has been found to yield two different, but much smaller exponents, one for non-painful electrical stimuli, 0.69, and another, 0.26, for painful electrical stimuli.

Contextual Effects

If the ratio scaling approach is valid, then the response scales, that is, the exponents and intercepts for each sensory modality, should be relatively invariant. However, Baird and Noma (1978) review a number of studies in which the expected invariance is not found under a variety of conditions, including different instructions, changes in modulus intensity, whether judgments are a multiple or a fraction of the modulus intensity, and differences in the range of intensities of the comparison stimuli. The influence of these non-sensory variables is termed context effects. Even attitudinal differences can have an effect. Baird et al. (1972) demonstrated that giving different groups information about the experimenter's belief in the relation between line length and the number scale caused the exponents to vary from 0.31 to 1.71. If the exponent is a pure measure of sensory sensitivity, it should not vary with induced changes in the subject's attitude. Contextual effects are probably responsible for the wide individual differences found among observers. This variability makes it necessary to average the

responses of a large number of observers in order to obtain a stable exponent. Such average data are of no use in the clinic where treatment decisions must be tailored to each patient. It should be noted that the sensory decision theory yields a much purer measure of sensory function, $P(A)$ or d', because the context effects are segregated into the report bias parameter, B or Lx. The presence of context effects demonstrates that both the power law and its exponents are not pure measures of sensory function, at least when data are collected by standard ratio scaling procedures.

Context effects present a serious problem for ratio scaling methodology. Although in a formal sense as long as the scales fulfill the criterion that equal stimulus ratios produce equal response ratios, the definition of a ratio scale remains untouched. However, it may be asked, what good are the ratio scales if a different one is required for each measurement condition? As Baird and Noma (1978) point out, we would not be happy with a ruler whose scale values depended upon whether it was used to measure the length of a desk or the height of a chair. The ratio scales obtained from human observers lack the invariance that is found with physical units such as centimeters, watts, lumens, etc.

Multivariate Analysis (MVA): Multidimensional Scaling (MDS) and Cluster Analysis (CA)

MVA includes a variety of mathematical models and data collection procedures (Lattin et al. 2003). The MDS and CA models are of particular interest here. These models yield geometric representations that make it possible for an investigator to uncover the "hidden structure" of complex databases. Both MDS and CA analyze pairwise similarity judgments based on a 10-point scale from "not at all similar" to "extremely similar" or some other measure of association, e.g., correlations, to generate proximities among a set of stimulus objects as input. Stimulus objects may be words, physical stimuli, or concepts. The response data are organized into a half-matrix in which rows and columns correspond to stimulus objects and the cells contain some measure of similarity such as ratings on a similarity scale. MVS has two advantages over presently used methods for questionnaire construction. First, the descriptors are grouped objectively by MVS analysis of similarity judgments; thus, an investigator does not make a priori judgments about which descriptors belong to which sensory, affective, etc. group of items. Secondly, the similarity judgments are made by lay people who may very well view the similarity and grouping of the descriptors in a different way from the "experts." This is important because it is lay people who will be responding to the pain questionnaire. That experts and lay people differ was demonstrated by Clark et al. (1995) who, using cluster analysis of similarity responses by college students, found no evidence to support the homogeneous status claimed for descriptors in the MPQ major evaluative class; furthermore, many of the words assigned to each of the 22 subgroups of the MPQ were found to be dispersed among different clusters.

Multidimensional Scaling Models (MDS)

For the MDS models, analysis of the same similarity half-matrices described above yields the ▶ group stimulus space, a configuration of points (the stimulus objects) as on a map. The stimulus objects appear along specific dimensions, e.g., sensory, affective, and motivational, in continuous space at the interval level of measurement. A ▶ dimension is a characteristic that defines a point by its coordinates. Another advantage of some of these multidimensional models (e.g., individual difference scaling [INDSCAL]) is that they quantify individual differences by providing subject coordinates in the source or subject weight space. The coordinates locate each individual with respect to the dimensions in the group stimulus space. The location of the coordinates quantifies the ▶ saliency or relative importance of each dimension to each individual. The coordinates of each individual subject weight can be used to distinguish among subpopulations in the sample and can be correlated with individual scores on psychological, sensory, or other measures.

INDSCAL has led to a novel approach to quantifying separately the neurosensory and the attitudinal components of the pain rating response to calibrated physical stimuli; the technique also can be used to construct verbal category scales at the interval scale level of measurement (Clark 2003b). First, subjects made 120 pairwise similarity judgments to all possible pairings of eight electrical stimuli that ranged from non-noxious to noxious intensities and eight related pain, somatosensory, and affect descriptors. Analysis by INDSCAL yielded a one-dimensional solution with the physical and verbal stimulus objects interdigitated with respect to their subjective magnitude. These coordinate values (ordinate) were plotted against the scale of electrical intensities in milliwatts (mW). Two distinct log-log linear functions with different slopes and intercepts that confirmed Stevens' psychophysical power law were found. The lower intensity function with a slope of 0.69 ranged from the descriptor "slight sensation" and the 3 mW physical stimulus to "faint pain" and 63 mW, while the higher intensity function with a slope of 0.26 ranged from "upsetting" and 88 mW to "severe pain" and 235 mW. The finding of two functions demonstrates the superiority of the INDSCAL approach over the magnitude estimation method, which was found to yield only a single function over a comparable set of electrical stimulus intensities. Since both the verbal and physical stimulus functions are at the interval scale level of measurement, it is possible to quantitate precisely the perceived intensities of the verbal descriptors at the interval scale level of measurement.

The superiority of multidimensional scaling over magnitude estimation resides in differences between the judgment tasks. The judgment of similarity between two stimulus objects allows the participant, not the investigator, to determine the dimension(s) that she/he considers relevant. Furthermore, the task of judging the similarity between two stimulus objects that are simultaneously present is much more direct and simple than comparing the relative strength of a test stimulus to the remembered strength of a previously presented standard stimulus (modulus). It has been demonstrated that cancer patients find the somatosensory pain dimension to be more salient than do healthy controls. This and other applications of MDS using either pain descriptors or physical, electrical, and thermal stimuli as stimulus objects to determine the number and characteristics of pain dimensions are reviewed elsewhere (Clark 2003a).

Cluster Analysis

The half-matrix obtained from pairwise similarity judgments just described for MDS can also be analyzed by one of the cluster analysis models. Indeed, the two models often provide complementary information and it is recommended that the same data set be analyzed both ways. Cluster models represent the structure of a set of stimulus objects (descriptors) and subsets of clusters in discrete space, where each cluster corresponds to a meaningful, homogeneous feature. Hierarchical agglomerative models (e.g., average linkage between groups) yield non-overlapping hierarchically organized, nested clusters, i.e., each cluster can be subsumed as a member of a larger, more inclusive, supraordinate cluster. A set of these hierarchically ordered clusters plotted along a relative similarity axis is termed a dendrogram. Clark et al. (1995) analyzed similarity judgments made to 270 descriptors of sensory pain, negative and positive emotions, motivations, physical illness, etc.; they found 50 subclusters subsumed under 18 primary clusters. Obviously, the objective cluster analysis approach to questionnaire construction is superior to the subjective, a priori approach of pain experts. In a subsequent study (Clark 2003a), healthy male and female African-American, Euro-American, and Puerto Rican subjects sorted 189 descriptors into similar piles (pile sort technique) and then sequentially merged their piles based on similarity judgments until only two piles remained (merge technique). After removing redundant descriptors and descriptors that had different meanings among the six sex/ethnocultural groups, i.e., were located in different clusters, there remained a dendrogram containing 101 descriptors in 30 clusters

subsumed within three superclusters, somatosensory pain, negative emotions, and well-being. The 101-item multidimensional affect and pain survey (MAPS) is based on this dendrogram. Thus, unlike the other questionnaires, the MAPS is based on the structure of the empirically derived dendrogram that emerged from the participants' own views of their sensory-emotional pain space. Factor analysis has demonstrated the validity of MAPS and hence the advantage of the cluster analytic approach to test construction (Clark et al. 2003).

Conclusions

This review should make it clear that many procedures that have been demonstrated to be inadequate for the accurate assessment of pain and suffering are still in use, while procedures proven to be superior tend to be ignored. For example, long questionnaires are based on subjective, a priori judgments of the investigators rather than being constructed by cluster analysis of lay persons' similarity judgments; short questionnaires are based on the investigators' opinion, not objectively by item analysis of patients' ratings on long questionnaires; sensory sensitivity is still determined by the method of serial exploration, not by statistical or sensory decision theory; and multidimensional scaling is seldom used to determine which pain dimension is most relevant to various patient groups.

Acknowledgment This entry and much of the research by Clark and colleagues was supported by USPHS grants NINCDS-20248 and NICDR-12725 and by the Nathaniel Wharton Fund for Research and Teaching in Brain, Body and Behavior.

References

Baird, J. C., & Noma, E. (1978). *Fundamentals of scaling and psychophysics.* New York: Wiley.

Baird, J. C., Kriendler, M., & Jones, K. (1972). Generations of multiple ratio scales with a fixed stimulus attribute. *Percept Psychophysics, 9*, 399–403.

Clark, W. C. (2003a). Pain, emotion, and drug-induced subjective states: Applications of multivariate scaling. In G. Adelman & B. Smith (Eds.), *Encyclopedia of neuroscience* (3rd ed.). Amsterdam: Elsevier. CD-ROM.

Clark, W. C. (2003b). *Do women actually experience more pain than men or do they merely report more? A multidimensional scaling approach* (p. 446). Prague: European Federation of the International Association for the Study of Pain. Book of Abstracts.

Clark, W.C. (2003c) Comments on: Petzke et al. Increased pain sensitivity in fibromyalgia: Effects of stimulus type and mode of presentation, Pain, 105, 403–413; and the related editorial: Hyperalgesia versus response bias in fibromyalgia, Fillingim. Pain, 105, 385–386

Clark, W. C., Fletcher, J. D., & Janal, M. N. (1995). Hierarchical clusters of 270 pain/emotion descriptors: Toward a revision of the McGill Pain Questionnaire. In B. Bromm & J. Desmedt (Eds.), *Pain and the brain: From nociception to sensation* (pp. 319–330). New York: Raven.

Clark, W. C., Yang, J. C., Tsui, S. L., et al. (2002). Unidimensional pain rating scales: A multidimensional affect and pain survey (MAPS) analysis of what they really measure. *Pain, 98*, 241–247.

Clark, W. C., Kuhl, J. P., Keohan, M. L., et al. (2003). Factor analysis validates the cluster structure underlying the Multidimensional Affect and Pain Survey (MAPS), challenges the a priori classification of the descriptors in the McGill Pain Questionnaire (MPQ) and demonstrates sex differences in cancer patients. *Pain, 106*, 357–363.

Coste, J., Guillemin, F., Pouchot, J., et al. (1997). Methodological approaches to shortening composite measurement scales. *Journal of Clinical Epidemiology, 50*, 247–252.

Griswold, G. A., & Clark, W. C. (2005). Item analysis of cancer patient responses to the multidimensional affect and pain survey demonstrates high inter-item consistency and discriminability and determines the content of a short form. *The Journal of Pain, 6*, 67–74.

Grossberg, J. M., & Grant, B. F. (1978). Clinical psychophysics: Applications of ratio scaling and signal detection methods to research on pain, fear, drug and medical decision making. *Psychological Bulletin, 85*, 1154–1176.

Knotkova, H., Clark, W. C., Mokrejs, P., et al. (2004). What do ratings on unidimensional pain and emotion scales really mean? A Multidimensional Affect and Pain Survey (MAPS) analysis of responses by cancer patients. *Journal of Pain and Symptom Management, 28*, 19–27.

Kremer, E., Atkinson, J. H., & Ignelzi, R. J. (1981). Measurement of pain: Patient preference does not confound pain measurement. *Pain, 10*, 241–248.

Lattin, J. M., Carroll, J. D., & Green, P. E. (2003). *Analyzing multivariate data*. Belmont: Duxbury Press.

Seymour, R. A., JM, S., Charlton, J. E., & Phillips, M. E. (1985). An evaluation of length and end-phrase of visual analogue scales in dental pain. *Pain, 21*, 177–185.

Sonty, N., Chokhavatia, S., Griswold, G. A., et al. (2004). *Studies of patients with cancer, low back pain, and irritable bowel syndrome demonstrate that short forms of the Multidimensional Affect and Pain Survey (MAPS) and other questionnaires must be tailored to each patient group.* Vancouver: American Pain Society.

Stevens, S. S. (1975). *Psychophysics*. New York: Wiley.

Wright, K. D., Asmundson, J. G., & McCreary, D. R. (2001). Factorial validity of the short-form McGill pain questionnaire (SF-MPQ). *European Journal of Pain, 5*, 279–284.

Pain Measures

▶ Pain Assessment in Neonates

Pain Mechanisms in Endometriosis

Karen J. Berkley and Natalia Dmitrieva
Program in Neuroscience, Florida State
University, Tallahassee, FL, USA

Synonyms

Central sensitization; Chronic pelvic pain; Comorbid pain conditions; Dyschesia; Dysmenorrhea; Dyspareunia; Referred pain

Definition

Endometriosis is an estrogen-dependent inflammatory disease defined by the presence of endometrial growths in abnormal locations.

Characteristics

Although endometriosis can be symptomless (its ectopic growths discovered during surgery for an unrelated problem), it is often associated with distressing chronic pain symptoms and/or subfertility. This entry focuses on endometriosis-associated pain and current views about its

Pain Mechanisms in Endometriosis, Fig. 1 *Some features of a rat model of endometriosis (ENDO): association of the ectopic growths with the CNS.* As discussed in the text, transplanted pieces of uterine horn form cysts that develop rapidly over a 1–2-month period and attract sensory and sympathetic innervation that sprouts from fibers in the splanchnic nerve. Part (**a**) is a diagram to illustrate the location of the four uterine transplants on the mesenteric arteries (*dots*) and how and where the developed cysts interact with the central nervous system via sensory and sympathetic fibers. The part of the uterine horn from which the transplants had been removed is shown by the X in the left uterine horn. The *dashed line* between the cysts and the spinal cord indicates that, as shown by retrograde tracing using DiI dye, the sensory and sympathetic innervation is associated mainly with the T9–T11 segments. The somata of the fibers are located in T9–T11 dorsal root ganglia and the celiac ganglion. The insets are two photographs of established cysts in situ (*arrows*, *left*) and a single harvested cyst (*right*). Part (**b**) includes four photomicrographs of cyst innervation. The top photo is a section through the middle of a whole cyst, immunostained with an antibody to protein gene product 9.5 (PGP9.5, a marker for all neurons). The second photo is an enlargement of the hilus region to show PGP9.5-positive fibers entering into the cyst wall. The third photo is an enlargement of the cyst wall, immunostained with an antibody to calcitonin gene-related peptide (CGRP, a marker for sensory C-fibers). Note CGRP-positive fibers invading the endothelium (layer adjacent to the lumen). The fourth photo is another enlargement of the cyst wall, immunostained with an antibody to vesicular monoamine transporter 2 (VMAT2, a marker for sympathetic fibers). Note VMAT2-positive fibers in the serosal, myometrial, and endothelial layers (Figure components were modified and adapted from Giamberardino et al. 2002 (*inset* in **a**), and Berkley et al. 2005 (diagram in **a** and **b**).) Finally, note that in the shamENDO surgery, only fat obtained from the uterine hysterectomy is transplanted, and no cysts form

underlying mechanisms. A more comprehensive review of this topic can be found in Stratton and Berkley (2011).

The sentinel pain symptom in endometriosis is severe dysmenorrhea (pain with menstruation), but dyspareunia (coital pain, vaginal hyperalgesia) and nonmenstrual, chronic pelvic-abdominal muscle pain also occur. In addition, painful endometriosis frequently co-occurs with other painful disorders, such as interstitial cystitis/painful bladder syndrome, irritable bowel syndrome, temporomandibular joint disorder,

Pain Mechanisms in Endometriosis, Fig. 2 *Influence of the estrous cycle on vaginal nociception in rats after ENDO or shamENDO surgery.* Parts (**a**) and (**b**) ENDO surgery (**a**), but not shamENDO surgery (**b**), induces significant increases in vaginal nociception that vary with the estrous cycle in parallel with estrous changes in plasma estradiol levels. Estrous changes in estradiol levels are shown in part (**c**) (Adapted from Smith et al. (1975)). The data were adapted from the study reported in Cason et al. (2003) in which the probability of behavioral escape responses to different volumes of vaginal distention was assessed (see pictures in **d**) before and > 1 month after ENDO or shamENDO surgery in different estrous stages during the time of day illustrated by the bars in (**c**). Part (**d**) shows the behavioral setup. **D1** shows the rat in the testing box. **D2** shows the rat "escaping" a vaginal distention stimulus by placing its head in the white tube (to break a light beam). **D3** shows the stimulus instrument: a distendable balloon that was inserted into the rat's vaginal canal ((**d**) was modified and adapted from Temple et al. (1999))

migraine headaches, other types of headache, and more. Indeed, up to one third of women with severe dysmenorrhea/chronic pelvic pain who undergo surgery are found to have associated endometriosis (Howard 1993). Endometriosis is also considered an "estrogen-dependent" condition, mainly because its pain symptoms and the lesions can be reduced by hormonal treatments that produce a hypoestrogenic state.

Until recently, most efforts at understanding endometriosis-associated pain were focused on the ectopic growths (variously called implants,

lesions, and cysts). This focus continued despite the recognition that there is a poor correlation between the amount and location of the lesions and the severity and location of the pain (Fauconnier and Chapron 2005). When grossly examined, however, endometriosis is subdivided into three subtypes: superficial peritoneal, deeply infiltrating (DIE), and ovarian cystic (endometrioma; OEA). In the 1990s, clinicians began to notice that certain subtypes of lesions had different likelihoods of being associated with severe pain (Vercellini et al. 1996). Thus, OEA

Pain Mechanisms in Endometriosis, Fig. 3 *Effect of ENDO transplant surgery on peripheral afferent background activity and in spinal cord.* (**a**) is an example of a recording of action potentials of a single afferent fiber in a T10 dorsal root. This fiber was activated by electrical stimulation of the splanchnic nerve (see diagram in (**b**)). Such recordings were made in three surgical groups > 8 weeks after the surgery or colony housing: (1) ENDO, (2) shamENDO, or (3) Naïve (no surgery). After such single fibers were identified, the stimulation was stopped and the activity of the fiber recorded for at least 30 min to establish an activity rate (spikes/s), as shown by the two examples in the *inset* in (**c**), one from an ENDO and, the other from a Naïve rat. (**c**) Is a graph of the results in the three groups. Note that background activity is significantly greater in the ENDO rats than in the other two groups, which indicates peripheral sensitization of these fibers in ENDO rats. (**d**) provides a graph that compares the number of c-Fos immunopositive fibers in different parts of the dorsal horn of the L6/S1 segment of spinal cord in ENDO versus shamENDO rats, both groups > 8 weeks post-transplant. Note the increase in "background" c-Fos labeling in the ENDO relative to shamENDO rats (#, $p < 0.05$). This finding indicates that the ENDO surgery produces central consequences (i.e., central sensitization) in segments remote from the T9-11 segments that receive input from the cysts (Fig. 1)

lesions were least likely, and DIE lesions were most likely to be associated with pain. Soon afterward, it was noticed that the DIE lesions appeared to encroach on nerve fibers and be associated with nerve growth factor (Anaf et al. 2000, 2002).

About this same time, a rat model that had been validated mainly to study mechanisms of endometriosis-associated subfertility issues (Vernon and Wilson 1985) began to be used to study endometriosis-associated pain. This model involves autotransplanting pieces of uterine horn (ENDO) or fat (shamENDO) on abdominal mesenteric arteries. The uterine, but not the fat, transplants develop into vascularized cysts (Fig. 1a). The model was then validated for

Pain Mechanisms in Endometriosis, Fig. 4 *A diagram of how endometrial lesions in women can engage the nervous system to give rise to different types of pain associated with endometriosis and co-morbid conditions.* Part *1*: This part of the diagram depicts the usual laparoscopic view of pelvic organs (seen by inserting the laparoscope at the umbilicus looking towards the reproductive organs) in which a deeply infiltrating lesion on the left uterosacral ligament is expanded in the inset. Both peptidergic sensory (*arrows* pointing rostrally) and sympathetic nerve fibers (*arrows* pointing caudally) sprout axon branches (*red dashed lines*) from nerve fibers that innervate nearby blood vessels to innervate this lesion. Sensory fibers that sprouted new axons become sensitized

endometriosis-associated pain and is association with estradiol. Thus, ENDO, but not shamENDO, rats develop both vaginal hyperalgesia (Cason et al. 2003) and referred abdominal hyperalgesia (Nagabukuro and Berkley 2007) as well as comorbid symptoms (bladder hyperreflexia) (Morrison et al. 2006) and increased pain from a kidney stone (Giamberardino et al. 2002), all of which can occur in women with endometriosis (Stratton and Berkley 2011). Furthermore, the severity of the vaginal hyperalgesia changes during the rat's ovarian cycle (Cason et al. 2003), so that it is greatest in "proestrus" when estradiol levels are highest and lowest in "estrus" when estradiol levels are lowest (Fig. 2a).

It was soon found using this model that, along with becoming vascularized, the cysts developed their own nerve supply. This supply comprised branches of both sensory and sympathetic nerve fibers that were derived (i.e., had "sprouted") from existing nerves (splanchnic) that supplied structures near the cysts (Berkley et al. 2004). This sprouting created a mechanism in which

the central nervous system communicated directly via a two-way connection with the cysts (Fig. 1b). Such sprouting of sensory and sympathetic fibers into the cysts was soon observed in samples of endometrial lesions harvested from women (Berkley et al. 2005).

This discovery led to the obvious hypothesis that this innervation contributes to endometriosis-associated pain. In the rat model, support for this hypothesis comes from findings that there are ovarian hormonal influences on cyst innervation that parallel changes in ENDO-induced vaginal hyperalgesia. Thus, there are proestrous-to-estrous decreases in the density of sympathetic fibers and of growth factors in the cysts that can influence activity in the sensory fibers (Zhang et al. 2008). These decreases follow the proestrus-to-estrus decreases in hyperalgesia (Cason et al. 2003). Additional support comes from data showing that this sprouting leads to sensitization of the sensory afferents (Fig. 3a–c). In concordance, it was further found that these peripheral influences were

---

Pain Mechanisms in Endometriosis, Fig. 4 (continued) (*asterisk*). The extent of sensitization is dynamically-modulated by estradiol and sympathetic-sensory coupling. Part 2: Two-way connection between innervated lesions and spinal cord is concentrated within sacral segments of the pelvic region. Sensitized peripheral nerve fibers, in turn, sensitize spinal sacral segment neurons. This "central sensitization," shown by the *asterisk* in the sacral segment (see also Fig. 3d) can become independent of and is modulated differently from peripheral sensitization (see Fig. 3c). Part 3: Although input from peripheral afferent fibers to the spinal cord via their dorsal roots is concentrated in the segment associated with the body part the fibers innervate (sacral segments), branches of the fibers extend to other segments (*dashed lines* pointing rostrally). Normally, these dorsal root branches have minimal impact on neurons in other segments unless the fibers become sensitized. If, however, the fibers become sensitized, they can in turn sensitize neurons in the other segments. Such remote actions are depicted by *dashed* branches into the lumbar, thoracic and cervical spinal cord dorsal horn and the asterisks at those levels. Part 4: Normally, multiple intersegmental spinal connections exist to coordinate healthy bodily functions via excitatory and inhibitory synaptic connections, shown by *double-arrowed lines* between segments. This intersegmental communication can influence how central sensitization

modifies how neurons in remote segments process nociceptive and non-nociceptive sensory information ("remote central sensitization"), shown as asterisks in the lumbar, thoracic, and cervical segments. Together, actions in *Parts 3* and *4* can lead to increased nociception not only at sacral entry segments but also in any other segment. *Part 5:* Multiple connections exist that ascend from every level of the spinal cord to the brain (shown by *lines* projecting rostrally) and descend from the brain to the spinal cord (shown by *lines* projecting caudally). Thus, in health, input from the spinal cord engages neurons throughout the brain that themselves are interconnected via complex ascending and descending inhibitory/excitatory synapses. Input from sensitized spinal neurons can affect activity throughout the neuroaxis, altering normal processing of nociceptive and non-nociceptive information. Some brain regions that can be influenced are depicted by *asterisks*. Although asterisks are shown on the medial surface of the cortex, some influenced areas extend to parts of the lateral prefrontal, frontal, parietal lobes and within the temporal lobe (*dotted ellipses*). These influences can become independent of peripheral sensitization associated with the lesions' innervation (Part 1). Such actions provide mechanisms for different types of endometriosis-associated and co-morbid pain, not only in the pelvis, but also elsewhere in the body (This figure is reprinted, with permission, from Stratton and Berkley 2011) accompanied by effects in the CNS, specifically central sensitization in the L6/S1 segment of the spinal cord, which is associated with processing of information from the vaginal canal. This sensitization was evidenced by an increased level of background activity of neurons in the dorsal horn of this segment in ENDO rats compared with shamENDO rats (Fig. 3d). Together, findings from the rat model suggest that sensory and sympathetic innervation of the ectopic growths contributes to pain in endometriosis, likely via the effects of this new innervation on central nervous system functioning, specifically central sensitization. Despite these advances, however, it remains poorly understood what kind of molecular signals are released by developing cysts into the surrounding environment that attract sensory and sympathetic fibers. Even less is known about potential molecular changes that sensory and sympathetic neurons undergo during cyst innervation and establishment of hyperalgesia. Improving our understanding of these signals and neuronal mechanisms that respond and process them could be the key for clues for the development of improved preventive measures for endometriosis-associated pain.

As reviewed in detail (Stratton and Berkley 2011; Vincent 2011), approaches to improving our understanding of the mechanisms of pain in women with endometriosis have only very recently begun to change. The main change has been to alter the primary focus away from the ectopic growths as pain generators toward the pain itself. In doing so, an important factor that needs to be considered is that chronic pelvic pain, including that experienced by women with severe dysmenorrhea, is very common (Berkley and McAllister 2011) and can frequently occur in many women without endometriosis. Thus, it is important to recognize that mechanisms underlying chronic pelvic pain with and without endometriosis may share common features. Thus, one way to frame the question is to ask how mechanisms of chronic pelvic pain might be influenced by aspects of the ectopic growths.

One likely influence, prompted by findings from the rat model, is the sensory and sympathetic innervation of the ectopic growths.

Thus, whether or not a woman's ectopic growths are innervated and how her growths are innervated (i.e., what types of fibers) could be part of the explanation for the "enigma" of endometriosis, i.e., the wide array in women with endometriosis of different symptoms – ranging from none (i.e., "silent" endometriosis) to all-encompassing chronic pains – that do not correlate with the ectopic growths. A few recent clinical studies, therefore, involved assessments in women of the relationship between the innervation of their growths and the severity of their pain (Mechsner et al. 2009; Anaf et al. 2011; Wang et al. 2010). These few results so far support the relationship, leading to suggestions for new treatment approaches (Asante and Taylor 2011). Together, however, the change in focus to pain has moved the emphasis even further toward the real generator of the women's pains, which is always the central nervous system (Fig. 4). This emphasis, in turn, strongly encourages a multidisciplinary approach toward alleviation of pain in women with endometriosis (Giudice 2010; Stratton and Berkley 2011).

References

Anaf, V., Simon, P., El Nakadi, I., Fayt, I., Buxant, F., Simonart, T., Peny, M. O., & Noel, J. C. (2000). Relationship between endometriotic foci and nerves in rectovaginal endometriotic nodules. *Human Reproduction, 15*, 1744–1750.

Anaf, V., Simon, P., El Nakadi, I., Fayt, I., Simonart, T., Buxant, F., & Noel, J. C. (2002). Hyperalgesia, nerve infiltration and nerve growth factor expression in deep adenomyotic nodules, peritoneal and ovarian endometriosis. *Human Reproduction, 17*, 1895–1900.

Anaf, V., El Nakadi, I., De Moor, V., Chapron, C., Pistofidis, G., & Noel, J. C. (2011). Increased nerve density in deep infiltrating endometriotic nodules. *Gynecologic and Obstetric Investigation, 71*, 112–117.

Asante, A., & Taylor, R. N. (2011). Endometriosis: The role of neuroangiogenesis. *Annual Review of Physiology, 73*, 163–182.

Berkley, K. J., & McAllister, S. L. (2011). Don't dismiss dysmenorrhea! *Pain, 152*, 1940–1941.

Berkley, K. J., Dmitrieva, N., Curtis, K. S., & Papka, R. E. (2004). Innervation of ectopic endometrium in a rat model of endometriosis. *Proceedings of the National Academy of Sciences of United States of America, 101*, 11094–11098.

Berkley, K. J., Rapkin, A. J., & Papka, R. E. (2005). The pains of endometriosis. *Science, 10*(308), 1587–1589.

Cason, A., Samuelsen, C., & Berkley, K. J. (2003). Estrous changes in vaginal nociception in a rat model of endometriosis. *Hormones and Behavior, 44*, 123–131.

Fauconnier, A., & Chapron, C. (2005). Endometriosis and pelvic pain: Epidemiological evidence of the relationship and implications. *Human Reproduction Update, 11*, 595–606.

Giamberardino, M. A., Berkley, K. J., Affaitati, G., Lerza, R., Centurione, L., Lapenna, D., & Vecchiet, L. (2002). Influence of endometriosis on pain behaviors and muscle hyperalgesia induced by ureteral calculosis in female rats. *Pain, 95*, 247–257.

Giudice, L. C. (2010). Clinical practice. Endometriosis. *The New England Journal of Medicine, 362*, 2389–2398.

Howard, F. M. (1993). The role of laparoscopy in chronic pelvic pain: Promise and pitfalls. *Obstetrical and Gynecological Survey, 48*, 357–387.

Mechsner, S., Kaiser, A., Kopf, A., Gericke, C., Ebert, A., & Bartley, J. (2009). A pilot study to evaluate the clinical relevance of endometriosis-associated nerve fibers in peritoneal endometriotic lesions. *Fertility and Sterility, 92*, 1856–1861.

Morrison, T. C., Dmitrieva, N., Winnard, K. P., & Berkley, K. J. (2006). Opposing viscerovisceral effects of surgically induced endometriosis and a control abdominal surgery on the rat bladder. *Fertility and Sterility, 86*, 1067–1073.

Nagabukuro, H., & Berkley, K. J. (2007). Influence of endometriosis on visceromotor and cardiovascular responses induced by vaginal distention in the rat. *Pain, 132*(Suppl 1), S96–103.

Smith, M. S., Freeman, M. E., & Neill, J. D. (1975). The control of progesterone secretion during the estrous cycle and early pseudopregnancy in the rat: prolactin, gonadotropin and steroid levels associated with rescue of the corpus luteum of pseudopregnancy. *Endocrinology, 1975*(96), 219–226.

Stratton, P., & Berkley, K. J. (2011). Chronic pelvic pain and endometriosis: Translational evidence of the relationship and implications. *Human Reproduction Update, 17*, 327–346.

Temple, J., Bradshaw, H. B., Wood, E., & Berkley, K. J. (1999). Effects of hypogastric neurectomy on escape responses to uterine distention in the rat. *Pain, 6*, S13–20.

Vercellini, P., Trespidi, L., De Giorgi, O., Cortesi, I., Parazzini, F., & Crosignani, P. G. (1996). Endometriosis and pelvic pain: Relation to disease stage and localization. *Fertility and Sterility, 65*, 299–304.

Vernon, M. W., & Wilson, E. A. (1985). Studies on the surgical induction of endometriosis in the rat. *Fertility and Sterility, 44*, 684–694.

Vincent, K. (2011). Pelvic pain in women: Clinical and scientific aspects. *Current Opinion in Supportive and Palliative Care, 5*, 143–149.

Wang, Y. Y., Leng, J. H., Shi, J. H., Li, X. Y., & Lang, J. H. (2010). Relationship between pain and nerve fibers distribution in multiple endometriosis lesions. *Zhonghua Fu Chan Ke Za Zhi, 45*, 260–263.

Zhang, G., Dmitrieva, N., Liu, Y., McGinty, K. A., & Berkley, K. J. (2008). Endometriosis as a neurovascular condition: Estrous variations in innervation, vascularization, and growth factor content of ectopic endometrial cysts in the rat. *American Journal of Physiology – Regulatory, Integrative and Comparative Physiology, 294*, R162–171.

Pain Memory

Stephen Morley
Academic Unit of Psychiatry and Behavioral Sciences, University of Leeds, Leeds, UK

Synonyms

Recall; Recognition; Recollection

Definition

Human memory is an extensive and complex system. Memories are influenced by a variety of processes occurring at the time of encoding, the period of storage, and the context of retrieval. Many memories are consciously accessible through language, and it is accepted that memories are reconstructed and influenced by the characteristics of the to-be-remembered material and the conditions at the time of recall. There is extensive evidence for the influence of implicit memories (material of which the person is unaware) on behavior and experience.

Characteristics

Memory for pain has not been extensively or systematically studied, and it is only recently that studies of memory for pain have begun to appear. Many studies have been stimulated by important practical questions, concerning the ► accuracy and ► reliability of retrospective clinical judgements of pain and improvement as a result of treatment but have lacked a strong theoretical rationale. Other relevant questions concern the influence of memory for pain on health behavior, for example, willingness to attend dental investigations or other potentially painful medical screening or diagnostic procedures. This entry considers factors that influence memory for discrete events in which pain is a defining characteristic, memory of persistent (chronic) pain, and memories that are characterized by the somatosensory reexperiencing of pain.

Contrary to earlier speculation, there is good evidence that people have ready access to ▶ autobiographical memories in which the experience of pain is a central feature – an *event memory*, for example, Morley (1993). Studies requesting participants to retrieve vivid memories of events report that up to 20 % of such memories concern accidents or personal injury. When vivid memories are not specified, between 6 % and 12 % of memories retrieved to neutral (non-pain-related) cues are of painful experiences. The presentation of pain-related cues, for example, body parts, enhances recall of pain-related events by a factor of 2–3. Available data indicate that retrieval of pain-related memories in these studies is not influenced by the presence of current pain. Participants recall pain events ranging from trivial injury to more major trauma, for example, road traffic accidents. The accounts are often rich in information concerning the contextual setting in which the event occurred, and participants are frequently able to make judgments about the sensory-intensity and affective dimensions of the pain *experience memory*. In marked contrast to this, participants in these studies do not typically report the subjective recollection of the event, that is, they do not reexperience the sensory and affective qualities of the original event – a *somatosensory memory*. The biopsychological advantages of retrieving the event and its context but not the experience are readily apparent.

Due to the retrospective nature of most autobiographical studies, that is, there are no data on the original experience, it is not possible to determine which aspects of the original event

Pain Memory, Fig. 1 Time course of two pain events – plotting pain intensity against time. See the text for an explanation

influence the recalled experience of pain. Where these data are available, it has been possible to begin to answer this question. In Fig. 1, curve A shows an idealized time course of a single event in which pain is experienced. There are three general models of how this experience may be processed and encapsulated in memory (Fredrickson and Kahneman 1993). The additive model proposes that the intensity of each moment of pain is summated: extending the experience with a series of low-intensity moments will therefore increase the total painfulness and result in the recall of a greater pain experience. The simple averaging model averages the moments of pain giving each moment unit weight. This model predicts that the extension of the pain experience by a series of low-intensity moments will reduce the average pain experienced and predicts that recalled pain will be reduced. Under this model, a short high-intensity pain is predicted to be recalled as less aversive than a longer low-intensity pain. A weighted averaging model allocates weights to different moments (the sum of the weights is constrained to 1). Empirical studies indicate that the peak and end of the experience are strongly predictive of recalled experience (Redelmeier and Kahneman 1996). The model, therefore, weights these two features of the experience and assigns zero, or very small weight, to other moments. It implies the phenomenon of *duration neglect*, that is, that the duration of the painful event is relatively unimportant in determining recalled experience. Thus, in Fig. 1, the experience represented by curve B will be recalled as more painful than that represented by curve A, although B represents less total pain. The model predicts that interventions to reduce the amount of recalled pain and suffering should be targeted at reducing the peak experience and ensuring that the end of the event is characterized by a slow decline, rather than an abrupt cessation of pain and rather than by a premature attempt to eliminate pain completely. Applications of this model are pertinent where painful procedures cannot be avoided. Managing the individual's pain experience to conform to the peak-end requirements is predicted to reduce recalled suffering and to enhance later compliance to similar painful procedures. For example, Redelmeier et al. (2003) showed that extending the duration of colonoscopy at low pain intensity reduced the patients recalled less total pain. There is evidence that chronic pain patients also discount pain-free periods when asked to recall their usual level of pain over the past week. As a consequence, the recalled usual pain overestimates the average experienced pain (Stone et al. 2004), indicating that retrospective estimates of pain are poorly suited to evaluating treatments where intermittent pain relief might be expected.

A person's expectation of a painful event is also consistently associated with variation in recall accuracy. Evidence from clinical observational studies indicates an individual's expectation of pain experience is frequently more strongly correlated with recalled pain, than is the actual experience. In clinical studies this relationship may be moderated by the individual's characteristics. For example, the presence of high anxiety in the relationship between expectation and recall is marked, but low anxiety attenuates or eliminates the relationship (Erskine et al. 1990). Expectation has also been established as a significant influence on recall in experimental studies. Price et al. (1999) elegantly demonstrated that placebo effects, based on remembered pain, were significantly (3–4 times in this study) greater than the concurrently estimated effect, and the recalled placebo effects were strongly related to an individual's expected pain levels.

The person's state, both at the time of the to-be-remembered experience and at the time of recall, has an influence on recalled pain. Several studies of chronic pain patients report that at the time of recall, if patients' pain levels are lower than at the period about which they are asked to make a memory judgement, then the recalled pain level is less than that reported. Similarly, a relatively raised level of pain at the time of recall is associated with an overestimation of the original pain. Eich et al. (1985) called this an assimilation effect. Experimental studies in which pain levels at the time of recall have been manipulated confirm the effect and also indicate that the bias is present when other concurrent aspects of the pain experience are assessed, such as affect and drug use. In chronic pain patients, the relative importance of pain at the time of recall appears to increase, as the time since the original event increases (Haas et al. 2002), that is, the original state becomes less important. For acute pain, however, there is evidence that negative emotions at the time of the to-be-remembered event are the stronger mediator of recall as time increases (Gedney and Logan 2004). These complex findings require further theoretical development but may be partly understood in the context of how an individual combines their implicit theory of events (expectations of pain experience over time) and information about their current state, relative to their implicit theory to reconstruct the original event (Ross 1989).

Somatosensory memory has not been extensively investigated, but there are sufficient observations to confirm the existence of such memories. Vivid recollection of previously experienced pain, complete with the accompanying affect, can be evoked by direct brain stimulation (Lenz et al. 1997). Somatosensory memories have also been reported by patients with phantom limbs (Katz and Melzack 1990) and in PTSD sufferers (Salomons et al. 2003). Under normal circumstances, the somatosensory memory would appear to be "inhibited" and not accessible via verbal recall (cf. Brewin et al. 1996). Compromised neurobiological integrity or altered states of consciousness appear to be prerequisites for the experience of somatosensory memories. The phenomenon is not well understood.

References

Brewin, C. R., Dalgleish, T., & Joseph, S. (1996). A dual representation theory of posttraumatic stress disorder. *Psychological Review, 103*, 670–686.

Eich, E., Reeves, J. L., Jaeger, B., & Graff-Radford, S. B. (1985). Memory for pain: Relation between past and present pain intensity. *Pain, 23*, 375–380.

Erskine, A., Morley, S., & Pearce, S. (1990). Memory for pain: A review. *Pain, 41*, 255–265.

Fredrickson, B. L., & Kahneman, D. (1993). Duration neglect in retrospective evaluations of affective episodes. *Journal of Personality and Social Psychology, 65*, 45–55.

Gedney, J. J., & Logan, H. (2004). Memory for stress associated acute pain. *The Journal of Pain, 5*, 83–91.

Haas, M., Nyiendo, J., & Aickin, M. (2002). One-year trend in pain and disability relief recall in acute and chronic ambulatory low back pain patients. *Pain, 95*, 83–91.

Katz, J., & Melzack, R. (1990). Pain 'memories' in phantom limbs: Review and clinical observations. *Pain, 43*, 319–336.

Lenz, F. A., Gracely, R. H., Zirh, A. T., Romanoski, A. J., & Dougherty, P. M. (1997). The sensory-limbic model of pain memory. *Pain Forum, 6*, 22–31.

Morley, S. (1993). Vivid memory for 'Everyday' pains. *Pain, 55*, 55–62.

Price, D. D., Milling, L. S., Kirsch, I., Duff, A., Montgomery, G. H., & Nicholls, S. S. (1999). An analysis of factors that contribute to the magnitude of placebo analgesia in an experimental paradigm. *Pain, 83*, 147–156.

Redelmeier, D. A., & Kahneman, D. (1996). Patients' memories of painful medical treatments: Real-time and retrospective evaluations of Two minimally invasive procedures. *Pain, 66*, 3–8.

Redelmeier, D. A., Katz, J., & Kahneman, D. (2003). Memories of colonoscopy: A randomized trial. *Pain, 104*, 187–194.

Ross, M. (1989). Relation of implicit theories to the construction of personal histories. *Psychological Review, 96*, 341–357.

Salomons, T. V., Osterman, J. E., Gagliese, L., & Katz, J. (2003). Pain flashbacks in posttraumatic stress disorder. *The Clinical Journal of Pain, 20*, 83–87.

Stone, A. A., Broderick, J. E., Shiffman, S. S., & Schwartz, J. E. (2004). Understanding recall of weekly pain from a momentary assessment perspective: Absolute agreement, between- and within-person consistency, and judged change in weekly pain. *Pain, 107*, 61–69.

Pain Modulation

▶ Descending Modulation of Nociceptive Processing

Pain Modulatory Circuitry and Endogenous Pain Control

▶ Descending Circuitry, Molecular Mechanisms of Activity-Dependent Plasticity

Pain Modulatory Pathways

▶ Descending Circuitry, Opioids

Pain Nurse(s)

Definition

A pain nurse is a nurse specially trained in the knowledge and skills required to contribute to the specific care of patients with pain. Those skills include, but are not necessarily limited to: participation in a pain team; providing nursing care to patients with pain; explaining to ward nurses any special care that a patient with pain might require, and coordinating that care; providing instruction to patients in methods for the relief of pain; administering prescribed interventions for pain; assessing the response of patients to interventions; recognising any complications of treatment, and alerting the pain team to those complications.

Cross-References

▶ Acute Pain Team
▶ Postoperative Pain, Acute Pain Team

Pain of Recent Origin

▶ Acute Pain, Subacute Pain, and Chronic Pain

Pain Ombudsman-Nurses

Definition

Pain ombudsman-nurses are specially trained to make sure that patients receive adequate pain relief in the ward they are responsible for.

Cross-References

▶ Acute Pain Team
▶ Postoperative Pain, Acute Pain Team

Pain Paroxysms

Ron Amir[1], Marshall Devor[2] and
Z. Harry Rappaport[3]
[1]Institute of Life Sciences, The Hebrew
University of Jerusalem, Jerusalem, Israel
[2]Department of Cell & Developmental Biology,
Institute of Life Sciences and Center for Research
on Pain, The Hebrew University of Jerusalem,
Jerusalem, Givat Ram, Israel
[3]Department of Neurosurgery, Rabin Medical
Center Tel-Aviv University, Petah Tiqva, Israel

Synonyms

Fits of pain; Lancinating pain; Lightning pain; Pain attacks

Definition

A pain paroxysm is a well-localized excruciating pain that has a sudden onset, is usually brief and can occur spontaneously, with no obvious precipitating event, or in response to a trigger stimulus. The term "paroxysm" derives from the Greek – *paroxunein*: "to irritate, provoke, or excite (literally to sharpen excessively)".

Characteristics

Clinical Manifestation

Pain paroxysms are part of the clinical picture in a variety of neurological conditions, but especially in chronic painful neuropathies. ▶ Trigeminal neuralgia (TN) is the prototypical neuropathy that features pain paroxysms. Extremely intense, rapid onset, brief, well-localized attacks of pain in the face typically cause the patient to wince, and hence an alternative name for the condition, tic douloureux (French for "painful twitching (or wincing)"). Paroxysmal pain is also reported in other cranial nerve neuropathies such as glossopharyngeal neuralgia, and is not uncommon in chronic neuropathies of segmental nerves. These include post-traumatic neuropathies, postherpetic neuralgia, diabetic neuropathy, and a variety of painful mono- and polyneuropathies associated with metabolic disorders, toxins, radiation, and genetic mutations (Otto et al. 2003; Scadding & Koltzenburg 2006; Zakrzewska and Lopez 2006). Painful paroxysms, when they occur, are a prime diagnostic indicator of neuropathic pain (Bouhassira et al. 2005), although painful neuropathies may present as a steady intense pain without paroxysms, and painful paroxysms can be present in conditions that are not considered neuropathic (e.g., gout).

Pain paroxysms also characterize specific headache types of still unknown origin, including paroxysmal hemicrania, cluster headache, SUNCT (short-lasting, unilateral, neuralgiform headache attacks with conjunctival injection and tearing), and idiopathic stabbing headache (Pareja et al. 1996; Sharav and Benouliel 2008). In these conditions, the paroxysms tend to be of slower onset, and of longer duration, than in neuropathy. Finally, acute neuropathic pain can be paroxysmal in nature if a sudden, brief, pain-provoking stimulus is applied such as a sharp rap over the ulnar nerve at the elbow.

Sensory Characteristics

There is no formal criterion for the suddenness of onset, intensity, or brevity of a pain that would properly be called paroxysmal. In common usage, these parameters vary among painful conditions. Descriptions such as "shooting," "stabbing," "electric-shock-like," "lancinating," and "lightning" imply a rapid rise-time of the pain, less than a second. Terms such as "boring" and "throbbing" indicate slower onset times, perhaps of several seconds. Pain duration is typically a few seconds to a few minutes, although an attack may last much longer. Furthermore, attacks may repeat at frequent intervals or appear as exacerbations on top of a steady burning or cramping pain. In TN, for example, pain attacks sometimes repeat so frequently that one attack runs into the next, with the patient having the impression of a continuous, uninterrupted pain. It is more usual, however, for there to be hours, days, or even months between attacks.

Pain paroxysms may be spontaneous or evoked. In the case of cluster headache, for example, there is no obvious trigger for individual paroxysms, although a number of environmental factors may increase the likelihood of a series (i.e., a cluster) of attacks. Likewise, in TN, lightning attacks may arise spontaneously. However, in this condition, it is usually possible to identify a "trigger point" at which a very light touch consistently evokes a painful paroxysm (Kugelberg and Lindblom 1959). The pain is usually felt at the point of stimulation, although it often radiates beyond. Frequently, the touch needs to be repeated several times in order to evoke an attack. Moreover, the pain may follow the provoking stimulus by several seconds. In patients in whom the trigger spot is in a location that is subject to repeated unintended stimulation, such as within the oral cavity during chewing or swallowing, evoked paroxysms may appear to be spontaneous.

When a painful paroxysm is evoked from a trigger point, the pain almost always outlasts the stimulus by at least a second or two (aftersensation), and fills an area larger than that stimulated. The mismatch in time and space between the trigger stimulus and the pain indicates that triggering is not simply amplification, as when light touch to tender skin evokes a sensation of pain (tactile allodynia). Indeed, in the region of the trigger point in TN, the skin tends to show somewhat *reduced* sensitivity to touch (Nurmikko 1991). Rather than amplification, paroxysms appear to reflect an intrinsic pain source that is "set off" by the trigger stimulus or that "explodes" spontaneously when an internal threshold is reached. Consistent with this concept, evoked pain paroxysms tend to be followed by a "refractory period" that lasts for several minutes after an evoked attack. During the refractory period, intentional stimulation of the trigger point fails to evoke a second attack (Kugelberg and Lindblom 1959).

Mechanism

The pathophysiological substrate(s) of pain paroxysms is not known with certainty. However, the striking characteristics of the sensation, namely, sudden, intense, brief, and localized, constrain the likely possibilities. Sensory experience in a conscious brain reflects corresponding impulse activity. Accordingly, a pain paroxysm presumably reflects a brief, intense, and rapid-onset barrage of impulse activity in a large, localized population of neurons. In the case of pain paroxysms evoked by a sudden intense stimulus applied to a nerve passing over bone, or to inflamed tissue, there is little mystery concerning the origin of the impulse barrage. The challenge is to understand spontaneous paroxysms, and painful paroxysms evoked by weak stimuli. Naturally, a discussion of this subject tends to revolve around the mechanism(s) of pain in TN.

Received wisdom holds that the intense focal activity underlying lightning pains in TN originates in the central nervous system (CNS), presumably in the trigeminal brainstem (Fromm and Sessle 1991). According to this CNS theory, a process akin to a focal epileptic seizure accounts for the paroxysmal pain attack. The classic observation that the anticonvulsants diphenylhydantoin (phenytoin, Dilantin) and carbamazepine (Tegretol) are highly effective, at least at first, at controlling pain paroxysms in TN is consistent with the analogy between pain paroxysms and epileptic seizures. However, the evidence for a primary CNS etiology does not go much beyond that. Idiopathic TN is not associated with focal lesions in the trigeminal brainstem, or with electrographic signs of seizure foci in this area. However, in a small minority of cases, demyelinating plaques of multiple sclerosis (MS) along the descending tract of the trigeminal root, and space occupying brainstem tumors in this region, constitute a likely focal pathology. Moreover, despite the effectiveness of phenytoin and carbamazepine, TN is not responsive to a variety of other anticonvulsants that do have considerable efficacy in epilepsy such as barbiturates and benzodiazepines.

The weight of evidence shifted to theories based on peripheral nervous system (PNS) pathophysiology, perhaps with a secondarily induced CNS component (▶ central sensitization), with two key discoveries (Devor et al. 2002). The first was the accumulation of data that in most cases of TN, there is a specific, focal demyelinating lesion in the trigeminal root, a few millimeters distal to its point of entry into the pons (see Fig. 1). Pathology in the trigeminal root ganglion (TRG) has also been documented. Focal demyelination of the trigeminal root is caused by compression of the root by a small impinging blood vessel (artery, or sometimes vein). Surgical separation of the offending vessel from the compressed root by ▶ microvascular decompression (MVD) frequently provides definitive, long-term relief from pain attacks. The second observation that undermined the idea that TN is due to seizure-like activity in the CNS was that phenytoin, carbamazepine, gabapentin, and other anticonvulsant drugs effective in TN reduce abnormal hyperexcitability of injured primary sensory neurons. That is, they have a "membrane-stabilizing" effect in the PNS. This PNS effect is in addition to the CNS effect, and is specific to the analgesic anticonvulsants. Membrane stabilization is typically due to pharmacological block of voltage-sensitive Na^+ channels, proteins that play a key role in neuronal excitability in both the PNS and the CNS (Catterall 1987; Amir et al. 2006). In contrast, the anticonvulsant drugs that are not effective in TN, such as barbiturates, tend to act

Pain Paroxysms, Fig. 1 Fascicular demyelination in a biopsy specimen removed from a patient with trigeminal neuralgia. In contrast to healthy cranial nerve roots, large numbers of non-myelinated and demyelinated axons are in close apposition to one another without an intervening glial process (e.g., *arrows*). This anatomical arrangement is a potential substrate for synchronous recruitment among the axons, including ephaptic cross talk and crossed after discharge (Fried et al. 1993). Scalebar: 2 μm

synaptically by enhancing inhibitory (e.g., GABAergic) neurotransmission. There are no GABAergic (or other) synapses along the trigeminal root or in the TRG.

In 1994, Rappaport and Devor proposed the "TRG ▶ ignition hypothesis," a PNS-based model that integrates the tissue pathology known to be present in TN, the pathophysiology of injured hyperexcitable primary afferent neurons, and the pharmacology of peripherally acting membrane-stabilizing drugs. The ignition hypothesis proposes that pain paroxysms result from synchronous recruitment of ectopic firing in a large population of hyperexcitable TRG neurons (or their axons) that innervate the territory in which the pain is felt. The ectopic hyperexcitability of these neurons is due to a trigeminal root neuropathy caused by vascular compression of the root, or less frequently to other causes such as MS. Synchronous recruitment results from non-synaptic coupling mechanisms: chemically mediated cross excitation and/or ephaptic cross talk. In cross excitation, neurotransmitter molecules, and probably also K⁺ ions, are released into the extracellular space from active neurons. These diffuse toward neighboring neurons and excite them (Amir and Devor 1996). The chemical mediator(s) are not known with confidence, but ATP is a likely candidate (Gu et al. 2010). Satellite glial cells in the ganglion may also play a role (Suadicani et al. 2010). Non-synaptic chemically mediated coupling occurs in DRGs even in the absence of nerve injury. However, impulse activity in neighboring afferents is recruited only in the presence of neuropathy. Neuropathy enhances electrogenesis in injured afferents and, in particular, leads to the emergence of subthreshold oscillations (Amir et al. 1999, 2002; Pedroarena et al. 1999; Gu et al. 2010). In ephaptic cross talk, the coupling is not chemical, but electrical. Excitatory electrical current passes directly from the active neuron to closely apposed passive neighbors. This does not occur in sensory ganglia, but it does occur in injured nerve trunks (Seltzer and Devor 1979).

According to the ignition hypothesis, paroxysms begin with discharge in a small cluster of trigeminal neurons, presumably low threshold afferents, that begin to fire in response to cutaneous trigger-point stimulation, or perhaps spontaneously. Cell-to-cell coupling in the injured trigeminal root or ganglion (chemical and/or electrical) then "ignites" activity in passive neighboring neurons, including nociceptors (Amir and Devor 1996, 2000). This augmented activity ignites additional passive neighbors, and these ignite yet others. The resulting positive-feedback chain-reaction builds up rapidly into an intense, explosive crescendo. This is felt as a pain paroxysm. Since afferent neurons of all types become active simultaneously, something that otherwise occurs only with electrical stimulation, the felt sensation is electric-shock-like. The ectopic impulse activity underlying the pain often persists for some time after the end of the stimulus ("afterdischarge"; Lisney and Devor 1987), giving rise to after-sensation (Figs. 2 and 3). After a few seconds of massive firing, activity-evoked after-suppression develops

Pain Paroxysms, Fig. 2 Pathophysiological afterdischarge following mechanical stimulation. In this experiment, the sciatic nerve of a rat was injured, and some days later, electrophysiological recordings of impulse activity arising ectopically from the DRG were made. A weak, momentary mechanical stimulus to the surface of the exposed DRG (<1 s, *arrow*) yielded 30 s of accelerated impulse discharge

(Amir and Devor 1997). This accounts for the "refractory period" observed clinically after paroxysmal attacks in TN.

In a tour-de-force case study, Burchiel and Baumann (2004) managed to obtain specific evidence supporting this process. Specifically, during the course of intraoperative recordings from the TRG in a patient with TN, they showed that the type of light and brief stimulation of the trigger point that typically evoked pain paroxysms initiated a few seconds of abnormal, intense burst discharge. This is just the type of activity that is expected to underlie a paroxysm.

Like CNS seizure activity, the ignition mechanism is expected to be sensitive to membrane-stabilizing anticonvulsants. Drug action, however, is supposed to be in the PNS rather than the CNS. Synaptic-acting anticonvulsants are ineffective because the nerve and ganglion contain no synapses. However, voltage-sensitive Na⁺-channel blockers effectively suppress the oscillatory mechanism and membrane hyperexcitability (Amir et al. 1999; Pedroarena et al. 1999; Xing et al. 2001; Yang et al. 2005). This is the presumed basis for their therapeutic effectiveness in TN. Finally, conduction block in

Pain Paroxysms, Fig. 3 Pathophysiological can also be generated in the DRG by gentle rubbing of the skin. In this experiment, the sciatic nerve of a rat was only partly injured, sparing many axons that innervate hindpaw skin. Electrophysiological recordings of impulse activity originating ectopically in the DRG were made a few days later. Rubbing the paw skin activated somata of DRG neurons with spared paw afferents. This cross-excited neighboring hyperexcitable DRG neurons whose axons had been cut and increased their spontaneous firing rate (see Devor and Wall 1990)

some demyelinated axons may account for the *reduced* sensitivity of the region of the trigger point to light touch (Nurmikko 1991).

Segmental Paroxysms and Hyperpathia

It is important to note that although the ignition hypothesis was developed with TN in mind, it applies equally to other cranial nerve neuralgias, and also to pain paroxysms felt at spinal segmental levels (Devor 2006). Particularly at spinal levels, paroxysmal afterdischarge bursts may be triggered by factors other than stimulation of sensory endings in the skin, such as mechanical forces applied to the DRG or to sites of axonal injury (Figs. 2 and 3). Examples are nerve traction or compression during straight leg raising, as articulating vertebrae move against one another in the presence of tissue swelling, intervertebral disk herniation, or the presence of osteophytes.

Afterdischarge, cross excitation, and ephaptic cross talk are expected to give rise to additional symptoms in neuropathy, not just pain. For example, they are expected to cause non-painful sensations to persist after the end of the triggering stimulus (after-sensation) and to result in the spread of sensation along a limb from the point of stimulation. Cross excitation is also a potential basis for the spread of sensation along the limb (perhaps including "shooting" sensations). Both after-sensation and the spread of sensation are typical "hyperpathic" sensory abnormalities, known to occur in neuropathy (Noordenbos 1959).

Treatment

Pain paroxysms in TN respond to specific anti-convulsant medications, notably carbamazepine and gabapentin, and to a lesser extent phenytoin. Tricyclic antidepressants and systemically applied local anesthetic agents, including the antiarrhythmic mexiletine, may also be effective. These same drugs reduce pain paroxysms in a wide range of other neuropathic pain conditions (Scadding and Koltzenburg 2006; Zakrzewska and Lopez 2006), supporting the contention that all, or most, instances of lightning pain are due to the same underlying mechanism. Barbiturates and benzodiazepines, although effective anticonvulsants, do not relieve pain paroxysms. This is presumably because they act synaptically rather than as ▶ membrane stabilizers.

The effectiveness of drug therapy in TN frequently declines over time perhaps because of progression of the neuronal injury with resulting increase in ectopic neuronal hyperexcitability. In these patients, interventional approaches are used. These fall into two categories. The first involves percutaneous partial damage to the trigeminal root and ganglion using radiofrequency coagulation, retro-Gasserian glycerol injection, balloon compression, or focal irradiation (Gamma-knife). These procedures, which cause various degrees of facial hypoesthesia, probably work by reducing the critical mass of electrical activity available for pain ignition. The second approach is to reverse the pathology that caused abnormal hyperexcitability and neuronal coupling in the first place. In TN, this can be accomplished with MVD surgery. This eliminates the source of continuous nerve irritation and permits remyelination.

The corresponding approaches are also available for the treatment of pain paroxysms in segmental neuropathic pain. Membrane stabilizers are the first line of medical treatment. Interventional approaches include radiofrequency partial thermocoagulation of paraspinal nerves or DRGs, mobilization of neuromas to sites where they are less likely to be triggered by mechanical forces, and surgical release of entrapped nerves.

Cross-References

▶ Trigeminal and Cranial Nerve Neuralgias

References

Amir, R., & Devor, M. (1996). Chemically-mediated cross-excitation in rat dorsal root ganglia. *The Journal of Neuroscience, 16*, 4733–4741.

Amir, R., & Devor, M. (1997). Spike-evoked suppression and burst patterning in dorsal root ganglion neurons of the rat. *The Journal of Physiology (London), 501*, 183–196.

Amir, R., & Devor, M. (2000). Functional cross-excitation between afferent A- and C-neurons in dorsal root ganglia. *Neuroscience, 95*, 189–195.

Amir, R., Michaelis, M., & Devor, M. (1999). Membrane potential oscillations in dorsal root ganglion neurons: Role in normal electrogenesis and in neuropathic pain. *The Journal of Neuroscience, 19*, 8589–8596.

Amir, R., Michaelis, M., & Devor, M. (2002). Burst discharge in primary sensory neurons: Triggered by subthreshold oscillations, maintained by depolarizing afterpotentials. *The Journal of Neuroscience, 22*, 1187–1198.

Amir, R., Argoff, C. E., Bennett, G. J., Cummins, T. R., Durieux, M. E., Gerner, P., Gold, M. S., Porreca, F., & Strichartz, G. R. (2006). The role of sodium channels in chronic inflammatory and neuropathic pain. *The Journal of Pain, 7*, S1–S29.

Bouhassira, D., Attal, N., Alchaar, H., Boureau, F., Brochet, B., Bruxelle, J., Cunin, G., Fermanian, J., Ginies, P., Grun-Overdyking, A., Jafari-Schluep, H., Lanteri-Minet, M., Laurent, B., Mick, G., Serrie, A., Valade, D., & Vicaut, E. (2005). Comparison of pain syndromes associated with nervous or somatic lesions and development of a new neuropathic pain diagnostic questionnaire (DN4). *Pain, 114*, 29–36.

Burchiel, K. J., & Baumann, T. K. (2004). Pathophysiology of trigeminal neuralgia: New evidence from a trigeminal ganglion intraoperative microneurographic recording. Case report. *Journal of Neurosurgery, 101*, 872–873.

Catterall, W. A. (1987). Common modes of drug action on Na + channels: Local anaesthetics, antiarrhythmics and anticonvulsants. *Trends in Pharmacological Sciences, 8*, 57–65.

Devor, M. (2006). Response of nerves to injury in relation to neuropathic pain. In S. L. McMahon & M. Koltzenburg (Eds.), *Wall and Melzack's textbook of pain* (5th ed., pp. 905–927). London: Churchill Livingstone.

Devor, M., Amir, R., & Rappaport, Z. H. (2002). Pathophysiology of trigeminal neuralgia: The ignition hypothesis. *The Clinical Journal of Pain, 18*, 4–13.

Devor, M., & Wall, P.D. (1990). Cross excitation among dorsal root ganglion neurons in nerve injured and intact rats. *Journal of Neurophysiology, 64*, 1733–1746.

Fried, K., Govrin-Lippmann, R., & Devor, M. (1993). Close apposition among neighbouring axonal endings in a neuroma. *Journal of Neurocytology, 22*, 663–681.

Fromm, G., & Sessle, B. (1991). *Trigeminal neuralgia: Current concepts regarding pathogenesis and treatment* (pp. 1–230). Boston: Butterworth-Heinemann.

Gu, Y., Chen, Y., Zhang, X., Li, G. W., Wang, C., & Huang, L. Y. (2010). Neuronal soma-satellite glial cell interactions in sensory ganglia and the participation of purinergic receptors. *Neuron Glia Biology, 6*, 53–62.

Kugelberg, E., & Lindblom, U. (1959). The mechanism of the pain in trigeminal neuralgia. *Journal of Neurology, Neurosurgery & Psychiatry, 22*, 36–43.

Lisney, S. J. W., & Devor, M. (1987). Afterdischarge and interactions among fibers in damaged peripheral nerve in the rat. *Brain Research, 415*, 122–136.

Noordenbos, W. (1959). *Pain* (p. 182). Amsterdam: Elsevier.

Nurmikko, T. J. (1991). Altered cutaneous sensation in trigeminal neuralgia. *Archives of Neurology, 48*, 523–527.

Otto, M., Bak, S., Bach, F. W., Jensen, T. S., & Sindrup, S. H. (2003). Pain phenomena and possible mechanisms in patients with painful polyneuropathy. *Pain, 101*, 187–192.

Pareja, J. A., Ruiz, J., de Isla, C., al-Sabbah, H., & Espejo, J. (1996). Idiopathic stabbing headache (jabs and jolts syndrome). *Cephalalgia, 16*, 93–96.

Pedroarena, C. M., Pose, I. E., Yamuy, M., Chase, M. H., & Morales, F. R. (1999). Oscillatory membrane potential activity in the soma of a primary afferent neuron. *Journal of Neurophysiology, 82*, 1465–1476.

Rappaport, Z. H., & Devor, M. (1994). Trigeminal neuralgia: The role of self sustaining discharge in the trigeminal ganglion. *Pain, 56*, 127–138.

Scadding, J. W., & Koltzenburg, M. (2006). Painful peripheral neuropathies. In S. B. McMahon & M. Koltzenburg (Eds.), *Wall and Melzack's textbook of pain* (5th ed., pp. 973–999). London: Churchill Livingstone.

Seltzer, Z., & Devor, M. (1979). Ephaptic transmission in chronically damaged peripheral nerves. *Neurology, 29*, 1061–1064.

Sharav, Y., & Benouliel, R. (Eds.). (2008). *Orofacial pain and headache* (p. 440). New York: Blackwell/Mosby-Elsevier/Edinburgh.

Suadicani, S. O., Cherkas, P. S., Zuckerman, J., Smith, D. N., Spray, D. C., & Hanani, M. (2010). Bidirectional calcium signaling between satellite glial cells and neurons in cultured mouse trigeminal ganglia. *Neuron Glia Biology, 6*, 43–45.

Xing, J. L., Hu, S. J., Xu, H., Han, S., & Wan, Y. H. (2001). Subthreshold membrane oscillations underlying integer multiples firing from injured sensory neurons. *Neuroreport, 12*, 1311–1313.

Yang, R. H., Xing, J. L., Duan, J. H., & Hu, S. J. (2005). Effects of gabapentin on spontaneous discharges and subthreshold membrane potential oscillation of type A neurons in injured DRG. *Pain, 116*, 187–193.

Zakrzewska, J. M., & Lopez, B. C. (2006). Trigeminal and glossopharyngeal neuralgia. In S. L. McMahon & M. Koltzenburg (Eds.), *Wall and Melzack's textbook of pain* (5th ed., pp. 1001–1010). London: Churchill Livingstone.

Pain Pathway

Definition

A conceptualization of pain mechanisms, which suggests that a dedicated interconnected set of CNS loci exist whose activation gives rise to the experience of pain; compare with dynamic ensemble.

Cross-References

▶ Gynecological Pain, Neural Mechanisms

"Pain" Pathway

▶ Spinothalamic Tract Neurons, Visceral Input

Pain Prevention

▶ Preemptive or Preventive Analgesia and Central Sensitization in Postoperative Pain

Pain Processing in the Cingulate Cortex, Behavioral Studies in Humans

Philippe Goffaux, Guillaume Léonard and Serge Marchand
Faculty of Medicine, University of Sherbrooke, Sherbrooke, QC, Canada

Definition

The cingulate cortex (CC) is a cortical structure located in the medial portion of the cerebral hemispheres immediately above the corpus callosum (see Fig. 1). The CC can be functionally divided in an anterior (ACC) and posterior part (PCC) with an intermediate or medial part (MCC) that all play specific roles in pain perception, anticipation, and reaction (Vogt et al. 2003). The function of the CC in pain has been studied in humans using different approaches, ranging from single-cell recordings during surgery to psychophysical and brain imaging studies. This key structure is also believed to play a critical role in a wide variety of regulatory human behaviors (see Table 1).

Characteristics

Structural Specialization for Pain Perception
Pain perception is supported by multiple brain regions including regions that play a major role in the sensory component of pain (intensity, localization, duration) and in the affective component of pain (fear, attention, unpleasantness, etc.). Brain regions believed to play a primary role in subsuming sensory-discriminative

Pain Processing in the Cingulate Cortex, Behavioral Studies in Humans, Fig. 1 MRI scan highlighting the anterior cingulate cortex (*ACC*), midcingulate cortex, (*MCC*) and posterior cingulate cortex (*PCC*). Also shown is a nonexhaustive subdivision of the ACC including Brodmann areas 32 (dorsal ACC), 24 (ventral ACC), and 25 (subgenual ACC)

Pain Processing in the Cingulate Cortex, Behavioral Studies in Humans, Table 1 Non-nociceptive functions of the CC

Functions	Associated subregions
Autonomic arousal, emotional processing, and social interactions	ACC (ventral pregenual and subgenual gyri and anterior dorsomedial paracingulate region)
Movement and monitoring of performance and reward	MCC (cingulate sulcus) and ACC (dorsal section)
Visuospatial orientation and memory	PCC and retrosplenial areas

dimensions of pain include the primary and secondary somatosensory cortices (SI, SII). On the other hand, the affective component of pain is thought to involve the insular and cingulate cortices (Apkarian et al. 2005). Most of the brain imaging or electrophysiological studies published to date consistently obtain increased activity in the CC following the administration of painful stimuli (Apkarian et al. 2005). However, dressing a unified portrait of the different functions subsumed by the CC has not yet been accomplished and may, in fact, prove difficult, partly because the cingulate cortex is not a homogeneous structure. From a cytoarchitectural point of view, the cingulate cortex is composed of a number of different areas, each involved in

producing different behavioral outputs, be they pain-related or not (see Vogt et al. 2005 as well as Beckmann et al. 2009). Given the results obtained when measuring the cortical activity underlying pain- and non-pain-related cingulate functions, it appears clear that different regions of the cingulate cortex are activated along with other brain structures to produce a distributed pattern of cortical activity (Garcia-Larrea et al. 2003). Nowhere is this more evident than when looking at the distributed representation underlying nociceptive processing. The matrix of cortical activity observed when noxious stimuli are given suggests that there is no central "pain structure" in the human brain but rather a pain-related network involving thalamic, somatosensory and frontolimbic structures. Furthermore, of all the different subsections of the CC, it is the ACC which most consistently coactivates when pain is experienced (Porro et al. 1998). Our review will, therefore, include the entire CC, but we will pay particular attention to the role of the ACC in nociceptive processing and pain perception.

ACC-PCC Coactivation

Using brain activity recording techniques that have a good temporal resolution (electrical brain source analysis with current reconstruction imaging), Bromm (2004) proposed that when applying a phasic nociceptive stimulus (i.e., rapid onset and short duration), the PCC is activated first and plays a role in the evaluation of sensory-discriminative aspects of the stimuli, while the ACC is activated later with a role for the motivational-affective aspect of pain. Bromm argues that the early, sensory-discriminative role played by the PCC is largely missed by most neuroimaging studies because these studies use imaging techniques that have poor temporal resolution (e.g., PET and fMRI) and so cannot dissociate early and late CC activation. This methodological constraint prevents the measurement of fast, transient PCC activity tied to the onset of nociceptive stimuli. One remaining possibility, however, is that fast, transient PCC activity is tied to the orientation of the body in response to sensory stimuli (i.e., a skeletomotor response). The extent to which PCC activity depends on stimulus strength remains to be tested.

ACC Activity: Pain Perception or Cognitive Control?

Early on, lesion studies suggested a link between ACC activity and the processing of pain affect. In particular, it appeared that cingulotomies helped reduce emotional responses to pain (decreased distress, bother, and worry) rather than the perception of pain intensity per se (Bouckoms 1994). However, even if the emotional role played by the ACC is important, recent studies have demonstrated that the ACC is more complex than previously thought. In fact, lesions of the anterior internal capsule (responsible for the transport of subcortical input to the ACC) affected both sensory and emotional aspects of pain (Talbot et al. 1995). Moreover, recent neuroimaging studies reveal that the ACC plays an important role in both the localization and intensity coding of pain (Oshiro et al. 2009). These findings argue against the idea that specific aspects of pain are tied to specific cerebral structures. In other words, a given cortical structure may be particularly sensitive in mediating a specific dimension of pain, but it is unlikely to be uniquely involved. Direct recordings of the ACC by either

microelectrodes (Hutchison et al. 1999) or subdural electrodes (Lenz et al. 1998) do however suggest a functional dissociation between the processing of nociceptive and non-nociceptive afferents. These electrode recordings show a clear implication of the ACC during, and in anticipation of, nociceptive stimulation. However, no activity is recorded following the administration of non-nociceptive stimuli. Interestingly, stimulation of these same ACC neurons fails to provoke any pain, while intense sensations of either fear or pleasure have been reported after ACC stimulation (Allman et al. 2001). It is also possible to draw a parallel between the signal obtained through modern imaging techniques and the subjective experience of pain. The results obtained by the studies that have done this appear, at first glance, to be equivocal. While some authors find that activity in the ACC covaries with pain unpleasantness (Rainville et al. 1997), others find that ACC activity is related to accompanying pain-triggered attentional/anticipatory factors (Lorenz et al. 2003). A closer look at these studies suggests, however, that their results may not be so contradictory. Studies which indicate that the ACC plays a key role in processing pain-affect primarily show activity in the ventral section of the ACC (Brodmann area 24), while those that argue that the ACC processes pain-related anticipation and attention primarily show activity in the dorsal section of the ACC (Brodmann area 32). Dissociating the neural mechanisms underlying pain-related cognitive factors from those underlying affect-related factors will constitute the next challenge in our field.

ACC and Endogenous Pain Modulation Systems

Pain is not solely the result of the activation of brain structures by nociceptive afferences but the integration of different endogenous pain modulation systems at different levels of the central nervous system. Descending inhibitory systems from brainstem structures, including the periaqueductal gray (PAG), play a major role in pain modulation (Fields and Basbaum 1999). Connectivity between the PAG and

ACC has been demonstrated, suggesting an influence between ACC and PAG (An et al. 1998; Kong et al. 2010). Recent brain imaging studies have shown that distraction- and expectation-induced pain inhibition activates both the ACC and the PAG, arguing for a possible role of the ACC in the descending modulation of nociception by the PAG (Valet et al. 2004; Wager et al. 2004).

Is ACC Activity for Pain Dissociable from ACC Activity for Negative Affect?

As already outlined in Table 1, the ACC is known to play an important role in affective processing. The colocalization of nociceptive and emotional functions in the ACC (particularly in the ventral and pregenual portions of the ACC) could suggest that this region is responsible for the processing of unpleasant/fearful states, be they triggered by noxious events or not. It is important to point out, however, that even if the processing of negative emotions and the perception of painful events depend on the activity of slightly different ACC regions (or subregions), the presence of numerous connections between ACC areas suggests that negative affective states could influence the processing of pain (Margulies et al. 2007; Beckmann et al. 2009). Interestingly, interactions between pain and emotion have recently been demonstrated by Yoshino et al. (2010) who showed that ACC activity increased under sad emotional contexts (compared to happy and neutral contexts) and that this increase was linked to higher subjective pain ratings. The results of Yoshino et al. (2010) unveil the presence of interactions between nociceptive- and emotional-processing functions of the ACC and provide the neuroanatomical basis by which negative emotions and pain relate (Garcia-Cebrian et al. 2006; Wiech and Tracey 2009).

Empathy for Pain

A growing amount of research on human empathy has been published recently. Several of these studies have focused on the neural signature associated with empathy for pain. Using various functional imaging techniques, these studies confirm the existence of a core neuronal network,

activated when rehearsing the pain of others. This network includes, as one of its key players, the ACC (see Lamm et al. 2011). In fact, the ACC, along with the anterior insula, appears to be one of the most consistently activated structures. Importantly, empathy-evoked activity within the ACC and the anterior insula overlaps greatly with the activity evoked during the actual experience of pain. This shared representation of "my pain" and "your pain" is thought to constitute the core mechanism underlying empathy for pain in humans.

ACC as Context-Driven Integrator

It has recently been suggested that since the ACC harbors neuronal connections with brain regions that are also involved in action, attention, executive control, speech, and affect, it should not be surprising that this structure play an integral role in regulating and modulating human behavior as a function of what is contextually appropriate (Paus 2001). With respect to pain processing, the ACC may serve to integrate intensity signals with the subjective meaning attributed to (or expected from) these signals so as to produce appropriate pain-related reactions. These reactions certainly include escape behaviors and overt manifestations of displeasure but may also include covert responses such as endogenous changes in the regulation of autonomic function and pain modulation. As reported above, anatomical data supports such a role for the ACC as this structure sends projections to brainstem regions involved in endogenous pain modulation (An et al. 1998) and autonomic activity (Vilensky and van Hoesen 1981). These projections would also explain why exposure to a tonic noxious stimulus leads to a rise in heart rate and to the concurrent trigger of descending pain control mechanisms (Willer et al. 1989). Of the different cortical structures implicated in pain perception, the CC plays a very complex role encompassing the anticipation and modulation of pain. Its role in the motivational-affective component of pain and in the emotionally driven responses to potential or actual danger for the organism makes it a key cortical structure in the complex and subjective experience of pain.

References

Allman, J. M., Hakeem, A., Erwin, J. M., et al. (2001). The anterior cingulate cortex, the evolution of an interface between emotion and cognition. *Annals of the New York Academy of Sciences, 935,* 107–117.

An, X., Bandler, R., Ongur, D., et al. (1998). Prefrontal cortical projections to longitudinal columns in the midbrain periaqueductal gray in macaque monkeys. *The Journal of Comparative Neurology, 401,* 455–479.

Apkarian, A. V., Bushnell, M. C., Treede, R. D., & Zubieta, J. K. (2005). Human brain mechanisms of pain perception and regulation in health and disease. *European Journal of Pain, 9,* 463–484.

Beckmann, M., Johansen-Berg, H., & Rushworth, M. F. (2009). Connectivity based parcellation of human cingulate cortex and its relation to functional specialization. *Journal of Neuroscience, 29,* 1175–1190.

Bouckoms, A. J. (1994). Limbic surgery for pain. In P. D. Wall & R. Melzack (Eds.), *Textbook of pain* (3rd ed., pp. 1171–1187). Edinburgh: Churchill Livingstone.

Bromm, B. (2004). The involvement of the posterior cingulate gyrus in phasic pain processing of humans. *Neuroscience Letters, 361,* 245–249.

Fields, H., & Basbaum, A. (1999). Central nervous system mechanisms of pain modulation. In P. D. Wall & R. Melzack (Eds.), *Textbook of pain* (pp. 309–329). Edinburgh: Churchill Livingstone.

Garcia-Cebrian, A., Gandhi, P., Demyttenaere, K., & Peveler, R. (2006). The association of depression and painful physical symptoms–a review of the European literature. *European Psychiatry, 21,* 379–388.

Garcia-Larrea, L., Frot, M., & Valeriani, M. (2003). Brain generators of laser-evoked potentials: From dipoles to functional significance. *Neurophysiologie Clinique, 33,* 279–292.

Hutchison, W. D., Lozano, A. M., Kiss, Z. H. T., et al. (1999). Pain-related neurons in human cingulate cortex. *Nature Neuroscience, 2,* 403–405.

Kong, J., Tu, P. C., Zyloney, C., & Su, T. P. (2010). Intrinsic functional connectivity of the periaqueductal gray, a resting fMRI study. *Behavioural Brain Research, 211,* 215–219.

Lamm, C., Decety, J., & Singer, T. (2011). Meta-analytic evidence for common and distinct neural networks associated with directly experienced pain and empathy for pain. *NeuroImage, 54,* 2492–2502.

Lenz, F. A., Rios, M., Zirh, A., Chau, D., Krauss, G., & Lesser, R. P. (1998). Painful stimuli evoke potentials recorded over the human anterior cingulate gyrus. *Journal of Neurophysiology, 79,* 2231–2234.

Lorenz, J., Cross, D. J., Minoshima, S., et al. (2003). Keeping pain out of mind: The role of the dorsolateral prefrontal cortex in pain modulation. *Brain, 126,* 1079–1091.

Margulies, D. S., Kelly, A. M., Uddin, L. Q., Biswal, B. B., Castellanos, F. X., & Milham, M. P. (2007). Mapping the functional connectivity of anterior cingulate cortex. *NeuroImage, 37,* 579–588.

Oshiro, Y., Quevedo, A. S., McHaffie, J. G., Kraft, R. A., & Coghill, R. C. (2009). Brain mechanisms supporting discrimination of sensory features of pain: A new model. *Journal of Neuroscience, 29,* 14924–14931.

Paus, T. (2001). Primate anterior cingulate cortex: Where motor control, drive and cognition interface. *Nature Reviews Neuroscience, 2,* 417–424.

Porro, C. A., Cettolo, V., Pia Francescato, M., et al. (1998). Temporal and intensity coding of pain in the human cortex. *Journal of Neurophysiology, 80,* 3312–3320.

Rainville, P., Duncan, G. H., Price, D. D., et al. (1997). Pain affect coded in human anterior cingulate but not somatosensory cortex. *Science, 227,* 968–971.

Talbot, J. D., Villemure, J. G., Bushnell, M. C., et al. (1995). Evaluation of pain perception after anterior capsulotomy: A case report. *Somatosensory and Motor Research, 12,* 115–126.

Valet, M., Sprenger, T., Boecker, H., et al. (2004). Distraction modulates connectivity of the cingulo-frontal cortex and the midbrain during pain-an fMRI analysis. *Pain, 109,* 399–408.

Vilensky, J. A., & van Hoesen, G. W. (1981). Corticopontine projections from the cingulate cortex in the rhesus monkey. *Brain Research, 205,* 391–395.

Vogt, B. A., Berger, G. R., & Derbyshire, S. W. (2003). Structural and functional dichotomy of human midcingulate cortex. *European Journal of Neuroscience, 18,* 3134–3144.

Vogt, B. A., Vogt, L., Farber, N. B., & Bush, G. (2005). Architecture and neurocytology of monkey cingulate gyrus. *The Journal of Comparative Neurology, 485,* 218–239.

Wager, T. D., Rilling, J. K., Smith, E. E., Sokolik, A., Casey, K. L., Davidson, R. J., et al. (2004). Placebo-induced changes in FMRI in the anticipation and experience of pain. *Science, 303,* 1162–1167.

Wiech, K., & Tracey, I. (2009). The influence of negative emotions on pain: Behavioral effects and neural mechanisms. *NeuroImage, 47,* 987–994.

Willer, J. C., De Broucker, T., & Le Bars, D. (1989). Encoding of nociceptive thermal stimuli by diffuse noxious inhibitory controls in humans. *Journal of Neurophysiology, 62,* 1028–1038.

Yoshino, A., Okamoto, Y., Onoda, K., Yoshimura, S., Kunisato, Y., Demoto, Y., et al. (2010). Sadness enhances the experience of pain via neural activation in the anterior cingulate cortex and amygdala: An fMRI study. *NeuroImage, 50,* 1194–1201.

Pain Progression

▶ Transition from Acute to Chronic Pain

Pain Prone Personality

Definition

A hypothesized personality constellation, which is thought to render the individual more prone to the development of a pain disorder.

Cross-References

▶ Pain as a Cause of Psychiatric Illness
▶ Pain-Prone Patients

Pain Questionnaire

▶ Pain Measurement by Questionnaires, Psychophysical Procedures, and Multivariate Analysis

Pain Questionnaires

▶ Outcome Measures

Pain Rehabilitation

Definition

A rehabilitation program aimed at improving activity and participation of individuals with chronic nonmalignant pain. The program is usually based on behavioral-cognitive principles and is administered in group form over 3–6 weeks by teams consisting of a pain or rehabilitation physician, a behaviorally and cognitively trained psychologist, a physiotherapist, and a vocational councillor or social worker. An occupational therapist and a registered nurse may also contribute.

Cross-References

▶ Physical Medicine and Rehabilitation, Team-Oriented Approach

Pain Scale

Definition

A pain scale is a system of rating the level, magnitude, or intensity of pain in order to quantify the individual's pain experience. This may include categorical items such as cartoon faces, simple classifications, and the assignment of numbers to the categories that rank in magnitude with the category descriptors. Depending on their levels of measurement and how they have been developed, category scale items (and summary scores derived from them) may only provide a measure of the grade of pain, rather than an absolute and accurate measure of differences in the magnitude of pain (scores at the interval or ratio level).

Cross-References

▶ Pain Assessment in Neonates

Pain Schema

Definition

An organized cognitive structure containing information about a particular domain of experience related to pain. Pain schemas are said to vary in terms of their level of activation. Once activated by a particular stimulus situation, pain schema will contribute to the preferential processing of pain-related information.

Cross-References

▶ Psychology of Pain, Assessment of Cognitive Variables

Pain Schools

▶ Information and Psychoeducation in the Early Management of Persistent Pain

Pain Sense Organs

▶ Nociceptor, Categorization

Pain Signal

▶ Central Pain: Outcome Measures in Preclinical Research

Pain System

▶ Lateral Thalamic Pain-Related Cells in Humans

Pain Tension Cycle

Definition

Pain tension cycle refers to the observation that pain leads to reflexive increases in tension that will again increase pain levels. This process may lead to a vicious cycle.

Pain Therapy

▶ Biofeedback in the Treatment of Pain

Pain Threshold

Definition

Pain threshold is defined as the lowest stimulus intensity that reliably elicits the perception of pain. Attention as well as motivational and emotional factors are known to influence the pain threshold that can be measured in the laboratory. There are three methods that are typically used to determine an individual's pain threshold: (1) Method of limits: Stimulus intensity is continuously increased until the subject perceives the stimulus as painful. Typically, several ascending and descending series of stimuli are used in order to reduce confounding influences due to expectation or habituation. (2) Method of adjustment: The subject adjusts stimulus intensity until the subjective pain threshold is reached. (3) Method of constant stimuli: A set of stimuli of fixed intensities are presented several times. Pain threshold is defined as the stimulus intensity that evokes a pain report in 50 % of the trials.

Cross-References

▶ Hyperpathia, Assessment
▶ Hypervigilance and Attention to Pain
▶ Hypoesthesia, Assessment
▶ Modeling, Social Learning in Pain
▶ Pain in Humans, Electrical Stimulation (Skin, Muscle, and Viscera)
▶ Pain in Humans, Thresholds
▶ Pain Measurement by Questionnaires, Psychophysical Procedures, and Multivariate Analysis
▶ Psychology of Pain, Self-efficacy
▶ Somatic Pain
▶ Tourniquet Test

Pain Thresholds

▶ Pain in Humans, Thresholds

Pain Tolerance

Definition

Pain tolerance is defined as the maximum stimulus intensity, or the maximum time of continuous painful stimulation, that a subject is willing to endure.

Cross-References

▶ Hypervigilance and Attention to Pain
▶ Hypoesthesia, Assessment
▶ Modeling, Social Learning in Pain
▶ Psychology of Pain, Self-efficacy
▶ Tourniquet Test

Pain Tools

▶ Pain Assessment in Neonates

Pain Treatment in Pediatric Palliative Care

Brian Carter
University of Missouri at Kansas City Children's Mercy Hospital and Bioethics Center, Kansas City, MO, USA

Synonyms

End-of-life care; Life-limiting condition; Life-threatening condition; Palliation; Quality of life

Introduction

As children – from infancy through adolescence – may experience life-threatening or life-limiting conditions that prompt the provision of palliative care, pain may be an episodic or pervasive symptom requiring assessment and management throughout a disease trajectory, and particularly at the end of life. As explained elsewhere in this work, pain is not simply a singular, objective, physiological phenomenon.

Characteristics

While developmental considerations such as the immaturity of the neonatal nervous system relative to an older child or adolescent, or the changing pharmacokinetics/pharmacodynamics of analgesics with age and disease processes, bear attention, the nature of pain and the ethos of the healing arts to alleviate suffering require healthcare professionals to give it priority among symptom management and therapeutic interventions.

Children who receive palliative and supportive treatments while continuing cure-oriented and life-extending interventions may reside in general or pediatric hospital inpatient care units, intensive care units (ICUs), oncology and myelosuppressive wards, or be under care at home through home health services (Orsey et al. 2009). They may make intermittent visits to emergency departments, or have brief inpatient stays. Other children whose care has progressed to a prevailing palliative and supportive care paradigm may also be inpatients in hospitals, perhaps occupy ICU beds, or receive care at home via home hospice services. Relatively few such children occupy inpatient hospice beds in the United States, but this may be different in Europe and the United Kingdom.

The conditions for which these children are receiving care may allow for a classification of their anticipated pain. Some *causative* subtypes of pain encountered in pediatric palliative care include physical causes: soft tissue pain, bone pain (such as osteopenia or infiltrative processes), nerve pain, central pain, pain from muscle spasm, visceral "colic"; and behavioral subtypes such as a more comprehensive/"total pain" that includes psychic, spiritual, and interpersonal aspects of human pain (Grégoire and Frager 2006; Friedrichsdorf and Kang 2007). An inquiry might include: "Are there soft tissue, musculoskeletal, or surgical etiologies for pain in this child?" for the former, and attention to meaning-assignment and interpretation of pain in the context of role, relationship, responsibilities, and place in the latter.

The *temporal nature* of pain lends itself to inquiries and pain assessment that may affect management (Hain et al. 2011):

- Is the pain acute, episodic, recurrent, pervasive/continuous, or chronic?
- Is the pain exacerbated or alleviated by environmental stimuli and changes (e.g., heat/cold, light/dark, light touch/ pressure, olfactory sensory stimuli, noise or music/repetitive calming sound)?

Additionally, the *responsiveness to pharmacologic interventions* – particularly to opioids – may also aid the clinician in classifying certain pain conditions (Friedrichsdorf 2010; Hain et al. 2011). Here, the likely response to acetaminophen or nonsteroidal anti-inflammatory agents, simple opioids, enhanced potency/prolonged duration opioids, or the requisite addition of adjuncts such as anticonvulsants, glucocorticoids, antidepressants, and benzodiazepines helps to underscore the severity and potential course of pain in a child being managed by palliative care.

Finally, another manner in which to classify pain is through examination of the *pathophysiologic processes* involved (Hain et al. 2011). The nature of the pathologic condition may guide the clinician to interventions that might be most effective (e.g., opioids or adjuncts). Categories may include:

- *Nociceptive pain*: Unpleasant sensory experience associated with actual or potential tissue damage
- *Somatic pain*: Commonly thermal (heat/cold), mechanical (crush injury or tissue tearing) and chemical (alcohol in a cut).
- *Visceral pain*: Commonly deep, vague pain associated with the stretching of internal tissues (e.g., bowel distention), inflammation or poor circulation (ischemia); often accompanied by nausea and vomiting
- *Neuropathic pain*: Damage or disease affecting the nervous system as it relates to bodily

feelings (e.g., peripheral neuropathic pain is often described by the patient to be "burning," "tingling," "electrical," "stabbing," or "pins and needles").

- *Central-pain syndromes*: Caused by damage to the central nervous system (including the brain, brainstem, and spinal cord). Commonly caused by stroke, multiple sclerosis, tumors, epilepsy, brain or spinal cord trauma, or some genetic conditions).
- *Sympathetic-mediated pain*: The sympathetic portion of the autonomic nervous system transmits pain signals to the brain that can be constant and severe, often described as "burning" and originating in the hands/feet. Perhaps the best described condition is Chronic Regional Pain Syndrome (CRPS).

Diagnosing and Managing Pain

Young or Noncommunicative Children

Abnormal behavior – other than wakeful play, eating, and sleeping – that is often described as "irritability" may indicate that a child is in pain. It might be manifest as distress with touching/examining the child, or moving him/her. Alternatively, a child that is withdrawn from normal play and interaction might be trying to avoid pain associated with movement. This can be carried to the extent of the infant or child lying perfectly still. When these patterns of behavior are observed over recurrent or lengthy periods of time, they may be more reliable and warrant a therapeutic trial of analgesia – especially if reliable parents and ongoing caregivers report such behaviors over a continuum to healthcare professionals that see the child only intermittently – even the palliative care consultant (Collins et al. 2011).

In conducting a thorough assessment and embarking on an effective management plan for pain in children living with a life-threatening or limiting illness, interdisciplinary providers bring much to the table. There are histories and beliefs that may be best elaborated by a child life specialist, an expressive therapist, or a developmental or pain psychologist. Social workers and chaplains may also help to unearth patient, family, and community backgrounds that shape the frequency and content of a child's expression of his/her pain – or that might impede it. More exists to the assessment of pain and related symptoms for these children than the simple application of a well-recognized pain assessment tool or scale by the patient's nurse. And the palliative physician needs to direct the palliative care team's attention to meeting overarching goals of relieving pain and suffering (Grégoire and Frager 2006).

When any caregiver is wondering whether a child is likely to experience pain as a result of an activity, treatment, or other intervention, a simple technique is to ask: *Would I be in pain if it were me?* If pain is likely, steps to minimize it – be they behavioral (e.g., distraction) or preemptive topical analgesia and anesthetics – and subsequently treat it should be taken. This is true regardless of whether or not the child is capable of verbalizing pain.

Pain assessments (pre- and post-procedural) are necessary to determine the effectiveness of measures taken to prevent pain or provide appropriate analgesia. The distinction between the relief of pain, rather than anxiety or behavioral phenomena, is important to know. Measuring the effect of pain relief can be one way to evaluate the overall effectiveness of pediatric palliative care. But it may not be enough as the comprehensive goals of palliative care are to improve the quality of life of the patient and family. Therefore, some practical tools to look at a child's quality of life (inclusive of pain relief) need to be developed – and particularly adapted for children living with life-threatening and limiting conditions.

Treating Pain

World Health Organization's Pain Ladder

The World Health Organization (WHO) has set the standard for universal pain management in children receiving palliative care regardless of where they live. The WHO Pain Ladder, describing a step-wise approach to analgesia, is familiar to many clinicians (Friedrichsdorf and Kang 2007; Friedrichsdorf 2010; Hain et al. 2011). As pain increases from mild to moderate and on to severe, its management moves from simple analgesics (e.g., acetaminophen as needed) to scheduled analgesics (e.g., a fixed-dose combination opioid plus acetaminophen), and onward to the use of major opioids (e.g., morphine, hydromorphone, oxycodone, hydrocodone, or fentanyl) (Hewitt et al. 2008). When applying this step-wise approach, clinicians are encouraged to do so as guided by the child's needs, the right route of administration, and with regular dosing intervals. Additional corollaries exist:

- The oral route (or enteral via tube) of administration is preferred when possible. Alternative routes may be available, depending upon the medicine, and at times necessary (palliative care clinicians know well the need to be flexible or innovative), but for most children, the oral route remains preferred (Hewitt et al. 2008).

- At times, the analgesics used can be augmented with using an appropriate adjuvant medication. Certain adjuvants find value in palliative care – such as anticonvulsants or antidepressants for neuropathic pain, a number of nonsteroidal anti-inflammatory drugs (NSAIDs) for musculoskeletal pain, anticholinergic agents for smooth muscle spasm and colic, corticosteroids to reduce inflammation or edema, and skeletal muscle relaxants for musculoskeletal pain. While these agents are not pain relievers, they may be effective for particular types of pain (Friedrichsdorf and Kang 2007; Friedrichsdorf 2010).

- Scheduling analgesics – including opioids – improves their effectiveness. Hence, they should be given "around-the-clock" in treating moderate or severe pain. At the same time, it may be necessary to prescribe additional doses for breakthrough pain (Friedrichsdorf and Kang 2007; Friedrichsdorf 2010).

- Environmental and behavioral (non-pharmacologic) methods should always be employed in conjunction with medications when addressing pediatric pain in general,

Pain Treatment in Pediatric Palliative Care, Table 1 Integrative pain management opportunities for children (Adapted from Grégoire and Frager (2006))

Age	Comfort (Physical)	Comfort (Behavioral)	Distraction
Infants: 0–1 year	Repetitive/calm movement (e.g., rocking), swaddling, "Kangaroo care," pacifier, sucrose, calming touch or massage, reduced light & noise	Teach/support parents in learning & responding to cues and means of comforting infant	Auditory: soothing voice/music/song Oral/suckling: pacifier, Visual: bubbles, mobile
Preschool: 2–5 years	Rocking, cuddling, pacifier, sucrose, hot/cold packs acupressure, physical therapy, and TENS	Art & music therapy, therapeutic play, imagery, and hypnosis teaching/supporting parents	Auditory: familiar songs, music, stories, Visual: miming Activity: blow bubbles, have pet visit
Early School: 6–11 years	Rocking, cuddling, reduced light and noise, acupressure & acupuncture, massage Physical therapy, and TENS	Art, music, expressive therapy (including dance), imagery/hypnosis, bio-feedback, games, play-acting and stories, psychotherapy	Auditory: music/songs, reading stories Visual: miming, puppets Activity: blow bubbles, pet visits;
Adolescent: 12–18 years	Massage, hot/cold packs, acupressure/acupuncture, physical therapy/TENS, comfortable & familiar environment	Imagery/hypnosis; art, music & expressive therapies, mindfulness meditation & relaxation, psychotherapy	Auditory: music, stories Visual: videos Activity: games, pet visit

TENS = Transcutaneous electrical nerve stimulation

and palliative care in particular. Integrative therapies may also be of particular /benefit (Grégoire and Frager 2006; Friedrichsdorf and Kang 2007).

- As always, clinicians need to be aware of potential adverse drug events or medication side effects and interactions, as well as the particulars of developmental pharmacokinetics in children.

Among common side effects of opioid analgesics, the most prevalent is constipation (Friedrichsdorf and Kang 2007). In the face of either acute or chronic illness, this can be exacerbated by both poor fluid intake (dehydration) and the child's inactivity. It should be paramount for all clinicians prescribing opioids to children that an accompanying stool softener or bulk laxative be prescribed – and perhaps even a stimulant.

Other opioid-related side effects that may be common in adults appear to be less so in children but need to be given attention should they arise – such as nausea and vomiting. On the other hand, urinary retention and pruritus appear to occur more often in children than in adults. Consideration for ondansetron, oral naloxone, or methylnaltrexone may also be of benefit. When the need to increase a particular opioid results in

adverse effects that are unacceptable, clinicians should consider rotating to an alternative opioid. Respiratory depression, while often feared, is rarely reported in children (Friedrichsdorf and Kang 2007).

Investigations of both morphine and midazolam pharmacokinetics for infants and children on ECMO suggest varied increases in clearance for morphine and decreased clearance of midazolam while the child is most critically ill. Clinical improvement generally results in normalized clearance, but given individual variation (including that of hepatic glucuronidation), clinicians need to be careful in titrating dosing when such support may be withdrawn, even in a palliative care context.

Little has been published on the use of opioids in the specific context of pediatric palliative care (Grégoire and Frager 2006). Some children may clear morphine more quickly than adults do, but wide variability exists for morphine half-life across patients. Some reduction in an otherwise typical adult dose of morphine, based upon a child's weight, is appropriate and clinicians are reminded to be attentive to impairments in hepatic or renal function in children near the end of life. However, it is not uncommon for pediatric palliative care patients to have such opioid tolerance that they require dosages that are larger than "conventional" dosing – this may be especially true at the end of life (Grégoire and Frager 2006; Hewitt et al. 2008).

Among opioids and commonly used adjuvants, there is little pediatric research on specific agents and their side effects. Nonetheless, certain truths remain (Friedrichsdorf 2010). The incidence of seizures, and the availability of alternatives, should make meperidine an unacceptable opioid for common use in pediatric pain relief. Tramadol has received little attention in pediatric palliative care but has been found to be helpful in postoperative pain. Methadone is especially useful for treating neuropathic pain. Hydromorphone, which is similar to but more potent than morphine, is a good alternative when matters of fluid restriction might interfere with other high-dose opioids. Buprenorphine and fentanyl also provide parenteral opioids without the need for needles, and both are sufficiently different from morphine to be appropriate for opioid substitution. The pediatric safety and efficacy data for fentanyl are much greater at present than for buprenorphine (Table 1).

Summary

In summary, pediatric palliative care requires early and comprehensive attention to pain and symptom management, with clinicians willing to be flexible and innovative, yet attentive to unique developmental and disease-mediated phenomena that are important in safe practice. Integrative and multimodal therapies stand to make care for children with life-threatening and limiting conditions both possible and effective.

References

Collins, J. J., Berde, C. B., & Frost, J. A. (2011). Pain assessment and management. In J. Wolfe, P. Hinds, & B. Sourkes (Eds.), *Textbook of interdisciplinary pediatric palliative care* (pp. 284–299). Philadelphia: Elsevier/Saunders.

Friedrichsdorf, S. J. (2010). Pain management in children with advanced cancer and during end-of-life care. *Pediatric Hematology and Oncology, 227*, 257–261.

Friedrichsdorf, S. J., & Kang, T. I. (2007). The management of pain in children with life-limiting illnesses. *Pediatric Clinics of North America, 54*, 645–672.

Grégoire, M.-C., & Frager, G. (2006). Ensuring pain relief for children at the end of life. *Pain Research & Management, 11*(3), 163–171.

Hain, R., Zeltzer, L., Hellston, M., Cohen, S. O., Orloff, S. F., & Gray, D. (2011). Holistic management of symptoms. In B. Carter, M. Levetown, & S. Friebert (Eds.), *Palliative care for infants, children and adolescents* (2nd ed., pp. 244–274). Baltimore: Johns Hopkins university Press.

Hewitt, M., Goldman, A., Collins, G. S., Childs, M., Hain, R. (2008). Opioid use in palliative care of children and young people with cancer. *J Pediatr, 152*, 39–44.

Orsey, A. D., Belasco, J. B., Ellenberg, J. H., Schmitz, K. H., & Feudtner, C. (2009). Variation in receipt of opioids by pediatric oncology patients who died in children's hospitals. *Pediatric Blood & Cancer, 52*, 761–766.

Pain Treatment, Implantable Pumps for Drug Delivery

Frederick A. Lenz and Khan W. Li
Department of Neurosurgery, Johns Hopkins Hospital, Baltimore, MD, USA

Synonyms

Epidural drug pumps; Intracerebroventricular drug pumps; Intrathecal drug pumps; Intraventricular drug pumps; Subdural drug pumps

Definition

Drug delivery system which infuses substances into the cerebrospinal fluid in a controlled fashion for treatment of pain or spasticity.

Introduction

Recent technological advances have made possible the delivery of drugs directly into cerebrospinal fluid via implantable pumps. This method is currently utilized when other medical or surgical therapies have failed or have been associated with significant side effects (Smith et al. 2002). Patients with cancer have been studied most frequently, although this technique is often used in patients with non*malignant pain syndromes*. The most common analgesic used is morphine, which has been shown to be both safe and efficacious (Wang et al. 1979; Smith et al. 2002). The main advantage of *intrathecal* or *intraventricular* administration is the decrease in systemic side effects due to decreased blood levels, as compared to the levels required for pain control when morphine is given orally or intravenously (Smith et al. 2002).

Before the availability of implantable pumps, opioids were directly infused in a single bolus injection via lumbar puncture or via an Ommaya reservoir with a catheter in the lumbar CSF space. This method was time consuming and fraught with potential complications. Repeated lumbar punctures or reservoir injections led to an increased risk of meningitis and cerebrospinal fluid leak. In addition, bolus injections were associated with respiratory depression (Simpson 2003). With the introduction of implantable pumps for infusion of morphine, the delivery system was completely internalized allowing for continuous infusion of analgesics, which is both safer and more effective in obtaining pain relief (Simpson 2003; Penn 2003; Levy and Salzman 1997; Smith et al. 2002).

Characteristics

Routes of Administration
Implantable pumps can be used for intraspinal (intrathecal/subdural or epidural) or intraventricular drug delivery. Intrathecal or intraventricular administration is generally thought to be superior to the epidural route. Epidural administration utilizes higher doses, which increases the risk of adverse systemic effects, such as respiratory depression, and which increases the frequency with which the drug

reservoir must be refilled. In addition, there is the risk of catheter misplacement or migration into the subdural space. With the increased opioids dosages used with epidural pumps, this complication could lead to overdosage and life-threatening respiratory depression. On the other hand, misplacement of an intrathecal catheter or migration into the epidural space would produce only inadequate pain relief or symptoms of withdrawal, which can both be treated with oral opioids (Simpson 2003).

Ballantyne et al. reviewed studies comparing intraventricular and intrathecal morphine delivery in cancer patients (Ballantyne et al. 1996). They found that 73 % of patients with intraventricular morphine drug delivery achieved excellent pain control which was not significantly different from patients with intrathecal morphine (62 %) (Ballantyne et al. 1996). There were fewer technical problems with intraventricular therapy, though adverse effects, such as sedation and confusion, can be substantial.

Intraventricular delivery is believed to be as efficacious as intraventricular delivery but to be more invasive, and so is usually reserved for cases that fail to respond to intraspinal delivery or in cases of refractory craniofacial pain. A study of the safety and efficacy of daily intraventricular drug delivery via Ommaya reservoir produced adequate relief of pain from cancer in a series of 90 patients (Karavelis et al. 1996). Only 10 % of their patients reported less than 50 % relief of pain. Complications included two intracerebral hematomas and two cases of infection, while long-term sequelae were not reported. Overall, only 3 of the 90 intraventricular catheters failed (Karavelis et al. 1996).

The most commonly used avenue of analgesic delivery via implantable pumps, however, remains the intrathecal route. The technique of pump implantation has been well described previously (Levy and Salzman 1997). First, a lumbar puncture is performed between L2 and L5 to avoid the caudal end to the spinal cord, which lies at L1 or above. A catheter is then threaded through the needle in a rostral direction. Catheter placement in the mid or lower thoracic region is then confirmed with fluoroscopy. Placement of the catheter into the cervical region has been performed, but increases the risk of respiratory depression. Attention is then turned to creating a subcutaneous pocket for the implantable pump. The site of this pocket must be accessible for refilling of the pump's reservoir and for possible reprogramming. Once an adequate pocket is made, the pump is implanted and the intrathecal catheter is tunneled subcutaneously, connected to the pump and secured to prevent catheter migration.

Implantable Pumps for the Treatment of Cancer Pain

A controlled trial demonstrated that intrathecal analgesia in patients with refractory cancer pain was more effective than best medical management (Smith et al. 2002). There were significant improvements in pain control overall, reduced pain VAS, drug toxicities, and survival. A large prospective multicenter trial demonstrated the efficacy of intrathecal morphine for the treatment of patients with cancer pain. Improvements included significant decreases in VAS pain ratings, in systemic opiate intake, and in the opioids complication severity index (Rauck et al. 2003). The most common complications (sleep disorders, daytime drowsiness, constipation, nausea) and the most frequent complication (urinary retention) significantly improved after intrathecal analgesia. Most likely, these results reflect the opioids sparing effect when changing from systemic route of opioids delivery to intrathecal delivery.

In another study, intrathecal analgesia was given to nine patients with advanced pancreatic cancer (Gilmer-Hill et al. 1999a) (see also Gilmer-Hill et al. 1999b). All patients receiving intrathecal opioids experienced good to excellent pain relief on a Likert 5 point scale. Using the visual analogue scale, there was also a significant and clinically relevant decrease from 9/10 to 2/10. Importantly, all patients reported improvements in their quality of life and improved function in activities of daily living. No patient developed loss of pain control that could not be resolved with increased doses. There is broad interest in the use of pumps for neuropathic

pain, although efficacy of these therapies is not yet established (Hempenstall et al. 2005; Wu and Raja 2008).

Complications

Infection with resulting meningitis remains the most common complication, usually necessitating pump removal and long-term intravenous antibiotics. Cerebrospinal fluid leaks can usually be treated with flat bed rest or placement of an injection of the patient's blood into the epidural space, that is, epidural blood patch. In some cases, reexploration of the site of dural penetration is necessary. Catheter migration, disconnection, obstruction, or kinking along with battery failure are all common occurrences which require replacement of the catheter or pump. In the setting of strong magnetic fields, such as an MRI scan, pumps should be turned off or the reservoirs emptied.

Long-term catheter use can lead to the development of inflammatory mass at the tip of the catheter. Recently, a review of the literature described 41 cases, all in patients with morphine pumps (Coffey and Burchiel 2002). Fourteen of these patients suffered complete, irreversible spinal cord lesion at the level of the mass. Intrathecal medications, including morphine, may be more responsible than the catheter since this complication is not described with lumbar-peritoneal shunt catheters (Coffey and Burchiel 2002). Although it was initially believed that morphine was the causative agent, further reports for other opioids and non-opioids agents have been implicated. In experimental animal studies, it was found that all opioids, with the exception of fentanyl, were associated with granuloma formation (Yaksh Anesth). Patients must be warned of the occurrence of this rare but potentially devastating complication.

Transient respiratory depression can occur with morphine pumps at any site but is most common with intraventricular delivery (Ballantyne et al. 1996) and appears to be less common with less hydrophilic compounds like hydromorphone or sufentanil. Symptoms of withdrawal may also occur with pump depletion or malfunction, which can usually be ameliorated by oral opioids therapy. Other adverse effects are similar to those of systemic opioids administration: pruritus, hypotension, constipation, confusion, drowsiness, and urinary retention. Painful myoclonus is a rare, mysterious complication (Penn 2003).

Pump Designs

The most commonly used design is a simple continuous infusion pump, which is both effective and inexpensive. The rate of opioids infusion is adjusted by replacing the contents of the reservoir with differing drug concentrations. There are also pumps activated by placing pressure over a switch on the subcutaneous (patient-controlled) pump, which can then deliver bolus infusions or adjust the continuous rate. Programmable pumps can be reprogrammed with an external device to adjust the rate of infusion at preset times. While programmable and patient-controlled pumps may initially appear to be superior to simple continuous infusion pumps, they have some disadvantages: they are much more costly and they utilize more battery power which leads to more frequent pump changes. The programmable and patient-controlled pumps should, in general, be reserved for patients with nonmalignant pain who do not have a short life expectancy (Simpson 2003).

Emerging Strategies in the Use of Intrathecal Therapies for Pain

While most drug pumps implanted for the treatment of pain have used morphine infusions, a number of other drug classes have been employed. These include sodium channel antagonists, calcium channel antagonists, NMDA antagonists, GABA agonists, alpha-2 adrenergic agonists, acetylcholinesterase inhibitors, adenosine agonists, and somatostatin analogues (Dougherty and Staats 1999; Lorenz et al. 2002; Middleton et al. 1996). More recently, Ziconotide has been employed with significant improvements in VAS of pain, and with side effects including nausea, dizziness, confusion, urinary retention, and somnolence (Rauck et al. 2009).

Recent studies have also looked at combination therapy using such drugs as clonidine, benzodiazepines, and bupivacaine in addition to morphine (Rainov et al. 2001). In another kind of adjuvant therapy, baclofen pumps may be effective when added to the therapy of patients in whom implanted spinal cord stimulators have proven to be ineffective (Lind et al. 2004; Schechtmann et al. 2010).

Other novel approaches include the implantation of synthetic microspheres containing cells or transplantation of cells which elaborate particular neurotransmitters, and so function as biologic drug pumps. Trials have already been performed with human chromaffin cells transplanted into the subarachnoid space of patients with refractory cancer pain (Pappas et al. 1997).

References

Ballantyne, J. C., Carr, D. B., Berkey, C. S., Chalmers, T. C., & Mosteller, F. (1996). Comparative efficacy of epidural, subarachnoid, and intracerebroventricular opioids in patients with pain due to cancer. *Regional Anesthesia, 21*, 542–556.

Coffey, R. J., & Burchiel, K. (2002). Inflammatory mass lesions associated with intrathecal drug infusion catheters: Report and observations on 41 patients. *Neurosurg, 50*, 78–86.

Dougherty, P. M., & Staats, P. S. (1999). Intrathecal drug therapy for chronic pain: From basic science to clinical practice. *Anesthesiology, 91*, 1891–1918.

Gilmer-Hill, H. S., Boggan, J. E., Smith, K. A., Frey, C. F., Wagner, F. C., Jr., & Hein, L. J. (1999a). Intrathecal morphine delivered via subcutaneous pump for intractable pain in pancreatic cancer. *Surgical Neurology, 51*, 6–11.

Gilmer-Hill, H. S., Boggan, J. E., Smith, K. A., & Wagner, F. C., Jr. (1999b). Intrathecal morphine delivered via subcutaneous pump for intractable cancer pain: A review of the literature. *Surgical Neurology, 51*, 12–15.

Hempenstall, K., Nurmikko, T. J., Johnson, R. W., A'Hern, R. P., & Rice, A. S. (2005). Analgesic therapy in postherpetic neuralgia: A quantitative systematic review. *PLoS Medicine, 2*, e164.

Karavelis, A., Foroglou, G., Selviaridis, P., & Fountzilas, G. (1996). Intraventricular administration of morphine for control of intractable cancer pain in 90 patients. *Neurosurgery, 39*, 57–61.

Levy, R. M., & Salzman, D. (1997). Implanted drug delivery systems for control of chronic pain.

P

In R. B. North & R. M. Levy (Eds.), *Neurosurgical management of pain* (pp. 302–324). New York: Springer.

Lind, G., Meyerson, B. A., Winter, J., & Linderoth, B. (2004). Intrathecal baclofen as adjuvant therapy to enhance the effect of spinal cord stimulation in neuropathic pain: A pilot study. *European Journal of Pain, 8*, 377–383.

Lorenz, M., Hussein, S., & Verner, L. (2002). Continuous intraventricular clonidine infusion in controlled morphine withdrawal–case report. *Pain, 98*, 335–338.

Middleton, J. W., Siddall, P. J., Walker, S., Molloy, A. R., & Rutkowski, S. B. (1996). Intrathecal clonidine and baclofen in the management of spasticity and neuropathic pain following spinal cord injury: A case study. *Archives of Physical Medicine and Rehabilitation, 77*, 824–826.

Pappas, G. D., Lazorthes, Y., Bes, J. C., Tafani, M., & Winnie, A. P. (1997). Relief of intractable cancer pain by human chromaffin cell transplants: Experience at two medical centers. *Neurological Research, 19*, 71–77.

Penn, R. D. (2003). Intrathecal medication delivery. *Neurosurgery Clinics of North America, 14*, 381–387.

Rainov, N. G., Heidecke, V., & Burkert, W. (2001). Long-term intrathecal infusion of drug combinations for chronic back and leg pain. *Journal of Pain and Symptom Management, 22*, 862–871.

Rauck, R. L., Cherry, D., Boyer, M. F., Kosek, P., Dunn, J., & Alo, K. (2003). Long-term intrathecal opioid therapy with a patient-activated, implanted delivery system for the treatment of refractory cancer pain. *The Journal of Pain, 4*, 441–447.

Rauck, R. L., Wallace, M. S., Burton, A. W., Kapural, L., & North, J. M. (2009). Intrathecal ziconotide for neuropathic pain: A review. *Pain Practice, 9*, 327–337.

Schechtmann, G., Lind, G., Winter, J., Meyerson, B. A., & Linderoth, B. (2010). Intrathecal clonidine and baclofen enhance the pain-relieving effect of spinal cord stimulation: A comparative placebo-controlled, randomized trial. *Neurosurgery, 67*, 173–181.

Simpson, R. K., Jr. (2003). Mechanisms of action of intrathecal medications. *Neurosurgery Clinics of North America, 14*, 353–364.

Smith, T. J., Staats, P. S., Deer, T., Stearns, L. J., Rauck, R. L., Boortz-Marx, R. L., Buchser, E., Catala, E., Bryce, D. A., Coyne, P. J., & Pool, G. E. (2002). Randomized clinical trial of an implantable drug delivery system compared with comprehensive medical management for refractory cancer pain: Impact on pain, drug-related toxicity, and survival. *Journal of Clinical Oncology, 20*, 4040–4049.

Wang, J. K., Nauss, L. A., & Thomas, J. E. (1979). Pain relief by intrathecally applied morphine in man. *Anesthesiology, 50*, 149–151.

Wu, C. L., & Raja, S. N. (2008). An update on the treatment of postherpetic neuralgia. *The Journal of Pain, 9*, S19–S30.

Pain Treatment, Intracranial Ablative Procedures

Philip L. Gildenberg
Baylor Medical College, Houston, TX, USA

Synonyms

Cancer pain; Cingulotomy; Intracranial ablative procedures; Mesencephalic tractotomy; Mesencephalotomy; Pain; Stereotactic; Thalamotomy

Definition

Interruption, within the brain or brain stem, of pathways or structures concerned with pain intensity, unpleasantness, or suffering for relief of subacute or ▶ chronic pain.

Characteristics

In the early part of the twentieth century, the spinothalamic tract was interrupted at several sites where it lies close to the surface of the brain stem (White and Sweet 1969; Gybels and Sweet 1989). The descending tract of the trigeminal nucleus, which concerns pain perception specifically, was interrupted in the dorsolateral medulla to treat head, neck, and face pain. The spinothalamic pathway lies just below the pial surface, at the lateral edge of the tectum of the midbrain, and presented a target for ▶ ablation. The localization of the tracts on visualization of the surface of the brain stem, however, is uncertain, so there was a high incidence of unsuccessful analgesia or pain relief, and a high incidence of unwanted neurological side effects.

When human stereotactic surgery was invented in 1947 (Spiegel et al. 1947), it became possible to insert an electrode accurately into any desired cerebral anatomic structure. This structure could then be ablated by application of a coagulating current or injection of a toxic substance, such as alcohol. One goal of such surgery was to treat persistent pain. The safety and effectiveness of these procedures was dramatically improved with the introduction of modern imaging techniques, such as computerized tomography or magnetic resonance imaging.

Selection of the proper patient is an important factor in whether ▶ intracranial ablative procedures might successfully relieve pain. Pain that involves the body or extremities may be treated by a procedure involving the peripheral nerves or spinal cord. Intracranial procedures are more likely to be used in patients with pain involving the head and/or neck, the upper part of the body, such as the brachial plexus, or pain that is widespread throughout the body. The etiology of the pain must be clearly defined, and the severity of the pain is consistent with the etiology (Gildenberg 1996; Gildenberg and DeVaul 1985). Ordinarily, such surgery is reserved for persistent pain accompanying serious conditions, such as involvement of the head or neck with cancer, cancer that is widespread throughout the body, or pain that may persist after severe nerve injury, especially involving the face. Pain may recur months or years after an initially successful procedure. Consequently, ablative intracranial surgery is used to treat ▶ cancer pain more often than persistent pain of other etiologies. Patients should not be considered for ablative procedures unless significant psychological factors are ruled out or controlled.

It seemed logical that pain would disappear if the primary pain pathway, the spinothalamic tract, were interrupted in the spinal cord or brain stem (Gildenberg 1973). When it was first interrupted stereotactically in the midbrain, a procedure called mesencephalotomy, hemisensory loss of pain sensation or analgesia could be achieved throughout the entire contralateral body, face and head, often with good but temporary relief of pain. A high risk of postoperative central pain was observed. It was observed that if the mesencephalic central gray were also interrupted, pain relief was more certain and persisted longer (Nashold et al. 1969). Even if only the gray matter is interrupted, it may be

possible to achieve pain relief without loss of pin stick pain sensation or analgesia (Shieff and Nashold 1990).

Ablation of targets in the thalamus has also been used for the management of persistent pain. Much of the spinothalamic tract terminates in the brain stem, in structures that may project to the thalamus, but its influence projects to many areas of the thalamus. The spinoreticulothalamic pathways may project to the more medial thalamic nuclei, including the intralaminar and centromedian nuclei. High success rates for treatment of a broad range of conditions have been reported by ▶ lesions of these nuclei and posteriorly adjacent nuclei (Jeanmonod et al. 1994), although these high success rates have not been duplicated (Gybels and Sweet 1989). Ablative lesions in those parts of the thalamus may successfully relieve persistent pain, especially such cancer pain as Pancoast syndrome, from infiltration of the brachial plexus by lung cancer, or diffuse cancer pain (Spiegel et al. 1964). It is sometimes observed that patients who have successful pain relief from lesions in the intralaminar or centromedian nuclei may abruptly withdraw their narcotics with a minimum of withdrawal symptoms (Gildenberg and DeVaul 1982).

Intracranial procedures ablating parts of the ▶ limbic system are sometimes used to treat such psychiatric conditions as obsessive-compulsive disorder or ▶ intractable depression, if they cannot be managed with medications. A common target for such psychosurgery is the anterior portion of the cingulate bundle, as it wraps around the anterior end of the corpus callosum. Ablation of that same area has been successful in alleviating severe persistent pain, particularly that of cancer, and particularly if the cancer has caused severe emotional distress (Hassenbusch 1998).

References

Gildenberg, P. L. (1973). Percutaneous cervical cordotomy. *Clinical Neurosurgery, 21*, 246–256.
Gildenberg, P. L. (1996). General and psychological assessment of the pain patient. In G. T. Tindall, P. R. Cooper, & D. L. Barrow (Eds.), *The practice of neurosurgery* (pp. 2987–2996). Baltimore: Williams and Wilkins.

Gildenberg, P. L., & DeVaul, R. A. (1982). Management of chronic pain refractory to specific therapy. In J. R. Youmans (Ed.), *Neurological surgery* (pp. 3749–3768). Philadelphia: W.B. Saunders.

Gildenberg, P. L., & DeVaul, R. A. (1985). *The chronic pain patient. Evaluation and management.* Basel: Karger.

Gybels, J. M., & Sweet, W. H. (1989). *Neurosurgical treatment of persistent pain. Physiological and pathological mechanisms of human pain.* Basel: Karger.

Hassenbusch, S. J. (1998). Cingulotomy for cancer pain. In P. L. Gildenberg & R. R. Tasker (Eds.), *Stereotactic and functional neurosurgery* (pp. 1447–1451). New York: McGraw-Hill.

Jeanmonod, D., Magnin, M., & Morel, A. (1994). Chronic neurogenic pain and the medial thalamotomy. *Schweizerische Rundschau für Medizin Praxis, 83,* 702–707.

Nashold, B. S., Wilson, W. P., & Slaughter, D. G. (1969). Stereotaxic midbrain lesions for central dysesthesia and phantom pain. Preliminary report. *Journal of Neurosurgery, 30,* 116–126.

Shieff, C., & Nashold, B. S. J. (1990). Stereotactic mesencephalotomy. *Neurosurgery Clinics of North America, 1,* 825–839.

Spiegel, E. A., Wycis, H. T., Marks, M., & Lee, A. S. (1947). Stereotaxic apparatus for operations on the human brain. *Science, 106,* 349–350.

Spiegel, E. A., Wycis, H. T., Szekely, E. G., Gildenberg, P. L., & Zanes, C. (1964). Combined dorsomedial, intralaminar and basal thalamotomy for relief of so-called intractable pain. *The Journal of the International College of Surgeons, 42,* 160–168.

White, J. C., & Sweet, W. H. (1969). *Pain and the neurosurgeon. A forty year experience.* Springfield: Charles C Thomas.

Pain Treatment, Motor Cortex Stimulation

Jeffrey A. Brown
Department of Neurological Surgery, Wayne State University School of Medicine, Detroit, MI, USA

Synonyms

Cortical stimulation for relief of pain; Motor cortex stimulation

Definition

A technique for relief of neuropathic pain, involving stimulation of the somatotopically appropriate area of ► motor cortex through electrodes implanted in the epidural space (see ► Cortex and Pain).

Characteristics

Epidural motor cortex stimulation is a form of neuromodulation used to treat central and neuropathic pain syndromes, including pain from poststroke ► central pain syndromes, trigeminal nerve injury, postherpetic neuralgia, brachial plexus injury, spinal cord (► spinal cord stimulation) injury, and phantom limb pain.

In order to develop a treatment for thalamic pain syndrome more effective than chronic stimulation of the thalamic relay nucleus with deep brain electrodes (► Deep Brain Stimulation), Hirayama et al. investigated the effects of stimulation of various brain regions (Hirayama et al. 1990). They described complete, long-term inhibition of the burst hyperactivity recorded in thalamic neurons in cats after anterolateral cordotomy. Tsubokawa et al. then applied the technique clinically in seven patients with chronic, intractable, neuropathic pain and reported success of the technique in all patients studied (Tsubokawa et al. 1991). Tsubokawa then reviewed a series of 11 patients with central pain after putaminal or thalamic hemorrhage, treated with motor cortex stimulation for 2 years. There was greater than 80 % pain relief maintained in 5 of the 11 patients (Tsubokawa et al. 1993). Two years later, Meyerson et al. (1993) successfully relieved greater than 50 % of pain in all five patients in his series who had trigeminal neuropathic pain. Allodynia, dysesthesia, and hyperesthesia diminished during stimulation (Meyerson et al. 1993). Katayama et al. treated four patients with lateral medullary infarctions, whose pain relief from ventral posterior thalamic nuclear stimulation had ceased to be effective. These patients were then treated by motor cortex stimulation and reported 40–60 % pain relief (Katayama et al. 1994).

Patient Screening

Yamamoto et al. noted that successful pain relief by motor cortex stimulation could be predicted, if patients responded by at least 40 % pain relief to incremental infusions of intravenous thiamylol to a maximum dose of 250 mg, but not to morphine in doses of up to 18 mg given over 5 h (Yamamoto et al. 1997). Katayama et al. have observed that satisfactory pain relief can be obtained in 73 % of patients when they have no motor weakness. If weakness is moderate to severe, then only 15 % of patients will benefit from stimulation. In addition, if motor responses cannot be induced by stimulation, then only 8 % of patients will obtain pain relief (Katayama et al. 1998). The conclusion to be drawn is that motor cortex stimulation requires an intact motor system to be effective, but not an intact somatosensory system.

Mechanism of Effect

Positron emission tomography, performed by Garcia-Larrea et al. in patients being treated by motor cortex stimulation, shows elevations in cerebral blood flow in the ipsilateral thalamus, cingulate gyrus, orbitofrontal cortex, and midbrain (see ▶ Thalamus and Pain). The degree of pain relief correlates with increased blood flow in the cingulate gyrus. Nociceptive spinal reflexes are also attenuated by stimulation. In general terms, the cingulate gyrus is believed to be involved in the affective dimension of ▶ chronic pain (Garcia-Larrea et al. 1999, 2000).

Morbidity

Several series have reported seizures occurring, usually during implantation of the stimulator. The effect of motor cortex stimulation in primates has been studies by Bezard et al. However, although stimulation could cause reversible seizures, neither epilepsy nor a reduced threshold for kindling of seizure occurred (Bezard et al. 1999). Epidural hematomas have been associated with the surgery, but without neurological injury. Stimulator pocket infections and electrode wire fractures have occurred, which are known complications of the implantation of neurostimulators.

Facial Pain

Nguyen et al. reviewed a series of 12 patients with neuropathic ▶ facial pain, of whom 75 % achieved good to excellent pain relief. This is due to the cortical representation of the face being large and accessible over the convexity. Therefore, it is easier to treat facial pain when the electrodes are placed somatotopically over the correlating representational region of the pain on the motor cortex (Nguyen et al. 2000).

Central Pain Syndromes

In a recent series of patients with central pain syndromes, it was reported that 10 out of 14 patients (77 %) achieved substantial pain relief (Nguyen et al. 2000).

Motor Effects

Motor cortex stimulation has been observed to improve motor function independent of pain control. Improved thalamic hand syndrome, spasticity, action tremor, intention myoclonus, advanced Parkinson's symptoms, and recovery after stroke have been reported to occur with stimulation.

Surgical Technique

The surgical technique has been refined since the initial report of Tsubokawa. Treatment planning begins with an MRI or CT scan that is integrated into a neuronavigational system. The target is selected on the primary motor cortex, based on anatomic landmarks. These were first established by Nguyen et al. in 2000. Most often, the target for facial pain, for example, is selected anterior to the central sulcus, at the level of the inferior frontal sulcus, as seen on the sagittal MRI or CT. Patients are induced with general endotracheal anesthesia and need not be awakened for evaluation of pain relief during surgery. The central sulcus, at the level of hand function, may be identified using median nerve somatosensory evoked potentials and determination of N20 P20 phase shift. Electromyographic recordings during cortical stimulation of the target muscles in the region of pain are done to determine the site of maximal electromyographic response to stimulation. A four-plate electrode paddle array is sutured to the dura, parallel to the central sulcus, overlying the primary motor cortex. Often patients are found to have substantial pain relief from the intraoperative stimulation. Several days of trial stimulation are performed, after which the pulse generator is implanted, when pain relief is greater than 50 %. Subthreshold stimulation relieves pain. Stimulation is done at low frequency (40 Hz), usually low pulse width (90 ms) and low amplitudes (2–10 V). More difficult to treat pain syndromes (such as anesthesia dolorosa) will require higher energy delivery.

Conclusions

Motor cortex stimulation represents a paradigm change in the treatment of central and neuropathic pain syndromes and potentially the treatment of movement disorders and paresis associated with stroke (Brown 2003). The morbidity is low and pain relief significant in patients who are otherwise not amenable to medical or surgical treatment. Clinical trials, published after the introduction of the technique a dozen years ago, continue to show favorable pain relief in this difficult to treat population of patients.

References

Bezard, E., Boraud, T., Nguyen, J. P., Velasco, F., Keravel, Y., & Gross, C. (1999). Cortical stimulation and epileptic seizure: A study of the potential risk in primates. *Neurosurgery, 45*, 346–350.

Brown, J. (2003). Guest editorial. *Neurological Research, 25*, 115–117.

Garcia-Larrea, L., Peyron, R., Mertens, P., Gregoire, M. C., Lavenne, F., Le Bars, D., et al. (1999). Electrical stimulation of motor cortex for pain control: A combined PET-scan and electrophysiological study. *Pain, 83*, 259–273.

Garcia-Larrea, L., Peyron, R., Mertens, P., Laurent, B., Mauguiere, F., & Sindou, M. (2000). Functional imaging and neurophysiological assessment of spinal and brain therapeutic modulation in humans. *Archives of Medical Research, 31*, 248–257.

Hirayama, T., Tsubokawa, T., Katayama, Y., et al. (1990). Chronic change in activity of thalamic relay neurons following spinothalamic tractotomy in cat: Effect of motor cortex stimulation. *Pain, 5*, 273.

Katayama, Y., Tsubokawa, T., & Yamamoto, T. (1994). Chronic motor cortex stimulation for central deafferentation pain: Experience with bulbar pain secondary to wallenberg syndrome. *Stereotactic and Functional Neurosurgery, 62*, 295–299.

Katayama, Y., Fukaya, C., & Yamamoto, T. (1998). Post-stroke pain control by chronic motor cortex stimulation: Neurological characteristics predicting a favorable response. *Journal of Neurosurgery, 89*, 585–591.

Meyerson, B. A., Lindblom, U., Linderoth, B., Lind, G., & Herregodts, P. (1993). Motor cortex stimulation as treatment of trigeminal neuropathic pain. *Acta Neurochir Suppl (Wien), 58*, 150–153.

Nguyen, J. P., Lefaucher, J. P., Le Guerinel, C., Eizenbaum, J. F., Nakano, N., Carpentier, A., et al. (2000a). Motor cortex stimulation in the treatment of central and neuropathic pain. *Archives of Medical Research, 31*, 263–265.

Nguyen, J. P., Lefaucheur, J. P., Le Guerinel, C., Fontaine, D., Nakano, N., Sakka, L., et al. (2000b). Treatment of central and neuropathic facial pain by chronic stimulation of the motor cortex: Value of neuronavigation guidance systems for the localization of the motor cortex. *Neurochirurgie, 46*, 483–491.

Tsubokawa, T., Katayama, Y., Yamamoto, T., Hirayama, T., & Koyama, S. (1991). Treatment of thalamic pain by chronic motor cortex stimulation. *Pacing and Clinical Electrophysiology, 14*, 131–134.

Tsubokawa, T., Katayama, Y., Yamamoto, T., Hirayama, T., & Koyama, S. (1993). Chronic motor cortex stimulation in patients with thalamic pain. *Journal of Neurosurgery, 78*, 393–401.

Yamamoto, T., Katayama, Y., Hirayama, T., & Tsubokawa, T. (1997). Pharmacological classification of central post-stroke pain: Comparison with the results of chronic motor cortex stimulation therapy. *Pain, 72*, 5–12.

Pain Treatment: Spinal Cord Stimulation

Björn A. Meyerson
Department of Clinical Neuroscience, Section of Neurosurgery, Karolinska Institute/University Hospital, Stockholm, Sweden

Synonyms

Dorsal column stimulation (DCS); Epidural spinal electrical stimulation (ESES); Spinal cord stimulation (SCS)

Definition

▶ Activation of the spinal ▶ dorsal columns by electrical stimuli, delivered via an implanted spinal electrode and a pulse generator. This technique is applied as treatment of neuropathic pain, preferably of peripheral origin, of ▶ angina pectoris, and of ischemic extremity pain. Stimulation evokes paresthesias that must cover the area of pain to be effective, and it may also induce peripheral vasodilatation.

Characteristics

▶ Spinal cord stimulation (SCS) evolved as a direct clinical application of the gate-control theory with the idea to produce inhibition of nociception at the first spinal relay by antidromic activation of the low-threshold dorsal column fibers. The general conceptualization of the theory may still be valid as a base for the understanding of the mode of action of SCS. The theory originally applied to acute nociceptive pain and it is therefore a paradox that SCS is efficacious exclusively, or at least predominantly, for neuropathic forms of pain; relief of ischemic pain relies on different mechanisms and is in fact presumably secondary to other effects of the simulation not directly related to pain. It was not until rodent models of neuropathic pain were introduced in the mid-1990s in experimental studies on SCS that some insight into its mode of action was made possible (Meyerson et al. 1995; Meyerson and Linderoth 2003). SCS applied in such animals may attenuate mechanical and thermal hypersensitivity ("allodynia") as well as neuronal hyperexcitability in the spinal dorsal horn. These effects appear to be preferentially related to Aβ fiber function while C-fiber-mediated pain is less likely to be affected, which is concordant with clinical observations. The suppressive effects on signs of neuropathy recorded in experimental animals are associated with prominent changes in various transmitter systems. There is an increased release of gamma-aminobutyric acid (GABA) in the dorsal horn, while the release of excitatory amino acids (glutamate, aspartate) is concomitantly decreased. The crucial role of GABA is further evidenced by the finding that the SCS effects on allodynia may be enhanced and counteracted by the spinal administration of $GABA_B$ agonists and antagonists, respectively. There is also evidence that spinal adenosine-related mechanisms are involved. Furthermore, there is an augmented release of acetylcholine, and this effect is present only in animals responding to SCS. The cholinergic activation is mediated via muscarinic receptors. These findings mostly relate to segmental spinal mechanisms, but there are data indicating that part of the pain-inhibiting effect of SCS is mediated via spinal-supraspinal-spinal loops (El-Khoury et al. 2002). SCS is regularly accompanied by paresthesias as a result of orthodromic activation of the dorsal columns, but it has been demonstrated that part of the SCS effect can be retained also when the stimulation is applied rostral to a transection of these tracts. A further support for the involvement of supraspinal mechanisms is the SCS activation of descending serotonin-containing fibers originating from the rostral medulla (reviews, Meyerson and Linderoth 2003; Linderoth et al. 2009; Guan 2012; Linderoth and Meyerson 2013a) (Figs. 1 and 2).

Ischemic pain is chiefly a nociceptive form of pain though neuropathic pain components may coexist. The beneficial effect on ischemic extremity pain is assumed to result from the reduction of ischemia, primarily induced by a vasodilatatory effect (Linderoth and Foreman 1999). This can be produced by an antidromic activation of primary afferents with a perispheral release of ▶ calcitonin gene-related peptide (CGRP) (Croom et al. 1997). Another mechanism is the suppression of sympathetic activity, and experimental findings suggest that these two modes of action operate in concert (Wu et al. 2008).

In myocardial ischemia, the antiangina effect of SCS is probably due to a restoration of the oxygen consumption/supply balance. It has been demonstrated that the time taken for the onset of angina during exercise or heart pacing can be significantly increased by SCS, and the depression of the ST segment in the electrocardiogram is counteracted

Pain Treatment: Spinal Cord Stimulation, Fig. 1 Schematic illustration of how a percutaneously introduced quadripolar electrode in the epidural space may activate the spinal dorsal columns. The electrode is connected to a subcutaneously implanted pulse generator, "neuropacemaker," supplying electrical pulses which are set by an external remote control. The stimulation-induced paresthesias must cover the painful region in order to produce pain relief. Experimental data indicate that the suppression of pain takes place at the first synaptic relay in the dorsal horn and by the activation of pain inhibiting supraspinal loops

(Mannheimer et al. 1993). It has also been shown that during controlled exercise, SCS causes a reversion of myocardial lactate production to extraction. Increased or redistributed coronary blood flow, reduced myocardial oxygen demand, and depression of cardiac sympathetic activity as well as reduced excitability of the intrinsic cardiac nervous system have all been suggested as underlying mechanisms (Foreman et al. 2000; Souherland et al. 2007).

SCS applied for either neuropathic, peripheral ischemic, or cardiac (angina pectoris) pain is indicated only when pharmacotherapy or surgical treatment options have been exhausted. Good treatment compliance is required because the patient externally controls the pulse generator.

There are only a few randomized, prospective SCS studies, and placebo-controlled trials are not possible due to the presence of paresthesias. On the other hand, there is a large number of case studies of SCS applied for various forms of neuropathic pain, comprising relatively large patient cohorts with long-term follow-up, and the reported outcomes are highly concordant. In general, good results, defined as >50 % pain reduction, supplemented by assessment of analgesic medication and global satisfaction, are reported in 60–70 % of the patients (Linderoth and Meyerson 2009, 2013b). It is notable that the analgesic effect often remains for many years, even decades, and there is no general tendency towards tolerance and fatigue.

Neuropathic Pain

The most common SCS neuropathic pain indication is lumbosacral rhizopathy, which in the literature is commonly referred to as "failed back surgery syndrome." This diagnosis often represents a mixed, complex pain condition, which also comprises nociceptive pain components located in the lumbar region. This "low back pain" is less likely to respond to SCS, although it has been claimed that it can also be ameliorated by using special types of multipolar electrodes and dual channel pulse generators; others maintain that it is preferably the "radiating leg pain" that is amenable to SCS. A systematic review of 72 case series from 1995 to 2002 concerning SCS applied for "chronic back and leg pain" comprised 3,427 implanted patients. Sixty-two percent of them had experienced >50 % pain relief (Taylor et al. 2005). The review includes an RCT study (North et al. 2005) where patients with radicular pain after lumbosacral spine surgery were randomized to SCS or to reoperation. A total of 45 patients were available for a mean follow-up of about 3 years, and it appeared that SCS was more successful than reoperation (P < 0.01). Patients initially randomized to SCS were significantly less likely to cross over than those randomized to reoperation. This study was scored as being of a high evidence level. Recently, a prospective randomized controlled multicenter trial of the

Pain Treatment: Spinal Cord Stimulation, Fig. 2 A schematic diagram showing a lumbar slice of the spinal cord with SCS applied just rostrally to this level. The antidromic impulses generated in the dorsal columns activate inhibitory interneurons, among them some GABAergic that reduce the activation of the hyperexcitable second-order neurons, in particular, the WDR cells and the release of excitatory transmitters. Another major impulse path is orthodromic to the brain, activating circuitry in the brain stem, ultimately giving rise to descending impulses via the dorsolateral funiculi amplifying the inhibitory processes at the spinal level. *DC* dorsal column, *DLF* dorsolateral funiculus, *GABA* g-amino butyric acid, *Ach* acetylcholine, *Aden* adenosine, *5-HT* 5-hydroxytryptophan (serotonin), *NE* noradrenalin, *SCS* spinal cord stimulation, *STT* spinothalamic tract, *WDR* wide-dynamic-range neuron (Redrawn from Linderoth and Meyersons 2010)

effectiveness of SCS in 100 patients suffering from radiating leg pain (lumbosacral rhizopathy), predominating over the pain in the low back. They were randomized to SCS combined with conventional medical management (CMM) or to CMM alone. The final outcome was evaluated at 24 months follow-up. In the intention-to-treat analysis, 37 % of the SCS patients and 2 % of the CMM patients (P = 0.0031) achieved 50 % or more pain relief in the legs. Also with regard to back pain relief, quality of life, and functional capacity, the SCS patients compared favorably to the CMM group. Interestingly, between 6 and 12 months, 32 of the 48 CMM patients crossed over to SCS. If instead the patients who actually received SCS were compared to those who did not, the percentages became instead 47 % and 7 %, respectively (P = 0.02). Thus, SCS proved to be clearly more beneficial than the conventional therapy (Kumar et al. 2008).

The second most common indication is pain following peripheral nerve injury or disease.

This type of pain is regarded by many to be the prime neuropathic pain indication for SCS with the highest probability of obtaining satisfactory and long-lasting pain relief. Such pain may occur as a result of surgery or trauma and is often associated with abnormalities of cutaneous sensibility (allodynia), which may give rise to evoked pain. The injury may comprise a major nerve or small, peripheral nerve branches (sometimes presenting as "scar pain"). Pain due to peripheral nerve injury sometimes presents as complex regional pain syndrome (CPRS type 2), and this condition is also considered to be a good indication (Kumar et al. 1997).

There is relatively good evidence for the efficacy of SCS for CRPS type 1. Data from 25 case studies with a mean follow-up of 33 months have been analyzed, and 67 % of the patients were reported to have a satisfactory pain relief (Taylor et al. 2006). There is one intention-to-treat RCT study included. Although there was a clearly significant difference in favor of SCS treatment as compared to physical therapy alone at 2 years follow-up, this effect was no longer present after 5 years. Nonetheless, the great majority of the SCS patients maintained that they had much benefitted from the SCS treatment (Kemler at al. 2012).

Pain associated with complete deafferentation, central pain, and pain located in the axial parts of the trunk are less likely to respond (Linderoth and Meyerson 2013b).

Ischemic Extremity Pain

Beside certain forms of neuropathic pain, ischemic pain in peripheral arterial occlusive disease and angina pectoris are important indications for SCS. However, it should be emphasized that the mechanisms involved in the SCS relief of ischemic pain are much different from those operating in the effect on neuropathic pain.

SCS for the management of ischemic pain in the extremities (arteriosclerosis and vasospastic disorders) is, to date, practiced in relatively few centers. Provided the treatment is applied only with stringent selection criteria, the overall outcome is favorable, with 60–70 % of the patients enjoying substantial pain relief. There

is also a marked amelioration of claudication and an increased walking distance. Three RCTs have been performed, and besides pain relief, the possible limb salvage effect was evaluated (Ubbink and Vermeulen 2006). In a well-designed but not randomized study, it was reported that in patients with inoperable, stable, critical limb ischemia, there was a highly significant difference in limb survival between those treated with SCS as compared to controls. The selection of patients was partly based on $TcpO_2$ measures, which was also employed for outcome assessment (Amann et al. 2003).

Angina Pectoris

Intractable angina pectoris has emerged as an important indication for SCS, since successful outcomes are reported in up to 85 % of patients (TenVaarwerk et al. 1999). Outcome is defined in terms of daily number of anginal attacks, consumption of nitrates, visits to the emergency room, and quality of life. Furthermore, exercise tolerance is increased, and ischemia-related ECG changes (ST depression) are reduced, which is paralleled by a reversal of cardiac lactate production to extraction. There are several RCTs comparing SCS to conventional treatment confirming the remarkably favorable outcome of the former. On the basis of an extensive literature search, it was concluded that there is strong evidence that SCS produces symptomatic benefits in refractory angina pectoris (Börjesson et al. 2008; Andréll et al. 2010). There has been much concern about the risk that SCS might conceal critical coronary ischemia leading to infarction. Several studies, however, have provided evidence that the warning pain signal under these circumstances is not influenced by SCS.

Summary

SCS is a well-established, minimally invasive treatment that is lenient and well tolerated and may offer effective management of pain conditions for which there are no alternative treatments available. It has also been shown, in a number of studies, to be cost-effective.

References

Amann, W., Berg, P., Gersbach, P., et al. (2003). Spinal cord stimulation in the treatment of non-reconstructable stable critical leg ischaemia: Results of the European Peripheral Vascular Disease Outcome Study (SCS-EPOS). *European Journal of Vascular and Endovascular Surgery, 26*, 280–286.

Andréll, P., Yu, W., Gersbach, P., et al. (2010). Long-term effects of spinal cord stimulation on angina symptoms and quality of life in patients with refractory angina pectoris–results from the European Angina Registry Link Study (EARL). *Heart, 96*, 1132–1136.

Börjesson, M., Andrell, P., Lundberg, D., et al. (2008). Spinal cord stimulation in severe angina pectoris – a systematic review based on the Swedish Council on Technology assessment in health care report on long-standing pain. *Pain, 140*, 501–508.

Croom, J. E., Foreman, R. D., Chandler, M. J., & Barron, K. W. (1997). Cutaneous vasodilatation during dorsal column stimulation is mediated by dorsal roots and CGRP. *The American Journal of Physiology, 272*, 950–957.

El-Khoury, C., Hawwa, M., Baliki, S. F., Atweh, S. F., Jabbur, S. V., & Saadé, N. E. (2002). Attenuation of neuropathic pain by segmental and supraspinal activation of the dorsal column system in awake rats. *Neuroscience, 112*, 541–553.

Foreman, R. D., Linderoth, B., Ardell, J. L., Barron, K. W., Dejongste, M. J. L., Hull, S. S., Chandler, M. J., TerHhorst, G. J., & Armour, J. A. (2000). Modulation of intrinsic, cardiac neuronal activity by spinal cord stimulation (SCS) in the dog: Implications for the use of SCS in angina pectoris. *Cardiovascular Research, 47*, 367–375.

Guan, Y. (2012). Spinal cord stimulation: Neurophysiological and neurochemical mechanisms of action. *Current Pain and Headache Reports, 16*, 217–225.

Kemler, M. A., de Vet, H. C., Barendse, G. A., et al. (2012). Effect of spinal cord stimulation for chronic complex regional pain syndrome Type I: Five-year final follow-up of patients in a randomized controlled trial. *Journal of Neurosurgery, 108*, 292–298.

Kumar, K., Nath, R. K., & Toth, C. (1997). Spinal cord stimulation is effective in the management of reflex sympathetic dystrophy. *Neurosurgery, 40*, 503–508.

Kumar, K., Taylor, R. S., Jacques, L., Eldabe, S., Meglio, M., Molet, J., Thomson, S., et al. (2008). The effects of spinal cord stimulation in neuropathic pain are sustained: A 24-month follow-up of the prospective randomized controlled multicenter trial of the effectiveness of spinal cord stimulation. *Neurosurgery, 63*, 762–770.

Linderoth, B., & Foreman, R. D. (1999). Physiology of spinal cord stimulation: Review and update. *Neuromodulation, 2*, 150–164.

Linderoth, B., & Meyerson, B. A. (2009). Spinal cord stimulation. Techniques, indications and outcome. In A. M. Lozano, P. L. Gildenberg, & R. R. Tasker (Eds.), *Textbook of stereotactic and functional neurosurgery* (pp. 2305–2330). Berlin: Springer.

Linderoth, B., & Meyerson, B. A. (2010). Spinal cord stimulation: exploration of the physiological basis of a widely used therapy. *Anesthesiology, 113*, 1265–1267.

Linderoth, B., Meyerson, B. A. (2013a). Spinal cord stimulation: Mechanisms of action. In K. Burchiel (Ed.), *Surgical management of pain* (2nd ed.). New York: Thieme Medical Publishers. (In press).

Linderoth, B., Meyerson, B. A. (2013b). Spinal cord and brain stimulation. In: S. B. McMahon, M. Koltzenburg (Eds.), *Textbook of pain* (6th ed.). Elsevier. 2013a (In press).

Linderoth, B., Foreman, R. D., & Meyerson, B. A. (2009). Mechanisms of action of spinal cord stimulation. In A. M. Lozano, P. L. Gildenberg, & R. R. Tasker (Eds.), *Textbook of stereotactic and functional neurosurgery* (pp. 2331–2347). Berlin: Springer.

Mannheimer, C., Eliasson, T., Andersson, B., et al. (1993). Effects of spinal cord stimulation in angina pectoris induced by pacing and possible mechanisms of action. *British Medical Journal, 307*, 477–480.

Meyerson, B. A., & Linderoth, B. (2003). Spinal cord stimulation – Mechanisms of action in neuropathic and ischemic pain. In B. Simpson (Ed.), *Electrical stimulation and the relief of pain* (pp. 161–182). Amsterdam: Elsevier.

Meyerson, B. A., Ren, B., Herregodts, P., & Linderoth, B. (1995). Spinal cord stimulation in animal models of mononeuropathy: Effects on the withdrawal response and the flexor reflex. *Pain, 61*, 229–243.

North, R. B., Kidd, D. H., Farrokhi, F., et al. (2005). Spinal cord stimulation versus repeated lumbosacral spine surgery for chronic pain: A randomized, controlled trial. *Neurosurgery, 56*, 98–106.

Souherland, E. M., Milhorn, D. M., Foreman, R. D., Linderoth, B., DeJongste, M. J., Armour, J. A., et al. (2007). Pre-emptive, but not reactive, spinal cord stimulation mitigates transient ischemia induced myocardial infarction via cardiac adrenergic neurons. *American Journal of Physiology. Heart and Circulatory Physiology, 292*, H311–H317.

Taylor, R.S., Van Buyten, J.P., & Buchser, E. (2005). Spinal cord stimulation for chronic back and leg pain and failed back surgery syndrome: A systematic review and analysis of prognostic factors. *Spine, 30*, 152–160.

Taylor, R. S., Van Buyten, J. P., & Buchser, E. (2006). Spinal cord stimulation for complex regional pain syndrome: A systematic review of the clinical and cost-effectiveness literature and assessment of prognostic factors. *European Journal of Pain, 10*, 91–101.

TenVaarwerk, I. A., Jessurun, G. A., DeJongste, M. J., Andersen, C., Mannheimer, C., Eliasson, T., Tadema, W., & Staal, M. J. (1999). Clinical outcome of patients treated with spinal cord stimulation for therapeutically refractory angina pectoris. The working group on neurocardiology. *Heart, 82*, 82–88.

Ubbink, D. T., & Vermeulen, H. (2006). Spinal cord stimulation for critical leg ischemia: A review of effectiveness and optimal patient selection. *Journal of Pain and Symptom Management, 31*(Suppl 4), S30–S35.

Wu, M., Linderoth, B., & Foreman, R. D. (2008). Putative mechanisms behind effects of spinal cord stimulation on vascular diseases: A review of experimental studies. *Autonomic Neuroscience, 138*, 9–23.

Pain Treatment: Spinal Nerve Blocks

Richard B. North
The Johns Hopkins University, Baltimore,
MD, USA

Synonyms

Diagnostic block; Prognostic block; Spinal nerve block; Therapeutic block

Definition

Spinal injections of ▶ local anesthetic agents are used (Boas 1991) as an aid to diagnosis, as prognostic tests for ablative procedures, and (Mooney 1991) as therapy. They may be administered within, or adjacent to, the spinal column.

Characteristics

Temporary nerve blocks using local anesthetic are often employed in the diagnostic evaluation of spinal pain syndromes – for example, in establishing lumbar radiculopathy (Boas 1991; Bonica and Buckley 1990; Mooney 1991; Steindler and Luck 1938) or lumbar facet disease (Carette et al. 1991; North et al. 1994; Schwarzer et al. 1994) as a cause of low back and sciatic pain. Temporary relief of pain by reversible local anesthetic blocks, however, does not reliably predict the long-term results of permanent ablative or anatomic procedures.

The spinal nerves arise from anterior (motor) and posterior (sensory) rootlets shortly after they exit the spinal cord. They traverse the subdural space (filled with spinal fluid) enclosed by the dura and fuse prior to exiting this space. As they exit the spinal canal through the foramina between the pedicles of adjacent vertebrae, the sensory axons proceed into their cell bodies, which are located in the dorsal root ganglion (DRG). Thereafter, a branch is sent to the ▶ anterior primary ramus, which provides the nerve supply to the extremities (e.g., the brachial plexus) and the chest wall. A second branch is the ▶ posterior primary ramus, which supplies branches to the paraspinal joints, ligaments, discs, and skin. Commonly used spinal nerve block targets include the nerve root (at the level of the DRG) (Boas 1991) and the medial branch of the posterior primary ramus (Bonica and Buckley 1990). "Facet" blocks are also commonly performed at multiple adjacent levels. ▶ Referred pain due to spinal pathology, for example, sciatica due to nerve root involvement, often responds not only to blockade of these spinal structures but also to peripheral nerve blockade in the distribution of the pain, viz. (Mooney 1991) sciatic nerve block at the medial border of the sciatic notch in the buttock. A lumbar subcutaneous control injection (superficial to lumbosacral fascia and superficial to any "trigger point") is seldom effective (Steindler and Luck 1938).

We have reported previously that in patients whose sciatica is relieved completely by monoradicular lumbosacral blocks, ablating the primary afferent neurons by dorsal root ganglionectomy does not provide long-term pain relief (North et al. 1991). This is consistent with prior studies of dorsal rhizotomy, whose long-term results were predicted poorly by temporary diagnostic or prognostic blocks (Loeser 1972). Likewise, using lumbar medial branch posterior primary ramus blocks to screen patients for radiofrequency denervations, we have observed that the results of the former predict the long-term results of the latter imperfectly (North et al. 1994). The temporary clinical effects of reversible blocks are not maintained by permanent ablative procedures for chronic pain. Furthermore, diagnostic or prognostic blocks have failed to predict the outcome of anatomic procedures, such as lumbar fusion or decompression, in patients responding to facet joint or nerve root blocks (Dooley et al. 1988; Esses and Moro 1993).

Pain may be relieved by temporary blocks directed to areas of pain referral, as opposed to areas of underlying pathology. In one series of patients with a chief complaint of sciatica attributable in all cases to radiculopathy, not only paraspinal lumbosacral root blocks and medial branch posterior primary ramus blocks (at or proximal to the pathology) but also sciatic nerve blocks (distal or collateral to the pathology) produced temporary relief in a majority of patients; the specificity of diagnostic blocks was as low as 24 % (North et al. 1996).

Nonspecific relief of segmental pain by reversible blockade may be explained anatomically by branching of primary afferents and/or by convergence upon second-order neurons in the spinal cord of primary afferents, from the regions of injury and of referred pain (Roberts et al. 1989). As reflected by single-unit recordings from centrally projecting wide dynamic range neurons, nociceptive as well as non-painful sensory inputs have excitatory effects upon the same central pain-signaling neurons (Gillette et al. 1993). Pain relief may reflect blockade of tonic, normally non-painful input to central neurons, which is necessary to maintain their ability to signal pain (Owens et al. 1992; Willis 1993).

Double blocks, using anesthetic agents with different durations of action, have been advocated as a method to isolate physiologic, as opposed to functional or placebo responses, to identify a putative pain generator (Schwarzer et al. 1994). Longer-lasting relief with a longer-acting agent is interpreted as a specific response, indicating that the pain generator has been identified. Segmental physiologic effects may vary, however, with duration of drug action, confounding diagnosis by double blocks.

Lack of specificity may in fact be advantageous in therapeutic applications of local anesthetic blockade: blocking an area of pain referral may afford relief even in the absence of local pathology.

There are nonspecific effects not only in the spatial domain but also in the time domain: prolonged relief of pain may occur after local anesthetic blocks, substantially exceeding their pharmacologic duration of action (Arn et al. 1990).

References

Arn, S., Lindblom, U., Myerson, B. A., & Molander, C. (1990). Prolonged relief of neuralgia after regional anesthetic blocks. A call for further experimental and systematic clinical studies. *Pain, 43*, 287–297.

Boas, R. A. (1991). Nerve blocks in the diagnosis of low back pain. In J. D. Loeser (Ed.), *Low back pain, neurosurgery clinics of North America* (pp. 807–816). Philadelphia: W. B. Saunders.

Bonica, J. J., & Buckley, F. P. (1990). Regional analgesia with local anesthetics. In J. J. Bonica (Ed.), *The management of pain* (pp. 1883–1966). Philadelphia: Lea & Fibiger.

Carette, S., Marcoux, S., Truchon, R., Grondin, C., Gagnon, J., Allard, Y., & Latulippe, M. A. (1991). Controlled trial of corticosteroid injections into facet joints for chronic low back pain. *The New England Journal of Medicine, 325*, 1002–1007.

Dooley, J. F., McBroom, R. J., Taguchi, T., & Macnab, I. (1988). Nerve root infiltration in the diagnosis of radicular pain. *Spine, 13*, 79–83.

Esses, S. I., & Moro, J. K. (1993). The value of facet joint blocks in patient selection for lumbar fusion. *Spine, 18*, 185–190.

Gillette, R. G., Kramis, R. C., & Roberts, W. J. (1993). Characterization of spinal somatosensory neurons having receptive fields in lumbar tissues of cats. *Pain, 54*, 85–98.

Loeser, J. D. (1972). Dorsal rhizotomy for the relief of chronic pain. *Journal of Neurosurgery, 36*, 745–754.

Mooney, V. (1991). Injection studies: Role in pain definition. In J. W. Frymoyer, T. B. Ducker, N. M. Hadler, J. P. Kostuik, J. N. Weinstein, & T. S. Whitecloud (Eds.), *The adult spine: Principles and practice* (pp. 527–540). New York: Raven.

North, R. B., Kidd, D. H., Campbell, J. N., & Long, D. M. (1991). Dorsal root ganglionectomy for failed back surgery syndrome: A five year follow-up study. *Journal of Neurosurgery, 74*, 236–242.

North, R. B., Han, M., Zahurak, M., & Kidd, D. H. (1994). Radiofrequency lumbar facet denervation: Analysis of prognostic factors. *Pain, 57*, 77–83.

North, R. B., Kidd, D. H., Zahurak, M., & Piantadosi, S. (1996). Specificity of diagnostic nerve blocks: A prospective, randomized, study of sciatica due to lumbosacral spine disease. *Pain, 65*, 77–85.

Owens, C. M., Zhang, D., & Willis, W. D. (1992). Changes in the response state of primate spinothalamic cells caused by mechanical damage or activation of descending controls. *Journal of Neurophysiology, 67*, 1509–1527.

Roberts, W. J., Gillette, R. G., & Kramis, R. C. (1989). Somatosensory input from lumbar paraspinal tissues: Anatomical terminations and neuronal responses to mechanical and sympathetic stimuli. *Society for Neuroscience Abstrtract, 15*, 755.

Schwarzer, A. C., Aprill, C. N., Derby, R., Fortin, J., Kine, G., & Bogduk, N. (1994). The false-positive rate of uncontrolled diagnostic blocks of the lumbar zygapophysial joints. *Pain, 58*, 195–200.

Steindler, A., & Luck, J. V. (1938). Differential diagnosis of pain in the low back: Allocation of the source of pain by procaine hydrochloride method. *Journal of the American Medical Association, 110*, 106–113.

Willis, W. D. (1993). Mechanical allodynia: A role for sensitized nociceptive tract cells with convergent input from mechanoreceptors and nociceptors? *APS Journal, 2*, 23–33.

Pain, Psychiatry, and Ethics

Mark D. Sullivan
Psychiatry and Behavioral Sciences, University of Washington, Seattle, WA, USA

Synonyms

Mental pain; Pain disorder; Psychogenic pain; Somatization

Definition

Pain disorder associated with psychological factors now officially exists as a subtype of the psychiatric diagnosis of pain disorder in the American Psychiatric Association's Diagnostic and Statistical Manual, Fourth Edition (DSM-IV) (1994). This diagnosis requires that (a) pain in one or more anatomical sites is of sufficient severity to warrant clinical attention; (b) pain causes clinically significant distress or impairment in social, occupational, or other important areas of functioning; (c) psychological factors are judged to have an important role in the onset, severity, exacerbation, or maintenance of the pain; (d) the symptom of pain is not intentionally produced or feigned (as in factitious disorder or malingering); and (e) the pain is not better accounted for by a mood, anxiety, or psychotic disorder and does not meet criteria for dyspareunia.

Characteristics

In many common pain syndromes (e.g., low back pain, headache, fibromyalgia), it is difficult to identify the tissue pathology giving rise to symptoms. When a somatic cause for pain cannot be identified, many clinicians begin to seek psychological causes. The identification of what has classically been called "psychogenic pain" is a difficult, and perhaps impossible, task. "Pain disorder" is the current psychiatric diagnosis that most closely corresponds to the concept of psychogenic pain.

We are drawn to the concept of psychogenic pain because it fills the gaps left when our attempts to explain clinical pain, exclusively in terms of tissue pathology, fail. Positive criteria for the identification of psychogenic pain, mechanisms for the production of psychogenic pain, and specific therapies for psychogenic pain are lacking. Psychiatric diagnosis of many disorders, such as depression, can be very helpful to the clinician and patient by pointing toward specific effective therapies. However, the diagnosis of psychogenic pain too often only serves to stigmatize further the patient who experiences chronic pain.

The recent history of the psychiatric diagnosis of "psychogenic pain" is instructive. In 1980, DSM-III introduced a new diagnostic category for pain problems, "psychogenic pain disorder" (American Psychiatric Association 1980). To qualify, a patient needed to have severe and prolonged pain inconsistent with neuroanatomical distribution of pain receptors or without detectable organic etiology or pathophysiologic mechanism. Related organic pathology was allowed, but the pain had to be "grossly in excess" of what was expected on the basis of physical exam. Difficulties in establishing that pain was psychogenic led to changes in the diagnosis for DSM-IIIR (American Psychiatric Association 1987). In DSM-IIIR, the diagnosis was renamed "somatoform pain disorder" and three major changes were made in the diagnostic criteria. The requirements for etiologic psychological factors and lack of other contributing mental disorders were eliminated, and a requirement for "preoccupation with pain for at least 6 months" was added

(Stoudemire and Sandu 1987). In DSM-IIIR, therefore, somatoform pain disorder becomes purely a diagnosis of exclusion. The diagnosis is made when medical disorders are excluded in a patient "preoccupied" with pain.

The DSM-IV subcommittee on pain disorders found that, despite these changes, "somatoform pain disorder" was rarely used in research projects or clinical practice. They identified a number of reasons for this: (1) the meaning of "preoccupation with pain" is unclear, (2) whether pain exceeds that which is expected is difficult to determine, (3) the diagnosis does not apply to many patients disabled by pain where a medical condition is contributory, (4) the term "somatoform pain disorder" implies that this pain is somehow different from organic pain, and (5) acute pain of less than 6 months duration was excluded (King and Strain 1992).

The DSM-IV subcommittee has tried to devise a broader diagnostic grouping encompassing both acute and chronic pain problems. They wanted to have all the factors relevant to the onset or maintenance of the pain delineated and also to have a diagnostic category that does not require more training than the majority of DSM-IV users would be expected to have. These two requirements may not be compatible. Furthermore, no guidance is given in determining when psychological factors have a major role in pain or are considered important enough in the presence of a painful medical disorder to be coded as a separate mental disorder. Given the high rates of mood and anxiety disorders among disabled chronic pain patients, many patients most appropriate for the diagnosis would be excluded. Although depression and anxiety diagnoses point toward specific proven therapies, this is not true for pain disorder. The diagnosis thus continues covertly as a diagnosis of exclusion, with neither clear inclusion criteria nor implications for therapy.

Prospective studies have demonstrated a strong association between medically unexplained symptoms (pain and nonpain) and depressive and anxiety disorders (Gureje et al. 2001). Patients with chronic pain will often dismiss a depression diagnosis, stating that their depression is a "direct reaction" to their pain

problem. Psychiatry has long debated the value of distinguishing a "reactive" form of depression caused by adverse life events and an endogenous from of depression caused by biological and genetic factors (Frank et al. 1994). Life events are important in many depressive episodes, though they play a less important role in recurrent and very severe or melancholic or psychotic depressions (Fishbain et al. 1997). Currently, the only life event that will exclude someone from a depression diagnosis, who otherwise qualifies, is bereavement. Determining whether a depression is a "reasonable response" to life's stress may be very important to patients seeking to decrease the stigma of a depression diagnosis and has been of interest to pain investigators. It is not, however, important in deciding that treatment is necessary and appropriate. Indeed, there is no clarity in assessment to be gained from debating whether the depression caused the pain or the pain caused the depression. If patients meet the criteria for a depressive disorder, it is likely that they can benefit from appropriate treatment.

Psychiatric diagnosis and treatment of these depressive and anxiety disorders can add an essential and often neglected component to the conceptualization and treatment of chronic pain problems (Sullivan and Turk 2000). However, it is absolutely critical to avoid a dualistic model, which postulates that pain is either physical or mental in origin. This model alienates patients who feel blamed for their pain. It also is inconsistent with modern models of pain causation. Since the gate control theory of pain, multiple lines of evidence suggest that pain is a product of efferent as well as afferent activity in the nervous system. Tissue damage and nociception are neither necessary nor sufficient for pain (Sullivan 1998). Indeed, it is now widely recognized that the relationship between pain and nociception is highly complex and must be understood in terms of the situation of the organism as a whole. Attribution of pain to solely psychological causes is ethically suspect, because it is often used as justification for denial of medical treatment to persons suffering from such pain.

We are only beginning to understand the complexities of the relationship between pain and suffering. Pain usually, but not always, produces suffering. Suffering can, through somatization, produce pain (Katon et al. 2001).We have traditionally understood this suffering, as we have understood nociception, as arising from a form of pathology intrinsic to the sufferer, hence the traditional view that pain is due to either tissue pathology (nociception) or psychopathology (suffering). An alternative model that allows us to escape this dualism is to think of pain as having causes outside as well as inside the body (Sullivan 2001). For humans, social pathology can be as painful as tissue pathology (Eisenberger et al. 2003). Psychological trauma as well as physical trauma appears to contribute to many chronic pain problems. Insurance carriers and medical practitioners who exclude psychological trauma from coverage or treatment may deny patients' resources essential for recovery.

Psychologic care for patients with chronic pain should occur within the medical treatment setting whenever possible. This is the most effective way to reassure patients that the somatic elements of their problems are not neglected. It also allows integration of somatic and psychological treatments in the most effective manner.

References

American Psychiatric Association. (1980). *Diagnostic and statistical manual of mental disorders* (3rd ed.). Washington, DC: American Psychiatric Association Press.

American Psychiatric Association. (1987). *Diagnostic and statistical manual of mental disorders* (Rev. 3rd ed.). Washington, DC: American Psychiatric Association Press.

American Psychiatric Association. (1994). *Diagnostic and statistical manual of mental disorders* (4th ed.). Washington, DC: American Psychiatric Association Press.

Eisenberger, E. I., Lieberman, M. D., & Williams, K. D. (2003). Does rejection hurt? An fMRI study of social exclusion. *Science, 302*, 290–292.

Fishbain, D. A., Cutler, R., Rosomoff, H. L., et al. (1997). Chronic pain-associated depression: Antecedent or consequence of chronic pain? A review. *The Clinical Journal of Pain, 13*, 116–137.

Frank, E., Anderson, B., Reynolds, C. F., et al. (1994). Life events and the research diagnostic criteria endogenous subtype. *Archives of General Psychiatry, 51*, 519–524.

Gureje, O., Simon, G. E., & Von Korff, M. (2001). A cross-national study of the course of persistent pain in primary care. *Pain, 92*, 195–200.

Katon, W. J., Sullivan, M. D., & Walker, E. A. (2001). Medical symptoms without identified pathology: Relationship to psychiatric disorders, childhood and adult trauma, personality traits. *Annals of Internal Medicine, 134*, 917–925.

King, S. A., & Strain, J. J. (1992). Revising the category of somatoform pain disorder. *Hospital & Community Psychiatry, 43*, 217–219.

Stoudemire, A., & Sandu, J. (1987). Psychogenic/idiopathic pain syndromes. *General Hospital Psychiatry, 9*, 79–86.

Sullivan, M. (1998). The problem of pain in the clinicopathological method. *The Clinical Journal of Pain, 14*, 196–201.

Sullivan, M. D. (2001). Finding pain between minds and bodies. *The Clinical Journal of Pain, 17*, 146–156.

Sullivan, M. D., & Turk, D. C. (2000). Psychiatric disorders and psychogenic pain. In J. D. Loeser, S. Butler, & D. C. Turk (Eds.), *Bonica's management of pain* (3rd ed., pp. 483–500). Philadelphia: Williams and Wilkins.

Pain-Facilitatory Systems

Definition

Systems within the central nervous system whose normal function is to increase pain perception.

Cross-References

▶ Descending Modulation of Nociceptive Processing
▶ Pain-Modulatory Systems, History of Discovery

Painful Channelopathies

Stephen G. Waxman
Department of Neurology, Yale School of Medicine, New Haven, CT, USA

Definition

Channelopathies are an important class of disorders in which aberrant function of ▶ ion channels, or lack of function of ion channels, produces clinical signs and symptoms. Acquired transcriptional ▶ channelopathies are the result of dysregulated production of channel proteins, due to changes in transcription of their respective genes; although the ion channel proteins are in themselves normal, deployment of abnormal ensembles of channel proteins distorts the function of neurons. Genetic channelopathies, on the other hand, are the result of mutations in ion channel genes, which result in the absence of channels, channels that fail to function, or channels that function abnormally. Painful channelopathies are those in which patients experience clinically significant pain.

Characteristics

Given the crucial role of voltage-gated ion channels in the generation and propagation of action potentials, it is not surprising that channelopathies (acquired, in which there are abnormal patterns of expression of normal ion channel genes, and genetic, in which mutant genes produce abnormal channels) can produce pain syndromes. However, it has only been over the past decade that, with the advent of contemporary molecular techniques of analyses, it has been possible to identify voltage-gated ion channels as major contributors to pain syndromes and, in fact, to provide precise molecular identification of the channel isoforms that are involved.

One type of acquired painful channelopathy involves the upregulation of the $Na_v1.3$ sodium channel following nerve injury. Waxman et al. (1994) demonstrated that, following transection of the axons of ▶ dorsal root ganglion (DRG) neurons, there is a marked upregulation of mRNA for the $Na_v1.3$ sodium channel. This was subsequently shown to be due, at least in part, to interruption of access to peripheral pools of neurotrophins, including ▶ nerve growth factor (NGF) and glial-cell-line-derived neurotrophic factor (GDNF) (Black et al. 1997; Dib-Hajj et al. 1998; Boucher et al. 2000). Using immunocytochemistry, Black et al. (1999) demonstrated that, following axonal transection, there is an upregulation of $Na_v1.3$ protein as well as

mRNA, with deposition of abnormal aggregations of $Na_v1.3$ within the injured axonal tips within experimental neuromas. ▶ Patch-clamp studies demonstrated that, concomitant with the upregulation of $Na_v1.3$, a new TTX-sensitive sodium current, characterized by rapid recovery from inactivation (rapid repriming), emerges within axotomized DRG neurons; the rapid recovery from inactivation would be expected to poise these neurons to fire at inappropriately high frequencies (Cummins and Waxman 1997). Evidence for production of the rapidly repriming current by $Na_v1.3$ was initially inferential, based on the concurrent time courses of upregulation of $Na_v1.3$ and the rapidly repriming current. However, subsequent studies, in which $Na_v1.3$ was heterologously expressed in HEK cells and within DRG neurons, provided definitive evidence that the $Na_v1.3$ channel produces a rapidly repriming current (Figs. 1 and 2).

A more recent study has demonstrated that, following injury to peripheral nerves, $Na_v1.3$ is expressed not only within DRG neurons but also within secondary (and possible tertiary) nociceptive neurons in superficial laminae within the dorsal horn (Hains et al. 2004). Concomitant with expression of $Na_v1.3$ within these dorsal horn nociceptive neurons, they become hyperresponsive, firing spontaneously and at higher-than-normal frequencies in response to stimulation. One study has reported that $Na_v1.3$ antisense oligonucleotides, which penetrate the dorsal horn but not DRG, reverse the upregulation of $Na_v1.3$ within dorsal horn neurons, attenuate their abnormal hyperresponsive firing, and ameliorate the pain-related behavior that occurs in this experimental model (Hains et al. 2004). Another study, assaying another $Na_v1.3$ antisense construct, did not observe significant amelioration of pain (Lindia et al. 2005). Moreover, it has been reported that knock-out of $Na_v1.3$ does not abrogate neuropathic pain (Nassar et al. 2006); this result suggests either that $Na_v1.3$ is not an essential contribution to neuropathic pain or that compensatory changes such as upregulation of other sodium channel isoforms eclipse the effect of knocking out the channel. On the other hand, a more recent study,

Painful Channelopathies, Fig. 1 Na$_v$1.3 sodium channel immunostaining in control (**a**, **c**) and axotomized (**b**, **d**) DRG neurons in vitro. Control DRG neurons exhibit low levels of or nondetectable Na$_v$1.3 immunostaining, while somata of peripherally axotomized small DRG neurons display robust Na$_v$1.3 immunoreactivity. (**e**) Na$_v$1.3 sodium channel immunostaining in ligated and transected sciatic nerve. Note the increased Na$_v$1.3 immunostaining immediately proximal to the ligature constricting the sciatic nerve. Control sciatic nerve (not shown) does not exhibit Na$_v$1.3 immunoreactivity. Scale bar: 100 μm (Modified from Black et al. 1999)

using sh RNA for knock down, showed that Na$_v$1.3 is a significant contributor to neuropathic pain (Samod et al. 2012), so Na$_v$1.3 remains an attractive therapeutic target.

Hains et al. (2003) have also demonstrated upregulation of Na$_v$1.3 within DRG and dorsal horn nociceptive neurons following contusive spinal cord injury. This upregulation of Na$_v$1.3 is accompanied by pain-related behaviors. As with peripheral nerve injury, they observed that introduction of Na$_v$1.3 antisense, which attenuates the expression of Na$_v$1.3 within dorsal horn neurons, blunts their hyperexcitability and ameliorates the pain-related behaviors.

Na$^+$ channel α subunits are the most salient of the neuronal membrane channels, whose expression in primary sensory neurons is regulated in the event of nerve injury. However, a number of other ion channels that may affect electrical excitability in afferent neurons are known to be regulated by axotomy. These include the Na$^+$ channel β2 and β3 subunits and a variety

Painful Channelopathies, Fig. 2 Recovery from inactivation for TTX-sensitive currents is faster in DRG neurons following transection of their peripheral axons, due at least in part to upregulation of Na$_v$1.3. (**a**) and (**b**): family of TTX-sensitive current traces from control (**a**) and axotomized (**b**) DRG cells showing the rate of recovery from inactivation at −80 mV. Cells are prepulsed to −20 mV for 20 ms to inactivate all of the current and then returned to −80 mV for increasing recovery durations before the test pulse of −20 mV. Maximum pulse rate was 0.5 Hz. (**c**) Time course for recovery from inactivation at −80 mV for the currents shown in (**a**) and (**b**). Recovery is much faster in the axotomized neuron. Single exponential functions fitted to the data yield time constants of 160 ms for the control neuron and 41 ms for the axotomized neurons (Modified from Black et al. 1999)

of voltage-sensitive K$^+$ channels (including Kv1.1, 1.2, 1.3, 1.4, 2.1, 4.3, 9.3, the α2 δ subunit of L-type Ca^{2+} channel, and the hyperpolarization-activated, cyclic nucleotide-modulated (HCN) "pacemaker" channel (Ishikawa et al. 1999; Xiao et al. 2002; Wang et al. 2002; Chaplan et al. 2003; Yang et al. 2004a).

Recent studies on hereditary erythromelalgia (a rare autosomal dominant disease characterized by intermittent burning pain that is often triggered by warmth or exercise, associated with warmth and redness in the feet and hands, also termed erythermalgia) provided the first example of a genetic painful channelopathy in which the physiological basis for pain is understood. Yang

et al. (2004b) demonstrated, in two families with erythromelalgia, two mutations in the Na$_v$1.7 channel. This channel is known to be selectively expressed within DRG neurons and is present within their neurites, including the axon tips (Toledo-Aral et al. 1997). Cummins et al. (1998) demonstrated that the Na$_v$1.7 channel exhibits slow closed-state inactivation, a biophysical characteristic that endows it with responsiveness to slow, small depolarizations. Thus, the Na$_v$1.7 channel appears to be deployed close to sensory terminals and is tuned in a manner that poises it to amplify depolarizing stimuli such as generator potentials. Yang et al. (2004b) reported two mutations in Na$_v$1.7 in erythromelalgia, one located in the II/S5 segment (L858H) and the other in the loop region between II/S4 and II/S5 (I848T). Importantly, the leucine and isoleucine at these two sites are highly conserved within most voltage-gated sodium channels, and a mutation similar to the I848T mutation in $Na_v1.7$ has previously been reported in familial paramyotonia. The physiological effects of the I848T and L858H $Na_v1.7$ mutations in erythromelalgia were first functionally characterized by Cummins et al. (2004) who observed that they produce a negative shift in the voltage dependence of activation (making it easier to open the channels), slowed deactivation (keeping them open longer once activated), and an enhanced response to small depolarizing inputs similar to generator potentials; these changes in functional characteristics of the mutant channels confer hyperexcitability on DRG neurons in erythromelalgia. Since these initial discoveries, more than ten additional mutations in $Na_v1.7$ have been found and characterized in other families with erythromelalgia. These mutations increase the excitability of nociceptive DRG neurons (Waxman and Dib-Hajj 2005). A compendium of these mutations has been published by Dib-Hajj et al. (2010).

A second genetic channelopathy involving $Na_v1.7$ that causes pain, paroxysmal extreme pain disorder (PEPD), was described in 2006 (Fertleman et al. 2006). PEPD is clinically distinct from familial erythromelalgia and is characterized by severe pain in the rectal region, triggered by perineal stimulation, which often migrates to the area around the eye or jaw in adulthood. Like hereditary erythromelalgia, PEPD is caused by point mutations in $Na_v1.7$. However, the mutations that cause PEPD impair inactivation of the $Na_v1.7$ channel (Fertleman et al. 2006). A patient with a mixed clinical phenotype, with symptoms of both erythromelalgia and PEPD, has been described, and in this patient, the $Na_v1.7$ mutation produced a set of physical changes common to PEPD (impaired fast inactivation) and inherited erythromelalgia (hyperpolarized activation, slowed deactivation, and enhanced ramp currents) (Estacion et al. 2008).

Inherited erythromelalgia and PEPD are both produced by gain-of-function mutations, i.e., mutations that enhance the activity of $Na_v1.7$ channels. Loss-of-function mutations of $Na_v1.7$ have also been described in humans. Channelopathy-associated insensitivity to pain (CIP) is caused by autosomal recessive mutations of $Na_v1.7$ that truncate the channel protein or impair splicing signals so that functional channels are not produced (Cox et al. 2006; Ahmad et al. 2007; Goldberg et al. 2007). Patients with CIP present a remarkable picture of insensitivity to all types of painful stimuli; they display painless fractures, painless burns, and painless bites to the lips and tongue and report that they do not feel any form of pain, although other sensory modalities, except olfaction, appear normal.

Still another example of a genetic painful channelopathy is familial hemiplegic migraine (FHM). FHM is a rare autosomal dominant subtype of migraine with aura. Missense mutations in the human chromosome 19 CACNA1A gene, which codes for the p-/q-type calcium channel $Ca_v2.1$, have been found in some affected families (Kors et al. 2002). The detailed relation between the gene mutation and the pain phenotype FHM is not yet fully understood. Understanding of the physiological effects of these mutations will almost certainly contribute substantially to our understanding of the associated pain syndromes.

References

Ahmad, S., Dahllund, L., Eriksson, A. B., Hellgren, D., Karlsson, U., Lund, P.-E., Meijer, I. A., Meury, L., Mills, T., Moody, A., Morinville, A., Morten, J., O'Donnel, D., Raynoschek, C., Salter, H., Rouleau, G., & Krupp, J. (2007). A stop codon mutation in SCN9A causes lack of pain sensation. *Human Molecular Genetics, 16*, 2114–2121.

Black, J. A., Cummins, T. R., Plumpton, C., Chen, Y. H., Hormuzdiar, W., Clare, J. J., & Waxman, S. G. (1999). Upregulation of a silent sodium channel after peripheral, but not central, nerve injury in DRG neurons. *Journal of Neurophysiology, 82*, 2776–2785.

Black, J. A., Langworthy, K., Hinson, A. W., Dibb-Hajj, S. D., & Waxman, S. G. (1997). NGF has opposing effects on Na + channel III and SNS gene expression in spinal sensory neurons. *NeuroReport, 8*, 2331–2335.

Boucher, T. J., Okuse, K., Bennett, D. L., Munson, J. B., Wood, J. N., & McMahon, S. B. (2000). Potent analgesic effects of GDNF in neuropathic pain states. *Science, 290*, 124–127.

Chaplan, S. R., Guo, H.-Q., Lee, D. H., Luo, L., Liu, C., Kuei, C., Velumian, A. A., Butler, M. P., Bown, S. M., & Dubin, A. E. (2003). Neuronal hyperpolarization-activated pacemaker channels drive neuropathic pain. *Journal of Neuroscience, 23*, 1169–1178.

Cox, J. J., Reiman, F., Nicholas, A. K., Thornton, G., Roberts, E., Springell, K., Karbani, G., Jafri, H., Mannan, J., Raashid, Y., Al-Gazali, L., Hamamy, H., Valente, E. M., Gorman, S., Williams, R., McHale, D. P., Wood, J. N., Gribble, F. M., & Woods, C. G. (2006). An SCN9A channelopathy causes congenital inability to experience pain. *Nature, 444*, 894–898.

Cummins, T. R., Dib-Hajj, S. D., & Waxman, S. G. (2004). Electrophysiological properties of mutant Nav1.7 sodium channels in a painful inherited neuropathy. *Journal of Neuroscience, 24*, 8232–8236.

Cummins, T. R., Howe, J. R., & Waxman, S. G. (1998). Slow closed-state inactivation: A novel mechanism underlying ramp currents in cells expressing the hNE/PN1 sodium channel. *Journal of Neuroscience, 18*, 9606–9619.

Cummins, T. R., & Waxman, S. G. (1997). Down-regulation of tetrodotoxin-resistant sodium currents and up-regulation of a rapidly repriming tetrodotoxin-sensitive sodium current in small spinal sensory neurons following nerve injury. *Journal of Neuroscience, 17*, 3503–3514.

Dib-Hajj, S. D., Black, J. A., Cummins, T. R., Kenney, A. M., Kocsis, J. D., & Waxman, S. G. (1998). Rescue of Î ± −SNS sodium channel expression in small dorsal root ganglion neurons after axotomy by nerve growth factor in vivo. *Journal of Neurophysiology, 79*, 2668–2676.

Dib-Hajj, S. D., Cummins, T. R., Black, J. A., & Waxman, S. G. (2010). Sodium channels and pathological pain. *Annual Review of Neuroscience, 33*, 325–347.

Estacion, M., Dib-Hajj, S. D., Benke, P. J., Te Morsche, R. H., Eastman, E. M., Macala, L. J., Drenth, J. P., & Waxman, S. G. (2008). Nav1.7 Gain-of-function mutations as a continuum: A1632E displays physiological changes associated with erythromelalgia and paroxysmal extreme pain disorder mutations and produces symptoms of both disorders. *Journal of Neuroscience, 28*, 11079–11088.

Fertleman, C. R., Baker, M. D., Parker, K. A., Moffatt, S., Elmslie, F. V., Abrahamsen, B., Ostman, J., Klugbauer, N., Wood, J. N., Gardiner, R. M., & Rees, M. (2006). SCN9A mutations in paroxysmal extreme pain disorder: Allelic variants underlie distinct channel defects and phenotypes. *Neuron, 52*, 767–774.

Goldberg, Y. P., MacFarlane, J., MacDonald, M. L., Thompson, J., Dube, M. P., Mattice, M., Fraser, R., Young, C., Hossain, S., Pape, T., Payne, B., Radomski, C., Donaldson, G., Ives, E., Cox, J., Younghusband, H. B., Green, R., Duff, A., Boltshauser, E., Grinspan, G. A., Dimon, J. H., Sibley, B. G., Andria, G., Toscano, E.,

Kerdraon, J., Bowsher, D., Pimstone, S. N., Samuels, M. E., Sherrington, R., & Hayden, M. R. (2007). Loss-of-function mutations in the Nav1.7 gene underlie congenital indifference to pain in multiple human populations. *Clinical Genetics, 71*, 311–319.

Hains, B. C., Klein, J. P., Saab, C. Y., Craner, M. J., Black, J. A., & Waxman, S. G. (2003). Upregulation of sodium channel Nav1.3 and functional involvement in neuronal hyperexcitability associated with central neuropathic pain after spinal cord injury. *Journal of Neuroscience, 23*, 8881–8892.

Hains, B. C., Saab, C. Y., Klein, J. P., Craner, M. J., & Waxman, S. G. (2004). Altered sodium channel expression in second-order spinal sensory neurons contributes to pain after peripheral nerve injury. *Journal of Neuroscience, 24*, 4832–4840.

Ishikawa, K., Tanaka, M., Black, J., & Waxman, S. (1999). Changes in expression of voltage-gated potassium channels in dorsal root ganglion neurons following axotomy. *Muscle & Nerve, 22*, 502–507.

Kors, E. E., van den Maagdenberg, A. M., Plomp, J. J., Frants, R. R., & Ferrari, M. D. (2002). Calcium channel mutations and migraine. *Current Opinion in Neurology, 15*, 311–316.

Lindia, J. A., Kohler, M. G., Martin, W. J., & Abbadie, C. (2005). Relationship between sodium channel Nav1.3 expression and neuropathic pain. *Pain, 117*, 145–153.

Nassar, M. A., Baker, M. D., Levato, A., Ingram, R., Mallucci, G., McMahon, S. B., & Wood, J. N. (2006). Nerve injury induces robust allodynia and ectopic discharges in Nav1.3 null mutant mice. *Molecular Pain, 2*, 33.

Samad, O. A., Tan, A. M., Cheng, X., Foster, E., Dib-Hajj, S. D., Waxman, S. G. (2012). Virus-mediated sh RNA knockdown of Na$_v$1.3 in rat dorsal root ganglion attenuates nerve injury induced neuropathic pain. *Molecular Therapy*, in press.

Toledo-Aral, J. J., Moss, B. L., He, Z.-J., Koszowski, A. G., Whisenand, T., Levinson, S. R., Wolf, J. J., Silos-Santiago, I., Halegoua, S., & Mandel, G. (1997). Identification of PN1, a predominant voltage-dependent sodium channel expressed principally in peripheral neurons. *Proceedings of the National Academy of Sciences of the USA, 94*, 1527–1532.

Wang, H., Sun, H., Della Penna, K., Benz, R., Xu, J., Gerhold, D., Holder, D., & Koblan, K. (2002). Chronic neuropathic pain is accompanied by global changes in gene expression and shares pathobiology with neurodegenerative diseases. *Neuroscience, 114*, 529–546.

Waxman, S. G., & Dib-Hajj, S. D. (2005). Erythermalgia: Molecular basis for an inherited pain syndrome. *Trends in Molecular Medicine, 11*, 555–562.

Waxman, S. G., Kocsis, J. D., & Black, J. A. (1994). Type III sodium channel mRNA is expressed in embryonic but not adult spinal sensory neurons, and is re-expressed following axotomy. *Journal of Neurophysiology, 72*, 466–471.

Xiao, H., Huang, Q., Zhang, F., Bao, L., Lu, Y., Guo, C., Yang, L., Huang, W., Fu, G., Xu, S., Cheng, X., Yan, Q.,

Zhu, Z., Zhang, X., Chen, Z., Han, Z., & Zhang, X. (2002). Identification of gene expression profile of dorsal root ganglion in the rat peripheral axotomy model of neuropathic pain. *Proceedings of the National Academy of Sciences of the USA, 99,* 8360–8365.

Yang, E. K., Takimoto, K., Hayashi, Y., de Groat, W. C., & Yoshimura, N. (2004a). Altered expression of potassium channel subunit mRNA and alpha-dendrotoxin sensitivity of potassium currents in rat dorsal root ganglion neurons after axotomy. *Neuroscience, 123,* 867–874.

Yang, J., Wang, Y., Li, S., Xu, Z., Li, H., Ma, L., Fan, J., Bu, D., Liu, B., Fan, Z., Wu, G., Jin, J., Ding, B., Zhu, X., & Shen, Y. (2004b). Mutations in SCN9A, encoding a sodium channel alpha subunit, in patients with primary erythermalgia. *Journal of Medical Genetics, 41,* 171–174.

Painful Disk Syndrome

▶ Chronic Back Pain and Spinal Instability

Painful Neuropathy

▶ Neuropathic Pain, Diagnosis, Pathology, and Management

Painful Scars

A. Lee Dellon
Department of Plastic Surgery and Neurosurgery, Johns Hopkins University, Baltimore, USA
Dellon Institutes for Peripheral Nerve Surgery, Baltimore, MD, USA

Synonyms

Collateral sprouting; Cutaneous neuroma; Neuroma

Definition

A scar results whenever the skin is cut. Since all normal skin is innervated, cutting the skin results in the division of at least small intraepithelial nerve endings. Usually, the cut extends deeper than just the skin, going to some depth either as required by the surgeon or as a result of trauma, and this places actual nerve fibers in cutaneous nerves in harm's way. When a scar becomes painful, it is due to injury to either the nerve endings within the skin or in the immediate, subjacent tissues. The pain may be due to a true neuroma, entrapment of the nerve within the scar, or inflammatory mediators stimulating the nerve. The treatment of the painful scar is related to nonoperative attempts to disengage the nerve from the scar, treat the inflammatory mediators, and desensitize the skin or operative attempts to relocate the injured nerve away from the scar.

Characteristics

Whenever an incision is made into the skin, there is the risk of having a painful scar. Normal skin is innervated by both large and small fibers, which are both quickly and slowly adapting. The nerve endings are related either directly to the skin itself, as is the case of the C-fibers, or to dermal appendages, like hair follicles, as is the case of the A-beta fibers, or end in encapsulated structures like the Meissner or Pacinian corpuscle (Dellon 1981). Injury to the skin itself can create entrapment of these small nerve endings within the scar. Injury that extends to the deeper tissues can create entrapment of the cutaneous nerves that supply that skin territory. Subsequent stimulation of the scar will stimulate the injured nerves, sending neural impulses interpreted as pain.

The management of the painful scar must include mechanical and chemical means to manipulate the wound healing process, with the goal of disentangling the nerve endings or the cutaneous nerve from the scar. The initial approach is to have the patient apply a topical cream to the scar and massage the scar. A cream is preferred over an ointment, as the cream becomes absorbed. Vitamin E cream is traditional. Use of a cortisone-containing cream is preferred as it decreases inflammation, which, in itself, may be contributing to the perceived pain through mediators like the cytokines (Devor et al. 1985). If the scar remains painful after 6 weeks of topical medication and massage, then ultrasound with steroid iontophoresis, available at most physical therapy centers, is recommended. The ultrasound transmits heat energy deeply into the tissues, where it may render the collagen more likely to respond to the massage and stretching of the therapist. The steroid is carried transdermally by this technique (iontophoresis). If the scar remains painful, direct injection of the scar and subdermal region with a steroid compound is recommended, accompanied by massage. A water soluble steroid is preferred, since if injected directly into the nerve it will not induce further nerve injury (Mackinnon et al. 1982). Topical xylocaine anesthetic patches (Lidoderm patches or EMLA cream) can be effective in controlling the painful scar. In physical therapy, direct attempts to desensitize the skin can be introduced, such as fluidotherapy, which bombards the painful skin surface with small pellets driven by air, or the use of transcutaneous nerve stimulators, which overdrive the sensory stimuli to the spinal cord, can be tried (Mackinnon and Dellon 1988). ▶ Desensitization is a form of sensory reeducation applied for the treatment of pain and is successful by inducing a reorganization of the cortical maps (Dellon 1997).

If a scar remains painful after 6 months of attempts to desensitize it, then operative approaches can be considered. If just the scar itself is sensitive, suggesting that the mechanism for the pain is related just to the scar, and not to an underlying cutaneous nerve, then the scar can be excised and the skin sutured again, reinitiating the wound healing process.

If the scar is painful and there is a trigger spot in the scar that causes a distal tingling into the skin, but the distal skin has not lost sensation, then a cutaneous nerve is adherent to the scar, but a true neuroma of that nerve may not be present. In this situation, a scar excision and a neurolysis of the cutaneous nerve is the best operative approach.

If the scar is painful and there is an area of numbness or dysesthesia distal to the scar and there is a trigger point for the pain that generates a perception of tingling along the course of a known cutaneous nerve, then the surgical approach must include the concept of dealing with a true neuroma of the cutaneous nerve. It has been demonstrated that such a neuroma is mechanosensitive and is the origin of spontaneous neural impulses (Meyer et al. 1985). The neuroma must therefore be excised. The problem then becomes what to do with the proximal end of the nerve so that it does not create another painful neuroma. It has been demonstrated that when a sensory nerve is implanted into a normal muscle, this alteration in the microenvironment of the sensory nerve permits axonal sprouting, but that, in the absence of nerve growth factor in the immediate tissues about the spouts and in the absence of tension, a classic end-bulb neuroma does not form (Mackinnon et al. 1985). This approach has been utilized for both upper and lower extremity painful scars.

In the upper extremity, the most common painful scar is related to injury to the palmar cutaneous branch of the median nerve. The appropriate surgical strategy is to interrupt function of this cutaneous neuroma and implantation of the proximal end of the palmar cutaneous nerve into the pronator quadratus muscle. This muscle is proximal to the wrist joint and has a limited excursion (Evans and Dellon 1994). A similar problem has been described for the painful scar related to ▶ tarsal tunnel decompression. This pain in the proximal scar is due to a neuroma of the saphenous nerve, which must be resected and the nerve implanted into the soleus muscle (Kim and Dellon 2001a). This pain in the distal scar is due to a neuroma of a calcaneal nerve, which must be resected and the nerve implanted into the flexor hallucis longus muscle (Kim and Dellon 2001b).

In the clinical situation, where more than one nerve may innervate the painful scar, nerve blocks are essential to identify whether one or more than one cutaneous nerve is involved in the pain mechanism. The most common site for this confusion is a painful scar over the dorsal radial aspect of the wrist. In this location, scars occur not only from trauma but from elective surgery in which a cyst from the dorsal wrist capsule, a ganglion, is excised or tendonitis of the first dorsal extensor compartment is treated by tendon sheath incision. In 1985, the source of the failure to treat painful scars in this region was related to the presence of overlap between the radial sensory nerve and the lateral antebrachial cutaneous nerve in 75 % of people (Mackinnon and Dellon 1985). The nerve block must first be done over the anterolateral forearm, to block the lateral antebrachial cutaneous nerve, as a more distal block will affect both of the nerves. The surgical approach is to resect both of the involved nerves and implant them into the brachioradialis muscle (Mackinnon and Dellon 1987). This is done with the elbow in extension so that subsequent elbow extension will not pull on the nerves implanted into the muscle.

The concept of placing the proximal end of the peripheral nerve in a "quiet" area is the reason that thoracotomy incisions are often more painful than abdominal incisions; with every breath, the ribs go through a range of motion, causing movement of the intercostal nerves and stimulating a painful neural impulse.

In the lower extremity, the most common location of a painful scar is over the dorsum of the foot. This area exemplifies all the problems related to treatment of the painful scar. Foot wear, such as shoes or even sandals, will press upon the scar. There is no subcutaneous padding over the dorsum of the foot. The extensor tendons create movement, and the ankle range of motion will put tension on the cutaneous nerve. There is overlap of the saphenous nerve, superficial peroneal nerve, deep peroneal nerves, and sural nerves in virtually all the dorsal skin. Initial nerve blocks are essential to identifying the source(s) of the painful inputs from the scar. Scars can be either from trauma or elective incisions designed for ankle fusion, capsular reconstruction, tendon surgery, tumor excision, treatment of "Morton's neuroma," or bunion surgery. Pain related to an interdigital nerve in the foot has been mistakenly termed a neuroma, whereas it is actually chronic nerve compression and should be treated by neurolysis instead of nerve resection (Dellon 1992). After identification of the involved nerve, its function can be interrupted by surgery planned for the leg, rather than the foot or the

ankle. The deep and superficial peroneal nerves can be interrupted in the anterolateral leg, translocating the proximal nerve away from tension and into a motor environment (Dellon and Aszmann 1998).

► Collateral sprouting can be a source of pain related to the painful scar. When a cutaneous nerve is divided, the distal axon degenerates but the Schwann cell lives. The Schwann cell upregulates and produces nerve growth factor. The nerve growth factor, instead of acting as a guide for the original nerve to regenerate to its target organ, stimulates normal axons in the adjacent skin. The original nerve cannot respond, as it is either stuck in the original scar or has been relocated to a new site. The sprouting of normal nerve fibers, in response to nerve growth factor, can create the perception of pain as this new pattern of neural impulses reaches the postcentral gyrus. Previously documented only in amphibian models for the C- and A-delta fiber populations, collateral sprouting has now been documented in the human in relationship to A-beta fibers, with the use of the Pressure-Specified Sensory Device™ in patients with lower extremity nerve injury (Inbal et al. 1987; Dellon et al. 1996). Treatment of pain due to collateral sprouting requires the same desensitization approach as for the painful scar itself (Dellon 1997). For the lower extremity, the most successful and practical approach is simply to walk in a swimming pool for 15 to 20 min, three or four times per week, allowing the water to provide the sensory stimuli to the skin.

References

Dellon, A. L. (1981). *Evaluation of sensibility and re-education of sensation in the hand* (pp. 27–46). Baltimore: Williams & Wilkins.

Dellon, A. L. (1992). Treatment of Morton's neuroma as a nerve compression: The role for neurolysis. *Journal of the American Podiatric Medical Association, 82,* 399–402.

Dellon, A. L. (1997). *Somatosensory testing and rehabilitation*. Bethesda, MD: American Occupational Therapy Association.

Dellon, A. L., & Aszmann, O. C. (1998). Treatment of dorsal foot neuromas by translocation of nerves into anterolateral compartment. *Foot & Ankle, 19,* 300–303.

Dellon, A. L., Aszmann, O. C., & Muse, V. (1996). Collateral sprouting documentation using the PSSD. *Annals of Plastic Surgery, 37*, 520–525.

Devor, M., Govrin-Lippmann, R., & Raber, P. (1985). Corticosteroids suppress ectopic neural discharge originating in experimental neuromas. *Pain, 22*, 127–137.

Evans, G. R. D., & Dellon, A. L. (1994). Implantation of the palmar cutaneous branch of the median nerve into the pronator quadratus for treatment of painful neuroma. *Journal of Hand Surgery, 19A*, 203–206.

Inbal, R., Rousso, M., Ashur, H., Wall, P. D., & Devor, M. (1987). Collateral sprouting in skin and sensory recovery after nerve injury in Man. *Pain, 28*, 141–154.

Kim, J., & Dellon, A. L. (2001a). Tarsal tunnel incisional pain due to neuroma of the posterior branch of saphenous nerve. *Journal of the American Podiatric Medical Association, 91*, 109–113.

Kim, J., & Dellon, A. L. (2001b). Calcaneal neuroma: Diagnosis and treatment. *Foot and Ankle Internatational, 22*, 890–894.

Mackinnon, S. E., & Dellon, A. L. (1985). The overlap pattern of the lateral antebrachial cutaneous nerve and the superficial branch of the radial nerve. *Journal of Hand Surgery, 10A*, 522–526.

Mackinnon, S. E., & Dellon, A. L. (1987). The results of treatment of recurrent dorsal radial wrist neuromas. *Annals of Plastic Surgery, 19*, 54–61.

Mackinnon, S. E., & Dellon, A. L. (1988). *Surgery of the peripheral nerve* (pp. 455–491). New York: Thieme.

Mackinnon, S. E., Hudson, A. R., Gentilli, F., et al. (1982). Peripheral nerve injection study with steroid agents. *Plastic and Reconstructive Surgery, 69*, 482–490.

Mackinnon, S. E., Dellon, A. L., Hudson, A. R., et al. (1985). Alteration of neuroma formation produced by manipulation of neural environment in primates. *Plastic and Reconstructive Surgery, 76*, 345–352.

Meyer, R. A., Raja, S. N., Campbell, J. N., et al. (1985). Neural activity originating from a neuroma in the baboon. *Brain Research, 325*, 255–260.

Painful Tonic Seizure

Definition

Paroxysmal painful sensations irradiating from the spinal cord into the extremities.

Cross-References

▶ Central Pain in Multiple Sclerosis

Pain-Inhibitory System

Definition

This refers to systems within the central nervous system whose normal function is the inhibition of pain.

Cross-References

▶ Descending Modulation of Nociceptive Processing
▶ Pain-Modulatory Systems, History of Discovery
▶ Stimulation-Produced Analgesia

Pain-inhibitory Systems

▶ Pain-Modulatory Systems, History of Discovery

Painkillers

▶ Simple Analgesics

Painless Neuropathies

Paolo Marchettini[1] and Marco Lacerenza[2]
[1]Pain Medicine Center, Scientific Institute San Raffaele, Milan, Italy
[2]Pain Medicine Center, Casa di Cura S. Pio X, Opera San Camillo Foundation, Milan, Italy

Definition

Any disorder affecting any part of the peripheral nervous system not associated with pain.

Characteristics

Peripheral neuropathies are abnormalities in structure and function of motor, sensory, and autonomic neurons and/or their axons. In spite of the variety of injuries (compressive, genetic, infective, inflammatory, ischemic, metabolic, traumatic, toxic, or associated with neoplasm), the peripheral nervous system reactions are limited to four major pathologies: ▶ Wallerian degeneration, axonal degeneration, neuronopathy, and ▶ segmental demyelination. However, according to the specific agent causing the neuropathy, such standardized pathologic manifestations take place in different époques. Therefore, considering the neuropathy as a time frame, where symptoms are due to an evolving biochemical and morphological reaction to an injury, it can be stated that most neuropathies are painless within a defined time frame. For example, a traumatic nerve injury is usually extremely painful in the first days, while in the later, chronic phase it is painful in only 2.5–5 % of patients (Kline and Hudson 1995). In the case of painful diabetic neuropathy, there are reports of patients in whom severe neuropathic pain lasting months wore off to be followed by hypoesthesia and ▶ paresthesias (Brown et al. 1976; Yuen et al. 2001).

Most mild neuropathies remain undiagnosed just because they are painless. Such is the case of mild genetic neuropathies, painless traumatic and iatrogenic neuropathies, or subclinical metabolic neuropathies. Therefore, although many may not share this view, painless neuropathies are likely to outnumber the painful ones.

Most of the time, pain in peripheral neuropathy is a positive sensory symptom arising from diseased primary afferent units. An exception is the nerve trunk pain due to the physiological activation of nervi nervorum nociceptors. This may be the cause of part of the complaints of patients with carpal tunnel syndromes, Guillain-Barré syndrome, ▶ radiculopathies, or vasculitis.

Through the technique of ▶ microneurography and intraneural microstimulation on human subjects, it has been demonstrated that excitation of nociceptors elicits either a dull burning pain (C nociceptors) or a sharp pricking pain (A delta nociceptors), while paresthesias arise from ▶ ectopic nerve impulses generated in dysfunctional large myelinated fibers (Ochoa and Torebjörk 1980, 1989; Torebjörk and Ochoa 1980). Therefore, extrapolating from the experimental information to clinical practice, it may be expected that neuropathies affecting only motor neurons or large myelinated primary afferent neurons are painless, apart from neuropathies associated with nerve trunk pain.

Generally, Bell's palsy involving the facial nerve is painless as well as the neuropathies of the spinal accessory nerve (causing weakness and wasting of the trapezius and sternocleidomastoid muscles), and the neuropathies of the hypoglossal, XII cranial nerve. However, if these neuropathies are due to an acute inflammatory or traumatic disease involving nervi nervorum, they can be painful at the specific site where the injury is.

Less clear is the pathophysiology of idiopathic trigeminal sensory neuropathy, in which sensory loss can be transient or permanent, localized mainly in the II and III nerve divisions, sometimes with paresthesias, but painless (Blau et al. 1969).

Isolated neuropathies of motor branches of the brachial plexus can be painless. Painless infraspinatus atrophy due to suprascapular nerve entrapment is described in subjects overexerting the upper extremity, for example, volleyball players (Tengan et al. 1993).

Damage of the long thoracic nerve causes weakness and wasting of the serratus anterior muscle and winging of the scapula. Among the most frequent causes of this neuropathy are closed trauma, sporting activities, iatrogenic procedures, viral infections, idiopathic conditions, and metabolic disorders. Most of the time, the neuropathy is painless, and the shoulder pain described by some patients is due to the dislocation of the scapula during movements of the upper limb and therefore it is a nociceptive pain. In many other cases, the condition is completely symptom free and its recognition accidental.

The comprehensive group of motor neuronopathies is painless with the exception of poliomyelitis in the acute phase and post-polio syndrome (Bendz 1954; Klein et al. 2004).

In both of the above conditions, pain is not related to the motor neuronopathy, albeit to myelitis and musculoskeletal complications, respectively. Cramps may indeed be relevant in up to 50 % of patients with motor neuronopathy and represent a nociceptive pain. This also includes the different sporadic and genetic forms of amyotrophic lateral sclerosis (ALS), bulbar motor neuron syndromes, spinal muscular atrophy, focal motor neuron disease, and paraneoplastic lower motor neuron syndrome.

Diffuse motor neuropathies can be predominantly demyelinating or axonal, and the differential diagnosis of ALS or other untreatable conditions is crucial since some are treatable. Among the treatable conditions, multifocal motor neuropathy (MMN) is the most representative. MMN is more common in males and has an estimated prevalence of 1 per 100,000; it is primarily a disease of young adults and is initially characterized by progressive asymmetrical limb weakness and later by muscular atrophy. The diagnostic clue is the electrophysiological demonstration of persistent localized motor conduction blocks, with normal or near-normal motor and sensory conduction velocities. MMN is painless apart from cramps, fasciculations, and (rarely) paresthesias, which may be part of the clinical picture. Evidence for an autoimmune mechanism is substantial, and the disease is treatable with intravenous immunoglobulin (IVIg) and/or cyclophosphamide (Biessels et al. 1997).

Acute axonal motor neuropathy is more common in Japan, China, and third world countries than in the West. It is often anticipated by diarrhea, gastrointestinal, or upper respiratory symptoms as a landmark of *Campylobacter jejuni* or *Haemophilus influenzae* infections. Weakness of distal rather than proximal muscles is prominent between the 6th and the 12th day from the onset. Sensory symptoms are absent. The electrophysiology points to a predominantly axonal dysfunction, and IgG vs. GM1 ganglioside can be found in 40–50 % of cases, with the treatment of choice being IVIg. Moreover, there is a slowly progressive form of painless motor neuropathy with axonal electrophysiological features, involving asymmetrically distal arms more

commonly than legs, without symptoms or signs of sensory involvement. The condition may be treatable with IVIg or prednisone suggesting an immune-mediated etiology (Katz et al. 2002).

Predominantly myelinated afferent and efferent fibers can be affected simultaneously in sporadic and genetic forms of neuropathy in painless clinical pictures. For example, in the group of neuropathies associated with monoclonal gammopathy of undetermined significance (MGUS), pain is an uncommon symptom. The illness has a male preponderance and begins insidiously around the fifth–sixth decade of life with a distal symmetrical sensorimotor polyneuropathy. The lower limbs and the sensory fibers are affected earlier, and electrophysiological studies reveal demyelination (mostly in IgM MGUS patients) or mixed demyelinating and axonal damage. More than 50 % of IgM MGUS neuropathy patients show a monoclonal protein with reactivity against myelin-associated glycoprotein (MAG) and have a favorable prognosis even after several years of illness. Aggressive immune interventions are limited to patients with progressive disabling neuropathy (Suarez and Kelly 1993; Nobile-Orazio et al. 2000).

In the last decade, rapid advances in the description of several new genes implicated in hereditary neuropathies have shed light on the pathophysiology of nerve demyelination and axonal degeneration. Charcot-Marie-Tooth (CMT), also called hereditary motor and sensory neuropathy (HMSN), is the most common type of hereditary neuropathy and comprises a family of clinically and genetically heterogeneous inherited neuropathies resulting from mutations in more than 40 genes expressed in Schwann cells and neurons. The clinical pictures differ from one another for the age of onset, mode of transmission, severity of the illness, electrophysiological, and nerve biopsy features pointing to a predominantly demyelinating or axonal damage. Disturbing sensory symptoms are uncommon in CMT1 patients and pain, not accountable by skeletal abnormalities, is rare. Traditional clinical judgment based on correspondence between localization of positive and negative sensory phenomena, commonly associated with neuropathic pain and the descriptive characteristics of pain quality, considers neuropathic pain in CMT an unlikely phenomenon, the most common pain by far being of the musculoskeletal type. A paper in disagreement with this view (Carter et al. 1998) reported a high prevalence of neuropathic pain localized mainly in the low back, knees, and ankles in CMT patients. The authors themselves properly caution that the finding is exclusively based on the wording of the neuropathic questionnaire.

Hereditary neuropathy with liability to pressure palsies (HNPP) is an autosomal dominant disorder, characterized by recurrent sensorimotor painless neuropathies associated with nerve conduction abnormalities typically localized to segments of nerves liable to pressure palsies. The disorder is pathologically characterized by focal myelin thickening, named tomacula, which was the hallmark of HNPP when the genetic test was not available. The genetic abnormality is a 1.5 Mb deletion on chromosome 17p11.2, and the clinical picture can vary from recurrent invalidating transient mononeuropathies and/or brachial plexopathy to benign forms with only some paresthesias and subclinical electrophysiological abnormalities (Harding 1995; Berciano and Combarros 2003).

In the group of hereditary sensory and autonomic neuropathies (HSAN), there are rare forms of painless neuropathies. For example, HSAN IV is an autosomal recessive disorder (gene locus: chromosome 1q21–22) where congenital insensitivity to pain, anhidrosis, abnormal temperature control, and mild mental retardation are the clinical presentation. Selective loss of small myelinated and unmyelinated fibers is associated with normal tendon reflexes and sensory nerve conduction velocities (Nagasako et al. 2003).

In the realm of diabetic neuropathies, it is possible to find very different clinical pictures related to multifactorial pathogenetic mechanisms. A few of them are painless neuropathies or relatively painless conditions for at least part of the natural history. For example, some patients affected by distal symmetrical polyneuropathy do not complain of positive sensory symptoms or pain and may even present recurrent painless neurotrophic foot ulcers. The rare condition named diabetic pseudotabes is a large-fiber

Painless Neuropathies, Table 1 Neuropathies/neuronopathies in which pain due to nerve damage is absent or uncommon

Motor nerve mononeuropathies	
	Spinal accessory nerve
	Hypoglossal nerve
	Facial nerve (predominantly motor)
	Motor branches of the brachial plexus
Motor neuropathies	
	Sporadic
	Hereditary
	Diffuse
	Focal
	Paraneoplastic
Large-fiber neuropathies	
Predominantly motor	Demyelinating: multifocal motor neuropathy
	Axonal: acute form; chronic form
	Painless diabetic motor neuropathy
Sensorimotor	Demyelinating: MGUS-related polyneuropathy
	Large-fiber diabetic neuropathy
Hereditary large-fiber neuropathy	
	Demyelinating: HNPP; CMT
	Axonal: CMT
Hereditary small-fiber neuropathy	
	HSAN IV

MGUS monoclonal gammopathy of undetermined significance, *HNPP* hereditary neuropathy with liability to pressure palsies, *CMT* Charcot-Marie-Tooth disease, *HSAN* hereditary sensory and autonomic neuropathy

symmetric polyneuropathy associated with sensory ataxia but rarely with pain.

A recent paper (Garces-Sanchez et al. 2011) describes a painless predominantly motor variant of diabetic lumbosacral radiculoplexus neuropathy (DLRPN). The clinical features of the 23 patients described with painless DLRPN vary from typical DLRPN in that they are more insidious and symmetrical with slower evolution. Although ischemic injury and microvasculitis appear to be the cause of this painless neuropathy, the authors suggest that the slower evolution, in respect to the classic form of painful DLRPN, may explain the lack of pain.

Although the cause of pain in diabetic neuropathies is still unknown, it appears that the absence of vascular or nerve inflammation, the sparing of small myelinated and unmyelinated fibers, a sort of inhibition of the later phase of myelinated fiber regeneration, and the absence of axonal atrophy may be the ingredients making a diabetic neuropathy virtually painless (Brown et al. 1976; Britland et al. 1990) (Table 1).

Cross-References

▶ Wallerian Degeneration

References

Bendz, P. (1954). Pain associated with acute poliomyelitis; neurologic and therapeutic considerations. AMA American Journal of Diseases of Children, 88, 141–147.

Berciano, J., & Combarros, O. (2003). Hereditary neuropathies. Current Opinion in Neurology, 16, 613–622.

Biessels, G. J., Franssen, H., van den erg, L. H., Gibson, A., Kappelle, L. J., Venables, G. S., & Wokke, J. H. J. (1997). Multifocal motor neuropathy. Journal of Neurology, 244, 143–152.

Blau, J. N., Harris, M., & Kennet, S. (1969). Trigeminal sensory neuropathy. The New England Journal of Medicine, 281, 873–876.

Britland, S. T., Young, R. J., Sharma, A. K., & Clarke, F. (1990). Association of painful and painless diabetic polyneuropathy with different patterns of nerve fiber degeneration and regeneration. Diabetes, 39, 898–908.

Brown, M. J., Martin, J. R., & Asbury, A. K. (1976). Painful diabetic neuropathy. Archives of Neurology, 33, 164–171.

Carter, G. T., Jensen, M. P., Galer, B. S., Kraft, G. H., Crabtree, L. D., Beardsley, R. M., Abresch, R. T., & Bird, T. D. (1998). Neuropathic pain in Charcot-Marie-Tooth disease. Archives of Physical Medicine and Rehabilitation, 79, 1560–1564.

Garces-Sanchez, M., Laughlin, R. S., Dyck, P. J., Engelstad, J. K., Norell, J. E., & Dyck, P. J. (2011). Painless diabetic motor neuropathy: A variant of diabetic lumbosacral radiculoplexus neuropathy? Annals of Neurology, 69(6), 1043–1054.

Harding, A. E. (1995). From the syndrome of Charcot, Marie and tooth to disorder of peripheral myelin proteins. Brain, 118, 809–818.

Katz, J. S., Barohn, R. J., Kojan, S., Wolfe, G. I., Nations, S. P., Saperstein, D. S., & Amato, A. A. (2002). Axonal multifocal motor neuropathy without conduction block or other features of demyelination. Neurology, 58, 615–620.

Klein, M. G., Keenan, M. A., Esquenazi, A., Costello, R., & Polansky, M. (2004). Musculoskeletal pain in polio survivors and strength-matched controls. Archives of Physical Medicine and Rehabilitation, 85, 1679–1683.

Kline, D. G., & Hudson, A. R. (1995). Nerve injuries. Philadelphia: WB Saunders.

Nagasako, E. M., Oaklander, A. L., & Dworkin, R. H. (2003). Congenital insensitivity to pain: An update. Pain, 101, 213–219.

Nobile-Orazio, E., Meucci, N., Baldini, L., Di Troia, A., & Scarlato, G. (2000). Long-term prognosis of neuropathy associated with anti-MAG IgM M-proteins and its relationship to immune therapies. Brain, 123, 710–717.

Ochoa, J., & Torebjörk, H. E. (1980). Paraesthesiae from ectopic impulse generation in human sensory nerves. Brain, 103, 835–853.

Ochoa, J. L., & Torebjörk, H. E. (1989). Sensation evoked by intraneural microstimulation of C nociceptor fibres in human skin nerves. Journal of Physiology, 415, 583–599.

Suarez, G. A., & Kelly, J. J., Jr. (1993). Polyneuropathy associated with monoclonal gammopathy of undetermined significance: Further evidence that IgM-MGUS neuropathies are different than IgG-MGUS. Neurology, 43, 1304–1308.

Tengan, C. H., Oliveira, A. S., Kiymoto, B. H., Morita, M. P., De Medeiros, J. L., & Gabbai, A. A. (1993). Isolated and painless infraspinatus atrophy in top level volleyball players. Report of two cases and review of the literature. Arquivos de Neuropsiquiatria, 51, 125–129.

Torebjörk, H. E., & Ochoa, J. (1980). Specific sensation evoked by activity in single identified sensory units in man. Acta Physiologica Scandinavica, 110, 445–447.

Yuen, K. C. J., Day, J. L., Flannagant, D. W., & Rayman, G. (2001). Diabetic neuropathic cachexia and acute bilateral cataract formation following rapid glycaemic control in a newly diagnosed type 1 diabetic patient. Diabetic Medicine, 18, 854–857. Diabetes UK.

Pain-Modulatory Systems, History of Discovery

Gregory W. Terman
Department of Anesthesiology and the Graduate Program in Neurobiology and Behavior, University of Washington, Seattle, WA, USA

Synonyms

Pain-facilitatory systems; Pain-inhibitory systems

Definition

Activation of nociceptive pathways by tissue damage promotes the important ability of an organism to perceive pain and make a subsequent adaptive response. More surprising, teleologically, is the existence of endogenous pain-modulatory systems that function to inhibit or facilitate nociception and thus alter the fidelity of pain perception. Nonetheless, since early in the 1970s, much evidence has accumulated to support the existence of such pain-modulatory systems. Moreover, activation of ▶ pain-inhibitory systems has been demonstrated to subserve the analgesic effects of several new and age-old therapies, and ▶ pain-facilitatory systems have been found to mediate pain in a number of animal models of clinical pain.

Characteristics

The concept of descending pain modulation goes back to Sherrington in the early 1900s and was an important part of Melzack and Wall's gate control theory of pain in 1965 (Melzack and Wall 1965). However, there was little direct evidence for dedicated pain-modulatory systems until Reynolds, in 1969, reported that electrical stimulation of the midbrain ▶ periaqueductal gray (PAG) resulted in sufficient analgesia to perform laparotomy in rats. Liebeskind and his colleagues replicated this phenomenon and labeled it "stimulation-produced analgesia" (SPA) (see ▶ Stimulation-Produced Analgesia). Moreover, in a series of studies spanning two decades, they identified the anatomical, electrophysiological, and neurochemical substrates of at least one neural system descending from brain to spinal cord, whose normal function was the inhibition of pain. Many laboratories have since studied this endogenous pain-inhibitory system and have found it to be potent, specific for pain, dependent on pathways from PAG through rostral ventral medulla down the dorsolateral funiculus of the spinal cord, and effective in inhibiting pain transmission as early as the first nociceptive synapse between primary afferent and spinal dorsal horn (Basbaum and Fields 1978; Mayer and Price 1976).

Another important characteristic of endogenous pain-inhibitory systems, found in early studies, was that these systems, activated by SPA, might also subserve analgesia from opiate drugs (Mayer and Price 1976). The same brain regions from which SPA can be elicited are rich in opiate receptors and can produce analgesia when microinjected with opiates. Like opiate analgesia, tolerance develops to the analgesic effects of repeated brain stimulation, and cross-tolerance develops between opiate analgesia and SPA suggesting a common underlying mechanism. Finally, opiate antagonists such as naloxone can antagonize SPA (although this finding was not universal). The commonality between opiate and endogenous pain inhibition encouraged the search and eventual identification of the endogenous opioid peptides (Hughes et al. 1975). Additional neurochemicals implicated in mediating endogenous pain inhibition include serotonin, norepinephrine, acetylcholine acting at nicotinic receptors, and the endogenous cannabinoids. Each of these has been studied as potential new analgesic therapeutics, although thus far the clinical utility has been hampered by low efficacy or toxic side effects. Of more apparent clinical utility was the success of SPA in inhibiting pain in some patients. Unfortunately, there are still no long-term placebo-controlled studies demonstrating success of this invasive and complicated therapy in alleviating pain in humans.

One alternative approach to identifying clinical applications of our knowledge of endogenous pain-inhibitory systems is to identify less invasive approaches to activating these systems. Given the importance of nociception for survival, it has been widely assumed that pain-inhibitory systems would be most active during times of life-threatening stress. Enduring pain in order to escape a predator, for example, might provide an adaptive advantage to animals that could inhibit nociception at least temporarily. The missionary, David Livingstone, wrote of his mauling by a lion that he had a feeling "of dreaminess in which there was no sense of pain." Moreover, numerous battlefield stories of seriously wounded soldiers without apparent pain exist. Tissue damage is not required to activate pain-inhibitory systems, however, and animal models of these phenomena have been labeled "▶ stress-induced analgesia" (Akil et al. 1976). Electric shock, restraint, rotation, forced swim, and intruder threat can all produce analgesia in laboratory animals without producing identifiable tissue damage (Hayes et al. 1978). Like SPA, early studies of SIA differed in the reported involvement of opioid peptides. Timing, severity, and location of a single stressor were all eventually reported to differentially produce opioid and nonopioid SIA (Terman et al. 1984). Moreover, opioid SIA itself proved not to be a unitary phenomenon, although lesion and cross-tolerance studies have demonstrated an anatomical and neurochemical overlap (including overlap with opioid SPA). Some forms of opioid SIA, although depending on descending pain-inhibitory pathways, appear to be activated entirely by the physical properties of the stressor and can be elicited even in the surgically anesthetized animal (Terman et al. 1984). On the other hand, other forms of opioid SIA depend on learning mechanisms and can be blocked by decerebration, muscarinic antagonists, amygdala lesions, and pentobarbital anesthesia. Placing rats in a cage where footshock was previously received, but is not currently administered, can also elicit analgesia ("▶ Conditioned Analgesia"). Evidently, animals cannot only activate their intrinsic pain-inhibitory systems in the presence of stress, but they can also learn to activate them in anticipation of such stimuli. Unconditioned expectations also appear capable of activating opioid descending pain-inhibitory substrates. The scent of an animal's natural predator, for instance, can induce analgesia. In man, ▶ placebo analgesia is also dependent on expectation and conditioning and is mediated by endogenous opiates and associated with activity in brainstem regions which support SPA (Benedetti and Amanzio 1997). A specific example of learned opiate SIA is the phenomenon of "learned helplessness." Inescapable stress in the laboratory, on the battlefield, or, as in Livingstone's case, in the jungle can eventually lead to exhaustion of all escape behaviors. Maier and colleagues demonstrated that rats actually learn to stop attempting escape from prolonged

electric shock (termed "▶ Learned Helplessness") and that they become analgesic in these conditions (Maier et al. 1983). This form of opioid SIA may be similarly activated by other intense forms of physical activity. Indeed, some hypothesize that this form of opioid SIA may occur in long-distance runners worldwide and motivate them to continue running even long after the lions have stopped chasing them.

Do studies of stress-induced analgesia have any other human applications? In addition to exercise, there are a number of pain-inhibitory phenomena in man provoked by procedures that might be considered stressful, including electric shock, acupuncture, moxibustion, transcutaneous electrical nerve stimulation, and even sexual activity. Elucidating the neural basis of SIA might one day lead to the development of new approaches to pain management. Perhaps the greatest promise for such an advance lies in determining the neurochemistry of ▶ nonopioid analgesia. The similar efficacy of opioid and nonopioid SIA bodes well for development of a novel nonopioid analgesic that does not show tolerance or cross-tolerance to mu opiates like morphine. The neurochemistry of "nonopioid" SIA, however, is still poorly understood despite considerable study. Probably the most convincing results have demonstrated ▶ kappa(κ) opiate/opioid receptors (not technically "nonopiate") and/or ▶ NMDA glutamate receptors to be important in mediating such phenomena. It should be noted that NMDA receptors can also be pro-nociceptive, and thus reduction of SIA by NMDA antagonists could work by decreasing nociceptive input necessary for producing SIA. That is, they could have their effects on the afferent (activating) rather than the efferent side of a nonopiate pain-inhibitory system. Le Bars et al., in a series of electrophysiological studies, reported just such a phenomenon termed ▶ Diffuse Noxious Inhibitory Controls (DNIC) in the anesthetized rat (Villanueva and Le Bars 1995). Dorsal horn neuronal nociceptive responses are inhibited by noxious stimulation applied outside the neuron's receptive field, but not if given in the presence of low doses of morphine, presumably because morphine inhibits the very nociceptive input that elicits DNIC. Thus, discriminating between a drug's effects on ascending and descending mechanisms underlying SIA is important, particularly if one hopes to activate descending nociception inhibitory pathways without activating ascending nociceptive pathways, as would be clinically desirable. Moreover, the NMDA receptor story is a prime example of a change in emphasis in the pain modulation field, beginning in the 1990s, shifting away from studying pain-inhibitory systems and towards studying pain-facilitatory systems.

Until the late 1980s, virtually all small animal studies of nociceptive processes involved threshold changes in responses to some nociceptive stimulus (usually heat). This approach is likely to have grown out of a preconception that the central nervous system, in general, and nociceptive transmission circuitry, specifically, normally underwent little change during an organism's adult life and was essentially hardwired. Since that time an explosion of neuroscience research has demonstrated the amazing plasticity of the adult brain and spinal cord. Correspondingly, and perhaps in response to the studies of pain-inhibitory systems described above, preclinical pain researchers have developed and investigated numerous innovative pain tests and animal models of pain demonstrating that basic nociception is under complex modulatory influences. Much of this research has attempted to mimic clinically relevant pain states, such as inflammation or nerve injury. Studies of such models have shown that the mechanisms underlying ► inflammatory and ► neuropathic pain differ considerably and that both differ from more acute or "transient" pains represented by threshold studies in previously untreated animals. This has served to focus attention more than ever on pain modulation systems, due to overwhelming evidence that pain responses elicited by noxious stimuli are integrally dependent on the prior nociceptive experiences of an animal.

Following inflammation, a host of neurochemical changes take place in primary afferents that make them more responsive to noxious stimuli (► Hyperalgesia) or produce a nociceptive response to normally non-noxious stimuli (► Allodynia). These changes are termed ► peripheral sensitization. Changes also take place within the nervous system postsynaptic to the primary afferent, and these changes can also produce facilitations in nociceptive responses – termed "► central sensitization."

Although Hardy in 1950 suggested that hyperalgesia following tissue injury or inflammation was due to changes in the spinal cord dorsal horn (secondary hyperalgesia), it was the work of Woolf and colleagues more than 30 years later that dramatically consolidated this concept. In 1983, Woolf reported that tissue injury caused an increase in responses (measured by motor neuron responses) that persisted even after local anesthetic blockade of the damaged tissue, suggesting central mediation of this sensitivity (Woolf 1983). Central sensitization has now been demonstrated in a number of laboratories using a variety of response endpoints (threshold stimuli, response frequency, and receptor field size), eliciting stimuli (thermal, chemical, acute joint inflammation, and C-fiber electrical stimulation), and a number of inflammatory and neuropathic pain models.

Descending serotonergic influences are now clearly implicated in both facilitatory and inhibitory forms of pain modulation, with the facilitatory or inhibitory effects being due primarily to differential effects of different serotonergic receptor subtypes (of which there are many) or, instead, from differences in circuitry. Serotonin, for example, can activate both excitatory and inhibitory neural transmission in the spinal cord. Stimulation of serotonin-rich regions of the RVM can have excitatory or inhibitory effects on dorsal horn nociception at virtually overlapping stimulation sites dependent on the glutamate dose or electrical stimulation intensity administered, emphasizing the intermingling of nociceptive inhibitory and facilitatory cells in the brainstem as described by Fields.

Like serotonin, norepinephrine may also have opposite effects on nociception, depending on the specific receptor subtypes and the neural circuitry activated. Microinjection of alpha2-adrenergic agonists into the RVM can inhibit the tail-flick response and produce a long-lasting decrease in ▶ on-cell firing rate; however, alpha1-noradrenergic receptor agonists excite RVM on cells suggesting pain-facilitatory effects. Indeed, lesions of some noradrenergic inputs to RVM have been found to both potentiate morphine analgesia and inhibit sensitization, suggesting pro-nociceptive effects of norepinephrine.

Other neurochemicals implicated in descending nociceptive facilitation include substance P, thyrotropin-releasing hormone (TRH), and ▶ cholecystokinin (CCK). Descending nociceptive facilitatory systems may also work by potentiating or disinhibiting spinal cord facilitatory mechanisms including glutamate and/or nitric oxide. Indeed, CCKB receptors are coupled to the production of arachidonic acid and presumably its cyclooxygenase and lipoxygenase metabolites that modulate nociception in the spinal cord. Glial activation and cytokine release have also been implicated in descending nociceptive facilitation (Watkins et al. 1995).

In this entry, we have described what is known about specific pain-inhibitory and pain-facilitatory systems that descend from the brain to modulate nociception in the spinal cord. It is likely that similar modulation takes place at every rostral stop of the nociceptive message on its way to consciousness – though perhaps by different mechanisms. The pain patient's knowledge of descending pain modulation provides a psychological double-edged sword. The existence of inhibitory systems can provide hope for overcoming debilitating nociception. Descending facilitatory systems, however, provide fodder for fears of patients (or their doctors) that their pain is "all in their heads." Such psychological dilemmas (and their treatment) are outside the scope of this entry but are integral to a full understanding of pain and its modulation.

References

Akil, H., Madden, J., et al. (1976). Stress-induced increases in endogenous opioid peptides: Concurrent analgesia and its partial reversal by naloxone. In H. W. Kosterlitz (Ed.), *Opiates and endogenous opioid peptides* (pp. 63–70). Amsterdam: Elsevier.

Basbaum, A. I., & Fields, H. L. (1978). Endogenous pain control mechanisms: Review and hypothesis. *Annals of Neurology, 4*, 451–462.

Benedetti, F., & Amanzio, M. (1997). The neurobiology of placebo analgesia: From endogenous opioids to cholecystokinin. *Progress in Neurobiology, 52*, 109–125.

Hayes, R. L., Bennett, G. J., Newlon, P. G., & Mayer, D. J. (1978). Behavioral and physiological studies of non-narcotic analgesia in the rat elicited by certain environmental stimuli. *Brain Research, 155*, 69–90.

Hughes, J., Smith, T. W., Kosterlitz, H. W., Fothergill, L. A., Morgan, B. A., & Morris, H. R. (1975). Identification of two related pentapeptides from the brain with potent opiate agonist activity. *Nature, 258*, 577–580.

Maier, S. F., Sherman, J. E., Lewis, J. W., Terman, G. W., & Liebeskind, J. C. (1983). The opioid/nonopioid nature of stress-induced analgesia and learned helplessness. *Journal of Experimental Psychology. Animal Behavior Processes, 9*, 80–90.

Mayer, D. J., & Price, D. D. (1976). Central nervous system mechanisms of analgesia. *Pain, 2*, 379–404.

Melzack, R., & Wall, P. D. (1965). Pain mechanisms: A new theory. *Science, 150*, 971–979.

Terman, G. W., Shavit, Y., Lewis, J. W., Cannon, J. T., & Liebeskind, J. C. (1984). Intrinsic mechanisms of pain inhibition: Activation by stress. *Science, 226*, 1270–1277.

Villanueva, L., & Le Bars, D. (1995). The activation of bulbo-spinal controls by peripheral nociceptive inputs: Diffuse noxious inhibitory controls. *Biological Research, 28*, 113–125.

Watkins, L. R., Maier, S. F., & Goehler, L. E. (1995). Immune activation: The role of pro-inflammatory cytokines in inflammation, illness responses and pathological pain states. *Pain, 63*, 289–302.

Woolf, C. J. (1983). Evidence for a central component of post-injury pain hypersensitivity. *Nature, 306*, 686–688.

Pain-Prone Patients

Rolf H. Adler
University of Berne Medical School, Kehrsatz, Switzerland

Characteristics

Clinical observation reveals that there are people who seem to experience pain with unusual intensity and frequency. With peripheral lesions, they seem to suffer more pain than most people do, but often they suffer pain without any peripheral process. Among such patients, the presence or absence of a peripheral disorder is not well correlated with the presence or absence of pain. Indeed, we often find that the discovery of the lesion and its removal or cure does not alleviate the pain, which may persist or even recur at a later date. In other words, there are certain individuals, called "pain prone," among whom psychic factors play the primary role in the genesis of pain, in the absence as well as in the presence of peripheral lesions.

Clinical psychological studies of many pain-prone persons have by now provided a fairly good understanding of the determinants of this susceptibility to suffer pain (Szasz 1957; Engel 1951, 1959, 1970; Rangell 1953; Kolb 1954; Hart 1947). The key comes through understanding how pain may yield pleasure. It is pleasure in a relative sense, which is in place of something even more distressing. Beginning from a primitive protective system, pain evolves into a complex psychic mechanism, part of the system whereby man maintains himself in his environment. Both as a warning system and as a mechanism of defense, pain helps to avoid or ward off even more unpleasant feeling states or experiences and may even offer the means whereby certain gratifications can be achieved, albeit at a price. If this adaptive role of pain in the psychic economy can be understood, it can be comprehended how it is that certain persons actually seek pain, even to the extent of creating it as a purely psychic experience if no peripheral stimulus is available to evoke it.

"Pain prone patients repeatedly or chronically suffer from one or another painful disability, sometimes with and sometimes without any recognizable peripheral change. In their choice of pain as the symptom, a long-term background of guilt and/or a guilt-provoking situation precipitating pain can be expected. Some of these individuals are chronically depressive, pessimistic and gloomy people whose guilty, self-depreciating attitudes are readily apparent. Some seem to have suffered the most extraordinary number and variety of defeats, humiliations and unpleasant experiences. They drift into situations or submit to relationships in which they are hurt, beaten, defeated, humiliated and seem not to learn from experience... they conspicuously fail to exploit situations which should lead to successes... Even though they complain of pain, for them the pain is almost a comfort or an old friend. It is an adjustment, a way of adaptation, acquired through psychic experience" (Adler et al. 1989, 1997; see also Adler 1976).

Patients who "need to suffer pain" to maintain their mental equilibrium have certain characteristics and modes of behavior that can be detected in a single conversation. The best results are obtained with a history-taking technique that allows the patient to tell his story in his own words but nevertheless to be guided into areas which seem important. Once the "need to suffer pain" has been diagnosed, time can be saved and the patient can be spared unnecessary, costly, and sometimes hazardous investigations and surgery with its serious consequences. Of course, one must not forget that a patient who "needs to suffer pain" may also have a painful organic disease.

It is believed that with this type of patient, pain plays an important part in early childhood in controlling the mental equilibrium (see also Engel's 1959, hypothesis of pain-prone patients having lived through a distinct pattern of developmental psychosocial experiences). Consequently, therapy that is aimed at eliminating "pain" is a threat to the patient's mental stability. This results in a breakdown in the doctor-patient relationship or in exacerbation of the pain with the subsequent tragic chain reaction of investigations and operations. The aim of the doctor-patient relationship should be for the doctor to give support to the patient in his pain. Only after months or years of patiently building up a working relationship should one work on the background to the patient's complaints. In some cases, the doctor has to be content with protecting the patient from surgery. He has to realize that some of these patients are so ill that to aim to cure them would be roughly comparable to an internist expecting a diabetic patient's pancreas to start producing insulin again when treatment is progressing satisfactorily (for further literature on the relationship between childhood experiences and pain in adulthood, see Adler et al. 1989; Egle et al. 1991; Raphael et al. 2001).

The validity of the assumption of a meaningful correlation between certain childhood experiences and pain proneness in adulthood is based on the following prerequisites. The samples must be homogenous, e.g., pelvic pain, psychogenic pain, multiple personality and disorder based on DSM-III fulfill this request. The psychosocial factors (childhood traumata) must be present more frequently in the samples than in the control groups. These factors must antedate the studied disease, which is the case in the cited studies. The findings should be confirmed by other researchers, which many studies in different countries do. The correlation must prevail after statistical control of confounding factors, a requirement fulfilled by many of the above-mentioned studies. A dosage-effect relation should exist. The studied factors should be specific for the disease. Relative specificity exists since other diseases such as depression and substance abuse are also related to childhood traumatic experiences. The correlation should be biologically plausible. This seems to be the case, e.g., same location of pain in sick parents and the child and later patient. In summary, Engel's 1959 paper on "psychogenic pain and the pain prone patient" is justifiably called seminal. A physician's knowledge about the correlation between traumatic childhood experiences and ensuing proneness to suffer repeatedly from pain to which psychosocial factors contribute heavily goes a long way with regard to diagnosis and therapy.

References

Adler, R. H. (1976). The need to suffer pain. *CIBA Review, 7.*
Adler, R. H., Zlot, I. S., Humy, C., et al. (1989). Engel's psychogenic pain and the pain prone patient: A controlled retrospective clinical study. *Psychosomatic Medicine, 51,* 87–101.
Adler, H. R., Zamboni, P., Hofer, T., et al. (1997). How not to miss a somatic needle in the hay stack of chronic pain. *Journal of Psychosomatic Research, 42,* 499–506.
Egle, U. T., Kissinger, D., & Schwab, R. (1991). Parent-child relations as a predisposition for psychogenic pain syndrome in adulthood. A controlled retrospective study in relation to GL Engel's "pain proneness". *Psychotherapie, Psychosomatik, Medizinische Psychologie, 41,* 247–256.
Engel, G. L. (1951). Primary atypical facial neuralgia. A hysterical conversion symptom. *Psychosomatic Medicine, 13,* 375–396.
Engel, G. L. (1959). "Psychogenic" pain and the pain prone patient. *American Journal of Medicine, 26,* 899–918.
Engel, G. L. (1970). Conversion symptoms. In C. M. Mac Bryde & R. S. Blacklow (Eds.), *Signs and symptoms: Applied physiology and clinical interpretation* (5th ed.). Philadelphia: Lippincott. chap. 30.
Hart, H. (1947). Displacement, guilt and pain. *Psychoanalytic Review, 34,* 259–273.
Kolb, L. C. (1954). *The painful phantom. Psychology, physiology, and treatment. American lecture series* (p. 111). Springfield: Charles C. Thomas.
Rangell, L. (1953). Psychiatric aspects of pain. *Psychosomatic Medicine 15.*
Raphael, K. G., Wisdom, C. S., & Lange, G. (2001). Childhood victimization and pain in adulthood: A prospective investigation. *Pain, 92,* 283–93.
Szasz, T. (1957). *Pain and pleasure.* New York: Basic Books.

Pain-related Anxiety

▶ Fear and Pain

Pain-related Disability

▶ Complex Chronic Pain in Children, Interdisciplinary Treatment

Pain-Related Fear

Definition

Pain-related fear is a general term used to describe several forms of fear with respect to pain, such as fear of movement or fear of physical activities that usually increase pain, ultimately originating in a fear of progressive injury.

Cross-References

- ▶ Disability, Fear of Movement
- ▶ Fear and Pain
- ▶ Fear Reduction Through Exposure in Vivo
- ▶ Hypervigilance and Attention to Pain
- ▶ Muscle Pain, Fear-Avoidance Model

Pain-Related Impairment

- ▶ Impairment, Pain-Related

Pain-Relevant Communication

Definition

Pain-relevant communication refers to verbal and nonverbal exchanges in the context of a person's expressions of pain.

Cross-References

- ▶ Spouse, Role in Chronic Pain

Paleospinothalamic Tract

Synonyms

Anterior spinothalamic tract

Definition

The paleospinothalamic tract is the medial and phylogenetically older component of the spinothalamic tract. It is comprised of the axons of nociceptive-specific and wide dynamic range neurons. It projects to the medial thalamus and is responsible for the autonomic and emotional aspects of pain. It is sometimes called anterior spinothalamic tract.

Cross-References

- ▶ Acute Pain Mechanisms
- ▶ Parafascicular Nucleus, Pain Modulation
- ▶ Somatic Pain

Palliation

- ▶ Pain Treatment in Pediatric Palliative Care

Palliation of Upper GI Cancer

- ▶ Palliative Surgery in Cancer Pain Management

Palliative Care

Definition

Palliative care refers to care of the patient with active, progressive, far-advanced disease in a multidisciplinary approach, for which the focus of the care is to optimize quality of life.

Cross-References

- ▶ Cancer Pain, Palliative Care in Children

Palliative Care and Cancer Pain Management

- ▶ Cancer Pain Management, Interface Between Cancer Pain Management and Palliative Care

Palliative Care in Children

▸ Cancer Pain, Palliative Care in Children

Palliative Medicine

Definition

Palliative Medicine is that branch of medical practice whose focus is the study and management of patients with an illness, typically cancer, for which there is no cure or definitive treatment, but which is progressive and ultimately life-threatening, and causes symptoms and impairments that disable the patient. Persistent pain is a typical symptom but is not necessarily the only symptom or an essential one.

Cross-References

▸ Cancer Pain Management, Interface Between Cancer Pain Management and Palliative Care

Palliative Radiotherapy

Definition

Radiotherapy is the administration of ionising radiation for the treatment of disease. Palliative radiotherapy is the application of radiotherapy with the intention of relieving symptoms or impairment without the expectation that the disease will be cured.

Cross-References

▸ Cancer Pain Management Radiotherapy

Palliative Rehabilitation

▸ Cancer Pain Management, Rehabilitative Therapies

Palliative Surgery in Cancer Pain Management

Kevin C. Conlon[1] and Anne M. Heffernan[2]
[1]Trinity College Dublin, University of Dublin, Dublin, Ireland
[2]The Adelaide and Meath Hospital Incorporating The National Children's Hospital, Dublin, Ireland

Synonyms

Cancer pain management; Palliation of upper GI cancer

Definition

The primary aim of any palliative procedure is the relief of symptoms with resultant preservation or improvement in the patient's quality of life. Although palliative surgery in cancer management may, in its broadest sense, refer to any surgery that is non-curative, it is generally considered to encompass surgical intervention that is aimed primarily at the treatment of symptoms or complications of a tumor in patients in whom a curative option is not available. The decision to proceed with a palliative procedure is complex and includes recognition of the general medical condition of the patient, the extent and prognosis of the disease, knowledge of the natural history of the symptoms, potential durability of the proposed intervention and its impact on the expectancy and quality of life, potential alternative nonsurgical options, and the patient's wishes (Kane 1984; Miner and Karpeh 2004). The following section focuses on gastrointestinal malignancies to explore the issues further.

Characteristics

Clinical Features

Despite recent advances in the diagnosis, screening, and multidisciplinary management of

Palliative Surgery in Cancer Pain Management, Table 1 Prevalence of pain in advanced or terminal cancer of the upper gastrointestinal tract (4, 5)

Primary site of cancer	Patients with pain (%) Mean	Range
Esophagus	87	80–93
Pancreas	81	72–100
Liver/biliary	79	65–100
Stomach	78	67–93
Rectal	25	12–30

malignant disease, approximately 560,000 patients in the United States will die each year of cancer (Jemal et al. 2004). Gastrointestinal (GI) cancers such as gastric, pancreatic, and colonic account for 25 % of the total deaths. For these patients, the end of life is often spent dealing with symptoms and complications requiring palliative surgical intervention, such as abdominal or pelvic pain, sepsis, gastrointestinal obstruction, biliary obstruction, hemorrhage, intestinal perforation, and ascites.

The incidence of pain in those receiving active treatment ranges from 24 % to 60 % and approaches 69 % in those with advanced terminal cancer (Van den Beuken-van Everdingen et al. 2007). Pain is a particularly distressing symptom for patients with GI malignancies and is generally indicative of local and perineural invasion. Table 1 details the prevalence of pain in common advanced GI cancers (Bonica 1986; Moran et al. 1987).

The goal of effective pain management is to enable patients to have a good quality of life, attain a reasonable performance status, tolerate diagnostic and therapeutic procedures, and to die relatively free of pain while maintaining freedom of choice and minimizing side effects. Successful pain therapy begins with a comprehensive pain assessment, taking into account the individual patient's goals and priorities and their definition of pain and suffering. A treatment plan is formulated which should be a dynamic process, changing in response to the underlying progressive disease and perhaps involving multiple treatment interventions over time.

Guidelines originally formulated by the World Health Organization (WHO) in 1982, and

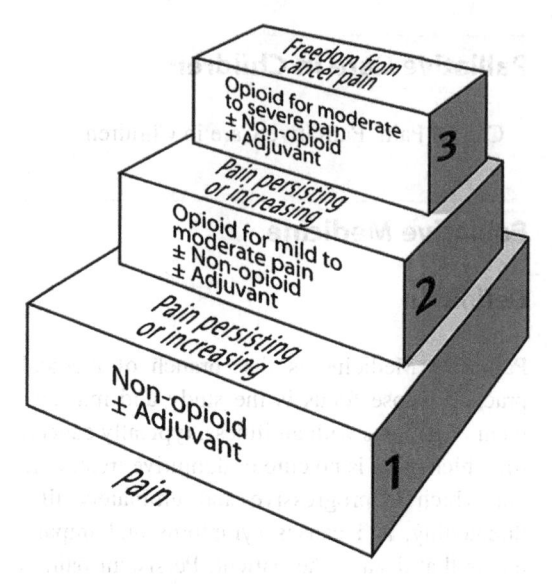

Palliative Surgery in Cancer Pain Management, Fig. 1 The WHO analgesic ladder (From World Health Organization 1996, with permission)

updated in 1996, have become the internationally accepted standard for the principles governing treatment of cancer pain. These guidelines focus on oral analgesic use as the mainstay of therapy (World Health Organization 1996; Schug et al. 1992).

The analgesic ladder is utilized (Fig. 1). Inadequate pain relief at one level results in a step up to the next level instead of changing to a drug on the same level. Non-opioids (paracetamol and nonsteroidal anti-inflammatory agents) are used for pain of mild to moderate intensity. If pain is not controlled, the addition of a so-called "weak" opioid (codeine or tramadol) to the non-opioid is reasonable. When the combination fails despite the use of appropriate doses, the opioid may be replaced by a stronger opioid (morphine sulfate, oxycodone, fentanyl). Any further increase in pain is then treated by an increase in the opioid dose. An argument has been made for the addition of a fourth step which would be intervention for management of pain crises. This is discussed below.

All steps of the analgesic ladder may be complemented by the use of co-analgesic drugs. These provide an alternative for patients who cannot be treated with conventional analgesics alone without encountering unacceptable morbidity. For example, drugs such as the tricyclic antidepressants (e.g., amitriptyline) and anticonvulsants (e.g., gabapentin and pregabalin) have been used successfully for neuropathic pain. Duloxetine, a new selective noradrenaline reuptake inhibitor, appears to have fewer side effects than the tricyclic antidepressants and appears useful in the management of neuropathic pain (Smith et al. 2011). Tapentadol is a new medication with a dual mechanism of action μ-opioid receptor agonism and noradrenaline reuptake inhibition, both mechanisms contributing to a reduction in analgesia. It is thought that its main use will be in patients with severe pain which also has a neuropathic element (Prommer EE 2010). Topical fentanyl patches are very useful in severe pain, and they have now very effective formulations for breakthrough pain in the form of lollipops or intranasal sprays (Mercandante et al. 2008). Corticosteroids may enhance analgesia in a variety of situations, including pain related to nerve compression or bony metastases. They may also improve mood, appetite, and general strength. When a pharmacological approach fails to relieve pain, interventional approaches to treatment are available, aimed at interrupting, destroying, or stimulating pain pathways. In addition to surgery, these interventional strategies include many types of neural blockade, neuraxial infusion, and other therapies. Their main pain relief benefit is in patients who have limited life expectancy.

Celiac sympathetic plexus blockade is the main interventional approach used very effectively in patients with life-limiting disease. This can be achieved either by intraoperative open or laparoscopic chemical splanchnicectomy, percutaneous or endoscopically directed celiac blockade, or thoracoscopic splanchnicectomy. Chemical splanchnicectomy typically involves the injection of 20 ml of 50 % alcohol around the base of the celiac artery. It has been demonstrated in patients with unresectable pancreatic cancer to show improved pain control, reduced narcotic usage, and reduced constipation (Lilliemoe et al. 1993; Yan and Myers 2007).

Complementary therapies may also have a role in achieving pain relief and are becoming increasingly popular with the general public. Massage, acupuncture, and aromatherapy have shown promise for the palliation of symptoms, but more rigorous examination is required before they can be considered an integral part of conventional pain management (Bardia et al. 2006).

Visceral pain may be secondary to gastrointestinal obstruction, which is not infrequent in patients with advanced disease. Dysphagia is the predominant symptom in 75 % of patients with esophageal carcinoma. For these patients, the choice of palliative therapy depends on tumor size, location, patient performance, status, and the experience available. In general, nonoperative approaches such as endoscopic dilation and/or stent placement, or laser ablation using the neodymium:yttrium-aluminum-garnet (Nd:YAG) laser are more commonly utilized than surgical procedures such as palliative resection, substernal bypass, or feeding enterostomy. While incomplete surgical resection may relieve dysphagia in 70–80 % of cases, median survival is short (7–9 months) and morbidity high (40–50 %) (Nash and Gerdes 2002).

The role of palliative resection as a palliative intervention is controversial in other GI sites as well. A recent retrospective review from Memorial Sloan-Kettering Cancer Center examined 307 patients who underwent a non-curative gastric resection (Miner et al. 2004). A palliative resection with residual gross disease was performed in 48 % of cases, specifically to relieve specific symptoms, control pain, or improve quality of life. Median survival was significantly reduced in this cohort compared to the non-palliative group (8.3 months vs. 13.5 months, $P < 0.001$). Only 24 % of patients required an additional palliative procedure during their remaining lifetime. However, many of these interventions were performed to treat new symptoms that arose after the initial palliative resection. The same group has subsequently attempted to better define the group of patients who would benefit from an aggressive surgical approach (Miner and Karpeh 2004). By using a portioned survival analysis, they defined three clinical health states: time with subjective toxic effects of treatment (TOX), time without symptoms requiring additional palliative procedures (TWiST), and time required in hospital for additional palliative procedures (REL). Patients with multiple sites of metastatic disease had decreased time in the TWiST state, suggesting that palliative resection may not be beneficial in this group.

For patients with adenocarcinoma of the pancreas in whom the primary tumor cannot be resected with negative margins, improvements in perioperative morbidity following pancreaticoduodenectomy have led some authors to propose an aggressive surgical approach (Lillemoe et al. 1996). Survival appears to be somewhat improved compared to palliative bypass. However, prospective data are lacking and this strategy is considered controversial.

Improvements in endoscopic techniques, coupled with the increasing use of laparoscopy to stage pancreatic cancer, have led many to question the role of surgery for the majority of patients who present with advanced pancreatic malignancies. Proponents of laparoscopic staging have argued that, due to the poor prognosis in patients with unresectable disease, avoidance of a laparotomy is beneficial with reduced hospital stay and less time for administration of adjuvant treatment. However, critics have argued that the procedure is of minimal benefit, as patients with unresectable disease will require a subsequent bypass procedure for biliary and/or gastric outlet obstruction. Historically, prophylactic bypasses have been supported by reports of the development of obstructive jaundice in as many as 70 % of patients and gastric outlet obstruction in up to 25 % of patients with unresectable pancreatic cancer. Although endoscopic techniques for the relief of obstructive jaundice have replaced surgery, the utility of surgery to palliate gastric and duodenal obstruction remains controversial.

A recent study examined the role of prophylactic gastric bypass in this disease, randomizing 87 patients who were thought to lack a significant risk for duodenal obstruction to receive either a retrocolic gastrojejunostomy or no gastrojejunostomy (Lillemoe et al. 1999). Late gastric

outlet obstruction requiring intervention developed in 8 of 43 patients (18.6 %) who did not receive a gastrojejunostomy at a median of 2 months from initial exploration, suggesting that prophylactic bypass is indicated in a significant proportion of patients. Similar data were recently published from Europe (Van Heek et al. 2003). However, a nonrandomized but prospective study from Memorial Sloan-Kettering appears to differ from the previously cited work (Espat et al. 1999). In this study, the outcome of 155 patients with laparoscopically staged, unresectable pancreatic cancer who did not undergo open enteric or biliary bypass at the time of initial laparoscopic staging was analyzed. A subsequent surgical bypass was only required in 3 % of patients, leading the authors to suggest that gastroenterostomy should be reserved for those who develop documented gastric outlet obstruction. These contrasting data indicate that further studies are required to determine the place of surgical palliation for this patient population.

In contrast to the controversy that exists in upper gastrointestinal cancer, surgical palliation is considered appropriate for both advanced colon and rectal cancer. Options include resection with restoration of intestinal continuity, abdominoperineal resection, or creation of a diverting colostomy, depending on the symptoms and clinical status of the patient. Each patient should be approached individually. Palliative resection may be indicated in a patient with minimal residual disease. However, in the patient with significant comorbidities, carcinomatosis, or extensive hepatic disease, a fecal diversion or no intervention may be more appropriate.

In summary, the role for surgical palliation for gastrointestinal cancer continues to evolve. Unfortunately, specific indications for palliative resection, surgical bypass, or endoscopic palliation remain unclear as well-designed prospective randomized trials are lacking. However, there is no doubt that selected patients do benefit from aggressive surgical intervention with improved symptomatic palliation and quality of life. Further prospective studies are required to identify the patient subgroups that would benefit from such an approach.

References

Bardia, A., Barton, D. L., Prokop, L. J., Bauer, B. A., & Moynihan, T. J. (2006). Efficacy of complementary and alternative medicine therapies in relieving cancer pain: A systematic review. *Journal of Clinical Oncology, 24*(34), 5457–5464.

Bonica, J. J. (1986). Cancer pain: Current status and future needs. In J. J. Bonica (Ed.), *The management of pain* (pp. 400–455). Philadelphia: Lea & Febiger.

Espat, N. J., Brennan, M. F., & Conlon, K. C. (1999). Patients with laparoscopically staged unresectable pancreatic adenocarcinoma do not require subsequent surgical biliary or gastric bypass. *Journal of the American College of Surgery, 188*, 649–655.

Jemal, A., Tiwari, R. C., Murray, T., et al. (2004). Cancer statistics 2004. *CA: A Cancer Journal for Clinicians, 54*, 8–29.

Kane, R. L., Wales, J., & Bernstein, L. (1984). A randomised controlled trial of hospice care. *Lancet, 1*, 890–894.

Lillemoe, K. D., Cameron, J. L., Kaufman, H. S., et al. (1993). Chemical splanchnicectomy in patients with unresectable pancreatic cancer: A prospective randomized trial. *Annals of Surgery, 23*, 447–455.

Lillemoe, K. D., Cameron, J. L., Yeo, C. J., et al. (1996). Pancreaticoduodenectomy: Does it have a role in palliation of pancreatic cancer? *Annals of Surgery, 223*, 718–725.

Lillemoe, K. D., Cameron, J. L., Hardacre, J. M., et al. (1999). Is prophylactic gastrojejunostomy indicated for unresectable periampullary cancer? A prospective randomized trial. *Annals of Surgery, 230*, 322–328.

Mercandante, S., Porzio, G., Ferrara, P., Fulfaro, F., Aielli, F., Verna, L., Villari, P., & Mangione, S. (2008). Sustained release oral morphine versus transdermal fentanyl and oral methadone in cancer pain management. *European Journal of Pain, 12*, 1040–1046.

Miner, T. J., & Karpeh, M. S. (2004). Gastrectomy for gastric cancer: Defining critical elements of patient selection and outcome assessment. *Surgical Oncology Clinics of North America, 13*, 455–466.

Miner, T. J., Jaques, D. P., Karpeh, M. S., et al. (2004). Defining palliative surgery in patients receiving noncurative resections for gastric cancer. *Journal of the American College of Surgeons, 198*, 1013–1021.

Moran, M. R., Rothenberger, D. A., Lahr, C. J., et al. (1987). Palliation for rectal cancer. *Archives of Surgery, 122*, 640–643.

Nash, C. L., & Gerdes, H. (2002). Methods of palliation of esophageal and gastric cancer. *Surgical Oncology Clinics of North America, 11*, 459–483.

Prommer, E. E. (2010). Tapentadol an initial analysis. *Journal of Opioid Management, 6*(3), 223–226.

Schug, S. A., Zech, D., & Grond, S. (1992). A long-term survey of morphine in cancer pain patients. *Journal of Pain and Symptom Management, 7*, 259–266.

Smith, E. M., Bakitas, M. A., Homel, P., Piehl, M., Kingman, L., Fadul, C. E., & Bookbinder, M. (2011). Preliminary assessment of a neuropathic pain treatment and referral algorithm for patients with cancer. *Journal of Pain and Symptom Management, 42*(6), 822–838.

Van den Beuken-van Everdingen, M. H., de Rijke, J. M., Kessels, A. G., Schouten, H. C., van Kleef, M., & Patijin, J. (2007). Prevalence of pain in patients with cancer. A systematic review of the past 40 years. *Annals of Oncology, 18*, 1437–1449.

Van Heek, N. T., De Castro, S. M., van Eijck, C. H., et al. (2003). The need for a prophylactic gastrojejunostomy for unresectable periampullary cancer: A prospective randomized multicenter trial with special focus on assessment of quality of life. *Annals of Surgery, 238*, 894–902.

World Health Organization. (1996). *Cancer pain relief and palliative care. Report of a WHO Expert Committee* (3rd ed.). Geneva: World Health Organization.

Yan, B. M., & Myers, R. P. (2007). Neurolytic celiac plexus block for pain control in unresectable pancreatic cancer. *The American Journal of Gastroenterology, 102*(2), 430–438.

Pancreatalgia

▶ Visceral Pain Model, Pancreatic Pain

Pancreatitis Pain

▶ Visceral Pain Model, Pancreatic Pain

Panic Attacks

Definition

These consist of repeated, unpredictable attacks of intense fear accompanied by severe anxiety symptoms in the body that may last from minutes to hours.

Cross-References

▶ Psychiatric Aspects of Visceral Pain

PAOD

▶ Peripheral Arterial Occlusive Disease

Papillae

Definition

Fungiform, foliate, and filiform papillae of the tongue, where a high concentration of nociceptive nerve endings can be found; they also host gustatory receptors.

Cross-References

▶ Nociception in Nose and Oral Mucosa

Papilledema

Definition

Swelling of the papilla of the optic nerve, due to high intracranial pressure.

Cross-References

▶ Headache Due to Low Cerebrospinal Fluid Pressure

PAR

▶ Proteinase-Activated Receptors

Parabrachial Hypothalamic and Amygdaloid Projections

Jean-François Bernard
INSERM UMR-894, UPMC Site
Pitié-Salpétrière, Paris, France

Definition

The parabrachial (PB) area is a group of neurons that surround the brachium conjunctivum (also named superior cerebellar peduncle). As shown in Fig. 1, the PB area is located in a dorsal and lateral region of the brain stem, at the level of junction between the pons (below the cerebellum) and the mesencephalon (below the inferior colliculus). Due to the PB localization, the coronal plane located at the level where the inferior colliculus (mesencephalic) merges with the pons, may be used as a plane of reference (Pr) (indicated with a dotted line in Fig. 1a). Thus, it is possible to identify a mesencephalic (mPB) and a pontine (pPB) portion, rostral and caudal to the Pr, respectively. From the Pr, mPB and pPB extend respectively ≈ 400 μm in a rostral direction and ≈ 800 μm in a caudal direction.

The PB is a complicated area, since in addition to the rostro-caudal division it includes at least ten subnuclei (Fig. 3a). However, it is important to consider only two major divisions: (1) The parabrachial lateral (PBl) (Fig. 1, dark gray) is dorsal and lateral to the bc. It is chiefly the nociceptive/visceral region that is presented here. (2) The parabrachial medial (PBm) (Fig. 1, white) is ventral and medial to the bc. It is chiefly a gustatory area, although even a small portion is linked functionally to the PBl.

Characteristics

Lamina I Afferent Projection to PB

Anatomical ▶ retrograde and ▶ anterograde axonal tracers have demonstrated that the PBl specifically receives a very extensive projection arising from the ▶ lamina I neurons of the trigeminal and spinal dorsal horn (Fig. 2) (Cechetto et al. 1985; Bernard et al. 1995). Lamina I neurons make up a key primary relay for nociceptive messages conveyed by Aδ and C fibers from periphery. Most of lamina I-PB neurons are nociceptive responsive, and a substantial proportion of them respond only to noxious stimuli (Bester et al. 2000). Smaller populations of lamina I neurons are sensitive to innocuous thermal stimuli (see ▶ Spinoparabrachial Tract). The PBl area also receives projections from spinal laminae IV, V, VIII

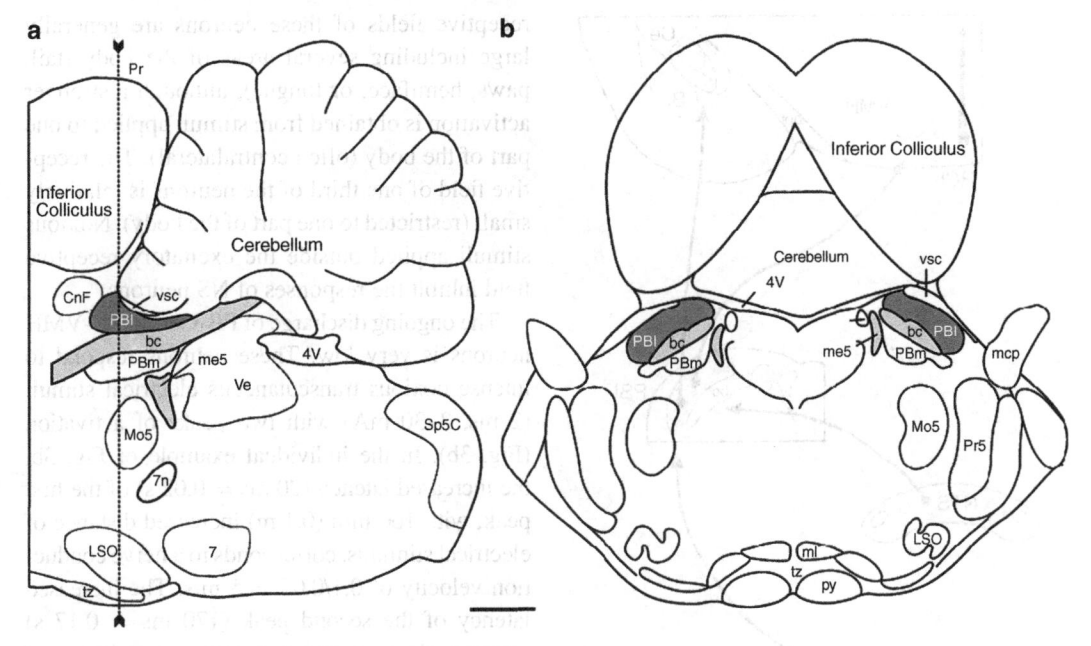

Parabrachial Hypothalamic and Amygdaloid Projections, Fig. 1 Diagrams indicating the location of the parabrachial area, lateral division (PBl). (**a**) Sagittal section of the brain stem of the rat (1.9 mm lateral); (**b**) Coronal section of the brain stem of the rat (−0.30 mm interaural) which corresponds to the level of the *dotted line* in (**a**). In both drawings, the PBl is in *dark gray*, and the brachium conjunctivum (bc) and the mesencephalic trigeminal tract (me5) are in *light gray*. Scale bar, 1 mm (**a** and **b**). Abbreviations: 4 V, fourth ventricle; 7, facial nucleus; 7n, facial nerve; CnF, cuneiformis nucleus; LC, locus coeruleus; LSO, lateral superior olive; mcp, medial cerebellar peduncle; ml, medial lemniscus; Mo5, motor trigeminal nucleus; PBm, parabrachial medial area; Pr, coronal plan of reference; Pr5, principal sensory trigeminal nucleus; py, pyramidal tract; Sp5C, spinal trigeminal nucleus, caudal; tz, trapezoid body; Ve, vestibular nuclei; vsc, ventral spinocerebellar tract

and the lateral spinal nucleus (see ▶ Spinoparabrachial Tract). However, this latter projection chiefly targets the medial most portion of the PBl (e.g., the internal lateral subnucleus), which does not project to the ▶ amygdala and the ▶ hypothalamus, and is thus not included in this report.

Solitary Afferent Projection to PB

The PBl receives a second very extensive projection arising from the nucleus of the solitary tract (Fig. 2) (Herbert et al. 1990). Although, not so directly involved in nociceptive processing as lamina I neurons, this input conveys, via the vagal nerve, visceral nociceptive and homeostatic (blood pressure, blood CO_2/pH, concentration of lipid and amino acid in duodenum, etc.) messages to the PBl.

Properties of PB-Ce and PB-Hypothalamic Neurons

The two primary projections of the PBl (with the exception of the internal lateral subnucleus) are the central nucleus of the amygdala (Ce), the ventromedial (VMH), and retrochiamatic (RCh) hypothalamic nuclei (Fig. 2). An extensive electrophysiological study has recorded numerous PBl neurons in rats under halothane anesthesia (0.5 %) in a nitrous oxide/oxygen mixture (Bernard and Besson 1990; Bester et al. 1995). These PB neurons were identified by antidromic electrical stimulation, applied within the Ce or the VMH/RCh. In agreement with anatomical studies, the locations of parabrachial neurons projecting to the Ce (PB-Ce) or to the VMH/RCh nuclei (PB-VMH) are clearly distinct. As shown in Fig. 3a, the PB-Ce neurons (black

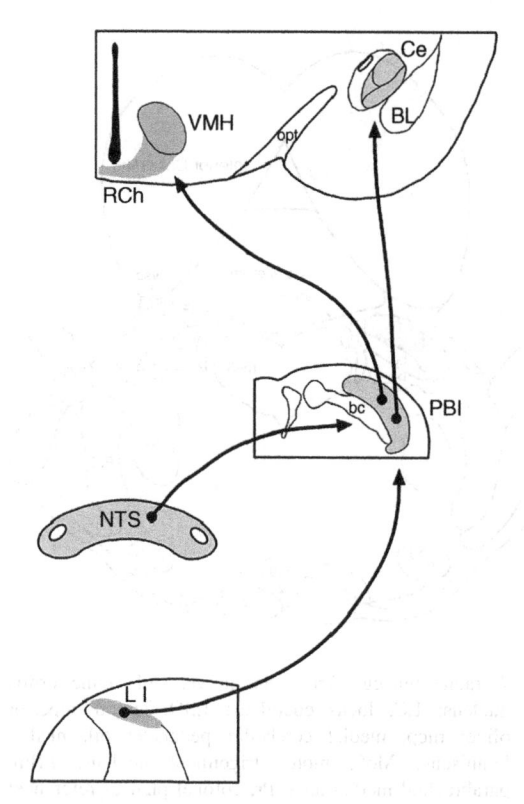

Parabrachial Hypothalamic and Amygdaloid Projections, Fig. 2 Summary diagram of the parabrachio-amygdaloid and parabrachio-hypothalamic pathways, with indication of the two primary inputs of the parabrachial lateral division (PBl). In *gray*: regions processing nociceptive and/or visceral messages. Abbreviations: bc, brachium conjunctivum; BL, basolateral amygdaloid nucleus; Ce, central nucleus of the amygdala; LI, lamina I of the dorsal horn of the spinal cord; NTS, nucleus of tract solitary; opt, optic tract; RCh, retrochiasmatic area; VMH, ventromedial nucleus of the hypothalamus

points) are located in the caudal "pontine" portion of the PBl (pPBl), whereas the PB-VMH neurons (stars) are located in the rostral "mesencephalic" portion of the PBl (mPBl).

A high proportion of PB-Ce (70 %) and PB-VMH (50 %) neurons are nociceptive specific (NS), the remainder being unresponsive. These NS neurons are strongly excited by noxious electrical (≥ 1 mA), mechanical (pinch, squeeze), and thermal (≥ 44 °C) cutaneous stimuli, whereas electrical (< 1 mA), tactile, innocuous pressure and thermal (< 44 °C) somatic stimuli are weakly or totally ineffective. The cutaneous

receptive fields of these neurons are generally large including several areas of the body (tail, paws, hemiface, or tongue), although a stronger activation is obtained from stimuli applied to one part of the body (often contralateral). The receptive field of one third of the neurons is relatively small (restricted to one part of the body). Noxious stimuli applied outside the excitatory receptive field inhibit the responses of NS neurons.

The ongoing discharge of PB-Ce and PB-VMH neurons is very low. These neurons respond to intense noxious transcutaneous electrical stimuli (2 ms, 3–30 mA) with two peaks of activation (Fig. 3b). In the individual example of Fig. 3b, the increased latency (20 ms = 0.02 s) of the first peak, with 100 mm (0.1 m) increased distance of electrical stimulus, corresponds to a nerve conduction velocity of 0.1/0.02 = 5 m/s. The increased latency of the second peak (170 ms = 0.17 s) corresponds to a nerve conduction velocity of 0.1/0.17 = 0.59 m/s. On average, the early and the late peak were triggered by the activation of peripheral fibers with conduction velocities in 8–20 m/s and 0.5–0.8 m/s range, for example, an activation of Aδ and C-fibers, respectively. Furthermore, the response due to C-fibers exhibits a wind-up phenomenon during repetitive stimulation (0.66 Hz).

PB-Ce and PB-VMH neurons respond to mechanical and/or thermal noxious stimuli with a strong and sustained but decreasing activation. These neurons exhibit a clear capacity to encode thermal stimuli in the noxious range. Figure 3c shows a PB neuron without ongoing activity, which does not respond to 42 °C applied to contralateral forepaw. On the other hand, the response appears at 44 °C and increases in parallel with temperature within the noxious range. The mean stimulus-response curve (discharge as a function of nociceptive temperature) increases progressively from a mean threshold around 44 °C up to a maximum of around 50/52 °C (Fig. 4d). The main change in the mean curve (steepest portion of the slope) is in 46–50 °C range (e.g., when the painful sensation becomes unbearable). Intravenous morphine induced a strong depression of the response elicited by noxious thermal stimuli in a dose-related and

Parabrachial Hypothalamic and Amygdaloid Projections, Fig. 3 (a) Location of nociceptive-specific neurons antidromically driven from the central nucleus of the amygdala (*black points*) or from the ventromedial hypothalamic nucleus (*stars*). Drawings of coronal sections of parabrachial region are presented from caudal to rostral, 300 μm apart. (**a** *1, 2*) pontine division of the PB area; (**a** *3, 4*) mesencephalic division of the parabrachial area. Brachium conjunctivum (bc) and mesencephalic trigeminal tract (me5) are in light gray. (**b**) Single sweep recording of the early and late peak of spike firing evoked by transcutaneous electrical stimulation of the base (*1*) and the tip (*2*) of the tail. Note the 20 ms and 170 ms increased latency of first and second discharge peak when the electrical stimulus is moved 100 mm apart from the base to the tip of tail. (**c**) Encoding properties of a nociceptive-specific parabrachio-amygdaloid neuron to thermal stimulation applied to the contralateral forelimb. (**d**) Mean stimulus–response curve of parabrachial neurons (discharge during 20 s) as a function of temperature applied on skin. Scale bar, 1 mm (**a**). Abbreviations: cl, central lateral; dl, dorsal lateral; el, external lateral; em, external medial; il, internal lateral; lcr, lateral crescent; m, medial; sl, superior lateral; vl, ventral lateral subnuclei of the parabrachial area.

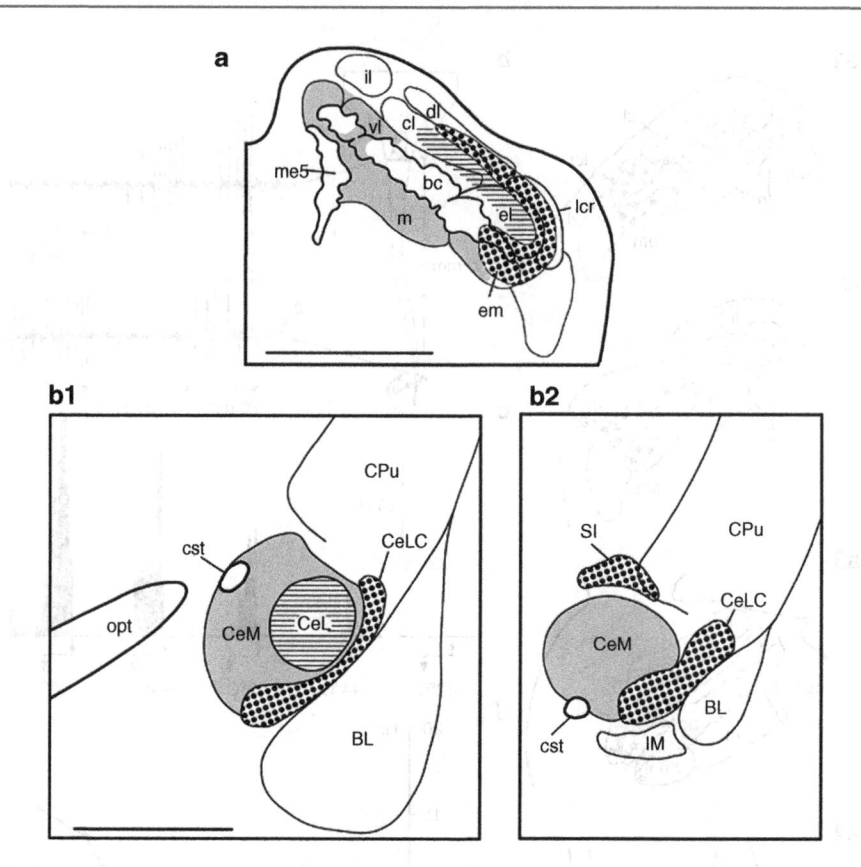

Parabrachial Hypothalamic and Amygdaloid Projections, Fig. 4 Summary diagram illustrating, in coronal sections, the projection pattern from the pontine parabrachial (pPB) area (**a**) to the caudal (**b1**) and rostral (**b2**) levels of the central nucleus of the amygdala (Ce). The pPB "nociceptive" area (*large dots*) projects upon the CeLC and the SI (*large dots*). pPB "nociceptive + visceral" area (horizontal hatching) projects upon the CeL (horizontal hatching). The pPB "gustatory" area (*gray*) projects upon the CeM (*gray*). Scale bars, 1 mm (**a** and **b**). Abbreviations: BL, basolateral amygdaloid nucleus; CeL, lateral subdivision of Ce; CeLC, lateral capsular subdivision of Ce; CeM, medial subdivision of Ce; cl, central lateral PB subnucleus; CPu, caudate-putamen; cst, commissural stria terminalis; dl, dorsal lateral PB subnucleus; el, external lateral PB subnucleus; em, external medial PB subnucleus; il, internal lateral PB subnucleus; IM, intercalated main amygdaloid nucleus; lcr, lateral crescent PB subnucleus; m, medial PB subnucleus; me5, mesencephalic trigeminal tract; opt, optic tract; PB, parabrachial area; SI, substantia innominata; sl, superior lateral PB subnucleus; vl, ventral lateral PB subnucleus

naloxone reversible fashion, with an ED50 = 1.8 mg/kg (Huang et al. 1993). PB neurons are also affected by visceral noxious stimuli (intraperitoneal injection of bradykinin and/or colorectal distension with mean threshold of 56 ± 24 mmHg) almost exclusively in the noxious range (Bernard et al. 1994). An additional small population of PB neurons responds specifically to cold. They encode linearly cold temperatures from an innocuous range (25–15 °C) up to a very low (−10 °C) and noxious cold (Menendez et al. 1996).

Projections upon the Amygdala

These projections were studied with retrograde (► WGA-HRP) (Moga et al. 1990) and antero-grade axonal tracer (► PHA-L) (Bernard et al. 1993). PB projections to the amygdala originate almost exclusively from the pPB and target chiefly the ipsilateral Ce.

1. The pPBl nociceptive and visceral areas centered on the external lateral subnucleus and including adjacent areas + the external medial subnucleus (in ventrolateral position) project primarily to the lateral capsular portion

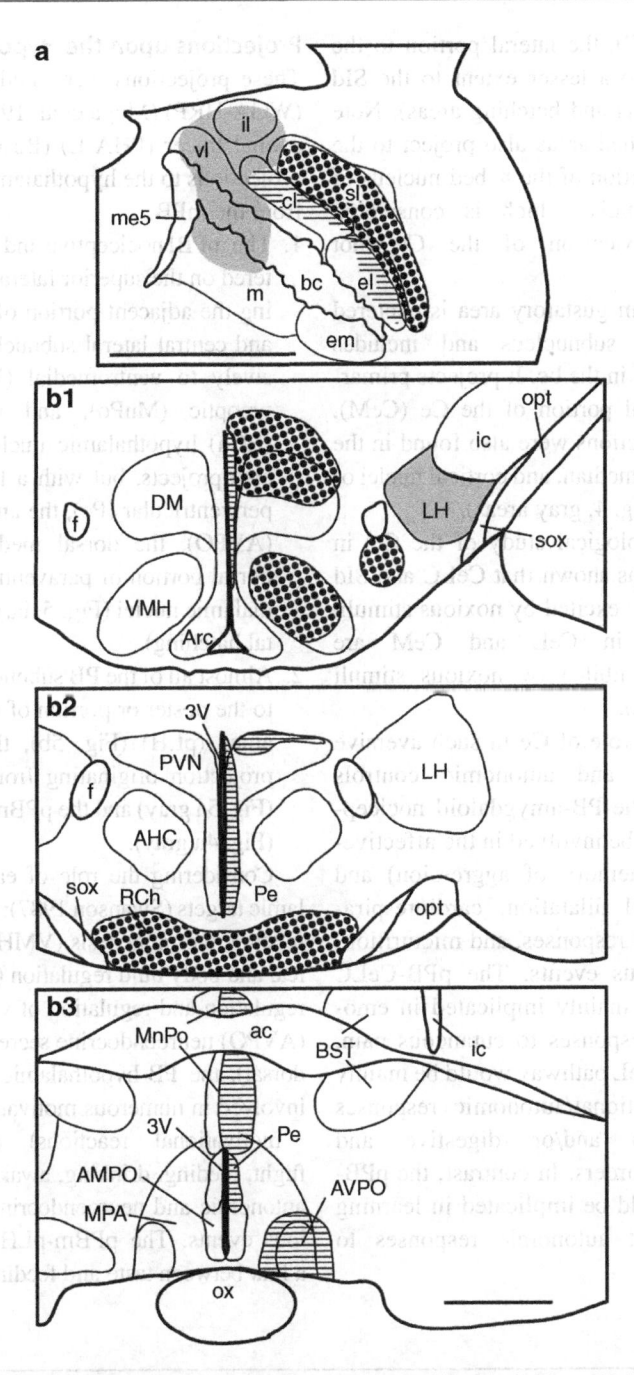

Parabrachial Hypothalamic and Amygdaloid Projections, Fig. 5 Summary diagram illustrating, in coronal sections, the projection pattern from the mesencephalic parabrachial (mPB) area (**a**) to the hypothalamus (caudal to rostral) (**b**, *1–3*). The mPB "nociceptive" area (*large dots*) projects upon the VMH, RCh, DM, and subfornical hypothalamic nuclei (*large dots*). The mPB "nociceptive + visceral" area (horizontal hatching) projects upon the Pe, PVN dorsal, MnPo, and AVPO hypothalamic nuclei (horizontal hatching). The mPB "gustatory" (*gray*) + pPB "gustatory" (see Fig. 4a, in *gray*) areas project upon the caudal portion of the lateral hypothalamus (*gray*). Scale bars, 1 mm (**a** and **b**). Abbreviations: 3 V, third ventricle; ac, anterior commissure; AHC, anterior hypothalamic area central; AMPO, anterior medial preoptic nucleus; Arc, arcuate nucleus; AVPO, anteroventral preoptic nucleus; BST, bed nucleus of stria terminals; cl, central lateral PB subnucleus; dl, dorsal lateral PB

of the Ce (CeLC), the lateral portion to the Ce (CeL), and to a lesser extent to the SId (Fig. 4, large dots and hatching areas). Note that the PB hatched areas also project to the lateral dorsal portion of the ▶ bed nucleus of the stria terminalis, which is considered a functional extension of the Ce (not illustrated).

2. The caudal pPBm gustatory area is centered on the medial subnucleus and includes the adjacent area in the bc. It projects primarily to the medial portion of the Ce (CeM). Additional projections were also found in the basolateral, basomedian, and cortical nuclei of the amygdala (Fig. 4, gray areas).

An electrophysiological study of the Ce, in anesthetized rats, has shown that CeLC and SId neurons are strongly excited by noxious stimuli, whereas neurons in CeL and CeM are unresponsive or inhibited by noxious stimuli (Bernard et al. 1992).

Considering the role of Ce in such aversive emotions as fear and autonomic controls (Aggleton 1992), the PB-amygdaloid nociceptive pathway could be involved in the affective-emotional (fear, memory of aggression) and ▶ autonomic (pupil dilatation, cardiorespiratory, adrenocortical responses, and micturition) reactions to noxious events. The pPB-CeLC pathway would be mainly implicated in emotional/autonomic responses to cutaneous pain, whereas the pPB-CeL pathway would be mainly implicated in emotional/autonomic responses to visceral pain and/or digestive and chemosensitive disorders. In contrast, the pPB-CeM pathway would be implicated in learning taste aversion and autonomic responses to feeding.

Projections upon the Hypothalamus

These projections were studied with retrograde (WGA-HRP) (Moga et al. 1990) and anterograde axonal tracer (PHA-L) (Bester et al. 1997). PB projections to the hypothalamus originate chiefly from the mPB.

1. The mPBl nociceptive and visceral region, centered on the superior lateral subnucleus including the adjacent portion of the external lateral and central lateral subnucleus, projects extensively to ventromedial (VMH), the median preoptic (MnPo), and the retrochiasmatic (RCh) hypothalamic nuclei. This mPBl area also projects, but with a lower density to the periventricular (Pe), the anteroventral preoptic (AVPO), the dorsal medial (DM), and the dorsal portion of paraventricular (PVN) hypothalamic nuclei (Fig. 5, large dots and horizontal hatching).

2. Almost all of the PB subnuclei project diffusely to the posterior portion of the lateral hypothalamus (pLH) (Fig. 5b), the more consistent projection originating from the medial mPB (Fig. 5a gray) and the pPBm "gustatory" region (Fig. 4a gray).

Considering the role of each of the hypothalamic targets (Swanson 1987): aggression, defense, feeding, thermogenesis (VMH, RCh, DM), chemical and body fluid regulation (MnPo, Pe), thermoregulation and regulation of vigilance/sleep states (AVPO) neuroendocrine secretions of CRH (PVN dorsal), the PB-hypothalamic pathway would be involved in numerous motivational behaviors (see ▶ motivational reactions) (aggression/defense, flight, feeding, drinking, awakening) as well as in autonomic and neuroendocrine responses to noxious events. The pPBm-pLH pathway could be a link between taste and feeding behaviors.

Parabrachial Hypothalamic and Amygdaloid Projections, Fig. 5 (continued) subnucleus; DM, dorsal medial nucleus; el, external lateral PB subnucleus; em, external medial PB subnucleus; f, fornix; ic, internal capsule; il, internal lateral PB subnucleus; lcr, lateral crescent PB subnucleus; LH, lateral hypothalamus; m, medial PB subnucleus; me5, mesencephalic trigeminal tract; MnPO, median preoptic nucleus; MPA, medial preoptic area; opt, optic tract; ox, optic chiasm; PB, parabrachial area; Pe, periventricular nucleus; pPB, pontine parabrachial area; RCh, retrochiasmatic area; sl, superior lateral subnucleus; sox, supraoptic decussation; vl, ventral lateral PB subnucleus; VMH, ventromedial hypothalamic nucleus

References

Aggleton, J. P. (1992). *The amygdala, neurobiological aspects of emotion, memory, and mental dysfunction*. New York: Wiley-Liss.

Bernard, J. F., & Besson, J. M. (1990). The spino (trigemino)-ponto-amygdaloid pathway: Electrophysiological evidence for an involvement in pain processes. *Journal of Neurophysiology, 63*, 473–490.

Bernard, J. F., Huang, G. F., & Besson, J. M. (1992). The nucleus centralis of the amygdala and the globus pallidus ventralis: Electrophysiological evidence for an involvement in pain processes. *Journal of Neurophysiology, 68*, 551–569.

Bernard, J. F., Aldén, M., & Besson, J. M. (1993). The organization of the efferent projections from the pontine parabrachial area to the amygdaloid complex: A phaseolus vulgaris leucoagglutinin (PHA-L) study in the rat. *The Journal of Comparative Neurology, 329*, 201–229.

Bernard, J. F., Huang, G. F., & Besson, J. M. (1994). The parabrachial area: Electrophysiological evidence for an involvement in visceral nociceptive processes. *Journal of Neurophysiology, 71*, 1646–1660.

Bernard, J. F., Dallel, R., Raboisson, P., Villanueva, L., & Le Bars, D. (1995). Organization of the efferent projections from the spinal cervical enlargement to the parabrachial area and periaqueductal grey: A PHA-L study in the rat. *The Journal of Comparative Neurology, 353*, 480–505.

Bester, H., Menendez, L., Besson, J. M., & Bernard, J. F. (1995). Spino (trigemino) parabrachiohypothalamic pathway: Electrophysiological evidence for an involvement in pain processes. *Journal of Neurophysiology, 73*, 568–585.

Bester, H., Besson, J. M., & Bernard, J. F. (1997). Organization of efferent projections from the parabrachial area to the hypothalamus: A phaseolus vulgaris-leucoagglutinin study in the rat. *The Journal of Comparative Neurology, 383*, 245–281.

Bester, H., Chapman, V., Besson, J. M., & Bernard, J. F. (2000). Physiological properties of the lamina I spinoparabrachial neurons in the rat. *Journal of Neurophysiology, 83*, 2239–2259.

Cechetto, D. F., Standaert, D. G., & Saper, C. B. (1985). Spinal and trigeminal dorsal horn projections to the parabrachial nucleus in the rat. *The Journal of Comparative Neurology, 240*, 153–160.

Herbert, H., Moga, M., & Saper, C. B. (1990). Connections of the parabrachial nucleus with the nucleus of the solitary tract and medullary reticular formation in the rat. *The Journal of Comparative Neurology, 293*, 540–580.

Huang, G. F., Besson, J. M., & Bernard, J. F. (1993). Morphine depresses the transmission of noxious messages in the spino (trigemino)-ponto-amygdaloid pathway. *European Journal of Pharmacology, 230*, 279–284.

Menendez, L., Bester, H., Besson, J. M., & Bernard, J. F. (1996). Parabrachial area: Electrophysiological evidence for an involvement in cold nociception. *Journal of Neurophysiology, 75*, 2099–2116.

Moga, M. M., Herbert, H., Hurley, K. M., Yasui, Y., Gray, T. S., & Saper, C. B. (1990). Organization of cortical, basal forebrain, and hypothalamic afferents to the parabrachial nucleus in the rat. *The Journal of Comparative Neurology, 295*, 624–661.

Swanson, L. W. (1987). The hypothalamus. In T. Hökfelt & L. W. Swanson (Eds.), *Handbook of chemical neuroanatomy* (Integrated systems of the CNS, part I, Vol. 5, pp. 1–124). Amsterdam/Oxford: Elsevier.

Parabrachial/PAG

Definition

Areas of the midbrain with important projections from lamina I of the spinal cord, which are thought to link to emotional aspects of pain and descending pathways back to the spinal cord.

Cross-References

▶ Opioids in the Spinal Cord and Central Sensitization

Paracetamol

Synonyms

Acetaminophen

Definition

Paracetamol is a nonacidic antipyretic analgesic with only weak anti-inflammatory activity. Paracetamol elicits a tissue-dependent inhibitory effect on the activity of the cyclooxygenase (COX) enzymes with a preference towards COX-2. With respect to its mode of action on COX, paracetamol has been suggested to act as a reducing agent within the peroxidase site of the

COX enzyme. However, other analgesic mechanisms appear likewise feasible.

Cross-References

▶ Cyclooxygenases in Biology and Disease
▶ Pharmacology of Cyclooxygenase Inhibitors
▶ Postoperative Pain, Acute Pain Management Principles
▶ Simple Analgesics

Paracetamol in Postoperative Pain

▶ Postoperative Pain, Paracetamol

Paracrine

Definition

A type of function in which a substance is synthesized in and released from endocrine cells to act locally on nearby cells of a different type and affects their function.

Cross-References

▶ Diencephalic Mast Cells

Paracrine Immune Cells

▶ Diencephalic Mast Cells

Paradoxical Cold Sensation

Definition

A sudden increase of the temperature above 45°C can temporarily induce an icy cold sensation. This phenomenon has been explained by the fact that sensitive cold receptors of the A-delta class can exhibit a short-timed bursting excitation during such a fast rising heat stimulus. The reason for the excitation by heat might be a co-localization of the cold transducing molecule TRPM8 with the heat activated TRPV1 in part of the cold fibers. However, the cold sensation requires a central projection of those units in the respective central pathway which mediates the cold sensation into the brain.

Cross-References

▶ Nociceptors, Cold Thermotransduction

Paradoxical Heat

Definition

A heat sensation (sometimes painful) that results, in particular, from sudden temperature decreases to very low temperatures. It can be explained by the excitation of afferent nerve fibers which feed into the central pathways for noxious cold and noxious heat.

Pathophysiology: Burning sensations upon exposition to cold are characteristic of some forms of neuropathies. These sensations should not be confounded with the intense burning sensations which occur regularly during rewarming of deeply cooled extremities.

Cross-References

▶ Threshold Determination Protocols

Paraesthesia

▶ Paresthesia

Parafascicular Nucleus (PF)

Definition

The parafascicular nucleus is part of the caudal group of intralaminar nuclei, which projects heavily to the striatum but also lightly to the cerebral cortex.

Cross-References

▶ Brainstem Subnucleus Reticularis Dorsalis Neuron
▶ Spinothalamocortical Projections to Ventromedial and Parafascicular Nuclei

Parafascicular Nucleus, Pain Modulation

Nachum Dafny and Erik Askenasy
Neurobiology and Anatomy, Health Science
Center, University of Texas, Houston, TX, USA

Synonyms

CM-PF complex; Intralaminar nuclei in pain

Introduction

The parafascicular nucleus (PF) and center median nucleus (CM) along with other intralaminar nuclei are considered higher integrative centers and major relay stations for sensory, cognition, and motor processes as well as in learning and memory function. They innervate different cortical and subcortical areas. These nuclei have been largely overlooked for a variety of reasons including their location deep within the brain making surgical manipulations both difficult and dangerous. Recently however, renewed interest concerning their role in pain, Parkinson, and Tourette modulation surfaced in

part because of the expansion of stereotactic neurosurgery together with deep brain stimulation. This chapter will focus on the intralaminar nuclei, mainly on the PF; however, due to its proximity and functional relationship to the CM, they will at times be discussed together. Because much of what is known about the intralaminar nuclei comes from animal studies, these findings will be presented as well.

Characteristics

The internal medullary lamina separates the thalamic medial nuclei from the lateral thalamic nuclei. The intralaminar thalamic nuclei can be divided into anterior (rostral) and posterior (caudal) groups. The caudal group consists of the PF and the CM and together with the central lateral nucleus (a rostral nucleus) comprises a majority of the intralaminar thalamic nuclei (Weigel and Krauss 2004). Posteriorly, the CM-PF complex is separated from the pretectum by a thin lamina. Medially, the CM merges with the PF, which lies just lateral to the anterior periaqueductal gray (Weigel and Krauss 2004). Nociceptive impulses reach the medially located intralaminar nuclei via the spinal cord by means of the diffusely projecting paleospinothalamic tract. It is likely that the affective and motivational components of pain (such as the desire to end, reduce, or escape from the noxious stimuli) are modulated at least in part by the effect of the intralaminar nuclei on the anterior cingulate cortex and other higher central nervous system areas (Taber et al. 2001). Conversely, the neospinothalamic tract projects to the lateral somatosensory thalamus (ventral posterior lateralis (VPL) and ventral posterior medialis (VPM)). These nuclei are involved in the sensory and discriminative aspects of pain and have projections to the primary somatosensory cortex (Young et al. 1995).

Parafasciculus Ascending and Descending

The main afferent projection to the PF nucleus is from the spinal cord, subcortical and cortical areas such as from the paleospinal thalamic and

archispinal thalamic tracts and the spinal trigeminal tract (Boivie 1979), periaqueductal gray
(Hamilton 1973), midbrain and brainstem reticular formation, superior colliculus, locus ceruleus,
raphe nuclei, tectum, cerebellum (Liu et al.
1993), pulvinar, striatum, globus pallidus, substantia
nigra, subthalamus, ventral midbrain, nucleus
accumbens, cingulate, prefrontal, and premotor
cortices (Sadikot and Rymar 2009), while
the main efferent projection from the PF nucleus
includes the primary somatosensory cortex,
prefrontal and frontal cortices, basolateral
amygdaloid complex, nucleus accumbens,
subthalamic nucleus, olfactory tubercle, as well
as parahippocampal cortex and hippocampus
(Berendse and Groenewegen 1991).

Types of Neurons
The vast majority of the information known
concerning different subsets of neurons within
the PF and their response to stimulation has
been determined from animal studies. These
investigations have shown that there are three
groups of neurons within the PF which can be
differentiated by their baseline electrical firing
patterns (Dafny et al. 1990; Liu et al. 1993;
Reyes-Vazquez et al. 1986). The first pattern
type I is characterized by a slow firing rate and
comprises approximately 60–70 % of the total
neuronal population. The second pattern type II
is characterized by spontaneous burst firing and
comprises approximately 10–20 % of the neuronal population. The third pattern type III is characterized by a fast firing rate and comprises
approximately 15–20 % of the total PF recorded
neuronal population. These three types of neurons are distributed randomly throughout the PF
(Liu et al. 1993).

Another way of grouping PF neurons is based
on their response to noxious stimuli. When painful stimuli such as tail pinch, sciatic nerve stimulation, or tail immersion in hot water were
administered to rats, type I and II neurons showed
a change in baseline discharge pattern as opposed
to type III neurons which did not show a change
from baseline activity (Dafny et al. 1990; Liu
et al. 1993; Reyes-Vazquez et al. 1986; Zhang
et al. 1997). Thus, type I and II neurons are

nociceptive reactive and can be classified as
either "nociceptive-on" (Fig. 1) or "nociceptive-
off" depending on whether an elevated or
depressed response to noxious stimuli occurs.
Type III neurons are "non-nociceptive" as they
do not respond to noxious stimuli. Further studies
have indicated that "nociceptive-on" neurons
comprise a majority of cells within the PF while
"non-nociceptive" cells comprise the minority
(see Table 1) (Dafny et al. 1990; Liu et al. 1993;
Reyes-Vazquez et al. 1986; Zhang et al. 1997).
This percentage correlates well with various studies which show that the majority of cells within
the PF respond to noxious stimuli (Dafny et al.
1990; Reyes-Vazquez et al. 1986).

Effect of Stimulation
In humans, pain can usually be controlled medically, and as such, surgical interventions for pain
alleviation (such as a medial thalamotomy) are
rarely used. In fact, surgery is usually considered
only as a last resort for patients with chronic
intractable pain refractory to medical treatment.
The lack of opportunity for intraoperative recording and obvious ethical considerations in designing studies has resulted in a paucity of
information regarding PF neuronal activity in
humans.

Intraoperative stimulation of the CM-PF complex in patients undergoing surgery for intractable pain has provided valuable information. Of
note, it is difficult to confirm electrode placement
in the intralaminar nuclei as they have no "signature" response to stimulation. Be that as it may,
stimulation of the CM-PF complex results in
unpleasant sensations such as a diffuse burning
(Sano et al. 1966) or warmth of the contralateral
body and a signal all over the body as if "I was
going to explode" (Mark et al. 1963). Pulling
of both sides of the face and shoulder (Rinaldi
et al. 1991); intense and painful cramping in the
head, neck, and contralateral arm; and a marked
increase in pain from which patients were initially complaining were also seen (Sano et al.
1966; Thoden et al. 1979). In animals, stimulation of the CM-PF complex results in aversive
pain-like responses (Rosenfeld and Holzman
1978). It was postulated that stimulation may

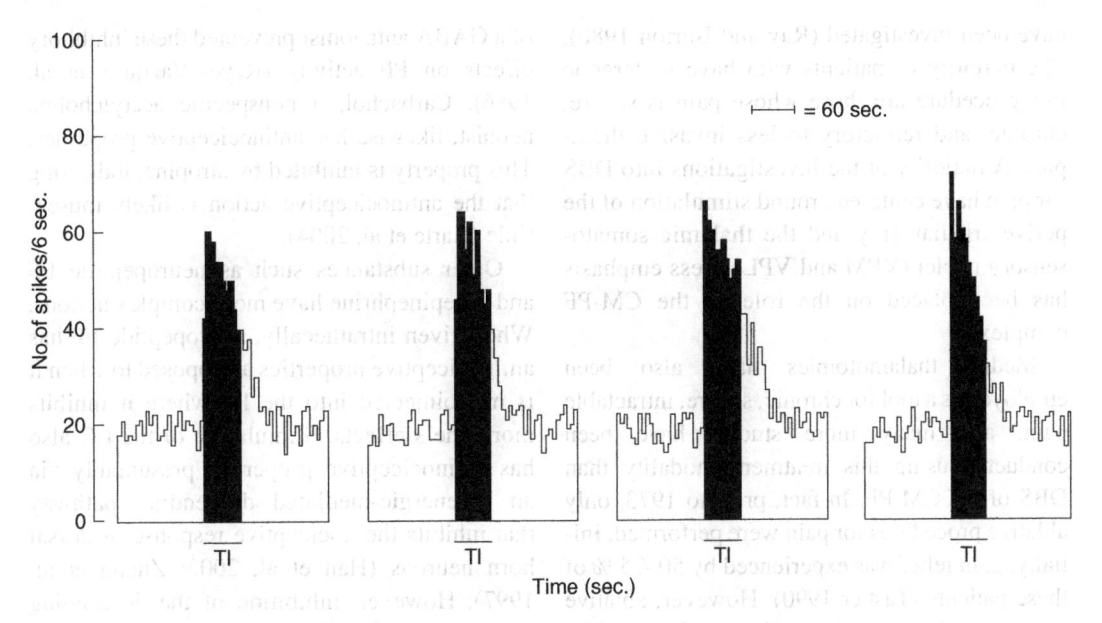

Parafascicular Nucleus, Pain Modulation, Fig. 1 The figure shows the effect of pain induced by tail immersion ('TI') in heated (50 °C) water bath for 60 s as indicated by the black horizontal bars in a representative single neuron recorded out from parafasciculus thalami. Four consecutive times were taken over a 20-min period showing an increase in neuronal firing rate following the TI as compared to baseline activity-unshaded bars, i.e., nociceptive on cells type I. The figure shows constant response to 50 °C hot water (pain)

cause both a diffuse unpleasant sensation and a state of awareness. Coupled, these sensations may be a motivating force to avoid the inciting stimulus. In addition, stimulation of the PF leads to activation of >80 % of periaqueductal gray (PAG) neurons. Thus, it is possible that the PF plays a significant role in ascending and descending pain modulation processes (Dafny et al. 1990).

Stimulation of Other Brain Nuclei Affecting PF Properties

In animals, electrical stimulation of various nuclei within the brain can also modify activity in the PF. Simulation of the locus ceruleus (LC) has antinociceptive properties as it produces a decrease in noxious-evoked impulses recorded from the PF (Han et al. 2002; Zhang et al. 1997). This effect was also noted upon stimulation of the CM or the substantia nigra stimulation. Stimulation of the dorsal raphe nucleus or cerebellar lateral nucleus results in an intensity-dependent suppression of impulses from "off" cells.

Parafascicular Nucleus, Pain Modulation, Table 1 Neuronal subpopulations within the PF (s/p stimulation)

Type	Percent	Change from baseline	Duration of change
I/"on"[a]	50–70 %	260–500 % increase	30–150 s
II/"off"[a]	20–30 %	60–80 % decrease	10–15 s
III/"non"[a]	10–20 %	No change	NA

[a]"on" indicates nociceptive-on neurons; "off" indicates nociceptive-off neurons; "non" indicates non-nociceptive neurons

However, low-intensity dorsal raphe or cerebellar lateral nucleus stimulation inhibits "on" cells while high-intensity stimulation was found to activate "on" cells (Dafny et al. 1990; Liu et al. 1993).

Effects of Deep Brain Stimulation and Lesioning

Since the early 1970s, practical clinical uses of deep brain stimulation (DBS) for pain relief have been investigated (Ray and Burton 1980). The majority of patients who have undergone this procedure are those whose pain is severe, chronic, and refractory to less invasive therapies. A majority of the investigations into DBS for pain have centered around stimulation of the periventricular gray and the thalamic somatosensory nuclei (VPM and VPL). Less emphasis has been placed on the role of the CM-PF complex.

Medial thalamotomies have also been employed as a tool for chronic, severe, intractable pain, and many more studies have been conducted using this treatment modality than DBS of the CM-PF. In fact, prior to 1973, only ablative procedures for pain were performed. Initially, pain relief was experienced by 50–75 % of these patients (Tasker 1990). However, ablative lesions are being used less frequently, possibly due to the high rate of recurrence of pain. Gamma knife has also been employed to perform medial thalamotomies. Results using radiofrequency ablation are similar to those seen using electrocautery (Ray and Burton 1980; Young et al. 1995).

More recently, it was reported that PF stimulation acted very powerfully to reduce tremors and contributed to the long-term management of L-Dopa-induced involuntary movements in Parkinson patients (Stefani et al. 2009), and in years to come, PF manipulations may have potential for treating movement disorders like Tourette syndrome and Parkinson's disease (Smith et al. 2009).

Effect of Drugs
Animal studies have investigated the effects of numerous drugs on PF activity. Examples of a few substances that appear to have antinociceptive properties include morphine (Fig. 2), gamma-aminobutyric acid (GABA), and acetylcholine. Morphine inhibits the increased activation of the PF in response to noxious stimuli (Fig. 2) and naloxone antagonized morphine's effects (Dafny et al. 1990; Reyes-Vazquez et al. 1986). Similarly, microinjections of GABA or GABA agonist into the PF also resulted in antinociception. Administration

of a GABA antagonist prevented these inhibitory effects on PF activity (Reyes-Vazquez et al. 1986). Carbachol, a nonspecific acetylcholine agonist, likewise has antinociceptive properties. This property is inhibited by atropine, indicating that the antinociceptive action is likely muscarinic (Harte et al. 2004).

Other substances such as neuropeptide FF and norepinephrine have more complex actions. When given intrathecally, neuropeptide FF has antinociceptive properties as opposed to when it is microinjected into the PF where it inhibits morphine's effects. Stimulation of the LC also has antinociceptive properties presumably via an adrenergic-mediated descending pathway that inhibits the nociceptive response of dorsal horn neurons (Han et al. 2002; Zhang et al. 1997). However, inhibition of the descending pathway via intrathecal phentolamine or dorsolateral funiculi transaction resulted in a facilitatory role of LC stimulation on PF nociceptive responses. This observation suggests that there are at least two pathways by which LC stimulation modulates nociception: a dominant descending inhibitory pathway to the spinal cord and an ascending predominantly facilitatory supraspinal pathway (Zhang et al. 1997). Additional studies have suggested that the adrenergic descending pathway does not directly activate dorsal horn neurons but that an enkephalinergic interneuron serves to connect the two.

Summary

In summary, the PF appears to be involved in the affective and motivational components of the pain response. The main afferent projections to the PF include the spinothalamic tracts, superior colliculi, and several brainstem nuclei. In turn, the PF sends significant projections to the caudate nucleus and prefrontal and premotor cortices (Weigel and Krauss 2004). Within the PF, three electrophysiological neuronal subpopulations can be classified by either their baseline firing pattern or their response to noxious stimuli. The "nociceptive-on" neurons comprise a majority of

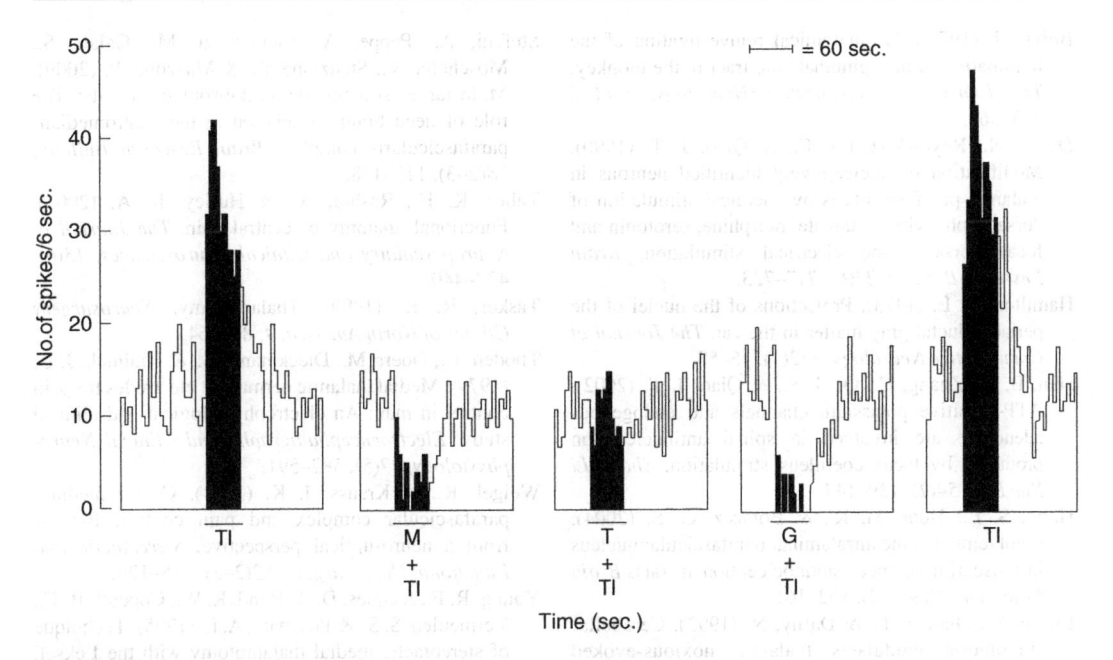

Parafascicular Nucleus, Pain Modulation, Fig. 2 The figure shows the effect of microiontophoretically (30 nA) morphine (M) on the effect of tail immersion (TI) stimulation and the effect of 30 nA microiontophoretically THIP (GABA agonist)-T and 5 min later 30 nA microiontophoretically GABA (G) applied together with pain stimulation (TI). The figure shows that morphine, GABA agonist, and GABA block to noxious input to PF nucleus. The figure shows that 10 min after the drug application, the PF cell responds to TI in a similar way to that prior to the drug application

neurons and as such the overall response of the PF to noxious stimuli is activation (Fig. 1). When the PF is stimulated intraoperatively, a variable response ranging from contractures and clonuses to a diffuse unpleasantness is seen (Sano et al. 1966). In animals, stimulation of the PF results in activation of the PAG which is intimately involved in the gate control theory of descending pain control. Therefore, a circuit can be theorized whereby painful stimuli transmitted to the PF via the spinothalamic tracts result in activation of the descending pain control pathway. Additionally, in animals stimulation of the locus ceruleus and substantia nigra decreases the response of the PF to noxious stimuli and may provide an explanation of the antinociceptive properties seen with norepinephrine and dopamine. As expected, administration of various substances to the PF either directly or intrathecally produces diverging results and must be examined individually.

Clearly, the role of the PF in ascending and descending pain modulation is complex and it is likely that dividing pain systems into the medial thalamus paleospinothalamic system (affective and motivational) versus the lateral thalamus neospinothalamic system (sensory and discriminatory) is an oversimplification (Weigel and Krauss 2004). Additionally, the PF has been postulated to play a role in attentional orienting to behaviorally significant events, seizure control, and various other behaviors. Continued investigations into the role of the PF in pain processing will further elucidate its function, increase our understanding of the physiology of pain, and may provide valuable insight into pain alleviating therapies.

References

Berendse, H. W., & Groenewegen, H. J. (1991). Restricted cortical termination fields of the midline and intralaminar thalamic nuclei in the rat. *Neuroscience*, *42*(1), 73–102.

Boivie, J. (1979). An anatomical reinvestigation of the termination of the spinothalamic tract in the monkey. *The Journal of Comparative Neurology, 186*(3), 343–369.

Dafny, N., Reyes-Vazquez, C., & Qiao, J. T. (1990). Modification of nociceptively identified neurons in thalamic parafascicularis by chemical stimulation of dorsal raphe with glutamate, morphine, serotonin and focal dorsal raphe electrical stimulation. *Brain Research Bulletin, 24*(6), 717–723.

Hamilton, B. L. (1973). Projections of the nuclei of the periaqueductal gray matter in the cat. *The Journal of Comparative Neurology, 152*(1), 45–58.

Han, B. F., Zhang, C., Qi, J. S., & Qiao, J. T. (2002). ATP-sensitive potassium channels and endogenous adenosine are involved in spinal antinociception produced by locus coeruleus stimulation. *Sheng Li Xue Bao, 54*(2), 139–144.

Harte, S. E., Hoot, M. R., & Borszcz, G. S. (2004). Involvement of the intralaminar parafascicular nucleus in muscarinic-induced antinociception in rats. *Brain Research, 1019*(1–2), 152–161.

Liu, F. Y., Qiao, J. T., & Dafny, N. (1993). Cerebellar stimulation modulates thalamic noxious-evoked responses. *Brain Research Bulletin, 30*(5–6), 529–534.

Mark, V. H., Ervin, F. R., & Yakovlev, P. I. (1963). Stereotactic thalamotomy: III. The verification of anatomical lesion sits in human thalamus. *Archives of Neurology Chicago, 8*, 528–538.

Ray, C. D., & Burton, C. V. (1980). Deep brain stimulation for severe, chronic pain. *Acta Neurochirurgica Supplement, Supplement (Wien), 30*, 289–293.

Reyes-Vazquez, C., Enna, S. J., & Dafny, N. (1986). The parafasciculus thalami as a site for mediating the antinociceptive response to GABAergic drugs. *Brain Research, 383*(1–2), 177–184.

Rinaldi, P. C., Young, R. F., Albe-Fessard, D., & Chodakiewitz, J. (1991). Spontaneous neuronal hyperactivity in the medial and intralaminar thalamic nuclei of patients with deafferentation pain. *Journal of Neurosurgery, 74*(3), 415–421.

Rosenfeld, J. P., & Holzman, B. S. (1978). Effects of morphine on medial thalamic and medial bulboreticular aversive stimulation thresholds. *Brain Research, 150*(2), 436–440.

Sadikot, A. F., & Rymar, V. V. (2009). The primate centromedian-parafascicular complex: Anatomical organization with a note on neuromodulation. *Brain Research Bulletin, 78*(2–3), 122–130.

Sano, K., Yoshioka, M., Ogashiwa, M., Ishijima, B., & Ohye, C. (1966). Thalamolaminotomy. A new operation for relief of intractable pain. *Confinia Neurologica, 27*, 63–66.

Smith, Y., Raju, D., Nanda, B., Pare, J. F., Galvan, A., & Wichmann, T. (2009). The thalamostriatal systems: Anatomical and functional organization in normal and parkinsonian states. *Brain Research Bulletin, 78*(2–3), 60–68.

Stefani, A., Peppe, A., Pierantozzi, M., Galati, S., Moschella, V., Stanzione, P., & Mazzone, P. (2009). Multi-target strategy for Parkinsonian patients: The role of deep brain stimulation in the centromedian-parafascicularis complex. *Brain Research Bulletin, 78*(2–3), 113–118.

Taber, K. H., Rashid, A., & Hurley, R. A. (2001). Functional anatomy of central pain. *The Journal of Neuropsychiatry and Clinical Neurosciences, 13*(4), 437–440.

Tasker, R. R. (1990). Thalamotomy. *Neurosurgery Clinics of North America, 4*, 841–64.

Thoden, U., Doerr, M., Dieckmann, G., & Krainick, J. U. (1979). Medial thalamic permanent electrodes for pain control in man: An electrophysiological and clinical study. *Electroencephalography and Clinical Neurophysiology, 47*(5), 582–591.

Weigel, R., & Krauss, J. K. (2004). Center median-parafascicular complex and pain control. Review from a neurosurgical perspective. *Stereotactic and Functional Neurosurgery, 82*(2–3), 115–126.

Young, R. F., Jacques, D. S., Rand, R. W., Copcutt, B. C., Vermeulen, S. S., & Posewitz, A. E. (1995). Technique of stereotactic medial thalamotomy with the Leksell Gamma Knife for treatment of chronic pain. *Neurological Research, 17*(1), 59–65.

Zhang, C., Yang, S. W., Guo, Y. G., Qiao, J. T., & Dafny, N. (1997). Locus coeruleus stimulation modulates the nociceptive response in parafascicular neurons: An analysis of descending and ascending pathways. *Brain Research Bulletin, 42*(4), 273–278.

Parahippocampal Region

Definition

The parahippocampal region is situated next to the hippocampal formation in the brain. Various parcellations of this region have been proposed. Here, we adhere to the definition used by Squire and colleagues: the perirhinal cortex, the entorhinal cortex, and the parahippocampal cortex comprise the parahippocampal region. The parahippocampal cortex, which forms part of the region, is not to be confused with the parahippocampal region itself.

Cross-References

▶ Hippocampus and Entorhinal Complex: Functional Imaging

Parallel Pain Processing

Definition

Parallel pain processing is the theoretical model of a dual pain system, in which the sensory-discriminative and affective-motivational dimensions of pain are simultaneously processed in two anatomically distinct neural circuits, termed the lateral pain system and the medial pain system, respectively.

Cross-References

▶ Thalamo-Amygdala Interactions and Pain

Paraneoplastic Syndrome

Definition

Paraneoplastic syndromes are disorders that can be painful, which occur with increased frequency in patients with cancer with etiologies that are incompletely understood. Paraneoplastic syndromes may precede the diagnosis of cancer and allow its earlier identification.

Cross-References

▶ Cancer Pain Management, Orthopedic Surgery

Paraphysiological Cramps

Definition

Characterized by the occurrence in healthy subjects during particular conditions such as pregnancy or exercising. Occasional cramps are those cramps that occasionally occur in healthy subjects, in the absence of exercising or pregnancy. In this case, a strong muscular contraction or a sustained abnormal posture may lead to muscle cramps.

Cross-References

▶ Muscular Cramps

Paraproteinemic Neuropathies

▶ Metabolic and Nutritional Neuropathies

Paraspinal

Definition

On the side of the spinal area.

Cross-References

▶ Lower Back Pain, Physical Examination

Parasylvian

Definition

In the vicinity of the Sylvian fissure.

Cross-References

▶ Secondary Somatosensory Cortex (S2) and Insula, Effect on Pain-Related Behavior in Animals and Humans

Parasympathetic Component

Definition

It is a group of visceral responses driven by the parasympathetic nervous system. This motor autonomic system is one of the two antagonistic visceral controls (the other being sympathetic). It includes schematically a first preganglionic neuron (cholinergic) located in the medulla oblongata (ambiguus nucleus, dorsal motor nucleus of the vagus) or in the sacral spinal cord. Parasympathetic axons of the first neurons travel within the cranial nerves (X, IX, VII, VIIbis, and III) and the pelvic nerve. The second neuron (cholinergic) is located in the nervous ganglia close to the viscera. It innervates the smooth muscles of the viscera and blood vessels, the heart, and the glandular tissues. For example, within the cardiovascular sphere, this system triggers a bradycardia and a decrease of blood pressure.

Cross-References

▶ Hypothalamus and Nociceptive Pathways

Paravertebral Muscle

▶ Lower Back Pain, Physical Examination

Parental Response

▶ Impact of Familial Factors on Children's Chronic Pain

Paresthesia

Turo J. Nurmikko
Department of Neurological Science, University of Liverpool, and Pain Research Institute, The Walton Centre NHS Foundation Trust, Liverpool, UK

Synonyms

Paraesthesia

Definition

An abnormal sensation, perceived in the skin, pleasant or unpleasant, and often described as if the skin is alive (pins and needles, tingling).

IASP definition: An abnormal sensation, whether spontaneous or evoked (Merskey and Bogduk 1994).

Characteristics

Paresthesia is a tingling or prickling sensation, often compounded by numbness, perceived in the skin or mucosa. It is described variably as pins and needles, skin crawling, electricity, or a limb going dead and is familiar to most people who have sat too long with their legs crossed or fallen asleep with their arm crooked under their head. In healthy people, paresthesias are usually transient, disappear quickly when the cause has been removed, and cause limited discomfort. In diseases affecting the nervous system, the symptoms usually persist and may become very annoying.

Existing pain nomenclature defines paresthesia as an abnormal cutaneous sensation, which is not painful or unpleasant (Merskey and Bogduk 1994). Painful or unpleasant sensation is referred to as dysesthesia. This dichotomy is artificial, and many hold it too puristic. In common clinical communication, the term "paresthesia" is used to describe all abnormal sensations in the skin whether unpleasant or not. The mechanisms that are responsible for the generation of paresthesia or dysesthesia are not likely to be very different.

Even in the healthy, paresthesia occurs commonly. Experiments using a tight cuff around one's arm for 20 min or more demonstrate that most people develop tingling during this time. They also experience tingling after the cuff is released, but it is then sharper and more intense and may become painful. Prolonged deep breathing (hyperventilation) also causes tingling in the hands, feet, and face in many people. Experiments using recordings in single nerves (microneurography) show increased activity in afferent nerve fibers supplying the area of the hand in which tingling is perceived (Mogyoros et al. 2000). Paresthesia and dysesthesia are therefore produced in the nerve, whether the cause is interruption of its blood supply, change in its chemical milieu, or some other disturbance. When a gentle electrical current is applied to the skin, e.g., by using a TENS machine, paresthesias are felt because of nerve ending activation. Warmth applied to the fingers and toes that have long been exposed to cold causes profuse tingling by the same mechanism.

Osteopaths and acupuncturists are familiar with tender spots or trigger points in muscles, which when pressed or massaged cause a tingling sensation usually referred away from the point. The most famous of such points is Ho-Ku, the classic acupuncture point in the webbing between the thumb and the index finger. Needling of the point causes tingling and numbing sensation, brought about by activation of nerve cells responsible for the sensation of wider areas than just the Ho-Ku point.

Permanent paresthesias may be a sign of a neurological disease. Practically any neurological condition affecting sensory tracts in the nervous system is capable of provoking paresthesia. Classical peripheral nervous system diseases include carpal tunnel syndrome, polyneuropathy, nerve injury, and shingles. The central nervous system may also be the origin of various paresthesias – they are frequently reported by patients with stroke, MS, and migraine. Intentional provocation of paresthesia is the cornerstone of several clinical tests, such as the ▶ Tinel signs and Phalen signs in ▶ carpal tunnel syndrome and Lhermitte's sign used in conjunction with MS. The location and quality of paresthesia tell little of the underlying cause, which can only be determined by neurological investigations.

What causes paresthesia and dysesthesia? Evidence from studies using microneurography – single nerve cell recordings in man from a pure sensory nerve of the arm or leg – points strongly to increased excitability of sensory axons and formation of ectopic impulses (Campero et al. 1998; Mogyoros et al. 2000). These come in abrupt onset, high-frequency bursts of action potentials in myelinated fibers (Campero et al. 1998).

The quality of paresthesia depends on which fibers are mostly active. Individual fibers when stimulated using weak currents give rise to specific sensations, such as pressure, vibration, and touch (Ochoa and Torebjork 1983; Vallbo et al. 1983). In an injured nerve, mechanical and electrical stimuli cause prolonged firing. Once filtered through the central nervous system, these impulse patterns are then interpreted as tingling by the brain. In the healthy nerve, the bursts originate at the site of nerve compression. In chronic neurological cases, bursts originate at the sites of ▶ demyelination and remyelination and regenerating nerve sprouts and in the dorsal root ganglion (Campero et al. 1998; Devor and Seltzer 1999; Mogyoros et al. 2000). A crucial ingredient for such alteration is accumulation of ▶ sodium channels. It is likely that many other factors contribute to this, either permanently or temporarily, including tissue ischemia, changes in blood gases and ion concentrations, and many substances released into the tissues during injury such as peptides, neurotrophins, histamine, bradykinin, prostaglandins, cytokines, and a host of other inflammatory mediators.

The ▶ axons in the central nervous system have also been shown to be capable of similar excessive discharge. Alterations in central myelin occur in MS, rendering surviving axons excitable. In spinal cord injury, increased burst activity is seen in the ▶ thalamus, while microstimulation of specific thalamic nuclei can give rise to different types of paresthesia (Ohara et al. 2004). Stimulation of the somatosensory cortex induces paresthesia in the corresponding part of the body as Penfield showed in his classical experiments during brain surgery for epilepsy. In migraine, moving paresthesia is associated with a wave of short-lasting excitation of the cortex, which precedes the better-known reduction of cortical activity (Leão's spreading cortical depression) (Spierings 2003).

There are no established guidelines for the treatment of paresthesia or dysesthesia. Intuitively, neurologists and pain specialists have used similar medication to that found useful in neuropathic pain as well as simple stimulation therapies and distraction. If chronic hyperventilation is the cause, training in proper breathing and relaxation programs is helpful. Whenever the underlying disease can be treated, e.g., release of a trapped nerve, most patients may expect amelioration of their paresthesia. In cases for which no curative treatment exists, annoying paresthesia may be controlled with antiepileptic and antidepressant medication. Interestingly, successful treatment of pain with ► spinal cord stimulation may be effective in controlling dysesthesia as well as ongoing pain, although in itself it produces clearly perceived tingling over the painful area. It is very unusual for patients in these circumstances to feel paresthesia as unpleasant.

Cross-References

► Cancer Pain Management, Neurosurgical Interventions
► Cervical Transforaminal Injection of Steroids
► Dysesthesia, Assessment
► Guillain-Barré Syndrome
► Metabolic and Nutritional Neuropathies
► Painless Neuropathies
► Thalamic Plasticity and Chronic Pain
► Threshold Determination Protocols

References

Campero, M., Serra, J., Marchettini, P., et al. (1998). Ectopic impulse generation and autoexcitation in single afferent myelinated fibers in patients with peripheral neuropathic and positive symptoms. *Muscle & Nerve, 21*, 1661–1667.
Devor, M., & Seltzer, Z. (1999). Pathophysiology of damaged nerves in relation to chronic pain. In P. D. Wall & R. Melzack (Eds.), *Textbook of pain* (4th ed., pp. 129–162). Edinburgh: Churchill Livingstone.
Merskey, H., & Bogduk, N. (1994). *Classification of chronic pain*. Seattle: IASP Press.
Mogyoros, I., Bostock, H., & Burke, D. (2000). Mechanisms of paresthesias arising from healthy axons. *Muscle & Nerve, 3*, 310–320.
Ochoa, J., & Torebjork, E. (1983). Sensations evoked by intraneural stimulation of single mechanoreceptor units innervating the human hand. *The Journal of Physiology, 342*, 633–654.
Ohara, S., Weiss, N., & Lenz, F. A. (2004). Microstimulation in the region of the human thalamic principal somatic sensory nucleus evokes sensations like those of mechanical stimulation and movement. *Journal of Neurophysiology, 91*, 736–745.
Spierings, E. (2003). Pathogenesis of migraine attack. *The Clinical Journal of Pain, 19*, 255–262.
Vallbo, A. B., Olsson, K. A., Westberg, K. G., et al. (1983). Microstimulation of single tactile afferents from the human hand. *Brain, 107*, 727–729.

Parietal Cortex

► PET and fMRI Imaging in Parietal Cortex (SI, SII, Inferior Parietal Cortex BA40)

Parity

Definition

Until a woman has delivered her first baby, she is said to be nulliparous or primiparous. When she has delivered she then becomes multiparous and the number of her completed pregnancies describes her parity.

Cross-References

► Postpartum Pain

Parkinson's Disease

Definition

It is a disease characterized by muscle tremor and difficulty in initiating and sustaining locomotion. The disease results from degeneration of dopamine-releasing neurons of the midbrain.

Cross-References

► Central Pain, Outcome Measures in Clinical Trials

Paroxysmal

Definition

An attack, recurrence, or intensification of disease that occurs repeatedly and suddenly. It also implies brevity, which may not always be the case.

Cross-References

▶ Trigeminal Neuralgia: Diagnosis and Treatment

Paroxysmal Hemicrania

Margarıta Sanchez-del-Rio
Neurology Department, Hospital Ruber Internacional, Madrid, Spain

Synonyms

Sjaastad's syndrome

Definition

Attacks of strictly unilateral pain usually in the orbital, supraorbital, and/or temporal region, with similar characteristics of pain and associated symptoms to those seen in ▶ cluster headache, but shorter-lasting (5–20 min) and more frequent (≥5 per 24 h). Occur more commonly in females and respond absolutely to indomethacin. See below for the IHS diagnostic criteria.

Paroxysmal (I.H.S. 3.2) Hemicrania Diagnostic Criteria

(A) At least 20 attacks fulfilling criteria B–D.
(B) Attacks of severe unilateral orbital, supraorbital, or temporal pain lasting 2–30 min.
(C) Headache is accompanied by at least one of the following:

1. Ipsilateral conjunctival injection and/or lacrimation
2. Ipsilateral nasal congestion and/or rhinorrhea
3. Ipsilateral eyelid edema
4. Ipsilateral forehead and facial sweating
5. Ipsilateral miosis and/or ptosis

(D) Attacks have a frequency above 5 per day for more than half of the time, although periods with lower frequency may occur.
(E) Attacks are prevented completely by therapeutic doses of indomethacin.[1]
(F) Not attributed to another disorder.[2]

Notes.

1. *In order to rule out incomplete response, indomethacin should be used in a dose of ≥150 mg daily orally or rectally or ≥100 mg by injection, but for maintenance, smaller doses are often sufficient.*
2. *History and physical and neurological examinations do not suggest any of the disorders listed in groups 5–12, or history and/or physical and/or neurological examinations do suggest such a disorder but it is ruled out by appropriate investigations or such a disorder is present but attacks do not occur for the first time in close temporal relation to the disorder.*

Episodic and Chronic Paroxysmal Hemicrania Diagnostic Criteria

Episodic Paroxysmal Hemicrania
1. Fulfils criteria for 3.2 paroxysmal hemicrania
2. At least two attack periods lasting 7–365 days and separated by pain-free remission periods of ≥1 month

Chronic Paroxysmal Hemicrania (CPH)
1. Fulfils criteria for 3.2 paroxysmal hemicrania.
2. Attacks recur over >1 year without remission periods or with remission periods lasting <1 month.

Probable Paroxysmal Hemicrania
1. Attacks fulfilling all but one of criteria A–E for 3.2 paroxysmal hemicrania
2. Not attributed to another disorder

Characteristics

Chronic Paroxysmal Hemicrania

▶ Chronic paroxysmal hemicrania (CPH) was first described by Sjaastad and Dale in 1974. It is a distinctive clinical syndrome that is part of a family of headache disorders referred to as the trigeminal autonomic cephalgias.

In contrast to cluster headache, in CPH there is a preponderance among women (3:1 female to male ratio). Onset is usually in adulthood with a mean age of 33 years, although childhood cases are reported. There is no family history of chronic paroxysmal headache, although migraine may occur in families. Pain is strictly unilateral and without side shift, although recently, it has been reported to alternate sides in 15 % of patients (Boes and Dodick 2002). Maximum pain intensity is located in ocular, temporal, maxilla, or frontal regions. Occasionally, pain may involve the nuchal, occipital, and retro-orbital areas. Pain is typically described as throbbing, pulsatile, or sharp, ranging from moderate to severe intensity. Although typically patients prefer to lie still, some prefer to pace as in cluster headache. The frequency of attacks is variable, ranging from 2 to 40× a day; the duration of most attacks ranges from 2 to 45 min. Patients may have soreness or tenderness in the interval between attacks, especially if the attacks are frequent. Attacks may also occur at night, awakening patients from sleep.

Associated symptoms during episodes of headache are ipsilateral lacrimation, conjunctival injection, nasal congestion, and rhinorrhea. Less frequently, there may be ipsilateral eyelid edema, mild miosis, photophobia, or nausea. There has been one reported case of dissociation between associated symptoms and pain (Pareja 1995). In a recent report, migrainous features were reported in 87 % of patients (Boes and Dodick 2002). There may be precipitating events such as mechanical stimulation, particularly with bending or rotating the head, pressure over the ipsilateral greater occipital nerve, or alcohol ingestion (Sjaastad et al. 1982, 1984). The pathophysiology is unknown and any presumption is based on the similarities with cluster headache.

The association of ▶ paroxysmal hemicrania with coexistent ▶ trigeminal neuralgia (CPH-tic syndrome) has recently been described (Caminero et al. 1998; Hannerz 1993). The importance of this observation is that both conditions require treatment. The pathophysiological significance of the association is not yet clear.

Episodic Paroxysmal Hemicrania

▶ Episodic paroxysmal hemicrania (EPH) was first described by Kudrow et al. (1987). Today there is sufficient clinical evidence to consider EPH as a subtype of paroxysmal hemicrania, based on the presence of symptomatic periods lasting from weeks to months separated by remission periods lasting more than 1 month. A seasonal pattern has been described (Veloso et al. 2001). While in CPH there is a female preponderance, EPH affects men and women practically equally (1:1.3 male to female ratio). EPH may be confused with episodic cluster headache but may be distinguished by the higher frequency and shorter duration of attacks.

Imaging

MRI studies of patients with CPH have been normal. Most cases are primary, but there have been cases secondary to gangliocytoma of the sella turcica, collagen vascular disease, cerebrovascular disease, Pancoast tumor, frontal lobe tumor, cavernous sinus meningioma, acoustic neurinoma, intracranial hypertension, or increased CSF pressure. Based on these findings, a complete work-up should initially include blood count, ESR, vasculitic investigations, and brain imaging.

Differential Diagnosis

The differential diagnoses of paroxysmal hemicrania include cluster headache, ▶ hemicrania continua, trigeminal neuralgia, and SUNCT.

Treatment

Indomethacin is the treatment of choice and a sine qua non criterion for diagnosis. The therapy should be initiated at 25 mg/8 h; if after

1 week there is no benefit, it should be increased to 50 mg/8 h. Complete resolution of the attack usually occurs within 1–2 days of initiating the effective dose. In order to rule out incomplete response, indomethacin should be used in a dose of ≥150 mg daily orally or rectally or ≥100 mg by injection, but for maintenance, smaller doses are often sufficient (Antonaci et al. 1998; Pareja et al. 2001). In CPH, long-term treatment is usually necessary, although long-lasting remissions have been reported following cessation of indomethacin (Sjaastad and Antonaci 1987). In EPH, indomethacin should be given for slightly longer than the typical headache phase and then gradually tapered. If there is not a complete response to indomethacin, the diagnosis should be reconsidered. Other agents that have demonstrated partial success in anecdotal cases are acetylsalicylic acid, verapamil, steroids, acetazolamide, piroxicam, and rocecoxib.

References

Antonaci, F., Pareja, J. A., Caminero, A. B., et al. (1998). Chronic paroxysmal hemicrania and hemicrania continua. Parenteral indomethacin: The 'indotest'. *Headache, 38*, 122–128.

Boes, C. J., & Dodick, D. W. (2002). Refining the clinical spectrum of chronic paroxysmal hemicrania: A review of 74 patients. *Headache, 42*, 699–708.

Caminero, A. B., Pareja, J. A., & Dobato, J. L. (1998). Chronic paroxysmal hemicrania-tic syndrome. *Cephalalgia, 18*, 159–161.

Hannerz, J. (1993). Trigeminal neuralgia with chronic paroxysmal hemicrania: The CPH-tic syndrome. *Cephalalgia, 13*, 361–364.

Kudrow, L., Esperanza, P., & Vijayan, N. (1987). Episodic paroxysmal hemicrania? *Cephalalgia, 7*, 197–201.

Pareja, J. A. (1995). Chronic paroxysmal hemicrania: Dissociation of the pain and autonomic features. *Headache, 35*, 111–113.

Pareja, J. A., Caminero, A. B., Franco, E., et al. (2001). Dose, efficacy and tolerability of long-term indomethacin treatment of chronic paroxysmal hemicrania and hemicrania continua. *Cephalalgia, 21*, 906–910.

Sjaastad, O., & Antonaci, F. (1987). Chronic paroxysmal hemicrania: A case report. Long-lasting remission in the chronic stage. *Cephalalgia, 7*, 203–205.

Sjaastad, O., & Dale, I. (1974). Evidence for a new (?) treatable headache entity. *Headache, 14*, 105–108.

Sjaastad, O., Russell, D., Saunte, C., et al. (1982). Chronic paroxysmal hemicrania. VI. Precipitation of attacks. Further studies on the precipitation mechanism. *Cephalalgia, 2*, 211–214.

Sjaastad, O., Saunte, C., & Graham, J. R. (1984). Chronic paroxysmal hemicrania. Mechanical precipitation of attacks: New cases and localization of trigger points. *Cephalalgia, 4*, 113–118.

Veloso, G. C., Kaup, A. O., Peres, M. F. P., et al. (2001). Episodic paroxysmal hemicrania with seasonal variation. *Arquivos de Neuro-Psiquiatria, 59*, 944–947.

Paroxysmal Hypertension

Definition

This is a sudden acute rise in blood pressure of at least 25 % of the diastolic value.

Cross-References

▶ Headache Due to Hypertension

Paroxysmal Hypertensive Headaches

Definition

These are severe generalized headaches caused by sudden severe rises in blood pressure to diastolic values of at least 25 % above normal/usual.

Cross-References

▶ Headache Due to Hypertension

Part of the Spinoreticular Tract

▶ Spinoparabrachial Tract

Partial Agonist

Definition

A partial agonist is a drug that binds to receptors, but is only partially effective.

Cross-References

▶ Pain Control in Children with Burns

Partial Dorsal Rhizotomy

▶ Dorsal Root Ganglion Radiofrequency

Partial Dorsal Root Ganglion Lesioning

▶ Dorsal Root Ganglion Radiofrequency

Partial Radiofrequency Dorsal Root Ganglion Lesion

▶ Dorsal Root Ganglion Radiofrequency

Partial Sciatic Nerve Ligation

Synonyms

PSL Model

Definition

Partial Sciatic Nerve Ligation (PSL Model) is a rodent model of peripheral nerve injury. In this model of neuropathic pain, the mid-thigh section of the sciatic nerve is surgically isolated and 33–50 % of the nerve thickness is tightly ligated with silk sutures.

Participation

Definition

It is described as an involvement in a life situation. It represents the societal perspective of functioning. Problems an individual may experience in involvement in life situations are participation restrictions. Limitations and restrictions are assessed against a generally accepted population standard; they reflect the discordance between the observed and the expected performance.

Cross-References

▶ Functioning and Disability Definitions
▶ Physical Medicine and Rehabilitation, Team-Oriented Approach

Participation Restrictions

Definition

Participation restrictions are problems an individual may experience in involvement in life situations.

Cross-References

▶ Physical Medicine and Rehabilitation, Team-Oriented Approach

Parvalbumin

Definition

Parvalbumin is a member of the EF-hand family calcium-binding proteins expressed in fast-firing neurons. They are small proteins and have three EF-hand domains, with two of them having calcium-binding capacity. Parvalbumin often exhibits a complementary distribution pattern with that of calbindin D-28 k in central neurons. In the monkey thalamus, antibodies against parvalbumin selectively label the rod domain of the medial ventroposterior nucleus of the thalamus, which is related to the trigeminothalamic projection from the principal sensory nucleus of the trigeminal nerve.

Cross-References

► Spinothalamic Terminations, Core and Matrix
► Thalamus, Receptive Fields, Projected Fields, Human
► Trigeminothalamic Tract Projections

Parvocellular Neurons

Definition

Neurons with soma of a small size located in the paraventricular hypothalamic nucleus. These neurons synthesize a number of releasing hormones that influence the secretion of pituitary anterior lobe hormones. These releasing hormones are transported from the hypothalamus to the pituitary gland via the venous hypophyseal portal system.

Cross-References

► Hypothalamus and Nociceptive Pathways

Parvocellular VP

Synonyms

VPpc

Definition

Parvocellular VP (VPpc) is the medial-most portion of VP_p, which includes the gustatory region of the thalamus.

Cross-References

► Thalamus, Visceral Representation

Passive

Definition

This refers to movement of a body part without using power generated from one's own muscle action.

Cross-References

► Cancer Pain Management, Orthopedic Surgery

Passive Avoidance

Definition

Passive avoidance is a behavioral test of the response of an experimental animal to a noxious stimulus. The animal is signaled when a stimulus is about to occur, and is allowed to avoid the stimulus by an action taken prior to the onset of the stimulus. Similar behaviors are thought to occur in humans.

Cross-References

► Visceral Pain Model, Lower Gastrointestinal Tract Pain

Passive Relaxation

► Relaxation in the Treatment of Pain

Passive Spinal Mobilization

Robert Gassin
Musculoskeletal Medicine Clinic, Frankston,
VIC, Australia

Synonyms

Maitland mobilization; Manipulation without impulse; Manipulation without thrust; Mobilization; Oscillatory mobilization

Definition

Passive spinal mobilization is a form of spinal manual therapy. It refers to the application of a rhythmic, oscillatory force to the spine with the aim of moving one or more spinal joints and increasing its range of motion. The procedure can be further qualified by specifying the region of the spine being mobilized, i.e., "lumbar" and "cervical" instead of "spinal."

Characteristics

Mechanism
The mechanism of action of passive spinal mobilization remains unknown. Several theories have been proposed to explain pain relief following the procedure. These include reduction of disc bulging or prolapse; correction of posterior joint dysfunction, breaking down or stretching adhesions in the zygapophysial joint capsules, freeing of an entrapped meniscoid in a zygapophysial joint; correction of spinal misalignment; muscle relaxation; stretching of the vertebral ligaments; normalization of local or regional reflex activity; activation of a central control mechanism; and placebo effect. None of these theories is strongly supported by scientific evidence.

Applications
The indications for the use of passive spinal mobilization depend on the individual practitioner's paradigm. Passive spinal mobilization is generally considered a means of improving the range of movement of a hypomobile spinal segment or region and of achieving pain relief.

Passive spinal mobilization is used as an isolated treatment or as part of a multimodal management strategy for the management of many musculoskeletal conditions.

Efficacy
Quality research on the efficacy of passive spinal mobilization is sparse. Interpretation of outcomes of trials is often difficult, as passive spinal mobilization is often used in conjunction with spinal manipulation or as part of a multimodal management approach. In the case of systematic reviews and meta-analyses, trials of spinal manipulation and passive spinal mobilization are often combined, making interpretation of results difficult.

Low Back Pain
There is little data on the efficacy of passive mobilization in the management of low back pain. Passive mobilization of the sacroiliac joint has been compared to flexion exercises, resulting in better function at 3 and 5 days (Delitto et al. 1992); and to massage, resulting in decreased sick-leave and analgesic consumption, but not pain, at 3 weeks (Wreje et al. 1992). In a recent randomized controlled trial, randomly selected passive spinal mobilization techniques were found to be as effective as therapist-selected ones for the treatment of low back pain (Chiradejnant et al. 2003).

Neck Pain
There have been several reviews describing studies on passive cervical spinal mobilization (Koes et al. 1991; Aker et al. 1996; Hurwitz et al. 1996; Kjellman et al. 1999; Harms-Ringdahl and Nachemson 2000; Gross et al. 2002; Gross et al. 2002b).

For the treatment of neck pain, most studies have combined passive spinal mobilization with other treatment modalities. Consequently, it is not possible to determine the contribution of individual treatment modalities to the overall outcome. In the most recent review, passive spinal mobilization alone was found to have similar effects to placebo, wait period, or control group, and appeared similar in benefit for pain relief to spinal manipulation. Multimodal care, including some combination of mobilizations and exercise, was found to be superior to control, other physical medicine methods, and rest (Gross et al. 2002). The latter conclusion is supported by at least one other review (Harms-Ringdahl and Nachemson 2000).

In a recent randomized controlled trial (Hoving et al. 2002), passive cervical mobilization was compared to exercise therapy and general practitioner care, in patients with at least 2 weeks of nonspecific neck pain. Treatment was provided over 6 weeks and outcomes measured at 7 weeks. Passive cervical mobilization scored consistently better than general practitioner care on most outcome measures but was not consistently and significantly better than exercise therapy.

Another trial compared the effectiveness of cervical spinal manipulation and passive cervical mobilization (Hurwitz et al. 2002). Mean reductions in pain and disability were similar in the manipulation and mobilization groups over 6 months. The authors concluded that cervical spinal manipulation and passive cervical mobilization yielded comparable clinical outcomes.

There is no evidence for the effectiveness of passive spinal mobilization as an isolated treatment in the management of whiplash. However, there are a number of studies that have demonstrated its effectiveness in combination with other treatment modalities (Mealy et al. 1986; McKinney et al. 1989; Provinciali et al. 1996). When compared with rest, analgesia, and a soft collar, mobilization combined with exercises, local heat, and analgesia provided statistically significant improvements in pain scores and range of movement at 4 and 8 weeks (Mealy et al. 1986). Mobilization combined with other interventions was more effective than rest and analgesia at 1 and 2 months after treatment, but not more effective than a program of home exercises (McKinney et al. 1989). At 1 year follow-up, significantly more patients who had been treated with home exercises were free of pain. Multimodal therapy, involving mobilization and a variety of other interventions, was more effective than

TENS, pulsed electromagnetic therapy, ultrasound, and calcic iontophoresis (Provinciali et al. 1996).

Headache of Cervical Origin

A small study found that a combination of oscillatory passive spinal mobilization and muscle energy technique was more effective than application of cold packs for the treatment of posttraumatic headache. Differences in outcome were statistically significant at 2 weeks, but not at 5 weeks posttreatment (Jensen et al. 1990).

A large study (Jull et al. 2002) found that mobilization therapy was more effective than general practitioner care in reducing the frequency and intensity of headaches of cervical origin, but was not more effective than exercise therapy or exercise therapy combined with mobilization therapy. Benefits achieved by either mobilization or exercise were sustained for 12 months.

Contraindications

Controversy exists regarding the contraindications to passive spinal mobilization. Nevertheless, it is generally agreed that in the presence of certain conditions, passive spinal mobilization should only be performed with extreme care, if at all. Listed below are such conditions (Gassin and Masters 2001):

Conditions in which passive spinal mobilization should be performed with extreme care or is contraindicated:

• Neoplasia (benign or malignant)
• Active infection
• Active inflammation
• Cauda equina syndrome
• Instability following trauma
• Active spondylolysis/spondylolisthesis
• Vertebral or rib fracture
• Osteoporosis
• Bleeding disorders
• Anticoagulation (warfarin therapy)
• Radicular pain
• Radiculopathy
• Vertebrobasilar insufficiency (cervical spine)
• Upper cervical spine in rheumatoid arthritis
• Abnormal spinal anatomy (e.g., upper cervical spine in Down's syndrome)

- Severe pain
- Severe distress
- Anxious patient
- Past adverse effect from passive spinal mobilization

Adverse Effects

Passive spinal mobilization is a safe procedure when performed by a qualified practitioner aware of high-risk situations. Most adverse effects following passive spinal mobilization are benign and of short duration. The most common is local discomfort. Seriously adverse effects have seldom been reported and are considered exceedingly rare. A report of a cerebrovascular accident following passive cervical mobilization is cited in a review of the risks and benefits of spinal manipulation (Di Fabio 1999).

References

Aker, P. D., Gross, A. R., Goldsmith, C. H., & Peloso, P. (1996). Conservative management of mechanical neck pain: Systematic overview and metaanalysis. *BMJ*, *313*, 1291–1296.

Chiradejnant, A., Maher, C. G., Latimer, J., & Stepkovitch, N. (2003). Efficacy of "therapist-selected" versus "randomly selected" mobilisation techniques for the treatment of Low back pain: A randomised controlled trial. *The Australian Journal of Physiotherapy, 49*, 233–241.

Delitto, A., Cibulka, M. T., Erhard, R. E., Bowling, R. W., & Tenhula, J. H. (1992). Evidence for use of an extension-mobilisation category in acute Low back syndrome: A prescriptive validation pilot study. *Physical Therapy, 73*, 216–223.

Di Fabio, R. P. (1999). Manipulation of the cervical spine: Risks and benefits. *Physical Therapy, 79*, 50–65.

Gassin, R., & Masters, S. (2001). Spinal manual therapy – the evidence. *Australasian Musculoskeletal Medicine, 6*, 26–31.

Gross, A. R., Kay, T., Hondras, M., Goldsmith, C., Haines, T., Peloso, P., Kennedy, C., & Hoving, J. (2002). Manual therapy for mechanical neck disorders: A systematic review. *Manual Therapy, 7*, 131–149.

Harms-Ringdahl, K., & Nachemson, A. (2000). Acute and subacute neck pain: Nonsurgical treatment. In A. Nachemson & E. Jonsson (Eds.), *Neck and back pain: The scientific evidence of causes, diagnosis, and treatment* (pp. 327–338). Philadelphia: Lippincott Williams and Wilkins.

Hoving, J. L., Koes, B. W., de Vet, H. C., van der Windt, D. A., Assendelft, W. J., van Mameren, H., Deville, W. L., Pool, J. J., Scholten, R. J., & Bouter, L. M. (2002). Manual therapy, physical therapy, or continued care by a general practitioner for patients with neck pain. A randomized, controlled trial. *Annals of Internal Medicine, 136*, 713–722.

Hurwitz, E. L., Aker, P. D., Adams, A. H., Meeker, W. C., Shekelle, P. G. (1996). Manipulation and mobilization of the cervical spine. A systematic review of the literature. *Spine, 21*(15), 1746–1759.

Hurwitz, E. L., Morgenstern, H., Harber, P., Kominski, G. F., Yu, F., & Adams, A. H. (2002). A randomized trial of chiropractic manipulation and mobilisation for patients with neck pain: Clinical outcomes from the UCLA neck-pain study. *American Journal of Public Health, 92*, 1634–1641.

Jensen, O. K., Nielsen, F. F., & Vosmar, L. (1990). An open study comparing manual therapy with the Use of cold packs in the treatment of post-traumatic headache. *Cephalalgia, 10*, 241–250.

Jull, G., Trott, P., Potter, H., Zito, G., Niere, K., Shirley, D., Emberson, J., Marschner, I., & Richardson, C. (2002). A randomized controlled trial of exercise and manipulative therapy for cervicogenic headache. *Spine, 27*, 1835–1843.

Kjellman, G. V., Skargren, E. I., & Oberg, B. E. (1999). A critical analysis of randomized clinical trials on neck pain and treatment efficacy. A review of the literature. *Scandinavian Journal of Rehabilitation Medicine, 31*, 139–152.

Koes, B. W., Assendelft, W. J., van der Heijden, G. J., Bouter, L. M., & Knipschild, P. G. (1991). Spinal manipulation and mobilization for back and neck pain: A blinded review. *BMJ, 2*, 1298–1303.

McKinney, L. A., Dornan, J. O., & Ryan, M. (1989). The role of physiotherapy in the management of acute neck sprains following road-traffic accidents. *Archives of Emergency Medicine, 6*, 27–33.

Mealy, K., Brennan, H., & Fenelon, G. C. (1986). Early mobilisation of acute whiplash injuries. *British Medical Journal, 292*, 656–657.

Provinciali, L., Baroni, M., Illuminati, L., & Ceravolo, G. (1996). Multimodal treatment to prevent the late whiplash syndrome. *Scandinavian Journal of Rehabilitation Medicine, 28*, 105–111.

Wreje, U., Nordgen, B., & Aberg, H. (1992). Treatment of pelvic joint dysfunction in primary care: A controlled study. *Scandinavian Journal of Primary Health Care, 10*, 310–315.

Patch-Clamp

Definition

An electrophysiological technique that makes it possible to record the electrical currents generated by single or multiple ion channels across cellular membranes. The technique is a refinement of the voltage clamp methodology, and has been crucial in elucidating the role of ion channels in membrane excitability and synaptic transmission. It was developed by Erwin Neher and Bert Sakmann. Both were awarded the Nobel Prize in Physiology or Medicine in 1991 for this splendid discovery.

Cross-References

▶ Painful Channelopathies

Pathological Fracture

Definition

A fracture due to weakening of the bone structure by pathological processes, such as tumors, infections, or other abnormal process.

Cross-References

▶ Cancer Pain Management, Orthopedic Surgery
▶ Cancer Pain Management Radiotherapy
▶ Chronic Low Back Pain: Definitions and Diagnosis

Pathophysiological Pain

Definition

Pathophysiological pain is that form of pain that results from abnormal processing within the peripheral or central nervous system, such as during hyperalgesia and/or allodynia, that results from nervous system disease.

Patient Education

Geoffrey Harding
Sandgate Spinal Medicine Clinic, Sandgate, QLD, Australia

Definition

Patient education is the provision to patients of information, in the form of booklets, pamphlets, or videotapes, on the biological basis for their pain, its effects, and how it can be treated, especially in terms of what the patient can do.

Characteristics

Rationale

Patient education was developed as a tool in the management of pain in the belief that patients who did not understand their condition responded less well to treatment. It was expected, therefore, that providing information that explained their condition would help them understand and would allay fears that they might have. Furthermore, patient education serves to inform patients of how they could contribute to their own rehabilitation instead of relying totally on interventions from others.

Instruments

Information can be delivered individually to patients (one-on-one) or in patient-groups (one-on-many or many-on-many). It can also be delivered to spouses or "significant others." Educative efforts can be directed at the general public in the form of community campaigns. The latter might involve the use of television, press, and radio.

Most of the educational material will be given in the form of informal drawings and diagrams, pamphlets, booklets, audiotapes, videotapes, and CD-ROM. These have been assigned the collective term "bibliotherapy" (Jones 2002).

Efficacy

Patient education seems to be of value for patients with arthritis. A meta-analysis (Superior-Cabuslay et al. 1996) found that patient education provides additional benefits that are 20–30 % as great as the effects of NSAID treatment for pain relief in osteoarthritis and rheumatoid arthritis, 40 % as great as NSAID treatment for improvement in functional ability in rheumatoid arthritis, and 60–80 % as great as NSAID treatment in reduction in tender joint counts. Others, however, contend that education is not of demonstrable benefit for patients with osteoarthritis of the knee (Lord et al. 1999), even though all practice guidelines for osteoarthritis recommend it (Pencharz et al. 2002).

For neck pain, a Cochrane Review (Gross et al. 2000) found only three studies that assessed patient education. It concluded that patient education was not shown to be beneficial in reducing pain.

For low back pain, trials have found patient education not to be effective, either in reducing pain or improving function (Roberts et al. 2002; Gilbert et al. 1985; Roland and Dixon 1989; Cherkin et al. 1996). An education booklet was no less effective for acute low back pain than chiropractic manipulation or McKenzie therapy (Cherkin et al. 1998). However, although not effective for pain, once established, education does have a modest effect in reducing absenteeism, when used as a preventative measure (Symonds et al. 1995).

More encouraging have been the results of mass-media campaigns. A media campaign (television commercials, aired in prime-time slots; radio and printed advertisements; outdoor billboards, posters, seminars; workplace visits and publicity articles), featuring medical experts and sporting and media personalities, was conducted in the State of Victoria, in Australia, over a 2-year period (Buchbinder et al. 2001). The campaign was backed up by the widespread distribution of "The Back Book" (Burton et al. 1996). Evaluation of the campaign showed a significant change in public attitudes to back pain.

Importantly, the major change was the de-medicalizing of the problem. Fewer patients felt that they needed to see a doctor. In addition, there were significant decreases in the number of compensable claims and the number of days on compensation during the period of the campaign, when compared with a "control" State (New South Wales) where no similar campaign was conducted. The change in attitude of the public was accompanied by a change in the attitude of doctors, a significant proportion of whom changed their beliefs from relying on passive interventions, to encouraging activation and self-help.

Indications

Despite the popularity of patient education among promoters of practice guidelines, there is little evidence to support the use of this intervention by individual practitioners. Although it may be of some benefit to patients with rheumatoid arthritis, its benefit in osteoarthritis is disputed. It has not been shown to be beneficial for reducing neck or back pain, but it may be of value as a preventative measure in reducing concerns about back pain and resultant absenteeism.

If education is to be of benefit, it seems more effective to conduct media campaigns that address the public as a whole; for this there is evidence (Buchbinder et al. 2001). The prospect is not only one of changing public attitudes but also that the public will drive changes in the attitudes of those who treat them.

Cross-References

▶ Psychological Treatment of Pain in Older Populations

References

Buchbinder, R., Jolley, D., & Wyatt, M. (2001). Breaking the back of back pain. Public policy initiatives directed towards managing the disability of back pain can be highly successful. Editorial. *The Medical Journal of Australia, 175*, 456–457.
Burton, A. K., Waddell, G., Burtt, R., & Blair, S. (1996). Patient educational material in the management of

Low back pain in primary care. *Bulletin of the Hospital for Joint Diseases, 55*, 138–141.

Cherkin, D. C., Deyo, R. A., Street, J. H., Hunt, M., & Barlow, W. (1996). Pitfalls of patient education. Limited success of a program for back pain in primary care. *Spine, 21*, 345–355.

Cherkin, D. C., Deyo, R. A., Battie, M., Street, J., & Barlow, W. (1998). A comparison of physical therapy, chiropractic manipulation, and provision of an educational booklet for the treatment of patients with low back pain. *The New England Journal of Medicine, 339*, 1021–1029.

Gilbert, J. R., Taylor, D. W., Hildebrand, A., & Evans, C. (1985). Clinical trial of common treatments for low back pain in family practice. *British Medical Journal, 291*, 791–794.

Gross, A. R., Aker, P. D., Goldsmith, C. H., & Peloso, P. (2000). Patient education for mechanical neck disorders. *Cochrane Database Systematic Review*, CD000962.

Jones, F. A. (2002). The role of bibliotherapy in health anxiety: An experimental study. *British Journal of Community Nursing, 7*, 498–504.

Lord, J., Victor, C., Littlejohn, P., Ross, F. M., & Axford, J. S. (1999). Economic evaluation of a primary care-based education programme for patients with osteoarthritis of the knee. *Health Technology Assessment, 3*, 1–55.

Pencharz, J. N., Grigoriadis, E., Jansz, G. F., & Bombardier, C. (2002). A critical appraisal of clinical practice guidelines for the treatment of lower-limb osteoarthritis. *Arthritis Research, 4*, 36–44.

Roberts, L., Little, P., Chapman, J., Cantrell, T., Pickering, R., & Langridge, J. (2002). General practitioner-supported leaflets may change back pain behaviour. *Spine, 27*(17), 1821–1828.

Roland, M., & Dixon, M. (1989). Randomized controlled trial of an educational booklet for patients presenting with back pain in general practice. *The Journal of the Royal College of General Practitioners, 39*, 244–246.

Superior-Cabuslay, E., Ward, M. M., & Lorig, K. R. (1996). Patient education interventions in osteoarthritis and rheumatoid arthritis: A meta-analytic comparison with non-steroidal anti-inflammatory drug treatment. *Arthritis Care and Research, 9*, 292–301.

Symonds, T. L., Burton, A. K., Tilotson, K. M., & Main, C. J. (1995). Absence resulting from low back trouble can be reduced by psychosocial intervention at the work place. *Spine, 20*, 2738–2745.

Patient Information

▶ Information and Psychoeducation in the Early Management of Persistent Pain

Patient-assisted Laparoscopy

▶ Chronic Pelvic Pain, Laparoscopic Pain Mapping

Patient-Centered

Definition

A way of approaching patient-therapist interactions where the goals, thoughts, and feelings of the patient are of primary importance.

Cross-References

▶ Chronic Pain, Patient-Therapist Interaction

Patient-centered Care

▶ Family-Centered Care

Patient-Controlled Analgesia

Synonyms

PCA

Definition

A method of analgesia that allows the patient to self-administer small incremental doses of an analgesic to treat pain under a controlled regimen. Nurse-administered intermittent opioid injection requires good staffing levels to minimize the delay between need and injection. Patient-controlled analgesia overcomes these logistical problems. The patient presses a button and receives a preset dose of opioid, from a microprocessor-controlled syringe pump or a disposable device connected to an intravenous or subcutaneous cannula. Safety is increased by the use of a lock-out interval, namely, the time from the end of the delivery of one dose until the device will respond to another patient demand. Patient-controlled analgesia produces a modest improvement in postoperative pain relief over a 24-h period compared with conventional analgesia. It is preferred by patients and is not associated with additional side effects. Patient-controlled epidural analgesia (PCEA) is a relatively new method of maintaining labor analgesia. There have been many studies performed that have compared the efficacy of PCEA with continuous epidural infusion (CEI). Patients who receive PCEA are less likely to require anesthetic interventions, require lower doses of local anesthetic, and have less motor block than those who receive CEI.

Cross-References

▶ Acute Pain in Children, Postoperative
▶ Alpha(α) 2-Adrenergic Agonists in Pain Treatment
▶ Analgesic Guidelines for Infants and Children
▶ Epidural Infusions in Acute Pain
▶ Postoperative Pain, Acute Pain Management Principles
▶ Postoperative Pain, Acute Pain Team
▶ Postoperative Pain, Appropriate Management
▶ Postoperative Pain, Data Gathering and Auditing
▶ Postoperative Pain, Preoperative Education
▶ Rest and Movement Pain

Patient-Controlled Epidural Analgesia

Definition

A technique used to provide continuous epidural analgesia. After an epidural catheter is placed, the patient is connected to a machine that continuously delivers medication and allows the patient to self-administer medication to supplement the analgesia. There is greater patient

satisfaction with this technique because it allows the patient to be in control of their doses. It is a safe technique because the physician sets limits on the machine to prevent over dosage.

Cross-References

▶ Analgesia During Labor and Delivery
▶ Postoperative Pain, Patient-Controlled Analgesia Devices, Epidural

Patient-controlled Intravenous Analgesia

▶ Postoperative Pain, Patient-Controlled Analgesia Devices, Parenteral

Patient-Therapist Interaction

Definition

The verbal and nonverbal interactions between healthcare providers and their patients or clients. These interactions may include, but are not necessarily limited to, communication and negotiations concerning patient history, diagnosis, and clinical care.

Cross-References

▶ Chronic Pain, Patient-Therapist Interaction

Pavlovian Conditioning

Synonyms

Classical conditioning

Definition

Pavlovian conditioning occurs if pairs of stimuli are presented together; a neutral stimulus by itself elicits the same response as the unconditioned stimulus. Other terms used are respondent conditioning and type S conditioning.

Cross-References

- ▶ Classical Conditioning
- ▶ Classical Conditioning and Chronic Pain Syndromes
- ▶ Fear Reduction Through Exposure in Vivo
- ▶ Respondent Conditioning of Chronic Pain

Paw Pressure Test

Definition

Application of constant pressure, through a blunt probe, to the dorsal aspect of the hind paw of a restrained rat, with the time for withdrawal of the leg being the latency of the test. This method is used to measure the threshold for mechanonociception.

Cross-References

- ▶ Thalamotomy, Pain Behavior in Animals

PCA

- ▶ Patient-Controlled Analgesia
- ▶ Postoperative Pain, Patient-Controlled Analgesia Devices, Parenteral

PCA Pump

Definition

Small, microprocessor-driven, programmable infusion devices for intravenous or epidural delivery of analgesics.

Cross-References

- ▶ Acute Pain in Children, Postoperative
- ▶ Postoperative Pain, Patient-Controlled Analgesia Devices, Parenteral

PCEA

- ▶ Postoperative Pain, Patient-Controlled Analgesia Devices, Epidural

PDPH

- ▶ Post-dural Puncture Headache

Pediatric Burns

- ▶ Pain Control in Children with Burns

Pediatric Chronic Pain Management

- ▶ Complex Chronic Pain in Children, Interdisciplinary Treatment

Pediatric Dosing Guidelines

- ▶ Analgesic Guidelines for Infants and Children

Pediatric Migraine

- ▶ Migraine, Childhood Syndromes

Pediatric Pain Management

- ▶ Psychological Treatment of Pain in Children

Pediatric Pain Measures

▶ Pain Assessment in Children

Pediatric Pain Scales

▶ Pain Assessment in Children

Pediatric Pain Treatment, Evolution

▶ Evolution of Pediatric Pain Treatment

Pediatric Pharmacological Interventions

▶ Acute Pain in Children, Procedural

Pediatric Postsurgical Pain

▶ Acute Pain in Children, Postoperative

Pediatric Procedural Analgesia

▶ Pain and Sedation of Children in the Emergency Setting

Pediatric Psychological Interventions

▶ Acute Pain in Children, Procedural

Pediatric Sedation

▶ Pain and Sedation of Children in the Emergency Setting

Peltier

▶ Pain in Humans, Thermal Stimulation (Skin, Muscle, Viscera), Laser, Peltier, Cold (Cold Pressure), Radiant, Contact

Pelvic Inflammatory Disease

▶ Chronic Pelvic Pain, Pelvic Inflammatory Disease and Adhesions
▶ Gynecological Pain, Neural Mechanisms

Pelvic Neurectomy

Definition

Surgical resection of the pelvic nerves.

Cross-References

▶ Visceral Pain Models, Female Reproductive Organ Pain

Pelvic Pain Syndrome

▶ Epidemiology of Chronic Pelvic Pain

Penile Block

Definition

Blockade of the paired dorsal penile nerves by subcutaneous infiltration with local anesthetic at the base of the penis, by subpubic injection of local anesthetic bilaterally beneath Scarpa's fascia, or by injection of local anesthetic at the 10:30 and 1:30 positions beneath Buck's fascia at the base of the penis.

Cross-References

▶ Acute Pain in Children, Postoperative

Peptides in Neuropathic Pain States

Frank Porreca[1], Carl-Olav Stiller[2] and
Todd W. Vanderah[3]
[1]Department of Pharmacology, University of
Arizona Health Sciences Center, Tucson,
AZ, USA
[2]Division of Clinical Pharmacology and
Department of Medicine, Karolinska University
Hospital, Stockholm, Sweden
[3]Department of Pharmacology, College of
Medicine University of Arizona, Tucson,
AZ, USA

Synonyms

Neuropeptides in Neuropathic Pain States

Definition

Several endogenous peptides have been implicated in neuropathic pain states. This section will address recent reports that suggest how some peptides may play a role in establishing and maintaining neuropathic pain. Studies have demonstrated that peptides and often their corresponding receptors are either upregulated or downregulated after damage to peripheral nerves. Here, we focus on a few selected neuropeptides which are upregulated during states of peripheral nerve injury and which may promote pain. All peptides mentioned are found within the somatosensory system but are not limited to this system.

Characteristics

Vasoactive Intestinal Peptide (VIP) and Pituitary Adenylate Cyclase-Activating Peptide (PACAP)

VIP is a 28-amino-acid polypeptide. PACAP occurs in two variants, a 38-amino-acid polypeptide and the C-terminally truncated form (PACAP-27). VIP and PACAP share 68 % amino acid homology and act at similar receptors, and hence, their roles in neuropathic pain will be summarized together.

The mRNA for PACAP in the L5 DRG dramatically increases within 2 days after ▶ chronic constriction injury (CCI) of the sciatic nerve and slowly reverses to basal levels within 1–2 weeks post-injury, whereas the mRNA for VIP in the L5 DRG dramatically increases within 1–2 weeks and remains elevated for long periods of time (42 days) post-injury (Nahin et al. 1994; Noguchi et al. 1989). In a rat model of diabetic neuropathy using streptozocin, the VIP content in the sciatic nerve was decreased; however, no change in VIP content was seen in the spinal cord as compared to control animals (Noda et al. 1990). In the CCI model, $VPAC_2$ receptor mRNA increases dramatically, whereas $VPAC_1$ receptor mRNA in laminae III and IV decreases. The level of mRNA for PAC_1 receptors does not change significantly (Inagaki et al. 1994; Dickinson et al. 1999). PACAP knockout mice do not develop allodynia or hyperalgesia after L5 spinal nerve lesion (Mabuchi et al. 2004).

Pharmacological studies in vivo have demonstrated an excitatory action of VIP and PACAP in either naïve or nerve-injured animals; however, antagonism at $VPAC_1$, $VPAC_2$, and PAC_1 receptors in nerve-injured animals showed negligible effects when applied to sustained brush-evoked activity of dorsal horn neurons ipsilateral to the injury. The $VPAC_1$ and PAC_1 antagonists significantly inhibited cold allodynia in the CCI animal. Antagonists for all three receptors inhibited mustard-oil-evoked thermal hypersensitivity (Dickinson et al. 1999). PACAP and VIP may play a role in establishing a state of neuropathy, and VIP may be important in maintaining an ongoing pain state.

Dynorphin

Elevated levels of spinal prodynorphin, a large precursor protein thought to be active as a 17-amino-acid polypeptide, serve to maintain enhanced pain states following nerve injury. Thus far, the binding site for ▶ dynorphin as an excitatory peptide has not been elucidated, although studies have shown an indirect pathway for modulating function at the NMDA receptor channel (Koetzner et al. 2004).

Peripheral nerve injury has been consistently associated with increased levels of immunoreactivity for dynorphin or mRNA for prodynorphin in neurons of the dorsal horn of the spinal cord (Malan et al. 2000). Peripheral nerve injury results in significant ipsilateral increases in immunoreactivity for dynorphin in laminae I–II and V–VII of the dorsal horn within 5 days of injury with peak elevations at day 10 (Malan et al. 2000). The time course of dynorphin upregulation after nerve injury was found to be consistent with the maintenance phase of neuropathic pain.

Mice not expressing spinal dynorphin (prodynorphin knockout mice) develop tactile and thermal hypersensitivities immediately after nerve injury, but these signs resolve spontaneously over the following 4–6 days, whereas wild-type mice expressing dynorphin upregulation after nerve injury maintain behavioral signs of neuropathic pain throughout the observation sessions (Gardell et al. 2004; Wang et al. 2001). Dynorphin is thought to promote ▶ spinal sensitization by directly or indirectly enhancing PGE_2 release, which may subsequently enhance release of excitatory transmitters from primary afferent terminals (Koetzner et al. 2004).

Dynorphin fragments not binding to kappa opioid receptors with high affinity, such as dynorphin 2–17 and 2–13 elicit effects similar to those of dynorphin in rats (Vanderah et al. 1996), yet this dynorphin-binding site has not been identified. Currently, it is thought that dynorphin plays a role in maintaining neuropathic pain.

Cholecystokinin (CCK)

CCK was first isolated from porcine intestine as a 33-amino-acid polypeptide; however, a shorter form of CCK (CCK-8) appears to act as a neurotransmitter in the central nervous system. In naïve animals, CCK is not found in the DRG or terminals of primary afferents of nonprimates but is detected in the superficial laminae of the spinal cord. Immunoreactivity for CCK is seen in CNS structures known to modulate nociception including the periaqueductal gray (PAG), the raphe nuclei, and the medullary reticular formation. CCK acts at two identified receptors, CCK_1 and CCK_2, with distribution in spinal and supraspinal regions, including regions of pain modulation. The ▶ axotomy model of neuropathic pain results in a slight increase in CCK in medium-diameter DRG neurons and a significant increase in mRNA for CCK in 30 % of all neurons of the DRG (see Hökfelt et al. 1994). Following complete section of the sciatic nerve, potassium-evoked release of CCK in the dorsal horn is slightly elevated 3 and 7 days after axotomy. The potassium-evoked CCK release in the cingulate cortex is markedly enhanced (4–6-fold) following axotomy of the sciatic nerve (Gustafsson et al. 2000). Recent studies using microdialysis in the rostral ventromedial medulla (RVM) have demonstrated a significant threefold increase in CCK-LI in L5/L6 spinal-nerve-ligated (▶ spinal nerve ligation (SNL) model) rats compared to sham-operated rats (Vanderah, unpublished observations). In addition, nerve injury results in a significant increase in mRNA for the CCK_2 receptor in medium- to large-diameter sensory fibers (Wank et al. 1992).

The microinjection of CCK into the RVM of normal rats resulted in time-dependent behavioral signs of pain, including an increased sensitivity to normally non-noxious mechanical stimuli and noxious thermal hyperalgesia. RVM administration of L365,260, a CCK_2 receptor antagonist, produced a reversible block of SNL-induced pain (Xie et al. 2005). CCK acts to promote nociception, since numerous studies have shown that CCK antagonists enhance morphine and/or endogenous opioid analgesia to mechanical or thermal stimuli. CCK is thought to play a role in maintaining neuropathic pain at the level of the spinal cord as well as at supraspinal regions, by driving descending facilitatory pathways (Xie et al. 2005).

Neuropeptide Y (NPY)

NPY, a 36-amino-acid polypeptide, has been implicated in modulating nociception. The pharmacological administration of NPY has resulted in both antinociception and nociception. Studies in animals with experimental neuropathies have shown a marked increase in NPY in large-diameter neurons (Zhang et al. 1993; Ossipov et al. 2002; Wakisaka et al. 1991).

Axotomy results in a significant increase in NPY expression in laminae III and IV, suggesting a role for medium and large DRG cells (Wakisaka et al. 1991). Following axotomy or spinal nerve ligation, NPY immunoreactivity increases significantly in fibers in the nucleus gracilis (Zhang et al. 1993; Ossipov et al. 2002). SNL results in a significant increase in NPY expression in the ipsilateral L5 and L6 DRG, in the nerve on the side proximal to the DRG, and in the ipsilateral side of the nucleus gracilis (Ossipov et al. 2002).

Microinjection studies of an NPY receptor antagonist into the nucleus gracilis, as well as NPY antiserum, resulted in a reversal of SNL mechanical hypersensitivities (Ossipov et al. 2002). Thus far using dose-selective antagonists, the NPY_1, but not the NPY_2, receptor has been identified in the nucleus gracilis as the receptor that NPY uses to promote mechanical hypersensitivity after peripheral nerve injury (Ossipov et al. 2002), although due to a lack of a highly selective antagonist, other NPY receptors (Y_3, Y_4, and Y_5) cannot be ruled out. NPY has been shown to produce antinociception in chemical and thermal-induced pain when given into either the spinal cord, nucleus raphe magnus, or periaqueductal gray.

References

Dickinson, T., Mitchell, R., Robberecht, P., et al. (1999). The role of VIP/PACAP receptor subtypes in spinal somatosensory processing in rats with an experimental peripheral mononeuropathy. Neuropharmacology, 38, 167–180.

Gardell, L. R., Ibrahim, M., Wang, R., et al. (2004). Mouse strains that lack spinal dynorphin upregulation after peripheral nerve injury do not develop neuropathic pain. Neuroscience, 123, 43–52.

Gustafsson, H., Stiller, C. O., & Brodin, E. (2000). Peripheral axotomy increases cholecystokinin release in the rat anterior cingulate cortex. Neuroreport, 11, 3345–3348.

Hökfelt, T., Morino, P., Verge, V., et al. (1994). CCK in cerebral cortex and at the spinal level. Annals of the New York Academy of Sciences, 713, 157–163.

Inagaki, N., Yoshida, H., Mizuta, M., et al. (1994). Cloning and functional characterization of a third pituitary adenylate cyclase-activating polypeptide receptor subtype expressed in insulin-secreting cells. Proceedings of the National Academy of Sciences, 91, 2679–2683.

Koetzner, L., Hua, X. Y., Lai, J., et al. (2004). Nonopioid actions of intrathecal dynorphin evoke spinal excitatory amino acid and prostaglandin E2 release mediated by cyclooxygenase-1 and −2. Journal of Neuroscience, 24, 1451–1458.

Mabuchi, T., Shintani, N., Matsumura, S., et al. (2004). Pituitary adenylate cyclase-activating polypeptide is required for the development of spinal sensitization and induction of neuropathic pain. Journal of Neuroscience, 24, 7283–7291.

Malan, T. P., Ossipov, M. H., Gardell, L. R., et al. (2000). Extraterritorial neuropathic pain correlates with multisegmental elevation of spinal dynorphin in nerve-injured rats. Pain, 86, 185–194.

Nahin, R. L., Ren, K., De Leon, M., & Ruda, M. (1994). Primary sensory neurons exhibit altered gene expression in a rat model of neuropathic pain. Pain, 58, 95–108.

Noda, K., Umeda, F., Ono, H., et al. (1990). Decreased VIP content in peripheral nerve from streptozocin-induced diabetic rats. Diabetes, 39, 608–612.

Noguchi, K., Senba, E., Morita, Y., et al. (1989). Prepro-VIP and preprotachykinin mRNAs in the rat dorsal root ganglion cells following peripheral axotomy. Brain Research. Molecular Brain Research, 6, 327–330.

Ossipov, M. H., Zhang, E. T., Carvajal, C., et al. (2002). Selective mediation of nerve injury-induced tactile hypersensitivity by neuropeptide Y. Journal of Neuroscience, 22, 9858–9867.

Vanderah, T. W., Laughlin, T., Lashbrook, J. M., et al. (1996). Single intrathecal injections of dynorphin A or des-Tyr-dynorphins produce long-lasting allodynia in rats: Blockade by MK-801 but not naloxone. Pain, 68, 275–281.

Wakisaka, S., Kajander, K. C., & Bennett, G. J. (1991). Increased neuropeptide Y (NPY)-like immunoreactivity in rat sensory neurons following peripheral axotomy. Neuroscience Letters, 124, 200–203.

Wang, Z., Gardell, L. R., Ossipov, M. H., et al. (2001). Pronociceptive actions of dynorphin maintain chronic neuropathic pain. Journal of Neuroscience, 21, 1779–1786.

Wank, S. A., Harkins, R., Jensen, R. T., Shapira, H., de Weerth, A., & Slattery, T. (1992). Purification, molecular cloning, and functional expression of the cholecystokinin receptor from rat pancreas. Proceedings of the National Academy of Sciences, 89, 3125–3129.

Xie, Y. Y., Herman, D. S., Stiller, C.-O., et al. (2005). Mediation of opioid-induced paradoxical pain and antinociceptive tolerance by cholecystokinin in the rostral ventromedial medulla. *Journal of Neuroscience, 25*, 409–416.

Zhang, X., Meister, B., Elde, R., et al. (1993). Large caliber primary afferent neurons projecting to the gracile nucleus express neuropeptide Y after sciatic nerve lesions: An immunohistochemical and in situ hybridization study in rats. *European Journal of Neuroscience, 5*, 1510–1519.

Percutaneous Electrical Nerve Stimulation

▶ Transcutaneous Electrical Nerve Stimulation Outcomes
▶ Transcutaneous Electrical Nerve Stimulation (TENS) in Treatment of Muscle Pain

Periaqueductal Gray

Synonyms

PAG

Definition

The Periaqueductal Gray is region of gray matter surrounding the aqueduct of Sylvius in the midbrain between the third and fourth ventricles.It has been implicated in several different functions including pain and analgesia, defensive behavior, fear, reproductive behavior and cardiovascular control.   It is part of a descending pain modulatory pathway that relays in the nucleus raphe magnus. Stimulation in the PAG in experimental animals generally reduces nociceptive behavior.

Cross-References

▶ Deep Brain Stimulation
▶ Descending Circuits in the Forebrain, Imaging
▶ Forebrain Modulation of the Periaqueductal Gray and Its Role in Pain

▶ GABA Mechanisms and Descending Inhibitory Mechanisms
▶ Nitrous Oxide Antinociception and Opioid Receptors
▶ Nociceptive Processing in the Nucleus Accumbens, Neurophysiology and Behavioral Studies
▶ Opioid Electrophysiology in PAG
▶ Opioids and Reflexes
▶ Peptides in Neuropathic Pain States
▶ Psychiatric Aspects of Visceral Pain
▶ Spinomesencephalic Tract
▶ Spinothalamic Tract Neurons, Descending Control by Brain Stem Neurons
▶ Spinothalamocortical Projections from SM
▶ Stimulation-Produced Analgesia

Periaqueductal Gray – Nucleus Raphe Magnus (PAG-NRM) Pathway

▶ Sex Differences in Descending Pain Modulatory Pathways

Periaqueductal Gray – Rostral Ventromedial Medulla (PAG-RVM) Pathway

▶ Sex Differences in Descending Pain Modulatory Pathways

Periaqueductal Grey

▶ Opioid Electrophysiology in PAG

Pericranial Muscles

Definition

The pericranial muscles, with importance in tension-type headache, are the neck muscles, chewing muscles, mimetic facial muscles, and muscles in the inner ear (tensor tympani, stapedius).

Cross-References

▶ Headache, Episodic Tension-Type

Peridural

▶ Epidural

Peridural Analgesia

▶ Epidural Analgesia

Peridural Block

▶ Epidural Block

Peridural Infusions

▶ Epidural Infusions in Acute Pain

Peridural Space

▶ Epidural Space

Perikarya

▶ Soma

Perineural Catheters

Definition

Plastic catheters placed near a peripheral nerve for application of numbing drug solutions.

Cross-References

▶ Cancer Pain Management, Anesthesiologic Interventions, and Neural Blockade

Perineural Injection

Definition

Local anesthetic injection around the nerve or into the sheath containing the nerve(s) either under direct surgical visualization or percutaneously.

Cross-References

▶ Acute Pain in Children, Postoperative

Perineuronal Basket

Definition

Following major peripheral nerve injuries with limited regeneration, some large-diameter sensory neurones within dorsal root ganglia that contain transected axons become surrounded by varicose nerve terminals of collateral branches of sympathetic and peptidergic sensory neurones. These terminals branch amongst the multiple layers of perisomatic glia that develop around these neurones. There is limited evidence of synaptic contact with neurone somata or functional responses to activation of small-diameter neurones.

Cross-References

▶ Sympathetic and Sensory Neurons after Nerve Lesions, Structural Basis for Interactions

Perineuronal Satellite Cells

▶ Satellite Glial Cells and Chronic Pain

Perineuronal Terminals

▶ Sympathetic and Sensory Neurons After Nerve Lesions, Structural Basis for Interactions

Period Prevalence

▶ Prevalence of Chronic Pain Disorders in Children

Periodic Disorders of Childhood that are Precursors to Migraine

▶ Migraine, Childhood Syndromes

Perioperative Nutritional Support

Definition

Perioperative nutritional support refers to perioperative enteral or parenteral nutritional supplementation of carbohydrates, proteins, vitamins, and minerals to improve the nutritional state of the patient.

Cross-References

▶ Postoperative Pain, Importance of Mobilization

References

Awad, S., & Lobo, D. N. (2011). What's new in perioperative nutritional support? *Current Opinion in Anaesthesiology, 24,* 339–348.

Peripheral and Central Neuronal Sensitization

▶ Neuropathic Pain in Children

Peripheral Arterial Occlusive Disease

Synonyms

PAOD

Definition

A condition caused by atherosclerotic narrowing of limb arteries, associated with activity-dependent pain and trophic disturbances.

Cross-References

▶ Vascular Neuropathies

Peripheral Glutamate Receptors

▶ Inflammation, Role of Peripheral Glutamate Receptors

Peripheral Mechanisms of Opioid Analgesia

▶ Opioids in the Periphery and Analgesia

Peripheral Modulation of Sensory Stimuli

▶ Perireceptor Elements

Peripheral Nerve Blocks

John Robinson
Pain Management Centre, Burwood Hospital,
Christchurch, New Zealand

Synonyms

Diagnostic blocks; Therapeutic blocks

Definition

Peripheral nerve blocks are procedures in which local anesthetic is injected onto a nerve in order to block conduction of nociceptive information along it and thereby temporarily relieve pain. Different types of block are distinguished by their purpose.

For ▶ diagnostic blocks the purpose is to obtain diagnostic information (Bogduk 2002). The objective of ▶ therapeutic blocks is to treat the pain. Prophylactic blocks are used to anesthetize a part of the body, so that other treatment can be applied to that part without aggravating the patient's pain. Prognostic block are used to predict the relief of pain, and any side effects, that might be expected from a treatment procedure in which the nerve will be destroyed.

Characteristics

Diagnostic Blocks
Principles
Diagnostic blocks are used to establish if a patient has a peripheral source of nociception and its location. If a patient's pain is not relieved by a block, a source of pain in the distribution of

the target nerve is excluded. If the block relieves the pain, the inference is that the nerve is responsible for mediating the patient's pain, and the source probably lies somewhere in the territory innervated by the nerve.

Applications
In patients with ▶ neuropathic pain, diagnostic blocks can be used to determine if a patient's pain, or other symptoms, are generated or sustained by activity in a peripheral nerve. Perhaps most critically, if peripheral nerve blocks do not relieve the patient's pain, a peripheral mechanism is excluded, whereupon a diagnosis of ▶ central pain becomes more likely. On the other hand, if a block relieves the pain, a peripheral contribution can be inferred. For example, if anesthetizing a neuroma relieves pain, then ectopic impulse generation from the neuroma can be deduced to be the basis of that pain.

An important caveat applies to patients in whom disinhibition is a mechanism of symptom production. In such patients, pain, ▶ hyperalgesia, and ▶ allodynia may arise, not because of nociceptive activity in damaged peripheral nerves but because of disinhibition of central connections of otherwise intact, normal nerves, whose receptive fields overlap into the territory of a damaged nerve. Under these conditions, relief of pain by blocking the overlapping nerves does indicate that they are responsible for mediating the patient's symptoms, but does not constitute evidence that the cause of pain lies in their territory.

In patients with ▶ somatic pain, diagnostic blocks can be used to identify sources of pain that cannot be detected by other means, such as medical imaging. If blocking a nerve completely relieves the patient's pain, a source somewhere in the territory of the nerve can be inferred. For defining the source of pain, such blocks become more accurate the more distally along a nerve they are applied, for under those conditions the number of possible sources is progressively less.

Diagnostic blocks can be applied in a similar manner for patients with ▶ visceral nociception and pain, but the number of accessible sites for blocks is small. Typically, they entail ▶ sympathetic blocks in which the celiac plexus or splanchnic nerves are anesthetized, in order to establish that the source of pain lies somewhere among the viscera of the upper abdomen.

Validity

For diagnostic blocks to be valid, they should be both target specific and controlled.

Target specificity means that the block must accurately anesthetize the target nerve and not unintentionally anesthetize some other structures that may be the cause of the patient's pain. For superficial and readily accessible nerves, target specificity can be achieved by limiting the injection of local anesthetic to the immediate vicinity of the target nerve. For deeper structures, which cannot be palpated, target specificity is achieved by using fluoroscopic guidance.

Controls are critical to the validity of diagnostic blocks (Bogduk 2002). Patients with pain may report relief for reasons other than the action of the local anesthetic administered. These constitute false-positive responses. Controls involve administering, on a separate occasion, an agent with an action different from that of the local anesthetic normally used for the block. This agent may be an inactive one, in which case the control is a ▶ placebo, or the agent may be another local anesthetic agent with a different duration of action (Bogduk 2002). The latter type is referred to as a comparative local anesthetic block. In either instance, the blocks should be conducted under double-blind conditions, so that neither the patient nor the physician knows what the expected effect is.

If placebo blocks are used, the criteria for a true-positive response would be complete relief of pain when the local anesthetic agent is administered, but no relief of pain when the placebo is administered. If the patient obtains relief when the placebo is administered, the response to the block cannot be considered positive.

For comparative blocks, typical agents are lignocaine, which has a short duration of action (1–4 h), and bupivacaine, which has a long duration of action (4–8 h). A true-positive response would be one in which the patient obtains

complete relief of pain each time that the nerve is blocked, but reports longer-lasting relief when the long-acting agent is used and short-lasting relief when the short-acting agent is used.

Utility

Peripheral nerve blocks are commonly practiced for the assessment of neuropathic pain. In this context, they are typically assumed to be valid. No studies have tested their validity by using controls (Bogduk 2002).

For the diagnosis of somatic pain, any nerve can be the target, depending on the site of the patient's pain (Buckley 2001). Controls have not typically been advocated for blocks of the trigeminal, intercostal, abdominal, or other major peripheral nerves. They are assumed to be valid, but this may not be correct (Bogduk 2002).

Only in the context of spinal pain have diagnostic blocks been tested and shown to have validity (Bogduk 2002). Diagnostic blocks of the medial branches of the dorsal rami, both in the cervical spine and in the lumbar spine, have target specificity, provided that needles are placed correctly under fluoroscopic control (Bogduk 2002; Barnsley and Bogduk 1993; Dreyfuss et al. 1997; Kaplan et al. 1998); they are diagnostically valid if performed under controlled conditions (Bogduk 2002; Barnsley et al. 1993; Lord et al. 1995; Schwarzer et al. 1994); and they are used to pinpoint the zygapophysial joints as the source of a patient's pain (see ▶ cervical medial branch blocks, ▶ lumbar medial branch blocks).

Therapeutic Blocks

Single or repeated peripheral nerve blocks are sometimes undertaken as a treatment for neuropathic pain. The expectation is that once nociceptive activity has been suppressed by the block, it may not return or may take some time to return after the local anesthetic has worn off. Although anecdotally such blocks are reputed to work for some patients (Arner et al. 1990), there is no formal evidence that they do work by more than a ▶ placebo effect.

There is no evidence that repeated blocks for somatic pain are an effective and useful treatment. Once the local anesthetic wears off, pain recurs.

Prophylactic Blocks

Prophylactic blocks are typically used in the treatment of patients with complex regional pain syndromes, general aspects. These patents may not be able to tolerate the affected body part to be touched, due to hyperesthesia, hyperalgesia, or allodynia. In order to permit physical therapy to be applied, peripheral nerve blocks can be used to anesthetize the affected part. A common example is the use of brachial plexus blocks or axillary blocks to anesthetize the upper limb. Although promoted as an adjunct in the management of complex regional pain syndromes, the efficacy of such blocks has not been vindicated by controlled trials (Azad et al. 2000; Lamacraft et al. 1997).

Prognostic Blocks

Prognostic blocks were classically applied as a test for surgical neurectomy. It was believed that if blocking a nerve would relieve the patient's pain temporarily, cutting or avulsing that nerve should secure permanent relief. Clinical experience showed that this was not the case. Despite nerves having been cut, pain recurred in many, if not most, patients. Consequently, surgical neurectomy has been all but abandoned as a treatment for pain, and prognostic blocks are no longer used for this purpose.

One explanation for the failure of neurectomy in the past may be that physicians did not understand and recognize the potential for patients with central pain to have apparently positive responses to diagnostic blocks. If symptoms are relieved because blocks anesthetize overlapping, but normal, nerves, neurectomy will not achieve lasting relief, for the mechanism of pain lies not in the cut nerve but in the central nervous system. Indeed, under such conditions, neurectomy is likely to worsen the patient's condition by increasing the number of damaged nerves, without at all addressing the mechanism of pain.

Prognostic blocks, however, have been vindicated as a predictive test for radiofrequency ▶ lumbar medial branch neurotomy and ▶ cervical medial branch neurotomy. If patients obtain complete relief following medial branch block, they can expect to obtain complete and lasting relief if the same nerve is coagulated (Lord et al. 1996; McDonald et al. 1999; Dreyfuss et al. 2000).

Cross-References

▶ Postoperative Pain, Regional Blocks

References

Arner, S., Lindblom, U., Meyerson, B. A., & Molander, C. (1990). Prolonged relief of neuralgia after regional anaesthetic blocks. A call for further experimental and systematic clinical studies. *Pain, 43*, 287–297.

Azad, S. C., Beyer, A., Romer, A. W., Galle-Rod, A., Peter, K., & Schops, P. (2000). Continuous axillary brachial plexus analgesia with low dose morphine in patients with complex regional pain syndromes. *European Journal of Anaesthesiology, 17*, 185–188.

Barnsley, L., & Bogduk, N. (1993). Medial branch blocks are specific for the diagnosis of cervical zygapophysial joint pain. *Regional Anesthesia, 18*, 343–350.

Barnsley, L., Lord, S., & Bogduk, N. (1993). Comparative local anaesthetic blocks in the diagnosis of cervical zygapophysial joints pain. *Pain, 55*, 99–106.

Bogduk, N. (2002). Diagnostic nerve blocks in chronic pain. *Best Practice & Research. Clinical Anaesthesiology, 16*, 565–578.

Buckley, F. P. (2001). Regional anesthesia with local anesthetics. In J. D. Loeser (Ed.), *Bonica's management of pain* (3rd ed., pp. 1893–1952). Philadelphia: Lippincott Williams & Wilkins.

Dreyfuss, P., Schwarzer, A. C., Lau, P., & Bogduk, N. (1997). Specificity of lumbar medial branch and L5 dorsal ramus blocks: A computed tomographic study. *Spine, 22*, 895–902.

Dreyfuss, P., Halbrook, B., Pauza, K., Joshi, A., McLarty, J., & Bogduk, N. (2000). Efficacy and validity of radiofrequency neurotomy for chronic lumbar zygapophysial joint pain. *Spine, 25*, 1270–1277.

Kaplan, M., Dreyfuss, P., Halbrook, B., & Bogduk, N. (1998). The ability of lumbar medial branch blocks to anesthetize the zygapophysial joint. *Spine, 23*, 1847–1852.

Lamacraft, G., Molloy, A. R., & Cousins, M. J. (1997). Peripheral nerve blockade and chronic pain management. *Pain Reviews, 4*, 122–147.

Lord, S. M., Barnsley, L., & Bogduk, N. (1995). The utility of comparative local anaesthetic blocks versus placebo-controlled blocks for the diagnosis of cervical zygapophysial joint pain. *The Clinical Journal of Pain, 11*, 208–213.

Lord, S. M., Barnsley, L., Wallis, B. J., McDonald, G. J., & Bogduk, N. (1996). Percutaneous radio-frequency neurotomy for chronic cervical zygapophysial-joint pain. *The New England Journal of Medicine, 335*, 1721–1726.

McDonald, G., Lord, S. M., & Bogduk, N. (1999). Long-term follow-up of cervical radiofrequency neurotomy for chronic neck pain. *Neurosurgery, 45*, 61–68.

Schwarzer, A. C., Aprill, C. N., Derby, R., Fortin, J., Kine, G., & Bogduk, N. (1994). The false-positive rate of uncontrolled diagnostic blocks of the lumbar zygapophysial joints. *Pain, 58*, 195–200.

Peripheral Nerve Injury

▶ Cathepsin and Microglia

Peripheral Neuralgia

▶ Peripheral Neuropathic Pain

Peripheral Neurogenic Pain

▶ Peripheral Neuropathic Pain

Peripheral Neuropathic Pain

Marshall Devor
Department of Cell & Developmental Biology, Institute of Life Sciences and Center for Research on Pain, The Hebrew University of Jerusalem, Jerusalem, Givat Ram, Israel

Synonyms

Nerve pain; Pain in ganglionopathy; Pain in radiculopathy; Peripheral neuralgia; Peripheral neurogenic pain

Definition

"Neuropathic pain" is defined by the International Association for the Study of Pain (IASP, 2011 update) as pain caused by a lesion or disease of the somatosensory nervous system (http://www.iasp-pain.org/Content/NavigationMenu/GeneralResourceLinks/PainDefinitions/default.htm#Peripheralneuropathicpain). This entry deals with disorders of the peripheral somatosensory system. ▶ Central (neuropathic) pain is reviewed in ▶ central pain in this Encyclopedia. A significant change introduced since the most recently published version of the IASP Pain Taxonomy (Merskey and Bogduk 1994) was deletion of the words "or dysfunction" from the definition. The revised definition requires demonstration of a lesion or a disease that satisfies established neurological diagnostic criteria. With this change certain diagnoses, such as chronic regional pain syndrome (CRPS type1, formerly known as reflex sympathetic dystrophy (RSD)), must be relegated to the category of "possibly neuropathic." Despite the fact that CRPS1 is widely presumed to be neuropathic in origin no identifiable neural lesion has been linked to it so far.

Neuropathic pain needs to be distinguished from normal (or "nociceptive") pain and from pain due to inflammation of innervated tissue. In nociceptive and inflammatory pain, the neural impulses that signal pain arise from sensory endings as part of the normal functioning of the pain detection system. Neuropathic pain results from abnormal neural discharge arising in a pain system that has been damaged. Pain due to demonstrable inflammation of a peripheral nerve or ganglion *is* properly considered peripheral neuropathic pain.

Characteristics

Precipitating Events
Peripheral neuropathic pain may result from any type of neural damage or disease, physical, chemical, biological, or metabolic that has the effect of inducing pathology in a peripheral nerve

(neuropathy), sensory or autonomic ganglion (ganglionopathy), or dorsal root (radiculopathy). If damage occurs suddenly, the initial impact may evoke pain sensation due to injury discharge. In general, however, peripheral neuropathic pain is not an immediate effect of the insult, but results from secondary pathophysiological changes that develop over time in the peripheral or central nervous systems (PNS, CNS), or both. Damage or disease affecting peripheral nerves induces two main types of pathological change, often in combination. (1) Various degrees of injury to the myelin sheath that surrounds many axons (dysmyelination and demyelination). Disruption of the Schwann cell sheath that surrounds small diameter nonmyelinated axons might also be of importance. (2) Various degrees of injury to axons, from minor axonopathy to frank transection (axotomy). More subtle disruptions may also occur, such as in the endoneurial matrix; but these ultimately cause pain by affecting nerve fibers and their glial sheath. While it is obvious why damage or disease that disrupts the ability of axons to propagate nerve impulses from peripheral tissues into the CNS may cause "negative" sensory (and motor) abnormalities such as hypoesthesia and numbness (or weakness/ataxia), it is not obvious why neural pathology causes "positive" sensory abnormalities such as dysesthesias and pain. Providing an explaining is the focus of this entry.

The neural pathology that brings about pain is usually due to one of the following precipitating events: trauma, infection, inflammation, metabolic abnormalities, malnutrition, vascular abnormalities, neurotoxins, radiation, inherited mutations, or autoimmune attack. Some typical examples and associated diagnostic entities are listed in Table 1. Many of these are discussed in more detail in focused entries in this Encyclopedia. Sometimes neuropathy arises spontaneously without an identified cause. This is "idiopathic" neuropathy.

Individual Variability

Pain in neuropathy is notoriously variable from patient to patient even when the precipitating injuries or diseases are essentially identical. This variability has traditionally been attributed to environmental factors such as upbringing and cultural norms, personality, habitual activity, use of prosthetics, drugs, and even diet. Evidence has accumulated in recent years, however, which points to genetic predisposition as being responsible for roughly half of the overall variability (Mogil 2004, 2012). A search for genetic polymorphisms that affect pain susceptibility is currently under way. Considerable progress has already been made at defining rare mutations that cause painful (and painless) peripheral neuropathies and a start has been made at defining common polymorphisms that predispose to acquiring neurological diseases that may be painful ("disease susceptibility genes"). Much less has been done so far to identify "pain susceptibility genes," genes whose alleles determine the amount of pain different individuals will suffer in the presence of the same neural pathology (e.g., Nissenbaum et al. 2010).

Individual differences can be quite extreme. For example, following amputation of a limb, with transection of all its major nerve trunks, about a third of individuals report frequent and often severe pain in their sump and/or phantom limb, while another third report little or no pain (Sherman et al. 1984). Likewise, while everyday cuts and bruises damage small cutaneous nerve branches with no long-term consequences, a papercut or accidental puncture of a small cutaneous fascicles with a fine sterile needle during venipuncture occasionally cascade into catastrophic chronic neuropathic pain (notably CRPS1). Indeed, the key distinguishing element in the definitions of CRPS1 and CRPS2 is that in CRPS1 there is no demonstrable nerve lesion. Partly for this reason there is a minority view that CRPS1 is not a neurological abnormality at all, but rather a psychiatric ("functional") one (Verdugo et al. 2004). Nonetheless, despite the absence of a clear precipitating lesion, most experts presume that CRPS1 is a neuropathic condition.

Abnormalities of the PNS fill a broad spectrum, and some are subtle. It is objectively difficult to rule out minor pathology or neuronal rearrangements at the molecular level.

Peripheral Neuropathic Pain, Table 1 Some categories of neuropathy often associated with chronic pain

Cross-references to other entries	Precipitating event	Some resulting conditions/diagnoses
▶ Brachial Plexus Avulsion and Surgery in the Dorsal Root Entry Zone (DREZ) ▶ Plexus Injuries and Deafferentation Pain ▶ Neuroma Pain ▶ Painful Scars	Nerve trauma, e.g., due to penetrating injuries, amputation, bone fractures	Stump pain, phantom limb pain, CRPS2 (causalgia)
▶ Iatrogenic Causes of Neuropathy	Ionizing radiation	Neuropathy from radiation burns, radiation therapy for cancer
▶ Painful Scars ▶ Iatrogenic Causes of Neuropathy	Iatrogenic nerve injury (e.g., surgery, radiotherapy, chemotherapy)	Scar pain, post-thoracotomy pain, post-mastectomy pain, accidental intraneural injection, paclitaxel neuropathy
▶ Thoracic Outlet Syndrome ▶ Entrapment Neuropathies, Carpal Tunnel Syndrome	Nerve compression due to entrapment	Carpal tunnel, ulnar nerve entrapment, foraminal (lateral) stenosis
▶ Radiculopathies, ▶ Trigeminal and Cranial Nerve Neuralgias ▶ Cancer Pain, Animal Models ▶ Sciatica	Nerve compression by: tumor, artery, edema, herniated intervertebral disc	Many cancer pains, trigeminal neuralgia, radicular pain (sciatica)
▶ Viral Neuropathies ▶ Hansen's Disease	Parasitic, bacterial, or viral infection	Nerve abscess, Hanson's disease (leprosy), postherpetic neuralgia, HIV/AIDS neuropathy, Lyme disease
▶ Inflammatory Neuritis	Inflammation (sterile)	Acute and chronic demyelinating neuropathy, dorsal root ganglionitis, tumor infiltration of nerves, or innervated tissues
▶ Small Fiber Neuropathies ▶ Vascular Neuropathies ▶ Diabetic Neuropathies ▶ Metabolic and Nutritional Neuropathies	Metabolic or nutritional distress, ischemia	Diabetic neuropathy, sickle cell anemia, occlusive/ischemic angiopathy, alcoholic neuropathy, vitamin B12 deficiency
▶ Toxic Neuropathies	Environmental toxins	Lead-, mercury-, ethanol-, or acrylamide-induced neuropathy
▶ Painful Channelopathies ▶ Hereditary Neuropathies	Hereditary neuropathies	Fabry's disease, some Marie-Charcot-Tooth neuropathies, familial erythromelalgia
▶ Small Fiber Neuropathies ▶ Ganglionopathies ▶ Guillain-Barré Syndrome	Autoimmune, idiopathic	Guillain-Barré, small fiber neuropathies, paraproteinemia, Sjögren's syndrome

Diagnostic electromyography and nerve conduction studies are insensitive to subtle damage, especially in small diameter slowly conducting axons and even nerve and skin punch biopsies, which may reveal small fiber loss, do not necessarily reveal a definitive cause of pain. One should therefore keep an open mind about the possibility of chronic pains of uncertain origin being neuropathic. This includes ones in deep somatic tissues and viscera where nerve injury and disease is particularly difficult to document

(Freeman 2005). The variable and unsystematic relation between precipitating events and pain report make neuropathic pain, especially at its boundaries, difficult to diagnose and difficult to treat.

Symptoms and Signs

Loss of function (negative symptoms) is expected following nerve injury. But positive sensory changes such as spontaneous or stimulus evoked parentheses (tingling, pins, and needles), dysesthesias (odd unpleasant sensations, not frankly painful), and frank ongoing or stimulus evoked pain are a priori unexpected. Nonetheless, they represent the main burden of neuropathy both for the patient and for the investigator seeking theoretical understanding and improved avenues for treatment. Ongoing pain (a positive symptom) may be present even when the limb is completely numb (a negative symptom). This is anesthesia dolorosa. Three categories of pain need to be distinguished.

1. *Spontaneous pain* (ongoing pain) is present at rest when no (intentional) stimulus is applied. Ongoing neuropathic pain can be truly independent of any stimulus or associated with skin temperature, gut motility, or internal factors such as blood chemistry, autonomic nervous system activity, or hormones. It is a bad habit to say "pain" when "spontaneous pain" is meant.

2. *Evoked pain* occurs on stimulation of the skin or other accessible tissues such as the oral or nasal mucosa. By current IASP usage (Merskey and Bogduk 1994), pain evoked by stimuli that are normally painless (touch, warm, cool, dilute chemicals) is called "allodynia" (tactile allodynia, heat allodynia, etc.). When a normally painful stimulus evokes more intense pain than expected, this is "hyperalgesia." Allodynia and hyperalgesia arise in inflammatory conditions and also in neuropathy. In circles where the term allodynia has not (yet) been adopted, hyperalgesia continues to mean excess pain from all types of stimuli, weak or strong. In the animal research literature, reduced threshold to mechanical stimulation is almost universally and correctly called tactile allodynia, while reduced threshold to cutaneous heating is widely called heat hyperalgesia, in violation of the IASP definition. When heat pain threshold is reduced pain occurs in response to normally nonpainful warming. This should be referred to as heat allodynia.

3. *Pain on movement and deep palpation* occurs on weight bearing or weight transfer (e.g., walking, bending), or on deep palpation of muscles, tendons, nerve trunks, or viscera.

When the effect is localized, one may refer to "tender spots," "trigger points," or clinical signs such as neuroma pain or the ▶ Tinel sign. This does not necessarily indicate neuropathy. It might simply reflect mechanical allodynia or hyperalgesia of the deep tissues, perhaps due to inflammation. However, it might also reflect ectopic mechanosensitivity at a location where damage has occurred along the course of a nerve.

When the source of pain is not at the surface it is often difficult to analyze, resulting in ambiguity. There are many recognized neuropathic pain conditions associated with injury or disease of cutaneous nerves. In contrast, there are few recognized neuropathic diagnoses associated with visceral organs or musculoskeletal structures. This could be due to some fundamental difference in the response of cutaneous versus deep nerves to injury or disease. But it is much more likely due to our failure to recognize some deep pains for what they are, neuropathic pains.

The Sensory Quality of Neuropathic Pain
Some of the words that people with neural damage or disease use to describe their sensory experience are generic. Others, however, are characteristic of neuropathy in general or even of a particular neuropathic pain syndrome (Bouhassira et al. 2004). Burning pain, for example, is very common in ▶ postherpetic neuralgia, small fiber peripheral neuropathies, and traumatic nerve injuries ("▶ causalgia," now called CRPS2, derives from the root caustic = burning). But it also results from acute burns. The feelings of intense, bone chilling cold and cold allodynia are more indicative of neuropathy. And descriptions such as persistent tingling, pins and needles, or electric shock-like paroxysms are fairly strong indicators of nerve involvement. Sometimes horrible sensations are reported that defy description, even by articulate patients ... pain unlike anything we normally experience. For lack of words, such sensations are put under the grab bag heading "dysesthesia."

A constellation of pain descriptors particular, and perhaps unique, to neuropathic pain conditions is "▶ hyperpathia." Hyperpathia refers to sensations with odd variations in time and space. For example, a gentle tap on the back of the hand may feel dull, as if felt through a boxing glove. However, with repeated tapping (say, once or twice a second for 10–20 s) the sensation "winds up," becoming stronger and stronger until it reaches a painful crescendo. Hyperpathic sensations also spread in space. For example, a localized touch on the hand might cause a stinging sensation that spreads up into the arm. ► Hyperpathia is different from "► referred pain," where stimulation at one point is felt somewhere else.

Anesthesia Dolorosa and Phantom Limb Pain

Special mention should be made of neuropathic pain felt in parts of the body that are completely numb, e.g., due to brachial plexus avulsion (► anesthesia dolorosa) or that no longer exists following amputation (phantom limb pain). Such pains can have all of the characteristics of other neuropathic pains (burning, cramping, electric shock-like, steady or waxing and waning) and may be exacerbated by factors such as palpation of the stump, urination, emotional upset, or cold weather. Phantom limb pain should not be confused with stump pain, which is (postsurgical) inflammatory or neuropathic pain felt in the stump.

Neuropathic Pain Mechanisms

Neuropathic pain is fundamentally a paradox. Like cutting a telephone cable, injuring a nerve ought to make the line go dead yielding only negative symptoms (numbness). Why, then, does neuropathy trigger paresthesias, dysesthesias, and pain? A related issue concerns diversity. Different neuropathic pain diagnoses are triggered by different precipitating events and present with different clinical pictures and natural histories. Does each have its own pain mechanism or is "peripheral neuropathic pain" a single disease with a variety of manifestations? Both dilemmas are rooted in the misconception that nerves are like copper telephone cables. Axons are not wires, but live protoplasmic

extensions of specialized cells, neurons. We need to understand how these cells respond to demyelination and axonopathy and how these changes affect the interaction of sensory neurons with their PNS and CNS neighbors.

Injury and Disease Alter Neuronal Phenotype

Depending on the nature and severity of the neuronal injury or disease, the distal part of the axon including the sensory ending may retract from the tissue it innervates, or undergo anterograde (Wallerian) degeneration. Severing a nerve or some of its axons always causes the part of the axon distal to the lesion to degenerate. But the proximal part of the axon, the DRG cell body, and connections with the CNS usually survive (Tandrup et al. 2000), and these can become a source of pain. Signaling in the injured neuron is drastically altered, however.

Neural signaling takes two forms, rapid electrical impulse traffic (measured in meters per second), and relatively slow axoplasmic transport of molecules (measured in centimeters per day). Electrical impulses convey moment-to-moment sensory information. Transported molecules, in contrast, may carry signals independent of impulse traffic.

Transported molecules fill a variety of roles, structural (e.g., cytoskeletal elements, membranes), sensory response (e.g., transducer molecules, receptors), neurotransmission (e.g., ion channels and other "molecules of excitability," the exocytosis machinery, neurotransmitter molecules themselves), and trophic regulation (e.g., neurotrophins, nerve cell adhesion molecules (NCAMs)). It is likely that spike activity, integrated over minutes, hours or days, affects both the incorporation of these transported molecules into the axon membrane (at peripheral sensory endings and central synaptic terminals) and the release of signaling molecules from the central axon terminal into the CNS.

Nerve injury affects both electrical signals and axoplasmic transport, resulting in numerous changes in nerve cell functioning, i.e., it causes changes in neuronal "phenotype" (► retrograde cellular changes after nerve injury). Probably, the most important change is the emergence of electrical *hyper*excitability and consequent abnormal impulse traffic. This is the basis for paresthesias, dysesthesias, and pain. Nerve injury also triggers a variety of changes in spinal processing of afferent nerve signals, many of which result in abnormal CNS amplification. The umbrella term for the latter is "central sensitization" ("▶ central sensitization"). Central sensitization goes beyond signal amplification. In its presence impulses carried by low threshold Aβ touch afferents can yield a sensation of pain, not just strong touch (tactile allodynia, "Aβ pain"). In this way, both injured and nearby noninjured sensory neurons contribute to neuropathic pain sensation (Campbell et al. 1988; Torebjork et al. 1992).

The cascade of events that lead to neuropathic pain begins with the injured primary sensory neuron. Axonal transection blocks the normal flow of neurotrophic signaling molecules between the periphery and the sensory cell body. This triggers a change in the quantity of various proteins synthesized ("expressed") by the cell body and exported to both peripheral and central axon endings. Some molecules start to be expressed in excess ("upregulation of gene expression"), while the synthesis of others is reduced ("downregulation"). It is now clear that several thousand genes are regulated in this way, a significant fraction of all of the genes expressed in sensory neurons (Hammer et al. 2010; Michaelevski et al. 2010; Persson et al. 2009)! A major challenge is to determine which of these numerous changes in gene expression are responsible for the most important functional changes in neuronal phenotype (Devor 2001).

In addition to changes in gene expression in the cell soma, the delivery ("▶ trafficking") of transported molecules is disrupted. Probably the most significant consequence of altered trafficking is the accumulation or depletion of molecules of excitability at sites of axonal injury, including zones of demyelination, swollen terminal endbulbs at the cut axon end, retraction bulbs (e.g., in axons that are dying back) and outgrowing sprouts (Fig. 1). The best-documented example is the accumulation of Na+ channels in neuroma endings (Devor et al. 1993). As for

Peripheral Neuropathic Pain, Fig. 1 Immunolabeling shows the accumulation of Na+ channels at the chronic cut end of injured axons (for details see (Devor et al. 1989)

demyelination, while it is not clear to what extent this alters gene expression, it does affect channel trafficking. In intact axons, Na+ channels are largely excluded from the axonal membrane under myelin. When myelin is stripped off the axon, Na+ channels accumulate locally. The ectopic accumulation of Na+ channels in neuroma endings and patches of demyelination gives rise to local electrical hyperexcitability and ectopic impulse discharge (Devor 2013).

Finally, neuropathy-induced phenotypic change may enhance electrical excitability by changing the current carrying ability of ion channels, and their kinetics. This could result from alternative splicing of channel subunits or altered stoichiometry in the assembled channel complex. Likewise, proinflammatory mediators such as prostaglandins, cytokines and trophic factors, and even co-neurotransmitters, can affect channel kinetics in this way (usually by activating protein kinases PKA and PKC (Gold et al. 1996; Natura et al. 2005; Hudmon et al. 2008)). Note that inflammatory mediators might enhance neuropathic discharge in the periphery in two ways, by exciting (depolarizing) injured neurons and by increasing their fundamental excitability (▶ inflammatory neuritis). If this occurs at a distance from the sensory ending, or in the presence of neuropathy, the resulting pain is neuropathic rather than inflammatory.

Spontaneous Ectopic Discharge

Altered expression, trafficking, and kinetics of ion channels renders many afferents prone to excess generation of electrical impulses, either spontaneously or in response to physical, thermal, or chemical stimuli (Fig. 2). Neurons that previously generated impulses only at their peripheral sensory ending in response to an adequate stimulus (touch, pinch, etc.) now begin to generate impulses at ectopic pacemaker sites. Ectopically generated impulses are conveyed along the dorsal roots into the CNS where they drive central pain signaling neurons. A broad variety of evidence from studies in animals and humans indicates that spontaneous ectopic discharge gives rise to spontaneous paresthesias, dysesthesias, and pain in neuropathy, including anesthesia dolorosa and phantom limb sensation (Devor 2013; Nordin et al. 1984; Serra et al. 2012).

It is natural to presume that the sensory quality evoked by spontaneous ectopic discharge bears a relation to the type of afferent involved. For example, thermal sensibility to heat and cold stimuli is normally due to the activation of thermosensitive C and Aδ afferents. Spontaneous burning pain in neuropathy would thus be due to spontaneous ectopic activity in injured thermal nociceptors. This inference could be wrong, however, given the major distortion wrought by central sensitization on the relation between afferent fiber type and resulting sensation (note Aβ pain). In principle, CNS pathways that signal hot, and are normally activated by thermal nociceptors, might now be driven by ectopic activity in nonthermal afferents at normal body temperature. Bizarre dysesthetic sensation presumably results from atypical mixtures of afferent drive. Electric shocks, for example, activate all afferent types in a nerve simultaneously. This is probably the type of ectopic activity that underlies electric shock-like paroxysms in neuropathy (Devor et al. 2002).

Injury evoked ectopic hyperexcitability is not simply a matter of lowered threshold for repetitive impulse discharge. Rather, it reflects a qualitative change in neuronal phenotype. Most intact sensory neurons are incapable of generating sustained impulse discharge in mid-nerve

Peripheral Neuropathic Pain, Fig. 2 Recordings (R) from afferent axons associated with an injured nerve frequently show abnormal ongoing tonic and burst discharge that originates ectopically at the nerve injury site and associated DRGs

or within sensory ganglia, even in the presence of strong sustained depolarizing stimuli. They are designed to fire exclusively in response to stimuli applied to sensory endings in skin, muscle, etc. Neuropathic alterations in neuronal phenotype actually creates repetitive firing capability at ectopic locations, de novo. Repetitive firing capability ("pacemaker capability") is a result of electrical resonance that emerges in the cell membrane of the injured afferent neurons, and subthreshold oscillations that develop as a result (Amir et al. 1999; Devor 2013). Affected neurons become responsive to slow onset and sustained depolarizing stimuli, which previously did not evoke discharge (Fig. 3). Some begin to fire spontaneously at resting membrane potential.

Pain on Movement and Deep Palpation

Gentle percussion over sites of nerve injury, areas of entrapment, or neuromas, for example, typically evokes an intense stabbing or electric shock-like sensation. This is the Tinel sign. In the event of injuries that leave the nerve in continuity (e.g., a crush or freeze lesion) a second Tinel may be evoked further distally, at the farthest position reached by regenerating sprouts. This is due to ectopic mechanosensitivity that develops at sites of nerve injury and

Peripheral Neuropathic Pain, Fig. 3 Ectopic burst discharge in neuropathy is driven by subthreshold membrane potential oscillations. (a) Shows the experimental setup for intracellular recording from rat DRG and an example of burst discharge with the associated distribution of interspike intervals. (b) Illustrates a neuron in which subthreshold oscillations were present at resting membrane potential (−58 mV), but with no spiking. When the cell was gradually depolarized, peaks of oscillatory sinusoids began to trigger ectopic burst discharge

demyelination, and in axonal sprouts. When pathology is disseminated along a nerve due to dying back and/or sprouting, odd sensations may be evoked by tapping along the entire course of major nerves (e.g., in diabetic neuropathy). Ectopic mechanosensitivity is also a feature of spinal roots entrapped by stenosis in the spinal canal or root foramina ("lateral stenosis"),

usually as a result of herniation of intervertebral disks. This is a major factor in pain evoked on movement, straight leg raising, and deep palpation in a variety of chronic musculoskeletal disorders.

For example, Kuslich et al. (1991) exposed deep paraspinal tissues in patients with radiculopathy and sciatica pain using local anesthetics applied sequentially to exposed surgical planes. Gentle palpation of the exposed nerve root and ganglion sheaths reproduced the patients' usual sciatica pain. Stimulation of other tissues such as muscles, tendons, and the annulus fibrosus of intervertebral discs did not do so. Interestingly, momentary mechanical probing of ectopic pacemaker sites frequently evokes discharge that long outlasts the stimulus itself. Such "afterdischarge" is the likely explanation of aftersensations and trigger-point pain. Ectopic mechanosensitivity is likely to develop at locations where small nerve branches cross through fascial planes, under tendons or are otherwise at risk of being locally pinched (▶ neuropathic pain, joint and muscle origin).

Abnormal Response to Other Stimuli

In addition to mechanical force, injured neurons also develop abnormal sensitivity to other depolarizing stimuli. Notable among these are ectopic responses to circulating catecholamines and noradrenalin released from nearby (injured) postganglionic sympathetic axons. Ectopic adrenosensitivity of sensory neurons yields sympathetic-sensory coupling, an important substrate of sympathetically maintained chronic pain states (SMP). Afferent response to inflammatory mediators such as TNFα, IL-1β, and prostaglandins is a second example of ectopic chemosensitivity (Shubayev et al. 2010). Abnormal discharge may also arise from temperature changes, ischemia, hypoxia, hypoglycemia, and other conditions capable of locally depolarizing afferent neurons at sites at which they have developed local resonance and ectopic pacemaker capability (Devor 2013).

Allodynia

Aside from spontaneous pain, tactile allodynia is the most irksome sensory abnormality in neuropathy. The simplest explanation is reduced response threshold in nociceptive afferents, the "excitable nociceptor hypothesis." However, there is precious little evidence that fibers that were originally mechano-nociceptors ever come to respond to the very weak tactile stimuli that typically evoke allodynia in animals or patients

(see references in Banik and Brennan 2008). Rather, tactile allodynia appears to be an abnormal sensory response to impulse activity in low threshold mechanosensitive Aβ afferents. Aβ afferents normally signal touch and vibration sense, but in neuropathy (and inflammation) they can evoke Aβ pain (Campbell et al. 1988; Torebjork et al. 1992) due to central sensitization. We must assume that central sensitization also causes ectopic spontaneous activity in Aβ fibers and Aβ fiber activity evoked ectopically by applied stimuli to be amplified in the CNS and felt as painful (▶ central changes after peripheral nerve injury).

Hyperpathia

Although the causes of hyperpathic sensory peculiarities are not known for certain they are probably a result of pathophysiological cross-excitation among injured sensory neurons, coupled with central sensitization. Injured sensory neurons communicate with one another through ▶ ephaptic (electrical) coupling and through a novel nonsynaptic, neurotransmitter-mediated (paracrine) mechanism, axonal and DRG "cross-excitation" (Amir and Devor 1996; Lisney and Devor 1987). Neuron to neuron crosstalk of this sort could also account for windup and for hyperpathic spread of sensation from the site of stimulation, and for pain paroxysms. The likely mechanism is incremental buildup and spread of the paracrine mediator within the nerve or DRG on repeated stimulation (see ▶ Pain Paroxysms).

The Relation of PNS Injury to Central Sensitization

Some authors limit use of the term "central sensitization" to functional changes that are dependent on ongoing nociceptive afferent activity and that reverse rapidly when the impulse activity is blocked (Ji et al. 2003). The best-known of many mechanism of this sort involves the recruitment of NMDA-type glutamate receptors on postsynaptic neurons in the dorsal horn. Other investigators use a broader definition that encompasses all of the many CNS changes that tend to increase spinal gain (amplification), whether or not they are labile or closely linked to impulse traffic. Numerous such changes have been reported, ▶ Central changes after peripheral nerve injury (Devor) including altered expression of neuromodulatory peptides in primary sensory neurons ("phenotypic switching") and release from their central afferent terminals, spinal disinhibition due to loss of inhibitory interneurons containing GABA, glycine, taurine, and/or endogenous opiates, altered gene expression and consequent hyperexcitability of intrinsic spinal neurons, denervation supersensitivity, afferent terminal sprouting, release by "activated" microglia and astrocytes of proinflammatory cytokines, upregulation of postsynaptic transcription factors and transmembrane signaling molecules (e.g., pERK, CREB, MAPK), suppression of brainstem descending inhibition and augmentation of brainstem descending facilitation, and many others. All of these changes are triggered by nerve injury (and some by peripheral inflammation).

In most instances, little is known about the mechanism by which the injury triggers the central change. There are three fundamental possibilities: (1) depolarization of postsynaptic neurons due to impulse traffic per se, (2) other actions of neurotransmitters, neuromodulators, cytokines, and additional neuroactive substances that are released within the spinal cord as a result of impulse traffic and that are capable of inducing trophic or metabolic changes in postsynaptic neurons or local glia, and (3) trophic effects of primary afferent terminals on dorsal horn neurons and glia that are mediated by cell-surface molecules, molecules that are not released into the extracellular space.

In healthy animals and people, when central sensitization is induced by acute noxious stimuli the neuroactive substances responsible for the CNS change are associated with nociceptive C fibers (and perhaps Aδ fibers) and are specific to these afferent types. Activity in Aβ afferents, no matter how intense, does not normally induce central sensitization or trigger pain. However, at least in the early stages of neuropathic pain in animal models, almost all of the ectopic firing present is in Aβ afferents (Liu et al. 2000).

How, then, does the neuropathy induce central sensitization? One proposed explanation is that central sensitization in neuropathy is due to ectopic spontaneous discharge that occurs at extremely low firing rates in C fibers that have not been injured directly, but that share peripheral nerve trunks and peripheral tissues with the damaged and degenerating injured afferents (Wu et al. 2001). The idea is that inflammatory molecules associated with the degeneration sensitize these "undamaged" C-afferents in the periphery. Another source of nociceptive input is the small number of injured C- (and Aδ) fibers that do develop ectopia soon after transection. Finally, it is possible that, in the presence of chronic inflammation and neuropathy, Aβ afferents *acquire* the ability to trigger central sensitization even though they are not normally able to do so. Specifically, this might occur when inflamed or axotomized Aβ afferents begin to synthesize and release neuroactive substances (e.g., substance P, CGRP, BDNF) that are normally present only in nociceptors. These are the very mediators that are thought to trigger central sensitization upon C-fiber activation (Devor 2013; Molander et al. 1994; Nitzan-Luques et al. 2011). If so, ectopic activity in Aβ afferents make two contributions to neuropathic pain: (1) it triggers and maintains a central hyperexcitability state and (2) it augments the normal Aβ input that undergoes central amplification.

Deafferentation Pain
In the context of PNS-triggered CNS changes, it is important to compare the frequently confused terms "neurectomy" and "▶ deafferentation." Nerve injury denervates peripheral tissue, but in adults most DRG cell somata survive transection of their axons for quite a long time (Tandrup et al. 2000). Therefore, impulses generated in the nerve end neuroma and associated DRGs continue to be able to activate the CNS (Wall and Devor 1981) and to evoke sensory experience (e.g., the Tinel sign). Deafferentation (by dorsal rhizotomy, root avulsion, or ganglionectomy) in contrast, immediately disconnects the periphery from the CNS and rapidly leads to the degeneration of central afferent terminals. Thus, even though the sensory

Peripheral Neuropathic Pain, Fig. 4 The distinction between "deafferentation" and "denervation." Dorsal rhizotomy prevents all peripheral impulse traffic from entering the spinal cord along the affected spinal segment. The cord has been deafferented. Transecting peripheral nerves disconnects the spinal cord from peripherally innervated tissues, but it does not prevent impulses that originate ectopically at the nerve injury site or the associated DRGs from entering the cord

neuron and its sensory endings survive, impulse activity generated in the sensory ending, or ectopically along the nerve or in the DRG, is no longer able to access the CNS and activate dorsal horn neurons (Fig. 4). A distinctive darkening of central terminals visible in electron micrographs of nerve injured animals and inappropriately termed "degeneration atrophy" (Knyihar-Csillik et al. 1989) has led some observers to presume that peripheral axotomy is equivalent to deafferentation by rhizotomy. This is clearly incorrect (Bailey and Ribeiro-Da-Silva 2006).

Both neurectomy and deafferentation can trigger neuropathic pain, but for different reasons and hence with different consequences for treatment. After neurectomy the primary pain mechanisms are ectopia in PNS and central sensitization as discussed above. Therapeutic modalities can target PNS or CNS targets, or both. The causes of deafferentation pain, in contrast, are almost entirely unknown, although they surely reside primarily in the CNS. Commonly offered explanations are "denervation supersensitivity" and

release of spinal cord circuitry from inhibition. Treatment modalities need to address central targets.

Interestingly, minor nerve injuries occasionally produce devastating chronic pain conditions (e.g., CRPS). In contrast, modest or diffuse deafferentation does not. C2 dorsal root ganglionectomy, for example, is frequently performed for the relief of severe headache and multi-segmental *partial* dorsal rhizotomy is routinely performed for relief of painful spasticity in children with cerebral palsy. Neither treatment provokes deafferentation pain (Sindou et al. 1986; White and Sweet 1969). Peripheral nerve injury does trigger diffuse loss of a fraction of axotomized DRG neurons through retrograde cell death, but the loss is not massive. For this reason, deafferentation probably does not contribute substantially to pain in most peripheral neuropathies. Possible exceptions, where deafferentation can be substantial, include late stage postherpetic neuralgia and tabes dorsalis. Overall, mechanisms of deafferentation pain, like central pain in general, remain a major research frontier.

Treatment Modalities

Impressive progress in understanding peripheral neuropathic pain mechanisms has not yet been translated into improved therapeutic agencies. The problem is not a dearth of drug targets, but the massive investment required to vet each one. On the plus side, we do have a better idea as to why certain empirically effective treatment modalities work.

Systemic Drugs

The first line analgesics with proven efficacy in the relief of neuropathic pain include (certain) anticonvulsants, (certain) antidepressants, (systemic) local anesthetics, and (certain) antiarrhythmics (McQuay and Moore 1998; Finnerup et al. 2010; http://papas.cochrane.org/our-reviews). As the names imply these drugs are "adjuvants," not developed originally as pain relievers. On the face of it they appear to fit into highly diverse drug families, but there is a common denominator. All of these drugs are "membrane stabilizers"; they reduce membrane excitability and hence suppress ectopic neural discharge (Catterall 1987). Not all anticonvulsants are effective. The ones that are, e.g., carbamazepine, phenytoin, and gabapentin, act on ion channels and suppress ectopia. In contrast, anticonvulsants that act by synaptic modulation (e.g.,barbiturates and benzodiazepines) are ineffective. Likewise for antidepressants. Tricyclics, and to a lesser extent SNRIs (serotonin and norepinephrine-selective reuptake inhibitors), are highly effective. Although known as inhibitors or catecholamine reuptake they are also powerful Na+ channel blockers and hence membrane stabilizers. SSRI antidepressants (selective serotonin reuptake inhibitors), in contrast, are ineffective at stabilizing membranes or at relieving pain (e.g., Dick et al. 2007). Local anesthetics are all membrane stabilizers as are corticosteroids (in addition to being anti-inflammatory). Each of the effective drug families suppresses ectopia originating in the PNS. This reduces abnormal afferent drive as well as reversing central sensitization and hence tactile allodynia. They kill two birds with one stone. The fact that each tends to have efficacy for diverse neuropathic pain diagnoses is a strong argument that these conditions share a common neural mechanism, ectopic hyperexcitability.

Unfortunately, the effectiveness of the adjuvant drugs is usually limited by the presence of difficult or intolerable side effects, sedation, vertigo, and nausea. All of the effective adjuvants share this drawback, if to varying degrees. The reason is that, just as they suppress the firing of afferent neurons, they also alter the firing of neurons in the brain. Indeed, their ability to directly suppress discharge in CNS pain processing networks probably contributes to a certain extent to their efficacy as analgesics (Woolf and Wiesenfeld-Hallin 1985). Since most of the desired analgesic action is in the PNS, while the unwanted side effects occur mostly in the CNS, there is an obvious avenue for improving therapeutic profile. Specifically, one could target drug delivery to PNS pain generators (e.g., DRGs), or reduce drug permeation through the blood–brain

barrier (while preserving PNS activity). Neither approach has been adequately exploited to date.

In addition to drugs with primary actions in the PNS, attempts have been made to directly suppress mechanisms of central sensitization in the CNS. A popular drug target is the NMDA-type glutamate receptor. Unfortunately, these centrally acting drugs also tend to have unwanted central side effects. But since their primary site of action is in the CNS, novel approaches will be needed to obtain therapeutic selectivity. Opioid analgesics, which act within the CNS by suppressing already amplified pain signals, are no longer viewed as ineffective in neuropathy (Rowbotham et al. 2003). But many experienced clinicians still believe that, dose for dose, they are less effective for neuropathic pain than for nociceptive or inflammatory pain of equivalent intensity. The same holds true for simple pain relievers such as aspirin and other NSAIDs, acetaminophen (paracetamol) and COX2 inhibitors.

The topical application of local anesthetics and other membrane stabilizers (e.g., ketamine and amitriptyline) also has some efficacy, especially for tactile allodynia. This is probably due to two effects. First, tactile allodynia requires responsive Aβ afferents, and their responsiveness is suppressed by the topical drugs. In addition, during the early stages of afferent dying back in some painful conditions, the afferent drive that underlies spontaneous burning pain and that maintains central sensitization may derive from nociceptor endings that are still in the skin. Drugs that suppress this ectopic activity (membrane stabilizers), or that destroy or desensitize the nociceptor endings themselves (notably topical capsaicin) can be effective. Block of afferent conduction in the peripheral nerve or plexus blocks can also be effective, as can spinal (foraminal, epidural or intrathecal) block. However, this requires interventional approaches that are technically demanding and carry risk. Worse still, sustaining the block over time is difficult. As a result, these methods are mostly used in extremis.

Physical modalities such as TENS (transcutaneous electrical nerve stimulation) and spinal cord stimulation (SCS) are often used with the intent of recruiting endogenous pain control mechanisms. To the extent that these modalities work, the effect is paradoxical. TENS and SCS preferentially activate Aβ fibers. In patients with tactile allodynia, this ought to provoke pain, not relieve it! Finally, surgical procedures may sometimes be used to reduce neural irritation and promote recovery. Examples are the release of a nerve from entrapping connective tissue (e.g., at the carpal tunnel) or by microvascular decompression in trigeminal neuralgia. Surgery in selected patients can also reduce mechanical stimulation of ectopic pacemaker sites, for example, mobilization of a stump neuroma to a location less likely to be subject to mechanical compression during weight bearing. Surgical procedures aimed at severing central pain transmission pathways have largely been abandoned, although the possibility of neuromodulation by surface or deep brain stimulation is still being explored.

Transition to Chronicity

There is a long held belief that unrelieved pain can "burn" itself into the CNS just as over eons a torrential river can gouge a canyon in solid rock (pain "centralization"). The presumption is that persistent pain, per se, induces a change in biological pain mechanisms rendering treatment by conventional analgesic modalities ineffective (Kalso 1997). This belief probably arose from anecdotes of pains felt in a limb prior to amputation, persisting in the phantom. More recently, the idea of centralization has become associated with CRPS1 on the grounds that patients with chronic CRPS1 pain are so difficult to treat.

As laid out above, neural trauma and disease can induce permanent CNS changes that contribute to chronic pain ▶ Central changes after peripheral nerve injury (Devor). However, the concept that pain alone can do the same thing, in the absence of nerve pathology, should be regarded with skepticism for several reasons. First, amputees relate similar anecdotes concerning the continued presence of rings, bunions, and other nonpainful features of the limb prior to amputation. Perhaps, dysesthesias resulting from nerve injury-induced ectopia is

being misinterpreted as persisting sensory experience. Second, centralization is not evident in situations where an obvious peripheral source of pain is present, for example, labor pain or pain from a kidney stone or an arthritic hip. These pains vanish without a trace when the peripheral noxious source is ultimately removed by delivery, passage of the stone, or joint replacement. Indeed, peripheral nerve block with local anesthetic drugs almost always eliminates pain for the duration of the drug's action. Where pain, per se, *can* cause permanent mischief is by inducing a downward spiral of psychosocial deterioration and financial distress in the chronic pain sufferer.

Perspective

Chronic peripheral neuropathic pain constitutes a significant burden for the sufferer, his/her family, carers, employers, and society at large. Whereas a generation ago very little was known about its underlying causes, this is no longer the case. A biological basis is now in place for understanding many aspects of the problem at the systems, cellular, and molecular levels. The new knowledge has not yet been translated into more effective treatment modalities. However, numerous promising targets and strategies have been identified in recent years and they will inevitably pay off in the form of better therapeutic options in the future.

Cross-References

▶ Neuropathic Pain Model, Tail Nerve Transection Model

References

Amir, R., & Devor, M. (1996). Chemically-mediated cross-excitation in rat dorsal root ganglia. *Journal of Neuroscience, 16*, 4733–4741.

Amir, R., Michaelis, M., & Devor, M. (1999). Membrane potential oscillations in dorsal root ganglion neurons: Role in normal electrogenesis and in neuropathic pain. *Journal of Neuroscience, 19*, 8589–8596.

Bailey, A. L., & Ribeiro-Da-Silva, A. (2006). Transient loss of terminals from non-peptidergic nociceptive fibers in the substantia gelatinosa of spinal cord following chronic construction injury of the sciatic nerve. *Neuroscience, 138*, 675–690.

Banik, R. K., & Brennan, T. J. (2008). Sensitization of primary afferents to mechanical and heat stimuli after incision in a novel in vitro mouse glabrous skin-nerve preparation. *Pain, 138*, 380–391.

Bouhassira, D., Attal, N., Fermanian, J., et al. (2004). Development and validation of the neuropathic pain symptom inventory. *Pain, 108*, 248–257.

Campbell, J. N., Raja, S. N., Meyer, R. A., et al. (1988). Myelinated afferents signal the hyperalgesia associated with nerve injury. *Pain, 32*, 89–94.

Catterall, W. A. (1987). Common modes of drug action on Na+ channels: Local anaesthetics, antiarrhythmics and anticonvulsants. *Trends in Pharmacological Sciences, 8*, 57–65.

Devor, M. (2001). Neuropathic pain: What do we do with all these theories? *Acta Anesthesiologica Scandinavica, 45*, 1121–1127.

Devor, M. (2013). Neuropathic pain: pathophysiological response of nerves to injury. In S. L. McMahon & M. Koltzenburg (Eds.), *Wall and Melzack's textbook of pain* (6th ed.). London: Churchill Livingstone.

Devor, M., Amir, R., & Rappaport, Z. H. (2002). Pathophysiology of trigeminal neuralgia: The ignition hypothesis. *The Clinical Journal of Pain, 18*, 413.

Devor, M., Govrin-Lippmann, R., & Angelides, K. (1993). Na$^+$ channel immuno-localization in peripheral mammalian axons and changes following nerve injury and neuroma formation. *Journal of Neuroscience, 13*, 1976–1992.

Devor, M., Keller, C. H., Deerinck, T., et al. (1989). Na+ channel accumulation on axolemma of afferents in nerve end neuromas in apteronotus. *Neuroscience Letters, 102*, 149–154.

Dick, I. E., Brochu, R. M., Purohit, Y., Kaczorowski, G. J., Martin, W. J., & Priest, B. T. (2007). Sodium channel blockade may contribute to the analgesic efficacy of antidepressants. *The Journal of Pain, 8*, 315–324.

Finnerup, N. B., Sindrup, S. H., & Jensen, T. S. (2010). The evidence for pharmacological treatment of neuropathic pain. *Pain, 150*, 573–581.

Freeman, R. (2005). Autonomic peripheral neuropathy. *Lancet, 365*, 1259–1270.

Gold, M. S., Reichling, D. B., Shuster, M. J., et al. (1996). Hyperalgesic agents increase a tetrodotoxin-resistant Na+ current in nociceptors. *Proceedings of the National Academy of Sciences, 93*, 1108–1112.

Hammer, P., Banck, M. S., Amberg, R., Wang, C., Petznick, G., Luo, S., Khrebtukova, I., Schroth, G. P., Beyerlein, P., & Beutler, A. S. (2010). mRNA-seq with agnostic splice site discovery for nervous system transcriptomics tested in chronic pain. *Genome Research, 20*, 847–860.

Hudmon, A., Choi, J. S., Tyrrell, L., Black, J. A., Rush, A. M., Waxman, S. G., & Dib-Hajj, S. D. (2008). Phosphorylation of sodium channel Na(v)1.8 by p38 mitogen-activated protein kinase increases current density in dorsal root ganglion neurons. *Journal of Neuroscience, 28*, 3190–3201.

Ji, R. R., Kohno, T., Moore, K. A., et al. (2003). Central sensitization and LTP: Do pain and memory share similar mechanisms? *Trends in Neurosciences, 26*, 696–705.

Kalso, E. (1997). Prevention of chronicity. In T. S. Jensen, J. A. Turner, & Z. A. Wiesenfeld-Hallin (Eds.), *Proceedings of the 8th World Congress on pain* (Progress in pain research and management, Vol. 8, pp. 215–230). Seattle: IASP Press.

Knyihar-Csillik, E., Rakic, P., & Csillik, B. (1989). Transneuronal degeneration atrophy in the Rolando substance of the primate spinal cord evoked by axotomy-induced transganglionic degenerative atrophy of central primary sensory terminals. *Cell and Tissue Research, 258*, 515–525.

Kuslich, S., Ulstro, C., & Michael, C. (1991). The tissue origin of low back pain and sciatica. *Orthopedic Clinic of North America, 22*, 181–187.

Lisney, S. J. W., & Devor, M. (1987). Afterdischarge and interactions among fibers in damaged peripheral nerve in the rat. *Brain Research, 415*, 122–136.

Liu, X., Eschenfelder, S., Blenk, K.-H., et al. (2000). Spontaneous activity of axotomized afferent neurons after L5 spinal nerve injury in rats. *Pain, 84*, 309–318.

McQuay, M., & Moore, R. A. (1998). *An evidence-based resource for pain relief* (p. 264). Oxford: Oxford University Press.

Merskey, H., & Bogduk, N. (1994). *Classification of chronic pain. Descriptions of chronic pain syndromes and definitions of pain terms* (pp. 40–43). Seattle: IASP Press.

Michaelevski, I., Segal Ruder, Y., Rozenbaum, M., Medzihradszky, K. F., Shalem, O., Coppola, G., Horn-Saban, S., Ben Yaakov, K., Dagan, S. Y., Rishal, I., Geschwind, D. H., Pilpel, Y., Burlingame, A. L., & Fainzilber, M. (2010). Signaling to transcription networks in the neuronal retrograde injury response. *Science Signaling, 3*(130), ra53.

Mogil, J. S. (2004). *The genetics of pain, progress in pain research and management* (Vol. 28, p. 349). Seattle: IASP Press.

Mogil, J. S. (2012). Pain genetics: Past, present and future. *Trends in Genetics, 28*, 258–266.

Molander, C., Hongpaisan, J., & Persson, J. K. (1994). Distribution of c-fos expressing dorsal horn neurons after electrical stimulation of low threshold sensory fibers in the chronically injured sciatic nerve. *Brain Research, 644*, 74–82.

Natura, G., von Banchet, G. S., & Schaible, H. G. (2005). Calcitonin gene-related peptide enhances TTX-resistant sodium currents in cultured dorsal root ganglion neurons from adult rats. *Pain, 116*, 194–204.

P

Nissenbaum, J., Devor, M., Seltzer, Z., Gebauer, M., Michaelis, M., Tal, M., Dorfman, R., Abitbul-Yarkoni, M., Lu, Y., Elahipanah, T., delCanho, S., Minert, A., Fried, K., Persson, A. K., Shpigler, H., Shabo, E., Yakir, B., Pisante, A., & Darvasi, A. (2010). Susceptibility to chronic pain following nerve injury is genetically affected by CACNG2. *Genome Research, 20*, 1180–1190.

Nitzan-Luques, A., Devor, M., & Tal, M. (2011). Genotype-selective phenotypic switch in primary afferent neurons contributes to neuropathic pain. *Pain, 152*, 2413–2426.

Nordin, M., Nystrom, B., Wallin, U., et al. (1984). Ectopic sensory discharges and paresthesiae in patients with disorders of peripheral nerves, dorsal roots and dorsal columns. *Pain, 20*, 231–245.

Persson, A. K., Gebauer, M., Jordan, S., Metz-Weidmann, C., Schulte, A. M., Schneider, H. C., Ding-Pfennigdorff, D., Thun, J., Xu, X. J., Wiesenfeld-Hallin, Z., Darvasi, A., Fried, K., & Devor, M. (2009). Correlational analysis for identifying genes whose regulation contributes to chronic neuropathic pain. *Molecular Pain, 5*, 7.

Rowbotham, M. C., Twilling, L., Davies, P. S., et al. (2003). Oral opioid therapy for chronic peripheral and central neuropathic pain. *The New England Journal of Medicine, 348*, 1223–1232.

Serra, J., Bostock, H., Sola, R., Aleu, J., Garcia, E., Cokic, B., Navarro, X., & Quiles, C. (2012). Microneurographic identification of spontaneous activity in C-nociceptors in neuropathic pain states in humans and rats. *Pain, 153*, 42–55.

Sherman, R. A., Sherman, C. J., & Parker, L. (1984). Chronic phantom and stump pain among American veterans: Results of a survey. *Pain, 18*, 83–95.

Shubayev, V. I., Kato, K., & Myers, R. R. (2010). Cytokines in pain, Ch.8. In L. Kruger & A. R. Light (Eds.), *In translational pain research: From mouse to man.* Boca Raton: CRC Press.

Sindou, M., Mifsud, J. J., Boisson, D., & Goutelle, A. (1986). Selective posterior rhizotomy in the dorsal root entry zone for treatment of hyperspasticity and pain in the hemiplegic upper limb. *Neurosurgery, 18*, 587–595.

Tandrup, T., Woolf, C. J., & Coggeshall, R. E. (2000). Delayed loss of small dorsal root ganglion cells after transection of the rat sciatic nerve. *The Journal of Comparative Neurology, 422*, 172–180.

Torebjork, H., Lundberg, L., & LaMotte, R. (1992). Central changes in processing of mechanoreceptive input in capsaicin-induced secondary hyperalgesia in humans. *Journal of Physiology, 448*, 765–780.

Verdugo, R. J., Bell, L. A., Campero, M., et al. (2004). Spectrum of cutaneous hyperalgesias/allodynias in neuropathic pain patients. *Acta Neurologica Scandinavica, 110*, 368–376.

Wall, P., & Devor, M. (1981). The effect of peripheral nerve injury on dorsal root potentials and on the transmission of afferent signals into the spinal cord. *Brain Research, 209*, 95–111.

White, J., & Sweet, W. (1969). *Pain and the neurosurgeon* (pp. 93, 123–256). Springfield: Thomas.

Woolf, C. J., & Wiesenfeld-Hallin, Z. (1985). The systemic administration of local anaesthetics produces a selective depression of C-afferent fibre evoked activity in the spinal cord. *Pain, 23*, 361–374.

Wu, G., Ringkamp, M., Hartke, T.V., et al. (2001). Early onset of spontaneous activity in uninjured C-fiber nociceptors after injury to neighboring nerve fibers. *J Neurosci, 21*:RC140 1–5.

Peripheral Neuropathies

Definition

Peripheral neuropathies are a variety of neurological disorders that affect sensory, motor, and/or autonomic fibers within one (mononeuropathy) or many (polyneuropathy) peripheral nerves. Peripheral neuropathies result in abnormal nerve functioning. The partial or complete failure of nerves to carry information normally to and from the brain and spinal cord results in the loss of sensation, motor control, and/or dysautonomia (negative symptoms). The development of neuronal hyperexcitability due to neuropathy can cause abnormal sensations including pain, abnormal muscle movements, and/or abnormal autonomic responses such as hyperhydrosis (positive symptoms). Peripheral neuropathies can have various etiologies such as trauma, diabetes, alcoholism, viral infection, toxin exposure, metabolic abnormalities, vitamin deficiency, and adverse effects of certain drugs, among others.

Cross-References

▶ Cancer Pain, Assessment in the Cognitively Impaired
▶ Experimental Pain in Children

Peripheral Opioid Analgesia

▶ Opioids in the Periphery and Analgesia

Peripheral Opioid Receptors (OR)

Definition

In addition to the central nervous system, three types of opioid receptors (μ, δ, and κ) are expressed in peripheral sensory neurons. Their activation by locally applied opioid agonists results in appropriate signal transduction and may contribute to inhibition of the generation and transmission of painful stimuli. Activation of these receptors can also contribute to adverse effects of opioids including constipation.

Cross-References

▶ Opioids and Inflammatory Pain

Peripheral Sensitization

Definition

Inflammatory mediators; intense, repeated, or prolonged noxious stimulation; or both can increase the excitability of nociceptors. The sensitized nociceptors exhibit a lowered threshold for activation, increased responsiveness to a given stimulus, enlargement of the receptive fields, and an increased rate of firing. Sleeping or silent nociceptors, usually not responsive to any stimulus or only responsive to stimuli with a very intense strength, get responsive and react like polymodal nociceptors. Many different mediators, including histamine or bradykinin as well as agents not traditionally considered to be mediators but released during inflammation (like H^+ ions), can cause sensitization of nociceptors.

Cross-References

▶ Arthritis Model, Kaolin-Carrageenan-Induced Arthritis (Knee)
▶ Cancer Pain, Animal Models

▶ Cytokines, Effects on Nociceptors
▶ Metabotropic Glutamate Receptors in Spinal Nociceptive Processing
▶ Nick Model of Cutaneous Pain and Hyperalgesia
▶ Spinothalamic Tract Neurons, Role of Nitric Oxide

Peripheral Sensitization in Muscle and Joint

▶ Sensitization of Muscular and Articular Nociceptors

Peripheral Vascular Disease

Definition

A clinical syndrome of ischemia of the extremities mostly due to arteriosclerosis leading to spontaneous or exercise-related pain in the extremities, often treated by bypass surgery and rarely by spinal cord stimulation.

Cross-References

▶ Pain Treatment: Spinal Cord Stimulation
▶ Sciatica

Periphery of the Ventral Posterior Complex (VPp)

Definition

A region surrounding VPL and VPM. Neurons in this region are small, and many respond to noxious stimuli. The region includes VPI.

Cross-References

▶ Thalamus, Nociceptive Cells in VPI, Cat and Rat

Perireceptor Elements

Ana Gomis
Instituto de Neurociencias de Alicante,
Universidad Miguel Hernández-CSIC,
San Juan de Alicante, Spain

Synonyms

Peripheral modulation of sensory stimuli;
Perireceptor events

Definition

These are the elements or events interposed
between the stimulus and the sensor element
involved in the stimulus-sensory receptor
coupling process. These structures may act as
filters for the stimulating energy, modulating the
transmission of mechanical forces or the accessibility of chemicals to the transducing area of the
receptor membrane.

Characteristics

Sensory transduction takes place in specialized
portions of sensory receptor cells which differ
from each other in their ability to respond
preferentially to a particular form of energy.
The primary transducer or detector is the
place where transformation of the stimulus
into an electrical signal begins and corresponds
to a molecular entity, located in the cell
membrane that is modified by the stimulus.
These include receptor molecules for chemicals,
mechanosensitive transduction channels, transduction molecules for thermal energy, etc. The
detector is usually linked to *perireceptor structures* that convey the stimulus to the detector,
and/or modify some of its characteristics.

A classical example of the participation of
perireceptor elements in chemical transduction
is provided by the mucus surrounding chemoreceptor cells in the olfactory system. There,

chemical signals, ► odorant, or ► pheromones,
are perceived by highly specialized sensory
cells, the olfactory receptors cells. In contrast to
aquatic animals that smell water-soluble odorants, terrestrial animals smell small, volatile lipophilic molecules; these molecules must pass
through an aqueous medium (mucus, sensillum
lymph) in order to reach the chemosensory membrane of the receptor neurons. Rheological properties of the mucus, i.e., diffusion parameters,
viscosity, mucociliary transport, are major determinants of odorant access. Moreover, the mucus
contains several different kinds of odorant-
binding proteins that probably assist in the delivery of some odorants to the receptor proteins, or
in their subsequent removal. After diffusion
through the mucus, the odors bind to one of the
large family of odorant receptor proteins (Buck
and Axel 1991).

In mechanoreceptor cells, receptor molecules
located in the transducing region of the
membrane are sensitive to mechanical forces.
Mammals have several types of morphologically distinct cells that sense mechanical signals,
such as hair cells of the ► cochlea or
mechanosensory neurons in ► sensory ganglia.
Some of these cells may have structures around
the transducing area which play an essential role
in modulating the receptor response. This is the
case of accessory cells in the ► Merkel,
► Ruffini, ► Meissner, or Pacinian receptor endings. For example, in the Pacinian corpuscle, the
nerve ending is surrounded by a gelatinous
structure formed from plasma membrane,
wrapped round the nerve much like the layers
of an onion. Pressure applied to the corpuscle
pushes on the layers of this structure and is
ultimately detected by the membrane of the
nerve, located deep within. The many layers of
membrane wrapped around the nerve act as a
viscoelastic filter so that sustained pressure is
transmitted to the transducing membrane as
brief and transient deformation at the onset and
the end of the stimulus.

Additionally, many mechanosensory cells
have distinctive cytoskeletal structures including,
microtubules, cilia, or actin-based filaments bundles (Sukharev and Corey 2004) as well as

Perireceptor Elements, Fig. 1 Effects of the elastoviscous solutions of different molecular weights (Synvisc: 6×10^6; Orthovisc: 1.3×10^6; Hyalgan: 0.7×10^6) on the movement-evoked impulse activity of fine primary afferents innervating normal and inflamed knee joints. The graphs in (**a, b, c**) display the averaged total number of evoked impulses of primary afferent units over the period of time shown on the abscissa and

extracellular features associated with the supposed sites of transduction. Evidence from genetics, genomics, and electrophysiological studies has suggested that mechanotransduction takes place in a mechanosensory complex formed by an ion channel coupled to extracellular and cytoskeletal proteins, and with modulating proteins (Du et al. 1996; Walker et al. 2000). Thus, proteins in the mechanosensory complex transferring mechanical energy to the ion channel also act, to some extent, as perireceptor elements.

In nociceptive endings, the influence of perireceptor events in the transmission and modulation of the stimulus is largely ignored. Nociceptors are naked nerve terminals embedded in an intercellular matrix that contains collagen fibrils and proteoglycans. In contrast with low threshold mechanoreceptors, they are devoid of specialized structures that would act as filters for the transmission of mechanical forces to the receptive sites, (Zelena 1994). Receptor sites in nociceptive terminals appear to be discontinuous patches of bare ▶ axolemma, which are separated from the adjacent tissue only by the basal lamina of the nerve fiber. Thus, the receptive plasma membrane seems to be particularly exposed to the direct action of stimuli. However, the role played by collagenous fibrils and viscous intercellular matrix surrounding nociceptive nerve endings and by proteins of the mechanosensory apparatus in the modulation of noxious stimuli is still undefined. Genetics screens for mutations in *Drosophila* revealed *painless*, a TRPA (ankyrin-repeat transient ▶ receptor potential) subfamily member. *Painless* is required for both thermal and mechanical nociception, but not for sensing light touch (Tracey et al. 2003). Indeed, the closest vertebrate homolog of *painless*, the TRPA1 channel, is expressed in a subset of

pain-sensing neurons of the mouse (Story et al. 2003). There is increasing evidence of a role for TRPA1 in noxious mechanosensation (Vilceanu and Stucky 2010; Anderson et al. 2009; Eid et al. 2008).

As mentioned above, it has been speculated that ▶ viscoelastic elements in the extracellular compartment and in the cytoskeletal domain of the receptor membrane play a role in the transmission of force, the determination of sensitivity, and the adaptation properties of mechanosensory channels (Hamill and McBride 1996). Therefore, the possibility exists that the viscoelastic characteristics of the extracellular matrix influence the functional behavior of mechanosensory channel in nociceptors. Another relevant property of nociceptors terminals, i.e., their sensitivity to exogenous and endogenous chemical agents, may be equally influenced by pericellular mechanisms that also are incompletely known. As was the case for the transmission of mechanical forces, the extracellular matrix located around nociceptive endings may modify the effect of chemicals by interfering with the diffusion and/or the accessibility of molecules to the membrane surface.

Joint nociceptors are an example of the modulation of impulse activity by extracellular perireceptor elements. In joints, hyaluronic acid, a ▶ glycosaminoglycan that fills the intercellular space between collagen fibrillar network surrounding the cells, vessels, and neural elements of the solid tissues of the joint, has been proposed to act as an elastoviscous buffer, reducing the transmission of mechanical stretch to the transducing membrane of nerve endings (Balazs 1982). In favor of this possibility is the observation that spontaneous and movement-evoked nerve impulse activity of joint nociceptors decreases in normal and inflamed joints after intra-articular injection of an elastoviscous

---

Perireceptor Elements, Fig. 1 (continued) expressed as the mean ± SEM percent of the average control responses. The *arrows* indicate the time when the test substance was injected into the knee joint. (**d**) Mean evoked impulse activity under control conditions and at different times after the injection of the test substances is expressed as percentages of the control response, which was taken as 100 %. Each column is the average of three consecutive recordings taken before the designated times. *Indicates significance at $P < 0.05$; *vertical bars* are SEM

a Cell-attached configuration

Pressure transducer

valve

negative pressure

stimulus

b

Barth's medium

I (pA)

−8mmHg/2s

Synvisc ®

I (pA)

23mm Hg/2s

⟨I⟩ Normalized

negative pressure (mmHg)

● Control Barth's medium, n=16
□ Hylan A 0.8% 6*10^6MW, n=9
△ Synvisc®, n=12
○ Hylan A 0.8% 96000MW, n=6

c Outside-out configuration

d

Compartment A Barth's medium	Compartment B Barth's medium
−60mmHg/3s	−60mmHg/3s
Compartment A Barth's medium	Compartment B HyA<0.1 M MW
−60mmHg/3s	−60mmHg/3s
Compartment A Barth's medium	Compartment B Synvisc®
−35mmHg/3s	−85mmHg/3s

Perireceptor Elements, Fig. 2 (a) Cell-attached configuration. A whole oocyte immersed in the solution. (b) *Left panel*, one example of currents recorded in control conditions and in presence of Synvisc. *Right panel*, Normalized averaged current against negative pressure applied in different conditions. (c) Recording chamber for excised patches. (d) Examples of currents from excised patches exposed to control solution or hylans with different elastoviscosities

solution of sodium hyaluronan (Pozo et al. 1997). The same inhibition was obtained with this substance in ▶ osteoarthritic pain in humans (Balazs and Denlinger 1984). In contrast, injection of non-elastoviscous solutions of low molecular mass hyaluronan did not cause a decrease of joint nociceptor activity in the cat or of joint pain in humans (Pozo et al. 1997). Moreover, high molecular weight hyaluronan solutions were more effective than lower molecular weight hyaluronan solutions in reducing the mean impulse frequency of movement-evoked discharges in pain nerve fibers in both, intact and acutely inflamed (arthritic) knee joints of the rat (Gomis et al. 2004 and Fig. 1).

Hyaluronan act as an elastoviscous filter for the transmission of force to mechanosensory channels in the cell membrane of *Xenopus laevis* ▶ oocytes "in vitro." There, elastoviscous solutions of high molecular weight reduced the response to membrane stretch of native mechanosensory channels in whole oocytes (cell-attached recording) and in isolated membrane patches (out-side out), while non-elastoviscous solutions were without effect (de la Peña et al. 2002 and Fig. 2).

Thus, it is possible that hyaluronan normally present in the surroundings of nociceptive nerve terminals acts as a perireceptor element for the modulation of the tension transmitted to mechanosensory channels in nociceptor endings of the joints and perhaps also in other tissues. Additionally, glycosaminoglycans and proteoglycans that are bound to the plasma membrane of nerve terminals and extend into the extracellular space may modify the surface charge and limit by electrostatic binding the diffusion of molecules such as inflammatory mediators to nociceptive nerve endings. Moreover, because of their ability to bind significant amounts of calcium, glycosaminoglycans may also act as an ionic buffer for this ion (Fukami 1986). In sensory fibers of the median articular nerve of guinea pig knee, it has been shown that the enhanced nerve impulse activity observed 24 h and seven days after surgical partial meniscectomy plus the transection of the lateral cruciate ligament was significantly reduced by highly elastoviscous sodium hyaluronan but also by its degraded, low molecular form, suggesting that this molecule acted not only as a mechanical filter but also as a chemical buffer for ▶ inflammatory mediators (Gomis et al. 2009). Still, the nature and the influence of perireceptor mechanisms in the modulation of ▶ sensory transduction in peripheral nociceptors remain to be defined.

References

Anderson, D. A., Gentry, C., Moss, S., et al. (2009). Clioquinol and pyrithione activate TRPA1 by increasing intracellular Zn^{2+}. *Proceedings of National Academy of Sciences USA, 106*, 8374–8379.

Balazs, E. A. (1982). The physical properties of synovial fluid and the special role of hyaluronic acid. In A. J. Helfet (Ed.), *Disorders of the knee* (pp. 61–74). Philadelphia: Lippincott Co.

Balazs, E. A., & Denlinger, J. L. (1984). The role of hyaluronic acid in arthritis and its therapeutic use. In J. C. Peyron (Ed.), *Osteoarthritis: Current clinical and fundamental problems* (pp. 165–174). Paris: Ciba Geigy.

Buck, L., & Axel, R. (1991). A novel multigene family may encode odorant receptors: A molecular basis for odor recognition. *Cell, 65*, 175–187.

De la Peña, E., Sala, S., Rovira, J. C., et al. (2002). Elastoviscous substances with analgesic effects on joint pain reduce stretch-activated ion channels activity in vitro. *Pain, 99*, 501–508.

Du, H., Gu, G., William, C. M., et al. (1996). Extracellular proteins needed for *C. elegans* mechanosensation. *Neuron, 16*, 183–94.

Eid, S. R., Crown, E. D., Moore, E. L., et al. (2008). HC-030031, a TRPA1 selective antagonist, attenuates inflammatory- and neuropathy-induced mechanical hypersensitivity. *Molecular Pain, 4*, 48.

Fukami, Y. (1986). Studies of capsule and capsular space of cat muscle spindles. *The Journal of Physiology, 376*, 281–297.

Gomis, A., Pawlak, M., Balazs, E. A., et al. (2004). Effects of different molecular weight elastoviscous hyaluronan solutions on articular nociceptive afferents. *Arthritis and Rheumatism, 50*, 314–326.

Gomis, A., Miralles, A., Schmidt, R. F., et al. (2009). Intra-articular injections of hyaluronan solutions of different elastoviscosity reduce nociceptive nerve activity in a model of osteoarthritic knee joint of the guinea pig. *Osteoarthritis and Cartilage, 17*, 798–804.

Hamill, O. P., & McBride, D. W., Jr. (1996). A supramolecular complex underlying touch sensitivity. *Trends in Neurosciences, 19*, 258–61.

Pozo, M. A., Balazs, E. A., & Belmonte, C. (1997). Reduction of sensory responses to passive movements of inflamed knee joints by hylan, a hyaluronan derivative. *Experimental Brain Research, 116*, 3–9.

Story, G. M., Peier, A. M., Reeve, A. J., et al. (2003). ANKTM1, a TRP-like channel expressed in nociceptive neurons, is activated by cold temperatures. *Cell, 112*, 819–829.

Sukharev, S., &Corey, D. P. (2004) Mechanosensitive channels: Multiplicity of families and gating paradigms. *Science STKE, 219*, re4.

Tracey, W. D., Jr., Wilson, R. I., Laurent, G., et al. (2003). Painless, a Drosophila gene essential for nociception. *Cell, 113*, 261–273.

Vilceanu, D., & Stucky, C. L. (2010). TRPA1 mediates mechanical currents in the plasma membrane of mouse sensory neurons. *PLos One, 5*, e12177.

Walker, R. G., Willingham, A. T., & Zuker, C. S. (2000). Drosophila mechanosensory transduction channel. *Science, 287*, 2229–2234.

Zelena, J. (1994). *Nerves and mechanoreceptors*. London: Chapman & Hall.

Perireceptor Events

▶ Perireceptor Elements

Perispinal Infusions

▶ Epidural Infusions in Acute Pain

Periventricular Gray

Definition

A gray matter structure at the most medial and posterior and inferior aspect of the wall of the third ventricle, just anterior to the aqueduct and the rostral extension of the periaqueductal gray. This is the region sometimes stimulated with deep brain stimulation electrodes in pain patients for stimulation-produced analgesia.

Cross-References

▶ Deep Brain Stimulation

Permanent Partial Disability

▶ Impairment Rating, Ambiguity
▶ Impairment Rating, Ambiguity, IAIABC System

Permanent Partial Impairment

▶ Impairment Rating, Ambiguity
▶ Impairment Rating, Ambiguity, IAIABC System

Permanent Restrictions

Definition

Activities or job tasks that an injured worker should never perform again.

Cross-References

▶ Pain in the Workplace, Risk Factors for Chronicity, Job Demands

Persistent Acute Pain

▶ Chronic Postsurgical Pain (CPSP)

Persistent Musculoskeletal Pain

▶ Chronic Musculoskeletal Pain

Persistent Pain

▶ Psychiatric Aspects of Pain and Dentistry

Persistent Pain Following Herpes Zoster (*shingles, varicella zoster virus, VZV*)

▶ Postherpetic Neuralgia
▶ Postherpetic Neuralgia, Pharmacological and Nonpharmacological Treatment Options

Persistent Postsurgical Pain

▶ Chronic Postsurgical Pain (CPSP)
▶ Chronic Postsurgical Pain Syndromes

Persisting Pain

▶ Acute Pain, Subacute Pain, and Chronic Pain

Person

Definition

A person is any human being with the possibility of self-awareness, which does not necessarily have to be actual, but can even be potential, as in the case of premature babies and severely ill patients.

Cross-References

▶ Ethics of Pain in the Newborn Human

Personal Factors

Definition

Personal factors are the particular background of an individual's life and living, and comprise features of the individual that are not part of a health condition or health states.

Cross-References

▶ Functioning and Disability Definitions

Personality

▶ Pain in the Workplace, Risk Factors for Chronicity, and Psychosocial Factors

Personality and Pain

Jennifer A. Haythornthwaite
Psychiatry & Behavioral Sciences, Johns Hopkins University, Baltimore, MD, USA

Synonyms

Disposition; Individual difference; Temperament; Trait

Definition

A dynamic system of behavioral, cognitive, emotional, and/or interpersonal processes that interact with situational characteristics and show consistency over time and situations.

Characteristics

Individuals come to the experience of pain, either acute or chronic, with personality characteristics – habitual styles of interpreting and responding to internal and external experiences – that influence the response to pain and adaptation when pain becomes chronic. Early work in the study of pain proposed a "pain-prone personality," the idea that some personality traits caused pain and pain-related disability. With increasing knowledge of the factors that are involved in the experience of pain, this early premise has grown in complexity and subtlety and evolved to conceptualize personality characteristics as vulnerabilities to respond to pain with certain affective, behavioral, cognitive, and interpersonal processes that promote the chronicity of pain or pain-related disability. One of the most useful conceptual models for understanding the role of personality in pain is the "diathesis-stress" model (Dersh et al. 2002). In this model, personality dimensions are "diatheses" – risk factors or vulnerabilities – that influence how the individual responds to the "stress" of pain. For example, an individual who experiences distress in the form of somatic symptoms – headaches or stomach aches – may respond to the stress of pain with hypervigilance and preoccupation with the pain. Such a response may inhibit effective coping strategies such as distraction, leading to prolonged pain and increased risk for disability. Alternatively, an individual prone to perceive threat and experience negative moods may respond to the stress of pain with catastrophizing thoughts about its meaning, controllability, and impact (Goubert et al. 2004).

Modern personality theory attempts to synthesize *dispositional* approaches, which focus on fixed, stable traits that have broad impact across situations, with *process* approaches, which focus on cognitive/emotional/behavioral/social processes that are activated by situations (Mischel and Shoda 1998). These newer approaches to personality, referred to as a cognitive-affective processing system, focus on mental experiences that are characteristically activated in distinctive patterns by specific internal and external experiences. This model nicely complements the diathesis-stress model, providing a framework for integrating the extensive information available about cognitive processes involved in the experience of pain and pain-related disability. For example, recent work identifying personality dimensions that activate cognitive processes in response to pain (Goubert et al. 2004) is consistent with this newer approach to personality that emphasizes stable *if...then* behavioral responses to situations (Mischel and Shoda, 1998).

Although an explicit assumption of this model is that personality precedes the experience of pain, most studies measure personality after the individual has already experienced pain. Since extensive work with the Minnesota Multiphasic Personality Inventory (MMPI) indicates that the experience of pain can influence the measurement of personality (Vendrig 2000) and that changes in pain with treatment correlate with changes in test scores (Fishbain et al. 2006), caution is warranted in interpreting many of the studies of personality factors and pain. Examples of important personality risk factors for pain include somatization, neuroticism or negative affectivity, and anxiety sensitivity, each of which is reviewed below. These personality dimensions are to be distinguished from personality disorders as psychopathology, which are discussed separately.

Hypochondriasis and Somatization

Hypochondriasis is a personality dimension characterized by preoccupation with the fear that one has a serious illness despite medical reassurance. Somatization is characterized by preoccupation with bodily sensations as expressions of distress, which is often expressed as illness behavior in the form of interrupting daily activities because of symptoms or seeking

treatment for these symptoms. Both hypochondriasis and somatization are characterized by preoccupation with symptoms that are typically minor, and both involve somatic amplification – sensitivity to or magnification of everyday somatic experiences. Somatization, which is more frequently studied in chronic pain, increases risk for developing chronic pain (Dersh et al. 2002), contributes to the chronicity of low back pain (Pincus et al. 2002), in part explains the high rates of psychological distress and depression that occur when people develop chronic pain (McBeth et al. 2002), and influences health-care use (Dersh et al. 2002) and disability outcomes (Laisne et al. 2012). In addition to being correlated with pain intensity, somatization and hypochondriasis improve following pain treatment (Fishbain et al. 2009).

Neuroticism or Negative Affectivity

Neuroticism is a personality dimension characterized by the propensity to experience negative emotions such as anxiety, depression, and irritability. Individuals high on neuroticism are hypervigilant to threatening internal and external cues and report greater negative mood, greater distress in response to stressful experiences, more somatic complaints, and poorer health. Some data indicate a correlation between neuroticism and reports of pain, although a detailed analysis of this relationship suggests that neuroticism may be related to ratings of pain unpleasantness more so than pain intensity (Wade et al. 1992). In the context of chronic pain, neuroticism as a personality diathesis influences cognitive appraisals of pain and coping with pain, as well as reports of mood (Affleck et al. 1992). In particular, neuroticism activates catastrophizing and fearful concerns about injury (Leeuw et al. 2007), thereby lowering the threshold for perceiving pain as threatening (Goubert et al. 2004).

Anxiety Sensitivity

Anxiety sensitivity is a personality dimension characterized by the fear that somatic anxiety symptoms (e.g., rapid heart rate) signal danger or illness – e.g., an impending heart attack (i.e., fear of fear). The role of anxiety in determining adaptation to pain, both acute and chronic, has stimulated interest in understanding personality vulnerabilities that increase risk for pain-related anxiety, fear of pain, and the resulting escape and avoidance responses often observed in highly fearful individuals. Anxiety sensitivity is a personality diathesis that predisposes individuals to experience pain-related anxiety and fear of pain (Asmundson et al. 1999), and negative mood more generally (Hadjistavropoulos et al. 2004). Anxiety sensitivity is also associated with greater pain and disability in clinical samples, as well as lower pain threshold in nonclinical samples (Ocanez et al. 2010).

Personality Disorders and Psychopathology

Personality Disorders

Personality disorders, including paranoid, obsessive-compulsive, borderline, avoidant, and passive-aggressive personality are more common in patients with chronic pain than in the general population or in other medical populations (Dersh et al. 2002). Given that much of the research on this topic measures individuals who are seeking treatment for chronic pain that has not responded to multiple treatments, the known higher rates of psychiatric comorbidity among patients seeking care for chronic painful conditions suggest caution in interpreting any causal relationship between personality disorder and the development of chronic pain. A recent epidemiologically based sample of adults assessed using structured interviews for personality disorders demonstrated a longitudinal relationship between personality disorder features and reports of pain, even after controlling for depression (Powers and Oltmanns 2012). This large, community-based study also found an association between personality disorder features and health-care utilization and medication use. The presence of a personality disorder, particularly borderline disorder, can complicate treatment and may require special intervention (Weisberg, 2000).

MMPI

Early work with the MMPI attempted to distinguish between organic and psychogenic pain,

an effort that has not been pursued in recent years as a result of increased knowledge of the complexities of emotional, cognitive, behavioral, and social factors that influence the expression of pain. The MMPI has been used to predict various outcomes, ranging from which personality type will develop pain to which type will respond, or not respond, to treatment, primarily in back pain (Vendrig 2000). While there is a strong tradition of using MMPI scales to aid in treatment planning, such as when patients are referred for psychological evaluation prior to back surgery (Epker and Block 2001), more recent systematic reviews of the available evidence suggest that depression may be the only consistent predictor of spinal cord stimulator therapy outcomes (Sparkes et al. 2010).

Summary

Individuals arrive at the experience of pain with important differences in temperament, chronic mood, and personal skills that activate cognitive, affective, behavioral, and interpersonal responses to pain. These responses influence the experience of pain and can contribute to the persistence of pain and pain-related disability, as well as influence how the individual interacts with the health-care system and adheres to treatments. Since the term "personality" often assumes trait-like or stable characteristics, it is critical to realize that many of the personality constructs measured in the scientific literature correlate with pain and improve with treatment and thus demonstrate mutability.

References

Affleck, G., Tennen, H., Urrows, S., & Higgins, P. (1992). Neutroticism and the pain-mood relation in rheumatoid arthritis: Insights from a perspective daily study. *Journal of Consulting and Clinical Psychology, 60,* 119–126.

Asmundson, G. J., Norton, P. J., & Veloso, F. (1999). Anxiety sensitivity and fear of pain in patients with recurring headaches. *Behaviour Research and Therapy, 37,* 703–713.

Dersh, J., Polatin, P. B., & Gatchel, R. J. (2002). Chronic pain and psychopathology: Research findings and theoretical considerations. *Psychosomatic Medicine, 64*, 773–786.

Epker, J., & Block, A. R. (2001). Presurgical psychological screening in back pain patients: A review. *The Clinical Journal of Pain, 17*, 200–205.

Fishbain, D. A., Coe, B., Cutler, R. B., Lewis, J., Rosomoff, H. L., & Rosomoff, R. S. (2006). Chronic pain and the measurement of personality: Do states influence traits? *Pain Medicine, 7*, 509–529.

Fishbain, D. A., Lewis, J. E., Gao, J., Cole, B., & Rosomoff, R. S. (2009). Is chronic pain associated with somatization/hypochondriasis? An evidence-based structured review. *Pain Practice, 9*, 449–467.

Goubert, L., Crombez, G., & Van Damme, S. (2004). The role of neuroticism, pain catastrophizing and pain-related fear in vigilance to pain: A structural equations approach. *Pain, 107*, 234–241.

Hadjistavropoulos, H. D., Asmundson, G. J., & Kowalyk, K. M. (2004). Measures of anxiety: Is there a difference in their ability to predict functioning at three-month follow-up among pain patients? *European Journal of Pain, 8*, 1–11.

Laisne, F., Lecomte, C., & Corbiere, M. (2012). Biopsychosocial predictors of prognosis in musculoskeletal disorders: A systematic review of the literature (corrected and republished). *Disability and Rehabilitation, 34*, 1912–1941.

Leeuw, M., Goossens, M., Linton, S. J., Crombez, G., Boersma, K., & Vlaeyen, J. (2007). The fear-avoidance model of musculoskeletal pain: Current state of scientific evidence. *Journal of Behavioral Medicine, 30*, 77–94.

McBeth, J., Macfarlane, G. J., & Silman, A. J. (2002). Does chronic pain predict future psychological distress? *Pain, 96*, 239–245.

Mischel, W., & Shoda, Y. (1998). Reconciling processing dynamics and personality dispositions. *Annual Review of Psychology, 49*, 229–258.

Ocanez, K., McHugh, R. K., & Otto, M. W. (2010). A meta-analytic review of the association between anxiety sensitivity and pain. *Depression and Anxiety, 27*, 760–767.

Pincus, T., Burton, A. K., Vogel, S., & Field, A. P. (2002). A systematic review of psychological factors as predictors of chronicity/disability in prospective cohorts of low back pain. *Spine, 27*, E109–E120.

Powers, A. D., & Oltmanns, T. F. (2012). Personality disorders and physical health: A longitudinal examination of physical functioning, healthcare utilization, and health-related behaviors in middle-aged adults. *Journal of Person Disorders, 26*, 524–538.

Sparkes, E., Raphael, J. H., Duarte, R. V., LeMarchand, K., Jackson, C., & Ashford, R. L. (2010). A systematic literature review of psychological characteristics as determinants of outcome for spinal cord stimulation therapy. *Pain, 150*, 284–289.

Vendrig, A. A. (2000). The Minnesota multiphasic personality inventory and chronic pain: A conceptual analysis of a long-standing but complicated relationship. *Clinical Psychology Review, 20*, 533–559.

Wade, J. B., Dougherty, L. M., Hart, R. P., Rafii, A., & Price, D. D. (1992). A canonical correlation analysis of the influence of neuroticism and extraversion on chronic pain, suffering, and pain behavior. *Pain, 51*, 67–73.

Weisberg, J. N. (2000). Personality and personality disorders in chronic pain. *Current Review of Pain, 4*, 60–70.

Personality Disorder

Definition

A pervasive well-established pattern of thinking and behavior that leads to distress for the individual and/or dysfunctional interpersonal relationships.

Cross-References

▶ Pain as a Cause of Psychiatric Illness

Personality Disorders and Pain

Michael R. Bond
University of Glasgow, Glasgow, UK

Synonyms

Psychosocial maladjustment and pain

Definition

1. Personality disorder is characterized by personality traits that are pervasive, inflexible, and maladaptive. They cause significant functional impairment and/or subjective distress.

2. Personality disorder is an enduring pattern of inner experience and behavior that deviates from the expectations of the individual's culture.
3. The onset of personality disorder occurs in adolescence or early adult life. The disorders are stable over time.
4. There are ten types of personality disorder, each of which has specific diagnostic criteria (DSM-IV-TR, 2000). Examples include paranoid, schizoid, antisocial, borderline, histrionic, dependent, and narcissistic.

Characteristics

A study of 200 individuals with chronic low back pain by Polatin et al. (1993) revealed that at some time in their lives, 54 % of the subjects had suffered from a major depressive disorder, 94 % had a history of substance abuse, and 95 % a history of anxiety disorder. At the time of assessment, 59 % had at least one psychiatric disorder, most often depression, but also substance abuse and/or an anxiety disorder. The criteria for a personality disorder were met by 51 % of the population studied, a surprisingly high figure. However, recent studies have not detected particular combinations of abnormal personality characteristics that give rise to greater vulnerability to chronic pain and disability. Fyer et al. (1988) reported that the presence of antisocial, borderline, and paranoid personality disorders as co-diagnoses in pain patients was associated with a poorer prognosis for recovery and a higher relapse rate than among individuals in pain but without personality disorders. Weisburg and Keefe (1997) reported that the presence of pathological personality traits is associated with reduced ability to cope with pain, a longer period in pain for any given condition, interference with treatment, and a poorer prognosis. It is clear, therefore, that both personality disorders and major psychiatric illnesses increase vulnerability to major stresses and are associated with greater levels of pain and disability, a longer duration of symptoms, and a poorer response to treatment. Finally, the presence of pain, especially chronic pain, intensifies the severity of preexisting personality traits.

References

American Psychiatric Association. (2000). *Diagnostic and statistical manual of mental disorders, 4th Revision. DSM-IV-TR*. Washington, DC: American Psychiatric Association.

Fyer, M. R., Francis, A. J., Sullivan, T., et al. (1988). Comorbidity of borderline personality disorder. *Archives of General Psychiatry, 45*, 348–352.

Polatin, P. B., Kinney, R. K., Gatchel, R. A., et al. (1993). Psychiatric illness and chronic low back pain. *Spine, 18*, 66–71.

Weisburg, J. N., & Keefe, F. J. (1997). Personality disorders in the chronic pain population: Basic concepts, empirical findings and clinical implications. *Pain Forum, 6*, 1–9.

Personally Relevant Stressful Situations

Definition

These are stress situations that are personally unique and meaningful compared to general stressors such as painful stimulation or mental arithmetic.

Cross-References

▶ Respondent Conditioning of Chronic Pain

Persuasion

Definition

The use of techniques such as information provision or argument in order to influence an individual's standing on a particular issue.

Cross-References

▶ Psychology of Pain, Self-Efficacy

Pervasiveness of Low Back Pain

▶ Low Back Pain, Epidemiology

PES

▶ TENS, Mechanisms of Action

PET

▶ Positron Emission Tomography

PET and fMRI Imaging

▶ PET and fMRI Imaging in Parietal Cortex (SI, SII, Inferior Parietal Cortex BA40)

PET and fMRI Imaging in Parietal Cortex (SI, SII, Inferior Parietal Cortex BA40)

Gary H. Duncan[1,2], Marie-Claire Albanese[3] and Mina Khoshnejad[4]
[1]Département de stomatologie, Faculté de médecine dentaire, Université de Montréal, Montreal, QC, Canada
[2]Department of Neurology and Neurosurgery, McGill University, Montreal, QC, Canada
[3]Department of Psychology, McGill University, Montreal, QC, Canada
[4]Department of Neuroscience, University of Montreal, Montreal, QC, Canada

Synonyms

fMRI imaging and PET in parietal cortex; Parietal cortex; PET and fMRI imaging

Definition

Parietal cortex approximates the superior middle third of the ▶ cerebral cortex, occupying a portion of the two ▶ hemispheres situated below the crown of the head. The parietal cortex is positioned in front of the occipital lobe and behind the frontal lobe. The human parietal cortex can be subdivided into the postcentral and posterior parietal regions; the postcentral gyrus contains the primary ▶ somatosensory cortex or SI (Brodmann areas 3, 1, and 2), while the posterior parietal region is made up of the superior (Brodmann areas 5 and 7) and inferior parietal lobules (Brodmann areas 39 and 40), which are separated by the intraparietal sulcus (Fig. 1). Medial and underneath the parietal cortex, in the superior bank of the lateral sulcus, lies the parietal operculum, defined functionally as the secondary somatosensory cortex or SII.

Characteristics

Brain imaging studies, using positron emission tomography (PET) or functional magnetic resonance imaging (fMRI), have consistently revealed activation in the parietal cortex following the presentation of noxious stimuli. A number of well-controlled studies now indicate that this parietal activity is directly associated with the human perception of acute experimental pain, regardless of the stimulus modality. For example, painful cutaneous thermal stimuli (Chen et al. 2002; Coghill et al. 2001, 2003; Strigo et al. 2003; Talbot et al. 1991), visceral stimuli (Strigo et al. 2003; Verne et al. 2003), and transcutaneous electrical nerve stimuli (Downar et al. 2003) reliably activate a common network of brain areas in the parietal lobe ▶ contralateral to the side of the stimulation. These regions of the parietal cortex (SI, SII, and inferior parietal lobule) appear to be preferentially related to processing the sensory-discriminative aspects of painful stimuli, such as quality, intensity, spatial, and temporal characteristics, rather than affective aspects of pain such as pain unpleasantness (Benuzzi et al. 2008; Hofbauer et al. 2001;

PET and fMRI Imaging in Parietal Cortex (SI, SII, Inferior Parietal Cortex BA40), Fig. 1 Schematic depiction of the lateral surface of the (*left*) cerebral hemisphere, which is subdivided into four lobes: frontal, parietal, temporal, and occipital. The central sulcus forms the anterior border of the parietal lobe, and the lateral sulcus makes the delineation between the parietal and temporal lobes. The parietal lobe is comprised of several anatomical areas, namely, the primary somatosensory cortex (*SI*), the secondary somatosensory cortex (*SII*), the superior lobule (Brodmann areas (*BA*) 5 and 7), and the inferior parietal lobule (BA 39 and 40)

Rainville et al. 1997) or empathy for pain (Bernhardt and Singer 2012; Lamm et al. 2011; Valentini 2010; Zaki et al. 2007; Jackson et al. 2005; Singer et al. 2004). However, it is now recognized that the conscious appreciation of pain sensation may involve cortical areas beyond the parietal cortex (Oshiro et al. 2007, 2009). In addition, recent studies argue that activation in the "pain network" is not specifically related to nociception and may function more generally as a salience detection network (Legrain et al. 2011; Mouraux et al. 2011) responsible for detecting potentially harmful events for the body's integrity regardless of stimulus modality.

The organization of functional units within the somatosensory cortex is a matter of dispute, especially concerning whether there is a parallel or serial mode of processing among the various units. Relatively few studies have directly examined the integration of information between the somatic sensory regions. Evidence for a serial mode of processing comes from studies showing an increased complexity in receptive field properties along the postcentral gyrus (Duffy and Burchfiel 1971; Iwamura et al. 1994; Iwamura 1998; Inui et al. 2004); likewise, a serial form of processing is suggested by evoked potential (Kitamura et al. 1997; Kakigi et al. 2000; Hu et al. 2012) and ablation studies indicate that responsiveness of SI is reduced or abolished by surgical ablation of SII and vice versa (Garraghty et al. 1990; Burton et al. 1990; Pons et al. 1992). However, one fMRI study (Liang et al. 2011) using dynamical causal modeling argues for a parallel input from thalamus to SI and SII. Likewise, one MEG study (Ploner et al. 2009) found no causal influence of SI on SII in the context of pain and thus argues for a parallel processing of noxious information.

Although activation by peripheral stimuli is predominantly observed in contralateral SI and SII, ipsilateral activation is also often observed, but in weaker magnitude. The nature of concomitant ipsilateral activations is debated; this ipsilateral activation might be due to interhemispheric integration through callosal fibers or ipsilateral pathways. A few fMRI studies looking at pain-evoked activation in callosotomized and acallosal patients (Fabri et al. 2002, 2005; Duquette et al. 2008) showed bilateral activation in SII in both healthy subjects and patient groups. Although the results from studies involving patients imply potential independence of the pain system from callosal connections, they may not be generalized to normal subjects as the activation in ipsilateral sites in the patients might be due to plasticity secondary to cortical damage.

It is important to remember that SI and SII are also involved in tactile perception, but the time courses of tactile- and pain-evoked activation show remarkable differences (Chen et al. 2002). Moreover tactile stimuli only evoke contralateral activation in SI and SII in callosotomized patients, which shows that interhemispheric transfer of non-nociceptive information depends more heavily on the integrity of the corpus callosum, particularly the posterior region (Fabri et al. 2005).

Although pain-evoked activation in SII has consistently been reported in imaging studies, there are inconsistencies regarding the activation of SI, especially in early studies (Jones et al. 1991; Apkarian et al. 1992; Timmermann et al. 2001). Such discrepancies have been attributed to uncontrolled but significant modulation of SI activity by cognitive factors like attention and suggestions (Bushnell et al. 1999) – i.e., when attention is directed away from a painful stimulus, the activity in SI is significantly reduced. Interestingly, changes in the perceived pain intensity produced through hypnotic suggestions (Rainville et al. 1999) correlated with activity in SI.

Additional evidence for the importance of the parietal cortex in the processing of intensity and location of painful stimuli comes from various case studies investigating changes in pain perception associated with lesions in SI and SII. For example, in a case report of an ischemic lesion to the right SI and SII, the patient was unable to perceive the intensity of painful heat stimuli applied to the left hand but complained that the stimuli were "clearly unpleasant" (Ploner et al. 1999). Moreover, a number of case studies have shown that lesions in SII lead to abnormalities in pain perception such as elevated mechanical and heat-evoked pain threshold (Greenspan and Winfield 1992; Greenspan et al. 1999) or episodic pain (Potagas et al. 1997) contralateral to the perceptual deficits.

In addition to the generally observed pain-related parietal activity contralateral to the area of the skin receiving the noxious stimulation, several studies (Coghill et al. 2001; Symonds et al. 2006) have demonstrated activation in the right inferior parietal cortex (Brodmann area 40) following cutaneous thermal pain, regardless of which arm received the noxious stimulation. This kind of functional lateralization in the right posterior parietal cortex is also important for the influence of attention on somatosensory processing. In fact, unilateral neglect, in which patients are not aware of the left side of their body or somatosensory stimuli applied to it, results only from a lesion in this area of the right hemisphere; a lesion of the homologous area of the left parietal lobe never results in unilateral neglect of either side of the body.

A similar network of cortical regions appears to be involved in the conscious experience of abnormal pain sensitivity, such as mechanical ▶ allodynia – a symptom of ▶ neuropathic pain. Experimental mechanical allodynia can be elicited in healthy human subjects by gently stroking with a soft brush an area of skin that has been pretreated with topical or ▶ intradermal injection of ▶ capsaicin. Using such models, brush-evoked allodynia has been associated with activation in contralateral SI, SII, and Brodmann area 40 of the inferior parietal cortex (Maihöfner et al. 2004), as well as ipsilateral SII (Iadarola et al. 1998; Maihöfner et al. 2004) and ipsilateral inferior parietal cortex (Iadarola et al. 1998). Moreover a recent meta-analysis (Lanz et al. 2011) showed that bilateral SII activation has a higher likelihood in experimental allodynia compared to normal evoked pain.

Patients who suffer from chronic pain conditions, such as fibromyalgia and chronic low back pain (CLBP), show more extensive patterns of activation in the parietal cortex. Various fMRI studies (Gracely et al. 2002; Giesecke et al. 2004) using experimental pressure pain, showed that fibromyalgia and CLBP patients exhibited an elevated level of activation in contralateral SI and bilateral SII concomitant with a higher level of reported pain, compared to that observed in the control group of healthy subjects. It also appears that poststroke central chronic pain is associated with decreased binding sites for endogenous opioids in contralateral SII and inferior parietal cortices (Willoch et al. 2004), which is consistent with the hypothesis that endogenous opioids take part in the inhibition of neural circuits subserving pain perception. Therefore, this lowered number of opioid binding sites in regions of the parietal cortex, which are directly involved in the sensory-discriminative aspects of pain, may contribute to sustained levels of pain associated with stroke.

Summary

In conclusion, functional neuroimaging data acquired with either ► PET or ► fMRI demonstrate that regions within the parietal cortex take part in processing the sensory-discriminative aspects of painful stimuli. Converging evidence from studies of healthy subjects involving either experimental pain or experimental models of chronic pain, as well as studies of chronic pain patients, suggests that activation in contralateral SI, SII, and inferior parietal cortices plays a large role in the conscious experience of pain.

References

Apkarian, A. V., Stea, R. A., Manglos, S. H., Szeverenyi, N. M., King, R. B., & Thomas, F. D. (1992). Persistent pain inhibits contralateral somatosensory cortical activity in humans. *Neuroscience Letters, 140*, 141–147.

Benuzzi, F., Lui, F., Duzzi, D., Nichelli, P. F., & Porro, C. A. (2008). Does it look painful or disgusting? Ask your parietal and cingulate cortex. *Journal of Neuroscience, 28*, 923–931.

Bernhardt, B. C., & Singer, T. (2012). The neural basis of empathy. *Annual Review of Neuroscience, 35*, 1–23.

Burton, H., Sathian, K., & Dian-hua, S. (1990). Altered responses to cutaneous stimuli in the second somatosensory cortex following lesions of the postcentral gyrus in infant and juvenile macaques. *The Journal of Comparative Neurology, 291*, 395–414.

Bushnell, M. C., Duncan, G. H., Hofbauer, R. K., et al. (1999). Pain perception: Is there a role for primary somatosensory cortex? *Proceedings of the National Academy of Sciences of the United States of America, 96*, 7705–7709.

Chen, J. I., Ha, B., Bushnell, M. C., et al. (2002). Differentiating noxious- and innocuous-related activation of human somatosensory cortices using temporal analysis of fMRI. *Journal of Neurophysiology, 88*, 464–474.

Coghill, R. C., Gilron, I., & Iadarola, M. J. (2001). Hemispheric lateralization of somatosensory processing. *Journal of Neurophysiology, 85*, 2602–2612.

Coghill, R. C., McHaffie, J. G., & Yen, Y. F. (2003). Neural correlates of interindividual differences in the subjective experience of pain. *Proceedings of the National Academy of Sciences of the United States of America, 100*, 8538–8542.

Downar, J., Mikulis, D. J., & Davis, K. D. (2003). Neural correlates of the prolonged salience of painful stimulation. *NeuroImage, 20*, 1540–1551.

Duffy, F. H., & Burchfiel, J. L. (1971). Somatosensory system: Organizational hierarchy from single units in monkey area 5. *Science, 172*, 273–275.

Duquette, M., Rainville, P., Alary, F., Lassonde, M., & Lepore, F. (2008). Ipsilateral cortical representation of tactile and painful information in acallosal and callosotomized subjects. *Neuropsychologia, 46*, 2274–2279.

Fabri, M., Polonara, G., Quattrini, A., & Salvolini, U. (2002). Mechanical noxious stimuli cause bilateral activation of parietal operculum in callosotomized subjects. *Cerebral Cortex, 12*, 446–451.

Fabri, M., Del Pesce, M., Paggi, A., Polonara, G., Bartolini, M., Salvolini, U., & Manzoni, T. (2005). Contribution of posterior corpus callosum to the interhemispheric transfer of tactile information. *Brain Research. Cognitive Brain Research, 24*, 73–80.

Garraghty, P. E., Pons, T. P., & Kaas, J. H. (1990). Ablations of areas 3b (SI proper) and 3a of somatosensory cortex in marmosets deactivate the second and parietal ventral somatosensory areas. *Somatosensory & Motor Research, 7*, 125–135.

Giesecke, T., Gracely, R. H., Grant, M. A., et al. (2004). Evidence of augmented central pain processing in idiopathic chronic low back pain. *Arthritis and Rheumatism, 50*, 613–623.

Gracely, R. H., et al. (2002). Functional magnetic resonance imaging evidence of augmented pain processing in fibromyalgia. *Arthritis and Rheumatism, 46,* 1333–1343.

Greenspan, J. D., & Winfield, J. A. (1992). Reversible pain and tactile deficits associated with a cerebral tumor compressing the posterior insula and parietal operculum. *Pain, 50,* 29–39.

Greenspan, J. D., Lee, R. R., & Lenz, F. A. (1999). Pain sensitivity alterations as a function of lesion location in the parasylvian cortex. *Pain, 81,* 273–282.

Hofbauer, R. K., Rainville, P., Duncan, G. H., et al. (2001). Cortical representation of the sensory dimension of pain. *Journal of Neurophysiology, 86,* 402–411.

Hu, L., Zhang, Z. G., & Hu, Y. (2012). A time-varying source connectivity approach to reveal human somatosensory information processing. *NeuroImage, 62,* 217–228.

Iadarola, M. J., Berman, K. F., Zeffiro, T. A., et al. (1998). Neural activation during acute capsaicin-evoked pain and allodynia assessed with PET. *Brain, 121,* 931–947.

Inui, K., Wang, X., Tamura, Y., Kaneoke, Y., & Kakigi, R. (2004). Serial processing in the human somatosensory system. *Cerebral Cortex, 14,* 851–857.

Iwamura, Y. (1998). Hierarchical somatosensory processing. *Current Opinion in Neurobiology, 8,* 522–528.

Iwamura, Y., Iriki, A., & Tanaka, M. (1994). Bilateral hand representation in the postcentral somatosensory cortex. *Nature, 369,* 554–556.

Jackson, P. L., Meltzoff, A. N., & Decety, J. (2005). How do we perceive the pain of others? A window into the neural processes involved in empathy. *NeuroImage, 24,* 771–779.

Jones, A. K., Brown, W. D., Friston, K. J., Qi, L. Y., & Frackowiak, R. S. (1991). Cortical and subcortical localization of response to pain in man using positron emission tomography. *Proceedings of the Royal Society B: Biological Sciences, 244,* 39–44.

Kakigi, R., Hoshiyama, M., Shimojo, M., Naka, D., Yamasaki, H., Watanabe, S., et al. (2000). The somatosensory evoked magnetic fields. *Progress in Neurobiology, 61,* 495–523.

Kitamura, Y., Kakigi, R., Hoshiyama, M., Koyama, S., Watanabe, S., & Shimojo, M. (1997). Pain-related somatosensory evoked magnetic fields following lower limb stimulation. *Journal of the Neurological Sciences, 145,* 187–194.

Lamm, C., Decety, J., & Singer, T. (2011). Meta-analytic evidence for common and distinct neural networks associated with directly experienced pain and empathy for pain. *NeuroImage, 54,* 2492–2502.

Lanz, S., Seifert, F., & Maihofner, C. (2011). Brain activity associated with pain, hyperalgesia and allodynia: An ALE meta-analysis. *Journal of Neural Transmission, 118,* 1139–1154.

Legrain, V., et al. (2011). The pain matrix reloaded: A salience detection system for the body. *Progress in Neurobiology, 93,* 111–124.

Liang, M., Mouraux, A., & Iannetti, G. D. (2011). Parallel processing of nociceptive and non-nociceptive somatosensory information in the human primary and secondary somatosensory cortices: evidence from dynamic causal modeling of functional magnetic resonance imaging data. *Journal of Neuroscience, 31*(24), 8976–8985.

Maihöfner, C., Schmelz, M., Forster, C., et al. (2004). Neural activation during experimental allodynia: A functional magnetic resonance imaging study. *European Journal of Neuroscience, 19,* 3211–3218.

Mouraux, A., et al. (2011). A multisensory investigation of the functional significance of the "pain matrix". *NeuroImage, 54,* 2237–2249.

Oshiro, Y., Quevedo, A. S., McHaffie, J. G., Kraft, R. A., & Coghill, R. C. (2007). Brain mechanisms supporting spatial discrimination of pain. *Journal of Neuroscience, 27,* 3388–3394.

Oshiro, Y., Quevedo, A. S., McHaffie, J. G., Kraft, R. A., & Coghill, R. C. (2009). Brain mechanisms supporting discrimination of sensory features of pain: A new model. *Journal of Neuroscience, 29,* 14924–14931.

Ploner, M., Freund, H. J., & Schnitzler, A. (1999). Pain affect without pain sensation in a patient with a postcentral lesion. *Pain, 81,* 211–214.

Ploner, M., Schoffelen, J. M., Schnitzler, A., & Gross, J. (2009). Functional integration within the human pain system as revealed by granger causality. *Human Brain Mapping, 30,* 4025–4032.

Pons, T. P., Garraghty, P. E., & Mishkin, M. (1992). Serial and parallel processing of tactual information in somatosensory cortex of rhesus monkeys. *Journal of Neurophysiology, 68,* 518–527.

Potagas, C., Avdelidis, D., Singounas, E., Missir, O., & Aessopos, A. (1997). Episodic pain associated with a tumor in the parietal operculum: A case report and literature review. *Pain, 72,* 201–208.

Rainville, P., Duncan, G. H., Price, D. D., et al. (1997). Pain affect encoded in human anterior cingulate but not somatosensory cortex. *Science, 277,* 968–971.

Rainville, P., Carrier, B., Hofbauer, R. K., Bushnell, M. C., & Duncan, G. H. (1999). Dissociation of sensory and affective dimensions of pain using hypnotic modulation. *Pain, 82,* 159–171.

Singer, T., Seymour, B., O'Doherty, J., et al. (2004). Empathy for pain involves the affective but not sensory components of pain. *Science, 303,* 1157–1162.

Strigo, I. A., Duncan, G. H., Boivin, M., & Bushnell, M. C. (2003). Differentiation of visceral and cutaneous pain in the human brain. *Journal of Neurophysiology, 89,* 3294–3303.

Symonds, L. L., et al. (2006). Right-lateralized pain processing in the human cortex: An FMRI study. *Journal of Neurophysiology, 95,* 3823–3830.

Talbot, J. D., Marrett, S., Evans, A. C., et al. (1991). Multiple representations of pain in human cerebral cortex. *Science, 251*, 1355–1358.

Timmermann, L., Ploner, M., Haucke, K., Schmitz, F., Baltissen, R., & Schnitzler, A. (2001). Differential coding of pain intensity in the human primary and secondary somatosensory cortex. *Journal of Neurophysiology, 86*, 1499–1503.

Valentini, E. (2010). The role of anterior insula and anterior cingulate in empathy for pain. *Journal of Neurophysiology, 104*, 584–586.

Verne, G. N., Himes, N. C., Robinson, M. E., et al. (2003). Central representation of visceral and cutaneous hypersensitivity in the irritable bowel syndrome. *Pain, 103*, 99–110.

Willoch, F., Schindler, F., Wester, H. J., et al. (2004). Central poststroke pain and reduced opioid receptor binding within pain processing circuitries: A [11C] diprenorphine PET study. *Pain, 108*, 213–220.

Zaki, J., Ochsner, K. N., Hanelin, J., Wager, T. D., & Mackey, S. C. (2007). Different circuits for different pain: Patterns of functional connectivity reveal distinct networks for processing pain in self and others. *Social Neuroscience, 2*, 276–291.

Pethidine

▶ Postoperative Pain, Pethidine

Petrissage

▶ Massage and Pain Relief Prospects

PGE$_2$

Definition

PGE$_2$ is one of many lipid mediators that are formed both by neurons and pro-inflammatory cells following activation of PLA2. Prostaglandins activate a number of G-protein-coupled receptors. Subsequent phosphorylation of ligand- and voltage-gated channels leads to enhanced nociceptor excitability.

Cross-References

▶ Nociceptors in the Orofacial Region (Skin/Mucosa)

PGH Synthase

▶ NSAIDs, COX-Independent Actions

PGHS

▶ NSAIDs, Mode of Action

PGHS-1

▶ COX-1 and COX-2 in Pain

PGHS-2

▶ COX-1 and COX-2 in Pain

PHA-L

Definition

Phaseolus vulgaris leucoagglutinin. A protein of the lectin group extracted from the red bean, which has a high affinity for the soma of neurons and migrates in an axonal anterograde (orthodromic) direction. It produces Golgi-like labeling of the axonal projections.

Cross-References

▶ Parabrachial Hypothalamic and Amygdaloid Projections

Phantom Limb Pain, Treatment

Herta Flor[1] and Ralf Baron[2]
[1]Department of Cognitive and Clinical
Neuroscience, Central Institute of Mental
Health, Medical Faculty Mannheim,
Heidelberg University, Mannheim, Germany
[2]Department of Neurological Pain Research and
Therapy, Christian Albrechts University Kiel,
Kiel, Germany

Synonyms

Treatment of neuropathic pain; Treatment of phantom pain

Definition

Phantom limb pain or phantom pain is defined as pain in a body part that is no longer present. It may be related to a certain position or movement of the phantom and may be elicited or exacerbated by a range of physical (e.g., changes in weather or pressure on the residual limb) and psychological factors (e.g., emotional stress). It seems to be more intense in the distal portions of the phantom and may have a number of different qualities such as stabbing, throbbing, burning, or cramping. Phantom limb pain is often confused with pain in the area adjacent to the amputated body part. This phenomenon is referred to as ▶ residual limb pain or stump pain and is usually positively correlated with phantom limb pain. In addition, ▶ postamputation pain at the site of the wound must be distinguished from pain in the residual limb and phantom limb pain, which may all co-occur in the early phase after amputation. Finally, it may be useful to assess acute and chronic ▶ preamputation pain, which was found to be related to the incidence, type, and severity of phantom limb pain in the phase following amputation.

Phantom limb pain is commonly classified as neuropathic pain and is assumed to be related to damage of peripheral neurons. Although phantom limb pain is more common after the amputation of an arm or leg, it may also occur after the surgical removal of other body parts such as a breast, the rectum, the penis, the testicles, an eye, the tongue, or the teeth. Both peripheral and central factors have been discussed as determinants of phantom limb pain. Psychological factors do not seem to contribute to the etiology of the problem but may rather affect the course and the severity of the pain. The general view today is that of multiple changes along the neuraxis contributing to the experience of phantom limb pain.

Characteristics

Several studies, including large surveys of amputees, have shown that most treatments for phantom limb pain are ineffective and fail to consider the mechanisms underlying the production of the pain (Flor 2002; Nikolajsen and Jensen 2006; Sherman et al. 1997). Most studies are uncontrolled short-term assessments of small samples of phantom limb pain patients. The maximum benefit reported from a host of treatments such as local anesthesia, sympathectomy, dorsal root entry zone lesions, cordotomy and rhizotomy, neurostimulation methods, or pharmacological interventions such as anticonvulsants, barbiturates, antidepressants, neuroleptics, and muscle relaxants seems to be around 30 % (Sherman et al. 1997). This does not exceed the placebo effect reported in other studies.

Pharmacological interventions include a host of agents, and although tricyclic antidepressants and sodium channel blockers have been indicated as treatments of choice for neuropathic pain (Nikolajsen and Jensen 2006), there are very few controlled studies for phantom limb pain. A study involving patients with phantom limb and residual limb pain found no support for the efficacy of amitriptyline (Robinson et al. 2004). Controlled studies have also been performed for opioids (Huse et al. 2001; Wu et al. 2002), calcitonin (Jaeger and Maier 1992), ketamine (Nikolajsen et al. 1996), dextromethorphan (Ben Abraham et al. 2003), and gabapentin

(Abbass 2012), all of which were found to effectively reduce phantom limb pain. Memantine, also an NMDA receptor antagonist like ketamine was, however, not effective (e.g., Maier et al. 2003), although animal studies suggest that ▶ cortical reorganization can be prevented and reversed by the use of NMDA receptor antagonists or GABA agonists. A recent Cochrane review (Alviar et al. 2011) showed that opioids, NMDA receptor antagonists, anticonvulsants, antidepressants, calcitonin, and anesthetics yielded unclear results and that morphine, gabapentin, and ketamine demonstrated trends towards short-term positive effects whereas memantine and amitriptyline were shown to have no effect.

Mechanism-based treatments are rare but have been shown to be effective in a few small but mostly uncontrolled studies. Lidocaine was found to reduce the phantom limb pain of patients with neuromas in two controlled studies (e.g., Chabal et al. 1992). In one controlled study, ▶ transcutaneous electrical nerve stimulation (TENS) yielded a small effect on phantom limb pain (Katz and Melzack 1991). ▶ Biofeedback (see also ▶ Biofeedback in the Treatment of Pain) treatments resulting in vasodilatation of the residual limb, or decreased muscle tension in the residual limb, helped to reduce phantom limb pain and seemed promising in patients where peripheral factors contributed to the pain (Sherman et al. 1997). Based on findings from brain imaging, changes in cortical reorganization might influence phantom limb pain (Flor et al. 2006). Animal work on stimulation-induced plasticity would suggest that extensive behaviorally relevant (but not passive) stimulation of a body part leads to an expansion of its representation zone. Thus, the use of a myoelectric prosthesis might be one method to influence phantom limb pain. It was shown that intensive use of a myoelectric prosthesis was positively correlated with both reduced phantom limb pain and reduced cortical reorganization (Lotze et al. 1999). When cortical reorganization was partialed out, the relationship between prosthesis use and reduced phantom limb pain was no

longer significant suggesting that cortical reorganization mediates this relationship. An alternative approach in patients where prosthesis use is not viable is the application of behaviorally relevant stimulation. A two-week training that consisted of a ▶ sensory discrimination training of electric stimuli to the stump for two hours per day led to significant improvements in phantom limb pain and a significant reversal of cortical reorganization(e.g., Flor et al. 2001). A control group of patients who received standard medical treatment and general psychological counseling in this time period did not show similar changes in cortical reorganization and phantom limb pain. The basic idea of the treatment was to provide input into the amputation zone, and thus undo the reorganizational changes that occurred subsequent to the amputation. Ramachandran and Rogers-Ramachandran (1996), who employed a virtual reality box to train patients to move the phantom limb and reduce phantom limb pain, described another behaviorally oriented approach. A mirror was placed in a box and the patients inserted both his or her intact arm and the phantom. The patient was then asked to look at the mirror image of the intact arm, which is perceived as an intact hand where the phantom used to be. The patients are then asked to make symmetric movements with both hands, thus suggesting real movement from the lost arm to the brain. This procedure seems to reestablish control over the phantom and to reduce phantom limb pain as shown in initial controlled studies (e.g., Chan et al. 2007). In addition, graded motor imagery that involves a combination of laterality recognition, imagery and mirror therapy, or motor imagery alone seem to be effective in phantom limb pain (Moseley and Flor 2012). In summary, controlled trials of effective treatment of phantom limb pain are rare.

Treatments that have been shown to be effective in controlled trials include opioids, calcitonin, ketamine, dextromethorphan, gabapentin, TENS, imagery, mirror training, motor imagery, and sensory discrimination training. For a review see Foell et al. (2011),

Flor (2008), and Weeks et al. (2010). Uncontrolled studies have reported the reduction of phantom limb pain by the use of lidocaine, biofeedback, prosthesis training, as well as motor cortex stimulation. Due to the paucity of controlled studies, it seems reasonable to base the treatment of phantom pain on recommendations for neuropathic pain in general, such as antidepressant medication other than tricyclics and calcium channel modulators including gabapentin. In addition, opioids, calcitonin, ketamine, dextromethorphan, TENS, imagery, mirror training or sensory discrimination training might be applied.

Prevention of Phantom Limb Pain

Katz and Melzack (1990) emphasized that there are somatosensory pain memories that may be revived after an amputation and lead to phantom limb pain. They have also noted that implicit and explicit memory components can be differentiated, both of which contribute to the experience of phantom limbs and phantom limb pain. They therefore suggested that both memory components needed to be targeted in preemptive analgesic trials destined to prevent the onset of phantom limb pain, i.e., that both general and spinal anesthesia were needed.

Preemptive analgesia refers to the attempt to prevent chronic pain by early intervention before acute pain occurs, e.g., before and during surgery. Based on the data on sensitization of spinal neurons by afferent barrage, it has been suggested that general anesthesia should be complemented by peripheral anesthesia, thus preventing peripheral nociceptive input from reaching the spinal cord and higher centers. However, preemptive analgesia that included both general and spinal anesthesia has not consistently been efficacious in preventing the onset of phantom limb pain (see Ypsilantis and Tang 2010). Whereas several studies reported a reduction of the incidence of phantom limb pain when additional epidural anesthesia was used in the pre- and perioperative stage, one well-controlled study failed to find a beneficial effect on phantom limb pain

(Nikolajsen et al. 1997). A preexisting pain memory that has already led to central and especially cortical changes would not necessarily be affected by a short-term elimination of afferent barrage. Thus, it is possible that peripheral analgesia would eliminate new but not preexisting central changes in the perioperative phase. Here, NMDA-antagonists as well as GABA agonists might be beneficial to prevent both central reorganization and phantom limb pain. A placebo-controlled double-blind randomized study (Wiech et al. 2001) that used the NMDA receptor antagonist memantine, in addition to brachial plexus anesthesia in patients undergoing traumatic amputations of individual fingers or a hand, found that the active drug significantly reduced the incidence of phantom limb pain one year after the surgery from 72 % to 20 %, whereas placebo failed to show a similar effect.

The development of more powerful treatments for phantom limb pain needs controlled treatment outcome, prospective, and double-blind placebo-controlled outcome research. Only then will effective evidence-based interventions be available.

References

Abbass, K. (2012). Efficacy of gabapentin for treatment of adults with phantom. *The Annals of Pharmacotherapy, 46*(12), 1707–1711.

Alviar, M. J., Hale, T., & Dungca, M. (2011). Pharmacologic interventions for treating phantom limb pain. *Cochrane Database System Review* (12), CD006380.

Ben Abraham, R., Marouani, N., & Weinbroum, A. A. (2003). Dextromethorphan mitigates phantom pain in cancer amputees. *Annals of Surgical Oncology, 10*(3), 268–274.

Chabal, C., Jacobson, L., Russell, L. C., & Burchiel, K. J. (1992). Pain response to perineuromal injection of normal saline, epinephrine, and lidocaine in humans. *Pain, 49*(1), 9–12.

Chan, B. L., Witt, R., Charrow, A. P., Magee, A., Howard, R., Pasquina, P. F., et al. (2007). Mirror therapy for phantom limb pain. *The New England Journal of Medicine, 357*(21), 2206–2207.

Flor, H. (2002). Phantom-limb pain: Characteristics, causes, and treatment. *Lancet Neurology, 1*(3), 182–189.

Flor, H. (2008). Maladaptive plasticity, memory for pain and phantom limb pain: Review and suggestions for new therapies. *Expert Review of Neurotherapeutics*, 8(5), 809–818.

Flor, H., Denke, C., Schäfer, M., & Grüsser, S. (2001). Effect of sensory discrimination training on cortical reorganisation and phantom limb pain. *The Lancet*, 357(9270), 1763–1764.

Flor, H., Nikolajsen, L., & Jensen, T. S. (2006). Phantom limb pain: A case of maladaptive CNS plasticity? *Nature Reviews. Neuroscience*, 7(11), 873–881.

Foell, J., Bekrater-Bodmann, R., Flor, H., & Cole, J. (2011). Phantom limb pain after lower limb trauma: Origins and treatments. *The International Journal of Lower Extremity Wounds*, 10(4), 224–235.

Huse, E., Larbig, W., Flor, H., & Birbaumer, N. (2001). The effect of opioids on phantom limb pain and cortical reorganization. *Pain*, 90(1–2), 47–55.

Jaeger, H., & Maier, C. (1992). Calcitonin in phantom limb pain: A double-blind study. *Pain*, 48, 21–27.

Katz, J., & Melzack, R. (1990). Pain "memories" in phantom limbs: Review and clinical observations. *Pain*, 43(3), 319–336.

Katz, J., & Melzack, R. (1991). Auricular transcutaneous electrical nerve stimulation (TENS) reduces phantom limb pain. *Journal of Pain and Symptom Management*, 6(2), 73–83.

Lotze, M., Grodd, W., Birbaumer, N., Erb, M., Huse, E., & Flor, H. (1999). Does use of a myoelectric prostesis prevent cortical reorganisation and phantom limb pain? *Nature Neuroscience*, 2(6), 501–502.

Maier, C., Dertwinkel, R., Mansourian, N., Hosbach, I., Schwenkreis, P., Senne, I., et al. (2003). Efficacy of the NMDA-receptor antagonist memantine in patients with chronic phantom limb pain – Results of a randomized double-blinded, placebo-controlled trial. *Pain*, 103(3), 277–283.

Moseley, G. L., & Flor, H. (2012). Targeting cortical representations in the treatment of chronic pain: A review. *Neurorehabilitation and Neural Repair*, 26(6), 646–652.

Nikolajsen, L., & Jensen, T. S. (2006). Phantom limb. In S. McMahon & M. Koltzenburg (Eds.), *Wall and Melzack's textbook of pain* (5th ed., pp. 961–971). London: Churchill-Livingstone.

Nikolajsen, L., Hansen, C. L., Nielsen, J., Keller, J., Arendt-Nielsen, L., & Jensen, T. S. (1996). The effect of ketamine on phantom pain: A central neuropathic disorder maintained by peripheral input. *Pain*, 67(1), 69–77.

Nikolajsen, L., Ilkjaer, S., Christensen, J. H., Kroner, K., & Jensen, T. S. (1997). Randomised trial of epidural bupivacaine and morphine in prevention of stump and phantom pain in lower-limb amputation. *The Lancet*, 350(9088), 1353–1357.

Ramachandran, V. S., & Rogers-Ramachandran, D. (1996). Synaesthesia in phantom limbs induced with mirrors. *Proceedings of the Royal Society B: Biological Sciences*, 263(1369), 377–386.

Robinson, L. R., Czerniecki, J. M., Ehde, D. M., Edwards, W. T., Judish, D. A., Goldberg, M. L., et al. (2004). Trial of amitriptyline for relief of pain in amputees: Results of a randomized controlled study. *Archives of Physical Medicine and Rehabilitation*, 85(1), 1–6.

Sherman, R. A., Devor, M., Jones, C., Katz, J., & Marbach, J. (Eds.). (1997). *Phantom limb pain – Plenum series in behavioral psychophysiology and medicine*. New York: Plenum.

Weeks, S. R., Anderson-Barnes, V. C., & Tsao, J. W. (2010). Phantom limb pain: Theories and therapies. *The Neurologist*, 16(5), 277–286.

Wiech, K., Preissl, H., Kiefer, T., Töpfner, S., Pauli, P., Grillon, C., et al. (2001). Prevention of phantom limb pain and cortical reorganization in the early phase after amputation in humans. *Society for Neuroscience Abstracts*, 163, 169.

Wu, C. L., Tella, P., Staats, P. S., Vaslav, R., Kazim, D. A., Wesselmann, U., et al. (2002). Analgesic effects of intravenous lidocaine and morphine on postamputation pain: A randomized double-blind, active placebo-controlled, crossover trial. *Anesthesiology*, 96(4), 841–848.

Ypsilantis, E., & Tang, T. Y. (2010). Pre-emptive analgesia for chronic limb pain after amputation for peripheral vascular disease: A systematic review. *Annals of Vascular Surgery*, 24(8), 1139–1146.

Phantom Sensation

Definition

Phantom sensation is a normal phenomenon after limb amputation and is experienced by virtually all amputees. The patient feels that the amputated limb is still present, and he/she may have vivid sensations in the missing limb, including feelings of movement and posture.

Cross-References

▶ Postoperative Pain, Postamputation Pain, Treatment, and Prevention

Phantom Tooth Pain

▶ Atypical Facial Pain: Etiology, Pathogenesis, and Management

Pharmacodynamics

Definition

Pharmacodynamics refers to the end-organ effects of drugs, i.e., "what the drug does to the body."

Cross-References

▶ Opioids in Geriatric Application

Pharmacogenetics

Definition

The study of the genetic factors influencing variable drug response, due to either variable pharmacodynamic or pharmacokinetic factors. The genetically variable *CYP* genes, producing P450 enzymes involved in drug metabolism, are the best known examples. This concept is also referred to as *pharmacogenomics*, with the latter term suggesting the simultaneous consideration of more than one gene.

Cross-References

▶ Opioid Analgesia, Strain Differences

Pharmacokinetic Profile of Fentanyl

Definition

Pharmacokinetic profile of fentanyl is best described by a three-compartment model.

Cross-References

▶ Postoperative Pain, Fentanyl

Pharmacokinetics

Definition

Pharmacokinetics refers to the effects of physiological processes of the body on drugs, including absorption, distribution, metabolism, and excretion (elimination, clearance), i.e., "what the body does to the drug."

Cross-References

▶ Opioids in Geriatric Application
▶ Pharmacology of Cyclooxygenase Inhibitors

Pharmacokinetics of the NSAIDs

▶ NSAIDs, Pharmacokinetics

Pharmacological Interventions

Definition

Used to prevent and manage pain and distress in infants. May include, but is not limited to, the use of opioids (e.g., morphine, fentanyl), oral analgesics (acetaminophen) and sedation (benzodiazepines, chloral hydrate), local anesthetics (lidocaine), and systemic anesthetics (ketamine). The decision to prescribe analgesia and sedation should be based on the drug, the clinical situation and setting, the person administering the drug, the physiological stability of the newborn, invasiveness of the procedure, the expected duration of drug effect, and the observed duration of pain. Subsequent doses should be modified based on multiple factors, including the cause of the pain, previous response, clinical condition, concomitant drug use, and the known properties of the sedative and analgesic drugs administered.

Cross-References

▶ Pain Assessment in Neonates

Pharmacology of Cyclooxygenase Inhibitors

Burkhard Hinz[1] and Kay Brune[2]
[1]Institute of Toxicology and Pharmacology, University of Rostock, Rostock, Germany
[2]Department of Experimental and Clinical Pharmacology and Toxicology, Friedrich Alexander University Erlangen-Nürnberg, Erlangen, Germany

Synonyms

Acetaminophen; Cyclooxygenase (COX) inhibitors; Nonsteroidal anti-inflammatory drugs (NSAIDs); Pyrazolinones; Selective COX-2 inhibitors

Definition

Cyclooxygenase (COX) inhibitors are a group of heterogenous substances including traditional nonsteroidal anti-inflammatory drugs (NSAIDs), selective COX-2 inhibitors, acetaminophen, and the pyrazolinones. This entry addresses the pharmacology of all of these drugs with particular emphasis on their rational use based on their diverse pharmacokinetic characteristics and adverse drug reaction profiles.

Characteristics

Mode of Action and Side Effects of Cyclooxygenase Inhibitors

In 1971, Vane showed that the anti-inflammatory action of nonsteroidal anti-inflammatory drugs (NSAIDs) rests in their ability to inhibit the activity of the cyclooxygenase (COX) enzyme, which in turn results in a diminished synthesis of proinflammatory prostaglandins (Vane 1971). The COX enzyme catalyzes the first step of the synthesis of prostanoids by converting arachidonic acid into prostaglandin H_2, which is the common substrate for specific prostaglandin

synthases. The enzyme is bifunctional, with fatty-acid COX activity (catalyzing the conversion of arachidonic acid to prostaglandin G_2) and prostaglandin hydroperoxidase activity (catalyzing the conversion of prostaglandin G_2 to prostaglandin H_2). In the early 1990s, COX was demonstrated to exist as two distinct isoforms (Masferrer et al. 1990; Xie et al. 1991). COX-1 is constitutively expressed as a "housekeeping" enzyme in nearly all tissues and mediates physiological responses (e.g., cytoprotection of the stomach, platelet aggregation). On the other hand, COX-2 expressed by cells that are involved in inflammation (e.g., macrophages, monocytes, synoviocytes) has emerged as the isoform that is primarily responsible for the synthesis of prostanoids involved in pathological processes, such as acute and chronic inflammatory states. The expression of the COX-2 enzyme is regulated by a broad spectrum of mediators involved in inflammation such as lipopolysaccharide, proinflammatory cytokines, and growth factors. Glucocorticoids have been reported to inhibit COX-2 expression (Masferrer et al. 1990).

Side Effects Associated with COX-1 Inhibition

All conventional NSAIDs interfere with the enzymatic activity of both COX-1 and COX-2 at therapeutic doses (Patrignani et al. 1997). In fact, many of the side effects of NSAIDs (e.g., gastrointestinal ulceration and bleeding, platelet dysfunctions) are due to a suppression of COX-1-derived prostanoids. COX-1 inhibition also confers hypersensitivity to aspirin and other chemically unrelated NSAIDs in 5–20 % of patients with chronic asthma and in an unknown fraction of patients with chronic urticaria-angioedema. Here, inhibition of COX-1 leads to activation of the lipoxygenase pathway and production of cysteinyl leukotrienes that induce bronchospasm and nasal obstruction. Asthmatic patients who are intolerant to NSAIDs produce low levels of bronchodilatatory prostaglandin E_2 (probably because of a lack of COX-2) and have increased levels of leukotriene C4 synthase and reduced levels of metabolites (lipoxins) released through the transcellular metabolism of arachidonic acid (Picado 2006).

Analgesic Action Dependent on COX-2 Inhibition

Inhibition of COX-2-derived prostanoids facilitates the anti-inflammatory, analgesic, and antipyretic effects of NSAIDs. Prostaglandins regulate the sensitivity of so-called polymodal nociceptors that are present in nearly all tissues. A significant proportion of these nociceptors cannot be easily activated by physiological stimuli such as mild pressure or some increase of temperature (Schaible and Schmidt 1988). However, following tissue trauma and subsequent release of prostaglandins, "silent" polymodal nociceptors become excitable to pressure, temperature changes, and tissue acidosis (Neugebauer et al. 1995). This process results in a phenomenon called hyperalgesia – in some instances allodynia. Prostaglandin E_2 and other inflammatory mediators facilitate the activation of tetrodotoxin-resistant Na^+ channels in dorsal root ganglion neurons (Akopian et al. 1996; England et al. 1996; Gold et al. 1996). Another important target of protein kinase A-mediated phosphorylation is the transient receptor potential vanilloid 1 (TRPV1), a nonselective cation channel of sensory neurons involved in the sensation of temperature and inflammatory pain (Lopshire and Nicol 1997; Caterina et al. 2000; Davis et al. 2000). TRPV1 responds to temperature above 40 °C and to noxious stimuli including capsaicin, the pungent component of chili peppers, and extracellular acidification. On the basis of this mechanism, prostaglandins produced during inflammatory states may significantly increase the excitability of nociceptive nerve fibers, including reactivity to temperatures below 40 °C (i.e., body temperature), thereby contributing to the activation of "sleeping" nociceptors and the development of burning pain. As such, it appears reasonable that at least a part of the peripheral antinociceptive action of COX inhibitors arises from prevention of this peripheral sensitization. Moreover, prostaglandins also act in the central nervous system to produce central hyperalgesia. Experimental data suggest that both acidic and nonacidic COX inhibitors antagonize central hyperalgesia in the dorsal horn of the spinal cord by modulating the glutamatergic signal transfer from nociceptive C fibers to secondary neurons, which propagate the signals to the higher centers of the central nervous system. Some COX-2 is expressed constitutively in the dorsal horn of the spinal cord and becomes upregulated briefly after a trauma, such as damage to a limb, in the corresponding sensory segments of the spinal cord (Beiche et al. 1996). The induction of spinal cord COX-2 expression may facilitate transmission of the nociceptive input. In line with a role of COX-2 in central pain perception, selective COX-2 inhibition has been reported to suppress inflammation-induced prostaglandin levels in cerebrospinal fluid, whereas selective inhibition of COX-1 was inactive in this regard (Smith et al. 1998). These observations were substantiated by findings showing a widespread induction of COX-2 expression in spinal cord neurons and in other regions of the central nervous system following peripheral inflammation (Samad et al. 2001). Several mechanisms have been proposed to underlie the facilitatory action of prostaglandin E_2 on central pain sensation. Accordingly, prostaglandin E_2 at relatively high concentrations has been shown to directly depolarize wide dynamic range neurons in the deep dorsal horn (Baba et al. 2001). In another study, prostaglandin E_2 at significantly lower concentrations reduced the inhibitory tone of the neurotransmitter glycine onto neurons in the superficial layers of the dorsal horn (Ahmadi et al. 2002) by phosphorylation of the specific glycine receptor subtype GlyR $\alpha3$ (Harvey et al. 2004), thereby causing a disinhibition of spinal nociceptive transmission. With respect to the underlying signaling element, prostaglandin E_2 receptors of the EP_2 receptor subtype have been identified to confer spinal inflammatory hyperalgesia (Reinold et al. 2005).

Side Effects Associated with COX-2 Inhibition

The hypothesis that selective inhibition of COX-2 might have therapeutic actions similar to those of NSAIDs, but without causing the unwanted side effects elicited by COX-1 inhibition, was the rationale for the development of selective COX-2 inhibitors. However, the simple concept of COX-2 being an exclusively proinflammatory and inducible enzyme cannot be sustained. Accordingly, COX-2 has also been shown to be expressed under basal conditions in organs including the ovary, uterus, brain, spinal cord, kidney, cartilage, bone, and even the gut, suggesting that this isozyme may play a more complex physiological role than previously recognized (for review, see Hinz and Brune 2002).

In agreement with this view, a permanent blockade of COX-2-dependent prostaglandins, including prostacyclin, is the currently most plausible explanation for the cardiovascular hazard conferred by long-term use of selective and nonselective COX-2 inhibitors. Prostacyclin, which is suppressed by over 60 % by both NSAIDs and selective COX-2 inhibitors (McAdam et al. 1999), is not only a potent inhibitor of platelet aggregation but also interferes with processes leading to hypertension, atherogenesis, and cardiac dysfunction. In this context, changes in arterial blood pressure have been proposed to underlie the long-term cardiovascular side effects of both NSAIDs and COX-2 inhibitors. The involvement of COX-2 in human renal function is supported by numerous clinical studies that showed that COX-2 inhibitors, similar to NSAIDs, can cause peripheral edema, hypertension, and exacerbation of preexisting hypertension by inhibiting water and salt excretion by the kidneys (Catella-Lawson et al. 1999; Whelton et al. 2000; Schwartz et al. 2002). These observations are of major importance given that relatively small changes in blood pressure could have a significant impact on cardiovascular events. In patients with osteoarthritis, increases in systolic blood pressure of 1–5 mmHg have been associated with 7,100–35,700 additional ischemic heart disease and stroke events over 1 year (Singh et al. 2003).

It has been suggested that both degree and time course of intravascular COX-2 inhibition might determine the differential profile of cardiovascular side effects associated with NSAIDs and COX-2 inhibitors (Hinz et al. 2006; Garcia Rodriguez et al. 2008; Hinz and Brune 2008). Claims that NSAIDs inhibit COX-1 thereby conferring cardioprotection proved wrong since platelet COX-1 activity has to be suppressed >95 % to translate into inhibition of platelet aggregation (Reilly and FitzGerald 1987). Such a complete COX-1 inhibition over the whole dosing interval is only achieved by low-dose aspirin and in some individuals by high-dose naproxen (500 mg twice daily) (Reilly and Fitz-Gerald 1987; Capone et al. 2004). Other NSAIDs such as ibuprofen and diclofenac suppress COX1 (>95 %) only at peak plasma concentrations which is consistent with an increased cardiovascular risk with high-dose diclofenac and ibuprofen, but not with naproxen (Kearney et al. 2006).

COX Inhibition in the Whole Blood Assay

Among the experimental systems developed to study COX-2 selectivity of drugs, the human whole blood assay (Patrignani et al. 1994) has emerged as the gold standard. Whole blood assays are based on the synthesis of coagulation-induced thromboxane B_2 as an index for COX-1 and on lipopolysaccharide-induced prostaglandin E_2 as an index of COX-2 activity. Ex vivo COX-2 inhibition in the whole blood assay by approximately 80 % is expected to result in analgesia due to the fact that the analgesic therapeutic plasma concentration of a COX inhibitor correlates with its inhibitory concentration $(IC)_{80}$ (the concentration that leads to 80 % inhibition) on COX-2 in the human whole blood assay (Huntjens et al. 2005). On the other hand, only a >95 % suppression of thromboxane B_2 formation translates into clinically relevant inhibition of platelet aggregation (Reilly and FitzGerald 1987), thus predicting whether a drug might cause cardioprotection (and bleeding) or not. In addition, ex vivo COX-2 inhibition measurements from whole blood have been suggested as a potential surrogate to estimate cardiovascular risk (Hinz et al. 2006; Garcia Rodriguez et al. 2008; Hinz and Brune 2008).

Nonsteroidal Anti-inflammatory Drugs (Acidic Antipyretic Analgesics)

In the 1960s, Winter developed an assay to search for drugs with anti-inflammatory activity (Winter et al. 1962). All analgesic/antipyretic/anti-inflammatory substances that survived the test of experimental pharmacology and subsequent clinical trials turned out to be acids with a high degree of lipophilic-hydrophilic polarity, similar pK_a values, and a high degree of plasma protein binding (for review, see Brune and Lanz 1985; Hinz et al. 2000; Brune et al. 2010).

Later it turned out that high concentrations of these acidic compounds are reached in the bloodstream, liver, spleen, and bone marrow (due to high protein binding and an open endothelial layer of the vasculature) but also in body compartments with acidic extracellular pH values (Brune et al. 1976). The latter type of compartments includes the inflamed tissue, the wall of the upper gastrointestinal tract, and the collecting ducts of the kidneys. Concerning gastrointestinal toxicity, there are, in fact, at least two major components contributing to NSAID's ulcerogenic action in the stomach, namely, a topical irritant effect on the epithelium and the ability to suppress prostaglandin synthesis (McCormack and Brune 1987; Price and Fletcher 1990; Wallace 1997; Somasundaram et al. 1997). Topical irritant properties are confined to acidic NSAIDs which accumulate in gastric epithelial cells because of the phenomenon of "ion trapping" (Brune et al. 1980).

Apart from aspirin, all of these compounds differ in their potency, i.e., the single dose necessary to achieve a certain degree of effect that ranges from a few milligrams (e.g., lornoxicam) to about 1 g (e.g., salicylic acid). They also differ in their pharmacokinetic characteristics. Interestingly, all traditional NSAIDs lack a relevant degree of COX-2 selectivity (Patrignani et al. 1997). The key characteristics of the most important NSAIDs are compiled in Table 1 (most data are from Brune and Lanz (1985)). This table also contains the data on aspirin which differs in many respects from the other NSAIDs and is therefore discussed in detail (see section "Compounds of Special Interest"). Otherwise, the drugs can be categorized in four different groups.

NSAIDs with Low Potency and Short Elimination Half-Life

The prototype is ibuprofen. Other drugs of this group are salicylates and mefenamic acid. The latter does not appear to offer major advantages. By contrast, this and other fenamates are rather toxic at overdosage (central nervous system).

Depending on its galenic formulation, fast or slow absorption of ibuprofen may be achieved. A fast absorption of ibuprofen was observed following administration of the respective lysine salt (Geisslinger et al. 1993). The bioavailability of ibuprofen is close to 100 %, and the elimination is always fast even in patients suffering from mild or severe impairment of liver or kidney function (Brune and Lanz 1985). Ibuprofen is used as single doses ranging from 200 mg to 1 g. A maximum dose of 3.2 g per day (United States) or 2.4 g (Europe) for rheumatoid arthritis is possible. At low doses, ibuprofen appears particularly useful for the treatment of acute occasional inflammatory pain. High doses of ibuprofen may also be administered, although with less benefit, for the treatment of chronic rheumatic diseases. Remarkably, at high doses, the otherwise harmless compound has been shown to result in an increased incidence of gastrointestinal side effects (Silverstein et al. 2000). In some countries, ibuprofen is also administered as the pure S-enantiomer, which comprises the active entity of the racemic mixture in terms of COX inhibition. On the other hand, a substantial conversion of the less-potent COX inhibitor R-ibuprofen (comprises 50 % of the usual racemic mixture) into the active S-enantiomer has been observed following administration of the racemic mixture (Rudy et al. 1991).

NSAIDs with High Potency and Short Elimination Half-Life

These drugs are predominantly prescribed for the treatment of rheumatic (arthritic) pain. The most widely used compound of this group is diclofenac. Other drugs of this group are lornoxicam, flurbiprofen, and indomethacin (very potent) but also ketoprofen and fenoprofen (less active). All of the latter drugs show a high oral bioavailability and good effectiveness but also a relatively high risk of unwanted drug effects (Henry et al. 1996).

Diclofenac appears to be less active on COX-1 as compared to COX-2 (Patrignani et al. 1997; Hinz et al. 2003). This is taken as a reason for the relatively low incidence of gastrointestinal side effects of diclofenac (Henry et al. 1996).

Pharmacology of Cyclooxygenase Inhibitors, Table 1 Physicochemical and pharmacological data of acidic antipyretic analgesics

Pharmacokinetic/chemical subclasses	pK_a	Binding to plasma proteins	Oral bioavailability	t_{max}[a]	$t_{1/2}$[b]	Single dose (maximal daily dose) for adults
Low potency/short elimination half-life						
Salicylates						
Aspirin	3.5	50–70 %	~50 % dose dependent	~15 min	~15 min	0.05–1 g[c] (~6 g)
Salicylic acid	3.0	80–95 % dose dependent	80–100 %	0.5–2 h	2.5–4.5 h dose dependent	0.5–1 g (~6 g)
2-Arylpropionic acids						
Ibuprofen	4.4	99 %	100 %	0.5–2 h	2 h	200–800 mg (2.4 g)
Anthranilic acids						
Mefenamic acid	4.2	90 %	70 %	2–4 h	1–2 h	250–500 mg (1.5 g)
High potency/short elimination half-life						
2-Arylpropionic acids						
Flurbiprofen	4.2	>99 %	No data	1.5–3 h	2.5–4(−8) h	50–100 mg (200 mg)
Ketoprofen	5.3	99 %	~90 %	1–2 h	2–4 h	25–100 mg (200 mg)
Aryl-/heteroarylacetic acids						
Diclofenac	3.9	99.7 %	~50 % dose dependent	1–12 h[d] very variable	1–2 h	25–75 mg (150 mg)
Indometacin	4.5	99 %	~100 %	0.5–2 h	2–3(−11) h[e] very variable	25–75 mg (200 mg)
Oxicams						
Lornoxicam	4.7	99 %	~100 %	0.5–2 h	4–10 h	4–12 mg (16 mg)
Intermediate potency/intermediate elimination half-life						
Salicylates						
Diflunisal	3.3	98–99 %	80–100 %	2–3 h	8–12 h dose dependent	250–500 mg (1 g)
2-Arylpropionic acids						
Naproxen	4.2	99 %	90–100 %	2–4 h	12–15 h[e]	250–500 mg (1.25 g)
Arylacetic acids						
6-Methoxy-2-naphthyl-acetic acid (active metabolite of nabumetone)	4.2	99 %	20–50 %	3–6 h	20–24 h	0.5–1 g (1.5 g)
High potency/long elimination half-life						
Oxicams						
Piroxicam	5.9	99 %	~100 %	3–5 h	14–160 h[e]	20–40 mg; initial: 40 mg
Tenoxicam	5.3	99 %	~100 %	0.5–2 h	25–175 h[e]	20–40 mg; initial: 40 mg
Meloxicam	4.08	99.5 %	89 %	7–8 h	20 h[d]	7.5–15 mg

[a]Time to reach maximum plasma concentration after oral administration
[b]Terminal half-life of elimination
[c]Single dose for inhibition of thrombocyte aggregation: 50–100 mg; single analgesic dose: 0.5–1 g
[d]Monolithic acid-resistant tablet or similar galenic form
[e]Enterohepatic circulation

The limitations of diclofenac result from its usual formulation (monolithic acid-resistant coated dragée or tablet). In fact, retention of such formulations in the stomach for hours or even days may cause retarded absorption of the active ingredient (Brune and Lanz 1985). Moreover, diclofenac has a considerable first-pass metabolism that causes its limited (about 50 %) oral bioavailability. Consequently, a lack of therapeutic effect may require adaptation of the dosage or change of the drug. New formulations (microencapsulations, salts, etc.) remedy some of these deficits (Hinz et al. 2005). The slightly higher incidence of liver toxicity associated with diclofenac may result from the high degree of first-pass metabolism, but other interpretations appear feasible.

NSAIDs with Intermediate Potency and Intermediate Elimination Half-Life

The third group is intermediate in potency and speed of elimination and comprises drugs such as naproxen and diflunisal. Because of its slow absorption, diflunisal is rarely used anymore.

The use of naproxen has been associated with a potential cardioprotective effect. Indeed, evidence suggests that continuous and regular administration of naproxen 500 mg administered twice daily can affect platelet COX-1 activity and subsequent platelet aggregation throughout the dosing interval in some, but not all patients (Capone et al. 2004). In line with this notion, a meta-analysis of 138 randomized trials (Kearney et al. 2006) concluded that the incidence of serious vascular events is similar between a COX-2 inhibitor and any non-naproxen COX inhibitor and that the risk of naproxen is in the placebo range. On the other hand, ulcer bleeds are seen more frequently with naproxen than with ibuprofen, which does not cause a long-lasting platelet inhibition (Goldstein et al. 2004).

NSAIDs with High Potency and Long Elimination Half-Life

This group consists of the oxicams (meloxicam, piroxicam, and tenoxicam). Oxicams are characterized by a high degree of enterohepatic circulation, slow metabolism, and slow elimination (Brune and Lanz 1985). Because of its long half-life (days), oxicams do not represent drugs of first choice for the treatment of acute pain of short duration. The main indication of the oxicams is inflammatory pain that persists for days, i.e., pain resulting from cancer (bone metastases) or chronic polyarthritis. The high potency and long persistence in the body may be the reason for the somewhat higher incidence of serious adverse drug effects in the gastrointestinal tract and in the kidney observed in the presence of these drugs (Henry et al. 1996).

Compounds of Special Interest

Aspirin, the prototype NSAID, deserves special discussion. This drug irreversibly inactivates both COX-1 (highly effective) and COX-2 (less effective) by acetylating an active-site serine. Consequently, this covalent modification interferes with the binding of arachidonic acid at the COX active site. Most cells compensate the enzyme loss due to acetylation by aspirin via de novo synthesis of this enzyme. However, as platelets are unable to generate fresh enzyme, a single dose of aspirin may suppress platelet COX-1-dependent thromboxane synthesis for the whole lifetime of platelets (8–11 days) until new platelets are formed.

Following oral administration, aspirin is substantially cleaved before, during, and shortly after absorption to yield salicylic acid. Consequently, the oral bioavailability is low and the plasma half-life of aspirin is only about 15 min. Aspirin may be used as a solution (effervescent) or as a (lysine) salt, allowing very fast absorption, distribution, and pain relief. Aspirin may cause bleeding from existing ulcers due to its long-lasting antiplatelet effect and topical irritation of the gastrointestinal mucosa (Kimmey 2004). The inevitable irritation of the gastric mucosa may be acceptable in otherwise healthy patients. Aspirin should not be used in pregnant women (premature bleeding, closure of ductus arteriosus) or children before puberty (Reye's syndrome) in addition to the contraindications pertinent to all NSAIDs.

When low doses of aspirin ($\leq$100 mg) are administered, aspirin acetylates the COX-1 isozyme of platelets presystemically in the portal circulation before aspirin is deacetylated to salicylate in the liver. By contrast, COX-2-dependent synthesis of vasodilatory and antithrombotic prostacyclin by vascular endothelial cells outside the gut is not altered by low-dose aspirin. The reason for this phenomenon lies in the rapid cleavage of aspirin, leaving little if any unmetabolized aspirin after primary liver passage. Thus, low-dose aspirin has its only indication in the prevention of thrombotic and embolic events.

Another problem concerns the use of low-dose aspirin together with other COX-2-selective or nonselective NSAIDs. In this context, it has been shown that the combination of low-dose aspirin with COX-2-selective inhibitors may abrogate the gastrointestinal-sparing effects of the latter compounds (Silverstein et al. 2000; Schnitzer et al. 2004). Moreover, ibuprofen and naproxen (the latter at higher than nonprescription doses) can interfere with the antiplatelet activity of low-dose aspirin when they are co-administered (Catella-Lawson et al. 2001; Capone et al. 2005). The underlying mechanism might be a competitive inhibition at the acetylation site of platelet COX-1. The clinical implication of this interaction is unclear. It is, however, potentially important because the cardioprotective effect of aspirin, when used for secondary prevention of myocardial infarction, could be decreased or negated if NSAIDs are used too. In this context, a small epidemiological study of survivors of myocardial infarction suggested that concurrent ibuprofen but not diclofenac undermined the efficacy of aspirin in preventing a second myocardial infarction (MacDonald and Wei 2003). Therefore, recommendations by the FDA (U.S. Food and Drug Administration 2006) advise patients who use immediate-release aspirin and require concurrent ibuprofen to take a single dose of 400 mg ibuprofen at least 30 min or longer after aspirin ingestion or more than 8 h before aspirin ingestion.

Selective COX-2 Inhibitors

Selective COX-2 inhibitors, also referred to as coxibs, have been developed as substances with therapeutic actions similar to those of NSAIDs but without causing gastrointestinal side effects. By definition, a substance may be regarded as a selective COX-2 inhibitor if it causes no clinically meaningful COX-1 inhibition at maximal therapeutic doses. Selective COX-2 inhibitors currently used in rheumatology are the sulfonamide celecoxib and the methylsulfone etoricoxib. Of these, celecoxib is the only COX-2 inhibitor available in the USA. Differences in physicochemical characteristics are reflected in different pharmacokinetic behavior (Table 2; for review, see Brune and Hinz 2004). Due to its very poor solubility, the absorption of celecoxib is relatively slow and incomplete; this compound undergoes considerable first-pass metabolism (20 –60 % oral bioavailability). Both factors limit celecoxib's utility for treatment of acute pain. In addition, its rate of elimination ($t_{1/2}$ ~6–12 h) appears to be highly variable (Werner et al. 2002, 2003; Hinz et al. 2006). In addition, celecoxib has been shown to inhibit the metabolism of the CYP2D6 substrate metoprolol, a widely used β-blocker (Werner et al. 2003). Etoricoxib is eliminated from the body slowly ($t_{1/2}$ ~20–26 h) and is absorbed at a fast rate, which appears to cause its fast onset of action. In fact, peak plasma concentrations of etoricoxib are reached within 1 h of administration in both healthy volunteers and patients who had undergone hip surgery (Renner et al. 2010).

The incidence of gastrointestinal side effects by selective COX-2 inhibitors as compared with conventional NSAIDs was tested in three large phase III clinical trials on a total of 35,000 patients. In the Gastrointestinal Outcomes Research (VIGOR) study (Bombardier et al. 2000) and in the Therapeutic Arthritis Research and Gastrointestinal Event Trial (TARGET) (Catella-Lawson et al. 2001), rofecoxib and lumiracoxib (both drugs were subsequently withdrawn due to cardiovascular or hepatotoxic side effects) were found to decrease the risk of confirmed gastrointestinal events (including ulcerations, bleedings, and perforations) associated with traditional NSAIDs by more than 50 %. In the Celecoxib Long-term Arthritis Safety Study

Pharmacology of Cyclooxygenase Inhibitors, Table 2 Physicochemical and pharmacological data of selective COX-2 inhibitors

	COX-1/ COX-2 ratio[a]	Binding to plasma proteins	V_d[b]	Oral bioavailability	t_{max}[c]	$t_{1/2}$[d]	Primary metabolism[e] (cytochrome P450 enzymes)	Recommended daily dose for adults
Sulfonamides								
Celecoxib	30	~97 %	455 l	20–60 %	2–4 h	6–12 h	Oxidation (CYP2C9, 3A4)[e]	200 mg (1 × 200 mg or 2 × 100 mg) for osteoarthrosis 200 mg (2 × 100 mg)–400 mg (2 × 200 mg) for rheumatoid arthritis
Methylsulfons								
Etoricoxib	344	~92 %	120 l	100 %	~1 h	20–26 h	Oxidation to 6'-hydroxymethyletoricoxib (major role: CYP3A4; ancillary role: CYP2C9, 2D6, 1A2)	30–60 mg for osteoarthrosis 90 mg for rheumatoid arthritis 120 mg for acute gouty arthritis

[a]Ratio of IC_{50} values (IC_{50} COX-1/IC_{50} COX-2) in the human whole blood assay
[b]Volume of distribution
[c]Time to reach maximum plasma concentration after oral administration
[d]Terminal half-life of elimination
[e]Compounds may inhibit CYP2D6

(CLASS), however, a significant beneficial effect of celecoxib was only evident when the definition of upper gastrointestinal endpoints was expanded to include symptomatic ulcers (Silverstein et al. 2000). Moreover, outcomes of the first 6 months were published instead of the complete 1-year data of this study.

Indeed, the risk of peptic ulcers in high-risk patients taking a COX inhibitor can be significantly reduced by concomitant administration of proton-pump inhibitors (Chan et al. 2002; Lai et al. 2005). This combination, however, does not provide protection against damage caused by COX inhibitors in the lower gastrointestinal tract. Accordingly, a double-blind, placebo-controlled trial using capsule endoscopy revealed celecoxib to be associated with considerably fewer small bowel mucosal breaks than naproxen plus omeprazole (Goldstein et al. 2005).

These data are supported by the CONDOR (Celecoxib versus omeprazole and diclofenac in patients with osteoarthritis and rheumatoid arthritis) trial that was performed on patients with osteoarthritis or rheumatoid arthritis at increased gastrointestinal risk receiving either celecoxib 200 mg twice a day or diclofenac slow release 75 mg twice a day plus omeprazole 20 mg once a day. This study has shown a lower risk of clinical outcomes throughout the gastrointestinal tract in patients treated with celecoxib than in those receiving diclofenac plus omeprazole (Chan et al. 2010).

In accordance with a decisive role of COX-1 in aspirin-induced asthma, COX-2 inhibitors are well tolerated by aspirin-sensitive asthmatic patients in several reexposure studies (Dahlen et al. 2001; Stevenson and Simon 2001; Szczeklik et al. 2001; Woessner et al. 2002).

However, these findings are as yet not seen as proof, and the product information of all COX-2 inhibitors still regards aspirin-induced asthma as a contraindication.

COX-2 inhibitors have been associated with an increased incidence of cardiovascular side effects. In fact, in placebo-controlled randomized clinical studies, rofecoxib and celecoxib were shown to increase the incidence of myocardial infarctions and other cardiovascular reactions after a prolonged period of treatment (Bresalier et al. 2005; Solomon et al. 2005). With respect to the original purpose of these studies (i.e., adeno-matous polyposis prevention study with rofecoxib [APPROVE] and adenoma prevention with celecoxib [APC]), both trials demonstrated a significant reduction in new adenoma formation associated with the use of COX-2 inhibitors in patients with a previous history of colorectal carcinomas. Moreover, in high-risk patients, short-term treatment with valdecoxib or parecoxib was associated with an increased num-ber of severe thromboembolic events (Nussmeier et al. 2005). In consequence to these observa-tions, rofecoxib (Vioxx) and valdecoxib (Bextra) were withdrawn from the market, and regulatory bodies were prompted to request changes in the labeling of both selective and nonselective COX inhibitors, including those available for over-the-counter use (U.S. Food and Drug Administration 2005). To minimize this risk, the respective sub-stances should be taken at the lowest effective dose for the shortest possible duration of treat-ment (European Medicines Agency 2005a, b).

In contrast to COX-2 inhibitors, no placebo-controlled randomized trial was designed to define the cardiovascular risk of NSAIDs. How-ever, a meta-analysis of 138 randomized trials (Kearney et al. 2006) concluded that the inci-dence of serious vascular events is similar between a COX-2 inhibitor and any non-naproxen NSAID and that the risk of naproxen is in the placebo range. The summary rate ratio for vascular events, compared with placebo, was 0.92 for naproxen, 1.51 for ibuprofen, and 1.63 for diclofenac. Furthermore, population-based nested case–control studies have shown an increased risk of myocardial infarction associated

with the current use of both COX-2 inhibitors and traditional NSAIDs (Hippisley-Cox and Coupland 2005; Singh et al. 2005). Finally, com-parable rates of thrombotic cardiovascular events have been reported for the highly selective COX-2 inhibitor etoricoxib and the traditional NSAID diclofenac in the MEDAL study program (Cannon et al. 2006), which involved ~35,000 patients with osteoarthritis or rheumatoid arthritis.

Acetaminophen

Acetaminophen (INN, paracetamol) possesses weak anti-inflammatory but analgesic and anti-pyretic activity. Acetaminophen is one of the most widely used over-the-counter antipyretic and analgesic drugs worldwide. It is recommended as first-line therapy for pain asso-ciated with osteoarthrosis (Schnitzer and Ameri-can College of Rheumatology 2002). Typical further indications of acetaminophen are fever and pain occurring in the context of viral infec-tions as well as headache. Acetaminophen is also used in children, but despite its somewhat lower toxicity in juvenile patients, fatalities due to involuntary overdosage have been reported.

Despite its improved gastrointestinal safety profile as compared to NSAIDs, acetaminophen is unique in causing a dose-dependent hepato-toxic effect: A small proportion of acetamino-phen is metabolized to the highly toxic nucleophilic N-acetyl-benzoquinoneimine that is usually inactivated by reaction with sulfhydryl groups in glutathione. However, following inges-tion of large doses of acetaminophen, hepatic glutathione is depleted, resulting in covalent binding of N-acetyl-benzoquinoneimine to DNA and structural proteins in parenchymal cells (e.g., in liver and kidney; for review, see Seeff et al. 1986). Under these circumstances, dose-depen-dent, potentially fatal hepatic necrosis may occur. When detected early, overdosage can be antago-nized within the first 12 h after intake of acetaminophen by administration of N-acetylcysteine that regenerates detoxifying mechanisms by replenishing hepatic glutathione stores. Accordingly, acetaminophen should not be given to patients with seriously impaired

Pharmacology of Cyclooxygenase Inhibitors, Table 3 Physicochemical and pharmacological data of acetaminophen and pyrazolinone derivatives

Chemical/pharmacological class	Binding to plasma proteins	Oral bioavailability	t_{max}[a]	$t_{1/2}$[b]	Single dose (maximal daily dose) for adults
Aniline derivatives					
Acetaminophen (INN, paracetamol)	5–50 % dose dependent	70–100 % dose dependent	0.5–1.5 h	1.5–2.5 h	0.5–1 g (4 g)
Pyrazolinone derivatives					
Phenazone	<10 %	~100 % dose dependent	0.5–2 h	11–12 h	0.5–1 g (4 g)
Propyphenazone	~10 %	~100 % dose dependent	0.5–1.5 h	1–2.5 h	0.5–1 g (4 g)
Dipyrone (INN, metamizol[c])	<20 %	–	–	–	0.5–1 g (4 g)
4-Methylaminophenazone[d]	58 %	~100 %	1–2 h	2–4 h	–
4-Aminophenazone[d]	48 %	–	–	4–5.5 h	–

[a]Time to reach maximum plasma concentration after oral administration
[b]Terminal half-life of elimination
[c]Noraminopyrinemethansulfonate-Na
[d]Metabolites of metamizol

liver function. As a matter of fact, excessive doses of acetaminophen are the most common cause of acute liver failure (Schilling et al. 2010). Remarkably, nearly half of acetaminophen-associated cases are caused by unintentional overdose due to taking multiple products containing this drug (Schilling et al. 2010). In addition, there is evidence that acetaminophen may still cause transient liver enzyme elevations and possibly hepatotoxicity at maximum therapeutic doses (i.e., 4 g daily), particularly in patients with risk factors (e.g., chronic alcohol use, malnutrition, concurrent use of inducers of the cytochrome P450 enzyme) (Watkins et al. 2006; Krahenbuhl et al. 2007; Larson 2007; Schwartz et al.). Accordingly, too short or insufficient pain relief associated with acetaminophen's fast elimination (Table 3) or limited potency should be considered as an additional factor leading to overdosages with this drug.

Although discovered over 100 years ago and extensively used for over 50 years, the mode of action of acetaminophen is still a matter of debate. For more than three decades, it was commonly stated that acetaminophen acts centrally and is at best a weak inhibitor of prostaglandin synthesis by COX-1 and COX-2 (Botting 2000). This concept is based on early work by Flower and Vane (Flower and Vane 1972) who showed that PG production in brain is 10 times more sensitive to inhibition by acetaminophen than that in spleen. Instead, several data published during the past few years suggest a tissue-dependent inhibitory effect of acetaminophen on the activity of both central and peripheral COX enzymes. Whereas NSAIDs and COX-2 inhibitors inhibit COX by competing with arachidonic acid for entering the COX reaction (Loll et al. 1996; Gierse et al. 1999), acetaminophen has been suggested to act as a reducing agent within the peroxidase site. In brief, acetaminophen quenches a protoporphyrin radical cation. The latter generates the tyrosine radical in the COX site which is responsible for catalyzing oxygenation of arachidonic acid (Ouellet and Percival 2001; Boutaud et al. 2002). In view of the fact that hydroperoxides oxidize the porphyrin within the peroxidase site, COX inhibition by acetaminophen is hampered by high peroxide levels. Therefore, high extracellular levels of peroxide in the inflamed tissue may also explain why acetaminophen does not suppress inflammation associated with rheumatoid arthritis (Boardman and Hart 1967;

Ring et al. 1974). On the other side, it is noteworthy that acetaminophen decreases tissue swelling following oral surgery in humans, with activity very similar to that of ibuprofen (Skjelbred and Lokken 1979; Bjornsson et al. 2003). Thus, the notion that acetaminophen has only weak anti-inflammatory properties rather than it causes no anti-inflammatory action at all appears to be more favorable.

In addition, a recent investigation indicates a preferential COX-2 inhibition by acetaminophen in the ex vivo human whole blood assay. Accordingly, oral administration of 1,000 mg acetaminophen to human volunteers was shown to inhibit blood monocyte COX-2 by more than 80 % (Hinz et al. 2008). On the other hand, a COX-1 blockade relevant for inhibition of platelet function (>95 %) was not achieved (Hinz et al. 2008). In this study, acetaminophen displayed approximately fourfold selectivity for inhibition of COX-2 both in vitro and in vivo (Hinz et al. 2008). With respect to the above mentioned "peroxide theory," further experiments revealed that acetaminophen elicits the most pronounced COX-2 inhibition in human whole blood when compared to the recombinant enzyme or freshly prepared human monocytes (Hinz et al. 2008), thus confirming the dependence of its potency as COX-2 inhibitor on the oxidant/antioxidant status of the surrounding system. As a matter of fact, human plasma comprising various enzymatic and nonenzymatic antioxidant components may provide favorable conditions in this respect. Collectively, preferential COX-2 inhibition may explain why short-term administration of acetaminophen at recommended single doses elicits no measurable toxic effect on the gastrointestinal tract (Garcia Rodriguez and Hernandez-Diaz 2001a), does not inhibit platelet function (Mielke 1981), and provokes less bronchoconstriction in aspirin-sensitive asthmatics than NSAIDs (Jenkins et al. 2004).

On the other hand, these findings raise concerns in particular to the hitherto proposed cardiovascular safety of the drug and the poorly investigated gastrointestinal safety under conditions of long-term treatment.

In contrast to the inflamed tissue, endothelial cells possess low levels of peroxide making an undisturbed COX-2 inhibition by acetaminophen possible (Boutaud et al. 2002). In line with this notion, acetaminophen was found to inhibit COX activity in human endothelial cells (O'Brien et al. 1993) and to diminish the urinary excretion of a stable prostacyclin metabolite in man (Green et al. 1989; O'Brien et al. 1993) In addition, epidemiological data found that regular consumption of acetaminophen is associated with a significantly higher relative risk for development of hypertension compared with no use (Forman et al. 2005) and that a frequent consumption of acetaminophen has nearly the same risk for major cardiovascular events as NSAIDs (Chan et al. 2006). This epidemiological data are supported by a recently published randomized, double-blind, placebo-controlled, crossover study demonstrating that acetaminophen induces a significant increase in ambulatory blood pressure in patients with coronary artery disease (Sudano et al. 2010). Collectively, these data raise concerns on the hitherto presumed cardiovascular safety of acetaminophen making further randomized trials necessary.

A favorable gastrointestinal tolerability of acetaminophen as compared to NSAIDs has been shown in a 7-day randomized trial, where healthy volunteers receiving acetaminophen 4 g per day showed no significant difference from the placebo group in terms of gastric mucosal injuries (Lanza et al. 1998). Unfortunately, no long-term endoscopy study has been performed with acetaminophen so far. However, observational studies suggest an increased gastrointestinal risk when higher doses of acetaminophen are taken for a longer period of time (Garcia Rodriguez and Hernandez-Diaz 2001a) or when acetaminophen is administered in combination with NSAIDs (Garcia Rodriguez and Hernandez-Diaz 2001b; Rahme et al. 2008). Side effects by an acetaminophen/NSAID combination were also addressed in a recently published randomized, active controlled trial on almost 900 participants with chronic knee pain most of them fulfilling ACR criteria for knee osteoarthritis (Doherty et al. 2011). On the basis of hemoglobin levels measured, the authors state that acetaminophen 3 g/day may cause similar degrees of blood loss as ibuprofen 1.2 mg/day and that the combination of the two appears to be additive or even synergistic in terms of the number of individuals with a greater than 2 g/dl decrease in hemoglobin (Doherty et al. 2011). Further investigations on acetaminophen's long-term gastrointestinal impact are therefore strongly advised.

Pyrazolinone Derivatives

Phenazone, propyphenazone, and dipyrone (INN, metamizol, water-soluble prodrug of 4-methylaminophenazone) are the predominantly used antipyretic analgesics in many countries (Latin America, many countries in Asia, Eastern and Central Europe). Pharmacokinetic key data are presented in Table 3. All nonacidic phenazone derivatives are devoid of gastrointestinal and (acute) renal toxicity.

Pharmacological aspects of dipyrone deserve special discussion. The drug is available in oral, rectal, and injectable forms. Dipyrone is indicated for severe pain conditions, in particular for those associated with smooth muscle spasm or colics affecting the gastrointestinal, biliary, or urinary tracts. Moreover, dipyrone is useful for treatment of cancer pain and migraine as well as fever refractory to other treatments.

Dipyrone has been accused of causing agranulocytosis. Although there appears to be a statistically significant link, the incidence is extremely rare (one case per million treatment periods) (Anonymous 1986; Kaufman et al. 1991; Ibáñez et al. 2005). However, a persisting debate concerning dipyrone's risk of agranulocytosis hampered further clinical investigations with the drug. Accordingly, there are no clinical studies addressing dipyrone's impact on chronic non-cancer pain (Hackenthal 1997). Likewise, animal studies suggesting a minor anti-inflammatory action of dipyrone as compared to NSAIDs (Brune and Alpermann 1983; Tatsuo et al. 1994) have not been readdressed in humans. Dipyrone has been shown to elicit a fast and pronounced inhibition of COX-2 in human blood monocytes of probands treated with the drug at pharmacological doses

(Hinz et al. 2007), supporting the view that a significant portion of dipyrone's analgesic action may be due to peripheral mechanisms. Moreover, dipyrone caused a pronounced COX-1 inhibition which was similar to that observed with various traditional NSAIDs (Hinz et al. 2007). Given the gastrointestinal safety of dipyrone shown in epidemiological studies (Laporte and Carne 1987; Laporte et al. 1991) and its apparently contradictory impact on COX-1, it was suggested that physicochemical factors (lack of acidity) rather than differences in COX-1 inhibition may be responsible for dipyrone's favorable gastric tolerability as compared to acidic NSAIDs.

References

Ahmadi, S., Lippross, S., Neuhuber, W. L., & Zeilhofer, H. U. (2002). PGE2 selectively blocks inhibitory glycinergic neurotransmission onto rat superficial dorsal horn neurons. *Nature Neuroscience, 5*, 34–40.

Akopian, A. N., Sivilotti, L., & Wood, J. N. (1996). A tetrodotoxin-resistant voltage-gated sodium channel expressed by sensory neurons. *Nature, 379*, 257–262.

Anonymous. (1986). The International Agranulocytosis and Aplastic Anemia Study. Risks of agranulocytosis and aplastic anemia. A first report of their relation to drug use with special reference to analgesics. *JAMA: The Journal of the American Medical Association, 256*, 1749–1757.

Baba, H., Kohno, T., Moore, K. A., & Woolf, C. J. (2001). Direct activation of rat spinal dorsal horn neurons by prostaglandin E2. *Journal of Neuroscience, 21*, 1750–1756.

Beiche, F., Scheuerer, S., Brune, K., Geisslinger, G., & Goppelt-Struebe, M. (1996). Up-regulation of cyclooxygenase-2 mRNA in the rat spinal cord following peripheral inflammation. *FEBS Letters, 390*, 165–169.

Bjornsson, G. A., Haanaes, H. R., & Skoglund, L. A. (2003). A randomized, double-blind crossover trial of paracetamol 1000 mg four times daily vs ibuprofen 600 mg: Effect on swelling and other postoperative events after third molar surgery. *British Journal of Clinical Pharmacology, 55*, 405–412.

Boardman, P. L., & Hart, F. D. (1967). Clinical measurement of the anti-inflammatory effects of salicylates in rheumatoid arthritis. *British Medical Journal, 4*, 264–268.

Bombardier, C., Laine, L., Reicin, A., Shapiro, D., Burgos-Vargas, R., Davis, B., Day, R., Ferraz, M. B., Hawkey, C. J., Hochberg, M. C., Kvien, T. K., Schnitzer, T. J., & Group, V. S. (2000). Comparison of upper gastrointestinal toxicity of rofecoxib and naproxen in patients with rheumatoid arthritis. VIGOR Study Group. *The New England Journal of Medicine, 343*, 1520–1528.

Botting, R. M. (2000). Mechanism of action of acetaminophen: Is there a cyclooxygenase 3? *Clinical Infectious Diseases, 31*, S202–S210.

Boutaud, O., Aronoff, D. M., Richardson, J. H., Marnett, L. J., & Oates, J. A. (2002). Determinants of the cellular specificity of acetaminophen as an inhibitor of prostaglandin H2 synthases. *Proceedings of the National Academy of Sciences of the United States of America, 99*, 7130–7135.

Bresalier, R. S., Sandler, R. S., Quan, H., Bolognese, J. A., Oxenius, B., Horgan, K., Lines, C., Riddell, R., Morton, D., Lanas, A., Konstam, M. A., Baron, J. A., & Adenomatous Polyp Prevention on Vioxx Trial I. (2005). Cardiovascular events associated with rofecoxib in a colorectal adenoma chemoprevention trial. *The New England Journal of Medicine, 352*, 1092–1102.

Brune, K., & Alpermann, H. (1983). Non-acidic pyrazoles: Inhibition of prostaglandin production, carrageenan oedema and yeast fever. *Agents and Actions, 13*, 360–363.

Brune, K., & Hinz, B. (2004). Selective cyclooxygenase-2 inhibitors: Similarities and differences. *Scandinavian Journal of Rheumatology, 33*, 1–6.

Brune, K., & Lanz, R. (1985). Pharmacokinetics of nonsteroidal anti-inflammatory drugs. In M. A. Bray, I. L. Bonta, & M. J. Parnham (Eds.), *Handbook of inflammation. The Pharmacology of Inflammation* (Vol. 5, pp. 413–449). Amsterdam: Elsevier.

Brune, K., Glatt, M., & Graf, P. (1976). Mechanisms of action of anti-inflammatory drugs. *General Pharmacology, 7*, 27–33.

Brune, K., Rainsford, K. D., & Schweitzer, A. (1980). Biodistribution of mild analgesics. *British Journal of Clinical Pharmacology, 10*, 279S–284S.

Brune, K., Renner, B., & Hinz, B. (2010). Using pharmacokinetic principles to optimize pain therapy. *Nature Reviews. Rheumatology, 6*, 589–598.

Cannon, C. P., Curtis, S. P., FitzGerald, G. A., Krum, H., Kaur, A., Bolognese, J. A., Reicin, A. S., Bombardier, C., Weinblatt, M. E., van der Heijde, D., Erdmann, E., Laine, L., & Committee, M. S. (2006). Cardiovascular outcomes with etoricoxib and diclofenac in patients with osteoarthritis and rheumatoid arthritis in the Multinational Etoricoxib and Diclofenac Arthritis Long-term (MEDAL) programme: A randomised comparison. *Lancet, 368*, 1771–1781.

Capone, M. L., Tacconelli, S., Sciulli, M. G., Grana, M., Ricciotti, E., Minuz, P., Di Gregorio, P., Merciaro, G., Patrono, C., & Patrignani, P. (2004). Clinical pharmacology of platelet, monocyte, and vascular cyclooxygenase inhibition by naproxen and low-dose aspirin in healthy subjects. *Circulation, 109*, 1468–1471.

Capone, M. L., Sciulli, M. G., Tacconelli, S., Grana, M., Ricciotti, E., Renda, G., Di Gregorio, P., Merciaro, G., & Patrignani, P. (2005). Pharmacodynamic interaction of naproxen with low-dose aspirin in healthy subjects. *Journal of the American College of Cardiology, 45*, 1295–1301.

Catella-Lawson, F., McAdam, B., Morrison, B. W., Kapoor, S., Kujubu, D., Antes, L., Lasseter, K. C., Quan, H., Gertz, B. J., & FitzGerald, G. A. (1999). Effects of specific inhibition of cyclooxygenase-2 on sodium balance, hemodynamics, and vasoactive eicosanoids. *The Journal of Pharmacology and Experimental Therapeutics, 289*, 735–741.

Catella-Lawson, F., Reilly, M. P., Kapoor, S. C., Cucchiara, A. J., DeMarco, S., Tournier, B., Vyas, S. N., & FitzGerald, G. A. (2001). Cyclooxygenase inhibitors and the antiplatelet effects of aspirin. *The New England Journal of Medicine, 345*, 1809–1817.

Caterina, M. J., Leffler, A., Malmberg, A. B., Martin, W. J., Trafton, J., Petersen-Zeitz, K. R., Koltzenburg, M., Basbaum, A. I., & Julius, D. (2000). Impaired nociception and pain sensation in mice lacking the capsaicin receptor. *Science, 288*, 306–313.

Chan, F. K., Hung, L. C., Suen, B. Y., Wu, J. C., Lee, K. C., Leung, V. K., Hui, A. J., To, K. F., Leung, W. K., Wong, V. W., Chung, S. C., & Sung, J. J. (2002). Celecoxib versus diclofenac and omeprazole in reducing the risk of recurrent ulcer bleeding in patients with arthritis. *The New England Journal of Medicine, 347*, 2104–2110.

Chan, A. T., Manson, J. E., Albert, C. M., Chae, C. U., Rexrode, K. M., Curhan, G. C., Rimm, E. B., Willett, W. C., & Fuchs, C. S. (2006). Nonsteroidal antiinflammatory drugs, acetaminophen, and the risk of cardiovascular events. *Circulation, 113*, 1578–1587.

Chan, F. K., Lanas, A., Scheiman, J., Berger, M. F., Nguyen, H., & Goldstein, J. L. (2010). Celecoxib versus omeprazole and diclofenac in patients with osteoarthritis and rheumatoid arthritis (CONDOR): A randomised trial. *Lancet, 376*, 173–179.

Dahlen, B., Szczeklik, A., Murray, J. J., & Celecoxib in Aspirin-Intolerant Asthma Study Group. (2001). Celecoxib in patients with asthma and aspirin intolerance. The Celecoxib in Aspirin-Intolerant Asthma Study Group. *The New England Journal of Medicine, 344*, 142.

Davis, J. B., Gray, J., Gunthorpe, M. J., Hatcher, J. P., Davey, P. T., Overend, P., Harries, M. H., Latcham, J., Clapham, C., Atkinson, K., Hughes, S. A., Rance, K., Grau, E., Harper, A. J., Pugh, P. L., Rogers, D. C., Bingham, S., Randall, A., & Sheardown, S. A. (2000). Vanilloid receptor-1 is essential for inflammatory thermal hyperalgesia. *Nature, 405*, 183–187.

Doherty, M., Hawkey, C., Goulder, M., Gibb, I., Hill, N., Aspley, S., & Reader, S. (2011). A randomised controlled trial of ibuprofen, paracetamol or a combination tablet of ibuprofen/paracetamol in community-derived people with knee pain. *Annals of the Rheumatic Diseases, 70*, 1534–1541.

England, S., Bevan, S., & Docherty, R. J. (1996). PGE2 modulates the tetrodotoxin-resistant sodium current in neonatal rat dorsal root ganglion neurones via the cyclic AMP-protein kinase A cascade. *Journal of Physiology -London, 495,* 429–440.

European Medicines Agency. (2005a). *Public statement: European medicines agency concludes action on COX2 inhibitors.* London, 27 June 2005.

European Medicines Agency. (2005b). *EMEA, Press release: European Medicines Agency update on non-selective NSAIDs.* London, October 17, 2005.

Flower, R. J., & Vane, J. R. (1972). Inhibition of prostaglandin synthetase in brain explains the anti-pyretic activity of paracetamol (4-acetamidophenol). *Nature, 240,* 410–411.

Forman, J. P., Stampfer, M. J., & Curhan, G. C. (2005). Non-narcotic analgesic dose and risk of incident hypertension in US women. *Hypertension, 46,* 500–507.

Garcia Rodriguez, L. A., & Hernandez-Diaz, S. (2001a). Relative risk of upper gastrointestinal complications among users of acetaminophen and nonsteroidal anti-inflammatory drugs. *Epidemiology, 12,* 570–576.

Garcia Rodriguez, L. A., & Hernandez-Diaz, S. (2001b). The risk of upper gastrointestinal complications associated with nonsteroidal anti-inflammatory drugs, glucocorticoids, acetaminophen, and combinations of these agents. *Arthritis Research & Therapy, 3,* 98–101.

Garcia Rodriguez, L. A., Tacconelli, S., & Patrignani, P. (2008). Role of dose potency in the prediction of risk of myocardial infarction associated with nonsteroidal anti-inflammatory drugs in the general population. *Journal of the American College of Cardiology, 52,* 1628–1636.

Geisslinger, G., Menzel, S., Wissel, K., & Brune, K. (1993). Single dose pharmacokinetics of different formulations of ibuprofen and aspirin. *Drug Investigation, 5,* 238–242.

Gierse, J. K., Koboldt, C. M., Walker, M. C., Seibert, K., & Isakson, P. C. (1999). Kinetic basis for selective inhibition of cyclo-oxygenases. *Biochemical Journal, 339,* 607–614.

Gold, M. S., Reichling, D. B., Shuster, M. J., & Levine, J. D. (1996). Hyperalgesic agents increase a tetrodotoxin-resistant Na+ current in nociceptors. *Proceedings of the National Academy of Sciences of the United States of America, 93,* 1108–1112.

Goldstein, J. L., Eisen, G. M., Agrawal, N., Stenson, W. F., Kent, J. D., & Verburg, K. M. (2004). Reduced incidence of upper gastrointestinal ulcer complications with the COX-2 selective inhibitor, valdecoxib. *Alimentary Pharmacology & Therapeutics, 20,* 527–538.

Goldstein, J. L., Eisen, G. M., Lewis, B., Gralnek, I. M., Zlotnick, S., Fort, J. G., & Investigators. (2005). Video capsule endoscopy to prospectively assess small bowel injury with celecoxib, naproxen plus omeprazole, and placebo. *Clinical Gastroenterology and Hepatology, 3,* 133–141.

Green, K., Drvota, V., & Vesterqvist, O. (1989). Pronounced reduction of in vivo prostacyclin synthesis in humans by acetaminophen (paracetamol). *Prostaglandins, 37,* 311–315.

Hackenthal, E. (1997). Paracetamol and metamizol in the treatment of chronic pain syndromes. *Schmerz, 11,* 269–275.

Harvey, R. J., Depner, U. B., Wassle, H., Ahmadi, S., Heindl, C., Reinold, H., Smart, T. G., Harvey, K., Schutz, B., Abo-Salem, O. M., Zimmer, A., Poisbeau, P., Welzl, H., Wolfer, D. P., Betz, H., Zeilhofer, H. U., & Muller, U. (2004). GlyR alpha3: An essential target for spinal PGE2-mediated inflammatory pain sensitization. *Science, 304,* 884–887.

Henry, D., Lim, L. L., Garcia Rodriguez, L. A., Perez Gutthann, S., Carson, J. L., Griffin, M., Savage, R., Logan, R., Moride, Y., Hawkey, C., Hill, S., & Fries, J. T. (1996). Variability in risk of gastrointestinal complications with individual non-steroidal anti-inflammatory drugs: Results of a collaborative meta-analysis. *British Medical Journal, 312,* 1563–1566.

Hinz, B., & Brune, K. (2002). Cyclooxygenase-2–10 years later. *Journal of Pharmacology and Experimental Therapeutics, 300,* 367–375.

Hinz, B., & Brune, K. (2008). Can drug removals involving cyclooxygenase-2 inhibitors be avoided? A plea for human pharmacology. *Trends in Pharmacological Sciences, 29,* 391–397.

Hinz, B., Dorn, C. P., Shen, T. Y., & Brune, K. (2000). Anti-inflammatory-antirheumatic drugs. In J. L. McGuire (Ed.), *Pharmaceuticals – Classes, therapeutic agents, areas of application* (Vol. 4, pp. 1671–1711). Weinheim: Wiley.

Hinz, B., Rau, T., Auge, D., Werner, U., Ramer, R., Rietbrock, S., & Brune, K. (2003). Aceclofenac spares cyclooxygenase 1 as a result of limited but sustained biotransformation to diclofenac. *Clinical Pharmacology and Therapeutics, 74,* 222–235.

Hinz, B., Chevts, J., Renner, B., Wuttke, H., Rau, T., Schmidt, A., Szelenyi, I., Brune, K., & Werner, U. (2005). Bioavailability of diclofenac potassium at low doses. *British Journal of Clinical Pharmacology, 59,* 80–84.

Hinz, B., Dormann, H., & Brune, K. (2006). More pronounced inhibition of cyclooxygenase 2, increase in blood pressure, and reduction of heart rate by treatment with diclofenac compared with celecoxib and rofecoxib. *Arthritis and Rheumatism, 54,* 282–291.

Hinz, B., Cheremina, O., Bachmakov, J., Renner, B., Zolk, O., Fromm, M. F., & Brune, K. (2007). Dipyrone elicits substantial inhibition of peripheral cyclooxygenases in humans: New insights into the pharmacology of an old analgesic. *The FASEB Journal, 21,* 2343–2351.

Hinz, B., Cheremina, O., & Brune, K. (2008). Acetaminophen (paracetamol) is a selective cyclooxygenase-2 inhibitor in man. *The FASEB Journal, 22,* 383–390.

Hippisley-Cox, J., & Coupland, C. (2005). Risk of myocardial infarction in patients taking cyclo-oxygenase-2 inhibitors or conventional non-steroidal anti-inflammatory drugs: Population based nested case–control analysis. *British Medical Journal, 330,* 1366.

Huntjens, D. R., Danhof, M., & Della Pasqua, O. E. (2005). Pharmacokinetic-pharmacodynamic correlations and biomarkers in the development of COX-2 inhibitors. *Rheumatology, 44*, 846–859.

Ibanez, L., Vidal, X., Ballarin, E., & Laporte, J. R. (2005). Agranulocytosis associated with dipyrone (metamizol). *European Journal of Clinical Pharmacology, 60*, 821–829.

Jenkins, C., Costello, J., & Hodge, L. (2004). Systematic review of prevalence of aspirin induced asthma and its implications for clinical practice. *British Medical Journal, 328*, 434.

Kaufman, D. W., Kelly, J. P., Levy, M., & Shapiro, S. (1991). *The drug etiology of agranulocytosis an aplastic anemia. Monographs in epidemiology and biostatistics* (Vol. 18). New York: Oxford University Press.

Kearney, P. M., Baigent, C., Godwin, J., Halls, H., Emberson, J. R., & Patrono, C. (2006). Do selective cyclo-oxygenase-2 inhibitors and traditional non-steroidal anti-inflammatory drugs increase the risk of atherothrombosis? Meta-analysis of randomised trials. *British Medical Journal, 332*, 1302–1308.

Kimmey, M. B. (2004). Cardioprotective effects and gastrointestinal risks of aspirin: Maintaining the delicate balance. *The American Journal of Medicine, 117*, 72S–78S.

Krahenbuhl, S., Brauchli, Y., Kummer, O., Bodmer, M., Trendelenburg, M., Drewe, J., & Haschke, M. (2007). Acute liver failure in two patients with regular alcohol consumption ingesting paracetamol at therapeutic dosage. *Digestion, 75*, 232–237.

Lai, K. C., Chu, K. M., Hui, W. M., Wong, B. C., Hu, W. H., Wong, W. M., Chan, A. O., Wong, J., & Lam, S. K. (2005). Celecoxib compared with lansoprazole and naproxen to prevent gastrointestinal ulcer complications. *The American Journal of Medicine, 118*, 1271–1278.

Lanza, F. L., Codispoti, J. R., & Nelson, E. B. (1998). An endoscopic comparison of gastroduodenal injury with over-the-counter doses of ketoprofen and acetaminophen. *The American Journal of Gastroenterology, 93*, 1051–1054.

Laporte, J. R., & Carne, X. (1987). Blood dyscrasias and the relative safety of non-narcotic analgesics. *Lancet, 1*, 809.

Laporte, J. R., Carne, X., Vidal, X., Moreno, V., & Juan, J. (1991). Upper gastrointestinal bleeding in relation to previous use of analgesics and non-steroidal anti-inflammatory drugs. Catalan Countries Study on Upper Gastrointestinal Bleeding. *Lancet, 337*, 85–89.

Larson, A. M. (2007). Acetaminophen hepatotoxicity. *Clinics in Liver Disease, 11*, 525–548.

Loll, P. J., Picot, D., Ekabo, O., & Garavito, R. M. (1996). Synthesis and use of iodinated nonsteroidal antiinflammatory drug analogs as crystallographic probes of the prostaglandin H2 synthase cyclooxygenase active site. *Biochemistry, 35*, 7330–7340.

Lopshire, J. C., & Nicol, G. D. (1997). Activation and recovery of the PGE2-mediated sensitization of the capsaicin response in rat sensory neurons. *Journal of Neurophysiology, 78*, 3154–3164.

MacDonald, T. M., & Wei, L. (2003). Effect of ibuprofen on cardioprotective effect of aspirin. *Lancet, 361*, 573–574.

Masferrer, J. L., Zweifel, B. S., Seibert, K., & Needleman, P. (1990). Selective regulation of cellular cyclooxygenase by dexamethasone and endotoxin in mice. *The Journal of Clinical Investigation, 86*, 1375–1379.

McAdam, B. F., Catella-Lawson, F., Mardini, I. A., Kapoor, S., Lawson, J. A., & FitzGerald, G. A. (1999). Systemic biosynthesis of prostacyclin by cyclooxygenase (COX)-2: The human pharmacology of a selective inhibitor of COX-2. *Proceedings of the National Academy of Sciences of the United States of America, 96*, 272–277.

McCormack, K., & Brune, K. (1987). Classical absorption theory and the development of gastric mucosal damage associated with the non-steroidal anti-inflammatory drugs. *Archives of Toxicology, 60*, 261–269.

Mielke, C. H., Jr. (1981). Comparative effects of aspirin and acetaminophen on hemostasis. *Archives of Internal Medicine, 141*, 305–310.

Neugebauer, V., Geisslinger, G., Rumenapp, P., Weiretter, F., Szelenyi, I., Brune, K., & Schaible, H. G. (1995). Antinociceptive effects of R(-)- and S (+)-flurbiprofen on rat spinal dorsal horn neurons rendered hyperexcitable by an acute knee joint inflammation. *Journal of Pharmacology and Experimental Therapeutics, 275*, 618–628.

Nussmeier, N. A., Whelton, A. A., Brown, M. T., Langford, R. M., Hoeft, A., Parlow, J. L., Boyce, S. W., & Verburg, K. M. (2005). Complications of the COX-2 inhibitors parecoxib and valdecoxib after cardiac surgery. *The New England Journal of Medicine, 352*, 1081–1091.

O'Brien, W. F., Krammer, J., O'Leary, T. D., & Mastrogiannis, D. S. (1993). The effect of acetaminophen on prostacyclin production in pregnant women. *American Journal of Obstetrics and Gynecology, 168*, 1164–1169.

Ouellet, M., & Percival, M. D. (2001). Mechanism of acetaminophen inhibition of cyclooxygenase isoforms. *Archives of Biochemistry and Biophysics, 387*, 273–280.

Patrignani, P., Panara, M. R., Greco, A., Fusco, O., Natoli, C., Iacobelli, S., Cipollone, F., Ganci, A., Creminon, C., Maclouf, J., et al. (1994). Biochemical and pharmacological characterization of the cyclooxygenase activity of human blood prostaglandin endoperoxide synthases. *Journal of Pharmacology and Experimental Therapeutics, 271*, 1705–1712.

Patrignani, P., Panara, M. R., Sciulli, M. G., Santini, G., Renda, G., & Patrono, C. (1997). Differential inhibition of human prostaglandin endoperoxide synthase-1 and -2 by nonsteroidal anti-inflammatory drugs. *Journal of Physiology and Pharmacology, 48*, 623–631.

Picado, C. (2006). Mechanisms of aspirin sensitivity. *Current Allergy and Asthma Reports, 6*, 198–202.

Price, A. H., & Fletcher, M. (1990). Mechanisms of NSAID-induced gastroenteropathy. *Drugs, 40*, 1–11.

Rahme, E., Barkun, A., Nedjar, H., Gaugris, S., & Watson, D. (2008). Hospitalizations for upper and lower GI events associated with traditional NSAIDs and acetaminophen among the elderly in Quebec, Canada. *The American Journal of Gastroenterology, 103*, 872–882.

Reilly, I. A., & FitzGerald, G. A. (1987). Inhibition of thromboxane formation in vivo and ex vivo: Implications for therapy with platelet inhibitory drugs. *Blood, 69*, 180–186.

Reinold, H., Ahmadi, S., Depner, U. B., Layh, B., Heindl, C., Hamza, M., Pahl, A., Brune, K., Narumiya, S., Muller, U., & Zeilhofer, H. U. (2005). Spinal inflammatory hyperalgesia is mediated by prostaglandin E receptors of the EP2 subtype. *The Journal of Clinical Investigation, 115*, 673–679.

Renner, B., Zacher, J., Buvanendran, A., Walter, G., Strauss, J., & Brune, K. (2010). Absorption and distribution of etoricoxib in plasma, CSF, and wound tissue in patients following hip surgery–A pilot study. *Naunyn-Schmiedeberg's Archives of Pharmacology, 381*, 127–136.

Ring, E. F., Collins, A. J., Bacon, P. A., & Cosh, J. A. (1974). Quantitation of thermography in arthritis using multi-isothermal analysis. II. Effect of nonsteroidal anti-inflammatory therapy on the thermographic index. *Annals of the Rheumatic Diseases, 33*, 353–356.

Rudy, A. C., Knight, P. M., Brater, D. C., & Hall, S. D. (1991). Stereoselective metabolism of ibuprofen in humans: Administration of R-, S- and racemic ibuprofen. *Journal of Pharmacology and Experimental Therapeutics, 259*, 1133–1139.

Samad, T. A., Moore, K. A., Sapirstein, A., Billet, S., Allchorne, A., Poole, S., Bonventre, J. V., & Woolf, C. J. (2001). Interleukin-1beta-mediated induction of Cox-2 in the CNS contributes to inflammatory pain hypersensitivity. *Nature, 410*, 471–475.

Schaible, H. G., & Schmidt, R. F. (1988). Time course of mechanosensitivity changes in articular afferents during a developing experimental arthritis. *Journal of Neurophysiology, 60*, 2180–2195.

Schilling, A., Corey, R., Leonard, M., & Eghtesad, B. (2010). Acetaminophen: Old drug, new warnings. *Cleveland Clinic Journal of Medicine, 77*, 19–27.

Schnitzer, T. J., & American College of Rheumatology. (2002). Update of ACR guidelines for osteoarthritis: Role of the coxibs. *Journal of Pain and Symptom Management, 23*, S24–S30.

Schnitzer, T. J., Burmester, G. R., Mysler, E., Hochberg, M. C., Doherty, M., Ehrsam, E., Gitton, X., Krammer, G., Mellein, B., Matchaba, P., Gimona, A., Hawkey, C. J., & Group, T. S. (2004). Comparison of lumiracoxib with naproxen and ibuprofen in the Therapeutic Arthritis Research and Gastrointestinal Event Trial (TARGET), reduction in ulcer complications: Randomised controlled trial. *Lancet, 364*, 665–674.

Schwartz, J., Stravitz, T., Lee, W. M., & American Association for the Study of Liver Disease Study Group. *AASLD position on acetaminophen.* www.aasld.org/about/publicpolicy/Documents/Public%2520Policy%2520Documents/AcetaminophenPosition.pdf

Schwartz, J. I., Vandormael, K., Malice, M. P., Kalyani, R. N., Lasseter, K. C., Holmes, G. B., Gertz, B. J., Gottesdiener, K. M., Laurenzi, M., Redfern, K. J., & Brune, K. (2002). Comparison of rofecoxib, celecoxib, and naproxen on renal function in elderly subjects receiving a normal-salt diet. *Clinical Pharmacology and Therapeutics, 72*, 50–61.

Seeff, L. B., Cuccherini, B. A., Zimmerman, H. J., Adler, E., & Benjamin, S. B. (1986). Acetaminophen hepatotoxicity in alcoholics. A therapeutic misadventure. *Annals of Internal Medicine, 104*, 399–404.

Silverstein, F. E., Faich, G., Goldstein, J. L., Simon, L. S., Pincus, T., Whelton, A., Makuch, R., Eisen, G., Agrawal, N. M., Stenson, W. F., Burr, A. M., Zhao, W. W., Kent, J. D., Lefkowith, J. B., Verburg, K. M., & Geis, G. S. (2000). Gastrointestinal toxicity with celecoxib vs nonsteroidal anti-inflammatory drugs for osteoarthritis and rheumatoid arthritis: The CLASS study: A randomized controlled trial. Celecoxib Long-term Arthritis Safety Study. *JAMA: The Journal of the American Medical Association, 284*, 1247–1255.

Singh, G., Miller, J. D., Huse, D. M., Pettitt, D., D'Agostino, R. B., & Russell, M. W. (2003). Consequences of increased systolic blood pressure in patients with osteoarthritis and rheumatoid arthritis. *The Journal of Rheumatology, 30*, 714–719.

Singh, G., Mithal, A., Triadafilopoulus, G. (2005). Both selective COX2 inhibitors and non-selective NSAIDs increase the risk of acute myocardial infarction in patients with arthritis: Selectivity is with the patient, not with the drug class. *Annual European Congress of Rheumatology, Vienna*, June 8–11 2005; (Abstract OP0091).

Skjelbred, P., & Lokken, P. (1979). Paracetamol versus placebo: Effects on post-operative course. *European Journal of Clinical Pharmacology, 15*, 27–33.

Smith, C. J., Zhang, Y., Koboldt, C. M., Muhammad, J., Zweifel, B. S., Shaffer, A., Talley, J. J., Masferrer, J. L., Seibert, K., & Isakson, P. C. (1998). Pharmacological analysis of cyclooxygenase-1 in inflammation. *Proceedings of the National Academy of Sciences of the United States of America, 95*, 13313–13318.

Solomon, S. D., McMurray, J. J., Pfeffer, M. A., Wittes, J., Fowler, R., Finn, P., Anderson, W. F., Zauber, A., Hawk, E., Bertagnolli, M., & Adenoma Prevention with Celecoxib Study I. (2005). Cardiovascular risk associated with celecoxib in a clinical trial for colorectal adenoma prevention. *The New England Journal of Medicine, 352*, 1071–1080.

Somasundaram, S., Rafi, S., Hayllar, J., Sigthorsson, G., Jacob, M., Price, A. B., Macpherson, A., Mahmod, T., Scott, D., Wrigglesworth, J. M., & Bjarnason, I. (1997). Mitochondrial damage: A possible mechanism of the "topical" phase of NSAID induced injury to the rat intestine. *Gut, 41*, 344–353.

Stevenson, D. D., & Simon, R. A. (2001). Lack of cross-reactivity between rofecoxib and aspirin in aspirin-sensitive patients with asthma. *The Journal of Allergy and Clinical Immunology, 108*, 47–51.

Sudano, I., Flammer, A. J., Periat, D., Enseleit, F., Hermann, M., Wolfrum, M., Hirt, A., Kaiser, P., Hurlimann, D., Neidhart, M., Gay, S., Holzmeister, J., Nussberger, J., Mocharla, P., Landmesser, U., Haile, S. R., Corti, R., Vanhoutte, P. M., Luscher, T. F., Noll, G., & Ruschitzka, F. (2010). Acetaminophen increases blood pressure in patients with coronary artery disease. *Circulation, 122*, 1789–1796.

Szczeklik, A., Nizankowska, E., Bochenek, G., Nagraba, K., Mejza, F., & Swierczynska, M. (2001). Safety of a specific COX-2 inhibitor in aspirin-induced asthma. *Clinical and Experimental Allergy, 31*, 219–225.

Tatsuo, M. A., Carvalho, W. M., Silva, C. V., Miranda, A. E., Ferreira, S. H., & Francischi, J. N. (1994). Analgesic and antiinflammatory effects of dipyrone in rat adjuvant arthritis model. *Inflammation, 18*, 399–405.

U.S. Food and Drug Administration. (2005). *FDA announces series of changes to the class of marketed Non-Steroidal Anti-Inflammatory Drugs (NSAIDs)*. Retrieved April 7, 2005, from http://www.fda.gov/NewsEvents/Newsroom/PressAnnouncements/2005/ucm108427.htm

U.S. Food and Drug Administration. (2006). *FDA Science Paper: Concomitant use of ibuprofen and aspirin: Potential for attenuation of the anti-platelet effect of aspirin*. Retrieved August 9, 2006, from http://www.fda.gov/cder/drug/infopage/ibuprofen/default.htm

Vane, J. R. (1971). Inhibition of prostaglandin synthesis as a mechanism of action for aspirin-like drugs. *Nature, 231*, 232–235.

Wallace, J. L. (1997). Nonsteroidal anti-inflammatory drugs and gastroenteropathy: The second hundred years. *Gastroenterology, 112*, 1000–1016.

Watkins, P. B., Kaplowitz, N., Slattery, J. T., Colonese, C. R., Colucci, S. V., Stewart, P. W., & Harris, S. C. (2006). Aminotransferase elevations in healthy adults receiving 4 grams of acetaminophen daily: A randomized controlled trial. *JAMA: The Journal of the American Medical Association, 296*, 87–93.

Werner, U., Werner, D., Pahl, A., Mundkowski, R., Gillich, M., & Brune, K. (2002). Investigation of the pharmacokinetics of celecoxib by liquid chromatography-mass spectrometry. *Biomedical Chromatography, 16*, 56–60.

Werner, U., Werner, D., Rau, T., Fromm, M. F., Hinz, B., & Brune, K. (2003). Celecoxib inhibits metabolism of cytochrome P450 2D6 substrate metoprolol in humans. *Clinical Pharmacology and Therapeutics, 74*, 130–137.

Whelton, A., Maurath, C. J., Verburg, K. M., & Geis, G. S. (2000). Renal safety and tolerability of celecoxib, a novel cyclooxygenase-2 inhibitor. *American Journal of Therapeutics, 7*, 159–175.

Winter, C. A., Risley, E. A., & Nuss, G. W. (1962). Carrageenin-induced edema in hind paw of the rat as an assay for antiiflammatory drugs. *Proceedings of the Society for Experimental Biology and Medicine, 111*, 544–547.

Woessner, K. M., Simon, R. A., & Stevenson, D. D. (2002). The safety of celecoxib in patients with aspirin-sensitive asthma. *Arthritis and Rheumatism, 46*, 2201–2206.

Xie, W. L., Chipman, J. G., Robertson, D. L., Erikson, R. L., & Simmons, D. L. (1991). Expression of a mitogen-responsive gene encoding prostaglandin synthase is regulated by mRNA splicing. *Proceedings of the National Academy of Sciences of the United States of America, 88*, 2692–2696.

Pharmacotherapy

Definition

Pharmacotherapy is the treatment of disease or symptoms by the administration of drugs.

Cross-References

▶ Preventive Migraine Therapy

Phase-1 Reactions

Definition

Phase-1 reactions are the reactions of biotransformation, where the drug molecule is transformed by oxidation, reduction, or hydrolysis.

Cross-References

▶ NSAIDs, Pharmacokinetics

Phase-2 Reactions

Definition

Phase-2 reactions are the reactions of biotransformation where the drug molecule is conjugated with different molecules, such as activated glucuronic acid, sulfate, amino acids, oligopeptides, and acetic acid.

Cross-References

▶ NSAIDs, Pharmacokinetics

Phenadone

▶ Postoperative Pain, Methadone

Phenotype

Definition

The phenotype describes the actual appearance of an individual, which depends on the interaction between the different alleles of genes and between the genotype and the environment.

Cross-References

▶ NSAIDs, Pharmacogenetics

Phentolamine Test

Definition

The phentolamine test is a test for SMP, in which phentolamine or placebo is administered systemically in a double-blinded fashion and a positive result is a decrease in pain.

Cross-References

▶ Complex Regional Pain Syndrome and the Sympathetic Nervous System
▶ Sympathetically Maintained Pain

Pheromones

Definition

A chemical secreted by an animal, especially an insect, that influences the behavior or development of others of the same species, often functioning as an attractant of the opposite sex.

Cross-References

▶ Perireceptor Elements

PHN

▶ Postherpetic Neuralgia
▶ Postherpetic Neuralgia, Pharmacological and Nonpharmacological Treatment Options

Phonophobia

Definition

Hypersensitivity to sound.

Cross-References

▶ Clinical Migraine with Aura
▶ Headache, Episodic Tension-Type
▶ Hemicrania Continua

Phosphatases and Glia

Alfonso Romero-Sandoval
Department of Pharmaceuticals and
Administrative Sciences, Presbyterian College
School of Pharmacy, Clinton, SC, USA

Synonyms

3CH134, ERP, PTPN10 in mouse; CL100 in rat. DUSP6 or MKP3: Pyst1 in human; Dual-specificity phosphatases (DUSP); DUSP1 or MKP-1: CL100, VH1, PTPN10 in human; Mitogen-activated protein kinase (MAPK); Mitogen-activated protein kinase phosphatases (MKP); VH6 in rat

Definition

Phosphatases are molecules that regulate cellular function by inducing dephosphorylation. Protein phosphatases possess differential substrate specificity; thus, they are classified as protein serine/threonine phosphatases (PS/TPases), protein tyrosine phosphatases (PTPases), or dual-specificity phosphatases (DUSP, protein tyrosine/serine/threonine phosphatases). Many DUSPs act as MAPK phosphatases (MKPs), which play a key role in the inactivation of different MAPK isoforms, and therefore modulate innate and adaptive immune effector functions.

Characteristics

MKPs are the major regulators of MAPK phosphorylation in immune cells in the periphery. However, the role of MKPs in the CNS and their functional implications in pain states are unknown.

Thus far, there are 16 identified dual-specificity MKPs (Jeffrey et al. 2007) that are structurally and functionally different from most of the other protein phosphatases (Theodosiou and Ashworth 2002). Some MKPs seem to be expressed in a tissue- or cell-specific fashion, and therefore, MKPs may possess individual roles that regulate specific cellular responses in certain tissue or cell types. The dephosphorylation induced by MKPs is a direct process, which makes MKPs 100–1,000 times more potent than the actions of kinases. This supports the idea that MKPs rather than kinases are responsible for the level of MAPK phosphorylation.

This entry focuses on the best characterized phosphatases that have been described in glial cells and that deactivate MAPKs, MKP-1 and MKP-3. These MKPs have differential cell localization and are substrate-specific. Thus, MKP-1 is localized in the nucleus and preferentially dephosphorylates p-p38 and p-JNK over p-ERK. MKP-3 is localized in the cytoplasm and preferentially dephosphorylates p-ERK (Farooq and Zhou 2004; Jeffrey et al. 2007). MKPs inactivate MAPKs through dephosphorylation of threonine and/or tyrosine residues (phosphorylated form of the TxY motif in the activated loop of MAPKs). MKPs possess a conserved signature motif, HCXXXXXR (where X is any amino acid), that is located in the active site loop of MKP structure. This signature motif together with a highly conserved aspartate residue (located in general acid loop) are essential for the dephosphorylation activity on MAPKs (Farooq and Zhou 2004; Jeffrey et al. 2007).

Most MKPs are inducible genes that respond individually in a differential way to certain stimuli. In contrast to MKP-3 that requires some time to be induced, MKP-1 is an immediate early gene. The expression or activation of MKP is molecularly driven by the activated form of their substrate. The phosphorylated form of ERK-1 is able to phosphorylate MKP-1, which prevents its proteosomal degradation by conferring stabilization of MKP-1 protein (Brondello et al. 1999). Conversely, the catalytic activity of MKP-3 is enhanced 25-fold once it is coupled with its substrate p-ERK-2 (Camps et al. 1998; Fjeld et al. 2000). Thus, MAPK dephosphorylation is a consequence of MKP stabilization and enhanced activity induced by an inhibitory feedback loop (Theodosiou and Ashworth 2002; Farooq and Zhou 2004; Ducruet et al. 2005). Likewise, the phosphorylated form of ERK2, JNK1, or p38 facilitates the catalytic activity of MKP-1 (Slack et al. 2001). Therefore, MKPs are molecules that can fine-tune in a very efficient fashion the activation of MAPKs. Also, it is logical to hypothesize that a failure in the enhancement or MKP expression or activity can result in a long-lasting phosphorylation of MAPKs.

In the central nervous system, MAPK activation occurs in neuronal and glial cells, and in some cases, primarily and time dependently in glial cells. For example, spinal p38 phosphorylation is found mainly in microglia following paw incision or peripheral nerve injury; spinal ERK phosphorylation occurs in microglia or astrocytes at different times following peripheral nerve injury; and JNK phosphorylation occurs preferentially in astrocytes following peripheral nerve injury. The prevention of phosphorylation, inhibition, or dephosphorylation of these MAPK isoforms induces anti-allodynic effects, which demonstrates the pivotal role of these glial molecular pathways in the generation and maintenance of mechanical hypersensitivity in rodent acute and chronic pain models. Since MKPs are the major regulators of MAPK phosphorylation, and altered MKPs may participate in the pathophysiology of several diseases (Ducruet et al. 2005), it is logical to hypothesize that MKPs may also regulate these glial molecular pathways in pain states.

Recent studies have shown that microglia cells express MKP-1 and MKP-3, and that these phosphatases inactivate MAPKs by dephosphorylation, which results in a reduction of pro-inflammatory molecule production (MAPK-dependent or downstream effectors). We have demonstrated that the LPS-induced ERK phosphorylation in primary microglia was blocked by the cannabinoid receptor type 2 (CB2 receptor), JWH015, via MKP-3, rather than MKP-1 induction. This enhancement of microglial MKP expression and subsequent ERK dephosphorylation resulted in a reduction of microglial migration toward ADP and

TNF alpha production (Romero-Sandoval et al. 2009). This change from a pro- to anti-inflammatory phenotype caused by MKPs in primary microglial cells has also been observed in the immortalized microglia cell line, BV-2 cells using the endocannabinoid anandamide (Eljaschewitsch et al. 2006). The anti-inflammatory agent, dexamethasone, has been shown to reduce monocyte chemoattractant protein-1 (MCP-1) via MKP-1 induction and MAPK dephosphorylation in microglial cells (Zhou et al. 2007). Likewise, it has been demonstrated in in vitro cultures that astrocytes express MKP-1, which modulates the production of MCP-1 (Lee et al. 2008). Based on these results, a role of glial MKP-1 and MKP-3 in the mechanisms of pain states is supported, since spinal TNF and astrocytic MCP-1 contribute to the generation of central sensitization and the maintenance of chronic pain in rodent models (Zhang et al. 2008; Gao et al. 2009, 2010).

Interestingly, MKP-1 expressing microglia has shown to be neuroprotective in organotypic brain slice cultures after glutamate damage via the NMDA receptor (Eljaschewitsch et al. 2006). This suggests that the functional role of these phosphatases is through glial-neuronal communication. Additionally, MKP-1 and MKP-3 (as well as MKP-2) is over-expressed in conjuction with MAPK phosphorylation in satellite glial cells of trigeminal ganglia in a rodent model of inflammatory pain (Freeman et al. 2008). Whether this association of MKPs and MAPKs in satellite cells is functionally relevant has yet to be demonstrated.

MKP-1 has been observed in invading microglial cells in inflammatory lesions of brains of patients with multiple sclerosis (Eljaschewitsch et al. 2006), a neurological disease that causes chronic pain. Whether these findings are related to a form of multiple sclerosis with lower or higher levels of pain has not been described thus far. However, these findings indicate that human brain glial cells express MKPs, suggesting a potential functional role in human disease.

The implication of MKPs in the pathophysiology of pain states seems logical. In fact, it has

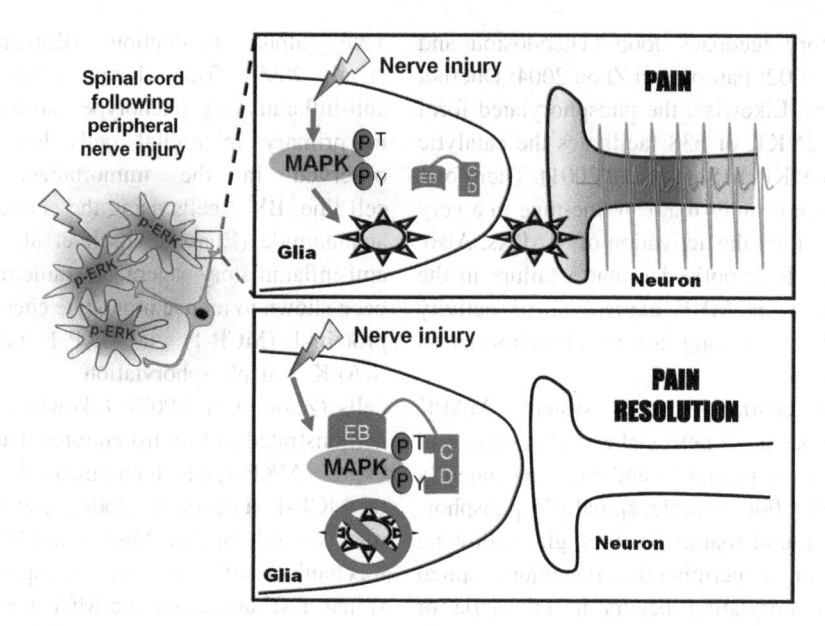

Phosphatases and Glia, Fig. 1 In homeostatic conditions, MKPs remain in a low-activity state, but upon MAPK binding MKPs are activated. Following peripheral nerve injury MAPKs are phosphorylated in spinal cord glial cells, orchestrating the production of pro-algesic factors that upon release sensitize spinal cord neurons, which contributes to the maintenance of pain. In these pathological conditions, MKPs are not activated and therefore, MAPK phosphorylation is not limited. To restore physiological conditions, MKPs bind MAPKs through their binding sides (*EB*: *ERK* binding side in the figure), which induces conformational changes and catalytic activation. Upon activation, spinal glial MKP dephosphorylates the MAPK through their catalytic domain (*CD* in the figure), inhibiting the production of pro-algesic molecules and reducing the subsequent spinal cord neuronal sensitization, which contributes to the resolution of pain

been shown that these molecules modulate nociceptive-related behavior in a rat model of neuropathic pain (Ndong et al. 2012; Landry et al. 2012). Accordingly, some drugs with analgesic properties induce an anti-inflammatory phenotype in glial cells via MKP induction (namely endocannabinoids, synthetic cannabinoid receptor type 2 agonists, or steroids). Interestingly, these drugs also modulate glial responses in spinal cord in in vivo experiments. As mentioned previously, glial MKP induction results in the dephosphorylation of MAPKs and the subsequent reduction of pro-algesic factor production, such as TNF or MCP-1. Since both, TNF and MCP-1 have been shown to participate in spinal neuronal sensitization and the maintenance of chronic pain in rodent models, the reduction of these factors by the induction of MKPs would result in a reduction of neuronal sensitization and pain-related behaviors (Fig. 1). Whether MKPs are potential targets to develop novel pharmacological strategies to treat or prevent chronic pain conditions is yet to be confirmed. The function of phosphatases in the mechanisms of pain or of the transition from acute to chronic pain is an unexplored but intriguing possibility that deserves further investigation.

References

Brondello, J. M., Pouyssegur, J., & McKenzie, F. R. (1999). Reduced MAP kinase phosphatase-1 degradation after p42/p44MAPK-dependent phosphorylation. *Science, 286,* 2514–2517.

Camps, M., Nichols, A., Gillieron, C., Antonsson, B., Muda, M., Chabert, C., Boschert, U., & Arkinstall, S. (1998). Catalytic activation of the phosphatase MKP-3 by ERK2 mitogen-activated protein kinase. *Science, 280,* 1262–1265.

Ducruet, A. P., Vogt, A., Wipf, P., & Lazo, J. S. (2005). Dual specificity protein phosphatases: Therapeutic targets for cancer and Alzheimer's disease. *Annual Review of Pharmacology and Toxicology, 45,* 725–750.

Eljaschewitsch, E., Witting, A., Mawrin, C., Lee, T., Schmidt, P. M., Wolf, S., Hoertnagl, H., Raine, C. S., Schneider-Stock, R., Nitsch, R., & Ullrich, O. (2006). The endocannabinoid anandamide protects neurons during CNS inflammation by induction of MKP-1 in microglial cells. *Neuron, 49*, 67–79.

Farooq, A., & Zhou, M. M. (2004). Structure and regulation of MAPK phosphatases. *Cellular Signalling, 16*, 769–779.

Fjeld, C. C., Rice, A. E., Kim, Y., Gee, K. R., & Denu, J. M. (2000). Mechanistic basis for catalytic activation of mitogen-activated protein kinase phosphatase 3 by extracellular signal-regulated kinase. *Journal of Biological Chemistry, 275*, 6749–6757.

Freeman, S. E., Patil, V. V., & Durham, P. L. (2008). Nitric oxide-proton stimulation of trigeminal ganglion neurons increases mitogen-activated protein kinase and phosphatase expression in neurons and satellite glial cells. *Neuroscience, 157*, 542–555.

Gao, Y. J., Zhang, L., Samad, O. A., Suter, M. R., Yasuhiko, K., Xu, Z. Z., Park, J. Y., Lind, A. L., Ma, Q., & Ji, R. R. (2009). JNK-induced MCP-1 production in spinal cord astrocytes contributes to central sensitization and neuropathic pain. *Journal of Neuroscience, 29*, 4096–4108.

Gao, Y. J., Zhang, L., & Ji, R. R. (2010). Spinal injection of TNF-alpha-activated astrocytes produces persistent pain symptom mechanical allodynia by releasing monocyte chemoattractant protein-1. *Glia, 58*, 1871–1880.

Jeffrey, K. L., Camps, M., Rommel, C., & Mackay, C. R. (2007). Targeting dual-specificity phosphatases: Manipulating MAP kinase signalling and immune responses. *Nature Reviews. Drug Discovery, 6*, 391–403.

Landry, R.P., Martinez, E., DeLeo, J.A., & Romero-Sandoval, E.A. (2012). Spinal cannabinoid receptor type 2 agonist reduces mechanical allodynia and induces mitogen-activated protein kinase phosphatases in a rat model of neuropathic pain. *Journal of Pain, 13*, 836–848.

Lee, J. H., Woo, J. H., Woo, S. U., Kim, K. S., Park, S. M., Joe, E. H., & Jou, I. (2008). The 15-deoxy-delta 12,14-prostaglandin J2 suppresses monocyte chemoattractant protein-1 expression in IFN-gamma-stimulated astrocytes through induction of MAPK phosphatase-1. *Journal of Immunology, 181*, 8642–8649.

Ndong, C., Landry, R.P., DeLeo, J.A., & Romero-Sandoval, E.A. (2012). Mitogen activated protein kinase phosphatase-1 prevents the development of tactile sensitivity in a rodent model of neuropathic pain. *Molecular Pain, 8*, 34.

Romero-Sandoval, E. A., Horvath, R., Landry, R. P., & DeLeo, J. A. (2009). Cannabinoid receptor type 2 activation induces a microglial anti-inflammatory phenotype and reduces migration via MKP induction and ERK dephosphorylation. *Molecular Pain, 5*, 25.

Slack, D. N., Seternes, O. M., Gabrielsen, M., & Keyse, S. M. (2001). Distinct binding determinants for ERK2/p38alpha and JNK map kinases mediate catalytic activation and substrate selectivity of map kinase phosphatase-1. *Journal of Biological Chemistry, 276*, 16491–16500.

Theodosiou, A., & Ashworth, A. (2002). MAP kinase phosphatases. *Genome Biology, 3*:REVIEWS3009.

Zhang, Y. H., Chi, X. X., & Nicol, G. D. (2008). Brain-derived neurotrophic factor enhances the excitability of rat sensory neurons through activation of the p75 neurotrophin receptor and the sphingomyelin pathway. *The Journal of Physiology, 586*, 3113–3127.

Zhou, Y., Ling, E. A., & Dheen, S. T. (2007). Dexamethasone suppresses monocyte chemoattractant protein-1 production via mitogen activated protein kinase phosphatase-1 dependent inhibition of Jun N-terminal kinase and p38 mitogen-activated protein kinase in activated rat microglia. *Journal of Neurochemistry, 102*, 667–678.

Photochemical Model

▶ Spinal Cord Injury Pain Model, Ischemia Model

Photophobia

Definition

Hypersensitivity to light.

Cross-References

▶ Clinical Migraine with Aura
▶ Headache, Episodic Tension-Type
▶ Hemicrania Continua

Photosensitivity

Definition

Photosensitivity of the skin: typical symptom of systemic lupus erythematosus (SLE).

Cross-References

▶ Headache Due to Arteritis

Physeptone

▶ Postoperative Pain, Methadone

Physical Abuse

Definition

Hitting, shaking, throwing, poisoning, burning or scalding, drowning, suffocating, or otherwise causing physical harm to another person.

Cross-References

▶ Chronic Pelvic Pain, Physical and Sexual Abuse

Physical Activity

Definition

Physical activity is defined as any bodily movement produced by skeletal muscles that requires energy expenditure (WHO). Physical inactivity has been identified as the fourth leading risk factor for global mortality. Regular moderate-intensity physical activity – such as walking, cycling, or participating in sports – has significant benefits for health. Chronic pain often leads to physical inactivity due to fear of progressive injury.

Cross-References

▶ Physical Exercise

Physical Agent

Definition

A physical agent is a form of acoustic, aqueous, electrical, mechanical, thermal, or light energy, which is applied to tissues in a systematic way to alter physiologic processes for therapeutic purposes.

The goal of use may include heating (hot packs, the diathermies), cooling (ice therapy), healing (e.g., low-intensity laser therapy), muscle stimulation (electrical stimulation), and pain control (e.g., TENS) as well as the introduction of pharmaceutical agents though the skin (iontophoresis). Many agents are thought to produce more than one effect (such as pain relief and the promotion of healing). Although the thermal properties of the heating and cooling agents are well established, these agents may also have nonthermal effects.

Cross-References

▶ Chronic Pain in Children, Physical Medicine and Rehabilitation
▶ Therapeutic Heat, Microwaves, and Cold

Physical Capacity Evaluation

▶ Disability, Functional Capacity Evaluations

Physical Conditioning Programs

Jean-Baptiste Fassier[1], Frederieke Geraldine Schaafsma[2] and Schonstein Eva[3]
[1]Department of Occupational Health and Medicine, UMRESTTE-Université Claude Bernard Lyon 1, Lyon, France
[2]Department of Public and Occupational Health, EMGO+ Institute/VU Medical Center, Amsterdam, MB, The Netherlands
[3]RehabCo, Fyshwick, ACT, Australia

Synonyms

Functional restoration/exercise programs; Work conditioning; Work hardening

Definition

Physical conditioning programs (PCPs) aim for return-to-work or improvement in work status for workers performing modified duties. Such programs either simulate or duplicate work and/or functional tasks in a safe, supervised environment. These tasks are structured and progressively graded to increase psychological, physical, and emotional tolerance and improve endurance and work feasibility (Lechner 1994). Work hardening programs are individualized, work-oriented activities that involve clients in simulated or actual work tasks. Work conditioning is a program with an emphasis on physical conditioning that addresses the issues of strength, endurance, flexibility, motor control, and cardiopulmonary function (Lechner 1994). Functional restoration refers to any intervention aimed at restoring a reasonable functional level for activities of daily living, including work (Bendix et al. 1996).

The goals of PCPs and the procedures used in them overlap with those of other treatment programs, such as patient care management, multidisciplinary treatments, pain clinics, standard medical care, or physiotherapy, which aim to reduce symptoms, pain intensity, use of medications, and health services, and to increase global improvement and quality of life (Guzman et al. 2001, 2002); physiological outcomes, such as range of motion and spinal flexibility (Hayden et al. 2005); or behavioral outcomes such as anxiety, depression, and cognition (Ostelo et al. 2005). But PCPs are distinguished from other programs by their principal outcome measured, which is time between injury and return to permanent, full-time duties.

Characteristics

Introduction to Cochrane Systematic Reviews

A systematic review of the literature (Schaafsma et al. 2010) conducted in the frame of the Cochrane collaboration recently updated a previous review (Schonstein et al. 2003a, b) about the effectiveness of physical conditioning

programs for workers with back pain. This update focused exclusively on workers with back pain and on work status outcomes and therefore excluded neck pain and secondary outcomes such as functional status and physiological outcomes, which were included in the original review. The definition of physical conditioning programs was refined so that the review was limited to studies with interventions that had a stated relationship with the workplace, a focus on job demands, and measured work outcomes. Only randomised controlled trials (RCT) and cluster RCTs were included in this review.

Characteristics of Study Population

The population included in this review comprised male and female adults (>16 years) with work disability related to back pain and who took part in physical conditioning programs. Back pain was defined as pain in the thoracic, lumbar, or gluteal region, or a combination, with or without radiation to the lower extremities. Studies with at least 50 % of workers with back pain were included. Work disability was defined as being on full or partial sick leave, or not being able to perform adequately at work due to back pain. All workers who were accepted into physical conditioning programs, whether they had acute (duration of symptoms less than 6 weeks), subacute (duration of symptoms more than 6 but less than 12 weeks), or chronic back pain (duration of symptoms more than 12 weeks), met the inclusion criteria. Studies with nonworkers, or workers with specific diagnoses such as infection, neoplasm, metastasis, osteoporosis, rheumatoid arthritis, fracture, inflammatory processes, or other conditions for which valid diagnoses had been demonstrated, were excluded.

Characteristics of Physical Conditioning Programs

Studies on physical conditioning programs were included only when they included the following three key elements:

- Exercises specifically designed to restore an individual's systemic, neurological, musculoskeletal (strength, endurance, movement, flexibility and motor control), cardiopulmonary function, or a combination;

- An explicit statement that the goal of the program was improvement of work status;
- A stated relationship between the intervention and functional job demands.

In addition to these three key elements, physical conditioning programs could include components such as operant conditioning behavioral approach, pain management, back pain education, advice on return-to-work or a workplace visit. The delivery of physical conditioning programs could involve multidisciplinary teams or individual health professionals. In addition, they could be delivered in a one-to-one fashion or in a group situation.

Two types of programs were differentiated depending on their intensity. Light physical conditioning programs included the three key elements and were delivered in fewer than five sessions (of 1 h) or were described by the primary study author as a light intervention program. Intense physical conditioning programs included the three key elements and were delivered in more than five sessions or were delivered on a full-time basis for more than 2 weeks.

Outcomes of Interest

Three different types of work-status outcomes were considered in this review: time between intervention and return-to-work, return-to-work status in terms of "at work" or "off work," and time on light or modified duties. When studies expressed time to return-to-work as mean numbers of days off work, this outcome was converted into standardized mean difference (SMD) between the intervention group and the control group. Rate of return-to-work was expressed as odd ratios (ORs). When studies could be pooled, ORs were converted into SMDs. Pooling of studies was only considered when statistical heterogeneity was less than moderate (I^2 less than 60 %).

In accordance with the original review, a mean saving of 10 sick days per year was considered the smallest effect that would be clinically worthwhile, based on the assumption of the cost of providing physical conditioning programs.

For dichotomous outcomes, an OR of 0.65 was used as the smallest clinically worthwhile effect. This corresponded to a number needed to treat (NNT) of 10 when the baseline rate (prevalence of the event measured (on/off work) was about 40 %, based on the two comparisons from the original review.

Quality Assessment of Included Studies

The criteria recommended by the Cochrane Back Review Group (Furlan et al. 2009) were used to assess the risk of bias of the selected RCTs. Each of the criteria was scored "yes," "no," or "unsure," according to the operationalization of the criteria. Following the recommendations of the Cochrane Back Review Group (Furlan et al. 2009), studies were rated as having a "low risk of bias" when at least 6 of the 12 CBRG criteria were met and the study had no serious flaws. Serious flaws were inadequate concealment of treatment allocation, inadequate blinding of outcome assessors, a large drop-out rate, or statistically significant and clinically important baseline differences not accounted for in the analyses. Studies with serious flaws, or those in which fewer than six of the criteria were met, were rated as having a "high risk of bias."

Overall Quality of the Evidence

The overall quality of the evidence for each outcome was assessed using an adapted GRADE approach (Atkins et al. 2004) as recommended by the Cochrane Back Review Group. The quality of the evidence for a specific outcome was based on the study design, limitations of the study, consistency, directness, precision of results, and publication bias. The GRADE group recommends four levels of evidence; the Cochrane Back Review Group recommends the addition of the fifth:

- High-quality evidence: there are consistent findings among 75 % of RCTs with low risk of bias that are generalizable to the population in question. There are sufficient data, with narrow confidence intervals. There is no known or suspected publication bias. Further research is unlikely to change either the estimate or confidence in the results.

- Moderate quality evidence: one of the domains is not met. Further research is likely to have an important impact on our confidence in the estimate of effect and may change the estimate.
- Low-quality evidence: two of the domains are not met. Further research is very likely to have an important impact on confidence in the estimate of effect and is likely to change the estimate.
- Very-low-quality evidence: three of the domains are not met. We are very uncertain about the estimate.
- No evidence: no RCTs are identified that address this outcome.

Overall Effectiveness of PCPs

Thirty-seven references, reporting on 23 RCTs (3,676 workers) were included (Altmaier et al. 1992; Lindström et al. 1992; Mitchell and Carmen 1994; Faas et al. 1995; Bendix et al. 1996, 1997, 2000 Corey et al. 1996; Loisel et al. 1997; Jensen et al. 2001; Skouen et al. 2002; Gatchel et al. 2003; Karjalainen et al. 2003; Storheim et al. 2003; van den Hout et al. 2003; Jousset et al. 2004; Staal et al. 2004; Kool et al. 2005; Meyer et al. 2005; Wright et al. 2005; Heymans et al. 2006; Steenstra et al. 2006; Roche et al. 2007). The main characteristics of the included studies with their population and intervention are given in Table 1.

The results highlighted the wide variation in the content and duration of interventions that are currently labeled as physical conditioning, work conditioning, work hardening, or functional restoration or exercise programs and the variations in the types of assessment of outcomes used in clinical trials investigating the effectiveness of these interventions. These variations limited the degree to which the results of trials could be compared and/or pooled. In addition, variations in the timing of the outcome assessments, subjects studied, limited methodological quality, inadequate reporting of results, and small sample sizes limited the conclusions that could be drawn regarding the efficacy of physical conditioning programs.

The overall effectiveness of PCPs when compared to other interventions is given in Tables 2 and 3 which detail the results according to the low back pain phase (acute, subacute, and chronic).

In workers with acute back pain, there was no effect of physical conditioning programs in reducing sickness absence. In workers with subacute back pain, conflicting results were found but subgroup analysis showed a positive effect in favor of interventions with workplace involvement. The overall effectiveness of physical conditioning programs at the acute and subacute phase of low back pain is given in Table 2.

In workers with chronic back pain, pooled results of five studies showed a small effect on sickness absence at long-term follow-up (SMD: -0.18 (95 % CI: -0.37 to 0.00)). In workers with chronic back pain, physical conditioning programs were compared to other exercise therapy in six studies, with conflicting results. The addition of cognitive-behavioral therapy to physical conditioning programs was not more effective than the physical conditioning alone. The overall effectiveness of physical conditioning programs at the chronic phase of low back pain is given in Table 3.

Discussion

Implications for practice

For workers with acute back pain, a physical conditioning program is not effective in reducing sickness absence duration. Light physical conditioning programs do not reduce sickness absence in workers with subacute nor chronic back pain. There is conflicting evidence regarding the effectiveness of intense physical conditioning program versus usual care for workers with subacute back pain. It might be that including workplace visits or execution of the intervention at the workplace is the component that renders a physical conditioning program effective. For workers with chronic back pain, there is moderate quality evidence that intense physical conditioning programs have a small but significant effect on sickness absence compared to care as usual. There is conflicting evidence on the effect of intense physical conditioning programs versus exercise therapy for workers with chronic back pain.

Physical Conditioning Programs, Table 1 Characteristics of the included studies

Population	
Population size	
Fewer than 100(6)	Altmaier et al. (1992); Bendix et al. (1996); Jousset et al. (2004); Meyer et al. (2005); Storheim et al. (2003); Wright (2005)
Between 100 and 200 (11)	Bendix et al. (1997, 2000); Gatchell (2003); Karjalainen et al. (2003); Kool et al. (2005); Lindstrom et al. (1992); Loisel et al. (1997); Roche et al. (2007); Staal et al. (2004); Steenstra et al. (2006); van den Hout et al. (2003)
More than 200 (6)	Corey et al. (1996); Faas et al. (1995); Heymans et al. (2006); Jensen et al. (2001); Mitchell and Carmen (1994) Skouen et al. (2002)
Low back pain phase	
Acute (3)	Faas et al. (1995); Gatchel et al. (2003); Wright (2005)
Subacute (8)	Heymans et al. (2006); Karjalainen et al. (2003); Kool et al. (2005); Lindstrom et al. (1992); Loisel et al. (1997); Staal et al. (2004); Steenstra et al. (2006); Storheim et al. (2003)
Chronic (12)	Altmaier et al. (1992); Bendix et al. (1996, 1997, 2000); Corey et al. (1996); Jensen et al. (2001); Jousset et al. (2004); Meyer et al. (2005); Mitchell and Carmen (1994); Roche et al. (2007); Skouen et al. (2002); van den Hout et al. (2003)
Work disability	
Sick leave (full/part time) (13)	Altmaier et al. (1992); Corey et al. (1996); Heymans et al. (2006); Jensen et al. (2001); Lindstrom et al. (1992); Loisel et al. (1997); Meyer et al. (2005); Mitchell and Carmen (1994); Staal et al. (2004); Steenstra et al. (2006); Storheim et al. (2003); van den Hout et al. (2003); Wright (2005)
Mix of sick leave/decreased ability (10)	Bendix et al. (1996, 1997, 2000); Faas et al. (1995); Gatchell et al. (2003); Jousset et al. (2004); Karjalainen et al. (2003); Kool et al. (2005); Roche et al. (2007); Skouen et al. (2002)
Intervention	
Intensity	
Light (5)	Faas et al. (1995); Heymans et al. (2006); Karjalainen et al. (2003); Skouen et al. (2002); Wright (2005)
Intense (18)	Altmaier et al. (1992); Bendix et al. (1996, 1997, 2000); Corey et al. (1996); Heymans et al. (2006); Jensen et al. (2001); Jousset et al. (2004); Kool et al. (2005); Lindstrom et al. (1992); Loisel et al. (1997); Meyer et al. (2005); Mitchell and Carmen (1994); Roche et al. (2007); Staal et al. (2004); Steenstra et al. (2006); Storheim et al. (2003); van den Hout et al. (2003)
Multidisciplinarity	
Yes (17)	Altmaier et al. (1992); Bendix et al. (1996, 1997, 2000); Corey et al. (1996); Gatchell et al. (2003); Jensen et al. (2001); Jousset et al. (2004); Karjalainen et al. (2003); Kool et al. (2005); Loisel et al. (1997); Meyer et al. (2005); Roche et al. (2007); Staal et al. (2004); van den Hout et al. (2003); Skouen et al. (2002); Wright (2005)
No (5)	Faas et al. (1995); Heymans et al. (2006); Lindstrom et al. (1992); Steenstra et al. (2006); Storheim et al. (2003)
Unclear (1)	Mitchell and Carmen (1994)
Other components	
Operant conditioning behavioral approach (15)	Altmaier et al. (1992); Bendix et al. (1996, 1997, 2000); Corey et al. (1996); Heymans et al. (2006); Jensen et al. (2001); Lindstrom et al. (1992); Loisel et al. (1997); Meyer et al. (2005); Mitchell and Carmen (1994); Skouen et al. (2002); Staal et al. (2004); Steenstra et al. (2006); van den Hout et al. (2003)
Occupational training or ergonomic advice (15)	Bendix et al. (1996, 1997, 2000); Gatchell et al. (2003); Heymans et al. (2006); Jensen et al. (2001); Jousset et al. (2004); Kool et al. (2005); Loisel et al. (1997); Meyer et al. (2005); Roche et al. (2007); Skouen et al. (2002); Storheim et al. (2003); van den Hout et al. (2003); Wright et al. (2005)

(continued)

Physical Conditioning Programs, Table 1 (continued)

Return to work advice (9)	Heymans et al. (2006); Jensen et al. (2001); Karjalainen et al. (2003); Lindstrom et al. (1992); Loisel et al. (1997); Meyer et al. (2005); Staal et al. (2004); Steenstra et al. (2006); van den Hout et al. (2003)
Workplace visit (6)	Jensen et al. (2001); Karjalainen et al. (2003); Lindstrom et al. (1992); Loisel et al. (1997); Meyer et al. (2005); van den Hout et al. (2003)
Study	
Design	
RCT (22)	All studies but Loisel et al. (1997)
Cluster RCT (1)	Loisel et al. (1997)
Risk of bias	
Low (13)	Faas et al. (1995); Heymans et al. (2006); Jensen et al. (2001); Karjalainen et al. (2003); Kool et al. (2005); Loisel et al. (1997); Meyer et al. (2005); Skouen et al. (2002); Staal et al. (2004); Steenstra et al. (2006); Storheim et al. (2003); van den Hout et al. (2003); Wright (2005)
High (10)	Altmaier et al. (1992); Bendix et al. (1996, 1997, 2000); Corey et al. (1996); Gatchell (2003); Jousset et al. (2004); Lindström et al. (1992); Mitchell and Carmen (1994); Roche et al. (2007)

Implications for Research

In the recently updated Cochrane review, a subgroup analysis of interventions that were done at the workplace or included a workplace visit seems to point toward a positive effect when a workplace intervention is included in an intense physical conditioning program. The effectiveness of interventions in or closely associated with the workplace is supported by current literature (Anema et al. 2007; van Oostrom et al. 2009). It is argued that workplace-based interventions can change both workers' and supervisors' perceptions about workers' return-to-work capabilities and opportunities (Anema et al. 2007). Workplace interventions may have different objectives ranging from gathering information in order to reproduce work demands in a clinical setting to gradually exposing workers to the demands of the real work environment, or permanently reducing the demands of the work situation (Durand et al. 2007). However, it is not known which components of the workplace interventions are necessary in order to reach effectiveness, nor the mechanisms by which these components might be effective. A meta-regression based on this Cochrane review could not identify any factor that explained the variation in outcomes between studies (Schaafsma et al. 2011). In order to overcome this black box phenomenon, it recommended that efforts be made to better describe the components of the workplace interventions (who, what, when, where, how) and to develop process outcomes representing the multidimensional effects produced in the workplace. The development of program theory within an evaluative research perspective (Durand et al. 2003) is necessary in order to better define and understand the workplace components of physical conditioning programs.

Recent high-quality primary studies have been published that bring new evidence supporting the effectiveness of an integrated care program for low back pain at the chronic phase (Lambeek et al. 2007, 2009, 2010a, b). The program consisted of a workplace intervention based on participatory ergonomics and a graded activity program based on cognitive behavioral principles. Integrated care was effective on return to work (hazard ratio 1.9, 95 % confidence interval 1.2–2.8, $P = 0.004$). After 12 months, patients in the integrated care group improved significantly more on functional status compared with patients in the usual care group ($P = 0.01$). The cost-benefit analysis showed that every £1 invested in integrated care would return an estimated £26. The net societal benefit of integrated care compared with usual care was £5,744. Based on these new findings, the conclusion of the Cochrane review described here is likely to be modified after its next update.

Physical Conditioning Programs, Table 2 Overall effectiveness of physical conditioning programs at the acute and subacute phase

Type of PCP	Effectiveness compared to other intervention	Follow-up	References	Overall level of evidence	
At the acute phase					
Light PCPs	Not more effective than usual care	At short-term FU	Wright (2005)	LQ	From two heterogeneous trials
		At long-term FU	Faas et al. (1995)		
Intense PCPs	Not more effective than usual care	At long-term FU	Gatchel et al. (2003)	VLQ	From one trial
At the subacute phase					
Light PCPs	Not more effective than usual care	At intermediate-term FU	Heymans et al. (2006)	LQ	From two heterogeneous trials
		At long-term FU	Karjalainen et al. (2003)		
		At very long-term FU	Karjalainen et al. (2003)		
Intense PCPs	More effective than pain-centered treatment	At short-term FU	Kool et al. (2005)	LQ	From one trial
		At long-term FU			
Intense PCPs	Not more effective than a brief cognitive intervention	At short-term FU	Storheim et al. (2003)	LQ	From one trial
Intense PCPs *without workplace component*	Not more effective than usual care	At short-term FU	Storheim et al. (2003)	LQ	From one trial
		At intermediate-term FU	Heymans et al. (2006)	Conflicting evidence	
			Steenstra et al. (2006)		
		At long-term FU	Steenstra et al. (2006)		
Intense PCPs *with a workplace component*	More effective than usual care	At intermediate-term FU	Staal et al. (2004)		
		At long-term FU	Lindström et al. (1992); Lindstrom et al. (1992)		
			Loisel et al. (1997)		
			Staal et al. (2004)		
Intense PCPs	More effective than usual care	At very long-term FU	Lindström et al. (1992); Lindstrom et al. (1992)	MQ	From two trials
			Staal et al. (2004)		
Intense PCPs	More effective than light PCPs	At intermediate-term FU	Heymans et al. (2006)	LQ	From one trial

FU: follow up; VLQ: very low quality; LQ: low quality; MQ: medium quality

However, new studies will be needed to ascertain the effectiveness of PCPs with a workplace component for low back pain at the chronic phase.

Eventually, new studies are needed in the field of implementation research. Although effective interventions have been developed, uptake of research findings by clinicians and healthcare organizations is low (Loisel et al. 2005), a discrepancy known as the "knowledge-to-action gap" (Grol 2001; Haines et al. 2004). It is currently recommended to perform a context analysis prior to implementing a complex or innovative intervention, in order to identify barriers and facilitators (Grol and Grimshaw 2003; Baker et al. 2010). An evidence-based conceptual framework has been elaborated in order to allow the identification of such barriers and facilitators before implementing

Physical Conditioning Programs, Table 3 Overall effectiveness of physical conditioning programs at the chronic phase

Type of PCP	Effectiveness compared to other intervention	Follow-up	References	Overall level of evidence	
At the chronic phase					
Intense PCPs	More effective than usual care	At short-term FU	Bendix et al. (1996)	VLQ	From one trial
	More effective than usual care but not clinically worthwhile[a]	At long-term FU	Mitchell and Carmen (1994)	MQ	From five trials
			Bendix et al. (1996)		
			Corey et al. (1996)		
			Jensen et al. (2001)		
			Skouen et al. (2002)		
	Not more effective than usual care	At very long-term FU	Bendix et al. (1996)	MQ	From three trials
			Jensen et al. (2001)		
			Skouen et al. (2002)		
Intense PCPs	Not more effective than exercise programs	At short-term FU	Roche et al. (2007)	VLQ	From one trial[b]
		At intermediate-term FU	Jousset et al. (2004)	VLQ	From one trial[b]
			Meyer et al. (2005)		
	Face conflicting evidence as regard their effectiveness when compared to exercise programs	At long-term FU	Bendix et al. (1997, 2000)	Conflicting evidence	
	More effective than exercise therapy	At very long-term FU	Bendix et al. (1997)	VLQ	From one trial
Intense PCPs	Not more effective than light PCPs	At long-term FU	Skouen et al. (2002)	LQ	From one trial
		At very long-term FU	Skouen et al. (2002)	LQ	From one trial
Intense PCPs	More effective than cognitive behavioral therapy	At long-term FU	Bendix et al. (1997)	LQ/VLQ	From two trials
			Jensen et al. (2001)		
	Face conflicting evidence as regard their effectiveness when compared to cognitive behavioral therapy	At very long-term FU		Conflicting evidence	

[a]The size of the effect is not clinically worthwhile with a SMD of -0.18 (95 % CI: -0.37 to 0.00).
[b]Two references pertained to the same study Jousset et al. (2004), Roche et al. (2007); one reference Meyer (2005) was a pilot study for a larger trial

return-to-work interventions (Fassier et al. 2011). Further feasibility studies should be conducted to document the construct validity and applicability of this conceptual framework.

Summary

The review of the evidence about the effectiveness of physical conditioning programs showed the absence of high-quality evidence. The overall quality of the evidence is ranging from very low quality to medium quality. It is therefore likely that further research will have an important impact on the confidence in the estimate of effect and may change the current state of the evidence.

References

Altmaier, E. M., Lehmann, T. R., Russell, D. W., Weinstein, J. N., & Kao, C. F. (1992). The effectiveness of psychological interventions for the rehabilitation of low back pain: A randomized controlled trial evaluation. *Pain, 3*, 329–335.

Anema, J. R., Steenstra, I. A., Bongers, P., de Vet, H. C. W., Knol, D. L., Loisel, P., & van Mechelen, W. (2007). Multidisciplinary rehabilitation for subacute low back pain: Graded activity or workplace intervention or both?: A randomized controlled trial. *Spine, 3*, 291–298.

Atkins, D., Best, D., Briss, P., Eccles, M., & Falck-Ytter, Y. (2004). GRADE working group. Grading quality of evidence and strength of recommendations. *BMJ, 328*, 1490–1494.

Baker, R., Camosso-Stefinovic, J., Gillies, C., Shaw, E. J., Cheater, F., Flottorp, S., & Robertson, N. (2010). Tailored interventions to overcome identified barriers to change: Effects on professional practice and health care outcomes [Systematic Review]. *Cochrane Database Systematic Reviews* (3).

Bendix, A. F., Bendix, T., Vaegter, K., Lund, C., Frolund, L., & Holm, L. (1996). Multidisciplinary intensive treatment for chronic low back pain: A randomized, prospective study. *Cleveland Clinic Journal of Medicine, 1*, 62–69.

Bendix, A. F., Bendix, T., Lund, C., Kirkbak, S., & Ostenfeld, S. (1997). Comparison of three intensive programs for chronic low back pain patients: A prospective, randomized, observer-blinded study with 1 year follow-up. *Scandinavian Journal of Rehabilitation Medicine, 2*, 81–89.

Bendix, T., Bendix, A., Labriola, M., Haestrup, C., & Ebbehoj, N. (2000). Functional restoration versus outpatient physical training in chronic low back pain: A randomized comparative study. *Spine, 19*, 2494–2500.

Corey, D., Koepfler, L., Etlin, D., & Day, H. (1996). A limited functional restoration program for injured workers: A randomized trial. *Journal of Occupational Rehabilitation, 4*, 239–249.

Durand, M. J., Vachon, B., Loisel, P., & Berthelette, D. (2003). Constructing the program impact theory for an evidence-based work rehabilitation program for workers with low back pain. *Work, 3*, 233–242.

Durand, M. J., Vézina, N., Loisel, P., Baril, R., Richard, M. C., & Diallo, B. (2007). Workplace interventions for workers with musculoskeletal disabilities: A descriptive review of content. *Journal of Occupational Rehabilitation, 1*, 123–136.

Faas, A., van Eijk, J. T., Chavannes, A. W., & Gubbels, J. W. (1995). A randomized trial of exercise therapy in patients with acute low back pain. Efficacy on sickness absence. *Spine, 8*, 941–947.

Fassier, J. B., Durand, M. J., & Loisel, P. (2011). 2nd place, PREMUS best paper competition: Implementing return-to-work interventions for workers with low-back pain - a conceptual framework to identify barriers and facilitators. *Scandinavian Journal of Work & Environmental Health, 2*, 99–108.

Furlan, A. D., Pennick, V., Bombardier, C., van Tulder, M., & Editorial Board, C. B. R. G. (2009). 2009 updated method guidelines for systematic reviews in the cochrane back review group. *Spine, 18*, 1929–1941.

Gatchel, R. J., Polatin, P. B., Noe, C., Gardea, M., Pulliam, C., & Thompson, J. (2003). Treatment- and cost-effectiveness of early intervention for acute low-back pain patients: A 1 year prospective study. *Journal of Occupational Rehabilitation, 1*, 1–9.

Grol, R. (2001). Successes and failures in the implementation of evidence-based guidelines for clinical practice. *Medical Care, 8*(Suppl 2), II46–II54.

Grol, R., & Grimshaw, J. (2003). From best evidence to best practice: Effective implementation of change in patients' care. *The Lancet, 9391*, 1225–1230.

Guzman, J., Esmail, R., Karjalainen, K., Malmivaara, A., Irvin, E., & Bombardier, C. (2001). Multidisciplinary rehabilitation for chronic low back pain: Systematic review. *BMJ, 7301*, 1511–1516.

Guzman, J., Esmail, R., Karjalainen, K., Malmivaara, A., Irvin, E., & Bombardier, C. (2002). Multidiciplinary biopsycho-social rehabilitation for chronic low back pain. *Cochrane Database Systematic Reviews* (1).

Haines, A., Kuruvilla, S., & Borchert, M. (2004). Bridging the implementation gap between knowledge and action for health. *Bulletin of the World Health Organization, 10*, 724–733.

Hayden, J. A., van Tulder, M. W., Malmivaara, A., & Koes, B. (2005). Exercise treatment for treatment of non-specific low back pain. *Cochrane Database Systematic Reviews* (3).

Heymans, M. W., de Vet, H. C., Bongers, P. M., Knol, D. L., Koes, B. W., & van Mechelen, W. (2006). The effectiveness of high-intensity versus low-intensity back schools in an occupational setting: A pragmatic randomized controlled trial. *Spine, 10*, 1075–1082.

Jensen, I., Bergström, G., Ljungquist, T., Bodin, L., & Nygren, A. (2001). A randomized controlled component analysis of a behavioral medicine rehabilitation program for chronic spinal pain: Are the effects dependent on gender? *Pain, 1–2*, 65–78.

Jousset, N., Fanello, S., Bontoux, L., Dubus, V., Billabert, C., Vielle, B., Roquelaure, Y., Penneau-Fontbonne, D., & Richard, I. (2004). Effects of functional restoration versus 3 h per week physical therapy: A randomized controlled study. *Spine, 5*, 487–493.

Karjalainen, K., Malmivaara, A., Pohjolainen, T., Hurri, H., Mutanen, P., Rissanen, P., Pahkajarvi, H., Levon, H., Karpoff, H., & Roine, R. (2003). Mini-intervention for subacute low back pain: A randomized controlled trial. *Spine, 6*, 533–540.

Kool, J. P., Oesch, P. R., Bachmann, S., Knuesel, O., Dierkes, J. G., Russo, M., de Bie, R. A., & van den Brandt, P. A. (2005). Increasing days at work using function-centered rehabilitation in nonacute nonspecific low back pain: A randomized controlled trial. *Archives of Physical Medicine and Rehabilitation, 5*, 857–864.

Lambeek, L. C., Anema, J. R., van Royen, B. J., Buijs, P. C., Wuisman, P. I., van Tulder, M. W., & van Mechelen, W. (2007). Multidisciplinary outpatient care program for patients with chronic low back pain: Design of a randomized controlled trial and cost-effectiveness study [ISRCTN28478651]. *BMC Public Health, 7*, 254.

Lambeek, L. C., van Mechelen, W., Buijs, P. C., Loisel, P., Anema, J. R. (2009). An integrated care program to prevent work disability due to chronic low back pain: A process evaluation within a randomized controlled trial. *BMC Musculoskeletal Disorders, 10*, 147.

Lambeek, L. C., Bosmans, J. E., Van Royen, B. J., Van Tulder, M. W., Van Mechelen, W., & Anema, J. R. (2010a). Effect of integrated care for sick listed patients with chronic low back pain: Economic evaluation alongside a randomised controlled trial. *BMJ, 341*, c:6414v.

Lambeek, L. C., van Mechelen, W., Knol, D. L., Loisel, P., & Anema, J. R. (2010b). Randomised controlled trial of integrated care to reduce disability from chronic low back pain in working and private life. *Bristish Medical Journal, 340*, c:1035.

Lechner, D. (1994). Work hardening and work conditioning interventions: Do they affect disability? *Physical Therapy, 4*, 471–493.

Lindström, I., Ohlund, C., Eek, C., Wallin, L., Peterson, L., Fordyce, W. E., & Nachemson, A. L. (1992). The effect of graded activity on patients with subacute low back pain: A randomized prospective clinical study with an operant-conditioning behavioral approach...Including commentary by Nelson RM with author response. *Physical Therapy, 4*, 279–293.

Lindstrom, I., Ohlund, C., Eek, C., Wallin, L., Peterson, L. E., & Nachemson, A. (1992). Mobility, strength, and fitness after a graded activity program for patients with subacute low back pain. A randomized prospective clinical study with a behavioral therapy approach. *Spine, 6*, 641–652.

Loisel, P., Abenhaim, L., Durand, P., Esdaile, J. M., Suissa, S., Gosselin, L., Simard, R., Turcotte, J., & Lemaire, J. (1997). A population-based, randomized clinical trial on back pain management. *Spine, 24*, 2911–2918.

Loisel, P., Buchbinder, R., Hazard, R., Keller, R., Scheel, I., van Tulder, M., & Webster, B. (2005). Prevention of work disability due to musculoskeletal disorders: The challenge of implementing evidence. *Journal of Occupational Rehabilitation, 4*, 507–524.

Meyer, K., Fransen, J., Huwiler, H., Uebelhart, D., & Klipstein, A. (2005). Feasibility and results of a randomised pilot-study of a work rehabilitation programme. *Journal of Back and Musculoskeletal Rehabilitation, 3–4*, 67–78.

Mitchell, R. I., & Carmen, G. M. (1994). The functional restoration approach to the treatment of chronic pain in patients with soft tissue and back injuries. *Spine, 6*, 633–642.

Ostelo, R., van Tulder, M., Vlaeyen, J., Linton, S., Morley, S., & Assendelft, W. (2005). Behavioural treatment for chronic low-back pain. *Cochrane Database Systematic Reviews* (1).

Roche, G., Ponthieux, A., Parot-Shinkel, E., Jousset, N., Bontoux, L., Dubus, V., Penneau-Fontbonne, D., Roquelaure, Y., Legrand, E., Colin, D., Richard, I., & Fanello, S. (2007). Comparison of a functional restoration program with active individual physical therapy for patients with chronic low back pain: A randomized controlled trial. *Archives of Physical Medicine and Rehabilitation, 10*, 1229–1235.

Schaafsma, F., Schonstein, E., Ojajärvi, A., & Verbeek, J. H. (2011). Physical conditioning programs for improving work outcomes among workers with back pain. *Scandinavian Journal of Work & Environmental Health, 1*, 1–5.

Schaafsma, F., Schonstein, E., Whelan, K., Ulvestad, E., Kenny, D., Verbeek, J. (2010). Physical conditioning programs for improving work outcomes in workers with back pain. *Cochrane Database Systematic Reviews* (1).

Schonstein, E., Kenny, D., Keating, J., Koes, B., & Herbert, R. D. (2003a). Physical conditioning programs for workers with back and neck pain: A cochrane systematic review. *Spine, 19*, E391–E395.

Schonstein, E., Kenny, D. T., Keating, J., & Koes, B. W. (2003b). Work conditioning, work hardening and functional restoration for workers with back and neck pain. *Cochrane Database Systematic Reviews* (1).

Skouen, J. S., Grasdal, A. L., Haldorsen, E. M., Ursin, H., Skouen, J. S., Grasdal, A. L., Haldorsen, E. M. H., & Ursin, H. (2002). Relative cost-effectiveness of extensive and light multidisciplinary treatment programs versus treatment as usual for patients with chronic low back pain on long-term sick leave: Randomized controlled study. *Spine, 9*, 901–909. discussion 909–10.

Staal, J. B., Hlobil, H., Twisk, J. W. R., Smid, T., Koke, A. J. A., & van Mechelen, W. (2004). Graded activity for low back pain in occupational health care: A randomized, controlled trial. *Annals of Internal Medicine, 2*, 77–84.

Steenstra, I. A., Anema, J. R., Bongers, P. M., de Vet, H. C., Knol, D. L., & van Mechelen, W. (2006).

The effectiveness of graded activity for low back pain in occupational healthcare. *Occupational and Environmental Medicine, 11*, 718–725.

Storheim, K., Brox, J. I., Holm, I., Koller, A. K., Bo, K., Storheim, K., Brox, J. I., Holm, I., Koller, A. K., & Bo, K. (2003). Intensive group training versus cognitive intervention in sub-acute low back pain: Short-term results of a single-blind randomized controlled trial. *Journal of Rehabilitation Medicine, 3*, 132–140.

van den Hout, J. H. C., Vlaeyen, J. W. S., Heuts, P. H. T. G., Zijlema, J. H. L., & Wijnen, J. A. G. (2003). Secondary prevention of work-related disability in nonspecific low back pain: Does problem-solving therapy help? A randomized clinical trial. *The Clinical Journal of Pain, 2*, 87–96.

van Oostrom, S. H., Driessen M. T., de Vet, H. C. W., Franche, R. L., Schonstein, E., Loisel, P., van Mechelen, W., & Anema, J. R. (2009). Workplace interventions for preventing work disability. *Cochrane Database Systematic Reviews* (2).

Wright, A., Lloyd-Davies, A., Williams, S., Ellis, R., & Strike, P. (2005). Individual active treatment combined with group exercise for acute and subacute low back pain. *Spine, 11*, 1235–41.

Physical Deconditioning

Definition

Physical deconditioning refers to the reduction in flexibility, strength, and endurance resulting from the restriction of muscular activity.

Cross-References

▶ Cognitive-Behavioral Perspective of Pain

Physical Dependence

Definition

Physical dependence is an adaptive state characterized by intense physical disturbances (withdrawal or abstinence syndrome characteristic of the particular drug, i.e., diarrhea, sweating, restlessness, etc.), after opioid cessation or blockade of receptors with an opioid antagonist) when repeated drug administration is suspended.

Cross-References

▶ Opiates During Development
▶ Opioid Therapy in Cancer Patients with Substance Abuse Disorders, Management
▶ Psychiatric Aspects of the Management of Cancer Pain

Physical Exercise

Kaisa Mannerkorpi
Department of Rheumatology and Inflammation Research, Institute of Medicine, University of Gothenburg, Göteborg, Sweden

Synonyms

Fitness training; Functioning; Physical performance capacity

Definition

▶ Physical activity is defined here as all spontaneous or regular everyday physical activity that consumes energy, including walking or cycling to work and working in the garden.

▶ Physical exercise is planned physical activity that consumes energy and increases the breathing rate and perspiration.

Characteristics

Many patients with ▶ chronic pain (IASP 1994) have reduced physical performance capacity, which may be associated with their primary impairment and pathological processes but also with a reduced overall physical activity level. A person with ▶ pain disorder may refrain from physical activity due to the anticipation of the aggravation of pain, which in turn will reduce his/her cardiovascular and muscle function, with a subsequent negative impact on the impairments, functional limitations, and other aspects of health. Multiple personal and environmental factors can contribute to a reduced level of activity; they include distress, reduced confidence in the ability to manage the disease, and disrupted relations with social fields that previously provided an opportunity for regular physical activity.

A growing body of evidence indicates that a sedentary lifestyle and a low level of physical activity are associated not only with poor fitness but also with the risk of coronary diseases, osteoporosis, diabetes mellitus, and so on. Exercise as a form of treatment for patients with chronic pain disorders can therefore have various goals. Specific exercises can be designed to restore identified dysfunction in the musculoskeletal system, such as muscle weakness, while general whole body exercise or aerobic exercise can aim to improve overall function, health, and mood or prevent secondary impairments and diseases associated with inactivity.

This entry will focus on the effects of exercise evaluated in randomized controlled trials (RCT). The results of the studies are discussed separately in patients with chronic low back pain (CLBP), rheumatoid arthritis (RA), and ▶ fibromyalgia (FM).

Chronic Low Back Pain (CLBP)

During the course of life, 70–85 % of individuals will experience low back pain, but most of them recover within a few weeks. Low back pain is considered to be chronic when it has lasted for more than 3 months. Active treatments are being increasingly advocated and recent research indicates that exercise plays an important role in the management of CLBP.

An RCT compared 0.5-h sessions of individual physiotherapy, 1-h sessions of muscle reconditioning on training devices, and 1-h sessions of group low-impact aerobic exercise (Mannion et al. 2001). All the three patient groups reported a reduction in pain intensity, pain frequency, and disability after the 3-month treatment period, with no differences between

them. The reduction in pain intensity and frequency was maintained in all three groups during the 1-year follow-up, while the reduction in disability was maintained in the strength training and aerobic exercise groups but not in those receiving individual physiotherapy (Mannion et al. 2001).

Moreover, a review (Liddle et al. 2004) comprising 16 RCTs of exercise, published between 1990 and 2001, found that exercise was beneficial for patients with CLBP, regardless of the chosen exercise. In this review, strengthening exercise was found to be the most common type of exercise, followed by flexibility, aerobic exercise, and multimodal programs. The strengthening exercises focused predominantly on the muscles in the lumbar spine, followed by the muscles in the abdomen and legs. Most of the studies had evaluated the effects on back-specific function and work disability, and improvements were found when comparing the intervention group with an untreated control group or a group receiving the advice of a physician, but not when comparing an exercise program with another type of program (Liddle et al. 2004). To summarize, exercise appears to be beneficial for patients with CLBP, regardless of the type of exercise that is chosen.

Exercise in Rheumatoid Arthritis (RA)

Rheumatoid arthritis (RA) is a chronic systemic inflammatory disease. The clinical picture of RA is dominated by pain, stiffness, fatigue, reduced range of motion, and muscle weakness. A reduction in muscle strength and aerobic capacity has been documented in several studies, and exercise has therefore been regarded as an important part of the rehabilitation of patients with RA. Several studies have indicated that patients with RA can engage in physical exercise without exacerbating their symptoms or increasing the inflammatory process. Furthermore, a 2-year study revealed that long-term, high-intensity exercise does not contribute to the progress of joint destruction in RA (de Jong et al. 2004).

A review, comprising 15 RCTs of exercise in RA, published between 1989 and 2000, found evidence that exercise improves aerobic capacity (six of the total of nine studies incorporating aerobic exercise) and muscle strength (eight of the total of 14 studies incorporating strengthening exercises) in patients with RA (Stenström and Minor 2003). Pain intensity was found to have decreased in three studies. Based on the results of this review, the authors suggest that both aerobic exercise and strengthening exercise for RA patients should be of moderate to hard intensity, which means 60–85 % of maximum heart rate in aerobic exercise versus 50–80 % of a maximum voluntary contraction in strength training.

Exercise in Fibromyalgia (FM)

Fibromyalgia (FM) is characterized by long-lasting, widespread pain (▶ Chronic Widespread Pain) and generalized tenderness (Wolfe et al. 1990), often accompanied by fatigue and other symptoms. Most patients with FM report activity limitations in daily life, and reduced physical performance capacity has been reported in several studies. Most researchers in the field suggest that aberrant physiological pain-processing mechanisms, together with multiple psychological and environmental factors, interact in the development of FM.

A systematic review (Mannerkorpi and Iversen 2003) comprising 27 RCTs of physical exercise for patients with FM and related syndromes, published between 1988 and 2002, concluded that exercise was beneficial for patients with FM and related syndromes. Moderate- to high-intensity aerobic exercise twice a week was found to improve aerobic capacity and reduce muscle tenderness (McGain et al. 1988), but these outcomes were not found in all the reviewed studies. The reasons for these inconsistent results may be related to the differences in the exercise programs and/or the patients' baseline physical capacity. For example, a study aiming to evaluate the effects of high-intensity aerobic exercise found that the patients that had been recruited to the study were not able to manage the planned level of exercise (van Santen et al. 2002). A meta-analysis of aerobic exercise performed at least 20 min a day, 27–3 days a week, showed evidence for positive effects on global well-being, physical function, pain and

tender points in FM. In this meta-analysis, no restriction was made on intensity of exercise. Standardized mean difference (SMD) was 0.49 for wellbeing; 0.66 for physical function; 0.65 for pain and 0.23 for tender points, implying that aerobic exercise can be recommended for patients with FM.

Walking is aerobic exercise that can be performed at varying intensities, and it appears to be an alternative for those patients with FM who are unable to participate in exercise programs of higher intensity. Studies of walking have indicated improvements in functioning, symptoms, and mood in sedentary people with FM (Mannerkorpi and Iversen 2003), something that was also reported in a study comprising sedentary women with FM who participated in a supervised walking program three times a week for 20 weeks (Valim et al. 2003).

Exercise in a temperate pool can be another alternative for sedentary patients with chronic pain, as physical properties such as buoyancy and temperature facilitate training in water and alleviate pain and stiffness, while viscosity provides the resistance required in aerobic and strengthening exercise. Pool exercise programs evaluated in patients with FM have been of varying length, ranging from 6 weeks to 6 months, and of varying intensity, ranging from low to moderate to high intensity of exercise (Mannerkorpi and Iversen 2003). Improvements have been observed in physical performance capacity and distress in several studies of pool exercise.

Learning how to modify the exercises to their own limitations may be important for patients with chronic pain, disabilities, and distress. This approach was applied in a study evaluating the effects of a 6-month pool exercise program combined with a six-session educational program in patients with FM (Mannerkorpi et al. 2000). The exercise program included aerobic, endurance, flexibility, relaxation, and body awareness exercises, and during the first weeks, the exercises focused on teaching patients how to perform the exercises properly and how to modify the exercises to match their threshold of pain and fatigue. The treatment group improved in terms of physical performance capacity (6-min walking test) and ratings of FM symptoms and distress, when compared with the control group. A long-term follow-up found that improvements in walking capacity and pain ratings lasted for 2 years (Mannerkorpi et al. 2002), probably because the patients had continued to exercise after completing the trial.

Few studies have independently evaluated the effects of strength training in FM. It appears that patients with FM can manage strength training when it is started at low loads and progresses individually. In one study, the intervention group exercised twice a week, starting at 40–60 % of their one-repetition maximum voluntary contraction and gradually increased the intensity to 60–80 % of their maximum (Häkkinen et al. 2001). Improvements were found in muscle strength, mood, and muscle firing patterns.

To summarize, most patients with FM are able to manage low-intensity exercise, while it is suggested that high-intensity exercise should be undertaken with care. Due to the large variability in functioning and symptom severity in patients with FM, it is recommended that exercise prescriptions should be individualized.

Exercise, Pain, and Mood

The primary goal of exercise as a treatment for chronic pain patients is to improve or maintain physical function and reduce disability. Most exercise studies have documented improvements in function following exercise, and this result is probably easiest to attain in sedentary patients with a low baseline level of functioning. Although pain reduction is desirable in patients with chronic pain, only a few studies have reported improvements in pain following exercise. Improvements in pain intensity and frequency have been reported in CLBP patients (Mannion et al. 2001), while studies of FM have reported improvements in mood (Häkkinen et al. 2001; Mannerkorpi et al. 2000; Valim et al. 2003).

Compliance

Motivating patients with chronic pain to start and continue regular physical activity and exercise is

crucial in order to improve function and maintain the gains. Adherence to exercise appears to improve by informing the patients about the risks and benefits of exercise and by adjusting the exercise to match individual limitations. Education and collaborative discussions about the patient's preferences and obstacles that prevent him/her from exercising might enhance adherence to planned exercise programs. Furthermore, many patients with chronic pain appreciate exercising in a supervised group, which also provides the social support that is frequently needed for regular exercise.

References

Busch, A. J., Barber, K. A., Overend, T. J., Peloso, P. M. J., & Schachter, C. I. (2007). Exercise for treating fibromyalgia syndrome. *Cochrane Database of Systematic Reviews*, (4). Art. No.: CD003786. doi:10.1002/14651858.CD003786.pub2

de Jong, Z., Munneke, M., Zwinderman, A., et al. (2004). Long term high intensity exercise and damage of small joints in rheumatoid arthritis. *Annals of the Rheumatic Diseases, 63*, 1399–1405.

Häkkinen, A., Häkkinen, K., Hannonen, P., et al. (2001). Strength training induced adaptations in neuromuscular function of premenopausal women with fibromyalgia: Comparison with healthy women. *Annals of the Rheumatic Diseases, 60*, 21–26.

IASP. (1994). *Classification of chronic pain*. Seattle: International Association for the Study of Pain.

Liddle, S., Baxter, G., & Gracey, J. (2004). Exercise and chronic low back pain: What works? *Pain, 107*, 176–190.

Mannerkorpi, K., & Iversen, M. (2003). Physical exercise in fibromyalgia and related syndromes. *Best Practice & Research. Clinical Rheumatology, 17*, 629–647.

Mannerkorpi, K., Nyberg, B., Ahlmén, M., et al. (2000). Pool exercise combined with an education program for patients with fibromyalgia syndrome. *Journal of Rheumatology, 27*, 2473–2481.

Mannerkorpi, K., Ekdahl, C., & Ahlmen, M. (2002). Six- and 24-month follow-up of pool exercise therapy and education for patients with fibromyalgia. *Scandinavian Journal of Rheumatology, 31*, 306–310.

Mannion, A., Muntener, M., Taimela, S., et al. (2001). Comparison of three active therapies for chronic low back pain: Results of a randomized clinical trial with one-year follow-up. *Rheumatology, 40*, 772–778.

McGain, G., Bell, D., Mai, F., et al. (1988). A controlled study of the effects of a supervised cardiovascular fitness training program on the manifestations of primary fibromyalgia. *Arthritis and Rheumatism, 31*, 1135–1141.

Stenström, C., & Minor, M. (2003). Evidence for the benefit of aerobic and strengthening exercise in rheumatoid arthritis. *Arthritis and Rheumatism, 49,* 428–434.

Valim, V., Oliveira, L., Suda, A., et al. (2003). Aerobic fitness effects in fibromyalgia. *Journal of Rheumatology, 30,* 2473–2481.

van Santen, M., Bolwijn, P., Verstappen, F., et al. (2002). A randomized clinical trial comparing fitness and biofeedback training versus basic treatment in patients with fibromyalgia. *Journal of Rheumatology, 29,* 575–581.

Wolfe, F., Smythe, H., Yunus, M., et al. (1990). The American college of rheumatology 1990 criteria for the classification of fibromyalgia. Report of the multi-center criteria committee. *Arthritis and Rheumatism, 33,* 160–172.

Physical Medicine and Rehabilitation

Bengt H. Sjölund
Department of Community Medicine and
Rehabilitation Medicine, Umea University,
Umea, Sweden

Introduction

Physical medicine has ancient roots in the medical field, being part of the more than 3,000-year-old traditional Chinese medicine as well as being known to have been utilized by the Romans. It is an umbrella notion containing diagnostic and above all diverse therapeutic procedures of nonsurgical and non-pharmacological character. Examples are the manual (mechanical) manipulation of musculoskeletal structures and the delivery of heat, cold, massage, dry needling, and, since the nineteenth century, application of electric currents or electromagnetic radiation to soft tissue, muscles, tendons, or joints. Furthermore, so-called balneology, i.e., the use of baths or irrigation of parts of the body with water, sometimes with specific mineral or other (e.g., iodine or sulfur) content, is traditionally included in physical medicine. These treatments were originally given in specific "health-promoting" environments, such as the ▶ spa (Latin: *salus per aqua* – health through water) or "Kurort" organizations, a custom that has lingered on in several middle European countries but is nowadays mostly found outside medical practice. The main indications for physical medicine treatments were usually the relief of chronic illness or symptoms such as aches or pain.

Rehabilitation medicine, on the other hand, originated in British experiences of patients with spinal cord injuries during the Second World War, where Dr. Ludwig (later Sir Ludwig) Guttmann in Stoke Mandeville, UK, pioneered in developing reliable rehabilitation programs for persons with paraplegia and tetraplegia (Bedbrook 1981). Rehabilitation literally means "redressing" (Latin *habitat* – dress). Apart from crisis intervention, these programs put strong emphasis on conservative treatment of the spinal fractures, on prophylactic treatment to avoid pressure sores, on care to empty the urinary bladder at regular intervals, and on how to train adequate techniques for breathing and coughing when having high injuries. Further, they teach effective means to develop techniques of independently taking care of personal hygiene as well as transferring between beds, chairs, and wheelchairs, including the effective handling of transport vehicles. Special efforts are made to counter autonomic dysfunction and metabolic deregulation and to make sexual relations as well as paternity or maternity possible.

After the post-acute rehabilitation, a lifelong follow-up ensues. The result is that the remaining number of life years has increased from 1 to 2 years after the injury even in young persons, to today's normal life span for a paraplegic, usually at an independent level and a moderately reduced life span for the tetraplegic, as a rule partly dependent.

To fulfill the many needs of persons in rehabilitation, it was recognized early on that many different health professions as well as the physician and the nurse were necessary in the daily work. This led to the development of rehabilitation teams, supervised by the physician and consisting also of nurses, physiotherapists, occupational therapists, social workers, psychologists, and sometimes vocational specialists and speech therapists. Somatic rehabilitation has gradually been expanded to include patients with disabilities from stroke, from neurological disease, from traumatic brain injury, from multi-trauma, from musculoskeletal conditions including amputations, and from chronic pain. Generally speaking, rehabilitation can be considered as a re-adaptive process, where the disabled person adapts his/her set of values to a different, more restricted life situation (Dijk 2001).

In the mid-twentieth century, the two medical areas merged in the medical specialty physical medicine and rehabilitation (now usually physical and rehabilitation medicine; PRM), "concerned with the promotion of physical and cognitive functioning, activities (including behavior), participation (including quality of life) and modifying personal and environmental factors" as presently defined (UEMS 2003) by the PRM Section of the Union Europeénne des Medecins Specialistes (UEMS).

An important part of the work in PRM is therefore disability assessment, usually performed with the ▶ team approach where health professionals jointly assess the patients. These assessments were greatly aided by the publication of the International Classification of Impairments, Disabilities, and Handicaps (ICIDH) by the World Health Organization in 1980. This classification distinguished between dysfunction at the organ level (impairment), the person level (disability), and the life situation level (handicap) rather than focusing on the underlying pathology and on the diagnosis of a disease. Through further international development, the ICIDH was replaced by the context sensitive International Classification of Functioning, Disability and Health (ICF; WHO 2001). Established in the ICF, contextual (environmental and personal) factors in addition to the impairments are crucial in assessing the functioning on the person level (activities) as well as on the life situation level (participation).

Even if patients in chronic pain have been treated at length with physical medicine approaches in rehabilitation centers since the 1920s in the Western world, it was only with the pioneering cooperation in the 1960s between professionals from the anesthesiology department and from the rehabilitation medicine department in the world's first pain clinic at the University of Washington, Seattle, USA (Bonica 1990; Fordyce 1976) that the multidisciplinary team approach from PRM was introduced into the work with patients in chronic pain.

For assessment, special instruments or questionnaires have been adapted and developed to help the team assess not only the pain but also the person and his or her perceived disability. The Minnesota Multiphasic Personality Inventory (MMPI) (Cox et al. 1978), the McGill Multidimensional Pain Questionnaire (MMPQ) (Melzack 1975), the Disability Rating Index (Salén et al. 1994), the Oswestry Low Back Pain Disability Questionnaire (Fairbank et al. 1980), the Multidimensional Pain Inventory (MPI) (Rudy et al. 1995), and various life satisfaction and health-related quality of life questionnaires (e.g., SF-36 and Lisat-11) (Melin et al. 2003) are among the most commonly used to provide supplementary information.

As regards a closer analysis of the mechanisms of chronic pain, which should be considered as a sensory impairment (Sjölund 2003) causing activity limitations and participation restrictions according to the ICF, this area still suffers the same shortcomings as in many other sectors of pain management; i.e., the exact mechanism of chronic musculoskeletal pain is usually not known. For example, whereas one-third of a large Western adult population reports chronic pain from the musculoskeletal system (Bergman et al. 2001), only 0.5 % fulfill the criteria of rheumatoid arthritis (Simonsson et al. 1999) and another 1.3 % those of fibromyalgia (Lindell et al. 2000). Most people on sick leave with back pain do not show signs of columnar disease or injury (Uden 1994). Additionally, it has been amply demonstrated that most disc herniations that can be demonstrated with MRT are asymptomatic (Savage and Roberts 1997; Stadnik et al. 1998). Instead, it may well be that many of these persons have pain due to a disturbed function in the sensory part of the central nervous system (dysfunctional pain; Johansen et al. 1999; Sjölund 1994, 2006).

Characteristics

Role of PRM in Pain Management

Three important roles for PRM can be discerned in modern pain management. First, even if the origin of pain is often uncertain, the PRM physician is trained to evaluate impairments, whether they are motor, cognitive, or, as is the case with longstanding pain, sensory in nature. In addition, he/she will assess if the patient is medically suitable for rehabilitation, as well as the rehabilitation potential of the patient. Further, he/she is trained to communicate a disabling condition and its prognosis to the patient and the significant other.

Second, some of the physical medicine modalities have been found to withstand the test of evidence-based medicine for the alleviation of various pain conditions, mostly those referred to the musculoskeletal system. Such techniques as well as pharmacological agents, singly or in combination, are commonly used by the PRM specialist to administer symptomatic impairment therapy.

Third, the PRM physician coordinates and facilitates the activities of the multi- or interdisciplinary team in the multiprofessional pain rehabilitation programs. This is carried through not only by mastering adequate rehabilitation strategies but also by promoting encouragement, motivation, and confidence during the rehabilitation process and by taking full medical responsibility for the team actions.

Physical Medicine Modalities

Some forms of ▶ spinal manipulation, characteristics techniques, as well as ▶ ultrasound for some painful conditions and injections, usually of steroids (▶ steroid injections) and local anesthetics or mixtures thereof, show a beneficial effect in chronic musculoskeletal pain but only short to intermediate term, as does ▶ dry needling, i.e., the use of an injection (or similar) needle for mechanical stimulation of superficial or deep tissue. Mechanical elongation of musculoskeletal tissues by force application, such as ▶ lumbar traction and ▶ stretching, has less

convincing effects. The same seems to be true for different forms of shortwave and microwave therapy (▶ therapeutic heat, microwaves, and cold) as well as for the delivery of heat, cold, ▶ massage and pain relief prospects, and low-level ▶ laser therapy. In general, the multitude of methods of single therapist-performed ▶ physiotherapy can be questioned for anything but very short lasting effects in chronic pain. On the other hand, ▶ physical exercise programs have recently been shown to have a beneficial effect, even in fibromyalgia.

Likewise, needle ▶ acupuncture, where the ▶ acupuncture mechanisms have become increasingly clarified during the last decades, has been found to have demonstrable positive long-term outcomes, e.g., in low back pain, even if there is a need for further randomized controlled studies. For ▶ transcutaneous electrical nerve stimulation (TENS), the classical hypothesis of induced activity in large mechanoceptive fibers closing a spinal gate is insufficient to explain the clinical observations of hour-long after effects, but new data on its mechanisms indicate a combination of spinal and supraspinal actions. There are systematic reviews indicating positive outcomes in chronic pain, but the picture is less clear regarding many other forms of pain. Here too, new studies are badly needed. An interesting new somatic stimulation technique, ▶ cutaneous field stimulation (CFS), holds promising results in cases where TENS is not effective.

Generally speaking, however, randomized controlled studies that include reasonably large homogenous patient groups ($n > 60$) are hitherto rarely available for physical treatment modalities, no doubt due both to the difficulties of blinding and to the almost impossible task of financing studies of sufficient size without the support of large organizations like the pharmaceutical industry. Therefore, many of the current recommendations for such therapeutic modalities are still based on best practice consensus or on single authorities rather than on large randomized controlled trials. It is a vital need in this field to develop more adequate test

Physical Medicine and Rehabilitation,
Fig. 1 PRM-related management of chronic pain

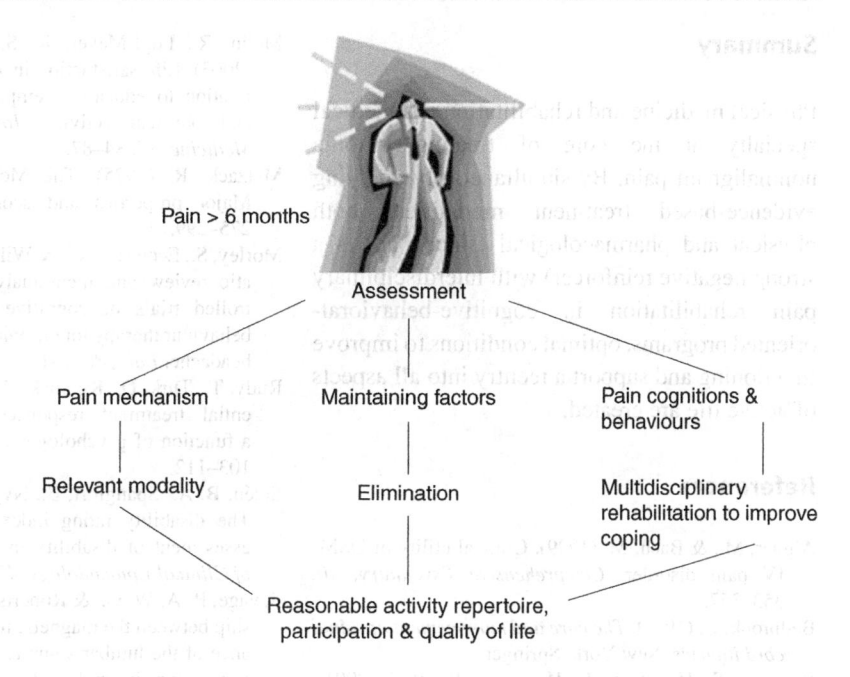

Pain > 6 months

Assessment

Pain mechanism   Maintaining factors   Pain cognitions & behaviours

Relevant modality   Elimination   Multidisciplinary rehabilitation to improve coping

Reasonable activity repertoire, participation & quality of life

paradigms for physical modalities, especially for long-term evaluation in patients with chronic pain.

Team-Oriented Rehabilitation
When it comes to multiprofessional rehabilitation programs, ► interdisciplinary pain rehabilitation has emerged as a successful integrated therapy, well supported by a number of RCTs as well as by systematic reviews (Flor et al. 1992; Morley et al. 1999; Henschke et al. 2010). However, the exact characterization of what is being treated is still lacking and has been denoted pain behaviors (Fordyce 1976), pain disorder (DSM-IV) (Aigner and Bach 1999), "chronic pain syndrome" (Black 1975), or dysfunctional MPI profile (Turk and Rudy 1988), as well as fear avoidance (Crombez et al. 1999). As pointed out previously (Sjölund 2003), the activity limitations and participation restrictions caused by chronic pain considered as a sensory impairment in conjunction with individual contextual factors in ICF terms probably describe these conditions most clearly at present.

The program should be based on behavioral-cognitive psychology to be effective but should also contain activity-related components, such as ► training by quota, ► body awareness therapies, and ► relaxation training, usually led by physiotherapists and/or occupational therapists. Education/information on the nature of chronic pain and on principles for the management of chronic as opposed to acute pain has been found to be an essential program component and should be given by the physician in charge (► Information and Psychoeducation in the Early Management of Persistent Pain). ► Coordination exercises may be of help in whiplash-associated disorders, and ergonomic counseling is usually relevant in all forms of spinal and upper extremity pain conditions.

Most importantly, interdisciplinary pain rehabilitation should be accompanied by ► vocational counseling to support a reentry into all aspects of active life. Furthermore, it is probably an advantage to combine evidence-based treatment modalities, both physical and pharmacological, with the multiprofessional cognitive-behavioral programs (Fig. 1), since pain is a strong negative reinforcer, and minimizing such a signal would diminish the learning of further pain behaviors.

Summary

Physical medicine and rehabilitation is a medical specialty at the core of treating chronic nonmalignant pain. By simultaneously applying evidence-based treatment modalities, both physical and pharmacological (since pain is a strong negative reinforcer) with interdisciplinary pain rehabilitation in cognitive-behavioral-oriented programs, optimal conditions to improve functioning and support a reentry into all aspects of active life are created.

References

Aigner, M., & Bach, M. (1999). Clinical utility of DSM-IV pain disorder. *Comprehensive Psychiatry, 40*, 353–357.

Bedbrook, G. (1981). *The care and management of spinal cord injuries*. New York: Springer.

Bergman, S., Herrstrom, P., Hogstrom, K., et al. (2001). Chronic musculoskeletal pain, prevalence rates, and sociodemographic associations in a Swedish population study. *Journal of Rheumatology, 28*, 1369–1377.

Black, R. G. (1975). The chronic pain syndrome. *The Surgical Clinics of North America, 55*, 4.

Bonica, J. (Ed.). (1990). *The management of pain* (2nd ed., pp. I–II). Philadelphia: Lea & Fabiger.

Cox, G., Chapman, C. R., & Black, R. G. (1978). The MMPI in chronic pain. *Journal of Behavioral Medicine, 1*, 437–444.

Crombez, G., Vlaeyen, J. W., Heuts, P. H., et al. (1999). Pain-related fear is more disabling than pain itself: Evidence on the role of pain-related fear in chronic back pain disability. *Pain, 80*, 329–339.

Fairbank, J., Couper, J., Davies, J. B., et al. (1980). The oswestry low back pain disability questionnaire. *Physiotherapy, 66*, 271–273.

Flor, H., Fydrich, T., & Turk, D. C. (1992). Efficacy of multidisciplinary pain treatment centers: A meta-analytic review. *Pain, 49*, 221–230.

Fordyce, W. B. (1976). *Behavioural methods in chronic pain and illness*. St. Louis: CV Mosby.

Henschke, N., Ostelo, R. W., van Tulder, M. W., Vlaeyen, J. W., Morley, S., Assendelft, W. J., & Main, C. J. (2010). Behavioural treatment for chronic low back pain. *Cochrane Database Systems Review*, (7): CD002014. doi: 10.1002/14651858.CD002014.pu.

Johansen, M., Graven-Nielsen, T., Olesen, A. S., et al. (1999). Generalised muscular hyperalgesia in chronic whiplash syndrome. *Pain, 83*, 229–234.

Lindell, L., Bergman, S., Petersson, I. F., et al. (2000). Prevalence of fibromyalgia and chronic widespread pain. *Scandinavian Journal of Primary Health Care, 18*, 149–153.

Melin, R., Fugl-Meyer, K. S., & Fugl-Meyer, A. R. (2003). Life satisfaction in 18-64-year-old Swedes in relation to education, employment situation, health and physical activity. *Journal of Rehabilitation Medicine, 35*, 84–87.

Melzack, R. (1975). The McGill pain questionnaire: Major properties and scoring methods. *Pain, 1*, 275–299.

Morley, S., Eccleston, C., & Williams, A. (1999). Systematic review and meta-analysis of randomized controlled trials of cognitive behaviour therapy and behaviour therapy for chronic pain in adults, excluding headache. *Pain, 80*, 1–13.

Rudy, T., Turk, D., Kubinski, J. A., et al. (1995). Differential treatment responses of TMD patients as a function of psychological characteristics. *Pain, 61*, 103–112.

Salén, B. A., Spangfort, E., Nygren, Å. L., et al. (1994). The disability rating index: An instrument for the assessment of disability in clinical settings. *Journal of Clinical Epidemiology, 47*, 1423–1435.

Savage, R. A. W. G., & Roberts, N. (1997). The relationship between the magnetic resonance imaging appearance of the lumbar spine and low back pain, age and occupation in males. *European Spine Journal, 6*, 106–114.

Simonsson, M., Bergman, S., Jacobsson, L. T., et al. (1999). The prevalence of rheumatoid arthritis in Sweden. *Scandinavian Journal of Rheumatology, 28*, 340–343.

Sjölund, B. (1994). Chronic pain in society -a case for chronic pain as a dysfunctional state? *Quality of Life Research, 3*, 5–9.

Sjölund, B. H. (2003). Current and future treatment strategies for chronic pain. In H. Ring & N. Soroker (Eds.), *Advances in physical and rehabilitation medicine* (pp. 307–313). Bologna: Monduzzi Editore.

Sjölund, B. H. (2006). Dysfunctional pain. In: R. F. Schmidt, & W. Willis (Eds.), *The encyclopaedia of pain*. Springer, Berlin.

Stadnik, T., Lee, R., Coen, H., et al. (1998). Annular tears and disk herniation: Prevalence and contrast enhancement on MR images in the absence of low back pain or sciatica. *Radiology, 205*, 49–55.

Turk, D. C., & Rudy, T. (1988). Toward an empirically devised taxonomy of chronic pain patients: Integration of psychology assessment data. *Journal of Consulting and Clinical Psychology, 56*, 233–238.

Uden, A. (1994). Specified diagnosis in 532 cases of back pain. *Quality of Life Research, 3*, 33–34.

UEMS, (2003). Definition of PRM. www.euro-prm.org

van Dijk, A. J. (2001). *On rehabilitation medicine. A theory-oriented contribution to assessment of functioning and individual experience* (pp. 1–280). Delft: Eburon Publishers.

WHO. (2001). *International classification of functioning, Disability and Health* (pp. 1–299). Geneva: World Health Organization.

Physical Medicine and Rehabilitation, Team-Oriented Approach

Bengt H. Sjölund
Department of Community Medicine and
Rehabilitation Medicine, Umea University,
Umea, Sweden

Synonyms

Group-oriented practice; Interdisciplinary; Interprofessional approach; Multidisciplinary; Multiprofessional; Team approach

Definition

The team-oriented approach is carried out by a group of healthcare professionals from different disciplines who share common values and objectives and provide a disabled person with holistic and comprehensive medical care, using the combined expertise of multiple caregivers. The team is the cardinal working unit in modern rehabilitation medicine, the medical specialty that primarily works with patients with ▶ impairments, ▶ activity limitations, and ▶ participation restrictions (ICF 2001).

Characteristics

In the rehabilitation process, the objective is to support the patient to reach the fullest possible physical, psychological, social, educational, vocational, and recreational goals that are consistent with existing impairments, contextual limitations, and life plans. The patient and his/her significant other work together with the rehabilitation team to develop realistic and multidimensional goals for the rehabilitation process and agree upon a ▶ rehabilitation plan to achieve these goals and to improve life satisfaction/quality of life (cf. Loisel et al. 2003).

The medical specialist supervising and carrying the responsibility for the rehabilitation process is the physical and rehabilitation medicine physician who has 3–5 years of residency training in this and related fields, focusing on conducting an effective team approach as well as on stabilizing and expanding the medical platform for the rehabilitation efforts, i.e., minimizing the impairment or dysfunction of (usually) the nervous and/or the musculoskeletal organ systems. In this context it should be remembered that persistent pain can be considered as a sensory impairment (Sjölund 2003), necessitating in-depth knowledge of pain analysis and management by the team physician in a ▶ pain rehabilitation team. He/she should further have a broad knowledge of the various rehabilitation treatments and strategies available, as well as of the profile and skills of the different allied health professions in pain rehabilitation (Table 1).

Consultations in traditional health care usually work according to the medical model, where a physician attends to the patient's needs. If another discipline is considered necessary to assess or treat the patient, that professional is consulted by the attending physician, either generally or more specifically. The consultant physician or allied health professional would usually discuss possible additional needs with the attending physician before major new diagnostic or therapeutic measures are undertaken. The medical model provides a clear chain of responsibility but may result in multiple professionals performing multiple tasks with little or no coordination (King et al. 1998). It should be remembered, however, that a prescription of specific training exercises or activity training can be as specific as a prescription of drugs and should never be changed in a major way without consulting the attending physician.

The ▶ multidisciplinary team model implies that organized interactions occur between the patient and not only the attending physician but also with other physicians and/or allied healthcare professionals, usually but not always from the same unit. It retains the medical approach in that the team is attending physician controlled and most of the communication is vertical, i.e., between the attending physician

Physical Medicine and Rehabilitation, Team-Oriented Approach, Table 1 Main allied health professions for pain rehabilitation, given in order of importance

Profession	Profile	Typical services
Psychologist	Cognitive-behavioral	Assess personality style, coping and problem-solving skills
		Conduct psycho-educational group sessions on the following subjects:
		Cognitive restructuring
		Coping self-statements
		Relaxation
		Assertiveness training
		Communication training
		Maintenance of gains
		Coping with flare-ups
Physical therapist	Restoration of motor function and awareness	Exercise, body awareness training, teach functional employment skills, e.g., by training by quota
	Administer pain relief	Offer various physical modalities such as TENS, acupuncture, heat, traction
Occupational therapist	Activity training	Assess activity limitations
		Train self-care to independence
		Train home management skills. Teach adaptive and ergonomic techniques to compensate for functional deficits
		Evaluate environmental barriers
		Explore and train vocational and recreational activities
		Educate the family to maintain patient independence
Vocational counselor	Attain realistic vocational goals	Evaluate vocational interests and skills
		Counsel on return to work, sometimes with adjustment of work environment
		Organize job-related activities
		Act as liaison and as mentor
		Provide support to employer
Social worker	Interact with family and insurers	Evaluate total living situation including family, finances, and employment history
		Coordinate funding resources
		Assess vocational barriers
		Facilitate discharge planning
		Provide emotional support
Rehabilitation nurse	Medical support	Nursing of gross physical impairment
		Case management when appropriate

and the other professionals. Assessments and treatments by different professionals in the team are usually independent of each other and a common strategy need not be shared, unless by the authority of the attending physician. Team conferences are usually conducted to report assessment results and progress from the various professionals to the attending physician. Coordination may suffer (Rothberg 1981).

The ▶ interdisciplinary team model differs from the previous two approaches in that the expected principle is group decision making in assessment, rehabilitation planning, and treatment. The patient is considered the most important member of the team, both in planning and in decision making. Common strategies of rehabilitation are used, where the different professionals contribute important components. The group shares the responsibility for the rehabilitation carried out and the team conference, led by any of the team members, is the important forum for lateral communication and for decision making.

This concept allows a free exchange of ideas and may benefit from group synergies in problem solving but can be time-consuming. The team members need training in the team process, which is rarely given during undergraduate studies. Coordination is facilitated. However, in a medical setting the medicolegal responsibility is still carried by the attending physician. Upcoming conflicts between different views should be resolved in team meetings. This is helped by communication ability and commitments by the different team members (Given and Simmons 1977).

The transdisciplinary team model is going a step further in that the various team members cross into each other's professional areas, in assessment, planning, or treatment (Hoffman 1990). This model has been said to originate from cost-reduction efforts or from lack of health professionals in underfinanced health care or social care and occurs so far mostly in educational and vocational rehabilitation settings. The exchange of information between different professionals as well as the consistency of information to patients have been reported to be highly valued in this approach, even if the therapy given has a lower quality. It is very doubtful whether this model can develop in medically oriented rehabilitation programs where the different professionals act upon highly specialized competences under government-regulated licensing qualifications. If the physician is not a co-therapist, such teams may cross medicolegal borders with unfortunate consequences (King et al. 1998).

One difficulty is inherent in the notion "disciplinary," which indicates either different medical disciplines (e.g., orthopedic surgery, psychiatry, neurology) or various professions (physician, ▶ physiotherapist, psychologist, ▶ occupational therapist, etc.). It is in fact more appropriate to reserve the notion multi-/interdisciplinary for teams containing disciplines within a profession and to use multi-/interprofessional for teams containing several professions (cf. UEMS PRM Section).

What then characterizes an effective team? There are many suggestions and statements, especially from the management literature (e.g., Leigh and Maynard 1995), but in fact little research has been performed in this field. However, there is a consensus that a coordinated team care approach seems more effective than fragmentary care for patients with long-term illness or pain (Flor et al. 1992; Keith 1991; Loisel et al. 2003). Among organization theorists, McGregor's 11 characteristics are often used to denote an effective work team (Characteristics of an effective team, McGregor 1960). In particular, a common objective is an exceedingly important factor, in pain rehabilitation consisting of a common strategy to assess and resolve the ▶ activity limitations and ▶ participation restrictions of a certain individual.

Characteristics of an Effective Team (Modified from McGregor 1960)

1. Informal and comfortable atmosphere.
2. Focused discussions where all participate.
3. The objective is well understood and accepted by all members of the group.
4. The members listen to each other and feel free to voice even unorthodox but creative ideas.
5. There may be disagreement but the group seeks to resolve rather than suppress it. Dominance by the majority or by the minority is avoided. A genuine disagreement may be taken up for reconsideration at a later moment.
6. Most decisions are reached by a consensus but there is a minimum of formal voting.
7. Criticism is frequent, frank, and constructive.
8. Team members are free to express their feelings and their ideas. There are few hidden agendas.
9. When action is decided, clear assignments are made and accepted.
10. The chairman of the group does not dominate it and the leadership shifts from time to time.
11. The group is self-conscious about its operations and regularly examines how well it performs.

In practice, the various team members combine their skills to achieve these goals and to produce a reasonable and efficient rehabilitation plan in cooperation with the patient and his/her significant other. This plan should be a written document, where overriding goals should be broken down into stepwise attainable smaller goals, with time frames, measures to be undertaken, and therapist involvement clearly stated. In particular, the role of the team physician is to perform an adequate sensory impairment analysis; to communicate the results in clear layman language to the patient; to treat the pain symptomatically, bearing in mind the long-term efficacy of the modality chosen; and to facilitate the rehabilitation process by educating the patient and supporting the actions of the other team members.

As for the patient, he/she is expected to make a transition from the passive observer role common in acute medical care to a more active participatory role in the team, emphasizing both autonomy and responsibility for his/her own rehabilitation. This requires not only information but also education of the patient and family by the health professionals.

Summary

The ▶ interdisciplinary team model is the preferred working strategy in rehabilitation, where the expected principle is group decision making in assessment, rehabilitation planning, and treatment. The patient is considered the most important member of the team, both in planning and in decision making. Common strategies of rehabilitation are used, where the different professionals contribute important components. The group shares the responsibility for the rehabilitation carried out and the team conference, led by any of the team members, is the important forum for lateral communication and for decision making. This concept allows a free exchange of ideas and may benefit from group synergies in problem solving but can be time-consuming. Coordination is facilitated. However, in a medical setting, the medicolegal responsibility is still carried by the attending physician.

References

Flor, H., Fydrich, T., & Turk, D. C. (1992). Efficacy of multidisciplinary pain treatment centers: A meta-analytic review. *Pain, 49*, 221–230.

Given, B., & Simmons, S. (1977). The interdisciplinary health care team: Fact or fiction? *Nursing Forum, 16*, 165–183.

Hoffman, L. P. (1990). Transdisciplinary team model: An alternative for speech language pathologists. *Texas Journal of Audiology and Speech-Language Pathology, 16*, 3–6.

ICF (2001). International classification of Functioning, disability and health. World Health Organization, Geneva, pp 1–299.

Keith, R. A. (1991). The comprehensive treatment team in rehabilitation. *Archives of Physical Medicine and Rehabilitation, 72*, 269–274.

King, J. C., Nelson, T. R., Heye, M. L., et al. (1998). Prescriptions, referrals, order writing and the rehabilitation team function. In J. A. DeLisa & B. M. Gans (Eds.), *Rehabilitation medicine. Principle and practice* (3rd ed., pp. 269–285). Philadelphia/New York: Lippincott Raven.

Leigh, A., & Maynard, M. (1995). *Leading your team - how to involve and inspire teams* (pp. 1–217). Great Britain: Nicholas Brealey Publishing.

Loisel, P., Durand, M. J., Diallo, B., et al. (2003). From evidence to community practice in work rehabilitation: The Quebec experience. *The Clinical Journal of Pain, 19*, 105–113.

McGregor, D. (Ed.). (1960). *The human side of enterprise* (pp. 232–235). New York: McGraw Hill.

Rothberg, J. S. (1981). The rehabilitation team: Future direction. *Archives of Physical Medicine and Rehabilitation, 62*, 407–410.

Sjölund, B. H. (2003). Current and future strategies for chronic pain. In N. Soroker & H. Ring (Eds.), *Advances in physical and rehabilitation medicine* (pp. 307–313). Bologna: Monduzzi Editore.

Physical Performance Capacity

▶ Physical Exercise

Physical Therapy

▶ Chronic Pain in Children, Physical Medicine and Rehabilitation

▶ Exercise

Physiological Pain Measure

Definition

Indirect measure of pain in which we infer presence or intensity of pain by noting changes from baseline levels for physiological parameters such as heart rate or respiration rate.

Cross-References

▶ Pain Assessment in Children

Physiological Pathways

▶ Stress and Pain

Physiotherapist

Definition

A physiotherapist is an allied healthcare professional working with motor control and musculoskeletal function, primarily treating problems on the impairment level. The profession was developed and recognized during the late nineteenth century and got specialized education from the early twentieth century, promoted by the need to treat disabled soldiers after the world wars. The education is academic in more than half of western countries, usually at the bachelor's level. In the pain field, there is a shift from passive (massage, manipulation, modalities) to active (exercise, circuit training, training by quota, body awareness therapy including range of motion and balance) physiotherapy.

Cross-References

▶ Physical Medicine and Rehabilitation, Team-Oriented Approach

Physiotherapy

Christoph Gutenbrunner
Department of Physical Medicine and Rehabilitation, Hanover Medical School, Hanover, Germany

Definition

Physiotherapy includes a wide range of therapeutic exercises carried out by physiotherapists, including active and passive movements as well as complex therapy concepts, aiming at the improvement of disturbed functions and recovery from diseases. Many physiotherapeutic techniques can be used for rehabilitation and prevention. Physiotherapy includes diagnostic measures but does not aim at the diagnosis of a disease, rather at the use of physiotherapeutic techniques and the follow-up of the therapy.

Characteristics

Methods
Basic techniques of physiotherapy are as follows:
1. Passive movements of joints and rest techniques aiming at the prevention of contractures or the improvement of the mobility of the joint in question. Other passive techniques are extensions and tractions. In manual therapy, special translatory passive movements are used in order to remove articular blocs.
2. Active movements of joints or functional units of the locomotor system in order to facilitate the respective movements. Active resisted movements are used for training purposes too, e.g., to train muscles or to improve the mechanical properties of ligaments, tendons, and bones.
3. Complex movements to initiate or practice complex locomotor functions and to improve coordinative capabilities. In order to use the facilitation effects of the buoyancy and the resistance caused by the viscosity, complex movements are often carried out in water pools.

4. Practicing special activities such as transferring oneself, walking, and climbing stairs. In this field there are some overlaps with occupational therapy which is, however, more oriented to everyday activities.
5. Cardiovascular training including endurance training of the cardiopulmonary system and training techniques aiming at an improvement in the peripheral blood flow.
6. Breathing exercises aiming at an improvement of respiratory functions or at the prevention of pulmonary complications of longer-lasting immobilization phases.
7. Other mechanical stimuli such as pressure or mechanical stimulation (e.g., deep friction) aiming at reactions of the tissue (e.g., stimulation of muscle blood flow) or at reflex responses (e.g., stimulation of tendon receptors).
8. Use of technical equipment in order to relieve affected joints or to support functions or activities (e.g., walking aids) and to stimulate sensorimotor functions (Pezzi ball) as well as for the training of muscles and the cardiopulmonary system (so-called medical training).
9. Relaxation therapies, e.g., using the mechanism of postisometric muscle relaxation or respiration techniques.

In addition, the personal interaction between therapist and patient is obviously a relevant factor in physiotherapy.

Besides these techniques in physiotherapy, many complex concepts and schools exist, most of them based on special theories of the function of the locomotor system and the effects of special techniques. These concepts and schools are mostly described by a single therapist or physician and many of them lack scientific evaluation. Some of them are only in use in single countries or regions. A system of the concepts and schools in physiotherapy has been produced by Gutenbrunner and Weimann (2004), differentiating between the following:

1. Organ-specific exercise concepts (e.g., respiration therapy, scoliosis therapy (Lehnert and Schroth), functional spine therapy (Laabs))
2. Training therapies (isokinetic muscle training, medical training therapy)
3. Manual medicine (chiropractic, manual medicine (Maitland concept), osteopathy)
4. Neurophysiological concepts (developmental neurobiological based concept (Bobath), developmental kinesiology (Vojta), sensomotoric facilitation (Janda), proprioceptive neuromuscular facilitation (Kabat and others))
5. Concepts based on special techniques (hippotherapy, functional movements (Klein-Vogelbach), aquatic exercise, loop (sling) table therapy)
6. Relaxation therapy and body-oriented psychotherapies (progressive muscle relaxation (Jacobson), functional integration (Feldenkrais), perceptive-cognitive therapy model (Affolter))

The concepts mentioned are of course only examples. The list of concepts is not complete nor are the concepts mentioned especially recommended.

Mechanisms

The application of physiotherapeutic techniques can induce a variety of reactions, most of which are therapeutically relevant. The main principles of these reactions refer to a system of adaptive capabilities of the organism (Gutenbrunner 2004):

1. Relief of strain or protection is a principle enabling the organism's healing and recovery processes. Physiologically speaking, immobilization leads to de-adaptation (muscle atrophy, calcium loss from bone, etc.). Therefore, only partial immobilization should be carried out. This can be effective, e.g., if muscle tenseness is caused by an increased afferent input from receptors of tendons or joint capsules. In physiotherapy, tractions aiming mainly at a relief of pressure on the joints and special rest techniques aiming at the prevention of contractures are of relevance.
2. Inhibition and facilitation are mechanisms that influence the proprioceptors of muscles and joint structures in order to facilitate or to reduce pathological muscle activity. The interneurons are located in the spinal cord.

Inhibition and facilitation mechanisms are of relevance mostly in neurological conditions.

3. Habituation, on the contrary, describes a mechanism controlled by the formatio reticularis. Habituation reduces (within minutes) autonomous reactions to different stimuli (increase of blood pressure or heart rate). Probably pain perception can also be influenced by habituation.

4. Sensomotoric adaptation describes repetitive exercises of specific movements that can induce coordinative capabilities. Probably peripheral (functional changes in the synapses of the motor cells), sprouting and central (activation of sleeping synapses, neuroplasticity) mechanisms are involved. Recently it has been shown that neuroplasticity can be induced by physiotherapeutic techniques (for an overview, see Rolnik 2004). Besides physical exercises, mental exercises can also improve coordinative performance.

5. Functional adaptation is a process that can be induced by repeated dynamic exercises; however, its development needs several weeks of time. It is triggered by hormonal mechanisms (pituitary-adrenocortical axis). It may lead to normalization of different autonomic functions (positive cross adaptations) and can be of some significance in the treatment of chronic pain, because pain perception is closely related to autonomic arousal.

6. Trophic and plastic adaptation describes the possibility of inducing growth of cells or intracellular organelles by repeated application of strain, e.g., muscle training, increase in the resistance of ligaments, or structural adaptations of the bones after repeated mechanical stress. Such processes are triggered by specific hormone systems (growth hormone, erythropoietin, etc.). The therapeutic significance is obvious in all states of weakness of structures of the locomotor system. Its relevance has been proven in several painful diseases such as low back pain or osteoporosis (for an overview, see Gutenbrunner 2004).

7. Changes of behavior and psychic adjustments can be interpreted as cortical adaptations. There is no doubt that movements can influence central nervous functions, e.g., strenuous work reduces pain perception, probably triggered by hormonal (opioid-like peptides as well as serotonin, adrenalin, and noradrenalin) or neural mechanisms. It has been proven that serial dynamic exercise may reduce depression and anxiety, which certainly is relevant in chronic pain syndromes. Physiotherapy can additionally play an important role in behavioral therapy concepts.

Effects in Pain Syndromes

In pain syndromes physiotherapy can be useful because of the following principles:

1. Diminished dysfunction of affected organs (joints, muscles, etc.)
2. Inhibiting pain perception by stimulation of sensory afferents according to the gate-control theory
3. Avoiding painful movements by an improvement of the quality of movement, e.g., by muscular training or improvement of coordinative skills
4. Normalization of autonomic nervous functions and reduction of depression by use of functional adaptive processes

For these reasons physiotherapy is indicated in all pain syndromes in which dysfunction of the locomotor system is involved. The active therapy should start as early as possible. Other indications justified by the reduction of depression and normalization of autonomic functions are other chronic pain syndromes with comorbidities and a marked reduction of quality of life. In most cases, physiotherapy is applied in combination with other treatments, such as analgesic drugs, other physical therapies, psychotherapy, and educational programs.

These indications are supported by clinical experience. However, the evidence in controlled clinical trials is weak. One reason for this is that there are only a few controlled studies using pain as the main outcome parameter. Secondly, many studies use the (recommended) strategy with a polypragmatic (rehabilitative) approach. Nevertheless – looking into the literature of the last few years (Medline 2001–2003) – there are some studies confirming the effectiveness of physiotherapy in many painful syndromes, e.g., low back pain (Pengel et al. 2002), cervical spine syndromes (Persson and Lilja 2001), cervicogenic headache (Jull et al. 2002), femoropatellar pain syndromes (Cowan et al. 2003; Crossley et al. 2002), shoulder pain (Hay et al. 2003), and craniomandibular dysfunction (Oh et al. 2002). Complex regimens were proven to have positive results in low back pain (Moseley 2002; Schonstein et al. 2003) and fibromyalgia (Gustafsson et al. 2002). For some techniques, controlled trials have shown negative results (Hemmila et al. 2002; Smidt et al. 2002) and some meta-analyses also show weak or missing evidence (Brosseau et al. 2002; Goodyear-Smith and Arroll 2001).

Summarizing the evidence from an expert point of view, it can be stated that implementing physiotherapy in pain therapy and rehabilitation programs is useful, although the scientific evidence is still weak. Therefore, further prospective controlled studies are needed.

References

Brosseau, L., Casimiro, L., Milne, S., et al. (2002). Deep transverse friction massage for treating tendinitis. *Cochrane Database of Systematic Reviews, 4,* CD003528.

Cowan, S. M., Bennell, K. L., Hodges, P. W., et al. (2003). Simultaneous feedforward recruitment of the vasti in untrained postural tasks can be restored by physical therapy. *Journal of Orthopaedic Research, 21,* 553–558.

Crossley, K., Bennell, K., Green, S., et al. (2002). Physical therapy for patellofemoral pain: A randomized, double-blinded, placebo-controlled trial. *The American Journal of Sports Medicine, 30,* 857–865.

Goodyear-Smith, F., & Arroll, B. (2001). Rehabilitation after arthroscopic meniscectomy: A critical review of the clinical trials. *International Orthopaedics, 24,* 350–363.

Gustafsson, M., Ekholm, J., & Broman, L. (2002). Effects of a multiprofessional rehabilitation programme for patients with fibromyalgia syndrome. *Journal of Rehabilitation Medicine, 34,* 119–127.

Gutenbrunner, C. (2004). Therapieprinzipien. In C. Gutenbrunner & G. Weimann (Eds.), *Krankengymnastische methoden und konzepte* (pp. 10–24). Berlin/Heidelberg/New York: Springer.

Gutenbrunner, C., & Weimann, G. (2004). *Krankengymnastische methoden und konzepte.* Berlin/Heidelberg/New York: Springer.

Hay, E. M., Thomas, E., Paterson, S. M., et al. (2003). A pragmatic randomised controlled trial of local corticosteroid injection and physiotherapy for the treatment of new episodes of unilateral shoulder pain in primary care. *Annals of the Rheumatic Diseases, 62,* 394–399.

Hemmila, H. M., Keinanen-Kiukaanniemi, S. M., Levoska, S., et al. (2002). Long-term effectiveness of bone-setting, light exercise therapy, and physiotherapy for prolonged back pain: A randomized controlled trial. *Journal of Manipulative and Physiological Therapeutics, 25,* 99–104.

Jull, G., Trott, P., Potter, H., et al. (2002). A randomized controlled trial of exercise and manipulative therapy for cervicogenic headache. *Spine, 27,* 1835–1843.

Moseley, L. (2002). Combined physiotherapy and education is efficacious for chronic low back pain. *The Australian Journal of Physiotherapy, 48,* 297–302.

Oh, D. W., Kim, K. S., & Lee, G. W. (2002). The effect of physiotherapy on post-temporomandibular joint surgery patients. *Journal of Oral Rehabilitation, 29,* 441–446.

Pengel, H. M., Maher, C. G., & Refshauge, K. M. (2002). Systematic review of conservative interventions for subacute low back pain. *Clinical Rehabilitation, 16,* 811–820.

Persson, L. C., & Lilja, A. (2001). Pain, coping, emotional state and physical function in patients with chronic radicular neck pain. A comparison between patients treated with surgery, physiotherapy or neck collar–a blinded, prospective randomized study. *Disability and Rehabilitation, 23,* 325–335.

Rolnik, J. D. (2004). Nervensystem. In C. Gutenbrunner & G. Weimann (Eds.), *Krankengymnastische methoden und konzepte* (pp. 54–55). Berlin/Heidelberg/New York: Springer.

Schonstein, E., Kenny, D. T., & Keating, J et al. (2003). Work conditioning, work hardening and functional restoration for workers with back and neck pain. *Cochrane Database Systematic Reviews.* CD001822.

Smidt, N., van der Windt, D. A., Assendelft, W. J., et al. (2002). Corticosteroid injections, physiotherapy, or a wait-and-see policy for lateral epicondylitis: A randomised controlled trial. *Lancet, 359,* 657–662.

Pin-Prick Pain

▶ First and Second Pain Assessment (First Pain, Pricking Pain, Pin-Prick Pain, Second Pain, Burning Pain)

PIRFT

Definition

Percutaneous intradiscal radiofrequency thermocoagulation, where a radiofrequency lesion is made in the nucleus pulposus, causes thermal fibrosis, which reduces nociceptive input from a painful intervertebral disc.

Cross-References

▶ Discogenic Back Pain

Piriformis

Definition

One of the groin muscles involved in hip movement.

Cross-References

▶ Sacroiliac Joint Pain

Pituitary Adenylate Cyclase-Activating Peptide

Definition

A polypeptide present in the pituitary and areas of the nervous system that modulates pain sensation as well as other neuronal processes.

Cross-References

▶ Peptides in Neuropathic Pain States

Pituitary Gland

Definition

The pituitary gland (hypophysis) is a neuroglandular organ located in the sella turcica just below the base of the brain (the hypothalamus). It includes two portions: (1) the anterior hypophysis (adenohypophysis) that is glandular and includes three parts, the anterior, intermedialis, and tuberalis, and (2) the posterior hypophysis (neurohypophysis) that is directly linked up to the hypothalamus via the pituitary stem which contains the axons of the hypothalamic magnocellular neurons.

Cross-References

▶ Hypothalamus and Nociceptive Pathways

PKA

▶ Protein Kinase A

PKC

▶ Protein Kinase C
▶ TRPV1 Modulation by PKC

Placebo

Scott Masters
Caloundra Spinal and Sports Medicine Centre,
Caloundra, QLD, Australia

Synonyms

Dummy; Inactive agent; Inert agent; Sham

Definition

Finding an adequate definition of ▶ placebo has proved a challenge. All definitions deal with the paradox of an intervention producing a beneficial effect when it is not supposed to do so. The paradox arises when an intervention has some effects that are due to a known physiological mechanism, and others whose mechanism is unknown. Respectively, these are referred to as the specific and nonspecific effects of a treatment. The paradox is resolved if and once the mechanisms of the nonspecific effects become known. However, while so long as its mechanism remains unknown, any effect that is not due to the specific, intended effects of a treatment is referred to as a ▶ placebo effect. Any beneficial effect that cannot be attributed to the specific, intended effects is referred to as a placebo response.

In its purist form, a placebo is an agent with no (known) therapeutic effect or "an intervention designed to simulate medical therapy" (Brody 2000) but which has no specific therapeutic effect.

The opposite of a placebo effect is a ▶ nocebo effect, in which harm occurs for no apparent reason (see ▶ nocebo).

Characteristics

The term placebo (from the Latin: "I shall please") was first used medically in the late 1700s. It described medicines adapted more to please patients rather than cure them. The use of placebos was widespread well into the first half of the twentieth century. In a 1981 hospital study, physicians and nurses admitted to using placebo administration in up to 80 % of patient contacts, mainly for analgesia (Gray and Flynn 1981).

The placebo effect is a specific phenomenon that needs to be distinguished from others with which it has been confused. It is not natural history, spontaneous recovery, regression to the mean, or the "Hawthorne effect."

The natural history of some conditions is that recovery occurs in due time, because of healing or resolution of the original problem that caused pain. The relief that occurs is a property of the disorder itself and neither implies nor requires a placebo effect. The two are distinguished by comparing the outcomes of a group treated with placebo and a group having no treatment.

Regression to the mean is a statistical phenomenon that occurs when measurements are taken. It refers to the tendency of measurements to be inflated at the commencement of an experiment but to return to average levels as the experiment continues. Patients with chronic pain are more likely to present to a doctor, or to enrol in studies, when their pain is worse than usual. In time, that pain settles to lesser, and more typical, average levels, regardless of the treatment implemented. The observed improvement is not due to a placebo effect; it is a reflection of normal fluctuations in reported levels of pain.

The "Hawthorne effect" is a phenomenon in which a subject's performance changes because they recognize that they are being studied. The change occurs consciously or subconsciously as the subject tries to guess and produce the behavior that they believe the investigator is looking for. The essential feature is that the subject believes that they are being judged in some way. A "Hawthorne effect" may contaminate a placebo effect, if the patient believes their role is to please the investigator. However, the essence of the placebo effect is that it is a response that the subject reports as due to an intervention, under open conditions, with no hidden agendas.

Mechanism

Four main theories are used to explain the placebo effect (Brody 2000; Peck and Coleman 1991).

According to *classic conditioning theory*, the placebo response is a conditioned response to features of the treatment setting (doctor's attire and equipment, pills, etc.). Relief occurs because going to the doctors in the past has been associated with relief.

The *response expectancy theory* maintains that patients respond favorably to treatment because they expect that treatment will relieve their pain.

The *meaning model* suggests three factors are important to maximize the placebo response. Firstly, the patient must feel listened to and must receive a valid coherent explanation for their illness. Secondly, the patient needs to feel care and compassion from their treatment environment. Lastly, the patient should feel empowered. The meaning model implies that patients amplify their pain because of fears about what is causing it and its implications. When these fears are eliminated, the pain is reduced to tolerable or trivial levels and emerges as a therapeutic effect.

Cognitive dissonance theory states that the holding of two or more beliefs, which are psychologically inconsistent, creates tension in the individual's psyche. This motivates the individual to reduce this inconsistency. For placebo treatments, the knowledge that a treatment has been administered, that the healer has said the treatment would work, and that only very ill people do not improve creates a dissonance if the patient does not improve. To reduce the likelihood of this anxiety-provoking dissonance, the patient alters their perception of their symptoms and their illness.

Biochemical Mediators

Modern research has demonstrated that placebo responses are not a psychological phenomenon or a sign of aberrant behavior. There is compelling evidence that endogenous opioids are involved in the placebo response to analgesia (Benedetti and Amanzio 1997; Ter Riet et al. 1998; Amanzio and Benedetti 1999; Amanzio et al. 2001). The placebo effect is reversible by naloxone; blockade of cholecystokinin receptors potentiates placebo analgesia; and placebo analgesia can mimic the respiratory depression side effect of morphine, and this side effect is reversible by naloxone. Catecholamines and cortisol have also been implicated as important mediators of the placebo effect (Brody 2000).

Response Rate

A figure of 35 % is often quoted as if it is a standard measure of the incidence of placebo response rates, but it is inaccurate and misleading. It is no more than the average incidence of placebo responses in the 15 papers that Beecher reviewed in his 1955 article: "The Powerful Placebo" (Beecher 1955). It is more realistic to recognize that the incidence of placebo responses can vary from 0 % to 100 %, depending on the disease, the setting, the administrator of the placebo, and numerous other factors (Voudouris et al. 1989).

Contrary to popular opinion, it is not the psychologically weak or the malingerer who responds to placebo; and there is no such entity as a "placebo responder." It has been shown that, under different circumstances, the same individual can be both a responder and a nonresponder (Peck and Coleman 1991; Voudouris et al. 1989).

Clinical Application

Brody has suggested that practice style and clinician behavior can be modified to potentiate the placebo effect in an ethical manner (Brody 2000). Areas for modification include:

Sustained Partnership

Developing a relationship with patients is often what separates primary practice from tertiary care. Six elements have been linked to superior health outcomes (Fig. 1).

Mastery

Empowering patients is an important part of the recovery process. Mastery encourages the patient to change from a passive instrument, who is totally dependent on the "healer," to a person ready to take control of their health issues by regaining control. The patient is encouraged to express their desires about management and verbalize their hidden thoughts about the reason for their problem and possible solutions.

Story

Many presentations to primary care are undifferentiated and remain so despite a barrage of investigations. Patients can feel abandoned if they are dismissed after a fruitless search for meaningful pathology. This is a common experience for people with chronic pain. If their pain story can be reconstructed in a way they can

Placebo, Fig. 1 Ethical attitudes and behaviors that serve to enhance placebo responses

DOCTOR	PATIENT
• expressess interest in the total person • provides ongoing temporal care • adaptive to patient's idiosyncrasies • avoids cookbook medicine	• feels they are in a caring, sensitive and empathetic environment • views the doctor as reliable and trustworthy

BOTH
• share in decision-making

understand, other management issues (e.g., physical therapy, vocational training, and medication use) are simpler to implement.

Similarly for acute pain, patients do much better if they are expressly told that they are receiving a strong painkiller, than if their treatment is covert (whereby they have no chance of a story). An elegant study on postoperative patients illustrates this difference (Voudouris et al. 1989). Patients were given buprenorphine on demand for 3 consecutive days along with a basal infusion of saline. Three different stories were told about the saline infusion. The first group was told nothing (natural history); the second was told that the saline could be either a powerful analgesic or a placebo (double-blind); and the third group were all told it was a potent analgesic (deceptive administration). The reduction in buprenorphine use in the double-blind group was 20 % compared with the natural history group and 33 % for the deceptive administration group.

These results are consistent with other hidden-injection studies (Amanzio et al. 2001). Patients unaware that they are receiving pain-killing injections get significantly less analgesic response than those who are made aware. This reinforces the importance of making certain that patients understand what their analgesics are for, how to use them properly, and to comprehend on their level how the analgesics will work in their body.

Importantly, the above scenarios involve complementing the use of active medication with placebo. Placebo used solely as therapy is not innocuous. It can delay patients seeking more effective treatments, it can reduce the response to active therapy in the future, it can add to the cost of treatment, and it may result in a nocebo effect. Also, by being a constant reminder of illness, it may prevent people from functioning to their otherwise full capacity.

Summary

Placebo treatment is not the same as no treatment. Normal treatment is the sum of natural history, placebo effect, and medical treatment. In pain medicine, it has been clearly shown that utilizing the placebo response can decrease the amount of analgesic needed and improve response to analgesics. Health workers would be well served to familiarize themselves with the ethical ways of harnessing the placebo effect.

Cross-References

▶ Amygdala, Pain Processing and Behavior in Animals
▶ Descending Circuits in the Forebrain, Imaging
▶ Functional Imaging of Cutaneous Pain
▶ Lumbar Traction
▶ Opioids in Geriatric Application

References

Amanzio, M., & Benedetti, F. (1999). Neuropharmacological dissection of placebo analgesia: Expectation-activated opioid systems versus conditioning activated specific subsystems. *Journal of Neuroscience, 19,* 484–494.

Amanzio, M., Pollo, A., Maggi, G., & Benedetti, F. (2001). Response variability to analgesics: A role for Non-specific activation of endogenous opioids. *Pain, 90,* 205–215.

Beecher, H. K. (1955). The powerful placebo. *The Journal of the American Medical Association, 159*, 1602–1606.

Benedetti, F., & Amanzio, M. (1997). The neurobiology of placebo analgesia: From endogenous opioids to cholecystekinin. *Progress in Neurobiology, 52*, 109–125.

Brody, H. (2000). The placebo response. *Journal of Family Practice, 49*, 649–654.

Gray, G., & Flynn, P. (1981). A survey of placebo use in a general hospital. *General Hospital Psychiatry, 3*, 199–203.

Peck, C., & Coleman, G. (1991). Implications of placebo theory for clinical research and practice in pain management. *Theoretical Medicine, 12*, 247–270.

Ter Riet, G., De Craen, A. J. M., De Boer, A., & Kessels, A. G. (1998). Is placebo analgesia mediated by endogenous opioids? A systematic review. *Pain, 6*, 273–275.

Voudouris, N. J., Peck, C. L., & Coleman, G. (1989). Conditioned response models of placebo phenomena: Further support. *Pain, 38*, 109–116.

Placebo Analgesia

Definition

This term can be applied either in reference to what a physician administers or what a patient with pain experiences.

When administered by a physician, placebo analgesia is an intervention intended to produce relief of pain through a placebo effect and not through any more specific effect of the intervention. Placebo analgesia is most typically administered in the course of controlled trials of interventions for pain.

When experienced by a patient, placebo analgesia refers to the relief of pain as a result of a placebo effect of an intervention.

Cross-References

▶ Hypnotic Analgesia
▶ Pain-Modulatory Systems, History of Discovery
▶ Placebo
▶ Placebo Analgesia and Descending Opioid Modulation

Placebo Analgesia and Descending Opioid Modulation

Fabrizio Benedetti
Department of Neuroscience, Clinical and Applied Physiology Program, University of Turin Medical School, Turin, Italy

Synonyms

Placebo analgesic effect; Placebo analgesic response

Definition

Placebo analgesia is the reduction or the disappearance of pain, when an inert treatment (the placebo) is administered to a subject who is told that it is a painkiller. The placebo effect, so far considered a nuisance in clinical research when a new analgesic treatment has to be tested, has now become a target of scientific investigation, to better understand the physiological and neurobiological mechanisms that link a complex mental activity to different functions of the body. Usually, in clinical research, the term placebo effect refers to any improvement in the condition of a group of subjects that has received a placebo treatment. Conversely, the term placebo response refers to the change in an individual caused by a placebo manipulation. However, these two terms are sometimes used interchangeably.

Characteristics

Methodological Aspects

The identification of a placebo effect is not easy from a methodological point of view, and its study is full of drawbacks and pitfalls. In fact, the positive effects following the administration of a placebo can be due to many factors, such as spontaneous remission, regression to the mean, symptom detection ambiguity, and biases (Benedetti 2008a). All these phenomena must be ruled out through adequate control groups. For example, spontaneous remission can be discarded by means of a no-treatment group, which gives us information about the natural course of a symptom. Regression to the mean, a statistical phenomenon due to selection biases at the enrolment in a clinical trial, can be controlled by using appropriate selection criteria. Symptom detection ambiguity and biases can be avoided by using objective physiological measurements. It is also important to rule out the possible effects of co-interventions. For example, the mechanical insertion of a needle for the injection of an inert substance may per se induce analgesia, thus leading to erroneous interpretations.

When all these phenomena are ruled out and the correct methodological approach is used, substantial placebo effects can be detected which are mediated by psychophysiological mechanisms worthy of scientific inquiry. Therefore, it is this psychological component that represents the real placebo effect.

Mechanisms

The placebo effect is basically a context effect, whereby the psychosocial context (e.g., the therapist's words, the sight of medical personnel and apparatuses, and other sensory inputs) around the medical intervention and the patient plays a crucial role. Today we know that the context may produce a therapeutic effect through at least two mechanisms: conscious anticipatory processes and unconscious conditioning mechanisms (Benedetti 2008b; Enck et al. 2008; Price et al. 2008). In the first case, expectation and anticipation of clinical benefit may affect the response to a therapy. In the second case, contextual cues (e.g., color and shape of a pill) may act as conditioned stimuli that, after repeated associations with an unconditioned stimulus (the pharmacological agent inside the pill), are capable alone of inducing a clinical improvement. In the case of pain and Parkinson's disease, it has been shown that conscious expectancies play a crucial role, even though a conditioning procedure is performed, whereas placebo responses in the immune and endocrine system are mediated by unconscious classical conditioning.

The neural mechanisms underlying the placebo effect are only partially understood, and most of our knowledge comes from the field of pain and analgesia, although Parkinson's disease, immune and endocrine responses, and depression have more recently emerged as interesting models. In each of these conditions, different mechanisms seem to take place, so that we cannot talk of a single placebo effect but many (Benedetti 2008a).

As far as pain is concerned, there is now compelling experimental evidence that the endogenous opioid systems play an important role in some circumstances. There are several lines of evidence indicating that placebo analgesia is mediated by a descending pain-modulating circuit which uses endogenous opioids as neuromodulators. This evidence comes from a combination of imaging and pharmacological studies. By using positron emission tomography (PET), it was found that some cortical and subcortical regions are affected by both a placebo and the opioid agonist remifentanil, thus indicating a related mechanism in placebo-induced and opioid-induced analgesia (Petrovic et al. 2002). In particular, the administration of a placebo affects the rostral anterior cingulate cortex (rACC), the orbitofrontal cortex (OrbC), and the brainstem. Moreover, there is a significant covariation in activity between the rACC and the lower pons/medulla at the level of the rostral ventromedial medulla (RVM) and a subsignificant covariation between the rACC and the periacqueductal gray (PAG), thus suggesting that the descending rACC/PAG/RVM pain-modulating circuit is involved in placebo analgesia. In a different study with functional magnetic resonance imaging (fMRI), it was shown that placebo administration produces a decrease of activity in many regions involved in pain transmission, such as the thalamus and the insula (Wager et al. 2004; Price et al. 2008).

A clear-cut involvement of endogenous opioids is shown by several pharmacological studies in which placebo analgesia is antagonized by the opioid antagonist naloxone. In addition, it has been shown that the endogenous opioid systems have a somatotopic organization, since local placebo analgesic responses in different parts of the body can be blocked selectively by naloxone (Benedetti 2008a). By using in vivo receptor binding techniques, placebos have been found to induce the activation of mu opioid receptors in different brain areas, like the dorsolateral prefrontal cortex, nucleus accumbens, insula, and rACC. It is also worth noting that opioids are activated together with dopamine, e.g., in the nucleus accumbens, thus indicating a complex interaction between different neurotransmitters (Scott et al. 2008).

The placebo-activated endogenous opioids do not act only on pain transmission but on the respiratory centers as well, since a naloxone-reversible placebo respiratory depressant effect has been described. Likewise, a reduction of beta-adrenergic sympathetic system activity, which is blocked by naloxone, has been found during placebo analgesia. These findings indicate that the placebo-activated opioid systems have a broad range of action, influencing pain, respiration, and the autonomic nervous system (Colloca and Benedetti 2005). The placebo-activated endogenous opioids have also been shown to interact with endogenous substances that are involved in pain transmission. In fact, on the basis of the anti-opioid action of cholecystokinin (CCK), CCK-antagonists have been shown to enhance placebo analgesia, thus suggesting that the placebo-activated opioid systems are counteracted by CCK during a placebo procedure (Colloca and Benedetti 2005).

It is important to point out that some types of placebo analgesia appear to be insensitive to naloxone. For example, if a placebo is given after repeated administrations (preconditioning) of the non-opioid painkiller ketorolac, the placebo analgesic response is not blocked by naloxone.

The release of endogenous substances following a placebo procedure is a phenomenon which is not confined to the field of pain, but it is also present in other conditions, such as Parkinson's disease. As occurs with pain, in this case patients are given an inert substance and are told that it is an anti-Parkinson drug that produces an improvement in their motor performance. A study used PET imaging in order to assess the competition between endogenous dopamine and ^{11}C-raclopride for D_2/D_3 receptors, a method that allows identification of endogenous dopamine release (de la Fuente-Fernandez et al. 2001). This study found that placebo-induced expectation of motor improvement activates endogenous dopamine in both the dorsal and ventral striatum, a region involved in reward mechanisms, which suggests that the placebo effect may be conceptualized as a form of reward. Placebo administration in Parkinson patients has also been found to affect the activity of neurons in the subthalamic nucleus, substantia nigra, and thalamus (Benedetti et al. 2004, 2009).

The study of the placebo effect in conditions other than pain is certainly useful to better understand how placebos work in pain itself. For example, the involvement of expectation mechanisms, anxiety reduction, classical and Pavlovian conditioning in the immune and endocrine systems, as well as the study of other conditions such as neuropsychiatric disorders and physical performance may help find similarities and differences across a variety of medical conditions and therapeutic interventions (Finniss et al. 2010).

Nocebo Hyperalgesia

The nocebo effect, or response, is a placebo effect in the opposite direction. Mainly due to ethical constraints, the nocebo effect has not been investigated in detail, as has been done for the placebo effect. In fact, a nocebo procedure is per se stressful and anxiogenic, as it induces negative expectations of clinical worsening. For example, the administration of an inert substance along with verbal suggestions of pain increase may induce a hyperalgesic effect. In this case, anticipatory anxiety plays a fundamental role. Nocebo hyperalgesia has been found to be blocked by proglumide, a nonspecific CCK-A/CCK-B receptor antagonist. This suggests that expectation-induced hyperalgesia is mediated, at least in part, by CCK (Enck et al. 2008). These effects of proglumide are not antagonized by naloxone, thus endogenous opioids are not involved.

Dopamine also plays a role in the nocebo hyperalgesic effect. In fact, whereas placebo

Placebo Analgesia and Descending Opioid Modulation, Fig. 1 Cascade of events that may occur during a placebo procedure. Pain is inhibited by a descending inhibitory network involving the rostral anterior cingulate cortex (*rACC*), the orbitofrontal cortex (*OrbC*), the periacqueductal gray (*PAG*), and the rostral ventromedial medulla (*RVM*). Endogenous opioids inhibit pain through this descending network and/or other mechanisms. The respiratory centers may be inhibited by opioid mechanisms as well. The β-adrenergic sympathetic system is also inhibited during placebo analgesia, although the underlying mechanism is not known (reduction of the pain itself and/or direct action of endogenous opioids). Non-opioid mechanisms are also involved. Cholecystokinin (*CCK*) counteracts the effects of the endogenous opioids, thus antagonizing placebo analgesia. Placebos may also affect serotonin-dependent hormone secretion, like growth hormone and cortisol

administration induces the activation of dopamine in the nucleus accumbens, the administration of a nocebo is associated with a deactivation of dopamine (Scott et al. 2008).

Clinical Implications
According to the classical methodology of randomized controlled trials, any drug must be compared with a placebo in order to assess its effectiveness. If the patients who take the drug show a larger clinical improvement than the patients who take the placebo, the drug is considered to be effective. However, in light of the recent advances in placebo research, some caution is necessary in the interpretation of some clinical trials. In fact, by considering the complex

cascade of biochemical events that take place after placebo administration, any drug that is tested in a clinical trial may interfere with these placebo/expectation-activated mechanisms, thus confounding the interpretation of the outcomes (Colloca and Benedetti 2005). As we have no a priori knowledge of which substances act on placebo-activated endogenous opioids, or dopamine, or serotonin – and indeed almost all drugs, e.g., analgesics and opioids, might interfere with these neurotransmitters – one way to eliminate this possible pharmacological interference is to make the placebo-activated biochemical pathways, so to speak, "silent." This can be achieved by the hidden administration of drugs (Colloca et al. 2004).

In fact, it is possible to eliminate the placebo (psychological) component and analyze the pharmacodynamic effects of a treatment, free of any psychological contamination, by administering drugs covertly. In this way, the biochemical events that are triggered by a placebo procedure can be eliminated. To do this, drugs are administered through hidden infusions by machines, so that the patient is not made aware that a medical therapy is being carried out. A hidden drug infusion can be performed through a computer-controlled infusion pump that is preprogrammed to deliver the drug at the desired time. It is crucial that the patient does not know that any drug is being injected, so that he or she does not expect anything. The computer-controlled infusion pump can deliver a drug automatically, without a doctor or nurse in the room, and without the patient being aware that a treatment has been started.

The analysis of different treatments, either pharmacological or not, in different conditions has shown that an open (expected) therapy, which is carried out in full view of the patient, is more effective than a hidden one (unexpected). Whereas the hidden injection represents the real pharmacodynamic effect of the drug, free of any psychological contamination, the open injection represents the sum of the pharmacodynamic effect plus the psychological component of the treatment. The latter can be considered to represent the placebo component of the therapy, even though it cannot be called placebo effect, as no placebo has been given. Therefore, by using hidden administration of drugs, it is possible to study the placebo effect without the administration of any placebo.

The influence of psychological factors on drug action is also shown by the balanced-placebo design (Benedetti 2008a). In this design, four groups of patients (1) receive the drug and are told it is a drug, (2) get the drug and are told it is a placebo, (3) get a placebo and are told it is a placebo, and (4) get a placebo and are told it is a drug. This design is particularly interesting because it indicates that verbally induced expectations can modulate the therapeutic outcome, both in the placebo group and in the active treatment group. It has been used in many conditions, such as alcohol research, smoking, and amphetamine effects (Benedetti 2008a) (Fig. 1).

References

Benedetti, F. (2008a). *Placebo effects: Understanding the mechanisms in health and disease*. New York: Oxford University Press.
Benedetti, F. (2008b). Mechanisms of placebo and placebo-related effects across diseases and treatments. *Annual Review of Pharmacology and Toxicology, 48*, 33–60.
Benedetti, F., Colloca, L., Torre, E., Lanotte, M., Melcarne, A., Pesare, M., Bergamasco, B., & Lopiano, L. (2004). Placebo-responsive Parkinson patients show decreased activity in single neurons of subthalamic nucleus. *Nature Neuroscience, 7*, 587–588.
Benedetti, F., Lanotte, M., Colloca, L., Ducati, A., Zibetti, M., & Lopiano, L. (2009). Electrophysiological properties of thalamic, subthalamic and nigral neurons during the anti-parkinsonian placebo response. *The Journal of Physiology, 587*, 3869–3883.
Colloca, L., & Benedetti, F. (2005). Placebos and painkillers: Is mind as real as matter? *Nature Reviews Neuroscience, 6*, 545–552.
Colloca, L., Lopiano, L., Lanotte, M., & Benedetti, F. (2004). Overt versus covert treatment for pain, anxiety and Parkinson's disease. *Lancet Neurology, 3*, 679–684.
de la Fuente-Fernandez, R., Ruth, T. J., Sossi, V., Schulzer, M., Calne, D. B., & Stoessl, A. J. (2001). Expectation and dopamine release: Mechanism of the placebo effect in Parkinson's disease. *Science, 293*, 1164–1166.
Enck, P., Benedetti, F., & Schedlowski, M. (2008). New insights into the placebo and nocebo responses. *Neuron, 59*, 195–206.
Finniss, D. G., Kaptchuk, T. J., Miller, F., & Benedetti, F. (2010). Biological, clinical, and ethical advances of placebo effects. *Lancet, 375*, 686–695.
Petrovic, P., Kalso, E., Petersson, K. M., & Ingvar, M. (2002). Placebo and opioid analgesia – Imaging a shared neuronal network. *Science, 295*, 1737–1740.
Price, D. D., Finniss, D. G., & Benedetti, F. (2008). A comprehensive review of the placebo effect: Recent advances and current thought. *Annual Review of Psychology, 59*, 565–590.
Scott, D. J., Stohler, C. S., Egnatuk, C. M., Wang, H., Koeppe, R. A., & Zubieta, J. K. (2008). Placebo and nocebo effects are defined by opposite opioid and dopaminergic responses. *Archives of General Psychiatry, 65*, 220–231.
Wager, T. D., Rilling, J. K., Smith, E. E., Sokolik, A., Casey, K. L., Davidson, R. J., Kosslyn, S. M., Rose, R. M., & Cohen, J. D. (2004). Placebo-induced changes in fMRI in the anticipation and experience of pain. *Science, 303*, 1162–1166.

Placebo Analgesic Effect

▶ Placebo Analgesia and Descending Opioid Modulation

Placebo Analgesic Response

▶ Placebo Analgesia and Descending Opioid Modulation

Placebo Effect

Definition

A placebo effect is the presumed, intangible influence that an intervention may have on a patient to relieve pain, or other symptoms, to a degree greater than that which can be expected from the known, or conventional, physiological effects of the intervention on the symptom being treated. The total effect of an intervention may be the combination of both its physiological effects and placebo effect. An absolute placebo effect occurs when the intervention has no known physiological effect.

Cross-References

▶ Placebo

Plain Radiography

Peter Lau
Department of Clinical Research, Royal
Newcastle Hospital, University of Newcastle,
Newcastle, NSW, Australia

Synonyms

Computed radiography; Digital radiography; Plain X-ray; Radiography

Definition

Plain radiography is a means of obtaining a picture of internal structures by passing X-rays through them and recording the shadows cast by these structures.

Characteristics

X-rays are a form of electromagnetic radiation that has the property of being able to pass through biological tissues. However, different tissues absorb the radiation to different extents, and not all the X-rays beamed at a body will emerge. Effectively, the tissues in the body cast shadows. If these shadows are recorded, an image can be developed of the tissues through which the X-rays have been beamed.

The classical means of obtaining an image have been to record the shadows of X-rays on a photographic film. The image is obtained by placing the film in a cassette on the opposite side of the body from that which the X-rays enter.

Modern techniques use electronic recording instead of photographic film. A detector placed behind the body records the density of radiation emerging. The captured information is stored in digital form (digital radiography), which can be processed by a computer in order to generate a photographic image (computed radiography). Apart from producing better quality images, digital radiography also allows data to be archived and distributed electronically.

Irrespective of how the images are obtained, plain radiographs are based on a single principle. Different tissues absorb X-rays to different extents and cast shadows of different densities.

Bones, or other tissues that contain calcium, absorb X-rays greatly and cast dense shadows. Photographically, they appear white, because the photographic emulsion in line with bones is relatively unexposed to radiation and remains chemically unaltered. Gas does not absorb X-rays and casts no shadows. Photographically it appears black (Fig. 1).

Tissues containing water or protein absorb X-rays to various, but small, extents. They appear

Plain Radiography, Fig. 1 A plain radiograph of the abdomen, illustrating the different radiographic densities of various tissues. The bones of the ribs (*R*), vertebrae (*V*), and pelvis (*P*) appear *white*. Gas (*G*) in the intestinal tract appears *black*. The liver (*L*) and the muscles (*M*) of the posterior abdominal wall appear *gray*

as various shades of gray. In standard radiography, using film-based images, these tissues are poorly resolved.

Plain radiography produces an incomplete image of internal structures. It produces a two-dimensional representation, or shadow, of a three-dimensional object. The image, therefore, needs to be carefully interpreted.

The advantage of plain radiography is that it is inexpensive and available almost everywhere. However, in the context of pain medicine, plain radiography has limited utility.

Contrast Media

Hollow organs, tubular structures, and spaces, which are normally invisible to X-rays, can be rendered visible by injecting into them a radiographic dye. The dyes contain iodine or barium, which absorb X-rays and cast a dense radiographic shadow. When injected into a cavity, the dye changes the contrast of the cavity by changing it from invisible to visible. For this reason, they are referred to as contrast media.

Utility

The cardinal utility of plain radiography is to demonstrate the status of bones. Radiographs can show the location, shape, internal architecture, and density of bones. When focused on joints, X-rays can demonstrate the proximity and relative position of the bones forming the joint and the shape of the ends of these bones (Fig. 2).

Fractures

The foremost application of plain radiography is in the detection of fractures. It is, therefore, critical in the investigation of acute pain following trauma. In cases of major trauma, where fracture is clinically obvious or highly likely, the utility of plain radiography is unquestioned.

However, when the trauma is not major or when fractures are unlikely to be the cause of pain, there has been a tendency for physicians to overuse radiographs. Radiographs are taken out of habit, just in case, or to protect physicians from accusations of negligence – the so-called defensive medicine.

Fractures are rarely the cause of acute low back pain. Contemporary guidelines deny the utility of plain radiography for the investigation of patients with back pain (Australian Acute Musculoskeletal Pain Guidelines Group 2003; Bogduk and McGuirk 2002). They recommend plain radiography only in patients with risk factors for fracture, such as major trauma or minor trauma in elderly patients, osteoporosis, use of corticosteroids, or the possibility of a pathological fracture. Otherwise, guidelines recommend medical imaging only if there are clinical indicators of a possible serious cause for the pain; but even so, magnetic resonance imaging, rather than plain radiography, is the modality of choice (Australian Acute Musculoskeletal Pain Guidelines Group 2003; Bogduk and McGuirk 2002).

In patients with idiopathic neck pain, the chances of fracture being the cause are essentially

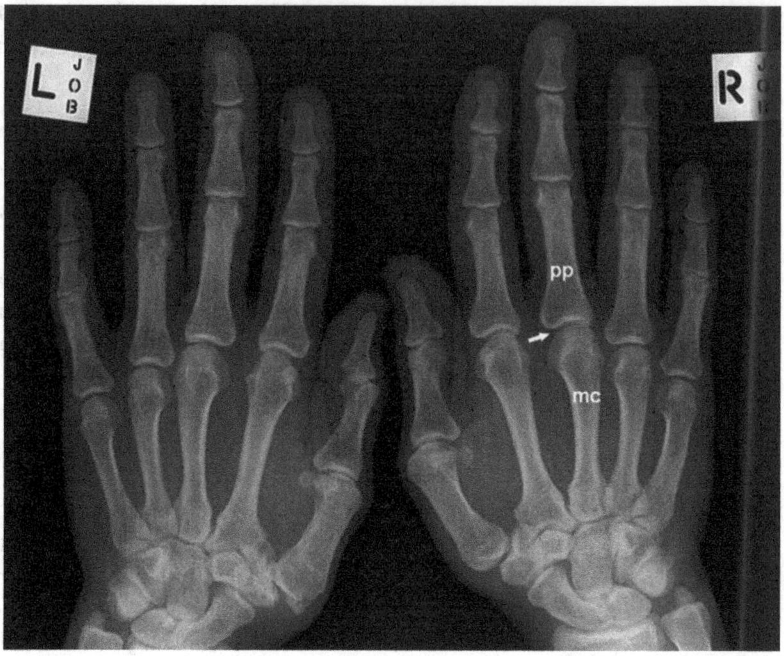

Plain Radiography,
Fig. 2 A plain radiograph
of a pair of hands,
illustrating how the
structure of bones and
joints can be demonstrated.
For example, the proximal
phalanx (*pp*) and the
metacarpal bone of the third
finger are labeled and the
arrow points to the
metacarpophalangeal joint

nil (Bogduk 1999; Heller et al. 1983; Johnson and
Lucas 1997). Even in patients with neck pain
after trauma, the prevalence of fractures is only
about 3 % (Bogduk 1999). In order to minimize
the use of plain radiographs in the pursuit of
cervical fractures, the Canadian C-spine rule has
been formulated and validated (Stiell et al. 2001)
(Fig. 3).

Similarly, in patients presenting with ankle
pain or knee pain, the overuse of plain radiographs can be safely reduced by following the
Ottawa ankle rule (Stiell et al. 1995) and the
Ottawa (Stiell et al. 1997) or Pittsburgh (Seaberg
and Jackson 1994; Seaberg et al. 1998) knee rules
(Table 1).

Infection and Tumors
Infections and tumors of the bone may manifest
with different forms of periosteal reaction or by
erosion of the bone. These can be detected by
plain radiography but tend to occur late in the
development of bone disease. Their early stages
are better detected by computerized tomography
(CT) or ► magnetic resonance imaging (MRI).

These modalities also provide better resolution
of later stages of tumors and infection.

Arthropathy
Although plain radiography can produce images
of joints, these are rarely diagnostic. Joint infections or inflammatory monoarthropathies are
diagnosed by aspiration and laboratory testing
of the aspirate. Polyarthropathies, such as rheumatoid arthritis and the various spondyloarthropathies, are diagnosed from their clinical
features. The utility of plain radiography is limited to confirming the joint pathology and grading
its severity in terms of erosions and loss of joint
space.

Degenerative Joint Disease
Plain radiography has, perhaps, been most
overused and wrongly interpreted in the context
of the so-called degenerative joint disease or
osteoarthritis. Although radiography can reveal
abnormalities, such as loss of joint space,
subchondral sclerosis, subchondral cysts, and
osteophytes, these bear an inconsistent, and

Plain Radiography,
Fig. 3 The Canadian
C-spine rule (Stiell et al.
2001)

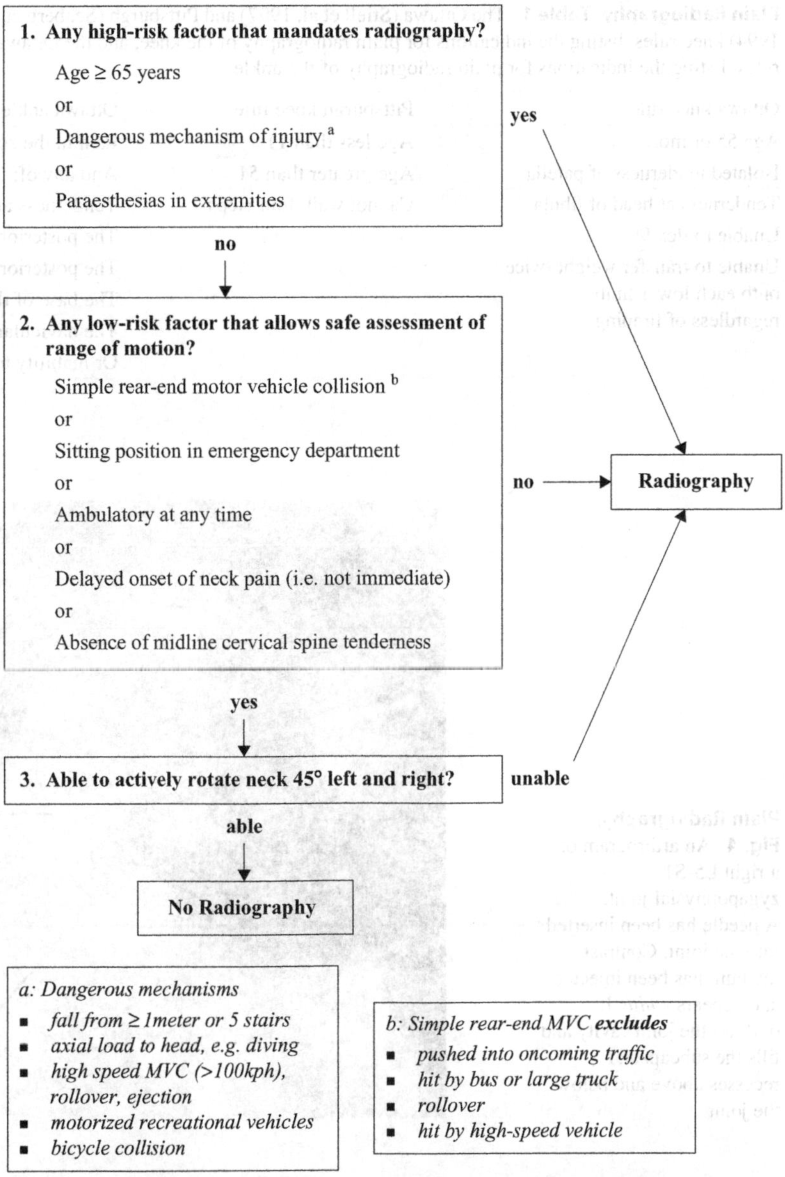

1. Any high-risk factor that mandates radiography?

Age ≥ 65 years

or

Dangerous mechanism of injury [a]

or

Paraesthesias in extremities

→ yes

↓ no

2. Any low-risk factor that allows safe assessment of range of motion?

Simple rear-end motor vehicle collision [b]

or

Sitting position in emergency department

or

Ambulatory at any time

or

Delayed onset of neck pain (i.e. not immediate)

or

Absence of midline cervical spine tenderness

no → **Radiography**

↓ yes

3. Able to actively rotate neck 45° left and right?   unable

↓ able

No Radiography

a: Dangerous mechanisms
- *fall from ≥ 1meter or 5 stairs*
- *axial load to head, e.g. diving*
- *high speed MVC (>100kph), rollover, ejection*
- *motorized recreational vehicles*
- *bicycle collision*

*b: Simple rear-end MVC **excludes***
- *pushed into oncoming traffic*
- *hit by bus or large truck*
- *rollover*
- *hit by high-speed vehicle*

sometimes negative, relationship to pain. They are normal age changes rather than a disease that causes pain.

Most patients with joint pain have minimal or no radiographic changes in their joints (Lawrence et al. 1996). Conversely, not all patients with radiographic osteoarthritis have pain (Lawrence et al. 1996). Only some 60–80 % of patients with severe radiographic changes in the hip or knee complain of pain. This figure drops to 30 % in

patients with radiographic changes in the hands. Under these conditions, plain radiography does not allow a diagnosis of the cause of pain to be made.

Spondylosis is very common in patients who do not have back pain. Consequently, there is only a marginal association between the presence of pain and the presence of degenerative changes in the lumbar spine (van Tulder et al. 1997). The association is too weak to allow degenerative

Plain Radiography, Table 1 The Ottawa (Stiell et al. 1997) and Pittsburgh (Seaberg et al. 1998; Seaberg and Jackson 1994) knee rules, listing the indications for plain radiography of the knee, and the Ottawa (Lawrence et al. 1996) ankle rules, listing the indications for plain radiography of the ankle

Ottawa knee rule	Pittsburgh knee rule	Ottawa ankle rule
Age 55 or more	Age less than 11	Pain in the zone of either malleolus
Isolated tenderness of patella	Age greater than 51	And any of:
Tenderness at head of fibula	Cannot walk four steps	Tenderness at:
Unable to flex 90°		The posterior edge of the lateral malleolus
Unable to transfer weight twice onto each lower limb, regardless of limping		The posterior edge of the medial malleolus
		The base of the fifth metatarsal
		The navicular bone
		Or inability to walk four steps

Plain Radiography, Fig. 4 An arthrogram of a right L5-S1 zygapophysial joint. A needle has been inserted into the joint. Contrast medium has been injected and appears *white*. It outlines the joint cavity and fills the subcapsular recesses above and below the joint

joint disease to be the diagnosis of back pain. Similarly, an association between spondylosis and neck pain is all but lacking (Australian Acute Musculoskeletal Pain Guidelines Group 2003; Bogduk 1999; Heller et al. 1983). Consequently, neck pain cannot be attributed to degenerative joint disease.

Structural Abnormalities
It had been the fashion to attribute low back pain to conditions such as spondylolysis and spondylolisthesis. However, epidemiological surveys have shown that these conditions, in adults, are just as common in patients with no pain as in patients with low back pain (van Tulder et al. 1997). They are incidental findings and cannot be regarded as the cause of pain.

Implications
Physicians have typically misused and overused plain radiography in the pursuit of a cause of pain. The utility of plain radiography is essentially

Plain Radiography, Fig. 5 Examples of contrast medium used to ensure the accuracy of nerve blocks. (**a**) Right L5 medial branch block. The contrast medium

remains in the correct target region for the nerve. (**b**) Left L4 spinal nerve block. The contrast medium outlines the course of the nerve and its roots

limited to the identification of fractures, which are uncommon, or even rare, causes of pain, other than in patients with severe trauma. Plain radiographs have no diagnostic utility for the pursuit of degenerative joint disease because this cannot be held to be the cause of pain, even when evident.

Fluoroscopy

Fluoroscopy is an adaptation of plain radiography. It involves capturing continuous images on a television screen as X-rays are continuously transmitted through the body. Continuous screening allows movements and events occurring within the body to be monitored and recorded. In radiology at large, fluoroscopy has been used to study the flow of contrast medium through blood vessels and through the urinary, genital, and gastrointestinal tracts. In pain medicine, the greatest utility of fluoroscopy is to guide injections into deep structures.

Prior to injecting a diagnostic or therapeutic agent, contrast medium can be injected in order to demonstrate where the injectate will flow. For intra-articular injections, contrast medium can be injected into the cavity of the target joint (Fig. 4).

For nerve blocks, contrast medium can be injected onto the target site where the nerve lies, in order to ensure that the injectate covers the target but also does not flow to affect other structures (Fig. 5).

In this regard, fluoroscopy itself is not a diagnostic procedure. It is a tool that uses modified, plain radiography to ensure the accuracy and safety of other diagnostic and therapeutic procedures.

References

Australian Acute Musculoskeletal Pain Guidelines Group. (2003). Evidence-based management of acute musculoskeletal pain. Brisbane: Australian Academic Press. Online. http://www.nhmrc.gov.au. Accessed 21 Sept 2004.

Bogduk, N. (1999). The neck. *Baillière's Clinical Rheumatology, 13*, 261–285.

Bogduk, N., & McGuirk, B. (2002). *Medical management of acute and chronic low back pain. An evidence-based approach*. Amsterdam: Elsevier.

Heller, C. A., Stanley, P., Lewis-Jones, B., et al. (1983). Value of x ray examinations of the cervical spine. *BMJ, 287*, 1276–1278.

Johnson, M. J., & Lucas, G. L. (1997). Value of cervical spine radiographs as a screening tool. *Clinical Orthopaedics, 340*, 102–108.

Lawrence, J. S., Bremner, J. M., & Bier, F. (1996). Osteoarthrosis: Prevalence in the population and relationship between symptoms and X-ray changes. *Annals of the Rheumatic Diseases, 25*, 1–24.

Seaberg, D. C., & Jackson, R. (1994). Clinical decision rule for knee radiographs. *The American Journal of Emergency Medicine, 12*, 541–543.

Seaberg, D. C., Yealy, D. M., Lukens, T., et al. (1998). Multicenter comparison of two clinical decision rules for the use of radiography in acute, high-risk knee injuries. *Annals of Emergency Medicine, 32*, 8–13.

Stiell, I., Wells, G., Laupacis, A., et al. (1995). Multicentre trial to introduce the Ottawa ankle rules for use of radiography in acute ankle injuries. *BMJ, 311*, 594–597.

Stiell, I. G., Wells, G. A., Hoag, R. H., et al. (1997). Implementation of the Ottawa knee rule for the use of radiography in acute knee injuries. *The Journal of the American Medical Association, 278*, 2075–2079.

Stiell, I. G., Wells, G. A., Vandemheen, K. L., et al. (2001). The Canadian C-spine rule for radiography in alert and stable trauma patients. *The Journal of the American Medical Association, 286*, 1841–1848.

van Tulder, M. W., Assendelft, W. J. J., Koes, B. W., et al. (1997). Spinal radiographic findings and nonspecific low back pain. A systematic review of observational studies. *Spine, 22*, 427–434.

Plain X-Ray

▶ Plain Radiography

Plantar Incision Model

▶ Nick Model of Cutaneous Pain and Hyperalgesia

Plantar Test

▶ Thermal Nociception Test

Plasma Concentration Versus Time Profiles

▶ NSAIDs, Pharmacokinetics

Plasma Extravasation

Definition

The leakage of plasma from blood vessels during acute inflammation associated with tissue injury. Proinflammatory mediators (PGE2, histamine, bradykinin, SP, CGRP) act on vasculature to induce vasodilation and increase endothelial cell gaps. This permits the movement of plasma into the extravascular space to form regions of edema (swelling) that surround wounds. Plasma extravasation is partly mediated by antidromic activation of primary afferent C-fibers which produce neurogenic inflammation.

Cross-References

▶ Formalin Test
▶ Inflammation, Modulation by Peripheral Cannabinoid Receptors
▶ Neurogenic Inflammation and Sympathetic Nervous System
▶ Nociceptors in the Orofacial Region (Skin/Mucosa)

Plasticity

Definition

Plasticity is the ability of central nervous system tissue to change due to a variety of causes, including psychological factors.

Changes can be in connectivity, inputs, and/or response properties of neurons.

Cross-References

▶ Cancer Pain, Palliative Care in Children
▶ Sensitization of Visceral Nociceptors
▶ Thalamic Plasticity and Chronic Pain
▶ Thalamus, Receptive Fields, Projected Fields, Human

Pleomorphic Vesicles

Definition

Pleomorphic vesicles are contained in axon terminals, a mixture of round, oval, and flattened synaptic vesicles.

Cross-References

▶ Trigeminal Brainstem Nuclear Complex, Anatomy

Plexopathy

Definition

Plexopathy is a term used to refer to the condition of a nerve plexus that has been widely injured, typically by ionising radiation in the treatment of cancer. It may present anywhere from 6 months to 20 years after radiation treatment. It is not usually a painful condition but at times can be extremely painful. Pain is typically referred to the shoulder, lateral arm, and thumb and fore finger, because the upper trunk of the brachial plexus is primarily involved. Lumbar plexopathy is uncommon.

Cross-References

▶ Cancer Pain Management, Neurosurgical Interventions
▶ Plexus Injuries and Deafferentation Pain

Plexus Injuries and Deafferentation Pain

Praveen Anand
Peripheral Neuropathy Unit, Imperial College London, London, UK

Synonyms

Anesthesia dolorosa due to plexus avulsion; Brachial plexus avulsion injury; Deafferentation; OBP; Obstetric brachial palsy; Plexopathy

Definition

Plexopathy refers to any lesion of the brachial or lumbosacral nerve plexus including trauma, infection, inflammation, ischemia, infiltration by or pressure from tumors, and post-radiation injury. Such lesions lead to pain syndromes that share the frequency and characteristics of other nerve lesions. Traction injuries, particularly of the ▶ brachial plexus, may result in ▶ avulsion of dorsal and ventral spinal roots from the spinal cord, which is frequently associated with severe, persistent pain due to ▶ deafferentation ("deafferentation pain"), and in some cases also because of injury to the dorsal spinal cord. The brachial plexus is the region of peripheral nervous system where the lowest four cervical and first thoracic spinal nerve roots (C5–T1) unite and branch, in the lower part of the neck and behind the clavicle. Nerve fibers in the plexus, as they proceed distally, form the main nerves to the upper limb – the median, ulnar, musculocutaneous, radial, and circumflex nerves. The lumbosacral plexus is formed by the union of lumbar and sacral spinal nerve roots (mainly L2–S3) as they traverse the posterior wall of the abdomen and pelvis. The main tributary branches emerge as the femoral, obturator, sciatic, and pudendal nerves.

Characteristics

Background and Anatomy

There are distinctive aspects of plexus injuries, which may involve both central and distal axons of sensory neurons at different spinal root levels (Fig. 1). Chronic pain has been estimated to occur in about 5 % of patients with peripheral nerve injury (Sunderland 1993), and this also applies to patients with plexus injuries where the lesion lies within the plexus, distal to the dorsal root ganglion. However, the majority of patients with spinal root avulsion injury, i.e., injury proximal to the dorsal root ganglion, report severe chronic pain at some point in the course of their condition (Wynn-Parry 1980; Birch et al. 1998). The degree of pain is related to the number of avulsed roots (Berman et al. 1998). Pain may occur from the time of injury, or within days, and is often intractable, lasting from months to years, or even decades.

In brachial plexus avulsion injury, there is intradural rupture of the spinal roots, which leads to degeneration of somatic efferent and pre-ganglionic autonomic efferent fibers, and central processes of sensory fibers; however, the distal axons of somatic afferents remain intact. The latter are electrophysiologically "functional", and continue to mediate cutaneous axon-reflex vasodilatation (Bonney and Gilliatt 1958). The transition from central to peripheral nervous system structures takes place in the transitional region, near to where the rootlets enter the spinal cord. CNS cells extrude into the rootlets in a cone-shaped arrangement: Thus, "each transitional region can be subdivided into an axial central nervous system compartment and a surrounding peripheral nervous system compartment" (Berthold et al. 1993). The axons traverse this region; the blood vessels that serve the nerve fibers do not. In traction lesions of the adult brachial plexus, the rupture is usually peripheral to the transitional zone. However, some avulsion injuries cause damage central to the transitional zone, which may result in a spinal cord lesion and "central pain."

The pain following spinal root avulsion injuries is more severe and persistent than injuries distal to the dorsal root ganglion, in accordance with its central origin. The avulsion injury-related pain is characteristically constant, described as crushing or burning, usually felt in the hand of patients with brachial plexus injury, with intermittent excruciating paroxysms of pain shooting down the arm. In complete brachial plexus avulsion injury (i.e., avulsions of C5–T1), the painful limb is insensate – hence, this pain is termed anesthesia dolorosa. As the pain after avulsion injury is related to the number of roots avulsed, the pain may be mild or even absent when there is only one root injured, particularly if it is an intradural injury of a dorsal root with no direct lesion of the spinal cord. This may explain why surgical dorsal rhizotomy has been practiced to produce pain relief, usually for a short duration such as in severe tumor-related pain. In approximately 30 % of patients with avulsion injuries, there is allodynia to mechanical and thermal stimuli between the injured and intact dermatomes (i.e., border-zone), prior to any surgical nerve repair, but this is usually clinically mild and often not noticed by the patient, and may require quantitative sensory testing to reveal it. A few patients may demonstrate more severe allodynia and hyperalgesia after spontaneous or postsurgical sensory recovery, with features similar to partial nerve injury.

Pain Mechanisms

It is now well established that peripheral nerve injury leads to a number of peripheral and central changes. Distinct mechanisms are involved in pain following spinal cord root avulsion. There is evidence that pain experienced after brachial plexus avulsion injury parallels the generation of abnormal activity within the dorsal horn of the spinal cord. Following dorsal rhizotomy in cats, increased burst activity is seen in dorsal horn cells in as early as 14 days (Loeser and Ward 1967). Continuous high-frequency activity has also been reported in the cat dorsal

Plexus Injuries and Deafferentation Pain, Fig. 1 Schematic representation of brachial plexus injuries. DRG, dorsal root ganglion; VR, ventral root; SC, spinal cord; G, graft; TR, transfer. (a) Before repair: (*i*) intact root, (*ii*) rupture, (*iii*) avulsion. (b) After repair: (*i*) intact root, (*ii*) rupture repaired by grafting, (*iii*) avulsion repaired by nerve transfer (from Berman et al. 1998).

horn and, significantly, this type of firing is characteristic when the denervation is due to avulsion rather than rhizotomy (Ovelmen-Levitt 1988), in accordance with clinical observations. Ischemic or hemorrhagic damage to the spinal cord, and death of inhibitory interneurons, may contribute. A high proportion of root avulsion patients report at least transient pain relief following dorsal root entry zone (DREZ) lesioning (Nashold et al. 1983), implicating this region in avulsion pain.

The usual current practice for repairing spinal cord root avulsion injuries is by transferring an intact neighboring nerve to the distal stump of the damaged nerve, in an effort to facilitate axonal regeneration and restore some motor and sensory function (Fig. 1). Clinical observations have indicated that successful surgical repair is associated with relief of avulsion pain, even in patients where repair was delayed for years and there was no prospect of motor recovery (Berman et al. 1996). In a prospective study, a correlation and close temporal relationship was observed between reduction in pain and functional recovery (Berman et al. 1998). It was concluded that

nerve repair can reduce pain from spinal root avulsions, and that the mechanism involves successful regeneration with restoration of peripheral inputs (e.g., from muscle), or central connections.

The findings are different after brachial plexus injuries in neonates (Anand and Birch 2002). Results of repair in the obstetric brachial plexus lesion are often disappointing for motor function, presumably related to the relative delay of surgery and consequent retrograde degeneration of motor neurons, a process which proceeds much more rapidly in neonates than in adults. Results of sensory recovery are very much better, demonstrated by quantitative sensory testing. While recovery of function after spinal root avulsion was demonstrably related to surgery, there were remarkable differences from adults, including excellent restoration of sensory function in children, and evidence of exquisite CNS plasticity, i.e., perfect localization of restored sensation in avulsed spinal root dermatomes, now presumably routed via nerves that had been transferred from a distant spinal region. Sensory recovery exceeded motor or cholinergic sympathetic recovery. Remarkably, there was no evidence of long-term chronic pain or other neuropathic syndromes, although pain was reported normally to external stimuli in unaffected regions. We proposed that differences in neonates that account for their lack of long-term chronic pain after spinal root avulsion injury are related to later maturation of injured fibers, and greater CNS plasticity (e.g., sprouting or maturation of descending inhibitory tracts).

Treatment

The clinician must always bear in mind that the sooner the distal axonal segment is reconnected to the proximal segment and cell body, the better the result will be, both for regeneration and preventing the development of chronic pain. The general treatment options are similar to other traumatic neuropathies, but pain may be more intractable after brachial plexus avulsion injury. There are few randomized placebo-controlled double-blind clinical trials of pain relief in patients with brachial plexus injury alone (Berman et al. 2004). Tricyclic antidepressants and anticonvulsants are widely used, in combination with opioids, cannabinoids, and other treatments. The latter include transcutaneous electrical nerve stimulation (TENS), ▶ acupuncture, spinal cord stimulation, nerve transfers (e.g., intercostal to ulnar nerve), ▶ DREZ lesions, and behavioral therapies.

References

Anand, P., & Birch, R. (2002). Restoration of sensory function and lack of long-term chronic pain syndromes after brachial plexus injury in human neonates. *Brain, 125*, 113–122.
Berman, J., Taggart, M., Anand, P., & Birch, R. (1996). Late intercostal nerve transfer relieves the pain of preganglionic injury to the brachial plexus. *The Journal of Bone & Joint Surgery, 78*, 759–760.
Berman, J. S., Birch, R., & Anand, P. (1998). Pain following human brachial plexus injury with spinal cord root avulsion and the effect of surgery. *Pain, 75*, 199–207.
Berman, J. S., Symonds, C., & Birch, R. (2004). Efficacy of two cannabis based medicinal extracts for relief of central neuropathic pain from brachial plexus avulsion: Results of a randomised controlled trial. *Pain, 112*, 299–306.
Berthold, C. H., Carlstedt, T., & Corneliuson, O. (1993). Anatomy of the mature transitional zone. In P. J. Dyck, P. K. Thomas, J. W. Griffin, P. A. Low, & J. F. Poduslo (Eds.), *Peripheral neuropathy* (3rd ed., pp. 75–80). Philadelphia: WB Saunders.
Birch, R., Bonney, G., & Wynn Parry, C. B. (1998). *Surgical disorders of the peripheral nerves* (pp. 467–490). London: Churchill Livingstone.
Bonney, G., & Gilliatt, R. W. (1958). Sensory nerve conduction after traction lesion of the brachial plexus. *Proceedings of the Royal Society of Medicine, 51*, 365–367.
Loeser, J. D., & Ward, A. A. (1967). Some effects of deafferentation on neurons of the cat spinal cord. *Archives of Neurology, 17*, 629–636.
Nashold, B. S., Ostdahl, R. H., Bullitt, E., Friedman, A., Brophy, B. (1983). Dorsal root entry zone lesions: A new neurosurgical therapy for deafferentation pain. In: Bonica, J. J. (Ed.), *Advanced Pain Research Therapy, 5*:739–750.
Ovelmen-Levitt, J. (1988). Abnormal physiology of the dorsal horn as related the deafferentation syndrome. *Applied Neurophysiology, 51*, 104–116.
Sunderland, S. (1993). *Nerves and nerve injuries*. London: Churchill Livingstone.
Wynn-Parry, C. B. (1980). Pain in avulsion lesions of the brachial plexus. *Pain, 9*, 41–53.

PLIF

Definition

Posterior lumbar interbody fusion – graft/cages placed between the vertebral bodies from posterior approach, retracting the thecal side to one side.

Cross-References

▶ Spinal Fusion for Chronic Back Pain

PMS

▶ Premenstrual Syndrome

Pneumothorax

Synonyms

PNX

Definition

An accumulation of air in the pleural space that results in complete or partial collapse of a lobe or entire lung. It can occur spontaneously or be secondary to underlying lung disease, chest trauma, mechanical ventilation, or perforated esophagus.

Cross-References

▶ Postoperative Pain, Regional Blocks

PNOC (*Homo sapiens*)

▶ Orphanin FQ

PNX

▶ Pneumothorax

PO

▶ Posterior Nucleus

Point Prevalence

Definition

Point prevalence refers to the rates of a specific disorder at a single point in time.

Cross-References

▶ Depression and Pain
▶ Prevalence of Chronic Pain Disorders in Children

Polyarteritis Enterica

▶ Reiter's Syndrome

Polyarthritis

Definition

Inflammation of several or many joints. Can have different etiologies (e.g., infection, autoimmune disease) and is often characterized by severe pain.

Cross-References

▶ Sciatica

Polymodal Nociceptor

▶ Nociceptors in the Orofacial Region (Skin/Mucosa)

Polymodal Nociceptor(s)

Definition

Sensory receptors in the skin and other organs with thinly myelinated (Aδ) or unmyelinated (C) afferent nerve fibers, which are activated by noxious stimuli of different modalities such as noxious (painful) mechanical stimuli, heat, lowering of pH, or different chemical mediators which typically appear in lesioned or inflamed tissues (e.g., bradykinin, prostaglandins, histamine, and others).

Cross-References

- ▶ Mechano-Insensitive C-Fibers, Biophysics
- ▶ Nick Model of Cutaneous Pain and Hyperalgesia
- ▶ Nociceptors in the Orofacial Region (Meningeal/Cerebrovascular)
- ▶ Polymodal Nociceptors, Heat Transduction
- ▶ Sensitization of Visceral Nociceptors
- ▶ Spinothalamocortical Projections to Ventromedial and Parafascicular Nuclei

---

Polymodal Nociceptors

- ▶ Histology of Nociceptors
- ▶ Polymodal Nociceptors, Heat Transduction

Polymodal Nociceptors, Heat Transduction

Carolina Roza
Biología de Sistemas (Unidad Fisiología),
Universidad de Alcalá, Alcalá de Henares,
Madrid, Spain

Synonyms

A delta(δ)-mechanoheat receptor; Capsaicin receptors; C-heat receptor; C-mechanoheat receptor; Mechanoheat nociceptor; Polymodal nociceptors; Thermal transduction

Definition

Sensory process by which noxious heat is transduced into propagated nerve impulses at the peripheral endings of nociceptive afferents. The impulse activity in these fibers may lead to the generation of subjective sensations of pain in humans.

Characteristics

Transduction of noxious heat takes place in two types of slowly conducting primary afferents: C- and Aδ-fibers. Although a group of C-fibers in human skin has been described that presented only responses to noxious heat (Schmelz et al. 2000b),

Polymodal Nociceptors, Heat Transduction, Fig. 1 (a) Responses of the cloned TRPV1 and TRPV2 expressed in Xenopus oocytes to capsaicin (1 μM), protons (pH 4), and noxious heat (from 23 °C to 55 °C in 15 s). Whereas cells expressing TRPV1 responded to all pain-producing stimuli, those expressing TRPV2 exhibited responses only to noxious heat. (b) TRPV1 and TRPV2 expressed in HEK293 cells showed different heat response thresholds (~42 °C and ~53 °C, respectively; final temperature 54 °C). Adapted from (Caterina et al. 1999)

a

Thermal Hyperalgesia

b

Polymodal Nociceptors, Heat Transduction, Fig. 2 TRPV1$^{-/-}$ mice showed reduced thermal hyperalgesia induced by inflammation evoked by (**a**) mustard oil (10 %) painted at the planter hindpaw and (**b**) by subplantar injection of carrageen (0.025 mls, 2 %). These results demonstrate a role for TRPV1 in the development of inflammation-induced heat hyperalgesia (Adapted from Caterina et al. (2000) and Davis et al. (2000), respectively)

studies in animals have demonstrated that both types of fibers also typically respond to mechanical forces of noxious intensity and hence have been categorized as mechanoheat nociceptors (CMH and AMH, respectively) (Campbell and Meyer 1996). In addition, when explored, these fibers usually also respond to various chemicals (Davis et al. 1993), and in particular to capsaicin, which also excites heat-sensitive fibers of viscera, muscle, and the cornea of the eye (Belmonte and Gallar 1996; Cervero 1994; Mense 1996). Therefore, it is likely that CMHs and AMHs belong to the group of polymodal nociceptors.

CMHs

They are present with similar heat thresholds in hairy and glabrous skin. CMHs respond both to heat and mechanical stimuli. Their heat threshold is over 38 °C but below 50 °C. After reaching their threshold, their response increases monotonically with the stimulus intensity, and this is in parallel with the heat sensation threshold and the intensity-response curve of pain evoked by heat in humans (LaMotte and Campbell 1978). With repeated stimulation, their response

decreases showing ▶ fatigue. In addition, they are activated only briefly after capsaicin application (Campbell and Meyer 1996).

CHs

In human skin, 45 % of the C units are mechanoheat afferents; however, a small percentage (4 %) responds exclusively to heating, with thresholds between 45 °C and 48 °C. In contrast with CMH, CH exhibit prolonged bursting discharges for minutes ($\geq$10) following injection of capsaicin. It is likely that this bursting discharge pattern induces the waxing and waning pain characteristic of capsaicin injections. It has been proposed that CH fibers are responsible for the development of ▶ flare and ▶ hyperalgesia after exposure to noxious heat (Schmelz et al. 2000a).

AMHs

According to their response to heat, fibers within this group have been divided into two subpopulations, type I AMH (or AMH I) and type II AMH (or AMH II) (Treede et al. 1995). Type II AMH fibers are only present in hairy skin; they have a heat threshold of ~46 °C; the peak response is reached early after the stimulus, and adaptation occurs over the next seconds. They show desensitization after a burn injury. With regard to their heat response, these fibers behave like CMH but with higher heat pain thresholds.

Type I AMHs are present in both hairy and glabrous skin and have heat thresholds over 53 °C; their response to heating winds up with time, reaching a peak late in the stimulus. They become sensitized after a burn injury. Type I AMHs have been proposed to be the high-threshold mechanonociceptors (or HTM) described by others, that also have the ability of transducing intense, noxious heat. These fibers probably contribute to heat pain evoked by a sustained, high-intensity heat stimulus (Campbell and Meyer 1996).

Most AMH units described in human hairy skin are similar to AMH type II, but some data suggest that AMH I units also exist in the glabrous skin of the human hand. A-δ fibers are less abundant than C-fibers in human skin nerves. Microneurography experiments additionally indicate that in human hairy skin, the proportion of heat-sensitive elements is smaller in the A-δ fibers group than among C-fibers. In humans, heat thresholds of AMH type II and CHM units were in the same range of 36–49 °C.

Cellular and Molecular Mechanisms

The transducer elements for mechanical forces and heat in A-δ and C polymodal afferents appear to be located in the same receptor ending (Treede et al. 1990) but are represented by separate molecular entities. This is because exposure of single nerve endings to capsaicin inactivates their responsiveness to heat and acid, but not to mechanical forces (Belmonte et al. 1991). Cesare and McNaughton (1996) first showed that heating over 42 °C evoked an inward current in the soma of a subpopulation of DRG neurons. The cloning of two members of the TRP ("transient receptor potential") cation channel family has settled the molecular basis underlying the transduction of noxious heat in the population of heat-sensitive neurons (Fig. 1) (Caterina et al. 1997, 1999). TRPV1 is a channel protein with a heat threshold of 43 °C that also responds to capsaicin and to

Polymodal Nociceptors, Heat Transduction, Fig. 3 Two different assays were used to analyze the withdrawal responses to noxious heat in two different TRPV1 mutant mice strains. (a) In the TRPV1 mutant mice constructed by Davis and co-workers, there were no significant differences between both genotypes in the hot-plate test at the two temperatures tested. (b) In the mice constructed by Caterina and co-workers, TRPV1$^{-/-}$ mice showed impaired responses to heat stimulation but only when noxious heat reached intensities far above TRPV1 threshold (~43 °C), as seen in both the hot-plate and tail immersion tests. Adapted from Davis et al. (2000) and Caterina et al. (2000)

protons, all pain-producing stimuli (Tominaga et al. 1998). The most compelling evidence that TRPV1 was the primary heat-responsive channel was obtained in dissociated neurons from TRPV1$^{-/-}$ sensory ganglia of mice, where heat currents were absent or greatly attenuated (Caterina et al. 2000; Davis et al. 2000). TRPV1 is expressed by both peptidergic and non-peptidergic primary afferents with small-diameter cell bodies and has been proposed as the molecular substrate for the primary heat response of CMH fibers. CMH and type II AMH fire at the same heat threshold that gates isolated TRPV1 channels. However, in contrast with the behavior of the channel, CMH and type II AMH fibers do not sensitize to heat. Thus, the contribution of TRPV1 to heat responses in this type of fiber is not clearly defined, because the absence of TRPV1 abolished the inflammation-induced thermal hyperalgesia possibly mediated by CH and type I AMH fibers (Fig. 2) (Caterina et al. 2000; Davis et al. 2000).

TRPV2 has 49 % identity with TRPV1, is not gated by capsaicin, and is expressed in primary afferents with cell bodies of all diameters, as well as in motoneurons and in a number of nonneural tissues (Caterina et al. 1999). Due to its high heat threshold ($\sim$52 °C), this channel may be the molecular substrate of the response of type I AMH fibers to strong heat and of the residual sensitivity to heat of TRPV1$^{-/-}$ mice (Fig. 3) (Caterina et al. 2000).

In spite of these advances in knowledge, the available information only partially explains the cellular and molecular basis of heat transduction. For instance, it has recently been described that nociceptors lacking TRPV1 and TRPV2 have normal heat responses (Woodbury et al. 2004), implying the existence of an additional and still ignored heat transduction mechanism. This may contribute to the variable responses to heat of the different classes of nociceptors.

References

Belmonte, C., & Gallar, J. (1996). Corneal Nociceptors. In C. Belmonte, & F. Cervero (Eds.), *Neurobiology of nociceptors* (pp. 147–183). Oxford: Oxford University Press.

Belmonte, C., Gallar, J., Pozo, M. A., et al. (1991). Excitation by irritant chemical substances of sensory afferent units in the cat's cornea. *The Journal of Physiology, 437*, 709–725.

Campbell, J. N., & Meyer, R. A. (1996). Cutaneous nociceptors. In C. Belmonte & F. Cervero (Eds.), *Neurobiology of nociceptors* (pp. 117–145). Oxford: Oxford University Press.

Caterina, M. J., Leffler, A., Malmberg, A. B., et al. (2000). Impaired nociception and pain sensation in mice lacking the capsaicin receptor. *Science, 288*, 306–313.

Caterina, M. J., Rosen, T. A., Tominaga, M., et al. (1999). A capsaicin-receptor homologue with a high threshold for noxious heat. *Nature, 398*, 436–441.

Caterina, M. J., Schumacher, M. A., Tominaga, M., et al. (1997). The capsaicin receptor: A heat-activated ion channel in the pain pathway. *Nature, 389*, 816–824.

Cervero, F. (1994). Sensory innervation of the viscera: Peripheral basis of visceral pain. *Physiological Reviews, 74*, 95–138.

Cesare, P., & McNaughton, P. (1996). A novel heat-activated current in nociceptive neurons and its sensitization by bradykinin. *Proceedings of National Academy of Science USA, 93*, 15435–15439.

Davis, J. B., Gray, J., Gunthorpe, M. J., et al. (2000). Vanilloid receptor-1 is essential for inflammatory thermal hyperalgesia. *Nature, 405*, 183–187.

Davis, K. D., Meyer, R. A., & Campbell, J. N. (1993). Chemosensitivity and sensitization of nociceptive afferents that innervate the hairy skin of monkey. *Journal of Neurophysiology, 69*, 1071–1081.

LaMotte, R. H., & Campbell, J. N. (1978). Comparison of responses of warm and nociceptive C-fiber afferents in monkey with human judgments of thermal pain. *Journal of Neurophysiology, 41*, 509–528.

Mense, S. (1996). Nociceptors in skeletal muscle and their reaction to pathological tissue changes. In C. Belmonte & F. Cervero (Eds.), *Neurobiology of nociceptors* (pp. 184–201). Oxford: Oxford University Press.

Schmelz, M., Michael, K., Weidner, C., et al. (2000a). Which nerve fibers mediate the axon reflex flare in human skin? *Neuroreport, 11*, 645–648.

Schmelz, M., Schmid, R., & Handwerker, H. O. (2000b). Encoding of burning pain from capsaicin-treated human skin in two categories of unmyelinated nerve fibres. *Brain, 123*, 560–571.

Tominaga, M., Caterina, M. J., Malmberg, A. B., et al. (1998). The cloned capsaicin receptor integrates multiple pain-producing stimuli. *Neuron, 21*, 531–543.

Treede, R. D., Meyer, R. A., & Campbell, J. N. (1990). Comparison of heat and mechanical receptive fields of cutaneous C-fiber nociceptors in monkey. *Journal of Neurophysiology, 64*, 1502–1513.

Treede, R. D., Meyer, R. A., & Raja, S. N. (1995). Evidence for Two different heat transduction mechanisms in nociceptive primary afferents innervating monkey skin. *The Journal of Physiology, 483*, 747–758.

Woodbury, C. J., Zwick, M., Wang, S., et al. (2004). Nociceptors lacking TRPV1 and TRPV2 have normal heat responses. *Journal of Neuroscience, 24*, 6410–6415.

Polymodal Sensory Neurons

Definition

Polymodal sensory neurons are primary afferent neurons that are activated by thermal, chemical, and mechanical stimuli; see also polymodal nociceptor.

Cross-References

▶ TRPV1 Receptor, Species Variability

Polymodality

Definition

Polymodality is the capacity of a neuron to respond to more than one type of stimulus: mechanical, thermal, or chemical.

Cross-References

▶ Mechanonociceptors

Polymorphism

Definition

A polymorphism is a genetic variant that appears in at least 1 % of a population.

Cross-References

▶ Capsaicin Receptor
▶ Thalamus, Clinical Pain, Human Imaging

Polymorphonuclear Neutrophil Granulocytes (PMN)

▶ Neutrophils in Inflammatory Pain

Polymyalgia Rheumatica

Definition

Polymyalgia rheumatica is a suspected autoimmune disease occurring in patients over the age of 50 with symptoms of muscle pain, most often in the shoulder and neck region but also in muscles of the trunk and pelvic girdle, and signs of inflammation in blood tests: elevated ESR and CRP.

Cross-References

▶ Cancer Pain, Assessment in the Cognitively Impaired
▶ Muscle Pain in Systemic Inflammation (Polymyalgia Rheumatica, Giant Cell Arteritis, Rheumatoid Arthritis)

Polymyositis

Definition

Polymyositis is an inflammatory muscle disease that only affects the muscles. The development of symptoms is subacute, always starting in proximal muscles. Muscle wasting is prominent in chronic forms. Muscle pain sometimes occurs. Muscle biopsy is diagnostic.

Cross-References

▶ Myositis

Polyol Pathway

Definition

The metabolism of glucose to sorbitol by the enzyme aldose reductase and the subsequent metabolism of sorbitol to fructose by sorbitol dehydrogenase. The enzymes of the polyol pathway are found in tissues that develop disorders during diabetes, and metabolism of excess glucose by the polyol pathway is considered to be an early etiologic mechanism in many diabetic complications. Inhibitors of aldose reductase have been extensively studied as a potential treatment for diabetic neuropathy and other complications. However, while these aldose reductase inhibitors have protective effects in many animal models of diabetic neuropathy, their clinical efficacy remains to be proven.

Cross-References

▶ Neuropathic Pain Model, Diabetic Neuropathy Model

Polyps

Definition

Placebo-controlled trials have been performed in patients with colorectal adenomas to assess if selective COX-2 inhibitors have any beneficial effect in this condition.

Cross-References

▶ Cyclooxygenases in Biology and Disease

Polyradiculopathy

Definition

Polyradiculopathy refers to pathologic involvement of the nerve roots.

Cross-References

▶ Viral Neuropathies

Polysomnography

Definition

The study of sleep patterns by the simultaneous recording of the electroencephalogram (EEG) and video monitoring of the patient, together with the recording of other biological data.

Cross-References

▶ Hypnic Headache

POm

▶ Medial Part of the Posterior Complex

Pons

Definition

Typically means bridge in Latin, but this reference is to the area of the brainstem between the medulla and midbrain; this area contains several neuronal structures, including parts of the trigeminal sensory nuclear complex and the reticular formation, as well as the trigeminal motor nucleus and the facial nucleus.

Cross-References

▶ Trigeminal Neuralgia: Diagnosis and Treatment

Poor Sleep Quality

▶ Orofacial Pain, Sleep Disturbance

Portacatheters

Definition

Portacatheters are surgically placed ports that are most often connected to a catheter placed in the internal jugular vein. These ports enable the medical team to easily administer repeat medications.

Cross-References

▶ Analgesic Guidelines for Infants and Children

Positioning

Definition

Data show that positioning is a key element in facilitating coping with acute pain and invasive procedures. Parents or caregivers provide a supportive hold to maximize comfort, safety, and efficiency. The supine position should be avoided unless contraindicated.

Cross-References

▶ Acute Pain Management in Infants

Positive Reinforcement

Definition

An event that increases the likelihood that the behavior that preceded it will be repeated. Often, but not always, positive reinforcement is perceived as pleasurable.

Cross-References

▶ Impact of Familial Factors on Children's Chronic Pain
▶ Motivational Aspects of Pain
▶ Operant Perspective of Pain

Positive Sensory Phenomenon

Definition

A positive sensory phenomenon is a clinical sign that is interpreted by the patient as less than when compared to normal bodily function and experiences.

Cross-References

▶ Hypoalgesia, Assessment
▶ Hypoesthesia, Assessment

Positron Emission Tomography

Synonyms

PET

Definition

Positron emission tomography (PET) is an imaging technique that uses radiolabeled compounds to study physiologic processes in vivo. In cerebral activation studies, ^{15}O-water is used to study regional cerebral blood flow. Cerebral regional increases in positron emissions during a condition of interest compared to baseline reflect increases in regional blood flow, which is a marker of neuronal activation. PET is often used to measure changes in regions of activity within the brain during an experimental task, thus implicating functional involvement of these areas in task performance. PET is also heavily used in clinical oncology (for medical imaging of tumors and to search for metastases).

Cross-References

- ▶ Amygdala, Functional Imaging
- ▶ Amygdala, Pain Processing and Behavior in Animals
- ▶ Cingulate Cortex, Functional Imaging
- ▶ Descending Circuits in the Forebrain, Imaging
- ▶ Human Thalamic Response to Experimental Pain (Neuroimaging)
- ▶ Pain Processing in the Cingulate Cortex, Behavioral Studies in Humans
- ▶ PET and fMRI Imaging in Parietal Cortex (SI, SII, Inferior Parietal Cortex BA40)
- ▶ Thalamus and Visceral Pain Processing (Human Imaging)
- ▶ Thalamus, Clinical Pain, Human Imaging
- ▶ Thalamus, Clinical Visceral Pain, Human Imaging

Positron Emission Tomography (PET)

- ▶ Amygdala, Functional Imaging

Possible Contributors

- ▶ Pain in the Workplace, Risk Factors for Chronicity, Job Demands

Postamputation Pain

Definition

Pain following amputation. It includes stump pain and phantom pain.

Cross-References

- ▶ Phantom Limb Pain, Treatment
- ▶ Postoperative Pain, Postamputation Pain, Treatment, and Prevention

Post-dural Puncture Headache

Synonyms

PDPH

Definition

A headache that occurs following spinal anesthesia. The etiology of the headache is secondary to a decrease in cerebrospinal (CSF) fluid leading to reduced support for intracranial structures. Traction on meningeal vessels and nerves, in the sitting or standing position, leads to the classic postural headache. Additionally, because of the loss of CSF, there is compensatory dilation of intracranial blood vessels that also causes headache.

Cross-References

- ▶ Analgesia During Labor and Delivery

Posterior Column

- ▶ Spinal Dorsal Horn Pathways, Dorsal Column (Visceral)

Posterior Column Neurons

- ▶ Postsynaptic Dorsal Column Neurons, Responses to Visceral Input

Posterior Horn

- ▶ Spinal Dorsal Horn Pathways, Dorsal Column (Visceral)

Posterior Intervertebral Joint

▶ Zygapophyseal Joint

Posterior Nucleus

Synonyms

PO

Definition

A transitional diencephalic zone of the thalamus consisting of complex and varied cells lying caudal to the ventral posterolateral nucleus, medial to the rostral part of the pulvinar, and dorsal to the medial geniculate body. It starts at the dorsolateral aspect of the suprageniculate nucleus and expands laterally to the medial geniculate complex, and anteriorly ventral to VPL, and becomes continuous with VPI.

Cross-References

▶ Lateral Thalamic Pain-Related Cells in Humans
▶ Spinothalamic Terminations, Core and Matrix
▶ Thalamic Nuclei Involved in Pain, Human and Monkey
▶ Thalamus, Visceral Representation

Posterior Part of the Ventromedial Nucleus

Synonyms

VMpo

Definition

The posterior part of the ventromedial nucleus (VMpo) is a region that is claimed to be the main spinothalamic termination site for lamina I neurons, especially in the monkey and human.

Cross-References

▶ Human Thalamic Response to Experimental Pain (Neuroimaging)

Posterior Primary Ramus

Definition

Posterior branch of a spinal nerve that supplies branches to the paraspinal joints, ligaments, discs, and skin.

Cross-References

▶ Pain Treatment: Spinal Nerve Blocks

Post-excitatory Effects of C-Fibers

▶ Mechano-Insensitive C-Fibers, Biophysics

Postherpetic Neuralgia

Synonyms

Persistent pain following herpes zoster (*shingles, varicella zoster virus, VZV*); PHN

Definition

Postherpetic neuralgia is the paroxysmal pain extending along the course of one or more nerves following herpes zoster (acute shingles). Definitions of this are arbitrary but include pain at 1, 3, or 6 months.

Cross-References

▶ Cancer Pain, Assessment in the Cognitively Impaired
▶ Neuralgia, Assessment
▶ Opioids in Geriatric Application
▶ Postherpetic Neuralgia, Etiology, Pathogenesis, and Management
▶ Postherpetic Neuralgia, Pharmacological and Nonpharmacological Treatment Options

---

Postherpetic Neuralgia, Etiology, Pathogenesis, and Management

Peter N. Watson
Department of Medicine, University of Toronto, Toronto, ON, Canada

Synonyms

Herpes zoster pain; Shingles pain; Zoster-associated pain

Definition

▶ Postherpetic neuralgia (PHN), as very generally defined, is pain along the course of a nerve following the characteristic acute segmental rash of ▶ herpes zoster (HZ). A more specific definition is pain persisting after the rash has healed (usually 1 month). Some investigators studying therapeutic approaches have chosen longer periods of 2, 3, or even 6 months after the disease onset. ▶ Zoster-associated pain (ZAP) is a recent term originating with studies of antiviral agents, which includes all pain beginning before, at onset, and after the rash, and does not discriminate between the acute pain of HZ and the chronic pain of PHN. It is more useful for the study of antiviral agents.

Characteristics

Incidence, Natural History, and Demographics

HZ is the most common neurological illness (Kurtzke 1984), and PHN is its most frequent complication. The incidence of PHN (defined as pain persisting for more than 1 month) has been variously estimated from 9 % to 14 % (Watson et al. 1991). Gradual improvement occurs in some patients after that time (Watson et al. 1991). The incidence of PHN is directly related to age. About 50 % at age 60 years and nearly 75 % at age 70 years with HZ have PHN 1 month or more following the rash (Watson et al. 1991). The dermatomes most commonly affected are the ophthalmic division of the fifth cranial nerve and the midthoracic segments.

Pathology and Pathogenesis

There is evidence from pathological studies of both peripheral and central abnormalities in PHN (Watson et al. 1988). The pathogenesis of PHN remains unknown. Ectopic generators of abnormal discharges may occur at different locations along a damaged sensory nerve, from the regenerating sprout to the dorsal root ganglion. The discovery of pathologic change in the dorsal horn (Watson et al. 1988) in some patients with PHN supports the concept of a possible derangement with hypersensitive neurons in the central nervous system.

Clinical Features

When the acute rash has healed, the affected skin often exhibits a reddish, purple, or brownish hue. As this subsides, pale scarring often remains. The scarred areas are usually at least hypoesthetic, and yet the skin often exhibits marked skin pain on tactile stimulation (▶ allodynia), or increased pain to noxious stimulation (▶ hyperalgesia), or an increased sensitivity to touch (▶ hyperesthesia). A steady burning or aching may occur and also a paroxysmal, lancinating pain. The examination of the affected, scarred skin often reveals a loss of sensation to pinprick, temperature, and touch over a wider area than the scars and an even wider area of sensitive or painful skin. The sensitive skin may paradoxically include the anesthetic areas, where it is elicited by stroking or skin traction between thumb and forefinger, an effect that may be caused by summation on hypersensitive central neurons with expanded receptive fields.

Prevention

A vaccine has been approved for the prevention of herpes zoster and postherpetic neuralgia in those over age 50 who are not immunocompromised. It is a live attenuated vaccine with injection site reactions as the commonest side effect which are usually mild. It reduces herpes zoster by 50 % and PHN by two-thirds (Oxman et al. 2005). A recent article answers frequently asked questions about this important approach (Shapiro et al. 2011). The other preventive avenue is to treat herpes zoster early and aggressively. It is important that antiviral therapy with valaciclovir or famciclovir be initiated within 72 h of the onset of pain or rash in order to inhibit viral replication. It appears that there is a modest reduction in ZAP and PHN (but not severe PHN) with this approach. Another measure that should be considered is the early, aggressive use of analgesics including opioids to prevent the development of sensitization of the nervous system. It is possible that regional anesthesia of the painful area may help to resolve acute pain and prevent PHN, although this is not scientifically proven. It has been suggested that early antidepressant therapy, at the stage of herpes zoster, may help to prevent persistent pain (Bowsher 1996). Gabapentin and pregabalin may be employed at this time in a similar and safe fashion, but this is not scientifically proven at this time.

The Treatment of Established Postherpetic Neuralgia

Due to space constraints, complete references for this section are not given but may be obtained from the author (CPNW) or a book chapter which, although published some years ago, is adequate for this purpose (Watson and Gershon 2001). Randomized controlled trials (RCTs) have repeatedly shown that the antidepressant amitriptyline relieves PHN. These studies have demonstrated that most patients are not depressed, and pain relief occurs without a change in depression ratings. This analgesia occurred at lower doses than usually used to treat depression (median 50–75 mg/day). Amitriptyline may have limitations in the long term because of side effects and the fact that good relief occurs in only about one-half of the patients. A good maxim is to start low (10–25 mgm daily qhs) and go slow (increase every 7 days or so). One of the effects of this drug is to potentiate both serotonin and noradrenaline in the central nervous system.

Experience with serotonergic agents (clomipramine, trazodone, nefazodone, fluoxetine, zimeldine) in PHN has been disappointing. Desipramine, a more selective norepinephrine reuptake inhibitor, has been shown to be more effective than placebo in PHN, and pain relief with this drug as well was found not to be mediated by mood elevation. An RCT comparing maprotiline (noradrenergic) with amitriptyline found that although both were effective, amitriptyline was more so. A comparison of nortriptyline (more noradrenergic) with amitriptyline showed about equal efficacy for both drugs, with less severe side effects with nortriptyline.

The anticonvulsant gabapentin has been proven superior to placebo in PHN, with 43 % of the patients reporting at least moderate improvement (versus 12 % with placebo). As it appears to have fewer side effects than antidepressants, it has been suggested as a first-line treatment for this condition. A dose of 3,500 mg/day was aimed for in the trial above, and many patients achieved that. A study reports favorable results with pregabalin (Dworkin et al. 2003). The efficacy of this drug seems comparable to gabapentin (see Table 1).

An RCT of the opioid compound-sustained release oxycodone has shown that 58 % of the subjects experienced at least moderate improvement in pain versus 18 % on placebo. An RCT has shown that although opioids and antidepressants (nortriptyline, desipramine) relieve PHN, opioids may be more effective (Raja et al. 2002).

Postherpetic Neuralgia, Etiology, Pathogenesis, and Management, Table 1 Number needed to treat (NNT) data in some neuropathic pain conditions

Drug	Condition				Comments
	Postherpetic neuralgia	Diabetic neuropathy	Painful neuropathy	Central pain	
Antidepressants					
McQuay et al. (1996)	2.3	3.0		1.7	Systematic review
Sindrup and Jensen (1999)	2.3	2.4		1.7	Review
Collins et al. (2000)	2.1	3.4			Systematic review
Imipramine			2.7		RCT
Venlafaxine			5.2		RCT
Sindrup et al. (2003)					
Gabapentin					
Sindrup and Jensen (1999)	3.2	3.7			Systematic review
Pregabalin					
Dworkin et al. (2003)	3.4				RCT
Oxycodone					
Watson et al. (1998)	2.5				RCT
		2.6			RCT
Tramadol					
Sindrup and Jensen (1999)			3.4		Systematic review
Boureau et al. (2003)	4.76				RCT

Tramadol was more effective than placebo in an RCT (Boureau et al. 2003).

Although there is a long tradition of combining an antidepressant with a neuroleptic for PHN and other neuropathic pain, clinical experience has indicated no additional analgesic effects, and an RCT now supports no benefit (Graff-Radford et al. 2000).

Topical therapies for PHN fall into three categories: capsaicin preparations; aspirin/non-steroidal, anti-inflammatory preparations; and local anesthetics.

Capsaicin remains the most controversial of the topical agents. Several hundred patients have now participated in controlled and uncontrolled clinical trials, and the two concentrations have been commercially available and in widespread use for several years. Yet, even in larger-scale studies, there is a wide divergence in reported results, with some investigators reporting no benefit in well-designed, controlled studies. Others have reported success rates for capsaicin that seem extraordinarily high for a condition long thought to be among the most intractable of chronic pain syndromes. Crushed aspirin in either chloroform or ethyl ether is internationally available but awkward to use. Lidocaine has been formulated into adhesive patches (Lidoderm®) with a soft, woven backing that can be applied to cover the area of pain. A randomized, placebo-controlled trial has demonstrated the benefit of these patches (Watson and Gershon 2001).

Kotani et al. (2000) reported the results of an RCT of an intrathecal corticosteroid (methylprednisolone), given with lidocaine, in patients with PHN. More than 90 % of the patients who received this reported excellent or good pain relief at 4 weeks and at 1 and 2 years, as compared with only approximately 6 % of those who received only lidocaine and approximately 4 % of those who received no treatment. During the 2 years of this trial, no critical adverse effects, such as arachnoiditis or neurotoxic effects from the methylprednisolone, were noted. To date (12 years since), there has been no replication of this approach. Use of this technique in larger numbers of patients, and observation during longer follow-up periods, will be important to identify any occurrences of these potentially severe side effects. Also preservative-free preparations are difficult to obtain, and recent reports of fungal meningitis and abscesses merit caution here. Surgical procedures are not useful for most patients with PHN (Watson and Gershon 2001).

How Effective Are Pharmacological Agents in Clinical Practice?

The ► number needed to treat (NNT) has been suggested to convey the clinical meaningfulness of a trial (Table 1). This evaluation is a description of an arbitrary therapeutic effect for a desired outcome of 50 % improvement or more. It describes the difference between an intervention such as a drug and a control treatment. It is expressed as the number of patients required to be treated for a favorable response. Antidepressants and oxycodone may have an advantage in relieving pain, according to NNT data; however, their side-effect profile may make them less satisfactory to patients. Caution is required in comparing NNT figures, because of differences in trial designs, study numbers, and data analyses.

A Summary of Practical Guidelines for the Prevention and Treatment of Acute Herpes Zoster and Postherpetic Neuralgia

The main advance since the previous article is the zoster prevention vaccine which at this time is key in controlling this disease. The aggressive treatment of acute zoster is critical for acute pain management but may not prevent severe long-lasting PHN.

For established PHN (neuropathic pain persisting more than 3 months after HZ), the most consistently effective agents appear to be the older antidepressant drugs, and several RCTs in PHN support this approach. These indicate that pain may be taken from moderate or severe to mild in about one-half to two-thirds of the

patients. Therapy may commence with nortriptyline (less severe side effects than amitriptyline) in a dose of 10 mg before bed in those over 65 years and with 25 mg in those 65 or under. The dose is increased by similar increments in a single bedtime dose every 7–10 days, until relief is obtained or intolerable side effects supervene. If these fail, then amitriptyline and then another noradrenergic agent, such as desipramine, are reasonable choices. Occasional patients failing these may benefit from a serotonergic drug, such as an SSRI, but no RCT has been done in PHN, and these agents are not useful for many individuals.

An alternative approach, which may also be regarded as first line, is gabapentin increased slowly to as much as 3,600 mg/day (1,200 × 3). A trial-and-error approach in refractory patients may also include the anticonvulsants carbamazepine, oxcarbazepine, phenytoin, clonazepam, lamotrigine, topiramate, or valproate. Resistant cases may require opioids on an as needed and/or round-the-clock basis. A variety of long-acting opioids are available. The dose can be increased to satisfactory relief or unacceptable side effects. A lidocaine patch has been shown to be useful by RCT. ► Transcutaneous electrical nerve stimulation (TENS) may be worth trying. Electrode placement, frequency, intensity, and duration of stimulation are a matter of trial and error. Some patients may benefit from nerve blocks that, if efficacious, may be repeated at appropriate intervals. "Scientifically based data regarding the efficacy of nerve blocks for either prevention or long-term treatment of PHN are not available," to quote from Perry Fine's superb review (Watson and Gershon 2001), and the interested reader is referred there for a comprehensive review of this subject. At least 30 % of our patients remain totally refractory or unsatisfactorily relieved, and our approach with those is to see them regularly and try any new or older approach that seems reasonable and safe. Approximately 50 % of patients, including even those with pain of long duration, will improve over the years with one-half of these being on no treatment (Watson et al. 1991).

References

Boureau, F., Legallicier, P., Kabir-Ahmadi, M., et al. (2003). Tramadol in postherpetic neuralgia: A randomized, double-blind, placebo-controlled trial. *Pain, 104*, 323–331.

Bowsher, D. (1996). Postherpetic neuralgia and its treatment: A retrospective survey of 191 patients. *Journal of Pain and Symptom Management, 12*, 327–331.

Collins, S. L., Moore, R. A., & McQuay, H. J. (2000). Antidepressants and anticonvulsants for diabetic neuropathy and postherpetic neuralgia: A quantitative systematic review. *Journal of Pain and Symptom Management, 20*, 449–458.

Dworkin, R. H., Dworkin, R. H., Corbin, A. E., et al. (2003). Pregabalin for the treatment of postherpetic neuralgia: A randomized, placebo-controlled trial. *Neurology, 60*, 1274–1283.

Graff-Radford, S. B., Shaw, L. R., & Naliboff, B. N. (2000). Amitriptyline and fluphenazine in postherpetic neuralgia. *The Clinical Journal of Pain, 16*, 188–192.

Kotani, N., Kushikata, T., Hashimoto, H., et al. (2000). Intrathecal methylprednisolone for intractable postherpetic neuralgia. *The New England Journal of Medicine, 343*, 1515–1519.

Kurtzke, J. R. (1984). Neuroepidemiology. *Annals of Neurology, 16*, 265–277.

McQuay, H. J., et al. (1996). A systematic review of antidepressants in neuropathic pain. *Pain, 68*, 217–227.

Oxman, M. N., Levin, M. D., Johnson, J. R., et al. (2005). A vaccine to prevent postherpetic neuralgia in older adults. *The New England Journal of Medicine, 352*, 2271–2282.

Raja, S. N., Haythornthwaite, J. A., Pappagallo, M., et al. (2002). Opioids versus antidepressants in postherpetic neuralgia: A randomized placebo-controlled trial. *Neurology, 59*, 1015–1021.

Shapiro, M., Kvern, B., Watson, C. P. N., et al. (2011). Update on herpes on herpes zoster vaccination: A family practitioners guide. *Canadian Family Physician, 57*, 1127–1132.

Sindrup, S. H., & Jensen, T. S. (1999). Efficacy of pharmacological treatment of neuropathic pain: An update and effect related to mechanism of drug action. *Pain, 83*, 389–400.

Sindrup, S. H., Bach, F. W., Madsen, E., et al. (2003). Venlafaxine versus imipramine in painful polyneuropathy. *Neurology, 60*, 1284–1289.

Watson, C. P. N., & Gershon, A. A. (2001). *Herpes zoster and postherpetic neuralgia* (Pain Research and Management 2nd ed., Vol. 11). Amsterdam Netherlands: Elsevier.

Watson, C. P. N., et al. (1988). Postherpetic neuralgia: Postmortem analysis of a case. *Pain, 35*, 129–138.

Watson, C. P. N., et al. (1998). Oxycodone relieves neuropathic pain: A randomized controlled trial in postherpetic neuralgia. *Neurology, 50*, 1837–1841.

Watson, C. P. N., Watt, V. R., Chipman, M., et al. (1991). The prognosis with postherpetic neuralgia. *Pain, 46*, 195–199.

Postherpetic Neuralgia, Pharmacological and Nonpharmacological Treatment Options

Miles Belgrade[1], Anne K. Bertelsen[2] and Trevor Anderson[3]
[1]Fairview Pain Management Center, University of Minnesota Medical Center, Minneapolis, MN, USA
[2]Neurology, Aleris Hospital, Bergen, Norway
[3]Fairview Pain & Palliative Care, University of Minnesota Medical Center, Minneapolis, MN, USA

Synonyms

Persistent pain following herpes zoster (*shingles, varicella zoster virus, VZV*); PHN

Definition

▶ Postherpetic neuralgia is pain that persists following acute ▶ herpes zoster, or shingles. The time point at which acute herpes zoster pain becomes postherpetic neuralgia has been the focus of disagreement in the medical literature. Nonetheless, one definition is pain persisting for more than three months after the crusting of skin lesions following an acute attack of herpes zoster (Bowsher 1997; Desmond et al. 2002).

Characteristics

Postherpetic neuralgia (PHN) affects up to 70 % of people over 60 after herpes zoster (HZ). It is difficult to treat and is the most common complication following HZ.

Herpes zoster is caused by reactivation of the varicella zoster virus (VZV), which can lay dormant in the dorsal root ganglia of the cranial or spinal nerves for decades after chickenpox. PHN may arise from abnormal activity in damaged or dysfunctional peripheral sensory neurons caused by the herpes zoster infection (Rowbotham et al. 1998). The pain in PHN is typically described as constant burning, throbbing or aching, or intermittent stabbing or shooting paroxysms and may extend beyond the margins of the original rash. Cutaneous allodynia is a characteristic feature of PHN that occurs in the affected dermatomes. Paresthesias and dysesthesias may be present in the affected dermatome. Sensory loss to thermal, tactile, sharp, or vibratory stimuli can occur.

There is evidence from pathological studies of both peripheral and central abnormalities in PHN (Watson et al. 1988). The exact pathogenesis of PHN remains unknown. Ectopic generators of abnormal discharges may occur at different locations along a damaged sensory nerve, from the regenerating sprout to the dorsal root ganglion. The discovery of pathological change in the dorsal horn (Watson et al. 1988) in some patients with PHN supports the concept of a possible derangement with hypersensitive neurons in the central nervous system. One important mechanism causing PHN is thought to be the accumulation of voltage-gated sodium channels along the damaged peripheral sensory neurons. This will result in a lower activation threshold, more intense response to stimuli and spontaneous ectopic discharges along the axons. These peripheral changes subsequently lead to hyperexcitability of central neurons in the dorsal horn, a state called central sensitization. Allodynia and spontaneous pain may then be manifested by augmented and spontaneous activation of nociceptive neurons (C- and A-delta fibers), thereby maintaining central sensitization. Other possible causes include peripheral sensitization, increased adrenergic activation of C fibers, neuromas, abnormal sprouting, and collateral innervation.

Vaccines

There are currently two vaccines that may affect the development of PHN. The varicella zoster vaccine is given mostly to children to prevent chickenpox. The vaccine does not prevent the development of HZ, and there is currently no conclusive evidence to support the suspicion that it decreases the incidence and severity of HZ. This is only due to the relatively recent introduction of widespread vaccination, and we can reasonably expect more conclusive evidence for the vaccine's effect on HZ and PHN over the next few decades.

In a clinical trial involving thousands of adults 60 years old or older, the shingles vaccine (Zostavax) reduced the risk of shingles by 51 % and the risk of postherpetic neuralgia by 67 %. While the vaccine was most effective in people 60–69 years old, it also provided some protection for older groups. The vaccine appears to be effective for at least six years. The CDC recommends that everyone 60 and over receive the vaccine. It can be given at any time, even after an episode of HZ, but is generally accepted that the vaccine should be given after resolution of the shingles rash (CDC 2011).

Pharmacological Management of Postherpetic Neuralgia

These recommendations are tiered as described and then classified for effectiveness according to the American Academy of Family Physicians level of evidence classification system (American Family Physician 2012):

Level A – Effectiveness on PHN supported by randomized controlled trial or meta-analysis

Level B – Effectiveness on PHN supported by well-designed clinical trial(s) or strong evidence not sufficient to meet level A

Level C – Effectiveness on PHN supported by consensus viewpoint or expert opinion

Tier 1

Tier 1 – Supported by the majority of literature as effective (>50 % pain reduction) first-line treatments for PHN:

Gabapentin (Neurontin) – 900–3600 mg daily divided into three doses. Side effects include dizziness, somnolence, and "cloudy" thinking. Well tolerated with slow titration and few drug interactions (Level A, FDA approved).

Pregabalin (Lyrica) – 150–600 mg daily divided into 2 doses. Side effects include dizziness and somnolence. Generally well tolerated with few drug interactions (Level A, FDA approved).

Lidocaine patch 5 % (Lidoderm) – Use up to three patches over intact, affected skin for 12 h on and 12 h off. The most common side effect is local irritation (Level A, FDA approved).

Tricyclic antidepressants – nortriptyline (25–75 mg qHS) and amitriptyline (10–75 mg qHS). Side effects include arrhythmia, orthostatic hypotension, sedation, confusion, and anticholinergic symptoms. Avoid in elderly and those with heart disease. Nortriptyline is generally accepted as having a milder side effect profile (Level A).

Tier 2

Tier 2 – Shown to be effective but either to a lesser degree than Tier 1 and/or with more significant side effects:

Capsaicin topical – 0.025–0.075 % 3–4× per day. Apply only to intact skin and use gloves or wash hands thoroughly after application. Poor tolerance of the burning sensation with treatment reduces compliance so thorough education is required (Level A, FDA approved).

Capsaicin patch (Qutenza 8 % patch) – Apply patch to affected area for 1 h. This should be performed by an experienced provider and a clinic. Reapply in 3 months if necessary. Topical lidocaine should be applied to the area first to minimize pain from the application (Level A, FDA approved).

Tramadol – 50–100 mg q4–6 h PRN. Generally well tolerated but nausea, constipation, and dizziness are the most common side effects. These can be minimized with a titrating schedule (Level A).

Serotonin and norepinephrine reuptake inhibitors (SNRIs) – Duloxetine (Cymbalta, 60 mg daily) and venlafaxine (Effexor, 25–37.5 mg BID–TID). The most common side effects include nausea, somnolence, and sexual dysfunction in men (Level B).

Tier 3

Tier 3 – Shown to be effective but side effects, better options, and/or minimal to no effect over

placebo limit their usefulness to cases when other options have been ruled out:

Opioids – Oxycodone 20 mg per day divided BID–QID. Side effects include sedation, dizziness, and constipation. Higher addiction potential than other treatments and withdrawal can occur (Level A).

Nongabapentinoid Anticonvulsants – Depakote, Carbamazepine, etc. (Level B, Selph et al. 2011).

Topical NSAIDS – Ketoprofen gel, etc. (Level C).

Not adequately tested or shown to be ineffective for PHN: acyclovir, codeine, ibuprofen, acupuncture, lorazepam, buspirone, memantine, dextromethorphan, topical diclofenac, epidural lidocaine or methylprednisolone, IV lidocaine, and triptans (Hempenstall et al. 2005).

Treatments that are inadequately studied but have shown some good preliminary effect include topical gabapentin and botulinum toxin.

Postherpetic neuralgia is unresponsive to common analgesics like nonsteroidal anti-inflammatory drugs. To date, no single treatment guarantees good efficacy in all patients. Different treatment options are discussed in more detail below.

Anticonvulsant Agents

Gabapentin (Neurontin) has been shown to be effective in the treatment of PHN (Argoff 2011; Chandra et al. 2006; Hempenstall et al. 2005; Parsons et al. 2004; Selph et al. 2011; Zin et al. 2010) and was approved by the US FDA in May 2002 for this indication. Structurally, gabapentin is an analogue of GABA, but pharmacologically it appears to have no direct effect on GABA uptake or metabolism. Its mechanism of action appears to involve binding to the alpha$_2$delta subunit of voltage-dependent calcium channels (Backonja 2000), resulting in inhibition of calcium influx with subsequent reduced release of excitatory neurotransmitters from presynaptic terminals and thus reduced nociceptive transmission. Gabapentin seems to be most effective at or above doses of 1,800 and up to 3,600 mg per day (Backonja and Glanzman 2003; Stacey and Glanzman 2003). Doses under 900 mg per day have shown limited effect in the treatment of postherpetic neuralgia. However, the initial daily dose should not be higher than 900 mg divided into three doses, and doses should be reduced and the dosing interval increased in patients with renal failure. Gabapentin is usually well tolerated, but dizziness, somnolence, and mental clouding may occur.

Pregabalin (Lyrica), another analogue of GABA, was approved by the US FDA in September 2004 for the management of PHN. It has been shown to have a pain-relieving effect similar to that of gabapentin (Argoff 2011; Baron et al. 2009; Hempenstall et al. 2005; Selph et al. 2011). Pregabalin, which also binds to the alpha$_2$delta subunit of voltage-dependent calcium channels, has a more linear bioavailability than gabapentin, and lower doses are therefore needed to achieve pain relief. This increases its therapeutic index with respect to gabapentin and should lead to fewer dose-related side effects. The therapeutic dose range for pregabalin is between 150 mg and 600 mg per day, divided into two or three doses (Dworkin et al. 2003). The initial daily dose should be 150 mg divided into two doses.

Adding a tricyclic antidepressant or lidocaine patch as a combination therapy with a gabapentinoid has been shown to be more effective than monotherapy (Level A, B) (Argoff 2011).

Tricyclic Antidepressants

Tricyclic antidepressants (e.g., amitriptyline, nortriptyline, and desipramine) were until recently the most commonly used medications in the treatment of PHN, and, interestingly, they are found to be strong sodium channel modulators (Pancrazio et al. 1998), in addition to their well-known effect on reuptake of serotonin and noradrenaline. The sodium channel-blocking effect is presumed to be the pain-relieving mechanism of these drugs in the case of neuropathic pain. The cumulative efficacy reported from trials suggests that about one-third of patients will achieve a 50 % reduction in neuropathic pain (Argoff 2011; Chandra et al. 2006; Hempenstall et al. 2005; Selph et al. 2011), and their analgesic

effect appears to be independent of their antidepressant effect. However, the benefits are often outweighed by intolerable side effects, especially among the elderly. These drugs should be given in appropriate doses and several should be tried before concluding that there is no response. The effect will often be delayed for up to four weeks. It is recommended to use desipramine or nortriptyline in the elderly to avoid the more frequent side effects (anticholinergic and orthostatic hypotension) reported with the use of amitriptyline. These medications are typically taken at night due to their tendency to cause somnolence. This can be a positive side effect if sleep augmentation is also desired.

Topical Analgesics

Lidocaine 5 % patch (Lidoderm) has been shown in several trials to significantly reduce pain intensity, allodynia, and heat and cold sensitivity in patients suffering from PHN (Argoff 2011; Baron et al. 2009). The most common adverse reactions are localized erythema or edema. Systemic reactions are rare, but CNS excitation and/or depression have been reported, as well as cardiovascular symptoms such as bradycardia and hypotension. Lidocaine should be applied to intact skin only.

Capsaicin cream (Zostrix), the active component in hot chili peppers, applied topically four to five times a day for 4 weeks is effective in PHN treatment and is approved by the US Food and Drug Administration (FDA) for this indication (Kost and Straus 1996). Although this has been shown to significantly reduce PHN (Argoff 2011; Hempenstall et al. 2005), initial burning of the skin upon application is an unpleasant side effect that leads to discontinuation of treatment in more than 30 % of patients. Capsaicin is thought to affect pain transmission by depleting the neural stores of substance P, a pain-modulating substance, in nociceptive fibers and/or promoting localized, temporary denervation of the epidermis. The capsaicin cream contains 0.025–0.075 % capsaicin and should be applied to intact skin only. Patients should wash their hands thoroughly after applying capsaicin cream in order to prevent inadvertent contact with other areas.

Capsaicin 8 % patch (Qutenza) uses an extremely potent dose of capsaicin as a one-time application. Due to the noxious nature of concentrated capsaicin, this is completed in a clinical setting under the supervision of an experienced practitioner. The patch is applied to the affected area for 1 h. There is typically significant pain associated with this treatment and topical lidocaine is used to lessen the unpleasant burning sensation. The treatment has been shown to be effective in reducing PHN for up to 3 months (Argoff 2011).

Opioid Analgesics

Although opioids are considered a last resort for control of neuropathic pain, there is data suggesting efficacy and tolerability of oxycodone and other opioids for this condition (Argoff 2011; Hempenstall et al. 2005; Watson 2003; Zin et al. 2010). They are best considered as a short-term adjuvant treatment for PHN significant enough to require hospitalization.

Tramadol is a weak mu-agonist and a serotonin and norepinephrine reuptake inhibitor. This combination of actions probably contributes to the strong neuropathic pain control action of tramadol as compared to other opioids (Hempenstall et al. 2005; Hollingshead et al. 2006). Tramadol doses should be started low and titrated slowly to reduce side effects such as nausea, constipation, and male sexual dysfunction. The side effect profile and abuse potential of tramadol is typically lower than that of traditional opioids.

Serotonin-Norepinephrine Reuptake Inhibitors (SNRIs)

Duloxetine (Cymbalta) and venlafaxine (Effexor) are antidepressant medications that have been shown to reduce PHN (Argoff 2011; Selph et al. 2011). The norepinephrine reuptake inhibition is thought to be a strong contributing factor as selective serotonin reuptake inhibitors (SSRI) have not been shown to be equally effective. SNRIs may be considered a first-line treatment for a PHN patient who also suffers from untreated depression or as an alternative to an SSRI they are already on.

Steroids

Steroids through various routes have shown weak (oral, IV), moderate (epidural), and strong (intrathecal) evidence of effectiveness in treating PHN but side effects and/or complex administration limit their usefulness in most cases (Hempenstall et al. 2005). Intrathecal steroids were shown to be very effective in one study (Level A, Kotani et al. 2000), but serious complications that can arise from intrathecal steroids make routine use of this treatment impractical (Argoff 2011).

Regional Anesthetic Techniques

Sensory Nerve Block

Nerve root involvement is a frequent characteristic of PHN, but the pain-relieving effect of sensory nerve blocks is limited and short-lived. There are some reports of success with this pain therapy in the early stage of the disease, although coincidental spontaneous resolution cannot be excluded.

Sympathetic Nerve Blocks

Sympathetic blocks are sometimes helpful in alleviating pain (Argoff 2011), although results are often temporary and may only be obtainable in patients with neuralgia of less than 2 months' duration (Wu et al. 2000). Sympathetic blocks of the stellate ganglion and trigeminal branch blocks are often used to treat trigeminal zoster.

Neurolytic Blocks

Neurolytic blocks may be considered in patients with PHN when other blocks have not given the patient significant relief. They should only be done after a prognostic block has shown that an effective block of the appropriate area can be achieved. Neurolytic agents are used in cases of prolonged destruction of the nerves.

These blocks include ethyl alcohol 50 % in aqueous solution, absolute alcohol 95 % in aqueous solution, and phenol 6 %. Ethyl alcohol causes a higher incidence of neuritis than phenol. The duration of effects may vary from days to years, but usually ranges from 2 to 6 months. Ammonium compounds can also be used for peripheral nerve blockade. Pain relief follows selective destruction of unmyelinated C fibers by the ammonium ion: e.g., a solution of ammonium sulfate 10 %, in lidocaine 1 %, or ammonium chloride 15 %. The duration of action ranges from 4 to 24 weeks. Cryoanalgesia has also been used as a means of producing long-term blocks with unpredictable results.

Complications that may result from any nerve block procedure include pain, local hemorrhage, infection, needle soreness, and numbness.

Other Nonpharmacological Therapies

Ice and Other Cold Therapies

Ice is applied to the skin for 2–3 min several times a day, starting with the least sensitive skin area and approaching the most sensitive area. Ethyl chloride, or other cold spray, can also be used. Fluid is then sprayed twice over the entire painful area, and the evaporation cools the area. This treatment may relieve pain for varying lengths of time. As soon as the pain returns to near its former intensity, the treatment is repeated. Two to three sets of spray per day are often necessary to relieve the pain in responsive patients.

Nerve Stimulation Therapies

Transcutaneous electrical nerve stimulation (TENS) has been used in an attempt to relieve intractable pain of PHN. Although the success rate is low, relief can be sufficient in individual cases to permit a return to normal activities without analgesic therapy.

Acupuncture

Due to the nature of acupuncture treatment, rigorous randomized, controlled, double-blind trials are challenging to design and carry out. There is evidence that pain relief has been obtained in patients with PHN using acupuncture (Level B, Hempenstall et al. 2005).

Psychological Consultation and Treatments

In patients suffering from PHN, emotional stability is almost always affected. Therefore, it is especially important to treat the whole patient and not just an area of the skin. Severe depression is seen in more than 50 % of these patients, and

suicide is commonly considered in those with long-term intractable pain. Consultation by a psychologist or clinical social worker who is experienced in pain management is a valuable adjunct to drug therapy.

Training the patient in stress management and relaxation techniques is important. Anxiety and stress can exacerbate and prolong the pain. By practicing these techniques, some patients are able to control pain to some degree. In some patients, the pain-tension-anxiety cycle can convert acute pain symptoms into a chronic condition. Often, no matter what is done to treat these patients, the pain is not relieved unless the stress factors are also removed.

Surgery

Surgery is the last resort for the treatment of severe intractable postherpetic neuralgia.

Dorsal column stimulators with an implanted electrode placed over the dorsal columns of the spinal cord have been used with some success (Argoff 2011).

Deep brain stimulators, which are patient-activated, have been applied to the mesencephalic medial lemniscus to block the pain-conducting systems and stimulate endorphin secretion, which then produces analgesia through an agonist action at opioid receptors. Good pain relief has been achieved in many patients with this therapy.

Cordotomy is often initially very effective in relieving pain; however, in most cases, the pain eventually returns. Early return has been blamed on failure to ablate all the nerves in the pathways, which resume function after the swelling has deceased.

References

American Family Physician (2012). Levels of Evidence in AFP. http://www.aafp.org/online/en/home/publications/journals/afp/afplevels.html

Argoff, C. E. (2011). Review of current guidelines on the care of postherpetic neuralgia. *Postgraduate Medicine, 123*(5), 134–142.

Backonja, M. M. (2000). Anticonvulsants (antineuropathics) for neuropathic pain syndrome. *The Clinical Journal of Pain, 16*, 67–72.

Backonja, M., & Glanzman, R. L. (2003). Gabapentin dosing for neuropathic pain: Evidence from randomized, placebo-controlled clinical trials. *Clinical Therapeutics, 25*, 81–104.

Baron, R., Mayoral, V., Leijon, G., et al. (2009). Efficacy and safety of 5 % lidocaine (lignocaine) medicated plaster in comparison with pregabalin in patients with postherpetic neuralgia and diabetic polyneuropathy: Interim analysis from an open-label, two-stage adaptive, randomized, controlled trial. *Clinical Drug Investigation, 29*(4), 231–241.

Bowsher, D. (1997). The management of postherpetic neuralgia. *Postgraduate Medical Journal, 73*, 623–629.

CDC. (2011). Shingles Vaccination: What You Need to Know. http://www.cdc.gov/vaccines/vpd-vac/shingles/vacc-need-know.htm

Chandra, K., Shafiq, N., Pandhi, P., et al. (2006). Gabapentin versus nortriptyline in post-herpetic neuralgia patients: A randomized, double-blind clinical trial–the GONIP trial. *International Journal of Clinical Pharmacology and Therapeutics, 44*(8), 358–363.

Desmond, R. A., Weiss, H. L., Arani, R. B., et al. (2002). Clinical applications for change-point analysis of herpes zoster pain. *Journal of Pain and Symptom Management, 23*, 510–516.

Dworkin, R. H., Corbin, A. E., Young, J. P., Jr., et al. (2003). Pregabalin for the treatment of postherpetic neuralgia: A randomized, placebo-controlled trial. *Neurology, 60*, 1274–1283.

Hempenstall, K., Nurmikko, T. J., Johnson, R. W., et al. (2005). Analgesic therapy in postherpetic neuralgia: A quantitative systematic review. *PLoS Medicine, 2*(7), e164.

Hollingshead, J., Dühmke, R. M., & Cornblath, D. R. (2006). Tramadol for neuropathic pain. *Cochrane Database of Systematic Reviews, 19*(3), CD003726.

Kost, R. G., & Straus, S. E. (1996). Postherpetic neuralgia-pathogenesis, treatment, and prevention. *The New England Journal of Medicine, 335*, 32–42.

Kotani, N., Kushikata, T., Hashimoto, H., et al. (2000). Intrathecal methylprednisolone for intractable postherpetic neuralgia. *The New England Journal of Medicine, 343*, 1514–1519.

Pancrazio, J. J., Kamatchi, G. L., & Roscoe, A. K. (1998). Lynch C 3rd. Inhibition of neuronal Na$^+$ channels by antidepressant drugs. *Journal of Pharmacology and Experimental Therapeutics, 284*, 208–214.

Parsons, B., Tive, L., & Huang, S. (2004). Gabapentin: A pooled analysis of adverse events from three clinical trials in patients with postherpetic neuralgia. *The American Journal of Geriatric Pharmacotherapy, 2*(3), 157–162.

Rowbotham, M., Harden, N., Stacey, B., et al. (1998). Gabapentin for the treatment of postherpetic neuralgia: A randomized controlled trial. *Journal of the American Medical Association, 280*, 1837–1842.

Selph, S., Carson, S., Fu, R., et al. (2011). *Drug class review: Neuropathic pain: Final update 1 report [Internet].* Portland (OR): Oregon Health & Science University.

Stacey, B. R., & Glanzman, R. L. (2003). Use of gabapentin for postherpetic neuralgia: Results of two randomized, placebo-controlled studies. *Clinical Therapeutics, 25*, 2597–2608.

Watson, C. P. N. (2003). Postherpetic neuralgia. In A. S. Rice, C. A. Warfield, D. Justins, et al. (Eds.), *Clinical pain management - chronic pain* (p. 453). London: Arnold Press.

Watson, C. P. N., et al. (1988). Postherpetic neuralgia: Postmortem analysis of a case. *Pain, 35*, 129–138.

Wu, C. L., Marsh, A., & Dworkin, R. H. (2000). The role of sympathetic nerve blocks in herpes zoster and postherpetic neuralgia. *Pain, 87*, 121–129.

Zin, C. S., Nissen, L. M., O'Callaghan, J. P., et al. (2010). A randomized, controlled trial of oxycodone versus placebo in patients with postherpetic neuralgia and painful diabetic neuropathy treated with pregabalin. *The Journal of Pain, 11*(5), 462–471.

Post-hypnotic Suggestion

Definition

These are suggestions given within the hypnotic session, which then take effect outside of formal hypnosis. The suggestions may be diffuse and general, simply to encourage a sense of well-being or self-control, or may be highly specific and linked to a post-hypnotic cue. The cue may be a word, image, or action which when exhibited elicits the post-hypnotic response. For example, a subject may be taught to look at a hand in a particular way, which then induces the return.

Cross-References

▶ Therapy of Pain, Hypnosis

Postictal Headache

▶ Post-seizure Headache

Postinfarctional Sclerodactylia

▶ CRPS, Evidence-Based Treatment

Postinfective Polyneuritis

▶ Guillain-Barré Syndrome

Post-Lumbar Puncture Headache

▶ Headache Due to Low Cerebrospinal Fluid Pressure

Postnatal Pain

▶ Postpartum Pain

Postoperative Pain Management

Definition

The treatment of postoperative pain.

Cross-References

▶ Postoperative Pain, Thoracic and Cardiac Surgery
▶ Postoperative Pain, Tramadol
▶ Postoperative Pain, Transition from Parenteral to Oral Drugs

Postoperative Pain Model

▶ Nick Model of Cutaneous Pain and Hyperalgesia

Postoperative Pain, Acute Neuropathic Pain

Michael J. Cousins[1] and Philip J. Siddall[2]
[1]Department of Anesthesia and Pain Management, Royal North Shore Hospital University of Sydney, Sydney, NSW, Australia
[2]Department of Pain Management, Greenwich Hospital, University of Sydney, Sydney, NSW, Australia

Definition

Neuropathic pain is defined as pain caused by a lesion or disease of the somatosensory nervous system (Jensen et al. 2011). In the context of postoperative pain, it refers to pain that occurs in the acute postoperative setting following trauma to the peripheral or central nervous system.

Characteristics

Prevalence

It has been reported that 1–3 % of patients experience neuropathic pain in the acute postoperative period (Hayes et al. 2002). In the long term, it has been reported that approximately 15–25 % of people may experience persistent neuropathic pain following procedures such as inguinal hernia repair (Callesen et al. 1999) and mastectomy (Jung et al. 2003). For surgery involving damage to major nerve trunks such as limb amputation, the prevalence is even higher (Nikolajsen and Jensen 2001). People with nervous system trauma such as brachial plexus avulsion or spinal cord injury also have a high incidence of acute neuropathic pain postoperatively. Failure to recognize acute postoperative neuropathic pain may result in inappropriate treatment and poorly controlled pain, with a subsequent adverse effect on postoperative outcomes. It has also been suggested that unrecognized or inadequately treated acute neuropathic pain may be associated with an increased likelihood of persistent neuropathic pain (Callesen et al. 1999).

Diagnosis

Neuropathic pain can usually be identified on the basis of history and clinical examination. Surgery or trauma that is known to result in damage to major nerve trunks, such as amputation or brachial plexus avulsion, should lead to a high index of suspicion for the presence of postoperative neuropathic pain. It is usually stated that descriptors such as shooting, stabbing, electric, and burning are suggestive of neuropathic pain. However, these descriptors are not exclusive and a diagnosis based on pain descriptors alone is often difficult. Diagnosis is therefore assisted by other symptoms of nerve involvement, such as dysesthesia, paresthesia, and numbness in the region corresponding to or adjacent to the site of pain. The pain may also be paroxysmal and, in contrast to nociceptive pain, may be spontaneous with no relation to position or movement. However, shooting electric sensations in a radicular pattern, which are aggravated by movement, may also be suggestive of nerve root impingement or irritation.

Clinical examination is also helpful. Once again, it is often stated that increased responsiveness to either thermal or mechanical stimuli (▶ Allodynia and ▶ Hyperalgesia) in a region is indicative of neuropathic pain. However, these features are often present with nociceptive pain, particularly where there is ▶ central sensitization following tissue trauma. Therefore, these signs are not diagnostic for neuropathic pain. A diagnosis of neuropathic pain is more likely in the presence of signs of reduced sensory function. This loss of sensation may be indicative of nervous system damage, which is a prerequisite for the diagnosis of neuropathic pain. The presence of vasomotor and sudomotor changes, in addition to the features described above, is indicative of a ▶ complex regional pain syndrome (Wasner et al. 2003).

The development of questionnaires such as the LANSS, Neuropathic Pain Questionnaire, DN4, and pain DETECT may be useful adjuncts for the diagnosis of neuropathic pain. Further examination with techniques such as quantitative sensory testing may provide greater precision in identifying the nature of the neurological lesion. Imaging may also be appropriate in some acute postoperative settings to assist with the diagnosis of neuropathic pain. For example, a person with radicular pain following spinal trauma may require magnetic resonance imaging to determine whether there is impingement of a nerve root that may require surgical treatment.

Mechanisms

Surgery is often carried out in those with trauma to the nervous system. This trauma may give rise to postoperative neuropathic pain. Surgery itself inevitably results in damage to small nerves and some procedures, such as amputation, deliberately sever or lesion larger nerves. Poor positioning during anesthesia may also result in damage to vulnerable nerves or the brachial plexus. Compromise in blood supply to the spinal cord may lead to ischemic damage. Any of these situations may lead to complete or partial deafferentation, leading to the development of neuropathic pain. Why some people develop neuropathic pain after surgery and not others does not solely appear to be due to physical or mechanical causes. There is increasing evidence for the involvement of genetic factors in the development of neuropathic pain (Tegeder et al. 2006).

The mechanisms of neuropathic pain are dealt with elsewhere but include:

1. Generation of ectopic impulses (Devor et al. 1992). Although generally too early for neuroma formation, acute nerve injury may be associated with the generation of ectopic impulses in primary afferents. These may arise from compression on the dorsal root ganglion following trauma and occur through an increased expression and activation of sodium and calcium channels (Rogawski and Loscher 2004).
2. Increased excitation of central neurons. An afferent barrage following acute nerve injury may induce a central neuronal hyperexcitability due to increased glutamate release and activation of N-methyl D-aspartate (NMDA), non-NMDA, and metabotropic glutamate receptors with subsequent ▶ central sensitization.
3. Decreased inhibition of central neurons. A reduction in local or descending inhibitory

processes involving γ-aminobutyric acid (GABA), glycine, opioids, serotonin, and noradrenaline may occur, as a result of either direct nervous system damage or secondary changes. This loss of inhibition will again lead to an increased excitability of central neurons.

4. Glial activation. This may lead to the release of proinflammatory cytokines and other factors that contribute further to central sensitization.

5. Altered sympathetic function. The mechanisms of complex regional pain syndromes are poorly understood, but broadly, there is a central hyperreflexic state affecting somatosensory, vasomotor, sudomotor, and somatomotor responses (Wasner et al. 2003).

Treatment

It has been suggested that preemptive analgesia may be useful in reducing the incidence of neuropathic pain postoperatively. Several agents and approaches have been used. These include the systemic administration of opioids and NMDA antagonists and epidural administration of opioids, local anesthetics, and clonidine (Power and Barratt 1999). Peripheral nerve block may also be beneficial. Although there is theoretical support for the use of preemptive analgesia, there is still no clear evidence for an improvement in long-term outcomes (Moiniche et al. 2002).

There are few studies that specifically examine the effectiveness of treatments of acute postoperative neuropathic pain. It is traditionally stated that opioids are relatively ineffective for the treatment of neuropathic pain. However, there is increasing evidence that they may be effective if an appropriate dose is used. Several randomized controlled studies, using intravenous administration of opioids for the treatment of a number of neuropathic pain states, demonstrate the efficacy of intravenous opioids in the treatment of neuropathic pain when compared with placebo (Attal et al. 2002). Tramadol may also be an option because of its serotonergic and noradrenergic effects, which may provide an advantage in the treatment of neuropathic pain.

Local anesthetics such as lidocaine (lignocaine) are also widely used for the treatment of acute postoperative neuropathic pain. One of the actions of local anesthetics is to produce sodium channel blockade. This reduces the amount of ectopic impulses generated by activity at these receptors. Relatively low concentrations of local anesthetic are required to reduce ectopic neural activity in damaged nerves (Tanelian and MacIver 1991). Parenteral administration of local anesthetic provides a dose that does not block normal nerve conduction but leads to a reduction in ectopic impulse generation. Therefore, local anesthetics are administered parenterally by intravenous or subcutaneous infusion. Topical application is also possible, although it is not clear whether adequate doses can be obtained rapidly to provide pain relief in this situation.

As mentioned above, nerve injury may lead to an afferent barrage with release of glutamate and activation of NMDA receptors resulting in central neuronal hyperexcitability. NMDA receptor antagonists such as ketamine have been used widely as treatment for acute postoperative neuropathic pain (Stubhaug and Breivik 1995). Administration is generally an infusion via the intravenous or subcutaneous route. One of the main problems with the use of ketamine is the occurrence of disturbing side effects such as hallucinations, although benzodiazepines may help reduce these symptoms. Careful monitoring can help to minimize the rate of occurrence of these side effects. However, they can be distressing to the person when they do occur.

Although intravenous and subcutaneous administration of drugs such as opioids, lidocaine, and ketamine is feasible in the inpatient setting, oral administration of a number of drugs is also possible and may be preferred. Unfortunately, most studies suggest that these oral agents are less effective than drugs used systemically. Most of the evidence for the effectiveness of opioids for the treatment of neuropathic pain comes from studies using acute intravenous administration. Local anesthetics are not available in oral form, and the local anesthetic congener mexiletine does not appear to be as effective as lidocaine in reducing neuropathic pain. Similarly, ketamine is difficult to administer orally, and other NMDA antagonists available for administration via the oral route, such as dextromethorphan, are also not as effective.

Several agents that are widely used for the treatment of persistent neuropathic pain are not available for parenteral administration but are available in the oral form. These include the tricyclic antidepressants such as amitriptyline and nortriptyline and anticonvulsants such as carbamazepine, valproate, lamotrigine, gabapentin, and pregabalin. There is some evidence that gabapentin and pregabalin may be effective in the postoperative setting and reduce analgesic requirements. However, the data are very limited and a Cochrane review found insufficient evidence to support the use of pregabalin in the postoperative period (Moore et al. 2010). Side effects of anticonvulsants and tricyclic antidepressants such as sedation, dizziness, postural hypotension, paralytic ileus, and urinary retention are also of concern postoperatively. Most evidence suggests that parenteral administration of agents such as lidocaine is just as effective, if not more effective, than tricyclic antidepressants and anticonvulsants for the treatment of neuropathic pain. Therefore, unless there is a reason to avoid parenteral administration, it may be preferable to use this route, at least initially, to obtain satisfactory symptomatic relief.

Epidural administration of agents such as morphine, clonidine, and local anesthetics may be useful for the control of acute neuropathic pain. This approach has the advantage of more specific targeting of changes, such as spinal cord sensitization and abnormal activity in primary afferents.

Complex regional pain syndromes are treated with a number of approaches including intravenous regional administration of agents such as guanethidine, sympathetic ganglion blockade, epidural administration of clonidine, and exercise of the affected limb. Unfortunately, there is little evidence for the effectiveness of any of these agents, and systematic review of the effectiveness of guanethidine found it was no different from placebo (McQuay et al. 1996).

References

Attal, N., Guirimand, F., Brasseur, L., et al. (2002). Effects of IV morphine in central pain – a randomized placebo-controlled study. *Neurology, 58*, 554–563.

Callesen, T., Bech, K., & Kehlet, H. (1999). Prospective study of chronic pain after groin hernia repair. *British Journal of Surgery, 86*, 1528–1531.

Devor, M., Wall, P. D., & Catalan, N. (1992). Systemic lidocaine silences ectopic neuroma and DRG discharge without blocking nerve conduction. *Pain, 48*, 261–268.

Hayes, C., Browne, S., Lantry, G., et al. (2002). Neuropathic pain in the acute pain service: A prospective survey. *Acute Pain, 4*, 45–48.

Jensen, T. S., Baron, R., Haanpää, M., et al. (2011). A new definition of neuropathic pain. *Pain, 152*, 2204–2205.

Jung, B. F., Ahrendt, G. M., Oaklander, A. L., et al. (2003). Neuropathic pain following breast cancer surgery: Proposed classification and research update. *Pain, 104*, 1–13.

McQuay, H., Carroll, D., & Moore, A. (1996). Variation in the placebo effect in randomised controlled trials of analgesics: all is as blind as it seems. *Pain, 64*, 331–335.

Moiniche, S., Kehlet, H., & Dahl, J. B. (2002). A qualitative and quantitative systematic review of pre-emptive analgesia for postoperative pain relief – the role of timing of analgesia. *Anesthesiology, 96*, 725–741.

Moore, A. R., Straube, S., Wiffen, P. J., et al. (2010). Pregabalin for acute and chronic pain in adults. *Cochrane Database of Systematic Reviews, 11*, 11.

Nikolajsen, L., & Jensen, T. S. (2001). Phantom limb pain. *British Journal of Anaesthesia, 87*, 107–116.

Power, I., & Barratt, S. (1999). Analgesic agents for the postoperative period. Nonopioids. *The Surgical Clinics of North America, 79*, 275–295.

Rogawski, M. A., & Loscher, W. (2004). The neurobiology of antiepileptic drugs for the treatment of nonepileptic conditions. *Nature Medicine, 10*, 685–692.

Stubhaug, A., & Breivik, H. (1995). Post-operative analgesic trials: Some important issues. *Bailliere's Clininical Anaesthesiology, 9*, 554–582.

Tanelian, D. I., & MacIver, M. B. (1991). Analgesic concentrations of lidocaine suppress tonic A-delta and C fiber discharges produced by acute injury. *Anesthesiology, 74*, 934–936.

Tegeder, I., Costigan, M., Griffin, R. S., et al. (2006). GTP cyclohydrolase and tetrahydrobiopterin regulate pain sensitivity and persistence. *Nature Medicine, 12*, 1269–1277.

Wasner, G., Schattschneider, J., Binder, A., et al. (2003). Complex regional pain syndrome-diagnostic, mechanisms, CNS involvement and therapy. *Spinal Cord, 41*, 61–75.

Postoperative Pain, Acute Pain Management Principles

Harald Breivik[1], Else K. B. Hals[2] and
Audun Stubhaug[1]
[1]Department of Pain Management and Research,
Oslo University Hospital, Rikshospitalet and
University of Oslo, Oslo, Norway
[2]University of Oslo, Faculty of Dentistry, Oslo,
Norway

Definition

Acute pain is the normal alarm function of experiencing pain when tissue is damaged by trauma, acute disease, or by the surgeon's knife. Postoperative pain is common after surgery, varying from mild to very severe; resting pain depending on degree of tissue injury; and dynamic pain is more intense with movements when the operation was close to joints, thorax, or upper abdomen. Indication for surgery (cancer or benign disease), outcome of surgery, and patient's mental state and mood, as well as genetic factors, influence severity of the postoperative pain experience. Effective relief of acute pain improves quality of life and reduces risk of postoperative cardiopulmonary, gastrointestinal, and infectious complications, possibly also the risk of chronic pain after surgery. Acute pain is most intense during the first few days after injury but can last until tissue injury is completely healed. If bothersome pain is still present 6 weeks after surgery, the pain may continue into a chronic pain condition in 1–10 % of cases (Romundstad et al. 2006; Romundstad 2007).

Introduction

Acute and postoperative pain has been a focus of interest for anesthesiologists for many decades. Anesthesiologists are experts at removing pain from the patient's conscious awareness during surgery and trauma by administering general anesthesia (patient is unconscious) and/or by administering central blockades (spinal, epidural, or paravertebral anesthesia) or by local anesthetic blocks of major nerve plexa or specific peripheral nerves. In the postanesthesia care units (PACU), anesthesiologists keep pain under control by continuing regional anesthesia techniques or by intravenous administration of potent analgesic drugs (NSAIDs, paracetamol (acetaminophen), and opioids). When patients are returned to surgical wards or are discharged from the outpatient clinic or day-care unit, responsibility for relief of postoperative pain may be transferred to an acute pain team, to the surgeon, or to the patients' general physician. The quality of postoperative pain relief, after the responsibility of the anesthesiologist ends, varies from excellent to poor quality (Breivik and Stubhaug 2008). Dentists are also experts at relieving severe acute pain and many advances in pharmacological management of acute pain have been developed with the "dental pain model" (Løkken and Skjelbred 1980; Skjelbred and Løkken 1982; Breivik et al. 1999).

Characteristics

Pharmacological Relief of Acute Postoperative Pain
Acute postoperative pain management should start with a ▶ nonopioid analgesic drug for mild pain; paracetamol is first line drug (Løkken and Skjelbred 1980); add a drug from another class of nonopioid analgesics for mild-to-moderate pain (Breivik et al. 1999), before adding an ▶ opioid analgesic for more severe pain (Breivik 2000). Local and regional analgesic techniques are most effective for severe pain provoked by movements, so-called dynamic pain (Breivik 2000).

Nonopioid analgesics in common use are the following: paracetamol, nonsteroidal anti-inflammatory drugs (NSAIDs and COX-2 inhibitors (coxibs)), metamizole (dipyrone, not available in Northwestern Europe due to serious bone marrow failure), nefopam (mostly used in Southern Europe), and glucocorticoids (Skjelbred and Løkken 1982; Romundstad 2007).

First, fully exploit paracetamol (acetaminophen) as the nonopioid drug with the least adverse effects when used in appropriate doses orally, rectally, or by i.v. infusion. Paracetamol is often underdosed for fear of liver damage (Korpela and Olkkola 1999). For acute pain, a safe dosing regimen is as follows (Breivik 2000; Breivik et al. 2007):

- Patients above 50 kg: start dose 1.5–2 g, followed by 1 g three to four times per 24 h for up to 2 weeks, thereafter not more than three grams per 24 h
- For children above 1 year of age and patients up to 49 kg: start dose 40 mg/kg, followed by 20 mg/kg three to four times per 24 h for up to 1 week, thereafter 15 mg/kg up to four times per 24 h

Then, when more potent pain relief is needed, take advantage of the additive analgesic effect of a ► nonsteroidal anti-inflammatory drug (NSAID or a coxib) when coadministered with paracetamol (Breivik et al. 1999).

However, NSAIDs are contraindicated when patients have:
- Actual or potential bleeding problems
- History of gastrointestinal ulceration
- Aspirin-sensitive asthma
- Renal impairment
- Hypovolemia
- Hyperkalemia
- Diuretic medication
- Preeclampsia
- Severe liver dysfunction
- Circulatory failure

NSAIDs should be used with caution in:
- Elderly patients (above 65–70 years), who often have significant renal impairment
- Diabetic patients (renal impairment likely)
- Patients with widespread vascular disease
- Patients on ACE inhibitors, beta-blockers, cyclosporin, and methotrexate

Except for bleeding and gastrointestinal ulcer disease, these are contraindications for coxibs as well.

Glucocorticoid drugs have prolonged analgesic effects with minimal risk of adverse effects when administered on the day of acute trauma or surgery (Skjelbred and Løkken 1982; Romundstad et al. 2004, Romundstad 2007), and they reduce postoperative nausea and fatigue as well (Holte and Kehlet 2002; Aasboe et al. 1998). For several years, we have administered routinely for adult patients 16 mg of dexamethasone or 80 mg of methylprednisolone (for children 0.2 mg/kg dexamethasone), half of this dose for less extensive surgery. Note that this dose is given only once during or soon after surgery, and the effects on pain persist for at least three days (Romundstad et al. 2004; Romundstad 2007).

Weak ► opioid analgesic drugs, such as codeine and tramadol, are often added to the nonopioid analgesics with some increased pain relief but with increased risk of adverse effects such as nausea, dizziness, sedation, and prolongation of postoperative ileus (Stubhaug et al. 1995; Bjune et al. 1996; Breivik et al. 1999).

A potent opioid is often required in addition to paracetamol (with or without an NSAID or a coxib) for more severe pain. A potent opioid like ► morphine, or ► oxycodone, is then administered either by mouth, rectally, or by injection. The administration of the opioid analgesic is usually under the control of a ward nurse who is guided by standing orders on dosing and monitoring of effects and adverse effects, respiratory depression in particular (Breivik et al. 2007). For day-case surgery, patients should be given the necessary drugs for the first 2–3 days after surgery, as well as a prescription for analgesic drugs that may be needed thereafter.

► Patient-controlled analgesia (PCA), i.e., the intravenous administration of a potent opioid in small bolus doses. This is controlled by the patient who is allowed to push a bolus-release button on a PCA pump whenever he/she needs more pain relief. An alternative to an i.v. PCA pump is to let the patient self-administer a limited number of tablets of a potent opioid (after day-case surgery). The basis for PCA is that pain is subjective, and only the patient can know how much pain-relieving medication is needed and how much subjective pain relief and adverse effects result from an analgesic drug dose. PCA devices must be constructed so that no patient can have an accidental overdose (Breivik et al. 2007).

Local Anesthetics, Nerve Blocks, and Neuraxial Analgesia

Most severe pain occurs on movement after major surgery of the thorax, upper abdomen, and after orthopedic surgery close to joints and on weight-bearing parts of the skeleton (dynamic pain). Whereas pain at rest may be controlled by increasing the potency of pharmacologic analgesia stepwise, as described above, pain provoked by deep breathing, coughing, moving a limb, or getting out of bed and walking around is not well relieved by analgesic drugs. Such pain is best relieved by local anesthetic drugs given as a peripheral nerve block, epidural, or spinal (subarachnoid) coadministration of a local anesthetic, an opioid, and an adrenergic drug, preferably adrenaline (epinephrine) (Capdevila et al. 1999; Breivik 2000; Niemi 2004; Breivik et al. 2007; Breivik 2008; Carli and Schricker 2009; Breivik 2012).

Acute Pain Team

For management of acute postoperative pain to be optimal, an acute pain team (APT) or ▶ acute pain service (APS) is necessary: An APT is organized in a hospital to prevent and treat acute pain after surgery, trauma, acute pain caused by other medical conditions, and even labor pain. An APT is most often based in a department of anesthesiology, the pain clinic of the hospital, or one of the clinical departments having many patients with acute pain problems. Basic to any acute pain team is a staff of well-trained and dedicated nurses, active during daytime working hours, or up to full 24 h 7 days per week (Breivik et al. 2007). The pain nurses must have an ▶ anesthesiologist attached to the APT on a part- or full-time basis, with night and weekend assistance with epidural catheter and

PCA-pump problems from the on-call anesthesia residents, or from a dedicated 24 h pain team in some major hospitals. Surgeons and other clinical specialists are associated with the pain team in various ways, again depending on the type of patients and pain conditions taken care of by the APT. An extremely important duty of any APT/APS is, in collaboration with surgeons, ward nurses, and administrators, to develop pain relief regimens for painful procedures and operations carried out in the hospital (Breivik et al. 2007; Macintyre and Scott 2009).

A well-planned postoperative pain management program will ensure that:

- All patients are offered optimal pharmacological pain relief with appropriate combinations of paracetamol and NSAIDs, when tolerated.
- Ward nurses have standing orders to take responsibility for appropriate dose of opioids by intramuscular or intravenous injections.
- Ward nurses are trained by the APT/APS to administer intravenous patient-controlled opioid analgesia with PCA pumps.
- Patients at increased risk of developing postoperative respiratory and cardiac complications, due to unrelieved dynamic pain after major thoracic or abdominal surgery, are offered optimal ▶ thoracic epidural analgesia (Niemi 2004; Breivik 2008, 2012). Paravertebral blocks may be an alternative, but it is not more effective and not safer than optimal epidural analgesia (Norum and Breivik 2010; Fagenholz et al. 2012).
- A dedicated acute pain team is established for effective and safe management of pain in all surgical patients. Continuous daytime coverage by pain nurses with anesthesiologist backup is a minimum.
- There is an ongoing educational program for ward nurses, surgeons, anesthesiologists, and patients.
- Robust monitoring of pain, pain relief, and any side effects (sedation and ▶ respiratory depression whenever opioids are administered) is continuous and results recorded at least every 4 h.

This will "make pain visible," focus attention on inadequate pain relief, and is a powerful tool in ongoing audit and continuous quality improvement of pain relief (Breivik et al. 2007)

- There is a high index of suspicion for early signs of intraspinal bleeding or infection and vigilant monitoring of leg weakness and any new backache, coupled with high preparedness for rapid verification of diagnosis (MRI or CT scan) and surgical removal of a blood clot or abscess, should any of these very rare complications to ▶ epidural analgesia happen (Breivik et al. 2007; Breivik et al. 2010).

- The risks are minimized and benefits are maximized for patients in terms of more comfort, less pain, and more gain in improved rehabilitation and outcome after surgery (Breivik et al. 2007).

Summary

The principles of acute pain management are to fully exploit paracetamol (acetaminophen), add a nonsteroidal anti-inflammatory analgesic drug (NSAID) or a glucocorticosteroid, and for more severe pain, administer an opioid in addition to paracetamol (with or without an NSAID), and a glucocorticoid. After surgery of the thorax, upper abdomen, and after orthopedic surgery close to joints and on weight-bearing parts of the skeleton, severe pains provoked by movement are best relieved by peripheral nerve blocks, epidural, or spinal analgesia. Relieving such dynamic pain facilitates coughing and mobilization and reduces cardiorespiratory complications after surgery. An acute pain team (APT) with dedicated nurse(s) will facilitate education of ward nurses, doctors, and patients, making implementation of improved pain-relieving methods possible, as well as ongoing audit and continuous quality improvement of postoperative pain management.

References

Aasboe, V., Ræder, J. C., & Grøgaard, B. (1998). Betamethasone reduces postoperative pain and nausea after ambulatory surgery. *Anesthesia and Analgesia, 87*, 319–323.

Bjune, K., Stubhaug, A., Dodgson, M. S., & Breivik, H. (1996). Additive analgesic effect of codeine and paracetamol can be detected in strong, but not moderate, pain after Caesarean section. Baseline pain-intensity is a determinant of assay-sensitivity in a postoperative analgesic trial. *Acta Anaesthesiologica Scandinavica, 40*, 399–407. http://www.ncbi.nlm.nih.gov/pubmed/8738682.

Breivik, H. (2000). High-tech versus low-tech approaches to postoperative pain management. In M. Devor, M. C. Robotham, & Z. Wiesenafeld-Hallin (Eds.), *Proceedings of the 9th world congress on pain* (pp. 787–807). Seattle: IASP Press.

Breivik, H. (2008). Epidural analgesia for acute pain after surgery and during labor, including patient-controlled epidural analgesia. In H. Breivik, W. Campbell, & M. K. Nicholas (Eds.), *Clinical pain management. Practice and procedures* (pp. 311–321). London: Hodder Arnold.

Breivik, H. (2012). Local anaesthetic blocks and epidurals. In S. B. McMahon, & M. Kolzenburg (Eds.), *Wall and Melzack's textbook of pain* (6th ed.) Chapter 38. London: Elsevier Churchill Livingstone.

Breivik, E. K., Barkvoll, P., & Skovlund, E. (1999). Combining diclofenac with acetaminophen or acetaminophen-codeine after oral surgery: A randomized, double-blind single-dose study. *Clinical Pharmacology and Therapeutics, 66*, 625–635.

Breivik, H., Curatolo, M., Niemi, G., Haugtom, H., & Kvarstein, G. (2007). How to implement an acute pain service: An update. In H. Breivik & M. Shipley (Eds.), *Pain best practice and research compendium* (pp. 255–270). London: Elsevier.

Breivik, H., & Stubhaug, A. (2008). Management of acute postoperative pain: Still a long way to go! *Pain, 137*, 233–234.

Breivik, H., Bang, U., Jalonen, J., Vigfússon, G., Alahuhta, S., & Lagerkranser, M. (2010). Nordic guidelines for neuraxial blocks in disturbed haemostasis from the Scandinavian society of anaesthesiology and intensive care medicine. *Acta Anaesthesiologica Scandinavica, 54*, 16–41.

Capdevila, X., Barthelet, Y., Biboulet, P., et al. (1999). Effects of perioperative analgesic technique on the surgical outcome and duration of rehabilitation after major knee surgery. *Anesthesiology, 91*, 8–15.

Carli, F., & Schricker, T. (2009). Modification of metabolic responses to surgery by neural blockade. In M. J. Cousins, D. B. Carr, T. T. Horlocker, & P. O. Bridenbaugh (Eds.), *Cousins and Bridenbaugh's neural blockade in clinical anaesthesia and pain medicine* (4th ed., pp. 133–143). Philadelphia: Wolters Kluwer/Lippincott Williams & Wilkins.

Fagenholz, P. J., Bowler, G. M. R., Carnochan, F. M., & Walker, W. S. (2012). Case report. Systemic local anaesthetic toxicity from continuous thoracic paravertebral block. *British Journal of Anaesthesia, 109*, 260–262.

Holte, K., & Kehlet, H. (2002). Perioperative single-dose glucocorticoid administration: Pathophysiologic effects and clinical implications. *Journal of the American College of Surgeons, 195*, 694–712.

Korpela, R., & Olkkola, K. T. (1999). Paracetamol – misused good Old drug? *Acta Anaesthesiologica Scandinavica, 43*, 245–247.

Løkken, P., & Skjelbred, P. (1980). Analgesic and anti-inflammatory effects of paracetamol evaluated by bilateral oral surgery. *British Journal of Clinical Pharmacology, 10*, 253–260.

Macintyre, P. E., & Scott, D. A. (2009). Acute pain management and acute pain services. In M. J. Cousins, D. B. Carr, T. T. Horlocker, & P. O. Bridenbaugh (Eds.), *Cousins' and Bridenbaugh's neural blockade in clinical anaesthesia and pain medicine* (4th ed., pp. 1036–1062). Philadelphia: Wolters Kluwer/Lippincott Williams & Wilkins.

Niemi, G. (2004). Optimizing postoperative epidural analgesia. PhD-thesis, Faculty of Medicine, University of Oslo, Unipub, Oslo.

Norum, H., & Breivik, H. A. (2010). Systematic review of comparative studies indicates that paravertebral block is neither superior nor safer than epidural analgesia for pain after thoracotomy. *Scandinavian Jounal of Pain, 1*, 12–23.

Romundstad, L. (2007). Effects of glucocorticoids and nonsteroidal anti-inflammatory drugs on postoperative and experimental pain. PhD-thesis, Faculty of Medicine, University of Oslo, Unipub, Oslo.

Romundstad, L., Breivik, H., Niemi, G., et al. (2004). Methylprednisolone intravenously 1 day after surgery has sustained analgesic and opioid-sparing effects. *Acta Anaesthesiologica Scandinavica, 48*, 1223–1231.

Romundstad, L., Breivik, H., Roald, H., et al. (2006). Chronic pain and sensory changes after augmentation mammoplasty: long term effects of preincisional administration of methylprednisolone. *Pain, 124*, 92–99.

Skjelbred, P., & Løkken, P. (1982). Postoperative pain and inflammatory reactions reduced by injection of a corticosteroid. A controlled trial in bilateral oral surgery. *European Journal of clinical Pharmacology, 21*, 391–396.

Stubhaug, A., Grimstad, J., & Breivik, H. (1995). Lack of analgesic effect of 50 and 100 mg oral tramadol after orthopaedic surgery: a randomized, double-blind, placebo and standard active drug comparison. *Pain, 62*, 111–118.

Postoperative Pain, Acute Pain Team

Harald Breivik[1], Kari Sørensen[2] and Audun Stubhaug[1]
[1]Department of Pain Management and Research, Oslo University Hospital, Rikshospitalet and University of Oslo, Oslo, Norway
[2]Department of Pain Management and Research, Oslo University Hospital, Rikshospitalet, Oslo, Norway

Definition

An ▶ acute pain team (APT) is organized in a hospital to prevent and treat acute pain after surgery, trauma, acute pain caused by other medical conditions, and labor pain. APT can improve pain relief for patients at high risk of severe acute pain, e.g., after extensive surgery, in patients on chronic opioid-therapy, patients with medical co-morbidities that increase their risk of analgesia-related complications or side effects. Children and elderly patients often need special focus on effective and safe pain relief. An APT is based in a department of anesthesiology, the pain clinic of the hospital, or one of the clinical departments that has patients with acute pain problems. For optimal management of acute postoperative pain, an APT is necessary for continuously ongoing education of ward nurses, doctors, and patients for implementing advanced pain-relieving methods, for effective and safe pain relief in high risk, young and old patients, for audit, and for continuous quality improvement in the hospital-wide management of acute pain.

Introduction

Advanced and highly effective methods of acute pain management, such as intravenous patient-controlled analgesia and prolonged epidural analgesia on surgical wards made it necessary to organize acute pain teams. Members of an APT have the necessary special skills and knowledge required for safe and effective patient care, and they organize and train other healthcare personnel in appropriate pain management. Without well-functioning APT, these advanced, often high-tech pain-relieving methods are neither effective nor safe for patients.

Characteristics

Depending on the size of the hospital, an acute pain team must have a staff of well-trained and dedicated pain nurses, varying from one to several, covering only daytime hours or up to full 24 h coverage 7 days per week. The ▶ pain nurses must be associated with a part-time or full-time ▶ anesthesiologist, depending on the size of the hospital, number of postoperative pain patients, and types of pain-relieving techniques offered. Surgeons and other clinical specialists are associated with the pain team in various ways, depending on the type of patients and pain conditions taken care of by the APT (Breivik et al. 2007; Counsell et al. 2008; Macintyre and Scott 2009). It is important to realize that a well organized APT will save resources by reducing postoperative complications and need for ICU-treatment: An acute pain service is cost-effective (Brodner et al. 2000; Breivik and Stubhaug 2008).

Why an Acute Pain Team Is Necessary
Epidural anesthesia had a renaissance about 30 years ago, and it was obvious that prolonging ▶ epidural analgesia into the immediate postoperative period and beyond could more effectively control ▶ dynamic pain after major surgery than parenteral opioids. It also soon became apparent that ▶ thoracic epidural analgesia with local anesthetic infusions had beneficial effects on postoperative complications such as thromboembolism, gastrointestinal immobility, and negative nitrogen balance (for review, see Breivik et al. 2007). Episodes of hypotension,

pronounced ▶ local anesthetic motor blockade causing leg weakness, and inability to spontaneously empty the urinary bladder were all frequent adverse effects of bolus injections or continuous epidural infusions of relatively high concentrations of local anesthetic, especially when epidural catheters were placed in the lumbar segments of the epidural space. When ▶ neuraxial morphine administration entered epidural analgesia practice, respiratory depression, nausea, and pruritus were often quite bothersome and sometimes even dangerous adverse effects (Stenseth et al. 1985).

These adverse effects and some dramatic case reports on ▶ delayed severe respiratory depression when opioids were administered epidurally caused restrictions on where and how epidural analgesia could be used for postoperative pain. This effective technique became strictly reserved for ▶ high-dependency or intensive care units. Attempts to move patients to the surgical wards with these epidural analgesia techniques were aborted when such complications occurred. It was, of course, not possible to keep many patients in the recovery ward or in the high-dependency unit solely for the purpose of giving pain relief with epidural analgesia techniques.

Thus, fear of side effects and complications, and the practice of limiting epidural analgesia to high-dependency nursing units, caused lack of confidence, skepticism, and mistrust among surgeons and ward nurses. When we started to plan major improvements in postoperative pain relief for most patients after major surgery, by offering to implement epidural analgesia and patient-controlled parenteral opioid analgesia on surgical wards, it was evident that we first needed to overcome these obstacles to wider use of the most effective pain-relieving methods available (Breivik 1993).

It was quite obvious to us that, in order to harvest the many potential benefits from optimal and prolonged epidural analgesia, we had to have a system that would enable nurses to care for patients with epidural analgesia on the surgical wards. Our early experience soon led to the conclusion that an acute pain organization was needed (Breivik 1993).

An acute pain team of well-trained nurses and dedicated anesthesiologist(s) must be responsible for:

- Fine-tuning an optimal thoracic epidural analgesia regimen
- Implementing safe patient-controlled intravenous opioid analgesia (PCA)
- Training programs for ward nurses in acute pain management with focus on pain assessment, handling and monitoring pain pumps, monitoring effects and observing and treating any adverse effects, including documentation of pain management
- Advise step-down program from advanced, neuraxial or intravenous pain-treatment to oral analgesic drugs. Prevent or treat withdrawal symptoms
- Teaching anesthesiologists in the operating room to place epidural catheters at optimal segmental spinal cord level
- Teaching and motivating on-call anesthesiologists to help solve epidural and peripheral nerve catheter and pain pump problems during nights and weekends
- Upgrading guidelines and upgrading ▶ traditional pharmacological pain relief with nonopioid and opioid analgesics in order to *improve postoperative pain relief for all surgical patients*

How to Implement an Acute Pain Team
Motivation for Change
Ward nurses, surgeons, anesthesiologists, and hospital administrators must be motivated and agree that traditional postoperative pain management can and should be improved and that this is important for patients' well-being, satisfaction with the hospital, and also that the outcome of surgery can benefit by optimal postoperative pain management.

Who Must Join the APT and How to Start
Initially, one full-time pain nurse and one anesthesiologist, able to spend most of his/her working day on the APT, establish procedures and regimens for epidural analgesia and

patient-controlled intravenous opioid analgesia (PCA) in the recovery room. Gradually, they spread knowledge and practice of these basic techniques to regular wards, one ward at a time.

Initially, the anesthesiologist and the pain nurse closely follow every patient, but with education and training of the ward nurses, they gradually take responsibility for monitoring and titration of epidural analgesia, as well as PCA. Soon, nurses on the surgical wards, as well as surgeons, realize that a well-tailored epidural analgesia can make the patient much more comfortable and satisfied with the postoperative pain relief. Nursing and mobilization of the patients become easier. Enthusiasm for the services of the APT grows.

The first reports of acute pain service teams came from the United States, and Ready et al. (1988) reported on the development of independent anesthesiologist-based postoperative pain management service in 1988. European models have been based on APT nurses, with focus on education and organization of regimens based on "pain ombudsman-nurses" in each ward, and on help from the responsible operating room anesthesiologists (Breivik 1993; Rawal and Berggren 1994; Breivik et al. 1995; Breivik 1995; Breivik et al. 2007).

Important Success Factors for Implementing an Effective Acute Postoperative Pain Team
The ultimate goal of an APT is to reduce unnecessary suffering from unrelieved pain in all surgical patients and to exploit potential benefits of optimal pain relief on postoperative complications and outcome of surgery (Breivik 2000; Rawal 1999; Breivik 2012). This is a major project in most hospitals, and the traditional obstacles to change are definitely present when an APT is introduced (Powel et al. 2008). The following are some factors that are important in order to overcome such obstacles:

- Well-trained and dedicated pain nurses supervised by an anesthesiologist.
- Anesthesiologist able to spend much of the working day outside the operating room, with the APT.

- APT is available continuously during working daytime.
- Ongoing educational program for all health personnel involved in the care of surgical patients.
- Information to patients (also in writing) before and after surgery.
- Optimal epidural analgesia for severe dynamic pain, prolonged for 3–7 days after major surgery (Breivik et al. 2007; Breivik 2008; Breivik 2012).
- Offer selected peripheral nerve blocks with continuous catheter techniques.
- Note that paravertebral block is a block of spinal nerves and also a central neuraxial block with the same adverse effects as spinal or epidural blocks (Norum and Breivik 2010). Paravertebral blocks need high doses of local anaesthetic drugs to be effective, with a real risk of serious local anesthetic toxicity (Fagenholz et al. 2012).
- Use upgraded, optimal pharmacological pain relief in all patients with paracetamol plus an NSAID (or COX-2 inhibitor) as base, plus parenteral, or oral opioid when needed (Breivik 2000; Breivik et al. 2007).
- Robust routine for monitoring and documenting pain-relieving effect, adverse effects, early symptoms of possible catastrophic complications from respiratory depression, infection, or bleeding in spinal canal (Breivik et al. 1995; Breivik et al. 2010; Breivik 2012).
- With increasing use of continuous peripheral nerve and paravertebral blocks, serious local anesthetic toxicity is again a real risk (Norum and Breivik 2010; Fagenholz et al. 2012).
- High preparedness for handling potentially dangerous complications from bleeding or infection in the spinal canal, as well as local anesthetic toxicity (Breivik et al. 2010; Norum and Breivik 2010; Fagenholz et al. 2012; Breivik 2012).
- Ongoing audit and continuous quality improvement of postoperative pain management (Breivik et al. 2007; Breivik et al. 2010).

References

Breivik, H. (1993). Recommendations for foundation of a hospital-wide postoperative pain service – a European view. *Pain Digest, 3*, 27–30.

Breivik, H. (1995). Benefits, risks and economics of postoperative pain management programmes. *Bailliere's Clinical Anaesthesiology, 9*, 403–422.

Breivik, H. (2000). High-tech versus low-tech approaches to postoperative pain management. In M. Devor, M. C. Robotham, & Z. Wiesenafeld-Hallin (Eds.), Proceedings of the 9th World Congress on pain. *Progress in Pain Research and Management, 16*, 787–807.

Breivik, H. (2008). Epidural analgesia. In H. Breivik, W. Campbell, & M. K. Nicholas (Eds.), *Clinical pain management. Practice and procedures* (pp. 311–321). London: Hodder Arnold.

Breivik, H. (2012). Local anaesthetic blocks and epidurals. In S. B. McMahon, & M. Kolzenburg (Eds.), *Wall and Melzack's Textbook of Pain* (6th ed.). London: Elsevier Churchill Livingstone, Chapter 38.

Breivik, H., Högström, H., Niemi, G., et al. (1995). Safe and effective post-operative pain-relief: Introduction and continuous quality-improvement of comprehensive post-operative pain management programmes. *Bailliere's Clinical Anaesthesiology, 9*, 423–460.

Breivik, H., Curatolo, M., Niemi, G., Haugtom, H., & Kvarstein, G. (2007). How to implement an acute pain service: An update. In H. Breivik & M. Shipley (Eds.), *Pain best practice and research compendium* (pp. 255–270). London: Elsevier.

Breivik, H., Bang, U., Jalonen, J., Vigfússon, G., Alahuhta, S., & Lagerkranser, M. (2010). Nordic guidelines for neuraxial blocks in disturbed haemostasis from the Scandinavian society of anaesthesiology and intensive care medicine. *Acta Anaesthesiologica Scandinavica, 54*, 16–41.

Breivik, H., & Stubhaug, A. (2008). Management of acute postoperative pain: Still a long way to go! *Pain, 137*, 233–234.

Brodner, G., Mertes, N., Buerkle H., Marcus, M. A., & Van Aken, H. (2000). Acute pain management: Analysis, implications and consequences after prospective experience with 6349 surgical patients. *European Journal of Anaesthesiology, 17*, 566–575.

Counsell, D., Macintyre, P. E., & Breivik, H. (2008). Organization and role of acute pain services. In H. Breivik, W. Campbell, & M. K. Nicholas (Eds.), *Clinical pain management. Practice and procedures* (pp. 579–603). London: Hodder Arnold.

Fagenholz, P. J., Bowler, G. M. R., Carnochan, F.M., & Walker, W. S. (2012). Case report: Systemic local anaesthetic toxicity from continuous thoracic paravertebral block. *British Journal of Anaesthesia, 109*, 260–262.

Macintyre, P. E., & Scott, D. A. (2009). Acute pain management and acute pain services. In M. J. Cousins, D. B. Carr, T. T. Horlocker, & P. O. Bridenbaugh (Eds.), *Cousins' and Bridenbaugh's neural blockade*

in clinical anaesthesia and pain medicine (4th ed.,
pp. 1036–1062). Philadelphia: Wolters Kluwer/
Lippincott Williams & Wilkins.

Norum, H., & Breivik, H. A. (2010). Systematic review of
comparative studies indicates that paravertebral block
is neither superior nor safer than epidural analgesia for
pain after thoracotomy. Scandinavian Jounal of Pain,
1, 12–23.

Powel, A. E., Davis, H. T. O., Bannister, J., & Macrae,
W. A. (2008). Acute pain services and organizational
change. In H. Breivik, W. Campbell, & M. K. Nicholas
(Eds.), Clinical pain management. Practice and
procedures (pp. 604–618). London: Hodder Arnold.

Rawal, N. (1999). 10 Years of acute pain services –
achievements and challenges. Regional Anesthesia
and Pain Medicine, 24, 68–73.

Rawal, N., & Berggren, L. (1994). Organization of acute
pain service: A Low-cost model. Pain, 57, 117–123.

Ready, L. B., Oden, R., Chadwick, H. S., et al. (1988).
Development of an anaesthesiology-based postoperative pain management service. Anesthesiology, 68,
100–106.

Stenseth, R., Sellevold, O., & Breivik, H. (1985). Epidural
morphine for postoperative pain: Experience with
1085 patients. Acta Anaesthesiologica Scandinavica,
29, 146–156.

Postoperative Pain, Acute Presentation of Complex Regional Pain Syndrome

Suellen M. Walker
University of Sydney Pain Management and
Research Centre, Royal North Shore Hospital,
NSW, Australia

Synonyms

CRPS type I: Algodystrophy; Posttraumatic
pain syndrome; Reflex sympathetic dystrophy;
Shoulder-hand syndrome; Sudeck's atrophy
CRPS type II: Causalgia

Definition

▶ Complex regional pain syndrome (CRPS) type
I has the following features (Stanton-Hicks
et al. 1995):
1. CRPS is usually preceded by a noxious event.
   This is not an essential diagnostic feature, as
   a precipitating injury cannot be identified in
   approximately 10 % of cases (Cepeda
   et al. 2002).
2. Spontaneous pain or allodynia/hyperalgesia
   occurs in a regional distribution, not limited
   to a single nerve territory, and the degree
   of pain is disproportionate in severity to the
   inciting event.
3. There is, or has been, evidence of edema,
   changes in skin blood flow, or abnormality of
   sudomotor activity in the region of the pain.
4. Other reasons for the symptoms and differential diagnoses have been excluded.

The criteria for CRPS type II are the same as
type I, but the syndrome develops in association
with a definable major nerve lesion.

Characteristics

CRPS is a variable constellation of symptoms
and signs, which may include pain and
sensory changes, autonomic dysfunction, trophic changes, motor impairment, and psychological changes. CRPS most commonly follows
injuries involving the limbs but may also
develop in association with visceral diseases
and central nervous system lesions. Recurrent
episodes of CRPS occur in 1030 % of patients,
and migratory symptoms involving multiple
extremities have also been reported (Walker
and Cousins 1997).

Pain

Pain is an essential criterion for CRPS and is
disproportionate in duration, severity, and distribution to the expected clinical course of the
inciting event. Pain is often described as deep
and diffuse in distribution and tearing, burning,
or stinging in character and may be exacerbated
by movement or position (e.g., when the limb is
dependent). Sensory symptoms are common,
particularly hyperalgesia and allodynia to light
touch, thermal stimulation, deep pressure, or
joint movement (Birklein et al. 2000).

Autonomic Dysfunction

Abnormalities of skin blood flow lead to changes
in skin temperature and color, ranging from warm
and red to cold, pale, and cyanotic. Edema and

increased or decreased sweating may occur, with side-to-side asymmetry between affected and non-affected limbs.

Trophic Changes

Trophic changes such as muscle wasting, thin shiny skin, coarse hair, and thickened nails are not invariably present. Joint stiffness may occur due to functional motor disturbances and/or trophic changes in joints and tendons. Periarticular osteoporosis occurs in a small percentage of cases.

Motor Impairment

Disturbances of motor function including weakness, tremor, decreased range of movement, and dystonia are common (Birklein et al. 2000) but were not included in the original criteria for CRPS. Voluntary guarding may be seen as patients attempt to protect their hyperalgesic limb, with range of movement limited by pain.

Psychological Issues

Psychological factors play a role in all forms of pain and dysfunction, as they influence patients' beliefs, expectations, incentive for recovery, coping mechanisms, and stress tolerance. Behavioral responses are particularly important in patients with CRPS, as immobilization and overprotection of the affected limb may produce or exacerbate demineralization, vasomotor changes, edema, and trophic changes. There is no evidence from current series that a particular psychological profile predisposes to CRPS or is specific to the diagnosis of CRPS, when compared to other patients with chronic pain. Therefore, most authors conclude that behavioral and emotional changes are a result of chronic pain rather than a cause of CRPS (Cepeda et al. 2002).

Diagnostic Criteria

Revision of the taxonomy and introduction of the term CRPS by an IASP Taskforce aimed to establish a non-mechanistic description with uniform terminology and diagnostic criteria (Stanton-Hicks et al. 1995). The diagnosis is based on clinical symptoms and signs. However, it has

been suggested that the original criteria have high sensitivity but low specificity, resulting in overdiagnosis of CRPS (Bruehl et al. 1999; Galer et al. 1998). Improved differentiation between neuropathic pain and CRPS has been achieved by inclusion of motor neglect signs (Galer et al. 1998) or by an algorithm that requires at least two sign categories and four symptom categories (Bruehl et al. 1999). These expanded criteria have not been extensively evaluated or validated (Cepeda et al. 2002). The lack of uniform diagnostic criteria in clinical studies limits our ability to assess the prevalence of the condition or the efficacy of different treatments. Unfortunately, even since the introduction of the IASP criteria, clinical studies still lack homogeneity, with diagnostic features being vaguely or inaccurately applied (Reinders et al. 2002).

Acute Presentation: Stages Versus Severity

Three stages of "RSD" were traditionally described: an early phase with acute pain and prominent sympathetic changes, a dystrophic stage with ongoing pain and motor/trophic changes, and finally, a late atrophic stage. Prospective studies following fracture or hand injury show gradual resolution of symptoms in the majority of patients, with relatively few experiencing increasingly problematic stages (Field and Atkins 1997; Bruehl et al. 2002). Progressive stages of symptoms and signs could not be demonstrated by a retrospective patient questionnaire (Galer et al. 2000) or by physicians' evaluation (Bruehl et al. 2002). Therefore, it has been suggested that different subtypes of severity, rather than stages, of CRPS exist: (1) a relatively limited syndrome with vasomotor signs predominating, (2) a relatively limited syndrome with neuropathic pain/sensory abnormalities predominating, and (3) a florid CRPS syndrome similar to "classic RSD" descriptions (Bruehl et al. 2002). However, all patients in this series were retrospectively assessed and had symptoms for an average of 24 months. Different subtypes of acute presentation have not been prospectively confirmed.

Pain (100 % of patients) and edema (78–88 %) are prominent features in the acute presentation of CRPS (Schurmann et al. 1999; Birklein et al. 2000). Other features include tremor (45 %), changes in skin color (71 %), and increased sweating (29 %) (Schurmann et al. 1999). In prospective studies, objective differences in skin temperature occur in 28–44 % of patients, and an increase rather than a decrease in skin temperature is more likely early in the disease (Birklein et al. 1998, Birklein et al. 2000, Schurmann et al. 1999). The reverse has been reported in a retrospective survey, with a greater proportion of patients reporting cold skin early in the disease (Galer et al. 2000), but may be influenced by the patients' memory of symptoms months to years later.

Differentiating Acute CRPS from Trauma

Identification of CRPS in patients with acute injury can be difficult, as many of the clinical diagnostic features of CRPS occur with trauma and inflammation. In prospective studies following Colles' fracture, symptoms and signs attributable to CRPS (pain, swelling, stiffness, and color change) can be identified in 24 % of patients, with a further 28 % having borderline symptoms (Field and Atkins 1997). Although differences may be apparent between groups of patients, clear differentiating features may be difficult to distinguish in individual patients (Schurmann et al. 1999). As a result, it may be several weeks before symptoms can no longer be explained by the severity of the initial injury, leading to some inherent delay in diagnosis. A high index of suspicion is required to identify CRPS if symptoms are not typical or severe. Comparison of patients with early CRPS and those with uncomplicated trauma or surgery of the upper limb (Birklein et al. 2001; Schurmann et al. 1999) found that pain, hyperalgesia, and edema were present in both groups. Increases in skin temperature, confirmed by ▶ thermography, were also present in approximately one-third of patients following either trauma or CRPS. Although different mechanisms for the increase in blood flow are likely (release of vasodilatory mediators from peripheral nerve endings and inflammatory cells following trauma, and inhibition of sympathetically

mediated vasoconstriction in CRPS) (Birklein et al. 2001), the presence or absence of skin temperature changes does not aid differentiation between the two groups. However, the presence of motor signs and increased sweating was specific to patients with CRPS. Tests of sympathetic reactivity showing impairment of sympathetic vasoconstrictor reflexes and/or hyperhidrosis (Birklein et al. 2001; Schurmann et al. 1999) have also been suggested as aids in the early differentiation of uncomplicated trauma and CRPS.

Investigations and Sympathetic Function Tests

The diagnostic criteria for CRPS are based on clinical symptoms and signs, and there is no "gold standard" diagnostic test. Radiological investigations, ▶ quantitative sensory testing, tests of ▶ sudomotor function, and measurement of skin temperature and regional blood flow have all been utilized to support, but do not confirm, the diagnosis of CRPS (Cepeda et al. 2002; Walker and Cousins 1997). Physician diagnosis of edema, skin temperature changes, and restricted movement is in strong agreement with objective measures of hand volume, infrared readings of skin temperature, and measurement of joint range of motion with a goniometer (Oerlemans et al. 1999). Therefore, the main role of objective tests may be in research settings or to more precisely evaluate the severity of signs and response to treatment. Patients experiencing pain relief following a sympathetic block are said to have a component of ▶ sympathetically maintained pain (SMP). However, the proportion of SMP can vary with time and between individuals, and there is no data about the specificity or sensitivity of ▶ sympathetic blockade as a diagnostic tool for CRPS (Cepeda et al. 2002).

Treatment

Early recognition and treatment are commonly believed to give the best chance of a successful outcome in patients with CRPS. However, identification of specific CRPS symptoms may be difficult early in the course of injury, and a delay of several months before diagnosis is still common. Prospective studies report some natural resolution of CRPS symptoms in the 3–12 months following Colles' fracture (Field and Atkins 1997), but the ability of treatment to hasten resolution has not been evaluated in prospective controlled trials.

Despite a diverse range of treatment options, there is limited scientific evidence to support many treatments currently used for CRPS (Cepeda et al. 2002; Kingery 1997). Despite introduction of the IASP criteria, clinical trials lack homogeneity in inclusion criteria and outcome evaluation. Relatively few blinded placebo-controlled trials have been conducted, and interpretation of the results is often hampered by small sample sizes, which increases the likelihood of a false-positive result (Cepeda et al. 2002).

Clinical consensus supports facilitating mobilization and activity with physical therapy to preserve and recover limb function and minimize secondary effects due to disuse, but no controlled trials have evaluated the impact of physical therapy on the natural history of CRPS (Cepeda et al. 2002). Controlled trials support the use of bisphosphonates, oral corticosteroids, and epidural clonidine, but the small sample size makes it difficult to evaluate precisely the extent of the treatment effect. Antidepressant and anticonvulsant drugs have been shown to be effective in patients with neuropathic pain, but trials have not specifically evaluated effects in CRPS patients. Beneficial effects of gabapentin are currently reported only in uncontrolled case series. Several studies have failed to identify a beneficial effect following intravenous regional block with guanethidine (Cepeda et al. 2002; Kingery 1997).

A systematic review of local anesthetic sympathetic blockade in CRPS has recently been conducted (Cepeda et al. 2002). Only three randomized controlled trials enrolling more than ten patients were identified, and the majority of reports were case series of poor quality. Of the 1,144 patients included in 29 studies, 29 % achieved a full response (decrease of pain intensity to below 25 % of preblock value or response described as "great," "dramatic," "complete," "absolute," or "excellent"), 41 % obtained a partial response, and in 32 % there was either no or only a minimal response. The authors noted that heterogeneity in the studies limited the interpretation of the results but concluded "soft evidence supports the therapeutic role of sympathetic blockade" (Cepeda et al. 2002). Further well-designed controlled trials are required.

References

Birklein, F., Kunzel, W., & Sieweke, N. (2001). Despite clinical similarities there are significant differences between acute limb trauma and complex regional pain syndrome I (CRPS I). *Pain, 93*, 165–171.

Birklein, F., Riedl, B., Claus, D., et al. (1998). Pattern of autonomic dysfunction in time course of complex regional pain syndrome. *Clinical Autonomic Research, 8*, 79–85.

Birklein, F., Riedl, B., Sieweke, N., et al. (2000). Neurological findings in complex regional pain syndromes – analysis of 145 cases. *Acta Neurologica Scandinavica, 101*, 262–269.

Bruehl, S., Harden, R. N., Galer, B. S., et al. (1999). External validation of IASP diagnostic criteria for complex regional pain syndrome and proposed research diagnostic criteria. International association for the study of pain. *Pain, 81*, 147–154.

Bruehl, S., Harden, R. N., Galer, B. S., et al. (2002). Complex regional pain syndrome: Are there distinct subtypes and sequential stages of the syndrome? *Pain, 95*, 119–124.

Cepeda, M. S., Lau, J., & Carr, D. B. (2002). Defining the therapeutic role of local anesthetic sympathetic blockade in complex regional pain syndrome: A narrative and systematic review. *The Clinical Journal of Pain, 18*, 216–233.

Field, J., & Atkins, R. M. (1997). Algodystrophy is an early complication of colles' fracture. What are the implications? *Journal Hand Surgical (Br), 22*, 178–182.

Galer, B. S., Bruehl, S., & Harden, R. N. (1998). IASP diagnostic criteria for complex regional pain syndrome: A preliminary empirical validation study. International association for the study of pain. *Clinical Journal Pain, 14*, 48–54.

Galer, B. S., Henderson, J., Perander, J., et al. (2000). Course of symptoms and quality-of-life measurement in complex regional pain syndrome: A pilot survey. *Journal of Pain and Symptom Management, 20*, 286–292.

Kingery, W. S. (1997). A critical review of controlled clinical trials for peripheral neuropathic pain and complex regional pain syndromes. *Pain, 73*, 123–139.

Oerlemans, H. M., Oostendorp, R. A., de Boo, T., et al. (1999). Signs and symptoms in complex regional pain syndrome type I/reflex sympathetic dystrophy:

Judgment of the physician versus objective measurement. *The Clinical Journal of Pain, 15*, 224–232.

Reinders, M. F., Geertzen, J. H., & Dijkstra, P. U. (2002). Complex regional pain syndrome type I: Use of the international association for the study of pain diagnostic criteria defined in 1994. *The Clinical Journal of Pain, 18*, 207–215.

Schurmann, M., Gradl, G., Andress, H. J., et al. (1999). Assessment of peripheral sympathetic nervous function for diagnosing early post-traumatic complex regional pain syndrome type I. *Pain, 80*, 149–159.

Stanton-Hicks, M., Janig, W., Hassenbusch, S., et al. (1995). Reflex sympathetic dystrophy: Changing concepts and taxonomy. *Pain, 63*, 127–133.

Walker, S. M., & Cousins, M. J. (1997). Complex regional pain syndromes: Including "reflex sympathetic dystrophy" and "causalgia". *Anaesthesia and Intensive Care, 25*, 113–125.

Postoperative Pain, Adverse Events (Associated with Acute Pain Management)

Stephan A. Schug
School of Medicine and Pharmacology,
Royal Perth Hospital, Faculty of Medicine,
Dentistry & Health Sciences, The University of
Western Australia, UWA Anaesthesiology,
Perth, Australia

Synonyms

Complications; Medical misadventures; Medical mishaps; Side effects

Definition

A medical adverse event is defined by the Food and Drug Administration of the USA (FDA) as any undesirable experience by a patient associated with a medical intervention. Such an adverse event is defined as a serious adverse event (SAE) when the patient outcome is either death or disability, the event is life threatening or potentially life threatening, or its sequelae require hospitalization or prolongation of hospital stay or lead to disability and congenital abnormality or require intervention to prevent permanent impairment or damage.

Characteristics

In the early days of Acute Pain Services (APS) (Ready et al. 1988), there was underlying concern among those involved in the care of the perioperative patient, that the benefits of such an approach must be balanced by additional risks associated with newer pain management techniques (Rowbotham 1992; Spittal 1992). Several large studies of morbidity and mortality associated with acute pain management techniques have been able to quantify this risk, demonstrating the excellent safety profiles of these techniques for use on the general surgical ward (e.g., Schug and Torrie 1993; Flisberg et al. 2003); these two large studies, examining a combined total of more than 5,500 surgical patients under the care of an APS, reported only one serious complication resulting in morbidity, and there was no mortality related to pain management techniques.

Respiratory Depression

Opioid administration is always linked to the risk of respiratory depression, but this risk is often overestimated.

Continuous morphine infusion has the highest rate of respiratory depression of around 1.65 % (Schug and Torrie 1993). This technique lacks the inherent safety of patient feedback provided by PCA and is thus only recommended for use in specific circumstances such as palliative care and patients requiring intravenous opiates who are unable to use any type of PCA device. There are new innovative PCA devices on the market, which do not require manual dexterity to operate, e.g., breath- or voice-activated devices, and this should widen the spectrum of patients able to use an analgesic technique with a "feedback" component.

PCA techniques without background infusions carry an extremely small risk of respiratory depression of between 0.1 % and 0.8 %

(Macintyre 2001). PCA devices used with an added background infusion have a respiratory depression rate almost as high as that of continuous opioid infusions (1.1–3.9 %) and are only recommended in specific circumstances such as patients with preoperative morphine requirements (Hansen et al. 1991). In a large audit of PCA use, of the 6 % of patients who experienced hypoxemia and the 2 % who experienced respiratory depression, virtually all had one of three risk factors: bolus dose greater than 1 mg morphine, age greater than 65 years, and intra-abdominal surgery (Sidebotham et al. 1997). A further risk factor is activation of the PCA device other than by the patient (parents, other visitors); the device should be clearly labelled for patient use only (Macintyre 2001).

Epidural opioids also carry a risk of respiratory depression due to systemic or CSF absorption of the agent. The incidence of respiratory depression has been reported in numerous studies involving data collection from the APS and lies between 0.24 % and 1.6 % (Wheatley et al. 2001). It is a dose-dependent effect and less common with more lipophilic opioids such as fentanyl. Respiratory depression is often associated with sedation.

Nausea and Vomiting

The incidence of nausea among patients with PCA techniques is highest on day 1 at 28 % and declines over the following 2 days to 14.3 % and then 4.7 % (Sidebotham et al. 1997). A large audit of 6,000 patients receiving PCA analgesia revealed that only five patients discontinued their PCA device because of nausea (Sidebotham et al. 1997). Nausea should be screened for among these patients and treated aggressively, as nausea and vomiting are one of the adverse effects most disliked by patients.

Nausea and vomiting in the patient receiving an epidural infusion may be due to drug effects, e.g., opioid-local anesthetic mixtures, or to hypotension secondary to sympatholysis. The overall frequency is similar to that associated with PCA techniques.

Sedation

This occurs in up to 30 % of patients on PCA infusions. Many patients dislike this adverse effect and underuse their PCA for this reason, choosing to remain alert in preference to becoming completely pain-free, but sedated (Sidebotham et al. 1997).

Addiction and Inappropriate Opioid Self-Administration

There is no evidence that this is a problem in any of the large surveys performed. Previous and current drug users have been safely managed using both regional and intravenous pain management techniques. It is important to remember that background opiate requirements will need to be continued over the perioperative period to avoid acute withdrawal states (Chapman and Hill 1989).

Drug and Programming Errors

These are a rare cause of adverse events when using pain management techniques. Premixed bags of epidural solutions and PCA syringes reduce the risk of drug error. Epidural hubs should be incompatible with ordinary syringes and giving sets. The programmed device should be routinely double-checked before being connected to a patient.

Neurological Injury After Regional Analgesia

The insertion of catheters near to neural structures carries a potential risk of neurological trauma. Several large published series of patients receiving epidural catheters have revealed a reassuring safety record. The rarity of the event makes it difficult to quantify it exactly, but the best available evidence points to an incidence of serious neurological sequelae between of 0.006 % and 0.005 %

(Kane 1981; Aromaa et al. 1997). The onset of neurological injury related to the use of epidural analgesia or anesthesia is not always a causal relationship; there are multiple reports in the literature of neurological disorder being wrongly blamed on the use of epidural techniques (Wheatley et al. 2001; Lovstad et al. 1999). Regional nerve catheter techniques also have an excellent safety record in this respect (Schug and Torrie 1993).

Local Anesthetic Toxicity During Regional Analgesia

The risk of local anesthetic toxicity from regional or peripheral nerve blocks is related to inadvertent misplacement of the catheter intravenously or subsequent migration of the catheter into a vascular space. Other potential causes are drug or programming errors, as discussed above, and the use of bolus doses of local anesthetic resulting in higher peak plasma levels than similar dosages given by infusion. The incidence of toxicity from racemic bupivacaine, notably convulsions, as a result of high peak plasma levels is 0.01–0.12 % (Brown et al. 1995). Overall, the newer enantiomer specific local anesthetics ropivacaine and levobupivacaine exhibit less cardiotoxicity than racemic bupivacaine and are preferable from a safety point of view (Stewart and Kellett 2003).

Hypotension Associated with Epidural Analgesia

Orthostatic hypotension occurs in about 0.7–3 % of patients receiving epidural analgesia (Wheatley et al. 2001). It is important to exclude intrathecal catheter migration as a cause of this by aspirating before administering a bolus dose of local anesthetic (Flisberg et al. 2003). The more stable block levels achieved by using a continuous infusion technique are probably advantageous in this regard.

Inadvertent Dural Puncture During Epidural Catheter Insertion

Dural puncture occurs in 0.32–1.23 % of epidural insertions and can result in a disabling post-dural-puncture headache (PDPH) (Wheatley et al. 2001).

Epidural Space Infection After Epidural Analgesia

Short-term catheter insertion is rarely associated with infection (Schug and Torrie 1993). A meta-analysis of 915 case reports of epidural abscesses identified that the majority was spontaneous, and only 5.5 % were related to epidural anesthesia and analgesia (Reihsaus et al. 2000). The overall incidence is estimated at somewhere between 1 in 2,000 and 1 in 7,000 (Wheatley et al. 2001). Associations include longer period of catheterization, immunocompromised patients, and perioperative anticoagulant therapy. Full aseptic precautions, including the use of gown and face mask, are strongly recommended during insertion.

Epidural Hematoma After Epidural Analgesia

This is another very rare complication defying accurate assessment of risk; nevertheless, the consequences can be catastrophic as a spinal hematoma not detected and treated early results in irreversible paraplegia. The best estimate of neurological dysfunction after hematoma formation associated with epidural anesthesia is 1 in 150,000 cases (Flisberg et al. 2003). The risk is clearly related to coagulopathies and, more relevant, the use of anticoagulation.

Inadequate Pain Management

Inadequate treatment effects are an adverse effect in itself. Its occurrence necessitating dose adjustments has been shown to be more common in patients receiving epidural analgesia than PCA techniques (Flisberg et al. 2003).

References

Aromaa, U., Lahdensuu, M., & Cozanitis, D. A. (1997). Severe complications associated with epidural and spinal anaesthesias in Finland 1987–1993. A study based on patient insurance claims. *Acta Anaesthesiologica Scandinavica, 41*, 445–452.

Brown, D. L., Ransom, D. M., Hall, J. A., et al. (1995). Regional anesthesia and local anesthetic-induced systemic toxicity: Seizure frequency and accompanying cardiovascular changes. *Anesthesia and Analgesia, 81*, 321–328.

Chapman, C. R., & Hill, H. F. (1989). Prolonged morphine self-administration and addiction liability. Evaluation of two theories in a bone marrow transplant unit. *Cancer, 63*, 1636–1644.

Flisberg, P., Rudin, A., Linner, R., et al. (2003). Pain relief and safety after major surgery. A prospective study of epidural and intravenous analgesia in 2696 patients. *Acta Anaesthesiologica Scandinavica, 47*, 457–465.

Hansen, L. A., Noyes, M. A., & Lehman, M. E. (1991). Evaluation of patient-controlled analgesia (PCA) versus PCA plus continuous infusion in postoperative cancer patients. *Journal of Pain and Symptom Management, 6*, 4–14.

Kane, R. E. (1981). Neurologic deficits following epidural or spinal anesthesia. *Anesthesia and Analgesia, 60*, 150–161.

Lovstad, R. Z., Steen, P. A., & Forsman, M. (1999). Paraplegia after thoracotomy – not caused by the epidural catheter. *Acta Anaesthesiologica Scandinavica, 43*, 230–232.

Macintyre, P. E. (2001). Safety and efficacy of patient-controlled analgesia. *British Journal of Anaesthesia, 87*, 36–46.

Ready, L. B., Oden, R., Chadwick, H. S., et al. (1988). Development of an anesthesiology-based postoperative pain management service. *Anesthesiology, 68*, 100–106.

Reihsaus, E., Waldbaur, H., & Seeling, W. (2000). Spinal epidural abscess: A meta-analysis of 915 patients. *Neurosurgical Review, 23*, 175–204.

Rowbotham, D. J. (1992). The development and safe use of patient-controlled analgesia. *British Journal Anaesthesia, 68*, 331–332.

Schug, S. A., & Torrie, J. J. (1993). Safety assessment of postoperative pain management by an acute pain service. *Pain, 55*, 387–391.

Sidebotham, D., Dijkhuizen, M. R., & Schug, S. A. (1997). The safety and utilization of patient-controlled analgesia. *Journal of Pain and Symptom Management, 14*, 202–209.

Spittal, M. J. (1992). Epidural morphine in a small community hospital: A false sense of security. *Anesthesia and Analgesia, 74*, 468–470.

Stewart, J., & Kellett, N. (2003). The central nervous system and cardiovascular effects of levobupivacaine and ropivacaine in healthy volunteers. *Anesthesia and Analgesia, 97*, 412–416.

Wheatley, R. G., Schug, S. A., & Watson, D. (2001). Safety and efficacy of postoperative epidural analgesia. *British Journal of Anaesthesia, 87*, 47–61.

Postoperative Pain, Anticonvulsant Medications

Ross MacPherson
Department of Anesthesia and Pain Management, Royal North Shore Hospital University of Sydney, Sydney, NSW, Australia

Synonyms

Anticonvulsant medication; Antiepileptic agents

Characteristics

A number of different anticonvulsant medications have found a place in the management of neuropathic pain states. Despite being a somewhat heterogeneous group, their targets of action often share similarities with those involved in pain transmission.

Anticonvulsants generally act at three main sites – the sodium or calcium channels or the GABA (A) receptor. Other agents, such as vigabatrin promote GABAminergic transmission by attenuating the rate of GABA breakdown by transaminases.

At the sodium channel, they attenuate neuronal conduction by maintaining the sodium channel in the inactivated state. This is similar to the mode of action of local anesthetic agents.

Another important target is the GABA receptor. As well as being an important site to suppress epileptic activity in the brain, it has been implicated in modulating the pain pathway at a number of sites (Wilcox 1999). Presynaptic GABA (B) receptors, through coupling to Ca channels have an important inhibitory role, while postsynaptically, GABA (A) receptors that act as chloride-sensitive ligand-gated ion channels can hyperpolarize dorsal horn neurones (Tremone-Lucas et al. 2000).

Although it has been known for some time that anticonvulsants are useful in the management of neuropathic pain, their role was initially limited to some degree by their side effect profile (Gourlay 1999). For example, McQuay et al. (1995) found that while agents such as carbamazepine and phenytoin were effective in the management of a range of pain states, the NNT (number needed to treat) to achieve pain relief and the appearance of minor adverse effects were very similar. In the case of diabetic neuropathy, for example, NNTs were 2.5 and 3.1, trigeminal neuralgia 2.6 and 3.4, and migraine prophylaxis 1.6 and 2.4, respectively. The implication from these data was that effective doses for pain management often brought significant CNS side effects, which many patients found particularly unpleasant and, hence, has limited the usefulness of these agents.

In examining the use of anticonvulsant agents in pain management, it is perhaps easiest to look at the drugs from a temporal viewpoint. Carbamazepine and phenytoin are older agents that have been available for over 20 years, while newer antiepileptic agents that have been successfully used in pain management include lamotrigine and gabapentin. There is also a group of very new antiepileptic agents such as topiramate, oxcarbazepine, and levetriacetam that have yet to be fully evaluated with regard to their application in the management of neuropathic pain states (Perucca 2001; Smith 2001).

Carbamazepine was one of the earliest drugs used for neuropathic pain and has been widely studied, although many of these studies suffer from either small numbers or lack of appropriate controls (Rowbotham 2002). Nevertheless, it has been shown to be effective, especially in trigeminal neuralgia and diabetic neuropathy (Campbell et al. 1966; Rull et al. 1969).

As mentioned above, associated CNS effects are bothersome, and include sedation, visual disturbances, and ataxia. In addition to these effects, carbamazepine is well known to cause a range of other serious, although rare adverse effects such as leukopenia, aplastic anemia, and hepatic failure. Furthermore, the enzyme inducing effect of carbamazepine is considerable, resulting in a long list of possible drug interactions. The evidence for the effectiveness of phenytoin is weak, and probably has little place in neuropathic pain management.

However, newer agents such as lamotrigine, gabapentin, and pregabalin now provide at least equal efficacy when compared with earlier agents, and with a superior side effect profile.

One recent advance has been the use of anticonvulsant medications not simply for the management of neuropathic pain, but as adjuncts in nociceptive pain control, especially in joint replacement protocols (Ménigaux et al. 2005). In pain management, the anticonvulsants cannot be considered a homogeneous group. They have different mechanisms of action and demonstrate varying efficacies in the management of different neuropathic pain states. More detailed information will be found under the entries for individual agents.

References

Campbell, F. G., Graham, J. G., & Zilkha, K. L. (1966). Clinical trials of carbamazepine in trigeminal neuralgia. *Journal of Neurology, Neurosurgery, and Psychiatry, 29*, 265–267.

Gourlay, G. K. (1999). Clinical pharmacology of the treatment of chronic non-cancer pain. In M. Devor (Ed.), *Pain – an updated review. Refresher course syllabus 9th world congress on pain* (pp. 433–442). Vienna.

McQuay, H. J., Carroll, D., Jadad, A. R., et al. (1995). Anti-convulsant drugs for the management of pain: A systematic review. *BMJ, 311*, 1047–1052.

Ménigaux, C., Adam, F., Guignard, B., Sessler, D. I., & Chauvin, M. (2005). Preoperative gabapentin decreases anxiety and improves early functional recovery from knee surgery. *Anesthesia and Analgesia, 100*(5), 1394–1399.

Perucca, E. (2001). Clinical pharmacology and therapeutic use of new antiepileptic drugs. *Fundamentals and Clinical Pharmacology, 15*, 405–417.

Rowbotham, M. C. (2002). Neuropathic pain: From basic science to evidence based treatment. In J. Dostrovsky (Ed.), *Pain – an updated review. Refresher course syllabus 10th world congress on pain* (pp. 165–176). San Diego.

Rull, J. A., Quibrera, R., et al. (1969). Symptomatic treatment of peripheral diabetic neuropathy with carbamazepine: A double-blind crossover trial. *Diabetologia, 5*, 215–218.

Smith, P. E. (2001). Clinical recommendations for oxcarbazepine. *Seizure, 91*, 87–91.

Tremone-Lucas, I. W., Megeff, C., & Backonja, M. M. (2000). Anti-convulsants for neuropathic pain syndromes: Mechanisms of action and place in therapy. *Drugs, 60*(5), 1029–1052.

Wilcox, G. L. (1999). Pharmacology of pain and analgesia. In M. Devor (Ed.), *Pain – an updated review. Refresher course syllabus 9th world congress on pain* (pp. 573–591). Vienna.

Postoperative Pain, Antidepressants

Ross MacPherson
Department of Anesthesia and Pain Management, Royal North Shore Hospital University of Sydney, Sydney, NSW, Australia

Synonyms

Antidepressant drugs

Characteristics

It was recognized that antidepressant drugs, specifically the tricyclic antidepressants, also had analgesic effects, more than 20 years ago (Watson et al. 1982) which were quite distinct from their mood-altering properties. For example, when it came to the use of antidepressants in pain management, the dose needed was much smaller than that needed to treat depression, and the time to onset of therapeutic effect was faster (Sindrup et al. 1992).

While there have been a number of trials examining the effectiveness of TCAs in chronic pain, there is almost no data available on their use in acute neuropathic pain states, despite this being an major problem in pain management.

The principal mode of action of the tricyclics, at least in the treatment of depression, is inhibition of reuptake of either noradrenaline or serotonin (or both) at the neuronal synapse. This is also clearly important in the drugs analgesic effect, as such an action will facilitate descending inhibitory pathways terminating at the spinal cord. Specifically, their action is probably via the α2-receptor site (Gray et al. 1999). However, as Rowbotham points out in his excellent review (Rowbotham 2002), this may be far from the whole story. Tricyclic antidepressants have also been shown to have significant activity at the sodium channel (Gerner et al. 2001), a property that they have in common with the antiepileptic agents (q. v.), and they may also have antagonistic effects at the NMDA receptor. Some workers have proposed that the antidepressants may have activity at the opioid receptor (Pick et al. 1992), although the importance of this mechanism is unclear.

Much has been written of the role of antidepressants in neuropathic pain, and these agents certainly have their supporters, but it is perhaps true to say that their role in neuropathic pain management has been slightly eclipsed by the antiepileptic agents.

Rowbotham questions the logic behind this, as he points out that the majority of clinical trials carried out have found them to be efficacious. The limiting factor, of course, has been the development of annoying side effects, which can range from just being a simple nuisance, such as dry mouth to hypotension – a response that can be associated with significant morbidity. Many of these adverse effects are as a result of the drugs' anticholinergic properties, a fact that must be borne in mind when prescribing the drug to patients with prostatism, glaucoma, or cardiac arrhythmias. When initiating treatment with TCAs, especially in the elderly, the lowest practicable dose would be employed. There has been a suggestion that nortryptyline has a more favorable side effect profile than amitryptiline and should be used in preference. It has also been shown to be useful when used in combination with gabapentin (Gilron et al. 2009).

Where the agents involved have significant sedative properties, it is usually prescribed as a single nighttime dose to take advantage of this effect. In the small doses used in pain management (e.g., amitriptyline 25–75 mg), the effect is usually transient.

However, not all populations have been equally responsive, as some studies involving TCAs in patients with either cancer pain (Anonymous 2002; Mercadante et al. 2002), spinal cord injury–induced pain (Cardenas et al. 2002), or fibromyalgia (Heymann et al. 2001) have not produced such positive results.

Furthermore, some studies have demonstrated that other medication, such as antiepileptics, can often be effective in cases where TCAs have failed (Anonymous 2002; Dallocchio et al. 2000). Occasional trials have examined the effectiveness of combinations of the two agents, i.e., TCAs and antiepileptic agents (Heugham and Sawynok 2002).

In contradistinction to the antiepileptic agents, the newer antidepressants, such as the selective serotonin reuptake inhibitors, have not provided any great advance. In fact, most studies have shown agents such as fluoxetine, paroxetine, and sertraline have generally been found to be either little better than placebo, or inferior to amitriptyline (Saarto and Wiffen 2007).

If it is decided to use TCAs for the management of neuropathic pain states, Rowbotham suggests a number of guidelines. Firstly, he suggests that if a TCA fails to produce pain relief having given an effective dose, other agents are unlikely to succeed; however, if a TCA proves effective, but side effects are a barrier, another antidepressant should be tried. Despite the fact that evidence for the effectiveness of the newer non-tricyclics is less than encouraging (Watson et al. 1982; Max et al. 1992; Gourlay 1999), he also suggests commencing treatment with the newer antidepressant agents, even though solid evidence for this is lacking.

McQuay et al. (1996) has examined the evidence for the use of antidepressant medications using diabetic neuropathy as the study condition, the NNT for pain relief and minor side effects (for all antidepressants) were 3 and 2.8, respectively.

For patients with postherpetic neuralgia, the data were 2.3 and 6, and for atypical and central pain 2.8 and 1.7, respectively.

In summary, there is a strong body of evidence to suggest that in many cases, antidepressant medications are at least as useful as other groups in the treatment of neuropathic pain states. However, should such an approach fail, then other drug groups should certainly be used.

References

Anonymous. (2002). Gabapentin: New indication. In *Postherpetic neuralgia when Amitriptyline fails. Prescrire international* (Vol. 11, pp. 111–112).

Cardenas, D. D., Warms, C. A., Turner, J. A., et al. (2002). Efficacy of amitriptyline for relief of pain in spinal cord injury: Results of a randomized controlled trial. *Pain, 96*, 365–373.

Dallocchio, C., Buffa, C., Mazzarello, P., et al. (2000). Gabapentin vs. Amitriptyline in painful diabetic neuropathy: An open-label pilot study. *Journal of Pain and Symptom Management, 20*, 280–285.

Gerner, P., Mujtaba, M., Sinnott, C. J., et al. (2001). Amitriptyline versus bupivacaine in Rat sciatic nerve blockade. *Anesthesiology, 94*, 661–667.

Gilron, I., Bailey, J. M., Tu, D., Holden, R. R., Jackson, A. C., & Houlden, R. L. (2009). Nortriptyline and gabapentin, alone and in combination for neuropathic pain: A double-blind, randomised controlled crossover trial. *The Lancet, 374*, 1252–1261.

Gourlay, G. (1999). Clinical pharmacology of the treatment of chronic non-cancer pain. In M. Devor (Ed.), *Pain – an updated review. Refresher course syllabus 9th world congress on pain*, (pp. 433–442). Vienna.

Gray, A. M., Pache, D. M., & Sewell, R. D. (1999). Do alpha2-adrenoceptors play an integral role in the antinociceptive mechanism of action of antidepressant compounds? *European Journal of Pharmacology, 378*, 161–168.

Heugham, C. E., & Sawynok, J. (2002). The interaction between gabapentin and amitriptyline in the Rat formalin test after systemic administration. *Anesthesia and Analgesia, 94*, 975–980.

Heymann, R. E., Helfenstein, M., & Feldman, D. (2001). A double-blind, randomized, controlled study of amitriptyline. Nortriptyline and placebo in patients with fibromyalgia. An analysis of outcome measures. *Clinical and Experimental Rheumatology, 19*, 697–702.

Max, M. B., Lynch, S. A., Muir, J., et al. (1992). Effects of desipramine, amitriptyline and fluoxetine on pain in diabetic neuropathy. *The New England Journal of Medicine, 326*, 1250–1256.

McQuay, H. J., Tramer, M., Nye, B. A., et al. (1996). A systematic review of antidepressants in neuropathic pain. *Pain, 68*, 217–227.

Mercadante, S., Arcuri, E., Tirelli, W., et al. (2002). Amitriptyline in neuropathic cancer pain in patients on morphine therapy: A randomized placebo-controlled, double-blind crossover study. *Tumori, 88*, 239–242.

Pick, C. G., Paul, D., Eison, M. S., et al. (1992). Potentiation of opioid analgesia by the antidepressant nefazodone. *European Journal of Pharmacology, 211*, 375–381.

Rowbotham, M. C. (2002). Neuropathic pain: from basic science to evidence based treatment. In J. Dostrovsky (Ed.), *Pain – an updated review. Refresher course syllabus 10th world congress on pain* (pp. 165–176). San Diego.

Saarto, T., & Wiffen, P. J. (2007). Antidepressants for neuropathic pain. *Cochrane Database of Systematic Reviews 4*.

Sindrup, S. H., Brosen, K., & Gram, L. F. (1992). Antidepressants in pain treatment: Antidepressant or analgesic effect? *Clinical Neuropharmacology, 15*, 636A–637A.

Watson, C. P., Evans, R. J., Reed, K., et al. (1982). Amitriptyline versus placebo in post-herpetic neuralgia. *Neurology, 32*, 671–673.

Postoperative Pain, Appropriate Management

Paul M. Murphy
Department of Anaesthesia and Pain Management, Royal North Shore Hospital, St. Leonards, NSW, Australia

Synonyms

Management of postoperative pain

Definition

Even minor surgery is associated with tissue damage and the potential for significant pain in the postoperative period. Pain experienced in this setting is typically nociceptive in nature; however, 1–3 % of referrals to an acute postoperative pain service have been reported as being neuropathic in nature (see ▶ neuropathic pain) (Hayes et al. 2002).

Characteristics

Assessment

Pain is undoubtedly a personal and subjective experience. Individuals undergoing a similar operative procedure may experience vastly different levels of pain. This may be dependent upon a number of factors including previous experiences, cultural factors, coping strategies, age, and gender. Importantly, greater than expected pain reported in the postoperative setting must always be considered as a potential indicator of surgical complications such as infection, compartment syndrome, or hematoma formation. Similarly an alteration in the character of the pain may indicate an underlying complication (Breivik 1993). It is well established that a poor correlation exists between patient and staff estimations of pain intensity (Coulling 2005). A self-report pain-rating tool is therefore an essential component of adequate postoperative pain management:

1. Categorical rating scale: verbal descriptors are utilized to rate the patients' pain, e.g., mild, moderate, severe, and worst.
2. Verbal numeric rating scale: the patient rates their pain from "none" (0) to "worst" (10).
3. Visual analogue scale (VAS): an unnumbered 10 cm line ranges from "no pain" to "worst pain." Numbering is avoided in order to prevent cueing of the patients' response. The distance marked along the line from the "no pain" (0) point is the VAS score.

Good correlation exists between VNRS and VAS; however, once a scoring system is established for an individual patient, it should be maintained in order to monitor progress (Averbuch and Katzper 2004). In addition to monitoring the patient's pain score, it is also important to assess adequately and monitor any potential side effects including sedation, nausea and vomiting, respiratory depression, and ▶ pruritus. Special care must be taken to use appropriate assessment tools in groups with special needs or communication difficulties such as children, the cognitively impaired, and ethnic groups unable to speak the dominant language (Morrison and Siu 2000).

Pharmacological Management of Postoperative Pain

A number of pharmacological agents may be utilized either in isolation or in a multimodal manner to achieve good postoperative pain control:

1. Opioids. In the majority of cases, opioid analgesics form the cornerstone of effective pain control. In the initial postoperative period, potent μ-▶ receptor agonists such as meperidine, morphine, or fentanyl may be administered via the parenteral route (SC/IM/IV) (Jensen et al. 2004). ▶ Patient-controlled analgesia (PCA) systems are also frequently utilized. These systems allow the patient to self-administer a predetermined dose of analgesic agent (opioid). A "lockout" period (commonly 5–6 min) is utilized in order to reduce the potential for overdose. Certain ystems allow a continuous or intermittent background infusion in addition to the patient-controlled boluses. There is ongoing conflict as to whether PCA systems result in improved pain control and reduced opioid consumption relative to "by the clock" or "on demand" administration; however, it is widely accepted that PCA systems are associated with a high degree of patient satisfaction (Ballantyne et al. 1993). Opioids have been administered perioperatively via the ▶ intrathecal and ▶ epidural routes in obstetric and orthopedic settings, with good analgesic benefit demonstrated for 24 h postoperatively (Murphy et al. 2003).

2. Nonsteroidal anti-inflammatory drugs (NSAIDs) are useful in the management of mild to moderate pain, particularly if there is an inflammatory component. They may also be utilized as adjuncts to opioids where they display ▶ synergistic analgesic properties. These agents are available in oral, rectal, intramuscular (IM), and intravenous (IV) formulations (Miranda et al. 2004).

3. Paracetamol is an important agent in the management of postoperative pain. It may be used alone in the management of mild to moderate pain or as an adjunct to opioids in the management of severe pain (Remy et al. 2005). It may be administered via the oral, rectal, or more recently IV route as proparacetamol.

Neuroaxial Blockade

Intrathecal and epidural analgesic techniques have been used with great success in the acute postoperative period (Rodgers et al. 2000). Techniques such as this significantly reduce the patients' requirement for systemic opioid administration and thus the potential for opioid-associated side effects. These techniques should be practiced and managed by an anesthesiologist. The route of administration (intrathecal versus epidural), location (thoracic versus lumbar), and the agents utilized must be decided on an individual basis considering the procedure to be performed, the medical status of the patient, and the "risk-benefit" ratio associated with the procedure. ▶ Neuroaxial analgesic techniques have the potential for serious adverse complications, and therefore, it is important that a stringent management plan including close observation of the patient is maintained in the postoperative period. The main agents administered via this route are local anesthetics and opioids. These agents display synergistic properties when administered via the neuroaxial route. Recently interest has developed in other potentially useful agents that may be administered by these routes including ▶ clonidine and ▶ midazolam (Tucker et al. 2004).

Regional Analgesic Techniques

A number of peripheral nerve blocks and local anesthetic infusions may be utilized to provide postoperative analgesia and to reduce systemic opioid requirements. Many of these techniques may incorporate a limited PCA function, which can be particularly useful in the management of "incident" pain in the rehabilitation setting. Commonly utilized techniques include continuous brachial plexus local anesthetic infusion and continuous femoral nerve analgesia (Ilfeld et al. 2005).

Additional Techniques

Other techniques that have been demonstrated to be of benefit in the management of acute and postoperative pain include cognitive behavioral techniques, relaxation therapy, ▶ TENS, and physical therapies including hot and cold packs and deep breathing exercises (Rakel and Frantz 2003).

References

Averbuch, M., & Katzper, M. (2004). Assessment of visual analog versus categorical scale for measurement of osteoarthritis pain. *Journal of Clinical Pharmacology, 44*, 368–372.

Ballantyne, J. C., Carr, D. B., Chalmers, T. C., et al. (1993). Postoperative patient controlled analgesia: Meta-analysis of initial randomized controlled trials. *Journal of Clinical Anesthesia, 5*, 182–193.

Breivik, H. (1993). Recommendations for foundation of a hospital-wide post-operative pain service — a European view. *Pain Digest, 3*, 27–30.

Coulling, S. (2005). Nurses' and doctors' knowledge of pain after surgery. *Nursing Standard, 19*, 41–49.

Hayes, C., Browne, S., Lantry, G., et al. (2002). Neuropathic pain in the acute pain service: A prospective survey. *Acute Pain, 4*, 45–48.

Ilfeld, B. M., Morey, T. E., Thannikary, L. J., et al. (2005). Clonidine added to a continuous interscalene ropivacaine perineural infusion to improve postoperative analgesia: A randomized, double-blind, controlled study. *Anesthesia and Analgesia, 100*, 1172–1178.

Jensen, M. P., Mendoza, T., & Hanna, D. B. (2004). The analgesic effects that underlie patient satisfaction with treatment. *Pain, 110*, 480–487.

Miranda, H. F., Silva, E., & Pinardi, G. (2004). Synergy between the anti-nociceptive effects of morphine and NSAIDs. *Canadian Journal of Physiology and Pharmacology, 82*, 331–338.

Morrison, R. S., & Siu, A. L. (2000). A comparison of pain and its treatment in advanced dementia and cognitively intact patients with hip fracture. *Journal of Pain and Symptom Management, 19*, 240–248.

Murphy, P. M., Stack, D., Kinirons, B., et al. (2003). Optimizing the dose of intrathecal morphine in older patients undergoing hip arthroplasty. *Anesthesia and Analgesia, 97*, 1709–1715.

Rakel, B., & Frantz, R. (2003). Effectiveness of transcutaneous electrical nerve stimulation on postoperative pain with movement. *The Journal of Pain, 4*, 455–464.

Remy, C., Marret, E., & Bonnet, F. (2005). Effects of acetaminophen on morphine side-effects and consumption after major surgery: Meta-analysis of randomized controlled trials. *British Journal of Anaesthesia, 94*, 505–513.

Rodgers, A., Walker, N., Schug, S., et al. (2000). Reduction of post operative mortality and morbidity with epidural or spinal anaesthesia: Results from overview of randomized trials. *British Medical Journal, 321*, 1493–1497.

Tucker, A. P., Mezzatesta, J., Nadeson, R., et al. (2004). Intrathecal midazolam II: Combination with intrathecal fentanyl for labor pain. *Anesthesia and Analgesia, 98*, 1521–1527.

Postoperative Pain, Compartment Syndrome

Stephan A. Schug[1] and Emma L. Brandon[2]
[1]School of Medicine and Pharmacology, Royal Perth Hospital, Faculty of Medicine, Dentistry & Health Sciences, The University of Western Australia, UWA Anaesthesiology, Perth, Australia
[2]Anaesthesiology and Pain Medicine, University of Western Australia Royal Perth Hospital, Perth, WA, Australia

Synonyms

Compartmental syndrome

Definition

Compartment syndrome is a condition in which an increase in the pressure within a closed, non-expansile muscle compartment reduces the level of capillary perfusion to less than that required for the viability of tissues. The resultant ischemia can have permanent and disabling sequelae; it is a limb and potentially life-threatening syndrome, which will lead to muscle necrosis with permanent functional impairment and renal failure if not treated. It was first described in 1872 by the German surgeon R. von Volkmann, whose name is associated with this condition (Volkmann 1882).

Characteristics

Compartment syndromes are characteristically seen following trauma to, or arterial reconstruction of, an extremity (Mubarak and Hargens 1983). However, they have also been described following the use of a tourniquet, surgical procedures requiring lithotomy, and as a result of drug abuse or medication injection (Franc-Law 2000), but can even occur with simple physical activity. In most cases of compartment syndrome, it is clear which factors led to the compromise of local blood flow, for example, a fracture of the lower third of the tibia, peripheral vascular disease, or a tight plaster cast. The factors that compromise local blood flow in the lithotomy position are not so obvious, and the term well leg compartment syndrome (WLCS) has recently been advocated for this situation (Heppenstall and Tan 1999). The possibilities include arterial hypoperfusion, venous obstruction, or an increase in compartment pressure resulting from the weight of the limb in the stirrups.

Regardless of the etiology, pathological elevation of intracompartmental pressures impedes capillary perfusion and may rapidly progress to muscle ischemia, metabolic acidosis, and rhabdomyolysis with subsequent release of metabolic toxins. Myoglobinuric renal failure, multiorgan failure, and possible death may ensue. Amputation of the limb is occasionally required and there is also the risk of long-term neuromuscular deficit of the limb and altered gait due to long-term ischemic contracture or nerve injury secondary to ischemia.

Pathophysiology

Possible explanations regarding the pathophysiology of compartment syndrome have developed with time, from Rowland's theory in 1910 of a direct plasma leak after ischemia to current concepts focusing on the ischemia-reperfusion theory of cellular injury (Rowlands 1910). Ischemia leads to the depletion of intracellular energy stores, causing cellular swelling, edema, and an increase in venular pressure due to venous outflow obstruction with a resultant increase in capillary pressure. Intracompartmental pressure eventually equals capillary pressure, stopping nutrient blood flow and leading to tissue infarction.

Diagnosis

The diagnosis of compartment syndrome requires a high index of clinical suspicion, and the typical postoperative presentation includes leg pain out of proportion to clinical findings. The earliest signs are often subtle and most often neurological because the tissues most sensitive to hypoxia are nonmyelinated type C sensory fibers. There may be swelling of the affected limb, pain with passive stretch of the muscles within the compartment, and paresthesias, and pulselessness may also be present. However, because compartment syndrome occurs at interstitial pressures below systolic arterial pressure, the presence of a pulse does not exclude the presence of a compartment syndrome. The condition may be misdiagnosed as a deep vein thrombosis or neurapraxia.

Compartment pressures may be measured invasively. However, there is still no universal agreement at what intracompartmental pressure one should consider fasciotomy. It has been argued that intracompartmental pressure monitoring should only be used to confirm clinical findings.

Management of Compartment Syndrome

Treatment involves correction of metabolic acidosis, treatment of renal failure, and decompressive fasciotomies of each compartment. If fasciotomy is not performed within 12 h of ischemia, there is little recovery of neuromuscular function.

Analgesia and Compartment Syndrome

Providing good intra- and postoperative analgesia is the goal of every anesthetist. However, much controversy exists about the most suitable analgesia for patients undergoing orthopedic/vascular extremity surgery or surgery requiring a prolonged lithotomy position, that is, those patients at risk of developing a compartment syndrome postoperatively.

The use of epidural analgesia has provoked the most concern. This relates to two main issues: the role of epidural analgesia in contributing to compartment syndrome and the risk of epidural analgesia masking the symptoms of compartment syndrome. It has been suggested that sympathetic blockade induced by the use of epidural local anesthetics causes increased blood flow to a limb and that this may contribute to a rise in intracompartmental pressure (Price et al. 1996). In contrast, epidurally administered opioids do not abolish the normal vasoconstrictor response, but may reduce afferent nociceptor input to the sympathetic nervous system by a direct effect on the peripheral vasculature, thereby potentially increasing blood flow to an extremity (Cousins and Glynn 1980).

The use of epidural analgesia for postoperative pain produces a sensory block of the lower limb. There is much debate about whether this masks the signs of compartment syndrome. Strecker et al. reported a case of compartment syndrome after osteocutaneous free fibula transfer, in which postoperative pain was managed with 0.125 % epidural bupivacaine for 4 days (Strecker et al. 1986). After discontinuation of the epidural, the patient complained of dull pain, but no treatment was considered for a further 3 days.

Morrow et al. (1994) described a patient who underwent bilateral closed intramedullary femoral rodding for femoral shaft fractures and received continuous epidural bupivacaine and fentanyl postoperatively. Thirteen hours after surgery, it was noted that the patient had numbness and paresis of the leg. Tibial pressures were elevated and decompressive fasciotomies were undertaken.

Tang and Chiu (2000) described a case of compartment syndrome following a total knee replacement, in which they claim continuous epidural analgesia masked the pain (Tang and Chiu 2000).

However, there have been many cases of compartment syndrome diagnosed successfully during epidural blockade (Dunwoody et al. 1997; Beerle and Rose 1993; Turnbull and Mills 2001): Beerle and Rose administered bupivacaine and fentanyl via the epidural route, which did not mask the signs of compartment syndrome due to the lithotomy position (Beerle and Rose 1993). Turnbull and Turnbull described three cases in which compartment syndrome developed after surgery in the Lloyd-Davies position, in which all three patients reported severe bilateral calf pain in the recovery room

despite functioning thoracic epidurals (Turnbull and Mills 2001).

Epidural local anesthetics act by blocking nerve transmission in the somatic and sympathetic fibers as they pass through the epidural space. The degree of neural blockade is dose and concentration dependent, with the lowest concentration causing sympathetic blockade and increasing concentrations producing analgesia, cutaneous anesthesia, and finally motor blockade. Epidural opioids exert their analgesic effect by binding to opioid receptors in the dorsal horn of the spinal cord and selectively blocking nociceptive stimulation, thereby producing analgesia. Unlike local anesthetics, they do not produce cutaneous anesthesia or motor blockade. It is common practice to combine low-dose local anesthetics and opioids in order to reduce the side effects of each agent.

In the light of these differences, several authors have advocated omitting the local anesthetic component and just delivering epidural opioids on their own with the aim of maintaining the sensation of skin and deep tissues, and thereby making a compartment syndrome easier to diagnose. However, the use of fentanyl as the sole agent in an epidural has been shown to mask a compartment syndrome in a patient undergoing bilateral corrective osteotomies (Price et al. 1996).

In summary, epidural analgesia is unlikely to mask the signs of complete compartment syndrome, although it may delay the diagnosis in patients with early compartment syndrome. However, quality and frequency of monitoring are important factors in preventing the syndrome being masked by epidural analgesia.

Postoperative epidural analgesia is not the only suspect when considering the association between analgesia and delayed diagnosis of compartment syndrome. Harrington et al. report the case of a 53-year-old male who underwent intramedullary nailing of a tibial fracture who subsequently developed compartment syndrome (Harrington et al. 2000). This went unrecognized until the patient went back to theater for wound inspection 36 h later. Initial postoperative pain relief was managed with a morphine patient-controlled analgesia (PCA) device. The authors concluded that nursing staff were not alerted to the increasing morphine consumption because use of the PCA reduces nurse-patient contact. They suggest the use of more sophisticated PCA pumps, which allows a specified 4 h maximum dose to be set. In this way, the patient with increasing analgesic requirements would be forced to seek nursing attention earlier, and this would allow earlier appropriate management. Other methods of postoperative analgesia have also been implicated in a delay in diagnosing compartment syndrome, including peripheral nerve blocks and spinals.

A recent survey of a group of orthopedic surgeons and anesthetists in the UK showed that significant differences exist regarding the preferred choice of postoperative analgesia in clinical situations, which have previously been shown to be associated with a high risk of compartment syndrome (Raghuram et al. 2004). Both surgeons and anesthetists favored the use of the PCA, but a higher proportion of anesthetists preferred the use of epidurals and nerve block compared to the orthopedic surgeons. The authors propose that management of postoperative pain in patients at risk of compartment syndrome should be along the lines of traditional on-demand intramuscular morphine, so that increasing analgesic requirements must be reviewed by the nursing staff prior to administration of the next dose. They also suggest that the use of local and regional nerve blocks, without intracompartmental pressure monitoring, should be contraindicated in these high-risk patients. However, these views are not widely supported by the evidence available.

Conclusion

Whatever mode of analgesia is used in patients at high risk of developing compartment syndrome, a high index of clinical suspicion, careful detailed examination, and compartmental pressure monitoring remain the cornerstone for diagnosis and treatment. Pain management should be multidisciplinary, and close liaison is necessary between surgeons and anesthetists when planning a postoperative analgesic regimen.

References

Beerle, B. J., & Rose, R. J. (1993). Lower extremity compartment syndrome from prolonged lithotomy position not masked by epidural bupivacaine and fentanyl. *Regional Anesthesia, 18,* 189–190.

Cousins, M. J., & Glynn, C. J. (1980). New Horizons. In M. J. Cousins & P. O. Bridenbough (Eds.), *Neural blockade in clinical anaesthesia and management of pain* (pp. 699–719). Philadelphia: JB Lippincott.

Dunwoody, J. M., Reichert, C. C., & Brown, K. L. (1997). Compartment syndrome associated with bupivacaine and fentanyl epidural analgesia in pediatric orthopaedics. *Journal of Pediatric Orthopedics, 17,* 285–288.

Franc-Law, J. M. (2000). Poisoning induced acute atraumatic compartment syndrome. *American Journal Emergency Medicine, 18,* 616–621.

Harrington, P., et al. (2000). Acute compartment syndrome masked by intravenous morphine from a patient controlled analgesia pump. *Injury, 31,* 387–389.

Heppenstall, B., & Tan, V. (1999). Well leg compartment syndrome. *Lancet, 354,* 970.

Morrow, B. C., Mawhinney, I. N., & Elliott, J. R. (1994). Tibial compartment syndrome complicating closed femoral nailing: Diagnosis delayed by an epidural analgesic technique — case report. *The Journal of Trauma, 37,* 867–868.

Mubarak, S. J., & Hargens, A. R. (1983). Acute compartment syndromes. *The Surgical Clinics of North America, 63,* 539–565.

Price, C., Ribeiro, J., & Kinnebrew, T. (1996). Compartment syndromes associated with postoperative epidural analgesia. A case report. *Journal Bone Joint Surgery American, 78,* 597–599.

Raghuram, T., et al. (2004). Differences in attitudes to analgesia in postoperative limb surgery put patients at risk of compartment syndrome. *Injury, 35,* 290–295.

Rowlands, R. P. (1910). Volkmann's contracture. *Guy's Hospital Gazette, 24,* 87.

Strecker, W. B., Wood, M. B., & Bieber, E. J. (1986). Compartment syndrome masked by epidural anaesthesia for postoperative pain. Report of a case. *Journal Bone Joint Surgery American, 68,* 447–448.

Tang, W. M., & Chiu, K. Y. (2000). Silent compartment syndrome complicating total knee arthroplasty: Continuous epidural anaesthesia masked the pain. *The Journal of Arthroplasty, 15,* 241–243.

Turnbull, D., & Mills, G. H. (2001). Compartment syndrome associated with the Lloyd Davies position: Three case reports and review of the literature. *Anaesthesia, 56,* 980–982.

Volkmann, R. (1882). Krankheiten der bewegungsorgane. In F. von Pitha & T. Billroth (Eds.), *Handbuch der allgemeinen und speziellen chirurgie* (Vol. 2, pp. 234–920). Stuttgart: Enke.

Postoperative Pain, COX-2 Inhibitors

Ian Power[1] and Lesley A. Colvin[2]
[1]Critical Care and Pain Medicine, The University of Edinburgh, Edinburgh, UK
[2]Department of Anaesthesia, Critical Care and Pain Medicine Western General Hospital, Edinburgh, UK

Synonyms

COX-2 inhibitors; Cyclooxygenase-2 inhibitors; Coxibs

Definition

COX-2 inhibitors are anti-inflammatory analgesics that act by specific inhibition of one of the isoenzymes involved in prostaglandin synthesis, cyclooxygenase 2 (COX-2). These drugs were developed as an alternative to traditional NSAIDs (▶ NSAIDs, Survey), with the aim of maintaining or improving efficacy and reducing adverse effects.

Characteristics

Mechanism of Action
The COX enzymes, of which there are several variants, are involved in prostaglandin production from ▶ arachidonic acid (see Fig. 1). Standard NSAIDs act by inhibiting activity of cyclooxygenase-1 and cyclooxygenase-2 enzymes (COX-1 and COX-2), otherwise known as prostaglandin endoperoxide synthases (PGHS). There is about 60 % homology of the amino acid sequence between COX-1 and COX-2, with both enzymes having very similar active sites and catalytic properties, although COX-2 has a larger potential binding site due to a secondary internal pocket. COX-2 is degraded more rapidly and has a shorter half-life than COX-1.

COX-1 is expressed constitutively in a wide range of tissues and is thought to be involved in many physiologic "housekeeping" functions. The prostanoids produced by COX-1 are functionally active in many areas, including the gastrointestinal tract, kidney, lung, and cardiovascular systems. In contrast, the functional COX-2 enzyme is normally only found in a few restricted tissues such as the brain, renal cortex, and tracheal epithelium, although COX-2 mRNA is widely distributed. In response to specific stimuli, the expression of the COX-2 isoenzyme is induced or upregulated. The prostaglandins subsequently produced play a key role in inflammatory processes and pain. Prostaglandin E2, in particular, plays an important role in ▶ inflammation, pyresis, and ▶ hyperalgesia. The role of COX-2 in pyrexia may depend on the etiology of the pyrexia. COX-2 may be involved in lipopolysaccharide-induced pyrexia, although COX-1 may be more important in fever occurring without infection. A variety of other stimuli found in the perioperative period have been shown to induce COX-2, including cytokines and growth factors (Masferrer et al. 1999).

Although COX-2 inhibitors act mainly on COX-2, they may have variable activity on the COX-1 enzyme. A range of assays, including in vitro whole blood assays, have been used to give an indication of the selectivity of different agents acting on the COX-1 and COX-2 enzymes (see Table 1 (Baigent and Patrono 2003)). These may not be clinically relevant, as there can be significant variability in individual plasma levels with the same dose, as well as variability in actual enzyme inhibition.

Efficacy
There is a large body of evidence of the efficacy of traditional NSAIDs in the treatment of both acute and chronic pain. Used alone, they are effective in mild to moderate postoperative pain and have an opioid sparing effect when used as part of a ▶ balanced analgesic regime in the treatment of moderate to severe postoperative pain. Although there is evidence of clinical efficacy and safety, not all COX-2 inhibitors are

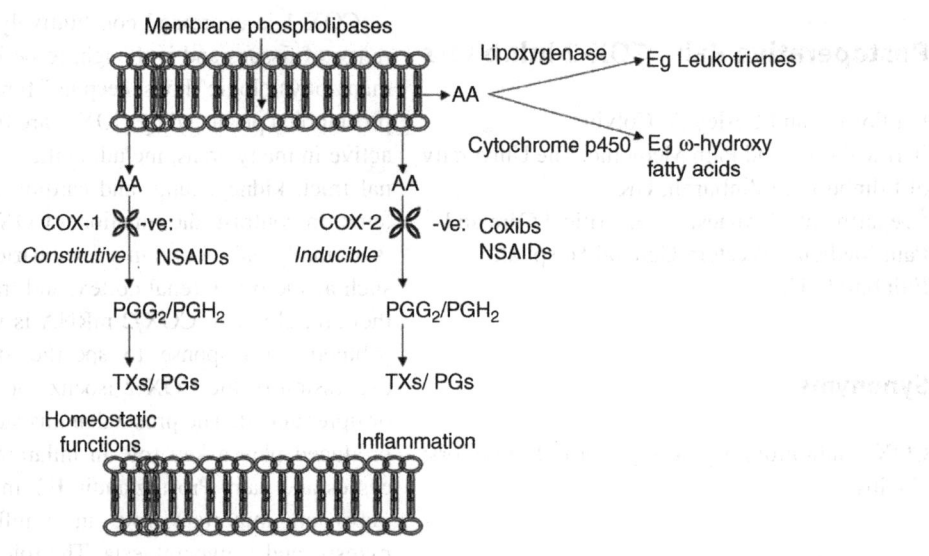

Postoperative Pain, COX-2 Inhibitors, Fig. 1 Free arachidonic acid (AA) is produced from membrane-bound phospholipids and may be metabolized by several different pathways into a variety of substances. The enzymes involved include lipoxygenase, cytochrome p450, and cyclooxygenases 1 and 2 (COX-1 and COX-2). COX-1 is involved in the production of a wide range of prostaglandins and thromboxanes that are essential for many homeostatic processes. COX-2 is induced by a variety of mediators, such as cytokines and growth factors, which are central to inflammatory processes. Nonsteroidal anti-inflammatory drugs (NSAIDs) inhibit both enzymes whereas COX-2 inhibitors are selective for COX-2

Postoperative Pain, COX-2 Inhibitors, Table 1 Relative selectivity of a range of anti-inflammatory agents for COX-1 and COX-2 enzymes, as measured using human whole blood assays in vitro using platelet COX-1 and monocyte COX-2. IC_{50} ratio = 50 % inhibitory concentration (modified from Baigent and Patrono 2003)

Inhibitor	COX-1:COX-2 (IC_{50} ratio)
Rofecoxib	267
Celecoxib	30
Diclofenac	29
Meloxicam	18
Indomethacin	1.9
Acetaminophen	1.6
Naproxen	0.7
Ibuprofen	0.5

licensed for the treatment of acute postoperative pain (rofecoxib and parecoxib are licensed). Ideally, any new agent would have to demonstrate efficacy similar to that of conventional ▶ NSAIDs but with reduced adverse effects, in particular, on the gastrointestinal tract and platelet function.

There is level one evidence (systematic reviews/meta-analyses) of the efficacy of COX-2 inhibitors for chronic pain conditions, such as rheumatoid arthritis and osteoarthritis, compared to both placebo or active comparator (usually a standard NSAID) (Deeks et al. 2002). There have also been studies demonstrating efficacy in dysmenorrhea and acute gout.

There is now an increasing amount of evidence that COX-2 inhibitors are at least as effective as standard NSAIDs in the treatment of postoperative pain. Clinical studies have looked at dental pain, orthopedic procedures including major spinal and joint replacement surgery, gynecological and cardiac surgery. To date there are several systematic reviews that demonstrate efficacy of specific COX-2 inhibitors in the acute postoperative setting (Barden et al. 2004a, b, c). These have been summarized in Table 2, along with some other agents for comparison. The ▶ number-needed-to-treat (NNT) is defined as the number of patients needed to receive the active agent (e.g., COX-2 inhibitor) for one patient to achieve a 50 % reduction in the level

Postoperative Pain, COX-2 Inhibitors, Table 2   Relative efficacy of different COX-2 inhibitors (Compiled from The Cochrane Library of Systemic Reviews/Bandolier) (Barden et al. 2004a)

Drug used	No of trials in systematic review	Type of surgery	Number of patients receiving active treatment	Number-needed-to-treat (NNT) (95 % CI)
Valdecoxib 20–40 mg	4	Dental	380	20 mg: 1.7 (1.4–2.0) 40 mg: 1.6 (1.4–1.8)
Celecoxib 400 mg	4	Dental; orthopedic	704	1.9 (1.6–2.3)
Etoricoxib 60–180 mg	3	Dental	824	60 mg: 2.0 (1.6–2.8) 120 mg 1.5 (1.3–1.7)
Rofecoxib 50 mg	7	Dental; orthopedic	667	2.2 (1.9–2.4)
Diclofenac 100 mg	7		618	2.3 (2.0–2.7)
Ibuprofen 400 mg	52		6,358	2.4 (2.3–2.6)
Morphine 10 mg (intramuscular)	20		696	2.9 (2.6–3.6)
Paracetamol 1,000 mg	46		2,530	3.8 (3.4–4.4)
Tramadol 100 mg	18		1,486	4.6 (3.4–4.4)

of their pain compared with placebo over a 4–6 h period. The NNTs for currently available COX-2 inhibitors are similar to that for diclofenac and better than that for tramadol or morphine.

More recently produced agents, such as valdecoxib and etoricoxib, may have an improved analgesic effect compared to older agents, with a good safety profile. New developments include agents that can be given both parenterally and orally, with obvious benefits for the perioperative management of pain. Some of these agents have very fast onset and a long duration of action.

Safety

Despite the considerable evidence for analgesic efficacy of standard NSAIDs, for postoperative use there is also considerable evidence regarding their potential for harm (Tramer et al. 2000). NSAIDs have been found to cause more deaths than asthma, multiple myeloma, cervical cancer, and Hodgkin's disease. As selective COX-2 inhibitors should not interfere with the physiological functions of COX-1 produced prostanoids, there should be an improved safety profile with these agents.

Most of the studies have looked at adverse events in the chronic pain setting with prolonged use of the agents, but generally at lower doses than those used for acute pain. The systematic

reviews of COX-2 inhibitors for efficacy in treating acute pain (see Table 2) have generally found no significant increase in adverse events compared to placebo, but information is limited and further study is required.

Gastrointestinal

One of the commonest side effects with NSAIDs is gastrointestinal toxicity. Approximately 1 in 1,200 patients on chronic NSAID treatment (>2 months) will die from related gastroduodenal complications (Tramer et al. 2000). The COX-1 isoenzyme is the predominant cyclooxygenase found in the gastric mucosa. The prostanoids produced here help to protect the gastric mucosa by reducing acid secretion, stimulating mucous secretion, increasing production of mucosal phospholipids and bicarbonate, and regulating mucosal blood flow. As COX-2 inhibitors do not interfere with these processes, a reduction in gastrointestinal complications compared to standard NSAIDs would be expected.

A reduction in morbidity and mortality from upper GI complications has been shown in several large studies of chronic pain populations as well as healthy volunteers. Studies have looked at several parameters including fecal red blood cell loss, endoscopic studies, and incidence of upper gastrointestinal clinical events. The majority of these studies demonstrated a significant reduction in GI bleeds, perforation, and peptic ulcer disease found in comparison to NSAIDs (Rostom et al. 2004). Using the strictest criteria of endoscopically proven ulcer, combining data from 3 large-scale trials (Simon et al. 1999; Laine et al. 1999; Hawkey et al. 2000), no significant difference in incidence rate was found after 12 weeks of treatment between placebo and all doses of coxibs (~6 %), compared to conventional NSAIDs with a significantly higher incidence rate (28 %).

Perioperatively, the use of higher doses than those used for chronic pain, combined with risk factors such as frail, elderly, or dehydrated patients may increase the risk of acute GI events, although there is no evidence of this to date. Studies in the acute setting have mainly focused on efficacy as a primary outcome measure, but none of these studies found any severe adverse GI events. There have been several studies using injectable parecoxib (a prodrug of valdecoxib), looking specifically at acute upper GI complications that found no endoscopically proven ulcers, compared to up to 23 % of patients receiving ketorolac (Stoltz et al. 2002).

Cardiovascular

Some of the large-scale studies of use of COX-2 inhibitors in chronic conditions have found an increased risk of cardiovascular events, mainly an increased risk of myocardial infarction (NHS Centre for Reviews and Dissemination 2004). The VIGOR study in particular, where patients on low-dose aspirin were excluded, found an increased risk of MI for those patients on rofecoxib compared to naproxen (Bombardier et al. 2002). Several explanations have been postulated including a protective effect of NSAIDs in reducing platelet aggregation, similar to low-dose aspirin. Alternatively, there may be an alteration in endothelial prostaglandin balance by COX-2 selective inhibition with increased thromboxane A2 production relative to prostacyclin.

In acute postoperative pain, one study found adverse effects of COX-2 inhibitors, when parecoxib followed by valdecoxib was used in patients undergoing CABG surgery for up to 2 weeks afterwards. An increase in cerebrovascular accidents, renal dysfunction, and sternal wound problems was found (Ott et al. 2003). Several studies have found an increase in thrombotic events resulting in the withdrawal of some of these agents. The risks and benefits of COX-2 selective agents remain the subject of considerable debate (Garner et al. 2005; Jones and Power 2005).

Renal

COX-2 is normally found in the renal cortex and is therefore inhibited by both conventional NSAIDs and COX-2 inhibitors. There is the potential for peripheral edema and hypertension, as well as direct effects on renal excretory function with oliguria and decreased creatinine clearance. Although this has so far not been clinically significant in trials on COX-2 inhibitors, it may become so in patients with a range of comorbidities that would have been excluded from clinical trials. Studies have not found any greater risk for these problems with COX-2 inhibitors compared to standard NSAIDs.

Hematological

Thromboxane production, requiring COX-1 activity, is important in the platelet aggregation necessary for hemostasis. With standard NSAIDs, there is the possibility of increased bleeding risk due to inhibition of platelet cyclooxygenase, which is mainly COX-1. This is of relevance, both in terms of surgical complications but also in the use of central regional techniques, such as epidural blocks. COX-2 inhibitors seem to have little effect on platelet activity or bleeding time, which may be beneficial in the perioperative situation.

Other

Respiratory problems have been reported, and as COX-2 is expressed in respiratory epithelium, their safety in susceptible asthmatics is not clear. Other reported side effects include nausea, constipation, dental infection, and congestive cardiac failure.

Conclusion

COX-2 inhibitors have been shown to be effective in the treatment of postoperative pain, with particular advantages in terms of improved GI safety compared to standard NSAIDs. As the evidence base for their clinical use increases, it appears that some of the newer agents may be more effective than previously available NSAIDs. COX-2 inhibitors provide a valuable addition to balanced postoperative pain management, with fast onset and long duration of action. Associated thrombotic events have restricted the critical use of COX-2 inhibition.

References

Baigent, C., & Patrono, C. (2003). Selective cyclooxygenase 2 inhibitors, aspirin and cardiovascular disease. *Arthritis and Rheumatism, 48*, 12–20.

Barden, J., Edwards, J, Moore, R., et al. (2004a). Single dose oral rofecoxib for postoperative pain. *Cochrane Database of Systematic Reviews 1*.

Barden, J., Edwards, J., McQuay, H., et al. (2004b). Single Dose Oral Celecoxib for Postoperative Pain. *Cochrane Database of Systematic Reviews 1*.

Barden, J., Edwards, J., Moore, R., et al. (2004c). Single Dose Oral Etoricoxib for Postoperative Pain. *Cochrane Database of Systematic Reviews 1*.

Bombardier, C., Laine, L., Reicin, A., et al. (2002). Comparison of upper gastrointestinal toxicity of rofecoxib and naproxen in patients with rheumatoid arthritis. VIGOR study group. *The New England Journal of Medicine, 343*, 1520–1528.

Deeks, J. J., Smith, L. A., & Bradley, M. D. (2002). Efficacy, tolerability, and upper gastrointestinal safety of celecoxib for treatment of osteoarthritis and rheumatoid arthritis: Systematic review of randomised controlled trials. *British Medical Journal, 325*, 619.

Garner, S., Fodan, D., Frankish, R., Maxwell, L. (2005). Rofecoxib for osteoarthritis. *Cochrane Database of Systematic Reviews 1*.

Hawkey, C., Laine, L., Simon, T., et al. (2000). Comparison of the effect of rofecoxib (A Cyclooxygenase 2 Inhibitor), ibuprofen, and placebo on the gastroduodenal mucosa of patients with osteoarthritis: A randomized, double-blind, placebo-controlled trial. The rofecoxib osteoarthritis endoscopy multinational study group. *Arthritis and Rheumatism, 43*, 370–377.

Jones, S. F., & Power, I. (2005). Editorial I: Postoperative NSAIDs and COX-2 inhibitors: Cardiovascular risks and benefits. *British Journal of Anaesthesia, 95*, 281–284.

Laine, L., Harper, S., Simon, T., et al. (1999). A randomized trial comparing the effect of rofecoxib, a cyclooxygenase 2-specific inhibitor, with that of ibuprofen on the gastroduodenal mucosa of patients with osteoarthritis. Rofecoxib Osteoarthritis endoscopy study group. *Gastroenterology, 117*, 776–783.

Masferrer, J. L., Koki, A., & Seibert, K. (1999). COX-2 inhibitors. A new class of antiangiogenic agents. *Annals of the New York Academy of Sciences, 889*, 84–86.

NHS Centre for Reviews and Dissemination. (2004). Risk of Cardiovascular Events Associated with Selective COX-2 Inhibitors. *Database of Abstracts of Reviews of Effectiveness Issue 1*.

Ott, E., Nussmeier, N. A., Duke, P. C., et al. (2003). Efficacy and safety of the cyclooxygenase 2 inhibitors parecoxib and valdecoxib in patients undergoing coronary artery bypass surgery. *The Journal of Thoracic and Cardiovascular Surgery, 125*, 1481–1492.

Rostom, A., Dube, C., Boucher, M., et al. (2004). Adverse gastrointestinal effects of COX-2 inhibitors for inflammatory diseases. *Cochrane Database of Systematic Reviews, 1*.

Simon, L. S., Weaver, A. L., Graham, D. Y., et al. (1999). Anti-inflammatory and upper gastrointestinal effects of celecoxib in rheumatoid arthritis — A randomized controlled trial. *Journal of the American Medical Association, 282*, 1921–1928.

Stoltz, R. R., Harris, S. I., Kuss, M. E., et al. (2002). Upper GI mucosal effects of parecoxib sodium in healthy elderly subjects. *American Journal of Gastroenterology, 97*, 65–71.

Tramer, M. R., Moore, R. A., Reynolds, D. J. M., et al. (2000). Quantitative estimation of rare adverse events which follow a biological progression: A new model applied to chronic NSAID use. *Pain, 85*, 169–182.

Postoperative Pain, Data Gathering and Auditing

Kersi Taraporewalla
Royal Brisbane and Womens' Hospital,
University of Queensland, Brisbane, QLD,
Australia

Synonyms

Acute pain database; Audit; Data auditing/collection/gathering/management/security; Data collection; Data gathering; Data management

Definition

▶ Data gathering consists of information collected by acute pain services, or their equivalents, in the day-to-day running of the service for purposes of patient care, improvement, auditing, comparison, or cost-benefit analysis.

Characteristics

Introduction

A database is more than a collection of information or data. It is a tool to perform useful functions for its users. Rather than merely store data, database management requires three vital components, each of which requires a critical question to be answered when a database is set up (Allen 2002):

- It needs to be a representation of a portion of the real world it is created from. The critical question is how much data needs to be collected to produce a clear picture of the portion of the world it represents.
- The information stored needs to be set out in a logically coherent manner, with inherent meaning. This element needs careful planning and forethought as to how different pieces of the data will be linked to present a realistic set rather than a distorted view.
- Specific purposes need to be addressed in creation of the database. The critical question is to what use is the information retrieved required, and what sorts of information will need to be collected to allow this to happen.

The purposes determine how the data needs to be analyzed and manipulated. This could vary from a simple count of one field or element to complex queries and reports across fields with Boolean operators.

Common purposes reported for the postoperative pain database (POPDB) have been auditing and quality assurance activities. Early databases, using paper records, tended to be simple and flat, partly due to the limitations of data retrieval. Improvement in computer technology has seen the development of relational databases, run by single operators. While the present databases represent a significant advance in retrieval, they have limitations due to the purposes intended. Significant advances in database technology can now be introduced including the use of wireless area network technology, computerized record keeping, and online analytic processing to allow for improved clinical decision making at point of treatment. This also requires the use of data warehouses for the sharing of data from multiple institutions and the pooling of data to allow for creation of standards of care and to improve audit comparisons.

For a database to have the power to improve performance, several issues need to be considered.

Culture of Data Collection

The clinical work of optimizing postoperative analgesia with minimal or no adverse effects is viewed as the primary aim of postoperative pain management units (such as acute pain services APS). The extra effort required in collecting data requires a culture that recognizes quality assurance of clinical routines and also improves postoperative pain management (Elmasri and Mavathe 2000).

Numerous reasons have been put forward for failure to collect data by APS units. These include:

- Cost of collection and management of a database given limited budgets
- Lack of personnel resources and support to collect and enter data
- Lack of institutional or departmental support
- Lack of expertise in database management or computer technology
- Previous failures at attempts in collecting data
- Issues of confidentiality of data and its security
- Medicolegal problems arising from deficiencies in management identified by the database
- Lack of equipment

These limitations have been partly addressed by the development of commercial databases for acute pain management; however, most have failed in being unable to generate adequate reporting support.

While all of the above reasons have a genuine origin, they can be overcome to produce databases with significant power to improve APS units and reduce patient discomfort.

Purposes of Data Collection

Data that are collected and stored but not used are costly and wasteful. The purposes that the POPDB is to be used for needs to be clarified at the design stage, so as to identify the types of data that need to be collected. The following uses of POPDB have been reported:

- Quantifying efficacy and safety of APS (Hammond et al. 2000)
- Providing evidence of quality of care (Kuwahara et al. 2000)
- Monitoring performance of APS
- Assessing the effects of training in postoperative pain care
- Allowing decision systems in the creation of protocols

Once data fields have been determined, consideration needs to be addressed as to how the data will be represented, entered, validated, and retrieved.

Methods of Data Collection

Early methods of data collection consisted of separate paper record forms derived from ward notes (McLeod 2002). Paper records have numerous drawbacks. They are retrospective, tedious, require repetitive data entry, and make statistics difficult. Systems using this form have reported achieving only 60 % completion (Sanders and Michel 2002). However, there are advantages to paper-based POPDB. They are easy to administer and tend not to need a large amount of capital outlay.

The development of relational databases has considerably added power to the POPDB. Data can now be categorized and stored with appropriate prompts for the user. Statistic manipulations become easier, and the total amount of data that can be stored is considerably greater. The types of tables that can be created include modalities used, equipment used, personnel involved, and patient demographics. The data tables can be generated to make intrinsic

sense as a stand-alone portion of the total database. The development of smaller-sized computers, as well as handheld devices such as personal digital assistants, allows for data to be entered at point of care delivery (Stomberg et al. 2003). This has a significant advantage in having data prospectively collected and analyzed and may also increase the chances of having data that are complete, consistent, and accurate.

Who Collects the Data

Past experience has shown that enthusiasts are most likely to collect data in a consistent manner, partly due to factors discussed in the culture of data collection. The advent of the pain nurse provides an ideal opportunity to perform regular audits by data entry (Tan and Wong 2003). However, as APS units grow in size, it becomes vital that all members be involved in data entry. This has significant implications for training of staff involved in the management of postoperative pain.

Data Needed to Be Collected

Enthusiasm for data collection tends to lead to collection of as much data as possible, in case the data might be needed in the future or become important, and the perception that it is better to collect data prospectively. There are a significant number of problems in collecting everything. These include increasing the time required to collect the data, staff annoyance at extra work for no return leading to poor data entry in fields not being analyzed, large databases slowing down computer performance, increased likelihood of data entry errors, and increased chance of confusion in queries setup for reporting. As a general principle, the purpose of the POPDB should dictate the data collected.

Authors reporting on quality assurance and workload activities have collected the following types of data:

- Patient demographics – name, age, sex, ASA status, etc.
- APS techniques and drugs used – ▶ PCA, epidural, regional blocks, etc.
- Adverse effects of techniques and drugs – nausea and vomiting, sedation, respiratory depression, and motor weakness

- Assessment of pain
- Personnel involved
- Equipment used
- Duration of treatment – time involved in care

There are two factors that have impaired how retrieved information is used by the wider community. There is no agreement as to what constitutes the minimal data set that POPDB should collect so as to gain meaningful comparisons from reports generated (Elmasri and Mavathe 2000; Turner and Halliwell 1999). Further, there is no uniform set of definitions and scoring systems, e.g., sedation, nausea, and pain intensity, making comparisons and standard setting difficult. These are urgently required for future developments and advances in postoperative pain care, as well as POPDB functions.

Data stored in POPDB should be in fixed coded systems rather than free text format. Free text formats create difficulties in analyzing due to spelling errors, abbreviations used, loose terminology, and the lack of clear definitions. Coded or fixed entry formats have the advantages of retrieval and checks for data quality, but need careful attention in the planning stage to allow for all possible values that may arise.

Quality of Data

The value of reports generated from a POPDB is dependent on the quality of data stored in it. This quality can be appreciated in terms of completeness, validation, error handling, and audit trails.

Incomplete forms and tables stored in the POPDB represent lost values that cannot be analyzed. Data entry rates reported at 60 % significantly impair the meaning that can be generated and introduce biases to interpretations made from reports produced. Staff averse to the ideas of data collection are more likely to be incomplete in their data entry. Data need to be checked for completeness at time of entry, and periodically thereafter. The former can be performed by a software approach that checks for the minimal fields that need to be completed before the form can be closed and stored in the database. Periodic checking of data quality requires a database manager to be employed to check fields for completeness and correction of

errors of omission from the patient notes. Changes made to the database after initial entry need to have an automatically generated timestamp to allow for audit trails.

Validity of data implies that data entered have sufficient meaning to be interpretable. Hence, a name field should not contain numbers, and a low pulse rate needs to have a range (e.g., <45) defined. Validation of data entries can be significantly improved if the user is given a pick list of fixed entries that can be used. This list needs to be complete and of limited size so as to be able to group and categorize the data in meaningful ways. Thus, a rare complication can be stored as "other complication" with an associated text entry field, and a drug list of local anesthetic agents can be generated from the limited number of drugs regularly used by the unit. Pick lists can become more useful when data entry can be generated by simple means such as pointers, touch screen, or using the first few letters of the choices. Validation can also be improved by incorporation of a software check at time of entry. It is important to realize that such validation does not check the truth of the data entered; rather, it eliminates nonsense entries into fields by accident.

Validity also implies consistency of what information is collected. A report from one institution produced a correlation coefficient of 0.4 for pain scores and 0.02 for sedation scores between ward nurses and APS staff (Wigfull and Welchew 1999). This suggests that audits need to validate their data prior to generating reports.

How the POPDB handles errors is of significant importance. Errors of omission can be checked for and corrected. The time of correction needs to be registered. Online correction of omission needs to be handled with diplomacy so as not to irritate the human user. Errors of commission are more difficult to manage. Some advances can be made in checking for logical errors, such as entering local anesthetic or sensory levels when PCA is used, or the entry of inappropriate infusion rates for continuous epidural infusions.

The use of default values in data fields has advantages and problems. Default values speed data entry and reduce the workload on the user. Care needs to be taken as to the values chosen for default, e.g., a pain score ▶ default value of 99 can be used to check that data have not been entered rather than a default value of 0. Default values that are also possible data entries can lead to errors at the time of retrieval, because of the interpretation ascribed to the collected results for the data field. Default values can also be used to check for errors of omission.

An audit trail is an important part of any database. It needs to be automatically generated and allow for validation and data entry checking and can explain differences in reports produced at different times. Database entries should not be used to generate patient notes; rather, data entries should be created simultaneously with patient notes. The patient notes could be produced as a report from the database. However, if the data fields are altered after patient records are generated, all reports should show the data trail of when the data was altered.

Data Security

There are two opposing factors at play. Data need to be stored in an easily accessible format for retrieval and transfer to other databases to become part of a larger set. However, sensitive patient information needs to be handled for confidentiality. Security of information in electronic form remains a problem for the computer industry, and ingenious solutions, such as encryption, have been formulated. The backwash of this problem is the lack of trust in providing information for storage by physicians. As a principle, shared information regarding patients should not be able to identify individual patients. This can be managed by creating appropriate tables that can be shared using numerous security limits for the sensitive data.

Audit Reports

The author has seen numerous proposed POPDB put for commercial use, with many failing to generate reports that can be produced by the user on an ad hoc basis. Most force reports perceived by the creators of the software as being important for an APS, rather than letting the APS users determine what information is of

value. No POPDB should be used unless the reports that can be generated are credible and serve the purpose the POPDB was set up to do.

Audits of activities of APS can be expressed in terms of:

- Efficiency – e.g., time to achieve satisfactory pain relief, number of pumps idle
- Effectiveness – in terms of no. of patients with satisfactory pain relief, complication rates, costs incurred in providing the service, and early discharge rate
- Resource utilization – costs, drugs used, and equipment used and needed
- Quality of service – as compared with standards set, other institutions, or nationally accepted rates

Certain points need to be addressed in reports generated by audits. When prevalence or incidences of events are reported, they need to be expressed in terms of incidence per 100 or 1,000 cases. Thus, both the numerator and denominator in terms of sample size need to be stated.

Interpretation needs to be provided in context for results obtained. Thus, a low complication rate for epidural analgesic techniques may be due to low rate of use, intense monitoring, or choice of drugs, etc. Explanations generated by members of the APS who understand the constructs of the figures are more likely to reflect reality.

The purpose of this form of activity is to enable suitable comparisons. The adoption of national audits, which produce a national average as well as a range for quality measures, such as epidural failure rates, allow for realistic achievable targets to be set for improvement activities (Wigfull and Welchew 1999).

Future of Data Collection

Considerable advances can be gained in postoperative patient care, in terms of research and evidence-based practice, on the basis of an individual unit's data collection activity. With pooling of data from multiple time periods and institutions, clinical decision making of the appropriate choice of drug or technique for an individual can be enhanced. This sets the stage for improved care. The POPDB of the future is likely to be organized so as to produce continuous performance indicators of the quality of care (Kuwahara et al. 2000).

However, a culture that encourages data collection and incorporation of technological advances in computing will be required before it can be established.

References

Allen, H. (2002). Using personal digital assistants for pain management. *Techniques in Regional Anesthesia and Pain Management, 6*, 158–164.

Elmasri, R., Mavathe, S. (2000). Chap 1: Databases and database users. In: *Fundamentals of database systems*, 3rd edn. (pp 3–21). Addison-Wesley.

Hammond, E. J., Veltman, M., Turner, G., et al. (2000). The development of a performance indicator to objectively monitor the quality of care provided by an acute pain team. *Anaesthesia and Intensive Care, 28*, 293–299.

Kuwahara, B., Klassen, K., Goresky, G., et al. (2000). The acute pain service at the Alberta Children's Hospital. *Clinical Pediatric Anesthesia, 6*, 105–115.

McLeod, G. (2002). The use of audit in acute pain services: A UK perspective. *Acute Pain, 4*, 57–64.

Sanders, M. K., & Michel, M. Z. M. (2002). Acute pain services – how effective are we? *Anaesthesia, 57*, 27.

Stomberg, M. W., Loerentzen, P., Joelsson, H., et al. (2003). Postoperative pain management on surgical wards – impact of database documentation of anesthesia organized services. *Pain Management Nursing, 4*, 155–164.

Tan, C. H., & Wong, W. S. H. A. (2003). Baseline audit of the efficacy and safety of an acute pain service (APS) for caesarean section patients. *Acute Pain, 4*, 99–104.

Turner, G., & Halliwell, R. (1999). Data collection by acute pain services in Australia and New Zealand. *Anaesthesia and Intensive Care, 27*, 632–635.

Wigfull, J., & Welchew, E. A. (1999). Acute pain service audit. *Anaesthesia, 54*, 299.

Postoperative Pain, Epidural Infusions

Edward A. Shipton
Department of Anaesthesia, Christchurch School of Medicine and Health Sciences, University of Otago, Christchurch, New Zealand

Synonyms

Extradural infusions

Definition

The epidural route of drug delivery has achieved widespread use for surgical and obstetrical analgesia (Shang and Gan 2003). Epidural analgesia is beneficial to patients in terms of improving pain control and reducing side effects (Taenzer and Clark 2010). Epidural blockade is usually performed in the thoracic and lumbar regions, or caudally to involve the sacral nerve distributions.

Epidural analgesia can be provided by a number of pharmacological agents administered into different levels of the epidural space for a wide variety of operations. Regardless of analgesic agent used, location of catheter, type of surgery, and type or time of pain assessment, the epidural route provides better pain relief than parenteral opioid administration (Guay 2006; Marret et al. 2007; Nishimori et al. 2006; Wu et al. 2005).

Introduction

Epidural analgesia is a well-established technique that has commonly been regarded as the gold standard in postoperative pain management. Epidural analgesia remains the gold standard for pain relief in labour because there are currently no good alternatives. However, newer, evidence-based outcome data show that the benefits of epidural analgesia are not as significant as previously believed (Rawal 2012).

Continuous Infusions

Epidural analgesia by the continuous administration of pharmacological agents into the epidural space (via an indwelling catheter) has become widely used in the management of acute pain after surgery or trauma in adults, children, and pregnant patients.

Continuous infusion remains the technique of choice when there is a prolonged operation or intense postoperative pain. Epidural analgesia provides superior pain relief and attenuates the stress response to surgery and pain, particularly when used in the form of continuous infusion both during and after surgery (Pyati and Gan 2007).

In children of all ages, continuous epidural infusions provide effective postoperative analgesia and are safe, provided experienced practitioners use appropriate doses and equipment, with adequate monitoring and management of complications (Macintyre 2010a).

Continuous epidural infusion of local anesthetics can reduce the stress response to surgery (Holte and Kehlet 2002) and allow rapid mobilization after major surgical procedures. A recent meta-analysis showed epidural analgesia with local anesthetics to result in a reduction in a range of adverse outcomes including the following: reduced rate of arrhythmias; earlier extubation; reduced intensive care unit stay; reduced stress hormone, cortisol, and glucose concentrations; and reduced incidence of renal failure, (Guay 2006). Although there are data suggesting that improved postoperative analgesia leads to better patient outcomes, there is insufficient evidence to support subsequent improvements in patient-centred outcomes such as quality of life and quality of recovery (Liu and Wu 2007). Epidural analgesia does not shorten the duration of hospital stay after colorectal surgery (Marret et al. 2007).

Characteristics

Local anesthetics and opioids are still the pharmacological agents more widely used epidurally (Congedo et al. 2009). Regarding the local anesthetics, the most recent literature focuses on the new enantiomers, ropivacaine and levobupivacaine, the efficacy of which is similar to that of bupivacaine with a reduced risk of cardiotoxicity (Congedo et al. 2009). Debate in the last few years has centered on the physicochemical properties of morphine and of the more recent lipophilic agents, fentanyl and sufentanil, in order to explain the main differences in efficacy and safety. The alpha (2)-adrenergic agonists, clonidine and dexmedetomidine as well as epinephrine (or adrenaline), in thoracic epidurals provide neuraxial analgesia via alpha-adrenergic receptors; these are being used increasingly as adjuvants to local anesthetics and

opioids. Ketamine, neostigmine, midazolam, and adenosine are the more recent studied drugs for epidural use; these are still under investigation and are not part of routine clinical practice (Congedo et al. 2009; Schug et al. 2006). Using adjuvants in the epidural space reduces the total dose of local anesthetic needed.

Most investigators compare infusions of bupivacaine or levobupivacaine at 0.1% or 0.125% with ropivacaine 0.2%, which removes any imbalance in comparative potency (Macintyre 2010b). Ropivacaine 0.2% and levobupivacaine 0.125% provided similar analgesia with similar adverse effects and no motor block when infused via thoracic epidural catheters for lung surgery (De Cosmo et al. 2008). Dose, not concentration or volume of local anesthetic infused, is relevant (Mendola et al. 2009; Dernedde et al. 2008).

In a comparison between ropivacaine 0.2%, bupivacaine 0.125%, and lignocaine 0.5%, the regression of sensory blockade under continuous infusion was least with ropivacaine (Kanai et al. 2007). Epidural patient-controlled analgesia with a mixture of low-concentration levobupivacaine (0.0625% or 0.1%) plus fentanyl, with basal infusion, has been extensively used for postoperative analgesia (Lin et al. 2010).

Recommended multimodal mixtures for epidural infusions consist of low concentrations of a local anesthetic (e.g., bupivacaine 1 mg/ml, ropivacaine 2 mg/ml, levobupivacaine 1 mg/ml), a lipophilic opioid (e.g., fentanyl 2 ug/ml, sufentanil 0.5 ug/ml), and an adrenergic agonist adrenaline (2 ug/ml) (Breivik 2002). Loading doses are usually given at the start of the infusion and the infusion rate titrated. Multimodal mixtures result in overall risks being lower for adverse effects. After major abdominal and thoracic surgery, pain during coughing is mild when the three components are used together (Breivik 2002).

Subanesthetic doses of local anesthetic inhibit excitatory synaptic mechanisms in the dorsal horn. The addition of opioids improves the quality of postoperative epidural analgesia, acting on the pre- and postsynaptic opioid receptors. Adrenaline acts on the pre- and postsynaptic alpha-2 receptors and has an analgesic effect of its own (dose-related), although it does not improve epidural analgesia at the lumbar level. It does not impair blood flow to the spinal cord and reduces systemic absorption and adverse effects of opioids (e.g., decreases pruritus) and of local anesthetics. The minimally effective concentration of epinephrine (adrenaline) to maintain relief of dynamic pain is approximately 1.5 ug/ml (Niemi and Breivik 2003).

Other additions to an epidural infusion include clonidine (10–20 ug/h in adults or 0.08–0.12 ug/kg/h in children), or a dose of the S+ enantiomer of ketamine (0.5 mg/kg) to reduce opioid consumption and improve analgesia (Cucchiaro et al. 2003; Ivani et al. 2003).

A meta-analysis comparing patient-controlled epidural analgesia (PCEA), continuous epidural infusions (CEI), and IV patient-controlled analgesia (PCA) opioids after surgery showed that both forms of epidural analgesia (with the exception of hydrophilic opioid-only epidural regimens) provided better pain relief with rest and with activity than PCA opioids (Wu et al. 2005). Conflicting results show that analgesia with a continuous epidural infusion to be superior to PCEA (Wu et al. 2005) or the reverse (Nightingale et al. 2007). Dose-only PCEA resulted in less total epidural dose used compared with the other modalities (Vallejo et al. 2007).

When compared to intravenous patient-controlled analgesia, epidural analgesia results in better analgesia and reduced opioid requirements (Standl et al. 2003). The magnitude of pain relief must be weighed against the frequency of adverse events. The incidence of serious complications is low (Ng and Goh 2002). Epidural analgesia using continuous infusion or patient-controlled administration of local anesthetic-opioid mixtures is safe on general hospital wards, if supervised by an anesthesia-based pain service with 24-h medical staff cover and is monitored by well-trained nursing staff (Macintyre 2010c). There is a high incidence of minor adverse effects especially during the first 48 h.

Local anesthetic may result in hypotension, motor blockade, and urinary retention if the catheter is in the low thoracic or thoracolumbar area (Breivik 2002). The hypotension depends on the dose of local anesthetic used and whether the patient is hypovolemic or not.

Epidural opioids may be associated with distressing side effects like nausea, vomiting, pruritus, urinary retention, gastrointestinal immotility, and respiratory depression (Ghai et al. 2009). Pruritus is more likely with epidural analgesia than with PCA opioids (Werawatganon and Charuluxanun 2005). Small doses or a continuous infusion of naloxone (0.25 ug/kg/h) or nalmefene (15–25 ug) are effective in reducing opioid-related adverse effects and paradoxically reduce postoperative opioid requirements as well (Shang and Gan 2003).

In children, continuous epidural infusion for postoperative pain is satisfactory. No major adverse effects occurred in children nursed on regular wards (Kampe et al. 2003). Epidural infusions of ropivacaine were effective and safe in neonates as well (Bösenberg et al. 2005).

Good knowledge of anatomic and physiologic differences between adults and children is necessary; new techniques such as nerve mapping and/or ultrasound can improve the success of blocks and reduce the risks (Ivani and Mossetti 2008).

Robust monitoring is, however, important during the early postoperative period as well as the involvement of the Acute Pain Service in the management of side effects, especially inadequate analgesia (e.g., catheter displacement, unilateral block, and tachyphylaxis). Complications related to catheter placement can occur, such as the potential for epidural hematoma with anticoagulation, technical issues with the infusion pumps, and the time and labor necessary to place, adjust, and manage catheters and infusions (Nagle and Gerancher 2007).

Severe neurological injury is rare. The incidence of transient neuropathy (radiculopathy) after epidural anesthesia was estimated to be 2.19:10,000. The risk of permanent neurological injury was less, and the incidences reported in the studies included in this review ranged from 0 to 7.6:10,000 (Brull et al. 2007). The rates of paraplegia and cauda equina syndrome associated with epidural anesthesia were estimated to be 0.09:10 000 and 0.23:10,000, respectively (Brull et al. 2007).

The Swedish case series quoted above puts the overall risk of epidural hematoma after epidural blockade at 1 in 10,300 (Moen et al. 2004). Leg weakness is a critical monitor of spinal cord health (Christie and McCabe 2007). Immediate diagnosis and decompression increases the likelihood of partial or good neurological recovery. The incidence of epidural abscess formation is rare. Methicillin-resistant Staphylococcus aureus is the predominant pathogen (Christie and McCabe 2007). Chlorhexidine-impregnated dressings of epidural catheters in comparison to placebo or povidone-iodine-impregnated dressings reduce the incidence of catheter colonization (Ho and Litton 2006).

The insertion of epidural needles or catheters should generally be avoided in patients with coagulation disorders or in those who are fully anticoagulated. Similarly, the use of low molecular weight heparin for deep vein thrombosis prophylaxis may be associated with a higher risk of epidural hematoma particularly in the higher dose ranges.

A thoracic epidural with local anesthetic is more beneficial for patients at high risk of cardiac or pulmonary complications after thoracic or abdominal surgery (Breivik 2002). It dilates stenotic coronary arteries, increases myocardial oxygen supply, decreases myocardial oxygen consumption, and decreases myocardial ischemic events and postoperative myocardial infarction (if extended for more than 24 h).

Beneficial effects on pulmonary outcome occur by promoting the postoperative recovery of diaphragmatic contractility that returns the functional residual capacity/closing capacity to normal. This has direct implications on the length of postoperative mechanical ventilation (de Leon-Casasola 2003). It improves lung function and oxygenation and decreases acute respiratory failure. Epidural analgesia protects against pneumonia following abdominal or thoracic surgery (Pöpping et al. 2008). The addition of enforced early mobilization may represent a further means of improving pulmonary outcome.

The addition of an epidural local anesthetic to an opioid shortens the time of intestinal paralysis after surgery. It improves gastrointestinal motility, decreases gastrointestinal complications, and improves bowel recovery after abdominal surgery; it is not associated with increased risk of anastomotic leakage after bowel surgery; it reduces renal insufficiency (Macintyre 2010c).

To derive the full protective cardiac effect from a perioperative epidural technique, the catheter must be inserted in the high thoracic region.

High thoracic epidural analgesia used for coronary artery bypass graft surgery reduces postoperative pain, risk of dysrhythmias, pulmonary complications, and time to extubation when compared with IV opioid analgesia (Macintyre 2010c).

Whereas lumbar epidural analgesia with local anesthetic dilates the arteries of the lower part of the body, it constricts the coronary arteries, decreases myocardial oxygen supply, causes leg weakness and urinary retention, and does not improve gastrointestinal function (Breivik 2002). Thoracic level neuraxial blockade has also been associated with decreased intraoperative blood loss and with fewer thromboembolic events (Chen et al. 2009).

The role of postoperative epidural analgesia on postoperative morbidity is controversial. Nevertheless, a substantial evidence base now exists to show that neuraxial blockade reduces postoperative morbidity and mortality in high-risk patients. Thoracic level neuraxial blockade has also been associated overall reduction of morbidity (Rodgers et al. 2000). The local anesthetic must be administered both during and after surgery in a continuous fashion for at least 72 h (de Leon-Casasola 2003) or 4–7 days after major thoracic or abdominal surgery (Breivik 2002).

Evidence is mounting that continuous epidural analgesia has a positive influence on protein and glucose metabolism, leading to preservation of muscle protein, a decrease in whole body protein catabolism, an improved nutritional substrate, and amelioration of insulin sensitivity (Carli et al. 2002). The combination of thoracic epidural analgesia with local anesthetics and nutritional support leads to preservation of total body protein after upper abdominal surgery (Lattermann et al. 2007).

Epidural analgesia in a multimodal analgesic regimen results in significantly less deterioration in postoperative functional status (lower pain and fatigue scores, earlier mobilization and return of gastrointestinal function) and health-related quality of life at 6-week follow-up (Wu and Raja 2002). Research into new delivery strategies, such as mandatory-programmed intermittent boluses and computerized feedback dosing, is ongoing (Halpern and Carvalho 2009). Although the efficacy of pain relief with epidural infusions can be outstanding, and there are some benefits in decreased cardiovascular and pulmonary morbidity in high-risk patients undergoing major open vascular or cardiac surgery, the use of epidural techniques is generally decreasing (Rawal 2012).

References

Bösenberg, A. T., Thomas, J., Cronje, L., et al. (2005). Pharmacokinetics and efficacy of ropivacaine for continuous epidural infusion in neonates and infants. *Paediatric Anaesthesia, 15*(9), 739–749.

Breivik, H. (2002). Postoperative pain: Toward optimal pharmacological and epidural analgesia. In M. A. Giamberardino (Ed.), *Pain 2002 - an updated review* (pp. 337–349). Seattle: IASP Press.

Brull, R., McCartney, C. J., Chan, V. W., et al. (2007). Neurological complications after regional anesthesia: contemporary estimates of risk. *Anesthesia and Analgesia, 104*(4), 965–974.

Carli, F., Mayo, N., Klubien, K., et al. (2002). Epidural analgesia enhances functional exercise capacity and health-related quality of life after colonic surgery: Results of a randomized trial. *Anesthesiology, 97*(3), 540–549.

Chen, L. M., Weinberg, V. K., Chen, C., et al. (2009). Perioperative outcomes comparing patient controlled epidural versus intravenous analgesia in gynecologic oncology surgery. *Gynecologic Oncology, 115*(3), 357–361.

Christie, I. W., & McCabe, S. (2007). Major complications of epidural analgesia after surgery: Results of a six-year survey. *Anaesthesia, 62*(4), 335–341.

Congedo, E., Sgreccia, M., & De Cosmo, G. (2009). New drugs for epidural analgesia. *Current Drug Targets, 10*(8), 696–706.

Cucchiaro, G., Dagher, C., Baujard, C., et al. (2003). Side effects of postoperative epidural analgesia in children: A randomized study comparing morphine and clonidine. *Paediatric Anaesthesia, 13*(4), 318–323.

De Cosmo, G., Congedo, E., Lai, C., et al. (2008). Ropivacaine vs. levobupivacaine combined with sufentanil for epidural analgesia after lung surgery. *European Journal of Anaesthesiology, 25*(12), 1020–1025.

de Leon-Casasola, O. A. (2003). When it comes to outcome, we need to define what a perioperative epidural technique is. *Anesthesia and Analgesia, 96*, 315–318.

Dernedde, M., Stadler, M., Taviaux, N., et al. (2008). Postoperative patient-controlled thoracic epidural analgesia: Importance of dose compared to volume or concentration. *Anaesthesia and Intensive Care, 36*(6), 814–821.

Ghai, B., Makkar, J. K., Chari, P., & Rao, K. L. (2009). Addition of midazolam to continuous postoperative epidural bupivacaine infusion reduces requirement for rescue analgesia in children undergoing upper abdominal and flank surgery. *Journal of Clinical Anesthesia, 21*(2), 113–119.

Guay, J. (2006). The benefits of adding epidural analgesia to general anesthesia: A meta analysis. *Journal of Anesthesia, 20*(4), 335–340.

Halpern, S. H., & Carvalho, B. (2009). Patient-controlled epidural analgesia for labor. *Anesthesia and Analgesia, 108*(3), 921–928.

Ho, K. M., & Litton, E. (2006). Use of chlorhexidine-impregnated dressing to prevent vascular and epidural catheter colonization and infection: A meta-analysis. *Journal of Antimicrobial Chemotherapy, 58*(2), 281–287.

Holte, K., & Kehlet, H. (2002). Effect of postoperative epidural analgesia on surgical outcome. *Minerva Anestesiologica, 68*(4), 157–161.

Ivani, G., & Mossetti, V. (2008). Regional anesthesia for postoperative pain control in children: Focus on continuous central and perineural infusions. *Paediatric Drugs, 10*(2), 107–114.

Ivani, G., Vercellino, C., & Tonetti, F. (2003). Ketamine: A new look to an old drug. *Minerva Anestesiologica, 69*(5), 468–471.

Kampe, S., Kiencke, P., Delis, A., et al. (2003a). The continuous epidural infusion of ropivacaine 0.1% with 0.5 microg x mL(−1) sufentanil provides effective postoperative analgesia after total hip replacement: A pilot study. *Canadian Journal of Anaesthesia, 50*(6), 580–585.

Kanai, A., Osawa, S., Suzuki, A., et al. (2007). Regression of sensory and motor blockade, and analgesia during continuous epidural infusion of ropivacaine and fentanyl in comparison with other local anesthetics. *Pain Medicine, 8*(7), 546–553.

Lattermann, R., Wykes, L., Eberhart, L., et al. (2007). A randomized controlled trial of the anticatabolic effect of epidural analgesia and hypocaloric glucose. *Regional Anesthesia and Pain Medicine, 32*(3), 227–232.

Liu, S. S., & Wu, C. L. (2007). The effect of analgesic technique on postoperative patient-reported outcomes including analgesia: A systematic review. *Anesthesia and Analgesia, 105*(3), 789–808.

Macintyre, P. E. (Ed.). (2010a). The paediatric patient. In *Acute pain management: Scientific evidence* (3rd ed., pp 339–384). Melbourne: Australian and New Zealand College of Anaesthetists.

Macintyre, P. E. (Ed.). (2010b). PCA, Regional and other local analgesia techniques. In *Acute pain management: Scientific evidence* (3rd ed., pp 175–224). Melbourne: Australian and New Zealand College of Anaesthetists.

Macintyre, P. E. (Ed.). (2010c). Regionally and locally administered analgesic drugs. In *Acute pain management: Scientific evidence* (3rd ed., pp 129–151). Melbourne: Australian and New Zealand College of Anaesthetists.

Marret, E., Remy, C., & Bonnet, F. (2007). Postoperative pain forum group. Meta-analysis of epidural analgesia versus parenteral opioid analgesia after colorectal surgery. *British Journal of Surgery, 94*(6), 665–673.

Mendola, C., Ferrante, D., Oldani, E., et al. (2009). Thoracic epidural analgesia in post-thoracotomy patients: Comparison of three different concentrations of levobupivacaine and sufentanil. *British Journal of Anaesthesia, 102*(3), 418–423.

Moen, V., Dahlgren, N., & Irestedt, L. (2004). Severe neurological complications after central neuraxial blockades in Sweden 1990–1999. *Anesthesiology, 101*(4), 950–959.

Nagle, P. C., & Gerancher, J. C. (2007). DepoDur® (extended-release epidural morphine): A review of an old drug in a new vehicle. *Techniques in Regional Anesthesia and Pain Management, 11*(1), 9–18.

Ng, J. M., & Goh, M. H. (2002). Problems related to epidural analgesia for postoperative pain control. *Annals of the Academy of Medicine, Singapore, 31*(4), 509–515.

Niemi, G., & Breivik, H. (2003). The minimally effective concentration of adrenaline in a low-concentration thoracic epidural analgesic infusion of bupivacaine, fentanyl and adrenaline after major surgery. A randomized, double-blind, dose-finding study. *Acta Anaesthesiologica Scandinavica, 47*(4), 439–450.

Nightingale, J. J., Knight, M. V., Higgins, B., et al. (2007). Randomized, double-blind comparison of patient controlled epidural infusion vs nurse-administered epidural infusion for postoperative analgesia in patients undergoing colonic resection. *British Journal of Anaesthesia, 98*(3), 380–384.

Nishimori, M., Ballantyne, J. C., & Low, J. H. (2006). Epidural pain relief versus systemic opioid-based pain relief or abdominal aortic surgery. *Cochrane Database of Systematic Reviews, 3*, CD005059.

Pöpping, D. M., Elia, N., Marret, E., Remy, C., & Tramèr, M. R. (2008). Protective effects of epidural analgesia on pulmonary complications after abdominal and thoracic surgery: A meta-analysis. *Archives of Surgery, 143*(10), 990–999.

Pyati, S., & Gan, T. J. (2007). Perioperative pain management. *CNS Drugs, 21*(3), 185–211.

Rawal, N. (2012). Epidural technique for postoperative pain: gold standard no more? *Regional Anesthesia and Pain Medicine, 37*(3), 310–317.

Rodgers, A., Walker, N., Schug, S., et al. (2000). Reduction of postoperative mortality and morbidity with epidural or spinal anaesthesia: Results from overview of randomised trials. *BMJ, 321*(7275), 1493.

Schug, S. A., Saunders, D., Kurowski, I., & Paech, M. J. (2006). Neuraxial drug administration: A review of treatment options for anaesthesia and analgesia. *CNS Drugs, 20*(11), 917–933.

Shang, A. B., & Gan, T. J. (2003). Optimising postoperative pain management in the ambulatory patient. *Drugs, 63*(9), 855–867.

Standl, T., Burmeister, M. A., Ohnesorge, H., Wilhelm, S., Striepke, M., Gottschalk, A., Horn, E. P., & Schulte Am Esch, J. (2003). Patient-controlled epidural analgesia reduces analgesic requirements compared to continuous epidural infusion after major abdominal surgery. *Canadian Journal of Anaesthesia, 50*(3), 258–264.

Taenzer, A. H., & Clark, C. (2010). Efficacy of postoperative epidural analgesia in adolescent scoliosis surgery: A meta-analysis. *Paediatric Anaesthesia, 20*(2), 135–143.

Vallejo, M. C., Ramesh, V., Phelps, A. L., et al. (2007). Epidural labor analgesia: Continuous infusion versus patient controlled epidural analgesia with background infusion versus without a background infusion. *The Journal of Pain, 8*(12), 970–975.

Werawatganon, T., & Charuluxanun, S. (2005). Patient controlled intravenous opioid analgesia versus continuous epidural analgesia for pain after intra-abdominal surgery. *Cochrane Database of Systematic Reviews, 25*(1), CD004088.

Wu, C. L., & Raja, S. N. (2002). Optimising postoperative analgesia. *Anesthesiology, 97*(3), 533–534.

Wu, C. L., Cohen, S. R., Richman, J. M., et al. (2005). Efficacy of postoperative patient-controlled and continuous infusion epidural analgesia versus intravenous patient-controlled analgesia with opioids: A meta-analysis. *Anesthesiology, 103*(5), 1079–1088.

Postoperative Pain, Fentanyl

Peter C. A. Kam
Anaesthesia, Central Clinical School,
The University of Sydney, Sydney, Australia

Synonyms

Actiq® oral transmucosal fentanyl citrate; Chemical name: N-Phenyl-N-(1-(2-phenylethyl)-4-piperidinyl-procainamide; Durogesic® – transdermal patch; Fentanyl; Sublimaze®

Definition

Fentanyl is a potent and rapid-onset synthetic opioid that is a phenylpiperidine derivative.

Characteristics

Fentanyl is a derivative of 4-anilinoperidine and is structurally related to pethidine. It is commercially formulated as a citrate, a white crystalline solid that is readily soluble in acidic aqueous solutions. It has a pK_a of 8.4, and each ml of aqueous solution contains 0.05 mg fentanyl base (0.0785 mg of the citrate). It is a highly lipophilic drug and produces analgesia, respiratory depression, vagal effects, emesis, constipation, and physical dependence.

Pharmacology

Fentanyl is an opioid analgesic that binds predominantly to mu (μ) receptors. Its primary therapeutic actions are analgesia and sedation. Mood alterations, euphoria, and dysphoria are common. Fentanyl also depresses the respiration and cough and may cause nausea and vomiting directly by stimulating the chemoreceptor trigger zone. A dose of fentanyl 100 μg is equivalent in analgesic activity to morphine 10 mg. Fentanyl appears to cause less vomiting than either morphine or pethidine.

Pharmacodynamics

Analgesia

Minimum effective analgesia serum concentrations of fentanyl in opioid naïve patients range from 0.3 to 1.5 ng/ml, and adverse effects increase in frequency at serum concentrations above 2 ng/ml. Loss of consciousness occurs at mean serum concentrations of approximately 34 ng/ml.

Clinically significant histamine release rarely occurs with fentanyl. Fentanyl preserves cardiac stability and attenuates stress-related endocrine changes at higher doses. Decreased sensitivity to CO_2 stimulation follows fentanyl administrations, and the duration and degree of respiratory depression is dose related. The peak respiratory depressant effect of a single intravenous fentanyl dose occurs 5–15 min following injection.

Pharmacokinetics

The pharmacokinetics of fentanyl is described by a three-compartment model, with a distribution half-life of 1.7 min, a redistribution half-life of 13 min, and an elimination half-life of 219 min (McLain and Hug 1980; Duthie et al. 1986). The volume of distribution of fentanyl is 4 l/kg and clearance is 57 l/h.

The onset of action is almost immediate, but maximal analgesic effects occur after several minutes. The average duration of action following a single intravenous dose of 100 μg is 30–60 min.

At physiological pH, 80 % fentanyl is protein bound, up to a concentration of 500 μg/ml. Fentanyl accumulates in skeletal muscle and fat and is slowly released into the blood. Fentanyl is metabolized in the liver by cytochrome CYP3A4 isoenzyme. It has a high first-pass clearance (N-dealkylation), and about 75 % of an intravenous dose is excreted in the urine primarily as metabolites, norfentanyl, and OH-prop-norfentanyl and less than 10 % as unchanged drug.

After a single bolus dose of fentanyl (2 μg/kg), the duration of effective analgesia is relatively short because of drug redistribution. With constant steady-state infusion of fentanyl, the fentanyl has a long steady half-time, which is the result of the large conductance ratio associated with the slow compartment of the steady-state model. In a simulated pharmacokinetics mode of steady-state fentanyl infusion, the context-sensitive half-time of fentanyl was approximately 20 min after 1 h infusion, and this increased to approximately 180 min after 4 h infusion and 280 min after 8 h infusion (Hughes et al. 1992). Intravenous fentanyl has been used for patient-controlled analgesia with a bolus dose of 1–2 μg/kg and lockout interval of 5 min (Nikkola et al. 1997).

Epidural and intrathecal administration of fentanyl, in combination with local anesthetics, is widely used for postoperative analgesia and obstetric analgesia. Most studies indicate that combination therapy reduces dose requirements of local anesthetics when they are administered as single drugs. This dose reduction is associated with reduced local anesthetic-related side effects (hypotension and motor block) and little (sedation) or no (vomiting or pruritus) reduction in opioid-related adverse effects (Walker et al. 2002). Epidural fentanyl (25–50 µg) produces a consistent and intense analgesia, but the duration is short, although this can be overcome by administering it as a continuous infusion. It is common practice to combine low concentrations of fentanyl (2 µg/ml) with a local anesthetic solution. Addition of fentanyl to spinal anesthetics produces synergistic analgesia for somatic and visceral pain. Doses added to spinal anesthetic range from 10 to 20 µg. There is very little risk of early respiratory depression and urinary retention, although the incidence of pruritus is increased.

Transmucosal Fentanyl Formulation
It is also available as a lollipop preparation (transmucosal preparation), which has been successfully used in children (Feld et al. 1989). The use of oral ▶ transmucosal fentanyl citrate in the management of acute and postoperative pain has not been adequately evaluated, and it is currently not recommended for analgesia in this setting. It has been used for breakthrough analgesia in opioid-tolerant patients with cancer. A formulation of transmucosal fentanyl system (▶ Actiq®) produced more significant pain relief at 15, 30, 45, and 60 min following administration (over a recommended 15 min) in opioid-tolerant cancer patients. It is contraindicated in acute and postoperative patients, because life-threatening respiratory depression can occur at any dose in patients not taking chronic opiates (Payne et al. 2001).

Transdermal Fentanyl
Transdermal fentanyl (▶ Durogesic®) is a transdermal patch system that continuously delivers fentanyl for 72 h (Jeal and Benfield 1997). It is a transparent unit that comprises of a protective liner and 4 functional layers. These layers include a backing of clear polyester film; a drug reservoir of fentanyl and alcohol gelled with

hydroxyethyl cellulose; an ethylene-vinylacetate copolymer membrane that controls the rate of delivery of fentanyl; and a silicone-adhesive layer. The approximate analgesic potency ratio of transdermally administered fentanyl to parenteral morphine ranges from 1:20 to 1:30 in opioid-naïve patients in acute pain.

Release of fentanyl from the patch is determined by the copolymer release membrane and the diffusion of fentanyl delivered to the systemic circulation across average skin (25, 50, 75, or 100 µg/h fentanyl). The small amount of alcohol in the patch enhances dry form through the copolymer release membrane and increases the permeability of the skin for diffusion of fentanyl. After initial application of a fentanyl patch, serum fentanyl concentrations increase gradually and peak serum fentanyl concentrations occur between 24 and 72 h. After repeated 72 h applications, serum concentrations reach a steady state that is maintained during subsequent applications of similar dose patches. Since elderly cachectic or debilitated patients have altered pharmacokinetics due to poor fat stores, muscle wasting, and altered clearance, they should not be started on doses greater than 25 µg/l, unless they have previously been taking another opioid equivalent or greater than 135 µg oral morphine per day. The terminal half-life after removal of the patch is 13–25 h. These properties make it unsuitable for acute pain (8nyl can be). Clinically significant doses of fentanyl can be administered transdermally by iontophoresis (currents 1–2 mA) for delivery periods of 2 h, but more research is required.

Adverse Reactions
The major adverse reactions associated with fentanyl are respiratory depression, apnea, muscle rigidity, and bradycardia. Respiratory depression is more common if an intravenous dose is given rapidly and can be immediately reversed by naloxone. Muscle rigidity can be associated with reduced thoracic or chest wall compliance and/or apnea. Bradycardia is caused by stimulation of the nucleus ambiguus and can be controlled by atropine. Itching and spasm of the sphincter of Oddi are other adverse effects.

Uses

Fentanyl can be used as an analgesic (2–10 μg/kg) or anesthetic (20–100 μg/kg) and can be administered intravenously, intramuscularly, epidurally, intrathecally, via mucous membranes, or transdermally. Transdermal fentanyl is extremely useful for chronic cancer pain when the oral route cannot be used.

References

Duthie, D. J., McLaren, A. D., & Nimmo, W. S. (1986). Pharmacokinetics of fentanyl during constant rate IV infusion for the relief of pain after surgery. *British Journal of Anaesthesia, 58*, 950–956.

Feld, L. H., Champeau, M. W., Van Steenis, C. A., et al. (1989). Preanaesthetic medication in children: A comparison of oral transmucosal fentanyl citrate versus placebo. *Anesthesiology, 71*, 374–377.

Hughes, M. A., Glass, P. S. A., & Jacobs, J. R. (1992). Context-sensitive half-time in multi-compartment pharmacokinetic models for intravenous anaesthetic drugs. *Anesthesiology, 76*, 334–341.

Jeal, W., & Benfield, P. (1997). Transdermal fentanyl: A review of its pharmacological properties and therapeutic efficacy in pain control. *Drugs, 53*, 109–138.

McLain, D. A., & Hug, C. C. (1980). Intravenous fentanyl pharmacokinetics. *Clinical Pharmacology and Therapeutics, 28*, 106–114.

Nikkola, E. M., Ekblad, U. U., Kero, P. O., et al. (1997). Intravenous fentanyl PCA during labour. *Canadian Journal of Anaesthesia, 44*, 1248–1255.

Payne, R., Coluzzi, P., Hart, L., et al. (2001). Long-term safety of oral transmucosal fentanyl citrate for breakthrough cancer pain. *Journal of Pain and Symptom Management, 22*, 575–583.

Walker, S. M., Goudas, L. C., Cousins, M. J., et al. (2002). Combination spinal analgesic chemotherapy: A systematic review. *Anesthesia and Analgesia, 95*, 674–715.

Postoperative Pain, Gabapentin

Ross MacPherson
Department of Anesthesia and Pain Management, Royal North Shore Hospital University of Sydney, Sydney, NSW, Australia

Synonyms

Aminomethyl-cyclohexane-acetic acid; Antiepileptic; Gabapentin

Characteristics

Gabapentin, a structural analogue of GABA, was initially synthesized in the late 1970s and developed during the following decade as an anticonvulsant (Gilron 2002). Since then, it has been in trials for a wide variety of uses including psychiatric illness (Malek-Ahmadi 2003), postmenopausal flushes (Guttuso et al. 2003a), chemotherapy-induced nausea (Guttuso et al. 2003b), and a number of different pain states. This entry will concentrate only on its use in pain medicine. Despite the structural similarity between these two moieties, it has been constantly shown that gabapentin has no direct GABAergic action, nor does it affect GABA uptake or metabolism (Backonja 2002). Its action may be through an effect on glutamate function, or alternatively, through alpha 2-delta calcium channel binding (Taylor et al. 1998).

Two large trials have confirmed the effectiveness of gabapentin in the treatment of diabetic neuropathy (Backonja et al. 1998), neuropathic pain states (Serpell and Neuropathic Pain Study Group 2002), and postherpetic neuralgia (Rowbotham et al. 1998). The drug appeared to be well tolerated, although transient dizziness and somnolence were reported through the various studies. Dosing with gabapentin needs to be increased gradually and titrated to effect. Backonja and Glanzman (2003), in reviewing a number of studies, have suggested that dosage should be gradually stepped up to 900 mg/day and thereafter increased up to 1,800 mg/day. This should be sufficient for most patients, although some may need further titration to 3,600 mg/day.

In a recent interesting study, Dirks et al. (2002) gave patients undergoing mastectomy a single oral dose of gabapentin (1,200 mg) orally 1 h prior to surgery. They found that this simple strategy resulted in a substantial reduction in postoperative morphine consumption and in movement related pain, confirming the results of previous work (Eckhardt et al. 2003).

Lamotrigine, another of the newer antiepileptic agents, is a phenyltriazine derivative. While it has been in trials in the treatment of a number of pain states including HIV-associated neuropathy (Simpson et al. 2000), trigeminal neuralgia

(Zakrzewska et al. 1997), spinal cord injury pain (Finnerup et al. 2002), and neuropathic pain states (Sandner-Kiesling et al. 2002; McCleane 1999), results have been conflicting. In at least one of these studies, there was a range of significantly adverse effects reported, including ataxia, constipation, somnolence, and diplopia. However, this has not been a consistent finding (Sandner-Kiesling et al. 2002).

Anticonvulsants continue to be used for a wide range of neuropathic and chronic pain states, despite the fact that in many cases, supporting evidence has been lacking. While controlled trials have been conducted into diabetic neuropathy and trigeminal neuralgia, large-scale studies of the effectiveness of anticonvulsants in the management of other pain states are lacking (Backonja 2002). However, a number of small trials examining the role of gabapentin in conditions such as pain associated with spinal cord injury, phantom limb pain (Bone et al. 2002), and central pain syndromes (Putzke et al. 2002; To et al. 2002; Tai et al. 2002) have been performed with encouraging results.

On this background, there are, however some critics of gabapentin in neuropathic pain states, some suggesting that many agents may improve pain simply by an indirect effect on mood and sleep (Bashford 1999).

In acute pain states, there have been a number of trails that have reported gabapentin administration resulted in improved postoperative pain coupled with reduced opioid consumption (McIntyre et al. 2010).

With regard to dosage, the initial dose, which can be as low as 100 mg t.d.s., should be titrated upward daily, or second daily, as circumstances permit. The upper limits of gabapentin use differ between centers, with some advocating a maximum dose of 3.6 g/day. An upper dose of 1.8 g/day should be tolerated by most patients.

References

Backonja, M. M. (2002). Use of anticonvulsants for the treatment of neuropathic pain. *Neurology, 59*, S14–S17.
Backonja, M., & Glanzman, R. L. (2003). Gabapentin dosing for neuropathic pain: Evidence from randomized, placebo-controlled clinical trials. *Clinical Therapeutics, 25*, 81–104.
Backonja, M., Beydoun, A., Edwards, K. R., et al. (1998). Gabapentin for the symptomatic treatment of painful neuropathy in patients with diabetes mellitus: A randomized controlled trial. *JAMA: The Journal of the American Medical Association, 280*, 1831–1836.
Bashford, G. M. (1999). The use of anti-convulsants for neuropathic pain. *Australian Prescriber, 22*, 140–141.
Bone, M., Critchley, P., & Buggy, D. J. (2002). Gabapentin in postamputation phantom limb pain: A randomized, double-blind, placebo-controlled, cross-over study. *Regional Anesthesia and Pain Medicine, 27*, 481–486.
Dirks, J., Fredensbor, B., & Domspaard, J. (2002). A randomized study of the effects of single-dose gabapentin versus placebo on postoperative pain and morphine consumption after mastectomy. *Anesthesiology, 97*, 560–564.
Eckhardt, K., Ammon, S., Hofmann, U., et al. (2003). Gabapentin enhances the analgesic effect of morphine in healthy volunteers. *Anesthesia and Analgesia, 91*, 185–191.
Finnerup, N. B., Sindrup, S. H., Bach, F. W., et al. (2002). Lamotrigine in spinal cord injury pain: A randomized controlled trial. *Pain, 96*, 375–383.
Gilron, I. M. (2002). Is gabapentin a "Broad-spectrum" analgesic? *Anesthesiology, 97*, 537–539.
Guttuso, T., Jr., Roscoe, J., & Griggs, J. (2003a). Effect of gabapentin on nausea induced chemotherapy in patients with breast cancer. *Lancet, 361*, 1703–1705.
Guttuso, T., Jr., Kurlan, R., McDermott, M. P., et al. (2003b). Gabapentin's effects on hot flashes in postmenopausal women: A randomized controlled trial. *Obstetrics and Gynecology, 101*, 337–345.
Malek-Ahmadi, P. (2003). Gabapentin and posttraumatic stress disorder. *The Annals of Pharmacotherapy, 37*, 664–666.
McCleane, G. (1999). 200 mg daily of lamotrigine has no analgesic effect in neuropathic pain: A randomized, double-blind placebo controlled trial. *Pain, 83*, 105–107.
McIntyre, P. E., Schug, S. A., Scott, D. A., Visser, E. J., Walker, S. M., & Working group of the Australian and New Zealand College of Anaesthetists and FAculty of Pain Medicine. (2010). *Acute pain management: Scientific guidelines* (3rd ed.). Melbourne: ANZCA & FPM.
Putzke, J. D., Richards, J. S., Kezar, L., et al. (2002). Long-term use of gabapentin for treatment of pain after traumatic spinal cord injury. *The Clinical Journal of Pain, 18*, 116–121.
Rowbotham, M., Harden, N., Stacey, B., et al. (1998). Gabapentin for the treatment of postherpetic neuralgia: A randomized controlled trial. *JAMA: The Journal of the American Medical Association, 280*, 1837–1842.
Sandner-Kiesling, A., Rumpold Seitliner, G., Dorn, C., et al. (2002). Lamotrigine monotherapy for control of

neuralgia after nerve section. *Acta Anaesthesiologica Scandinavica, 46*, 1261–1264.

Serpell, M. G., & Neuropathic Pain Study Group. (2002). Gabapentin in neuropathic pain syndromes: A randomised, double-blind, placebo-controlled trial. *Pain, 99*, 557–566.

Simpson, D. M., Onley, R., McArthur, J. C., et al. (2000). A placebo-controlled trial of lamotrigine for painful HIV-associated neuropathy. *Neurology, 54*, 2115–2119.

Tai, Q., Kirshblum, S., Chen, B., et al. (2002). Gabapentin in the treatment of neuropathic pain after spinal cord injury: A prospective, randomized, double-blind, crossover trial. *The Journal of Spinal Cord Medicine, 25*, 100–105.

Taylor, C. P., Gee, N. S., Su, T. Z., et al. (1998). A summary of mechanistic hypotheses of gabapentin pharmacology. *Epilepsy Research, 29*, 233–249.

To, T. P., Lim, T. C., Hill, S. T., et al. (2002). Gabapentin for neuropathic pain following spinal cord injury. *Spinal Cord, 40*, 282–285.

Zakrzewska, J., Chardhry, Z., Nurmikko, T. J., et al. (1997). Lamotrigine (lamictal) in refractory trigeminal neuralgia: Results from a double-blind placebo-controlled crossover trial. *Pain, 73*, 223–230.

Postoperative Pain, Hydromorphone

Ross MacPherson
Department of Anesthesia and Pain Management, Royal North Shore Hospital University of Sydney, Sydney, NSW, Australia

Synonyms

Dihydromorphinone; Dilaudid; Hydromorphone

Definition

A hydrogenated ketone derivative of morphine.

Characteristics

Hydromorphone has had something of a dual life. First synthesized in the early part of last century, it was introduced into clinical practice by Krehl in 1926, and gained immediate popularity as a potent analgesic and an alternative to morphine.

Interest in hydromorphone continued for the next 50 years (Hanna et al. 1962) after which its popularity started to wane.

It remained more of a second-line agent, often prescribed for the treatment of chronic nonproductive cough, until the last 10 years or so, during which time the drug has undergone a type of renaissance. There are a number of reasons for this.

The development of Pain Medicine as a speciality probably initiated a reexamination of many older drugs, to try to evaluate whether they had a place in modern therapy. This was made timelier by the discovery of the concept of ▶ opioid rotation – where the substitution of one opioid for another, especially in patients with chronic pain states, often resulted in a relative reduction in both dosage and side effects (Mercadante 1999).

On a milligram-for-milligram basis, hydromorphone is clearly more potent than morphine, although there is some variation. Most studies have found the hydromorphone:morphine equianalgesic dose ratio to be in the range of 5:1 to 7:1, although some studies have found the ratio to be in the order of 10:1 (Jain et al. 1989). Hydromorphone is then significantly more potent than morphine and caution needs to be exercised in its prescribing, especially in opioid-naive patients, where it is most likely not the drug of first choice in any case.

Quigley and Wiffen (2003) have recently reviewed the studies looking at hydromorphone in both acute and chronic pain states. They make the point that only 43 studies had actually been conducted between 1966 and 2000, and that many of these were poorly controlled and had very small patient numbers. Their overall assessment of the data available was that there was little difference between hydromorphone and other opioids with regard to analgesic efficacy and side-effect profile. Some studies have suggested that troublesome pruritus, so often a complication of morphine use, may be less with hydromorphone (Katcher and Walsh 1999; Chaplan et al. 1992; Goodarzi 1999).

Despite this, hydromorphone does have some interesting pharmacological features that set it apart to some degree from other agents.

Its potency and favorable solubility – it is about 8 times more soluble in water than morphine and 15 times more soluble than fentanyl – have made it a popular agent for use in implanted intrathecal or epidural pumps, and many studies have concentrated on this aspect of its use (Chaplan et al. 1992; Drakeford et al. 1991; Gaeta et al. 1995).

The drug is well absorbed orally, however, the substantial first-pass effect, as seen with all opioids reduces its oral bioavailability to about 35–60 % and a short half life which, when given orally, usually necessitates multiple daily dosing. However, a new sustained release dosage form, incorporating an osmotic pump system, provides the promise of sustained analgesia following a single oral dose (Angst et al. 2001).

Metabolism of hydromorphone does not result in the production of any metabolites with analgesic efficacy.

However, a number of metabolites are produced, at least one of which has been implicated in the precipitation of adverse effects.

The neuroexcitatory effects of opioids have been well described (Bruera and Neumann 1999). Although any member of the opioid group can precipitate a neurotoxic response, morphine and hydromorphone have been particularly implicated. These responses are thought to be due, at least in part, to accumulation of excitatory metabolites – in this case morphine-3-glucuronide or hydromorphone-3-glucuronide, respectively (Wright et al. 2001; Armstrong and Cozza 2003). Despite having a more polar structure than the parent molecule, they can cross the blood-brain barrier and exert significant effects (Smith 2000). Both agents rely on renal excretion and can accumulate in patients with impaired renal function; such effects should be watched for (McIntyre et al. 2010).

Lastly, some researchers are looking at novel routes of hydromorphone administration. At least one series has been conducted looking at the absorption of intranasal hydromorphone with encouraging results (Coda et al. 2003). In this study, following the administration of a single intranasal dose of either 1 or 2 mg of hydromorphone, the authors found that absorption was rapid, and overall bioavailability was in the order of 50 %. The drug has also been administered by inhalation (Sarhill et al. 2000).

A long-acting product has recently become available, given on a once-daily basis. Although clearly this will be of more use in the management of chronic pain, a number of sources have pointed out the high potency of hydromorphone products and the consequent risk of toxicity in inadvertent overdose.

In summary, hydromorphone has established itself as a useful addition to the analgesic repertoire. It is most useful in opioid rotation, in patient-controlled analgesia devices, and in intrathecal use.

References

Angst, M., Drover, D., Lotsch, J., et al. (2001). Pharmacodynamics of orally administered sustained-release hydromorphone in humans. *Anesthesiology, 94*, 63–73.

Armstrong, S. C., & Cozza, K. L. (2003). Pharmacokinetic drug interactions of morphine, codeine, and their derivatives: Theory and clinical reality, part I. *Psychosomatics, 44*, 167–171.

Bruera, E., & Neumann, C. M. (1999). Cancer pain. In M. Devor (Ed.), *Pain – an updated review. Refresher course syllabus 9th world congress on pain* (pp. 25–35). Vienna.

Chaplan, S., et al. (1992). Morphine and hydromorphone epidural analgesia. *Anesthesiology, 77*, 1090–1094.

Coda, B. A., Rudy, A. C., Archer, S. M., et al. (2003). Pharmacokinetics and bioavailability of single-dose intranasal hydromorphone hydrochloride in healthy volunteers. *Anesthesia and Analgesia, 97*, 117–123.

Drakeford, M. K., Pettine, K. A., Brookshire, L., et al. (1991). Spinal narcotics for postoperative analgesia in total joint arthroplasty. *Journal of Bone & Joint Surgery, 73*, 424–428.

Gaeta, R. R., Marcario, A., Brodsky, J. B., et al. (1995). Pain outcomes after thoracotomy: Lumbar epidural hydromorphone versus intrapleural bupivacaine. *Journal of Cardiothoracic and Vascular Anesthesia, 9*, 534–537.

Goodarzi, M. (1999). Comparison of epidural morphine, hydromorphone and fentanyl for postoperative pain control in children undergoing orthopaedic surgery. *Paediatric Anaesthesia, 9*, 419–422.

Hanna, C., Mazuzan, J., & Abajian, J. (1962). An evaluation of dihydromorphinone in treating postoperative pain. *Anesthesia and Analgesia, 11*, 39–44.

Jain, A. K., McMahon, F. G., Reder, R., et al. (1989). A placebo-controlled study of an oral solution of 5 and 10 mg of hydromorphone hydrochloride in post-operative pain. *Clinical Pharmacology and Therapeutics, 45*, 175.

Katcher, J., & Walsh, D. (1999). Opioid-induced itching: Morphine sulfate and hydromorphone hydrochloride. *Journal of Pain and Symptom Management, 17*, 70–72.

Krehl, L. (1926). Aerztliche Erfahrungen mit Dilaudid (medical experiences with Dilaudid). *Munchener Medizinische Wochenschrift, 73*, 596.

McIntyre, P. E., Schug, S. A., Scott, D. A., Visser, E. J., Walker, S. M., & Working Group of the Australian and New Zealand College of Anaesthetists and Faculty of Pain Medicine. (2010). *Acute pain management: Scientific guidelines* (3rd ed.). Melbourne: ANZCA & FPM.

Mercadante, S. (1999). Opioid rotation for cancer pain: Rationale and clinical aspects. *Cancer, 86*, 1856–1866.

Quigley, C., & Wiffen, P. (2003). A systematic review of hydromorphone in acute and chronic pain. *Journal of Pain and Symptom Management, 25*, 169–178.

Sarhill, N., Walsh, D., Khawam, E., et al. (2000). Nebulized hydromorphone for dyspnea in hospice care of advanced cancer. *The American Journal of Hospice & Palliative Care, 17*, 389–391.

Smith, M. T. (2000). Neuroexcitatory effects of morphine and hydromorphone: Evidence implicating the 3-glucuronide metabolites. *Clinical and Experimental Pharmacology and Physiology, 27*, 524–528.

Wright, A. W., Mather, L. E., & Smith, M. T. (2001). Hydromorphone-3-glucuronide: A more potent neuro-excitant than its structural analogue, morphine-3-glucuronide. *Life Sciences, 69*, 409–420.

Postoperative Pain, Importance of Mobilization

Peter C. A. Kam
Anaesthesia, Central Clinical School, The
University of Sydney, Sydney, Australia

Synonyms

Accelerated recovery programs; Enforced mobilization; Fast-track surgery; Mobilization following surgery

Definition

There is no formal definition of early mobilization following surgery. However, it may be defined as early resumption of "normal" physical activity of the patient following a surgical procedure. The term "normal physical activity" usually refers to normal "activities of daily living," such as getting out of bed, walking (initially with support), sitting, and personal toilet care (Harwood and Ebrahim 2002).

Characteristics

Traditionally, postoperative hospitalization of patients is undertaken so that they can be observed and treated for surgical or anesthetic complications that may arise and that they recover to an adequate level of self-care before discharge. Traditional postoperative care includes bed rest, which is undesirable because it predisposes to weakness and loss of muscle tissue, thromboembolic and pulmonary complications, and orthostatic circulatory disturbances (Harper and Lyles 1988).

There is no scientific basis for restrictions of ▶ activities of daily living in the postoperative period (Kehlet and Wilmore 2002). All efforts should be undertaken to enforce postoperative mobilization, which is possible with effective pain relief, education, and nutritional support. Therefore, new approaches are directed towards evaluating new perioperative therapies to reduce postoperative morbidity, shorten convalescence, decrease the length of hospital stay and resources, and reduce economic hospital costs.

The methods used to facilitate early postoperative mobilization include potential improvements in the preoperative preparation of patient, intraoperative strategies, and revising postoperative care regimens.

Preoperative Period

Patient education of the postoperative care plan can modify the patient's response to the operative experience. Effective postoperative analgesia can be achieved by good preoperative instruction of patients on the various methods of providing postoperative analgesia (Egbert et al. 1964). Informed patients experience significantly less pain and require less analgesia in the postoperative period. ▶ Preoperative education also aids coping, reduces preoperative anxiety, and can enhance postsurgical recovery (Daltroy et al. 1988; Klafta and Roizer 1966).

Enhanced recovery derived from early mobilization can be facilitated by preoperative nutritional support. In severely malnourished patients who suffer >15 % weight loss, ▶ preoperative nutritional support for 7–10 days effectively reduces postoperative complications (Veteran's Affairs Total Parenteral Nutrition Study Group 1991). Micronutrient (vitamin and minerals) deficiencies are common in elderly patients. Evaluation and preoperative nutritional supplementation may be beneficial in selected patients by reducing fatigue and improving wound healing.

Postoperative confusional states delay postoperative recovery. Patients who abuse alcohol, even in the absence of alcohol-related organ dysfunction, experience higher morbidity and prolonged recovery after surgery (Tonnesen and Kehlet 1999). Patients who use recreational drugs or substances should be weaned off these agents to avoid withdrawal symptoms. In smokers, nicotine patches applied postoperatively reduce nicotine withdrawal in the postoperative period, which may cause postoperative confusional states.

Intraoperative Period

Multimodal analgesia and anesthesia regimens that include regional anesthesia appear to facilitate early mobilization of surgical patients. The concept of ▶ multimodal or ▶ balanced analgesia takes advantage of additive or synergistic effects of the combination of multiple drugs and techniques. This results in lower doses of the individual drugs and will enable a concomitant reduction in the side effects. The reduced side effects experienced by the patient may be brought about by the differences in the side effect profile of the drugs (Kehlet 1997).

Intraoperative infiltration of local anesthesia into the surgical field can allow early discharge and mobilization following inguinal hernia surgery with minimal side effects. Urinary retention is eliminated with pronounced economic

benefits (Callesen et al. 1998). Similar results have been reported with paravertebral block anesthesia for surgery on the breast for cancer (Coveney et al. 1998). The use of simple and safe local anesthetic infiltration techniques in minor operations should be encouraged (Moiniche et al. 1998).

Regional anesthesia, utilizing local anesthetics, is the most effective method to attenuate metabolic response (increased cortisol, catecholamine, and glucogen levels; insulin resistance; and negative nitrogen balance) to surgery (Liu et al. 1995). These inhibitory effects on the catabolic responses are most pronounced when regional anesthesia is employed as continuous epidural analgesia for 24–48 h postoperatively. Total afferent neural blockade is achieved by spinal or epidural anesthesia in lower body procedures. However, the inflammatory responses (e.g., increased acute-phase proteins, IL-6) are not attenuated by central neural blockade. Postoperative ileus, which is associated with intra-abdominal procedures, is shortened by several days by continuous thoracic epidural analgesia with local anesthetic (+ small doses of opioids). This is not achieved after lumbar epidural local anesthetic or opioid techniques. A meta-analysis demonstrated significant reduction in pulmonary complications (−30 %), pulmonary embolism (50 %), myocardial infarction (30 %), ileus (2 days), and blood loss (∼30 %) in patients after upper and lower abdominal procedures who received continuous epidural local anesthetic analgesia (Ballantyne et al. 1998).

The use of minimal invasive abdominal surgical techniques can also facilitate early mobilization and discharge following surgery. Inflammatory and immune mediators are reduced by minimally invasive surgical techniques. Most studies report less pain, shorter hospital stay, and reduced morbidity after laparoscopic surgery.

The use of modern anesthetic agents that cause less nausea and vomiting may aid in the implementation of early mobilization programs. Propofol and total intravenous anesthesia techniques can facilitate more rapid recovery after anesthesia and early mobilization of patients in the postoperative period.

Postoperative Care

Postoperative recovery or convalescence depends on several factors of which pain, fatigue, rehabilitation, and specific surgical factors are most important. Early postoperative fatigue may arise from sleep deprivation resulting from noise in hospitals, drugs (opioids, benzodiazepines), and inflammatory response (IL-6). Late fatigue may occur several weeks after surgery, and this may be related to loss of weight and muscle mass and associated weakness.

Effective Analgesia

Effective treatment of postoperative analgesia allows early mobilization. This can be achieved by multimodal or balanced analgesia. Multimodal analgesia takes advantage of additive or synergistic effects by combining agents that act at different mechanisms and/or sites. There is concomitant reduction of side effects, because of the reduced doses of the individual drugs and differences between drugs in side effect profiles. The concomitant administration of paracetamol and nonsteroidal analgesic drugs with opioids can reduce total opioid consumption. This can decrease the adverse effects of the opioids, such as constipation, to facilitate early feeding and recovery of metabolic function of the patient. It is an important component of enhanced recovery after surgery (ERAS) protocols (Lassen et al. 2009).

▶ Early mobilization after major procedures can only be effectively achieved by using epidural or local anesthetic techniques. The use of systemic analgesics (such as opioids or NSAIDS) alone does not reliably facilitate early mobilization. The use of simple and safe local anesthetic infiltration techniques in minor operations aids early mobilization by providing good analgesia (Moiniche et al. 1998). Continuous epidural techniques are more effective for more severe pain. In a randomized trial where patients who underwent elective colonic resection received either ▶ patient-controlled analgesia with morphine or thoracic epidural analgesia with bupivacaine and fentanyl, the patients in the epidural group had lower postoperative pain and fatigue scores, which allowed them to mobilize to a greater extent and eat more. The superior quality of analgesia provided by thoracic epidural analgesia had a positive impact on out-of-bed mobilization, recovery of bowel function, and intake of food, with long-lasting effects on functional exercise capacity and health-related quality of life after colonic surgery (Carli et al. 2002). ▶ Effective analgesia during continuous passive motion after major knee surgery, achieved with continuous epidural infusion and continuous femoral block, improved early postoperative rehabilitation and hastened convalescence (Capdevila et al. 1999).

Early Enteral Feeding

Early enteral (oral) feeding reduces catabolism and limits loss of muscle function and postoperative fatigue. Epidural analgesia and more effective treatment of nausea and vomiting and postoperative ileus can facilitate early enteral feeding and early mobilization in the postoperative period (Silk and Green 1998).

Postoperative Hypoxemia

Early postoperative hypoxemia is a result of anesthesia-related causes. Late postoperative hypoxemia can occur as an episodic event (related to REM-sleep disturbances) or as a persistent phenomenon (due to pulmonary complications and the supine position). Postoperative sleep disturbances and episodic hypoxemia can be reduced by the use of regional local anesthetic techniques and reduced use of opioids. Postoperative pulmonary function is most effectively improved by continuous epidural analgesia with local anesthetics.

A multimodal interventional approach is therefore essential for an effective postoperative rehabilitation program. Therefore, early mobilization of the postsurgical patient requires organization to successfully interface between the patient, the surgeon, the anesthetist, the surgical nurse, and the physiotherapist. Such programs require nurse specialization, specialized wards or units, optimal use of regional anesthetic techniques and multimodal analgesia, opioid-free or opioid-reduced analgesia, and comprehensive revision of traditional concepts of postoperative surgical care.

References

Ballantyne, J. C., Carr, D. B., & De Ferranti, S. (1998). The comparative effects of postoperative analgesic therapies on pulmonary outcome: Cumulative meta-analysis of randomised controlled trials. *Anesthesia and Analgesia, 86,* 598–612.

Callesen, T., Bech, K., & Kehlet, H. (1998). The feasibility, safety and cost of infiltration anaesthesia for hernia repair. *Anaesthesia, 53,* 31–35.

Capdevila, X., Barthelet, Y., Biboulet, P., et al. (1999). Effects of perioperative analgesic technique on the surgical outcome and duration of rehabilitation after major knee surgery. *Anesthesiology, 91,* 8–15.

Carli, F., Mayo, N., Klubien, K., et al. (2002). Epidural analgesia enhances functional exercise capacity and health related quality of life after colonic surgery. *Anesthesiology, 97,* 540–549.

Coveney, E., Wetz, C. R., Greengass, R., et al. (1998). Use of paravertebral block anaesthesia in the surgical management of breast cancer: Experience in 156 cases. *Annals of Surgery, 227,* 496–501.

Daltroy, L. H., Morlino, C. I., Eaton, H. M., et al. (1988). Preoperative education for total hip and knee replacement patients. *Arthritis Care and Research, 11,* 469–478.

Egbert, L. D., Bart, G. E., Welch, C. E., et al. (1964). Reduction of postoperative pain by encouragement and instruction of patients. *The New England Journal of Medicine, 207,* 824–827.

Harper, C. M., & Lyles, Y. M. (1988). Physiology and complications of bed rest. *Journal of American Geriatrics Society, 36,* 1047–1054.

Harwood, R. H., & Ebrahim, S. (2002). The validity, reliability and responsiveness of the Nottingham extended activities of daily living scale in patients undergoing total hip replacement. *Disability and Rehabilitation, 24,* 371–377.

Kehlet, H. (1997). Multimodal approach to control postoperative pathophysiology and rehabilitation. *British Journal of Anaesthesia, 78,* 606–617.

Kehlet, H., & Wilmore, D. W. (2002). Multimodal strategies to improve surgical outcome. *American Journal of Surgery, 183,* 630–641.

Klafta, J. M., & Roizer, M. F. (1966). Current understanding of patients attitudes toward and preparation for anaesthetic: A review. *Anesthesia and Analgesia, 83,* 1314–1321.

Lassen, K., et al. (2009). Consensus review of optimal perioperative care in colorectal surgery. *Arch Surg, 144,* 961–969.

Liu, S. S., Carpenter, R. L., & Neal, J. M. (1995). Epidural anaesthesia and analgesia, their role in postoperative outcome. *Anesthesiology, 82,* 1474–1506.

Moiniche, S., Mikkelson, S., Wettersev, J., et al. (1998). A qualitative systematic review of incisional local anaesthesia for postoperative pain relief after abdominal operations. *British Journal of Anaesthesia, 81,* 377–383.

Silk, B. B. A., & Green, C. J. (1998). Perioperative nutrition: Parenteral versus enteral. *Current Opinion in Clinical Nutrition and Metabolic Care, 1,* 21–27.

Tonnesen, H., & Kehlet, H. (1999). Preoperative alcoholism and postoperative morbidity. *British Journal of Surgery, 86,* 867–874.

Veteran's Affairs Total Parenteral Nutrition Study Group. (1991). Perioperative total parenteral nutrition in surgical patients. *New England Journal of Medicine, 325,* 525–532.

Postoperative Pain, Intrathecal Drug Administration

Edward A. Shipton
Department of Anaesthesia, Christchurch School of Medicine and Health Sciences, University of Otago, Christchurch, New Zealand

Synonyms

Subarachnoid drug administration

Definition

Deep to the ▶ arachnoid membrane is the subarachnoid or ▶ intrathecal space, containing cerebrospinal fluid, brain, spinal cord above the level of L1-2, and the cauda equina (the lumbar and sacral nerve roots) below L1-2. The dural sac ends at S2. For the management of postoperative pain management, multimodal combinations of drugs (usually local anesthetics and opioids) are inserted into the intrathecal space. The intrathecal route has the advantages of simplicity, reliability, and low-dose requirements. It is characterized by rapid onset and offset, easy administration, minimal expense, and minimal adverse effects or complications (Shipton 1999). The use of intrathecal opioids has profoundly changed the quality of spinal anesthesia, with improved analgesia, a reduction in local anesthetic requirements, and shorter duration of motor blockade.

Characteristics

Local Anesthetics

Local anesthetics have been placed in the intrathecal space for approximately 100 years. Local anesthetics given intrathecally provide only short-term postoperative analgesia. Different characteristics of patients and local anesthetic formulations will influence the spread of spinal anesthesia. The height (level) and the duration of sensory analgesia are important clinically. The important characteristics of intrathecal local anesthetics in determining this are the density, the dose, the concentration, and the volume of local anesthetic injected (Shipton 1999). The predictability of the spread of spinal anesthesia can be improved by adjusting both the baricity of the solution and the position of the patient during the intrathecal local anesthetic injection.

The use of short-acting local anesthetic agents such as mepivacaine, 2-chloroprocaine, and articaine permits rapid onset of intrathecal anesthesia with early recovery profiles (O'Donnell and Iohom 2008). Lignocaine (5 % in 7.5 % dextrose) provides a short- to intermediate-acting intrathecal drug. Tetracaine (0.5 % in 5 % dextrose) and bupivacaine (0.75 % or 0.5 % in 8.5 % dextrose) provide intermediate- to long-duration neural blockade. Increasingly, the clinical isobaric forms of bupivacaine (0.5 % and 0.75 %) are being used. Ropivacaine (0.5 % and 0.75 %) is as effective as bupivacaine, but with a more rapid postoperative recovery of sensory and motor functions (McNamee et al. 2002). There are many possible choices of local anesthetics for outpatient spinal anesthesia including procaine, mepivacaine, prilocaine, and small doses of lignocaine (20–25 mg) or bupivacaine (5 mg) combined with an intrathecal opioid (fentanyl 20 mcg).

Since its introduction in 1948, hyperbaric 5 % lignocaine (lidocaine) with its predictable onset and limited duration of action has been used for millions of spinal anesthetics. Concern about the use of spinal lignocaine began in 1991 with published reports of cauda equina syndrome after continuous spinal anesthesia. Large concentrations of local anesthetics administered intrathecally increased glutamate concentrations in the cerebrospinal fluid (Yamashita et al. 2003). The margin of safety may be smallest with lignocaine.

Opioids

Intrathecal opioid disposition depends on the lipid solubility of the individual drug. Intrathecal opioids penetrate the spinal cord and bind to nonspecific sites within the white matter as well as to specific receptors within the dorsal horn (Rathmell et al. 2005). Drug within the spinal cord eventually reaches the plasma compartment through venous uptake.

Opioids also penetrate the dura mater to enter the epidural space. In the epidural space, the opioid is sequestered in fat and enters the plasma compartment via venous uptake. The sum of these multiple avenues for drug disposition results in the clinical characteristics observed (Rathmell et al. 2005). Any drug given intrathecally rapidly redistributes within the CSF; opioid is detectable in the cisterna magna after lumbar intrathecal administration within 30 min accounting for the small incidence of respiratory depression observed immediately after lumbar intrathecal injection.

The lipid solubility of opioids largely determines the speed of onset and duration of intrathecal analgesia; hydrophilic drugs such as morphine have a slower onset of action and longer half-lives in cerebrospinal fluid with greater dorsal horn bioavailability and greater cephalad migration compared with lipophilic opioids such as fentanyl (Bernards et al. 2003).

Although the toxicity of the most common preservatives appears to be small, benzyl alcohol and the parabens have been implicated as the cause of neurotoxicity after intrathecal administration (Hetherington and Doooley 2000). Preservative-free morphine, fentanyl, and sufentanil have shown no neurotoxicity at clinical intrathecal doses. Intrathecal techniques clearly produce reliable analgesia in patients undergoing cardiac surgery. Additional potential benefits include stress response attenuation and thoracic cardiac sympathectomy (Chaney 2006).

Morphine

Morphine, a more hydrophilic opioid, has limited binding to fat within the epidural space and limited binding to nonspecific receptors within the spinal cord white matter. Transfer to the systemic circulation is likewise slower than the lipophilic drugs (Rathmell et al. 2005). Concentrations within the cerebrospinal fluid (CSF) decline more slowly than similar doses of lipophilic drugs, accounting for the greater degree of rostral spread, delayed respiratory depression, and extensive dermatomal analgesia.

Preservative-free morphine in doses from 100 to 500 ug provides excellent postoperative analgesia (Shipton 1999). Doses <100 μg often suffice for pain control after Caesarean delivery, whereas doses in the area of 500 μg may be required for more extensive surgery, such as thoracotomy or open abdominal aortic aneurysm repair (Rathmell et al. 2005). Intrathecal morphine offers improved analgesia and opioid sparing for up to 24 h especially following abdominal surgery (Meylan et al. 2009).

Following gynecological surgery, intrathecal morphine (200 mcg) gives optimal analgesia. Increasing the dose shows no more analgesic efficacy but increases the number of pruritic patients requiring treatment. A meta-analysis showed that after coronary artery bypass surgery, intrathecal morphine reduced systemic morphine use and global pain scores, but increased pruritus; there were no significant effects on mortality, myocardial infarction, dysrhythmias, nausea and vomiting, or time to tracheal extubation (Liu et al. 2004).

Intrathecal morphine (150 ug) when added to intrathecal isobaric ropivacaine (0.5 %, 15 mg) provided sufficient postoperative analgesia for Caesarean delivery with shorter motor duration of motor block (Ogun et al. 2003). When combined with low-dose bupivacaine for Caesarean section, 100 mcg intrathecal morphine produced analgesia comparable with doses as high as 400 mcg, with significantly less pruritus (Girgin et al. 2008).

Lipophilic Opioids

The lipophilic opioids rapidly traverse the dura where they are sequestered in the epidural fat and enter the systemic circulation; they also rapidly penetrate the spinal cord where they bind to both nonspecific sites within the white matter as well as dorsal horn receptors and eventually enter the systemic circulation as they are cleared from the spinal cord (Rathmell et al. 2005). This rapid transfer from the CSF to both spinal cord and the epidural fat accounts for the rapid onset and the prompt decline in CSF levels of opioid; this accounts for their minimal rostral spread, shorter duration and more segmental action, and lack of delayed respiratory depression; vascular uptake accounts for the limited duration of analgesia of lipophilic opioids (Rathmell et al. 2005).

The adverse effect profiles of intrathecal lipophilic opioids are now well characterized and appear less troublesome than intrathecal morphine. Lipophilic opioid/local anesthetic combinations permit more rapid motor recovery (Shipton 1999). Intrathecal fentanyl and sufentanil allow clinicians to use smaller doses of spinal local anesthetic, yet still provide excellent anesthesia for surgical procedures without producing significant adverse effects (Axelsson and Gupta 2009). In summary, the addition of small doses of lipophilic opioids during spinal anesthesia for ambulatory procedures can produce more rapid onset and better quality surgical block, and lead to more rapid recovery of motor function and allow for earlier discharge after surgery (Rathmell et al. 2005).

The addition of intrathecal fentanyl (15 mcg) to hyperbaric bupivacaine (0.5 %) in parturients undergoing Caesarean section improves the quality of anesthesia. The combination of the intrathecal and intravenous routes (with fentanyl) provides better hemodynamic stability and a less pronounced stress response, as reflected by 24-h urinary cortisol excretion.

Sufentanil (5 mcg) in combination with mepivacaine provided more complete and prolonged analgesia than mepivacaine alone after Caesarean section (Meininger et al. 2003). When compared to intrathecal morphine, intrathecal sufentanil provides shorter postoperative analgesia (mean 6.3 h) than intrathecal morphine (mean 19.5 h) with no difference in side effects (Karaman et al. 2006). In summary, intrathecal fentanyl and sufentanil provide a rapid onset of analgesia (10–15 min) with a short duration of action (2–5 h) (Rathmell et al. 2005).

Meperidine (pethidine) has local anesthetic properties in addition to its opioid properties. It has been used as the sole intrathecal agent for anesthesia but has no real advantages over lignocaine (Urmey 2003) owing to the frequent association with nausea and vomiting. Detailed dose-response studies examining analgesic effects at smaller intrathecal doses are not available (Rathmell et al. 2005).

Other Opioids

For Caesarean section, the Effective Dose 95 for intrathecal diamorphine is 0.4 mg (Saravanan et al. 2003). Combining intrathecal diamorphine with hyperbaric bupivacaine gives superior analgesia than bupivacaine alone.

Hydromorphone is a potent analgesic, with dose-related clinical effects, and an adverse effect profile similar to that of other mu-opioid receptor agonists (Quigley 2002). The limited available data suggest that intrathecal hydromorphone 50–100 mcg produces analgesia and side effects similar to 100–200 mcg of intrathecal morphine with a similar duration of action (Rathmell et al. 2005).

Adjuvant Drugs

A variety of adjuvant drugs (many not licensed for use as spinal analgesic agents) have been used with intrathecal analgesics, including clonidine, ketamine, neostigmine, and midazolam. An example of this is the intrathecal administration of bupivacaine (15 mg), clonidine (75 mcg), and morphine (0.2 mg) for radical prostate surgery (Brown et al. 2004). Another example is the combination of intrathecal morphine (500 mcg) plus fentanyl 150 mcg after liver resection that improved immediate analgesia and decreased morphine requirements compared to IV patient-controlled analgesia (PCA) (Roy et al. 2006).

Adrenaline

Adrenaline acts on the pre- and postsynaptic alpha-2 receptors, has an analgesic effect of its own (dose related), does not impair blood flow to the spinal cord, reduces systemic absorption, and reduces adverse effects of opioids (e.g., pruritis) and local anesthetics. Adding the vasoconstrictor adrenaline (0.1–0.6 mg) to spinal local anesthetics, intensifies and extends the duration of sensory and motor block in a dose-dependent fashion (Kito et al. 1998). In the lumbar intrathecal space, adrenaline (epinephrine) had no effect on the pharmacokinetics of alfentanil, fentanyl, or sufentanil, but it increased the area under the concentration-time curve of morphine and decreased its elimination half-life (Bernards et al. 2003).

Clonidine

The well-demonstrated opioid-sparing effect of alpha-2 agonists may be one of the most promising properties, particularly in those patients who are at high risk for developing postoperative respiratory depression and for ambulatory procedures. Postoperative analgesia has been shown to be effective when intrathecal clonidine is used either alone or as an adjunct to local anesthetic agents. However, when it is injected intrathecally with morphine, the results are mixed (Chan et al. 2010).

A meta-analysis concluded that adjunctive use of clonidine (15–150 mcg) to intrathecal local anesthetics prolonged the postoperative pain-free period; the long-term analgesic profile of intrathecal clonidine remains unknown (Elia et al. 2008).

In coronary artery bypass surgery, intrathecal clonidine 100 mcg and 0.5 mg morphine resulted in lower pain scores and opioid consumption in the first 24 h postoperatively compared to intrathecal morphine 0.5 mg alone (Nader et al. 2009). In radical prostatectomies, intrathecal clonidine added to morphine significantly prolonged the time until the first PCA dose, and significantly decreased PCA morphine consumption compared to intrathecal morphine alone (Andrieu et al. 2009).

In inguinal hernia repairs, the addition of clonidine (15 mcg) to hyperbaric bupivacaine (6 mg) increases the spread of analgesia, prolongs the time to first analgesic request, and decreases postoperative pain, as compared to bupivacaine alone. In total knee replacement surgery, intrathecal clonidine (25 or 75 mcg) combined with intrathecal morphine (250 mcg) provides superior analgesia as compared to intrathecal morphine alone (Sites et al. 2003). Similar analgesic benefits were shown for post-Caesarean section patients who received intrathecal injection of clonidine plus morphine (Neves et al. 2006).

Small doses of intrathecal clonidine (30 mcg), combined with local anesthetics and opioids, prolong labor analgesia. Hypotension can occur and must be promptly treated by ephedrine to avoid fetal side effects (Roelants 2006).

Dexmedetomidine

Dexmedetomidine has rarely been given intrathecally, presumably because of the possible neurotoxic effects associated with epidural administration (Konakci et al. 2008). When clonidine or dexmedetomidine was added to intrathecal local anesthetics, the regression of sensory, motor block increased dose dependently, and postoperative analgesia was prolonged (Axelsson and Gupta 2009). The potency of intrathecal clonidine: dexmedetomidine seems to be 10:1 (Axelsson and Gupta 2009).

Fadolmidine

The effect of fadolmidine, the newest alpha-2 agonist in postoperative analgesia is yet to be determined, as the drug is undergoing Phase I clinical trials at present. It is a full agonist of all subtypes of alpha-2 adrenoreceptors and does not cross the blood-brain barrier (Chan et al. 2010). Its use seems promising for analgesia in animal studies, especially when administered intrathecally (Chan et al. 2010). Novel combinations of intrathecal analgesics such as clonidine and gabapentin deserve further study (Joshi et al. 2008).

Midazolam

Intrathecal midazolam binds with gamma aminobutyric acid-A receptors in the spinal cord leading to an analgesic effect. Clinical studies suggest that intrathecal midazolam might reduce nausea and vomiting as well when used as an adjunct to other spinal medications (Ho and Ismail 2008). However, the potential neurotoxic effect of intrathecal midazolam remains a concern. A recent meta-analysis showed that adding intrathecal midazolam to other spinal medications reduced the incidence of nausea and vomiting, and delayed the time to request for rescue analgesia. Intrathecal midazolam did not affect the duration of motor blockade (Ho and Ismail 2008). The incidence of neurological symptoms after intrathecal midazolam was uncommon (Ho and Ismail 2008).

Neostigmine

In the perioperative and the peripartum settings, a meta-analysis found that intrathecal neostigmine increased the incidence of nausea and vomiting, bradycardia (requiring intravenous atropine), anxiety, agitation, and restlessness (Ho et al. 2005). It improved the overall 24-h Visual Analogue Scale score, delayed the time of first request for rescue analgesia, and reduced the total number of rescue injections of nonsteroidal anti-inflammatory drug within the first 24 h. It did not affect the duration of motor blockade or the total amount of ephedrine required (Ho et al. 2005). Adding intrathecal neostigmine to other spinal medications improves perioperative analgesia marginally when compared with placebo. Disadvantages include a high incidence of adverse events, mainly nausea and vomiting; these limit the clinical usefulness of this route of administration and outweigh any minor improvement in analgesia achieved (Habib and Gan 2006).

Nonsteroidal Anti-inflammatory Drugs

Spinal prostaglandin synthesis has been implicated in acute pain processes, and in the generation and maintenance of central sensitization. The intrathecal injection of cyclooxygenase inhibitors produces antinociception and reduces hypersensitivity in animals. Spinal cylcooxygenase activity may, however, not contribute to chronic or postoperative pain. The intrathecal injection of the cyclooxygenase inhibitor, ketorolac (0.25–2 mg) in healthy volunteers has shown no serious adverse effects (Eisenach et al. 2002). However, intrathecal ketorolac did not relieve chronic pain or extend anesthesia or analgesia of intrathecal bupivacaine administered at the beginning of surgery (Eisenach et al. 2010).

Adenosine

Investigations have been performed in patients with acute perioperative pain treated with intrathecal adenosine (Hayashida et al. 2005). Intrathecal adenosine (0.25–2 mg) provides analgesia without affecting blood pressure, heart rate, end-tidal carbon dioxide, or neurological function. Minor effects such as headache and back pain, however, have been reported. By administering sufficient doses of adenosine compounds during surgery, however, significant and long-lasting perioperative pain relief can be achieved via central A1 receptor-mediated antinociceptive/analgesic actions, as well as via peripheral A2a or A3 receptor-mediated antiinflammatory actions (Hayashida et al. 2005).

Sameridine

Sameridine (25 mg) is a new compound with both local anesthetic and opioid properties for intrathecal administration. It has been tested to provide anesthesia for surgery and extended postoperative analgesia (Modalen et al. 2003).

Obstetrics

A small dose of opioid delivered into the cerebrospinal fluid provides almost immediate relief from labor pain without motor block or significant haemodynamic perturbation, and with minimal risks to the mother and fetus (Minty et al. 2007). Considerable research has focused on the optimum dose of opioids when delivered intrathecally, with or without adjuncts. Fentanyl and sufentanil have emerged as the most useful. Bupivacaine is added to provide anesthesia and prolong the duration of analgesia, although this tends to increase the likelihood of motor blockade of the lower extremities. A combination of 2.5 mg of bupivacaine, 25 mcg of fentanyl, and 250 mcg of morphine intrathecally usually provides a 4-h window of acceptable analgesia for patients without complications not anticipating protracted labor.

The combined spinal-epidural (CSE) technique has been developed to counteract the short duration of effect (90–180 min) of intrathecal opioids for labor analgesia. Spinal anesthesia (hyperbaric bupivacaine 12 mg with intrathecal fentanyl 10 mcg and morphine 200 mcg) provided analgesia in unplanned Caesarean sections (Carvalho et al. 2010).

Children

The use of spinal anesthesia in children has been primarily limited to situations in which general anesthesia was considered to pose an excessive risk. Intrathecal bupivacaine is used in high-risk infants as analgesia persists for several hours after the return of motor function. The ex-premature infant and the neurologically impaired child account for the majority of spinal anesthetics used today (Lederhaas 2003). After spinal fusion in children, low-dose intrathecal morphine (2–5 ug/kg) supplemented by intravenous patient-controlled analgesia (PCA) morphine provides better analgesia than PCA or morphine alone. With intrathecal analgesia for spinal surgery, clinical data offer evidence of other benefits, including decreased intraoperative blood loss and quicker return of gastrointestinal function (Tobias 2004).

Adverse Effects

Too high a dose of local anesthetic may result in hypotension, motor blockade, and urinary retention (Shipton 1999). With high sympathetic blockade, sudden bradycardia and cardiac arrest may occur. Mepivacaine and lignocaine have each been associated with transient neurological symptoms following intrathecal administration. This has stimulated development of alternative agents and combinations of local anesthetics and opioids (Urmey 2003).

With opioids, adverse effects increase in proportion to the dose administered, and may be more common if the opioid is administered intrathecally. Adverse effects of opioids include pruritus, nausea and vomiting, urinary retention, and respiratory depression. Nausea and vomiting remains the main adverse effect of intrathecal morphine (Rawal 1999). Adequate postoperative analgesia with fewer adverse effects may be obtained with significantly less morphine, although at lower doses there is not a clear dose-response relationship for some adverse effects or for pain relief (Meylan et al. 2009). The lowest effective dose should be used.

A meta-analysis showed a greater risk of respiratory depression and of nausea and vomiting doses >200 mcg of intrathecal morphine (Gehling and Tryba 2009). Doses in excess of 300 mcg produce nausea, vomiting, pruritus, and urinary retention in most recipients (Rathmell et al. 2005). Indeed, there appears to be a ceiling analgesic effect for intrathecal morphine above which the risk of side effects outweighs the benefits of improved analgesia.

Although rare, respiratory depression continues to be a major problem of intrathecal opioids (Rawal 1999). Respiratory depression occurs in up to 1.2–7.6 % of patients given intrathecal morphine (Meylan et al. 2009). Intrathecal morphine doses of 300 mcg or more increase the risk of respiratory depression. In patients following major surgery, the incidence of respiratory depression and pruritus increased with intrathecal morphine over control treatment (IV PCA morphine) (Meylan et al. 2009). In combination with local anesthetic for analgesia (after Caesarean section), the rate of respiratory depression with intrathecal opioids (all opioids and all doses) is low (Dahl et al. 1999). For example, of 1,915 patients, 0.26 % developed bradypnea associated with 0.15 mg intrathecal morphine. The incidence of severe bradypnea requiring naloxone was 0.052 % (Kato et al. 2008). Risk factors for the development of respiratory depression include large doses, concomitant use of additional opioids and/or sedatives, administration in opioid-naïve patients, and age >65 years (Etches et al. 1989). The most reliable clinical sign of significant respiratory depression appears to be a depressed level of consciousness.

None of the currently available opioids are completely safe; however, extensive international experience has shown that patients receiving spinal opioids for postoperative analgesia can be safely nursed on regular wards, provided that trained personnel and appropriate guidelines are available.

High-dose intrathecal opioids combined with postoperative IV naloxone infusion provided excellent analgesia for radical prostate surgery. IV naloxone infusion appeared to control opioid side effects without diminishing the analgesia (Rebel et al. 2009). Following noninjurious intervals of spinal ischemia, intrathecal morphine has been found to potentiate motor dysfunction.

The rapid onset of analgesia with intrathecally administered agents must be balanced against the added risks of dural puncture. Development of small-gauge, pencil-point needles are responsible for the success of outpatient spinal anesthesia with acceptable rates (0–2 %) of postdural puncture headache (PDPH) (Urmey 2003). In high-risk patients (e.g., age <50 years, postpartum, large-gauge-needle puncture), patients should be offered early epidural blood patch (within 24–48 h of dural puncture).

Nausea and vomiting induced by neuraxial opioids may be a systemic effect, particularly with lipophilic opioids, or may be the result of cephalad migration of drug in the CSF and subsequent interaction with opioid receptors located in the area postrema (Rathmell et al. 2005). Postoperative nausea and vomiting (PONV) is also common after intrathecal morphine with incidences reported at 30 % and 24 %, respectively, although this rate was similar to IV PCA morphine (Meylan et al. 2009). In contrast, fentanyl and sufentanil have been associated with little or no nausea or vomiting after intrathecal administration of a single dose (Lo et al. 1999). After gynecological surgery, the combination of ondansetron with either dexamethasone or droperidol has a better antiemetic effect (Sanchez-Ledesma et al. 2002); dexamethasone plus droperidol seems superior to either alone (Wu et al. 2007).

Pruritus caused by stimulation of spinal and supraspinal mu-opioid receptors is significantly higher than that for patients receiving intrathecal morphine as compared to those receiving IV PCA morphine (Meylan et al. 2009). Opioid antagonists (naloxone, nalbuphine) seem to be the most potent antipruritic drugs, but they also decrease analgesia, which limits their usage (Reich and Szepietowski 2010). Other agents used include ondansetron, propofol, and antihistamines. To date, the data are conflicting, and only limited studies have been performed to confirm their efficacy (Reich and Szepietowski 2010).

Urinary retention is common after administration of intrathecal opioids. This adverse effect can be observed soon after the intrathecal injection of morphine and lasts for 14–16 h, regardless of the dose used (Rawal et al. 1983). Urinary retention is infrequent with use of lipophilic intrathecal opioids (Rathmell et al. 2005). Naloxone and nalbuphine can reverse urodynamic changes after neuraxial morphine.

Of major concern is the development of intrathecal and epidural bleeding producing spinal cord or cauda equina compression and paresis (Shipton 1999). Coagulopathies or anticoagulant therapy form the predominant risk factors, whereas low-dose heparin thromboprophylaxis, nonsteroidal anti-inflammatory drugs, and low-dose aspirin are rarely associated with spinal bleeding complications. The use of higher doses of low-molecular-weight heparin for thromboprophylaxis may be associated with a higher risk of epidural hematoma.

References

Andrieu, G., Roth, B., & Ousmane, L. (2009). The efficacy of intrathecal morphine with or without clonidine for postoperative analgesia after radical prostatectomy. *Anesthesia and Analgesia, 108*, 1954–1957.

Axelsson, K., & Gupta, A. (2009). Local anaesthetic adjuvants: Neuraxial versus peripheral nerve block. *Current Opinion in Anaesthesiology, 22*(5), 649–654.

Bernards, C. M., Shen, D. D., Sterling, E. S., et al. (2003). Epidural, cerebrospinal fluid, and plasma pharmacokinetics of epidural opioids (part 2): Effect of epinephrine. *Anesthesiology, 99*(2), 466–475.

Brown, D. R., Hofer, R. E., Patterson, D. E., et al. (2004). Intrathecal anesthesia and recovery from radical prostatectomy: A prospective, randomized, controlled trial. *Anesthesiology, 100*(4), 926–934.

Carvalho, B., Coleman, L., Saxena, A., et al. (2010). Analgesic requirements and postoperative recovery after scheduled compared to unplanned cesarean delivery: A retrospective chart review. *International Journal of Obstetric Anesthesia, 19*(1), 10–15.

Chan, A. K., Cheung, C. W., & Chong, Y. K. (2010). Alpha-2 agonists in acute pain management. *Expert Opinion on Pharmacotherapy, 11*, 2849–2868.

Chaney, M. A. (2006). Intrathecal and epidural anesthesia and analgesia for cardiac surgery. *Anesthesia and Analgesia, 102*(1), 45–64.

Dahl, J. B., Jeppesen, I. S., Jørgensen, H., et al. (1999). Intraoperative and postoperative analgesic efficacy and adverse effects of intrathecal opioids in patients undergoing cesarean section with spinal anesthesia: A qualitative and quantitative systematic review of randomized controlled trials. *Anesthesiology, 91*(6), 1919–1927.

Eisenach, J. C., Curry, R., Hood, D. D., & Yaksh, T. L. (2002). Phase I safety assessment of intrathecal ketorolac. *Pain, 99*(3), 599–604.

Eisenach, J. C., Curry, R., Rauck, R., et al. (2010). Role of spinal cyclooxygenase in human postoperative and chronic pain. *Anesthesiology, 112*(5), 1225–1233.

Elia, N., Culebras, X., Mazza, C., et al. (2008). Clonidine as an adjuvant to intrathecal local anesthetics for surgery: Systematic review of randomized trials. *Regional Anesthesia and Pain Medicine, 33*, 159–167.

Etches, R. C., Sandler, A. N., & Daley, M. D. (1989). Respiratory depression and spinal opioids. *Canadian Journal of Anaesthesia, 36*, 165–185.

Gehling, M., & Tryba, M. (2009). Risks and side-effects of intrathecal morphine combined with spinal anaesthesia: A meta-analysis. *Anaesthesia, 64*(6), 643–651.

Girgin, N. K., Gurbet, A., Turker, G., et al. (2008). Intrathecal morphine in anesthesia for cesarean delivery: Dose–response relationship for combinations of low-dose intrathecal morphine and spinal bupivacaine. *Journal of Clinical Anesthesia, 20*(3), 180–185.

Habib, A. S., & Gan, T. J. (2006). Use of neostigmine in the management of acute postoperative pain and labour pain: a review. *Central Nervous System Drugs, 20*(10), 821–839.

Hayashida, M., Fukuda, K., & Fukunaga, A. (2005). Clinical application of adenosine and ATP for pain control. *Journal of Anesthesia, 19*(3), 225–235.

Hetherington, N. J., & Dooley, M. J. (2000). Potential for patient harm from intrathecal administration of preserved solutions. *The Medical Journal of Australia, 173*, 141–143.

Ho, K. M., & Ismail, H. (2008). Use of intrathecal midazolam to improve perioperative analgesia: A meta-analysis. *Anaesthesia and Intensive Care, 36*(3), 365–373.

Ho, K. M., Ismail, H., Lee, K. C., & Branch, R. (2005). Use of intrathecal neostigmine as an adjunct to other spinal medications in perioperative and peripartum analgesia: A meta-analysis. *Anaesthesia and Intensive Care, 33*(1), 41–53.

Joshi, G. P., Bonnet, F., Shah, R., et al. (2008). Is intrathecal morphine of benefit to patients undergoing cardiac surgery? *Anesthesia and Analgesia, 107*(3), 1026–1040.

Karaman, S., Kocabas, S., Uyar, M., et al. (2006). The effects of sufentanil or morphine added to hyperbaric bupivacaine in spinal anaesthesia for caesarean section. *European Journal of Anaesthesiology, 23*(4), 285–291.

Kato, R., Shimamoto, H., Terui, K., et al. (2008). Delayed respiratory depression associated with 0.15 mg intrathecal morphine for cesarean section: A review of 1915 cases. *Journal of Anesthesia, 22*(2), 112–116.

Kito, K., Kato, H., Shibata, M., et al. (1998). The effect of varied doses of epinephrine on duration of lidocaine spinal anesthesia in the thoracic and lumbosacral dermatomes. *Anesthesia and Analgesia, 86*, 1018–1022.

Konakci, S., Adanir, T., Yilmaz, G., & Rezanko, T. (2008). The efficacy and neurotoxicity of dexmedetomidine administered via the epidural route. *European Journal of Anaesthesiology, 25*, 403–409.

Lederhaas, G. (2003). Spinal anaesthesia in paediatrics. *Best Practice & Research. Clinical Anaesthesiology, 17*(3), 365–376.

Liu, S. S., Block, B. M., & Wu, C. L. (2004). Effects of perioperative central neuraxial analgesia on outcome after coronary artery bypass surgery: A meta-analysis. *Anesthesiology, 101*(1), 153–161.

Lo, W. K., Chong, J. L., & Chen, L. H. (1999). Combined spinal epidural for labour analgesia: Duration, efficacy and side effects of adding sufentanil or fentanyl to bupivacaine intrathecally vs plain bupivacaine. *Singapore Medical Journal, 40*, 639–643.

McNamee, D. A., McClelland, A. M., Scott, S., et al. (2002). Spinal anaesthesia: Comparison of plain ropivacaine 5 mg ml(−1) with bupivacaine 5 mg ml (−1) for major orthopaedic surgery. *British Journal of Anaesthesia, 89*(5), 702–706.

Meininger, D., Byhahn, C., Kessler, P., et al. (2003). Intrathecal fentanyl, sufentanil, or placebo combined with hyperbaric mepivacaine 2% for parturients undergoing elective cesarean delivery. *Anesthesia and Analgesia, 96*(3), 852–858.

Meylan, N., Elia, N., Lysakowski, C., & Tramèr, M. R. (2009). Benefit and risk of intrathecal morphine without local anaesthetic in patients undergoing major surgery: Meta-analysis of randomized trials. *British Journal of Anaesthesia, 102*(2), 156–167.

Minty, R. G., Kelly, L., Minty, A., & Hammett, D. C. (2007). Single-dose intrathecal analgesia to control labour pain: is it a useful alternative to epidural analgesia? *Canadian Family Physician, 53*(3), 437–442.

Modalen, A. O., Westman, L., Arlander, E., et al. (2003). Hypercarbic and hypoxic ventilatory responses after intrathecal administration of bupivacaine and sameridine. *Anesthesia and Analgesia, 96*(2), 570–575.

Nader, N. D., Li, C. M., & Dosluoglu, H. H. (2009). Adjuvant therapy with intrathecal clonidine improves postoperative pain in patients undergoing coronary artery bypass graft. *The Clinical Journal of Pain, 25*, 101–106.

Neves, J. F., Monteiro, G. A., & Almeida, J. R. (2006). Postoperative analgesia for cesarean section: Does the addition of clonidine to subarachnoid morphine

improve the quality of the analgesia? *Revista Brasileira de Anestesiologia, 56*, 370–376.

O'Donnell, B. D., & Iohom, G. (2008). Regional anesthesia techniques for ambulatory orthopedic surgery. *Current Opinion in Anaesthesiology, 21*(6), 723–728.

Ogun, C. O., Kirgiz, E. N., Duman, A., Okesli, S., & Akyurek, C. (2003). Comparison of intrathecal isobaric bupivacaine-morphine and ropivacaine-morphine for Caesarean delivery. *British Journal of Anaesthesia, 90*(5), 659–664.

Quigley, C. (2002). Hydromorphone for acute and chronic pain. *Cochrane Database of Systematic Reviews*, (1), CD003447.

Rathmell, J. P., Lair, T. R., & Nauman, B. (2005). The role of intrathecal drugs in the treatment of acute pain. *Anesthesia and Analgesia, 101*(5 Suppl.), S30–S43.

Rawal, N. (1999). Epidural and spinal agents for postoperative analgesia. *The Surgical Clinics of North America, 79*(2), 313–344.

Rawal, N., Mollefors, K., Axelsson, K., et al. (1983). An experimental study of urodynamic effects of epidural morphine and of naloxone reversal. *Anesthesia and Analgesia, 62*, 641–647.

Rebel, A., Sloan, P., & Andrykowski, M. (2009). Postoperative analgesia after radical prostatectomy with high-dose intrathecal morphine and intravenous naloxone: A retrospective review. *Journal of Opioid Management, 5*(6), 331–339.

Reich, A., & Szepietowski, J. C. (2010). Opioid-induced pruritus: An update. *Clinical and Experimental Dermatology, 35*(1), 2–6.

Roelants, F. (2006). The use of neuraxial adjuvant drugs (neostigmine, clonidine) in obstetrics. *Current Opinion in Anaesthesiology, 19*(3), 233–237.

Roy, J. D., Massicotte, L., Sassine, M. P., et al. (2006). A comparison of intrathecal morphine/fentanyl and patient-controlled analgesia with patient-controlled analgesia alone for analgesia after liver resection. *Anesthesia and Analgesia, 103*(4), 990–994.

Sanchez-Ledesma, M. J., López-Olaondo, L., Pueyo, F. J., et al. (2002). A comparison of three antiemetic combinations for the prevention of postoperative nausea and vomiting. *Anesthesia and Analgesia, 95*(6), 1590–1595.

Saravanan, S., Robinson, A. P., Qayoum Dar, A., Columb, M. O., & Lyons, G. R. (2003). Minimum dose of intrathecal diamorphine required to prevent intraoperative supplementation of spinal anaesthesia for Caesarean section. *British Journal of Anaesthesia, 91*(3), 368–372.

Shipton, E. A. (1999). Central neural blockade. In E. A. Shipton (Ed.), *Pain – acute and chronic* (pp. 177–209). London: Arnold.

Sites, B. D., Beach, M., Biggs, R., et al. (2003). Intrathecal clonidine added to a bupivacaine-morphine spinal anesthetic improves postoperative analgesia for total knee arthroplasty. *Anesthesia and Analgesia, 96*(4), 1083–1088.

P

Tobias, J. D. (2004). A review of intrathecal and epidural analgesia after spinal surgery in children. *Anesthesia and Analgesia, 98*(4), 956–965.

Urmey, W. F. (2003). Spinal anaesthesia for outpatient surgery. *Best Practice & Research. Clinical Anaesthesiology, 17*(3), 335–346.

Wu, J. I., Lo, Y., Chia, Y. Y., Liu, K., et al. (2007). Prevention of postoperative nausea and vomiting after intrathecal morphine for Cesarean section: A randomized comparison of dexamethasone, droperidol, and a combination. *International Journal of Obstetric Anesthesia, 16*(2), 122–127.

Yamashita, A., Matsumoto, M., Matsumoto, S., Itoh, M., Kawai, K., & Sakabe, T. (2003). A comparison of the neurotoxic effects on the spinal cord of tetracaine, lidocaine, bupivacaine, and ropivacaine administered intrathecally in rabbits. *Anesthesia and Analgesia, 97*(2), 512–519.

Postoperative Pain, Joint Replacement

Gavin Pattullo
Department of Anesthesia and Pain Management, Royal North Shore Hospital, University of Sydney, St. Leonards, NSW, Australia

Introduction

Total knee arthroplasty (TKA) has become one of the most frequently performed major surgeries in the western world. This status quo is likely to be maintained under the dual influences of an aging population and the present obesity epidemic. Owing to the large patient numbers involved it is possible to achieve considerable population gains, when measured as both quality of life improvements and financially, with even modest improvements in efficiency. TKA has resultantly come under a great deal of scrutiny along the lines of enhanced recovery after surgery programs. Surgeons aim for patient mobilization within 24 h of surgery and a stay in the acute-care hospital of 5 days or less. Analgesia regimens need to effectively match and enable these goals.

Characteristics

TKA analgesia regimens used at major centers are all consistent in their emphasis on multimodal analgesia; however, the use of interventional techniques may vary – largely reflecting institutional expertise.

Components of any TKA analgesia regimen that warrant considerations are discussed below :

General or Neuraxial Anesthesia

TKA can be performed under spinal anesthesia using local anesthetic (e.g., bupivacaine 0.5 %, heavy or isobaric, 2–3 mL) with a short-acting lipid-soluble opioid (e.g., fentanyl 10–20 mcg). Patient tolerance of the operating room environment is improved with the addition of anesthetic agents to produce sedation – the depth of which is determined by institutional practice. Benefits of utilizing spinal over general anesthesia include reduced postoperative nausea and vomiting (lessening interruption to calorific intake and allowing greater success with oral analgesia) and a lowering of wound infection rates (Macfarlane et al. 2009; Chang et al. 2010). Administration of general anesthesia is favored by many clinicians simply because of overwhelming patient preference coupled with the enviable safety record of modern general anesthesia.

Epidural Analgesia

The use of lumbar epidural analgesia (LEA) has declined rapidly since being the favored technique at many centers through the 1980s to 1990s. Coercive factors in this change are multiple, but the greatest by far has been the concern for epidural hematoma that was exacerbated by the arrival of low-molecular-weight heparins in the 1990s. Much of this fear has since been allayed through better understanding and education on the need to strictly apply minimum time intervals between the LMWH dosing and any epidural needling or catheter manipulations. Anesthetists familiar with anticoagulant guidelines for neuraxial blockade are increasingly finding themselves abandoning plans for LEA owing to the use in some centers of potent and poorly reversible chemical anticoagulant practices in the early postoperative period. Coincident to the concern for hematoma has been a heightened awareness of the frequent coexistence of lumbar spinal canal stenosis in the arthritic population undergoing TKA and the role this plays in the pathogenesis of spinal cord injury. Stenosis confines the vertebral canal and results in a physical state of low compliance within the epidural space. Subsequently, there is a smaller volume available to extraneous substances (such as catheters and analgesic infusions) to fill before pressure in the epidural space is pathologically raised. This state may be further worsened by the presence of hematoma – even including those of such small volume that they would not ordinarily be significant. Any elevated epidural pressure is readily transferred through the meninges to exert an extramural pressure on the spinal cord vasculature. If this pressure approaches or exceeds the intravascular pressure of a spinal artery or vein, the resulting ischemia of the spinal cord may cause devastating neurological injury. Further contributing to the decline in LEA use has been the popularization of rehabilitation programs emphasizing early mobilization (precluded by LEA-induced lower limb weakness), and the goal of enhanced recovery programs for early demedicalization of recovering patients by eliminating the need for intravenous drips and urinary catheters (precluded in the presence of LEA).

Intrathecal Morphine

An analgesic benefit ascribed to the selective spinal effect of IT morphine (in doses of up to 300 mcg) has been demonstrated in the absence of peripheral nerve blockade. However, Rathmell and coworkers found the improvement in analgesia was not accompanied by the expected reduction in opioid use via a PCA, so questioning the benefit of routine use of IT morphine in TKA (Rathmell et al. 2003).

Femoral Nerve Block

Effective analgesia for TKA can be achieved through selective blockade of the lower lumbar

plexus as in a femoral nerve block (Salinas et al. 2006). The anterior "femoral nerve block" approach, in comparison to the possible alternative of a posterior "lumbar plexus" approach offers the advantage of greater simplicity with an easier and shorter trajectory from skin to neural structure, so lessening the chance of misadventure or failure – two factors that bedevil the posterior approach. Anesthetists experienced in regional anesthesia point out the need to also block nociceptive inputs traveling in the obturator and sciatic nerves in order to achieve complete anesthesia of the operative site. However, others have questioned the clinical analgesic significance of such a thorough blockade. Analgesia with single-dose femoral nerve block can be expected to last 8–14 h. Prolongation of local anesthetic's benefits can be achieved with the use of continuous infusions via catheters placed at the time of the initial dose. Early studies using catheters aimed to be near the femoral nerve or, alternatively merely placed under the fascia iliaca, have reported unreliable analgesia as a consequence of frequent malpositioning of the catheter tip (Capdevila et al. 2002). More recently, high rates of successful analgesia have been reported using placement technique of a stimulating introducer needle targeting a quadriceps motor response endpoint combined with an emphasis on shorter catheter feeds limited to 5 cm (Barrington et al. 2008). Successful sensory blockade with catheters has a problematic downside of being achieved at the expense of quadriceps paresis. This not only prevents ambulatory physiotherapy but also worryingly increases the rate of patient falls and the potential for serious damage to the operative site (Ilfeld et al. 2010). Femoral nerve blockade carries a small risk of neural injury probably best estimated with data from a large prospective study at a rate of 1 in 2,500 (Barrington et al. 2009). Although rare, fear of such a detrimental complication, sometimes accompanied by apparent clustering of higher rates of neural injury, has seen blanket abandonment of femoral nerve block in some centers. The move away from femoral nerve blockade has driven the popularization of alternative sites for local anesthetic placement such as with local infiltration analgesia (see below).

Local Infiltration Analgesia

Impressive analgesia can be achieved through the simple technique of high volume, low concentration local anesthetic infiltration of the pericapsular structures and incision site (Kehlet and Andersen 2011; Otte 2009). There is as yet little data on the optimal concentration and volume of local anesthetic to use, or whether additives are beneficial. A solution in common use consists of 100 mL of 0.2 % ropivacaine to which is added adrenaline 5 mcg/mL. The operating surgeon undertakes infiltration of the solution during the latter half of the operation. Meticulous attention is necessary when injecting to ensure all nociceptive structures are captured while taking care to avoid larger neural structures in close proximity, such as the common peroneal nerve. Although an analgesic benefit of LIA compared with placebo has been demonstrated, any comparative studies with femoral nerve blockade are awaited.

Opioid

With the immense nociceptor stimulation of TKA, the need for a strong opioid during the postoperative period is mostly unavoidable. Opioid is readily delivered via a patient–controlled analgesia device – at least for the initial phase. Enhanced recovery protocols emphasize the early transition from the PCA to oral opioid to enable unencumbered mobilization of the patient as well as to prevent analgesia failure resulting from loss of IV access. Through the implementation of multimodal analgesia regimens that emphasize correctly placed peripheral local anesthetic, many authors and clinicians report the ability to successfully avoid the need for a PCA altogether (Hebl et al. 2005). One option for oral opioid use is to deliver the bulk of expected oral opioid requirement as an extended release formulation, e.g., oxycodone-controlled release 10–20 mg every 8–12 hourly (for up to 5 days). An immediate release formulation, e.g., oxycodone 5–15 mg third hourly is then ordered on an as-required basis to cover any unmet analgesic need.

When considering the use of empiric dosing of extended release opioid, the clinician must exercise particular care to avoid neurodepression in the opioid naïve patient.

Adjunct Analgesia

A number of agents provide an essential contribution to the success of multimodal analgesia in TKA:

Paracetamol (acetaminophen). For maximal benefit, the recommended dosing of 1 g four times a day should be administered regularly. Reduce dosing in elderly (> 80 years) and those at increased risk of hepatotoxicity. Limit dosing to 5–7 days.

NSAIDs/COX-2. Orthopedic surgeon resistance to the use of NSAIDs/COX-2 due to concern for delay in bone healing is commonly encountered. This anxiety appears to be unnecessary when these agents are administered for only a few days, as only the extended use over 30 days has been implicated in impaired bone healing. Of additional reassurance is the very large number of patients who have received these agents following orthopedic injuries and for whom impairment of bone healing has not been noted. Prescriptions for NSAID/COX-2 should emphasize dosing cessation after 3–4 days in order to limit side effects.

Gabapentin. Dosing of gabapentin at 300 mg three times a day is well tolerated with few side effects and substantially reduces opioid requirements post TKA (Clarke et al. 2009). Dose reduction may warrant consideration for the elderly or frail patient. Administration is commonly continued up to the fifth postoperative day. Administration as a preoperative dose (600 mg) may be beneficial, but it is not essential and can lead to disconcerting sedation in the recovery period when combined with general anesthesia. Pregabalin has also been studied in this setting. One study demonstrated the perioperative use of pregabalin for 14 days eliminated chronic knee pain at 6 months – compared to the placebo group with a rate of 5 % (Buvanendran et al. 2010). Early postoperative sedation and confusion were reported more frequently in the pregabalin-treated patients.

Analgesic Considerations for Major Arthroplasty at Other Sites

For surgery at other anatomical sites, a multimodal regimen encompassing similar concepts to the above can be successfully implemented. Total hip arthroplasty (THA) protocols can include use of regional anesthesia with posterior lumbar plexus block. The modest analgesic benefit of this block should be weighed against the risk of needle entry into nearby vulnerable vascular structures, with potential for life-threatening consequences. This consideration is particularly salient given that hip surgery patients (in common with TKA) are frequently obese so making correct placement of the needle tip very difficult. Patients undergoing shoulder arthroplasty can be exceedingly well analgesed with the provision of interscalene block. Lower local anesthetic volumes of 5–10 mL are increasingly favored for the accompanying reduction in phrenic nerve palsy (Riazi et al. 2008). Prolongation of analgesia can be provided with the use of catheter techniques or the addition of clonidine to the local anesthetic. Clonidine increases analgesia by up to 2 h. The ideal dose of clonidine is unclear, however limiting dose to 150 mcg or 2 mcg/kg, and possibly even lower, will help minimize troublesome side effects of hypotension and postoperative sedation (Popping et al. 2009). As most shoulder surgery is performed in the upright or beach-chair position, the anesthetist wishing to use clonidine must be especially vigilant for, and manage, any intraoperative hypotension to prevent cerebral hypoperfusion.

References

Barrington, M. J., Olive, D. J., McCutcheon, C. A., Scarff, C., Said, S., Kluger, R., et al. (2008). Stimulating catheters for continuous femoral nerve blockade after total knee arthroplasty: a randomized, controlled, double-blinded trial. *Anesth Analg, 106*(4), 1316–1321.

Barrington, M. J., Watts, S. A., Gledhill, S. R., Thomas, R. D., Said, S. A., Snyder, G. L., et al. (2009). Preliminary results of the Australasian Regional Anaesthesia Collaboration: a prospective audit of more than 7000 peripheral nerve and plexus blocks for neurologic and other complications. *Reg Anesth Pain Med, 34*(6), 534–541.

Buvanendran, A., Kroin, J. S., Della Valle, C. J., Kari, M., Moric, M., & Tuman, K. J. (2010). Perioperative oral pregabalin reduces chronic pain after total knee arthroplasty: a prospective, randomized, controlled trial. *Anesth Analg, 110*(1), 199–207.

Capdevila, X., Biboulet, P., Morau, D., Bernard, N., Deschodt, J., Lopez, S., et al. (2002). Continuous three-in-one block for postoperative pain after lower limb orthopedic surgery: where do the catheters go? *Anesthesia & Analgesia, 94*(4), 1001–1006.table of contents.

Chang, C. C., Lin, H. C., & Lin, H. W. (2010). Anesthetic management and surgical site infections in total hip or knee replacement: a population-based study. *Anesthesiology, 113*(2), 279–284.

Clarke, H., Pereira, S., Kennedy, D., Gilron, I., Katz, J., Gollish, J., et al. (2009). Gabapentin decreases morphine consumption and improves functional recovery following total knee arthroplasty. *Pain Res Manag, 14*(3), 217–222.

Hebl, J. R., Kopp, S. L., Ali, M. H., Horlocker, T. T., Dilger, J. A., Lennon, R. L., et al. (2005). A comprehensive anesthesia protocol that emphasizes peripheral nerve blockade for total knee and total hip arthroplasty. *J Bone Joint Surg Am, 87*(Suppl 2), 63–70.

Ilfeld, B. M., Duke, K. B., & Donohue, M. C. (2010). The association between lower extremity continuous peripheral nerve blocks and patient falls after knee and hip arthroplasty. *Anesth Analg, 111*(6), 1552–1554.

Kehlet, H., & Andersen, L. O. (2011). Local infiltration analgesia in joint replacement: the evidence and recommendations for clinical practice. *Acta Anaesthesiol Scand, 55*(7), 778–784.

Macfarlane, A. J., Prasad, G. A., Chan, V. W., & Brull, R. (2009). Does regional anesthesia improve outcome after total knee arthroplasty? *Clin Orthop Relat Res, 467*(9), 2379–2402.

Otte, K. S. (2009). Local infiltration analgesia in total knee arthroplasty. *Acute Pain, 10*, 111–116.

Popping, D. M., Elia, N., Marret, E., Wenk, M., & Tramer, M. R. (2009). Clonidine as an adjuvant to local anesthetics for peripheral nerve and plexus blocks: a meta-analysis of randomized trials. *Anesthesiology, 111*(2), 406–415.

Rathmell, J. P., Pino, C. A., Taylor, R., Patrin, T., & Viani, B. A. (2003). Intrathecal morphine for postoperative analgesia: a randomized, controlled, dose-ranging study after hip and knee arthroplasty. *Anesth Analg, 97*(5), 1452–1457.

Riazi, S., Carmichael, N., Awad, I., Holtby, R. M., & McCartney, C. J. (2008). Effect of local anaesthetic volume (20 vs 5 ml) on the efficacy and respiratory consequences of ultrasound-guided interscalene brachial plexus block. *Br J Anaesth, 101*(4), 549–556.

Salinas, F. V., Liu, S. S., & Mulroy, M. F. (2006). The effect of single-injection femoral nerve block versus continuous femoral nerve block after total knee arthroplasty on hospital length of stay and long-term functional recovery within an established clinical pathway. *Anesth Analg, 102*(4), 1234–1239.

www.medicine.ox.ac.uk/bandolier/booth/painpag/wisdom/NSB.pdf

www.postoppain.org. Website containing regularly updated evidence based guidelines for analgesia following a number of surgery types, including knee and hip arthroplasty.

Postoperative Pain, Ketamine

Ross MacPherson
Department of Anesthesia and Pain Management,
Royal North Shore Hospital University of
Sydney, Sydney, NSW, Australia

Synonyms

Chloro-phenyl-2-methylaminocyclohexanone-hydrochloride; Ketalar; Ketamine

Definition

Ketamine is a general anesthetic and analgesic agent, which is a novel structure unrelated to other classes of anesthetic agents. It is a phencyclidine derivative.

Characteristics

Activity

Ketamine has wide-ranging activities at a number of discrete sites within the CNS (Krystal et al. 1994). From an anesthetic viewpoint, its main site of action is most likely the NMDA receptors that are most densely localized in the cerebral cortex and hippocampus. The NMDA receptor comprises of an ion channel, which when blocked attenuates the capacity of excitatory amino acids, notably glutamate, from opening the ion channel and increasing calcium permeability. Ketamine binds to the receptor in a noncompetitive manner.

Other receptors at which ketamine is known to have activity include mu- and sigma-opioid receptors, as well as nicotinic muscarinic and cholinergic receptors within the CNS and PNS. In addition to all these effects at receptor sites, ketamine has also been shown to inhibit reuptake of a range of neurotransmitters including noradrenaline, dopamine, and serotonin. It is also a weak inhibitor of acetylcholinesterase.

Ketamine as an Anesthetic Agent

Since ketamine was originally intended to be used as a general anesthetic agent, it is important to review this aspect of the drugs pharmacological profile. Administered in a dose of 1–2 mg/kg I.V., it provides a period of unconsciousness of up to 15 min, although this can be prolonged by administration as a continuous I. V. infusion (15–30 μg/kg/min). The state of unconsciousness produced by ketamine has been designated "dissociative anesthesia," which differs from that produced by other anesthetic agents, insomuch as many reflexes such as breathing and upper airway protective reflexes are maintained.

Ketamine as an Analgesic Agent

Ketamine's analgesic properties were recognized soon after the drug's release. This again was unusual, as most anesthetic induction agents either have no analgesic properties or are mildly analgesic. Yet it took almost 20 years before this aspect of ketamine's pharmacological profile was specifically exploited. In a comprehensive review of the literature, Schmid et al. (1999) suggested that there are three dosage ranges that must be considered with regard to ketamine. The first is the dose employed when the drug is used as an anesthetic agent, outlined above. The second is the dose needed for ketamine to have analgesic or antihyperalgesic effects in its own right – so-called low dose ketamine, usually in the range of an I.V. or epidural bolus of 1 mg/kg, or continuous I.V. administration in the order of 20 μg/kg/min or less. Lastly, the authors suggest we need to consider even lower doses than this, where ketamine has no analgesic effects on its own, but may act synergistically with other agents such as opioids.

Ketamine as a Local Anesthetic Agent

As ketamine gained popularity as an analgesic, studies were undertaken to investigate its efficacy when administered by routes of administration other than the parenteral routes.

Laboratory investigations have confirmed that ketamine possesses significant sodium channel blocking properties, sufficient to explain the "local anesthetic" effect of the drug when given intrathecally (Reckziegel et al. 2002; Haeseler et al. 2003). Indeed, animal studies of intrathecal ketamine have demonstrated promising results, although there are still some concerns that either ketamine itself, or some of the additives found in the nonpreservative free product, may result in spinal cord damage. Others have reported use of ketamine (or its enantiomers) without adverse effects (Sator-Katzenschlager et al. 2001), claiming that its use is without sequelae.

Studies where ketamine has been administered as the sole agent by the epidural route have failed to demonstrate any significant analgesic effects (Peat et al. 1989; Wong et al. 1997), although when combined with morphine it does exert an opioid-sparing effect with improved analgesia (Chia et al. 1998). Similar results have been seen when epidural morphine has been given in combination with intravenous ketamine (Aida et al. 2000).

Postoperative Pain

There have been a large number of reports of low-dose ketamine used in a variety of different surgical procedures. Results have generally shown a morphine-sparing effect (Menigaux et al. 2001; Weinbroum 2003), but this has not been a consistent finding (Reeves et al. 2001). There have also been some trials looking at the preemptive effects of ketamine. Since the whole subject of the efficacy of preemptive analgesia is under question, it is no surprise that again results have been conflicting (Holthusen et al. 2002; White 2002).

Novel Routes of Administration

Since ketamine is often "snorted" by drug abusers, this suggested that the drug has appreciable absorption when administered by the intranasal route, and studies are currently

being undertaken to assess the suitability of this method of administration. The oral route of administration has also been reported to have been successful in some individual case reports (Kannan et al. 2002), having been used successfully in the management of cancer-related neuropathic pain. This is perhaps unusual since it is known that the oral bioavailability of ketamine is low with sublingual and oral bioavailabilities of ketamine having been reported as being 23 % and 32 %, respectively (Schug et al. 2006).

There are also a number of trials currently being conducted on the efficacy of ketamine in a topical formulation (of between 0.25 % and 1.5 %) in the treatment of complex regional pain syndrome Type I (Ushida et al. 2002), and early results are encouraging. As has been pointed out by Persson (2010) apart from oral and topical administration, the drug has also been administered by intramuscular, rectal, transdermal, intranasal, epidural, and intrathecal routes with varying degrees of success.

Adverse Effects

The adverse effects of ketamine when used as a general anesthetic have been well described. While the drug has effects on most physiological systems, it is the cardiovascular and psychological effects that have received the most attention. Pronounced cardiovascular stimulation, brought about through either central sympathetic stimulation or inhibition of reuptake of peripheral catecholamines, results in an increase in mean blood pressure, heart rate, and cardiac index. This may be beneficial in the anesthetic management of critically ill patients in whom hypotension is to be avoided. The effect of ketamine on the cardiovascular system must not be underestimated. In a recent study (Sigtermans et al. 2009), S(+) ketamine was infused over 2 h eventually reaching a maximum rate of 1.2 mg/kg/min over 15 min. Over the dose range, the researchers found a 40–50 % increase in cardiac output. Interestingly the researchers also found a significant sex difference in ketamine metabolism, with the elimination of ketamine being 20 % greater in women, resulting in higher plasma levels in the male subjects.

The psychological effects of the drug are also well known, and indeed have probably limited the use of the drug in anesthetic practice. Adults, much more so than children, are prone to a range of disturbing and bizarre dreams following ketamine use, which may become recurring. Numerous strategies have been employed to try and limit these effects, including pretreatment with either benzodiazepines or droperidol, and allowing patients to emerge from ketamine anesthesia without undue stimulation. It is during emergence that many of these effects are particularly prominent, including auditory and visual hallucinations, disorientation, restlessness, and confusion.

There are two important features relating to the psychomimetic effects of ketamine. The first is that they are dose dependent. The second is that they are thought to be induced to a much greater extent by the $R(-)$ isomer of ketamine.

In the acute pain setting, infusions of low-dose ketamine, usually given by the subcutaneous route, have often been used to control acute pain poorly responsive to opioids, although it has been pointed out that solid evidence for this is lacking (McIntyre et al. 2010).

There have been a number of studies specifically investigating the risks of psychomimetic events in patients receiving low-dose ketamine for analgesia. In his review of these reports, Schmid et al. (1999) summarized the results as suggesting that I.V. low-dose ketamine (at a dose of <2.5 µg/kg leading to a blood level of <50 ng/ml) does not cause hallucinations or impairment of cognitive function. However, at higher levels, even using dosages that might be said to be in the "low-dose" range, psychologically adverse effects can be demonstrated (Krystal et al. 1994); however, blood levels, where measured, are usually in the region of 200 ng/ml.

Abuse Potential

It is important for clinicians to be aware of the abuse potential of ketamine. Despite having been relatively freely available for over 40 years, it is only recently that the drug has been utilized for nontherapeutic uses. Due to the drug's extensive use in veterinary anesthetic practice, its ease of availability probably led to its illicit use; however, with awareness of the drug's abuse potential, authorities in most countries have restricted its availability, which has made supply through hospitals and other medical facilities more difficult. This is important, as diversion of ketamine from legitimate sources was the main source of illicit supply, since the drug is difficult to manufacture from other products (Smith et al. 2002). However, we should not let the fact that ketamine, in some sectors, become a drug of abuse limit its use in appropriate medical settings (Akporehwe et al. 2006). There is no compelling evidence that the use of ketamine as part of medically supervised pain management leads in any way to ongoing abuse.

References

Aida, S., Yakamura, T., Baba, H., et al. (2000). Preemptive analgesia by intravenous low-dose ketamine and epidural morphine in gastrectomy: A randomized double-blind study. *Anesthesiology, 92,* 1624–1630.

Akporehwe, N. A., Wilkinson, P. R., Quibell, R., & Akporehwe, K. A. (2006). Ketamine: A misunderstood analgesic? *BMJ, 332,* 1466.

Chia, Y. Y., Liu, K., Liu, Y. C., et al. (1998). Adding ketamine in a multimodal patient-controlled epidural regimen reduces postoperative pain and analgesic consumption. *Anesthesia and Analgesia, 86,* 1245–1249.

Haeseler, G., Tetzlaff, D., Bfler, J., et al. (2003). Blockage of voltage-operated neuronal and skeletal muscle sodium channels by S(+)- and R(-)-ketamine. *Anesthesia and Analgesia, 96,* 1019–1026.

Holthusen, H., Backhaus, P., Boeminghaus, F., et al. (2002). Preemptive analgesia: No relevant advantage of preoperative compared with postoperative intravenous administration of morphine, ketamine, and clonidine in patients undergoing transperitoneal tumour nephrectomy. *Regional Anesthesia and Pain Medicine, 27,* 249–253.

Kannan, T. R., Saxena, A., Bhatnagar, S., et al. (2002). Oral ketamine as an adjuvant to oral morphine for neuropathic pain in cancer patients. *Journal of Pain and Symptom Management, 23,* 60–65.

Krystal, J., Karper, L., Seibyl, J., et al. (1994). Subanaesthetic effects of the noncompetive NMDA antagonist, ketamine, in humans: Psychotomimetic, perceptual, cognitive, and neuroendocrine responses. *Archives of General Psychiatry, 51,* 199–214.

McIntyre, P. E., Schug, S. A., Scott, D. A., Visser, E. J., Walker, S. M., & Working group of the Australian and

New Zealand College of Anaesthetists and FAculty of Pain Medicine. (2010). *Acute pain management: Scientific guidelines* (3rd ed.). Melbourne: ANZCA & FPM.

Menigaux, C., Guignard, F. D., et al. (2001). Intraoperative small-dose ketamine enhances analgesia after outpatient knee arthroscopy. *Anesthesia and Analgesia, 93*, 606–612.

Peat, S. J., Bras, P., & Hanna, M. H. (1989). A double-blind comparison of epidural ketamine and diamorphine for postoperative analgesia. *Anaesthesia, 44*, 555–558.

Persson, J. (2010). Wherefore ketamine? *Current Opinion in Anaesthesiology, 23*(4), 455–460.

Reckziegel, G., Friederich, P., & Urban, B. (2002). Ketamine effects on human neuronal Na+channels. *European Journal of Anaesthesiology, 19*, 634–640.

Reeves, M., Lindholm, D., Myles, P., et al. (2001). Adding ketamine to morphine for patient-controlled analgesia after major abdominal surgery: A double-blinded, randomized controlled trial. *Anesthesia and Analgesia, 93*, 116–120.

Sator-Katzenschlager, S., Deusch, E., Maier, P., et al. (2001). The long-term antinociceptive effect of intrathecal S(+)-ketamine in a patient with established morphine tolerance. *Anesthesia and Analgesia, 93*, 1032–1034.

Schmid, R. L., Sandler, A. N., & Katz, J. (1999). Use and efficacy of low-dose ketamine in the management of acute postoperative pain: A review of current techniques and outcomes. *Pain, 82*, 111–125.

Schug, S. A., Page-Sharp, M., & Ilett, K. (2006). Bioavailability of ketamine after oral or sublingual administration. *Anaesthesia and Intensive Care, 34*(4), 529.

Sigtermans, M. J., van Hilten, J. J., Bauer, M. C., et al. (2009). Ketamine produces effective and long-term pain relief in patients with complex regional pain syndrome type 1. *Pain, 145*(3), 304–311.

Smith, K. M., Larive, L. L., & Romanelli, F. (2002). Club drugs: Methylene-dioxy-methamphetamine, flunitrazepam, ketamine hydrochloride, and [gamma]-hydroxybutyrate. *American Journal of Health-System & Pharmacy, 59*, 1067–1076.

Ushida, T., Tani, T., Kanbara, T., et al. (2002). Analgesic effects of ketamine ointment in patients with complex regional pain syndrome type 1. *Regional Anesthesia and Pain Medicine, 27*, 524–528.

Weinbroum, A. (2003). A single small dose of postoperative ketamine provides rapid and sustained improvement in morphine analgesia in the presence of morphine-resistant pain. *Anesthesia and Analgesia, 96*, 789–795.

White, P. (2002). The role of non-opioid analgesic techniques in the management of pain after ambulatory surgery. *Anesthesia and Analgesia, 94*, 577–585.

Wong, C. S., Lu, C. C., Cherng, C. H., et al. (1997). Pre-emptive analgesia with ketamine morphine and epidural lidocaine prior to total knee replacement. *Canadian Journal of Anaesthesia, 44*, 31–37.

Postoperative Pain, Lignocaine

Ross MacPherson
Department of Anesthesia and Pain Management, Royal North Shore Hospital University of Sydney, Sydney, NSW, Australia

Synonyms

Lidocaine

Definition

Local anesthetic of the amide type; Membrane-stabilizing drug.

Characteristics

Lignocaine has a number of uses in anesthesia and pain medicine. It is perhaps most commonly used as a local anesthetic agent either in local or regional anesthesia or in epidural or spinal blockade. However, it is also given parenterally in the management of neuropathic pain states.

The first reports of the use of parenteral lignocaine for analgesic purposes were in 1961 (Bartlett and Hutaserani 1961), although it was not until about 20 years later that it began to be seriously investigated for this purpose (Mao and Chen 2000).

Since then lignocaine has been extensively employed both as a diagnostic and a therapeutic tool in the management of neuropathic pain by a variety of means (Bath et al. 1990). Although it has mainly been administered by the intravenous or subcutaneous routes, impressive results have also been achieved when the drug has been used by the topical route.

Where lignocaine is used parenterally, in most studies the drug was administered by the intravenous route (Kastrup et al. 1987), but the subcutaneous route has also been employed (Devulder et al. 1993). The route of administration does not seem to be important, so long as a plasma concentration of between 2 and 5 micrograms/ml is achieved (Rowbotham et al. 1995), and onset of relief of symptoms, if it is to occur, generally starts within 60 min.

In the many studies reported, lignocaine was administered either as a bolus or as a bolus followed by a continuous infusion. Bolus doses have been in the range of 1.5–5 mg/kg and infusions 2 mg/kg/h. As dosage increases, so naturally does the incidence of adverse effects. Serial measurements of plasma lignocaine levels are important, if treatment is to be continued over any period of time.

Lignocaine therapy has been employed in the treatment of a wide variety of pain states, including central and peripheral neuropathic pain states, postherpetic neuralgia, and neuropathic cancer pain. While there have been good results in many of these studies, it does appear that systemic lignocaine is most effective in treating pain resulting from peripheral nerve injury (Galer et al. 1993). A recent study has also shown its effectiveness in the management of refractory complex regional pain syndrome (Schwartzman et al. 2009).

More recently, lignocaine in high concentration (5 %) in a topical patch product has been used to control local pain from conditions such as postherpetic neuralgia (Galer et al. 2002). While this approach does not produce plasma levels comparable to those achieved with parenteral administration, the results are encouraging. In another study (Devers and Galer 2000), the authors were able to demonstrate significant pain relief using the lignocaine patch, in a group of patients whose pain had hitherto been unrelieved by all standard analgesic agents.

Clearly, the use of parenteral lignocaine will be limited to the hospital population. However, the response of patients to lignocaine infusion has been used as a diagnostic tool, in an effort to identify those patients whose pain state is likely to respond to other membrane-stabilizing agents. Patients are usually administered a single dose of between 2 and 5 mg/kg of lignocaine intravenously (with or without control placebo) and the response to the infusion recorded (Galer et al. 1996).

In general most studies have focussed on the use of lignocaine in neuropathic pain states; however, a recent review by Vigneault et al. (2011) examined 29 studies where intravenous lignocaine was used as a means of controlling postoperative pain. The general consensus was that treatment with intravenous lignocaine in many studies reduced pain both at rest and during movement and also resulted in lower opioid use and reduced nausea and vomiting. However, as the authors point out, toxic levels were reported in some studies, and the safety of this form of management needs to be born in mind.

While many workers use the so-called lignocaine test to predict response to other agents, Mao and Chen (2000) have pointed out that data supporting the use of the test is somewhat anecdotal. More studies need to be carried out to truly identify its validity.

References

Bartlett, E. E., & Hutaserani, O. (1961). Xylocaine for the relief of postoperative pain. *Anesthesia and Analgesia, 40*, 296–304.

Bath, F. W., Jensen, T. S., Kastrup, J., et al. (1990). The effect of intravenous lidocaine on nociceptive processing in diabetic neuropathy. *Pain, 40*, 29–34.

Devers, A., & Galer, B. S. (2000). Topical lidocaine patch relieves a variety of neuropathic pain conditions: An open-label study. *The Clinical Journal of Pain, 16*, 205–208.

Devulder, J. E., Ghys, L., Dhondt, W., et al. (1993). Neuropathic pain in a cancer patient responding to subcutaneously administered lignocaine. *The Clinical Journal of Pain, 9*, 220–223.

Galer, B. S., Miller, K. V., & Rowbotham, M. C. (1993). Responses to intravenous lidocaine infusion differs based on clinical diagnosis and site of nervous system injury. *Neurology, 43*, 1233–1235.

Galer, B. S., Harle, J., & Rowbotham, M. C. (1996). Responsive to intravenous lidocaine infusion predicts subsequent response to oral mexiletine: A prospective study. *Journal of Pain and Symptom Management, 12*, 161–167.

Galer, B. S., Jensen, M. P., Ma, T., et al. (2002). The lidocaine patch 5 % effectively treats all neuropathic pain qualities: Results of a randomized, double-blind, vehicle-controlled, 3-week efficacy study with use of the neuropathic pain scale. *The Clinical Journal of Pain, 18*, 297–301.

Kastrup, J., Petersen, P., Dejgard, A., et al. (1987). Intravenous lidocaine infusion – a new treatment of chronic painful diabetic neuropathy. *Pain, 28*, 75.

Mao, J., & Chen, L. (2000). Systemic lidocaine for neuropathic pain relief. *Pain, 87*, 7–17.

Rowbotham, M. C., Davies, P. S., & Fields, H. L. (1995). Topical lidocaine gel relieves postherpetic neuralgia. *Annals of Neurology, 37*, 246–253.

Schwartzman, R. J., Patel, M., Grothusen, J. R., & Alexander, G. A. (2009). Efficacy of 5-day continuous lidocaine infusion for the treatment of refractory complex regional pain syndrome. *Pain Medicine, 10*(2), 401–412.

Vigneault, L., et al. (2011). Perioperative intravenous lidocaine infusion for postoperative pain control: A meta-analysis of randomized controlled trials. *Canadian Journal of Anaesthesia, 58*(1), 22–37.

Postoperative Pain, Local Anesthetics

Laurence E. Mather
Department of Anaesthesia and Pain Management, University of Sydney at Royal North Shore Hospital, St. Leonards, Sydney, NSW, Australia

Synonyms

Local analgesics

Definition

A local anesthetic is a chemical substance that, when applied in the region of a neuronal structure in sufficient concentration, produces reversible blockade of axonal conduction without depolarization, and produces loss of sensation in the region innervated. These drugs produce neural blockade, otherwise referred to as local anesthesia, regional anesthesia, conduction anesthesia, or analgesia, depending on how they are used. A local anesthetic should be capable of producing a differential neural blockade, that is, sensory without significant motor neural blockade, and have an acceptable difference between beneficial and toxic doses. The latter clause is important because of its contemporary significance: Achieving greater safety in local anesthetics,

especially of the long-acting variety, has been the major driving force in the medicinal chemistry and pharmacology of these drugs for many decades. Although many drugs have local anesthetic properties, only a few satisfy these requirements and are used clinically for that purpose. Conversely, local anesthetics have a variety of pharmacological actions, some of which are exploited therapeutically and others of which are avoided assiduously.

Some substances, and/or formulations of some substances, have been introduced/used that produced, either intentionally or unintentionally, irreversible local anesthesia. Such agents are neurolytic and, if intentionally neurolytic, still may have a small amount of use in the management of chronic pain.

Local anesthetics may be applied with great precision to the neuraxis to produce neurally selective central blockade of a range of spinal nerves or peripherally to produce blockade of a plexus or defined nerves, or they may be infiltrated to produce generalized blockade of nerve endings and axons in a more diffuse region. They may also be applied topically to produce surface anesthesia in the mucous membranes or the skin.

Introduction

Provision of a pain-free state in a conscious, cooperative patient undergoing an operative or otherwise painful procedure has been the goal of humankind since time immemorial. However, it was not until the local anesthetic properties of the natural product cocaine were discovered over a century ago and combined with the developing knowledge of neuroanatomy and neurophysiology that reliable techniques of neural blockade/locoregional anesthesia/analgesia could be used to benefit patients.

Contemporary local anesthetics (Table 1) work by reversibly blocking neural impulses in excitable membranes, mainly by interfering with the function of voltage-gated sodium (Na^+) channels, thereby preventing neural impulse propagation. In the presence of local anesthetics, Na^+ channels in the neural membrane are less likely

Postoperative Pain, Local Anesthetics, Table 1 Physicochemical properties of local anesthetics (Mather and Tucker 2008)

Agent	Mol. wt	pK$_a$(25 °C)	n-octanol; pH 7.4; 25 °C distribution coefficient[a]	Plasma protein binding (%)	Aqueous solubility (mg HCl/ml; pH 7.37, 37 °C)
Esters					
Procaine	236	9.0	1.7	5.8	–
Chloroprocaine	271	9.3	9.0	–	–
Tetracaine (amethocaine)	264	8.6	221	76	1.4
Amides					
Prilocaine[a]	220	8.0	25	55	–
Lidocaine (lignocaine)	234	7.8	43	64	24
Mepivacaine[a]	246	7.9	21	77	15
Ropivacaine	274	8.2	115	90	–
Bupivacaine[a] (and levobupivacaine)	288	8.2	346	95	0.83
Etidocaine[a]	276	7.9	800	94	–

[a]Designates a chiral caine used as a racemate

to open in response to a stimulating depolarization, and the resulting transmembrane Na$^+$ current is decreased. At sufficiently high local anesthetic concentrations, enough channels are impaired to prevent impulse generation, thereby producing neural blockade. Overall, local anesthetics are not highly receptor selective and thus also block other ion channels (Strichartz et al 2008); To some extent, they also expand membrane volumes, thereby disrupting ion channels, and this is thought to augment ion channel blockade. A modulated receptor hypothesis has been developed to account for the fact that these substances inhibit depolarization of stimulated channels (phasic block) more than resting channels (tonic block).

Local anesthetics should be viewed in the light of the unique and positive aspects of clinical neural blockade. These include (1) complete afferent neural blockade, potentially resulting in complete relief of pain and abolition of the stress response; (2) efferent neural blockade, resulting in increased blood flow to the region of neural blockade; (3) ease of assessment of dose requirements by objective assessment of sensory, sympathetic, and/or motor blockade in the appropriate area; and (4) preclusion of unwanted side effects, such as nausea, ventilatory depression, and gastric stasis, frequently associated with conventional forms of pain control, such as opioid pharmacotherapy.

In clinical use, local anesthetics work as expected and produce local anesthesia limited, mainly, by the skill of the anesthesiologist in placing the right dose of local anesthetic agent, at the right place and at the right time. The right dose is derived from the right combination of concentration and volume. Higher concentrations generally lead to relatively greater motor block, lower concentrations generally lead to relatively greater sensory block, and larger volumes generally lead to greater spread of block. This is essentially as expected from consideration of concentration gradient diffusion and bulk flow factors in relation to the thickness and length of the relevant neural membrane sheath.

Overall, the concept of local anesthetic potency is important as it relates to the minimum effective dose for any given procedure, as well as the potential for toxicity. Clinical aspects of neural blockade, however, are beyond the scope of this brief entry. Local anesthetic potency is typically assessed experimentally by examining drug concentrations that abolish membrane currents in isolated frog or rat sciatic nerve preparations ex vivo or concentrations that produce nerve blocks in laboratory rodents in vivo

(Strichartz and Ritchie 1987; Kanai, et al. 1999; Sinnott and Strichartz 2003; Muguruma et al. 2006). In both laboratory and clinical research, the relative potency of local anesthetic agents is often specified in terms of their "equi-effective anesthetic concentrations," their "maximum recommended doses," or even their "equi-toxic doses" (Rosenberg et al. 2004). Equi-effectiveness is sometimes judged by proportions of subjects having similar neural blockade, or in frequency (or amount) of supplemental analgesic used (or demanded) by patients, in some or other procedure. However, it is a fairly inexact standard in that, with reasonably similar agents being tested in the typically small cohorts of most clinical research studies, with the characteristically high variability found among patients, most studies are more likely to fail to detect a difference between agents (or doses or concentrations, etc.) rather than show that there is no difference between agents. The observed relative potency of the agents, in reality, derives from an amalgam of the structural, stereochemical, and physicochemical properties in that these properties govern the fraction of a dose reaching the relevant receptor, as well as the agent's affinity for the receptor, and its efficacy once it has bound to the receptor. Relative potency is a truly complex issue, but in clinical research it is normally reduced to a simple measure (with usually unknown uncertainty) and then applied to some measure of relative risk from chosen doses of different local anesthetics.

As pointed out below, contemporary local anesthetics have formulaic molecular structures – an aromatic head, an amino tail, and a linking chain – that derive from deconstruction and reconstruction of cocaine, with various functional group modifications from aralkyl substitutions on each component. These substitutions govern the physicochemical properties. Also as pointed out below, physicochemical properties are very important to local anesthetics and are the key to analyzing the differences between otherwise chemically similar agents. Local anesthetics for injection are prepared as slightly acidic aqueous solutions. Their pKa regulates the degree of ionization so that the (protonated, charged) conjugate acid form vastly

exceeds that of the relatively water-insoluble local anesthetic base form. They diffuse through the nerve membrane as the base, but their conjugate acid is required within the axoplasm for receptor block (Hemmings and Greengard 2010). Latency of neural blockade may be reduced by alkalinizing the solution for injection, typically with sodium bicarbonate. However, if too much is added, the local anesthetic base may precipitate from solution – and this is related to the water solubility and structural hydrophobicity or, its obverse, the lipophilicity. Lipophilicity is a fundamental physicochemical property and is assessed as a partition or distribution coefficient of agent between an organic solvent (such as n-octanol) and water (or an aqueous buffer) under laboratory conditions. Greater lipophilicity promotes the local distribution over systemic absorption, thus affecting the potency for neural blockade, local toxicity, and duration of action, as well as the potency for the toxicity of the local anesthetic, once systemically absorbed, essentially all in rank order consistency. The actions and side effects are essentially mediated via the kinetics of binding to and, more importantly, unbinding from their ion channel receptors.

Safe and successful use of local anesthesia requires that effects remain local (Mather 2010). The ability of local anesthetics to block Na^+ and other ion channels in other regional membranes, especially the heart and brain, also leads to their systemic actions. These are generally, but not exclusively, toxic effects in other excitable tissues especially the heart and brain. A useful cardiac antiarrhythmic effect and even an anticonvulsant effect, resultants of initial intrinsic membrane stabilizing actions, occur at lower concentrations. However, the well-known problem is that local anesthetics can achieve sufficient circulating concentrations to cause serious toxicity, particularly CNS excitotoxicity and convulsions along with myocardial depression, electromechanical dissociation, and malignant arrhythmias. These agents can also exert local toxicity, including neurotoxicity and myonecrosis, as well as other, but rarer, forms of systemic toxicity, including allergy and methemoglobinemia.

Most toxicity is explicable in terms of the chemistry and physicochemical properties of the agents. Notwithstanding, we now know that all local anesthetics have a variety of direct and indirect pharmacological actions at subcellular, cellular, organ, and whole-body levels, apart from any untoward physiological effects of neural blockade. Most of these actions are unwanted and are considered toxic, but most pass undetected in patients, particularly where only small doses are used. These side effects can manifest clinically when large amounts of drug are rapidly absorbed from a site of application or, even worse, when the drug is accidentally delivered into a blood vessel during a procedure for neural blockade. The latter is a particularly threatening outcome in which the body is presented with the highest plasma drug concentrations almost immediately after injection so that the probability of toxicity is greatest within seconds to minutes after injection. Under these circumstances, the patient will normally fare better if the local anesthetic is one with a lower intrinsic toxicity. However, the most important pharmacological differences between the various local anesthetic agents can be revealed only by preclinical or experimental studies where their effects can be studied in an isolated cell or organ systems or where intentionally toxic doses can be studied in vivo. Over the past 2 decades, for example, essentially all experimental pharmacological studies indicate that both levobupivacaine and ropivacaine would less likely than bupivacaine to cause serious toxicity at anesthetically equipotent doses, that is, they have a greater margin of safety than bupivacaine, and such pharmacological considerations may be important for the anesthesiologist's selection of a local anesthetic agent. Local anesthetic intoxication is a relatively rare clinical event – but it can have serious consequences (Barrington et al. 2009; Neal et al. 2010; Drasner 2010; Di Gregorio et al. 2010). However, near misses undoubtedly pass undetected, and apart from prudence and/or wisdom after the event, it is not easy to assess the value of choosing an agent with a greater margin of safety.

As far as is being disclosed in the public domain, there does not appear to be further novel molecular development along the lines of conventional local anesthetics. Various local anesthetic applications of natural and semisynthetic biotoxins have appeared sporadically for decades, but there are not yet prominent developments. Pharmaceutical manipulation therefore remains an important area of research, especially for increasing the duration of neural blockade with the existing agents. The simplest method for making local anesthetics longer acting is to administer them by an infusion catheter method, and various novel catheter designs have been introduced for this purpose. However, shorter-acting local anesthetics can have the problem of tachyphylaxis when administered repeatedly, and the potential for toxicity through systemic accumulation of any local anesthetic has to be considered, although the consequences may be more serious for the longer-acting agents. The most familiar pharmacological method for increasing the duration of action is to add a vasoconstrictor to the local anesthetic solution, thereby decreasing the clearance of the agent from the deposited region. This method has been used for over 100 years, with advantages and disadvantages that are well known to the anesthesiologist. Various pharmaceutical methods continue to be developed, whereby the local anesthetic agent is combined in a pharmaceutical preparation to release the agent slowly and thereby prolong its neural residence time. Older methods are based on incorporation of the local anesthetic agent into an oily or a viscous preparation with a greater risk of local toxicity; newer methods are based on incorporation into a biodegradable polymer, encapsulation into liposomes, or complexation with a cyclodextrin. These certainly favor local residence over systemic absorption as well, but the risks from intravascular injection appear to not yet have been studied.

Characteristics

Local anesthesia for attenuating pain from surgical and other physical procedures is a very old concept (Liljestrand 1971). Three main methods have been used:
1. Refrigeration. Avicenna (980–1030) used snow or ice for refrigeration to alleviate pain,

a forerunner of present-day ethyl chloride spray that was introduced in the early twentieth century.

2. Pressure. Nerve compression is an ancient method of alleviating pain, especially for amputations. Paré (1510–1590), a military surgeon, recorded the use of pressure for amputations and found that it produced a better surgical field, hemostasis, and diminished pain. Esmarch (1823–1909), also a military surgeon, followed this idea by using a rubber tube as a compression bandage above the site of operation to produce hemostasis, a technique that was subsequently adopted to reduce the systemic absorption of local anesthetics.

3. True local anesthetics as we now know them. The leaves of the coca bush spring from ancient Peru. Although the extraction and preparation of the pure alkaloid cocaine from coca leaves derive from a nineteenth-century golden age of European chemistry, cocaine was originally identified as a stimulant, like caffeine, and not identified with anesthesia. Its use as a possible local anesthetic came from an observation of von Anrep in 1880, who, after personally observing its anesthetizing properties on cutaneous nerves, tongue, and pupil, recommended such a use. The story of Koller in 1884, trying it as an anesthetic for ophthalmic surgery, is illustrious in anesthesia history; its subsequent use for anesthesia of other mucous membranes followed rapidly. Alongside these experiments were those with its use for producing conduction block. Such experiments then enabled great progress to be made in neurophysiology and have continued to do so through use of chemical variants of the local anesthetics to act as probes. From a clinical perspective, the administration of cocaine represents a somewhat dramatic watershed in the history of anesthesia, because it enabled a new realm of surgery to be performed on conscious patients having anesthesia only of the required part of the body, at a time when operating conditions were still fairly primitive, general anesthetic agents were still in their infancy, and muscle relaxants had not been discovered (Brown and Fink 2008).

Final elucidation of the chemical structure of cocaine did not occur until 1924, but by then, a host of synthetic substitutes had already been prepared and tried, based upon the anesthetic portion ("anesthesiophore") of the cocaine molecule. "**Ester caines**": The introduction of procaine (Novocaine) by Einhorn in 1905 heralded the next generation of local anesthetics having greater safety than cocaine and, importantly, lacking its addictive/psychostimulant properties. As the cocaine anesthesiophore contained an aromatic group, an amine, and an ester linking group, the ensuing generation of synthetic local anesthetics, such as procaine, were also aromatic amines and contained an ester group (usually as p-aminobenzoic acid esters). "**Amide caines**": Apart from cinchocaine (dibucaine, introduced in 1924), which contains a carbamoyl (reversed amide) linkage, the next and present generation of synthetic amino local anesthetics evolved in the mid-1940s, with an amide linkage in place of the isosteric ester from the prototype lidocaine (lignocaine, Xylocaine, which remains in widespread clinical use). These have a major advantage of chemical stability to hydrolysis compared to esters and have largely replaced them. Moreover, allergic reactions derive from chemical structure specificity, and the preponderance of clinical allergic reactions, estimated to be 1% of all local anesthetic adverse reactions, were provoked by "ester caines" probably mediated through their p-aminobenzoic acid metabolites. "**Chiral caines**": Cocaine is a chiral substance; being an enzymically synthesized natural product, it is enantiopure. Further evolution of the "amide caines" took place during the 1990s: Driven by attempts to reduce the systemic toxicity of the long-acting synthetic local anesthetic, bupivacaine (racemic), when the enantiopure agents, ropivacaine and levobupivacaine (both S-configuration), were introduced as substitutes. Of the commonly used drugs, lidocaine, tetracaine, and the procaine family are not chiral local anesthetics.

Like all drugs, the pathway to clinical acceptance of local anesthetics is dominated by toxicity testing, as mentioned above. They first have to survive laboratory testing for tissue toxicity – a primary screen – a local anesthetic that causes leakiness of cell membranes is of no value. Next, they have to have no unexpected or disproportionately high systemic toxicity – and this is critical. Although administered locally for local effects, these agents will be absorbed systemically with a time course that closely relates to the vascularity of the administration site and any maneuvers made to reduce their rate of systemic absorption, such as use of physical or chemical hemostasis with vasoconstrictors (Mather and Tucker 2008). Both tissue and systemic toxicity are determined by the physicochemical properties of the molecule (see Table 1). Two properties predominate: Extent of ionization as determined by the pK_a of the amine group and lipophilicity of its various substituents, especially of the amino group (although two local anesthetics, benzocaine and butamben, have survived from early days without having an ionizable amine group). The clinically useful local anesthetics have a pK_a value within a reasonably narrow range: Those for the surviving "ester caines" (around 9) are higher than for "amide caines" (around 8). The extent of ionization is important, because it bears an inverse relationship to the speed of onset of effect. Lipophilicity is related to the extent of ionization and is important because it bears a direct relationship to the duration of effect through promotion of local distribution into fatty tissues and thereby local residence time.

Prolonging the duration of action of local anesthetics has had an interesting history (Liljestrand 1971; Brown and Fink 2008). Among the early investigations, Corning, in 1885, found that effect duration of cocaine injected subcutaneously could be prolonged by a tourniquet applied above the site of injection; Braun, in 1903, found that the addition of "suprarenal extract" could delay the absorption of procaine, and thus, he founded the first "chemical tourniquet" – a practice that survives today, usually with adrenaline as the vasoconstrictor (so that the commercial products are generally offered as "plain" or "with adrenaline"); Hoffman and Kochmann, in 1912, found that the procaine effect on isolated frog nerves could be prolonged by potassium phosphate – the precursor of knowledge about the importance of K^+ and Na^+ flux in axonal transmission; in 1928, Yeomans et al. used procaine in oil, and Pitkin used a procaine-alcohol-mucilage ("Spinocaine") – the neurolytic properties of oil and alcohol would certainly have prolonged the block! Chemical modifications (and concurrent physicochemical properties), such as through simple substituents, converted procaine into amethocaine (tetracaine) and through homologous series converted mepivacaine into bupivacaine, thereby prolonging duration and increasing toxicity! Unfortunately, overall, the same molecular changes increase both activity and toxicity.

In the 1950s, the quest for new local anesthetics produced a family of N-alkyl piperidine 2,6-xylidides, which helped to reveal the significance of the physicochemical properties (Åberg, Dhuner and Sydnes 1977). It was found that increasing their N-n-alkyl carbon chain gave increased lipophilicity as expected, along with increased tissue toxicity, and a bell-shaped relationship with both duration of neural blockade and systemic toxicity: This is interpreted in terms of extracellular-intracellular membrane permeability having an optimal value for axons; for vital organs, notably heart and brain; and for causing direct effects. In both latter cases, the relevant maxima occurred at C4 or C5 alkyl chain length (see Fig. 1). From this family, the N-methyl homologue, mepivacaine, was selected for clinical development on the basis of its favorable local anesthetic activity, combined with relatively low tissue and systemic toxicity, in animals. The others were deemed unsuitable because of local tissue or systemic toxicity. However, it was soon realized that when the doses were scaled to equi-toxicity in animal tests, a relatively longer duration neural blockade could be obtained; the N-n-butyl derivative, bupivacaine, was tested in the 1960s and soon found acceptance in clinical anesthesiology. Since then, despite a variety of new conventional amino amide type local anesthetics, for example, etidocaine, or new molecular forms, for example, biotoxins and cyclizing lidocaine analogues, being tested and/or introduced for producing long duration of action, bupivacaine has become the de facto standard long-acting local anesthetic.

Postoperative Pain, Local Anesthetics, Fig. 1 Relationships between tissue toxicity, systemic toxicity, and duration of neural blockade for a homologous series of racemic N-alkyl piperidine 2,6-xylidides (Data from Åberg et al. 1977)

Many other properties correlate with lipophilicity. Indeed, classification by lipophilicity is as functionally relevant as classifying these molecules by chemical structures, although the classification would range only between moderately (e.g., lidocaine, mepivacaine) and highly lipophilic (e.g., ropivacaine, bupivacaine). Importantly, binding to plasma and tissue proteins also correlates directly: These properties are significant in drug distribution effects and are believed to play a role, for example, in ameliorating the transplacental distribution when these drugs are used for obstetric anesthesia and/or analgesia.

Although of long-standing concern, the modern-day appreciation of local anesthetic systemic toxicity and the subsequent quest for safer long-acting local anesthetics arose from a brief report in 1979, of sudden onset of convulsions and ventricular fibrillation from etidocaine used for caudal anesthesia in a 31 year-old healthy fit male (Prentiss 1979). This report was followed the same year by an editorial containing a small case report series in which bupivacaine used epidurally, brachial plexus blocks, intravenous regional anesthesia (Bier's block), or caudal block, was believed to be causative in patients undergoing "sudden cardiovascular collapse" from which resuscitation was "difficult" (Albright 1979). A variety of other reports followed implicating bupivacaine and its then competitor, etidocaine, as having disproportionately high cardiotoxicity compared to shorter-acting agents. It had been known from the early days that local anesthetics caused CNS toxicity (frank seizures) and that local anesthetics could cause myocardial depression, but the heart was thought to be more resistant to conduction toxicity from clinical doses. Through preclinical nonhuman and human studies, with intravenous doses of possible replacement drugs, it was learned that all local anesthetics caused similar CNS toxicity in parallel to local anesthetic potency (Mather 2010). The symptomatology in humans could be graded through prodromal features including circumoral numbness (a local sensory effect), light-headedness, tinnitus, visual disturbances, slurring of speech, irrational conversation, and muscle irritability; more serious effects such as seizures, coma, apnea, respiratory arrest, and finally myocardial failure have been studied in experimental animals. Moreover, cardiac studies with in vitro and in vivo models in animals have shown that all local anesthetics are myocardial depressants, again with dose-related potency that parallels local anesthetic potency and with doses that apparently cause no CNS effects. A progression to fatality has been found with decreased left ventricular dP/dt_{max}, increased left ventricular end diastolic pressure, and increased pulmonary artery pressure, progressing to cardiac failure, with hypotension, and abrupt-onset malignant arrhythmias. In all of this, the role of CNS and sympathetic nervous system stimulation seems paramount; the onset of seizures causes a reversal of the myocardial depressant effects but predisposes toward malignant arrhythmias (Mather and Chang 2001; Copeland et al 2008).

Ropivacaine and levobupivacaine, the drugs intended as replacements for bupivacaine, are enantiopure: both are of the S-configuration. Enantioselectivity in CNS models indicates a greater toxicity of the R-enantiomers. In electrophysiological and cardiac contractile function models, it has been found that, with cell channel blockade, for example, firing frequency, $R > S$ enantioselectivity is modest at Na^+ channels, high at certain K^+ channels, and moderate at Ca^{2+} channels. In isolated heart muscle, for example, papillary muscle, $R > S$ enantioselectivity occurs with faster and more potent block of inactivated channels. Other models do not show significant enantioselectivity, for example, disturbances of membrane Na^+/Ca^{2+} pump, inhibition of c-AMP production, and disturbances of mitochondrial energy transduction (Mather and Chang 2001; Groban 2003; Butterworth 2010). Whole-body toxicity studies have consistently found that the S-enantiomers of these drugs produce the same toxicity as either the racemic or R-enantiomers but require greater doses to do so. As they have essentially the same anesthetic potency, the S-enantiomers thus have a greater margin of safety, and this is the rationale for their introduction into clinical practice (Mather and Chang 2001).

Local anesthetics present the reverse side of conventional ▶ pharmacokinetics: Drug absorption is bad, and minimal systemic bioavailability is good (Strichartz 2008; Mather and Tucker 2008). Blood-borne drugs will be delivered to all parts of the body in proportion to the distribution of cardiac output such that, in clinical use, systemic toxicity is well related to drug plasma concentrations, as far as these mirror drug concentrations in well-perfused vital organs. Toxicity occurs because of overdose, unexpectedly rapid absorption, or inadvertent intravascular injection. The role of dose in toxicity is clear, but dosing guidelines are usually couched in terms of dose/body weight. However, while

a small person probably should receive a smaller dose than a large person, the use of weight as a linearizing factor is questionable. Absorption rate is related to vascularity and the surface area of the administration site. All local anesthetics are vasodilators at the concentrations used for neural blockade; hence, the use of vasoconstrictors, in general, reduces the rate of absorption from the administration site, thereby prolonging the duration and reducing the risk of systemic toxicity. The S-enantiomers vasodilate relatively less than the R-enantiomers, so there is, usually, less gain in their being used with adrenaline.

The pharmacokinetics associated with drug systemic absorption from neural blockade can be broadly generalized by several points. (1) A systemically absorbed drug has a similar blood concentration profile to that after intramuscular injection. Biphasic absorption is the rule. This can be interpreted as a "portion" of the dose being absorbed reasonably rapidly with a half-life of around 5–10 min, generating the "peak" arterial blood concentration at around 10 min after injection. The remaining "portion," presumably that distributed into fatty tissues, is absorbed more slowly with a half-life measured in many tens of minutes, thereby sustaining the blood drug concentrations compared to intravenous drug administration. (2) Most pharmacokinetic studies are based upon venous (typically antecubital fossa) blood sampling, and these find peak venous blood concentrations at around 30–60 min after administration. The discrepancy occurs because of drug equilibration in the tissues being drained. However, due to circulatory readjustments to central neural blockade, concomitant medications, etc., venous profiles can vary markedly. (3) The magnitude of the peak drug concentration in arterial blood is usually considerably greater than in venous blood. (4) Due to binding to plasma proteins, plasma drug concentrations are usually greater than blood concentrations, with the discrepancy being greater for the more lipophilic agents. Toxicity, should it occur, will usually occur close to the time of maximum arterial blood drug concentrations. If the drug is administered repeatedly or by infusion, the late-onset

Postoperative Pain, Local Anesthetics, Fig. 2 Drug concentrations after epidural administration in healthy volunteer subjects, showing the time courses in arterial and venous blood samples (Data from Tucker and Mather 1979)

toxicity becomes a concern. Toxicity will then take longer to recede, because the body burden of drug is high (see Fig. 2).

The purpose of knowing "toxic dosage" is, of course, to be able to avoid it. However, the question of toxic dosage needs to be rephrased as knowing the maximum dose that could be used for the neural blockade procedure, while still precluding a significant risk of causing toxicity. To help this, tables of maximum recommended doses of local anesthetics have been produced over the years. Rosenberg et al. (2004) reviewed the issue of currently recommended maximum doses of local anesthetic agents. They asserted that there is no scientific justification nor clinical relevance for the now-dated maximum recommended doses of local anesthetic agents; instead, they proposed that only clinically adequate and safe dose ranges, which account for the site of local anesthetic injection as well as patient comorbidities, are appropriate and that the doses for specific blocks should be modified individually according to patient characteristics.

As pointed out above, the most sinister outcome of neural blockade is that of rapid intravascular injection (Mather et al. 2005). In this case, the body is presented with the highest plasma drug concentrations almost immediately after injection, so that the probability of CNS and/or cardiac toxicity is greatest. Thus, the effects are likely to be particularly severe; however, they are likely to be short lived. However, the systemic effects of the longer-acting and more lipophilic agents are relatively more serious, and probably longer lasting, than those of the shorter-acting agents, despite the overall general similarity in their pharmacokinetic profiles. The essential difference lies in the ability of the longer-acting agents to bind with greater affinity to the ion channel receptors, predisposing and causing a greater probability of fatal cardiac arrhythmias. CNS excitotoxicity is serious enough and can be readily arrested by prompt use of an intravenous anesthetic agent and oxygenation support. Cardiotoxicity remains a serious problem. A survey of American academic regional anesthesiologists indicated that there was no real consensus for treatment of local anesthetic cardiotoxicity (Corcoran et al. 2006). Indeed, there is no strong evidence-based pharmacological method yet developed, even in animals, but oxygenation and cardiac support remain paramount (Moore 2007). Experiments in laboratory animals suggest that lipid infusion (20% Intralipid™) may reduce the cardiotoxic potency of bupivacaine and provide a means of preventing or arresting these effects of bupivacaine (Weinberg et al. 1998, 2003). There are still many unanswered questions about the mechanisms of lipid infusion, but data so far appear encouraging (Rosenblatt et al. 2006). It is not yet clear whether the lipid formulation achieves its desired effect by direct inotropic support or by acting as a "sink" for absorbing bupivacaine from the circulation and/or tissues or by a combination of effects (Weinberg et al. 2006). The applicability to toxicity from other local anesthetics is not yet established and its mechanisms of action still remain controversial (Buckenmaier 2012; Litonius et al 2012a, b). The principles of treatment of cardiotoxicity are still based on general principles of resuscitation, although recent guidelines are indicating that lipid infusion should be readily available (Neal et al. 2010; Weinberg 2010).

Summary

Local anesthetics are among the most versatile of drugs for the management of pain, given the application of the right dose at the right time. Most of their applications, however, require the skill, if not artistry, of an experienced anesthesiologist to place them correctly – where they can provide anesthesia with or without motor blockade or analgesia with or without anesthesia, using the best combination of volume and concentration. However, all local anesthetics, absorbed into or placed in the systemic circulation, can cause systemic toxic effects, where the heart and brain are particularly at risk. In laboratory studies, of the three main contemporary long-acting local anesthetic agents, ropivacaine has been consistently found the least toxic, levobupivacaine intermediate, and bupivacaine the most toxic – all are relatively more toxic than their shorter-acting counterparts, and ropivacaine is somewhat less potent than levobupivacaine and bupivacaine in many

clinical bioassays. By and large, fatal intoxication from bupivacaine is more likely to be caused by ventricular fibrillation, and fatal intoxication from ropivacaine is more likely to be caused by a lidocaine-like electromechanical dissociation or "pump failure." Levobupivacaine fits in between in most measures of toxicity. However, these are generalizations and there are no absolutes in any of this, and death can result from either contractile failure or conduction abnormalities from any of the local anesthetics. The reality is that both ropivacaine and levobupivacaine push the dose-response curves to the right, that is, they are somewhat less potent than bupivacaine in producing their toxic effects, whether CNS excitotoxicity or propensity to cause lethal arrhythmias, but they still can cause the same outcome as bupivacaine. The relative potency for causing CNS and cardiotoxicity appears to run parallel with local anesthetic potency, being linked through lipophilicity and receptor affinity.

All local anesthetics can, and undoubtedly do, cause some systemic effects from their systemic absorption from nerve blocks, but this is rarely monitored. It is not frequently problematic clinically and tends to occur at a more or less predictable time after local anesthetic administration, with successful neural blockade. It does become problematic clinically if the agent is absorbed unexpectedly rapidly, or if a dose is excessive, or if the recipient of the drug is especially susceptible (Rosenberg et al. 2004; Drasner 2010). If the dose is infused, then the effects may occur expectedly and may take a long time to dissipate as the body load of agent may be quite substantial. However, a more fearsome side effect of local anesthetics is acute toxicity due to accidental intravascular administration, occasionally intra-arterial, more commonly intravenous. In such a situation, even conservative dosage recommendations may be problematic because the local anesthetic dose required for most nerve blocks could produce acute toxicity if injected intravascularly (Rosenberg et al. 2004; Di Gregiorio et al. 2010; Mulroy and Hejtmanek 2010). Given that the possibility of accidental intravascular injection is not dependent on the local anesthetic agent selected, it behooves the anesthesiologist to select the local anesthetic best suited to the procedure to be performed and with least tendency to produce toxicity in the event of an accident. The production of the anesthetically reliable "chiral" long-acting local anesthetic agents, ropivacaine and levobupivacaine, has given the anesthesiologist a safer choice (Zink and Graf 2008; Leone et al. 2008). And, it is appearing that intravenous lipid preparations may be the best pharmacological treatment of such intoxication on the horizon should serious systemic cardiotoxicity occur (Weinberg et al. 2010).

Things have not changed much since the early days of cocaine. Good local anesthesia still can be wonderful, and bad local anesthesia still can be lethal.

References

Åberg, G., Dhuner, K.-G., & Sydnes, G. (1977). Studies on the duration of local anaesthesia: Structure/activity relationships in a series of homologous local anaesthetics. *Acta Pharmacologica et Toxicologica, 41*, 432–443.

Albright, G. A. (1979). Cardiac arrest following regional anesthesia with etidocaine or bupivacaine (editorial). *Anesthesiology, 51*, 285–287.

Barrington, M. J., Watts, S. A., Gledhill, S. R., Thomas, R. D., Said, S. A., Snyder, G. L., Tay, V. S., & Jamrozik, K. (2009). Preliminary results of the Australasian regional anaesthesia collaboration: A prospective audit of more than 7000 peripheral nerve and plexus blocks for neurologic and other complications. *Regional Anesthesia and Pain Medicine, 34*, 534–541.

Brown, D. L., & Fink, B. R. (2008). The history of regional anesthesia. In M. J. Cousins & P. O. Bridenbaugh (Eds.), *Neural blockade* (4th ed., pp. 1–23). Philadelphia: Lippincott.

Buckenmaier, C. C., Capacchione, J., Mielke, A. R., Bina, S., Shields, C., Kwon, K. H., McKnight, G., Fish, D. A., Bedocs, P. (2012). The effect of lipid emulsion infusion on postmortem ropivacaine concentrations in swine: endeavoring to comprehend a soldier's death. *Anesthesia and Analgesia, 114*, 894–900.

Butterworth, J. F. (2010). Models and mechanisms of local anesthetic cardiac toxicity. A review. *Regional Anesthesia and Pain Medicine, 35*, 167–176.

Copeland, S. E., Ladd, L. A., Gu, X.-Q., & Mather, L. E. (2008). Effects of general anesthesia on the central nervous and cardiovascular system toxicity of local anesthetics. *Anesthesia and Analgesia, 106*, 1429–1439.

Corcoran, W., Butterworth, J., Weller, R. S., Beck, J. C., Gerancher, J. C., Houle, T. T., & Groban, L. (2006). Local anesthetic-induced cardiac toxicity: A survey of contemporary practice strategies among academic anesthesiology departments. *Anesthesia and Analgesia, 103*, 1322–1326.

Di Gregorio, G., Neal, J. M., Rosenquist, R. W., & Weinberg, G. L. (2010). Clinical presentation of local anesthetic systemic toxicity. *Regional Anesthesia and Pain Medicine, 35*, 181–187.

Drasner, K. (2010). Local anesthetic systemic toxicity. A historical perspective. *Regional Anesthesia and Pain Medicine, 35*, 162–166.

Groban, L. (2003). Central nervous system and cardiac effects from long-acting amide local anesthetic toxicity in the intact animal model. *Regional Anesthesia and Pain Medicine, 28*, 3–11.

Hemmings, H. C., & Greengard, P. (2010). Positively active - how local anesthetics work. *Anesthesiology, 113*, 250–252.

Kanai, Y., Tateyama, S., Nakamura, T., Kasaba, T., & Takasaki, M. (1999). Effects of levobupivacaine, bupivacaine, and ropivacaine on tail-flick response and motor function in rats following epidural or intrathecal administration. *Regional Anesthesia and Pain Medicine, 24*, 444–452.

Leone, S., Di Cianni, S., Casati, A., & Fanelli, G. (2008). Pharmacology, toxicology, and clinical use of new long acting local anesthetics, ropivacaine and levobupivacaine. *Acta Bio-Medica, 79*, 92–105.

Liljestrand, G. (1971). The historical development of local anesthesia. In P. Lechat (Ed.), *Local anesthetics* (International encyclopedia of pharmacology and therapeutics, Vol. 1, pp. 1–38). New York: Pergamon Press.

Litonius, E. S., Niiya, T., Neuvonen, P. J., Rosenberg, P. H. (2012a). Intravenous lipid emulsion only minimally influences bupivacaine and mepivacaine distribution in plasma and does not enhance recovery from intoxication in pigs. *Anesthesia and Analgesia, 114*, 901–906.

Litonius, E., Tarkkila, P., Neuvonen, P. J., Rosenberg, P. H. (2012b). Effect of intravenous lipid emulsion on bupivacaine plasma concentration in humans. *Anaesthesia, 67*, 600–605.

Mather, L. E. (2010). The acute toxicity of local anesthetics. *Expert Opinion on Drug Metabolism & Toxicology, 6*, 1313–1332.

Mather, L. E., & Chang, D.-H. T. (2001). Cardiotoxicity of local anaesthetics. *Drugs, 61*, 333–343.

Mather, L. E., & Tucker, G. T. (2008). Properties, absorption and disposition of local anesthetics. In M. J. Cousins, D. B. Carr, T. T. Horlocker, & P. O. Bridenbaugh (Eds.), *Neural blockade* (4th ed., pp. 48–95). Philadelphia: Wolters Kluwer/Lippincott, Williams & Wilkins.

Mather, L. E., Copeland, S. E., & Ladd, L. A. (2005). Acute toxicity of local anesthetics: Underlying pharmacokinetic and pharmacodynamic concepts. *Regional Anesthesia and Pain Medicine, 30*, 553–566.

Moore, D. C. (2007). Lipid rescue from bupivacaine cardiac arrest: A result of failure to ventilate and maintain cardiac perfusion? *Anesthesiology, 106*, 636–637.

Muguruma, T., Sakura, S., Kirihara, Y., & Saito, Y. (2006). Comparative somatic and visceral antinociception and neurotoxicity of intrathecal bupivacaine, levobupivacaine, and dextrobupivacaine in rats. *Anesthesiology, 104*, 1249–1256.

Mulroy, M. M. F., & Hejtmanek, M. R. (2010). Prevention of local anesthetic systemic toxicity. *Regional Anesthesia and Pain Medicine, 35*, 177–180.

Neal, J. M., Bernards, C. M., Butterworth, J. F., IV, Di Gregorio, G., Drasner, K., Hejtmanek, M. R., Mulroy, M. F., Rosenquist, R. W., & Weinberg, G. L. (2010). ASRA practice advisory on local anesthetic sysemic toxicity. *Regional Anesthesia and Pain Medicine, 35*, 152–161.

Prentiss, J. E. (1979). Cardiac arrest following caudal anesthesia. *Anesthesiology, 50*, 51–53.

Rosenberg, P. H., Veering, B. T., & Urmey, W. F. (2004). Maximum recommended doses of local anesthetics: A multifactorial concept. *Regional Anesthesia and Pain Medicine, 29*, 564–575.

Rosenblatt, M. A., Abel, M., Fischer, G. W., Itzkovich, C. J., & Eisenkraft, J. B. (2006). Successful Use of a 20% lipid emulsion to resuscitate a patient after a presumed bupivacaine-related cardiac arrest. *Anesthesiology, 105*, 217–218.

Sinnott, C. J., & Strichartz, G. R. (2003). Levobupivacaine versus ropivacaine for sciatic nerve block in the rat. *Regional Anesthesia and Pain Medicine, 28*, 294–303.

Strichartz, G. R. (2008). Novel ideas of local anaesthetic actions on various ion channels to ameliorate postoperative pain. *British Journal of Anaesthesia, 101* (1), 45–47.

Strichartz, G. R., & Ritchie, J. M. (1987). The actions of local anesthetics on Ion channels of excitable tissues. In G. R. Strichartz (Ed.), *Local anesthetics, handbook of experimental pharmacology* (Vol. 81, pp. 21–52). Berlin: Springer.

Strichartz, G. R., Pastjin, E., & Sugimoto, K. (2008). Neural physiology and local anesthetic action. In M. J. Cousins & P. O. Bridenbaugh (Eds.), *Neural blockade* (4th ed., pp. 26–47). Philadelphia: Lippincott.

Tucker, G. T., & Mather, L. E. (1979). Clinical pharmacokinetics of local anaesthetic agents. *Clinical Pharmacokinetics, 4*, 241–278.

Weinberg, G. (2010). Treatment of local anesthetic systemic toxicity (LAST). *Regional Anesthesia and Pain Medicine, 35*, 188–189.

Weinberg, G., Ripper, R., Feinstein, D. L., & Hoffman, W. (2003). Lipid emulsion infusion rescues dogs from bupivacaine-induced cardiac toxicity. *Regional Anesthesia and Pain Medicine, 28*, 198–202.

Weinberg, G. L., Ripper, R., Murphy, P., Edelman, L. B., Hoffman, W., Strichartz, G., & Feinstein, D. L. (2006). Lipid infusion accelerates removal of bupivacaine and recovery from bupivacaine toxicity in the isolated rat heart. *Regional Anesthesia and Pain Medicine, 31*, 296–303.

Weinberg G. L., VadeBoncouer T., Ramaraju G. A., Garcia-Amaro M. F., Cwik M. J. (1998). Pretreatment or resuscitation with a lipid infusion shifts the dose-response to bupivacaine-induced asystole in rats. *Anesthesiology, 88*, 1071–1075.

Zink, W., & Graf, B. M. (2008). The toxicity of local anesthetics: The place of ropivacaine and levobupivacaine. *Current Opinion in Anaesthesiology, 21*, 645–650.

Postoperative Pain, Membrane-Stabilizing Agents

Ross MacPherson
Department of Anesthesia and Pain Management, Royal North Shore Hospital University of Sydney, Sydney, NSW, Australia

Synonyms

Membrane-stabilizing agents

Definition

The term "membrane stabilization" refers to the suppression or attenuation of the transmission of neuronal impulses, usually via a reduction in ionic fluxes through sodium channels (Khodorova et al. 2001). The most common membrane-stabilizing drugs are lignocaine, mexiletine, and flecainide.

Characteristics

Membrane-stabilizing agents have multiple pharmacological functions. All have significant anti-arrhythmic activity and are primarily used for this purpose, while their utility as analgesics has been a relatively late discovery (Mao and Chen 2000).

At a cellular level, much research has been carried out examining the different subgroups of sodium channels involved in the development of neuropathic pain (Strichartz et al. 2002). Divided into ▶ tetrodotoxin-sensitive and tetrodotoxin-resistant channels, it is the latter group that seems to be involved in the generation and conduction of nociceptive neuronal discharges (Bau et al. 2001). Recent studies have identified further subgroups within each of these divisions (Porreca et al. 1999).

Lignocaine has multiple uses. While it is most widely used as a local anesthetic agent, its membrane-stabilizing properties can be exploited through its administration systemically, either by the subcutaneous or intravenous routes. When used in this way, the lignocaine infusion can be used in both the diagnosis and treatment of neuropathic pain. Flecainide and mexiletine can be administered orally.

Unfortunately, two barriers limit the use of membrane-stabilizing drugs in acute postoperative pain. The first is that well controlled trials demonstrating efficacy are lacking and secondly the adverse effect profile is significant and the risk of toxicity, even when used at "therpautic doses" is considerable.

References

Bau, M. F., Dreimann, M., Olschewski, A., et al. (2001). Effect of drugs used for neuropathic pain management on tetrodotoxin-resistant Na$^+$ currents in rat sensory neurons. *Anesthesiology, 94*, 137–144.

Khodorova, A., Meissner, K., Leeson, S., et al. (2001). Lidocaine electively blocks abnormal impulses arising from noninactivating Na channels. *Muscle & Nerve, 24*, 634–637.

Mao, J., & Chen, L. L. (2000). Systemic lidocaine for neuropathic pain relief. *Pain, 87*(1), 7–17.

Porreca, F., Lai, J., Bian, D., et al. (1999). Comparison of the potential role of the tetrodotoxin-insensitive sodium channels, PN3/SNS and NaN/SNS2, in rat models of chronic pain. *Proceedings of National Academics of Sciences USA, 96*, 7640–7644.

Strichartz, G. R., Zhou, Z., Sinnott, C., et al. (2002). Therapeutic concentrations of local anaesthetics unveil the potential role of sodium channels in neuropathic pain. *Novartis Foundation Symposium, 241*, 189–201.

Postoperative Pain, Methadone

Ross MacPherson
Department of Anesthesia and Pain Management, Royal North Shore Hospital University of Sydney, Sydney, NSW, Australia

Synonyms

Amidine; Amidine hydrochloride; (+/−)-6-Dimethylamino-4,4-Diphenylheptane-3-One-Hydrochloride; Methadone; Phenadone; Physeptone

Definition

Synthetic opioid of the diphenylpropylamine class.

Characteristics

Methadone has a unique place among opioid analgesics, by virtue of its somewhat unusual pharmacokinetic profile and its range of clinical applications, and its role in acute pain management is undergoing an expansion. Methadone is a truly synthetic opioid, its usefulness as an analgesic first being noted by German chemists during the 1940s as a result of the systematic investigation of compounds possessing pethidine-like structures.

In terms of its pharmacological profile, while it shares many of the attributes common to members of the opioid group, in general there are some important differences (Garrido and Troconiz 1999). The first is its impressive oral bioavailability. Most opioids are well known for their poor and erratic absorption following oral dosing, yet methadone has a bioavailability variously quoted as being from 66 % to 100 % (Kristensen et al. 1996). It can be detected in the plasma within 30 min after oral dosing, with peak plasma levels generally achieved within 4–6 h.

Secondly, is its variable and prolonged duration of action, which has been measured at anything from between 12 and 150 h following a single dose.

This has important clinical applications. For one thing, these data suggest significant intersubject variation in dosage requirements, stressing the importance of individualizing drug regimes. Many reasons for this variability have been suggested, with variations in both cytochrome P450 3A (CYP3A) and CYP2D6 activity and binding to alpha 1 glycoprotein being cited as critical (Boulton et al. 2001). Such variability means that the time required to reach steady-state levels can be as long as a week, or as short as a day, implying that methadone is not always a suitable choice for the first-line treatment of acute pain states. Likewise, the prescriber must be alert for the development of delayed toxicity, which can present even after cessation of treatment because of drug accumulation and deposition in extravascular sites. Furthermore, the fact that methadone is so reliant on P450 cytochromes for metabolism, it must be remembered that a wide variety of other drugs can enhance (e.g., carbamazepine, phenytoin) or inhibit (SSRIs, antifungal agents) methadone metabolism (McIntyre et al. 2010).

The interesting pharmacokinetic profile of methadone also allows it to be administered on a once-daily basis for the management of abstinence syndrome in opioid-dependent patients (Mattick et al. 2002). It is also particularly useful in ▶ opioid rotation programs, because it seems that only partial tolerance occurs between methadone and other full opioid agonists.

Patients taking methadone as part of an opioid replacement program can present particular problems, and these can be difficult to address, as there have been no high-quality trials to assess the efficacy of any particular strategy. However, one set of general guidelines (Alford et al. 2006) suggests that patients taking methadone as management of opioid withdrawal do not gain any significant analgesic effect from the drug. Their usual dose should be continued throughout the period of acute pain, which should be managed in the usual way, with short-acting opioids as needed.

The activity of methadone extends beyond that at the mu-opioid receptor, as it has been shown to also possess antagonist action at NMDA receptor sites. Although this has been known for some time (Ebert et al. 1998), it is still a matter of debate as to the degree to which this activity contributes to its analgesic effect (Chizh et al. 2000).

Since drugs with NMDA antagonist properties have a clear role in the management of neuropathic pain, it is not surprising that there have been reports of methadone being successfully used to treat such conditions (Bruera and Sweeney 2002; Makin and Ellershaw 1998).

The majority of literature dealing with the analgesic properties of methadone is concerned with its application in either chronic pain or cancer pain states (Ripamonti and Bianchi 2002; Bidlack et al. 2000). However, there is increasing interest in its use in acute pain states (Shir et al. 2001), especially where there may be an element of neuropathic pain, or of tolerance to other opioids such as morphine (Sartain and Mitchell 2002). Indeed, despite the somewhat unfavorable pharmacokinetic parameters outlined above, intravenous methadone by either bolus or P.C.A has been used in difficult cases to assist in the management of intractable pain states (Sabatowski et al. 2002; Fitzgibbon and Ready 1997). Although results have been favorable, methadone used by this route may result in high plasma levels, which is not without risk, as cases of serious cardiac arrhythmias been reported in some patients (Krantz et al. 2002). Other research has focused on the administration of methadone via the epidural route in the management of acute postoperative pain. In one study (Prieto-Alvarez et al. 2002), methadone was administered epidurally either as a continuous infusion or as an intermittent bolus. Due to the long half-life of the drug, bolus administration is needed only every eight hours. Results were favorable with good pain relief in both groups of patients who had undergone either abdominal or lower limb surgery. However, because of its long time to reach steady state, there are probably better alternatives for the acute management of severe pain. However, for ongoing acute pain, it has a definite role.

As with many other commonly used analgesics such as ketamine, methadone is currently presented as a racemic mixture, with individual enantiomers displaying significant differences. The levorotatory enantiomer (R-methadone) displays greater mu-receptor affinity and subsequently has enhanced analgesic potency, assessed as between 8 and 50 times more when compared to S-methadone. However, both enantiomers bind to the NMDA receptors. Likewise the desired therapeutic effects of the drug, such as analgesia and suppression of withdrawal syndrome, lie almost entirely with the R-isomer (Olsen et al. 1976). Again, as with ketamine, researchers have recently succeeded in producing preparations of the active isomer, which may herald an expanded role for methadone in the future.

In summary, methadone is a potent opioid agonist with a favorable pharmacokinetic profile that allows for oral administration on a once- or twice-daily basis. It also has antagonist activity at NMDA receptor which might contribute to its analgesic activity, especially in neuropathic pain states. Due to limited cross-tolerance between methadone and other opioids, it can be of use in the management of nociceptive pain states where patients have become tolerant to other opioids.

References

Alford, D. P., Compton, P., & Samat, J. H. (2006). Acute pain management for patients receiving maintenance methadone or buprenorphine therapy. *Annals of Internal Medicine, 144*(2), 127–134.

Bidlack, J. M., McLaughlin, J. P., & Wentland, M. P. (2000). Partial opioids. Medications for the treatment of pain and drug abuse. *Annals of the New York Academy of Sciences, 909,* 1–11.

Boulton, D. W., Arnaud, P., & DeVane, C. L. (2001). Pharmacokinetics and pharmacodynamics of methadone enantiomers after a single oral dose of racemate. *Clinical Pharmacology & Therapeutics, 70,* 48–57.

Bruera, E., & Sweeney, C. (2002). Methadone use in cancer patients with pain: A review. *Journal of Palliative Medicine, 5,* 127–138.

Chizh, B. A., Schults, H., Scheede, M., et al. (2000). The N-methyl-D-asparate antagonistic and opioid components of d-methadone antinociception in the Rat spinal cord. *Neuroscience Letters, 296,* 117–120.

Ebert, B., Thorkildsen, C., Anderson, S., et al. (1998). Opioid analgesics as noncompetive N-methyl-D-asparate (NMDA) anatagonists. *Biochemical Pharmacology, 56,* 553–559.

Fitzgibbon, D. R., & Ready, L. B. (1997). Intravenous high-dose methadone administered by patient controlled analgesia and continuous infusion for the treatment of cancer pain refractory to high-dose morphine. *Pain, 73,* 259–261.

Garrido, M. J., & Troconiz, I. F. (1999). Methadone: A review of its pharmacokinetic/pharmacodynamic properties. *Journal of Pharmacological and Toxicological Methods, 42,* 61–66.

Krantz, M. J., Lewkowiez, L., Hays, H., et al. (2002). Torsade de Pointes associated with very high dose methadone. *Annals of Internal Medicine, 137,* 501–504.

Kristensen, K., Blemmer, T., Angelo, H. R., et al. (1996). Stereoselective pharmacokinetics of methadone in chronic pain patients. *Therapeutic Drug Monitoring, 18,* 221–227.

Makin, M. K., & Ellershaw, J. E. (1998). Substitution of another opioid for morphine: Methadone can be used to manage neuropathic pain related to cancer. *British Medical Journal, 317,* 81.

Mattick, R. P., et al. (2002). Methadone maintenance therapy versus no opioid replacement therapy for opioid dependence. *Cochrane Database of Systematic Reviews,* (4), CD002209.

McIntyre, P. E., Schug, S. A., Scott, D. A., Visser, E. J., Walker, S. M., & Working group of the Australian and New Zealand College of Anaesthetists and FAculty of Pain Medicine. (2010). *Acute pain management: Scientific guidelines* (3rd ed.). Melbourne: ANZCA & FPM.

Olsen, G., Wendel, H. A., Livermore, J. D., Leger, R. M., Lynn, R. K., & Gerber, N. (1976). Clinical effects and pharmacokinetics of racaemic methadone and its optical isomers. *Clinical Pharmacology & Therapeutics, 21,* 147–157.

Prieto-Alvarez, P., Tello-Galindo, I., Cuenca-Pena, J., et al. (2002). Continuous epidural infusions of racaemic methadone results in effective postoperative analgesia and Low plasma concentrations. *Canadian Journal of Anesthesia, 49,* 25–31.

Ripamonti, C., & Bianchi, M. (2002). The use of methadone for cancer pain. *Hematology/Oncology Clinics of North America, 16,* 543–555.

Sabatowski, R., Kasper, S. M., & Radbruch, L. (2002). Patient-controlled analgesia with intravenous L-methadone in a child with cancer pain refractory to high-dose morphine. *Journal of Pain and Symptom Management, 23,* 3–5.

Sartain, J. B., & Mitchell, S. J. (2002). Successful use of oral methadone after failure of intravenous morphine and ketamine. *Anaesthesia and Intensive Care, 30,* 487–489.

Shir, Y., Rosen, G., Zelden, A., et al. (2001). Methadone is safe for treating hospitalized patients with severe pain. *Canadian Journal of Anesthesia, 48,* 1109–1113.

Postoperative Pain, Mexiletine

Ross MacPherson
Department of Anesthesia and Pain Management,
Royal North Shore Hospital University of
Sydney, Sydney, NSW, Australia

Synonyms

Methyl-2-(2,6-xylyloxy)-ethylamine-hydrochloride; Mexiletine/Mexitil; Mexitil

Definition

Membrane-stabilizing and antiarrhythmic agent
(Class IB).

Characteristics

Mexiletine, an antiarrhythmic agent structurally
related to lignocaine, has been the most
commonly used oral agent of the membrane-
stabilizing group, although still generally much
less popular than antidepressant and anticonvulsant agents. As a result, the number of studies
involving mexiletine is small, and the evidence
supporting its use is somewhat less impressive
than drugs in other categories. Indeed some
studies have failed to demonstrate any significant
effects on neuropathic pain symptoms (Wallace
et al. 2000; Chiou-Tan et al. 1996).

Furthermore, these studies have produced
mixed results, with some favoring the drug's
use (Dejgard et al. 1988; Lindstrom and
Lindblom 1987), but others suggesting that
even in high doses, it has minimal effectiveness
(Wallace et al. 2000; Ando et al. 2000). The main
concerns are with side effects, both cardiac and
noncardiac, which can be common (Chabal et al.
1992; Jarvis and Coukell 1998; Kreeger and
Hammill 1987). Patients in whom the use of
mexiletine is proposed need to undergo cardio-
logical assessment and have serial ECG
examinations performed prior to commencement
of therapy (Portenoy 1999), and any symptom
suggestive of being related to cardiac abnormality needs to be investigated. Noncardiac adverse
effects include upper gastrointestinal discomfort,
which is common and often limits treatment
(Rowbotham 2002), and tremor. As the range of
agents available for neuropathic pain management has expanded, the role of mexilitine, with
its potentially significant side effects as declined
significantly, with other less toxic agents being
preferred.

References

Ando, K., Wallace, M., Braun, J., et al. (2000). Effect of
    oral mexiletine on capsaicin-induced allodynia and
    hyperalgesia: A double-blind, placebo-controlled,
    crossover study. *Regional Anesthesia and Pain
    Medicine, 25,* 1–2.
Chabal, C., Jacobson, L., Mariano, A., et al. (1992).
    The Use of oral mexiletine for the treatment of pain
    after peripheral nerve injury. *Anesthesiology, 76,*
    513–517.
Chiou-Tan, F. Y., Tuel, S. M., Johnson, J. C., et al. (1996).
    Effect of mexiletine on spinal cord injury dysesthetic
    pain. *American Journal of Physical Medicine &
    Rehabilitation, 75,* 84–87.
Dejgard, A., Petersen, P., & Kastrup, J. (1988). Mexiletine
    for treatment of chronic painful diabetic neuropathy.
    The Lancet, 1, 9–11.
Jarvis, B., & Coukell, A. J. (1998). Mexiletine: A review
    of its therapeutic use in painful diabetic neuropathy.
    Drugs, 56, 691–707.
Kreeger, W., & Hammill, S. C. (1987). New antiarrhythmic
    drugs: Tocainide, mexiletine, flecainide, encainide, and
    amiodarone. *Mayo Clinic Proceedings, 62,* 1033–1050.
Lindstrom, P., & Lindblom, U. (1987). The analgesic
    effect of tocainide in trigeminal neuralgia. *Pain, 28,*
    45–50.
Portenoy, R. K. (1999). Opioid and adjuvant analgesics.
    In M. Devor (Ed.), *Pain – an updated review.*
    Refresher course syllabus 9th world congress on pain
    (pp. 1–18). Vienna: IASP Press.
Rowbotham, M. C. (2002). Neuropathic pain: From basic
    science to evidence based treatment. In J. Dostrovsky
    (Ed.), *Pain - an updated review.* Refresher
    Course Syllabus 10th World Congress on Pain
    (pp. 165–176). San Diego: IASP Press.
Wallace, M., Magnuson, S., & Ridgeway, B. (2000).
    Efficacy of oral mexiletine for neuropathic pain with
    allodynia: A double-blind, placebo-controlled, cross-
    over study. *Regional Anesthesia and Pain Medicine,
    25,* 459–467.

Postoperative Pain, Morphine

Peter C. A. Kam
Anaesthesia, Central Clinical School,
The University of Sydney, Sydney, Australia

Synonyms

Kapanol® controlled-release pellets within a capsule; Morphine; Morphine sulfate; MS Contin® controlled (slow)-release morphine sulfate; Ordine®-morphine-hydrochloride

Definition

A strong (potent) naturally occurring opiate.

Characteristics

▶ Morphine is the most commonly used strong opioid of choice for acute pain. Other strong opioids are used mainly when morphine is not readily available or when the patient experiences intolerable adverse effects with morphine (McQuay 1999; Moore and McQuay 1998).

Chemistry

Morphine is a naturally occurring phenanthrene derivative. It is poorly lipid soluble due to the 2 hydroxyl moieties. The OH group at C3 enhances van der Waal forces that bind the aromatic phenol ring to the opioid receptor.

Morphine is formulated as tablets, suspensions, and as slow-release capsules and granules in a wide range of strengths. The parenteral preparation may be given subcutaneously, intramuscularly, and intravenously. Morphine is the principal alkaloid of opium and constitutes 9–17 % of opium.

Pharmacodynamics

Morphine is a μ-agonist, binding to receptors in the brain, spinal cord, and other tissues. Following its binding to μ-receptors in the brain (periaqueductal gray, hippocampus) and the spinal cord (lamina 1), morphine produces analgesia – particularly effective for visceral pain. Sedation and drowsiness are frequently observed with therapeutic doses. In the absence of pain, morphine sometimes produces dysphoria (unpleasant sensation of fear and anxiety). Nausea and vomiting occur via stimulation of serotonin and dopamine receptors in the chemoreceptor trigger zone by morphine. Morphine may induce mild bradycardia due to vagal stimulation and hypotension secondary to histamine release and a decrease in sympathetic tone. Morphine constricts the sphincters of the gastrointestinal tract, resulting in constipation. Pruritus, most marked following epidural or intrathecal administration, does not appear to be associated with histamine release. Morphine inhibits the release of ACTH, prolactin, and gonadotrophic hormones. ADH secretion is increased. The tone of bladder detrusor and sphincter of the bladder is increased, and this precipitates urinary retention.

Pharmacokinetics

Morphine is a weak base ($pK_a - 8.0$). When administered orally, morphine is ionized in the acidic gastric contents, and absorption only occurs when it becomes unionized in the alkaline environment of the small bowel. It undergoes extensive first-pass metabolism, and its oral bioavailability is 25–30 %. Its peak effects occur after 10 and 30 min following intravenous and intramuscular injection, respectively, lasting 3–4 h. Unlike most other opioids, morphine is relatively water soluble (Hull 1991; Aitkenhead et al. 1984; Dahlstrom et al. 1982).

Morphine undergoes extensive hepatic biotransformation by phase II reactions to morphine-3-glucuronide (70 %) and morphine-6-glucuronide (5–10 %), and the remainder undergoes sulfation. Morphine-3-glucuronide has no analgesic effects and is possibly a μ-antagonist. Morphine-6-glucuronide is pharmacologically active and is 13 times more potent than morphine. The plasma concentration of morphine-6-glucuronide exceeds that of the parent drug (by a factor of 9) after 30 min. of intravenous administration. It crosses the blood–brain barrier and contributes to the analgesic effects of morphine. Both metabolites are excreted renally and accumulate in renal failure.

There are two slow-release formulations, ▶ Kapanol® and MS Contin®. Kapanol is a formulation of morphine prepared as polymer-coated sustained-release pellets contained in a capsule. Following oral administration, the rate of absorption of morphine is slower. A single 50 mg oral dose of Kapanol results in a mean peak plasma concentration of 8.1 ng/ml at 8.5 h. A steady state is achieved within 2 days when Kapanol is given on a fixed dosing regimen. Kapanol®, when given on a 12 hourly dosing regimen, results in a lower mean peak plasma morphine concentration and higher mean trough plasma morphine concentration than the same total dose of morphine solution given at 4 hourly dosing schedules (Gourlay et al. 1997; Gourlay 1998). The reduced fluctuation in plasma morphine concentration may reduce the incidence of breakthrough pain and adverse effects.

▶ MS Contin® is a controlled (slow)-release tablet formulation of morphine. MS Contin tablets produce peak morphine concentrations 4–5 h post dose, and therapeutic concentrations persist for about 12 h.

Clinical Efficacy

A systematic review of a single IM dose of morphine 10 mg for moderate to severe postoperative pain reported ▶ numbers needed to treat (NNT) of 2.9. This means that one out of every three patients treated achieved more than 50 % pain relief with morphine. Increasing the dose of morphine would lower the NNT. Therefore, the key to morphine use is to titrate the dose to analgesic effect in individual patients.

Morphine is given by intramuscular or subcutaneous injection in doses of 0.1–0.2 mg/kg body weight every 3–6 h. Lower doses are used in children and elderly patients and should be titrated to effect. When a rapid onset of action is desirable, morphine can be administered intravenously with caution in a dose of 0.05–1.0 mg/kg over 5–15 min with close monitoring. Morphine

can also be administered by continuous intravenous infusion using a controlled rate infusion pump or a syringe driver. The dose of morphine should be titrated according to the patient's analgesic requirements and previous opioid usage. Most adult patients, with no previous exposure to opioid usage, will require a morphine infusion at a rate of 0.5–2.0 mg/h after adequate analgesia has been achieved with a loading dose. In children, an infusion at a rate of 0.01–0.05 mg/kg/h provides adequate analgesia.

▶ Patient-controlled analgesia using morphine as the agent to provide postoperative pain relief is a common method used for postoperative analgesia. This allows the patient to titrate the amount of morphine they require for adequate pain relief against sedation or other side effects. For most opioid-naïve adult's demands, doses of 0.5–2.0 mg, with a lockout interval of 5 min, are adequate via PCA. In some patients, a background continuous infusion at a basal rate (usually 1 mg/h) may be required to provide adequate analgesia, but this is associated with a higher incidence of adverse effects, and therefore requires close monitoring.

When morphine is administered to patients with chronic pain associated with cancer, the dose of morphine must be individualized according to the response and tolerance of patient. The dose of morphine should be reduced in poor-risk patients, in very old or very young patients, and in patients receiving other depressants. Oral morphine should be used in preference to parenteral morphine whenever it is possible to achieve analgesia by this route. When patients are converted from parenteral morphine to oral morphine, the dose of oral morphine should be increased, because about 60 % of oral morphine is metabolized in the liver by the first-pass effect (Moore and McQuay 1998).

Conversion of the Slow-Release Formulation

Based on single-dose studies, parenteral morphine 10 mg is equipotent to oral morphine 60 mg. In chronic use this ratio does not apply, and the ratio of parenteral morphine 10 mg to oral morphine 30 mg may be more appropriate. However, it is best to assume that the parenteral to oral potency ratio is high when converting from a slow-release morphine (e.g., Kapanol) to parenteral morphine and estimate the parenteral morphine based on 1:6 ratio. The total daily oral morphine dose is estimated, and should then be divided into two Kapanol or MS Contin doses given every 12 h. Kapanol is not bioequivalent to other controlled-release morphine formulations and has reduced fluctuation in dose-adjusted plasma morphine levels (Gourlay et al. 1997).

Opioid analgesic agents do not relieve dysesthetic pain, postherpetic neuralgia, and some forms of headache.

Morphine may have a prolonged and cumulative effect in patients with impaired renal function. In these patients, the active metabolite morphine-6-glucuronide can accumulate so that analgesia may last for 6–8 h (Aitkenhead et al. 1984).

References

Aitkenhead, A. R., Vater, M., Achola, K., et al. (1984). Pharmacokinetics of single dose IV morphine in normal volunteers and patients with end stage renal failure. *British Journal of Anaesthesia, 56,* 813–819.

Dahlstrom, B., Tamsen, A., Paalzow, L., et al. (1982). Patient-controlled analgesic therapy, part IV: Pharmacokinetics and plasma concentrations of morphine. *Clinical Pharmacokinetics, 7,* 266–279.

Gourlay, G. K. (1998). Sustained relief of chronic pain. Pharmacokinetics of sustained release morphine. *Clinical Pharmacokinetics, 35,* 173–190.

Gourlay, G. K., Cherry, D. A., Onley, M. M., et al. (1997). Pharmacokinetics and pharmacodynamics of twenty-four hourly kapanol compared to twelve hourly MS contin in the treatment of severe cancer pain. *Pain, 69,* 295–302.

Hull, C. J. (1991). *Pharmacokinetics for anaesthesia* (pp. 290–295). Oxford: Butterworth-Heinemann.

Kalso, E., Heiskanen, H., Rantio, M., et al. (1996). Epidural and subcutaneous morphine in the management of cancer pain: A double cross over study. *Pain, 67,* 443–449.

McQuay, H. (1999). Opioids in pain management. *Lancet, 353,* 2229–2232.

McQuay, H. J., & Moore, A. (1998). *An evidence-based resource of pain relief.* Oxford: Oxford University Press.

Moore, R. A., & McQuay, H. (1998). Peak plasma concentrations after oral morphine – A systematic review. *Journal of Pain and Symptom Management, 16,* 388–402.

Postoperative Pain, Nonsteroidal Anti-Inflammatory Drugs

Ian Power[1] and Lesley A. Colvin[2]
[1]Critical Care and Pain Medicine,
The University of Edinburgh, Edinburgh, UK
[2]Department of Anaesthesia, Critical Care and Pain Medicine Western General Hospital, Edinburgh, UK

Synonyms

Cyclooxygenase inhibitors; NSAID

Definition

In the twentieth century many chemical substances, now known as the ▶ NSAIDs, Survey or NSAIDs, were developed with the same anti-inflammatory, analgesic and antipyretic effects of aspirin. The introduction a few years ago of injectable preparations of the NSAIDs ketorolac, diclofenac, ketoprofen, and tenoxicam delivered perioperative analgesia free from opioid disadvantages of respiratory depression, sedation, nausea and vomiting, gastrointestinal stasis, or abuse potential. Extensive clinical investigation and use of NSAIDs have confirmed that they are effective postoperative analgesics, although significant contraindications and adverse effects limit this use.

Characteristics

General

NSAIDs have a spectrum of analgesic, anti-inflammatory, and antipyretic effects, but those used for perioperative pain relief produce marked analgesia with little anti-inflammatory action.

NSAID absorption is rapid by all routes of administration, whether enteral or by injection. NSAIDs are highly protein bound and have low volumes of distribution of the order of 0.1 l/kg and the unbound fraction is active. Consequently,

Postoperative Pain, Nonsteroidal Anti-Inflammatory Drugs, Fig. 1 Eicosanoid production

NSAIDs can potentiate the effects of other highly protein-bound drugs by displacing them from protein-binding sites (oral anticoagulants, oral hypoglycemics, sulfonamides, and anticonvulsants).

Mechanism of Action

Many NSAID and aspirin effects can be explained by inhibition of prostaglandin synthesis in peripheral tissues, nerves, and the central nervous system. However, the NSAIDs and aspirin may have other mechanisms of action independent of any effect on prostaglandins, including effects on basic cellular and neuronal processes (McCormack 1994).

Von Euler first described "prostaglandins" as locally active tissue agents that produce smooth muscle contraction. Many prostaglandins are now recognized, and they are all based on a 20-carbon chain molecule. Hence, they are one family of the "eicosanoids" (from "eicosa," Greek for "twenty"), oxygenated metabolites of arachidonic acid, and other polyunsaturated fatty acids that include leukotrienes. The basal rate of prostaglandin production is low and is regulated by tissue stimuli or trauma that activates phospholipases to release arachidonic acid. Prostaglandins are then produced by the enzyme prostaglandin endoperoxide (PGH) synthase, which has both cyclooxygenase and hydroperoxidase sites. Two subtypes of cyclo-oxygenase enzyme have been identified:

the "constitutional" COX-1 and the "inducible" COX-2 (see Fig. 1).

Prostaglandins have many physiological functions including gastric mucosal protection, renal tubular function and vasodilation, and bronchodilation; endothelial prostacyclin produces vasodilation and prevents platelet adhesion; platelet thromboxane produces aggregation and vessel spasm. Such physiological roles are mainly regulated by COX-1 and are the basis for many of the adverse effects associated with NSAID use. Tissue damage induces COX-2 production with the production of prostaglandins that produce pain and inflammation. COX-2 may be "constitutive" in some tissues, including the kidney.

NSAIDs, like aspirin, are "non-selective" cyclo-oxygenase inhibitors that inhibit both COX-1 and COX-2, unlike the newer "COX-2 inhibitors" that have been developed to inhibit selectively the inducible form. Paracetamol may work by inhibiting a further subtype "COX-3" in the central nervous system (see Fig. 2).

Nonselective COX-1 and COX-2 inhibition confers both the analgesic and the adverse effect (peptic ulceration, renal impairment, bleeding, and aspirin-induced asthma) profile of NSAIDs.

Efficacy

The efficacy of single-dose NSAIDs as postoperative analgesics has been examined and

Postoperative Pain, Nonsteroidal Anti-Inflammatory Drugs,
Fig. 2 NSAID mechanism of action

```
                Phospholipids                    ┌──────────┐
                                                 │ Steroids │
                                                 └──────────┘
                                                    │ −
                     │            ◄─ ─ ─ ─ ─ ─ ─ ─ ─ ─ ─ ─ ─ ─ ─  Phospholipase
                     ▼
                Arachidonic acid

                     │            ◄─ ─ ─ ─ ─ ─ ─ ─ ─ ─ ─ ─ ─ ─ ─  Cyclo-oxygenase
                     ▼                              ▲ −
                Endoperoxides                  ┌────────┐
                                               │ NSAID  │
                     │                         └────────┘
            ┌────────┴────────┐
            ▼                 ▼
       Prostaglandins    Thromboxanes
```

confirmed by many studies. Using the concept of ► number needed to treat (NNT), obtained from meta-analyses of well-designed studies of postoperative analgesia, McQuay and Moore from Oxford University have enabled comparison of the efficacy of NSAIDs with other types of analgesics for acute pain relief after surgery (www.jr2.ox.ac.uk/bandolier/booth/painpag/). From such work, the NNT of ketorolac 10 mg is 2.6, diclofenac 50 mg 2.3, and ibuprofen 400 mg 2.4. For comparison, the NNT of morphine 10 mg intramuscularly is 2.9 and codeine 60 mg 16.7.

Individual studies have confirmed that NSAIDs, such as ketorolac, are effective postoperative analgesics (Power et al. 1990; NHMRC 1999). When given in combination with opioids, NSAIDs produce better analgesia and reduce opioid consumption by 25–50 %. Of the NSAIDs available, most experience has been gained in surgical patients with ketorolac, which has proven to be an effective postoperative analgesic (Gillis and Brogden 1997). To improve safety, ketorolac doses were revised in the United Kingdom after the drug was launched, and the current recommended dose in adults younger than 65 years is 10 mg every 4–6 h, with a maximum daily dose of 40 mg. In adults over 65 years of age, the recommended dose is still 10 mg, but the dosing interval is lengthened to 6–8 h, with a daily maximum of 30 mg. NSAIDs are insufficient for sole use for the relief of very severe pain

immediately after major surgery, although they are very useful analgesic adjuncts, improving pain relief while reducing opioid requirements. They are of considerable value in day case surgery, where they can be given by injection at the time of surgery and then continued in tablet form when the patient is discharged from hospital, often avoiding the need for any opioid therapy.

Safety
NSAIDs possess undesirable effects, because of their mechanism of action, as prostaglandins are local tissue hormones regulating function, and interference with their synthesis can produce problems. NSAID side effects are more common with long-term use, but in the perioperative period, the main concerns are the potential of producing peptic ulceration, interference with platelets, renal impairment, and bronchospasm in individuals who have aspirin-induced asthma. In general, the risk and severity of NSAID-associated side effects is increased in the elderly population.

Platelet Function
Platelet cyclooxygenase is essential for the production of the cyclic endoperoxides and thromboxane A_2, which mediate the primary hemostatic response to vessel injury by producing vasoconstriction and platelet aggregation. Unlike aspirin, which acetylates cyclooxygenase irreversibly, NSAIDs inhibit platelet

cyclooxygenase in a reversible fashion. Studies have confirmed that single doses of NSAID, such as ketorolac and diclofenac, do inhibit platelet function (prolong skin bleeding time and inhibit platelet function in vitro) (Power et al. 1998), but do not tend to increase surgical blood loss in normal patients (Moiniche et al. 2003). However, the presence of a bleeding diathesis or administration of anticoagulants may increase the risk of significant surgical blood loss upon NSAID administration.

Renal Function

In the kidney prostaglandins have many physiological roles, including the maintenance of renal blood flow and glomerular filtration rate in the presence of circulating vasoconstrictors, regulation of tubular electrolyte handling, and modulation of the actions of renal hormones. The adverse renal effects of chronic NSAID use are common and well recognized. In some clinical conditions high circulating concentrations of the vasoconstrictors renin, angiotensin, norepinephrine, and vasopressin increase production of intrarenal vasodilators including prostacyclin, and renal function can be sensitive to NSAIDs. Diclofenac has been shown to affect renal function in the immediate postoperative period after major surgery (Power et al. 1992), and administration of other potential nephrotoxins, such as gentamicin, increased the renal effect of ketorolac (Jaquenod et al. 1998). However, in clinical practice, with careful patient selection and monitoring, the incidence of NSAID-induced renal impairment is low in the perioperative period (Lee et al. 2004).

Aspirin-Induced Asthma

Precipitation of bronchospasm is a recognized phenomenon in individuals with asthma, chronic rhinitis, and nasal polyps. Such "aspirin-induced asthma" affects 10–15 % of asthmatics and can be severe, and there is a cross sensitivity with NSAIDs (Szczeklik and Stevenson 2003). A history of aspirin-induced asthma is a contraindication to NSAID use after surgery, although there is no reason to avoid NSAIDs in other asthmatics. The mechanism of this is unclear, but the reaction increases with the potency of a drug as a cyclooxygenase inhibitor. One hypothesis is that cyclooxygenase inhibition increases arachidonic acid availability for production of inflammatory leukotrienes by lipooxygenase pathways. The syndrome is less common in children, and the susceptibility may be revealed in adult life by a viral illness.

Peptic Ulceration

The gastric and duodenal epithelia have various protective mechanisms against acid and enzyme attack, and many of these involve prostaglandin production. Chronic NSAID use is associated with peptic ulceration and bleeding (Henry et al. 1996), and the latter may be exacerbated by the anti-platelet effect. One meta-analysis has estimated the risk associated with NSAIDs of perforations, ulcers, and bleeds, finding that the pooled relative risk from nine cohort studies, comprising over 750,000 person-years of exposure, was 2.7 (95 % Cl: 2.1, 3.5) (Ofman et al. 2002). Acute gastroduodenal damage and bleeding can occur with short-term NSAID use in the perioperative period. One post-marketing surveillance study of ketorolac found that the risk of gastrointestinal hemorrhage was small but larger and more clinically important when ketorolac was used in higher doses, in older subjects, and for more than 5 days (Strom et al. 1996).

Other

Prostaglandin production has been shown to be important in animal models of bone healing, but there is little research evidence that any inhibitory effect of NSAIDs is important clinically (Harder and An 2003).

Contraindications

The presence, in patients, of a bleeding diathesis, peptic ulceration, significant renal impairment, peptic ulceration, and aspirin-induced asthma should be respected as contraindications to NSAID administration.

Conclusion

NSAIDs are powerful analgesics that can be used to relieve postoperative pain effectively and minimize opioid requirements. NSAIDs can be recommended as an integral component of multimodal analgesia in combination with paracetamol and opioids, but contraindications effectively prevent their use in many surgical patients (NHMRC 1999).

References

Gillis, J. C., & Brogden, R. N. (1997). Ketorolac. A reappraisal of its pharmacodynamic and pharmacokinetic properties and therapeutic use in pain management. *Drugs, 53*, 139–188.

Harder, A. T., & An, Y. H. H. (2003). The mechanisms of the inhibitory effects of nonsteroidal anti-inflammatory drugs on bone healing: A concise review. *Journal of Clinical Pharmacology, 43*, 807–815.

Henry, D., Lim, L. L., et al. (1996). Variability in risk of gastrointestinal complications with individual non-steroidal anti-inflammatory drugs – results of a collaborative meta-analysis. *BMJ, 312*, 1563–1566.

Jaquenod, M., Ronnedh, C., et al. (1998). Factors influencing ketorolac-associated perioperative renal dysfunction. *Anesthesia and Analgesia, 86*, 1090–1097.

Lee, A. C. M., Craig, J. C., Knight, J. F. et al. (2004). Effects of nonsteroidal anti-inflammatory drugs on post-operative renal function in normal adults (Cochrane Review). In *The cochrane library*. Chichester, UK: John Wiley & Sons.

McCormack, K. (1994). Non-steroidal anti-inflammatory drugs and spinal nociceptive processing. *Pain, 59*, 9–43.

Moiniche, S., Romsing, J., et al. (2003). Nonsteroidal anti-inflammatory drugs and the risk of operative site bleeding after tonsillectomy: A quantitative systematic review. *Anesthesia and Analgesia, 96*, 68–77.

NHMRC. (1999). *Acute pain management: Scientific evidence*. Canberra: Australian National Health and Medical Research Council.

Ofman, J. J., MacLean, C. H., et al. (2002). A meta-analysis of severe upper gastrointestinal complications of nonsteroidal anti-inflammatory drugs. *The Journal of Rheumatology, 29*, 804–812.

Power, I., Noble, W., et al. (1990). Comparison of I.M. ketorolac trometamol and morphine sulphate for pain relief after cholecystectomy. *British Journal of Anaesthesia, 65*, 448–455.

Power, I., Cumming, A. D., et al. (1992). Effect of diclofenac on renal function and prostacyclin generation after surgery. *British Journal of Anaesthesia, 69*, 451–456.

Power, I., Noble, W., et al. (1998). Effects of ketorolac and dextran-70 alone and in combination on haemostasis. *Acta Anaesthesiologica Scandinavica, 42*, 982–986.

Strom, B. L., Berlin, J. A., et al. (1996). Parenteral ketorolac and risk of gastrointestinal and operative site bleeding. A postmarketing surveillance study. *Journal of the American Medical Association, 275*, 376–382.

Szczeklik, A., & Stevenson, D. D. (2003). Aspirin-induced asthma: Advances in pathogenesis, diagnosis, and management. *Journal of Allergy and Clinical Immunology, 111*, 913–921.

Postoperative Pain, Opioids

Peter C. A. Kam
Anaesthesia, Central Clinical School,
The University of Sydney, Sydney, Australia

Synonyms

Opiate; Opioid

Definition

An ▶ opioid is any compound (endogenous or exogenous – naturally occurring or synthetic) that has pharmacological activity at an ▶ opioid receptor. The term "opiate" describes a substance that is a naturally occurring opioid.

Characteristics

Opioid Receptor

In 1976, Martin and colleagues first classified the opioid receptors as mu (μ), kappa (κ), and sigma (σ), based on the specificity of three agents acting at these receptors, i.e., morphine, ketocyclazocine, and SKF 10047, respectively. Two further opioid receptors were described, delta (δ) and epsilon (ε) receptors, based on their responses to enkephalins and epsilon β-endorphins, respectively. However, later studies demonstrated conclusively that sigma and epsilon receptors were not opioid receptors (Shook et al. 1988). From the 1980s, only three opioid receptors, μ, δ, and κ,

Postoperative Pain, Opioids, Table 1 Classification of opioid receptors

Conventional nomenclature	Previous nomenclature	Proposed nomenclature
Delta	OP1	DOR
Kappa	OP2	KOR
Mu	OP3	MOR
Orphan	ORL1	NOR

were established. The opioid receptors were cloned in the 1990s (Thompson et al. 1993 and Yasuda et al. 1993).

In 1996, the International Union of Pharmacology (IUPHAR) recommended that the Greek alphabets μ, δ, and κ should be replaced by OP_3, OP_1, and OP_2, respectively (numbered in the order in which they were cloned) (Dhawan et al. 1996). In 1994, the ▶ nociceptin/orphanin FQ receptor was discovered. This accidentally cloned new opioid receptor was called orphan receptor because no ligand was known at that time. An endogenous opioid called orphanin F/Q or nociceptin was discovered. Nociceptin F/Q is a heptadecapeptide and shows similarity to dynorphin A (Calo et al. 2000).

Latest guidelines from IUPHAR show that the opioid receptors should be called ▶ MOR, ▶ DOR, ▶ KOR, or NOR (see Table 1).

Location of Opioid Receptors

Opioid receptors synthesized in sensory neuron cell bodies are found throughout the central nervous system. Mu and kappa receptors are located in the cerebral cortex, amygdala, hippocampus, thalamus, mesencephalon, pons, medulla, and spinal cord. Kappa receptors are also present in the hypothalamus, while the delta receptors are present in the telencephalon and the spinal cord.

In the spinal cord, opioid receptors are found at the terminal zones of C-fibers, primarily in lamina I. OP_3 (mu) receptors predominate in the spinal cord ($\mu = 70\%$, $\delta = 24\%$, $\kappa = 6\%$ in the rat).

Peripherally, opioid receptors have been identified in peripheral tissue. They are silent until activated by inflammatory mediators

(Stein 1993). Opioid receptors have also been identified on the surface of immune cells such as τ- and β-lymphocytes, monocytes, and macrophages.

Stimulation of presynaptic OP_1 (δ) and OP_3 (μ) receptors results in hyperpolarization of the nerve terminal with reduced excitatory neurotransmitter release, mediated by inhibition of voltage-gated calcium channels. Activation of postsynaptic opioid receptors activates potassium channels, enhancing the outward flow of potassium, thereby hyperpolarizing the membrane and inhibiting the response of the postsynaptic membrane to neurotransmitters. The action of morphine on opioid receptors is mediated by inhibitory G-protein, coupled to adenylyl cyclase, thus reducing cAMP formation.

Classification of Opioid Drugs

Traditionally, opioids have been classified as weak, intermediate, and strong. Codeine is considered a "weak" opioid; partial μ-agonists or mixed agonist – antagonists e.g., buprenorphine, nalbaphine as "intermediate" and pure agonists (e.g., morphine) as "strong" opioids. Opioids can also be classified according to their chemical structures: morphinans (e.g., morphine, codeine), phenylpiperidines (meperidine, fentanyl), diphenylpropylamine (e.g., methadone, dextropropoxyphene), and esters (e.g., remifentanil). The best way of classifying opioids is by utilizing a functional approach: pure agonists (e.g., morphine, fentanyl), partial agonists (e.g., buprenorphine), agonist-antagonists (e.g., pentazocine), and mixed action (e.g., meperidine – opioid agonist with anticholinergic and membrane-stabilizing properties; tramadol – weak μ-agonist; and inhibition of reuptake of norepinephrine and 5 HT) (see Table 2).

Pharmacodynamics

The opioid effects on the central nervous system include analgesia, sedation, respiratory depression, miosis, nausea, and vomiting.

Analgesia

There are two distinct anatomical sites, supraspinal and spinal, which are responsible

Postoperative Pain, Opioids, Table 2 Bioavailability of commonly used opioids

Opioid	Bioavailability (%)
Morphine	25
Meperidine	52
Codeine	50
Oxycodone	60
Hydromorphone	25
Methadone	80

for opioid-mediated analgesia. Systemically administered opioids produce both spinal and supraspinal analgesia. Activation of supraspinal opioid receptors, mainly μ-receptors, transits descending impulses that inhibit spinal nociceptive neurons and reflexes to produce analgesia. The periaqueductal gray is a major site for supraspinal opioid analgesia. The neurotransmitters in the spinal cord that mediate the effects of descending inhibitory pathways are norepinephrine and 5HT. All three types of opioid receptor are present on the peripheral terminals of sensory nerves, but their activation requires an inflammatory reaction. Euphoria often occurs, and dysphoria is associated with κ agonists. Muscle rigidity appears to be associated with central activation of the μ-receptor, whereas supraspinal δ and κ receptors may attenuate this effect. Rigidity involving the thoracic and abdominal muscles can interfere with breathing. It may be related to opioid-mediated inhibition of dopamine release in the striatum. Involuntary movements (catatonic) of the limbs are frequently observed in patients receiving high doses of opioids.

Gastrointestinal Tract

Significant concentrations of μ-opioid receptors are present in the gastrointestinal tract. Opioids increase intestinal tone, decrease peristalsis, and delay gastric emptying resulting in ileus or constipation. Opioids also cause spasm of the sphincter of Oddi and increase bile duct tone, resulting in increased common bile duct pressure and decreased bile production and flow. Nausea and vomiting is mediated by stimulation of opioid receptors within the chemoreceptor trigger zone in the area postrema of the medulla.

Respiratory System

Opioids depress both the central (medullary) and peripheral chemoreceptors. The CO_2 response curve is shifted to the right and the slope decreased, with a shift of the apneic threshold. The effect on the medullary chemoreceptor involves inhibition of acetylcholine release. The hypoxic drive is also depressed by opioids. Pain may counteract the depressant effects of opioids.

Tolerance and Dependence

▶ Tolerance can be defined as a reduction in response to the same dose of drug after repeated administration. The exact mechanism underlying tolerance is not known, but several mechanisms appear to be involved. Acute (desensitization) and chronic tolerance may involve the uncoupling of the receptor from the G-protein-adenyl cyclase cascade. Chronic receptor activation may cause compensatory increase in adenylyl cyclase activity and increased cellular cAMP concentration, which induces cAMP phosphodiesterase and increased cAMP degradation. The stimulatory effects of the opioids may activate phosphokinase C and desensitize the K^+ channels normally opened by opioids.

▶ Dependence is a state in which an abstinence syndrome may occur following abrupt withdrawal, dose reduction, or administration of an antagonist. ▶ Addiction is a behavioral state characterized by compulsive self-administration of a drug, continuously or periodically, in order to experience its psychic effects and sometimes to avoid the discomfort of its absence, often obtaining supply by deceptive or illegal means. Addiction rarely develops in patients receiving opioids for severe pain. The fear of addiction is a common cause of under-prescription and under-use of strong opioid analgesics, and this is unfounded and unnecessary.

Opioid Hyperexcitability and Rotation

In patients who experience ▶ opioid hyperexcitability (whole body hyperalgesia and allodynia, abdominal muscle spasms, and symmetrical jerking of legs), methadone is preferred. The exact mechanism of opioid hyperexcitability is not known, but it has been suggested that

Postoperative Pain, Opioids, Table 3 Pharmacokinetics of opioids

Drug	Volume of distribution L/kg	Clearance Ml/min/kg	Elimination half-life (h)
Morphine	3.5	15	3
Meperidine	4	12	4
Codeine	2.6	11	3
Oxycodone	2.6	10	3.7
Hydromorphine	4.1	22	3.1
Fentanyl	4.0	13	3.5
Remifentanil	0.4	40	0.1
Alfentanil	0.8	6	1.6
Methadone	3.8	1.4	35
Buprenorphine	7	70	2.5

metabolites such as morphine-3-glucuronide, normorphine, and hydromorphone-3-glucuronide may be involved. When opioid-related side effects occur before satisfactory analgesia is achieved, the first opioid is ceased and a new opioid is introduced. This treatment strategy is known as "▶ opioid rotation," and the change to an alternative drug can yield a better balance between analgesia and side effects.

Pharmacokinetics of Opioids
The bioavailability of commonly used opioids tends to be poor and variable. The bioavailability of methadone, however, is 80 %, but it is not suitable for the use of acute pain because of its prolonged half-life.

The pharmacokinetics of the opioids is summarized in Table 3.

Routes of Administration
Oral, sublingual, rectal, parenteral, and spinal (epidural and intrathecal) routes of administration of opioids are employed.

For acute pain the parenteral route is usually used, whereas the oral route is the norm for cancer patients who can swallow.

Sublingual Route
Buprenorphine is given sublingually because of considerable increase in bioavailability compared to the oral route (55 % vs. 15 % after oral administration).

Rectal Route
Rectal administration of opioids bypasses the portal circulation to a variable extent, according to the amount absorbed into the middle and inferior rectal veins (systemic circulation), compared with the superior rectal vein (portal circulation), resulting in variable rectal bioavailability ranging from 70 % to 120 %.

Transdermal Route
Passive diffusion of highly lipid-soluble opioids like fentanyl is possible. Fentanyl is delivered via patches that contain a drug reservoir and a rate-controlling membrane. The system is convenient and comfortable but does not permit rapid titration of dose. Therefore, the transdermal route is not suitable for acute pain, but is useful for patients with stable pain who cannot take oral medications.

Parenteral Routes
Opioids can be administered parenterally by subcutaneous, intramuscular, and intravenous routes, either via continuous infusions or intermittent injections. Traditionally, IM opioids are prescribed 4 hourly PRN, but a reluctance to give opioids more frequently is a major factor for the lack of efficacy of IM regimen. Although the subcutaneous route is often used for opioid administration for cancer pain management, it is also used for acute pain management. Morphine is the opioid commonly used for intermittent SC injection.

Continuous and intermittent intravenous opioid administration is also used in acute pain management. Intermittent intravenous administration of opioids results in large variations in blood concentrations of the drug and is therefore not an effective way to achieve good analgesia. If sustained analgesia is to be achieved without side effects, smaller doses more often produce less variability in blood concentrations of the drug with better analgesia.

To avoid the peaks and troughs in the blood concentrations of opioids, constant intravenous infusions of opioid are used. However, a loading dose is required. There is considerable delay in matching the amount of opioid delivered to the amount of opioid actually needed. There is a risk that blood levels of the drug or its active metabolite may continue to rise after analgesia is achieved. As such, ▶ patient-controlled analgesia (PCA) is frequently used. PCA has been associated with better pain relief and greater intermittent opioid injection. Smaller and frequent intravenous bolus doses that are received whenever pain becomes uncomfortable enable individual titration of analgesia and maintenance of opioid blood concentrations within the analgesic range. This overcomes inter-patient variability in opioid requirements. As acute pain intensity is rarely constant, PCA enables the amount of opioid delivered to be rapidly titrated to effect and against dose-related side effects.

Spinal Opioids

The epidural and intrathecal routes of administration of opioids allow smaller doses, prolonged duration of action, and minimal side effects. The main aim is to achieve selective spinal analgesia. Intrathecal opioids can provide prolonged analgesia. In cardiac surgical patients, analgesia for about 36 h was achieved with lumbar intrathecal 2–4 mg morphine.

Highly lipid-soluble opioids are less effective or potent when administered intrathecally because of the diffusion of the opioid from the subarachnoid space and their sites of action in the dorsal horn. A method of converting from conventional parenteral routes to intrathecal administration is to use 1 % of the former total daily dose. Lower doses should be used if the opioid is combined with a local anesthetic. Delayed respiratory depression may occur with the non-lipophilic opioids.

References

Calo, G., Guerrini, R., Rizzi, A., et al. (2000). Pharmacology of nociceptin and its receptor: A novel therapeutic target. *British Journal of Pharmacology, 129*, 1261–1283.

Dhawan, B. N., Celsselin, F., & Raghbir, R. (1996). International union of pharmacology. Classification of opioid receptors. *Pharmacological Reviews, 12*, 567–592.

Shook, J. E., Kazmierski, W., Wire, W. S., et al. (1988). Opioid receptor selectivity of Î²-endorphin in vitro and in vivo: Âµ, Î′ and Îµ receptors. *Journal of Pharmacology and Experimental Therapeutics, 246*, 1018–1025.

Stein, C. (1993). Peripheral mechanisms of opioid analgesia. *Anesthesia and Analgesia, 76*, 182–191.

Thompson, R. C., Mansour, A., Akil, A., et al. (1993). Cloning and pharmacological characterisation of a rat Î¼-opioid receptor. *Neuron, 11*, 903–911.

Yasuda, K., Raynor, K., Kong, et al. (1993). Cloning and functional comparison of kappa opioid and delta opioid receptors from mouse brain. *Proceedings of the National Academy of Sciences of the United States of America, 90*, 6736–6740.

Postoperative Pain, Oxycodone

Peter C. A. Kam
Anaesthesia, Central Clinical School,
The University of Sydney, Sydney, Australia

Synonyms

Endone®; Oxycodone; Oxynorm®; Roxicodone®

Definition

▶ Oxycodone is a semi-synthetic opioid full agonist with μ- and κ-activity.

Characteristics

▶ Oxycodone (14-hydroxy-7,8-dihydrocodeinone) is a semisynthetic derivative of the opium alkaloid thebaine. It is a full opioid agonist, acting on OP_3 (μ) and OP_2 (κ) receptors.

Pharmacodynamics

The analgesic effect of oxycodone is mediated by the parent compound and not by oxymorphone, a potent minor metabolite (Kaiko 1997). The affinity of oxycodone for the μ-receptor is 1 : 10–1 : 40 that of morphine and four times that of meperidine (pethidine). Oxycodone also binds the OP_2 or kappa receptor, which contributes to its analgesic, dysphoric, and sedative effects (Poyhoa et al. 1993).

Pharmacokinetics

Oxycodone has an elimination half-life of 2.6–3.2 h, a volume of distribution at steady state of 2.6 l/kg, and plasma protein binding of 38–45 %. The plasma clearance of oxycodone is 0.8 l/min. The oral bioavailability of oxycodone is approximately 60 %, which is almost double that of morphine (Leow et al. 1992). Maximal plasma concentrations are achieved at 1–1.5 h after an oral dose of oxycodone. Oral oxycodone undergoes ▶ low first-pass metabolism and is among those opioids possessing a short ▶ elimination half-life. These properties make it possible for oral oxycodone to provide for both a relatively short-time to steady-state pain control and less variation in bioavailability and drug effects.

Metabolism

Oxycodone is extensively metabolized in the liver by demethylation to noroxycodone and oxymorphone, and then these are conjugated to glucuronides. The major circulating metabolite is noroxycodone, with an AUC ratio of 0.6 relative to that of oxycodone. Noroxycodone is a considerably weaker analgesic than oxycodone. Oxymorphone is present in only low concentrations in the plasma, although it is a potent analgesic. Therefore, the ▶ analgesic effect of oxycodone is mainly mediated by the parent compound. The formation of oxymorphone is mediated by CYP2D6.

Excretion

Oxycodone and its metabolites are excreted primarily by the kidney. The excretory products of oxycodone in the urine include free oxycodone (19 %), conjugated oxycodone (50 %), conjugated oxymorphone (14 %), and free and conjugated noroxycodone. The plasma concentrations of oxycodone are 15 % greater in elderly patients compared with young subjects. Female subjects have higher plasma concentration (~25 %) than males.

In patients with impaired renal functions (~creatinine clearance <60 ml/min), peak plasma oxycodone and noroxycodone are 50 % and 20 % higher than in normal subjects, respectively. This is associated with increased sedation. The T $1/2$ β oxycodone in patients with renal impairment increases by 1 h. In patients with mild-to-moderate hepatic dysfunction, peak plasma oxycodone and noroxycodone concentrations are 50 % and 20 % higher, respectively, than in normal subjects, and the T $1/2$ β of oxycodone is increased by 2–3 h.

OxyContin Controlled-Release Formulation

Although immediately released oxycodone has a high oral bioavailability and a short elimination half-life, it has to be administered frequently to maintain plasma concentrations within the therapeutic range. This may be convenient in patients with persistent pain conditions.

▶ OxyContin is a controlled-release formulation of oxycodone that provides controlled release/delivery of oxycodone over 12 h, with an onset of analgesia within 1 h. Convenient oral dosing of oxycodone every 12 h to provide sustained analgesia can therefore be achieved.

A dual matrix (Agro Contin System) that uses 2 different types of retarding polymers is utilized to achieve measured/controlled release of the active drug. Two hydrophobic molecules are used containing an acrylic polymer. Gastrointestinal fluids dissolve the oxycodone tablet surface and expose the hydrophobic/acrylic matrix, releasing initial quantities of oxycodone. The active oxycodone then diffuses through the hydrophobic/acrylic matrix, becoming available for prolonged absorption. This formulation enables a biphasic absorption pattern to be achieved: an initial rapid absorption with an onset of action of 1 h, followed by a slow release and absorption phase that assures effective blood concentrations of oxycodone over 12 h.

Pharmacokinetics of OxyContin

OxyContin has 2 half-lives of absorption consistent with a biphasic absorption pattern. The absorption of oxycodone from OxyContin follows a biexponential process, with a rapid phase of T $1/2$ absorption $= 37$ min and a slow phase of T 1/2 absorption component accounting for 38 % of the available dose and the slow phase accounting for 63 % of the available dose.

Studies comparing the pharmacokinetic profile of OxyContin with that of immediate-release oxycodone show that the bioavailability was comparable for the 2 preparations and OxyContin yielded about half the Cmax, with about twice the time to Tmax. Steady-state plasma oxycodone concentrations were achieved after 24–36 h of dosing with q12h OxyContin tablets (Kaiko et al. 1996).

Clinical Efficacy

Oxycodone has been widely used in many countries to treat moderate-to-severe postoperative pain (Kalso et al. 1991), as well as for chronic cancer and non-cancer pain by oral, rectal, and subcutaneous routes. In a randomized double-blind study, Kalso et al. compared the efficacy of IV oxycodone with that of IV morphine in patients after abdominal surgery following opioid free inhalational anesthesia. Pain relief was achieved faster (28 min vs. 46 min) and lasted longer (39 min vs. 27 min) with oxycodone than morphine. Morphine caused more sedation than oxycodone.

A controlled-release formulation of oxycodone in patients undergoing anterior cruciate ligament repair on an ambulatory basis provided significant analgesic benefit, with less side effects compared with fixed dose or as needed oxycodone regimens (Reuben et al. 1999).

Kalso and Vaino also demonstrated that both oxycodone and morphine effectively relieved cancer pain with both IV and oral administration in a double-blind crossover trial. A higher bioavailability for oxycodone was demonstrated by the mean daily consumption of the opioids: oxycodone 150 mg orally vs. morphine 204 mg orally. Morphine caused more nausea and hallucinations compared with oxycodone. A multicenter randomized double-blind parallel-group study of patients with chronic cancer pain compared the effectiveness and safety of controlled-release oxycodone tablets with immediate-release oxycodone. There were no significant differences in effective analgesia

and the incidence of adverse events. The study showed that controlled-release oxycodone administered every 12 h could provide equally effective analgesia as immediate-release oxycodone administered 4 times at the same total daily dose and, in addition, offered the benefit of twice-daily dosing (Parris et al. 1998).

Dose titration can be accomplished as readily with oral controlled-release oxycodone as with immediate-release oxycodone, in patients with chronic malignant or nonmalignant pain of moderate-to-severe intensity (Saltzman et al. 1999).

Systematic reviews indicate that the ▶ NNT of oxycodone 15 mg is 2.3. The effect of oxycodone in the treatment of postherpetic neuralgia has been evaluated in a long-term trial. Oxycodone relieved steady pain, brief pain, and allodynia.

The NNT of oxycodone for postherpetic neuralgia was 2.5 (Watson and Babul 1998).

Adverse Effects

The adverse effects of oxycodone are typical of opioid agonists. The most common adverse drug reactions during oxycodone therapy are constipation, nausea, dizziness, headache, dry mouth, confusion, sweating, and drowsiness.

References

Kaiko, R. F. (1997). Pharmacokinetics and pharmacodynamics of controlled release opioids. *Acta Anaesthesiologica Scandinavica, 41,* 166–174.

Kaiko, R. F., Benziger, D. P., Fitzmartin, R. D., et al. (1996). Pharmacodynamic-pharmacokinetic relationships of controlled release oxycodone. *Clinical Pharmacology and Therapeutics, 59,* 52–61.

Kalso, E., Poyhia, R., Annela, P., et al. (1991). Intravenous morphine and oxycodone for pain after abdominal surgery. *Acta Anesthesiologica Scandinavica, 35,* 642–646.

Leow, K. P., Smith, M. T., Williams, B., et al. (1992). Single dose and steady state of oxycodone in patients with cancer. *Clinical Pharmacology and Therapeutics, 52,* 487–495.

Parris, W. C. V., Johnson, B. W., Croghan, M. K., et al. (1998). The use of controlled – release oxycodone for the treatment of chronic cancer pain: A randomised, double blind study. *Journal of Pain and Symptom Management, 16,* 205–211.

Poyhoa, R., Vainio, A., & Kalso, E. (1993). A review of oxycodone's clinical pharmacokinetics and pharmacodynamics. *Journal of Pain and Symptom Management*, 8, 63–67.

Reuben, S. S., Connelly, N. R., & Maciolek, H. (1999). Post operative analgesia with controlled release oxycodone for outpatient anterior cruciate ligament surgery. *Anesthesia and Analgesia, 88*, 1286–1291.

Saltzman, R. T., Roberts, M. S., Wild, J., et al. (1999). Can a controlled release oral dose form of oxycodone be used as readily as an immediate release form for the purpose of titrating to stable pain control? *Journal of Pain Management, 18*, 271–279.

Watson, C. P. N., & Babul, N. (1998). Efficacy of oxycodone in neuropathic pain: A randomised trial in postherpetic neuralgia. *Neurology, 50*, 1837–1841.

Postoperative Pain, Paracetamol

Ian Power[1] and Lesley A. Colvin[2]
[1]Critical Care and Pain Medicine, The University of Edinburgh, Edinburgh, UK
[2]Department of Anaesthesia, Critical Care and Pain Medicine Western General Hospital, Edinburgh, UK

Synonyms

Acetaminophen; Paracetamol in postoperative pain

Definition

Paracetamol is the remaining para-aminophenol used in clinical practice for analgesia, being the active metabolite of the earlier, more toxic, drugs acetanilide and phenacetin. Paracetamol can be given orally, rectally, or parentally; has little anti-inflammatory activity; and is an effective analgesic and antipyretic (Fig. 1).

Characteristics

General

Paracetamol is absorbed rapidly from the small intestine after oral administration and can be given rectally, and parenteral preparations have

Postoperative Pain, Paracetamol, Fig. 1 Paracetamol

recently been introduced into clinical practice (Bannwarth and Pehourcq 2003). Paracetamol has lower protein binding (hence less potential drug interactions) than NSAIDs and a higher volume of distribution than NSAIDs. The recommended paracetamol dose in adults is 0.5–1 g oral or rectal, every 3–6 h when necessary, with maximum of 6 g a day in divided doses for acute use and 4 g a day for chronic use.

Mechanism of Action

The mechanism of action of paracetamol is not well understood, but it may act by inhibiting prostaglandin synthesis in the central nervous system, with little peripheral nervous system effect. Unlike morphine, paracetamol has no known endogenous binding sites and unlike the NSAIDs apparently does not inhibit peripheral cyclooxygenase activity. There is growing evidence of a central antinociceptive effect of paracetamol. Possible mechanisms include central COX-2 inhibition; the existence of a cyclooxygenase, COX-3, that is selectively susceptible to paracetamol; and modulation of descending serotonergic pathways that suppress spinal cord nociceptive transmission (Bonnefont et al. 2003; Botting 2003; Warner and Mitchell 2002). Paracetamol has also been shown to prevent prostaglandin production at the cellular transcriptional level, independent of cyclooxygenase activity (Mancini et al. 2003). A new potent nitric oxide-releasing version of paracetamol (Moore and Marshall 2003) (nitroxyparacetamol or nitroacetaminophen) has anti-inflammatory and analgesic properties, with a described mechanism of action in the spinal cord, which may differ from that of paracetamol (Moore and Marshall 2003; Romero-Sandoval et al. 2003).

Efficacy

The efficacy of single-dose paracetamol as a postoperative analgesic has been examined and confirmed by many studies. Using the concept of number needed to treat (NNT), obtained from meta-analyses of well-designed studies of postoperative analgesia, McQuay and Moore from Oxford University have enabled comparison of the efficacy of paracetamol with other types of analgesics for acute pain relief after surgery (www.jr2.ox.ac.uk/bandolier/booth/painpag/). From such work, the NNT of paracetamol is 3.8 (3.4–4.4) (Barden et al. 2004). For comparison, the NNT of morphine 10 mg intramuscularly is 2.9, ibuprofen 400 mg 2.4, and codeine 60 mg 16.7. The combination of paracetamol 1,000 mg plus codeine 60 mg has a NNT of 2.2. Paracetamol is therefore an effective postoperative analgesic, with a potency slightly less than that of a standard dose of morphine or of the nonsteroidal anti-inflammatory drugs (Barden et al. 2004).

Paracetamol is an effective adjunct to opioid analgesia, requirements for the latter typically being reduced by 20–30 % when combined with a regular regime of oral or rectal paracetamol (Romsing et al. 2002). One study has demonstrated that the use of oral paracetamol in addition to PCA morphine lowered pain scores, shortened the duration of PCA use, and improved patient satisfaction (Schug et al. 1998). However, recent work has questioned whether the usual dose of rectal paracetamol, 1 g 6-hourly, is too low for maximal opioid-sparing benefit as this produces a suboptimal plasma paracetamol concentration (Kvalsvik et al. 2003). The addition of an NSAID to paracetamol also improves efficacy (Hyllested et al. 2002; Romsing et al. 2002). Paracetamol can be recommended an integral component of multimodal analgesia in combination with NSAIDs and opioids.

Intravenous paracetamol is an effective analgesic after surgery (Hernandez-Palazon et al. 2001), is as effective as morphine, and is better tolerated after dental surgery (Van Aken et al. 2004), although there is some evidence of a ceiling effect (Hahn et al. 2003).

Safety

In general, ▶ acetaminophen has fewer side effects than the NSAIDs and can be used when the latter are contraindicated (e.g., asthma, peptic ulcers).

Hepatic

A small amount of ingested paracetamol undergoes cytochrome P450-mediated hydroxylation producing a reactive toxic metabolite that is normally rendered harmless by conjugation with liver glutathione and renal excretion as mercapturic derivatives. With larger doses, the rate of formation of the reactive metabolite exceeds that of glutathione conjugation, and the reactive metabolite combines with hepatocellular macromolecules, resulting in cell death. The clinical picture is of centrilobular hepatocellular necrosis, occasionally with acute renal tubular necrosis. Doses of more than 150 mg/kg taken within 24 h may result in severe liver damage, hypoglycemia, and acute tubular necrosis. Individuals taking enzyme-inducing agents are more likely to develop hepatotoxicity. Early signs include nausea and vomiting, followed by right subcostal pain and tenderness the next day. Hepatic damage is maximal 3–4 days after ingestion and may lead to death. Treatment consists of gastric emptying and the specific antidotes methionine and acetylcysteine, which offer effective protection up to 10–12 h after ingestion. Acetylcysteine is effective to 24 h and perhaps beyond. The plasma paracetamol concentration related to time from ingestion indicates the risk of liver damage (acetylcysteine is given if the plasma paracetamol concentration is greater than 200 mg/l at 4 h and 6.25 mg/l at 24 h after ingestion). The molecular mechanisms of paracetamol hepatotoxicity are being researched actively, in the hope that advances will make it possible to prevent liver failure (Prescott 2003).

Care is required to avoid inadvertent overdoses in children where the dose changes with age (e.g., 6–12 years 250–500 mg 6-hourly) and when combination preparations containing paracetamol are given in error together with the parent drug to excess. There is encouraging evidence that nitroxyparacetamol may have less hepatic toxicity than paracetamol (Futter et al. 2001).

Conclusion

Paracetamol is an effective postoperative analgesic that should be used as a fundamental component of multimodal analgesia (NHMRC 1999). In normal doses paracetamol is safe, but with overdose, liver toxicity may be severe. In the perioperative period when oral intake is not possible, the introduction of parenteral preparations may expedite the use of paracetamol.

References

Bannwarth, B., & Pehourcq, F. (2003). Pharmacological rationale for the clinical use of paracetamol: Pharmacokinetic and pharmacodynamic issues. *Drugs, 63*, 5–13.

Barden, J. E. J., Moore, A., & McQuay, H. (2004). *Single dose oral paracetamol (acetaminophen) for postoperative pain (Cochrane review)*. Chichester: The Cochrane Library/Wiley.

Bonnefont, J., Courade, J. P., Allouie, A., et al. (2003). Mechanism of the antinociceptive effect of paracetamol. *Drugs, 63*, 1–4.

Botting, R. (2003). COX-1 and COX-3 inhibitors. *Thrombosis Research, 110*, 269–272.

Futter, L. E., Al-Swayeh, O. A., & Moore, P. K. (2001). A comparison of the effect of nitroparacetamol and paracetamol on liver injury. *British Journal of Pharmacology, 132*, 10–12.

Hahn, T. W., Mogensen, T., Lund, C., et al. (2003). Analgesic effect of i.v. paracetamol: Possible ceiling effect of paracetamol in postoperative pain. *Acta Anaesthesiologica Scandinavica, 47*, 138–145.

Hernandez-Palazon, J., Tortosa, J. A., Martinez-Lage, J. F., et al. (2001). Intravenous administration of propacetamol reduces morphine consumption after spinal fusion surgery. *Anesthesia and Analgesia, 92*, 1473–1476.

Hyllested, M., Jones, S., & Perdersen, J. L. (2002). Comparative effect of paracetamol, NSAIDs or their combination in postoperative pain management: A qualitative review. *British Journal of Anaesthesia, 88*, 199–214.

Kvalsvik, O., Borchgrevink, P. C., Hagen, L., et al. (2003). Randomized, double-blind, placebo-controlled study of the effect of rectal paracetamol on morphine consumption after abdominal hysterectomy. *Acta Anaesthesiologica Scandinavica, 47*, 451–456.

Mancini, F., Landolfi, C., Muzio, M., et al. (2003). Acetaminophen down-regulates interleukin-1 beta-induced nuclear factor-kappa B nuclear translocation in a human astrocytic cell line. *Neuroscience Letters, 353*, 79–82.

Moore, P. K., & Marshall, M. (2003). Nitric oxide releasing acetaminophen (nitroacetaminophen). *Digestive and Liver Disease, 35*, 49–60.

NHMRC. (1999). *Acute pain management: Scientific evidence*. Canberra: National Health and Medical Research Council.

Prescott, L. F. (2003). Future perspectives with paracetamol. *Drugs, 63*, 51–56.

Romero-Sandoval, E. A., Del Soldato, P., & Herrero, J. F. (2003). The effects of sham and full spinalization on the antinociceptive effects of NCX-701 (nitroparacetamol) in monoarthritic rats. *Neuropharmacology, 45*, 412–419.

Romsing, J., Moiniche, S., & Dahl, J. B. (2002). Rectal and parenteral paracetamol, and paracetamol in combination with NSAIDs, for postoperative analgesia. *British Journal of Anaesthesia, 88*, 215–226.

Schug, S. A., Sidebotham, D. A., McGuinnety, M., et al. (1998). Acetaminophen as an adjunct to morphine by patient-controlled analgesia in the management of acute postoperative pain. *Anesthesia and Analgesia, 87*, 368–372.

Van Aken, H., Thys, L., Veekman, L., et al. (2004). Assessing analgesia in single and repeated administrations of propacetamol for postoperative pain: Comparison with morphine after dental surgery. *Anesthesia and Analgesia, 98*, 159–165.

Warner, T. D., & Mitchell, J. A. (2002). Cyclooxygenase-3 (COX-3): Filling in the gaps toward a COX continuum? *Proceedings of the National Academy of Sciences USA, 99*, 13371–13373.

Postoperative Pain, Pathophysiological Changes in Cardiovascular Function in Response to Acute Pain

Stephan A. Schug
School of Medicine and Pharmacology,
Royal Perth Hospital, Faculty of Medicine,
Dentistry & Health Sciences, The University of
Western Australia, UWA Anaesthesiology,
Perth, Australia

Synonyms

Cardiac stress response; Reactive tachycardia

Definition

Cardiovascular responses are part of the manifestations of the acute stress response exhibited by an organism in response to injury, surgery, and acute pain; acute pain is typically associated with a neuroendocrine stress response that is proportional to the pain intensity. The sympathetic activation increases efferent sympathetic tone to all viscera and releases catecholamines from the adrenal medulla. This hormonal response results from increased sympathetic tone and hypothalamically mediated reflexes (Aantaa and Scheinin 1993).

The major catecholamines involved are epinephrine and norepinephrine (Desborough 2000). The primary precursor for both of these hormones is tyrosine, which is available from both the diet and from the synthesis of phenylalanine (Desborough 2000). Epinephrine is the methylated product of norepinephrine (Aantaa and Scheinin 1993).

Characteristics

The cardiovascular effects of the acute trauma and pain process are often very prominent and include hypertension, tachycardia, enhanced myocardial irritability, and increased systemic vascular resistance (Desborough 2000).

Cardiac output increases in most people, but in people with compromised left ventricular function, the cardiac output may decrease. The increase in cardiac output will result in an increase in myocardial oxygen demand, precipitating myocardial ischemia in those with coronary artery disease.

Sympathetic fibers are widely distributed throughout the heart. The cardiac sympathetic fibers originate in the thoracic spinal cord (T1–T4) and travel to the heart, initially through the cervical ganglia (stellate), then as cardiac nerves (Aantaa and Scheinin 1993).

Norepinephrine release causes positive chronotropic, dromotropic, and inotropic effects,

primarily through activation of the beta-1 adrenergic receptors. Beta-2 adrenergic receptors are fewer in number and are mainly found in the atria; activation increases the heart rate and, to a lesser extent, contractility. Alpha-1 adrenergic receptors have a positive inotropic effect (Millar 2000).

Parasympathetic fibers primarily innervate the atria and conducting tissues. Acetylcholine acts on specific cardiac muscarinic receptors (M2) to produce negative chronotropic, dromotropic, and inotropic effects (Morgan et al. 2002).

Cardiac autonomic innervation has an apparent sidedness, as the right sympathetic and right vagus nerves primarily affect the sinoatrial node, while the left sympathetic and vagus principally affect the atrioventricular node. Vagal effects frequently have a rapid onset and resolution, while sympathetic influences generally have a gradual onset and dissipation.

Effects on Heart Rate

Cardiac output is generally proportional to heart rate. Heart rate is an intrinsic function of the sinoatrial node, but is modified by autonomic, humoral, and local factors.

Enhanced vagal activity slows the heart rate via stimulation of M2 cholinergic receptors, while enhanced sympathetic activity increases the heart rate, mainly by activation of beta-1 adrenergic receptors and, to a lesser extent, beta-2 adrenergic receptors. The beta-1 adrenergic receptor responds to the circulating catecholamine release as part of the stress response.

Direct stimulation of beta-1 receptors by epinephrine raises the cardiac output and oxygen demand by increasing the contractility and heart rate (increased rate of spontaneous phase 4 depolarization) and systolic blood pressure, although beta-2-mediated vasodilatation in skeletal muscle may lower diastolic blood pressure.

Effects on Stroke Volume

Three major factors, preload, afterload, and contractility, normally determine stroke volume.

Preload is the muscle fiber length prior to contraction, while afterload is the resistance against which the ventricle must contract. Contractility is an intrinsic property of the cardiac muscle that is related to the force of contraction but is independent of both the preload and afterload (Morgan et al. 2002). Heterometric autoregulation results in an increase in stretch of a muscle myocyte, thereby producing an increase in the tension developed, and hence influencing the ventricular function (Millar 2000). This is part of Starling's law of the heart (Millar 2000).

Experimentally, maximally developed tension in cardiac myocytes occurs at a resting sarcomere length of 2.0–2.3 μm, which allows optimum crossbridge formation between the contractile proteins that make up the thick and thin filaments (Millar 2000). Clinically, this translates to peak ventricular output occurring at filling pressures of 10–12 mm Hg.

Beta receptor occupancy predominantly activates adenyl cyclase within the cardiac myocytes to increase the conversion of adenosine triphosphate to cyclic adenosine monophosphate (cAMP). The cAMP facilitates the intracellular shift of calcium which increases inositol phosphate (IP3 and IP4). This in turn has positive chronotropic (increased heart rate), dromotropic (increased conduction), and inotropic (increased contractility) effects.

Effects on Systemic Vascular Resistance

Arteriolar tone is the principal determination of the systemic vascular resistance. Although both the sympathetic and parasympathetic systems can exert important influences on the circulation, autonomic control of the vasculature is primarily sympathetic. Sympathetic outflow to the circulation passes out of the spinal cord at all thoracic and the first two lumbar segments (Ganong 2001). These fibers reach the blood vessels via specific autonomic nerves or by travelling along the spinal nerves. Sympathetic fibers innervate all parts of the vasculature except for the capillaries. Their principal function is to regulate the vascular tone. Variation of arterial vascular tone serves

to regulate the blood pressure and the distribution of blood flow to the various organs, while variations in venous tone alter venous return to the heart.

The vasculature has sympathetic vasoconstrictors and vasodilator fibers, but the former are more important physiologically in most tissue beds. Sympathetically induced vasoconstriction (via alpha-1 adrenergic receptors) can be potent in skeletal muscles, kidneys, gut, and skin; it is the least active in the brain and heart (Ganong 2001).

The vasomotor center in the reticular formation of the medulla and the lower pons controls vascular tone and autonomic influences on the heart. Distinct vasoconstrictor and vasodilator centers have been identified. The anterolateral column mediates vasoconstriction. The vasomotor center is also responsible for the adrenal secretion of the catecholamines, as well as the enhancement of cardiac automaticity and contractility. Vasodilatory centers, which are located in the lower medulla, are also adrenergic but function by projecting inhibitory fibers upwards to the vasoconstrictor areas (Ganong 2001).

Overall, the effects of stress due to injury, surgery, and acute pain on cardiovascular function are potentially harmful, in particular to patients at high risk such as those with coronary heart disease. Providing a blockade to this stress response and acute pain stimuli has the potential to reduce postoperative cardiovascular complications.

References

Aantaa, R., & Scheinin, M. (1993). Alpha-adrenergic agents in anaesthesia. *Acta Anaesthesiologica Scandinavica, 37*, 1–16.

Desborough, J. P. (2000). The stress response to trauma and surgery. *British Journal of Anaesthesia, 85*, 109–117.

Ganong, W. F. (2001). *Review of medical physiology* (20th ed.). New York: McGraw Hill.

Millar, R. D. (2000). *Anaesthesia* (5th ed.). Edinburgh: Churchill Livingstone.

Morgan, G. E., Mikhail, M. S., Murray, M. J., & Larson, C. P. (2002). *Clinical anesthesiology 3rd edn*. New York: McGraw Hill.

Postoperative Pain, Pathophysiological Changes in Metabolism in Response to Acute Pain

Stephan A. Schug
School of Medicine and Pharmacology,
Royal Perth Hospital, Faculty of Medicine,
Dentistry & Health Sciences, The University of
Western Australia, UWA Anaesthesiology,
Perth, Australia

Synonyms

Catabolism; Destructive metabolism; Stress metabolism

Definition

Acute pain, trauma, and surgery are typically associated with a neuroendocrine stress response that is proportional to the severity of the injury and the pain intensity. Sympathetic activation increases efferent sympathetic tone to all viscera and releases catecholamines from the adrenal medulla (Ganong 2001).

The hormonal response results from increased sympathetic tone and hypothalamically mediated reflexes (Desborough 2000), increases in catabolic hormones (catecholamines, cortisol, and glucagon), and a decrease in anabolic hormones (insulin and testosterone). The net effect is a negative nitrogen balance, carbohydrate intolerance, and increased lipolysis. The increase in cortisol, renin, aldosterone, angiotensin, and antidiuretic hormone results in sodium retention, water retention, and a secondary expansion of the extracellular space (UKPD Group 1998).

Characteristics

Carbohydrate Metabolism

Blood glucose concentration increases as part of the response to stress and pain stimulus. Cortisol and catecholamines facilitate glucose production as a result of increased hepatic glycogenolysis and gluconeogenesis (Bent et al. 1978). In addition, insulin resistance develops with a reduction in the peripheral utilization of glucose (Bent et al. 1978).

Therefore, processes of normal glucose maintenance and homeostasis are ineffective in the perioperative period. Hyperglycemia persists because catabolic hormones promote glucose production and there is a relative lack of insulin, together with a peripheral insulin resistance (Cuthbertson 1932).

Protein Metabolism

Protein catabolism is stimulated by an increased cortisol concentration (Desborough and Hall 1993). Muscle proteins, as well as visceral proteins, are catabolized to release constituent amino acids. These amino acids are either catabolized for energy or utilized in the liver to form new proteins, in particular the so-called "acute phase" proteins (Colley et al. 1983). Constituent amino acids are also metabolized by the liver to other substrates such as glucose, fatty acids, and ketone bodies (Bent et al. 1978). Urinary nitrogen can be measured as an indirect measure of protein catabolism.

This catabolism results in marked weight loss and, even more relevant, muscle wasting in patients after major surgery and burns, thus impairing rehabilitation.

Fat Metabolism

Lipolysis converts the triglycerides in the fat stores of the body into glycerol and fatty acids. This process is stimulated by catecholamines, growth hormone, and cortisol; insulin on the other hand inhibits this process. The free fatty acids are utilized for the process of gluconeogenesis in the liver (Bent et al. 1978).

Water and Electrolyte Balance

Preservation of adequate body fluid volume is also influenced by the numerous hormonal changes. Arginine vasopressin, released by the posterior pituitary gland, results in water retention, and this, as a consequence, results in the production of both concentrated urine and a relative oliguria (Desborough and Hall 1993).

The sympathetic stimulation also increases the production of renin from the cell of the juxtaglomerular apparatus, which in turn stimulates the production of angiotensin 2 (Desborough 2000). Angiotensin 2 promotes the release of aldosterone leading to reabsorption of sodium and water, further promoting the retention of water.

Overall, the catabolic changes of the metabolism and the disturbance of the water and electrolyte balance impair postoperative recovery and cause postoperative complications. Provision of multimodal analgesia and a multidisciplinary rehabilitative approach to the postoperative period aim to reduce these detrimental changes (Kehlet 1997).

References

Bent, J. M., Paterson, J. L., Mashiter, K., et al. (1978). Effects of high dose fentanyl anaesthesia on the established metabolic and endocrine response to surgery. *Anaesthesia, 39*, 19–23.

Colley, C. M., Fleck, A., Goode, A. W., et al. (1983). Early time course of the acute phase protein response in man. *Journal of Clinical Pathology, 36*, 203–207.

Cuthbertson, D. P. (1932). Observation on the disturbance of metabolism produced by injury to the limb. *Quarterly Journal of Medicine, 1*, 233–246.

Desborough, J. P. (2000). The stress response to trauma and surgery. *British Journal of Anaesthesia, 85*, 109–117.

Desborough, J. P., & Hall, G. M. (1993). Endocrine responses to surgery. In L. Kaufman (Ed.), *Anaesthesia review* (Vol. 10, pp. 131–148). Edinburgh: Churchill Livingstone.

Ganong, W. F. (2001). *Textbook of medical physiology* (20th ed.). New York: McGraw Hill.

Kehlet, H. (1997). Multimodal approach to control postoperative pathophysiology and rehabilitation. *British Journal of Anaesthesia, 78*, 606–617.

UKPD Group. (1998). Intensive blood glucose control with sulphonylureas or insulin compared with conventional treatment and risk of complications in patients with type 2 diabetes (UKPDS 33). UK prospective diabetes study. *Lancet, 352*, 837–853.

Postoperative Pain, Pathophysiological Changes in Neuroendocrine Function in Response to Acute Pain

Stephan A. Schug
School of Medicine and Pharmacology, Royal Perth Hospital, Faculty of Medicine, Dentistry & Health Sciences, The University of Western Australia, UWA Anaesthesiology, Perth, Australia

Synonyms

General adaptation syndrome; Hypothalamic-pituitary-adrenal axis response; Stress response

Definition

The neuroendocrine response is part of the stress response that is mediated as a reaction to acute pain. The response is mediated via the afferents from the site of primary noxious stimulation and relies primarily on the hypothalamic-pituitary-adrenal axis. This response has a role in the preservation of an organism in the face of an acute injury.

Characteristics

Acute pain is characterized by the increased secretion of pituitary hormones and the activation of the sympathetic nervous system (Desborough 2000). Hypothalamic activation of the autonomic nervous system results in increased secretion of catecholamines, epinephrine, and norepinephrine from the adrenal medulla.

This increased sympathetic activity also results in the cardiovascular effects of tachycardia and hypertension (Desborough 2000).

The following relevant changes in hormone secretion occur in response to acute pain and surgery (Desborough 2000; Millar 2000; Ganong 2001; Desborough and Hall 1989).

The hypothalamus releases stimulating factors, which in turn stimulate the anterior pituitary gland to synthesize the corticotrophins. Therefore, the release of adrenocorticotrophic hormone (ACTH), thyroid-stimulating hormone (TSH), growth hormone, and prolactin increases. Arginine vasopressin (AVP) is the major hormone produced by the posterior pituitary gland; it is a major antidiuretic hormone that influences water balance in the kidneys.

ACTH then stimulates the adrenal cortex to produce increased levels of glucocorticoids, resulting in increased levels of cortisol. Cortisol has a role in carbohydrate metabolism and promotes protein breakdown.

Growth hormone has a major role in growth regulation and has some of its action mediated through the insulin-like growth factors. Growth hormone also promotes protein breakdown, promotes lipolysis, and has an anti-insulin-like effect. Plasma levels of growth hormone also influence glycogenolysis.

The pancreas often decreases insulin production, while increased amounts of glucagon are released. Insulin is the major anabolic hormone. It is a polypeptide, with two chains linked by two disulfide bridges. Insulin is synthesized and secreted by the beta cells of the pancreas. Insulin in normal situations is secreted after food intake, when the blood glucose and amino acid concentration increases. Insulin, as part of its action, promotes the uptake of glucose into muscles and adipose tissues and facilitates the conversion of glucose into glycogen and triglycerides. Insulin also stimulates the formation of glycogen from glucose in the liver and inhibits protein catabolism and lipolysis.

Acute pain and surgery causes a reduction in the production of insulin, which in turn results in the failure of the secretion to match the catabolic and hyperglycemic response. The normal insulin responsive cells also fail to respond appropriately to the reduced levels of insulin, resulting in the "insulin resistance" that occurs in the perioperative period.

Glucagon production in the alpha cells of the pancreas increases, and this promotes hepatic

glycogenolysis. Hepatic gluconeogenesis also increases, but this response is not a major contributor to the hyperglycemic state seen in the perioperative period.

TSH, released from the anterior pituitary, stimulates the thyroid gland to synthesize and release thyroxine (T4) and triiodothyronine (T3) into the circulation. Triiodothyronine is three to five times more metabolically active than thyroxine.

These hormones are bound extensively to their binding proteins: albumin, thyroxine-binding prealbumin, and thyroid-binding globulin. The free thyroid hormones in the plasma are metabolically active, and the very low concentration of the free T3 and T4 that are present in the circulation are in equilibrium with the bound hormones in the plasma and tissues. Thyroid hormones increase the oxygen consumption of most of the metabolically active tissues of the body. The brain tissues and the spleen cells are not under the influence of the thyroid hormones. The metabolic rate and heat production are also increased as a direct influence of the thyroid hormones. The other actions of the thyroid hormones include an increase in carbohydrate absorption from the gut, stimulation of the central and peripheral nervous system, and its influence on growth and development.

Like the thyroid hormones, the catecholamines also increase the metabolic rate and stimulate the nervous system. The thyroid hormones increase the number and affinity of the beta-adrenoreceptors in the heart and also increase the sensitivity of the heart to the actions of the catecholamines.

Last, but not least, beta-endorphin and prolactin levels are also increased during acute pain and surgery. Beta-endorphin is an endogenous opioid peptide and has a role in the perception of pain.

Follicle-stimulating hormone (FSH) and luteinizing hormone (LH) do not change markedly during surgery and acute pain. The gonadotrophin secretion and release may be reduced in situations of acute pain, but their significance in this situation remains undetermined.

References

Desborough, J. P. (2000). The stress response to trauma and surgery. *BJA, 85*, 109–117.

Desborough, J. P., & Hall, G. M. (1989). Modification of the hormonal and metabolic response to surgery by narcotics and general anaesthesia. *Clinical Anaesthesiology, 3*, 317–334.

Ganong, W. F. (2001). *Review of medical physiology* (20th ed.). New York: McGraw Hill.

Millar, R. D. (2000). *Textbook of anaesthesia* (5th ed.). Edinburgh: Churchill Livingstone.

Postoperative Pain, Patient-Controlled Analgesia Devices, Epidural

Stephan A. Schug[1] and Per Flisberg[2]
[1]School of Medicine and Pharmacology, Royal Perth Hospital, Faculty of Medicine, Dentistry & Health Sciences, The University of Western Australia, UWA Anaesthesiology, Perth, Australia
[2]Department of Anaesthesia and Intensive Care, Lund University Hospital, Lund, Sweden

Synonyms

On-demand epidural analgesia; Patient-controlled epidural analgesia; PCEA

Definition

▶ Patient-controlled epidural analgesia (PCEA) is an analgesic technique which transfers the concept of classical PCA (parenteral) to the provision of epidural analgesia. It allows the individual patient to adjust the epidural block height and density by self-administering the analgesic medications via a pump into the epidural space. The pump is usually an electronic or mechanical, reusable or disposable device, delivering a metered dose of drug in response to the patient trigger. The device can be programmed in a variety of ways to provide only on-demand doses or to also administer a continuous background infusion. Pumps need a mechanism that ensures that the selected dose cannot be given too frequently; thus, a lock-out time interval can be programmed.

Characteristics

Epidural analgesia in general offers improved analgesia compared to intramuscular analgesia and intravenous patient-controlled analgesia (Dolin et al. 2002). Since epidural analgesia is not a single entity, there are difficulties when interpreting the available data on the efficacy of PCEA. PCEA can be provided with a variety of medications, each on its own and in combination. The catheter can be inserted at different levels of the epidural space, and the technique can be provided for a large variety of different operations.

The introduction of PCEA technique has individualized and improved the patients' epidural analgesia, by giving patients control over their own quality of analgesia. Studies have shown that patient-controlled epidural analgesia leads to reductions in pain intensity, less consumption of drugs, and superior patient satisfaction (Lubenow et al. 1994; Liu et al. 1998; Silvasti and Pitkanen 2001).

The Issue of Background Infusions

The best regimen for PCEA after major surgery has not yet been fully clarified; however, data support the use of an additional background infusion, in contrast to parenteral PCA.

The addition of a background infusion when using a local anesthetic-opioid combination epidurally was found to be superior compared to PCEA alone in patients undergoing gastrectomy (Komatsu et al. 1998). The same investigators have also shown that a nighttime infusion added to PCEA following gastrectomy decreases the incidence of postoperative pain associated with coughing during the night and improves sleep compared with PCEA alone (Komatsu et al. 2001). In addition, a large prospective study in patients undergoing major abdominal surgery found that PCEA combined with background infusion was superior regarding dynamic postoperative pain scores compared with intravenous PCA (Flisberg et al. 2003).

PCEA for Labor Pain

In labor analgesia, the PCEA concept became a widely accepted technique early on. PCEA in labor analgesia offers several advantages over conventional techniques of pain relief including reduced drug consumption, improved analgesia, and thereby superior patient satisfaction, reducing unscheduled clinician interventions and has been proven to be safe (Ferrante et al. 1994; Gambling et al. 1988; Paech 1991; Halpern and Carvalho 2009). In a recent study, it was shown that even ultra-low doses of an opioid combined with a local anesthetic administered via PCEA reduced the dose required, compared with continuous epidural infusion (Ledin et al. 2003). When PCEA is used in labor, a diluted local anesthetic combined with the lowest clinically effective concentration of lipophilic opioid should be used in order to avoid unnecessary motor block and pruritus (Halpern and Carvalho 2009).

A meta-analysis compared PCEA with continuous infusion in labor analgesia; it confirmed that patients having PCEA are less likely to require analgesic interventions and have less disturbance of motor function than those receiving continuous epidural infusion (Van der Vyver et al. 2002).

Economic Aspects of PCEA

No randomized-controlled trial has so far been published comparing the cost of PCEA versus intravenous PCA. Recently, a retrospective study examined the costs involved with the use of postoperative PCEA. It concluded that the treatment length was the main cost driver, both for drugs and for staff costs (Schuster et al. 2004). A comparison of the cost-effectiveness of PCEA versus intrathecal morphine has also been performed (Vercauteren et al. 2002); it found that PCEA resulted in better pain relief and caused less nausea and vomiting but was more expensive to provide than intrathecal morphine.

Safety Aspects of PCEA

In order to achieve effective and safe analgesia with PCEA, the patient needs some introductory education about the technique and has to understand the basic principle. This has become evident in a study where several patients had difficulties in operating the PCEA device (Silvasti and Pitkanen 2001). However, if used correctly, PCEA results in improved mental status in elderly patients (Mann et al. 2000), compared with parenteral PCA.

A number of large surveys have shown that PCEA can be provided safely on general hospital wards, as long as certain safety requirements are maintained and supervision by an anesthesia-based pain service is offered (Flisberg et al. 2003; Liu et al. 1998).

Future Perspectives

New techniques and new software regarding epidural delivery systems such as computer-integrated PCEA and programmed intermittent epidural boluses (PIEB) are under development (Lim et al. 2006; Sia et al. 2007). Computer-integrated PCEA uses an algorithm which adjusts the background infusion in response to the number of patient-demanded doses. The programmed intermittent epidural module system, in comparison, delivers bolus doses of local anesthetic at programmed intervals. Neither the computer-integrated PCEA nor the programmed intermittent epidural boluses are currently commercially available but may be incorporated in future epidural delivery systems.

References

Dolin, S. J., Cashman, J. N., & Bland, J. M. (2002). Effectiveness of acute postoperative pain management: I evidence from published data. *British Journal of Anaesthesia, 89,* 409–423.

Ferrante, F. M., Rosinia, F. A., Gordon, C., et al. (1994). The role of continuous background infusions in patient-controlled epidural analgesia for labor and delivery. *Anesthesia and Analgesia, 79,* 80–84.

Flisberg, P., Rudin, A., Linner, R., et al. (2003). Pain relief and safety after major surgery. A prospective study of epidural and intravenous analgesia in 2,696 patients. *Acta Anaesthesiologica Scandinavica, 47,* 457–465.

Gambling, D. R., Yu, P., Cole, C., et al. (1988). A comparative study of patient controlled epidural analgesia (PCEA) and continuous infusion epidural analgesia (CIEA) during labour. *Canadian Journal of Anaesthesia, 35*, 249–254.

Halpern, S. H., & Carvalho, B. (2009). Patient-controlled epidural analgesia for labor. *Anesthesia and Analgesia, 108*, 921–928.

Komatsu, H., Matsumoto, S., Mitsuhata, H., et al. (1998). Comparison of patient-controlled epidural analgesia with and without background infusion after gastrectomy. *Anesthesia and Analgesia, 87*, 907–910.

Komatsu, H., Matsumoto, S., & Mitsuhata, H. (2001). Comparison of patient-controlled epidural analgesia with and without night-time infusion following gastrectomy. *British Journal of Anaesthesia, 87*, 633–635.

Ledin, E. S., Gentele, C., & Olofsson, C. H. (2003). PCEA compared to continuous epidural infusion in an ultra-low-dose regimen for labor pain relief: A randomized study. *Acta Anaesthesiologica Scandinavica, 47*, 1085–1090.

Lim, Y., Sia, A. T., & Ocampo, C. E. (2006). Comparison of computer integrated patient controlled epidural analgesia vs. conventional patient controlled epidural analgesia for pain relief in labour. *Anaesthesia, 61*, 339–344.

Liu, S. S., Allen, H. W., & Olsson, G. L. (1998). Patient-controlled epidural analgesia with bupivacaine and fentanyl on hospital wards: Prospective experience with 1,030 surgical patients. *Anesthesiology, 92*, 433–441.

Lubenow, T. R., Tanck, E. N., Hopkins, E. M., et al. (1994). Comparison of patient-assisted epidural analgesia with continuous-in-fusion epidural analgesia for postoperative patients. *Regional Anesthesia, 19*, 206–221.

Mann, C., Pouzeratte, Y., Boccara, G., et al. (2000). Comparison of intravenous or epidural patient-controlled analgesia in elderly after major abdominal surgery. *Anesthesiology, 92*, 433–441.

Paech, M. J. (1991). Epidural analgesia in labour: Constant infusion plus patient – controlled boluses. *Anaesth Intens Care, 19*, 32–39.

Schuster, M., Gottschalk, A., Freitag, M., et al. (2004). Cost drivers in patient-controlled epidural analgesia for postoperative pain management after major surgery. *Anesthesia and Analgesia, 98*, 708–713.

Sia, A. T., Lim, Y., & Ocampo, C. (2007). A comparison of a basal infusion with automated mandatory boluses in parturient-controlled epidural analgesia during labor. *Anesthesia and Analgesia, 104*, 673–678.

Silvasti, M., & Pitkanen, M. (2001). Patient-controlled epidural analgesia versus continuous epidural analgesia after total knee arthroplasty. *Acta Anaesthesiologica Scandinavica, 45*, 472–476.

Van der Vyver, M., Halpern, S., & Joseph, G. (2002). Patient-controlled epidural analgesia versus continuous infusion for labour analgesia: A meta-analysis. *British Journal of Anaesthesia, 89*, 459–645.

Vercauteren, M., Vereecken, K., Malfa, M., et al. (2002). Cost-effectiveness of analgesia after caesarean section. A comparison of intrathecal morphine and epidural PCA. *Acta Anaesthesiologica Scandinavica, 46*, 85–89.

Postoperative Pain, Patient-Controlled Analgesia Devices, Parenteral

Stephan A. Schug
School of Medicine and Pharmacology, Royal Perth Hospital, Faculty of Medicine, Dentistry & Health Sciences, The University of Western Australia, UWA Anaesthesiology, Perth, Australia

Synonyms

On-demand analgesia; On-demand analgesia computer; Patient-controlled intravenous analgesia; PCA

Definition

A PCA device is usually an electronic or mechanical pump, which delivers a metered dose or volume of drug in response to a patient trigger. It is most often given to patients who undergo moderate or major surgery and are expected to be in moderate to severe pain postoperatively. The patient can therefore determine when and how much analgesia they receive within the prescribed and programmed limits of the device. Historically, the devices were used only in the postoperative setting, using intravenous opioid. Increasingly they are used in other clinical settings, via other parenteral routes and with non-opioid drugs.

Characteristics

▶ Patient-controlled analgesia is a concept of pain relief, which takes the subjective nature of pain and the wide interindividual variability of opioid requirements into consideration. The concept was first used by Sechzer to measure pain in 1968 (Sechzer 1968).

PCA devices have been used since the 1980s, primarily using intravenous opioids. Most of the evidence on efficacy, adverse effects, education, and safety is based on opioid usage.

Devices can be separated into two types of pumps. An electronic pump allows the operator to set bolus volume, lockout period, time-dose limit, and background infusions. Mechanical pumps are simple devices that allow a fixed volume of liquid drug to be delivered, the empty chamber filling up over a period of 5–6 min, effectively producing a mechanical lockout period.

PCA Versus Conventional Analgesia

In 1993, Ballantyne et al. performed a meta-analysis of the randomized controlled trials comparing outcomes of PCA and intramuscular (i.m.) analgesia (Ballantyne et al. 1993). Significantly greater analgesic efficacy was seen using PCA, though the magnitude of the pain score reduction was small. More recent studies comparing PCA with traditional methods of opioid analgesia have produced contradictory results. Some show significantly improved analgesia with PCA, while others show no difference. In a later review, Macintyre suggested that at least some of the trials showing no difference involved patients who had a higher level of postoperative nursing care than a general ward setting and possibly fast access to conventional analgesia (Macintyre 2001). A later meta-analysis of 32 trials in 2001 concluded that opioid via PCA provided better efficacy than conventional opioid analgesia, with no real difference in the amount of opioids consumed (Walder et al. 2001). A more recent Cochrane Review confirmed this and found that PCA provided better pain control and greater patient satisfaction than conventional parenteral "as-needed" analgesia (Hudcova et al. 2006).

PCA has also been used effectively in settings other than the postoperative, i.e., in bone marrow transplantation, burns, and cancer pain. Many trials comparing the effectiveness of new drugs

also use PCAs to measure the drug's analgesic ability to decrease PCA opioid requirements, "the opioid-sparing effect."

Dose

The PCA delivers its bolus dose in response to a patient trigger. This is usually pressing a button or lever but can be by blowing into a mouthpiece. The optimal amount of bolus dose has been investigated. Owen et al. found that the optimal dose for morphine is 1 mg, with lower doses associated with ineffective analgesia and higher doses with increased incidence of respiratory depression (Owen et al. 1989). To increase safety, many hospitals have a policy of keeping the concentration of drug used in the device fixed. Variable bolus PCA devices have been investigated but did not offer any increased patient satisfaction or lead to decreased complications (Love et al. 1996). There seems to be no advantage in setting hourly or 4-hourly dose limits on the electronic devices (Sidebotham et al. 1997). There is wide variation in patients' opioid requirements, and there have been several reports of respiratory depression at doses well inside the standard set limits.

Lockout Interval

This is the time between a bolus delivery and the next valid delivery request. Within this latent period, a patient is "locked out" from further access to the drug regardless of the number of times the request button is pressed. Electronic pumps may allow a variable lockout period. Ideally, the time period would reflect the time it would take to achieve full effect from the drug delivered. In practice, for most intravenous opioids, the lockout period is set to 5–10 min (Badner et al. 1996).

Continuous Infusions

Most electronic PCA devices can deliver continuous (background) infusions on top of the demand doses. However, studies in both adults and children have been unable to show improved pain relief and have shown an increased incidence of respiratory depression (Badner et al. 1996). Prescribing a background infusion may also increase programming errors. However, it may be suitable for opioid-tolerant patients (on long-term opioids for cancer or chronic pain or on maintenance programs) or those who report waking up in severe pain.

Drug

There seems little evidence to suggest major differences in efficacy or side effects between the most commonly used PCA opioids (Woodhouse et al. 1999). However, patients who experience particular side effects from one drug may benefit from switching to another at an equipotent dose ("► opioid rotation"). Pethidine use in PCAs should be avoided because of the complications associated with accumulation of the metabolite, norpethidine. Side effects ranging from anxiety to seizure activity have been reported. This may occur within the first 24 h of commencing the PCA, even in the absence of renal impairment.

Patient Factors

Patients from 4 years old to over 90 have used PCA effectively. Since successful use requires reasonable cognitive function, it may be an unsuitable analgesic plan for patients who are confused or who have dementia. Opioid requirements can be reduced in the elderly, and reduction of the bolus dose may be required. In opioid-tolerant patients, a background infusion and higher bolus dose may be indicated. In all cases, appropriate patient selection and education is essential to improve compliance, satisfaction, and safety. Psychological factors also influence usage (Perry et al. 1994).

Patient Outcome

Patient satisfaction scores are often higher in those using PCA compared with conventional analgesia (Ballantyne et al. 1993). This is attributed to improved feelings of autonomy, less anxiety, and lower postoperative pain scores. However, in a more recent study, PCA was shown to be valued as a way to avoid the difficulty of disclosing pain or securing pain relief within the usual nurse-patient relationship (Taylor et al. 1996).

Adverse Effects

Respiratory depression is the most concerning potential side effect. Several audits have reported an incidence of 0.1–0.8 %, although with a background infusion, this rises to between 1.1 % and 3.9 % (Macintyre 2001). Respiratory rate seems to correlate poorly with opioid-related respiratory depression; the best marker seems to be sedation score. However, some studies have suggested that respiratory depression associated with i. m. analgesia is even more common. This is probably associated with higher plasma concentrations as the initial bolus is absorbed.

There have been several case reports of malfunction of both electronic and mechanical pumps (Baird and Schug 1996). Manufacturers have since altered much of the associated hardware and software. Anti-reflux valves are now recommended for use with all pumps. Triggering by patients, relatives, or other staff has also been reported. Operator error in the programming of pumps or in preparation of content and concentration of syringes has been a reasonably common safety issue. Hospital protocol and staff education on the use of PCA should decrease these incidences.

Several cases have been reported where the PCA device was thought to be "masking" the signs of myocardial infarction, pulmonary embolus, urinary retention, and compartment syndrome. Concern has been expressed that a patient with pain from a new medical or surgical problem could control the pain with PCA so effectively that the cause may remain undiagnosed. However, most PCA treatment prescriptions incorporate regular clinical assessments of pain scores and amount of drug used. These are done on up to an hourly basis. An unexpected change in the rate of analgesic use, nature of pain, or its intensity should trigger careful monitoring and investigation.

Other Routes and Drugs

PCA has been used successfully via the intranasal (Toussaint et al. 2000), subcutaneous, and oral route. An interesting new concept is the transdermal electrotransport of fentanyl via a patient-controlled transdermal system using iontophoresis (Chelly et al. 2004); this might become commercially available in the near future.

Overall PCA is a very effective and relatively safe method of postoperative analgesia and compares well to conventional analgesia. Having a standard protocol, and then tailoring it to suit the individual patient by a single specialist team, will increase efficacy and safety.

References

Badner, N. H., Doyle, J. A., Smith, M. H., et al. (1996). Effect of varying intravenous patient-controlled analgesia dose and lockout interval while maintaining a constant hourly maximum dose. *Journal of Clinical Anesthesia, 8*, 382–385.
Baird, M. B., & Schug, S. A. (1996). Safety aspects of postoperative pain relief. *Pain Digest, 6*, 219–225.
Ballantyne, J. C., Carr, D. B., Chalmers, T. C., et al. (1993). Post-operative patient-controlled analgesia: Meta-analyses of initial randomized control trials. *Journal of Clinical Anesthesia, 5*, 182–193.
Chelly, J. E., Grass, J., Houseman, T. W., et al. (2004). The safety and efficacy of a fentanyl patient-controlled transdermal system for acute postoperative analgesia: A multicenter, placebo-controlled trial. *Anesthesia and Analgesia, 98*, 427–433.
Hudcova, J., McNicol, E., Quah, C., Lau, J., & Carr, D. B. (2006). Patient controlled opioid analgesia versus conventional opioid analgesia for postoperative pain. *Cochrane Database of Systematic Reviews, 4*, CD003348.
Love, D. R., Owen, H., Ilsley, A. H., et al. (1996). A comparison of variable-dose patient-controlled analgesia. *Anesthesia and Analgesia, 83*, 1060–1064.
Macintyre, P. E. (2001). Safety and efficacy of patient-controlled analgesia. *British Journal of Anaesthesia, 87*, 36–46.
Owen, H., Plummer, J. L., Armstrong, I., et al. (1989). Variables of patient-controlled analgesia. 1. Bolus size. *Anaesthesia, 44*, 401–406.
Perry, F., Parker, R. K., White, P. F., et al. (1994). Role of psychological factors in postoperative pain control and recovery with patient-controlled analgesia. *The Clinical Journal of Pain, 10*, 57–63.
Sechzer, P. H. (1968). Objective measurement of pain. *Anesthesiology, 29*, 209–213.
Sidebotham, D., Dijkhuizen, M. R., & Schug, S. A. (1997). The safety and utilization of patient-controlled analgesia. *Journal of Pain and Symptom Management, 14*, 202–209.
Taylor, N., Hall, G. M., & Salmon, P. (1996). Is patient-controlled analgesia controlled by the patient? *Social Science & Medicine, 43*, 1137–1143.
Toussaint, S., Maidl, J., Schwagmeier, R., & Striebel, H. W. (2000). Patient-controlled intranasal analgesia: effective alternative to intravenous PCA for postoperative pain relief. *Canadian Journal of Anaesthesia, 47* (4), 299–302.
Walder, B., Schafer, M., Henzi, I., et al. (2001). Efficacy and safety of patient-controlled opioid analgesia for acute postoperative pain. A quantitative systematic review. *Acta Anaesthesiologica Scandinavica, 45*, 795–804.
Woodhouse, A., Ward, M., & Mather, L. (1999). Inter-subject variability in post-operative patient-controlled analgesia (PCA): Is the patient equally satisfied with morphine, pethidine and fentanyl? *Pain, 80*, 545–553.

Postoperative Pain, Persistent Acute Pain

Edward A. Shipton
Department of Anaesthesia, Christchurch School of Medicine and Health Sciences, University of Otago, Christchurch, New Zealand

Synonyms

Chronic pain; Subacute pain

Definition

Persistent acute postoperative pain is pain in the location of the surgery that persists beyond the usual course of an acute injury (surgery) and is different from that suffered preoperatively.

Characteristics

Normally acute surgical pain declines over the first few days after surgery, but in some instances the pain can persist and become both chronic and severe. Every chronic pain was once acute. Few studies have defined chronic postsurgical pain. In general, chronic postsurgical pain (CPSP) has been defined as pathological pain that persists for longer than two months postsurgery (Macrae and Davies 1999).

Risk Factors

The transition of acute postoperative pain to chronic postsurgical pain is a complex and a poorly understood developmental process. Surgical, psychosocial, socio-environmental, and patient-related factors (Clarke et al. 2010) and the known polymorphisms in human genes are involved in perpetuating the pain.

Genes

The genetic regulation of pain has been derived from animal models developed mainly in mice. Several hereditary syndromes of complete or almost complete insensitivity to pain have been identified and include channelopathy-associated pain insensitivity, of which the most likely candidate gene is the α-subunit of the voltage-gated sodium channel known as Na(v)1.7 (Tremblay and Hamet 2010). Five hereditary sensory and autonomic neuropathy syndromes have been described. Variable pain sensitivity in the general population has been linked to common variants of the μ-opioid receptor and of the catecholamine-O-methyltransferase genes; this potentially leads to increased opioid tonus (Tremblay and Hamet 2010). Variants of the guanosine triphosphate cyclohydrolase 1/dopa-responsive dystonia gene appear to regulate nociception. Other candidate genes are the transient receptor potential cation channel, subfamily 5 member 1, gene, and the melanocortin-1 receptor gene (Tremblay and Hamet 2010).

Genetic variations can occur during uptake, transport, at the effector site, and during the metabolism and excretion of a drug. Candidate genes for predicting opioid efficacy are drug-metabolizing enzymes and transporters – including cytochrome P450, uridine 5′-diphosphate-glucuronosyl-transferases, and adenosine triphosphate-binding cassette transporters that are involved in opioid metabolism (Tremblay and Hamet 2010). Genetic analysis of receptors, of drug transporters, and of metabolizing enzymes may be needed to establish the effective doses of each drug in the individual patient to prevent adverse effects and achieve pain relief in a shorter period of time; this may prevent acute pain from becoming chronic.

Preoperative Risk Factors

Both postoperative pain and the risk for the development of chronic postoperative pain are determined by preoperative, intraoperative, and postoperative factors. The intensity of suprathreshold heat pain (i.e., pain beyond patient threshold) has been most consistently shown to correlate with postoperative pain (Abrishami et al. 2011).

Patients with preoperative pain related (or not related) to the surgical site are significantly at risk to develop CPSP (Gerbershagen et al. 2009, 2010). Studies have reported that female gender (Clarke et al. 2010) and younger age (Clarke et al. 2010; Katz et al. 2005) predict intense acute postoperative pain.

High preoperative anxiety is associated with increased acute postoperative pain (Clarke et al. 2010; Katz et al. 2008). High catastrophizing levels are associated with increased pain severity, increased incidence of development of chronic pain, and poorer quality of life after surgery (Khan et al. 2011). Long-term fear is associated with an increased risk of more postoperative pain (Sommer et al. 2010). Social and environmental factors are associated with persistent postsurgical pain; these include stressful life events, solicitous responding from significant others, and social support (Nikolajsen and Minella 2009).

Intraoperative Risk Factors

Intraoperative risk factors that are associated with an increased likelihood of developing chronic postsurgical pain include (Katz and Seltzer 2009) the following: site (e.g., thoracotomy, sternotomy, mastectomy, major limb amputation); extent of surgery and incision type (open versus laparoscopic approach) (Liem et al. 2003); increased duration of surgery (Kalso et al. 2001); a low (compared to a high) volume surgical unit (Tasmuth et al. 1999); pericostal (vs. intracostal) stitches; and intraoperative nerve damage (Liem et al. 2003). Damage to nerves occurs by surgical section, compression (in a suture or by clips), stretching, ischemia, or infection (Power et al. 2010).

Postoperative Risk Factors

Red flags, such as infection, bleeding, organ rupture, and the onset of a compartment syndrome, need to be excluded. Postoperative risk factors include unrelieved and severe pain (Aasvang et al. 2010), high postoperative use of analgesics, surgery performed in a previously injured area, radiation therapy and chemotherapy, and repeated surgery in the same area.

Mechanisms

Progression from acute to chronic pain involves altered pain processing. One of the triggers of pain chronicity is thought to be the amount of injury discharge produced intraoperatively and its impact on the central nervous system (Nikolajsen and Minella 2009). Chemical, mechanical, and thermal receptors, along with leukocytes and macrophages, determine the intensity, location, and duration of noxious events (Voscopoulos and Lema 2010). Noxious stimuli are transduced to the dorsal horn of the spinal cord, where amino acid and peptide transmitters activate second-order neurones. Spinal neurones then transmit signals to the brain. Persistent, intense pain activates secondary mechanisms both at the periphery and within the central nervous system that cause allodynia, hyperalgesia, and hyperpathia. Continued input from the perioperative noxious injury barrage maintains central sensitization and secondary hyperalgesia, amplifies postoperative pain, and contributes to chronic pain (De Kock 2009). These changes begin in the periphery with upregulation of cyclooxygenase-2 and interleukin-1β-sensitizing first-order neurones occur at the periphery; this leads to eventual sensitization of the second-order spinal neurones by activating N-methyl-d-aspartic acid channels and signaling microglia to alter neuronal cytoarchitecture (Voscopoulos and Lema 2010). Throughout these processes, prostaglandins, endocannabinoids, ion-specific channels, and scavenger cells all play a key role in the transformation of acute to chronic pain. Recently, a mechanism of neuronal plasticity in primary afferent nociceptive nerve fibers (nociceptors)

has been identified by which an acute inflammatory insult or environmental stressor can trigger long-lasting hypersensitivity of nociceptors to inflammatory cytokines (Reichling and Levine 2009). Hyperalgesic priming depends on the epsilon isoform of protein kinase C (PKC epsilon) and a switch in intracellular signaling pathways that mediate cytokine-induced nociceptor hyperexcitability. Studies in the rat model confirm a switch to a G (i)-activated and PKC epsilon-dependent signaling pathway in primary mechanical hyperalgesia that is induced by stress or inflammation (Dina et al. 2009).

Prevention and Management

The single best approach to persistent acute postoperative pain is to prevent it. Standardized pain evaluation and treatment protocols are strongly recommended. Psychosocial screening questionnaires that predict the psychosocial factors in patients that make them high risk for developing acute persistent pain should be used (Browne et al. 2011). Patients with high risk factors for developing persisting postsurgical pain should be marked and followed up after discharge. Preoperatively education of both the patient and clinical staff about the procedure and intended pain management should be provided. A thorough explanation of the surgery and the expected outcome should be given to the patient. Patient attitudes and concerns should be addressed. The patient and family should be provided with a written workable, effective, and safe pain management program for use at home. This is particularly important for parents of infants and children undergoing surgical procedures on a day-surgery basis.

Operative procedures that cause severe pain should be identified. Thoracic, major limb amputation, and spinal surgeries are the most painful procedures. The prevention of nerve and tissue damage during surgery is probably the most important preventive factor (Nikolajsen and Minella 2009). The least painful surgical approach (e.g., keyhole) with acceptable exposure should be used.

Multimodal pharmacological analgesia in addition to afferent neural blockade (neuraxial, plexus, peripheral nerve, wound infiltration, and wound catheters) should be instituted. Early improved outcomes are achieved by reducing opioid-related adverse effects. Rational multimodal analgesic drug combinations (e.g., NSAIDs, cyclooxygenase-2 inhibitors, paracetamol, gabapentinoids, clonidine, ketamine, and local and regional anesthetic techniques) are supplemented by opioid analgesics. An optimal procedure-specific analgesic regimen (see PROSPECT Web site: www.postoppain.org) should be integrated into an enhanced recovery care program (Abraham and Albayati 2011). Above all, analgesia should be continued well into the postoperative period.

Future

Perioperative identification of patients who are at high risk for developing chronic pain after surgery is a major goal for the future. To identify the contribution of relevant determinants and modulators of postoperative pain, large prospective-controlled clinical studies are urgently required that include measurement and documentation of all risk factors and procedure-specific aspects.

References

Aasvang, E. K., Gmaehle, E., Hansen, J. B., et al. (2010). Predictive risk factors for persistent postherniotomy pain. *Anesthesiology, 112*(4), 957–969.

Abraham, N., & Albayati, S. (2011). Enhanced recovery after surgery programs hasten recovery after colorectal resections. *World Journal of Gastrointestinal Surgery, 3*(1), 1–6.

Abrishami, A., Chan, J., Chung, F., & Wong, J. (2011). Preoperative pain sensitivity and its correlation with postoperative pain and analgesic consumption: A qualitative systematic review. *Anesthesiology, 114*(2), 445–457.

Browne, A. L., Andrews, R., Schug, S. A., & Wood, F. (2011). Persistent pain outcomes and patient satisfaction with pain management after burn injury. *The Clinical Journal of Pain, 27*(2), 136–145.

Clarke, H., Kay, J., Mitsakakis, N., & Katz, J. (2010). Acute pain after total hip arthroplasty does not predict the development of chronic postsurgical pain 6 months later. *Journal of Anesthesia, 24*(4), 537–543.

De Kock, M. (2009). Expanding our horizons: Transition of acute postoperative pain to persistent pain and

establishment of chronic post surgical pain services. *Anesthesiology, 111*(3), 461–463.

Dina, O. A., Khasar, S. G., Gear, R. W., & Levine, J. D. (2009). Activation of Gi induces mechanical hyperalgesia post stress or inflammation. *Neuroscience, 160*(2), 501–507.

Gerbershagen, H. J., Dagtekin, O., Gaertner, J., et al. (2010). Preoperative chronic pain in radical prostatectomy patients: preliminary evidence for enhanced susceptibility to surgically induced pain. *European Journal of Anaesthesiology, 27*(5), 448–454.

Gerbershagen, H. J., Ozgür, E., Dagtekin, O., et al. (2009). Preoperative pain as a risk factor for chronic postsurgical pain – six month follow-up after radical prostatectomy. *European Journal of Pain, 13*(10), 1054–1061.

Kalso, E., Mennander, S., Tasmuth, T., & Nilsson, E. (2001). Chronic post-sternotomy pain. *Acta Anaesthesiologica Scandinavica, 45*(8), 935–939.

Katz, J., Buis, T., & Cohen, L. (2008). Locked out and still knocking: Predictors of excessive demands for postoperative intravenous patient-controlled analgesia. *Canadian Journal of Anaesthesia, 55*, 88–99.

Katz, J., Poleshuck, E. L., Andrus, C. H., et al. (2005). Risk factors for acute pain and its persistence following breast cancer surgery. *Pain, 119*, 16–25.

Katz, J., & Seltzer, Z. (2009). Transition from acute to chronic postsurgical pain: Risk factors and protective factors. *Expert Review of Neurotherapeutics, 9*(5), 723–744.

Khan, R. S., Ahmed, K., Blakeway, E., Skapinakis, P., Nihoyannopoulos, L., Macleod, K., Sevdalis, N., Ashrafian, H., Platt, M., Darzi, A., & Athanasiou, T. (2011). Catastrophizing: A predictive factor for postoperative pain. *American Journal of Surgery, 201*(1), 122–131.

Liem, M. S., van Duyn, E. B., van der Graaf, Y., et al. (2003). Recurrences after conventional anterior and laparoscopic inguinal hernia repair: A randomized comparison. *Annals of Surgery, 237*(1), 136–141.

Macrae, W., & Davies, H. (1999). Chronic postsurgical pain. In I. K. Crombie, S. Linton, P. Croft, M. Von Knorff, & L. Leresche (Eds.), *The epidemiology of chronic pain* (pp. 125–142). Washington: IASP Press.

Nikolajsen, L., & Minella, C. E. (2009). Acute postoperative pain as a risk factor for chronic pain after surgery. *European Journal of Pain, Supplements, 3*(2), 29–32.

Power, I., McCormack, J. G., & Myles, P. S. (2010). Regional anaesthesia and pain management. *Anaesthesia, 65*(Suppl. 1), 38–47.

Reichling, D. B., & Levine, J. D. (2009). Critical role of nociceptor plasticity in chronic pain. *Trends in Neurosciences, 32*(12), 611–618.

Sommer, M., de Rijke, J. M., van Kleef, M., Kessels, A. G., Peters, M. L., Geurts, J. W., Patijn, J., Gramke, H. F., & Marcus, M. A. (2010). Predictors of acute postoperative pain after elective surgery. *The Clinical Journal of Pain, 26*(2), 87–94.

Tasmuth, T., Blomqvist, C., & Kalso, E. (1999). Chronic post-treatment symptoms in patients with breast cancer operated in different surgical units. *European Journal of Surgical Oncology, 25*(1), 38–43.

Tremblay, J., & Hamet, P. (2010). Genetics of pain, opioids, and opioid responsiveness. *Metabolism, 59* (Suppl 1), S5–S8.

Voscopoulos, C., & Lema, M. (2010). When does acute pain become chronic? *British Journal of Anaesthesia, 105*(Suppl 1), 69–85.

Postoperative Pain, Pethidine

Ross MacPherson
Department of Anesthesia and Pain Management, Royal North Shore Hospital University of Sydney, Sydney, NSW, Australia

Synonyms

Ethyl 1-methyl-4-phenylpiperidine-4-carboxylate hydrochloride; Meperidine; Operidine; Pethidine

Characteristics

Meperidine has no real place in contemporary pain management, and it has been eliminated from many hospitals, although usage in American institutions remains high.

Iatrogenic meperidine dependence is just one of the problems associated with this particular agent, whose continuing use is coming under increasing scrutiny.

The adverse effects of meperidine are well known and have been reported for years. In addition to having a risk of dependence higher than that seen with other opioids, it has a short duration of action, produces, even at recommended doses nor meperidine, a potentially toxic metabolite that is analgesic and epileptognic (Anon 1993), and can contribute to important drug interactions (Kredo and Onia 2005). Meperidine, when delivered by patient-controlled analgesia devices, is particularly dangerous and associated with a high incidence of serious adverse effects (Seifert and Kennedy 2004; Stone et al. 1973; McHugh 1999).

Meperidine has an undeserved reputation for being some sort of "improved analgesic" with no basis in fact. The reason for this, in part, lies in its history. Following its synthesis in 1939, meperidine was released for clinical use 8 years later. At that stage, the only other opioid readily available was morphine, whose spectrum of adverse effects was well known, and the hope was that somehow this new opioid would offer substantial advantages over morphine, and perhaps even replace it.

As outlined in the recent review of meperidine pharmacology by Latta et al. (2002), this initial enthusiasm for meperidine was fueled by a number of early studies that purported to demonstrate that it indeed possessed superior analgesic activity to morphine, had less side effects, and in addition possessed a beneficial effect on relaxation of smooth muscle. Despite the fact that most of these claims were refuted almost as quickly as they appeared, for example, papers showing a lack of smooth muscle relaxing effect appeared in 1947, the die was cast, and the myths surrounding the enhanced clinical efficacy of meperidine have persisted into the twenty-first century.

In many countries, prescribing of meperidine is now actively discouraged. There are considerable grounds well questioning its ongoing use in all settings. Likewise in the USA, where meperidine usage has always been substantial (Seifert and Kennedy 2004), its continued use, especially in postoperative pain management, has been questioned (O'Conner et al. 2005; Vermeulen et al. 1997; Beckwith et al. 2002).

In the hospital setting, an enterprising start has been made by reducing or eliminating its use in hospital emergency departments (Kaye et al. 2005). The result of this has been very positive with emergency physicians having no difficulty in pain management using other agents and at the same time reducing the time having to be spent dealing with meperidine-seeking individuals.

The work by (Shipton 2006), in particular, is an excellent example of scientific clarity in which the author points out in half a dozen pages the range problems associated with meperidine and the reasons why we should not be used it. The critical review by Latta et al. (2002) and the position statement by independent groups on meperidine (Davis et al. 2004) are also highly recommended.

Meperidine has a long tradition of use in obstetrics based on the myth that it does not cause respiratory depression and has no effects on the fetus (Latta et al. 2002). Recent studies have shown that not only are effects on the fetus significant, but can last for as long as six weeks after delivery (Kunhert et al. 1980). If more evidence was needed, one should consider the recent study by Olofsson et al.(1996), who has shown that not only meperidine but other opioids do not seem to actually provide any significant analgesia in labor at all, but rather simply sedate the patient! Furthermore, patients given meperidine have a significant incidence of adverse effects such as nausea, vomiting, and dizziness (Watts 2004). With regard to its use in gastroenterology, long-term prescribers will likewise find the truth somewhat unpalatable. The sphincter of Oddi is sensitive to all opioids at equianalgesic doses, and meperidine is no more effective in the management of biliary colic than any other opioid (Hubbard and Wolfe 2003), or NSAID for that matter (Dula et al. 2001).

Historically, meperidine has also been seen as the drug of choice in the management of shivering associated with range of conditions such as recovery from general anesthesia, following spinal or epidural analgesia, or as a result of blood transfusion or chemotherapy. However, a range of recent reports has shown that this is not a response unique to meperidine, and indeed all opioids possess the capacity to suppress shivering. Furthermore a number of non-opioid agents such as tramadol and ketamine have been shown to be are at least effective, or even superior in management of shivering (Matthews et al. 2002; Bhatnagar et al. 2001; Tsai and Chu 2001; Sagir et al. 2007).

References

Anon. (1993). Pethidine and convulsions. *Australian Adverse Drug Reactions Bulletin, 16*(3), 1–2.

Beckwith, M. C., Fox, E. R., & Chandramouli, J. (2002). Removing meperidine from the health system formulary – frequently asked questions. *Journal of Pain & Palliative Care Pharmacotherapy, 16*(3), 45–59.

Bhatnagar, S., Saxena, A., Kannan, T. R., Punj, J., Panigrohi, M., & Mishra, S. (2001). Tramadol for postoperative shivering; a double blind comparison with pethidine. *Anaesthesia and Intensive Care, 29*, 149–154.

Davis, S., Dowsett, R., & Graudins, A. (2004). Use of pethidine for pain management in the Emergency Department. A position Statement of the NSW Therapeutic Assessment Group.

Dula, D. J., Anderson, R., & Wood, G. C. (2001). A prospective study comparing i.m. ketorolac with i.m. meperidine in the treatment of acute biliary colic. *The Journal of Emergency Medicine, 20*, 121–124.

Hubbard, G. P., & Wolfe, K. R. (2003). Meperidine misuse in a patient with sphincter of Oddi dysfunction. *The Annals of Pharmacotherapy, 37*, 534–537.

Kaye, K. I., Welch, S. A., Graudins, L. V., Graudins, A., Rotem, T., Davies, S. R., & Day, R. O. (2005). Pethidine in emergency departments: Promoting evidenced-based prescribing. *The Medical Journal of Australia, 183*(3), 129–133.

Kredo, T., & Onia, R. (2005). Pethidine-does familiarity or evidence perpetuates its use? *South African Medical Journal, 95*, 100–101.

Kunhert, B. R., Kunhert, P. M., Philipson, E. H., et al. (1980). Meperidine disposition in mother, neonate and non-pregnant female. *Clinical Pharmacology & Therapeutics, 27*, 478–491.

Latta, K. S., Ginsberg, B., & Barkin, R. (2002). Meperidine: A critical review. *American Journal of Therapeutics, 9*, 53–68.

Matthews, S., Al Mulla, A., Varghese, P. K., Radmin, K., & Mumtaz, A. (2002). Postanaesthetic shivering – a new look at tramadol. *Anaesthesia, 57*(4), 394–398.

McHugh, G. J. (1999). Norpethidine accumulation and generalised seizure during pethidine patient-controlled analgesia. *Anaesthesia and Intensive Care, 27*(3), 289–291.

O'Conner, A. B., Lang, V. J., & Quill, T. E. (2005). Eliminating analgesic meperidine use with a supported formulary restriction. *American Journal of Medicine, 118*(8), 885–889.

Olofsson, C. H., Ekblom, A., Ekman-Ordeberg, G., Hjelm, A., & Irestedt, L. (1996). Lack of analgesic effect of systemically administered morphine or pethidine on labour pain. *BJOG: An International Journal of Obstetrics and Gynaecology, 102*(10), 968–972.

Sagir, O., Gulhas, N., Toprak, H., Yucel, A., Begec, Z., & Ersoy, O. (2007). Control of shivering during regional anaesthesia: Prophylactic ketamine and graniserton. *Acta Anaesthesiologica Scandinavica, 51*(1), 44–49.

Seifert, C. F., & Kennedy, S. (2004). Meperidine is alive and well in the new millennium: Evaluation of meperidine usage patterns and frequency of adverse drug reactions. *Pharmacotherapy, 24*, 776–783.

Shipton, E. (2006). Should New Zealand continues signing up to the pethidine protocol? *Journal of the*

New Zealand Medical Association, 119(1230), U1875. http://www.nzma.org.nz/journal/1-19-1230/1875/

Stone, P. A., Macintyre, P. E., & Jarvis, D. A. (1973). Norpethidine toxicity and patient controlled analgesia. *British Journal of Anaesthesia, 71*, 738–740.

Tsai, Y.-C., & Chu, K.-S. (2001). A comparison of tramadol, amtitripyline and meperidine for postepidural anesthetic shivering in parturients. *Anesthesia and Analgesia, 93*, 1288–1292.

Vermeulen, L. C., Bollinger, K. A., Anonopoulos, J., Meek, P. D., Goshman, L. M., & Poelz, P. A. (1997). Multifaceted approach to medication use policy development: The restriction of meperidine. *Pharmacy Practice Management Quarterly, 16*(4), 66–75.

Watts, R. W. (2004). Does pethidine still have a place in the management of labour pain? *Aust Prescr, 27*, 34–35.

Postoperative Pain, Postamputation Pain, Treatment, and Prevention

Lone Nikolajsen
Department of Anaesthesiology, Danish Pain
Research Center Aarhus University Hospital,
Aarhus, Denmark

Synonyms

Postamputation pain

Definition

Phantom pain may occur following amputation of other body parts than limbs, but this entry will focus on clinical characteristics, mechanisms, treatment, and possible preventive measures of phantom pain after limb amputation.

The phantom complex includes three elements:

- Phantom pain: painful sensations referred to the missing limb
- ▶ Phantom sensation: any sensation of the missing limb, except pain
- Stump pain: pain localized to the amputation stump

There is an overlap between these elements, and in the same individual, phantom pain, phantom sensations, and ▶ stump pain often coexist.

Characteristics

Clinical Characteristics Including Mechanisms

Sixty to eighty percent of patients experience phantom pain following limb amputation. The incidence does not seem to be influenced by age in adults, gender, and side of amputation, level of amputation, or cause (civilian versus traumatic) of amputation. Phantom pain is less frequent in young children and congenital amputees. Prospective studies in patients amputated mainly because of peripheral vascular disease have shown that the onset of phantom pain is usually within the first week after amputation, but case reports have suggested that the onset may be delayed for months or even years. The number of patients with severe phantom pain is probably in the range of 5–10 %. Pain is intermittent in most amputees; only few are in constant pain. Episodes of pain attacks are most often reported to occur daily or at daily or weekly intervals. Phantom pain is primarily localized to the distal parts of the missing limb. In upper limb amputees, pain is normally felt in the fingers and palm of the hand, and in lower limb amputees, pain is generally experienced in the toes, foot, or ankle. The reason for this clear and vivid phantom experience of distal limb parts is not clear but may be explained by the larger cortical representation by the hand and foot, as opposed to the lesser representation of the more proximal parts of the limb.

Experimental and clinical studies indicate that a series of mechanisms are involved in generating phantom pains and that these include elements in the periphery, the spinal cord, brainstem, thalamus, and cerebral cortex. It is likely that the first events occur in the periphery, which subsequently generates a cascade of events that sweep more centrally, finally recruiting cortical brain structures. The latter may be responsible for the complex and vivid sensation that characterizes certain phantom pain sensations (for review on clinical characteristics and mechanisms, see Nikolajsen 2012).

Treatment

Phantom pain may be very difficult to treat, and – despite much research in the area – there is little evidence from randomized trials to guide clinicians with treatment. A systematic literature search (Medline 1966–99) was performed recently in order to determine the optimal management of phantom pain. The authors identified 186 articles, but after exclusion of letters, reviews, descriptive trials without intervention, case reports, and trials with major methodological errors, only 12 articles were left for review (Halbert et al. 2002). Since then a few well-designed studies have been published. Until more clinical data become available, guidelines in analogy with treatment regimens used for other ▶ neuropathic pain conditions are probably the best approximation.

A large number of randomized controlled trials have shown a beneficial effect of ▶ tricyclic antidepressants and ▶ sodium channel blockers in different neuropathic pain conditions. Only few controlled data are available for phantom pain, but the drugs are generally believed to be effective, at least in some patients (Wilder-Smith et al. 2005). The effect of ▶ gabapentin was examined in a well-designed crossover study including 19 patients with phantom pain. After 6 weeks of treatment, gabapentin at a daily dose of 2,400 mg was better than placebo in reducing phantom pain (Bone et al. 2002). A similar effect of gabapentin was described in an open study. Permanent phantom pain should not be accepted until ▶ opioids have been tried. Opioids can probably be used safely for several years with a limited risk of dependence. In a randomized, double-blind, cross-study with active placebo, 31 amputees received a 40-min. infusion of lidocaine, morphine, or diphenhydramine. Compared with placebo, morphine reduced both stump and phantom pain, whereas lidocaine only decreased stump pain (Wu et al. 2002). In another placebo-controlled, crossover study including 12 patients, a significant reduction in phantom pain was found during treatment with oral morphine (Huse et al. 2001). Case reports have suggested that methadone may reduce phantom pain. The effect of ▶ NMDA receptor antagonists have been examined in different studies. Ketamine administered intravenously reduced pain in amputees with stump and phantom pain (Nikolajsen et al. 1996). However, two other well-designed studies found no effect of memantine, an NMDA receptor antagonist available for oral use, at doses of 20 and 30 mg, respectively. In both studies, memantine was administered in a blinded, placebo-controlled crossover fashion to patients with stump and phantom pain (Nikolajsen et al. 2000; Maier et al. 2003). Calcitonin is probably not effective in established phantom pain (Eichenberger et al. 2008). A large number of other treatments, for example, beta-blockers, mexiletine (the oral congener of lidocaine), topical application of capsaicin, intrathecal opioids, various anesthetic blocks, and injection of botulinum toxin, have been claimed to be effective in phantom pain, but none of them have proven to be effective in well-controlled trials with a sufficient number of patients.

Physical therapy involving massage, manipulation, and passive movements may prevent trophic changes and vascular congestion in the stump. Other treatments such as transcutaneous electrical nerve stimulation (TENS), acupuncture, ultrasound, and hypnosis may, in some cases, have a beneficial effect on stump and phantom pain. Others have used visual feedback with a mirror, but the results are not consistent (Ramachandran and Rogers-Ramachandran 2000; Brodie et al. 2007). It has also been demonstrated that sensory discrimination training, obtained by applying stimuli at the stump, reduced pain in five upper limb amputees (Flor et al. 2001). The advantage of most of the above-mentioned nonmedical treatments is the absence of side effects and complications and the fact that the treatment can be easily repeated. However, most of these studies are uncontrolled observations.

Surgery on amputation neuromas and more extensive amputation have previously played important roles in the treatment of stump and phantom pain. Today stump revision is probably only performed in cases of obvious stump pathology, and in properly healed stumps, there is almost never any indication for proximal extension of the amputation because of pain. Spinal cord stimulation may relieve postamputation pain but is most effective in the treatment of stump pain. The results of other invasive techniques such as ▶ dorsal root entry zone lesions (DREZ), sympathectomy, and ▶ cordotomy have generally been unfavorable, and most of them have been abandoned today. Surgery may produce short-term pain relief, but pain often reappears.

Prevention

Some early studies showed that preamputation increases the risk of subsequent phantom pain and that phantom pains in some cases are a replicate of pain experienced before the amputation.

The above observations led to the theory that preamputation pain creates an imprint in memorizing structures of the central nervous system and that such an imprint could be responsible for persistent pain after amputation. Inspired by this, Bach et al. (1988) carried out the first study on the prevention of phantom pain: 25 patients were allocated by means of the year of their birth to either epidural pain treatment 72 h before the amputation (B: 11 patients) or conventional analgesics (C: 14 patients). All patients had spinal or epidural analgesia for the amputation, and both groups received conventional analgesics to treat postoperative pain. Blinding was not described. After 6 months, the incidence of phantom pain was lower among patients who had received the preoperative epidural blockade. Since then a few trials have examined the short-term and long-term impact of regional analgesia (epidural and nerve blocks) on phantom pain. Some of the studies, but not all, have confirmed that regional analgesia may be effective in reducing postamputation pain. Unfortunately, some of the trials are of very poor methodological quality (see Nikolajsen 2012 for details). Nikolajsen et al. (1997) carried out a randomized, double-blind, and placebo-controlled study, in which 60 patients scheduled for lower ▶ limb amputation were randomly assigned into one of two groups: a blockade group that received epidural bupivacaine and morphine before the amputation and during the operation (29 patients) and a control group that received epidural saline and oral/intramuscular morphine (31 patients). Both groups had general anesthesia for the amputation, and all patients received epidural analgesics for postoperative pain management. Patients were interviewed about preamputation pain on the day before the amputation and about stump and phantom pain after 1 week and 3, 6, and 12 months. Median duration of preoperative epidural blockade (blockade group) was 18 h. After 1 week, the percentage of patients with phantom pain was 51.9 in the blockade group and 55.6 in the control group. Subsequently, the figures were (blockade/control): at 3 months, 82.4/50; at 6 months, 81.3/55; and at 12 months, 75/68.8. Intensity of stump and phantom pain and consumption of opioids were also similar in both groups, at all four postoperative interviews. So according to this study, it is not possible to prevent phantom pain by a preoperative epidural blockade. The aim of preemptive treatment is to avert spinal sensitization by blocking, in advance, the cascade of intraneuronal responses that take place after peripheral nerve injury. A true preemptive approach is probably not possible in patients scheduled for amputation. Many have suffered from ischaemic pain for months or years and are likely to present with preexisting neuronal hyperexcitability. It cannot be excluded that a preoperative epidural blockade for a longer period (weeks?) would prevent phantom pain from developing. However, this would be very inconvenient from a practical point of view, as the decision to amputate is often not taken until the day before or even on the same day as surgery.

Summary

In conclusion, epidural blockade is effective in the treatment of preoperative ischaemic pain and postoperative stump pain, but at present no studies of sufficient methodological quality have provided evidence that preoperative epidural blockade has any beneficial effect in preventing phantom pain. It cannot be excluded that other approaches may be effective. For example, it was suggested recently in an open nonrandomized

Postoperative Pain, Preemptive or Preventive Analgesia

study including 71 lower limb amputees that postoperative treatment with a perineural catheter for a median period of 30 days could reduce the incidence of phantom pain (Borghi et al. 2010).

References

Bach, S., Noreng, M. F., & Tjéllden, N. U. (1988). Phantom limb pain in amputees during the first 12 months following limb amputation after preoperative lumbar epidural blockade. *Pain, 33*, 297–301.

Bone, M., Critchley, P., & Buggy, D. J. (2002). Gabapentin in postamputation phantom limb pain: A randomized, double-blind, placebo controlled, cross-over study. *Regional Anesthesia and Pain Medicine, 27*, 481–486.

Borghi, B., D'Addabbo, M., White, P. F., Gallerani, P., Toccaceli, L., Raffaeli, W., Tognu, A., Fabbri, N., & Mercuri, M. (2010). The use of prolonged peripheral neural blockade after lower extremity amputation: The effect on symptoms associated with phantom limb syndrome. *Anesthesia and Analgesia, 111*, 1308–1315.

Brodie, E. E., Whyte, A., & Niven, C. A. (2007). Analgesia through the looking-glass? A randomized controlled trial investigating the effect of viewing a 'virtual' limb upon phantom limb pain, sensation and movement. *European Journal of Pain, 11*, 428–436.

Eichenberger, U., Neff, F., Sveticic, G., Bjorgo, S., Petersen-Felix, S., Arendt-Nielsen, L., & Curatolo, M. (2008). Chronic phantom limb pain: The effects of calcitonin, ketamine, and their combination on pain and sensory thresholds. *Anesthesia and Analgesia, 106*, 1265–1273.

Flor, H., Denke, C., Schaefer, M., et al. (2001). Effect of sensory discrimination training on cortical reorganisation and phantom limb pain. *The Lancet, 357*, 1763–1764.

Halbert, J., Crotty, M., & Cameron, I. D. (2002). Evidence for the optimal management of acute and chronic phantom pain: A systematic review. *Clinical Journal of Pain, 18*, 84–92.

Huse, E., Larbig, W., Flor, H., et al. (2001). The effect of opioids on phantom limb pain and cortical reorganization. *Pain, 90*, 47–55.

Maier, C., Dertwinkel, R., Mansourian, N., et al. (2003). Efficacy of the NMDA-receptor antagonist memantine in patients with chronic phantom limb pain – results of a randomized double-blinded, placebo-controlled trial. *Pain, 103*, 277–283.

Nikolajsen, L. (2012). Postamputation pain: Studies on mechanisms. *Danish Medical Journal, 59*(10), B4527.

Nikolajsen, L., Hansen, C. L., Nielsen, J., et al. (1996). The effect of ketamine on phantom pain: A central neuropathic disorder maintained by peripheral input. *Pain, 67*, 69–77.

Nikolajsen, L., Ilkjær, S., Krøner, K., et al. (1997). Randomised trial of epidural bupivacaine and morphine in prevention of stump and phantom pain in lower-limb amputation. *Lancet, 350*, 1353–1357.

Nikolajsen, L., Gottrup, H., Kristensen, A. G. D., et al. (2000). Memantine (a N-methyl D-aspartate receptor antagonist) in the treatment of neuropathic pain following amputation or surgery: A randomised, double-blind, cross-over study. *Anesthesia and Analgesia, 91*, 960–966.

Ramachandran, V. S., & Rogers-Ramachandran, D. (2000). Phantom limbs and neural plasticity. *Archives of Neurology, 57*, 317–320.

Wilder-Smith, C. H., Hill, L. T., & Laurent, S. (2005). Postamputation pain and sensory changes in treatment-naive patients: Characteristics and responses to treatment with tramadol, amitriptyline, and placebo. *Anesthesiology, 103*, 619–628.

Wu, C. L., Tella, P., Staats, P. S., et al. (2002). Analgesic effects of intravenous lidocaine and morphine on postamputation pain. *Anesthesiology, 96*, 841–848.

Postoperative Pain, Preemptive or Preventive Analgesia

Jorgen B. Dahl
Department of Anaesthesiology, Glostrup University Hospital, Herlev, Denmark

Synonyms

Prevention of postoperative pain; Preventive analgesia; Protective analgesia

Definition

"▶ Preemptive analgesia" is a treatment that starts before surgery, in order to prevent or reduce the establishment of ▶ sensitization of dorsal horn neurons caused by tissue injury. Sensitization of dorsal horn neurons is supposed to amplify postoperative pain.

Characteristics

The idea of preemptive analgesia was addressed by Crile in 1913 (Crile 1913) and further elaborated by Patrick Wall in 1988 (Wall 1988). The concept was originally based on experimental observations demonstrating that a short barrage of nociceptor afferent input to the dorsal horn neurons results in an activity-dependent hyperexcitability, which may outlast the stimulus – a phenomenon now termed "heterosynaptic ▶ central sensitization" (Woolf and Salter 2006). Wall suggested that it might be preferable (and more efficacious) to prevent pain and the consequences of a noxious input to the CNS (i.e., dorsal horn sensitization), rather than to treat pain when this sensitization is already established (Wall 1988).

Central Sensitization and "Pain Memory"

Central sensitization includes an altered processing of both nociceptive signals and innocuous input/stimulus, tactile impulses from myelinated afferents, leading to painful sensations when these afferents are activated. As suggested by Woolf and Salter, central sensitization manifests itself in the clinic as ▶ clinical pain, with hyperalgesia, allodynia, spread of sensitization, and persistent pain (Woolf and Salter 2006). The neurophysiological and biochemical mechanisms of these alterations are complex and not fully understood (Woolf and Salter 2006), but include an increase in synaptic efficacy mediated by neurokinin and N-methyl-d-aspartic acid (NMDA)-receptor mechanisms (Woolf and Salter 2006). Since central sensitization may outlast the stimuli that triggered the alterations in the first place, it may be considered as a "pain memory."

Clinical Studies

The idea of "prevention of postoperative pain," published by Wall in 1988 (Wall 1988), resulted in substantial clinical interest. The interpretation and clinical implementation of the concept however has varied and provoked substantial controversy (Kissin 2005). The basic assumption is that the surgical injury leads to sensitization and hyperexcitability of dorsal horn neurons in the spinal cord, which in turn amplify postoperative pain. "Preemptive analgesia" aims at

preventing the central alterations since experimental studies have demonstrated that this may be possible with pre-injury measures, whereas the same measures applied post-injury are less effective.

As pointed out by Kissin, at least two approaches have been used to reveal preemptive analgesia (Kissin 2005). One has been to compare the pain relieving effect of identical analgesic regimens administered before versus after the surgical incision or procedure. The other has been to compare the effect of a presurgical administered analgesic with no treatment and subsequently to evaluate the analgesic effect beyond the expected presence of the drug in the biophase (Kissin 2005). Outcome measures of these trials have varied. In most studies, pain intensity and/or requirements for supplemental analgesics have been the primary outcomes, whereas in other studies allodynia and hyperalgesia surrounding the surgical wound have been the major focus of interest.

Evidence from Controlled Clinical Trials of Identical Analgesic Treatments, Initiated Before Versus After the Surgical Incision or Procedure

Recently, two systematic reviews of the clinical literature have examined the effects of identical analgesic treatments initiated before versus after surgical incision, with contrasting results (Møiniche et al. 2002; Ong et al. 2005). In a review of 80 controlled trials with different analgesic regimens, Møiniche et al. were not able to demonstrate any substantial clinical benefits of preemptive analgesia with NSAIDs, iv-opioids, iv-ketamine, dextromethorphan, peripheral local anesthetics, or epidural or caudal analgesia (Møiniche et al. 2002).

In the most recent meta-analysis, Ong et al. investigated the effect of five types of analgesic interventions (epidural analgesia, local anesthetic wound infiltration, systemic NMDA-receptor antagonists, systemic NSAIDs, and systemic opioids) on three outcome variables, postoperative pain scores, analgesic consumption, and time to first rescue analgesia, in

66 clinical studies (Ong et al. 2005). The overall conclusion from this latter meta-analysis was that preemptive epidural analgesia resulted in consistent improvements in all three outcome variables, whereas preemptive local anesthetic wound infiltration and NSAID administration reduced analgesic consumption and increased time to first rescue analgesia, without effect on postoperative pain scores. The least proof of efficacy was found with systemic NMDA antagonist and opioid administration (Ong et al. 2005). Reasons for the contrasting results between these meta-analyses were discussed by Ong et al. (2005). These reasons may include differences in methodology and included data, as well as differences in included studies.

Evidence from Controlled Clinical Trials of Presurgical Versus No Treatment

It has been argued that not only the surgical incision but also other noxious intraoperative and postoperative stimuli may trigger central sensitization (Dierking et al. 1992). As a result, both pre- and postoperative administration of analgesics may reduce central sensitization and thus decrease postoperative pain intensity (McCartney et al. 2004). Consequently, it may not be possible to demonstrate preemptive analgesic effects with studies of identical analgesic treatments, initiated before versus after surgery (McCartney et al. 2004). A different approach is to study effects of an analgesic treatment initiated before surgical incision on pain and/or analgesic consumption beyond the expected pharmacological duration of action relative to an untreated or placebo control ("preventive analgesia") (McCartney et al. 2004).

McCartney et al. conducted a systematic review of the clinical literature to determine the extent to which NMDA antagonists have yielded analgesic effects beyond five half-lives when given during the perioperative period (McCartney et al. 2004). The results of this meta-analysis demonstrated that ketamine and dextromethorphan produced a significant preventive analgesic benefit in 14 of 24 and in 8 of 12 studies, respectively, whereas none of 4 studies examining

magnesium demonstrated a preventive analgesic benefit (McCartney et al. 2004).

Effects of Preemptive Analgesia on the Development of Chronic Pain

It has been assumed that preemptive analgesia may reduce the risk of developing chronic postoperative pain, including phantom limb pain. This assumption may be supported by data suggesting that patients with high-intensity acute postoperative pain scores also have a higher risk of developing a chronic pain state (Perkins and Kehlet 2000). One trial has investigated the effect of identical pre- versus post-incisional treatment on long-term pain (Obata et al. 1999). Results showed that the percentage of patients with pain at 6 months postoperatively was significantly reduced (Obata et al. 1999). Unfortunately, the preliminary promising results with preemptive analgesia in patients undergoing limb amputation were not confirmed by a recent meta-analysis (Halbert et al. 2002).

Other Approaches to Reduce Central Sensitization

Antihyperalgesics such as ketamine and dextromethorphan reduce the injury-induced hyperexcitability of the central nervous system, but have no effect on the nociceptive input per se. Ketamine and dextromethorphan are both NMDA-receptor antagonists. The effect of ketamine on postoperative pain has been examined in a number of recent meta-analyses (Elia and Tramer 2005; Bell et al. 2005).

The overall conclusion from these reviews is that ketamine has a modest opioid sparing effect after surgery (Elia and Tramer 2005; Bell et al. 2005). Adverse effects are mild or absent with low-dose regimens, and ketamine may reduce opioid-related side effects such as nausea and vomiting (Bell et al. 2005). The data should however be interpreted with caution, since published original studies with ketamine are very heterogeneous and the results of the different meta-analyses cannot be translated into any specific administration regimen with ketamine (Elia and Tramer 2005; Bell et al. 2005).

In a systematic review examining the effects of dextromethorphan in postoperative pain, results showed that dextromethorphan did not reduce the postoperative pain score with a clinical significant magnitude. Time to first analgesic request was significantly prolonged in most comparisons, and significant, albeit modest, reductions in supplemental opioid consumption were observed in a majority of studies with parenteral administration and in about half of the studies with oral administration (Duedahl et al. 2006).

The anticonvulsant gabapentin is widely used for treatment of chronic pain, and experimental studies have demonstrated antihyperalgesic effects of gabapentin in models involving central neuronal sensitization, without affecting acute pain transmission. A substantial number of clinical studies investigating the effect of gabapentin on postoperative pain have recently been published (Dahl et al. 2004). The results of these studies have demonstrated promising effects on both postoperative pain and opioid requirements. So far, side effects have been few and not disturbing (Dahl et al. 2004), but further studies are obviously needed.

In summary, ▶ timing of analgesia per se may have some effect on postoperative pain, but results are conflicting. One meta-analysis has demonstrated analgesic effects of NMDA antagonists beyond five half-lives when administered during the perioperative period, which may point to a preemptive effect. Antihyperalgesics such as ketamine and dextromethorphan have been demonstrated to have moderate effects on postoperative opioid requirements. Gabapentin has demonstrated promising effects in postoperative pain treatment. In future studies, attention should be directed toward protective analgesia (Møiniche et al. 2002; Dahl et al. 2004) aimed at the prevention of pain hypersensitivity through intensive and prolonged, multimodal analgesic ("protective") interventions. Furthermore, antihyperalgesic drugs and methods should be further developed and evaluated in clinical trials of postoperative pain.

References

Bell, R. F., Dahl, J. B., Moore, A., et al. (2005). Peri-operative ketamine for acute post-operative pain: A quantitative and qualitative systematic review. *Acta Anaesthesiologica Scandinavica, 49,* 1405–1428.

Crile, G. W. (1913). The kinetic theory of shock and its prevention through association. *Lancet, 185,* 7–16.

Dahl, J. B., Mathiesen, O., & Møiniche, S. (2004). 'Protective premedication': An option with gabapentin and related drugs? *Acta Anaesthesiologica Scandinavica, 48,* 1130–1136.

Dierking, G., Dahl, J. B., Kanstrup, J., et al. (1992). The effect of pre- versus postoperative inguinal field block on postoperative pain following herniorrhaphy. *British Journal of Anaesthesia, 68,* 344–348.

Duedahl, T. H., Rømsing, J., Møiniche, S., et al. (2006). A qualitative systematic review of dextromethorphan in postoperative pain. *Acta Anaesthesiologica Scandinavica, 50,* 1–13.

Elia, N., & Tramer, M. R. (2005). Ketamine and postoperative pain-a quantitative systematic review of randomised trials. *Pain, 113,* 61–70.

Halbert, J., Crotty, M., & Cameron, I. D. (2002). Evidence for the optimal management of acute and chronic phantom pain: A systematic review. *The Clinical Journal of Pain, 18,* 84–92.

Kissin, I. (2005). Preemptive analgesia at the crossroad. *Anesthesia and Analgesia, 100,* 754–756.

McCartney, C. J., Sinha, A., & Katz, J. (2004). A qualitative systematic review of the role of N-methyl-D-aspartate receptor antagonists in preventive analgesia. *Anesthesia and Analgesia, 98,* 1385–1400.

Møiniche, S., Kehlet, H., & Dahl, J. B. (2002). A qualitative and quantitative systematic review of preemptive analgesia for postoperative pain relief – the role of timing of analgesia. *Anesthesiology, 96,* 725–741.

Obata, H., Saito, S., Fujita, N., et al. (1999). Epidural block with mepivacaine before surgery reduces long-term post-thoracotomy pain. *Canadian Journal of Anesthesia, 46,* 1127–1132.

Ong, C. K.-S., Lirk, P., Seymour, R. A., et al. (2005). The efficacy of preemptive analgesia for acute postoperative pain management: A meta-analysis. *Anesthesia and Analgesia, 100,* 757–773.

Perkins, F., & Kehlet, H. (2000). Chronic pain as an outcome of surgery. A review of predictive factors. *Anesthesiology, 93,* 1123–1133.

Wall, P. D. (1988). The prevention of postoperative pain. *Pain, 33,* 289–290.

Woolf, C. J., & Salter, M. W. (2006). Postoperative pain and its management. In S. B. McMahon & M. Koltzenburg (Eds.), *Wall and Melzack's textbook of pain* (pp. 91–105). Edinburgh, NY: Churchill Livingstone.

Postoperative Pain, Pregabalin

Gavin Pattullo
Department of Anaesthesia and Pain
Management, Royal North Shore Hospital,
University of Sydney, St. Leonards, NSW,
Australia

Introduction

Since its release in 2004, pregabalin has followed gabapentin into use for a wide range of pain conditions. Pregabalin and gabapentin together comprise a group of drugs termed the "gabapentinoids." The naming of the gabapentinoids is somewhat deceptive, as the clinical actions of both are thought to predominately result from binding to the alpha-2 delta subunit of voltage-sensitive calcium channels to reduce neuronal activity (Taylor 2009). The analgesic effect of pregabalin is largely ascribed to anti-hyperalgesic and anti-allodynic actions. In addition, pregabalin's anxiolytic properties can be beneficial for some patients at the time of surgery.

Characteristics

Pharmacology

Pregabalin possesses favorable pharmacokinetic qualities, including high oral bioavailability, absence of a deactivating metabolic pathway, and renal excretion. Dosing is twice daily, usually commencing at 100–150 mg per day followed by a buildup to the target dose of 450–600 mg/day over a period of days. Dose-limiting side effects most commonly include sedation and dizziness, both of which can be more pronounced in the elderly so necessitating a caution in maximum dosing for this age group. Pregabalin is often observed to be better tolerated with fewer side effects in the elderly than is gabapentin. Patients with renal impairment require a lowering of maximum target dose.

Efficacy for Postoperative Pain

Preoperative pregabalin dosing (up to 600 mg) has been demonstrated to have considerable opioid-sparing effect for some surgery types – the most notable being major orthopedic surgery (Engelman and Cateloy 2011). Pregabalin's actions in countering allodynia and hyperalgesia are thought to underlie this efficacy against acute nociceptive pain – since both processes are features of acute nociceptive pain (Tiippana et al. 2007). Owing to this mechanism of action, most research conducted on pregabalin has focused on prevention of central sensitizing processes by incorporating the preoperative administration of pregabalin, with limited duration of use into the postoperative period. Importantly, the opioid-sparing effect of pregabalin is often not paralleled by a reduction in reported pain scores (Zhang et al. 2011). Therefore, the most impressive clinical response with pregabalin is seen when administered as an adjunct to opioid administration, and notably when the opioid dose would otherwise be significant. This contrasts starkly to pregabalin's use as a sole agent in acute pain, or even in combination with a NSAID/COX2, where the observed clinical response can be unimpressive. Owing to a more established use of gabapentin (and hence a greater volume of research into its use in acute pain), many clinicians still have a preference for gabapentin as their first-line gabapentinoid adjunct agent – reserving pregabalin for those intolerant of gabapentin.

Prevention of Chronic Pain

A small number of studies in limited surgery types have reported a reduction in chronic postsurgical pain with the perioperative use of pregabalin (Clarke et al. 2012). Further research is required in this area before routine use of pregabalin for prevention of chronic postsurgical pain is recommended.

Clinical Use

In clinical practice, dosing of pregabalin for acute pain is undertaken in a different manner to that of use in chronic pain. Introduction is at the target dose – rather than a gradual escalation of dose – so as to avoid any loss of therapeutic advantage. Careful titration of dose during introduction should still be utilized for the elderly and those with renal impairment. There is no consensus of opinion as to the ideal duration of pregabalin use in acute pain. A clinical regimen of use for the period while opioid demand is at its greatest is perhaps the most logical. Limiting dose to just the first few days is also the simplest to administer. Following a short period of pregabalin use of only 3–5 days, most individuals can safely have their dose abruptly curtailed. Gradual tapering may be considered for patients with coincident epilepsy, or for those receiving treatment for a prolonged period of time.

Side Fffects

Pregabalin's most commonly encountered side effects of dizziness and sedation can be lessened by limiting dose to the lower range of less than 300 mg/day. The unique combination of a high preoperative pregabalin dose with a general anesthetic can result in disconcerting levels of sedation in the PACU. Many clinicians therefore prefer to avoid this combination by delaying pregabalin administration until after surgery, or by using a preoperative dose only when regional anesthesia is planned. Based on extrapolation of research into gabapentin use postoperatively, it is not unreasonable to expect an opioid-sparing effect when commencement of pregabalin is delayed until the postoperative period (Clarke et al. 2009).

References

Clarke, H., Pereira, S., Kennedy, D., Gilron, I., Katz, J., Gollish, J., et al. (2009). Gabapentin decreases morphine consumption and improves functional recovery following total knee arthroplasty. *Pain Research & Management, 14*(3), 217–222.

Clarke, H., Bonin, R. P., Orser, B. A., Englesakis, M., Wijeysundera, D. N., & Katz, J. (2012). The prevention of chronic postsurgical pain using gabapentin and pregabalin: A combined systematic review and meta-analysis. *Anesthesia and Analgesia, 115,* 428–442.

Engelman, E., & Cateloy, F. (2011). Efficacy and safety of perioperative pregabalin for post-operative pain: A meta-analysis of randomized-controlled trials. *Acta Anaesthesiologica Scandinavica, 55*, 927–43.

Taylor, C. P. (2009). Mechanisms of analgesia by gabapentin and pregabalin–calcium channel alpha2-delta [Cavalpha2-delta] ligands. *Pain, 142* (1–2), 13–16.

Tiippana, E. M., Hamunen, K., Kontinen, V. K., & Kalso, E. (2007). Do surgical patients benefit from perioperative gabapentin/pregabalin? A systematic review of efficacy and safety. *Anesthesia and Analgesia, 104*(6), 1545–1556.

Zhang, J., Ho, K. Y., & Wang, Y. (2011). Efficacy of pregabalin in acute postoperative pain: A meta-analysis. *British Journal of Anaesthesia, 106*(4), 454–462.

Postoperative Pain, Preoperative Education

Edward A. Shipton
Department of Anaesthesia, Christchurch School of Medicine and Health Sciences, University of Otago, Christchurch, New Zealand

Synonyms

Preoperative patient education; Preoperative teaching

Definition

Pain is a sensory and emotional experience that is influenced by physiological, sensory, affective, cognitive, sociocultural, and behavioral factors. Research in postoperative pain management indicates room for improvement, especially in the area of patient education (Chung 1999). Patients who undergo surgery experience acute psychological distress in the preoperative period. It is essential to ensure patients are adequately assessed, well educated, and prepared both clinically and psychologically for the event. Since the 1960s, many primary studies and several ▶ meta-analyses have been conducted to assess the effectiveness of education for patients undergoing surgery. Many have demonstrated the beneficial impact that preoperative education exerts on the postoperative recovery of patients (Mordiffi et al. 2003). These studies are often in discreet patient cohorts with little attention paid to generalizing the results (Hobbs 2002). The delivery is often based on the healthcare providers' view of what information should be included. It is also clear that patients' coping behavior varies considerably and strongly influences the usefulness of providing detailed preoperative information.

Characteristics

The delivery is often based on healthcare providers' view of what information should be included. It is also clear that patients' coping behavior varies considerably and strongly influences the usefulness of providing detailed preoperative information. The amount of care the patient will need, who will care for them, and the pain management strategies needed are important issues to address before surgery.

Current patient education is influenced by shorter hospital stays and by patients' growing expectations and knowledge, and all these are placing new demands on patient education (Johansson et al. 2005). It is not currently apparent, which is the most effective method of delivering preoperative information, or at what stage of the preoperative phase is the optimum time to deliver such important information (Walker 2007). Preoperative interview is an important tool to receive and give information concerning postoperative pain management (Walker 2007).

Preoperative patient education should recognize that different patients have various misconceptions, expectations, and needs. Multiple modes may be required to increase knowledge for informed consent and decrease patient anxiety (Lee and Gin 2005). The use of a multimedia health educational program remains effective in reducing preoperative anxiety and postoperative pain in patients undergoing elective laparoscopic cholecystectomy (Stergiopoulou et al. 2006).

The concept of preoperative education can have many implications from a prepared program with specified content, approaches, and measurable outcomes to an informal method that takes into account the patient's perceptions, beliefs, learning styles, and organizational constraints (Whyte and Grant 2005). The preoperative teaching process is best approached as a team effort (Whyte and Grant 2005). Establishing a seamless preoperative process that anticipates all of the patient's clinical needs helps achieve program goals and outcomes, reduces costly delays, and improves patient and family satisfaction. Careful planning, teamwork, and feedback from patients will lead to successful implementation of a preoperative educational program (Kruzik 2009). The additional goal of obtaining informed consent has become an ingrained component to the current physician-patient relationship (Whyte and Grant 2005).

Anxiety
Independent risk factors for preoperative anxiety include history of cancer and smoking, psychiatric disorders, negative future perception, moderate to intense depressive symptoms, high trait anxiety, moderate to intense pain, medium surgery, female gender, American Society of Anesthesiologists' Status Category III, and having received up to and more than 12 years of education (Caumo et al. 2001).

Preoperative instruction has been demonstrated to have benefit with regard to patient anxiety, postoperative pain, postoperative complications, and length of hospitalization. Preoperative anxiety correlated with high postoperative anxiety, increased postoperative pain and analgesic requirements, decreased patient satisfaction, and prolonged hospital stay. Female patients had higher preoperative anxiety than males.

Patient education decreases preoperative anxiety and pain (Giraudet-Le Quintrec et al. 2003). But for patients with high anxiety levels and poor coping skills, excessive information can exacerbate anxiety and pain. The question remains as to what type of preprocedural preparation works best to reduce anxiety? There may also be beneficial effects when preoperative education is tailored according to anxiety, or targeted at those most in need of support (McDonald et al. 2004). Patient's sociodemographic and psychological characteristics and type of surgery need to be considered for identifying patients at risk of experiencing high anxiety both before and after surgery so that psychological support and clinical management can be tailored to alleviate their anxiety.

Fears

The most requested information from patients is related to anxiety-creating factors such as pain and postoperative symptoms after surgery. Many patients fear and expect postoperative pain. This can affect pain perception and analgesic requirements. Patients place more emphasis on the duration of surgery, with less priority being given to questions relating to complications of anesthesia (Moores and Pace 2003).

However, fears of brain damage, death, and intraoperative awareness associated with general anesthesia remain prevalent, suggesting that preoperative education of patients should also address these. Preoperative education will help to ensure that patients no longer arrive at the operating theater frightened and unaware of what will happen to them (Garretson 2004). Education can further decrease the proportion of surgical patients who fear addiction to pain medication.

Preadmission

Preadmission clinics offer an excellent means for patient education. The overall goal of preadmission testing programs is to ensure patient preparedness while increasing quality health care and overall customer satisfaction. In pediatric cardiac fast-track surgery, parents overestimate the time it takes for several key milestones in the fast-track process to occur. Preadmission education will allow parents/caregivers to be more involved in their child's care (Jawahar and Scarisbrick 2009).

An effective preadmission program is a major investment in a patient's recovery. Patients begin to formulate their expectations of the postoperative hospitalization during the preadmission program. The challenge is to better understand the factors patients consider important when formulating judgments about the quality of preadmission education. The preadmission clinic gives an opportunity to educate patients about pain control (including patient-controlled analgesia), adverse effects (such as nausea and vomiting), and answer any questions they might have.

Method

Preoperative education or presurgical preparation involves conveying developmentally appropriate, detailed information regarding the surgical process (Frisch et al. 2010). Preoperative education improves patient or carer knowledge of pain and encourages a more positive attitude toward pain relief (Cheung et al. 2007).

Patient information should include procedural information (current and reflect best practice), sensory information (what the patient can expect), and physiological coping information. In addition to pain and mobility information, patient education should reflect knowledge of what can be expected at 3 months after surgery, including expected and potential improvements in mobility, pain, and ambulation and in nonphysical dimensions, such as sleep, home management, and social interaction.

The increasing multiculturalism and diversity of the surgical population challenges the ability of the preoperative educator to convey information effectively. There is often a lack of patient focus when delivering such information, and it needs repetition. The information should be realistic. Education should include treatment goals, options available for monitoring and treating acute pain, and likely benefits and probability of success (Macintyre and Schug 2007; Counsell et al. 2008).

There is a need to communicate inadequate analgesia and adverse effects and address fears of opioid addiction (Macintyre and Ready 2002). When warranted, education should include the use of patient-controlled analgesia (either intravenously, subcutaneously, or epidurally) and regional anesthesia catheters. Implementation of an acute pain service may improve pain relief and reduce the incidence of side effects (Macintyre 2010). Staff education and the use of guidelines improve pain assessment and pain relief (Macintyre 2010).

But what are the optimal methods for educating patients? Preoperative preparation programs for patient or carer education may take a number of forms. These may consist of hospital tours, play therapy, informational videos, surgical brochures, medical role-play, and/or phone interviews (Frisch et al. 2010). Written information given to patients prior to the interview is better than verbal information given at the time of the interview (Binhas et al. 2008; Walker 2007). Structured preoperative education may be better than routine information (Jawahar and Scarisbrick 2009).

The most common methods are the use of booklets or short videos and specialist one-on-one education (Macintyre 2010). Patients' ability to recall potential risks is significantly increased by an educational intervention using printed material (Chan et al. 2002). A preadmission educational booklet for coronary bypass surgery patients resulted in fewer concerns about asking for help and taking analgesia (Watt-Watson et al. 2000). Overall, patients prefer direct information from their healthcare professional than via the mail.

The use of video teaching gives better reported pain control and satisfaction after invasive procedures (Duhalde et al. 2002). Video education of patients with a whiplash injury reduces the incidence of persistent pain (Oliveira et al. 2006). Education using a video about patient-controlled analgesia (PCA) improved both patient knowledge and pain relief (Chen et al. 2005) For example, using videotapes decreased anxiety and stress, measured in terms of urinary cortisol excretion and intraoperative systolic blood pressure increase, in patients undergoing hip replacement surgery and prepared them to cope better with postoperative pain. Cognitive interventions such as distraction and reappraisal have been used to improve the care of women having an elective hysterectomy (Cheung et al. 2003).

Which group of patients would most likely benefit from more information is an important question that remains relatively untouched. Whether the preoperative contact, supplemented by videotape or in-hospital, on-demand television programming, or computer networks, such as the World Wide Web or home television, is the most effective and practical means for this education remains to be seen? Nevertheless, a culturally sensitive and acceptable plan of care should be developed that establishes trust, anticipates needs, and preserves the integrity of the patients (Deyirmenjian et al. 2006). This can be achieved through patient-centered communication skills used by the health workers, those that would empower the patients and help them to become autonomous and active participants (Deyirmenjian et al. 2006).

Measures

The most common outcome measures related to pain, knowledge, anxiety, exercises, and length of stay and the least common to self-efficacy and empowerment (Johansson et al. 2005). A variety of tools can be used to assess the effectiveness of preoperative education. These include the SF-36 Health Status questionnaire and the General Well-Being questionnaire that emphasize the outcome of medical care as the patient sees it; the State-Trait Anxiety Inventory, the Hospital Anxiety and Depression Scale, and the Montgomery-Asberg Depression Rating Scale to measure anxiety and depression; as well as pain measurement tools such as the visual analogue pain scale and the McGill Pain Questionnaire Short Form.

Children

Developmentally appropriate presurgical educational programs and parental involvement in the surgical experience can help alleviate the anxiety of both children and parents during the pediatric surgical experience (Frisch et al. 2010). Examples of age-appropriate educational programs for infants and toddlers include using children's books and puppet shows to help children become comfortable with the preoperative environment. Preschool and school-aged children would benefit from medical play, tour of the surgical area, and picture books (Felder-Puig et al. 2003; Hatava et al. 2000; Justus et al. 2006). Preop education programs provided by nurse practitioners are beneficial in decreasing the anxiety state among children and parents prior to surgery (Frisch et al. 2010). Benefits to patients and families were an increase in patient satisfaction as well (Varughese et al. 2006).

Preoperatively, parents are more anxious when their child is less than 1 year of age or when it is the child's first surgery. Postoperative pain is negatively related to parents' provision of surgery-relevant information during the preoperative period. With patient-controlled analgesia, preoperative education on its use provides children and parents with invaluable information, alleviating their concerns about getting "hooked on drugs," overdosing, adverse effects, and being able to get pain relief when needed (Kotzer et al. 1998). In adolescents, cognitive-behavioral interventions designed to prepare them for surgery should be tailored to individual factors and developmental needs, especially the adolescents' preoperative anxiety level and age. Preoperative education (information plus coping skills) is effective in reducing postoperative anxiety in adolescents with high preoperative anxiety (LaMontagne et al. 2003).

Results
Since the 1960s, many primary studies and several meta-analyses have been conducted to assess the effectiveness of education for patients undergoing surgery. A meta-analysis of 20 studies assessing variation in length of hospital stay and pain outcomes in relation to age, ethnicity, gender, and education showed the positive moderate effects of preoperative teaching on these outcomes (Guruge and Sidani 2002). However, there is little evidence to support the use of preoperative education over and above standard care to improve postoperative outcomes in patients undergoing hip or knee replacement surgery (McDonald et al. 2004).

Research findings therefore remain contradictory with respect to the effect of preoperative information on postoperative pain and patient satisfaction. There is still a lack of systematic evidence on the quality and effectiveness of patient education (Johansson et al. 2005). Despite inconsistencies in the evidence, the studies all point to the fact that preoperative education is beneficial to adult surgical patients (Oshodi 2007).

Other advantages of patient education include good patient rapport and communication. This is associated with a lower incidence of malpractice litigation. Liability reduction may be more likely to arise from the educational model (Brenner et al. 2009). Patients taught to assess their pain will have more control over the dose and delivery of analgesic drugs, regardless of the analgesic technique used (Macintyre and Ready 2002).

A preoperative teaching interview guide exploring five dimensions of preoperative information (situational/procedural information, sensation/discomfort information, patient role information, psychosocial support, and skills training), found a correlation between the amount of preoperative teaching received and the value thereof (Bernier et al. 2003). Being able to introduce an effective smoking intervention program 6–8 weeks before surgery will reduce postoperative morbidity.

In coronary artery bypass surgery, patient education was more effective if the content was individualized and given in a combination of media on an individual basis and in more than one session (Fredericks et al. 2009). For orthopedic patients, preoperative education can help to reduce postoperative pain medication and increase self-efficacy (Johansson et al. 2005). In a study of patients with pain presenting to an emergency department, those shown educational videos or printed brochures had greater decreases in self-reported pain than those given no education (Marco et al. 2006). After preoperative education, a group of hysterectomy patients ambulated significantly earlier than those without teaching (Oetker-Black et al. 2003).

Preoperative teaching at least provides the surgical patient with pertinent information concerning the surgical process and the intended surgical procedure, as well as anticipated patient behaviors (e.g., anxiety, fear), expected sensations, and probable outcomes (Kruzik 2009). However, research does indicate that the provision of good-quality preoperative information facilitates patients' active involvement in their care and therefore may contribute to an overall increase in satisfaction (Walker 2007).

Future

Well-designed, methodologically sound research is needed into the outcomes of patient education. Surgical services should be reengineered to include preadmission assessment, education, and care after discharge. Preoperative education should be individually tailored and include individualized discharge analgesic packages (Chung 1999).

References

Bernier, M. J., Sanares, D. C., Owen, S. V., et al. (2003). Preoperative teaching received and valued in a day surgery setting. *AORN, 77*(3), 563–572.

Binhas, M., Roudot-Thoraval, F., Thominet, D., et al. (2008). Impact of written information describing postoperative pain management on patient agreement with proposed treatment. *European Journal of Anaesthesiology, 25*(11), 884–890.

Brenner, L. H., Brenner, A. T., & Horowitz, D. (2009). Beyond informed consent: Educating the patient. *Clinical Orthopaedics and Related Research, 467*(2), 348–351.

Brown, V. (2006). Preparing a patient information leaflet. *Journal of Perioperative Practice, 16*(11), 540–545.

Caumo, W., Schmidt, A. P., Schneider, C. N., et al. (2001). Risk factors for preoperative anxiety in adults. *Acta Anaesthesiologica Scandinavica, 45*(3), 298–307.

Chan, Y., Irish, J. C., Wood, S. J., et al. (2002). Patient education and informed consent in head and neck surgery. *Archives of Otolaryngology – Head & Neck Surgery, 128*(11), 1269–1274.

Chen, H. H., Yeh, M. L., & Yang, H. J. (2005). Testing the impact of a multimedia video CD of patient-controlled analgesia on pain knowledge and pain relief in patients receiving surgery. *International Journal of Medical Informatics, 74*(6), 437–445.

Cheung, L. H., Callaghan, P., & Chang, A. M. (2003). A controlled trial of psycho-educational interventions in preparing Chinese women for elective hysterectomy. *International Journal of Nursing Studies, 40*(2), 207–216.

Cheung, A., Finegan, B. A., Torok-Both, C., Donnelly-Warner, N., & Lujic, J. (2007). A patient information booklet about anesthesiology improves preoperative patient education. *Canadian Journal of Anaesthesia, 54*(5), 355–360.

Chung, F. (1999). Postoperative pain control in ambulatory surgery. *The Surgical Clinics of North America, 79*(2), 401–430.

Counsell, D., Macintyre, P. E., & Breivik, H. (2008). Organisation and role of acute pain services. In H. Breivik, W. I. Campbell, & M. K. Nicholas (Eds.), *Clinical pain management: Practice and procedures.* London: Hodder Arnold.

Deyirmenjian, M., Karam, N., & Salameh, P. (2006). Preoperative patient education for open-heart patients: A source of anxiety? *Patient Education and Counseling, 62*(1), 111–117.

Duhalde, A., Boucheny, D., & Dougados, M. (2002). Effects of video information on preoperative anxiety level and tolerability of joint lavage in knee osteoarthritis. *Arthritis and Rheumatism, 47*(4), 380–382.

Felder-Puig, R., Maksys, A., Noestlinger, C., et al. (2003). Using a children's book to prepare children and parents for elective ENT surgery: Results of a randomized clinical trial. *International Journal of Pediatric Otorhinolaryngology, 67*(1), 35–41.

Fredericks, S., Ibrahim, S., & Puri, R. (2009). Coronary artery bypass graft surgery patient education: A systematic review. *Progress in Cardiovascular Nursing, 24*(4), 162–168.

Frisch, A. M., Johnson, A., Timmons, S., & Weatherford, C. (2010). Nurse practitioner role in preparing families for pediatric outpatient surgery. *Pediatric Nursing, 36*(1), 41–47.

Garretson, S. (2004). Benefits of pre-operative information programmes. *Nursing Standard, 18*(47), 33–37.

Giraudet-Le Quintrec, J. S., Coste, J., Vastel, L., et al. (2003). Positive effect of patient education for hip surgery: A randomized trial. *Clinical Orthopaedics, 414*, 112–120.

Guruge, S., & Sidani, S. (2002). Effects of demographic characteristics on preoperative teaching outcomes: A meta-analysis. *Canadian Journal of Nursing Research, 34*(4), 25–33.

Hatava, P., Olsson, G. L., & Lagerkranser, M. (2000). Preoperative psychological preparation for children undergoing ENT operations: A comparison of two methods. *Paediatric Anaesthesia, 10*(5), 477–486.

Hobbs, F. D. (2002). Does pre-operative education of patients improve outcomes? The impact of pre-operative education on recovery following coronary artery bypass surgery: a randomized controlled clinical trial. *European Heart Journal, 23*(8), 600–601.

Jawahar, K., & Scarisbrick, A. A. (2009). Parental perceptions in pediatric cardiac fast-track surgery. *AORN Journal, 89*(4), 725–731.

Johansson, K., Nuutila, L., Virtanen, H., et al. (2005). Preoperative education for orthopaedic patients: Systematic review. *Journal of Advanced Nursing, 50*(2), 212–223.

Justus, R., Wyles, D., Wilson, J., et al. (2006). Preparing children and families for surgery: Mount Sinai's multidisciplinary perspective. *Pediatric Nursing, 32*(1), 35–43.

Kotzer, A. M., Coy, J., & LeClaire, A. D. (1998). The effectiveness of a standardized educational program for children using patient-controlled analgesia. *Journal of the Society of Pediatric Nurses, 3*(3), 117–126.

Kruzik, N. (2009). Benefits of preoperative education for adult elective surgery patients. *AORN Journal, 90*(3), 381–387.

LaMontagne, L. L., Hepworth, J. T., Cohen, F., & Salisbury, M. H. (2003). Cognitive-behavioral intervention effects on adolescents' anxiety and pain following spinal fusion surgery. *Nursing Research, 52*(3), 183–190.

Lee, A., & Gin, T. (2005). Educating patients about anaesthesia: Effect of various modes on patients' knowledge, anxiety and satisfaction. *Current Opinion in Anaesthesiology, 18*(2), 205–208.

Macintyre, P. (Ed) (2010). Provision of safe and effective acute pain management. In *Acute pain management: Scientific evidence*, (3rd edn, pp. 47–56). Melbourne: Australian and New Zealand College of Anaesthetists.

Macintyre, P. E., & Ready, L. B. (2002). *Acute pain management* (2nd ed., pp. 225–231). London: WB Saunders.

Macintyre, P. E., & Schug, S. A. (2007). *Acute pain management: A practical guide*. London: Elsevier.

Marco, C. A., Marco, A. P., Plewa, M. C., et al. (2006). The verbal numeric pain scale: Effects of patient education on self-reports of pain. *Academic Emergency Medicine, 13*(8), 853–859.

McDonald, S., Hetrick, S., & Green, S. (2004). Preoperative education for hip or knee replacement. *Cochrane Database of Systematic Reviews, 1*, CD003526.

Moores, A., & Pace, N. A. (2003). The information requested by patients prior to giving consent to anaesthesia. *Anaesthesia, 58*(7), 703–706.

Mordiffi, S. Z., Tan, S. P., & Wong, M. K. (2003). Information provided to surgical patients versus information needed. *AORN Journal, 77*(3), 546–549.

Oetker-Black, S. L., Jones, S., Estok, P., et al. (2003). Preoperative teaching and hysterectomy outcomes. *AORN Journal, 77*(6), 1215–1218. 1221–1231.

Oliveira, A., Gevirtz, R., & Hubbard, D. (2006). A psycho-educational video used in the emergency department provides effective treatment for whiplash injuries. *Spine, 31*(15), 1652–1657.

Oshodi, T. O. (2007). The impact of preoperative education on postoperative pain. Part 2. *British Journal of Nursing, 16*, 790–797.

Stergiopoulou, A., Birbas, K., Katostaras, T., et al. (2006). The effect of a multimedia health educational program on the postoperative recovery of patients undergoing laparoscopic cholecystectomy. *Studies in Health Technology and Informatics, 124*, 920–925.

Varughese, A. M., Byczkowski, T. L., Wittkugel, E. P., et al. (2006). Impact of a nurse practitioner-assisted preoperative assessment program on quality. *Paediatric Anaesthesia, 16*(7), 723–733.

Walker, J. A. (2007). What is the effect of preoperative information on patient satisfaction? *British Journal of Nursing, 16*(1), 27–32.

Watt-Watson, J., Stevens, B., Costello, J., et al. (2000). Impact of preoperative education on pain management outcomes after coronary artery bypass graft surgery: A pilot. *Canadian Journal of Nursing Research, 31*(4), 41–56.

Whyte, R. I., & Grant, P. D. (2005). Preoperative patient education in thoracic surgery. *Thoracic Surgery Clinics, 15*(2), 195–201.

Postoperative Pain, Regional Blocks

Spencer S. Liu
Virginia Mason Medical Center, Seattle, WA, USA

Synonyms

Peripheral nerve blocks; Regional anesthesia; Regional blocks

Definition

The chemical blockade of inputs in response to noxious mechanical stimuli originating from various anatomical groupings of afferent nerve fibers and their distal innervations. The goal of the regional blockade is to provide analgesia and rapid functional recovery to patients during the perioperative period while simultaneously taking steps to utilize correct technique and local anesthetic agents so as to reduce potential side effects associated with these blocks.

Characteristics

The numerous regional blocks available to treat postoperative pain can best be understood by classifying each into anatomical subheadings and addressing the different afferent innervations they are designed to suppress. Each block will be

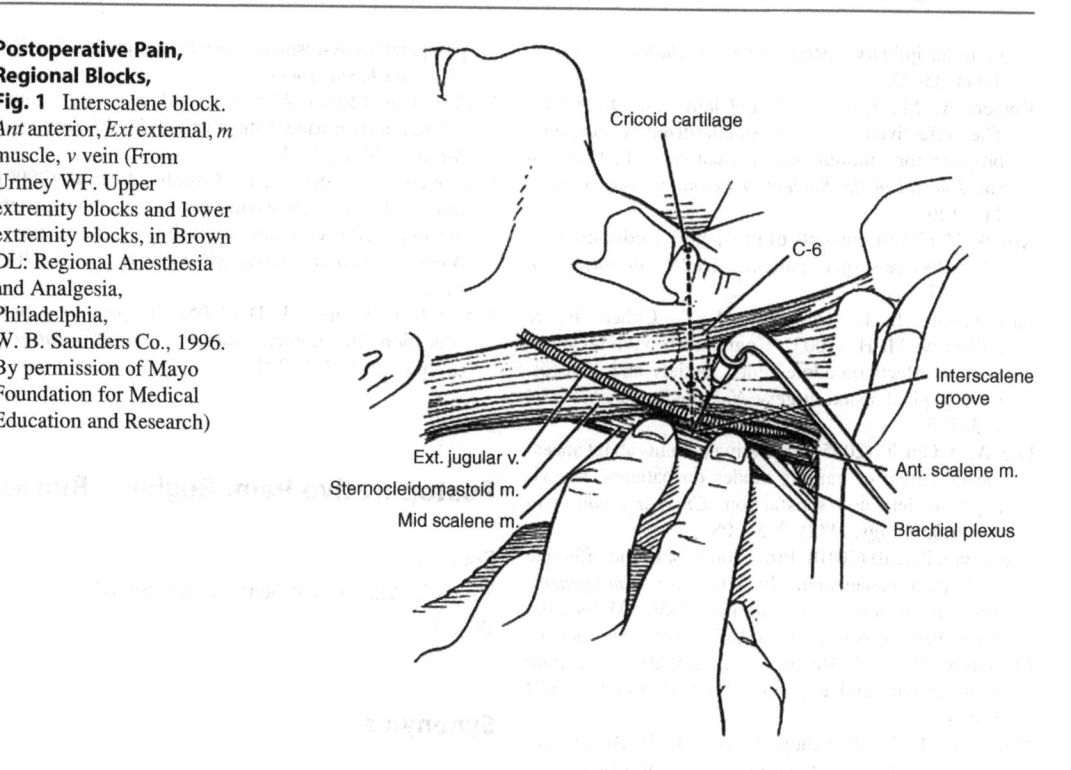

Postoperative Pain, Regional Blocks, Fig. 1 Interscalene block. *Ant* anterior, *Ext* external, *m* muscle, *v* vein (From Urmey WF. Upper extremity blocks and lower extremity blocks, in Brown DL: Regional Anesthesia and Analgesia, Philadelphia, W. B. Saunders Co., 1996. By permission of Mayo Foundation for Medical Education and Research)

described with respect to the most typical indications, complications, and technique associated with their use, with appropriate local anesthetic agent selection and overall contraindication from regional blocks to follow.

The Upper Extremity

Brachial Plexus

Indications for ▶ brachial plexus blockade are procedures involving the shoulder or upper extremity. Examples of these types of surgeries include acromioplasty rotator cuff repair and humeral, hand, and forearm procedures. Common techniques available for brachial plexus blockade are the ▶ interscalene approach and the axillary approach.

Interscalene Approach

Often utilized for open shoulder surgeries and surgeries of the proximal upper extremity, the interscalene approach can be used for the placement of continuous interscalene catheters or single-shot local anesthetic dosing. Continuous interscalene analgesia has been shown to reduce patient requirements for postoperative

opioids, provide better analgesia, reduce opioid-related side effects, and give better patient satisfaction for at least the first 48 h following surgery, when compared with IV ▶ patient-controlled analgesia in prospective, randomized, controlled trials (Liu and Salinas 2003; Ilfeld et al. 2003).

Potential complications to the interscalene approach include inadvertent injection into the vertebral artery, epidural and subarachnoid space, phrenic and recurrent laryngeal nerve block, block of the ipsilateral sympathetic chain, and ▶ pneumothorax (Liu and Salinas 2003).

Technique involves needle entry in the interscalene groove, typically utilizing a needle capable of electrical peripheral nerve stimulation for localization of the brachial plexus via muscle twitch and either a single-shot injection of a local anesthetic agent or the insertion of catheters 5–10 cm into the brachial plexus sheath for continuous anesthetic infusion (Mulroy 2002) (Fig. 1).

Axillary Approach

Often employed for more distal upper extremity procedures involving the hand and forearm, the

**Postoperative Pain,
Regional Blocks,**
Fig. 2 Illustration of
anesthesiologist and patient
for axillary block. Patient's
arm bent at 90°.
Anesthesiologist's fingers
straddle artery (*a*); humeral
head is shown and
facilitates plexus fixation. *n*
nerve (From Urmey
WF. Upper extremity
blocks and lower extremity
blocks, in Brown DL:
Regional Anesthesia and
Analgesia, Philadelphia,
W. B. Saunders Co., 1996.
By permission of Mayo
Foundation for Medical
Education and Research)

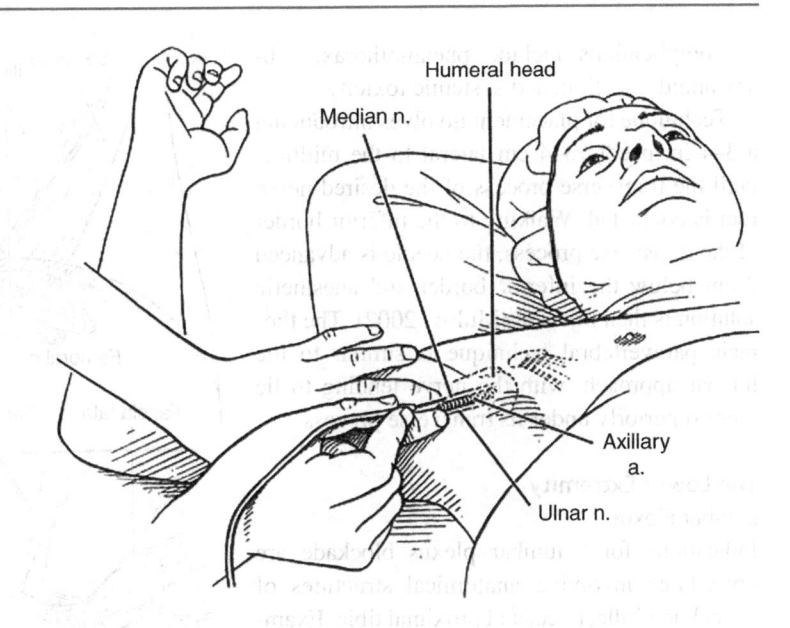

**Postoperative Pain,
Regional Blocks,
Fig. 2** Illustration of
anesthesiologist and patient
for axillary block. Patient's
arm bent at 90°.
Anesthesiologist's fingers
straddle artery (*a*); humeral
head is shown and
facilitates plexus fixation. *n*
nerve (From Urmey
WF. Upper extremity
blocks and lower extremity
blocks, in Brown DL:
Regional Anesthesia and
Analgesia, Philadelphia,
W. B. Saunders Co., 1996.
By permission of Mayo
Foundation for Medical
Education and Research)

benefit of using continuous axillary brachial plexus block over PCA has not been established as for continuous interscalene analgesia (Liu and Salinas 2003).

A complication unique to this technique is axillary artery hematoma.

Techniques for the axillary approach are numerous and involve a periarterial single-shot injection of local anesthetics or continuous catheter placement via the use of nerve stimulators, fluoroscopic, and ultrasound-guided needle placement (Mulroy 2002) (Fig. 2).

The Thorax and Abdomen

Intercostal Approach

Indicated for abdominal and chest-wall procedures, this block can offer an alternative to spinal or epidural anesthesia. Bilateral blockade of the 6th through to the 12th intercostal nerves provides sensory anesthesia for the thoracic and abdominal wall in these respective dermatomes, i.e., from the xiphoid to the pubis, and has been shown to be weakly associated with improved pulmonary outcome measures in postoperative recovery studies (Mulroy 2002; Ballantyne et al. 1998).

Complications include pneumothorax, airway obstruction secondary to sedation in the prone position during placement, systemic toxicity, and hypotension.

Technique is typically the posterior approach. The classic posterior approach inserts the needle at the inferior portion of the body of the rib, walking the needle tip inferiorly off the body of the rib. A continuous technique has been described, inserting a catheter into the ▶ intercostal space to provide analgesia to several levels via spread of injected solutions (Mulroy 2002).

Paravertebral Approach

Indicated for blocking peripheral nerves as they exit from the intervertebral foramina along the thoracic and lumbar vertebrae. Particularly useful for blocking lumbar roots, which do not have a convenient rib to mark their peripheral course, this block may be employed for procedures involving the chest (i.e., breast procedures) or abdominal wall as well as lower abdominal and upper leg procedures (i.e., inguinal hernia repair). Paravertebral blocks provide ipsilateral somatic and sympathetic nerve blockade in multiple contiguous thoracic dermatomes above and below the site of injection and are effective in treating acute and chronic pain of unilateral origin from the chest and abdomen (Karmakar 2001). Useful as an alternative to epidural or intercostal anesthesia, this method has a decreased accuracy of placement associated with its use, with a 10 % failure rate reported in several series (Mulroy 2002).

Complications include pneumothorax, subarachnoid injection, and systemic toxicity.

Technique for placement involves introducing a 3–4 in. needle 3–4 cm lateral to the midline, until the transverse process of the desired nerve root is contacted. Walking to the inferior border of the transverse process, the needle is advanced 2 cm below the inferior border and anesthetic solution is then injected (Mulroy 2002). The thoracic paravertebral technique is similar to the lumbar approach, with the nerve tending to lie more superiorly under its transverse process.

The Lower Extremity

Lumbar Plexus

Indications for ▶ lumbar plexus blockade are procedures involving anatomical structures of the pelvic girdle, knee, and proximal tibia. Examples of these types of surgeries are total hip replacement (THR), knee arthroscopy, knee replacement, ACL repair, and repair of tibia plateau fracture. Common techniques available for lumbar plexus blockade include the femoral nerve sheath approach and the ▶ psoas compartment approach.

Femoral Nerve Sheath Approach

Indicated in a majority of open knee procedures, continuous femoral analgesia for total knee replacement has been shown to provide comparable or improved analgesia, with fewer side effects, for at least the first 48 h following surgery, faster short-term functional recovery of knee flexion, accelerated physical rehab, and reduction in hospital stay and total rehab when compared to IV PCA in prospective clinical trials. Data also supports continuous femoral analgesia after total hip replacement, with significant morphine-sparing effects resulting in decreased incidence of nausea, vomiting, pruritus, and sedation versus IV PCA.

Complications arising from femoral nerve sheath blocks include inadvertent injection of the femoral vein/artery, femoral nerve injury, and infection.

The technique most commonly employed in the placement of femoral nerve sheath single-shot injections of local anesthetics or continuous

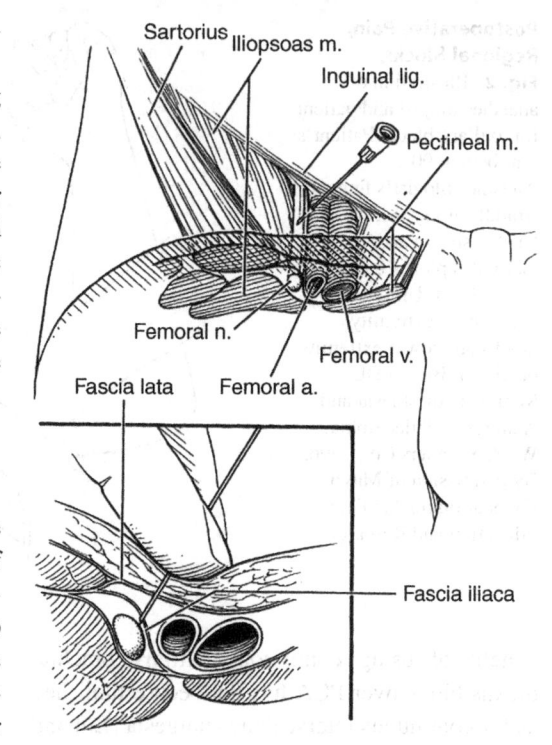

Postoperative Pain, Regional Blocks, Fig. 3 Anteroposterior (**a**) and cross-sectional (**b**) views of the femoral nerve (*n.*), which lies between the fascia lata and the iliac fascia, lateral to the femoral artery (*a*) and vein (*v*). *lig* ligament, *m* muscle (From Urmey WF. Upper extremity blocks and lower extremity blocks, in Brown DL: Regional Anesthesia and Analgesia, Philadelphia, W. B. Saunders Co., 1996. By permission of Mayo Foundation for Medical Education and Research)

catheter infusions involves perivascular isolation of femoral nerve twitch with nerve stimulators. Catheters are typically inserted 10–15 cm into the femoral sheath to maximize cephalad proximity to the lumbar plexus (Fig. 3).

Psoas Compartment Approach

Indicated for procedures that require a more reliable block of the obturator nerve (e.g., total hip replacement (THR)), continuous psoas compartment analgesia has been shown to provide improved analgesia and patient satisfaction in THR/hip fracture patients when compared to IV PCA (Liu and Salinas 2002).

Complications are similar to FIC and femoral nerve sheath blocks with infection, neurologic, and vascular injuries being the most common.

The technique of psoas compartment block involves placing the patient in the lateral decubitus position with the operative extremity up. The needle insertion point is 3 cm caudal and 5 cm lateral to the L4 spinous process. A 15 cm stimulating needle is advanced perpendicular to the skin and directed slightly midline until the L5 transverse process is encountered. The needle is then redirected slightly cephalad, sliding past the superior aspect of the L5 transverse process and advanced until stimulation of the quadriceps muscle is elicited (Mulroy 2002).

Sciatic Nerve

Indications for sciatic nerve block primarily involve open procedures of the foot and distal lower extremity. When compared to IV PCA, continuous sciatic nerve block has been shown to provide superior analgesia with significantly reduced incidence of nausea/vomiting, urinary retention, and sedation. After moderately painful orthopedic surgery of the lower extremity, ropivacaine infusion using a portable mechanical pump and a popliteal sciatic perineural catheter at home also decreased pain, opioid use and related side effects, and sleep disturbances and improved overall satisfaction (Ilfeld et al. 2002).

The incidence of complications associated with sciatic nerve block is very low and similar to those listed for lumbar plexus blocks with infectious, vascular, and neurologic injury having the highest frequency.

There are two techniques available in blocking the sciatic nerve, the posterior popliteal approach and the lateral approach. The posterior popliteal approach is performed with the patient in the prone position, with the benefits over PCA as described above. This approach is often associated with decreased catheter durability and function, given its close proximity to the knee joint. The lateral approach can be performed in a supine patient, a position that has been anecdotally shown to be advantageous in allowing a more secure placement of the catheter between the vastus lateralis and biceps femoris, away from the mobile knee joint. There have been no clinical trials, however, to determine which approach is optimal for catheter placement (Liu and Salinas 2002).

Drugs for Continuous Perineural Anesthesia

Lidocaine, bupivacaine, and ropivacaine have all been used in conjunction with continuous infusions, PCA, or a combination of background infusions and PCA boluses, for prolonged plexus analgesia. The longer-acting amino-amides bupivacaine, levobupivacaine, and ropivacaine, being the most commonly used agents, are frequently mixed with adjuvants such as epinephrine, clonidine, and opioids, offering the ability to spare local anesthetic volume, reduce motor and sensory blocks, and improve the quality of analgesia. In specific agent selection, there is insufficient data to determine an optimal analgesic solution. Both levobupivacaine and ropivacaine, however, have less CNS and cardiovascular toxicity than bupivacaine and may be safer (Ladd et al. 2002).

Contraindications

Common contraindications include infection at the block site and allergy to analgesics. Peripheral nerve blocks have generally been considered safe in the anticoagulated patient. However, compressive neuropathies have been described in anticoagulated patients, secondary to post-block perineural hematoma formation. Placement of peripheral nerve catheters during a central neuraxial or general anesthesia to improve patient comfort, although not contraindicated, may lead to potential neurologic injury due to inadvertent intraneural injection. Use of a nerve stimulator does not guarantee avoidance of this type of neurologic injury (Liu and Salinas 2002).

References

Ballantyne, J. C., Carr, D. B., deFerranti, S., et al. (1998). The comparative effects of postoperative analgesic therapies on pulmonary outcome: Cumulative meta-analyses of randomized, controlled trials. *Anesthesia and Analgesia, 86*, 598–612.

Ilfeld, B. M., Morey, T. E., Wang, R. D., et al. (2002). Continuous popliteal sciatic nerve block for postoperative pain control at home: A randomized, double-blinded, placebo-controlled study. *Anesthesiology, 97*, 959–965.

Ilfeld, B. M., Morey, T. E., Wright, T. W., et al. (2003). Continuous interscalene brachial plexus block for postoperative pain control at home: A randomized, double-blinded, placebo-controlled study. *Anesthesia and Analgesia, 96*, 1089–1095.

Karmakar, M. (2001). Thoracic paravertebral block. *Anesthesiology, 95*, 771–780.

Ladd, L., Chang, D., Wilson, K. A., et al. (2002). Effects of CNS site-directed carotid arterial infusions of bupivacaine, levobupivacaine, and ropivacaine in sheep. *Anesthesiology, 97*, 418–428.

Liu, S., & Salinas, F. (2003). Continuous plexus and peripheral nerve blocks for postoperative analgesia. *Anesthesia and Analgesia, 96*, 262–272.

Mulroy, M. (2002). *Regional anesthesia: An illustrated procedural guide* (3rd ed.). Philadelphia: Lippincott Williams & Wilkins.

Postoperative Pain, Thoracic and Cardiac Surgery

Elmar Berendes[1] and Thomas Hachenberg[2]
[1]Klinik für Anästhesiologie, Operative Intensivmedizin und Schmerztherapie, Klinikum Krefeld, Krefeld, Germany
[2]Clinic for Anaesthesiology and Intensive Therapy, Otto-von-Guericke University, Magdeburg, Germany

Synonyms

Acute postoperative pain therapy; Cardiac surgery; High thoracic epidural anesthesia; Postoperative pain management; Thoracic epidural anesthesia; Thoracic surgery

Definition

Pain influences patients' morbidity and mortality postoperatively after cardiac and thoracic surgical procedures. Adequate perioperative analgesia is essential to avoid pulmonary and cardiac complications. ▶ Postoperative pain management must be aggressive to relieve severe dynamic pain, e.g., when the patient is coughing and needs robust routines that will discover early symptoms and signs of potentially serious complications. ▶ Thoracic epidural anesthesia provides excellent analgesia, improves left ventricular function, and reduces perioperative pulmonary complications and myocardial ischemia. Thus, optimal use of pharmacological analgesia, combined with a well-tailored perioperative thoracic epidural anesthesia, can markedly reduce complications in patients at high risk of developing postoperative respiratory infections and cardiac ischemic events.

Characteristics

Pulmonary and Cardiac Function After Thoracic and Cardiac Surgery

Sternotomy, and in particular lateral thoracotomy, result in postoperative pain and impair respiratory mechanics and coughing. Both affect the retraction forces of the chest wall. After sternotomy, thoracic compliance decreases, while compliance of the lungs increases. These mechanisms may enhance the development of atelectasis, even if tidal volume and airway pressure remain unchanged during mechanical ventilation. The opened pleura further increases the compliance of the respiratory system. The chest wall and pleura compartment offer significant impedance to lung expansion after sternotomy and rib extraction, unless the pleura is opened (Hachenberg and Pfeiffer 1999; Barnas et al. 1996). The deflation of the lung during one lung ventilation invariably results in an increased shunt, impaired oxygenation, and carbon dioxide elimination. After thoracic surgical procedures, the following major problems result in an impaired pulmonary function: (1) decreased lung volume (e.g., atelectasis, resection of lung tissue, pleural effusion, and thoracic restriction), (2) impaired ventilation (decreased functional residual capacity, dysfunction of diaphragm, dysfunction of intercostal muscles, increased airway resistance), and (3) impaired gas exchange (atelectasis, lung edema, decreased cardiac output and decreased minute ventilation).

Cardiac and thoracic surgical procedures are further associated with exaggerated perioperative adrenergic stimulation, resulting in hormonal response and systemic inflammation. These pathophysiological changes can lead to perioperative myocardial ischemia or other life-threatening cardiac events. The incidence of perioperative myocardial infarction is 0.2 % in patients without and 4.0 % in patients with preexisting coronary artery disease (Ashton et al. 1993). Severe myocardial ischemia appears most frequently during the first 48 h after major surgery (Metzler et al. 1997). Thus, perioperative pain management should focus on effective perioperative pain relief and a reduction of pulmonary and cardiac complications.

Effects of Thoracic Epidural Anesthesia on Pulmonary and Myocardial Function

The effects of thoracic epidural anesthesia on lung volume, respiratory mechanics, and pulmonary gas exchange depend on the extent of segmental regional blockade and sympathicolysis. Theoretically, thoracic epidural anesthesia may lead to an alteration of intrathoracic blood volume, lung volume and pulmonary vasotone by sympathicolysis, paralysis of intercostal muscles, decreased volume of thoracic cavity, and increased thoracic and abdominal compliance. However, the results of prospective, randomized clinical trials postulate beneficial effects of thoracic epidural anesthesia, and no study has reported a deterioration of pulmonary function by thoracic epidural anesthesia during and after ▶ thoracic surgery (Hachenberg and Pfeiffer 1999). The excellent analgesia after thoracic surgery, by contrast, results in lower incidence of respiratory morbidity (Ballantyne et al. 1998).

An activation of myocardial sympathetic nerves in ▶ cardiac surgery can result in myocardial ischemia, especially in patients with coronary artery disease (Nabel et al. 1988). Consequently, a reversible cardiac sympathectomy by ▶ high thoracic epidural anesthesia has anti-ischemic effects owing to the blockade of efferent sympathetic fibers (Olausson et al. 1997). Clinical studies of high thoracic epidural anesthesia imply beneficial effects for the perioperative management of patients who underwent coronary artery bypass grafting. The sympathetic blockade by high thoracic epidural anesthesia results in depression of the endocrine

perioperative stress response (Loick et al. 1999), an improvement of global systolic and diastolic left ventricular function (Schmidt et al. 2000), and a reduction of new wall motion abnormalities (Berendes et al. 2003) during and after coronary artery bypass grafting. These effects of high thoracic epidural anesthesia may improve the long-term outcome after myocardial revascularization. Figure 1 shows the effects of high thoracic epidural anesthesia on postoperative and global left ventricular function, cardiac troponin I concentrations, and afterload after coronary artery bypass grafting.

Specific Problems of High Thoracic Epidural Anesthesia for Coronary Artery Bypass Grafting

Several studies have reported the advantages of an additional high thoracic epidural anesthesia such as earlier weaning from mechanical ventilation, decreased catecholamine response, reduction of perioperative myocardial ischemia, improved renal and pulmonary outcome, and better pain control. However, there are some problems limiting the routine use of high thoracic epidural anesthesia for coronary artery bypass grafting. Since the neurological risk of high thoracic epidural anesthesia for this procedure remains inadequately defined, placement of the epidural catheter 1 day prior to surgery is mandatory to minimize the risk of epidural hematoma because of intraoperative systemic anticoagulation. Moreover, in many cases, pain control may be not effective for chest tubes and venous graft preparation. Thus, an additional systemic analgesia with opioids and/or non-opioid analgesics is routinely necessary.

Clinical Implications of Thoracic Epidural Anesthesia for Thoracic and Cardiac Surgery

In thoracic, and especially cardiac, surgical procedures, an effective cardiomyocyte functional protection during a perioperative insult stress requires a balanced preservation vs. blunting of β-adrenergic signaling. Adrenergic sensitization optimizes post-ischemic functional recovery, while desensitization protects against intraoperative oxygen supply/demand

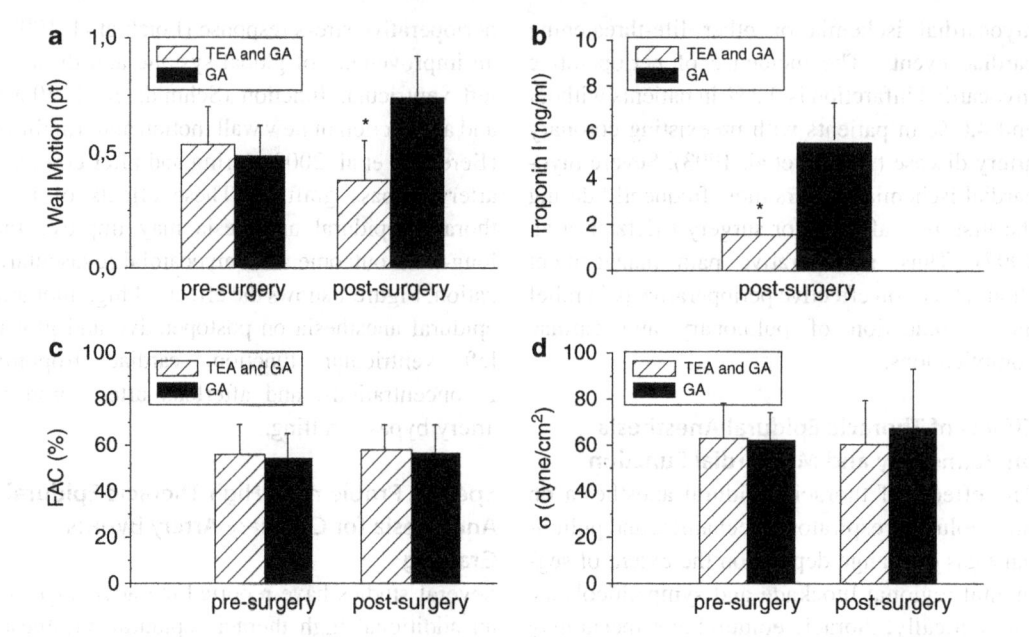

Postoperative Pain, Thoracic and Cardiac Surgery, Fig. 1 Patients' global and regional left ventricular function and afterload. Pre- and postoperative values of (**a**) global left ventricular wall motion index, (**c**) left ventricular fractional area change (FAC), and (**d**) left ventricular end-systolic meridional wall stress (σ) as well as (**b**) postoperative concentrations of cardiac troponin I in patients receiving general anesthesia (GA) or general anesthesia and high thoracic epidural anesthesia (TEA). Mean $\pm$ SD, * $P < .05$ (Berendes et al. 2003)

imbalance. The deleterious effects of adrenergic stimulation on β-adrenergic receptor density and β-adrenergic receptor coupling and the positive effects of thoracic epidural anesthesia have been well documented (Lefkowitz et al. 2000). Thus, attenuation of the stress response via blockade and/or selective anesthetic regimens, i.e., thoracic epidural anesthesia, has the potential to prevent adverse events. Consequently, a multimodal approach to reduce patients' stress response and improve recovery, consisting of intraoperative general anesthesia, thoracic epidural anesthesia, postoperative patient-controlled epidural anesthesia, early extubation, early oral nutrition, and enforced mobilization, is the most appropriate regimen, particularly for patients undergoing thoracic and cardiac surgical procedures (Brodner et al. 2001). Other local anesthetic techniques such as intercostal nerve block, intrapleural local anesthesia, or wound infiltration are less effective than thoracic epidural anesthesia and have not been introduced in clinical practice for routine use (Ballantyne et al. 1998).

Systemic Analgesia After Thoracic and Cardiac Surgery

Systemic postoperative pain management after thoracic and cardiac surgery is mainly based on intravenous opioids when thoracic epidural anesthesia is not possible. Opioids are potent analgesics; their use, however, may cause a variety of adverse effects, such as excessive sedation, respiratory depression, biliary spasm, depression of gastrointestinal motility, and postoperative nausea and vomiting. Some systemic side effects can be blunted by adequately tailored opioid analgesia, given in time and not on demand, combined with a non-opioid analgesic like propacetamol or metamizol (Hyllested et al. 2002). Nonsteroidal anti-inflammatory drugs effectively reduce opioid consumption after thoracic surgery (Perttunen et al. 1999); however, the fear of potentially serious side effects, such as predisposition to increased postoperative bleeding and deterioration of renal function, means that these drugs are not widely used, especially after cardiac surgical procedures performed with cardiopulmonary bypass. Thus, recent studies have focused on the reduction of opioid consumption by paracetamol or propacetamol (Lahtinen et al. 2002; Avellanda et al. 2000) showing different results. In the study of Lahtinen et al. (2002), an adjunctive therapy with propacetamol neither decreased cumulative opioid consumption nor reduced adverse effects after coronary bypass grafting, while Avellanda et al. (2000) found that pain scores were significantly decreased with a single additional propacetamol infusion after cardiac surgery. The combination of nonsteroidal anti-inflammatory drugs and paracetamol may confer additional analgesic efficacy compared with paracetamol alone, and the published data suggest that paracetamol may enhance analgesia when added to a nonsteroidal anti-inflammatory drug, compared with nonsteroidal anti-inflammatory drugs alone (Hyllested et al. 2002).

A good interdisciplinary acute pain service and/or well-designed algorithm for postoperative pain therapy implemented in routine postoperative care is essential, especially for thoracic and cardiac surgical procedures.

References

Ashton, C. M., Petersen, N. J., Wray, N. P., et al. (1993). The incidence of perioperative myocardial infarction in men undergoing non-cardiac surgery. *Annals of Internal Medicine, 118*, 504–510.

Avellanda, C., GÂ3mez, A., Martos, F., et al. (2000). The effect of a single intravenous dose of metamizol (2 g), ketorolac (30 mg) and propacetamol (1 g) on hemodynamic parameters and postoperative pain after heart surgery. *European Journal of Anaesthesiology, 17*, 85–89.

Badner, N. H., Knill, R. L., Brown, J. E., et al. (1998). Myocardial infarction after non-cardiac surgery. *Anesthesiology, 88*, 572–578.

Ballantyne, J. C., Carr, D. B., de Ferranti, D., et al. (1998). The comparative effects of postoperative analgesic therapies on pulmonary outcome: Cumulative meta-analyses of randomised, controlled trials. *Anesthesia and Analgesia, 86*, 598–612.

Barnas, G. M., Gilbert, T. B., Watson, R. J., et al. (1996). Respiratory mechanics in the open chest: Effects of parietal pleurae. *Respiratory Physiology, 104*, 63–70.

Berendes, E., Schmidt, C., Van Aken, H., et al. (2003). Reversible cardiac sympathectomy by high thoracic epidural anesthesia improves regional left ventricular function in patients undergoing coronary artery bypass grafting. *Archives of Surgery, 138*, 1283–1290.

Brodner, G., Van Aken, H., Hertle, L., et al. (2001). Multimodal perioperative management – combining thoracic epidural analgesia, forced mobilization, and oral nutrition – reduces hormonal and metabolic stress and improves convalescence after major urologic surgery. *Anesthesia and Analgesia, 92*, 1594–1600.

Hachenberg, T., & Pfeiffer, B. (1999). Use of thoracic epidural anaesthesia for thoracic surgery and its effect on pulmonary function. *Best Practice & Research. Clinical Anaesthesiology, 13*, 57–72.

Hyllested, M., Jones, S., Pedersen, J. L., et al. (2002). Comparative effect of paracetamol, NSAIDs, or their combination in postoperative pain management: A qualitative review. *British Journal of Anaesthesia, 88*, 199–214.

Lahtinen, P., Kokki, H., Hendolin, H., et al. (2002). Propacetamol as adjunctive treatment for postoperative pain after cardiac surgery. *Anesthesia and Analgesia, 95*, 813–819.

Lefkowitz, R. J., Rockman, H. A., & Koch, W. J. (2000). Catecholamines, cardiac Î2-adrenergic receptors, and heart failure. *Circulation, 101*, 1634–1637.

Loick, H. M., Schmidt, C., Van Aken, H., et al. (1999). Modulation of immune and stress response by high thoracic epidural anesthesia or clonidine in patients undergoing cardiopulmonary bypass grafting. *Anesthesia and Analgesia, 88*, 701–709.

Nabel, E. G., Ganz, P., Gordon, J. B., et al. (1988). Dilation of normal and constriction of atherosclerotic coronary arteries caused by the cold pressure test. *Circulation, 77*, 43–52.

Olausson, K., Magnusdottir, H., Lurje, L., et al. (1997). Anti-ischemic effects and anti-anginal effects of thoracic epidural anesthesia versus those of conventional therapy in the treatment of severe refractory unstable angina pectoris. *Circulation, 96*, 2178–2182.

Perttunen, K., Nilsson, E., & Kalso, E. (1999). I. V. Diclofenac and ketorolac for pain after thoracoscopic surgery. *British Journal of Anaesthesia, 82*, 221–227.

Schmidt, C., Wirtz, S., Van Aken, H., et al. (2000). Effect of high thoracic epidural anesthesia on global left ventricular function in patients with coronary artery disease. *Anesthesiology, 93*, A 656.

Postoperative Pain, Tramadol

Peter C. A. Kam
Anaesthesia, Central Clinical School,
The University of Sydney, Sydney, Australia

Synonyms

Tramal®

Definition

▶ Tramadol is a synthetic, centrally acting analgesic agent that is structurally related to codeine and morphine. It has a methyl substitution on the phenol moiety, and this accounts for its affinity for opioid receptors. It is a racemic mixture and the two enantiomers function in a complementary manner to enhance analgesic actions.

Characteristics

Pharmacodynamic Properties
The overall mechanisms of the analgesic actions of tramadol are mediated by modulation of opioid and monoamine descending inhibitory pathways mediated by the raphe nucleus, periaqueductal grey, locus coeruleus, and reticulospinal projections (Lee et al. 1993). Tramadol acts as an opioid agonist and inhibits the reuptake of norepinephrine and serotonin. The (+)-enantiomer of tramadol has selective mu (μ) agonist actions. The affinity of the (+)-enantiomer for the μ-receptor is ∼6,000 times less than that of morphine and ten times less than codeine. It binds weakly to kappa and delta receptors. O-desmethyl (MI) tramadol metabolite formed during extensive hepatic metabolism of tramadol is pharmacologically active, with a ∼200-fold higher affinity for the μ-receptor compared with racemic tramadol. The (+)-enantiomer is also a potent inhibitor of 5-HT reuptake. Thirty percent of the analgesic actions of tramadol are antagonized by naloxone.

The non-opioid analgesic actions of tramadol are mediated by its actions on monoamine systems, where it inhibits the reuptake of norepinephrine and serotonin. The (−) enantiomer is a more effective inhibitor of 5-HT reuptake and also increases its release by autoactivation (Bamigbade and Langford 1998). O-desmethyl tramadol (MI) metabolite may have an important role in mediating the analgesic actions of tramadol. As it is metabolized by liver cytochrome P450 (CYP) 2D6 sparteine oxygenase, tramadol may have reduced analgesic effects in patients deficient in CYP 2D6 enzyme (∼8 % Caucasian population).

In double-blind controlled studies, the analgesic effects of oral tramadol (100 mg) peaked at 1–2 h and lasted for 3–6 h after drug administration. ▶ Intravenous tramadol (2 mg/kg) provided similar antinociceptive effects to that of pethidine (1 mg/kg). The minimum effective serum concentration (MEC) of tramadol is 20.2–986.3 μg/l (median 287.7 μg/l), with the MEC for the MI metabolite being 0.9–190.5 μg/l (median 36.2 μg/l).

Respiratory Effects
In clinical studies in postoperative patients using equianalgesic dose (potency ratio 10:1 for tramadol/morphine), tramadol had less effect on the respiratory center compared with morphine.

Gastrointestinal Motility
Unlike opioids, which impair gastrointestinal motility, tramadol generally has no significant effect on gastrointestinal function. Tramadol does not prolong gastric emptying. In volunteers receiving oral tramadol, median oro-cecal transit times were similar to those receiving a placebo, although there was a trend towards an increase in total colonic transit time.

Central Nervous System
Tramadol has been shown to cause idiopathic seizures in clinical use with an incidence of <10 %. Significant dose-related activation of EEG variables (spectral edge, median power frequency, delta power, and alpha/delta ratio) is produced by tramadol (Vaughan et al. 2000). However, tramadol has no effects on the depth of anesthesia, as measured by auditory-evoked response, when it is used for intraoperative analgesia during surgery.

Pharmacokinetics
Tramadol is rapidly absorbed with an oral bioavailability of 70 % after a single dose administration and ∼90 % after multiple doses (due to saturated first-pass hepatic metabolism). A Cmax of 308 μg/l is reached in 1–6 h after a single oral 100 mg dose. Multiple doses result in higher Cmax values for tramadol and its MI metabolite. Tramadol has a large volume of

distribution (~260 l), indicating high tissue affinity of the drug. It is ~20 % bound to plasma proteins. Tramadol crosses the placental barrier with a foetal/maternal ratio of 0:8.

Tramadol undergoes first-pass hepatic metabolism via two metabolic pathways, mainly involving isoenzyme CDYP2D6 and, to a lesser extent, CYP3A. Tramadol is metabolized by CYP2D6 to O-desmethyl tramadol (MI), which is active (binds to μ-receptor). In patients who are deficient in CYP2D6 isoenzyme, inactive N-desmethyl tramadol is produced. The O- and N-desmethylated compounds are conjugated (sulfation or glucuronidation).

The elimination kinetics of tramadol follows a two-compartmental model, with a terminal elimination half life of ~5.5 h. The T 1/2 β of its MI metabolite is approximately 6.6–6.9 h. Tramadol and its metabolites are excreted primarily via the kidneys (90 %), with the remaining (10 %) eliminated in the feces. Therefore, the T 1/2 β is prolonged in patients with impaired renal and hepatic function (10.8 and 13.3 h, respectively).

Clinical Efficacy

Earlier studies on the anesthetic efficacy of oral or parenteral tramadol in acute postoperative pain management showed that intravenous tramadol 50–150 mg was equivalent in analgesic efficacy to morphine 5–15 mg in the treatment of moderate, but not in severe, pain. The peak analgesic effect of intramuscular tramadol occurred within 1–2 h and lasted for 5–6 h. In a meta-analysis of 3,453 postoperative patients, it was found that the analgesic efficacy of tramadol 75–100 mg was similar to that of codeine 60 mg, and combinations of aspirin 650 mg plus codeine 60 mg, and paracetamol 650 mg plus dextropropoxyphene 100 mg (Moore and McQuay 1997). The NNT of tramadol 50 mg is 8.3, that of tramadol 100 mg is 4.8, and that of tramadol 150 mg is 2.4, indicating a dose-dependent response.

Tramadol (dose titrated to response) reduced severe postoperative pain associated with abdominal, orthopedic, and cardiac surgery by 46–57 % after 4–6 h, compared with a 69 % reduction with morphine (Gritti et al. 1998; Lanzetta et al. 1998; Bloch et al. 2002).

With nurse-controlled analgesia, the overall analgesic efficacy with tramadol was comparable to that of equipotent doses of parenteral morphine. Comparative studies, comparing the analgesic efficacy of tramadol with the NSAIDS, suggest that tramadol is a more effective analgesic for pain following orthopedic surgery compared with ketorolac. Investigators reported that the analgesic efficacy of tramadol was excellent in 79 % patients, compared with 58 % in patients receiving ketorolac (Lanzetta et al. 1998). Tramadol provided longer mean pain-free intervals between doses than that achieved with ketorolac.

Two double-blind comparative studies reported that continuous intravenous infusions of tramadol (12 mg/h) achieved analgesic efficacy similar to that obtained with intermittent boluses of tramadol 50 mg (Hartjen et al. 1996). However, there was an increased consumption of tramadol with continuous infusion compared with intermittent bolus doses (223.5 vs. 176.6 mg after 6–8 h and 449 vs. 201 mg after 24 h), although this was not associated with increased side effects.

Patient-Controlled Analgesia (PCA)

Comparative studies have shown that tramadol provides effective analgesia over a 24–48-h period using PCA for moderate to severe postoperative pain. The analgesic efficacy of tramadol was similar to that achieved with morphine PCA (Pang et al. 1999; Stamer et al. 1997). In these studies, a loading dose (100–150 mg or 3 mg/kg) followed by a bolus dose (16–30 mg, or 0.3 mg/kg) was used.

A randomized double-blind controlled study comparing the analgesic efficacy of tramadol infusion with epidural morphine for post-thoracotomy pain showed that IV tramadol (150 mg bolus followed by continuous infusion 450 mg/24 h) was as effective as epidural morphine (2 mg loading dose, followed by infusion 0.2 mg/h). This technique utilizing tramadol avoids the necessity of placement of a thoracic epidural catheter (Bloch et al. 2002).

Pediatric Use

Oral tramadol 1.5 mg/kg combined with oral midazolam (0.5 mg/kg; maximum dose 7.5 mg) provided effective postoperative analgesia in children undergoing extraction of six or more teeth, without delay in recovery (Roelofse and Payne 1999).

Two recent studies investigated the analgesic effects of caudal tramadol in children undergoing hypospadia surgery. In one study involving 90 children, it was concluded that the caudal tramadol (2 mg/kg), or its concomitant use with caudal bupivacaine, did not provide additional analgesia. In the children who received caudal tramadol, pain scores at 3 h postoperative were significantly higher.

Intraoperative Analgesia

Intraoperative tramadol provided analgesia similar to that of equipotent doses of morphine; potency ratio of morphine to tramadol is 1:11.23.

There were early concerns that tramadol may predispose to intraoperative awareness because of its stimulatory effects on EEG of patients receiving tramadol. However, recent studies using tramadol in combination with potent volatile or intravenous anesthetics have not shown any increase in intraoperative awareness in patients undergoing surgery (Coetzee et al. 1996).

Day Surgery

Its lack of sedative and respiratory depressant effects makes tramadol an attractive analgesic agent. Studies have reported that perioperative tramadol provided better analgesia than fentanyl in combination with codeine/paracetamol or ketorolac. In a large multicentre study, oral tramadol (100 mg) after discharge, in combination with intravenous tramadol (100 mg pre-, intra-, or postoperative doses), provided better analgesia than a combination of intravenous fentanyl (100 μg) and postoperative codeine (16 mg)/paracetamol (1 Gm) tablets (Bamigkade et al. 1998).

Neuropathic Pain

The treatment of pain in polyneuropathy is often difficult. The tricyclic antidepressants and anticonvulsants are the main drugs used in the treatment of pain in polyneuropathy, with NNT of 2.6 for tricyclic antidepressants, 2.5 for anticonvulsant sodium channel blockers, and 4.1

for gabapentin. However, tramadol has been shown to be effective for pain in polyneuropathy with an NNT of 2.4 (Sindrup 2000).

Side Effects

The most common side effects of tramadol in clinical and postmarketing surveillance studies of oral or parenteral doses of tramadol were nausea (6.1 %), dizziness (4.6 %) drowsiness (2.4 %), sweating (1.9 %), vomiting (1.7 %), and dry mouth (1.6 %).

There are no differences in the incidence of nausea and vomiting in patients receiving tramadol compared with other opioids. In a randomized double-blind parallel group study involving patients undergoing laparoscopic surgery who received intraoperative intravenous tramadol 100 mg or morphine 10 mg, the incidence of vomiting (30 %) and nausea (15–16 %) was similar in both groups. The incidence of seizures is estimated to be <1 %. The risk of epilepsy is increased in patients receiving pro-convulsant drugs, neuroleptic agents, and monoamine oxidase inhibitors or patients with epilepsy.

Drug Interactions

Drugs acting on hepatic CYP3A and CYP2D6 isoenzymes can affect the pharmacokinetics of tramadol. Carbamazepine induces hepatic enzymes. Concomitant administration of tramadol with carbamazepine results in a 50 % decrease in T $1/2$ β of tramadol as a result of induction of the metabolism of tramadol. Cimetidine, a potent inhibitor of cytochrome P450 isoenzymes, prolongs the T $1/2$ β of tramadol and its M_i metabolite.

Ondansetron is a 5-HT$_3$ receptor antagonist and an antiemetic agent commonly used for postanesthetic nausea and vomiting. When administered to a patient receiving tramadol for analgesia, ondansetron reduced the overall analgesic effect of tramadol by approximately 25 %, probably by blocking spinal 5-HT$_3$ receptors (Arcioni et al. 2002).

There have been case reports of enhanced effects of warfarin in a patient receiving concomitant warfarin and tramadol. Tramadol is not recommended in patients who are receiving monoamine oxidase inhibitors or the serotonin reuptake inhibitors or within 2 weeks of their withdrawal. Administration of tramadol and these antidepressant drugs predisposes to the serotonin syndrome.

Recommended dosage of oral or parenteral tramadol in adults <75 year is 50–100 mg every 4–6 h with a maximum of 400–600 mg/day.

References

Arcioni, R., della Rocca, M., Romano, S., et al. (2002). Ondansetron inhibits the analgesic effects of tramadol: A possible 5HT$_3$ spinal involvement in acute pain in humans. *Anesthesia and Analgesia, 94*, 1553–1557.

Bamigbade, T. A., & Langford, R. M. (1998). Tramadol hydrochloride: An overview of current use. *Hospital Medicine, 59*, 373–376.

Bamigkade, T. A., Langford, R. M., Blower, A. L., et al. (1998). Pain control in day surgery: Tramadol vs. standard analgesia. *British Journal of Anaesthesia, 80*, 558–559.

Bloch, M. B., Dyer, R. A., Heijke, S. A., et al. (2002). Tramadol infusion for post-thoracotomy pain relief: A placebo controlled comparison with epidural morphine. *Anesthesia and Analgesia, 94*, 523–528.

Coetzee, J. F., Maritz, J. S., & du Toit, J. C. (1996). Effect of tramadol on depth of anaesthesia. *British Journal of Anaesthesia, 76*, 415–418.

Gritti, G., Verri, M., Launo, C., et al. (1998). Multi centre trial comparing tramadol and morphine for pain after abdominal surgery. *Drugs under Experimental and Clinical Research, 24*, 9–16.

Hartjen, K., Fisher, M. V., Mewes, R., et al. (1996). Preventive tramadol infusion in comparison with bolus application on demand during the early postoperative period. *Anaesthetist, 45*, 538–544.

Lanzetta, A., Vizzardi, M., Letzia, G., et al. (1998). Intramuscular tramadol versus ketorolac in patients with orthopedic and traumatologic postoperative pain: A comparative multicenter trial. *Current Therapeutic Research, Clinical and Experimental, 59*, 39–47.

Lee, C. R., McTavish, D., & Sorkin, E. M. (1993). Tramadol a preliminary review of its pharmacodynamic and pharmacokinetic properties, and therapeutic potential in acute and chronic pain states. *Drugs, 46*, 313–340.

Moore, R. A., & McQuay, H. J. (1997). Single-patient data meta-analysis of 3453 postoperative patients: Oral tramadol versus placebo, codeine and combination analgesics. *Pain, 69*, 287–294.

Pang, W. W., Mok, M. S., Lin, C. H., et al. (1999). Comparison of patient-controlled analgesia (PCA) with tramadol or morphine. *Canadian Journal of Anaesthesia, 46*, 1030–1035.

Roelofse, J. A., & Payne, K. A. (1999). Oral tramadol analgesic efficacy in children following multiple dental extractions. *European Journal of Anaesthesiology, 16*, 441–447.

Sindrup, S. H., & Jensen, T. S. (2000). Pharmacologic treatment of pain in polyneuropathy. *Neurology, 55*, 915–920.

Stamer, V. M., Maier, C., Grond, S., et al. (1997). Tramadol in the management of postoperative pain: A double-blind, placebo- and active drug-controlled study. *European Journal of Anaesthesiology, 14*, 646–654.

Vaughan, D. J., Shinner, G., Thornton, C., et al. (2000). Effect of tramadol on electroencephalographic and auditory evoked variables during light anaesthesia. *British Journal of Anaesthesia, 85*, 705–707.

Postoperative Pain, Transition from Parenteral to Oral Drugs

Stephan A. Schug[1] and Emma L. Brandon[2]
[1]School of Medicine and Pharmacology, Royal Perth Hospital, Faculty of Medicine, Dentistry & Health Sciences, The University of Western Australia, UWA Anaesthesiology, Perth, Australia
[2]Anaesthesiology and Pain Medicine, University of Western Australia Royal Perth Hospital, Perth, WA, Australia

Synonyms

Transition from parenteral to oral analgesic drugs

Definition

The change from intravenous or intramuscular administration of analgesic drugs to the oral route.

Characteristics

Oral analgesia is the route of choice for all patients where the gastrointestinal tract is intact, motility is adequate, and the patient is not being starved (NHMRC 1998). In acute postoperative pain, these conditions are frequently not met, and for severe pain, the delay between administration and onset of effect makes the oral route less appealing. Nonetheless, it is still the commonest route for drug administration and the mainstay for ambulant patients.

In those patients where parenteral routes of administration have been chosen in the first instance, the issue of transition to oral medication is quite complex and must be planned with some idea of the natural history of the pain in mind.

A UK Audit Commission report in 1997 highlighted some deficiencies in postoperative analgesia (Audit Commission 1997). The commission commented in particular on the period between immediate postoperative care, usually coordinated by an acute pain service, and discharge on simple oral analgesia. Smith and Power (Smith and Power 1998) coined the term "analgesic gap" to describe increased pain in this period and pointed out that responsibility for covering the gap was usually delegated to the junior surgical team. They also referred to the lack of research effort in this area, which remains a problem.

The benefits of ▶ multimodal analgesia have been known since around 1990 (Kehlet and Dahl 1993). The use of combinations of local anesthesia, NSAIDs, paracetamol, other non-opioid drugs, and opioids to improve pain relief and reduce side-effect intensity is widespread among acute pain services. One interpretation of the Audit Commission report is that this practice is not continued appropriately once the initial severe pain has been controlled. Possible reasons for this include the presence of contraindications to the use of NSAID, and reluctance to use oral opioids after parenteral treatment is ceased. Although oral opioids are often said to be difficult to use for the immediate treatment of postoperative pain because of impaired gastric emptying due to both drug and surgical factors, there are reports of success as well as failure in this setting (Derbyshire et al. 1985; Pearl et al. 2002). Once GIT function has begun to return to normal, however, there is no evidence to suggest that parenteral routes offer any advantage. A neuropathic basis should be considered for pain that is poorly responsive to opioids, and appropriate adjunctive agents either substituted or added to the regimen.

Three groups of patients will be considered:
1. Those where the pain intensity is expected to decline rapidly
2. Those where the decline in pain intensity will occur over days to a week
3. Those where pain is thought to be stable but ongoing for longer periods of time

Patients Where Pain Intensity Is Expected to Decline Rapidly

Examples of such patients include those undergoing laparoscopic surgery, minor orthopedics, or superficial surgery of other types. Treatment for such patients should be based on simple analgesics (NSAIDs, paracetamol) (NHMRC 1998), but provision should be made for opioid supplementation with a rapidly acting formulation if required. As gastric emptying is usually rapidly reestablished in such patients, the oral route is to be preferred, as fear of needles is a known cause of poor analgesia in prn regimens.

Patients Where the Decline in Pain Intensity Will Occur Over Days to a Week

These patients are probably the highest risk group for experiencing the analgesic gap. Again, multimodal analgesia should be the key to ongoing therapy, with regular oral NSAID/paracetamol forming the cornerstone. Despite this, there will be a number of patients where pain relief will deteriorate if parenteral opioids are simply ceased (Ng et al. 2000). The requirement for strong opioids can be judged from the pattern and amount of opioid use prior to cessation of parenteral therapy. This is easiest if the parenteral modality has been PCA but may still be ascertained for infusions, by examination of rate variation and bolus requirement in response to various maneuvers such as dressing changes. Patients requiring more than two doses per hour may be assumed to have a need for ongoing opioids. In such patients, this can achieved by the method described by Smith and Power (Smith and Power 1998), using a baseline of slow-release opioid, with fast-acting oral opioid available on demand. As they point out, this regimen provides a background level of analgesia well below the threshold for respiratory depression, but adequate for control of rest pain, with the demand doses available for control of dynamic pain.

The use of controlled-release oxycodone following PCA for major abdominal, gynecological, or orthopedic surgery has been studied by Ginsberg (Ginsberg et al. 2003). The authors suggest the use of a conversion factor of 1.2X the IV morphine daily dose to predict the daily dose of controlled-release oxycodone. As the tablets are only available in certain sizes, doses were usually adjusted down to the nearest tablet strength available. Note the contrast with the conversion factor of 3 that is usually quoted.

The length of time that slow-release opioids are required will vary with the severity of the pain but may be quite extended after painful surgery such as thoracotomy and orthopedics. In the study of Ginsberg et al. (Ginsberg et al. 2003), the average duration of oral opioid therapy was 4 days for abdominal and gynecological surgery and 5 days for orthopedics; however, one third of the 189 patients remained on opioids at the conclusion of the 7-day study period. Pain control was maintained at four or less on a 11-point numerical scale throughout the 12-h dose period, but most patients required supplemental doses of rapidly acting oxycodone (62 % on day 1 and 38 % on day 7).

The safety of the technique was also evaluated. Most patients were commenced on oral controlled-release treatment on the first postoperative day, and the incidence of paralytic ileus was reported as 2.1 % (4/189). There were also four patients with low oxygen saturations, three of whom had low baseline readings.

Alternatively, the dose of slow-release drug may also be estimated from the daily dose of fast-acting drug, where this is prescribed alone in the first instance. Whichever method is used, the kinetics of the slow-release drug must be respected and adequate time allowed to reach steady state. Where PCA is being used, the controlled-release drugs may be safely commenced the night before it is planned to cease PCA.

Patients Where Pain Is Thought to Be Stable but Ongoing for Longer Periods of Time

Examples of such patients include burn patients, multi-trauma patients, and some painful medical problems such as calciphylaxis. Typical patterns of pain would include a stable background component present throughout the whole day, with possible periods of increased intensity associated with dressings or other clinical activities. In this group, conversion from parenteral to oral using the conversion factors usually quoted for chronic pain may be more appropriate. Fast-acting agents may be required to cover breakthrough pain and can also be used to estimate dosage changes, either as a result of tolerance or increases in pain intensity.

A conversion chart such as Table 1 (NHMRC 1998) should be used to guide dosage adjustment, but care must be exercised when changing drugs because cross-tolerance is often incomplete, and doses lower than the equianalgesic dose are recommended, with dosage titrated to clinical effect. As mentioned above, it is usual practice to adjust doses down to the nearest appropriate tablet size.

Choice of Drugs

There is little evidence on which to base the decision as to which oral drug is best to use. The four commonest options, morphine, oxycodone, methadone, and tramadol, will be considered.

Morphine

Morphine has a long history of use via the oral route and is probably still the commonest oral opioid. It is available in several controlled-release preparations designed for once or twice daily dosing. The exact kinetics of these varies and should be considered in prescribing and interpreting responses to doses. It is also available in rapidly acting syrup formulation, which provides capacity for treatment of those with problems swallowing tablets and for supplemental dosing. Care should be exercised, particularly in those with impaired renal function, because of accumulation of the glucuronide metabolites, both of which are pharmacologically active.

Postoperative Pain, Transition from Parenteral to Oral Drugs, Table 1 Transition from parenteral to oral

Agonists	Approximate equianalgesic dose	
	Parenteral	Oral
Morphine	10 mg	30 mg
Oxycodone	15 mg	20–30 mg
Fentanyl	100 mcg	N/A
Hydromorphone	1.5 mg	7.5 mg
Methadone	10 mg	20 mg
Codeine	130 mg	200 mg
Dextropropoxyphene		130 mg
Tramadol	100 mg	????
Partial agonists		
Buprenorphine	0.4 mg	0.4–0.8 mg (sublingual)

Morphine-3-glucuronide may be responsible for central excitation seen with chronic dosing, while morphine-6-glucuronide may produce respiratory depression. Both have half-lives significantly longer than the parent drug.

Oxycodone

Oxycodone (Davis et al. 2003) has recently become available in a variety of dosages in controlled-release form. Rapidly acting tablets are available for supplemental doses. It has a 60 % oral bioavailability, and the rapid onset and sustained duration of effect make it a viable alternative to morphine. The main metabolite, noroxycodone, is inactive, although 10 % of the dose is converted to oxymorphone, which has 14X the analgesic potency of oxycodone. Since both are renally excreted, doses should be reduced in renal failure. It also has more activity at ▶ kappa opioid receptors, which may offer some advantage where there is felt to be a neuropathic component to the pain, but may mean that there is a lower conversion ratio for patients converting from oxycodone to morphine than those switching from morphine to oxycodone.

Methadone

Methadone (Mancini et al. 2000) is not available in controlled-release formulation, and both absorption and elimination are highly variable.

In its favor it is cheap, has a high lipid solubility and bioavailability, and can usually be dosed twice daily for pain management. It has no active metabolites and does not accumulate in chronic renal failure. It is active at both mu and ▶ delta opioid receptors. It would probably have a very limited role because of its variable kinetics were it not for its ▶ NMDA antagonist properties, which give it a potential advantage over the other opioids for the treatment of ▶ neuropathic pain. Care must be taken to avoid toxicity due to increasing duration of effect with chronic dosing.

Tramadol

Tramadol (Scott and Perry 2000; Shipton 2000) also has unusual characteristics. Only 30 % of its analgesic activity is attributed to ▶ mu(μ)-opioid receptor agonism, the remainder being due to enhancement of ▶ noradrenergic and serotonergic inhibitory pathways. This results in analgesia associated with low risk of respiratory depression and generally with low levels of sedation. Unfortunately, this is balanced by a strong emetic effect. Care must also be taken to avoid drug interactions with other drugs producing raised serotonin levels and those reducing the seizure threshold (particularly SSRIs and MAOIs). It is available in both controlled-release and immediate-onset capsules. Dose reduction is required in renal failure. Due to its alternative mechanism of action, tramadol may be the drug of choice where low mu activity is particularly desirable, e.g., patients where sedation and reduced GIT motility are problematic, although high doses may be required to produce sufficient analgesia in severe pain.

References

Audit Commission. (1997). *Anaesthesia under examination*. London: Audit Commission.

Davis, M. P., Varga, J., Dickerson, D., et al. (2003). Normal-release and controlled-release oxycodone: Pharmacokinetics, pharmacodynamics, and controversy. *Supportive Care in Cancer, 11*, 84–92.

Derbyshire, D. R., Bell, A., Parry, P. A., et al. (1985). Morphine sulphate slow release. Comparison with I.M. Morphine for postoperative analgesia. *British Journal of Anaesthesia, 57*, 858–865.

Ginsberg, B., Sinatra, R. S., Adler, L. J., et al. (2003). Conversion to oral controlled-release oxycodone from intravenous opioid analgesic in the postoperative setting. *Pain Medicine, 4*, 31–38.

Kehlet, H., & Dahl, J. B. (1993). The value of "multimodal" or "balanced analgesia" in postoperative pain treatment. *Anesthesia and Analgesia, 77*, 1048–1056.

Mancini, I., Lossignol, D. A., & Body, J. J. (2000). Opioid switch to oral methadone in cancer pain. *Current Opinion in Oncology, 12*, 308–313.

Ng, A., Hall, F., Atkinson, A., Kong, K., et al. (2000). Bridging the analgesic gap. *Acute Pain, 3*, 172–180.

NHMRC. (1998). *Acute pain management: Scientific evidence*. Canberra: AusInfo.

Pearl, M. L., McCauley, D. L., Thompson, J., et al. (2002). A randomized controlled trial of early oral analgesia in gynecologic oncology patients undergoing intra-abdominal surgery. *Obstetrics and Gynecology, 99*, 704–708.

Scott, L. J., & Perry, C. M. (2000). Tramadol: A review of its use in perioperative pain. *Drugs, 60*, 139–176.

Shipton, E. A. (2000). Tramadol – present and future. *Anaesthesia and Intensive Care, 28*, 363–374.

Smith, G., & Power, I. (1998). Audit and bridging the analgesic gap. *Anaesthesia, 53*, 521–522.

Postoperative Pain, Venous Thromboembolism

Stephan A. Schug
School of Medicine and Pharmacology, Royal Perth Hospital, Faculty of Medicine, Dentistry & Health Sciences, The University of Western Australia, UWA Anaesthesiology, Perth, Australia

Synonyms

Deep vein thrombosis (DVT); Embolism; Pulmonary embolism (PE); Red thrombi; Vein thrombosis

Definition

Venous thrombi (red thrombi) are blood clots in a vein, principally erythrocytes trapped in a fine fibrin mesh with few platelets. The less common arterial or white thrombi are large aggregates of platelets trapped in fibrin strands with very few erythrocytes. Unregulated activation of the hemostatic system may cause thrombosis. Certain patients can be identified as at-high-risk; these may have acquired or inherited factors. A venous thrombus may be friable and dislodge as an embolus, which may obstruct blood flow to critical organs such as the lung (pulmonary embolism).

Characteristics

Causation

The triad of thrombogenesis, first described by R. Virchow in 1858 (Virchow's triad), identified three factors causal of venous thrombosis: stasis of blood, endothelial injury, and hypercoagulability.

In the postoperative setting, bed rest, immobilization, and inactivity due to postoperative pain may cause stasis of blood flow and protect activated procoagulants from circulating inhibitors and fibrinolysins and liver clearance. In addition, postoperative plasma concentrations of clotting factors rise. Vessel wall damage can be due to direct trauma, catheters, infection, or external compression. This tissue damage evokes a stress hormone response, leading to activation of cytokines, adhesion molecules, and coagulation factors.

The final result of these processes is a hypercoagulable state, which contributes to the multifactorial pathogenesis of venous thromboembolism. Once initiated, coagulation is the dominant process and produces retrograde and prograde thrombosis. Thrombosis may be friable and therefore dislodge and, thus, embolize. Such embolization leads to occlusion of blood flow to critical organs, in particular the lungs in the form of pulmonary embolism (PE).

However, the presence of deep vein thrombosis (DVT) does not reliably predict the sequelae of PE, death, or post-thrombotic complications (Handoll et al. 2002). The outcome of the DVT may be complete resolution without ill effects or, in the worst case, cause pulmonary embolism and thereby even death. DVT prevalence in the community is 0.2 % but rises to 30–40 % after surgery for hip fractures; 0.9 % of deaths in hospital are due to PE.

Methods of Reduction

The incidence of postoperative venous thromboembolism can be reduced by preventative measures. These can be mechanical devices such as thromboembolic deterrent stockings (TEDS), which are either elastic compression stockings (ECS) or graduated compression stockings (GCS), as well as pneumatic leg compression devices (muscle pumps). Furthermore, there are pharmacological methods including low-dose aspirin, warfarin, heparin (unfractionated (UFH) or low-molecular-weight (LMWH) heparin), and dextran.

Effective pain relief can modify the surgical stress response and enable mobilization, thus reducing the predisposition to the hypercoagulable state. Techniques of neuraxial blockade (epidural or spinal anesthesia) have also been shown to be very effective in reducing venous thromboembolism in some situations.

The efficacy of these methods has been assessed in detail. Graduated compression stockings significantly reduce the occurrence of deep vein thrombosis in hospitalized moderate- and high-risk postoperative patients of all specialities (DVT incidence, 13 % in treatment group, 27 % in control group) (Sachdeva et al. 2010). The data suggest that GCS are most effective when combined with another method of thromboprophylaxis (sequential compression + dextran 70 + subcutaneous heparin: DVT incidence, 2 % in treatment group, 15 % in control group) (Sachdeva et al. 2010).

Following surgery for hip fractures, physical methods of pneumatic compression showed a significant decrease in DVT and PE incidence. However, current data are insufficient to determine their effect on fatal PE or mortality. Physical methods can be as effective as injectable anticoagulants in this patient group, but compliance remains a problem (Handoll et al. 2002).

In colorectal surgery, a combination of GCS and low-dose heparin was found to be more effective than the heparin alone for DVT prevention (Wille-Jørgensen et al. 2003).

Low-dose aspirin shows favorable results in surgery for hip fractures (Handoll et al. 2002), although evidence indicates little effect on the venous side of the circulation (Sachdeva et al. 2010). Recent guidelines propose physical methods and/or aspirin for thromboprophylaxis in post-surgery hip fracture patients (Handoll et al. 2002). It is suggested that injectable anticoagulation be reserved for high-risk patients and those with contraindications to physical methods post-surgery for hip fractures (Handoll et al. 2002).

Unfractionated heparin greatly reduces the incidence of DVT and appears to reduce the mortality from PE in general and orthopedic surgery (Handoll et al. 2002). In lower limb joint replacement, low-molecular-weight heparin is more effective than both unfractionated heparin and warfarin (Choi et al. 2003).

Both unfractionated and low-molecular-weight heparin can be used as effective thromboprophylaxis after colorectal surgery (Wille-Jørgensen et al. 2003).

Data on prevention of fatal PE is inadequate to make conclusions on treatment with heparin (Handoll et al. 2002).

Effects of Neuraxial Blockade on Venous Thromboembolism

There are several physiological effects of neuraxial blockade that provide a rationale for an expected reduction in venothrombotic sequelae (Steele et al. 1991). As described above, hypercoagulability manifested by increased fibrinogen and platelet activity has been implicated in the cause of postoperative venous thrombosis (Tuman et al. 1991). Neural blockade attenuates this response by its effect on afferent and efferent neural transmission and by local anesthetic drug effects (Kehlet 1998).

The effect on postoperative neuroendocrine response partially explains the difference in platelet function between general anesthesia and continuous epidural blockade, as neuraxial blockade substantially reduces the surgical stress response (Steele et al. 1991; Tuman et al. 1991; Kehlet 1998). Other favorable physiological changes also occur. The sympathetic blockade increases blood flow to lower extremities (Steele et al. 1991; Kehlet 1998), in particular in atherosclerotic patients (Tuman et al. 1991). Furthermore, the venous emptying rate and venous capacity are increased (Tuman et al. 1991).

Both local anesthetic solutions alone and local anesthetic in combination with epidural opiates are more effective than epidural opioids alone at reducing the hypercoagulable state (Kehlet 1998). Local anesthetic causes an inhibition of platelet aggregation in vitro, and this effect may also occur in vivo (Steele et al. 1991). It is also thought that local anesthetic may block leukocyte locomotion and, by preventing these cells from adhering to or invading venous walls, preserve endothelial structure (Kehlet 1998). Neuronal blockade may inhibit the increase in platelet aggregation seen postoperatively; thromboelastography demonstrates an increased coagulation inhibition in continuous epidural blockade (Tuman et al. 1991; Kehlet 1998).

Most clinical trials into the effect of neuraxial blockade have been too small to detect a significant effect on mortality and major life threatening events. This is mostly due to the low incidence of such events and the lack of screening for complications. Most evidence comes from systematic reviews of the relevant randomized trials, although single studies in orthopedic surgery demonstrate a reduced incidence in deep venous thrombosis (e.g., Jorgensen et al. 1991). In a systematic review, an overall reduction in mortality was found in postoperative patients treated with neuraxial blockade; one less death per 100 patients was seen in the 30 days postoperatively if treated with a neuraxial blockade (Rodgers et al. 2000). This might be partially due to the effect of neuraxial blockade on thromboembolic complications, as the incidence of deep venous thrombosis and pulmonary embolism is reduced by 50 % (Rodgers et al. 2000).

Conclusion

Good postoperative pain relief is essential to enable ambulation and physiotherapy but, on its own, is not sufficient to reduce the incidence of venous thromboembolism as it is often a result of multiple causative factors. Thus, the approach to prophylaxis needs to be multimodal and appropriate to the individual risk and may include beside good analgesia, the use of neuraxial blockade, as well as mechanical and pharmacological measures.

References

Choi, P. T., Bhandari, M., Scott, J., Douketis, J. (2003). Epidural analgesia for pain relief following hip or knee replacement. *Cochrane Database Systematic Reviews*, (3):CD003071.

Handoll, H. H., Farrar, M. J., McBirnie, J., Tytherleigh-Strong, G., Milne, A. A., & Gillespie, W. J. (2002). Heparin, low molecular weight heparin and physical methods for preventing deep vein thrombosis and pulmonary embolism following surgery for hip fractures. *Cochrane Database Systematic Reviews*, (4):CD000305.

Jorgensen, L. N., Rasmussen, L. S., Nielsen, P. T., Leffers, A., & Albrecht-Beste, E. (1991). Antithrombotic efficacy of continuous extradural analgesia after knee replacement. *British Journal of Anaesthesia, 66*, 8–12.

Kehlet, H. (1998). Modification of responses to surgery by neural blockade. In M. J. Cousins & B. P. Bridenbough (Eds.), *Neural blockade in clinical anaesthesia and management of Pain* (pp. 145–146). Philadelphia: Lippincott-Raven, 158–159.

Rodgers, A., Walker, N., Schug, S., McKee, A., Kehlet, H., van Zundert, A., Sage, D., Futter, M., Saville, G., Clark, T., & MacMahon, S. (2000). Reduction of postoperative mortality and morbidity with epidural or spinal anaesthesia: Results from overview of randomised trials. *British Medical Journal, 321*, 1–12.

Sachdeva, A., Dalton, M., Amaragiri, S. V., Lees, T. (2010). Elastic compression stockings for prevention of deep vein thrombosis. *Cochrane Database Systematic Review*, (7), CD001484. doi:10.1002/14651858. CD001484.pub2. Review. PMID: 20614425.

Steele, S. M., Slaughter, T. F., Greenberg, C. S., & Reves, J. G. (1991). Epidural anesthesia and analgesia: Implications for perioperative coagulability. *Anesthesia and Analgesia, 73*, 683–685.

Tuman, K. J., McCarthy, R. J., March, R. J., DeLaria, G. A., Patel, R. V., & Ivankovich, A. D. (1991). Effects of epidural anesthesia and analgesia on coagulation and outcome after major vascular surgery. *Anesthesia and Analgesia, 73*, 696–704.

Wille-Jørgensen, P., Rasmussen, M. S., Andersen, B. R., Borly, L. (2003). Heparins and mechanical methods for thromboprophylaxis in colorectal surgery. *Cochrane Database Systematic Reviews*, (4): CD001217.

Postpartum Pain

Nicola Beale
Nuffield Department of Anaesthetics, Oxford
University Hospitals NHS Trust, UK

Synonyms

Afterpains; Postnatal pain

Introduction

The world birth rate is estimated at 19.3 per 1,000 population, with a world population that exceeded seven billion in 2012. Of these 133 million deliveries, approximately 18.5 million are by caesarean section.

A recent study by Eisenach and colleagues demonstrated that the vast majority of women (96 %) experience pain in the immediate aftermath of delivery, and their study went on to describe persisting pain at 6 months in 1.8 % of women and in 0.3 % of women at 12 months. (Eisenach et al. 2013) These figures are lower than previous studies: Kainu described persisting pain 1 year post delivery in 19 % of women who had undergone caesarean section and in 10 % of women who had delivered vaginally (Kainu et al. 2010), whereas Nikolajsen described significant daily pain 10 months following caesarean section in 5.9 % of women (Nikolajsen et al. 2004). The presence of pain has long since been recognized with not only increased morbidity but long-term detrimental psychological effects (Carr and Goudas 1999) which can affect both the mother and baby. When compared with other surgical procedures, the existence of persistent pain after caesarean section is lower (Kehlet et al. 2006), but because pain after vaginal delivery also exists and caesarean section is such a commonly performed procedure, this represents an enormous number of women worldwide and a potentially enormous burden on society.

In the latter stages of pregnancy and in labor, higher circulating levels of sex steroid hormones and endogenous neuropeptides such as endorphins confer a protective effect raising pain threshold levels. After delivery these hormone levels drop rapidly, contributing to increasing pain levels.

Another hormone, oxytocin, is released from the posterior pituitary during the second and third stages of labor causing uterine contractions which are responsible for delivery of the fetus and placenta. Following delivery, maternal suckling causes further oxytocin release that triggers uterine contractions and pain (commonly referred to as afterpains) as the uterus involutes. Afterpains are more common with increasing parity. Conversely, though, oxytocin has been proposed as one of the underlying mechanisms that promote maternal protection from developing persisting pain after childbirth (Gutierrez et al. 2013).

Common causes of postpartum pain are pelvic girdle and lower back pain, headache, and perineal and abdominal pain, and these are discussed.

Characteristics

Pregnancy-Related Pelvic Girdle Pain (PPGP) and Pregnancy-Related Low Back Pain (PLBP)

Over the years there have been differences in the description and terminology relating to pelvic girdle pain, which makes comparative analysis of literature on the subject difficult. Since Vleeming's paper in 2007, there has been more consistency, using a definition of pelvic irdle pain that is localized from the level of the posterior iliac crest and gluteal fold over the anterior and posterior elements of the bony pelvis while excluding gynecological or urological causes (Vleeming et al. 2007). Low back pain however can occur between the 12th rib and gluteal fold.

PPGP can develop at any time in pregnancy. Using the above definition and prospective studies only, a review by Kanakaris (Kanakaris et al. 2011) showed that the incidence of the condition is 16–25 % in pregnancy, but only 5–8.5 % at 2 years postpartum, with the majority of symptoms resolving by 6 months. This review quoted predictive factors for PPGP in the longer term as being previous lower back pain (LBP), pelvic girdle pain or pelvic injury, and a history of strenuous work. A recent study has also shown that higher somatization antenatally and at 6 weeks postpartum, increasing fetal weight (previously thought not to be related), higher pain scores at 6 weeks postpartum, and requiring more than 8 h sleep or bed rest at 30 weeks gestation were predictive of pain at 12 weeks after delivery, whereas bed rest (up to 3–4 days postpartum) was associated with less long-term PPGP (Stomp-van den Berg et al. 2012). This study did not contact the mothers immediately post delivery though, and 25 % of the women experiencing PPGP at 12 weeks did not actually have symptoms at 6 weeks postnatally.

If a woman has experienced low back pain and/or PGP in one pregnancy, it does tend to recur in future pregnancies, but does not appear to predict pain without pregnancy.

Ongoing low back pain is not associated with regional anesthetic techniques (Loughnan et al. 2002; Russell et al. 1996).

Treating the pain in the postpartum period is most effective using individualized exercise regimes, focusing on stabilization of the pelvic joints, and oral analgesia using paracetamol and nonsteroidal anti-inflammatories. Surgery to stabilize the pelvis has been used but generally as an end-stage treatment, when other techniques have failed.

Headache

Headache is a common phenomenon in the postpartum period, but establishing the true incidence and determining the actual diagnosis can be difficult.

Goldszmidt's prospective study reported that 39 % of women in a study population of 985 experienced symptoms of headache or neck/shoulder stiffness within a week of delivery, with the majority suffering either primary tension headaches or migraine (Goldszmidt et al. 2005). The median time to presentation of symptoms was 2 days, with a median duration of only 4 h. 4 % of women described their symptoms as incapacitating. In this study risk factors for developing headaches were sustaining an inadvertent dural puncture, increasing age, multiparity, prepregnancy headaches, and a shorter pushing time in the second stage.

Levels of estrogen influence migraine, with symptoms often improving in pregnancy as circulating levels of estrogen (and endorphins) increase, but relapsing after delivery. Sances interviewed 49 women with a known history of migraine 1 week and 1 month postpartum and demonstrated recurrence in symptoms in 34 % and 69 % of responders at respective time intervals (Sances et al. 2003). Parity bore no effect on these symptoms, but breastfeeding was shown to be protective. This protective effect was also seen in a Japanese study, where 63 % of women with known migraine described return of symptoms at 1 month postpartum, with an increase in this figure to 87.5 % at 1 year (Hoshiyama et al. 2012). This study saw a difference in age with over 30s being more susceptible to recurrence, which has been previously described.

Turner and colleagues identified a lower rate of all headache in parturients, with 3.7 % of women experiencing symptoms at 72 h post delivery and 4.5 % at 8 weeks (Turner et al. 2012). The risk factors for each time interval differed, with experience of headache prior to pregnancy *not* associated with pain at 72 h, but associated with pain at 8 weeks. Regional analgesia/anesthesia techniques were also implicated in headache at 72 h, with a threefold increase in likelihood of headache in those who had received a spinal or epidural. As acknowledged by the authors, this study was a secondary analysis of a larger study (Eisenach et al. 2008), in which the international Classification of Headache Disorders, 2nd Edition (ICHD-II) (International Headache Society 2004) was not used, which may affect reporting of specific pre- and post pregnancy symptoms. There was also potential recall bias at reporting at 72 h and limited return from participants in two of the four participating centers at the 8-week follow-up.

Postdural puncture headache (PDPH) is of particular concern to the anesthetist due to the iatrogenic nature of the condition. It occurs when cerebrospinal fluid (CSF) leaks out through the dura following breach of the membrane, either deliberately with spinal procedures or following inadvertent dural puncture (IADP) with an epidural needle. Subsequent low CSF pressure causes symptoms of headache (often frontal or occipital) that worsen after sitting or standing, but improve with lying down, and are sometimes associated with tinnitus, photophobia, neck stiffness, or nausea. These symptoms are frequently considered to be severe, increasingly so with increasing size of needle. They typically develop within 5 days of the causative procedure, and resolution occurs either spontaneously within a week or after an epidural blood patch (EBP) according to the ICHD criteria (International Headache Society 2004), although if left untreated they have been shown to persist for much longer (Sprigge and Harper 2008). The incidence of IADP with epidural is often quoted as 1 %, with 75 % of women requiring EBP for incapacitating symptoms. PDPH after spinal anesthetic occurs less frequently due to the smaller needle gauge used in obstetric anesthesia (25–27 G) and the atraumatic needle tips. There is currently little data to support the use of an intrathecal catheter following recognized IADP, although Heesen suggests that this practice reduces the need for epidural blood patch, although it does not reduce the incidence of PDPH (Heesen et al. 2013).

It is interesting that in Goldszmidt's study while a confirmed diagnosis of PDPH accounted for nearly 5 % of those with a headache, it was the actual cause of low pressure headache in only 21 % of women with those symptoms (Goldszmidt et al. 2005), which highlights the conundrum that is presented to clinicians when assessing women in the postpartum period.

It is also important to recognize that while most headaches in the postpartum period are primary headaches or attributed to regional anesthetic techniques, there are a significant number of rare yet serious pathologies that need to be excluded from one's differential diagnosis, which are summarized by Klein (Klein and Loder 2010) (Table 1). This is especially pertinent in the

Postpartum Pain, Table 1 Differential diagnosis of postpartum headache

Primary causes
Migraine
Tension
Cluster
Orgasmic
Secondary causes
Postdural puncture headache
Preeclampsia/eclampsia
Ischemic stroke
Hemorrhagic stroke
Hypertension
Aneurysm/arteriovascular malformation rupture
Subarachnoid hemorrhage
Cerebral venous/arterial thrombosis
Reversible vasoconstrictive syndromes
Subdural hematoma
Pituitary infarction
Vasculitis
Arterial dissection
Benign intracranial hypertension
Sinusitis
Meningitis
Tumor
Drugs

context of a prothrombotic population who may or may not develop a hypertensive disorder of pregnancy. Concerning features include new neurological symptoms associated with headache, a change in the nature of a preexisting headache, uncontrollable vomiting, change in mental status, no relief from analgesic agents, or concurrent medical complaints, e.g., coagulation disorder (Klein and Loder 2010). Seeking further neurological evaluation with imaging early is beneficial in these circumstances.

Perineal Pain

The incidence of perineal pain in the first postpartum day has been reported as 75 % of women with an intact perineum, 97 % of those who had undergone episiotomy, and 100 % of those who had sustained a third- or fourth-degree tear (Macarthur and Macarthur 2004). Six weeks post delivery, 7 % of study participants still had pain, but this group contained no patients who had an intact perineum at delivery and 20 % of those who had sustained maximal perineal damage.

A smaller study by Paterson showed that 17.5 % of all patients had ongoing pelvic or genital pain at 1 year after delivery, and 26 % had experienced equivalent pain in the interim year between delivery and follow-up, but it had resolved (Paterson et al. 2009). The sample size was too small to comment on the effect of mode of delivery, but chronic pain was a factor in the development of pelvic pain.

In Eisenach's study the only patients experiencing pain at 1 year were those who had undergone vaginal delivery(Eisenach et al. 2013), a group in which pain tended to be localized in the pelvis. The pain interfered with normal activities of daily living, and all of these women had high depression scores.

The Cochrane Collaboration is undertaking a multistage review to look at efficacy of drugs in the first 4 weeks postpartum – paracetamol (Chou et al. 2010) and rectal nonsteroidal antiinflammatories (Hedayati et al. 2003) are effective in the early post delivery period. They have also published reviews demonstrating reduced postpartum pain when using continuous and absorbable sutures to repair perineal tears or episiotomies (Kettle et al. 2010, 2012).

Abdominal Pain

Abdominal pain is seen more commonly following caesarean rather than vaginal delivery.

Papers that looked at both caesarean and vaginal delivery (Eisenach et al. 2008, 2013 and Kainu et al. 2010) tend to agree that caesarean delivery is associated with abdominal or scar pain, compared with vaginal delivery when the pain tends to be located in the perineum or pelvis.

Nikolajsen's paper asked specifically about "pain in the operated area" (Nikolajsen et al. 2004), which was exacerbated by carrying heavy bags or the baby in 75 % of those experiencing pain at 1 year.

Surgical technique has been shown to influence the prevalence of post caesarean abdominal pain. Methods to reduce this include reducing the length of abdominal incision (reducing damage to ilioinguinal and iliohypogastric nerves) (Luijendijk et al. 1997), avoiding uterine exteriorization (Nafisi 2007), and restricting uterine closure to one layer (although with the potential risk of subsequent rupture) (Dodd et al. 2008).

Acute versus Chronic Pain

Eisenach's investigation in 2008 was primarily examining whether caesarean section increased the risk of persistent pain after childbirth, but a secondary outcome was to look at the presence of acute pain after both caesarean and vaginal delivery and to assess whether severity of pain in the acute phase predicted persistent pain development or depression at 8 weeks postpartum (Eisenach et al. 2008). In this study there was a definite link between acute severe post delivery pain and development of persisting pain and depression, but no relationship between mode of delivery and pain. In the longer-term study, there were insufficient patient numbers with pain to establish predictive factors (Eisenach et al. 2013).

Kainu identified a previous history of chronic pain, chronic disease and acute post delivery pain as risk factors for the development of long-term pain (International Headache Society 2004).

Nikolajsen established a link between the presence of pain elsewhere at the time of interview, general anesthesia, and recall of severe acute postoperative pain with the development of persisting pain attributed to caesarean (Nikolajsen et al. 2004).

The repeated link between acute severe pain and ongoing pain correlates with Kehlet's previous work on acute pain after other surgical procedures (Kehlet et al. 2006), which should be borne in mind when considering this large opulation group.

The greatest concern of a woman around the time of caesarean section is feeling pain, yet despite this there is a reluctance in women to take analgesics, and a subsequent increase in their tolerance to pain intensity, in order to prevent adverse effects on the newborn (Carvalho et al. 2005).

In the future there is much scope for developing improved surgical, anesthetic, educational, and psychological techniques in order to improve a woman's experience of pain associated with childbirth; however, it must be remembered that these strategies must be balanced against her expectations.

References

Carr, D. B., & Goudas, L. C. (1999). Acute pain. *The Lancet, 353*, 2051–2058.

Carvalho, B., Cohen, S. E., Lipman, S. S., Fuller, A., Mathusamy, A. D., & Macario, A. (2005). Patient preferences for anesthesia outcomes associated with cesarean delivery. *Anesthesia and Analgesia, 101*, 1182–1187.

Chou, D., Abalos, E., Gyte, G. M. L., & Gülmezoglu, A. M. (2010). Paracetamol/acetaminophen (single administration) for perineal pain in the early postpartum period. *Cochrane Database of Systematic Reviews*, (3). Art. No.: CD008407.

Dodd, J. M., Anderson, E. R., & Gates, S. (2008). Surgical techniques for uterine incision and uterine closure at the time of caesarean section. *Cochrane Database Systematic Reviews*, CD004732.

Eisenach, J. C., Pan, P. H., Smiley, R., Lavand'homme, P., Landau, R., & Houle, T. T. (2008). Severity of acute pain after childbirth, but not type of delivery, predicts persistent pain and postpartum depression. *Pain, 140*, 87–94.

Eisenach, J. C., Pan, P., Smiley, R. M., Lavand'homme, P., Landau, R., & Houle, T. T. (2013). Resolution of pain after childbirth. *Anesthesiology, 118*(1), 143–151.

Goldszmidt, E., Kern, R., Chaput, A., & Macarthur, A. (2005). The incidence and etiology of postpartum headaches: A prospective cohort study. *Canadian Journal of Anesthesia, 52*, 971–977.

Gutierrez, S., Liu, B., Hayashida, K. D. V. M., Houle, T., & Eisenach, J. (2013). Reversal of peripheral nerve injury-induced hypersensitivity in the postpartum period: Role of spinal oxytocin. *Anesthesiology, 118*(1), 152–159.

Hedayati, H., Parsons, J., & Crowther, C. A. (2003). Rectal analgesia for pain from perineal trauma following childbirth. *Cochrane Database of Systematic Reviews*, (3). Art. No.: CD003931.

Heesen, M., Klohr, S., Rossaint, R., Walters, M., Straube, S., & van de Velde, M. (2013). Insertion of an intrathecal catheter following accidental dural puncture: A meta-analysis. *International Journal of Obstetric Anesthesia, 22*, 26–30.

Hoshiyama, E., Tatsumoto, M., Iwanami, H., Akihiro, S., Watanabe, H., Inaba, N., & Hirata, K. (2012). Postpartum migraine: A long-term prospective study. *Internal Medicine, 51*, 3119–3123.

International Headache Society. (2004). The international classification of headache disorders, 2nd ed. *Cephalalgia, 24*(Suppl. 1), S1–S151.

Kainu, J. P., Sarvela, J., Tippana, E., Halmesmaki, E., & Korttila, K. T. (2010). Persistent pain after caesarean section and vaginal birth: A cohort study. *International Journal of Obstetric Anesthesia, 19*, 4–9.

Kanakaris, N. K., Roberts, C. S., & Giannoudis, P. V. (2011). Pregnancy-related pelvic girdle pain: An update. *BMC Medicine, 9*, 15.

Kehlet, H., Jensen, T. S., & Woolf, C. J. (2006). Persistent postsurgical pain: Risk factors and prevention. *Lancet, 367*, 1618–1625.

Kettle, C., Dowswell, T., & Ismail, K. M. K. (2010). Absorbable suture materials for primary repair of episiotomy and second degree tears. *Cochrane Database of Systematic Reviews*, (6). Art. No.: CD000006.

Kettle, C., Dowswell, T., & Ismail, K. M. K. (2012). Continuous and interrupted suturing techniques for repair of episiotomy or second-degree tears. *Cochrane Database of Systematic Reviews*, (11). Art. No.: CD000947.

Klein, A. M., & Loder, E. (2010). Postpartum headache. *International Journal of Obstetric Anesthesia, 19*, 422–430.

Loughnan, B. A., Carli, F., Romney, M., Dore, C. J., & Gordon, H. (2002). Epidural analgesia and backache: A randomized controlled comparison with intramuscular meperidine for analgesia during labour. *British Journal of Anaesthesia, 89*, 466–472.

Luijendijk, R. W., Jeekel, J., & Storm, R. K. (1997). The low transverse pfannenstiel incision and the prevalence of incisional hernia and nerve entrapment. *Annals of Surgery, 225*, 365–369.

Macarthur, A. J., & Macarthur, C. (2004). Incidence, severity and determinants of perineal pain after vaginal delivery: A prospective cohort study. *American Journal of Obstetrics and Gynecology, 191*, 1199–1204.

Nafisi, S. (2007). Influence of uterine exteriorization versus in situ repair on post-cesarean maternal pain: A randomized trial. *International Journal of Obstetric Anesthesia, 16*, 135–138.

Nikolajsen, L., Sørensen, H. C., Jensen, T. S., & Kehlet, H. (2004). Chronic pain following caesarean section. *Acta Anaesthesiologica Scandinavica, 48*, 111–116.

Paterson, L. Q. P., Davis, S. N. P., Khalife, S., Amsel, R., & Binik, Y. M. (2009). Persistent genital and pelvic pain after childbirth. *The Journal of Sexual Medicine, 6*, 215–221.

Russell, R., Dundas, R., & Reynolds, F. (1996). Long term backache after childbirth: Prospective search for causative factors. *BMJ, 312*, 1384–1388.

Sances, G., Granella, F., & Nappi, R. E. (2003). Course of migraine during pregnancy and postpartum: A prospective study. *Cephalalgia, 23*, 197–205.

Sprigge, J. S., & Harper, S. J. (2008). Accidental dural puncture and post dural puncture headache in obstetric anaesthesia: Presentation and management: A 23-year survey in a district general hospital. *Anaesthesia, 63*, 36–43.

Stomp-van den Berg, S. G. M., Hendriksen, I. J. M., Bruinvels, D. J., Twisk, J. W. R., van Mechelen, W., & van Poppel, M. N. M. (2012). Predictors for post-partum pelvic girdle pain in working women: The Mom@work cohort study. *Pain, 153*, 2370–2379.

Turner, D., Smitherman, T. A., Eisenach, J. C., Penzien, D. B., & Houle, T. T. (2012). Predictors of headache before, during and after pregnancy: A cohort study. *Headache, 52*, 348–362.

Vleeming, A., Albert, H. B., Östgaard, H. C., Stuge, B., & Sturesson, B. (2007). European guidelines for the diagnosis and treatment of pelvic girdle pain. *European Spine Journal, 17*, 794–819.

Post-seizure Headache

Hans-Christoph Diener[1] and Miguel J. A. Lainez-Andres[2]
[1]Department of Neurology, University Hospital Essen, University Duisburg-Essen, Essen, Germany
[2]Department of Neurology, University of Valencia, Valencia, Spain

Synonyms

Postictal headache; Seizure-associated headache

Definition

Headache begins after the seizure. The temporal relationship between seizure and headache is not clearly defined. A time interval of 1 h has recently been suggested (Leniger et al. 2001). In cases in which headache is the sole ictal epileptic manifestation is called ictal epileptic headache (Belcastro et al. 2011).

Characteristics

Epidemiology and Clinical Features

The comorbidity of epilepsy and migraine is well established. The frequency of epilepsy among patients with migraine is higher than in the general population, and the incidence of migraine is higher in patients with epilepsy

(Lipton et al. 1994). The incidence of post-seizure headache (PSH) varies between 34 % and 43 % (Förderreuther et al. 2002; Leniger et al. 2001). Higher PSH frequencies of up to 51 % reported by Schon and Blau (1987) cannot be compared because the IHS criteria were not applied in that study. Of patients with PSH, 60–66 % experienced seizures, always combined with headache (Förderreuther et al. 2002; Leniger et al. 2001). In 62.5 % of 47 patients with PSH, the headache lasted longer than 4 h (Förderreuther et al. 2002). Leniger et al. (2001) reported an average of 18.6 h duration and an average pain intensity of 6.1 on the visual analogue scale in 115 patients with PSH. Thus, PSH is a frequent, long-lasting, and severe symptom accompanying epileptic seizures. Phonophobia, photophobia, hemicrania, throbbing pain quality, and nausea are typical symptoms observed in 41–72 % of patients with PSH (Leniger et al. 2001). The pathophysiology of the two conditions was discussed in a review paper by Papetti et al. (2013)

Subtypes of PSH

According to the IHS criteria, PSH could be predominantly classified as a migraine-like headache in 34–56 % of patients, followed by tension-type headache in 34–37 % of patients (Förderreuther et al. 2002; Leniger et al. 2001). PSH could not be classified in 8–21 % of patients (Förderreuther et al. 2002; Leniger et al. 2001). The PSH subtypes were not significantly associated with a certain seizure type or an epilepsy syndrome (focal versus generalized epilepsy syndrome) (Leniger et al. 2001).

Migraine-Like PSH

A history of migraine was significantly associated with the occurrence of migraine-like PSH (Förderreuther et al. 2002; Leniger et al. 2001). So far, the linkage of migraine headache and epileptic seizures has not been fully understood. It may be unspecific or based on the pathophysiological mechanism of ▶ spreading depression. Next to the predominance of migraine-like headache in PSH, epidemiological data revealed an increased prevalence of (seizure independent) migraine in patients with epilepsy and vice versa (Andermann and Andermann 1987; Leniger and Diener 1999; Ottmann and Lipton 1994). If spreading depression plays a pivotal role in these comorbid conditions, migraine with aura may be overrepresented. However, only a minority of patients with migraine-like PSH (16 %) complained about a migraine headache with aura symptoms (Leniger et al. 2001). In contrast, patients with migraine attacks occurring independently of epileptic seizures were found to have significantly more migraine with aura as compared to patients with migraine alone (Leniger et al. 2003b). The latter finding supports the hypothesis of spreading depression as an underlying substrate in comorbid conditions. It can be speculated that in patients with migraine-like PSH, the seizure itself represents or covers the aura symptoms, so that a decreased frequency of migraine-like PSH with aura symptoms can be observed (Leniger et al. 2003). Thus, the predominance of migraine-like PSH is not associated with a specific epilepsy syndrome or seizure type, but the linkage of migraine and epilepsy may be based on spreading depression as an expression of altered brain state with ▶ neuronal hyperexcitability (Leniger et al. 2003).

Therapy

There are no recommendations for the treatment of PSH. Förderreuther et al. (2002) reported a self-medication with acetylsalicylic acid and nonsteroidal antirheumatics in 30 % of their patients with PSH. Triptans may have good efficacy in migraine-like PSH (Förderreuther et al. 2002; Jacob et al. 1996; Ogunyemi and Adams 1993). A number of ▶ anticonvulsant (agent) (e.g., valproate, gabapentin, and topiramate) have demonstrated some efficacy in migraine prophylaxis (Freitag et al. 2002; Hering and Kuritzky 1992; Klapper et al. 1997; Jensen et al. 1994; Mathew et al. 1995; Mathew et al. 2001; Mathew et al. 2002; Von Seggern et al. 2002). However, there are no data showing that these anticonvulsants are also effective in the treatment of migraine-like PSH.

References

Andermann, E., & Andermann, F. (1987). Migraine-epilepsy relationships: Epidemiological and genetic aspects. In F. Andermann & E. Lugaresi (Eds.), *Migraine and epilepsy* (pp. 281–291). Boston: Butterworths.

Belcastro, V., Striano, P., Kasteleijn-Nolst Trenite, D. G., Villa, M. P., & Parisi, P. (2011). Migralepsy, hemicrania epileptica, post-ictal headache and "ictal epileptic headache": A proposal for terminology and classification revision. *The Journal of Headache and Pain, 12*(3), 289–294. Epub 2011/03/02.

Förderreuther, S., Henkel, A., Noachtar, S., et al. (2002). Headache associated with epileptic seizures: Epidemiology and clinical characteristics. *Headache, 42,* 649–655.

Freitag, F. G., Collins, S. D., Carlson, H. A., et al. (2002). A randomized trial of divalproex sodium extended-release tablets in migraine prophylaxis. *Neurology, 58,* 1652–1659.

Hering, R., & Kuritzky, A. (1992). Sodium valproate in the prophylactic treatment of migraine: A double-blind study versus placebo. *Cephalalgia, 12,* 81–84.

Jacob, J., Goadsby, P. J., & Duncan, J. S. (1996). Use of sumatriptan in post-ictal migraine headache. *Neurology, 47,* 1104.

Jensen, R., Brinck, T., & Olesen, J. (1994). Sodium valproate has a prophylactic effect in migraine without aura: A triple-blind, placebo-crossover study. *Neurology, 44,* 647–651.

Klapper, J., & On behalf of the Divalproex Sodium in Migraine Prophylaxis Study Group. (1997). Divalproex sodium in migraine prophylaxis: A dose-controlled study. *Cephalalgia, 17,* 103–108.

Leniger, T., & Diener, H. C. (1999). Migräne und Epilepsie – ein Zusammenhang? *Akt Neurol, 26,* 116–120.

Leniger, T., Isbruch, K., von den Driesch, S., et al. (2001). Seizure-associated headache in epilepsy. *Epilepsia, 42,* 1176–1179.

Leniger, T., Diener, H. C., & Hufnagel, A. (2003a). Erhöhte cerebrale Erregbarkeit und Spreading depression – Ursachen für eine Komorbidität von Epilepsie und Migräne? *Nervenarzt, 74,* 869–874.

Leniger, T., von den Driesch, S., Isbruch, K., et al. (2003b). Clinical characteristics of patients with comorbidity of migraine and epilepsy. *Headache, 43,* 672–677.

Lipton, R. B., Ottman, R., Ehrenberg, B. L., & Hauser, W. A. (1994). Comorbidity of migraine: The connection between migraine and epilepsy. *Neurology, 44,* 28–32.

Mathew, N. T., Saper, J. R., Silberstein, S. D., et al. (1995). Migraine prophylaxis with divalproex. *Archives of Neurology, 52,* 281–286.

Mathew, N. T., Rapoport, A., Saper, J., et al. (2001). Efficacy of gabapentin in migraine prophylaxis. *Headache, 41,* 119–128.

Mathew, N. T., Kailasam, J., & Meadors, L. (2002). Prophylaxis of migraine, transformed migraine, and cluster headache with topiramate. *Headache, 42,* 796–803.

Ogunyemi, A., & Adams, D. (1993). Migraine-like symptoms triggered by occipital lone seizures: Response to sumatriptan. *Canadian Journal of Neurological Sciences, 25,* 151–153.

Ottmann, R., & Lipton, R. B. (1994). Comorbidity of migraine and epilepsy. *Neurology, 144,* 2105–2110.

Papetti, L., Nicita, F., Parisi, P., Spalice, A., Villa, M. P., Kasteleijn-Nolst Trenite, D. G. (2013). "Headache and epilepsy" – How are they connected? *Epilepsy and Behavior, 26*(3), 386–393.

Schon, F., & Blau, N. J. (1987). Post-epileptic headache and migraine. *Journal of Neurology, Neurosurgery, and Psychiatry, 50,* 1148–1152.

Von Seggern, R. L., Mannix, L. K., & Adelman, J. U. (2002). Efficacy of topiramate in migraine prophylaxis: A retrospective chart analysis. *Headache, 42,* 804–809.

Poststroke Central Pain

Definition

A ▶ central pain condition caused by a stroke. Central pain is a neurological condition caused by a lesion or disease of the central somatosensory nervous system. See also ▶ central poststroke pain.

Cross-References

▶ Thalamic Plasticity and Chronic Pain

Poststroke Pain Model, Cortical Pain (Injection of Picrotoxin)

Jean-Louis Oliveras
Inserm, Paris, France

Definition

Producing "pain-like" behavioral reactions in animals in which the cortical GABAergic

control, a powerful inhibitory process potentially disturbed in various poststroke situations including epilepsy, is removed.

Characteristics

After microinjection of picrotoxin (2 μg/ 0.2 μl/3 min) at somato-motor (SmI) cortical level in the rat (rear paw region) (Fig. 1), remarkable reproducible and specific behavioral reactions do occur, for at least 1 h, together with cortical events, such as electrocorticogram (ECoG) continuous high amplitude spike and waves complexes, and less frequent and irregular long duration bursts.

These "pain-like" manifestations, observed contralaterally to the microinjection site, are not in phase with the ECoG spikes and bursts (Fig. 2), except the "tremor" reaction (see below), and consist of the following:

1. Lifting off the floor, fast withdrawal of the paw towards the abdomen, followed by placing the paw onto the floor, after a variable time from very fast to several seconds, in which the paw stays retracted.
2. Licking the paw, palm, and digits, varying from very briefly to a few seconds, sometimes ending with biting the tip of the digits.
3. "Turn-in paw," which consists of turning the foot in a supine position on the floor or against

Poststroke Pain Model, Cortical Pain (Injection of Picrotoxin), Fig. 1 Cortical localization of the effective "pain-like" reactions producing sites in the rat after picrotoxin microinjection. Relative to the vibrissae region (*open circles*), which provoke minor effects, *black circles*, mainly located to layer VI, indicate "pain-like" reactions restricted to the hind paw

Poststroke Pain Model, Cortical Pain (Injection of Picrotoxin), Fig. 2 ECoG isolated spikes and behavioral manifestations after microinjection of picrotoxin into SmI. From this example without any burst (see text), it is clear that the behavioral reactions are not in phase with the ECoG, even for the "lifting." The *triangle* indicates biting of the tip of the digits (the symbol "Cx 32" designates the animal number)

the flank, frequently grasping the digits inside the palm and staying in this posture for a while.

4. Tremor, probably another manifestation of seizure, as very fast trembling of the paw or the whole limb, sometimes associated with lifting, and particularly noticeable in the digits. This response can be prolonged for as long as several seconds.

In addition to these primary reactions, there are other manifestations, such as "limping" and "shaking," which are invariably associated with lifting or tremor. Episodes of "neglected paw" are frequently noted, as the animal suddenly loses its paw and limb placement reflexes, or is unable to use them for walking properly, or lying on the ipsilateral side with the "paw on the tail," or intense rearing and urination. Vocalizations are rarely noted.

These "pain-like" manifestations resemble those observed after formalin administration into the paw (Abbott et al. 1995; Olivéras and Montagne-Clavel 1996) and are not apparently due to indirect "peripheral effect" (Montagne-Clavel 1996), i.e., as possible cramps provoked by excessive cortical centrifugal motor excitation (see, e.g., Richardson 1987). Except the tremor, all these reactions can be quantitatively modulated by pharmacological agents directly administered into the rat SmI cortex, such as naloxone,

a specific opiate antagonist (Richardson 1987) or cholinergic antagonists (Montagne-Clavel and Olivéras 1997).

All these observations serve as a simple "model" for poststroke central pain, especially in the case of seizure, "possibly painful" (Kryzhanovski et al. 1992; Lenz et al. 1989; Mauguière and Gourgeon 1978; Nair et al. 2001; Scholtz et al. 1999; Wilkinson 1973; Yamashiro et al. 1991), and strongly emphasize potent common deficit mechanisms between central pain and epilepsy, such as a loss of somatosensory and somato-motor cortical GABA control (Dykes et al. 1984, Li and Prince 2002). Technically, this model requires simple procedures, such as classical microinjection technique and ECoG chronic recordings in awakened animals, and a good sense of observation (see details in Olivéras and Montagne-Clavel 1996).

References

Abbott, F. V., Franklin, K. B. J., & Westbrook, R. F. (1995). The formalin test: Scoring properties of the first and second phases of the pain responses in rats. *Pain, 60*, 91–102.

Dykes, R. W., Landry, P., Metherate, R., & Hicks, T. P. (1984). Functional role of GABA in the cat primary somato-sensory cortex: Shaping receptive fields of cortical neurons. *Journal of Neurophysiology, 52*, 1066–1093.

Kryzhanovski, G. N., Reshetnyak, V. K., Igonkina, S. I., & Zinkevich, V. A. (1992). Epileptiform activity in the somato-sensory cortex of rats with trigeminal neuralgia. *Pathol Physiol Gen Pathol, 114*, 126–128.

Lenz, F. A., Kwan, H. C., Dostrovsky, J. O., & Tasker, R. R. (1989). Characteristics of the bursting pattern of action that occurs in the thalamus of patients with central pain. *Brain Research, 496*, 357–360.

Li, H., & Prince, D. A. (2002). Synaptic activity in chronically injured, epileptogenic sensory-motor neocortex. *Journal of Neurophysiology, 88*, 2–12.

Mauguière, F., & Gourgeon, J. (1978). Somato-sensory epilepsy: A review of 127 cases. *Brain, 101*, 307–332.

Montagne-Clavel, J., & Olivéras, J.-L. (1997). Cholinergic modulation of the picrotoxin-induced "pain-like" symptoms at somato-motor cortical level in the rat. *Experimental Brain Research, 117*, 362–368.

Nair, D. R., Najm, I., Bulacio, J., & Luders, H. (2001). Painful auras in focal epilepsy. *Neurology, 57*, 700–7002.

Olivéras, J.-L., & Montagne-Clavel, J. (1996). Picrotoxin produces a "central" pain-like syndrome when microinjected into the somato-motor cortex of the rat. *Physiology and Behavior, 60*, 1425–1434.

Richardson, D. E. (1987). Does epileptic pain really exist? *Applied Neurophysiology, 50*, 365–368.

Scholtz, J., Vieregge, P., & Moser, A. (1999). Central pain as a manifestation of partial epileptic seizures. *Pain, 80*, 445–450.

Wilkinson, H. A. (1973). Epileptic pain, an uncommon manifestation with localizing value. *Neurology, 23*, 518.

Yamashiro, K., Iwayama, K., Kurihara, M., Mori, K., Tasker, R. R., & Albe-fessard, D. (1991). Neurons with epileptiform discharge in the central nervous system and chronic pain. *Acta Neurochirurgica Supplement (Wien), 52*, 130–132.

Zilles, K. (1985). *The cortex of the rat: A stereotaxic atlas.* Berlin: Springer.

Poststroke Pain Model, Thalamic Pain (Lesion)

Nayef E. Saadé and Suhayl J. Jabbur
Anatomy, Cell Biology and Physiology, Faculty of Medicine, American University of Beirut, Beirut, Lebanon

Definition

Blockage of the blood supply to a brain center, as with other organs, often leads to cell death in the area due to asphyxia. This pathology in the brain is commonly known as stroke. Patients suffering from stroke involving parts of the thalamus may complain from motor and sensory disturbances referred, in general, to the opposite side of the body (Tasker et al.1991). The most frequent and paradoxical manifestations are spontaneous pain, ► allodynia (painful sensation induced by innocuous or nonpainful stimulation), and ► hyperalgesia (increased reactivity to a painful stimulus). This pathology was first referred to as the "thalamic syndrome" (Dejerine and Roussy 1906). This syndrome was later shown to be one aspect of a pathological entity known as ► central pain, resulting from injuries or trauma at various levels of the central nervous system (Bowsher 1996). Several observations based on clinical and/or animal experimentation shed more light on the origin of this syndrome.

Characteristics

Roussy (1907) was among the first to attempt to reproduce the thalamic syndrome pathology in experimental animals, including monkeys, dogs, and cats. He defined two main approaches, the open, bloody method (méthode sanglante) and the blind method (méthode aveugle) (Roussy 1907). The first was based on a surgical approach to produce lesions in the posterior thalamus through the roof of the third ventricle of the brain. The second (blind) was based on an electrical lesion through electrodes placed in the brain (as illustrated in Fig. 1).

About a century after the studies of Roussy, modern techniques for placement of controlled lesions in the thalamus still use the blind method, but with more refined tools. Based on the established three-dimensional (stereotaxic, see ► stereotaxy) coordinates of an animal's brain and the use of the Horsley-Clark apparatus (► Horsley-Clarke Apparatus or Stereotaxic Frame), punctate and size-controlled lesions can be placed in different parts of the thalamus as in other areas of the brain.

Figure 2a represents the penetration points on the dorsal aspect of the skull to reach the different nuclear groups of the rat thalamus, based on

Poststroke Pain Model, Thalamic Pain (Lesion), Fig. 1 Blind lesion from the original drawings by Roussy (1907). These show the bipolar platinum electrodes (**a**), their presumed trajectory in the brain (**b**), and a transverse brain section illustrating the extent of the lesion (**c**)

Poststroke Pain Model, Thalamic Pain (Lesion), Fig. 2 Stereotaxic placement of lesions in the thalamus. Panel (**a**) shows a dorsal view of the exposed surface of the skull illustrating the penetration points used to produce lesions in the thalamus. These points correspond to the coordinates taken from Paxinos and Watson's Atlas (Paxinos and Watson 1986). Panel (**b**) represents a transverse section of the rat's brain illustrating a lesion in the lateral thalamus induced by kainic acid injection. Panel (**c**) shows another section in a different rat illustrating an electrolytic lesion placed in the medial thalamus. Both lesions are indicated by *arrows*

data extracted from the Paxinos and Watson atlas (Paxinos and Watson 1986). Lesions can be placed either in the lateral (posterior to bregma −2 to −4 mm, lateral 2.5–3.5 mm, vertical 5.25–6.25 mm) or the medial (posterior to bregma −1.5 to −4 mm, lateral 0.2–0.75 mm, vertical 5.25–6.25 mm) thalamic nuclear groups by using either electrical current or chemical substances (Fig. 2b and c). For this purpose, the head of the anesthetized rat is placed in the stereotaxic frame, the bony skull is exposed by skin incision, and small holes are drilled in the bone to allow the penetration of electrodes or pipettes (Fig. 2).

For the electrolytic lesion, coaxial bipolar metal electrodes, insulated except for about 50 μm from their tips (inter-tip distance 0.1–0.2 mm) and fixed on a micromanipulator are introduced stereotaxically into a nucleus of the thalamus. Constant current (20–30 μA) is injected for a period of 20–50 s. This current has been shown to produce cellular death by electrolysis. The volume and extent of the lesion depend on the duration and the intensity of the injected current. For lesions that involve the rostrocaudal extent of a thalamic nuclear group, more than one penetration and current injection are needed (Saadé et al. 1999). This technique can produce graded lesions in different parts of the thalamus and allow for the observation of their effects on the general sensorimotor behavior of the animal. However, it presents the inconvenience of nonselective destruction of local neurons in addition to passing fibers in the area that are not necessarily involved in the activities of the lesioned center.

A more selective local lesion, restricted to neuronal cell bodies, is based on the injection of excitatory amino acids, such as kainic or ibotenic acids. Excitotoxic lesion of neuronal cell bodies is attributed to the excessive activation of neurons leading to the accumulation of calcium ions inside the cells (the mitochondria, in particular) that leads to ▶ osmolysis (Schinder et al. 1996; Wall et al. 1988). Solutions of ibotenic (0.03 M) or kainic acid (0.01 M) are injected through a micropipette (tip diameter 50–75 μm) or a Hamilton syringe (5 μl) fixed to a microinjector and placed stereotaxically in a thalamic nucleus. The injection is performed slowly over a period of 2–3 min using a manual or motorized microinjector. The size and the extent of the lesion depend on the injected volume (0.25–0.75 μl) and the location and number of injections in a part (lateral or medial) of the thalamus. The initial excitatory effects of the injections are not, in general, observed, since they occur during a short period following the injection when the animal is under deep anesthesia. In both types (electrolytic or chemical) of lesions, however, the stable effects are observed 2–3 days after the recovery of the animal from anesthesia and the acute effects of surgery. These effects can be assessed by observing the variations in the thresholds of the acute phasic pain tests (paw pressure, paw withdrawal, tail flick, hot plate, etc.) or the temporal evolution of the formalin test considered as a model for acute tonic pain (Saadé et al. 1999; Saadé et al. 1997). The effects of either lateral or medial thalamic lesions in a rat model of mononeuropathy are also under investigation.

A rat model of intracerebral hemorrhage, based on the injection of the proteolytic enzyme of collagenase, has been described by Rosenberg et al. in 1990. This consists in injecting few microliters of the enzyme that can disrupt the collagen of the basal lamina of the brain blood vessels. This results in extensive bleeding and brain edema that develop over few hours, leading to necrotic permanent changes. The size of stroke depends on the volume and concentration of the injected collagenase. This model has been used recently by Wasserman and Koeberle (2009) to produce hemorrhagic stroke in the ventrobasal complex of the rat thalamus. Rats subjected to thalamic hemorrhage elicited hypersensitivity to mechanical and thermal stimuli, applied to the contralateral limb, within 1 week and over a period of several weeks.

Another method for experimental block consists in reversible and selective anesthetic-block of various thalamic nuclear groups by injecting small volumes of 0.5–1 μl of lidocaine through stereotaxically implanted cannulas. Long-lasting anesthesia can be obtained by implanting miniosmotic pumps that can deliver lidocaine at a rate of 0.5 μl per hour over a period of several days. The effects of such selective block can be observed on the various tests of nociception and/or on allodynia and hyperalgesia in a rat model of mononeuropathy (Saadé et al. 2006). This model presents the advantages of producing selective and reversible block of a part of the thalamus, but cannot be compared to pathological situations resulting from ischemic damage or injury to the thalamus.

Finally, it is worth noting that other techniques such as surgical ablation or embolization of the arterial supply to the thalamus are still practiced in experimental animals. The extent of the resulting lesion, however, is not as well circumscribed as those of localized electrolytic or chemical lesions.

References

Bowsher, D. (1996). Central pain: Clinical and physiological characteristics. *Journal of Neurology, Neurosurgery & Psychiatry, 61*, 69.

Dejerine, J., & Roussy, G. (1906). Le syndrome thalamique. *Revista de Neurologia, 14*, 521–532.

Paxinos, G., & Watson, C. (1986). *The rat brain in stereotaxic coordinates*. London: Academic.

Roussy, G. (1907). *La Couche Optique*. Steinheil: Le syndrome thalamique.

Saadé, N. E., Atweh, S. F., Bahuth, N., et al. (1997). Augmentation of nociceptive reflexes and chronic deafferentation pain by chemical lesions of either striatal dopaminergic terminals or midbrain dopaminergic neurons. *Brain Research, 751*, 1–12.

Saadé, N. E., Kafrouni, A. I., Saab, C. Y., et al. (1999). Chronic thalamotomy increases pain-related behavior in rats. *Pain, 83*, 401–409.

Saadé, N. E., Al Amin, H., AbdelBaki, S., Safieh-Garabedian, B., Atweh, S. F., & Jabbur, S. J. (2006). Transient attenuation of neuropathic manifestations in rats following lesion or reversible block of the lateral thalamic somatosensory nuclei. *Experimental Neurology, 197*, 157–166.

Schinder, A. F., Olson, E. C., Spitzer, N. C., et al. (1996). Mitochondrial dysfunction is a primary event in glutamate neurotoxicity. *Journal of Neuroscience, 16*, 6125–6133.

Tasker, R. R., Decarvalho, G., & Dostrovsky, J. O. (1991). The history of central pain syndromes, with observations concerning pathophysiology and treatment. In K. L. Casey (Ed.), *Pain and central nervous system disease: The central pain syndromes* (pp. 31–58). New York: Raven.

Wall, P. D., Bery, J., & Saadé, N. E. (1988). Effects of lesion to rat spinal cord lamina cell projection pathways on reactions to acute and chronic noxious stimuli. *Pain, 35*, 327–339.

Wasserman, J. K., & Koeberle, P. D. (2009). Development and characterization of a hemorrhagic rat model of central post-stroke pain. *Neuroscience, 161*(1), 173–183.

Postsurgical Nerve Injury

▶ Iatrogenic Causes of Neuropathy

Postsynaptic Dorsal Column Neurons

Synonyms

PSDC

Definition

Postsynaptic dorsal column neurons are cells that are found predominantly in lamina III–V of the spinal cord, projecting to the dorsal column nuclei. While cells in the lumbosacral enlargement project to the gracile nucleus (GN), those in the cervical enlargement project to the cuneate nucleus. PSDC neurons receive convergent inputs from high-threshold muscle, skin, and joint afferents, as well as low-threshold muscle and cutaneous afferents. Some receive visceral input.

Cross-References

▶ Nociceptive Circuitry in the Spinal Cord
▶ Opioids in the Spinal Cord and Modulation of Ascending Pathways (N. gracilis)
▶ Postsynaptic Dorsal Column Projection, Functional Characteristics
▶ Spinal Ascending Pathways, Colon, Urinary Bladder, and Uterus
▶ Spinal Dorsal Horn Pathways, Dorsal Column (Visceral)

Postsynaptic Dorsal Column Neurons, Responses to Visceral Input

Elie D. Al-Chaer
Center for Pain Research, University of Arkansas for Medical Sciences, Little Rock, AR, USA

Synonyms

Dorsal horn neurons; Posterior column neurons; Second-order neurons; Visceral nociceptive tracts

Definition

A postsynaptic dorsal column neuron is a spinal cord neuron whose cell body and dendrites are located in the dorsal horn or central gray matter of the spinal cord, and whose axon projects in the dorsal column to terminate in the dorsal column nuclei of the medulla oblongata, usually on the side ipsilateral to the cell body. The axons of some postsynaptic dorsal column neurons convey visceral nociceptive information through the dorsal column to the dorsal column nuclei. Postsynaptic dorsal column neurons can be identified electrophysiologically by antidromic stimulation of the dorsal column or the dorsal column nuclei, or histologically by injecting a retrograde tracer into the dorsal column nuclei.

Characteristics

Postsynaptic dorsal column neurons are a class of sensory projection neurons. They form the post-synaptic component of the dorsal column, one of several different sensory pathways that transmit signals of touch, ▶ kinesthesia, and two-point discrimination from the level of the spinal cord to the brain. Evolving clinical and experimental evidence has confirmed a role of the dorsal column in visceral pain that involves largely postsynaptic dorsal column neurons (Al-Chaer et al. 1996a, b; 1999; Hirshberg et al. 1996). The nociceptive signals depend on input to the spinal cord from peripheral nociceptors in the viscera. The signals are processed in the spinal cord and are then transmitted to the ipsilateral dorsal column nuclei (Al-Chaer et al. 1996b, 1997a) and, after a synaptic relay, to the contralateral ventrobasal complex of the thalamus (Al-Chaer et al. 1996a, 1998). Clinical observations have shown that a ▶ midline myelotomy, severing axons in the human posterior columns, can help attenuate otherwise intractable visceral pain (Hirshberg et al. 1996), whereas stimulation of the posterior columns in patients with severe ▶ irritable bowel syndrome produces immediate increased intensity of abdominal pain (Malcolm et al. 2001). These observations concur with experimental data obtained in animals, which show that severing the axons of the dorsal columns in rats or monkeys reduces the responses to ▶ colorectal distension (CRD) of neurons in the ventral posterolateral nucleus of the thalamus, and of neurons in the dorsal column nuclei,

Postsynaptic Dorsal Column Neurons, Responses to Visceral Input, Fig. 1 In (**a**) are shown the responses of a PSDC cell to cutaneous stimuli as well as the blood pressure traces obtained during stimulation. The cell was located in the dorsal commissural region at S1 (**e**), was antidromically activated from the nucleus gracilis (**d**) and had a cutaneous receptive field over the rump area (**c**). (**b**) shows traces of antidromic activation (*left*), collision (middle trace; *dot* indicates expected time of antidromic spike), and high frequency following (*right*). (**f**) shows the responses of the cells to CRD and the corresponding changes in blood pressure (top traces). (**g**) shows a trace of the cellular spike

particularly in nucleus gracilis (Al-Chaer et al. 1996a, b, 1997a, 1998).

The dorsal column nuclei are known to receive both a direct sensory input from primary afferent fibers (Kuo and De Groat 1985; Giuffrida and Rustioni 1992), and an indirect input from a large number of postsynaptic dorsal column neurons located in the ipsilateral dorsal horn and the central gray matter of the spinal cord. The cells of origin of the postsynaptic dorsal column pathway were mapped at different levels of the spinal cord using retrograde (Giesler et al. 1984), anterograde (Cliffer and Giesler 1989), and degeneration tracing techniques (Rustioni 1973). The projections are somatotopic; the cuneate nucleus receives more projections from the thoracic and cervical segments of the cord and the gracile nucleus receives more projections from the lumbar and sacral segments (Campbell et al. 1974). Iontophoretic injections of an anterograde tracer (*Phaseolus vulgaris* leucoagglutinin, PHA-L) in the rat's sacral cord label dorsal column fibers near the midline terminating within the most medial part of the ipsilateral nucleus gracilis (Cliffer and Giesler 1989), whereas injections in the midthoracic spinal cord label fibers near the dorsal intermediate septum, and that terminate in the border area between the nucleus gracilis and the nucleus cuneatus (Wang et al. 1999). Investigation of viscerosensitive postsynaptic dorsal column

Postsynaptic Dorsal Column Neurons, Responses to Visceral Input, Fig. 2 (a) Location of PSDC cell in the dorsal commissural gray at S1. (b) Traces of antidromic activation of the cell from the cervical FG as well as of collision of the antidromic spike with orthodromic action potential (middle trace, *the dot* indicates the expected time of the antidromic spike). (c) Receptive field of the cell extending from perineal area to base of tail and upper aspect of hind limb. (d) and (e) Responses of the cell to cutaneous stimuli (d) and to graded CRD (e) before and after morphine, after naloxone and after CNQX

neurons in rats and monkeys, using retrograde labeling and ▶ Fos protein expression in response to visceral stimulation, showed them to be located mainly in the gray matter near the central canal (Lamina X) (Hirshberg et al. 1996). These observations concurred with the location of postsynaptic dorsal column neurons isolated electrophysiologically using antidromic activation.

The role of the postsynaptic dorsal column pathway in somatosensory processing is not clear, but evidence shows that postsynaptic dorsal column neurons receive nociceptive input from the viscera, and relay signals of visceral nociception to the dorsal column nuclei.

Microdialysis administration of morphine or of a non-N-methyl-D-aspartate antagonist (CNQX) into the sacral spinal cord, after restricting the afferent input from the colon onto the sacral cord to the pelvic nerves by sectioning the hypogastric nerves, can block visceral nociceptive responses recorded from neurons of the nucleus gracilis. The ability of morphine or CNQX to block the visceral responses in the nucleus gracilis is consistent with a synaptic action, presumably at synapses of viscerosensory afferents onto postsynaptic dorsal column neurons (Al-Chaer et al. 1996b).

Recordings from individual antidromically identified postsynaptic dorsal column neurons in rats showed them to be activated by graded intensities of colorectal distension spanning the innocuous and noxious range (Fig. 1), and confirmed that their responses to visceral stimuli could in fact be blocked by microdialysis of morphine or CNQX (Fig. 2). Furthermore, viscerosensitive postsynaptic dorsal column neurons in the sacral cord receive convergent somatic input from the skin over the hind paws, the hind legs, the rump, and the perineal area, and respond mainly to innocuous cutaneous stimuli, but also mildly to noxious cutaneous pinch (Al-Chaer et al. 1996b).

Responses of viscerosensitive postsynaptic dorsal column neurons to CRD sensitize following colon inflammation with mustard oil. The sensitization is characterized by a decrease in threshold and an increase in firing rate in response to CRD and an attenuation of the responses to cutaneous stimuli (Al-Chaer et al. 1997b).

References

Al-Chaer, E. D., Lawand, N. B., Westlund, K. N., & Willis, W. D. (1996a). Visceral nociceptive input into the ventral posterolateral nucleus of the thalamus: A new function of the dorsal column. *Journal of Neurophysiology, 76*, 2661–2674.

Al-Chaer, E. D., Lawand, N. B., Westlund, K. N., & Willis, W. D. (1996b). Pelvic visceral input into the nucleus gracilis is largely mediated by the postsynaptic dorsal column pathway. *Journal of Neurophysiology, 76*, 2675–2690.

Al-Chaer, E. D., Westlund, K. N., & Willis, W. D. (1997a). The nucleus gracilis: An integrator for visceral and somatic information. *Journal of Neurophysiology, 78*, 521–527.

Al-Chaer, E. D., Westlund, K. N., & Willis, W. D. (1997b). Effects of colon inflammation on the responses of postsynaptic dorsal column cells to visceral and cutaneous stimulation. *Neuro Report, 8*, 3267–3273.

Al-Chaer, E. D., Feng, Y., & Willis, W. D. (1998). A role for the dorsal column in nociceptive visceral input into the thalamus of primates. *Journal of Neurophysiology, 79*, 3143–3150.

Al-Chaer, E. D., Feng, Y., & Willis, W. D. (1999). A comparative study of viscerosomatic input onto postsynaptic dorsal column and spinothalamic tract neurons in the primate. *Journal of Neurophysiology, 82*, 1876–1882.

Campbell, S. K., Parker, T. D., & Welker, W. (1974). Somatotopic organization of the external cuneate nucleus in albino rats. *Brain Research, 77*, 1–23.

Cliffer, K. D., & Giesler, G. J. (1989). Postsynaptic dorsal column pathway of the rat. III. Distribution of ascending afferent fibers. *Journal of Neuroscience, 9*, 3146–3168.

Giesler, G. J., Nahin, R. L., & Madsen, A. M. (1984). Postsynaptic dorsal column pathway of the rat. I. Anatomical studies. *Journal of Neurophysiology, 51*, 260–275.

Giuffrida, R., & Rustioni, A. (1992). Dorsal root ganglion neurons projecting to the dorsal column nuclei of rats. *The Journal of Comparative Neurology, 316*, 206–220.

Hirshberg, R. M., Al-Chaer, E. D., Lawand, N. B., Westlund, K. N., & Willis, W. D. (1996). Is there a pathway in the dorsal funiculus that signals visceral pain? *Pain, 67*, 291–305.

Kuo, D. C., & De Groat, W. C. (1985). Primary afferent projections of the major splanchnic nerve to the spinal cord and the nucleus gracilis of the cat. *The Journal of Comparative Neurology, 231*, 421–434.

Malcolm, A., Phillips, S. F., Kellow, J. E., & Cousins, M. J. (2001). Direct clinical evidence for spinal hyperalgesia in a patient with irritable bowel syndrome. *American Journal of Gastroenterology, 96*, 2427–2431.

Rustioni, A. (1973). Non-primary afferents to the nucleus gracilis from the lumbar cord of the cat. *Brain Research, 51*, 81–95.

Wang, C. C., Willis, W. D., & Westlund, K. N. (1999). Ascending projections from the area of the spinal cord around the central canal: A *phaseolus vulgaris* leucoagglutinin study in rats. *The Journal of Comparative Neurology, 415*, 341–367.

Postsynaptic Dorsal Column Pathway (See Postsynaptic Dorsal Column Neurons)

▶ Spinal Ascending Pathways, Colon, Urinary Bladder, and Uterus

Postsynaptic Dorsal Column Projection, Anatomical Organization

Karin Westlund High
Department of Physiology, University of Kentucky, Lexington, KY, USA

Synonyms

Cutaneous mechanical sensation; Ipsilateral second order sensory relay; PSDC projection; Visceral pain

Definition

Postsynaptic dorsal column (PSDC) cells are second-order ▶ sensory relay neurons in the spinal cord. The PSDC cells receive cutaneous mechanical and/or noxious visceral information from primary afferent nerves in the periphery. The information is relayed by PSDC axons ascending uncrossed through the dorsal column that innervate the ▶ dorsal column nuclei in the caudal medulla. The largest numbers of PSDC cells are located in laminae III and IV, primarily transmitting information about innocuous cutaneous mechanical input. The PSDC cells located

in medial laminae III–VII and X transmit information about noxious input received from afferent fibers innervating inflamed or distended visceral organs.

Characteristics

The traditional view of the dorsal column as solely a direct pathway for primary afferent fiber transmission of discriminative cutaneous and deep sensory stimuli has been expanded in light of the many clinical and basic studies describing transmission of information about visceral pain through axons in the dorsal column midline. In addition to ▶ myelinated and unmyelinated primary afferent nerve fibers, the dorsal column is composed of second-order fibers arising from postsynaptic neurons in the spinal gray matter (Fig. 1). The long, ascending tracts in the dorsal column terminate in the dorsal column nuclei. The uncrossed dorsal column fibers are topographically arranged in both human and other animals. Second-order fibers arising from PSDC cells in lumbosacral levels of the spinal cord travel medially in the ▶ fasciculus gracilis to the ▶ gracile nucleus. Those arising from cervicothoracic levels of the spinal cord travel in the ▶ fasciculus cuneatus to terminate in the ▶ cuneate nucleus.

Postsynaptic Dorsal Column Projection in Animals

The existence of ascending non-primary afferent fibers in the dorsal columns was first demonstrated in cats (Petit 1973; Rustioni 1973) and has been well studied in several species. The information comes originally from both physiologic localization studies and from retrograde and anterograde labeling studies in rats (Rustioni et al. 1979; Bennett et al. 1983; Giesler et al. 1984; Enevoldson and Gordon 1989; Palecek et al. 2003a, b; Wang et al. 1999). The cells of origin of the second-order pathway terminating in the dorsal column nucleus are referred to as postsynaptic dorsal column neurons. These neurons have been described as being primarily localized in laminae III and IV of the spinal gray matter,

**Primary Afferent and Postsynaptic Dorsal Column
Projections in the Dorsal Column**

Nuc. gracilis

Nuc. cuneatus

Medulla

Spinal nuc of N. V

Neuron II (internal
arcuate fibers)

Dorsal root ganglion
cell Neuron I

Decussation of medial
lemniscus fibers

Pacinian
corpuscle   C VIII

Fasciculus cuneatus

Unencapsulated
joint receptor   T IV

Fasciculus gracilis

Golgi-Mazzoni
corpuscle   L III

Meissner's
corpuscle   S IV

* Postsynaptic Dorsal Column Neurons

**Postsynaptic Dorsal Column Projection, Anatomical
Organization, Fig. 1** Sensory information is transduced by specialized receptor endings in the periphery.
The information is relayed by primary afferent nerves
(1) directly to the dorsal column nuclei in the dorsal
midline of the caudal medulla or (2) to the spinal cord
gray matter. The cutaneous mechanical information is
sent through ascending projections in the dorsal
column from postsynaptic dorsal column cells (*stars*) in
laminae III and IV. Information about noxious input in
visceral structures is also relayed through the dorsal
column in projection fibers arising from postsynaptic
neurons in lamina X. Dorsal column projections are
somatotopically arranged with information from the
lower body situated medially in the dorsal column
while the upper body is represented in the lateral portions
of the dorsal column (Modified from Truex and
Carpenter 1969)

but PSDC neurons are also located medially in
deeper laminae V–VII and around the central
canal in lamina X (Giesler et al. 1984; Palecek
et al. 2003a, b). The PSDC neurons convey visceral pain information sending their axons along
the dorsal column midline (Al-Chaer et al. 1996;
Hirshberg et al. 1996; Wang et al. 1999; Willis
et al. 1999; Palecek et al. 2002).

**Laminae III and IV Postsynaptic Dorsal
Column Neurons**

Postsynaptic dorsal column neurons form a cell
column throughout the length of the spinal cord,
but the PSDC neurons in laminae III–IV are
localized in highest density in the cervical and
lumbar enlargements (Giesler et al. 1984;

Hirshberg et al. 1996; Palecek et al. 2003). Noxious somatic and visceral mechanical stimulation
induces Fos expression in the numerous laminae
III and IV postsynaptic dorsal column neurons.

PSDC neurons receive peripheral input about
noxious mechanosensation; however, transmission of this information is not believed to be the
major function of the postsynaptic dorsal column
pathway (Giesler and Cliffer 1985). For example,
in an unanesthetized, decerebrated, spinalized rat
preparation, 64 % of PSDC cells responded only
to innocuous mechanical stimuli, while only
36 % responded to noxious pinch. Rather, an
abundance of studies have shown that a major
function of the PSDC neurons is transmission of
visceral pain. This was demonstrated in terminal

cancer patients receiving midline dorsal column lesions (Gildenberg and Hirshberg 1984; Hirshberg et al. 1996; Nauta et al. 1997, 2000). Many subsequent clinical case reports and basic studies in monkeys and rats have confirmed this role for the PSDC pathway in the midline dorsal column (Al-Chaer et al. 1996; Wang et al. 1999; Willis et al. 1999; Vilela Filho et al. 2001; Becker et al. 2002; Hwang et al. 2004; Francisco et al. 2006). In anesthetized preparations, noxious visceral input activated PSDC neurons and the neurons in the medullary gracile nucleus where the ascending PSDC neurons terminate (Al-Chaer et al. 1996, 1997; Houghton et al. 2001). Eleven percent of the neurons in the dorsal column nuclei that were isolated using somatic inputs were excited and 7 % were inhibited.

Medial Laminae III–VII Postsynaptic Dorsal Column Neurons

Figure 2 summarizes what is now understood about the anatomy of the postsynaptic dorsal column pathway as it relates to visceral pain. Retrograde labeling of axons in the dorsal columns of the cervical spinal cord demonstrated the presence of a large population of postsynaptic dorsal column neurons in spinal cord lamina III–IV, as well as neurons scattered in the medial deep dorsal horn and the area around the central canal (Hirshberg et al. 1996; Palecek et al. 2002, 2003). It is known that visceral primary afferent fibers project to this region (Morgan et al. 1981; Neuhuber 1982; Kuo et al. 1983, 1984; Sugiura et al. 1989). After ureter distension, among the PSDC neurons identified by retrograde labeling from the dorsal column nuclei, those located medially in laminae III–VII (and X) expressed Fos, in spinal levels T4–L6, and a small percentage expressed NK-1 (Palecek et al. 2003). Axons of PSDC neurons in the medial lumbosacral cord traveled near the dorsal column midline and terminated in the medial gracile nucleus (Cliffer and Giesler 1989; Wang et al. 1999). Similarly, medial PSDC neurons in thoracic spinal cord sent their axons to terminal sites in cuneate nucleus and ascend in a position near the intermediate septum in the dorsal column (Wang et al. 1999). Thus, there is anatomic evidence of

Postsynaptic Dorsal Column Projection, Anatomical Organization, Fig. 2 Distribution of PSDC neurons (a) retrogradely labeled from the dorsal column nuclei with fluorescent-labeled dextran tracers. (b) Fos expression was evident in some of the PSDC cells after distention of the ureter. The retrogradely labeled PSDC neurons were found in laminae III–IV and in the vicinity of the central canal (Modified from Palecek et al. 2003)

a postsynaptic dorsal column pathway that originates from a part of the spinal cord gray matter that is known to contain neurons that respond to somatic and/or visceral stimuli (Nahin et al. 1983; Honda 1985; Honda and Perl 1985; Ness and Gebhardt 1987, 1988). Anterograde tracing has shown that other axons arising from PSDC neurons continue on to innervate the ventral brainstem, lateral pons, periaqueductal gray, and the medial and central lateral nuclei of the thalamus after small anterograde tracer injections (Wang et al. 1999).

Lamina X Postsynaptic Dorsal Column Neurons

Other postsynaptic dorsal column neurons are identified in lamina X around the central canal

a

b

Postsynaptic Dorsal Column Projection, Anatomical Organization, Fig. 3 (a) Distribution of physiologically characterized somatic cutaneous C-fiber afferent nerve terminal endings from guinea pig. Somatic terminations are discretely confined to laminae I and II. (b) Terminations of a single visceral C-afferent fiber. The highly branched terminal endings innervate the superficial dorsal horn broadly as well as deeper dorsal horn and in the vicinity of the central canal on both sides of the spinal cord. The nerve endings were identified after stimulating and filling the celiac ganglia (Modified from Sugiura et al. 1989)

that relay noxious visceral input to the dorsal column nuclei (Al-Chaer et al. 1996; Honda 1985; Honda and Lee 1985). While cutaneous primary afferents typically end discretely in lamina I, visceral afferent input is received throughout the dorsal horn, including lamina X around the central canal, and likely contributes to the diffuse character of visceral pain (Fig. 3) (Sugiura et al. 1989). Both somatic and visceral inputs have been shown functionally to converge on lamina X PSDC neurons and are then relayed to the dorsal column nuclei (Honda 1985; Honda and Lee 1985). The PSDC cells in lamina X have been shown to receive terminals of primary afferent fibers containing ▶ vasoactive intestinal polypeptide, a peptide found in abundance in visceral afferent fibers, as well as fibers containing somatostatin, substance P, leu-enkephalin, and serotonin (Honda and Lee 1985). While lamina X PSDC neurons typically remain silent during low-intensity visceral stimulation, responses to intense noxious visceral input evokes activity (Al-Chaer et al. 1997; Houghton et al. 2001). Inflammation of visceral organs increases Fos expression and background activity of neurons in PSDC. Fos expression and retrograde labeling from the gracile nucleus in PSDC neurons in the vicinity of the central canal are greatest in spinal levels receiving visceral inputs, including lower thoracic and sacral cord levels (Hirshberg et al. 1996; Willis et al. 1999; Palecek et al. 2002, 2003).

PSDC axons travel in the dorsal column on the same side, as opposed to the crossed spinothalamic tract (Fig. 4). The route taken by the fibers of lamina X postsynaptic neurons as they leave the gray matter in rats is to travel for several segments in the position predicted by earlier neurosurgical successes in patients with intractable pain. As they travel rostrally, the fibers are elevated dorsally along the midline septum by the addition of other postsynaptic dorsal column fibers that enter the dorsal column at each successive spinal segment. Those second-order PSDC axons receiving information about viscera in the lower body (pelvic) are pushed to the midline and eventually form a small wedge situated at the dorsal medial edge at upper cervical level to enter the dorsal column nuclei (Wang et al. 1996; Hirshberg et al. 1996). PSDC axons carrying information about thoracic viscera travel somatotopically along the intermediate septum between the gracile and cuneate fasciculi. Thoracic PSDC neurons transmit mechanical cardiac

P

Postsynaptic Dorsal Column Projection, Anatomical Organization, Fig. 4 Spinothalamic tract neurons in laminae I, IV–VII, and X primarily receive information about somatic pain from afferent fibers innervating the skin and deep tissues. The crossed spinothalamic tract axons travel in the dorsolateral and ventrolateral white matter. Information about visceral pain is conveyed by afferent fibers associated anatomically with the autonomic nervous system and terminates centrally in the dorsal horn or brainstem nucleus tractus solitarius (vagus nerve afferents). Non-visceral afferent fibers have their cell bodies located in dorsal root ganglia as do many visceral afferents. Because each internal organ receives input from two nerves, other visceral afferents have cell bodies in nodose and jugular ganglia (not illustrated). PSDC neurons in laminae III–VII and X receive information about visceral pain from visceral afferents with cell bodies in dorsal root ganglia. Their uncrossed axons ascend in the dorsal column midline to the dorsal column nuclei

information, but they play a minimal role in cardiac nociception (Goodman-Keiser et al. 2010).

Postsynaptic Dorsal Column Projection in Humans

Clinical reports primarily in the 1970s reported hundreds of neurosurgical cases in which unexpected and immediate postoperative relief of pain, primarily intractable cancer pain, was relieved after lesions in the central spinal cord. These lesions, designed to cut the crossing fibers of the spinothalamic tract segmentally (Figs. 4 and 5), were created by passing a fine surgical instrument longitudinally along the midline. The results were in most cases quite successful in providing widespread alleviation of pain with only a girdle of cutaneous ▶ analgesia and temperature loss, without the complications seen with anterolateral lesion of the spinothalamic tract or dorsal root section. It was never

previously understood why this type of lesion removed cancer pain in regions below the lesion as well as segmentally. Thus, the practice was mostly abandoned. Further refinement of the neurosurgical technique by some neurosurgeons, however, yielded remarkably positive results for pain relief below the level of the lesion (i.e., making a shallow punctuate lesion of the dorsal column axons immediately adjacent to the midline septum; Hitchcock 1970; Gildenberg and Hirshberg 1984; Nauta et al. 1997). The anatomical basis for pain relief in patients after punctate midline myelotomy was resolved based on animal studies reporting that visceral nociceptive information is transmitted by PSDC axons in the dorsal columns (Hirshberg et al. 1996; Wang et al. 1999; Willis et al. 1999). Surgical incisions in the spinal cord midline that had aimed either at crossing spinothalamic tract axons would necessarily

**VISCERAL PAIN INPUT PATHWAYS
IN SPINAL CORD**

Postsynaptic Dorsal
Column Pathway

To Thalamus

To Dorsal Column Nuclei

Spinothalamic
Tract

Anterior:
1. Commissure
2. Spinal Artery
3. Median Sulcus

1. 2. 3.

Punctate Midline
Myelotomy

Posterior
Spinal
Arteries

Postsynaptic Dorsal Column Projection, Anatomical Organization, Fig. 5 Interruption of the dorsal column midline relieves visceral pain. Postsynaptic dorsal column neurons in laminae III–VII and X receive information about visceral pain and send their uncrossed axons in the dorsal column midline to the dorsal column nuclei. Spinothalamic tract neurons in laminae I, IV–VII, and X primarily receive information about non-visceral, somatic pain and send their crossed axons to the thalamus. The single punctate surgical midline myelotomy disrupts the ascending axons of the postsynaptic dorsal column neurons below the lesion without other significant neurological effects. The histological sections from three patients confirmed the lesions in the dorsal column midline (See Nauta et al. 2000)

have passed through and eliminated the ascending postsynaptic dorsal column axons (Wang et al. 1999). As the axons travel rostrally, the fibers are elevated dorsally and pushed along the midline septum in the position of the surgical punctate midline myelotomy. The midline myelotomy has relieved pelvic cancer pain for many terminal cancer patients without other significant neurological effects (Nauta et al. 1997; Vilela Filho et al. 2001; Becker et al. 2002; Hwang et al. 2004; Francisco et al. 2006). Some terminal cancer patients with midline surgical lesions survived 3–4 years without return of their intractable visceral pain.

Kim and Kwon (2000) reported a series of high thoracic spinal cord lesions of which one was particularly successful. This lesion placed at the lateral edge of the gracile fasciculus at the dorsal root entry zone eliminated stomach cancer pain, while midline lesions were less effective in eliminating visceral pain arising from thoracic levels. Additional surgical studies are needed to clarify the effectiveness of a lesion placed off

midline between the gracile and cuneate nuclei, perhaps at the C7 segment where the septum is oriented vertically or at the dorsal root entry zone at the C8–T1 border. However, clinical reports imply that human spinal cord is organized similarly to that reported in animal studies (i.e., carrying visceral nociceptive information to the brain as a component of the postsynaptic dorsal column system).

References

Al-Chaer, E. D., Lawand, N. B., Westlund, K. N., et al. (1996a). Pelvic visceral input into the nucleus gracilis is largely mediated by the postsynaptic dorsal column pathway. *Journal of Neurophysiology, 76*, 2675–2690.

Al-Chaer, E. D., Westlund, K. N., & Willis, W. D. (1996b). Potentiation of thalamic responses to colorectal distension by visceral inflammation. *NeuroReport, 7*(1), 635–1639.

Al-Chaer, E. D., Westlund, K. N., & Willis, W. D. (1997). Sensitization of postsynaptic dorsal column neuronal responses by colon inflammation. *NeuroReport, 8*, 3267–3273.

Becker, R., Gatscher, S., Sure, U., & Bertalanffy, H. (2002). The punctate midline myelotomy concept for visceral cancer pain control–case report and review of the literature. *Acta Neurochirurgica. Supplement, 79,* 77–78.

Bennett, G. J., Seltzer, Z., Lu, G. W., Nishikawa, N., & Dubner, R. (1983). The cells of origin of the dorsal column postsynaptic projection in the lumbosacral enlargements of cats and monkeys. *Somatosensory Research, 1,* 131–149.

Brown, A. G., Brown, P. B., Fyffe, R. E., et al. (1983). Receptive field organization and response properties of spinal neurones with axons ascending the dorsal columns in the cat. *The Journal of Physiology, 337,* 575–588.

Enevoldson, T. P., & Gordon, G. (1989). Postsynaptic dorsal column neurons in the cat: A study with retrograde transport of horseradish peroxidase. *Experimental Brain Research, 75,* 611–620.

Francisco, A. N., Lobão, C. A., Sassaki, V. S., Garbossa, M. C., & Aguiar, L. R. (2006). Punctate midline myelotomy for the treatment of oncologic visceral pain: Analysis of three cases. *Arquivos de Neuro-Psiquiatria, 64,* 446–450.

Giesler, G. J., Jr., & Cliffer, K. D. (1985). Postsynaptic dorsal column pathway of the rat. II. Evidence against an important role in nociception. *Brain Research, 326,* 347–356.

Giesler, G. J., Jr., Nahin, R. L., & Madsen, A. M. (1984). Postsynaptic dorsal column pathway of the rat. I. Anatomical studies. *Journal of Neurophysiology, 51,* 260–275.

Gildenberg, P. L., & Hirshberg, R. M. (1984). Limited myelotomy for the treatment of intractable cancer pain. *Journal of Neurology, Neurosurgery, and Psychiatry, 47,* 94–96.

Goodman-Keiser, M. D., Qin, C., Thompson, A. M., & Foreman, R. D. (2010). Upper thoracic postsynaptic dorsal column neurons conduct cardiac mechanoreceptive information, but not cardiac chemical nociception in rats. *Brain Research, 1366,* 71–84.

Hirshberg, R. M., Al Chaer, E. D., Lawand, N. B., et al. (1996). Is there a pathway in the posterior funiculus that signals visceral pain? *Pain, 67,* 291–305.

Hitchcock, E. (1970). Stereotactic cervical myelotomy. *Journal of Neurology, Neurosurgery & Psychiatry, 33,* 224–230.

Honda, C. N. (1985). Visceral and somatic afferent convergence onto neurons near the central canal in the sacral spinal cord of the cat. *Journal of Neurophysiology, 53,* 1059–1078.

Honda, C. N., & Lee, C. L. (1985). Immunohistochemistry of synaptic input and functional characterizations of neurons near the spinal central canal. *Brain Research, 343,* 120–128.

Honda, C. N., & Perl, E. R. (1985). Functional and morphological features of neurons in the midline region of the caudal spinal cord of the cat. *Brain Research, 340,* 285–295.

Houghton, A. K., Wang, C. C., & Westlund, K. N. (2001). Do nociceptive signals from the pancreas travel in the dorsal column? *Pain, 89,* 207–220.

Hwang, S. L., Lin, C. L., Lieu, A. S., Kuo, T. H., Yu, K. L., Ou-Yang, F., Wang, S. N., Lee, K. T., & Howng, S. L. (2004). Punctate midline myelotomy for intractable visceral pain caused by hepatobiliary or pancreatic cancer. *Journal of Pain and Symptom Management, 27,* 79–84.

Kim, Y. S., & Kwon, S. J. (2000). High thoracic midline dorsal column myelotomy for severe visceral pain due to advanced stomach cancer. *Neurosurgery, 46,* 85–90.

Kuo, D. C., Nadelhaft, I., Hisamitsu, T., & de Groat, W. C. (1983). Segmental distribution and central projections of renal afferent fibers in the cat studied by transganglionic transport of horseradish peroxidase. *The Journal of Comparative Neurology, 216,* 162–174.

Kuo, D. C., Oravitz, J. J., Eskay, R., & de Groat, W. C. (1984). Substance P in renal afferent perikarya identified by retrograde transport of fluorescent dye. *Brain Research, 323,* 168–171.

Morgan, C., Nadelhaft, I., & de Groat, W. C. (1981). The distribution of visceral primary afferents from the pelvic nerve to Lissauer's tract and the spinal gray matter and its relationship to the sacral parasympathetic nucleus. *The Journal of Comparative Neurology, 201,* 415–440.

Nahin, R. L., Madsen, A. M., & Giesler, G. J., Jr. (1983). Anatomical and physiological studies of the gray matter surrounding the spinal cord central canal. *The Journal of Comparative Neurology, 220,* 321–335.

Nauta, H. J., Hewitt, E., Westlund, K. N., et al. (1997). Surgical interruption of a midline dorsal column visceral pain pathway. Case report and review of the literature. *Journal of Neurosurgery, 86,* 538–542.

Nauta, H. J., Soukup, V. M., Fabian, R. H., Lin, J. T., Grady, J. J., Williams, C. G., Campbell, G. A., Westlund, K. N., & Willis, W. D., Jr. (2000). Punctate midline myelotomy for the relief of visceral cancer pain. *Journal of Neurosurgery, 92,* 125–130.

Ness, T. J. (2000). Evidence for ascending visceral nociceptive information in the dorsal midline and lateral spinal cord. *Pain, 87,* 83–88.

Ness, T. J., & Gebhardt, G. F. (1987). Characterization of neuronal responses to noxious visceral and somatic stimuli in the medial lumbosacral spinal cord of the rat. *Journal of Neurophysiology, 57,* 1867–1892.

Ness, T. J., & Gebhart, G. F. (1988). Characterization of neurons responsive to noxious colorectal distension in the T13-L2 spinal cord of the rat. *Journal of Neurophysiology, 60,* 1419–1438.

Neuhuber, W. (1982). The central projections of visceral primary afferent neurons of the inferior mesenteric plexus and hypogastric nerve and the location of the related sensory and preganglionic sympathetic cell bodies in the rat. *Anatomy and Embryology (Berl), 164,* 413–425.

Palecek, J., Paleckova, V., & Willis, W. D. (2003a). Fos expression in spinothalamic and postsynaptic dorsal column neurons following noxious visceral and cutaneous stimuli. *Pain, 104*, 249–257.

Palecek, J., Paleckova, V., & Willis, W. D. (2003b). Postsynaptic dorsal column neurons express NK1 receptors following colon inflammation. *Neuroscience, 116*, 565–572.

Petit, D. (1972). Postsynaptic fibres in the dorsal columns and their relay in the nucleus gracilis. *Brain Research, 48*, 380–384.

Raslan, A. M., Cetas, J. S., McCartney, S., & Burchiel, K. J. (2011). Destructive procedures for control of cancer pain: The case for cordotomy. *Journal of Neurosurgery, 114*, 155–170.

Rustioni, A. (1973). Non-primary afferents to the nucleus gracilis from the lumbar cord of the cat. *Brain Research, 51*, 81–95.

Rustioni, A., Hayes, N. L., & O'Neill, S. (1979). Dorsal column nuclei and ascending spinal afferents in macaques. *Brain, 102*, 95–125.

Sikandar, S., & Dickenson, A. H. (2011). Pregabalin modulation of spinal and brainstem visceral nociceptive processing. *Pain, 152*, 2312–2322.

Sugiura, Y., Terui, N., & Hosoya, Y. (1989). Difference in distribution of central terminals between visceral and somatic unmyelinated (C) primary afferent fibers. *Journal of Neurophysiology, 62*, 834–840.

Traub, R. J., Herdegen, T., & Gebhart, G. F. (1993). Differential expression of c-fos and c-jun in two regions of the rat spinal cord following noxious colorectal distention. *Neuroscience Letter, 160*, 121–125.

Traub, R. J., Stitt, S., & Gebhart, G. F. (1995). Attenuation of c-Fos expression in the rat lumbosacral spinal cord by morphine or tramadol following noxious colorectal distention. *Brain Research, 701*, 175–182.

Truex, R. C., & Carpenter, M. B. (1969). *Human neuroanatomy* (6th ed.). Baltimore: Williams and Wilkins.

Vilela Filho, O., Araujo, M. R., Florencio, R. S., Silva, M. A., & Silveira, M. T. (2001). CT-guided percutaneous punctate midline myelotomy for the treatment of intractable visceral pain: A technical note. *Stereotactic and Functional Neurosurgery, 77*, 177–182.

Wang, C. C., Willis, W. D., & Westlund, K. N. (1999). Ascending projections from the area around the spinal cord central canal: A Phaseolus vulgaris leucoagglutinin study in rats. *The Journal of Comparative Neurology, 415*, 341–367.

Willis, W. D., Al Chaer, E. D., Quast, M. J., et al. (1999). A visceral pain pathway in the dorsal column of the spinal cord. *Proceedings of the National Academy of Sciences of the United States of America, 96*, 7675–7679.

Zhai, Q. Z., & Traub, R. J. (1999). The NMDA receptor antagonist MK-801 attenuates c-Fos expression in the lumbosacral spinal cord following repetitive noxious and non-noxious colorectal distention. *Pain, 83*, 321–329.

Postsynaptic Dorsal Column Projection, Functional Characteristics

Karin N. Westlund
Department of Physiology, University of Kentucky, Lexington, KY, USA

Synonyms

Dorsal horn; Second-order cutaneous mechanical input; Visceral pain pathway

Definition

Second-order neurons in the deep layers of the spinal cord dorsal horn receive cutaneous mechanical and/or visceral information that is relayed through the ► dorsal column to the ► dorsal column nuclei in the medulla. The ascending axons of these second-order neurons are referred to as the postsynaptic dorsal column pathway. The ► postsynaptic dorsal column (PSDC) cells located in laminae III and IV primarily transmit noxious cutaneous mechanical input, while the PSDC cells medially in laminae III, VII, and X transmit noxious visceral pain input.

Characteristics

The dorsal column has previously been known solely as a direct pathway for primary afferent axonal transmission of ► epicritic information (e.g., discriminative responses to cutaneous and deep sensory stimuli). The direct presynaptic primary afferent nerve fibers provide discriminative, site-specific input to dorsal column nuclei in the caudal medulla about light touch, vibration, and limb position in space. In addition, it is now known that the dorsal column is also a conduit for uncrossed second-order axons of postsynaptic dorsal column neurons located in the deep dorsal horn. The functional characteristics of the dorsal column inputs have been expanded in light of the many animal and clinical studies describing postsynaptic dorsal column neurons that relay information to the dorsal column nuclei about non-noxious mechanoreception and noxious visceral pain. The latter includes reports of pelvic cancer pain relief after discrete midline dorsal column lesions (Nauta et al. 2000; Raslan et al. 2011). The ▶ epicritic information is carried by large, heavily myelinated fibers whereas second-order axons of postsynaptic dorsal column neurons include medium-sized myelinated axons, based on conduction velocities in rats of 38.3 ms^{-1} on average (range 22–61 ms^{-1}) (Rustioni 1973; Cervero 1983; Giesler and Cliffer 1985).

Postsynaptic dorsal column (PSDC) neurons are distributed as a column through the length of the spinal cord. Response properties of PSDC neurons have been characterized by Brown and colleagues (Brown et al. 1983). PSDC cells have resting discharges characterized as single impulses or busts of impulses at short intervals separated by longer, irregular periods. Receptive fields were identified as being either excitatory, with clearly defined zones of high and low thresholds which were sometimes discontinuous, or they were inhibitory (40 %). Inhibitory fields were either small if adjacent to an excitatory field or were large and separated from an excitatory field. Almost all PSDC cells received convergent inputs from low- and/or high-threshold cutaneous and sub-cutaneous mechanoreceptors (▶ Cutaneous Mechanoreception, ▶ Subcutaneous Mechano-reception) from regions tested, which included both hairy and glabrous skin. Intense visceral stimulation also activates PTSD neurons in the resting state (Honda 1985). PSDC neurons are likely among the "silent neurons" that upon sensitization induced by inflammation produce a sustained increase in background firing in response to convergent inputs from cutaneous, deep, and visceral tissues (Cervero 1983; Ness and Gebhardt 1987; Schaible et al. 1987; Al-Chaer et al. 1996a, b, 1997; Feng et al. 1998).

Laminae III and IV Postsynaptic Dorsal Column Neurons

In mammals, including monkeys and humans, PSDC neurons are primarily found in medial laminae III–VI, at cervical levels and more laterally in lumbar levels throughout the length of the spinal cord (Rustioni et al. 1979). PSDC cells specifically identified in Laminae III–IV are responsive to mechanical input following cutaneous stimulation (Giesler and Cliffer 1985). Background discharge in PSDC neurons consists of short high-frequency bursts separated by longer intervals. In unanesthetized, decerebrated animals, 64 % of the PSDC cells responded only to innocuous cutaneous input, while the remainder responded best to strong mechanical stimuli. Input can be received from more than one afferent fiber and can include response to more than one modality (Uddenberg 1968). For example, a neuron may respond to hair movement, pressure, and/or pinch stimulation of the skin. Neuronal responses to different types of input vary in their discharge pattern in agreement with input type (e.g., hair movement produces rapidly adapting responses whereas maintained tactile contact produces slowly adapting responses; Angaut-Petit 1975). Inhibition of PSDC neuron responses to peripheral input can be achieved by stimulation of other peripheral nerves, demonstrating complex interactions within this system (Brown et al. 1983). Thus, deep dorsal horn PSDC neurons have an integra-tive function.

Lamina X Postsynaptic Dorsal Column Neurons

Intense noxious visceral input activates lamina X PSDC neurons (Honda 1985). Responses to less intense noxious visceral input are evoked in lamina X PSDC neurons in a state of sensitization induced by inflammation of visceral organs. A state of sensitization induced by inflammation of a visceral organ produces a sustained increase in background firing and increased responsive-ness in a greater number of PSDC cells (Al-Chaer et al. 1996a, b, 1997; Ness and Gebhardt 1987; Feng et al. 1998; Wang et al. 1999). For example, inflammation of the colon with mustard oil or

Postsynaptic Dorsal Column Projection, Functional Characteristics, Fig. 1 (a) Effects of dorsal column (*DC*) and ventrolateral (*VLC*) lesions at T10 in rats (*n* = 20) on responses to mechanical stimulation of the skin (Brush, Press, and Pinch) and to graded intensities of colorectal distension (*CRD*; 20, 40, 60, and 80 mmHg) in **b** (responses were normalized). (**c**) Effects of microdialysis administration into the sacral spinal cord of drugs that block nociceptive synaptic transmission on responses of gracile neurons (*n* = 10) to graded intensities of CRD. The drugs included morphine and CNQX (6-cyano-7-nitroquinoxaline-2,3-dione). Naloxone was given systemically to reverse the action of morphine. In some animals (*n* = 3), a lesion was placed in the DC. This confirms receipt of noxious visceral input from PSDC neurons through the dorsal column. *indicates a significant change (**a** and **b** is reproduced with permission from Al-Chaer et al. 1996a; **c** is reproduced from Al-Chaer et al. 1996b (Copyright 1996, The American Physiological Society))

of the pancreas with bradykinin has been shown to lower the threshold of PSDC cells and increase responses of visceroreceptive (▶ Visceroreception) lamina X neurons (Al-Chaer et al. 1997; Willis et al. 1999). Both excitatory and inhibitory responses can be recorded in PSDC cells in response to noxious

visceral input. Visceral nociceptive responses are recorded in neurons of the dorsal column nuclei and thalamus as a consequence (Al-Chaer et al. 1997; Willis et al. 1999; Wang and Westlund 2001; Houghton et al. 2001). Lesions of the dorsal column or spinal morphine dramatically reduce responses to graded colon distension

Postsynaptic Dorsal Column Projection, Functional Characteristics, Fig. 2 Responses of a cell in the dorsal column nucleus to (**a**) cutaneous stimuli (brush, *BR*; press, *PR*; pinch, *PI*) and (**b**) noxious chemical activation by bradykinin applied to pancreatic afferent fibers. The accompanying change in blood pressure (mmHg) is also shown. (**c**) This cell had a cutaneous receptive over the back of the thorax. (**d**) The cell was located at the lateral edge of the gracile nucleus (*GR*) (Reprinted from Wang and Westlund (2001) NeuroReport 12:2527–2530)

after sensitization (Fig. 1). Inflammation of visceral organs also increases Fos expression (Palecek et al. 2002, 2003). Increased activation recorded from neurons in the dorsal column nuclei and thalamus after noxious visceral stimulation is dramatically reduced by dorsal column lesions, while response to cutaneous pinch persists (Al-Chaer et al. 1996a, b; Feng et al. 1998; Houghton et al. 2001). Consequences of noxious visceral stimulation on exploratory behaviors are reduced by surgical lesions in the dorsal columns (Houghton et al. 1997, 2001).

Both somatic and visceral inputs have been shown to converge onto the majority of the lamina X PSDC neurons (Honda 1985; Honda and Perl 1985). Those excited by noxious input are sometimes inhibited by innocuous inputs and vice versa. In addition to direct input from ▸ mechanosensitive visceral afferents, convergent somatic and visceral inputs are relayed by PSDC neurons in lamina X through the dorsal columns. Visceral nociceptive responses are recorded in up to 50 % of cells in the dorsal column nuclei (Fig. 2) (Rustioni 1973; Angaut-Petit 1975; Cliffer and Giesler 1989; Berkeley and Hubscher 1995). PSDC cells in lamina X have been shown to receive input from afferent fibers containing vasoactive intestinal polypeptide as well as from fibers containing somatostatin, substance P, leu-enkephalin, and serotonin (Honda and Lee 1985), some of which may be from descending brainstem projections providing further complex interactions within this system.

References

Al-Chaer, E. D., Lawand, N. B., Westlund, K. N., et al. (1996a). Pelvic visceral input into the nucleus gracilis is largely mediated by the postsynaptic dorsal column pathway. *Journal of Neurophysiology, 76,* 2675–2690.

Al-Chaer, E. D., Lawand, N. B., Westlund, K. N., & Willis, W. D. (1996b). Visceral nociceptive input into the ventral posterolateral nucleus of the thalamus: A new function for the dorsal column pathway. *Journal of Neurophysiology, 76,* 2661–2674.

Al-Chaer, E. D., Westlund, K. N., & Willis, W. D. (1997). Sensitization of postsynaptic dorsal column neuronal responses by colon inflammation. *Neuroreport, 8,* 3267–3273.

Angaut-Petit, D. (1975). The dorsal column system: I. Existence of long ascending postsynaptic fibres in the cat's fasciculus gracilis. *Experimental Brain Research, 22,* 457–470.

Berkeley, K. J., & Hubscher, C. H. (1995). Are there separate central nervous system pathways for touch and pain? *Nature Medicine, 1,* 766–773.

Brown, A. G., Brown, P. B., Fyffe, R. E., et al. (1983). Receptive field organization and response properties of spinal neurones with axons ascending the dorsal columns in the cat. *The Journal of Physiology, 337,* 575–588.

Cervero, F. (1983). Somatic and visceral inputs to the thoracic spinal cord of the cat: Effects of noxious stimulation of the biliary system. *The Journal of Physiology, 337,* 51–67.

Cliffer, K. D., & Giesler, G. J., Jr. (1989). Postsynaptic dorsal column pathway of the rat. III. Distribution of ascending afferent fibers. *The Journal of Neuroscience, 9,* 3146–3168.

Feng, Y., Cui, M., Al-Chaer, E. D., & Willis, W. D. (1998). Epigastric antinociception by cervical dorsal column lesions in rats. *Anesthesiology, 89,* 411–420.

Giesler, G. J., Jr., & Cliffer, K. D. (1985). Postsynaptic dorsal column pathway of the rat. II. Evidence against an important role in nociception. *Brain Research, 326,* 347–356.

Honda, C. N. (1985). Visceral and somatic afferent convergence onto neurons near the central canal in the sacral spinal cord of the cat. *Journal of Neurophysiology, 53,* 1059–1078.

Honda, C. N., & Lee, C. L. (1985). Immunohistochemistry of synaptic input and functional characterizations of neurons near the spinal central canal. *Brain Research, 343,* 120–128.

Honda, C. N., & Perl, E. R. (1985). Functional and morphological features of neurons in the midline region of the caudal spinal cord of the cat. *Brain Research, 340,* 285–295.

Houghton, A. K., Kadura, S., & Westlund, K. N. (1997). Dorsal column lesions reverse the reduction of homecage activity in rats with pancreatitis. *Neuroreport, 8,* 3795–3800.

Houghton, A. K., Wang, C. C., & Westlund, K. N. (2001). Do nociceptive signals from the pancreas travel in the dorsal column? *Pain, 89,* 207–220.

Nauta, H. J., Soukup, V. M., Fabian, R. H., Lin, J. T., Grady, J. J., Williams, C. G., Campbell, G. A., Westlund, K. N., & Willis, W. D., Jr. (2000). Punctate midline myelotomy for the relief of visceral cancer pain. *Journal of Neurosurgery, 92,* 125–130.

Ness, T. J., & Gebhardt, G. F. (1987). Characterization of neuronal responses to noxious visceral and somatic stimuli in the medial lumbosacral spinal cord of the rat. *Journal of Neurophysiology, 57,* 1867–1892.

Palecek, J., Paleckova, V., & Willis, W. D. (2002). The roles of pathways in the spinal cord lateral and dorsal funiculi in signaling nociceptive somatic and visceral stimuli in rats. *Pain, 96,* 297–307.

Palecek, J., Paleckova, V., & Willis, W. D. (2003). Fos expression in spinothalamic and postsynaptic dorsal column neurons following noxious visceral and cutaneous stimuli. *Pain, 104,* 249–257.

Raslan, A. M., Cetas, J. S., McCartney, S., & Burchiel, K. J. (2011). Destructive procedures for control of cancer pain: The case for cordotomy. *Journal of Neurosurgery, 114,* 155–170.

Rustioni, A. (1973). Non-primary afferents to the nucleus gracilis from the lumbar cord of the cat. *Brain Research, 51,* 81–95.

Rustioni, A., Hayes, N. L., & O'Neill, S. (1979). Dorsal column nuclei and ascending spinal afferents in macaques. *Brain, 102,* 95–125.

Schaible, H. G., Schmidt, R. F., & Willis, W. D. (1987). Convergent inputs from articular, cutaneous and muscle receptors onto ascending tract cells in the cat spinal cord. *Experimental Brain Research, 66,* 479–488.

Uddenberg, N. (1968). Functional organization of long, second-order afferents in the dorsal funiculus. *Experiencia, 15,* 441–442.

Wang, C. C., & Westlund, K. N. (2001). Responses of rat dorsal column neurons to pancreatic nociceptive stimulation. *Neuroreport, 12,* 2527–2530.

Wang, C. C., Willis, W. D., & Westlund, K. N. (1999). Ascending projections from the area around the spinal cord central canal: A PHA-L study in rats. *The Journal of Comparative Neurology, 415,* 341–367.

Willis, W. D., Al Chaer, E. D., Quast, M. J., et al. (1999). A visceral pain pathway in the dorsal column of the spinal cord. *Proceedings of the National Academy of Sciences of the United States of America, 96,* 7675–7679.

Postsynaptic Inhibition

Definition

Postsynaptic inhibition is a form of synaptic activation by an inhibitory transmitter in which the responses of a postsynaptic neuron are reduced. Mechanisms of postsynaptic inhibition include the opening of potassium or chloride channels, resulting in an increased membrane conductance and often a hyperpolarization.

Cross-References

▶ GABA Mechanisms and Descending Inhibitory Mechanisms

Postsynaptic Opioid Receptors

▶ Opioid Receptors at Postsynaptic Sites

Posttraumatic Dystrophy

▶ CRPS, Evidence-Based Treatment

Posttraumatic Headache

▶ Headache, Acute Posttraumatic

Posttraumatic Neuralgia

Definition

Pain syndrome that occurs after a traumatic nerve injury. A more appropriate term is posttraumatic peripheral neuropathic pain.

Cross-References

▶ Neuralgia, Assessment

Posttraumatic Pain Syndrome

▶ Postoperative Pain, Acute Presentation of Complex Regional Pain Syndrome

Postural Position

Synonyms

Rest position

Definition

Postural or rest positions are habitual mandibular positions that are assumed when the person is relaxed, with no voluntary activation of the jaw-closing muscles. They are determined in part by the viscoelasticity of the muscles and by the contractile elements of the muscles.

Cross-References

▶ Orofacial Pain
▶ Orofacial Pain, Movement Disorders

Posture

Definition

Posture refers to the position of a joint or complex of joints in the body, which is maintained for a specific unit of time. A posture is always associated with static loading.

Cross-References

▶ Ergonomic Counseling

Potassium Sensitivity Testing

Definition

Intravesical infusion of a potassium solution that often elicits pain in patients with interstitial cystitis.

Cross-References

► Interstitial Cystitis and Chronic Pelvic Pain

Potentially damaging stimulus

► Muscle Nociceptors, Neurochemistry

Potent Opioid Analgesic

Definition

Morphine-like pain-relieving drug that can relieve very intense pain by administering appropriate doses intravenously, epidurally, or directly into the cerebrospinal fluid around the spinal cord.

Cross-References

► Postoperative Pain, Acute Pain Management Principles

Pourfour du Petit Syndrome

► CRPS, Evidence-Based Treatment

Pourpre Myelotomy

► Midline Myelotomy

Power Law

Definition

Power law refers to a stimulus-response function wherein equal stimulus ratios result in equal response ratios. In psychophysics, the law is expressed as $\Psi = k(S-S_0)^X$.

Cross-References

► Pain in Humans, Psychophysical Law
► Spinothalamic Tract Neurons, in Deep Dorsal Horn

Power of Ultrasound

Definition

The power is total energy per unit time.

Cross-References

► Ultrasound Therapy of Pain from the Musculoskeletal System

PPI

► Present Pain Intensity Subscale

Preadmission Clinics

Definition

Preadmission clinics are used to carry out a preanesthetic evaluation. Goals include creating a doctor-patient relationship, identifying the present surgical problem and the medical conditions that coexist, mounting a management strategy for perioperative anesthetic care, and obtaining informed consent for the anesthetic plan. The consultation is recorded in the patient's records. Anesthetic options with their risks and benefits are discussed. The preadmission clinic corrects conditions as far as possible to diminish patient anxiety and decrease perioperative morbidity and mortality.

Cross-References

▶ Postoperative Pain, Preoperative Education

Preamputation Pain

Definition

Preamputation pain refers to pain prior to the amputation that is often related to later phantom limb pain.

Cross-References

▶ Phantom Limb Pain, Treatment

Precontemplation

Definition

A readiness to change stage in which a person sees no need and has no intention of changing a behavior that others (not the person) might see as problematic (e.g., problem drinking, smoking, avoiding exercise).

Cross-References

▶ Motivational Aspects of Pain

Predentin

Definition

The deepest thin layer of dentin that is not mineralized.

Cross-References

▶ Nociceptors in the Dental Pulp

Preemptive Analgesia

Preemptive analgesia is a treatment that is initiated before and is operational during the tissue injury (e.g., a surgical procedure), in order to reduce the physiological consequences of nociceptive transmission provoked by the procedure. The hypothesis behind this is that the transmission of noxious afferent input from the periphery to the spinal cord during the induction of a tissue injury induces a prolonged state of central sensitization, which amplifies subsequent input from the injury and leads to heightened pain. By interrupting the transmission of noxious inputs to the spinal cord during the injury, a preemptive approach is suggested to prevent the establishment of central sensitization, resulting in reduced pain intensity and lower analgesic requirements, even after the analgesic effects of the (preemptive) agents have worn off.

Cross-References

▶ Formalin Test
▶ Nick Model of Cutaneous Pain and Hyperalgesia
▶ Preemptive or Preventive Analgesia and Central Sensitization in Postoperative Pain

Preemptive or Preventive Analgesia and Central Sensitization in Postoperative Pain

Stephan A. Schug[1] and Emma L. Brandon[2]
[1]School of Medicine and Pharmacology, Royal Perth Hospital, Faculty of Medicine, Dentistry & Health Sciences, The University of Western Australia, UWA Anaesthesiology, Perth, Australia
[2]Anaesthesiology and Pain Medicine, University of Western Australia Royal Perth Hospital, Perth, WA, Australia

Synonyms

Pain prevention; Prophylactic pain treatment

Definition

Preemptive Analgesia

This term was used for the first time by P. Wall in an editorial in pain in 1988. He wrote "that we should consider ... the possibility that pre-emptive preoperative analgesia has prolonged effects which long outlast the presence of the drugs" (Wall 1988). Preemptive analgesia is now commonly defined as a preoperative analgesic treatment that is more effective than the identical treatment administered after incision or surgery (as evidenced by reduced pain or analgesic consumption or both). The only difference is the timing of administration.

Preventive Analgesia

In contrast, the definition of preventive analgesia focuses on duration of effect. Postoperative pain and/or analgesic consumption is reduced relative to another treatment, to a placebo treatment, or to no treatment, as long as the effect is observed at a point in time that exceeds the expected duration of action of the target agent. The intervention may or may not be initiated before surgery. For example, a preventive effect is present if postoperative administration of a target analgesic agent, but not a placebo, results in reduced postoperative pain or analgesic consumption after the effects of the target agent have worn off (Nguyen et al. 2001; Reuben et al. 2001).

Characteristics

Although the concept that pain is more difficult to treat late rather than early in its course is not new, the scientific basis of this phenomenon was not realized until Woolf and others elucidated mechanisms for the effect in the spinal cord in the early 1980s (Woolf 1983). The demonstration of preemptive analgesia in animals (Woolf 1983) and the unravelling of spinal cord mechanisms by scientists led to the hypothesis that analgesia prior to surgery may enhance postoperative pain management (Wall 1988).

However, randomized controlled trials and their meta-analysis have failed to show clinically useful benefits of timing of analgesia when comparing pre-incisional and post-incisional interventions (Moiniche et al. 2002; Eisenach 2000; Abdi et al. 2000). Reasons for this include problems with definitions, the design of the studies, or no clinical benefit of preemptive analgesia.

The almost exclusive focus on the narrow definition of preemptive analgesia had the unintended effect of diverting attention away from certain clinically significant findings because they did not conform to what had become the accepted definition. The concept of preemptive analgesia has now evolved and has led to progress in our understanding of the mechanisms that contribute to acute postoperative pain.

In addition to skin incision, central and peripheral sensitization is now believed to be triggered by other factors including preoperative noxious inputs and pain, the extent of intraoperative tissue injury and postoperative inflammation, and ectopic neural activity in the case of postsurgical nerve injury. Hence, the focus has shifted from the timing of a single intervention to the concept of "preventive" analgesia (Kissin 1994). Each of the aforementioned factors is a potential target for a preventive approach.

A systematic review of 27 studies identified evidence for a benefit of preventive analgesia, with 60 % of studies finding reduced pain or analgesic consumption beyond the clinical duration of action of the analgesic intervention (Katz and McCartney 2002). Of the six preventive studies that examined the use of NMDA antagonists, four found positive preventive effects. Similarly in a most recent other systematic review of NMDA antagonists, 14 of 24 ketamine studies and 8 of 12 dextromethorphan studies report a preventive analgesic effect (McCartney et al. 2004). This may reflect their role in reducing central sensitization by their actions at the NMDA receptor-ion channel complex or by reducing acute opioid tolerance (Mao et al. 1995). Preventive analgesic effects have also been shown with clonidine, gabapentin, local anesthetic, NSAIDs, and morphine (Katz and McCartney 2002).

Preventive analgesia may even reduce the risk of developing chronic postoperative pain states; while studies investigating the role of neural blockade in preemptive analgesia have found no effect, the use of a local anesthetic regime throughout the perioperative period has been shown to provide effective and prolonged analgesia (e.g., Reuben et al. 2001; Fischer et al. 2000; Gottschalk et al. 1998; Obata et al. 1999; Senturk et al. 2002). Reuben et al. demonstrated a significant reduction in pain 24 h after post-incisional administration of 5 mg morphine injected intramuscularly or infiltrated directly into the exposed cancellous bone and bone marrow cavity in patients undergoing iliac bone harvest (Reuben et al. 2001). Remarkably, the incidence of pain 1 year later was significantly lower in the iliac infiltration group (5 %) than in the intramuscular injection group (37 %).

Several studies have shown an effect of perioperative epidural analgesia for thoracotomies with a significant reduction of pain at 6 months postoperatively (Obata et al. 1999; Senturk et al. 2002). Similarly, Gottschalk et al. found prolonged benefit (up to 9.5 weeks) after pre- and intraoperative epidural analgesia for radical prostatectomy (Gottschalk et al. 1998). More trials are required to confirm this hypothesis.

In conclusion, the previous concept of preempting nociception by providing analgesia before surgical incision should be broadened to the concept of preempting central excitatory changes by providing effective analgesia, specific antagonists to the excitatory process, and/or complete neural blockade throughout the various stages of injury, i.e., by providing preventive analgesia.

References

Abdi, S., Lee, D. H., Park, S. K., et al. (2000). Lack of pre-emptive analgesic effects of local anaesthetics on neuropathic pain. *British Journal of Anaesthesia, 85*, 620–623.

Eisenach, J. C. (2000). Pre-emptive hyperalgesia, not analgesia? *Anesthesiology, 92*, 308–309.

Fischer, S., Troidi, H., MacLean, A. A., et al. (2000). Prospective double blind randomised study of a new regimen of pre-emptive analgesia for inguinal hernia repair: Evaluation of postoperative pain course. *European Journal of Surgery, 166*, 545–551.

Gottschalk, A., Smith, D. S., Jobes, D. R., et al. (1998). Pre-emptive epidural analgesia and recovery from radical prostatectomy. A randomised controlled trial. *Journal of the American Medical Association, 279*, 1076–1082.

Katz, J., & McCartney, C. J. (2002). Current status of pre-emptive analgesia. *Current Opinion in Anaesthesiology, 15*, 435–441.

Kissin, I. (1994). Pre-emptive analgesia: Terminology and clinical relevance. *Anesthesia and Analgesia, 79*, 809–810.

Mao, J., Price, D. D., & Mayer, D. J. (1995). Mechanisms of hyperalgesia and morphine tolerance: A current review of their possible interactions. *Pain, 62*, 259–274.

McCartney, C. J. L., Sinha, A., & Katz, J. (2004). A qualitative systematic review of the role of n-methyl-d-aspartate receptor antagonists in preventive analgesia. *Anesthesia and Analgesia, 98*, 1385–1400.

Moiniche, S., Kehlet, H., & Dahl, J. B. (2002). A qualitative and quantitative systematic review of pre-emptive analgesia for post-operative pain relief: The role of timing of analgesia. *Anesthesiology, 96*, 725–741.

Nguyen, A., Girard, F., Boudreault, D., et al. (2001). Scalp nerve blocks decrease the severity of pain after craniotomy. *Anesthesia and Analgesia, 93*, 1272–1276.

Obata, H., Saito, S., Fujita, N., et al. (1999). Epidural block with mepivacaine before surgery reduces long-term post-thoracotomy pain. *Canadian Journal of Anaesthesia, 46*, 1127–1132.

Reuben, S. S., Vieira, P., Faruqi, S., et al. (2001). Local administration of morphine for analgesia after iliac bone graft harvest. *Anesthesiology, 95*, 390–394.

Senturk, M., Ozcan, P. E., Talu, G. K., et al. (2002). The effects of three different analgesia techniques on long-term postthoracotomy pain. *Anesthesia and Analgesia, 94*, 11–15.

Wall, P. D. (1988). The prevention of post-operative pain. *Pain, 33*, 289–290.

Woolf, C. J. (1983). Evidence for a central component of post-injury pain hypersensitivity. *Nature, 308*, 686–688.

Preferential Processing

Definition

A systematic bias for the processing of information related to a specific domain of experience.

Cross-References

▶ Psychology of Pain, Assessment of Cognitive Variables

Prefrontal Cortex, Effects on Pain-Related Behavior

Jürgen Lorenz
Faculty of Life Science, Competence Center Gesundheit, University of Applied Science Hamburg, Hamburg, Germany

Definition

The prefrontal cortex (PFC) is a ▶ neocortical brain region of the frontal lobe that is farthest developed in humans. As indicated by human functional brain imaging, it plays a major role in cognitive, attentional, and emotional processing of painful stimuli and recruitment of endogenous pain control.

Characteristics

The PFC is considered to be important for higher cortical functions that characterize the flexibility of human behavior, the ability to control attention, and the richness of intellectual and emotional competence. PFC comprises orbital and medial (Brodman areas 10, 11, 13, 14), ventrolateral (Brodman areas 12, 45), dorsolateral (Brodman area 46), and mid-dorsal (Brodman area 9) regions of the frontal lobe and has manifold, mostly reciprocal, connections to visual, auditory, somatosensory, olfactory, gustatory, visceral, and motor cortical areas. As part of neuronal circuits, the PFC is linked to subcortical structures such as the basal ganglia (important for motivation and initiation of movement) and ▶ thalamus, and limbic structures (important for emotional behavior) such as ▶ cingulate cortex, ▶ amygdala, hypothalamus, and brain stem.

The PFC is activated in several, yet not all functional, imaging studies using experimental painful stimuli or clinical pain states. It typically fails to show a clear pain-related stimulus-response function (Coghill et al. 1999), which led to the assumption that PFC activity mainly relates to the engagement of attention during pain processing (Peyron et al. 1999). On the other hand, it has long been assumed that higher cortical functions represented by the PFC participate in endogenous pain control. The biological significance of endogenous pain control, primarily governed by endorphins (the body's own morphine-like substances), is generally seen in the context of behavioral conflicts in which the subject needs to disengage from pain in order to fight or escape at the presence of body injury. Analogous human life situations are sporting competition or combat, during which a subject may fail to be aware of even severe tissue damage, which becomes painful when the victim releases engagement in these activities.

A particular part of the PFC, the dorsolateral prefrontal cortex (DLPFC), is important for continuous monitoring of the external world, maintenance of information in short-term memory, and governing efficient performance control in the presence of distracting or conflicting stimuli (Bunge et al. 2001; Sakai et al. 2002). In turn, orbital and medial portions of the prefrontal cortex (OMPFC) are known to be important for mood and emotional behavior, e.g., when guided

a

b

Prefrontal Cortex, Effects on Pain-Related Behavior, Fig. 1 Surface-rendered and sagittal views of statistical (Z-score) maps of O^{15}-water positron emission tomography (*PET*) scans from 14 volunteers during contact heat stimulation of their left volar forearm which was sensitized by a topical solution of capsaicin. The maps illustrate significant activation of the dorsolateral prefrontal cortex (*DLPFC*), the orbital and medial prefrontal cortex (*OMPFC*), the perigenual anterior cingulate cortex (*perig ACC*), the medial thalamus, and the dorsomedial midbrain.

DLPFC activity showed a negative correlation with the visual analogue score of unpleasantness (*upper right diagram*). When dividing scans according to a median-half-split of left DLPFC activity, scans during low DLPFC activity (*open circles*) yielded significantly greater correlations than scans with high left DLPFC (*filled circles*) activity between midbrain and medial thalamic activations. This result suggests a modulation of synaptic connectivity within the afferent medial thalamic pathway by which the affective reaction to the painful stimulation is diminished

by cues of reward or punishment, as well as for visceral and autonomic homeostasis related to eating and drinking behavior (Price et al. 1996). Lorenz et al. (2002, 2003) confirmed differential responses of DLPFC and OMPFC during pain. They used O^{15}-water injections in healthy volunteers, to scan cerebral blood flow responses by ▶ positron emission tomography (PET), following contact heat stimuli applied upon normal and capsaicin-treated skins of the volar forearm. They found strong activation of the bilateral DLPFC and OMPFC during capsaicin-induced heat allodynia (Fig. 1). Whereas DLPFC exhibited a negative, OMPFC yielded a positive relationship to the unpleasantness of perceived pain. Furthermore, the interregional correlation of activity between midbrain and medial thalamus was significantly reduced dependent on the amount of DLPFC activity, which could indicate a top-down mode of inhibition of effective synaptic connectivity between brain stem and medial thalamus, causing reduced pain unpleasantness. Consistent with this view, the perigenual cingulate cortex (perig ACC), as well as the bilateral OMPFC which receives strong input from the medial thalamus, exhibited a clear reduction of positive relationship to pain unpleasantness during scans of high compared to low DLPFC activity.

The interaction of DLPFC with critical nodes of the descending inhibitory pathways such as the perigenual ACC and the midbrain, both brain regions that are richly provided with opioid receptors, may therefore be particularly important for disengagement from the prepotent pain behavior to pursue other behavioral goals. This is consistent with the notion that lateral PFC exerts cognitive control during the processing of emotionally salient stimuli (Goel and Dolan 2003), especially when the emotional content of task stimuli is task-irrelevant and distracting (Bishop et al. 2004). Lorenz and Bromm (1997) found that the ability to maintain performance accuracy in a short-term memory task during experimental painful muscle ischemia positively correlates with frontal lobe activity measured by event-related brain potentials of the electroencephalogram elicited by the memory stimuli. Disengagement from pain is also a prominent feature of placebo analgesia. Bingel et al. (2006) and Wager et al. (2004) provided consistent evidence that DLPFC engages anticipatory and cognitive processes that mediate placebo analgesia through descending pathways involving cingulate cortex and brain stem. Goel and Dolan (2003) argued that the reciprocity of neural responses within lateral and ventral medial PFC

cortex reflects the degree to which decision making is guided by respective rational or emotional reasoning. During acute pain inflicting the intact body, such as when touching a hot plate, immediate withdrawal is of pivotal importance, resulting in activation of brain areas like the premotor and motor cortex, dorsal basal ganglia, and cerebellum (Casey et al. 2001) which allow rapid and spatially coordinated motor reactions. Thus, whereas acute pain orchestrates a habitual pain behavior, ongoing pain states caused by tissue damage requires executive attentional and behavioral control governed by the DLPFC.

Recent clinical studies lend further support to the association of the DLPFC with pain suppression. Apkarian et al. (2004) observed a reduction of the gray matter density, determined by morphometry of magnetic resonance scans in bilateral DLPFC and right thalamus in chronic back pain patients, which was strongly related to pain characteristics. There is evidence that decreased gray matter density in DLPFC is the consequence and not the cause of chronic pain based on the observation that chronic osteoarthritis pain can normalize after surgical hip replacement (Rodriguez-Raecke et al. 2009). In agreement with the suggested role of DLPFC in pain control, high-frequency transcranial magnetic stimulation (rTMS) over left DLPFC was able to ameliorate chronic migraine (Brighina et al. 2004). This is consistent with earlier studies in animals where electrical stimulation of fiber connections of the prefrontal cortex to the midbrain mediated antinociceptive effects in rodents (Hardy and Haigler 1985). In summary, collective evidence suggests that the PFC represents a brain substrate for the human ability to actively cope with pain.

References

Apkarian, A. V., Sosa, Y., Sonty, S., et al. (2004). Chronic back pain is associated with decreased prefrontal and thalamic gray matter density. *Journal of Neuroscience, 24,* 10410–10415.

Bingel, U., Lorenz, J., Schoell, E., Weiller, C., & Büchel, C. (2006). Mechanisms of placebo analgesia: rACC recruitment of a subcortical antinociceptive network. *Pain, 120,* 8–15.

Bishop, S., Duncan, J., Brett, M., et al. (2004). Prefrontal cortical function and anxiety: Controlling attention to threat-related stimuli. *Nature Neuroscience, 7,* 184–188.

Brighina, F., Piazza, A., Vitello, G., et al. (2004). rTMS of the prefrontal cortex in the treatment of chronic migraine: A pilot study. *Journal of the Neurological Sciences, 227,* 67–71.

Bunge, S. A., Ochsner, K. N., Desmond, J. E., et al. (2001). Prefrontal regions involved in keeping information in and out of mind. *Brain, 124,* 2074–2086.

Casey, K. L., Morrow, T. J., Lorenz, J., et al. (2001). Temporal and spatial dynamics of human forebrain activity during heat pain: Analysis by positron emission tomography. *Journal of Neurophysiology, 85,* 951–959.

Coghill, R. C., Sang, C. N., Maisog, J. M., et al. (1999). Pain intensity processing within the human brain: A bilateral, distributed mechanism. *Journal of Neurophysiology, 82,* 1934–1943.

Goel, V., & Dolan, R. J. (2003). Reciprocal neural response within lateral and ventral medial prefrontal cortex during hot and cold reasoning. *NeuroImage, 20,* 2314–2321.

Hardy, S. G. P., & Haigler, H. J. (1985). Prefrontal influences upon the midbrain: A possible route for pain modulation. *Brain Research, 339,* 285–293.

Lorenz, J., & Bromm, B. (1997). Event-related potential correlates of interference between cognitive performance and tonic experimental pain. *Psychophysiology, 34,* 436–445.

Lorenz, J., Cross, D. J., Minoshima, S., et al. (2002). A unique representation of heat allodynia in the human brain. *Neuron, 35,* 383–393.

Lorenz, J., Cross, D. J., Minoshima, S., et al. (2003). Keeping pain out of mind: The role of the dorsolateral prefrontal cortex in pain modulation. *Brain, 126,* 1079–1091.

Peyron, R., Garcia-Larrea, L., Groegoire, M.-C., et al. (1999). Haemodynamic brain responses to acute pain in humans. Sensory and attentional networks. *Brain, 122,* 1765–1780.

Price, J. L., Carmichael, S. T., Drevets, W. C., et al. (1996). Networks related to the orbital and medial prefrontal cortex; a substrate for emotional behavior. In G. Holstege, R. Bandler, & C. B. Saper (Eds.), *Progress in Brain Research* (Vol. 107, pp. 523–536). Amsterdam: Elsevier.

Rodriguez-Raecke, R., Niemeier, A., Ihle, K., Ruether, W., & May, A. (2009). Brain gray matter decrease in chronic pain is the consequence and not the cause of pain. *Journal of Neuroscience, 29,* 13746–13750.

Sakai, K., Rowe, J. B., & Pasingham, R. E. (2002). Active Maintenance in prefrontal area 46 creates distractor-resistant memory. *Nature Neuroscience, 5,* 479–484.

Wager, T. D., Rilling, J. K., Smith, E. E., Sokolik, A., Casey, K. L., et al. (2004). Placebo-induced changes in FMRI in the anticipation and experience of pain. *Science, 303,* 1162–1167.

Premenstrual Dysphoric Disorder

▶ Premenstrual Syndrome

Premenstrual Syndrome

Judith A. Mikacich and Andrea J. Rapkin
Department of Obstetrics and Gynecology,
University of California Los Angeles Medical
Center, Los Angeles, CA, USA

Synonyms

PMS; Premenstrual dysphoric disorder; Premenstrual tension

Definition

Premenstrual syndrome (PMS) is defined as the presence of disabling physical and psychological symptoms during the premenstrual phase, with relief soon after the onset of menses. Formal diagnosis of PMS is made when one or more physical or psychological symptoms are present for 5 days prior to menses, with relief by the cessation of menstrual flow, in each of the three previous cycles (ACOG 2000). Prominent physical complaints include bloating, mastalgia, and increased appetite. Psychological symptoms include irritability, anxiety, mood swings, and depression. Cognitive changes including confusion and poor concentration are also reported. Behaviorally, social withdrawal and increased arguing are typically seen. The symptoms of PMS begin after ovulation and can persist through the first 4–5 days of the follicular phase of the menstrual cycle. A symptom-free interval should be present from the end of menses until the approximate time of ovulation. Symptoms should be severe enough to result in some functional impairment. Other medical or psychological diagnoses that might also explain the symptoms must be excluded. Diagnostic criteria

Premenstrual Syndrome, Table 1 Diagnostic criteria for premenstrual syndrome (PMS). PMS is diagnosed when the patient prospectively documents at least one of the following affective and somatic symptoms during the 5 days before menses for three menstrual cycles. Symptoms are of sufficient severity to impact social or economic performance. Symptoms abate during the first 5 days of the menstrual cycle and do not recur until that would explain the symptoms. There is no concomitant pharmacological therapy, hormone ingestion, or drug or alcohol abuse (Adapted from ACOG 2000)

American college of obstetricians and gynecologists	
Affective symptoms	Somatic symptoms
Depression	Breast tenderness
Angry outbursts	Abdominal bloating
Irritability	Headache
Anxiety	Swelling of extremities
Confusion	
Social withdrawal	

for PMS according to the American College of Obstetrics and Gynecology (ACOG) are listed in Table 1 (ACOG 2000, reaffirmed 2010).

Premenstrual dysphoric disorder (PMDD) is a more severe form of premenstrual disorder with a focus on the affective symptoms (Table 1) (American Psychiatric Association 2000). To meet criteria for PMDD, women must experience at least five symptoms, including one or more moderately severe to disabling premenstrual mood symptom during most of the preceding 12 menstrual cycles. Impairment from the symptoms is an important prerequisite for the PMDD diagnosis.

For PMS or PMDD, symptoms must be confirmed by prospective documentation for at least two cycles in the absence of concomitant pharmacological therapy, hormone ingestion, or drug or alcohol intake (Table 2).

Characteristics

An estimated 50–80% of menstruating women experience some physical and/or psychological symptoms during the premenstrual period. In approximately 3–5% of cases, symptoms are of sufficient severity to disrupt social or psychological functioning (Pearlstein and Steiner 2008;

Premenstrual Syndrome, Table 2 DSM-IV criteria for premenstrual dysphoric disorder (PMDD), American Psychiatric Association (Adapted from the American Psychiatric Association 2000)

DSM-IV criteria for premenstrual dysphoric disorder (PMDD) American Psychiatric Association
PMDD is diagnosed when, for most of the preceding twelve cycles, the patient meets the following criteria:
1. Experiences five or more symptoms, including at least one core symptom
Markedly depressed mood, hopelessness, self-deprecating thoughts[a]
Marked anxiety, tension[a]
Marked affective lability[a]
Persistent and marked anger or irritability[a]
Decreased interest in usual activities
Subjective sense of difficulty in concentrating
Subjective sense of being out of control
Lethargy, easy fatigability
Marked change in appetite
Hypersomnia or insomnia
Other physical symptoms, such as breast tenderness, headache, bloating
2. Reports symptoms during the last week of the luteal phase, with remission within a few days of onset of menses
3. Documents absence of symptoms during the week following menses
4. Demonstrates marked interference of symptoms with work, school, or usual social activities and relationships
5. Symptoms are not an exacerbation of another disorder
6. Prospective daily ratings confirm three of the above criteria during at least two consecutive symptomatic menstrual cycles

[a]Core symptom

Rapkin and Winer 2009). A recent study noted a 6.3% prevalence of PMDD among a community-based sample of American women; other estimates have suggested a prevalence of 3–8% of women (Soares et al. 2001). However, up to 19% of reproductive women are recognized to be "near threshold" cases that do not meet the strict DSM-IV criteria for PMDD (Freeman 2005).

Diagnosis

- A clinical diagnosis of PMS or PMDD relies on the patient's daily recording of symptoms for at least two successive menstrual cycles. Cyclical symptoms confined to the latter half of each menstrual cycle and resolving by the end of menses, as well as a clearly symptom-free interval from menses until ovulation, are critical to the diagnosis of PMS/PMDD. A validated prospective rating scale, the Daily Record of Severity of Problems (RSP) can be downloaded at www.pmdd.factsforhealth.org/drsp_month/pdf.

- Consistent postmenstrual timing of symptoms suggests an underlying affective or anxiety disorder, as does the presence of physical symptoms relating only to appetite, pain, energy, and sleep and not including breast tenderness and/or bloating.
- A complete history and physical examination should be performed, in order to exclude other diagnoses. The differential diagnosis of PMS includes various conditions subject to menstrual magnification or premenstrual exacerbation, including depressive disorders, generalized anxiety disorder, panic disorder, hypothyroidism, irritable bowel syndrome, endometriosis, anemia, chronic fatigue syndrome, fibromyalgia, systemic lupus erythematosus, perimenopause, drug or alcohol abuse, and domestic violence.
- In women with PMS or PMDD, the lifetime prevalence of major depression varies between 30% and 76%. The premenstrual disorders also frequently are comorbid with anxiety disorders and postpartum depression.

Etiology

The pathophysiology underlying the premenstrual disorders is not known. Estrogen, progesterone, testosterone, prolactin, and cortisol concentration are not abnormal in women with premenstrual symptoms. Two prominent hypotheses with theoretical evidence include the dysregulation of serotonergic activity and/or abnormalities of gamma-aminobutyric acid subunit A (GABA$_A$) receptor functioning (Birzniece et al. 2006). Decreased serotonin transmission in the brain is thought to contribute to depressed mood, irritability, anger and aggression, poor impulse control, and increased carbohydrate craving. Most, but not all, serotonergic parameters that have been evaluated are altered only during the symptomatic luteal phase in women with PMS (Rapkin and Mikacich 2001). Furthermore, serotonergic antidepressants are effective for the treatment of PMS/PMDD, but agents that augment norepinephrine or dopamine are not useful.

The inhibitory neurotransmitter GABA$_A$ is considered a primary regulator of mood and cognitive functioning. The pharmacological effect is modulated by various hormonal and other chemical modulators that bind to and secondarily alter the subunit configuration of the GABA$_A$ receptor. Allopregnanolone is a neuroactive steroid produced by the ovary, the adrenal gland, and *de novo* in the brain, as a metabolite of both cholesterol and progesterone. Similar to benzodiazepines and barbiturates, allopregnanolone binds to the chloride channel of the GABA$_A$ receptor and has potent anxiolytic and anticonvulsant effects. Rodent studies of exposure to and withdrawal from progesterone (after metabolism to allopregnanolone) reveal decreased sensitivity of GABA$_A$ receptors to GABA agonists, resulting in insensitivity to endogenous anxiolytic neurosteroids or to benzodiazepines. The decreased sensitivity has been attributed to alterations in GABA$_A$ receptor subunit composition (Smith et al. 1998; Maguire et al. 2005). Women with PMS/PMDD may also be less sensitive to the anxiolytic, sedating effects of progesterone metabolites and other neurosteroids during the luteal phase (Sundstrom et al. 2003). Diminished levels of or sensitivity to allopregnanolone and/or altered GABAergic transmission could contribute to symptoms of anxiety, irritability, and depression in women with PMS/PMDD (Rapkin et al. 1997). Additionally, women with premenstrual mood disorders may not be able to mount an appropriate increase in neuroactive steroids in response to stress (Girdler et al. 2001). Fluoxetine and paroxetine, two selective serotonin reuptake inhibitors (SSRIs), increase brain allopregnanolone content in rats and possibly in individuals with depression. The increase in allopregnanolone concentrations in the brain acting via GABA receptors may contribute to the rapid anxiolytic and antidepressant clinical action of SSRIs.

The interaction between ovarian sex steroids, the GABA$_A$ receptor complex, and/or the central serotonergic system differs in women with PMS/PMDD compared with asymptomatic women or individuals with an affective or anxiety disorder. This conclusion is based on the consistent findings in women with premenstrual disorders of (1) diminished serotonergic activity, mostly limited to the luteal phase; (2) rapid therapeutic response to the administration of serotonergic antidepressants such as selective serotonin reuptake inhibitors; and (3) alterations in neurosteroid concentrations or response to neurosteroid exposure.

Neuroimaging studies utilizing functional magnetic resonance imaging (fMRI) and positron emission tomography (PET) have also demonstrated neurobiological differences between asymptomatic women and those with PMDD. Functional MRI results in women with PMDD show a tendency toward enhanced negative emotional processing, diminished positive emotional processing, and diminished impulse control via the prefrontal cortex during the premenstrual phase (Protopopescu et al. 2008). Functional MRI studies also suggest that reward processing in the brain is influenced by menstrual cycle changes in gonadal steroids (Dreher et al. 2007). A recent PET study revealed increased cerebellar activity in the late luteal phase when compared with follicular phase in women with PMDD and asymptomatic women. This increased activity was paralleled by worsening of mood, implicating cerebellar dysfunction in PMDD (Rapkin et al. 2011). The cerebellum has a heavy concentration of GABA$_A$ receptors and increased cerebellar activity may reflect lack of appropriate GABA-mediated inhibition.

Treatment

Selective serotonin reuptake inhibitors (SSRIs) are considered the treatment of choice in PMS/PMDD and can be given daily (continuous regimen) or during the luteal phase of the menstrual cycle (intermittent regimen). Randomized controlled trials demonstrate that SSRIs are highly effective in treating affective, functional, behavioral, and even the physical premenstrual

symptoms. All of the SSRIs that have been studied, including fluoxetine; paroxetine; sertraline; fluvoxamine; citalopram; escitalopram; the serotonin-norepinephrine reuptake inhibitors (SNRIs), such as venlafaxine; and the serotonergic tricyclic antidepressant clomipramine, have all proven effective (Brown et al 2009; Cunningham et al 2009). Response is usually seen within the first few days of exposure. Intermittent treatment with SSRIs during the luteal phase, for the 14 days before menses, has also proven effective for about 60% of woman with PMDD (Halbreich et al. 2002). Doses are generally lower for the treatment of PMS/PMDD than for treatment of unipolar depression. Fluoxetine is effective at 10–20 mg/day, sertraline at 50–150 mg/day, and paroxetine continuous release at 12.5–25 mg/day. These three agents received FDA approval for PMDD. Venlafaxine at 75–112.5 mg/day used intermittently during the luteal phase resulted in significantly improved symptom scores and was well tolerated (Cohen et al. 2004).

Escitalopram at 20 mg/day was superior to 10 mg/day when used intermittently during the luteal phase. In this study, irritability, depressed mood, tension, and affective lability were decreased by 90% (Eriksson et al. 2008). Symptom-onset dosing, generally 7 days before the onset of menses, is a newer dosing strategy with promising early results. Treatment with citalopram at 10–20 mg/day beginning at the onset of symptoms was associated with significant improvement in symptom scores (Ravindran et al. 2007).

In addition to improvement in mood, other aspects of PMS/PMDD are relieved by treatment with SSRIs. SSRIs reduce the most common physical symptoms, such as bloating and breast tenderness, associated with PMS (Steiner et al. 1995). Intermittent luteal phase dosing of sertraline failed to demonstrate significant improvement in physical symptoms, but did reduce complaints of premenstrual cognitive disturbance, increased appetite, increased sleep, and lethargy (Halbreich et al. 2002). Higher doses and continuous daily therapy may be necessary for treatment of the physical symptoms.

Failure of PMDD symptoms to respond to intermittent therapy should prompt a change to continuous daily administration. If there has been inadequate response after 1–2 cycles, the dose can be increased. Lack of response should prompt switch to another SSRI or a SNRI as other drugs in a particular class may still be effective. Side effects of SSRIs and SNRIs can include gastrointestinal disturbance, headache, jitteriness, fatigue, sexual dysfunction, and weight gain.

Oral contraceptive pills (OCs) have not proven uniformly beneficial in PMS/PMDD. This is a conundrum, since ovulation is the trigger for PMS symptoms and OCs inhibit ovulation. This failure of efficacy of most OCs for PMS treatment may be related to an adverse effect of the progestagen in the particular OC on GABA receptor functioning (Pluchino et al. 2009).

A newer OC containing the progestin drospirenone, a spironolactone derivative with antiandrogenic and antimineralocorticoid effects and with a shorter pill-free interval, has shown efficacy in the treatment of PMDD in at least two studies (Yonkers et al. 2005; Pearlstein et al. 2005; Breech and Braverman 2010). In a multicenter, double-blind, placebo-controlled crossover study of women with PMDD, the OC formulation containing drospirenone 3 mg and ethinyl estradiol 20 µg in a 24/4 regimen (24 active pills followed by four placebo pills) resulted in a mean decrease in Daily Record of Severity of Problems (DRSP) scores that was significantly superior to placebo. Greatest improvements were in breast tenderness, swelling, bloating, headache, and muscle pain. Mood was also significantly improved and quality of life in the treatment group showed a twofold improvement. The same OC was found to improve health-related quality of life and general well-being in women with PMS (Borenstein et al. 2003). While a large placebo effect was noted in some studies, the drospirenone/EE 24/4 regimen OC reduced premenstrual physical and emotional symptoms in women with PMDD (Lopez et al. 2009). Continuous daily OC administration, without a break in active pills, is also under investigation for the treatment of PMS/PMDD.

The gonadotropin-releasing hormone (GnRH) agonist creates a temporary chemical oophrectomy by downregulating the production of gonadotropins and of ovarian sex steroids to menopausal levels. GnRH agonists provide symptomatic relief in a majority of women with PMS (Wyatt et al. 2004). Bone density and cardiovascular and vaginal health can theoretically be maintained and vasomotor symptoms prevented with add back hormonal therapy using estrogen, progestin (oral or intrauterine), or tibolone; however, long-term studies are lacking.

Bilateral oophorectomy with hormone add back is also effective for severe PMS and PMDD. This approach should be reserved for women with severe PMS/PMDD where other methods have failed and reproduction has been completed or if a hysterectomy is planned for concomitant pathology, such as endometriosis. In these cases, it is helpful to ensure that 6 months of treatment with a GnRH agonist with estrogen or estrogen/progestin add back has been successful.

High doses of exogenous 17 β-estradiol 100–200 mcg (one or two 100-mcg patches) used continuously will inhibit ovulation in many cases and has shown efficacy for PMS (Smith et al. 1995). This approach requires a progestin (norethindrone 5 mg orally for 7 days per month or a levonorgestrel-containing intrauterine device have been studied) to protect the endometrium from the development of hyperplasia related to unopposed estrogen. Nausea and breast tenderness are less common with the 100-mcg dose of estradiol, and with transdermal administration, the risk of venous thromboembolism is theoretically lower than the oral route. The intrauterine route produces negligible systemic progestin concentrations, which is preferable as for many women, as adequate doses of oral progestagens can reinstate PMS-like symptoms.

Several complementary and alternative medicine approaches have demonstrated some efficacy in premenstrual disorders, although studies were not well controlled. These include light therapy, certain herbal and nutritional supplements, including Vitex agnus castus (Chasteberry) and calcium 600 mg twice daily. Exercise and mind-body approaches, such as relaxation or cognitive-behavioral therapy (CBT) (Girman et al. 2003), can also be effective.

References

American College of Obstetricians and Gynecologists (ACOG). (2000, reaffirmed 2010). ACOG practice bulletin: Premenstrual syndrome. *Compendium of Selected Publications, 15*, 1–9.

American Psychiatric Association. (2000). *Diagnostic and statistical manual of mental disorder-IV* (4th ed.). Washington, DC: American Psychiatric Association.

Birzniece, V. B. T., Johansson, I. M., Lindblad, C., Lundgren, P., Lofgren, M., Olsson, T., Ragagnin, G., Taube, M., Turkmen, S., Wahlstrom, G., Wang, M. D., Wihlback, A. C., & Zhu, D. (2006). Neuroactive steroid effects on cognitive functions with a focus on the serotonin and GABA systems. *Brian Research Review, 51*(2), 212–239.

Borenstein, J., Yu, H. T., Wade, S., Chiou, C. F., & Rapkin, A. (2003). Effect of an oral contraceptive containing ethinyl estradiol and drospirenone on premenstrual symptomatology and health-related quality of life. *The Journal of Reproductive Medicine, 48*(2), 79–85.

Breech, L. L., & Braverman, P. K. (2010). Safety, efficacy, actions, and patient acceptability of drospirenone/ethinyl estradiol contraceptive pills in the treatment of premenstrual dysphoric disorder. *International Journal of Women's Health, 1*, 85–95.

Brown, J., O'Brien, P. M. S., Marjoribanks, J., & Wyatt, K. (2009). Selective serotonin reuptake inhibitors for premenstrual syndrome. *Cochrane Database of Systematic Reviews, 15*(2), CD001396.

Cohen, L. S., Soares, C. H., Lyster, A., Cassano, P., Brandes, M., & Leblanc, G. A. (2004). Efficacy and tolerability of premenstrual use of venlafaxine (flexible dose) in treatment of premenstrual dysphoric disorder. *Journal of Clinical Psychopharmacology, 24*(5), 540–543.

Cunningham, J., Yonkers, K. A., O'Brien, S., & Weiksson, E. (2009). Update on research and treatment of premenstrual dysphoric disorder. *Harvard Review of Psychiatry, 17*(2), 120–137.

Dreher, J., Schmidt, P., Kohn, P., Furman, D., Rubinow, D., & Berman, K. (2007). Menstrual, cycle phase modulates reward-related neural function in women. *PNAS, 104*, 2465–2470.

Eriksson, E., Ekman, A., Sinclair, S., Sorvik, K., Ysander, C., Mattson, U. B., & Nissbrandr, H. (2008). Escitalopram administered in the luteal phase exerts a marked and dose-dependent effect in premenstrual dysphoric disorder. *Journal of Clinical Psychopharmacology, 28*(2), 195–202.

Freeman, E. W. (2005). Effects of antidepressants in quality of life in women with premenstrual dysphoric disorder. *PharmacoEconomics, 23*(5), 433–444.

Girdler, S. S., Straneva, P. A., Light, K. C., Pedersen, C. A., & Morrow, A. L. (2001). Allopregnanolone levels reactivity to mental stress in premenstrual dysphoric disorder. *Biological Psychiatry, 49*, 788–797.

Girman, A., Lee, R., & Kligler, B. (2003). An integrative medicine approach to premenstrual syndrome. *American Journal of Obstetrics and Gynecology, 188*, S56–S65.

Halbreich, U., Bergeron, R., Yonkers, K. A., Freeman, E., Stout, A. L., & Cohen, L. (2002). Efficacy of intermittent, luteal phase sertraline treatment of premenstrual dysphoric disorder. *Obstetrics and Gynecology, 100*(6), 1219–1229.

Lopez, L. M., Kaptein, A. A., & Helmerhorst, F. M. (2009). Oral contraceptives containing drospirenone for premenstrual syndrome. *Cochrane Database of Systematic Reviews, 15*(2), CD006586.

Maguire, J., Stell, B., Rafizadeh, M., & Mody, I. (2005). Ovarian cycle-linked changes in $GABA_A$ receptors mediating tonic inhibition alter seizure susceptibility and anxiety. *Nature Neuroscience, 8*(6), 797–804.

Pearlstein, T., & Steiner, M. (2008). Premenstrual dysphoric disorder: Burden of illness and treatment update. *Journal of Psychiatry & Neuroscience, 33*(4), 291–301.

Pearlstein, T. B., Bachmann, G. A., Zacur, H. A., & Yonkers, K. A. (2005). Treatment of premenstrual dysphoric disorder with a new drospirenone-containing oral contraceptive formulation. *Contraception, 72*, 414–421.

Pluchino, N., Cubeddu, A., Giannini, A., Merlini, S., Cela, V., Angioni, S., & Genazazani, A. R. (2009). Progestogens and brain: An update. *Maturitas, 62*(4), 349–355.

Protopopescu, Z., Tuescher, O., Pan, H., Epstein, J., Root, J., Chang, L., Altemus, M., Polanecsky, M., McEwen, B., Stern, E., & Silbersweig, D. (2008). Toward a functional neuroanatomy of premenstrual dysphoric disorder. *Journal of Affective Disorders, 108*(1–2), 87–94.

Rapkin, A. J., & Mikacich, J. A. (2001). Premenstrual syndrome: Gynaecology or psychiatry? *Reproductive Medicine Review, 9*, 223–239.

Rapkin, A. J., & Winer, S. A. (2009). Premenstrual syndrome and premenstrual dysphoric disorder: Quality of life and burden of illness. *Expert Review of Pharmacoeconomics & Outcomes Research, 9*(2), 157–170.

Rapkin, A. J., Morgan, M., Goldman, L., Brann, D. W., Simone, D., & Mahesh, V. B. (1997). Progesterone metabolite allopregnanolone in women with premenstrual syndrome. *Obstetrics and Gynecology, 90*(5), 709–714.

Rapkin, A., Berman, S., Mandelkern, N., Silverman, D., Morgan, M., & London, E. (2011). Neuroimaging evidence of cerebellar involvement in premenstrual dysphoric disorder. *Biological Psychiatry, 69*, 374–380.

Ravindran, L. N., Woods, S. A., Steiner, M., & Ravindran, A. V. (2007). Symptom-onset dosing with citalopram in the treatment of premenstrual dysphoric disorder (PMDD): A case series. *Archieves of Women's Mental Health, 10*(3), 125–127.

Smith, R. N., Studd, J. W., Zamblera, D., & Holland, E. F. (1995). A randomised comparison over 8 months of

100 micrograms and 200 micrograms twice weekly doses of transdermal oestradiol in the treatment of severe premenstrual syndrome. *British Journal of Obstetrics and Gnaecology, 102*(6), 475–484.

Smith, S. S., Gong, Q. H., Hsu, F. C., Markowitz, R. S., ffrench-Mullen, M. H., & Li, X. (1998). GABA_A receptor alpha-4 subunit suppression prevents withdrawal properties of an endogenous steroid. *Nature, 392*, 926–929.

Soares, C. N., Cohen, L. S., Otto, M. W., & Harlow, B. L. (2001). Characteristics of women with premenstrual dysphoric disorder (PMDD) who did or did not report history of depression: A preliminary report from the Harvard Study of Moods and Cycles. *Journal of Women's Health and Gender-Based Medicine, 10*(9), 873–878.

Steiner, M., Steinberg, S., Stewart, D., Carter, D., Berger, C., Reid, R., Grover, D., & Streiner, D. (1995). Fluoxetine in the treatment of premenstrual dysphoria. Canadian fluoxetine/premenstrual dysphoria collaborative study group. *The New England Journal of Medicine, 332*(23), 1529–1534.

Sundstrom, P. I., Smith, S., & Gulinello, M. (2003). GABA receptors, progesterone and premenstrual dysphoric disorder. *Archieves of Women's Health, 6*(1), 23–41.

Wyatt, K. M., Dimmock, P. W., Ismail, K. M. K., Jones, P. W., & O'Brien, P. M. S. (2004). The effectiveness of GnRHa with and without "add-back" therapy in treating premenstrual syndrome: A media analysis. *BJOG: An International Journal of Obstetrics and Gynaecology, 111*, 585–593.

Yonkers, K. A., Brown, C., & Pearlstein, T. B. (2005). Efficacy of a new low-dose oral contraceptive with drosperinone in premenstrual dysphoric disorder. *Obstetrics and Gynecology, 106*, 493–501.

Premenstrual Tension

▶ Premenstrual Syndrome

Premorbid Functioning

Definition

Premorbid functioning refers to the state of existence before the occurrence of physical disease or emotional illness.

Cross-References

▶ Pain Inventories

Preoperative Education

Definition

Preoperative education includes instructions concerning techniques of postoperative analgesia, mobilization, and respiratory care (coughing and other physiotherapy maneuvers). This aids coping and reduces perioperative anxiety.

Cross-References

▶ Postoperative Pain, Importance of Mobilization

Preoperative Patient Education

▶ Postoperative Pain, Preoperative Education

Preoperative Teaching

▶ Postoperative Pain, Preoperative Education

Prepaid Care

Definition

A fee paid in advance by an employer and/or employee to cover all anticipated health-care costs.

Cross-References

▶ Disability Management in Managed Care System

Preparation

Definition

A readiness to change stage in which a person states an intention to change soon and in fact may be making initial steps towards changing behavior.

Cross-References

▶ Motivational Aspects of Pain
▶ Psychological Treatment in Acute Pain

Preparation Programs

▶ Psychological Treatment of Pain in Children

Prepuce Clitoris

Definition

This is the hood of the clitoris, comparable to the male foreskin, which may completely or incompletely cover the glans.

Cross-References

▶ Clitoral Pain

Present Pain Intensity Subscale

Synonyms

PPI

Definition

Zero to five scale, with word anchors for the subjective measurement of pain intensity.

Cross-References

▶ Cancer Pain, Assessment in the Cognitively Impaired

Pressing/Tightening

Definition

Pain of a constant quality often compared to an iron band around the head.

Cross-References

▶ Headache, Episodic Tension-Type

Pressure Algometer

Definition

A pressure algometer is a device to measure the detection threshold or tolerance threshold of pressure-induced pain.

Cross-References

▶ Headache, Episodic Tension-Type

Presynaptic

Definition

Events occurring in the axonal terminal before the neurotransmitter crosses into the synapse.

Cross-References

▶ Amygdala, Pain Processing and Behavior in Animals

Presynaptic Inhibition

Definition

Presynaptic inhibition is a form of synaptic inhibition in which the release of a transmitter, in response to the invasion of the action potential into the synaptic terminals, is reduced by the activation of an axo-axonic GABAergic synapse. In such cases, this is the result of a depolarization of primary afferent terminals by activation of GABA$_A$ receptors. In other cases, transmitter release is reduced because of a reduction in the Ca++ influx required for transmitter release. An example is the presynaptic inhibition produced by GABA action on GABA$_B$ receptors on presynaptic endings. Presynaptic inhibition is known to occur on primary afferent terminals that drive spinal cord projection neurons.

Cross-References

▶ Central Changes After Peripheral Nerve Injury
▶ GABA and Glycine in Spinal Nociceptive Processing
▶ GABA Mechanisms and Descending Inhibitory Mechanisms

Presynaptic Reuptake

Definition

Pumping of neurotransmitter from synaptic cleft back into the presynaptic neuron.

Cross-References

▶ Antidepressants in Neuropathic Pain

Prevalence

Definition

The number of events in a given population at a designated time, e.g., the proportion of individuals in the population with a disease at a particular time. Prevalence is determined by the average incidence and average duration of disease.

Cross-References

▶ Low Back Pain, Epidemiology
▶ Migraine Epidemiology
▶ Pain in the Workplace, Risk Factors for Chronicity, Job Demands
▶ Prevalence of Chronic Pain Disorders in Children
▶ Psychiatric Aspects of the Epidemiology of Pain
▶ Psychological Aspects of Pain in Women

Prevalence of Chronic Pain Disorders in Children

Sara King
Faculty of Education (School Psychology),
Mount Saint Vincent University, Halifax, NS,
canada

Synonyms

Childhood chronic pain; Chronic pain prevalence; Epidemiology; Period prevalence; Point prevalence

Definition

The International Association for the Study of Pain (IASP) defines chronic pain as pain lasting more than 3 months (Merskey et al. 1994). Chronic pain can be either recurrent (i.e., periods of pain interspersed with pain-free periods) or persistent (i.e., pain that does not remit). Prevalence refers to the total number of cases of a particular condition at a given time. Lifetime prevalence refers to the number of individuals in a given population who have, in their lifetime, experienced a particular condition. The current review will present aggregated prevalence rates as well as correlates (i.e., age, sex, and psychosocial functioning) of pediatric chronic headache, abdominal pain, back pain, musculoskeletal pain, multiple pains, and other pains.

Introduction

Pediatric chronic pain is a serious developmental health concern that has been shown to interfere significantly with the daily functioning of affected children (Palermo 2000; Palermo and Chambers 2005). For example, children challenged by chronic or recurrent pain often miss school, withdraw from social and extracurricular activities, and have been shown to be at risk of developing internalizing symptoms, such as anxiety and depression as a result of their pain condition (McGrath et al. 1986; Oddson et al. 2006; Palermo 2000). Given the consequences of pediatric chronic pain, it is important for researchers and clinicians to understand the epidemiology of this condition for the purposes of treatment development and implementation. However, despite the importance of understanding chronic pain in children, epidemiological studies of the condition have yielded mixed findings and large variability in prevalence rates. Until recently, the most current comprehensive review of pediatric pain epidemiology was published over 20 years ago (Goodman and McGrath 1991). In their 1991 review, Goodman and McGrath identified several

methodological limitations of the existing studies and made suggestions for the improvement of epidemiological studies of pediatric chronic pain. An updated review of the epidemiology of pediatric chronic pain was published in 2011 (King et al. 2011) with the goal of updating the Goodman and McGrath paper and providing a systematic examination of epidemiological studies of pediatric chronic and recurrent pain. The goal of the current review is to provide an abbreviated version of the King et al. paper and to summarize the current state of the literature in the area of pediatric chronic pain.

Characteristics

Forty-one epidemiological studies of pediatric chronic pain were published between 1991 and 2009. Sample sizes ranged from 495 to 20,745, with the majority of studies including more than 1,000 participants. Of these studies, there was a great deal of variability with respect to definitions of chronic pain and the reporting periods used to assess prevalence rates. Specifically, reports of prevalence ranged from point prevalence to lifetime prevalence, with many studies reporting monthly and weekly prevalence rates. Participants ranged in age from less than 1 year (Perquin et al. 2000) to 18 years (Barea et al. 1996; Perquin et al. 2000; Rhee 2000; Rhee et al. 2005; Stanford et al. 2008; Zapata et al. 2006). Prevalence rates varied substantially, ranging from 4 % for abdominal pain, musculoskeletal pain, and multiple pains to 88 % for other pains. King et al. (2011) note that prevalence rates were typically higher among girls and generally increased with age in both boys and girls. Finally, using a set of quality criteria developed specifically for the purposes of the review, King et al. (2011) found that the majority of the studies did not meet criteria.

Prevalence Estimates (1991–2009)

Key electronic databases (i.e., EMBASE, Medline, CINAHL, and PsycINFO) were

Prevalence of Chronic Pain Disorders in Children, Table 1 Studies of headache prevalence

Prevalence estimate	Authors/country	Number of participants	Age
Yearly	Abu-Arefeh and Russell (1994, United Kingdom)	1,754	5–15
Yearly	Barea et al. (1996; Brazil)	538	10–18
Yearly	Rhee (2000; United States)	6,072	11–18
6-monthly	Sillanpaa et al. (1996; Finland)	1,436	7
3-monthly	Fendrich et al. (2007; Germany)	3,072	12–15
Monthly	Bugdayci et al. (2005; Turkey)	5,562	7–10
Monthly	Carlsson (1996; Sweden)	1,297	7–16
Weekly	Bandell-Hoekstra et al. (2001; Netherlands)	2,358	10–17
Weekly	Brun Sundblad et al. (2007; Sweden)	1,908	9–15
Weekly	Kristjansdottir (1993; Iceland)	2,140	11–12, 15–16
Weekly	Petersen (2003; Sweden)	1,121	6–13
Weekly	Stanford (2008; Canada)	2,488	10–18

Prevalence of Chronic Pain Disorders in Children, Table 2 Studies of abdominal pain prevalence

Prevalence estimate	Authors/country	Number of participants	Age
≥3 months	Boey and Yap (1999; Malaysia)	148	11–12
≥3 months	Boey et al. (2000; Malaysia)	1,549	11–16
≥3 months	Ostkirchen et al. (2006; Germany)	555	5–7
≥3 months	Ramchandani et al. (2005; United Kingdom)	13,971	2–6
Weekly	Brun Sundblad et al. (2007; Sweden)	1,908	9–15
Weekly	Kristjansdottir (1996; Iceland)	2,162	11–12, 15–16
Weekly	Petersen (2003; Sweden)	1,155	6–13
Weekly	Stanford et al. (2008; Canada)	2,488	10–18

searched for articles published between 1991 and 2009 using the following search terms: child, adolescent, chronic pain, pediatric pain, recurrent pain, prevalence, incidence, pain epidemiology, headache, abdominal pain, musculoskeletal pain, and back pain. This search resulted in a total of 185 citation abstracts, of which 41 were deemed to meet the inclusion criteria for the review. The most frequently studied type of chronic pain in children and adolescents was headache, with 12 studies examining prevalence (Stanford et al. 2008; Brun Sundblad et al. 2007; Fendrich et al. 2007; Bugdayci et al. 2005; Petersen et al. 2003; Bandell-Hoekstra et al. 2001; Rhee 2000; Barea et al. 1996; Carlsson 1996; Sillanpaa and Anttila 1996; Abu-Arefeh and Russell 1994; Kristjansdottir and Wahlberg 1993) (see Table 1). A total of 29,746 participants were included in studies of headache, and prevalence rates ranged from 8 % (weekly) to 83 % (yearly). In general, headaches were found to be more common in girls and prevalence rates increased with increasing age. Prevalence of abdominal pain was also frequently studied (see Table 2), with nine studies examining this pain condition (Stanford et al. 2008; Brun Sundblad et al. 2007; Ostkirchen et al. 2006; Ramchandani et al. 2005; Oh et al. 2004; Petersen et al. 2003; Boey et al. 2000; Boey and Yap 1999; Kristjansdottir 1996b). A total of 27,526 participants were included in studies of abdominal pain, and prevalence rates ranged from 4 % (≥3 month) to 53 % (weekly). As with headache, abdominal pain was more common among girls, but findings were mixed with respect to age differences in abdominal pain prevalence rates. A total of four studies examined prevalence rates of back pain in children and adolescents (Stanford et al. 2008; Petersen et al. 2003; Watson et al. 2002; Kristjansdottir 1996a),

Prevalence of Chronic Pain Disorders in Children, Table 3 Studies of back pain prevalence

Prevaelnce estimate	Authors/country	Number of participants	Age
Monthly	Watson et al. (2002; United Kingdom)	1,446	11–14
Weekly	Kristjansdottir (1996; Iceland)	2,173	11–12, 15–16
Weekly	Petersen et al. (2003; Sweden)	1,155	8–13
Weekly	Stanford et al. (2008; Canada)	2,488	10–18

Prevalence of Chronic Pain Disorders in Children, Table 4 Studies of musculoskeletal pain prevalence

Prevalence estimate	Authors/country	Number of participants	Age
6-monthly	Zapata et al. (2006; Brazil)	791	10–18
Monthly	Mikkelsson et al. (1997; Finland)	1,626	9–11
Weekly	Brun Sundblad et al. (2007; Sweden)	1,908	9–15
Weekly	Mikkelsson et al. (1997; Finland)	1,626	9–11
Weekly	Smedbraten et al. (1998; Norway)	569	10–15
Point prevalence	Vahasarja (1995; Finland)	856	9–10, 14–15

Prevalence of Chronic Pain Disorders in Children, Table 5 Studies of combined pain prevalence

Prevalence estimate	Authors/country	Number of participants	Age
Yearly	Rhee et al. (2005; United States)	20,745	11–18
Monthly	Kristjansdottir (1997; Iceland)	2,173	11–12, 15–16
Weekly	Bakoula et al. (2006; Greece)	7,925	7
Weekly	Stanford et al. (2008; Canada)	2,488	10–18
Point prevalence	Ostkirchen et al. (2006; Germany)	555	5–7

for a total of 7,262 participants (see Table 3). Prevalence rates for back pain ranged from 9 % (weekly or "at least weekly") to 24 % (monthly). Sex differences were not as clear with respect to back pain compared to other pain types; however, the studies that identified sex differences (Stanford et al. 2008; Watson et al. 2002) found higher prevalence rates in girls as compared to boys. Prevalence rates of back pain increased with increasing age. A total of five studies examined musculoskeletal/limb pain in children and adolescents (Brun Sundblad et al. 2007; Zapata et al. 2006; Smedbraten et al. 1998; Mikkelsson et al. 1997; Vahasarja 1995) (see Table 4). A total of 5,750 participants were included in these studies, and prevalence rates ranged from 4 % (point prevalence) to 40 % (6 month). The majority of these studies found that this type of pain is more common in girls than in boys; however, one large-scale study (Rhee et al. 2005)

found that boys report higher rates of musculoskeletal pain than girls. Prevalence rates of musculoskeletal pain increased with age. Five studies, yielding a total of 33,886 participants examined multiple/combined pain (Stanford et al. 2008; Bakoula et al. 2006; Ostkirchen et al. 2006; Rhee et al. 2005; Kristjansdottir 1997) (see Table 5). Pain type varied widely in these studies; therefore, prevalence rates were also highly variable. The prevalence of multiple/combined pains ranged from 4 % (weekly) to 78 % (monthly). Sex differences in prevalence rates for multiple/combined pains were generally consistent with the pattern observed in other types of pain, namely, this type of pain is more prevalent in girls than in boys. Findings with respect to age differences in prevalence rates were less clear, likely due to the wide variety of pain combinations that were included in these studies.

Prevalence of Chronic Pain Disorders in Children, Table 6 Studies of other/general pain prevalence

Prevalence estimate	Authors/country	Number of participants	Age
3-monthly	Perquin et al. (2000; Netherlands)	5,424	0–18
Monthly	van Dijk et al. (2008; Canada)	495	9–13
Point prevalence	Barajas et al. (2001; Spain)	571	6–15
Point prevalence	Huguet and Miro (2008; Spain)	561	8–16
Point prevalence	van Dijk et al. (2006; Canada)	495	9–13

Finally, five studies reported prevalence rates for general (i.e., nonspecific) pain in children and adolescents (Huguet and Miro 2008; van Dijk et al. 2006, 2008; Barajas et al. 2001; Perquin et al. 2000) (see Table 6). A total of 7,051 participants were included in studies of general pain, and prevalence rates ranged from 5 % (point prevalence) to 88 % (3 month). General pain was generally found to be more prevalent in girls as compared to boys. With respect to age differences in prevalence rates of general pain, findings were mixed; however, one study (Barajas et al. 2001) noted an interesting sex x age interaction such that higher prevalence rates were observed in girls aged 6–7 years and in boys aged −10 years.

In their seminal review of the epidemiology of pain in children, Goodman and McGrath (1991) identified several methodological limitations of the studies they reviewed. These limitations made it difficult to accurately assess prevalence rates of chronic pain in children and, as such, these authors provided several recommendations to strengthen further epidemiological studies in this area. Given these recommendations, King et al. (2011) developed a set of criteria with which to examine the quality of recent epidemiological studies of chronic and recurrent pain in children. Examples of quality criteria employed by King et al. (2011) included use of an unbiased sample, use of a large sample, operational definition of pain, clearly defined recall period, and presentation of demographic information (see King et al. (2011) for a full list of quality criteria). Generally, findings from the King et al. (2011) study indicated that the quality of the majority of studies included in this systematic review was low to moderate. Areas

of strength included randomized sampling procedures, high response rates, and providing information regarding the types of questions used to assess pain in children. Areas of weakness included using convenience samples, failing to include analysis of nonresponders, and omission of demographic information of participants such as age, sex, and socioeconomic status (SES).

Impact of Pain on Functioning

Pediatric chronic and recurrent pain has been described as an important developmental health concern that has the potential to affect various domains of child and family quality of life and functioning (Palermo 2000; Palermo and Chambers 2005; Perquin et al. 2000). Perquin et al. (2000) note that one third of children continue to experience significant pain at 2-year follow-up. They also found that pain impacted health-related factors such as medication use, health care use, and health status. Various authors have described the academic and psychosocial consequences of chronic and recurrent pain (e.g., Simons et al. 2010; Logan et al. 2009; Oddson et al. 2006; Palermo 2000; McGrath et al. 1986). For example, children who experience chronic and recurrent pain are at risk for disrupted peer relationships and withdrawal from social activities (Forgeron et al. 2010; McGrath et al. 1986), academic consequences such as school failure and decline in grades (Logan et al. 2009), disrupted family functioning (Palermo 2000), and development of internalizing disorders such as anxiety and depression (Oddson et al. 2006). In a comprehensive systematic review, King et al. (2011) found that many types of chronic and recurrent pain in children (especially headache) were associated with low socioeconomic status (SES) in families of children who experience pain. It should be noted, however, that many studies of chronic and recurrent pain in children did not systematically examine psychosocial functioning as a primary focus of the study (see Forgeron et al. 2010 for a review), meaning that the precise psychosocial sequelae of pain conditions are not entirely clear. It is imperative that future studies systematically and reliably examine the psychosocial factors associated with pain conditions so that interventions can be more streamlined to specific pain types and developmental stages.

Summary

According to a recent systematic review of epidemiological studies of chronic and recurrent pain in children (King et al. 2011), prevalence rates vary widely and are largely dependent on the reporting period used in the original study. The most commonly studied and reported pain type was headache, with an estimated median prevalence rate of 23 %. Abdominal pain, back pain, musculoskeletal pain, and various other pains and pain combinations were also reported frequently, with prevalence rates ranging widely, depending on study methodology. Girls generally reported experiencing more pain than boys and pain complaints generally increased with increasing age, although abdominal pain was noted to be common in younger children. Given the increase in pain complaints with increasing age, it is important to identify pain and intervene at a young age to decrease the likelihood of developing a chronic pain condition that extends into adolescence and adulthood. Conducting more longitudinal pain research will be key in determining the relation between childhood pain complaints and adolescent/adult chronic pain conditions.

There is a dearth of research examining the psychosocial and sociodemographic factors (e.g., SES, psychological/mental health functioning, parental factors, and peer factors) in pediatric chronic and recurrent pain. Indeed,

Forgeron et al. (2010) note that psychosocial factors are often included as an afterthought in studies of pediatric pain and are assessed in a very rudimentary fashion. Future studies should consider assessing psychosocial and sociodemographic factors in a more systematic and prescribed fashion to facilitate understanding of the contribution and impact of these variables to the development and maintenance of pain, as well as the impact of pain on these factors. King et al. (2011) state that more focus on these factors may lead to earlier identification of children and families who may benefit from medical or psychological interventions for pain. The goal in delivering more effective and targeted interventions should be to prevent the onset of chronic pain conditions in children and adolescents.

There is little consistency in existing studies of pediatric pain epidemiology with respect to operational definitions of pain, accurate reporting of pain (i.e., intensity, frequency, and duration), and inclusion of psychosocial and sociodemographic factors related to pain. This is perhaps the area most in need of improvement, as recommendations to increase study quality and consistency were made by Goodman and McGrath in their 1991 review. Very little progress has been made in this area; it is imperative to design epidemiological studies with these quality criteria in mind to facilitate combination of studies to provide accurate prevalence rates and to improve developmental models of chronic and recurrent pain.

In summary, despite limitations in the current body of literature, it is clear that chronic and recurrent pain is highly prevalent in pediatric populations and that this condition has far-reaching implications for functioning. Future epidemiological studies in this area should consider applying consistent quality criteria, as well as inclusion of systematic assessment of psychosocial and sociodemographic variables. Continued focus in the area of pediatric chronic and recurrent pain may result in increased quality of life, decreased rates of pain-related disability, and improved health and psychosocial outcomes for children and families challenged by chronic and recurrent pain.

Acknowledgment The author wishes to acknowledge the support of the Canadian Child Health Clinician Scientist Program, a Canadian Institutes of Health Research Strategic Initiative in partnership with SickKids Foundation, Child and Family Research Institute BC, Women and Children's Health Research Institute (Alberta), and Manitoba Institute of Child Health.

References

Abu-Arefeh, I., & Russell, G. (1994). Prevalence of headache and migraine in schoolchildren. *BMJ (Clinical Research Ed.), 309*(6957), 765–769.

Bakoula, C., Kapi, A., Veltsista, A., Kavadias, G., & Kolaitis, G. (2006). Prevalence of recurrent complaints of pain among Greek schoolchildren and associated factors: A population-based study. *Acta Paediatrica (Oslo, Norway: 1992), 95*(8), 947–951. doi:10.1080/08035250600684453.

Bandell-Hoekstra, I. E., Abu-Saad, H. H., Passchier, J., Frederiks, C. M., Feron, F. J., & Knipschild, P. (2001). Prevalence and characteristics of headache in Dutch schoolchildren. *European Journal of Pain (London, England), 5*(2), 145–153. doi:10.1053/eujp.2001.0234.

Barajas, C., Bosch, F., & Banos, J. (2001). A pilot survey of pain prevalence in schoolchildren. *The Pain Clinic, 13*(2), 95–102.

Barea, L. M., Tannhauser, M., & Rotta, N. T. (1996). An epidemiologic study of headache among children and adolescents of southern Brazil. *Cephalalgia: An International Journal of Headache, 16*(8), 545–549. discussion 523.

Boey, C. C., & Yap, S. B. (1999). An epidemiological survey of recurrent abdominal pain in a rural malay school. *Journal of Paediatrics and Child Health, 35*(3), 303–305.

Boey, C. C., Yap, S., & Goh, K. L. (2000). The prevalence of recurrent abdominal pain in 11- to 16-year-old Malaysian schoolchildren. *Journal of Paediatrics and Child Health, 36*(2), 114–116.

Brun Sundblad, G. M., Saartok, T., & Engstrom, L. M. (2007). Prevalence and co-occurrence of self-rated pain and perceived health in school-children: Age and gender differences. *European Journal of Pain (London, England), 11*(2), 171–180. doi:10.1016/j.ejpain.2006.02.006.

Bugdayci, R., Ozge, A., Sasmaz, T., Kurt, A. O., Kaleagasi, H., Karakelle, A., & Siva, A. (2005). Prevalence and factors affecting headache in Turkish schoolchildren. *Pediatrics International: Official Journal of the Japan Pediatric Society, 47*(3), 316–322. doi:10.1111/j.1442-200x.2005.02051.x.

Carlsson, J. (1996). Prevalence of headache in schoolchildren: Relation to family and school factors. *Acta Paediatrica (Oslo, Norway: 1992), 85*(6), 692–696.

Fendrich, K., Vennemann, M., Pfaffenrath, V., Evers, S., May, A., Berger, K., & Hoffmann, W. (2007). Headache prevalence among adolescents–the German DMKG headache study. *Cephalalgia: An International Journal of Headache, 27*(4), 347–354. doi:10.1111/j.1468-2982.2007.01289.x.

Forgeron, P. A., King, S., Stinson, J. N., McGrath, P. A., MacDonald, A. J., & Chambers, C. T. (2010). Social functioning and peer relationships in children and adolescents with chronic pain: A systematic review. *Pain Research & Management, 15*(1), 27–41.

Goodman, J. E., & McGrath, P. J. (1991). The epidemiology of pain in children and adolescents: A review. *Pain, 46*(3), 247–264. doi:10.1016/0304-3959(91)90108-A.

Huguet, A., & Miro, J. (2008). The severity of chronic pediatric pain: An epidemiological study. *The Journal of Pain: Official Journal of the American Pain Society, 9*(3), 226–236. doi:10.1016/j.jpain.2007.10.015.

King, S., Chambers, C. T., Huguet, A., MacNevin, R. C., McGrath, P. J., Parker, L., & MacDonald, A. J. (2011). The epidemiology of chronic pain in children and adolescents revisited: A systematic review. *Pain, 152*(12), 2729–2738.

Kristjansdottir, G. (1996a). Prevalence of self-reported back pain in school children: A study of sociodemographic differences. *European Journal of Pediatrics, 155*(11), 984–986.

Kristjansdottir, G. (1996b). Sociodemographic differences in the prevalence of self-reported stomach pain in school children. *European Journal of Pediatrics, 155*(11), 981–983.

Kristjansdottir, G. (1997). Prevalence of pain combinations and overall pain: A study of headache, stomach pain and back pain among school-children. *Scandinavian Journal of Social Medicine, 25*(1), 58–63.

Kristjansdottir, G., & Wahlberg, V. (1993). Sociodemographic differences in the prevalence of self-reported headache in icelandic school-children. *Headache, 33*(7), 376–380.

Logan, D. E., Simons, L. E., & Kaczynski, K. J. (2009). School functioning in adolescents with chronic pain: The role of depressive symptoms in school impairment. *Journal of Pediatric Psychology, 34*(8), 882–892. doi:10.1093/jpepsy/jsn143.

McGrath, P. J., Dunn-Gier, J., Cunningham, S. J., Brunette, R., D'Astous, J., Humphreys, P., & Keene, D. (1986). Psychological guidelines for helping children cope with chronic benign intractable pain. *The Clinical Journal of Pain, 1*, 229–233.

Merskey, H., Bogduk, N., & International Association for the Study of Pain. (1994). *Classification of chronic pain: Descriptions of chronic pain syndromes and definitions of pain terms.* Seattle: IASP Press.

Mikkelsson, M., Salminen, J. J., & Kautiainen, H. (1997). Non-specific musculoskeletal pain in preadolescents. Prevalence and 1-year persistence. *Pain, 73*(1), 29–35.

Oddson, B. E., Clancy, C. A., & McGrath, P. J. (2006). The role of pain in reduced quality of life and depressive symptomology in children with spina bifida. *The Clinical Journal of Pain, 22*(9), 784–789.

Oh, M. C., Aw, M. M., Chan, Y. H., Tan, L. Z., & Quak, S. H. (2004). Epidemiology of recurrent abdominal pain among singaporean adolescents. *Annals of the Academy of Medicine, Singapore, 33*(Suppl. 5), S10–S11.

Ostkirchen, G. G., Andler, F., Hammer, F., Pohler, K. D., Snyder-Schendel, E., Werner, N. K., & Diener, H. C. (2006). Prevalences of primary headache symptoms at school-entry: A population-based epidemiological survey of preschool children in Germany. *The Journal of Headache and Pain, 7*(5), 331–340. doi:10.1007/s10194-006-0325-z.

Palermo, T. M. (2000). Impact of recurrent and chronic pain on child and family daily functining: A critical review of the literature. *Journal of Developmental and Behavioral Pediatrics, 21*, 58–69.

Palermo, T. M., & Chambers, C. T. (2005). Parent and family factors in pediatric chronic pain and disability: An integrative approach. *Pain, 119*, 1–4.

Perquin, C. W., Hazebroek-Kampschreur, A. A. J. M., Hunfeld, J. A. M., Bohnen, A. M., van Suijlekom-Smit, L. W. A., Passchier, J., & van der Wouden, J. C. (2000). Pain in children and adolescents: A common experience. *Pain, 87*, 51–58.

Petersen, S., Bergstrom, E., & Brulin, C. (2003). High prevalence of tiredness and pain in young schoolchildren. *Scandinavian Journal of Public Health, 31*(5), 367–374.

Ramchandani, P. G., Hotopf, M., Sandhu, B., Stein, A., & ALSPAC Study Team. (2005). The epidemiology of recurrent abdominal pain from 2 to 6 years of age: Results of a large, population-based study. *Pediatrics, 116*(1), 46–50. doi:10.1542/peds.2004-1854.

Rhee, H. (2000). Prevalence and predictors of headaches in US adolescents. *Headache, 40*(7), 528–538.

Rhee, H., Miles, M. S., Halpern, C. T., & Holditch-Davis, D. (2005). Prevalence of recurrent physical symptoms in U.S. adolescents. *Pediatric Nursing, 31*(4), 314–319, 350.

Sillanpaa, M., & Anttila, P. (1996). Increasing prevalence of headache in 7-year-old schoolchildren. *Headache, 36*(8), 466–470.

Simons, L. E., Logan, D. E., Chastain, L., & Stein, M. (2010). The relation of social functioning to school impairment among adolescents with chronic pain. *The Clinical Journal of Pain, 26*(1), 16–22. doi:10.1097/AJP.0b013e3181b511c2.

Smedbraten, B. K., Natvig, B., Rutle, O., & Bruusgaard, D. (1998). Self-reported bodily pain in schoolchildren. *Scandinavian Journal of Rhuematology, 27*, 273–276.

Stanford, E. A., Chambers, C. T., Biesanz, J. C., & Chen, E. (2008). The frequency, trajectories and predictors of adolescent recurrent pain: A population-based approach. *Pain, 138*(1), 11–21. doi:10.1016/j.pain.2007.10.032.

Vahasarja, V. (1995). Prevalence of chronic knee pain in children and adolescents in northern Finland. *Acta Paediatrica (Oslo, Norway: 1992), 84*(7), 803–805.

van Dijk, A., McGrath, P. A., Pickett, W., & VanDenKerkhof, E. G. (2006). Pain prevalence in nine- to 13-year-old schoolchildren. *Pain Research & Management: The Journal of the Canadian Pain Society (Journal De La Societe Canadienne Pour Le Traitement De La Douleur), 11*(4), 234–240.

van Dijk, A., McGrath, P. A., Pickett, W., & Van Den Kerkhof, E. G. (2008). Pain and self-reported health in canadian children. *Pain Research & Management: The Journal of the Canadian Pain Society (Journal De La Societe Canadienne Pour Le Traitement De La Douleur), 13*(5), 407–411.

Watson, K. D., Papageorgiou, A. C., Jones, G. T., Taylor, S., Symmons, D. P., Silman, A. J., & Macfarlane, G. J. (2002). Low back pain in schoolchildren: Occurrence and characteristics. *Pain, 97*(1–2), 87–92.

Zapata, A. L., Moraes, A. J., Leone, C., Doria-Filho, U., & Silva, C. A. (2006). Pain and musculoskeletal pain syndromes in adolescents. *The Journal of Adolescent Health: Official Publication of the Society for Adolescent Medicine, 38*(6), 769–771. doi:10.1016/j.jadohealth.2005.05.018.

Prevalence of Low Back Pain

▶ Low Back Pain, Epidemiology

Preventative Rehabilitation

▶ Cancer Pain Management, Rehabilitative Therapies

Prevention of Postoperative Pain

▶ Postoperative Pain, Preemptive or Preventive Analgesia

Preventive Analgesia

▶ Postoperative Pain, Preemptive or Preventive Analgesia

Preventive Migraine Therapy

Stephen D. Silberstein
Jefferson Medical College, Thomas Jefferson
University and Jefferson Headache Center,
Thomas Jefferson University Hospital,
Philadelphia, PA, USA

Definition

▶ Migraine is a primary episodic headache
disorder characterized by various combinations
of neurologic, gastrointestinal, and autonomic
changes. Diagnosis is based on the headache's
characteristics and associated symptoms
(Silberstein et al. 2001). The International Headache Society's diagnostic criteria for headache
disorders were recently revised and provide
criteria for a total of seven subtypes of migraine
(Headache Classification Committee 2013). This
entry relied on the Technical Reports of the
Agency for Healthcare Policy and Research
(Goslin et al. 1994; Gray et al. 1999) and the US
Headache Consortium Guidelines (Ramadan
et al. 1999). It was updated by the practice parameter of the AAN (Silberstein et al. 2012).

Characteristics

Treatment

Migraine treatment begins with making
a diagnosis (Silberstein et al. 2001), explaining it
to the patient, and developing a treatment plan that
considers coincidental or comorbid conditions.
Headache calendars record headache duration,
severity, and treatment response. ▶ Comorbidity
indicates an association between two disorders
that is more than coincidental. Conditions that
occur in migraineurs with a higher prevalence
than would be expected include stroke, epilepsy,
Raynaud's syndrome, and affective disorders,
which include depression, mania, anxiety, and
panic disorder. Possible associations include
essential tremor, mitral valve prolapse, and irritable bowel syndrome (Silberstein 2004).

▶ Pharmacotherapy may be acute (abortive)
or preventive (prophylactic), and patients may
require both approaches. Acute treatment
attempts to reverse or stop the progression of
a headache once it has started. It is appropriate
for most attacks and should be limited to 2–3 days
a week. ▶ Preventive therapy is designed to
reduce attack frequency and severity.

Preventive Treatment

Preventive medications reduce attack frequency,
duration, or severity (Silberstein and Goadsby
2004; Silberstein et al. 2001). According to the
US Headache Consortium Guidelines (Ramadan
et al. 1999), indications for preventive treatment
include:

- Migraine that significantly interferes with the
  patient's daily routine despite acute treatment
- Failure of, contraindication to, or troublesome
  adverse events (AEs) from acute medications
- Acute medication overuse
- Very frequent headaches (>2/week) (risk of
  medication overuse)
- Patient preference
- Special circumstances, such as hemiplegic
  migraine or attacks with a risk of permanent
  neurologic injury

Preventive medication groups include
beta-adrenergic blockers, anticonvulsants,
▶ antidepressants, ▶ calcium-channel antagonists, ▶ serotonin antagonists, and nonsteroidal
anti-inflammatory drugs. Choice is based on
efficacy, AEs, and coexistent and comorbid
conditions. The medication is started at a low
dose and increased slowly until therapeutic
effects develop or the ceiling dose is reached.
A full therapeutic trial may take 2–6 months.
Acute headache medications should not be
overused. If headaches are well controlled, medication can be tapered and discontinued. Dose
reduction may provide a better risk-to-benefit
ratio. Women of childbearing potential should
be on adequate contraception.

Behavioral and psychological interventions
used for prevention include ▶ relaxation training,
thermal ▶ biofeedback combined with relaxation
training, electromyography biofeedback, and
cognitive-behavioral therapy (Silberstein 2004).

These interventions are effective as monotherapy but are more effective when used in conjunction with pharmacologic management.

Medication

▶ Beta(β) blockers, ▶ propranolol, metopropolol, and ▶ timolol are effective preventive drugs; ▶ nadolol and atenolol are probably effective (Silberstein et al. 2012). Their relative efficacy has not been established; choice is based on beta selectivity, convenience, AEs, and patients' reactions (Silberstein et al. 2001). Beta blockers can produce behavioral AEs, such as drowsiness, fatigue, lethargy, sleep disorders, nightmares, depression, memory disturbance, and hallucinations; they should be avoided when patients are depressed. Decreased exercise tolerance limits their use by athletes. Less common AEs include impotence, orthostatic hypotension, and bradycardia. Beta blockers are useful for patients with angina or hypertension. They are relatively contraindicated for patients with congestive heart failure, asthma, Raynaud's disease, and insulin-dependent diabetes.

Antidepressants

▶ Amitriptyline (a tricyclic antidepressant) and venlafaxine (a serotonin-norepinephrine reuptake inhibitor) are the only antidepressants with fairly consistent support for efficacy (Silberstein et al. 2012). AEs of ▶ amitriptyline include increased appetite, weight gain, dry mouth, and sedation; cardiac toxicity and orthostatic hypotension occasionally occur (Saper et al. 1994). There has been one positive trial for fluoxetine. Sexual dysfunction is a common AE of fluoxetine and venlafaxine. Antidepressants are especially useful for patients with comorbid depression and anxiety disorders.

Calcium-Channel Blockers

The Agency for Healthcare Policy and Research analyzed 45 controlled trials (Gray et al. 1999). Flunarizine was effective. Nicardipine, nifedipine, nimodipine, and verapamil have mixed results and their efficacy is uncertain (Silberstein et al. 2012). Verapamil's most common AE is constipation. Flunarizine is the most effective drug of this class, but it is not available everywhere. AEs include Parkinsonism, depression, and weight gain.

Anticonvulsant Medications

Divalproex sodium (500–1,000 mg) and sodium valproate are effective, as is the extended release formulation (Silberstein et al. 2012). The most frequent AEs were nausea (42 %), alopecia (31 %), tremor (28 %), asthenia (25 %), dyspepsia (25 %), somnolence (25 %), and weight gain (19 %) (Silberstein and Collins 1999). Hepatotoxicity and pancreatitis are the most serious AEs, but irreversible hepatic dysfunction is extremely rare in adults. Baseline liver function studies should be obtained, but adults on monotherapy probably do not need follow-up studies. Divalproex carries a high risk of congenital abnormality.

▶ Gabapentin (1,800–2,400 mg) efficacy is uncertain (Silberstein et al. 2012). The most common AEs were dizziness or giddiness and drowsiness.

▶ Topiramate, a D-fructose derivative, is effective. It has been associated with weight loss, not weight gain. In two large, double-blind, placebo-controlled, multicenter trials, topiramate, both 100 and 200 mg, was effective in reducing migraine attack frequency by 50 % in half of the patients (Brandes et al. 2004; Silberstein et al. 2004). Dropouts due to AEs were common in the topiramate groups, but did not affect statistical significance (Silberstein et al. 2012).

Divalproex and topiramate are useful in patients with epilepsy, anxiety disorder, or manic-depressive illness. They can be used in patients with depression, Raynaud's disease, asthma, and diabetes, circumventing the contraindications to beta blockers.

Serotonin Antagonists

Methysergide is effective (Gray et al. 1999; Silberstein 1998). AEs include transient muscle aching, claudication, abdominal distress, nausea, weight gain, and hallucinations. The major complication is rare (1/2,500) retroperitoneal, pulmonary, or endocardial fibrosis (Silberstein 1998).

To prevent this, a 4-week medication-free interval is recommended after 6 months of continuous treatment (Silberstein 1998; Silberstein et al. 2001). Methysergide is unavailable in many parts of the world.

Pizotifen, a benzocycloheptathiophene derivative, is effective (Ramadan et al. 1999). AEs include drowsiness, increased appetite, and weight gain. It is not available in the USA.

Natural Products

Petasites hybridus root (butterbur), a perennial shrub, is effective for migraine prevention. The most common AE was belching (Holland, et al. 2012). Feverfew (*Tanacetum parthenium*) is a medicinal herb which is probably effective. Riboflavin (400 mg) is also probably effective. A standardized extract (75 mg bid) was effective in a double-blind, placebo-controlled study. The most common AE was belching (Silberstein 2004).

Newer Treatments

Botulinum toxin type A (Botox^R 0, 25, or 75 U) showed promising results in one placebo-controlled, double-blind trial. It was injected into glabellar, frontalis, and temporalis muscles. The 25 U treatment group was significantly better than the placebo group in reducing mean frequency of moderate to severe migraines during days 31–60, incidence of 50 % reduction in all migraine at days 61–90, and reduction in all migraine at days 61–90 (Silberstein et al. 2000).

Preventive Migraine Therapy

See Table 1.

Summary

Migraine is an extremely common neurobiologic headache disorder that is due to increased CNS excitability. It ranks among the world's most disabling medical illnesses. Diagnosis is based on the headache's characteristics and associated symptoms. The economic and societal impact of migraine is substantial. It impacts on sufferers' quality of life and impairs work, social activities, and family life. Increased headache frequency is an

Preventive Migraine Therapy, Table 1 Choices of preventive treatment in migraine: influence of comorbid conditions

Drug	Relative contraindication	Relative indication
Beta blockers	Asthma, depression, CHF, Raynaud's disease, diabetes	HTN, angina
Antiserotonin Pizotifen	Obesity	
Methysergide	Angina, PVD	Orthostatic hypotension
Calcium-channel blockers		
Verapamil	Constipation, hypotension	Migraine with aura, HTN, angina, asthma
Flunarizine	Parkinson's	Hypertension, FHM
Antidepressants		
TCAs	Mania, urinary retention, heart block	Other pain disorders, depression, anxiety disorders, insomnia
SSRIs/SNRIs	Mania	Depression, OCD
Anticonvulsants		
Divalproex/ valproate	Liver disease, bleeding disorders	Mania, epilepsy, anxiety disorders
Topiramate	Kidney stones	Mania, epilepsy, anxiety disorders
NSAIDs		
Naproxen	Ulcer disease, gastritis	Arthritis, other pain disorders

CHF Congestive heart failure, *OCD* Obsessive compulsive disorder, *HTN* Hypertension, *PVD* Peripheral vascular disease, *MAOIs* Monoamine oxidase inhibitors, *SNRI* Serotonin-norepinephrine reuptake inhibitor, *SSRI* Serotonin specific reuptake inhibitor, *NSAIDs* Nonsteroidal anti-inflammatory drugs, *TCA* Tricyclic antidepressants

indication for preventive treatment. Preventive treatment decreases migraine frequency and improves quality of life. More treatments are being developed, which provides hope to the many sufferers who are still uncontrolled.

References

Brandes, J. L., Saper, J. R., Diamond, M., et al. (2004). Topiramate for migraine prevention: A randomized controlled trial. *JAMA: The Journal of the American Medical Association, 291*, 965–973.

Gray, R. N., Goslin, R. E., & McCrory, D. C. (1999). *Drug treatments for the prevention of migraine headache.* Prepared for the Agency for Health Care Policy and Research, Contract No. 290-94-2025. Available from the National Technical Information Service, Accession No. 127953, 1999.

Headache Classification Committee. (2013). *The International Classification of Headache Disorders* (3rd ed.). *Cephalalgia.* (in press).

Holland, S., Silberstein, S. D., Freitag, F., et al. (2012). Evidence-based guideline update: NSAIDs and other complementary treatments for episodic migraine prevention in adults. *Neurology, 78*, 1346.

Ramadan, N. M., Silberstein, S. D., & Freitag, F. G. (1999). Evidence-based guidelines of the pharmacological management for prevention of migraine for the primary care provider. *Neurology.* www.neurology. org.

Saper, J. R., Silberstein, S. D., & Lake, A. E. (1994). Double-blind trial of fluoxetine: Chronic daily headache and migraine. *Headache, 34*, 497–502.

Silberstein, S. D. (1998). Methysergide. *Cephalalgia, 18*, 421–435.

Silberstein, S. D. (2004). Migraine. *Lancet, 363*, 391.

Silberstein, S. D., & Collins, S. D. (1999). Safety of divalproex sodium in migraine prophylaxis: An open-label, long-term study (for the long-term safety of depakote in headache prophylaxis study group). *Headache, 39*, 633–643.

Silberstein, S. D., Mathew, N., & Saper, J. (2000). Botulinum toxin type A as a migraine preventive treatment. *Headache, 40*, 445–450.

Silberstein, S. D., Saper, J. R., & Freitag, F. (2001). Migraine: Diagnosis and treatment. In S. D. Silberstein, R. B. Lipton, & D. J. Dalessio (Eds.), *Wolff's headache and other head pain* (7th ed., pp. 121–237). New York: Oxford University Press.

Silberstein, S. D., Neto, W., Schmitt, J., et al. (2004). Topiramate in the prevention of migraine headache: A randomized, double-blind, placebo-controlled, multiple-dose study. For the MIGR-001 study group. *Archives of Neurology, 61*, 490–495.

Silberstein, S. D., Holland, S., Freitag, F., Dodick, D. W., Argoff, C., & Ashman, E. (2012). Evidence-based guideline update: Pharmacologic treatment for episodic migraine prevention in adults. *Neurology, 78*, 1337–1345.

Preventive Therapy

Definition

Designed to reduce headache attack frequency and severity.

Previously Used Diagnostic Labels

▶ Temporomandibular Joint Disorders

Pricking Pain

▶ First and Second Pain Assessment (First Pain, Pricking Pain, Pin-Prick Pain, Second Pain, Burning Pain)

Primary Afferent

▶ Nociceptors, Action Potentials, and Post-Firing Excitability Changes
▶ Prostaglandins, Spinal Effects

Primary Afferents of the Urinary Bladder

Definition

Thinly myelinated (Aδ) and unmyelinated (C) fibers with nerve endings in the suburothelial lamina propria and smooth muscle layers of the bladder.

Cross-References

▶ Opioids and Bladder Pain/Function

Primary Afferents/Neurons

Definition

Primary afferents are sensory neurons (axons or nerve fibers) in the peripheral nervous system that transduce information about mechanical, thermal, and chemical states of the body and transmit it to sites in the central nervous system. Those that innervate the trunk and limbs have their cell bodies in dorsal root ganglia. The cell body gives rise to a peripheral process, which terminates in tissue (e.g., skin, muscle), and a central process that enters the spinal cord. Primary afferents are highly specialized such that separate populations transmit information about different types of innocuous mechanical, innocuous thermal, and noxious information. At the first synapse of the afferent pathway, afferent input and temporal and spatial processing of the afferent input take place. Enhanced sensitivity of the central processing of afferent input can contribute to inflammatory hyperalgesia (central sensitization).

Cross-References

► Encoding of Noxious Information in the
  Spinal Cord
► Mechanonociceptors
► Neutrophils in Inflammatory Pain
► NGF, Regulation During Inflammation
► Nociceptive Circuitry in the Spinal Cord
► Nociceptor(s)
► Opioids in the Periphery and Analgesia
► Postsynaptic Dorsal Column Projection,
  Anatomical Organization
► Primary Afferents/Neurons
► Prostaglandins, Spinal Effects
► Vagal Input and Descending Modulation

Primary and Secondary Hyperalgesia

Definition

Hyperalgesia means increased sensitivity to painful or to normally non-painful stimulation. There are two definitions of "primary" and "secondary" hyperalgesia. (1) According to one definition, a hyperalgesia is primary if it occurs in an inflamed body area or – in mononeuropathies – in the innervation territory of a lesioned nerve. Secondary are hyperalgesias occurring outside such an area. (2) According to another definition,

a hyperalgesia is primary when the reason is sensitization of nociceptors and secondary when it is due to alterations of central synaptic transmission. For practical reasons, the first type of definition should be preferred, since it is purely descriptive.

In the case of inflammatory pain, primary hyperalgesia is usually due to sensitization of the primary afferent nociceptors. However, altered central nervous synaptic transmission may contribute. There are several forms of primary hyperalgesia: to tonic pressure (pressure hyperalgesia), to pin prick (pin prick hyperalgesia), and to warming (heat hyperalgesia). One form of mechanical hyperalgesia, the hyperalgesia to light touch stimuli, for example, a swab of cotton wool (also called "Allodynia," see the respective definition), may occur in the primary zone of inflammatory processes, but also in a secondary zone (see below). There is good evidence that this type of hyperalgesia is due to plastic changes in the central synaptic transmission. If the second type of definition is chosen, "touch hyperalgesia" and also "pin prick hyperalgesia" are always secondary, independent of the site of occurrence.

Secondary hyperalgesia occurs in inflammatory pain outside the inflamed area, sometimes also in other tissues (e.g., the skin above an inflamed knee joint, intestinal pain projected into Head zones). This type of hyperalgesia is due to changes in the central synaptic transmission, in particular increased receptive fields of central transmission neurons. Two forms of secondary hyperalgesia have been extensively characterized in the literature: "touch hyperalgesia," in which sensitive mechanosensors mediate the pain when at the same time spontaneously active nociceptive C-fibers provide a barrage of input to the central pain transmission pathways, and "pin prick hyperalgesia," which is mediated by nociceptors. It is highly controversial if secondary heat hyperalgesia occurs.

Cross-References

► Quantitative Sensory Testing
► Spinothalamic Neuron

Primary and Secondary Zones

Definition

The primary zone describes the area of skin directly affected by the inflammatory injury. The secondary zone is the area surrounding the primary zone which, as a consequence of the injury at the primary zone, has altered sensory properties. Sensory changes in the primary zone may be due to ▶ peripheral sensitization or ▶ central sensitization. Sensory changes in the secondary zone are usually assumed to be exclusively caused by ▶ central sensitization.

Cross-References

▶ Quantitative Thermal Sensory Testing of Inflamed Skin

Primary Anorgasmia

Definition

This means that the individual has never been able to achieve an orgasm.

Cross-References

▶ Clitoral Pain

Primary Cells (RVM)

Definition

Primary cells are a class of RVM neurons defined in vitro by lack of response to mu-opioid agonists. Primary cells presumably correspond to off-cells and neutral cells defined in vivo.

Cross-References

▶ Opiates, Rostral Ventromedial Medulla, and Descending Control

Primary Cough Headache

James W. Lance
Institute of Neurological Sciences, University of New South Wales, Sydney, Australia

Synonyms

Benign cough headache; Valsalva maneuver headache

Definition

Headache precipitated by coughing in the absence of any intracranial disorder.

Characteristics

The headache starts suddenly as a sharp pain on coughing, straining, or other forms of ▶ Valsalva maneuver and not at other times. It usually lasts less than 1 min but may sometimes persist for 30 min or longer.

Organic causes of cough headache, such as intermittent obstruction of the cerebrospinal pathway, internal carotid or ▶ vertebrobasilar arterial disease, and cerebral ▶ aneurysms, must be excluded by imaging.

Primary cough headache mainly affects patients older than 40 years of age, is usually bilateral, and is responsive to indomethacin.

Clinical Reports

In 1932, Tinel described four patients with severe head pain on coughing and various Valsalva-like maneuvers (Ekbom 1986). Sir Charles Symonds reported 21 patients with cough headache in whom no intracranial disease became apparent, together with 6 patients whose headaches resulted from basilar impression caused by Paget's disease or space-occupying lesions in the posterior fossa (Symonds 1956). He concluded that there was a syndrome of benign cough headache as it improved or disappeared in 15 patients during the follow-up period of 18 months to 12 years. Ages ranged from 37 to 77 years (mean 55 years). Of the 21 patients, 18 were men.

Rooke (1968) reported cough headache as a form of exertional headache in presenting the histories of 103 patients, whose headaches were precipitated by running, bending, coughing, sneezing, lifting, or straining. During the follow-up period of 3 years or more, structural lesions such as platybasia, ▶ Chiari type I malformation, subdural hematoma, and cerebral or cerebellar tumors came to light in 10 patients. The remaining 93 patients fell into the primary cough or exertional headache category. Thirty were free of headache in 5 years, and 73 were improved or free of headache after 10 years. Men were affected more than women in the ratio 4:1. Some patients reported that their cough headache followed a respiratory infection. Like Symonds (1956), Rooke noted that some patients improved after the removal of abscessed teeth.

Cough headache is best considered as an entity separate from exertional headache. Pascual et al. (1996) reported that the average age of their patients with cough headache was 43 years older than their patients with exertional headache. They analyzed their experience with 72 patients whose headaches were precipitated by coughing (30), physical exercise (28), or sexual activity (14). Of the patients with cough headache, 17 were secondary to Chiari type I malformations. The remaining 13 patients, 10 men and 3 women, were diagnosed as "benign cough headaches." Their ages ranged between 44 and 81 years (mean 67 ± 11 years). Headache was also brought on by a sudden Valsalva maneuver in four subjects but never by physical exertion. Headache was bilateral in 12 cases and unilateral in one. Doppler ultrasound showed no evidence of carotid disease in this case. Indomethacin 75 mg daily was effective in the six patients for whom it was prescribed but not

for patients with Chiari type I malformation. The tendency to benign cough headache persisted for 2 months to 2 years.

An atypical cough headache in a 57-year-old man was unilateral and recurred in bouts of 30–60 days reminiscent of episodic cluster headache (Perini and Toso 1998). The headaches involved one supraorbital region, were very severe, and lasted for 1–10 min without any autonomic features. The pain was eased by a nonsteroidal anti-inflammatory agent, nimesulide.

Symptomatic Cases

The most common secondary cough headaches are those caused by Chiari type I malformation in which a tongue of cerebellum projects through the foramen magnum and provides a valve-like obstruction to the flow of cerebrospinal fluid (CSF) on coughing. Patients with this Chiari type I malformation are significantly younger than patients with primary cough headache. Pascual et al. (1996) found that the age of onset in the secondary cases varied from 15 to 64 years (mean 39 ± 14 years). His patients complained of occipital and suboccipital pain precipitated by laughing, weight lifting, or sudden changes in posture as well as by coughing. All patients later developed posterior fossa symptoms or signs. Eight patients underwent operation with improvement of headache in seven. Ertsey and Jelencsik (2000) reported a 48-year-old woman with Chiari type I malformation whose cough headaches were suppressed in 3–4 days by indomethacin 25 mg, but this does not reflect the general experience.

Cough headache was the presenting feature of a cerebral aneurysm in a report by Smith and Messing (1993). Their patient described a severe right temporal ache on coughing or bending lasting 1–5 min, not relieved by indomethacin. After 24 days the pain became continuous, and a right third cranial nerve palsy developed. She was found to have a posterior communicating artery aneurysm. Her headache settled postoperatively. Cough headache has also been reported as the presenting symptom of carotid stenosis persisting for 1 year before other neurological symptoms appeared (Britton and Guiloff 1988). The headache disappeared after the artery occluded, causing a hemiparetic stroke.

Pathophysiology of Cough Headache

Williams (1976) studied the relationship between CSF pressure in the cisterna magna and lumbar sac during coughing. At first, lumbar pressure was greater than cisternal pressure, then the pressure gradient reversed. He postulated that any obstruction that interfered with the downward rebound pulsation at the foramen magnum could cause cough headache. He later reported two patients whose cerebellar tonsils extended below the foramen magnum and who presented with cough headache (Williams 1980). During a Valsalva maneuver, he recorded a marked craniospinal pressure differential in the rebound phase. After the cerebellar tonsils were decompressed, the steep pressure gradient on coughing was eliminated and the cough headache disappeared.

Whether this explanation accounts for primary cough headache is uncertain. Coughing increases intrathoracic and intra-abdominal pressure, which is transmitted to the epidural veins, causing a pressure wave to move rostrally in the CSF. Temporary impaction of the cerebellar tonsils during the relaxation phase after coughing even in the absence of a Chiari type I malformation remains the most plausible mechanism of primary cough headache.

The possibility of venous distension being sufficient in itself to cause headache seems unlikely. Lance (1991) described a man with a large goiter whose face became purple when he lifted his arms above his head. As the veins in his neck distended, he developed a sudden headache. The headache vanished after the goiter was removed. This circumstance is exceptional.

The Valsalva maneuver does not usually initiate headache, and some source of venous sensitization would have to be proposed to account for primary cough headache of venous origin. "Benign Valsalva manoeuvre-related headache" has been put forward as a more inclusive descriptive term (Calandre et al. 1996), but "benign" or "primary cough headache" is more succinct and will be hard to displace.

Treatment

Mathew (1981) reported two patients who improved with indomethacin 50 mg 3 × daily, one of whom had not responded to propranolol or methysergide. Raskin (1995) followed 30 patients with primary cough headache for 3 years or more. Out of 16 patients treated with indomethacin 50–20 mg daily, 10 lost their headache completely, 4 were moderately improved, and 2 did not respond.

A separate group of 14 patients was subjected to lumbar puncture with 40 ml of CSF being removed. Six patients improved immediately and three improved gradually, but the procedure was ineffective in eight patients, six of whom later responded to indomethacin. Raskin commented that indomethacin and lumbar puncture may both be effective by lowering intracranial pressure.

Boes et al. (2002) stated that they usually combined indomethacin with a protein pump inhibitor for long-term use. They mentioned that some patients have improved with acetazolamide and methysergide.

Summary

The International Headache Society Diagnostic Criteria for the diagnosis of primary cough headache are as follows:

1. Headache of sudden onset lasts from 1 s to 30 min.
2. Headache brought on by coughing, straining, and/or Valsalva maneuver.
3. The same headache does not occur without coughing or straining.
4. Not attributed to any other disorder.

The condition appears to be caused by a pulse of increased intracranial pressure and relieved by treatments that reduce CSF pressure. It is imperative to exclude cough headache caused by structural lesions, such as Chiari type I malformation, by MRI before the diagnosis is made. The syndrome has been the subject of a comprehensive review by Boes et al. (2002). Primary cough headache is best regarded as an entity separate from primary exertional headache, which affects a younger age group. The prognosis is for gradual improvement over a period of 2 months to 12 years. The symptom is usually controlled, at least partially, by indomethacin.

References

Boes, C. J., Matharu, M. S., & Goadsby, P. J. (2002). Benign cough headache. *Cephalalgia, 22*, 772–779.

Britton, T. C., & Guiloff, R. J. (1988). Carotid artery disease presenting as cough headache. *Lancet, 1*, 1406–1407.

Calandre, L., Hernandez-Lain, A., & Lopez-Valdes, E. (1996). Benign Valsalva's manoeuvre-related headache: An MRI study of six cases. *Headache, 36*, 251–253.

Ekbom, K. (1986). Cough headache. In F. C. Rose (Ed.), *Headache. Handbook of clinical neurology* (Vol. 4, pp. 367–376). Amsterdam: Elsevier.

Ertsey, C., & Jelencsik, I. (2000). Cough headache associated with Chiari type I malformation: Responsiveness to indomethacin. *Cephalalgia, 20*, 518–520.

Lance, W. (1991). Solved and unsolved headache problems. *Headache, 31*, 439–445.

Mathew, N. T. (1981). Indomethacin responsive headache syndrome. *Headache, 21*, 147–150.

Pascual, P., Iglesias, F., Oterino, A., et al. (1996). Cough, exertional and sexual headache. *Neurology, 4*, 1520–1524.

Perini, F., & Toso, V. (1998). Benign cough "cluster" headache. *Cephalalgia, 18*, 493–494.

Raskin, N. H. (1995). The cough headache syndrome: Treatment. *Neurology, 45*, 1784.

Rooke, E. (1968). Benign exertional headache. *The Medical Clinics of North America, 52*, 801–808.

Smith, W. S., & Messing, R. O. (1993). Cerebral aneurysm presenting as cough headache. *Headache, 33*, 203–204.

Symonds, C. (1956). Cough headache. *Brain, 79*, 557–568.

Williams, B. (1976). Cerebrospinal fluid pressure changes in response to coughing. *Brain, 99*, 331–346.

Williams, B. (1980). Cough headache due to craniospinal pressure dissociation. *Archives of Neurology, 37*, 226–230.

Primary Dysmenorrhea

Definition

Primary dysmenorrhea is menstrual pain, beginning at or shortly after onset of menstrual bleeding, without associated pelvic pathology.

Cross-References

▶ Dyspareunia and Vaginismus

Primary Exertional Headache

Andreas Straube
Department of Neurology, Ludwig Maximilians University, Munich, Germany

Synonyms

Bath headache; Diver's headache; Effort headache; Sports headache; Weight-lifter's headache

Definition

The headache is precipitated by any form of exercise. Recognized subvarieties include "weight-lifter's headache, diver's headache, effort headache, etc." The differentiation from orgasmic headache and forms of thunderclap headache is somewhat unclear. Symptomatic cases have always to be excluded before the diagnosis primary exertional headache can be done.

Introduction

Primary exertional headache is classified in the 2nd IHS classification under paragraph 4 – Other primary headaches.

Characteristics

Since a patient with primary exertional headache (pEH) seldom goes to the doctor, there are only very few epidemiological studies or reports on treatment options for pEH.

The new classification of the International Headache Society (IHS) provides the following description and diagnostic criteria for pEH.

Diagnostic Criteria

1. Pain is bilateral, throbbing on onset, and short or long lasting (5 min–48 h).
2. It specifically occurs during or immediately after physical exercise requiring exertion.

3. There is a close temporal relationship between pain and physical exercise, particularly in hot weather or at high altitude.
4. The headache is prevented by avoiding excessive exertion.
5. It cannot be attributed to any other disorder.

When acute onset of this headache type first occurs, the possibility of a subarachnoid hemorrhage must be excluded. A further differential diagnosis is thunderclap headache, which often occurs also after exertion or bathing. The primary exertional headache is to differentiate from the primary cough headache whereby Valsalva-like maneuvers may be also play a role in exertional headache. In cough-related headache, 50–60 % is classified as symptomatic (e.g., Arnold Chiari syndrome). In some patients, pEH is prevented by ingesting ergotamine tartrate. Indomethacin is also reported to be effective in the majority of such cases. Exercise-induced migraine is classified under migraine.

Epidemiology

According to the Vågå, Norway, study of headache, 12.3 % of all individually interviewed inhabitants of this community (ages 18–65 years old) reported having had pEH. There was a slight preponderance of females (F/M: 1.38) and the age of onset usually occurred before the third decade. This is supported by Pascual et al. (1996), who noted that the onset of pEH occurs at an earlier age than cough headache. In a population of college students, about 35 % reported about experience with sport- and exercise-related headache (Williams and Nukada 1994). In a recent survey of about 2,000 adolescents (age 13–15 years) the prevalence of pEH was 30 %, females were more often affected and the prevalence decreased with increasing age. Migraineurs reported also more often on pEH (Chen et al. 2009). Exertional headache is thought to be experience by nearly 4 % of weight lifters. However, since a ▶ Valsalva maneuver is often performed during weight lifting, it is not clear if this is really the case (Powell 1982; Rooke 1968). Older studies reported a definite relationship between benign ▶ sexual activity headache and pEH. A study on 45 patients with sexual

activity headache reported that 60 % had also experienced pEH and 47 % also had a personal history of migraine (Silbert et al. 1991). Contrary to benign sexual activity headache, which is seldom a persisting problem, two thirds of 93 patients with pEH still complained of it 5 years after their first experience with it (Rooke 1968). Seldom may an orgasmic headache convert into a benign exertional headache (Brilla and Evers 2005).

Clinical Characteristics

Most authors describe pEH as an abrupt, bilateral, severe headache that lasts for minutes to hours. Sometimes it has a pulsating character that later becomes dull. Some patients also have vomiting. Only two studies analyzed the headache duration and intensity in more detail (Sjaastad and Bakketeig 2002). The headache was found to last minimally 5 min and maximally 24 h and had a mild to moderate intensity in the majority of the patients. The same study also reported that brief and long-lasting attacks occurred equally often, and that the longer lasting attacks had more migraine-like features. Both patient groups frequently complained of ▶ jab-like and jolt-like headaches (Sjaastad and Bakketeig 2003). Chen et al. (2009) described that in about 80 % of their adolescent patients, the headache lasted less than 1 h was in about 50 % bilateral and in 60 % pulsating. In patients with a history of migraine, the exertional headache had more features of migraine. Only 14 % took painkillers in the course of pEH.

Differential Diagnosis

The more limited form of pEH must be differentiated from the sudden onset headache due to a closed glottis, which elevates the intrathoracic and intracranial pressure. Increased intracranial pressure is more related to the cough headache. Other forms of exertional headache are caused by stimulation of cold-sensitive trigeminal afferences (ice cream headache) or are due to pressure (e.g., goggle headache in divers; Cheshire and Ott 2001). Especially in cases of sudden onset headache, other causes of symptomatic forms have to be excluded, such as subarachnoid hemorrhage (IHS 6.2.2), mass lesions, ▶ spontaneous CSF leaks, or an anomaly of the posterior fosse (Buzzi et al. 2003; Green 2001; Mokri 2002). Rarely an acute cardial ischemia can cause exertional-like headache, often only lasting for minutes (Lance and Lambros 1998; Lipton et al. 1997; Wei and Wang 2008). The most important differential diagnosis is the so-called primary thunderclap headache (ISH 4.6). Primary thunderclap headache is more and more attributed to the reversible cerebral angiopathy syndrome and is often triggered by exertion and bathing (Chen et al. 2006). Typically, the patients describe this headache as an exploding headache, which can show multiple recurrences in a period of 2–4 weeks. The IHS also recognizes sexual activity headache as a distinct diagnosis, although it is still not clear if this more slowly developing headache shares some common features with pEH. The activation of a migraine by exertion also has to be differentiated from those forms of headaches that are elicited by exertion.

Pathophysiology

There is no generally accepted pathophysiological explanation for exertional headache. The acute-onset exertional headache, seen in weight lifters, seems to be related to an acute increase in intrathoracic pressure. This increased pressure in turn causes an increase of pressure in the cranial veins, which results in the increased intracranial pressure regularly seen during the monitoring of patients with brain edema. However, it is still unclear why the Valsalva maneuver does not elicit headache in normal subjects. Some authors discuss if abnormalities of the cranial sinus are a predisposition for the development of an exertional headache (Donnet et al. 2008). A second explanation is that exertion causes an inappropriate reaction of the cerebral vasculature in some patients (Lane and Gulevich 2002). Using stress transcranial Doppler to evaluate the myogenic mechanism of the cerebral vessels, Heckmann et al. (1997) detected signs in two patients that cerebral vascular autoregulation is disturbed during ergometer stress; no constriction occurs during systemic blood pressure increase.

Another mechanism under discussion is the retention of carbon dioxide. Especially in divers, this causes a relaxation of the cerebrovascular smooth muscles (Cheshire and Ott 2001). The finding of an asymmetric bifrontal hypoperfusion in one subject with a postexercise-provoked headache (Basoglu et al. 1996) is in line with the findings in primary thunderclap headache where segmental vasospasms of intracranial arteries can be seen (Ducros et al. 2007; Chen et al. 2006). There are now case reports on patients who had sexual activity headache with identified cerebral ▶ arterial spasms during angiography (Silbert et al. 1989). Thus, it will be to show in the future if short lasting segmental vasospasms may also occur in exertional headache.

Today, a final pathophysiological concept of exertional headache is impossible to draw. Nevertheless, it is highly probable that the disturbed ▶ autoregulation of the cerebral vessels that results in relaxation or more likely segmental vasoconstriction under increased sympathetic activity (stress) best explains the mechanisms underlying the headache. The migraine-like features point to a secondary activation of the trigeminovascular system. Secondary forms of the exertional headache (e.g., cardiac exertional headache) may be caused by coactivation of the autonomic systems and secondary relaxation/vasoconstriction of the cerebral vessels.

Treatment

Controlled studies on the treatment of pEH are not available. Most reviews as well as case reports draw attention to the therapeutic benefit of indomethacin, a strong COX 1 and COX 2 inhibitor. Indomethacin reduces the intracranial pressure in pseudotumor cerebri (Förderreuther and Straube 2000), probably by having a vasoconstrictory effect on the cerebral vessels. In general, initial doses of 25 mg p.o. twice daily are recommended, but some patients require higher dosages (up to 75–100 mg p.o. twice daily). Another infrequently used alternative is DHE, which is limited to subcutaneous administration. Since propranolol may also be helpful for sexual activity headache, trials with the beta-blocker are called for. There are few reports on the benefit of triptans in preventing exertional headaches (Green 2001), but acute cardial ischemia has to be excluded.

Summary

Primary exertional headache has to be always differentiated from symptomatic cases and in the future it may be possible that several different subtypes concerning the underlying pathophysiology are differentiated.

References

Basoglu, T., Ozbenli, T., Bernay, I., et al. (1996). Demonstration of frontal hypoperfusion in benign exertional headache by Technetium-99 m-HMPAO SPECT. *Journal of Nuclear Medicine, 37*, 1172–1174.

Brilla, R., & Evers, S. (2005). A patient with orgasmic headaches converting to concurrent orgasmic and benign exertional headaches. *Cephalalgia, 25*(12), 1182–3.

Buzzi, M. G., Formisano, R., Colonnese, C., et al. (2003). Chiari-associated exertional, cough, and sneeze headache responsive to medical therapy. *Headache, 43*, 404–406.

Chen, S. P., Fuh, J. L., Lirng, J. F., Chang, F. C., & Wang, S. J. (2006). Recurrent primary thunderclap headache and benign CNS angiopathy: Spectra of the same disorder? *Neurology, 67*(12), 2164–9.

Chen, S. P., Fuh, J. L., Lu, S. R., & Wang, S. J. (2009). Exertional headache–a survey of 1963 adolescents. *Cephalalgia, 29*(4), 401–7.

Cheshire, W. P., & Ott, M. C. (2001). Headache in divers. *Headache, 41*, 235–247.

Donnet, A., Dufour, H., Levrier, O., Metellus, P. (2008). Exertional headache: a new venous disease. *Cephalalgia. 28*(11):1201–1203.

Ducros, A., Boukobza, M., Porcher, R., Sarov, M., Valade, D., & Bousser, M. G. (2007). The clinical and radiological spectrum of reversible cerebral vasoconstriction syndrome. A prospective series of 67 patients. *Brain, 130*(Pt. 12), 3091–3101.

Förderreuther, S., & Straube, A. (2000). Indomethacin reduces CSF pressure in pseudotumor cerebri. *Neurology, 55*, 1043–1045.

Green, M. W. (2001). A spectrum of exertional headaches. *The Medical Clinics of North America, 85*, 1085–1092.

Heckmann, J. G., Hilz, M. J., Muck-Weymann, M., et al. (1997). Benign exertional headache / benign sexual headache: A disorder of myogenic cerebrovascular autoregulation? *Headache, 37*, 597–598.

Lance, J. W., & Lambros, J. (1998). Unilateral exertional headache as a symptom of cardiac ischemia. *Headache, 38*, 315–316.

Lane, J. C., & Gulevich, S. (2002). Exertional, cough, and sexual headaches. *Current Treatment Options in Neurology, 4*, 375–381.

Lipton, R. B., Lowenkopf, T., Bajwa, Z. H., et al. (1997). Cardiac cephalagia: A treatable form of exertional headache. *Neurology, 49*, 813–816.

Mokri, B. (2002). Spontaneous CSF leaks mimicking benign exertional headaches. *Cephalalgia, 22*, 780–783.

Pascual, J., Iglesias, F., Oterino, A., et al. (1996). Cough, exertional, and sexual headaches: An analysis of 72 benign and symptomatic cases. *Neurology, 46*, 1520–1524.

Powell, B. (1982). Weight lifter's cephalgia. *Annals of Emergency Medicine, 11*, 449–451.

Rooke, E. D. (1968). Benign exertional headache. *The Medical Clinics of North America, 52*, 801–808.

Silbert, P. L., Hankey, G. J., Prentice, D. A., et al. (1989). Angiographically demonstrated arterial spasm in case of benign sexual headache and benign exertional headache. *Australian and New Zealand Journal of Medicine, 19*, 466–468.

Silbert, P. L., Edis, R. H., Stewart-Wynne, E. G., et al. (1991). Benign vascular sexual headache and exertional headache: Interrelationships and long term prognosis. *Journal of Neurology, Neurosurgery, and Psychiatry, 54*, 417–421.

Sjaastad, O., & Bakketeig, L. S. (2002). Exertional headache I. Vågå study of headache epidemiology. *Cephalalgia, 22*, 784–790.

Sjaastad, O., & Bakketeig, L. S. (2003). Prolonged benign exertional headache. The Vågå study of headache epidemiology. *Headache, 43*, 611–615.

Wei, J. H., & Wang, H. F. (2008). Cardiac cephalalgia: Case reports and review. *Cephalalgia, 28*(8), 892–6.

Williams, S. J., & Nukada, H. (1994). Sport and exercise headache: Part 1 Prevalence among university students. *British Journal of Sports Medicine, 28*, 90–95.

Primary Hyperalgesia

Definition

See "▶ Primary and Secondary Hyperalgesia"

Cross-References

▶ Descending Modulation and Persistent Pain
▶ Hyperalgesia

▶ Muscle Pain Model, Inflammatory Agents-Induced

▶ Nick Model of Cutaneous Pain and Hyperalgesia

▶ Nocifensive Behaviors, Muscle and Joint

▶ Opioids and Muscle Pain

▶ Spinothalamic Neuron

▶ UV-Induced Erythema

Primary Motor Cortex

Definition

A part of the frontal lobe cerebral cortex, confined to the gyrus in front of the central sulcus (Brodmann's area 4). This area has a lower threshold of excitability to induce muscle movement compared with other motor areas and contributes a large number of fibers to the corticospinal and corticobulbar tracts. It is the only cortical area to contain the large Betz's neurons.

Cross-References

▶ Motor Cortex, Effect on Pain-Related Behavior

Primary Somatosensory Cortex (S1), Effect on Pain-Related Behavior in Humans

Markus Ploner[1] and Alfons Schnitzler[2]
[1]Department of Neurology, Technische Universität München, Munich, Germany
[2]Department of Neurology, Heinrich Heine University, Düsseldorf, Germany

Definition

The primary somatosensory cortex (S1) is a functionally defined part of the somatosensory

and ▶ nociceptive system. Anatomically, S1 occupies the postcentral cortex and consists of four architectonic fields each containing a somatotopic representation of the contralateral body half. Afferent projections to S1 originate mainly from the ventroposterior nucleus of the thalamus. Physiological, functional imaging and lesion studies provide converging evidence for an association between S1 and the ▶ sensory-discriminative aspect of pain perception.

Characteristics

The behavioral effects of cortical lesions determined early concepts of the relevance of the cerebral cortex for the perception of pain. In the first half of the twentieth century, the view that pain has "little, if any, true cortical representation" predominated (Penfield and Boldrey 1937). This notion was mainly based on Henry Head's observation that "pure cortical lesions cause no increase or decrease of sensibility to measured painful stimuli" (Head and Holmes 1911, p. 154) and on the fact that electrical stimulations of S1 only rarely elicited painful sensations (Penfield and Boldrey 1937). In the second half of the twentieth century, an opposing view emerged. Various lesion studies contradicted Head's findings by clearly demonstrating alterations of pain perception due to lesions of the S1 region (reviewed in Kenshalo and Willis 1991; Sweet 1982).

Russell (1945) and Marshall (1951) presented two seminal studies in patients with traumatic head injury involving the postcentral cortex. Both authors observed patients with a loss of all forms of somatosensory sensations including a loss of pain perception. In these patients, lesions were located in the more anterior parts of the postcentral cortex (the left panel of Fig. 1 shows the lesion documentation of Russell (1945) in this type of patient). By contrast, more extended lesions located in the posterior parts of postcentral cortex yielded deficits in position sense and two-point discrimination, but no deficits in pain perception (Fig. 1, middle panel). Additionally, in a few cases of both series, hyperpathia

Primary Somatosensory Cortex (S1), Effect on Pain-Related Behavior in Humans, Fig. 1 Locations of lesions causing permanent deficits in all forms of somatosensory sensation including pain (*left*), in discriminative sensations excluding pain (*middle*), and of lesions causing hyperpathia (*right*). Figures, depth of wounds in centimeters; *D* depressed fracture without dural tear, *L* leg, *A* arm, *H* hand, *F* face (From Russell (1945))

was observed (Fig. 1, right panel). Thus, Marshall and Russell both concluded that the postcentral cortex comprising S1 is concerned with the appreciation of pain but that the effects of postcentral lesions on pain perception are largely inconsistent.

This side-by-side hypo-, norm-, and hyperalgesia after postcentral lesions may partly reflect that S1 participates not only in the mediation but also in the modulation of pain perception (Bushnell et al. 1999). Accordingly, surgical excision of S1 has at least transiently been successful in alleviating phantom limb and central pain conditions (White and Sweet 1969).

However, the inconsistency of the observed effects of S1 lesions might also be due to methodological reasons. Despite the very good clinical documentation of the perceptual lesion effects, information about lesion extent or presence of additional lesions was often scant before high-resolution structural brain imaging became available. Furthermore, clinical examination of pain perception was carried out with nonselective painful stimuli that yield a coactivation of nociceptive and tactile afferents and result in an interference between touch and pain perception. Moreover, the complex and multidimensional nature of pain may have not been sufficiently appreciated, so that the different sensory, ▶ cognitive aspects of pain and affective aspects of pain perception could not be disentangled.

Conceptual and methodological advances now allow more subtle investigations of lesion effects on pain perception. Recently, the case of a patient with an ischemic lesion of the postcentral cortex was reported (Ploner et al. 1999) (Fig. 2). This patient suffered from a complete loss of tactile and thermal sensation on the contralateral hand. However, application of selective pain stimuli to the affected hand yielded a clearly unpleasant feeling emerging from an ill-localized and extended area somewhere between fingertips and shoulder that he wanted to avoid. The fully eloquent patient was completely unable to further describe quality, localization, and intensity of the perceived stimulus nor did he choose the given description "pain" for the perceived sensation. This pattern of deficits sheds light on some basic functional aspects of the role of S1 in pain perception. The postcentral lesion has resulted in a dissociation between different aspects of pain perception with a loss of sensory-discriminative capabilities while sparing pain affect. This indicates a crucial role of S1 for the sensory-discriminative aspect of pain perception, which is in accordance with anatomical, neurophysiological, and functional imaging data from the past decades (reviewed in Duncan and Albanese 2003; Schnitzler and Ploner 2000). In addition, the fact of a perceptual dissociation of pain perception strongly suggests that the different aspects of pain perception are organized in parallel.

In conclusion, lesion studies indicate that S1 constitutes an essential part of a parallel-organized cortical network of pain perception.

Primary Somatosensory Cortex (S1), Effect on Pain-Related Behavior in Humans, Fig. 2 Magnetic resonance images of an ischemic lesion of the right postcentral cortex causing a loss of pain sensation with preservation of pain affect (From Ploner et al. (1999))

Within this network, S1 is particularly associated with sensory-discriminative capabilities of pain perception and may also be involved in the modulation of pain perception.

References

Bushnell, M. C., Duncan, G. H., Hofbauer, R. K., et al. (1999). Pain perception: Is there a role for primary somatosensory cortex? *Proceedings of the National Academy of Sciences of the United States of America, 96*, 7705–7709.

Duncan, G. H., & Albanese, M. C. (2003). Is there a role for the parietal lobes in the perception of pain? *Advances in Neurology, 93*, 69–86.

Head, H., & Holmes, G. (1911). Sensory disturbances from cerebral lesions. *Brain, 34*, 102–254.

Kenshalo, D. R., & Willis, W. D. (1991). The role of the cerebral cortex in pain sensation. In A. Peters & E. G. Jones (Eds.), *Cerebral cortex* (Vol. 9, pp. 153–212). New York: Plenum Press.

Marshall, J. (1951). Sensory disturbances in cortical wounds with special reference to pain. *Journal of Neurology, Neurosurgery, and Psychiatry, 14*, 187–204.

Penfield, W., & Boldrey, E. (1937). Somatic motor and sensory representation in the cerebral cortex of man as studied by electrical stimulation. *Brain, 60*, 389–443.

Ploner, M., Freund, H. J., & Schnitzler, A. (1999). Pain affect without pain sensation in a patient with a postcentral lesion. *Pain, 81*, 211–214.

Russell, W. R. (1945). Transient disturbances following gunshot wounds of the head. *Brain, 68*, 79–97.

Schnitzler, A., & Ploner, M. (2000). Neurophysiology and functional neuroanatomy of pain perception. *Journal of Clinical Neurophysiology, 17*, 592–603.

Sweet, W. H. (1982). Cerebral localization of pain. In R. A. Thompson & J. Green (Eds.), *New perspectives in cerebral localization* (pp. 205–242). New York: Raven.

White, J. C., & Sweet, W. H. (1969). *Pain and the neurosurgeon. A forty-year experience*. Springfield: Charles C Thomas.

Primary Somatosensory Cortex (SI)

Definition

The majority of neurons originating in the ventroposterior nucleus of the thalamus project to the primary somatosensory cortex (SI), while the remaining fibers project to the secondary somatosensory cortex (SII) in the posterior parietal lobe.

Cross-References

► Opioid Receptor Localization
► Spinothalamic Input: Cells of Origin (Monkey)

Primary Stabbing Headache

James W. Lance
Institute of Neurological Sciences, University of
New South Wales, Sydney, Australia

Synonyms

Ice-pick pains; Idiopathic stabbing headache
(ISH); Jabs and jolts syndrome; Ophthalmodynia
periodica

Definition

Localized stabs of pain in the head that occur
spontaneously in the absence of organic disease
of underlying structures or of the cranial nerves.

Characteristics

Primary stabbing headaches (PSH) are pains
occurring as a single stab or series of stabs
confined to the head, predominantly in the
orbit, temple, or parietal area. Each stab lasts
from a fraction of a second to 3 s, and pains
recur irregularly, from a single stab occasionally
to many each day. There are no accompanying
symptoms or signs. Such pains are more common in patients subject to migraine, cluster, or
tension headache than in the headache-free
population.

The briefest of stabs, previously known as
▶ ice-pick pains, usually respond to indomethacin. The more prolonged stabs, previously called
the "jabs and jolts syndrome" are less responsive
to medication. Ophthalmodynia periodica is
a stabbing pain limited to one eye.

Clinical Reports

Sharp stabbing pains in the head resembling a jab
from a nail or a needle were described by Raskin
and Schwartz (1980) as "ice-pick" pains. They
found that 42 out of 100 migraine patients had
experienced this symptom, compared with only 3
out of 100 headache-free control subjects.

The most common sites affected were the
orbits or temples. The pains were often associated
with the times of recurrence of migraine
headache. They commented that similar pains
could occur with temporal arteritis.

Drummond and Lance (1984) obtained a history of ice-pick pains in 200 out of 530 patients
with recurrent migraine and tension headache. In
the 92 patients in whom the site of ice-pick pains
had been recorded, it coincided with the site
affected by the patient's customary headaches in
37, 19 unilateral and 18 bilateral. Stabbing pains
have also been reported in conjunction with cluster
headaches (Ekbom 1975). Sudden stabbing
pains in one eye are termed "ophthalmodynia
periodica." Over 60 % of patients with this syndrome are migraine sufferers (Lansche 1964).

The "jabs and jolts syndrome" was first
described by Sjaastad et al. (1979). These are
sharp knifelike pains lasting less than 1 min in
patients with migraine, tension, and cluster
headache as well as in chronic paroxysmal
hemicrania (▶ CPH). They were called "jabs
and jolts" to distinguish them from episodes of
CPH which last more than 3 min (Sjaastad 1992).
Pareja et al. (1999) considered the differential
diagnosis of idiopathic (primary) stabbing
headache from ▶ trigeminal neuralgia affecting
the first division and SUNCT (short-lasting unilateral neuralgiform headache attacks with conjunctival injection and tearing) syndrome.
SUNCT is confined to one periocular area and is
associated with autonomic features not present in
PSH, which is multifocal and variable in location.
SUNCT lasts more than 5 s and commonly
continues for 1 min or more. It affects men
more than women in contrast to PSH, which
mainly affects women.

Sjaastad et al. (2001) conducted a large-scale
study of headache epidemiology in 1,838 adults
living in Vågå, Norway, and found that 35.2 %
claimed to have experienced primary stabbing
headaches, a much higher proportion than found
in previous studies. The female:male ratio was
about 1.5, whereas it was 6.6 in an earlier report

by Pareja et al. (1996). The mean age of onset was 28 in the Norwegian series and 47 in the Spanish series. The primary stabbing headaches were associated with another form of headache in half of the patients reported by Pareja et al. (1996). Contrary to the experience in adults, PSH in children is not usually accompanied by other forms of headache (Soriani et al. 1996). Martins et al. (1995) reported recurrent stabbing pains in bouts they termed "status," outside the area innervated by the trigeminal nerve in the retro-auricular, parietal, and occipital regions in six patients. Attacks were daily, recurred every minute, and subsided after 1 week. None were associated with migraine. A recent publication from the Vågå study (Sjaastad et al. 2003) also pointed out that stabbing pains were not necessarily confined to the head. They were reported throughout the body in 4 patients, in the face in 3, and in the neck in 12.

Pathophysiology

The stabbing quality of these pains suggests a paroxysmal neuronal discharge comparable with trigeminal neuralgia. The localization of pains to the habitual site of migraine or cluster headache suggests the possibility of a defect in the brain's ► endogenous pain control pathway that permits the spontaneous synchronous discharge of neurones receiving impulses from the area to which the head pain is referred.

Treatment

Mathew (1981) reported the response of five patients to indomethacin 50 mg three times daily but not to aspirin or placebo. Twenty patients in whom PSH accompanied atypical headaches responded well to indomethacin (Medina and Diamond 1981), which has remained the treatment of choice. The bout of frequent stabs ("status") described by Martins et al. (1995) also settled promptly with indomethacin. Provesan et al. (2002) described PSH developing after a stroke in three patients, all of whom improved with the COX-2 inhibitor celecoxib. Sjaastad (1992) reported that the "jabs and jolts syndrome" responded only partially to indomethacin or not at all.

Summary

It remains to be determined whether brief, stabbing ice-pick pains have the same pathophysiology as the longer-lasting "jabs and jolts syndrome" and form part of a spectrum linking up with SUNCT and CPH, although devoid of their autonomic features. The frequent response to indomethacin in ice-pick pains and CPH is a striking clinical difference from "jabs and jolts" and SUNCT, which respond poorly if at all.

In view of the gross discrepancy in the reported prevalence of PSH in different communities, further studies are required to clarify this and to determine whether the reported association of PSH with other headaches is greater than could be accounted for by chance.

Cross-References

► Hemicrania Continua

References

Drummond, P. D., & Lance, J. W. (1984). Neurovascular disturbances in headache patients. *Clinical and Experimental Neurology, 20*, 93–99.

Ekbom, K. (1975). Some observations on pain in cluster headache. *Headache, 14*, 219–225.

Lansche, R. K. (1964). Ophthalmodynia periodica. *Headache, 4*, 247–249.

Martins, I. P., Parreira, B., & Costa, I. (1995). Extratrigeminal ice-pick status. *Headache, 35*, 107–110.

Mathew, N. T. (1981). Indomethacin responsive headache syndrome. *Headache, 21*, 147–150.

Medina, J. L., & Diamond, S. (1981). Cluster headache variant: Spectrum of a new headache syndrome. *Archives of Neurology, 38*, 705–709.

Pareja, J. A., Ruiz, J., de Isla, C., et al. (1996). Idiopathic stabbing headache (jabs and jolts syndrome). *Cephalalgia, 16*, 93–96.

Pareja, J. A., Kruszewski, P., & Caminero, A. B. (1999). SUNCT syndrome versus idiopathic stabbing headache (jabs and jolts syndrome). *Cephalalgia, 19* (Suppl. 25), 46–48.

Provesan, E. J., Zukerman, E., Kowacs, P. A., et al. (2002). COX-2 inhibitor for the treatment of idiopathic stabbing headache secondary to cerebrovascular diseases. *Cephalalgia, 22*, 197–200.

Raskin, N. H., & Schwartz, R. K. (1980). Icepick-like pain. *Neurology, 30*, 203–205.

Sjaastad, O. (1992). *Cluster headache syndrome* (pp. 310–312). London: WB Saunders.

Sjaastad, O., Egge, K., Hørven, I., et al. (1979). Chronic paroxysmal hemicrania V. Mechanical precipitation of attacks. *Headache, 19,* 31–36.

Sjaastad, O., Pettersen, H., & Bakketeig, L. S. (2001). The Vågå study; epidemiology of headache. *Cephalalgia, 21,* 207–215.

Sjaastad, O., Pettersen, H., & Bakketeig, L. S. (2003). Extracephalic jabs/idiopathic stabs. Vågå study of headache epidemiology. *Cephalalgia, 23,* 50–54.

Soriani, G., Battistella, P. A., Arnaldi, C., et al. (1996). Juvenile idiopathic stabbing headache. *Headache, 36,* 565–567.

Principalism

▶ Ethics of Pain, Culture, and Ethnicity

PRM Specialist

Definition

PRM specialist is a physician with 3–5 years of postgraduate training in Physical and Rehabilitation Medicine, carrying final medical responsibility for team actions.

Cross-References

▶ Physical Medicine and Rehabilitation, Team-Oriented Approach

PRN

▶ Pro Re Nata

Pro Re Nata

Synonyms

PRN

Definition

According to circumstances; as necessary.

Cross-References

▶ Assessment of Pain Behaviors
▶ Operant Treatment of Chronic Pain
▶ Psychology of Pain and Psychological Treatment

Probability

Definition

The chance or likelihood that a given event will occur.

Cross-References

▶ Pain in the Workplace, Risk Factors for Chronicity, and Demographics

Probable Cause

Definition

A reasonable ground for supposing that something will predispose toward a particular outcome.

Cross-References

▶ Pain in the Workplace, Risk Factors for Chronicity, and Demographics

Procedural Distress

Definition

The aggregate of pain and anxiety that induces a stress response to an invasive procedure.

Cross-References

▶ Cancer Pain, Assessment in Children

Prognosis

Definition

The prognosis is a measure of the probability of a favorable or unfavorable outcome of a given condition.

Cross-References

▶ Psychiatric Aspects of the Epidemiology of Pain

Prognostic Block

▶ Pain Treatment: Spinal Nerve Blocks

Programmed Cell Death

▶ Apoptosis

Progressive Adjustive

▶ Traction

Progressive Muscle Relaxation

Definition

Progressive muscle relaxation is a relaxation technique in which patients are trained to systematically tense and relax specific muscle groups.

Benefits are achieved as individuals become more aware of the presence of muscle tension in their bodies and as they learn to reduce this tension.

Cross-References

▶ Coping and Pain
▶ Psychological Treatment in Acute Pain
▶ Relaxation in the Treatment of Pain

Proinflammatory

▶ Proinflammatory Cytokines

Proinflammatory Cytokines

Julie Wieseler, Steven F. Maier and
Linda R. Watkins
Department of Psychology and Neuroscience,
and The Center for Neuroscience, University of
Colorado at Boulder, Boulder, CO, USA

Synonyms

Cytokines; Inflammatory cytokines; Proinflammatory

Definition

A family of proteins, including tumor necrosis factor (TNF), interleukin-1 (IL1), and (often but not always) IL6. "Proinflammatory" refers to their classical function in the peripheral immune system, where they enhance inflammation by orchestrating the early immune response to damage and infection.

Characteristics

Peripheral Tissues

Proinflammatory cytokines are released by activated immune cells at sites of infection, inflammation, and damage. Their release can enhance nociceptive signaling in skin, muscle, and joints. A wide array of cell types at these sites can release proinflammatory cytokines including macrophages, neutrophils, mast cells, fibroblasts, chondrocytes, keratinocytes, and endothelial cells. In addition, mast cells can further contribute to pain signaling by cytokines as they can release enzymes that cleave extracellular pools of inactive proinflammatory cytokines to their active forms (for a review, see Watkins et al. 1999). It is now well accepted that not only do dorsal root ganglion neurons express mRNAs for proinflammatory cytokine receptors but also these receptors are expressed at sensory nerve terminals of these DRG neurons (for a review, see Watkins et al. 2007; Watkins and Maier 2002).

In addition to activating nociceptive neurons that project directly to the spinal cord, peripheral proinflammatory cytokines can also enhance pain as a natural consequence of immune-to-brain communication (for a review, see Maier and Watkins 1998; Watkins and Maier 2000). Immune-to-brain communication refers to a pathway activated by localized (such as at sites of infection or inflammation) or blood-borne proinflammatory cytokines. These activate sensory neurons (e.g., vagal afferents) and/or *circumventricular brain structures* (e.g., area postrema), respectively (for a review, see Watkins et al. 1995). This communication pathway relays to the brain that a peripheral immune challenge has occurred. Subsequently, specific brain and brain-to-spinal cord pathways become activated, causing *glial activation* and proinflammatory cytokine release. This, in turn, results in the expression of well-characterized sickness responses. Sickness responses refer to a brain-orchestrated constellation of changes aimed at enhancing the survival of the host. These include a wide array of symptoms such as fever, increases in sleep, decreases in food and

water intake, and decreases in social behavior, as well as increases in pain to formerly innocuous stimuli (sickness-induced hyperalgesia) (for a review, see Maier and Watkins 1998; Watkins and Maier 2000). Whether hyperalgesia continues throughout sickness, versus being replaced by hypoalgesia at later time points, is a subject of debate.

Peripheral Nerves

Whereas sensory neurons are classically believed to be activated by stimuli in their terminal receptive fields, immune cells can exert excitatory influences over sensory neuronal function at mid-axonal sites as well. Cells resident within peripheral nerves that can produce proinflammatory cytokines include dendritic cells, endothelial cells, fibroblasts, macrophages, mast cells, and Schwann cells. In addition, under inflammatory, infectious, or traumatic conditions, the blood-nerve barrier breaks down. This allows immune cells (predominantly macrophages and neutrophils) to be recruited out of the bloodstream and into the nerve, where they contribute to local proinflammatory cytokine production (for a review, see Watkins et al. 2007; Watkins and Maier 2002).

Involvement of immune cells in pain from nerve fiber inflammation without damaging the nerve has been studied predominantly in two animal models: *Aplysia* and rats. In the simpler nervous system of *Aplysia*, it is known that activated immunocytes (immune-like cells of *Aplysia*) release TNF-like and IL1-like proinflammatory cytokines. These alter ion channels in *Aplysia* neurons, causing hyperexcitability. Hyperexcitability of injured *Aplysia* nerves is greater in the presence of activated immunocytes, an effect mimicked by application of IL1 (for a review, see Clatworthy 1999). Similarly in rats, neuropathy is linked to the activity of macrophages recruited to the site of injury. Notably, simply delaying the *recruitment* of macrophages to the injury site delays the development of neuropathic pain, while attracting activated immune cells to the injury site enhances the development of neuropathic pain (for a review, see Myers et al. 1999, 2006). Indeed, simply attracting and activating immune cells to an otherwise healthy sciatic nerve induces pain facilitation. Of the various neuroactive substances released by activated immune cells, TNF, IL1, and IL6 have received the most experimental support for their involvement in neuropathic pain of both traumatic and inflammatory origin. Blockade of TNF, IL1, or IL6 reduces neuropathic pain. For example, TNF injected into the sciatic nerve induces pain facilitation and TNF applied topically to the sciatic nerve evokes *ectopic activity* in single primary afferent nociceptive fibers (for a review, see Watkins et al. 2007; Watkins and Maier 2002).

Although it is clear that proinflammatory cytokines can induce pain, how they do this from mid-axonal sites is controversial (for a review, see Watkins and Maier 2002). Proinflammatory cytokine receptors are expressed by dorsal root ganglia neuronal cell bodies, but whether or not functional receptors are expressed along the course of the peripheral nerves is unknown. Of the proinflammatory cytokines, TNF has been the most intensively studied with regard to its effects on neuroexcitability. It rapidly alters peripheral nerve activity, suggesting that it may exert direct effects on axonal function. Membrane modeling studies suggest that TNF can dimerize and insert through cell membranes, forming a pore-like structure. Formation of TNF "pores" is facilitated under the acidic conditions that occur at inflammation sites. Intriguingly, the TNF "pores" have been reported to form voltage-dependent sodium channels. Other lines of evidence suggest that TNF enhances nerve excitability by increasing the membrane conductance of endogenous sodium and calcium channels. Like TNF, IL1 also rapidly increases neuronal excitation and produces long-lasting increases in conductance of sodium and calcium channels. Thus, multiple mechanisms may account for proinflammatory cytokine actions on peripheral nerves.

In addition to direct effects on axonal excitability, proinflammatory cytokines may serve as retrogradely transported signals that influence gene activation in intact dorsal root ganglia somata. TNF and IL6 have each been reported to be retrogradely transported from the site of peripheral nerve injury to the dorsal root ganglia. Intriguingly, the transport of TNF after nerve trauma appears to be bidirectional. That is, after nerve injury, medium and large dorsal root ganglion neurons produce TNF and anterogradely transport it to the injury site (Schafers et al. 2002).

Dorsal Root Ganglia

Proinflammatory cytokines can be produced by a variety of cell types in dorsal root ganglia, including glially derived *satellite cells*, macrophages, endothelial cells, and neurons. Under conditions of neuropathy, immune cells migrate from the systemic circulation into the dorsal root ganglia. In response to peripheral nerve injury, satellite cells (and presumably other immune cells) are activated and release proinflammatory cytokines in the dorsal root ganglia.

Dorsal root ganglia neurons can also be affected by proinflammatory cytokines produced by herniated discs. Herniated discs spontaneously produce TNF, IL1, and IL6, among other inflammatory products. In addition, exposure of nerve roots to herniated disc substances (nucleus pulposus) induces elevations in dorsal root ganglion expression of proinflammatory cytokine mRNAs, as well as pain facilitation. Topical application of TNF to dorsal root ganglia and proximal roots induces spontaneous activity, possibly via binding to TNF receptors expressed by neuronal somata. Applying TNF to nerve roots causes activation of Schwann cells, local production of TNF, and recruitment of macrophages, just as does application of nucleus pulposus. Moreover, blocking TNF activity delays the development of pain facilitation in response to applied nucleus pulposus (for a review, see Watkins and Maier 2002).

Spinal Cord

Spinal proinflammatory cytokines have been attracting increasing interest (for a review, see DeLeo et al. 2004; DeLeo and Yezierski 2001; Watkins and Maier 2003; Watkins et al. 2001). Whereas TNF transport from dorsal root ganglia to spinal cord has been reported after peripheral nerve injury (Shubayev and Myers 2002), it remains unknown whether this TNF can be released so to affect spinal cord function. If this were true, it would be a novel source of spinal cord neuroexcitation by proinflammatory cytokines. What is clear is that the reverse is true; namely, proinflammatory cytokines in spinal cord CSF can be retrogradely transported to dorsal root ganglia where they affect gene expression of sensory neurons.

The predominant view is that proinflammatory cytokines, likely of glial origin, are important in the induction and maintenance of pain facilitation. In some models, recruitment of immune cells to spinal cord may also contribute to alterations in pain responsivity (for a review, see DeLeo et al. 2004; DeLeo and Yezierski 2001). Intrathecal delivery of IL1 and IL6 has each been reported to enhance pain responsivity. Rapid glial activation and rapid elevations of proinflammatory cytokines have been observed in spinal cord and surrounding CSF, the latter observation implying cytokine release. Blockade of spinal TNF, IL1, and IL6 has each been reported to prevent and/or reverse pain facilitation induced by inflammation of peripheral tissues, inflammation and trauma to either peripheral or spinal nerves, and spinal cord glial activation in response to viral gp120. Importantly, glial involvement in pain facilitation is not restricted to the time of injury. Rather, recent data from several laboratories point to prolonged glial activation, lasting many weeks. For example, intrathecal blockade of IL1, even 2 months after traumatic neuropathy, still rapidly and completely reverses neuropathic pain (Watkins and Maier 2003). Prolonged upregulation of proinflammatory cytokines for many weeks after tissue or nerve injury has also been reported. Such observations are striking in that they imply that spinal cord glial activation and proinflammatory cytokines may be important contributors to perseverative pain states.

It is not yet clear whether glial activation and proinflammatory cytokine release is linked to spinal cord neurodegeneration. Several laboratories have now reported neurodegeneration in spinal cord dorsal horn following peripheral nerve injury. It is intriguing that available evidence suggests that a positive feedback cycle may exist wherein neural excitation triggers glial activation, which leads to the release of neuroexcitatory (and potentially neurotoxic) substances from glia, which leads to further neuroexcitation. If this cycle is sufficiently potent or prolonged, neuronal death could occur. Indeed, a number of substances released from activated glia (including proinflammatory cytokines) have been linked to neurodegeneration in a variety of diseases, such as Alzheimer's, Parkinson's, prion disease, multiple sclerosis, and AIDS dementia. If glia contribute to neurodegeneration in spinal cord, this could in turn further activate glia, because neurons normally keep glia in an unactivated state via contact inhibition. Dying and dead neurons could thereby induce a state of perseverative hyperexcitability in nearby glia. While clearly speculative for spinal cord, the implications of this inference are intriguing.

Brain

There is a growing literature implicating brain proinflammatory cytokines in the modulation of head pain (for a review, see Sessle 2011). Several laboratories have now reported that *intracerebroventricular* administration of TNF, IL1, and IL6 induces pain facilitation and intracerebroventricular IL1 enhances nociceptive neuronal responses in trigeminal nucleus caudalis (Oka and Hori 1999). A complex dose response function has been reported for intracerebroventricular IL1, wherein low doses (picograms) induce hyperalgesia and high doses (nanograms) induce hypoalgesia. Likewise, intracerebroventricular antagonists of TNF and IL1 block pain facilitation induced by intraperitoneal inflammation.

Microinjections of proinflammatory cytokines into discrete brain regions reveal marked heterogeneity of effects. For example, IL1 microinjection into the preoptic area produces pain facilitation. In contrast, IL1 microinjections into the ventromedial hypothalamus, centromedial thalamus, and gelatinosus nucleus of the thalamus produce hypoalgesia. Such differential effects of proinflammatory cytokines in diverse brain regions may, at least in part, explain the complex dose response function reported after intraventricular administration.

Lastly, there is a small but growing literature suggesting that proinflammatory cytokine levels increase in brain following manipulations that induce pain facilitation. Increases in IL1 in glial cells of the hypothalamus have been reported following subcutaneous formalin. Following peripheral neuropathy, neuronal TNF increases in hippocampus and in the region of the locus coeruleus, and blockade of hippocampal TNF blocks the development of neuropathic pain. Very recently, subcutaneous complete Freund's adjuvant was found to produce nonhomogeneous increases in glial activation markers in brain, with greater activation in midbrain and thalamic and cerebral cortex, compared to pons or medulla (Raghavendra et al. 2004). Notably, mRNA for TNF, IL1, and IL6 were all increased in each of these brain regions (Raghavendra et al. 2004).

Thus, while little is yet known about supraspinal actions of proinflammatory cytokines in pain modulation, these initial studies clearly suggest that proinflammatory cytokine regulation of pain within the central nervous system includes supraspinal as well as spinal sites of action.

Implications for Pain Control

The animal data are clear that immune- and glia-derived products can dramatically amplify pain. It is no longer a question of whether they do, at least in laboratory animals. It is now a matter of how to focus drug discovery to take advantage of this knowledge to test whether clinical pain, which is resistant to currently available drugs, may be resolved by using compounds that target immune and/or glial cells.

One issue that immediately arises is whether systemic drug delivery is desirable. One issue with systemic delivery is that drugs that will control immune/glial contributions to pain would also be expected to be generally *immuno-suppressive*. A second issue with systemic delivery is that, unless the blood-nerve barrier is disrupted, it may be difficult for many compounds to reach therapeutic levels in nerve. A third issue is that by the time drugs are administered, damage to peripheral nerves may have already occurred. While treatment may prevent further immune-derived damage, the damage already present may be sufficient to maintain the pain. The last issue is that blood-brain barrier permeability is a double-edged sword for immune/glial active compounds. If a drug does not cross the blood-brain barrier, it cannot directly control the glial activation contributing to pain facilitation. On the other hand, if a drug does cross the blood-brain barrier, it would disrupt glial function in brain as well as spinal cord. As glia are now implicated as regulators of wide-ranging phenomena, such as learning and memory, hormone regulation, thermoregulation, and so forth, pervasive alterations in glial function require careful consideration before implementation.

One alternative approach would be to specifically target glia within the spinal cord, via direct administration to the cerebrospinal fluid space surrounding the spinal cord. As glia normally serve very important functions, one simply wants to control their activation rather than to aim to destroy these cells. From studies of animal models, a variety of compounds have now been identified which can prevent and/or reverse exaggerated pain states of diverse etiologies. These include drugs that disrupt glial activation (e.g., minocycline), proinflammatory cytokine antagonists (e.g., infliximab, etanercept, anakinra), drugs that disrupt proinflammatory cytokine synthesis (e.g., propentofylline, thalidomide and congeners), or drugs that disrupt both proinflammatory cytokine signaling and synthesis (e.g., leflunomide, methotrexate, p38 *MAP kinase* inhibitors, the antiinflammatory cytokine interleukin-10). For detailed discussion of these compounds with regard to pain control, readers are referred to the review by Watkins and Maier (2003).

Conclusions

Over the past decade, an increasing and convincing literature has linked proinflammatory cytokines with pain. Proinflammatory cytokines generated by both peripheral immune cells and central (primarily) glia are clearly involved in pain sensitization. Peripherally, proinflammatory cytokines increase excitability at peripheral nerve terminals as well as mid-axonally along nerve bundles. They signal the brain both via sensory neuron-to-spinal cord and immune-to-brain circuitries. Centrally, to date, most evidence of glially driven pain facilitation has accrued in spinal cord. While the study of glial activation in brain is still in its infancy, it is clear that an intriguing story will emerge there as well.

Cross-References

▶ Cord Glial Activation
▶ Cytokines, Regulation in Inflammation

References

Clatworthy, A. L. (1999). Evolutionary perspectives of cytokines in pain. In L. R. Watkins & S. F. Maier (Eds.), *Cytokines and pain* (pp. 21–38). Basel: Birkhauser.

DeLeo, J. A., & Yezierski, R. P. (2001). The role of neuroinflammation and neuroimmune activation in persistent pain. *Pain, 2001*, 1–6.

DeLeo, J. A., Tanga, F. Y., & Tawfik, V. L. (2004). Neuroimmune activation and neuroinflammation in chronic pain and opioid tolerance/hyperalgesia. *The Neuroscientist, 10*, 40–52.

Maier, S. F., & Watkins, L. R. (1998). Cytokines for psychologists: Implications of bi-directional immune-to-brain communication for understanding behavior, mood, and cognition. *Psychological Review, 105*, 83–107.

Myers, R. R., Wagner, R., & Sorkin, L. S. (1999). Hyperalgesic actions of cytokines on peripheral nerves. In L. R. Watkins & S. F. Maier (Eds.), *Cytokines and pain* (pp. 133–158). Basel: Birkause.

Myers, R. R., Campana, W. M., & Shubayev, V. I. (2006). The role of neuroinflammation in neuropathic pain: Mechanisms and therapeutic targets. *Drug Discovery Today, 11*(1–2), 8–20.

Oka, T., & Hori, T. (1999). Brain cytokines and pain. In L. R. Watkins & S. F. Maier (Eds.), *Cytokines and pain* (pp. 183–204). Basel: Birkhauser.

Raghavendra, V., Tanga, F. Y., & DeLeo, J. A. (2004). Complete Freunds' adjuvant-induced peripheral inflammation evokes glial activation and proinflammatory cytokine expression in the CNS. *European Journal of Neuroscience, 20*(2), 467–473.

Schafers, M., Geis, C., Brors, D., Yaksh, T. L., & Sommer, C. (2002). Anterograde transport of tumor necrosis factor-alpha in the intact and injured rat sciatic nerve. *Journal of Neuroscience, 22*, 536–545.

Sessle, B. J. (2011). Peripheral and central mechanisms of orofacial inflammatory pain. *International Review of Neurobiology, 97*, 179–206.

Shubayev, V. I., & Myers, R. R. (2002). Anterograde TNF alpha transport from rat dorsal root ganglion to spinal cord and injured sciatic nerve. *Neuroscience Letters, 320*, 99–101.

Watkins, L. R., & Maier, S. F. (2000). The pain of being sick: Implications of immune-to-brain communication for understanding pain. *Annual Review of Psychology, 51*, 29–57.

Watkins, L. R., & Maier, S. F. (2002). Beyond neurons: Evidence that immune and glial cells contribute to pathological pain states. *Physiological Reviews, 82*, 981–1011.

Watkins, L. R., & Maier, S. F. (2003). Glia: A novel drug discovery target for clinical pain. *Nature Reviews. Drug Discovery, 2*, 973–985.

Watkins, L. R., Maier, S. F., & Goehler, L. E. (1995). Cytokine-to-brain communication: A review and analysis of alternative mechanisms. *Life Sciences, 57*, 1011–1026.

Watkins, L. R., Hansen, M. K., Nguyen, K. T., Lee, J. E., & Maier, S. F. (1999). Dynamic regulation of the proinflammatory cytokine, interleukin-1beta: Molecular biology for non-molecular biologists. *Life Sciences, 65*, 449–481.

Watkins, L. R., Milligan, E. D., & Maier, S. F. (2001). Glial activation: A driving force for pathological pain. *Trends in Neuroscience, 24*, 450–455.

Watkins, L. R., Hutchinson, M. R., Milligan, E. D., & Maier, S. F. (2007). "Listening" and "talking" to neurons: Implications of immune activation for pain control and increasing the efficacy of opioids. *Brain Research Reviews, 56*(1), 148–169.

Projection Neurons

Definition

Projection neurons of the spinal cord are those with axons that travel rostrally to reach the brain.

Cross-References

▶ Nociceptive Circuitry in the Spinal Cord
▶ Opioid Receptors at Postsynaptic Sites

Projective Field (PF)

Definition

A projective field is an area of the body in which a sensation is projected due to stimulation of a neuron.

Cross-References

▶ Central Pain, Human Studies of Physiology
▶ Thalamus, Receptive Fields, Projected Fields, Human

Prolactin

Definition

Prolactin is one of the gonadotrophic hormones secreted by the anterior pituitary that stimulates the gonads to release hormones, thereby regulating sexual and reproductive function. Other gonadotrophic hormones include luteinizing hormone and follicle-stimulating hormone.

Cross-References

▶ Cancer Pain Management, Opioid Side Effects, Endocrine Changes, and Sexual Dysfunction

Proliferant

Definition

An injection solution used in prolotherapy that aims to induce inflammation, the release of tissue growth factors, and the subsequent deposition of collagen fibers.

Cross-References

▶ Prolotherapy

Prolonged Acute Pain

▶ Chronic Postsurgical Pain (CPSP)

Prolotherapy

Michael Yelland
School of Medicine, Logan campus, Griffith University, Logan, Australia

Synonyms

Reconstructive therapy; Sclerotherapy

Definition

The term ▶ prolotherapy derives from the Latin word "proles" meaning "growth" or "offspring." It refers to a treatment that involves the injection of chemicals designed to strengthen weakened ligaments, based on the premise that these ligaments are a source of instability and chronic pain.

Characteristics

Mechanism
Prolotherapy is based on the premise that much chronic musculoskeletal pain and disability is the result of ligamentous laxity or weakness and that the correction of this laxity will result in a reduction in pain and disability. The evidence for a direct link between ligamentous laxity and pain/disability is weak. Other mechanisms of action, such as central modulation of pain perception and denervation of painful structures by the phenol component of prolotherapy solutions (Klein et al. 1993), are just two other possibilities.

The purported mechanism is that "▶ proliferant" solutions strengthen weakened ligaments and tendinous attachments (entheses) by inducing inflammation, the release of tissue growth factors, and the subsequent deposition of collagen fibers. In osteoarthritis, proliferant solutions are believed to stimulate cartilage growth and repair. There is some evidence from animal and human studies that the proliferant solutions can induce thickening of ligaments and increase chondrocyte activity in cartilage (Hackett et al. 1993; Reeves and Hassanein 2000a; Klein et al. 1989). However, dextrose (glucose) prolotherapy has not been shown to have any effects on the laxity or mechanical properties of injured ligaments over control injections in a rat model (Jensen et al. 2008).

There are three major classes of proliferants: the ▶ irritants (phenol, guaiacol, and tannic acid), the ▶ chemotactics (sodium morrhuate), and the ▶ osmotics (dextrose, glycerine, and zinc sulfate) (Banks 1991). There is some overlap in their actions. Irritants act by triggering cellular trauma and by attracting macrophages, which ingest them and secrete polypeptide growth factors. Chemotactics attract inflammatory cells. The osmotics cause an osmotic shock to cells, leading to the release of proinflammatory substances.

By far the most common proliferant used in current prolotherapy practice is hypertonic dextrose in concentrations ranging from 10 % to 25 %. It is combined with a local anesthetic, usually lidocaine (lignocaine) in a concentration of 0.2–0.5 %. In knee osteoarthritis, it has been proposed that hypertonic dextrose acts by stimulating chondrogenesis with enlargement of articular cartilage and repair of full thickness joint cartilage defects (Reeves and Hassanein 2000a).

Technique

Injections into ligaments are performed through an anesthetized wheal of skin over each site, after first contacting bone to confirm their position. Approximately 0.5–3 ml of solution is infiltrated at each site, depending on the size of the structures treated. Intravenous analgesia and sedation is often given with more extensive injection regimens. For joint pain and osteoarthritis, the joint cavity is injected along with the tender ligaments, although in some protocols, the joint cavity alone is injected (Reeves and Hassanein 2000a). Analgesics, heat, and general activity are recommended for post-injection pain and stiffness, but anti-inflammatory medications are proscribed in order to avoid negating the inflammatory effect of the injections.

Applications

Prolotherapists treat patients with pain of mechanical origin, on the assumption that the pain is of ligamentous or tenoperiosteal in origin. This is based on a history, which excludes nonmechanical causes, and an examination, which finds tenderness in ligaments consistent with the distribution of the pain. Maps relating the putative structures to the distribution of pain have been drawn by Hackett (Hackett et al. 1993) and are based on observations of pain responses to needling of ligaments and entheses in thousands of patients.

Treatments are commonly given at 1–2-week intervals, but intervals of up to 2 months have been described. Improvement in symptoms can occur with as little as one treatment, but more typically it occurs slowly over several treatments. Typically, six treatments are given for low back pain; five to six for the knee, hip, and shoulder; four to five for the neck or upper back pain; and three for thumb and finger joints. After the initial course of treatment, follow-up treatments may be given if the response has been inadequate or a relapse occurs.

Vitamin and mineral supplements (including vitamin C, zinc, and manganese) are given in some protocols, in the hope that they will facilitate ligament healing (Dhillon 1997). Exercises may also be added to improve healing. Some protocols emphasize the importance of resuming activities that had been reduced or ceased due to pain (Klein et al. 1993; Ongley et al. 1987).

Efficacy

Low Back Pain

Four trials of prolotherapy for nonspecific chronic low back pain have been performed (Table 1). The first three used a dextrose/

Prolotherapy, Table 1 Comparison of efficacy data for the four prolotherapy trials on the treatment of chronic low back pain. Note that pain and disability scales for each study have different ranges that are outlined in footnotes to the table

	Ongley et al.		Klein et al.		Dechow et al.		Yelland et al.	
	Proliferant (n = 40)	Control (n = 41)	Proliferant (n = 39)	Control (n = 40)	Proliferant (n = 36)	Control (n = 38)	Proliferant (n = 54)	Control (n = 56)
>50 % reduction in pain/disability at 6 months	88 % (disability scores)	39 % (disability scores)	77 %	53 %	Not reported	Not reported	50 % (pain) 49 % (disability)	46 % (pain) 32 % (disability)
Mean pain (CI) at inception	3.8 (3.4–4.2)	4.0 (3.6–4.4)	4.9 (4.5–5.3)	4.6 (4.2–4.9)	5.4 (3.8–7.0)	5.3 (3.7–6.9)	5.2 (4.7–5.7)	5.5 (4.9–6.1)
Mean pain (CI) at 6 months	1.5 (1.1–1.9)	3.1 (2.5–3.7)	2.3 (1.8–2.8)	2.9 (2.3–3.4)	5.2 (3.4–7.0)	4.4 (2.6–6.2)	3.1 (2.4–3.8)	3.4 (2.7–4.1)
Mean pain (CI) at 12 months	–	–	–	–	–	–	3.3 (2.5–4.1)	3.7 (2.9–4.4)
Mean disability (CI) at inception	11.5 (9.7–14.1)	11.8 (10.0–13.6)	9.4 (8.3–10.5)	8.3 (7.2–9.3)	35 (17–53)	34 (16–52)	13.7 (12.3–15.0)	14.3 (13.1–15.5)
Mean disability (CI) at 6 months	3.4 (2.0–4.9)	8.3 (6.1–10.5)	4.0 (2.9–5.2)	4.4 (3.1–5.6)	36 (18–54)	35 (17–53)	7.9 (6.0–10.0)	9.3 (7.7–10.9)
Mean disability (CI) at 12 months	–	–	–	–	–	–	8.0 (6.0–10.0)	9.8 (8.1–11.5)
Reduction in pain means at 6 months	60 %	23 %	53 %	38 %	4 %	17 %	39 %	38 %
Reduction in disability means at 6 months	70 %	30 %	57 %	47 %	3 %	–3 %	42 %	35 %
Reduction in pain means at 12 months	–	–	–	–	–	–	36 %	35 %
Reduction in disability means at 12 months	–	–	–	–	–	–	42 %	31 %

(a) VAS pain scale, range 0–8 cm; (b) VAS pain scale, range 0–7.5 cm; (c) VAS pain scale, range 0–10 cm; (d) combination of Roland-Morris and Waddell's disability questionnaires, range 0–33; (e) Oswestry disability scale, range 0–50; (f) modified Roland-Morris questionnaire, range 0–23

glycerine/phenol/lignocaine proliferant (Klein et al. 1993; Ongley et al. 1987; Dechow et al. 1999). The fourth used a simpler dextrose/lignocaine proliferant (Yelland et al. 2004). All trials included only patients with pain for longer than 6 months, which had failed prior treatments.

In the first trial (Ongley et al. 1987), the proliferant group had an initial manipulation under local anesthesia, while the saline controls had a sham manipulation under a lower dose of local anesthesia. All participants had six injection treatments and did exercises for 6 months. The proportion of participants with greater than 50 % reductions in pain/disability scores at 6 months was significantly higher in the proliferant group (0.88) than in the saline controls (0.39). These results, however, have been construed as evidence of efficacy of manipulation rather than prolotherapy (van Tulder et al. 1997) and as the cumulative neurolytic effect of phenol on treated ligaments (Klein et al. 1993).

Prompted by these uncertainties, Klein et al. (1993) repeated the first trial, testing only the phenol/glycerine/dextrose (sclerosant) components of the solution by having lidocaine controls, and compared proliferant solution with lidocaine controls (Klein et al. 1993). Both groups showed improvements in mean pain or disability scores at 6 months, but the differences between groups did not quite reach statistical significance. However, the proportion with greater than 50 % reductions in pain or disability scores at 6 months was significantly higher in the proliferant group (0.77) than in the lidocaine controls (0.53).

The third trial (Dechow et al. 1999) gave only three injection treatments, with much lower volumes of solution than in the other trials. It used a lidocaine control group. No change in mean pain or disability occurred in either group over 6 months.

The fourth trial (Yelland et al. 2004), used a dextrose/lidocaine solution and six to eight injection treatments. The proportion at 6 months the proportion of patients with greater than 50 % reduction in pain at six months was 0.50, compared with 0.46 for saline controls. The proportions for greater than 50 % reduction in disability were 0.46 and 0.32, respectively. Neither of these differed significantly. Responses at 6 months were maintained to 24 months. This trial also showed no additive benefit from lumbar flexion-extension exercises.

Poorer results with prolotherapy for chronic low back pain have been noted in a clinical trial in patients with hyperirritable foci in the muscles and ligaments (Klein et al. 1993) and anecdotally in bedridden patients, cigarette smokers, the malnourished, alcohol and drug abusers, post-lumbar surgery patients, and those with "true fibromyalgia" (Klein and Eek 1997).

Osteoarthritis of the Knee

There have been two trials of prolotherapy for chronic osteoarthritis of the knee. The first compared two monthly intra-articular injections of dextrose/lidocaine with lidocaine alone in 38 knees. Both groups showed significant improvements in pain, subjective swelling, and goniometric knee measures at 6 months. However, only the dextrose/lidocaine groups showed a reduction in the number of knee buckling episodes. A superior response in the dextrose/lidocaine group only came to light by combining measures in a multivariate analysis ($p = 0.015$). The improvements in pain were similar to those seen with medications such as NSAIDs, but improvements in flexion range were greater.

More recently a high quality trial of 89 patients with knee osteoarthritis randomized to dextrose/lidocaine prolotherapy, saline injections, or at-home exercises control showed prolotherapy is statistically and clinically superior to blinded saline injections and at-home exercises on Western Ontario and McMaster Universities Osteoarthritis Index and Knee Pain Scale from 9 to 52 weeks (Rabago et al. 2011).

Osteoarthritis of the Hand

A small trial ligament prolotherapy for chronic osteoarthritis of the hand found that pain at rest and with grip improved more in the dextrose/lidocaine group than in the lidocaine group, at 6 months, but not by a significant margin. Improvement in pain with movement of fingers and the finger flexion range improved more significantly in the dextrose/lidocaine group (Reeves and Hassanein 2000b).

Lateral Epicondylalgia

There are three pilot size trials of prolotherapy for the treatment of lateral epicondylalgia (tennis elbow) with mixed results. The first comparing dextrose/morrhuate/lidocaine injections with saline injections in 24 participants showed significantly greater reduction in pain scores at 16 and 52 weeks and significantly better improvements in grip strength at 16 weeks (Scarpone et al. 2008).

The next trial with 31 participants randomized to dextrose/lidocaine prolotherapy, dextrose/ morrhuate/lidocaine prolotherapy, or waitlist control showed dextrose/lidocaine prolotherapy was statistically superior to waitlist in Patient-Rated Tennis Elbow Evaluation at 8 weeks and in function at 16 weeks but not in pain at either follow-up point (Rabago et al. 2013).

Finally, a pilot trial of 24 patients with lateral epicondylalgia randomized to two treatments of dextrose/lidocaine prolotherapy or corticosteroid/lidocaine showed no significant differences between groups for changes in pain or the Disabilities of Arm, Shoulder, and Hand scores over 6 months (Carayannopoulos et al. 2011). Notably the number of injections for both treatments was limited to only two per participant to reflect common, safe practice for steroid injections. However, two injection treatments would be regarded by prolotherapists as an inadequate course of prolotherapy (Yelland et al. 2012).

Wrist

A trial of 39 patients with chronic wrist pain randomized to dextrose/lidocaine prolotherapy vs. a lidocaine control showed a statistically but not clinically significant differences in Patient-Rated Wrist Evaluation scores in favor of prolotherapy at 12 months (Hooper et al. 2011).

Knee

A trial of 51 boys with Osgood-Schlatters disease randomized to dextrose/lidocaine prolotherapy, lidocaine injections, or usual care showed significantly better return to sport rates for prolotherapy and lidocaine injections than usual care at 3 months and for prolotherapy than lidocaine injections and usual care at 12 months (Topol et al. 2011).

Achilles Tendon

A trial of 43 participants with painful midportion Achilles tendinosis randomized to prolotherapy, eccentric loading exercises (ELE), or a combination of both showed prolotherapy, and particularly ELE combined with prolotherapy gives more rapid clinical improvements than ELE alone, but outcomes at 12 months are similar (Yelland et al. 2011).

Safety

In the trials of prolotherapy for chronic low back pain, transient post-injection soreness and stiffness occurred in the majority of patients. Significant post-injection headache, suggestive of lumbar puncture, occurred in 2–4 %. There were no long-term sequelae in any of these trials. The phenol component of the dextrose/glycerine/ phenol/lidocaine solution raises the most concern because of its potential neurotoxic effect. At the recommended concentration of 1.25 %, phenol is unlikely to damage nerves, but significant complications have been reported due to an error in its concentration in other settings (Reeves and Baker 1992).

Adverse events in knee and hand prolotherapy include temporary discomfort and stiffness, a flare in pain requiring corticosteroid injection, and, in the fingers, a transient restriction of venous return due to pressure effects (Reeves and Hassanein 2000a, b).

Pneumothorax and nonlife-threatening allergic reactions have also been reported following prolotherapy (Dorman 1993).

Indications

Advocates of prolotherapy recommend its use for patients with chronic pain that have been unresponsive to other treatments and for which a ligament or tendon source is suspected. It is not indicated for ▶ radicular pain or ▶ neuropathic pain.

Prolotherapy may play a role in the treatment of knee and hand osteoarthritis where there is some evidence for its efficacy. There may be a role for it with recalcitrant low back pain and lateral epicondylalgia, where there is conflicting evidence for its efficacy. It may speed recovery in Osgood-Schlatters disease and Achilles tendinosis.

The effect of the components of treatment other than the ligament/tendon injections, including the doctor-patient relationship, skin wheals, the physical trauma caused by the needles themselves, and the vitamin and mineral supplements used, has still to be studied. Nevertheless, substantial proportions of patients with chronic musculoskeletal pain do benefit. Although contentious, prolotherapy may, therefore, constitute a viable alternative for patients not responsive to other interventions.

References

Banks, A. R. (1991). A rationale for prolotherapy. *Journal of Orthopaedic Medicine, 13*(3), 54–59.

Carayannopoulos, A., Borg-Stein, J., Sokolof, J., Meleger, A., & Rosenberg, D. (2011). Prolotherapy versus corticosteroid injections for the treatment of lateral epicondylosis: A randomized controlled trial. *PM R, 3*(8), 706–715. Epub 2011/08/30.

Dechow, E., Davies, R. K., Carr, A. J., & Thompson, P. W. (1999). A randomized, double-blind, placebo-controlled trial of sclerosing injections in patients with chronic low back pain. *Rheumatology, 38*, 1255–1259.

Dhillon, G. S. (1997). Prolotherapy in lumbo-pelvic pain. *Australian Musculoskeletal Medicine, 2*(2), 17–19.

Dorman, T. (1993). Prolotherapy: A survey. *Journal of Orthopaedic Medicine, 15*, 49.

Hackett, G. S., Hemwall, G. A., & Montgomery, G. A. (1993). *Ligament and tendon relaxation treated by prolotherapy* (5th ed., pp. 14–15). Oak Park: Hemwall G.A.

Hooper, R. A., Hildebrand, K., Faris, P., Westaway, M., & Freiheit, E. (2011). Randomized controlled trial for the treatment of chronic doral wrist pain with dextrose prolotherapy. *International Musculoskeletal Medicine, 33*(3), 100–106.

Jensen, K. T., Rabago, D. P., Best, T. M., Patterson, J. J., & Vanderby, R., Jr. (2008). Response of knee ligaments to prolotherapy in a rat injury model. *The American Journal of Sports Medicine, 36*(7), 1347–1357.

Klein, R. G., & Eek, B. C. J. (1997). Prolotherapy: An alternative approach to managing low back pain. *The Journal of Musculoskeletal Medicine, 14*(5), 45–49.

Klein, R. G., Dorman, T. A., & Johnson, C. E. (1989). Proliferant injections for low back pain: Histologic changes of injected ligaments and objective measurements of lumbar spine mobility before and after treatment. *The Journal of Neurological and Orthopaedic Medicine and Surgery, 10*(2), 123–126.

Klein, R. G., Eek, B. C., DeLong, W. B., & Mooney, V. (1993). A randomized double-blind trial of dextrose-glycerine-phenol injections for chronic, low back pain. *Journal of Spinal Disorders, 6*(1), 23–33.

Ongley, M. J., Klein, R. G., Dorman, T. A., Eek, B. C., & Hubert, L. J. (1987). A new approach to the treatment of chronic low back pain. *Lancet, 2*(8551), 143–146.

Rabago, D., Miller, D., Zgierska, A., Mundt, M., Kijowski, R., Belling, J., et al. (2011). Dextrose prolotherapy for knee osteoarthritis: Results of a randomized controlled trial. *Osteoarthritis and Cartilage, 19*(S1), S142–S143.

Rabago, D., Lee, K., Ryan, M. S., Chourasia, A. O., Sesto, A. E., Zgierska, A., et al. (2012). The efficacy of prolotherapy for lateral epicondylosis: A pilot-level randomized controlled trial. American College of Sports Medicine annual conference, May 29–June 2, San Francisco. Abstract number 3822vbCrLf.

Rabago, D., Lee, K. S., Ryan, M., Chourasia, A. O., Sesto, M. E., Zgierska, A. E., Kijowski, R., Grettie, J., Wilson, J. J., & Miller, D. (2013). Hypertonic dextrose and morrhuate sodium injections (prolotherapy) for lateral epicondylosis (tennis elbow): Results of a single-blind, pilot-level randomized controlled trial. *American Journal of Physical Medicine & Rehabilitation*. E-pub ahead of print PMID: 23291605; Jan 3, 2013.

Reeves, K. D., & Baker, A. (1992). Mixed somatic peripheral phenol nerve block for painful or intractable spasticity: A review of 30 years of use. *American Journal of Pain Management, 2*, 205–210.

Reeves, K. D., & Hassanein, K. (2000a). Randomized, prospective, placebo-controlled double-blind study of dextrose prolotherapy for osteoarthritic thumb and finger (DIP, PIP, and trapeziometacarpal) joints: Evidence of clinical efficacy. *Journal of Alternative and Complementary Medicine, 6*(4), 311–320.

Reeves, K. D., & Hassanein, K. (2000b). Randomized prospective double-blind placebo-controlled study of dextrose prolotherapy for knee osteoarthritis with or without ACL laxity. *Alternative Therapies in Health and Medicine, 6*(2), 68–70,2–4,7–80.

Scarpone, M., Rabago, D. P., Zgierska, A., Arbogast, G., & Snell, E. (2008). The efficacy of prolotherapy for lateral epicondylosis: A pilot study. *Clinical Journal of Sport Medicine, 18*(3), 248–254.

Topol, G. A., Podesta, L. A., Reeves, K. D., Raya, M. F., Fullerton, B. D., & Yeh, H. W. (2011). Hyperosmolar dextrose injection for recalcitrant Osgood-Schlatter disease. *Pediatrics, 128*(5), e1121–e1128. Epub 2011/10/05.

van Tulder, M. W., Koes, B. W., & Bouter, L. M. (1997). Conservative treatment of acute and chronic nonspecific low back pain. A systematic review of randomized controlled trials of the most common interventions. *Spine, 22*(18), 2128–2156.

Yelland, M. J., Glasziou, P. P., Bogduk, N., Schluter, P., & McKernon, M. (2004). Prolotherapy injections, saline injections, and exercises for chronic low-back pain: A randomized trial. *Spine, 29*(1), 9–16.

Yelland, M. J., Sweeting, K. R., Lyftogt, J. A., Ng, S. K., Scuffham, P. A., & Evans, K. A. (2011). Prolotherapy injections and eccentric loading exercises for painful Achilles tendinosis: A randomised trial. *British Journal of Sports Medicine, 45*(5), 421–428.

Yelland, M. J., Bissett, L., Rabago, D., & Patterson, J. (2012). Can double blinding distort results in randomized controlled trials? *PM R, 4*(4), 322–323; author reply 3. Epub 2012/05/01.

Promoter

Definition

A region of DNA, upstream from the transcription start site, which contains binding sites for RNA polymerase and a number of proteins that regulate the rate of transcription of the adjacent gene.

Cross-References

▶ NSAIDs and Cancer

Prophylactic Pain Treatment

▶ Preemptive or Preventive Analgesia and Central Sensitization in Postoperative Pain

Propofol

Definition

Propofol is an intravenous anesthetic acting, at least partly, via $GABA_A$ receptors.

Cross-References

▶ GABA and Glycine in Spinal Nociceptive Processing

Propranolol

Definition

A beta-blocker.

Proprioception

Definition

Sensory information from the muscles, tendons, or joints about limb position, movement, and position of that area of the body to the brain.

Cross-References

▶ Cordotomy Effects on Humans and Animal Models
▶ Trigeminal Brainstem Nuclear Complex, Anatomy
▶ Trigeminal Neuralgia: Diagnosis and Treatment

Propriospinal Neurons

Definition

Neurons with cell bodies in the spinal cord and axonal projections with sites of termination in other spinal cord sites.

Cross-References

▶ Spinal Ascending Pathways, Colon, Urinary Bladder, and Uterus

Prospective Observational Cohort Study

Definition

A study that starts with the present condition of a population of individuals and observes them into the future.

Cross-References

▶ Low Back Pain, Epidemiology

Prostacyclin

Definition

A member of the prostaglandin family, also called PGI2. It is synthesized from arachidonic acid by cyclooxygenase. Nonsteroidal anti-inflammatory drug inhibit the formation of PGI2 by blocking cyclooxygenase. Prostacyclin exerts proinflammatory actions and is particularly active in preventing of platelet aggregation.

Cross-References

▶ NSAIDs and Their Indications

Prostadynia

Definition

Perineal pain, often felt in the region of the prostate when sitting, typical of prostatitis-induced pain, but without infection, and often caused by myofascial trigger points in the pelvic floor musculature.

Cross-References

▶ Chronic Pelvic Pain, Musculoskeletal Syndromes

Prostaglandin Endoperoxide Synthase

▶ Cyclooxygenases in Biology and Disease

Prostaglandin Endoperoxide Synthase 1

▶ COX-1 and COX-2 in Pain

Prostaglandin Endoperoxide Synthase 2

▶ COX-1 and COX-2 in Pain

Prostaglandin H Synthase

▶ NSAIDs, COX-Independent Actions
▶ NSAIDs, Mode of Action

Prostaglandins

Definition

Prostaglandins are endogenous mediators that are synthesized from arachidonic acid. These hormone-like, lipid-soluble regulatory molecules have a variety of effects in the body, including potentiation of the effects of painful mediators, the production of fever, smooth muscle contraction and relaxation, kidney regulation, neuronal activity, and reproduction. Prostaglandins are also involved in the processes of inflammation.

Cross-References

- ▶ Cancer Pain, Animal Models
- ▶ COX-1 and COX-2 in Pain
- ▶ NSAIDs and Their Indications
- ▶ NSAID-Induced Lesions of the Gastrointestinal Tract
- ▶ Prostaglandins, Spinal Effects

Prostaglandins in Inflammatory Nociceptor Sensitization

- ▶ Inflammatory Nociceptor Sensitization, Prostaglandins and Leukotrienes

Prostaglandins, Spinal Effects

Hans-Georg Schaible[1] and Horacio Vanegas[2]
[1]Department of Physiology - Neurophysiology, Friedrich-Schiller-University of Jena, Jena, Germany
[2]Instituto Venezolano de Investigaciones Cientificas (IVIC), Caracas, Venezuela

Synonyms

Autacoids; COX; Cyclooxygenases; Dorsal horn; Dorsal root ganglion; Eicosanoids; Primary afferent; Prostanoids; Spinal cord

Definition

Prostaglandins (PGs) are derivatives of arachidonic acid, a 20-carbon fatty acid which generally originates from cell membranes by the action of the phospholipase A_2 enzyme family. Arachidonic acid is acted upon by enzymes commonly called cyclooxygenases (COX-1 and COX-2) to yield PGH_2, which is subsequently transformed into PGD_2, PGE_2, $PGF_{2\alpha}$, and PGI_2 (prostacyclin) by terminal synthases. Spinal PGs are key players in the processing and transmission of pain signals, particularly during inflammation of peripheral tissues.

Introduction

In the spinal cord, PGs may be synthesized and released by the incoming primary afferents as well as by intrinsic spinal neurons, glial cells, and vascular cells. Primary afferents and intrinsic spinal cells are in turn a target for PG actions. Therefore, the effects of spinal PGs have been studied on primary afferent neurons in the dorsal root ganglion (DRG) as well as on the spinal cord itself (for reviews, see, e.g., Kawabata 2011; Svensson and Yaksh 2002; Vanegas and Schaible 2001).

Under basal conditions, both COX-1 mRNA and COX-2 mRNA are present in DRG neurons of small to medium size, a class which includes the nociceptive afferents. In the spinal cord, COX-1 and -2 mRNA as well as COX-1 and -2 protein are expressed during basal conditions. In particular, COX-2 has been found in neurons of all laminae and in white matter glial cells. Basal release of PGD_2, PGE_2, $PGF_{2\alpha}$, and PGI_2 occurs in DRGs and spinal cord.

In addition, COX-2 is an inducible enzyme. Acute and chronic peripheral inflammation, damage to peripheral nerves, spinal cord injury, and the spinal action of glutamatergic agonists, dynorphin, interleukins, or nuclear factor kappa B, all significantly increase the expression of COX-2 and the release of PGE_2 and PGI_2 (Ghilardi et al. 2004; Koetzner et al. 2004; Kunori et al. 2011; Lee et al. 2004; Maihöfner et al. 2000; O'Rielly and Loomis 2006; Samad et al. 2001; Svensson and Yaksh 2002; Tegeder et al. 2001; Vanegas and Schaible 2001; Yaksh et al. 2001). As mentioned above, these PGs may arise both from primary afferents and from spinal intrinsic neurons and glial cells. Also, COX-2 is induced in inflammatory cells such as macrophages, mast cells, microglia, and monocytes by interleukin-1β, tumor necrosis factor α, nerve growth factor, the monocyte chemoattractant protein 1 (MCP1 or CCL2), and reactive oxygen species (Ma and Quirion 2008). Upon damage to neural tissue, the extracellular signal-regulated kinase 1/2 (ERK1/2) in microglia is phosphorylated, thus leading to PGE_2 release (Zhao et al. 2007). Most of these mediators constitute reciprocal signals between neurons and glial cells. Even glutamate, via the NMDA receptor, can induce COX-2 upregulation (Li et al. 2009), and genetically induced deficiency in PGE_2 synthesis impairs activation of microglia and NMDA receptors (Kunori et al. 2011). Furthermore, activation of descending pathways from the rostral ventromedial medulla that facilitate neuropathic pain mechanisms results in a COX-mediated elevation of PGE_2 release in the spinal cord (Marshall et al. 2012).

Once released into the intercellular space, PGs bind to specific membrane receptors on primary afferent fibers and intrinsic spinal cells (Vanegas and Schaible 2001). The receptors for PGD, PGE, PGF, and PGI are respectively called DP, EP, FP, and IP receptors. There are four general types of EP receptors, i.e., EP1, EP2, EP3 with subtypes EP3α, EP3β, and EP3γ, and EP4 (Kawabata 2011). All PG receptors are G-protein-coupled molecules with seven transmembrane domains. Activation of most PG receptors triggers intracellular signals which lead to increases of intracellular calcium (EP1), cAMP (EP2, EP4 and IP), and inositol phosphates (EP3, FP and IP). These actions lead to cellular activation. In contrast, EP3α and EP3ß receptors mediate a decrease in cAMP and, consequently, cellular inhibition. The EP3γ receptor may have both stimulatory and inhibitory effects on adenylyl cyclase.

Induction of COX-2 expression and the subsequent increase in PG synthesis and release are critical factors in the development of spinal cord hyperexcitability (central sensitization) that results from peripheral inflammation and trauma, and significantly contribute to the resulting hyperalgesia. This is supported, in general, by the following findings (Vanegas and Schaible 2001): (a) induction of COX-2 in primary afferents and spinal cord has been found in models of inflammation and trauma; (b) PGs are released in the spinal cord in models of inflammation and trauma; (c) PGs induce neuronal hyperexcitability and release of neuroactive compounds in DRG cells and spinal cord; (d) application of PGs to the spinal cord in freely moving animals induces hyperalgesia and allodynia, which are typical of central sensitization; (e) there is close correlation between manipulations that induce spinal release of PGs and manipulations that induce hyperalgesia and allodynia; and (f) spinal administration of COX inhibitors reduces manifestations of central sensitization in various experimental models and in humans.

Interestingly, PGD_2 may have a protective action against neurotoxicity (Grill et al. 2008; Liang et al. 2005).

Characteristics

PG Effects on Ion Currents, Release of Mediators, and Synaptic Transmission

PGs can have a direct effect upon neurons, and also facilitate the effect of other mediators (Ebersberger et al. 2011; Vanegas and Schaible 2001). Primary nociceptive afferents express both TTX-sensitive and TTX-resistant sodium currents, in addition to other voltage-dependent ion currents. PGE_2 enhances TTX-resistant sodium currents, and PGE_2 and PGI_2 diminish potassium currents. As a consequence, primary nociceptive afferents show a decreased threshold and an increased firing rate. These effects are mediated by the cAMP-protein kinase A (PKA) cascade. PGs also facilitate a voltage-gated calcium (inward) current. The PG-induced increase in primary afferent excitability, firing rate, and calcium entry may be the reason why PGs increase the release of glutamate, aspartate, substance P, CGRP, and nitric oxide in the spinal cord. Additionally, PGE_2 increases the expression of the NK1 receptor for substance P in primary afferent neurons, also by way of the cAMP-PKA cascade (Segond von Banchet et al. 2003; Vanegas and Schaible 2001). As a consequence, a larger proportion of these neurons show an increase in intracellular calcium when exposed to substance P. In neurons of laminae II–III, where nociceptive afferents abundantly terminate, PGE_2 application induces an increase in glutamatergic miniature excitatory postsynaptic currents (EPSCs), which indicates an enhancement of spontaneous glutamate release from presynaptic terminals (Minami et al. 1999). PGE_2, and to some extent PGF_{2a}, also induces an increase in the glutamate-dependent EPSCs evoked by electrical stimulation of presynaptic axons, which indicates a sensitization of pre- and/or postsynaptic elements. In agreement with behavioral experiments (see below), neonatal treatment with capsaicin, which destroys nociceptive primary afferents, reduces the sensitizing effect of PGE_2 upon evoked EPSCs. All these effects of PGs upon primary afferents certainly contribute to the enhancement of spinal excitability and the resulting hyperalgesia during peripheral inflammation and trauma.

PGs also have direct effects upon spinal postsynaptic neurons. The EP2 receptor has been located in laminae IV–VI of the dorsal horn (Vanegas and Schaible 2001) and, through this receptor, PGE_2 excites neurons located here because it induces an inward sodium current (Baba et al. 2001). Furthermore, PGE_2 via the EP2 receptor reduces glycinergic synaptic inhibition in the dorsal horn during peripheral inflammation (Reinold et al. 2005) but not during nerve damage (Hösl et al. 2006). Both the increase in excitation and the decrease in inhibition must of course contribute to hyperalgesia. In addition, several PGs may facilitate the action of other neuroactive substances (Vanegas and Schaible 2001). For example, PGs facilitate bradykinin-induced release of substance P and CGRP. This is not surprising since activation of PKA is a common factor in the effect of PGs and bradykinin, and since induction of substance P and CGRP release by bradykinin is at least partly mediated by PG synthesis. PGs also facilitate calcium inflow through the transient receptor potential type 1 (TRPV1) or "capsaicin" channel, also via PKA (Kawabata 2011). This is probably why PGs facilitate capsaicin-induced release of substance P and CGRP in the spinal cord, a phenomenon which may lead to nociceptive behavior in freely moving animals and is exacerbated during peripheral inflammation.

All these effects of PGs lead to an enhanced elicitation of pain signals and an increase in their synaptic transfer in the spinal cord.

PG Effects on Responses to Natural Stimuli

Action potential discharges by spinal nociceptive neurons constitute the drive for segmental and suprasegmental nociceptive reflexes as well as for the thalamocortical transmission of pain signals. A study of such discharges is therefore essential for understanding the influence of PGs upon the mechanisms of pain and hyperalgesia. PGE_2 induces an increase in the responses of spinal nociceptive neurons to innocuous and noxious stimulation of, for example, the knee joint (Vasquez et al. 2001). This mimics inflammation-induced spinal hyperexcitability (Schaible and Grubb 1993) and fits in with the proposed role of PGs in central sensitization (see above). The facilitatory influence of PGE_2, however, is subject to plasticity and seems to be important only for the initiation of spinal sensitization. Once the inflammation of a peripheral tissue is fully developed, PGE_2 application to the spinal cord has negligible effects upon the enhanced neuronal discharges to innocuous or noxious stimulation. This may in part be related to downregulation of PG receptors. Indeed PGE_2 binding to membranes obtained from the spinal cord of animals with inflammation is 30 % reduced, and exposing DRG neurons to PGE_2 for 72 h decreases PGE_2 binding. But peripheral inflammation also brings about a change in the relative role of the various PG receptor types. Before inflammation, spinal application of EP1, EP2, and EP4 agonists increases spinal neuronal firing to innocuous and noxious stimulation, whereas $EP3\alpha$ receptor agonists have no effect (Bär et al. 2004). However, when the inflammation is fully developed, EP2 and EP4 agonists have no effect, like PGE_2, and EP3 agonists now exert an inhibitory effect upon hyperalgesia.

PGD_2 is the most abundant PG in the central nervous system (Vanegas and Schaible 2001). Its receptors, DP1 and DP2, are G-protein-coupled membrane molecules. Like PGE_2, application of PGD_2 to the spinal cord at high doses sensitizes spinal nociceptive neurons to mechanical stimulation of the normal knee joint (Telleria-Diaz et al. 2008). This effect may result from an increase in substance P and CGRP release from primary afferents. However, PGD_2, via the DP1 receptor, reduces responses of spinal nociceptive neurons to stimulation of an inflamed knee joint. PGD_2 also reduces the PGE_2-induced increase in responses to stimulation of a non-inflamed knee. During inflammation, PGD_2 therefore counteracts the facilitatory action of PGE_2 on central sensitization (Telleria-Diaz et al. 2008; Vanegas and Schaible 2001), an effect that might be due to an activation of DP1 receptors on inhibitory spinal interneurons (Minami et al. 1997).

In sum, the actions of COX, PGs, and COX inhibitors prior to inflammation may be significantly changed during inflammation, a fact that has considerable importance for the treatment of inflammatory pain.

In general, intrathecal PG administration to normal and freely moving rats or mice has elicited an increase in the behavioral responses to noxious stimuli (hyperalgesia) as well as aversive and/or aggressive responses to touch (allodynia) (Vanegas and Schaible 2001). This fits in with the effects of PGs upon nociceptive afferents and spinal neurons (see above). Somatic hyperalgesia was evidenced, for example, as a decrease in the time it takes for an animal to withdraw from contact with a hot object. Visceral hyperalgesia was evidenced, for example, as an increase in the number of contractions of the abdominal wall (writhing) elicited by intraperitoneal injection of acetic acid.

Intrathecal PGE_2 and PGD_2 elicit hyperalgesia and allodynia, and both phenomena are strongly attenuated in animals that have been neonatally injected with capsaicin and are thus deficient in primary nociceptive afferents. Therefore, the effects of PGE_2 and PGD_2 might partially be exerted upon nociceptive afferents. EP1, EP3, and DP1 receptors in primary afferents, and EP2 receptors in spinal neurons might be involved in these effects (Ebersberger et al. 2011; Telleria-Diaz et al. 2008; Vanegas and Schaible 2001).

Summary

Given their role in hyperalgesia, particularly during inflammation, the effects of PGs in the spinal cord have been intensively investigated, as have also been the spinal actions of COX inhibitors. There is general agreement that PGs significantly contribute to the increase in spinal neuronal excitability caused by inflammation and trauma, and that this in turn is a fundamental cause of pain and hyperalgesia. The extent of glial cell involvement in this hyperexcitability is becoming increasingly obvious. The effects of spinal PGs during induction of hyperexcitability may radically change once this process is fully established, and this must have important consequences for the interpretation and treatment of hyperalgesia.

References

Baba, H., Kohno, T., Moore, K. A., & Woolf, C. J. (2001). Direct activation of rat spinal dorsal horn neurons by prostaglandin E_2. *Journal of Neuroscience, 21*, 1750–1756.

Bär, K.-J., Natura, G., Telleria-Diaz, A., Teschner, P., Vogel, R., Vasquez, E., Schaible, H.-G., & Ebersberger, A. (2004). Changes in the effect of spinal prostaglandin E_2 during inflammation: Prostaglandin E (EP1-EP4) receptors in spinal nociceptive processing of input from the normal or inflamed knee joint. *Journal of Neuroscience, 24*, 642–651.

Ebersberger, A., Natura, G., Eitner, A., Halbhuber, K.-J., Rost, R., & Schaible, H.-G. (2011). Effects of prostaglandin D_2 on tetrodotoxin-resistant Na^+ currents in DRG neurons of adult rats. *Pain, 152*, 1114–1126.

Ghilardi, J. R., Svensson, C. I., Rogers, S. D., Yaksh, T. L., & Mantyh, P. W. (2004). Constitutive spinal cyclooxygenase-2 participates in the initiation of tissue injury-induced hyperalgesia. *Journal of Neuroscience, 24*, 2727–2732.

Grill, M., Heinemann, A., Hoefler, G., Peskar, B. A., & Schuligoi, R. (2008). Effect of endotoxin treatment on the expression and localization of spinal cyclooxygenase, prostaglandin synthases, and PGD2 receptors. *Journal of Neurochemistry, 104*, 1345–1357.

Hösl, K., Reinold, H., Harvey, R. J., Müller, U., Narumiya, S., & Zeilhofer, H. U. (2006). Spinal prostaglandin E receptors of the EP2 subtype and the glycine receptor a3 subunit, which mediate central inflammatory hyperalgesia, do not contribute to pain after peripheral nerve injury or formalin injection. *Pain, 126*, 46–53.

Kawabata, A. (2011). Prostaglandin E2 and pain –An update. *Biological and Pharmaceutical Bulletin, 34*, 1170–1173.

Koetzner, L., Hua, X.-Y., Lai, J., Porreca, F., & Yaksh, T. (2004). Nonopioid actions of intrathecal dynorphin evoke spinal excitatory amino acid and prostaglandin E_2 release mediated by cyclooxygenase-1 and -2. *Journal of Neuroscience, 24*, 1451–1458.

Kunori, S., Matsumura, S., Okuda-Ashitaka, E., Katano, T., Audoly, L. P., Urade, Y., & Ito, S. (2011). A novel role of prostaglandin E2 in neuropathic pain: Blocade of microglial migration in the spinal cord. *Glia, 59*, 208–218.

Lee, K.-M., Kang, B.-S., Lee, H.-L., Son, S.-J., Hwang, S.-H., Kim, D.-S., Park, J.-S., & Cho, H.-J. (2004). Spinal NF-kB activation induces COX-2 upregulation and contributes to inflammatory pain hypersensitivity. *European Journal of Neuroscience, 19*, 3375–3381.

Li, S. Q., Xing, Y. L., Chen, W. N., Yue, S. L., Li, L., & Li, W. B. (2009). Activation of NMDA receptor is associated with up-regulation of COX-2 expression in the spinal dorsal horn during nociceptive inputs in rats. *Neurochemical Research, 34*, 1451–1453.

Liang, X., Wu, L.-J., Hand, T., & Andreasson, K. (2005). Prostaglandin D2 mediates neuronal protection via the DP1 receptor. *Journal of Neurochemistry, 92*, 477–486.

Ma, W., & Quirion, R. (2008). Does COX2-dependent PGE_2 play a role in neuropathic pain? *Neuroscience Letters, 437*, 165–169.

Maihöfner, C., Tegeder, I., Euchenhofer, C., Dewitt, D., Brune, K., Bang, R., Neuhuber, W., & Geisslinger, G. (2000). Localization and regulation of cyclo-oxygenase-1 and -2 and neuronal nitric oxide synthase in mouse spinal cord. *Neuroscience, 101*, 1093–1108.

Marshall, T. M., Herman, D. S., Largent-Milnes, T. M., Badghisi, H., Zuber, K., Holt, S. C., Lai, J., Porreca, F., & Vanderah, T. W. (2012). Activation of descending pain-facilitatory pathways from the rostral ventromedial medulla by cholecystokinin elicits release of prostaglandin-E2 in the spinal cord. *Pain, 153*, 86–94.

Minami, T., Okuda-Ashitaka, E., Nishizawa, M., Mori, H., & Ito, S. (1997). Inhibition of nociceptin-induced allodynia in conscious mice by prostaglandin D_2. *British Journal of Pharmacology, 122*, 605–610.

Minami, T., Okuda-Ashitaka, E., Hori, Y., Sakuma, S., Sugimoto, T., Sakimura, K., Mishina, M., & Ito, S. (1999). Involvement of primary afferent C-fibres in touch-evoked pain (allodynia) induced by prostaglandin E_2. *European Journal of Neuroscience, 11*, 1849–1856.

O'Rielly, D. D., & Loomis, C. W. (2006). Increased expression of cycooxygenase and nitric oxide isoforms, and exaggerated sensitivity to prostaglandin E_2, in the rat lumbar spinal cord 3 days after L5-L6 spinal nerve ligation. *Anesthesiology, 104*, 328–337.

Reinold, H., Ahmadi, S., Depner, U. B., Layh, B., Heindl, C., Hamza, M., Pahl, A., Brune, K., Narumiya, S., Müller, U., & Zeilhofer, H. U. (2005). Spinal inflammatory hyperalgesia is mediated by prostaglandin

E receptors of the EP2 type. *The Journal of Clinical Investigation, 115*, 673–679.

Samad, T., Moore, K. A., Sapirstein, A., Billet, S., Allchorne, A., Poole, S., Bonventre, J. V., & Woolf, C. J. (2001). Interleukin-1b-mediated induction of COX-2 in the CNS contributes to inflammatory pain hypersensitivity. *Nature (London), 410*, 471–475.

Schaible, H.-G., & Grubb, B. D. (1993). Afferent and spinal mechanisms of joint pain. *Pain, 55*, 5–54.

Segond von Banchet, G., Scholze, A., & Schaible, H.-G. (2003). Prostaglandin E_2 increases the expression of the neurokinin₁ receptor in adult sensory neurones in culture: A novel role for prostaglandins. *British Journal of Pharmacology, 139*, 672–680.

Svensson, C. I., & Yaksh, T. L. (2002). The spinal phospholipase-prostanoid cascade in nociceptive processing. *Annual Review of Pharmacology and Toxicology, 42*, 553–583.

Tegeder, I., Niederberger, E., Vetter, G., Bräutigam, L., & Geisslinger, G. (2001). Effects of selective COX-1 and -2 inhibition on formalin-evoked nociceptive behavior and prostaglandin E_2 release in the spinal cord. *Journal of Neurochemistry, 79*, 777–786.

Telleria-Diaz, A., Ebersberger, A., Vasquez, E., Schache, F., Kahlenbach, J., & Schaible, H.-G. (2008). Different effects of spinally applied prostaglandin D_2 on responses of dorsal horn neurons with knee input in normal rats and in rats with acute knee inflammation. *Neuroscience, 156*, 184–192.

Vanegas, H., & Schaible, H.-G. (2001). Prostaglandins and cyclooxygenases in the spinal cord. *Progress in Neurobiology, 64*, 327–363.

Vasquez, E., Bär, K.-J., Ebersberger, A., Klein, B., Vanegas, H., & Schaible, H.-G. (2001). Spinal prostaglandins are involved in the development but not the maintenance of inflammation-induced spinal hyperexcitability. *Journal of Neuroscience, 21*, 9001–9008.

Yaksh, T. L., Dirig, D. M., Conway, C. M., Svensson, C., Luo, Z. D., & Isakson, P. C. (2001). The acute antihyperalgesic action of nonsteroidal, antiinflammatory drugs and release of spinal prostaglandin E_2 is mediated by the inhibition of constitutive spinal cyclooxygenase-2 (COX-2) but not COX-1. *Journal of Neuroscience, 21*, 5847–5853.

Zhao, P., Waxman, S. G., & Hains, B. C. (2007). Extracellular signal-regulated kinase-regulated microglianeuron signaling by prostaglandin E_2 contributes to pain after spinal cord injury. *Journal of Neuroscience, 27*, 2357–2368.

Prostanoids

Definition

Prostanoids are generated from arachidonic acid in response to a wide range of different stimuli.

Once arachidonic acid has been liberated, cyclooxygenases catalyze the formation of the cyclic endoperoxides, prostaglandin (PG)G2 and PGH2, by oxygenation and subsequent cyclization of arachidonic acid. The type of prostanoid that is produced from PGH2 is largely dependent on the enzymes present in the individual cells. The term prostanoid encompasses related compounds such as the prostaglandins (PGD2, PGE2, PGF2alpha), thromboxane A2 (TXA2), and prostacyclin (PGI2).

Cross-References

▶ COX-1 and COX-2 in Pain
▶ Neutrophils in Inflammatory Pain
▶ Prostaglandins, Spinal Effects

Protective Analgesia

▶ Postoperative Pain, Preemptive or Preventive Analgesia

Protein Extravasation

▶ Nociceptor, Axonal Branching

Protein Kinase A

Synonyms

PKA

Definition

Cyclic AMP (cAMP), which is generated by adenylate cyclase (AC) activation, activates the Protein Kinase A (PKA) by binding to the regulatory subunits. PKA has been implicated in persistent pain mechanisms in both the peripheral and central nervous systems. The sensitization of primary afferent neurons that occurs in the setting of inflammation has been shown to involve cAMP- and PKA-dependent mechanisms.

Cross-References

▶ ERK Regulation in Sensory Neurons during Inflammation

Protein Kinase C

Synonyms

PKC

Definition

Protein Kinase C (PKC) is activated by the intracellular lipid second messengers phosphatidylserine and diacylglycerol (DAG) and to some extent by Ca^{2+}. There is substantial evidence supporting a role for PKC expressed in dorsal horn neurons in regulating pain hypersensitivity in a number of different pain models. The ε-isoform of PKC plays a major role in peripheral sensitization of the peripheral terminals of nociceptors.

Cross-References

▶ ERK Regulation in Sensory Neurons during Inflammation
▶ TRPV1 Modulation by p2Y Receptors
▶ TRPV1 Modulation by PKC

Proteinase-Activated Receptors

Synonyms

PAR

Definition

Proteinase-activated receptors (PAR) are seven-transmembrane-spanning receptors coupled to G-proteins. Proteinases hydrolyze, at a specific cleavage site, the extracellular N-terminus of the receptor, whereby a new N-terminus is exposed that acts as a tethered ligand, which binds intramolecularly to initiate cellular signals. Four PARs have been cloned, of which PAR-1 and PAR-2 are localized on peptidergic afferents. Activation of PAR-2 by trypsin or mast-cell tryptase is an important mechanism to stimulate and sensitize nociceptive afferents.

Protein-Calorie Malnutrition

Definition

PCM stems from the inadequate intake of carbohydrate, protein, and fat to meet metabolic requirements and/or the reduced absorption of macronutrients. PCM in cancer results from multiple factors most often associated with anorexia, cachexia, and the early satiety sensation frequently experienced. These factors range from altered tastes to a physical inability to ingest or digest food, leading to reduced nutrient intake.

Cross-References

▶ Cancer Pain, Evaluation of Relevant Comorbidities and Impact

Proteoglycan

Definition

A protein containing one or more covalently linked and usually sulfated polysaccharides called glycosaminoglycans.

Cross-References

▶ Diencephalic Mast Cells

Provocation Discogram

▶ Cervical Discography

Provocation Discography

▶ Lumbar Discography

Provocative

Definition

Discography: injection of a vertebral disc leading to reproducible pain.

Cross-References

▶ Cervical Discography
▶ Chronic Low Back Pain: Definitions and Diagnosis
▶ Lumbar Discography

Provocative Discography

▶ Cervical Discography
▶ Chronic Low Back Pain: Definitions and Diagnosis
▶ Lumbar Discography

Provocative Maneuver

▶ Lower Back Pain, Physical Examination

Provoked Vestibulodynia

▶ Dyspareunia and Vaginismus

Proximal Diabetic Neuropathy

▶ Diabetic Neuropathies

Pruriception

Definition

The sensory modality of itch or behavior related
to itch (from the Latin word *prurire*, to itch).

Cross-References

▶ Alloknesis and Allodynia

Pruriceptive

Definition

See "▶ Pruriception".

Cross-References

▶ Allodynia and Alloknesis

Pruritic Stimulus

Definition

A stimulus that produces a sensation of itch.

Cross-References

▶ Spinothalamic Tract Neurons, Central
Sensitization

Pruritus

Definition

Pruritus refers to itching, taken from the Latin
word *prurire*, to itch.

Cross-References

▶ Cutaneous Field Stimulation
▶ Itch
▶ Itch/Itch Fibers
▶ Postoperative Pain, Appropriate Management
▶ Responses of Spinothalamic Tract Neurons to
  Pruritic Stimuli

PSDC

▶ Postsynaptic Dorsal Column Neurons

PSDC Projection

▶ Postsynaptic Dorsal Column Projection,
Anatomical Organization

Pseudaffective

▶ Nocifensive Behaviors, Gastrointestinal Tract

Pseudaffective Response

Definition

An alternative spelling is "pseudaffective." A term coined by Sherrington in 1890s referring to reflex responses suggestive of emotional responses to painful stimuli. Initially characterized in decerebrate cats and these responses included cardiovascular reflexes, visceromotor reflexes, flexion-withdrawal reflexes, and vocalization. Examples include changes in heart rate, blood pressure, respiratory rate, and activation of sweat glands.

Cross-References

▶ Nocifensive Behaviors, Gastrointestinal Tract
▶ Nocifensive Behaviors of the Urinary Bladder
▶ Visceral Pain Model, Urinary Bladder Pain (Irritants or Distension)

Pseudoaddiction

Definition

Pseudoaddiction is a term coined for such patients who seek the drug (usually opiates or analgesics) in the presence of inadequate pain relief and who show no such preoccupation once the dose is escalated and comfort has been reached. May accompany an addictive disorder.

Cross-References

▶ Cancer Pain, Evaluation of Relevant Comorbidities and Impact
▶ Psychiatric Aspects of the Management of Cancer Pain
▶ Visceral Pain Model, Lower Gastrointestinal Tract Pain

Pseudoaffective

▶ Nocifensive Behaviors, Gastrointestinal Tract

Pseudoaffective Response

▶ Nocifensive Behaviors of the Urinary Bladder

Pseudomigraine with Lymphocytic Pleocytosis

▶ Transient Headache and CSF Lymphocytosis

Pseudoptosis

Definition

Pseudoptosis refers to decreased palpebral width that is not paretic in nature.

Cross-References

▶ Sunct Syndrome

Pseudotumor Cerebri

Definition

Pseudotumor cerebri is a neurologic manifestation of Behçet's disease.

Cross-References

▶ Headache Due to Arteritis

PSL Model

▶ Partial Sciatic Nerve Ligation

Psoas Compartment

Definition

The psoas compartment is a potential space in the psoas muscle through which the lumbar plexus travels.

Cross-References

▶ Postoperative Pain, Regional Blocks

Psoriatic Arthritis

Definition

Psoriatic arthritis is an inflammatory arthritis that develops in about 5–8% of patients with psoriasis, a disease that is characterized by sharply demarcated inflammatory plaques on the skin.

Cross-References

▶ NSAIDs and Their Indications
▶ Sacroiliac Joint Pain

Psychalgia (ICD-10)

▶ Psychiatric Aspects of the Epidemiology of Pain

Psychiatric Aspects of Litigation and Pain

George Mendelson
Department of Psychological Medicine,
Monash University, Caulfield, VIC, Australia

Definition

The Concept of "Compensation Neurosis"

The concept of "neurosis" has been traditionally invoked with reference to personal injury litigants who complain of physical symptoms for which there is no objectively demonstrable organic basis. Very frequently, such symptoms include the complaint of persistent pain.

Terms such as "compensation neurosis" and "accident neurosis" were introduced to refer to this type of clinical presentation, with the expectation that all the symptoms complained of will resolve following the finalization of litigation or cessation of compensation payments. However, there is no empirical support for the view either that there is a specific diagnosis of "compensation neurosis" or that after litigation is finalized the individual will return to his or her pre-accident level of functioning and resume work (Mendelson 1995).

The Relationship Between Compensation Status and Persistent Pain

There are three approaches that can be used to examine the possible relationship between compensation and chronic pain, and these will be briefly reviewed:

1. Comparison of the described pain experience of litigants (or those receiving compensation payments) with that of pain patients with similar complaints who are not involved in litigation and/or not receiving compensation benefits
2. Comparison of treatment response between groups of pain patient with similar clinical syndromes, where members of one group are

involved in litigation or receive compensation, and the others do not

3. Examination of the presentation and duration of chronic pain complaints under different systems of compensation or access to common law litigation

Compensation Status and Pain Description

Studies of the impact of compensation benefits on the described characteristics of pain have generally compared patients with chronic pain receiving compensation and those who had no such claims; the two groups are compared on pain measures such as duration, intensity, locus of pain, pain description, and the quality of pain. Most results of such comparisons have not shown significant differences between the compensation and non-compensation groups on any clinical characteristic of pain (Mendelson 1984; Melzack et al. 1985).

A minority of studies, however, have found that involvement in litigation does influence pain reports. In one study comparing patients with "whiplash" involved in litigation with similar patients whose litigation had been finalized, the authors concluded that "the rather robust effect of litigation status on pain reports" could be attributed "to the potential mediational role of the stress of litigation."

Similar comparisons of patients with chronic pain have been undertaken on measures of illness behavior, using the Illness Behaviour Questionnaire (IBQ). No differences between compensation and no-compensation groups were found in the mean group scores on the scales of the IBQ or on the Conscious Exaggeration Scale of the IBQ which, it had been asserted, can detect deception with the object of financial gain among chronic pain patients involved in litigation (Mendelson 1987).

The Effect of Compensation on Pain Treatment Outcome

Studies of the effect of compensation on treatment outcome of chronic pain have been reviewed elsewhere (Mendelson 1983) and were in general agreement that patients in receipt of compensation and/or involved in litigation had a worse prognosis than those with similar complaints who were neither in receipt of such payments nor involved in litigation. It is also generally considered that litigants and those receiving compensation payments show reduced motivation for effective treatment.

In one study, those injured at work who were receiving pain-contingent benefits were away from work for a mean of 14.2 months, as compared with 4.9 months for those with similar complaints not in receipt of compensation payments. The authors concluded that "the financial rewards of compensation" were responsible for the prolonged work absence of those injured at work.

Comparisons of workers' compensation and nonworkers' compensation patients, who had undergone surgery, showed that residual symptoms were significantly more common in those receiving compensation benefits. Several studies have also found that there was no relationship between physical status and work disability status following surgery, and it was postulated that receipt of compensation was the cause of failure to return to work.

In a population-based epidemiological study, Blyth et al. (2003) found that a significant proportion of patients with chronic pain were able to work, indicating that complete relief of pain may not be an essential therapeutic target. The authors also found that litigation, mainly work related, for chronic pain was strongly associated with higher levels of pain-related disability, even after taking into account other factors associated with poor functional outcomes.

Although most published studies have reported that the receipt of compensation has a negative effect on treatment outcome and prolongs work disability, it has been demonstrated that specific treatment programs can reduce the likelihood of progression to pain chronicity following work injuries. It has been shown that a treatment program for low back pain, which incorporates behavioral methods, is more effective than traditional management. Thus, the adverse effect of compensation on the outcome of treatment for acute pain following work injuries can be modified by specific programs that stress early mobilization and return to the workplace, appropriate activity and exercises, and avoidance of other factors that might promote "learned pain behaviour" (Tyrer 1986).

The Compensation System and Chronic Pain

Several studies have compared the impact of compensation on chronic pain complaints by examining data from New Zealand, where in 1974 both the traditional workers' compensation scheme and the right to sue in civil courts at common law for personal injuries were abolished, and comparing the New Zealand experience with the traditional common law systems and workers' compensation schemes in the United States and Australia. In the USA and Australia, personal injury claims can be litigated in court, with the plaintiff seeking damages from the wrongdoer, and similarly allow an injured worker to sue the employer if negligence can be established as having caused or contributed to the injury.

These studies showed that in New Zealand a significantly smaller proportion of those injured in accidents had chronic pain, and similarly, the period of work incapacity was significantly shorter in the New Zealand sample.

The relationship between societal expectations concerning illness and illness behavior was explored by Balla (1982) in an analysis of the "late whiplash syndrome," as manifested in Australia and in Singapore. A prospective study of Singapore patients with an acute whiplash injury, whose "original acute complaints of pain in the head and neck region were the same as seen in Australia," showed that the symptoms resolved and that the patients did not develop chronic neck pain.

On the basis of these findings, Balla postulated that the "late whiplash syndrome" is a "culturally constructed illness behaviour based on indigenous categories and social structural determinants." He also noted that in Australia such a condition had been accepted as compensable by the courts and that this provided a positive reinforcement for "illness behaviour" following the acute phase of "whiplash."

Characteristics

Chronic Pain, Litigation, and Psychiatric Diagnosis

When litigants, and/or compensation recipients, complain of chronic pain or disability that cannot be explained on the basis of objectively demonstrable organic abnormality, they are often referred for a psychiatric assessment. The question is then posed as to whether the individual is suffering from a mental disorder that could "explain" the complaint of persistent pain and the alleged resultant disability.

The "classification of chronic pain" published by the International Association for the Study of Pain (Merskey and Bogduk 1994) recognizes "pain of psychological origin" (Code I-16) among the group of "relatively generalized syndromes." "Pain of psychological origin" is further classified as arising through four different mechanisms: through muscle tension; delusional or hallucinatory pain; hysterical, conversion, or hypochondriacal pain; and pain associated with depression.

If the patient has pain associated with a clinically significant psychiatric illness, such as major depression or psychosis, then specific treatment of the mental disorder is required; with appropriate treatment, the mental illness and pain should resolve. More commonly, the patient with chronic pain might have some symptoms that overlap those found in mental disorders, such as sleep disturbance or dysphoria, but the nature of the psychiatric symptoms elicited and findings on mental state examination do not warrant the formal diagnosis of a specific mental disorder such as a depressive illness, a generalized anxiety disorder, or a psychotic condition.

In the absence of a specific mental disorder that could explain the presence of pain not attributable to an objectively demonstrable organic lesion, or pain that is disproportionate to the extent of any objectively demonstrable pathology or evidence of continuing physical injury, clinicians sometimes use a variety of terms such as psychogenic pain disorder, somatoform pain disorder, pain disorder (American Psychiatric Association (2000), or learned pain behavior (Tyrer 1986).

The most recent edition of the Diagnostic and Statistical Manual of Mental Disorders (American Psychiatric Association 2000) includes the diagnosis of "Pain Disorder," with three sub-types: "Pain Disorder Associated with Psychological Factors," "Pain Disorder Associated with Both Psychological Factors and a General Medical Condition," and "Pain Disorder Associated with a General Medical Condition." The latter "*is not considered a mental disorder and is coded on Axis III*. It is listed in this section to facilitate differential diagnosis" (italics in original).

The diagnosis of "Pain Disorder Associated with Psychological Factors" requires that "psychological factors are judged to have the major role in the onset, severity, exacerbation, or maintenance of the pain (If a general medical condition is present, it does not have a major role in the onset, severity, exacerbation, or maintenance of the pain). This type of Pain Disorder is not diagnosed if criteria are also met for Somatization Disorder."

The diagnosis of "Pain Disorder Associated with Both Psychological Factors and a General Medical Condition" is made if "both psychological factors and a general medical condition are judged to have important roles in the onset, severity, exacerbation, or maintenance of the pain." The relevant general medical condition or anatomical site of the pain is to be coded on Axis III.

In the assessment of personal injury litigants or compensation recipients with chronic pain, it is uncommon for such a patient to acknowledge the presence of any pre-injury emotional problems or psychological conflict that could "explain" the persistence of disproportionate pain and resultant disability following the injury. In the absence of any understandable intrapsychic basis for the "psychogenesis" of chronic pain in such patients, it is usually postulated that the pain persists beyond the usual duration of healing because of environmental factors that reinforce the pain complaints and pain behavior.

In the context of the evaluation of a patient with chronic pain – where the pain cannot be attributed to a physical lesion – in a medicolegal setting, it is inaccurate to make the diagnosis of Pain Disorder, which is a mental disorder, unless there is some specific evidence of preexisting intrapsychic conflict or psychological disturbance. In such a situation, the presentation of a somatic complaint (i.e., pain) allows the individual to avoid acknowledging and dealing with the conflict and thus also avoid the experience of anxiety that this would evoke (primary gain).

For that reason, the term "learned pain behaviour" (Tyrer 1986) might provide a more accurate description of the psychological mechanisms that lead to the persistence of pain following an injury in compensable circumstances.

Tyrer has described factors that promote pain complaints and pain behavior, such as sympathetic attention from those of importance to the individual, or permission to avoid unpleasant activities. Block et al. (1980) have demonstrated empirically that patients with "sympathetic spouses" showed more marked pain behavior when they were told that the spouse was observing the interview through a one-way mirror.

In general, environmental factors that might reinforce pain behavior include the availability of medications prescribed on a "pain-contingent" basis as well as, in some cases, availability or entitlement to "pain-contingent" compensation benefits. Such factors, which tend to promote pain complaints and pain behavior, have been termed "secondary gains."

Finally, "tertiary gain," which might also perpetuate pain complaints, is that which is obtained by others in the patient's environment who obtain some form of benefit from the patient's continuing sick role. Tertiary gain may accrue to spouses, children, and relatives of the patient, as well as to that individual's medical and paramedical attendants and lawyers. It has been commented that a family that is "fearful, suspicious, or highly competitive" might undermine a rehabilitation program, and the same authors have emphasized the need to identify such families and design intervention strategies, which will minimize the risk that such a family may adversely affect the patient's progress towards recovery.

Bayer (1985) has provided an illuminating discussion of the mechanisms of self-deception in patients with pain of nonorganic origin, discussing patterns of abnormal illness behavior shown by such patients in relation to cognitive dissonance theory. This approach also helps to explain the mental processes which lead to disproportionate disability among individuals who have minimal, or no, discernible physical basis for the complaint of pain.

Fibromyalgia following trauma might be considered a special example of Pain Disorder Associated with Both Psychological Factors and a General Medical Condition. In an editorial, Wolfe (2000) described the case vignette of a patient who had been diagnosed with fibromyalgia by two "of the best fibromyalgia experts in the US (well known authors and clinicians)." Subsequent surveillance and a "detailed review of medical and nonmedical records" caused the clinicians to review their diagnosis and to conclude "that the patient was malingering." Wolfe wrote that "observations such as this call into question the validity and reliability of the fibromyalgia evaluations in the medicolegal setting."

Iatrogenic factors have also been implicated in contributing to the entrenchment of pain complaints in the absence of an organic disorder, in particular where the diagnosis of fibromyalgia had been made.

In the area of psychiatric evaluation of chronic pain patients in receipt of, or seeking, compensation also lies the question frequently asked of the psychiatrist by solicitors or insurance agents, namely, "does this person have real pain or is he/she malingering." The recent introduction of technological equipment in an attempt to assess physical impairment "objectively" has given rise to claims that such equipment can be used to "catch malingerers." This is discussed in the next section.

Alleged Malingering of Chronic Pain

Complaints of chronic pain in the setting of personal injury litigation not infrequently give rise to the allegation that the plaintiff is either exaggerating the severity of the pain and the resultant alleged disability or malingering. It needs to be emphasized that malingering is neither a medical nor a psychiatric diagnosis. Malingering is fraud within a medical context, and thus, it is an act that is properly assessed by a tribunal or a court of law.

The range of contemporary clinical settings in which malingering may need to be considered has been illustrated in the current edition of the Diagnostic and Statistical Manual of Mental Disorders (DSM-IV-TR) (American Psychiatric Association 2000), which describes malingering as the intentional production of false or grossly exaggerated physical or psychological symptoms, motivated by external incentives such as avoiding military duty, avoiding work, obtaining financial compensation, evading criminal prosecution, or obtaining drugs.

It must be noted that DSM-IV-TR does not include malingering among the other diagnosable mental disorders, but rather places it among various "conditions" that might be "a focus of clinical attention." In DSM-IV-TR, malingering is given the code V65.2.

In the current tenth edition of the International Classification of Diseases (ICD-10), malingering similarly is not included among the diagnosable mental and behavioral disorders, but is listed among "factors influencing health status and contact with health services." In ICD-10, malingering (conscious simulation) is given the code Z76.5 and includes "persons feigning illness with obvious motivation."

A number of studies have reported the views of clinicians about the rate of malingering. There has been no study that has reported the incidence of malingering based on legal findings. Some authors have noted that the views of examiners are influenced by whether the medicolegal assessment is requested on behalf of the plaintiff or the defendant. In reading such articles, it is important to keep in mind that, despite their titles, they are not studies of the frequency and characteristics of pain malingering as determined by the "triers of fact" (i.e., the courts), but only surveys of views (and possibly prejudices) about malingering.

A variety of methods have been suggested to identify putative malingerers complaining of persistent pain. These can be conveniently grouped into four main types, namely, questionnaires, clinical examination, facial expression, and testing with mechanical devices. Other attempts to detect malingering have included differential spinal blocks, thermography, and the parenteral administration of agents such as Amytal and Pentothal.

There have been a number of published studies describing questionnaires that purport to be able to detect malingered pain. Such questionnaires are usually constructed by asking volunteers or patients to simulate and pretend that they have pain or to exaggerate the pain that they do experience. Some authors have claimed that a subset of pain descriptors is able to differentiate between volunteers and back pain patients instructed to exaggerate the severity of the pain they experienced.

The research paradigm used in such studies has been criticized and described as the "paradox in asking subjects to *comply* to instructions to *fake* in order to study subjects who *fake* when ordered to *comply*."

The introduction of spinal dynamometry, which allows measurement of the range of motion and strength of exertion, led to it being called "a spinal lie detector." However, Jayson (1992) has shown that, in relation to low back pain, dynamometric techniques that may indicate that a submaximal effort has been made cannot be used as "the final arbiter" of malingering.

Similar considerations apply to the use of the Jamar dynamometer to measure grip strength and to the use of other mechanical devices as indicators of alleged malingering.

It has also been suggested that nondermatomal sensory deficits ("glove and stocking" anesthesia) in association with chronic pain are indicative of malingering (or, as an alternative, that the pain does not have an organic basis). Others have argued, however, that nondermatomal somatosensory deficits might have a central basis.

Conclusions

Although it is often alleged that compensation and litigation invariably cause exaggeration of pain complaints and prolong disability, the available empirical evidence does not support such a simplistic view. While there are studies that support the view that compensation and litigation often have an adverse effect on treatment response in chronic pain, this effect can be modified by work and by the provision of specific treatment programs.

The view that financial considerations are the main determinants of behavior by injured workers ignores the extensive research that has shown the importance of occupational and other psychosocial factors.

The prevalence of chronic pain complaints among personal injury litigants is positively correlated with compensation systems that provide pain-contingent benefits and allow litigation at common law, but there is no consistent evidence to support the view that – as a group – chronic pain patients involved in litigation or receiving compensation describe their pain as more severe or more distressing than similar patients with pain which is not compensable.

There is some evidence that the rate and duration of claims are influenced by the level of compensation payments, but the research in this area is meager and does not allow for any firm conclusions.

There is no support for the view that litigants and/or compensation recipients with chronic pain suffer from or have a specific psychiatric "condition" such as "compensation neurosis" or "accident neurosis." There is also no support for the view that litigants with chronic pain are significantly more psychologically disturbed than similar patients not claiming compensation (Mendelson 1984). These findings contradict the view that patients involved in litigation develop the putative psychiatric condition termed "accident neurosis" or "compensation neurosis," as these patients could not be distinguished on psychometric testing from those not seeking compensation. Some authors have specifically commented on the deleterious effect of using such labels, which have no clinical validity.

It needs to be emphasized to clinicians and lawyers alike that malingering is not a diagnosis and that there is no valid clinical methodology that can detect purported "malingering" of pain. Malingering is an act, and it is fraud in relation to one's health status, that is, in a medical context. Whether or not a person is malingering should, therefore, be determined by the trier of fact, be it a judge, a jury, or a tribunal.

Numerous studies have demonstrated that the assessment of chronic pain patients needs to take into consideration societal expectations and that attitudes to illness and illness behavior must also be considered. The impact of the compensation system and of litigation on the process of becoming a chronic pain patient has important social policy implications that thus far have been generally neglected by governments and by policy-makers.

The various findings reviewed above suggest that the effect of compensation on chronic pain complaints and disability can best be conceptualized as occurring at an unconscious rather than a conscious level, as do other secondary gains that tend to perpetuate psychogenic symptoms. These include personality, cultural factors, interpersonal dynamics, occupational factors, and societal factors (see list below). For that reason, the simplistic notion of "compensation neurosis" as a specific clinical condition that is reversible once the claim is finalized is invalid. Emotional factors that promote psychogenic symptoms also lead to other changes in the individual's level of functioning which, once established, perpetuate the symptoms after litigation has been finalized.

Factors that influence outcome of compensable injuries, including chronic pain:

- Personality:
  - Childhood emotional deprivation
  - Hypochondriacal traits
  - Unmet dependency needs
- Demographic factors:
  - Age
  - Education level
- Cultural factors:
  - Illness behavior
  - Folk belief concerning illness
- Interpersonal dynamics:
  - Within the family
  - Social milieu
- Occupational factors:
  - Job dissatisfaction
  - Work stress

- Type of work performed
- Employer/supervisor attitude
- Psychological reaction:
  - Alteration in self-concept and body image
  - Personality disorganization
  - Regression
  - Development of a mental disorder
- Physical factors:
  - Presence and nature of physical injury or disease
  - Extent of residual organic impairment
- Economic factors:
  - Compensation payments
  - Job security/level of unemployment
  - Litigation
  - Level of wages
- Societal factors:
  - Acceptance of complaint as work or injury related
  - Expectations concerning prognosis
  - Availability of rehabilitation
- Therapeutic variables:
  - Appropriate early pain management
  - Treatment of specific mental disorder if present
  - Early intervention to develop a vocational rehabilitation plan

This emphasizes the need to carefully evaluate each litigant with chronic pain to assess the relative importance of compensation and other factors and to avoid stereotyping or labelling litigants as universally motivated by potential financial gains.

In the diagnostic evaluation of patients with chronic pain seeking compensation, in the absence of an objectively demonstrable organic lesion or evidence of continuing physical injury that could explain the persistence of pain, diagnoses such as "psychogenic pain disorder" are problematic.

This issue has also caused difficulty for courts that have to make decisions concerning entitlement to benefits, and contradictory judgments have been reported.

A legal solution to this dilemma had been adopted by the Ontario Workers' Compensation Appeals Tribunal (1987). In a lengthy decision on a pension application by a worker who had sustained some injuries when a wall collapsed, but whose subsequent chronic pain complaints were described as predominantly psychogenic, the Tribunal held, in effect, that if pain that was considered to be psychogenic persisted following a documented physical injury that had resolved, the claimant would be entitled to benefits. It was noted that in the Tribunal's view, "a non-organic element in the chronic condition is compensable" and that "the Tribunal determined that the injuries resulting from the fall were a significant contributing factor in the development of both the organic and the non-organic components of the worker's chronic pain condition."

References

American Psychiatric Association. (2000). *Diagnostic and statistical manual of mental disorders* (4th ed.). Washington, DC: Text Revision APA.

Balla, J. I. (1982). The late whiplash syndrome: A study of an illness in Australia and Singapore. *Culture, Medicine and Psychiatry, 6*, 191–210.

Bayer, T. L. (1985). Weaving a tangled web: The psychology of deception and self deception in psychogenic pain. *Social Science & Medicine, 20*, 517–527.

Block, A. R., Kremer, E. F., & Gaylor, M. (1980). Behavioral treatment of chronic pain: The spouse as a discriminative cue for pain behavior. *Pain, 9*, 243–252.

Blyth, F. M., March, L. M., Nicholas, M. K., & Cousins, M. J. (2003). Chronic pain, work performance and litigation. *Pain, 103*, 41–47.

Jayson, M. I. V. (1992). Trauma, back pain, malingering, and compensation. *BMJ, 305*, 7–8.

Melzack, R., Katz, J., & Jeans, M. E. (1985). The role of compensation in chronic pain: Analysis using a new method of scoring the McGill pain questionnaire. *Pain, 23*, 101–112.

Mendelson, G. (1983). The effect of compensation and litigation on disability following compensable injuries. *American Journal of Forensic Psychiatry, 4*, 97–112.

Mendelson, G. (1984). Compensation, pain complaints, and psychological disturbance. *Pain, 20*, 169–177.

Mendelson, G. (1987). Measurement of conscious symptom exaggeration by questionnaire: A clinical study. *Journal of Psychosomatic Research, 31*, 703–711.

Mendelson, G. (1995). "Compensation neurosis" revisited: Outcome studies of the effects of litigation. *Journal of Psychosomatic Research, 39*, 695–706.

Merskey, H., & Bogduk, N. (1994). *Classification of chronic pain: Descriptions of chronic pain syndromes and definitions of pain terms* (2nd ed.). Seattle: IASP Press.

Ontario Workers' Compensation Appeals Tribunal (1987) Decision no. 915. Research Publications Department, WCAT, Toronto.

Tyrer, S. P. (1986). Learned pain behaviour. *BMJ, 292*, 1–2.

Wolfe, F. (2000). For example in not evidence: Fibromyalgia and the law. *Journal of Rheumatology, 27*, 1115–1116.

Psychiatric Aspects of Multimodal Treatment for Pain

▶ Multimodal Rehabilitation Treatment and Psychiatric Aspects of Multimodal Treatment for Pain

Psychiatric Aspects of Pain and Dentistry

Charlotte Feinmann
Eastman Dental Institute London, University College London, London, UK

Synonyms

Orofacial pain; Persistent pain; Temporomandibular joint

Definition

There are four recognizable symptom complexes of persistent orofacial pains, which may, however, coexist: temporomandibular disorder (TMD, myofascial face pain), persistent facial pain (atypical facial neuralgia), persistent odontalgia (atypical and phantom tooth pain), and burning mouth syndrome (oral dysaesthesia).

Characteristics

Persistent physical symptoms are common, disabling, and economically important. Less than

10 % are found to have a physically identifiable cause yet they account for 22 % of primary care consultations and 25–70 % of acute outpatient activity; 30 % have an associated mental health problem, and the cost of irrelevant investigation of these patients is enormous. In the general population, these symptoms are mild and transient but a substantial subsection of patients have more persistent disabling symptoms and may meet the criteria for one or more recognizable functional somatic syndromes such as irritable bowel syndrome or persistent facial pain. Persistent orofacial pain is an ill-understood group of conditions, which may affect the whole of the mouth and face. Unfortunately, descriptions of disorders and treatment tend to be influenced by the background of the specialist assessing the patient. Thus, patients who see maxillofacial surgeons have symptoms described in terms of clicking, sticking, and locking of the temporomandibular joint (TMJ) and pain in the associated musculature. ENT surgeons may retain Costen's outdated notion that the pain is due to missing molar teeth and may refer on to the maxillofacial surgeons or restorative dental specialists. Despite advice from the National Institute of Health that "there is no evidence linking occlusal abnormalities with pain," patients continue to have pain treated by occlusal adjustment by ill-informed practitioners, often leading to more problems for patients (NIH 1996).

Marcfarlane et al. (2002) have found that 26 % of a UK population complained of orofacial pain, but only 46 % of those sought help. An obvious explanation for why some individuals choose to consult while others do not is the severity of their symptoms. Sadly, epidemiological studies of facial pain have tended to ignore intensity and frequency of symptoms and recorded only their presence or absence.

Patients may complain of pain, clicking, or locking in the temporomandibular joint and surrounding muscles. TMJ pain is often associated with disorders of function; the patients may complain of locking of the joint, clicking, and sticking. Eating and other oral activities such as kissing may be difficult, and on examination the jaw may deviate to one side.

Traditional signs of temporomandibular disorders (*TMD*) have included temporomandibular joint sounds, mandibular motility, and pain elicited on palpation of the joint and associated musculature. The prevalence of joint sounds and the wide range of mandibular motility in healthy populations, together with the subjectivity of mobility assessment and pain report on palpation, mean that these signs must be interpreted with considerable caution. Also, joint sounds may remain unchanged, despite perceived improvement of TMD after treatment. There is growing support for a distinction between muscle-related ("myogenous") TMD and joint-related ("arthrogenous") TMD, and hence, pain on palpation may be the most useful clinical measure.

The publication of Research Diagnostic Criteria for TMD (RDC/TMD) has provided an opportunity for standardized evaluation of TMD2. While the distinction between myogenous and arthrogenous complaint may be valid, the inclusion of non-pain signs such as joint sounds and jaw opening, as suggested in the RDC/TMD, appears inappropriate, and self-reported disability factors may be more relevant.

Current diagnoses include tension headache, migraine, neckache, temporomandibular disorder (temporomandibular joint pain dysfunction syndrome, facial arthromyalgia), and persistent facial pain or *atypical facial pain*. These pains appear to arise from blood vessels, muscles, and joint capsules rather than sensory nerve branches, as in trigeminal neuralgia. Artificial distinctions in clinical presentation lead patients to different specialists providing different treatments, including dentists, neurologists, otorhinologists, osteopaths, chiropractors, and psychiatrists, with little collaboration between the specialties.

There are additional important problems concerning the recognition and definition of underlying psychiatric disturbances. Emotional disturbance, when present, is often mild and of brief duration, and psychiatric classification has proved an inadequate measure to capture the problem.

Atypical facial pain is no longer included in the International Association for the Study of Pain's "Classification of Chronic Pain." The term originated in order to distinguish the condition from "typical" trigeminal neuralgia, since the pain neither follows the distribution of the peripheral nerve nor responds to antiepileptic agents. However, the categorization of individuals with similar pain histories into a diagnostic pigeon-hole labeled "atypical" is self-contradictory, and the definition of a condition by what it is not, rather than what it is, is unsatisfactory. A better term might be "persistent facial pain" (PFP), since the defining characteristics are longevity and site, as distinct from TMD, which affects the jaw rather than the (mid-) face and intraoral pains.

The pain is usually a continuous dull ache with intermittent severe episodes, primarily affecting areas of the face other than joints and muscles of mastication, such as the zygomatic maxilla. Pain may be bilateral and will often have been present for several years. Analgesics are ineffective.

Persistent (atypical) odontalgia (PO) has a similar character, but is localized to one or more premolar or molar teeth, simulating pulpitis. There may be a history of inappropriate dental treatment including extraction and subsequent recurrence of symptoms, apparently from another tooth. Dentists must resist the temptation to over-investigate and repeat tests carried out by others. The term "chronic" is used in preference to "atypical" for reasons described above.

Positron emission tomography has demonstrated an increased contralateral cingulate cortex activity (and decreased prefrontal cortex activity), in response to both heat and nociceptive stimuli, in CAFP patients relative to controls. This is suggestive of an exaggerated perception of pain in response to peripheral stimuli (Sarlain and Greenspain 2003).

Patients may complain of a burning tongue; occasionally the gingivae and lips are also involved, or the denture-bearing areas of the hard palate and lower alveolus, making the wearing of a denture impossible.

The patient is often middle aged and female, but can be of any age or sex. The picture is invariably one of a burning sensation that gradually increases in severity and frequency, until the patient suffers it for the greater part of every day.

It is not present on waking but very quickly builds up to its maximum intensity for that day. It may be unilateral or bilateral. The most striking feature of the condition is that it is usually relieved by eating or drinking, and patients may chew gum or other food in order to seek relief. Organic pains of the tongue are made worse by eating and drinking, and this distinction may be crucial in making the diagnosis.

Denture intolerance due to discomfort or pain in the absence of any mucosal or bony lesion is a comparable problem. The diagnosis can be made on the many sets of dentures carried by the patient.

A common presentation is with posturing of the jaw and bizarre complaints such as an inability to speak clearly.

A full history and appropriate investigations should eliminate any deficiencies of serum iron; vitamins B12, B1, B2, and B6; and folate. A localized diabetic neuropathy can be excluded by blood and urine glucose estimations.

The persistent "nasty" taste (dysgeusia), usually with an obsessional concern of having halitosis, and the dry mouth sensation may occur separately or together with a burning tongue. This patient may also complain of dysphagia. Examination invariably reveals no gross oral, pharyngeal, or antral sepsis. Where the complaint is principally that of a dry mouth, there may be adequate or overtly reduced salivary flow. However, sialography and salivary gland biopsies and stimulated flow rates to exclude Sjögren's syndrome are normal. Further, concern in patients over halitosis is not uncommon. Variations include an illusion of sand in saliva, excess saliva, or excess mucus, which cannot be objectively substantiated.

These complaints are often a manifestation of stress. Recognizable causes are professional worries, bereavement, loneliness, and curiously enough, in women, the premature retirement of a successful husband (perhaps a symbolic bereavement). The patient may also show symptoms of agitation, early waking, and loss of appetite and libido. There is occasionally a firmly held delusional explanation such as amalgam fillings, "hyperacidity," or impaired pancreatic or hepatic function, which may require referral to a psychologist.

In phantom bite (Marbach 1978), patients complain of continuous discomfort because their teeth do not meet correctly.

They are often intensively involved with, and superficially knowledgeable about, details of dental anatomy, physiology, and restorative dentistry. At consultation they frequently present numerous radiographs of the teeth, casts, splints, old crowns, and pictures of themselves when perfect. A careful history will reveal a long succession of attempted treatment plans by a variety of dentists who have not been successful. The spouse may be involved in the delusion (a folie à deux). Dentists must be cautious in trying to provide any treatment for these patients.

It is possible that the problem is a disturbance in kinesthesia (position sense), particularly as this can be produced experimentally when vibration is applied to muscle tendons. Prolonged dental treatment, especially occlusal equilibration and the replacement of natural tooth contour, may have a similar effect in obsessional individuals. Continued attempts to "equilibrate" the occlusion do not help.

If there is no clinical evidence of an iatrogenic grossly deranged occlusion, the condition should be considered to be an obsessional or *psychotic* problem. If this is not recognized, the patient will eventually be rendered cuspless, then edentulous, incapable of tolerating false dentures and invariably acquiring other intractable syndromes.

Initiation and Diagnosis

Many persistent facial pain patients specifically relate the onset of their symptoms to dental treatment itself. Other frequently reported precipitating factors include infections, toxins, and stress, such as following bereavement. However, once initiated, the patient may inadvertently exacerbate, and thereby maintain, the pain problem through his or her own actions. For example, a proportion of patients completely avoid movement of the jaw, which eventually results in muscular atrophy and greater joint stiffness. Others compulsively stretch and hyperextend the jaw numerous times each day, provoking local

irritation. Frequent prodding and touching the painful areas of the face, teeth, or gums is also common in facial pain patients and is also likely to irritate already sensitive muscles and nerves. The role of fear-avoidance tendency to avoid fear painful activity is useful for patients to understand. Sarlain and Greenspain (2003) have described a generalized hyperexcitability in the central processing of the nociceptive cortex as the pathophysiological basis of the TMJ pain.

Underpinning these behaviors is the patient's mood state. High levels of anxiety – related to concerns such as "will the pain worsen?" and "is there an undetected malignant cause?" – increase the perception of pain, as does significantly depressed mood.

Associated Symptoms: Physical

It is crucial for an understanding of these patients to realize that headache and facial pain are not the patient's exclusive problems. Nor are these symptoms mutually exclusive, as they may occur sequentially or simultaneously in any patient. Eighty percent of patients will complain, if prompted, of other recurrent symptoms such as persistent neck and low back pain, migraine, pruritic skin disturbances, irritable bowel, or dysfunctional uterine bleeding. The prevalence of these symptoms is much higher than in the general population, indicating that facial pain and headache are features of a more generalized bodily distress disorder. Specialists unfamiliar with the variety of the disorders, however, may not elicit the range of symptoms. There does seem to be an adequate muscular correlate to account for the pain of chronic daily headache or persistent facial pain, and there is no satisfactory explanation about the etiology of facial pain. The combination of a number of disorders has been termed a "vulnerable neurochemistry" or a pain-prone patient; however, appreciation of *central sensitization* and the phenomenon of *wind up* has improved our understanding of the pathophysiology of pain.

Persistent Symptoms

Persistent symptoms and functional syndromes pose a major challenge to medicine: they are

Psychiatric Aspects of Pain and Dentistry, Table 1 Chronic syndromes by specialty

Gastroenterology	Irritable bowel syndrome, non-ulcer dyspepsia
Gynecology	Premenstrual syndrome, chronic pelvic pain
Rheumatology	Fibromyalgia
Cardiology	Atypical or non-cardiac chest pain
Respiratory medicine	Hyperventilation syndrome
Infectious disease	Chronic (postviral) fatigue syndrome
Neurology	Tension headache
Dentistry	Temporomandibular joint dysfunction, atypical facial pain
Ear, nose, and throat	Globus syndrome
Allergy	Multiple chemistry sensitivity

common and persistent and are associated with significant distress, disability, and unnecessary expenditure of medical resources. In UK primary care, somatic symptoms and syndromes account for 20 % of consultations. Among medical outpatients, somatic complaints accounted for 35 % of new referrals in a UK study. Even among medical inpatients, a substantial proportion has complaints that are found to be functional. The prevalence of emotional distress and disorder in patients who attend hospital with unexplained syndromes (such as irritable bowel syndrome) is higher than in patients with comparable medical conditions (such as inflammatory bowel disease), and many such patients are severely disabled.

Wessley et al. (1999) suggest that each medical specialty has defined its own syndrome or syndromes in terms of symptoms that relate to their organ or interest (see Table 1).

Patients seek help from doctors for symptoms, and doctors diagnose diseases to explain them. Symptoms are the patient's subjective experience of changes in his or her body; diseases are objectively observable abnormalities in the body. Difficulties arise when the doctor can find no objective changes to explain the patient's subjective experience. The symptoms are then referred to as medically unexplained or functional.

Wessely and colleagues postulate that: "the existence of specific somatic syndromes is largely an artefact of medical specialisation. That is to say, the differentiation of specific syndromes reflects the tendency of specialists to focus on only those symptoms pertinent to their specialty, rather than any real differences between patients." Treating the symptoms in terms of severity, number of symptoms, associated psychological distress, and functional disturbance will in the future prevent harm done to patients by idiosyncratic isolated specialists. The terms persistent symptoms and functional syndromes are preferred as they are less pejorative and more likely to engage the patients in treatment.

A "diathesis-stress" framework has been proposed to explain the high comorbidity between chronic pain and depression. This approach encourages identification of vulnerability factors in the individual as well as investigation into the nature of the stressor and may be a useful theoretical basis from which to advance the study of depression in persistent facial pain.

Management

The treatment offered to a persistent facial pain sufferer will be determined by the specialist understanding of the clinician to whom he/she is referred. A specialist must eliminate possible dental causes with clinical and radiographic investigation, but extensive investigation can lead to patients being off-loaded from specialist to specialist in search of a diagnosis and feeling ill understood, over-investigated, and dissatisfied. Evidence suggests that it may be more helpful to assess patients in terms of disability and coping strategies, rather than pain intensity per se (Feinmann and Madland 2001). It is crucial that physicians and patients reach an agreement about how to manage the pain. The mainstays of treatment are counseling, cognitive behavior therapy, and antidepressant medication.

It is important to acknowledge the genuine anger that patients may have about previous treatment and to attempt to explain the interaction between peripheral and central perception of pain before attempting treatment.

Psychiatric Aspects of Pain and Dentistry, Table 2

Evidence-based treatment
Antidepressants
Cognitive behavior therapy
Relaxation
Hypnotherapy
Reattribution

The evidence-based treatments are antidepressants, cognitive therapy, hypnosis, and relaxation (see Table 2).

Pain relief with tricyclic agents appears to be independent of the antidepressant effect of the drugs. These drugs are considered to act by altering the sensory discrimination component of pain. The possibility of interference with serotonin reuptake in the brainstem has been proposed. A study of 181 (Harrison et al. 1999) patients treated with either fluoxetine or cognitive behavior therapy, especially targeted at facial pain, found that fluoxetine reduced pain at 3 months compared to placebo, and these improvements were maintained when the drug ceased. Dworkin et al. (2002) has also confirmed the effectiveness of CBT in facial pain. The decision as to whether a tricyclic or an SSRI is needed is really dependent on which side effects, such as sedation profile, the patient requires, and general treatment needs to be continued for at least 6 months and may be for as long as 3 years to maintain improvement. Antidepressants are used independently of their antidepressant effect and probably have a central pain relieving effect.

Both relaxation treatment and hypnosis have been shown to be effective in the management of facial pain. Patients who respond least well to treatment are non-anxious somatizers, those who insist on a physical cause for pain, and those with odd dysfunctional health beliefs, lots of unsuccessful surgery, and no life event before onset of pain. There is some evidence that these patients respond to intensive psychotherapy or extended CBT. The physical and psychological functioning of patients with chronic disabling TMJ pain is substantially associated with the patient's beliefs; in particular, stopping patients catastrophizing about disability was shown to be crucial for pain relief (Dworkin et al. 2002).

Counseling and Changing the Agenda

Antidepressants should not be used in isolation. A pragmatic treatment approach must be adopted, in which the diagnosis is first established by a careful history that identifies other idiopathic pains and seeks out stressful life events. Many patients will respond to reattribution or "change of agenda" of their physical symptoms to an emotional cause by an explanation of amplifying normal bodily sensations.

Summary of Management

1. For successful management it is important to identify not only the features of the joint pain but also chronic pain disorders elsewhere, for example, the head, neck, back, abdomen, and pelvis.
2. Identify predisposing adverse life events or a history of an emotional or psychiatric disturbance.
3. Emphasize to the patient that the pain is real rather than imaginary, arising in "cramped" muscles and dilated blood vessels as a response to emotional stress, and that drug therapy is not being used to treat depression but has a direct effect in relieving the painful muscles and blood vessels.
4. Tricyclic antidepressants such as nortriptyline 10–100 mg, TCAs, or selective serotonin reuptake inhibitors (SSRIs) such as escitalopram 5 mg daily and venlafaxine 37.5 mg bd should be prescribed at night in gradually increasing doses, combined with regular reviews at 3–6-week intervals to provide reassurance and to achieve drug compliance. Little or no response at 12 weeks indicates a need for further investigation.
5. Dental treatment should be confined to essential problems such as sensitive carious cavities, pulpal inflammation, or marked occlusal disturbances, usually where there is a gross deficiency of functional teeth.

References

Dworkin, S., Turner, J., & Mand, L. (2002). A randomised clinical trial of CBT. *Journal of Orofacial Pain, 16*(14), 259–276.

Feinmann, C., & Madland, G. (2001). Chronic facial pain – A multidisciplinary problem. *Journal of Neurology, Neurosurgery & Psychiatry, 6*, 7126–7129.

Harrison, S., Glover, L., & Feinmann, C. (1999). A comparison of antidepressant medication alone and in combination with cognitive behaviour therapy for chronic idiopathic facial pain. In T. Jensen & I. Turner (Eds.), *Proceedings of the 8th world congress on pain* (Progress in pain research and management, Vol. 8). Seattle: IASP Press.

Marbach, J. (1978). Phantom bite syndrome. *The American Journal of Psychiatry, 135–4*, 476–479.

Marcfarlane, T., Blinkhourn, A., Davies, R., et al. (2002). Orofacial pain in the community. Prevalence and associated impact. *Community Dentists and Oral Epidemiology, 30*, 50–60.

NIH. (1996). *Technology assessment statement management of temporomandibular disorders*. Washington, DC: Author.

Sarlain, E., & Greenspain, J. D. (2003). Evidence for generalised hyperalgesia in temporomandibular disorders. *Pain, 102*, 221–226.

Wessley, S., Nimnuan, C., & Sharpe, M. (1999). Functional somatic syndromes: One or many. *The Lancet, 354*, 936–939.

Psychiatric Aspects of the Epidemiology of Pain

Eldon Tunks[1], Joan Crook[2] and Robin Weir[2]
[1]Hamilton Health Sciences Corporation, Hamilton, ON, Canada
[2]School of Nursing, Faculty of Health Sciences, McMaster University, Hamilton, ON, Canada

Synonyms

Chronic fibromyalgia; Chronic pain disorder (DSM-IV); Chronic pain, persistent pain; Chronic widespread pain; Dysthymia, chronic depression; Fibrositis; Opioid addiction; Opioid dependency; Psychalgia (ICD-10); Somatoform pain disorder (DSM-III R and ICD-10)

Definition

▶ Prevalence: the frequency at one point in time of a condition in a population.

▶ Risk: a measure of the probability that a given condition will arise.

▶ Prognosis: a measure of the probability of a favorable or unfavorable outcome of a given condition.

In ▶ epidemiology, the units of observation are generally samples of people drawn from populations of interest to the research. Epidemiology is concerned with comparing one sample with another, and inferring relevance to the larger population from which the sample is drawn. Although classically epidemiology has been understood to be the study of the characteristics of larger populations, such as in a census or transcultural studies, epidemiological methods are applied to a great variety of scientific inquiry, including the study of natural history of illnesses, the comparison of clinical or social policy interventions or natural events that affect health outcomes, studies of risk factors predicting new illnesses, factors affecting prognosis, and other applications, most of which is beyond the scope of this brief discussion.

Characteristics

In epidemiological methods, procedures are established to detect "cases" of interest. The operational definitions used will have a significant effect on the characteristics of the "cases" studied, and on the results. For example, some definitions in the earliest epidemiological studies of "chronic pain" used selection criteria that were specific for constant and prolonged pain, while others used criteria that detected constant and prolonged pain but also intermittent pain, and hence the prevalence reports varied with the definitions used in the methodology. (The estimates of chronic pain prevalence, using various methodologies, have ranged from 10 % to 55 %.) Recently, with the development of standardized diagnostic criteria and instruments, some epidemiological studies have assessed the prevalence of specific subgroups of pain disorders such as migraine, chronic fibromyalgia, temporomandibular joint pain, or chronic low back pain. In reading any epidemiological studies, it is essential to look first at the methods section and the sampling methods, before reading the results.

We have confined ourselves here to studies of various populations that either have had, or who were at risk for, both psychological (psychiatric) and chronic pain problems. The hypotheses underlying many of these studies were the following three:

1. Chronic pain is a symptom of psychological maladjustment, for example, a form of somatization.

2. Chronic pain causes ▶ depression (or anxiety), or depression (or anxiety) causes pain.

3. Pain and depression (or anxiety) are separate phenomena that may be associated for various reasons, including common risk factors, selection factors that bring them to clinical attention, or common constitutional predispositions.

There is no reason to believe that these hypotheses must be mutually exclusive, but our brief review is concerned with the second and third hypothesis.

Epidemiologists have studied these hypotheses through the following questions:

1. What is the prevalence of chronic pain associated with psychiatric ▶ comorbidity?

2. How does comorbidity affect the prognosis and/or the clinical course of chronic pain or depression?

3. How does pain influence the risk of developing psychiatric conditions?

4. How do psychiatric conditions influence the risk of developing pain disorders?

A few such studies and their findings are presented here for illustration.

What Is the Prevalence of Chronic Pain Associated with Psychiatric Comorbidity?

Crook et al. (1986) conducted a standardized telephone survey of adults in 393 randomly selected

households registered with a group family practice, and separately 62 files representing all patients who had attended a university outpatient pain clinic within the previous 6 months. "Persistent pain" was defined as having both "pain always or usually present" and "noteworthy pain within the previous 2 weeks" Psychological and psychosocial problems were assessed by standardized interview questions identifying physical function, pain on activities, emotional status, attitude, pain-related behavior, and social consequences of long-term pain. In both community and specialty clinic samples, similar ranges of problems were found of pain in the neck, upper or lower back, upper or lower limbs, or headache, and demographic variables were similar across the two samples. Significant differences were found between groups, however, with regard to a higher prevalence of work-related accidents, more constant pain and greater reported disability, greater psychosocial difficulties and distress, in the specialty pain clinic group. Patients referred to tertiary-level specialty pain clinics are not necessarily characteristic of patients in the community. Patients are more likely to be selected for referral if there are significant psychosocial or functional problems.

Breslau and Davis (1992) used standardized structured interview schedules to identify cases with lifetime diagnoses of migraine and/or major depression or anxiety disorders, in a random sample of 1,200 drawn from young adults in a large HMO. At baseline, 12.9 % of the sample met the criteria for migraine at any point in their lives. Those with migraine were found to have a threefold to sixfold increased prevalence of anxiety or mood disorders, compared to those who did not have migraine.

Ohayon and Schatzberg (2003) conducted a cross-section telephone survey of a random sample of 18,980 subjects, representative of the general populations of the UK, Germany, Italy, Portugal, and Spain. The prevalence of persistent pain (more than 6 months duration), of non-painful physical conditions, and the relationship with "major depression" were determined.

The prevalence of at least one chronic pain condition was 17.1 %, and of at least one depressive symptom was 16.5 %. Chronic pain occurred 4 times more often in those with "major depression." There was an even stronger association with "major depression" if both chronic pain and another medical condition coexisted. Constant pain increased the probability of having "major depression" (▶ Adjusted Odds Ratio 1.6).

Based on the above studies, one can conclude that chronic pain disorders in general, and specific chronically painful physical conditions, are associated with increased prevalence of psychiatric diagnoses. This prevalence of psychological comorbidity is further increased by ▶ selection factors operative in the referral of chronic pain patients to tertiary specialty pain clinics.

Do Chronic Pain and Psychological Comorbidity Affect the Prognosis and/or the Clinical Course of Chronic Pain or Depression?

Crook et al. (1989) followed up the community and specialty pain clinic patients studied in their initial survey, using the same definitions and methods. 77 % were reinterviewed. After 2 years, it was found that 13 % of the persistent pain sufferers from the specialty pain clinic, and 36 % of the persistent pain sufferers from the community family practice group no longer reported that pain was a problem. Of those who suffered ongoing pain, on average they reported that their pain had become more intermittent, that psychological distress factors improved, and that their use of health services decreased. However, the emotional factors that distinguished the pain clinic group from family practice group were still an important distinguishing factor, and likely played an important role in the chronic morbidity and worse overall prognosis of these patients. A standardized and validated inventory of "coping" was employed in the same follow-up study (Crook et al. 1988). Multiple regression analysis found that "adversarialness" explains some of the variance in patients' adjustment to chronic pain. The more that patients were adversarial as a way of coping, the more depressed and anxious they

tended to be, the more they assessed their pain as serious, and felt a lack of ability to control, prevent, or lessen their pain. The "ability to mobilize environmental resources" explained a small amount of variance with regard to sexual relations and social consequences of the pain.

Weir et al. (1992) conducted an historical cohort survey of 571 patients referred to the specialty pain clinic, through chart review and mailed questionnaire, which was followed up by a standardized telephone interview including demographic and economic and utilization variables and validated measures (assessing meaning of illness, social support, and psychosocial adjustment to illness). Eighty-one percent of the sample were "fair to poorly adjusted psychosocially," and 56 % of these fell in the "poorly adjusted" range. Compared with other specialty clinic samples, the prevalence of those who were poorly adjusted was greater among pain service patients. It was found that individuals who lacked intimate and caring social relationships, and whose pain problem had a high impact on their lives, and were low in perceived resilience, were unemployed, and in constant pain, and perceived the burden of the pain as affecting all aspects of their lives, were the individuals who were most likely to be poorly adjusted. On the positive side, it was found that patients who had used specialty pain clinic services generated proportionately less costs in the use of other health services, when compared with those who had not used the specialty pain clinic services: this provides evidence for the efficacy and social benefits of multidisciplinary services for these patients with comorbid chronic pain and psychological problems.

Livingston et al. (2000), in a survey of elderly patients with physical disability and psychiatric morbidity, found that those who developed depression became more likely to consult with their doctors. It appeared that elderly subjects with physical limitations who had become depressed were more likely to complain to their doctors about an illness, but only two subjects in their sample reported consultation because of depression, suggesting that depression may be underreported or unnoticed in this group of patients with comorbid pain, physical limitation, and depression.

MacFarlane et al. (1996) reported assessment of 141 subjects initially defined with "chronic widespread pain" "regional pain" or "no pain" after a follow-up period of a median of 27 months. Of those with chronic widespread pain initially, 35 % still had chronic widespread pain at follow-up, 50 % had regional pain, and 15 % no pain at follow-up. Of those initially with regional pain, 65 % still had regional pain, 19 % chronic widespread pain, and 16 % no pain at follow-up. They were more likely to have widespread pain at follow-up if at original assessment they had any of the following: 11 or more tender points on examination, a high score on a measure of psychological distress and fatigue, or if they had problems with sleep, micturition, abdominal pain, or headache. Any of the above symptoms increased the odds of non-improvement by at least threefold, although because of sample size only the increased odds associated with sleep problems were statistically significant.

Wells et al. (1992) surveyed the course of depression over a 2-year period in 626 adult outpatients suffering with major depressive disorders or depressive symptoms. The course of depression was assessed over the two follow-up years using a validated structured telephone interview. Initial functional impairment was greater in those with ▶ dysthymia or "double depression." By 2 years, the initial level of functional impairment and the initial depression severity both accounted for much of the variance in final depression outcomes. Impaired functional status and well-being, associated with bodily pain, made a significant contribution to depressive symptoms at the end of the second follow-up year. This indicates that comorbid dysthymia and pain confer a guarded prognosis. Note also that other epidemiology studies have shown that in chronic pain populations, dysthymia is more prevalent than expected.

Magni et al. (1994) studied, at an 8-year interval, a sample of 2,324 noninstitutionalized adults, who were interviewed for presence or absence of musculoskeletal pain and depression. For those who had chronic pain at baseline, there were no socio-demographic variables that significantly predicted the persistence of pain at 8 years of follow-up. For patients who were depressed at baseline, the only significant predictive variable for persistent depression at 8 years was unemployment.

These findings are important in several respects. With regard to prognosis, comorbid chronic pain and psychological factors confer more guarded outcomes, both with respect to psychiatric outcomes as well as with respect to pain and functional outcomes. Given that patients with comorbid psychological and chronic pain disorders are more likely to be referred to tertiary specialty chronic pain centers, it is important that chronic pain management teams be equipped with the necessary resources to deal with psychological and psychosocial complications and functional impairments, and not just with pain relief techniques. Finally, general community rehabilitation clinics dealing with chronic pain need also to be alert to those patients with higher loading on psychosocial factors, in order to identify and provide timely multidisciplinary care to these individuals with a potentially worse prognosis.

Does Chronic Pain Influence the Risk of Developing Psychiatric Conditions?

Livingston et al. (2000) conducted a community survey of psychiatric morbidity in people 65 years old or more, and selected those for follow-up who needed significant daily help with ADL due to physical ill health, but who did not initially have a psychiatric disorder (depression, anxiety, or dementia). After 3 years, 77 of the initial 141 subjects were able to be reinterviewed. Ten percent became seriously depressed, and a total of 24 % became mildly, moderately, or severely depressed. Frequency of pain was significantly associated with onset of depression. Consultation for chronic illness was not significantly associated with development of depression.

Patten (2001) studied data from a Canadian national population health survey, with follow-up 2 years later. "Major Depression" was assessed using a standardized interview for depression, and self-reported long-term medical diagnoses (providing that the diagnosis was made by a health professional and the condition was expected to last more than 6 months). The diagnoses surveyed included migraine, arthritis or rheumatism, back problems, and other painful or non-painful medical conditions. After adjusting for age and gender, the association between any long-term medical condition and developing "major depression" was highly significant. With increasing age, long-term medical conditions became less influential in predicting new onset "major depression." Female gender was another factor predictive of new onset "major depression."

Do Psychiatric Conditions Influence the Risk of Developing Pain Disorders?

Croft et al. (1995) studied a sample drawn from 4,501 adults registered in two family practices. A standardized postal questionnaire was used to define psychological distress (depression and anxiety). New episodes of low back pain were identified by monitoring primary care attendances and by the second postal survey after 12 months. The odds of developing low back pain among those with the highest scores on the psychological measure were 2.4 times greater than the odds of low back pain in those with the lowest psychological scores, taking into account age and gender.

Bidirectional Influence of Pain and Psychological Problems as Risk Factors

Breslau et al. (1994) followed a random sample of 1,007 young adults (median age 26 years) over a follow-up of 3.5 years. Those who initially had migraine but not major depression had a Relative Risk for developing "major depression" of 3.2, compared to individuals with no migraine or depression at baseline. Those who initially had a diagnosis of "major depression" but not migraine, on follow-up at 3.5 years had Relative Risk for developing migraine of 3.1, compared to individuals with no migraine or depression at baseline.

Gureje et al. (2001) reported on results of the WHO Study of Psychological Problems in General Health Care, involving 15 centers and 14 countries. Primary care patients aged 18 to 65 were sampled, and stratified according to scores on the General Health Questionnaire. Interviews were conducted most commonly in patients' homes, and diagnoses were made according to ICD-10 criteria. The disability was rated according to an occupational disability schedule, and treating physicians completed ratings of medical illness severity. Patients with significant psychological or somatic symptoms, and a random sample of the remainder, were followed up at 12 months (a total of 3,197). The only demographic factor significantly associated with non-recovery was greater age (over 40), while the presence of anxiety or depression at baseline was a marginally significant predictor of non-recovery. Non-recovery was also associated with low back pain (78.3 %), headache (72.3 %), joint pain (71.3 %), limb pain (63.9 %), chest pain (59.2 %), abdominal pain (53.5 %), and other pains (34.1 %). With regard to new onset of pain during the study period, the age of greater than 40 was the only demographic factor significantly associated, and the clinical factors predicting the onset were anxiety or depression, interviewer-rated occupational role disability, and fair or poor self-rated health. After adjusting for other variables, occupational role disability at baseline was a significant predictor of onset of persistent pain (▶ Odds Ratio of 3.63), with baseline anxiety or depression contributing further risk (Odds Ratio 2.18). After adjusting for other variables, occupational role disability at baseline was a significant predictor of psychological disorder (Odds Ratio 3.23), and the effect of baseline pain disorder contributed a further risk of psychological disorder (Odds Ratio 2.33). This demonstrates that there was a bidirectional influence of risk factors between psychological disorder and pain disorder, but with occupational role disability at baseline conferring greater variance for the risk of either chronic pain or psychiatric disorder.

Hotopf et al. (1998), in a population-based birth cohort study, at 36 and 43 years follow-up, found that psychiatric disorder at 36 years increased the odds of reporting physical symptoms 7 years later, by threefold to sevenfold, and likewise physical symptoms at 36 years predicted psychological symptoms 7 years later.

Magni et al. (1994) studied, at an 8-year interval, a sample of 2,324 noninstitutionalized adults, who were interviewed for presence or absence of musculoskeletal pain and depression, using standardized measures. Among those who were free of pain at baseline, depressive symptoms significantly predicted the development of chronic musculoskeletal pain at 8 years, with an Odds Ratio of 2.14, although depression accounted for only a small percentage of the variance. Among those who were free of depression at baseline, chronic pain predicted the onset of depression at 8 years, with an Odds Ratio of 2.85. Other predictive factors for depression included lower education, coming from rural or small urban locations, unemployment, and female gender.

What Psychological/Psychiatric Conditions Affect the Course and Prognosis of Chronic Pain?

Tunks et al. (2000) reported a review of the extensive epidemiological literature concerning the course and prognosis of chronic pain resulting from musculoskeletal injury. Several prospective studies have demonstrated that there is a higher level of persistent pain and functional disability than previously assumed. Initial return to work after injury marks a return to stable employment for less than half of injured workers. The majority of the workers who initially return to work incur periods of further work absence related to the original injury. The probability of eventual successful return to work decreases with the number of months off the job.

In long-term follow-up of low back patients who have completed pain clinic treatment, only a minority have completely favorable outcomes on all adjustment measures at 5 years, and for most the outcomes are variable, and are not consistent over time, with a large percentage never presenting a completely successful picture at any time.

Work factors associated with a higher prevalence of back pain include heavy, unpleasant, and dangerous work, higher age and lower wage, and especially previous pain sick-listings. Some of the factors associated with prolonged work absence include difficult working conditions, physical limitations experienced while performing work, and job dissatisfaction. Some factors improving occupational prognosis include the possibility of changing occupations, and the ability to modify the job or make a better fit between the worker's capacity and the job requirements. The number of times a worker attempts work reentry increases the likelihood of eventual success. In one study, those who were not working at the time of treatment onset had poorer outcomes than those who still worked.

Significant pain factors influencing prognosis include the number of painful sites, pain intensity and the number of disability days, fatigue, and pain behavior.

Significant psychological factors affecting prognosis include comorbid depression, anxiety, or dysthymia, and psychosocial maladjustment. There is no test which prospectively identifies individuals who will have an unfavorable outcome, but on average, those with higher levels of impairment tend to have the highest number of painful sites, the highest number of reported functional limitations, and the greatest pain behavior. On average, similar individuals during follow-up tend to demonstrate greater interference in occupational, social, and family roles, and higher levels of subjective distress including pain, insomnia, and fatigue.

References

Breslau, N., & Davis, G. C. (1992). Migraine, major depression and panic disorder: A prospective epidemiologic study of young adults. *Cephalalgia, 12,* 85–90.

Breslau, N., Davis, G. C., Schultz, L. R., & Peterson, E. L. (1994). Migraine and major depression: A longitudinal study. *Headache, 34,* 387–393.

Croft, P. R., Papageorgiou, A. C., Ferry, S., Thomas, E., Jayson, M. I., & Silman, A. J. (1995). Psychologic distress and low back pain. Evidence from a prospective study in the general population. *Spine, 20*, 2731–2737.

Crook, J., Tunks, E., Rideout, E., & Browne, G. (1986). Epidemiologic comparison of persistent pain sufferers in a specialty clinic and in the community. *Archives of Physical Medicine and Rehabilitation, 67*, 451–455.

Crook, J., Tunks, E., Kalaher, S., & Roberts, J. (1988). Coping with persistent pain: A comparison of persistent pain sufferers in a specialty pain clinic and in a family practice clinic. *Pain, 34*, 175–184.

Crook, J., Weir, R., & Tunks, E. (1989). An epidemiological follow-up survey of persistent pain sufferers in a group family practice and specialty pain clinic. *Pain, 36*, 49–61.

Gureje, O., Simon, G. E., & Von Korff, M. (2001). A cross-national study of the course of persistent pain in primary care. *Pain, 92*, 195–200.

Hotopf, M., Mayou, R., Wadsworth, M., & Wessely, S. (1998). Temporal relationship between physical symptoms and psychiatric disorder. Results from a national birth cohort. *The British Journal of Psychiatry, 173*, 255–261.

Livingston, G., Watkin, V., Milne, B., Manela, M. V., & Katona, C. (2000). Who becomes depressed? The Islington community study of older people. *Journal of Affective Disorders, 58*, 125–133.

Macfarlane, G. J., Thomas, E., Papageorgiou, A. C., Schollum, J., Croft, P. R., & Silman, A. J. (1996). The natural history of chronic pain in the community: A better prognosis than in the clinic? *The Journal of Rheumatology, 23*, 1617–1620.

Magni, G., Moreschi, C., Rigatti-Luchini, S., & Merskey, H. (1994). Prospective study on the relationship between depressive symptoms and chronic musculoskeletal pain. *Pain, 56*, 289–297.

Ohayon, M. M., & Schatzberg, A. F. (2003). Using chronic pain to predict depressive morbidity in the general population. *Archives of General Psychiatry, 60*, 39–47.

Patten, S. B. (2001). Long-term medical conditions and major depression in a Canadian population study at waves 1 and 2. *Journal of Affective Disorders, 63*, 35–41.

Tunks, E., Crook, J., & Crook, M. (2000). The natural history and effective treatment of chronic pain from musculoskeletal injury. In T. Sullivan (Ed.), *Injury and the new world of work* (pp. 219–245). Vancouver: UBC Press.

Weir, R., Browne, G. B., Tunks, E., Gafni, A., & Roberts, J. (1992). A profile of users of specialty pain clinic services: Predictors of use and cost estimates. *Journal of Clinical Epidemiology, 45*, 1399–1415.

Wells, K. B., Burnam, M. A., Rogers, W., Hays, R., & Camp, P. (1992). The course of depression in adult outpatients. *Archives of General Psychiatry, 49*, 788–794.

Psychiatric Aspects of the Management of Cancer Pain

Santosh K. Chaturvedi
Psychiatry, National Institute of Mental Health & Neurosciences, Bangalore, India

Synonyms

Psychiatric management of cancer pain; Psychological management of cancer pain; Psychosocial management of cancer pain

Introduction

Cancer pain is a common symptom in cancer patients. It has been often noted that psychiatric symptoms are under-recognized and undertreated in cancer patients, including those with cancer pain. Cancer pain causes distress and emotional, psychological, behavioral, and cognitive symptoms, many of which need psychiatric evaluation and management.

Definition

▶ Cancer pain has been described as a somatopsychic experience, with its intensity depending both on the extent of tissue damage and on the patient's psychological state. In patients with cancer, the pain will be exacerbated by associated physical symptoms and by ▶ psychological factors like mood disturbances and social factors like responses of partners and caregivers (Sutton et al. 2002; Keefe et al. 2003). Psychiatric management of pain is the use of psychiatric methods in the assessment and relief of cancer pain.

Characteristics

The management of cancer pain depends on a comprehensive assessment that characterizes the symptom in terms of phenomenology and pathogenesis, assesses the relation between the pain and the disease, and clarifies the impact of the pain and comorbid conditions on the patient's ▶ quality of life. This assessment requires an approach that explores the many dimensions of pain and other features of cancer (Portenoy and Lesage 1999). In this context, the psychiatric aspect of cancer pain assessment and management needs to be addressed.

Psychiatric Management of Cancer Pain

In order to address the issues related to psychiatric aspects of cancer pain, it is important to understand the different causes and varieties of pain syndromes in cancer patients. According to Twycross (1997) about three quarters of patients with advanced cancer experience pain. Most of them have multiple pains. Causes of pain fall into four broad categories: the cancer itself, related to the cancer with or without debility, related to treatment, and concurrent disorder. From a neuropathological perspective, pain is either nociceptive or neuropathic. Whatever the cause, pain is a "somatopsychic" experience (Twycross 1997). In addition, there are patients with pain from cancer, patients with a history of drug ▶ addiction, and dying patients with cancer-related pain. Pain in cancer patients is associated with many psychological antecedents and reactions that not only aggravate pain due to progressive disease or its treatment but also cause ▶ psychogenic pain in a cancer patient. Though cancer pain purely due to psychological factors may be rare, there can be a ▶ somatoform pain disorder or atypical somatoform disorder presenting as cancer pain or an ▶ abnormal illness behavior aggravating the cancer pain and interfering in its relief (Chaturvedi et al. 2006). Somatic symptom burden has been reported to be high in cancer patients with pain and/or depression. The strong association between prevalence of somatic symptoms in cancer and disability indicates the importance of recognizing and managing somatic symptoms in improving the quality of life and functional status of cancer patients (Kroenke et al. 2010a).

Psychological Factors and Psychological Symptoms in Cancer Pain Patients

Psychological factors play an important part in chronic cancer-related pain patients. The sense of hopelessness and fear of impending death may add to and exaggerate the pain, which in turn contributes to the overall suffering of the patient. The chronicity of cancer pain is associated with a series of psychological signs, i.e., disturbances in sleep, reduction in appetite, impaired concentration, irritability, and with signs of depressive disorder. Higher levels of pain intensity in cancer patients have been found to be associated with mood disturbances like ▶ depression, frustration, anger, exhaustion, maladaptive behavioral coping responses, beliefs that pain is related to cancer progression, greater life stress, feelings of helplessness, and hopelessness (Speigel et al. 1994; Poulas et al. 2001; Sela et al. 2002; Okano et al. 2001). In a recent study, depression and pain together were reported in 44 % of cancer patients (Kroenke et al. 2010b). Depression and pain are prevalent and disabling, commonly co-occur, and have additive adverse effects, emphasizing the need for enhanced detection and management of this disabling symptom dyad (Kroenke et al. 2010b).

A recent critical review of the literature by Zaza and Baine (2002) examined the evidence for an association between chronic cancer pain and psychological distress, social support, and coping and found that, based on several criteria, the reviewed studies confirmed that the evidence was strong for psychological distress, moderate for social support, and inconclusive for coping. This review, thus, emphasized that chronic pain assessment should include routine screening for psychological distress.

Most patients with cancer who experience chronic pain also develop other physical and psychological symptoms. Studies have shown that pain, fatigue, and psychological distress are the most common symptoms in patients with cancer (Portenoy et al. 1994). A broad psychiatric assessment of symptoms is an essential part of the management of cancer patients. Fear of pain is one of the most common and dreaded fears

P

among cancer patients and is an important factor in causing depression in cancer patients. Patients with pain-related emotional states (including anger and frustration) and negative moods reported higher levels of cancer pain (Sela et al. 2002; Greenwood et al. 2003).

Cancer pain has been reported to be associated with delirium, dementia, and adjustment disorders, besides depressive disorders (Ogawa et al. 2010).

Addiction and Fear of Addiction

Many health-care workers are concerned that there is a significant risk of addiction when using opioids for cancer pain. Most surveys have shown that patients or physicians, except when there is a worsening of pain, usually do not escalate pain medication. It has also been shown that tolerance to pain relief does not occur as frequently as it occurs for the other side effects of opiates. Secondly, tolerance or ▶ physical dependence alone should not be criteria for dependence or "addiction." Addiction should only be considered if there is a psychological and behavioral syndrome characterized by craving, compulsive use, aberrant drug seeking behavior (such as reuse or falsification of prescriptions, use of street drugs), use of the drug for other effects, unsupervised escalation of drugs and contact with several doctors for the same, and relapses after withdrawal. Cancer patients should also be evaluated for pain relief, as patients with inadequate pain relief may also show "undue preoccupation" about the drug. "Pseudo addiction" is a term coined for such patients who seek the drug in the presence of inadequate pain relief and who show no such preoccupation once the dose is escalated and comfort reached (Weissman and Haddox 1989).

Issues Related to Psychiatric Management

The psychiatric management of cancer pain would consist of elicitation of features of depression, ▶ anxiety, somatic focusing, hypochondriasis, and abnormal illness behavior and treating them by appropriate methods.

Treatment of Psychiatric Disorders in Cancer Patients with Pain

The cancer patient with pain has an enhanced risk of developing the common psychiatric disorders in cancer. In terms of frequency, depression, anxiety, or mixed symptoms of anxiety and depression are the most common problems. These need to be evaluated for their severity and persistence and treated appropriately with a combination of psychopharmacological and psychosocial methods.

Use of Psychotropics for Cancer Pain Management

The response of cancer pain to tricyclic antidepressants (TCAs) is usually good (Panerai et al. 1991). Relatively low dosages of tricyclic antidepressants given at night are often effective and benefit in reducing pain, normal sleep is restored, fatigue is diminished, and the patient is able to engage more energetically in daily activities. The exact mechanism of antidepressant effects in cancer pain is not fully understood. The antinociceptive action of these agents seems to be independent of any beneficial effect on depression or mood. Although the co-administration of TCAs may increase plasma morphine concentrations, any potentiation of morphine analgesia is thought not to be due to an increased bioavailability of the opiate, but to an intrinsic analgesic effect of antidepressants. On this basis, the use of antidepressants in combination with opioids for the treatment of cancer pain is suitable when a component of deafferentation is present or when there is concomitant depressive illness (Panerai et al. 1991). The interaction between the cognitive component (perception of nociceptive stimuli) and the affective component of pain is reflected in the overlap between the actions of antinociceptive agents (analgesics) and mood altering drugs (▶ Psychotropics). Anxiety, depression, fear, and sleeplessness may all respond to psychotropic drugs, and this may result in reduction in pain or a greater ability to cope with it. Selective serotonin reuptake inhibitors (SSRIs) may also be useful in the management of cancer pain, without significant adverse drug reactions. It is thought that these agents might be more analgesic than other antidepressants, because serotoninergic neurons are involved in the response to noxious stimulation through pathways in the midbrain. However, clinical trials have not convincingly confirmed their superiority in pain relief. Phenothiazines and butyrophenones in low doses also act as adjuncts in cancer pain relief. Psychotropics are also known to potentiate narcotic analgesia.

Psychological Treatments, Communication, and Counseling

A number of psychological treatment approaches are available for cancer pain patients depending upon the presence of contributing factors. Family counseling, marital counseling, individual psychotherapy, cognitive therapy, and supportive therapy may be advised wherever necessary. ▶ Psychoeducational interventions on pain in adults with cancer were found to be beneficial in a meta-analysis (Devine 2003). These are not a substitute for analgesics, but they may serve as adjuvant therapy. The intervention must be acceptable to the patients and not too burdensome for cancer patients in pain to use (Devine 2003).

Behavioral Intervention for Cancer Pain

▶ Behavior therapies are also found to be useful. Relaxation exercises, yoga, meditation, and biofeedback may help in pain relief in cancer patients. Pharmacological and behavioral therapies may be combined for some patients. Multidisciplinary pain clinics can provide a vast range of treatments for cancer pain patients. Reasonably strong evidence exists documenting beneficial effects of relaxation-based cognitive-behavioral interventions, education about analgesic usage, and supportive counseling (Devine 2003).

Maintaining Quality of Life

Suffering due to cancer pain can be an important reason for poor quality of life. Psychological symptoms and psychiatric morbidity can aggravate the suffering from cancer pain and lead to poor quality of life. Pain relief is therefore very important for enhancing the quality of life. Multidisciplinary approaches, with due attention to psychological and psychiatric factors, prove to be most effective. Where most approaches have failed, it is important to encourage and motivate the patient in the process of teaching him to "live with pain."

Summary

This entry deals with the psychiatric aspects of management of cancer pain. Many challenging psychiatric symptoms, like depression, dementia, delirium, and distress, coexist with cancer pain. It is important to appreciate this association, to effectively manage both psychiatric symptoms and cancer pain. The difficulties in management of cancer pain, due to fear of addiction, are discussed. To improve the quality of life of cancer patients, it is essential to manage the psychiatric disorders existing with cancer pain.

Acknowledgement I acknowledge the inspiration by the late Professor Peter Maguire for guiding me through my involvement in cancer and cancer pain.

P

References

Chaturvedi, S. K., Maguire, P., & Somashekar, B. S. (2006). Somatization in cancer. *International Review of Psychiatry, 18*, 49–54.

Devine, E. C. (2003). Meta-analysis of the effect of psycho educational interventions on pain in adults with cancer. *Oncology Nursing Forum, 30*, 75–89.

Greenwood, K. A., Thurston, R., Rumble, M., Waters, S. J., & Keefe, F. J. (2003). Anger and persistent pain: Current status and future directions. *Pain, 103*, 1–5.

Keefe, F. J., Ahles, T. A., Porter, L. S., Sutton, L. M., McBride, C. M., Pope, M. S., McKinstry, E. T., Furstenberg, C. P., Dalton, J., & Baucom, D. H. (2003). The self-efficacy of family caregivers for helping cancer patients manage pain at end of life. *Pain, 103*, 157–162.

Kroenke, K., Zhong, X., Theobald, D., Wu, J., Tu, W., & Carpenter, J. S. (2010a). Somatic symptoms in cancer patients with pain and/or depression prevalence, disability, and health care use. *Archives of Internal Medicine, 170*, 1686–1694.

Kroenke, K., Theobald, D., Wu, J., Loza, J. K., Carpenter, J. S., & Tu, W. (2010b). The association of depression and pain with health related quality of life, disability, and health care use in cancer patients. *Journal of Pain and Symptom Management, 40*, 327–341.

Ogawa, A., Shimuzu, K., Akuzuki, N., & Uchitomi, Y. (2010). Involvement of a psychiatric consultation service in a palliative care team at a Japanese cancer center hospital. *Japanese Journal of Clinical Oncology, 40*, 1139–1146.

Okano, Y., Okamura, H., Watanabe, T., Narabayashi, M., Katsumata, N., Ando, M., Adachi, I., Kazuma, K., Akechi, T., & Uchitomi, Y. (2001). Mental adjustment to first recurrence and correlated factors in patients with breast cancer. *Breast Cancer Research and Treatment, 67*, 255–262.

Panerai, A. E., Bianchi, M., Sacerdote, P., Ripamonti, C., Ventafridda, V., & De Conno, F. (1991). Antidepressants in cancer pain. *Journal of Palliative Care, 7*, 42–44.

Portenoy, R. K., & Lesage, P. (1999). Management of cancer pain. *Lancet, 353*, 1695–1700.

Portenoy, R. K., Thaler, H. T., Kornblith, A. B., Lepore, J. M., Friedlander-Klar, H., Coyle, N., Smart-Curley, T., Kemeny, N., Norton, L., Hoskins, W., et al. (1994). Symptom prevalence, characteristics and distress in cancer population. *Quality of Life Research, 3*, 183–189.

Poulas, A. R., Gertz, M. A., Pankratz, V. S., & Post-White, J. (2001). Pain, mood disturbance, and quality of life in patients with multiple myeloma. *Oncology Nursing Forum, 28*, 1163–1171.

Sela, R., Bruera, L., Conner-Spady, B., Cumming, C., & Wallen, C. (2002). Sensory and affective dimensions of advanced cancer pain. *Psycho-Oncology, 11*, 23–34.

Speigel, D., Sands, S., & Koopman, D. (1994). Pain and depression in patients with cancer. *Cancer, 74*, 2570–2578.

Sutton, L. M., Porter, L. S., & Keefe, F. J. (2002). Cancer pain at the end of life: A biopsychosocial perspective. *Pain, 99*, 5–10.

Twycross, R. G. (1997). *Introducing palliative care.* (2nd edn). Radcliffe Medical Press Oxford.

Weissman, D. E., & Haddox, J. D. (1989). Opioid pseudo-addiction – an iatrogenic syndrome. *Pain, 36*, 363–366.

Zaza, C., & Baine, N. (2002). Cancer pain and psychosocial factors: Critical review of the literature. *Journal of Pain and Symptom Management, 24*, 526–542.

Psychiatric Aspects of Visceral Pain

Ann Wigham[1] and Stephen P. Tyrer[2]
[1]McCoull Clinic, Prudhoe Hospital, Northumberland, UK
[2]Royal Victoria Infirmary, University of Newcastle upon Tyne, Newcastle-on-Tyne, UK

Synonyms

Functional aspects of visceral pain; Nonorganic aspects of visceral pain

Definition

The internal organs of the body, closely related to or contained within the pleural, pericardial, or peritoneal cavities, are termed the viscera. Visceral pain is pain that is caused by activation of pain receptors from infiltration, compression, extension, or stretching of the thoracic, abdominal, or pelvic viscera. Psychiatric, in the context of this chapter, refers to the effects of the mind and illnesses affecting the mind, on the perception of pain arising from the viscera. The word functional refers to symptoms or disorders which, after appropriate medical assessment, cannot be explained by a defined medical disease.

Characteristics

The general characteristics of pain due to visceral pathology are that it is:
1. Diffuse and poorly localized with referral to somatic structures
2. Referred to the somatic structures in the body wall
3. Often accompanied by ▶ autonomic and motor reflexes
4. Not necessarily associated with injury (Cervero and Laird 1999)

Pain is not evoked from all viscera, as many solid viscera are not innervated by "sensory" receptors. However, pains from the lining membranes of hollow organs such as the colon are very sensitive. The combined inputs from somatic and visceral afferent fibers from these structures converge onto spinal cord neurons enhancing their excitability. Visceral ► hyperalgesia results through the process of what is known as ► central sensitization (Roza et al. 1998). Although visceral pain is referred to ► cutaneous areas, subjects are able to distinguish pain arising from a cutaneous or visceral origin, with visceral pain being perceived as more unpleasant (Strigo et al. 2002).

Stimuli that induce visceral pain are distension, ischemia, and inflammation. However, unlike somatic pain, the severity of the pain experienced does not always correlate with the seriousness of the condition leading to the pain.

Neuroanatomy of Visceral Pain

It is only recently, with the development of functional neuroimaging methods, that the anatomical correlates of visceral pain have been able to be investigated in human subjects. There are two distinct classes of nociceptive sensory receptors that innervate internal organs. The first class respond to mechanical stimuli that are perceived as unpleasant, the so-called high-threshold receptors. The second class, low-threshold receptors, respond to a wide range of stimuli from innocuous to noxious (Sengupta and Gebhart 1994).

The cell bodies of visceral afferent neurons are situated in the dorsal root ganglia of the thoracic and lumbar nerves, the afferent fibers from which are conveyed in the sympathetic trunk and then to the viscera by the cardiac, pulmonary, and splanchnic nerves. The majority of these fibers terminate in the ► substantia gelatinosa. From here, the connections ascend to the inferior solitary nucleus, the midbrain, the medial forebrain bundle, and the hypothalamus, via the limbic system to the thalamus, and then to

the cerebral cortex to evaluate and localize the pain. In addition, the ► periaqueductal gray matter (PAG) in the midbrain transmits descending opiate-mediated pain inhibitory messages back to the spinal cord by way of the brain stem raphe nuclei as well as conveying information rostrally to the amygdala.

Experimentally applied visceral pain to normal subjects has been shown to activate the anterior cingulate cortex (ACC) (Strigo et al. 2003), part of the limbic lobe that comprises a high density of opiate receptors. The ACC has extensive connections with PAG, a region that is involved with diminishment of pain. There is some evidence that this descending antinociceptive system is defective in visceral pain (Naliboff et al. 2001). The known relationship between intense emotional feelings and visceral sensations, such as nausea and feelings of chest discomfort, can be explained on the basis of these pathways.

Association with Functional Somatic Symptoms

Reports of patients with abdominal pain in the absence of organic findings were described 120 years ago. The conditions from which these individuals suffered were described as the visceral neuroses. The etiology of these illnesses included (Ryle 1939):

- Peripheral or central irritation from a previous injury
- "Inborn sensitivity" traits
- Apprehension that the symptoms are due to organic disease
- Hereditary predisposition
- Lack of manifestation of organic disease under congenial conditions but which become apparent under stress

The beliefs of these early authors have been borne out by subsequent work. Inflammation of the esophagus reduces the threshold of viscerosomatic spinal neurons to noxious stimuli in cats (Garrison et al. 1992) and in healthy volunteers (Sarkar et al. 2000). In this second study, volunteers exposed to acid infusion into the lower esophagus subsequently had a lower pain threshold in the upper esophagus after signs of inflammation had resolved. Furthermore, patients with noncardiac chest pain in the same investigation had a lower resting esophageal pain threshold than healthy controls, and this threshold fell further and for longer than the controls following acid infusion. This study illustrates both hypersensitivity of these visceral neurons and secondary ▶ allodynia in the distribution of the same nerves, independently of the site of the stimulus, the phenomenon of ▶ central sensitization (Roza et al. 1998).

The genetic contribution to these disorders has only been investigated in detail in IBS. Studies have shown that this syndrome is twice as common in monozygotic twins as in dizygotic pairs (Levy et al. 2001). These authors posit evidence showing that these disorders are related to both inborn factors and social learning. It has been proposed that the symptoms described in functional somatic disorders are not greatly different from each other (Wessely et al. 1999) and that the manifestation of one disorder rather than another results from excessive attention to one symptom dependent on familial experience (Levy et al. 2001).

The ▶ hypochondriacal fears of patients who have functional somatic disorders may be related to the distress that these disorders cause (Wessely et al. 1999).

Relationship of Psychiatric Issues in Disorders Leading to Visceral Pain

Depression and anxiety are the illnesses most frequently found in patients with chronic pain attending pain clinics (Tyrer et al. 1989; Fishbain et al. 1998). Anxiety has been found to decrease pain threshold and tolerance (Cornwall and Donderi 1988), and depression is also associated with poorer treatment outcomes. ▶ Panic attacks are particularly associated with subjective reports of symptoms referable to the viscera. The symptoms occurring during panic attacks are similar to those occurring in animals during fearful response to conditioned stimuli. ▶ Functional imaging has demonstrated activation of the

same anatomical structures in the limbic system, in particular the amygdala, during panic attacks as those occurring during visceral pain (Gorman et al. 2000). Autonomic activation is found in both conditions.

Disorders that give rise to visceral pain in which there is lack of evidence of organic findings are now considered under the rubric of somatoform disorders. These cover a wide variety of disorders in which there is repeated presentation of physical symptoms together with persistent requests for medical investigations, in spite of negative findings and reassurance by doctors that the symptoms have no physical basis. The condition that is most often found in patients with visceral pain is somatoform autonomic dysfunction (F45.3 in the ICD-10 classification). Such autonomic symptoms are unsurprising in view of the fact that autonomic nerves innervate the viscera. Details of the most common conditions that fall under this category are given below.

Noncardiac Chest Pain

Pain in the upper chest region is common. In a comprehensive population-based birth cohort study, chest pain was reported in 17 % of respondents at the age of 36, with a prevalence of exertional chest pain of 1 % (Hotopf et al. 1999). There was a significant relationship between psychiatric disorder and the presence of chest pain, and this was particularly high for exertional chest pain. Furthermore, with increasing severity of the psychiatric disorder, the odds of reporting chest pain increased. The exertional chest pain reported was not due to cardiac disease; only one subject who reported exertional chest pain had evidence of cardiac illness. As in most studies involving medical conditions, the most prominent psychiatric diagnoses were anxiety and depression. This study suggests that if a child has relatives with physical illnesses or suffers from fatigue, this primes the person to attend to normal physiological sensations in a different way.

The relationship between chest pain, psychiatric morbidity, and the presence of coronary disease in older patients shows a lesser trend for an increase in psychiatric morbidity in those with nonorganic findings. Evidence of psychiatric illness is higher in this group. A large study did not reveal any difference in psychiatric morbidity in those with positive angiographic findings than in those without, but 25 % of subjects showed evidence of a psychiatric disorder (Valkamo et al. 2001).

Irritable Bowel Syndrome
The symptoms of irritable bowel syndrome consist of pain or discomfort for 12 weeks of the previous 12 months associated with relief with defecation and abnormal frequency of defecation. The prevalence of this disorder varies from 3 % to 22 % according to the criteria used (Talley and Spiller 2002). Children with persistent abdominal pain are more likely to suffer from psychiatric disorders in adulthood, but there is debate about whether they are especially prone to physical symptoms once the psychiatric disorder is controlled (Hotopf et al. 1998; Blanchard and Scharff 2002). As with most of the functional somatic disorders, there is female predominance. It is widely thought that psychological distress is a major factor in the use of health care by adults with irritable bowel syndrome and patients attending medical clinics have high levels of generalized anxiety, depression, and hypochondriasis. However, studies have found that those with similar symptoms who do not seek health care for their symptoms are indistinguishable from controls without the disorder (Drossman et al. 1988). There is a relationship between severity of pain as opposed to other symptoms and referral to health care (Koloski et al. 2001). Although acute stress probably aggravates the disorder, more important chronic sustained stresses such as separation and bereavement are likely to be important in contributing to persistence of symptoms.

Treatment options are centered on the principle that the patient is largely responsible for care. Tricyclic antidepressants, ▶ cognitive behavior therapy, and hypnotherapy have all been reported to be superior to standard care in patients.

Uterine and Pelvic Pain
Nociceptive stimulation of the uterine cervix by balloon dilation leads to pain in the hypogastric and low back regions, in the same areas in which pain is experienced during labor and menstruation (Bajaj et al. 2002). The gynecological symptom that is associated with the highest frequency of psychiatric disorder is abdominal or pelvic pain (Bixo et al. 2001). Patients with these symptoms had a more than threefold increased risk of having an anxiety disorder. Although it has been proposed that women with chronic pelvic pain have a higher incidence of sexual abuse in the past and/or sexual or relationship difficulties at the time of presentation, a recent comprehensive study using a cohort design in over a thousand patients showed that physically and sexually abused individuals were not at risk for increased pain symptoms (Raphael et al. 2001). Previous positive findings in this area are likely to have been related to selection bias.

Conclusion
The prevalence of psychiatric disorders in patients with visceral pain is greater than with other forms of somatic or neuropathic pains. The involvement of the anterior cingulate cortex in the limbic system in visceral pain states helps to explain this. The increased distress that these symptoms cause may also reflect the hyperalgesia resulting from central sensitization. The autonomic accompaniments of visceral pain, which occur in anxiety-provoking situations, further reinforce the feeling of threat that arises. Cognitive behavioral therapy is an advantage in visceral pain conditions, but less attention has been paid to this form of treatment than in other painful conditions. Explanation and education about visceral pain mechanisms should be more widely employed.

References
Bajaj, P., Drewes, A. M., Gregersen, H., et al. (2002). Controlled dilatation of the uterine cervix – an experimental visceral pain model. *Pain, 99*, 433–442.

Bixo, M., Sundstrom-Poromaa, I., Bjorn, I., & Astrom, M. (2001). Patients with psychiatric disorders in gynecological practice. *American Journal of Obstetrics and Gynecology, 185*, 396–402.

Blanchard, E., & Scharff, L. (2002). Psychosocial aspects of assessment and treatment of irritable bowel syndrome in adults and recurrent abdominal pain in children. *Journal of Consulting and Clinical Psychology, 70*, 725–738.

Cervero, F., & Laird, J. M. (1999). Visceral pain. *Lancet, 353*, 2145–2148.

Cornwall, A., & Donderi, D. C. (1988). The effect of experimental induced anxiety on the experience of pressure pain. *Pain, 35*, 105–113.

Drossman, D., McKee, D. C., Sandler, R. S., et al. (1988). Psychosocial factors in the irritable bowel syndrome. A multivariate study of patients and nonpatients with irritable bowel syndrome. *Gastroenterology, 95*, 701–708.

Fishbain, D. A., Cutler, R., & Rosomoff, H. (1998). Comorbid psychiatric disorders in chronic pain patients. *Pain Clinical, 11*, 79–87.

Garrison, D., Chandler, M. J., & Foreman, R. D. (1992). Viscerosomatic convergence onto feline spinal neurons from esophagus, heart and somatic fields: Effects of inflammation. *Pain, 49*, 373–382.

Gorman, J. M., Kent, J. M., Sullivan, G. M., et al. (2000). Neuroanatomical hypothesis of panic disorder, revised. *The American Journal of Psychiatry, 157*, 493–505.

Hotopf, M., Carr, S., Mayou, R., et al. (1998). Why do children have chronic abdominal pain, and what happens to them when they grow up? Population based cohort study. *BMJ, 316*, 1196–1200.

Hotopf, M., Mayou, R., & Wadsworth, M. (1999). Psychosocial and developmental antecedents of chest pain in young adults. *Psychosomatic Medicine, 61*, 861–867.

Koloski, N., Talley, N. J., Boyce, P. M., et al. (2001). Predictors of health care seeking for irritable bowel syndrome and nonulcer dyspepsia: A critical review of the literature on symptom and psychosocial factors. *American Journal of Gastroenterology, 96*, 1340–1349.

Levy, R., Jones, K. R., Whitehead, W. E., et al. (2001). Irritable bowel syndrome in twins: Heredity and social learning both contribute to etiology. *Gastroenterology, 121*, 799–804.

Naliboff, B. D., Derbyshire, S. W., Munakata, J., et al. (2001). Cerebral activation in patients with irritable bowel syndrome and control subjects during rectosigmoid stimulation. *Psychosomatic Medicine, 63*, 365–375.

Raphael, K., Widom, C. S., & Lange, G. (2001). Childhood victimization and pain in adulthood: A prospective investigation. *Pain, 92*, 283–293.

Roza, C., Laird, J. M. A., & Cervero, F. (1998). Spinal mechanisms underlying persistent pain and referred hyperalgesia in rats with an experimental ureteric stone. *Journal of Neurophysiology, 79*, 1603–1612.

Ryle, J. A. (1939). Visceral neuroses. *Lancet, II*, 297–301.

Sarkar, S., Aziz, Q., Woolf, C. J., et al. (2000). Contribution of central sensitisation to the development of non-cardiac chest pain. *Lancet, 356*, 1154–1159.

Sengupta, J. N., & Gebhart, G. F. (1994). Characterization of mechanosensitive pelvic nerve afferent fibres innervating the colon of the rat. *Journal of Neurophysiology, 71*, 2046–2060.

Strigo, I. A., Bushnell, M. C., Boivin, M., et al. (2002). Psychophysical analysis of visceral and cutaneous pain in human subjects. *Pain, 97*, 235–246.

Strigo, I. A., Duncan, G., Boivin, M., et al. (2003). Differentiation of visceral and cutaneous pain in the human brain. *Journal of Neurophysiology, 89*, 3294–3303.

Talley, N., & Spiller, R. (2002). Irritable bowel syndrome: A little understood organic bowel disease? *Lancet, 360*, 555–564.

Tyrer, S. P., Capon, N., Peterson, D. N., Charlton, J. E., & Thompson, J. W. (1989). The detection of psychiatric illness and psychological handicaps in a British pain clinic population. *Pain, 36*, 63–74.

Valkamo, M., Hintikka, J., Niskanen, L., et al. (2001). Psychiatric morbidity and the presence and absence of angiographic coronary disease in patients with chest pain. *Acta Psychiatrica Scandinavica, 104*, 391–396.

Wessely, S., Nimnuan, C., & Sharpe, M. (1999). Functional somatic symptoms: One or many? *Lancet, 354*, 936–939.

Psychiatric Disorder

Definition

Psychiatric disorder is a term usually reserved for conditions that have met defined diagnostic criteria in a system of psychiatric classification.

Cross-References

▶ Pain as a Cause of Psychiatric Illness

Psychiatric Evaluation

▶ Disability Assessment, Psychological/Psychiatric Evaluation

Psychiatric Illness

▶ Pain as a Cause of Psychiatric Illness

Psychiatric Management of Cancer Pain

▶ Psychiatric Aspects of the Management of Cancer Pain

Psychiatry and Pain

Michael R. Bond
University of Glasgow, Glasgow, UK

Definition

The use of psychiatric diagnoses is one aspect of the multimodal system used for the assessment of individuals in pain, and especially those in chronic pain. This system represents a movement away from a single, linear, or "cause and effect" model of pain, to the use of a biopsychosocial model which has been one of the most important developments in the field of pain research and management in the past 35 years. Psychiatric diagnoses provide a structured framework for emotional and physical symptoms, taking into account social factors and they serve as a basis for the treatment and prognosis of certain painful disorders.

The use of psychiatric diagnoses is in accordance with the definition of pain developed in 1979 by the Taxonomy Committee of the International Association for the Study of Pain (IASP). It states that pain is, "an unpleasant, sensory, and emotional experience associated with actual or potential tissue damage, or described in terms of such damage." Pain is a subjective experience and the use of the term, and behavior associated with it, is learned in

childhood but shaped throughout life by additional experiences and the social contexts in which it occurs.

Several hypotheses have been proposed to explain why certain emotional changes take place as an aspect of pain experiences and all involve psychological mechanisms, which have been the subject of extensive research in the past three decades, and are discussed elsewhere in this volume. (see Flor, 'Psychophysiologic Model of Pain' and Aamir and Large 'Pain' as a cause of Psychiatric Illness', in this volume). The reader might also refer to the publication by Price and Bushnell (2004) who sought to explain the link between noxious stimulation and the short- and long-term changes in emotion and behavior produced via intermediary cognitive processes but these mechanisms do not explain, however, why different psychiatric disorders develop in different individuals.

People experience mood changes that occur in all painful disorders, but their intensity and duration is often relatively minor and falls short of a formal psychiatric condition. Suffering undoubtedly occurs, however, and in these circumstances its relief is a necessary part of management. Acute pain gives rise to feelings of both anxiety and depression, but primarily to anxiety, whereas those in chronic pain develop more complex changes which include feelings of depression and anxiety which may proceed to become a psychiatric disorder. For example, Aamir and Large review publications dealing with the link between chronic pain and psychiatric diagnoses and conclude that the best estimate of any psychiatric disorder between the ages of 18 and 54 years diagnosed by DSM III-R is 21 %. The rate for a major depressive episode is 6.5 % which is slightly higher than in the general population. Aamir and Large conclude with the quote "Hence there is a definite association between a psychiatric disorder and pain." To enable sense to be made of such changes, clinicians need a frame of reference, a systematic diagnostic index, and the one most often used in pain research and management is the Diagnostic and Statistical Manual of the American Psychiatric Association in its fourth and revised version (American Psychiatric Association 2002). DSM IV-TR diagnoses have five axes. The first axis is for the primary diagnosis, for example, a major depressive disorder, and the second axis for any personality disorder detected. The third axis allows reporting of current general medical conditions that are relevant, the fourth enables a description of psychosocial and environmental problems, and the fifth an overall global assessment of function. The first three axes are particularly important, but quite often consideration of personality disorder is neglected, even though a high rate of such a disorder exists in chronic pain patients. Aamir and Large note the incidence of a personality disorder of one form or another is as high as 51 % of patients with chronic pain. Although this has relevance to the development and management of the primary condition none of the personality disorders predict specifically the development of chronicity of pain. For further information, and about mood and anxiety disorders, the reader is referred to Gallagher and Verma (2004).

Considerable effort has been put into identifying the extent to which psychiatric disorders occur as comorbidity in populations of pain patients. There is a wide variation in the levels of incidence and prevalence described in the literature and this is reviewed in this volume by Tunks, Crook, and Weir, who, in addition, review the relationship between certain aspects of mental disorder and pain. Further consideration of epidemiology is covered in the contribution from Aamir and Large. Despite the variations that occur, as stated earlier, it is clear that there is a definite increase in the number of individuals who develop psychiatric disorders when physically ill compared with a healthy population, and that the number increases further when chronic pain is present. It is known also that the extent to which psychiatric symptoms occur increases with age when depression, if present, will be central to the maintenance of the pain-frailty relationship, in fact, more so than a pain-related medical diagnosis. It should be noted that a significant proportion of chronic pain patients have a higher incidence of psychiatric disorder, for example, depression, anxiety, or a personality disorder before the onset of pain than is the case in the general population. The effect of pain itself, fear of pain, and the nature of the condition are important in the generation of psychiatric disorders and this is mentioned in entries on cancer pain by Chaturvedi, on visceral pain by Tyrer and Wigham, and on facial pain by Feinmann.

Mood disorders during painful illness, or post injury, are associated with increased levels of pain and lowered tolerance for it. Mood changes make rehabilitation more difficult and prolong recovery time leading to extended pain-related disability (Hall et al. 2011). For these reasons also their detection and treatment is important. Psychiatric disorders that occur include major depressive disorder, general anxiety disorder, posttraumatic stress disorder, personality disorders, and adjustment disorder – a condition that develops when a patient does not adapt emotionally and socially to a chronic physical disorder including those that are painful.

It is important to be aware also that depressive illnesses and certain forms of anxiety disorder have pain as a symptom, for example, pain is present in approximately half of all patients who develop depressive disorders.

A wide range of factors need to be taken into account in an analysis of pain disorders and their management. This is illustrated in the contribution to this volume by Hapidou who deals with psychological aspects of pain in women. She indicates that the differences in pain experiences, including mood changes, behavior, and responses to treatment, do exist between women and men.

The social consequences of chronic pain have emotional effects and this is a further relatively neglected issue. Writing about social dislocation and chronic pain, Roy (2001) drew attention to the fact that loss of ability to work, loss of income, loss of status within and outside the family, and forced role changes, are all factors which tend to promote or exacerbate mood change and other psychiatric disorders. Therefore, social aspects of the biopsychosocial model are significant and they have implications for the management of pain problems and, in particular, the processes of rehabilitation and return to work.

It is readily appreciated that painful conditions result in changes of mood. Equally, we are aware that individuals with anxieties in daily life may develop pain, for example, headache or backache, although there is no known underlying tissue injury. This is a normal psychological process known as somatization and it is almost always transient and treatable. In a more powerful form it is the basis for a group of psychiatric disorders known as the somatoform disorders, some of which include pain as a symptom. In these conditions anxiety is focused somatically by an unconscious process and, therefore, there will be evidence of background emotional problems, or distress, not linked by the patient to his or her physical complaints and disability, both of which will be out of proportion to any physical pathology if present. It is a matter for a doctor, or other health professional, to decide the extent to which a discrepancy exists, whether it can be dealt with simply by advice and reassurance, or whether it constitutes one of the somatoform group disorders and needs more specialist attention. The two somatoform disorders, in which pain features prominently and most often, are hypochondriasis and pain disorder. A third condition, somatization disorder, is less common.

The concept of the sick role is fundamental to the understanding of somatoform disorders. The behavior shown in the sick role is known as illness behavior and it is important when assessing a person with a somatization disorder. At times, it is significantly out of keeping with the level of physical abnormality demonstrated and is associated with a level of disability far greater than would be expected. These issues are addressed briefly in this volume by Pilowsky who deals with hypochondriasis. In this condition, the key features are an increase in somatic focusing and the sufferer's fears or beliefs that they have a serious disease as a cause for pain, for example, cancer or heart disease. In some,

hypochondriasis is a personality trait and the individual is well aware that their fears may be groundless but, nevertheless, seek reassurance. At other times the beliefs are held with greater conviction and may even reach delusional proportions. Care must be taken with patients who have marked hypochondriasis because, on occasions, this is a manifestation of a major depressive disorder.

An unconscious link between the concerns or anxieties of everyday life, and the presence of pain, is characteristic of the condition of pain disorder. The diagnosis is made only when comorbid anxiety or depression is absent. Most often the condition comes to light after exhaustive clinical investigations have failed to reveal a physical cause for pain and after the patient has been labeled "psychiatric." The difficulties in making the diagnosis, other than by exclusion, are described by Mark Sullivan (see page x). The fact that the definition of this condition has changed three times in the DSM series since 1980 underlies his comments. He makes the important point that the route the patient is taken down by doctors and the labeling that takes place during that process, frustrates and angers the sufferer who usually rejects attempts to involve the use of psychological terms which to them implies that their pain is not "genuine." This raises difficulties in management once the diagnosis has been made, although the development of multimodal clinics and management programs has eased the problem.

Fishbain's entry on multimodal rehabilitation treatment and its psychiatric aspects outlines clearly why such facilities are needed for the treatment of complex pain problems and the role within them of psychiatric methods. The reasons for an integrated approach are evident and such programs have been found to be superior to the use of single therapies. Fishbain comments, however, that the reason why such packages work is not fully understood. From the patient's perspective the multimodal approach is often much more acceptable than single therapy, especially if that is psychologically based because, for obvious reasons, blending physical, psychological, and social techniques, allows the therapy team to present to the patient what seems a more reasonable and acceptable set of explanations and reasons for therapies needed.

Nonpsychiatrists play an important part in multimodal treatment programs. Psychiatrists share techniques based on cognitive and behavioral concepts with psychologists, nurse therapists, and others and, in addition, have the vital role of directing the use of psychotropic drugs and, in particular, those of the antidepressant group for the treatment of psychiatric disorders and pain. Originally developed to combat depressive disorders, the antidepressants of the tricyclic group were noted, as long ago as the 1960s, to have an independent analgesic effect at doses lower than those used in the treatment of depression. The value of tricyclics as analgesics and, in particular, nortriptyline, amitriptyline, and imipramine, has been established for some years. There is little evidence of potency in the treatment of acute pain, but some suggest that amitriptyline may be of value in rheumatic disorders. The greatest success is gained in the treatment of pains due to nerve injury, or neuropathic pain, e.g., PHN and diabetic neuropathy. Evidence to date indicates that "mixed" noradrenergic and serotonergic drugs (amitriptyline and imipramine) are more effective than single noradrenergic drugs such as desipramine; nortriptyline has advantages in use because it produces fewer adverse effects than amitriptyline. The serotonergic drugs, paroxitine, clomipramine, and others, are much less effective. The tricyclic group appears equal in effect to the anticonvulsant drug, Gabapentin, widely used currently in the treatment of neuropathic pain. Decisions about the selection of individual drugs are based upon their effectiveness and, also, taking into consideration adverse effects they may cause.

The pain specialist, most often an anesthetist, but perhaps a psychologist or occasionally a psychiatrist, may be asked to act as an expert witness in cases of post-injury litigation. The complexities of such problems are described in the entry by Mendelson and can be demanding

of witnesses who require extensive knowledge of this field and the ability to express opinions clearly. Many claimants have a history of a painful injury with persistent post-injury pain and disability with, in a significant number, marked changes in emotions. Key questions put to medical experts, first in reports and perhaps later in Court by lawyers for the insurer, will at some point focus upon the extent to which the accident has caused the level of pain and disability of which the litigant complains, and any emotional and social difficulties that have developed. At times it is implied that the claimant's evidence is a fabrication and a means of gaining financially. In these circumstances the term malingering may be used by lawyers but, as Mendelson points out, this is not a medical or psychiatric diagnosis but a word used to describe a fraud within a medical context. As such, it is for lawyers to determine whether or not it is present. Lawyers tend to seek definitive psychiatric diagnoses when emotional factors form a major part of the litigant's problem. Some, but not all, do have specific disorders, for example, a major depressive disorder, or posttraumatic stress disorder, due to the effects of the accident or injury, but the terms accident neurosis, or compensation neurosis, should not be used because they do not have clinical validity. Finally, although there is no clear evidence that the severity of pain among litigants is any greater than in non-litigants, but the often stressful process of seeking compensation may well contribute to the chronicity of the condition, albeit it an unconscious rather than a conscious level. In some cases, the behavior shown appears to be exaggerated but it may reflect the individuals heightened emotion resulting from the stress of the examination. Once established over a period of several years, symptoms including pain often continue after litigation has been settled and, as such, are difficult to treat.

To conclude, the discipline of psychiatry has an important role in the evaluation and management of pain, both when determined by physical causes and when it is a symptom of a mental disorder.

References

American Psychiatric Association. (2002). *Diagnosis and statistical manual of mental disorders (fourth revision) DSM IV-TR*. Washington, DC: American Psychiatric Association.

Gallagher, R. N., & Verma, S. (2004). Mood and anxiety disorders in chronic pain. In R. H. Dworkin & W. S. Breitbart (Eds.), *Psychosocial aspects of pain: A handbook for health care providers*. Seattle: IASP Press.

Hall, A. M., Kamper, S. J., Maher, C. G., Latimer, J., Ferrerira, M. L., & Nicholas, M. K. (2011). Symptoms of depression and stress mediate the effect of pain on disability. *Pain, 152*, 1044–1057.

Price, D. D., & Bushnell, M. C. (2004). Overview of pain dimensions and their psychological modulation. In D. D. Price & D. C. Bushnell (Eds.), *Psychological methods of pain control: Basic science and clinical perspectives* (Progress in pain research and management, Vol. 29). Seattle: IASP Press.

Roy, R. (2001). *Social relations and chronic pain*. New York: Kluver/Plenum.

Psychobiological Model

▶ Chronic Pain, Diathesis-Stress Model of

Psychoeducation

▶ Information and Psychoeducation in the Early Management of Persistent Pain
▶ Psychological Treatment in Acute Pain

Psychoeducational Intervention

Definition

A technique that uses a didactic approach, including lectures and discussions, to teach coping skills to the individual.

Cross-References

▶ Psychiatric Aspects of the Management of Cancer Pain

Psychogenic

Definition

Having a psychological origin.

Cross-References

▶ Chronic Back Pain and Spinal Instability

Psychogenic Headache

▶ Headache, Episodic Tension-Type

Psychogenic Pain

Definition

This term is controversial since pain is a psycho-physiological phenomenon that always has psychological and physiological components and the term psychogenic suggests a dichotomy that is no longer viewed appropriate.

Cross-References

▶ Pain, Psychiatry, and Ethics
▶ Psychiatric Aspects of the Management of Cancer Pain

Psychogenic Rheumatism

▶ Fibromyalgia

Psychological [Behavioral Health] Assessment of Pain

Dennis C. Turk
Department of Anesthesiology & Pain Research, University of Washington, Seattle, WA, USA

Synonyms

Assessment; Evaluation; Screening

Definition

A thorough psychological assessment typically involves an interview with the patient (and, if possible, ► significant others), observation of the patient's (and significant other's) behaviors, and completion of a standardized set of self-report questionnaires. The overall objectives of assessment are to determine the extent to which cognitive (► cognitive factors), emotional, and behavioral factors are exacerbating the pain experience, interfering with physical and emotional functioning, or impeding response to treatment and rehabilitation, and maintaining disability. The information obtained should assist in treatment planning, specifically the matching of treatment (► Treatment Matching) components to the needs of individual patients, and modification of the treatment plan as needed.

Characteristics

In some settings, such as hospitals, health professionals are asked to conduct bedside patient evaluations or provide a pain consultation service for clinicians treating patients with complicated symptoms or on rehabilitation units. Under these circumstances, a brief psychological ► screening may be all that is feasible. In those instances, patients are routinely queried as to pain severity, location, characteristics, and modifiers (e.g., movement, coughing). In addition, when feasible, patients should be asked about the impact of pain on their activities (e.g., socializing, eating, ambulating), current and past treatments for pain, and patients' expectations for pain relief. In addition, behavioral manifestations of pain should be observed (e.g., grimacing, protective body postures, moaning), extent and changes in these (when patient observation over time) should be noted. Regular reassessment of patients attends to alterations (improvement or deterioration) in symptom severity, physical functioning, emotional functioning, and social functioning (Flor and Turk 2011; Turk et al. 2009).

A thorough psychological evaluation will reveal aspects of the patient's history that are relevant to the current situation. For example, the behavioral health professional will gather information about psychological disorders, substance abuse or dependence, vocational difficulties, and family role models for chronic illness. In terms of current status, topics covered include recent life stresses, vocational, emotional, social, and physical functioning, and sleep patterns. The purpose of the evaluation is to determine whether historical or current factors are influencing the way the patient perceives, copes, and adapts to pain.

A central component of a psychological evaluation is the interview. A number of topics, roughly fitting within ten general areas, are covered in the interviews:

Experience of Pain and Related Symptoms

Location and description of pain (e.g., "sharp," "burning")

Onset and progression

Pattern of symptoms (e.g., symptoms worse at certain times of day or following activity or stress)

Exacerbating and relieving factors (e.g., exercise, relaxation, stress, massage). "What makes your pain worse?" "What makes you pain better?"

Perception of cause (e.g., trauma, virus, stress)

What have they been told about their symptoms and condition? Do they understand and believe that what they have been told is accurate?

Sleep habits (e.g., difficulty falling asleep or maintaining sleep, sleep hygiene)

Thoughts, feelings, and behaviors that precede, accompany, and follow fluctuations in symptoms

Treatments Received and Currently Receiving

Medication (prescribed and over-the-counter). How helpful have these been?

Pattern of medication use (as needed or time contingent), changes in quantity or schedule

Physical modalities (e.g., physical therapy). How helpful have these been?

Exercise (e.g., Do they participate in a regular exercise routine? Is there evidence of deactivation and avoidance of activity due to fear of pain or exacerbation of injury?). Has the pattern changed (increased, decreased)?

Complementary and alternative (e.g., chiropractic manipulation, counseling). How helpful have these been?

Which treatments have they found the most helpful?

Compliance/adherence with recommendations of health-care providers

Attitudes toward previous health-care providers

Compensation/Litigation

Current disability status (e.g., receiving or seeking disability, amount, percentage of former job income, expected duration of support)

Current or planned litigation (e.g., "Have you hired an attorney?")

Activity

Typical daily routine ("How much time do you spend sitting, standing, walking, lying down?")

Changes in activities and responsibilities (both positive and obligatory) due to symptoms ("What activities did you use to engage in prior to your symptoms?" "How has this changed since your symptoms began?")

Changes in significant other's activities and responsibilities due to patient's symptoms

What does the patient do when pain is not bothering him or her?

Activities that patient avoids because of symptoms

Activities continued despite symptoms

Responses by Patient and Significant Others

Patient's behavior when pain increases or flares up ("What do you do when your pain is bothering you?" "Can others tell when your pain is bothering you?" "How do they know?")

Significant others' responses to behavioral expressions of pain ("What do your significant others do when they can tell your pain is bothering you?" "Are you satisfied with their responses?")

Significant other's response when patient is active ("How does your significant other respond to your engaging in activities?")

Impact of symptoms on interpersonal, family, marital, and sexual relations (e.g., changes in desire, frequency, or enjoyment)

Pattern of activity and pacing of activity (can use activity diaries that ask patients to record their pattern of daily activities [time spent sitting, standing, walking, and reclining] for several days or weeks)

Coping

How does the patient try to cope with his or her symptoms? (e.g., "What do you do when your pain worsens?" "How helpful are these efforts?"). Does patient view himself or herself as having any role in symptom management? "What role?"

Current life stresses

Pleasant activities ("What did you enjoy doing before your pain onset?" "What do you currently enjoy doing?")

Educational and Vocational History

Level of education completed (any special training)

Work history

How long at the most recent job?

How satisfied with the most recent job and supervisor?

What they like least about the most recent job?

Would they like to return to the most recent job? If not, what type of work would they like?

Current work status (including homemaking activities)

Vocational and avocational plans

Social History

Relationships with family or origin

History of pain or disability in family members

History of or current physical, emotional, and sexual abuse. Was the patient a witness to abuse of someone else?

Marital history and current status

Quality of current marital and family relations

Alcohol and Substance Use

History of substance abuse in family members

Current and history of alcohol use (quantity, frequency)

History and current use of illicit psychoactive drugs

History and current use of prescribed psychoactive medications

Consider the CAGE questions as a quick screen for alcohol dependence (Mayfield et al. 1987). Depending on response, consider other instruments for alcohol and substance abuse (Allen and Litten 1998).

Psychological Dysfunction

Current psychological symptoms/diagnosis (depression including suicidal ideation, anxiety disorders, somatization, posttraumatic stress disorder). Depending on responses, consider using standardized questionnaire (e.g., Hospital Anxiety and Depression Scale [HADS], Zigmond and Snaith 1983; PHQ-9, Kroenke et al. 2001) and conducting formal structured clinical interview for DSM (SCID) (American Psychiatric Association 1997).

Is the patient currently receiving treatment for psychological symptoms? If yes, what treatments (e.g., psychotherapy or psychiatric medications). How helpful?

History of psychiatric disorders and treatment including family counseling

Family history of psychiatric disorders

Concerns and Expectations

Patient concerns/fears (e.g., Does the patient believe he/she has serious physical problems that have not been identified? Or that symptoms will become progressively worse and patient will become more disabled and more dependent? Does the patient worry that he/she will be told the symptoms are all psychological?)

Explanatory models ("What have you been told is the cause of your symptoms?" "Does this explanation make sense?" "What do you think is the cause of your pain now?")

Expectations regarding the future and regarding treatment (will get better, worse, never change)

Attitude toward rehabilitation versus "cure"

Treatment Goals

Behavioral health professionals are interested in how patients experience their pain, what types of things exacerbate or alleviate the symptoms and what thoughts and feelings they have about their pain.

In addition to asking about the severity of pain, patients are asked about the location and changing (spreading) of pain, the characteristics of pain (e.g., burning, aching), the effect of pain on activities, and what they do when their pain is particularly severe as well as how they typically control their pain. Patients are asked about their sleep, specifically if they have any difficulty initiating or maintaining sleep? They may be asked about what treatments they have tried in the past and are using presently, how effective these have been.

The patient's social history is important. The psychologist inquires about the patient's background including family structure, marital (social) status, socioeconomic status, educational background, health history, and presence of anyone in the family with a history of chronic pain.

When patients with persistent pain seek compensation for lost wages or are involved in litigation, these processes can add an additional layer of distress. Behavioral Health practitioners attempt to determine whether compensation or litigation statuses might inadvertently be contributing to and maintaining the patients' symptoms.

Antecedents and consequences of pain symptoms and associated behaviors (▶ Antecedents and Consequences of Behavior) can potentially shape future experiences and behaviors (Fordyce 1976). Behavioral health practitioners use information to formulate hypotheses about what behavioral factors in a person's life may serve to maintain or exacerbate the pain experience. It is helpful to gather this information through interviews with patients and significant others, together as well as separately. During conjoint interviews, the behavioral health practitioner observes interactions between the significant others, and responses by significant others to patients expressions of pain and suffering.

Symptoms of chronic pain are extremely distressing and many times there is no cure or treatment capable of substantially reducing all symptoms (Boothby et al. 1999). At the present time, rehabilitation, including improvement in emotional functioning, physical functioning, and quality of life, is the goal. Rehabilitation in spite of pain is a daunting task even for patients with ample ▶ coping skills.

People who feel that they have a number of successful methods for coping with pain may suffer less than those who behave and feel helpless, hopeless, and demoralized. Thus, assessments focus on identifying factors that *exacerbate* and *ameliorate* the pain experience. The behavioral health professional not only focuses on deficits and weaknesses in coping efforts and coping repertoire, but also strengths.

Characteristics of jobs are important in chronic pain. The behavioral health professional inquires about the patient's current and previous work history; the nature of the work; physical demands required; how the patient gets along with coworkers, supervisors, and employees; and whether the patient believes that he or she will be able to return to their previous occupation. Questions about whether the patient liked his or her previous job and wishes to return to the same or a related job are discussed along with other plans they have for the future.

History of and current alcohol and substance use are important topics to be addressed during a psychological assessment. Moreover, a significant percentage of people with chronic pain are prescribed one or more analgesic medications, with a substantial percentage receiving prescriptions for opioid medication (Clark 2002; Hojsted and Sjogren 2007). A behavioral health professional may use a structured interview such as the SCID (American Psychiatric Association 1997) to determine if the patient meets the criteria for substance abuse or dependence. In addition, a psychological assessment will include determination of whether patients have a prior history of psychiatric illness, and whether they are currently being treated for psychological problems. The SCID-I and SCID-II (1997) can be used to determine whether the patient has any Axis I (primary psychiatric diagnosis) or Axis II (personality disorder) DSM-IV diagnoses (American Psychiatric Association 1994). In addition to the use of interviews, psychologists may make use of a number of self-report measures to assess psychological status and emotional distress (DeGood and Cook 2011).

Patients should be asked about their beliefs and expectations about the future of their pain problem (DeGood and Cook 2011). Are they convinced that they will not be "fixed" unless they have a surgery? What would they do if their pain were eliminated? These questions are meant not only to assess the patient's thoughts (beliefs, expectations, attitudes) surrounding their pain problem, but also to assess whether the patient has considered that rehabilitation is possible. To what extent have they internalized the disability role? Are they expecting to improve?

Observation of patients' behaviors (ambulation, body postures, facial expressions) can occur while they are being escorted to interview, during the interview, and when exiting interview (observation checklists are available to assist in assessing pain behaviors, Keefe et al. 2011). Observation of significant others' responses to patients can occur at the same time.

A large number of psychological instruments have been used to assess domains relevant to patients with chronic pain (Turk and Melzack 2011). Data gathered from measures not specifically developed or standardized (normed) on a chronic pain sample should be interpreted with caution, as their medical condition may influence some of the responses (Turk et al. 2009). Still, standardized assessment measures can provide an alternate source of information about areas that appear to be influencing patients' adaptation to their pain and their response to treatment.

Once areas of concern are identified from the evaluation, it is important to develop a plan for how to assess progress. Health-care providers should look for signs that indicate the patients' psychosocial, physical, and behavioral functioning have improved or declined. Patients may be asked to complete diaries in which they report (daily, several times a day) the activities they performed (e.g., number of hours sitting, standing, walking), their mood (e.g., fear, anxiety, depression), medication usage, thoughts, use of coping strategies, and sleep quality (Turk et al. 2004).

The high levels of emotional distress, disability, and reduced quality of life noted in many chronic pain patients suggests that psychological screening is essential; in the majority of cases, a thorough psychological evaluation is desirable. Psychosocial assessment allows health-care professionals to tailor treatment to meet individual needs and preferences (Turk 1990). A comprehensive assessment is a complex task, involving an exploration of a broad range of areas. The importance of behavioral health professionals in the assessment and treatment of chronic pain has been accepted by a number of agencies and governmental bodies in the United States, Canada, and England (e.g., United States Veterans Administration; U.S. Social Security Administration, Ontario Workplace Safety and Insurance Board). In fact, for multidisciplinary treatment programs to be certified, the Commission on the Accreditation of Rehabilitation Facilities in the United States requires involvement of psychologists in the treatment.

References

American Psychiatric Association. (1994). *Diagnostic and statistical manual of mental disorders* (4th ed., Text Revision). Washington, DC: American Psychiatric Association Press.

American Psychiatric Association. (1997). *User's Guide for the structured clinical interview for DSM-IV axis I disorders SCID-1: Clinician version.* Washington, DC: American Psychiatric Press.

Clark, J. D. (2002). Chronic pain prevalence and analgesic prescribing in a general medical population. *Journal of Pain and Symptom Management, 23,* 131–137.

DeGood, D. E., & Cook, A. J. (2011). Psychosocial assessment: Comprehensive measures and measures specific to pain beliefs and coping. In D. C. Turk & R. Melzack (Eds.), *Handbook of pain assessment* (3rd ed., pp. 67–97). New York: Guilford.

Flor, H., & Turk, D. C. (2011). *Chronic pain: An integrated biobehavioral approach.* Seattle: IASP Press.

Fordyce, W. E. (1976). *Behavioral methods for chronic pain and illness.* St. Louis: Mosby.

Hojsted, J., & Sjogren, P. (2007). An update on the role of opioids in the management of chronic pain of nonmalignant origin. *Current Opinion in Anaesthesiology, 20,* 451–455.

Keefe, F. J., Somers, T. J., Williams, D. A., & Smith, S. J. (2011). Assessment of pain behaviors. In D. C. Turk & R. Melzack (Eds.), *Handbook of pain assessment* (3rd ed., pp. 134–150). New York: Guilford.

Kroenke, K., Spitzer, R., & Williams, J. W. (2001). The PHQ-9: Validity of a brief depression measure. *Journal of General Internal Medicine, 16,* 601–616.

Turk, D. C. (1990). Customizing treatment for chronic pain patients: Who, what, and why. *The Clinical Journal of Pain, 6,* 255–270.

Turk, D. C., & Melzack, R. (2011). *Handbook of pain assessment* (3rd ed.). New York: Guilford.

Turk, D. C., Monarch, E. S., & Williams, A. R. (2004). Pain Assessment. In T. Hadjistavropoulos & K. D. Craig (Eds.), *Pain: Psychological perspectives* (pp. 209–244). Mahwah: Erlbaum.

Zigmond, A. S., & Snaith, R. P. (1983). The Hospital Anxiety and Depression Scale. *Acta Psychiatrica Scandinavica, 67,* 361–370.

Psychological Aspects of Pain

▶ Psychology of Pain, Assessment of Cognitive Variables

Psychological Aspects of Pain in Women

Eleni G. Hapidou
Department of Psychiatry and Behavioural
Neurosciences, McMaster University and
Hamilton Health Sciences Chronic Pain
Management Unit, Chedoke Hospital,
Hamilton, ON, Canada

Definition

Considerable evidence from ▶ epidemiology, the clinic and the laboratory suggested that women might experience pain differently than men (Berkley and Holcroft 1999; Fillingim et al. 2003; Riley et al. 1998; Rollman 2003; Unruh 1996). Women were reported to be more sensitive to exogenous and endogenous painful stimulation, to report greater clinical pain with the same pathology or tissue injury as men, to have a wider pain distribution and a greater likelihood to develop chronic pain after trauma, and to use analgesics more than men (Berkley and Holcroft 1999; Unruh 1996; Fillingim et al. 2003). Moreover, women's pain as part of reproductive life was found to be unparalleled to anything in men. Because of documented differences between women and men, the National Institute of Health mandated sex/gender differential designs and analyses in behavioral and biomedical research (Rollman 2003). More recent evidence has disputed experimental pain differences between women and men and has also questioned the usefulness of the laboratory differences (Racine et al. 2012).

Characteristics

Epidemiology of Pain in Women

Women have a higher ▶ prevalence of migraine; trigeminal neuralgia; causalgia; neuralgia; neck, shoulder, knee, and abdominal pain; pain from osteoporosis; ▶ irritable bowel syndrome (IBS); interstitial cystitis; rheumatoid arthritis;

▶ fibromyalgia; carpal tunnel syndrome; and ▶ temporomandibular disorder (TMD) (Fillingim et al. 2003; Unruh 1996; Yunus 2002). Prevalence patterns vary according to condition and across the life span.

Clinical Pain

There is an even higher female predominance of several clinical pain syndromes. Prevalence rates for TMD, IBS, migraine, and fibromyalgia are two to nine times higher in women than in men (Yunus 2002). Some studies also show that women report more chronic back pain, pain causing activity limitations, osteoarthritis, rheumatoid arthritis, pain from multiple sclerosis, colonoscopy, oral surgery, dental pain, and orthopedic surgery, but findings for chronic pain severity are less consistent (Fillingim 2002).

Pain Sensitivity in the Laboratory

Laboratory pain in women has been widely investigated in the context of sex/gender differences and the menstrual cycle in both pain-free women and in women with clinical pain syndromes. Studies using a variety of noxious stimuli (e.g., mechanical, ischemic, and the ▶ cold pressor task (CPT)) and measuring a variety of responses (pain threshold, tolerance, pain ratings) in different body systems (cardiovascular, neuromuscular, cerebral, autonomic nervous system) show increased pain responsiveness in women (Fillingim 2002). The magnitude of gender differences varies across pain induction methods, with effects being relatively consistent and robust with pressure and cold pressor pains but rather contradictory with thermal and electrical pains (Fillingim 2002; Riley et al. 1998). Discrepancies in ▶ effect sizes are also evident between experimental and clinical studies with the latter obtaining smaller or minimal ones (Myers et al. 2003).

Factors Mediating Gender Differences/Proposed Mechanisms

Biological, psychological, and sociocultural factors have all been implicated to explain women's differential sensitivity to pain. Specific mechanisms include body size, gonadal hormones, brain function, anxiety, cardiovascular reactivity/ blood pressure, cortisol, gender role expectancies, self-efficacy, and learning history regarding pain behavior (Derbyshire et al. 2002; Fillingim et al. 2003; Riley et al. 1998). Only some of the main factors mentioned are reviewed here.

The Influence of the Menstrual Cycle/ Reproductive Hormone Status

Gonadal hormones can alter ▶ nociception both peripherally and centrally (Aloisi 2003). Although not consistent for all types of noxious stimuli, on average, pain threshold and tolerance are highest in the follicular and lowest in the luteal phase of the menstrual cycle (Riley et al. 1999). Effect size across all stimulus modalities is about 0.40. The fact that certain pain syndromes such as TMD, migraine, myofascial pain, autoimmune diseases, and irritable bowel syndrome (IBS) are more prevalent in women during their reproductive years suggests the role of gonadal hormones in the development and pathogenesis of these problems.

Central Processing

Neuroimaging findings from positron emission tomography (PET) suggest that women are different from men in terms of brain activation patterns in response to noxious heat (Derbyshire et al. 2002; Paulson et al. 1998). Greater cerebral flow was shown in the anterior insula, thalamus and prefrontal left hemisphere (Paulson et al. 1998), and the perigenual cingulate cortex (Derbyshire et al. 2002) in women. These structures are also implicated in affective processing (Derbyshire et al. 2002). Women with IBS are also different from men in response to a visceral stressor. They showed greater activation in the ventromedial prefrontal cortex, right anterior cingulate cortex, and left amygdala (Naliboff et al. 2003).

Gender Roles/Expectations

Women differ from men in terms of their expectations of the typical male and female response to pain (Robinson et al. 2001). When the effect of gender role expectations on laboratory pain was assessed with the gender role expectation

of pain (GREP) scale in a large number of university students, women were rated as more willing to report pain, as more sensitive to pain, and as less enduring of pain than men. Men rated their own endurance as higher than that of the typical man. Willingness to report pain accounted for 46 % of the variance, 1.8 standard deviations above the mean. Self-efficacy beliefs are another factor that could account for the lower tolerance and the higher pain intensity ratings by women in the CPT (Jackson et al. 2002). Women, who were told that men do better in tolerating ischemic pain, had the lowest systolic blood pressure reactivity and reported lower perceived ability to tolerate the painful task (Fillingim et al. 2002).

Emotional Factors

Women tend to be more worried and irritated about pain (Unruh 1996) and have more negative cognitions and emotions, such as ▶ catastrophizing about pain, than men (Keefe et al. 2000). They show a much higher prevalence of somatic depression (appetite, sleep disturbance, and fatigue), but not of pure depression, than men, and this is also associated with high rates of pain (Silverstein 2002). Anxiety plays a role in women's increased sensitivity to laboratory pain as compared to men, in reporting clinical pain and symptoms, and in predicting poorer outcome of pain relief procedures (blocks and injections) for chronic pain (Edwards et al. 2003). Preoperative anxiety and depression in women undergoing radical mastectomy is associated with higher postoperative pain and analgesic requirements (Ozalp et al. 2003). Pregnant parous and nulliparous women tolerated less pain in the CPT and experienced more severe pain if they feared pain of labor (Saisto et al. 2001). In a population study in Sweden, older women's (60–74 years) higher risk for migraine was associated with a history of major depression, higher levels of susceptibility to stress, and somatic trait anxiety (Mattsson and Ekselius 2002).

Depression and anxiety are more common in women with TMD and fibromyalgia (Yunus 2002). Women with chronic pain differ from men in terms of anger management style and hostility (Burns et al. 1996). Women who express hostile anger characterized by cynicism and resentment may suffer greater pain than women with less hostile anger. Women who express non-hostile anger report the best adjustment to chronic pain (lowest pain, highest activity).

Women responded more positively than men to pleasant odors in terms of reducing their perceived pain intensity and unpleasantness of thermal stimuli (Marchand and Arsenault 2002). Odor-related brain activity was also higher in women than men in functional magnetic resonance imaging (fMRI) (Yousem et al. 1999). The pleasant odors that may reduce pain perception could have implications for alternative therapies, such as the use of aromatherapy in pain management.

Trauma/Abuse and Pain

A trauma history that includes physical and sexual abuse has been associated with poor health status, the development of chronic pain such as pelvic pain in women, other heterogeneous pain problems, catastrophizing, psychological distress, functional disability, and lifetime surgeries (Seville et al. 2003). Women with a history of trauma from non-clinical or primary care samples are no different from men in terms of pain and distress, but emotional distress occurs more in traumatized men with chronic pain.

Coping Styles

Pain coping instructions have different effects on women and men in the experimental setting. When instructed to focus on the sensory aspects of pain in the CPT, pain-free women reported more negative pain responses – lower pain threshold and tolerance and higher sensory pain ratings than did men (Keogh and Henderfeldt 2002). When instructed to focus on the emotional aspects of pain, they reported higher affective pain ratings. In dealing with clinical pain, women tend to use a greater range of pain coping strategies than men, including alternative pain therapies (Berkley and Holcroft 1999). Women with osteoarthritis and rheumatoid arthritis tend to use emotion-focused strategies (venting emotions, redefinition, seeking spiritual comfort, and

emotional support) more than problem-focused strategies (attempting pain reduction, relaxation, and distraction), which are more often used by men (Affleck et al. 1999). Women with chronic pain who view themselves as responsible for their health status tend to use more adaptive coping strategies than those who rely more on medical professionals (Buckelew et al. 1990).

Clinicians' Responses to Women's Symptoms and Pain

Gender biases and expectations seem to influence health-care professionals' responses to pain. Male and female physicians treat chronic neck pain differently in women and men (Hamberg et al. 2002). Three times more women than men are referred to a mind/body clinic (Nakao et al. 2001), and this approximates normative patterns for women. Women with TMD are treated surgically more often than men (Marbach et al. 1997). Physicians seem to prescribe opioids to men with affective distress and to women with functional disability (Fillingim et al. 2003).

Treatment Outcome

Improved long-term behavioral medicine rehabilitation outcomes (less chance of obtaining early retirement and better quality of life) were found for women with chronic pain as compared to untreated controls (Jensen et al. 2001). However, there is a paucity of research on differential treatment outcome for chronic pain based on gender. Most research exists with respect to pharmacological outcome.

Analgesic Responses

Pharmacological interventions act differentially on women in both clinical and experimental pain conditions, but the pattern of results is not consistent across drugs or testing methods (Fillingim 2002). Also, the practical implications are not yet known. Results are more consistent for opioids, which produce greater analgesia for women (Craft 2003). Women respond better to mixed ► mu- and kappa-opioids such as pentazocine, nalbuphine, and butorphanol; experience longer analgesia than men with higher doses of nalbuphine; and have greater morphine analgesia than men (Craft 2003). Morphine, meperidine, and hydromorphone, all mu-agonists, produced equipotent analgesia in women for pressure pain but greater analgesia for women in the CPT (Zacny 2002). There are also gender differences in anesthetic drug effects (Pley et al. 2003). Whereas animal studies on sex differences in opioid analgesia typically show opposite results from studies in humans, it was recently reported that the melanocortin 1 receptor (Mc1r) gene (in humans expressed in red hair and fair skin) is responsible for altered kappa-opioid analgesia in both humans and animals (Mogil et al. 2003). Women with two variant MC1R alleles had a greater analgesic response to pentazocine than did men and non-redhead women, and "this represents an example of a direct translation of a pharmacogenetic finding from mouse to human" (p. 4867). Mechanisms suggested to account for sex differences in opioid analgesia include drug metabolism, genotype, density of receptors, gonadal steroids, and psychosocial factors such as anxiety and expectancies (Craft 2003; Fillingim 2002). Sex differences in opioid analgesia need to be examined in models of acute inflammatory and chronic pain, as well in long-acting opioids (Craft 2003). Other analgesics such as anti-inflammatories need to be examined as well (Fillingim 2002).

References

Affleck, G., Tennen, H., Keefe, F. J., et al. (1999). Everyday life with osteoarthritis or rheumatoid arthritis: Independent effects of disease and gender on daily pain, mood, and coping. *Pain, 83*, 601–609.

Aloisi, A. M. (2003). Gonadal hormones and sex differences in pain reactivity. *The Clinical Journal of Pain, 19*, 168–174.

Berkley, K. J., & Holcroft, A. (1999). Sex and gender differences in pain. In P. Wall & R. Melzack (Eds.), *Textbook of pain* (4th ed., pp. 951–965). London: Churchill Livingstone.

Buckelew, S. P., Shutty, M. S., Jr., Hewett, J., et al. (1990). Health locus of control, gender differences and adjustment to persistent pain. *Pain, 42*, 287–294.

Burns, J. B., Johnson, B. J., Mahoney, N., et al. (1996). Anger management style, hostility and spouse responses: Gender differences in predictors of adjustment among chronic pain patients. *Pain, 64*, 445–453.

Craft, R. M. (2003). Sex differences in opioid analgesia: "From mouse to man". *The Clinical Journal of Pain, 19*, 175–186.

Derbyshire, S. W., Nichols, T. E., Firestone, L., et al. (2002). Gender differences in patterns of cerebral activation during equal experience of painful laser stimulation. *The Journal of Pain, 3*, 401–411.

Edwards, R., Augustson, E., & Fillingim, R. (2003). Differential relationships between anxiety and treatment-associated pain reduction among male and female chronic pain patients. *The Clinical Journal of Pain, 19*, 208–216.

Fillingim, R. B. (2002). Sex differences in analgesic responses: evidence from experimental pain models. *European Journal of Anaesthesiology, 19*(Suppl 26), 16–24.

Fillingim, R. B., Browning, A. D., Powell, T., et al. (2002). Sex differences in perceptual and cardiovascular responses to pain: The influence of a perceived ability manipulation. *Pain, 3*, 439–445.

Fillingim, R. B., Doleys, D. M., Edwards, R. R., et al. (2003). Clinical characteristics of chronic back pain as a function of gender and oral opioid use. *Spine, 28*, 143–150.

Hamberg, K., Risberg, G., Johansson, E. E., et al. (2002). Gender bias in physician's management of neck pain: A study of the answers in a Swedish national examination. *Journal of Women's Health & Gender-Based Medicine, 11*, 653–666.

Jackson, T., Iezzi, T., Gunderson, J., et al. (2002). Gender differences in pain perception: The mediating role of self-efficacy beliefs. *Sex Roles, 47*, 561–568.

Jensen, I. B., Bergstom, G., Ljungquist, T., et al. (2001). A randomized controlled component analysis of a behavioral medicine rehabilitation program for chronic spinal pain: Are the effects dependent on gender? *Pain, 91*, 65–78.

Keefe, F. J., Lefebvre, J. C., Egert, J. R., et al. (2000). The relationship of gender to pain, pain behavior and disability in osteoarthritis patients: The role of catastrophizing. *Pain, 87*, 325–334.

Keogh, E., & Henderfeldt, M. (2002). Gender, coping and the perception of pain. *Pain, 97*, 195–201.

Marbach, J. J., Ballard, G. T., Frankel, M. R., et al. (1997). Patterns of TMJ surgery: Evidence of sex differences. *Journal of the American Dental Association, 128*, 609–614.

Marchand, S., & Arsenault, P. (2002). Odors modulate pain perception. A gender-specific effect. *Physiology and Behavioral, 76*, 251–256.

Mattsson, P., & Ekselius, L. (2002). Migraine, major depression, panic disorder, and personality traits in women aged 40–74 years: A population-based study. *Cephalalgia, 22*, 543–551.

Mogil, J. S., Wilson, S. G., Chesler, E. J., et al. (2003). The melanocortin-1 receptor gene mediates female-specific mechanisms of analgesia in mice and humans. *Neuroscience, 100*, 4867–4872.

Myers, C. D., Riley, J. L., 3rd, & Robinson, M. E. (2003). Psychosocial contributions to sex-correlated differences in pain. *The Clinical Journal of Pain, 19*, 225–232.

Nakao, M., Fricchione, G., Zuttermeister, P. C., et al. (2001). Effects of gender and marital status on somatic symptoms of patients attending a mind/body medicine clinic. *Behavioral Medicine, 26*, 159–168.

Naliboff, B. D., Berman, S., Chang, L., et al. (2003). Sex-related differences in IBS patients: Central processing of visceral stimuli. *Gastroenterology, 124*, 1975–1977.

Ozalp, G., Sarioglu, R., Tuncel, G., et al. (2003). Preoperative emotional states in patients with breast cancer and postoperative pain. *Acta Anaesthesiologica Scandinavica, 47*, 26–29.

Paulson, P. E., Minoshima, S., Morrow, T. J., & Casey, K. L. (1998). Gender differences in pain perception and patterns of cerebral activation during noxious heat stimulation in humans. *Pain, 76*, 223–229.

Pley, H., Spigset, O., Kharasch, E. D., et al. (2003). Gender differences in drug effects: Implications for anesthesiologists. *Acta Anaesthesiologica Scandinavica, 47*, 241–259.

Racine, M., Tousignant-Laflamme, Y., Kloda, L. A., Dion, D., Dupuis, G., & Choinière, M. (2012). A systematic literature review of 10 years of research on sex/gender and experimental pain perception – Part 1: Are there really differences between women and men? *Pain, 153*, 602–618.

Riley, J. L., III, Robinson, M. E., Wise, E. A., et al. (1998). Sex differences in the perception of noxious experimental stimuli: A meta-analysis. *Pain, 74*, 181–187.

Riley, J. L., III, Robinson, M. E., Wise, E. A., et al. (1999). A meta-analytic review of pain perception across the menstrual cycle. *Pain, 81*, 225–235.

Robinson, M. E., Riley, J. L., Myers, C., et al. (2001). Gender role expectations of pain: Relationship to sex differences in pain. *Pain, 2*, 251–257.

Rollman, G. B. (2003). Sex makes a difference: Experimental and clinical pain responses. *The Clinical Journal of Pain, 19*, 204–207.

Saisto, T., Kaaja, R., Ylikorkala, O., et al. (2001). Reduced pain tolerance during and after pregnancy in women suffering from fear of labor. *Pain, 93*, 123–127.

Seville, J. L., Ahles, T. A., Wasson, J. H., et al. (2003). Ongoing distress from emotional trauma is related to pain, mood, and physical function in a primary care population. *Journal of Pain and Symptom Management, 25*, 256–263.

Silverstein, B. (2002). Gender differences in the prevalence of somatic versus pure depression: A replication. *The American Journal of Psychiatry, 159*, 1051–1052.

Unruh, A. M. (1996). Gender variations in clinical pain experience. *Pain, 65*, 123–167.

Yousem, D. M., Maldjian, J. A., Siddiqi, F., et al. (1999). Gender effects on odor-stimulated functional magnetic resonance imaging. *Brain Research, 818*, 480–487.

Yunus, M. B. (2002). Gender differences in fibromyalgia and other related syndromes. *The Journal of Gender-Specific Medicine, 5*, 42–47.

Zacny, J. P. (2002). Gender differences in opioid analgesia in human volunteers: Cold pressor and mechanical pain (CPDD abstract). *NIDA Research Monograph, 182*, 22–23.

Psychological Consequences of Pain

▶ Pain as a Cause of Psychiatric Illness

Psychological Evaluation

▶ Disability Assessment, Psychological/Psychiatric Evaluation

Psychological Factors

Definition

Factors related to life events, stress, emotions, behaviors, and other events in the internal or external environment that have an impact on the individual's state of mind.

Cross-References

▶ Pain in the Workplace, Risk Factors for Chronicity, and Psychosocial Factors
▶ Psychiatric Aspects of the Management of Cancer Pain

Psychological Management of Cancer Pain

▶ Psychiatric Aspects of the Management of Cancer Pain

Psychological Pain Measures

Definition

Direct measure of pain in which individuals describe the intensity, location, and quality of their pain experience using a broad array of subjective rating scales such as visual analogue scales, adjective checklists, and numerical scales. In a broader sense, this also relates to measures of pain-related variables such as interference related to pain or psychological distress.

Psychological Predictors of Chronicity

Michael K. Nicholas
Royal North Shore Hospital, University of Sydney, St. Leonards, NSW, Australia

Synonyms

Psychological risk factors; Psychosocial predictors or risk factors

Definition

Beliefs, responses, mood states, personality characteristics, and environmental interactions that increase the likelihood that pain will persist beyond the acute stage and that pain will be more disabling.

Characteristics

Personality Theories
Early attempts to predict the development of chronic pain and associated ▶ disability based on psychological factors focused mainly on personality theories, traumatic childhood experiences, or preexisting psychopathology (for reviews, see Asghari and Nicholas 1999;

Gamsa 1994; Roy 1985; Weisberg and Keefe 1999). However, these propositions are unproven, and the support that is available is mostly derived from retrospective or cross-sectional analyses. In recent years researchers have employed stronger research methods, especially prospective research designs, but methodological problems still persist. These are mostly related to differences between studies over variables such as sample composition and size, relevant outcomes, measures employed, and time covered (e.g., Linton 2000; Pincus et al. 2002).

Prospective Studies
Several systematic reviews of prospective studies have been published to date. Overall, despite their differences, they provide consistent evidence that psychological and social/environmental factors contribute to the progression from acute pain to chronic, disabling pain.

In an early review of 26 prospective studies relating to the development of back and neck pain, Linton (2000) found a consistent relationship between some psychological factors and the onset of pain, the transition from acute to chronic pain, and the development of associated disability. These factors included beliefs, emotional reactions, and behaviors. Truchon and Fillion (2000) confirmed Linton's finding that when psychological variables were studied in combination with medical or clinical variables (e.g., diagnosis and clinical tests, like straight leg raising), the psychological variables were more powerful predictors of future low back pain (LBP) and disability problems.

Additional systematic reviews of prospective studies following recent injury (or the onset of pain) have provided broad support for the thesis that psychological factors contribute to the progression from acute pain to chronic, disabling pain (Crook et al. 2002; Pincus et al. 2006; Steenstra et al. 2005; Waddell et al. 2003). The most cautious findings were reported by Pincus et al. (2006), possibly due to their more stringent quality criteria and their focus on studies relating to the fear-avoidance model. Nevertheless, they did find consistent supporting evidence for the influence of distress/depression as an obstacle to recovery from recent-onset LBP, but not for fear-avoidance beliefs or catastrophizing, but they did recommend that in order that for an adequately test of the fear-avoidance hypothesis, a more specific measure of avoidance behavior was required.

Steenstra et al. (2005) systematic review of prospective studies of prognostic factors for duration of sick leave after onset of low back pain examined studies that selected only workers who had had less than 6 weeks of LBP-related sick leave. Steenstra et al. reported that specific LBP, higher disability levels, older age, female gender, more social dysfunction and more social isolation, heavier work, and receiving higher compensation were significant predictors for a longer duration of sick leave.

A narrative review by Sullivan et al. (2005) found evidence for both psychological factors (fear, beliefs in severity of health conditions, catastrophizing, and poor problem solving) and work-related factors (low return to work expectancies and lack of confidence in performing work activities). They also found evidence that heightened pain severity and more severe depressive symptoms contributed to the *transition* to chronicity. Since publication of these reviews, additional evidence has accumulated that psychological factors can be predictive of persisting pain and disability or nonreturn to work in workers with acute or subacute back pain (Carragee et al. 2005; Gheldof et al. 2005; Grotle et al. 2005, 2007; Turner et al. 2006; Henschke et al. 2008).

Fear-Avoidance Model

The fear-avoidance model has provided a theoretical framework for a key group of psychosocial factors (Vlaeyen et al. 1995). In this model fear of pain or re-injury and consequent-activity avoidance has been considered prognostic for future disability. While there is much empirical evidence in support of the model (Leeuw et al. 2007), conflicting evidence has also been reported (Pincus et al. 2006). A possible explanation for these discrepancies has been reported by Boersma and Linton (2006) and Sieben et al. (2005) who found that

while pain-related fear made only a minor contribution to the persistence of pain and disability overall, there seem to be subgroups of people for whom it is a significant contributor. These authors cautioned that simply identifying psychological factors in isolation from contextual factors is likely to be insufficient for accurate prediction of pain-related disability. Nevertheless, such findings do reinforce views that multiple interacting factors are likely to influence the development of pain-related disability and these may vary from sample to sample as well as between individuals (Labriola et al. 2006).

Most recently, Vlaeyen and Linton (2012) provided a further update on the concept of fear avoidance. They noted that two main conclusions could be drawn from the research to date: (1) While pain may be intrinsically threatening, its threat value can vary across contexts and individuals; (2) While protective responding (e.g., avoidance) may be adaptive initially, it may worsen the problem in the long run.

Vlaeyen and Linton highlighted the role of both extero- and interoceptive stimuli that could become conditioned stimuli for predicting pain and noted that interoceptive stimuli may actually become better predictors due to their proximity and ease of access to the person. Along with other interoceptive stimuli, they argued that proprioceptive stimuli were particularly salient stimuli in the case of musculoskeletal pain and pointed to recent experimental findings that showed that simple paring of electric shock and certain hand movements could readily come to elicit fear of movement-related pain in healthy individuals.

As a whole, these reviews provide strong evidence that a number of psychological and social (i.e., environmental) variables play a significant role in the progression of pain experience from acute to chronic and the associated development of disability. There is also evidence that when both psychosocial and medical/clinical variables are measured, the psychosocial variables are stronger predictors of future pain and disability problems. Nevertheless, what started out as quite a simple idea, "fear avoidance" has become more elaborate and complicated with many interacting facets. However, empirical support for associations between variables within the models has also grown, and in many ways, fitting it all together has become increasingly complicated.

Diathesis-Stress Model

While the study of psychosocial risk factors has provided important information, it remains the case that, even when exposed to many of these risk factors, a large proportion of people with acute or subacute pain do not develop chronic pain and associated disability. As Linton (2002) has pointed out, these psychosocial factors should be termed "risk factors" rather than "predictors," as they only increase the risk of a person developing chronic, disabling pain. Some have argued that this observation may be explained by the concept of variance in vulnerability or predisposition to respond to certain stressors, also known as a stress-diathesis model (see ▶ Diathesis-Stress Model of Chronic Pain) (e.g., Turk 2002). Other explanations have suggested that the impact of a persisting pain depends on dynamic processes, such as learning, which would include the person's interactions with their pain, injury, and environment (e.g., Linton 2002).

Flor and Turk's (2012) description of a diathesis-stress model in relation to fibromyalgia syndrome incorporated potential predisposing psychological factors such as ▶ anxiety sensitivity, general fearful appraisals of bodily sensations, and previous experiences with pain. He argued that a person with heightened anxiety sensitivity, fearful appraisals of bodily sensations, and previous troublesome experiences with pain might be more likely to respond to the onset of a new pain with fear, overly alarmist thought processes (▶ catastrophizing), and low confidence in their ability to cope with the pain (low self-efficacy beliefs), which could lead to their engaging in avoidance or escape behaviors (from the pain or expected pain). These responses, in turn, could result in greater disability than would be the case if these responses had been different. Several studies of personality in relation to pain have taken a similar perspective, with personality characteristics acting as a modulator of a person's responses to pain, rather than a causal influence (for reviews, see Asghari and Nicholas 1999; Weisberg and Keefe 1999).

Linton (2002) explored the possible role of processes, like learning (conditioning), within the context of an evolving interplay of psychological risk factors, somatic factors (e.g., nature of the injury), and environmental factors (e.g., work demands) that could lead to the development of chronic pain and disability over an extended period. Some of these factors, like small lifestyle changes to minimize pain, might seem quite innocuous at the time, but over an extended period they could have important ramifications for adjustment, especially given the recurrent nature of musculoskeletal pain and the associated opportunities for learning. Linton argued that at the level of the individual person, many aspects of the development of chronic pain and disability were likely to involve events that were salient or critical to the individual at a certain time, but not at all times, and not necessarily across all injured people in a regular or linear manner. Identifying these risk factors at the individual level is likely to require careful assessment (and time), but a short (self-report) screening instrument developed by Linton and Halldén (1998) has proved to be useful in this context (e.g., Boersma and Linton 2002). The most recent review of this topic was reported by Nicholas et al. (2011), and this included mention of available brief measures as well as evidence for the role of these psychosocial factors in the development of chronic disabling pain.

Summary

Overall, the evidence that psychological factors can contribute to the development of chronic pain and associated disability is strong. There is also good evidence that other somatic and environmental factors can also play a role in this process. Rather than see these factors as predictors of chronic pain and disability, it may be more accurate and useful to view them as risk factors for future pain-related problems. The strength of the findings in relation to psychological and environmental interactions has potentially significant implications for the treatment/management of acute/subacute back and neck pain particularly, as well as the prevention of their progression to chronic pain and associated disability. For example, as many of the psychological and environmental risk factors (e.g., fears, beliefs, responses, and work demands) are amenable to change, it suggests that interventions aimed at modifying these factors could greatly limit the progression and impact of chronic pain.

References

Asghari, A., & Nicholas, M. K. (1999). Personality and adjustment to pain. *Pain Reviews, 6*, 85–97.

Boersma, K., & Linton, S. J. (2002). Early assessment of psychological factors: The Orebro screening questionnaire for pain. In S. J. Linton (Ed.), *New avenues for the prevention of chronic musculoskeletal pain and disability, pain research and clinical management* (Vol. 12, pp. 205–213). Amsterdam: Elsevier.

Boersma, K., & Linton, S. J. (2006). Psychological processes underlying the development of a chronic pain problem: A prospective study of the relationship between profiles in the fear-avoidance model and disability. *The Clinical Journal of Pain, 22*, 160–166.

Carragee, E. J., Alamin, T. F., Miller, J. L., & Carragee, J. M. (2005). Discographic, MRI and psychosocial determinants of low back pain disability and remission: A prospective study in subjects with benign persistent back pain. *The Spine Journal, 5*, 24–35.

Crook, J., Milner, R., Schultz, I. Z., & Stringer, B. (2002). Determinants of occupational disability following a low back injury: a critical review of the literature. *Journal of Occupational Rehabilitation, 12*, 277–295.

Flor, H., & Turk, D. C. (2012). *Chronic pain: An integrated biobehavioral approach*. Seattle: IASP Press.

Gamsa, A. (1994). The role of psychological factors in chronic pain. Part I. A Half Century of Study. *Pain, 57*, 5–15.

Gheldof, E. L., Vinck, J., Vlaeyen, J. W. S., Hidding, A., & Crombez, G. (2005). The differential role of pain, work characteristics and pain-related fear in explaining back pain and sick leave in occupational settings. *Pain, 113*, 71–81.

Grotle, M., Brox, J. I., Veierød, M. B., Glomsrød, B., Lønn, J. H., & Vøllestad, N. K. (2005). Clinical course and prognostic factors in acute low back pain: Patients consulting primary care for the first time. *Spine, 30*, 976–982.

Grotle, M., Brox, J. I., Glomsrød, B., Lønn, J. H., & Vøllestad, N. K. (2007). Prognostic factors in first-time care seekers due to acute low back pain. *European Journal of Pain, 11*, 290–298.

Henschke, N., Maher, C. G., Refshauge, K. M., et al. (2008). Prognosis in patients with recent onset low back pain in Australian primary care: Inception cohort study. *British Medical Journal, 337*, a171.

Labriola, M., Lund, T., Christensen, K. B., & Kristensen, T. S. (2006). Multilevel analysis of individual and contextual factors as predictors of return to work. *Journal of Occupational and Environmental Medicine, 48*(11), 1181–1188.

Leeuw, M., Goossens, M. E. J. B., Linton, S. J., et al. (2007). The fear-avoidance model of musculoskeletal pain: State of the scientific evidence. *Journal of Behavioral Medicine, 30*, 77–94.

Linton, S. J. (2000). A review of psychological risk factors in back and neck pain. *Spine, 25*, 1148–1156.

Linton, S. J. (2002). Why does chronic pain develop? A behavioural approach. In S. J. Linton (Ed.), *New avenues for the prevention of chronic musculoskeletal pain and disability* (Pain research and clinical management, Vol. 12, pp. 67–82). Amsterdam: Elsevier.

Linton, S. J., & Halldén, K. (1998). Can we screen for problematic back pain? A screening questionnaire for predicting outcome in acute and subacute back pain. *The Clinical Journal of Pain, 14*, 209–215.

Nicholas, M. K., Linton, S. J., Watson, P. J., & Main, C. J. (2011). The early identification and management of psychological risk factors (yellow flags) in patients with low back pain: A reappraisal. *Physical Therapy, 11*(91), 737–753.

Pincus, T., Burton, A. K., Vogel, S., et al. (2002). A systematic review of psychological factors as predictors of chronicity/disability in prospective cohorts of Low back pain. *Spine, 27*, 109–120.

Pincus, T., Vogel, S., Burton, A. K., Santos, R., & Field, A. P. (2006). Fear avoidance and prognosis in back pain: A systematic review and synthesis of current evidence. *Arthritis and Rheumatism, 54*(12), 3999–4010.

Roy, R. (1985). Engel's pain-prone disorder patients: 25 years after. *Psychotherapy and Psychosomatics, 43*, 126–135.

Sieben, J. M., Vlaeyen, J. W. S., Portegeijs, P. J. M., Verbunt, J. A., VanRiet-Rutgers, S., Kester, A. D. M., Von Korff, M., Arntz, A., & Knottnerus, J. A. (2005). A longitudinal study on the predictive validity of the fear-avoidance model in low back pain. *Pain, 117*, 162–170.

Steenstra, I. A., Verbeek, J. H., Heymans, M. W., & Bongers, P. M. (2005). Prognostic factors for duration of sick leave in patients sick listed with acute low back pain: A systematic review of the literature. *Occupational and Environmental Medicine, 62*, 851–860.

Sullivan, M. J., Feuerstein, M., Gatchel, R., et al. (2005). Integrating psychosocial and behavioral interventions to achieve optimal rehabilitation outcomes. *Journal of Occupational Rehabilitation, 15*, 475–489.

Truchon, M., & Fillion, L. (2000). Biopsychosocial determinants of chronic disability and Low-back pain:

P

A review. *Journal of Occupational Rehabilitation, 10*, 117–142.

Turk, D. C. (2002). A diathesis-stress model of chronic pain and disability following traumatic injury. *Pain Research & Management, 7*, 9–19.

Turner, J. A., Franklin, G., Fulton-Kehoe, D., et al. (2006). Worker recovery expectations and fear-avoidance predict work disability in a population-based workers' compensation back pain sample. *Spine, 31*, 682–689.

Vlaeyen, J. W., & Linton, S. J. (2012). Fear-avoidance model of chronic musculoskeletal pain: 12 years on. *Pain, 153*, 1144–1147.

Vlaeyen, J. W., Kole-Snijders, A. M., Boeren, R. G., et al. (1995). Fear of movement/(Re)injury in chronic Low back pain and its relation to behavioral performance. *Pain, 62*, 363–372.

Waddell, G., Burton, A. K., & Main, C. J. (2003). Screening to identify people at risk of long-term incapacity for work. London: Royal Society of Medicine Press.

Weisberg, J. N., & Keefe, F. J. (1999). Personality, individual differences, and psychopathology in chronic pain. In R. J. Gatchel & D. C. Turk (Eds.), *Psychosocial factors in pain: Critical perspectives* (pp. 56–73). New York: Guilford Press.

Psychological Risk Factors

▶ Psychological Predictors of Chronicity

Psychological Strategies

Definition

Strategies relating to the modification of emotion, cognition, or behavior of the individual by psychological methods.

Cross-References

▶ Acute Pain in Children, Procedural

Psychological Treatment

▶ Psychological Treatment of Chronic Pain, Prediction of Outcome

Psychological Treatment in Acute Pain

Barbara A. Hastie
Community Dentistry & Behavioral Science,
University of Florida College of Dentistry,
Gainesville, FL, USA

Synonyms

Behavioral; Cognitive; Cognitive restructuring; Distraction; Hypnosis; Mood; Preparation; Psychoeducation; Relaxation; Sensory focusing

Definition

Psychological treatments encompass all non-pharmacological, nonsurgical interventions in which the patient actively participates to alter his/her affect, behavior, and/or cognitions.

Characteristics

Psychological factors are intricately related in the biopsychosocial model of pain and influence an individual's experience of pain. Despite this, practitioners are not always aware of the utility and benefits of psychological interventions to augment pain treatment, especially in acute pain settings. As a result, acute pain management is suboptimal, and this negative pain experience can cause lasting effects on the individual. Myriad cognitive, affective, and behavioral approaches for coping with acute pain have been developed, and despite the distinctions, these treatments overlap substantially with one another. In general, given the limited temporal nature and intensity of acute pain, the most effective and parsimonious psychological treatment techniques are those that are brief, require limited to no training, and are highly portable, particularly to increase compliance and efficacy.

Education and Preparation

"Psychoeducation" is a common intervention for acute pain, and it involves providing patients with information about the cause and course of acute pain either from trauma (i.e., whiplash) or medical procedure such as surgery. It entails actively engaging them in the garnering of knowledge of what to expect and the procedures and after painful procedures. Preparation is a related intervention often used in presurgical settings to engage patients and bolster their psychological resources to cope with impending acute post-procedural pain. It is particularly useful in elective surgeries that allow for more time between preparation and the actual procedure thereby allowing for building a person's coping repertoire and also for practice of the techniques by the patient (Fredericks et al. 2010). Education and preparation can substantially lower pre-procedural anxiety and alter negative expectation (Oshodi 2007). Such techniques can aid patient decision-making, may help prevent disability from acute trauma-induced pain, and potentially reduce transition to chronic pain in certain acute pain condition (Meeus et al. 2012).

Relaxation

Abundant research demonstrates the benefits of relaxation techniques in reducing acute and chronic pain though the efficacy of the techniques varies across individuals. An additional benefit is that there are several types of relaxation techniques that can be modified and applicable on an individual and setting-specific basis. Acute pain causes a "stress response" which includes increases in muscle tension, blood pressure, heart and respiratory rates, blood glucose levels, and blood coagulation. Conversely, relaxation techniques augment action of the parasympathetic nervous system, essentially causing an antagonistic effect to such excitation. Additionally, the tenets of ▶ psychophysiologic model of pain assert that muscle tension may intensify pain at an injury site, and thus, a main goal of relaxation training is to prevent such exacerbation. Though there are many relaxation practices, the most common techniques used with acute pain and medical population for inducing the "relaxation response" include deep breathing, systematic relaxing of individual muscle groups (i.e., ▶ progressive muscle relaxation), or visualizing pleasant scenes (i.e., ▶ guided imagery). Despite decades of research, no specific relaxation technique ranks superior to others, although individual preferences and effectiveness vary across patients. Audio recordings of relaxation techniques are as effective as clinician-directed interventions and may offer significant time and cost savings. In general, relaxation has been rated as one of the most effective intervention techniques and is used with increasing frequency across multiple populations ranging from pediatrics (Tsao and Zeltzer 2005) to geriatrics (Ardery et al. 2003) and across acute pain conditions (Schaffer and Yucha 2004). Additionally, relaxation may be an effective alternative or adjunct for individuals who have limitations of pharmacotherapy or may otherwise be medically compromised.

Hypnosis

Hypnosis is essentially a modified state of consciousness during which particular suggestions for improved health are given. Hypnotic intervention usually entails two components of induction and suggestion. The initial induction phase involves fixed attention deepening to a dissociative state with concurrent suggestion that deepen the "trancelike" state. Once induction is achieved, it is followed by specific suggestions for reduced pain and analgesia. A series of literature reviews on hypnosis for pain have consistently revealed that hypnotic analgesia decreases pain compared to no treatment and is superior for pain reduction to non-hypnotic non-pharmacological interventions (i.e., education, supportive therapy), and hypnosis is comparable to other techniques that have hypnotic qualities (such as progressive muscle relaxation) but is not surpassed in efficacy by those other psychological treatments but may be more effective in some populations. For acute pain specifically, hypnosis is more effective for planned procedures as opposed to trauma but consistently shows efficacy for pain reduction, reduced anxiety as well as reduced hospital stay, and better recovery across multiple acute pain populations including children as well as for labor pain, burn wound dressing change, planned invasive surgical procedures, and cancer and related treatment side effects (Wood and Bioy 2008; Neron and Stephenson 2007). In fact, 89 % of surgical patients benefited from hypnosis by reducing pain, psychological distress, physiological disruption, and shorter hospital stay (Montgomery et al. 2002). Historically, hypnosis required significant time and professional commitment, but more recent studies have shown efficacy in limited interventions and self-hypnosis training of patients.

Distraction

Distraction techniques for acute pain reduction are straightforward and easily implemented. Distraction for pain reduction is based on the antithesis that paying attention to painful stimuli reliably increases reports of pain; hence, diverting attention away from the pain should be an effective pain-reducing technique. Multiple distraction techniques have been reported as effective in analgesic effects including standard cognitive distraction, virtual reality and video games, and music (Villarreal et al. 2012). The main goal of all of the distraction techniques is to elicit selective attention away from the acute pain, and efficacy is often based on individual preference and cognitive style. While distraction techniques can be highly effective across ages and venues, they may not be effective for individuals with an attentional bias toward pain and may actually increase pain (Van Ryckeghem et al. 2012). Yet, overall, active and interactive distraction is shown to be more effective than passive distraction (i.e., music, nature sounds), and most techniques are portable, easily implementable and modifiable, and have shown high efficacy in reducing acute pain in children and adults.

Sensory Focusing

Sensory focusing is in direct opposition to distraction and is based on the tenet of a limited capacity model of attentional allocation meaning focusing attentional resources to specific sensation (i.e., pain) limits available resources to other modalities. It is also based on the notion that adopting an emotional schema of pain may intensify its effect, whereas adopting a more objective-sensation-based schema, such as sensory focusing, might help reduce pain. Essentially, sensory focusing techniques assert that aligning one's attention closely to somatic sensations such as pain may reduce distress by providing self-regulatory information, increasing perceptions of control while essentially indirectly minimizing one's ability to interpret the meaning of the painful sensations. Limited studies exist to evaluate the efficacy of sensory focusing in clinical populations and those have been limited primarily to dental populations and burn patients. In the latter, individuals trained in sensory focusing reported greater pain relief compared to a music distraction control group, and they remembered less pain than a usual care comparison group (Haythornthwaite et al. 2001). In experimental settings of acute pain, sensory focusing was effective in the presence of high fear of pain (Roelofs et al. 2004). Efficacy of sensory focusing may be gender-specific whereby sensory focusing may not be effective in females regardless of affective distress, but it may be more effective for males when anxiety is high (Thompson et al. 2011). Notably, since distraction has proven effective in reducing acute pain of milder intensity in other contexts, it may behoove healthcare providers to reserve sensory focusing interventions for moderate to severe pain while being mindful of anxiety level, gender, and also being careful to match the intervention to the patients' preferences (i.e., focusing on pain vs. ignoring pain) to increase the efficacy of the intervention in conjunction with individual cognitive style, preference, and pain.

Cognitive Restructuring

Cognitive restructuring is an action-oriented intervention that focuses on the role of thoughts, attitudes, and beliefs in determining responses to pain. Related interventions challenge negative self-talk and maladaptive patterns of coping, such as ▶ catastrophizing (e.g., "the pain is terrible, it overwhelms me and I feel it's never going to get any better"). Working with the patient, such statements are replaced with more balanced cognitions that emphasize control, reduce negative emotions, and encourage adaptive coping (i.e., "although it hurts, it won't last forever and I can keep going"). The techniques are based on the premise that many of the negative thoughts and cognitions are automatic and not challenged for their veracity but have strong valence with regard to pain coping. Thus, additional specific skills related to cognitive restructuring include monitoring thoughts (via a daily diary and then employing the statement replacement technique above), thought-stopping, and generating accurate cognitions. Such interventions, including restructuring, stress management, and panic management via treatment of cognitions, have shown positive effects in reducing pain and emotional distress related to invasive medical procedures (Blacker et al. 2012), improve short- and long-term outcomes, as well as improve immune function postsurgically. Moreover, it is postulated that early psychological interventions, especially cognitive restructuring, may reduce transition to chronic pain following acute injury (Mavros et al. 2011; Katz and Seltzer 2009).

Social Support

While most of the literature on social support in pain has been conducted among individuals with chronic or life-threatening pain conditions, recent work reveals the utility of social connection even in acute pain settings. Notably, a recent large study showed that among patients undergoing total knee replacement surgery, those with strong social support had shorter hospital stays and overall better outcomes (Theiss et al. 2011). Support was measured by the presence of family or friend during preoperative classes, in the preoperative holding area, and/or during the last physical therapy session. In another large study of surgical recovery, social "connectedness" and support improved outcomes, including reducing pain (Mitchinson et al. 2008). Similarly, among postoperative patients, protective factors for better recovery include optimism, hope, and social support, whereas increased psychological distress including disruptive social support such as marital hostility complicated recovery and lengthened postsurgical hospital stay (Mavros et al. 2011; Mitchinson et al. 2008). Moreover, even in experimental acute pain settings, either active or passive social support reduced pain ratings and increased tolerance compared to those who underwent the procedures alone. Likewise, animal studies have shown the importance of social connectedness in pain reduction, and it would seem logical that such emotional influence would translate into the human realm of acute pain and coping.

Memory-Influencing Strategies

Memory is an ever-changing fluid process, continually modified by experiences, entrenched by learning, and subject to multiple biases. Acute pain in the presence of negative emotions can exacerbate memories of painful or unpleasant procedures and can negatively affect current real-time pain by lowering tolerance and increasing perception of intensity (Gedney and Logan 2004). Thus, emotional arousal may influence how memory for pain is encoded, processed, and retrieved. Similarly, prior painful experiences can modulate expectations of pain, influence current and retrospective reports of pain (Terry et al. 2008), and may also predispose a person to chronic pain through changes in neural networks. Importantly, learning and memory of pain can result in plastic changes of the central nervous system along with other related maladaptive changes (Flor 2012). Consequently, since learning is modifiable, treatment of negative memories of pain should focus on extinction of aversive pain memories with strong behavioral components to mitigate the long-term deleterious influences. While a brief post-procedure discussion session reduced anticipatory distress and discomfort during subsequent procedures among children, among adults, it may be beneficial to add a brief period of minimally painful stimulation to the end of a painful procedure to take advantage of the concept that the most recent events are best remembered (i.e., "recency effect"). An example was shown among adults who underwent a colonoscopy and who were exposed to several additional minutes of mild stimulation (i.e., the tip of the colonoscope was left just barely inserted) and who reported less unpleasantness, remembered less pain, and were more likely to return for a repeat scope compared to individuals who underwent the same procedure during which greater pain was reported during the last several minutes of the colonoscopy (Redelmeier et al. 2003). Hence, in clinical situations that require repeated painful procedures (e.g., dressing changes for burns, biopsies for cancer detection, surgical radiology), it may be useful to aim for as minimally unpleasant a memory as possible mainly to work toward relearning more positive memories and extinguishing negative ones related to the painful procedures. In addition to classical behavioral strategies, this could involve combination non-pharmacological therapies such as distraction coupled with music, pleasant aroma therapy, social support, and other aspects to elicit other senses with positive memory-influencing strategies.

Humor and Positive Mood Induction

Most reports of humor or "laughter therapy" have been in the context of cancer populations, but recent report demonstrates positive effects on immune function, pain tolerance, and decreased stress response across patient populations (MacDonald 2004). An older review (Matz and Brown 1998) revealed the positive influences of humor on pain management and asserted "humor assessment guidelines" and treatment recommendations for inclusion of humor in pain protocols. Recently, clowns have resulted in reduced preoperative anxiety and distress among children (Fernandes and Arriaga 2010). In experimental settings, positive mood using humorous films or pleasant images reliably increases pain tolerance and decreases pain ratings (review by Martin 2001). Humor and positive mood induction have not been systematically studied as an

intervention for pain, although reports on benefits to immune function and mood have been reported. Likewise, many healthcare professionals may engage in lighthearted or humorous interaction before or during painful procedures. Indeed, humor may help ease pain, may show the human side of healthcare professionals, and may improve the overall coping for the patient and the provider. Further investigation is warranted as to the efficacy on acute pain reduction and long-term outcome and affect modulation.

Positive Reinforcement

Positive reinforcement in the context of acute pain management entails providing tangible rewards for any positive behavior and any adaptive coping efforts during the procedure. It has been utilized mainly in pediatric populations undergoing painful procedures including interventional radiology (Alexander 2012) but also with other potentially painful interventions. The main tenet is to allow for pain to be viewed as a challenge and not as a punishment. To be most effective, rewards should be given immediately after the procedure has been completed. This is built on the theoretical principle that relief from pain will be associated with rewarding qualities (Becker et al. 2012). This technique also aligns with extinction of negative memory of pain, reduction of pain-related fear, and instilling more positive pain-related learning. Moreover, neuroimaging studies have shown that relief and reward are associated with similar brain regions, thus further supporting the concept that termination of aversive stimulation has rewarding qualities of its own, thus advocating for effective acute pain management in the immediate post-procedural settings. Nonetheless, motivational pathways (i.e., monetary, relief of pain) can influence valence of reinforcement and should be age and individual appropriate. In general, positive reinforcement technique is relatively easy to implement, and it can be effectively combined with other non-pharmacological psychological interventions (i.e., relaxation) with immediate individually meaningful rewards for employing the techniques to manage pain.

Treatment Combinations

Though the above psychological interventions were differentiated as distinct, in actual clinical applications, most techniques are delivered as treatment packages. To date, there is a paucity of research evaluating the efficacy of one technique over others or in the optimal combination of interventions for reducing acute pain. Yet mounting evidence reveals compelling results on the positive benefits of psychological interventions for acute pain especially in recovery from invasive surgical procedures (Mavros et al. 2011). It is also asserted that the interaction of psychological (viz., neural) networks with physiological aspects of pain can have a synergistic effect in terms of pain intensity and presumably in relief. Psychological interventions have also been used as adjuvant therapies to augment the efficacy of pharmacological treatment. Notably, in some postoperative populations, patients receiving various types of relaxation or distraction along with postoperative patient-controlled opioid analgesia (PCA) have reported less pain than those who receive PCA alone. Among children, while independent techniques aided in lowering pain and distress, combined cognitive-behavioral interventions were most effective in reduced reports of pain (Chambers et al. 2009). Given the nature and treatment venues of acute pain, cognitive-behavioral treatments (CBT) as delineated above may need to be modified for the location, procedure, and patient-specific characteristics (i.e., age, gender, cognition), but conceivably, they are interactive and synergistic in their effects not just on addressing emotional components related to acute pain but also the unpleasant nociceptive stimulus and related CNS effects of acute pain. Given these facts, it would behoove healthcare providers to be mindful of the potential benefits of implementing and/or referring their patients for psychological interventions to help address acute pain particularly among those where the procedure is known ahead giving time to train, learn, and practice techniques. Yet, even with onetime interventions, psychological approaches have proven efficacious as both primary and adjunctive treatments in this reduction of acute pain and therefore should become standard in their integration with multimodal pain treatment.

Summary

Compelling evidence supports psychological factors and acute pain as highly consistent risk factors for the development of chronic pain and disability. Indeed, it is posited that risk for transition to chronic pain following an acute pain episode may be mediated by cognitive and behavioral factors, such as anxiety, mood, expectations, and cognitions, which are highly amenable to intervention (Katz and Seltzer 2009). While there is burgeoning research in the field of cancer-related and chronic pain and the influence and interaction of psychological interventions and combination of pharmacological and non-pharmacological treatments, there remains the need for more standardized assessment and clinical trials in various acute pain settings. Since all chronic pain was once acute, targeting risk factors as well as determining the most effective non-pharmacological psychological interventions may improve short-term outcomes of acute pain and may also mitigate transition to chronic pain through the increased focus and implementation to the psychological domain of the biopsychosocial model of pain management. Though not addressed in this section, it is important to note that employing psychological interventions concomitantly with standard pain treatment often lead to short-term benefits (reduced acute pain and faster recovery) and long-term improved outcomes (presumably reduced incidence of chronic pain). Yet, importantly, there is the societal aspect that heralds the need for effective acute pain management as cost savings for institutions and on the macroeconomic societal level, to effectively reduce the public health crisis of pain. Lastly, most, if not all, of the psychological interventions are learnable and portable, and it is important to bear in mind the need to effectively match the patient to the intervention while making the patient more of an active participant in their care and bolstering their repertoire of pain management skills.

References

Alexander, M. (2012). Managing patient stress in pediatric radiology. *Radiologic Technology, 83*, 549–560.

Ardery, G., Herr, K. A., Titler, M. G., Sorofman, B. A., & Schmitt, M. B. (2003). Assessing and managing acute pain in older adults: A research base to guide practice. *Medsurg Nursing, 12*, 7–18.

Becker, S., Gandhi, W., & Schweinhardt, P. (2012). Cerebral interactions of pain and reward and their relevance for chronic pain. *Neuroscience Letters, 520*, 182–187.

Blacker, K. J., Herbert, J. D., Forman, E. M., & Kounios, J. (2012). Acceptance-versus change-based pain management: The role of psychological acceptance. *Behavior Modification, 36*, 37–48.

Chambers, C. T., Taddio, A., Uman, L. S., & McMurtry, C. M. (2009). Psychological interventions for reducing pain and distress during routine childhood immunizations: A systematic review. *Clinical Therapeutics, 31*(2), S77–S103. doi:10.1016/j.clinthera.2009.07.023.

Fernandes, S. C., & Arriaga, P. (2010). The effects of clown intervention on worries and emotional responses in children undergoing surgery. *Journal of Health Psychology, 15*, 405–415.

Flor, H. (2012). New developments in the understanding and management of persistent pain. *Current Opinion in Psychiatry, 25*, 109–113.

Fredericks, S., Guruge, S., Sidani, S., & Wan, T. (2010). Postoperative patient education: A systematic review. *Clinical Nursing Research, 19*, 144–164.

Gedney, J. J., & Logan, H. (2004). Memory for stress-associated acute pain. *The Journal of Pain, 5*, 83–91.

Haythornthwaite, J. A., Lawrence, J. W., & Fauerbach, J. A. (2001). Brief cognitive interventions for burn pain. *Annals of Behavioral Medicine, 23*, 42–49.

Katz, J., & Seltzer, Z. (2009). Transition from acute to chronic postsurgical pain: Risk factors and protective factors. *Expert Review of Neurotherapeutics, 9*, 723–744.

MacDonald, C. M. (2004). A chuckle a day keeps the doctor away: Therapeutic humor and laughter. *Journal of Psychosocial Nursing and Mental Health Services, 42*, 18–25.

Martin, R. A. (2001). Humor, laughter, and physical health: Methodological issues and research findings. *Psychological Bulletin, 127*, 504–519.

Matz, A., & Brown, S. T. (1998). Humor and pain management. A review of current literature. *Journal of Holistic Nursing, 16*, 68–75.

Mavros, M. N., Athanasiou, S., Gkegkes, I. D., Polyzos, K. A., Peppas, G., & Falagas, M. E. (2011). Do psychological variables affect early surgical recovery? *PLoS One, 6*, e20306.

Meeus, M., Nijs, J., Hamers, V., Ickmans, K., & Oosterwijck, J. V. (2012). The efficacy of patient education in whiplash associated disorders: A systematic review. *Pain Physician, 15*, 351–361.

Mitchinson, A. R., Kim, H. M., Geisser, M., Rosenberg, J. M., & Hinshaw, D. B. (2008). Social connectedness and patient recovery after major operations. *Journal of the American College of Surgeons, 206*, 292–300.

Montgomery, G. H., David, D., Winkel, G., Silverstein, J. H., & Bovbjerg, D. H. (2002). The effectiveness of adjunctive hypnosis with surgical patients: A meta-analysis. *Anesthesia and Analgesia, 94*, 1639–1645. Table.

Neron, S., & Stephenson, R. (2007). Effectiveness of hypnotherapy with cancer patients' trajectory: Emesis, acute pain, and analgesia and anxiolysis in procedures. *International Journal of Clinical and Experimental Hypnosis, 55*, 336–354.

Oshodi, T. O. (2007). The impact of preoperative education on postoperative pain. Part 2. *British Journal of Nursing, 16*, 790–797.

Redelmeier, D. A., Katz, J., & Kahneman, D. (2003). Memories of colonoscopy: A randomized trial. *Pain, 104*, 187–194.

Roelofs, J., Peters, M. L., van der Zijden, M., & Vlaeyen, J. W. (2004). Does fear of pain moderate the effects of sensory focusing and distraction on cold pressor pain in pain-free individuals? *The Journal of Pain, 5*, 250–256.

Schaffer, S. D., & Yucha, C. B. (2004). Relaxation & pain management: The relaxation response can play a role in managing chronic and acute pain. *The American Journal of Nursing, 104*(75–9), 81.

Terry, R. H., Niven, C. A., Brodie, E. E., Jones, R. B., & Prowse, M. A. (2008). Memory for pain? A comparison of nonexperiential estimates and patients' reports of the quality and intensity of postoperative pain. *The Journal of Pain, 9*, 342–349.

Theiss, M. M., Ellison, M. W., Tea, C. G., Warner, J. F., Silver, R. M., & Murphy, V. J. (2011). The connection between strong social support and joint replacement outcomes. *Orthopedics, 34*, 357.

Thompson, T., Keogh, E., & French, C. C. (2011). Sensory focusing versus distraction and pain: Moderating effects of anxiety sensitivity in males and females. *The Journal of Pain, 12*, 849–858.

Tsao, J. C., & Zeltzer, L. K. (2005). Complementary and alternative medicine approaches for pediatric pain: A review of the state-of-the-science. *Evidence-Based Complementary and Alternative Medicine, 2*, 149–159.

Van Ryckeghem, D. M., Crombez, G., Van, H. L., & Van, D. S. (2012). Attentional bias towards pain-related information diminishes the efficacy of distraction. *Pain, 153*, 2345–2351.

Villarreal, E. A., Brattico, E., Vase, L., Ostergaard, L., & Vuust, P. (2012). Superior analgesic effect of an active distraction versus pleasant unfamiliar sounds and music: the influence of emotion and cognitive style. *PLoS One, 7*: e29397.

Wood, C., & Bioy, A. (2008). Hypnosis and pain in children. *Journal of Pain and Symptom Management, 35*, 437–446.

Psychological Treatment of Chronic Pain, Prediction of Outcome

Robert J. Gatchel
Department of Psychology, The University of Texas, Arlington, TX, USA

Synonyms

Biopsychosocial model; Chronic pain; Psychological treatment; Treatment outcome

Definition

The biopsychosocial model of pain has greatly increased our ability to evaluate more comprehensively the intricate interaction among biological, psychological, and social factors that all contribute to the unique individual experience of pain. Adhering to such a model, in turn, leads to a better method of predicting pain treatment outcomes.

Characteristics

Current efforts to study pain conditions have broadened from antiquated, unidimensional biomedical reductionist models to more heuristic, multidimensional models of pain that take into account physiological, psychological, and social variables. This biopsychosocial perspective has greatly increased our ability to evaluate more comprehensively factors that can contribute to a better method of predicting treatment outcomes. Of course, when reviewing this area, it is important to keep in mind that treatment response predictors will vary from one form of pain syndrome to another. To date, the following pain conditions have received the greatest amount of empirical attention: chronic low back pain, headache, irritable bowel syndrome, rheumatoid arthritis, upper-extremity disorders, and temporomandibular joint disorders. Each of these disorders will be discussed below. A more comprehensive discussion of many of these pain conditions can be found in Turk and Gatchel (2002) and Gatchel and Oordt (2003).

Chronic Low Back Pain

There have been a number of studies linking psychosocial factors to treatment outcome. In an early study, Elkayam et al. (1996) found that the presence of personality disorders was highly correlated with poor outcome after a multidisciplinary treatment program, as were psychosocial stressors such as divorced marital status and unemployment. Hildebrandt et al. (1997) found that the most important variable in predicting a successful treatment outcome after a functional restoration and behavioral support program was the reduction in the patient's subjective feelings of disability. Rainville and colleagues (1997) found that compensation involvement had an adverse effect on self-reported pain and disability, both before and after a spine rehabilitation program. Such compensation issues have been indicated in numerous other studies with chronic pain patients. Finally, Harkapaa et al. (1996) reported that a patient's optimistic expectations and locus of control beliefs significantly predicted increased functional capacity in a group of patients who underwent an intensive multidisciplinary treatment program. Thus, taken as a whole, these earlier studies generally suggested that the following psychosocial variables appear to be predictors of response to treatment: presence of personality disorders; psychosocial stressors, such as divorced marital status and employment status; objective feelings of disability; optimistic expectations; locus of control beliefs; and compensation status. Moreover, a study by Pulliam et al. (2001) found psychosocial differences (such as coping skills) between those acute low back pain patients who were at higher risk for developing CLBP. Subsequent studies reported similar findings (Gatchel et al. 2003; Whitfill et al. 2010).

Headache

Gatchel and colleagues (1985) found that scores on the Millon Behavioral Health Inventory (MBHI)

significantly predicted response to a multidisciplinary treatment program for chronic headache patients. Response to treatment was measured by the number of daily headaches, duration of headaches, intensity of headaches, and all medications taken. Of all the scales on the MBHI, the emotional vulnerability scale was found to be consistently related to outcome on all four longitudinal treatment measures. In another study, Osterhaus et al. (1993) evaluated the outcome of a combined behavioral treatment consisting of relaxation training and temperature biofeedback, as well as cognitive training, in school children suffering from migraine headaches. Predictors of positive outcome were found to be gender, headache history, age, and psychosomatic complaints before training. Finally, Lipchick and colleagues (2002) have noted that a variety of qualitative and analytic reviews strongly indicate that cognitive-behavioral treatment yields a 40–60 % reduction in headaches due to it addressing a host of biopsychosocial factors.

Irritable Bowel Syndrome (IBS)
Blanchard and colleagues have conducted the most extensive biopsychosocial research with IBS patients. In an early study, Blanchard et al. (1988) found three potentially important predictors of biobehavioral treatment outcome: trait anxiety; frequency of symptom-free days at baseline; and gender. More specifically, having lower initial levels of trait anxiety, being male and having more episodic symptoms (i.e., more symptom-free days) were associated with successful outcomes. However, subsequently, Blanchard et al. (1992) found that neither trait anxiety, gender, nor baseline symptom-free days predicted outcome. The only variable that significantly predicted outcome was the presence or absence of an Axis I disorder as measured by the Anxiety Disorders Interview Schedule-Revised. They found that the absence of an Axis I disorder predicted more than twice the likelihood of successful outcome from treatment. Subsequently, based on the idea that IBS patients are very anxious and overaroused, and that they cannot adaptively solve their problems,

Blanchard (2001) developed a comprehensive biopsychosocial assessment-treatment program for IBS.

Rheumatoid Arthritis
There is currently a paucity of research evaluating which patients are most likely to have a positive response to treatment. In one study, Young (1992) reported no clear delineation of patient demographic characteristics associated with better outcomes of patients undergoing cognitive-behavioral treatment. Radojevik et al. (1992), in contrast, reported that spousal participation in treatment tended to enhance outcome. Finally, Sinclair and Wallston (2001) found that, in a cohort of women with rheumatoid arthritis who received brief cognitive-behavioral intervention, personal coping resources and maladaptive and adaptive pain coping behaviors were predictors of improvement. Overall, in the case of most chronic pain conditions, a comprehensive biopsychosocial treatment program will work best to deal with the various factors involved in this disorder. Obviously, additional clinical research in this area is greatly needed.

Upper Extremity Disorders
Work-related upper extremity musculoskeletal disorders have become more prevalent during the last decade. A prospective cohort study investigating the effects of psychosocial factors in long-term employment status of a sample of such chronic patients who completed an interdisciplinary functional restoration program was conducted by Burton et al. (1997). Overall, the results revealed that patients, who were older, Caucasian, had a current diagnosis of an anxiety disorder and whose perceptions of their disability deteriorated from pre- to post-treatment evaluations, were significantly less likely to have returned to work at 1-year follow-up. Thus, again, psychosocial variables influence the successful rehabilitation of these pain patients. Moreover, a comprehensive biopsychosocial treatment program was also found to produce significant improvements in patients with work-related upper extremity disorders (Mayer et al. 1999).

Temporomandibular Joint Disorder

There had been few early studies evaluating those psychosocial variables that serve to predict response to treatment. Schwartz et al. (1979) found that those patients who responded less positively to conservative treatment, relative to those who responded well, had greater elevations on several scales of the MMPI (hypochondriasis, depression, hysteria, psychopathic deviation, psychoasthenia, and schizophrenia). Such elevations subsequently were interpreted to suggest a tendency toward excessive pain sensitivity and agitated depression in TMD patients unresponsive to treatment (Block 1996). A study by Fricton and Olsen (1996) evaluated chronic TMD patients prior to entering a multidisciplinary treatment program. Overall, these authors concluded that symptoms of depression were the best predictors of response to treatment.

Rudy and colleagues (1995) were the first to utilize the Multidimensional Pain Inventory (MPI) to classify TMD patients within three psychosocial-based coping subgroups and then evaluate their differential response to treatment. Findings indicated that, as an overall group, patients improved significantly, but comparisons across subgroups revealed differential patterns of response to treatment. The most notable changes occurred with the subgroup characterized by the greatest amount of psychological distress (dysfunctional copers). This subgroup demonstrated significantly greater increases on measures of pain intensity, perceived impact of TMD symptoms on their lives, depression and negative thoughts, compared with subgroups characterized by greater personal problems (interpersonally distressed copers) and those patients who appeared to be less disabled by TMD (adaptive copers). However, this was interpreted as an "initial levels" effect because the dysfunctional patients had a greater range to improve. Subsequently, Kight et al. (1999) evaluated multiple psychosocial variables in an attempt to predict response to treatment in a group of TMD patients treated with cognitive therapy, biofeedback, or a combination of both. The most robust predictor of poor response to treatment was coping ability as measured by the MPI.

Those subjects who were categorized as either dysfunctional or interpersonally distressed copers displayed significantly poorer response to treatment. Epker and Gatchel (2000) and Wright et al. (2004) also found that those TMD patients who had a high risk for developing chronic TMD had poorer coping resources (measured by different indices) than those who did not develop chronicity.

Summary and Conclusions

As a whole, various studies have delineated variables that appear to hold promise for differentiating between those patients who respond and those who do not respond to a particular treatment program. For the cognitive-behavioral treatment of chronic pain in general, McCracken and Turk (2002) found that decreased negative emotional responses to pain, decreased perceptions of disability, and increased orientation toward self-management during the course of treatment predicted a favorable treatment outcome. Indeed, Gatchel and Okifuji (2006) conducted an extensive review, which clearly revealed that comprehensive biopsychosocial treatment programs produce the most therapeutic- and cost-effective outcomes for a wide array of chronic pain conditions, relative to traditional treatment methods. It should also be kept in mind that, even for the same pain condition, there may be a different array of predictor variables, depending upon the specific type of treatment program administered. Such response specificity should always be remembered when comparing results across studies. Also, one must be aware of the "pain homogeneity" myth. That is to say, not all pain conditions respond similarly to a particular treatment program. This was especially highlighted in the studies evaluating the different coping subgroups identified by the MPI. As Turk and Gatchel (1999) have pointed out, future research should be directed at "...not whether a treatment is effective but rather what treatment components delivered in what way, and when, produce the most successful outcomes for individual pain sufferers with what set of characteristics" (p. 490).

References

Blanchard, E. B., Schwarz, S. P., & Neff, D. F. (1988). Two-year follow-up of behavioral treatment of irritable bowel syndrome. *Behavior Therapy, 19,* 67–73.

Blanchard, E. B., Schwartz, S. P., Suls, J. M., et al. (1992). Two controlled evaluations of multicomponent psychological treatment of irritable bowel syndrome. *Behaviour Research and Therapy, 30,* 175–189.

Blanchard, E. B. (2001). *Irritable bowel syndrome: Psychological assessment and treatment.* Washington, DC: American Psychological Association.

Block, A. R. (1996). *Presurgical psychological screening in chronic pain syndromes.* Hillsdale: Erlbaum.

Burton, K., Polatin, P. B., & Gatchel, R. J. (1997). Psychosocial factors and the rehabilitation of patients with chronic work-related upper extremity disorders. *Journal of Occupational Rehabilitation, 7,* 139–153.

Elkayam, O., Ben Itzhak, S., Avrahami, E., et al. (1996). Multidisciplinary approach to chronic back pain: Prognostic elements of the outcome. *Clinical and Experimental Rheumatology, 14,* 281–288.

Epker, J., & Gatchel, R. J. (2000). Coping profile differences in the biopsychosocial functioning of TMD patients. *Psychosomatic Medicine, 62,* 69–75.

Fricton, J. R., & Olsen, T. (1996). Predictors of outcome for treatment of temporomandibular disorders. *Journal of Orofacial Pain, 10,* 54–65.

Gatchel, R. J., Deckel, A. W., Weinberg, N., et al. (1985). The utility of the Millon behavioral health inventory in the study of chronic headaches. *Headache, 25,* 49–54.

Gatchel, R. J., & Okifuji, A. (2006). Evidence-based scientific data documenting the treatment- and cost-effectiveness of comprehensive pain programs for chronic nonmalignant pain. *The Journal of Pain, 7*(11), 779–793.

Gatchel, R. J., & Oordt, M. S. (2003). *Clinical health psychology and primary care: Practical advice and clinical guidance for successful collaboration.* Washington, DC: American Psychological Association.

Gatchel, R. J., Polatin, P. B., Noe, C., Gardea, M., Pulliam, C., & Thompson, J. (2003). Treatment- and cost-effectiveness of early intervention for acute low-back pain patients: A one-year prospective study. *Journal of Occupational Rehabilitation, 13*(1), 1–9.

Harkapaa, K., Jarvikoski, A., & Estlander, A. M. (1996). Health optimism and control beliefs as predictors for treatment outcome of a multimodal back treatment program. *Psychology & Health, 12,* 123–134.

Hildebrandt, J., Pfingsten, M., Saur, P., et al. (1997). Prediction of success from a multidisciplinary treatment program for chronic low back pain. *Spine, 22,* 990–1001.

Kight, M., Gatchel, R. J., & Wesley, L. (1999). Temporomandibular disorders: An example of the interface between psychological and physical diagnoses. *Health Psychology, 18,* 177–182.

Lipchik, G. L., Holroyd, K. A., & Nash, J. M. (2002). Cognitive-behavior management of recurrent headache disorders: A minimal-therapist-contact approach. In D. C. Turk & R. J. Gatchel (Eds.), *Psychological approaches to pain management: A practitioner's handbook* (2nd ed.). New York: Guilford.

Mayer, T., Gatchel, R. J., Polatin, P., & Evans, T. (1999). Outcomes comparison of treatment for chronic disabling work-related upper extremity disorders. *Journal of Occupational and Environmental Medicine, 41,* 761–770.

McCracken, L. M., & Turk, D. C. (2002). Behavioral and cognitive-behavioral treatment for chronic pain. *Spine, 27,* 2564–2573.

Osterhaus, S. O., Passchier, J., van der Helm-Hylkema, H., et al. (1993). Effects of behavioral psychophysiological treatment on schoolchildren with migraine in a nonclinical setting: Predictors and process variables. *Journal of Pediatric Psychology, 18,* 697–715.

Pulliam, C. B., Gatchel, R. J., & Gardea, M. A. (2001). Psychosocial differences in high risk versus low risk acute low-back pain patients. *Journal of Occupational Rehabilitation, 11*(1), 43–52.

Radojevik, V., Nicassio, P. M., & Weisman, M. H. (1992). Behavioral intervention with and without family support for rheumatoid arthritis. *Behavior Therapy, 23,* 13–20.

Rainville, J., Sobel, J., Hartigan, C., et al. (1997). The effect of compensation involvement of the reporting of pain and disability by patients referred for rehabilitation of chronic low back pain. *Spine, 22,* 2016–2024.

Rudy, T. E., Turk, D. C., Kubinski, J. A., et al. (1995). Differential treatment response of TMD patients as a function of psychological characteristics. *Pain, 61,* 103–112.

Schwartz, R. A., Greene, C. S., & Laskin, D. M. (1979). Personality characteristics of patients with myofascial pain-dysfunction (MPD) syndrome unresponsive to conventional therapy. *Journal of Dental Research, 58,* 1435–1439.

Sinclair, V. G., & Wallston, K. A. (2001). Predictors of improvement in a cognitive-behavioral intervention for women with rheumatoid arthritis. *Annals of Behavioral Medicine, 23,* 291–297.

Turk, D. C., & Gatchel, R. J. (1999). Psychosocial factors and pain: Revolution and evolution. In R. J. Gatchel & D. C. Turk (Eds.), *Psychosocial factors in pain: Critical perspectives.* New York: Guilford.

Turk, D. C., & Gatchel, R. J. (2002). *Psychological approaches to pain management: A practitioner's handbook.* New York: Guilford.

Whitfill, T., Haggard, R., Bierner, S. M., Pransky, G., Hassett, R. G., & Gatchel, R. J. (2010). Early intervention options for acute low back pain patients: A randomized clinical trial with one-year follow-up outcomes. *Journal of Occupational Rehabilitation, 20,* 256–263.

Wrigth, A., Gatchel, R. J., Wildenstein, L., Riggs, R., Buschang, P. H., & Ellis, E. (2004). Biopsychosocial differences in high-risk vs low-risk acute TMD pain-related patients. *Journal of the American Dental Association, 135*, 474–483.

Young, L. D. (1992). Psychological factors in rheumatoid arthritis. *Journal of Consulting and Clinical Psychology, 60*, 619–627.

Psychological Treatment of Headache

Frank Andrasik[1], Beverly E. Thorn[2] and Joshua C. Eyer[2]
[1]Department of Psychology, Applied Psychophysiology and Biofeedback, The University of Memphis, Memphis, TN, USA
[2]Department of Psychology, University of Alabama, Tuscaloosa, AL, USA

Synonyms

Behavioral treatment; Biofeedback; Cognitive-behavioral treatment; Nondrug treatment; Non-pharmacological treatment; Psychosocial interventions; Relaxation; Stress management

Definition

Psychological therapies are ▶ psychosocial interventions applied to the management of headache as adjunctive or alternative procedures to medication. Psychological headache therapies with demonstrated efficacy include behavioral treatments (e.g., relaxation techniques) and cognitive treatments, generally accompanied by pain education. Frequently, these components are combined under the rubric of cognitive-behavioral therapy (CBT). Relaxation, ▶ biofeedback, and behavioral pacing strategies are most often categorized as behavioral components of CBT, with the greatest bulk of the headache treatment outcome research having utilized relaxation and biofeedback. Accumulating research also supports mindfulness meditation as a relaxation technique that promotes greater stress management while fostering positive cognitive and affective changes. The two major cognitive components of CBT are ▶ cognitive restructuring and cognitive coping-skills training. Cognitive restructuring includes the identification and modification of dysfunctional thought patterns, typically dysfunctional pain-related cognitions about headaches. ▶ Cognitive coping training for pain includes, but is not limited to, attention diversion strategies, reinterpreting pain sensations, calming self-statements, and imagery techniques. Treatment outcome research indicates that cognitive-behavioral therapies effectively decrease headache frequency and intensity, decrease psychological distress and dysfunction, and increase adaptive coping responses and sense of personal control over headaches (Andrasik 2007; Blanchard and Andrasik 1985). Motivational interviewing can be a useful adjunctive technique for fostering greater change (Miller and Rollnick 2002).

Characteristics

Efficacy

CBT is currently the only empirically supported treatment for chronic headache and has become the standard psychosocial intervention for headache pain. With a primary focus on headache prevention, CBT has proven effective at reducing the occurrence of headaches by approximately 40–50 % (Andrasik 2007; Penzien et al. 2002), although benefits to headache severity are also common. The benefits of psychosocial therapies have been effectively extended beyond adult patients to pediatric as well as elderly populations (Blanchard and Diamond 1996). These interventions not only reduce pain, but also increase adaptive coping responses, self-efficacy, and physical functioning, and decrease dysfunctional cognitions (Holroyd 2002). Additionally, psychological treatment of headache is cost-effective relative to biomedical approaches. Measures of patient acceptance of psychological treatment, patient attrition, and long-term maintenance of improvements indicate that patients are willing and able to engage in and benefit from CBT. In particular, long-term maintenance of psychological headache treatment appears to be much better (5 years and longer) than with drug treatment. It is hypothesized that psychological treatments teach patients certain cognitive and/or behavioral pain coping skills that they can subsequently utilize following treatment, whereas drug treatments do not foster such skills acquisition (Blanchard and Diamond 1996). Neither generic psychotherapy (e.g., insight-oriented therapy) nor hypnotherapy has been shown to be more effective than control or placebo conditions for the management of headache (Holroyd 2002; Holroyd and Lipchik 1999). However, newer psychotherapeutic approaches, such as Acceptance and Commitment Therapy and Mindfulness-Based Cognitive Therapy, have shown early signs of clinical efficacy.

Efficacy for psychological treatments is often evaluated by comparing those receiving the treatment to participants who monitor their symptoms while waiting for treatment (wait-list comparison groups) or to participants who receive credible attention-placebo treatment (often educational information regarding headaches and/or supportive psychotherapy). A common methodological difficulty associated with the extant treatment outcome research is small sample size, which limits the statistical power necessary to identify subtle but important treatment effects. Nonetheless, meta-analyses, and several studies using adequate sample sizes (i.e., over 30 subjects per cell) allow certain generalizations to be made. First, combinations of electromyographic (EMG) biofeedback and relaxation instruction were superior to placebo treatment for tension headaches. Second, including cognitive therapy with relaxation or relaxation plus EMG biofeedback for tension headaches added incremental utility to treatment gains. Third, for migraine headache, thermal biofeedback to induce hand warming accompanied by relaxation instruction was superior to placebo. Findings regarding the incremental advantage of adding a cognitive component to treatment for migraine are mixed. Nonetheless, the US Headache Consortium Guidelines advised that relaxation, thermal

biofeedback plus relaxation, and cognitive-behavioral therapy are all appropriate treatments for the prevention of migraine (Silberstein and Rosenberg 2000). The most current meta-analysis provides continued support for the efficacy of various biofeedback approaches, not just for pain outcome but for significant change with respect to other key variables: improvements in self-efficacy and symptoms of anxiety and depression (Nestoriuc et al. 2008).

Therapeutic Mechanisms

A crucial issue in all intervention research is identifying the mechanism by which a treatment achieves therapeutic effects. Behavioral techniques may reduce pain by altering underlying physiological processes related to stress and headache type, whereas cognitive techniques rely on altering cognitive patterns to foster greater coping, usually by changing the way a patient thinks about his or her pain. Research into the mechanisms associated with psychosocial headache treatments contributed to the reconceptualization of previously accepted pathophysiological models for the most common headache types: tension headache (prolonged pericranial muscle contraction) and migraine headache (vasospasm of cranial blood vessels). Data from electromyographic biofeedback (EMG-BF) studies for tension headache demonstrated that the actual change in EMG level is not associated with treatment success. Rather, patients' belief in the success of the training and related increases in self-efficacy and internal ► locus of control predicted improvements. In similar studies examining temperature biofeedback (TBF) for migraine, feedback indicating successful performance of TBF, independent of actual success in hand warming, predicted improvements in migraine among those with some improvement (Holroyd 2002; Seng and Holroyd 2010). Additionally, patients who attribute improvements following therapy to their own efforts demonstrate better long-term outcome than patients who attribute improvement to the interventions of health-care providers (Spinhoven et al. 1992). Such findings have led to the proposal of a cognitive-attributional model as the mechanism for psychosocial treatment success. The most recent research regarding psychophysiological mechanisms of headache (as well as other pain disorders) points to a likely central nervous system dysfunction, i.e., ▶ central sensitization, as maintaining both types of headache (Holroyd 2002). Thus, it is likely that these treatments affect central mechanisms of headache as well.

Areas of Ongoing Research
Difficult-to-Treat Headache Types
The discussion above pertains primarily to typical forms of headache. As research progresses, certain forms of headache have been shown to be particularly difficult to treat. These include patients experiencing medication-overuse headache, cluster headache, posttraumatic headache, or headaches accompanied by comorbid conditions. For these headache types, intensive, comprehensive, multidisciplinary treatments may be needed. Although menstrual migraine is common, the limited number of available studies has produced equivocal findings. Further research is also warranted here (Andrasik 2004; Andrasik et al. in press).

Cognitive-Attributional Emphases
It is possible that certain patient characteristics contribute to the determination of the relative success of psychological treatment for headache. One patient characteristic with particular relevance is dysfunctional thinking. In the pain literature, dysfunctional thinking has often been referred to as ▶ catastrophizing. Headache sufferers have been shown to have higher levels of catastrophizing than pain-free controls, and pain-related catastrophizing is a robust predictor of treatment outcome (Hassinger et al. 1999; Ukestad and Wittrock 1996). Treatments specifically targeted at reducing catastrophizing are being developed, examined, and promulgated (Thorn et al. 2002, 2007 Thorn 2004).

Mindfulness Meditation Techniques
A new direction in chronic headache pain treatment is instruction in mindfulness meditation techniques. Mindfulness meditation is a therapeutic, self-regulation technique that builds cognitive skills for managing stress, attention, and emotion through an accepting, nonjudgmental focus on one's present, moment-to-moment experience as it occurs. Mindfulness-Based Stress Reduction, a formalized group intervention that teaches mindfulness skills, has been found efficacious for chronic pain conditions, although the effects for headache were small-moderate (Rosenzweig et al. 2010). More complex interventions that feature mindfulness meditation as a prominent behavioral technique are also showing promise. A recent test of Mindfulness-Based Cognitive Therapy, an intentional melding of mindfulness meditation and cognitive components, has shown efficacy for migraine compared to a passive control (Day et al. 2012), and in another study Acceptance and Commitment Therapy (ACT), a psychotherapeutic approach that includes training in mindfulness, showed equivalent benefits to CBT for patients with chronic headache (Wetherell et al. 2011). Consequently, mindfulness meditation, and perhaps ACT, appears to be a positive direction for future investigation.

Application in Primary Practice Settings
Although the evidence clearly supports the efficacy of CBT for headache management, the effectiveness of such interventions applied in primary practice settings and/or neurology clinics needs to be established (Buse and Andrasik 2009). Promising evidence for the effectiveness of home-based and/or limited contact therapies for self-management of headache may provide a practical vehicle through which the implementation of CBT procedures in medical practice settings can be generalized. Novel public health interventions within the school system or communities at large or via internet-based therapies are presently in their infancy with regards to exploring their treatment utility and applicability (Andrasik 2004, Andrasik et al. 2012; Holroyd 2002; Penzien et al. 2002).

Integration with Drug Therapy
Clinical trials that include direct comparisons of drug treatment vs. psychological treatment for headache are few in number. The most extensive

investigation to date showed a quicker response to medication, with the behavioral treatment achieving a similar level of benefit over time for individuals experiencing chronic tension-type headache. Combining both treatments revealed the greatest effect (Holroyd et al. 2001, 2010). CBT investigations aimed directly at increasing the integration of ▶ behavioral self-management skills into everyday life (Hoodin et al. 2000), as well as increasing adherence to proper medication scheduling or for medication withdrawal (Andrasik 2004; Blanchard and Diamond 1996; Holroyd 2002; Nicholson et al. 2011; Penzien et al. 2002), will be useful for the future.

References

Andrasik, F. (2004). Behavioral treatment of migraine: Current status and future directions. *Expert Review of Neurotherapeutics, 4*, 89–99.

Andrasik, F. (2007). What does the evidence show? Efficacy of behavioural treatments for recurrent headaches in adults. *Neurological Sciences, 28*, S70–S77.

Andrasik, F. (2012). Behavioral treatment of headaches: Extending the reach. *Neurological Sciences, 33* (Suppl. 1), 127–130.

Andrasik, F., Sollars, C. M., Walch, S. E., & Buse, D. C. (2012). Headaches. In I. B. Weiner, A. M. Nezu, C. M. Nezu, & P. A. Geller (Eds.), *Handbook of psychology, Vol. 9: Health psychology* (2nd ed.). Hoboken, NJ: Wiley.

Blanchard, E. B., & Andrasik, F. (1985). *Management of chronic headaches: A psychological approach.* New York: Pergamon.

Blanchard, E. B., & Diamond, S. (1996). Psychological treatment of benign headache disorders. *Professional Psychology: Research and Practice, 27*, 541–547.

Buse, D. C., & Andrasik, F. (2009). Headaches in primary care. In R. A. Di Tomasso, H. Morris, & B. A. Golden (Eds.), *Handbook of cognitive behavioral approaches in primary care* (pp. 653–675). New York: Springer.

Day, M., Thorn, B. E., Ward, L. C., Rubin, N., Hickman, S., Scogin, F., & Kilgo, G. (2012). *Mindfulness-based cognitive therapy for the treatment of headache pain: A pilot study.* Manuscript submitted for publication. (Editorial decision on revision pending).

Hassinger, H. J., Semenchuk, E. M., & O'Brien, W. H. (1999). Appraisal and coping responses to pain and stress in migraine headache sufferers. *Journal of Behavioral Medicine, 22*, 327–340.

Holroyd, K. A. (2002). Assessment and psychological management of recurrent headache disorders. *Journal of Consulting and Clinical Psychology, 70*, 656–677.

Holroyd, K. A., & Lipchik, G. L. (1999). Psychological management of recurrent headache disorders: Progress and prospects. In R. J. Gatchel & D. C. Turk (Eds.), *Psychosocial factors in pain: Critical perspectives* (pp. 193–212). New York: Guilford Press.

Holroyd, K. A., O'Donnell, F. J., Stensland, M., Lipchik, G. L., Cordingley, G. E., & Carlson, B. W. (2001). Management of chronic tension-type headache with tricyclic antidepressant medication, stress management therapy, and their combination: A randomized controlled trial. *Journal of the American Medical Association, 285*, 2208–2215.

Holroyd, K. A., Cottrell, C. K., O'Donnell, F. J., Cordingley, G. E., Drew, J. B., Carlson, B. W., & Himawan, L. (2010). Effect of preventive (beta blocker) treatment, behavioural migraine management or their combination on outcomes of optimized acute treatment in frequent migraine: Randomised controlled trial. *British Medical Journal, 341*, c4871.

Hoodin, F., Brines, B. J., Lake, A. E., III, Wilson, J., & Saper, J. R. (2000). Behavioral self-management in an inpatient headache treatment unit: Increasing adherence and relationship to changes in affective distress. *Headache, 40*, 377–383.

Miller, W. R., & Rollnick, S. (2002). *Motivational interviewing: Preparing people for change* (2nd ed.). New York: Guilford Press.

Nestoriuc, Y., Martin, A., Rief, W., & Andrasik, F. (2008). Biofeedback treatment for headache disorders: A comprehensive efficacy review. *Applied Psychophysiology and Biofeedback, 33*, 125–140.

Nicholson, R. A., Buse, D. C., Andrasik, F., & Lipton, R. B. (2011). Nonpharmacologic treatments for migraine and tension-type headache: How to choose and when to use. *Current Treatment Options in Neurology, 13*, 28–40.

Penzien, D. B., Rains, J. C., & Andrasik, F. (2002). Behavioral management of recurrent headache: Three decades of experience and empiricism. *Applied Psychophysiology and Biofeedback, 27*, 163–181.

Rosenzweig, S., Greeson, J. M., Reibel, D. K., Green, J. S., Jasser, S. A., & Beasley, D. (2010). Mindfulness-based stress reduction for chronic pain conditions: Variation in treatment outcomes and role of home meditation practice. *Journal of Psychosomatic Research, 68*, 29–36.

Seng, E. K., & Holroyd, K. A. (2010). Dynamics of changes in self-efficacy and locus of control expectancies in the behavioral and drug treatment of severe migraine. *Annals of Behavioral Medicine, 40*, 235–247.

Silberstein, S. D., & Rosenberg, J. (2000). Multispecialty consensus on diagnosis and treatment of headache. *Neurology, 54*, 1553–1554.

Spinhoven, P., Lenssen, A. C. G., Van Dyck, R., & Zitman, F. G. (1992). Autogenic training and self-hypnosis in the control of tension headache. *General Hospital Psychiatry, 14*, 408–415.

Thorn, B. E. (2004). *Cognitive therapy for chronic pain: A step-by-step guide*. New York: Guilford Press.

Thorn, B. E., Boothby, J. L., & Sullivan, M. J. L. (2002). Targeted treatment of catastrophizing for the management of chronic pain. *Cognitive and Behavioral Practice, 9*, 127–138.

Thorn, B. E., Pence, L. B., Ward, L. C., Kilgo, G., Clements, K. L., Cross, T. H., & Tsui, P. W. (2007). A randomized clinical trial of cognitive behavioral treatment targeted at the reduction of catastrophizing in chronic headache sufferers. *The Journal of Pain, 68*, 938–949.

Ukestad, L. K., & Wittrock, D. A. (1996). Pain perception and coping in female tension headache sufferers and headache-free controls. *Health Psychology, 15*, 65–68.

Wetherell, J. L., Afari, N., Rutledge, T., Sorrell, J. T., Stoddard, J. A., Petkus, A. J., & Hampton Atkinson, J. (2011). A randomized, controlled trial of acceptance and commitment therapy and cognitive-behavioral therapy for chronic pain. *Pain, 152*, 2098–2107.

Psychological Treatment of Pain in Children

Christiane Hermann
Department of Clinical Psychology, Justus-Liebig University Giessen, Giessen, Germany

Synonyms

Acceptance-based treatment; Biofeedback; Cognitive-behavioral therapy; Internet therapy; Multimodal treatment; Non-pharmacological interventions; Operant treatment; Pediatric pain management; Preparation programs; Relaxation

Definition

Children and adolescents experience pain from a number of different sources. Pediatric pain problems that have been the target of psychological interventions are procedure-related acute pains (e.g., lumbar puncture), disease-related chronic pain (e.g., rheumatoid arthritis, cancer, sickle cell disease), and chronic pain of benign or unknown origin (e.g., migraine, tension-type headache, abdominal pain, widespread musculoskeletal pain). A variety of psychological

treatments for pediatric pain have been developed, mostly using well-established psychological interventions for adults as a model. These include (1) cognitive-behavioral interventions that focus on improving pain and stress coping skills or, more recently, that focus on pain acceptance and pursuit of valued goals, (2) relaxation techniques such as progressive muscle relaxation or autogenic training, (3) biofeedback, and (4) operant treatment approaches with the aim to reduce pain behavior. For infants undergoing painful procedures, specific psychological interventions have been developed that either focus on altering the psychosocial context of the procedure or rely on behavioral techniques.

Introduction

Psychological treatments for pediatric pain can be subdivided into two broad categories: (a) interventions for acute procedure-related pain and (b) interventions for chronic pain regardless of its origin. Although both types of interventions rely on cognitive and behavioral techniques, they differ with regard to design, goal, treatment delivery, and the target population.

Characteristics

Procedure-Related Pain

Children at all ages are faced with pain due to medical procedures (e.g., immunization, injection, blood taking, surgery, lumbar puncture, bone marrow aspiration), especially when they undergo treatment for an illness or injury. Preterm or ill-born newborns are exposed to many invasive procedures per day when being treated in an intensive care unit. Analgesics are routinely given for major painful events such as surgery. For other more minor painful events such as blood taking or injections, however, a number of non-pharmacological interventions have been developed. Several recent meta-analyses have summarized the available evidence for their efficacy based on ▶ randomized controlled trials (RCTs). For neonates, non-nutritive sucking (e.g., a pacifier inducing sucking behavior), swaddling (i.e., firmly wrapping the infant in a blanket to prevent excessive limb movements) or facilitated tucking (i.e., holding the infant in a bended position by placing the caregiver's hands on the head and the lower limbs), kangaroo care (i.e., placing the infant skin-to-skin on the caregiver's chest during the procedure) and administration of sucrose are effective in reducing the pain response to procedures such as blood taking, injections, endotracheal suctioning, and, in preterms only, diaper change and weighing (Pillai Riddell et al. 2011; Stevens et al. 2010; routine immunization: Shah et al. 2009). Yet, the obtained effects vary considerably across studies and not all of these behavioral interventions have been systematically studied in preterm and full-term neonates and in older infants.

In the early 1980s, due to the demonstrated success of surgery preparation programs, efforts were made to develop programs that would specifically help children between the ages of about 2 and 18 years to prepare for and cope with painful medical procedures. The vast majority of these programs rely on ▶ cognitive-behavioral treatment (CBT) techniques to reduce procedure-related pain and distress. Cognitive interventions aim at identifying and modifying negative thoughts about the procedure and one's own coping abilities. Behavioral techniques target modifying the behavior by applying learning principles. A variety of interventions have been evaluated: (1) cognitive (e.g., counting, non-procedure-related talk, virtual reality) and behavioral ▶ distraction (e.g., games, playing with toys, solving puzzles); (2) cognitive interventions such as thought stopping, coping self-statements and suggestions (i.e., providing a verbal or non-verbal cue suggesting that the intervention will reduce pain and/or distress), preparation for and education about the procedure, and reframing negative memories of previous procedures; (3) imagery of a pleasant situation or experience; (4) ▶ hypnosis; and (5) ▶ relaxation such as progressive muscle relaxation or breathing exercises (e.g., blowing a party blower; pretending to be a tire). These techniques can be taught and practiced using modeling, positive reinforcement, rehearsal of coping skills and desensitization. For example, films telling the story of a peer model coping well with the procedure have been used. Or, children are reinforced (e.g., praising, small rewards) for using coping skills and for holding still. Role-plays have been utilized for a gradual exposure to the feared procedure (desensitization) and as a means for rehearsing the learned coping behavior. For example, the child first learns about the procedure and then performs the procedure using a doll while being encouraged to apply the learned coping skills prior to the actual procedure. Treatment protocols often include active coaching (i.e., prompting the child to use coping skills) either by the parent, a nurse, or a psychologist during the actual medical procedure. Indeed, in some of the programs it is the parents or staff rather than the child him/herself who are trained to coach the child in using the cognitive and behavioral strategies.

Preparation programs are brief and, due to the specific situational demands, involve only 1–2 up to 4 sessions that are delivered within a short period of time (0–2 days) prior to the procedure. The majority of programs have been evaluated for immunization, injections, blood taking, bone marrow aspiration, and lumbar puncture. Due to their developmental stage, younger children rely more on primary coping (i.e., attempts to alter the situation) and use less secondary coping strategies (i.e., attempts to adjust oneself to the situation). Due to their cognitive development, young children cannot be expected to remember and initiate the use of coping skills, hence active coaching and prompting by an adult is important. There is good evidence that parental anxiety and parental distress-promoting behavior (e.g., criticism, reassurance, empathetic comments) contribute to the child's pain and distress (e.g., Mahoney et al. 2010). Yet, the findings regarding a potentially detrimental effect of parental presence on children's distress and pain are equivocal (Piira et al. 2005). Clearly, parents themselves perceive their presence as helpful which may help to reduce their distress and increase their feeling of being in control. Similarly, almost all children perceive the presence of parents as most helpful. Training the parent may not only be beneficial because it facilitates the child's active coaching, but it can also promote the transfer of the learned skills to subsequent procedure sessions (e.g., Kazak et al. 1998).

Studies on the efficacy of psychological interventions for needle-related pain in children and adolescents have relied on self-report (pain intensity, distress), observer report, behavioral observation, and physiological measures (heart rate, cortisol, oxygen saturation, etc.). Across most outcome measures and across most procedures, distraction, combined cognitive-behavioral programs, and hypnosis emerged as the most efficacious interventions (Uman et al. 2008). For other psychological interventions, there is insufficient data to determine their efficacy. The combination of CBT and adequate pharmacological intervention (analgesics, anesthetics, and sedatives) may be somewhat more effective in reducing child and parent distress than pharmacological intervention alone (Kazak et al. 1998). Yet, overall, there is insufficient data on such combined treatments. While CBT is effective in reducing distress during an actual procedure, little is known about the maintenance and generalization of treatment gains. Initial findings in healthy children undergoing routine vaccination suggest that distraction may be particularly effective in preventing the formation of pain-related aversive memories, even if being equivalent to topical anesthetics in reducing actual pain and anxiety during the procedure (Cohen 2001).

Chronic Pain
A substantial number of children and adolescents suffer from chronic pain of 3 months or longer. Despite the rather common occurrence of pediatric chronic pain, conclusions with regard to its severity and to pediatric pain as a health problem are difficult to draw. Many children cope quite well and do not experience heightened levels of functional disability. Yet, chronic pain during childhood may increase the risk of a lifetime of chronic pain and/or the risk of mental disorders. Treatments for chronic pediatric pain aim at decreasing the frequency and intensity of pain and at reducing pain-related disability by improving the child's coping skills, thus reducing the impact of chronic pain on the child's personal growth and development. Even though data are still lacking, such interventions may also be important as a preventive measure for reducing the risk of developing a life-long pain problem.

▶ Progressive muscle relaxation (PMR) is one of the most frequently used interventions for pediatric pain either alone or as part of CBT. Typically, PMR is administered by a therapist on an individual basis or in groups. CBT packages for chronic pediatric pain are highly similar to adult programs and usually comprise four major components: (1) education about the pain; (2) learning of cognitive and behavioral pain coping skills such as imagery, distraction, and relaxation; (3) stress management, that is, identifying and coping with stressful situations using thought stopping, cognitive restructuring, assertiveness, and problem solving; and (4) relapse prevention. More recently, first attempts have been made to adapt adult pain treatment programs that are based on the acceptance and commitment therapy approach (McCracken 2005). Rather than focusing on controlling pain and modifying cognitions, this approach aims at helping patients to pursue personally meaningful goals in their life despite pain. A first RCT in children and adolescents showed that an acceptance-based treatment was more efficacious than a standard multimodal program (Wicksell et al. 2009).

Biofeedback is one of the best evaluated psychological interventions for pediatric chronic pain. Skin temperature or thermal ▶ biofeedback (BFB) (i.e., volitional hand-warming) and ▶ EMG-BFB from the m. frontalis have most frequently been used in the treatment of childhood headache. For migraine, additional forms of BFB such as vasoconstriction training, that is, learned voluntary constriction of extracranial blood vessels, and a specific form of EEG-BFB, that is, BFB of the ▶ contingent negative variation (CNV), a slow cortical potential that is enhanced in migraine have also been evaluated (Hermann 2011).

CBT and relaxation training for chronic pediatric pain have been evaluated both in individual and group formats. In addition, self-help formats have been developed. For example, a self-help booklet for children with headache is available since many years. Such self-help programs are increasingly offered via Internet (e.g., Trautmann and Kröner-Herwig 2009).

The role of parents in the treatment of pediatric pain still awaits a more systematic evaluation (Hermann 2011). The ▶ operant model of chronic pain postulates that chronic pain is maintained by operant reinforcement of pain behaviors such as verbal and nonverbal expressions of pain, guarding, resting, and reduced activity. Hence, parents' response may contribute to pain maintenance. Based on the operant model, brief parent trainings (1–3 sessions) have been added to CBT of biofeedback protocols. Such parent trainings focus on principles of ▶ contingency management with the aim to minimize positively (e.g., paying attention) or negatively reinforcing (e.g., excuse from daily chores) parental responses to the child's pain. Moreover, parents are taught to encourage their children to practice and use pain coping skills and maintain normal daily activities during pain episodes. Recently, an Internet-delivered family CBT (i.e., CBT including a parent training) was shown to be effective in reducing pain activity and pain-related disability in adolescents suffering from various recurrent pains (e.g., headache, abdominal pain, leg pain; Palermo et al. 2009). Yet, parent behavior changed also in the control condition which makes it difficult to determine the specific impact of the parent training.

Psychological programs typically address children and adolescents between the ages of 8 and 18 years. All available psychological treatments for chronic pediatric pain require the child to engage in active coping and self-initiate coping attempts, thus not being a promising approach in very young children. According to several meta-analyses and systematic reviews (Palermo et al. 2010; Trautmann et al. 2006), psychological interventions yield large effect sizes in pain reduction in childhood ▶ headache, abdominal pain, and ▶ fibromyalgia. Functional disability and psychosocial adjustment were only slightly improved. This may at least partially be accounted for by

a lack of studies assessing this outcome variable. According to the recommendations of the Initiative on Methods, Measurement, and Pain Assessment in Clinical Trials in pediatric pain (PedIMMPACT, McGrath et al. 2008), pain intensity, physical and emotional functioning, role functioning, sleep, adverse events, and satisfaction with treatment represent core outcomes that need to be assessed. Future studies will allow to estimate treatment effects on outcome dimensions beyond pain itself. Conclusions with regard to the differential efficacy of psychological interventions are limited since direct comparisons between treatments are rare. Moreover, few data on long-term outcome (>6 months) are available. It also needs to be emphasized that the efficacy of psychological treatments has mostly been demonstrated for pediatric headache with very few studies focusing on other pain problems. Psychological interventions are increasingly offered as part of multimodal inpatient and outpatient treatment programs (e.g., Hechler et al. 2009). In most programs, participants receive medical treatment, psychological interventions, and physical therapy. Some programs entail additional services such as occupational therapy, consultations by social workers, and schooling by teachers in the clinic.

Summary

Psychological treatments both for acute procedural as well as chronic pain are effective. Yet, rather little is known about their differential efficacy. Aside from determining what works best for which type of pediatric pain problem, at what age and for whom, it will be the greatest challenge to make such treatments available for those who need it.

References

Cohen, L. L. (2001). Children's expectations and memories of acute distress: Short- and long-term efficacy of pain management interventions. *Journal of Pediatric Psychology, 26*, 367–374.
Hechler, T., Dobe, M., Kosfelder, J., Damschen, U., Hubner, B., Blankenburg, M., et al. (2009).

Effectiveness of a 3-week multimodal inpatient pain treatment for adolescents suffering from chronic pain: Statistical and clinical significance. *The Clinical Journal of Pain, 25*, 156–166.

Hermann, C. (2011). Psychological interventions for chronic pediatric pain: State of the art, current developments and open questions. *Pain Management, 1*, 473–483.

Kazak, A. E., Penati, B., Brophy, P., & Himelstein, B. (1998). Pharmacologic and psychologic interventions for procedural pain. *Pediatrics, 102*, 59–66.

Mahoney, L., Ayers, S., & Seddon, P. (2010). The association between parent's and healthcare professional's behavior and children's coping and distress during venepuncture. *Journal of Pediatric Psychology, 35*, 985–995.

McCracken, L. M. (2005). *Contextual cognitive-behavioral therapy for chronic pain*. Seattle: IASP Press.

McGrath, P. J., Walco, G. A., Turk, D. C., Dworkin, R. H., Brown, M. T., Davidson, K., et al. (2008). Core outcome domains and measures for pediatric acute and chronic/recurrent pain clinical trials: PedIMMPACT recommendations. *The Journal of Pain, 9*, 771–783.

Palermo, T. M., Wilson, A. C., Peters, M., Lewandowski, A., & Somhegyi, H. (2009). Randomized controlled trial of an internet-delivered family cognitive-behavioral therapy intervention for children and adolescents with chronic pain. *Pain, 146*, 205–213.

Palermo, T. M., Eccleston, C., Lewandowski, A. S., Williams, A. C. D. C., & Morley, S. (2010). Randomized controlled trials of psychological therapies for management of chronic pain in children and adolescents: An updated meta-analytic review. *Pain, 148*, 387–397.

Piira, T., Sugiura, T., Champion, G. D., Donnelly, N., & Cole, A. S. (2005). The role of parental presence in the context of children's medical procedures: A systematic review. *Child: Care, Health and Development, 31*, 233–243.

Pillai Riddell, R. R., Racine, N. M., Turcotte, K., Uman, L. S., Horton, R. E., Din Osmun, L. et al. (2011). Non-pharmacological management of infant and young child procedural pain (Review). *Cochrane database of systematic reviews*, CD006275.

Shah, V., Taddio, A., & Rieder, M. J. (2009). Effectiveness and tolerability of pharmacologic and combined interventions for reducing injection pain during routine childhood immunizations: Systematic review and meta-analyses. *Clinical Therapeutics, 31*(Suppl. 2), S104–S151.

Stevens, B., Yamada, J., & Ohlsson, A. (2010). Sucrose for analgesia in newborn infants undergoing painful procedures. *Cochrane database of systematic reviews*, CD001069.

Trautmann, E., & Kröner-Herwig, B. (2009). A randomized controlled trial of internet-based self-help training for recurrent headache in childhood and adolescence. *Behaviour Research and Therapy, 48*, 28–37.

Trautmann, E., Lackschewitz, H., & Kroner-Herwig, B. (2006). Psychological treatment of recurrent headache in children and adolescents–a meta-analysis. *Cephalalgia, 26*, 1411–1426.

Uman, L. S., Chambers, C. T., McGrath, P. J., & Kisely, S. (2008). A systematic review of randomized controlled trials examining psychological interventions for needle-related procedural pain and distress in children and adolescents: An abbreviated Cochrane review. *Journal of Pediatric Psychology, 33*, 842–854.

Wicksell, R. K., Melin, L., Lekander, M., & Olsson, G. L. (2009). Evaluating the effectiveness of exposure and acceptance strategies to improve functioning and quality of life in longstanding pediatric pain–a randomized controlled trial. *Pain, 141*, 248–257.

Psychological Treatment of Pain in Older Populations

Stephen Gibson[1] and Steven Savvas[2]
[1]National Ageing Research Institute Department of Medicine, University of Melbourne, Parkville, VIC, Australia
[2]Biomedical Division, National Ageing Research Institute, Melbourne, VIC, Australia

Synonyms

Cognitive-behavior therapy; Psychological treatment of pain in the elderly

Definition

The world's population is ageing and as the post-World War II baby boomer generation is now entering their seventh decade of life, the number of persons aged over 65 years is expected to triple by 2050 and will constitute more than one third of the total world population (U. S. Census Bureau 2012). There is no universally accepted definition of "old age," although chronological convention sets the demarcation point at 65 years. However, it is important to note that chronological age is an imperfect marker of biological age and that variation in physical, social, mental, and functional abilities is larger between individuals within any age cohort than it is across the adult life span. For this reason, the so-called elderly cannot be viewed as a homogenous group. Rather, the focus should be on the individual. In some cases the person may be frail, with cognitive impairment and multiple physical disabilities, whereas in many others of the same chronological age, there will be few or no obvious signs of senescent decline.

▶ Psychological treatments refer to a broad range of therapies which aim to improve the self-management of pain and its associated impacts (see Flor's ▶ Psychology of Pain and Psychological Treatment for detailed description). Many different psychological approaches have been used to manage pain including self-regulatory control (biofeedback, relaxation, stress reduction, hypnosis), operant behavioral treatments (utilizing principles of reinforcement of healthy/adaptive behaviors and extinction of maladaptive behaviors), and cognitive-behavioral therapy (cognitive restructure of maladaptive beliefs and appraisals, improved coping skills, promoting acceptance of pain, dealing with fear avoidance of activity, goal setting and behavioral activity attainment, problem solving, stress reduction, communication skills) among others. Rather than attempting to reiterate the conceptual frameworks and fundamental principles of the various psychological approaches to pain management in older persons, this entry will describe studies of psychological pain management in older persons and elaborate on age-related considerations in the delivery of psychological therapies.

Introduction

Chronic or persistent pain is very common in ▶ older persons. Epidemiologic studies reveal considerable variation in the prevalence estimates of persistent pain in older adults (25–70 %), although the median pain prevalence approaches 40–50 % of the community-dwelling older population (Helme and Gibson 2001). Pain in institutional settings, such as nursing homes, is even more common and has been reported in up to 80 % of residents (Takai et al. 2010). The high prevalence of persistent pain likely reflects the increasing disease burden in persons of advanced age and particularly the increasing presence of degenerative joint disease (i.e., osteoarthritis, degenerative spinal disorders).

Given the extremely high prevalence of persistent pain in the older segments of the population, one might expect a plethora of studies in the aged examining pharmacological and nonpharmacological approaches to the management of pain. Unfortunately, this is not the case and comprehensive information to help guide clinical practice is currently lacking. Nonetheless, there are a number of reasons as to why psychological interventions may be particularly valuable in the management of pain in older adults. The use of typical analgesic medications may be more problematic in older persons due to the increased risk of ▶ drug-drug and drug-disease interactions. The typical 70-year-old living in the community has three to four comorbid medical conditions and takes seven to eight different medications. As a result, drug-drug interactions are likely to be more common and may even restrict pharmacological options entirely. There is also an increased risk of adverse drug reactions in the elderly, including renal impairment, cardiovascular and gastrointestinal effects, falls, and cognitive impairment (Fine 2004). The elderly have a more external locus of control (Gibson and Helme 2000) and are often characterized as being more passive receivers of health interventions. The active, self-management approaches that are featured in many psychological therapies are likely to be more efficacious in older persons as there is greater room to gain from this previously underutilized approach. Finally, the loss of social support networks and spousal assistance through natural attrition over time can greatly reduce the level of support that is available to those of very advanced age. Psychological therapies conducted within the context of a small group setting can often ameliorate this loss and allow for much needed social interaction with all attendant benefits. In aggregate, these considerations provide a strong rationale for the more frequent and widespread use of psychological therapies in older persons with persistent, bothersome pain.

Characteristics

Psychological Management of Persistent Pain in Older Persons

Despite the very high prevalence of persistent pain in older segments of the community, there have been relatively few studies to examine the use and efficacy of psychological treatments. A couple of uncontrolled comparative studies of younger and older adults have been reported. In one of the first attempts, Middaugh and colleagues (1988) examined the treatment response in a group of 20 younger patients (29–48 years) and 17 older patients (55–70 years). All patients received 10 × 2 h training sessions including ▶ cognitive-behavioral therapy (CBT), coupled with ▶ relaxation/biofeedback. All patients improved and there was no overall difference in the magnitude of improvement between younger and older adults with respect to self-rated pain, activity levels, and medication use. Older adults showed a greater reduction in health service utilization (93 % vs. 7.5 %). There were no posttreatment changes in mood in either group and the authors conclude that older patients benefit at least as much as younger patients when exposed to this multidisciplinary intervention. Cutler et al. (1994) reported similar findings with a CBT program delivered at a multidisciplinary pain center. Three age cohorts were compared: geriatric (n = 153, 65 + years), middle-aged (n = 126, 45–64 years), and younger adults (n = 191, 21–44 years). Geriatric patients were distinctly different from younger cohorts at baseline, but all age groups showed a similar magnitude of posttreatment improvement in 37 of 43 measures, including self-rated pain severity, mood, functional status, and behavioral variables (bend forwards, bend backwards, twisting, stretching, etc.).

Studies on the treatment response to psychological therapies have also been undertaken exclusively in geriatric populations. A 6-week CBT group program offered as part of multidisciplinary pain management demonstrated that 96 % of older patients (n = 125) improved on pain severity, mood disturbance, or activity levels (Helme et al. 1996). The treatment program included attention to coping skills, restructure of maladaptive thoughts, relaxation, problem solving and operant behavioral reinforcement of healthy behaviors, and achieving preset functional goals. More than 65 % of the sample was over 70 years and 25 % exceeded 80 years of age. Finally, Reid et al. (2003) examined the use of CBT in older adults (mean age = 77 years) with chronic low back pain. Ten weekly individual sessions were administered by a psychologist. Compliance with treatment protocols was excellent and there was a significant posttreatment reduction in pain severity and disability.

Randomized Controlled Trials

The vast majority of empirical studies using randomized controlled designs relates to CBT. Indeed, the literature on mind-body therapies, operant behavior therapy, or specific subtypes of psychotherapy (i.e., hypnosis, acceptance and commitment therapy, fear avoidance exposure therapy) when used in older adults with persistent pain is largely lacking. Marone and Greco (2007) undertook a systematic review of mind-body therapies in older adults including ▶ biofeedback, ▶ progressive muscle relaxation, ▶ guided imagery, ▶ hypnosis, tai chi, qi gong, and yoga. While more than 20 studies were identified, only two focused on adults greater than 65 years, although several included adults over 50 years. The overall conclusion was that mind-body therapies are feasible in older persons, are likely to be safe, are potentially useful in reducing pain and its impacts, but are currently lacking sufficient evidence of treatment efficacy. In persons over 65 years, only meditation (Morone et al. 2008) and guided imagery (Baird and Sands 2004) have been assessed and the quality of these studies was low.

Nine randomized controlled trials have examined the efficacy of CBT for pain management in older adults. Several other studies comprise a large age range usually with some older adults, but these studies have not been included unless the mean age of the sample exceeds 64 years. In two studies of older adults with osteoarthritis knee pain (n = 99, mean age 64.5 years), Keefe and colleagues (1990a, 1990b) demonstrated the efficacy of cognitive-behavioral pain coping skills training when compared to patient education or standard medical care control groups. Patients received 10 weekly sessions each of 90-min duration and were assessed pre- and posttreatment and at 6 months follow-up. Findings demonstrate a significant reduction in pain severity and psychological disturbance, but not in physical disability in the coping skills group (Keefe et al. 1990a), and these improvements were maintained at 6 months (Keefe et al. 1990b). Another study of coping skills training in older adults with chronic knee pain (n = 69, mean age = 69 years) attempted to match the training to the preferred coping style of the patient (Fry and Wong 1991). Patients were classified as predominately problem-focused (problem solving, relaxation, cognitive restructure) or emotion-focused (hope and faith, social support) copers and then randomized to a matching intervention or a mixed approach with both coping styles. All patients improved in pain and emotional distress following treatment, but the matched intervention group showed the greatest magnitude of improvement and this was maintained at 24 weeks follow-up.

The use of CBT-based pain management has also been examined in older nursing home residents (n = 22, mean age = 80 years). Ten weekly group sessions lasting 60–90 min each were undertaken onsite at the nursing home and included education and reconceptualization of pain, relaxation, guided imagery, goal setting, and coping skills training. When compared to an attention/support control group, the CBT program produced significantly greater reductions in pain intensity and disability, although medication use and depression scores remained unchanged in either group. Several of the participants had mild cognitive impairment but adherence to the treatment was excellent with over 80 % of sessions attended. These findings provide clear support for CBT in frail older adults even in the presence of mild cognitive impairment. Particularly noteworthy is the fact that CBT could be successfully delivered in a nursing home setting because this highly dependent and vulnerable group rarely has access to this treatment modality and yet may have the greatest level of need.

Ersek et al. (2003) examined a CBT-based self-management program in a community retirement village (n = 45, mean age = 82 years). The intervention involved 90-min weekly group sessions delivered over 8 weeks and involving education on the definition and mechanisms of pain, pharmacological and nonpharmacological treatments, communication with health professionals, relaxation, goal setting, and problem solving. Individualized pain management goals were selected by participants and assistance was provided to achieve these goals. When compared to a control group who received an educational booklet, the self-management intervention showed significantly greater posttreatment improvements in pain and physical functioning, but not pain-related beliefs, depression, or interference. A larger (n = 218, mean age = 82 years), more recent study (Ersek et al. 2008) failed to replicate these findings despite a significantly greater use of relaxation and exercise/stretching in the intervention group. The authors speculate that a more heterogeneous patient group (neuropathic and nociceptive pain) may have contributed to the negative findings. Nonetheless, this study is one of the largest randomized trials in adults of advanced age and the results raise some concerns regarding patient selection, the exact content of the CBT program, and how it should be delivered.

A recent pilot study of a psychologist-led group program for older adults (n = 21, mean age = 72 years) provides further preliminary support for CBT. While few statistically significant treatment effects were observed (perhaps due to the small sample size), all measures trended in the expected direction, and significant improvement was found in measures of self-efficacy, pain-related impairment, and quality-of-life (Andersson et al. 2012). Compared to a waitlist control, providing 8 individual treatment sessions with a psychologist using a cognitive-behavioral orientation resulted in fewer maladaptive beliefs about pain and greater use relaxation in older persons (n = 95, mean age = 75 years) with persistent musculoskeletal pain (Green et al. 2009). All patients (controls and intervention) showed a reduction in pain severity and the improvement in beliefs and relaxation coping skills in the intervention group were maintained at 3 months follow-up. Self- reported disability and mood were not measured. Finally, a CBT program for the treatment of insomnia in older persons with osteoarthritis pain (n = 51, mean age = 69 years) demonstrated significant reductions in pain severity (Vitiello et al. 2009). It is suggested that CBT directed at treating the impact of persistent pain (in this case insomnia) may have a useful role in managing pain even though instruction in pain control strategies was not part of the treatment program. Such generalized, nonspecific treatment effects emphasize the multidimensional relevance of CBT and the potential importance of monitoring treatment outcome effects across a broad range of physical, psychosocial, and functional domains.

Psychological Approaches to the Management of Pain: Special Considerations in Older Persons

When considering the use of psychological therapies in older adults, it is worthwhile to explore how standard treatment approaches might need to be modified in order to better meet the needs of this population (Gibson et al. 1996; Helme and Gibson 1998; Waters et al. 2005). Perhaps one of the most fundamental considerations in tailoring psychological therapy for older persons relates to the specific goals of treatment. Goals for functional restoration might include vocational rehabilitation and resumption of all physical activities in younger adults, whereas an increase in recreational pastimes, spiritual and social interaction, as well as maintenance of functional independence in the home environment are likely to be of greater priority for the older person. Goals for improved communication skills and the specific types of maladaptive beliefs to be targeted in CBT may also vary with age (see below). Reductions in medication use may be more difficult in the patient with multiple comorbid medical conditions, and when seeking to improve disability or mood state, it is important to recognize that persistent pain may be only one of several other potentially more significant medical (i.e., frailty,

falls, cognitive impairment) and social (bereavement, loss of independence) contributors.

Many older adults will suffer from sensory impairments, including hearing loss, visual problems, decreased sensitivity to touch, and proprioceptive input. Such sensory deficits can potentially interfere with various aspects of psychological management unless appropriately assessed and the treatment modified to take these issues into account. For instance, activity pacing may be more difficult with reduced proprioceptive feedback; handout materials may be more difficult to read unless the print size and contrast sensitivity is increased, while hearing loss can interfere with verbal instructions and counselling sessions unless the volume and level of background noise are adjusted. The presence of other comorbid illnesses might also require some consideration. The average 70-year-old living in the community will have three to four different disease states requiring active medical management. Dementia alone affects about 1.5 % of adults aged 65–70, but the rate doubles every 5 years thereafter, peaking at over 40 % prevalence in those in the tenth decade of life. A slower pace of group education and counselling sessions might be needed to accommodate such comorbidities. The duration of sessions might need to be reduced and the use of less abstract content with more repetition of cognitive-behavioral instructions may be required. Medical comorbidity can reduce the ability to travel and therefore access outpatient- and office-based psychological treatment programs. Greater acceptance and adherence to group-based programs may result from providing treatment within community settings or at a residential retirement village (Ersek et al. 2003, 2008). Severe cognitive impairment is a contraindication for CBT, although programs have been shown to be effective in those with mild to moderate cognitive impairment (Cook et al. 1998).

Some beliefs and attitudes about pain are known to vary as a function of age. Older adults have been reported to be more accepting of pain and may misattribute mild aches and joint pains to the normal ageing process, rather than a consequence of a treatable disease (see Gibson 2005 for review). Stoic attitudes (hide pain, carry on despite pain) to the presence of pain are more frequently endorsed by persons of advanced age (Yong et al. 2003); a higher conviction in finding a medical cure (Gibson 2003) and a more external locus of control over pain (chance or belief in powerful others) have been reported (Gibson and Helme 2000). This may be important given that patients with an internal locus (one's own actions control pain) are known to be more successful with self-help, active management approaches, and this is often a fundamental principle of CBT programs. Elements of psychological therapy that focus on cognitive restructuring of thoughts should consider these age differences in pain beliefs and attitudes and possibly target such beliefs in addition to other well-recognized maladaptive thoughts (i.e., catastrophizing).

Summary

The evidence from both controlled and uncontrolled evaluations of psychological therapies for the management of persistent pain suggest that it is well accepted by older adults and can lead to significant improvements. The treatment outcomes in specific domains (pain, mood, disability, beliefs, etc.) are somewhat mixed between studies particularly with respect to change in mood and functional disability, although all but one study demonstrated an improvement in self-rated pain severity. There are considerable variations in the exact type of CBT offered, the duration of treatment, and how it was delivered (individual vs group sessions). Moreover, many of the CBT programs were administered as part of a multidisciplinary pain program rather than as the sole therapy. This multimodal, multidisciplinary approach to pain management may be particularly indicated for older adults given the high levels of medical morbidity and complex functional and psychosocial impacts of multimorbid pain. However, the use of a combination, multimodal therapy makes it difficult to evaluate the specific merit of psychological components of therapy. Despite these considerations, the weight of evidence does indicate a clear role for the inclusion of psychological approaches in the management of bothersome persistent pain in older adults, and contrary to negative, ageist stereotypes, such therapies are often efficacious.

References

Andersson, G., Johansson, C., Nordlander, A., & Asmundson, G. J. (2012). Chronic pain in older adults: A controlled pilot trial of a brief cognitive-behavioural group treatment. *Behavioural and Cognitive Psychotherapy, 40*(2), 239–244.

Baird, C. L., & Sands, L. (2004). A pilot study of the effectiveness of guided imagery with progressive muscle relaxation to reduce chronic pain and mobility difficulties of osteoarthritis. *Pain Management Nursing, 5*(3), 97–104.

Cook, A. J. (1998). Cognitive-behavioral pain management for elderly nursing home residents. *The Journals of Gerontology. Series B, Psychological Sciences and Social Sciences, 53*(1), P51–P59.

Cutler, R. B., Fishbain, D. A., Rosomoff, R. S., & Rosomoff, H. L. (1994). Outcomes in treatment of pain in geriatric and younger age groups. *Archives of Physical Medicine and Rehabilitation, 75*(4), 457–464.

Ersek, M., Turner, J. A., McCurry, S. M., Gibbons, L., & Kraybill, B. M. (2003). Efficacy of a self-management group intervention for elderly persons with chronic pain. *The Clinical Journal of Pain, 19*(3), 156–167.

Ersek, M., Turner, J. A., Cain, K. C., & Kemp, C. A. (2008). Results of a randomized controlled trial to examine the efficacy of a chronic pain self-management group for older adults. *Pain, 138*(1), 29–40.

Fine, P. G. (2004). Pharmacological management of persistent pain in older patients. *The Clinical Journal of Pain, 20*(4), 220–226.

Fry, P. S., & Wong, P. T. P. (1991). Pain management training in the elderly: Matching interventions with the subjects' coping styles. *Stress Medicine, 7*, 93–98.

Gibson, S. J. (2003). Pain and ageing. In J. O. Dostrovsky, D. B. Carr, & M. Koltzenburg (Eds.), *Proceedings of 10th world congress on pain* (pp. 767–790). Seattle: IASP Press.

Gibson, S. J. (2005). Age differences in psychological factors related to pain perception and report. In S. J. Gibson & D. K. Wiener (Eds.), *Pain in older adults* (pp. 87–110). Seattle: IASP press.

Gibson, S. J., & Helme, R. D. (2000). Cognitive factors and the experience of pain and suffering in older persons. *Pain, 85*(3), 375–383.

Gibson, S. J., Farrell, M. J., Katz, B., & Helme, R. D. (1996). Multidisciplinary management of chronic nonmalignant pain in older adults. In B. A. Ferrell & B. R. Ferrell (Eds.), *Pain in the elderly* (pp. 91–101). Seattle: IASP press.

Green, S. M., Hadjistavropoulos, T., Hadjistavropoulos, H., Martin, R., & Sharpe, D. (2009). A controlled investigation of a cognitive behavioural pain management program for older adults. *Behavioural and Cognitive Psychotherapy, 37*(2), 221–226.

Helme, R. D., & Gibson, S. J. (1998). Measurement and management of pain in older people. *Australia Journal Ageing, 17*(1), 5–10.

Helme, R. D., & Gibson, S. J. (2001). The epidemiology of pain in elderly people. *Clinics in Geriatric Medicine, 17*(3), 417–431.

Helme, R. D., Katz, B., Gibson, S. J., Bradbeer, M., Farrell, M., Neufeld, M., & Corran, T. (1996). Multidisciplinary pain clinics for older people. Do they have a role? *Clinical Geriatrie Medicine, 12*(3), 563–582.

Keefe, F. J., Caldwell, D. S., Williams, D. A., Gil, K. M., Mitchell, D., Robinson, C., Martinez, S., Nunley, J., Beckham, J. C., & Helms, M. (1990a). Pain coping skills training in the management of osteoarthritic knee pain: A comparative study. *Behavior Therapy, 21*, 49–62.

Keefe, F. J., Caldwell, D. S., Williams, D. A., Gil, K. M., Mitchell, D., Robinson, C., Martinez, S., Nunley, J., Beckham, J. C., & Helms, M. (1990b). Pain coping skills training in the management of osteoarthritic knee pain-II: Follow-up results. *Behavior Therapy, 21*, 435–447.

Middaugh, S. J., Levin, R. B., Kee, W. G., Barchiesi, F. D., & Roberts, J. M. (1988). Chronic pain: Its treatment in geriatric and younger patients. *Archives of Physical Medicine and Rehabilitation, 69*(12), 1021–1026.

Morone, N. E., & Greco, C. M. (2007). Mind-body interventions for chronic pain in older adults: A structured review. *Pain Medicine, 8*(4), 359–375.

Morone, N. E., Greco, C. M., & Weiner, D. K. (2008). Mindfulness meditation for the treatment of chronic low back pain in older adults: A randomised controlled pilot study. *Pain, 134*(3), 310–319.

Reid, M. C., Otis, J., Barry, L. C., & Kerns, R. D. (2003). Cognitive-behavioral therapy for chronic low back pain in older persons: A preliminary study. *Pain Medicine, 4*(3), 223–230.

Takai, Y., Yamamoto-Mitani, N., Okamoto, Y., Koyama, K., & Honda, A. (2010). Literature review of pain prevalence among older residents of nursing homes. *Pain Management Nursing, 11*(4), 209–223.

U. S. Census Bureau, D. I. S. (n.d.). (2012). National Population Projections: Summary Tables. Retrieved 18 February 2013, from http://www.census.gov/population/projections/data/national/2012/summarytables.html.

Vitiello, M. V., Rybarczyk, B., Von Korff, M., & Stepanski, E. J. (2009). Cognitive behavioural therapy for insomnia improves sleep and decreases pain in older adults with co-morbid insomnia and osteoarthritis. *Journal of Clinical Sleep Medicine, 5*(4), 355–362.

Waters, S. J., Woodward, J. T., & Keefe, F. J. (2005). Cognitive-behavioral therapy for pain in older adults. In S. J. Gibson & D. K. Wiener (Eds.), *Pain in older adults* (pp. 87–110). Seattle: IASP Publications.

Yong, H.-H., Bell, R., Workman, B., & Gibson, S. J. (2003). Psychometric properties of the Pain Attitudes Questionnaire (revised) in adult patients with chronic pain. *Pain, 104*(3), 673–681.

Psychological Treatment of Pain in the Elderly

▶ Psychological Treatment of Pain in Older Populations

Psychology of Pain and Psychological Treatment

Herta Flor[1] and Niels Birbaumer[2]
[1]Department of Cognitive and Clinical Neuroscience, Central Institute of Mental Health, Medical Faculty Mannheim, Heidelberg University, Mannheim, Germany
[2]Institute of Medical Psychology and Behavioral Neurobiology, University of Tübingen, Tübingen, Germany

Characteristics

The Role of Learning in Chronic Pain

Both associative and nonassociative learning (see also ▶ Modeling, Social Learning in Pain) processes as well as social learning have been found to be of fundamental significance in the development of chronic pain. The repeated application of painful stimuli leads to reduced responsivity, that is, ▶ habituation to the painful stimulation. In many states of chronic pain, ▶ sensitization (see also ▶ Psychology of Pain, Sensitization, Habituation, and Pain) rather than habituation occurs due to changes at both the level of the receptor and the central nervous system (Woolf and Mannion 1999). Sensory information accelerates habituation and reduces activation caused by surprise, insecurity, and threat. This mechanism may underlie the effects reported in a large number of studies that support the positive results of preparatory information prior to acutely painful procedures, such as surgery or bone marrow aspiration.

The most influential model of psychological factors in chronic pain was W. E. Fordyce's assumption that chronic pain can develop and be maintained due to ▶ operant conditioning (see also ▶ Operant Treatment of Chronic Pain; ▶ Operant Perspective of Pain) of pain behaviors, that is, overt expressions of pain. Fordyce (1976) postulated that acute pain behaviors such as limping or moaning may come under the control of external contingencies of reinforcement and thus develop into a chronic pain problem. Positive reinforcement of pain behaviors (see also ▶ Motivational Aspects of Pain) (e.g., by attention and the expression of sympathy), negative reinforcement of pain behaviors (see also ▶ Motivational Aspects of Pain) (e.g., the reduction of pain through the intake of medication or by the cessation of activity), as well as a lack of reinforcement of healthy behaviors (see also ▶ Motivational Aspects of Pain) could provoke chronicity in the absence of nociceptive input. Thus, ▶ pain behaviors (see also ▶ Motivational Aspects of Pain) – originally elicited by nociceptive input – may, over time, occur in response to environmental contingencies. This model has generated much research that has not only confirmed the original assumptions made by Fordyce but has also shown that in addition to pain behaviors, the subjective experience of pain and physiological responses related to pain are subject to reinforcement learning. A special role has been assigned to significant others of chronic pain patients, who have high reinforcement potential. When solicitous, that is, pain-reinforcing spouses were present, several studies found that the patients were more pain sensitive during acute pain tests than when the spouses were absent. The patients with nonsolicitous spouses did not differ in the spouse present or absent conditions. These studies suggest that spouses can serve as discriminative stimuli for the display of pain behaviors by chronic pain patients, including their reports

of pain intensity (Flor et al. 1995a). Health care providers may also become discriminative cues influencing patients' responses. Solicitous responses by significant others can also lead to increased physiological pain responses in the patients, whereas focusing on healthy behaviors by significant others can have positive effects on the pain experience.

Of equal importance is the operant conditioning related to the intake of pain medication. Patients are often told by their physicians or by well-meaning family members that they should not continue to take analgesic medication unless the pain increases to a point where it becomes intolerable (referred to ▶ prn from the Latin "take as needed"). When pain medication is taken at that stage, both pharmacological and behavioral factors can contribute to the development of misuse of medication and in severe cases even drug dependence. If analgesic medication is taken only at peak pain levels, the effect of the medication is less potent and patients cycle between high and low levels of medication, which facilitates the development of dependence. In addition, medication intake is negatively reinforcing, since the intake of medication ends an aversive state (pain). Subsequently, the pain-reducing behavior (use of analgesics) increases in frequency. Thus, both pharmacotherapists and behavioral psychologists recommend that analgesic medication should not be taken on a pain contingent but rather in a time contingent fashion adapted to the specific pain level of the patient and the half-life of the drug.

The negative reinforcement of activity levels is an important process in the development of disability. A specific activity, for example, walking, is performed until pain sets in, at which point the activity is interrupted and replaced by rest. Subsequently, the pain will be reduced. This reduction of an aversive state – pain – negatively reinforces the cessation of activity. As was the case with the intake of analgesic medication, the cessation of activity has to be made dependent on amount of activity achieved – quota based (e.g., number of stairs climbed, distance walked), rather than on the amount of pain. Thus, the pain-reinforcing quality of rest is eliminated. This formulation supports the strategy of encouraging patients to perform activities to meet a specific quota and not until pain is perceived as overwhelming.

The respondent conditioning (see also ▶ Respondent Conditioning of Chronic Pain) model postulates that numerous formerly neutral cues can be associated with the experience of pain and can – over time – themselves elicit responses that lead to an increased pain response and create the experience of pain in the absence of nociceptive input. To illustrate the proposed process from a respondent conditioning perspective, the patient may have learned to associate increases in muscle tension with all kinds of stimuli that were originally associated with nociceptive stimulation. Thus, sitting, walking, bending, or even thoughts about these movements may elicit anticipatory ▶ anxiety (see also ▶ Depression and Pain, ▶ Fear and Pain) and increases in muscle tension. This fear of movement or "▶ kinesiophobia" (see also ▶ Fear and Pain) has been discussed as an important factor in the maintenance and exacerbation of chronic pain (e.g., Vlaeyen and Linton 2000, 2012). Subsequently, patients may display maladaptive responses to any number of stimuli and reduce the frequency of performance of many activities other than those that initially reduced pain. Thus, although the original association between injury and pain results in anxiety regarding movement, over time the anxiety may lead to increased muscle tension and pain even if the nociceptive stimuli are no longer present. In addition, stress situations can increase muscle tension levels and cause sympathetic activation and may thus reinforce the process. Many patients have reported that an acute pain problem evolved into chronic pain at a time where personal stressors co-occurred with the pain. ▶ Stress (see also ▶ Stress and Pain) situations may serve as additional US and also as CS for muscle tension increases, increased sympathetic activation, and subsequently pain. It has been shown that pain perception can be increased or decreased by classically conditioning positive or negative stimuli to pain (Flor and Turk 2011).

Nonoccurrence of pain is a powerful reinforcer for reduction of movement. Thus, the original respondent conditioning may be complemented by an ▶ operant process (see also ▶ Operant Treatment of Chronic Pain; ▶ Operant Perspective of Pain), whereby the nociceptive stimuli need no longer be present for the avoidance behavior to occur. People who suffer from acute back pain, regardless of the cause, may adopt specific behaviors (e.g., limping) to avoid pain, and they may never obtain "corrective feedback" because they fail to perform more natural movements and learn that they may not induce pain. Reduction in physical activity may subsequently result in muscle atrophy and increased disability. In this manner, the physical abnormalities proposed by biomechanical models of pain may actually be secondary to changes in behavior initiated through learning. Chronic pain patients tend to focus their attention on impending pain and subsequently avoid many types of activity, thus fostering the development of disability and depression. The release of endogenous opioids – the body's own analgesic system – may also be influenced by respondent conditioning (cf. Flor et al. 2002) as well as brain responses related to the experience of pain (cf. Schneider et al. 2004). Moreover, fear of pain and subsequent avoidance of activity is one of the best predictors (see also ▶ Psychological Predictors of Chronicity) of subsequent chronicity (see also ▶ Psychological Predictors of Chronicity) (Asmundson et al. 1999). Conditioned ▶ fear of pain and movement may lead to the avoidance of many behaviors, including motor behaviors.

Response acquisition through the observation of others or ▶ social learning (see also ▶ Modeling, Social Learning in Pain) is an essential mechanism of learning new patterns of behavior. Children acquire attitudes about health and health care, the perception and interpretation of symptoms, and physiological processes from their parents and social environment. They also learn appropriate responses to injury and disease and thus may be more or less likely to ignore or overrespond to normal bodily sensations they experience. The culturally acquired perception and interpretation of symptoms determines how people deal with illness. The observation of others in pain is an event that captivates attention. This attention may have survival value, may help to avoid experiencing more pain, and may help to learn what to do about acute pain.

Modeling (see also ▶ Modeling, Social Learning in Pain) probably plays a part in the phenomenon of "pain-prone families" families with a significantly elevated occurrence of pain problems (Levy et al. 2007). It was, for example, reported that children show the same pain syndromes that their parents currently have, rather than the pain problems their parents had in their own childhood. The large cultural variations in pain expression are also important. It is likely that social factors play a much larger role in pain perception and pain coping than previously thought and that they are especially important in the response of children to pain (Hadjistav-ropoulos et al. 2011; Vervoort et al. 2011). In common clinical practice, the acquisition or extinction of pain-related behavior by means of modeling has received little attention. However, there are occasional indications for the role of modeling in treating pain problems in children on burn units and in the treatment of postoperative pain.

Despite the great deal of data available on the modification of experimentally induced pain behavior by means of modeling in normal (healthy) subjects, there are still few experimental results concerning chronic pain patients. Nor are there any longitudinal studies of the development of pain syndromes in "pain-prone families." Further investigation of modeling and social learning as a factor in the development of chronic pain disorders is necessary.

Cognitive Factors and Pain

Cognitive-behavioral models of chronic pain emphasize that the evaluation of the pain experience by the patient greatly determines that amount of pain that is experienced as well as its negative consequences (Flor and Turk 2006, 2011; Turk et al. 1983). General assumptions that characterize the cognitive-behavioral perspective are as follows: (1) people are active

processors of information and not passive reactors; (2) thoughts (e.g., appraisals, expectancies) can elicit or modulate mood, affect physiological processes, influence the environment, and serve as an impetus for behavior (conversely, mood, physiology, environmental factors, and behavior can influence thought processes); (3) behavior is reciprocally determined by the person and environmental factors; (4) people can learn more adaptive ways of thinking, feeling, and behaving; and (5) people are capable and should be involved as active agents in change of maladaptive thoughts, feelings, and behaviors.

From the cognitive-behavioral perspective (see also ▶ Cognitive-Behavioral Perspective of Pain), people suffering from chronic pain are viewed as having negative expectations about their own ability to control certain motor skills, such as performing specific physical activities (e.g., climbing stairs, lifting objects) that are attributed to one overwhelming factor, namely, a chronic pain syndrome. Moreover, chronic pain patients tend to believe that they have a limited ability to exert any control over their pain. Such negative, maladaptive appraisals of the situation and personal efficacy may reinforce the experience of demoralization, inactivity, and overreaction to nociceptive stimulation. Specifically, too much attention to pain may be associated with a state of hypervigilance (see ▶ Hypervigilance and Attention to Pain) that increases the impact of pain and leads to unnecessary suffering (Van Damme et al. 2010).

A great deal of research has been directed toward identifying cognitive factors that contribute to pain and disability. Numerous studies have been conducted to examine the importance of cognitive components of the pain. These have consistently demonstrated that patients' attitudes, beliefs, expectancies about their plight, their coping resources, and the health care system affect their reports of pain, activity, disability, and response to treatment (e.g., Wiech et al. 2008).

Pain, when interpreted as signifying ongoing tissue damage or a progressive disease, seems to produce considerably more suffering and behavioral dysfunction than if it is viewed as being the result of a stable problem that is expected to improve. A number of studies have used experimental pain stimuli and demonstrated that the conviction of personal control (see also ► Psychological Treatment of Headache) can ameliorate the experience of experimentally induced nociception. Moreover, the type of thoughts employed during exposure to painful stimulation has been related to pain tolerance and pain intensity ratings. ► Catastrophizing (see also essay ► catastrophizing) thoughts have been associated with lower pain tolerance and higher ratings of pain intensity. In contrast, coping thoughts have been related to higher pain tolerance and lower pain intensity ratings.

Certain beliefs may lead to maladaptive ► coping (see also ► Fear and Pain, ► Motivational Aspects of Pain), increased suffering, and greater disability. Patients who believe their pain is likely to persist may be passive in their coping efforts and fail to make use of available strategies, even when in their repertoire, to cope with pain. Patients who consider their pain to be an "unexplainable mystery" may negatively evaluate their own abilities to control or decrease pain and are less likely to rate their coping strategies as effective in controlling and decreasing pain.

Once beliefs and expectancies (cognitive schemata) about a disease are formed, they become stable and are very difficult to modify. Patients tend to avoid experiences that could invalidate their beliefs and they guide their behavior in accordance with these beliefs, even in a situation where the belief is no longer valid (no corrective feedback is received to discredit this belief). For example, feeling some muscular pain following activity may be caused by lack of muscle strength and general deconditioning and not by additional tissue damage.

Self-regulation of pain and its impact depends upon the person's specific ways of dealing with pain, adjusting to pain, and reducing or minimizing pain and distress caused by pain – their coping strategies. Coping is assumed to be manifested by spontaneously employed purposeful and intentional acts, and it can be assessed in terms of overt and covert behaviors. Overt, behavioral coping strategies include rest, medication, and use of relaxation. Covert coping strategies include various means of distracting oneself from pain, reassuring oneself that the pain will diminish, seeking information, and problem solving. Studies have found active coping strategies (efforts to function in spite of pain or to distract oneself from pain such as activity, ignoring pain) to be associated with adaptive functioning and passive coping strategies (depending on others for help in pain control and restricted activities) to be related to greater pain and depression. However, beyond this, there is no evidence supporting the greater effectiveness of one active coping strategy compared to any other (Fernandez and Turk 1989). Specific coping strategies need not always be adaptive or maladaptive. It seems more likely that different strategies will be more effective than others for some people at some times, but not necessarily for all people all of the time or even the same person at different times. One of the most important factors for improved coping with pain as well as the success of psychological treatment seems to be a sense of mastery over the goals one seeks to accomplish in the sense of increased self-efficacy (Keefe et al. 2004).

Pain and Affect

The affective factors associated with pain include many different emotions, but they are primarily negative in quality. ► Anxiety (see also ► Depression and Pain; ► Fear and Pain) and ► depression (see also ► Depression and Pain; ► Fear and Pain) have received the greatest amount of attention in chronic pain patients; however anger has recently received considerable interest as an important emotion in chronic pain patients. Research suggests that from 40 % to 50 % of chronic pain patients suffer from depression (Gracely et al. 2012; Romano and Turner 1985). There have been extensive and fruitless debates concerning the causal relationship between depression and pain. In the majority of cases, depression appears to be the patients' reaction to their plight (Fishbain et al. 1997). The presence of depression is closely related to the feelings of loss of control and helplessness often associated with pain.

Several investigators have also found a close association between fear of pain and dysfunctional coping. In addition, high comorbidity between anxiety disorders and pain seems to be present. Muscular hyperreactivity to stress seems to be closely associated with fear of pain (see also ► Fear and Pain) (e.g., Geisser et al. 2004) and exposure therapy (see also ► Fear Reduction Through Exposure In Vivo) has been suggested as a new treatment option. Anger (see also ► Anger and Pain) has been widely observed in individuals with chronic pain and there is conflicting evidence if anger in or anger out styles are related to higher pain sensitivity (e.g., Bruehl et al. 2006; Greenwood et al. 2003). Bruehl et al. (2009) proposed that a greater propensity for anger out behavior is linked with more intense pain perception, and they suggested an opioid dysfunction hypothesis in which inadequate endogenous opioid inhibitory activity in several pain-regulating brain regions is thought to contribute to links between trait anger out and pain. Anger and hostility are closely associated with pain in persons with spinal cord problems. Frustrations related to persistence of symptoms, limited information on etiology, and repeated treatment failures, along with anger toward employers, the insurance, the health care system, family members, and themselves, also contribute to the general dysphoric mood of these patients. The impact of anger and frustration on exacerbation of pain and treatment acceptance has not received adequate attention. It would be reasonable to expect that the presence of anger may serve as an aggravating factor, associated with increasing autonomic arousal and blocking of motivation and acceptance of treatments oriented toward rehabilitation and disability management rather than cure, which are often the only treatments available for chronic pain.

Biobehavioral Perspective

A biobehavioral or a ► diathesis stress (see also ► Diathesis-Stress Model of Chronic Pain) model of chronic pain needs to consider the factors discussed above and their mutual interrelationships in the explanation of chronic pain. The existence of a physiological disposition or diathesis is one important component. This predisposition is related to a reduced threshold for noxious stimulation and can be determined by genetic factors or acquired through early learning experiences. For example, Mogil (2012) summarized research that showed that large genetic variations in individual pain sensitivity exist. Very impressive evidence for the role of early traumatic experience comes from the work of Anand et al. (1999) who showed that minor noxious experience in neonate rats leads to dramatic alterations (sensitization) in nociceptive processing in the adult organism. These studies were also done in humans suggesting similar long-term effects (Hermann et al. 2006). A further component of the biobehavioral model is a response stereotypy of a particular bodily system, such as exaggerated muscular responses of the lower back muscle to stress and pain that is based both on the diathesis and on aversive experiences present at the time of the development of the response. These aversive stimuli may include personal or work stress or problematic occupational conditions and will lead not only to painful responses but also to avoidance behaviors and associated maladaptive cognitive and affective processes. The cognitive evaluation of these external or internal stimuli is of great importance in the pain response as discussed above.

Memory for Pain

An important maintaining factor in this chronicity process is the development of pain memories (see also ► Fear and Pain). These pain-related memories may be explicit or implicit and may subsequently guide the patient's experience and behaviors (Erskine et al. 1990). For example, pain patients have a tendency to remember preferentially negative and pain-related life events and show a deficit in the retrieval of positive memories. The experience of chronic pain also leads to the development of somatosensory pain memories, for example, an expanded representation of the affected body part in primary somatosensory cortex. This expanded cortical representation is accompanied by increased sensitivity to both painful and nonpainful stimuli and may be further enhanced by learning processes or attention to painful stimulation. An even more dramatic example of a learned memory for pain has been found in phantom limb pain patients (Flor et al. 1995b). In upper extremity amputees, the magnitude of the phantom limb pain was found to be proportional to the amount of reorganization in primary somatosensory cortex, namely, the shift of the cortical mouth representation into the area where the amputated limb was formerly represented. The brain obviously maintains a memory of the former input to the deafferented area and subsequently stimulation stemming from areas adjacent to the deafferented zone elicits sensations and pain in the now absent limb. Phantom sensations and cortical reorganization are absent in congenital amputees. The focus of the biobehavioral perspective is thus on the patient and not just on the symptoms or the underlying pathology and this focus also requires that the treatment of the patients be tailored not only to medical factors, but incorporate psychosocial variables that may often be predominant in states of chronic pain.

Psychological Assessment of Pain and Its Consequences

As noted above, pain is viewed as a complex response with verbal-subjective, motor-behavioral, and physiological components that need to be assessed in a multifactorial model (Turk and Melzack 2011). A comprehensive psychological assessment (see also ▶ Psychophysiological Assessment of Pain; ▶ Psychological [Behavioral Health] Assessment of Pain; ▶ Psychology of Pain, Assessment of Cognitive Variables) of pain includes the measurement of the intensity and quality of pain, as well as its consequences in many areas of life and a thorough analysis of pain-eliciting and pain-maintaining factors. Pain coping strategies and pain-related cognitions give important information about how patients deal with pain and how this may lead to chronicity. In view of the high comorbidity of pain and depression, the assessment of depression is important in addition to the measurement of anger and anxiety related to pain. Pain behaviors are important for the analysis of pain-reinforcing

factors and can be assessed by observation checklists or video recordings with subsequent behavioral analyses. The amount of disability should also be determined as well as the patients' general activity levels. Psychophysiological assessments (▶ Psychophysiological Assessment of Pain; ▶ Assessment of Pain Behaviors) become more common as the cost of equipment as well as the complexity of the assessments are decreasing due to new technical developments. General measures of arousal such as heart rate and skin conductance as well as more specific pain-related variables such as muscle tension levels or responses to stress have been assessed. Turk and Rudy (1988) proposed a multiaxial pain assessment based on the Multidimensional Pain Inventory (Kerns et al. 1985) that includes patients' estimates of and responses to pain, significant other responses to pain behaviors, and activity levels. Cluster analyses were used to identify various patient subgroups that were similar across different diagnoses and that reflected psychosocial patient characteristics.

Psychological Treatment of Chronic Pain

Biofeedback and Relaxation

▶ Biofeedback (see also ▶ Biofeedback in the Treatment of Pain) refers to the modification of a bodily process (e.g., skin temperature, muscle tension) that normally does not reach consciousness (see ▶ Consciousness and Pain) by making the bodily process perceptible to the patient. The physiological signal is measured and amplified and fed back to the patients by the use of a computer that translates variations in bodily processes into visual or auditory or tactile signals. Seeing or hearing their blood pressure or muscle tension enables persons to self-regulate it. The most common type of biofeedback for chronic pain is muscle tension or electromyographic (EMG) biofeedback, which was found to be effective for several chronic musculoskeletal pain syndromes (e.g., Flor et al. 1992; Kröner-Herwig 2011). For migraine headache (see ▶ Psychological Treatment of Headache), temperature, blood flow in the temporal artery, or slow cortical potentials have been fed back with good results (Penzien et al. 2002). Similar good results are available for Raynaud's disease with respect to temperature feedback (e.g., Kranitz and Lehrer 2004). Other respondent methods are various types of relaxation training (see also ▶ Relaxation in the Treatment of Pain), among which progressive muscle relaxation seems to be especially suited for the treatment of chronic musculoskeletal pain (e.g., Ostelo et al. 2005).

Operant Behavioral Treatment
Patients who show high levels of pain behaviors and are very incapacitated by their pain should profit from ▶ operant behavioral treatment (see also ▶ Cognitive-Behavior Treatment of Pain). The goals of this treatment are increases in activity levels and healthy behaviors related to work, leisure time, and the family, as well as medication reduction and change in the behavior of significant others (cf. Fordyce 1976). The overall goal is to reduce disability by reducing pain and increasing healthy behaviors. This approach has been found to be effective with patients with chronic back pain as well as other pain syndromes.

Cognitive-Behavioral Treatment of Chronic Pain
The cognitive-behavioral (see ▶ Cognitive-Behavior Treatment of Pain; ▶ Cognitive-Behavioral Perspective of Pain) model of chronic pain emphasizes the role of cognitive, affective, and behavioral factors on the development and maintenance of chronic pain. The cognitive-behavioral perspective that focuses on the important role of assigning more responsibility to the patients can be applied to any type of treatment. The central tenet of cognitive-behavioral treatment is to reduce feelings of helplessness and uncontrollability and to establish a sense of control over pain in the patients. This is achieved by the modification of pain-eliciting and pain-maintaining behaviors, cognitions, and emotions. The CB approach teaches patients various techniques to deal effectively with episodes of pain. Pain-related cognitions are changes produced by cognitive restructuring and pain coping strategies, such as attention diversion, self-hypnosis (see also ▶ Therapy of Pain, Hypnosis), and use of imagery or relaxation that increase

self-efficacy (see also ▶ Psychology of Pain, Self-Efficacy). A variant of cognitive-behavioral therapy is fear exposure training for chronic pain or acceptance and commitment therapy that address specifically fear of pain and maladaptive coping and have shown initial positive results (Vlaeyen et al. 2012). Several studies (Kröner-Herwig 2011) have examined the efficacy (see ▶ Psychology of Pain, Self-efficacy) of cognitive-behavioral pain management, which must be considered as a very effective treatment of chronic pain (e.g., Morley 2011).

Important Aspects of Psychological Treatment of Chronic Pain
Psychological treatment of chronic pain is usually performed in an interdisciplinary setting (see ▶ Motivational Aspects of Pain) that includes medical interventions, physiotherapy, and social measures that are often combined in a multimodal approach. The problem with multimodal approaches is that some parts of the treatments may counteract each other and that it is difficult to assess the contribution of the individual components. Rather than combining an array of diverse intervention strategies, it might be more fruitful to aim for a differential indication of various treatment components based on the pain-related characteristics of the patients. For example, Turk et al. (1998) found that persons characterized by high levels of dysfunction responded better to an interdisciplinary pain treatment program for fibromyalgia than those who were interpersonally distressed. The prediction of outcome (see ▶ Psychological Treatment of Chronic Pain, Prediction of Outcome) in psychological pain treatment would greatly benefit from such an indication-based approach.

Another important aspect of psychological pain management is the motivation (see ▶ Motivational Aspects of Pain) of the patients for a psychological approach. This is often difficult since they may be concerned that the referral to a mental health professional implies that their pain is "not real," they are exaggerating, the pain they feel is really "all in their head," or their pain is a psychological and therefore not a physical problem. Furthermore, many pain sufferers fear that a referral for psychological intervention implies that they can no longer be helped by the traditional health care system and that they are being abandoned as "hopeless cases." They may view the referral as requiring that they prove that they do have legitimate reasons for their reported symptoms. These people usually believe that psychological assessment is not relevant to their problem, when they know that there must be a known physical basis for their symptoms. The patient may believe that cure of the disease or elimination of the symptoms or physical limitations is all that is required or ask why they are being referred to a psychologist or psychiatrist. This requires a motivational phase prior to treatment that familiarizes the patients with the multidimensional view of chronic pain and motivates them to view a psychological approach as a chance to alter their attitude toward their pain and a first step toward improvement as well as an optimal patient-therapist interaction (see ▶ Chronic Pain, Patient-Therapist Interaction) that is characterized by an emphasis of the self-management skills of the patient. Psychological pain treatments are also effective in children (see ▶ Psychological Treatment of Pain in Children) (Eccleston et al. 2003) and older populations (see ▶ Psychological Treatment of Pain in Older Populations) (Karp et al. 2008) and can be used throughout the life span. To what extent spouses (see ▶ Spouse, Role in Chronic Pain) or significant others should be included in psychological pain treatment is a matter of controversy; the literature is not consistent on this point (van Lankveld et al. 2004). Moreover, gender (see ▶ Gender and Pain) aspects have been addressed with respect to pain sensitivity, but are virtually unexplored with respect to the efficacy of psychological pain treatment. Psychological treatment methods have also been successfully applied to the treatment of acute pain (see ▶ Psychological Treatment in Acute Pain) states (e.g., Luebbert et al. 2001), where they can be employed to prevent chronicity (see ▶ Chronicity, Prevention) and the development of persistent pain memories (e.g., Linton et al. 2005).

References

Anand, K. J., Coskun, V., Thrivikraman, K. V., Nemeroff, C. B., & Plotsky, P. M. (1999). Long-term behavioral effects of repetitive pain in neonatal rat pups. *Physiology and Behavior, 66*(4), 627–637.
Asmundson, G. J., Norton, P. J., & Norton, G. R. (1999). Beyond pain: The role of fear and avoidance in chronicity. *Clinical Psychology Review, 19*(1), 97–119.
Bruehl, S., Chung, O. Y., & Burns, J. W. (2006). Anger expression and pain: An overview of findings and possible mechanisms. *Journal of Behavioral Medicine, 29*(6), 593–606.
Bruehl, S., Burns, J. W., Chung, O. Y., & Chont, M. (2009). Pain-related effects of trait anger expression: Neural substrates and the role of endogenous opioid mechanisms. *Neuroscience and Biobehavioral Reviews, 33*(3), 475–491.
Eccleston, C., Yorke, L., Morley, S., Williams, A. C., & Mastroyannopoulou, K. (2003). Psychological therapies for the management of chronic and recurrent pain in children and adolescents. *Cochrane Database of Systematic Reviews, 1,* CD003968.
Erskine, A., Morley, S., & Pearce, S. (1990). Memory for pain: A review. *Pain, 41*(3), 255–265.
Fernandez, E., & Turk, D. C. (1989). The utility of cognitive coping strategies for altering pain perception: A meta-analysis. *Pain, 38*(2), 123–135.
Fishbain, D. A., Cutler, R., Rosomoff, H. L., & Rosomoff, R. S. (1997). Chronic pain-associated depression: Antecedent or consequence of chronic pain? A review. *The Clinical Journal of Pain, 13*(2), 116–137.
Flor, H., & Turk, D. C. (2006). Cognitive and learning aspects. In S. B. McMahon & M. Koltzenburg (Eds.), *Wall and Melzack's textbook of pain* (5th ed., pp. 241–258). London: Elsevier.
Flor, H., & Turk, D. C. (2011). *Chronic pain: An integrated biobehavioral approach.* Seattle: IASP Press.
Flor, H., Schugens, M. M., & Birbaumer, N. (1992). Discrimination of muscle tension in chronic pain patients and healthy controls. *Biofeedback and Self-Regulation, 17*(3), 165–177.
Flor, H., Breitenstein, C., Birbaumer, N., & Fürst, M. (1995a). A psychophysiological analysis of spouse solicitousness towards pain behaviors, spouse interaction, and pain perception. *Behavior Therapy, 26,* 255–272.
Flor, H., Elbert, T., Knecht, S., Wienbruch, C., Pantev, C., Birbaumer, N., et al. (1995b). Phantom-limb pain as a perceptual correlate of cortical reorganization following arm amputation. *Nature, 375*(6531), 482–484.
Flor, H., Birbaumer, N., Schulz, R., Grüsser, S. M., & Mucha, R. F. (2002). Pavlovian conditioning of opioid and nonopioid pain inhibitory mechanisms in humans. *European Journal of Pain, 6*(5), 395–402.
Fordyce, W. E. (1976). *Behavioral methods for chronic pain and illness.* St Louis: Mosby.

Geisser, M. E., Haig, A. J., Wallbom, A. S., & Wiggert, E. A. (2004). Pain-related fear, lumbar flexion, and dynamic EMG among persons with chronic musculoskeletal low back pain. *The Clinical Journal of Pain, 20*(2), 61–69.

Gracely, R. H., Ceko, M., & Bushnell, M. C. (2012). Fibromyalgia and depression. *Pain Research Treatment, 2012*, 486590.

Greenwood, K. A., Thurston, R., Rumble, M., Waters, S. J., & Keefe, F. J. (2003). Anger and persistent pain: Current status and future directions. *Pain, 103* (1–2), 1–5.

Hadjistavropoulos, T., Craig, K. D., Duck, S., Cano, A., Goubert, L., Jackson, P. L., et al. (2011). A biopsychosocial formulation of pain communication. *Psychological Bulletin, 137*(6), 910–939.

Hermann, C., Hohmeister, J., Demirakca, S., Zohsel, K., & Flor, H. (2006). Long-term alteration of pain sensitivity in school-aged children with early pain experiences. *Pain, 125*(3), 278–285.

Karp, J. F., Shega, J. W., Morone, N. E., & Weiner, D. K. (2008). Advances in understanding the mechanisms and management of persistent pain in older adults. *British Journal of Anaesthesia, 101*(1), 111–120.

Keefe, F. J., Rumble, M. E., Scipio, C. D., Giordano, L. A., & Perri, L. M. (2004). Psychological aspects of persistent pain: current state of the science. *Journal of Pain, 5*(4), 195–211.

Kerns, R. D., Turk, D. C., & Rudy, T. E. (1985). The West Haven-Yale Multidimensional Pain Inventory (WHYMPI). *Pain, 23*(4), 345–356.

Kranitz, L., & Lehrer, P. (2004). Biofeedback applications in the treatment of cardiovascular diseases. *Cardiology in Review, 12*(3), 177–181.

Kröner-Herwig, B. (2011). Psychological treatments for pediatric headache. *Expert Review of Neurotherapeutics, 11*(3), 403–410.

Levy, R. L., Langer, S. L., & Whitehead, W. E. (2007). Social learning contributions to the etiology and treatment of functional abdominal pain and inflammatory bowel disease in children and adults. *World Journal of Gastroenterology, 13*(17), 2397–2403.

Linton, S. J., Boersma, K., Jansson, M., Svard, L., & Botvalde, M. (2005). The effects of cognitive-behavioral and physical therapy preventive interventions on pain-related sick leave: A randomized controlled trial. *The Clinical Journal of Pain, 21*(2), 109–119.

Luebbert, K., Dahme, B., & Hasenbring, M. (2001). The effectiveness of relaxation training in reducing treatment-related symptoms and improving emotional adjustment in acute non-surgical cancer treatment: A meta-analytical review. *Psycho-Oncology, 10*(6), 490–502.

Mogil, J. S. (2012). Pain genetics: Past, present and future. *Trends in Genetics, 28*(6), 258–266.

Morley, S. (2011). Efficacy and effectiveness of cognitive behaviour therapy for chronic pain: Progress and some challenges. *Pain, 152*(3 Suppl.), S99–S106.

Ostelo, R. W., van Tulder, M. W., Vlaeyen, J. W., Linton, S. J., Morley, S. J., & Assendelft, W. J. (2005). Behavioural treatment for chronic low-back pain. *Cochrane Database of Systematic Reviews, 1*, CD002014.

Penzien, D. B., Rains, J. C., & Andrasik, F. (2002). Behavioral management of recurrent headache: three decades of experience and empiricism. *Applied Psychophysiology and Biofeedback, 27*(2), 163–181.

Romano, J. M., & Turner, J. A. (1985). Chronic pain and depression: Does the evidence support a relationship? *Psychological Bulletin, 97*(1), 18–34.

Schneider, C., Palomba, D., & Flor, H. (2004). Pavlovian conditioning of muscular responses in chronic pain patients: Central and peripheral correlates. *Pain, 112*(3), 239–247.

Turk, D. C., & Melzack, R. (Eds.). (2011). *Handbook of pain assessment* (3rd ed.). New York: Guilford Press.

Turk, D. C., & Rudy, T. E. (1988). Toward an empirically derived taxonomy of chronic pain patients: Integration of psychological assessment data. *Journal of Consulting and Clinical Psychology, 56*(2), 233–238.

Turk, D. C., Meichenbaum, D. H., & Genest, M. (1983). *Pain and behavioral medicine: A cognitive-behavioral approach.* New York: Guilford Press.

Turk, D. C., Okifuji, A., Sinclair, J. D., & Starz, T. W. (1998). Interdisciplinary treatment for fibromyalgia syndrome: Clinical and statistical significance. *Arthritis Care and Research, 11*(3), 186–195.

Van Damme, S., Legrain, V., Vogt, J., & Crombez, G. (2010). Keeping pain in mind: A motivational account of attention to pain. *Neuroscience and Biobehavioral Reviews, 34*(2), 204–213.

van Lankveld, W., van Helmond, T., Naring, G., de Rooij, D. J., & van den Hoogen, F. (2004). Partner participation in cognitive-behavioral self-management group treatment for patients with rheumatoid arthritis. *Journal of Rheumatology, 31*(9), 1738–1745.

Vervoort, T., Caes, L., Trost, Z., Sullivan, M., Vangronsveld, K., & Goubert, L. (2011). Social modulation of facial pain display in high-catastrophizing children: An observational study in schoolchildren and their parents. *Pain, 152*(7), 1591–1599.

Vlaeyen, J. W., & Linton, S. J. (2000). Fear-avoidance and its consequences in chronic musculoskeletal pain: A state of the art. *Pain, 85*(3), 317–332.

Vlaeyen, J. W., & Linton, S. J. (2012). Fear-avoidance model of chronic musculoskeletal pain: 12 years on. *Pain, 153*(6), 1144–1147.

Vlaeyen, J. W., Morley, S., Linton, S. J., Boersma, K., & de Jong, J. (2012). *Pain-related fear: Exposure-based treatment of chronic pain.* Seattle: IASP Press.

Wiech, K., Ploner, M., & Tracey, I. (2008). Neurocognitive aspects of pain perception. *Trends in Cognitive Sciences, 12*(8), 306–313.

Woolf, C. J., & Mannion, R. J. (1999). Neuropathic pain: Aetiology, symptoms, mechanisms, and management. *Lancet, 353*(9168), 1959–1964.

Psychology of Pain, Assessment of Cognitive Variables

Michael J. L. Sullivan and Whitney Scott
Department of Psychology, McGill University,
Montreal, QC, Canada

Synonyms

Cognitive assessment; Evaluation of pain-related thought; Psychological aspects of pain

Definition

Cognitive assessment refers to a variety of procedures designed to elucidate the content or processes of thought implicated in the pain experience. Cognitive assessment in pain research may address stable internal factors such as beliefs or situationally determined or transient factors such as automatic thoughts, ▶ expectancies, or ▶ coping strategies. Assessments that use self-report questionnaires are typically aimed at assessing thought content (e.g., whether an individual endorses a specific pain-related belief). Assessments that involve performance on ▶ cognitive tasks are typically aimed at the analysis of thought process (e.g., vigilance, bias, accessibility).

Characteristics

Background

With increased awareness of the important role of psychological factors as determinants of pain experience, clinicians and researchers have been faced with the challenge of reliably and validly assessing pain-related cognitions. A variety of self-report questionnaires have been developed to assess the content of cognitions associated with pain. Questionnaires are relatively easy to administer and score, thus facilitating their inclusion in research protocols. Self-report instruments have been used extensively to assess coping, ▶ appraisals, expectancies, and beliefs about pain. However, self-report questionnaires are restricted to the study of consciously accessible thought content. Questionnaires are also susceptible to willful distortion, which can be a significant drawback if they are administered under conditions where there might be incentives (e.g., compensation) for certain forms of self-presentation (e.g., disability).

A number of cognitive performance paradigms have also been developed to assess thought processes associated with pain (Pincus and Morley 2001; Keogh et al. 2001; Grumm et al. 2008). Paradigms that assess the amount of time required to make decisions about different aspects of pain stimuli, the interpretation of ambiguous pain-related stimuli, memory of pain stimuli, and cognitive interference due to pain have been useful in the study of cognitive dimensions such as vigilance, accessibility, and ▶ preferential processing (Crombez et al. 1998; Pincus and Morley 2001).

Due to the multitude of assessment instruments and procedures that have been developed, it is not possible to provide a comprehensive review of these within the space limitations of this essay. The review below is selective by necessity and focuses on areas of high research activity and current theoretical relevance.

Cognitive Assessment of Pain-Related Thought Content

Coping

The assessment of coping dominated much of the research on the psychology of pain in the 1980s. This work emerged from the view that pain symptoms could be viewed as a physical and psychological stressor and that the manner in which individuals coped with their pain would have implications for physical and emotional outcomes. Numerous investigations have used the coping strategies questionnaire (CSQ; Rosenstiel and Keefe 1983) to assess coping strategies for pain. The CSQ consists of 7 coping subscales: coping self-statements, praying or hoping, ignoring pain sensations, reinterpreting pain sensations, increasing behavioral activities, and catastrophizing. Subscale scores have been used to examine the correlates of coping in research, as well as in clinical practice as a means of identifying targets of intervention.

A different approach to the assessment of coping and adaptation is reflected in the Multidimensional Pain Inventory (MPI; Kerns et al. 1985), which integrates medical, psychosocial, and behavioral data. The MPI is based on an empirically derived typology and classifies individuals as (1) adaptive copers, (2) interpersonally distressed, or (3) dysfunctional. The MPI has been used to predict health outcomes in patients with persistent pain conditions and to match patients with appropriate interventions (Turk and Rudy 1990; Soderlund and Denison 2006).

Self-efficacy

In pain research, ▶ self-efficacy has been described as individuals' confidence in their ability to control or manage various aspects of health conditions associated with pain (Lorig et al. 1989). A number of scales have been developed to assess self-efficacy for pain and illness-related symptoms (Miles et al. 2011). Some measures are tailored to specific pain-related conditions (e.g., the Arthritis Self-Efficacy Scale; Lorig et al. 1989), while others are intended for a wide range of pain conditions (The Pain Self-Efficacy Scale; Nicholas et al. 1992).

Pain Catastrophizing

A variety of assessment instruments have been developed to assess catastrophic thinking associated with pain. The catastrophizing subscale of the coping strategies questionnaire (Rosenstiel and Keefe 1983) and the ▶ Pain Catastrophizing Scale (Sullivan et al. 1995) have been used extensively. On these questionnaires, individuals are asked to indicate the frequency with which they experience different catastrophic thoughts in relation to their pain (i.e., "It's terrible and it's never going to get any better").

Pain Acceptance

The role of acceptance of chronic pain has gained increasing attention in the literature of the management of chronic pain. The chronic pain acceptance questionnaire (CPAQ;

McCracken et al. 2004) is one instrument that has been developed to assess pain-related acceptance. On the CPAQ, individuals rate their agreement with statements regarding their willingness to engage in life activities despite pain. Research indicates that the CPAQ predicts a number of pain-related outcomes (McCracken and Eccleston 2005).

Pain/Health Beliefs

Health beliefs have been discussed as a construct central to successful adjustment to persistent pain conditions. Numerous investigations have addressed the role of control beliefs with the Multidimensional Health Locus of Control Scale (MHLC; Wallston et al. 1978). Another frequently used measure of pain beliefs is the survey of pain attitudes (SOPA; Jensen et al. 1994). The SOPA assesses 7 domains of beliefs hypothesized to be important to pain outcomes. One- and two-item measures of pain beliefs have recently been developed in order to facilitate their inclusion in clinical and research protocols (Jensen et al. 2003).

Expectancies

Expectancies for pain have been discussed as a significant determinant of pain experience (Kirsch 1985). A distinction has been drawn between "response expectancies" and "behavioral outcome expectancies" (Bandura 1983). Predictions about non-volitional responses (e.g., pain, sleep, emotional arousal) are referred to as "response expectancies." Behavioral outcome expectancies refer to individuals' estimates of the probability of occurrence of a given behavioral outcome that is under volitional control. It has been suggested that expectancies might represent the final common pathway of several pain-related psychological factors (Sullivan et al. 2011). Typically, single-item measures are used to assess an individual's expectancies for pain in relation to specific clinical or experimental procedures.

Cognitive Assessment of Pain-Related Thought Process

A modified Stroop procedure has been used in a number of studies to assess the selective processing of pain-related stimuli. In this procedure, participants are presented with words printed in different colors, and their task is to state the color of the word as quickly as possible. Certain words will tend to "grab the attention" of the participant such that color-naming latency will be increased. It has been suggested that chronic pain patients might possess a "▶ pain schema" that will draw their attention to pain-related stimuli and consequently will have longer color-naming reaction times for pain words (Pincus and Morley 2001).

An implicit pain association task (pain IAT) has been used to assess automatic processing of pain-related information in relation to the self (Grumm et al. 2008). The pain IAT measures the speed at which participants sort "self" words with "pain" and "pain-free" words. Participants' reaction time is assumed to reflect the strength of the association between pain and self in memory, or a "pain-self schema." Chronic pain patients have been shown to have stronger pain-self associations on the IAT than pain-free controls (Grumm et al. 2008).

The dot probe procedure has also been used to address questions concerning patient's ▶ pain processing of pain-related stimuli. In this task, two words are presented on a screen, and one is then replaced by a dot. The participant's task is to identify as quickly as possible the location of the dot. If the word that remains on the screen is a word that is attention grabbing, latency to identify the location of the dot will be increased. There are indications that chronic pain patients with high levels of fear of pain have difficulty disengaging their attention from pain-related words (Keogh et al. 2001).

Primary task paradigms and ▶ thought suppression procedures have been used to address the degree of attentional control individuals have over pain-related information (Crombez et al. 1998; Sullivan et al. 2001). Preferential processing of pain-related information has been addressed with procedures assessing differential memory for pain and non-pain information and the interpretation of ambiguous words (Pincus and Morley 2001).

References

Bandura, A. (1983). Reflections on self-efficacy. *Advances in Behaviour Research and Therapy, 1*, 237–269.

Crombez, G., Eccleston, C., Baeyens, F., et al. (1998). When somatic information threatens, catastrophic thinking enhances attentional interference. *Pain, 74*, 230–237.

Grumm, M., Erbe, K., von Collani, G., et al. (2008). Automatic processing of pain: The change of implicit pain associations after psychotherapy. *Behaviour Research and Therapy, 46*, 701–714.

Jensen, M. P., Romano, J. M., Turner, J. A., et al. (1994). Relationship of pain-specific beliefs to chronic pain adjustment. *Pain, 57*, 301–309.

Jensen, M. P., Keefe, F. J., Lefebvre, J. C., et al. (2003). One- and two-item measures of pain beliefs and coping strategies. *Pain, 104*, 453–469.

Keogh, E., Ellery, D., Hunt, C., et al. (2001). Selective attentional bias for pain-related stimuli amongst pain fearful individuals. *Pain, 91*, 91–100.

Kerns, R. D., Turk, D. C., & Rudy, T. E. (1985). The west haven yale multidimensional pain inventory (WHYMPI). *Pain, 23*, 345–356.

Kirsch, I. (1985). Response expectancy as a determinant of experience and behavior. *American Psychologist, 40*, 1189–1202.

Lorig, K., Chastain, R., Ung, E., et al. (1989). Development and evaluation of a scale to measure perceived self-efficacy in people with arthritis. *Arthritis and Rheumatism, 32*, 37–44.

McCracken, L. M., & Eccleston, C. (2005). A prospective study of acceptance of pain and patient functioning with chronic pain. *Pain, 118*, 164–169.

McCracken, L. M., Vowles, K. E., & Eccleston, C. (2004). Acceptance of chronic pain: Component analysis and a revised assessment method. *Pain, 107*, 159–166.

Miles, C. L., Pincus, T., Carnes, D., & Taylor, S. J. C. (2011). Measuring pain self-efficacy. *The Clinical Journal of Pain, 27*, 461–470.

Nicholas, M. K., Wilson, P. H., & Goyen, J. (1992). Comparison of cognitive-behavioral group treatment and an alternative non-psychological treatment for chronic low back pain. *Pain, 48*, 339–347.

Pincus, T., & Morley, S. (2001). Cognitive processing bias in chronic pain: A review and integration. *Psychological Bulletin, 127*, 599–617.

Rosenstiel, A. K., & Keefe, F. J. (1983). The use of coping strategies in chronic low back pain patients: Relationship to patient characteristics and current adjustment. *Pain, 17*, 33–44.

Soderlund, A., & Denison, E. (2006). Classification of patients with whiplash associated disorders (WAD): Reliable and valid subgroups based on the multidimensional pain inventory (MPI-S). *European Journal of Pain, 10*, 113–119.

Sullivan, M. J. L., Bishop, S., & Pivik, J. (1995). The pain catastrophizing scale: Development and validation. *Psychological Assessment, 7*, 524–532.

Sullivan, M. J. L., Thorn, B., Haythornthwaite, J. A., et al. (2001). Theoretical perspectives on the relation between catastrophizing and pain. *The Clinical Journal of Pain, 17*, 52–64.

Sullivan, M. J. L., Tanzer, M., Reardon, G., Amirault, D., Dunbar, M., & Stanish, W. D. (2011). The role of pre-surgical expectancies in predicting pain and function one year following total knee arthroplasty. *Pain, 152*, 2287–2293.

Turk, D. C., & Rudy, T. E. (1990). The robustness of an empirically derived taxonomy of chronic pain patients. *Pain, 43*, 27–35.

Wallston, K. A., Wallston, B. S., & DeVellis, R. (1978). Development of the multidimensional locus of control (MHLC) scales. *Health Education Monographs, 6*, 161–170.

Psychology of Pain, Efficacy of Cognitive-Behavioral Treatment

Stephen Morley
Academic Unit of Psychiatry and Behavioral Sciences, University of Leeds, Leeds, UK

Synonyms

Treatment outcome research

Definition

Efficacy of psychological treatments (mainly ▶ cognitive-behavioral therapy) is established through randomized controlled trials. Absolute efficacy – comparison of treated groups with untreated groups – has been established using statistical criteria. There is some evidence of relative ▶ efficacy, i.e., that cognitive-behavioral treatments are relatively more efficacious than a range of other treatments. Little is known about the efficacy of specific treatment components, and the criteria for establishing clinical criteria for evaluating these treatments have yet to be formalized.

Characteristics

Psychological treatments for pain are broadly cognitive-behavioral in orientation. They contain multiple components and have been designed to address the problem of chronic pain rather than acute pain. The cognitive-behavioral analysis of chronic pain has generally eschewed differences between various medically defined diagnostic groups, and focused on the common psychological consequences of the experience of chronic pain *per se*. Most cognitive-behavioral treatments are made up of multiple components, i.e., different therapeutic procedures, which are packaged together and delivered to groups of patients by multidisciplinary teams, e.g., psychologist, physical therapist, nurse, anesthesiologist. The content of treatment packages varies considerably but a common element is the inclusion of a significant educational component, strategies to socialize patients into active participation in the treatment, and an emphasis on practical skills to be rehearsed and practiced. Treatments have been continuously developed over the last 50 years and have drawn inspiration from several sources including research in basic behavioral science and astute clinical observations that have then been empirically tested (Morley 2011). Treatments may include procedures developed from ▶ operant analyses of pain, which focus on alteration of observable behavior (Fordyce 1976), procedures such as ▶ relaxation and ▶ biofeedback, stress management (coping skills) strategies (Keefe et al. 2002), and a range of cognitive procedures for which the central focus of therapy are changes in a person's appraisal, and increases in self-management of pain and its associated consequences (Thorn 2004; Flor and Turk 2011). Typical components of a broad based cognitive-behavioral program are graded behavioral activities; the development of cognitive strategies to modify attention to, and appraisal of, pain; strategies to manage adverse mood states; and strategies to modify social contingencies (including family and spouse) that may reinforce undesirable pain-related behaviors. Recent developments in treatment include the application of mindfulness procedures that have their roots in ancient Buddhist philosophy and Acceptance and Commitment Therapy (ACT), which has developed from the radical behaviorist analysis of the function of language (Hayes et al. 2001).

Psychology of Pain, Efficacy of Cognitive-Behavioral Treatment, Table 1 Measurement domains used in the evaluation of psychological treatments for chronic pain conditions (After Morley et al. 1999)

Domain	Definition
Pain experience	Measures of subjective pain experience captured by ratings of intensity, sensation, and unpleasantness
Mood and affect	Measures of mood and affective state (but not trait predispositions): these measures were divided into measures of depression and measures of other states – predominantly anxiety
Cognitive coping and appraisal	Measure of coping strategies and appraisals used in attempts to manage pain: these measures were divided into measures of negative coping, i.e., variables known to be correlated with poor adjustment (catastrophizing), and measures of positive coping, i.e., associated with good adjustment
Pain behavior	Observable behavior associated with pain: behavioral expression concerns behavior that signals the presence of pain, e.g., guarding, whereas behavior activity indicates behavioral activation. It is assumed that effective treatment results in the decrease of behavioral expression and increase of behavioral activity
Social role interference	These measures assess the impact of pain on the ability of the sufferer to function in a variety of social roles ranging from simple self-care to more complex family, work, and social activities. Effective treatment should result in the increase in social role activity – decreased interference
Biological and fitness	Assessments of biological functioning and physical fitness
Use of health care	Use of a variety of health-care facilities ranging from medication to clinic visits
Miscellaneous	This category contains a wide range of measures not otherwise accommodated in the above definitions. It included pain drawings and a variety of personality diagnostics and idiographic measures

The intended therapeutic outcomes are changes in behavioral activity, effective use of coping strategies, improved mood, and reduction in inappropriate medication and health costs. With regard to the evaluation of ACT, researchers have also sought to determine that patients have made a fundamental shift in the processes presumed to underlie successful treatment (McCracken 2005). Finally, a treatment (graded exposure) based on the fear-avoidance model of pain has been developed. In contrast to the majority of cognitive-behavioral treatments, graded exposure has a relatively specific remit (to change pain-related fearful behavior) and the treatment procedure is relatively constrained and not multi-componential (Leeuw et al. 2007).

The measurement of outcomes for cognitive-behavioral treatment has been broad based and in Table 1 shows the heterogeneity of the measures and one schema for organizing them.

The absolute efficacy of treatment programs has been established through randomized controlled trials of comparing treatment with no-treatment (waiting list) control groups (Wampold 2001). Comparisons across trials and

measurement domains may be made by combining ▶ effect-sizes in a ▶ meta-analysis, which were computed by comparing posttreatment scores of treated and untreated groups. Early meta-analyses of trials (Morley et al. 1999) suggested that the median effect-sizes across all domains of measurement were around 0.5 units. More recent analyses (Hoffman et al. 2007; Eccleston et al. 2009) indicate that with the addition of more trials the effect may have shrunk to around 0.2 units. Possible reasons for this are: the exclusion of trials with small numbers which are likely to provide biased estimates; improvement in the design and implementation (quality) of trials and the consequent reduction of bias (see Fig. 1); and weaker implementation of treatment, both reduced time and less skilled implementation. At present, it is not possible to provide a definitive verdict on these alternatives.

Figure 1 shows plots of the quality scores for randomized controlled trials between 1985 and 2009 for two parameters, the quality of trial design (includes characteristics such as randomization and blinding of assessors) and the quality of treatment implementation (use of manuals,

Psychology of Pain, Efficacy of Cognitive-Behavioral Treatment, Fig. 1 Plots of the quality scores for randomized controlled trials between 1985 and 2009 for, the quality of trial design and the quality of treatment implementation

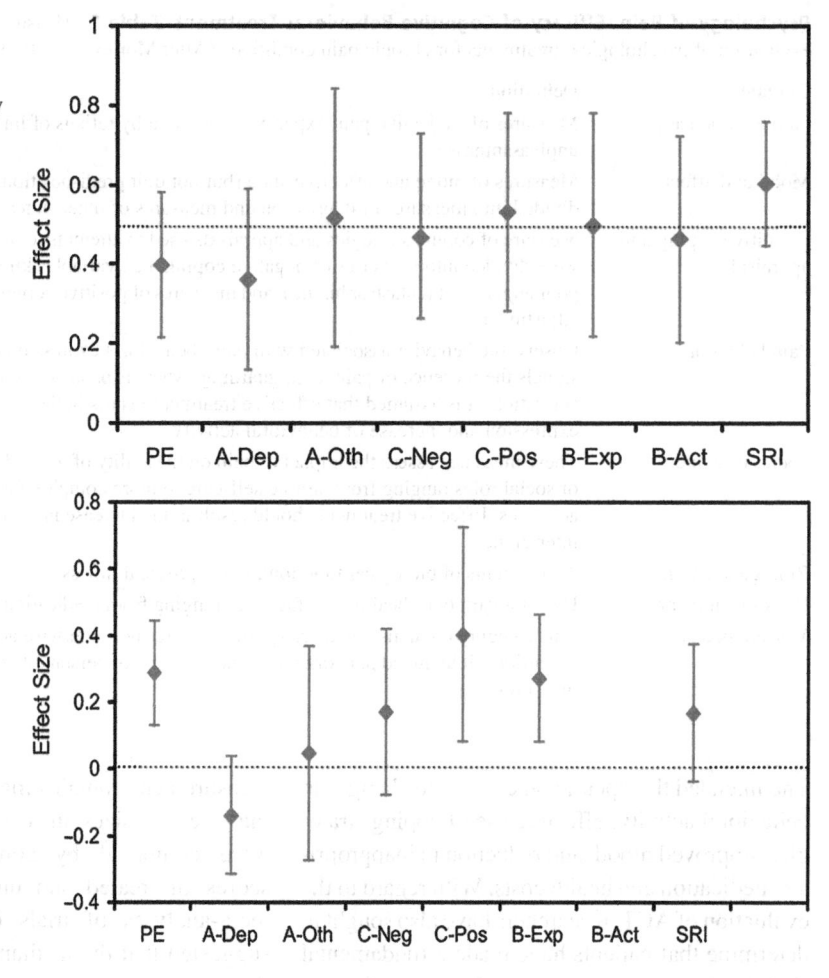

skilled therapists). Data on the trials is reported in Eccleston et al. (2009) and the ratings were made using a standardized scale (Yates et al. 2005). The data show that whereas the quality of study design has improved over the years (significant B in regression analysis), the quality of treatment implementation has not.

Cognitive-behavioral treatments have also been applied to children with pain, but absolute efficacy has only established for children with headache, abdominal pain, and fibromyalgia (Palermo et al. 2010).

Considerable progress in establishing the efficacy of psychological treatments for chronic pain has been made and several current developments will enhance our understanding. These are:
1. The development of outcomes based on significant clinically important criteria. At present,

the predominant evaluation of trials is determined by the statistical difference between means from measures on a continuous scale. In many cases, we do not know if the means encompass clinically important gains. One approach to this problem is to use information about the distribution of scores in functional and dysfunction samples to determine cut scores that can be used to categorize patients as clinically improved or not improved (Jacobson et al. 1999). The advantage of this method is that it may also be used to evaluate uncontrolled studies of treatment implementation (Morley et al. 2008). Other nonstatistical definitions of clinically important gain require further investigation, including patient-determined outcomes (Robinson et al. 2005; Thorne and Morley 2009).

2. The development of outcomes whose value is understood by clinicians, researchers, and patients. The majority of outcomes are determined by researcher-clinicians (Turk et al 2003). Recent analyses show that many outcomes valued by patients are not assessed in trials (Turk et al. 2008; Beale et al. 2011).
3. Ensuring the quality (integrity and competence of delivery) of psychological treatments. Figure 1 shows that this aspect of study efficacy requires further attention and development.

Thus far, it has not been possible to establish the efficacy of specific cognitive-behavioral treatments for chronic pain, i.e., whether the processes invoked by the treatment rationale are necessary for therapeutic change, or whether changes are due to other processes not invoked by cognitive-behavioral theory. This is a considerable challenge within the general arena of evaluating psychological therapies (Wampold 2001), and the complex, multicomponential nature of contemporary psychological therapy for pain suggests that this will be a substantial task. Resolving this issue might be enhanced by the development of more precisely articulated formulations of chronic pain. The consideration of possible subgroups of patients defined by psychological principles, rather than medical diagnoses, such as the fear-avoidance model (Vlaeyen et al. 2001; Vlaeyen and Linton 2000), is a recent development that may encourage a further generation of well designed trials to determine the absolute, relative, and specific efficacy of psychological treatments for pain.

References

Beale, M., Cella, M., & Williams, A. (2011). Comparing patients' and clinician-researchers' outcome choice for psychological treatment of chronic pain. *Pain, 152*(10), 2283–3386.

Eccleston, C., Williams, A. C. deC., & Morley, S. (2009). Psychological therapies for the management of chronic pain (excluding headache) in adults. *Cochrane Database of Systematic Reviews,* (2).

Flor, H., & Turk, D. C. (2011). *Chronic pain: An integrated biobehavioral approach.* Seattle: IASP Press.

Fordyce, W. E. (1976). *Behavioral methods for chronic pain and illness.* St Louis: Mosby.

Hayes, S. C., Barnes-Holmes, D., & Roche, B. (2001). *Relational frame theory: A post-Skinnerian account of human language and cognition.* New York: Kluwer.

Hoffman, B. M., Papas, R. K., Chatkoff, D. K., & Kerns, R. D. (2007). Meta-analysis of psychological interventions for chronic low back pain. *Health Psychology, 26*(1), 1–9.

Jacobson, N. S., Roberts, L. J., Berns, S. B., & McGlinchey, J. B. (1999). Methods for defining and determining the clinical significance of treatment effects: Description, application, and alternatives. *Journal of Consulting and Clinical Psychology, 67*(3), 300–307.

Keefe, F. J., Beaupré, P. M., Gil, K. M., Rumble, M. E., & Aspnes, A. K. (2002). Group therapy with patients with chronic pain. In D. C. Turk & R. J. Gatchel (Eds.), *Psychological approaches to pain management: A practitioner's handbook* (pp. 234–255). New York: Guilford.

Leeuw, M., Goossens, M. E., Linton, S. J., Crombez, G., Boersma, K., & Vlaeyen, J. W. (2007). The fear-avoidance model of musculoskeletal pain: Current state of scientific evidence. *Journal of Behavioral Medicine, 30*(1), 77–94.

McCracken, L. M. (2005). *Contextual cognitive-behavioral therapy for chronic pain.* Seattle: IASP Press.

Morley, S. (2011). Efficacy and effectiveness of cognitive behaviour therapy for chronic pain: Progress and some challenges. *Pain, 152*(3 (Suppl. 1)), 99–106.

Morley, S., Eccleston, C., & Williams, A. (1999). Systematic review and meta-analysis of randomized controlled trials of cognitive behaviour therapy and behaviour therapy for chronic pain in adults, excluding headache. *Pain, 80*(1–2), 1–13.

Morley, S., Williams, A. C. C., & Hussain, S. (2008). Estimating the clinical effectiveness of cognitive behavioural therapy in the clinic: Evaluation of a CBT informed pain management programme. *Pain, 137*(3), 670–680.

Palermo, T. M., Eccleston, C., Lewandowski, A. S., Williams, A. C., & Morley, S. (2010). Randomized controlled trials of psychological therapies for management of chronic pain in children and adolescents: An updated meta-analytic review. *Pain, 148*(3), 387–397.

Robinson, M. E., Brown, J. L., George, S. Z., Edwards, P. S., Atchison, J. W., Hirsh, A. T., Waxenberg, L. B., Wittmer, V., & Fillingim, R. B. (2005). Multidimensional success criteria and expectations for treatment of chronic pain: The patient perspective. *Pain Medicine, 6*(5), 336–345.

Thorn, B. E. (2004). *Cognitive therapy for chronic pain.* New York: Guilford.

Thorne, F. M., & Morley, S. (2009). Prospective judgments of acceptable outcomes for pain, interference and activity: Patient-determined outcome criteria. *Pain, 144*(2/3), 262–269.

Turk, D. C., Dworkin, R. H., Allen, R. R., Bellamy, N., Brandenburg, N., Carr, D. B., Cleeland, C., Dionne, R., Farrar, J. T., & Galer, B. S. (2003). Core outcome domains for chronic pain clinical trials: IMMPACT recommendations. *Pain, 106*(3), 337–345.

Turk, D. C., Dworkin, R. H., Revicki, D., Harding, G., Burke, L. B., Cella, D., Cleeland, C. S., Cowan, P., Farrar, J. T., Hertz, S., Max, M. B., & Rappaport, B. A. (2008). Identifying important outcome domains for chronic pain clinical trials: An IMMPACT survey of people with pain. *Pain, 137*(2), 276–285.

Vlaeyen, J. W., Linton, S. J. (2000). Fear-avoidance and its consequences in chronic musculoskeletal pain: a state of the art. *Pain, 85*(3):317–32.

Vlaeyen, J. W. S., De Jong, J., Geilen, M., Heuts, P. H. T. G., et al. (2001). Graded exposure in vivo in the treatment of pain-related fear: A replicated single-case experimental design in four patients with chronic low back pain. *Behaviour Research and Therapy 39*(2):151–166.

Wampold, B. E. (2001). *The great psychotherapy debate: Models, methods, and findings.* Mahwah: Lawrence Erlbaum Associates.

Yates, S., Morley, S., Eccleston, C., & Williams, A. C. C. (2005). A scale for rating the quality of trials of psychological trials for pain. *Pain, 117*, 314–325.

Psychology of Pain, Self-efficacy

Michael J. L. Sullivan
Department of Psychology, McGill University,
Montreal, QC, Canada

Synonyms

Confidence in coping abilities; Outcome expectancies; Self-efficacy

Definition

▶ Self-efficacy was originally defined as "belief in one's capabilities to organize and execute the course of action required to produce given attainments" (Bandura 1977). In research addressing the role of self-efficacy in the context of pain, self-efficacy has been described as the degree of confidence individuals have about their ability to control or manage various aspects of their pain-related health conditions (Lorig et al. 1989). In pain research, self-efficacy has been operationally defined in terms of one's *overall* confidence in the ability to deal with symptoms, stresses, or limitations associated with a pain condition, or in terms of the ability to manage, control, or decrease *specific* components of symptoms or disability (e.g., ability to decrease pain, ability to complete certain activities in spite of pain) (Lackner et al. 1996; Shelby et al. 2008).

Characteristics

Self-efficacy and Pain-Related Outcomes

Self-efficacy has been shown to be associated with a number of pain-related outcomes such as pain intensity, disability, and depression (Foster et al. 2010; Lorig et al. 1989; Asghari and Nicholas 2001). A relation between self-efficacy and pain-related outcomes has been reported in patients with acute pain symptoms (Kaplan et al. 1996), patients with chronic pain (Sarda et al. 2009), and in response to experimentally induced pain (Bandura et al. 1987). Considerable research has addressed the relation between self-efficacy and pain-related outcomes in patients with arthritic conditions (e.g., rheumatoid arthritis, osteoarthritis, fibromyalgia; Lorig et al. 1989, Keefe et al. 1997; Shelby et al. 2008).

Theoretical Perspectives on Self-efficacy

Self-efficacy is a particular form of expectancy, that is, a probabilistic judgment about the future occurrence of a specific outcome. Bandura (1977) distinguishes between two forms of expectancies related to behavioral outcomes: outcome expectancies and efficacy expectancies. Outcome expectancies refer to individuals' estimates that a given behavior will result in a given outcome. Efficacy expectancies refer to the confidence individuals have that they possess the ability to successfully execute the behavior required to yield a given outcome. Under conditions where individuals possess the necessary skills for execution of a particular behavior, and when adequate incentives are in place, efficacy expectancies are said to be a major determinant of individuals' activity choices and the effort they will expend to attain desired outcomes (Bandura 1977).

According to Bandura, self-efficacy expectancies develop as a result of a number of experiential and ▶ social learning factors. Past experience with success or failure in producing specific outcomes will influence individuals' confidence to produce such outcomes in the future. ▶ Vicarious experience, exposure to information, and the attributions made for outcomes can also influence the development of self-efficacy expectancies (Bandura 1977).

As discussed by Bandura (1977), self-efficacy can be construed as a final common pathway of a variety of psychological influences on behavior. A number of investigations have provided support for the role of self-efficacy as a mediator of the relation between various psychological factors and adverse pain outcomes. Shelby et al. (2008) found that self-efficacy mediated the relation between catastrophizing and pain in patients with osteoarthritis. One investigation revealed that self-efficacy mediated the relation between pain and disability in patients with low back pain (Costa et al. 2011). Foster et al. (2010) reported that self-efficacy made unique contributions to the persistence of low back pain, even when controlling for catastrophizing, fear of pain, and depression.

Bandura (1977) suggests that while self-efficacy may ▶ mediate various health outcomes, as a function of dynamic feedback processes, self-efficacy can also influence its cognitive and motivational antecedents. High levels of self-efficacy may increase individuals' investment in their coping efforts, thereby increasing the probability of positive outcome and, in turn, contributing to the maintenance or enhancement of self-efficacy (Sarda et al. 2009).

Mechanisms of Action: Self-efficacy and Pain

Considerable attention has been given to the role of coping strategies in the development of self-efficacy expectancies. Keefe et al. (1997) reported that certain pain coping strategies were associated with varying levels of self-efficacy.

For example, ignoring pain sensations and coping self-statements were associated with higher self-efficacy. High levels of pain catastrophizing have been associated with lower levels of self-efficacy (Sullivan et al. 2001). Litt et al. (2009) found that daily fluctuations in self-efficacy contributed significant variance in the prospective prediction of pain in patients with temporomandibular disorder.

A study by Bandura et al. (1987) provided evidence of a link between self-efficacy and endogenous opiates. In this study, subjects were taught a variety of cognitive strategies to control pain prior to participation in an experimental pain procedure (i.e., Cold Pressor Test). Subjects were asked to rate their confidence in their ability to control pain. Training in the use of cognitive pain control strategies enhanced self-efficacy, which in turn increased ▶ pain tolerance. However, the administration of naloxone (i.e., an opioid antagonist) interfered with the pain tolerance enhancing effects of increased self-efficacy.

Mechanism of Action: Self-efficacy and Disability

According to Bandura (1977), efficacy expectancies determine how much effort people will expend and how long they will persist in the face of obstacles and aversive experiences (p. 194). The proposed influence of self-efficacy on behavioral persistence suggests a possible relation between self-efficacy and pain-related disability. Consistent with this account, there is growing evidence that self-efficacy expectancies contribute to the severity of disability associated with chronic pain (Sarda et al. 2009; Shelby et al. 2008; Lackner et al. 1996).

The term functional self-efficacy has been used to describe patients' confidence in their ability to carry out specific physical maneuvers that might be associated with pain (Lackner et al. 1996). Lower levels of functional self-efficacy are thought to impact on disability by limiting the range of activities the patient will undertake, or by reducing the effort invested in activity. Rejeski et al. (2001) found that low levels of functional self-efficacy predicted declining levels of physical function over a 30-month period in older adults with knee pain.

Assessment

Self-efficacy judgments are considered to be most predictive of proximal behavior (Bandura 1977). As such, the bulk of research on self-efficacy and behavioral outcomes has made use of single item scales, worded in a fashion directly relevant to the behavior of interest (Shelby et al. 2008). A number of scales have also been developed to assess a "generalized" self-efficacy orientation to dealing with pain and illness-related symptoms (Miles et al. 2011). The Arthritis Self-Efficacy Scale (Lorig et al. 1989) has been used in numerous studies to assess individuals' confidence in their ability to perform various behaviors aimed at controlling pain and arthritis symptoms. Similar measures have been developed to assess self-efficacy beliefs in chronic pain patients not restricted to arthritis (Nicholas et al. 1992, Anderson et al. 1995). A measure of functional self-efficacy was developed by Barry et al. (2003). On this scale, respondents are asked to rate their confidence in their ability to carry out ten different activities of daily living.

Treatment

There are no intervention programs that have been designed to specifically target self-efficacy expectancies. However, several components of pain management programs are considered to be conducive to the enhancement of self-efficacy (Nicholas et al. 1992). Strategies considered to enhance self-efficacy include ▶ mastery experience, ▶ role modeling, ▶ persuasion, and reinterpretation of physiological and affective states (Bandura 1977). These strategies, either alone or in combination, have been incorporated in several interventions aimed at assisting individuals in managing distress and disability associated with persistent pain conditions (Nicholas et al. 1992). There are indications that the increases in self-efficacy observed following participation in cognitive-behavioral interventions may be one of the mechanisms through which positive treatment outcomes are achieved (Litt et al. 2009).

References

Anderson, K. O., Dowds, B. N., Pelletz, R. E., & Edwards, W. T. (1995). Development and initial validation of a scale to measure self-efficacy beliefs in patients with chronic pain. *Pain, 63*, 77–84.

Asghari, A., & Nicholas, M. K. (2001). Pain self-efficacy beliefs and pain behavior: A prospective study. *Pain, 94*, 85–100.

Bandura, A. (1977). *Self-efficacy: The exercise of control.* New York: Freeman.

Bandura, A., O'Leary, A., Taylor, C. B., Gauthier, J., et al. (1987). Perceived self-efficacy and pain control: Opioid and non-opioid mechanism. *Journal of Personality and Social Psychology, 53*, 563–571.

Barry, L. C., Guo, Z., Kerns, R. D., Duong, B. D., & Reid, M. C. (2003). Functional self-efficacy and pain-related disability among older veterans with chronic pain in a primary care setting. *Pain, 104*, 131–137.

Costa, LdCM., Maher, C. G., McAuley, J. H., Hancock, M. J., & Smeets, R. J. E. M. (2011). Self-efficacy is more important than fear of movement in mediating the relationship between pain and disability in chronic low back pain. *European Journal of Pain, 15*, 213–219.

Foster, N. E., Thomas, E., Bishop, A., Dunn, K. M., & Main, C. J. (2010). Distinctiveness of psychological obstacles to recovery in low back pain patients in primary care. *Pain, 148*, 398–406.

Kaplan, G. W., Wurtele, S. K., & Gillis, D. (1996). Maximal effort during functional capacities evaluations: An evaluation of psychological factors. *Archives of Physical Medicine and Rehabilitation, 77*, 161–174.

Keefe, F. J., Kashikar-Zuck, S., Robinson, E., Salley, A., Beaupre, P., Caldwell, D., Baucom, D., & Haythornthwaite, J. (1997). Pain coping strategies that predict patients' and spouses' ratings of patients' self-efficacy. *Pain, 73*, 191–199.

Lackner, J. M., Carosella, A., & Feuerstein, M. (1996). Pain expectancies, pain and functional self-efficacy expectancies as determinants of disability in patients with chronic low back pain disorders. *Journal of Consulting and Clinical Psychology, 64*, 212–220.

Litt, M. D., Shafer, D. M., Ibanez, C. R., Kreutzer, D. L., & Tawfik-Yonkers, Z. (2009). Momentary pain and coping in temporomandibular disorder pain: Exploring mechanisms of cognitive behavioral treatment for chronic pain. *Pain, 145*, 160–168.

Lorig, K., Chastain, R., Ung, E., Shoor, S., & Holman, H. R. (1989). Development and evaluation of a scale to measure perceived self-efficacy in people with arthritis. *Arthritis and Rheumatism, 32*, 37–44.

Miles, C. L., Pincus, T., Carnes, D., & Taylor, S. J. C. (2011). Measuring pain self-efficacy. *The Clinical Journal of Pain, 27*, 461–470.

Nicholas, M. K., Wilson, P. H., & Goyen, J. (1992). Comparison of a cognitive behavioural group treatment and an alternative non-psychological treatment for chronic low back pain patients. *Pain, 4*, 339–347.

Rejeski, W. J., Miller, M. E., Foy, C., Messier, S., & Rapp, S. (2001). Self-efficacy and the progression of functional limitations and self-reported disability in older adults with knee pain. *Journal of Gerontology, 56*, S261–S265.

Sarda, J., Nicholas, M. K., Asghari, A., & Pimenta, C. A. M. (2009). The contribution of self-efficacy and depression to disability and work status in chronic pain patients: A comparison of Australian and Brazilian samples. *European Journal of Pain, 13*, 189–195.

Shelby, R. A., Somers, T. J., Keefe, F. J., Pells, J. J., Dixon, K. E., & Blumenthal, J. A. (2008). Domain specific self-efficacy mediates the impact of pain catastrophizing on pain and disability in overweight and obese osteoarthritis patients. *The Journal of Pain, 9*, 912–919.

Sullivan, M. J. L., Thorn, B., Haythornthwaite, J. A., Keefe, F. J., Martin, M., Bradley, L. A., & Lefebvre, J. C. (2001). Theoretical perspectives on the relation between catastrophizing and pain. *The Clinical Journal of Pain, 17*, 52–64.

Psychology of Pain, Sensitization, Habituation, and Pain

Michael K. Nicholas
Royal North Shore Hospital, University of Sydney, St. Leonards, NSW, Australia

Synonyms

Desensitization; Habituation; Hyperalgesia; Hypersensitivity; Sensitization

Definition

Sensitization in relation to ► pain occurs when a person's threshold for responding to a stimulus as noxious (painful) is reduced. This is analogous to allodynia (experiencing a normally non-noxious stimulus as painful) or ► hyperalgesia (raised pain report to a previously experienced noxious stimulus). Habituation or desensitization in relation to pain may be seen when greater tolerance for persisting pain (of normal severity for that person) is reported or the person reports their pain as not as troubling as it had been. Habituation should be evident in the presence of

the pain and not when pain is absent. Thus, when a pain-relieving intervention has been administered, habituation cannot be assessed until the pain severity has returned to previous levels.

Characteristics

There is strong evidence that certain chemicals and nerve injuries can induce hyperalgesia and allodynia phenomena (e.g., Hood et al. 2003) and that central (CNS) changes or mechanisms, especially central sensitization, appear to underlie these phenomena when they persist (see Woolf 2011, for a recent review). At the same time, there is growing evidence that the psychological processes associated with learning and attentional mechanisms may also contribute to these common phenomena of chronic pain (see Apkarian et al. 2011, for a recent review that is relevant to this topic). This evidence has important implications for potential treatments.

Nonassociative Learning Mechanisms

In contrast to the principle of learning processes normally considered in relation to pain, (i.e., operant (or instrumental) learning and classical (or Pavlovian) conditioning), sensitization and habituation are regarded as nonassociative forms of learning (Flor and Turk 2012). However, that does not mean they are independent of reinforcement processes. For example, there is evidence that intrinsic reinforcement processes can enhance pain sensitivity (Becker et al. 2008). In contrast to extrinsic reinforcement (e.g., external feedback like verbal praise following a behavior, such as a pain report), Becker et al. proposed that intrinsic reinforcement (i.e., negative reinforcement internal to the nociceptive system, e.g., by pain reduction) could initiate a process of perceptual discrimination learning leading to sustained hypersensitivity (to noxious stimulation). In this process, the perception of pain could provoke a behavioral response (e.g., to change position), resulting in pain reduction (i.e., intrinsic reinforcement). Such reinforcement may cause the behavioral response to become more likely when pain is perceived again. Over time, the pain may become a discriminative signal for the performance of the reinforced behavior (the conditioned response). Eventually, Becker et al. (2008) argued progressively weaker nociceptive signals might serve as discriminative signals for the performance of the conditioned behavior. This paradigm provides a model of how acute changes in clinical pain could lead to prolonged hypersensitivity and as a result to gradual immobility by learning to avoid pain.

It is still unclear how long these effects might last and whether they can account for sustained changes in pain sensitivity and whether these mechanisms are different in clinical pain patients. But it is consistent with this paradigm that some differences between healthy people and chronic pain patients in relation to responses to noxious stimuli have been found, with healthy controls tending to habituate while chronic pain patients tend to become more sensitized (e.g., Colloca et al. 2006; Kleinbohl et al. 1999). Flor et al. (2004) also found that, relative to healthy control subjects and episodic headache patients, chronic back pain patients tended to show sensitization to both painful (intracutaneous electrical) stimulation and innocuous stimulation. It was especially notable that the chronic back pain patients considered stimuli to be intolerable when they were reported by the healthy controls as non-painful. Flor and Turk (2012) suggested these effects may be related to implicit operant conditioning (via intrinsic reinforcement) and that when sensitization was evident in a patient, treatment should be directed to reducing this sensitization.

Associative Learning Mechanisms

Birbaumer et al. (1995) reviewed evidence supporting the presence of learning processes in the development and maintenance of chronic pain. This included evidence of both classical and instrumental (operant) learning responses, reflected in stereotypy of certain muscle groups to personally relevant stressful situations relative to control conditions, as well as conditioning of muscle and pain responses to previously neutral tones and slides in healthy subjects. Birbaumer et al. also cited evidence of learned facilitation

for pain-related information by the primary cortical projection areas in chronic pain patients. Research since then has added further support to these early findings.

Flor et al. (2002) demonstrated that operant learning of both increased and decreased pain responses could be demonstrated under controlled conditions and was also associated with measurable cortical responses. This study showed slower extinction rates of learned cortical responses among chronic back pain patients than normal controls.

Most recently, Apkarian et al. (2011) summarized much of the laboratory research work using classical (Pavlovian) conditioning that used pain (usually from electrical shocks) as the unconditioned stimulus (UCS) with which various events were associated. These studies had demonstrated that single painful stimuli can be learned and remembered for months. Significantly, Apkarian et al. pointed out that once the associated learning occurred, the extinction of these associations required repeated exposure of the organism to the conditioned stimulus in the absence of the painful event (UCS). They concluded that these studies indicated that chronic pain could be considered as a state of continuous learning (via associations), coupled with reduced opportunity for forgetting (as the UCS is rarely absent). This analysis presents a clear challenge to pain treatments.

Leeuw et al. (2007) reviewed the evidence supporting a role for fear of pain contributing to the development of avoidance responses and, ultimately, disability in many people with persisting pain. Consistent with the fear-avoidance model of chronic post-injury pain described by Vlaeyen and Linton (2000), they found that negative appraisals (catastrophizing) of internal and external stimuli, negative affectivity, and ► anxiety sensitivity could contribute to the development of pain-related fear and how that, in turn, could lead to escape and avoidance behaviors, as well as ► hypervigilance to internal and external illness information, muscular reactivity, as well as physical disuse and behavioral changes. From another perspective, while the repeated exposure to a stimulus (e.g., pain) might be expected to lead to ▶ habituation (to that stimulus), these findings might also suggest that this is less likely in those people who develop pain-related fear, escape and avoidance responses, as well as hypervigilance to such internal stimuli. These findings can be considered in relation to the laboratory studies described by Apakarian et al. (2011).

Laboratory evidence of habituation to a repeated experience of (experimentally induced) pain suggests that it does happen to some extent, but not completely (Crombez et al. 1997), and it is unstable, being readily reversed in the presence of a stressor or when the pain becomes unpredictable, perhaps consistent with Apkarian et al.'s (2011) observations.

Predisposing personality attributes have been proposed to explain why some individuals and not others develop fear of pain and subsequent difficulties. Turk and Okifuji (2002) have referred to these constructs within the rubric of a diathesis-stress model (see Flor and Turk 2012, for an example of this model applied to fibromyalgia syndrome), whereby a person's predisposing attributes only emerge under certain conditions of stress (e.g., the presence of pain). For example, Asmundson (1999) suggested that anxiety sensitivity might predispose individuals to developing a fear of pain and indirectly predict ▶ fear/avoidance behavior. Similar arguments have been posited for the personality construct of neuroticism (Asghari and Nicholas 1999).

It has been argued that one of the crucial functions of anxiety is concerned with prioritizing the selection and processing of information (Aldrich et al. 2000). By logical extension, it might be expected that people in chronic pain who are also anxious may attend to pain-related information at the expense of other, less personally relevant information. There is certainly some evidence that chronic pain patients exhibit selective attention to pain information relative to other, non-pain, information and that such individuals may also have trouble disengaging their attention from their pain, especially if they engage in frequent catastrophizing (Van Damme et al. 2006). However, the results of studies on attention bias in pain patients have not always

been supportive of this perspective (Asmundson and Hadjistavropoulos 2007). Haggman et al. (2010), for example, found that relative to health control participants, both acute and chronic pain patients displayed attentional biases towards sensory pain words only, and not words reflecting threat, and that the level of pain-related fear was not related to attentional biases. It has long been observed that increased attention to pain sensations increases the intensity of experienced pain (Bushnell et al. 2004), but clearly, factors other than attention contribute to this effect as well.

Summary

The evidence summarized here suggests that the psychological processes associated with learning (both associative and nonassociative), attention, affect, and thought processes in relation to pain can account for many of the typical features of pain sensitization phenomena. This evidence has a number of possible treatment implications for people in distressing or disabling persisting pain. For example, if pain-related fear is a potential driver of escape and avoidance behavior, then interventions, such as graded exposure to the feared stimulus, that have proven successful in people with other fears (like phobias for certain places) ought to be useful for people with pain-related fears and associated disability. To date, there is preliminary evidence that encouraging chronic low back pain patients with high levels of fear of pain to engage in activities they would normally avoid, such as falling (onto a foam rubber mattress), can help to reduce such fears, catastrophic beliefs, and pain-related disability (de Jong et al. 2005). In the context of pain patients who are highly sensitized or hypervigilant to the experience of pain, this intervention could be described as a form of desensitization or habituation. These findings suggest that if pain patients can have their fears or distressing beliefs about their pain desensitized by direct experience, as opposed to attempts to reassure them, they might more easily limit the intrusive or activity limiting effects of their pain.

In addition to graded behavioral exposure tasks, recent anxiety research would also suggest that pain patients could be desensitized or habituated to their pain by encouraging their repeated (for prolonged periods), deliberate experiencing of their pain, as opposed to their trying to avoid or escape from it by attention-distraction or behavioral change (e.g., Forsyth and Eifert 1998). There is some evidence for this effect in laboratory studies (Van Bockstaele et al. 2010), but only preliminary evidence from clinical trials (Flink et al. 2009). Similarly, laboratory research on the role of intrinsic reinforcement (Becker et al. 2008) suggests that inverse operant training, described as habituation training by Holzl et al. (2005), may contribute to the prevention of hypersensitivity to nociceptive stimuli and reduce existing hypersensitivity. However, application of this procedure in clinical trials is still to be demonstrated.

References

Aldrich, S., Eccleston, C., & Crombez, G. (2000). Worrying about chronic pain: Vigilance to threat and misdirected problem solving. Behaviour Research and Therapy, 38, 457–470.

Apkarian, A. V., Hashmi, J. A., & Baliki, M. N. (2011). Pain and the brain: Specificity and plasticity of the brain in clinical chronic pain. Pain, 152, S49–S64.

Asghari, A., & Nicholas, M. K. (1999). Personality and adjustment to pain. Pain Reviews, 6, 85–97.

Asmundson, G. J. G. (1999). Anxiety sensitivity and chronic pain: Empirical findings, clinical implications, and future directions. In S. Taylor (Ed.), Anxiety sensitivity: Theory, research, and treatment of the fear of anxiety (pp. 269–285). Mahwah: Erlbaum.

Asmundson, G. J. G., & Hadjistavropoulos, H. D. (2007). Is high fear of pain associated with attentional biases for pain-related or general threat? A categorical reanalysis. The Journal of Pain, 8, 11–18.

Becker, S., Kleinbohl, D., Klossika, I., & Holzl, I. (2008). Operant conditioning of enhanced pain sensitivity by heat-pain titration. Pain, 140, 104–114.

Birbaumer, N., Flor, H., Lutzenberger, W., & Elbert, T. (1995). The corticalization of chronic pain. In B. Bromm & J. E. Desmeldt (Eds.), Pain and the brain: From nociception to cognition (Advances in pain research and therapy, Vol. 22, pp. 331–343). New York: Raven.

Bushnell, M. C., Villemure, C., & Duncan, G. H. (2004). Psychophysical and neurophysiological studies of pain modulation by attention. In D. D. Bushnell (Ed.), Psychological methods of pain control: Basic science and clinical perspectives. Seattle: IASP Press.

Colloca, L., Benedetti, F., & Pollo, A. (2006). Repeatability of autonomic responses to pain anticipation and pain stimulation. European Journal of Pain, 10, 101–118.

Crombez, G., Eccleston, C., Baeyens, F., & Eelen, P. (1997). Habituation and the interference of pain with task performance. Pain, 70, 149–154.

de Jong, J. R., Vlaeyen, J. W., Onghena, P., Cuypers, C., den Hollander, M., & Ruijgrok, J. (2005). Reduction of pain-related fear in complex regional pain syndrome type I: The application of graded exposure in vivo. Pain, 116(3), 264–275.

Flink, I., Nicholas, M. K., Boersma, K., & Linton, S. J. (2009). Reducing the threat value of chronic pain: A preliminary replicated single-case study of interoceptive exposure versus distraction in six individuals with chronic back pain. Behaviour Research and Therapy, 47, 721–728.

Flor, H., & Turk, D. C. (2012). Chronic pain: An integrated biobehavioral approach. Seattle: IASP Press.

Flor, H., Knost, B., & Birbaumer, N. (2002). The role of operant conditioning in chronic pain: An experimental investigation. Pain, 95, 111–118.

Flor, H., Diers, M., & Birbaumer, N. (2004). Peripheral and electrocortical responses to painful and nonpainful stimulation in chronic pain patients, tension headache patients, and healthy controls. Neuroscience Letters, 361, 147–150.

Forsyth, J. P., & Eifert, G. H. (1998). Phobic anxiety and panic: An integrative behavioural account of their origin and treatment. In J. J. Plaud & G. H. Eifert (Eds.), From behavior theory to behavior therapy (pp. 40–67). Boston: Allyn and Bacon.

Haggman, S., Sharpe, L., Nicholas, M., & Refshauge, K. (2010). The nature of attentional biases towards sensory pain words in acute and chronic pain patients. The Journal of Pain, 11(11), 1136–1145.

Holzl, R., Kleinbohl, D., & Huse, E. (2005). Implicit operant learning of pain sensitization. Pain, 115, 12–20.

Hood, D. D., Curry, R., & Eisenach, J. C. (2003). Intravenous remifentanil produces withdrawal hyperalgesia in volunteers with capsaicin-induced hyperalgesia. Anaesthesia and Analgesia, 97, 810–815.

Kleinbohl, D., Holzl, R., Moltner, A., Rommel, C., Weber, C., & Osswald, P. M. (1999). Psychophysical measures of sensitization to tonic heat discriminate chronic pain patients. Pain, 81, 35–43.

Leeuw, M., Goossens, M. E. J. B., Linton, S. J., et al. (2007). The fear-avoidance model of musculoskeletal pain: State of the scientific evidence. Journal of Behavioral Medicine, 30, 77–94.

Turk, D. C., & Okifuji, A. (2002). Psychological factors in chronic pain: Evolution and revolution. Journal of Consulting and Clinical Psychology, 70(3), 678–690. doi:10.1037/0022-006X.70.3.678.

Van Bockstaele, B., Verschuere, B., DeHouwer, J., & Crombez, G. (2010). On the costs and benefits of directing attention towards or away from threat-related stimuli: A classical conditioning experiment. *Behaviour Research and Therapy, 48*, 692–697.

Van Damme, S., Crombez, G., Eccleston, C., & Koster, E. H. W. (2006). Hypervigilance to learned pain signals: A componential analysis. *The Journal of Pain, 7*(5), 346–357.

Vlaeyen, J. W. S., & Linton, S. J. (2000). Fear-avoidance and its consequences in chronic musculoskeletal pain: a state of the art. *Pain, 85*, 317–332.

Woolf, C. (2011). Central sensitization: Implications for the diagnosis and treatment of pain. *Pain, 152*, S2–S15.

Psychometric Function

Definition

A psychometric function describes the relationship between the probability $P_b(a)$ that a stimulus "a" is judged as exceeding a fixed standard stimulus "b," from the viewpoint of some sensory attribute in a discrimination experiment. This term is also used in a detection experiment, when it describes the probability $P_b(a)$ of detecting a stimulus "a" embedded in some "background noise" denoted by "b." When a psychometric function is graphed (i.e., $P_b(a)$ as a function of stimulus intensity "a") it typically exhibits an S-shaped curve.

Cross-References

▶ Pain Evaluation, Psychophysical Methods

Psychometrics

Definition

Psychometrics is the branch of psychology that deals with the design, administration, and interpretation of quantitative tests for the measurement of psychological variables.

Cross-References

▶ Pain Inventories

Psychomotor Physiotherapy

Definition

A branch within physiotherapy that focuses on the body-mind relationship, originally developed in Norway in the late 1940s. Through body awareness techniques, the aim is to improve and normalize the muscular control and help the patient to become aware of how the mind and body interact. Different techniques, ranging from stimulating massage to active exercise, are part of the therapy sessions. Common for all approaches is to teach the patient to register tension and to notice the difference between contracted and relaxed muscles. Furthermore, the therapy also focuses on how respiration can influence muscle tension and guarded movements.

Cross-References

▶ Body Awareness Therapies

Psychomyogenic Headache

▶ Headache, Episodic Tension-Type

Psychopathology

Definition

A broad term to describe abnormal psychological functioning leading to emotional suffering for the individual and/or dysfunctional interpersonal relationships.

Cross-References

▶ Pain as a Cause of Psychiatric Illness

Psychopathology Caused by Pain

▶ Pain as a Cause of Psychiatric Illness

Psychophysical Law

▶ Pain in Humans, Psychophysical Law

Psychophysical Procedures

Stefan Lautenbacher
Physiologische Psychologie, Otto-Friedrich-
Universität, Bamberg, Germany

Introduction

Experimental pain measurement involves two separate issues, namely, the standardized activation of the nociceptive system and the assessment of the evoked responses. Especially, the latter is the domain, which requires the use of psychophysical procedures. Psychophysical procedures are used to determine the quantitative relationship between noxious stimulus and subjective pain experience. These procedures can be applied in the laboratory for basic studies (e.g., experimental study of central hyperexcitability or preclinical screening of drug efficacy) but also in the clinic to characterize patients with sensory dysfunctions and/or pain (e.g., QST in patients with neuropathic pain). An exclusively psychophysical approach to the measurement of pain perception would certainly not be adequate because it stresses the sensory-discriminative side and neglects emotional, motivational, and cognitive features of pain.

Characteristics

Assessment of Pain Responses

Psychophysical procedures require first the experimentally controlled way of pain induction. Numerous experimental methods to induce pain have been developed. So far, low as well as high correlations have been found between different methods indicating that different information can be gathered. The level of correlation is dependent on which methods of pain induction are compared (Janal et al. 1994; Neddermeyer et al. 2008; Neziri et al. 2011). When selecting the method of pain induction, it is therefore appropriate not to assume simply that the various types of experimental pain are equivalent. It is rather advisable either to examine pain perception using several pain induction techniques or to explicitly deduce the appropriate pain induction method from the question under investigation. The most frequently used pain induction methods are derived from the application of mechanical, thermal, electrical, or chemical stimuli (for a an overview see Arendt-Nielsen and Lautenbacher 2004).

Second, the assessment of the evoked responses follows. Psychophysical determinations can roughly be divided into response-dependent and stimulus-dependent methods (see reviews by Arendt-Nielsen and Lautenbacher 2004, Chapman et al. 1985). The response-dependent methods are constructed by a series of fixed stimulus intensities and a rating of each stimulus. The stimulus-dependent methods are based on adjustment of the stimulus intensity until a predefined sensory criterion, typically a threshold, is reached. The stimulus intensity required to reach the threshold is recorded in physical units.

Stimulus-response functions are more informative than a threshold determination alone because suprathreshold response characteristics can be derived from the data. All psychophysical tests require awake and alert individuals, who fully understand the instructions given and are fully capable of cooperating during testing. The choice of the suitable method is therefore dependent both on the precision of a given method and on the given diagnostic or scientific target of an investigation; the latter guides the assessment to the aspect of pain perception of interest.

Pain and Pain Tolerance Thresholds

The determination of pain and pain tolerance threshold has most often been used in experimental pain research. There are several ways of assessing pain and pain tolerance thresholds: the method of constant stimuli, the method of adjustment, the method of limits, etc. They all look for the least amount of physical energy necessary to elicit pain in the case of pain threshold or for the most amount of physical energy still tolerable for an individual in the case of pain tolerance threshold (Arendt-Nielsen and Lautenbacher 2004; Gracely 2005).

The thresholds have been criticized because they are said to confound perceptual and motivational as well as emotional dimensions. This holds especially for the pain tolerance threshold (Chapman et al. 1985). It is true that the thresholds are determined by multiple factors and are not capable of separating these factors. However, the thresholds share this problem with other experimental pain parameters. An ideal parameter in this respect still awaits development. A further objection against the thresholds has been raised because of its selective mapping of the whole pain range. In fact, they only demonstrate where the pain range starts (pain threshold) and where it ends (pain tolerance thresholds). There is no information about pain perception between these range delimiters.

Despite these limitations of the pain and pain tolerance thresholds, they are still of great value because they are reliably, easily, and simply assessed. A further advantage results from the limited cognitive demands that are imposed during threshold assessment upon the individuals, who have to judge perceptions only as "painful" or as "non-painful."

Much less common has been the use of the drug intervention threshold. This threshold is defined as the physical energy of a stimulus that lets a subject want to take an analgesic. The clinical usefulness seems to be extraordinary at the first glance. However, the use of analgesics is determined by so many factors besides the intensity of pain that the meaningfulness of the drug intervention threshold appears doubtful. This method can only be used in pain patients and not in experimental studies on healthy volunteers.

The perception or detection thresholds are also of relevance in this context although they do in fact not measure painful sensations. These thresholds are defined as the minimal physical energy of a stimulus necessary to evoke a first sensation or to allow the detection of a stimulus. Some experimental pain assessment algorithms require the expression of the pain threshold, the pain tolerance threshold, or of other experimental pain parameter relative to the perception threshold in order to take the general somatosensitivity into account (e.g., Rollman and Harris 1987; Komiyama and de Laat 2005). Such an approach is especially useful when the pain threshold is very close to the perception threshold as in some conditions with neuropathic pain. The disturbance of somatosensitivity, in which already slight pressure and moderate heat lead to pain, is called allodynia. However, it is not wise to make no difference between perception and pain threshold. This has been occasionally done by giving the argument that only nociceptors exist in certain tissues, for example, in the tooth pulp, and that, therefore, only painful sensations are possible (e.g., Ghione et al. 1988). However, psychophysical thresholds can be defined and measured only by use of sensory criteria and cannot rely on physiological assumptions.

Staircase or tracking procedures were designed to assess thresholds not only once but repeatedly over time (Gracely 2005; Lautenbacher et al. 1989). A common feature of all of these procedures is the attempt of keeping the stimulation close to the threshold according to the subject's responses in the preceding trials. This approach is especially useful to assess the time course of analgesic treatments because the analgesic effects can be monitored continuously without making the subjects aware of any change in pain perception because the subjective sensations are kept on the same level (Gracely 2005).

Rating Scales

There are many variants of pain rating scales, which can be classified roughly into verbal and numerical scales as well as methods of direct scaling.

A verbal scale is normally divided into several categories, which are labeled, for example, with "no pain," "slight pain," "moderate pain," and "strong pain." A subject is asked to express her/his pain perception by the use of these categories. Her/his answers are quantified afterward in case of a 4-point rating scale by assigning a number from 1 to 4 according to the category chosen. The advantage of this approach is that it is easy to be understood by the subject and does not require a skilled investigator. However, the assumed equidistance between categories has not been proven for many of these scales. Therefore, ordinal scale properties ought to be taken for granted instead of interval scale properties (Chapman et al. 1985; Jensen and Karoly 2001). These limitations can be avoided by calibrating the psychometric distance between categories in advance. The verbal descriptor scales are examples of such empirically validated scales. The stability of distances between categories across studies requires a universal pain language, which is not very likely existent even within a given linguistic area (Gracely 2005; Herr et al. 2004).

The numerical scales use verbal anchors such as "no pain" and "worst pain imaginable" at best at the starting and end points of the scale to define the scale range. The categories themselves are represented only by numbers. It is tempting to assume that the numbers guarantee the equidistance between categories. Since the use of numbers by the patients or the subjects and not the numbers themselves determines the psychometric distances between categories, numerical scales are not a simple solution to this problem. Numerical scales are like verbal scale easy to be explained and understood, facts that have determined their frequent application.

Variants of the methods of direct scaling are magnitude estimation and cross-modality matching. Magnitude estimation requires the subjects to assign numbers proportionally to the intensity of pain perception. At the beginning, the numbers can either be chosen completely arbitrarily by the subjects, or the investigator introduces certain values. Thus, the subject is instructed to assign a 10 to the first stimulus or a 50 to a "barely painful stimulus." Furthermore, certain stimulus intensities can be used as references, which are assigned certain scale values. Such preset scale values are called modulus and are thought to homogenize the use of a scale. The metric properties of such scales can be outstanding as long as a trained and skilled subject is the rater. However, the high cognitive demands associated with such scales prevent the use in populations with low education and cognitive disabilities.

Cross-modality matching requires the subject to express the intensity of pain perception in a second sensory modality (Chapman et al. 1985; Gracely 2005). A well-known example is the handgrip method, in which a pressure-sensitive grip is to be compressed proportionally to the pain intensity.

Another widely used form of cross-modality matching is the visual analog scale (VAS), which has horizontal or vertical lines of 10 or 15 cm length and scale anchors at the beginning like "no pain" and at the end like "worst pain imaginable." The subject is required to express the intensity of pain perception by indicating a line length subjectively proportional to the perception (Jensen and Karoly 2001). The VAS is said to be highly sensitive to treatment effects. However, the cognitive demands in using a VAS are too high for a good proportion of the patients under investigation (Williamson and Hoggart 2005).

The direct scaling methods are said to produce scales with interval or even ratio qualities (Price et al. 1994). Such metric qualities are, however, no unalterable features because they are dependent on the rating behavior of the given population under investigation. It is possible that the same scale allows for ratio quality when used by pain-free undergraduates and does not exceed ordinal scale quality in chronic pain patients with a low education.

Multidimensional Assessment of Pain

The multidimensionality of pain has been often claimed theoretically but has not been acknowledged as often in practice by use of appropriate instruments (Gracely 2005; Wade et al. 1996). The use of two separate scales (mainly visual analog scales or verbal descriptor scales) for assessment of pain intensity and pain unpleasantness has become a kind of a standard with a limited distribution so far. The two dimensions have been proven to be sufficiently independent to justify their separate assessment (Jensen and Karoly 2001).

One of the first, who tried to assess pain by a multidimensional approach, was Melzack and his colleagues (Melzack 1975). The result was the McGill pain questionnaire (MPQ) with a sensory, an affective, an evaluative, and "miscellaneous" pain dimension. The psychometric quality of the MPQ has been repeatedly challenged, but it is still in use. Whereas many clinical pain conditions have been quantified by means of the MPQ, it has infrequently been used to assess experimental pain. Multidimensional comparisons of clinical and experimental pain are, however, badly needed to understand the subjective quality of pain.

A further approach consists of the determination of the dimensionality of pain itself without using an a priori set of dimensions. The method of multidimensional scaling allows statistically finding the dimensionality of a given pain by using similarity estimates for series of perceptions of this pain under systemically varied conditions (e.g., different intensities, different durations, different areas, etc.). The number of distinct dimensions of pain is, by that, not a prerequisite but a result. The resulting dimensions are, however, not necessarily straightforward in their physiological or psychological meaning (Clark et al. 1989).

Signal Detection Theory

The introduction of signal detection theory (SDT) into the domain of experimental pain assessment was at first accompanied by great enthusiasm. One held the belief that the SDT might solve a long-standing problem in experimental pain assessment, which was the reliable separation of the sensory pain component from motivational, emotional, and cognitive influences. However, the influential critic of Rollman (1977) led to the conclusion that the SDT is not as useful in the pain domain as in other sensory modalities. The reasons for that are as follows: The classical SDT paradigm requires a subject to detect a weak stimulus (signal) on the background of sensory noise. However, pain perception cannot be analyzed by using weak stimuli for obvious reasons. Because of that, the adaptation of the classical SDT paradigm to experimental pain research requires the subjects to discriminate strong, painful stimuli from barely weaker, sometimes still painful stimuli. Discrimination from sensory noise alone would be too easy and, by that, uninformative. Accordingly, the SDT of pain requires discrimination and not detection tasks.

As long as the SDT of pain is dealt as an approach to investigate discrimination and as long as the SDT parameters for the discrimination ability (d', P(A), etc.) and for the response bias (b, B, L_X, etc.) are not overloaded by unverified assumptions, the SDT can enrich the methodology of experimental pain research. However, such a theoretically cautious attitude toward the use of SDT has unfortunately been given up repeatedly, which has led to more confusion than insight and some fruitless debates between proponents and opponents (Chapman et al. 1985; Rollman 1977, 1992).

Furthermore, the SDT paradigms ask for long series of pain stimuli under constant conditions, which constitute a prerequisite that is difficult to be met. The study of patients does often not allow for sessions of long durations. Sensitization, adaptation, or habituation prevents constant conditions in long series of stimuli. Such limitations have led to a decline in the use of SDT in experimental pain research.

References

Arendt-Nielsen, L., & Lautenbacher, S. (2004). Assessment of pain perception. In S. Lautenbacher & R. B. Fillingim (Eds.), *Pathophysiology of pain perception* (pp. 25–42). New York: Plenum/Kluver.

Chapman, C. R., Casey, K. L., Dubner, R., et al. (1985). Pain measurement: An overview. *Pain, 22,* 1–31.

Clark, W. C., Ferrer-Brechner, T., Janal, M. N., et al. (1989). The dimensions of pain: A multidimensional scaling comparison of cancer patients and healthy volunteers. *Pain, 37,* 23–32.

Ghione, S., Rosa, C., Mezzasalma, L., et al. (1988). Arterial hypertension is associated with hypalgesia in humans. *Hypertension, 12*, 491–497.

Gracely, R. H. (2005). Studies of experimental human pain. In S. McMahon & M. Koltzenburg (Eds.), *Wall and Melzack's textbook of pain* (5th ed., pp. 267–289). Philadelphia: Elsevier/Churchill Livingston.

Herr, K. A., Spratt, K., Mobily, P. R., et al. (2004). Pain intensity assessment in older adults: Use of experimental pain to compare psychometric properties and usability of selected pain scales with younger adults. *The Clinical Journal of Pain, 20*, 207–219.

Janal, M. N., Glusman, M., Kuhl, J. P., et al. (1994). On the absence of correlation between responses to noxious heat, cold, electrical and ischemic stimulation. *Pain, 58*, 403–411.

Jensen, M. P., & Karoly, P. (2001). Self-report scales and procedures for assessing pain in adults. In D. C. Turk & R. Melzack (Eds.), *Handbook of pain assessment* (pp. 15–34). New York: Guilford Press.

Komiyama, O., & De Laat, A. (2005). Tactile and pain thresholds in the intra- and extra-oral regions of symptom-free subjects. *Pain, 115*, 308–315.

Lautenbacher, S., Galfe, G., Hölzl, R., et al. (1989). Threshold tracking for assessment of long-term adaptation and sensitization in pain perception. *Perceptual and Motor Skills, 69*, 579–589.

Melzack, R. (1975). The McGill pain questionnaire: Major properties and scoring methods. *Pain, 1*, 277–299.

Neddermeyer, T. J., Flühr, K., & Lötsch, J. (2008). Principle components analysis of pain thresholds to thermal, electrical, and mechanical stimuli suggests a predominant common source of variance. *Pain, 138*, 286–291.

Neziri, A. Y., Curatolo, M., Nüesch, E., et al. (2011). Factor analysis of responses to thermal, electrical, and mechanical painful stimuli supports the importance of multi-modal pain assessment. *Pain, 152*, 1146–1155.

Price, D. D., Bush, F. M., Long, S., et al. (1994). A comparison of pain measurement characteristics of mechanical visual analogue and simple numerical rating scales. *Pain, 56*, 217–226.

Rollman, G. B. (1977). Signal detection theory measurement of pain: A review and critique. *Pain, 3*, 187–211.

Rollman, G. B. (1992). Cognitive effects in pain and pain judgments. In D. Algom (Ed.), *Psychophysical approaches to cognition* (pp. 515–574). Amsterdam: North-Holland.

Rollman, G. B., & Harris, G. (1987). The detectability, discriminability, and perceived magnitude of painful electrical shock. *Perception & Psychophysics, 42*, 257–268.

Wade, J. B., Dougherty, L. M., Archer, C. R., et al. (1996). Assessing the stages of pain processing: A multivariate analytical approach. *Pain, 68*, 157–167.

Williamson, A., & Hoggart, B. (2005). Pain: A review of three commonly used pain rating scales. *Journal of Clinical Nursing, 14*, 798–804.

Psychophysical Testing

▶ Quantitative Sensory Testing

Psychophysics

Definition

Psychophysics is the branch of psychology concerned with the relationships of physical stimuli to sensory processes of organisms, including detection, discriminability, sensations, perceptions, and hedonic judgments. It describes the relationship between physical stimuli and the perceptions they evoke.

Cross-References

▶ Pain in Humans, Psychophysical Law
▶ Pain in Humans, Thresholds
▶ Quantitative Thermal Sensory Testing of Inflamed Skin
▶ Secondary Somatosensory Cortex (S2) and Insula, Effect on Pain-Related Behavior in Animals and Humans
▶ Spinothalamic Tract Neurons, in Deep Dorsal Horn

Psychophysiologic Model of Pain

Definition

A model in which pain is conceptualized as resulting from both psychological and physiological components that interact. For example, pain can be viewed as part of a stress response, which includes muscle tensing or activation of other physiologic mechanisms that contribute to the onset or exacerbation of pain. The model postulates a series of positive-feedback loops such that pain results in increased muscle tension, which exacerbates pain, etc.

Cross-References

▶ Coping and Pain
▶ Psychological Treatment in Acute Pain

Psychophysiological Assessment

Definition

A procedure designed to identify the specific physiological dysfunction relevant to a particular pain condition in order to guide treatment efforts and gauge treatment progress.

Cross-References

▶ Biofeedback in the Treatment of Pain
▶ Psychophysiological Assessment of Pain

Psychophysiological Assessment of Pain

Frank Andrasik[1] and Beverly E. Thorn[2]
[1]Department of Psychology, Applied Psychophysiology and Biofeedback,
The University of Memphis, Memphis, TN, USA
[2]Department of Psychology, University of Alabama, Tuscaloosa, AL, USA

Synonyms

Psychophysiological stress profile; Stress profile

Definition

▶ Psychophysiological assessment is designed to identify the physiological dysfunction or response modalities assumed to be relevant to the pain condition and to do so under varied stimulus conditions, psychological and physical, which mimic work and rest (reclining, bending, stooping, lifting, working a keyboard, simulated stressors, etc.) in order to guide treatment efforts and gauge progress. Our comments are restricted to peripheral measures, particularly muscle tension, as these have garnered the greatest attention by researchers and clinicians. Readers seeking more extended discussion of peripheral, and the less commonly utilized central, measures of physiology are referred to Flor (2001), Cacioppo et al. (2000), Peek (2003), and Stern et al. (2001).

Common Measures Used in Pain Assessment
Electromyography (EMG)
The EMG signal arises from the electrochemical changes that occur when a muscle contracts, and it is monitored by placing a series of electrodes along the muscle fibers (to assess the muscle action potentials associated with the ion exchange across the membrane of the muscles). Various factors affect this signal, including sensor composition, size, and placement; distance between sensors; adiposity; and bandpass or recording range. Sensors are typically placed either along the forehead (when measures of general tension are of interest) or at specific sites of interest (e.g., frontal-posterior neck placement for tension-type headache (Hudzynski and Lawrence 1988), masseter and temporalis muscles for temporomandibular pain (Crider and Glaros 1999)).

Other Measures of Interest
Other measures may be of interest for conditions that arise from vascular problems (e.g., migraine or Raynaud's disease) or dysfunction of the sympathetic nervous system (e.g., complex regional pain syndromes, phantom pain). Measures used include the following: blood flow/volume, skin temperature, heart rate, blood pressure, and skin conductance.

Work with phantom limb pain illustrates the value of matching the psychophysiological response to pain type (Sherman 1997). Pain described as burning, throbbing, and tingling is associated with decreased temperature in the stump, while pain described as cramping is preceded by and associated with EMG changes. Targeting feedback accordingly leads to the greatest outcome.

Characteristics

Flor (2001) identified the six functions, utility, and advantages of psychophysiological data collection as it is commonly performed:

1. Demonstrate the role of psychological factors in maladaptive physiological functioning.
2. Provide a necessary prerequisite or justification for the use of biofeedback and related psychophysiological therapies.
3. Facilitate tailoring treatments to individual patients.
4. Make it possible for therapists and researchers to document efficacy, generalization, and transfer of treatment objectively.
5. Identify predictors of treatment response.
6. Serve as a source of motivation (e.g., as patients realize they are able to influence bodily processes by their own thoughts, emotions, and actions, their feelings of helplessness decrease, and they become more open to trying psychological approaches).

The key components of a psychophysiological assessment are reviewed below (and are discussed more fully in Flor (2001) and Arena and Schwartz (2003), among others).

Adaptation

An ▶ adaptation phase is included for three chief reasons:

1. To allow patients to become familiar with the setting and recording procedure
2. To minimize pre-session effects (rushing to the appointment, temperature and humidity differences between office and outdoors)
3. To permit habituation of the orienting response and response stability to occur

Although the importance of completing a pre-baseline period is widely acknowledged, little research has been conducted to help identify key parameters of adaptation or the amount of time needed to achieve stability. Most, but not all, individuals will achieve stability within 5–20 min (some individuals, though, are not fully adapted even after a 60 min session). Practitioners are encouraged to extend this period until some stability is achieved for the key responses of interest (defined as minimal to no fluctuation within

a specified time period). Patients are instructed merely to sit quietly during this period and to refrain from any conscious efforts to relax or alter their physiology.

Baseline

Once adaptation has occurred, the clinician will need to collect some type of baseline data. The baseline data serve as the basis of comparison for subsequent assessment phases and as the basis for gauging progress within and across future treatment sessions. Again, there are no definitive data to document the optimal approach (should eyes be open or closed? should the patient be fully reclined or sitting upright? should conditions be neutral or designed to promote relaxation?) or the desired duration of baseline data collection. In practice, the ▶ baseline phase typically ranges from 1 to 5 min, which should be adequate for providing a representative sample of the patient's typical responses during resting states.

Clinicians (and researchers) often collect a second type of baseline, which is intended to assess preexisting abilities to regulate physiology. When the goal of biofeedback is generalized relaxation, it is useful to collect a second baseline during which the patient is instructed in a manner similar to this: "I would now like to see what happens when you try to relax as deeply as you can. Use whatever means you believe will be helpful. Please let me know when you are as relaxed as possible."

It was once thought that elevated resting levels of muscle tension might be a unique characteristic of patients experiencing chronic pain. A review of 60 psychophysiological investigations conducted with headache, back, and TMD patients found minimal support for this notion (Flor and Turk 1989). Research on this topic, however, is compounded by questions about measurement reliability/stability (Flor 2001; Arena and Schwartz 2003).

Reactivity

The third component examines psychophysiological activity in response to simulated or clinician- or experimenter- induced stressors, which are designed to be personally relevant or to approximate conditions that rival real-world events that are associated with pain onset, exacerbation, or perpetuation. Again, there is no standard empirically validated approach. Some examples of commonly used stimulus conditions are:

- Negative imagery, wherein a patient concentrates on a personally relevant unpleasant situation (the details of which have been obtained during the intake interview, such as a stressful encounter with a coworker or boss)
- Cold exposure (e.g., Raynaud's) or cold pressor test (as a general physical stressor)
- Movement, such as sitting, rising, bending, stooping, or walking
- Load bearing, such as lifting or carrying an object
- Operation of a keyboard

Although baseline differences for EMG have not been found to be reliable in characterizing pain disorders, symptom-specific responses to stimuli have been found for certain pain conditions on a more consistent basis (see Flor 2001 for review).

Recovery

Another component involves ▶ recovery or return to baseline. If multiple stressful stimuli are presented to a patient, then a post-stress recovery period is recommended after each stimulus presentation. This phase continues until the patient's physiology returns to a value close to that observed prior to stimulus presentation (often responses do not fully return to their starting values).

The above components constitute the basic approach to psychophysiological assessment. The final components mentioned here are less common in practice (but they are gaining in acceptance) and may be very useful as well.

Muscle Scanning

Cram (1990) developed a procedure for quickly sampling EMG activity from multiple sites, in a manner that does not require multiple separate recording channels (only two channels are needed). This approach is made possible by the use of two handheld "post" electrodes, which are used to obtain brief (around 2 s per site) sequential bilateral recordings while the patient

is sitting or moving. A normative database helps the therapist determine if any readings are abnormally high or low and if any asymmetries (right-versus left-side differences) exist, as these may be suggestive of bracing or favoring of a position or posture. The goal of biofeedback is to return the aberrant readings to a more normal state. Although this type of approach seems straightforward at first blush, in actuality it is more complex. A number of factors can influence the readings obtained, including the angle and force by which the sensors are applied, the amount of adipose tissue present (fat acts as an insulator and dampens the signal), and the degree to which the sensors are placed in a similar location to that used for the norming sample (plus other variables that affect EMG in general).

Muscle Discrimination

Some have speculated that an inability to perceive bodily states accurately may be a factor in maintaining chronic pain. Flor and colleagues found that patients with chronic pain were unable to perceive muscle tension levels accurately, both in the affected and nonaffected muscles, and that, when exposed to tasks requiring production of muscle tension, these patients overestimated physical symptoms, rated the task as more aversive, and reported greater pain (Flor et al. 1999). These findings point to a heightened sensitivity.

Flor (2001) describes a procedure that can be easily used to assess muscle discrimination abilities in a clinical setting:

- Present the patient with a bar of varying height on a monitor.
- Instruct the patient to tense a muscle to the level reflected in the height of the bar.
- Vary the bar height from low to high.
- Correlate the EMG readings obtained with the actual heights of the bars.
- Define as "good" discrimination abilities correlation coefficients ≥.80.
- Define as "bad" or poor discrimination abilities correlation coefficients ≤.50.

Trigger Points

Some researchers have turned their attention to the psychophysiological model of Travell and

Simons (1983), who postulated that a large percentage of chronic muscle pain resulted from trigger points. Hubbard (1996) provides the rationale for this approach below:

- Muscle tension and pain are sympathetically mediated hyperactivity of the muscle stretch receptors or the muscle spindles.
- Muscle spindles, which are scattered throughout the muscle belly (hundreds within the trapezius muscle), are encapsulated organs that contain their own muscle fibers.
- Although traditionally viewed as a stretch sensor, the muscle spindle is now recognized to be a pain and pressure sensor, and an organ that can be activated by sympathetic stimulation; thus, the pain associated with trigger points arises in the spindle capsule.

Support for this model comes from studies where careful needle electrode placements have detected high levels of EMG activity in the trigger point itself, but data collected from adjacent non-tender sites just 1 cm away are relatively silent (Hubbard 1993). Further, when exposed to a stressful stimulus, EMG activity increases at the trigger point but not at the adjacent site (McNulty et al. 1994). This work provides further evidence of the link between behavioral and emotional factors and mechanisms of muscle pain.

Summary

Flor (2001) has listed a number of recommendations for conducting psychophysiological assessment with pain patients, which serves as a good summary:

- Use multiaxial classification of patients, to identify specific somatic and psychosocial characteristics of the patients.
- If possible, use normative data from controls.
- Control for pain status (i.e., test in a pain-free and a painful state, if possible).
- Control for medication (i.e., make sure patient has not taken analgesic or psychotropic medication for several days, if possible).
- Use sites both proximal and distal to the painful site.
- Make sure that the measures selected are relevant for the specific type of pain being studied (e.g., temperature recordings for Raynaud's syndrome rather than EMG levels).
- Use ecologically valid methods of stress induction (i.e., use self-selected stressors; test stressfulness by assessing subjective stress rating, heart rate, or skin conductance levels).
- Use sufficiently long adaptation phases and baselines.
- Use a syndrome-specific and a general autonomic measure.

References

Arena, J. G., & Schwartz, M. S. (2003). Psychophysiological assessment and biofeedback baselines for the front-line clinician: A primer. In M. S. Schwartz & F. Andrasik (Eds.), *Biofeedback: A practitioner's guide* (3rd ed., pp. 128–158). New York: Guilford Press.

Cacioppo, J. T., Tassinary, L. G., & Bernston, G. G. (2000). *Handbook of psychophysiology* (2nd ed.). Cambridge, MA: Cambridge University Press.

Cram, J. R. (1990). EMG muscle scanning and diagnostic manual for surface recordings. In J. R. Cram & Associates (Eds.), *Clinical EMG for surface recordings, vol 2* (pp. 1–141). Nevada City: Clinical resources.

Crider, A. B., & Glaros, A. G. (1999). A meta-analysis of EMG biofeedback treatment of temporomandibular disorders. *Journal of Orofacial Pain, 13*, 29–37.

Flor, H. (2001). Psychophysiological assessment of the patient with chronic pain. In D. C. Turk & R. Melzack (Eds.), *Handbook of pain assessment* (2nd ed., pp. 76–96). New York: Guilford.

Flor, H., Fürst, M., & Birbaumer, N. (1999). Deficient discrimination of EMG levels and overestimation of perceived tension in chronic pain patients. *Applied Psychophysiology and Biofeedback, 24*, 55–66.

Flor, H., & Turk, D. C. (1989). Psychophysiology of chronic pain: Do chronic pain patients exhibit symptom-specific psychophysiological responses? *Psychological Bulletin, 105*, 219–259.

Hubbard, D. (1996). Chronic and recurrent muscle pain: Pathophysiology and treatment, and review of pharmacologic studies. *Journal of Musculoskeletal Pain, 4*, 123–143.

Hubbard, D., & Berkoff, G. (1993). Myofascial trigger points show spontaneous EMG activity. *Spine, 18*, 1803–1807.

Hudzynski, L. G., & Lawrence, G. S. (1988). Significance of EMG surface electrode placement models and headache findings. *Headache, 28*, 30–35.

McNulty, E., Gevirtz, R., Hubbard, D., & Berkoff, G. (1994). Needle electromyographic evaluation of trigger point response to a psychological stressor. *Psychophysiology, 31*, 313–316.

Peek, C. J. (2003). A primer of biofeedback instrumentation. In M. S. Schwartz & F. Andrasik (Eds.), *Biofeedback: A practitioner's guide* (3rd ed., pp. 43–87). New York: Guilford Press.

Sherman, R. (1997). *Phantom pain*. New York: Plenum Press.

Stern, R. M., Ray, W. J., & Quigley, K. S. (2001). *Psychophysiological recording* (2nd ed.). Oxford: Oxford University Press.

Travell, J., & Simons, D. (1983). *Myofascial pain and dysfunction: The trigger point manual*. New York: Williams & Wilkins.

Psychophysiological Reactivity

Definition

Reactivity of psychophysiological systems (for instance, muscle tone, heart rate, sweating) in reaction to fear-eliciting situations, such as certain physical activities in case of chronic low back pain patients with fear of movement/(re)injury.

Cross-References

▶ Disability, Fear of Movement

Psychophysiological Stress Profile

▶ Psychophysiological Assessment of Pain

Psychosocial Factors

Definition

Thoughts, emotions, and contextual factors that may have an influence on behavior and physiological reactivity.

Cross-References

▶ Impact of Familial Factors on Children's Chronic Pain
▶ Multiaxial Assessment of Pain
▶ Stress and Pain

Psychosocial Interventions

Definition

Psychosocial interventions are non-pharmacological treatments that, while recognizing biological contributions to pain, focus on individual psychological factors (e.g., cognitions, beliefs) as well as social contextual factors (e.g., family system).

Cross-References

▶ Psychological Treatment of Headache

Psychosocial Maladjustment and Pain

▶ Personality Disorders and Pain

Psychosocial Management of Cancer Pain

▶ Psychiatric Aspects of the Management of Cancer Pain

Psychosocial Obstacles to Recovery

▶ Yellow Flags

Psychosocial Predictors or Risk Factors

▶ Psychological Predictors of Chronicity

Psychosocial Risk Factors

▶ Yellow Flags

Psychosocial Treatments for Pain

Definition

Psychosocial treatments for pain include interventions that target maladaptive thoughts and behaviors that contribute to the maintenance and exacerbation of pain symptoms. Cognitive-behavioral therapy (CBT) involves questioning negative thought patterns using a Socratic approach (e.g., what is the evidence that one's pain will never end?), as well as direct behavioral modifications (e.g., pacing oneself during pain-free periods to avoid overexertion). In children, psychosocial treatments for pain often involve the parent(s) or the larger family unit.

Cross-References

▶ Experimental Pain in Children

Psychosocioeconomic

Definition

The combination of psychological, sociological, and economic factors affecting an individual's surrounding environment.

Cross-References

▶ Lower Back Pain, Physical Examination

Psychosocioeconomic Syndrome

▶ Lower Back Pain, Physical Examination

Psychosomatic and Psychiatric Physiotherapy

▶ Body Awareness Therapies

Psychotherapy

Definition

Psychotherapy refers to the treatment of psychological, emotional, or behavioral disorders through interpersonal communications and methods of behavioral changes between the patient and a trained counselor or therapist.

Cross-References

▶ Complex Chronic Pain in Children, Interdisciplinary Treatment

Psychotic

Definition

Refers to a state of mind with impaired reality testing characterized by delusions and hallucinations. It also often includes an illogical thought process characterized by a loosening of the associations between concepts.

Cross-References

▶ Psychiatric Aspects of Pain and Dentistry

Psychotropics

Definition

Drugs that affect psychological or emotional states, behavior, or experience.

Cross-References

▶ Psychiatric Aspects of the Management of Cancer Pain

Ptgs

▶ Cyclooxygenases in Biology and Disease

Ptgs1

▶ COX-1 and COX-2 in Pain

Ptgs2

▶ COX-1 and COX-2 in Pain

PTHA

▶ Headache, Acute Posttraumatic

Ptosis

Definition

Ptosis refers to drooping of the upper eyelid. During acute bouts of cluster headache and during exacerbations of the hemicrania continua, the eyelid on the side of the pain often droops.

Cross-References

▶ Hemicrania Continua

PTPN10

▶ Phosphatases and Glia

Pudendal Nerve

Definition

The pudendal nerve arises from three sacral roots (S-2 to S-4) and leaves the pelvis through the greater sciatic foramen to reach the deep gluteal region and bend around the sacrospinous ligament. After entering Alcock's canal through the lesser sciatic foramen, the nerve courses within a duplication of the muscle fascia along the obturator internus muscle and divides into several branches supplying the anal and urethral sphincters, the pelvic floor muscles, and the anal, perineal, and genital sensitivity.

Cross-References

▶ Pudendal Neuralgia in Women

Pudendal Nerve Entrapment

▶ Pudendal Neuralgia in Women

Pudendal Neuralgia

Definition

Pudendal nerve entrapment may be diagnosed clinically. Computed tomography-guided steroid nerve blocks may provide temporary or long-term relief. A prolonged pudendal nerve distal motor latency on electrodiagnostic testing may confirm the diagnosis. Surgery frequently finds the pudendal nerve flattened at the ischial spine level or in the pudendal canal of Alcock in contact with the sharp inferior border of the sacrospinous ligament. After surgical decompression significant relief of pain is frequently obtained. Pudendal nerve entrapment should be considered in the differential diagnosis of chronic urogenital or anorectal pain, particularly if the pain is aggravated by sitting or if there is a history of bicycle riding.

Cross-References

► Clitoral Pain
► Pudendal Neuralgia in Women

Pudendal Neuralgia in Women

Roger Robert[1] and Jean-Jacques Labat[2]
[1]Neurosurgery, Hotel Dieu, Nantes, CHU, France
[2]Neurology and Rehabilitation, Clinique
Urologique, Nantes, France

Synonyms

Alcock's canal syndrome; Pudendal nerve entrapment

Definition

Pudendal neuralgia was first described by G Amarenco in 1987 who identified the condition in cyclists. He hypothesized compression in the Alcock's canal (Amarenco et al. 1987). Later, anatomical and surgical studies showed the possibility of pudendal nerve entrapment at several levels in the gluteal or perineal areas (Robert et al. 1998).

Pudendal neuralgia (Hibner et al. 2010) is a chronic, painful, debilitating condition that occurs in both men and women, although it is suspected that about two-thirds of those with the disease are women. The main symptom is pain in the area of the pudendal nerve when pressure is applied. The intense pain can be triggered by a sitting position. It is a neuropathic-type pain which patients describe as severe tooth pain, burning, knifelike or aching, stabbing, pinching, twisting, and even numbness. It is not a pelvic pain.

Mostly, the beginning is insidious, but a lot of patients believe that the starting point of the pain is from a pelvi-perineal surgery.

Introduction

What is the pudendal nerve?

The pudendal nerve arises mostly from the S3 root, sometimes from S2 and S4. Its course is first in the presacral area, then it goes laterally, penetrates the great ischiatic overture under the piriformis muscle medially to the ischiatic nerve. In the gluteal region, it crosses behind the distal insertion of the sacrospinal ligament, then goes medially to enter the pudendal tunnel in the inner aspect of the ischiatic tuberosity. In its perineal course, the nerve is in an infra-levatorian space and is situated in an undoubling of the internal obturator muscle fascia named Alcock's tunnel. Its branches perforate this fascia to reach their target zones. The motor branches supply the external striated sphincters of the anus and mainly of the urethra. The sensitive fibers supply the skin of the anal region, the intermediate perineum, the clitoris, and the labia. Four potential entrapment zones can be described: under the piriformis canal, between the sacro-tuberal and the sacrospinous ligaments which cross each other, at the entry of the Alcock's tunnel by the falciform process of the sacro-tuberal ligament, and in the pudendal canal, among the undoubling of the internal obturator fascia (Alcock's tunnel). In the seated position on cadavers, we can see the ascension of the ischiorectal fat tissue which leads to compression of the nerve trunk at the levels mentioned above.

Characteristics

Diagnosis

Pudendal neuralgia is difficult to diagnose. It can be confused but also associated with vulvodynia, piriformis syndrome, levator ani syndrome, sacroiliac joint dysfunction, painful bladder syndrome (interstitial cystitis), etc.

A working party has validated a set of simple internationally recognized diagnostic criteria: Nantes criteria (Labat et al. 2008).

The most important element in making a diagnosis is taking a careful clinical history. There may be trauma or injury to the nerve from childbirth, accident, surgery, or disease. A thorough vaginal/rectal physical examination is crucial. The five essential diagnostic criteria are:

Pain in the anatomical area of the pudendal nerve. Worsened by sitting.

The patient is not usually woken at night by the pain.

No objective sensory loss on clinical examination.

Positive anesthetic pudendal nerve block.

There are also complementary diagnostic criteria which include the words used to describe the pain such as allodynia, rectal or vaginal foreign body sensation, worsening of pain during the day, pain that may be relieved sitting on a toilet seat, pain mainly on one side, pain triggered by bowel movements, and obvious pain on applying pressure on the ischial spine. Exclusion criteria include exclusively coccyx, gluteal, pubic or hypogastric pain, itching, and imaging abnormalities able to account for the pain.

Pudendal Neuralgia Is Associated with Pelvi-perineal Dysfunction

Pudendal neuralgia is often associated with reflex contractures of the perineum, with functional dysuria or anorectal dyschezia and central sensitization phenomena explaining pollakiuria or painful bladder syndrome and a frequent dyspareunia, pain after sexual intercourse. Myofascial pain in the piriformis muscle or obturator internal muscle can explain buttock pain, irradiations into ischiatic or obturator nerve territories.

Pudendal Neuralgia Is a Clinical Diagnosis

Anomalies found during implementation of electrophysiological studies have linked this to a pudendal neuropathy. But these tests lack sensitivity because they only investigate large motor fibers and may not detect selective lesions of small sensory fibers. Furthermore, due to the postural nature of the pain, a neurological lesion may not always be present in the context of intermittent conflict. Lastly, the tests are not specific because of persistent floor neuropathy which could be due to delivery or terminal constipation (Lefaucheur et al. 2007).

Medical imaging is not contributive to the positive diagnosis of pudendal neuralgia, but can be useful to exclude other diagnoses (Figs. 1 and 2). However, imaging may reveal an intercurrent disease clearly unrelated to the neuralgia, for which treatment will not modify the course of the neuralgia. The finding of arachnoid cysts remains a difficult problem, but these cysts are classically considered to be asymptomatic. Imaging is therefore essential, whenever the clinical features do not strictly meet the diagnostic criteria and will be decisive when it demonstrates a lesion able to account for the pain (especially a nerve tumor).

Thus, diagnosis of pudendal neuralgia is primarily a clinical diagnosis.

Etiology

Pudendal neuralgia can appear among cyclists, but not in all; it is just another risk factor. They often complained in the past of pain or numbness after a race but which did not last.

Operative findings reveal anatomical variants of the course of the pudendal nerve and its relations with ligaments, for example, passage of the nerve through the sacrospinous ligament,

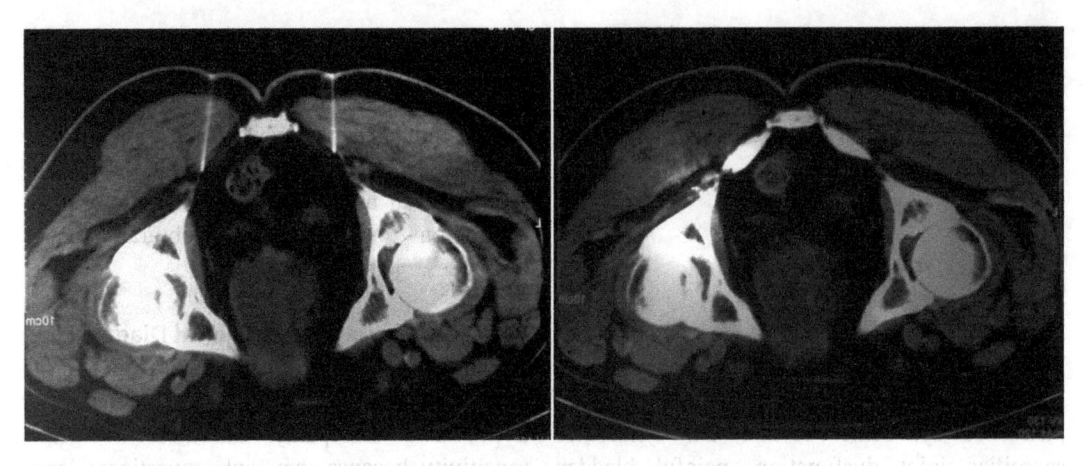

Pudendal Neuralgia in Women, Fig. 1 Infiltration of the sacrospinous ligament

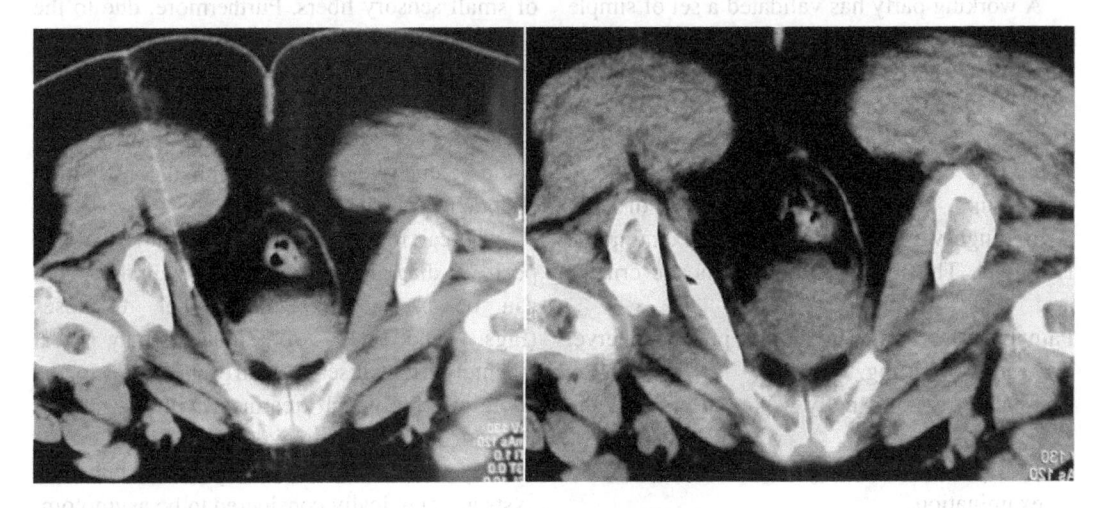

Pudendal Neuralgia in Women, Fig. 2 Infiltration of the pudendal canal

requiring particular caution during surgery, as these variants cannot be detected preoperatively.

The direct trauma of the nerve is only after sacrospinous ligament fixation procedure. The position on the operating table can also be a reason for indirect trauma; this can be seen after fracture table traction. These patients had pudendal sensitive and motor deficiency with sexual dysfunction getting better after a few weeks.

If in these cases, we can understand that the nerve has been touched, the pain can appear after any operation concerning pelvic and perineal areas, without any explanation. The operation

triggers off the pain which gets worse and worse. How can we explain this? Could it be a sympathetic local reaction? We know that many nerve entrapments can remain without symptoms for a long time and are only revealed after an event. Every operation diminishes the threshold of the perception of pain. Preexisting pudendal conflict could therefore be revealed by metameric sensitization.

Importance of Anesthetic Blocks in the Diagnosis

Infiltrations have an essential diagnostic role. Another essential point is their importance in the context of the doctor-patient relationship. Pudendal nerve blocks can be useful for diagnostic using a short term analgesic which even transiently relieves the patient's pain. This confirms a nonexclusively psychiatric origin of the pain and reinforces the patient's confidence.

CT-guided infiltration is clearly the reference technique (McDonald and Spigos 2000), allowing precise procedures at various levels and retrospective evaluation (due to the concomitant use of radiopaque solution) of the technical quality of the infiltration. Two types of CT-guided pudendal nerve infiltration are possible:

Infiltration of the Sacrospinous Ligament
  The patient in the prone position, the needle is introduced into the medial extremity of the ischial spine into the sacrospinous ligament.

Infiltration of the Pudendal Canal
  A CT slice is acquired through the middle of the obturator foramen, and the pelvic part of the obturator internus muscle is identified. The needle is introduced tangentially into the fascia on the medial surface of this muscle. Ideally, the solution molds the medial surface of the muscle without any intrapelvic diffusion.
  Several techniques have been also proposed with or without *using a neurostimulator*: *transperineal infiltration*, *ultrasound-guided infiltration*, and *fluoroscopy-guided infiltration*.
  An immediately positive block in the sitting position is a strong argument in favor of pudendal neuropathy, but does not exclude the possibility of associated inferior cluneal neuralgia. An immediately negative block is a strong argument against the diagnosis of pudendal neuralgia by pudendal nerve entrapment.

Treatment

Minor Measures

Adaptation of the sitting position can provide a certain amount of comfort that must not be neglected. An inflatable ring, a shape memory cushion, a toilet seat, or a chair without gluteal weightbearing (weight on the knees) can be proposed.

Change of defecation times: pain is often considerably worsened after defecation.

Whenever possible, changing the patient's bowel habit so that he/she defecates shortly before going to bed can reduce daytime pain and therefore the overall quantity of pain.

Patients generally try many types of cushions, sometimes making their own cut from foam and covering it with fabric.

Medication

Few studies have assessed the real efficacy of the drug treatments proposed for neuropathic pain. The usual range of drugs (antidepressants, antiepileptics, tramadol) may be prescribed, but they are limited by their adverse effects, their relative efficacy, and the need to evaluate the therapeutic gain obtained. As for any type of chronic pain, only the patient can say whether or not treatment is effective and well tolerated. The choice of treatment must therefore be decided in collaboration with the patient.

Transcutaneous Electrical Nerve Stimulation (TENS)

Transcutaneous electrical nerve stimulation can be useful. The zone of stimulation most frequently effective is the medial retromalleolar space (innervated by the S2 nerve root). The advantages of this technique, apart from its intrinsic efficacy, are its low cost, its relative safety, and the need for the patient's active participation in the treatment.

Physical Therapy

Contraction of the piriformis muscle, the internal obturator muscle, and the levator ani muscle may accompany pudendal neuralgia. Physical therapy should be started early. However, it is the exact opposite of classical physiotherapy following childbirth or for urinary incontinence. A physiotherapist should look at posture, the way the patient stands and moves as it is possible for the pelvis to be misaligned.

Physiotherapy

Pelvic myofascial syndromes and minor intervertebral disorders of the thoracolumbar junction (Maigne syndrome) are often associated, either as a cause or a consequence of pudendal neuralgia. Physiotherapy allows improvement or even resolution of the symptoms in 60 % of patients with pudendal neuralgia accompanied by pelvic myofascial syndromes (piriformis, obturator, levator ani, deep transverse perineal, rectus femoris, psoas muscles) or minor intervertebral disorders of the thoracolumbar junction. The safety, low cost, and relative simplicity (gentle stretching) of the technique make it a first-line treatment modality. The only real limitation is the number of physiotherapists trained in these techniques, although they are easy to learn. Finally, the presence of an associated fibromyalgia syndrome may be an indication for relaxing physiotherapy based on massage and balneotherapy. However, it is the exact opposite of classical physiotherapy following childbirth or for urinary incontinence.

Pudendal Nerve Blocks

A multicenter, randomized, double-blind, 201-patient study has demonstrated the lack of efficacy of corticosteroids during infiltration. Only 14 % of patients have a gain greater than 30 % of their pain 3 months after pudendal infiltration whether or not corticosteroids associated with local anesthetics.

Surgery

Where a patient has intractable pain that does not respond to other treatments, pudendal nerve decompression surgery is an option. A randomized controlled trial of the results of the transgluteal surgical approach has been published (Robert et al. 2005). At 12 months 71.4 % of the surgical group compared with 13.3 % of the nonsurgical group were improved (Robert et al. 2005). Older patients have a poorer response to nerve surgery. Effective postsurgical pain management and follow-up is generally needed after surgery. There are other surgical techniques for which there are currently no randomized trials.

The surgical technique, which remains the reference technique, consists of performing surgical release of the pudendal nerve from the infrapiriformis foramen to Alcock's canal via a transgluteal approach. This surgical procedure is safe.

Spinal Cord Stimulation: Neuromodulation

Conventional spinal cord stimulation has limited efficacy in pudendal neuralgia, possibly because of poor patient selection (TENS responders?). However, several favorable results have been reported for retrograde stimulation performed after neurological and psychiatric assessment and with a clearly defined therapeutic contract. It is obviously only proposed in the case of proven or highly probable failure of surgery (very elderly subjects, atypical clinical features, etc.).

References

Amarenco, G., Lanoe, Y., Perrigot, M., & Goudal, H. (1987). Un nouveau syndrome canalaire: La compression du nerf pudendal dans le canal d'Alcock ou paralysie périnéale des cyclistes. *Presse Médicale, 16*, 399.

Hibner, M., Desai, N., Robertson, L. J., & Nour, M. (2010). Pudendal neuralgia. *Journal of Minimally Invasive Gynecology, 17*(2), 148–153.

Labat, J. J., Riant, T., Robert, R., Amarenco, G., Lefaucheur, J. P., & Rigaud, J. (2008). Diagnostic criteria for pudendal neuralgia by pudendal nerve entrapment (Nantes criteria). *Neurourology and Urodynamics, 27*, 306–310.

Lefaucheur, J. P., Labat, J. J., Amarenco, G., et al. (2007). What is the place of electroneuromyographic studies in the diagnosis and management of pudendal neuralgia related to entrapment syndrome? *Neurophysiologie Clinique, 37*, 223–228.

McDonald, J. S., & Spigos, D. G. (2000). Computed tomography guided pudendal block for treatment of pelvic pain due to pudendal neuropathy. *Obstetrics and Gynecology, 95*, 306–309.

Robert, R., Prat-Pradal, D., Labat, J. J., et al. (1998). Anatomic basis of chronic perineal pain: Role of the pudendal nerve. *Surgical and Radiologic Anatomy, 20*, 93–98.

Robert, R., Labat, J. J., Bensignor, M., et al. (2005). Decompression and transposition of the pudendal nerve in pudendal neuralgia: A randomized controlled trial and long-term evaluation. *European Urology, 47*, 403–408.

Puerperium

Definition

The term puerperium refers to the first six weeks after delivery. It is during this time that the reproductive organs are returning to their nonpregnant condition following labor and delivery.

Cross-References

► Postpartum Pain

PUF Scale

Definition

Pelvic Pain, Urgency, and Frequency Patient Symptom Scale, utilized for quantifying the degree of symptoms associated with interstitial cystitis.

Cross-References

► Interstitial Cystitis and Chronic Pelvic Pain

Pulmonary Embolism (PE)

► Postoperative Pain, Venous Thromboembolism

Pulpal Nociceptors

► Nociceptors in the Dental Pulp

Pulpalgia

► Dental Pain, Etiology, Pathogenesis, and Management

Pulpitis

Definition

This is a state of inflammation of the dental pulp. In its acute stage it is associated with severe attacks of paroxysmal pain, and is exacerbated by cold or hot foods. Most cases are due to penetration of a carious lesion into the pulp chamber.

Cross-References

► Atypical Facial Pain: Etiology, Pathogenesis, and Management
► Dental Pain, Etiology, Pathogenesis, and Management
► Substance P Regulation in Inflammation

Pulsed Electromagnetic Therapy

► Modalities

Punch Skin Biopsies

Definition

A newer, minimally invasive method of observing and quantitating small sensory axons in which small skin punches (typically 3 mm in diameter) are removed, sectioned, and immunolabeled. The procedure can be performed at various locations on the body and repeated to monitor disease progress or the effectiveness of treatment. The method does not provide information about motor or large myelinated axons, demyelination, or vasculitis. Serious complications are nil.

Cross-References

► Diabetic Neuropathies

Punctate Allodynia

Definition

Punctate stimuli such as a pinprick evoke pain.

Cross-References

▶ Hyperpathia, Assessment

Punctate Hyperalgesia

Definition

Punctate hyperalgesia is the increase in pain above normal levels perceived to mechanical stimulation produced by a small contact diameter probe such as a von Frey hair.

Cross-References

▶ Quantitative Thermal Sensory Testing of Inflamed Skin

Punctate Midline Myelotomy

▶ Midline Myelotomy

Punishment

Definition

Punishment has no useful role to play in contingency management and should be avoided.

Cross-References

▶ Training by Quotas

Purine Receptor Targets in the Treatment of Neuropathic Pain

Lucy Vulchanova
Department of Veterinary and Biomedical Sciences, University of Minnesota, St. Paul, MN, USA

Synonyms

Nucleotide receptors; Purinergic receptors

Definition

Purine nucleotides, such as ▶ adenosine $5'$ triphosphate (ATP), adenosine diphosphate (ADP), and uridine di- and triphosphate (UDP and UTP), as well as the ATP metabolite adenosine, have been recognized as ubiquitous intercellular messengers. Based on their preferred ligands and functional characteristics, purine receptors are classified as follows: (a) ▶ P_1 receptors – activated by adenosine, seven transmembrane, G protein-coupled; (b) ▶ P_2X receptors – activated by ATP, cation channels; and (c) ▶ P_2Y receptors – activated by ATP, ADP, UTP or UDP, seven transmembrane, G protein-coupled. Therapeutic approaches that lead to activation of adenosine receptors in spinal cord and inhibition of P_2X receptors in sensory neurons and spinal cord are under development as potential treatments of neuropathic pain.

Characteristics

Modulation of Adenosine Signaling for the Treatment of Neuropathic Pain

Four subtypes of P_1 receptors have been identified: A_1, A_{2A}, A_{2B}, and A_3 (for review see Ribeiro et al. 2003). In general, A_1 and A_3 have inhibitory actions, whereas A_{2A} and A_{2B} have stimulatory actions. In the central nervous system (CNS), high levels of A_1 are present in the cortex, cerebellum, hippocampus, and dorsal horn of spinal cord. The expression of A_{2A} is prominent in the basal ganglia. Modulation of adenosine signaling at A_1 and A_{2A} in the CNS has therapeutic potential for the treatment of a number of disorders including Alzheimer's disease, Parkinson's disease, and chronic pain. Activation of peripheral adenosine receptors is associated with effects such as hypotension, bradycardia, hypothermia, and bronchial constriction, which limit the systemic use of adenosine agonists.

In different experimental paradigms, spinal, peripheral, and systemic modulation of adenosine signaling has shown potential for the treatment of neuropathic pain (Zylka 2011). The analgesic effects of adenosine appear to be mediated predominantly by A_1 receptors in dorsal horn of spinal cord; however, recent evidence also suggests involvement of A_3 (Chen et al. 2012). One of the strategies developed to exploit adenosine analgesia has been the design of A_1 allosteric modulators that enhance the effects of endogenous adenosine without producing side effects in its absence. Intrathecal and systemic administration of the allosteric modulator T62 was able to reduce mechanical hypersensitivity in the spinal nerve ligation model of neuropathic pain, but its chronic use was associated with the development of ► tolerance. A_{2A} receptors have been implicated in microglial responses to nerve injury. However, the contributions of A_{2A} to the mechanisms generating nerve injury-induced hypersensitivity remain unclear at present since constitutive deletion of A_{2A} prevented the establishment of persistent pain after nerve injury (Bura et al. 2008), while in a rat model of neuropathic pain, A_{2A} agonists reversed established allodynia (Loram et al. 2009).

P2X Receptors

P_2X receptors are cation channels activated by ATP (for review see North 2002). The seven cloned subunits (P_2X_{1-7}) each have two transmembrane domains with intracellular amino- and carboxy-termini and a large extracellular loop. Studies on their stoichiometry suggest that individual channels are comprised of three subunits. When expressed in ► transfected cells, all subunits but P_2X_6 can form functional ► homomeric

channels. Co-expression of different subunits has also suggested the existence of ► heteromeric channels ($P_2X_{2/3}$, $P_2X_{1/5}$, $P_2X_{2/6}$, $P_2X_{4/6}$, $P_2X_{4/7}$). P_2X receptors are widely expressed in CNS neurons, where they mediate fast excitatory transmission in response to synaptically released ATP, as well as in glial cells (e.g., microglial P_2X_4 and P_2X_7). P_2X receptors are also present in autonomic and sensory ganglia of the peripheral nervous system as well as in nonneuronal tissues such as smooth muscle.

P2X Receptors in Sensory Neurons as Targets for Neuropathic Pain

Six (P_2X_{1-6}) of the seven cloned P_2X receptor subunits are expressed in sensory neurons at varying levels, P_2X_3 being the most abundant (for example see Collo et al. 1996). Application of ATP to dissociated dorsal root ganglion (DRG) neurons activates three types of inward currents: (1) transient, rapidly desensitizing; (2) sustained, slowly desensitizing; and (3) a combination of transient and sustained. It is believed that the transient, rapidly desensitizing currents are mediated by homomeric P_2X_3 channels, whereas some of the sustained currents are mediated by heteromeric $P_2X_{2/3}$ channels (North 2002; Tsuzuki et al. 2003). Cells that respond to ATP with a combination of transient and sustained currents are thought to express both homo- and heteromeric channels composed of P_2X_2 and P_2X_3. Other homo- or heteromeric channel configurations are also likely to contribute to sustained currents in DRG neurons (North 2002; Tsuzuki et al. 2003).

Considerable research effort has been directed at the P_2X_3 subunit, whose expression is largely restricted to sensory neurons. Immunohistochemical studies have demonstrated that in DRG P_2X_3 is present predominantly in small, neuropeptide-poor neurons, many of which also express other channels known to play a role in nociception and chronic pain such as the capsaicin receptor VR1/TRPV1 (reviewed in North 2003). Following nerve injury, P_2X_3 expression is differentially modulated in injured and uninjured primary afferent neurons. Although the pattern of changes is different depending on the neuropathic pain model used, the modulation of P_2X_3 levels is consistent with a role of P_2X_3 in nerve injury-induced hypersensitivity.

The ability of ATP agonists to excite nociceptive nerve fibers and the high levels of P_2X_3 expression in presumed nociceptors have led to the identification of the P_2X_3 receptor subunit as a pronociceptive molecule whose inhibition may reduce pain (reviewed in North 2003). Knockdown of P_2X_3 expression by ► antisense oligonucleotide (ASO) treatment has been instrumental in the functional validation of P_2X_3 as a therapeutic target for neuropathic pain. ASO inhibited nerve injury-induced mechanical hypersensitivity in two different rodent models of neuropathic pain. In the ► spinal nerve ligation model, there was approximately 50% inhibition of mechanical ► allodynia after ASO treatment. In the partial nerve ligation model, ASO treatment resulted in approximately 30% reversal of mechanical ► hyperalgesia but was ineffective in inhibiting mechanical allodynia. In both studies, the expression of P_2X_3 was inhibited by approximately 50%.

Knockdown of P_2X_3 has also been achieved using siRNA (Hemmings-Mieszczak et al. 2003; Dorn et al. 2004). ASO and siRNA work through different mechanisms, and in vitro studies have suggested that siRNA are more potent and have longer-lasting effects. In the partial nerve ligation model, ► siRNA was slightly more effective than ASO in attenuating nerve injury-induced mechanical allodynia and hyperalgesia (Dorn et al. 2004). However, the reduction in P_2X_3 protein expression by siRNA treatment was comparable to that achieved by ASO treatment. Interestingly, in transfected cells, simultaneous targeting of two different regions of the P_2X_3 mRNA by ASO and siRNA resulted in greater inhibition of P_2X_3 expression than either of the two treatments alone (Hemmings-Mieszczak et al. 2003), suggesting that a siRNA- and ASO-based combinatorial approach to P_2X_3 knockdown may have an enhanced analgesic effect. A systematic comparison of the two technologies in vivo will aid substantially their evaluation and optimization for basic research and clinical use as novel therapeutic agents (Holmlund 2003).

The relative contribution of P_2X_3 homomeric and $P_2X_{2/3}$ heteromeric channels to neuropathic pain is poorly understood because the P_2X_3 antagonist A-317491 as well as ASO and siRNA affect both types of channels. It has been suggested that heteromeric $P_2X_{2/3}$ channels mediate mechanical allodynia following intraplantar injection of the ATP analog $\alpha\beta$-methylene-ATP (Tsuda et al. 2000). Therefore, it is possible that the mechanical hypersensitivity in neuropathic pain depends on heteromeric channels rather than homomeric P_2X_3. The relative contribution of heteromeric channels is particularly relevant to ASO or siRNA approaches that target P_2X_3. For example, if P_2X_3 is required for assembly and/or trafficking of the heteromeric channels, P_2X_3 knockdown would result in functional inactivation of both homomeric and heteromeric channels. Conversely, in the absence of P_2X_3, P_2X_2 may form aberrant homomeric or heteromeric channels. In this case, simultaneous targeting of both subunits may be more effective in attenuating nerve injury-induced mechanical hypersensitivity.

The therapeutic potential of P_2X_3 inhibition has led to efforts toward development of selective P_2X_3 antagonists. A-317491 is a selective non-nucleotide blocker of P_2X_3 and $P_2X_{2/3}$ channels (reviewed in North 2003). Administered systemically, the drug was effective in inhibiting hypersensitivity in several models of chronic pain. A different P_2X_3 and $P_2X_{2/3}$ antagonist, AF-353, was recently reported to attenuate bone cancer pain (Gever et al. 2010; Kaan et al. 2010).

Microglial P_2 Receptors as Targets for Neuropathic Pain

Microglia are activated in the ipsilateral dorsal horn of spinal cord after peripheral nerve injury. Several P2 receptors are expressed in microglia. Among those, signaling through P_2X_4, P_2X_7, and P_2Y_{12} in activated microglia has been implicated in the spinal mechanisms of nerve injury-induced hypersensitivity in rodent models (reviewed in (Jarvis 2010; Trang et al. 2012)).

The expression of P_2X_4 receptors is upregulated in activated microglia with a time course that parallels the development of mechanical hypersensitivity (Tsuda et al. 2003).

Microglial P_2X_4 upregulation is likely dependent on the release of signaling molecules, such as CCL21, from injured sensory neurons (Biber et al. 2011). Mechanical allodynia was substantially reduced after P_2X_4 inhibition or ASO-mediated knockdown (Tsuda et al. 2003) as well as in mice lacking the P_2X_4 receptor (Ulmann et al. 2008; Tsuda et al. 2009). Moreover, intrathecal injection of ATP-primed activated microglia induced allodynia in a P_2X_4-dependent manner (Tsuda et al. 2003; Coull et al. 2005). Taken together, these results indicate that the P_2X_4 receptor in microglia is both necessary and sufficient for the induction of nerve injury-evoked mechanical hypersensitivity. P_2X_4 receptors mediate microglial release of BDNF in a p38 MAPK-dependent manner, and BDNF signaling through the TrkB receptor contributes to the mechanisms of spinal neuroplasticity after nerve injury (Coull et al. 2005; Ulmann et al. 2008; Trang et al. 2009).

The P_2X_7 receptor also represents a target for neuropathic pain treatment. Both genetic deletion and pharmacological inhibition of P_2X_7 attenuate nerve injury-induced hypersensitivity (Chessell et al. 2005; Honore et al. 2006; Perez-Medrano et al. 2009). Microglial P_2X_7 is likely to participate in the spinal mechanisms of nerve injury-induced hypersensitivity through release of the proinflammatory cytokine IL1β (Kawasaki et al. 2008; Mingam et al. 2008; Clark et al. 2010). However, it is possible that P_2X_7 receptors expressed in hematopoietic cells in the periphery and in Schwann cells also contribute to the analgesic effects of systemic P_2X_7 inhibition and the anti-allodynic effects of constitutive P_2X_7 gene deletion (Shamash et al. 2002).

Finally, several lines of evidence suggest that the P_2Y_{12} receptor participates in the induction of neuropathic pain. P_2Y_{12} is upregulated in spinal microglia after nerve injury and its pharmacological inhibition; ASO knockdown or genetic deletion attenuates the development of hypersensitivity as well as microglial p38 MAPK activation (Kobayashi et al. 2008; Tozaki-Saitoh et al. 2008; Ando et al. 2010); conversely, intrathecal administration of P_2Y_{12} agonist results in microglial p38 MAPK activation and mechanical allodynia (Kobayashi et al. 2008).

P_2Y Receptors in Sensory Neurons

P_2Y receptors are ▶ G-protein-coupled purinergic receptors. The agonist preferences of these receptors vary between different subtypes as well as between different species homologs of the same subtype (for review see von Kugelgen and Wetter 2000). In addition to ATP, some P_2Y receptors are activated by adenosine diphosphate (ADP) or by uridine di- or triphosphate (UTP or UDP). P_2Y receptor activation generally has stimulatory effects through Ca^{2+} release from intracellular stores (i.e., P_2Y_1), although some P_2Y receptors have also been shown to couple to inhibitory pathways (i.e., P_2Y_{12} and P_2Y_{13}). Sensory neurons express high levels of P_2Y_1 and P_2Y_2 receptors, and activation of these receptors increases their excitability (Yousuf et al. 2011). P_2Y_2 is present in nociceptors, and analysis of P_2Y_2 KO mice suggests that it is involved in thermal nociception (Stucky et al. 2004; Malin et al. 2008). Sensory neurons also express inhibitory P_2Y receptors, whose activation reduced inflammatory hyperalgesia (Malin and Molliver 2010). Although the role of P2Y receptors in neuropathic pain has not been explored directly, these findings suggest that P_2Y_2 may play a role in nociceptive signaling as well as in sensory neuron plasticity underlying chronic pain states.

References

Ando, R. D., Mehesz, B., Gyires, K., Illes, P., & Sperlagh, B. (2010). A comparative analysis of the activity of ligands acting at P2X and P2Y receptor subtypes in models of neuropathic, acute and inflammatory pain. *British Journal of Pharmacology, 159*, 1106–1117.

Biber, K., Tsuda, M., Tozaki-Saitoh, H., Tsukamoto, K., Toyomitsu, E., Masuda, T., Boddeke, H., & Inoue, K. (2011). Neuronal CCL21 up-regulates microglia P2X4 expression and initiates neuropathic pain development. *The EMBO Journal, 30*, 1864–1873.

Bura, S. A., Nadal, X., Ledent, C., Maldonado, R., & Valverde, O. (2008). A 2A adenosine receptor regulates glia proliferation and pain after peripheral nerve injury. *Pain, 140*, 95–103.

Chen, Z., Janes, K., Chen, C., Doyle, T., Bryant, L., Tosh, D. K., Jacobson, K. A., & Salvemini, D. (2012). Controlling murine and rat chronic pain through A3 adenosine receptor activation. *The FASEB Journal, 26*(5), 1855–1865.

Chessell, I. P., Hatcher, J. P., Bountra, C., Michel, A. D., Hughes, J. P., Green, P., Egerton, J., Murfin, M., Richardson, J., Peck, W. L., Grahames, C. B., Casula, M. A., Yiangou, Y., Birch, R., Anand, P., & Buell, G. N. (2005). Disruption of the P2X7 purinoceptor gene abolishes chronic inflammatory and neuropathic pain. *Pain, 114*, 386–396.

Clark, A. K., Staniland, A. A., Marchand, F., Kaan, T. K., McMahon, S. B., & Malcangio, M. (2010). P2X7-dependent release of interleukin-1beta and nociception in the spinal cord following lipopolysaccharide. *The Journal of Neuroscience, 30*, 573–582.

Collo, G., North, R. A., Kawashima, E., Merlo-Pich, E., Neidhart, S., Surprenant, A., & Buell, G. (1996). Cloning OF P2X5 and P2X6 receptors and the distribution and properties of an extended family of ATP-gated ion channels. *The Journal of Neuroscience, 16*, 2495–2507.

Coull, J. A., Beggs, S., Boudreau, D., Boivin, D., Tsuda, M., Inoue, K., Gravel, C., Salter, M. W., & De Koninck, Y. (2005). BDNF from microglia causes the shift in neuronal anion gradient underlying neuropathic pain. *Nature, 438*, 1017–1021.

Dorn, G., Patel, S., Wotherspoon, G., Hemmings-Mieszczak, M., Barclay, J., Natt, F. J., Martin, P., Bevan, S., Fox, A., Ganju, P., Wishart, W., & Hall, J. (2004). siRNA relieves chronic neuropathic pain. *Nucleic Acids Research, 32*, e49.

Gever, J. R., Soto, R., Henningsen, R. A., Martin, R. S., Hackos, D. H., Panicker, S., Rubas, W., Oglesby, I. B., Dillon, M. P., Milla, M. E., Burnstock, G., & Ford, A. P. (2010). AF-353, a novel, potent and orally bioavailable P2X3/P2X2/3 receptor antagonist. *British Journal of Pharmacology, 160*, 1387–1398.

Hemmings-Mieszczak, M., Dorn, G., Natt, F. J., Hall, J., & Wishart, W. L. (2003). Independent combinatorial effect of antisense oligonucleotides and RNAi-mediated specific inhibition of the recombinant rat P2X3 receptor. *Nucleic Acids Research, 31*, 2117–2126.

Holmlund, J. T. (2003). Applying antisense technology: Affinitak and other antisense oligonucleotides in clinical development. *Annals of the New York Academy of Sciences, 1002*, 244–251.

Honore, P., Donnelly-Roberts, D., Namovic, M. T., Hsieh, G., Zhu, C. Z., Mikusa, J. P., Hernandez, G., Zhong, C., Gauvin, D. M., Chandran, P., Harris, R., Medrano, A. P., Carroll, W., Marsh, K., Sullivan, J. P., Faltynek, C. R., & Jarvis, M. F. (2006). A-740003 [N-(1-{[(cyanoimino)(5-quinolinylamino) methyl] amino}-2,2-dimethylpropyl)-2-(3,4-dimethoxyphenyl) acetamide], a novel and selective P2X7 receptor antagonist, dose-dependently reduces neuropathic pain in the rat. *The Journal of Pharmacology and Experimental Therapeutics, 319*, 1376–1385.

Jarvis, M. F. (2010). The neural-glial purinergic receptor ensemble in chronic pain states. *Trends in Neurosciences, 33*, 48–57.

Kaan, T. K., Yip, P. K., Patel, S., Davies, M., Marchand, F., Cockayne, D. A., Nunn, P. A., Dickenson, A. H., Ford, A. P., Zhong, Y., Malcangio, M., & McMahon, S. B. (2010). Systemic blockade of P2X3 and P2X2/3 receptors attenuates bone cancer pain behaviour in rats. *Brain, 133*, 2549–2564.

Kawasaki, Y., Xu, Z. Z., Wang, X., Park, J. Y., Zhuang, Z. Y., Tan, P. H., Gao, Y. J., Roy, K., Corfas, G., Lo, E. H., & Ji, R. R. (2008). Distinct roles of matrix metalloproteases in the early- and late-phase development of neuropathic pain. *Nature Medicine, 14*, 331–336.

Kobayashi, K., Yamanaka, H., Fukuoka, T., Dai, Y., Obata, K., & Noguchi, K. (2008). P2Y12 receptor upregulation in activated microglia is a gateway of p38 signaling and neuropathic pain. *The Journal of Neuroscience, 28*, 2892–2902.

Loram, L. C., Harrison, J. A., Sloane, E. M., Hutchinson, M. R., Sholar, P., Taylor, F. R., Berkelhammer, D., Coats, B. D., Poole, S., Milligan, E. D., Maier, S. F., Rieger, J., & Watkins, L. R. (2009). Enduring reversal of neuropathic pain by a single intrathecal injection of adenosine 2A receptor agonists: A novel therapy for neuropathic pain. *The Journal of Neuroscience, 29*, 14015–14025.

Malin, S. A., & Molliver, D. C. (2010). Gi- and Gq-coupled ADP (P2Y) receptors act in opposition to modulate nociceptive signaling and inflammatory pain behavior. *Molecular Pain, 6*, 21.

Malin, S. A., Davis, B. M., Koerber, H. R., Reynolds, I. J., Albers, K. M., & Molliver, D. C. (2008). Thermal nociception and TRPV1 function are attenuated in mice lacking the nucleotide receptor P2Y2. *Pain, 138*, 484–496.

Mingam, R., De Smedt, V., Amedee, T., Bluthe, R. M., Kelley, K. W., Dantzer, R., & Laye, S. (2008). In vitro and in vivo evidence for a role of the P2X7 receptor in the release of IL-1 beta in the murine brain. *Brain, Behavior, and Immunity, 22*, 234–244.

North, R. A. (2002). Molecular physiology of P2X receptors. *Physiological Reviews, 82*, 1013–1067.

North, R. A. (2003). The P2X3 subunit: A molecular target in pain therapeutics. *Current Opinion in Investigational Drugs, 4*, 833–840.

Perez-Medrano, A., Donnelly-Roberts, D. L., Honore, P., Hsieh, G. C., Namovic, M. T., Peddi, S., Shuai, Q., Wang, Y., Faltynek, C. R., Jarvis, M. F., & Carroll, W. A. (2009). Discovery and biological evaluation of novel cyanoguanidine P2X(7) antagonists with analgesic activity in a rat model of neuropathic pain. *Journal of Medicinal Chemistry, 52*, 3366–3376.

Ribeiro, J. A., Sebastião, A. M., & de Mendonça, A. (2003). Adenosine receptors in the nervous system: pathophysiological implications. *Progress in Neurobiology, 68*, 377–392.

Shamash, S., Reichert, F., & Rotshenker, S. (2002). The cytokine network of Wallerian degeneration: Tumor necrosis factor-alpha, interleukin-1alpha, and interleukin-1beta. *The Journal of Neuroscience, 22*, 3052–3060.

Stucky, C. L., Medler, K. A., & Molliver, D. C. (2004). The P2Y agonist UTP activates cutaneous afferent fibers. *Pain, 109*, 36–44.

Tozaki-Saitoh, H., Tsuda, M., Miyata, H., Ueda, K., Kohsaka, S., & Inoue, K. (2008). P2Y12 receptors in spinal microglia are required for neuropathic pain after peripheral nerve injury. *The Journal of Neuroscience, 28*, 4949–4956.

Trang, T., Beggs, S., Wan, X., & Salter, M. W. (2009). P2X4-receptor-mediated synthesis and release of brain-derived neurotrophic factor in microglia is dependent on calcium and p38-mitogen-activated protein kinase activation. *The Journal of Neuroscience, 29*, 3518–3528.

Trang, T., Beggs, S., & Salter, M. W. (2012). ATP receptors gate microglia signaling in neuropathic pain. *Experimental Neurology, 234*, 354–361.

Tsuda, M., Koizumi, S., Kita, A., Shigemoto, Y., Ueno, S., & Inoue, K. (2000). Mechanical allodynia caused by intraplantar injection of P2X receptor agonist in rats: Involvement of heteromeric P2X2/3 receptor signaling in capsaicin-insensitive primary afferent neurons. *The Journal of Neuroscience, 20*, 90.

Tsuda, M., Shigemoto-Mogami, Y., Koizumi, S., Mizokoshi, A., Kohsaka, S., Salter, M. W., & Inoue, K. (2003). P2X4 receptors induced in spinal microglia gate tactile allodynia after nerve injury. *Nature, 424*, 778–783.

Tsuda, M., Kuboyama, K., Inoue, T., Nagata, K., Tozaki-Saitoh, H., & Inoue, K. (2009). Behavioral phenotypes of mice lacking purinergic P2X4 receptors in acute and chronic pain assays. *Molecular Pain, 5*, 28.

Tsuzuki, K., Ase, A., Seguela, P., Nakatsuka, T., Wang, C. Y., She, J. X., & Gu, J. G. (2003). TNP-ATP-resistant P2X ionic current on the central terminals and somata of rat primary sensory neurons. *Journal of Neurophysiology, 89*, 3235–3242.

Ulmann, L., Hatcher, J. P., Hughes, J. P., Chaumont, S., Green, P. J., Conquet, F., Buell, G. N., Reeve, A. J., Chessell, I. P., & Rassendren, F. (2008). Up-regulation of P2X4 receptors in spinal microglia after peripheral nerve injury mediates BDNF release and neuropathic pain. *The Journal of Neuroscience, 28*, 11263–11268.

von Kugelgen, I., & Wetter, A. (2000). Molecular pharmacology of P2Y-receptors. *Naunyn-Schmiedeberg's Archives of Pharmacology, 362*, 310–323.

Yousuf, A., Klinger, F., Schicker, K., & Boehm, S. (2011). Nucleotides control the excitability of sensory neurons via two P2Y receptors and a bifurcated signaling cascade. *Pain, 152*, 1899–1908.

Zylka, M. J. (2011). Pain-relieving prospects for adenosine receptors and ectonucleotidases. *Trends in Molecular Medicine, 17*, 188–196.

Purinergic Receptors

Definition

Purinergic receptor molecules bind adenosine triphosphate (ATP) and its metabolites. They are grouped in two large classes with many subgroups, namely, P2X receptors that control ion channels and P2Y receptors that are coupled to a G protein and induce changes in the intracellular cascade of second messengers. The P2X3 receptor is assumed to be specific for nociceptive endings.

Cross-References

▶ Muscle Pain Model, Ischemia-Induced and Hypertonic Saline-Induced
▶ Nociceptors in the Orofacial Region (Skin/Mucosa)
▶ Purine Receptor Targets in the Treatment of Neuropathic Pain

Putamen

Definition

Putamen is a part of the striatum, a component of the basal ganglia, which receives afferents from motor and somatosensory cortical areas, substantia nigra, and the centromedian thalamic nucleus. It projects efferents to the globus pallidus, which in turn projects to the thalamus and to cortical motor areas.

Cross-References

▶ Nociceptive Processing in the Nucleus Accumbens, Neurophysiology and Behavioral Studies

Pyeloplasty

Definition

A surgical operation to relieve an obstruction at the junction of the pelvis of the kidney and the ureter, often involving a refashioning of the junction. Such an obstruction in infants is often diagnosed prenatally by the presence of hydronephrosis (accumulation of fluid in the kidney) observed by routine ultrasound scan during pregnancy.

Cross-References

▶ Infant Pain Mechanisms

Pyramidal Cells

Definition

The pyramidal cells are the major neuronal type in the hippocampus. The pyramidal cells of field CA1 have basal dendrites that extend into the stratum oriens and apical dendrites that extend to the stratum lacunosum-moleculare. The axons of CA1 pyramidal cells project in a topographic fashion to the adjacent subiculum and to other cortical and subcortical areas including the pre- and infralimbic cortex, the lateral septal area, and the amygdala.

Cross-References

▶ Nociceptive Processing in the Hippocampus and Entorhinal Cortex, Neurophysiology and Pharmacology

Pyramidal Neuron

Definition

A pyramidal neuron is a specific type of cortical cell that sends its axon to other brain areas.

Cross-References

▶ Corticothalamic and Thalamocortical Interactions

Pyrazolinones

▶ Pharmacology of Cyclooxygenase Inhibitors

Pyst1

▶ Phosphatases and Glia

Encyclopedia of Pain

Gerald F. Gebhart • Robert F. Schmidt

Editors

Encyclopedia of Pain

Second Edition

Volume 6

Q–S

With 795 Figures and 217 Tables

Springer Reference

Editors
Gerald F. Gebhart
Department of Anesthesiology
Center for Pain Research
University of Pittsburgh
Pittsburgh, PA, USA

Robert F. Schmidt
Physiologisches Institut der
Universität Würzburg
Würzburg, Germany

ISBN 978-3-642-28752-7   ISBN 978-3-642-28753-4 (eBook)
ISBN 978-3-642-28754-1 (print and electronic bundle)
DOI 10.1007/978-3-642-28753-4
Springer Heidelberg New York Dordrecht London

Library of Congress Control Number: 2013951551

Printed on acid-free paper

Springer is part of Springer Science+Business Media (www.springer.com)

Preface to the Second Edition

As was pointed out in the Preface to the first edition, pain is the principal reason individuals seek medical and dental attention. Fortunately, acute tissue insults associated with pain – and typically inflammation – readily resolve with appropriate treatment and care. Chronic pain states, in contrast, are notoriously difficult to manage and are commonly associated with comorbidities (e.g., depression, cross-organ sensitization) and reduced quality of life. Indeed, chronic pain has come to be considered by some as a disease in its own right. This edition of the *Encyclopedia of Pain* updates the content of the 1st edition and introduces new topics consistent with advances in our knowledge of underlying mechanisms of pain. Thus, new content on channelopathies, expanded and updated material on the role of glia and immune-competent cell interactions with nociceptive neurons, and advances in imaging and changes in brain structure in chronic pain states are included in this edition.

The cardinal objective of research on pain is the translation of knowledge between the laboratory and the clinic. The present prevailing refrain is "bench to bedside," but in fact many ideas for research in pain derive from clinical observations and the absence of knowledge to explain clinical pain states. In the recent past, basic and clinical science research has seen the application of imaging techniques to visualize areas of the brain that are involved in both the sensory and affective aspects of pain, cloning of pain-related channels and receptors, and the application of elegant genetic manipulations that selectively "label" functionally distinct dorsal root ganglion sensory and spinal neurons, two-photon imaging, and optogenetic approaches.

Accordingly, much is new in this second edition of the *Encyclopedia of Pain*. The number of authors who have contributed to this edition is close to 800. In addition to a print edition, the publisher, Springer-Verlag, will produce an electronic version that will make comprehensive searches easy to manage. The electronic version, to be made available on the online publishing platform SpringerReference, can be updated regularly to ensure that the content remains current.

October 2013
Gerald F. Gebhart
Pittsburgh, PA, USA

Robert F. Schmidt
Würzburg, Germany

Preface to the First Edition

As all medical students know, pain is the most common reason for a person to consult a physician. Under ordinary circumstances, acute pain has a useful, protective function. It discourages the individual from activities that aggravate the pain, allowing faster recovery from tissue damage. The physician can often tell from the nature of the pain what its source is. In most cases, treatment of the underlying condition resolves the pain. By contrast, children born with congenital insensitivity to pain suffer repeated physical damage and die young (see Sweet WH (1981) Pain 10:275).

Pain resulting from difficult to treat or untreatable conditions can become persistent. Chronic pain "never has a biologic function but is a malefic force that often imposes severe emotional, physical, economic, and social stresses on the patient and on the family..." (Bonica JJ (1990) The Management of Pain, vol 1, 2nd edn. Lea & Febiger, Philadelphia, p 19). Chronic pain can be considered a disease in its own right.

Pain is a complex phenomenon. It has been defined by the Taxonomy Committee of the International Association for the Study of Pain as "An unpleasant emotional and sensory experience associated with actual or potential tissue damage, or described in terms of such damage" (Merskey H and Bogduk N (1994) Classification of Chronic Pain, 2nd edn. IASP Press, Seattle). It is often ongoing, but in some cases it may be evoked by stimuli. Hyperalgesia occurs when there is an increase in pain intensity in response to stimuli that are normally painful. Allodynia is pain that is evoked by stimuli that are normally non-painful.

Acute pain is generally attributable to the activation of primary efferent neurons called nociceptors (Sherrington CS (1906) The Integrative Action of the Nervous System. Yale University Press, New Haven; 2nd edn, 1947). These sensory nerve fibers have high thresholds and respond to strong stimuli that threaten or cause injury to tissues of the body. Chronic pain may result from continuous or repeated activation of nociceptors, as in some forms of cancer or in chronic inflammatory states, such as arthritis.

However, chronic pain can also be produced by damage to nervous tissue. If peripheral nerves are injured, peripheral neuropathic pain may develop. Damage to certain parts of the central nervous system may result in central neuropathic pain. Examples of conditions that can cause central neuropathic pain include spinal cord injury, cerebrovascular accidents, and multiple sclerosis.

Research on pain in humans has been an important clinical topic for many years. Basic science studies were relatively few in number until experimental work on pain accelerated following detailed descriptions of peripheral nociceptors and central nociceptive neurons that were made in the 1960s and 1970s, by the discovery of the endogenous opioid compounds and the descending pain control systems in the 1970s and the application of modern imaging techniques to visualize areas of the brain that are affected by pain in the 1990s. Accompanying these advances has been the development of a number of animal models of human pain states, with the goal of using these to examine pain mechanisms and also to test analgesic drugs or non-pharmacologic interventions that might prove useful for the treatment of pain in humans. Basic research on pain now emphasizes multidisciplinary approaches, including behavioral testing, electrophysiology, and the application of many of the techniques of modern cell and molecular biology, including the use of transgenic animals.

The *Encyclopedia of Pain* is meant to provide a source of information that spans contemporary basic and clinical research on pain and pain therapy. It should be useful not only to researchers in these fields but also to practicing physicians and other health care professionals and to health care educators and administrators. The work is subdivided into 35 fields, and the Field Editor of each of these describes the areas covered in each in a brief review entry. The topics included in a field are the subject of a series of short essays, accompanied by key words, definitions, illustrations, and a list of significant references. The number of authors who have contributed to the encyclopedia exceeds 550. The plan of the publisher, Springer-Verlag, is to produce both print and electronic versions of this encyclopedia. Numerous links within the electronic version should make comprehensive searches easy to manage. The electronic version will be updated at sufficiently short intervals to ensure that the content remains current.

The editors thank the staff at Springer-Verlag who have provided oversight for this project, including Rolf Lange, Thomas Mager, Claudia Lange, Natasja Sheriff, and Michaela Bilic. Working with these outstanding individuals has been a pleasure.

July 2006
ROBERT F. SCHMIDT   WILLIAM D. WILLIS
Würzburg, Germany   Galveston, Texas, USA

About the Editors

Gerald F. Gebhart received his PhD from The University of Iowa at Iowa City, Iowa, USA, after which he completed a 2-year fellowship at the Université de Montréal, Montréal, Canada, under the direction of Herbert H. Jasper. He began his academic career in the Department of Pharmacology at The University of Iowa in 1973, where he developed a productive research program and led a Pain Interest Group. His career includes a sabbatical year in Heidelberg, Germany (1981–1982), and leadership as head of the Department of Pharmacology from 1996 to 2006. In 2006, Dr. Gebhart was named director of the Center for Pain Research, Department of Anesthesiology, the University of Pittsburgh at Pittsburgh, Pennsylvania, USA. Dr. Gebhart has served the pain community throughout his career. He was elected to the Board of Directors of the American Pain Society (1991–1994) and subsequently elected president of the APS (1997). In 2000, he became the founding editor of *The Journal of Pain*, the official journal of the APS, serving as editor-in-chief until 2010. Dr. Gebhart also served on the Council and later as president-elect, president, and past president of the International Association for the Study of Pain (2005–2012).

Dr. Gebhart is best known for his research contributions in descending modulation of pain and mechanisms of visceral pain, and was designated in 2004 by Thomson ISI as a Highly Cited Researcher (Neuroscience). His scientific contributions have been recognized with several awards, including a 5-year Bristol Myers-Squibb Award for Excellence in Pain Research (1989–1994), a 10-year MERIT Award from the National Institutes of Health (1993–2003),

the Frederick W.L. Kerr Award from the American Pain Society (1994), the Purdue Pharma Prize for Pain Research (2004), the Janssen Award in Gastroenterology (2005), the Founders Award from the American Academy of Pain Medicine (2006), and the Patrick D. Wall Award from the British Pain Society (2012).

Robert F. Schmidt studied medicine at Heidelberg University (Germany). He graduated in 1959 and was promoted to Dr.med. in the same year. After residencies in clinical disciplines, Dr. Schmidt went to the John Curtin School of Medical Research of the Australian National University in Canberra, ACT, where he worked under the head of the Department of Neurophysiology, Sir John C. Eccles. He received his Ph.D. in 1963, the same year Sir John was awarded the Nobel Prize for Physiology and Medicine.

Returning to Heidelberg Medical School, Dr. Schmidt continued his academic career. He completed his habilitation in 1964, becoming a tenured lecturer (Wissenschaftlicher Rat) in 1966 and an associate professor in 1970. After an 8-month sabbatical with Sir John at the State University of New York, Dr. Schmidt became professor and director of the Department of Physiology at the University of Kiel, Germany. In 1982, he moved to the University of Würzburg to take the same position in the Department of Physiology. In 2000, Dr. Schmidt became professor emeritus at this department.

The research work of Dr. Schmidt and his associates in both Kiel and Würzburg focused on the neurophysiology of fine sensory afferents and their connections and functions in the spinal cord. In Würzburg, he concentrated on the processes leading to the changes in response characteristics of nociceptive afferents (i.e., peripheral sensitization) during inflammation of joints, discovering in the process silent nociceptors, which have subsequently been described in all vertebrates studied, including humans. In addition, Dr. Schmidt has edited, coedited, authored, and coauthored numerous textbooks and monographs, most of them having several editions and translations.

In recognition of his research contributions, Dr. Schmidt received many prizes, awards, and honors, including a D.Sc. honoris causa from the University of New South Wales, Sydney, Australia, and an Honorary Professorship from the University of Tübingen, Germany. He was elected to membership in the Akademie der Wissenschaften und der Literatur, Mainz, Germany. Dr. Schmidt is an Honorary Member of the Colombian Association for the Study of Pain, the Japan Physiological Society, the German Society for the Study of Pain, the German Pain Association, the International Association for the Study of Pain (IASP), and the German Physiological Society (in which he served as president in 1979). In 2000, he was awarded the Bundesverdienstkreuz 1st Class (Order of Merit of the German Federal Republic).

In recognition of his research contributions, Dr. Schmidt received many prizes, awards, and honors, including a D.Sc. honoris causa from the University of New South Wales, Sydney, Australia, and an Honorary Professorship from the University of Tübingen, Germany. He was elected to membership in the Akademie der Wissenschaften und der Literatur, Mainz, Germany. Dr. Schmidt is an Honorary Member of the Colombian Association for the Study of Pain, the Japan Physiological Society, the German Society for the Study of Pain (DGSS), the German Physiological Society, and the German Pharmacological Society (in which he served as president in 1979). In 2000, he was awarded the Bundesverdienstkreuz 1st class (order of Merit of the German Federal Republic).

Section Editors

Ebtesam Ahmed College of Pharmacy and Health Sciences, St. John's University, Department of Pain Medicine and Palliative Care, Beth Israel Medical Center, New York, NY, USA

A. Vania Apkarian Department of Physiology, Northwestern University Feinberg School of Medicine, Chicago, IL, USA

Lars Arendt-Nielsen Center for Sensory-Motor Interaction, Department of Health Science and Technology, School of Medicine, Aalborg University, Aalborg, Denmark

Carlos Belmonte Instituto de Neurociencias, Universidad Miguel Hernandez-CSIC, San Juan de Alicante, Spain

Niels Birbaumer Institute of Medical Psychology and Behavioral Neurobiology, University of Tübingen, Tübingen, Germany

Nikolai Bogduk University of Newcastle, Newcastle Bone & Joint Institute, Royal Newcastle Centre, Newcastle, NSW, Australia

Jin Mo Chung Department of Neuroscience and Cell Biology, University Chair in Neuroscience, University of Texas Medical Branch, Galveston, TX, USA

Joyce DeLeo Dartmouth Hitchcock Medical Center, Dartmouth Medical School, Hanover, NH, USA

Marshall Devor Department of Cell & Developmental Biology, Institute of Life Sciences and Center for Research on Pain, The Hebrew University of Jerusalem, Jerusalem, Givat Ram, Israel

Hans-Christoph Diener Department of Neurology, University Hospital Essen, University Duisburg-Essen, Essen, Germany

Jonathan O. Dostrovsky Department of Physiology, University of Toronto, Toronto, ON, Canada

Ronald Dubner Department of Neural and Pain Sciences, University of Maryland Dental School, Baltimore, MD, USA

Michel Y. Dubois NYU Medical School, NYU Langone Medical Centers, Center for Research & Treatment of Pain, NYU Ambulatory Care Center, New York, NY, USA

Nanna Brix Finnerup Aarhus University Hospital, Danish Pain Research Center, Aarhus, Denmark

Herta Flor Department of Cognitive and Clinical Neuroscience, Central Institute of Mental Health, Medical Faculty Mannheim, Heidelberg University, Mannheim, Germany

Gerald F. Gebhart Department of Anesthesiology, Center for Pain Research, University of Pittsburgh, Pittsburgh, PA, USA

Maria Adele Giamberardino Pathophysiology of Pain Laboratory and Centre for Fibromyalgia and Musculoskeletal Pain, Department of Medicine and Science of Aging, University of Chieti, Chieti, Italy

Glenn J. Giesler Department of Neuroscience, University of Minnesota, Minneapolis, MN, USA

Peter Goadsby Headache Center, Department of Neurology, University of California, San Francisco, San Francisco, CA, USA

Michael S. Gold Department of Anesthesiology, Center for Pain Research, University of Pittsburgh, Pittsburgh, PA, USA

Hermann O. Handwerker Department of Physiology and Pathophysiology, University of Erlangen/Nürnberg, Erlangen, Germany

Burkhard Hinz Institute of Toxicology and Pharmacology, University of Rostock, Rostock, Germany

Wilfrid Jaenig Physiologisches Institut, Christian-Albrechts-Universität zu
Kiel, Kiel, Germany

Michaela Kress Department für Physiologie und Medizinische Physik,
Medizinische Universität Innsbruck, Innsbruck, Austria

Frederick A. Lenz Department of Neurosurgery, Johns Hopkins Hospital, Baltimore, MD, USA

Tonya M. Palermo Seattle Children's Hospital Research Institute, CW8-6 - Child Health, Behavior and Development, University of Washington, Seattle, WA, USA

Frank Porreca Department of Pharmacology, University of Arizona Health Sciences Center, Tucson, AZ, USA

Russell K. Portenoy Department of Pain Medicine and Palliative Care, Beth Israel Medical Center, New York, NY, USA

James P. Robinson Department of Anesthesiology & Pain Medicine, University of Washington, Seattle, WA, USA

Hans-Georg Schaible Department of Physiology - Neurophysiology, Friedrich-Schiller-University of Jena, Jena, Germany

Robert F. Schmidt Physiologisches Institut der Universität Würzburg, Würzburg, Germany

Stephan A. Schug School of Medicine and Pharmacology, Royal Perth Hospital, Faculty of Medicine, Dentistry & Health Sciences, The University of Western Australia, UWA Anaesthesiology, Perth, Australia

Barry J. Sessle Faculty of Dentistry, University of Toronto, Toronto, ON, Canada

Bengt H. Sjölund Department of Community Medicine and Rehabilitation Medicine, Umea University, Umea, Sweden

Mark D. Sullivan Psychiatry and Behavioral Sciences, University of Washington, Seattle, WA, USA

Irene Tracey Pain Imaging Neuroscience (PaIN) Group, Department of Physiology, Anatomy and Genetics, Oxford University, Oxford, UK

Dennis C. Turk Department of Anesthesiology & Pain Research, University of Washington, Seattle, WA, USA

Félix Viana Instituto de Neurociencias, Universidad Miguel Hernandez-CSIC, San Juan de Alicante, Alicante, Spain

Katy Vincent Nuffield Department of Obstetrics and Gynaecology, The Women's Centre, John Radcliffe Hospital, University of Oxford, Oxford, UK

Gary A. Walco Department of Anesthesiology and Pain Medicine, University of Washington School of Medicine, Seattle Children's Hospital, Seattle, WA, USA

George L. Wilcox Department of Neuroscience, Pharmacology and Dermatology, University of Minnesota Medical School, Minneapolis, MN, USA

Katja Wiech Nuffield Department of Clinical Neurosciences, John Radcliffe Hospital, Oxford, UK

George L. Wilcox Department of Neuroscience, Pharmacology and Dermatology, University of Minnesota Medical School, Minneapolis, MN, USA

Katja Wiech Nuffield Department of Clinical Neurosciences, John Radcliffe Hospital, Oxford, UK

Contributors

Tipu Aamir The Auckland Regional Pain Service, Auckland Hospital, Auckland, New Zealand

Catherine Abbadie Department of Pharmacology, Merck Research Laboratories, Rahway, NJ, USA

Frances V. Abbott Departments of Psychiatry and Psychology, McGill University, Montreal, QC, Canada

Heather Adams Department of Surgery, McGill University Health Centre, Montreal, QC, Canada

Mary O. Adedoyin Department of Anatomy, University of California San Francisco, San Francisco, CA, USA

Rolf H. Adler University of Berne Medical School, Kehrsatz, Switzerland

Claire D. Advokat Department of Psychology, Louisiana State University, Baton Rouge, LA, USA

Reto M. Agosti Headache Center Hirlsanden, Zurich, Switzerland

Ebtesam Ahmed College of Pharmacy and Health Sciences, St. John's University, Department of Pain Medicine and Palliative Care, Beth Israel Medical Center, New York, NY, USA

Marie-Claire Albanese Department of Psychology, McGill University, Montreal, QC, Canada

Kathryn M. Albers Department of Neurobiology, University of Pittsburgh School of Medicine, Pittsburgh, PA, USA

Phillip J. Albrecht Center for Neuropharmacology and Neuroscience, Albany Medical College, Albany, NY, USA

Elie D. Al-Chaer Center for Pain Research, University of Arkansas for Medical Sciences, Little Rock, AR, USA

Terry Altilio Department of Pain Medicine and Palliative Care, Beth Israel Medical Center, New York, USA

Rainer Amann Medical University Graz, Graz, Austria

Ron Amir Institute of Life Sciences, The Hebrew University of Jerusalem, Jerusalem, Israel

Praveen Anand Peripheral Neuropathy Unit, Imperial College London, London, UK

Karen O. Anderson The University of Texas MD Anderson Cancer Center, Houston, TX, USA

Ole K. Andersen Department of Health Science & Technology, Aalborg University, Aalborg, Denmark

Stanley W. Anderson Department of Neurosurgery, Johns Hopkins Hospital, Johns Hopkins Medical Institutions, Johns Hopkins University, Baltimore, MD, USA

Trevor Anderson Fairview Pain & Palliative Care, University of Minnesota Medical Center, Minneapolis, MN, USA

Karl-Erik Andersson Wake Forest Institute for Regenerative Medicine, Wake Forest University School of Medicine, Winston Salem, USA

Frank Andrasik Department of Psychology, Applied Psychophysiology and Biofeedback, The University of Memphis, Memphis, TN, USA

A. Vania Apkarian Department of Physiology, Northwestern University Feinberg School of Medicine, Chicago, IL, USA

Lars Arendt-Nielsen Center for Sensory-Motor Interaction, Department of Health Science and Technology, School of Medicine, Aalborg University, Aalborg, Denmark

Kirsten Arndt CNS Pharmacology, Boehringer Ingelheim Pharma GmbH & Co. KG, Biberach, Germany

Charles-Daniel Arreto Université Paris Descartes, INSERM UMR 894, France

Erik Askenasy Neurobiology and Anatomy, Health Science Center, University of Texas, Houston, TX, USA

J. H. Atkinson Psychiatry Service, VA San Diego Healthcare System and Department of Psychiatry, University of California, San Diego, CA, USA

Nadine Attal Centre d'Evaluation et de Traitement de la Douleur, Hôpital Ambroise Paré, INSERM U 987 APHP, Boulogne-Billancourt, France

Christoph Aufenberg Functional Neurosurgery, Neurosurgical Clinic University Hospital, Zürich, Switzerland

Beate Averbeck Sanofi-Aventis Deutschland GmbH, Frankfurt, Germany

Qasim Aziz Section of GI Sciences, Hope Hospital University of Manchester, Salford, UK

Cathrine Baastrup Danish Pain Research Center, Aarhus University Aarhus University Hospital, Aarhus, Denmark

F. W. Bach Danish Pain Research Center, Department of Neurology, Aarhus University Hospital, Aarhus, Denmark

S. K. Back Neuroscience Research Institute and Department of Physiology, Korea University, College of Medicine, Seoul, Republic of Korea

Misha-Miroslav Backonja Department of Neurology, University of Wisconsin-Madison, Madison, WI, USA

Banani Banerjee Division of Gastroenterology and Hepatology and Pediatric Gastroenterology, Medical College of Wisconsin, Milwaukee, WI, USA

John D. Banja Department of Rehabilitation Medicine, Center for Ethics Emory University, Atlanta, GA, USA

Carsten Bantel Department of Surgery & Cancer? Anaesthetics Section, Imperial College of Science and Medicine Chelsea and Westminster Campus, London, UK

Andrew Paul Baranowski The Pain Management Centre, National Hospital for Neurology and Neurosurgery, UCL Hospitals NHS Foundation Trust, London, UK

Alex Barling Department of Neurology, Birmingham Muscle and Nerve Centre Queen Elizabeth Hospital, Birmingham, UK

Ralf Baron Department of Neurological Pain Research and Therapy, Christian Albrechts University Kiel, Kiel, Germany

Gordon A. Barr Department of Anesthesiology and Critical Care Medicine, Children's Hospital of Philadelphia University of Pennsylvania, Philadelphia, PA, USA

Emily J. Bartley Department of Psychology, The University of Tulsa, Tulsa, OK, USA

Jeffrey R. Basford Department of Physical Medicine and Rehabilitation, Mayo Clinic and Mayo Foundation, Rochester, NY, USA

Michele C. Battie Department of Physical Therapy, University of Alberta, Edmonton, AB, Canada

Ulf Baumgartner Chair of Neurophysiology, Center for Biomedicine and Medical Technology Mannheim, Heidelberg University, Mannheim, Germany

Nicola Beale Nuffield Department of Anaesthetics, Oxford University Hospitals NHS Trust, UK

Alain Beaudet Montreal Neurological Institute, McGill University, Montreal, QC, Canada

Lino Becerra P.A.I.N. Group, Department of Anesthesiology, Perioperative and Pain Medicine, Boston Children's Hospital Department of Radiology Massachusetts General Hospital Center for Pain and the Brain, Harvard Medicals, Boston, MA, USA

Yaakov Beilin Department of Anesthesiology, and Obstetrics, Mount Sinai School of Medicine, New York, NY, USA

Alvin J. Beitz Department of Veterinary and Biomedical Sciences, University of Minnesota, St. Paul, MN, USA

Miles Belgrade Fairview Pain Management Center, University of Minnesota Medical Center, Minneapolis, MN, USA

Carlo Valerio Bellieni Neonatal Intensive Care Unit, University Hospital, Siena, Italy

Carlos Belmonte Instituto de Neurociencias, Universidad Miguel Hernandez-CSIC, San Juan de Alicante, Spain

Allan J. Belzberg Department of Neurosurgery, Johns Hopkins School of Medicine, Baltimore, MD, USA

Matt Bender Department of Neurosurgery, Johns Hopkins Hospital, Baltimore, MA, USA

Fabrizio Benedetti Department of Neuroscience, Clinical and Applied Physiology Program, University of Turin Medical School, Turin, Italy

David L. H. Bennett Wolfson CARD, Kings College London, London, UK

Nuffield Department of Clinical Neurosciences, John Radclife Hospital, The University of Oxford, Oxford, UK

Robert Bennett Department of Medicine, Oregon Health and Science University, Portland, USA

Edward C. Benzel Department of Neurosurgery, Center for Spine Health Neurological Institute Cleveland Clinic, Cleveland, OH, USA

David A. Bereiter Department of Diagnostic and Biological Sciences, University of Minnesota School of Dentistry, Minneapolis, MN, USA

Elmar Berendes Klinik für Anästhesiologie, Operative Intensivmedizin und Schmerztherapie, Klinikum Krefeld, Krefeld, Germany

Karen J. Berkley Program in Neuroscience, Florida State University, Tallahassee, FL, USA

Peter Berlit Department of Neurology, Alfried Krupp Hospital, Essen, Germany

Jean-François Bernard INSERM UMR-894, UPMC Site Pitié-Salpêtrière, Paris, France

Anne K. Bertelsen Neurology, Aleris Hospital, Bergen, Norway

Jennifer Bickel Integrative Pain Management, Children's Mercy Hospitals and Clinics, University of Missouri-Kansas City School of Medicine, Kansas City, MO, USA

Marcelo E. Bigal Department of Neurology and Montefiore Headache Unit, Albert Einstein College of Medicine, Bronx, NY, USA

Yitzchak M. Binik McGill University, Montreal, QC, Canada

Niels Birbaumer Institute of Medical Psychology and Behavioral Neurobiology, University of Tübingen, Tübingen, Germany

Frank Birklein Department of Neurology, University of Mainz, Mainz, Germany

Kathryn A. Birnie Department of Psychology, Dalhousie University, Halifax, NS, Canada

Thomas Bishop Centre for Neuroscience, King's College London, London, UK

Henning Bliddal The Parker Institute, Frederiksberg Hospital, Copenhagen, Denmark

Nikolai Bogduk University of Newcastle, Newcastle Bone & Joint Institute, Royal Newcastle Centre, Newcastle, NSW, Australia

Jorgen Boivie Department of Neurology, University Hospital, Linköping, Sweden

Michael R. Bond University of Glasgow, Glasgow, UK

Richard L. Boortz-Marx Division of Pain Medicine, Department of Anesthesiology, University of North Carolina, Chapel Hill, North Carolina, USA

James M. Borowczyk Department of Orthopaedics and Musculoskeletal Medicine, University of Otago, Christchurch, Christchurch, New Zealand

David Borsook P.A.I.N. Group, Department of Anesthesiology, Perioperative and Pain Medicine, Boston Children's Hospital Department of Radiology Massachusetts General Hospital Center for Pain and the Brain, Harvard Medicals, Boston, MA, USA

George S. Borszcz Department of Psychology, Wayne State University, Detroit, USA

Jasenka Borzan Department of Anesthesiology and Critical Care Medicine, Johns Hopkins University, Baltimore, USA

Regina M. Botting London, UK

Didier Bouhassira Centre d'Evaluation et de Traitement de la Douleur, Hôpital Ambroise Paré, INSERM U 987 APHP, Boulogne-Billancourt, France

Laurence Bourgeais INSERM UMR 894, Centre de Psychiatrie et Neurosciences, Paris, France

David Bowsher Pain Research Institute, University Hospital Aintree, Liverpool, UK

Emma L. Brandon Anaesthesiology and Pain Medicine, University of Western Australia Royal Perth Hospital, Perth, WA, Australia

Jennifer Brawn P.A.I.N. Group, Center for Pain and the Brain, Boston Children's Hospital, Harvard Medical School, Waltham, MA, USA

Harald Breivik Department of Pain Management and Research, Oslo University Hospital, Rikshospitalet and University of Oslo, Oslo, Norway

Kori L. Brewer Department of Emergency Medicine, Brody School of Medicine East Carolina University, Greenville, NC, USA

Christopher R. Brigham Brigham and Associates, Inc., Portland, USA

Abigail Broderick Nuffield Department of Obstetrics and Gynaecology, University of Oxford, Oxford, UK

Jeffrey A. Brown Department of Neurological Surgery, Wayne State University School of Medicine, Detroit, MI, USA

Stephen C. Brown Department of Anesthesia, The Hospital for Sick Children, Toronto, ON, Canada

Eduardo Bruera Department of Palliative Care and Rehabilitation Medicine, The University of Texas M. D. Anderson Cancer Center, Houston, TX, USA

Pablo Brumovsky Faculty of Biomedical Sciences, Austral University, Buenos Aires, Argentina

Consejo Nacional de Investigaciones Científicas y Técnicas (CONICET), Buenos Aires, Argentina

Kay Brune Department of Experimental and Clinical Pharmacology and Toxicology, Friedrich Alexander University Erlangen-Nürnberg, Erlangen, Germany

Maria Burian Department of General, Visceral and Transplantation Surgery, University of Heidelberg, Heidelberg, Germany

Dan Buskila Rheumatic Disease Unit, Soroka Medical Center, Beer Sheva, Israel

Catherine M. Cahill Anesthesiology & Perioperative Care, University of California, Irvine, Irvine, CA, USA

Brian E. Cairns Faculty of Pharmaceutical Sciences, University of British Columbia, Vancouver, BC, Canada

Nigel A. Calcutt Department of Pathology, University of California San Diego, La Jolla, CA, USA

Margarita Calvo Wolfson CARD, Kings College London, London, UK

James N. Campbell Johns Hopkins University School of Medicine, Baltimore, MD, USA

Lauren Campbell Department of Psychology, York University, Toronto, ON, Canada

Ling Cao Department of Biomedical Sciences, College of Osteopathic Medicine, University of New England, ME, USA

Adam Carinci Department of Anesthesia, Critical Care and Pain Medicine, Massachusetts General Hospital Harvard Medical School, Boston, MA, USA

Giancarlo Carli Department of Physiology, University of Siena, Siena, Italy

Christer P. O. Carlsson Rehabilitation Clinic, Lunds University Hospital, Lund, Sweden

Susan M. Carlton Neuroscience and Cell Biology, University of Texas Medical Branch, Galveston, TX, USA

Richard Carr Clinic for Anaesthesiology, Heidelberg University, Mannheim, Baden-Wuerttemberg, Germany

Benjamin S. Carson Department of Neurosurgery, Johns Hopkins Hospital, Baltimore, MA, USA

Brian Carter University of Missouri at Kansas City Children's Mercy Hospital and Bioethics Center, Kansas City, MO, USA

Kenneth L. Casey Department of Neurology, University of Michigan Medical Center, Ann Arbor, MI, USA

Christine T. Chambers Departments of Pediatrics and Psychology, Dalhousie University, Halifax, NS, Canada

Sandra Chaplan Janssen Research and Development, LLC, San Diego, CA, USA

C. Richard Chapman Department of Anesthesiology, University of Utah School of Medicine, Salt Lake City, UT, USA

Santosh K. Chaturvedi Psychiatry, National Institute of Mental Health & Neurosciences, Bangalore, India

Andrew C. N. Chen Human Brain Mapping and Cortical Imaging Laboratory, Aalborg University, Aalborg, Denmark

Hsinlin Thomas Cheng Department of Neurology, University of Michigan, Ann Arbor, MI, USA

Pavel Cherkas Faculty of Dentistry, University of Toronto, Toronto, ON, Canada

Andrea Cheville Department of Physical Medicine and Rehabilitation, Mayo Clinic, Rochester, MN, USA

Chen Yu Chiang Faculty of Dentistry, University of Toronto, Toronto, ON, Canada

N. L. Chillingworth School of Physiology and Pharmacology, University of Bristol, Bristol, UK

Edward Chow Department of Radiation Oncology, Sunnybrook Health Sciences Centre Odette Cancer Centre (T2-182), Toronto, ON, Canada

Julie A. Christianson Department of Anatomy and Cell Biology, University of Kansas Medical Center, Kansas City, KS, USA

MacDonald J. Christie Discipline of Pharmacology, University of Sydney, Sydney, Australia

Jin Mo Chung Department of Neuroscience and Cell Biology, University Chair in Neuroscience, University of Texas Medical Branch, Galveston, TX, USA

Kyungsoon Chung Department of Neuroscience and Cell Biology, University of Texas Medical Branch, Galveston, TX, USA

Anna K. Clark Wolfson Centre for Age Related Diseases, King's College London, London, UK

W. Crawford Clark College of Physicians and Surgeons, Columbia University, New York, USA

James F. Cleary Pain & Policy Studies Group, UW Carbone Cancer Center, Madison, WI, USA

John De Clercq Department of Rehabilitation Sciences and Physiotherapy, Ghent University Hospital, Ghent, Belgium

Jacqui Clinch Royal National Hospital for Rheumatic Diseases NHS Foundation Trust, Bath, UK

Terence J. Coderre Departments of Anesthesia, Neurology & Neurosurgery and Psychology, McGill University, Montreal, QC, Canada

Robert Coghill Department of Neurobiology and Anatomy, Wake Forest University School of Medicine, Winston-Salem, NC, USA

Lindsey Cohen Psychology, Georgia State University, Atlanta, GA, USA

Alan L. Colledge Utah Labor Commission, International Association of Industrial Accident Boards and Commissions Central Utah Medical Clinic, South Salt Lake City, UT, USA

Beverly J. Collett University Hospitals of Leicester, Leicester, UK

Lesley A. Colvin Department of Anaesthesia, Critical Care and Pain Medicine Western General Hospital, Edinburgh, UK

Kevin C. Conlon Trinity College Dublin, University of Dublin, Dublin, Ireland

Mark Connelly Integrative Pain Management, Children's Mercy Hospitals and Clinics, University of Missouri-Kansas City School of Medicine, Kansas City, MO, USA

Brian Y. Cooper Department of Oral and Maxillofacial Surgery, University of Florida, Gainesville, FL, USA

Christine Cordle University Hospitals of Leicester, Leicester, UK

Laura A. Cousins Psychology, Georgia State University, Atlanta, GA, USA

Michael J. Cousins Department of Anesthesia and Pain Management, Royal North Shore Hospital University of Sydney, Sydney, NSW, Australia

Verne C. Cox Department of Psychology, University of Texas at Arlington, Arlington, TX, USA

George A. Crabill Department of Neurosurgery, Johns Hopkins Hospital, Baltimore, MA, USA

Rebecca M. Craft Psychology, Washington State University, Pullman, WA, USA

Kenneth D. Craig Department of Psychology, University of British Columbia, Vancouver, BC, Canada

Richard Crevenna Medizinische Universität Wien, Wien, Austria

Geert Crombez Department of Experimental Clinical and Health Psychology, Ghent University, Ghent, Belgium

Joan Crook School of Nursing, Faculty of Health Sciences, McMaster University, Hamilton, ON, Canada

Giorgio Cruccu Department of Neurological Sciences, La Sapienza University, Rome, Italy

Michele Curatolo Department of Anesthesiology, Division of Pain Therapy, University Hospital of Bern Inselspital, Bern, Switzerland

Nachum Dafny Neurobiology and Anatomy, Health Science Center, University of Texas, Houston, TX, USA

Jorgen B. Dahl Department of Anaesthesiology, Glostrup University Hospital, Herlev, Denmark

Yi Dai Department of Anatomy and Neuroscience, Hyogo College of Medicine, Hyogo, Japan

Stefaan Van Damme Department of Experimental Clinical and Health Psychology, Ghent University, Ghent, Belgium

Alexandre F. M. DaSilva Biologic & Materials Sciences, University of Michigan Dental School, MI, USA

Steve Davidson Pain Center & Department of Anesthesiology, Washington University School of Medicine, MO, USA

Brian Davis Department of Neurobiology, University of Pittsburgh School of Medicine, Pittsburgh, PA, USA

Karen D. Davis Brain, Imaging and Behaviour – Systems Neuroscience, Toronto Western Research Institute, Toronto, ON, Canada

Department of Surgery, University of Toronto, Toronto, ON, Canada

Richard O. Day Department of Clinical Pharmacology and Toxicology, St Vincent's Hospital, Sydney, NSW, Australia

Isabelle Decosterd Pain Center, University Hospital Center and University of Lausanne, Lausanne, Switzerland

Ruth Defrin Physical Therapy, Faculty of Medicine, Tel-Aviv University, Tel-Aviv, Israel

Antoon F. C. De Laat Department of Oral/Maxillofacial Surgery, Catholic University of Leuven, Leuven, Belgium

Joyce DeLeo Dartmouth Hitchcock Medical Center, Dartmouth Medical School, Hanover, NH, USA

A. Lee Dellon Department of Plastic Surgery and Neurosurgery, Johns Hopkins University, Baltimore, USA

Dellon Institutes for Peripheral Nerve Surgery, Baltimore, MD, USA

Marshall Devor Department of Cell & Developmental Biology, Institute of Life Sciences and Center for Research on Pain, The Hebrew University of Jerusalem, Jerusalem, Givat Ram, Israel

Perry Dhaliwal Department of Neurosurgery, Cleveland Clinic Spine Institute, Cleveland, OH, USA

David Diamant Neurological and Spinal Surgery, Lincoln, NE, USA

Anthony H. Dickenson Department of Pharmacology, University College London, London, UK

Hans-Christoph Diener Department of Neurology, University Hospital Essen, University Duisburg-Essen, Essen, Germany

Natalia Dmitrieva Program in Neuroscience, Florida State University, Tallahassee, FL, USA

L. F. Donaldson School of Life Sciences, University of Nottingham, Nottingham, UK

Henri Doods CNS Pharmacology, Boehringer Ingelheim Pharma GmbH & Co. KG, Biberach, Germany

Victoria Dorf Social Security Administration, Baltimore, MD, USA

Michael J. Dorsi Department of Neurosurgery, Johns Hopkins School of Medicine, Baltimore, MD, USA

Jonathan O. Dostrovsky Department of Physiology, University of Toronto, Toronto, ON, Canada
Faculty of Dentistry, University of Toronto, Toronto, ON, Canada

Robyn Drach Fairview Pain Management Center, Minneapolis, MN, USA

Ron Dressler Utah Labor Commission, International Association of Industrial Accident Boards and Commissions Central Utah Medical Clinic, South Salt Lake City, UT, USA

Paul Dreyfuss Department of Rehabilitation Medicine, University of Washington, Seattle, USA

Peter D. Drummond School of Psychology, Murdoch University, Perth, Australia

Ronald Dubner Department of Neural and Pain Sciences, University of Maryland Dental School, Baltimore, MD, USA

Anne Ducros Headache Emergency Department, INSERM, Paris, France

Gary H. Duncan Département de stomatologie, Faculté de médecine dentaire, Université de Montréal, Montreal, QC, Canada

Department of Neurology and Neurosurgery, McGill University, Montreal, QC, Canada

Mary J. Eaton Research Service (151), Miami VA Health Care System, Miami VAMC, Miami, FL, USA

Andrea Ebersberger Department of Physiology, Friedrich Schiller University of Jena, Jena, Germany

Christopher Eccleston Pain Management Unit, University of Bath, Bath, UK

John Edmeads Sunnybrook and Woman's College Health Sciences Centre, University of Toronto, Toronto, Canada

Edzard Ernst Complementary Medicine, Peninsula Medical School Universities of Exeter and Plymouth, Exeter, UK

Reuben Escorpizo Swiss Paraplegic Research, Nottwil, Switzerland

Department of Health Sciences and Health Policy, University of Lucerne, Lucerne, Switzerland

ICF Research Branch in cooperation with the World Health Organization (WHO) Collaborating Centre for the Family of International Classifications in Germany, Nottwil, Switzerland

Schonstein Eva RehabCo, Fyshwick, ACT, Australia

Joshua C. Eyer Department of Psychology, University of Alabama, Tuscaloosa, AL, USA

Marie Faaborg-Andersen McGill University, Montreal, QC, Canada

Keri L. Fakata College of Pharmacy and Pain Management Center, University of Utah, Salt Lake City, USA

Marie Fallon Department of Oncology, Edinburgh Cancer Center University of Edinburgh, Edinburgh, UK

Melissa Farmer Department of Physiology / Feinberg School of Medicine, Northwestern University, Chicago, IL, USA

Samantha R. Fashler Department of Psychology, University of British Columbia, Vancouver, BC, Canada

Jean-Baptiste Fassier Department of Occupational Health and Medicine, UMRESTTE-Université Claude Bernard Lyon 1, Lyon, France

Charlotte Feinmann Eastman Dental Institute London, University College London, London, UK

Elizabeth Roy Felix The Miami Project to Cure Paralysis, University of Miami Miller School of Medicine, Miami, FL, USA

Bin Feng Center for Pain Research, Department of Anesthesiology, School of Medicine, University of Pittsburgh, Pittsburgh, Pennsylvania, USA

Yi Feng Department of Anesthesiology, Peking University People's Hospital, Peking, China

Andres Fernandez Department of Neurosurgery, Johns Hopkins Hospital, Baltimore, MD, USA

Antonio Vicente Ferrer-Montiel Institute of Molecular and Cell Biology, University Miguel Hernández, Elche, Alicante, Spain

Perry G. Fine Pain Research Center, School of Medicine, University of Utah, Salt Lake City, UT, USA

David J. Fink Department of Neurology, University of Michigan and VA Ann Arbor Healthcare System, Ann Arbor, MI, USA

Nanna Brix Finnerup Aarhus University Hospital, Danish Pain Research Center, Aarhus, Denmark

David A. Fishbain Department of Psychiatry, University of Miami, Miami, FL, USA

Maria Fitzgerald Neuroscience, Physiology & Pharmacology, University College London, London, UK

Per Flisberg Department of Anaesthesia and Intensive Care, Lund University Hospital, Lund, Sweden

Herta Flor Department of Cognitive and Clinical Neuroscience, Central Institute of Mental Health, Medical Faculty Mannheim, Heidelberg University, Mannheim, Germany

Christopher M. Flores Drug Discovery, Johnson and Johnson Pharmaceutical Research and Development, Spring House, PA, USA

Wilbert E. Fordyce Department of Rehabilitation, University of Washington School of Medicine, Seattle, WA, USA

Robert D. Foreman Department of Physiology, College of Medicine, University of Oklahoma Health Sciences Center, Oklahoma City, OK, USA

Gary M. Franklin Department of Environmental and Occupational Health Sciences, University of Washington, Seattle, WA, USA

Department of Health Services, University of Washington, Seattle, WA, USA

Jan Fridén Department of Hand Surgery, Sahlgrenska University Hospital, Gothenborg, Sweden

Justin Friedlander The Arthur Smith Institute for Urology, North Shore-LIJ Healthcare System, Hofstra North Shore LIJ School of Medicine, New Hyde Park, NY, USA

Allan H. Friedman Division of Neurosurgery, Duke University Medical Center, Durham, NC, USA

Perry N. Fuchs Department of Psychology, University of Texas at Arlington, Arlington, TX, USA

Deborah Fulton-Kehoe Department of Environmental and Occupational Health Sciences, University of Washington, Seattle, WA, USA

Andrea D. Furlan Institute for Work & Health, Toronto, ON, Canada

Vasco Galhardo Departamento de Biologia Experimental, Faculdade de Medicina, Universidade do Porto, Portugal

Andreas R. Gantenbein RehaClinic Bad Zurzach, Bad Zurzach, Switzerland

I. Garonzik Department of Neurosurgery, Johns Hopkins Medical Institutions, Baltimore, MD, USA

Robert Gassin Musculoskeletal Medicine Clinic, Frankston, VIC, Australia

Robert J. Gatchel Department of Psychology, The University of Texas, Arlington, TX, USA

Gerald F. Gebhart Department of Anesthesiology, Center for Pain Research, University of Pittsburgh, Pittsburgh, PA, USA

Gerd Geisslinger Department of Clinical Pharmacology, University Frankfurt am Main, Frankfurt, Germany

Robert D. Gerwin Department of Neurology, Johns Hopkins University, Bethesda, MD, USA

Maria Adele Giamberardino Pathophysiology of Pain Laboratory and Centre for Fibromyalgia and Musculoskeletal Pain, Department of Medicine and Science of Aging, University of Chieti, Chieti, Italy

Stephen Gibson National Ageing Research Institute Department of Medicine, University of Melbourne, Parkville, VIC, Australia

Glenn J. Giesler Department of Neuroscience, University of Minnesota, Minneapolis, MN, USA

Philip L. Gildenberg Baylor Medical College, Houston, TX, USA

Myra Glajchen Department of Pain Medicine and Palliative Care, Beth Israel Medical Center Albert Einstein College of Medicine, New York, NY, USA

Hartmut Göbel Kiel Migraine and Headache Center, Kiel Pain Center, Kiel, Germany

Peter Goadsby Headache Center, Department of Neurology, University of California, San Francisco, San Francisco, CA, USA

Philippe Goffaux Faculty of Medicine, University of Sherbrooke, Sherbrooke, QC, Canada

Ana Gomis Instituto de Neurociencias de Alicante, Universidad Miguel Hernández-CSIC, San Juan de Alicante, Spain

Gilbert R. Gonzales Department of Neurology, Memorial Sloan-Kettering Cancer Center, New York, USA

Allan Gordon Wasser Pain Management Centre, Mount Sinai Hospital University of Toronto, Toronto, ON, Canada

Liesbet Goubert Department of Experimental Clinical and Health Psychology, Ghent University, Ghent, Belgium

Jayantilal Govind Pain Clinic, Department of Anaesthesia, University of New South Wales, Sydney, NSW, Australia

Richard H. Gracely Departments of Medicine-Rheumatology and Neurology, University of Michigan Health System, Ann Arbor, MI, USA

Garry G. Graham Department of Pharmacology, School of Medical Sciences, University of New South Wales, Sydney, NSW, Australia

David K. Grandy Department of Physiology and Pharmacology, Oregon Health and Science University, Portland, OR, USA

Thomas Graven-Nielsen Laboratory for Musculoskeletal Pain and Motor Control, Center for Sensory-Motor Interaction, Aalborg University, Aalborg, Denmark

Barry Green Pierce Laboratory, Yale School of Medicine, New Haven, CT, USA

Paul G. Green Oral & Maxillofacial Surgery, University of California San Francisco, San Francisco, CA, USA

Joel D. Greenspan Department of Neural and Pain Sciences, University of Maryland School of Dentistry, Baltimore, MD, USA

John W. Griffin Department of Neurology, Johns Hopkins University School of Medicine, Baltimore, MD, USA

Cornelius Groenwald Department of Anesthesiology and Pain Medicine, Seattle Children's Hospital, University of Washington School of Medicine, Seattle, WA, USA

Douglas P. Gross Department of Physical Therapy, University of Alberta WCB-Alberta Millard Health, Edmonton, AB, Canada

Sabine Grösch Institute of Clinical Pharmacology, Clinical Center of the Goethe University, Frankfurt, Germany

Blair D. Grubb Department of Cell Physiology and Pharmacology, University of Leicester, Leicester, UK

Ruth Eckstein Grunau Pediatrics Dept, University of British Columbia Developmental Neurosciences & Child Health, Child & Family Research Institute, Vancouver, BC, Canada

Christoph Gutenbrunner Department of Physical Medicine and Rehabilitation, Hanover Medical School, Hanover, Germany

Maija Haanpää Department of Neurosurgery, Helsinki University Central Hospital, Helsinki, Finland

Sherri Haas Fairview Pain Management Center, Minneapolis, MN, USA

Ute Habel Department of Psychiatry and Psychotherapy, RWTH Aachen University, Aachen, Germany

Thomas Hachenberg Clinic for Anaesthesiology and Intensive Therapy, Otto-von-Guericke University, Magdeburg, Germany

Nouchine Hadjikhani Martinos Center for Biomedical Imaging, Massachusetts General Hospital Harvard Medical School, Charlestown, MA, USA

Ole Haegg Department of Orthopedic Surgery, Sahlgren University Hospital, Goteborg, Sweden

Bryan C. Hains Department of Neurology, Yale University School of Medicine, West Haven, CT, USA

Else K. B. Hals University of Oslo, Faculty of Dentistry, Oslo, Norway

Darryl T. Hamamoto Diagnostic and Biological Sciences, University of Minnesota, Minneapolis, MN, USA

Donna L. Hammond Department of Anesthesia, University of Iowa, Iowa City, IA, USA

Menachem Hanani Hadassah-Hebrew University Medical Center Hebrew University Medical School, Jerusalem, Israel

Hermann O. Handwerker Department of Physiology and Pathophysiology, University of Erlangen/Nürnberg, Erlangen, Germany

George M. Hanna Department of Anesthesia, Critical Care and Pain Medicine, Massachusetts General Hospital Harvard Medical School, Boston, MA, USA

Jing-Xia Hao Department of Physiology and Pharmacology, Karolinska Institutet, Stockholm, Sweden

Eleni G. Hapidou Department of Psychiatry and Behavioural Neurosciences, McMaster University and Hamilton Health Sciences Chronic Pain Management Unit, Chedoke Hospital, Hamilton, ON, Canada

Geoffrey Harding Sandgate Spinal Medicine Clinic, Sandgate, QLD, Australia

Louise Harding Hospitals NHS Foundation Trust, University College London, London, UK

Kenneth M. Hargreaves Department of Endodontics, University of Texas Health Science Center at San Antonio, San Antonio, USA

D. Paul Harries Pain Treatment Center, Samaritan Hospital Lexington, Lexington, KY, USA

Denise Harrison Nursing, Children's Hospital of Eastern Ontario (CHEO), University of Ottawa, Ottawa, ON, Canada

Steven Harte Anesthesiology and Internal Medicine/Rheumatology, University of Michigan VA Ann Arbor Healthcare System, Ann Arbor, MI, USA

Barbara A. Hastie Community Dentistry & Behavioral Science, University of Florida College of Dentistry, Gainesville, FL, USA

Jennifer A. Haythornthwaite Psychiatry & Behavioral Sciences, Johns Hopkins University, Baltimore, MD, USA

Heinz-Joachim Häbler FH Bonn-Rhein-Sieg, Rheinbach, Germany

John H. Healey Orthopaedic Service, Department of Surgery, Memorial Sloan-Kettering Cancer Center Weill Medical College of Cornell University, New York, NY, USA

Alicia A. Heapy VA Connecticut Healthcare System, Westhaven, CT, USA

Anne M. Heffernan The Adelaide and Meath Hospital Incorporating The National Children's Hospital, Dublin, Ireland

Geert Van der Heijden Julius Center, Department of Epidemiology, University Medical Center Utrecht, Utrecht, The Netherlands

Mary M. Heinricher Departments of Neurological Surgery and Behavioral Neuroscience, Oregon Health & Science University, Portland, USA

Robert D. Helme Department of Medicine, Royal Melbourne Hospital, University of Melbourne, Melbourne, Australia

Kim Helsen Research Group on Health Psychology, KU Leuven Katholieke Universiteit Leuven, Leuven, Belgium

Department of Experimental–Clinical and Health Psychology, Ghent University, Ghent, Belgium

Amin S. Herati The Arthur Smith Institute for Urology, North Shore-LIJ Healthcare System, Hofstra North Shore LIJ School of Medicine, New Hyde Park, NY, USA

Robert D. Herbert Neuroscience Research Australia, Sydney, Australia

Christiane Hermann Department of Clinical Psychology, Justus-Liebig University Giessen, Giessen, Germany

Juan F. Herrero Department of Physiology, Faculty of Medicine University of Alcala, Alcalá de Henares Madrid, Spain

Aki J. Hietaharju Department of Neurology and Rehabilitation, Pain Clinic Tampere University Hospital, Tampere, Finland

Karin Westlund High Department of Physiology, University of Kentucky, Lexington, KY, USA

Marita Hilliges Dalarna University, Halmstad, Sweden

Max J. Hilz University of Erlangen-Nuremberg, Erlangen, Germany

Burkhard Hinz Institute of Toxicology and Pharmacology, University of Rostock, Rostock, Germany

Anthony R. Hobson Section of GI Sciences, Hope Hospital University of Manchester, Salford, UK

Edward Högestätt Department of Laboratory Medicine, Lund University, Division of Clinical Chemistry and Pharmacology, Lund, Sweden

Tomas Hökfelt Department of Neuroscience, Karolinska Institute, Stockholm, Sweden

Sarah V. Holdridge Department of Pharmacology and Toxicology, Queen's University, Kinston, Canada

Kjell Hole University of Bergen, Bergen, Norway

Graham R. Holland Department of Cariology, University of Michigan, Ann Arbor, MI, USA

Chang-Zern Hong Department of Physical Medicine and Rehabilitation, University of California, Irvine, CA, USA

Minoru Hoshiyama Department of Integrative Physiology, National Institute for Physiological Sciences, Okazaki, Japan

Fred M. Howard University of Rochester, Rochester, NY, USA

Sung-Tsang Hsieh Department of Anatomy and Cell Biology, Department of Neurology, National Taiwan University Hospital, National Taiwan University, Taipei, Taiwan

James W. Hu Faculty of Dentistry, University of Toronto, Toronto, Canada

Claire E. Hulsebosch Spinal Cord Injury Research, University of Texas Medical Branch, Galveston, TX, USA

Thomas Hummel Department of Otorhinolaryngology, Technical University of Dresden Medical School, Dresden, Germany

Mark R. Hutchinson School of Medical Sciences, University of Adelaide, Adelaide, SA, Australia

Charles E. Inturrisi Department of Pharmacology, Weill Cornell Medical College, New York, USA

Koji Inui Department of Integrative Physiology, National Institute for Physiological Sciences, Okazaki, Japan

Koichi Iwata Department of Physiology, Nihon University School of Dentistry, Chiyoda-ku, Japan

Suhayl J. Jabbur Anatomy, Cell Biology and Physiology, Faculty of Medicine, American University of Beirut, Beirut, Lebanon

Neuroscience Program, Faculty of Medicine, American University of Beirut, Beirut, Lebanon

Nora Janjan University of Texas M. D. Anderson Cancer Center, Houston, TX, USA

Luc Jasmin Department of Anatomy, University of California San Francisco, San Francisco, CA, USA

Anne Jaumees Royal North Shore Hospital, St Leonards, NSW, Australia

Wilfrid Jaenig Physiologisches Institut, Christian-Albrechts-Universität zu Kiel, Kiel, Germany

Daniel Jeanmonod Functional Neurosurgery, The Center of Ultrasound Functional Neurosurgery, Solothurn, Switzerland

Andrew Jeffreys Director Department of Anaesthesia & Pain Medicine, Western Health, Melbourne, VIC, Australia

Mark P. Jensen Department of Rehabilitation Medicine, University of Washington, Seattle, WA, USA

Troels Staehelin Jensen Danish Pain Research Center, Department of Neurology, Aarhus University Hospital, Aarhus, Denmark

Kurt L. Johnson Department of Rehabilitation Medicine, University of Washington, Seattle, WA, USA

Mark I. Johnson Faculty of Health and Social Sciences, Leeds Pallium Research Group, Leeds Metropolitan University, Leeds, UK

Mark Johnston Hibiscus Coast Highway, Orewa, New Zealand

Edward Jones Center for Neuroscience, University of California, Davies, CA, USA

Jeroen R. De Jong Department of Rehabilitation, University Hospital Maastricht, Maastricht, The Netherlands

Sven-Eric Jordt Department of Pharmacology, Yale University School of Medicine, New Haven, CT, USA

Ellen Jørum The Laboratory of Clinical Neurophysiology, Rikshospitalet University, Oslo, Norway

Stefan Just CNS Pharmacology, Boehringer Ingelheim Pharma GmbH & Co. KG, Biberach, Germany

Timothy K. Y. Kaan Department of Anatomy, University of California San Francisco, San Francisco, CA, USA

Noufissa Kabli Department of Pharmacology and Toxicology, Queen's University, Kinston, Canada

Ryusuke Kakigi Department of Integrative Physiology, National Institute for Physiological Sciences, Okazaki, Japan

Peter C. A. Kam Anaesthesia, Central Clinical School, The University of Sydney, Sydney, Australia

Giresh Kanji The Sports and Pain Clinic, Wellington, New Zealand

Joel Katz Department of Psychology, McGill University, Montreal, QC, Canada

Valerie Kayser NSERM U677 91Bd de l'Hôpital, Paris, France

Francis J. Keefe Pain Prevention and Treatment Research, Department of Psychiatry and Behavioral Sciences, Duke University Medical Center, Durham, NC, USA

Lois J. Kehl Department of Anesthesiology, University of Minnesota, Minneapolis, MN, USA

Maureen Kelley Department of Pediatrics, Bioethics Division, University of Washington School of Medicine, WA, USA

Seattle Children's Hospital, Seattle, WA, USA

Stephen Kennedy Nuffield Department of Obstetrics and Gynaecology, University of Oxford, Oxford, UK

Edmund Keogh Department of Psychology, University of Bath, Bath, UK

Robert D. Kerns VA Connecticut Healthcare System, Westhaven, CT, USA

Sanjay Khanna Department of Physiology, National University of Singapore, Singapore

Kok Eng Khor Department of Pain Management, Prince of Wales Hospital University of New South Wales, Randwick, NSW, Australia

Mina Khoshnejad Department of Neuroscience, University of Montreal, Montreal, QC, Canada

H. J. Kim Department of Life Science, Yonsei University, Wonju, Republic of Korea

Sara King Faculty of Education (School Psychology), Mount Saint Vincent University, Halifax, NS, Canada

Wade King Department of Clinical Research, Royal Newcastle Centre University of Newcastle, Newcastle, NSW, Australia

K. L. Kirsh Behavioral Medicine, The Pain Treatment Center of the Blue grass, Lexington, KY, USA

Henriette Klit Danish Pain Research Center, Aarhus University Hospital, Aarhus C, Denmark

Ricardo J. Komotar Department of Neurological Surgery, Neurological Institute Columbia University, New York, NY, USA

Sungtae Koo Division of Meridian and Structural Medicine, School of Korean Medicine, Pusan National University, Yangsan, Korea

Rhonda K. Kotarinos Oakbrook Terrace, IL, USA

Katalin J. Kovacs Department of Veterinary and Biomedical Sciences, University of Minnesota, St. Paul, MN, USA

Malgorzata Krajnik Chair of Palliative Care, Nicolaus Copernicus University in Torun, Collegium Medicum in Bydgoszcz, Bydgoszcz, Poland

Elliot Krane Anesthesia and Pediatrics, Stanford University School of Medicine Lucile Packard Children's Hospital at Stanford, Stanford, CA, USA

Ajit Krishnaney Department of Neurosurgery, Cleveland Clinic, Cleveland, OH, USA

Greg Krohm International Association of Industrial Accident Boards and Commissions (IAIABC), Utah Valley University, USA

Lawrence Kruger UCLA Geffen School of Medicine, Los Angeles, CA, USA

M. M. ter Kuile Department of Gynecology, Leiden University Medical Center, Leiden, The Netherlands

Promil Kukreja Department of Anesthesiology, University of Alabama at Birmingham, Birmingham, AL, USA

Alice Kvåle Department of Public Health and Primary Health Care, Faculty of Medicine and Dentistry University of Bergen, Bergen, Norway

Gunnvald Kvarstein Department of Pain Management and Pain Research, Oslo University Hospital, Rikshospitalet, Oslo, Norway

Jean-Jacques Labat Neurology and Rehabilitation, Clinique Urologique, Nantes, France

Gaston Labrecque Faculty of Pharmacy, University of Laval, Quebec City, QC, Canada

Marco Lacerenza Pain Medicine Center, Casa di Cura S. Pio X, Opera San Camillo Foundation, Milan, Italy

Uri Ladabaum Gastroenterology & Hepatology, Stanford Cancer Institute, Redwood City, CA, USA

Marie Andree Lahaie McGill University, Montreal, QC, Canada

Miguel J. A. Lainez-Andres Department of Neurology, University of Valencia, Valencia, Spain

Robert H. LaMotte Department of Anesthesiology, Yale School of Medicine, New Haven, CT, USA

James W. Lance Institute of Neurological Sciences, University of New South Wales, Sydney, Australia

Robert G. Large The Auckland Regional Pain Service, Auckland Hospital, Auckland, New Zealand

Alice A. Larson Department of Veterinary and Biomedical Sciences, University of Minnesota, St. Paul, MN, USA

Roberta Lattanzi Department of Physiology and Pharmacology, Sapienza University of Rome, Rome, Italy

Peter Lau Department of Clinical Research, Royal Newcastle Hospital, University of Newcastle, Newcastle, NSW, Australia

Stefan A. Laufer Institute of Pharmacy, Eberhard-Karls University of Tuebingen, Tuebingen, Germany

Stefan Lautenbacher Physiologische Psychologie, Otto-Friedrich-Universität, Bamberg, Germany

Gilles J. Lavigne Facultés de Médecine Dentaire et de Médecine, Université de Montréal, Hopital Sacre Coeur de Montreal. Surgery-trauma dept, Montréal, QC, Canada

Emily Law Center for Child Health, Behavior, and Development, Seattle Children's Research Institute, Seattle, WA, USA

Peter G. Lawlor Our Lady's Hospice, Medical Department, Dublin, Ireland

Michel Lazdunski Institut de Pharmacologie Moléculaire et Cellulaire (IPMC) CNRS/UNS UMR7275, Valbonne-Sophia Antipolis, France

Vincenzo Di Lazzaro Institute of Neurology, Campus Biomedico University, Rome, Italy

Alyssa Lebel Boston Children's Hospital, Boston, MA, USA

Frank Lee Johns Hopkins University, Baltimore, MD, USA

Maaike Leeuw Department of Medical, Clinical and Experimental Psychology, Maastricht University, Maastricht, The Netherlands

Siri Graff Leknes Research Fellow, Psykologisk institutt, Oslo, Norway

Frederick A. Lenz Department of Neurosurgery, Johns Hopkins Hospital, Baltimore, MD, USA

Guillaume Léonard Faculty of Medicine, University of Sherbrooke, Sherbrooke, QC, Canada

Pauline Lesage Department of Pain Medicine and Palliative Care, Beth Israel Medical Center, New York, USA

Isobel Lever Neural Plasticity Unit, Institute of Child Health University College London, London, UK

Khan W. Li Department of Neurosurgery, Johns Hopkins Hospital, Baltimore, MD, USA

Alan R. Light Department of Anesthesiology, University of Utah, Salt Lake City, UT, USA

Michael Lim Department of Neurosurgery and Department of Neurology, Johns Hopkins Hospital, Baltimore, MA, USA

Deolinda Lima Department of Experimental Biology, Faculdade de Medicina da Universidade do Porto, Porto, Portugal

Volker Limmroth Department of Neurology, University of Essen, Essen, Germany

Eric Lingueglia Institut de Pharmacologie Moléculaire et Cellulaire (IPMC) CNRS/UNS UMR7275, Valbonne-Sophia Antipolis, France

Steven J. Linton Department of Occupational and Environmental Medicine, Orebro Medical Center, Orebro, Sweden

Arthur G. Lipman Departments of Pharmscotherpay and Anesthesiology and Pain Management Center, University of Utah, Salt Lake City, UT, USA

Richard B. Lipton Departments of Neurology, Epidemiology and Population Health, Albert Einstein College of Medicine, Bronx, NY, USA

Diana Lisi Department of Psychology, York University, Toronto, ON, Canada

Kenneth M. Little Division of Neurosurgery, Duke University Medical Center, Durham, NC, USA

Edward Littleton Queen Elizabeth Hospital, Birmingham, UK

Spencer S. Liu Virginia Mason Medical Center, Seattle, WA, USA

Anne Elisabeth Ljunggren Section for Physiotherapy Science, Department of Public Health and Primary Health, Faculty of Medicine, University of Bergen, Bergen, Norway

John D. Loeser Department of Neurological Surgery, University of Washington, Seattle, WA, USA

Joern Loetsch Institute for Clinical Pharmacology, Pharmaceutical Center Frankfurt, Johann Wolfgang Goethe University, Frankfurt, Germany

Patrick Loisel Disability Prevention Research and Training Centre, Université de Sherbrooke, Longueuil, Canada

Elisa Lopez-Dolado Servicio de Rehabilitacion, National Hospital of Paraplèjicos SESCAM, Toledo, Spain

Jürgen Lorenz Faculty of Life Science, Competence Center Gesundheit, University of Applied Science Hamburg, Hamburg, Germany

Dayna R. Loyd US Army Institute of Surgical Research, Fort Sam Houston, TX, USA

Daniel Lubelski Cleveland Clinic Lerner College of Medicine, Cleveland, OH, USA

Stephen Lutz Blanchard Valley Regional Cancer Center, Findlay, OH, USA

Bruce Lynn Department of Physiology, University College London, London, UK

Ross MacPherson Department of Anesthesia and Pain Management, Royal North Shore Hospital University of Sydney, Sydney, NSW, Australia

Christophe Maes Manager Innovatie AZ Alma, Belgium

Michel Magnin INSERM, Neurological Hospital, Lyon, France

Bryan Mahony Department of Anesthesiology, Ohio State University Medical Center, Mount Sinai Hospital, New York, NY, USA

Steven F. Maier Department of Psychology and Neuroscience, and The Center for Neuroscience, University of Colorado at Boulder, Boulder, CO, USA

Thorsten J. Maier Institute of Pharmaceutical Chemistry, Goethe University, Frankfurt, Germany

Chris J. Main Keele University, Keele, Staffordshire, UK

Marzia Malcangio Wolfson Centre for Age Related Diseases, King's College London, London, UK

Paolo L. Manfredi Department of Neurology, Memorial Sloan-Kettering Cancer Center, New York, USA

Kaisa Mannerkorpi Department of Rheumatology and Inflammation Research, Institute of Medicine, University of Gothenburg, Göteborg, Sweden

Barton H. Manning Amgen Inc., Thousand Oaks, CA, USA

Patrick W. Mantyh Department of Pharmacology, The University of Arizona, AZ, USA

Jianren Mao Department of Anesthesia and Critical Care, Harvard Medical School, Boston, MA, USA

Serge Marchand Faculty of Medicine, University of Sherbrooke, Sherbrooke, QC, Canada

Paolo Marchettini Pain Medicine Center, Scientific Institute San Raffaele, Milan, Italy

Sarah R. Martin Psychology, Georgia State University, Atlanta, GA, USA

Peggy Mason Department of Neurobiology, Pharmacology and Physiology, University of Chicago, Chicago, IL, USA

Scott Masters Caloundra Spinal and Sports Medicine Centre, Caloundra, QLD, Australia

Laurence E. Mather Department of Anaesthesia and Pain Management, University of Sydney at Royal North Shore Hospital, St. Leonards, Sydney, NSW, Australia

Bill McCarberg Pain Services, Kaiser Permanente, San Diego, USA

Lance McCracken Health Psychology/Psychology Department, King's College London, London, UK

Kathleen McGinn Anesthesiology & Pain Medicine, Seattle Children's Hospital, Seattle, WA, USA

Patricia A. McGrath The Hospital for Sick Children Pain Research Center, Toronto, ON, Canada

Patrick McGrath Departments of Psychology, Pediatrics and Psychiatry, Dalhousie University, Canada

Greg McIntosh Canadian Back Institute, Toronto, ON, Canada

Peter McIntyre Department of Pharmacology, University of Melbourne, Melbourne, Australia

Elspeth M. McLachlan Neuroscience Research Australia, University of New South Wales, Randwick, NSW, Australia

University of Glasgow, Glasgow, UK

Peter A. McNaughton Department of Pharmacology, University of Cambridge, Cambridge, UK

J. Mark Melhorn Section of Orthopaedics, Department of Surgery, University of Kansas School of Medicine, Wichita, KS, USA

Ronald Melzack Department of Psychology, McGill University, Montreal, QC, Canada

Lorne M. Mendell Department of Neurobiology and Behavior, Stony Brook University, Stony Brook, NY, USA

George Mendelson Department of Psychological Medicine, Monash University, Caulfield, VIC, Australia

Siegfried Mense Neuroanatomy, Heidelberg University, Medical Faculty Mannheim, CBTM, Mannheim, Germany

Daniel Menétrey Biomédicale, Paris, France

Sebastiano Mercadante Pain Relief & Palliative Care, La Maddalena Cancer Center University of Palermo, Palermo, Italy

Susan Mercer School of Biomedical Sciences, The University of Queensland, Brisbane, Australia

Karl Messlinger University of Erlangen-Nürnberg, Erlangen, Germany

Richard A. Meyer Department of Neurosurgergy, School of Medicine Johns Hopkins University, Baltimore, MD, USA

Walter J. Meyer III Shriners Hospitals for Children, Shriners Burns Hospital, Galveston, TX, USA

Department of Psychiatry, University of Texas Medical Branch, Galveston, TX, USA

Björn A. Meyerson Department of Clinical Neuroscience, Section of Neurosurgery, Karolinska Institute/University Hospital, Stockholm, Sweden

Martin Michaelis Sanofi-Aventis Deutschland GmbH, Frankurt am Main, Germany

Joseph Mignogna Michael E. DeBakey VA Medical Center, Houston VA HSR&D Center of Excellence (152), Houston, TX, USA

Judith A. Mikacich Department of Obstetrics and Gynecology, University of California Los Angeles Medical Center, Los Angeles, CA, USA

Kensaku Miki Department of Integrative Physiology, National Institute for Physiological Sciences, Okazaki, Japan

Vjekoslav Miletic School of Medicine and Public Health, University of Wisconsin, Madison, WI, USA

Scott R. Miller Memorial Sloan-Kettering Cancer Center, New York, USA

Erin D. Milligan Department of Neurosciences, University of New Mexico, Albuquerque, NM, USA

Michael S. Minett Molecular Nociception Group, Wolfson Institute for Biomedical Research, University College London, London, WC, UK

J. Mocco Department of Neurological Surgery, Neurological Institute Columbia University, New York, NY, USA

Jeffrey S. Mogil Department of Psychology, McGill University, Montreal, QC, Canada

Harvey Moldofsky Sleep Disorders Clinic of the Centre for Sleep and Chronobiology, Toronto, ON, Canada

Robert M. Moldwin The Arthur Smith Institute for Urology, North Shore-LIJ Healthcare System, Hofstra North Shore LIJ School of Medicine, New Hyde Park, NY, USA

Derek C. Molliver Department of Neurobiology, University of Pittsburgh School of Medicine, Pittsburgh, PA, USA

Robert D. Mootz Washington State Department of Labor and Industries, Olympia, WA, USA

Laura Mordillo-Mateos FENNSI Group, Hospital Nacional de Parapléjicos, Toledo, Spain

Anne Morel Laboratory for Functional Neurosurgery, Neurosurgical Clinic University Hospital Zürich, Zürich, Switzerland

Anne Morinville Montreal Neurological Institute, McGill University, Montreal, QC, Canada

Stephen Morley Academic Unit of Psychiatry and Behavioral Sciences, University of Leeds, Leeds, UK

David B. Morris University of Virginia, Charlottesville, VA, USA

G. Lorimer Moseley Sansom Institute for Health Research, University of South Australia Neuroscience Research Australia, Adelaide, Australia

Joachim Mossner Medical Clinic and Policlinic II, University Clinical Center Leipzig AöR, Leipzig, Germany

Eric A. Moulton P.A.I.N. Group, Anesthesiology, Boston Children's Hospital, Harvard Medical School, Waltham, MA, USA

Francis Githae Muriithi Department of Obstetrics and Gynaecology, Aga Khan University, Nairobi, Kenya

Anne Z. Murphy Neuroscience Institute, Georgia State University, Atlanta, GA, USA

Paul M. Murphy Department of Anaesthesia and Pain Management, Royal North Shore Hospital, St. Leonards, NSW, Australia

Alison Murray Foothills Medical Center, Calgary, AB, Canada

H. S. Na Neuroscience Research Institute and Department of Physiology, Korea University, College of Medicine, Seoul, Republic of Korea

Daisuke Naka Department of Integrative Physiology, National Institute for Physiological Sciences, Okazaki, Japan

Yoshio Nakamura Pain Research Center, Department of Anesthesiology University of Utah School of Medicine, Salt Lake City, UT, USA

Barbara Namer Department of Physiology & Pathophysiology, FAU, Erlangen/Nuremberg, Bavaria, Germany

Matti V. O. Närhi Department of Biomedicine/Physiology, Department of Dentistry, University of Eastern Finland, Institute of Medicine, Kuopio, Finland

H. J. W. Nauta University of Texas Medical Branch, Galveston, TX, USA

Lucia Negri Department of Physiology and Pharmacology, Sapienza University of Rome, Rome, Italy

Gary Nesbit Department of Radiology, Oregon Health and Science University, Portland, OR, USA

Timothy Ness Department of Anesthesiology, University of Alabama at Birmingham, Birmingham, AL, USA

Volker Neugebauer Department of Neuroscience & Cell Biology, University of Texas Medical Branch, Galveston, TX, USA

Lawrence C. Newman Albert Einstein College of Medicine, New York, NY, USA

Michael K. Nicholas Royal North Shore Hospital, University of Sydney, St. Leonards, NSW, Australia

Ellen Niederberger Pharmazentrum frankfurt, Institute of Clinical Pharmacology, Goethe-University Hospital, Frankfurt, Germany

Lone Nikolajsen Department of Anaesthesiology, Danish Pain Research Center Aarhus University Hospital, Aarhus, Denmark

Katie Nixdorf Comprehensive Pain Center, Oregon Health and Science University, Portland, OR, USA

Donald R. Nixdorf Department of Diagnostic and Biological Sciences, School of Dentistry University of Minnesota, Minneapolis, MN, USA

Carl E. Noe University of Texas Southwestern Medical Center, Lubbock, TX, USA

Koichi Noguchi Department of Anatomy and Neuroscience, Hyogo College of Medicine, Hyogo, Japan

Richard B. North The Johns Hopkins University, Baltimore, MD, USA

Turo J. Nurmikko Department of Neurological Science, University of Liverpool, and Pain Research Institute, The Walton Centre NHS Foundation Trust, Liverpool, UK

Anne Louise Oaklander Nerve Injury Unit, Departments of Neurology and Pathology, Massachusetts General Hospital Harvard Medical School, Boston, MA, USA

Koichi Obata Department of Anatomy and Neuroscience, Hyogo College of Medicine, Hyogo, Japan

Tim F. Oberlander Division of Developmental Pediatrics, University of British Columbia, Vancouver, BC, Canada

Bruno Oertel Institute for Clinical Pharmacology, Pharmaceutical Center Frankfurt, Johann Wolfgang Goethe University, Frankfurt, Germany

Andrea C. Ogbonna Wolfson Centre for Age Related Diseases, King's College London, London, UK

Alfred T. Ogden Department of Neurological Surgery, Neurological Institute Columbia University, New York, NY, USA

S. Ohara Department of Neurosurgery, Johns Hopkins Hospital, Johns Hopkins Medical Institutions, Johns Hopkins University, Baltimore, MD, USA

Peter T. Ohara Department of Anatomy, University of California San Francisco, San Francisco, CA, USA

Akiko Okifuji Department of Anesthesiology, University of Utah, Salt Lake City, UT, USA

Jean-Louis Oliveras Inserm, Paris, France

Antonio Oliviero FENNSI Group, National Hospital of Paraplégicos, Toledo, Spain

Sean O'Mahony Palliative medicine, Rush University Medical Center, Chicago, IL, USA

Peregrine B. Osborne Dept Anatomy & Neuroscience, University of Melbourne, Melbourne, VIC, Australia

Michael H. Ossipov Department of Pharmacology, University of Arizona College of Medicine, Tucson, AZ, USA

Tonya M. Palermo Seattle Children's Hospital Research Institute, CW8-6 - Child Health, Behavior and Development, University of Washington, Seattle, WA, USA

Shreela Palit Department of Psychology, The University of Tulsa, Tulsa, OK, USA

Juan A. Pareja Department of Neurology, Fundación Hospital Alcorcón, Madrid, Spain

Steven D. Passik Psychiatry and Anesthesiology, Vanderbilt University, Nashville, TN, USA

Gavril W. Pasternak Laboratory of Molecular Neuropharmacology, Memorial Sloan-Kettering Cancer Center, New York, USA

S. H. Patel Department of Neurosurgery, Johns Hopkins Hospital, Baltimore, MD, USA

Gavin Pattullo Department of Anesthesia and Pain Management, Royal North Shore Hospital, University of Sydney, St. Leonards, NSW, Australia

Kevin Pauza Tyler Spine and Joint Hospital, Tyler, TX, USA

Eric M. Pearlman Pediatrics, Mercer University School of Medicine, The Children's Hospital at Memorial University Medical Center, Savannah, USA

Elvira De la Pena Instituto de Neurociencias, Universidad Miguel Hernández-CSIC, Alicante, Spain

Yuan Bo Peng Department of Psychology, University of Texas at Arlington, Arlington, TX, USA

Jose Pereira Foothills Medical Center, Calgary, AB, Canada

Edward Perl Department of Cell and Molecular Physiology, School of Medicine University of North Carolina, Chapel Hill, NC, USA

Tiffany Grace Perry Center for Spine Health, The Cleveland Clinic Foundation, Cleveland, OH, USA

Marie Pertin Pain Center, University Hospital Center and University of Lausanne, Lausanne, Switzerland

Antti Pertovaara Department of Physiology, Institute of Biomedicine University of Helsinki, Helsinki, Finland

Peter Peter Department of Neurobiology and Anatomy, University of Rochester Medical Center, Rochester, NY, USA

Lisa M. Peters Department of Anesthesiology and Pain Medicine, Seattle Children's Hospital, Seattle, WA, USA

Pramit Phal Department of Radiology, Oregon Health and Science University, Portland, OR, USA

Rebecca Pillai Riddell Department of Psychology, York University, Toronto, ON, Canada

Department of Psychiatry, Hospital for Sick Children, Toronto, ON, Canada

Issy Pilowsky University of Adelaide, Adelaide, Australia

Leon Plaghki Institute of Neuroscience (IoNS), Université Catholique de Louvain, Brussels, Belgium

Markus Ploner Department of Neurology, Technische Universität München, Munich, Germany

Esther M. Pogatzki-Zahn Department of Anesthesiology and Intensive Care Medicine, University Muenster, Muenster, Germany

Michael Polydefkis Department of Neurology, Johns Hopkins University School of Medicine, Baltimore, MD, USA

James D. Pomonis Algos Preclinical Services, Roseville, MN, USA

Dieter Pongratz Friedrich Baur Institute, Ludwig Maximilians University, Munich, Germany

Hayley Poole Hadley House, Leicester General Hospital, Leicester, UK

Frank Porreca Department of Pharmacology, University of Arizona Health Sciences Center, Tucson, AZ, USA

Russell K. Portenoy Department of Pain Medicine and Palliative Care, Beth Israel Medical Center, New York, NY, USA

Ian Power Critical Care and Pain Medicine, The University of Edinburgh, Edinburgh, UK

Donald D. Price Oral Surgery and Neuroscience, McKnight Brain Institute University of Florida, Gainesville, FL, USA

Daryl Pullman Medical Ethics, Memorial University of Newfoundland, St. John's, NL, Canada

Yunhai Qiu Department of Integrative Physiology, National Institute for Physiological Sciences, Okazaki, Japan

Michael Quittan Department of Physical Medicine and Rehabilitation, Kaiser-Franz-Joseph Hospital, Vienna, Austria

Raymond M. Quock Department of Pharmaceutical Sciences, Washington State University College of Pharmacy, Pullman, WA, USA

Jennifer Rabbitts Department of Anesthesiology and Pain Medicine, Seattle Children's Hospital, University of Washington School of Medicine, Seattle, WA, USA

Nicole Racine Psychology, York University, Toronto, ON, Canada

Gabor B. Racz Texas Tech University Health Sciences Center, Lubbock, TX, USA

Lukas Radbruch Department of Palliative Medicine, University of Bonn, Bonn, Germany

Robert B. Raffa Department of Pharmaceutical Sciences, Temple University School of Pharmacy, Philadelphia, PA, USA

Vasudeva Raghavendra Algos Therapeutics Inc., Saint Paul, MN, USA

Pierre Rainville Department of Stomatology, Faculty of Dental Medicine, University of Montreal, Montreal, QC, Canada

Srinivasa N. Raja Department of Anesthesiology and Critical Care Medicine, Johns Hopkins University, Baltimore, USA

Alan Randich University of Alabama, Birmingham, AL, USA

Andrea J. Rapkin Department of Obstetrics and Gynecology, University of California Los Angeles Medical Center, Los Angeles, CA, USA

Z. Harry Rappaport Department of Neurosurgery, Rabin Medical Center Tel-Aviv University, Petah Tiqva, Israel

Matthew N. Rasband Department of Molecular and Cellular Biology, Baylor College of Medicine, Farmington, Houston TX, USA

University of Connecticut Health Center, Farmington, CT, USA

Douglas Rasmusson Department of Physiology and Biophysics, Dalhousie University, Halifax, Canada

James P. Rathmell Department of Anesthesiology, University of Vermont College of Medicine, Burlington, USA

K. Ravishankar The Headache and Migraine Clinic, Jaslok Hospital and Research Centre Lilavati Hospital and Research Centre, Mumbai, India

Ke Ren Department of Neural and Pain Sciences, University of Maryland Dental School, Baltimore, MD, USA

Seth Resnick Silver Hill Hospital, New York University School of Medicine, USA

Cielito Reyes-Gibby The University of Texas MD Anderson Cancer Center, Houston, TX, USA

Jamie L. Rhudy Department of Psychology, The University of Tulsa, Tulsa, OK, USA

Alfredo Ribeiro-da-Silva Department of Pharmacology and Therapeutics, McGill University, Montreal, QC, Canada

Frank L. Rice Center for Neuropharmacology and Neuroscience, Albany Medical College, Albany, NY, USA

Matthias Ringkamp Department of Neurosurgergy, School of Medicine Johns Hopkins University, Baltimore, MD, USA

Pierre J. M. Rivière Ferring Research Institute, San Diego, CA, USA

Claude Robert Université Paris Descartes, INSERM UMR 894, France

Roger Robert Neurosurgery, Hotel Dieu, Nantes, CHU, France

James P. Robinson Department of Anesthesiology & Pain Medicine, University of Washington, Seattle, WA, USA

John Robinson Pain Management Centre, Burwood Hospital, Christchurch, New Zealand

Alfonso Romero-Sandoval Department of Pharmaceuticals and Administrative Sciences, Presbyterian College School of Pharmacy, Clinton, SC, USA

David Roselt Aberdovy Clinic, Bundaberg, QLD, Australia

Jackie Ross Early Pregnancy Unit, King's College Hospital Early Pregnancy Unit, London, UK

Shlomo Rotshenker Department of Medical Neurobiology, IMRIC, Faculty of Medicine Hebrew University of Jerusalem, Jerusalem, Israel

Paul C. Rousseau Medical University of South Carolina, Charleston, SC, USA

Ranjan Roy Department of Clinical Health Psychology, Faculty of Medicine, University of Manitoba, Winnipeg, MB, Canada

Carolina Roza Biología de Sistemas (Unidad Fisiología), Universidad de Alcalá, Alcalá de Henares, Madrid, Spain

Todd D. Rozen Department of Neurology, Geisinger Health System, Wilkes-Barre, PA, USA

Roman Rukwied Department of Anaesthesiology and Intensive Care Medicine, Faculty of Clinical Medicine Mannheim University of Heidelberg, Heidelberg, Germany

Irwin Jon Russell Division of Clinical Immunology and Rheumatology, Department of Medicine, The University of Texas Health Science Center, San Antonio, TX, USA

Nayef E. Saadé Anatomy, Cell Biology and Physiology, Faculty of Medicine, American University of Beirut, Beirut, Lebanon

Mary Ann C. Sabino Oral and Maxillofacial Surgery, Hennepin County Medical Center University of Minnesota, Minneapolis, MN, USA

Thomas E. Salt Institute of Ophthalmology, University College London, London, UK

A. Samdani Department of Neurosurgery, Johns Hopkins Medical Institutions, Baltimore, MD, USA

Yasmin Sana King's College Hospital, London, UK

Margarita Sanchez-del-Rio Neurology Department, Hospital Ruber Internacional, Madrid, Spain

Jürgen Sandkühler Department of Neurophysiology, Center for Brain Research Medical University of Vienna, Vienna, Austria

Peter S. Sándor RehaClinic, Cantonal Hospital Baden, Baden, Switzerland

Johannes Sarnthein Functional Neurosurgery, Neurosurgical Clinic University Hospital, Zürich, Switzerland

Susanne K. Sauer Department of Physiology and Pathophysiology, University of Erlangen, Erlangen, Germany

Steven Savvas Biomedical Division, National Ageing Research Institute, Melbourne, VIC, Australia

Jana Sawynok Department of Pharmacology, Dalhousie University, Halifax, Canada

John W. Scadding The National Hospital for Neurology and Neurosurgery, London, UK

Frederieke Geraldine Schaafsma Department of Public and Occupational Health, EMGO+ Institute/VU Medical Center, Amsterdam, MB, The Netherlands

Hans-Georg Schaible Department of Physiology - Neurophysiology, Friedrich-Schiller-University of Jena, Jena, Germany

Steven S. Scherer Department of Neurology, The Perelman School of Medicine at the University of Pennsylvania, Philadelphia, PA, USA

Michael Schäfer Department of Anesthesiology and Intensive Care Medicine, Charité University, Berlin, Germany

Martin Schmelz Department of Anesthesiology in Mannheim, University of Heidelberg, Mannheim, Germany

Robert F. Schmidt Physiologisches Institut der Universität Würzburg, Würzburg, Germany

Suzan Schneeweiss The Hospital for Sick Children, Toronto, ON, Canada

Frank Schneider Department of Psychiatry and Psychotherapy, RWTH Aachen University, Aachen, Germany

Alfons Schnitzler Department of Neurology, Heinrich Heine University, Düsseldorf, Germany

Jean Schoenen Department of Neurology, University of Liège, CHR Citadelle, Belgium

Department of Neuroanatomy, University of Liège, CHR Citadelle, Belgium

Jens Schouenborg Neuronano Research Center, Lund University, Lund, Sweden

Stephan A. Schug School of Medicine and Pharmacology, Royal Perth Hospital, Faculty of Medicine, Dentistry & Health Sciences, The University of Western Australia, UWA Anaesthesiology, Perth, Australia

Patrick Schuss Department of Neurosurgery, University-Hospital Bonn, Bonn, Germany

Anthony C. Schwarzer Faculty of Medicine and Health Sciences, The University of Newcastle, Newcastle, Australia

Whitney Scott Department of Psychology, McGill University, Montreal, QC, Canada

Laura C. Seidman Pediatric Pain Program, Department of Pediatrics David Geffen School of Medicine at UCLA, Los Angeles, CA, USA

Ze'ev Seltzer Faculty of Dentistry, University of Toronto Centre for the Study of Pain, Toronto, ON, Canada

Peter Selwyn Albert Einstein College of Medicine, Bronx, NY, USA

Patrick B. Senatus Department of Neurological Surgery, Neurological Institute Columbia University, New York, NY, USA

Jyoti N. Sengupta Division of Gastroenterology and Hepatology and Pediatric Gastroenterology, Medical College of Wisconsin, Milwaukee, WI, USA

Mariano Serrao Department of Neurology and Otorinolaringoiatry, University of Rome La Sapienza, Rome, Italy

Barry J. Sessle Faculty of Dentistry, University of Toronto, Toronto, ON, Canada

Faculty of Dentistry, University of Toronto, Toronto, ON, Canada

Virginia S. Seybold Department of Neuroscience, University of Minnesota, Minneapolis, MN, USA

T. Shah Royal Perth Hospital and University of Western Australia, Crawley, WA, Australia

Yair Sharav Department of Oral Medicine, School of Dental Medicine, Hebrew University-Hadassah, Jerusalem, Israel

S. Murray Sherman Department of Neurobiology, University of Chicago, Chicago, IL, USA

Yoshio Shigenaga Department of Oral Anatomy and Neurobiology, Osaka University, Osaka, Japan

Susan Shin Neurology, Mount Sinai School of Medicine, New York, NY, USA

Edward A. Shipton Department of Anaesthesia, Christchurch School of Medicine and Health Sciences, University of Otago, Christchurch, New Zealand

Yoram Shir The Alan Edwards Pain Management Unit, McGill University Health Centre, Montreal, QC, Canada

Peter Shrager Department of Neurobiology and Anatomy, University of Rochester Medical Center, Rochester, NY, USA

Marc J. Shulman VA Connecticut Healthcare System, Yale University, New Haven, CT, USA

David M. Sibell Comprehensive Pain Center/Spine Center, Oregon Health and Science University, Portland, OR, USA

Philip J. Siddall Department of Pain Management, Greenwich Hospital, University of Sydney, Sydney, NSW, Australia

Stephen D. Silberstein Jefferson Medical College, Thomas Jefferson University and Jefferson Headache Center, Thomas Jefferson University Hospital, Philadelphia, PA, USA

Donald A. Simone Diagnostic and Biological Sciences, University of Minnesota, Minneapolis, MN, USA

David G. Simons Department of Rehabilitation Medicine, Emory University, Atlanta, GA, USA

David Simpson Clinical Neurophysiology Laboratories, Mount Sinai Medical Center, New York, NY, USA

Marc Sindou Department of Neurosurgery, Hopital Neurologique Pierre Wertheimer, University of Lyon 1, Lyon, France

Soeren H. Sindrup Department of Neurology, Odense University Hospital, Odense, Denmark

Bengt H. Sjölund Department of Community Medicine and Rehabilitation Medicine, Umea University, Umea, Sweden

Carsten Skarke Institute for Translational Medicine and Therapeutics (ITMAT), University of Pennsylvania Perelman School of Medicine, Philadelphia, PA, USA

Paul A. Sloan Department of Anesthesiology, University of Kentucky, Lexington, KY, USA

Kathleen A. Sluka Physical Therapy and Rehabilitation Science Graduate Program, University of Iowa, Iowa City, IA, USA

Seymour Solomon Montefiore Medical Center, Albert Einstein College of Medicine, Bronx, NY, USA

Claudia Sommer Department of Neurology, University of Würzburg, Würzburg, Germany

Linda S. Sorkin Occupational Therapy and Rehabilitation, Anesthesia Research Laboratories, University of California, San Diego, CA, USA

Ingrid Söderback Uppsala University, Uppsala, Sweden

Kari Sørensen Department of Pain Management and Research, Oslo University Hospital, Rikshospitalet, Oslo, Norway

Michael Spaeth Friedrich-Baur-Institute, University of Munich, Munich, Germany

Peregrin O. Spielholz Washington State Department of Labor and Industries, SHARP Program, Olympia, WA, USA

Peter S. Staats Division of Pain Medicine, The Johns Hopkins University, Baltimore, MD, USA

Brett R. Stacey Comprehensive Pain Center, Oregon Health and Science University, Portland, OR, USA

Wendy M. Stein San Diego Hospice and Palliative Care, UCSD, San Diego, CA, USA

Christoph Stein Department of Anesthesiology and Intensive Care Medicine, Freie Universitaet Berlin Charité Campus Benjamin Franklin, Berlin, Germany

Michael P. Steinmetz Department of Neurosurgery, MetroHealth Medical Center, Case Western Reserve University School of Medicine, Cleveland, OH, USA

Jair Stern Functional Neurosurgery, Neurosurgical Clinic University Hospital, Zürich, Switzerland

Bonnie J. Stevens University of Toronto and The Hospital for Sick Children, Toronto, ON, Canada

Carl-Olav Stiller Division of Clinical Pharmacology and Department of Medicine, Karolinska University Hospital, Stockholm, Sweden

Jennifer N. Stinson The Hospital for Sick Children, Toronto, ON, Canada

Henry Stockbridge University of Washington School of Public Health and Community Medicine, Seattle, WA, USA

Christian S. Stohler Baltimore College of Dental Surgery, University of Maryland, Baltimore, MD, USA

Laura Stone Faculty of Dentistry, Alan Edwards Centre for Research on Pain, Montreal, QC, Canada

William Stones Department of Obstetrics and Gynaecology, Aga Khan University, Nairobi, Kenya

Andreas Straube Department of Neurology, Ludwig Maximilians University, Munich, Germany

Jenny Strong University of Queensland, Brisbane, QLD, Australia

Audun Stubhaug Department of Pain Management and Research, Oslo University Hospital, Rikshospitalet and University of Oslo, Oslo, Norway

Gerold Stucki Swiss Paraplegic Research, Nottwil, Switzerland

Department of Health Sciences and Health Policy, University of Lucerne, Lucerne, Switzerland

ICF Research Branch in cooperation with the World Health Organization (WHO) Collaborating Centre for the Family of International Classifications in Germany, Nottwil, Switzerland

Cheryl L. Stucky Department of Cell Biology Neurobiology and Anatomy, Medical College of Wisconsin, Milwaukee, WI, USA

Mathias Sturzenegger Department of Neurology, University Hospital, Inselspital, University of Bern, Bern, Switzerland

Yasuo Sugiura School of Human Care Study, Nagoya University of Arts and Sciences, Nissin City, Aichi, Japan

Mark D. Sullivan Psychiatry and Behavioral Sciences, University of Washington, Seattle, WA, USA

Michael J. L. Sullivan Department of Psychology, McGill University, Montreal, QC, Canada

Rie Suzuki Department of Pharmacology, University College London, London, UK

Kristina B. Svendsen Danish Pain Research Center, Aarhus University Hospital, Aarhus, Denmark

Peter Svensson Clinical Oral Physiology, Aarhus University, Aarhus, Denmark

Peter Szucs Spinal Neuronal Networks, Instituto de Biologia Molecular e Celular - IBMC, Porto, Portugal

A. Taghva Department of Neurosurgery, Johns Hopkins Hospital, Baltimore, MD, USA

Kimberson Cochien Tanco Department of Palliative Care and Rehabilitation Medicine, The University of Texas M. D. Anderson Cancer Center, Houston, TX, USA

Kersi Taraporewalla Royal Brisbane and Womens' Hospital, University of Queensland, Brisbane, QLD, Australia

Irmgard Tegeder Pharmazentrum frankfurt, Institute of Clinical Pharmacology, Goethe-University Hospital, Frankfurt, Germany

Astrid Juhl Terkelsen Danish Pain Research Center, Department of Neurology, Aarhus University Hospital, Aarhus, Denmark

Gregory W. Terman Department of Anesthesiology and the Graduate Program in Neurobiology and Behavior, University of Washington, Seattle, WA, USA

Alison Thomas Lismore, NSW, Australia

Benjamin Thomas Chelsea & Westminster Hospital Foundation Trust, London, UK

Jacob Thomas School of Medical Sciences, University of Adelaide, Adelaide, SA, Australia

Miles Thompson Goldsmiths, University of London, London, UK

Beverly E. Thorn Department of Psychology, University of Alabama, Tuscaloosa, AL, USA

Cindy L. Thurston-Stanfield Department of Biomedical Sciences, University of South Alabama, Mobile, AL, USA

Arne Tjølsen Department of Bomedicine, University of Bergen, Haukeland University Hospital, Bergen, Norway

Andrew J. Todd Institute of Neuroscience and Psychology, College of Medical, Veterinary and Life Sciences, University of Glasgow, Glasgow, UK

Makoto Tominaga Department of Cellular and Molecular Physiology, Mie University School of Medicine, Tsu Mie, Japan

Victor Tortorici Instituto Venezolano de Investigaciones Científicas (IVIC), Caracas, Venezuela

Irene Tracey Pain Imaging Neuroscience (PaIN) Group, Department of Physiology, Anatomy and Genetics, Oxford University, Oxford, UK

Diep Tuan Tran Department of Integrative Physiology, National Institute for Physiological Sciences, Okazaki, Japan

Rolf-Detlef Treede Institute of Physiology and Pathophysiology, Johannes Gutenberg University, Mainz, Germany

Jennie C. I. Tsao Pediatric Pain Program, Department of Pediatrics David Geffen School of Medicine at UCLA, Los Angeles, CA, USA

Lawrence Tsen Department of Anesthesiology, Perioperative and Pain Medicine, Brigham and Women's Hospital Harvard Medical School, Boston, MA, USA

Jeanna Tsenter Department of Physical Medicine and Rehabilitation, Hadassah University Hospital, Jerusalem, Israel

Maurits Van Tulder Institute for Research in Extramural Medicine, Free University Amsterdam, Amsterdam, The Netherlands

Eldon Tunks Hamilton Health Sciences Corporation, Hamilton, ON, Canada

Susan Tupper Pediatrics, University of Saskatchewan, Saskatoon, SK, Canada

Dennis C. Turk Department of Anesthesiology & Pain Research, University of Washington, Seattle, WA, USA

Judith A. Turner Department of Psychiatry and Behavioral Sciences, University of Washington, Seattle, WA, USA

Department of Rehabilitation Medicine, University of Washington, Seattle, WA, USA

Stephen P. Tyrer Royal Victoria Infirmary, University of Newcastle upon Tyne, Newcastle-on-Tyne, UK

Alexander Tzabazis Department of Anesthesia, Stanford University, Stanford, CA, USA

Kene Ugokwe Department of Neurosurgery, St Elizabeth Health Center, Youngstown, OH, USA

Lindsay S. Uman Department of Psychology, IWK Health Centre, Halifax, NS, Canada

Christiane Vahle-Hinz Department of Neurophysiology and Pathophysiology, University Medical Center Hamburg-Eppendorf, Hamburg, Germany

Muthukumar Vaidyaraman Division of Pain Medicine, The Johns Hopkins University, Baltimore, MD, USA

Todd W. Vanderah Department of Pharmacology, College of Medicine University of Arizona, Tucson, AZ, USA

Guy Vanderstraeten Department of Rehabilitation Sciences and Physiotherapy, Ghent University Hospital, Ghent, Belgium

Horacio Vanegas Instituto Venezolano de Investigaciones Científicas (IVIC), Caracas, Venezuela

Jean-Jacques Vatine Outpatient and Research Division, Reuth Medical Center, Sackler Faculty of Medicine, Tel Aviv University, Tel Aviv, Israel

Leigh M. Vaughan Medical University of South Carolina, Charleston, SC, USA

Linda K. Vaughn Department of Biomedical Sciences, Marquette University, Milwaukee, WI, USA

Enrique Vazquez Instituto Venezolano de Investigaciones Científicas (IVIC), Caracas, Venezuela

Louis Vera-Portocarrero Department of Pharmacology, College of Medicine University of Arizona, Tucson, AZ, USA

Paul Verrills Metropolitan Spinal Clinic, Prahran, VIC, Australia

Tine Vervoort Department of Experimental Clinical and Health Psychology, Ghent University, Ghent, Belgium

Félix Viana Instituto de Neurociencias, Universidad Miguel Hernández-CSIC, San Juan de Alicante, Alicante, Spain

Charles J. Vierck Jr. Department of Neuroscience and McKnight Brain Institute, University of Florida College of Medicine, Gainesville, FL, USA

Luis Villanueva INSERM UMR 894, Centre de Psychiatrie et Neurosciences, Paris, France

Marcelo Villar Faculty of Biomedical Sciences, Austral University, Buenos Aires, Argentina

Katy Vincent Nuffield Department of Obstetrics and Gynaecology, The Women's Centre, John Radcliffe Hospital, University of Oxford, Oxford, UK

David Vivian Metro Spinal Clinic, South Caulfield, VIC, Australia

Johan W. S. Vlaeyen Research Group Health Psychology, Catholic University of Leuven, Leuven, Belgium

Department of Clinical Psychological Science, Maastricht University, Maastricht, The Netherlands

Stéphanie Volders Research Group on Health Psychology, Leuven, Belgium

Ernest Volinn Pain Research Center, University of Utah, Salt Lake City, UT, USA

Lucy Vulchanova Department of Veterinary and Biomedical Sciences, University of Minnesota, St. Paul, MN, USA

Paul W. Wacnik Neuromodulation Research, Medtronic Inc., Minneapolis, MN, USA

Gordon Waddell University of Glasgow, Glasgow, UK

Gary A. Walco Department of Anesthesiology and Pain Medicine, University of Washington School of Medicine, Seattle Children's Hospital, Seattle, WA, USA

Suellen M. Walker University of Sydney Pain Management and Research Centre, Royal North Shore Hospital, NSW, Australia

Victoria C. J. Wallace Department of Anaesthetics, Pain Medicine and Intensive Care Chelsea and Westminster Hospital Campus, London, UK

Xiaohong Wang Department of Integrative Physiology, National Institute for Physiological Sciences, Okazaki, Japan

Gunnar Wasner Department of Neurological Pain Research and Therapy, Neurological Clinic, Kiel, Germany

Shoko Watanabe Department of Integrative Physiology, National Institute for Physiological Sciences, Okazaki, Japan

Linda R. Watkins Department of Psychology and Neuroscience, and The Center for Neuroscience, University of Colorado at Boulder, Boulder, CO, USA

Peter N. Watson Department of Medicine, University of Toronto, Toronto, ON, Canada

James Watt North Shore Hospital, Auckland, New Zealand

Stephen G. Waxman Department of Neurology, Yale School of Medicine, New Haven, CT, USA

Inge Wegner Julius Center, Department of Epidemiology, University Medical Center Utrecht, Utrecht, The Netherlands

Feng Wei Department of Neural and Pain Sciences, University of Maryland Dental School, Baltimore, MD, USA

Christian Weidner Department of Physiology and Experimental Pathophysiology, University Erlangen, Erlangen, Germany

P. T. M. Weijenborg Department of Gynecology, Leiden University Medical Center, Leiden, The Netherlands

Robin Weir School of Nursing, Faculty of Health Sciences, McMaster University, Hamilton, ON, Canada

Nirit Weiss Department of Neurosurgery, Mount Sinai Medical Center, New York, NY, USA

Ursula Wesselmann Departments of Neurology and Anesthesiology / Division of Pain Management, University of Alabama at Birmingham, Birmingham, AL, USA

Dagmar Westerling Department of Anesthesiology, Central Hospital of Kristianstad, Kristianstad, Sweden

Karin N. Westlund Department of Physiology, University of Kentucky, Lexington, KY, USA

Thomas M. Wickizer Health Services Management and Policy, Ohio State University, Colombus, OH, USA

Eva Widerström-Noga The Miami Project to Cure Paralysis, Department of Neurological Surgery, University of Miami Miller, School of Medicine, Miami, FL, USA

Julie Wieseler Department of Psychology and Neuroscience, and The Center for Neuroscience, University of Colorado at Boulder, Boulder, CO, USA

Zsuzsanna Wiesenfeld-Hallin Karolinska Institute, Stockholm, Sweden

Ann Wigham McCoull Clinic, Prudhoe Hospital, Northumberland, UK

George L. Wilcox Department of Neuroscience, Pharmacology and Dermatology, University of Minnesota Medical School, Minneapolis, MN, USA

Oliver H. G. Wilder-Smith Pain Centre, Department of Anaesthesiology, Radboud University Nijmegen Medical Centre, Nijmegen, The Netherlands

Kjersti Wilhelmsen Section for Physiotherapy Science, Department of Public Health and Primary Health, Faculty of Medicine, University of Bergen, Bergen, Norway

Victor Wilk Brighton Spinal Group, Brighton, VIC, Australia

W. A. Williams Royal Perth Hospital and University of Western Australia, Perth, WA, Australia

Sara E. Williams Division of Pediatric Gastroenterology, Medical College of Wisconsin, Milwaukee, WI, USA

William D. Willis Department of Neuroscience and Cell Biology, University of Texas Medical Branch, Galveston, TX, USA

John B. Winer Department of Neurology, Birmingham Muscle and Nerve Centre Queen Elizabeth Hospital, Birmingham, UK

Christopher J. Winfree Department of Neurological Surgery, Neurological Institute Columbia University, New York, NY, USA

Janet Winter Novartis Institute for Medical Science, London, UK

Alain Woda Faculté de Chirurgie Dentaire, University Clermont-Ferrand, Clermont-Ferrand, France

John N. Wood Molecular Nociception Group, Wolfson Institute for Biomedical Research, University College London, London, WC, UK

Lee Woodson Shriners Hospitals for Children, Shriners Burns Hospital, Galveston, TX, USA

Department of Anesthesia, University of Texas Medical Branch, Galveston, TX, USA

Anava A. Wren Pain Prevention and Treatment Research, Department of Psychiatry and Behavioral Sciences, Duke University Medical Center, Durham, NC, USA

Xiao-Jun Xu Department of Physiology and Pharmacology, Section of Integrative Pain Rsearch Karolinska Institutet, Stockholm, Sweden

Hiroshi Yamasaki Department of Integrative Physiology, National Institute for Physiological Sciences, Okazaki, Japan

Michael Yelland School of Medicine, Logan campus, Griffith University, Logan, Australia

Sriram Yennurajalingam Department of Palliative Care and Rehabilitation Medicine, The University of Texas M. D. Anderson Cancer Center, Houston, TX, USA

David C. Yeomans Department of Anesthesia, Stanford University, Stanford, CA, USA

Robert P. Yezierski Comprehensive Center for Pain Research and Department of Neuroscience, The McKnight Brain Institute University of Florida, Gainesville, FL, USA

Way Yin Department of Anesthesiology, University of Washington, Seattle, WA, USA

Atsushi Yoshida Department of Oral Anatomy and Neurobiology, Osaka University, Osaka, Japan

Joanna M. Zakrzewska Eastman Dental Hospital UCLH NHS Foundation Trust, London, UK

Jenna E. Zauk St. George's University, School of Medicine, Grenada, West Indies

Hanns Ulrich Zeilhofer Institute for Pharmacology and Toxicology, University of Zurich, and Swiss Federal Institute of Technology (ETH) Zurich, Zürich, Switzerland

Lonnie K. Zeltzer Pediatric Pain Program, Department of Pediatrics David Geffen School of Medicine at UCLA, Los Angeles, CA, USA

Giovambattista Zeppetella St. Clare Hospice, Hastingwood, UK

Xijing Zhang Department of Neuroscience, University of Minnesota, Minneapolis, MN, USA

Min Zhuo Department of Physiology, University of Toronto, Toronto, ON, Canada

C. Zöllner Department of Anesthesiology and Intensive Care Medicine, Charité University, Berlin, Germany

Krina Zondervan Genetic and Genomic Epidemiology Unit, Nuffield Dept of Obstetrics and Gynaecology, Wellcome Trust Centre for Human Genetics, Oxford, UK

Peter Zygmunt Department of Laboratory Medicine, Lund University, Division of Clinical Chemistry and Pharmacology, Lund, Sweden

Zbigniew Zylicz Hildegard Hospiz, Basel Switzerland, Basel, Switzerland

David C. Yeomans Department of Anesthesia, Stanford University, Stanford, CA, USA

Robert P. Yezierski Comprehensive Center for Pain Research and Department of Neuroscience, The McKnight Brain Institute, University of Florida, Gainesville, FL, USA

Wei Yin Department of Anesthesiology, University of Washington, Seattle, WA, USA

Atsushi Yoshida Department of Oral Anatomy and Neurobiology, Osaka University, Osaka, Japan

Joanna M. Zakrzewska Eastman Dental Hospital UCLH NHS Foundation Trust, London, UK

Jonas E. Zault St. George's University, School of Medicine, Grenada, West Indies

Hanns Ulrich Zeilhofer Institute for Pharmacology and Toxicology, University of Zurich, and Swiss Federal Institute of Technology, Zurich, Switzerland

Leonie K. Zeltzer Pediatric Pain Program, Department of Pediatrics, David Geffen School of Medicine at UCLA, Los Angeles, CA, USA

Cleoniki Zoppi-Fella St. Clare Hospice, Hastingwood, UK

Xijun Zhang Department of Neuroscience, University of Minnesota, Minneapolis, MN, USA

Min Zhuo Department of Physiology, University of Toronto, Toronto, ON, Canada

C. Zöllner Department of Anesthesiology and Intensive Care Medicine, Charité Universität, Berlin, Germany

Irina Zezierska Genetic and Genomic Epidemiology Unit, P. Rill Cox Department of Obstetrics and Gynaecology, Wellcome Trust Centre for Human Genetics, Oxford, UK

Peter Zygmunt Department of Laboratory Medicine, Lund University, Division of Clinical Chemistry and Pharmacology, Lund, Sweden

Zbigniew Zylicz Hildegard Hospiz, Basel, Switzerland, Basel, Switzerland

Q

QOL

▶ Quality of Life

QST

▶ Quantitative Sensory Testing
▶ Quantitative Sensory Tests

QTL

▶ Quantitative Trait Locus Mapping

Quadratus Lumborum

Definition

Quadratus lumborum is the muscle involved in hip flexion and balancing postural distortion.

Cross-References

▶ Sacroiliac Joint Pain

Quality of Life

Synonyms

QOL

Definition

Quality of life (QOL) is defined as an individual's perception of their position in life, in the context of the culture and value systems in which they live and in relation to their goals, expectations, standards, and concerns. It is a broad concept, incorporating the person's physical health, level of independence, social relationships, personal beliefs, and their relationship to salient features of the environment in a complex way. QOL includes both positive and negative dimensions embedded in sociocultural and environmental contexts.

It has also been defined as "the subjective satisfaction expressed or experienced by an individual in his physical, mental and social situation," "the capacity of an individual to realize his life plans," and "the difference at a particular period in time between the hopes and expectations of the individual's present experience."

Cross-References

▶ Cancer Pain Management, Interface Between Cancer Pain Management and Palliative Care

G.F. Gebhart, R.F. Schmidt (eds.), *Encyclopedia of Pain*, DOI 10.1007/978-3-642-28753-4,
© Springer-Verlag Berlin Heidelberg 2013

- ▶ Central Pain, Outcome Measures in Clinical Trials
- ▶ Pain Treatment in Pediatric Palliative Care
- ▶ Psychiatric Aspects of the Management of Cancer Pain

Quantal Release Probability

Definition

Quantal release probability refers to the likelihood of vesicular release events of quanta of neurotransmitter occurring in a population of nerve terminals or a single nerve terminal, usually during invasion of an action potential. Drugs that have presynaptic actions on nerve terminals either increase or decrease quantal release probability.

Cross-References

- ▶ Opioid Electrophysiology in PAG

Quantification of Pain

- ▶ Central Pain: Outcome Measures in Preclinical Research

Quantitative Sensory Testing

Oliver H. G. Wilder-Smith
Pain Centre, Department of Anaesthesiology, Radboud University Nijmegen Medical Centre, Nijmegen, The Netherlands

Synonyms

Psychophysical Testing; QST

Definition

In the context of human pain, quantitative sensory testing (QST) is generally taken to involve the application of a variety of nociceptive stimuli to the body and the elicitation of voluntary verbal or nonverbal responses defining the subject's experience of the stimulus (Zaslansky and Yarnitsky 1998; Kosek and Ordeberg 2000; Wilder-Smith 2000a; Greenspan 2001; Graven-Nielsen and Arendt-Nielsen 2002). The responses to a given test stimulation depend on the entire neuraxis, thereby limiting its localizing value. QST should be distinguished from other forms of formal sensory testing entailing the direct measure (e.g., via electrical or magnetic signals) of involuntary responses to stimulation (e.g., nociceptive flexion reflexes, evoked or event-related potentials) (Dotson 1997).

QST explores nervous system sensory processing under standardized conditions in an intact organism. The methodology can be either stimulus dependent (e.g., threshold determination) or response dependent (e.g., generating a stimulus–response curve using pain intensity visual analogue scoring). Furthermore, QST can be either simple, i.e., involving only the response to test stimuli, or conditioned, i.e., involving the use of a conditioning stimulus to probe how pain processing is modulated.

Introduction

The historical origin of this method lies in the field of neurology and clinical neurophysiology, where it has been extensively used in research and clinical contexts since the 1980s. In this framework, QST has been used for the study and diagnosis of many types of disorders of the nervous system, with a particular emphasis on the sensory assessment of small fiber function in the context of neuropathies (for reviews, see Zaslansky and Yarnitsky 1998; Siao and Cros 2003).

The use of QST in the field of pain medicine – on which this contribution will concentrate – is more recent, and experience less extensive, than in the context of neurology. In contrast to neurological QST, which concentrates more on sensory fiber (dys)function, pain QST is interested in diagnosing alterations in pain processing such as increased (hyperalgesia) or decreased (hypoalgesia) sensitivity to pain (Wilder-Smith and Arendt-Nielsen 2006; Arendt-Nielsen and Yarnitsky 2009). Thus, pain QST seeks to answer scientifically and clinically relevant questions such as "is central sensitization present in this patient" or "is this patient able to generate effective inhibitory modulation of painful input."

Characteristics

Assessment and Induction Techniques

Two major types of responses (or endpoints) are typically used in QST. The first (stimulus dependence) entails the determination of thresholds (e.g., ▶ pain detection or pain tolerance threshold), while the second (response dependence) involves rating the magnitude of the pain (e.g., by ▶ visual analogue scale) produced by the given stimulus. The use of thresholds has the advantage of providing an easily defined – and thus generally reliable and reproducible – endpoint for the subject. Pain magnitude rating, while more difficult to teach and thus potentially less reliable, has the advantage that ▶ suprathreshold stimuli can be used. These are important for studying true nociceptive processing, frequently not reliably or well reflected using threshold or subthreshold stimuli. For optimal results, subjects must undergo some preliminary QST training in order to be able to detect or rate painful stimuli consistently and reliably, and the stimuli must be applied under standardized conditions (Zaslansky and Yarnitsky 1998; Rommel et al. 2001; Graven-Nielsen and Arendt-Nielsen 2002; Wilder-Smith et al. 2003; Arendt-Nielsen and Yarnitsky 2009). Under such conditions, QST results can be expected to remain relatively stable over days or weeks in normal subjects (Siao and Cros 2003). It must, however, be remembered that many factors influence QST results, including the technical characteristics of the devices and the testing algorithms (protocols) used.

As for all neurophysiological measures, QST is subject to large interindividual variability, which may in part be genetically determined (Flores and Mogil 2001; Lacroix-Fralish and Mogil 2009). This, combined with the present lack of large normal value databases, may make it difficult to reliably determine whether a given QST measurement is "normal" or "abnormal." As a result, methods using a given individual to provide his or her own normal measures are preferred for the analysis of QST. Examples of such methods include the determination of baseline values before predictable nociceptive events such as surgery, or comparison of affected limbs (or sides) with their normal opposite (Wilder-Smith et al. 2003; Rolke et al. 2006).

Choice of the type, characteristics, and location of the defined QST stimulus are important, as the options thus provided open the possibility of testing many different aspects of nociceptive processing. Stimulation using physiological stimuli (e.g., pressure, temperature, chemical) generally provides a means of including peripheral nociceptors in the analysis of nervous system processing. In contrast, electrical stimulation is considered to more or less bypass peripheral nociceptors and may thus be more indicative of central aspects of nociceptive processing (Zaslansky and Yarnitsky 1998). Dermatomal stimulation stimulates the most terminal fine nerve fibers (A delta or C fibers), a part of the nervous system bypassed if direct stimulation of larger sensory nerves (e.g., sural nerve) is used.

The most frequently stimulated tissue for QST is the skin, a superficial somatic structure, due to its easy accessibility. However, stimulation of other tissues, such as visceral (e.g., esophagus) or deep (e.g., muscle) structures, provides further different but fundamentally important types of topographical information (Arendt-Nielsen and Yarnitsky 2009).

The topography of changes in nervous system processing can also provide useful information on underlying processes. Thus, QST measurements performed at different sites may, e.g., provide the basis for differentiating between generalized (e.g., supraspinal) and segmental (e.g., spinal) changes in sensory processing (Wilder-Smith et al. 1996), or to differentiate changes in referred pain areas from those in the primary painful site (Kosek and Hansson 2003). A further example of topographical differences reflecting differing mechanisms is the differences in QST results in areas of ▶ primary and secondary hyperalgesia associated with a surgical wound, where the former reflects mainly local and peripheral nociceptive mechanisms and the latter, central mechanisms (Treede et al. 1992; Brennan 1999).

Simple QST involves single test stimuli applied to provide a basic assessment of sensitivity to nociceptive stimuli. Stimuli maybe summated either in time or space to provide more dynamic information on aspects of nociceptive processing relevant to clinical pain states such as postoperative pain or chronic pain diseases (e.g., ▶ windup, central sensitization) (Arendt-Nielsen et al. 1994, 1997). Conditioned QST is a further development and involves the investigation of changes in pain sensitivity induced by a variety of nociceptive conditioning stimuli, thus permitting insight into how the system defends itself against nociceptive challenge – and how vulnerable it may be to such a challenge. A classic example is the conditioned pain modulation (CPM) paradigm (Yarnitsky et al. 2010), in which a tonic painful conditioning stimulus (e.g., ▶ ice water bucket test) is applied to elicit diffuse noxious inhibitory controls (DNIC), a form of descending inhibitory modulation (Pud et al. 2009).

Use of QST in Practice

QST can be used in three major ways in the field of human pain and nociception:

1. To study the baseline processing of nociceptive stimuli by the (healthy) nervous system and how the nervous system modulates nociceptive input
2. To study how processing and modulation of nociceptive stimuli by the nervous system is altered by nociceptive processes or pain diseases (▶ nociceptive neuroplasticity) and their progression (e.g., development of chronic pain)

3. To study how pharmacological or therapeutic interventions affect nociceptive processing and modulation in health and disease

In the field of pain, QST is suitable for use for both research and clinical purposes. Typical research applications include the experimental exploration of nociceptive processing in health, the investigation of pain mechanisms in disease, and the study of how therapeutic interventions (both pharmaceutical and non-pharmaceutical) affect nociceptive processing. On the clinical side, QST provides information for diagnostic, prognostic, and therapeutic purposes. In pain diagnostics, the insight QST provides into nociceptive mechanisms makes it a basis for ▶ mechanism-based approaches to pain and nociception. Understanding the baseline sensitivity of the nervous system to nociception, and how nociceptive input is modulated, may provide clues as to the vulnerability of the person to future nociceptive challenge (e.g., surgery, trauma). On the pain therapeutic side, using QST to make nociceptive neuroplasticity visible can help achieve a mechanism-based choice of treatment method (e.g., drugs, physical stimulation), as well as being suitable for the documentation and study of treatment effects.

Examples

In the context of acute perioperative pain, QST has proven particularly valuable in providing insight into postoperative nociceptive neuroplasticity, its mechanisms, and the impact of various analgesic or antinociceptive therapeutic interventions (Wilder-Smith et al. 1996, 2003; Wilder-Smith 2000b). For example, using change in electric pain tolerance thresholds compared to preoperative baseline, it has been shown that surgery under "non-analgesic" anesthesia (unsupplemented isoflurane/nitrous oxide) is followed by long-lasting spinal sensitization, manifesting as generalized hyperalgesia 5 days after surgery (see Fig. 1). Another study has demonstrated that preemptive analgesia, with a single dose of fentanyl before back surgery lasting less than an hour, is able to prevent the appearance of such generalized hyperalgesia (central sensitization) 5 days postoperatively, while the similar

Quantitative Sensory Testing, Fig. 1 (a) Generalized neuroplasticity after back surgery: Time course of change in pain tolerance thresholds to transcutaneous electrical stimulation (mA, change is vs. preoperative baseline value) at site distant to surgery. (b) Segmental neuroplasticity after back surgery: Time course of change in normalized thresholds to transcutaneous electrical stimulation (ratio; i.e., threshold close to surgery divided by threshold distant to surgery, change is vs. preoperative baseline value). Bar values are means, thin lines are standard deviations. X-axis gives time after end of surgery in hours (h) or days (d).* = $P < 0.05$ vs. preoperative baseline value (Figure adapted from Wilder-Smith et al. 1996)

application of ketorolac is not (Wilder-Smith et al. 2003) (see Fig. 2).

Serial QST measurements have also proven useful in understanding why certain patients develop chronic pain after surgery, while others do not. In this context, it has been shown that after abdominal surgery, the patients who show persisting pain 6 months after surgery are those in whom secondary incisional and generalized hyperalgesia (a sign of rostral spread of central sensitization) persists after surgery (Wilder-Smith et al. 2010) (see Fig. 3). Furthermore, vulnerability to the development of chronic pain 6 months postoperatively is increased by the presence of poor inhibitory descending modulation preoperatively (Wilder-Smith et al. 2010), a finding also confirmed in other studies (Yarnitsky et al. 2008). Further studies have shown that better perioperative analgesia is not only associated with smaller areas of peri-incisional hyperalgesia in the early postoperative period but also links to less persistent pain up to 1 year later (Lavand'homme et al. 2005).

QST is now also starting to prove valuable in understanding the pathological neuroplasticity associated with chronic pain conditions (Kosek

Quantitative Sensory Testing, Fig. 2 Change at 5 days postoperatively vs. preoperative baseline value of pain tolerance threshold to transcutaneous electrical stimulation at site distant to surgery for patients undergoing back surgery and receiving placebo (saline i.v.), ketorolac (30 mg i.v.) or fentanyl (3 µg/kg i.v.) supplementation for isoflurane/nitrous oxide/oxygen general anesthesia. Bar values (mA) are means, standard deviations omitted for clarity.* = $P < 0.05$ vs. preoperative baseline value (Figure adapted from Wilder-Smith et al. 2003)

Quantitative Sensory Testing, Fig. 3 Percent change in electric pain tolerance thresholds (ePTT) 1 day to 6 months after abdominal surgery at peri-incisional (abdomen) and distant (leg) sites in patients with (yes) or without (no) chronic pain 6 months postoperatively. Values are means and 95 % confidence intervals (Adapted from Wilder-Smith et al. 2010)

and Ordeberg 2000; Rommel et al. 2001). An example of such use is given here in the context of CRPS type 1, where patients suffering from active, symptomatic disease had significantly lower pressure pain thresholds in the affected upper extremity as compared to the unaffected one, while there was no difference between the two extremities in patients with inactive, asymptomatic disease (Vaneker et al. 2005) (see Fig. 3).

Summary

Quantitative sensory testing is a useful method of assessing and quantifying changes in pain processing in health or disease or due to therapeutic interventions. From a research point of view, it can provide valuable insights into the mechanisms underlying pain disorders and their treatment; for clinical use, it is a simple-to-use bedside technique providing useful information concerning the prognosis, diagnosis, and therapeutics of acute and chronic pain diseases and their progression.

References

Arendt-Nielsen, L., & Yarnitsky, D. (2009). Experimental and clinical applications of quantitative sensory testing applied to skin, muscles and viscera. *The Journal of Pain, 10*, 556–572.

Arendt-Nielsen, L., Brennum, J., Sindrup, S., et al. (1994). Electrophysiological and psychophysical quantification of temporal summation in the human nociceptive system. *European Journal of Applied Physiology and Occupational Physiology, 68*, 266–273.

Arendt-Nielsen, L., Graven-Nielsen, T., Svensson, P., et al. (1997). Temporal summation in muscles and referred pain areas: An experimental human study. *Muscle & Nerve, 20*, 1311–1313.

Brennan, T. J. (1999). Postoperative models of nociception. *ILAR Journal, 40*, 129–136.

Dotson, R. M. (1997). Clinical neurophysiology laboratory tests to assess the nociceptive system in humans. *Journal of Clinical Neurophysiology, 14*, 32–45.

Flores, C. M., & Mogil, J. S. (2001). The pharmacogenetics of analgesia: Toward a genetically-based approach to pain management. *Pharmacogenomics, 2*, 177–194.

Graven-Nielsen, T., & Arendt-Nielsen, L. (2002). Peripheral and central sensitization in musculoskeletal pain disorders: An experimental approach. *Current Rheumatology Reports, 4*, 313–321.

Greenspan, J. D. (2001). Quantitative assessment of neuropathic pain. *Current Pain and Headache Reports, 5*, 107–113.

Kosek, E., & Hansson, P. (2003). Perceptual integration of intramuscular electrical stimulation in the focal and the referred pain area in healthy humans. *Pain, 105*, 125–131.

Kosek, E., & Ordeberg, G. (2000). Abnormalities of somatosensory perception in patients with painful osteoarthritis normalize following successful treatment. *European Journal of Pain, 4*, 229–238.

Lacroix-Fralish, M. L., & Mogil, J. S. (2009). Progress in genetic studies of pain and analgesia. *Annual Review of Pharmacology and Toxicology, 49*, 97–121.

Lavand'homme, P., De Kock, M., & Waterloos, H. (2005). Intraoperative epidural analgesia combined with ketamine provides effective preventive analgesia in patients undergoing major digestive surgery. *Anesthesiology, 103*, 813–820.

Pud, D., Granovsky, Y., & Yarnitsky, D. (2009). The methodology of experimentally induced diffuse noxious inhibitory control (DNIC)-like effect in humans. *Pain, 144*, 16–19.

Rolke, R., Baron, R., Maier, C., Tölle, T. R., Treede, R. D., Beyer, A., Binder, A., Birbaumer, N., Birklein, F., Bötefür, I. C., Braune, S., Flor, H., Huge, V., Klug, R., Landwehrmeyer, G. B., Magerl, W., Maihöfner, C., Rolko, C., Schaub, C., Scherens, A., Sprenger, T., Valet, M., & Wasserka, B. (2006). Quantitative sensory testing in the German research network on neuropathic pain (DFNS): Standardized protocol and reference values. *Pain, 123*, 231–243.

Rommel, O., Malin, J. P., Zenz, M., & Janig, W. (2001). Quantitative sensory testing, neurophysiological and psychological examination in patients with complex regional pain syndrome and hemisensory deficits. *Pain, 93*, 279–293.

Siao, P., & Cros, D. P. (2003). Quantitative sensory testing. *Physical Medicine and Rehabilitation Clinics of North America, 14*, 261–286.

Treede, R. D., Meyer, R. A., Raja, S. N., & Campbell, J. N. (1992). Peripheral and central mechanisms of cutaneous hyperalgesia. *Progress in Neurobiology, 38*, 397–421.

Vaneker, M., Wilder-Smith, O. H., et al. (2005). Patients initially diagnosed as 'warm' or 'cold' CRPS 1 show differences in central sensory processing some 8 years after diagnosis: A quantitative sensory testing study. *Pain, 115*, 204–211.

Wilder-Smith, O. H. (2000a). Changes in sensory processing after surgical nociception. *Current Review of Pain, 4*, 234–241.

Wilder-Smith, O. H. (2000b). Pre-emptive analgesia and surgical pain. *Progress in Brain Research, 129*, 505–524.

Wilder-Smith, O. H., & Arendt-Nielsen, L. (2006). Post-operative hyperalgesia: Its clinical importance and relevance. *Anesthesiology, 104*, 601–607.

Wilder-Smith, O. H., Tassonyi, E., Senly, C., et al. (1996). Surgical pain is followed not only by spinal sensitization but also by supraspinal antinociception. *British Journal of Anaesthesia, 76*, 816–821.

Wilder-Smith, O. H., Tassonyi, E., Crul, B. J., et al. (2003). Quantitative sensory testing and human surgery: Effects of analgesic management on postoperative neuroplasticity. *Anesthesiology, 98*, 1214–1222.

Wilder-Smith, O. H., Schreyer, T., Scheffer, G. J., & Arendt-Nielsen, L. (2010). Patients with chronic pain after abdominal surgery show less preoperative endogenous pain inhibition and more postoperative hyperalgesia: A pilot study. *Journal of Pain & Palliative Care Pharmacotherapy, 24*, 119–128.

Yarnitsky, D., Crispel, Y., Eisenberg, E., Granovsky, Y., Ben-Nun, A., Sprecher, E., Best, L. A., & Granot, M. (2008). Prediction of chronic post-operative pain: Pre-operative DNIC testing identifies patients at risk. *Pain, 138*, 22–28.

Yarnitsky, D., Arendt-Nielsen, L., Bouhassira, D., Edwards, R. R., Fillingim, R. B., Granot, M., Hansson, P., Lautenbacher, S., Marchand, S., & Wilder-Smith, O. (2010). Recommendations on terminology and practice of psychophysical DNIC testing. *European Journal of Pain, 14*, 339.

Zaslansky, R., & Yarnitsky, D. (1998). Clinical applications of quantitative sensory testing (QST). *Journal of Neurological Sciences, 153*, 215–238.

Quantitative Sensory Tests

Synonyms

QST

Definition

Quantitative sensory tests (QSTs) are techniques employed to measure the intensity of stimuli needed to produce specific sensory perceptions. They quantify nervous system input-response relations, allowing detection and quantification of nociceptive neuroplasticity. Quantitative sensory tests provide insight into nociceptive mechanisms, with the potential to provide the diagnostics for mechanism-based approaches to perioperative nociception and pain management. Quantitative sensory tests are used to evaluate a sensory detection threshold or other sensory responses from suprathreshold stimulation. The common physical stimuli are (1) touch pressure, (2) vibration, and (3) coolness, warmth, cold pain, and heat pain. Thermal somatosensory testing allows the clinician to test small nerve fibers. In this technique, thresholds for warmth, cold, heat-induced pain, and cold-induced pain are quantitatively measured and then compared to age-matched normal population values. A deviation from the normal range can indicate the existence of peripheral nerve disease. Vibratory sensory testing enables the clinician to test large nerve fibers and is useful as an aid in early diagnosis, follow-up, and therapy assessment. It measures sensory thresholds for vibration. These thresholds deviate from the normal range in diabetic neuropathy and peripheral nerve diseases. In a QST, the subject must be able to comprehend what is being asked by the test, be alert and not taking mind-altering medications, and not be biased to a certain test outcome.

Cross-References

► Diagnosis and Assessment of Clinical Characteristics of Central Pain
► Postoperative Pain, Acute Presentation of Complex Regional Pain Syndrome
► Postoperative Pain, Persistent Acute Pain

Quantitative Thermal Sensory Testing of Inflamed Skin

Louise Harding
Hospitals NHS Foundation Trust, University College London, London, UK

Definition

Psychophysical tests using hot and cold stimuli (or heat) to investigate inflammation evoked changes in small fiber pathways.

Characteristics

Quantitative thermal sensory testing gives an indirect measurement of the somatosensory pathways involved in communicating thermal signals. Thermal stimuli are detected and transduced by Aδ and C primary afferent fibers. Aδ cool fibers transmit the sensation of cold, and Aδ nociceptors relay fast sharp thermal pain. Warm C fibers convey innocuous warming signals, and C nociceptors mediate burning sensations. Through studying the temperature-sensation relationship of both innocuous and noxious

thermal sensations, sensory testing can be used to investigate changes in activity of these small fiber pathways in experimental and clinical conditions.

The methods used to investigate thermal sensation have changed over the years, as more and more sophisticated assessment tools have evolved. Many investigators use electronically controlled contact thermodes, which are held against the skin and programmed to heat or cool following specific operator set parameters. Alternatively, if only a heat stimulus is required, xenon lamps, argon, or CO_2 lasers can be used (Arendt-Nielsen and Bjerring 1988; Sumikura et al. 2003). These radiant heat methods cannot be manipulated and controlled as readily as contact thermodes, but as they do not touch the skin, they have the advantage of giving heat without also activating mechanosensitive fibers. The subject uses feedback systems, rating scales, and questionnaires to report thermal sensation information. By manipulating the rate, amplitude, and duration of thermal stimulus, a wealth of information about thermal sensibility can be recorded.

Inflamed skin, as anyone who has tried to take a hot shower with sunburn will know, is much more sensitive to heat. Using human models of inflammatory pain, quantitative thermal sensory testing can be used to characterize this heat hypersensitivity. The most prominent change in thermal sensitivity that develops in inflamed skin is a decrease in the heat pain threshold, i.e., the temperature at which a hot stimulus produces a sensation of pain. Heat pain thresholds can be reduced by as much as 8 °C in inflamed skin, from an average heat pain threshold of 43 °C to a normally lukewarm 36 °C. This reduction in heat pain threshold can be readily assessed using contact thermodes and a testing algorithm called the method of limits (Yarnitsky and Ochoa 1990). The essence of the test is as follows: the thermode is held against the skin at a neutral temperature and warmed at an operator-controlled rate. Subjects are requested to press a button the instance they feel the thermode stimulus change from "hot" to "pain." When the button is pressed, a feedback system switches off

Quantitative Thermal Sensory Testing of Inflamed Skin, Table 1 Heat pain thresholds in human models of experimental inflammation

Model	Heat pain threshold (°C)				
	Burn[a]	Capsaicin[b]	UV[c]	Freeze[b]	Mustard oil[d]
Baseline	44.5	42.5	44.5	42.5	42.1
Inflamed skin	39.7	33.7	36.7	37.1	36.5
Change in threshold	−4.8	−8.8	−7.8	−5.4	−5.6

[a]Pedersen and Kehlet 1998b
[b]Kilo et al. 1994
[c]Benrath et al. 2001
[d]Schmelz et al. 1996

the heating, and the arresting temperature is recorded as the heat pain threshold. The method is quick and simple with high repeatability. Decreases in threshold, which have been shown in a number of different experimental models of inflammatory pain, are shown in Table 1. The method used to evoke inflammation will affect the change in threshold. For instance, capsaicin, which directly sensitizes the heat responsive ion channels, tends to produce a bigger decrease in heat pain threshold than tissue damage models where sensitization is indirect, evoked by injury-released endogenous inflammatory mediators sensitizing heat-responsive ion channels.

The development of heat pain sensitization appears to follow a similar time frame to the development of other features of inflammation such as pain and erythema. This has been shown in the burn and capsaicin models, where decreased heat pain thresholds are evident within 5 min of application of the inflammatory stimulus. Heat pain thresholds can remain reduced for the duration of the inflammatory response, and generally return to pre-inflammation levels with the resolution of skin reddening. Some drop in heat pain threshold may, however, still be evident for many hours after blood flow has returned to normal (Benrath et al. 2001).

Inflammation produces changes in thermal sensitivity not only within the inflamed area but also in surrounding skin, in an area termed the secondary zone. Absolute heat pain thresholds are not changed in this area, but there is an increase in the perceived intensity of supra-pain threshold heat stimuli (Pedersen and Kehlet 1998a; Yucel et al. 2002). This change in perceived intensity can again be characterized with simple quantitative thermal sensory testing. Typically, a contact thermode or radiant heat source is programmed to give a supra-heat pain threshold stimulus, and the subject is instructed to rate the pain intensity of the stimulus using simple visual analogue scales (VAS) or 0–10 ordinal pain intensity scales. This type of assessment method is relatively crude, and consequently, only large changes in sensation intensity tend to be detected. Integrated electronic VAS systems, which capture intensity scores every tenth of a second, are currently available with many commercial contact thermode assessment machines. These greatly increase the sensitivity of the test, such that subtle changes in stimulus intensity can be identified. Electronic intensity scales also allow real-time intensity measures with changing temperature, thus enabling precise elucidation of stimulus response function of human perception of thermal stimuli.

While decreased heat pain thresholds are generally accepted as a peripheral event, heat hyperalgesia in skin surrounding the inflamed area is thought, by many, to be a central phenomena. Pharmacological studies showing differential modulation of sensory changes in the primary and secondary zones have helped clarify differences in mechanisms (Ilkjaer et al. 1997; Werner et al. 2001), although drug studies of secondary heat hyperalgesia are currently lacking. It has been suggested that heat hyperalgesia and secondary punctate hyperalgesia share common mechanisms. Like heat hyperalgesia, punctate hyperalgesia is characterized by an increase in the sharp pricking pain quality and is also thought to be predominantly Aδ mediated (Ziegler et al. 1999). Further research is required before full characterization of increased heat sensitivity in inflamed skin is complete.

References

Arendt-Nielsen, L., & Bjerring, P. (1988). Reaction times to painless and painful CO_2 and argon laser stimulation. *European Journal of Applied Physiology and Occupational Physiology, 58*, 266–273.

Benrath, J., Gillardon, F., & Zimmermann, M. (2001). Differential time courses of skin blood flow and hyperalgesia in the human sunburn reaction following ultraviolet irradiation of the skin. *European Journal of Pain, 5*, 155–167.

Ilkjaer, S., Dirks, J., Brennum, J., Wernberg, M., & Dahl, J. B. (1997). Effect of systemic N-methyl-D-aspartate receptor antagonist dextromethorphan on primary and secondary hyperalgesia in humans. *British Journal of Anaesthesia, 79*, 600–605.

Kilo, S., Schmelz, M., Koltzenburg, M., & Handwerker, H. O. (1994). Different patterns of hyperalgesia induced by experimental inflammation in human skin. *Brain, 117*, 385–396.

Pedersen, J. L., & Kehlet, H. (1998a). Secondary hyperalgesia to heat stimuli after burn injury in man. *Pain, 76*, 377–384.

Pedersen, J. L., & Kehlet, H. (1998b). Hyperalgesia in a human model of acute inflammatory pain: A methodological study. *Pain, 74*, 139–151.

Schmelz, M., Schmidt, R., Ringkamp, M., Forster, C., Handwerker, H. O., & Torebjörk, H. E. (1996). Limitation of sensitization to injured parts of receptive fields in human skin C-nociceptors. *Experimental Brain Research, 109*, 141–147.

Sumikura, H., Andersen, O. K., Drewes, A. M., & Arendt-Nielsen, L. (2003). Spatial and temporal profiles of flare and hyperalgesia after intradermal capsaicin. *Pain, 105*, 285–291.

Werner, M. U., Perkins, F. M., Holte, K., Pedersen, J. L., & Kehlet, H. (2001). Effects of gabapentin in acute inflammatory pain in humans. *Regional Anesthesia and Pain Medicine, 26*, 322–328.

Yarnitsky, D., & Ochoa, J. L. (1990). Studies of heat pain sensation in man: Perception thresholds, rate of stimulus rise and reaction time. *Pain, 40*, 85–91.

Yucel, A., Andersen, O. K., Nielsen, J., & Arendt Nielsen, L. (2002). Heat hyperalgesia in humans: Assessed by different stimulus temperature profiles. *European Journal of Pain, 6*, 357–364.

Ziegler, E. A., Magerl, W., Meyer, R. A., & Treede, R. D. (1999). Secondary hyperalgesia to punctate mechanical stimuli. Central sensitization to a-fibre nociceptor input. *Brain, 122*, 2245–2257.

Quantitative Trait Locus Mapping

Synonyms

QTL

Definition

Genetic linkage mapping as applied to quantitatively varying traits. In QTL mapping, a genetically segregating population (e.g., an *F2 intercross* or backcross) is "phenotyped" at a trait of interest and then "genotyped" at polymorphic DNA markers (e.g., microsatellites or single-nucleotide polymorphisms) of known genomic location. Statistical evidence of co-inheritance between phenotypic values and genotype at a marker or series of markers is sought, called *genetic linkage*. Once genetic linkage between the phenotype and marker genotype is established, one can infer the presence of a controlling gene or genes in the genomic vicinity of the marker.

Cross-References

▶ Opioid Analgesia, Strain Differences
▶ Single-Nucleotide Polymorphisms

Quotas

▶ Training by Quotas

R

α₂-Receptor Agonists

▶ Alpha(α) 2-Adrenergic Agonists in Pain Treatment

Race

▶ Ethics of Pain, Culture, and Ethnicity

Radial Arm Maze

Definition

A behavioral test used to measure spatial and working memory.

Radiant Heat

Definition

In the tail-flick test, the nociceptive stimulus is often a light beam focused on the tip of the tail.

Cross-References

▶ Tail-Flick Test

Radiant Heat Test

▶ Thermal Nociception Test

Radiant Stimulation

▶ Pain in Humans, Thermal Stimulation (Skin, Muscle, Viscera), Laser, Peltier, Cold (Cold Pressure), Radiant, Contact

Radiculalgia

▶ Radicular Pain, Diagnosis

Radicular Leg Pain

▶ Sciatica

Radicular Nerve Root Pain

Definition

Pain radiating from the back (usually from a slipped disk) to a distal part of the leg (or arm).

G.F. Gebhart, R.F. Schmidt (eds.), *Encyclopedia of Pain*, DOI 10.1007/978-3-642-28753-4,
© Springer-Verlag Berlin Heidelberg 2013

Cross-References

▶ Cancer Pain Management, Anesthesiologic Interventions, and Neural Blockade

Radicular Pain

Definition

Pain in a dynatomal distribution (the distribution of referred symptoms caused by spinal nerve root irritation), which may resemble the distribution of classic dermatomal maps for cervical nerve roots, but is not infrequently provoked outside of the distribution of these classic dermatomal maps.

Cross-References

▶ Cervical Transforaminal Injection of Steroids
▶ Radicular Pain, Diagnosis

Radicular Pain, Diagnosis

Jayantilal Govind
Pain Clinic, Department of Anaesthesia, University of New South Wales, Sydney, NSW, Australia

Synonyms

Brachialgia; Radiculalgia; Sciatica

Definition

Radicular pain is a form of pain caused by irritation of the sensory root or the dorsal root ganglion (DRG) of a spinal nerve.

Radicular pain is not nociceptive pain, for the neural activity arises from the dorsal root and not from stimulation of peripheral nerve endings.

Therefore, it is not synonymous with ▶ somatic pain or ▶ somatic referred pain and needs to be distinguished from them.

Nor is radicular pain synonymous with radiculopathy. Whereas radicular pain is caused by the generation of ectopic impulses, radiculopathy is caused by the blocking of conduction along sensory and motor axons and is characterized by loss of nerve function (Merskey and Bogduk 1994).

Characteristics

Pathophysiology
The introduction of computerized tomography (CT) and ▶ magnetic resonance imaging (MRI) has brought into question the belief that a simple mass effect adequately accounts for the mechanism of lumbar radicular pain. Studies have shown that in patients whose symptoms of sciatica had resolved, still showed the same mass effect on serial CTs, while conversely disc herniations evident on CT or MRI may not be associated with either low back or lumbar radicular pain (Bogduk and Govind 1999). In animal studies, compression of a normal nerve root causes only a momentary discharge, too brief to account for radicular pain (Howe et al. 1977).

Ectopic impulses evoked by the compression of an uninjured DRG produce a train of sustained discharges mediated via the A-δ, C, and the A-β fibers (Howe et al. 1977), which may account for the distinctive quality of radicular pain – its shooting or electric nature. Hence, a mechanical basis for radicular pain can be accommodated, provided that the dorsal root ganglion is regarded as the source of pain.

However, experiments in laboratory animals (Howe et al. 1977) and in postsurgical patients (Smith and Wright 1959; Kuslich et al. 1991) have shown that radicular pain can be generated from nerve roots, provided that they have been previously damaged. Thus, for lumbar radicular pain, it is feasible that two distinctive but not mutually exclusive mechanisms prevail.

On the one hand, nerve roots that are chronically compressed may become sensitized to mechanical stimulation. While the exact mechanism is not known, the pathogenesis might include partial damage to the axon, neuroma in continuity, or focal demyelination, intraneural edema, and impaired microcirculation, each of which is capable of generating ectopic impulses from the affected nerve (Devor 1996).

The second explanation invokes a chemically mediated noncellular inflammatory process – a type of "chemical radiculitis." Studies have shown the nucleus pulposus to be inflammatogenic and leukotactic (Olmarker et al. 1995); that injecting phospholipase A_2 (PLA2) into rat sciatic nerve produces demyelination, axonal injuries, and increased macrophage phagocytic activity; and that the resulting mechanical hyperalgesia correlates with PLA2 immunoreactivity (Kawakami et al. 1996). In the absence of nerve compression, applying nucleus pulposus in the epidural space causes a delay in the nerve conduction velocity of nerve roots, and methylprednisolone injected intravenously within 20 h prevents this reduction in conduction velocity (Olmarker et al. 1994).

The evidence would suggest that, unlike the DRG, for the dorsal nerve root to be a source of radicular pain due to disc herniation, an injury either mechanically or chemically induced is a prerequisite.

With respect to cervical radicular pain, there are no equivalent experimental data. One study showed that applying a sufficiently strong stimulus to a normal dorsal nerve root is always followed by a peripheral radiation of pain, but gentle stimulation of dorsal roots previously affected by compressive lesions evoked a sensation of pain or paresthesia (Fryckholm 1951). Herniated cervical intervertebral discs have been shown to produce nitric oxide, metalloproteinases, interleukin-6, and prostaglandin E_2 (Kang et al. 1995), but their role as possible mediators of nerve root inflammation has not been fully ascertained.

Clinical Features

Radicular pain has certain distinctive qualitative features. It is not dull or achy in quality. Where radicular pain has been produced experimentally, in all instances it has been perceived as a shooting or lancinating pain (Smith and Wright 1959; Kuslich et al. 1991), and unlike somatic pain, radicular pain has both a cutaneous and deep quality. Radicular pain can be episodic, recurrent, or paroxysmal.

In the case of lumbar and sacral radicular pain, the pain is felt in the lower limb when L4, L5, S1, or S2 nerve roots are involved. Pain is typically felt along the back of the thigh, into the leg, and into the foot. Pain does not follow the corresponding dermatome. Lumbar radicular pain travels through the lower limb along a narrow band usually not more than 5–8 cm wide (Smith and Wright 1959). While it may be felt throughout the entire length of the lower limb, it is more commonly experienced below the knee than above the knee.

With respect to the cervical spine, typical patterns of radicular pain have been mapped (Slipman et al. 1998), but cervical radicular pain and cervical somatic referred pain have similar distributions. Hence, the pattern of pain cannot be used to determine its cause or mechanism. Pain over the shoulder girdle and upper arm can be either somatic referred pain or radicular pain, but pain in the forearm and hand is unlikely to be somatic referred pain and is more likely to be radicular in nature (Bogduk 1999). Pain that radiates into the upper limb, and is shooting electric in quality, is bound to be radicular in origin.

Etiology

A large number of lesions have been reported to cause radicular pain (Bogduk and Govind 1999; Bogduk 1999). Systematically, any abnormality of the vertebral column, or any space-occupying lesion of spinal tissues, that impinges on a spinal nerve or its roots may cause radicular pain. These include osteophytes, cysts, and tumors of the vertebral column; cysts and tumors of the spinal cord, nerve roots, and their dural sheaths; tumors of epidural fat and blood vessels; and infections, infestations, or metastatic deposits in the vertebral column or vertebral canal. However, herniation of an intervertebral disc is the single most common cause, followed by foraminal stenosis due to osteoarthrosis of the zygapophysial joints.

The differential diagnosis includes intrinsic disorders of the dorsal root or its axons, such as diabetic neuropathy, postherpetic neuralgia, and various, rare dorsal root ganglionopathies.

Diagnosis

Radicular pain must be distinguished from somatic referred pain. Although not an absolute distinction, radicular pain is sharp, shooting, or lancinating in quality, and topographically it must involve a region beyond the spine. Imaging studies, including computerized tomography and magnetic resonance imaging, may assist in determining both the putative cause and the segmental level. Features of radiculopathy including absent reflexes, weakness, numbness, or muscle wasting may be evident clinically. Electrophysiological tests for the investigation of patients with acute radicular pain are generally not helpful (Bogduk and Govind 1999; Bogduk 1999).

Treatment

Physicians often prescribe what is commonly referred to as "conservative therapy" for radicular pain. Typically, this consists of exercises, traction, analgesics, and other measures, such as collars for cervical radicular pain. Multiple studies have shown that these interventions confer no benefit greater than that of the natural history of the condition (Bogduk and Govind 1999; Bogduk 1999).

Traditionally, the mainstay for treatment of radicular pain has been surgery, in the form of laminectomy and microdiscectomy for lumbar radicular pain and foraminotomy for cervical radicular pain. Studies have shown that surgery does not achieve better long-term outcomes than conservative therapy, but surgery has the distinct advantage of providing prompt relief when used for patients with severe pain that does not respond to other measures (Bogduk and Govind 1999; Bogduk 1999).

Other interventions that have been advocated and tested for radicular pain include epidural injection of steroids and ▶ transforaminal injection of steroids.

Acknowledgement Dr. Jayantilal Govind passed away on June 20, 2009, at age 86. We thank Prof. Nik Bogduk for the update of his entry.

References

Bogduk, N. (1999). *Medical management of acute cervical radicular pain: An evidence-based approach.* Newcastle: Newcastle Bone and Joint Institute.

Bogduk, N., & Govind, J. (1999). *Medical management of acute lumbar radicular pain: An evidence-based approach.* Newcastle: Newcastle Bone and Joint Institute.

Devor, M. (1996). Pain arising from nerve root and dorsal root ganglion. In J. N. Weinstein & S. L. Gordon (Eds.), *Low back pain: A scientific and clinical overview* (pp. 187–208). Rosemont, IL: American Academy of Orthopaedic Surgeons.

Fryckholm, R. (1951). Cervical nerve root compression resulting from disc degeneration and root-sleeve fibrosis: A clinical investigation. *Acta Chirurgica Scandinavica Supplement, 160,* 10149.

Howe, J. F., Loeser, J. D., & Calvin, W. H. (1977). Mechanosensitivity of dorsal root ganglia and chronically injured axons: A physiological basis for the radicular pain of nerve root compression. *Pain, 3,* 25–41.

Kang, J. D., Georgescu, H. I., McIntyre-Larkin, L., Stanovic-Racic, M., & Evans, C. H. (1995). Herniated cervical intervertebral discs spontaneously produce matrix metalloproteinases, nitric oxide, interleukin-6 and prostaglandin E_2. *Spine, 22,* 2373–2378.

Kawakami, M., Tamaki, T., Weinstein, J. N., Hashizume, H., Nishi, I., & Meller, S. T. (1996). Pathomechanism of pain-related behaviour produced by allografts of the intervertebral disc in the rat. *Spine, 21,* 2101–2107.

Kuslich, S. D., Ulstrom, C. L., & Michael, C. J. (1991). The tissue origin of low back pain and sciatica: A report of pain response to tissue stimulation during operations on the lumbar spine using local anaesthesia. *Orthopedic Clinics of North America, 22,* 181–187.

Merskey, H., & Bogduk, N. (Eds.). (1994). *Classification of chronic pain. Description of chronic pain syndromes and definitions of pain terms* (2nd ed.). Seattle: IASP Press.

Olmarker, K., Blomquist, J., Stromberg, J., Nanmark, U., Thomsen, P., & Rydevik, B. (1995). Inflammatogenic properties of nucleus pulposus. *Spine, 20,* 665–669.

Olmarker, K., Byrod, G., Cornefjord, M., Nordberg, B., & Rydevik, B. (1994). Effects of methylprednisolone on nucleus pulposus induced nerve root injury. *Spine, 19,* 1803–1808.

Slipman, C. W., Plastaras, C. T., Palmitier, R. A., Huston, C. W., & Sterenfeld, E. B. (1998). Symptom provocation of fluoroscopically guided cervical nerve root stimulation: Are dynatomal maps identical to derma-tomal maps? *Spine*, *23*, 2235–2242.

Smith, M. J., & Wright, V. (1959). Sciatica and the intervertebral disc. An experimental study. *Journal of Bone & Joint Surgery, 40A*, 1401–1418.

Radiculopathies

Paolo Marchettini

Pain Medicine Center, Scientific Institute San Raffaele, Milan, Italy

Synonyms

Back pain; Cauda equina; Lumbago; Neck pain; Root disease; Sciatica

Definition

▶ Radiculopathy is a pathological condition affecting the spinal roots of peripheral nerves (see ▶ spinal root disease). The term has no implication for the mechanism causing damage to the root, which may be mechanical compression/ischemia, inflammation, or primary or metastatic tumor. Spondyloarthritis is by far the most common cause of radiculopathy. Radiculopathy, independent of the cause, is, in the majority of cases, a painful medical condition. Besides pain, root injury may also cause weakness and sensory disorder. Weakness is obvious in the limbs, while abnormal sensation may also be detected in the chest and abdomen. Patients with radiculopathy tend to seek the opinion of a specialist in orthopedic surgery, neurology, or neurosurgery, or, when pain prevails, of a pain specialist or an anesthesiologist.

Characteristics

Clinical Features
Pain and related sensory motor symptoms in radiculopathy may appear suddenly (acute) or develop slowly with progressive increase,

followed by improvement and relapse (chronic). Acute radiculopathy is almost exclusively caused by sudden compression of a root by a herniated intervertebral disk. In some cases, the root is squeezed inside the radicular pocket by a disk fragment herniated into the foramina. The compressed root undergoes ischemia and also swelling that worsens the ischemia. Moderate compression causes venous congestion ultimately responsible for the swelling. The periradicular veins and meningeal veins have no valves (Baxter plexus) and minimal increases in vein pressure worsen periradicular vein congestion ultimately responsible for pain worsening. Pain in radiculopathy is usually reported to be very intense, interfering with common simple daily activities and sleep.

Pain is both nociceptive, originating from the ruptured ligament, the perineural sheath, and the meninges, and neuropathic, originating from ectopic impulses generated by mechanical compression and ischemia of nociceptive afferents (small myelinated and myelinated). The nociceptive pain originates at root level and spreads above or below along the spine. The neuropathic pain radiates into the innervation territory of the root. Sometimes, nociceptive pain might radiate into the thigh (pseudoradicular pain). The most reliable clinical test of root involvement in such cases is the Lasègue sign. Quite often, pain in lumbosacral radiculopathy is worsened by maneuvers that increase spinal fluid or vein pressure, such as sneezing, coughing, and voiding. Such maneuvers more rarely increase neck and upper limb pain in cervical radiculopathy. Transient mechanical compression of the cord causes sensations of an "electric shower" known as the Lhermitte sign. Lhermitte's sign, which is usually not painful, may be provoked by forced extension or flexion of the neck. Lateral rotation of the head in the opposing direction of an elevated arm may cause root stretching in cervical radiculopathy (Spurling sign). Elevation of a lower limb kept straight in the supine position causes lumbosacral root stretching (Lasègue sign). Spurling and Lasègue signs are commonly painful, evoking both local pain (in the neck or back) and radicular pain. The single root stimulated by such maneuvers may be readily identified by the selective anatomical distribution of the radicular dysesthesias. Neuropathic sensory symptoms and neuropathic pain are projected into the root's anatomical territory. This anatomical territory consists of the cutaneous and also the bone and muscle structures innervated by that root (territory of positive sensation). It is a territory wider than the area of cutaneous hypesthesia that may ensue following complete root section (territory of negative sensation) (Marchettini 1993; Marchettini and Ochoa 1994). Many patients are good "witnesses" and clearly perceive and describe the difference between the pain originating from ligaments (purely nociceptive) and the pain originating from root compression (nociceptive and neuropathic).

The typical medical history of radiculopathy in disk herniation is of preceding neck or back pain for many days or even months, which over time becomes very intense and more focal. The back pain suddenly disappears and is followed within hours by the radicular pain. Radicular pain combines nociceptive neck or back pain of various qualities, with neuropathic pain projected into the limb. The ▶ neck pain, which is due to meningeal and cervical root compression, radiates over a wide area in the neck, and is often referred into the scapular area. The equivalent back pain radiates into the buttocks. The neuropathic radicular pain combines large fiber positive symptoms (tingling and numbness) and small fiber positive symptoms (pins and needles from small myelinated fibers, burning, deep soreness, squeezing, and cramp-like from unmyelinated fibers). Axonal degeneration is not confined to the site of root compression, but extends to the primary sensory neurons within the dorsal root ganglion as a result of the axon reaction. Therefore, sensory loss may be long lasting even after root decompression.

When the disk pinches the root, any movement elongating it worsens the pain. This leads to reflex paraspinal muscle contraction to avoid movement and to the search for a comfortable position reducing radicular stretching. This position is usually neck and back hyperextension and limb flexion, particularly the lower limb. To further avoid radicular stretching, patients tend to sleep on their side. In the ▶ cauda equina syndrome, short distance walking evokes symptoms of neuropathic pain and paresthesias in the legs and buttocks. Curiously, biking or walking, or any lower limb activity practiced while bending forward, may be symptom-free. Probably, this happens because tilting forward reduces intraspinal pressure and interferes less with radicular blood flow. Rare alternative mechanical causes of single segment radiculopathy or cauda equina syndrome are epidural varices (Genevay et al. 2002), synovial cysts (Yarde et al. 1995), or even perineural cysts filled with cerebrospinal fluid (Tarlov 1938). Primary root tumors are schwannomas and neurinomas. Spinal roots may also be compressed by bone metastasis or, more rarely, invaded by leptomeningeal metastasis. The latter event is becoming more common because of spinal fluid sequestration of metastatic cells in patients surviving systemic chemotherapy.

History

Until the mid part of the last century, intervertebral disk herniation, which is the most common cause of radiculopathy, was hard to distinguish among the multiple other nerve root diseases caused by tuberculosis, syphilis, and cancer. Radiculopathy caused by disk compression was often recognized late when it reached the paralyzing state and thus carried a poor general prognosis. Nevertheless, in the late 1940s, Guillain and Barré polyradiculoneuropathy, which is a rare condition, was considered the most common "reversible" or "curable" root disease (Henri Roger quoted by Bénard 1966). Historical note, Cotugno, a master of the Salernitan School, is classically credited with having been the first to describe the characteristic distribution of radicular pain in the leg that in 1764 he named "▶ sciatica" (ischias postica or morbus ischiaticus). However, it is not until 1934 that Mixter and Barr (quoted by Parisien and Ball 1998) established a relationship between rupture of the intervertebral disk and lumbosacral radiculopathy, an observation that took about two additional decades to gain wider recognition. This knowledge led to enthusiastic, aggressive root decompression through surgical laminectomy that consequently produced a new iatrogenic entity, the "failed back surgery syndrome". The second most common cause of radiculopathy in spondyloarthritis, the intermittent claudication of the cauda equina, was recognized later by Blau and Logue in 1961.

Anatomy

Nerve roots are made of four to five voluminous fascicles that ramify distally and give rise to the peripheral nerve plexus and peripheral nerves. Radicular fascicles already possess anatomical segregation, each one hosting fibers of well-defined peripheral territories. Thus, partial root compression affecting one or two fascicles may mimic a peripheral nerve injury. Radicular fascicles are separated by epineural septa and embedded in the connective tissue of the perineurium. From the radicular pocket inward into the spinal canal, the perineurium fuses with the dura mater and is surrounded by the arachnoid that keeps the nerve root bathed in cerebrospinal fluid. Meninges and perineurium are sensitive structures, innervated by nociceptive nerve endings. As a consequence, inflammation and trauma of nerve roots always cause pain of mixed quality with irritation of the meninges and perineurium (nociceptive pain) and of radicular axons (neuropathic pain). Cervical and thoracic roots cross the intervertebral foramen corresponding to their anatomical level, while the lumbar and sacral roots show a progressive downward shift, due to the termination of the conus medullaris approximately at the first lumbar vertebral body. As a consequence, the cauda equina's lumbar and sacral roots first travel downward within the vertebral canal and then reach the intervertebral foramen by taking off sideways with a variable angle (mean 40° for the lumbar and 23° for the sacral roots). Because of this, they are more exposed to ischemia than thoracic or cervical roots. At the L4–L5

intervertebral level, the L5 root is situated anterolaterally, displacing the S1 root laterally, while the lower sacral roots are positioned dorsally. Therefore, the most common disk protrusion, the L4–L5, may hit the L5 or S1 root depending on the median or lateral side of its bulging or herniation. Nerve roots receive a vascular supply from peripheral and central sources. However, they are not served by a regional segmental supply. In the case of the cauda equina, ischemia is a considerable risk because the majority of the arteries are "end" arteries without effective anastomotic connections. The neural structure of the cauda equina occupies about half (44 %) of the cross-sectional area of the spinal space. Constriction of the spinal canal increases the spinal fluid pressure on the roots by about 50 mmHg for each third reduction from normal. As a consequence, narrowing of the spinal canal severely affects the blood supply to the cauda equina.

Diagnostic Work-Up

Plain radiographs have been the traditional first-choice imaging technique, and many clinicians still routinely prescribe them in back pain patients. It is well recognized, however, that particularly in the early stages of disk disease or other vertebral diseases, they have a low yield (McNally et al. 2001). Plain radiographs may show indirect signs of disk degeneration through reduction of the intervertebral space and osteoarthritic changes. However, such signs appear only in the advanced stages of spondyloarthritis. Their main utility is to rule out more worrisome causes of back pain and radiculopathy such as infection or metastasis. A loss of about 80 % of vertebral bone may be required before a destructive lesion can be visualized with this method. In addition, plain radiographs cannot define the cause of nerve root compression. CT scanning allows proper examination of horizontal planes and explores roots and disks that are best visualized when degenerative changes from gas formation and calcification begin to appear. The gas, also known as vacuum phenomenon, was first observed by Fick in 1910; in 1942, Knutsson described its radiographic characteristics and it was first analyzed by Ford and coworkers (1977), who reported that the gas was 90–95 % nitrogen. CT scans, however, may completely miss early disk pathology. MRI, eventually with limited time-saving protocols, is becoming the method of first choice to explore radiculopathy.

One has to be aware that imaging techniques cannot predict clinical outcome. In the case of CT scans, for example, it has been shown that none of the features of disk herniation has any significant correlation with the pain prognosis. On the contrary, larger herniation or the presence of free disk fragments may be more common in patients who go on to have good outcomes (Beauvais et al. 2003). The same probably applies to MRI.

Another prognostic approach, electromyography, does not reveal all structural or clinical radiculopathy. However, its sensitivity may reach 92 % (Dillingham and Dasher 2002). In lumbosacral radiculopathy, electromyographic screening of four to five leg and paraspinal muscles may provide valuable information in 89 % of cases of root disease. Electromyography is indicated when differential diagnosis is considered. The most common examples are polyradiculoneuropathy (Guillain-Barré), plexopathy, multineuropathy, and the overall assessment of the patient to examine peripheral nerve function when systemic diseases such as severe diabetes or uremia coexist. Electromyography is also recommended as a presurgical assessment before performing laminectomy, particularly in the presence of muscle weakness.

Therapy

The majority of patients heal spontaneously without needing invasive procedures. Pain control is the only medical need in most cases. However, the clinical course in radiculopathy is quite variable. Although symptoms usually improve within 2 weeks, in a conspicuous minority of patients, pain continues for months and at times years. In the first 2 months, about 60 % of the patients have a marked improvement in back and leg pain. By 1 year, one third are still complaining of back or

leg pain. Due to the frequent spontaneous recovery, there is controversy on the effectiveness of physical therapy, exercise, and pharmacological treatment. When conservative treatment fails, transforaminal corticosteroid administration is recommended (level of evidence 2B+) in subacute lumbosacral radicular pain below L3 level. Radiofrequency treatment of the spinal ganglion (DRG) can be considered (2C+) in chronic lumbosacral radiculopathy. Refractory radicular pain could also be treated (2B+) with adhesiolysis and epiduroscopy. Surgery is usually decided upon within the first year. Large and migrated disk herniations decrease spontaneously more than disk protrusions or small contained herniations. Morphologic changes (reduction in protrusion or even hernia reentry) may be observed; however, they usually follow rather than anticipate the clinical improvement. Patients resistant to therapy and affected by failed back surgery syndrome are candidate for spinal cord stimulation (2A+).

In addition, clinical improvement may be quite remarkable in the absence of morphological improvement. New minimally invasive percutaneous techniques have been proposed to reduce the risk of chronic arachnoiditis following back surgery. However, there are no clinical trials of percutaneous diskectomy technique to provide definite evidence supporting the efficacy or advantages of the procedure. The success rate is 41 % for percutaneous diskectomy versus 40 % for the conventional open method (Haines et al. 2002). There is controversy regarding the use of epidural steroids. The only agreement regarding its efficacy is in decreasing the symptoms of the acute phase (Boswell et al. 2005). There is no clear evidence that epidural steroids alter the natural history of radiculopathy or the morphology of herniation.

Cross-References

▶ Cervical Transforaminal Injection of Steroids
▶ Pain Treatment: Spinal Cord Stimulation
▶ Whiplash

References

Bénard, H. (1966). Eulogy of henri roger (1860–1946). *Bulletin de l'Académie Nationale de Médecine, 150*(32), 651–656.

Beauvais, C., Wybier, M., Chazerain, P., et al. (2003). Prognostic value of early computed tomography in radiculopathy due to lumbar intervertebral disk herniation. A prospective study. *Joint Bone Spine, 70*, 134–139.

Boswell, M. V., Shah, R. V., Everett, C. R., et al. (2005). Interventional techniques in the management of chronic spinal pain: Evidence-based practice guidelines. *Pain Physician, 8*, 1–47.

Dillingham, T. R., & Dasher, K. J. (2002). The lumbosacral electromyographic screen: Revisiting a classic paper. *Clinical Neurophysiology, 111*, 2219–2222.

Ford, L. T., Gilula, L. A., Murphy, W. A., & Gado, M. (1977). Analysis of gas in vacuum lumbar disc. *American Journal of Roentgenology, 128*, 1056–1057.

Genevay, S., Palazzo, E., Huten, D., et al. (2002). Lumboradiculopathy due to epidural varices: Two case reports and a review of the literature. *Joint, Bone, Spine, 69*, 214–217.

Haines, S. J., Jordan, N., Boen, J. R., et al. (2002). Discectomy strategies for lumbar disc herniation: Results of the LAPDOG trial. *Journal of Clinical Neuroscience, 9*, 411–417.

Marchettini, P. (1993). Muscle pain: Animal and human experimental and clinical studies. *Muscle & Nerve, 16*, 1033–1039.

Marchettini, P., & Ochoa, J. L. (1994). The clinical implications of referred muscle pain sensation. *American Pain Society Journal, 3*, 10–12.

McNally, E. G., Wilson, D. J., & Ostlere, S. J. (2001). Limited magnetic resonance imaging in low back pain instead of plain radiographs: Experience with first 1000 cases. *Clinical Radiology, 56*, 922–925.

Parisien, R. C., & Ball, P. A. (1998). William Jason mixter (1880–1958). Ushering in the "dynasty of the disc". *Spine, 23*, 2363–2366.

Tarlov, I. M. (1938). Perineural cysts of the spinal root. *Archives of Neurology & Psychiatry, 40*, 1067–1074.

Yarde, W. L., Arnold, P. M., Kepes, J. J., et al. (1995). Synovial cysts of the lumbar spine: Diagnosis, surgical management, and pathogenesis report of eight cases. *Surgical Neurology, 43*, 459–465.

Radiculopathy

▶ Lower Back Pain, Physical Examination

Radiofrequency

Definition

An electric current oscillating at high frequencies which generates heat and causes lesions when delivered in a focused manner, e.g., through an electrode.

Cross-References

▶ Facet Joint Pain

Radiofrequency Ablation

Definition

The use of high-frequency current to generate heat and thus destroy tissue.

Cross-References

▶ Cancer Pain Management, Neurosurgical Interventions
▶ Facet Joint Procedures for Chronic Back Pain

Radiofrequency Denervation

Definition

Procedure by which a nerve is coagulated by heating the tip of an electrode to 80 °C, with consequent loss of nociceptive transmission.

Cross-References

▶ Whiplash

Radiofrequency Lesion

▶ Radiofrequency Neurotomy, Electrophysiological Principles

Radiofrequency Neurotomy

▶ Lumbar Medial Branch Blocks

Radiofrequency Neurotomy, Electrophysiological Principles

Jayantilal Govind
Pain Clinic, Department of Anaesthesia,
University of New South Wales, Sydney, NSW,
Australia

Synonyms

Heat lesion; Radiofrequency lesion; RF; Thermal neuroablation; Thermocoagulation

Definition

Radiofrequency (RF) neurotomy is a means of treating pain, by which the nerves from a source of pain are coagulated using an electrode, through which a high-frequency electrical current is passed, in order to heat the tissues immediately surrounding the tip of the electrode.

Characteristics

RF neurotomy is not electrocautery. It is the creation of a targeted, therapeutic, thermal lesion achieved by passing a low-energy, high-frequency, alternating current (100,000–500,000 Hz) between the small surface area on the uninsulated tip of an (active) electrode and the large surface area of a ground plate (the dispersive electrode) that is applied to a remote area of the body. The concentration of current in tissues around the active electrode achieves sufficient density to denature those tissues by heating them.

In pain medicine, RF neurotomy is used to coagulate peripheral nerves, in order to block nociceptive information from a source that is responsible for a patient's pain. Similar techniques are used to produce lesions in the central nervous system for the relief of ▶ central pain.

Electrophysiological Principles

Figure 1 illustrates the pattern of the electric current lines in the body between the active and dispersive electrode. The current leaving the large area of the dispersive plate is concentrated onto the small area of the exposed tip of the electrode.

The alternating current produced by the generator passes through the body and agitates charged molecules in the tissues, thereby heating them. Where the current converges and concentrates towards the active tip of the electrode, current density is greater and heating is greater.

Heating occurs as a result of agitation of charged molecules in the tissues and the resultant friction between them. Thus, the tissue adjacent to the electrode becomes the source of heat, not the electrode itself. In proportion to the current density, temperatures are higher closer to the electrode. Accordingly, around the tip of the electrode, isotherms can be depicted (Fig. 2).

If tissues are heated sufficiently, they coagulate. The volume of coagulated tissue assumes the shape of a "prolate spheroid" (Organ 1976), whose long axis is formed along the uninsulated tip of the electrode (Fig. 3).

Creation and Size of Lesion

The principal factors that govern temperature equilibrium between the electrode and the heated tissue are distance from the tip (r), current intensity (I), and duration (t) of application (Lord and Bogduk 2002). Tissue heating (T) varies with the square of the current intensity (I^2); it decreases rapidly away from the tip by a factor of $1/r^4$ and increases with the duration

Radiofrequency Neurotomy, Electrophysiological Principles, Fig. 1 The electric field produced by a radiofrequency (*RF*) generator, between a ground plate on the body surface and an electrode introduced into the body. The field lines converge on the tip of the electrode

Radiofrequency Neurotomy, Electrophysiological Principles, Fig. 2 The isotherms of a radiofrequency current. As the electric field (*E*) oscillates, tissues are heated in proportion to the density of the field lines. Around the tip of the electrode, a temperature gradient is established, with temperatures highest at the surface of the electrode, but progressively less away from the electrode. The isotherms represent regions with the same temperature. The 65° isotherm is labelled

of application. Mathematically, the relationship is (Lord and Bogduk 2002)

$$T \propto \left[\frac{I^2}{r^4}\right] \times \log(Kt)$$

where K is a constant.

Sufficient time must be allowed for some of the heat to be transmitted to the periphery of the lesion volume. Optimal current should generate a lesion of a maximum volume for the

size of the electrode tip. The rapid application of high current, however, may result in solidification and char formation, which in turn will impede current flow and heating. Due to the increased resistance, lesion size becomes submaximal.

As the temperature of the electrode reaches 60 °C, coagulation commences on the surface of the electrode tip (Lord and Bogduk 2002; Lord et al. 1988; Bogduk et al. 1987). The volume of coagulated tissue expands as the temperature reaches 80 °C. At this temperature, the lesion is initially about 60 % of maximal achievable size. If temperature is maintained at 80 °C for 30 s, the lesion increases to 85 % size. After 60 s, it reaches 94 % maximal size. The growth of the lesion asymptotes at 90 s. Maintaining the current for longer than this does not achieve appreciable increases in the size of the lesion (Bogduk et al. 1987). Accordingly, the optimal duration for the maximal coagulation lies between 60 and 90 s.

Temperature monitoring is critical to safety. During coagulation, the temperature should be increased slowly from body temperature to 80 °C. A rate of 1 °C per second is recommended. This slow increase allows patients to report any untoward sensations and allows the operator to abort the lesion before any appreciable damage is done. Maintaining the temperature under 85 °C not only ensures creating a lesion of maximal volume but also avoids the risk of boiling, charring, and gas or steam formation, which can occur when higher temperatures are applied

Radiofrequency Neurotomy, Electrophysiological Principles, Fig. 3 The shape of a radiofrequency lesion. Tissues are heated in a more radial direction from the tip of the electrode than in a distal direction. On average, the effective radius (r) of the lesion is about twice the diameter, or less, of the electrode's width (ew)

Radiofrequency Neurotomy, Electrophysiological Principles, Fig. 3 The shape of a radiofrequency lesion. Tissues are heated in a more radial direction from the tip of the electrode than in a distal direction. On average, the effective radius (r) of the lesion is about twice the diameter, or less, of the electrode's width (ew)

and which compromise both the safety and the efficacy of the lesion made.

During coagulation, impedance should remain reasonably constant as power and temperature are increased. Erratic behavior of the impedance and fluctuation of temperature indicate either faulty equipment or connections, or dissipation of heat into tissues such as blood or other fluids.

Electrodes with larger exposed tips produce larger lesions. Longer exposed tips produce elongated, elliptical lesions (Organ 1976; Alberts et al. 1966). Coagulation occurs principally in a radial direction perpendicular to the long axis of the electrode, i.e., sideways (Lord and Bogduk 2002; Lord et al. 1988; Bogduk et al. 1987). Depending on the shape of the point of the electrode, relatively less coagulation occurs distal to the point.

The dimensions of lesions generated are proportional to the size of the electrode used and can be normalized and expressed in terms of electrode widths (Lord et al. 1988). In a radial direction, large-diameter (1.6 mm) electrodes coagulate tissues up to 1.6 ± 0.3 (mean $\pm$ sd) electrode widths away from the surface of the electrode. Distally, tissues up to 0.4 ± 0.2 electrode widths from the tip of the electrode are coagulated. With electrodes of smaller diameter (0.7 mm), the radial range is 2.3 ± 0.4 electrode widths, and the distal range is 1.4 ± 0.4 electrode widths.

These figures indicate that electrodes must be placed accurately, i.e., very close to the target nerve, in order to ensure that it is coagulated. Since the electrode-width of smaller electrodes is less than 1 mm, the electrode will, on average, coagulate nerves within 2 mm of its surface. However, there is a 17.5 % chance of failing to coagulate the nerve if it is not within 1.3 mm (mean size of lesion minus one sd) of the electrode. Larger electrodes coagulate tissues within 2.6 mm from their surface, with a one standard deviation lower limit of 2 mm.

Electrodes placed perpendicular to the target nerve may miss coagulating the nerve completely (Fig. 4a). For optimal coagulation, the electrode must be placed parallel to the nerve (Lord and Bogduk 2002; Lord et al. 1988; Bogduk et al. 1987). Creating multiple, parallel lesions compensates for possible inaccuracies in placing the electrode exactly on the target nerve. However, to be effective those lesions must be centered no further than one electrode-width from another. At greater displacements, nerves may escape coagulation because of the circular, cross-sectional shape of the lesions generated (Fig. 4b).

Physiology

By coagulating neural tissue, RF neurotomy creates a mechanical barrier to the transmission of nociceptive traffic. The denatured proteins in the nerve are incapable of transmitting action potentials across the denatured zone.

Earlier views that radiofrequency neurotomy selectively destroys $A\delta$ and C fibers (Letcher and Goldring 1968) have not been corroborated. Experiments have established that radiofrequency neurotomy results in indiscriminate destruction of

Radiofrequency Neurotomy, Electrophysiological Principles, Fig. 4 The problems of matching a radiofrequency electrode (*e*) and its lesion to the target nerve (*n*). (**a**) Top view and (**b**) transverse view of an electrode placed perpendicularly onto a nerve. Although the electrode rests on the nerve, the lesion it makes is largely proximal to the nerve, for little lesion is produced distal to its tip. As a result, the nerve may escape complete coagulation. (**c**) Top view and (**d**) transverse view of electrodes placed parallel to the target nerve. Since the electrodes coagulate transversely, the nerve is more likely to be incorporated into the lesion. (**e**) Top view and (**f**) transverse view of an electrode placed in two positions on a target nerve. If the two placements are not more than two electrode widths apart, their lesions overlap and maximize the likelihood of the nerve being coagulated thoroughly

small unmyelinated, small myelinated, and large myelinated fibers (Smith et al. 1981), including alpha motor neurones (Dreyfuss et al. 2000).

Recovery from RF neurotomy is not a simple matter of nerve regeneration. Before regeneration can occur, the coagulated proteins have to be removed and replaced and the affected segment of nerve reconstituted. This takes considerably longer time than conventional regeneration of a nerve after it has been cut.

Indications
The classical indication for RF neurotomy is ▶ trigeminal neuralgia. Otherwise, RF neurotomy can, in principle, be applied to any peripheral nerve that has been shown to be responsible for mediating a patient's pain. However, the procedure has only been validated in the context of certain forms of spinal pain.

Pain arising from the lumbar or cervical zygapophysial joints can be treated by performing RF neurotomy of the nerves that innervate those joints. The cardinal indication for this therapy is complete relief of pain following controlled, ▶ lumbar medial branch blocks or ▶ cervical medial branch blocks, respectively.

Experimental applications include lumbar discogenic pain and certain forms of sacroiliac joint pain (Lord and Bogduk 2002).

Side Effects and Complications
In principle, the side effects of RF neurotomy are due to the loss of function of the nerve that is coagulated. If the nerve has a cutaneous distribution, numbness will occur in that distribution. Hyperesthesia may complicate this denervation. If the nerve has a major proprioceptive function, impaired proprioception should be expected. When motor fibers are coagulated, muscle denervation will occur.

The generic complications that may occur are those associated with any percutaneous procedure: infection, hematoma, allergy, and unintended damage to structures adjacent to the target nerve. Potential complications specific to RF neurotomy are electrical in nature and include burns resulting from damaged insulation of the electrode or from incorrect application of the dispersive plate. In particular, alternatives to plate electrodes, such as spinal needles, should not be used as the ground electrode. In that event, current concentrates around the needle and will cause burns along its entire length.

Efficacy

To be consistent with the rationale for the procedure, complete relief of pain should be the standard outcome. If diagnostic blocks produce complete relief, so should RF. If complete relief is not achieved, either the patient was incorrectly selected or the procedure may have been technically imperfect.

Specific data, from observational studies and from controlled studies, are available for ▶ cervical medial branch neurotomy, for ▶ lumbar medial branch neurotomy, and for trigeminal neurotomy. Early data are available for sacroiliac neurotomy. These are provided in the sections dealing with these procedures. For other applications of RF neurotomy, comparable data are not available.

Acknowledgement Dr. Jayantilal Govind passed away on June 20, 2009, at age 86. We thank Prof. Nik Bogduk for the update of his entry.

References

Alberts, W. W., Wright, E. W., Feinstein, B., & von Bonin, G. (1966). Experimental radiofrequency brain lesion size as a function of physical parameters. *Journal of Neurosurgery, 25*, 421–423.

Bogduk, N., MacIntosh, J., & Marsland, A. (1987). Technical limitations to the efficacy of radiofrequency neurotomy for spinal pain. *Pain, 20*, 529–535.

Dreyfuss, P., Halbrook, B., Pauza, K., Joshi, A., McLarty, J., & Bogduk, N. (2000). Efficacy and validity of radiofrequency neurotomy for chronic lumbar zygapophyseal joint pain. *Spine, 25*, 1270–1277.

Letcher, F. S., & Goldring, S. (1968). The effect of radiofrequency current and heat on peripheral nerve action potential in the Cat. *Journal of Neurosurgical Sciences, 29*, 42–47.

Lord, S. M., & Bogduk, N. (2002). Radiofrequency procedures in chronic pain. Best practice and research. *Clinical Anaesthesiology, 16*, 597–617.

Lord, S. M., McDonald, G. J., & Bogduk, N. (1988). Percutaneous radiofrequency neurotomy of the cervical medial branches: A validated treatment for cervical zygapophyseal joint pain. *Neurosurgery Quarterly, 8*, 288–308.

Organ, L. W. (1976). Electrophysiologic principles of radiofrequency lesion making. *Applied Neurophysiology, 39*, 69–76.

Smith, H. P., McWhorter, J. M., & Challa, V. R. (1981). Radiofrequency neurolysis in a clinical model: Neuropathological correlation. *Journal of Neurosurgery, 55*, 246–253.

Radiofrequency Rhizotomy

Definition

Treatment of TN by an injury produced by a radiofrequency current, applied through a needle placed in the space around the sensory (Gasserian) ganglion containing the cell bodies of the sensory fibers in the trigeminal nerve.

Cross-References

▶ Trigeminal, Glossopharyngeal, and Geniculate Neuralgias

Radiography

▶ Plain Radiography

Radioisotope

Definition

An isotope that is radioactive, i.e., one having an unstable nucleus, which gives it the property of decay by one or more of several processes.

Cross-References

▶ Cancer Pain Management Radiotherapy

Radioligand Binding

Definition

A method for detecting the distribution of receptors in tissue sections, based on the use of radiolabeled compounds that bind specifically to the receptor. These can then be revealed with autoradiography.

Cross-References

▶ Opioid Receptors at Postsynaptic Sites

Radiopharmaceuticals

Definition

Radioisotopic medications that emit radiation for diagnostic or therapeutic purposes including bone cancer pain management.

Cross-References

▶ Adjuvant Analgesics in Management of Cancer-Related Bone Pain

Radiosurgical Rhizotomy

Definition

Treatment of TN by a mild injury produced by irradiation of the intracranial portion of the trigeminal nerve using focused, MRI-guided radiation, using either cobalt (gamma ray) or a linear accelerator as a source.

Cross-References

▶ Trigeminal, Glossopharyngeal, and Geniculate Neuralgias

Radiotherapy

Definition

Treatment of disease by ionizing radiation.

Cross-References

▶ Adjuvant Analgesics in Management of Cancer-Related Bone Pain

Raeder's Paratrigeminal Syndrome

Definition

A combination of pain, ipsilateral oculosympathetic defect, and ipsilateral trigeminal dysfunction; a paratrigeminal oculosympathetic syndrome that indicates pathology in the middle cranial fossa ipsilateral to the pain.

Cross-References

▶ Headache Due to Dissection

Ramus

Definition

Ramus refers to a branch or a projecting part.

Cross-References

▶ Facet Joint Pain

Randall-Selitto Paw Pressure Test

Valerie Kayser
NSERM U677 91Bd de l'Hôpital, Paris, France

Definition

Rat-appropriate mechanical assay based on the use of short-duration stimuli (in the order of seconds). In the course of this pain test, a pressure of increasing intensity is applied to a punctiform area on the hindpaw or, far less commonly, on the tail. The monitored reactions range from paw withdrawal reflexes (the rat withdraws its paw) to more complex organized unlearned behaviors (escape or vocalization). The measured parameter is the threshold (weight in grams) for the appearance of a given behavior. Tests using constant pressure have been abandoned progressively for those applying gradually increasing pressures.

Characteristics

The Randall-Selitto paw pressure test (Randall and Selitto 1957) is a sensitive assay, able to show effects for the different classes of antinociceptive agents, at doses comparable to those used for analgesics in humans (refs. in Guilbaud et al. 1999). Its predictability is strongly related to situations in which one is trying to understand the basic mechanisms underlying pain and in terms of identifying analgesic molecules. Results obtained with this assay are reproducible not only within the same laboratory but also between different laboratories. Its advantage is a precise stimulus application (e.g., right vs. left paw). Further, the test allows simultaneous analysis of the ▶ threshold of response to noxious stimulation of a spinal reflex (▶ withdrawal of the limb), on the one hand, and a centrally processed reaction (▶ vocalization), on the other (Kayser and Christensen 2000). The method has been helpful especially for assessing mechanical hypersensitivity under varied experimental conditions, such as during different phases in localized inflammation (Kayser and Guilbaud 1987; Kayser et al. 1998). It requires, however, restraining of the rat by hand. Thus, the animals cannot control the intensity or duration of the stimulus. From a physiological point of view, it seems essential that three parameters in this test can be controlled with some precision by the experimenter: the intensity, the duration, and the surface area of stimulation. These three parameters determine the global quantity of nociceptive information that will be carried to the central

nervous system by the peripheral nervous system. In healthy rats, this type of mechanical stimulation has a certain number of disadvantages. Firstly, repetition of the stimulus can produce a diminution, or conversely an increase, in the sensitivity of the stimulated part of the body; in the latter case, this carries the risk that the tissues may be altered by inflammatory reactions that could call into question the validity of repeated tests. Secondly, there is the necessity of applying relatively high pressures. Lastly, there is a non-negligible level of variability of the responses. With the aim of improving the sensitivity of the test, Randall and Selitto (1957) proposed comparing the thresholds observed with a healthy paw and with an inflamed paw, based on the principle that inflammation increases the sensitivity to pain and that this increased sensitivity is susceptible to modification by analgesics. The inflammation was induced beforehand by a subcutaneous injection into the area to be stimulated of substances such as croton oil, beer yeast, or carrageenin, the last of these being the most commonly used today. Even though it was found that the sensitivity of the method was improved, it was to the detriment of its specificity, because two different pharmacological effects, analgesic and anti-inflammatory, could be confused. However, a comparison in the same animal of responses triggered from a healthy and an inflamed paw allows this problem to be overcome: nonsteroidal anti-inflammatory drugs are inactive on the former, but do increase the lowered vocalization threshold when pressure is applied to the latter (Winter and Flataker 1965). The mechanical stimulation used differs notably from the application of von Frey filaments, often almost revered by neurologists, but which has the disadvantage of activating low-threshold mechanoreceptors as well as nociceptors. There are also technical difficulties in applying mechanical stimuli in freely moving rats.

In practice, as illustrated in Fig. 1, the paw or tail (Kayser et al. 1996) is jammed between a plane surface and a blunt point mounted on top of a system of cogwheels with a cursor that can be displaced along the length of a graduated beam. These devices permit the application of

Randall-Selitto Paw Pressure Test, Fig. 1 Mechanical stimulation is made using the procedure of Randall and Selitto (1957), with a Ugo Basile analgesimeter: a linearly increasing force is applied via a dome-shaped plastic tip ($\varnothing$ = 1 mm) onto the surface of the hindpaw. The force causing paw withdrawal is quoted, as well as that inducing vocalization (audible squeak)

constantly increasing pressure and the interruption of the assay when the threshold is reached. When the pressure increases, one can see successively the reflex withdrawal of the paw, a more complex movement whereby the animal tries to release its trapped limb, then a sort of struggle (animal tries to escape, not always observed), and thereafter vocalization (an audible squeak) (Winter and Flataker 1965). The stimulus intensity necessary to elicit paw withdrawal from mechanical stimulation, struggle reaction, or vocal response from the animal is determined. If the first of these monitored reactions is undoubtedly a proper spinal reflex, although under influence of descending supraspinal inputs, the last two clearly involve supraspinal structures (Winter and Flataker 1965). It was firmly established in experiments in which the effect of interruption of the anterolateral quadrant of the rat cervical spinal cord, as well as partial lesion of the lateral ventrobasal complex of the thalamus has been determined, using paw withdrawal and vocalization threshold to paw pressure as the endpoints. Both lesions led to a large enhancement of the vocalization threshold to paw pressure but did not alter the paw withdrawal threshold (Kayser et al. 1985). Other studies have shown that systemic

low doses of morphine, that strongly depressed neuronal thalamic but not spinal responses to noxious stimuli in the rat, produced marked effects in the first but not in the second of these two tests (Kayser and Guilbaud 1990; Kayser 1994).

In our laboratory, a mechanical stimulus was applied using the Ugo Basile analgesimeter (Apelex). This instrument, controlled by a pedal, generates a linearly increasing pressure applied through a dome-shaped plastic tip (1 mm diameter) onto the dorsal surface of the paw. The rat is gently held by the trunk, leaving the head and limbs freely exposed. Testing sessions are conducted in a quiet room. Rats are randomly assigned to groups of five or six for a given series of tests and are not acclimatized to the test situation beforehand. Force is applied until the rat withdraws its limb (withdrawal threshold to paw pressure) and/or squeaks (vocalization threshold to paw pressure). The withdrawal reflex that occurs before vocalization is prevented by smoothly holding the rat's hindpaw in position under the pusher until vocalization. The two paws are tested consecutively in each rat. The sequence of sides is alternated between the animals to prevent "order" effects. For each rat, a control threshold (mean of two consecutive stable thresholds expressed in grams) is determined before injecting the drug. After drug administration, nociceptive pressure thresholds are measured every 5 or 10 min, until they have returned to the level of the control values. Thus, each rat is its own control. Results are then expressed in grams or as a percentage of the control values. Based on the principle that the application of the stimulus must not produce lesions, one often defines a limit for how long the animal should be exposed to the stimulus (the cutoff level; around 600 g, Fletcher et al. 1997). This limit is absolutely necessary when the intensity of the stimulus is increasing for the prevention of skin damage. In some cases, the threshold of another reaction, ▶ struggle, is noted, but this response, although considered by some authors as reliable (see discussion in Winter and Flataker 1965), is difficult to assess and therefore appears unsuitable for pharmacological studies.

In the paw withdrawal threshold test, the pressure for a brisk motor response is used as the endpoint. This simple reflex measure permits the animal to have control over stimulus magnitude and thus ensures that the animal can control the level of pain. Reflex responses, however, suffer from a number of limitations as measures of pain behavior. They are a measure of reflex activity and not pain sensation. Changes in reflex activity can result from alterations in motor as well as sensory processing. In addition, drugs that affect motoneuron or muscle function will alter reflex latencies in a manner similar to analgesic drugs. Vocalization is another commonly used reaction to painful stimuli. Using the vocalization threshold to paw pressure test, antinociceptive effects of analgesic compounds have been elicited using low doses, devoid of obvious side effects such as drowsiness or locomotor deficit (Christensen et al. 1998). Thus, the test may be useful when assessing the activity of newly designed analgesic substances (Kayser et al. 2003). This approach also appears to be a good tool to determine the optimal use of peripheral versus systemic administration of analgesic compounds (refs. in Kayser and Guilbaud 1994; Guilbaud et al. 1999). Finally, the effects of compounds on the two nociceptive thresholds can be systematically compared in order to explore the relative contribution of spinal versus supraspinal mechanisms to their antinociceptive effects (Kayser and Christensen 2000; Aubel et al. 2004).

References

Aubel, B., Kayser, V., Mauborgne, A., Farre, A., Hamon, M., & Bourgoin, S. (2004). Antihyperalgesic effects of cizolirtine in diabetic rats: Behavioral and biochemical studies. *Pain, 110*, 22–32.

Christensen, D., Idänpään-Heikkilä, J. J., Guilbaud, G., & Kayser, V. (1998). The antinociceptive effect of combined systemic administration of morphine and the glycine/NMDA receptor antagonist, (+)-HA966 in a rat model of peripheral neuropathy. *British Journal of Pharmacology, 125*, 1641–1650.

Fletcher, D., Kayser, V., & Guilbaud, G. (1997). The influence of the timing of bupivacaine infiltration induced by two carrageenin injections seven days apart. *Pain, 69*, 303–309.

Guilbaud, G., Kayser, V., Perrot, S., & Keita, H. (1999). Antinociceptive effect of opioid substances in different models of inflammatory pain. In E. Kalso, H. McQuay, & Z. Wiesenfeld-Hallin (Eds.), *Opioid sensitivity of chronic noncancer pain, progress in pain research and management* (Vol. 14, pp. 201–223). Seattle: IASP Press.

Kayser, V. (1994). Endogenous opioid systems involved in the modulation of pain: Behavioural studies in rat models of persistent hyperalgesia. In G. F. Gebhart, D. L. Hammond, & T. S. Jensen (Eds.), *Progress in pain research and management, proceedings of the 7th world congress on pain* (Vol. II, pp. 553–568). Seattle: IASP Press.

Kayser, V., & Christensen, D. (2000). Antinociceptive effect of systemic gabapentin in mononeuropathic rats, depends on stimulus characteristics and level of test integration. *Pain, 88*, 53–60.

Kayser, V., & Guilbaud, G. (1987). Local and remote modifications of nociceptive sensitivity during carrageenin-induced inflammation in the rat. *Pain, 28*, 99–108.

Kayser, V., & Guilbaud, G. (1990). Differential effects of various doses of morphine and naloxone on two nociceptive test thresholds in arthritic and normal rats. *Pain, 41*, 353–363.

Kayser, V., & Guilbaud, G. (1994). Peripheral aspects of opioid activity: Studies in animals. In J. M. Besson, G. Guilbaud, & H. Ollat (Eds.), *Peripheral neurons in nociception: Physiopharmacological aspects* (pp. 137–156). Paris: John Libbey Eurotext.

Kayser, V., Peschanski, M., & Guilbaud, G. (1985). Neuronal loss in the ventrobasal complex of the rat thalamus alters behavioural responses to noxious stimulation. In M. J. Field (Ed.), *Advances in pain research and therapy* (Vol. 9, pp. 277–284). New York: Raven.

Kayser, V., Berkley, K. J., Keita, H., Gautron, M., & Guilbaud, G. (1996). Estrous and sex variations in vocalization thresholds to hindpaw and tail pressure stimulation in the rat. *Brain Research, 742*, 352–354.

Kayser, V., Idänpän-Heikkilä, J. J., & Guilbaud, G. (1998). Sensitization of the nervous system, induced by two successive hindpaw inflammations, is suppressed by a local anesthetic. *Brain Research, 794*, 19–27.

Kayser, V., Farré, A., Hamon, M., & Bourgoin, S. (2003). Effects of the novel analgesic, cizolirtine, in a rat model of neuropathic pain. *Pain, 104*, 169–177.

Randall, L. O., & Selitto, J. J. (1957). A method for measurement of analgesic activity on inflamed tissue. *Archives Internationales de Pharmacodynamie et de thérapie, CXI*, 409–419.

Winter, C. A., & Flataker, L. (1965). Reaction thresholds to pressure in edematous hindpaws of rats and responses to analgesic drugs. *The Journal of Pharmacology and Experimental Therapeutics, 150*, 165–171.

Randomized

Definition

Randomized allocation to treatment, which requires the use of a list of random numbers usually generated by a computer program, so that each successive subject has an equal chance of being assigned to each treatment or treatment sequence.

Cross-References

▶ Antidepressants in Neuropathic Pain
▶ Central Pain, Pharmacological Treatments

Randomized Controlled Trial

Synonyms

RCT

Definition

An epidemiologic experiment in which subjects in a population are randomly allocated into groups to receive an experimental intervention or not.

Cross-References

▶ Lumbar Traction

RAP

▶ Recurrent Abdominal Pain

Raphe Nuclei

Definition

The major source of serotonin production in the brainstem. Neurons in these midline nuclei have descending projections to spinal cord as well as ascending projections. The Nucleus Raphe Magnus (NRM) which is largely congruent with the Rostral Ventromedial Medulla (RVM) is the best known of these nuclei in terms of pain in view of its involvement in descending modulation of pain.

Cross-References

▶ Molecular Contributions to the Mechanism of Central Pain

Rapid Eye Movement Sleep

Synonyms

REM sleep

Definition

The rapid eye movement (REM) phase of sleep is associated with dreaming, nocturnal emissions, enuresis, and the onset of some headaches that may awaken the subject.

Cross-References

▶ Hypnic Headache

Rapidly Adapting Responses

Definition

Rapidly adapting responses refer to the fast acclimation of peripheral afferents to incoming information about peripheral stimuli that are responsible for the perception of localized movement along the skin, for example. The stimulus usually produces only a short-lasting response.

Cross-References

► Postsynaptic Dorsal Column Projection, Functional Characteristics

RAS

► Reticular Activating System

Rat Tail Model

► Neuropathic Pain Model, Tail Nerve Transection Model

Rating Impairment Due to Pain in a Workers' Compensation System

Henry Stockbridge
University of Washington School of Public Health and Community Medicine, Seattle, WA, USA

Synonyms

Disability rating; Evaluation of permanent impairment; Impact on activities of daily living; Impairment due to pain; Impairment rating; Workers' compensation

Characteristics

Introduction

Workers' compensation systems around the world have many similarities and even more differences. In the USA, each state has its own laws governing many aspects of workers' compensation, ranging from broad, overarching principles to minute details affecting day-to-day operations, including calculation of the monetary award for impairment due to pain (Barth and Niss 1999; Cocchiarella and Lord 2002, personal communication 2003).

A key function of workers' compensation systems is the provision of a monetary award, to compensate injured workers for any permanent impairment resulting from work-related conditions. Healthcare providers are often given the responsibility of quantifying the impairment, so that adjudicators can calculate the appropriate monetary payment. In many cases, quantifying impairment is easy. For example, an uncomplicated amputation of a digit can be straightforward. However, quantifying impairment can also be highly complex. When pain contributes to impairment, the process of quantifying impairment can require special knowledge and skills.

The distinction between the two terms "disability" and "impairment" is important for an understanding of the differences among workers' compensation systems, especially with respect to the issue of impairment due to pain. The terms are defined slightly differently in different workers' compensation systems, but "disability" generally refers to the inability to perform a specific task or job, while impairment is the loss of function of an organ or part of the body (Medical Examiners' Handbook 2000).

For example, if a classical pianist and a truck driver both lose a finger, both would have the same impairment and receive the same award, if the law is written to compensate impairment. However, their disabilities would be different, because the truck driver would be able to continue in his or her job, and the pianist would not. If the law is written to compensate disability rather than impairment, the pianist would receive a significantly higher monetary award.

Although disability and impairment can be distinguished conceptually, the distinctions between them become ambiguous in certain types of disorders, such as mental illnesses.

Also, workers' compensation systems often contribute to the ambiguities surrounding the terms by using them interchangeably.

This entry will illustrate the range of approaches to the problem of rating impairment/disability associated with pain by describing two very different systems: the workers' compensation systems of California and Washington State.

California Workers' Compensation System

In the current system in California, monetary awards for permanent residuals from work-related injuries are based on disability rather than impairment. The role of pain in the rating process is understood more easily in light of this legal underpinning. In 1917, California's Workers' Compensation Insurance and Safety Act added a provision that, when rating permanent disability, "consideration be given to the diminished ability of such an injured employee to compete in the open labor market." In 1951, the California State Senate decided that this language "imposed the obligation to make... due allowance for such disabling subjective factors as, for example, pain, discomfort, and psychiatric or mental disturbances, provided, of course, the subjective factor or factors be considered to be of a permanent nature" (Physician's Guide 2001, personal communication 2003).

In fact, the California Constitution (Article XIV, Section 4) states that a complete system of workers' compensation must include "... adequate provisions for the comfort, health and safety and general welfare of any and all workers ... to the extent relieving from the consequences of any injury or death incurred or sustained by workers in the course of their employment...."

Instructions for healthcare providers on this issue are presented in the Physician's Guide: Medical Practice in the California Workers' Compensation System (Physician's Guide 2001). The Physician's Guide is explicit in stating that pain should be considered in the rating process. It says, "Some impairments can be defined in purely medical terms and can be objectively measured.... Impairment also applies to less easily measured functions such as emotional stability, loss of concentration, or pain that causes a handicap in performing activities of work or daily living."

The Physician's Guide goes further, actually to require consideration of pain. It says: "There are four categories of information, called factors of disability, that must be covered in a disability evaluation: objective factors; subjective factors; work restrictions; loss of pre-injury capacity." It also states: "Although a genuine human experience, pain is not readily measurable or objectively validated. Consequently, the evaluation, management, and disability evaluation of pain disorders depend on the assessment of an individual's verbal and non-verbal pain behavior. This presumes a reasonable degree of expertise and competence in the field.... Although fraud and malingering are uncommon, over-elaboration, symptom magnification, and embellishment often characterize pain complaints. The credibility of the injured worker must be assessed.... The disability rating of pain should address ... the extent to which the pain interferes with the performance of work related tasks.... Subjective factors can be the only ratable factor for an injured worker...."

The Physician's Guide instructs the doctor to list subjective complaints in one section of the medical-legal report using the worker's own words. The doctor then describes subjective disability in a separate section, to express the doctor's judgment as to the intensity and frequency of pain. This judgment is based on complaints, diagnosis, physical exam findings, test results, history, credibility assessment, etc. An example of a complete description of subjective factors is: "Occasional minimal pain at rest, increasing to intermittent slight to moderate pain on repetitive fingering activities, such as key-stroking continuously for over 30 min." The doctor's role is to provide a narrative description, which is translated by a state disability evaluator into a numerical rating using defined techniques.

The bottom line for the injured worker is the monetary award, which is based on the worker's percentage of permanent disability. For example, in California, a worker with 100 % permanent total disability may receive $602 per week for life. The California system permits even 100 % disability awards to be made essentially, entirely on the basis of subjective disability (although this is rare). An example is a 100 % disability award for constant, severe back pain.

Washington State Workers' Compensation System

In contrast to California, Washington State law requires that monetary awards for permanent residuals from work-related injuries be based on impairment rather than disability (although the two terms often seem to be used interchangeably in Washington workers' compensation). Impairment is rated using primarily two systems: the Washington State Category Rating System (CRS) and the American Medical Association Guides to the Evaluation of Permanent Impairment (the Guides) (American Medical Association Guides to the Evaluation of Permanent Impairment 2001). The latter is generally used for conditions of the upper and lower extremities, hearing loss, and loss of vision. The former is used for most other conditions (disorders of the spine, digestive tract, respiratory tract, and so forth). The process and method of rating impairment is described in some detail in a publication of the Washington State Department of Labor and Industries, the Medical Examiners' Handbook (2000).

Until 2001, neither of these two systems provided a method for quantifying impairment due solely to pain in the absence of objective findings. Each of these systems included consideration of pain within the context of individual organ systems, such as with peripheral neuropathies. However, they did not provide a method for quantifying pain in excess of what was already included in those organ systems.

In 2001, the American Medical Association published the Fifth Edition of the Guides, which included an entire chapter, Chap. 18, providing a method for rating impairment associated with pain.

When the Fifth Edition was published, it was necessary to determine whether the new method in Chap. 18 conflicted with existing laws contained in the Washington State Industrial Insurance Act.

In particular, in Washington State, there is a strong emphasis on objective findings to support medical opinions and adjudicative decisions. Since pain is by definition subjective, it was necessary to determine if Chap. 18 conflicted with this and other aspects of the legal foundation of the system.

After considerable legal analysis, it was determined that Chap. 18 of the Fifth Edition of the AMA's Guides to the Evaluation of Permanent Impairment could not be used to calculate awards for permanent impairment under Washington's Industrial Insurance Act. The decision was published in Provider Bulletin 02–12, "Rating Permanent Impairment," (2002), which includes the rationale saying, in part:

"This rule is in response to a new chapter in the AMA's Guides to the Evaluation of Permanent Impairment, Chap. 18, which proposes to rate pain in excess of what is already considered in other chapters of that document. Chapter 18 fails to segregate a worker's subjective areas of pain already taken into consideration in the Category rating system... and in the organ and body systems of the AMA's Guides to the Evaluation of Permanent Impairment. This will result in overlapping PPD awards, a double recovery, for the same impairment. Further, there have been no studies performed demonstrating that the methods outlined in the new chapter are valid or reliable. The American Medical Association itself has issued two subsequent publications highlighting the confusion and lack of clarity surrounding Chap. 18. In addition, Chap. 18 of the Guides conflicts with existing law, and may result in similarly situated injured workers receiving different PPD awards for the same degree of impairment.... The amount of the permanent partial disability award is not dependent upon or influenced by the economic impact of the occupational injury or disease on an individual worker. Rather, Washington's Industrial Insurance Act requires that permanent partial disability be established primarily by objective physical or clinical findings establishing a loss of function.... The impairment caused by the worker's pain complaints is already taken into consideration in the categories and in the organ and body system ratings in the AMA Guides.... Chap. 18 of the 5th Edition of AMA Guides to the Evaluation of Permanent Impairment cannot be used to calculate awards for permanent partial disability under Washington's Industrial Insurance Act."

Summary

The California and Washington workers' compensation systems have developed very different approaches to the problem of rating impairment associated with pain. In California, providers are required to evaluate subjective factors such as pain, and subjective factors can be the only ratable factor for an injured worker. While California does provide a method for characterizing pain, any such characterization is necessarily a product of judgment, rather than objective measurement, and that judgment is inherently imprecise, affected by predisposition, perspective, prior experience, and other factors. Also, the Physician's Guide states that the credibility of the injured worker must be assessed, which can pose a number of challenges for many providers. While rare, the California system even permits 100 % disability awards to be made essentially entirely on the basis of subjective reports of a worker. In contrast, in Washington State, there is a strong emphasis on objective findings to support medical opinions and adjudicative decisions. The positions reached by the two states regarding the rating of pain reflect differences in the overall legal frameworks that the states have developed for workers' compensation. For example, California law stipulates that permanent awards for an injured worker should take into account his or her ability to work (i.e., disability), whereas Washington State compensation law indicates (though with some ambiguity) that awards should be made on the basis of impairment. Systems developed to address this issue will most likely continue to change as interest in this topic grows and further research is conducted.

References

American Medical Association. (2001). *Guides to the evaluation of permanent impairment.* Chicago: AMA Press.

Barth, P. S., & Niss, M. (1999). Permanent partial disability benefit: Interstate differences. Workers Compensation Research Institute. WC-99-2. September.

Cocchiarella, L., & Lord, S. J. (2002). *Master the AMA guides fifth.* Chicago: AMA Press; Chicago Physician's Guide (2001) *Medical practice in the California workers' compensation system* (3rd edn). State of California, Department of Industrial Relations, Industrial Medical Council.

Medical examiners' handbook. (2000, September). Olympia: Washington State Department of Labor and Industries. American college of occupational and environmental medicine. http://usability.lni.wa.gov/IPUB/252-001-000.pdf.

Personal communication. (2003). State of California, Department of Industrial Relations.

Personal communication. (2003). Madison: International Association of Industrial Accident Boards and Commissions (IAIABC).

Rating Permanent Impairment. (2002). Washington State Department of Labor and Industries. Provider Bulletin 02-12.

The Physician's guide to Medical practice in the California workers. (2001). http://www.dir.ca.gov/dwc/medicalunit/toc.pdf. Accessed 18 June 2013.

Ratio Scale

Definition

Ratio scale is a scale that yields an accurate measurement of true ratios or proportions of magnitude and has a true zero point.

Cross-References

▶ Pain in Humans, Psychophysical Law
▶ Pain Measurement by Questionnaires, Psychophysical Procedures, and Multivariate Analysis

Ratio Scaling

▶ Pain Measurement by Questionnaires, Psychophysical Procedures, and Multivariate Analysis

Raynaud's Syndrome

Definition

Raynaud's syndrome refers to a set of symptoms characteristic of peripheral vascular disease, caused by an inappropriate response of the peripheral arteries in reaction to environmental stimuli (i.e., usually to the cold). Common symptoms of Raynaud's in the affected parts (e.g., toes, fingers) include the color change of first white, then blue/red and icy cold hands and feet.

Cross-References

▶ Experimental Pain in Children

Raynaud's Vibration White Finger Syndrome

Definition

Disorder characterized by blanching of the fingers and pain when exposed to cold. It can be naturally occurring or the result of prolonged exposure to hand-arm vibration, particularly noted in colder environments.

Cross-References

▶ Ergonomics Essay
▶ Raynaud's Syndrome

RCT

▶ Randomized Controlled Trial

Reactive Arthritis

▶ Reiter's Syndrome

Reactive Tachycardia

▶ Postoperative Pain, Pathophysiological Changes in Cardiovascular Function in Response to Acute Pain

Reactivity

Definition

A measure of psychophysiological response to a stressful stimulus.

Cross-References

▶ Psychophysiological Assessment of Pain

Real-Time PCR

Definition

Real-time polymerase chain reaction assays allow fluorescent detection and quantification of nucleic acid sequences as they accumulate in a PCR reaction with time.

Cross-References

▶ Nerve Growth Factor Overexpressing Mice as Models of Inflammatory Pain

Reassurance and Activation

▶ Activation/Reassurance

Recall

Definition

A procedure in which the person remembering generates a report of a memory either in response to a cue (cued recall) or in response to a general instruction to recall remembered events (free recall).

Cross-References

▶ Pain Memory
▶ Recognition

Receiver Operating Characteristics

Synonyms

ROC

Definition

A graphic method for assessing the ability of a test to discriminate between disease and health. The size of the area under the curve is a measure of this ability.

Cross-References

▶ Oswestry Disability Index

Receptive Field (RF)

Definition

The receptive field of a somatosensory neuron is the body region where a stimulus of a given type elicits a change in the firing of that neuron. For example, excitatory receptive fields are body regions where application of a stimulus produces increases in discharge frequencies of neurons, while inhibitory receptive fields are regions where stimuli produce decreases in discharge frequencies of neurons. Some types of neurons may have separate receptive fields for noxious and innocuous information, and portions of these receptive fields may be both excitatory and inhibitory. Some cells have both somatic and visceral receptive fields. Receptive fields vary in size according to cell type and stage of development. The size of dorsal horn cell peripheral cutaneous receptive fields has been shown to be larger in the newborn rat but decreases with age.

Cross-References

▶ Allodynia (Clinical, Experimental)
▶ Amygdala, Pain Processing and Behavior in Animals
▶ Central Pain, Human Studies of Physiology
▶ Encoding of Noxious Information in the Spinal Cord
▶ Exogenous Muscle Pain
▶ Hyperalgesia
▶ Infant Pain Mechanisms
▶ Insular Cortex, Neurophysiology, and Functional Imaging of Nociceptive Processing
▶ Mechano-Insensitive C-Fibers, Biophysics
▶ Mechanonociceptors

Receptor

Definition

Receptors are specialized proteins located in their neuronal membrane that bind neurotransmitters, neuropeptides, and drugs. Binding of a transmitter or neuropeptide to its receptor induces a response in the neuron or target cell associated with the receptor. Also used to refer to cells or nerve endings specialized in sensory transduction.

Cross-References

▶ Descending Circuitry, Transmitters and Receptors
▶ Mechanonociceptors

Receptor Affinity

Definition

Receptor affinity refers to the ability of a ligand to bind a receptor. The affinity of the various nonradioactive compounds is evaluated by displacement curves for the binding of a radioactive-ligand endowed with high affinity for the receptor. It is expressed as the inhibition constant (K_i) of the binding of the radioactive-selective ligand calculated from the concentration of peptide required to displace 50 % of the labeled ligand (IC_{50}), by means of the equation $K_i = IC_{50}/[(2L/L_0) + (L/K_D) - 1]$, where L is the free radioligand concentration in equilibrium with IC_{50} of unlabeled ligand, L_0 is the free radioligand concentration in the absence of competing ligand, and K_D is the equilibrium dissociation constant of the radioligand.

Cross-References

▶ Opioid Peptides from the Amphibian Skin

Receptor Potential

Definition

The membrane potential change elicited in receptor neurons during sensory transduction.

Cross-References

▶ Nociceptor Generator Potential
▶ Perireceptor Elements

Receptor Selectivity

Definition

Receptor selectivity refers to the ratio between the affinities for two different receptor types. For example, the mu/delta selectivity of a compound is the ratio between the mu-opioid receptor affinity and the delta-opioid receptor affinity of such a compound.

Cross-References

▶ Opioid Peptides from the Amphibian Skin

Receptors for These Neurotransmitters That Are Expressed by the Neurons

▶ Spinothalamic Tract Neurons: Neurotransmitter Inputs and Receptors

Receptors in the Descending Circuitry

▶ Descending Circuitry, Transmitters and Receptors

Recist

Definition

Recist is an acronym for the Response Evaluation Criteria in Solid Tumors, the evidence based validating assessment of treatment effects of reducing tumor volume.

Cross-References

▶ Cancer Pain Management, Chemotherapy

Recognition

Definition

A procedure in which the person remembering reports whether an event (stimulus) that is currently presented to them has occurred in the past. Under most conditions recognition memory produces more remembered material than memory retrieved under recall conditions, where the materials have to be actively retrieved.

Cross-References

▶ Pain Memory
▶ Recall

Recognizing Issues Contributing to Pain Secondary to Burn Injury and Its Treatment

▶ Pain Control in Children with Burns

Recollection

Definition

Recollective memory occurs when an individual recalls a specific episode from their past experience. A defining feature of recollective memory is that in addition to remembering that an event occurred, the rememberer has some phenomenal experience of the remembered event. This is often in the form of a visual or auditory image or some affective experience.

Cross-References

▶ Pain Memory

Reconstructive Therapy

▶ Prolotherapy

Recording of Pain Behavior

▶ Assessment of Pain Behaviors

Recovery

Definition

A period designed to allow physiological responses to a return to a normal or near-normal state.

Cross-References

▶ Psychophysiological Assessment of Pain

Recovery Cycles

▶ Nociceptors and Activity-Dependent Changes in Axonal Conduction Velocity

Recruitment to an Injury Site

Definition

In the case of immune cells, this refers to attracting immune cells out of the bloodstream and into the tissue at a site of injury. Immune cell recruitment occurs in response to chemical signals generated at the injury site, which attracts and activates immune cells to respond.

Cross-References

▶ Proinflammatory Cytokines

Rectangular Pulses

Definition

Rectangular electrical pulses are current pulses with a square shape. They are characterized by the amplitude and their duration.

Cross-References

▶ Pain in Humans, Electrical Stimulation (Skin, Muscle, and Viscera)

Recurrent Abdominal Pain

Synonyms

RAP

Definition

Recurrent abdominal pain (RAP) was initially defined as three or more bouts of pain, severe enough to affect activities over a period of at least 3 months, in school-age children. Subsequently, this definition of RAP has been expanded to include any child with recurrent abdominal pain of any etiology, including inflammatory bowel disease, metabolic disorders, and irritable bowel syndrome, as examples.

Cross-References

▶ Experimental Pain in Children
▶ Functional Abdominal Pain in Children

Recurrent Abdominal Pain (RAP)

▶ Functional Abdominal Pain in Children

Recurrent Pelvic Pain

▶ Epidemiology of Chronic Pelvic Pain

Red Flags

Definition

Red flags are presenting signs and symptoms considered indicative of possible spinal pathology or of the need for an urgent surgical evaluation. These "risk factors" for serious pathology or disease became incorporated into screening tools, recommended for use in primary care by clinicians to identify those patients in whom an urgent specialist opinion was indicated. Assessment of these risk factors was included within a new set of clinical guidelines for the management of acute low back pain, reviewed in Waddell (1998). Increasingly, however, the problem of low back disability has become recognized as a biopsychosocial phenomenon, and research reviews have demonstrated the powerful influence of psychological factors (Linton 2000) and social factors (Waddell et al. 2002) as predictors of disability.

Cross-References

▶ Yellow Flags

References

Linton, S. J. (2002). Why does chronic pain develop? A behavioural approach. In: Linton, S. J. (ed). Pain Research and Clinical Management, Volume 12. Elsevier Science: Amsterdam.
Waddell, G., (1998). The Back Pain Revolution. Churchill Livingstone, Edinburgh.
Waddell, G., Aylward, M., Sawney, P. (2002). Back Pain, Incapacity for Work, and Social Security Benefits. London: The Royal Society of Medicine Press.

Red Thrombi

▶ Postoperative Pain, Venous Thromboembolism

Reentrant Excitation

Definition

Excitation due to a positive feedback signal.

Cross-References

▶ Corticothalamic and Thalamocortical Interactions

Referred Abdominal Muscle Pain Model

▶ Visceral Pain Models, Female Reproductive Organ Pain

Referred Hyperalgesia

Synonyms

Secondary hyperalgesia

Definition

Hyperalgesia of the somatic body wall tissues in an area of referred pain, i.e., an area other than that in which the noxious stimulation takes place (Head, 1983, Brain 16:1). Referred hyperalgesia from viscera may involve both superficial (skin, subcutis) and deep tissues (muscle), but is most often confined to the muscle.

In humans with colics from calculosis of one upper urinary tract, hyperalgesia of the oblique musculature of the ipsilateral lumbar region is revealed by a significant decrease in the pain threshold to pressure and electrical muscle stimulation with respect to the contralateral side and normal control values (Vecchiet et al. 1989).

Similarly, in patients with biliary colics from gallbladder calculosis, hyperalgesia of the rectus abdominis muscle at the cystic point (upper right abdominal quadrant) is revealed by a significant decrease in the same thresholds with respect to the contralateral symmetrical site and normal values (Giamberardino et al. 2005).

Cross-References

▶ Gynecological Pain, Neural Mechanisms
▶ Muscle Pain, Referred Pain
▶ Visceral Pain Model, Kidney Stone Pain

References

Giamberardino, M. A., Affaitati, G., Lerza, R., Lapenna, D., Costantini, R., & Vecchiet, L. (2005). Relationship between pain symptoms and referred sensory and trophic changes in patients with gallbladder pathology. *Pain, 114*, 239–49.
Vecchiet, L., Giamberardino, M. A., Dragani, L., Albe-Fessard, D. (1989). Pain from renal/ureteral calculosis: evaluation of sensory thresholds in the lumbar area. *Pain, 36*, 289–95.

Referred Muscle Pain

Definition

Referred pain from one muscle to another is related to the ▶ central sensitization of nociceptive sensory neurons in the spinal cord. The pattern is similar to the "meridian connections" of acupuncture points.

Cross-References

▶ Dry Needling
▶ Muscle Pain, Referred Pain
▶ Referred Muscle Pain, Assessment

Referred Muscle Pain, Assessment

Thomas Graven-Nielsen
Laboratory for Musculoskeletal Pain and Motor Control, Center for Sensory-Motor Interaction, Aalborg University, Aalborg, Denmark

Definition

Pain perceived at a site adjacent to or at a distance from the site of origin (local pain, see Fig. 1). To distinguish between "referred pain" and "spread of pain," referred pain is typically restricted to areas outside the local pain area.

Characteristics

Referred pain has been known and described for more than a century and has been used extensively as a diagnostic tool in the clinic. Mainly musculoskeletal and visceral pain conditions are accompanied by local and/or referred pain. Originally, the term "referred tenderness and pain" was used.

In contrast to the sharp and localized characteristics of cutaneous pain, muscle pain is described as aching with diffuse and referred localization. Kellgren (1938) was one of the pioneers to study experimentally the characteristics of muscle pain and the actual locations of referred pain to selective stimulation of specific muscle groups. Several theories regarding the appearance of referred pain have been suggested (Graven-Nielsen 2006), but no firm neurophysiologic-based explanations for referred pain exist. It has been shown that ▶ wide dynamic range neurons, as well as nociceptive-specific neurons in the spinal cord and in the brain stem, receive convergent afferent input from the skin, muscles, joints, and viscera. This may cause misinterpretation of the afferent information coming from muscle afferents when reaching higher levels in the central nervous system and hence be one reason for the diffuse and referred characteristics.

Referred pain is probably a combination of central processing and peripheral input, as it is possible to induce referred pain to limbs with complete sensory loss due to an anesthetic block (Kellgren 1938). However, the involvement of peripheral input from the referred pain area is not clear, as anesthetizing this area shows inhibitory or no effects on the referred pain intensity. ▶ Central sensitization may be involved in the generation of referred pain. A complex network of extensive collateral synaptic connections for each muscle afferent fiber onto multiple dorsal horn neurons is assumed (Mense and Simons 2001). Under normal conditions, afferent fibers have fully functional synaptic connections with dorsal horn neurons, as well as latent synaptic connections to other neurons within the same region of the spinal cord. Following ongoing strong noxious input, latent synaptic connections become operational, thereby allowing for convergence of input from more than one source. Animal studies show a development of new and/or expansion of existing ▶ receptive fields by a noxious muscle stimulus. Recordings from a dorsal horn neuron, with a receptive field located in the biceps femoris muscle, show new receptive fields in the tibialis anterior muscle and at the foot after intramuscular (IM) injection of bradykinin into the tibialis anterior muscle (Hoheisel et al. 1993). In the context of referred pain, the unmasking of new receptive fields due to central sensitization could mediate referred pain (Graven-Nielsen 2006). A number of studies have found that the area of the referred pain correlated with the intensity of the muscle pain (Laursen et al. 1997). Moreover, the appearance of referred pain is delayed (20–40 s) compared to the local muscle pain (Graven-Nielsen et al. 1997; Laursen et al. 1997), indicating that a time-dependent process, like the unmasking of new synaptic connections, is involved in the neural mediation of referred pain. Referred pain has been suggested to be the phenomenon of ▶ secondary hyperalgesia in deep tissue.

Substantial clinical knowledge exists on the patterns of referred muscle pain from various skeletal muscles and after activation of trigger points in myofascial pain patients. Similarly, the pattern of referred pain evoked by exogenous muscle pain has been thoroughly mapped, initially by Kellgren (1938) and later by others (Fig. 1). Referral of muscle pain is typically described as a sensation from deep structures (Graven-Nielsen et al. 2002), in contrast to visceral referred pain that is both superficially and deeply located. The pattern and size of referral seem to be changed in chronic musculoskeletal pain conditions; e.g., fibromyalgia patients experience stronger pain and larger referred areas after exogenous muscle pain (hypertonic saline) compared to matched controls (Sörensen et al. 1998). Interestingly, these manifestations were present in lower limb muscles where the patients typically do not experience ongoing pain. Normally, pain from the tibialis anterior is projected distally to the ankle and only rarely proximally (Fig. 1). In the fibromyalgia patients, substantial proximal spread of the experimentally induced referred pain areas was found (Fig. 2). This corresponds to basic neurophysiological experiments in rats, where muscle nociception caused a proximal sensitization of dorsal horn neurons (Mense and Simons 2001). Enlarged referred pain areas in pain patients suggest that the efficacy of central processing is increased (central sensitization, Arendt-Nielsen and Graven-Nielsen 2008; Graven-Nielsen and Arendt-Nielsen 2010). Moreover, the expanded referred pain areas in fibromyalgia patients were partly inhibited by ketamine (an N-methyl-D-aspartate receptor antagonist) targeting central sensitization (Graven-Nielsen et al. 2000). Extended referred pain areas from the tibialis anterior muscle have also been shown in patients suffering from chronic whiplash pain; the extended areas of referred pain were also found when stimulating the neck/shoulder region (Johansen et al. 1999). Similarly, in patients with temporomandibular pain disorders, enlarged pain areas were found after experimental masseter muscle

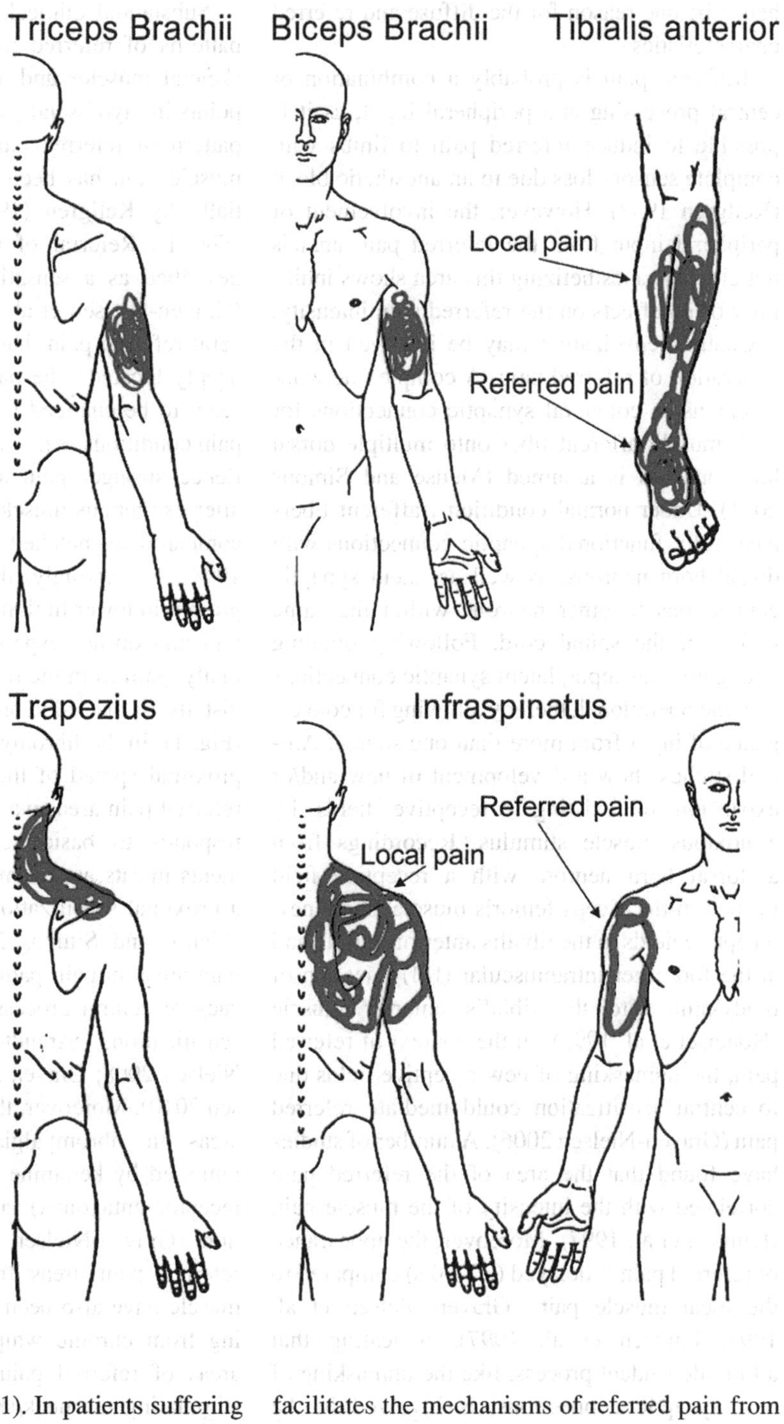

Referred Muscle Pain, Assessment,
Fig. 1 Distribution of experimentally induced local and referred pain. Hypertonic saline (0.5–1 ml, 5.8 %) was injected into musculus triceps brachii, biceps brachii, tibialis anterior, trapezius, and infraspinatus. The subjects ($n = 9$–15) outlined the area of pain. Saline-induced pain in tibialis anterior and infraspinatus showed distinct referred pain areas (not included in the local pain area), whereas the other muscles showed more localized pain around the injection site

pain (Svensson et al. 2001). In patients suffering from chronic ostereoarthritic knee pain (Bajaj et al. 2001) and low back pain (O'Neill et al. 2007), extended areas of saline-induced referred pain have been found. This shows that noxious joint input to the central nervous system facilitates the mechanisms of referred pain from muscle, possibly due to central sensitization.

Referred Hyperalgesia
Central sensitization facilitates the mechanism for referred pain. Likewise, central sensitization

Control   Whiplash   Fibromyalgia

Referred Muscle Pain, Assessment, Fig. 2 Typical distribution of experimentally induced local and referred pain in healthy subjects and musculoskeletal pain patients (fibromyalgia and whiplash). Hypertonic saline was injected into musculus tibialis anterior. Expanded pain referrals were characteristically seen in the musculoskeletal pain patients

might be involved in hyperalgesia at sites distant from the pain locus, as referred or widespread hyperalgesia. Referred hyperalgesia can be present in areas with or without referred pain. In referred pain areas, Kellgren (1938) found tenderness to pressure, but some of the later studies have been unable to reproduce this finding. Skin sensitivity in the referred pain area has been reported depended on the stimulus modality tested (Graven-Nielsen et al. 1997; Leffler et al. 2000). Increased sensitivity to electrical cutaneous stimulation, and decreased sensitivity to radiant heat or pinprick stimulation, has been reported in referred pain areas (Graven-Nielsen et al. 1997; Leffler et al. 2000). Moreover, hyperalgesia to pressure was found distal to the referred pain area, produced by experimental pain induced in the tibialis anterior muscle

(Graven-Nielsen et al. 2002). This suggests involvement of summation between muscle afferents and the somatosensory afferents from the hyperalgesic area, eventually facilitated by central sensitization. In clinical studies, a modality-specific and tissue-specific change of the somatosensory function in referred pain areas has been reported (Leffler et al. 2003), in line with the experimental findings.

Central sensitization of dorsal horn or brain stem neurons, initiated by nociceptive activity from muscles, may explain the expansion of pain with referral to other areas and probably hyperalgesia in these areas. However, facilitated neurons cannot account for the decreased sensation to certain sensory stimuli in the referred area. Descending inhibitory control of the dorsal horn neurons may explain the decreased response to additional noxious stimuli in the referred pain area and therefore mask any eventual increase in somatosensory sensitivity in referred pain areas (Graven-Nielsen et al. 2002).

References

Arendt-Nielsen, L., & Graven-Nielsen, T. (2008). Translational aspects of musculoskeletal pain: From animals to patients. In T. Graven-Nielsen, L. Arendt-Nielsen, & S. Mense (Eds.), *Fundamentals of musculoskeletal pain* (pp. 347–366). Seattle: IASP Press.

Bajaj, P., Graven-Nielsen, T., & Arendt-Nielsen, L. (2001). Osteoarthritis and its association with muscle hyperalgesia: An experimental controlled study. *Pain, 93*, 107–114.

Graven-Nielsen, T. (2006). Fundamentals of muscle pain, referred pain, and deep tissue hyperalgesia. *Scandinavian Journal of Rheumatology, 35*(Suppl. 122), (5), 1–43.

Graven-Nielsen T, Arendt-Nielsen L. (2010). Assessment of mechanisms in localized and widespread musculoskeletal pain. *Nat Rev Rheumatol, 6*:599–606.

Graven-Nielsen, T., Arendt-Nielsen, L., Svensson, P., et al. (1997). Stimulus-response functions in areas with experimentally induced referred muscle pain – a psychophysical study. *Brain Research, 744*, 121–128.

Graven-Nielsen, T., Aspegren, K. S., Henriksson, K. G., et al. (2000). Ketamine reduces muscle pain, temporal summation, and referred pain in fibromyalgia patients. *Pain, 85*, 483–491.

Graven-Nielsen, T., Gibson, S. J., Laursen, R. J., Svensson, P., & Arendt-Nielsen, L. (2002). Opioid-Insensitive hypoalgesia to mechanical stimuli at sites ipsilateral and contralateral to experimental muscle pain in human volunteers. *Experimental Brain Research, 146*, 213–222.

Hoheisel, U., Mense, S., Simons, D. G., et al. (1993). Appearance of New receptive fields in Rat dorsal horn neurons following noxious stimulation of skeletal muscle: A model for referral of muscle pain? *Neuroscience Letters, 153*, 9–12.

Johansen, M. K., Graven-Nielsen, T., Schou, O. A., et al. (1999). Generalised muscular hyperalgesia in chronic whiplash syndrome. *Pain, 83*, 229–234.

Kellgren, J. H. (1938). Observations on referred pain arising from muscle. *Clinical Science, 3*, 175–190.

Laursen, R. J., Graven-Nielsen, T., Jensen, T. S., et al. (1997). Quantification of local and referred pain in humans induced by intramuscular electrical stimulation. *European Journal of Pain, 1*, 105–113.

Leffler, A. S., Kosek, E., & Hansson, P. (2000). Injection of hypertonic saline into musculus infraspinatus resulted in referred pain and sensory disturbances in the ipsilateral upper arm. *European Journal of Pain, 4*, 73–82.

Leffler, A. S., Hansson, P., & Kosek, E. (2003). Somatosensory perception in patients suffering from long-term trapezius myalgia at the site overlying the most painful part of the muscle and in an area of pain referral. *European Journal of Pain, 7*, 267–276.

Mense, S., & Simons, D. G. (2001). *Muscle pain. Understanding its nature, diagnosis, and treatment.* Philadelphia: Lippincott Williams & Wilkins.

O'Neill, S., Manniche, C., Graven-Nielsen, T., & Arendt-Nielsen, L. (2007). Generalized deep-tissue hyperalgesia in patients with chronic low-back pain. *European Journal of Pain, 11*, 415–420.

Sörensen, J., Graven-Nielsen, T., Henriksson, K. G., et al. (1998). Hyperexcitability in fibromyalgia. *Journal of Rheumatology, 25*, 152–155.

Svensson, P., List, T., & Hector, G. (2001). Analysis of stimulus-evoked pain in patients with myofascial temporomandibular pain disorders. *Pain, 92*, 399–409.

spinal) second order neurons from visceral organs and the corresponding myotome or dermatome. Common examples of referred pain include pain down the left arm when a patient is having a heart attack, or pain in the flank/groin with a urinary colic. Referred pain has been known and described for more than a century, and has been used extensively as a diagnostic tool in the clinic.

Cross-References

▶ Cancer Pain
▶ Cancer Pain, Goals of a Comprehensive Assessment
▶ Exogenous Muscle Pain
▶ Gynecological Pain, Neural Mechanisms
▶ Muscle Pain, Referred Pain
▶ Nociceptors in the Orofacial Region (Meningeal/Cerebrovascular)
▶ Opioids and Muscle Pain
▶ Pain Mechanisms in Endometriosis
▶ Pain Treatment: Spinal Nerve Blocks
▶ Peripheral Neuropathic Pain
▶ Referred Hyperalgesia
▶ Somatic Referred Pain
▶ Spinothalamic Tract Neurons, Visceral Input
▶ Thalamus, Visceral Representation
▶ Visceral Nociception and Pain
▶ Visceral Pain and Nociception
▶ Visceral Referred Pain

Referred Pain

Definition

A sensation of pain arising from a body region remote from the site of the stimulus and afferent activation, most frequently appearing (with or without hyperalgesia) in muscle or skin of a somatic region as a result of visceral nociceptive activation. Referred pain is usually explained by convergent afferent input onto central (usually

Reflective Listening

Definition

Reflective listening involves making statements that are meant to clarify and restate what the other person is saying. The usual intent of reflective listening is to build rapport and allow the other person to explore his or her thoughts and feelings about the issue being discussed in further depth.

Cross-References

▶ Chronic Pain, Patient-Therapist Interaction

Reflex Neurovascular Dystrophy (Unconventional Taxonomy Used by a Few Clinicians)

▶ CRPS-1 in Children

Reflex Sympathetic Dystrophy

▶ Complex Regional Pain Syndrome and the Sympathetic Nervous System
▶ Complex Regional Pain Syndromes, Clinical Aspects
▶ CRPS-1 in Children
▶ Neuropathic Pain Models, CRPS-I Neuropathy Model
▶ Postoperative Pain, Acute Presentation of Complex Regional Pain Syndrome

Reflex Sympathetic Dystrophy (RSD)

▶ Complex Regional Pain Syndrome
▶ CRPS, Evidence-Based Treatment

Reflexes and Opioids

▶ Opioids and Reflexes

Refractory Period

Definition

The period of time following the end of a painful attack during which no stimulus, however strong, precipitates a further attack. In cells, it refers to the time during which an excitable membrane does not respond to a stimulus that normally generates a response.

Cross-References

▶ Molecular Contributions to the Mechanism of Central Pain
▶ SUNCT Syndrome

Regional Anesthesia

Definition

Various regional anesthetic techniques exist to promote postoperative pain relief. Regional anesthesia techniques can block or reduce pain anywhere in the body for several hours to several days, depending on the technique used. Techniques used include neuraxial local anesthetic techniques, peripheral nerve blocks and regional plexus blocks with local anesthetic, and wound infiltration with local anesthetics. The infiltration of wounds with local anesthetics, besides providing analgesia, appears to reduce the local inflammatory response to trauma or surgery. Local infiltration may help reduce the upregulation of peripheral nociceptors that manifests as hypersensitivity to stimuli. Local anesthetic can also be administered intra-articularly (knee and shoulder), or on bone wounds (iliac crest graft sites). Catheter techniques can be used to prolong the duration of peripheral nerve blocks, and are either infused continuously or by bolus administration with patient-controlled pumps. With the advent of disposable infusion pumps, home discharge with a catheter and pump in place is possible. Advantages include better postoperative pain control, reduced blood loss, and decreased rates of deep venous thrombosis. If given as the sole anesthetic technique intraoperatively, it allows patient involvement and avoids the common adverse effects of general anesthesia.

Cross-References

▶ Cancer Pain Management, Anesthesiologic Interventions, and Neural Blockade
▶ Multimodal Analgesia in Postoperative Pain
▶ Postoperative Pain, Regional Blocks

Regional Blocks

▶ Postoperative Pain, Regional Blocks

Regional Heterogeneity

Definition

Regional heterogeneity, as it applies to glia, refers to the fact that glia in various CNS sites are not the same. These cells express receptors dependent on their specific microenvironments. Regional heterogeneity exists in the spinal cord, as glia within the dorsal horn exhibit higher glutamate transporter expression and higher fractalkine receptor expression than other regions of spinal cord.

Cross-References

▶ Cord Glial Activation

Regional Infusion

▶ Intravenous Infusions, Regional and Systemic

Regression to the Mean

Definition

A statistical phenomenon that assumes that individuals tend to receive their initial pain assessment when pain is near its greatest intensity, and that their pain level is likely to be lower when they return for a second assessment.

Cross-References

▶ Placebo Analgesia and Descending Opioid Modulation

Rehabilitation

Definition

The World Health Organization (WHO) has defined rehabilitation as "a process aimed at enabling disabled persons to reach and maintain their optimal physical, sensory, intellectual, psychological and social functional levels. Rehabilitation provides disabled people with the tools they need to attain independence and self-determination." Thus, the aim of rehabilitation has traditionally been seen as facilitating the normalization of human functioning after injury and disease or due to congenital defects. It is usually said to comprise a number of coordinated measures of medical, psychological, social, educational, and vocational nature. The Convention of the Rights of Persons with Disabilities, in its article 26, calls for "appropriate measures /-/ to enable persons with disabilities to attain and maintain their maximum independence, full physical, mental, social and vocational ability, and full inclusion and participation in all aspects of life" (United Nations, 2006).

Cross-References

▶ Chronic Pain in Children, Physical Medicine and Rehabilitation
▶ Compensation, Disability, and Pain in the Workplace
▶ Disability, Effect of Physician Communication

Rehabilitation Plan

Definition

A rehabilitation plan is a written document specifying stepwise goals to improve functions, activities, and participation in disabled persons including necessary contextual modifications. Ideally, the document is written together with and approved by the disabled person. It should specify interventions, responsible therapist, and time frames.

Cross-References

▶ Physical Medicine and Rehabilitation, Team-Oriented Approach

Reinforcers

Definition

Reinforcers are defined in a circular fashion. If a consequence systematically following an operant leads to strengthening that operant, it is, by definition, a reinforcer. If it fails to have such an effect when observed over time, it is not a reinforcer for that activity, for that person, in that context.

Cross-References

▶ Malingering, Primary and Secondary Gain
▶ Training by Quotas

Reiter's Syndrome

Synonyms

Arthritis urethritica; Polyarteritis enterica; Reactive arthritis; Venereal arthritis

Definition

Reiter's syndrome, an autoimmune disorder, is the combination of three seemingly unlinked symptoms: inflammatory arthritis of large joints, inflammation of the eyes (conjunctivitis and uveitis), and urethritis. It is named after Hans Reiter, a German military physician, who in 1916 described the disease in a World War I soldier who had recovered from a bout of diarrhea. It is also known as arthritis urethritica, venereal arthritis, reactive arthritis, and polyarteritis enterica.

Cross-References

▶ Chronic Low Back Pain: Definitions and Diagnosis
▶ Sacroiliac Joint Pain

Relationship

▶ Chronic Pain, Patient-Therapist Interaction

Relative Potency

Definition

Relative potency is determined by comparing graded dose-response curves for two drugs (A and B). From these curves the dose required to produce 50 % of the maximum response can be estimated, and the ratio of the dose of A/B that produces an equivalent response is found.

Cross-References

▶ Opioid Rotation

Relaxation

Definition

A group of therapeutic procedures that aim to teach individuals how to induce a state of reduced musculoskeletal activity. The induction of relaxation is accompanied by quiescence in visceral activity and brain states associated with mental calmness and a positive mood (see also ▶ Relaxation Therapy).

Cross-References

- ▶ Psychological Treatment in Acute Pain
- ▶ Psychological Treatment of Headache
- ▶ Psychological Treatment of Pain in Children
- ▶ Psychology of Pain, Efficacy of Cognitive-Behavioral Treatment
- ▶ Therapy of Pain, Hypnosis

Relaxation in the Treatment of Pain

Beverly E. Thorn[1], Joseph Mignogna[2] and Frank Andrasik[3]
[1]Department of Psychology, University of Alabama, Tuscaloosa, AL, USA
[2]Michael E. DeBakey VA Medical Center, Houston VA HSR&D Center of Excellence (152), Houston, TX, USA
[3]Department of Psychology, Applied Psychophysiology and Biofeedback, The University of Memphis, Memphis, TN, USA

Synonyms

Autogenic training; Diaphragmatic breathing; Guided imagery; Hypnotic relaxation; Meditation; Passive relaxation; Progressive muscle relaxation

Definition

▶ Relaxation interventions are categorized as psychological therapies, psychophysiological therapies, and/or behavioral therapies. Relaxation therapies involve a systematic process of teaching patients to become aware of their physiological responses and to gain a sense of physiological and cognitive/emotional calm or serenity. Successful relaxation is thought to decrease overall sympathetic nervous system arousal, which is associated with physiological concomitants (e.g., reductions in muscle tension, heart rate, blood pressure, and respiration) as well as cognitive/affective components (e.g., a sense of tranquillity, reduced anxiety, and distress). Relaxation can be achieved through a variety of means, including specific muscle relaxation or breathing exercises, biofeedback-assisted relaxation, ▶ hypnotic relaxation, ▶ meditation, or mental imagery. Relaxation therapies are often used in conjunction with other psychological interventions, such as cognitive stress-coping therapy and other cognitive-behavioral approaches. Relaxation interventions are widely utilized with both acute and chronic pain management, and there is a well-established evidence base for their use (Andrasik 1986, 2004; Arena and Blanchard 1996).

Characteristics

Types of Relaxation Therapy

The most frequently used form of relaxation therapy is ▶ progressive muscle relaxation (PMR), whereby patients are taught to discriminate between tensed and relaxed muscles by volitionally tensing, and then relaxing, a prescribed set of muscle groups throughout the body (Bernstein and Borkovec 1973; Bernstein and Carlson 1993; Jacobsen 1938; Lehrer 1982). This form of relaxation is often used as the sole relaxation intervention or as a first component of a multi-session relaxation protocol. PMR is thought to help patients discriminate between tensed and relaxed muscles, thus facilitating the process of learning to achieve reductions in muscle tension. Patients with previous injuries, pain, or strain in any particular muscle group are instructed to omit that group of muscles from the tension-relaxation protocol.

Another common form of relaxation therapy is ▶ guided mental imagery, whereby a patient is taught to create a mental "image" of a relaxing setting (the image may include auditory, olfactory, tactile, and even gustatory imagery, as well as visual imagery; Rime and Andrasik 2011). The most commonly employed guided imagery involves asking the patient to imagine a peaceful and pleasant scene or place in great detail. Patients are encouraged to include in their image as many of the sensory aspects of the scene as possible; although visual imagery is most predominant, supplementing the visual image with the other senses (e.g., auditory or tactile imagery) is thought to enhance the depth of the experience.

A third form of relaxation therapy involves ▶ autogenic training, whereby various verbal cues associated with physiological aspects of the relaxation response (e.g., "my hands are heavy and warm") are repeated slowly by the participant (Schultz and Luthe 1969). Autogenic relaxation training is also called cue-assisted relaxation since it involves self-verbalized cues associated with decreased sympathetic nervous system arousal. Cue-assisted relaxation can also involve a simpler format, in which a verbal cue (e.g., "relax") is repeated in the mind of the patients as they continue to relax and breathe slowly and deeply.

Meditative relaxation is another form of relaxation that is most commonly practiced in two forms, transcendental meditation (TM) and mindfulness-based meditation. In TM, a person brings awareness to an assigned word or phrase that can be spiritual or secular (e.g., "God is love" or "I am happy") that is silently and continuously repeated while a person remains in a comfortable seated position (Morone and Greco 2007; Wachholtz and Pargament 2005). Alternatively, mindfulness-based meditation is commonly

taught in conjunction with the Mindfulness-Based Stress Reduction program (MBSR). Mindfulness "simply...involves bringing non-judgemental moment-to-moment awareness to thoughts, sensations, or emotions as they arise," and is practiced in different forms, namely, "sitting meditation, walking meditation, loving-kindness meditation, or the body scan" (p. 366, Morone and Greco 2007). MBSR programs are generally administered over at least 8 weeks and include daily practice between sessions (Morone and Greco 2007); however, recent findings support the efficacy of a condensed 3 day mindfulness meditation training program at improving mindfulness skills, as well as reducing pain ratings, pain sensitivity, and state anxiety (Zeidan et al. 2010).

Most forms of relaxation include instruction in the use of ▶ diaphragmatic breathing. Diaphragmatic breathing involves teaching the patient to draw air more deeply into the lungs by moving the diaphragm downward toward the umbilicus and thus causing a notable expansion of the abdomen upon inspiration. In addition to coupling diaphragmatic breathing with other forms of relaxation during periods of practice, patients are often instructed to use diaphragmatic breathing as a brief intervention tool in response to stressors as they arise during their daily routine.

Many relaxation protocols, especially when they involve more than a single training session, utilize deepening strategies to help patients increase their depth of relaxation. Deepening strategies often involve counting slowly, each number representing a deeper level of relaxation. In some cases, the patient is instructed to visualize descending a very secure staircase, and each step represents a deeper level of relaxation. These deepening strategies share similarities with hypnosis, although they may or may not be labeled hypnosis. In general, similar treatment outcomes are reported in the literature when self-hypnosis was compared to other "traditional" relaxation procedures, like PMR and autogenic training in treating chronic pain (Jensen and Patterson 2006). Castel and colleagues (2007) compared

the effects of hypnosis with relaxation suggestions to hypnosis with analgesia suggestions and
to a relaxation procedure not involving hypnosis
on patients with fibromyalgia. They found that
hypnosis with analgesia suggestions had a greater
effect on pain intensity and on the sensory (but
not affective) components of pain in comparison
to hypnosis with relaxation suggestions. Additionally, there was no difference in the effect of
hypnosis with relaxation suggestions compared
to the relaxation alone on pain intensity, sensory
components of pain, and affective components
of pain.

An example of a multi-session relaxation
instruction regimen is taken from Andrasik
(1986) and is briefly summarized here. Following
an education phase where the client is informed
about relaxation procedures, instruction begins
with progressive muscle relaxation in which the
patient tenses, then relaxes, a relatively large
number of muscle groups (e.g., 18 specific muscle groups), followed by a deepening strategy.
A second session incorporates imagery training
into the protocol. Subsequent sessions include
discrimination training, where patients are taught
various gradations of tension and relaxation in
order to understand that the state of muscle tension or relaxation is a continuous process, rather
than a dichotomous, either-or state. Later sessions decrease the number of muscle groups that
are contracted and then relaxed, which shortens
the procedure and makes it more easily "portable," and relaxation by recall is introduced, where
the patient attempts relaxation without going
through the tension-relaxation cycles, but rather
by focusing on the process of relaxation and
recalling the previous instruction. Final sessions
incorporate cue-assisted relaxation (using diaphragmatic breathing and often using a word
cue to signal relaxation, such as "relax").

Although relaxation treatments most commonly lead to positive outcomes, a small number
of individuals can experience initial negative
outcomes (muscle cramps, disturbing sensory,
cognitive or emotional reactions) and/or encounter other problems that impact adherence and
practice. Schwartz, Schwartz, and Monastra
(2003) provide a detailed discussion of these
potential problems along with solutions for
addressing them.

Efficacy

The efficacy of relaxation treatments has most
frequently been evaluated by comparing one
form of relaxation training to another or comparing relaxation training to other forms of credible
treatment. For example, in a study comparing
progressive muscle relaxation to autogenic relaxation training in low back pain patients, both
forms of relaxation were found to increase feelings of relaxation (Rohrmann et al. 2001). Other
efficacy investigations have compared individuals receiving relaxation instruction to waiting
list control comparison groups or to participants
who receive credible attention placebo treatment
(often education information regarding pain or
supportive psychotherapy). For example, in
a study comparing relaxation and guided imagery
for individuals in rehabilitation for anterior cruciate ligament reconstruction, patients were compared to controls (no treatment) or placebo
treatment groups (attention, encouragement, and
support). In this study, those patients receiving
guided imagery/relaxation training reported significantly less pain and reinjury anxiety at
24 week post-surgery than patients in the placebo
group or the no treatment condition (Cupal and
Brewer 2001).

Another example of a well-controlled efficacy
study of relaxation assessed self-reported pain of
bone marrow transplant patients after one of four
treatments: relaxation/imagery training alone,
combined cognitive-behavioral treatment (CBT)
with imagery/relaxation training, medical treatment as usual, or therapist support (Syrjala et al.
1995). In this clinical trial, relaxation/imagery
training alone or combined relaxation/imagery/
CBT resulted in lower pain reports compared to
the treatment as usual or therapist support groups,
although the addition of CBT did not add incremental value.

Kwekkeboom and Gretarsdottir (2006)
revealed support for relaxation interventions in
reducing pain in 8 out of 15 studies identified in
their systematic review of the literature. They
cited support for the efficacy of PMR in reducing

arthritis pain (receiving the greatest support), chronic low back pain, and pregnancy-related leg pain. Additionally, support was cited for jaw relaxation (an intervention focused on bringing the mouth, throat, and face into a relaxed position to bring about a reduction in muscle tension throughout the body), passive systemic relaxation (an intervention focused on bodily sensations during relaxation), and autogenic training in reducing postoperative pain. Importantly, significant reductions in pain were only observed in studies that employed relaxation interventions over the course of multiple occasions across several days or weeks and not following a single session. In another structured review of the literature, Morone and Greco (2007) identified and reviewed studies of mind-body interventions for chronic pain among older adults, although a shortage of relevant published studies focusing on older adults led the authors to also include studies with younger adults. These authors found support for using PMR with guided imagery in reducing pain and improving function among older adults with osteoarthritis pain. Also, limited support was provided for meditation in improving the functioning or coping ability of older adults with osteoarthritis or low back pain. Additionally, among the studies not concerned with age-related benefits of relaxation interventions, PMR and hypnosis were associated with reductions in pain. While these two systematic reviews of the literature provided support for a number of relaxation interventions for pain, both reviews also highlight the need for larger controlled clinical trials (Kwekkeboom and Gretarsdottir 2006; Morone and Greco 2007).

Therapeutic Mechanism
The specific mechanism responsible for the success of relaxation in pain management is not known. There is no evidence that clearly demonstrates the superiority of one type of relaxation training over other types, even for specific pain disorders. It is, however, widely observed that relaxation instruction elicits general reductions in muscle tension.

Flor and colleagues (1999) have shown that patients with chronic pain are unable to perceive muscle tension levels accurately (in either affected or nonaffected muscles) and, further, when exposed to tasks requiring production of muscle tension, they overestimate physical symptoms, rate the task as more aversive, and report greater pain. Their findings indicate a heightened sensitivity and suggest that relaxation strategies may lead to increased discrimination accuracy with a concomitant reduction in sensitivity.

Additionally, it is commonly observed that relaxation is associated with reductions in heart rate and respiratory activity and physiological responses thought to moderate ▶ sympathetic nervous system arousal (Good et al. 1999). Thus, the success of the various forms of relaxation instruction may be more related to an overall reduction in levels of sympathetic nervous system arousal and perhaps a reduction in the cognitive and affective factors that exacerbate pain. Emery and colleagues (2008) proposed and tested one mechanism to account for how PMR affects pain, specifically its impact on descending inhibition of nociception. These authors utilized the nociceptive flexion reflex (NFR) as an objective measure of nociceptive response and a polysynaptic spinal reflex that can be elicited by electrical stimulation to the foot and recorded by electromyography (EMG) of the hamstring muscles. A brief PMR intervention for college students without chronic pain resulted in PMR condition participants experiencing an increased NFR threshold while no change was observed in the control condition. While subject ratings of pain did not change during the study for the PMR or control condition participants, participants in the PMR condition reported decreases in stress following the brief intervention.

Relaxation interventions, such as PMR, meditation, and guided imagery, have been shown to elicit the relaxation response (RR; Benson and Goodale 1981), and the RR has demonstrated its clinical benefit in treating health conditions exacerbated or caused by stress, including pain (Caudill et al. 1991). RR is associated with numerous physiological changes, including decreased consumption of oxygen; reductions in carbon dioxide elimination; reductions in blood pressure, heart rate, and respiration rate; changes in the cortical and subcortical regions of the brain; and pronounced low-frequency heart rate oscillations. Additionally, recent evidence supports the idea that RR causes changes in gene expression in both short- and long-term RR practitioners, and these changes in gene expression may lead to cellular activity that counteracts damage following chronic psychological stress (Dusek et al. 2008).

References

Andrasik, F. (1986). Relaxation and biofeedback for chronic headaches. In A. D. Holzman & D. C. Turk (Eds.), *Pain management: A handbook of treatment approaches* (pp. 213–239). NY: Pergamon Press.

Andrasik, F. (2004). The essence of biofeedback, relaxation, and hypnosis. In R. H. Dworkin & W. S. Breitbart (Eds.), *Psychosocial aspects of pain: A handbook for health care providers* (pp. 285–305). Seattle: IASP Press.

Arena, J. G., & Blanchard, E. B. (1996). Biofeedback and relaxation therapy for chronic pain disorders. In R. J. Gatchel & D. C. Turk (Eds.), *Psychological approaches to pain management: A Practitioner's handbook* (pp. 179–230). New York: Guilford Press.

Benson, H., & Goodale, I. L. (1981). The relaxation response: Your inborn capacity to counteract the harmful effects of stress. *The Journal of the Florida Medical Association, 68*(4), 265–267.

Bernstein, D. A., & Borkovec, T. D. (1973). *Progressive relaxation training.* Champaign: Research Press.

Bernstein, D. A., & Carlson, C. R. (1993). Progressive relaxation: Abbreviated methods. In P. M. Lehrer & R. L. Woolfolk (Eds.), *Principles and practice of stress management* (2nd ed.). New York: Guilford Press.

Castel, A., Perez, M., Sala, J., Padrol, A., & Rull, M. (2007). Effect of hypnotic suggestion on fibromyalgic pain: Comparison between hypnosis and relaxation. *European Journal of Pain, 11*(4), 463–468.

Caudill, M., Schnable, R., Zuttermeister, P., Benson, H., & Friedman, R. (1991). Decreased clinic use by chronic pain patients: Response to behavioral medicine intervention. *The Clinical Journal of Pain, 7*(4), 305–310.

Cupal, D. D., & Brewer, B. W. (2001). Effects of relaxation and guided imagery on knee strength, reinjury anxiety, and pain following anterior cruciate ligament reconstruction. *Rehabilitation Psychology, 46*, 28–43.

Dusek, J. A., Otu, H. H., Wohlhueter, A. L., Bhasin, M., et al. (2008). Genomic counter-stress changes induced by the relaxation response. *PLoS One, 3*(7), e2576.

Emery, C. F., France, C. R., Harris, J., Norman, G., & Vanarsdalen, C. (2008). Effects of progressive muscle relaxation training on nociceptive flexion reflex threshold in healthy young adults: A randomized trial. *Pain, 138*(2), 375–379.

Flor, H., Fürst, M., & Birbaumer, N. (1999). Deficient discrimination of EMG levels and overestimation of perceived tension in chronic pain patients. *Applied Psychophysiology and Biofeedback, 24*, 55–66.

Good, M., Stanton-Hicks, M., Grass, J. A., Anderson, G., Choi, C., Schoolmeesters, L., & Salman, A. (1999). Relief of postoperative pain with Jaw relaxation, music, and their combination. *Pain, 81*, 163–172.

Jacobsen, E. (1938). *Progressive relaxation.* Chicago: University of Chicago Press.

Jensen, M., & Patterson, D. R. (2006). Hypnotic treatment of chronic pain. *Journal of Behavioral Medicine, 29*(1), 95–124.

Kwekkeboom, K. L., & Gretarsdottir, E. (2006). Systematic review of relaxation interventions for pain. *Journal of Nursing Scholarship, 38*(3), 269–277.

Lehrer, P. M. (1982). How to relax and how not to relax: A reevaluation of the work of Edmund Jacobson-I. *Behaviour Research and Therapy, 20*, 417–428.

Morone, N. E., & Greco, C. M. (2007). Mind-body interventions for chronic pain in older adults: A structured review. *Pain Medicine, 8*(4), 359–375.

Rime, C., & Andrasik, F. (2011). Relaxation techniques and guided imagery. In S. D. Waldman (Ed.), *Pain management* (2nd ed., pp. 967–975). Philadelphia: Saunders/Elsevier.

Rorhmann, S., Hopf, M., Hennig, J., & Netter, P. (2001). Psychobiological effects of autogenic training and progressive muscle relaxation in patient with back pain, multiple sclerosis (MS) and healthy individuals. *Zeitschrift fuer Klinische Psychologie, Psychiatrie und Psychotherpie, 49*, 373–387.

Schultz, J. H., & Luthe, W. (1969). *Autogenic training* (Vol. 1). New York: Grune and Stratton.

Schwartz, M. S., Schwartz, N. M., & Monastra, V. J. (2003). Problems with relaxation and biofeedback-assisted relaxation and guidelines for management. In M. S. Schwartz & F. Andrasik (Eds.), *Biofeedback: A Practitioner's guide* (3rd ed., pp. 251–264). New York: Guilford Press.

Syrjala, K. L., Donaldson, G. W., Davis, M. W., Kippes, M. E., et al. (1995). Relaxation and imagery and cognitive-behavioral training reduce pain during cancer treatment: A controlled clinical trial. *Pain, 63*, 189–198.

Wachholtz, A. B., & Pargament, K. I. (2005). Is spirituality a critical ingredient of meditation? Comparing the effects of spiritual meditation, secular meditation, and relaxation on spiritual, psychological, cardiac, and pain outcomes. *Journal of Behavioral Medicine, 28*(4), 369–384.

Zeidan, F., Gordon, N. S., Merchant, J., & Goolkasian, P. (2010). The effects of brief mindfulness meditation training on experimentally induced pain. *The Journal of Pain, 11*(3), 199–209.

Relaxation Interventions

Definition

A systematic process of teaching patients to become aware of their physiological responses and to gain a sense of physiological and cognitive/emotional calm or serenity.

Cross-References

▶ Psychological Treatment of Headache

Relaxation Therapy

Definition

Relaxation therapy is standardized instructions or meditative techniques used to elicit a relaxation response, characterized by a generalized decrease in the sympathetic nervous system and metabolic activity and an altered state of consciousness, described as a subjective experience of well-being. The techniques include autogenic training, transcendental meditation, yoga, and Jacobson's method of progressive relaxation.

Cross-References

▶ Biofeedback in the Treatment of Pain
▶ Body Awareness Therapies
▶ Complex Chronic Pain in Children, Interdisciplinary Treatment
▶ Relaxation
▶ Relaxation in the Treatment of Pain
▶ Relaxation Training

Relaxation Training

Bengt H. Sjölund
Department of Community Medicine and
Rehabilitation Medicine, Umea University,
Umea, Sweden

Synonyms

Autogenic training; Stress management

Definition

A relaxation procedure, taught to patients during a number of individual or group sessions, from which a psychophysiologically determined relaxation response should preferably be elicited. It is often used as one component of a pain rehabilitation program (e.g., Turner-Stokes et al. 2003).

Characteristics

Systematic training in relaxation techniques for continued self-use has been reported to have beneficial results in patients with chronic temporomandibular disorder, both on pain severity and life interference from pain (Carlson et al. 2001), immediately after intervention and at a 26-week follow-up. There were also significant decreases in affective distress, somatization, obsessive-compulsive symptoms, tender point sensitivity, awareness of tooth contact, and sleep dysfunction.

Similarly, in an evaluation of the effectiveness of a meditation-based stress reduction program on fibromyalgia, 77 patients meeting the 1990 criteria of the American College of Rheumatology for fibromyalgia took part in a 10-week group outpatient program with a carefully defined treatment approach. The mean scores of all the patients completing the program showed improvement, but 51 % showed moderate to marked improvement and were considered as responders, indicating that a meditation-based stress reduction program is indeed effective for patients with fibromyalgia (Kaplan et al. 1993).

In a study of the relative effectiveness of EMG biofeedback, applied relaxation training and a combined procedure versus a wait list control for the management of chronic, upper extremity cumulative trauma disorder, patients in all three treatment conditions showed significant short-term reductions in pain and psychopathology in comparison to the wait list group (Spence et al. 1995). There was some relapse on measures of depression, anxiety, and pain beliefs for treated patients during a 6-month follow-up period, although the measures remained significantly below pretreatment levels for most outcome indices and self-monitored pain continued to decrease. Interestingly, the greatest short-term treatment benefits on measures of pain, distress, interference in daily living, depression, and anxiety were shown by patients receiving applied relaxation training (cf. Stuckey et al. 1986), whereas by the follow-up, the differences between the treatment groups were no longer evident.

In children with chronic tension-type headache, a controlled study has shown that headache activity in the children treated with relaxation training was significantly more reduced than among those in the no treatment control group at posttreatment as well as the 6-month follow-up. At these evaluations, 69 % and 73 %, respectively, of the pupils treated with relaxation had achieved a clinically significant headache improvement (at least a 50 % improvement) as compared to 8 % and 27 %, respectively, of the pupils in the no treatment control group (Larsson and Carlsson 1996). In adults, stress management therapy consisting of relaxation training and cognitive coping has been found to produce larger (but modest) reductions in tension-type headache activity, analgesic medication use, and headache-related disability than placebo, but antidepressant medication yielded more rapid improvements in headache activity (Holroyd et al. 2001). Combined therapy was, however, found more likely to produce clinically significant reductions in headache index scores than monotherapy.

In a recent systematic review, positive effects (medium range) of autogenic training and of autogenic training versus control in a meta-analysis of at least 3 studies were found for tension headache/migraine, somatoform pain disorder (unspecified type), anxiety disorders, mild to moderate depression/dysthymia, and functional sleep disorders (Stetter and Kupper 2002), demonstrating that this technique can be of considerable value as part of a multidisciplinary pain rehabilitation program. In a recent Cochrane review on behavioral treatment of chronic low back pain (Ostelo et al. 2004), moderate evidence (2 trials, 39 people) was found in favor of progressive relaxation for a large positive effect on pain and behavioral outcomes, but only short-term.

References

Carlson, C. R., Bertrand, P. M., Ehrlich, A. D., et al. (2001). Physical self-regulation training for the management of temporomandibular disorders. *Journal of Orofacial Pain, 15*, 47–55.

Holroyd, K. A., O'Donnell, F. J., Stensland, M., et al. (2001). Management of chronic tension-type headache with tricyclic antidepressant medication, stress management therapy, and their combination: A randomized controlled trial. *Journal of the American Medical Association, 285*, 2208–2215.

Kaplan, K. H., Goldenberg, D. L., & Galvin-Nadeau, M. (1993). The impact of a meditation-based stress reduction program on fibromyalgia. *General Hospital Psychiatry, 15*, 284–289.

Larsson, B., & Carlsson, J. (1996). A school-based, nurse-administered relaxation training for children with chronic tension-type headache. *Journal of Pediatric Psychology, 21*, 603–614.

Ostelo, R. W. J. G, Tulder, M. W. van., Vlaeyen, J. W. S., et al. (2004). Behavioural treatment for chronic low-back pain. *The Cochrane Database of Systematic Reviews*, (3). Art. No. CD002014. doi: 10.1002/14651858.CD002014.pub2.

Spence, S. H., Sharpe, L., Newton-John, T., et al. (1995). Effect of EMG biofeedback compared to applied relaxation training with chronic, upper extremity cumulative trauma disorders. *Pain, 63*, 199–206.

Stetter, F., & Kupper, S. (2002). Autogenic training: A meta-analysis of clinical outcome studies. *Applied Psychophysiology and Biofeedback, 27*, 45–98.

Stuckey, S. J., Jacobs, A., & Goldfarb, J. (1986). EMG biofeedback training, relaxation training, and placebo for the relief of chronic back pain. *Perceptual and Motor Skills, 63*, 1023–1036.

Turner-Stokes, L., Erkeller-Yuksel, F., Miles, A., et al. (2003). Outpatient cognitive behavioral pain management programs: A randomized comparison of a group-based multidisciplinary versus an individual therapy model. *Archives of Physical Medicine and Rehabilitation, 84*, 781–788.

Reliability, Table 1 A contingency table from which the reliability of a diagnostic test can be derived

Observer One	Observer Two	
	Positive	Negative
Positive	a	b
Negative	c	d

Relaxation-Induced Anxiety

Definition

A sudden increase in anxiety during deep relaxation that can range from mild to moderate intensity and that can approach the level of a minor panic episode.

Cross-References

▶ Biofeedback in the Treatment of Pain
▶ Relaxation in the Treatment of Pain

Release of CCK

▶ Opioid-Induced Release of CCK

Reliability

David Vivian
Metro Spinal Clinic, South Caulfield,
VIC, Australia

Definition

▶ Reliability is the extent to which two observers obtain the same results, when using the same diagnostic test, on the same sample of patients.

Characteristics

Reliability is determined by having two observers independently apply the test on the same patients,

and recording the results in a contingency table (Table 1).

In such a table, the total number (N) of patients in the sample is a + b + c + d; "a" constitutes the number of patients in whom both observers found that the test was positive; "d" is the number of patients in whom both observers found that the test was negative; "b" is the number of patients in whom observer one found the test to be positive, but observer two found it to be negative; and "c" is the number of patients in whom observer one found the test to be negative, but observer two found it to be positive. Such a table shows that the observers agreed that the test was positive in "a" cases, and agreed that the test was negative in "d" cases. Their apparent rate of agreement is $[(a + d) / N] \times 100\%$.

The apparent rate of agreement, however, is not an accurate measure of how good the test is, for it incorporates agreement that might have occurred by chance alone. The true agreement is provided by discounting the apparent agreement for agreement by chance. This is illustrated in Fig. 1.

This figure illustrates that if there is not complete agreement in all cases, the observed agreement (Po) is less than complete agreement by the extent to which there is disagreement. The figure also illustrates that complete agreement consists of two hypothetical parts: the agreement expected by chance alone (Pe), and the range available for possible agreement beyond chance alone (1-Pe).

Meanwhile, the observed agreement (Po) also consists of one portion due to agreement by chance alone (Pe), and another portion that is the agreement beyond chance alone (Po-Pe). In determining the reliability of the test, credit is not given to the observers for that agreement which they achieved by chance alone. Their true

Reliability, Fig. 1 The distinction between observed agreement and agreement beyond chance

complete agreement in all cases (P = 1)	
observed agreement (Po)	disagreement
expected agreement by chance alone (Pe)	available range of agreement beyond chance alone (1 − Pe)
expected agreement by chance alone (Pe)	observed agreement beyond chance alone (Po − Pe)

skill, and the strength of the test, is determined by how well they agreed beyond chance alone.

That is established by measuring how far their observed agreement extends into the range of agreement available beyond chance alone, and is expressed by a statistic called kappa, where:

$$\kappa = (Po - Pe) \,/\, (1 - Pe).$$

In words, kappa is the extent to which the observed agreement (discounted for chance) fills the range of possible agreement available (also discounted for chance).

An example to help understand this concept is a multiple choice examination. If there are 100 questions each with four choices, a candidate could, on average, answer one in four questions correctly simply by guessing. By chance alone, they could score 25 %. A candidate, who scores 65 %, has not demonstrated a proficiency of 65 %, for 25 of those 65 marks could have been gained by chance alone. Their true skill is demonstrated by how well they performed beyond chance alone. Accordingly, their raw score (65) is discounted by the number of questions that they could have answered correctly by chance alone (25), which yields 40. Similarly, the total number of questions (100) is discounted by the number of questions that might have been answered correctly by chance alone (25), to yield the available number of questions that might have been answered correctly beyond chance alone (75). The true skill of the candidate is then the proportion of the available number of questions (75) that the candidate correctly answered (40),

Reliability, Table 2 A contingency table from which agreement can be derived

Observer One	Observer Two		Totals
	Positive	Negative	
Positive	a	b	a + b
Negative	c	d	c + d
Totals	a + c	b + d	N = a + b + c + d

i.e., 40/75 = 53 %. The true skill of the candidate, discounted for chance, is 53 %, not the raw score of 65 %.

Not obvious in the determination of reliability is how the expected agreement by chance alone should be estimated. This is based on what the two observers find, on average, which is shown by the sums of the columns and rows of the contingency table (Table 2).

This table shows that, overall, observer one recorded positive findings in (a + b) cases. On average, therefore, this observer would record positive findings in (a + b)/N of all cases presented to him. Meanwhile, observer two found (a + c) cases presented to him to be positive. If these cases were presented to observer one, the number that observer one would be expected to record as positive, by chance alone, would be (a + c) × [(a + b) / N].

Similarly, the proportion of cases that observer one recorded as negative is (c + d)/N. The number of cases that observer two found to be negative is (b + d). If these cases were presented to observer one, the number that would be expected to be recorded as negative, by chance alone, would be (b + d) × [(c + d) / N].

These derived numbers provide an estimate of the agreement expected by chance alone (Pe), i.e.,

$$Pe = [(a + c)(a + b)/N + (b + d)(c + d)/N] / N$$

This is different from the observed agreement (Po), viz.

$$Po = (a + d)/N.$$

From these equations, Po, Pe, (Po-Pe), and (1-Pe) can be calculated, and hence, kappa can be calculated. Readers interested in further reading about kappa can consult the original literature (Cohen 1960) or various other educational resources (Sackett et al. 1991; Bogduk 1998).

For practical purposes, however, readers should understand that once calculated, kappa can range in value from 0 to 1, or from 0 to -1. Negative values of kappa indicate abject disagreement, which occurs only for very unreliable tests. For most tests, the values are positive. A value of 0 indicates no agreement beyond chance alone. A value of 1 indicates complete, i.e., perfect agreement.

Values between 0 and 1 can be accorded a range of verbal descriptors (Table 3). The kappa values indicate quantitatively, and the descriptors indicate qualitatively, just how reliable the diagnostic test is.

If practitioners use diagnostic tests with high kappa scores, they can be confident that the reliability of the test is good. However, if the kappa score is low, practitioners should realize that the diagnostic test does not work well; that someone else using the same test on the same patient is not likely to obtain the same result. Consequently, they have grounds to question or doubt the result that they have obtained.

The kappa score does not, and cannot, determine which of two observers is correct. It measures only how consistent any two observers are, or might expect to be. However, if consistency is lacking, the test or how it is used is defective. In essence, diagnostic tests with low kappa scores just do not work well enough to be reliable. Their results are little better than guessing, and do not reflect professional proficiency.

Reliability, Table 3 Verbal translations of kappa scores

Kappa Value	Descriptor
0.8–1.0	Very good
0.6–0.8	Good
0.4–0.6	Moderate
0.2–0.4	Slight
0.0–0.2	Poor

Cross-References

▶ Multiaxial Assessment of Pain
▶ Oswestry Disability Index
▶ Pain Assessment in Children
▶ Pain Assessment in Neonates

References

Bogduk, N. (1998). Truth in musculoskeletal medicine II. Truth in diagnosis: Reliability. *Australasian Musculoskeletal Medicine, 3,* 21–23.
Cohen, J. (1960). A coefficient of agreement for nominal scales. *Educational and Psychological Measurement, 20,* 37–46.
Sackett, D. L., Haynes, R. B., Guyatt, G. H., & Tugwell, P. (1991). *Clinical epidemiology. A basic science for clinical medicine* (2nd ed., pp. 119–139). Boston: Little, Brown.

REM Sleep

▶ Rapid Eye Movement Sleep

Remote Management of Pediatric Pain

Tonya M. Palermo
Seattle Children's Hospital Research Institute,
CW8-6 -Child Health, Behavior and Development,
University of Washington, Seattle, WA, USA

Synonyms

Computer applications; Distance treatment; eHealth; Internet; Telehealth; Telemedicine

Definition

Remote management of pain refers to pain interventions utilized outside of the clinic/hospital setting to reach children in their homes or communities. Interventions are typically delivered using some form of technology, such as the Internet, or may rely on other media such as telephone counseling or use of written self-help materials. There are several related terms including self-management, minimal therapist contact, distance treatment, telehealth, and eHealth that also may refer to remote pain management. eHealth refers specifically to the application of interactive and communication technologies to improve or enable health and health care in children, adolescents, and families (Palermo and Wilson 2009). Most typically, remote management of pain includes monitoring, counseling, and/or delivery of behavioral and cognitive-behavioral therapy (CBT) interventions.

Introduction

Many children's hospitals serve large geographic regions. There are often major barriers for families to access specialized pain services due to this geographical distance separating children and families from qualified specialists. As early as the 1980s, clinicians and researchers recognized the advantages of moving interventions out of the clinic and into the homes of children and families. The primary purpose of remote pain management was to address difficulties with access; to contain costs of family travel and therapist time; and to better allocate clinical resources.

There has been limited consideration of remote management for acute pediatric pain. Some exceptions to this are for pediatric surgery. Telephone and telehealth services have been used for pediatric surgical consultations and follow-up. More commonly, remote management has been used to deliver interventions to children who require longer-term pain management such as children and families coping with chronic pain. The type of intervention that has received the most attention to date is the delivery of psychological interventions remotely. This is a particular need as the majority of youth with chronic pain do not receive psychological interventions. For example, in one study of health service patterns among youth with chronic pain, while the majority of youth had used medications for pain, only 2.8 % of children and adolescents received any psychological treatment services (Perquin, et al. 2001).

Characteristics

The earliest forms of remote management relied on low levels of technology including written self-help manuals, portable biofeedback monitors, and relaxation audiotapes used by the child for self-guided behavioral intervention for pain (e.g., Burke and Andrasik 1989; McGrath et al. 1992). Primarily, children and adolescents with headache disorders were the focus of this early work. The results of this work were very promising, indicating both intervention efficacy and cost-effectiveness. Subsequently, as technological advances became available to the masses, intervention delivery options expanded to personal computers via CD-ROM applications and then via Internet interventions. Most recently, remote pain management has expanded to the use of mobile communications (e.g., cell phones) Rosser & Eccleston (2011).

Five specific intervention modalities that have been utilized in pediatric pain populations are reviewed below, including the use of written materials/booklets, telephone counseling and telehealth, CD-ROM interventions, Internet Interventions, and mobile communications.

Types of Interventions Utilized
Booklets/Written Manuals/Tapes
Remote treatment utilizing written materials and audiocassette tapes rely on the child or parent reading and working through manualized behavioral treatment programs. For example, McGrath and colleagues (1992) used a self-help manual and relaxation audiotape to teach CBT to teens with recurrent headaches. Several investigators

explored the use of home biofeedback monitoring in youth with headache. Youth were sent home with a portable temperature monitor to practice skills taught in a single clinic session. These studies demonstrated not only that self-administered treatments were feasible but that they also produced equal or better results as equivalent therapist-led in-office treatments (Burke and Andrasik 1989; McGrath et al. 1992; Kroener-Herwig and Denecke 2002). Importantly, in a study examining the costs of treatment, the remote self-administered treatment was also found to be three times more cost-effective compared to the equivalent in-office treatment (Larsson et al. 1987). In a subgroup analysis conducted as part of a meta-analytic review of psychological treatments for chronic pediatric pain, Palermo and colleagues (2010) reported that home-based treatments had equivalent effect sizes for pain reduction as clinic-based treatments.

Telephone and Telemedicine

Telephone counseling refers to the provision of support and therapeutic interventions via a health care professional or trained nonprofessional delivered on the phone to the child and/or parent. Telephone counseling may serve as the only medium by which pain intervention is delivered, or telephone counseling may be adjunctive to other therapeutic materials delivered via booklet or computer. For example, Hicks and colleagues (2006), in a randomized trial with children and adolescents with recurrent headache or abdominal pain, used telephone counseling to support an educational website to deliver CBT.

Telephone and telehealth services have been used for pediatric surgical consultations and follow-up, finding that it is an effective method for addressing postoperative pain concerns, and that clinicians and patients are highly satisfied with the experience (e.g., Czarnecki et al. 2007; Shivji et al. 2011). There are ongoing clinical programs providing telemedicine consultations for pain problems, but limited publications have resulted at this time related to the experience or of patient outcome.

CD-ROM Intervention Programs

Educational and behavioral intervention programs that are delivered via CD-ROMs have the advantage of using interactive media to engage children in psychological treatments while being cheap and portable. Investigators have also developed their own CD-ROM intervention programs to provide remote pain management. For example, Connelly and colleagues (2006) developed and evaluated a CD-ROM pain management program (called Headstrong), which focused on relaxation training for children aged 7–12 years with recurrent headache. They found significant improvements in headache activity (less intense and frequent pain) in children who received the 4-week intervention versus children receiving standard care. The primary disadvantages of CD-ROMs are the inability to update information or to communicate with the user in real time, and increased difficulty in dissemination in comparison to publicly available resources such as the Internet. Thus, at this point in time, CD-ROM intervention programs have been replaced with other technologies.

Internet Interventions

The Internet is a medium with a potential to reach a large number of people at a low cost. It makes two-way communication possible, which facilitates the development of cost-effective interventions. This mode of presentation has several advantages including allowing children and their parents private access to treatment information at any time, thereby increasing treatment convenience as well as helping to decrease costs, and takes advantage of a mode of communication, the computer, that is often highly appealing to children, and increasingly available in the majority of households in developed countries. The expressive graphic, video, and audio capabilities of the computer/Internet make it an ideal tool for use with children.

Internet interventions are typically behaviorally based treatments that have been transformed for delivery via the Internet. The International Society for Research on Internet Interventions further describes these interventions as characteristically highly structured, self-guided, based on effective face-to-face interventions, personalized to the user, interactive, enhanced by multimedia features, and tailored to provide feedback and follow-up. These features distinguish Internet interventions from other websites that solely provide health information content, but are not interactive, personalized, or designed to achieve behavior change (Ritterband et al. 2003).

A recent systematic review summarized the evidence base for internet interventions for chronic pain in both adult and pediatric populations (Bender et al. 2011), finding that most cognitive-behavioral treatment (CBT) studies showed an improvement in pain and activity limitations. In this review, five studies targeted children and adolescents with chronic pain. Three examples are provided here to illustrate the development of this research area.

Palermo and colleagues (2009) reported on the findings of internet family CBT compared to standard care in 48 adolescents aged 11–17 years with musculoskeletal, abdominal, or headache pain. Significant group differences in pain and disability were found at posttreatment in that children receiving the Internet treatment had greater reductions in pain and disability in comparison to the standard care group. Pain and disability continued to decline at 3-month follow-up. This investigatory team is currently evaluating the Internet intervention in a much larger trial with 300 youth recruited from around the USA and Canada.

In a trial conducted in Germany, 65 youth with recurrent headache were randomized to two different online programs (CBT or relaxation) or to an educational control group (Trautmann and Kroner-Herwig 2010). Findings were somewhat mixed in that all groups improved, but there were significant differences in responders (50 % reduction in headache frequency) at posttreatment in the CBT group compared to the other two groups.

Stinson and colleagues (2010) conducted a randomized controlled trial with adolescents with juvenile idiopathic arthritis (JIA) to teach self-management skills more broadly. In this trial, 46 adolescents with JIA were randomized to online self-management or to standard care. Findings were also mixed; although there were no group differences on health-related quality of life or stress, there was a significant decrease in average weekly pain intensity.

Mobile Communications

Cell phones and text messaging have been used to deliver interventions for a range of health conditions including for health promotion as well as symptom management (Rosser et al. 2009).

A recent review found 111 smartphone applications (or apps) that were aimed at pain education or pain management. However, at this time, there are no published examples of the use of apps for pediatric pain management, although there are studies in progress in this area. The ability of cell phones to access internet-based resources is promising, however, and is included here as a potential modality for the delivery of remote pain management. In current clinical practice, there are many apps in use to teach behavioral skills to children such as relaxation and deep breathing.

Summary

Remote pain interventions are the largest treatment advance to make evidence-based pain interventions more accessible to children and families. Over time, newer forms of technology have been used to offer self-administered treatments for pain in children. Barriers to evidence-based pediatric pain care are a major issue for children living all over the world and remote pain management could be a promising solution.

The field of remote pain management is rapidly developing, and there are as yet few published examples for certain methods of technology delivery. Internet interventions have received the most research attention to date, with published examples for several different pediatric chronic pain conditions with promising findings for pain reduction. Telemedicine, while in widespread use clinically for many pediatric health conditions, has not yet been evaluated in pediatric pain. Smartphone applications, while readily available for pain education or management, have also not yet been evaluated.

There remain issues to address related to the high cost and effort in initial development of internet and mobile applications, and there are not yet precedences for widespread dissemination of such programs. Individual differences in previous experience with computer technology may impact the efficacy of remote intervention using these modalities, and there remain some gaps in access along sociodemographic lines. Nevertheless, remote pain management holds promise for greatly reducing disparities in pain care by enhancing access to evidence-based pain interventions.

References

Bender, J. L., Radhakrishnan, A., Diorio, C., Englesakis, M., & Jadad, A. R. (2011). Can pain be managed through the Internet? A systematic review of randomized controlled trials. *Pain, 152*, 1740–1750.

Burke, E. J., & Andrasik, F. (1989). Home vs clinic-based biofeedback treatment for pediatric migraine: Results of treatment through one-year follow-up. *Headache, 29*, 434–440.

Connelly, M., Rapoff, M. A., Thompson, N., & Connelly, W. (2006). Headstrong: A pilot study of a CD-ROM intervention for recurrent pediatric headache. *Journal of Pediatric Psychology, 31*, 737–747.

Czarnecki, M. L., Garwood, M. M., & Weisman, S. J. (2007). Advanced practice nurse-directed telephone management of acute pain following pediatric spinal fusion surgery. *Journal for Specialists in Pediatric Nursing, 12*, 159–169.

Hicks, C. L., von Baeyer, C. L., & McGrath, P. J. (2006). Online psychological treatment for pediatric recurrent pain: A randomized evaluation. *Journal of Pediatric Psychology, 31*, 724–736.

Kroener-Herwig, B., & Denecke, H. (2002). Cognitive-behavioral therapy of pediatric headache: Are there differences in efficacy between a therapist-administered group training and a self-help format? *Journal of Psychosomatic Research, 53*, 1107–1114.

Larsson, B., Daleflod, B., Hakansson, L., & Melin, L. (1987). Therapist assisted versus self-help relaxation treatment of chronic headaches in adolescents: A school-based intervention. *Journal of Child Psychology and Psychiatry, 28*, 127–136.

McGrath, P. J., Humphreys, P., Keene, D., Goodman, J. T., Lascelles, M. A., Cunningham, S. J., & Firestone, P. (1992). The efficacy and efficiency of a self-administered treatment for adolescent migraine. *Pain, 49*, 321–324.

Palermo, T. M., & Wilson, A. C. (2009). eHealth applications in pediatric psychology. In M. Roberts &

R. Steele (Eds.), *Handbook of pediatric psychology* (4th ed., pp. 227–237). New York: Guilford Press.

Palermo, T. M., Wilson, A. C., Peters, M., Lewandowski, A., & Somhegyi, H. (2009). Randomized controlled trial of an Internet-delivered family cognitive-behavioral therapy intervention for children and adolescents with chronic pain. *Pain, 146*, 205–213.

Palermo, T. M., Eccleston, C., Lewandowski, A. S., Williams, A. C., & Morley, S. (2010). Randomized controlled trials of psychological therapies for management of chronic pain in children and adolescents: An updated meta-analytic review. *Pain, 148*, 387–397.

Perquin, C. W., Hunfeld, J. A., Hazebroek-Kampschreur, A. A., van Suijlekom-Smit, L. W., Passchier, J., Koes, B. W., & van der Wouden, J. C. (2001). Insights in the use of health care services in chronic benign pain in childhood and adolescence. *Pain, 94*, 205–213.

Ritterband, L. M., Gonder-Frederick, L. A., Cox, D. J., Clifton, A. D., West, R. W., & Borowitz, S. M. (2003). Internet interventions: In review, in use, and into the future. *Professional Psychology: Research and Practice, 34*, 527–534.

Rosser, B. A., & Eccleston, C. (2011). Smartphone applications for pain management. *Journal of Telemedicine and Telecare, 17*, 308–312.

Rosser, B. A., Vowles, K. E., Keogh, E., Eccleston, C., & Mountain, G. A. (2009). Technologically-assisted behavior change: A systematic review of studies of novel technologies for the management of chronic illness. *Journal of Telemedicine and Telecare, 15*, 327–338.

Shivji, S., Metcalfe, P., Khan, A., & Bratu, I. (2011). Pediatric surgery telehealth: Patient and clinician satisfaction. *Pediatric Surgery International, 27*, 523–526.

Stinson, J., McGrath, P. J., Hodnett, E. D., et al. (2010). An Internet-based self-management program with telephone support for adolescents with arthritis: A pilot randomized controlled trial. *Journal of Rheumatology, 37*, 1944–1952.

Trautmann, E., & Kroner-Herwig, B. (2010). A randomized controlled trial of Internet-based self-help training for recurrent headache in childhood and adolescence. *Behaviour Research and Therapy, 48*, 28–37.

Renin

Definition

Renin is a proteolytic enzyme secreted by specific kidney cells. It catalyzes the formation of angiotensin and thus affects blood pressure.

Cross-References

▶ NSAIDs, Adverse Effects

Repetitive Strain Injuries

Synonyms

RSI

Definition

A descriptive term used to label pain associated with repetitive activities.

Cross-References

▶ Disability, Upper Extremity

Repetitive Transcranial Magnetic Stimulation

Synonyms

rTMS

Definition

This form of stimulation uses a rapidly (approximately 10 Hz) changing magnetic field to stimulate the nervous system for the treatment of pain, movement and psychiatric disorders.

Cross-References

▶ Stimulation Treatments of Central Pain

Repetitiveness of Movement

Definition

This term is used for repetitive movements with a cycle time of at least 1 per 30 s.

Cross-References

▶ Ergonomic Counseling

Repriming

Definition

Repriming refers to the resetting of the electrical status of a discharged neuron that permits re-firing.

Cross-References

▶ Molecular Contributions to the Mechanism of Central Pain

Rescue Analgesic

Definition

Rescue analgesic agents are medications prescribed in addition to regularly scheduled analgesic medications, which are intended to be taken during episodes of pain not controlled by a patient's scheduled analgesic regimen.

Cross-References

▶ Evoked and Movement-Related Neuropathic Pain

Residual Functional Capacity

Synonyms

RFC

Definition

A person's remaining capacity to perform work-related physical and mental activities. The assessment of RFC is a determination reserved for the Commissioner of Social Security.

Cross-References

▶ Disability Evaluation in the Social Security Administration

Residual Limb Pain

Definition

Also called stump pain, refers to pain in the portion of the body adjacent to the amputation line.

Cross-References

▶ Pain Assessment in Children
▶ Phantom Limb Pain, Treatment

Resiniferatoxin

Synonyms

RTX

Definition

An investigational drug for the treatment of interstitial cystitis that appears to desensitize urothelial nerve fibers.

Cross-References

▶ Interstitial Cystitis and Chronic Pelvic Pain

Resistance

Definition

Patient behaviors (such as arguing, changing the subject, interrupting, denying a problem) that indicate the patient is not interested in changing his or her behavior.

Cross-References

▶ Chronic Pain, Patient-Therapist Interaction

Respiratory Centers

Definition

Brainstem centers that specifically regulate ventilation.

Cross-References

▶ Placebo Analgesia and Descending Opioid Modulation

Respiratory Depression

Definition

Severe slowing or arrest of spontaneous breathing caused by morphine and morphine-like drugs.

Cross-References

▶ Postoperative Pain, Acute Pain Management Principles

Respondent Conditioning

▶ Classical Conditioning and Chronic Pain Syndromes

Respondent Conditioning of Chronic Pain

Herta Flor
Department of Cognitive and Clinical
Neuroscience, Central Institute of Mental Health,
Medical Faculty Mannheim, Heidelberg
University, Mannheim, Germany

Synonyms

Classical conditioning; Pavlovian conditioning

Definition

Classical conditioning refers to a type of learning, where a neutral or conditioned stimulus (CS) is repeatedly paired with a biologically significant or unconditioned stimulus (US) and comes to elicit a conditioned response (CR) that is often, but not always, similar to the original response to the US, the unconditioned response (UR).

Characteristics

Gentry and Bernal (1977) were the first to describe a respondent model of the development of chronic pain. They suggested that acute pain (the unconditioned stimulus, US) that is associated with sympathetic activation and increased generalized ▶ muscle tension (the unconditioned response, UR) may evolve into a chronic pain problem through a process of classical conditioning. In their model, the frequent association of innocuous stimuli (conditioned stimuli, CS) such as a certain environment or

a certain body position with acute pain states may elicit ▶ fear of pain (see also ▶ Fear and Pain), sympathetic activation, and increased muscle tension (the conditioned response, CR) to these previously neutral stimuli. Gentry and Bernal further suggested that this process of conditioning might lead to a ▶ pain-tension cycle, which may maintain the chronic pain problem independent of the original tissue damage. Thus, the pain resulting from this process may be a purely ▶ musculoskeletal type of pain completely unrelated to the original cause of pain.

Linton et al. (1985) have further elaborated on the respondent conditioning perspective of chronic pain and have emphasized that there is a wide range of stimuli that may serve as CSs and USs. They point out that both direct and indirect noxious stimuli (e.g., pain originating from a herniated disk, carrying something heavy with a problem back) may be important USs. Whereas the US may be related to an injury, the UR usually manifests itself as activation of the sympathetic nervous system, as anxiety and muscle tension increases. The CS related to the US and the CR is not pain initially, but can be pain-provoking over time. Linton et al. (1985) state that there is no evidence of classical conditioning of neurological pain, but only of anxiety and related physiological activation. This elevated anxiety may then lead to heightened sensitivity to noxious stimuli (Asmundson et al. 1999; Vlaeyen and Linton 2000). Whether muscle tension produced by muscle contraction leads to pain is dependent on (1) the amount of muscle contraction, (2) the duration of the contraction, and (3) the individual vulnerability (predispositional factors such as prior injury or personality characteristics). The pain produced by muscle tension may thus not be the same as the original pain; however, patients (and physicians) may not be able to discriminate between the two.

There is evidence on the role of respondent conditioning in chronic pain patients. For example, patients who suffered from upper back pain and healthy controls received unconditioned stimuli (painful electric stimulation) to the forearm, and a picture of a dead body (thought to have a high belongingness with pain) served as the CS+ (followed by shock most of the time), and picture of a rabbit (positive cue unrelated to pain) served as the CS− (never followed by shock). The chronic back pain patients already showed anticipatory high muscle tension in the arm during the preconditioning phase, when the dead body was never followed by pain. In the learning phase, they displayed an increase of the muscular response in the arm close to where the painful stimulus was applied and, in addition, had conditioned responses in the trapezius muscle (Schneider et al. 2004; Vlaeyen et al. 1999). These results suggest that a conditioning process may already have taken place in chronic pain patients and that the often observed heightened stress reactivity in these patients (Flor et al. 1992; Moulton and Spence 1992) may be a result of classical conditioning that is quite specific to the site of pain. Klinger et al. (2010) also showed both enhanced unconditioned and conditioned muscular responses in patients with chronic back pain and tension-type headache. These data lend credence to the assumption that classically conditioned pain responses may not only be a consequence of the occurrence of an acute pain episode but may also play a role in the evolution of chronic pain states. The role of high levels of muscle tension in the development of pain has also been substantiated by stress-induction studies in chronic pain patients. For example, syndrome-specific (i.e., lower back and facial pain) alterations in muscle tension in response to stress were observed in patients with chronic low back pain or ▶ temporomandibular pain and dysfunction. Subjects who suffered from chronic temporomandibular pain showed increases in masseter muscle tension when exposed to personally relevant stress episodes; in contrast, subjects who suffered from chronic back pain selectively tensed their lumbar erector spinae muscles. Healthy controls showed high cardiovascular responses but not elevated muscular reactivity to stress. Skin conductance levels were equally elevated in all groups when stress or pain, as compared to neutral imagery, was employed. The observation that patients respond selectively to ▶ personally relevant stressful situations but not to general stressors suggests the presence of classically conditioned responses rather than unconditioned responses in these patients (see Flor et al. 1992, 1991; Glombiewski et al. 2008; Moulton and Spence 1992). The specificity of the response to the situation from the patients' lives support the learning (see also ▶ Modeling, Social Learning in Pain) view.

To illustrate the proposed process, from a respondent conditioning perspective, the patient may have learned to associate increases in muscle tension with all kinds of stimuli that were originally associated with nociceptive stimulation. Thus, sitting, walking, bending, or even thoughts about these movements may elicit ▶ anticipatory anxiety and increases in muscle tension. This fear of movement or "▶ kinesiophobia" (see also ▶ Fear and Pain) has been discussed as an important factor in the maintenance and exacerbation of chronic pain (e.g., Asmundson et al. 1997; Flor and Turk, 2011; Vlaeyen and Linton 2012). Subsequently, patients may display maladaptive responses to any number of stimuli and reduce the frequency of performance of many activities other than those that initially reduced pain. This process is referred to as stimulus generalization. The anticipatory anxiety related to movement may act as a conditioned stimulus for muscle tension (conditioned response) that may be maintained after the original unconditioned stimulus (e.g., injury) and unconditioned response (pain and muscle tension) have subsided. Thus, although the original association between injury and pain results in anxiety regarding movement, over time the anxiety may lead to increased muscle tension and pain, even if the nociceptive stimuli are no longer present. In addition, stress situations can increase muscle tension levels and cause sympathetic activation and may, thus, reinforce this process. Many of our patients have reported that an acute pain problem evolved into chronic pain at a time where personal stressors co-occurred with the pain. Stress situations may serve as additional US and also as CS for muscle tension increases, increased sympathetic activation, and subsequently pain.

Nonoccurrence of pain is a powerful reinforcer for the reduction of movement. Thus, the original respondent conditioning may be complemented by an *operant process* (see also ▶ Operant Treatment of Chronic Pain; ▶ Operant Perspective of Pain), whereby the nociceptive stimuli need no longer be present for the avoidance behavior to occur. People who suffer from acute back pain, regardless of the cause, may adopt specific behaviors (e.g., limping) to avoid pain, and they may never obtain "corrective feedback" because they fail to perform more natural movements and fail to learn that they may not induce pain (Vlaeyen et al. 1995). Reduction in physical activity may subsequently result in muscle atrophy and increased disability. In this manner, the physical abnormalities proposed by biomechanical models of pain (cf. Dolce and Raczynski 1985) may actually be secondary to changes in behavior initiated through learning. Similarly, Lethem et al. (1983) have emphasized that chronic pain patients tend to focus their attention on impending pain and subsequently avoid many types of activity, thus fostering the development of disability and depression.

As the pain symptoms persist, more and more situations may elicit anxiety and anticipation of pain. Depression, pain, and dependence on medication may follow, further intensifying the pain-tension cycle. Thus, psychological expectations may lead to modified behavior that, in turn, produces physical changes leading to still further physical deconditioning. In the case of chronic pain, the anticipation of, or prevention of, pain may be sufficient for the long-term maintenance of the avoidance behaviors. These observations add support to the importance of active physical therapy, with patients progressively increasing their activity levels despite fear of injury and discomfort associated with renewed use of deconditioned muscles. Disconfirmation of the expected and feared outcomes by exposure may help to desensitize patients and will serve to reinforce performance of additional activities that had previously been feared and avoided. The importance of operant learning factors will be further discussed in the following section. The role of fear of pain and movement is described in another essay.

In addition to the anticipation of pain and to associated physiological processes, the subjective evaluation of pain may also be modified by its association with affective variables such as positive, negative, or neutral emotional states. Wunsch et al. (2003) showed that aversive slides paired with a painful stimulus lead to higher pain intensity ratings of the same painful stimulus than when appetitive or neutral slides preceded the painful stimulus. They demonstrated that the conditioning changed the evaluation of the painful stimulus without the participants' awareness and concluded that pain is the result of complex cognitive-emotional interactions that need to be considered when the experience of pain is evaluated. In a study in healthy controls, Diesch and Flor (2007) showed that the use of pain as an unconditioned stimulus, and non-painful tactile stimuli as conditioned stimuli, leads to a fast acquisition of conditioned muscle tension increases, as well as an expansion of the representation of the CS that signals pain in primary somatosensory cortex. In addition, the presence of pain, no matter if the CS was paired with the US or not, leads to a more aversive evaluation of the non-painful tactile sensation that served as CS. These data suggest that the mere presence of pain leads to a more aversive evaluation of any type of non-painful bodily sensation. This might explain why chronic pain patients frequently complain about a host of physical symptoms and are often classified as suffering from ▶ somatization disorder.

Meulders et al (2011, 2013) used movement as a conditioned stimulus and showed that its association with a painful unconditioned stimulus can lead to enhanced fear of movement, slowed movement, and enhanced startle responses as an indication of enhanced emotional processing of a normally innocuous movement. They also found enhanced generalization to other movements when contextual conditioning using unpredictable pain was used suggesting that this might be important for widespread pain conditions. This approach has also been used in the treatment of pain (Meulders and Vlaeyen 2012) and is relevant for exposure treatment for chronic pain (see ▶ Disability, Fear of Movement).

Finally, several animal studies and studies in humans have shown that pain-inhibiting descending systems, both opioid and non-opioid mediated, can be classically conditioned and thus be influenced by learning (e.g., Flor et al. 2002). For example, the ticktock of a clock (CS) was systematically combined with the application of stress (mental arithmetic and noise, US), which leads to a stress-induced hypoalgesia or ▶ analgesia (SIA). After several pairings of the CS and the US, the ticktock of the clock alone elicited a hypoalgesic response that could be partially reversed by naloxone, an opiate antagonist. This suggests that biochemical variables involved in the transmission of nociception as well as ▶ antinociception are also influenced by learning. It was also shown that SIA activates similar descending inhibitory mechanisms as, for example, ▶ placebos, and that this involves brain regions such as the rostral anterior cingulate cortex and the periaqueductal grey (Yilmaz et al. 2010). These learning processes could be used to enhance analgesic processes in states of chronic pain.

These results support the contribution of respondent factors in the maintenance and exacerbation of chronic pain syndromes. They also suggest that interventions designed to alter high levels of muscle tension, and specifically stress reactivity as well as the cognitive and emotional evaluation of pain, might be of great value in the treatment of chronic pain patients.

References

Asmundson, G. J., Norton, G. R., & Allerdings, M. D. (1997). Fear and avoidance in dysfunctional chronic back pain patients. *Pain, 69*(3), 231–236.

Asmundson, G. J., Norton, P. J., & Norton, G. R. (1999). Beyond pain: The role of fear and avoidance in chronicity. *Clinical Psychology Review, 19*(1), 97–119.

Diesch, E., & Flor, H. (2007). Alteration in the response properties of primary somatosensory cortex related to differential aversive Pavlovian conditioning. *Pain, 131*(1–2), 171–180.

Dolce, J. J., & Raczynski, J. M. (1985). Neuromuscular activity and electromyography in painful backs: Psychological and biomechanical models in assessment and treatment. *Psychological Bulletin, 97*(3), 502–520.

Flor, H., Birbaumer, N., Schugens, M. M., & Lutzenberger, W. (1992). Symptom-specific psychophysiological responses in chronic pain patients. *Psychophysiology, 29*(4), 452–460.

Flor, H., Birbaumer, N., Schulte, W., & Roos, R. (1991). Stress-related electromyographic responses in patients with chronic temporomandibular pain. *Pain, 46*(2), 145–152.

Flor, H., Birbaumer, N., Schulz, R., Grüsser, S. M., & Mucha, R. F. (2002). Pavlovian conditioning of opioid and nonopioid pain inhibitory mechanisms in humans. *European Journal of Pain, 6*(5), 395–402.

Flor, H., & Turk, D. C. (2011). *Chronic pain: An integrated biobehavioral approach.* Seattle: IASP Press.

Gentry, W. D., & Bernal, G. A. A. (1977). Chronic pain. In R. Williams & W. D. Gentry (Eds.), *Behavioral approaches to medical treatment* (pp. 173–182). Cambridge: Ballinger.

Glombiewski, J. A., Tersek, J., & Rief, W. (2008). Muscular reactivity and specificity in chronic back pain patients. *Psychosomatic Medicine, 70*(1), 125–131.

Klinger, R., Matter, N., Kothe, R., Dahme, B., Hofmann, U. G., & Krug, F. (2010). Unconditioned and conditioned muscular responses in patients with chronic back pain and chronic tension-type headaches and in healthy controls. *Pain, 150*(1), 66–74.

Lethem, J., Slade, P. D., Troup, J. D. G., & Bentley, G. (1983). Outline of a fear-avoidance model of exaggerated pain perception. *Behaviour Research and Therapy, 21*(4), 401–408.

Linton, S. J., Melin, L., & Götestam, K. G. (1985). Behavioral analysis of chronic pain and its management. In M. Hersen, A. Bellack, & M. Eisler (Eds.), *Progress in behavior modification* (pp. 1–42). New York: Academic Press.

Meulders, A., Vansteenwegen, D., & Vlaeyen, J. W. (2011). The acquisition of fear of movement-related pain and associative learning: A novel pain-relevant human fear conditioning paradigm. *Pain, 152*(11), 2460–2469.

Meulders, A., & Vlaeyen, J. W. (2012). Reduction of fear of movement-related pain and pain-related anxiety: An associative learning approach using a voluntary movement paradigm. *Pain, 153*(7), 1504–1513.

Meulders, A., & Vlaeyen, J. W. (2013). The acquisition and generalization of cued and contextual pain-related fear: An experimental study using a voluntary movement paradigm. *Pain, 154*(2), 272–282.

Moulton, B., & Spence, S. H. (1992). Site-specific muscle hyper-reactivity in musicians with occupational upper limb pain. *Behaviour Research and Therapy, 30*(4), 375–386.

Schneider, C., Palomba, D., & Flor, H. (2004). Pavlovian conditioning of muscular responses in chronic pain patients: Central and peripheral correlates. *Pain, 112*(3), 239–247.

Vlaeyen, J. W., Kole-Snijders, A. M., Boeren, R. G., & van Eek, H. (1995). Fear of movement/(re)injury in chronic low back pain and its relation to behavioral performance. *Pain, 62*(3), 363–372.

Vlaeyen, J. W., & Linton, S. J. (2000). Fear-avoidance and its consequences in chronic musculoskeletal pain: A state of the art. *Pain, 85*(3), 317–332.

Vlaeyen, J. W., & Linton, S. J. (2012). Fear-avoidance model of chronic musculoskeletal pain: 12 years on. *Pain, 153*(6), 1144–1147.

Vlaeyen, J. W., Seelen, H. A. M., Peters, M., de Jong, P., Aretz, E., Beisiegel, E., et al. (1999). Fear of movement/(re)injury and muscular reactivity in chronic low back pain patients: An experimental investigation. *Pain, 82*(3), 297–304.

Wunsch, A., Philippot, P., & Plaghki, L. (2003). Affective associative learning modifies the sensory perception of nociceptive stimuli without participant's awareness. *Pain, 102*(1–2), 27–38.

Yilmaz, P., Diers, M., Diener, S., Rance, M., Wessa, M., & Flor, H. (2010). Brain correlates of stress-induced analgesia. *Pain, 151*(2), 522–529.

Respondents

Definition

Activity elicted by stimuli to which one reflex responds, hence the term "respondent." The light of a room suddenly brightens, eliciting a reflexive pupillary response. At the sight of food to a hungry person, reflexy elicts salivation. These are autonomically mediated.

Cross-References

▶ Training by Quotas

Response Prevention

Definition

Response prevention consists of gradually refraining from safety behaviors (checking body parts, escape from and avoiding pain-increasing activities) that lead to a temporary reduction of fear.

Cross-References

▶ Fear Reduction Through Exposure In Vivo

Response to Analgesics

Definition

Different pains respond in different ways to analgesics. Inflammatory pain is thought to be most sensitive to opioids. Neuropathic pain is less sensitive to opioids, although some opioids like methadone, levorphanol, and fentanyl may have more effect than morphine. Bone pain is less sensitive to opioids, but the effect of opioids increases after administration of NSAIDs. Muscle cramp may be insensitive to opioids or may even be exacerbated by it. Analyzing complex pain histories and responses to different analgesics should be investigated. The effect of analgesics should always be seen in the context of the adverse effects.

Cross-References

▶ Cancer Pain

Response-dependent Method

▶ Pain Evaluation, Psychophysical Methods

Responses of Spinothalamic Tract Neurons to Pruritic Stimuli

Steve Davidson
Pain Center & Department of Anesthesiology, Washington University School of Medicine, MO, USA

Synonyms

Dorsal horn; Itch; Pruritus; Scratch; Second-order/ascending pruriceptive pathway; Spinal cord

Definition

Spinothalamic tract (STT) neurons located in the dorsal horn of the spinal cord receive direct and indirect inputs from primary afferent sensory neurons, including those responsive to pruritic stimuli. Spinothalamic tract axons decussate and then ascend within the ventrolateral funiculus to terminate mainly in several nuclei of the ventrobasal and posterior thalamus, providing a route for the transmission of pruritic information from the spinal cord to the brain.

Characteristics

The perception of itch induced by a cutaneous pruritogen requires that sensory information ascend to the brain because lesion of the anterolateral funiculus of the spinal cord, where several sensory tracts ascend, abolishes the perception of itch from regions of the skin below and contralateral to the lesion (Bickford 1938; White and Sweet 1969). An early study showed that a cutaneously applied non-histaminergic pruritogen (cowhage) could evoke action potentials from individual axons of wide dynamic range neurons that were located in the ventrolateral funiculus of cat spinal cord (Wei and Tuckett 1991). Among the ascending tracts of the ventrolateral funiculus, the STT is well established to be involved in the transmission of information related to noxious and innocuous thermal, mechanical, and chemical stimuli. The STT has also been identified as an important route for transmitting pruriceptive information to the brain (Andrew and Craig 2001). Spinothalamic tract neurons were identified by antidromic activation from the thalamus of anesthetized cats and histamine was administered to the skin by iontophoresis. This produced activation in a subset of mechanically insensitive lamina I STT neurons in a manner consistent with the time course of histamine-evoked itch in humans. These histamine-responsive STT neurons exhibited almost no spontaneous activity, received inputs from slowly conducting primary afferent fibers, and propagated ascending action potentials to the thalamus with a conduction velocity that was significantly slower than mechanically sensitive STT neurons. The encoding of pruritic information in the spinal cord remains enigmatic however, because spinal neurons that respond to pruritic stimuli also respond to stimuli that elicit painful sensations, when tested. In vivo experiments in anesthetized nonhuman primates demonstrate that STT neurons recorded from the marginal zone and deep dorsal horn can respond to histamine and are also responsive to capsaicin, heat, and can be categorized by mechanical sensitivity as wide dynamic range or high threshold type (Simone et al. 2004). Histamine-responsive wide dynamic range neurons exhibit an increased response to punctuate mechanical stimuli after the application of histamine, suggesting a possible mechanism for hyperknesis involving sensitization of pruriceptive STT neurons.

Histaminergic and Non-histaminergic Itch

Clinically relevant pruritus is often not controlled by antihistamines indicating the existence of histamine-independent pathways for itch (Davidson and Giesler 2010). Both histamine and non-histaminergic pruritogens activate primate STT neurons; however, histamine and the non-histaminergic pruritogen cowhage were found to activate separate populations of STT neurons (Davidson et al. 2007). About 30 % of antidromically identified primate STT neurons responded to one pruritogen or the other, while the remaining 70 % of STT neurons responded to noxious stimuli but not to either pruritogen. Histamine-responsive STT neurons had significantly slower conduction velocities than those responsive to cowhage. Pruritoceptive neurons were recorded in marginal zone and also within the deep dorsal horn and their axons projected to the ventrobasal and posterior nuclei of the thalamus. Neurons in both the histamine-responsive subset and cowhage-responsive subset of STT neurons could be classified as high threshold or wide dynamic range type, nearly all responded to capsaicin, and most were activated by thermal stimulation demonstrating that they were all polymodal.

Spinal Inhibition of Itch

Scratching the skin can reduce the perception of itch and scratching also reliably evokes activity in naïve STT neurons. Histamine-responsive STT neurons responded to scratching by firing action potentials while in the naive state. However, during an ongoing response to histamine, scratching the skin reduced the histamine-evoked firing in these cells. In contrast, capsaicin-evoked firing in the same histamine-responsive STT neurons was not reduced by scratching, indicating a state-dependent mechanism for the inhibition of the itch signal in the spinal cord (Davidson et al. 2009).

References

Andrew, D., & Craig, A. D. (2001). Spinothalamic lamina I neurons selectively sensitive to histamine: A central pathway for itch. *Nature Neuroscience, 4*, 72–77.

Bickford, R. G. (1938). Experiments relating to the itch sensation, its peripheral mechanism and central pathways. *Clinical Science, 3*, 377–386.

Davidson, S., & Giesler, G. J., Jr. (2010). The multiple pathways for itch and their interactions with pain. *Trends in Neurosciences, 33*, 550–588.

Davidson, S., Zhang, X., Yoon, C. H., Khasabov, S. G., Simone, D. A., & Giesler, G. J., Jr. (2007). The itch-producing agents histamine and cowhage activate separate populations of primate spinothalamic tract neurons. *The Journal of Neuroscience, 27*, 10007–10014.

Davidson, S., Zhang, X., Khasabov, S. G., Simone, D. A., & Giesler, G. J., Jr. (2009). Relief of itch by scratching: State-dependent inhibition of primate spinothalamic tract neurons. *Nature Neuroscience, 12*, 544–546.

Simone, D. A., Zhang, X., Li, J., Zhang, J. M., Honda, C. N., LaMotte, R. H., & Giesler, G. J., Jr. (2004). Comparison of responses of primate spinothalamic neurons to pruritic and algogenic stimuli. *Journal of Neurophysiology, 91*, 213–222.

Wei, J. Y., & Tuckett, R. P. (1991). Response of cat ventrolateral spinal axons to an itch-producing stimulus (cowhage). *Somatosensory & Motor Research, 8*, 227–239.

White, J. C., & Sweet, W. H. (1969). *Pain and the neurosurgeon.* Springfield: Charles C. Thomas.

Responsive Neurostimulation

Definition

A treatment of epilepsy in which stimulation is delivered to the brain in response to a detected event which predicts the onset of a seizure in that patient.

Cross-References

▶ Pain Treatment: Spinal Cord Stimulation

Responsiveness

Definition

Responsiveness refers to a measure of the ability of an outcome instrument used to detect a score change.

Cross-References

▶ Oswestry Disability Index

Rest and Movement Pain

Paul M. Murphy
Department of Anaesthesia and Pain Management, Royal North Shore Hospital, St. Leonards, NSW, Australia

Definition

In a 1990 study of ▶ breakthrough pain in cancer sufferers, Portenoy and Hagen proposed a standard definition of breakthrough pain. It was defined as a transient increase in the intensity of moderate or severe pain, occurring in the presence of well-established baseline pain.

Breakthrough pain was further characterized as being of rapid in onset (typically less than 3 min) and of short duration (median 30 min) (Portenoy and Hagen 1990). This entry paved the way for the systematic study of breakthrough pain, resulting in the 2002 definition of breakthrough pain by the American Pain Society as

> Intermittent exacerbations of pain that can occur spontaneously or in relation to specific activity; pain that increases above the level of pain addressed by the ongoing analgesic; includes incident pain and end-of-dose failure.

Characteristics

In the acute setting, breakthrough pain is predominantly somatic but may also be visceral or occasionally neuropathic (see ► Visceral Nociception and Pain, ► neuropathic pain, ► somatic pain).

Potential Etiology
1. Breakthrough pain due to voluntary or involuntary movement is referred to as ► incident pain.
2. End-of-dose failure.
3. Idiopathic (Caraceni and Portenoy 1999; McQuay and Jadad 1994; Portenoy et al. 1999).

Management of Movement-Associated or Incident Pain
In order to achieve the optimal balance between adequate analgesia and side effect profile, so-called rescue medication for breakthrough pain should be individualized and carefully titrated (Patt and Ellison 1998).
1. Breakthrough pain is typically managed with short-acting oral immediate-release opioids, given as required. In many institutions, it is standard practice to use the same opioid for the treatment of baseline and breakthrough pain, but in different formulations, such as long-acting morphine for baseline pain and short-acting morphine for breakthrough pain. Some sources recommend the use of different opioids for the management of baseline and breakthrough pain, suggesting that as it allows for the administration of lower doses of each

agent, dose-dependent side effects may be minimized. Unfortunately, many of the opioids used in the management of postoperative incident/breakthrough pain, such as morphine, hydromorphone, and oxycodone, are not very ► lipophilic, and thus, oral formulations of these agents are relatively poorly absorbed. The poor absorption kinetics of such agents may render them less than perfect in managing breakthrough pain because of the delayed onset of analgesia (Simmonds 1999).
2. Another strategy that may be implemented is the prophylactic treatment of breakthrough pain using higher dosages of baseline opioids. The sometimes significantly higher opioid doses utilized in these strategies may be associated with an increased risk of adverse effects (Bruera 1999).
3. ► Patient-controlled analgesia (PCA) systems provide the patient with a considerable degree of autonomy in meeting their analgesic needs. Protocols typically allow the patient to self-administer a predetermined dose of analgesic agent (morphine, fentanyl) once a "lockout" period (commonly 5–6 min) has elapsed. More advanced protocols allow for background drug infusions to run in parallel with bolus administration, thus providing baseline analgesic control. A potential benefit of a background infusion is that it prevents exacerbations in pain following a period of sleep, during which time the patient has not administered drug boluses. Evidence in this regard is contradictory, with some authors suggesting that this type of protocol does not improve analgesic outcome but simply increases the administered drug dose. Typically, it has been suggested that PCA systems are associated with both improved pain control and reduced opioid consumption relative to "on demand" IM or IV administration by nursing staff, although recent evidence is controversial. Irrespective of the conflicting data with regard to pain control and drug utilization, it is widely accepted that PCA systems are associated with significantly enhanced patient satisfaction in the management of both rest and movement/incident pain (Upton et al. 1997).

References

Bruera, E. (1999). Management of breakthrough pain due to cancer – The Simmonds article reviewed. *Oncology, 13*, 1110–1113.

Caraceni, A., & Portenoy, R. K. (1999). An international survey of cancer pain characteristics and syndromes. *Pain, 82*, 263–274.

McQuay, H. J., & Jadad, A. R. (1994). Incident pain. *Cancer Surveys, 21*, 17–24.

Patt, R. B., & Ellison, N. M. (1998). Breakthrough pain in cancer patients: Characteristics, prevalence and treatment. *Oncology, 12*, 1035–1052.

Portenoy, R. K., & Hagen, N. A. (1990). Breakthrough pain: Definition, prevalence and characteristics. *Pain, 41*, 273–281.

Portenoy, R. K., Payne, D., & Jacobsen, P. (1999). Breakthrough pain: Characteristics and impact in patients with cancer pain. *Pain, 81*, 129–134.

Simmonds, M. A. (1999). Management of breakthrough pain due to cancer. *Oncology, 13*, 1103–1114.

Upton, R. M., Semple, T. J., & Macintyre, P. E. (1997). Pharmacokinetic optimisation of opioid treatment in acute pain therapy. *Clinical Pharmacokinetics, 33*, 225–244.

Rest Position

▶ Postural Position

Resting State Network

Definition

PET and fMRI studies have suggested that the resting brain has a default mode of internal (possibly conceptual) processing. The activation of the areas involved in the resting state network during baseline recording in functional imaging may mask out the activation of these areas in a task-baseline comparison. That the default-mode network includes the hippocampus suggests that episodic memory may be incorporated in default-mode cognitive processing.

Cross-References

▶ Hippocampus and Entorhinal Complex: Functional Imaging

Restless Legs Syndrome

Michael Polydefkis
Department of Neurology, Johns Hopkins University School of Medicine, Baltimore, MD, USA

Synonyms

RLS

Definition

Restless legs syndrome (RLS) is characterized by a desire to move the limbs, usually associated with unpleasant sensations in lower extremities, less often in the arms.

Characteristics

RLS symptoms are worse or exclusively present during periods of rest in the evening or nighttime with at least partial or temporary relief by activity (Allen et al. 2003). RLS interferes with sleep quality leading to marked daytime fatigue and impaired quality of life. Though scientific interest in RLS has recently dramatically increased, the etiology and pathophysiology remain unknown. While RLS generally falls within the domain of movement disorders, there is increasing evidence for investigators to view RLS from a neuropathic pain perspective.

RLS was originally described in 1945 by Karl Ekbom (1945) who reported two forms of the syndrome: one characterized by ▶ dysesthesias, "asthenia crurum paresthetica," and one by pain, "asthenia crurum dolorosa." Adjectives used by Ekbom to describe patient symptoms included many neuropathic descriptors such as burning, tingling, or nerve vibrations. In part, this has prompted speculation of a causal association with peripheral neuropathy. Indeed, there is a loose correlation between painful neuropathies and RLS in terms of evening/nighttime worsening of symptoms, sleep interference, and body topography. Several prominent causes of painful neuropathy including diabetes (O'Hare et al. 1994), cryoglobulinemia (Gemignani et al. 1997), uremia, and amyloid (Salvi et al. 1990) have all been associated with RLS, supporting such a link. While the incidence of RLS among patients with neuropathy has been reported as similar to the general population at 5.2 % (Rutkove et al. 1996), this investigation did not include evaluation of small-caliber unmyelinated nerve fibers. This distinction may be important, as RLS is observed in many patients with small fiber neuropathy (Polydefkis et al. 1999), and has prompted some to perceive RLS among the spectrum of painful symptoms experienced by these patients. Furthermore, many patients with primary RLS have subclinical small fiber abnormalities, suggesting that abnormal peripheral sensation may play a role in RLS pathophysiology (Polydefkis et al. 2000). In addition, other forms of peripheral nerve injury such as radiculopathy (Walters et al. 1996) and spinal stenosis (LaBan et al. 1990) have also been associated with RLS.

RLS and painful neuropathies also share many similarities in treatment options. Opiates, antiepileptic agents, and the $\alpha2$-agonists all have well-accepted efficacy for both conditions. The diurnal pattern of RLS was noted to have an inverse relationship to endogenous dopamine levels, and a clinical response to dopaminergic agents emerged as a hallmark for RLS. Interestingly, dopamine also plays a role in pain processing, with high levels of dopaminergic activity in the striatum reducing neuronal encephalin content, leading to a compensatory rise in μ-opioid receptor expression. A role for dopamine as an inhibitor of pain perception is supported by its efficacy in relieving the pain associated with breast cancer and bone metastases, herpes zoster (Kernbaum and Hauchecorne 1981), painful Parkinson's disease (Quinn et al. 1986), and diabetic neuropathy.

Recent studies have suggested that ► central sensitization may exist in RLS (Stiasny-Kolster et al. 2004). Enhanced spinal reflexes are a feature of central sensitization in both animal models and human subjects and have also been documented to occur in RLS (Bara-Jimenez et al. 2000). The presence of profound static ► mechanical hyperalgesia, a hallmark of central sensitization, in RLS patients (Stiasny-Kolster et al. 2004) provides further evidence. Sensitization of dorsal horn neurons to mechanical stimuli by conditioning stimulation of nociceptive C fiber afferents is well described (Simone et al. 1991) and could explain the association between peripheral nerve abnormalities, particularly painful neuropathies and RLS. In contrast to typical neuropathic pain syndromes, the absence of ► dynamic mechanical hyperalgesia (allodynia) (Stiasny-Kolster et al. 2004) in RLS may explain why central sensitization of the nociceptive system in RLS has been underappreciated.

Viewing RLS from a neuropathic pain perspective has the opportunity to foster new treatment strategies. Glutamate and substance P are important in the induction of central sensitization and imply that NMDA antagonists and NK1 receptor antagonists may have a future as RLS treatments. In support of this is the report of the NMDA receptor antagonist amantadine improving RLS symptoms. Furthermore, incorporating central sensitization into the pathophysiology of RLS can explain why antinociceptive agents such as epilepsy medications are efficacious in RLS.

In summary, pain as a symptom of RLS has been underappreciated. There is compelling clinical evidence that the two share many common features. Approaching RLS from such a viewpoint offers the potential to improve our pathophysiologic understanding of the disease, as well as offers novel treatment strategies.

References

Allen, R. P., Kushida, C. A., Atkinson, M. J., et al. (2003). Factor analysis of the international restless legs syndrome study Group's scale for restless legs severity. *Sleep Medicine, 4*, 133–135.

Bara-Jimenez, W., Aksu, M., Graham, B., et al. (2000). Periodic limb movements in sleep: State-dependent excitability of the spinal flexor reflex. *Neurology, 54*, 1609–1616.

Ekbom, K. (1945). Restless legs. *Acta Medica Scandinavica, 158*, 1–123.

Gemignani, F., Marbini, A., Di Giovanni, G., et al. (1997). Cryoglobulinaemic neuropathy manifesting with restless legs syndrome. *Journal of Neurological Sciences, 152*, 218–223.

Kernbaum, S., & Hauchecorne, J. (1981). Administration of levodopa for relief of herpes zoster pain. *JAMA: The Journal of the American Medical Association, 246*, 132–134.

LaBan, M. M., Viola, S. L., Femminineo, A. F., et al. (1990). Restless Legs syndrome associated with diminished cardiopulmonary compliance and lumbar spinal stenosis – a motor concomitant of "Vesper's curse". *Archives of Physical Medicine and Rehabilitation, 71*, 384–388.

O'Hare, J. A., Abuaisha, F., & Geoghegan, M. (1994). Prevalence and forms of neuropathic morbidity in 800 diabetics. *Irish Journal of Medical Science, 163*, 132–135.

Polydefkis, M., Hauer, P., et al. (1999). Small fiber sensory neuropathy in restless legs syndrome. *Neurology, 52*(Suppl. 2), A411.

Polydefkis, M., Allen, R. P., Hauer, P., et al. (2000). Subclinical sensory neuropathy in late-onset restless legs syndrome. *Neurology, 55*, 1115–1121.

Quinn, N. P., Koller, W. C., Lang, A. E., et al. (1986). Painful Parkinson's disease. *Lancet, 1*, 1366–1369.

Rutkove, S. B., Matheson, J. K., & Logigian, E. L. (1996). Restless legs syndrome in patients with polyneuropathy. *Muscle & Nerve, 19*, 670–672.

Salvi, F., Montagna, P., Plasmati, R., et al. (1990). Restless legs syndrome and nocturnal myoclonus: Initial clinical manifestation of familial amyloid polyneuropathy. *Journal of Neurology, Neurosurgery & Psychiatry, 53*, 522–525.

Simone, D. A., Sorkin, L. S., Oh, U., et al. (1991). Neurogenic Hyperalgesia: Central neural correlates in responses of spinothalamic tract neurons. *Journal of Neurophysiology, 66*, 228–246.

Stiasny-Kolster, K., Magerl, W., Oertel, W. H., et al. (2004). Static Mechanical hyperalgesia without dynamic tactile allodynia in patients with restless legs syndrome. *Brain, 127*, 773–782.

Walters, A. S., Wagner, M., & Hening, W. A. (1996). Periodic Limb movements as the initial manifestation of restless legs syndrome triggered by lumbosacral radiculopathy. *Sleep, 19*, 825–826.

Restorative Rehabilitation

▶ Cancer Pain Management, Rehabilitative Therapies

Reticular Activating System

Synonyms

RAS

Definition

The system of cells of the reticular formation of the medulla oblongata that receive collaterals from the ascending sensory pathways and project to higher centers; they control the overall degree of central nervous system activity, including wakefulness, attentiveness, and sleep; abbreviated RAS.

Cross-References

▶ Somatic Pain

Reticular Formation

Definition

Reticular formation is the central core within the brain stem, consisting of a short diffusely projecting web of neurons located in the brainstem, receiving many inputs and projecting widely throughout the central nervous system. It receives also nociceptive inputs via the spinoreticular pathways and collaterals of the spinothalamic tract and some of its neurons respond to noxious stimuli.

Cross-References

▶ Pain Treatment, Intracranial Ablative Procedures
▶ Spinothalamocortical Projections to Ventromedial and Parafascicular Nuclei

Reticular Nucleus Inputs from Cortex

▶ Corticothalamic and Thalamocortical Interactions

Retinal Slip

Definition

Movements of images across the retina. Retinal slip is considered to be an error signal necessary for vestibular compensation. The error signal is brought on by a combination of visual fixation and head movements. The brain tries to minimize the error signal by increasing the gain of the vestibulo-ocular reflexes.

Cross-References

▶ Coordination Exercises in the Treatment of Cervical Dizziness

Retrograde

Definition

Signals and transport of molecules from the periphery (axonal nerve endings) along the axoplasm toward the cell body.

Cross-References

▶ Retrograde Cellular Changes after Nerve Injury
▶ Substance P Regulation in Inflammation

Retrograde Axonal Tracer

Definition

A substance (protein, enzyme) that is injected at the level of axonal endings. It is incorporated

within the endings, then conveyed in a retrograde (antidromic) direction within the axon back to the soma. The tracer is generally colored by a histochemical reaction, with or without an earlier immune amplification reaction; see also ▶ Retrograde Tracing Technique.

Cross-References

▶ Parabrachial Hypothalamic and Amygdaloid Projections

Retrograde Cellular Changes After Nerve Injury

Pablo Brumovsky[1,2], Marcelo Villar[1] and Tomas Hökfelt[3]
[1]Faculty of Biomedical Sciences, Austral University, Buenos Aires, Argentina
[2]Consejo Nacional de Investigaciones Científicas y Técnicas (CONICET), Buenos Aires, Argentina
[3]Department of Neuroscience, Karolinska Institute, Stockholm, Sweden

Synonyms

Axotomy; Bennett model; Chung model; Gazelius model; Nerve injury; Nerve ligation; Other types of nerve injury; Seltzer model; Spared nerve injury

Definition

▶ Retrograde changes here refer to the alteration in expression of various molecules in the parent dorsal root ganglion (DRG, sensory) or autonomic neurons after any type of lesion of the peripheral axon (branch). This occurs in humans after various types of lesions, for example, surgery, and is mimicked in experimental animal models, mainly in attempts to understand mechanisms underlying ▶ neuropathic pain.

Characteristics

Nerve Injury and Neuropathic Pain

Nerve injury, such as transection of a nerve (axotomy), occurs under a wide variety of circumstances in real life, ranging from a simple accidental cut during cooking to more or less complicated surgical procedures in hospitals, amputation of a limb being the extreme. Early work on experimental animals by Wall, Devor, and colleagues showed that such a transection has profound consequences as recorded by electrophysiological methods, including massive and persistent spontaneous discharges in lumbar dorsal rootlets, possibly contributing to chronic neuropathic pain (Devor and Wall 1981; Tal et al. 1999).

Dramatic and robust chemical changes have been found to occur in ▶ DRGs after peripheral nerve injury (see Hökfelt et al. 1994), in fact, leading to a completely new neuronal phenotype, encompassing hundreds of regulatory molecules, including inter- and intracellular messengers, as shown in recent global gene array studies comparing lesioned and nonlesioned DRGs (see Costigan et al. 2002; Xiao et al. 2002). The purpose and consequences of these changes are beginning to be understood, with not only upregulated and/or de novo synthesized molecules possibly contributing to the survival and regeneration of the damaged neuron but also to the generation (pronociceptive) and/or attenuation (antinociceptive) of pain. It is, however, clear that changes affect not only neurons but also satellite and other nonneuronal cells in the DRGs (see Bradman et al. 2010) and around the lesion site. Moreover, the effect is not limited to the DRGs but also includes neurons and glia in subsequent relay stations, primarily the dorsal horn of the spinal cord (Yang et al. 2004; Watkins et al. 2007). Interestingly, similar chemical changes have been observed in a number of nerve injury models not involving a complete transection of the nerve (Koltzenburg 2005; Scholz and Woolf 2005; Hökfelt et al. 2005).

In this entry, we will focus on the changes seen after damage to the peripheral branch of sensory and sympathetic neurons. It should be noted that transection of the dorsal roots, that is, the central, afferent branches of the DRG neurons terminating in the dorsal horn, has been shown to spare DRG neurons from several of the neurochemical changes observed after injury to their peripheral nerve endings (see McGraw et al. 2005). However, in gene array studies in rat (Takeuchi et al. 2008) and human (Rabert et al. 2004) DRGs, changes in the expression of up to 91 genes were reported after dorsal root lesions, including the upregulation of the cyclooxygenase-2 (COX-2), involved in acute pain and inflammation (Takeuchi et al. 2008).

Sympathetic neurons, classically identified by their expression of the noradrenergic marker tyrosine hydroxylase (TH) but also capable of synthesizing a diversity of other neurotransmitters, including acetylcholine and neuropeptides, are also affected by axotomy of their postganglionic projections. Thus, downregulation of the neuropeptide tyrosine (NPY) or TH (Zigmond and Sun 1997; Landry et al. 2000; Navarro 2009) and upregulation of galanin (Lindh et al. 1993; Zigmond and Sun 1997; Landry et al. 2000; Navarro 2009), vasoactive intestinal peptide (VIP), and substance P (SP) (Zigmond and Sun 1997; Landry et al. 2000; Navarro 2009) have been shown in superior cervical ganglion neurons after axotomy of their postganglionic projections in rodents. Similar effects on some of the neurotransmitters mentioned above have also been shown in cat lumbar sympathetic chain (LSC) (Lindh et al. 1993). More recently, the unexpected de novo expression of the vesicular glutamate transporter type 2 ($VGLUT_2$) in a subpopulation of LSC neurons in the mouse has been demonstrated after axotomy of the pelvic nerve (Brumovsky et al. 2011). Many of these neurochemical changes could relate to neuropathic pain mechanisms after nerve injury by sympathetic-sensory coupling (Jänig 2003; Gibbs et al. 2008).

Do DRG Neurons Die After Peripheral Nerve Injury?

This issue has been much debated, but recent studies based on unbiased ▶ stereological techniques have shown that in the rat no significant loss of neurons occurs up until 4 weeks after the "standard" lesion (transection of the sciatic nerve at mid-thigh level) (Tandrup 2004; Welin et al. 2008). In contrast, after mid-thigh axotomy in the mouse, there is already a significant loss of neurons after 1 week (Shi et al. 2001), possibly due to the short distance between lesion and cell bodies in this small animal.

Transmitters/Neuropeptides in DRG Neurons

Today we know that the main and principal (classic) transmitter in most DRG neurons is glutamate, as supported by the abundant expression of VGLUTs, in particular $VGLUT_2$, essential for the incorporation of the amino acid into synaptic vesicles (see Brumovsky et al. 2007a, 2011). Glutamate thus acts on pre- and postsynaptic glutamate receptors of various classes in the dorsal horn (Salter 2005) and the periphery (Raab and Neuhuber 2007; Miller et al. 2011), including both ligand-gated ion channels and 7-transmembrane, G-protein-coupled (metabotropic) receptors (GPCRs). Other transmitter candidates are the nucleotide adenosine triphosphate (ATP) and the "gaseous" transmitter nitric oxide, as well as a number of neuropeptides such as substance P, calcitonin gene-related peptide (CGRP), ▶ somatostatin, galanin, and pituitary adenylate cyclase-activating peptide (PACAP) (Hökfelt et al. 1997). There is a second population of small DRG neurons, many of which apparently lack most of these neuropeptides, and they are characterized by expression of components of the receptor for the glial-derived ▶ neurotrophic factor (GDNF) and the isolectin B4 (IB4) from Griffonia simplicifolia I, which is a marker for many of these cells (McMahon and Priestley 2005). Finally, other, more recently described subpopulations of DRG neurons lack both neuropeptides or IB4 while expressing the noradrenergic marker tyrosine hydroxylase (Brumovsky et al. 2006), the neuropeptide tyrosine (NPY) type 2 receptor (Brumovsky et al. 2005), $VGLUT_1$ (Brumovsky et al. 2007b), or $VGLUT_3$ (Seal et al. 2009). This rich neuronal variety in DRGs, tightly associated to developmental and environmental cues, has become an important model for the understanding of neuronal diversification (Lallemend and Ernfors 2012).

Expression of Regulatory Molecules After Axotomy

Substance P and CGRP are downregulated in DRG neurons after peripheral axotomy, whereas vasoactive intestinal polypeptide (VIP), ▶ neuropeptide Y (NPY), ▶ galanin, PACAP, and the enzyme ▶ nitric oxide synthase are upregulated (Hökfelt et al. 1997) (Figs. 1 and 2). Receptors for peptides (e.g., the cholecystokinin B receptor) and classic transmitters (e.g., the α_{2A}-adrenoceptor) are also regulated (Fig. 1). Many of these changes are long-lasting and remain, if regeneration is efficiently prevented (Navarro 2009). A similar effect is also seen in several neuropathic pain models, that is, without a complete transection of the axons (see McMahon and Priestley 2005). In non-peptide DRG neurons several markers, including certain vanilloid receptors and purinoreceptors as well as some sodium channels, are upregulated and other molecules are downregulated (McMahon and Priestley 2005) (Fig. 1). For information on other GPCRs, ligand-gated receptors, and ion channels, see below and McMahon and Priestley (2005).

Classic Growth Factors

▶ Nerve growth factor (NGF), brain-derived neurotrophic factor (BDNF), and neurotrophin 3 (NT3) are the main members of the classic growth factor family and are in general target derived, acting on neuronal tyrosine kinase (trk) receptors (▶ trkA, trkB, and trkC) (see Ernsberger 2009). However, BDNF is, in fact, constitutively expressed in neurons, mainly in small, trkA-positive (NGF-sensitive) ones; packaged in large dense core vesicles (the same vesicles that store neuropeptides); transported into the dorsal horn of the spinal cord; and released, acting in a transmitter-like fashion (see Obata and Noguchi 2006). BDNF is mainly upregulated in medium-sized and large-diameter (trkB- and trkC-positive) neurons after axotomy. Also, members of the fibroblast growth factor (FGF) family and their receptors are present in

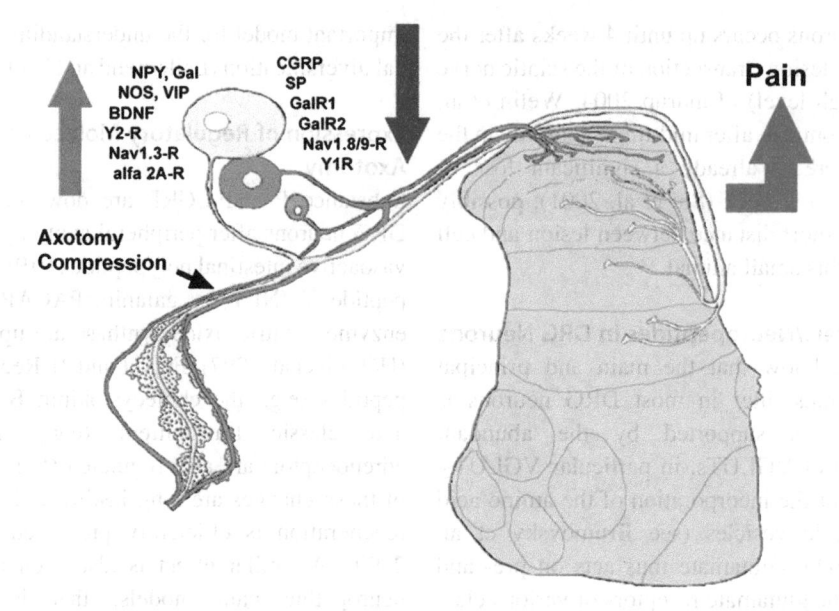

Retrograde Cellular Changes After Nerve Injury, Fig. 1 Schematic drawing showing changes in the expression of different neurotransmitters, receptors and channels in DRGs, and spinal cord of rat after axotomy or constriction of a peripheral nerve. Normally, small- and medium-sized neurons produce detectable levels of, for example, CGRP, substance P (SP), and GalR2 and Y1R receptors, while larger neurons express GalR1 and Y2R

receptors and some CGRP. Axotomy of the sciatic nerve induces dramatic changes in the expression of these molecules, some being upregulated (e.g., NPY, galanin (Gal), Nav 1.3R, Y2R) and others downregulated (CGRP, substance P, Y1R). In contrast, the expression of peptides and receptors in dorsal horn neurons largely seems to be unaffected by axotomy. Moreover, myelin basic protein shows degeneration distal to the sciatic nerve lesion

DRG neurons and regulated by nerve injury (Grothe et al. 2001).

Voltage-Gated Sodium Channels

Much attention is presently focused on voltage-gated sodium channels in DRG neurons and their role in chronic pain (Dib-Hajj et al. 2010; Kullman and Waxman 2010). Many subtypes are expressed in DRG neurons and are of ▶ tetrodotoxin (TTX)-sensitive or TTX-resistant subtypes. The former, especially $Na_v1.3$, are involved in the generation of ectopic discharges after injury and in maintaining neuropathic pain. Also, some subtypes are upregulated, while others are downregulated after nerve injury (see Dib-Hajj et al. 2010).

Global Gene Expression Studies

Global gene expression analyses have shown that hundreds of different types of coding genes change in sensory ganglia after nerve injury

(Costigan et al. 2002; Xiao et al. 2002). Thus, Xiao et al. (2002) examined 7,523 genes and expressed sequence tags (ESTs) of which, respectively, 122 and 51 were up and downregulated changed after axotomy. Costigan et al. (2002) categorized such genes into "ion channels," "neurotransmission," "vesicle trafficking," "apoptosis," "cytoskeleton," "immunologically related genes," and "unknown genes," the latter reflecting identification of many "novel" DRG genes. These changes are bidirectional in several categories.

Species Differences: Strain Matters

Most studies of this type have been carried out on rats and some also on mice. While in general the regulation of several molecules shows little qualitative differences between these two types of rodents, it has been recently shown that the expression of a number of pain molecules is differentially affected among different inbred

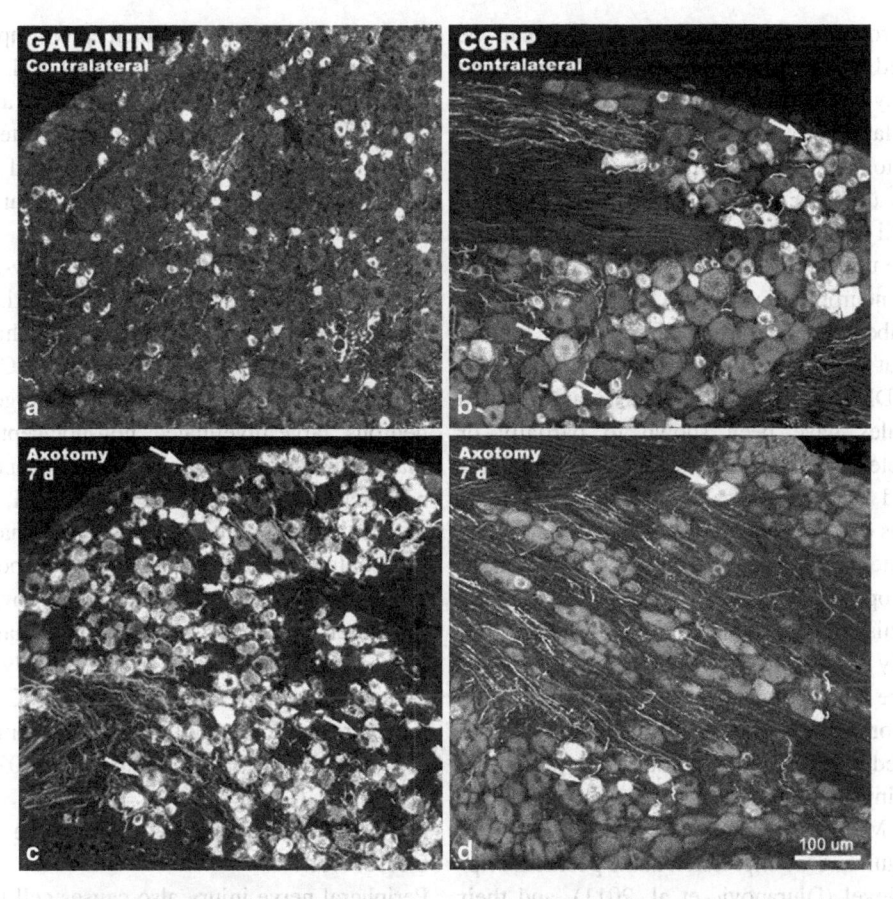

Retrograde Cellular Changes After Nerve Injury, Fig. 2 Fluorescence micrographs showing mouse DRG sections after incubation with galanin (**a, c**) or CGRP (**b, d**) antisera, contralateral (**a, b**) and ipsilateral (**c, d**) to a 7-day axotomy of the sciatic nerve at the mid-thigh level. Several galanin-immunoreactive small neurons are seen in contralateral DRGs (**a**) and many CGRP-IR neurons, ranging from small to large (*arrows* in **b**) in size. A dramatic upregulation of galanin is observed ipsilateral to the nerve lesion, in many small and some large DRG neurons (*arrow* in **c**). In contrast, the expression of CGRP is strongly downregulated, and only some large CGRP-IR neurons are still observed (*arrows* in **d**). Note that all figures have the same magnification

mouse strains (see Persson et al. 2010). Moreover, the influence of strain is not limited to the neurochemical expression in DRGs but also transcends to pain behavior (see Mogil 2012). Interestingly, clustering analysis suggests that molecules with similar phylogenetic origin and function are regulated similarly by peripheral nerve injury (see (Persson et al. 2010)). In monkey (Macaca mulatta), the upregulation of galanin is strong and NPY weak, and guinea pigs show the reverse situation (Zhang et al. 1993; Hökfelt et al. 1997). Human ganglia have been analyzed only to a limited degree, nevertheless showing that substance P, CGRP, and galanin are expressed in DRG neurons in proportions similar to those found in the monkey (Landry et al. 2003). Also the capsaicin receptor TRPV1 has been identified in human DRGs, spinal cord, and peripheral nerves and, in the latter, showing changes during peripheral neuropathies (Lauria et al. 2006).

Mechanisms Underlying Injury-Induced Phenotypic Changes

The various molecules seem to be regulated by "individual" mechanisms. The downregulation of the two excitatory peptides substance P and CGRP is a consequence of interrupted supply of

NGF from peripheral tissues, since it can be reversed by exogenous NGF (McMahon and Priestley 2005). The dramatic axotomy-induced upregulation of galanin is strongly influenced by the cytokine leukemia inhibitory factor (LIF) (Rao et al. 1993), with GDNF also being involved (Gardell et al. 2003). Other groups have shown that the upregulation of VIP is dependent on the ciliary neurotrophic factor (CNTF), just as LIF a member of the neuropoietic cytokine family, and that NPY upregulation involves both NGF and BDNF. Interestingly, GDNF and related molecules have been shown to partially or completely reverse many of the nerve injury-induced morphological and neurochemical changes described above, as well as to block the associated neuropathic pain state (Gardell et al. 2003), opening up new venues for understanding mechanisms underlying neuropathic pain and possibly its treatment.

More recently, the expression of other type of regulatory molecules, the microRNAs, has been analyzed in DRGs of rat or mouse after peripheral nerve injury (Zhou et al. 2011; Kusuda et al. 2011). MicroRNAs are small noncoding RNAs that regulate gene expression at the posttranscriptional level (Djuranovic et al. 2011), and their changes and action in sensory neurons after nerve injury have only begun to be studied. However, their capacity to influence intracellular signaling cascades and signal transduction suggests an important role in processes such as regeneration and pain. In fact, it has been observed that in different models of pain in mouse the regulation of mircroRNAs is differential (Kusuda et al. 2011).

Functional Consequences of Altered Gene Expression in the DRG

The regulation of molecules in DRG neurons could, among others, promote survival and regeneration and/or affect various sensory modalities, in particular pain. Based on studies of the nociceptive flexor reflex and other functional models, it has been concluded that the downregulation of excitatory CGRP and substance P attenuates pain transmission in the dorsal horn. In contrast, the increased galanin levels may serve to reduce the nociceptive input after nerve injury (see Xu et al. 2008) and may be important for regeneration (Wynick et al. 2001; Hobson et al. 2008). Also, upregulated NPY could both enhance regeneration and control pain threshold (see Brumovsky et al. 2007; Smith et al. 2007).

An interesting aspect of axotomy-induced regulation concerns the so-called sprouting paradigm, proposed to underlie neuropathic pain. Using the cholera toxin-β subunit (CTB) as a selective retrograde tracer for large DRG neurons, large myelinated, non-nociceptive primary afferents seem to sprout from deeper laminae into spinal laminae I and II, where nociceptive afferents normally terminate, and this could lead to neuropathic pain (Woolf et al. 1992). However, several groups have now shown that after peripheral nerve injury, small DRG neurons also take up CTB, presumably due to upregulation of the monoganglioside GM1, the receptor for CTB, and that no or only little sprouting seems to occur (Bao et al. 2002).

Nerve Injury-Induced Cell Loss in the Dorsal Horn

Peripheral nerve injury also causes cell loss and apoptotic cell death in the dorsal horn (Coggeshall et al. 2001; Lee et al. 2009), which can be blocked by treatment with an NMDA receptor antagonist (MK-801) (Azkue et al. 1998); Coggeshall et al. (2001) have shown that it is the release of glutamate from myelinated afferents (A-fibers) that is responsible for this cell death, possibly affecting inhibitory GABAergic mechanisms.

Changes in Spinal Glia After Peripheral Nerve Injury

It has become increasingly clear that spinal glia are also activated by peripheral nerve injury, especially if there is an inflammatory component, and this activation is associated with pain-related behavior (Zhuo et al. 2011; Calvo and Bennett 2012). Here pro-inflammatory cytokines and other molecules play important roles (Watkins et al. 2007; O'Callaghan and Miller 2010).

Summary

Peripheral nerve injury induces many changes in DRG and sympathetic neurons as well as in the spinal cord, as clearly demonstrated in a number of global gene expression analyses. Here, we could only deal with a limited number of the molecules involved. These changes may play a role in nociception and be important for survival/regeneration and may lead to novel therapeutic strategies.

Importantly, and in contrast, in inflammatory pain substance P and CGRP are upregulated in DRG neurons, without distinct effects on galanin, VIP, or NPY in these neurons. Instead, inflammation, but not peripheral axotomy, activates opioid and other peptides in local dorsal horn neurons (Dubner and Ruda 1992; Przewłocki and Przewłocki 2001). Thus, the nerve injury-induced changes are distinctly different from the regulatory events seen after peripheral inflammation, suggesting existence of separate defense systems for inflammatory and neuropathic pain, associated with, respectively, dorsal horn and DRG neurons (Hökfelt et al. 1997).

Acknowledgement This work was supported by the Swedish Research Council, the Marianne and Marcus Wallenberg Foundation, the Knut and Alice Wallenberg Foundation, funds from Karolinska Institutet, the International Association for the Study of Pain, an Unrestricted Bristol-Myers Squibb Neuroscience Grant, CONICET (Consejo Nacional de Investigaciones Científicas y Técnicas), and the Austral University. We thank Drs. Zsuzsanna Wiesenfeld-Hallin, Xiao-Jun Xu, and Zhang Xu for valuable support.

References

Azkue, J. J., Zimmermann, M., Hsieh, T. F., et al. (1998). Peripheral nerve insult induces NMDA receptor-mediated, delayed degeneration in spinal neurons. *The European Journal of Neuroscience, 10*, 2204–2206.

Bao, L., Wang, H. F., Cai, H. J., et al. (2002). Peripheral axotomy induces only very limited sprouting of coarse myelinated afferents into inner lamina II of rat spinal cord. *The European Journal of Neuroscience, 16*, 175–185.

Bradman, M. J., Arora, D. K., Morris, R., et al. (2010). How do the satellite glia cells of the dorsal root ganglia respond to stressed neurons? – nitric oxide saga from embryonic development to axonal injury in adulthood. *Neuron Glia Biology, 6*, 11–17.

Brumovsky, P., Stanic, D., Shuster, S., et al. (2005). Neuropeptide Y2 receptor protein is present in peptidergic and nonpeptidergic primary sensory neurons of the mouse. *The Journal of Comparative Neurology, 489*, 328–348.

Brumovsky, P., Villar, M. J., & Hökfelt, T. (2006). Tyrosine hydroxylase is expressed in a subpopulation of small dorsal root ganglion neurons in the adult mouse. *Experimental Neurology, 200*, 153–165.

Brumovsky, P., Shi, T. S., Landry, M., et al. (2007a). Neuropeptide tyrosine and pain. *Trends in Pharmacological Sciences, 25*, 93–102.

Brumovsky, P., Watanabe, M., & Hökfelt, T. (2007b). Expression of the vesicular glutamate transporters-1 and -2 in adult mouse dorsal root ganglia and spinal cord and their regulation by nerve injury. *Neuroscience, 147*, 469–490.

Brumovsky, P. R., Robinson, D. R., La, J. H., et al. (2011). Expression of vesicular glutamate transporters type 1 and 2 in sensory and autonomic neurons innervating the mouse colorectum. *The Journal of Comparative Neurology, 519*, 3346–3366.

Calvo, M., & Bennett, D. L. (2012). The mechanisms of microgliosis and pain following peripheral nerve injury. *Experimental Neurology, 234*, 271–282.

Coggeshall, R. E., Lekan, H. A., White, F. A., et al. (2001). A-fiber sensory input induces neuronal cell death in the dorsal horn of the adult rat spinal cord. *The Journal of Comparative Neurology, 435*, 276–282.

Costigan, M., Befort, K., Karchewski, L., et al. (2002). Replicate high-density rat genome oligonucleotide microarrays reveal hundreds of regulated genes in the dorsal root ganglion after peripheral nerve injury. *BMC Neuroscience, 3*, 16.

Devor, M., & Wall, P. D. (1981). Plasticity in the spinal cord sensory map following peripheral nerve injury in rats. *The Journal of Neuroscience, 1*, 679–684.

Dib-Hajj, S. D., Cummins, T. R., Black, J. A., et al. (2010). Sodium channels in normal and pathological pain. *Annual Review of Neuroscience, 33*, 325–347.

Djuranovic, S., Nahvi, A., & Green, R. (2011). A parsimonious model for gene regulation by miRNAs. *Science, 331*, 550–553.

Dubner, R., & Ruda, M. A. (1992). Activity-dependent neuronal plasticity following tissue injury and inflammation. *Trends in Neurosciences, 15*, 96–103.

Ernsberger, U. (2009). Role of neurotrophin signalling in the differentiation of neurons from dorsal root ganglia and sympathetic ganglia. *Cell and Tissue Research, 336*, 349–384.

Gardell, L. R., Wang, R., Ehrenfels, C., et al. (2003). Multiple actions of systemic artemin in experimental neuropathy. *Nature Medicine, 9*, 1383–1389.

Gibbs, G. F., Drummond, P. D., Finch, P. M., et al. (2008). Unravelling the pathophysiology of complex regional pain syndrome: Focus on sympathetically maintained pain. *Clinical and Experimental Pharmacology & Physiology, 35*, 717–724.

Grothe, C., Meisinger, C., & Claus, P. (2001). In vivo expression and localization of the fibroblast growth factor system in the intact and lesioned rat peripheral nerve and spinal ganglia. *The Journal of Comparative Neurology, 434*, 342–357.

Hobson, S. A., Bacon, A., Elliot-Hunt, C. R., et al. (2008). Galanin acts as a trophic factor to the central and peripheral nervous systems. *Cellular and Molecular Life Sciences, 65*, 1806–1812.

Hökfelt, T., Zhang, X., & Wiesenfeld-Hallin, Z. (1994). Messenger plasticity in primary sensory neurons following axotomy and its functional implications. *Trends in Neurosciences, 17*, 22–30.

Hökfelt, T., Zhang, X., Xu, Z.-Q., et al. (1997). Cellular and synaptic mechanisms in transition of pain from acute to chronic. In T. S. Jensen, J. A. Turner, & Z. Wiesenfeld-Hallin (Eds.), *Proceedings of the 8th World CongrPain, Prog Pain Res Management, vol 8* (pp. 133–153). Seattle: IASP Press.

Hökfelt, T., Zhang, X., Xu, X. J., & Wiesenfeld-Hallin, Z. S. (2005). Central consequences of peripheral nerve damage (Chapter 60). In S. McMahon & M. Koltzenburg (Eds.), *Textbook of pain* (5th ed., pp. 947–959). Churchill Livingstone: Elsevier.

Jänig, W. (2003). Relationship between pain and autonomic phenomena in headache and other pain conditions. *Cephalalgia, 23*(Suppl 1), 43–48.

Koltzenburg, M. (2005). Mechanisms of peripheral neuropathic pain. In S. Hunt & M. Koltzenburg (Eds.), *The neurobiology of pain* (pp. 115–147). Oxford: Oxford University Press.

Kullman, D. M., & Waxman, S. G. (2010). Neurological channelopathies: New insights into disease mechanisms and ion channel function. *The Journal of Physiology, 588*, 1823–1827.

Kusuda, R., Cadetti, F., Ravanelli, M. I., Sousa, T. A., Zanon, S., De Lucca, F. L., & Lucas, G. (2011). Differential expression of microRNAs in mouse pain models. *Molecular Pain, 7*, 17.

Lallemend, F., & Ernfors, P. (2012). Molecular interactions underlying the specification of sensory neurons. *Trends in Neurosciences, 35*, 373–381.

Landry, M., Holmberg, K., Zhang, X., & Hökfelt, T. (2000). Effect of axotomy on expression of NPY, galanin, and NPY Y1 and Y2 receptors in dorsal root ganglia and the superior cervical ganglion studied with double-labeling in situ hybridization and immunohistochemistry. *Experimental Neurology, 162*, 361–384.

Landry, M., Aman, K., Dostrovsky, J., et al. (2003). Galanin expression in adult human dorsal root ganglion neurons. *Neuroscience, 117*, 795–809.

Lauria, G., Morbin, M., & Lombardi, R. (2006). Expression of capsaicin receptor immunoreactivity in human peripheral nervous system and in painful neuropathies. *Journal of the Peripheral Nervous System, 11*, 262–271.

Lee, J. W., Siegel, S. M., & Oaklander, A. L. (2009). Effects of distal nerve injuries on dorsal-horn neurons and glia: Relationships between lesion size and mechanical hyperalgesia. *Neuroscience, 158*, 904–914.

Lindh, B., Risling, M., Remahl, S., et al. (1993). Peptide-immunoreactive neurons and nerve fibres in lumbosacral sympathetic ganglia: Selective elimination of a pathway-specific expression of immunoreactivities following sciatic nerve resection in kittens. *Neuroscience, 55*, 545–562.

McGraw, J., Gaudet, A. D., Oschipok, L. W., et al. (2005). Regulation of neuronal and glial galectin-1 expression by peripheral and central axotomy of rat primary afferent neurons. *Experimental Neurology, 195*, 103–114.

McMahon, S. B., & Priestley, J. V. (2005). Nociceptor plasticity. In S. P. Hunt & M. Koltzenburg (Eds.), *The neurobiology of pain* (pp. 35–64). Oxford: OUP.

Miller, K. E., Hoffman, E. M., Sutharshan, M., et al. (2011). Glutamate pharmacology and metabolism in peripheral primary afferents: Physiological and pathophysiological mechanisms. *Pharmacology & Therapeutics, 130*, 283–309.

Mogil, J. S. (2012). Pain genetics: Past, present and future. *Trends in Genetics, 28*, 258–266.

Navarro, X. (2009). Chapter 27: Neural plasticity after nerve injury and regeneration. *International Review of Neurobiology, 87*, 483–505.

Obata, K., & Noguchi, K. (2006). BDNF in sensory neurons and chronic pain. *Neuroscience Research, 55*, 1–10.

O'Callaghan, J. P., & Miller, D. B. (2010). Spinal glia and chronic pain. *Metabolism, 59*(Suppl 1), S21–S26.

Persson, A. K., Xu, X. J., Wiesenfeld-Hallin, Z., et al. (2010). Expression of DRG candidate pain molecules after nerve injury – a comparative study among five inbred mouse strains with contrasting pain phenotypes. *Journal of the Peripheral Nervous System, 15*, 26–39.

Przewłocki, R., & Przewłocki, B. (2001). Opioids in chronic pain. *European Journal of Pharmacology, 429*, 79–91.

Raab, M., & Neuhuber, W. L. (2007). Glutamatergic functions of primary afferent neurons with special emphasis on vagal afferents. *International Review of Cytology, 256*, 223–275.

Rabert, D., Xiao, Y., Yiangou, Y., et al. (2004). Plasticity of gene expression in injured human dorsal root ganglia revealed by GeneChip oligonucleotide microarrays. *Journal of Clinical Neuroscience, 2004*(11), 289–299.

Rao, M. S., Sun, Y., Escary, J. L., et al. (1993). Leukemia inhibitory factor mediates an injury response but not a target-directed developmental transmitter switch in sympathetic neurons. *Neuron, 11*, 1175–1185.

Salter, M. W. (2005). Cellular signalling pathways of spinal pain neuroplasticity as targets for analgesic development. *Current Topics in Medicinal Chemistry, 5*, 557–567.

Scholz, J., & Woolf, C. (2005). Mechanisms of neuropathic pain. In P. Marco (Ed.), *The neurological basis of pain* (pp. 71–95). New York: McGraw-Hill.

Seal, R. P., Wang, X., Guan, Y., et al. (2009). Injury-induced mechanical hypersensitivity requires C-low threshold mechanoreceptors. *Nature, 462*, 651–655.

Shi, T. J., Tandrup, T., Bergman, E., et al. (2001). Effect of peripheral nerve injury on dorsal root ganglion neurons in the C57 BL/6J mouse: Marked changes both in cell numbers and neuropeptide expression. *Neuroscience, 105*, 249–263.

Smith, P. A., Moran, T. D., Abdulla, F., et al. (2007). Spinal mechanisms of NPY analgesia. *Peptides, 28*, 464–474.

Takeuchi, H., Kawaguchi, S., Mizuno, S., et al. (2008). Gene expression profile of dorsal root ganglion in a lumbar radiculopathy model. *Spine, 33*, 2483–2488.

Tal, M., Wall, P. D., & Devor, M. (1999). Myelinated afferent fiber types that become spontaneously active and mechanosensitive following nerve transection in the rat. *Brain Research, 824*, 218–223.

Tandrup, T. (2004). Unbiased estimates of number and size of rat dorsal root ganglion cells in studies of structure and cell survival. *Journal of Neurocytology, 33*, 173–192.

Watkins, L. R., Hutchinson, M. R., Milligan, E. D., et al. (2007). "Listening" and "talking" to neurons: Implications of immune activation for pain control and increasing the efficacy of opioids. *Brain Research Reviews, 56*, 148–169.

Welin, D., Novikova, L. N., Wiberg, M., et al. (2008). Survival and regeneration of cutaneous and muscular afferent neurons after peripheral nerve injury in adult rats. *Experimental Brain Research, 186*, 315–323.

Woolf, C. J., Shortland, P., & Coggeshall, R. E. (1992). Peripheral nerve injury triggers central sprouting of myelinated afferents. *Nature, 355*, 75–78.

Wynick, D., Thompson, S. W., & McMahon, S. B. (2001). The role of galanin as a multi-functional neuropeptide in the nervous system. *Current Opinion in Pharmacology, 1*, 73–77.

Xiao, H. S., Huang, Q. H., Zhang, F. X., et al. (2002). Identification of gene expression profile of dorsal root ganglion in the rat peripheral axotomy model of neuropathic pain. *Proceedings of the National Academy of Sciences of the United States of America, 99*, 8360–8365.

Xu, X. J., Hökfelt, T., & Wiesenfeld-Hallin, Z. (2008). Galanin and spinal pain mechanisms: Where do we stand in 2008? *Cellular and Molecular Life Sciences, 65*, 1813.

Yang, L., Zhang, F. X., Huang, F., et al. (2004). Peripheral nerve injury induces trans-synaptic modification of channels, receptors and signal pathways in rat dorsal spinal cord. *The European Journal of Neuroscience, 19*, 871–883.

Zhang, X., Ju, G., Elde, R., et al. (1993). Effect of peripheral nerve cut on neuropeptides in dorsal root ganglia and the spinal cord of monkey with special reference to galanin. *Journal of Neurocytology, 22*, 342–381.

Zhou, S., Yu, B., Qian, T., Yao, D., et al. (2011). Early changes of microRNAs expression in the dorsal root ganglia following rat sciatic nerve transection. *Neuroscience Letters, 494*, 89–93.

Zhuo, M., Wu, G., & Wu, L. J. (2011). Neuronal and microglial mechanisms of neuropathic pain. *Molecular Brain, 4*, 31.

Zigmond, R. E., & Sun, Y. (1997). Regulation of neuropeptide expression in sympathetic neurons. Paracrine and retrograde influences. *Annals of the New York Academy of Sciences, 814*, 181–197.

Retrograde Labeling

▶ Retrograde Tracing Technique

Retrograde Tracing Technique

Synonyms

Retrograde labeling

Definition

Retrograde tracing (retrograde labeling) is a neuroanatomical method used to determine the location of the cells of origin of a nervous system pathway. A tracer substance that will be taken up by synaptic terminals (and sometimes by axons) is injected into a region of interest, such as a central nervous system nucleus. The tracer is then conveyed retrogradely by axonal transport to the cell bodies, and often just the proximal dendrites of the neurons that give rise to the projection being labeled. After a suitable time has passed to allow for the uptake and transport of the tracer, the nervous tissue is fixed, sectioned, and the retrograde tracer located microscopically. In some cases, histological processing is required to visualize the tracer. In other cases, the tracer is fluorescent and can be identified using suitable optics without special processing.

Cross-References

▶ Spinohypothalamic Tract, Anatomical
  Organization, and Response Properties
▶ Spinothalamic Input: Cells of Origin (Monkey)
▶ Spinothalamic Neuron
▶ Trigeminal Brainstem Nuclear Complex,
  Anatomy

Retrograde Transport

Definition

Movement of proteins toward the cell body.

Cross-References

▶ Opioid Receptor Trafficking in Pain States

Retroperitoneal

Definition

Visceral pain that is referred to the somatic
structures outside the lining of the abdominal
cavity.

Cross-References

▶ Animal Models and Experimental Tests to
  Study Nociception and Pain
▶ Visceral Pain Model, Pancreatic Pain

Retrospective

Definition

Looking back at events that have already taken
place.

Cross-References

▶ Facet Joint Pain

Reversal Reaction

▶ Type 1 Reaction (Leprosy)

Reversible Posterior Leukoencephalopathy Syndrome

▶ Headache Due to Hypertension

Rexed's Lamina II

▶ Substantia Gelatinosa

Rexed's Laminae

Definition

Rexed's laminae is an architectural classification
of the structure of the spinal cord, based on the
cytological features of the neurons in different
regions of the gray substance described by the
Swedish Anatomist B. Rexed. It consists of nine
laminae (I–IX) that extend throughout the cord,
roughly paralleling the dorsal and ventral columns of the gray substance, and a tenth region
(lamina X) that surrounds the central canal and
consists of the dorsal and ventral commissures
and the central gelatinous substance. Laminae
I and II are often referred to as the superficial
part of the dorsal horn.

Cross-References

▶ DREZ Procedures
▶ Opioid Receptors at Postsynaptic Sites

▶ Spinothalamic Neuron
▶ Spinothalamic Tract Neurons, Glutamatergic Input
▶ Visceral Nociception and Pain
▶ Visceral Pain and Nociception

RF

▶ Radiofrequency Neurotomy, Electrophysiological Principles

RFC

▶ Residual Functional Capacity

RF-DRG

▶ Dorsal Root Ganglion Radiofrequency

Rheumatoid Arthritis

Definition

Rheumatoid arthritis is a chronic autoimmune disease in which mainly peripheral joints are inflamed. The principal characteristics are pain, swelling, and redness (due to increased bloodflow) of joints. There is also overgrowth of the synovial tissue by pannus. Marked destruction of cartilage and bone may occur result in deformities of joints. The disease affects about 1 % of the population. Pathologic processes may also be present outside joints.

Cross-References

▶ Muscle Pain in Systemic Inflammation (Polymyalgia Rheumatica, Giant Cell Arteritis, Rheumatoid Arthritis)
▶ NSAIDs and Their Indications

Rhinorrhea

Definition

Rhinorrhea refers to running of the nose. During acute bouts of cluster headache and during exacerbations of hemicrania continua, the nostril on the side of the pain often runs.

Cross-References

▶ Hemicrania Continua

Rhizotomy

Definition

Destruction of a nerve root within the subarachnoid space, such as dorsal rhizotomy in which the dorsal roots of the spinal cord are severed.

Cross-References

▶ Anesthesia Dolorosa Model, Autotomy
▶ Thalamic Plasticity and Chronic Pain

Risk

Definition

Risk is the likelihood, probability, or degree of probability of loss or injury.

Cross-References

▶ Pain in the Workplace, Risk Factors for Chronicity, and Demographics
▶ Psychiatric Aspects of the Epidemiology of Pain

Risk Factor

Definition

A behavior, environmental exposure, or inherent human characteristics that is associated with an increased probability of a particular health-related outcome (Oleckno 2002).

Cross-References

▶ Prevalence of Chronic Pain Disorders in Children

References

Oleckno, W. A. (2002). *Essential Epidemiology: Principles and Application*. Long Grove, IL: Waveland Press.

Risk Factors for Chronicity

▶ Pain in the Workplace, Risk Factors for Chronicity, and Workplace Factors
▶ Pain in the Workplace, Risk Factors for Chronicity, Job Demands

Risk for Development of a WMSD

Definition

A risk is determined by three factors: (1) the degree of exposure to the risk factors; (2) the likelihood that a ▶ work-related musculoskeletal disorder (WMSD) will develop on exposure to the risk factors; and (3) the severity of the damage.

Consequently, the longer a person is exposed to the risk factors, the higher the risk. The same also applies if the relation between the risk factors and the development of a WMSD is more likely. Finally, a risk is defined as higher if a more severe form of WMSD develops.

Cross-References

▶ Ergonomic Counseling

RLS

▶ Restless Legs Syndrome

ROC

▶ Receiver Operating Characteristics

Rolandic Sulcus

▶ Central Sulcus

Role In Inflammation

▶ Inflammation, Role of Peripheral Glutamate Receptors

Role Modeling

Definition

An instructional strategy involving the actual demonstration of a particular behavior.

Cross-References

▶ Psychology of Pain, Self-Efficacy

Role Plays

Definition

Role plays refer to practice situations in pain treatment, where desirable behaviors are practiced with the help of the therapist or group members.

Cross-References

▶ Operant Treatment of Chronic Pain

Root Canal Therapy

Definition

Root canal therapy is a procedure carried out by dentists in which the dental pulp is removed from the tooth (usually under local anesthesia) and replaced with an inert filling material.

Cross-References

▶ Dental Pain, Etiology, Pathogenesis, and Management

Root Disease

▶ Radiculopathies

Rostral Ventral Medulla

Synonyms

RVM

Definition

The rostral ventral medulla is a region of the brainstem near the pontomedullary junction, and includes the nucleus raphe magnus and the adjacent magnocellular reticular formation. Many neurons in the rostral ventral medulla project serotonergic axons to the spinal cord. Other neurotransmitters, including peptides, can also be contained in axons of neurons in this region.

Cross-References

▶ Nociceptive Processing in the Nucleus Accumbens, Neurophysiology and Behavioral Studies
▶ Pain-Modulatory Systems, History of Discovery
▶ Spinothalamic Tract Neurons, Descending Control by Brain Stem Neurons
▶ Stimulation-Produced Analgesia

Rostral Ventromedial Medulla

Peggy Mason
Department of Neurobiology, Pharmacology and Physiology, University of Chicago, Chicago, IL, USA

Synonyms

RVM

Ventromedial medulla (VMM), raphe magnus (RM) and the adjacent reticular nucleus, which is termed reticularis magnocellularis in human, sheep, cat, and rat, but is often termed reticularis gigantocellularis pars alpha in rat.

Definition

Rostral ventromedial medulla (RVM) contains serotonergic and non-serotonergic cells in the caudal raphe and the adjacent medullary reticular formation. RVM cells receive input from rostral sites such as the hypothalamus and its caudal extension the periaqueductal gray, which are important in modulating nociception as well as other physiological processes. RVM cells project to the spinal cord, particularly the superficial dorsal horn, the intermediolateral cell column that contains preganglionic sympathetic neurons, and the central canal region. Through these connections, RVM cells form one of the primary descending pathways involved in the modulation of nociceptive transmission and other spinal processes. RVM has been strongly implicated in contributing to opioid analgesia and to the hyperalgesia associated with some forms of chronic pain.

Characteristics

Neurons in RVM receive a strong direct projection from sites in the midbrain ▶ periaqueductal gray (PAG), especially the caudal ventrolateral portion of PAG. A monosynaptic, glutamatergic connection from PAG to RVM is critical to the antinociception evoked by PAG stimulation (Aimone and Gebhart 1986). RVM neurons also receive inputs from nuclei in the hypothalamus and from the amygdala. While such inputs are likely to activate nociceptive modulatory circuits in RVM under stressful conditions, little is known about how these circuits are engaged in the normal, uninjured animal.

Electrical stimulation throughout the RVM region mainly blocks withdrawals from noxious stimuli and greatly reduces the responses of dorsal horn cells to noxious stimuli (reviewed in Basbaum and Fields 1984). Microinjection of glutamate, an agonist at the most common excitatory receptor in the CNS, into the breadth of RVM produces similar antinociceptive effects. Thus, there exists within the RVM a neural element that when excited will produce antinociception. Further, the similarity of effects upon activation of cells in both the medial and lateral portions of RVM (Sandkuhler and Gebhart 1984) may be due to the large dendritic fields of

the non-serotonergic component neurons that extend mediolaterally throughout the region (Potrebic and Mason 1993). The functional and anatomical continuity within the RVM suggest that the entire region is a functional unit, albeit highly heterogeneous. For example, while the antinociceptive effects of stimulation in the midline and lateral regions of the RVM are phenomenologically similar, they are pharmacologically distinguishable (Barbaro et al. 1985).

Microinjection of a mu-opioid receptor agonist into RVM has similar effects as does glutamate microinjection, suggesting that the antinociceptive output cell of RVM is excited by mu-opioid receptor agonists. This is likely to occur through a disinhibitory mechanism, with mu-opioid receptor agonists inhibiting a tonically active GABAergic input to the antinociceptive output cell (reviewed in Fields et al. 1991). Endogenously, local neurons provide at least one of the major sources of opioid input to the RVM.

Inactivation of RVM greatly attenuates the antinociception evoked by either electrical stimulation of the PAG or opioid microinjection into the PAG. Thus, the descending antinociceptive pathway from PAG requires excitation of an RVM cell type. Neurons that are indirectly excited by mu-opioid receptor agonists are thought to be the antinociceptive output cells of the RVM. Inactivation of RVM also blocks the hyperalgesia associated with naloxone-precipitated withdrawal from morphine, leading to the idea that RVM modulates nociceptive transmission positively as well as negatively (Kaplan and Fields 1991). Bidirectional modulation is accomplished by two populations of RVM neurons, one that suppresses pain transmission and one that facilitates it. The latter group is comprised of neurons that are directly inhibited by morphine. Recently, Porreca and co-workers have greatly extended our understanding of RVM-mediated nociceptive facilitation. They demonstrated that ablation of opioid-binding neurons in RVM blocks the allodynia and hyperalgesia associated with neuropathic pain (Porreca et al. 2001). Thus, RVM's facilitation of nociception may contribute to behaviors associated with persistent pain.

Most RVM cells project caudally into the spinal cord (Skagerberg and Bjorklund 1985). Rostrally projecting RVM cells are few and are concentrated in the pontine portion of raphe magnus. Within the spinal cord, RVM axons travel primarily in the dorsolateral funiculus and target the superficial and deep dorsal horn, the intermediolateral cell column, and the central canal. The oligosynaptic projection from RVM neurons to most sympathetic targets suggests that RVM cells contribute to the control of many homeostatic functions as well as to sensory modulation. Indeed, evidence supports a role for RVM neurons in thermoregulation (Nakamura et al. 2004).

All of the serotonin in the spinal cord arises from the medulla. Most of the serotonin in the intermediate and dorsal gray comes from RVM as opposed to other caudal serotonergic cells in the cat; this has not been confirmed in the rat (Oliveras et al. 1977). While all or most of the serotonin in the dorsal horn arises from RVM neurons, only a minority of cells in RVM, about 15–25 %, contains serotonin. These serotonergic cells are anatomically, pharmacologically, and physiologically distinct from their non-serotonergic neighbors. Most of the serotonergic RVM cells project to the spinal cord, principally through thin unmyelinated fibers that descend in the dorsolateral funiculus. Although antinociception is sensitive to serotonin receptor antagonists, serotonergic cells are not activated and serotonin release is not increased by antinociceptive manipulations (Gao et al. 1998; Matos et al. 1992). It therefore appears that the sensitivity of antinociception to manipulations of serotonin stems from influences on the effects of other neurotransmitters released from non-serotonergic RVM cells.

Cross-References

► Descending Modulation and Persistent Pain
► Forebrain Modulation of the Periaqueductal Gray and Its Role in Pain
► Opiates, Rostral Ventromedial Medulla, and Descending Control

References

Aimone, L. D., & Gebhart, G. F. (1986). Stimulation-produced spinal inhibition from the midbrain in the rat is mediated by an excitatory amino acid neurotransmitter in the medial medulla. *The Journal of Neuroscience, 6*, 1803–1813.

Barbaro, N. M., Hammond, D. L., & Fields, H. L. (1985). Effects of intrathecally administered methysergide and yohimbine on microstimulation-produced antinociception in the rat. *Brain Research, 343*, 223–229.

Basbaum, A. I., & Fields, H. L. (1984). Endogenous pain control systems: Brainstem spinal pathways and endorphin circuitry. *Annual Review of Neuroscience, 7*, 309–338.

Fields, H. L., Heinricher, M. M., & Mason, P. (1991). Neurotransmitters in nociceptive modulatory circuits. *Annual Review of Neuroscience, 14*, 219–245.

Gao, K., Chen, D. O., Genzen, J. R., et al. (1998). Activation of serotonergic neurons in the raphe magnus is not necessary for morphine antinociception. *The Journal of Neuroscience, 18*, 1860–1868.

Kaplan, H., & Fields, H. L. (1991). Hyperalgesia during acute opioid abstinence: Evidence for a nociceptive facilitating function of the rostral ventromedial medulla. *The Journal of Neuroscience, 11*, 1433–1439.

Matos, F. F., Rollema, H., Brown, J. L., et al. (1992). Do opioids evoke the release of serotonin in the spinal cord? An in vivo microdialysis study of the regulation of extracellular serotonin in the rat. *Pain, 48*, 439–447.

Nakamura, K., Matsumura, K., Hubschle, T., et al. (2004). Identification of sympathetic premotor neurons in medullary raphe regions mediating fever and other thermoregulatory functions. *The Journal of Neuroscience, 24*, 5370–5380.

Oliveras, J. L., Bourgoin, S., Hery, F., et al. (1977). The topographical distribution of serotoninergic terminals in the spinal cord of the cat: Biochemical mapping by the combined use of microdissection and microassay procedures. *Brain Research, 138*, 393–406.

Porreca, F., Burgess, S. E., Gardell, L. R., et al. (2001). Inhibition of neuropathic pain by selective ablation of brainstem medullary cells expressing the mu-opioid receptor. *The Journal of Neuroscience, 21*, 5281–5288.

Potrebic, S. B., & Mason, P. (1993). Three-dimensional analysis of the dendritic domains of on- and off-cells in the rostral ventromedial medulla. *The Journal of Comparative Neurology, 337*, 83–93.

Sandkuhler, J., & Gebhart, G. F. (1984). Characterization of inhibition of a spinal nociceptive reflex by stimulation medially and laterally in the midbrain and medulla in the pentobarbital-anesthetized rat. *Brain Research, 305*, 67–76.

Skagerberg, G., & Bjorklund, A. (1985). Topographic principles in the spinal projections of serotonergic and non-serotonergic brainstem neurons in the rat. *Neuroscience, 15*, 445–480.

Rostral Ventromedial Medulla Cell Types

Peggy Mason
Department of Neurobiology, Pharmacology
and Physiology, University of Chicago, Chicago,
IL, USA

Synonyms

RVM cell types

Definition

Rostral ventral medulla (RVM) cells have been classified as on, off, neutral, and serotonergic. RVM contains both serotonergic and non-serotonergic cell types. In addition to the obvious neurochemical difference between these two cell types, there are anatomical, physiological, and functional differences.

Characteristics

Serotonergic cells comprise about 15 % of the cells in RVM. They have relatively few (3–5) mediolaterally oriented dendrites, which typically extend for less than 750 μm (Gao and Mason 1997). Most serotonergic RVM cells send unmyelinated axons through the dorsolateral funiculus into the spinal cord. The axons travel through and collateralize heavily in the ventrolateral medulla, a region that contains neurons involved in cardiovascular and respiratory control. Serotonergic cells are heterogeneous in neurochemistry, anatomy, and physiology. Most serotonergic cells contain at least one, and often several, co-transmitters, including excitatory and inhibitory amino acids and a plethora of neuropeptides (Bowker et al. 1982). The functional implications of serotonergic cell heterogeneity are unclear and represent a challenge for the future. Since decreasing the availability of serotonin or antagonizing the activation of

serotonergic receptors strongly attenuates antinociception, serotonergic RVM cells were originally thought to be the antinociceptive output cell of the RVM. However, serotonergic RVM cells are not activated by midbrain periaqueductal gray stimulation or opioid administration that evokes antinociception. Further, supraspinal opioid administration that causes analgesia does not consistently evoke serotonin release in the spinal cord. Serotonergic cells in RVM discharge slowly (<5 Hz) and steadily, never bursting or pausing for long periods of time (Auerbach et al. 1985). Serotonergic RVM cells fire at their highest rates when an animal is active and moving about, fire progressively more slowly as an animal proceeds from drowsy to slow-wave sleep, and are silent during rapid eye movement sleep. The discharge of serotonergic RVM cells in anesthetized animals resembles that of animals in a quiet waking state. Thus, the antagonism of RVM-induced antinociception by manipulations that decrease serotonin transmission is likely to be due to antagonism of the actions of tonically released serotonin. Within the dorsal horn, serotonin is likely to primarily function by presynaptically modulating the release of glutamate and other transmitters (Li and Zhuo 1998).

In 1983, Fields and colleagues introduced a physiological classification scheme for three classes of non-serotonergic RVM cells (Fields et al. 1983). This classification system, based on data from anesthetized rats, has proved heuristic and has greatly stimulated the pain modulation field. In essence, the classification system's utility stems from the ability to predict a cell's response to analgesic doses of opioids by its response to noxious cutaneous heat (Barbaro et al. 1986). "On" cells are excited by noxious cutaneous stimuli and inhibited by opiates and are thought to facilitate nociceptive transmission. "Off" cells are inhibited by noxious cutaneous stimuli, excited by opiates, and are thought to inhibit nociceptive transmission. On and Off cell discharge is correlated with increases and decreases, respectively, in cutaneous nociceptive sensitivity evoked by opioid receptor-mediated manipulations, including morphine administration and withdrawal from the same. These findings have implicated On and Off cells in the bidirectional opioid modulation of cutaneous heat-evoked nociceptive responses. Thus, RM neurons integrate brain stem and forebrain input to modulate dorsal horn circuitry, with On cells' facilitating and Off cells' inhibiting nociceptive transmission.

A third class of cells was originally termed neutral, because they failed to respond to noxious heating of the rat tail. These cells also failed to respond to analgesic manipulations, leading to great uncertainty as to the role, if any, of neutral cells in pain modulation. The picture has recently been complicated by clear findings by several investigators that neutral cells respond to noxious stimulation of other modalities and at other sites (Ellrich et al. 2000; Miki et al. 2002).

In the anesthetized rat, there are nonserotonergic cells that are excited or inhibited by noxious stimulation and thus resemble On and Off cells (Leung and Mason 1999; Oliveras et al. 1989). These cells respond similarly to a number of non-noxious stimuli, such as brush and even non-somatic stimuli such as clap. Thus, non-serotonergic RVM cells appear to respond to unexpected external stimulation, regardless of modality, and may therefore function to modify sensory processing at times of danger. Additionally, On and Off cells are active preferentially during waking and sleep, respectively. This pattern may contribute to the lack of responsiveness during sleep to noxious stimulation that is experienced as moderate pain during waking. There is an increase in On and Off cell activity after hindpaw inflammation (Miki et al. 2002). In addition, some neutral cells change their response profile to that of On or Off cells after hindpaw inflammation that persists for 10–24 h.

To date, the Fields' cell classification system explains far more of the existing data than any other. Yet, there are problems that need to be addressed. For instance, cells that are excited by noxious stimulation are not inhibited by morphine in the awake animal (Martin et al. 1992). Finally, as discussed above, RVM neurons have been implicated in a number of processes in addition to pain modulation. Yet, the role of On

and Off cells in these other processes is either unknown or controversial. As an example of the latter, two groups have investigated RVM cell responses to a thermoregulatory cold challenge. One group found that On cells were activated and the other that Off cells were activated (Rathner et al. 2001; Young and Dawson 1987). Understanding if and how individual RVM neurons contribute to multiple modulatory functions presents an exciting challenge for the future.

References

Auerbach, S., Fornal, C., & Jacobs, B. L. (1985). Response of serotonin-containing neurons in nucleus raphe magnus to morphine, noxious stimuli, and periaqueductal gray stimulation in freely moving cats. *Experimental Neurology, 88*, 609–628.

Barbaro, N. M., Heinricher, M. M., & Fields, H. L. (1986). Putative pain modulating neurons in the rostral ventral medulla: Reflex-related activity predicts effects of morphine. *Brain Research, 366*, 203–210.

Bowker, L. K. N., Sullivan, M. C., Wilber, J. F., et al. (1982). Transmitters of the raphe-spinal complex: Immunocytochemical studies. *Peptides, 3*, 291–298.

Ellrich, J., Ulucan, C., & Schnell, C. (2000). Are "neutral cells" in the rostral ventro-medial medulla subtypes of on- and off-cells? *Neurosciences Research, 38*, 419–423.

Fields, H. L., Bry, J., & Hentall, I. (1983). The activity of neurons in the rostral medulla of the rat during withdrawal from noxious heat. *The Journal of Neuroscience, 3*, 2545–2552.

Gao, K., & Mason, P. (1997). Somatodendritic morphology and axonal anatomy of intracellularly labeled serotonergic neurons in the rat medulla. *The Journal of Comparative Neurology, 389*, 309–328.

Leung, C. G., & Mason, P. (1999). Physiological properties of medullary raphe neurons during sleep and waking. *Journal of Neurophysiology, 81*, 584–595.

Li, P., & Zhuo, M. (1998). Silent glutamatergic synapses and nociception in mammalian spinal cord. *Nature, 393*, 695–698.

Martin, G., Montagne, C. J., & Oliveras, J. L. (1992). Involvement of ventromedial medulla "multimodal, multireceptive" neurons in opiate spinal descending control system: A single-unit study of the effect of morphine in the awake, freely moving Rat. *The Journal of Neuroscience, 12*, 1511–1522.

Miki, K., Zhou, Q. Q., Guo, W., et al. (2002). Changes in gene expression and neuronal phenotype in brain stem pain modulatory circuitry after inflammation. *Journal of Neurophysiology, 87*, 750–760.

Oliveras, J. L., Vos, B., Martin, G., et al. (1989). Electrophysiological properties of ventromedial medulla neurons in response to noxious and non-noxious stimuli in the awake, freely moving rat: A single-unit study. *Brain Research, 486*, 1–14.

Rathner, J. A., Owens, N. C., & McAllen, R. M. (2001). Cold-activated raphe-spinal neurons in rats. *The Journal of Physiology, 535*, 841–854.

Young, A. A., & Dawson, N. J. (1987). Static and dynamic response characteristics, receptive fields, and interaction with noxious input of midline medullary thermoresponsive neurons in the rat. *Journal of Neurophysiology, 57*, 1925–1936.

Roxicodone®

▶ Postoperative Pain, Oxycodone

RSD (Obsolete Taxonomy)

▶ CRPS-1 in Children

RSI

▶ Repetitive Strain Injuries

RSIs

▶ Disability, Upper Extremity

rTMS

▶ Repetitive Transcranial Magnetic Stimulation

RTX

▶ Resiniferatoxin

Ruffini Receptor

Definition

Spindle-like structures in the dermis and close to joints in which the terminals of a fast conducting myelinated afferent axon are situated. The structure serves as a slowly adapting mechanosensor particularly sensitive to shear forces in skin or joints. The transducer molecules are probably situated in the nerve terminals.

Cross-References

▶ Perireceptor Elements

Rumination

Definition

A tendency to fix attention on a particular symptom, situation, or outcome.

Cross-References

▶ Catastrophizing

RVM

▶ Descending Modulation of Nociceptive Transmission During Persistent Damage to Peripheral Tissues
▶ Rostral Ventral Medulla
▶ Rostral Ventromedial Medulla

RVM Cell Types

▶ Rostral Ventromedial Medulla Cell Types

RVM, Rostral Ventromedial Medulla

▶ Nucleus Raphe Magnus

S

S/N Ratio

▶ Signal-to-Noise Ratio

S2

▶ Nociceptive Processing in the Secondary Somatosensory Cortex

Sacral Hiatus

Definition

A normally occurring gap at the lower end of the sacrum. It is closed by the sacrococcygeal ligament and provides access to the sacral epidural space for administration of anesthetics.

Cross-References

▶ Epidural Infusions in Acute Pain

Sacroiliac Injury

▶ Sacroiliac Joint Pain

Sacroiliac Joint

Definition

The sacroiliac joint lies next to the spine and connects the sacrum (the triangular bone at the bottom of the spine) with the pelvis (iliac crest).

Cross-References

▶ Chronic Low Back Pain: Definitions and Diagnosis
▶ Lower Back Pain, Physical Examination

Sacroiliac Joint Blocks

Anthony C. Schwarzer
Faculty of Medicine and Health Sciences,
The University of Newcastle, Newcastle,
Australia

Synonyms

Intra-articular sacroiliac joint block; Intra-articular sacroiliac joint injection; Sacroiliac joint injection

Definition

Intra-articular injection of local anesthetic into the sacroiliac joint (SIJ) under controlled

G.F. Gebhart, R.F. Schmidt (eds.), *Encyclopedia of Pain*, DOI 10.1007/978-3-642-28753-4,
© Springer-Verlag Berlin Heidelberg 2013

conditions is a diagnostic test that is used to confirm or deny that the SIJ is the source of a patient's pain.

Characteristics

Principles

The explicit purpose of a SIJ block is to test if a patient's pain is relieved by anesthetizing the joint targeted. The block tests the hypothesis that perhaps the pain arises from the SIJ. If pain is not relieved, the target joint cannot be regarded as mediating the patient's pain, whereupon a new hypothesis about the source of pain is required. If pain is relieved, the response constitutes prima facie evidence that the targeted joint is the source of the patient's pain, but steps need to be taken to ensure that the observed response is not false-positive. A positive response occurs when there is complete relief of a defined region of pain. The block need not relieve every area of pain, for a diagnosis of SIJ pain does not preclude other sources giving rise to pain in other areas. Remaining areas of pain may be targeted separately, using other investigations.

Setting

The procedure is conducted using a C-arm fluoroscope, with the patient lying face down on a radiolucent table. No sedation is required. Needle placement and spread of contrast medium is documented with hard-copy films or an image on specialized paper.

Technique

The procedure is performed using a strict aseptic technique, with the physician using sterile gloves. The skin of the back is cleansed using an iodine-based solution, chlorhexidine, or an alcohol-based antiseptic. A 90 mm 25 gauge spinal needle is optimal. Local anesthetic agents, typically bupivacaine 0.5 % or lignocaine 2 %, are injected. Contrast medium is required to verify intra-articular placement of the needle.

The target point lies along the inferior, posterior aspect of the joint, approximately 1–2 cm

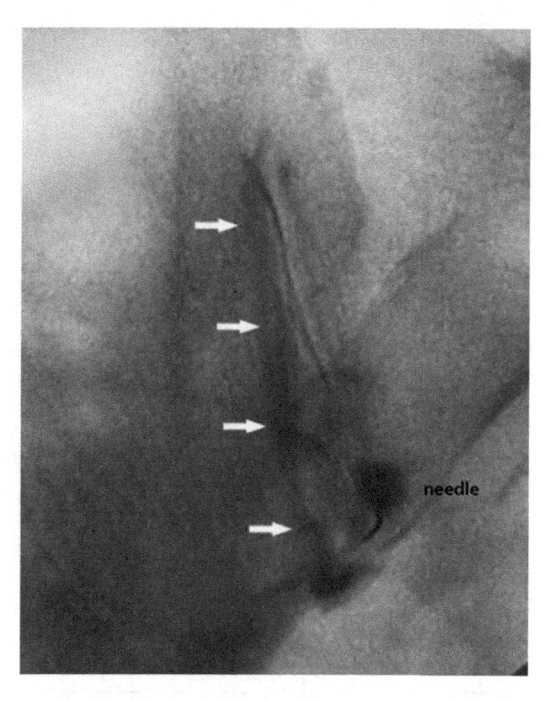

Sacroiliac Joint Blocks, Fig. 1 An AP radiograph of a sacroiliac joint arthrogram. The needle lies in the inferior end of the joint cavity. The *arrows* indicate contrast medium that has spread throughout the joint cavity

cephalad of its most inferior end. The C-arm of the fluoroscope is rotated until the medial cortical line of the medial silhouette of the SIJ is maximally crisp. Once the inferior, posterior margin of the joint has been isolated radiographically, the target point is selected. The puncture point directly overlies the target (Dreyfuss et al. 1994; Schwarzer et al. 1995).

The needle is directed through the puncture point towards the target point, but aiming initially at the sacrum. The needle is then withdrawn slightly and redirected towards the joint space a few millimeters cranial to its lower end. Once the needle has entered the joint space, intra-articular placement must be confirmed with an injection of a small volume (0.3–0.5 ml) of contrast medium. If the needle has been correctly placed, upon injection the contrast medium will outline the joint space (Fig. 1). Local anesthetic is then injected and continued until there is a firm resistive end point. No more than 1.5 ml of local anesthetic is used.

Complications

Like any invasive procedure, SIJ blocks carry the nominal risk of infection, bleeding, and allergic reaction.

Evaluation

After completion of the procedure, the response is evaluated. A validated pain measure such as the visual analogue pain scale is used. Information is obtained at half-hourly to hourly intervals for at least 4 h after the procedure. An independent assessor also records the response before, and half an hour after, the procedure. A response is considered positive if there is greater than 75 %, and ideally greater than 90 %, relief from the injection. The duration of relief must be consistent with the known duration of action of the local anesthetic used. If the patient has a positive response, then a second, or confirmatory, block is performed on a different day.

Application

SIJ blocks are used in patients with low back pain that is maximal below the topographic level of the L5 vertebra. The patient may or may not have somatic referred pain in the lower limb. The pain must have failed to resolve with conservative means and with the passage of time.

Validity

Intra-articular SIJ blocks have face validity. An injection of contrast medium followed by up to 1.5 ml of local anesthetic will remain within the joint and not anesthetize any other structure. A volume exceeding 2.5 ml will leak through the joint capsule, thereby anesthetizing structures posteriorly and/or anteriorly (Fortin et al. 1994a; Fortin and Tolchin 1993). If an injection of local anesthetic into the joint results in complete relief of the patient's usual pain, then there is evidence to suggest that the pain stems from the SIJ. However, this does not constitute proof that the SIJ is the source of pain. There is the risk of false-positive responses. Patients may report

relief of pain for reasons other than the action of the agent injected. In order to avoid or to reduce false-positive diagnoses based on the results of SIJ blocks, these blocks should be subjected to some form of control.

A compelling form of control is the injection of a placebo agent such as normal saline, but this would require informed consent, and patients might object to undergoing a sham procedure. An alternative form of control is to anesthetize the joint on separate occasions using agents with different durations of action. This form of control, known as comparative local anesthetic blocks, has been validated in the context of peripheral nerve blocks, but it has not been validated in the context of intra-articular blocks (Barnsley et al. 1993; Lord et al. 1995). It is not known whether local anesthetic agents exhibit the same duration of action when injected intra-articularly, as they do when injected into or onto more superficial tissues. Nevertheless, comparative blocks have some degree of concept validity and strict content validity. Their critical utility arises when repeat blocks fail to relieve the patient's pain. In that event, they demonstrate that the first response was false-positive and thereby reduce false-positive diagnoses based on single blocks.

An alternative idiom is the use of anatomical controls. A positive response to a SIJ block is more compelling if the patient has previously had negative responses to zygapophysial joint blocks and/ or lumbar discography (Schwarzer et al. 1995). Such patterns of response do not absolutely rule out the chance of a placebo response, but the fact that the patient has not exhibited a placebo response during earlier investigations intuitively increases the credibility of their response.

Using anatomical controls, Schwarzer et al. (1995) estimated that in patients with chronic low back pain, the source of pain could be traced to the SIJ in 13 % of cases (95 % confidence interval: 6–20 %). Maigne et al. (1996) used controlled diagnostic blocks and found a prevalence of 19 % (95 % CI: 9–29 %).

In the absence of any other valid, diagnostic tests, controlled SIJ blocks constitute the only available means of establishing a diagnosis of SIJ pain. There have been no studies of the validity of pain radiographs using controlled diagnostic blocks as the criterion standard. CT analysis of the SIJ has limited value as judged against diagnostic intra-articular SIJ injections due to poor sensitivity and specificity (Elgafy et al. 2001). There are no studies of the diagnostic validity of MRI for SIJ pain. Bone scintigraphy demonstrated a specificity of 90–100 %, but poor sensitivity (Slipman et al. 1996; Maigne et al. 1998). No specific features on history or physical examination features have been validated scientifically, which accurately identify a painful SIJ (Schwarzer et al. 1995). None of the commonly used items of history or physical examination tests that are purported to diagnose SIJ pain have been validated (Dreyfuss et al. 1994, 1996). Patterns of referred pain are not diagnostic of SIJ pain (Schwarzer et al. 1995). However, it has been established by several investigators that, in patients with SIJ pain, pain is always maximal below L5 (Schwarzer et al. 1995; Fortin et al. 1994a, b). Pain maximal above L5 is highly unlikely to be of SIJ origin.

Utility

SIJ blocks have diagnostic utility in that if and once positive, they identify the source of pain. SIJ blocks have limited therapeutic utility. There is no proven treatment for SIJ pain if and once it is diagnosed.

Injections of corticosteroids into the SIJ have been shown to be efficacious in the treatment of sacroiliitis, due to various spondyloarthropathies in most, but not all, controlled and observational outcome studies. These studies imply a therapeutic role for intra-articular SIJ corticosteroid injections when the diagnosis of true sacroiliitis has been established (Maugars et al. 1992, 1996). One retrospective, uncontrolled case series on the use of IA corticosteroids exists for those with documented idiopathic SIJ pain with promising results (Slipman et al. 2001). Uncontrolled, observational reports suggest some benefit from the use of extra-articular corticosteroids for presumed SIJ pain.

Indications

The cardinal indication for SIJ blocks is the need to know if a patient's pain is arising from a SIJ or not. SIJ blocks may be performed in the course of the systematic investigation of low back pain or when there is, a priori, a high index of suspicion that the joint is painful.

References

Barnsley, L., Lord, S., & Bogduk, N. (1993). Comparative local anaesthetic blocks in the diagnosis of cervical zygapophysial joints pain. *Pain, 55*, 99–106.
Dreyfuss, P., Dreyer, S., Griffin, J., Hoffman, J., & Walsh, N. (1994). Positive sacroiliac joint screening tests in asymptomatic adults. *Spine, 19*, 1138–1143.
Dreyfuss, P., Michaelsen, M., Pauza, K., McLarty, J., & Bogduk, N. (1996). The value of history and physical examination in diagnosing sacroiliac joint pain. *Spine, 21*, 2594–2602.
Elgafy, H., Semaan, H. B., Ebraheim, N. A., & Coombs, R. J. (2001). Computed tomography findings in patients with sacroiliac pain. *Clinical Orthopaedics, 382*, 112–118.
Fortin, J. D., & Tolchin, R. B. (1993). Sacroiliac provocation and arthrography. *Archives of Physical Medicine and Rehabilitation, 74*, 125–129.
Fortin, J. D., Aprill, C. N., Ponthieux, B., et al. (1994a). Sacroiliac joint: Pain referral maps upon applying a new injection/arthrography technique part II: Clinical evaluation. *Spine, 19*, 1483–1489.
Fortin, J. D., Dwyer, A. D., West, S., & Pier, J. (1994b). Sacroiliac joint: Pain referral maps upon applying a new injection/arthrography technique. Part I: Asymptomatic volunteers. *Spine, 19*, 1475–1482.
Lord, S. M., Barnsley, L., & Bogduk, N. (1995). The utility of comparative local anaesthetic blocks versus placebo-controlled blocks for the diagnosis of cervical zygapophysial joint pain. *The Clinical Journal of Pain, 11*, 208–213.
Maigne, J. Y., Aivaliklis, A., & Pfefer, F. (1996). Results of sacroiliac joint double block and value of sacroiliac pain provocation tests in 54 patients with low back pain. *Spine, 21*, 1889–1892.
Maigne, J. Y., Boulahdour, H., & Chatellier, G. (1998). Value of quantitative radionuclide bone scanning in the diagnosis of sacroiliac joint syndrome in 32 patients with low back pain. *European Spine Journal, 7*, 328–331.
Maugars, Y., Mathis, C., Vilon, P., & Prost, A. (1992). Corticosteroid injection of the sacroiliac joint in patients with seronegative spondyloarthropathy. *Arthritis and Rheumatism, 35*, 564–568.
Maugars, Y., Mathis, C., & Berthelot, J. (1996). Assessment of the efficacy of sacroiliac corticosteroid

injections in spondyloarthropathies: A double-blind study. *British Journal of Rheumatology, 35*, 767–770.

Schwarzer, A. C., Aprill, C. N., & Bogduk, N. (1995). The sacroiliac joint in chronic low back pain. *Spine, 20*, 31–37.

Slipman, C. W., Sterenfeld, E. B., Chou, L. H., Herzog, R., & Vresilovic, E. (1996). The value of radionuclide imaging in the diagnosis of sacroiliac joint syndrome. *Spine, 21*, 2251–2254.

Slipman, C. W., Lipetz, J. S., Plastaras, C. T., Jackson, H. B., Vresilovic, E. J., Lenrow, D. A., & Braverman, D. L. (2001). Fluoroscopically guided therapeutic sacroiliac joint injections for sacroiliac joint syndrome. *American Journal of Physical Medicine Rehabilitation, 80*, 425–432.

Sacroiliac Joint Denervation

▶ Sacroiliac Joint Radiofrequency Denervation

Sacroiliac Joint Injection

▶ Sacroiliac Joint Blocks

Sacroiliac Joint Neurotomy

▶ Sacroiliac Joint Radiofrequency Denervation

Sacroiliac Joint Pain

Jenna E. Zauk[1], Ricardo J. Komotar[2] and Christopher J. Winfree[3]
[1]St. George's University, School of Medicine, Grenada, West Indies
[2]Department of Neurological Surgery, Neurological Institute Columbia University, New York, NY, USA
[3]Department of Neurological Surgery, Neurological Institute Columbia University, New York, NY, USA

Synonyms

Lower back pain; Sacroiliac injury; Sacroiliitis

Definition

Sacroiliac (SI) joint pain is defined as back discomfort in the area of the SI joint, where the spine and pelvis meet (Tanner 1997). This pain may be the result of an underlying condition or trauma, leading to nerve damage and bone necrosis.

Characteristics

Background

Sacroiliac joint syndrome (SIJS) was a common diagnosis for low back pain in the early 1900s (Tanner 1997). In 1932, however, the discovery of vertebral disk herniation altered the diagnosis and management of back pain. Currently, SIJS is difficult to distinguish from other types of low back pain and, as a result, is often overlooked. Nevertheless, SIJS remains an important contributor to disability, accounting for nearly 20 % of chronic low back pain cases (Maigne et al. 1996).

Anatomy

The SI joints exist bilaterally and hinge the lower half of the skeleton by connecting the sacrum to the ilium (Bogduk 1995). Their articular surfaces are integrated and covered by a combination of hyaline cartilage and fibrocartilage. Not designed for motion, these joints commonly fuse with age and are difficult to examine. Although the SI joints typically move only 2–3 mm during exertion, they may be mobilized by certain therapeutic positional exercises. Limited movement enables the SI joints to absorb shock and act as fulcrums during ambulation (Bowen and Cassidy 1981).

Medical History

The most classic presentation of SIJS is medial buttock discomfort, with sacral sulcus pain nearly always being present (Fortin et al. 1994). A distinguishing feature for this syndrome is a lack of pain above the L-5 level (Fukui and Nosaka 2002). Patients may also complain of referred pain to the buttocks, hip, groin, abdomen, thigh, or calf. Standing for an extended period of time and stair climbing exacerbates the pain. Sacroiliitis can be due to a traumatic injury, infection, arthritis, or pregnancy.

Abnormal back posture is typical in patients with arthritis in the lumbopelvic region. A recent study (Chang 2009) demonstrates that sacroilliac joint problems can be associated with high anti-streptolysin O titers and postural imbalance. High levels of anti-streptolysin O (ASO titer > 116 IU/mL) cause an overstimulation of the sacroiliac joint, which in turn causes pain and lost control of upright posture. In this particular study, patients with undifferentiated arthritis and high ASO titers presented with higher sway velocity and intensity values (which were the measures of postural control) than in arthritic patients with low titer levels of ASO < 116 IU/mL (P < 0.0001). A small subject pool and confounding factors limit the study's validity: Only 84 subjects were tested (mean age: 23 years), and it is difficult to differentiate between SIJS and other types of low back pain that may cause similar problems (Chang 2009).

Physical Exam

Physicians need to rule out other potential causes of lower back pain, including disk disease, spinal stenosis, degenerative hip joint disease, and posterior facet syndrome (Bogduk 2004). With SIJS, full lumbar spine and hip range of motion are expected. Furthermore, neurological examination is usually normal, with negative nerve-root tension signs. Gluteus medius weakness may be found and is often combined with tightness in the piriformis muscle and quadratus lumborum (Calvillo et al. 2000). In addition to radiographs, springing the pelvis and percussion of the lower spine are two methods to assess the condition. Patients with sacroiliitis will experience sensitivity and pain (Peh 2009).

Diagnostic Testing

No single test accurately identifies SIJS. Diagnostic examination aims to detect abnormal movement in the SI joint and elicit pain by stressing adjacent structures, such as the lumbar spine, hip joint, and femoral or sciatic nerves (Cibulka and Koldehoff 1999). Of note, the thigh thrust and Gaenslen's sign have the greatest

predictive value and should be used in combination. Other diagnostic tests that may be implemented during physical exam include the distraction test, sacroiliac shear test, Patrick's test, and compression test.

Radiology

Conventional X-ray and computed tomography (CT) are rarely useful in evaluating possible SIJS. Bone scans, despite their high specificity, are not recommended for the diagnostic workup of SIJS due to their low sensitivity (Esdaile and Rosenthall 1983). Similarly, MRI has a high sensitivity and allows visualization of soft-tissue anatomy, yet it is of limited value due to its low specificity (Bigot et al. 1999). With this in mind, the gold standard for the diagnosis of SIJS is provocative injection under fluoroscopic guidance (Prather 2003). More specifically, this technique confirms the diagnosis of SIJS by inducing pain upon injection, followed by symptom relief with an anesthetic block. This is the most reliable method for establishing the diagnosis of SIJS and allows immediate radiographic interpretation. It is an invasive procedure, however, and should be used after less complicated and invasive therapeutic measures are exhausted.

Differential Diagnosis

Numerous conditions may mimic SIJS (Miller 1973). For instance, sacral stress fractures should be considered in athletes with chronic pain in this region. Spondyloarthropathies often begin with symptoms of sacroiliitis (Khan and van der Linden 1990). Ankylosing spondylitis and spondyloarthropathy associated with inflammatory bowel disease may also produce findings consistent with SIJS. In addition, psoriatic arthritis, Reiter's syndrome, myofascial pain, and osteitis condensans ilii are other potentially confounding diagnoses that may present with sacroiliitis.

Treatment Options

First-line treatment for sacroiliac joint disease should consist of conservative therapy. To begin with, patients should avoid any activities, including athletic endeavors, which aggravate symptoms. In addition, patients should be advised to take anti-inflammatory medications for pain relief. Irrespective of their analgesic effects, these medications are beneficial in reducing regional inflammatory processes that may cause symptoms. Therefore, patients should continue on these medications as long as they are efficacious.

Physical therapy is another key component of conservative treatment, as strengthening the musculature surrounding the SI joint may help increase flexibility and protect from further injury. Rehabilitation must address functional biomechanical deficits with exercise regimens that stabilize the SI joints against large shear stresses (Prather 2003). These targeted exercises may be used in combination with therapeutic joint mobilization to counter muscle disability or abnormal shear forces. Abnormal posture or leg-length discrepancy must also be treated, as this condition is associated with detrimental positioning of the sacrum. If leg-length discrepancy is detected on the physical exam, orthoses and heel lifts can help correct this deformity.

If conservative treatments fail, corticosteroid injections may be effective. This procedure delivers concentrated anti-inflammatory medication directly into the sacroiliac joint. As the SI joint lies deep within the pelvis, these injections are usually given under X-ray guidance in a hospital setting (Bellaiche 1995). Recent therapy for this pain involves sacroiliac joint injections that inject corticosteroids into or around the sacroiliac joint. To date, limited trials are available to thoroughly assess this form of treatment. A study in 2002 (n = 24) confirms greater progress in pain relief (pain index = 0.017, p = 0.047) after 1 month of treatment from a corticosteroid injection compared with an anesthetic injection (Luukkainen et al. 2002).

Conclusion

SIJS remains a diagnostic challenge. After ruling out other potential sources of pain, a combination of provocative maneuvers and diagnostic injections has proven reliable in the evaluation of SIJS. If conservative management consisting of physical therapy does not foster recovery,

fluoroscopically guided injection of analgesics into the SIJ should be considered for both diagnostic confirmation of the pain source and symptomatic improvement.

References

Bellaiche, L. (1995). Cortisone injection into the sacroiliac joint. *Annales de Radiologie, 38*, 192–195.

Bigot, J., Loeuille, D., Chary-Valckenaere, I., Pourel, J., Cao, M. M., & Blum, A. (1999). Determination of the best diagnostic criteria of sacroiliitis with MRI. *Journal de Radiologie, 80*, 1649–1657.

Bogduk, N. (1995). The anatomical basis for spinal pain syndromes. *Journal of Manipulative and Physiological Therapeutics, 18*, 603–605.

Bogduk, N. (2004). Management of chronic low back pain. *The Medical Journal of Australia, 180*, 79–83.

Bowen, V., & Cassidy, J. D. (1981). Macroscopic and microscopic anatomy of the sacroiliac joint from embryonic life until the eighth decade. *Spine, 6*, 620–628.

Calvillo, O., Skaribas, I., & Turnipseed, J. (2000). Anatomy and pathophysiology of the sacroiliac joint. *Current Review of Pain, 4*, 356–361.

Chang, C. C., Chu, H. Y., Chiang, S. L., Li, T. Y., & Chang, S. T. (2009). Impairment of static upright posture in subjects with undifferentiated arthritic in sacroilliac joint in conjuction with eleveation of streptococcal serology. *Journal of Back and Musculoskeletal Rehabilitation, 22(1)*:33–41.

Cibulka, M. T., & Koldehoff, R. (1999). Clinical usefulness of a cluster of sacroiliac joint tests in patients with and without low back pain. *The Journal of Orthopaedic and Sports Physical Therapy, 29*, 83–89. discussion 90–82.

Esdaile, J., & Rosenthall, L. (1983). Radionuclide joint imaging. *Comprehensive Therapy, 9*, 54–63.

Fortin, J. D., Aprill, C. N., Ponthieux, B., & Pier, J. (1994). Sacroiliac joint: Pain referral maps upon applying a new injection/arthrography technique. Part II: Clinical evaluation. *Spine, 19*, 1483–1489.

Fukui, S., & Nosaka, S. (2002). Pain patterns originating from the sacroiliac joints. *Journal of Anesthesia, 16*, 245–247.

Khan, M. A., & van der Linden, S. M. (1990). A wider spectrum of spondyloarthropathies. *Seminars in Arthritis and Rheumatism, 20*, 107–113.

Luukkainen, R. K., Wennerstrand, P. V., Kautiainen, H. H., Sanila, M. T., & Asikainen, E. L. (2002). Efficacy of periarticular corticosteroid treatment of the sacroiliac joint in nonspondylarthropathic patients with chronic low back pain in the region of the sacroiliac joint. *Clinical and Experimental Rheumatology, 20*, 52–54.

Maigne, J. Y., Aivaliklis, A., & Pfefer, F. (1996). Results of sacroiliac joint double block and value of sacroiliac pain provocation tests in 54 patients with low back pain. *Spine, 21*, 1889–1892.

Miller, D. S. (1973). Low back pain. Differential diagnosis determines treatment: Hospitalization enables more efficient treatment. *Medical Trial Technique Quarterly, 20*, 1–41.

Peh WC. March 25, 2009. http://emedicine.medscape.com/article/386639-overview

Prather, H. (2003). Sacroiliac joint pain: Practical management. *Clinical Journal of Sport Medicine, 13*, 252–255.

Tanner, J. (1997). Sacroiliac joint pain. *Spine, 22*, 1673–1674.

Sacroiliac Joint Radiofrequency

▶ Sacroiliac Joint Radiofrequency Denervation

Sacroiliac Joint Radiofrequency Denervation

Way Yin
Department of Anesthesiology, University of Washington, Seattle, WA, USA

Synonyms

Sacroiliac joint denervation; Sacroiliac joint neurotomy; Sacroiliac joint radiofrequency; Sensory stimulation-guided radiofrequency neurotomy; Stereotactic sacroiliac joint denervation

Definition

Sacroiliac joint radiofrequency denervation is a treatment for low back pain originating from the sacroiliac joint complex (SIJC). Radiofrequency (RF) current is channeled through percutaneous electrodes to coagulate sensory nerves from the SIJC. Coagulation of these sensory nerves disrupts the transmission of nociceptive signals from the SIJC, thereby relieving pain.

Characteristics

Mechanism

▶ Radiofrequency neurotomy achieves relief of pain by coagulating sensory nerves that are conducting nociceptive information from a painful structure (see ▶ Radiofrequency Neurotomy, Electrophysiological Principles).

Indications

Sacroiliac joint radiofrequency denervation is a structure-specific therapeutic intervention, specifically for the treatment of chronic back pain arising from the sacroiliac joint complex. SIJC RF should only be considered in patients who have demonstrable pain arising from the sacroiliac joint, and in whom other structural sources of low back pain have been eliminated (e.g., zygapophysial joints, intervertebral discs).

Demonstration of SIJC pain requires the patient to experience complete or near-complete relief of pain following controlled local anesthetic blocks of the SIJC. Response to a single anesthetic injection of the SIJC is insufficient to justify proceeding with denervation, as the rate of false-positive responses to single anesthetic injections may exceed 35 % (Barnsley et al. 1993; Schwarzer et al. 1994). SIJC RF should only be considered in the absence of specific contraindications to therapeutic intervention.

Technique

Any attempt to denervate the SIJC requires a detailed understanding of relevant sensory neuroanatomy. The primary sensory neuroanatomy of the SIJC is currently felt to arise from the dorsal rami of L5-S3 (Grob et al. 1995; Fortin et al. 1999; Willard et al. 1998). Numerous lateral branches of the segmental dorsal rami form a lateral dorsal sacral plexus, and branches from this plexus travel laterally in a complex and highly variable three-dimensional topography, to ultimately supply sensation to the deep interosseous ligaments and the sacroiliac joint proper. These nerves constitute the target for denervation of the SIJC.

Several techniques for denervation of the SIJC by radiofrequency neurotomy have been described

Sacroiliac Joint Radiofrequency Denervation, Fig. 1 Sacroiliac joint radiofrequency denervation, bipolar technique (Courtesy Kline MT)

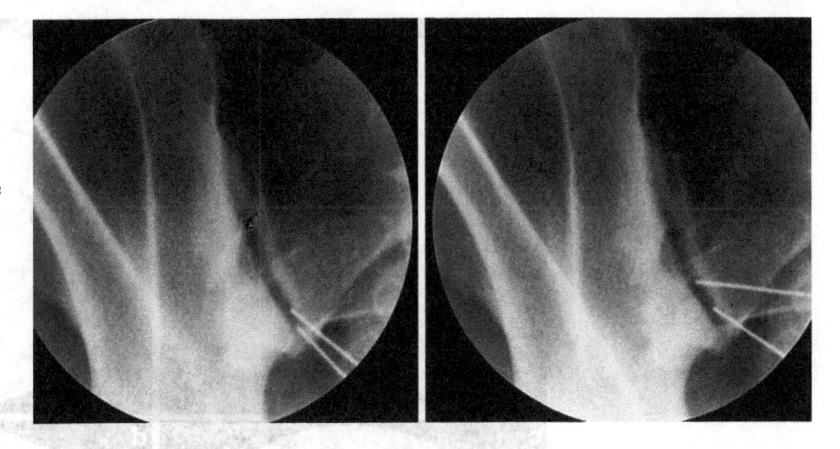

(Gevargez et al. 2002; Kline 1992; Lennard 2000; Ferrante et al. 2001; Yin et al. 2003), all of which utilize fluoroscopic (Lennard 2000; Ferrante et al. 2001; Kline 1992; Yin et al. 2003) or CT guidance (Gevargez et al. 2002).

The first technique described to denervate the SIJC is attributed to Kline (Kline 1992). He performed serial bipolar radiofrequency lesions along the medial aspect of the posterior ilium on the dorsal sacral plate (Fig. 1). Other techniques have used single lesions along the medial sacral ala and lateral aspect of the first dorsal sacral foramen (Gevargez et al. 2002), or along the lateral margins of the S1–S3 dorsal sacral foramina (Lennard 2000).

A technique based on the neuroanatomy of the dorsal sacral plexus and incorporating patient-blinded, sensory stimulation to select symptom-bearing lateral branch nerves has recently been described (Yin et al. 2003). This sensory stimulation-guided (SSG) technique is based on the known neuroanatomy of the dorsal sacral plexus and has produced the most favorable outcomes of all techniques to date (Yin et al. 2003).

The SSG technique involves inserting a needle electrode into the vicinity of each of the lateral branches of the sacral dorsal that may be mediating the patient's pain (Yin et al. 2003). A stimulating current is passed through the electrode, in an effort to reproduce the patient's pain. The patient is asked to report any responses to stimulation and must be able to discriminate between reproduction of concordant pain and non-painful paresthesia.

Reproduction of the patient's pain indicates that a nerve that mediates the pain has been located. Wherever a symptomatic nerve is identified with reproducible low-voltage sensory stimulation, it is thermocoagulated with the electrode.

Sensory mapping is undertaken systematically across the medial sacral ala and the lateral aspects of the dorsal sacral foramina, in order to find all nerves that may be involved (Fig. 2a–d). Symptomatic nerves are nearly always identified from L5 and S1, with progressively fewer nerves being found from S2 and S3 (Yin et al. 2003). Patients with referred pain in the lower extremity or groin associated with SIJC pain will typically experience relief of their referred pain following successful SIJC denervation, especially if concordant symptoms are elicited on lateral branch sensory mapping.

Efficacy

To date, the efficacy of SIJC RF is based on non-controlled prospective (Gevargez et al. 2002) and retrospective case series (Ferrante et al. 2001; Yin et al. 2003). No controlled or long-term (i.e., greater than 1 year) efficacy studies have been published to date.

If patients are properly selected, the duration of properly performed SIJC RF exceeds that of local anesthetic and steroid injections into the SIJC and, as published reports to date suggest, may provide efficacy beyond that provided by conservative, noninterventional means (Gevargez et al. 2002; Yin et al. 2003). Published

Sacroiliac Joint Radiofrequency Denervation,
Fig. 2 Sacroiliac joint radiofrequency denervation, sensory stimulation-guided technique, and typical electrode placement. (**a**) L5 lateral branch, (**b**) S1 lateral branch, (**c**) S2 lateral branch, (**d**) S3 lateral branch

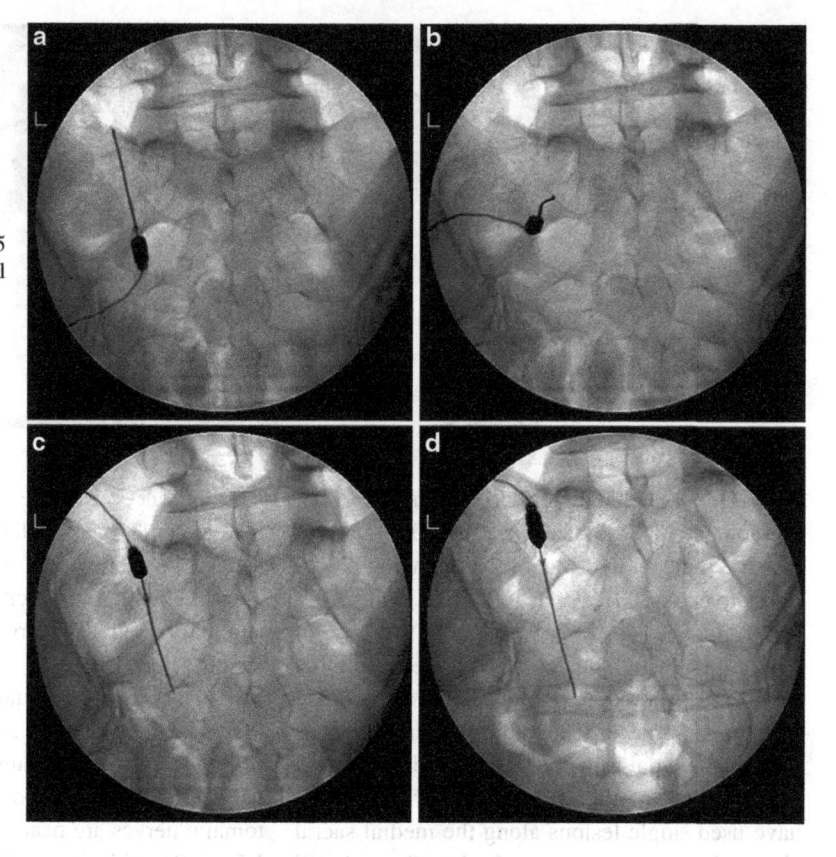

efficacy ranges from a low of 36.4 % of patients experiencing greater than 50 % reduction in pain for more than 6 months (Ferrante et al. 2001) to a high of 64 % of patients experiencing greater than 60 % relief for more than 6 months using the SSG technique (Yin et al. 2003). Some 36 % of patients experienced complete relief for more than 6 months (Yin et al. 2003).

Although controlled trials have still to be conducted, radiofrequency neurotomy is the only procedure for sacroiliac joint pain that has a documented efficacy. No other procedure has been shown to provide lasting relief, and none has been found to produce complete relief of pain.

Complications

Reported complications following SIJC RF by any technique are rare, and no mortality has been reported. Patients most commonly experience a period of postoperative discomfort that is self-limited. Patients will occasionally experience moderate-term localized dysesthesia or hypoesthesia in the skin of the ipsilateral buttock following SIJC RF denervation.

References

Barnsley, L., Lord, S., Wallis, B., & Bogduk, N. (1993). False-positive rates of cervical zygapophysial joint blocks. *The Clinical Journal of Pain, 9*, 124–130.

Ferrante, F. M., King, L. F., Roche, E. A., et al. (2001). Radiofrequency sacroiliac joint denervation for sacroiliac syndrome. *Regional Anesthesia and Pain Medicine, 26*, 137–142.

Fortin, J., Kissling, R., O'Connor, B., et al. (1999). Sacroiliac joint innervation and pain. *The American Journal of Orthopedics, 28*, 687–690.

Gevargez, A., Groenemeyer, D., Schirp, S., et al. (2002). CT-guided percutaneous radiofrequency denervation of the sacroiliac joint. *European Radiology, 12*, 1360–1365.

Grob, K., Neuhuber, W., & Kissling, R. (1995). Innervation of the sacroiliac joint of the human. *Zeitschrift für Rheumatologie, 54*, 117–122.

Kline, M. T. (1992). *Stereotactic radiofrequency lesions as part of the management of chronic pain*. Orlando: Paul M Deutsch.

Lennard, T. E. (2000). *Pain procedures in clinical practice* (2nd ed.). Philadelphia: Hanley & Belfus.

Schwarzer, A. C., Aprill, C. N., Derby, R., & Bogduk, N. (1994). The false-positive rate of uncontrolled diagnostic blocks of the lumbar zygapophysial joints. *Pain, 58*, 195–200.

Willard, F., Carreiro, J., & Manko, W. (1998). The long posterior interosseous ligament and the sacrococcygeal plexus. *Third interdisciplinary world congress on low back and pelvic pain*. Vienna.

Yin, W., Willard, F., Carreiro, J., et al. (2003). Sensory stimulation-guided sacroiliac joint radiofrequency neurotomy: Technique based on neuroanatomy of the dorsal sacral plexus. *Spine, 28*, 2419–2425.

Sacroiliitis

▶ Sacroiliac Joint Pain

Sacrum

Definition

The sacrum is the wedge-shaped bone consisting of five fused vertebrae forming the posterior part of the pelvis.

Cross-References

▶ Sacroiliac Joint Pain

Saliency

Definition

Saliency quantifies the relative importance of each dimension to an individual.

Cross-References

▶ Pain Measurement by Questionnaires, Psychophysical Procedures, and Multivariate Analysis

Salpingitis-Oophoritis-Peritonitis

▶ Chronic Pelvic Pain, Pelvic Inflammatory Disease and Adhesions

Sarcomere

Definition

A sarcomere is the basic contractile unit of striated muscle. Sarcomeres are attached to each other end-to-end at the Z line, where the actin molecules are anchored and where the myosin is anchored by titan. The sarcomere is essentially a constant-volume structure that can change length by a factor of nearly two in most muscles. Localized microscopic abnormal shortening of sarcomeres appears to be characteristic of myofascial trigger points.

Cross-References

▶ Myofascial Trigger Points

Satellite Cells

Synonyms

Satellite glia cells, SCG

Definition

Satellite cells are glial cells located in the sensory (dorsal root and trigeminal ganglia) and autonomic ganglia. Each neuron within the ganglion is encircled by a layer of satellite cells in close apposition. Satellite cells become activated in response to injury of "their" neuron's peripheral axon.

Cross-References

▶ Glial Cells in Orofacial Pathological Pain Mechanisms
▶ Proinflammatory Cytokines

Satellite Glia Cells, SCG

▶ Satellite Cells

Satellite Glial Cells

▶ Satellite Glial Cells and Chronic Pain

Satellite Glial Cells and Chronic Pain

Menachem Hanani
Hadassah-Hebrew University Medical Center
Hebrew University Medical School,
Jerusalem, Israel

Synonyms

Perineuronal satellite cells; Satellite glial cells

Definition

Satellite glial cells (SGCs) closely surround neuronal cell bodies in sensory ganglia and also in peripheral sympathetic and parasympathetic ganglia.

Introduction

Sensory ganglia contribute to pathological pain by generating abnormal (ectopic) firing (Devor 2006). Most of the research on sensory ganglia has focused on the neurons, whereas the role of glial cells in sensory ganglia received little attention. Glial activation (gliosis) is the most prominent pathologic change in glial cells, and there is considerable evidence that activation of microglia and astrocytes in the spinal cord is associated with chronic pain (Milligan and Watkins 2009; Tsuda et al. 2013; Gosselin et al. 2010). Satellite glial cells (SGCs) are the main

type of glial cells in sensory ganglia, and current research has shown that they undergo activation following peripheral injury and contribute to a variety of pain states. The unique arrangement of SGCs around sensory neurons and the absence of blood-nerve barrier in sensory ganglia make SGCs a preferred therapeutic target.

Terminology

Some authors use the term "satellite cells" for "satellite glial cells" (SGCs). This is a misnomer, as "satellite cells" are defined as the progenitor cells in striated muscles. Also, in the older literature, the term "capsule" is used to describe the SGC sheath around the sensory neurons, but this word should be reserved for the connective tissue layer covering the ganglia.

Characteristics

Structure of Satellite Glial Cells

Satellite glial cells share some functional and biochemical features with astrocytes. The main distinguishing morphological property of SGCs is that they completely surround the somata of sensory neurons, thus forming a functional unit that consists of a neuron and the SGC sheath around it (Fig. 1). Thus, SGCs are distinct from Schwann cells, which are associated with axons, rather than neuronal somata (except in special cases, such as the acoustic and vestibular ganglia). The number of SGCs around each neuron increases with the size of the neuron (Pannese 2010) and can be from one to over ten. Although the SGC sheath around the neuron is tight, it is not an absolute barrier, and large molecules, such as proteins, and even macrophages can reach the neuronal soma (Hanani 2005). The gap between SGCs and the neurons is about 20 nm, and therefore, the extracellular space of the neurons is very small. As SGCs are endowed with a variety of transport mechanism (e.g., for amino acid neurotransmitters), they can control the neuronal extracellular space and regulate the passage of substances from the space around the neuron-glia unit to the neuronal surface (Hanani 2005). The plasma membrane of sensory neurons is not

Satellite Glial Cells and Chronic Pain, Fig. 1 Low-magnification electron micrograph of mouse dorsal root ganglion showing the cell body of a sensory neuron (*N1*) enveloped by a SGC sheath, which is highlighted. The entire outer contour of the sheath is smooth and is completely separated from the sheaths encircling the adjacent nerve cell bodies (*N2–N4*) by the connective tissue space (*ct*). The small empty regions within the SGCs envelope are fine neuronal protrusions (see Pannese 2002) that increase the contact between the two cell types. *v* blood vessel. Scale bar, 4 μm

smooth, but there are numerous slender processes emerging from them that project into invaginations in the SGCs (Pannese 2002). This increases the surface area of the neuronal membrane and facilitates interactions between neurons and SGCs. Normally SGCs do not send processes outside the neuron-glia unit.

In most cases each sensory neuron is ensheathed with its own glial cover, but some neurons share a common glia sheath (Pannese 2010). These arrangements, which usually consist of 2–3 neurons, have been termed "clusters." The percentage of clusters declines with age, but they persist in adult and old animals (e.g., the cluster incidence in dorsal root ganglia in the adult is 5.6 % for rats and 4.3 % for rabbits). Within some of these clusters, the neurons are in direct contact, with no intervening SGCs, but in others, they remain separated from one another by a SGC sheath (Fig. 2). The functional significance of the clusters is not known, but the greater proximity of the neurons in them might promote

neuron-neuron interactions. The presence of a common SGC envelope may contribute to chemical interactions between neurons and SGCs.

Gap Junctions

Gap junctions are channels between cells, through which ions (e.g., Na^+, K^+) and molecules (with molecular weight of up to 1,000) can pass between the connected cells, enabling intercellular electrical and metabolic communications. Electron microscopy demonstrated gap junctions in SGCs (Pannese 2010), but direct evidence for gap junction-mediated coupling among SGCs was obtained by injecting a fluorescent dye into a single SGC and observing dye spread into neighboring SGCs – the dye coupling method (Hanani et al. 2002). There is immunohistochemical evidence for the presence in SGCs of connexins (Fig. 3a), which are the proteins that make up the gap junctions (Jasmin et al. 2010; Durham and Garrett 2010). This correlates with the functional observations on coupling. Aging is associated with an increase in dye coupling and in gap junction number in SGCs (Huang et al. 2006).

Identification of Satellite Glial Cells

The most prominent feature of SGCs is that they completely ensheathe the sensory neurons. However, identifying SGCs can pose difficulty because the sheath they form around the neurons may be at places too thin to be detected by light microscopy (Fig. 1). Therefore, confirmatory immunohistochemical labeling is recommended for identifying these cells. SGCs and Schwann cells share several marker proteins such as S100 and laminin, and therefore, these markers are not definitive. More selective SGC markers are glutamine synthetase (Hanani 2005; Jasmin et al. 2010; Hanani and Spray 2013), the glutamate transporter GLAST (Ohara et al. 2008), the small-conductance calcium-activated K^+ channel SK3 (Jasmin et al. 2010), or the receptor for the purinergic P2 receptor P2X7 (Fig. 3b). Using glial fibrillary acidic protein (GFAP) to identify SGCs is problematic because under resting conditions, SGCs express very low levels of GFAP, but after a variety of insults such as axotomy and

Satellite Glial Cells and Chronic Pain, Fig. 2 A cluster consisting of two neurons covered with a common sheath of SGCs. The asterisk denotes a SGC that was injected with the fluorescent dye Lucifer yellow. The dye passed to several other SGCs (*black triangles*), indicating that they are coupled by gap junctions. The *white triangle* indicates the neuronal nucleus. Scale bar, 20 μm

Satellite Glial Cells and Chronic Pain, Fig. 3 Immunohistochemical characterization of SGCs in sensory ganglia. (a) SGCs are labeled for glutamine synthetase and are seen to surround the neurons, which are not labeled and appear as empty spaces. The *bright dots* are gap junctions that were labeled for connexin 43. (b) Labeling of SGCs for the purinergic receptor P2X7. SGC nuclei are labeled with DAPI and are seen as *bright circles*. The neuronal nuclei were also labeled but are very faint. Neurons are not labeled

inflammation, GFAP expression is greatly increased. The increase in the number of neurons that are surrounded by GFAP-positive SGCs is used to quantify the extent of glial activation.

Physiological Properties of SGCs

As in astrocytes, the plasma membrane of SGCs is highly permeable to K^+ ions. Current–voltage curves for SGCs obtained using voltage clamp recordings in mouse trigeminal ganglia were nearly linear, consistent with the near absence of voltage-dependent ion channels and in being inexcitable cells, but a small component of delayed rectifying K^+ channels was present (Hanani 2005). Using immunohistochemistry, Vit et al. (2008) observed inward rectifying K^+ channels (type Kir4.1) in SGCs from rat trigeminal ganglia, which are believed to be important in regulating the extracellular K^+ concentration, especially during intense nerve activity. As inexcitable cells, SGCs do not conduct action potentials, but, again like astrocytes, they can maintain long-range communications by the propagation of calcium waves, which are mediated by gap junctions and P2 receptors (Suadicani et al. 2010).

Pharmacology

As glial cells do not conduct action potentials, their main means for short- and long-range communications is chemical signaling. Of major interest is the nature of neuron-SGC interactions, which are believed to be largely via chemical messengers. Therefore, learning about the pharmacology of SGCs is essential for understanding their functions. Like astrocytes, SGCs do not receive synapses, but possess membrane receptors for bioactive molecules, some of which have been studied functionally. Using calcium imaging, Weick et al. (2003) found that the pain mediator ATP induces an increase in the intracellular calcium concentration ($[Ca^{2+}]_{in}$) in SGCs. The response was largely mediated by activating purinergic P2Y receptors, which are metabotropic, and their activation leads to Ca^{2+} release from intracellular stores. The purinergic ionotropic P2X7 receptors are also present in SGCs (Gu et al. 2010; Villa et al. 2010), see Fig. 3b. As sensory neurons release ATP, it is very likely the purinergic signaling is of major importance for neuron-SGC interactions (Gu et al. 2010; Suadicani et al. 2010; Villa et al. 2010). For example, ATP released from sensory neurons activates P2X7 receptors in SGCs, which induces the release of the cytokine TGFα from the SGCs. In turn, the TGFα potentiates P2X3 receptor-mediated excitation of the neurons (Gu et al. 2010). However, the picture is more complicated, because ATP released from SGCs also activates P2Y1 receptors in the neurons, which reduces P2X3 receptors in the neurons (Gu et al. 2010). The mechanism for ATP release from SGCs is not understood and could be via pannexin1 channels.

Sensory neurons contain the peptide endothelin 1 (ET1), and SGCs display the ETB receptors. In a calcium imaging study on mouse trigeminal ganglia, ET1 raised $[Ca^{2+}]_{in}$ with a very low threshold (0.05 nM) (Feldman-Goriachnik and Hanani 2011), and therefore, this peptide might serve for neuron-SGC signaling. Functional evidence for the following receptors in SGCs has been reported: bradykinin (England et al. 2001) [but see Villa et al. (2010), who reported direct action of bradykinin on the neurons], tumor necrosis factor alpha (Gu et al. 2010), interleukin 1β (Takeda et al. 2009), and glutamate and GABA (Shoji et al. 2010).

Satellite Glial Cells in Tissue Culture

Cultures of sensory ganglia are used frequently to study the physiology and pharmacology of sensory neurons, and in particular of pain mechanisms. In contrast, the study of SGCs in culture has been relatively neglected. One potential pitfall with work on cultured SGCs is that they undergo major phenotypic changes in culture. There is evidence that they can differentiate into muscle cells, astrocytes, and neurons in culture; see Belzer et al. (2010) for references. In a study on trigeminal ganglion cultures where cells were observed continuously under a microscope, Belzer et al. (2010) have shown that after 24 h in culture, SGCs were migrating away from their parent neurons, and altered their morphology; they became elongated and assumed Schwann cell-like morphology. Thus, their main characteristic – tightly wrapping neuronal somata – was lost. Moreover, the SGC marker glutamine synthetase was nearly absent after 1–2 days in culture. Therefore, it is difficult to justify the identification of these cells as SGCs, and it is recommended to use the cultures at 1–2 days after culturing, when SGCs are in close proximity to the neurons. Such cultures are useful for investigating neuron-SGC interactions (Suadicani et al. 2010).

SGCs in Nerve Injury and Pain Models

The potential role of SGCs in chronic pain has received considerable attention because they undergo striking changes following peripheral nerve injury. These changes include:

I. A large upregulation of GFAP in SGCs, indicative of glial activation (Hanani 2005;

Satellite Glial Cells and Chronic Pain, Fig. 4 Dye coupling in sensory ganglia is augmented by peripheral inflammation. The fluorescent dye Lucifer yellow (*LY*) was injected into SGCs of S1 DRG ganglia of mice to determine the degree of gap junction-mediated coupling between the cells. Dinitrobenzoate sulfate (*DNBS*) was instilled into the large intestine to induce inflammation, and the ganglia were examined in vitro 10 days later. (**a**), LY-injected SGC is coupled to other SGCs only around the same neuron. This is the usual case in controls. (**b**), Dye coupling between SGCs around different neurons, observed in DNBS-treated mouse. The asterisks indicate the LY-injected SGCs. Scale bars, 20 μm

Jasmin et al. 2010). In a rat pain model based on spinal nerve ligation, the elevation of GFAP persisted for at least 56 days (Liu et al. 2012). Still, it is unlikely that the GFAP upregulation per se drives the other functional changes in SGCs that are believed to be associated with pathological pain.

The signals that trigger SGC activation are not known yet, but they are very likely to be chemical, as they appear to spread from affected neurons to non-affected ones. For example, a localized injury, like damaging the pulp in a single tooth, led to GFAP upregulation in all three divisions of the rat trigeminal ganglion (Stephenson and Byers 1995). Likewise, focal colonic inflammation in mice induced an increase in SGC coupling and an increase in neuronal excitability in a large population of neurons in DRG (Huang et al. 2010). Intense neuronal firing, which occurs after injury and can induce the released of signaling molecules from the neurons, is an important factor (though possibly not exclusive) in GFAP elevation (Xie et al. 2009).

II. A large increase (four to sevenfold) in coupling between SGCs encircling different neurons (Hanani 2012), see Fig. 4. These gap junctions are located in abnormal processes that grow from SGCs and form bridges between glial envelopes (Fig. 5). These observations were correlated with increased levels of connexins in SGCs in pain models (Durham and Garrett 2010; Jasmin et al. 2010). The increased coupling may be beneficial for extracellular buffering of K^+ during intense neuronal activity but apparently contributes to pain because of enhanced spread of excitatory influences in the ganglia, which may cause pain. In support of this conclusion, there are studies showing that blocking gap junctions in a variety of rodent pain models reduces allodynia (Durham and Garrett 2010; Huang et al. 2010; Jasmin et al. 2010; Hanani 2012).

III. Following peripheral injury, the pharmacology of SGC, like their morphology, is altered markedly. In an in vitro model of inflammation, incubation of SGCs with the pro-inflammatory mediator bradykinin upregulated SGC responses to purinergic agonists via the activation of P2Y receptors (Villa et al. 2010). Kushnir et al. (2011) showed that the sensitivity of SGCs of trigeminal ganglion to ATP increased 100-fold following focal skin inflammation or axotomy of the infraorbital nerve. Whereas P2Y receptors were dominant in control ganglia, 7 days after the induction of inflammation, the receptor population

Satellite Glial Cells and Chronic Pain, Fig. 5 Electron micrograph showing the effect of peripheral inflammation on SGCs. Neuritis of the sciatic nerve induced the formation of a bridge (*arrow*) connecting two SGCs (*s1, s2*), each is a part of the sheath around a different neuron (*N1, N2*). The inset is a high magnification of the boxed area, which is within the bridge formed by the two SGCs. The dark band (indicated by an *arrow*) is a gap junction. *Ct* connective tissue space, *v* blood vessel. Scale bar, 2 μm; *inset*: scale bar, 0.2 μm

switched to P2X (possibly P2X2, 4 and 5). It can be concluded that SGCs display high plasticity with respect to P2 receptors, similarly to spinal cord microglia (Tsuda et al. 2013).

P2X7 receptors, which are located in SGCs, but not in neurons, have been implicated in neuropathic pain. A study on rat DRG did not support this idea and indicated that P2X7 receptor stimulation of SGCs inhibited hyperalgesia that was induced by inflammation and nerve injury (Gu et al. 2010). Also, Kushnir et al. (2011) did not detect changes in P2X7 receptor expression following facial skin inflammation. Still, in cases where the balance of the opposing actions of P2X7 receptor stimulation tilts toward excitation, these receptors will contribute to hyperalgesia.

IV. SGCs, but not neurons, contain the K^+ channel Kir4.1. Silencing of Kir4.1 using RNA interference leads to spontaneous and evoked facial pain-like behavior in rats (Vit et al. 2008). Furthermore, the level of Kir4.1 in the trigeminal ganglion was reduced after chronic constriction injury of the infraorbital nerve, indicating that disrupting the K^+ buffering function of SGCs contributes to neuropathic pain. These conclusions are in agreement with a study showing that facial inflammation in rats reduced Kir4.1 immunoreactivity in SGCs in the trigeminal ganglion, which was also supported by in vivo electrical recordings from SGCs (Takeda et al. 2011). This impairment of glial K^+ homeostasis can contribute to trigeminal pain because elevated K^+ levels may increase neuronal excitability.

V. Following facial skin inflammation, the level of the cytokine interleukin 1β (IL-1β) in SGCs in the rat trigeminal ganglia is increased, and in parallel, IL-1 receptors in small diameter neurons were upregulated (Takeda et al. 2009). As IL-1β is excitatory to the neurons, these changes will lead neuronal excitation and pain. Indeed, in this model the number of Aδ neurons that fired spontaneously increased about twofold compared with controls. Applying the IL-1 receptor antagonist reduced neuronal firing and also reduced hyperalgesia, supporting the proposed role of IL-1β-mediated SGC-neuron interactions in inflammatory pain (Takeda et al. 2009).

SGCs as Therapeutic Target for Pain

The available information supports the idea that SGCs can be a suitable target for treating chronic pain. The location of sensory ganglia in the pain pathways, the close proximity of SGCs to the sensory neurons, and the lack of blood-ganglion barrier should be an important consideration in developing pain therapy. Below is a summary of developments in this area:

I. Gap Junctions

In a number of animal models of chronic pain, blocking the augmented gap junctions resulted in a reduction in pain behavior (Jasmin et al. 2010; Durham and Garrett 2010; Hanani 2012). It should be noted that in most cases, pain due to localized insult was studied (axotomy, nerve ligation, etc.). However, there is evidence that SGC coupling also increases following systemic insult such as in a model of chemotherapy-induced neuropathic pain (Warwick and Hanani 2013). The administration of gap junction blockers reduced pain in this model, as found in localized pain models.

II. Kir4.1 Channels

The presence of Kir4.1 channels in SGCs offers another therapeutic target. Silencing of Kir4.1 using RNA interference leads to spontaneous and evoked facial pain-like behavior in freely moving rats, and there is evidence that Kir4.1 channels are suppressed in pain models (Vit et al. 2008; Takeda et al. 2011). Thus, the selective opening of these channels can reduce firing in sensory neurons.

III. P2 Receptors

SGCs are endowed with P2 receptors, which are upregulated after peripheral injury (Kushnir et al. 2011). As these receptors mediate neuron-SGC communications, they may serve as therapeutic targets, but more research is needed because of the complex and opposing interactions in which they are involved. The precise subtype of P2 receptor that can be targeted for pain therapy has to be determined. Neurons have P2/3 excitatory receptors, and their blockade is an option.

IV. Other Targets

Blocking other mechanisms that mediate SGC-neuron communications may serve as therapeutic target. These include receptors for cytokines, such as Il-1β and TGFα, and also prostaglandins.

Acknowledgement Work done in the author's laboratory was supported by the European Community's 7th Framework Programme through the Marie Curie Initial Training Network Edu-GLIA, the Israel Cancer Association, the Israel Science Foundation, and the United States-Israel Binational Science Foundation.

References

Belzer, V., Shraer, N., & Hanani, M. (2010). Phenotypic changes in satellite glial cells in cultured trigeminal ganglia. *Neuron Glia Biology, 6*, 237–243.

Devor, M. (2006). Response to nerve injury in relation to neuropathic pain. In S. B. McMahon & M. Koltzenburg (Eds.), *Wall and Melzack's textbook of pain* (5th ed., pp. 905–927). Philadelphia: Elsevier Churchill Livingstone.

Durham, P. L., & Garrett, F. G. (2010). Emerging importance of neuron-satellite glia interactions within trigeminal ganglia in craniofacial pain. *The Open Pain Journal, 3*, 3–13.

England, S., Heblich, F., James, I. F., Robbins, J., & Docherty, R. J. (2001). Bradykinin evokes a Ca^{2+}-activated chloride current in non-neuronal cells isolated from neonatal rat dorsal root ganglia. *The Journal of Physiology, 530*, 395–403.

Feldman-Goriachnik, R., & Hanani, M. (2011). Functional study of endothelin B receptors in satellite glial cells in trigeminal ganglia. *Neuroreport, 22*, 465–469.

Gosselin, R. D., Suter, M. R., Ji, R. R., & Decosterd, I. (2010). Glial cells and chronic pain. *The Neuroscientist, 16*, 519–531.

Gu, Y., Chen, Y., Zhang, X., Li, G. W., Wang, C., & Huang, L. Y. (2010). Neuronal soma-satellite glial cell interactions in sensory ganglia and the participation of purinergic receptors. *Neuron Glia Biology, 6*, 53–62.

Hanani, M. (2005). Satellite glial cells in sensory ganglia: From form to function. *Brain Research. Brain Research Reviews, 48*, 457–476.

Hanani, M. (2012). Intercellular communication in sensory ganglia by purinergic receptors and gap junctions: Implications for chronic pain. *Brain Research, 1487*, 183–191.

Hanani, M., & Spray, D. C. (2013). Glial cells in autonomic and sensory ganglia. In H. Kettenmann & B. R. Ransom (Eds.), *Neuroglia* (pp. 122–133). New York: Oxford University Press.

Hanani, M., Huang, T. Y., Cherkas, P. S., Ledda, M., & Pannese, E. (2002). Glial cell plasticity in sensory ganglia induced by nerve damage. *Neuroscience, 114*, 279–283.

Huang, T. Y., Hanani, M., Ledda, M., De Palo, S., & Pannese, E. (2006). Aging is associated with an increase in dye coupling and in gap junction number in satellite glial cells of murine dorsal root ganglia. *Neuroscience, 137*, 1185–1192.

Huang, T. Y., Belzer, V., & Hanani, M. G. (2010). Gap junctions in dorsal root ganglia: Possible contribution to visceral pain. *European Journal of Pain, 14*, 49.e1–49.e11.

Jasmin, L., Vit, J. P., Bhargava, A., & Ohara, P. T. (2010). Can satellite glial cells be therapeutic targets for pain control? *Neuron Glia Biology, 6*, 63–71.

Kushnir, R., Cherkas, P. S., & Hanani, M. (2011). Peripheral inflammation upregulates P2X receptor expression in satellite glial cells of mouse trigeminal ganglia: A calcium imaging study. *Neuropharmacology, 61*, 739–746.

Liu, F. Y., Sun, Y. N., Wang, F. T., Li, Q., Su, L., Zhao, Z. F., Meng, X. L., Zhao, H., Wu, X., Sun, Q., Xing, G. G., & Wan, Y. (2012). Activation of satellite glial cells in lumbar dorsal root ganglia contributes to neuropathic pain after spinal nerve ligation. *Brain Research, 1427*, 65–77.

Milligan, E. D., & Watkins, L. R. (2009). Pathological and protective roles of glia in chronic pain. *Nature Reviews. Neuroscience, 10*, 23–36.

Ohara, P. T., Vit, J. P., Bhargava, A., & Jasmin, L. (2008). Evidence for a role of connexin 43 in trigeminal pain using RNA interference in vivo. *Journal of Neurophysiology, 100*, 3064–3073.

Pannese, E. (2002). Perikaryal surface specializations of neurons in sensory ganglia. *International Review of Cytology, 220*, 1–34.

Pannese, E. (2010). The structure of the perineuronal sheath of satellite glial cells (SGCs) in sensory ganglia. *Neuron Glia Biology, 6*, 3–10.

Shoji, Y., Yamaguchi-Yamada, M., & Yamamoto, Y. (2010). Glutamate- and GABA-mediated neuron-satellite cell interaction in nodose ganglia as revealed by intracellular calcium imaging. *Histochemistry and Cell Biology, 134*, 13–22.

Stephenson, J. L., & Byers, M. R. (1995). GFAP immunoreactivity in trigeminal ganglion satellite cells after tooth injury in rats. *Experimental Neurology, 131*, 11–22.

Suadicani, S. O., Cherkas, P. S., Zuckerman, J., Smith, D. N., Spray, D. C., & Hanani, M. (2010). Bidirectional calcium signaling between satellite glial cells and neurons in cultured mouse trigeminal ganglia. *Neuron Glia Biology, 6*, 43–51.

Takeda, M., Takahashi, M., & Matsumoto, S. (2009). Contribution of the activation of satellite glia in sensory ganglia to pathological pain. *Neuroscience and Biobehavioral Reviews, 33*, 784–792.

Takeda, M., Takahashi, M., Nasu, M., & Matsumoto, S. (2011). Peripheral inflammation suppresses inward rectifying potassium currents of satellite glial cells in the trigeminal ganglia. *Pain, 152*, 2147–2156.

Tsuda, M., Beggs, S., Salter, M. W., & Inoue, K. (2013). Microglia and intractable chronic pain. *Glia, 61*, 55–61.

Villa, G., Fumagalli, M., Verderio, C., Abbracchio, M. P., & Ceruti, S. (2010). Expression and contribution of satellite glial cells purinoceptors to pain transmission in sensory ganglia: An update. *Neuron Glia Biology, 6*, 31–42.

Vit, J. P., Ohara, P. T., Bhargava, A., Kelley, K., & Jasmin, L. (2008). Silencing the Kir4.1 potassium channel subunit in satellite glial cells of the rat trigeminal ganglion results in pain-like behavior in the absence of nerve injury. *The Journal of Neuroscience, 16*, 4161–4171.

Warwick, R. A., & Hanani, M. (2013). The contribution of satellite glial cells to chemotherapy-induced neuropathic pain. *European Journal of Pain*. doi:10.1002/j.1532-2149.2012.00219.x [Epub ahead of print].

Weick, M., Cherkas, P. S., Härtig, W., Pannicke, T., Uckermann, O., Bringmann, A., Tal, M., Reichenbach, A., & Hanani, M. (2003). P2 receptors in satellite glial cells in trigeminal ganglia of mice. *Neuroscience, 120*, 969–977.

Xie, W., Strong, J. A., & Zhang, J. M. (2009). Early blockade of injured primary sensory afferents reduces glial cell activation in two rat neuropathic pain models. *Neuroscience, 160*, 847–857.

Scalenus Anticus Syndrome

▶ Thoracic Outlet Syndrome

Scanning

Definition

A high and fast rate of switching attention to different stimuli/locations, in order to facilitate the detection of information.

Cross-References

▶ Hypervigilance and Attention to Pain

SCC

▶ Spinal Cord Compression

Schedules of Reinforcement

Definition

Reinforcement of a behavior can follow certain time schedules. For example, reinforcement can be continuous (for each response a reinforcer is given) or intermittent (a reinforcer is provided only after a certain number of behaviors). Various schedules of reinforcement (e.g., fixed, variable, interval, ratio) have been developed and empirically tested that have different influences on the acquisition of a response as well as on its maintenance.

Schema

Definition

A schema is a perceptual hypothesis that serves as the fundamental unit for constructing awareness.

Cross-References

▶ Consciousness and Pain

Schema Activation

Definition

The hypothesized engagement of an organized knowledge structure (i.e., pain schema), which leads to the preferential processing of schema-relevant information.

Cross-References

▶ Catastrophizing

Schemata

Definition

Schemata are fuzzy, preconscious, and dynamical patterns, roughly related to dynamically stable patterns in neural networks.

Cross-References

▶ Consciousness and Pain

Schrober's Test

▶ Lower Back Pain, Physical Examination

Schwann Cell

Definition

These cells form the sheath around axons in the peripheral nervous system; the sheath and axon together are called a nerve fiber. The fibers are either myelinated or unmyelinated. Myelinated fibers have Schwann cells wrapped around the axon as a spiral. The spiral (the myelin) consists of the membrane of a single Schwann cell from which the cytoplasm has been removed during the wrapping process. The Schwann cells of a single nerve fiber are separated from each other by nodes of Ranvier. Note that unmyelinated nerve fibers also have a sheath of Schwann cells. However, in this case the Schwann cell forms a single cell layer around the axon.

Cross-References

▶ Hansen's Disease
▶ Hereditary Neuropathies
▶ Sensitization of Muscular and Articular Nociceptors
▶ Toxic Neuropathies
▶ Wallerian Degeneration

Sciatic Inflammatory Neuritis

Synonyms

SIN

Definition

Inflammation of the sciatic nerve. An animal model of Sciatic Inflammatory Neuritis (SIN) is produced by causing a local inflammation in one sciatic nerve. This leads to enhanced responses to weak mechanical stimuli.

Cross-References

▶ Neuropathic Pain Model, Neuritis/Inflammatory Neuropathy

Sciatic Inflammatory Neuropathy

▶ Neuropathic Pain Model, Neuritis/Inflammatory Neuropathy

Sciatic Neuralgia

▶ Sciatica

Sciatic Neuropathy

▶ Sciatica

Sciatic Notch

▶ Lower Back Pain, Physical Examination

Sciatica

Gabor B. Racz[1] and Carl E. Noe[2]
[1]Texas Tech University Health Sciences Center, Lubbock, TX, USA
[2]University of Texas Southwestern Medical Center, Lubbock, TX, USA

Synonyms

Cotunnius disease; Lumbosacral radiculitis; Radicular leg pain; Sciatic neuralgia; Sciatic neuropathy

Definition

Sciatica is pain in the distribution of the sciatic nerve (buttocks, posterior thigh, and/or lower limb).

Characteristics

Sciatica is a clinical diagnosis, one made by history and physical examination as opposed to a radiographic diagnosis made by a test such as a magnetic resonance imaging (MRI) study. Pain with sciatica is within the territory of the sciatic nerve. The sciatic nerve is the largest peripheral nerve in humans and is formed by several branches of nerve roots passing out of the lower lumbar spine and sacrum. The sciatic nerve travels through the buttocks and into the backside of the thigh. Branches of the sciatic nerve continue into the lower leg and foot.

Physical examination findings include a positive straight leg-raising test (Lasègue's sign). Classically, the patient reclines and the examiner lifts the foot while the leg remains unbent at the knee. This has the effect of stretching the nerve fibers that comprise the sciatic nerve. Normally, there is enough slack in the system to prevent any symptoms, but if nerve root irritation is present, pain is exacerbated.

Acute Sciatica

Acute sciatica refers to sciatic pain of recent onset. Chronic sciatica is used to describe pain persisting beyond 6 month duration. Some patients have recurrent episodes of sciatica, sometimes associated with recurrent injuries and subsequent resolution.

Acute sciatica can be caused by a number of diagnoses which are more specific than the general clinical diagnosis of sciatica. Sudden onset of sciatica commonly occurs with an acute lumbar disk herniation. In this circumstance, the lumbar disk protrudes out of its normal position and irritates a nerve root, usually the fifth lumbar or first sacral, or another nerve root(s) which forms the sciatic nerve. As a result, the nerve root discharges, sending nerve impulses to the brain, which is experienced as pain from the body region where the irritated nerve fibers have receptors. Often the natural history of acute sciatica is favorable, resolving with time and analgesics. Oral or injected corticosteroids are common treatments. A minority of patients with acute sciatica progress to have surgery to decompress nerve roots. In addition to pain, patients with sciatica may experience loss of sensation, weakness, or loss of bladder and bowel function. These symptoms may push patients toward surgical treatment. Other spinal causes of acute sciatica include spondylolisthesis and spinal stenosis. Spondylolisthesis is a shifting of one vertebra out of normal alignment. Spondylolisthesis can be traumatic or degenerative. This can result in a narrowing of spaces within the spine where nerve roots pass. Spinal stenosis is a narrowing of canals inside the spine where nerves pass. Stenosis can involve the central spinal canal or nerve root foramina. Spinal stenosis is usually degenerative and caused by bone spurs and thickening of ligaments. These diagnoses are confirmed by x-rays or other imaging tests. Patients with these diagnoses are frequently treated with surgery.

Rare but important causes of acute sciatica include acute epidural abscess, epidural hematoma and spinal cord infarction, and conus medullaris syndrome.

Other causes of sciatica of more gradual onset include piriformis syndrome and peripheral neuropathy. Piriformis syndrome results from the sciatic nerve being compressed by the piriformis muscle. This muscle is a part of the musculature of the buttocks, near the point where the sciatic nerve passes out of the pelvis into the leg. Piriformis syndrome is frequently treated with exercises to stretch the piriformis muscle.

Peripheral neuropathy is commonly associated with diabetes and may also be a result of trauma to the sciatic nerve. Diabetic peripheral neuropathy is frequently treated with medications for neuropathic pain.

Tumors are less common causes of sciatica but can occur at any location along the nerve pathway involving the sciatic nerve.

Diseases and conditions in the pelvis can produce symptoms of sciatica, since the nerve roots form a plexus that lies in the pelvis.

Chronic Sciatica

Chronic sciatica can result from failed back surgery syndrome, arachnoiditis, and chronic lumbosacral radiculopathy. Failed back surgery syndrome is a nonspecific term used to classify patients who have persistent or recurrent symptoms, even after having back surgery. Sometimes patients have a definitive cause for symptoms, but often patients have pain that cannot be explained by conventional medical tests. Epidural adhesions may be associated with chronic sciatica. Epidurography is a newer test that is used to identify epidural adhesions, and ▶ lysis of adhesions is a newer procedure used to treat this problem.

Following lysis of adhesions, an exercise program, including the technique of "neural flossing," is recommended.

Arachnoiditis is inflammation of a lining of the nerve roots. This can result in scarring around nerve roots and permanent nerve damage.

Patients with chronic sciatica may benefit from rehabilitation treatment to improve physical and psychosocial function. Medications for neuropathic pain, opioids, antidepressants, and anticonvulsants, are also commonly used for symptomatic relief. Spinal cord stimulation is also used (Dubuisson 2003).

Differential Diagnosis

Conditions that mimic sciatica include referred pain from the sacroiliac and zygapophyseal (facet) joints (Wiener 1993). Referred pain is pain which is localized away from the area of pathology. Patients with coronary artery disease sometimes report left arm pain when their problem is a lack of blood flow to their heart. The innervation of the internal organs is less specific than the innervation of the face or hand, and sometimes the localization of the problem is inaccurate. Sacroiliitis and lumbar spondylosis (degenerative arthritis of the zygapophyseal joints) are sometimes associated with referred pain into the back of the leg and mistakenly diagnosed as sciatica. Hip arthritis can produce referred pain to the knee as well.

▶ Trochanteric bursitis is an inflammation of the bursa near the greater trochanter of the hip. This condition can produce pain in the thigh.

▶ Meralgia paresthetica is a pain syndrome involving the lateral femoral cutaneous nerve which results in pain in the lateral thigh.

▶ Herpes virus can infect nerve roots, including the nerve roots of the sciatic nerve. In the case of shingles (herpes zoster), a rash is usually present along the course of the infected nerve root.

▶ Polyarthritis involving joints in the back, hip, knee, and ankle can produce such widespread pain that even though pain is localized to joints, patients will describe it as "leg pain" and even draw their pain pattern as leg pain rather than separate joint pain.

▶ Hamstring muscle strain produces pain in the biceps femoris muscles in the back of the thigh. Stretching the muscle can produce pain as is found with a straight leg-raising maneuver.

Multiple sclerosis (see ▶ Central Pain in Multiple Sclerosis) sometimes manifests as radiating pain in a distribution similar to that of the sciatic nerve.

▶ Peripheral vascular disease can be a cause of leg pain at rest or with walking.

▶ Deep venous thrombosis (blood clot) commonly affects the lower leg and can present as a pain syndrome mimicking sciatica with primarily distal extremity pain.

Phantom pain may occur in the distribution of the sciatic nerve following amputation or congenital lower limb absence. Patients may experience pain described as shooting, spasms, or cramps. Patients may also experience the sensation of the limb being twisted or unusually positioned. The end of the nerve may form a neuroma which may be treated by excision or neuromodulation.

Cross-References

▶ Ergonomics Essay
▶ Facet Joint Pain
▶ Guillain-Barré Syndrome
▶ Radicular Pain, Diagnosis
▶ Radiculopathies

References

Dubuisson, D. (2003). Nerve root disorders and arachnoiditis. In R. Melzack & P. Wall (Eds.), *Handbook of pain management – A clinical companion to wall and Melzack's textbook of pain* (pp. 289–304). London: Elsevier.

Wiener, S. (1993). Acute unilateral upper and lower leg pain. In S. Wiener (Ed.), *Differential diagnosis of acute pain by body region* (pp. 559–565). New York: McGraw-Hill.

Sclerotherapy

▶ Prolotherapy

Scotoma

Definition

Scotoma refers to a blind spot, area of visual loss.

Cross-References

▶ Clinical Migraine with Aura

Scratch

▶ Responses of Spinothalamic Tract Neurons to Pruritic Stimuli

Screening

Definition

Screening refers to an initial assessment that serves as the basis for more in-depth and comprehensive evaluation.

Cross-References

▶ Psychological [Behavioral Health] Assessment of Pain

Second Messenger Cascades

Definition

Second messenger cascades are intracellular signaling pathways that depend on one or more second messengers, and that lead to the activation of a protein kinase. The protein kinase can then alter protein function by protein phosphorylation.

Cross-References

▶ Spinothalamic Tract Neurons, Role of Nitric Oxide

Second Order Neuron

Definition

There are three orders of neurons. The first-order neurons carry signals from the periphery to the spinal cord; the second-order neurons carry signals from the spinal cord to the thalamus; and the third-order neurons carry signals from the thalamus to the primary sensory cortex. Second-order neurons are generally located in the spinal cord or brainstem. These neurons together form a neuronal pathway.

Cross-References

▶ Postsynaptic Dorsal Column Neurons, Responses to Visceral Input
▶ Postsynaptic Dorsal Column Projection, Functional Characteristics

Second Pain Assessment

▶ First and Second Pain Assessment (First Pain, Pricking Pain, Pin-Prick Pain, Second Pain, Burning Pain)

Second Somatic Area

▶ Nociceptive Processing in the Secondary Somatosensory Cortex

Second Somatosensory Cortex

Synonyms

SII

Definition

The second somatosensory cortex (SII) is located at the upper bank of the Sylvian fissure, and is considered to play a role in pain perception. It receives thalamic input from the VPL and VPM nuclei and has a somatotopic organization. Its presence in humans has been confirmed by MEG.

Cross-References

▶ Magnetoencephalography in Assessment of Pain in Humans
▶ Nociceptive Processing in the Secondary Somatosensory Cortex
▶ Secondary Somatosensory Cortex (S2) and Insula, Effect on Pain-Related Behavior in Animals and Humans
▶ Spinothalamic Input: Cells of Origin (Monkey)

Secondary Cells (RVM)

Definition

Secondary cells are a class of RVM neurons defined in vitro by response to mu-opioid agonists. These neurons were originally proposed to be unresponsive to kappa agonists, but this characteristic has since been questioned. Secondary cells presumably correspond to on-cells recorded in vivo.

Cross-References

▶ Opiates, Rostral Ventromedial Medulla, and Descending Control

Secondary Dysmenorrhea

Definition

Menstrual pain arising in association with underlying pathology, such as leiomyomata, adenomyosis, or endometriosis.

Cross-References

▶ Dyspareunia and Vaginismus

Secondary Hyperalgesia

Definition

See "▶ Primary and Secondary Hyperalgesia"

Cross-References

▶ Allodynia
▶ Descending Circuits in the Forebrain, Imaging
▶ Descending Modulation and Persistent Pain
▶ Hyperalgesia
▶ Muscle Pain Model, Inflammatory Agents-Induced
▶ Nick Model of Cutaneous Pain and Hyperalgesia
▶ Nociceptors, Action Potentials, and Post-Firing Excitability Changes
▶ Opioids and Muscle Pain
▶ Referred Hyperalgesia
▶ Referred Muscle Pain, Assessment
▶ Spinothalamic Tract Neurons, Glutamatergic Input
▶ Spinothalamic Tract Neurons, Peptidergic Input
▶ UV-Induced Erythema

Secondary Losses

Definition

Losses that can occur from being a patient or being disabled.

Cross-References

▶ Malingering, Primary and Secondary Gain

Secondary Prevention

▶ Chronicity, Prevention
▶ Disability Prevention
▶ Disability, Effect of Physician Communication

Secondary Somatosensory Cortex

▶ Nociceptive Processing in the Secondary Somatosensory Cortex

Secondary Somatosensory Cortex (S2) and Insula, Effect on Pain-Related Behavior in Animals and Humans

Joel D. Greenspan[1] and Eric A. Moulton[2]
[1]Department of Neural and Pain Sciences, University of Maryland School of Dentistry, Baltimore, MD, USA
[2]P.A.I.N. Group, Anesthesiology, Boston Children's Hospital, Harvard Medical School, Waltham, MA, USA

Definition

The functions of the secondary somatosensory cortex (S2) and ▶ insula can be inferred by evaluating the perceptual or behavioral consequences of lesions (disruptions) or direct stimulation of these anatomical sites.

Characteristics

Lesions

In the current context, a lesion is a disruption of a brain region that adversely affects its function, most often by way of tissue injury. In animal studies, lesions are intentionally produced in a brain area of interest, while lesions in humans are "accidents of nature," most often resulting from pathologies such as strokes or tumors. Functional deficits associated with lesions are determined through behavioral and psychophysical evaluations. The study of the effects of lesions relies on the concept of functional modularity, which assumes that a biological process can be localized to a particular area in the brain. The drawbacks of using lesions to assess function are (1) that they may damage the passage of transmitting neural tracts along with the neural integration regions at the site of injury, thus altering neural function of a site remote from the physical disruption, (2) functional processes may be distributed, such that damage to a functionally relevant area may be compensated for by other involved regions, and (3) other parts of the cortex may change their function to adopt the functional properties of the damaged area, an example of neural plasticity.

Early studies on patient populations reported that ▶ parasylvian lesions could result in decreased pain and temperature sensation on the side of the body contralateral to the lesion (Biemond 1956; Davison and Schick 1935). Lesions in this area could also selectively mitigate the affective reaction to painful stimuli, without significantly altering thresholds for pain sensation (Berthier et al. 1988). ▶ Central pain sometimes accompanied deficits in pain and temperature discrimination associated with parasylvian lesions. More recent studies have attempted to determine whether the precise location of a lesion within the parasylvian cortex was correlated with specific functional deficits. Lesions to the posterior insula and adjacent parietal operculum (including S2) were associated with impaired pain and temperature detection, while lesions to the anterior insula did not show such sensory effects (Cereda et al. 2002;

Greenspan and Winfield 1992; Greenspan et al. 1999; Schmahmann and Leifer 1992). In one study (Greenspan and Winfield 1992), the sensory deficits were alleviated after the subject underwent surgery to extract a temporal lobe tumor that compressed posterior insula and S2. Note that large lesions in the insular or parietal cortex, rather than specific focal lesions within these areas, tend to reduce thermal pain sensitivity. In contrast, innocuous thermal perception is attenuated by small or large lesions in various areas of insular and parietal cortex (Veldhuijzen et al. 2010). This functional resilience of pain processing with focal lesions may be related to compensation by other brain structures, and the more distributed nature of nociception representation in the brain. Though discrete lesions of posterior insula versus S2 are rare in clinical cases, a group analysis indicated that lesions that include S2 alter pain thresholds, while posterior insular lesions do not (Greenspan et al. 1999). Instead, posterior insular lesions were associated with increased pain tolerance. These results suggest that S2 has a ▶ sensory-discriminative role in pain processing, while the neighboring insula is more related to ▶ affective-motivational processing.

The involvement of these parasylvian areas in pain processing is also supported by a primate study evaluating the effects of an accidentally induced cerebral compression of a portion of the posterior parietal cortex, including S2 and insula (Dong et al. 1996). The investigators found that escape behavior to noxious temperatures was nearly eliminated, while detection of temperature changes remained intact.

The insula and/or S2 may have a more complex modulatory role in pain processing. A recent study reported that pain intensity ratings of long duration suprathreshold temperatures were enhanced in two patients with extensive insular/parietal lesions, rather than showing deficits in pain perception (Starr et al. 2009).

Stimulation

Stimulation refers to the direct electrical and/or pharmacological stimulation of a particular cortical area to determine its functional relevance. An invasive procedure, cortical stimulation is commonly performed in animals, though it can also be performed on people with implanted electrodes or while undergoing certain neurosurgical procedures. The function of an area is inferred from evoked animal behavior or human subject report following the stimulation.

Electrical stimulation of the insula in monkeys does not elicit apparent ▶ nocifensive responses, though such studies were performed with the animals under anesthesia (Hoffman and Rasmussen 1953; Kaada et al. 1949). Instead, stimulation of the insula elicited autonomic reactions, including changes in blood pressure, heart rate, and respiration. Cardiovascular responses to insular stimulation have also been reported in humans (Oppenheimer et al. 1992).

Though early work in patients suggested that stimulation of the insula would not evoke pain (Penfield and Faulk 1955), a more recent study has reported that electrical insular stimulation can elicit painful sensations and reactions in humans (Ostrowsky et al. 2002). Using intracranial electrodes for stimulation, they found that stimulation of the posterior insula in 14 out of 43 patients elicited pain ranging from mild to intolerable. The quality of the evoked pain was described as burning, stinging, shocking, and/or disabling. Stimulation of the right posterior insula more frequently evoked pain (n = 12) than the left (n = 3). A subsequent study from the same group found evidence of a somatotopic representation of pain-evoking stimulations in the posterior insula (Mazzola et al. 2009).

Studies that have stimulated in the vicinity of S2 in humans have not evoked pain (Ostrowsky et al. 2002; Penfield and Jasper 1954). Interestingly, some evidence suggests that S2 stimulation can produce ▶ antinociception (Kuroda et al. 2001). Using a formalin model for chronic pain in unanesthetized rats, this group found that electrical stimulation of S2 along with inhibition of neuronal nitric oxide synthase could synergistically suppress spinal nociceptive neurons as well as behavioral reactions (Kuroda et al. 2001).

Overview

There are few pain-related studies involving lesions or direct stimulation of S2 or insula. Lesion study interpretations are limited by the location specificity of lesions in patients, and the potential for remote effects. The available evidence suggests that the posterior insula and parietal operculum including adjacent S2 are necessary for normal pain sensation. The posterior insula may have a more direct role in pain perception, particularly suprathreshold pain perception, than S2 but this distinction is only tentatively supported at this point. More work is needed to better elaborate the pain-related functionality of these brain regions.

References

Berthier, M., Starkstein, S., & Leiguarda, R. (1988). Asymbolia for pain: A sensory-limbic disconnection syndrome. *Annals of Neurology, 24*, 41–49.

Biemond, A. (1956). The conduction of pain above the level of the thalamus opticus. *Archives of Neurology and Psychiatry, 75*, 231–244.

Cereda, C., Ghika, J., Maeder, P., et al. (2002). Strokes restricted to the insular cortex. *Neurology, 59*, 1950–1955.

Davison, C., & Schick, W. (1935). Spontaneous pain and other subjective sensory disturbances. *Archives of Neurology and Psychiatry, 34*, 1204–1237.

Dong, W., Hayashi, T., Roberts, V., et al. (1996). Behavioral outcome of posterior parietal cortex injury in the monkey. *Pain, 64*, 579–587.

Greenspan, J., & Winfield, J. (1992). Reversible pain and tactile deficits associated with a cerebral tumor compressing the posterior insula and parietal operculum. *Pain, 50*, 29–39.

Greenspan, J., Lee, R., & Lenz, F. (1999). Pain sensitivity alterations as a function of lesion location in the parasylvian cortex. *Pain, 81*, 273–282.

Hoffman, B., & Rasmussen, T. (1953). Stimulation studies of insular cortex of *Macaca mulatta*. *Journal of Neurophysiology, 14*, 343.

Kaada, B., Pribam, K., & Epstein, J. (1949). Respiratory and vascular responses in monkeys from temporal pole, insula, orbital surface and cingulate gyrus. *Journal of Neurophysiology, 12*, 347–356.

Kuroda, R., Kawao, N., Yoshimura, H., et al. (2001). Secondary somatosensory cortex stimulation facilitates the antinociceptive effect of the NO synthase inhibitor through suppression of spinal nociceptive neurons in the rat. *Brain Research, 903*, 110–116.

Mazzola, L., Isnard, J., Peyron, R., Guenot, M., & Mauguiere, F. (2009). Somatotopic organization of pain responses to direct electrical stimulation of the insular cortex. *Pain, 146,* 99–104.

Oppenheimer, S., Gelb, A., & Girvin, J. (1992). Cardiovascular effects of human insular cortex stimulation. *Neurology, 42,* 1727–1732.

Ostrowsky, K., Magnin, M., Ryvlin, P., et al. (2002). Representation of pain and somatic sensation in the human insula: A study of responses to direct electrical cortical stimulation. *Cerebral Cortex, 12,* 376–385.

Penfield, W., & Faulk, M. (1955). The insula: Further observations on its function. *Brain, 78,* 445–470.

Penfield, W., & Jasper, H. (1954). *Epilepsy and the functional anatomy of the human brain.* Boston: Little Brown.

Schmahmann, J., & Leifer, D. (1992). Parietal pseudothalamic pain syndrome. Clinical features and anatomic correlates. *Archives of Neurology, 49,* 1032–1037.

Starr, C. J., Sawaki, L., Wittenberg, G. F., Burdette, J. H., Oshiro, Y., Quesvedo, A. S., & Coghill, R. C. (2009). Roles of the insular cortex in the modulation of pain: Insights from brain lesions. *Journal of Neuroscience, 29,* 2684–2694.

Veldhuijzen, D. S., Greenspan, J. D., Kim, J. H., & Lenz, F. A. (2010). Altered pain and thermal sensation in subjects with isolated parietal and insular cortical lesions. *European Journal of Pain, 14,* 535.e1.

Second-Generation Antidepressants

▶ Antidepressant Analgesics in Pain Management

Second-Order Cutaneous Mechanical Input

▶ Postsynaptic Dorsal Column Projection, Functional Characteristics

Second-Order Neurons

▶ Postsynaptic Dorsal Column Neurons, Responses to Visceral Input

Second-Order Subtype: Central Lateral Thalamotomy

▶ Thalamotomy for Human Pain Relief

Second-order/Ascending Pruriceptive Pathway

▶ Responses of Spinothalamic Tract Neurons to Pruritic Stimuli

Sedentary Work

Definition

Work performed primarily in the seated position without exposure to vibration.

Cross-References

▶ Pain in the Workplace, Risk Factors for Chronicity, Job Demands

Segmental Analgesia

Definition

Analgesic effect within a segment.

Cross-References

▶ Transcutaneous Electrical Nerve Stimulation Outcomes

Segmental Anti-Nociceptive Mechanisms

Definition

Neural circuitry that is located within the spinal cord that, when active, prevents the onward transmission of noxious information en route to higher centers in the brain.

Cross-References

▶ Transcutaneous Electrical Nerve Stimulation (TENS) in Treatment of Muscle Pain

Segmental Demyelination

Definition

Segmental demyelination is the loss of myelin with sparing of axonal continuity; it can be the result of injury of either the myelin sheath or Schwann cells. Demyelination can also be related to alterations in axonal caliber (secondary demyelination). Both axonal atrophy and swelling may cause myelin alterations over many consecutive internodes of individual fibers.

Cross-References

▶ Demyelination
▶ Painless Neuropathies

Seizure-Associated Headache

Definition

Headache which occurs before, during, or after seizure.

Cross-References

▶ Post-seizure Headache

Seizures

Definition

Frequent symptom in cerebral vasculitis; manifestation of systemic vasculitides and the isolated angiitis of the central nervous system.

Cross-References

▶ Headache Due to Arteritis

Selectins

Definition

Selectins are a family of transmembrane molecules, expressed on the surface of leukocytes and activated endothelial cells. The selectin family members, L-selectin, P-selectin, and E-selectin, are involved in the adhesion of leukocytes to activated endothelium.

Cross-References

▶ Cytokine Modulation of Opioid Action

Selection Factors

Definition

Selection factors are characteristics that increase the probability that the patient will be referred for treatment or that increase the probability that a subject will be chosen for an experiment.

Cross-References

▶ Psychiatric Aspects of the Epidemiology of Pain

Selective Anesthesia

Definition

Selective anesthesia is a technique sometimes used to localize a source of pain. Small areas or individual teeth are anesthetized sequentially until the pain disappears.

Cross-References

▶ Dental Pain, Etiology, Pathogenesis, and Management

S

Selective Attention

▶ Hypervigilance and Attention to Pain

Selective COX-2 Inhibitors

▶ NSAIDs and Their Indications
▶ Pharmacology of Cyclooxygenase Inhibitors

Selective Nerve Root Block

Definition

A pain management technique for selectively anesthetizing a single nerve root.

Cross-References

▶ Epidural Steroid Injections for Chronic Back Pain
▶ Spinal Cord Injury Pain

Selective Nerve Root Injection

Definition

This term is often used to refer to a local anesthetic block placed outside of the neuroforamen, and some practitioners have advocated for its use as a diagnostic tool for identifying the level of pain-producing pathology. However, placing even a small volume of injectate along the course of the spinal nerve within or near the neural foramen can result in epidural spread and analgesia at more than one spinal level. Placebo responses are also common. Thus, the usefulness of selective nerve root injection as a diagnostic tool is uncertain.

Cross-References

▶ Cervical Transforaminal Injection of Steroids
▶ Epidural Steroid Injections for Chronic Back Pain

Self-constructs

Definition

A term used in personal construct theory, which focuses on the constructs developed by individuals to explain and deal with their social world. Established chronic pain is often associated with self-constructs of being physically ill or impaired, with diminished control.

Cross-References

▶ Therapy of Pain, Hypnosis

Self-control

▶ Biofeedback in the Treatment of Pain

Self-Efficacy

Definition

Self-efficacy is a person's appraisal of his or her ability to engage in specific behaviors, ability to cope with a specific situation, or ability to cope with pain symptoms or stress in general.

Cross-References

▶ Assessment of Pain Behaviors
▶ Motivational Aspects of Pain
▶ Psychology of Pain, Assessment of Cognitive Variables
▶ Psychology of Pain, Self-efficacy
▶ Therapy of Pain, Hypnosis

Self-management

Definition

Self-management is an individual's willingness and motivation to exert effort on their own behalf to maximize their health or to minimize symptoms and disability. Self-management involves active efforts to control life and symptoms in contrast to passive dependency on others to provide treatment and assistance at all levels.

Cross-References

► Cognitive-Behavioral Perspective of Pain

Self-Organization

Definition

Self-organization is a process whereby a pattern at the global level of a system emerges solely from interactions among lower-level components of the system.

Cross-References

► Consciousness and Pain

Self-regulation

► Biofeedback in the Treatment of Pain

Self-reinforcement

Definition

Reinforcement that is not provided by others but the patient himself or herself.

Cross-References

► Operant Treatment of Chronic Pain

Self-Report Pain Measures

Definition

Self-report pain measures are instruments that rely on the patient's assessments of their pain intensity, pain affect, pain quality, and pain location, as well as constructs related to chronic pain.

Cross-References

► Pain Inventories

Self-traction

► Traction

Seltzer Model

► Retrograde Cellular Changes After Nerve Injury

Sense of Control

Definition

The belief that one is able to exert some control on life, problems, symptoms, and health. Perceptions of low levels of control contribute to passivity, dependency, low levels of persistence in the face of adversity, and ultimately a feeling of helplessness, emotional distress, and demoralization.

Cross-References

► Cognitive-Behavioral Perspective of Pain

Sensibility

▶ Ulceration, Prevention by Nerve Decompression

Sensitive Locus

Synonyms

Local-twitch-response locus; LTR locus

Definition

The sensory component of an MTrP locus is the sensitive locus, or local-twitch-response locus (LTR locus), which consists of sensitized receptors. From this locus, referred pain and LTR can be elicited by mechanical stimulation with a needle.

Cross-References

▶ Dry Needling

Sensitivity

Definition

Sensitivity is a statistical decision theory term equivalent to hit rate. It defines the probability that the test is positive when given to a group of patients with the disease.

Cross-References

▶ Sacroiliac Joint Pain

Sensitization

Definition

In the periphery, it describes the process by which the peripheral endings of nociceptive sensory neurons become more easily activated by stimuli. An increase in the excitability of the terminal membrane of peripheral nociceptive fibers reduces the amount of depolarization required to initiate an action potential discharge. Sensitization of a nociceptor is not caused by an unspecific lesion of the nociceptor, but by binding of sensitizing substances to specific receptor molecules in the membrane of the nociceptor. Typical sensitizing substances are prostaglandin E2 and bradykinin. In the CNS, it is a process by which nociceptive neurons become more responsive to peripheral stimulation. This process requires a series of steps that include release of certain neurotransmitters, activation of signal transduction pathways in the neurons that become sensitized, and phosphorylation of excitatory surface membrane receptors, such as glutamate receptors, on the affected central neurons.

Cross-References

▶ Allodynia (Clinical, Experimental)
▶ Clinical Migraine Without Aura
▶ Deafferentation Pain
▶ Hyperalgesia
▶ Infant Pain Mechanisms
▶ Inflammation, Modulation by Peripheral Cannabinoid Receptors
▶ Nociceptive Processing in the Amygdala: Neurophysiology and Neuropharmacology
▶ NSAIDs, Mode of Action
▶ Psychology of Pain, Sensitization, Habituation, and Pain
▶ Postoperative Pain, Preemptive or Preventive Analgesia
▶ Postsynaptic Dorsal Column Projection, Functional Characteristics
▶ Quantitative Thermal Sensory Testing of Inflamed Skin
▶ Sensitization of Muscular and Articular Nociceptors
▶ Sensitization of Visceral Nociceptors
▶ Spinothalamic Tract Neurons, Glutamatergic Input
▶ Thalamus, Dynamics of Nociception

Sensitization of Muscular and Articular Nociceptors

Hans-Georg Schaible[1] and Siegfried Mense[2]
[1]Department of Physiology - Neurophysiology, Friedrich-Schiller-University of Jena, Jena, Germany
[2]Neuroanatomy, Heidelberg University, Medical Faculty Mannheim, CBTM, Mannheim, Germany

Synonyms

Peripheral sensitization in muscle and joint

Definition

In intact tissue, nociceptors have no resting activity and exhibit a high threshold to mechanical stimuli. In addition, a nociceptor is supposed to encode the intensity of stimuli within the noxious range. During pathological processes in muscle and joint, nociceptive primary afferent neurons supplying these tissues are rendered active and hyperexcitable to mechanical stimuli. This process of sensitization of deep tissue afferents contributes to the dysesthesia and mechanical allodynia/hyperalgesia observed after various forms of deep tissue injury.

Characteristics

Ongoing Pain/Dysesthesias and Mechanical Allodynia/Hyperalgesia During Diseases or Overuse of Muscle and Joint

A number of different conditions can cause painful sensations from muscle: e.g., overuse during strenuous exercise, trauma (blow to a muscle or muscle tear), tonic or ischemic contractions, myofascial trigger points, and acute myositis. All these conditions are accompanied by the release of agents from muscle tissue that sensitize nociceptors (Mense 1993, 2003, 2009). Generally, muscle lesions first cause

tenderness (▶ allodynia, i.e., there is no spontaneous pain, but light mechanical stimuli or movements are painful), and later, spontaneous pain develops in addition to the tenderness.

Joint inflammation is characterized by hyperalgesia and persistent pain at rest which is usually dull and poorly localized. Noxious stimuli cause greater pain than normal, and pain is even evoked by mechanical stimuli whose intensity does not normally elicit pain (i.e., movements in the working range and gentle pressure, e.g., during palpation) (Schaible and Grubb 1993; Schaible 2006).

Mechanosensitivity Changes in Muscle Afferents During Pathophysiological Conditions

If a muscle contracts under ischemic conditions, severe pain develops within 1 or 2 min. During ischemia, a subpopulation of group IV units responds strongly to contractions, although these units are not or only minimally excited by contractions when the blood supply is intact (Mense and Stahnke 1983). Even muscle contractions at low workload can become painful, particularly if the work is monotonous and stressful. For instance, blood flow and muscle oxygenation were reduced after prolonged computer work (Cagnie et al. 2012). The resulting increase in mechanosensitivity and pain may be due to compounds that are released by ischemia such as bradykinin, prostaglandins, and protons.

Experimental muscle inflammation is the typical example of chemical sensitization of nociceptors by pro-inflammatory agents released from muscle tissue. One effect of the sensitization is an increased discharge rate of group III and IV fibers. The activity rises gradually from intact to acute (carrageenan-induced) and chronic myositis (Fig. 1b). The discharges of group IV units in inflamed muscle often are intermittent, with bursts separated by periods of complete silence (Berberich et al. 1988) (Fig. 1c, lower panel). The bursting pattern of activity is known to be particularly effective at central synapses. It is worth noting, however, that the mean discharge rate was relatively low and did not exceed 1 Hz in chronically inflamed muscle. Moreover, in

Sensitization of Muscular and Articular Nociceptors, Fig. 1 (a) Sensitization of a group IV afferent unit by serotonin (5-hydroxytryptamine, 5-HT) to bradykinin (Brad.). The substances were injected into the muscle artery. (b) Mean background (ongoing) activity in group IV muscle afferents in rats with intact, acutely inflamed, and chronically inflamed muscle. (c) Original recordings of single group IV fibers. The upper unit had irregular increased activity, whereas the unit in the lower panel exhibited bursting discharges

inflamed muscle, the mechanical threshold of muscle nociceptors drops markedly. This change expresses itself by a shift in the mechanosensitivity from high to low thresholds (Fig. 2b). In intact muscle, the high-threshold mechanosensitive (HTM) units are generally assumed to be nociceptive, whereas the low-threshold mechanosensitive (LTM) units may inform higher centers about the degree of muscle activity, i.e., they fulfill ergoreceptive or proprioceptive functions. In inflamed muscle, no functional classification based on threshold can be made: an LTM unit may be a sensitized nociceptor or an originally non-nociceptive ending. The sensitization of nociceptors is assumed to be the peripheral neuronal mechanism of tenderness of an inflamed or otherwise lesioned muscle (in addition to central nervous sensitization).

Data from group IV units in normal muscle show that low concentrations of inflammatory agents such as serotonin and PGE_2 sensitize free nerve endings without exciting them (Fig. 1a), whereas high concentrations are excitatory (Mense 1981). Therefore, when a muscle lesion develops and the concentrations of inflammatory substances rise, nociceptors are likely to be first sensitized and then activated. In patients, often a similar temporal sequence can often be observed; first there is tenderness (due to sensitization of nociceptors) followed by spontaneous pain (due to ongoing nociceptor activity). There are also cases where tenderness does not develop into spontaneous pain; a prominent example is delayed onset muscle soreness after unaccustomed exercise. In this case, the patients experience tenderness to pressure and pain during movements, but no pain during rest.

Nerve growth factor (NGF) had no immediate mechanical sensitizing action on muscle nociceptors when injected into the rat gastrocnemius-soleus muscle. There was just a

Sensitization of Muscular and Articular Nociceptors, Fig. 2 Changes in mechanical thresholds of group IV afferents following muscle inflammation. *Left panel*, data from intact muscle; *middle panel*, data from rats

with acute myositis (duration 2–8 h); *right panel*, data from chronic myositis (duration 12 days). The *bars* show the proportion of units with low (*open bars*) and high mechanical threshold (*filled bars*)

transient trend toward a higher mechanosensitivity. A significant NGF-induced sensitization took more than 1 day to develop (Hoheisel et al. 2005). However, the time-course of the sensitizing action of NGF appears to differ between muscles: In the rat masseter muscle, a faster NGF-induced sensitization has been reported to occur (Svensson et al. 2010). Injection of NGF into the masseter muscle in humans likewise led to a fast and long-lasting mechanical allodynia/hyperalgesia (Svensson et al. 2003). The NGF injection itself was not painful.

Not all substances released during muscle inflammation or other tissue lesions sensitize muscle nociceptors. The cytokine TNF-α and the neurotrophin brain-derived neurotrophic factor (BDNF) had a desensitizing effect on the mechanosensitivity of muscle nociceptors. The effect was of short duration and complete recovery occurred 8–10 minutes after i.m. injection (Hoheisel et al. 2005). Thus, it appears that in the very body periphery – at the nociceptive ending – there is a sort of balance between sensitizing and desensitizing factors and processes, at least for several minutes after onset of the lesion.

Interestingly, a sensitizing substance can be switched into a desensitizing one, if it acts

sequentially or for a longer time on a nociceptive ending. This process has been shown to be controlled by Ca^{++}/calmodulin-dependent protein kinase II (Hucho et al. 2012).

Mechanosensitivity Changes in Articular Afferent Fibers During Joint Inflammation

An important mechanism for heightened pain sensitivity in a joint is an increase in mechanosensitivity in joint afferents. Changes in the discharge properties have been studied in models of acute and chronic joint inflammation. As an acute model of inflammation, the kaolin/carrageenan model has been used. After injection of kaolin and carrageenan into the joint, inflammation with swelling and cellular infiltration develops within 1–3 h. This allows to record from neurons during development of inflammation, and directly document neuronal changes. As a model of chronic inflammation, Freund's adjuvant-induced inflammation of the joint has been used.

During development of acute inflammation, the discharge properties of articular afferents (see ▶ articular nociceptors) are significantly altered (Coggeshall et al. 1983; Schaible and Schmidt 1985, 1988). First, some low-threshold group II fibers with rapidly conducting axons and corpuscular sensory endings show transiently increased responses to innocuous and noxious joint movements in the initial hours of inflammation. These fibers do not develop ongoing discharges. Second, many low-threshold group III and IV fibers with free nerve endings that show responses to movements within the working range in the normal joint exhibit increased responses to these movements. Third, a large proportion of articular afferents that are in the normal joint only weakly activated by innocuous movements are sensitized, as well as high-threshold afferents that do not normally respond to innocuous movements are sensitized. After sensitization, they show pronounced responses to movements in the working range of the inflamed joint and enhanced responses to noxious movements. Fourth, neurons that have a receptive field but do not respond to movements of the normal joint develop responses to joint movements. Fifth, numerous initially mechanoinsensitive afferents (silent nociceptors) are sensitized and become mechanosensitive. Under normal conditions silent nociceptors do not respond to mechanical stimuli. However, in the inflammatory process, a receptive field becomes detectable, and the fibers start to respond to movements of the joint (Grigg et al. 1986; Schaible and Schmidt 1988). In addition, many units develop ongoing discharges in the resting position.

Figure 3 shows discharges of a group IV fiber from the inflamed knee joint of the cat. The dots in Fig. 3a mark the most sensitive spots at which the fiber was activated by pressure to the medial collateral ligament and to the capsule. The fiber was also activated by pressure to the shaded areas. Under normal conditions, nociceptive articular afferents are only activated from tiny spots. This fiber had substantial resting discharges (Fig. 3b) and was strongly activated by extension (Fig. 3c) and flexion (Fig. 3d) within the working range of the joint. Increased mechanosensitivity has also been found during chronic Freund's complete adjuvant-induced arthritis, suggesting that mechanical sensitization is an important neuronal basis for chronic persistent hyperalgesia of the inflamed joint (Guilbaud et al. 1985).

Sensitization of Muscular and Articular Nociceptors, Fig. 3 Discharges of a group IV fiber (conduction velocity 2.1 m/s) from the acutely inflamed knee joint of a cat. Inflammation was evoked by kaolin/carrageenan several hours before the recordings. (a) Receptive fields in the knee joint. *Dots* show most sensitive spots in the medial collateral ligament and the fibrous capsule. *Shaded areas* show sensitive areas in the medial and anterior aspect of the knee. (b) Resting discharges in the absence of mechanical stimulation. (c) Response to extension within the working range. (d) Response to flexion within the working range

Mechanisms of Sensitization of Muscular Nociceptors

As was to be expected, there is a marked interaction between the many substances released during a muscle lesion. A sensitizing action of serotonin and PGE_2 on the responses of group IV units to bradykinin has been described above. Conversely, acetylsalicylic acid, a blocker of PG synthesis, reduced responses to bradykinin, indicating that bradykinin releases PGs from muscle tissue and thus potentiates its own action (Mense 1982). In inflamed tissue, a neuroplastic change of the bradykinin receptor takes place: in intact tissue, normal tissue, bradykinin acts by binding to the B2 receptor, in inflamed tissue, a new receptor molecule for bradykinin (the B1 receptor) is synthesized by the dorsal root ganglion cell and mediates the bradykinin action. A more general mechanism for mechanical sensitization has been described for cutaneous and visceral receptors (Grant et al. 2007): During inflammation and other painful diseases, tissue proteases are released that cleave protease-activated receptor 2 (PAR2) on (nociceptive) nerve endings. PAR2 activates protein kinases in the ending which sensitize the mechanosensitive transient receptor potential vanilloid 4 (TRPV4) receptor. Another intracellular pathway leads from the NGF receptor TrkA via a mitogen-activated protein (MAP) kinase to sensitization of nociceptors (Hucho and Levine 2007). If these mechanisms also take place in muscle nociceptors, is unknown at present. The mechanisms involved in ischemic muscle pain and sensitization are still obscure. Data obtained in experiments in which muscles performed isometric contractions indicate that high interstitial concentrations of potassium ions or protons may be important factors (Hník et al. 1976). Potassium ions could depolarize nerve endings and protons may excite muscle nociceptors by binding to acid sensitive receptor molecules such as ASICs or TRPV1. Since ischemia is a very effective stimulus for the cleavage of bradykinin from plasma proteins (Nakahara 1971), the kinin is another agent that may sensitize or excite nociceptors during ischemic contractions. Recently, novel agents that are released in a pathologically altered muscle (protons, ATP, interleukin-6 (IL-6), tumor necrosis factor-α (TNF-α), BDNF, and NGF) have been tested for their excitatory and sensitizing action on single muscle group IV endings (Hoheisel et al. 2004, 2005).

During the first 10–15 minutes after i.m. injection of these agents, none sensitized the endings to noxious mechanical stimuli. However, TNF-α and BDNF had a desensitizing action, with the maximum effect occurring 6 min after injection.

Several lines of evidence (e.g., pharmacological, immunohistological) show that muscle group IV units possess both the high-affinity TrkA and low-affinity p75 receptors for NGF. For the NGF-induced mechanical sensitization, which takes longer than 15 minutes to develop, the TrkA receptor appears to be the important one (Svensson et al. 2010).

Mechanisms of Mechanical Sensitization of Articular Afferents

Key to changes of mechanosensitivity of sensory group III and IV fibers is their chemosensitivity. Mediators such as bradykinin, prostaglandins, serotonin, proinflammatory cytokines and others are able to excite and/or sensitize primary afferent neurons for mechanical and chemical stimuli (see ▶ Articular nociceptors). The application of these mediators can partly mimic the inflammation-evoked sensitization (see ▶ Articular nociceptors). A single injection of either TNF-α, interleukin-6, interleukin-1β, or interleunkin-17A into the joint causes a slowly developing hyperexcitability of group IV fibers to mechanical stimuli which persists for hours (see ▶ Articular nociceptors). Thus, cytokines may play a particular role in the long-term sensitization.

An important question is which compounds are able to reverse mechanical sensitization. Recordings of articular afferents have documented effects of nonsteroidal anti-inflammatory drugs (NSAIDs) in articular afferents. Both PGE_2 and PGI_2 cause ongoing discharges and/or sensitization to mechanical stimulation of the joint (cf. Schaible 2006). Conversely, NSAIDs such as aspirin and indomethacin reduce spontaneous discharges from acutely and chronically inflamed joints and attenuate the responses to mechanical stimulation (Heppelmann et al. 1986). Figure 4 shows the effects of systemic indomethacin on the discharges of a group III fiber (Fig. 4a) and a group IV fiber (Fig. 4b) from the inflamed knee joint. Indomethacin reduced both the movement-evoked responses and the resting discharges (note that there is an interval of about 1 h between the initial testing and the testing after indomethacin in Fig. 4a). The effect of indomethacin was reversed by subsequent intraarterial application of PGE_2 into the joint.

The reduction of inflammation-evoked activity by opioids, somatostatin and somatostatin receptor agonists, cannabinoids has been addressed in the chapter on articular nociceptors (see ▶ Articular nociceptors). A blocker of the sodium channel $Na_v1.8$, applied intraarterially close to the knee joint, reduced the firing of joint afferents to noxious rotation (but not of innocuous rotation) in the monoiodoacetate model of osteoarthritis, and it reduced the incapacitance and secondary hyperalgesia (Schuelert and McDougall 2012).

The neutralization of some cytokines was found to reduce inflammation-evoked

Sensitization of Muscular and Articular Nociceptors, Fig. 4 Effects of indomethacin and subsequent PGE_2 on the discharges of articular afferents from inflamed cat knee joint. (a) Group III unit with three receptive fields in the inflamed knee. The initial three *dots* show the responses to flexion of the inflamed knee, the *circles* the resting discharges in the 30 s before flexion. Both responses to flexion and resting discharges were significantly reduced approximately 60 min after indomethacin. The intraarterial injection of PGE_2 at increasing doses into the knee joint reversed the effect of indomethacin on resting discharges and flexion movements. (b) Group IV unit with a receptive field in the patellar ligament of the inflamed knee. The *symbols* show average resting discharges in three 10-min periods before indomethacin, a reduction in the resting discharges after indomethacin and the reversal of the effect by intraarterial injection of PGE_2 into the knee joint. *CV* conduction velocity, *T* threshold in electrical stimulation of joint nerve (From Heppelmann et al. 1986)

mechanical (and thermal) hyperalgesia in behavioral experiments using models of arthritis. A reduction of mechanical hyperalgesia by neutralization of TNF-α was observed in mice with collagen-induced arthritis (Inglis et al. 2007), in rats with antigen-induced arthritis (AIA) (Boettger et al. 2008), and in mice with K/BxN arthritis (Christianson et al. 2010). The injection of etanercept into the knee joint of rats at day 3 of AIA significantly reduced the responses of C-fibers to innocuous and noxious outward rotation of the inflamed knee within 1 hour after intraarticular injection, whereas it did not influence the responses of C-fibers to innocuous and noxious outward rotation of the normal knee joint (Boettger et al. 2008). In patients with rheumatoid arthritis, the application of infliximab reduced mechanical hyperalgesia on the first day after application, and at this time point the inflammatory process was not yet reduced (Hess et al. 2011). These data show the attenuation or reversal of mechanical sensitization by neutralization of TNF-α, suggesting that TNF-α acts on neuronal targets.

Intraarticular sgp130 pretreatment, which neutralizes IL-6/soluble IL-6 receptor complexes, attenuated the development of mechanical hyperalgesia in AIA, but systemic application of sgp130 after AIA onset only slightly reduced mechanical hyperalgesia (Boettger et al. 2010). Thus, the IL-6-induced sensitization is more difficult to reverse (see also Brenn et al. 2007).

Neutralization of IL-1ß in rat AIA reduced only thermal hyperalgesia but neither mechanical hyperalgesia nor the inflammatory process per se (Ebbinghaus et al. 2012). Neutralization of IL-17 in mice with AIA improved the guarding score and reduced secondary mechanical hyperalgesia (Richter et al. 2012).

In patients with moderate to severe osteoarthritis, antibodies against nerve growth factor (NGF) produced significant pain relief (Lane et al. 2010) raising the possibility that NGF sensitizes joint afferents. A soluble NGF receptor, TrkAd5, effectively suppressed pain in an experimental osteoarthritis in mice (McNamee et al. 2010).

A totally different approach to attenuate osteoarthritic pain is the intraarticular injection of hyaluronic acid (HA) preparations. The therapeutical benefit of HA was and is being discussed controversially (for review see Schaible 2012). However, the intraarticular injection of HA preparations reduced rapidly movement-evoked discharges of rat joint nociceptors (Gomis et al. 2009) thus directly showing an antinociceptive effect at least at a short time scale. In a rat model of joint pain, in which bradykinin and PGE_2 were repetitively administered intra-articularly, a single injection of HA preparations at the beginning provided a significant antinociceptive effect (Boettger et al. 2011). The mechanisms of action of HA preparations are unclear. HA may provide mechanical protection for the joint due to their viscous properties, or they may cover sensory endings so that these are protected against sensitizing inflammatory mediators. Alternatively, inflammatory mediators might be entrapped in HA through electrostatic interaction (Gomis et al. 2009).

References

Berberich, P., Hoheisel, U., & Mense, S. (1988). Effects of a carrageenan-induced myositis on the discharge properties of Group III and IV muscle receptors in the cat. *Journal of Neurophysiology, 59,* 1395–1409.

Boettger, M. K., Hensellek, S., Richter, F., et al. (2008). Antinociceptive effects of TNF-α neutralization in a rat model of antigen-induced arthritis. *Arthritis Rheumatism, 58,* 2368–2378.

Boettger, M. K., Leuchtweis, J., Kümmel, D., et al. (2010). Differential effects of locally and systemically administered soluble glycoprotein 130 on pain and inflammation in experimental arthritis. *Arthritis Research Therapy, 12,* R140.

Boettger, M. K., Kümmel, D., Harrison, A., et al. (2011). Evaluation of long-term antinociceptive properties of of stabilized hyaluronic acid preparation (NASHA) in an animal model of repetitive joint pain. *Arthritis Research Therapy, 13,* R110.

Brenn, D., Richter, F., & Schaible, H.-G. (2007). Sensitization of unmyelinated sensory fibers of the joint nerve for mechanical stimuli by interleukin-6 in the rat. An inflammatory mechanism of joint pain. *Arthritis Rheumatism, 56,* 351–359.

Cagnie, B., Dhooge, F., Van Akeleyen, J., Cools, A., Cambier, D., Danneels, L. (2012). Changes in microcirculation of the trapezius muscle during a prolonged computer task. *Eur J Appl Physiol, 3305*–3312.

Christianson, C. A., Corr, M., Firestein, G. S., et al. (2010). Characterization of the acute and persistent pain state present in K/BxN serum transfer arthritis. *Pain, 151,* 394–403.

Ebbinghaus, M., Uhlig, B., Richter, F., et al. (2012). The role of interleukin-1ß in arthritic pain: Main involvement in thermal but not in mechanical hyperalgesia in rat antigen-induced arthritis. *Arthritis Rheumatism, 64,* 3897–3907.

Gomis, A., Miralles, A., Schmidt, R. F., et al. (2009). Intra-articular injections of hyaluronan solutions of different elastoviscosity reduce nociceptive nerve activity in a model of osteoarthritic knee joint of the guinea pig. *Osteoarthritis Cartilage, 17,* 798–804.

Grant, A. D., Cottrell, G. S., Amadesi, S., et al. (2007). Protease-activated receptor 2 sensitizes the transient receptor potential vanilloid 4 ion channel to cause mechanical hyperalgesia in mice. *The Journal of Physiology, 578,* 715–733.

Grigg, P., Schaible, H.-G., & Schmidt, R. F. (1986). Mechanical sensitivity of group III and IV afferents from posterior articular nerve in normal and inflamed cat knee. *Journal of Neurophysiology, 55,* 635–643.

Guilbaud, G., Iggo, A., & Tegner, R. (1985). Sensory receptors in ankle joint capsules of normal and arthritic rats. *Experimental Brain Research, 58,* 29–40.

Heppelmann, B., Pfeffer, H.-P., Schaible, H.-G., et al. (1986). Effects of acetylsalicylic acid and indomethacin on single groups III and IV sensory units from acutely inflamed joints. *Pain, 26,* 337–351.

Hess, A., Axmann, R., Rech, J., et al. (2011). Blockade of TNF-α rapidly inhibits pain responses in the central nervous system. *Proceedings of the National Acadamy of Sciences USA, 108,* 3731–3736.

Hník, P., Holas, M., Krekule, I., et al. (1976). Work-induced potassium changes in skeletal muscle and effluent venous blood assessed by liquid ion-exchanger microelectrodes. *Pflugers Archive, 362,* 85–94.

Hoheisel, U., Reinöhl, J., Unger, T., et al. (2004). Acidic pH and capsaicin activate mechanosensitive group IV muscle receptors in the rat. *Pain, 110,* 149–157.

Hoheisel, U., Unger, T., & Mense, S. (2005). Excitatory and modulatory effects of inflammatory cytokines and neurotrophins on mechanosensitive group IV muscle afferents in the rat. *Pain, 114*, 168–176.

Hucho, T., & Levine, J. D. (2007). Signaling pathways in sensitization: Toward a nociceptor cell biology. *Neuron, 55*, 365–376.

Hucho, T., Suckow, V., Joseph, E. K. et al. (2012) Ca++/CaMKII switches nociceptor-sensitizing stimuli into desensitizing stimuli. *Journal of Neurochemistry, 123*, 589–601.

Inglis, J. J., Notley, C. A., Essex, D., et al. (2007). Collagen-induced arthritis as a model of hyperalgesia. Functional and cellular analysis of the analgesic actions of tumor necrosis factor blockade. *Arthritis Rheumatism, 56*, 4015–4023.

Lane, N. E., Schnitzer, T. J., Birbara, C. A., et al. (2010). Tanezumab for the treatment of pain from osteoarthritis of the knee. *New England Journal of Medicine, 363*, 1521–1531.

McNamee, K. E., Burleigh, A., Gompels, L. L., et al. (2010). Treatment of murine osteoarthritis with TrkAd5 reveals a pivotal role for nerv growth factor in non-inflammatory joint pain. *Pain, 149*, 386–392.

Mense, S. (1981). Sensitization of group IV muscle receptors to bradykinin by 5-hydroxytryptamine and prostaglandin E2. *Brain Research, 225*, 95–105.

Mense, S. (1982). Reduction of the bradykinin-induced activation of feline group III and IV muscle receptors by acetylsalicylic acid. *The Journal of Physiology, 326*, 269–283.

Mense, S. (1993). Nociception from skeletal muscle in relation to clinical muscle pain. *Pain, 54*, 241–289.

Mense, S. (2003). The pathogenesis of muscle pain. *Current Pain and Headache Reports, 7*, 419–425.

Mense, S. (2009). Algesic agents exciting muscle nociceptors. *Experimental Brain Research, 196*, 89–100.

Mense, S., & Stahnke, M. (1983). Responses in muscle afferent fibres of slow conduction velocity to contractions and ischaemia in the cat. *The Journal of Physiology, 342*, 383–397.

Nakahara, M. (1971). The effect of a tourniquet on the kinin-kininogen system in blood and muscle. *Thromb Diath Haemorrh, 26*, 264–274.

Richter, F., Natura, G., Ebbinghaus, M., et al. (2012). Interleukin-17 sensitizes joint nociceptors for mechanical stimuli and contributes to arthritic pain through neuronal IL-17 receptors in rodents. *Arthritis Rheumatism, 64*, 4125–4134.

Schaible, H.-G. (2006). Basic mechanisms of deep somatic pain. In S. B. McMahon & M. Koltzenburg (Eds.), *Textbook of pain*. Philadelphia: Elsevier/Churchill, Livingston.

Schaible, H.-G. (2012). Mechanisms of chronic pain in osteoarthritis. *Current Rheumatology Reports, 14*, 549–556.

Schaible, H.-G., & Grubb, B. D. (1993). Afferent and spinal mechanisms of joint pain. *Pain, 55*, 5–54.

Schaible, H.-G., & Schmidt, R. F. (1985). Effects of an experimental arthritis on the sensory properties of fine articular afferent units. *Journal of Neurophysiology, 54*, 1109–1122.

Schaible, H.-G., & Schmidt, R. F. (1988). Time course of mechanosensitivity changes in articular afferents during a developing experimental arthritis. *Journal of Neurophysiology, 60*, 2180–2195.

Schuelert, N., & McDougall, J. J. (2012). Involvement of $Na_v1.8$ sodium channels in the transduction of mechanical pain in a rodent model of osteoarthritis. *Arthritis Research Therapy, 14*, R5.

Svensson, P., Cairns, B. E., Wang, K., et al. (2003). Injection of nerve growth factor into human masseter muscle evokes long-lasting mechanical allodynia and hyperalgesia. *Pain, 104*, 241–247.

Svensson, P., Wang, M. W., Dong, X.D., et al. (2010). Human nerve growth factor sensitizes masseter muscle nociceptors in female rats. *Pain, 148*, 473–480.

Sensitization of the Nociceptive System

Definition

A process that leads to a decrease of the threshold of a certain stimulus (e.g., heat, pressure) that activates nociceptive neurons and elicits pain. Sensitization manifests in ► hyperalgesia, which is an increased response to nociceptive stimuli, and ► allodynia, which is a sensation of pain caused by normally non-painful stimuli such as touch and cold.

Cross-References

► COX-1 and COX-2 in Pain

Sensitization of Visceral Nociceptors

Gerald F. Gebhart[1] and Bin Feng[2]
[1]Department of Anesthesiology, Center for Pain Research, University of Pittsburgh, Pittsburgh, PA, USA
[2]Center for Pain Research, Department of Anesthesiology, School of Medicine, University of Pittsburgh, Pittsburgh, Pennsylvania, USA

Synonyms

Excitability; Hyperalgesia; Hypersensitivity; Plasticity; Sensitization; Visceral Nociceptors

Definition

Among all categories of cutaneous receptors (i.e., mechanoreceptors, thermoreceptors, chemoreceptors, and nociceptors), only nociceptors sensitize. Bessou and Perl (see Perl 1996 for review) described an enhancement of responsiveness of C-fiber ▶ polymodal nociceptors after exposure to noxious intensities of skin heating. This sensitization was manifest by an increase in discharge for a given stimulus intensity, associated with a lowering of the response threshold to heat stimulation. Stretch-sensitive visceral ▶ mechanonociceptors exhibit the same ability as cutaneous nociceptors to sensitize but with significant differences. Like some cutaneous nociceptors, most, if not all, visceral mechanonociceptors also respond to thermal (cold and/or hot) and/or chemical stimuli (e.g., see Su and Gebhart 1998). Unlike low-threshold mechanosensitive cutaneous receptors, low (as well as high)-threshold stretch-sensitive visceral mechanoreceptors sensitize, suggesting that most stretch-sensitive mechanosensitive visceral receptors can function and contribute to visceral nociception. Mechanically *in*sensitive colorectal afferents also exhibit the ability to acquire mechanosensitivity (an expression of sensitization) when exposed to inflammatory mediators (Feng and Gebhart 2011).

Characteristics

Sensitization represents a process by which neuron excitability is increased initially by events in the local environment of the nociceptive ending and, later, also by transcriptional events in the cell soma. Local events associated with sensitization include release of, synthesis of, or attraction of chemical mediators and modulators to the site of organ insult. The list of putative mediators/modulators of nociceptor sensitization is long and includes amines such as histamine and serotonin, peptides such as substance P, products of arachidonic acid metabolism, protons, cytokines, ATP, etc. The lumen of hollow organs also contains at various times secreted peptides, protons,

serotonin derived from enterochromaffin cells, potassium ions in the urinary bladder, bile salts, etc. Two points of emphasis are important. First, many so-called functional visceral disorders such as nonulcer dyspepsia, irritable bowel syndrome, and painful bladder syndrome exist in the absence of biochemical or morphological evidence of tissue inflammation or pathology. Second, many if not all of the chemicals identified above are normally present, either within the lumen of hollow organs or released into the lumen as part of normal organ function. Accordingly, pain and discomfort associated with functional disorders can arise by an increase in excitability of mechanoreceptor endings in organ tissue layers in the absence of frank organ insult (e.g., see Feng et al. 2012).

As described in *Visceral Mechanoreceptors*, approximately 75–80 % of stretch/tension-sensitive visceral afferents recorded in a variety of animal species, and innervating a variety of organs (e.g., stomach, gall bladder, urinary bladder, colon, uterus), have low thresholds for response; the remaining 20–25 % have thresholds for response in the noxious range. Both low- and high-threshold stretch/tension-sensitive visceral afferents are slowly adapting, and both encode stimulus intensity into the noxious range. Importantly, and unlike low-threshold cutaneous mechanoreceptors, most low-threshold stretch/tension-sensitive visceral receptors sensitize after experimental exposure of receptive endings to an inflammogen/inflammatory soup or after organ insult. Accordingly, because both low- and high-threshold stretch/tension-sensitive visceral receptors (1) encode into the noxious range and (2) sensitize, they are functionally distinct from low- and high-threshold mechanoreceptors in the cutaneous realm and thus are likely able to contribute to nociception after organ insult. Figure 1 illustrates responses of low- and high-threshold stretch-sensitive afferents innervating the mouse colorectum (pelvic nerve) before and after acute (and reversible) sensitization. The low-threshold fiber (left) exhibits a clear increase in response magnitude and also acquired spontaneous activity. The high-threshold pelvic (right) exhibits a similar increase in response magnitude,

Sensitization of Visceral Nociceptors, Fig. 1 Examples of sensitization by inflammatory soup (IS) of low-threshold (*left*) and high-threshold (*right*) stretch-sensitive pelvic nerve afferents innervating the mouse colorectum. Sensitization includes an increase in response magnitude and reduction in response threshold, both of which are apparent in the recordings. The ramped stretch stimulus (0–170 mN at 5 mN/s) is equivalent to ~45-mmHg colorectal distension (Feng and Gebhart 2011) and permits extrapolation of response threshold, illustrated by the arrowheads on the broken red lines (Examples from Feng and Gebhart, unpublished)

as well as a decrease in response threshold from 30 to 10 mmHg.

In addition to exhibiting the property of sensitization after acute exposure to inflammatory mediators, accumulating evidence reveals that stretch-sensitive colorectal afferents can exist in a persistent sensitized state, in conjunction with behavioral colorectal hypersensitivity, in the absence of organ inflammation. Figure 2 shows examples of muscular mucosal afferents studied 3 weeks after intracolonic instillation of saline or zymosan.

Sensitization has also been documented as an increase in neuronal excitability associated with alterations in voltage-gated and/or ligand-gated ion channels in visceral sensory neurons studied using whole-cell patch clamp techniques (e.g., Beyak 2010). Voltage-sensitive ion channels underlie both the generation of action potentials and repolarization/hyperpolarization of sensory neurons. After experimental organ insult (e.g., bladder inflammation), the excitability of bladder neurons in dorsal root ganglia is increased (Fig. 3). The increase in excitability (sensitization) is reflected by changes in voltage-sensitive or ligand-gated currents. For example, the whole-cell inward sodium current recovers more rapidly from inactivation and is significantly increased, principally contributed to by an increase in the tetrodotoxin-resistant sodium current (e.g., Bielefeldt et al. 2002a, b); in addition, the current density of a slowly inactivating outward A-type potassium current is significantly reduced (e.g., Dang et al. 2004). These changes in sodium and potassium currents after organ insult are wholly consistent with an increase in excitability that underlies the process of sensitization. Ligand-gated ion channels can similarly play a role in the process of sensitization, although direct evidence at present is limited. Several such channels/receptors are candidates as modulators of sensory neuron excitability: acid-sensing ion channels (ASICs), transient receptor potential vanilloid 1 (TRPV1) receptors, purinergic P2X channels, and proteinase-activated receptors (PARs), to name but a few. In conjunction with whole-cell patch clamp recordings of identified visceral sensory neurons, the use of receptor-selective agonists and antagonists, antisense oligonucleotides, and knockout

Sensitization of Visceral Nociceptors, Fig. 2 Examples of responses of pelvic nerve stretch-sensitive muscular-mucosal afferents studied in mice 3 weeks after intracolonic instillation of either saline or zymosan. Saline-treated mice did not develop colorectal hypersensitivity to balloon distension whereas zymosan-treated mice were hypersensitive at all times tested after treatment (not shown). The response of the muscular-mucosal afferent to circumferential stretch (0–170mN @ 5mN/s) taken from a saline-treated mouse (top record) did not differ from previous reports (Feng and Gebhart 2011) or from those recorded in naive mice. The response of the muscular-mucosal afferent from the zymosan-treated mouse, however, exhibits clear persistent sensitization: the response threshold is reduced and response magnitude is significantly increased. See Feng et al. 2012 for experimental details and grouped responses (Examples from Feng and Gebhart, unpublished)

Sensitization of Visceral Nociceptors, Fig. 3 Whole-cell patch clamp recordings (current clamp mode) of lumbosacral dorsal root ganglion (DRG) neurons innervating the rat urinary bladder. In bladder DRG neurons taken from mice treated with saline (control, *left*), there is no spontaneous activity and rheobase (the minimum injected current required to generate an action potential) was 160 nA. In bladder DRG neurons taken from mice treated every other day with cyclophosphamide (three doses of 100 mg/kg i.p. each), which is metabolized to a bladder irritant (acrolein), spontaneous activity was present in a significant proportion of neurons and rheobase was significantly reduced, in this example to 40 nA. When twice the rheobase current (2X) was injected, only one action potential was generated in this neuron taken from a control animal whereas when 2X rheobase current was injected into this neuron taken from a CYP-treated animal, the number of action potentials produced by current injection was significantly greater at a lower intensity of current, an index of sensitization (Taken from Dang et al. 2008)

or knockdown strategies have further established roles for these channels/receptors in either or both basal visceral nociceptive sensitivity or the process of sensitization. For example, ATP is released from colorectal epithelium and bladder urothelium during stretch or distension and can activate pelvic nerve afferent fibers (e.g., Rong et al. 2002), P2X2 and P2X3 receptors are increased in the urothelium of interstitial cystitis patients (Tempest et al. 2004), receptor subunit composition is changed in rats with bladder inflammation (e.g., Dang et al. 2008), and P2X3-deficient mice exhibit reduced pain behaviors in general and urinary bladder hyporeflexia (Cockayne et al. 2000). Similarly, TRPV1 is expressed in visceral organs, and colon (Jones et al. 2005) and bladder (Birder 2007) functions are significantly altered in TRPV1-deficient mice.

Similarly, roles for TRPV4 (Cenac et al. 2008), TRPA1 (Cattaruzza et al. 2010), and TRPM8 (Harrington et al. 2011) and interactions between TRPV1 and TRPA1 (Schwartz et al. 2011) in visceral hypersensitivity have been reported.

References

Beyak, M. J. (2010). Visceral afferents – determinants and modulation of excitability. *Autonomic Neuroscience, 153*, 69–78.

Bielefeldt, K., Ozaki, N., & Gebhart, G. F. (2002a). Experimental ulcers alter voltage-sensitive sodium currents in rat gastric sensory neurons. *Gastroenterology, 122*, 394–405.

Bielefeldt, K., Ozaki, N., & Gebhart, G. F. (2002b). Mild gastritis alters voltage-sensitive sodium currents in gastric sensory neurons. *Gastroenterology, 122,* 752–761.

Birder, L. A. (2007). TRPs in bladder diseases. *Biochimica et Biophysica Acta, 1772,* 879–884.

Cattaruzza, F., Spreadbury, I., Miranda-Morales, M., Grady, E. F., Vanner, S., & Bunnett, N. W. (2010). Transient receptor potential ankyrin- 1 has a major role in mediating visceral pain in mice. *American Journal of Physiology, 298,* G81–G91.

Cenac, N., Altier, C., Chapman, K., Liedtke, W., Zamponi, G., & Vergnolle, N. (2008). Transient receptor potential vanilloid-4 has a major role in visceral hypersensitivity symptoms. *Gastroenterology, 135,* 937–946.

Cockayne, D. A., Hamilton, S. G., Zhu, Q. M., et al. (2000). Urinary bladder hyporeflexia and reduced pain-related behaviour in P2X$_3$-deficient mice. *Nature, 407,* 1011–1015.

Dang, K., Bielefeldt, K., & Gebhart, G. F. (2004). Gastric ulcers reduce a-type potassium currents in rat gastric sensory ganglion neurons. *American Journal of Physiology, 286,* G573–G579.

Dang, K., Lamb, K., Cohen, M., Bielefeldt, K., & Gebhart, G. F. (2008). Cyclophosphamide-induced bladder inflammation sensitizes and enhances P2X receptor function in rat bladder sensory neurons. *Journal of Neurophysiology, 99,* 49–59.

Feng, B., & Gebhart, G. F. (2011). Characterization of silent afferents in the pelvic and splanchnic innervations of the mouse colorectum. *American Journal of Physiology, 300,* G170–G180.

Feng, B., La, J.-H., Schwartz, E. S., & Gebhart, G. F. (2012). Neural and neuro-immune mechanisms of visceral hypersensitivity in irritable bowel syndrome. *American Journal of Physiology, 302*(10), G1085–G1098.

Harrington, A. M., Hughes, P. A., Martin, C. M., Yang, J., Castro, J., Isaacs, N. J., Blackshaw, L. A., & Brierley, S. M. (2011). A novel role for TRPM8 in visceral afferent function. *Pain, 152,* 1459–1468.

Jones, R. C. W., III, Xu, L., & Gebhart, G. F. (2005). Mechanosensitivity of mouse colon afferent fibers and their sensitization by inflammatory mediators require TRPV1 and ASIC3. *Journal of Neuroscience, 25,* 10981–10989.

Perl, E. R. (1996). Cutaneous polymodal receptors: Characteristics and plasticity. In T. Kumazawa, L. Kruger, & K. Mizumura (Eds.), *Progress in brain research* (The polymodal receptor: A gateway to pathological pain, Vol. 113, pp. 21–37). Amsterdam: Elsevier.

Rong, W., Spyer, K. M., & Burnstock, G. (2002). Activation and sensitisation of low and high threshold afferent fibres mediated by P2X receptors in the mouse urinary bladder. *The Journal of Physiology, 541,* 591–600.

Schwartz, E. S., Christianson, J. A., Chen, X., La J-H Davis, B. M., Albers, K. M., & Gebhart, G. F. (2011). Synergistic role of TRPV1 and TRPA1 in pancreatic pain and inflammation. *Gastroenterology, 140,* 1283–1291.

Su, X., & Gebhart, G. F. (1998). Mechanosensitive pelvic nerve afferent fibers innervating the colon of the rat are polymodal in character. *Journal of Neurophysiology, 80,* 2632–2644.

Tempest, H. V., Dixon, A. K., Turner, W. H., et al. (2004). P2X And P2X receptor expression in human bladder urothelium and changes in interstitial cystitis. *BJU International, 93,* 1344–1348.

Sensory Decision Theory

▶ Pain Measurement by Questionnaires, Psychophysical Procedures, and Multivariate Analysis

Sensory Dimension of Pain

Definition

Dimension of pain characterized by the quality of the experience (e.g., burning, aching, or stinging), its spatial location, its temporal dynamics, and its subjective intensity.

Cross-References

▶ Cingulate Cortex, Functional Imaging

Sensory Discrimination

Definition

The ability to experience and recognize changes in the quality, duration, location, and intensity of stimuli.

Cross-References

▶ Pain in Humans, Sensory-Discriminative Aspects
▶ Spinohypothalamic Tract, Anatomical Organization, and Response Properties
▶ Spinothalamic Neuron

Sensory Endings in Joint Tissues

▶ Articular Afferents, Morphology

Sensory Focusing

▶ Psychological Treatment in Acute Pain

Sensory Ganglia

Definition

Cluster of neurons in the somatic peripheral nervous system that contain the cell bodies of sensory nerve axons. Sensory ganglia also have nonneuronal supporting cells.

Cross-References

▶ Perireceptor Elements

Sensory Ganglionectomy

▶ Dorsal Root Ganglionectomy and Dorsal Rhizotomy

Sensory Ganglionitis

▶ Ganglionopathies

Sensory Impairment

▶ Dysfunctional Pain and the International Classification of Function

Sensory Modalities

Definition

Sensory modalities are the basic types of sensory phenomena, such as vision, audition, somato-sensation, smell, and taste.

Cross-References

▶ Pain in Humans, Psychophysical Law

Sensory Motor Retuning

▶ Brain Based Interventions

Sensory Neuronopathy

▶ Ganglionopathies

Sensory Receptor

Definition

An organ having nerve endings (e.g., the skin, viscera, eye, nose, mouth) that respond to stimulation.

Cross-References

▶ Perireceptor Elements

Sensory Relay Neurons

Definition

Sensory relay neurons are projection neurons located between the spinal cord and the cerebral cortex relaying nociceptive messages to higher brain centers.

Cross-References

▶ Postsynaptic Dorsal Column Projection, Anatomical Organization

Sensory Rhizotomy

▶ Dorsal Root Ganglionectomy and Dorsal Rhizotomy

Sensory Saturation

Definition

Sensory saturation is a maneuver for non-pharmacologic analgesia. It implies administering oral sugar, massaging the baby, attracting its attention by speaking, etc.

Cross-References

▶ Ethics of Pain in the Newborn Human

Sensory Stimulation-Guided Radiofrequency Neurotomy

▶ Sacroiliac Joint Radiofrequency Denervation

Sensory Transduction

Definition

Process by which energy in the environment is converted into electrical signals by sensory receptors.

Cross-References

▶ Perireceptor Elements

Sensory-Discriminative Aspect of Pain

Definition

Certain aspects of pain sensation can be discriminated with accuracy. These include pain quality, location, intensity, and duration. Sensory discrimination of painful stimuli is thought to depend on the processing of nociceptive information by the part of the spinothalamic tract that synapses in the ventral posterior lateral nucleus, and by that nucleus and the cortical areas to which it projects, including SI and SII.

Cross-References

▶ Amygdala, Pain Processing and Behavior in Animals
▶ Hypnotic Analgesia
▶ Pain in Humans, Sensory-Discriminative Aspects
▶ Parafascicular Nucleus, Pain Modulation
▶ Secondary Somatosensory Cortex (S2) and Insula, Effect on Pain-Related Behavior in Animals and Humans
▶ Thalamo-Amygdala Interactions and Pain
▶ Thalamus, Nociceptive Cells in VPI, Cat and Rat

Sensory-Discriminative Aspects

▶ Pain in Humans, Sensory-Discriminative Aspects

SEPs

▶ Somatosensory Evoked Potentials

Sequential Relay

Definition

A sequential relay has been proposed that consists of (a) perforant path fibers of the entorhinal cortex that innervate dentate granule cells, (b) mossy fibers of dentate granule cells that innervate CA3 pyramidal cells, and (c) Schaffer collaterals of CA3 pyramidal cells that innervate CA1 pyramidal cells. Stimulation of perforant path fibers can lead to sequential activation of the other members of the relay.

Cross-References

▶ Nociceptive Processing in the Hippocampus and Entorhinal Cortex, Neurophysiology and Pharmacology

Serotonergic

Definition

Related to action of the serotonin, a biochemical neurotransmitter found primarily in the central nervous system, gastrointestinal tract, and platelets.

Cross-References

▶ Functional Abdominal Pain in Children
▶ Stimulation-Produced Analgesia

Serotonin

Synonyms

5-Hydroxytryptamine; 5-HT

Definition

Serotonin is a monoamine neurotransmitter synthesized in serotonergic neurons in the central nervous system and enterochromaffin cells in the gastrointestinal tract. In the central nervous system, serotonin is involved in the regulation of mood, sleep, emesis, sexuality, and appetite. It is thought to be significant in the biochemistry of depression, migraine, bipolar disorder, and anxiety. Serotonin is contained in brainstem (medulla and midbrain) neurons with descending projections to the spinal cord or ascending projections in the forebrain. Projections to spinothalamic and spinomesencephalic tract neurons are believed to have inhibitory and/or facilitatory influences on these projection neurons. Serotonin is also released by platelets and mast cells during the inflammatory response.

Cross-References

▶ Descending Circuitry, Transmitters and Receptors
▶ Fibromyalgia
▶ Molecular Contributions to the Mechanism of Central Pain
▶ Spinomesencephalic Tract
▶ Thalamic Neurotransmitters and Neuromodulators

Serotonin Antagonist

Definition

A substance that blocks the action of serotonin.

Serotonin Antagonist Reuptake Inhibitors (SARIs)

▶ Antidepressant Analgesics in Pain Management

Serotonin Blockers

Definition

Type 3 serotonin (5-hydroxytryptamine [5-HT3]) receptor blocker, class of antiemetic agents, e.g., dolasetron, granisetron, ondansetron, and tropisetron.

Cross-References

▶ Cancer Pain Management, Chemotherapy

Serotonin Selective Reuptake Inhibitors (SSRIs)

▶ Antidepressant Analgesics in Pain Management

Serotonin-Norepinephrine Reuptake Inhibitors (SNRIs)

▶ Antidepressant Analgesics in Pain Management

Serous Meningitis

▶ Headache in Aseptic Meningitis

Sex

Definition

Sex is the classification of living things, generally as male or female according to their reproductive organs and functions assigned by the chromosomal complement. In essence, sex refers to the biological component of being male or female.

Cross-References

▶ Gender and Pain
▶ Psychological Aspects of Pain in Women

Sex and Chronicity

▶ Pain in the Workplace, Risk Factors for Chronicity, and Demographics

Sex Differences in Descending Pain Modulatory Pathways

Dayna R. Loyd[1] and Anne Z. Murphy[2]
[1]US Army Institute of Surgical Research, Fort Sam Houston, TX, USA
[2]Neuroscience Institute, Georgia State University, Atlanta, GA, USA

Synonyms

Endogenous descending pain modulatory pathways; Descending inhibition; Descending inhibitory pathway; Periaqueductal gray – rostral ventromedial medulla (PAG-RVM) pathway; Periaqueductal gray – nucleus raphe magnus (PAG-NRM) pathway

Definition

The midbrain periaqueductal gray (PAG) and its descending projections to the rostral ventromedial medulla (RVM) have long been recognized as the key neural circuit mediating opioid-based analgesia. It is well established that administration of the opioid morphine produces greater analgesia in males compared to females, and recent studies indicate that this PAG-RVM pathway contributes to the sexually dimorphic actions of morphine. Many anatomical and physiological studies since the 1960s have characterized this descending pathway. However, the majority of these studies were conducted exclusively in males with the implicit assumption that the anatomy and physiology of this circuit was the same in females, and this now is known not to be the case. Recent anatomical and behavioral studies in the rat indicate that the PAG-RVM circuit is sexually dimorphic in its anatomical organization and in its activation during persistent inflammatory pain states.

Introduction

The midbrain periaqueductal gray (PAG), and its descending projections to the rostral ventromedial medulla (RVM) and spinal cord, comprises an essential neural substrate for both endogenous and exogenous opioid-mediated analgesia. The PAG contains a high density of mu opioid receptor (MOR) containing neurons, and microinjection of opioid antagonists into the PAG significantly attenuates the analgesic effects of systemic morphine. While it is well established that the PAG-RVM-spinal cord pathway is essential for the analgesic actions of morphine, these early studies were conducted exclusively in males. Only recently have studies begun including "sex" as an independent variable, and it is becoming increasingly clear that morphine does not produce the same degree of analgesia in males and females. Indeed, in models of persistent visceral or inflammatory pain, the ED50 for females is typically two-fold higher than the ED50 for males. To date, only a limited number of clinical studies have examined "gender" or "sex" as an independent variable, with no unequivocal finding reported. Recent studies examining the biological bases for the sexually dimorphic effects of morphine suggest that there are inherent differences in how the central nervous system of males and females responds to opiates.

Characteristics

The midbrain periaqueductal gray has been identified as a key neural structure in the endogenous

descending analgesia circuit (Basbaum and Fields 1979, 1984; Behbehani 1995). Electrical or chemical stimulation of the PAG inhibits both spinal and supraspinal nociceptive withdrawal reflexes, as well as dorsal horn neuronal responses to noxious peripheral stimulation. In addition to stimulation-produced analgesia (Reynolds 1969; Carstens et al. 1979; Basbaum and Fields 1979), the PAG is also a crucial neural substrate for opioid analgesia. Administration of morphine, or other mu opioid receptor agonists, into the PAG produces potent analgesia, which is blocked by central or systemic administration of the opioid antagonist naloxone (Jensen and Yaksh 1986).

There are no direct projections from the PAG to the dorsal horn of the spinal cord. Rather, the PAG provides direct and extensive input to the rostral ventromedial medulla (RVM). Bulbospinal fibers from the RVM descend within the dorsolateral funiculus, and terminate in the dorsal horn of the spinal cord at all rostrocaudal levels (Basbaum et al. 1978; Behbehani 1995). Similar to the PAG, stimulation of the RVM produces a powerful analgesic response, and the RVM is a critical site for morphine-induced analgesia (Le Bars et al. 1980). Lesions of the RVM or dorsolateral funiculus completely abolish PAG stimulation or opiate-induced analgesia, indicating that this PAG-RVM-spinal cord circuit is a critical neural pathway for stimulation-produced, or opiate-induced analgesia (Behbehani and Fields 1979).

Periaqueductal Gray
Rostral Ventromedial Medulla
From Periphery
Spinal Cord Dorsal Horn

Sex Differences in Descending Pain Modulatory Pathways, Fig. 1 Each *black dot* denotes one PAG neuron projecting to the RVM in male (*left*) versus female (*right*) rats across three representative rostrocaudal levels of the PAG (a-c)

It is becoming increasingly clear that males and females differ in both their sensitivities to pain (Fillingim 2000), as well as in the ability of opioids and other analgesic compounds to alleviate it (Craft 2003). For example, Krzanowska and Bodnar (1999) reported an intra-PAG morphine ED50 of 1.2 μg for male rats in comparison to >50 μg in estrus female rats. Similarly, in a model of chronic inflammatory pain, ED50s for systemic morphine in male rats was 5 mg/kg versus 9 mg/kg in cycling female rats (Wang et al. 2006). As the PAG-RVM circuit is an essential pathway by which morphine produces an analgesic response, sex differences in the anatomical organization and/or steroid regulation of the PAG-RVM pathway may account for sex-dependent differences in opioid analgesia (Loyd and Murphy 2009).

Anatomical Organization of the PAG-RVM Circuit

Neuroanatomical tract tracing techniques using the retrograde tracer fluorogold enable the examination of qualitative and/or quantitative differences in the projection from the PAG to the RVM in males and female rats (Fig. 1). The dorsomedial, lateral, and ventrolateral PAG heavily projects to the RVM in both male and female rats. While there are no qualitative sex differences in the overall distribution of PAG-RVM projection neurons, quantitatively, females have significantly more PAG neurons projecting to the RVM across the rostrocaudal axis of the PAG compared with males (Loyd and Murphy 2006). The average number of retrogradely labeled cells across the rostrocaudal extent of the PAG is greater by a third in female compared with male rats (Fig. 3a). The most prominent sex difference in retrograde labeling is seen within the lateral and ventrolateral regions of the PAG, areas known to contain a dense distribution of mu opioid receptors.

Steroid Modulation of the PAG-RVM Circuit

The PAG contains a large population of both estrogen (ERα) and androgen (AR) receptor–containing neurons (Murphy and Hoffman 1999, 2001). Indeed, this region contains the largest population of steroid receptors outside of the hypothalamus. ERα immunoreactive neurons are localized primarily within the dorsomedial and lateral PAG, and are highly co-distributed among PAG-RVM output neurons (Loyd and Murphy 2008).

Males have a significantly greater number of AR immunoreactive neurons localized within the dorsomedial, lateral, and ventrolateral PAG compared with females and the expression of ERα in the PAG is comparable between the sexes (Fig. 2). Both receptor types increase in density

Sex Differences in Descending Pain Modulatory Pathways, Fig. 2 Color photomicrograph showing a low (**a, c**) and high (**b, d**) power example of single- and double-labeled ERα and FG immunoreactive cells in the lateral PAG (bregma −8.00) of a male (**a, b**) and female rat (**c, d**). Scale bar = 100 μm for low power images; scale bar = 50 μm for high power images

along the rostrocaudal axis of the PAG with the highest expression localized within the caudal PAG. In addition, 30–37 % of PAG neurons projecting to the RVM express AR or ERα with the highest density in the lateral/ventrolateral PAG (Fig. 3d). The high density of steroid receptors localized on the PAG-RVM pathway may provide the biological basis for sex differences in morphine analgesia.

Activation of the PAG-RVM Circuit

In addition to the sexually dimorphic anatomical organization of the PAG, the activation of this circuit during either persistent inflammatory pain (Loyd and Murphy 2006) or systemic morphine administration is similarly dimorphic (Loyd et al. 2007). Figure 3b-c illustrates the percentage of PAG-RVM output neurons that also contained the Fos protein, a marker for neural activation, following CFA (Complete Freund's Adjuvant) injection into the hindpaw or systemic morphine administration (5 mg/kg). Twenty-four hours of inflammatory pain resulted in extensive Fos induction throughout the rostrocaudal axis of PAG. Interestingly, while there are a greater number of PAG-RVM output neurons present in

females versus males, few were activated during inflammatory pain or morphine. In males, over 40 % of PAG-RVM neurons expressed Fos following persistent inflammatory pain (Fig. 3b) or systemic morphine (Fig. 3c), while under 30 % were activated in females. Sex differences in the activation of the PAG-RVM circuit were most prevalent within the lateral and ventrolateral PAG. These areas are the critical PAG sites for either stimulation-produced or opioid-induced analgesia.

Sex Differences in Mu Opioid Receptor Expression in the PAG

The PAG contains a dense population of mu opioid receptor–containing neurons that project to the RVM. Males have a significantly higher density of mu opioid receptors (MORs) in the PAG in comparison to cycling females, with females in the proestrus stage having the lowest density of MORs (Fig. 4). In addition, morphine administered directly to the PAG produces a greater degree of analgesia in males compared with females. Site-directed lesions of PAG MORs (using dermorphin-saporin) dose-dependently reduce morphine analgesia in males

a Mean # PAG-RVM Neurons

♂ 71 93 ♀

96 125

b % Pain-Induced Activation

♂ 42% 29% ♀

44% 27%

c % Morphine-Induced Activation

♂ 46% 19% ♀

44% 19%

d % Expressing Steroid Receptor

AR 35% 37% ERα

30% 37%

Sex Differences in Descending Pain Modulatory Pathways, Fig. 3 (a) Average of the mean number of PAG cells retrogradely labeled from the RVM across the rostrocaudal axis of the PAG in males (*left*) and females (*right*). (b) Percentage of Fos-positive neurons that were retrogradely labeled from the RVM in males (*left*) and females (*right*) following 24 h of inflammation. (c) Average of the percentage of androgen (*left*) and estrogen (*right*) receptor-expressing periaqueductal cells retrogradely labeled from the RVM. (d) Percentage of Fos-positive neurons that were retrogradely labeled from the RVM in males (*left*) and females (*right*) following 24 h of inflammation and 1 h of morphine (5 mg/kg)

only (Loyd et al. 2009). Together, these results indicate that the PAG is both necessary and sufficient for the antinociceptive effects of morphine in males, and that sex differences in MOR expression provide a primary mechanism for the sexually dimorphic actions of morphine.

Summary

Research spanning four decades has shown that the PAG, and its descending projections to the RVM (and from there to the spinal cord dorsal horn), constitutes an essential neural circuit for opioid-based analgesia. During the last half of that period, numerous rodent and human studies have established sex differences in the analgesic effects of morphine. The anatomical and physiological characteristics of the PAG and its descending projections to the RVM are sexually dimorphic, with clear biological consequences in terms of morphine action. Sex differences, in both the anatomical and functional

Sex Differences in Descending Pain Modulatory Pathways, Fig. 4 Example photomicrographs of MOR expression in the ventrolateral PAG of male (a) versus female (b) rats and the mean density of expression across the female rat estrous cycle versus males (c). Also shown is paw withdrawal latency (% maximal possible effect) following MOR lesions (dermorphin-saporin injected into the PAG) versus controls (blank conjugated saporin) following a cumulative morphine dosing paradigm (d)

organization of the PAG-RVM circuit, may contribute to sexually dimorphic pain thresholds and opioid responsiveness.

Acknowledgments This work was supported by NIH grant DA16272.

References

Basbaum, A. I., & Fields, H. L. (1979). The origin of descending pathways in the dorsolateral funiculus of the spinal cord of the cat and rat: Further studies on the anatomy of pain modulation. *The Journal of Comparative Neurology, 187*, 513–531.

Basbaum, A. I., & Fields, H. L. (1984). Endogenous pain control systems: Brainstem spinal pathways and endorphin circuitry. *Annual Review of Neuroscience, 7*, 309–338.

Basbaum, A. I., Clanton, C. H., & Fields, H. L. (1978). Three bulbospinal pathways from the rostral medulla of the cat: An autoradiographic study of pain modulating systems. *The Journal of Comparative Neurology, 178*, 209–224.

Behbehani, M. M. (1995). Functional characteristics of the midbrain periaqueductal gray. *Progress in Neurobiology, 46*, 575–605.

Behbehani, M. M., & Fields, H. L. (1979). Evidence that an excitatory connection between the periaqueductal gray and nucleus raphe magnus mediates stimulation produced analgesia. *Brain Research, 170*, 85–93.

Carstens, E., Yokota, T., & Zimmermann, M. (1979). Inhibition of spinal neuronal responses to noxious skin heating by stimulation of mesencephalic periaqueductal gray in the cat. *Journal of Neurophysiology, 42*, 558–568.

Craft, R. M. (2003). Sex differences in opioid analgesia: "From mouse to man". *The Clinical Journal of Pain, 19*, 175–186.

Fillingim, R. (2000). Sex, gender, and pain: Women and men really are different. *Current Review of Pain, 4*, 24–30.

Jensen, T. S., & Yaksh, T. L. (1986). III. Comparison of the antinociceptive action of mu and delta opioid receptor ligands in the periaqueductal gray matter, medial and paramedial ventral medulla in the rat as studied by microinjection technique. *Brain Research, 372*, 301–312.

Krzanowska, E. K., & Bodnar, R. J. (1999). Morphine antinociception elicited from the ventrolateral periaqueductal gray is sensitive to sex and gonadectomy differences in rats. *Brain Research, 821*, 224–230.

Le Bars, D., Dickenson, A. H., & Besson, J. M. (1980). Microinjection of morphine within nucleus raphe magnus and dorsal horn neurone activities related to nociception in the rat. *Brain Research, 189*, 467–481.

Loyd, D. R., & Murphy, A. Z. (2006). Sex differences in the anatomical and functional organization of the periaqueductal gray-rostral ventromedial medullary pathway in the rat: A potential circuit mediating the sexually dimorphic actions of morphine. *The Journal of Comparative Neurology, 496*(5), 723–738.

Loyd, D. R., & Murphy, A. Z. (2008). Androgen and estrogen (alpha) receptor localization on periaqueductal gray neurons projecting to the rostral ventromedial medulla in the male and female rat. *Journal of Chemical Neurology, 36*(3–4), 216–226.

Loyd, D. R., & Murphy, A. Z. (2009). The role of the periaqueductal gray in the modulation of pain in males and females: are the anatomy and physiology really that different? *Neural Plasticity*, doi 10.1155/2009/462879.

Loyd, D. R., Morgan, M. M., & Murphy, A. Z. (2007). Morphine preferentially activates the periaqueductal gray-rostral ventromedial medullary pathway in the male rat: A potential mechanism for sex differences in antinociception. *Neuroscience, 147*(2), 456–468.

Loyd, D. R., Wang, X., & Murphy, A. Z. (2009). Sex differences in micro-opioid receptor expression in the rat midbrain periaqueductal gray are essential for eliciting sex differences in morphine analgesia. *Journal of Neuroscience, 28*(52), 14007–14017.

Murphy, A. Z., & Hoffman, G. E. (1999). Distribution of androgen and estrogen receptor containing neurons in the male rat periaqueductal gray. *Hormones and Behavior, 36*, 98–108.

Murphy, A. Z., & Hoffman, G. E. (2001). Distribution of gonadal steroid receptor-containing neurons in the preoptic-periaqueductal gray-brainstem pathway: a potential circuit for the initiation of male sexual behavior. *The Journal of Comparative Neurology, 438*(2), 191–212.

Reynolds, D. V. (1969). Surgery in the rat during electrical analgesia induced by focal brain stimulation. *Science, 164*, 444–445.

Wang, X., Traub, R. J., & Murphy, A. Z. (2006). Persistent pain model reveals sex difference in morphine potency. *American Journal of Physiology. Regulatory, Integrative and Comparative Physiology, 291*(2), R300–R306.

Sex Differences in Opioid Analgesia

Rebecca M. Craft
Psychology, Washington State University, Pullman, WA, USA

Synonyms

Gender differences in opioid analgesia (Hughes 2003)

Definition

Opioid analgesia that differs in potency, efficacy, time course, or some other variable, between males and females (defined anatomically or via sex chromosome).

Characteristics

Recent reviews of epidemiological, clinical, and experimental studies indicate that women suffer disproportionately more pain than men. A majority of experimental studies in humans report that women have lower pain thresholds and lower pain tolerance than men for many though not all types of noxious stimuli (Berkley 1997). Women also report more endogenous pains, and a number of painful clinical syndromes, such as migraine, temporomandibular disorder, rheumatoid arthritis, fibromyalgia, and irritable bowel syndrome, are more prevalent in women than in men (Berkley 1997; Unruh 1996; Fillingim 2000). The existence of sex differences in pain and the fact that the developmental profile of some types of pain clearly parallels reproductive function strongly suggest that gonadal steroid hormones significantly influence pain (Fillingim 2000; Riley et al. 1999; Craft et al. 2004).

Given the significant roles of endogenous opioids and opioid receptors in pain and analgesia, it is not surprising that sex differences in opioid analgesia have been observed. Greater opioid analgesia in women than men was initially reported in a series of studies conducted in dental pain patients (e.g., Gear et al. 1996, 2000). In these studies, the mixed-action opioid agonists pentazocine, nalbuphine, and butorphanol were found to be longer acting and/or more effective in women than in men. A subsequent review of sex difference studies in humans concluded that the finding of greater opioid analgesia in women than men was more likely to be observed (1) for morphine rather than for other opioids and (2) when analgesia was assessed over days (typically in studies in which morphine was self-administered via patient-controlled analgesia) rather than acutely (Niesters et al. 2010). In contrast to studies conducted in humans, rat studies conducted over the past 20 years often report that mu-opioid agonists such as morphine are more potent, and in some cases more efficacious, in *males* than in females (Craft 2003a). Several studies report no sex differences in opioid analgesia in rodents (particularly mice); however, almost none report greater opioid analgesia in females compared to males (e.g., Kest et al. 1999). The finding of greater opioid analgesia in male compared to female rats extends to opioids with mixed (mu/kappa) actions and to selective kappa opioid agonists (Craft 2003a). Sex differences in opioid analgesia do not appear to depend on the type of acute noxious stimulus used in the pain test (thermal, mechanical, electrical, chemical), although they do depend on the intensity of the noxious stimulus (particularly when lower efficacy agonists are examined) and the hormonal state of rats at the time of testing (Craft 2003a). For example, females in vaginal estrus (or ovariectomized females chronically replaced with estradiol) are typically the least sensitive to morphine analgesia (Craft et al. 2004).

The opposite sex difference in morphine analgesia observed in humans (females more sensitive) vs. rats (males more sensitive) may be explained by variables that typically differ between human and rat studies. For example, nearly all rat studies of sex differences in opioid analgesia have used models of acute pain in which the opioid is administered once by the experimenter; in contrast, most human studies that demonstrate sex differences have examined analgesia after patients self-administered opioid for hours to days. Thus, the greater analgesia observed in women could be due to lesser tolerance development in women compared to men. Lesser morphine tolerance development in females compared to males has been reported in rats (Craft et al. 1999). To address possible species differences that may be due simply to differential testing parameters, it will be important in future rodent studies to model the human condition more closely, by comparing self-administration of opioid analgesics in males vs. females postsurgically or under other conditions of chronic pain.

Mechanisms underlying sex differences in opioid analgesia are not yet well understood. Gonadal steroid hormones such as estradiol and testosterone have been shown to act both organizationally (during development) and activationally (in the adult) to influence sensitivity to opioid analgesia in male and female rats (Craft et al. 2004). Gonadal steroids may influence opioid analgesia by modulating opioid pharmacokinetics and/or pharmacodynamics. In rodents, gonadal hormones have been shown to influence opioid disposition and metabolism, as well as opioid receptor density and signal transduction, at least in some brain areas (Craft et al. 2004; Craft 2003a). In humans, there is little evidence for sex differences in opioid pharmacokinetics, but one group has demonstrated that brain mu-opioid receptor binding potential differs between men and women exposed to a painful stimulus (Zubieta et al. 2002) and that brain mu-opioid receptor binding potential is estradiol dependent in women (Smith et al. 2006). These findings suggest that sex differences in opioid receptor density and endogenous opioid release contribute to sex differences in both pain and opioid analgesic sensitivity.

The clinical impact of these findings is not yet known. Several labs have now demonstrated that genotype is a significant moderating factor in sex difference studies in rodents (Kest et al. 1999; Mogil et al. 2000, 2003; Cook et al. 2000), and this appears to be the case in humans as well (Mogil et al. 2003; Olsen et al. 2012). Of further potential clinical relevance, some rodent and human studies report that side effects of opioids such as respiratory depression, sedation, and euphoria also differ between the sexes (Craft 2003a). Because side effects often limit the therapeutic use of opioid analgesics, they will be important to consider in determining whether opioid analgesics should be used according to "sex-specific guidelines" in the future (Craft et al. 2004; Craft 2003b).

References

Berkley, K. J. (1997). Sex differences in pain. *Behavioral Brain Science, 20*, 371–380.

Cook, C. D., Barrett, A. C., Roach, E. L., Bowman, J. R., & Picker, M. J. (2000). Sex-related differences in the antinociceptive effects of opioids: Importance of rat genotype, nociceptive stimulus intensity, and efficacy at the mu opioid receptor. *Psychopharmacology, 150*, 430–442.

Craft, R. M. (2003a). Sex differences in opioid analgesia: "From mouse to man". *The Clinical Journal of Pain, 19*, 175–186.

Craft, R. M. (2003b). Sex differences in drug- and non-drug-induced analgesia. *Life Sciences, 72*, 2675–2688.

Craft, R. M., Stratmann, J. A., Bartok, R. E., Walpole, T. I., & King, S. J. (1999). Sex differences in development of morphine tolerance and dependence in the rat. *Psychopharmacology, 143*, 1–7.

Craft, R. M., Mogil, J. S., & Aloisi, A. M. (2004). Sex differences in pain and analgesia: The role of gonadal hormones. *European Journal of Pain, 8*, 397–411.

Fillingim, R. B. (Ed.). (2000). *Sex, gender, and pain.* Seattle: IASP Press.

Gear, R. W., Miaskowski, C., Gordon, N. C., Paul, S. M., Heller, P. H., & Levine, J. D. (1996). Kappa-opioids produce significantly greater analgesia in women than in men. *Nature Medicine, 2*, 1248–1250.

Gear, R. W., Miaskowski, C., Gordon, N. C., Paul, S. M., Heller, P. H., & Levine, J. D. (2000). Action of naloxone on gender-dependent analgesic and antianalgesic effects of nalbuphine in humans. *The Journal of Pain, 1*, 122–127.

Hughes, R. N. (2003). The categorization of male and female laboratory animals in terms of "gender". *Brain Research Bulletin, 60*, 189–190.

Kest, B., Wilson, S. G., & Mogil, J. S. (1999). Sex differences in supraspinal morphine analgesia are dependent on genotype. *Journal of Pharmacology and Experimental Therapeutics, 289*, 1370–1375.

Mogil, J. S., Chesler, E. J., Wilson, S. G., Juraska, J. M., & Sternberg, W. F. (2000). Sex differences in thermal nociception and morphine antinociception in rodents depend on genotype. *Neuroscience and Biobehavioral Reviews, 24*, 375–389.

Mogil, J. S., Wilson, S. G., Chesler, E. J., Rankin, A. L., Nemmani, K. V. S., Lariviere, W. R., Groce, M. K., Wallace, M. R., Kaplan, L., Staud, R., Ness, T. J., Glover, T. L., Stankova, M., Mayorov, A., Hruby, V. J., Grisel, J. E., & Fillingim, R. B. (2003). The melanocortin-1 receptor gene mediates female-specific mechanisms of analgesia in mice and humans. *Proceedings of the National Academy of Sciences, 100*, 4867–4872.

Niesters, M., Dahan, A., Kest, B., Zacny, J., Stijnen, T., Aarts, L., & Sarton, E. (2010). Do sex differences exist in opioid analgesia? A systematic review and meta-analysis of human experimental and clinical studies. *Pain, 151*, 61–68.

Olsen, M. B., Jacobsen, L. M., Schistad, E. I., Pedersen, L. M., Rygh, L. J., Røe, C., & Gjerstad, J. (2012). Pain intensity the first year after lumbar disc herniation is associated with the A118G polymorphism in the opioid receptor mu 1 gene: Evidence of a sex and genotype interaction. *The Journal of Neuroscience, 32*, 9831–9834.

Riley, J. L., Robinson, M. E., Wise, E. A., & Price, D. D. (1999). A meta-analytic review of pain perception across the menstrual cycle. *Pain, 81*, 225–235.

Smith, Y. R., Stohler, C. S., Nichols, T. E., Bueller, J. A., Koeppe, R. A., & Zubieta, J.-K. (2006). Pronociceptive and antinociceptive effects of estradiol through endogenous oppid neurotransmission in women. *The Journal of Neuroscience, 26*, 5777–5785.

Unruh, A. M. (1996). Gender variations in clinical pain experience. *Pain, 65*, 123–167.

Zubieta, J.-K., Smith, Y. R., Bueller, J. A., Xu, Y., Kilbourn, M. R., Jewett, D. M., Meyer, C. R., Koeppe, R. A., & Stohler, C. S. (2002). Mu-opioid receptor-mediated antinociceptive responses differ in men and women. *The Journal of Neuroscience, 22*, 5100–5107.

Sexual Abuse

Definition

Forcing or enticing another person to take part in sexual activities, whether or not they are aware of what is happening. The activities may involve physical contact, including penetrative (e.g., rape or buggery) and non-penetrative acts. They may include non-contact activities, such as involving another person in looking at, or in the production of, pornographic material or watching sexual activities or encouraging another person to behave in sexually inappropriate ways.

Cross-References

▶ Chronic Pelvic Pain, Physical and Sexual Abuse

Sexual Activity Headache

Definition

Headache associated with sexual activity may be primary, i.e., without any external cause and can be seen both with excitement and with orgasm. Orgasmic headache should always make one consider subarachnoid hemorrhage.

Cross-References

▶ Primary Exertional Headache

Sexual Dysfunctions

▶ Gynecological Pain and Sexual Functioning

Sexual Pain Disorder

▶ Dyspareunia and Vaginismus

Sexual Response

Definition

There are four phases involved in the sexual response: desire, excitement, orgasm, and

resolution. Desire is also referred to as libido. Excitement is a state of arousal that often occurs as a result of physical contact or emotional readiness. During excitement the body responds by increasing the heart rate, blood pressure, and respiratory rate. Blood flow increases to the genitals in both men and women. In women, this creates a state of engorgement in the labia and surrounding structures, as well as lubrication of the vagina. In men, this results in an erection. The third phase, orgasm, is also called climax, where muscles in the genitalia and pelvic region contract. In males this is accompanied by ejaculation. Finally, resolution is the return of the body to its original state.

Cross-References

▶ Cancer Pain Management, Opioid Side Effects, Endocrine Changes, and Sexual Dysfunction

SF-36 Health Status Questionnaire

Definition

The SF-36 is a multipurpose, short-form health survey with 36 questions. It consists of an eight-scale profile of functional health and well-being scores, psychometrically-based physical and mental health summary measures, and a preference-based health utility index. It is a generic measure, in contrast to one that targets a specific age, disease, or treatment group. The SF-36 has been used in surveys of general and specific populations, comparing the relative burden of diseases, and in differentiating the health benefits produced by a wide range of different treatments.

Cross-References

▶ Postoperative Pain, Preoperative Education

SFEMG

▶ Single-Fiber Electromyography

SF-MPQ

Definition

Short-Form McGill Pain Questionnaire.

Cross-References

▶ McGill Pain Questionnaire

SF-MPQ-2

▶ McGill Pain Questionnaire

Sham

▶ Placebo

Shear Stress

Definition

Shear stress is applied to vascular endothelial cells by the force of laminar blood flow along their surface. This induces COX-2 in the cells, which supplies endoperoxides for the synthesis of prostacyclin (PGI_2).

Cross-References

▶ Cyclooxygenases in Biology and Disease

Shingles Pain

▶ Postherpetic Neuralgia, Etiology, Pathogenesis, and Management

Short-Lasting Unilateral Neuralgiform Headache Attacks with Conjunctival Injection and Tearing

▶ SUNCT Syndrome

Shortwave Diathermy

▶ Modalities

Shoulder Pain

Definition

Nonspecific shoulder pain is a prominent feature of polymyalgia rheumatica.

Cross-References

▶ Muscle Pain in Systemic Inflammation (Polymyalgia Rheumatica, Giant Cell Arteritis, Rheumatoid Arthritis)

Shoulder-Hand Syndrome

▶ CRPS, Evidence-Based Treatment
▶ Postoperative Pain, Acute Presentation of Complex Regional Pain Syndrome

SIA

Definition

Stress-Induced Analgesia.

Cross-References

▶ Acupuncture Mechanisms

Sicca Syndrome

Definition

Dryness of the mouth and eyes may be a sign of Sjögren's syndrome, in which salivary glands are the main target for the autoimmune disease.

Cross-References

▶ Muscle Pain in Systemic Inflammation (Polymyalgia Rheumatica, Giant Cell Arteritis, Rheumatoid Arthritis)

Sick Role

Definition

A role legitimized by medical sanction, which by its "special status" relieves the patient of the usual demands and obligations and takes priority over other social roles (e.g., occupation, familial). In children, these behaviors may be learned or reinforced by parental or family responses.

Cross-References

▶ Impact of Familial Factors on Children's Chronic Pain
▶ Malingering, Primary and Secondary Gain
▶ Pain as a Cause of Psychiatric Illness

Sickle Cell Disease

Definition

Sickle cell disease is an inherited, chronic condition in which the red blood cells, which are normally disc shaped, become crescent shaped. As a result, the red blood cells function abnormally and cause small blood clots. These clots give rise to recurrent painful episodes called "sickle cell pain crises." Sickle cell disease occurs most commonly in African Americans or in people of Mediterranean descent. Its severity can range from mild to life threatening.

Cross-References

▶ Experimental Pain in Children

Side Effects

▶ Postoperative Pain, Adverse Events (Associated with Acute Pain Management)

Sign of Formication

▶ Tinel Sign

Signal Detection Analysis/Theory

Definition

Signal detection analysis is used to explain how decisions are made in uncertain or ambiguous situations, e.g., judgments concerning the presence or absence of a stimulus at threshold intensity. Psychophysics, i.e., the characterization of the relationship between a physical stimulus and its subjective perception, represents an important area of application for signal detection analysis. Signal detection theory assumes that a person's response to stimulation is influenced both by the ability to discriminate between stimuli and by a response bias, i.e., subjective criteria used to label a stimulus, for example, as painful versus not painful. Signal detection analysis provides methods on how to estimate these two response components based on the person's response pattern across trials.

Cross-References

▶ Modeling, Social Learning in Pain

Signaling

Definition

The process of conditioning an infant to anticipate the occurrence of a painful event based on the repeated pairing of a stimulus with the painful event.

Cross-References

▶ Acute Pain Management in Infants

Signaling Molecules of Thalamic Regions

▶ Thalamic Neurotransmitters and Neuromodulators

Signals from Injured Neurons that Modulate Microglia

Margarita Calvo[1] and David L. H. Bennett[1,2]
[1]Wolfson CARD, Kings College London, London, UK
[2]Nuffield Department of Clinical Neurosciences, John Radcliffe Hospital, The University of Oxford, Oxford, UK

Introduction

Neuronal Signals Regulating the Microglial Response to Peripheral Nerve Damage

Microglia are the resident macrophages and the only immune cells in the central nervous system (CNS). These cells are derived from myeloid precursor cells, which enter the developing CNS during embryogenesis (Ransohoff and Perry 2009; Ginhoux et al. 2010). Under the influence of the CNS microenvironment, microglia develop fine and long processes and protrusions which are continually surveying their microenvironment (Nimmerjahn et al. 2005). Local cell division at a very low level maintains the number of resident microglia in the brain of rodents (Lawson et al. 1992). When the tightly regulated CNS homeostasis is disturbed by any kind of insult, microglia rapidly change their morphology and gene expression profile (reviewed in Hanisch and Kettenmann 2007). This response of microglia (or "microgliosis") has a number of different components including microglial proliferation and migration to accumulate at regions of neuronal degeneration, phagocytosis of cellular debris, antigen presentation, adoption of an amoeboid morphology, and the release of a wide variety of pro-inflammatory molecules (reviewed in Calvo and Bennett 2012). Production of such pro-inflammatory substances including IL-1β, TNF-α, NO, BDNF, or IL-6, can further recruit microglia, activate astrocytes, and increase neuronal excitability (reviewed in Milligan and Watkins 2009). The presence of "reactive" microglia has been documented in almost every kind of CNS injury or disease (reviewed in Hanisch and Kettenmann 2007). Indeed, injury to a peripheral nerve results in a marked microgliosis within the dorsal horn of the spinal cord and this contributes to the development of neuropathic pain (see Fig. 1, Tsuda et al. 2003). In this entry, we focus on how signals derived from injured peripheral neurons drive this microglial response.

Characteristics

Microgliosis is Dependent on Primary Afferent Derived Injury Signals

Following peripheral nerve injury, the accumulation of reactive microglia into the site of injury is orchestrated through different signals which appear in the spinal cord microenvironment or which are released from neurons, astrocytes, and other immune cells. For instance, purines are believed to "leak" from injured neurons and cells, and through activation of P2X4, P2X7, P2Y6, and P2Y12 receptors in microglia can attract these cells and induce a pro-inflammatory

Signals from Injured Neurons that Modulate Microglia, Fig. 1 *Microgliosis in the spinal cord after L5 SNL.* The microglial response to nerve injury develops during the first week following injury. Here we show a representative section of the spinal cord of an animal three days after SNL that has been immunostained with Iba1 (a microglial marker). (**a**) An increase in microglial numbers is observed in the ipsilateral dorsal and ventral horn of the spinal cord of L5 SNL animals. (**b-c**) A closer view shows that microglia change their morphology from the uninjured state in which they have long processes and a small soma (termed surveying microglia, in **b**) to having retracted processes and a hypertrophic soma (termed an effector or activated morphology 3 days following L5 SNL, in c). Scale bars: **a** = 500 μm, in **b** and **c** = 50 μm

phenotype (reviewed in Tsuda et al. 2010 and in Beggs et al. 2007). The activation of the Toll-like receptors 2, 3, and 4 in microglia by a yet-to-be-identified endogenous ligand leads to cytokine release and further activation of these cells (Tanga et al. 2005; Obata et al. 2008; Kim et al. 2007). Interferon-γ (IFN-γ) probably secreted from infiltrating T cells acts in microglia upregulating the MHC-II antigen presenting pathway and thus further promoting CD4+ T-cell recruitment (Tsuda et al. 2009; Neumann 2001).

The complement cascade is also activated following peripheral nerve injury and its components could act as potent chemoattractants to microglia (Griffin et al. 2007).

Many heterogeneous signals therefore appear to contribute to microgliosis; however, evidence suggests that these are initially triggered by the injury response of damaged primary afferents. Nociceptive inputs from damaged or neighboring uninjured sensory neurons projecting to the dorsal horn appear to be critical in the development of spinal microgliosis. In a study using stereological techniques, the spatial distribution of microgliosis after spared nerve injury was carefully investigated. Accumulation of microglia mainly occurred at the central territory of peripherally axotomized afferents although it could also extend into areas of uninjured terminal afferents. Nevertheless, this microgliosis was always restricted to the central terminals of the intact branch of the sciatic nerve. The central terminals of adjacent but independent nerves (such as the saphenous) remained unaffected (Beggs and Salter 2007). An injury signal is therefore conveyed to the spinal cord by damaged primary afferents and neighboring uninjured afferents which share the same nerve trunk. Direct evidence that microgliosis occurs as a consequence of the altered functional state of sensory neurons and not just as a result of degeneration of nerve terminals is provided by comparing the effects of dorsal rhizotomy (i.e., transection of the dorsal root proximal to the DRG) to those of peripheral nerve axotomy. After dorsal rhizotomy microgliosis occurs but it is predominantly concentrated in the dorsal funiculus (and not the dorsal horn) at lumbar and thoracic levels, most likely as part of the immune response to Wallerian degeneration of the lesioned axons (Colburn et al. 1999). In contrast, after peripheral nerve axotomy, accumulation of reactive microglia is seen at the areas where injured and neighboring uninjured afferents fibers terminate (Beggs and Salter 2007) within the dorsal horn. It is very likely that either ongoing activity and subsequent neurotransmitter release from primary afferent terminals or transported factors from the cell body of these sensory neurons may be providing microglia with activation signals.

A number of studies have examined the relative contribution of activity in different classes of sensory neurons to microgliosis. Intense but nondamaging electrical stimulation of naïve C fibers is capable of inducing central sensitization and can elicit microgliosis in the areas where these fibers terminate. This C-fiber stimulation and subsequent microgliosis are sufficient to evoke significant and long-lasting changes in mechanical sensitivity (Hathway et al. 2009). The development of microgliosis following nerve injury is not solely dependent on C-fiber activity. Blockade of the peripheral input of TRPV1-positive fibers (Aδ and C fibers) was not enough to prevent nerve injury-induced spinal microgliosis. However, bupivacaine, which induces a complete sensory blockade, resulted in a significant inhibition of the microglial response. Therefore, the peripheral input from large myelinated fibers seems to play an important role in the development of microgliosis (Suter et al. 2009).

Neuroglial Signaling Molecules

In the introduction to this entry, we discussed some well-known signaling pathways which modulate microglial function; a number of these pathways will be further described in other entries of this section. In this entry, we focus on a number of recently described signaling pathways which enable direct communication between injured primary afferents and microglia. An emerging theme is that chemokines – a family of proteins structurally defined by the presence of four conserved cysteine residues and first described for their chemotactic activity on immune cells – provide a potent means of neuroglial signaling. A number of these pathways involve complex two-way relationships between injured neurons and microglia, and in many cases injury modulates the activity-dependent release of secreted proteins or promotes the cleavage of a transmembrane precursor.

Fractalkine

The chemokine fractalkine is expressed on neurons in the dorsal horn of the spinal cord and the CX3CR1 receptor for fractalkine is exclusively expressed by microglia in the spinal cord (Verge et al. 2004). Fractalkine expression and release from neurons is induced by nerve injury; the chemokine domain of fractalkine can be shed from neurons via the protease cathepsin S which is secreted by microglia in a P2X7-dependent way (Clark et al. 2007). In neuropathic dorsal horn slices ex vivo, noxious electrical stimulation of dorsal roots evokes soluble fractalkine release and high levels of soluble fractalkine are detected in the CSF of neuropathic animals

(Clark et al. 2009). The fractalkine receptor CX3CR1 causes phosphorylation of p38MAPK in microglia (Clark et al. 2007; Zhuang et al. 2007) and leads to IL-1β, IL-6, and nitric oxide release (Milligan et al. 2005). Furthermore, intrathecal administration of the chemokine domain of fractalkine results in thermal and mechanical hypersensitivity that can be prevented by the absence of the CX3CR1 receptor or by pretreatment with neutralizing antibodies against CX3CR1 or fractalkine itself (Milligan et al. 2005; Clark et al. 2007).

In summary, following peripheral nerve injury, spinal reactive microglia can release cathepsin S which enzymatically cleaves neuronal fractalkine. This soluble fractalkine binds microglial CX3CR1 receptors leading to a further release of pro-inflammatory mediators and establishing a positive feedback which may contribute to a chronic pain state. This pathway will be looked in more detail in another entry.

CCL2

The chemokine monocyte chemoattractant protein 1 (MCP-1, also known as CCL2) binds with high affinity to the chemokine receptor CCR2, a G-protein-coupled receptor, and stimulates monocyte migration. CCL2 is constitutively expressed in small and medium diameter size neurons in the dorsal root ganglia (DRG) under naïve conditions and following peripheral nerve injury it is upregulated and transported to the central terminals of primary afferents in the dorsal horn (Abbadie et al. 2003). CCR2 is normally expressed on cells of monocyte/macrophage lineage in the periphery and can be upregulated in spinal microglia by peripheral nerve injury. Exogenous spinal administration of CCL2 induces spinal microgliosis and mechanical allodynia, both effects which are dependent on CCR2 expression. Similarly, following peripheral nerve injury, CCR2 knockout mice fail to develop tactile allodynia (Zhang et al. 2007; reviewed in Abbadie et al. 2009). CCL2 is probably a key signaling molecule in chemoattraction of microglia within the spinal cord and may even be important in the recruitment of macrophages at the site of peripheral injury.

Metalloproteinase-9

Another neuronally derived molecule that may be important for the development of microgliosis is the metalloproteinase-9 (MMP-9). After peripheral nerve injury, MMP-9 shows a rapid and transient upregulation in injured DRG primary sensory neurons. Spinal MMP-9 injection resulted in increased expression of CD11b (microglial surface marker) and in p38MAPK phosphorylation and induced mechanical hypersensitivity in naïve animals. Following peripheral nerve injury, MMP-9 inhibition or knockdown of this protein decreased microgliosis and mechanical allodynia (Kawasaki et al. 2008). Substrates of MMP-9 that could recruit microglia into a proinflammatory phenotype are unclear but IL-1β is a potential candidate (Kawasaki et al. 2008). Indeed, spinal MMP-9 treatment in naïve animals increases the cleaved forms of IL-1β in the DRG. Therefore, the hypothesis is that neuronal MMP-9 is secreted after peripheral nerve injury and cleaves pro-inflammatory mediators such as IL-1β. These in turn promote microgliosis and increase the excitability of dorsal horn neurons leading to a hypersensitive state.

Neuregulin-1

NRG1 is one of a family of growth factors (*NRG1-4*) which has a key role in neural development, it can modulate synaptic plasticity and stimulate the proliferation, survival, and motility of a number of different cell types (reviewed in Mei and Xiong 2008). *Nrg1* is expressed in sensory neurons and undergoes alternative splicing and differential promoter use to produce at least 15 isoforms, which include both secreted and transmembrane forms (Esper et al. 2006). All *Nrg1* isoforms contain an epidermal growth factor (EGF)-like domain, which is critical for mediating biologic activity and binds to the tyrosine kinase receptors erbB3 and erbB4. These receptors subsequently heterodimerize with erbB2, which lacks a ligand-binding domain but is a key coreceptor in mediating signal transduction. ErbB3 lacks an active tyrosine kinase domain, and so its interaction with erbB2 is essential for signaling. ErbB4 can signal via homodimers but dimerizes preferentially with erbB2.

Microglia express the NRG1 receptors erbB2, 3, and 4. We have recently shown that NRG1 is a survival, proliferative, and chemotactic factor for microglia in vitro and in addition can promote the release of Il-1β from these cells. Consequent with the effects of NRG1 in cultured microglia, we and others have observed that intrathecal treatment of naive animals with NRG1 induced microglial proliferation, adoption of an effector morphology, and development of cold and mechanical pain-related hypersensitivity (Calvo et al. 2010; Lacroix-Fralish et al. 2008). Primary afferents express a number of NRG1 isoforms, and in culture systems active NRG1 can be released on activity and in response to neurotrophin treatment. We have found that in vivo following peripheral nerve injury, NRG1 is released from primary afferents in the dorsal horn and activates the erbB2 receptors in microglia. Phosphorylation of erbB2 leads to the activation of the MEK/ERK1/2 pathway, which triggers a mitotic response in microglia. Microglia accumulation at the site of injury in the spinal cord and the associated cytokine release leads to increased excitability of dorsal horn neurons and thus contributes to the generation of neuropathic pain. Blockade of erbB2 receptor or inhibition of the MEK/ERK1/2 pathway or sequestration of endogenous NRG1 following peripheral nerve injury have been shown to reduce microgliosis (Calvo et al. 2010, 2011). This reduction in microgliosis is accompanied by a reduction in mechanical and cold pain-related hypersensitivity following nerve injury.

Therefore, NRG1 secreted from injured primary afferent terminals is a key regulator of microgliosis in the dorsal horn in the context of peripheral axotomy (see Fig. 2).

Signals from Injured Neurons that Modulate Microglia, Fig. 2 *NRG1 secreted from terminal afferents after nerve injury promote microglia proliferation, chemotaxis, and secretion of cytokines.* After peripheral nerve injury, NRG1 is released from primary afferent terminals in the dorsal horn of the spinal cord. NRG1 binds the erbB3 or 4 receptor, which heterodimerize with the erbB2 receptors present on microglia. Phosphorylation of erbB2 leads to the activation of the MEK/ERK1/2 pathway, which triggers a proliferative response in microglia. NRG1 also induces chemoattraction and enhances the release of IL-1β from these cells. Microglia accumulation at the site of injury in the spinal cord and the associated cytokine release leads to increased excitability of dorsal horn neurons and thus contributes to the generation of neuropathic pain

Summary

Conclusions

Damage to a peripheral nerve triggers a microglial response in the spinal cord which contributes to the generation of neuropathic pain. Microglia in the spinal cord accumulate at the sites were sensory afferents from peripherally axotomized axons terminate and change their phenotype to secrete pro-inflammatory cytokines. These molecules contribute to create a hypersensitive state in spinal cord neurons which will manifest as pain-related behaviors in the injured animal. Many heterogeneous signals appear to contribute to microgliosis; however, evidence suggests that signals from damaged or neighboring uninjured sensory neurons appear to be critical in the development of spinal microgliosis. In this entry, we described some of the recently characterized forms of neuroglial communication. Fractalkine is cleaved from spinal cord neurons and increases the release of pro-inflammatory cytokines from microglia. CCL-2 is upregulated in DRG cells after peripheral axotomy and is an important signal for recruiting microglia apparently through chemoattraction of these cells. The metalloproteinase MMP-9 is also upregulated in DRG cells after axotomy and probably through cleavage, and activation of cytokines recruits microglia to an inflammatory response. NRG-1 is expressed in DRG neurons and following peripheral nerve injury is released by terminal afferents in the spinal cord and promotes microglial proliferation and cytokine release. Therefore, following nerve injury, signals from neurons are vital in orchestrating the microglial response, which plays a critical role in the generation of neuropathic pain. Targeting such molecules therapeutically will present a challenge because of their pleiotropic effects. For instance, NRG1 is not only important in establishing the microglial response to nerve injury but also is essential for effective nerve repair. The microglial response to injury may not always be deleterious. In certain model systems of CNS injury, these cells can be neuroprotective (Hanisch and Kettenmann 2007) and so depending on the specific activating conditions microglia may promote either harmful or beneficial outcomes. Treatments should therefore be aimed at modulating the microglial response toward a "healing anti-inflammatory" phenotype rather than completely ablating the function of these cells.

References

Abbadie, C., Bhangoo, S., De Koninck, Y., Malcangio, M., Melik-Parsadaniantz, S., & White, F. A. (2009). Chemokines and pain mechanisms. *Brain Research Reviews, 60*(1), 125–134.

Abbadie, C., Lindia, J. A., Cumiskey, A. M., Peterson, L. B., Mudgett, J. S., Bayne, E. K., DeMartino, J. A., MacIntyre, D. E., & Forrest, M. J. (2003). Impaired neuropathic pain responses in mice lacking the chemokine receptor CCR2. *Proceedings of the National Academy of Sciences of the United States of America, 100*, 7947–7952.

Beggs, S., & Salter, M. W. (2007). Stereological and somatotopic analysis of the spinal microglial response to peripheral nerve injury. *Brain, Behavior, and Immunity, 21*, 624–633.

Beggs, S., Trang, T., & Salter, M. (2007). The role of ATP and microglia in enhanced pain states. In J. A. DeLeo, L. S. Sorkin, & L. R. Watkins (Eds.), *Immune and glial regulation of pain* (pp. 283–296). Seattle: IASP Press.

Calvo, M., & Bennett, D. L. (2012). The mechanisms of microgliosis and pain following peripheral nerve injury. *Experimental Neurology, 234*(2), 271–282.

Calvo, M., Zhu, N., Grist, J., Ma, Z., Loeb, J. A., & Bennett, D. L. (2011). Following nerve injury neuregulin-1 drives microglial proliferation and neuropathic pain via the MEK/ERK pathway. *Glia, 59*(4), 554–568.

Calvo, M., Zhu, N., Tsantoulas, C., Ma, Z., Grist, J., Loeb, J. A., & Bennett, D. L. (2010). Neuregulin-ErbB signaling promotes microglial proliferation and chemotaxis contributing to microgliosis and pain after peripheral nerve injury. *Journal of Neuroscience, 30*(15), 5437–5450.

Clark, A. K., Yip, P. K., Grist, J., Gentry, C., Staniland, A. A., Marchand, F., Dehvari, M., Wotherspoon, G., Winter, J., Ullah, J., Bevan, S., & Malcangio, M. (2007). Inhibition of spinal microglial cathepsin S for the reversal of neuropathic pain. *Proceedings of the National Academy of Sciences of the United States of America, 104*, 10655–10660.

Clark, A. K., Yip, P. K., & Malcangio, M. (2009). The liberation of fractalkine in the dorsal horn requires microglial cathepsin S. *Journal of Neuroscience, 29*(21), 6945–6954.

Colburn, R. W., Rickman, A. J., & DeLeo, J. A. (1999). The effect of site and type of nerve injury on spinal glial activation and neuropathic pain behavior. *Experimental Neurology, 157*, 289–304.

Esper, R. M., Pankonin, M. S., & Loeb, J. A. (2006). Neuregulins: Versatile growth and differentiation factors in nervous system development and human disease. *Brain Research Reviews, 51*, 161–175.

Ginhoux, F., Greter, M., Leboeuf, M., Nandi, S., See, P., Gokhan, S., Mehler, M. F., Conway, S. J., Ng, L. G., Stanley, E. R., Samokhvalov, I. M., & Merad, M. (2010). Fate mapping analysis reveals that adult microglia derive from primitive macrophages. *Science, 330*(6005), 841–845.

Griffin, R. S., Costigan, M., Brenner, G. J., Ma, C. H., Scholz, J., Moss, A., Allchorne, A. J., Stahl, G. L., & Woolf, C. J. (2007). Complement induction in spinal cord microglia results in anaphylatoxin C5a-mediated pain hypersensitivity. *Journal of Neuroscience, 27*, 8699–8708.

Hanisch, U. K., & Kettenmann, H. (2007). Microglia: Active sensor and versatile effector cells in the normal and pathologic brain. *Nature Neuroscience, 10*, 1387–1394.

Hathway, G. J., Vega-Avelaira, D., Moss, A., Ingram, R., & Fitzgerald, M. (2009). Brief, low frequency stimulation of rat peripheral C-fibres evokes prolonged microglial-induced central sensitization in adults but not in neonates. *Pain, 144*, 110–118.

Kawasaki, Y., Xu, Z. Z., Wang, X., Park, J. Y., Zhuang, Z. Y., Tan, P. H., Gao, Y. J., Roy, K., Corfas, G., Lo, E. H., & Ji, R. R. (2008). Distinct roles of matrix metalloproteases in the early- and late-phase development of neuropathic pain. *Nature Medicine, 14*, 331–336.

Kim, D., Kim, M. A., Cho, I. H., Kim, M. S., Lee, S., Jo, E. K., Choi, S. Y., Park, K., Kim, J. S., Akira, S., Na, H. S., Oh, S. B., & Lee, S. J. (2007). A critical role of toll-like receptor 2 in nerve injury-induced spinal cord glial cell activation and pain hypersensitivity. *Journal of Biological Chemistry, 282*, 14975–14983.

Lacroix-Fralish, M. L., Tawfik, V. L., Nutile-McMenemy, N., & DeLeo, J. A. (2008). Neuregulin 1 is a pronociceptive cytokine that is regulated by progesterone in the spinal cord: Implications for sex specific pain modulation. *European Journal of Pain, 12*, 94–103.

Lawson, L. J., Perry, V. H., & Gordon, S. (1992). Turnover of resident microglia in the normal adult mouse brain. *Neuroscience, 48*, 405–415.

Mei, L., & Xiong, W. C. (2008). Neuregulin 1 in neural development, synaptic plasticity and schizophrenia. *Nature Reviews Neuroscience, 9*, 437–452.

Milligan, E. D., & Watkins, L. R. (2009). Pathological and protective roles of glia in chronic pain. *Nature Reviews Neuroscience, 10*, 23–36.

Milligan, E., Zapata, V., Schoeniger, D., Chacur, M., Green, P., Poole, S., Martin, D., Maier, S. F., & Watkins, L. R. (2005). An initial investigation of spinal mechanisms underlying pain enhancement induced by fractalkine, a neuronally released chemokine. *European Journal of Neuroscience, 22*, 2775–2782.

Neumann, H. (2001). Control of glial immune function by neurons. *Glia, 36*(2), 191–199.

Nimmerjahn, A., Kirchhoff, F., & Helmchen, F. (2005). Resting microglial cells are highly dynamic surveillants of brain parenchyma in vivo. *Science, 308*, 1314–1318.

Obata, K., Katsura, H., Miyoshi, K., Kondo, T., Yamanaka, H., Kobayashi, K., Dai, Y., Fukuoka, T., Akira, S., & Noguchi, K. (2008). Toll-like receptor 3 contributes to spinal glial activation and tactile allodynia after nerve injury. *Journal of Neurochemistry, 105*(6), 2249–2259.

Ransohoff, R. M., & Perry, V. H. (2009). Microglial physiology: Unique stimuli, specialized responses. *Annual Review of Immunology, 27*, 119–145.

Suter, M. R., Berta, T., Gao, Y. J., Decosterd, I., & Ji, R. R. (2009). Large A-fiber activity is required for microglial proliferation and p38 MAPK activation in the spinal cord: Different effects of resiniferatoxin and bupivacaine on spinal microglial changes after spared nerve injury. *Molecular Pain, 5*, 53.

Tanga, F. Y., Nutile-McMenemy, N., & DeLeo, J. A. (2005). The CNS role of Toll-like receptor 4 in innate neuroimmunity and painful neuropathy. *Proceedings of the National Academy of Sciences of the United States of America, 102*(16), 5856–5861.

Tsuda, M., Shigemoto-Mogami, Y., Koizumi, S., Mizokoshi, A., Kohsaka, S., Salter, M. W., & Inoue, K. (2003). P2X4 receptors induced in spinal microglia gate tactile allodynia after nerve injury. *Nature, 424*, 778–783.

Tsuda, M., Masuda, T., Kitano, J., Shimoyama, H., Tozaki-Saitoh, H., & Inoue, K. (2009). IFN-gamma receptor signaling mediates spinal microglia activation driving neuropathic pain. *Proceedings of the National Academy of Sciences of the United States of America, 106*, 8032–8037.

Tsuda, M., Tozaki-Saitoh, H., & Inoue, K. (2010). Pain and purinergic signaling. *Brain Research Reviews, 63* (1–2), 222–232.

Verge, G. M., Milligan, E. D., Maier, S. F., Watkins, L. R., Naeve, G. S., & Foster, A. C. (2004). Fractalkine (CX3CL1) and fractalkine receptor (CX3CR1) distribution in spinal cord and dorsal root ganglia under basal and neuropathic pain conditions. *European Journal of Neuroscience, 20*, 1150–1160.

Zhang, J., Shi, X. Q., Echeverry, S., Mogil, J. S., De, K. Y., & Rivest, S. (2007). Expression of CCR2 in both resident and bone marrow-derived microglia plays a critical role in neuropathic pain. *Journal of Neuroscience, 27*, 12396–12406.

Zhuang, Z. Y., Kawasaki, Y., Tan, P. H., Wen, Y. R., Huang, J., & Ji, R. R. (2007). Role of the CX3CR1/p38 MAPK pathway in spinal microglia for the development of neuropathic pain following nerve injury-induced cleavage of fractalkine. *Brain, Behavior, and Immunity, 21*, 642–651.

Signal-to-Noise Ratio

Synonyms

S/N ratio

Definition

After some kind of stimulation, both real signals and background noises are recorded. This is the ratio between them. If its value is high, it means a good recording.

Cross-References

▶ Magnetoencephalography in Assessment of Pain in Humans

Significant Others

Definition

Significant others are important persons in the social environment and most often include spouses or living partners, family members, neighbors, close friends, and even a health- care provider.

Cross-References

▶ Psychological [Behavioral Health] Assessment of Pain
▶ Spouse, Role in Chronic Pain

Significant Others' Responses to Pain

Definition

It has been shown that the responses of significant others such as the spouse play an important role in the development of chronic pain.

Cross-References

▶ Operant Treatment of Chronic Pain

SII

▶ Nociceptive Processing in the Secondary Somatosensory Cortex
▶ Second Somatosensory Cortex

SII/PV

▶ Nociceptive Processing in the Secondary Somatosensory Cortex

Sildenafil

Synonyms

Viagra

Definition

Sildenafil (Viagra) is indicated for the treatment of erectile dysfunction. Erection is mediated by the release of nitric oxide in the corpus cavernosum of the penis. Nitric oxide stimulates the enzyme guanylate cyclase, which increases levels of cGMP; it is this response that creates smooth muscle relaxation and blood flow. A substance called phosphodiesterase type 5 (PDE5) leads to the degradation of cGMP. Sildenafil increases the availability of cGMP. Randomized controlled clinical trials demonstrate improved sexual performance as measured by firmness of erection, ability to maintain erection after penetration, as well as satisfaction and enjoyment with intercourse. The usual recommended dose is 50 mg administered 30–60 min prior to intercourse, with maximum recommended dosing frequency once per day. Clinical trials regarding the use of sildenafil to treat opioid-induced sexual dysfunction, in men or women, have not been conducted.

Cross-References

▶ Cancer Pain Management, Opioid Side Effects, Endocrine Changes, and Sexual Dysfunction

Silent Myocardial Ischemia

Definition

Myocardial ischemia is the term for inadequate oxygen supply (from inadequate blood flow) to the heart muscle (myocardium). In silent myocardial ischemia, patients have myocardial ischemia as indicated by the marked ST segment changes of the electrocardiogram (ECG) without experiencing angina (chest discomfort).

Cross-References

▶ Thalamus and Visceral Pain Processing (Human Imaging)
▶ Thalamus, Clinical Visceral Pain, Human Imaging
▶ Visceral Pain Model, Angina Pain

Silent Nociceptor

Robert F. Schmidt
Physiologisches Institut der Universität
Würzburg, Würzburg, Germany

Synonyms

Mechanoinsensitive nociceptor; Sleeping nociceptor

Definition

▶ Nociceptors that are normally insensitive to even very intense mechanical stimuli but they become sensitized in the course of pathophysiological processes in the tissue to become responsive to (formerly) nonnoxious mechanical stimuli. Silent nociceptors form a widespread, if not a universal, component of the somatovisceral afferent system in mammals.

Characteristics

Investigations of single primary afferent units in sensory nerves or in dorsal roots usually reveal a certain proportion of afferent fibers that despite their electrical identification cannot be excited by adequate natural mechanical stimulation. Except for the ambiguity with respect to unmyelinated afferents (▶ C Fibers) in peripheral nerves (some 50 % of which are preganglionic autonomic efferents), the lack of responsiveness of such units is usually attributed to failure to find the peripheral receptive field of the unit or to inadequate or subthreshold natural stimulation. Usually, such units are not even mentioned in reports of such studies, except perhaps indirectly in the data on the total number of units isolated. The concept of silent nociceptors first emerged in a series of studies that explored the responses of fine articular primary afferent units to passive movements of the cat's knee joint (Coggeshall et al. 1983; Schaible and Schmidt 1983a, b 1985; Grigg et al. 1986). It was firmly established in experiments in which silent articular nociceptors were first identified and then afterward exposed to an experimental arthritis. As illustrated in Fig. 1, this inflammation sensitized the silent nociceptors to the point where many of them developed resting activity, as well as vigorous responses to movements in the working range of the joint (Schaible and Schmidt 1988a, b).

While the concept of mechanoinsensitive, i.e., silent nociceptors, in the articular nerves of the cat's knee joint emerged from the experiments quoted above, it became obvious that there is most probably a range of mechanical thresholds for articular fine afferents under both physiological and pathophysiological conditions. Under physiological conditions, the spectrum of mechanical thresholds is rather evenly distributed, whereas under pathophysiological conditions, such as in experimental arthritis, all units become sensitized and the spectrum of thresholds is skewed, i.e., most units can now be activated by stimulation which under normal conditions is completely ineffective (for details, see Schmidt et al. 1994).

Silent Nociceptor, Fig. 1 Effects of an experimental arthritis on a silent ("sleeping") Group IV (C-fiber) unit. The unit had no receptive field and no resting activity in the beginning of the recording session. (**a**) Time course of the development of resting discharges. Mean and standard deviation of resting discharges quantified during the indicated times (*bars*). The first spontaneous activity was recorded about 2 h after the intraarticular injection of kaolin (*K* and *red arrow*) and carrageenan (*C* and second *red arrow* in (**a**)). The receptive field (*RF* in (**a**)) shown in (**d**) (*red dot* near the patella) appeared about 100 min after the onset of inflammation. Response to movements (**b**) and resting activity (**c**) in the inflamed state at the indicated times after the kaolin application; *Ext.* extension, *IR* pronation (Modified from Schaible and Schmidt 1988a)

Silent nociceptors were also identified in cutaneous and visceral nerves. Silent nociceptors in the skin do not respond to noxious mechanical and thermal stimuli applied to normal skin (Ringkamp et al. 2001; Schmelz et al. 2000; Weidner et al. 1999). In the viscera silent nociceptors were described that lack mechanosensitivity under normal conditions but become mechanosensitive during inflammation (Cervero and Laird 1999; Häbler et al. 1988). Thus, silent primary afferent units may be a universal feature of the somatovisceral afferent system of mammals and presumably of other vertebrates. They seem to form a substantial subpopulation of the fine afferent fibers in the ▶ A-delta (Group III) and ▶ C fiber (Group IV) range with particularly slow conduction velocities. Interestingly, silent nociceptors have distinct axonal biophysical characteristics separating them from polymodal nociceptors (Orstavik et al. 2003; Weidner et al. 1999).

Many details of the functional characteristics and the possible afferent and efferent functions of silent nociceptors have still to be elucidated.

For instance, many of the mechanoinsensitive units can be activated by chemical stimulation, particularly by algesic substances such as bradykinin and capsaicin. Silent nociceptors of the skin show a particular long lasting response to algogenic chemical stimuli (Ringkamp et al. 2001; Schmelz et al. 2000; Weidner et al. 1999). Functionally silent nociceptors produce pronounced neurogenic inflammation in humans (Schmelz et al. 2000). Furthermore, they play a major role in initiating central sensitization (Klede et al. 2003).

Cross-References

▶ Arthritis Model, Kaolin-Carrageenan-Induced Arthritis (Knee)
▶ Articular Nociceptors

References

Cervero, F., Laird, J. M. (1999). Visceral pain. *Lancet, 353*, 2145–2148.

Coggeshall, R. E., Hong, K. A., Langford, L. A., et al. (1983). Discharge characteristics of fine medial articular afferents at rest and during passive movements of inflamed knee joints. *Brain Research, 272*, 185–188.

Grigg, P., Schaible, H.-G., & Schmidt, R. F. (1986). Mechanical sensitivity of group III and IV afferents from posterior articular nerve in normal and inflamed cat knee. *Journal of Neurophysiology, 55*, 635–643.

Häbler, H.-J., Jänig, W., Koltzenburg, M. (1988). A novel type of unmyelinated chemosensitive nociceptor in the acutely inflamed urinary bladder. *Agents Actions, 25*, 219–221.

Klede, M., Handwerker, H. O., Schmelz, M. (2003) Central origin of secondary mechanical hyperalgesia. *Journal of Neurophysiology, 90*, 353–359.

Orstavik, K., Weidner, C., Schmidt, R., Schmelz, M., Hilliges, M., Jørum, E., Handwerker, H. O., Torebjörk, H. E. (2003). Pathological C-fibres in patients with a chronic painful condition. *Brain, 126*, 567–578.

Ringkamp, M., Peng, Y. B., Wu, G., Hartke, T. V., Campbell, J. N., Meyer, R. A. (2001). Capsaicin responses in heat-sensitive and heat-insensitive A-fiber nociceptors. *Journal of Neuroscience, 21*, 4460–4468.

Schaible, H.-G., & Schmidt, R. F. (1983a). Activation of groups III and IV sensory units in medial articular nerve by local mechanical stimulation of knee joint. *Journal of Neurophysiology, 49*, 35–44.

Schaible, H.-G., & Schmidt, R. F. (1983b). Responses of fine medial articular nerve afferents to passive movements of knee joints. *Journal of Neurophysiology, 49*, 1118–1126.

Schaible, H.-G., & Schmidt, R. F. (1985). Effects of an experimental arthritis on the sensory properties of fine articular afferent units. *Journal of Neurophysiology, 54*, 1109–1122.

Schaible, H.-G., & Schmidt, R. F. (1988a). Direct observation of the sensitization of articular afferents during an experimental arthritis. In R. Dubner, G. F. Gebhardt, & M. R. Bond (Eds.), *Pain research and clinical management series, proceedings of the 5th world congress on pain* (Vol. III, pp. 44–50). Amsterdam: Elsevier Science.

Schaible, H.-G., & Schmidt, R. F. (1988b). Time course of mechanosensitivity changes in articular afferents during a developing experimental arthritis. *Journal of Neurophysiology, 60*, 2180–2195.

Schmelz, M., Schmidt, R., Handwerker, H. O., Torebjörk, H. E. (2000a) Encoding of burning pain from capsaicin-treated human skin in two categories of unmyelinated nerve fibres. *Brain, 123*, 560–571.

Schmelz, M., Michael, K., Weidner, C., Schmidt, R., Torebjörk, H. E., Handwerker, H. O. (2000b). Which nerve fibers mediate the axon reflex flare in human skin? *Neuroreport, 11*, 645–648.

Schmidt, R. F., Schaible, H.-G., Meßlinger, K., et al. (1994). Silent and active nociceptors: Structure, functions, and clinical implications. In G. F. Gebhart, D. L. Hammond, & T. S. Jensen (Eds.), *Proceedings of the 7th world congress on pain, progress in pain research and management* (Vol. 2, pp. 213–250). Seattle: IASP Press.

Weidner, C., Schmelz, M., Schmidt, R., Hansson, B., Handwerker, H. O., Torebjörk, H. E. (1999). Functional attributes discriminating mechano-insensitive and mechano-responsive C nociceptors in human skin. *Journal of Neuroscience, 19*, 10184–10190.

Silent Nociceptors

▶ Histology of Nociceptors

Simple Analgesic

▶ Cancer Pain Management, Principles of Opioid Therapy, Drug Selection

Simple Analgesics

Nikolai Bogduk
University of Newcastle, Newcastle Bone & Joint
Institute, Royal Newcastle Centre, Newcastle,
NSW, Australia

Synonyms

Acetaminophen; Over-the-counter analgesics;
Painkillers; Paracetamol

Definition

The term "simple analgesics" is used to imply
a class of drugs, designed to relieve pain, but
which are less potent and safer than ▶ opioids,
and which are not ▶ NSAIDs, Survey
(NSAIDs). In effect, the class of drugs is now
represented only by paracetamol, known also as
acetaminophen.

Characteristics

Paracetamol was first introduced in 1893 and
became commercially available during the
1950s (Prescott 2000). It has become the most
widely available and most widely used analgesic. Much of its attraction stems from its relative
safety. Although in large doses, either acutely or
cumulatively, it can be toxic, paracetamol is
substantially free of adverse effects when used
in recommended doses. For this reason, it rapidly became available to the public as "over-the-
counter" medication, i.e., not requiring
a prescription.

In its pure form, paracetamol is a relatively
weak analgesic, substantially less potent than
opioids. In order to provide greater analgesia,
preparations are available, called compound
analgesics, in which small but various doses of
opioids, such as codeine or dextropropoxyphene,
are combined with paracetamol.

Mechanism

The mechanism by which paracetamol exerts
its analgesic effects remains largely a mystery.
When it is used, the excretion of prostaglandin
metabolites is decreased, but paracetamol
does not reduce the synthesis of prostaglandins.
It has only a weak inhibitory effect on cyclooxygenase 1 and cyclooxygenase 2. It is suspected to
have an effect on cyclooxygenase 3 or a variant of
cyclooxygenase 2 (Botting 2000).

Application

Paracetamol is suitable for any and all forms of
mild to moderate pain and is used widely for
such pain. Its application is limited only by its
modest efficacy. It is perhaps most widely used
for short-term management of acute but self-
limiting pain, or intermittent pain; but it has also
become the drug of first choice for persistent or
chronic pain.

Efficacy

For a drug so widely used, and with such a long
history of use, there is surprisingly little literature
explicitly on the efficacy of paracetamol. Rather,
paracetamol has most often been studied as the
benchmark, or comparison, drug against which
other and newer analgesics have been tested.
Nevertheless, in recent years data on the efficacy
of paracetamol have emerged.

Paracetamol is effective for the relief of postoperative pain. For a 50 % reduction in pain, it
has an NNT (see ▶ Number Needed to Treat
(NNT)) of 4 (Barden et al. 2004). For the postoperative pain of dental extraction, it is more effective than placebo, and as effective as diclofenac
(Kubitzek et al. 2003).

For the relief of pain due to dysmenorrhea,
paracetamol is more effective than placebo and
no less effective than ▶ NSAIDs (Marjoribanks
et al. 2003). It has fewer side effects than NSAIDs.

For the relief of tension-type headache,
paracetamol is more effective than placebo, and
more effective than naproxen (Prior et al. 2002).
For the relief of acute attacks of migraine, it is as
effective, or slightly less effective, than various
triptans (Diener and Limmroth 2001).

Paracetamol has not been tested for the relief of spinal pain. There is conflicting evidence as to whether it is less effective than NSAIDs or not (Van Tulder et al. 2003).

It can be used for rheumatoid arthritis, but surveys show that both patients and physicians prefer to use NSAIDs for this condition (Wienecke and Gotzsche 2004). For the treatment of pain due to osteoarthritis, the division is less clear. For this condition, paracetamol is more effective than placebo but less effective than NSAIDs for the relief of pain (Towheed et al. 2003; Zhang et al. 2004), although equally effective to NSAIDs for improving function (Towheed et al. 2003). NSAIDs, however, carry a greater risk of gastrointestinal side effects (Towheed et al. 2003; Zhang et al. 2004). For this reason, paracetamol is recommended as a first-line drug for osteoarthritis (Towheed et al. 2003; Zhang et al. 2004).

Although compound analgesics were developed with the prospect of providing greater analgesia and, therefore, greater patient satisfaction, than paracetamol alone, the outcome data are disappointing. A meta-analysis of the literature on oral surgery and postoperative pain showed that codeine added only a 5 % increase in pain relief to that afforded by paracetamol alone (de Craen et al. 1996); but the addition of codeine significantly increased the risk of side effects, particularly with repeated use. Another meta-analyses found that, for 50 % relief of postoperative pain, the NNT of paracetamol 650 mg compounded with codeine 60 mg was 4: a value neither substantially nor statistically better than that of the NNT of paracetamol alone (4–5) (Moore et al. 2000). Compared to paracetamol alone, paracetamol compounded with codeine had an NNT of 7.7 (Moore et al. 2000). Paracetamol compounded with dextropropoxyphene has an NNT of 4.4 (Collins et al. 2000).

For the relief of back pain, compound analgesics containing codeine are not more effective than NSAIDs, but produce more side effects (Bogduk 2004). Compound analgesics containing dextropropoxyphene are no more effective than paracetamol with caffeine, or paracetamol alone (Bogduk 2004).

Cross-References

► Cancer Pain Management, Principles of Opioid Therapy, Drug Selection
► NSAIDs and Their Indications

References

Barden, J., Edwards, J., Moore, A. et al. (2004). Single dose oral paracetamol (acetaminophen) for postoperative pain. *Cochrane Database System Review*, CD004602.
Bogduk, N. (2004). Pharmacological alternatives for the alleviation of back pain. *Expert Opinion Phamacotherapy, 5*, 2091–2098.
Botting, R. M. (2000). Mechanism of action of acetaminophen: Is there a cyclooxygenase 3? *Clinical Infectious Diseases, 31*, 202–210.
Collins, S. L., Edwards, J. E., Moore, R. A. et al. (2000). Single dose dextropropoxyphene, alone and with paracetamol (acetaminophen), for postoperative pain. *Cochrane Database System Review*, CD001440.
de Craen, A. J. M., Di Giulio, G., Lampe-Schoenmaeckers, A. J. E. M., et al. (1996). Analgesic Efficacy and safety of paracetamol-codeine combinations versus paracetamol alone: A systematic review. *British Medical Journal, 313*, 321–325.
Diener, H. C., & Limmroth, V. (2001). Analgesics. *Current Medical Research and Opinion, 17*, 13–16.
Kubitzek, F., Ziegler, G., Gold, M. S., et al. (2003). Analgesic efficacy of Low-dose diclofenac versus paracetamol and placebo in postoperative dental pain. *Journal of Orofacial Pain, 17*, 237–244.
Marjoribanks, J., Proctor, M. L., & Farquhar, C. (2003). Nonsteroidal anti-inflammatory drugs for primary dysmenorrhoea. *Cochrane Database System Review*, CD001751.
Moore, A., Collins, S., Carroll, D. et al. (2000). Single Dose Paracetamol (Acetaminophen), With And Without Codeine, For Postoperative Pain. *Cochrane Database System Review*, CD001547.
Prescott, L. F. (2000). Paracetamol: Past, present, and future. *American Journal of Therapeutics, 7*, 143–147.
Prior, M. J., Cooper, K. M., May, L. G., et al. (2002). Efficacy and safety of acetaminophen and naproxen in the treatment of tension-type headache. A randomized, double-blind, placebo-controlled trial. *Cephalalgia, 22*, 740–748.
Towheed, T. E., Judd, M. J., Hochberg, M. C. et al. (2003). Acetaminophen for osteoarthritis. *Cochrane Database System Review*, CD004257.
Van Tulder, M. W., Scholten, R. J. P. M., Koes, B. W., et al. (2003). Non steroidal anti-inflammatory drugs for low back pain (Cochrane Review). In Cochrane Library, Issue 3. Update software, Oxford.

Wienecke, T., Gotzsche, P. C. (2004). Paracetamol versus nonsteroidal anti-inflammatory drugs for rheumatoid arthritis. *Cochrane Database Systems Review*, CD003789.

Zhang, W., Jones, A., & Doherty, M. (2004). Does paracetamol (acetaminophen) reduce the pain of osteoarthritis? a meta-analysis of randomised controlled trials. *Annals of the Rheumatic Diseases, 63*, 901–917.

Simple Nerve Blocks

Definition

Numbing of a single peripheral nerve.

Cross-References

▶ Cancer Pain Management, Anesthesiologic Interventions, and Neural Blockade

Simple Synapse

Definition

Synapse is composed of a complex of the axon terminal and dendrite or dendritic spine and is often seen on presynaptic axons.

Cross-References

▶ Morphology, Intraspinal Organization of Visceral Afferents

Simulated Job Tasks

Definition

Simulated job tasks are in contrast to real work tasks. These are performed in job-related environments that imitate the clients' regular occupation without producing products or services. The Revised Handbook for Analyzing Jobs is a useful starting point in designing simulated job tasks.

Cross-References

▶ Vocational Counseling

SIN

▶ Neuropathic Pain Model, Neuritis/Inflammatory Neuropathy
▶ Sciatic Inflammatory Neuritis

Single Interval Procedure

Definition

A single interval procedure is a statistical decision method in which a sensory decision is made after each stimulus presentation.

Single or Straight Form of Terminal Branches

Definition

Terminal branches form several patterns depending on the function of the afferent fiber. Somatic afferents from skin make complicated, concentrated nest-like terminations different from visceral afferents, which have single or straight terminal branches.

Cross-References

▶ Morphology, Intraspinal Organization of Visceral Afferents

Single Photon Emission Computed Tomography

Peter Lau
Department of Clinical Research, Royal
Newcastle Hospital, University of Newcastle,
Newcastle, NSW, Australia

Synonyms

SPECT

Definition

Single photon emission computed tomography is a means of obtaining images of the internal structure of the body by combining the technology of nuclear medicine with that of computerized tomography (CT). Virtual images of the body are synthesized by a computer, based on the radiation emitted from a radioactive chemical administered intravenously to the patient.

Characteristics

According to their metabolic activities and blood flow, different healthy and diseased tissues absorb radionuclides (radioactive isotopes) at different rates. When injected into the body, these radionuclides accumulate in different tissues to different extents. From those tissues, radioactive emissions will emerge, in proportion to the concentration of radionuclide absorbed by them. By detecting those emissions across various diameters of the body, an image can be synthesized on the basis of the size and location of the inferred source of radiation. Images can be displayed in sagittal, axial, or coronal planes.

SPECT scanning does not demonstrate tissues directly. Rather, it detects areas of increased metabolic activity or increased blood flow. The site of these abnormalities, however, can be deduced from a knowledge of the anatomy of the region being studied.

SPECT is a major advance on conventional nuclear medicine imaging because it provides high-resolution images of the region being studied. It converts the normally flat, two-dimensional image of a bone scan into multiple, axial, sagittal, and coronal slices. This allows any source of radiation to be localized accurately to specific structures present in the plane being studied (Murray 1994).

Applications

The application of SPECT in pain medicine is limited to those conditions that might be associated with pain and which themselves are characterized by increased blood flow.

Among musculoskeletal disorders, those that are associated with increased blood flow are stress reactions in bone (Fig. 1), acute fractures (Fig. 2), pseudoarthroses, and certain bone tumors. However, although several descriptive studies extol what SPECT can show, few studies have demonstrated the validity of SPECT quantitatively.

In the context of low back pain, a review of the literature (Littenberg et al. 1995) found 13 reports on the accuracy of SPECT. Only three studies provided a reasonable reference standard and allowed the calculation of sensitivity and specificity. The review concluded that SPECT is useful in detecting pseudoarthrosis after failed spinal fusion, evaluating young patients with back pain and distinguishing benign from malignant lesions in cancer patients.

Although pseudoarthrosis can be detected, there is no evidence as yet either that this abnormality is a cause of pain or that detecting it makes a difference to management. In young athletes with back pain, SPECT is far more sensitive than plain radiography for the detection of pars fractures. However, because SPECT shows increased blood flow, rather than the fracture itself, it lacks specificity. SPECT will be positive not only in patients with actual fractures but also in patients with stress reactions and incipient fractures. Supplementary CT is required to demonstrate the morphology of the abnormality detected by SPECT (Fig. 3).

Single Photon Emission Computed Tomography, Fig. 1 A SPECT scan showing stress reaction in the sacroiliac joints after lumbar spinal fusion. Both joints show intense uptake of radionuclides (*arrows*). (**a**) Axial view. (**b**) Coronal view (Images provided by courtesy of Dr John Booker of Hunter Imaging Group and Sonic Health, Newcastle, Australia)

Single Photon Emission Computed Tomography, Fig. 2 A SPECT scan of a vertebral compression fracture in the lower thoracic spine. (**a**) Coronal view. (**b**) Lateral view. (**c**) Axial view. (**d**) The accompanying bone scan (Images provided by courtesy of Dr John Booker of Hunter Imaging Group and Sonic Health, Newcastle, Australia)

Single Photon Emission Computed Tomography, Fig. 3 A SPECT scan of a bilateral pars stress fracture. The *arrows* point to the areas of increased uptake in the pars interarticularis. (**a**) Axial view. (**b**) Lateral view. (**c**) Coronal view. (**d**) The accompanying bone scan (Images provided by courtesy of Dr John Booker of Hunter Imaging Group and Sonic Health, Newcastle, Australia)

Osteoid osteoma accounts for approximately 10 % of benign bone tumors and frequently affects young adults in the second decade. Although more commonly found in the proximal femur and tibia, it can also affect the posterior elements of the spine. SPECT allows for localization of the uptake to the posterior element of the spine rather than the anterior body or the pedicles, which are sites of malignant primary and metastatic neoplasm. Therefore with solitary lesions, metastatic malignant disease should be suspected when the area of increased uptake extends from the vertebral body into the pedicle, whereas simultaneous involvement of the vertebral body and a portion of the posterior arch is most often caused by benign conditions. For other lesions associated with back pain, SPECT has not been sufficiently evaluated for either validity or cost-effectiveness.

In the evaluation of hip pain, SPECT can be used to distinguish activity from overlying or underlying tissues that otherwise obscures structures at the depth of interest. Thus, the hyperemia of the overlying soft tissues and the underlying acetabulum can be differentiated by SPECT from the abnormal activity of the femoral head in avascular necrosis.

Some practitioners have used SPECT to demonstrate the juxta-articular hyperemia that occurs in some patients with complex regional pain syndromes. While perhaps satisfying, this is a superfluous diagnostic investigation. The diagnosis of complex regional pain syndrome is entirely clinical and does not require SPECT for confirmation.

For certain neurological disorders associated with pain, SPECT has potential, but still not proven, applications. To date, it has been used in an exploratory manner in experimental settings. In patients with migraine, SPECT has been used to demonstrate neurogenic inflammation of the meninges (Pappagallo 1999). During migraine, antidromic activity in trigeminal afferents and orthodromic parasympathetic vasodilatory activity increase the permeability of vessels in the meninges to molecules such as albumin. SPECT is able to detect this exudation. In this regard, however, SPECT is not diagnostic. Migraine is diagnosed by its clinical features. Rather, SPECT is a potential tool by which to explore the mechanisms involved in this condition.

SPECT has been used to explore possible neurological correlates with somatization disorders. A small, initial study has suggested that somatization disorders may be associated with hypoperfusion of certain areas of the brain (Garcia-Campayo et al. 2001). It remains to be established, however, that these changes are specific to and biologically meaningful as the basis of somatization disorders.

References

Garcia-Campayo, J., Sanz-Carrilo, C., Baringo, T., et al. (2001). SPECT scan in somatization disorder patients: An exploratory study of eleven cases. *Australian and New Zealand Journal of Psychiatry, 35,* 359–363.

Littenberg, B., Siegel, A., Tosteson, A. N., et al. (1995). Clinical efficacy of SPECT bone imaging for low back pain. *Journal of Nuclear Medicine, 36,* 1707–1713.

Murray, I. P. C. (1994). Bone scintigraphy: The procedure and interpretation. In I. P. C. Murray & P. J. Lell (Eds.), *Nuclear medicine in clinical diagnosis and treatment* (Vol. 2, pp. 925–931). Edinburgh: Churchill Livingstone.

Pappagallo, M. (1999). The pain in migraine is not mainly in the brain. *Johns Hopkins Physicians Update, 11,* 2.

Single Unit Activity

Synonyms

SUA

Definition

A microelectrode in the extracellular space records action potentials that constitute the output signal of neurons. On the basis of their size and shape, the action potentials are assigned to individual putative neurons.

Cross-References

▶ Thalamotomy for Human Pain Relief

Single-Fiber Electromyography

Synonyms

SFEMG

Definition

SFEMG uses dedicated intramuscular electrodes to record action potentials from individual muscle fibers. SFEMG is the most sensitive in vivo method to detect dysfunctions of neuromuscular transmission. It is part of the diagnostic work-up of neuromuscular disorders, in particular myasthenia.

Cross-References

▶ Clinical Migraine Without Aura

Single-Nucleotide Polymorphisms

Synonyms

SNP

Definition

Substitution of one nucleotide for another somewhere in the genome of different individuals. These are the most common type of DNA variant, occurring on average once in every 1,000 base pairs. Single-nucleotide polymorphisms (SNPs) are being catalogued and mapped in the genomes of various species and are useful for both QTL mapping and association studies.

Cross-References

▶ Association Study
▶ NSAIDs, Pharmacogenetics
▶ Opioid Analgesia, Strain Differences
▶ Quantitative Trait Locus Mapping
▶ TRPV1 Receptor, Species Variability

Single-Unit Recording

Definition

Single-unit recording is a neurophysiological recording of a single neuron using a microelectrode.

Cross-References

▶ Insular Cortex, Neurophysiology, and Functional Imaging of Nociceptive Processing

Sinus Thrombosis

Definition

Sinus thrombosis is a neurologic manifestation of Behçet's disease.

Cross-Reference

▶ Headache Due to Arteritis

Sinus-Venous Thrombosis (SVT)

▶ Headache Due to Venous Sinus Thrombosis

siRNA

Definition

siRNA (small interfering RNA) are double-stranded RNA sequences, approximately 20 nucleotides in length, with characteristic 2–3 nucleotide 3' overhangs. Introduction of siRNA into cells activates a posttranscriptional gene-silencing mechanism known as RNA interference (RNAi). siRNA are a relatively novel tool for inhibition of the synthesis of a protein of interest.

Cross-References

▶ Purine Receptor Targets in the Treatment of Neuropathic Pain

Situational Assessment

Synonyms

Job-Site Evaluation

Definition

Situational assessment is undertaken in the ordinary work environment during a visit to the individual's ordinary workplace. It consists of systematic observation for assessing work-related behavior and occupational capabilities during the individual's performance of ordinary work tasks. The observations are video-recorded and used for ergonomic and/or job analyses.

Cross-References

▶ Vocational Counseling

Situational Factors

Definition

Situational factors are a multiplicity of ontogenetic factors, such as cognitive, behavioral emotional, and social elements, as well as broader contextual factors, such as family, institutional, and other social ecological elements, whose interplay greatly affects a child's pain experiences.

Cross-References

▶ Cancer Pain, Palliative Care in Children

Situational Pain Questionnaire

Definition

Situational pain questionnaire refers to the application of statistical decision theory to the construction of, and the analysis of, responses to a pain questionnaire.

Sjaastad's Syndrome

▶ Paroxysmal Hemicrania

Sjögren's Syndrome

Definition

Sjögren's syndrome is characterized by kerato-conjunctivitis sicca and symptomatic xerostomia (the sicca-syndrome), and associated with the detection of anti-Ro (SSA – 97 %) and anti-La (SSB – 78 %) autoantibodies. It is the most frequent of the autoimmune diseases characterized by sicca syndrome, and is often combined with rheumatoid arthritis.

Cross-References

▶ Headache Due to Arteritis
▶ Muscle Pain in Systemic Inflammation (Polymyalgia Rheumatica, Giant Cell Arteritis, Rheumatoid Arthritis)

Skin Conductance-Assisted Relaxation

Definition

A form of biofeedback that employs information about sweat gland activity in order to facilitate overall relaxation.

Cross-References

▶ Biofeedback in the Treatment of Pain

Skin Nociception

▶ Species Differences in Skin Nociception

Skin-Nerve Preparation

Definition

The skin-nerve preparation is an in vitro preparation consisting of a skin patch and its nerve supply. The skin is mounted in a chamber, external side down, and superfused with physiological solutions. The nerve is used for extracellular recordings of action potentials from single nerve fibers which can be characterized by conduction velocity and receptive properties. This preparation, originally introduced by P.W. Reeh, has been extensively used for studying nociceptors, in particular, from rodents.

Cross-References

▶ Nerve Growth Factor, Sensitizing Action on Nociceptors

Skin-to-Skin Contact

Definition

Also known as kangaroo care: an infant is placed on his/her mother's bare chest during a painful procedure or for soothing after a painful procedure.

Cross-References

▶ Acute Pain Management in Infants

SLE

▶ Systemic Lupus Erythematosus

SLEA

▶ Sublenticular Extended Amygdala

Sleep and Pain

Emily Law
Center for Child Health, Behavior, and Development, Seattle Children's Research Institute, Seattle, WA, USA

Synonyms

Chronic pain; Insomnia; Sleep disturbance; Sleep problems

Definition

Chronic pain impacts 25–35 % of youth (Perquin et al. 2000) and is commonly associated with disturbed sleep. Sleep difficulties among youth with chronic pain can include insufficient sleep duration, nighttime awakenings, insomnia symptoms, frequent daytime napping, poor perceived sleep quality, and poor sleep habits (Law et al. in press; Meltzer et al. 2005; Palermo et al. 2011; Tsai et al. 2008; Valrie et al. 2007).

Introduction

Sleep disturbances are common among healthy adolescents and include obtaining an insufficient amount of nightly sleep (i.e., <8 h), greater delays in the timing of sleep onset, longer sleep onset latencies (i.e., >30 min to fall asleep), an increasing discrepancy between weekday and weekend sleep patterns (i.e., later bedtimes and waketimes on weekends), and spending more time sleeping during the day (National Sleep Foundation 2006). Adolescents who frequently do not get enough sleep at night may show inconsistent and delayed sleep patterns, excessive daytime sleepiness, and higher rates of insomnia (Sadeh et al. 2002; Wolfson and Carskadon 1998). Shifts in sleep patterns can perpetuate a cycle of sleep difficulties into adulthood (Sadeh et al. 2002; Wolfson and Carskadon 1998).

Sleep is closely connected to the health and well-being of adolescents with medical conditions, and disturbed sleep in these populations have been associated with poor quality of life and increased risk for poor health outcomes (Lewandowski et al. 2011). Chronic pain is one such pediatric medical condition and has been associated with a variety of sleep problems.

Characteristics

Over 50 % of youth with chronic pain report frequent problems with insomnia, defined as difficulty falling or staying asleep (LaPlant et al. 2007; Palermo et al. 2011). Youth with chronic pain also commonly report delayed sleep onset, frequent night awakenings, reduced sleep duration, and increased daytime sleep (Law et al. in press; Meltzer et al. 2005; Palermo et al. 2011; Tsai et al. 2008). Sleep problems have been documented in groups of children and adolescents with specific pain conditions, including juvenile fibromyalgia (Roizenblatt et al. 1997), juvenile rheumatoid arthritis (Amos et al. 1997; Bloom et al. 2002), headache (Bruni et al. 1997; Miller et al. 2003), sickle cell disease (Valrie et al. 2007), and complex regional pain syndrome (Meltzer et al. 2005), as well as in samples of youth with mixed chronic pain conditions (Palermo et al. 2008, 2011; Law et al. in press).

There are limited data available on behavioral and psychosocial factors that may predict sleep disturbance in children and adolescents with chronic pain. Available research indicates that higher levels of cognitive presleep arousal (e.g., worries at bedtime) and depressive symptoms are associated with worse sleep problems in these youth (Murray et al. 2012; Palermo et al. 2011).

Impact of Comorbid Chronic Pain and Sleep Disturbance

Insomnia symptoms such as difficulty falling and staying asleep have been associated with greater activity limitations and poor health-related quality of life among youth with chronic pain (Palermo and Kiska 2005). Increased daytime sleep has also been associated with greater pain-related disability (Law et al. in press). The adverse effects of sleep disturbance in children and adolescents with chronic pain highlight the need for appropriate assessment and intervention for these youth.

Mechanisms of Pain and Sleep

While having chronic pain is related to sleep problems, conversely, disrupted sleep patterns may contribute to increased pain sensations (hyperalgesia) and decreased functioning among adolescents with chronic pain. Lewin and Dahl (1999) proposed a theoretical framework positing bidirectional relationships between pain and sleep, stating that pain sensations can increase the time it takes to fall asleep and contribute to nighttime awakenings, and poor sleep can subsequently increase pain sensitivity and lower pain tolerance. This model has been supported by studies of adults and adolescents with chronic pain, which show that pain predicts sleep disturbances and that sleep disturbances predict pain the following day (Affleck et al. 1996; Lewandowski et al. 2010; Roehrs and Roth 2005). The bidirectional relationship between sleep problems and pain sensations may lead to a sleep-pain cycle that worsens pain and increases functional limitations.

Sleep Assessment

Assessment of sleep disturbance in youth with chronic pain can be incorporated into a standard evaluation of youth with chronic pain that includes a physical exam as well as review of medical history, developmental history, family history, pain history, and psychosocial functioning. Given potential discrepancies between objective and subjective sleep assessments, multi-method approaches to assessing sleep are recommended (Mindell and Owens 2003).

Clinical interview. Sleep in youth with chronic pain can be assessed by having the family describe general sleep behaviors over the past several weeks. As recommended by Mindell and Owens (2003), the clinical interview should include assessment of bedtime behaviors (e.g., timing and consistency of bedtime, bedtime routine, timing and location of sleep onset), the sleep environment (e.g., bedroom sharing, bedroom location, presence of light/noise), sleep schedule (e.g., timing and duration of naps, weekday vs. weekend schedule), and daytime behaviors (e.g., timing of morning waking, consequences for difficulty waking, daytime sleepiness, timing and intensity of daytime activities).

Sleep diaries. Sleep diaries typically collect information regarding time to bed, sleep onset latency, number and duration of night awakenings, time of waking, time out of bed, total sleep time, and duration and time of sleep during the day. Sleep diaries are typically completed twice per day, in the morning upon waking (to report on nighttime sleep) and in the evening before going to bed (to report on daytime sleep). In clinical settings, 2 weeks of daily sleep diaries have been recommended to establish stable baseline (Mindell and Owens 2003). In research settings, sleep diaries have been collected using paper-and-pencil and electronic formats (Palermo et al. 2011).

Polysomnography. Polysomnography is an overnight sleep study that provides information about sleep architecture, breathing, limb/body movements, and awakenings during sleep. Polysomnography typically requires an overnight stay in a sleep laboratory, where a qualified technician uses video recording, piezoelectric belts, pulse oxometry, electroencephalogram (EEG), electromyogram (EMG), and electrocardiogram (ECG) monitoring to assess airflow, oxygen saturation, heart rate, and limb/body movements.

Actigraphy. An actiwatch is worn on the nondominant wrist and records sleep-wake patterns based on the presence/absence of movements detected by an omnidirectional sensor. Actigraphy provides estimates of total sleep in minutes, minutes awake after sleep onset, and sleep efficiency (the ratio of estimated total sleep time and total time spent in bed). An advantage of actigraphy over polysomnography is that it allows for unobtrusive assessment of sleep patterns in the child's home over an extended period of time. Actigraphy has been used to assess sleep patterns in adolescents with chronic pain (Law et al. in press; Palermo et al. 2007; Tsai et al. 2008) and has demonstrated up to 95 % agreement with traditional polysomnography recordings (Sadeh et al. 1995). Five to seven nights of actigraphy recordings are recommended to obtain reliable estimates of sleep patterns (Acebo et al. 1999).

Self-report measures. The Adolescent Sleep Wake Scale (ASWS) is a 28-item self-report measure of sleep quality that has been validated for use in 12-to-18-year-olds (LeBourgeois et al. 2005). The ASWS assesses five behavioral dimensions of sleep quality: going to bed, falling asleep, maintaining sleep, reinitiating sleep, and returning to wakefulness. *The Adolescent Sleep Hygiene Scale* (ASHS) is a 24-item measure of sleep hygiene (i.e., behaviors that facilitate and inhibit sleep, such as maintaining a regular bedtime routine, caffeine consumption, etc.) that has been validated for use in 12-to-18-year-olds (LeBourgeois et al. 2005). The ASHS assesses six domains of sleep hygiene: physiological, cognitive, emotional, sleep environment, substances, and sleep stability. The ASWS and ASHS have been used in research with samples of youth with chronic pain (Palermo et al. 2007, 2011) and have demonstrated adequate internal consistency.

Treatment of Comorbid Chronic Pain and Sleep

Research on the treatment of comorbid sleep disturbance in children and adolescents with chronic pain is extremely limited. In one study of children and adolescents with migraine, sleep hygiene education was associated with reduced frequency and duration of migraine attacks compared to youth who did not receive sleep hygiene education (Bruni et al. 1999). Although these findings are promising, this study was limited by small sample size and lack of assessment of changes in sleep in response to the intervention. Given the high rate of sleep problems among youth with chronic pain and their relationship to functional disability and quality of life, the development and evaluation of targeted sleep interventions in this population is paramount.

More progress has been made in the treatment of comorbid chronic pain and sleep disturbances in adult populations. Cognitive-behavioral therapy for insomnia (CBT-I) has been effectively used in adults with a variety of comorbid medical conditions (Perlis et al. 2005), including adults with chronic pain (Vitiello et al. 2009). Primary treatment strategies of CBT-I include cognitive strategies to reduce worries at bedtime, relaxation and deep breathing strategies, and stimulus control and sleep restriction strategies to increase associations between sleep and the sleep environment. Randomized clinical trials are needed to evaluate the effectiveness of CBT-I in children and adolescents with comorbid chronic pain and sleep disturbance.

Summary

Sleep problems are common among children and adolescents with chronic pain, and are associated with worse functional disability and poorer quality of life. Clinicians working with youth who have chronic pain should conduct a detailed assessment of sleep patterns and sleep quality using both subjective and objective measures when possible. Targeted sleep interventions may benefit children and adolescents with comorbid chronic pain and sleep disturbance, and research is needed to develop and evaluate such interventions in this population.

References

Acebo, C., Sadeh, A., Seifer, R., Tzischinsky, O., Wolfson, A. R., Hafer, A., & Carskadon, M. A. (1999). Estimating sleep patterns with activity monitoring in children and adolescents: How many nights are necessary for reliable measures? *Sleep, 22*(1), 95–103.

Affleck, G., Urrows, S., Tennen, H., Higgins, O., & Abeles, M. (1996). Sequential daily relations of sleep, pain intensity, and attention to pain among women with fibromyalgia. *Pain, 68,* 363–368.

Amos, C. E., Jr., Curry, M. R., Drutz, I. E., Frost, J. D., Jr., & Warren, R. W. (1997). Sleep disruption and daytime sleepiness in school-age children with JRA. *Arthritis Rheumatology, 40,* S244.

Bloom, B. J., Owens, J. A., McGiunn, M., Nobile, C., Shaeffer, L., & Alario, A. J. (2002). Sleep and its relationship to pain, dysfunction, and disease activity in juvenile rheumatoid arthritis. *Journal of Rheumatology, 29,* 169–173.

Bruni, O., Fabrizi, P., Ottaviano, S., Cortesi, F., Giannotti, F., & Guidetti, V. (1997). Prevalence of sleep disorders in childhood and adolescents with headache: A case-control study. *Cephalalgia, 17,* 492–498.

Bruni, O., Galli, F., & Guidetti, V. (1999). Sleep hygiene and migraine in children and adolescents. *Cephalalgia, 19*(Suppl. 25), 57–59.

LaPlant, M. M., Adams, B. S., Haftel, H. M., & Chervin, R. D. (2007). Insomnia and quality of life in children referred for limb pain. *Journal of Rheumatology, 34*(12), 2486–2490.

Law, E. F., Dufton, L., & Palermo, T. M. (in press). Brief report: Daytime and nighttime sleep patterns in adolescents with and without chronic pain. *Health Psychology.*

LeBourgeois, M. K., Giannotti, F., Cortesi, F., Wolfson, A. R., & Harsh, J. (2005). The relationship between reported sleep quality and sleep hygiene in Italian and American adolescents. *Pediatrics, 115*(1 Suppl.), 257–265.

Lewandowski, A. S., Palermo, T. M., De la Motte, S., & Fu, R. (2010). Temporal daily associations between pain and sleep in adolescents with chronic pain versus healthy adolescents. *Pain, 151*(1), 220–225.

Lewandowski, A. S., Ward, T. M., & Palermo, T. M. (2011). Sleep problems in children and adolescents with common medical conditions. *Pediatric Clinics of North America, 58,* 699–713.

Lewin, D. S., & Dahl, R. E. (1999). Importance of sleep in the management of pediatric pain. *Journal of Developmental and Behavioral Pediatrics, 20,* 244–252.

Meltzer, L. J., Logan, D. E., & Mindell, J. A. (2005). Sleep patterns in female adolescents with chronic musculoskeletal pain. *Behavioral Sleep Medicine, 3*(4), 193–208.

Miller, V. A., Palermo, T. M., Powers, S. W., Scher, M. S., & Hershey, A. D. (2003). Migraine headaches and sleep disturbances in children. *Headache, 43,* 362–368.

Mindell, J. A., & Owens, J. A. (2003). *A clinical guide to pediatric sleep: Diagnosis and management of sleep problems in children and adolescents.* Philadelphia: Lippincott Williams & Wilkins.

Murray, C. B., Murphy, L. K., Palermo, T. M., & Clarke, G. M. (2012). Brief report: Pain and sleep-wake disturbances in adolescents with depressive disorders. *Journal of Clinical Child and Adolescent Psychology, 41,* 482–490.

National Sleep Foundation. (2006). *2006 Sleep in America poll: Summary of findings.* Washington, DC: National Sleep Foundation.

Palermo, T. M., Fonareva, I., & Janosy, N. R. (2008). Sleep quality and efficiency in adolescents with chronic pain: Relationship with activity limitations and health related quality of life. *Behavioral Sleep Medicine*, 6, 234–250.

Palermo, T. M., & Kiska, R. (2005). Subjective sleep disturbances in adolescents with chronic pain: Relationship to daily functioning and quality of life. *The Journal of Pain*, 6(3), 201–207.

Palermo, T. M., Toliver-Sokol, M., Fonareva, I., & Koh, J. L. (2007). Objective and subjective assessment of sleep in adolescents with chronic pain compared to healthy adolescents. *The Clinical Journal of Pain*, 23(9), 812–820.

Palermo, T. M., Wilson, A. C., Lewandowski, A. S., Toliver-Sokol, M., & Murray, C. B. (2011). Behavioral and psychosocial factors associated with insomnia in adolescents with chronic pain. *Pain*, 152(1), 89–94. doi: 10.1016/j.pain.2010.09.035.

Perlis, M. L., Junquist, C., Smith, M. T., & Posner, D. (2005). *Cognitive behavioral treatment of insomnia*. New York: Springer.

Perquin, C. W., Hazebroek-Kampschreur, A. A., Hunfeld, J. A., Bohnen, A. M., van Suijlekom-Smit, L. W., Passchier, J., & van der Wouden, J. C. (2000). Pain in children and adolescents: A common experience. *Pain*, 87(1), 51–58.

Roehrs, T., & Roth, T. (2005). Sleep and pain: Interaction of two vital functions. *Seminars in Neurology*, 25, 106–116.

Roizenblatt, S., Tufik, S., Goldenberg, J., Pinto, L. R., Hilario, M. O., & Feldman, D. (1997). Juvenile fibromyalgia: Clinical and polysomnographic aspects. *Journal of Rheumatology*, 24(3), 579–585.

Sadeh, A., Hauri, P. J., Kripke, D. F., & Lavie, P. (1995). The role of actigraphy in the evaluation of sleep disorders. *Sleep*, 18, 288–302.

Sadeh, A., Gruber, R., & Raviv, A. (2002). Sleep, neurobehavioral functioning, and behavior problems in school-age children. *Child Development*, 73, 405–417.

Tsai, S. Y., Labyak, S. E., Richardson, L. P., Lentz, M. J., Brandt, P. A., Ward, T. M., & Landis, C. A. (2008). Actigraphic sleep and daytime naps in adolescent girls with chronic musculoskeletal pain. *Journal of Pediatric Psychology*, 33(3), 307–311.

Valrie, C. R., Gil, K. M., Redding-Lallinger, R., & Daeschner, C. (2007). Brief report: Sleep in children with sickle cell disease: An analysis of daily diaries utilizing multilevel models. *Journal of Pediatric Psychology*, 32(7), 857–861.

Vitiello, M. V., Rybarczyk, B., Von Korff, M., & Stepanski, E. J. (2009). Cognitive behavioral therapy for insomnia improves sleep and decreases pain in older adults with co-morbid insomnia and osteoarthritis. *Journal of Clinical Sleep Medicine*, 5(4), 355–362.

Wolfson, A. R., & Carskadon, M. A. (1998). Sleep schedules and daytime functioning in adolescents. *Child Development*, 69, 875–887.

Sleep Disturbance

▶ Orofacial Pain, Sleep Disturbance
▶ Sleep and Pain

Sleep Disturbances

▶ Pain in Humans, Sleep Disturbances

Sleep Fragmentation

Definition

Interruption of any sleep stage by isolated or repetitive events such as sleep stage shifts (deeper to lighter), micro-arousal or awakening, short duration of deep (St 3 and 4) sleep, or alpha EEG wave intrusion. As a result, EEG frequency is in the fast range, heart rate is increased, and muscle tone higher with occasional body movements. As a consequence, sleep continuity may be impaired and sleep complaints are frequent (e.g., unrefreshing).

Cross-References

▶ Orofacial Pain, Sleep Disturbance

Sleep Problems

▶ Sleep and Pain

Sleep Stage

Definition

A division of specific sleep periods into sleep stages (St): light sleep (St 1 and 2), deep sleep

(St 3 and 4), and rapid eye movement (REM). Each sleep stage has its own EEG characteristics (frequency change or large amplitude signal, such as K complex in stage 2), heart rate, and muscle tone, best demonstrated by polygraphic recording, which allows sleep stage scoring.

Cross-References

▶ Orofacial Pain, Sleep Disturbance

Sleeping Nociceptor

▶ Silent Nociceptor

Small Fiber Neuropathies

John W. Griffin
Department of Neurology, Johns Hopkins
University School of Medicine, Baltimore,
MD, USA

Definition

The small fiber neuropathies are a group of disorders of diverse etiologies that have predominant involvement of Aδ and C-fibers, with consequent impairment of nociception, variable degrees of autonomic dysfunction, and frequently neuropathic pain.

Characteristics

The "small fibers" of the PNS include the Aδ and C sensory fibers, the γ efferent fibers, and the preganglionic parasympathetic and postganglionic

John W. Griffin: deceased.

sympathetic C-fibers. In contrast, the large fibers of the PNS include the Aβ afferents from skin, muscle, and the internal organs, as well as the α motor neurons. Most disorders of the PNS are characterized by distally predominant loss of axons, a process often referred to as distal axonal degeneration or "dying back." Distal axonal degeneration affects long fibers before shorter ones, with the distal-most regions of the long axons underlying pathologic changes that closely resemble Wallerian degeneration following axonal injury. With time, the process moves proximally up the long fibers, and shorter fibers become affected. This process can affect large caliber fibers, producing numbness in the toes, loss of the tendon reflexes at the ankle, and weakness in the toes and ankles. When distal axonal degeneration affects small fibers predominantly, there can initially be a loss of protective and nociceptive sensibility in the toes and feet, autonomic features such as orthostatic hypotension, and often spontaneous stimulus-independent neuropathic pain.

Causes of Small Fiber Neuropathies

Neuropathies that can prominently or predominantly involve small fibers include diabetic polyneuropathy (Dyck and Dyck 1999); amyloid neuropathy, either heritable or associated with monoclonal gammopathies; the sensory neuropathy of AIDS; occasional patients with sarcoidosis; Fabry's disease, a sex-linked recessive disorder; some other types of inherited sensory neuropathy; and idiopathic painful polyneuropathy, a frequent problem that especially affects elderly individuals (Holland et al. 1998; Periquet et al. 1999).

The understanding of the relationship of SFN to diabetes is evolving. Symptomatic diabetic neuropathy usually occurs in the setting of longstanding, well-documented diabetes. However, recent studies have found that loss of small sensory fibers and autonomic dysfunction can occur early in the course of diabetes. Presentation with neuropathic pain can occur at the stage of impaired glucose tolerance, before frank diabetes supervenes (Singleton et al. 2001; Sumner et al. 2003).

Clinical Manifestations

The term "small fiber neuropathy" implies the triad of loss of protective sensibility, presence of spontaneous neuropathic pain, and autonomic dysfunction. The portion of the "small fiber" spectrum that an individual physician sees is defined by his or her specialty and patient population. To neurologists, the most frequent cause of presentation is neuropathic pain. To generalists and diabetologists, presenting problems may reflect painless injuries or autonomic dysfunction alone. Examples include diabetic patients who develop painless ulcers or foot fractures without other neurologic symptoms or diabetics who present with complaints of impotence in the absence of other symptoms.

Irrespective of the initial complaint, most patients will develop elevated pin and thermal thresholds in the feet and some elements of autonomic dysfunction. The feature that distinguishes small fiber neuropathies from large or mixed neuropathies is the absence of loss of tendon reflexes at the ankle, of joint position sensibility in the toes, or of strength in the intrinsic muscles of the feet and in the ankle dorsiflexors and evertors.

The neuropathic pain is usually greatest in the feet, and with time may affect the fingers. Various adjectives are used to describe the pain, with "burning" among the most frequent. There are often superimposed lightning pains, sudden "stabbing" or "electric" pains in one limb that last for seconds to a minute or so. Patients frequently complain that the pain is worse at night, presumably reflecting the lack of distracting stimuli. The affected extremities may have hypalgesia to touch and temperature, hyperalgesia or allodynia to these modalities, or any variation.

Occasional patients have widespread small fiber loss and pain, involving the face, trunk, and arms as well as the legs (Holland et al. 1998). Such patients may well have a sensory ganglionopathy or neuronopathy (see ▶ ganglionopathies), rather than a length-dependent process as the explanation for the widespread distribution of their involvement.

Laboratory Tests for SFN

Several autonomic function tests that have been extensively used have well-established norms for the whole age range. They are limited by the fact that they are not widely available. Measures of cardiac and blood pressure regulation are most widely available. These tests include heart period variability (sinus arrhythmia) with standardized breathing, the Valsalva response, and measures of blood pressure changes in going from lying to standing or on a tilt table (Diem et al. 2003).

Sudomotor (sweat) testing is useful (Diem et al. 2003; Kennedy 2002; Low 1990). This approach allows assessment of regional differences in the body, including testing of the distal leg, the first region affected in most nerve diseases. Qualitative testing with dyes that change color with moisture, such as quinizarin and starch/iodine, can be done anywhere but benefits from a controlled temperature/humidity room. Sweat production can be quantitated by Silastic molds or by the quantitative sudomotor axon reflex test (QSART) (Low 1990). QSART is a measure of sudomotor function and requires specialized machinery. This test has been extensively validated, but is not widely available.

As a test of small fiber sensory function, quantitative thermal testing has been validated for a variety of instruments but also has limited availability.

Skin biopsies can be used to study cutaneous nerve fibers, using selective immunocytochemical stains for specific C-fiber types (Herrmann et al. 1999; Holland et al. 1998; Kennedy and Wendelschafer-Crabb 1994). The technique is most highly developed for assessment of epidermal fibers. Most of these fibers are nociceptors. This technique offers the promise of being able to examine the same patient at multiple sites and at multiple times, creating a spatial and temporal picture of the degree of small fiber sensory loss. The use of punch skin biopsies with local anesthesia is well tolerated by patients. Normative data has been published, and the reproducibility and interobserver reliability are excellent

(McArthur et al. 1998). Abnormalities on skin biopsy and QSART have a high degree of concordance, suggesting that small fiber involvement typically affects both somatic sensory and autonomic postganglionic C-fibers, although there can be specificity of involvement by fiber population.

Nerve biopsy can also be used to assess small fibers (Herrmann et al. 1999) and has the advantage that large fibers can be assessed on the same specimen. However, there are several limitations. Only a few nerves, such as the sural, superficial peroneal, and some cutaneous nerves of the forearm are suitable for biopsy. Assessment of C-fibers requires electron microscopy, and the quantitation by EM morphometry is laborious. Only one site in time and space can be biopsied.

Pathogenesis

It is widely accepted that neuropathic pain can be initiated by sensitization, hyperexcitability, and/or spontaneous discharge in primary afferent neurons, leading to "▶ central sensitization" or facilitation. When neuropathic pain of peripheral origin occurs in small fiber neuropathies, are spontaneous discharges in the degenerating small fibers driving the process? Or are the responsible nerve fibers their intact neighbors? Where in the afferent neuron does the spontaneous discharge arise, in the perikaryon, the degenerating terminal, or in intact terminals? The mechanisms by which degeneration of small sensory fibers including nociceptors generate neuropathic pain are not understood. An attractive hypothesis is that a new population of ion channels is inserted into the excitable membranes of the responsible neurons. Evidence that this mechanism can apply in man comes from erythromelalgia, a disorder characterized by marked heat hyperalgesia. Microneurographic studies identified a decrease in conduction velocity and an increase in activity-dependent slowing in C-fibers (Orstavik et al. 2003). In addition, C-fibers that would normally be classified as mechanically insensitive afferents, based on their conduction properties, had

mechanosensitivity, suggesting that sensitization of nociceptors may play a role in this disease. Recently, mutations in the Na channel Na(v)1.7 have been identified (Yang et al. 2004).

Treatment

When possible, the treatment of the underlying disease should be treated. If there is loss of protective sensibility, protection from painless injury should be undertaken, as described in ▶ Diabetic Neuropathy Treatment. Education regarding the nature of neuropathic pain is often helpful; patients interpret the spontaneous pain as evidence of ongoing tissue damage. Understanding the pain as dysfunctional nerve fibers can be reassuring.

The first-line medications of the neuropathic pain associated with SFN are anticonvulsants, and gabapentin is most widely used (Sindrup and Jensen 2000). An analog, pregabalin, has had reported efficacy in neuropathic pain. No head-to-head comparison with gabapentin is available. Carbamazepine and Dilantin are alternatives for which some evidence of efficacy is available, and there are limited data surrounding lamotrigine (Sindrup and Jensen 2000).

There is increasing attention to serotonin-norepinephrine reuptake inhibitors, including duloxetine and venlafaxine. In older populations they are expected to provide a better safety profile compared to tricyclic antidepressants such as nortriptyline and amitriptyline.

Tramadol, an analgesic that acts through both opiate and non-opioid receptors, is a second-line medication. In patients with severe pain not responsive to other modalities, opiates may be required. Opiates can be effective in neuropathic pain states (Raja et al. 2002).

References

Diem, P., Laederach-Hofmann, K., Navarro, X., Mueller, B., Kennedy, W. R., & Robertson, R. P. (2003). Diagnosis of diabetic autonomic neuropathy: A multivariate approach. *European Journal of Clinical Investigation, 33*, 693–697.

Dyck, P. J. B., & Dyck, P. J. (1999). Diabetic polyneuropathy. In P. J. Dyck & P. K. Thomas (Eds.), *Diabetic neuropathy* (pp. 255–278). Philadelphia: W.B.Saunders.

Herrmann, D. N., Griffin, J. W., Hauer, P., Cornblath, D. R., & McArthur, J. C. (1999). Epidermal nerve fiber density and sural nerve morphometry in peripheral neuropathies. *Neurology, 53*, 1634–1640.

Holland, N. R., Crawford, T. O., Hauer, P., Cornblath, D. R., Griffin, J. W., & McArthur, J. C. (1998). Small-fiber sensory neuropathies: Clinical course and neuropathology of idiopathic cases. *Annals of Neurology, 44*, 47.

Kennedy, W. R. (2002). Usefulness of the silicon impression mold technique to evaluate sweating. *Clinical Autonomic Research, 12*, 9–10.

Kennedy, W. R., & Wendelschafer-Crabb, G. (1994). Quantification of nerve in skin biopsies from control and diabetic subjects. *Neurology, 44*, A275.

Low, P. A. (1990). The effect of aging on cardiac autonomic and post ganglionic sudomotor function. *Muscle & Nerve, 13*, 152–157.

McArthur, J. C., Stocks, A., Hauer, P., Cornblath, D. R., & Griffin, J. W. (1998). Epidermal nerve fiber density: Normative reference range and diagnostic efficiency. *Archives of Neurology, 55*, 1513–1520.

Orstavik, K., Weidner, C., Schmidt, R., Schmelz, M., Hilliges, M., Jorum, E., Handwerker, H., & Torebjork, E. (2003). Pathological C-fibres in patients with a chronic painful condition. *Brain, 126*, 567–578.

Periquet, M. I., Novak, V., Collins, M. P., Nagaraja, H. N., Erdem, S., Nash, S. M., Freimer, M. L., Sahenk, Z., Kissel, J. T., & Mendell, J. R. (1999). Painful sensory neuropathy: Prospective evaluation using skin biopsy. *Neurology, 53*, 1641–1647.

Raja, S. N., Haythornthwaite, J. A., Pappagallo, M., Clark, M. R., Travison, T. G., Sabeen, S., Royall, R. M., & Max, M. B. (2002). Opioids versus antidepressants in postherpetic neuralgia: A randomized, placebo-controlled trial. *Neurology, 59*, 1015–1021.

Sindrup, S. H., & Jensen, T. S. (2000). Pharmacologic treatment of pain in polyneuropathy. *Neurology, 55*, 915–920.

Singleton, J. R., Smith, A. G., & Bromberg, M. B. (2001). Painful sensory polyneuropathy associated with impaired glucose tolerance. *Muscle & Nerve, 24*, 1225–1228.

Sumner, C. J., Sheth, S., Griffin, J. W., Cornblath, D. R., & Polydefkis, M. (2003). The spectrum of neuropathy in diabetes and impaired glucose tolerance. *Neurology, 60*, 108–111.

Yang, Y., Wang, Y., Li, S., Xu, Z., Li, H., Ma, L., Fan, J., Bu, D., Liu, B., Fan, Z., Wu, G., Jin, J., Ding, B., Zhu, X., & Shen, Y. (2004). Mutations in SCN9A, encoding a sodium channel alpha subunit, in patients with primary erythermalgia. *Journal of Medical Genetics, 41*, 171–174.

Small Fibers

Definition

A collective term for small myelinated axons, unmyelinated axons, and small-diameter axons.

Cross-References

▶ Toxic Neuropathies

SMP

▶ Sympathetically Maintained Pain
▶ Sympathetically Maintained Pain, Clinical Pharmacological Tests

SNI Model

▶ Neuropathic Pain Model, Spared Nerve Injury

SNL

▶ Spinal Nerve Ligation Model

SNP

▶ Single-Nucleotide Polymorphisms

Social and Interpersonal Features of Pain

▶ Empathy and Pain

Social Dislocation and the Chronic Pain Patient

Ranjan Roy
Department of Clinical Health Psychology, Faculty of Medicine, University of Manitoba, Winnipeg, MB, Canada

Synonyms

Displacement; Disruption

Definition

Dislocation and loss of social roles associated with chronic pain are legion. They occur mainly in three areas, namely, work, family relations, and social life.

Characteristics

The magnitude of social upheaval caused by chronic pain disorders is significant for many patients. Generally, a pain patient seen at a pain clinic has failed to respond to the ministrations of modern medicine. She/he has lived with the pain for many months or even years by the time the patient arrives at a pain clinic. The descent of a person into chronic patienthood, and the new reality of this world, constitutes the central feature of this chapter. Many patients lose most of their valued roles, and the chronic sick role tends to loom large in their lives.

The World of Work
There can be little debate about the beneficial outcome of pain management programs, even in terms of patients returning to work following treatment. On the other hand, many patients do not. This job loss has profound consequences. At its simplest, it calls for a redefinition of self. A person's sense of who and what she/he is tends to be closely tied to what one does. "What do you do?" is probably one of the most oft asked

questions in social situations. Work is often at the center of one's self-image and constitutes the core identity. Hence, the loss of that identity for many patients poses an emotional and intellectual challenge of a magnitude that becomes the source of much grief and even depression. This process of redefinition of self, inevitably in the negative direction, and the associated sense of loss are the two binding themes of this chapter.

Loss of work, for many patients and their families, translates into loss of income, loss of social standing, and loss of friends and mates, and social isolation is not an unknown outcome. For many patients there is no greater insult than their inability to earn a living. Work for many is not a means to an end, but an end in itself. Job loss for health reasons that cannot always be medically substantiated, and challenged by employers, insurance companies, and workers compensation board, creates an untenable situation for the patient. Anger combined with humiliation and grief is not an uncommon presentation for these patients. Loss of the occupational role is extraordinarily undesirable and has consequences for every aspect of a person's life. However, the impact of unemployment that results from chronic pain has received scanty attention from researchers.

When unemployment is caused by accidents and health problems, then it carries additional burdens of coping with poor health as well as unemployment. Both chronic pain and unemployment are known to produce similar negative consequences. Both generate family conflict, depression, ▶ somatization disorders, social isolation, and loss of roles. Unemployment also contributes to family violence, family crisis, health problems, and suicide (Roy 2001). A major Canadian study involving 14,313 subjects revealed the far-reaching consequences of unemployment (D'Arcy and Siddique 1987). Significant health differences emerged between the employed and the unemployed: the unemployed showing a higher level of distress, short- and long-term disabilities, numerous health problems, and proportionately greater utilization of health services. Interaction effects of socioeconomic status and demographic variables showed an association of employment status and emotional health. The blue-collar workers were more prone to physical illness in contrast to white-collar workers, who seemed more prone to emotional distress. Low-income unemployed, who were the principal wage earners, were the most psychologically distressed. An inescapable truth is that the pain clinic population is well represented by this last group. It serves as a double-edged sword, because insufficient education is a known barrier to returning to work among individuals suffering from arthritis and musculoskeletal disorders. Fear of layoff and unemployment is common in workers with low-back strain injuries, resulting in lost time.

A high rate of unemployment among chronic pain patients was reported in a comparative study of headache and back pain patients that was designed to grade the severity of chronic pain (Stang et al. 1998). The sample consisted of 662 headache and 1,024 backache subjects. Over the study period of 3 years, 13 % of headache subjects and 18 % of back pain subjects were unable to obtain or keep full-time work due to their pain conditions. Among the employable, 12 % head and back pain subjects, respectively, were unemployed. Grading of chronic pain emerged as a significant factor in predicting disability, and the highest grade of chronic pain predictably showed the highest levels of affective distress. Unemployment was strongly associated with pain grade, which has great significance for the population attending pain clinics, whose pain levels are generally very high.

Studies investigating the consequences of the added burden of unemployment in the chronic pain population are few and far between. Jackson and associates (Jackson et al. 1998) investigated the complex nature of the effects of unemployment in a group of 83 chronic pain subjects and 88 healthy controls. After controlling for length of current unemployment and number of pain sites, several psychosocial measures, such as structured and purposeful time-use, perceived financial security, and social support from formal sources, were the most powerful predictors of emotional distress. Structured and purposeful

time-use emerged as the most significant predictor of emotional distress for both groups. This study addressed in a most comprehensive way some of the consequences of chronic pain and unemployment and their power to predict emotional distress.

▶ Depression, or at least many depressive symptoms, is ubiquitous in the chronic pain population. As previously noted, unemployment, on the basis of limited data, is high in this population. Averill and associates (Averill et al. 1996) investigated the impact of unemployment on depression in 300 patients randomly selected from 1,000 patients referred to a pain clinic. A remarkable 67 % of the subjects were unemployed. A statistically significant association was found between work-related variables and Beck Depression Inventory (BDI). Unequivocal support was found for an association between unemployment and increased depressive symptomatology. A Swedish study reported higher levels of depression in a group of unemployed women compared to an employed group (Hall and Johnson 1998). Their results showed that even after controlling for social support, stressful life-events, and marital status, depression on the basis of BDI was higher in the unemployed group. Pervasive feelings of social isolation, embarrassment, mild depression, and a generally low life satisfaction were reported in an investigation of 73 unemployed workers over the age of 50 (Rife and First 1989). Shame appears to be a common response to unemployment, associated with guilt, poorer mental health, and, not infrequently, clinical depression (Eales 1998).

A Norwegian investigation of the effects of long-term unemployment in a group of 270 subjects reported that the prevalence of depression, anxiety, and somatic illness was 4–20 times higher than a control group of employed persons at baseline (Classen et al. 1993). A striking finding was that the chances of reemployment were reduced by 70 % among those with a psychiatric diagnosis. This study is of special relevance to the chronic pain population, many of whom fall into the long-term unemployment category, and many of whom have major and minor psychiatric diagnoses.

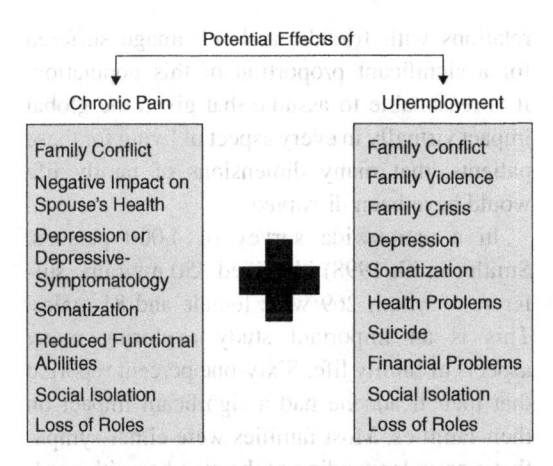

Social Dislocation and the Chronic Pain Patient, Fig. 1 Potential effects of chronic pain and unemployment

The double jeopardy of chronic pain and unemployment often creates an untenable situation for pain patients. Chronic pain and unemployment often create the same problems. Both give rise to family conflicts, depression and somatization, social isolation, financial difficulties, and loss of roles. Unemployment is associated with a rise in family violence and suicide. The level of vulnerability for chronic pain patients is significantly worsened by the problem of unemployment (see Fig. 1).

Family Relations

Family disruption caused by chronic and debilitating illness including chronic pain is considerable. Research into this important topic is extensive, all pointing in the direction that many and varied aspects of family functioning is placed under jeopardy by chronic pain in a family member. Research in one critical area, namely, what may constitute healthy functioning for a family with a chronic pain patient in its midst, is nonexistent. Should this resemble "normal" families as proposed by family researchers? Do these families learn to function effectively without necessarily emulating normal families?

The research literature reveals the diversity and complexity of family problems encountered by the chronic pain families. Findings are contradictory, and to date there is no adequate way of

judging what may be construed as optimum functioning for families with chronic patients in their midst. Family system-based studies have yielded very contradictory findings. Two family functioning scales, Family Environment Scale (FES) and Family Adaptability and Cohesion Evaluation Scale (FACES), are the most common measures of family functioning in the chronic pain literature. Studies vary widely in the specifics of what precise family function may be adversely affected by chronic pain. Findings based on both these scales, on the one hand, show extensive family dysfunction in almost every aspect of family life and, on the other hand, virtually normal family functioning (Roy 2001).

Prima facie, a review of the literature leads to a predictable conclusion, that no aspect of family functioning remains unaffected by chronic pain. Family roles, communication, capacity for expression of positive feelings, family cohesion, family's capacity for adapting to a person with chronic illness, and family organization all fall prey to chronic pain.

The following three studies investigated family functioning from a broad perspective and reported similar findings. One of earlier qualitative studies investigated family functioning of 12 back pain and 20 headache patients and their families (Roy 1989). The study was based on the author's personal knowledge of these patients and their families. Roy's assessment of these families was based on the McMaster Model of Family Functioning (MMFF). MMFF assesses the following dimensions of family functioning: (1) problem solving, (2) communication, (3) roles, (4) affective responsiveness, (5) affective involvement, and (6) behavior control.

The capacity to solve problems was compromised for 75 % of back pain families and 100 % of headache families; communication was at the pathological end of the continuum for 50 % of back pain and 75 % of the headache families. Direct and clear communication was supplanted by unsatisfactory communication patterns. Another incidental finding was the actual amount of communication between the family members and the patient declined.

Problems related to role functioning was pervasive. Nurturance and support related roles were ineffective for 75 % of back pain and 100 % of headache families. Children were the main victims of the failure of this critical aspect of family life. Marital and sexual gratification declined for 75 % of back pain and 60 % of headache families. Occupational and household roles were compromised for 50 % of the back pain group, but the headache group remained unaffected.

These families' capacity to express a wide range of emotions or their affective responsiveness was seriously compromised. Nearly 65 % of both groups expressed difficulties in this area. Not only did they fail to express the whole range of emotions, they were inclined to show their negative emotions more readily than positive feelings. Their capacity for empathic involvement was also hampered, and 83 % of back pain families and 60 % of headache families were found wanting. Finally, in the area of behavior control, which deals with family rules, 83 % and 80 % of back and head pain families, respectively, were encountering difficulties. This particular aspect of family life has special implications for families with children.

This was a qualitative study and the findings were based on extensive family interviews. The interviews were anchored to the MMFF. No specific family functioning instruments were used. The objective of this study was to develop a comprehensive picture of how families were affected in their day-to-day functioning when an adult, usually a parent, was afflicted with a chronic pain disorder.

One American study employed the survey method to assess the social and personal impact of headache in a community sample of headache sufferers in Kentucky (Kryst and Scherl 1994). A total of 647 persons were assessed for serious headache. The 12-month period of prevalence for all serious headaches was 13.4 %. A vast majority of these patients, 73.6 %, reported that headache had adversely affected at least one aspect of their lifestyle. Of these 20 % of men and 62 % of women reported negative effects on their family relations. In addition, efficiency at work, attending social events, capacity for planning ahead,

relations with friends, and self-image suffered for a significant proportion of this population. It is reasonable to assume that given the global impact virtually in every aspect of living for these patients, that many dimensions of family life would have been disrupted.

In a nationwide survey of 4,000 persons, Smith (Smith 1998) identified 350 migraine sufferers of whom 269 were female and 81 males. This is an important study exploring many aspects of family life. Sixty-one percent reported that their headache had a significant impact on their families. Most families were either sympathetic or understanding of the member with headache. Nevertheless, headache delayed or postponed household duties for nearly 79 % of the respondents, and another 64 % reported that activities with children and spouses were adversely affected.

The household activities delayed or postponed included 81 % delaying or postponing house cleaning and yard-work, 79 % laundry and shopping, 76 % cooking, 69 % activities with spouse, 62 % activities with children, and 18 % stayed in bed.

Social activities were either canceled or postponed, albeit by a smaller number of patients. One striking finding was that 61 % of the subjects had to give up parental care for children under 12. This included 61 % canceling plans for playing, helping with homework or spending time together. Sixty-six percent of the children kept quiet, 25 % of the children became confused, and another 17 % were hostile. This is an impressive catalogue of problems that affected younger children.

For the older children between the ages of 12 and 17, 87 % stopped playing music or engaged in any noisy activities, 61 % stopped asking questions or for help with homework, 42 % stopped inviting friends home and 34 % stopped visiting friends. However, children over 12 showed more understanding (87 %) and 42 % were helpful. Finally, 25 % of the migraine sufferers reported that their headaches had a negative effect on their relationship with their spouse or partner. For a full 24 % of the respondents, frequency and/or quality of sexual relations fell. However, only

5 % reported divorce and another 5 % cited head-ache as a cause for separation from a spouse or partner.

This study, painted with a broad brush, leaves little room for doubt that migraine produces sig-nificant problems for the families and that role-related activities are seriously compromised. By extension, an argument can be made that the affective aspects of family life were also adversely affected. The author concluded that ". . . the patients' family should not be ignored when treating the individual patient and that appropriate involvement of the family may be indicated."

These three studies taken together show the extent of the damage that chronic pain inflicts on families. Family roles, communication, marital relations, relations with children are all affected by parental chronic pain. One central issue that needs to be addressed is how to determine effec-tive family functioning in families with chronic pain patients. These survey-type studies present a more comprehensive picture of the problems encountered by chronic pain families. The other point of note is the high level of agreement between the studies about the pervasiveness of family problems. The findings of these three stud-ies are presented in the following tables (see Tables 1, 2, and 3).

Social Disruption

The loss of a support system for chronic pain patients, especially those seen in pain clinics, is sadly a common tale. Common sense dictates that persistent and severe pain narrows or eliminates social interaction. When job loss is added, the sense of isolation for some patients becomes almost as intolerable as the pain itself.

As a starting point, the nature of interaction between an individual and the social support sys-tem is explored. It is convenient to visualize the support system as informal, semiformal, and for-mal. Within the confines of the informal system, it is usual to have partners, close relatives, chil-dren, parents, grandparents, and perhaps close friends. The semiformal system represents work, church, stores, voluntary associations, family physicians, and so on. The formal systems

Social Dislocation and the Chronic Pain Patient, Table 1 Distribution of family problems in a group of chronic pain patients (Roy 1989). Subjects: n = 12 with chronic back pain, n = 20 with chronic headache

Family problems reported	Back pain	Headache
Problem solving	75 %	100 %
Communication	50 %	75 %
Role function (marital/sexual)	75 %	100 %
Household roles	50 %	Nil
Affective responsiveness	65 %	65 %
Absence of empathic involvement	83 %	60 %

Social Dislocation and the Chronic Pain Patient, Table 2 Distribution of family problems in a community sample of headache sufferers (Kryst and Scherl 1994). n = 647

At least one aspect of lifestyle affected	73.6 %
Family relations adversely affected: men	20 %
Family relations adversely affected: women	62 %

In addition
Efficiency at work
Attending social events
Capacity for planning ahead
Relations with friends
were issues for a majority of patients

Social Dislocation and the Chronic Pain Patient, Table 3 Distribution of family problems in a community sample of migraine sufferers (Smith 1998). n = 350, female = 269, male = 81

Negative impact on family	61 %
Postponed household duties	79 %
Postponed housecleaning and yard-work	81 %
Postponed laundry and shopping	79 %
Postponed cooking	76 %
Postponed activities with spouse	69 %
Postponed activities with children	62 %
Stayed in bed	18 %
Given up parental care on children under12	61 %
Negative impact on partner relationship	25 %
Decline in sexual relations	24 %

consist of medical services, workers compensa-tion board, unemployment insurance, Canada Pension, courts, police, and so on. Close listening to the patients reveals an interesting pattern in relation to the social system network. As their interaction with informal and semiformal systems shrink, their involvement with the formal systems tends to rise.

Social Support and Headache

The literature on the role of social support and headache is a collection of studies without benefit of a central theme, such as testing the buffering role of social support. The literature is more concerned with the lack of social support or the need for its availability. Martin and Theunissen (Martin and Theunissen 1993) compared 28 subjects with chronic headaches with two individually matched control groups in terms of life-events, stress, coping skills, and social support. The headache group scored significantly lower on social support compared to the control groups. No differences were found on life-events, stress, or coping between the groups. The authors suggested that clinicians should make some effort to mobilize social support, and that effective social support should be considered a desirable treatment goal.

Lack or ineffective use of social support was noted in a study of 87 undergraduates with recurrent tension headaches, compared to 177 healthy subjects in a matched control group (Holm et al. 1986). Headache subjects relied on ineffective coping strategies of avoidance and self-blame and made less use of social support than did normal controls. In another study involving Australian and American undergraduate students, a major finding was that the prevalence of headache was three to four times higher among American students than their Australian counterparts (Martin and Nathan 1987). A greater incidence of stressors and a deficient social support system accounted for much of the difference between the two groups. A study involving 62 adult chronic headache sufferers and 64 non-headache controls (Martin and Soon 1993) found that it was not just the ineffective use of social support that was problematic, but a lack of satisfaction with the available support. However, support measures did not reveal any linear relationship with chronicity of headache. They made a plea for greater attention from clinicians and researchers to the social dimension of headache.

Another study noted poorer social support in 24 female patients with cluster headache compared to randomly selected and age-matched migraine sufferers (Blomkvist et al. 1997). However, cluster headache patients were found to be significantly more positive as to their anticipated activities in the future compared to migraine patients. The consequences for the lack of social support were not addressed.

These studies on the whole failed to make a compelling case for the value of social support and relied on the proposition of the intrinsic merit of such support. Nevertheless, a couple of these studies made a strong plea that the social dimensions of headaches, including social support, should be an integral part of a clinical investigation.

Chronic Low-Back Pain (CLBP) and Social Support

A handful of studies investigated the influence of social support on CLBP. These studies generally confirmed the value of social support in coping with CLBP. Social support appears to have the capacity to enable CLBP patients to gain good pain control. That was one of the findings of a study involving 95 African-American CLBP patients (Klapow et al. 1995). Twenty-five subjects, categorized with chronic pain syndrome, reported greater life adversity, more reliance on passive-avoidant coping strategies, and a less satisfactory network. In contrast, subjects reporting good pain control reported less adversity, less reliance on passive-avoidant coping strategies, and a more satisfactory social support network. A third group comprised of 24 subjects presented a mixed picture, to the extent that they reported less life adversity, but more reliance on passive-avoidant coping strategies and more satisfactory social support network. Their adaptation to pain was positive. Two points are noteworthy: Firstly, a social support network was only one of the psychosocial factors predicting good pain control, and secondly, the concept of a social support network was not distinguished from intimate social support.

A Swedish study that looked beyond intimate social support and involved a nonclinical population of 90 registered nurses found that social support from coworkers, in addition to psychological demands, authority over decisions, and skill utilization, had a significant effect on the symptoms of low-back pain (Ahlberg-Hulten and Theorell 1995). However, neck and shoulder pain were only related to support at work. The latter finding was telling in terms of the role of social support in mitigating shoulder and neck pain. However, from a theoretical or clinical perspective, the reasons for social support being more effective with one kind of pain than another remain unclear.

In another Swedish study, involving 22,180 employees, 31 % reported neck pain and 39 % reported back pain (Linton 1990). Lifting, monotonous work tasks, vibration, and uncomfortable work postures were the most important ergonomic factors. Work content and social support were the critical psychosocial factors. The combination of a poor psychosocial work environment and exposure to one of the ergonomic variables produced the highest risk factors. In other words, inadequate social support emerged as a very critical factor in predicting high risk for back pain.

Trief and colleagues (Trief et al. 1995) investigated the role of social support in the mitigation of depression in 70 patients with CLBP. The depressed patients differed from the nondepressed patients in their perceived social support and family environment. A noteworthy finding was that depression did not appear to have any association with rehabilitation outcomes. Clearly, susceptibility to depression in CLBP patients is related to the availability of social support, and to that extent, this study furnishes further evidence for the stress-buffer model for social support.

It is virtually impossible to draw any general conclusions on the basis of a handful of rather varied studies on social support and CLBP. While these studies confirm the usefulness of social support in preventing depression or coping more effectively with pain, they vary widely in scope and methodology. From a clinical perspective, they lend further credence to the necessity of careful assessment of the patient's social support system. Yet, a common clinical observation that the social network for chronic pain patients tends to shrink over time has yet to be empirically confirmed (see Fig. 2).

Conclusion

Significant losses of valued social roles demanding a redefinition of one's sense of self or identity underpin the struggle for chronic pain patients. A patient's familiar assumptive world is turned upside down. This entry is a conscious attempt to emphasize the "social" in the biopsychosocial perspective of chronic pain. There is truly no debate about the significance of social factors in complicating medical disorders, and yet as a focus of research, the social aspects of chronic pain disorders still remains the least investigated. Not a single study is to be found on the global consequences of chronic pain. Job loss, disruption to family relationships, reduced physical capacity, increased social isolation for the patient combined with decline in spousal health, and disruptive behavior in children is not as exaggerated a scenario as one might be inclined to think. For many chronic pain patients, if pain was the only problem they had to deal with, life for them would be a great deal more tolerable. Unfortunately, chronic pain disorders have a way of imposing related and yet unforeseen problems that tend to make coping with pain much more onerous. A simple fact that depression and chronic pain together predict a poor outcome illustrates this point. Not infrequently, the root cause for depression is to be found in the altered circumstances of the patients.

Deliberate and well-planned social interventions to address specific problems need to be incorporated into the overall management and treatment of our patients. Great strides have been made in relation to psychological treatment of chronic pain, but that cannot be said for social interventions. While at a common sense level clinicians of every stripe fully appreciate the importance of the patient's environment, for reasons that remain unclear, their social problems are seldom addressed. On the other hand, when the social problems are assessed alongside the medical and psychological ones, patients become the main beneficiaries of such a comprehensive

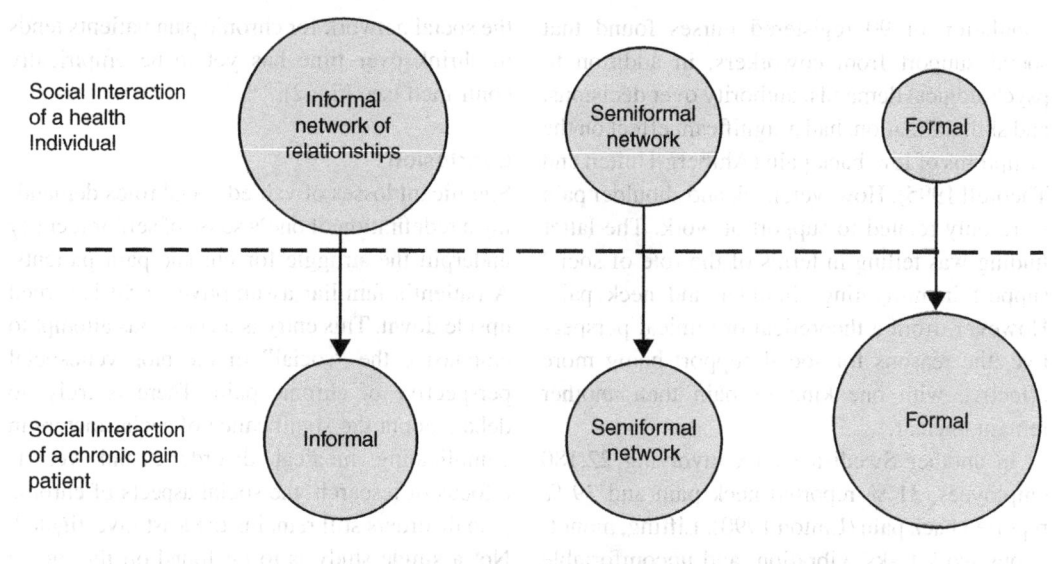

Social Interaction of a health Individual	Informal network of relationships	Semiformal network	Formal

Social Interaction of a chronic pain patient	Informal	Semiformal network	Formal

Examples of support systems

Informal systems	Semiformal systems	Formal systems
Spouses/partners	Workplace	Medical services
Children	Church	WCB
Parents	Volunteer organization	Insurance companies
Friends/relatives	Family physicians	Legal systems
Close friends		Canada pension
Neighbours		

Social Dislocation and the Chronic Pain Patient, Fig. 2 Representation of social network

approach. Research is rather silent on this topic. There is no empirical evidence to suggest that recognition and treatment of social problems alongside medical and psychological improves outcome. While researchers may want to sort that out, at a clinical level, the patient's need for help with social problems cannot and should not be ignored.

The way ahead is to pay heed to the social environment of the patient. Pain clinics must incorporate routine social assessment of each patient alongside medical and psychological. Finally, a case is presented that captures the essence of social dislocation.

Mr. Albert, in his mid-30s, was involved in a work-related accident, which had profound impact on every aspect of his life. He was married man with a baby daughter. He was in the construction business and had a reputation for his knowledge of commercial constructions. His wife worked, and together they generated a substantial

income. He enjoyed a very close relationship with his parents. He was a churchgoer, had an extensive network of friends, and belonged to a number of social organizations.

The accident, which occurred at a construction site, was initially not thought be serious. While lifting heavy material, he slipped and fell, sustaining bruises on his buttocks and back. He was off work for a few days, seemingly made a complete recovery and returned to his job. He soon discovered that he was experiencing severe back pain, especially while lifting. He persisted until his pain worsened to the point that he had to take time off work. His employer remained very sympathetic to Mr. Albert's situation. That was about to change.

Mr. Albert sought medical treatment, and after extensive radiological, orthopedic, and neurological investigations, he was pronounced fit. He was incredulous as his pain continued to worsen. He filed for worker's compensation, which was denied. As a starting point, he was left with no income and a very painful lower back. He was in the early stages of conflict with three key systems, medical, employer, and the worker's compensation board. His problems were just beginning. An immediate consequence of the action of these three organizations was a precipitous drop in his income. Thus, began his struggle on the domestic front as well as the world outside.

Regarding his trials and tribulations on the domestic front, a great reversal of fortune occurred along with an equally measurable change in his roles and attitudes. Sexual activities came to a halt. Mr. Albert assumed the role, albeit grudgingly, of a homemaker and babysitter, since they could no longer afford daycare. This family had no savings and they were forced into finding cheaper accommodation. All this time, Mr. Albert was not so gradually sinking into a chronic sick role. Level of tension between him and his wife became palpable, and his pain continued unabated.

Mr. Albert's search for a medical cure gained momentum. He saw several specialists and sub-specialists including a psychiatrist, but all the findings were negative. In the meantime, strained finances created serious tensions in family relationships. Social outings, family get-togethers, and even church and social activities were abandoned. All this happened in a matter of 6 months. His pain continued to be a mystery. Hopelessness and helplessness were the overpowering emotions he experienced, at the basis of which was a terrific loss of self-esteem as man, a father, a partner, and a provider. This is a summary of the information obtained during a routine psychosocial assessment at the point of Mr. Albert's admission to a pain clinic. His story is not unique among chronic pain sufferers.

References

Ahlberg-Hulten, G., & Theorell, T. (1995). Social support, job strain and musculo-skeletal pain among female health care personnel. *Scandinavian Journal of Work, Environment & Health, 21*, 435–439.

Averill, P., Novy, D., Nelson, D., et al. (1996). Correlates of depression in chronic pain patients: A comprehensive examination. *Pain, 65*, 93–100.

Blomkvist, V., Hannerz, J., Orth-Gomer, K., et al. (1997). Coping styles and social support in women suffering from cluster headaches or migraine. *Psychotherapy and Psychosomatics, 66*, 150–154.

Classen, B., Bjorndal, A., & Hjort, P. (1993). Health and Re-employment in a two-year follow-up of long-term unemployed. *Journal of Epidemiology and Community Health, 47*, 14–18.

D'Arcy, C., & Siddique, C. (1987). Unemployment and health: An analysis of Canada health survey@ data. Inter. *International Journal of Health Services, 15*, 609–635.

Eales, M. (1998). Depression and anxiety in unemployed men. *Psychological Medicine, 18*, 935–945.

Hall, E., & Johnson, J. (1998). Depression in unemployed Swedish women. *Social Science & Medicine, 27*, 134–135.

Holm, J., Holroyd, V., Hursey, K., et al. (1986). The role of stress in recurrent tension headache. *Headache, 26*, 160–167.

Jackson, T., Lezzi, A., Lafreniere, K., et al. (1998). Relations of employment status to emotional distress among chronic pain patients: A path analysis. *The Clinical Journal of Pain, 14*, 55–60.

Klapow, J., Slater, M., Patterson, T., et al. (1995). Psychosocial factors discriminate multidimensional clinical groups of chronic low-back patients. *Pain, 62*, 349–355.

Kryst, S., & Scherl, E. (1994). A population based survey of the social and personal impact of headache. *Headache, 34*, 344–350.

Linton, S. (1990). Risk-factors for neck and back pain in a working population in Sweden. *Work and Stress, 4*, 41–49.

Martin, P., & Nathan, P. (1987). Differential prevalence ratio for headaches: A function of stress and social support. *Headache, 27*, 292–333.

Martin, P., & Soon, K. (1993). The relationship between perceived stress, social support and chronic headache. *Headache, 33*, 307–314.

Martin, P., & Theunissen, C. (1993). The role of life-event stress, coping and social support in chronic headache. *Headache, 33*, 301–306.

Rife, J., & First, R. (1989). Discouraged older workers: An exploration study. *International Journal of Aging & Human Development, 29*, 195–203.

Roy, R. (1989). *Chronic pain and the family: A problem centered perspective*. New York: Human Sciences Press.

Roy, R. (2001). *Social relations and chronic pain*. New York: Kluwer Academic/Plenum Publishers.

Smith, R. (1998). Impact of migraine on the family. *Headache, 38*, 424–426.

Stang, P., Von Korff, M., & Galer, B. S. (1998). Reduced labor force participation among primary care patients with headache. *Journal of General Internal Medicine, 13*, 296–302.

Trief, P., Carnrike, C., & Drudge, O. (1995). Chronic pain and depression: Is social support relevant? *Psychological Reports, 76*, 27–236.

Social Learning

Definition

The process by which individuals acquire knowledge about different aspects of their social environment. This may include experience, behavioral consequences (e.g., reward vs. punishment), or observation. In short, it is a term used to describe the process by which new behaviors are learned through observation.

Cross-References

► Behavioral Therapies to Reduce Disability
► Psychology of Pain, Self-efficacy

Social Learning Theory

Definition

A psychological theory, proposed by Albert Bandura, in which the importance of observing and modeling the behaviors, attitudes, and emotional reactions of others is emphasized.

Cross-References

► Impact of Familial Factors on Children's Chronic Pain

Social Security Disability Insurance

Synonyms

SSDI

Definition

The Social Security Disability Insurance (SSDI) is a national program in the United States that is administered by the Social Security Administration. It provides disability benefits to disabled workers who are insured under the Social Security Act and have worked long enough to be eligible. It also provides disability benefits for children of insured workers who become disabled before age 22 and disabled widows or widowers and certain surviving divorced spouses of insured workers.

Cross-References

► Disability Evaluation in the Social Security Administration

Social Security Disability Insurance (United States)

► Disability Evaluation in the Social Security Administration

Social Support

Definition

Social support generally refers to the availability or provision of instrumental and/or emotional assistance from significant others.

Cross-References

► Spouse, Role in Chronic Pain

Socioeconomic Factors

► Pain in the Workplace, Risk Factors for Chronicity, and Workplace Factors

Sodium Channel

Definition

Large transmembrane proteins with a pore that selectively allows sodium ions to pass through the cell membrane. There are a multitude of different sodium channels in excitable and in non-excitable cells. For nerve cells, particularly important are voltage-dependent sodium channels which become permeable for sodium ions upon depolarization of the cell membrane. The terminals of nociceptors have particular sodium channels which are not found in other neurons. Particularly important are the channels NaV1.7 and NaV1.8. These and other channels have been structurally identified.

Cross-References

▶ Demyelination

Sodium Channel Blockers

Definition

Chemical substances that block either specifically particular types of sodium channels or a multitude of them. Clinically, sodium channel blockers have been used for treating very different diseases, e.g., neuropathic pain.

Cross-References

▶ Postoperative Pain, Postamputation Pain, Treatment, and Prevention

Sodium Pump

Definition

Upon excitation, sodium ions penetrate into nerve cells which have to be pumped out after termination of excitations. There are several types of sodium pumps. The most important is the "sodium-potassium ATPase" which pumps sodium ions out of the neuron in exchange for potassium ions which are brought into the cell, to maintain the equilibrium of extracellular-intracellular concentration gradients for these two ions. The pump requires energy which is supplied by ATP in a stoichiometric fashion: one ATP for exchange of 3 sodium ions against 2 potassium ions. Since more positive charges are brought out of the neuron, the pump is "electrogenic," i.e., it increases the hyperpolarization of the neuron.

Cross-References

▶ Nociceptors, Action Potentials, and Post-Firing Excitability Changes

Soft Tissue Manipulation

▶ Massage, Basic Considerations

Soft Tissue Pain Syndromes

Synonyms

STP syndromes

Definition

A generic term used to include painful medical conditions, such as bursitis, tendinitis, complex regional pain syndrome (CRPS), myofascial pain syndrome (MPS), and fibromyalgia syndrome (FMS), which substantially contribute to morbidity and physical dysfunction in the general population.

Cross-References

▶ Muscle Pain, Fibromyalgia Syndrome (Primary, Secondary)

Soft Tissue Rheumatism

▶ Myofascial Pain

Softener

Definition

A softener is a laxative drug whose principal mode of action is to make the bowel movement softer and therefore easier to pass.

Solicitous

Definition

Manifesting or expressing attentive care and protectiveness that is full of concern or fear.

Cross-References

▶ Assessment of Pain Behaviors

Solicitous Responses

Definition

Solicitous responses are defined as a set of positive responses (e.g., expressions of sympathy, taking over duties and responsibilities) delivered by significant others, including the spouse or family members, contingent on the display of pain behaviors.

Cross-References

▶ Spouse, Role in Chronic Pain

Solicitousness

▶ Spouse, Role in Chronic Pain

Soma

Synonyms

Perikarya

Definition

Soma refers to the cell body of the neurone, which accommodates its nucleus and the main part of its synthesis machinery such as the endoplasmic reticulum, ribosomes, Golgi complex, and mitochondria.

Cross-References

▶ Nociceptors, Action Potentials, and Post-Firing Excitability Changes
▶ Spinothalamic Tract Neurons, Morphology

Somatic Dysfunction

Definition

Somatic dysfunction is the impairment or altered functioning of the related elements of the musculoskeletal system. This is characterized by asymmetry of bony landmarks, altered articular mobility, and tissue texture abnormality.

Cross-References

▶ Chronic Pelvic Pain, Physical Therapy Approaches, and Myofascial Abnormalities

Somatic Pain

Paul M. Murphy
Department of Anaesthesia and Pain
Management, Royal North Shore Hospital,
St. Leonards, NSW, Australia

Definition

This term is derived from the Greek σομα, meaning body. In principle, somatic pain is pain evoked by nociceptive information arising from any of the tissues that constitute the structure of the body. These would include the bones, muscles, joints, ligaments, and tendons of the spine, the trunk, and the limbs, but technically they would include also the skull, the pachymeningeal coverings of the brain and spinal cord, and the teeth. More explicitly, somatic pain is used to distinguish pain that does not arise from the viscera, i.e., internal organs, of the body. Pain from those structures is referred to as ▶ visceral nociception and pain. Similarly, pain from the head or skull is referred to as ▶ headache, and pain from the teeth is ▶ dental pain. Consequently, somatic pain is effectively restricted to refer to pain arising from musculoskeletal structures, the limbs, the spine, the chest wall, and the abdominal wall.

Somatic pain is typically described as aching or stabbing. It is localized to the area of tissue injury or stimulation and may follow the distribution of a nerve root or peripheral nerve. A pivotal feature of somatic pain is that it is caused by the stimulation of the nerve endings of the peripheral nerves that innervate the tissues that are the source of pain. This feature distinguishes somatic pain from ▶ neuropathic pain, ▶ neuralgia, and ▶ radicular pain, in which the source of nociception lies in axons of the affected nerve. In this respect, somatic pain and visceral pain do not differ. Both are evoked by the stimulation of free nerve endings. In mechanism and clinical features, both are quite similar. They differ only with respect to the classification of the tissues from which they arise. Both somatic pain and visceral pain can be referred to remote sites. Although the mechanism of referral is similar, the clinical features of each differ (see ▶ Somatic Referred Pain and ▶ Visceral Referred Pain).

Characteristics

Background

▶ Specificity theory suggests that the ▶ somatosensory system processes sensory modalities by means of specific receptors and neuroanatomical pathways, although this view is less rigid than previously thought. As early as 1906, it was proposed that cutaneous receptors responding to noxious stimuli should be termed nociceptors (Sherrington 1906).

Nociceptors

Nociceptors in the skin and other deeper somatic tissues such as the periosteum are morphologically free nerve endings or simple receptor structures. A noxious stimulus activates the nociceptor, depolarizing the membrane via a variety of stimulus-specific ▶ transduction mechanisms. C polymodal nociceptors are the most numerous of somatic nociceptors and respond to a full range of mechanical, chemical, and thermal noxious stimuli (see ▶ Noxious Stimulus). These nociceptors are coupled to unmyelinated C fibers. Electrophysiological activity in these slow conduction C fibers is characteristically perceived as dull, burning pain. Faster conducting Aδ fibers are coupled to more selective thermal and mechanothermal receptors considered to be responsible for the perception of sharp or "stabbing" pain (Julius and Basbaum 2001).

Spinal Cord Integration

The majority of somatic nociceptive neurons enter the dorsal horn spinal cord at their segmental level. A proportion of fibers pass either rostrally or caudally in ▶ Lissauer's tract. Somatic primary afferent fibers terminate predominantly in laminae I (marginal zone) and II (▶ substantia gelatinosa) of the ▶ dorsal horn, where they synapse with projection neurons and excitatory/inhibitory interneurons. Some Aδ fibers penetrate more deeply into lamina V (Parent 1996).

Projection neurons are classified as follows:

1. Nociceptive specific (NS):
   - Respond exclusively to noxious stimuli
   - Small receptive field
   - Predominate in lamina I
   - Also in laminae II and V
2. Low-threshold neurons (LT):
   - Respond to innocuous stimuli only
   - Laminae III, IV
3. Wide dynamic range neurons (WDR):
   - Input from wide range of sensory afferents
   - Large receptive field
   - Predominate in lamina V (also in I)

The amino acids ▶ glutamate and ▶ aspartate are the primary ▶ neurotransmitters involved in spinal excitatory transmission. Fast postsynaptic potentials generated via the action of glutamate on AMPA receptors are primarily involved in nociceptive transmission. Prolonged C fiber activation facilitates glutamate-mediated activation of ▶ NMDA receptors and subsequent "▶ wind-up." The peptidergic neurotransmitters ▶ substance P and calcitonin G-related peptide (CGRP) are coproduced in glutamatergic neurons and released with afferent stimulation. These transmitters appear to play a neuromodulatory role facilitating the action of excitatory amino acids. A number of other molecules including glycine, ▶ GABA, somatostatin, endogenous opioids, and endocannabinoids play modulatory roles in spinal nociceptive transmission (Fürst 1999).

Supraspinal Somatic Pain Pathways

Nociceptive somatic input is relayed to higher cerebral centers via three main ascending pathways (Basbaum and Jessel 2000):

1. Spinothalamic path:
   - Originates in laminae I and V–VII
   - Composed of NS and WDR neuron axons
   - Projects to thalamus via lateral tracts (▶ neospinothalamic tract) and medial tracts (▶ paleospinothalamic tract)
   - Lateral tract passes to ventral posterior lateral (VDL) nucleus and subserves discriminative components of pain
   - Medial tract is responsible for autonomic and emotional components

- Additional fibers pass to ▶ reticular activating system (arousal response) and the periaqueductal gray matter (PAG) (▶ descending modulation)
2. Spinoreticular pathway:
   - Originates in laminae VII and VIII
   - Terminates predominantly on the medial medullary reticular formation
3. Spinomesencephalic pathway:
   - Originates in laminae I and V
   - Terminates in midbrain
   - Projects to mesencephalic PAG
   - Not essential for pain perception but appears to modulate afferent input

Cortical Representation of Somatic Pain

Multiple cortical areas are activated by nociceptive afferent input including the primary and secondary somatosensory cortex, the insula, the anterior cingulate cortex, and the prefrontal cortex. Pain is a multidimensional experience with sensory-discriminative and affective-motivational components. Advances in functional brain imaging have allowed further understanding of the putative role of cortical structures in the pain experience. The primary and secondary somatosensory cortices and the insula predominantly subserve pain intensity. The prefrontal cortex, right posterior cingulate cortex, and the brainstem/periventricular gray matter play important roles in interpreting pain intensity. The affective unpleasant sensation of pain appears to be subserved by the left anterior cingulate cortex, while the right anterior cingulate cortex, left thalamus, and frontal inferior cortex play important roles in somatic ▶ pain threshold (Treede et al. 1999).

Cross-References

- ▶ Cancer Pain Management, Treatment of Neuropathic Components
- ▶ Chronic Pelvic Pain, Musculoskeletal Syndromes
- ▶ Diagnosis of Pain, Epidural Blocks
- ▶ Rest and Movement Pain
- ▶ Taxonomy

References

Basbaum, A. I., & Jessel, T. M. (2000). The perception of pain. In E. R. Kandel, J. H. Schwartz, & T. M. Jessell (Eds.), *Principles of neural science* (pp. 472–491). New York: McGraw-Hill.

Fürst, S. (1999). Transmitters involved in antinociception in the spinal cord. *Brain Research Bulletin, 48*, 129–141.

Julius, D., & Basbaum, A. I. (2001). Molecular mechanisms of nociception. *Nature, 413*, 203–207.

Parent, A. (1996). *Carpenter's human neuroanatomy.* Baltimore: Williams & Wilkins.

Sherrington, C. S. (1906). *The integrative action of the nervous system.* New York: Scribner.

Treede, R.-D., Kenshalo, D. R., Gracely, R. H., et al. (1999). The cortical representation of pain. *Pain, 79*, 105–111.

Somatic Referred Pain

David Roselt
Aberdovy Clinic, Bundaberg, QLD, Australia

Synonyms

Referred pain

Definition

Referred pain is pain perceived in a region innervated by nerves other than those innervating the source of the pain (Bogduk 1987; Merskey and Bogduk 1994). Somatic referred pain is explicitly somatic pain that becomes referred. The term is used to distinguish referred pain that arises from the musculoskeletal tissues of the body from ▶ visceral referred pain.

Characteristics

The physiological basis of referred pain is convergence. Common neurons in the spinal cord and thalamus receive sensory input from different peripheral sites and relay this onto higher centers. Under those conditions, and in the absence of additional sensory input to clarify the situation, the brain is unable to identify the source of the pain accurately and attributes it erroneously to the entire area subtended by the common neurons (Bogduk 1987).

Somatic referred pain typically occurs when the source of pain lies in a deep musculoskeletal structure, from which the brain is unaccustomed to receiving nociceptive input. Ambiguity as to the source of information arises, either or both, because the painful structure is not densely innervated, and the central pathways along which the information is relayed are not highly organized somatotopically. In essence, the ambiguity is "built in" to the nervous system.

Nature

Much of our understanding of referred pain arose from clinical experiments by Kellgren in the late 1930s (Bogduk 2002; Kellgren 1938, 1939). He showed that noxious stimulation of muscle using injections of hypertonic saline produced local and referred pain that was diffuse and often perceived as being removed from the site of stimulation. In the limbs, muscle pain tended to be referred towards the joint upon which the muscle acted (Kellgren 1938).

When investigating spinal referred pain by injecting interspinous ligaments with hypertonic saline, Kellgren (1939) found that stimulation of thoracic segments produced referred pain in the posterior and anterior chest wall (Fig. 1). Injection into the cervical and lumbar segments produced referred pain in the upper and lower limbs, respectively (Figs. 2 and 3).

These experiments showed that spinal pain could arise from noxious stimulation of intrinsic structures of the vertebral column and produce pain referred to the trunk and limbs. This was in the face of the prevailing wisdom and contemporary assumptions (that persist today) that referred pain from the spine could only be caused by nerve root compression (Bogduk 2002).

Nerve root compression may cause ▶ radiculopathy, which is conduction block resulting in numbness and or weakness. It may cause ▶ radicular pain, which is pain arising from irritation of a spinal nerve or its roots. This pain

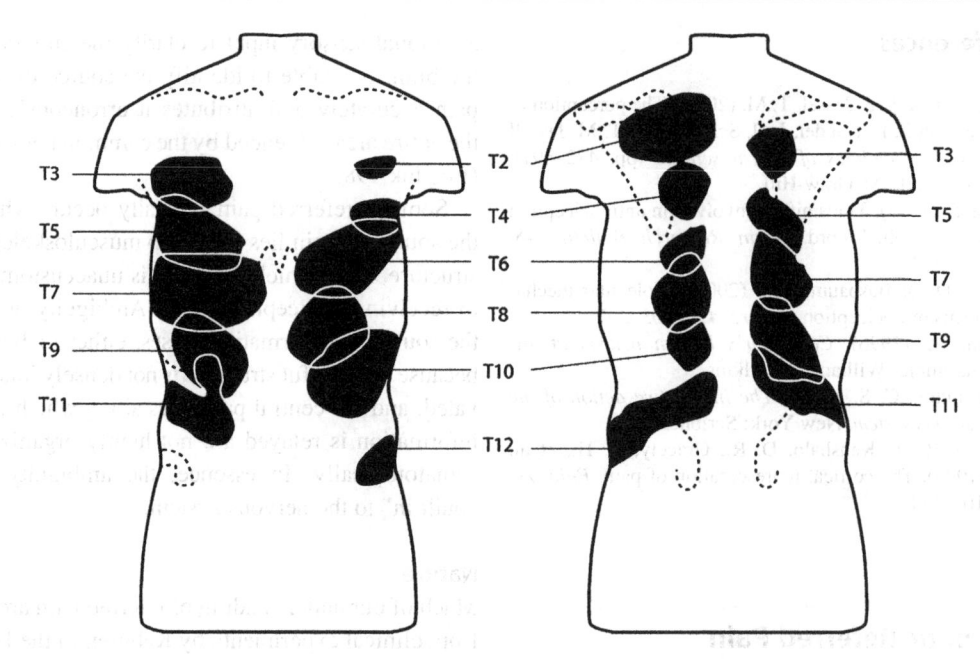

Somatic Referred Pain, Fig. 1 Patterns of referred pain produced by noxious stimulation of the thoracic interspinous ligaments at the segments indicated (Based on Kellgren (1939))

has a different quality and distribution from somatic referred pain and is due to ectopic activation of the nerve root axons or dorsal root ganglia, rather than stimulation of peripheral nociceptors (Bogduk 1987; Merskey and Bogduk 1994).

Somatic referred pain occurs in a segmental distribution, and such stimulation of successively lower spinal segments produces pain in successively more caudal regions of the trunk or limbs (Figs. 1, 2, and 3). This segmental pattern, however, does not correspond to dermatomes.

Kellgren's work was subsequently reproduced by others (Feinstein et al. 1954) who produced maps of referred pain similar to those of Kellgren. The patterns of distribution of pain from a given segment are not identical between studies, and differ between individuals. Nevertheless, all studies showed a segmental pattern.

Inman and Saunders (1944) published an influential paper that firmly introduced the concept of sclerotomes. They proposed that sclerotomes represented the segmental pattern of innervation of skeletal tissues and were different from dermatomes and myotomes. However,

although conceptually attractive, this concept is poorly founded.

Dermatomes are the areas of skin supplied by individual spinal nerves. They were first mapped by studying the zones of eruption of vesicles in herpes zoster, which provide a physical tracer of segmental nerves. Subsequently, they were corroborated by studies of the zones of sensory loss following dorsal rhizotomy (Bogduk 2002). In this regard, dermatomes have a tangible, anatomical substrate.

Myotomes are the groups of muscles supplied by a given spinal nerve. They were derived from mapping zones of weakness after segmental nerve injury and reinforced by mapping EMG activity in response to electric stimulation of segmental nerves (Bogduk 2002). Therefore, they, too, have an anatomical substrate.

In contrast, no anatomical substrate for sclerotomes has been established. No one has traced the segmental innervation of skeletal tissues. Sclerotomes were only inferred to exist simply because pain from deep, skeletal tissues seemed to follow a segmental pattern. The maps published by Inman and Saunders (1944) were

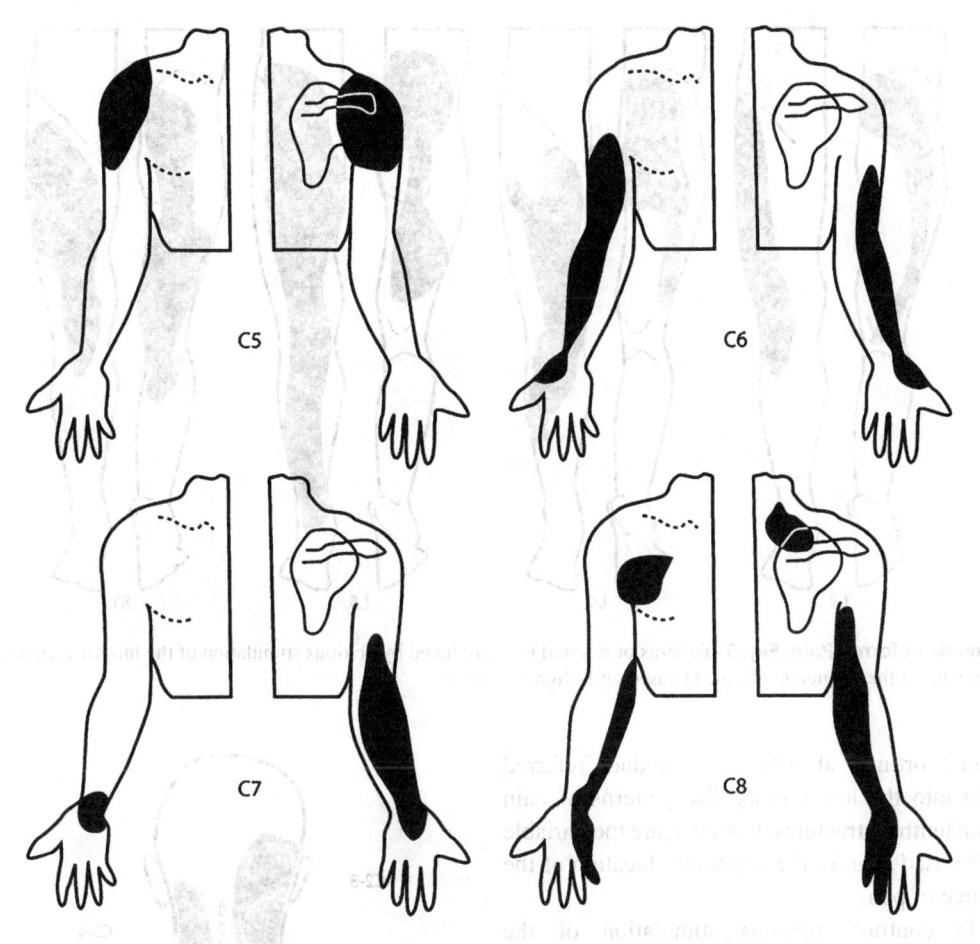

Somatic Referred Pain, Fig. 2 Patterns of referred pain produced by noxious stimulation of the cervical interspinous ligaments at the segments indicated (Based on Kellgren (1939))

idealized, and not based on published quantitative data (Bogduk 2002). They did not take into account the variance between individuals and between studies, in the patterns of somatic referred pain. This variance suggests that the purported pattern of segmental innervation of deep tissues is not consistent. Alternatively, and furthermore, pattern referral may be more due to differences in central connections of deep somatic afferents, than to patterns of peripheral innervation.

Kellgren's work was extended, when researchers from the mid-1970s onwards investigated patterns of referred pain from structures more likely to be clinically relevant than the interspinous ligaments.

Mooney and Robertson (1976) injected hypertonic saline into the lower lumbar zygapophysial joints in normal volunteers and produced low back pain with referral to the lower limbs. This knowledge was then applied to relieve similar clinical pain, by injecting local anesthetic and steroid into patient's lumbar zygapophysial joints.

Others have reproduced or emulated this study, using intra-articular injections and electrical stimulation, to confirm that the lumbar zygapophysial joints can be a source of referred pain into the buttock and lower limb (Bogduk and McGuirk 2001). Studies have also shown that noxious stimulation of lumbar intervertebral discs (O'Neill et al. 2002) and the sacroiliac

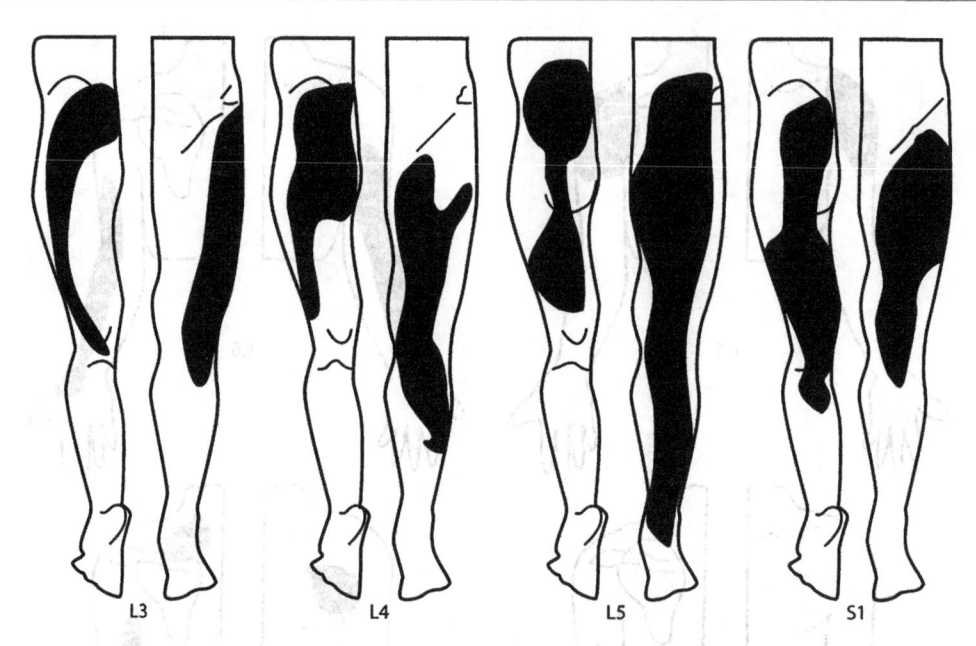

Somatic Referred Pain, Fig. 3 Patterns of referred pain produced by noxious stimulation of the lumbar interspinous ligaments at the segments indicated (Based on Kellgren (1939))

joint (Fortin et al. 1994) can produce referred pain into the lower limb. The patterns of pain from lumbar structures, however, are too variable to be used to infer the segmental location of the source of pain.

In contrast, noxious stimulation of the zygapophysial joints of the cervical spine causes pain in distinctive patterns, which can be reliably used to pinpoint the segmental source of pain (Bogduk 2003) (Fig. 4). According to their segmental location, these joints refer pain to the shoulder girdle or into the head (Bogduk 2003, 2004). The same segmental patterns apply to cervical intervertebral discs (Grubb and Kelly 2000).

Somatic referred pain can be associated with muscle spasm in the areas of referred pain. Mooney and Robertson (1976) found that referred pain from the lumbar zygapophysial joints was associated with involuntary activity in the hamstring muscles, and demonstrated this with EMG. This phenomenon was explored by Bogduk (1980), who replicated Kellgren's experiments. Referred pain produced by injecting the lower lumbar interspinous ligaments with hypertonic saline was associated with involuntary

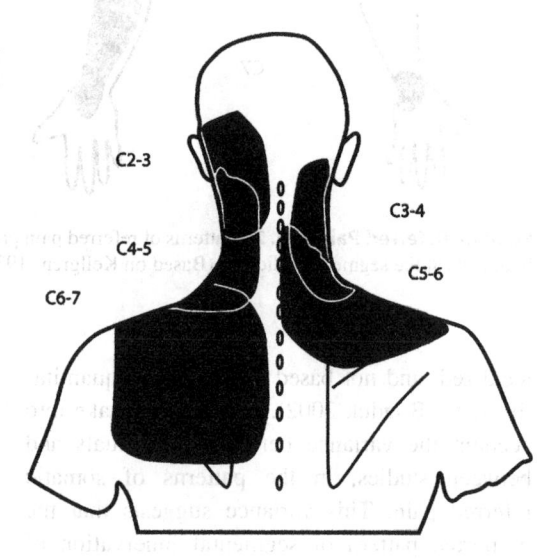

Somatic Referred Pain, Fig. 4 Patterns of referred pain produced by noxious stimulation of the cervical zygapophysial joint or the cervical intervertebral discs

muscle activity in the multifidi, tensor fasciae latae, and gluteus medius muscles. This activity started shortly after the onset of pain and disappeared as the pain passed over the next few minutes (Bogduk 1980).

Clinical Features

All the experimental studies on somatic referred pain have consistently shown that it has a characteristic quality. It is felt deeply, and is aching in quality, or like an expanding pressure. It expands to encompass a broad area. Once established, its location remains constant. Subjects or patients are aware of its distribution in a general sense; they can clearly identify the centroid of the distribution, but its boundaries are hard to define.

These features distinguish somatic referred pain from radicular pain and ▶ neuropathic pain, which differ in quality. Radicular pain is typically shooting or lancinating in quality and extends along a narrow band. Neuropathic pain is usually burning in quality and is associated with sensory disturbances. Somatic referred pain is not associated with neurological abnormalities.

Making the distinction between somatic referred pain and radicular or neuropathic pain is crucial, for the investigation and subsequent treatment of these different types of pain are distinctly different. Confusing or misrepresenting one for the other becomes a recipe for failure and misadventure.

References

Bogduk, N. (1980). Lumbar dorsal ramus syndrome. *The Medical Journal of Australia, 2,* 537–541.

Bogduk, N. (1987). *Clinical anatomy of the lumbar spine and sacrum* (3rd ed., pp. 187–213). Edinburgh: Churchill Livingstone.

Bogduk, N. (2002). The physiology of deep somatic pain. *Australasian Musculoskeletal Medicine, 7,* 6–15.

Bogduk, N. (2003). The anatomy and pathophysiology of neck pain. *Physical Medicine and Rehabilitation Clinics of North America, 14,* 455–472.

Bogduk, N. (2004). The neck and headaches. *Neurologic Clinics of North America, 22,* 151–171.

Bogduk, N., & McGuirk, B. (2001). *Medical management of acute and chronic low back pain. An evidence-based approach* (pp. 9–10). Amsterdam: Elsevier.

Feinstein, B., Langton, J. N. K., & Jameson, R. M. (1954). Experiments on pain referred from deep somatic tissues. *Journal of Bone and Joint Surgery, 35A,* 981–987.

Fortin, J. D., Dwyer, A. P., & West, S. (1994). Sacroiliac joint: Pain referral maps upon applying a new injection/arthrography technique: Part 1: Asymptomatic volunteers. *Spine, 19,* 1475–1482.

Grubb, S. A., & Kelly, C. K. (2000). Cervical discography: Clinical implications from 12 years of experience. *Spine, 25,* 1382–1389.

Inman, V. T., & Saunders, J. B. D. (1944). Referred pain from skeletal structures. *The Journal of Nervous and Mental Disease, 99,* 660–667.

Kellgren, J. H. (1938). Observations on referred pain arising from muscle. *Clinical Science, 3,* 175–190.

Kellgren, J. H. (1939). On the distribution of pain arising from deep somatic structures with charts of segmental pain areas. *Clinical Science, 4,* 35–46.

Merskey, H., & Bogduk, N. (1994). *Classification of chronic pain. Descriptions of chronic pain syndromes and definitions of pain terms* (2nd ed., pp. 11–16). Seattle: IASP Press.

Mooney, V., & Robertson, J. (1976). The facet syndrome. *Clinical Orthopaedics, 115,* 149–156.

O'Neill, C., Kurgansky, M., Derby, R., et al. (2002). Disc stimulation and patterns of referred pain. *Spine, 27,* 2776–2781.

Somatic Sensory and Intralaminar Nuclei of Thalamus

Definition

Nuclei that transmit input from the dorsal column nuclei and STT to the cortex.

Cross-References

▶ Deep Brain Stimulation

Somatic Sensory Nucleus of Human Thalamus

Definition

Relevant to the discussion of plasticity is the principle somatic sensory nucleus of the thalamus (ventral posterior – VP, also known as ventral caudal – Vc).

Cross-References

▶ Thalamus, Receptive Fields, Projected Fields, Human

Somatization

Definition

The tendency to use numerous physical symptoms to express emotional distress.

Cross-References

▶ Functional Abdominal Pain in Children
▶ Pain in the Workplace, Risk Factors for Chronicity, and Psychosocial Factors
▶ Pain, Psychiatry, and Ethics

Somatoform Disorders

Definition

A category of mental disorders in which individuals have physical symptoms that suggest a general medical condition, but are not fully explained by a general medical condition, by the direct effects of a substance, or by another mental disorder. The symptoms must cause clinically significant distress or impairment in social, occupational, or other areas of functioning.

Cross-References

▶ Hypochondriasis, Somatoform Disorders, and Abnormal Illness Behavior

Somatoform Pain Disorder

Definition

A term used in the Diagnostic and Statistical Manual of Mental Disorders, Third Edition of the American Psychiatric Association to describe a preoccupation with pain causing distress and disability, which is presumed to be generated by psychological factors.

Cross-References

▶ Pain as a Cause of Psychiatric Illness
▶ Psychiatric Aspects of the Management of Cancer Pain

Somatoform Pain Disorder (DSM-III R and ICD-10)

▶ Psychiatric Aspects of the Epidemiology of Pain

Somatosensory

Definition

Derived from the Greek word for "body," somatosensory input refers to sensory signals from all tissues of the body including skin, viscera, muscles, and joints. Somatic usually refers to input from body tissue other than viscera.

Cross-References

▶ Acute Pain Mechanisms
▶ Infant Pain Mechanisms
▶ Pain in Humans, Psychophysical Law
▶ PET and fMRI Imaging in Parietal Cortex (SI, SII, Inferior Parietal Cortex BA40)
▶ Somatic Pain

Somatosensory Evoked Potentials

Synonyms

SEPs

Definition

These are electrical signals generated by the nervous system following surface excitation (until now almost always using electrical stimuli) of mixed afferent fibers in peripheral nerves. These afferent signals can be recorded over the peripheral nerve, spinal cord, or over the somatosensory cortex using scalp electrodes, with various surface-recording montages (electrode locations) in humans.

Cross-References

▶ Diagnosis and Assessment of Clinical Characteristics of Central Pain
▶ Infant Pain Mechanisms
▶ Metabolic and Nutritional Neuropathies

Somatosensory Nervous System

Definition

The somatosensory nervous system is part of the nervous system that conveys sensations from the body, such as touch, position, pain, and temperature.

Cross-References

▶ Hypoalgesia, Assessment
▶ Hypoesthesia, Assessment

Somatosensory Pain Memories

Definition

Somatosensory pain memories refer to memory processes related to painful somatosensory stimulation. These can occur already at the level of

the spinal cord, but also involve the brain stem, the thalamus, and cortical as well as subcortical regions.

Cross-References

▶ Diathesis-Stress Model of Chronic Pain

Somatosensory Pathways

Definition

The neuronal pathways that bring sensory information from the periphery (skin, deep somatic tissues, and internal organs) to the brain. Each pathway has a number of levels of neural integration where sensory information may be processed.

Cross-References

▶ Quantitative Thermal Sensory Testing of Inflamed Skin

Somatostatin

Synonyms

Somatotropin

Definition

Somatostatin is a cyclic peptide of 14 amino acids. It was originally isolated from the hypothalamus but is expressed in a variety of neurons and endocrine cells. Its main action is inhibition of release of other mediators.

Cross-References

▶ Cancer Pain Management, Adjuvant Analgesics in Management of Pain Due to Bowel Obstruction
▶ Opioid Modulation of Nociceptive Afferents In Vivo

Somatotopic

Definition

The correspondence of information transmitted from receptors in regions of the body through respective nerve fibers to specific integrative functional regions and the cerebral cortex. For the motoric system, it is the other way round.

Cross-References

▶ Postsynaptic Dorsal Column Projection, Anatomical Organization

Somatotopic Organization

Definition

A central nervous system structure is said to have a somatotopic organization when there is a systematic spatial correspondence between the locations of neurons receiving somatosensory information from given parts of the body and particular parts of the nervous system structure. For example, the lateral part of the ventral posterior nucleus of the thalamus receives somatosensory information from the hindlimb and the trunk below the midthoracic level, whereas the medial part of this nucleus receives input from the forelimb and upper trunk. The motoric system is also organized in a somatotopic fashion.

Cross-References

▶ Spinothalamic Input: Cells of Origin (Monkey)
▶ Spinothalamic Tract Neurons, in Deep Dorsal Horn
▶ Trigeminal Brainstem Nuclear Complex, Anatomy

Somatotopical Location

Definition

Both the motor and somatosensory cortices receive, and give rise to, projections in an ordered fashion that mirrors the body.

Cross-References

▶ Motor Cortex, Effect on Pain-Related Behavior

Somatotopy

Definition

Relationship between particular brain and body regions. Neuronal activation in a brain region that is related to a specific sensory pathway is referred to as somatotopically organized.

Cross-References

▶ Trigeminothalamic Tract Projections

Somatotropin

▶ Somatostatin

Somatovisceral Convergence

Definition

The phenomenon or functional property that shows both somatic and visceral inputs converging on the same neurons in the dorsal horn. Stimuli in viscera often evoke pain at the somatic level, that is, in skin, subcutis, and muscle of the body wall in areas neuromerically connected to the affected organ(s). This is called referred pain, which is relevant to the somatovisceral convergent cells.

Cross-References

▶ Morphology, Intraspinal Organization of Visceral Afferents
▶ Muscle Pain, Referred Pain
▶ Referred Hyperalgesia
▶ Referred Pain

Sonography

▶ Ultrasound

Soul Pain

Definition

Deeper pain or anguish relating to spiritual or existential distress.

Cross-References

▶ Cancer Pain Management, Interface Between Cancer Pain Management and Palliative Care

Source Analysis

▶ Magnetoencephalography in Assessment of Pain in Humans

Spa

Definition

The term is derived from the Walloon word "espa" meaning fountain.

Cross-References

▶ Spa Treatment

Spa Therapy

▶ Spa Treatment

Spa Treatment

Dan Buskila
Rheumatic Disease Unit, Soroka Medical Center, Beer Sheva, Israel

Synonyms

Balneotherapy; Medical hydrology; Spa therapy; Thalassotherapy; Thermal therapy

Definition

Different modalities of treatment at a ▶ spa include balneotherapy (bathing in mineral water), ▶ mud therapy (heated mud packs applied to the body), ▶ hydrotherapy (immersion of the body in thermal water), ▶ thalassotherapy (bathing in the sea), ▶ heliotherapy (exposure to sun), and more (Sukenik et al. 1999).

The term balneotherapy comes from the Latin *balneum* (bath). This form of treatment was used since the Roman era but became popular for the management of arthritis and pain in Europe during the eighteenth century. Spa therapy or balneotherapy has been frequently and widely used in classical medicine in the management of various musculoskeletal conditions (Verhagen et al. 1997).

Characteristics

During the years, we have conducted several studies on the effect of balneotherapy in the ▶ Dead Sea area of Israel. The Dead Sea region is the major spa area in Israel for patients with arthritis. The unique climatic conditions in this area and the balneological therapy which is based primarily on mud packs and bathing in ▶ sulfur baths and Dead Sea water combine to alleviate the symptoms of pain and arthritis (Sukenik et al. 1990). The patients are treated, usually once a day, with salt baths heated to 35–37 °C for 20 min, and mud packs heated to 40–42 °C are applied to the body for 20 min. Similar studies from other parts of the world have used water at a temperature around 34 °C (Verhagen et al. 1997).

The mechanisms by which spa therapy alleviates pain and arthritis in musculoskeletal disorders are not fully understood. However thermal, chemical, and mechanical effects have been suggested. Mud therapy has been shown to influence chondrocyte activity in osteoarthritic patients by modulating the production of serum cytokines (interleukin 1) and by increasing insulin growth factor 1 and decreasing tumor necrosis factor alpha in the serum after 12 days of mud pack application (Bellometti et al. 1997). Beta-endorphin and stress hormones decreased significantly in patients with osteoarthritis undergoing thermal mud therapy (Pizzoferrato et al. 2000).

It has been suggested that thermal treatment reduces pain by reducing inflammation and therefore diminishes the cause of stress. Mud bath therapy has been shown to influence nitric oxide, myeloperoxidase, and glutathione peroxidase serum levels in arthritic patients (Bellometti et al. 2000). Recently it was shown that sulfur bath therapy could cause a reduction in oxidative stress, alterations in superoxide dismutase activity, and a tendency towards improvement of lipid levels (Ekmekcioglu et al. 2002). Mechanically, spa therapy may favorably affect muscle tone, joint mobility, and pain intensity, and immersion in spa water at 35 °C has been shown to induce diuresis and natriuresis (Sukenik et al. 1999).

The effectiveness of ▶ spa treatment has been evaluated in various musculoskeletal conditions and pain syndromes. The results of these studies are summarized by Sukenik et al. (1999) and Verhagen et al. (1997). Balneotherapy was found to be effective in the treatment of inflammatory arthritides and spondylarthropathies, namely, rheumatoid arthritis and psoriatic arthritis (Sukenik et al. 1990, 1994).

A statistically significant improvement in most of the clinical parameters assessed for the short period of 1 month was observed in patients with rheumatoid arthritis treated with Dead Sea bath salts (Sukenik et al. 1990). In another study, we found that treatment of psoriasis and psoriatic arthritis in the Dead Sea area is very efficacious and that the addition of balneotherapy can have additional beneficial effects on patients with psoriatic arthritis (Sukenik et al. 1994). Furthermore, it was shown that combined spa exercise therapy and standard treatment with drugs and weekly group physical therapy is more effective and shows favorable cost-effectiveness and cost-utility ratios compared with standard treatment alone in patients with ankylosing spondylitis (Van Tubergen et al. 2002).

We have evaluated the effectiveness of Dead Sea salts dissolved in bathtubs in a group of 26 patients suffering from osteoarthritis of the knees (Sukenik et al. 1995). A statistically significant improvement in patients' self-assessment of disease severity was found in the treatment group vs. a control group at the end of the study. Similar results have been reported on the effectiveness of Puspokladany thermal water on osteoarthritis of the knee joints (Szucs et al. 1989). The effectiveness of spa treatment has also been evaluated in pain syndromes. A randomized study has demonstrated that spa treatment was effective in patients with chronic low back pain, immediately and after 6 months, compared with a control group of patients who did not receive this therapy (Constant et al. 1995).

The fibromyalgia syndrome is a chronic disorder of widespread pain or stiffness in the muscles or joints, accompanied by tenderness on examination at specific predictable anatomic sites known as tender points. We found that balneotherapy was effective in alleviating the pain and other symptoms in patients with fibromyalgia (Buskila et al. 2001). Relief from the severity of pain, fatigue, stiffness, and anxiety were reported, especially in patients receiving sulfur baths, as compared with a control group.

Furthermore, we have shown that staying at the Dead Sea spa, in addition to balneotherapy, can transiently improve the quality of life of patients with fibromyalgia as measured by the short form-36 (SF-36) (Neumann et al. 2001). Interestingly, another study has reported meaningful improvements in the quality of life of patients with osteoarthritis treated with spa therapy in France (Guillemin et al. 2001). Although several studies suggest effectiveness of spa treatment in pain syndromes and rheumatic conditions, there still remains a controversy about this treatment modality. Verhagen et al. (1997) provided a systematic review of the literature analyzing the efficacy of balneotherapy in patients with arthritis. It was a concluded that because of methodological flaws, a conclusion about the efficacy of balneotherapy could not be provided from studies that were reviewed.

In conclusion, spa treatment has been shown to be effective in temporarily relieving pain and arthritis is a variety of pain syndromes and rheumatic conditions. However, well-conducted prospective randomized clinical trials are needed to strengthen these findings.

References

Bellometti, S., Cecchettin, M., & Galzigna, L. (1997). Mud pack therapy in osteoarthrosis. Changes in serum levels chondrocyte markers. *Clinica Chimica Acta, 268*, 101–106.

Bellometti, S., Poletto, M., Gregotti, C., et al. (2000). Mud bath therapy influences nitric oxide, myeloperoxidase and glutathione peroxidase serum levels in arthritic patients. *International Journal of Clinical Pharmacology Research, 20*, 69–80.

Buskila, D., Abu-Shakra, M., Neumann, L., et al. (2001). Balneotherapy for fibromyalgia at the Dead Sea. *Rheumatology International, 20*, 105–108.

Constant, F., Collin, J. F., & Guillemin, F. (1995). Effectiveness of the spa therapy in chronic low back pain: A randomized clinical trial. *The Journal of Rheumatology, 22*, 1315–1320.

Ekmekcioglu, C., Strauss-Blasche, G., Holzer, F., et al. (2002). Effect of sulfur baths on antioxidative defense systems, peroxidase concentrations and lipid levels in patients with degenerative osteoarthritis. *Forschende Komplementärmedizin und Klassische Naturheilkunde, 9*, 216–220.

Guillemin, F., Virion, J. M., Escudier, P., et al. (2001). Effect on osteoarthritis of spa therapy at Bourbonne-Les-Bain. *Joint, Bone, Spine, 68*, 499–503.

Neumann, L., Sukenik, S., Bolotin, A., et al. (2001). The effect of balneotherapy at the Dead Sea on the quality of life of patients with fibromyalgia. *Clinical Rheumatology, 20*, 15–19.

Pizzoferrato, A., Garzia, I., Cenni, E., et al. (2000). Beta-endorphin and stress hormones in patients affected by osteoarthritis undergoing thermal mud therapy. *Minerva Medica, 91*, 239–245.

Sukenik, S., Neumann, L., Buskila, D., et al. (1990). Dead Sea bath salts for the treatment of rheumatoid arthritis. *Clinical and Experimental Rheumatology, 8*, 353–357.

Sukenik, S., Giryes, H., Halevy, S., et al. (1994). Treatment of psoriatic arthritis at the Dead Sea. *The Journal of Rheumatology, 21*, 1305–1309.

Sukenik, S., Mayo, A., Neumann, L., et al. (1995). Dead Sea bath salts for osteoarthritis of the knees. *Harefuah, 129*, 100–103.

Sukenik, S., Flusser, D., & Abu-Shakra, M. (1999). The role of spa therapy in various rheumatic diseases. *Rheumatic Diseases Clinics of North America, 25*, 883–897.

Szucs, L., Ratka, L., & Lesko, T. (1989). Double-blind trial on the effectiveness of the Puspokladany thermal water on arthrosis of the knee joints. *Journal of the Royal Society of Health, 109*, 7–14.

Van Tubergen, A., Boonen, A., Landewe, R., et al. (2002). Cost effectiveness of combined spa exercise therapy in ankylosing spondylitis: A randomized controlled trial. *Arthritis and Rheumatism, 47*, 459–467.

Verhagen, A. P., de Vet, H. C. W., de Bie, R. A., et al. (1997). Taking baths: The efficacy of balneotherapy in patients with arthritis. A systematic review. *The Journal of Rheumatology, 24*, 1964–1971.

Spared Nerve Injury

▶ Retrograde Cellular Changes After Nerve Injury

Spared Nerve Injury Model

▶ Neuropathic Pain Model, Spared Nerve Injury

Spasm

Definition

Involuntary sudden muscle contraction. It is a generic term that does not necessarily indicate a muscle cramp. For example, a facial spasm cannot be included as a muscle cramp because it is not painful. Thus, a muscle cramp is simply a painful spasm.

Cross-References

▶ Muscular Cramps

Spasticity

Definition

Increased tone or rigidity of muscles as a result of abnormal upper motor neuron function. Spasticity may be differentiated from contractures by its improvement with peripheral nerve blocks or the use of botulinum toxin.

Cross-References

▶ Spinal Cord Injury Pain

Spatial Summation

Definition

Spatial summation is when progressively larger numbers of primary afferent (presynaptic) neurons are activated simultaneously, until sufficient neurotransmitter is released to activate an action potential in the spinal cord (postsynaptic) neuron. The term is also used to describe the increase in response that results from stimulation of larger areas or greater amounts of body tissue. Thus, pain increases in intensity as a function of this factor. In the context of nocifensive behaviors, spatial summation is the response that results from simultaneous activation of spatially separated visceral receptors. Because the viscera are generally less densely innervated than most other tissues (e.g., skin), production of nocifensive

behaviors (e.g., by distending a hollow organ) typically requires that a larger area of tissue is activated than would be necessary in other tissues. Thus, distension of hollow organs with longer balloons activates more visceral receptors and generally produces nocifensive behaviors at lower distending pressures.

Cross-References

▶ Exogenous Muscle Pain
▶ IB4-Positive Neurons, Role in Inflammatory Pain
▶ Nocifensive Behaviors, Gastrointestinal Tract
▶ Pain in Humans, Electrical Stimulation (Skin, Muscle, and Viscera)
▶ Pain in Humans, Psychophysical Law

Special Place Imagery

Definition

This is a very common use of guided imagery. The subject is invited to enter a "special place" in imagination, which has qualities of comfort, safety, and other desirable attributes. The intention is to engender feelings of comfort and safety that can lead to reductions in pain experience.

Cross-References

▶ Therapy of Pain, Hypnosis

Specialist Palliative Care

Definition

Comprehensive care provided by an interdisciplinary team of professionals who have received education and training in palliative care. In some countries, the professionals involved have the ability to become certified as specialists.

Cross-References

▶ Cancer Pain Management, Interface Between Cancer Pain Management and Palliative Care

Species Differences

▶ Species Differences in Skin Nociception

Species Differences in Skin Nociception

Bruce Lynn
Department of Physiology, University College London, London, UK

Synonyms

Skin nociception; Species differences

Definition

Differences (and similarities) in how different animals, from the simplest to the most complex, detect noxious stimuli at the body surface.

Characteristics

The detection of dangerous events or situations at the body surface is of great importance to the survival of an organism. Motile organisms can respond to this information and take avoiding action. It is therefore no surprise to find that, as a result of natural selection, most organisms possess excellent nociceptive systems at the

body surface. In this entry, the nociceptive systems of a relatively simple organism, the nematode *C. elegans* will be described. Then the nociceptors in mammalian skin will be looked at, and here the focus will be on the differences between species, especially those differences that scale with body size.

Caenorhabditis elegans

C. elegans is a small (1 mm long) ▶ nematode worm with a nervous system comprising only 302 neurons. The main sensory neurons linked to the body surface are shown in Fig. 1. There are sensitive mechanoreceptors (ALM, AVM, PLM) that respond to very light contact (<10 µN force) and trigger locomotor responses (Goodman and Schwarz 2003). After destruction of these neurons the worm still responds to stronger contacts (>100 µN force) with avoidance movements. These responses are triggered by the two PVD sensory neurons that therefore act like mechanical nociceptive neurons (Kaplan 1996; Goodman and Schwarz 2003). At the anterior end of the worm are two other neurons (ASH in Fig. 1) that respond to pressure and also to dangerous chemical states such as high osmotic pressure and acid pH (Goodman and Schwarz 2003). These neurons thus have a polymodal response profile.

A similar organization is seen in many species, i.e., separate specialized nociceptive sensory neurons that operate in parallel with specialized mechanoreceptors and, in some species, thermoreceptors. Within the nociceptors, there is a further specialization into two classes, one class "tuned" for mechanical noxious stimuli and one class with broad, polymodal, response properties.

As a further example, we can consider the rainbow trout. Recent studies of the trigeminal system in the trout have revealed the presence of ▶ polymodal nociceptors in large numbers (approximately 60 % of all high threshold, presumed nociceptive, neurons) (Sneddon 2003). Smaller numbers of ▶ mechanical nociceptors were also found, plus some mechano-heat units that did not respond to the irritant used in this study (acetic acid). Interestingly the trout

Species Differences in Skin Nociception, Fig. 1 Sensory cells in *C. elegans*. The ASH, FLP and OLQ neurons sense touch to the nose. The ASH cells in addition sense noxious chemicals and so have a polymodal response profile. AVM and ALM neurons sense light touch to the anterior region whereas PLM neurons sense light touch to the posterior region. PVD neurons sense only intense mechanical stimuli, i.e., they have a mechanical nociceptor profile. The anatomical abbreviations are *L* left, *R* right, *D* dorsal, and *V* ventral. The figure shows the left lateral side of the worm (From Kaplan and Horvitz (1993))

trigeminal system has few ▶ C-fibers (only 5 % of all fibers), so the nociceptors are predominantly ▶ A-fiber conducting.

Mammalian Skin

Four major classes of nociceptor have been described in mammalian skin: polymodal nociceptors, mechanical nociceptors, heat nociceptors, and sleeping or silent nociceptors. Properties are summarized in Table 1. Polymodal nociceptors with C-fiber axons and mechanical nociceptors with A-fiber axons are present in all species examined. Probably some insensitive "sleeping" units that only respond to irritant chemicals in uninflamed skin are also present in all species. However, heat nociceptors are not always found. They comprise 7–19 % of the C-nociceptor population in primate and pig. However, they are rare or absent in rabbit and rodent skin.

Conduction velocity varies with species, as does the proportion of A to C fiber nociceptors. In general, conduction velocities increase with body size. For example, A-mechanical nociceptors in the mouse conduct at an average of 7 m/s (Cain et al. 2001; Koltzenburg et al. 1997) while similar units in the pig average 20 m/s (Lynn et al. 1995). Similarly C-nociceptors conduct at an average of 0.7 m/s in the mouse but at 1.2 m/s in the pig. There are also small differences in the conduction velocities of different classes of nociceptor. Most

Species Differences in Skin Nociception, Table 1 Response profiles to key stimulus modalities of different classes of nociceptor in mammalian skin

Class of nociceptor	Main conduction velocity band	Stimulus		
		Strong mechanical	Strong heat	Irritant chemicals
Polymodal nociceptor	C	+	+	+
Mechanical nociceptor	A	+	0	0
Heat nociceptor	C	0	+	+
Sleeping nociceptor	C	(+)[a]	(+)[a]	+

[a]Sleeping (or silent) nociceptors show no responses to mechanical or heat stimuli in normal skin, but fire like polymodal nociceptors when the skin becomes inflamed

interesting is the finding of significant numbers of A-fiber nociceptors with polymodal properties in primate skin, whereas such units are uncommon in rodents.

The mechanical thresholds and receptive field areas of nociceptors vary with body size within the mammals. Average pressure thresholds of C-polymodal nociceptors, determined using fine ▶ von-Frey-type filaments, increase from 6 mN in the mouse to 70 mN in the pig. If pressure threshold is plotted against body mass, it appears that thresholds increase with the 0.3 power of body mass (Fig. 2, upper part). Interestingly this relation holds for both A-fiber mechanical nociceptors and for C-polymodal nociceptors.

Species Differences in Skin Nociception, Fig. 2 *Top.* Pressure thresholds versus body weight for seven mammalian species on the hairy skin of the limb. Slope of best fit line for C-polymodal nociceptors (C-PMN) is 0.29. A-Noci = A-mechanical nociceptor. C-mech = sensitive C-fiber mechanoreceptors. *Bottom.* Receptive field area of C-PMN in five species. *Dotted line*, best fit, slope 0.62 (Data from the following studies: mouse, Koltzenburg et al. (1997); rat, Lynn (1994); cat, Bessou and Perl (1969), Burgess and Perl (1967); rabbit, Fitzgerald (1978); monkey, Treede et al. (1995); pig, Lynn et al. (1995); man, Schmidt et al. (1997))

It appears to hold for C-fiber mechanoreceptors as well. Receptive field areas increase approximately with the 0.67 power of body mass, i.e., receptive field area increases roughly in proportion to total body surface area (Fig. 2, lower part).

Note that the changes in properties with body size described above can lead to problems of definition. A neuron with a pressure threshold at the most sensitive end of the nociceptor range in a mouse would probably be considered a mechanoreceptor in a large species such as the pig. A neuron with conduction velocity of 25 m/s is a fast fiber in the mouse but would be firmly in the A-delta, slow myelinated, band in larger mammalian species.

Molecular Evolution: Species Differences, but Molecular Similarities

Recent studies into the key proteins involved in sensory transduction in nociceptors have revealed some interesting similarities across the animal kingdom. Two families of ▶ ion channel proteins appear to play a role in nociception in all species, invertebrate and vertebrate, so far studied. These are the TRPv family and the ▶ ENaC/DEG family (Julius and Basbaum 2001). To illustrate this pattern, consider the TRPv family. The first mammalian member to be characterized was the ▶ capsaicin receptor (originally designated VR1, now designated TRPV1) that amazingly turns out to be a ligand-gated ion channel that also opens to noxious heat and low pH. Truly a polymodal channel! It plays an important role in noxious heat ▶ transduction and is expressed in most C-polymodal neurons. If expressed artificially in non-sensory cells, then these cells develop chemo- and heat sensitivity. Several other members of the TRP family are also expressed in small sensory ganglion cells. For example, TRPV2 (originally VRL1) is another heat-gated channel but lacks capsaicin sensitivity (Julius and Basbaum 2001).

In *C. elegans* a ▶ homologous gene, osm-9, is found in ASH cells and if it is missing, all sensory functions of the ASH cells are lost. However, unlike TRPV1, attempts to generate transduction by expressing the osm-9 gene in non-sensory cell lines have been unsuccessful. However, this may not indicate that osm-9 is not involved directly in transduction, but perhaps that other proteins are necessary to get a working channel. Also, noxious heat reactions of *C. elegans* are unaffected in mutants lacking osm-9, although there are several other TRPv proteins whose function has yet to be determined and which might therefore turn out to be involved in noxious heat transduction (Wittenburg and Baumeister 1999).

Summary

Specialized nociceptors sensing mechanical and thermal conditions at the body surface and reacting to noxious chemicals appear to be omnipresent in Metazoa.

Commonly, a subclass of nociceptor sensing just strong mechanical stress is found alongside another "polymodal" subclass sensing a range of noxious stimuli.

Within the vertebrates, nociceptor pressure thresholds scale with 1/3 power of body weight, while receptive field area scales with body surface area.

Despite many detailed differences between nociceptors in different species, they share a closely related family of molecular transduction proteins. Notable are the TRP family of sensory-gated ion channels that subserve mechano-, thermal-, and chemo-transduction in many species.

References

Bessou, P., & Perl, E. R. (1969). Response of cutaneous sensory units with unmyelinated fibers to noxious stimuli. *Journal of Neurophysiology, 32,* 1025–1043.

Burgess, P. R., & Perl, E. R. (1967). Myelinated afferent fibres responding specifically to noxious stimulation of the skin. *The Journal of Physiology, 190,* 541–562.

Cain, D. M., Khasabov, S. G., & Simone, D. A. (2001). Response properties of mechanoreceptors and nociceptors in mouse glabrous skin: An in vivo study. *Journal of Neurophysiology, 85,* 1561–1574.

Fitzgerald, M. (1978). *The sensitization of cutaneous nociceptors.* Ph.D. Thesis, University of London.

Goodman, M. B., & Schwarz, E. M. (2003). Transducing touch in *Caenorhabditis elegans. Annual Review of Physiology, 65,* 429–452.

Julius, D., & Basbaum, A. I. (2001). Molecular mechanisms of nociception. *Lancet Neurology, 413,* 203–210.

Kaplan, J. M. (1996). Sensory signalling in *Caenorhabditis elegans. Current Opinion in Neurobiology, 6,* 494–499.

Kaplan, J. M., & Horvitz, H. R. (1993). A dual mechanosensory and chemosensory neuron in *Caenorhabditis elegans. Proceedings of the National Academy of Sciences USA, 90,* 2227–2231.

Koltzenburg, M., Stucky, C. L., & Lewin, G. R. (1997). Receptive properties of mouse sensory neurons innervating hairy skin. *Journal of Neurophysiology, 78,* 1841–1850.

Lynn, B. (1994). The fibre composition of cutaneous nerves and the classification and response properties of cutaneous afferents, with particular reference to nociception. *Pain Reviews, 1,* 172–183.

Lynn, B., Faulstroh, K., & Pierau, F. K. (1995). The classification and properties of nociceptive afferent units from the skin of the anaesthetized pig. *The European Journal of Neuroscience, 7,* 431–437.

Schmidt, R., Schmelz, M., Ringkamp, M., et al. (1997). Innervation territories of mechanically activated C nociceptor units in human skin. *Journal of Neurophysiology, 78,* 2641–2648.

Sneddon, L. U. (2003). Trigeminal somatosensory innervation of the head of a teleost fish with particular reference to nociception. *Brain Research, 972,* 44–52.

Treede, R. D., Meyer, R. A., Raja, S. N., et al. (1995). Evidence for two different heat transduction mechanisms in nociceptive primary afferents innervating monkey skin. *The Journal of Physiology, 483,* 747–758.

Wittenburg, N., & Baumeister, R. (1999). Thermal avoidance in *Caenorhabditis elegans*: An approach to the study of nociception. *Proceedings of the National Academy of Sciences USA, 96,* 10477–10482.

Specificity

Definition

Specificity is a statistical decision term equivalent to correct rejection rate. The probability that the test will be negative among patients who do not have the disease.

Cross-References

▶ Sacroiliac Joint Pain

Specificity Theory

Definition

This term is often used in the context of pain mechanisms. It characterizes the hypothesis that pain is mediated by a specific subset of neurons in peripheral nerves (the nociceptors) and also specific projection neurons in the CNS ("labeled line" hypothesis). M.v. Frey who discovered the "pain points" in the skin in the late nineteenth century is regarded as the inaugurator of the specificity hypothesis. Contrary hypotheses are often summarized as "pattern theories." Generally, they assume that a particular pattern of excitation in the CNS is responsible for pain.

Cross-References

▶ Somatic Pain

SPECT

Definition

SPECT (Single Photon Emission Computed Tomography) is a method of functional brain imaging. The technique involves the injection of a radiopharmaceutical contrast agent into the subject; the distribution of the contrast agent within the brain is detected using gamma rays.

Cross-References

▶ Single Photon Emission Computed Tomography
▶ Thalamus, Clinical Pain, Human Imaging

Spike

Definition

Spike is a synonym for action potential.

Cross-References

▶ Nociceptors, Action Potentials, and Post-Firing Excitability Changes

Spike Bursting

▶ Thalamic Bursting Activity, Chronic Pain

Spike Bursts

Definition

Spike bursts are action potentials that comprise of the thalamic firing pattern during bursting mode.

Cross-References

▶ Central Pain, Human Studies of Physiology
▶ Thalamic Bursting Activity, Chronic Pain

Spike Initiation and Conduction in Nociceptors

▶ Nociceptors and Activity-Dependent Changes in Axonal Conduction Velocity

Spillover

Definition

Spillover refers to the diffusion of a synaptically released neurotransmitter out of the synaptic cleft to neighboring synapses or extrasynaptic receptors.

Cross-References

▶ GABA and Glycine in Spinal Nociceptive Processing

Spinal (Neuraxial) Opioid Analgesia

Definition

For selected patients, the spinal route of drug delivery may be used for opioid analgesics. Opioids are administered via catheters placed in the epidural or intrathecal (subarachnoid) space. The spinal route of administration allows the drug to be delivered in close proximity to opioid receptors in the dorsal horn of the spinal cord, often permitting effective analgesia at considerably lower doses than would be required if a systemic route of drug administration were used.

Spinal Analgesia

Definition

Pain relief obtained by drugs acting directly on the spinal cord after intrathecal administration,

e.g., morphine, local anesthetics, adrenaline (epinephrine), clonidine, or neostigmine.

Cross-References

▶ Postoperative Pain, Acute Pain Management Principles

Spinal Arthrodesis

Definition

Another term for spinal fusion.

Cross-References

▶ Spinal Fusion for Chronic Back Pain

Spinal Ascending Pathways

Definition

The neuronal tracts of spinal origin that project directly upward to supraspinal structures.

Cross-References

▶ Spinal Ascending Pathways, Colon, Urinary Bladder, and Uterus
▶ Spinothalamic Projections in Rat

Spinal Ascending Pathways, Colon, Urinary Bladder, and Uterus

Timothy Ness and Promil Kukreja
Department of Anesthesiology, University of Alabama at Birmingham, Birmingham, AL, USA

Synonyms

Postsynaptic dorsal column pathway (see postsynaptic dorsal column neurons); Spinal dorsal horn pathways, colon, urinary bladder, and uterus; Visceral spinohypothalamic tract (see spinohypothalamic neurons); Visceral spinomedullary tract (see spinomedullary neurons); Visceral spinomesencephalic tract (see spinomesencephalic neurons); Visceral spinoreticular tract; Visceral spinothalamic tract

Definition

Spinal dorsal horn neurons receiving excitatory inputs from pelvic viscera, which include the urinary bladder, the colon-rectum, and the uterus-vagina, frequently have axonal extensions projecting to supraspinal sites of termination (Fig. 1). These sites include the thalamus, the hypothalamus, the midbrain, the pons, and the medulla. These axons travel within the white matter of the spinal cord in two main sites: the anterolateral quadrants and the dorsal midline. Decussation of fibers to the contralateral anterolateral white matter occurs for axons projecting to the thalamus and most other brainstem sites, although many axons with sites of termination in "reticular" structures will remain in the ipsilateral anterolateral white matter. The spinothalamic track is classically considered to be the major "nociceptive" pathway, but recent findings have shown significant role of dorsal column pathways in visceral pain (Kim and Kwon 2000). Dorsal column pathways have sites of termination in the gracile nucleus of the medulla. Intraspinal pathways have also been demonstrated, as has extensive collateralization of axons with multiple sites of termination. Novel pathways for ascending visceral information may also include dorsolateral spinothalamic pathways related to lamina I neurons (McMahon et. al 1983).

Characteristics

General
Spinal dorsal horn neurons receiving excitatory visceral inputs from pelvic structures have been demonstrated to be present throughout the dorsal

Spinal Ascending Pathways, Colon, Urinary Bladder, and Uterus, Fig. 1 Schematic diagram of spinal cord cell body localization of neurons receiving pelvic organ visceral inputs (*filled circles*) and spinal white matter pathways discussed in the text (see also ▶ spinosolitary neurons; ▶ spinoparabrachial neurons)

horn, but with particular localization to laminae I, II, V, and X. Mapping of primary afferents from visceral structures has demonstrated sites of connectivity that are heaviest in these same laminae. Multisegmental connections of single primary afferents from visceral structures have been demonstrated to travel via lateral (e.g., Lissauer's tract) and medial tracts, prior to synaptic contact with multiple neurons in multiple laminae (Sugiura et al. 1993). Neurons throughout the dorsal horn receiving visceral inputs have been demonstrated to possess axonal extensions that project to distant (rostral and caudal) sites within the spinal cord and brainstem. No quantitative study has defined the precise percentage of spinal neurons with visceroceptive input that have axonal projections within any specific portion of spinal white matter, and information related to the importance of specific tracts on subsequent neuronal function has been extrapolated from studies utilizing spinal lesions to reduce or abolish behaviors, reflexes, or supraspinal neuronal responses. Inputs from some visceral structures such as the uterus to supraspinal sites such as the nucleus tractus solitarius have been demonstrated to travel by spinal as well as nonspinal pathways, which include the vagus nerve (Berkley and Hubscher 1995). The unmyelinated afferent fibers are not only for pain anymore based on the spinal termination patterns in the superficial dorsal horn (Ling et al. 2003).

Intraspinal (Propriospinal) Pathways

The spinal dorsal horn collectively executes the function of sensory information processing without any significant communications with other regions of spinal gray matter. However, within the spinal dorsal horn, the different subdivisions like rostro-caudal and mediolateral intensively communicate with each other through propriospinal pathways (Petko and Antal 2012). Multiple interconnections occur between spinal neurons. This has been particularly demonstrated in the case of neurons receiving visceral afferent input from pelvic structures. With afferent pathways traveling via the pelvic nerve and in association with sympathetic nerve pathways, the site of entry of afferent information for pelvic organs is split between the lower thoracic-upper lumbar and lower lumbar-sacral segments. McMahon and Morrison (1982) demonstrated clear connectivity between inputs via these two pathways by performing a cordotomy at the midlumbar spinal cord and demonstrating an abolition of excitatory responses to respective inputs. A more subtle white matter localization of the axonal extensions of ▶ propriospinal neurons has not been performed, but they are presumed to follow the paths of other propriospinal neurons, which include dorsally located white matter paths and some within-gray matter extensions. Collateral intraspinal extensions of ascending axons located within the ventrolateral white matter have also been demonstrated. Spinocervicothalamic pathways have not been implicated in visceroception.

Anterolateral White Matter

The traditional pain pathways are the spinothalamic and spinoreticular tracts (see ▶ spinoreticular neurons) located within the anterolateral white matter of the spinal cord. Both ipsilateral and contralateral localizations of axon pathways have been demonstrated en route to supraspinal sites, with the predominant path for afferents traveling to the ventrobasal thalamus residing on the contralateral side. Numerous neurons with axonal projections to the thalamus traversing the anterolateral white matter have been demonstrated to receive visceral inputs from pelvic organs (e.g., Milne et al. 1981). Despite these observations, experiments utilizing lesions of the anterolateral spinal white matter have demonstrated reduction/abolition of ventrobasal thalamic neuron responses to noxious cutaneous stimuli but failed to demonstrate a significant reduction in the responses to colorectal distension in rats and primates (Al Chaer et al. 1999; Hirshberg et al. 1996; Menetrey and DePommery 1991; Ness 2000). In contrast, ventrolateral medullary neuronal responses to colorectal distension and the evocation of cardiovascular and visceromotor reflexes to the same stimulus disappeared with ventrolateral white matter lesions (Ness 2000). Neurons with projections to other supraspinal sites such as the hypothalamus, parabrachial nucleus, mesencephalon, pons, and other medullary sites have been demonstrated to be excited by pelvic organ stimulation (e.g., Katter et al. 1996). It must be extrapolated that the axonal projections of this specific subset of neurons follow the same spinal white matter pathways as other spinal neurons with similar terminal sites of projection. Hence, the ipsilateral and contralateral ventrolateral white matter pathways are the presumed predominant pathway, with an unknown balance between ipsilateral and contralateral localizations.

Lamina I Spinothalamic Pathways

Lamina I region of spinal cord has become recognized as a part of important projection system relaying information from specific afferent receptors to rostral centers, including the thalamus, the midbrain, and the medulla (Andrew and Craig 2001). ▶ Spinothalamic neurons located within lamina I have been demonstrated to send axonal projections to the contralateral thalamus via a contralateral dorsolateral pathway. As lamina I neurons have also been demonstrated to receive pelvic visceral inputs, it is presumed that similar white matter projections exist for visceroceptive lamina I neurons.

Dorsal Column Pathways

One of the most exciting neurophysiological findings related to visceral pain is the demonstration of the existence of a spinal pathway in the midline of the dorsal columns for the rostral transmission of pelvic visceral nociception. Behavioral, electrophysiological, and immunohistochemical methods used in animal experiments showed that dorsal column lesion led to decreased activation of thalamic and gracile neurons by visceral stimuli and prevented potentiation of visceromotor reflexes evoked by visceral organ distention under inflammatory conditions (Palecek 2004). Discrete neurosurgical lesions of this portion of the spinal cord have been demonstrated to relieve cancer-related pain in patients with pelvic visceral pathology (Hirshberg et al. 1996; Nauta et al. 2000). Parallel studies in nonhuman animals have demonstrated that in this area of spinal white matter, there exist axons of postsynaptic dorsal horn neurons receiving noxious excitatory input from the colon, bladder, and/or uterus. Excitatory responses of neurons located in the medullary gracile nucleus or ventrobasal thalamus to noxious visceral stimuli are attenuated/abolished with lesions of the dorsal midline region of the spinal cord (Al Chaer et al. 1999; Berkley and Hubscher 1995; Hirshberg et al. 1996; Ness 2000). Inputs from the uterus to medullary gracile nucleus neurons have been demonstrated not to require dorsolateral funiculus pathways (Berkley and Hubscher 1995).

Summary

Spinal dorsal horn neurons excited by stimulation of the colon, urinary bladder, and/or uterus have axonal projections to numerous intraspinal and supraspinal sites. Unique to these visceral structures is the suggestion of a role for dorsal column pathways in nociceptive processing. Other pathways appear to be shared with somatic nociceptive pathways.

Acknowledgment Authors: Timothy J Ness MD, PhD (Department of Anesthesiology, University of Alabama at Birmingham). Promil Kukreja MD, PhD (Department of Anesthesiology, University of Alabama at Birmingham).

References

Al Chaer, E. D., Feng, Y., & Willis, W. D. (1999). Comparative study of viscerosomatic input onto postsynaptic dorsal column and spinothalamic tract neurons in the primate. *Journal of Neurophysiology, 82*, 1876–1882.

Andrew, D., & Craig, A. D. (2001). Spinothalamic lamina I neurons selectively responsive to cutaneous warming in cats. *The Journal of Physiology, 537*, 489–495.

Berkley, K. J., & Hubscher, C. H. (1995). Are there separate central nervous system pathways for touch and pain? *Nature Medicine, 1*, 766–773.

Hirshberg, R. M., Al-Chaer, E. D., Lawand, N. B., et al. (1996). Is there a pathway in the posterior funiculus that signals visceral pain? *Pain, 67*, 291–305.

Katter, J. T., Dado, R. J., Kostarczyk, E., et al. (1996). Spinothalamic and spinohypothalamic tract neurons in the sacral spinal cord of rats. II. Responses to cutaneous and visceral stimuli. *Journal of Neurophysiology, 75*, 2606–2628.

Kim, Y. S., & Kwon, S. J. (2000). High thoracic midline dorsal column myelotomy for severe visceral pain due to stomach cancer. *Neurosurgery, 46*, 85–90.

Ling, L. J., Honda, T., Shimada, Y., Ozaki, N., Shiraishi, Y., & Sugiura, Y. (2003). Central projection of unmyelinated (C) primary afferent fibers from gastrocnemius muscle in the guinea pig. *The Journal of Comparative Neurology, 461*, 140–150.

McMahon, S. B., & Morrison, J. F. B. (1982). Two groups of spinal interneurones that respond to stimulation of the abdominal viscera of the cat. *The Journal of Physiology, 322*, 21–34.

McMahon, S. B., & Wall, P. D. (1983). A system of rat spinal cord lamina I cells projecting through the contralateral dorsolateral funiculus. *The Journal of Comparative Neurology, 20*, 217–223.

Menetrey, D., & DePommery, J. (1991). Origins of spinal ascending pathways that reach central areas involved in visceroception and visceronociception in the rat. *European Journal of Neuroscience, 3*, 249–259.

Milne, R. J., Foreman, R. D., Giesler, G. J., Jr., et al. (1981). Convergence of cutaneous and pelvic visceral nociceptive inputs onto primate spinothalamic neurons. *Pain, 11*, 163–183.

Nauta, H. J., Soukup, V. M., Fabian, R. H., et al. (2000). Punctate midline myelotomy for the relief of visceral cancer pain. *Journal of Neurosurgery, 92*, 125–130.

Ness, T. J. (2000). Evidence for ascending visceral nociceptive information in the dorsal midline and lateral spinal cord. *Pain, 87*, 83–88.

Palecek, J. (2004). The role of dorsal columns pathway in visceral pain. *Physiological Research, 53*(Suppl I), S125–S130.

Petko, M., & Antal, M. (2012). Propriospinal pathways in the dorsal horn (laminae I-IV) of the rat lumbar spinal cord. *Brain Research Bulletin, 89*(1–2), 41–49.

Sugiura, Y., Terui, N., Hosoya, Y., et al. (1993). Quantitative analysis of central terminal projections of visceral and somatic unmyelinated (C) Primary afferent fibers in the guinea pig. *The Journal of Comparative Neurology, 15*, 315–325.

Spinal Block

Definition

Injection of local anesthetic into the spinal fluid of the lumbar spinal canal; rarely used for diagnostic blocks.

Cross-References

▶ Pain Treatment: Spinal Nerve Blocks

Spinal Cord

Definition

The spinal cord is part of the central nervous system and connects the brain to the rest of the body. The dorsal horn, which is the dorsal part of the gray matter, contains the cell bodies of sensory neurons.

Cross-References

▶ Opioid Receptor Localization
▶ Prostaglandins, Spinal Effects
▶ Responses of Spinothalamic Tract Neurons to Pruritic Stimuli

Spinal Cord Astrocyte

▶ Cord Glial Activation

Spinal Cord Compression

Synonyms

SCC

Definition

Spinal cord compression (SCC) is due to extrinsic compression of the cord from the epidural space secondary to extension of adjacent bony or soft tissue lesions. Less commonly, SCC can be a complication of intramedullary metastases or primary tumors (even benign). Of cancer patients, 5–10 % suffer SCC. Most cases of SCC are caused by extension of vertebral bone metastases to the epidural space. Metastases of cancer of breast, lung, and prostate are the most common causes of SCC. Pain precedes the neurological symptoms (e.g., paresis) in almost all cases; however, the diagnosis of SCC is always delayed until the onset of a full-blown neurological syndrome.

Cross-References

▶ Cancer Pain
▶ Cancer Pain Management Radiotherapy

Spinal Cord Injury

▶ Spinal Cord Injury Pain Model, Hemisection Model

Spinal Cord Injury Pain

D. Paul Harries
Pain Treatment Center, Samaritan Hospital
Lexington, Lexington, KY, USA

Characteristics

Incidence

Spinal cord injury (SCI) occurs at an annual incidence of between 20 and 40 cases per million of the populations. Of those affected, 69 % suffer from chronic pain, a third of these patients describing their pain as severe. Half of the patients describe primarily neuropathic symptoms, 25 % primarily non-neuropathic symptoms, and 25 % a mixture of pain types. Generally, ▶ neuropathic pain seems to be worse with distal lesions and in incomplete spinal cord injuries (▶ incomplete spinal cord injury) (Kirshblum 2002).

Pain in Relation to Level of Injury

Pain can occur at the site of injury, above the injury, and/or below the injury. Somatic and neuropathic pain can occur at each of these levels. It is helpful when evaluating a patient to try and think of pain origin in these terms. Formal classification has been suggested (Bryce and Ragnarsson 2000; Siddall et al. 1997).

Cranial to Injury

Probably the commonest source of somatic pain cranial to the injury is from the shoulder joint. This joint is particularly vulnerable to repeated trauma in the paraplegic using their arms to propel a wheelchair.

Neuropathic pain cranial to the injury is classically a result of peripheral nerve compression from wheelchair positioning and lack of padding and a result of poorly fitting orthotics. Prevention is the best approach. ▶ Entrapment neuropathies do occur at an increased rate, and treatment for these follows conventional guidelines for these syndromes once external compression has been eliminated.

At the Level of Injury

At the level of injury, pain can be somatic from associated injuries to bony structures and soft tissues; fusion hardware can contribute to this component of the pain.

Neuropathic pain can be as a result of either injury to the exiting root at the time of injury or surgery or from ongoing compression, the latter being potentially amenable to surgical decompression. Neuropathic pain at the level of injury can also arise from within the cord itself. ▶ Selective nerve root blocks may be helpful in making this distinction.

Caudad to Injury

Neuropathic pain distal to the level of injury is classically of central origin. Somatic pain can be as a result of acquired musculoskeletal problems and is particularly associated with poorly controlled ▶ spasticity.

Psychological Factors and Substance Abuse

It is estimated that 50 % of patients with SCI in the United States have a prior history of substance abuse. There are a similar percentage of patients with significant pre-morbid psychological/psychiatric issues. These clearly affect the patient's suffering after a SCI (Kennedy et al. 1995; Kirshblum 2002).

Other Factors Causing Pain After SCI

Poorly controlled spasticity is associated with increased pain. Spasticity may result in compromised wheelchair seating, and as a result postural imbalances may develop that increase pain. It is particularly important to evaluate a patient for contractures around the hip joint, as these may result in a cascade of seating-related pain generators.

▶ Syringomyelia can present with new onset of pain years after a SCI. To think of a syrinx is usually enough to make the diagnosis!

▶ Heterotopic ossification develops in up to 50 % of SCI patients; this can be painful. NSAIDs (with or without surgical resection) are the mainstay of treatment where tolerated.

Management of Particular SCI Pain Syndromes Medications

Gabapentin has become first-line therapy for neuropathic symptoms following SCI (Ahn et al. 2003; Levendoglu et al. 2004; Tai et al. 2002). Numerous studies have demonstrated its effectiveness. Carbamazepine and other AEDs have also been utilized. Baclofen and Tizanidine have been used as adjuvant therapy.

Spasticity-related pain is controlled using Baclofen, Tizanidine, benzodiazepines, or dantrolene.

Narcotics together with NSAIDs are the mainstay of pharmacologic management of somatic pain. Narcotic management can be very challenging given the 50 % rate of substance abuse.

Chronic shoulder pain in the SCI patient who uses their arms for propulsion can be very difficult to treat. Evaluation of scapular stabilizer muscle strength is important, and strengthening exercises may well be necessary. The trade-off between using a power chair with improvement in pain vs. loss of exercise and physiological fitness is a difficult balance.

▶ Autonomic dysreflexia is a syndrome that can occur in patients with an SCI above T6. It is characterized by an exaggerated autonomic response to what would normally represent a painful area (e.g., ingrown toenail) or even normal visceral distension (e.g., distended bladder) stimuli. Severe hypertension can occur, which left untreated can be fatal. Optimal treatment starts with removal of the responsible stimulus. Careful clinical evaluation is mandatory.

Interventional Pain Techniques

Selective nerve root blocks and ganglion radiofrequency lesions can be performed for segmental neuropathic pain. There are no good outcome studies for these procedures.

Facet joint denervation can be helpful in some patients with facet pain (see ▶ Facet Joint Procedures for Chronic Back Pain and ▶ Dorsal Root Ganglion Radiofrequency).

Intrathecal infusions of morphine and clonidine for pain, and Baclofen for spasticity, have been utilized with success in SCI (Siddall et al. 2000; Sjolund 2002).

Spinal cord stimulation for neuropathic pain after SCI has generally been disappointing, with the exception of pain following cauda equina lesions (Kirshblum 2002; Midha and Schmitt 1998).

Surgical Interventions

Dorsal root entry lesions have been used for severe neuropathic pain at the site of injury with limited success (Werhagen et al. 2004; Young 1990).

Deep brain stimulation has been used for chronic pain after SCI; however, there is no evidence to support its use (Kumar 1997).

Cross-References

▶ Functional Changes in Sensory Neurons Following Spinal Cord Injury in Central Pain
▶ Opioids and Reflexes
▶ Spinal Cord Injury Pain Model, Hemisection Model

References

Ahn, S. H., Park, H. W., Lee, B. S., et al. (2003). Gabapentin effect on neuropathic pain compared among patients with spinal cord injury and different durations of symptoms. *Spine, 28,* 341–347.
Bryce, T. N., & Ragnarsson, K. T. (2000). Pain after spinal cord injury. *Physical Medicine and Rehabilitation Clinics of North America, 11,* 157–168.
Kennedy, P., Lowe, R., Grey, N., et al. (1995). Traumatic spinal cord injury and psychological impact: A cross-sectional analysis of coping strategies. *The British Journal of Clinical Psychology, 34,* 627–639.
Kirshblum, S. (2002). *Spinal cord medicine* (pp. 389–408). Philadelphia: Lippincott Williams and Wilkins.
Kumar, K. (1997). Deep brain stimulation for intractable pain: A 15-year experience. *Neurosurgery, 40,* 736–747.
Levendoglu, F., Ogun, C. O., Ozerbil, O., et al. (2004). Gabapentin is a first line drug for the treatment of neuropathic pain in spinal cord injury. *Spine, 29,* 743–751.
Midha, M., & Schmitt, J. K. (1998). Epidural spinal cord stimulation for the control of spasticity in spinal cord injury patients lacks long-term efficacy and is not cost-effective. *Spinal Cord, 36,* 190–192.
Siddall, P. J., Taylor, D. A., & Cousins, M. J. (1997). Classification of pain following spinal cord injury. *Spinal Cord, 35,* 69–75.
Siddall, P. J., Molloy, A. R., Walker, S., et al. (2000). The efficacy of intrathecal morphine and clonidine in the

treatment of pain after spinal cord injury. *Anesthesia and Analgesia, 91*, 1493–1498.

Sjolund, B. H. (2002). Pain and rehabilitation after spinal cord injury: The case of sensory spasticity? *Brain Research. Brain Research Reviews, 40*, 250–256.

Tai, Q., Kirshblum, S., Chen, B., et al. (2002). Gabapentin in the treatment of neuropathic pain after spinal cord injury: A prospective, randomized, double-blind, crossover trial. *The Journal of Spinal Cord Medicine, 25*, 100–105.

Werhagen, L., Budh, C. N., Hultling, C., et al. (2004). Neuropathic pain after traumatic spinal cord injury – Relations to gender, spinal level, completeness, and age at the time of injury. *Spinal Cord, 42*, 665–673.

Young, R. F. (1990). Clinical experience with radiofrequency and laser DREZ lesions. *Journal of Neurosurgery, 72*, 715–720.

Spinal Cord Injury Pain Model, Contusion Injury Model

Philip J. Siddall
Department of Pain Management, Greenwich Hospital, University of Sydney, Sydney, NSW, Australia

Synonyms

Contusion injury model

Definition

The contusion injury model of spinal cord injury (SCI) pain refers to an animal model of SCI in which a spinal cord ▶ contusion is produced by dropping a small weight either directly or indirectly onto the exposed surface of the cord. This results in a motor dysfunction and in a proportion of animals is accompanied by altered behavioral responses to peripheral mechanical and thermal stimulation suggestive of pain.

Introduction

The contusion model of spinal cord injury is based on the Allen weight drop method and is sometimes referred to as the modified Allen weight drop method (Wrathall 1992). In humans, most spinal cord injuries occur with a relatively fast application of a force that results in compression and contusion to the spinal cord. The contusion model, therefore, was designed to mimic the impact forces that are sustained in the clinical setting. It has been in use for nearly a century to investigate the sequelae of SCI, particularly in studies that investigate prevention of secondary consequences of cord trauma, regeneration, and restoration of function. Although the contusion model has been used for some time in studies of SCI, most studies have focused on motor effects of SCI and paid little attention to sensory changes. It is only more recently that it has been used to investigate mechanisms of neuropathic pain following SCI (Siddall et al. 1995; Hubscher and Johnson 1999; Mills et al. 2000a; Lindsey et al. 2000).

Characteristics

Description of the Model

To produce a contusion injury, the animal is deeply anesthetized and a laminectomy is performed to expose the spinal cord, usually at the lower thoracic or upper lumbar levels. A small window is created over the dorsal surface of the spinal cord and the dura mater is left intact. The vertebral processes above and below are clamped to stabilize the spine. A guide tube containing a cylindrical weight (usually 10 g) is positioned directly over the exposed cord. The weight is then dropped from a fixed height (approximately 5–25 mm) in a guide tube by removing a pin inserted through holes at fixed intervals through the tube. Several devices are available that produce a standardized lesion such as the New York University (NYU) impactor. The weight drops though the guide tube and the tip, 2–2.5 mm diameter, impacts either directly onto the cord surface or against a plastic device with the same configuration resting on the cord surface.

Another technique for producing a contusion injury has also been described by Hubscher and Johnson (1999). Rather than using an impact device, a contusion injury is induced by rapidly compressing the cord using a probe connected to a displacement-controlled device.

Following the contusion injury, the wound is closed and the animal is allowed to recover from the anesthetic with antibiotic and analgesic cover. The contusion results in weakness of the hind limbs that is apparent immediately on recovery from anesthesia. The severity of hind limb weakness is proportional to the height of the weight drop. A 12.5–20 mm drop typically results in a transient hind limb paralysis immediately following injury. However, depending on the severity of the injury, most animals regain motor function with a gradual improvement over the 2–4 weeks following injury (Siddall et al. 1995; Hulsebosch et al. 2000; Lindsey et al. 2000).

Although motor function improves following the contusion injury, a proportion of rats develop mechanical and ► thermal allodynia following spinal cord contusion. This ► allodynia may be present almost immediately following the injury (Siddall et al. 1995) but may also develop later and persist for more than 10 weeks post injury (Lindsey et al. 2000). Interestingly, in the study by Lindsey et al. (2000) in which animals were assessed for 10 weeks following injury, after an initial increase in responsivity in the first 2 weeks, animals displayed a decrease in responsivity which then reappeared at 8–10 weeks following injury. This suggests a biphasic onset pattern of allodynia which has also been noted in the ischemic SCI pain model.

Allodynia and ► hyperalgesia are most commonly assessed by a change in vocalization threshold to peripheral stimuli but other responses such as orientation and escape behaviors may also be observed. ► Mechanical allodynia is evidenced by a decreased vocalization threshold to mechanical stimulation. This is usually assessed by the application of graded ► Von Frey monofilaments onto the skin surface, noting the caliber and force of the filament that results in consistent vocalization (Siddall et al. 1995; Mills et al. 2000a; Lindsey et al. 2000). Thermal allodynia or hyperalgesia may be assessed by application of a radiant heat stimulus (Mills et al. 2002) or by application of an ice probe (Lindsey et al. 2000). Stimuli are applied most commonly in the dermatomes close to the level of injury (trunk) (Siddall et al. 1995;

Hubscher and Johnson 1999; Hulsebosch et al. 2000; Lindsey et al. 2000) as well as the forelimbs and hind limbs (Hulsebosch et al. 2000; Mills et al. 2000b; Lindsey et al. 2000).

As well as these evoked responses, changes in behavior suggestive of spontaneous pain have also been observed. Animals exhibit a reduction in exploratory behaviors including rearing and other activities that were independent of locomotor ability and may be indicative of the presence of spontaneous neuropathic pain (Mills et al. 2001).

Behavioral changes suggestive of evoked pain appear to occur most commonly in those animals that have moderate injuries. Most studies suggest that an injury of between 12 and 25 mm results in the most consistent expression of allodynia (Siddall et al. 1995; Lindsey et al. 2000). Although the development of allodynia appears to be related to the severity of injury, it does not appear to correspond with the location of injury or the spinal cord structures that are affected.

Histologically, the extent of cell loss corresponds to the severity of injury. A 12–25 mm injury, which is most commonly used in SCI pain studies, results in cell loss which predominantly affects the spinal gray matter and dorsal columns. Cystic cavitation is evident in the central region of the spinal cord with a spared rim of white matter (Siddall et al. 1995; Lindsey et al. 2000). The lesion mainly involves the injured segment but extends rostrally and caudally for a distance of approximately 1 segment in each direction with a mean cavity length of 4.2 mm following a 25 mm injury (Lindsey et al. 2000).

Clinical Relevance

One of the issues confronting any animal model is its relevance to the clinical situation. The nature of the SCI following a contusive injury is very similar to the clinical situation in which most injuries are due to high impact trauma to the cord. Contusive trauma results in secondary biochemical changes, through to cell death and necrosis similar to that seen in the spinal cord of humans who have had a high impact SCI. A weight drop from a relatively low height (6–25 mm depending on the device used) results in a lesion predominantly affecting the dorsal columns and central gray matter with relative preservation of lateral and anterior fiber tracts. This is similar to the pathology evident in people with central cord injuries.

Allodynia and hyperalgesia typically occur and are strongest in the dermatomes close to the level of injury, although changes in responsiveness with stimulation of hind limbs and forelimbs have also been noted. Most studies have focused on relatively early (<4 weeks) changes. The combination of these characteristics (damage to central structures, early onset, allodynia in the segments close to the level of injury) suggests that the model mimics the acute segmental allodynia and hyperalgesia that is sometimes a feature of central cord injuries (Siddall et al. 1999). A later onset form of this type of pain may also be indicated by the findings of Lindsey et al. (2000) and is also apparent clinically (Siddall et al. 1999). Thus, the model appears to correspond well with a recognized pattern of ▶ at-level neuropathic pain that is observed following SCI (Siddall et al. 2002). This means that any observations on the underlying mechanisms and potential treatments are more likely to relate to at-level neuropathic SCI pain with allodynia and hyperalgesia.

Findings

The model has been used in a number of studies to investigate the mechanisms underlying neuropathic pain following spinal cord injury. Using this model, it has been demonstrated that animals with allodynia following a contusion injury have higher levels of basal and evoked neuronal activity in the spinal cord close to the level of injury as measured by ▶ Fos immunoreactivity (Siddall et al. 1999). Single-cell recordings in this model also demonstrate an increased responsiveness of some neurons in the spinal cord close to the damaged segment (Drew et al. 2001). It has also been demonstrated that an increase in neuronal responsiveness is not confined to spinal neurons. Neurons in the caudal brain stem show a decreased threshold to mechanical stimulation of the trunk (Hubscher and Johnson 1999). Evoked responses of thalamic neurons were also exaggerated in allodynic animals with a contusion injury with

a higher proportion of cells exhibiting ▶ afterdischarges (Gerke et al. 2003).

The receptor changes that may underlie these changes in behavior and neuronal responsiveness have been examined in a series of experiments investigating alterations in metabotropic glutamate expression following contusive SCI. These experiments indicate that contusion results in an increase in the group I metabotropic glutamate receptor (mGluR1) in the spinal cord (Mills et al. 2001) and that this increase in expression occurs on spinothalamic tract cells both just rostral to the site of the lesion and on cells in the cervical enlargement (Mills and Hulsebosch 2002).

Thus, findings from this model suggest that neuropathic pain following spinal cord injury is associated with abnormal responsiveness of spinal and supraspinal neurons with both an increased rate of firing and prolonged firing beyond the period of stimulation. The cellular mechanisms underlying this increased neuronal responsiveness are still unclear but may include receptor changes such as alterations in metabotropic glutamate receptors.

References

Drew, G. M., Siddall, P. J., & Duggan, A. W. (2001). Responses of spinal neurones to cutaneous and dorsal root stimuli following contusive spinal cord injury in the rat. *Brain Research, 893*, 59–69.

Gerke, M. B., Xu, L., Duggan, A. W., & Siddall, P. J. (2003). Ventrobasal thalamic neuronal activity in rats with mechanical allodynia following contusive spinal cord injury. *Neuroscience, 117*, 715–722.

Hubscher, C. H., & Johnson, R. D. (1999). Changes in neuronal receptive field characteristics in caudal brain stem following chronic spinal cord injury. *Journal of Neurotrauma, 16*, 533–541.

Hulsebosch, C. E., Xu, G. Y., Perez-Polo, J. R., Westlund, K. N., Taylor, C. P., & McAdoo, D. J. (2000). Rodent model of chronic central pain after spinal cord contusion injury and effects of gabapentin. *Journal of Neurotrauma, 17*, 1205–1217.

Lindsey, A. E., LoVerso, R. L., Tovar, C. A., Hill, C. E., Beattie, M. S., & Bresnahan, J. C. (2000). An analysis of changes in sensory thresholds to mild tactile and cold stimuli after experimental spinal cord injury in the rat. *Neurorehabilitation and Neural Repair, 14*, 287–300.

Mills, C. D., Xu, G. Y., Johnson, K. M., McAdoo, D. J., & Hulsebosch, C. E. (2000a). AIDA reduces glutamate release and attenuates mechanical allodynia after spinal cord injury. *Neuroreport, 11*, 3067–3070.

Mills, C. D., Fullwood, S. D., & Hulsebosch, C. E. (2000b). Changes in metabotropic glutamate receptor expression following spinal cord injury. *Experimental Neurology, 170*, 244–257.

Mills, C. D., Grady, J. J., & Hulsebosch, C. E. (2001). Changes in exploratory behavior as a measure of chronic central pain following spinal cord injury. *Journal of Neurotrauma, 18*, 1091–1105.

Mills, C. D., & Hulsebosch, C. E. (2002). Increased expression of metabotropic glutamate receptor subtype 1 on spinothalamic tract neurons following spinal cord injury in the rat. *Neuroscience Letters, 319*, 59–62.

Mills, C. D., Johnson, K. M., & Hulsebosch, C. E. (2002). Role of group II and group III metabotropic glutamate receptors in spinal cord injury. *Experimental Neurology, 173*, 153–167.

Siddall, P. J., Taylor, D. A., McClelland, J. M., Rutkowski, S. B., & Cousins, M. J. (1999b). Pain report and the relationship between physical factors and the development of pain in the first six months following spinal cord injury. *Pain, 81*, 187–197.

Siddall, P. J., Xu, C. L., & Cousins, M. J. (1995). Allodynia following traumatic spinal cord injury in the rat. *Neuroreport, 6*, 1241–1244.

Siddall, P. J., Xu, C. L., Floyd, N., & Keay, K. A. (1999a). C-fos expression in the spinal cord in rats exhibiting allodynia following contusive spinal cord injury. *Brain Research, 851*, 281–286.

Siddall, P. J., Yezierski, R. P., & Loeser, J. D. (2002). Taxonomy and epidemiology of spinal cord injury pain. In R. P. Yezierski & K. Burchiel (Eds.), *Spinal cord injury pain: Assessment, mechanisms, management, progress in pain research and management* (Vol. 23, pp. 9–23). Seattle: IASP Press.

Wrathall, J. R. (1992). Spinal cord injury models. *Journal of Neurotrauma, 9*, 129–134.

Spinal Cord Injury Pain Model, Cordotomy Model

Alan R. Light[1] and Charles J. Vierck Jr.[2]
[1]Department of Anesthesiology, University of Utah, Salt Lake City, UT, USA
[2]Department of Neuroscience and McKnight Brain Institute, University of Florida College of Medicine, Gainesville, FL, USA

Definition

▶ Cordotomy (originally ▶ anterolateral cordotomy – referring to the geometric term "chord" to indicate interruption of white matter) was developed as a surgical procedure to sever the spinothalamic tract at a spinal level above dermatomes to which extreme chronic pain is referred. However, physicians noted early on that while pain was initially reduced or eliminated contralateral and below the level of cordotomy, pain often recurred, even 6 months or more after the lesion. Other studies have reported a strong correlation between development of chronic pain after traumatic ▶ spinal cord injury (SCI) and interruption of the spinothalamic tract. This prompted investigations of evidence in animal models for recurrence of pain sensitivity following anterolateral spinal lesions, in order to understand mechanisms for chronic pain caudal to the level of the lesion following spinal cord injury.

Characteristics

Spinal cord injury produces loss of motor and sensory functions below the ▶ dermatomal level of the lesion, as a predictable consequence of interrupting long pathways descending from and ascending to the brain. If the spinal cord is transected (e.g., at a thoracic level), consciously initiated motor activities of the legs and sensations from stimulation of the lower limb are permanently lost. However, spontaneous sensations that are referred to the legs (perceived as if they originated in the legs) are commonly experienced after ▶ spinal transection, and these can be painful (Defrin et al. 2001; Finnerup et al. 2003).

Numerous investigations of human patients following incomplete spinal cord injuries or vascular strokes that affect ascending somatosensory pathways have shown that chronic pain can also develop with these partial lesions. Interestingly, in these patients, the body region from which the chronic, spontaneous pain originates often has reduced or no sensitivity to noxious stimuli or temperature, yet mild stimuli of these same regions can enhance or trigger the chronic pain (e.g., Boivie et al. 1989). These results implicate interruption of the lateral spinothalamic tract as necessary for development and maintenance of chronic, below-level pain.

Thus, paradoxically, interruption of the ascending pathways considered most important for ascending conduction of nociceptive input can result in chronic pain. However, spontaneous pain does not always develop after transection of the spinothalamic tract (see ▶ Cordotomy Effects on Humans and Animal Models). Therefore, this lesion may represent a necessary condition, but does not represent a sufficient condition for chronic, below-level pain. Possibly, the extent of a lesion that involves the spinothalamic tract (i.e., partial interruption, total interruption, or inclusion of another critical pathway) is the determining feature. Alternatively, excitatory influences at the lesion site could have important influences on rostral sites that have been partially deafferented by the spinothalamic lesion. These factors have been evaluated in laboratory animal models of spinal cord injury.

Two approaches have been used to study chronic pain in laboratory animals after spinal lesions. The first is observation of spontaneous behaviors that might be elicited by pain. For example, Levitt and Levitt (1981) observed episodes of self-injurious behavior directed towards regions caudal and contralateral to anterolateral cordotomy. The episodic nature of these behaviors is consistent with reports of pain episodes experienced by humans after spinal cord injury. However, the qualities of spontaneous sensations experienced by a laboratory animal cannot be deduced from self-injurious behavior, which could be elicited by non-noxious sensations such as itching or tickle. Animals might persist in such behavior, even when they severely injure the body part, because scratching or biting the region of sensory referral would not be painful (or would be minimally painful) after interruption of the spinothalamic tract.

The second approach to study the development of chronic pain caused by an incomplete spinal lesion such as cordotomy is to evaluate sensitivity to nociceptive stimulation before and after surgery. When chronic pain develops after cordotomy in humans, sensitivity to noxious stimulation recovers for stimulation within the region of pain referral and can be heightened outside the region of referral (e.g., ipsilateral to the lesion) (Nagaro et al. 2001).

Testing of operant escape from noxious stimulation in monkeys has demonstrated that restricted, superficial lesions of the anterolateral tract result in long-lasting contralateral hypoalgesia, similar to the effects of superficially located cordotomy in humans (Nathan and Smith 1979; Vierck et al. 1990). In contrast, medially extensive lesions often resulted in recovery of pain sensibility and even the development of hyperalgesia (Vierck et al. 1990). This result suggests that lesion extent is a critical feature of anterolateral cordotomy, a hypothesis that is difficult to evaluate in humans, because histological confirmation of lesion extent has rarely been possible. Because mechanisms of recovery of behavioral responses to nociceptive stimulation were difficult to study in monkeys, a rat model was developed to determine if similar results would be obtained (Vierck et al. 1995). This study demonstrated that, like monkeys and humans, rats with medially extensive anterolateral spinal lesions showed initial contralateral hypoalgesia, which recovered in some animals over time.

Overall, the monkey and rat studies showed a remarkable correspondence to human patients with cordotomy, in that animals with medially extensive lesions recovered pain sensitivity in approximately 50 % of the cases, very similar to the percentage of patients with cordotomy who have recurrence of their pain (White and Sweet 1969). Thus, another apparently paradoxical finding was confirmed: that larger, more medially extensive lesions resulted in the recurrence of pain, while smaller, more restricted lesions resulted in permanent pain reduction.

The explanation for less doing more, in terms of reducing pain following cordotomy, may lie in reactions to the injury in the spinal gray, rather than in the tracts lesioned. For example, Vierck and Light (1999) observed ipsilateral hyperalgesia when hemotoxic ischemia was intentionally introduced within the gray matter at the cordotomy lesion sites of rats (Fig. 1).

Other investigations of neuronal activity within spinal segments bordering a variety of spinal lesion models have shown conclusively that some neurons exhibit abnormal spontaneous activity and hyperexcitable responses to

Spinal Cord Injury Pain Model, Cordotomy Model, Fig. 1 Three examples of effects of spinal lesions on nociceptive sensitivity of rats. Each lesion cavity was filled with blood to produce hemotoxicity within the spinal gray matter. Each graph presents the speed of operant escape responses to electrocutaneous stimulation (ordinate) within successive postoperative testing sessions (abscissa; five sessions per week). (a) A large lesion involving both lateral white columns and the *gray* matter produced a bilateral reduction of somatosensory input and abnormal afterdischarge. The lesion models include lateral hemisection (Hains et al. 2003), contusion injury (Gerke et al. 2003), photochemical ischemia nociceptive sensitivity that recovered to preoperative levels (*dashed line*). (b) In contrast, a small superficial lesion of anterolateral white matter on the right produced a substantial and enduring contralateral reduction of escape speed, accompanied by an ipsilateral hypersensitivity. (c) A large unilateral lesion produced a moderate contralateral reduction of escape speed that episodically recovered to preoperative levels (*dashed line*). Ipsilateral sensitivity for this animal was increased, particularly early

(Hao et al. 1992), and excitotoxicity (Yezierski and Park 1993). These demonstrations of neuronal hyperexcitability are most clearly related mechanistically to ▶ dysesthesias and pain, which can be experienced within dermatomes bordering a spinal lesion (▶ at-level phenomena). In addition, they may provide evidence for a determining factor in development of below-level pain after interruption of the spinothalamic tract. Thus, it may be that ischemia/excitotoxicity establishes abnormal activity of sufficient magnitude in neurons bordering a spinal lesion that propagates via diffuse spinal projection systems and influences spontaneous and evoked activity of rostral projection sites, particularly including those partially deafferented, by interruption of the spinothalamic tract.

Evidence for propagation of excitatory influences along the propriospinal system of diffuse spinal projections has been provided by assessment of nociceptive reflexes and by psychophysical observations of humans after SCI. Excitotoxic injury to the spinal gray matter of rats enhances reflexes elicited by mechanical and thermal stimuli delivered to segments caudal to the lesion (Yezierski et al. 1998), and thoracic lateral hemisection of rats has this effect on forelimb and hindlimb reflex responses to mechanical and thermal stimulation (Christensen et al. 1996). In monkeys, SCI can transform the valence of intersegmental modulation from inhibitory to excitatory, using a condition-test paradigm (Vierck et al. 2002), and nociceptive flexion reflexes can recover from an initial depression following anterolateral cordotomy, similar to the return of contralateral pain sensitivity (Vierck et al. 1990). Thus, modulation of segmental spinal reflexes in laboratory animals is affected rostral and caudal to a spinal lesion, and this effect can change over time after SCI. Similarly, exaggeration of flexion reflexes caudal to SCI is well documented in humans as a component of the spastic syndrome (Dimitrijevic and Nathan 1967), and autonomic and somatic reflexes rostral to a lesion can be enhanced in humans (Calancie 1991; Curt et al. 1997). In addition, psychophysical tests of humans after SCI have shown that pain sensitivity can be exaggerated for stimulation rostral to a spinal lesion (Vierck et al. 2002).

Both laboratory animal and human studies have implicated combined contributions of spinal hyperexcitability and rostral deafferentation for establishment of chronic pain after SCI. Following anterolateral cordotomy and more extensive spinal lesions in rats, monkeys, and humans, abnormal spontaneous activity is present in a primary thalamic termination site of the spinothalamic tract (nucleus ventralis posterolateralis: VPL) (Weng et al. 2000). Clearly, sufficient spontaneous activity in a pain transmission system could produce sensations of pain, but the pattern or intensity of this activity that elicits pain is difficult to determine. An investigation with rats after contusion injury of the spinal cord, however, has shown that spontaneous and elicited activity in partially deafferented thalamic nucleus VPL is greater in animals with exaggerated reflex responses to at-level stimulation than for animals with comparable lesions but no reflex enhancement (Gerke et al. 2003). Also, a study of humans with incomplete spinal lesions has thoroughly characterized the sensitivity to below-level stimulation of those with and without chronic pain (Finnerup et al. 2003). Absence of below-level thermal sensitivity was present for all subjects, indicating that interruption of the spinothalamic tract was common to all lesions. However, regions of below-level allodynia were unique to subjects with pain, and the abnormally sensitive regions corresponded to locations of referred pain. These studies indicate that spinal hyperexcitability is a contributing factor to abnormal thalamic activity and chronic pain for incomplete spinal lesions. It remains to be determined if this is the case for complete spinal transections that result in below-level central pain.

Summary

In summary, interruption of the spinothalamic tract is a necessary condition for development of below-level pain following SCI. In addition, abnormal discharge within the gray matter bordering a lesion appears to be a necessary condition for development of increased pain sensitivity over time, an accompaniment of chronic pain following incomplete spinal lesions. Neither of these factors represents a sufficient condition for chronic pain or increased nociceptive sensitivity. Therefore, attenuation or prevention of abnormal activity within the spinal gray matter bordering a spinal lesion has the potential to reduce the probability that a chronic pain condition will become established.

References

Boivie, J., Leijon, G., & Johansson, I. (1989). Central post-stroke pain – A study of the mechanisms through analyses of the sensory abnormalities. *Pain, 37*, 173–185.

Calancie, B. (1991). Interlimb reflexes following cervical spinal cord injury in man. *Experimental Brain Research, 85*, 458–469.

Christensen, M. D., Everhart, A. W., Pickelman, J. T., & Hulsebosch, C. E. (1996). Mechanical and thermal allodynia in chronic central pain following spinal cord injury. *Pain, 68*, 97–107.

Curt, A., Nitsche, B., Rodic, B., Schurch, B., & Dietz, V. (1997). Assessment of autonomic dysreflexia in patients with spinal cord injury. *Journal of Neurology, Neurosurgery, and Psychiatry, 62*, 473–477.

Defrin, R., Ohry, A., Blumen, N., & Urca, G. (2001). Characterization of chronic pain and somatosensory function in spinal cord injury subjects. *Pain, 89*, 253–263.

Dimitrijevic, M. R., & Nathan, P. W. (1967). Studies of spasticity in man. I. Some features of spasticity. *Brain, 90*, 1–30.

Finnerup, N. B., Johannesen, I. L., Fuglsang-Frederiksen, A., Bach, F. W., & Jensen, T. S. (2003). Sensory function in spinal cord injury. Patients with and without central pain. *Brain, 126*, 57–70.

Gerke, M. B., Duggan, A. W., & Siddall, P. J. (2003). Thalamic neuronal activity in rats with mechanical allodynia following contusive spinal cord injury. *Neuroscience, 117*, 715–722.

Hains, B. C., Willis, W. D., & Hulsebosch, C. E. (2003). Temporal plasticity of dorsal horn somatosensory neurons after acute and chronic spinal cord hemisection in rat. *Brain Research, 970*, 238–241.

Hao, J. X., Xu, X.-J., Seiger, A., & Wiesenfeld-Hallin, Z. (1992). Transient spinal cord ischemia induces temporary hypersensitivity of dorsal horn wide dynamic range neurons to myelinated, but not unmyelinated fiber input. *Journal of Neurophysiology, 68*, 384–391.

Levitt, M., & Levitt, J. (1981). The deafferentation syndrome in monkeys: Dysesthesias of spinal origin. *Pain, 10*, 129–147.

Nagaro, T., Adachi, N., Tabo, E., Kimura, S. A. T., & Dote, K. (2001). New pain following cordotomy: Clinical features, mechanisms, and clinical importance. *Journal of Neurosurgery, 95*, 425–431.

Nathan, P. W., & Smith, M. C. (1979). Clinico-anatomical correlation in anterolateral cordotomy. *Advances in Pain Research and Therapy, 3*, 921–926.

Vierck, C. J., & Light, A. (1999). Effects of combined hemotoxic and anterolateral spinal lesions on nociceptive sensitivity. *Pain, 82*, 1–11.

Vierck, C. J., Jr., Greenspan, J. D., & Ritz, L. A. (1990). Long-term changes in purposive and reflexive responses to nociceptive stimulation in monkeys following anterolateral chordotomy. *Journal of Neuroscience, 10*, 2077–2095.

Vierck, C. J., Jr., Lee, C. L., Willcockson, H. H., Kitzmiller, A., DiRuggiero, D., Bullitt, E., & Light, A. R. (1995). Effects of anterolateral spinal lesions on escape responses of rats to hindpaw stimulation. *Somatosensory & Motor Research, 12*, 163–174.

Vierck, C. J. J., Cannon, R. L., Stevens, K. A., & Wirth, E. D. (2002). Mechanisms of increased pain sensitivity within dermatomes remote from an injured segment of the spinal cord. In R. P. Yezierski & K. J. Rurchiel (Eds.), *Spinal cord injury pain: Assessment, mechanisms, management* (pp. 155–173). Seattle: IASP Press.

Weng, H.-R., Lee, J., Lenz, F., Schwartz, A., Vierck, C., Rowland, L., & Dougherty, P. (2000). Functional plasticity in primate somatosensory thalamus following chronic lesion of the ventral lateral spinal cord. *Neuroscience, 101*, 393–401.

White, J. C., & Sweet, W. H. (1969). *Pain and the neurosurgeon: A forty-year experience*. Springfield: CC Thomas.

Yezierski, R. P., & Park, S.-H. (1993). The mechanosensitivity of spinal sensory neurons following intraspinal injections of quisqualic acid in the rat. *Neuroscience Letters, 157*, 115–119.

Yezierski, R. P., Liu, S., Ruenes, G. L., Kajander, K. J., & Brewer, K. L. (1998). Excitotoxic spinal cord injury: Behavioral and morphological characteristics of a central pain model. *Pain, 75*, 141–155.

Spinal Cord Injury Pain Model, Hemisection Model

Claire E. Hulsebosch
Spinal Cord Injury Research, University of Texas Medical Branch, Galveston, TX, USA

Synonyms

Central neuropathic pain; Chronic central pain models; Spinal cord injury

Definition

Spinal hemisection model of chronic central pain is a surgical lesion to the low thoracic spinal cord, between the dorsal root entry zones, in which the spinal cord is cut, unilaterally, from dorsal to ventral, in one spinal segment. The surgical cut includes unilateral lesions of the dorsal column system, dorsal corticospinal tract, Lissauer's tract, lateral and ventral funiculi (including the dorsal lateral fasciculus, spinal thalamic, spinal cerebellar, and ventral corticospinal tracts), and the gray matter on one side of the cord. Lesion sites vary from T8 to T13, to model paraplegia, and C2 or C3 to model quadriplegia, depending on the study. Care should be taken to avoid dorsal blood vessels to eliminate the confounding effect of additional ischemia combined with the surgical lesion. Mechanical allodynia develops over 2–3 weeks bilaterally in the dermatome of the segment of lesion as well as bilaterally in both the forelimbs and hindlimbs; while thermal allodynia (allodynia – since lower temperature than that of presurgery, evokes response after surgery) develops in these same regions but a week or so later than mechanical allodynia. The mechanical and thermal allodynia persists for the life of the animal (Christensen et al. 1996; Christensen and Hulsebosch 1997).

Characteristics

There are several advantages of the spinal hemisection (HS) model over other models of CCP (chronic central pain) after SCI (spinal cord injury): (1) HS avoids the variability of vascular-dependent lesions; (2) responses to a variety of somatic stimuli are examined, since pharmacological interventions may selectively alter one quality of sensation (e.g., thermal and not tactile); (3) changes in locomotion are assessed using the Basso, Beattie, and Bresnahan (BBB) open-field test scale developed in the laboratory of M. Basso to control for alterations in motor control that might affect the nociceptive tests; (4) most animals develop allodynia (only a subset (44 %) of the spinal ischemic rats, 50 %

of the quisqualate spinally injected rats, and 80 % of the spinal contused rats (Mills et al. 2001a, b) develop mechanical allodynia); (5) both spontaneous and evoked components of central neuropathic pain occur (Hulsebosch 2002a, b); (6) surgical ease and reproducibility (i.e., this is a relatively low technical approach that does not require an expensive contusion weight drop apparatus); (6) the surgical approach allows consistency in injury for ontological comparisons since the spinal cord increases in size as rodents age (Gwak et al. 2004); and finally, (7) while spinal contusion lesions may better parallel the injury profile described in human spinal cord injury (see Hulsebosch et al. 2000), we have determined empirically that the HS model has several advantages over the contusion model such as no "twice daily" bladder expressions for up to 2 weeks, no accompanying bladder infections, better hindlimb recovery for spontaneous and evoked testing, allowing below-level testing of "pain-like" behavior, and, importantly, more animals develop allodynia in this model than other models (Mills et al. 2001b; Hulsebosch 2002a, b).

Because of the discrete unilateral spinal lesion, this surgical approach has been used to study regeneration of tracts in response to neurotrophins or antibodies to neutralize growth inhibitors (Schnell et al. 2011), unilateral viral delivery of substances to attenuate development of allodynia (Peng et al. 2006), effect of discrete spinal lesion on brainstem, thalamic and cortical activity (Hubscher and Johnson 2002; Gwak et al. 2010; Yague et al. 2011), bilateral effects along the neuraxis from unilateral central injury (Gwak and Hulsebosch 2009), and effects on respiratory activity (Singh et al. 2012).

To approach this concept at the single cell level, if one focuses on a pain projection neuron, such as a spinothalamic tract neuron, input from primary afferents, interneurons, and descending inputs can either facilitate or inhibit the excitability of the projection neurons by receptor-mediated alterations in membrane potential. Equally important in projection neuron excitability is the presence of ion channels and a variety of second messenger and trans-synaptic signaling cascades. Thus, by pharmacological interventions using antagonists of excitatory receptors, or agonists of inhibitory receptors, or appropriate manipulation of ion channel permeability, it is possible to alter the hyperexcitability that is present after SCI (Hains et al. 2003). The spinal HS allows easy interpretation of whole animal behavior in response to pharmacological intervention since the hindlimbs are not compromised.

For example, it has been demonstrated by our laboratory and others that CGRP is a putative nociceptive transmitter. Thus, blocking CGRP receptor activation with a neutralizing peptide fragment that binds in the CGRP ligand binding site but does not activate the G-protein-coupled activation of adenyl cyclase would be predicted to attenuate nociception. In fact, this results in attenuation of evoked responses in the forelimbs and the hindlimbs that is consistent with mechanical and thermal allodynia (above-level and below-level pain) (Bennett et al. 2000a). Similarly, inhibition of excitatory amino acid (EAA) receptors by the NMDA receptor antagonist D-AP 5 and a non-NMDA receptor antagonist that is specific for the AMPA/kainate receptor (NBQX) (Bennett et al. 2000b) attenuates mechanical allodynia but has no effect on thermal allodynia in either the forelimbs or hindlimbs. Furthermore, application of exogenous catecholamines (Hains et al. 2000a) or serotonin (Hains et al. 2000b) attenuate the mechanical and thermal allodynia both above and below level, presumably subserving the descending inputs that provide tonic inhibition of projection neurons that were severed in the hemisection. In another example, examples of evoked and spontaneous allodynia behaviors are attenuated with a L-type Ca channel blocker (gabapentin; see Hulsebosch 2002a, b) in the rat spinal HS model after development of mechanical and thermal allodynia in both forelimbs and hindlimbs. Behavioral measures of both mechanical and thermal allodynia are described in details elsewhere (Christensen et al. 1996). It is important to note that the measures (frequency or latency) of brisk paw withdrawals are accompanied by active attention of the rat to the stimulus by head turning and stimulus attack,

whole body posturing to avoid a repeated stimulus, etc. in response to stimuli. The inclusion of these complex behaviors excludes simple hyperreflexia, which is a segmental response. Briefly, all the above interventions resulted in attenuation of both mechanical and thermal allodynia with the exception of the NMDA receptor antagonist (D-AP5) and the non-NMDA receptor antagonist (NBQX) that attenuated mechanical but not thermal allodynia, in both above-level and below-level subclassifications of CCP.

Summary

To summarize, there are many models of CCP after spinal cord injury and each have advantages and disadvantages. The above discussion focuses on the spinal hemisection model as a useful model of clinical CCP.

References

Bennett, A. D., Chastain, K. M., & Hulsebosch, C. E. (2000a). Alleviation of mechanical and thermal allodynia by CGRP$_{8-37}$ in a rodent model of chronic central pain. *Pain, 86*, 163–175.

Bennett, A. D., Everhart, A. W., & Hulsebosch, C. E. (2000b). Intrathecal administration of an NMDA or a non-NMDA receptor antagonist reduces mechanical but not thermal allodynia in a rodent model of chronic central pain after spinal cord injury. *Brain Research, 859*, 72–82.

Christensen, M. D., & Hulsebosch, C. E. (1997). Chronic central pain after spinal cord injury. *Journal of Neurotrauma, 14*, 517–531.

Christensen, A. D., Everhart, A. W., Pickelmann, J. T., & Hulsebosch, C. E. (1996). Mechanical and thermal allodynia in chronic central pain following spinal cord injury. *Pain, 68*, 97–107.

Gwak, Y. S., & Hulsebosch, C. E. (2009). Remote astrocytic and microglial activation modulates neuronal hyperexcitability and below-level neuropathic pain after spinal injury in rat. *Neuroscience, 161*, 895–903.

Gwak, Y. S., Hains, B. C., Johnson, K. M., & Hulsebosch, C. E. (2004). Effect of age at time of spinal cord injury on behavioral outcomes in rat. *Journal of Neurotrauma, 21*, 983–993.

Gwak, Y. S., Kim, H. K., Kim, H. Y., & Leem, J. W. (2010). Bilateral hyperexcitability of thalamic VPL neurons following unilateral spinal injury in rats. *The Journal of Physiological Sciences, 60*, 59–66.

Hains, B. C., Chastain, K. M., Everhart, A. W., McAdoo, D. J., & Hulsebosch, C. E. (2000a). Transplants of adrenal medullary chromaffin cells reduce forelimb and hindlimb allodynia in a rodent model of chronic central pain after spinal cord hemisection injury. *Experimental Neurology, 164*, 426–437.

Hains, B. C., Johnson, K. A., Eaton, M. J., & Hulsebosch, C. E. (2000b). Transplantation of immortalized serotonergic neurons attenuates chronic central pain after spinal hemisection injury in rat. *Neuroscience Abstracts, 26*, 2303.

Hains, B. C., Willis, W. D., & Hulsebosch, C. E. (2003). Serotonin receptors 5-HT$_{A1}$ and 5-HT$_3$ reduce hyperexcitability of dorsal horn neurons after chronic spinal cord hemisection injury in rat. *Brain Research, 149*, 174–186.

Hubscher, C. H., & Johnson, R. D. (2002). Differential effects of chronic spinal hemisection on somatic and visceral inputs to caudal brainstem. *Brain Research, 947*, 234–242.

Hulsebosch, C. E. (2002a). Pharmacology of chronic pain after spinal cord injury: Novel acute and chronic intervention strategies. In R. P. Yezierski & K. J. Burchiel (Eds.), *Spinal cord injury pain: Assessment, mechanisms, management* (Progress in pain research and management, Vol. 23, pp. 189–204). Seattle: IASP Press.

Hulsebosch, C. E. (2002b). Recent advances in pathophysiology and treatment of spinal cord injury. *Advances in Physiology Education, 26*, 238–255.

Hulsebosch, C. E., Hains, B. C., Waldrep, K., & Young, W. (2000). Bridging the gap: From discovery to clinical trials in spinal cord injury. *Journal of Neurotrauma, 17*, 1117–1128.

Mills, C. D., Grady, J. J., & Hulsebosch, C. E. (2001a). Changes in exploratory behavior as a measure of chronic central pain following spinal cord injury. *Journal of Neurotrauma, 18*, 1091–1105.

Mills, C. D., Hains, B. C., Johnson, K. M., & Hulsebosch, C. E. (2001b). Strain and model differences in behavioral outcomes after spinal cord injury in rat. *Journal of Neurotrauma, 18*, 743–756.

Peng, X.-M., Zhou, Z.-G., Glorioso, J. C., Fink, D. J., & Mata, M. (2006). Tumor necrosis factor-a contributes to below-level neuropathic pain after spinal cord injury. *Annals of Neurology, 59*, 843–851.

Schnell, L., Hunanyan, A. S., Bowers, W. J., Homer, P. J., Federoff, H. J., Gullo, M., Schwab, M. E., Mendell, L. M., & Arvanian, V. L. (2011). Combined delivery of nogo-a antibody, neurotrophin-3 and the NMDA-NR2d subunit establishes a functional 'detour' in the hemisected spinal cord. *The European Journal of Neuroscience, 34*, 1256–1267.

Singh, L. P., Devi, T. S., & Nantwi, K. D. (2012). Theophylline regulates inflammatory and neurotrophic factor signals in functional recovery after C2-hemisection in adult rats. *Experimental Neurology, 238*, 79–88.

Yague, J. G., Foffami, G., & Aguilar, J. (2011). Cortical hyperexcitability in response to preserved spinothalamic inputs immediately after spinal cord hemisection. *Experimental Neurology, 227*, 252–263.

Spinal Cord Injury Pain Model, Ischemia Model

Zsuzsanna Wiesenfeld-Hallin[1], Jing-Xia Hao[2] and Xiao-Jun Xu[3]
[1]Karolinska Institute, Stockholm, Sweden
[2]Department of Physiology and Pharmacology, Karolinska Institutet, Stockholm, Sweden
[3]Department of Physiology and Pharmacology, Section of Integrative Pain Research Karolinska Institutet, Stockholm, Sweden

Synonyms

Central pain model; Ischemia model; Photochemical model

Definition

A photochemical technique was developed to induce ischemia in nervous tissue (Dietrich et al. 1986), which results from a reaction between a photosensitizing dye circulating in the blood stream and a focused light source, producing singlet oxygen, leading to damage of the endothelial layer of the vessel. The developing thrombosis causes localized ischemia. Spinal cord injury can be produced in rats following intravenous injection of the red dye erythrosin B, and subsequent irradiation with an argon ion laser (Hao et al. 1991, 1992a). By varying the duration of laser irradiation, the extent of ischemia and consequent injury can be controlled (Hao et al. 1991, 1994). In this series of studies, ischemia was induced in lower thoracic to midlumbar segments in Sprague-Dawley rats.

Characteristics

Acute Pain-Like Behaviors, Its Underlying Physiological Mechanisms and Pharmacology

Within 24 h following spinal ischemia, about 90 % of rats develop hypersensitivity to

Spinal Cord Injury Pain Model, Ischemia Model, Fig. 1 Spinal cord injury pain models/ischemia model. Illustration of typical area exhibiting mechanical allodynia in an acute (**a**) and a chronic (**b**) allodynic rat. In (**a**) injury is at spinal segments L3-4. In (**b**) injury is at spinal segment T11-12. The vocalization threshold in the acutely alloydynic rat is 27.5 g in the *shaded* area and 13.0 g at *black* area. In the rat with chronic allodynia, the vocalization threshold in the *shaded* area is 2.6 g and in the *black* area 0.6 g. The normal vocalization threshold is 80 g

mechanical stimuli (Hao et al. 1991, 1992a). The acute allodynia is present in animals with both mild and severe ischemic injury. Following transient ischemia, motor deficits are minor and are accompanied by little or no morphological damage in the spinal cord at the light microscopic level (Hao et al. 1991, 1992a). A similar acute hypersensitivity follows severe ischemia, producing severe motor deficit, including bladder paralysis for about 1 week and extensive injury to the dorsal spinal cord. In these cases, the acute phase is often followed by a chronic phase of allodynia.

The behavioral abnormalities during the acute phase include mechanical hypersensitivity to brushing and pressure and cold hypersensitivity, but no increased responsiveness to heat. The allodynic area is usually fairly large, including the flanks and the whole back region (Fig. 1a).

To study the underlying physiological mechanisms of acute allodynia, the characteristics of wide dynamic range (WDR) neurons in the dorsal horn were compared in normal and allodynic animals (Hao et al. 1992b). The receptive field sizes and background activity were similar in the two groups. Cutaneous stimuli that activated both A and C afferents evoked a biphasic response, with a short latency A-fiber mediated and a long latency C-fiber-mediated response in normal animals. In allodynic animals, the stimulus evoked a prolonged burst, with no separation between A- and C-fiber-mediated activities. The overall response magnitude was increased, but the A-fiber-mediated response especially was much greater than normal. The response of the WDR neurons to mechanical stimulation with calibrated von Frey hairs in normal animals increased linearly with increased pressure. In allodynics, the cells had a significantly lower threshold than normal and the response magnitude increased exponentially, reaching maximal discharge frequencies at lower stimulus intensities than in normal animals. In contrast, the neuronal responses to noxious thermal stimuli applied with a CO_2 laser were indistinguishable in normal and allodynic animals. Thus, the responses of single neurons clearly reflected the behavioral abnormalities, where allodynia to mechanical, but not to heat, stimulation was observed.

The pharmacological basis of acute allodynia was also examined (Table 1).

Systemic or intrathecal administration of the GABA-B receptor agonist baclofen alleviated the allodynia and normalized the response pattern of WDR neurons recorded from allodynic rats, indicating an important role for activation of the GABA-B receptors in controlling the expression of mechanical allodynia. The GABA-A agonist muscimol had no effect. We have subsequently demonstrated that the number of dorsal horn neurons exhibiting GABA-like immunoreactivity was temporarily decreased after spinal cord

Spinal Cord Injury Pain Model, Ischemia Model, Table 1 Summary of the effect of drugs applied systemically or intrathecally on acute and chronic allodynia in spinally injured rats. The drug's mechanism of action and side effects are indicated in parentheses

Drug	Acute allodynia		Chronic allodynia	
	Systemic	Intrathecal	Systemic	Intrathecal
Morphine (μ-opioid agonist)	−	±	− (sedation)	+
DAMGO (μ-opioid agonist)	nt	nt	nt	+
CI977, U50488 (κ-opioid agonist)	nt	nt	− (sedation)	− (increased pain)
CI988 (CCK-B receptor antagonist)	nt	nt	+	nt
DPDPE (δ-opioid agonist)	nt	nt	nt	+
Clonidine (α-2 adrenergic agonist)	nt	±	− (sedation)	+
R-PIA (adenosine 1 receptor agonist)	nt	±	nt	+
Baclofen (GABA-B agonist)	+	+	−	nt
Muscimol,THIP (GABA-A agonist)	−	−	−	nt
Diazepam (benzodiazepine)	nt	nt	− (sedation)	nt
Tocainide, mexiletine (Na$^+$-channel blocker)	+	nt	+	nt
Carbamazepine (tricyclic antidepressant)	nt	nt	− (sedation)	nt
Gabapentin (Ca^{++} current inhibitor)	nt	nt	+	nt
NBQX (AMPA antagonist)	+	nt	− (sedation)	nt
MK-801, CGS 19755 (NMDA antagonist)	+	nt	± (sterotypy)	−
Dextromethorphan (NMDA antagonist)	nt	nt	+	nt
F 13640 (5-HT $_{A1}$ receptor agonist)	−	nt	+	nt

+ = good effect, ± = moderate effect, − = no effect, nt = not tested

ischemia, during the period when the acute allodynia was observed (Zhang et al. 1994). These results suggest that the acute allodynia is predominantly mediated by abnormal input from myelinated afferents. The sensory abnormality is primarily due to disinhibition, involving a loss of GABAergic presynaptic control of input via myelinated afferents, which may result from a high susceptibility of GABAergic neurons to excitotoxicity.

Chronic Allodynia, Its Underlying Physiological Mechanisms and Pharmacology

A chronic syndrome, with sensory abnormalities more severe than in the acute condition, develops in a subgroup of animals following severe, irreversible spinal cord ischemia (Xu et al. 1992). The main symptoms of this chronic syndrome are mechanical and cold allodynia referred to zones rostral to the damaged spinal cord segments (Fig. 1b).

Injury to dorsal gray matter is necessary for the development of chronic allodynia, which

appears after a delay of 2–8 weeks following the injury. Once present, the allodynia persists without signs of remission. Not all rats with a similar degree of spinal cord injury developed chronic allodynia, which may be due to different levels of endogenous inhibitory control exerted by opioids. Thus, systemic or intrathecal naloxone, or selective μ-opioid receptor antagonists, can trigger the appearance of typical allodynia-like responses in non-allodynic, spinally injured rats (Hao et al. 1998). Finally, the level of endogenous inhibitory control may be set by antiopioid systems, such as the neuropeptide cholecystokinin (CCK) (Xu et al. 1994).

Several abnormalities were found in the distribution and response characteristics of dorsal horn neurons in the chronic allodynic rats (Xu et al. 2002). In the allodynic rats, 17 % of the units versus 0 % in controls had no receptive field. Most of these units were located at or close to the lesioned spinal segment and they discharged spontaneously at high frequencies. Furthermore, allodynic rats showed a significant change in the relative proportion of low threshold (LT),

WDR, or high threshold (HT) neurons recorded, with fewer LT and more WDR cells in allodynic rats than in normal rats. The rate of ongoing activity of HT neurons was significantly higher, and the responses to mechanical stimulation were increased in allodynic rats. In allodynics, the WDR neurons responded with higher discharge rates to innocuous mechanical stimuli compared to controls, and the percentage of WDR and HT neurons showing afterdischarges to noxious pinch was also significantly increased. Finally, the proportion of WDR and HT neurons responding to innocuous cold stimulation increased from 53 % and 25 %, respectively, in control rats to 91 % and 83 % in allodynic animals. Neurons located up to three segments rostral to the lesion exhibited mechanical and cold hypersensitivity, which supports the presence of allodynia-like behavior to mechanical and cold stimuli applied in these dermatomes.

A large number of drugs have been tested in the chronic allodynic model (Table 1). Systemic or intrathecal morphine has a weak effect on acute allodynia, which may be related to a reduction in the density of μ-opioid receptors in the spinal cord after ischemia (Hao et al. 1991; Yu et al. 1999). In contrast, during chronic allodynia intrathecal, but not systemic, morphine was highly effective. The potency of the antiallodynic effect of intrathecal morphine was, however, reduced compared to its antinociceptive effect, and there was a rapid development of tolerance to the effects of intrathecal morphine in spinally injured rats. Thus it is unclear whether monotherapy with intrathecal morphine could provide long-term pain relief in patients with spinal cord injury pain, as tolerance development may present a problem.

Adenosine is an endogenous purine nucleotide which is extensively distributed intra- and extracellularly in the nervous system. Activation of spinal adenosine A1 receptors has been shown to produce antinociception. In a recent series of studies, we have shown that intrathecal R-phenylisopropyladenosine (R-PIA), an adenosine A1 receptor agonist, effectively alleviated chronic allodynia in spinally injured rats. The antiallodynic effect of R-PIA persisted considerably longer than that of morphine upon repetitive administration, and there was a synergistic interaction between R-PIA and morphine (von Heijne et al. 2000). Interestingly, i. t. R-PIA has only limited effect against acute allodynia, possibly indicating that the strong antiallodynic effect of R-PIA against chronic allodynia depends on structural and/or functional plasticity after injury.

There are several other interesting differences between the pharmacology of acute and chronic allodynia (Table 1). While baclofen is effective in treating acute allodynia (Hao et al. 1991), it is ineffective in chronic allodynia (Xu et al. 1992b), suggesting that GABAergic mechanisms may not play as important a role in the pathophysiology of chronic allodynia as they do immediately after injury. Moreover, acute allodynia is sensitive to blockade of the AMPA receptor for glutamate (Xu et al. 1993), but not the NMDA receptor, whereas the opposite is true for chronic allodynia (Hao and Xu 1996). Again, this suggests the involvement of central nervous system plasticity in the mechanisms of chronic pain, for which the activation of NMDA receptors may play an important role. Systemically applied local anesthetics, such as tocainide and mexiletine, are effective in both conditions (Xu et al. 1992a).

Two drugs were recently tested that may offer new approaches to treat central pain following spinal cord injury (Table 1). One is the highly selective 5-HT1A receptor agonist F 13640, which was effective in reversing chronic allodynia, possibly through the mechanism of inverse tolerance. The second is gabapentin, a drug that has been shown to be useful in treating some neuropathic pain conditions. Thus, the ischemic spinal injury model may be useful for testing the efficacy of potential analgesics for the treatment of neuropathic central pain.

References

Dietrich, W. D., Ginsgerg, M. D., Busto, R., & Watson, B. D. (1986). Photochemically induced cortical infarction in the rat. 1. Time course of hemodynamic consequences. *Journal of Cerebral Blood Flow and Metabolism, 6*, 184–194.

Hao, J.-X., & Xu, X.-J. (1996). Treatment of chronic allodynia-like symptoms in rats after spinal cord injury: Effects of systemic glutamate receptor antagonists. *Pain, 66,* 279–286.

Hao, J.-X., Xu, X.-J., Aldskogius, H., Seiger, Å., & Wiesenfeld-Hallin, Z. (1991). Allodynia-like effect in rat after ischemic spinal cord injury photochemically induced by laser irradiation. *Pain, 45,* 175–185.

Hao, J.-X., Xu, X.-J., Aldskogius, H., Seiger, Å., & Wiesenfeld-Hallin, Z. (1992a). Photochemically induced transient spinal ischemia induces behavioral hypersensitivity to mechanical and cold, but not to noxious heat, stimuli in the rat. *Experimental Neurology, 118,* 187–194.

Hao, J.-X., Xu, X.-J., Yu, Y.-X., Seiger, Å., & Wiesenfeld-Hallin, Z. (1992b). Transient spinal cord ischemia induces temporary hypersensitivity of dorsal horn wide dynamic range neurons to myelinated, but not unmyelinated, fiber input. *Journal of Neurophysiology, 68,* 384–391.

Hao, J.-X., Herregodts, P., Lind, G., Meyerson, B., Seiger, Å., & Wiesenfeld-Hallin, Z. (1994). Photochemically induced spinal cord ischaemia in rats: Assessment of blood flow by laser Doppler flowmetry. *Acta Physiologica Scandinavica, 151,* 209–215.

Hao, J.-X., Yu, W., & Xu, X.-J. (1998). Evidence that spinal endogenous opioidergic systems control the expression of chronic pain-related behaviors in spinally injured rats. *Experimental Brain Research, 118,* 259–268.

von Heijne, M., Hao, J.-X., Sollevi, A., Xu, X.-J., & Wiesenfeld-Hallin, Z. (2000). Marked enhancement of anti-allodynic effect by combined intrathecal administration of the adenosine a-1 receptor agonist R-phenylisopropyladenosine and morphine in a rat model of central pain. *Acta Anaesthesiologica Scandinavica, 44,* 665–671.

Xu, X.-J., Hao, J.-X., Aldskogius, H., Seiger, Å., & Wiesenfeld-Hallin, Z. (1992a). Chronic pain-related syndrome in rats after ischemic spinal cord lesion: A possible animal model for pain in patients with spinal cord injury. *Pain, 48,* 279–290.

Xu, X.-J., Hao, J.-X., Seiger, Å., Arnér, S., Lindblom, U., & Wiesenfeld-Hallin, Z. (1992b). Systemic mexiletine relieves chronic allodynia-like symptoms in rats with ischemic spinal cord injury. *Anesthesia and Analgesia, 74,* 652–694.

Xu, X.-J., Hao, J.-X., Seiger, Å., & Wiesenfeld-Hallin, Z. (1993). Systemic excitatory amino acid receptor antagonist of the a-amino-3-hydroxy-5-methyl-4-isoxazolepropionic acid (AMPA) receptor, but not of N-methyl-D-aspartate (NMDA) receptor, relieves the mechanical hypersensitivity in rats after transient spinal cord ischemia. *Journal of Pharmacology and Experimental Therapeutics, 267,* 140–144.

Xu, X.-J., Hao, J.-X., Seiger, Å., Hughes, J., Hökfelt, T., & Wiesenfeld-Hallin, Z. (1994). Chronic pain-related behaviors in spinally injured rats: Evidence for functional alterations of the endogenous cholecystokinin and opioid systems. *Pain, 56,* 271–277.

Xu, X.-J., Hao, J.-X., & Wiesenfeld-Hallin, Z. (2002). Physiological and pharmacological characterization of a Rat model of spinal cord injury pain after ischemic spinal cord injury. In R. Yezierski & K. Burchiel (Eds.), *Spinal cord injury pain* (pp. 175–187). Seattle: IASP Press.

Yu, W., Hao, J.-X., Xu, X.-J., Hökfelt, T., & Wiesenfeld-Hallin, Z. (1999). Spinal cord ischemia reduces Mu-opioid receptor in rats: Correlation with morphine insensitivity. *NeuroReport, 10,* 87–91.

Zhang, A.-L., Hao, J.-X., Seiger, Å., Xu, X.-J., Wiesenfeld-Hallin, Z., Grant, G., & Aldskogius, H. (1994). Decreased GABA immunoreactivity in spinal cord dorsal horn neurons after transient spinal cord ischemia in the rat. *Brain Research, 656,* 187–190.

Spinal Cord Injury, Central Neuropathic Pain

▶ Contusion Model of Chronic Central Pain After Spinal Cord Injury

Spinal Cord Injury, Excitotoxic Model

Kori L. Brewer[1] and Robert P. Yezierski[2]
[1]Department of Emergency Medicine, Brody School of Medicine East Carolina University, Greenville, NC, USA
[2]Comprehensive Center for Pain Research and Department of Neuroscience, The McKnight Brain Institute University of Florida, Gainesville, FL, USA

Synonyms

Excitotoxic model

Definition

While the loss of normal sensory and motor function is considered the most significant consequence of spinal cord injury (SCI), the development of chronic pain after injury strongly affects quality of life, and relief from correlative pain conditions is an important factor to the overall ability of patients to cope with SCI (van Leeuwen et al. 2012). The impact of SCI pain has been revealed in survey studies of this population. In a postal survey, 11 % of those responding reported pain, rather than loss of motor function, stopped them from working (Rose et al. 1988). A study by Widerstrom-Noga et al. (1999) found that the difficulty of dealing with chronic SCI pain was only surpassed by the decreased ability to walk or move, loss of sexual function, and decreased ability to control bladder and bowel function.

The recently developed International Spinal Cord Injury Pain (ISCIP) Classification defines the criteria for neuropathic pain associated with SCI. These criteria state that the pain must be at or below the neurological level of the injury, with the presence of allodynia or hyperalgesia within the pain distribution supporting the classification of SCI-related neuropathic pain (Bryce et al. 2012). This pain may be continuous or intermittent spontaneous pain, as well as stimulus-evoked (Finnerup and Baastrup 2012).

Research to identify the underlying mechanisms responsible for chronic pain after SCI focuses on the use of injury models that simulate pathological changes associated with human spinal injuries that lead to clinically relevant pain-related behaviors. Intramedullary injection of excitatory amino acid receptor agonists simulates the increase in endogenous excitatory amino acids known to occur after traumatic SCI, produces pathological and behavioral changes consistent with those observed after human SCI, and is a reliable method in which to study post-SCI pain states and their molecular correlates in rodents (Yezierski et al. 1998; Yezierski 2009). Simulating this chemical change that occurs during injury leads to neurochemical, inflammatory, and pathophysiological changes similar to those produced by traumatic or ischemic injuries, including cavitations of the spinal cord and altered response properties of surviving neurons (Yezierski and Park 1993; Yezierski et al. 1993). In addition, this excitotoxic SCI produces predictable pain behaviors including below-level

mechanical and thermal allodynia (Yezierski et al. 1998; Choi et al. 2011). These behavioral and pathophysiological events characterize the excitotoxic model of spinal injury.

Experimental Models of Spinal Cord Injury Pain

In recent years a number of experimental models have been used to study spinal cord injury (SCI) and the associated pain conditions (Nakae et al. 2011). All of these models have distinctive characteristics related to a critical component of the primary injury (e.g., trauma or ischemia). The photochemical (Xu et al. 2002) and hemisection (Hulsebosch 2002) models have provided important information regarding the pathophysiology and pharmacology of injury-induced sensory changes, including pain. The weight drop or contusion model, which is the oldest and most widely used model of SCI, has also been used in studies related to altered sensation following injury (Kerr and David 2007). The clip compression injury, which resembles a spinal contusion injury, also results in sensory alterations and pain (Bruce et al. 2002). A final approach in the study of injury-induced abnormal sensation involves the use of selected spinal lesions simulating the mechanical damage to ascending and descending pathways associated with spinal injury (Vierck and Light 2002). While most of these models are used primarily in rodents, some have been adapted for use in larger mammals as well as nonhuman primates (Nout et al. 2012). These different experimental strategies all produce pathological and/or behavioral changes that mimic those seen with human SCI, and, therefore, provide an opportunity to study the spinal and supraspinal mechanisms responsible for the condition of injury-induced pain. Examples of the pathological changes observed with different models of SCI compared to the human condition are shown in Fig. 1.

The Excitotoxic Model of Spinal Cord Injury

In recent years, substantial evidence has established glutamate as a major contributor to the tissue-damaging effects of SCI. For this reason, glutamate is viewed as one of several

Spinal Cord Injury, Excitotoxic Model, Fig. 1 Examples of histological sections from spinal cords that underwent contusion (weight drop), vasoconstriction (ischemic), ischemia (photochemical), and chemical (excitotoxic) injuries. The objective of each of these experimental models is to simulate different components of the human condition using features (e.g., trauma, vascular, chemical) known to be involved in producing primary and/or secondary tissue damage. Sections from the rat spinal cords were taken 3–6 weeks after injury, while the human sample was obtained 15 years after contusion injury (Reprinted with permission from Yezierski 2002)

putative chemical mediators contributing to a cellular cascade of secondary pathological changes following spinal injury. In an effort to evaluate the contribution of different glutamate receptor subtypes to the pathophysiology of SCI, a series of studies was undertaken in which NMDA and non-NMDA receptor agonists were microinjected into the rat spinal cord (Yezierski et al. 1993, 1998). In these studies the intraspinal injection of quisqualic acid (QUIS), a functional agonist of multiple glutamate receptor subtypes, resulted in neuronal death and the formation of spinal cavities; similar results have been obtained with injections of NMDA (Choi et al. 2011). Based on initial observations, the excitotoxic model of SCI was developed. The precision of the intramedullary injection technique makes it possible to evaluate the physiological and behavioral consequences of simulating small elevations in the concentration of specific glutamate agonists on specific cell populations.

This unique animal model allows for the study of sensory disorders resulting from selective gray matter loss after SCI and reliably results in the development of pain behaviors in both rats and mice (Abraham et al. 2004; Yezierski et al. 1998), including at-level (spontaneous overgrooming) and below-level (allodynia, hyperalgesia) pain syndromes. Excitotoxic SCI produces a pattern of neuronal loss and syrinx formation that mimics that seen after traumatic and ischemic SCI, without the resulting white matter loss, motor deficits, and spasticity below the level of injury observed in spinal contusion or transection models that complicate the study of sensory sequelae after injury. As with traumatic injury, these changes are, in part, mediated through the excessive stimulation of NMDA

Spinal Cord Injury, Excitotoxic Model, Fig. 2 Summary of the major components of the spinal *central injury cascade* that are believed to be responsible for the onset and progression of *at-level* and *below-level* pain following ischemic or traumatic spinal injury. Evidence supporting the basic concept of this cascade follows from results of clinical studies, as well as those obtained from the use of ischemic, lesion, contusion, and excitotoxic models of spinal cord injury. The four major components of the cascade (neurochemical, excitotoxicity, anatomical, and inflammation) are represented as being interactive and collectively lead to changes in the physiological state of spinal and supraspinal neurons. The end result of this cascade is the onset of clinical symptoms, e.g., allodynia, hyperalgesia, and pain. Abbreviations: *EAAs* excitatory amino acids, *Sub P* substance P, *cGMP* cyclic guanidine monophosphate, *NO* nitric oxide, *NF-kB* nuclear factor kappa B, *PKC* protein kinase C, *TNF* tumor necrosis factor, *IL-1β* interleukin-1β, *PLA₂* phospholipase A2, *NOS* nitric oxide synthase, *COX-2* cyclooxygenase-2, *RF* receptive field (Reprinted with permission from Yezierski 2000)

receptors as NMDA antagonists provide significant neuroprotection when administered immediately after QUIS (Liu et al. 1997). All of the pathophysiological, molecular, and behavioral consequences initiated by the transient increase in spinal levels of EAAs are also part of the sequelae associated with human SCI and similar to those described following experimental ischemic and traumatic injury in rodents.

The pathological findings seen with the excitotoxic model were initially believed to result solely from cellular events associated with excitotoxicity, i.e., overstimulation of glutamate receptors. More recent studies have revealed that contributions from other components of a *central injury cascade* associated with SCI are intricately involved. For example, studies of the inflammatory component of the QUIS injury model have shown that, following QUIS injections, there is an upregulation of mRNA for cytokines (IL-1β, TNF-α), COX-2, iNOS, and the death-inducing ligands TRAIL and CD-95 (Plunkett et al. 2001; Brewer and Nolan 2007). QUIS injury also activates the ERK1/2 and NF-κB signaling cascades, an effect that when blocked prevented the onset of injury-induced pain behaviors (Yu and Yezierski 2005). Therefore, the excitotoxic model, originally developed to study the temporal profile of pathological consequences following injury-induced elevations of EAAs, uncovered the interrelationship between excitotoxicity and other components of

Spinal Cord Injury, Excitotoxic Model, Fig. 3 Responses to mechanical and thermal stimuli delivered to the hind paws. Animals were pretested over a period of 7–10 days prior to receiving intraspinal injections of 125 mM quisqualic acid (QUIS). Animals received unilateral injections (1.2 μl) directed at the dorsal horn and intermediate gray (right side) in spinal segments ranging from T12 to L2. Postinjection testing commenced 10 days after surgery and continued for a period of 34 days. (a) Results of testing with mechanical stimuli delivered to the hind paws ipsilateral and contralateral to the side of QUIS injections. During postinjection testing, there was a significant decrease in stimulus intensity required to elicit withdrawal responses (bilaterally). Each point on the graph represents the mean threshold for all animals ($n = 10$) on each day of testing. Error bars represent standard errors around the mean. Stimulus intensity in grams is represented on the y-axis and days pre- and

postinjection on the x-axis. Statistical comparisons were made between the mean pre-injection baseline value and data obtained on each day of postinjection testing: *, triangle = $P < 0.05$ and **, double triangle = $P < 0.01$. (b) Results of the paw flick test assessing the sensitivity to a radiant heat stimulus delivered to the hind paws (same group of animals as tested in (a)). During postinjection testing, there was a significant decrease in withdrawal latencies to the thermal stimulus (bilaterally). Each point on the graph represents the mean of all trials ($n = 3$) for all animals ($n = 10$) on each day of testing. Error bars represent standard errors around the mean. Time in seconds is represented on the y-axis and days on the x-axis. Statistical comparisons were made between the mean pre-injection baseline value and data obtained on each day of postinjection testing: *, triangle = $P < 0.05$ and **, double triangle = $P < 0.01$ (Reprinted with permission from Yezierski et al. 1998)

a secondary injury cascade, as well as a link with the spinal mechanisms responsible for the onset of injury-induced abnormal sensation and pain-related behaviors (Fig. 2).

Characteristics

Behavioral Consequences of Excitotoxic Spinal Injury

Mechanical and Thermal Hypersensitivity

A significant observation following QUIS injections is the onset of mechanical and thermal allodynia below the level of injury (below-level pain). In the evaluation of mechanical thresholds, baseline responses (withdrawals) were elicited by stimulus intensities of 10–35 g (mean baseline

value 21.0 ± 9.8 g) applied to the glabrous skin of the hind paws. Following unilateral QUIS injections targeting spinal segments T12-L2, stimulus intensities were significantly lower (1.5–8.0 g) than pre-injection values (Fig. 3a). The same phenomenon of lowered thresholds to withdrawal from a thermal (heat) stimulus was demonstrated in the same animals (Fig. 3b). Hypersensitivity to both mechanical and thermal stimulation developed approximately 10–12 days post-QUIS injections and persisted throughout the 35-day post-injury evaluation period, underscoring the chronic nature of the behavioral effect (Yezierski et al. 1998). Because both tests involve stimulation of the hind paw, responses reflecting a hypersensitivity to thermal stimulation were observed in dermatomes remote from those represented by segments receiving QUIS injections (Yezierski et al. 1998). These evoked pain behaviors have been shown to develop after intramedullary injection of both NMDA and non-NMDA glutamate receptor agonists (Choi et al. 2011).

Excessive Grooming Behavior (Overgrooming)
Excessive grooming behavior is a spontaneous, self-directed behavior that begins 10–15 days post-injury. This behavior is characterized by excessive biting and scratching of the skin directed towards the dermatome associated with the level of spinal injury. This behavior continues with removal of hair and can progress to the extent where there is damage to the superficial and deep layers of skin. Because overgrooming behavior is a progressive condition, a classification scheme for different phases of this behavior was developed: (a) Class I: hair removal over contiguous portions of a dermatome; (b) Class II: extensive hair removal combined with signs of damage to the superficial layers of skin; (c) Class III: hair removal and damage to dermal layers of skin; and (d) Class IV: subcutaneous tissue damage (experiment terminated). Clinically, pain in paraplegia is also referred to peripheral dermatomes corresponding to spinal segments at or below the site of injury. Of special interest in QUIS-injected animals is the parallel between the delayed onset of excessive grooming and a similar temporal profile of central pain in patients with SCI. Injury-induced changes following QUIS injections are therefore believed to be part of a progression of altered sensations related to the clinical condition of central pain. Overgrooming behavior has been proposed as a model of at-level pain associated with SCI (Yezierski et al. 1998).

Excessive grooming behavior is correlated with a dorsal horn lesion in which the superficial laminae are spared, and is viewed as a variant of the well described "deafferentation autotomy" (Yezierski et al. 1998). Support for the conclusion that ectopic activity in the superficial dorsal horn may contribute to this behavior was demonstrated by the ability to affect overgrooming behavior by selectively eliminating different

regions of the gray matter. Figure 4a shows a cord where the superficial region and neck of the dorsal horn on the side of injection were eliminated, and this animal did not exhibit overgrooming behavior. By contrast, the pattern of neuronal loss in Fig. 4b included the neck of the dorsal horn, with sparing of the superficial region. This animal exhibited overgrooming behavior ipsilateral to the side of injury. This pattern of neuronal loss is referred to as "grooming-type damage." Examples of grooming-type damage are also shown in Fig. 4c–f.

A specific population of cells in the superficial dorsal horn that contribute to the generation of overgrooming behavior was defined by a study in which a conjugate of substance P and the ribosome-inactivating protein saporin was used to ablate neurons expressing the neurokinin-1 (NK-1) receptor in lamina I of the dorsal horn (Yezierski et al. 2004). When treated with this conjugate at the time of injury, animals undergoing excitotoxic SCI showed a significant decrease in the onset and severity of the behavior. When treatment was administered after overgrooming had developed, the behavior was again significantly reduced. This study suggests that the substrate for this specific at-level pain behavior includes the spinal neurons that express the NK-1 receptor in the superficial dorsal horn.

Efforts to systematically characterize events contributing to the onset and progression of excessive grooming have been carried out in three different strains of male rats (Gorman et al. 2001): Sprague–Dawley (SDM), Long Evans (LEM), and Wistar Furth (WFM) rats. Differences in grooming characteristics between male and female rats and the modulatory effects of female gonadal hormones were also evaluated in Sprague–Dawley females (SDF), bilaterally ovariectomized Sprague–Dawley females (OVX), and SDM's treated with either 17-β-estradiol (SDM-Est) or progesterone (SDM-Pro). The results showed that the development of overgrooming behavior in males and ovariectomized females is related to the rostrocaudal spread of a specific pattern of neuronal loss (see below). The onset, severity, and progression of

Spinal Cord Injury, Excitotoxic Model, Fig. 4 Patterns of neuronal loss following intraspinal injection of 125 mM quisqualic acid (QUIS). All injections were made on the right side of the spinal cord. (a) Neuronal loss throughout the dorsal horn following injection of 0.6 µl of QUIS at a depth of 300 µm in spinal segment L2 (survival period 30 days). (b) Grooming-type damage represented by neuronal loss in the neck of the dorsal horn (*arrows*) following injection of 0.6 µl of QUIS at a depth of 900 µm in spinal segment T13 (survival period 32 days). (c) Neuronal loss throughout the dorsal horn (ipsilateral to injection) and in the neck of the dorsal horn (*arrows*) contralateral to injection site (L4) where 1.2 µl of QUIS was injected at depths of 600 µm and 1,200 µl (survival period 18 days). The pattern of neuronal loss contralateral to injection represents grooming-type damage. (d) Bilateral grooming-type damage represented by neuronal loss in the dorsal horn partially sparing the superficial lamina (*arrowheads*) following injection of 1.2 µl of QUIS at depths of 600 µm and 1,200 µm in spinal segment T13 (survival period 34 days). (e) Bilateral grooming-type damage represented by neuronal loss throughout the neck of the dorsal horn following injection of 1.2 µl of QUIS at depths of 600 µm and 1,200 µm in spinal segment L1 (survival period 27 days). Note partial and complete sparing of the superficial lamina (*arrowheads*) ipsilateral and contralateral, respectively, to injection site. (f) Bilateral grooming-type damage represented by neuronal loss below the superficial lamina ipsilateral and contralateral following injection of 1.2 µl of QUIS at depths of 600 µm and 1,200 µm in spinal segment L3 (survival period 28 days). Note: sparing of superficial lamina (*arrowheads*) contralateral and ipsilateral to the site of injection. Scale bar in (e) equals 190 µm in **a–c, e–f**, and 320 µm in **d** (Reprinted with permission from Yezierski et al. 1998)

overgrooming in OVX females was similar to that found in SDMs. Furthermore, estradiol-treated SDMs developed severe overgrooming characterized by early onset, while progesterone treatment delayed the onset of grooming and attenuated its severity and progression. Strain-related differences were also observed with WFMs exhibiting more aggressive overgrooming than SDMs or LEMs. The results of this study showed that gender, strain, and gonadal hormones influence the onset and progression of overgrooming behavior following excitotoxic SCI (Gorman et al. 2001).

In conclusion there are three important similarities between excessive grooming behavior and the well-documented clinical condition of at-level pain in patients with spinal injury: (a) delayed onset, (b) spontaneous nature, and (c) dermatomal distribution relative to site of injury. The delayed onset of excessive grooming behavior suggests the neural mechanism is not simply an inhibitory release phenomenon, but instead requires significant changes (over time) in the functional state of spinal, and possibly supraspinal, sensory neurons. It is hypothesized that this behavior may be due to a loss of spinal inhibitory neurons, thus creating an imbalance between normal gating and biasing mechanisms within spinal and supraspinal somatosensory pathways (Yezierski 2009). Combined with a loss of segmental and/or supraspinal inhibitory influences, spinal neurons become hyperactive, and these focal pattern generators are responsible for producing paraesthetic and/or dysesthetic sensations referred to the affected dermatome.

Spatial Profile of Spinal Cord Damage Required for the Onset of Pain Behaviors Following Excitotoxic Spinal Injury

Initial studies evaluating the behavioral consequences of QUIS-induced spinal injury gave little attention to the longitudinal extent of neuronal loss that might be required to produce these behaviors. While it is acknowledged that the cellular and molecular events accompanying injury are undoubtedly important in inducing pain behaviors, the question was asked if the longitudinal extent over which neurons are affected by

the injury process is also important. To answer this question, an analysis of the longitudinal distribution of grooming-type damage was carried out in animals with and without overgrooming behavior. The results demonstrated a clear relationship between a critical extent of damage along the rostrocaudal axis of the cord and the onset of this spontaneous pain behavior. In animals with injury-induced grooming behavior, grooming-type damage was in excess of 5,000 μm. By contrast, damage in non-grooming animals was less than 4,000 μm. Based on these data, a hypothetical model of tissue damage versus pain behavior was proposed (Fig. 5). In this model there is a gradual progression of cord damage away from the injury epicenter. As the injury evolves, the extent of tissue damage exceeds a critical threshold, triggering the onset of pain behaviors. This model suggests that it is not only the secondary injury cascade that is responsible for the onset of central pain following spinal injury but also important is the longitudinal extent over which events in this cascade spread throughout the cord.

Several studies have provided support of this "neuroprotective hypothesis." When QUIS-injured animals were administered intraperitoneal injections of the NMDA antagonist and NOS inhibitor agmatine (Yu et al. 2003) at the time of QUIS injury, there was a significant delay in the onset of overgrooming behavior. Furthermore, following a 14-day treatment with agmatine after the onset of overgrooming, the final area of skin involvement targeted, and the severity of the behavior were significantly reduced. Neuroprotective effects similar to those with agmatine have also been obtained with the potent anti-inflammatory IL-10 (Yu et al. 2003; Abraham et al. 2004). These studies showed that IL-10 significantly reduced the onset time and area of overgrooming following QUIS injections. The fact that agmatine and IL-10 can be used effectively as preventive treatments for injury-induced pain behaviors underscores the importance of this intervention as a preemptive strategy of pain management for patients predisposed to progressive tissue damage in the spinal cord (e.g., syringomyelia, spinal cord injury).

Spinal Cord Injury, Excitotoxic Model, Fig. 5 Hypothetical progression of the *central injury cascade* from the epicenter (EPI) of an ischemic, traumatic, or excitotoxic insult to the spinal cord. In this model, the extent of injury and/or the area of cord influenced by different components of the injury cascade expand to include 2 ° and 3 ° areas of injury. If left untreated the amount of cord damage will continue to expand until it exceeds the threshold required for the onset of pain behavior (Reprinted with permission from Yezierski 2000)

Further studies related to preventing spinal injury pain behaviors evaluated using the immunosuppressant, cyclosporin A (CsA). In a "prevention protocol," rats received treatment with either saline or CsA thirty minutes postinjection of QUIS. In a "treatment protocol," rats received treatment with either saline or CsA beginning after the onset of QUIS-induced excessive grooming. The results of this study showed in the "prevention protocol" CsA delayed the onset of excessive grooming, reduced grooming area, reduced grooming severity, and reduced neuronal loss in the spinal cord compared to saline-treated animals (Yu et al. 2003). Treatment of excessive grooming behavior with CsA significantly reduced grooming area, grooming severity, and neuronal loss in the spinal cord compared to saline treatment. The conclusion from this study was that systemic administration of CsA significantly delayed the onset and reduced the progression of this spontaneous pain-related behavior. The results of the above studies have shown that there is a critical distance of neuronal loss along the longitudinal axis of the cord, which when exceeded, leads to the expression of a pain-related behavior. Interventions which limit the spread of neuronal loss result in the prevention or delayed onset of these behaviors. The above

studies support the "neuroprotective hypothesis" of SCI pain, and point to the possibility that using neuroprotective strategies, targeting specific components of the spinal injury cascade that interfere with death-inducing events may be useful in the prevention or treatment of pain conditions associated with SCI.

Functional Correlates of Behavioral Changes Following Excitotoxic Spinal Cord Injury
Evaluation of the neural correlate of QUIS-induced behavioral changes was undertaken by examining the functional properties of dorsal horn neurons, including cells belonging to the projection system from the spinal cord to the mesencephalon, i.e., spinomesencephalic tract (SMT). Supportive of a spinal mechanism for the evoked and spontaneous behavioral changes following QUIS injections was the finding that spinal wide dynamic range neurons adjacent to the injury site undergo significant functional changes, including a shift to the left in the stimulus–response function, an increase in the level of background activity, and an increase in the duration of afterdischarge, i.e., responses following removal of a stimulus (Yezierski and Park 1993). These changes appeared within 3–7 days of

Spinal Cord Injury, Excitotoxic Model, Fig. 6 Poststimulus time histograms showing the responses of two WDR neurons to graded intensities of mechanical stimuli. Cells were recorded in uninjected (**a**) and quisqualic acid-injected (**b**) animals. (**a**) Response profile of a WDR neuron recorded in the dorsal horn of spinal segment L3 (*inset*). Mechanical stimuli delivered to the receptive field on the foot included brush (*BR*), light pressure (*PR*), pinch (*PI*), and noxious squeeze (*SQ*). Note increased magnitude of response with increased intensity of stimulation. This cell had no spontaneous activity or afterdischarge response following removal of a stimulus. (**b**) Response profile of a WDR neuron recorded in the dorsal horn of spinal segment L4 (*inset*) caudal to an excitotoxic injury site located in L1-2. Note the presence of background activity, response magnitude to each stimulus condition (compared to cell in **a**), and the afterdischarge responses that continued following removal of each stimulus. Time in seconds is represented on the x-axis and spikes/s on the y-axis (Reprinted with permission from Yezierski 1996)

injury and were especially prevalent in animals following the onset of excessive grooming behavior. Afterdischarges lasting 5–15 min, and "windup" of background discharges with repeated stimulation, were novel characteristics of neurons in QUIS-injected animals (Fig. 6). The fact that these changes were observed in cells belonging to the SMT supports the notion of a possible supraspinal component to the observed pain behaviors. Changes in the response characteristics of spinal neurons, similar to those observed following QUIS injuries, have been reported for cells following ischemic and hemisection injury of the spinal cord. The increased excitability, bursting discharges, and long afterdischarge responses of neurons in QUIS-injected animals are reminiscent of the abnormal functional characteristics of neurons recorded in patients with chronic pain following SCI and present a potential target for elimination of this activity as a treatment for spontaneous pain.

Supraspinal Changes Associated with Excitotoxic Spinal Cord Injury

Efforts to evaluate supraspinal changes associated with excitotoxic injury were carried out, to evaluate whether injury-induced changes extend to sites remote from the site of injury. Changes at supraspinal sites could potentially contribute to the central mechanism responsible for the onset and progression of injury-induced pain behaviors, especially spontaneous pain behaviors (i.e., overgrooming). These studies included an evaluation of changes in forebrain blood flow (Morrow et al. 2000), opioid peptides (Abraham et al. 2001), and inflammatory cytokines (Brewer and Nolan 2007) in selected supraspinal sites. Morrow et al. (2000) demonstrated significant increases in regional cerebral blood flow in 7/22 supraspinal structures examined, including the arcuate nucleus, hindlimb region of S1 cortex, parietal cortex and the thalamic posterior, and ventral posterior medial and lateral nuclei. Work from Brewer and colleagues examined changes in peptidergic transmitter systems at spinal and supraspinal levels following excitotoxic SCI (Abraham et al. 2001). Preproenkephalin (PPE) and preprodynorphin (PPD) expression was shown to increase in cortical regions associated with nociceptive function at various time points following injury: PPE in the cingulate cortex and PPD in the parietal cortex. PPE expression in the anterior cingulate cortex and PPD expression in the contralateral parietal cortex were significantly higher in grooming versus non-grooming animals. All of these structures are associated with the processing of somatosensory information. The important conclusion from these reports is that following injury to the spinal cord, there are significant changes at selected supraspinal sites, including somatosensory structures putatively involved in pain processing. These findings of the supraspinal impact of spinal cord injury are consistent with a recent report in humans using diffusion tensor imaging, showing changes in regional brain anatomy in pain-related brain regions in SCI patients with pain (Gustin et al. 2010).

Peripheral Changes Associated with Excitotoxic Spinal Cord Injury

The behavioral manifestations of post-QUIS injection pain, including hypersensitivity to both mechanical and thermal stimuli, are similar to what has been described after peripheral nerve injury. While there is considerable data on the central consequences of peripheral nerve injury, there is little on the peripheral consequences of central nervous system (CNS) injury. Clinical evidence exists to show that CNS lesions from stroke can produce damage to peripheral afferents at the level innervated by the damaged CNS structures, often with related pain syndromes (Chemaili and Zhou 2007). This type of peripheral neuronal loss is frequently suggested to underlie the development of neuropathic pain.

Experimentally, QUIS-induced excitotoxic SCI leads to the loss of peripheral innervations of the skin at the level of injury (Brewer et al. 2008, Fig. 7). This pattern of denervation of the dermatome associated with the level of injury correlates with the spontaneous behavior of overgrooming as described above, providing an anatomical substrate for the behavior (Fig. 8).

Spinal Cord Injury, Excitotoxic Model, Fig. 7 Photomicrographs showing PGP95$^+$ fibers of sampled skin tissues (a–c; *arrow head*) and ATF3$^+$ cells from an ipsilesional DRG (d; *arrow*) 2 weeks post-QUIS injection. Dense PGP95-immunoreactive nerve fibers were observed in the epidermal and dermal layers from control skin (a). Non-groomed QUIS-injured skin contained fewer PGP95-immunoreactive nerve fibers (c), and groomed QUIS-lesioned skin contained almost no PGP95-immunoreactive nerve fibers (b). Presence of ATF3+ cells was observed from a DRG ipsilesional to grooming. *Epi* epidermis, *Der* dermis (Reproduced with permission from Brewer et al. 2008)

Spinal Cord Injury, Excitotoxic Model, Fig. 8 Mean axonal length in skin taken from unoperated rats ($n = 11$), rats injected with QUIS with no signs of gross pathology on the side ipsilateral to skin biopsy ($n = 5$), rats injected with QUIS with obvious pathology on the side ipsilateral to skin biopsy ($n = 20$), ungroomed regions after QUIS injection ($n = 7$), and skin targeted for overgrooming after QUIS injection ($n = 13$). All skin taken from QUIS-injected animals had reduced neurite density compared to unoperated controls. Skin that was overgroomed after QUIS injection had significantly reduced neurite density compared to ungroomed skin from QUIS-injected animals. *Significant difference from unoperated; # significant difference from ungroomed skin ($p < 0.01$) (Reproduced with permission from Brewer et al. 2008)

In addition, dorsal root ganglia (DRG) collected from QUIS-injected spinal cord levels contained neuronal labeling for ATF-3, suggesting that the injury creates an at-level neuronopathy (Brewer et al. 2008, Fig. 7). This peripheral nerve damage may be contributed to the sensory syndromes associated with excitotoxic SCI, suggesting that neuropathic pain after SCI may share a common mechanism with the pain that develops post-peripheral nerve injury and that the peripheral nervous system must be considered when developing treatment strategies for post-SCI pain.

Summary

In conclusion, the results of the excitotoxic model described above support the general scheme that, following spinal injury, there are neurochemical, anatomical, molecular, and physiological changes that collectively constitute a *central injury cascade* responsible for the development of clinical symptoms, i.e., altered sensory (and motor) function. It is noteworthy that many of the changes described following excitotoxic injury parallel descriptions of events thought to

be responsible for the development of neurogenic pain following peripheral nerve and tissue injury. Although there may be subtle differences in specific components of the central cascade responsible for neurogenic versus central pain, it is difficult to ignore the similarities between the end result, i.e., pain, and the involvement of neurochemical, excitotoxic, anatomical, and physiological changes proposed for both central and peripheral pain. These similarities suggest that the central responses brought on by peripheral injury, and which underlie the onset of neurogenic pain, should not be overlooked in the search for a central mechanism of pain following spinal injury.

Acknowledgement The work related to the excitotoxic model of spinal injury was supported by NS40096. The authors would like to thank the collaborative assistance of Drs. Shanliang Liu, Chen-Guang Yu, Jeffery Plunkett, and Laurel Gorman.

References

Abraham, K. E., McGinty, J. F., & Brewer, K. L. (2001). Spinal and supraspinal changes in opioid mRNA expression are related to the onset of pain behaviors following excitotoxic spinal cord injury. *Pain, 90*, 181–190.

Abraham, K. I., McMIllen, D., & Brewer, K. L. (2004). The effects of endogenous interleukin-10 in gray matter damage and development of pain behaviors following excitotoxic spinal cord injury in the mouse. *Neuroscience, 124*(4), 945–952.

Brewer, K. L., & Nolan, T. A. (2007). Spinal and supraspinal changes in tumor necrosis factor-alpha expression following excitotoxic spinal cord injury. *Journal of Molecular Neuroscience, 31*(1), 13–21.

Brewer, K. L., Lee, J. W., Downs, H., Oaklander, A. L., & Yezierski, R. P. (2008). Dermatomal scratching after intramedullary quisqualate injection: Correlation with cutaneous denervation. *The Journal of Pain, 9*(11), 999–1005.

Bruce, J. C., Oatway, M. A., & Weaver, L. C. (2002). Chronic pain after clip-compression injury of the rat spinal cord. *Experimental Neurology, 178*(1), 33–48.

Bryce, T. N., Biering-Sorensen, F., Finnerup, N. B., Cardenas, D. D., Defrin, R., et al. (2012). International spinal cord injury pain classification: Part I. Background and description. March 6-7, 2009. *Spinal Cord, 50*(6), 413–417.

Chemaili, K. R., & Zhou, L. (2007). Small fiber degeneration in post-stroke complex regional pain syndrome I. *Neurology, 69*, 316–317.

Choi, S. S., Hahm, K. D., Min, H. G., & Leem, J. G. (2011). Comparison of the spinal neuropathic pain induced by intraspinal injection of N-methyl-d-aspartate and quisqualate in rats. *Journal of Korean Neurosurgical Society, 50*(5), 420–425.

Finnerup, N. B., & Baastrup, C. (2012). Spinal cord injury pain: Mechanisms and management. *Current Pain and Headache Reports, 16*, 207–216.

Gorman, A. L., Yu, C. G., Sanchez, D., Ruenes, G. R., Daniels, L., & Yezierski, R. P. (2001). Conditions affecting the onset, severity, and progression of a spontaneous pain-like behavior after excitotoxic spinal cord injury. *The Journal of Pain, 2*, 229–240.

Gustin, S. M., Wrigley, P. J., Siddall, P. J., & Henderson, L. A. (2010). Brain anatomy changes associated with persistent neuropathic pain following spinal cord injury. *Cerebral Cortex, 20*(6), 1409–1419.

Hulsebosch, C. E. (2002). Pharmacology of chronic pain after spinal cord injury: Novel acute and chronic intervention strategies. In R. P. Yezierski & K. Burchiel (Eds.), *Spinal cord injury pain: Assessment, mechanisms, management* (pp. 189–204). Seattle: IASP Press.

Kerr, B. J., & David, S. (2007). Pain behaviors after spinal cord contusion injury in two commonly used mouse strains. *Experimental Neurology, 206*(2), 240–247.

Liu, S., Ruenes, G. L., & Yezierski, R. P. (1997). NMDA and non-NMDA receptor antagonists protect against excitotoxic injury in the rat spinal cord. *Brain Research, 756*(1–2), 160–167.

Morrow, T. J., Paulson, P. E., Brewer, K. L., Yezierski, R. P., & Casey, K. L. (2000). Chronic, selective forebrain responses to excitotoxic dorsal horn injury. *Experimental Neurology, 161*, 220–226.

Nakae, A., Nakai, K., Yan, K., Hosokawa, K., Shibata, M., & Mashimo, T. (2011). The animal model of spinal cord injury as an experimental pain model. *Journal of Biomedicine and Biotechnology, 2011*, 939023.

Nout, Y. S., Rosenzweig, E. S., Brock, J. H., Strand, S. C., Moseanko, R., Hawbecker, S., Zdunowski, S., Nielson, J. L., Roy, R. R., Courtine, G., Ferguson, A. R., Edgerton, V. R., Beattie, M. S., Bresnahan, J. C., & Tuszynski, M. H. (2012). Animal models of neurologic disorders: A nonhuman primate model of spinal cord injury. *Neurotherapeutics, 9*(2), 380–392.

Plunkett, J. A., Yu, C.-G., Bethea, J. R., & Yezierski, R. P. (2001). Effects of interleukin-10 (IL-10) on pain behavior and gene expression following excitotoxic spinal cord injury in the Rat. *Experimental Neurology, 169*, 144–154.

Rose, M., Robinson, J. E., Ellis, P., & Cole, J. D. (1988). Pain following spinal injury: Results from a postal survey. *Pain, 34*, 101–102.

Van Leeuwen, C. M., Post, M. W., van Asbeck, F. W., Bongers-Janssen, H. M., van der Woude, L. H., de Groot, S., & Lindeman, E. (2012). Life satisfaction in people with spinal cord injury during the first five years after discharge from inpatient rehabilitation. *Disability and Rehabilitation, 34*, 76–83.

Vierck, C. J., & Light, A. R. (2002). Assessment of pain sensitivity in dermatomes caudal to spinal cord injury in rats. In R. P. Yezierski & K. Burchiel (Eds.), *Spinal cord injury pain: Assessment, mechanisms, management* (pp. 137–154). Seattle: IASP Press.

Widerstrom-Noga, E. G., Cuevo, E. F., Broton, J. G., Duncan, R. C., & Yezierski, R. P. (1999). Perceived difficulty in dealing with consequences of SCI. *Archives of Physical Medicine and Rehabilitation, 80*, 580–586.

Xu, X.-J., Hao, J.-X., & Wiesenfeld-Hallin, Z. (2002). Physiological and pharmacological characterization of a Rat model of spinal cord injury pain after spinal ischemia. In R. P. Yezierski & K. Burchiel (Eds.), *Spinal cord injury pain: Assessment, mechanisms, management* (pp. 175–187). Seattle: IASP Press.

Yezierski, R. P. (1996). Pain following spinal cord injury: The clinical problem and experimental studies. *Pain, 68*(2–3), 185–194.

Yezierski, R. P. (2000). Pain following spinal cord injury: Pathophysiology and central mechanisms. *Progress in Brain Research, 129*, 429–449.

Yezierski, R. P. (2002). Pathophysiology and animal models of spinal cord injury pain. In R. P. Yezierski & K. Burchiel (Eds.), *Spinal cord injury pain: Assessment, mechanisms, management* (pp. 1117–136). Seattle: IASP Press.

Yezierski, R. P. (2009). Spinal cord injury pain: Spinal and supraspinal mechanisms. *Journal of Rehabilitation Research and Development, 46*(1), 95–108.

Yezierski, R. P., & Park, S. H. (1993). The mechanosensitivity of spinal sensory neurons following intraspinal injections of quisqualic acid in the Rat. *Neuroscience Letters, 157*, 115–119.

Yezierski, R. P., Santana, M., Park, S. H., & Madsen, P. W. (1993). Neuronal degeneration and spinal cavitation following intraspinal injections of quisqualic acid in the rat. *Journal of Neurotrauma, 19*(4), 445–456.

Yezierski, R. P., Liu, S., Ruenes, G. L., Kajander, K. J., & Brewer, K. L. (1998). Excitotoxic spinal cord injury: Behavioral and morphological characteristics of a central pain model. *Pain, 75*, 141–155.

Yezierski, R. P., Yu, C. G., Mantyh, P. W., Vierck, C. K., & Lappi, D. A. (2004). Spinal neurons involved in the generation of at-level pain following spinal injury in the rat. *Neuroscience Letters, 361*(1–3), 232–236.

Yu, C. G., & Yezierski, R. P. (2005). Activation of the ERK 1/2 signaling cascade by excitotoxic spinal cord injury. *Brain Research. Molecular Brain Research, 138*(2), 244–255.

Yu, C. G., Fairbanks, C. A., Wilcox, G. L., & Yezierski, R. P. (2003). Effects of agmatine, interleukin-10 and cyclosporin on spontaneous pain behavior following excitotoxic spinal cord injury in rats. *The Journal of Pain, 4*, 129–140.

Spinal Cord Laminae

Definition

Cytoarchitecturally differentiated regions of the spinal cord gray matter (described by Rexed).

Cross-References

▶ Stimulation-Produced Analgesia

Spinal Cord Microglia

▶ Cathepsin and Microglia

Spinal Cord Neurons Projecting to the Lateral Parabrachial Area

▶ Spinoparabrachial Tract Neurons, Anatomical Features

Spinal Cord Neurons Projecting to the Thalamus

▶ Spinothalamic Tract Neurons: Neurotransmitter Inputs and Receptors

Spinal Cord Nociception

▶ Encoding of Noxious Information in the Spinal Cord
▶ Metabotropic Glutamate Receptors in Spinal Nociceptive Processing

Spinal Cord Nociception and CGRP

▶ CGRP and Spinal Cord Nociception

Spinal Cord Nociception, Neurotrophins

Marzia Malcangio, Anna K. Clark and Andrea C. Ogbonna
Wolfson Centre for Age Related Diseases
King's College London, London, UK

Synonyms

Neurotrophins; Neurotrophins in spinal cord nociception; Spinal nociceptive processing

Definition

The neurotrophins are a family of polypeptides, namely, nerve growth factor (NGF), neurotrophin 3 (NT-3), NT-4, and brain-derived neurotrophic factor (BDNF). Peripheral sensory neurons require the neurotrophins produced by peripheral target tissue for survival during embryonic life. Then in adulthood, the

Spinal Cord Nociception, Neurotrophins, Fig. 1 Peripherally produced neurotrophins are retrogradely transported to the dorsal root ganglia by sensory neurons expressing Trk receptors. BDNF is also expressed by nociceptive neurons along with SP and CGRP

neurotrophins can influence the morphology, excitability, and synaptic plasticity of sensory neurons expressing their high-affinity receptors. These receptors are tropomyosin receptor kinases (Trks) coupled to a tyrosine kinase domain, TrkA, TrkB, and TrkC, and bind NGF, BDNF, and NT4/5 and NT-3, respectively.

Characteristics

The first pain synapse is formed between the central terminal of primary sensory neurons (presynaptic element) and second order neurons in the dorsal horn of spinal cord (postsynaptic element). The strength of this synapse is plastic and modifiable, and it is now well established that neurotrophic factors are strong regulators of synaptic efficacy.

Primary sensory neurons are pseudo-unipolar cells that have cell bodies in the dorsal root ganglia (DRG) and send their axons centrally to the dorsal horn of the spinal cord and peripherally to the skin and internal tissues. Sensory neurons can be broadly divided into two groups. The first group includes neurons with a large cell body diameter and large myelinated axons. These respond to low threshold innocuous stimuli

applied to peripheral tissues, and their central axons terminate in the deeper laminae of the spinal cord. These neurons mainly express TrkC and TrkB receptors and do not express peptides under normal circumstances. The second group includes neurons with a small cell body diameter and mostly unmyelinated axons. These cells are nearly all nociceptors and the central axons terminate in the superficial layers of the spinal cord. Many of these neurons contain peptides, and a subpopulation of these small diameter sensory neurons expresses the TrkA receptor for NGF and expresses neurotrophins (McMahon et al. 1994; Michael et al. 1997) (Fig. 1).

Neurotrophic factors can regulate spinal nociceptive processing at the first synapse in the dorsal horn in two main ways. Firstly, if synthesized and released by primary sensory neurons, they will directly modulate synaptic efficacy. Secondly, neurotrophins can indirectly modulate synaptic strength by regulating the expression and release of transmitters/modulators of the first pain synapse.

Nerve Growth Factor

NGF indirectly regulates the synaptic function of sensory neuron terminals in the dorsal horn by maintaining or stimulating the expression of excitatory peptides, receptors, and neurotrophins in TrkA-expressing neurons. The high-affinity receptor for NGF, TrkA, is expressed in the superficial laminae of the dorsal horn where it is found on the central terminals of some unmyelinated fibers that also express substance P (SP), calcitonin gene-related peptide (CGRP), and BDNF. However, NGF does not directly modulate the activity of the central terminals of sensory neurons (Malcangio et al. 1997). Intrathecally injected NGF does not modify the rat nociceptive threshold unless administered for 6–9 days, when it induces thermal hyperalgesia and increases sensory neuron transmitter content and consequently activity-induced release (Malcangio et al. 2000). In contrast, systemically administered NGF induces thermal hyperalgesia in rats very rapidly (Lewin and Mendell 1993). Endogenous NGF produced peripherally in the skin can sensitize the peripheral terminals of sensory neurons (e.g., in inflammation). Exogenous NGF induces hypersensitivity to thermal and mechanical noxious stimuli and sensitizes the response of sensory neurons to capsaicin. Thermal hyperalgesia develops quickly and is maintained for days. It is sympathetically maintained and depends on peripheral mechanisms involving degranulation of mast cells and spinal cord mechanisms involving the NMDA receptor as well as the regulation of the content/release of nociceptive neuromodulators from sensory neurons. NGF-induced mechanical hyperalgesia seems to be independent of mast cell degranulation or central NMDA receptor sites. Endogenous NGF levels rise in inflammatory conditions. Reduction in endogenous NGF levels following systemic treatment with TrkA-IgG, which sequesters peripherally produced NGF, induces hypoalgesia and reduces inflammatory pain (McMahon et al. 1995). However, endogenous NGF levels decrease in peripheral neuropathies (e.g., diabetic neuropathy), and NGF supplementation can be anti-hyperalgesic in neuropathic pain conditions (McMahon and Priestley 1995).

Neurotrophin 3

Unlike NGF, NT-3 can acutely modulate sensory neuron transmission in the dorsal horn, where NT-3 inhibits activity-induced neuropeptide release from sensory neuron terminals. This effect is mainly indirect since trkC receptors are not found in peptidergic neurons. NT-3 intrathecally or intraplantarly delivered does not modify thermal thresholds (Shu et al. 1999; Malcangio et al. 2000). The effects of systemic NT-3 on nociceptive thresholds include short-lasting thermal and mechanical hyperalgesia (Malcangio et al. 1997). However, activation of more than one Trk receptor by NT-3 makes it difficult to interpret the effect of this neurotrophin on sensory mechanisms.

Brain-Derived Neurotrophic Factor

BDNF is unique among the neurotrophins in that it is constitutively synthesized by sensory neurons. Unlike NGF, BDNF can directly modulate nociceptive processing in the dorsal horn (Pezet et al. 2002; Malcangio and Lessmann 2003).

Although peripherally produced BDNF can be retrogradely transported by sensory neurons expressing TrkB receptors, target-derived BDNF represents a small fraction of the total BDNF within the DRG. BDNF is found along with SP and CGRP in a subpopulation of sensory neurons that is sensitive to NGF (Michael et al. 1997). BDNF is synthesized in the dorsal root ganglia, packaged in vesicles and anterogradely transported to the central terminals of sensory neurons in the superficial laminae of the dorsal horn. Here BDNF can be released by certain noxious stimuli that produce bursting activity in nociceptive fibers (Lever et al. 2001) and binds its high-affinity TrkB receptors, functioning to modulate nociceptive transmission (Garraway et al. 2003; Lever et al. 2003) (Fig. 2). BDNF content in sensory neurons is upregulated in inflammatory states, and the activation of TrkB receptors in the dorsal horn by endogenous BDNF appears to contribute to the hyperalgesia associated with peripheral inflammation via facilitation of NMDA receptor activation (Kerr et al. 1999, Garraway et al. 2003). In neuropathic rat models, BDNF undergoes changes in injured and uninjured sensory neurons. Whether de novo expressed/upregulated endogenous BDNF contributes to spinal nociceptive processing has not been definitively proved (Malcangio and

Spinal Cord Nociception, Neurotrophins, Fig. 2 pERK labeling increases in mouse dorsal horn after application of BDNF-releasing stimulus to the attached dorsal roots. (**a**) Control tissue. (**b**) Stimulated tissue (300 pulses at 100 Hz, 10 mA, 0.5 ms). Scale bar = 122 μm. (**c**) In the dorsal horn, BDNF released by nociceptive neuron terminals binds trkB postsynaptically and contributes to synaptic plasticity. TrkB activation promotes activation of ERK (MAP kinase) directly and via NMDA receptor facilitation (Lever et al. 2003)

Lessmann, 2003). However, intrathecal injection of BDNF in neuropathic rats transiently reversed thermal hyperalgesia via release of GABA in the dorsal horn, and overexpression of BDNF in the spinal cord of neuropathic rats can alleviate chronic neuropathic pain (Eaton et al. 2002).

Therefore, BDNF may be pro- or antinociceptive depending on circumstances, e.g., inflammatory versus neuropathic pain conditions.

References

Eaton, M. J., Blits, B., Ruitenberg, M. J., et al. (2002). Amelioration of chronic neuropathic pain after partial nerve injury by adeno-associated viral (AAV) vector-mediated over-expression of BDNF in the rat spinal cord. *Gene Therapy, 9*, 1387–1395.

Garraway, S. M., Petruska, J. C., & Mendell, L. M. (2003). BDNF sensitizes the responses of lamina II neurons to high threshold primary afferent inputs. *European Journal of Neuroscience, 18*, 2467–2476.

Kerr, B. J., Bradbury, E. J., Bennett, D. L. H., et al. (1999). Brain-derived neurotrophic factor modulates nociceptive sensory inputs and NMDA-evoked responses in the rat spinal cord. *Journal of Neuroscience, 19*, 5138–5148.

Lever, I. J., Bradbury, E. J., Cunningham, J. R., et al. (2001). Brain-derived neurotrophic factor is released in the dorsal horn by distinctive patterns of afferent fiber stimulation. *Journal of Neuroscience, 21*, 4469–4477.

Lever, I. J., Pezet, S., McMahon, S. B., et al. (2003). The signalling components of sensory fiber transmission involved in the activation of ERK MAP kinase in the mouse dorsal horn. *Molecular and Cellular Neuroscience, 24*, 259–270.

Lewin, G. R., & Mendell, L. M. (1993). Nerve growth factor and nociception. *Trends in Neurosciences, 16*, 353–359.

Malcangio, M., & Lessmann, V. (2003). A common thread for pain and memory synapses? Brain-derived neurotrophic factor and TrkB receptors. *Trends in Pharmacological Sciences, 24*, 116–121.

Malcangio, M., Garrett, N. E., Cruwys, S., et al. (1997). Nerve growth factor- and neurotrophin-3- induced changes in nociceptive threshold and the release of substance P from the rat isolated spinal cord. *Journal of Neuroscience, 17*, 8459–8467.

Malcangio, M., Ramer, M. S., Boucher, T. J., et al. (2000). Intrathecally injected neurotrophins and the release of substance P from the rat isolated spinal cord. *European Journal of Neuroscience, 12*, 139–144.

McMahon, S. B., & Priestley, J. V. (1995). Peripheral neuropathies and neurotrohic factors: Animal models and clinical perspectives. *Current Opinion in Neurobiology, 5*, 616–624.

McMahon, S. B., Armanini, M. P., Ling, L. H., et al. (1994). Expression and co-expression of Trk receptors in subpopulations of adult primary sensory neurones projecting to identified peripheral targets. *Neuron, 12*, 1161–1171.

McMahon, S. B., Bennett, D. L., Priestley, J. V., et al. (1995). The biological effects of endogenous NGF on adult sensory neurones revealed by a TrkA-IgG fusion molecule. *Nature Medicine, 1*, 774–780.

Michael, G. J., Averill, S., Nitkunan, A., et al. (1997). Nerve growth factor treatment increases brain-derived neurotrophic factor selectively in tyrosine kinase A expressing dorsal root ganglion cells and their central terminations in the dorsal horn. *Journal of Neuroscience, 17*, 8476–8490.

Pezet, S., Malcangio, M., & McMahon, S. B. (2002). BDNF: A neuromodulator in nociceptive pathways? *Brain Research Reviews, 40*, 240–249.

Shu, X.-Q., Llinas, A., & Mendell, L. M. (1999). Effects of TrkB and TrkC neurotrophin receptor agonists on thermal nociception: A behavioural and electrophysiological study. *Pain, 80*, 463–470.

Spinal Cord Segmental Level

Definition

The part of the spinal cord which supplies a segment of the body with sensory nerves (conducting, e.g., pain impulses).

Cross-References

▶ Cancer Pain Management, Anesthesiologic Interventions, and Neural Blockade

Spinal Cord Stimulation

Definition

Chronic electrical stimulation of the spinal cord through an epidural electrode attached to an implanted pulse generator, a common minimally invasive technique for treatment of chronic pain.

Cross-References

▶ Cancer Pain Management, Anesthesiologic Interventions, Spinal Cord Stimulation, and Neuraxial Infusion
▶ Complex Regional Pain Syndrome and the Sympathetic Nervous System
▶ Deep Brain Stimulation
▶ Pain Treatment, Motor Cortex Stimulation
▶ Pain Treatment: Spinal Cord Stimulation

Spinal Cord Stimulation (SCS)

▶ Pain Treatment: Spinal Cord Stimulation

Spinal Dorsal Horn

Definition

Spinal structure mediating sensory inputs from the somatic distributions to the spinal cord.

Cross-References

▶ Calcium Channels in the Spinal Processing of Nociceptive Input
▶ DREZ Procedures
▶ NMDA Receptors in Spinal Nociceptive Processing

Spinal Dorsal Horn Pathways, Colon, Urinary Bladder, and Uterus

▶ Spinal Ascending Pathways, Colon, Urinary Bladder, and Uterus

Spinal Dorsal Horn Pathways, Dorsal Column (Visceral)

William D. Willis[1] and Karin Westlund High[2]
[1]Department of Neuroscience and Cell Biology, University of Texas Medical Branch, Galveston, TX, USA
[2]Department of Physiology, University of Kentucky, Lexington, KY, USA

Synonyms

Posterior column; Posterior horn; Visceral nociceptive tracts

Definition

The dorsal column visceral nociceptive pathway originates from postsynaptic dorsal column (PSDC) neurons (see ▶ Postsynaptic Dorsal Column Neurons). Although most previous work has emphasized the cutaneous input to PSDC neurons, it has now been shown that at least some of these neurons respond to noxious visceral stimuli as well as to cutaneous stimuli. The cell bodies of PSDC neurons are concentrated in the nucleus proprius of the spinal cord dorsal horn, but some are located in the central gray matter. The axons of some of these neurons convey visceral nociceptive information through the dorsal column to the dorsal column nuclei of the medulla oblongata. The dorsal column nuclei transmit visceral nociceptive information to the contralateral ventral posterior lateral (VPL) thalamic nucleus, which in turn relays the information to the somatosensory cerebral cortex.

Characteristics

A number of spinal cord ascending pathways convey visceral nociceptive information to the brain. These pathways include the spinothalamic, spinoreticular, spinoparabrachial, spinohypothalamic, spinoamygdalar, and other spinolimbic tracts, which ascend in the lateral and ventral funiculi (Willis and Coggeshall 2004) (see ▶ Spinal Ascending Pathways, Colon, Urinary Bladder, and Uterus). However, clinical observations have shown that midline lesions of the human posterior columns at a midthoracic level (Fig. 1) can relieve the pain of pelvic cancer and reduce the need for analgesic drugs (see ▶ Analgesics) (Hirshberg et al. 1996; Nauta et al. 2000; Kwon and Kim 2000). For example, the lesions shown in Fig. 1a–c were made in patients with cancer pain. The lesion in Fig. 1a was made at T10 in a patient with colon cancer (Hirshberg et al. 1996). After the surgery, this patient spent the remaining 3 months of his life pain-free, without the need for strong analgesics. The lesion in Fig. 1b was made at T7 in a patient with lung cancer. This was the least successful lesion in relieving cancer pain in a series of cases (Nauta et al. 2000), perhaps because the lesion was largely unilateral or because the lesion was not placed sufficiently rostrally in this patient, who had lung cancer. Pelvic cancer pain was relieved for a period of several years by the lesion shown in Fig. 1c in another patient. A technique for making a punctate midline myelotomy (Nauta et al. 2000) is shown in Fig. 1d (drawing at the left). The postsynaptic dorsal column pathway that was proposed to convey visceral nociceptive signals to the brain based on animal experiments (see below) is shown at the right in Fig. 1d.

The rationale for performing a midline myelotomy for pelvic cancer pain was the previous success of midline myelotomies performed by Hitchcock and other neurosurgeons at the upper cervical level in relieving visceral and other forms of pain (Gybels and Sweet 1989). However, placement of a midline myelotomy at a midthoracic level rostral to afferent input from pelvic viscera is a safer procedure (Hirshberg et al. 1996). The interruption of a nociceptive pathway in the posterior columns of humans could also account for neurosurgical reports that commissural myelotomies can produce pain relief in much more caudal parts of the body than would be suggested by the segmental levels of the commissural myelotomy (Gybels and Sweet 1989).

The presence of a visceral nociceptive path in the dorsal columns has now been demonstrated in

Spinal Dorsal Horn Pathways, Dorsal Column (Visceral), Fig. 1 (a–d) The dorsal column visceral pain pathway in humans. (a–c) Lesions made in the dorsal columns of patients in an effort to relieve cancer pain. (a) Lesion at T10 completely relieved the pain of colon cancer. (b) Lesion at T7 was only partially effective in relieving pain from lung cancer, perhaps because the lesion did not extend sufficiently laterally on one side. (c) A lesion that relieved pelvic cancer pain for several years after the surgery. (d) The different levels of the dorsal column visceral pain pathway, including visceral nociceptive afferents, spinal cord processing circuits, postsynaptic dorsal column neurons projecting through the dorsal column to the dorsal column nuclei, and nucleus gracilis neurons projecting through the medial lemniscus to the VPL thalamic nucleus (From Willis and Westlund 2004)

animal experiments. For example, in rats, the responses of neurons in the *ventral posterior lateral* (VPL) thalamic nucleus to noxious colorectal distension were found to be reduced by about 80 % following a lesion restricted to the dorsal columns (Fig. 2), whereas similar responses in other rats were reduced by only about 20 % after a lesion confined to the ventrolateral

quadrant of the spinal cord (Al-Chaer et al. 1996a). Visceral nociceptive responses could also be recorded from neurons of the gracile nucleus, and these were blocked by microdialysis administration of morphine or of the non-*N*-methyl-D-aspartate antagonist, CNQX, into the sacral spinal cord, after restricting the afferent input from the colon to the sacral cord to the

Spinal Dorsal Horn Pathways, Dorsal Column (Visceral), Fig. 2 (a–f) Reduction in the responses of a neuron in the VPL thalamic nucleus in a rat to cutaneous and visceral stimulation following successive lesions of the dorsal column (DC) and the ventrolateral column (VLC). (a) Recording site; (b) cutaneous receptive field; (c) maximum extent of lesions of DC and VLC; (d) responses to brush (*BR*), pressure (*PR*), and pinch (*PI*) stimuli applied to the receptive field; (e) responses to graded colorectal distention (20, 40, 60, and 80 mmHg); and (f) action potential of VPL neuron. The DC lesion eliminated the responses to BR and PR, but not PI, and it nearly eliminated the responses to colorectal distention. The VLC lesion eliminated the remaining responses. The action potential remained consistent, indicating that the recording was unchanged throughout the experiment (From Al-Chaer et al. 1996)

pelvic nerves by sectioning the hypogastric nerves prior to the experiment (Al-Chaer et al. 1996b). The ability of these drugs to block the visceral responses in the gracile nucleus is consistent with a synaptic action, presumably at synapses of visceral primary afferent neurons on PSDC neurons. Recordings from individual antidromically identified PSDC neurons confirmed that their responses to visceral stimuli were in fact blocked by microdialysis administration of morphine or CNQX into the sacral spinal cord. Furthermore, the responses of VPL neurons to colorectal distention were greatly reduced by a small lesion placed in the gracile nucleus (Al-Chaer et al. 1997). Thus, the visceral responses in the gracile and VPL nuclei depend on a synaptic relay in the sacral spinal cord and on subsequent transmission of visceral signals through the dorsal columns and dorsal column nuclei (Fig. 1d). Midthoracic dorsal column lesions were also observed to block changes in regional blood flow in the brains of monkeys at the level of the thalamus, using functional magnetic resonance imaging (Willis et al. 1999), indicating that there is a dorsal column visceral nociceptive pathway in primates, as well as in rats.

An anatomical study using the anterograde tracer, (see ▶ Anterograde Axonal Tracer

(Anterograde Labeling)) *Phaseolus vulgaris* leucoagglutinin, showed that PSDC neurons in the sacral spinal cord project their axons through the midline dorsal column to the gracile nucleus (Wang et al. 1999). However, the axons of PSDC neurons in the midthoracic spinal cord were found in the dorsal column, near the dorsal intermediate septum, between the gracile and cuneate fasciculi, and the axons terminated in the border area between the gracile and cuneate nuclei (Wang et al. 1999). Thus, the visceral component of the PSDC pathway has a viscerotropic organization, similar to the somatotopic organization of the cutaneous component of this pathway. Due to this viscerotropic organization, dorsal column lesions to reduce visceral pain arising from abdominal organs would need to be placed more laterally than the midline lesions that are used to alleviate pelvic visceral pain.

Evidence confirming that responses to noxious visceral stimuli applied to abdominal viscera are blocked by lesions of the dorsal columns placed at the border between the fasciculus gracilis and fasciculus cuneatus has been obtained in experiments using recordings from VPL neurons before, during, and after duodenal distention (Feng et al. 1998). In such animals, midline dorsal column lesions were ineffective. Similar laterally placed dorsal column lesions in rats were also needed to reduce the responses of VPL neurons to noxious stimulation of the pancreas using bradykinin applications (Houghton et al. 2001).

Behavioral studies have underlined the importance of the visceral pathway in the dorsal column for responses to noxious visceral stimuli. For example, dorsal column lesions reduce the writhing responses of rats to duodenal distention (Feng et al. 1998), as well as some of the changes in exploratory activity that result from noxious stimulation of the pancreas (Houghton et al. 2001). Furthermore, dorsal column lesions prevent or disrupt the visceromotor reflexes, recorded as electromyographic responses from the external abdominal oblique muscle in rats, following noxious colorectal distention in animals in which the colon had been inflamed by injection of mustard oil, but not in non-inflamed control animals (Palecek and Willis 2003). The effect of a combination of colon inflammation and colorectal distention in reducing exploratory behavior of rats is partially reversed by bilateral dorsal column lesions at an upper cervical level, but not by a lesion of the ventrolateral funiculus (Palecek et al. 2002).

Another approach that has been used to demonstrate the role of the PSDC pathway, as well as of the spinothalamic tract, in signaling nociceptive visceral activity is a demonstration of the expression of Fos protein in response to ureter distention (Palecek et al. 2003). Retrograde labeling from the nucleus gracilis and the contralateral VPL nucleus, respectively, was used to identify neurons of the PSDC pathway and of the spinothalamic tract. A substantial fraction of neurons belonging to each pathway was found to express Fos protein after noxious distention of the ureter (Fig. 3). The leftmost column of Fig. 3 shows drawings of histological sections taken at various levels of the spinal cord. The locations of retrogradely labeled PSDC neurons are plotted on the drawings. The second column of drawings shows the locations of PSDC neurons that were immunoreactive for Fos protein following distention of the ureter. Some of the Fos-labeled PSDC neurons were in the central gray matter of the spinal cord, an area known to process visceral information. Other Fos-positive PSDC neurons were in the nucleus proprius. Similarly, the rightmost column in Fig. 3 shows the locations of retrogradely labeled spinothalamic tract (STT) cells, and the third column shows those STT cells that were immunoreactive for Fos protein after ureter distention.

Although not rigorously proven as yet, there is a plausible explanation for why a dorsal column lesion can interrupt nociceptive behavior in response to visceral stimuli, despite the fact that many spinothalamic and other ascending tract cells also respond to noxious visceral stimuli. It is well established that visceral responses in the spinal cord depend on the activation of an

Spinal Dorsal Horn Pathways, Dorsal Column (Visceral), Fig. 3 Fos protein expression in postsynaptic dorsal column (PSDC) and spinothalamic tract (STT) neurons following noxious distention of the ureter in rats. PSDC and STT neurons were labeled retrogradely from the nucleus gracilis and thalamus, respectively. The filled circles in the leftmost column of drawings of spinal cord sections from Th4 to L6 indicate the locations of PSDC neurons that did not express Fos protein, and those in the next column of drawings show the positions of neurons that did express Fos protein after ureter distention. The third column shows STT cells that expressed Fos protein and the rightmost column those that did not (From Palecek et al. 2003)

excitatory pathway that descends from the brain stem to activate spinal cord visceroceptive neurons (Zhuo and Gebhart 2002; see also Willis and Coggeshall 2004). It seems likely that activity ascending in the dorsal column is not only transmitted by way of the dorsal column nuclei to the thalamus, but that this information also serves to activate this descending excitatory pathway (Palecek and Willis 2003). The brain stem excitation would normally help activate spinothalamic and other ascending tracts that convey visceral nociceptive information to higher centers. Interruption of the dorsal column could interfere with transmission, not only in the postsynaptic dorsal column path but also in the other ascending visceroceptive tracts, such as the spinothalamic tract.

References

Al-Chaer, E. D., Lawand, N. B., Westlund, K. N., & Willis, W. D. (1996a). Visceral nociceptive input into the ventral posterolateral nucleus of the thalamus: A new function for the dorsal column pathway. *Journal of Neurophysiology, 76,* 2661–2674.

Al-Chaer, E. D., Lawand, N. B., Westlund, K. N., & Willis, W. D. (1996b). Pelvic visceral input into the nucleus gracilis is largely mediated by the postsynaptic dorsal column pathway. *Journal of Neurophysiology*, *76*, 2675–2690.

Al-Chaer, E. D., Westlund, K. N., & Willis, W. D. (1997). Nucleus gracilis: An integrator for visceral and somatic information. *Journal of Neurophysiology, 78*, 521–527.

Feng, Y., Cui, M., Al-Chaer, E. D., & Willis, W. D. (1998). Epigastric antinociception by cervical dorsal column lesions in rats. *Anesthesiology, 89*, 411–420.

Gybels, J. M., & Sweet, W. H. (1989). *Neurosurgical treatment of persistent pain*. Basel: Karger.

Hirshberg, R. M., Al-Chaer, E. D., Lawand, N. B., Westlund, K. N., & Willis, W. D. (1996). Is there a pathway in the posterior funiculus that signals visceral pain? *Pain, 67*, 291–305.

Houghton, A. K., Wang, C. C., & Westlund, K. N. (2001). Do nociceptive signals from the pancreas travel in the dorsal column? *Pain, 89*, 207–220.

Kwon, S. J., & Kim, Y. S. (2000). High thoracic midline dorsal column myelotomy for severe visceral pain due to advanced stomach cancer. *Neurosurgery, 46*, 85–90.

Nauta, H. J. W., Soukup, V. M., Fabian, R. H., Lin, J. T., Grady, J. J., Williams, C. G. A. S., Campbell, G. A., Westlund, K. N., & Willis, W. D. (2000). Punctate midline myelotomy for the relief of visceral cancer pain. *Journal of Neurosurgery (Spine 1), 92*, 6989–6997.

Palecek, J., & Willis, W. D. (2003). The dorsal column pathway facilitates visceromotor responses to colorectal distention after colon inflammation in rats. *Pain, 104*, 501–507.

Palecek, J., Paleckova, V., & Willis, W. D. (2002). The roles of pathways in the spinal cord lateral and dorsal funiculi in signaling nociceptive somatic and visceral stimuli in rats. *Pain, 96*, 297–307.

Palecek, J., Paleckova, V., & Willis, W. D. (2003). Fos expression in spinothalamic and postsynaptic dorsal column neurons following noxious visceral and cutaneous stimuli. *Pain, 104*, 249–257.

Wang, C. C., Willis, W. D., & Westlund, K. N. (1999). Ascending projections from the area around the spinal cord central canal: A *Phaseolus vulgaris* leucoagglutinin study in rats. *The Journal of Comparative Neurology, 415*, 341–367.

Willis, W. D., & Coggeshall, R. E. (2004). *Sensory mechanisms of the spinal cord* (3rd ed.). New York: Kluwer Academic/Plenum Publishers.

Willis, W. D., & Westlund, K. N. (2004). Pain system. In G. Paxinos & J. K. Mai (Eds.), *The human nervous system* (2nd ed., pp. 1125–1170). San Diego: Elsevier Academic Press.

Willis, W. D., Al-Chaer, E. D., Quast, M. J., & Westlund, K. N. (1999). A visceral pain pathway in the dorsal column of the spinal cord. In: The neurobiology of pain, NAS colloquium. *Proceedings of the National Academy of Sciences of the United States of America, 96*, 7675–7679.

Zhuo, M., & Gebhart, G. F. (2002). Facilitation and attenuation of a visceral nociceptive reflex from the rostroventral medulla in the Rat. *Gastroenterology, 122*, 1007–1019.

Spinal Dorsal Horn Pathways, Muscle and Joint

Hans-Georg Schaible[1] and Siegfried Mense[2]
[1]Department of Physiology - Neurophysiology, Friedrich-Schiller-University of Jena, Jena, Germany
[2]Neuroanatomy, Heidelberg University, Medical Faculty Mannheim, CBTM, Mannheim, Germany

Synonyms

Deep somatic pain, muscle and joint pain

Definition

▶ *Deep somatic pain*: Pain from muscle, fascia, tendons, and joint capsules.

▶ *Receptive field*: Those regions of the body from which a second-order (or higher-order) neuron can be excited or inhibited.

Clinical Characteristics

In contrast to cutaneous pain, which is usually well localized, deep somatic pain is rather diffuse and has a stronger tendency to spread. Moreover, deep pain is often associated with more marked autonomic reactions such as sweating and an increase in heart rate.

Spinal Terminations
Primary Afferents from Muscle
Transganglionic labelling of afferent fibers from the cat gastrocnemius-soleus (GS) muscle has shown that the small fibers terminate in laminae I and IV–VI (Mense and Craig 1988). Intraaxonal staining of single group IV fibers from muscle in the guinea pig demonstrated spinal terminations mainly in laminae I and II (Ling et al. 2003). Thus, the superficial laminae - but also laminae of the deep dorsal horn - appear to be

the main targets of small-caliber muscle afferent fibers, many of which are nociceptive.

Primary Afferents from Joint

Articular nerves supplying the knee or elbow joint of rat, cat, and monkey project to several spinal segments. Staining of whole nerves with horseradish peroxidase showed dense projections of joint afferents to lamina I and to deep laminae IV, V, and VI (and VII) in cat, but other studies found projections mainly in laminae II and III (Schaible and Grubb 1993).

Dorsal Horn Laminae Containing Second-Order Neurons Involved in Muscle and Joint Pain

In the past, pain research was focused on cutaneous pain, and dorsal horn neurons that were described as mediating skin pain were not tested for additional receptive fields (RFs) in deep somatic tissues. In recent years, it became clear that most dorsal horn neurons – particularly in the deep dorsal horn – have convergent input from joint, muscle, and skin. Neurons with exclusive input from muscle or joint are extremely rare.

Second-Order Neurons Involved in Nociception from Muscle and Fascia

Lamina I: In cat, most of the lamina I cells with input from the GS muscle were found to respond only to noxious stimuli and were driven by group III fibers (Craig and Kniffki 1985). Interestingly, those cells that projected to the contralateral thalamus had no additional group IV-fiber input. However, neurons not projecting to the thalamus exhibited responses to electrical stimulation of group IV fibers. Apparently, lamina I is a rather specific target for muscle nociceptors with thin myelinated fibers.

Laminae IV–VI: In these laminae, a marked convergence from deep somatic tissues and skin is present, and all functional types of cells can be found: low-threshold mechanosensitive (LTM, presumably non-nociceptive), high-threshold mechanosensitive (HTM or nociceptive specific (NS), presumably nociceptive), and wide dynamic range or multireceptive (WDR or MR, presumably nociceptive) cells.

Under normal experimental conditions (i.e., in the absence of a deliberate experimental lesion) there are no dorsal horn neurons that receive input from muscle nociceptors, only. The neurons have a convergent input from several tissues such as muscle plus skin and from both nociceptive and non-nociceptive peripheral receptors (Hoheisel et al. 1990). Thus, these cells behave like WDR neurons.

Recently, the thoracolumbar fascia has attracted much interest as a potential pain source in patients with unspecific low back pain. The fascia in rats and humans has been found to possess a dense innervation including nociceptive and sympathetic fibers. The nociceptive input from fascia excites WDR neurons in the dorsal horn (Hoheisel et al. 2011) and evokes pain-related behavior in non-anesthetized animals when tested with noxious stimuli.

Spread of Muscle-Induced Excitation of Dorsal Horn Neurons Under Pathologic Conditions

Input from muscle is known to be more effective in inducing spinal neuroplastic changes than is input from the skin (Wall and Woolf 1984). In the rat, an experimental myositis induced marked changes in the connections between the afferents of the inflamed muscle and dorsal horn neurons (a lesion-induced "functional reorganization" of the spinal cord). One aspect of this reorganization was an expansion of the target region of the muscle nerve, i.e., input from inflamed muscle excited a larger population of spinal neurons than input from intact muscle (Koch et al. 1994; Fig. 1). This expansion was mainly due to activation of NMDA channels (binding glutamate) and NK1 receptors (binding substance P); AMPA/kainate receptors were not involved. The expansion is an expression of dorsal horn hyperexcitability (► central sensitization). Under the influence of nociceptive input from an inflamed muscle, formerly silent or ineffective synapses become effective, and the target area of the muscle nerve expands. Clinically, these findings could explain the subjective spread and referral of muscle pain.

Spinal Dorsal Horn Pathways, Muscle and Joint, Fig. 1 Expansion of the target area of a muscle nerve after an acute myositis (duration 8 h). The target area is that region of the dorsal horn in which neurons can be excited by standard electrical stimulation of the GS muscle nerves. *Dark shading*, target area of rats with intact muscle (control); *light shading*, target area of myositis animals. For further explanation, see text

Glia-Neuron Interaction

Since long ago it is known that the great majority of cells in the CNS are glial cells (astrocytes, microglia, oligodendrocytes). Only recently has the role of glial cells in pain been acknowledged (Watkins and Milligan 2001). In chronic myositis rats astrocytes exhibit an increased synthesis of the cytoskeletal protein GFAP and of the basic fibroblast factor (FGF-2) (Tenschert et al. 2004; Fig. 2a). Activation of central nervous glial cells by a peripheral lesion and the ensuing release of cytokines and other proinflammatory substances are assumed to be a major aspect of the transition to chronic pain. For instance, microglial cells are activated by a tissue or CNS lesion and sensitize dorsal horn neurons by releasing substances such as TNF-alpha (Marchand et al. 2009). Blocking microglia activation with minocycline abolishes the sensitization of dorsal horn neurons (Chacur et al. 2009). However, glial cells synthesize also agents (e.g., FGF-2) that inhibit neuronal activity (Blum et al. 2001; Fig. 2b).

Nerve Growth Factor (NGF) as a Sensitizing Substance

The neurotrophin NGF has been found to sensitize both peripheral nociceptors and dorsal horn neurons after intramuscular injection in man and rat (Svensson et al. 2003; Hoheisel et al. 2007). Interestingly, the NGF injection excited many muscle nociceptors but did not evoke pain (Hoheisel et al. 2005). However, after a few hours the injected muscle was allodynic/hyperalgesic and the dorsal horn neurons supplying the muscle were hyperexcitable. The puzzling finding that in spite of the excitation of muscle nociceptors NGF did not elicit pain can be

Spinal Dorsal Horn Pathways, Muscle and Joint, Fig. 2 Activation of astrocytes by a chronic myositis (a) and the effect of FGF-2 on neuronal activity (b). (a) Astrocytes were double-labeled with antibodies to glial fibrillary acidic protein (GFAP; *white* spiderlike structures) and basic fibroblast growth factor (FGF-2; *dark* nuclei of the astrocytes). (b) *Upper panel*, resting activity of a dorsal horn neuron during iontophoresis of FGF-2 and vehicle, respectively. *Lower panel*, recording of the iontophoresis current used for depositing the FGF-2 solution close to the neuron. FGF-2 was ejected at a concentration of 5 nM through a glass microelectrode with three barrels, one of which was used for recording of the neuron's activity

explained by the response behavior of the dorsal horn neurons: the neurons did not fire action potentials in response to NGF i.m. but reacted with subthreshold excitatory postsynaptic potentials (EPSPs; Hoheisel et al. 2007). These potentials are not transmitted to higher nociceptive centers, but have a marked sensitizing action at the spinal level.

Second-Order Neurons Involved in Nociception from Joint

The spinal cord contains neurons that respond to pressure onto to the knee joint but not to stimulation of the skin overlying the knee (Figure 3c). Typically, these neurons respond also to stimulation of adjacent muscles (Figure 3a). NS neurons with joint input are only excited by noxious pressure onto joint and/or muscle and twisting of the joint against resistance of the tissue. WDR neurons with joint input respond to light pressure onto the joint and to movements in the working range and show stronger responses to noxious stimuli (Figure 3b). Most WDR neurons exhibit convergent inputs from skin and deep tissue, whereas many NS neurons in the deep dorsal horn have only deep input. The cat spinal cord contains neurons with cell bodies in the ventral horn and axons in the spinoreticular tract, they are only driven by noxious stimulation of deep tissue (Schaible 2006; Schaible and Grubb 1993).

Inflammation-Evoked Hyperexcitability of Spinal Neurons with Joint Input

During knee inflammation, NS and WDR neurons develop pronounced hyperexcitability. Responses to innocuous and noxious mechanical stimulation of the inflamed knee increase, and NS neurons show a drop of threshold to the innocuous range. Responses to pressure onto non-inflamed muscles in the limb and to the ankle joint show a similar increase, and the receptive field often expands toward the paw. Thus, the spinal sensitization results from increased input from sensitized joint afferents and from an intraspinal increase of the gain (Schaible 2006; Schaible et al. 1987).

a

b

c

Spinal Dorsal Horn Pathways, Muscle and Joint, Fig. 3 Receptive field and responses of spinal cord neurons with input from the knee joint. (a) Receptive field of a neuron in the deep dorsal horn with input from the knee and adjacent muscles. (b) Responses of an ascending WDR neuron in lamina VIII of cat spinal cord with input from the knee joint, adjacent muscles, and skin over the ankle joint. The neuron responded weakly to innocuous movements of the knee (flexion, extension, outward rotation, OR) and showed pronounced responses to forced extension (f. ext) and noxious outward rotation (n.OR) of the knee. (c) Responses of a lamina I neuron in L7 with joint input in cat spinal cord. The neuron was activated by pressure applied to the lateral and medial side of the knee joint through the skin and by probing the exposed lateral and medial side of the knee but not by squeezing the skin overlying the joint

Central Sensitization Can Persist During Chronic Inflammation

During chronic inflammation, spinal neurons were more sensitive than in normal rats and had expanded receptive fields in deep tissue and skin (Menétrey and Besson 1982; Schaible and Grubb 1993). Numerous neurons in lamina I and in the deep dorsal and ventral horn of several segments expressed the activity marker c-Fos. Particularly during chronic inflammation, c-Fos was mainly elevated in the deep dorsal horn and only marginally in the superficial dorsal horn (Abbadie and Besson 1992; Menétrey et al. 1989). These changes in the spinal cord are likely to account for the pronounced forms of referred pain in the deep tissue that are induced by noxious stimulation of deep tissue in humans. For example, at advanced stages patients with osteoarthritis often report widespread pain beyond the OA joint (Imamura et al. 2008; Ordeberg 2009) and exhibit lower pressure pain thresholds in cutaneous and subcutaneous structures of the leg (Hendiani et al. 2003; Imamura et al. 2008). Patients with strong local hyperalgesia at the OA knee joint exhibit in addition higher pain summation scores upon repetitive stimulation of the OA knee which is another indicator for central sensitization in severe OA (Arendt-Nielsen et al. 2010).

Transmitters Involved in Activation and Sensitization of Spinal Neurons with Joint Input

Spinal application of antagonists at AMPA/kainate (non-NMDA) receptors reduced the responses of spinal neurons to innocuous and noxious pressure onto the knee joint, whereas NMDA receptor antagonists reduced only the responses to noxious pressure. During acute joint inflammation, the intraspinal release of glutamate is enhanced, and antagonists at AMPA/kainate and NMDA receptors prevented the development of hyperexcitability. These antagonists also reduced the responses of the neurons to joint stimulation after established inflammation. Thus, glutamate receptors play a key role in the generation and maintenance of inflammation-evoked spinal hyperexcitability.

Noxious but not innocuous compression of the normal joint enhanced the intraspinal release of substance P, neurokinin A, and CGRP. During acute inflammation, release of these mediators was even evoked by innocuous mechanical stimulation. Spinal application of antagonists at neurokinin 1, neurokinin 2, and CGRP receptors reduced the responses of spinal cord neurons to noxious compression of the normal joint and attenuated the inflammation-evoked hyperexcitability (Schaible 2006).

During joint inflammation, PGE_2 was released within the dorsal and ventral horn. Topical application of PGE_2 (Vasquez et al. 2001) or of EP_1, EP_2, and EP_4 receptor agonists (Bär et al. 2004) to the spinal cord surface facilitated the responses of spinal cord neurons to pressure onto the normal joint. Topical spinal application of the COX inhibitor indomethacin before inflammation attenuated the development of hyperexcitability. Thus, spinal PGs are also involved in the generation of inflammation-evoked spinal hyperexcitability (Vasquez et al. 2001). Similarly, a spinally applied specific IkB kinase inhibitor prevented spinal hyperexcitability (Ebersberger et al. 2006). However, once spinal hyperexcitability was established neither spinal indomethacin or diclofenac (Telleria-Diaz et al. 2010; Vasquez et al. 2001) nor the IkB kinase inhibitor (Ebersberger et al. 2006) reduced the enhanced responses of spinal cord neurons to mechanical stimulation of the inflamed knee joint. Thus the presence of spinal PGE_2 seems to be required for the generation but not for the maintenance of spinal hyperexitability. In fact, after

establishment of inflammation, neither PGE_2 nor the EP2 and EP4 agonist further increased responses to mechanical stimulation in rats with inflammation (Bär et al. 2004).

However, spinal application of a selective COX-2 inhibitor reduced the spinal responses to mechanical stimulation of the joint even after establishment of hyperexcitability; but this action was rather related to the inhibition of the breakdown of the spinal endocannabinoid 2-acylglycerol (2-AG, COX-2 metabolizes 2-AG) than to the inhibition of PG synthesis (Telleria-Diaz et al. 2010). PGD_2, another major PG in the CNS, dose-dependently reduced responses of spinal cord neurons to stimulation of the inflamed knee joint, whereas an antagonist at the DP1 receptor increased responses (Telleria-Diaz et al. 2008). Thus PGD_2 counteracts the sensitizing effects of PGE_2.

Spinal cord levels of the endocannabinoids 2-acylglycerol and anandamide were increased in MIA-treated rats, due to increased synthesis. Spinal application of both antagonists at the cannabinoid 1 and 2 receptor increased responses of neurons cutaneous stimulation suggesting tonic inhibition by endocannabinoids during OA (Sagar et al. 2010).

While spinal glia activation is common in models of peripheral nerve injury, it is not generally present in inflammation (Clark et al. 2007). In K/BxN serum transfer arthritis astrocyte and microglia activation was observed initially and microglia activation persisted. Because ATF3 immunreactivity appeared it was concluded that the inflammatory state showed a transition into a neuropathic pain state (Christianson et al. 2010).

References

Abbadie, C., & Besson, J. M. (1992). C-fos expression in rat lumbar spinal cord during the development of adjuvant-induced arthritis. *Neuroscience, 48,* 985–993.

Arendt-Nielsen, L., Nie, H., Laursen, M. B., et al. (2010). Sensitization in patients with painful knee osteoarthritis. *Pain, 149,* 573–581.

Bär, K.-J., Natura, G., Telleria-Diaz, A., et al. (2004). Changes in the effect of spinal prostaglandin E_2 during inflammation – Prostaglandin E (EP1-EP4) receptors in spinal nociceptive processing of input from the normal or inflamed knee joint. *Journal of Neuroscience*, *24*, 642–651.

Blum, T., Hoheisel, U., Unger, T., et al. (2001). Fibroblast growth factor-2 acutely influences the impulse activity of rat dorsal horn neurones. *Neuroscience Research*, *40*, 115–123.

Chacur, M., Lambertz, D., Hoheisel, U., et al. (2009). Role of spinal microglia in myositis-induced central sensitisation: an immunohistochemical and behavioural study in rats. *European Journal of Pain*, *13*, 915–923.

Christianson, C. A., Corr, M., Firestein, G. S., et al. (2010). Characterization of the acute and persistent pain state present in K/BxN serum transfer arthritis. *Pain*, *151*, 394–403.

Clark, A. K., Gentry, C., Bradbury, E. J., et al. (2007). Role of spinal microglia in rat models of peripheral nerve injury and inflammation. *European Journal of Pain*, *11*, 223–230.

Craig, A. D., & Kniffki, K. D. (1985). Spinothalamic lumbosacral lamina I cells responsive to skin and muscle stimulation in the cat. *The Journal of Physiology*, *365*, 197–221.

Ebersberger, A., Buchmann, M., Ritzeler, O., et al. (2006). The role of spinal nuclear factor-κB in spinal hyperexcitability. *Neuroreport*, *17*, 1615–1618.

Ferrington, D. G., Sorkin, L. S., Willis, W. D., Jr. (1986). Responses of spinothalamic tract cells in the cat cervical spinal cord to innocuous and graded noxious stimuli. *Somatosensory and Motor Research*, *3*, 339–358.

Hendiani, J. A., Westlund, K. N., Lawand, N., et al. (2003). Mechanical sensation and pain thresholds in patients with chronic arthropathies. *The Journal of Pain*, *4*, 203–211.

Hoheisel, U., Koch, K., & Mense, S. (1994). Functional reorganization in the rat dorsal horn during an experimental myositis. *Pain*, *59*, 111–118.

Hoheisel, U., & Mense, S. (1990). Response behaviour of cat dorsal horn neurones receiving input from skeletal muscle and other deep somatic tissues. *Journal of Physiology*, *426*, 265–280.

Hoheisel, U., Unger, T., & Mense S. (2005). Excitatory and modulatory effects of inflammatory cytokines and neurotrophins on mechanosensitive group IV muscle afferents in the rat. *Pain*, *114*, 168–176.

Hoheisel, U., Unger, T., & Mense, S. (2007). Sensitization of rat dorsal horn neurons by NGF-induced subthreshold potentials and low-frequency activation. A study employing intracellular recordings in vivo. *Brain Research*, *1169*, 34–43.

Hoheisel, U., Taguchi, T., Treede, R. D., & Mense, S. (2011). Nociceptive input from the rat thoracolumbar fascia to lumbar dorsal horn neurones. *European Journal of Pain*, *15*, 810–815.

Imamura, M., Imamura, S. T., Kaziyama, H. H. S., et al. (2008). Impact of nervous system hyperalgesia on pain, disability, and quality of life in patients with knee osteoarthritis: a controlled analysis. *Arthritis Rheumatism (Arthritis Care Research)*, *59*, 1424–1431.

Ling, L. J., Honda, T., Shimada, Y., et al. (2003). Central projection of unmyelinated (C) primary afferent fibers from gastrocnemius muscle in the guinea pig. *The Journal of Comparative Neurology*, *461*, 140–150.

Marchand, F., Tsantoulas, C., Singh, D., et al. (2009). Effects of Etanercept and Minocycline in a rat model of spinal cord injury. *European Journal of Pain*, *13*, 673–681.

Menétrey, D., & Besson, J. M. (1982). Electrophysiological characteristics of dorsal horn cells in rats with cutaneous inflammation. *Pain*, *13*, 343–364.

Menétrey, D., Gannon, A., Levine, J. D., et al. (1989). Expression of c-fos protein in interneurons and projection neurons of the rat spinal cord in response to noxious somatic, articular, and visceral stimulation. *The Journal of Comparative Neurology*, *285*, 177–195.

Mense, S., & Craig, A. D. (1988). Spinal and supraspinal termination of primary afferent fibers from the gastrocnemius-soleus muscle in the cat. *Neuroscience*, *26*, 1023–1035.

Ordeberg, G. (2009). Evidence of sensitization to pain in human osteoarthritis. In D. T. Felson, & H.-G. Schaible (Eds.), *Pain in osteoarthritis* (pp. 199–209). Hoboken, NJ: Wiley Blackwell.

Sagar, D. R., Staniaszek, L. E., Okine, B. N., et al. (2010). Tonic modulation of spinal hyperexcitability by the endocannabinoid receptor system in a rat model of osteoarthritis pain. *Arthritis Rheumatism*, *62*, 3666–3676.

Schaible, H. G. (2005). Basic mechanisms of deep somatic pain. In S. B. McMahon & M. Koltzenburg (Eds.), *Wall and Melzack's textbook of pain* (pp. 621–633). Philadelphia: Elsevier/Churchill Livingston.

Schaible, H. G., & Grubb, B. D. (1993). Afferent and spinal mechanisms of joint pain. *Pain*, *55*, 5–54.

Schaible, H. G., Schmidt, R. F., & Willis, W. D. (1987). Enhancement of the responses of ascending tract cells in the cat spinal cord by acute inflammation of the knee joint. *Experimental Brain Research*, *66*, 489–499.

Svensson, P., Cairns, B. E., Wang, K., et al. (2003). Injection of nerve growth factor into human masseter muscle evokes long-lasting mechanical allodynia and hyperalgesia. *Pain*, *104*, 241–247.

Telleria-Diaz, A., Ebersberger, A., Vasquez, E., et al. (2008). Different effects of spinally applied prostaglandin D_2 (PGD$_2$) on responses of dorsal horn neurons with knee input in normal rats and in rats with acute knee inflammation. *Neuroscience*, *156*, 184–192.

Telleria-Diaz, A., Schmidt, M., Kreusch, S., et al. (2010). Spinal antinociceptive effects of cyclooxygenase inhibition during inflammation: involvement of prostaglandins and endocannabinoids. *Pain*, *148*, 26–35.

Tenschert, S., Reinert, A., Hoheisel, U., et al. (2004). Effects of a chronic myositis on structural and functional features of spinal astrocytes in the rat. *Neuroscience Letters, 361*, 196–199.

Vasquez, E., Bär, K. J., Ebersberger, A., et al. (2001). Spinal prostaglandins are involved in the development but not the maintenance of inflammation-induced spinal hyperexcitability. *Journal of Neuroscience, 21*, 9001–9008.

Wall, P. D., & Woolf, C. J. (1984). Muscle but not cutaneous C-afferent input produces prolonged increases in the excitability of the flexion reflex in the rat. *The Journal of Physiology, 356*, 443–458.

Watkins, L. R., Milligan, E. D., & Maier, S. F. (2001). Spinal cord glia: New players in pain. *Pain, 93*, 201–205.

Spinal Flattened Neuron

Definition

Spinal lamina I neuron with a disk-shaped horizontal cell body and a roughly circular sparsely ramified aspiny dendritic tree confined to lamina I.

Cross-References

▶ Spinothalamic Tract Neurons, Morphology

Spinal Fusiform Neuron

Definition

Spinal lamina I neuron with a spindle-shaped cell body oriented rostrocaudally and a narrow longitudinal dendritic tree with numerous short pedicled dendritic spines.

Cross-References

▶ Spinothalamic Tract Neurons, Morphology

Spinal Fusion for Chronic Back Pain

Edward C. Benzel and Perry Dhaliwal
Department of Neurosurgery, Cleveland Clinic
Spine Institute, Cleveland, OH, USA

Synonyms

Low back pain; Spinal arthrodesis

Definition

Elimination of movement between various elements of spinal column at one or more segmental levels, with or without instrumentation, is termed spinal fusion or arthrodesis.
▶ Spinal arthrodesis may be a more appropriate term, since it involves removing the articular surfaces and securing bony union. The origin of spinal fusion stems from the vast orthopedic literature supporting fusion for painful joints. Hippocrates first observed fusion of the facet joints in a patient with spinal tuberculosis and described it as nature's attempt to halt the progression of the deformity. Hibbs is credited for the first spinal fusion performed; in 1911 he surgically fused the posterior spinal elements in young patients with spinal tuberculosis.

Introduction

Indications

Chronic back pain may be due to a multitude of degenerative spinal conditions including spondylolisthesis, lumbar spinal stenosis, degenerative scoliosis or pseudoarthrosis following a previous spinal fusion. However, there is also a large proportion of the population that will have non-specific low back pain that may not be attributable to the aforementioned causes (Deyo et al. 2006). Thus, the responsible physician must differentiate between the various causes of back pain and identify those patients that have pathology that may be amenable to a surgical intervention.

In general terms, spinal fusion for back pain is not advocated in the absence of associated leg pain. In a patient with imaging findings of spondylolisthesis, spinal stenosis, or degenerative scoliosis one must evaluate the patient for signs and symptoms consistent with radicular pain or neurogenic claudication. Several studies confirm the effectiveness of surgery under these circumstances (Weinstein et al. 2007; Zdeblick 1993). Though many patients with degenerative spinal conditions will present with both back and leg pain, there are some patients that will present with isolated back pain in the setting of degenerative spondylolisthesis, lumbar spinal stenosis or degenerative scoliosis. In these circumstances, the decision making becomes much more difficult. To evaluate these patients, one must begin with a full history and physical examination. The appropriate imaging investigations should be performed and conservative therapies should be exhausted prior to considering spinal surgery. Viable options for conservative therapy include weight loss, physical therapy, aerobic exercise, steroid injections, massage therapy, chiropractic manipulation, acupuncture and analgesics. In chronic pain patients, an often overlooked management strategy is that of cognitive behavioural therapy (Morley et al. 1999). Cognitive behavioural therapy, in combination with the aforementioned therapies, aims to rearrange patient's perceptions of pain and teach behavioural strategies to help patients better manage their pain. There is overwhelming evidence of its efficacy in the management of chronic pain and specifically as related to low back pain. One should also give particular consideration to a history of depression in any patient with back pain. It should be noted that outcomes from spinal surgery in the context of depression are unfavourable (Sinikallio et al. 2007, 2009) and it is of utmost importance to obtain appropriate psychological evaluation and treatment prior to considering surgery in such a patient.

In those patients that continue to have back pain despite an adequate trial of conservative therapy, spinal fusion may be considered in select circumstances. In patients with degenerative spondylolisthesis, some have suggested that back pain may be the result of abnormal motion at the spinal segment where the spondylolisthesis is seen. However, these patients will undoubtedly have degenerative spinal disease across other segments of the spine and thus the back pain may not be associated with the spondylolisthesis. For these reasons, the exact management of these patients is unclear. In light of the uncertainty, a more conservative approach to management of these patients may be warranted. Such an approach would aim to localize the pain to the segmental level where the spondylolisthesis is seen on imaging using facet joint injections. Relief of pain would suggest that fusion of that segment would also provide relief of back pain. Another strategy would be to obtain both supine and weight bearing x-rays to assess the degree of motion at the level of the spondylolisthesis. If there is an increase in the spondylolisthesis when the patient is standing, then spinal fusion may be warranted. If there is no motion across the segment, then the management becomes much more controversial. Some authors view spondylolisthesis as a manifestation of chronic movement across the level of interest and therefore would recommend fusion to attempt treatment of back pain and to prevent progression of the spondylolisthesis. Others may not perform spinal fusion if there is no clear motion seen across the involved segment.

In patients with degenerative scoliosis, possible indications for surgery include back pain, neurological symptoms, cosmesis and/or progression of the deformity. Given the difficulty and high morbidity associated with deformity correction, conservative measures to treat the patient's pain are of utmost importance. Though a full discussion around planning for deformity correction is beyond the scope of this chapter, suffice it to say that careful analysis of the patient's spinal anatomy is mandatory before embarking on any attempts to correct the patient's spinal deformity.

Investigations

Before embarking on a surgical fusion procedure, it is imperative that the surgeon establishes an anatomic explanation for the patient's symptoms. There are many different techniques that can assist in the decision making process including plain radiographs, magnetic resonance imaging (MRI) scans, computed tomography (CT) scans and facet joint injections.

Plain radiographs are useful to assess the patient's alignment, disc space height, and presence or absence of osteoporosis. Flexion and extension views (or supine and weight-bearing films) are used to help identify excessive segmental motion/instability. It should be noted that MRI and CT scans are most often performed in the supine position and weight bearing plain radiographs may demonstrate segmental motion that may not be seen if one were to rely on the findings of an MRI or CT scan alone. In non-traumatic settings, the extent to which this excessive motion represents instability is poorly defined (Nachemson 1991) but generally 4–5 mm of translation and 15° of relative angulation is considered instability in degenerative disease (Spratt et al. 1993).

Magnetic resonance imaging provides a very comprehensive evaluation of the patient's spinal anatomy. One can evaluate the spine for vertebral or disc pathology, compression of neural elements and other more subtle findings including soft tissue edema, foramina narrowing and facet joint degeneration. Given the sensitivity of MRIs in identifying changes in spinal anatomy, one must be very judicious in ensuring that the patient's symptoms are correlated to the findings on the imaging study. For example, MRI is very sensitive regarding degenerative changes in discs and the surrounding soft tissue structures, spawning the concept of "black disc disease." Black disc disease is a loss of bright signal on T2 weighted images due to age-related loss of water content in the disk. High intensity zones confined to the annulus fibrosus, annular tears of various configurations and various endplate abnormalities of degenerative disc disease (Modic et al. 1988) are identified on MRI. It is precarious to attribute low back pain to these MRI abnormalities, as these are found in a fairly high percentage of asymptomatic individuals. Borenstein et al. (2001) demonstrated that the presence of such MRI findings of disc degeneration in healthy subjects does not predict the subsequent development of lumbar spine symptomatology. Provocative discography has thus been proposed to identify the painful disc (see chapter on ▶ Discogenic Back Pain). The most useful aspect of the procedure in the context of determining levels for surgical fusion is the subjective pain response, with the reproduction of the patient's low back pain symptoms during the injection of the contrast agent representing a "concordant" pain response. However, the natural history of low back pain in patients with positive discography is uncertain. Smith et al. (1995) observed that approximately two thirds of patients with positive discography, who were considered operative candidates but did not have surgery, had satisfactory relief with non-operative therapy.

Computed tomography scans can also provide useful information in the evaluation of a patient with back pain. Firstly, CT scans are much better at visualization of bone structures as compared to MRI. This is particularly important in cases of isthmic spondylolisthesis where facet damage may be missed on MRI scans. Secondly, CT scans are particularly useful in the context of previous spinal fusion as MR images may be significantly distorted as a result of metal artifact. In such circumstances, CT myelograms may be the only imaging modality that would allow one to evaluate the neural elements and may significantly change surgical planning.

Another consideration is that of facet joint injections as an important diagnostic tool. As mentioned previously, in some patients with degenerative spondylolisthesis, it may be useful to perform a facet joint injection to evaluate a response. Should the patient find relief from his or her symptoms, then spinal fusion at that level may be warranted. Absence of a response from a facet joint injection may suggest that the spondylolisthesis may not be entirely responsible for the patient's pain syndrome.

From this description of various diagnostic tools, it is obvious that our ability to delineate the exact cause of back pain, and then to predict the outcome of surgical fusion based on these findings, is far from perfect. However, the surgeon can use the available imaging modalities to help narrow the aetiology of back pain and assist in the formulation of a management strategy for the patient.

Characteristics

As with any pathology in the spine, the basic surgical algorithm includes decision-making around the need for decompression, stabilization and fusion. Decompression may be performed directly or indirectly through distraction of the segments of the spine and opening the neural foramina. Stabilization is typically performed through the use of a variety of instrumentation which may include pedicle screws, translaminar screws, rods, hooks, and/or wires. Mobility across a spinal segment may be limited using posterolateral fusion or interbody fusion techniques. The full details of surgical technique are beyond the scope of this chapter, but a brief description of each fusion procedure, along with expected results, follows.

Postero-lateral fusion involves decortication of facet joints, transverse processes and pars-interarticularis, with the onlay of bone graft to encompass the exposed bony surfaces. Although this technique is very effective at fusing the spine, good to excellent relief of back pain was reported in only 39–57 % of the patients (France et al. 1999) (Zdeblick 1993). This would suggest that there is limited success for treatment of back pain with spinal fusion and that there are likely other factors that contribute the back pain that cannot be identified with imaging techniques alone.

Lumbar Interbody Fusion (LIF) thus became popular since it has multiple theoretical advantages over posterolateral fusion. Interbody fusion places the bone graft in the load bearing position of anterior and middle spinal columns (which support 80 % of spinal loads and provide 90 % of the osseous surface area), thereby enhancing the potential for fusion. This procedure also helps to restore the disc space height, lumbar lordosis, and coronal and sagittal balance of the spine, whereas a posterolateral fusion has

limited potential to do this (Mummaneni et al. 2004). As this procedure involves complete discectomy, it theoretically eliminates discogenic pain if the lumbar disc is the pain generator in that particular patient. Three different approaches for LIF are widely practiced:

- **ALIF**: Anterior Lumbar Interbody Fusion – where discectomy and fusion is performed by ventral approach and fusion is often supplemented with posterior instrumentation. Advantages include direct access for reconstruction of anterior spinal column, ability to improve sagittal balance, and ability to avoid paraspinal muscle trauma and denervation. Drawbacks include risk of deep venous thrombosis and vascular injury due to retraction of iliac vessels, retrograde ejaculation from hypogastric plexus injury, and muscular atony of abdominal wall leading to hernias.
- **PLIF**: Posterior Lumbar Interbody Fusion (Fig. 1a, b) – where a laminectomy is performed and thecal sac is retracted to access the disc space, and to perform discectomy and placement of interbody cages filled with bone graft. Simultaneous posterolateral fusion is performed to achieve circumferential fusion. Since this technique requires significant bilateral retraction of the thecal sac and nerve roots, risks include cerebrospinal fluid leakage, dysesthetic nerve root pain syndromes, nerve root injury, epidural fibrosis and potential injury to conus medullaris if performed above L3 level.
- **TLIF**: Transforaminal Lumbar Interbody Fusion - where hemilaminectomy and hemifacetectomy is performed and discectomy and bone graft/cages are placed through a posterior transforaminal approach, thus avoiding undue retraction of the thecal sac. Thus, it avoids all the risks associated with ALIF and TLIF, and can be used safely even above L3 level.
- Minimally Invasive Lumbar Fusion: the methods are associated with varying degrees of paraspinal muscle injury due to retraction induced intramuscular pressure and ischemia. Gejo et al. (1999) showed that the incidence of low back pain after open lumbar surgery was

Spinal Fusion for Chronic Back Pain, Fig. 1 (a) MRI of lumbar spine showing L5-S1 spondylolisthesis in a patient with mechanical back pain. (b) Post-operative x-ray of posterior lumbar interbody fusion (PLIF) – position of cages between the intervertebral bodies is identified by the radio-opaque markers

significantly increased in patients who had long muscle retraction times. Rantanen et al. (1993) demonstrated that patients with poor outcomes after lumbar surgery were more likely to have persistent pathologic changes in their paraspinal muscles. Minimally invasive lumbar fusion technology has evolved to obviate these risks associated with open surgeries. Foley et al. (2003) reported that in 39 patients, who had percutaneous minimally invasive lumbar fusion, two thirds had excellent outcome and one third had good outcome, as determined by the modified MacNab criteria, with 100 % solid fusion by radiographic criteria.

Outcomes after Lumbar Fusion: Spinal immobilization and fusion are obviously indicated for spinal instability due to trauma or tumor, but huge controversy exists about the role of fusion as treatment for low back pain secondary to degenerative disc disease and spondylolisthesis. There is a definitive trend towards increasing use of spinal instrumentation and fusion over the past 20 years, despite the lack of evidence that surgery can positively influence the course of degenerative spine disorders (Bono and Lee 2004). The reader should also be cognizant of the potential complications of invasive spinal surgery, occurring in up to 33 % of patients (Mayer and Korge 2002), which include implant failure, non-union, graft related morbidity, sexual dysfunction etc. A Cochrane review by Gibson et al. (Gibson and Waddell 2005) came to the following conclusions: (1) There is no acceptable evidence of the efficacy of any form of fusion for degenerative lumbar spondylosis, back pain or instability. (2) There is strong evidence that instrumented fusion may produce a higher rate of fusion, but does not improve clinical outcome. The same observations can be made about non-surgical therapeutic options like peridural injections, catheters, intradiscal electrothermal therapy (▶ IDET) etc. Until randomized prospective blinded clinical trials are performed, controversy will persist regarding spinal fusion. Suffice it to say, fusion for backache is not an alternative to expert rehabilitation that emphasizes both flexibility enhancement and core muscle strengthening with muscle retraining and excercises.

Summary

Lumbar spinal fusion is an effective surgical option for chronic back pain secondary to degenerative spinal disorders when a patient has failed optimal conservative and less invasive management. Randomized clinical trails are necessary to study the efficacy of spinal fusion on clinical outcomes. Emerging technology as an alternative to spinal fusion include total disc replacement with prosthetic discs (like SBIII Charite Disc or ProDisc II) in order to preserve the segmental motion of the involved spine.

References

Bono, C. M., & Lee, C. K. (2004). Critical analysis of trends in fusion for degenerative disc disease over the past 20 years: Influence of technique on fusion rate and clinical outcome. *Spine (Phila Pa 1976)*, *29*(4), 455–63. discussion Z5.

Borenstein, D. G., O'Mara, J. W., Jr., Boden, S. D., Lauerman, W. C., Jacobson, A., Platenberg, C., et al. (2001). The value of magnetic resonance imaging of the lumbar spine to predict low-back pain in asymptomatic subjects: A seven-year follow-up study. *The Journal of Bone and Joint Surgery American*, *83-A*(9), 1306–1311.

Deyo, R. A., Mirza, S. K., & Martin, B. I. (2006). Back pain prevalence and visit rates: Estimates from U.S. national surveys, 2002. *Spine (Phila Pa 1976)*, *31*(23), 2724–2727.

Foley, K. T., Holly, L. T., & Schwender, J. D. (2003). Minimally invasive lumbar fusion. *Spine (Phila Pa 1976)*, *28*(15 Suppl), S26–S35.

France, J. C., Yaszemski, M. J., Lauerman, W. C., Cain, J. E., Glover, J. M., Lawson, K. J., et al. (1999). A randomized prospective study of posterolateral lumbar fusion. Outcomes with and without pedicle screw instrumentation. *Spine (Phila Pa 1976)*, *24*(6), 553–560.

Gejo, R., Matsui, H., Kawaguchi, Y., Ishihara, H., & Tsuji, H. (1999). Serial changes in trunk muscle performance after posterior lumbar surgery. *Spine (Phila Pa 1976)*, *24*(10), 1023–1028.

Gibson, J. N. A., & Waddell, G. (2005). Surgery for degenerative lumbar spondylosis: Updated Cochrane Review. *Spine (Phila Pa 1976)*, *30*(20), 2312–2320.

Mayer, H. M., & Korge, A. (2002). Non-fusion technology in degenerative lumbar spinal disorders: Facts, questions, challenges. *European Spine Journal, 11* (Suppl 2), S85–S91.

Modic, M. T., Steinberg, P. M., Ross, J. S., Masaryk, T. J., & Carter, J. R. (1988). Degenerative disk disease: Assessment of changes in vertebral body marrow with MR imaging. *Radiology, 166*(1 Pt 1), 193–199.

Morley, S., Eccleston, C., & Williams, A. (1999). Systematic review and meta-analysis of randomized controlled trials of cognitive behaviour therapy and behaviour therapy for chronic pain in adults, excluding headache. *Pain, 80*(1–2), 1–13.

Mummaneni, P. V., Haid, R. W., & Rodts, G. E. (2004). Lumbar interbody fusion: State-of-the-art technical advances. Invited submission from the joint section meeting on disorders of the spine and peripheral nerves, March 2004. *Journal of Neurosurgery: Spine, 1*(1), 24–30.

Nachemson, A. L. (1991). Instability of the lumbar spine. Pathology, treatment, and clinical evaluation. *Neurosurgery Clinics of North America, 2*(4), 785–790.

Rantanen, J., Hurme, M., Falck, B., Alaranta, H., Nykvist, F., Lehto, M., et al. (1993). The lumbar multifidus muscle five years after surgery for a lumbar intervertebral disc herniation. *Spine (Phila Pa 1976), 18*(5), 568–574.

Sinikallio, S., Aalto, T., Airaksinen, O., Herno, A., Kröger, H., Savolainen, S., et al. (2007). Depression is associated with poorer outcome of lumbar spinal stenosis surgery. *European Spine Journal, 16*(7), 905–912.

Sinikallio, S., Aalto, T., Koivumaa-Honkanen, H., Airaksinen, O., Herno, A., Kröger, H., et al. (2009). Life dissatisfaction is associated with a poorer surgery outcome and depression among lumbar spinal stenosis patients: A 2-year prospective study. *European Spine Journal, 18*(8), 1187–1193.

Smith, S. E., Darden, B. V., Rhyne, A. L., & Wood, K. E. (1995). Outcome of unoperated discogram-positive low back pain. *Spine (Phila Pa 1976), 20*(18), 1997–2000. discussion 2000-1.

Spratt, K. F., Weinstein, J. N., Lehmann, T. R., Woody, J., & Sayre, H. (1993). Efficacy of flexion and extension treatments incorporating braces for low-back pain patients with retrodisplacement, spondylolisthesis, or normal sagittal translation. *Spine (Phila Pa 1976), 18*(13), 1839–1849.

Weinstein, J. N., Lurie, J. D., Tosteson, T. D., Hanscom, B., Tosteson, A. N. A., Blood, E. A., et al. (2007). Surgical versus nonsurgical treatment for lumbar degenerative spondylolisthesis. *The New England Journal of Medicine, 356*(22), 2257–2270.

Zdeblick, T. A. (1993). A prospective, randomized study of lumbar fusion. Preliminary results. *Spine (Phila Pa 1976), 18*(8), 983–991.

Spinal Injection

▶ Intrathecal Injection

Spinal Manipulation, Characteristics

Bengt H. Sjölund
Department of Community Medicine and Rehabilitation Medicine, Umea University, Umea, Sweden

Synonyms

Chiropractics; Osteopathy

Definition

Manual manipulation of near vertebral column structures to relieve pain.

Characteristics

Originally based on anecdotal nonscientific observations (for the "subluxation theory," see Jarvis 2001), spinal manipulative therapy has been used to a varying extent in the western world, both as a separate modality outside and inside traditional medical practice and as part of physical and rehabilitation medicine. As part of complementary medicine, it has received great popularity in the general population (Ernst et al. 2001).

The most common types of spinal manipulative therapy are high velocity, low amplitude thrust manipulation, low velocity, small or large amplitude mobilization, manual traction (see ▶ Lumbar Traction), and so-called craniosacral therapy. While there are controlled data that acute back pain may in fact be temporarily relieved by manipulation (e.g., Koes et al. 1992), the results in chronic back pain are ambiguous. Furthermore, to be clinically relevant, the long-term pain relief after a limited series of manipulation sessions should last at least 3–6 (12) months and be reproducible with additional spinal manipulation sessions at a later stage. In addition, the risk of serious side effects after initial or subsequent manipulations should be negligible.

At present there are several systematic reviews available on the effects of spinal manipulative therapy in chronic low back pain. Ferreira et al. (2002) included nine randomized controlled trials (RCTs) of mostly moderate quality in their review. The patients were adult and had low back pain of at least 3 months duration. The outcome measures in the studies had to include at least one of pain, disability, quality of life, adverse events, return to work, global perceived effect, or patient's satisfaction with therapy. The mean duration of pain was 28.1 months. The mean reduction in pain intensity was 7 (95 % CI 1–14) mm on a 100 mm VAS scale at 1 month when compared with placebo, and the disability score was reduced by 6 (95 CI 11–40) points on a 100-point disability questionnaire, indicating no relevant clinical effects from the therapy.

In a periodically updated Cochrane analysis of spinal manipulative therapy in low back pain, Assendelft et al. (2008) aimed to analyze comparative studies between spinal manipulative therapy and other therapies and to include recent RCTs up to September 2003. They have found 39 RCTs. In acute low back pain, the effect of spinal manipulation was superior to sham therapy with a pain VAS reduction of 10 (95 % CI 2–17) mm short term and 19 (95 % CI 3–35) mm after more than 6 weeks. The improvement at the same point in time on the Rivermead disability questionnaire was 2.6 (95 CI 0.5–4.8) points, i.e., of little clinical importance. Spinal manipulation was not more effective than general practice care, analgesics, physiotherapy in general, exercise, or back schools. The results for chronic low back pain were similar, and the authors conclude that there is at present no evidence that spinal manipulation is superior to standard treatments for acute or chronic low back pain.

The so-called craniosacral therapy is based on bizarre mechanistic speculations (see Hartman and Norton 2002) and has not been shown to relieve pain in controlled studies (e.g., Ventegodt et al. 2004).

An additional factor is the risk of side effects with manipulation, which is very low with lumbar manipulation but not with cervical manipulation, where many clinical reports have been published of vascular deficiency to structures of the posterior part of the brain. Recently, a case control study (Smith et al. 2003) of 150 patients with vertebral artery dissection and ischemic stroke or transient ischemic attacks compared with 300 matched controls could demonstrate that spinal manipulative therapy is an independent risk factor for vertebral artery dissection (odds ratio 6.62, 95 % CI 1.4–30). Hence, it can be considered entirely inappropriate to perform cervical spinal manipulative therapy for the symptomatic treatment of pain.

Summary

Spinal manipulation is not more effective than general practice care, analgesics, physiotherapy in general, exercise, or back schools.

References

Assendelft, W. J. J., Morton, S. C., Yu, E. I., et al. (2008). Spinal manipulative therapy for low back pain. *The Cochrane Database of Systematic Reviews, 1.* doi:10.1002/14651858.CD000447.pub2.

Ernst, E., Pittler, M. H., Eisenberg, D., et al. (2001). *The desktop guide to complementary and alternative medicine.* Edinburgh: Mosby.

Ferreira, M. F., Ferreira, P. H., Latimer, L., et al. (2002). Does spinal manipulative therapy help people with chronic low back pain? *Australian Journal of Physiotherapy, 48,* 277–284.

Hartman, S. E., & Norton, J. M. (2002). Craniosacral therapy is not medicine. *Physical Theraphy, 82,* 1146–1147.

Jarvis, W. T. (2001). Fact sheet on chiropractic. National Council against health fraud, USA, http://www.ncahf.org/articles/c-d/chiro.html

Koes, B. W., Bouter, L. M., van Mameren, H., et al. (1992). The effectiveness of manual therapy, physiotherapy and treatment by the general practitioner for non-specific back and neck complaints. *Spine, 17,* 28–36.

Smith, W. S., Johnston, S. C., Skalabrin, E. J., et al. (2003). Spinal manipulative therapy is an independent risk factor for vertebral artery dissection. *Neurology, 60,* 1424–1428.

Ventegodt, S., Merrick, J., Andersen, N. J., & Bendix, T. (2004). A combination of gestalt therapy, Rosen body work, and cranio sacral therapy did not help in chronic whiplash-associated disorders (WAD)–results of a randomized clinical trial. *Scientific World Journal, 4,* 1055–1068.

Spinal Manipulation, Pain Management

Robert Gassin
Musculoskeletal Medicine Clinic, Frankston, VIC, Australia

Synonyms

High-velocity thrust manipulation (HVTM); Mobilization with impulse; Mobilization with thrust

Definition

Spinal manipulation is a treatment that involves the application of a mechanical force to the spine, with the aim of moving one or more spinal joints beyond the physiological range and towards the anatomical limit of motion. Spinal manipulation is described as "long lever" when the force is applied some distance from the area where it is expected to have its effect and "short lever" when the force is applied to a specific spinal segment.

Characteristics

Mechanism

The mechanism of action of spinal manipulation remains unclear. The major theories include:

- The reduction of disc bulging or prolapse
- Stretching of adhesions in the zygapophyseal joint capsules or freeing of an entrapped meniscoid from a zygapophyseal joint
- Correction of spinal misalignment
- Inducing muscle relaxation
- Stretching of the vertebral ligaments
- Release of endogenous opioids
- Normalization of local or regional reflex activity
- Normalization of blood flow
- Normalization of CSF flow
- Activation of a central control mechanism
- A placebo effect

None of these theories are strongly supported by scientific evidence.

Applications

The indications for spinal manipulation depend on what the practitioner has been trained to regard as a "manipulable lesion," i.e., a condition that invites treatment by manipulation. The practitioner decides if the patient has such a lesion on the basis of medical history, musculoskeletal examination, and perhaps investigations such as plain radiography.

Spinal manipulation may be used as an isolated treatment or as part of a multimodal management strategy for the management of many musculoskeletal conditions.

Efficacy

Low Back Pain

An early systematic review found that patients with uncomplicated, acute low back pain, treated with spinal manipulation, had a 17 % greater chance of recovery at 3 weeks (Shekelle et al. 1992). No other attributable effect was demonstrated. Since that review, further studies have been published and have been subjected to a meta-analysis (Assendelft et al. 2003).

For patients with acute low back pain, spinal manipulation was superior only to sham therapy and therapies judged to be ineffective or even harmful. Spinal manipulation had no statistically or clinically significant advantage over general practitioner care, analgesics, physical therapy, exercises, or back school.

Results for patients with chronic low back pain were similar. Radiation of pain, study quality, profession of manipulator, and use of manipulation alone or in combination with other therapies did not affect these results.

The authors concluded that there is no evidence that spinal manipulation is superior to other standard treatments for patients with acute or chronic low back pain (Assendelft et al. 2003). This conclusion is consistent with that of previous systematic reviews (Koes et al. 1996; van Tulder et al. 1997).

Thoracic Pain

Scientific data on the efficacy of spinal manipulation for thoracic pain is sparse. A recent randomized controlled trial, comparing spinal manipulation to nonfunctional ultrasound, has demonstrated significant improvement in lateral flexion and reductions in numerical pain ratings at the conclusion of a 2- to 3-week treatment period. At 1-month follow-up, the changes were maintained but were no longer better than in the placebo group (Schiller 2001). These results suggest that spinal manipulation has greater short-term benefits than placebo treatment in the management of thoracic pain but offers no sustained benefit.

Neck Pain

The effectiveness of spinal manipulation for neck pain remains unproven. In one review (Hurwitz et al. 1996), the combination of three of the randomized controlled trials for patients with subacute or chronic neck pain showed an improvement on a 100 mm visual analogue scale of pain at 3 weeks of 12.6 mm (95 % confidence interval,-0.15, 25.5) for manipulation, compared with muscle relaxants or usual medical care. This suggests that cervical spinal manipulation, at best, provides marginal benefit for some patients with subacute and chronic neck pain.

The most recent systematic review found that manipulation used as a sole treatment consistently showed effects similar to those of placebo, wait period, or control (Gross et al. 2002). It was only if combined with exercises, as part of a multimodal treatment package, that there was any detectable benefit of manipulation for neck pain. However, in that context, the attributable effects of manipulation cannot be distinguished from those of exercise.

In a randomized trial of chiropractic manipulation and mobilization for patients with neck pain, both modalities were found to yield comparable clinical outcomes in terms of mean reductions in pain and disability (Hurwitz et al. 2002).

Cervicogenic Headaches

In one controlled trial, the active and control interventions consisted of 6 sessions over 3 weeks (Nilsson et al. 1997). The active intervention group received spinal manipulation at each session, whereas the control intervention group received deep friction massage to the posterior muscles of the shoulder girdle, the upper thoracic and lower cervical regions, plus treatment with a laser light in the upper cervical region. One week following the final treatment, statistically significant improvement was evident in the active intervention group compared to the control group, with respect to decreased use of analgesia (36 % vs. 0 %, $p = 0.04$), decreased number of headache hours per day (69 % vs. 37 %, $p = 0.03$), and decreased headache intensity per episode (36 % vs. 17 %, $p = 0.04$). Evidence of any longer-lasting effect was not provided.

Migraine

One controlled trial of spinal manipulation in the management of migraine consisted of 2 months of data collection (before treatment), 2 months of treatment, and a further 2 months of data collection (after treatment) (Tuchin et al. 2000).

The active treatment consisted of a maximum of 16 sessions of HVTM administered by a chiropractor. The control group received detuned interferential therapy. The average response of the treatment group showed statistically significant improvement in migraine frequency, duration, disability, and medication use, when compared to the control.

A prospective, randomized, parallel-group comparison of spinal manipulation, amitriptyline, and a combination of both therapies for the prophylaxis of migraine (Nelson et al. 1998) revealed clinically important, but not statistically significant, improvement in a headache index score, in all 3 study groups compared to baseline. The headache index score represents the weekly sum of the patient's daily headache pain score. The authors claim that the study shows that HVTM is as effective as a well-established and efficacious treatment, amitriptyline. These results suggest that spinal manipulation is an option for certain migraine patients who have failed trials of proven conventional treatments.

Contraindications

Although certain conditions are almost universally acknowledged as contraindications to spinal manipulation, others are seen by some and not others as contraindications (Gassin and Masters 2001).

Universally acknowledged contraindications to spinal manipulation include neoplasia (benign or malignant), active infection, active inflammation, neurological disorders, instability following trauma, acute spondylolysis/spondylolisthesis, and vertebral or rib fracture. Widely accepted contraindications to spinal manipulation are osteoporosis, bleeding disorders, anticoagulation therapy, radicular pain, radiculopathy, vertebrobasilar insufficiency, rheumatoid arthritis of the upper cervical spine, abnormal spinal anatomy, severe pain or distress, past adverse effect from spinal manipulation, and hypermobility (including pregnancy).

Adverse Effects

Most adverse effects following spinal manipulation are benign and of short duration. Local discomfort or pain is by far the most common, accounting for approximately two-thirds of the total adverse effects (Senstad et al. 1997). The possible adverse effects of spinal manipulation, ranked in order of decreasing incidence, are local discomfort, headache, fatigue, discomfort outside area of treatment, dizziness, nausea, intervertebral joint injury, triggering or exacerbation of neurological symptoms, vertebral and rib fracture, vertebral artery injury, carotid artery injury, and spinal cord injury.

Vertebral and internal carotid artery injuries results from manipulation of the cervical spine above the C6 spinal segment. Most published data suggest an incidence of 0.5–1 per 1,000,000 manipulations (Hurwitz et al. 1996; Haldeman et al. 2002).

In a review of case reports of accidents due to cervical spine manipulation published between 1925 and 1997, the most frequently reported injuries involved arterial dissection or spasm, lesions of the brain stem, and Wallenberg syndrome. Death occurred in 18 % of cases (Di Fabio 1999). The incidence of death following cervical spinal manipulation has been estimated at 0.3 per 1,000,000 (Hurwitz et al. 1996).

References

Assendelft, W. J., Morton, S. C., Yu, E. I., Suttorp, M. J., & Shekelle, P. G. (2003). Spinal manipulative therapy for low back pain. A meta-analysis of effectiveness relative to other therapies. *Annals of Internal Medicine, 138*, 871–881.

Di Fabio, R. P. (1999). Manipulation of the cervical spine: Risks and benefits. *Physical Therapy, 79*, 50–65.

Gassin, R., & Masters, S. (2001). Spinal manual therapy – The evidence. *Australasian Musculoskeletal Medicine, 6*, 26–31.

Gross, A. R., Kay, T., Hondras, M., Goldsmith, C., Haines, T., Peloso, P., Kennedy, C., & Hoving, J. (2002). Manual therapy for mechanical neck disorders: A systematic review. *Manual Therapy, 7*, 131–149.

Haldeman, S., Carey, P., Townsend, M., & Papadopoulos, C. (2002). Clinical perceptions of the risk of vertebral artery dissection after cervical manipulation: The effect of referral bias. *The Spine Journal, 2*, 334–342.

Hurwitz, E. L., Aker, P. D., Adams, A. H., Meeker, W. C., & Shekelle, P. G. (1996). Manipulation and mobilization of the cervical spine. A systematic review of the literature. *Spine, 21*, 1746–1759.

Hurwitz, E. L., Morgenstern, H., Harber, P., Kominski, G. F., Yu, F., & Adams, A. H. (2002). A randomized trial of chiropractic manipulation and mobilization for patients with neck pain: Clinical outcomes from the UCLA neck-pain study. *American Journal of Public Health, 92,* 1634–1641.

Koes, B. W., Assendelft, W. J., van der Heijden, G. J., & Bouter, L. M. (1996). Spinal manipulation for low back pain. An updated systematic review of randomized clinical trials. *Spine, 21,* 2860–2871.

Nelson, C. F., Bronfort, G., Evans, R., Boline, P., Goldsmith, C., & Anderson, A. V. (1998). The efficacy of spinal manipulation, Amitriptyline and the combination of both therapies for the prophylaxis of migraine headache. *Journal of Manipulative and Physiological Therapeutics, 21,* 511–519.

Nilsson, N., Christensen, H. W., & Hartvigsen, J. (1997). The effect of spinal manipulation in the treatment of cervicogenic headache. *Journal of Manipulative and Physiological Therapeutics, 20,* 326–330.

Schiller, L. (2001). Effectiveness of spinal manipulative therapy in the treatment of mechanical thoracic spinal pain. *Journal of Manipulative and Physiological Therapeutics, 24,* 394–401.

Senstad, O., Leboeuf-Yde, C., & Borchgrevink, C. (1997). Frequency and characteristics of side effects of spinal manipulative therapy. *Spine, 22,* 435–441.

Shekelle, P. G., Adams, A. H., Chassin, M. R., Hurwitz, E. L., & Brooks, R. H. (1992). Spinal manipulation for low-back pain. *Annals of Internal Medicine, 117,* 590–598.

Tuchin, P. J., Pollard, H., & Bonello, R. (2000). A randomized controlled trial of chiropractic spinal manipulative therapy for migraine. *Journal of Manipulative and Physiological Therapeutics, 23*(2), 91–95.

van Tulder, M. W., Koes, B. W., & Bouter, L. M. (1997). Conservative treatment of acute and chronic nonspecific low back pain. A systematic review of randomized controlled trials of the most common interventions. *Spine, 2,* 2128–2156.

Spinal Modulatory Impulses

Definition

Nerve impulses generated inside the spinal cord or at supraspinal centers that modulate the transmission of primary impulses through spinal projection neurons.

Cross-References

▶ Spinothalamic Tract Neurons, Morphology

Spinal Multipolar Neuron

Definition

Spinal lamina I neuron with an ovoid cell body and dendritic tree oriented rostrocaudally and ventrally, highly ramified near the cell body and covered by numerous multishaped, often long pedicled spines.

Cross-References

▶ Spinothalamic Tract Neurons, Morphology

Spinal Nerve Block

▶ Pain Treatment: Spinal Nerve Blocks

Spinal Nerve Ligation Model

Synonyms

Chung model; SNL

Definition

Spinal nerve ligation, also known as the "Chung" model, is currently one of the most frequently used models for nerve injury and neuropathic pain behavior in animals (rats/mice; Kim and Chung, 1992). The lumbar 5 spinal nerve is ligated with silk suture just distal to the L5 DRG. Usually, the spinal nerve is then cut just distal to the ligation. In some cases, the lumbar 6 spinal nerve is also ligated and cut. In non-injured animals, the distal axons of L4 and L5 neurons comingle in the sciatic nerve and innervate overlapping territories in the skin. After L5 spinal nerve ligation, the distal axons of the L5 nerve degenerate within the sciatic nerve. It is thought that inflammatory and immune processes involved in the degeneration of the distal L5 fibers then affect the adjacent "uninjured" L4 nerve fibers, causing functional changes in the L4 neurons. The advantage of this model, compared to other nerve injury models, is that all L5 DRG neurons are directly injured, whereas the adjacent L4 DRG neurons remain intact. This allows anatomical and functional analyses to be performed on injured and adjacent DRG neurons separately.

Cross-References

▶ Immunocytochemistry of Nociceptors
▶ Neuropathic Pain Model, Partial Sciatic Nerve Ligation Model
▶ Neuropathic Pain Model, Spinal Nerve Ligation Model
▶ Peptides in Neuropathic Pain States
▶ Purine Receptor Targets in the Treatment of Neuropathic Pain

Spinal Neurons Projecting to the Thalamus

▶ Spinothalamic Tract Neurons, Morphology

Spinal Nociceptive Processing

▶ Spinal Cord Nociception, Neurotrophins

Spinal Nociceptive Tail Flick Withdrawal Reflex

▶ Opioids and Reflexes

Spinal Nociceptive Transmission

▶ Calcium Channels in the Spinal Processing of Nociceptive Input

Spinal Pyramidal Neuron

Definition

Spinal lamina I neuron with a triangular prism-shaped cell body partly encased in the overlying white matter by its dorsal portion and a dendritic tree occupying a similarly shaped domain with branches oriented rostrocaudally, mediolaterally, and dorsally.

Cross-References

▶ Spinothalamic Tract Neurons, Morphology

Spinal Radicular Arteries

Definition

Small arterial branches that course through the intervertebral foramina adjacent to the exiting spinal nerve roots. In the cervical region, the radicular arteries arise from the vertebral artery and supply the spinal cord via the anterior and posterior spinal arteries.

Cross-References

▶ Cervical Transforaminal Injection of Steroids

Spinal Reflex

Definition

A reflex with the reflex center in the spinal cord. The nociceptive tail-flick reflex is a spinal reflex that is present after a transection of the spinal cord above the lumbar level. However, it has been suggested that for longer tail-flick latencies, supraspinal structures may be involved.

Cross-References

▶ Central Pain: Outcome Measures in Preclinical Research
▶ Tail-Flick Test

Spinal Root Disease

Definition

A radiculopathy caused by compression, inflammation, ischemia, or tumor.

Cross-References

▶ Radiculopathies

Spinal Segment

Definition

Morphological division of the spinal cord based on the dorsal root ganglia and ventral roots. The cervical, thoracic, lumbar, and sacral cord have 8, 12, 5, and 5 pairs of roots, respectively. They are arranged rostro-caudally, and their length or width is correlated with the nerve function innervating the peripheral tissue.

Cross-References

▶ Morphology, Intraspinal Organization of Visceral Afferents

Spinal Sensitization

Definition

Central sensitization refers to a state of increased synaptic efficacy, established in spinal dorsal

horn (DH) neurons following intense peripheral noxious stimulation, a tissue injury, inflammation, or nerve damage. ▶ Central sensitization is associated with spontaneous DH neuron activity, the recruitment of responses from neurons that normally only respond to low-intensity stimuli (i.e., altered neural connections), and expansion of DH neuron receptive fields.

Cross-References

▶ Nick Model of Cutaneous Pain and Hyperalgesia
▶ Peptides in Neuropathic Pain States

Spinal Stenosis

Definition

Narrowing of the anterior-posterior dimension of the spinal canal.

Cross-References

▶ Chronic Back Pain and Spinal Instability
▶ Sacroiliac Joint Pain

Spinal Subarachnoid Space

Definition

The space surrounding the spinal cord and containing the cerebrospinal fluid.

Cross-References

▶ Cancer Pain Management, Anesthesiologic Interventions, and Neural Blockade

Spinal Tract of the Trigeminal Nerve

Definition

The central processes of trigeminal ganglion cells that descend in the lateral brainstem to terminate in the caudal spinal trigeminal nucleus. The spinal tract of the trigeminal nerve gives off collaterals to provide input for trigeminal reflexes and convey nociceptive information from the head and face to the subnucleus caudalis.

Cross-References

▶ Trigeminothalamic Tract Projections

Spinal Transection

Definition

Spinal transection is a surgical procedure in which the spinal cord is severed.

Cross-References

▶ Opioids and Reflexes

Spinoannular

Definition

The projection from the spinal cord to the periaqueductal gray (PAG) is known as the spinoannular pathway.

Cross-References

▶ Spinomesencephalic Tract

Spinohypothalamic Neurons

Definition

Neurons with cell bodies in the spinal cord and axonal projections with sites of termination in the hypothalamus.

Cross-References

▶ Spinal Ascending Pathways, Colon, Urinary Bladder, and Uterus

---

Spinohypothalamic Tract, Anatomical Organization, and Response Properties

Xijing Zhang
Department of Neuroscience, University of Minnesota, Minneapolis, MN, USA

Definition

Neuron whose cell body and dendrites are located in the spinal cord, and whose axon projects to or through the hypothalamus, decussates in the posterior optic chiasm, turns posteriorly, and descends in the other side of the brain. The axon or collateral branches of the spinohypothalamic neuron terminate in both sides of the hypothalamus, thalamus, superior colliculus, and reticular formations in the midbrain, pons, and medulla (Fig. 1). The axons of spinohypothalamic neurons form a nociceptive tract that is responsible for conveying information to the brain that mainly contributes to the affective component of pain, including various reflexes, endocrine adjustments, and emotional changes caused by pain.

Characteristics

Spinohypothalamic neurons are a type of sensory projection neuron. The axons of

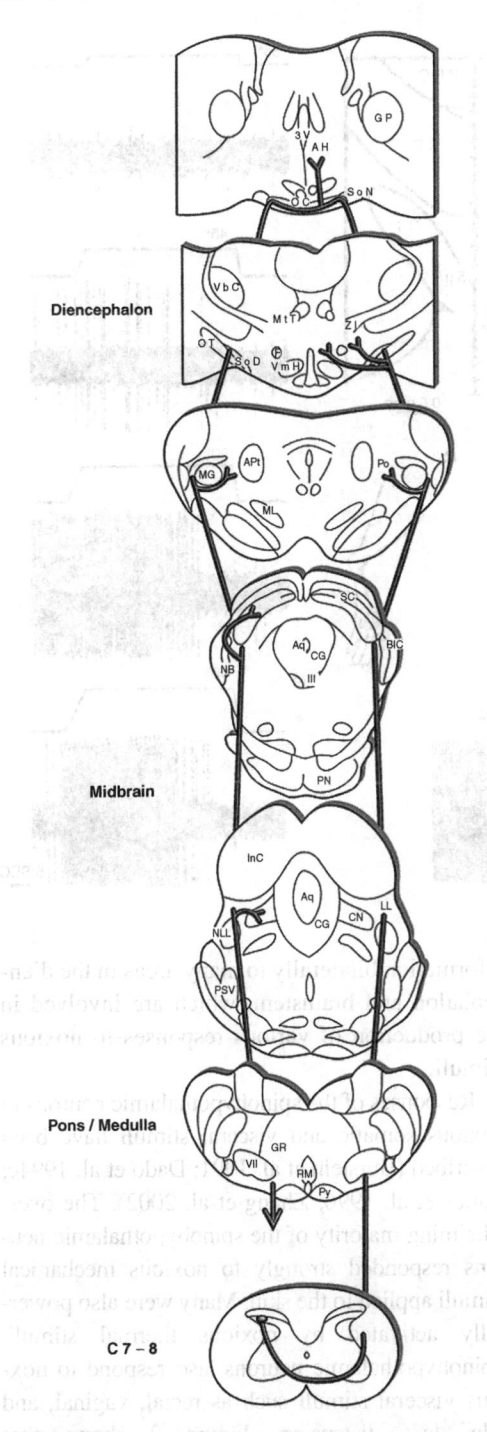

Spinohypothalamic Tract, Anatomical Organization, and Response Properties, Fig. 1 Schematic drawing of the spinohypothalamic neuron with its axon projecting trajectory in the brain. From Zhang et al. (1995). Abbreviations: *3V* third ventricle; *AH* anterior hypothalamus area; *Apt* anterior pretectal nucleus; *Aq* aqueduct; *BIC*

spinohypothalamic neurons form the spinohypothalamic tract, which is one of several sensory pathways that transmit nociceptive signals from the spinal cord to the brain (Giesler et al. 1994). Locations of the spinohypothalamic neurons in the spinal cord have been determined using ▸ retrograde labeling (Burstein et al. 1990) and ▸ antidromic activation techniques (Burstein et al. 1991; Zhang et al. 1995). Spinohypothalamic neurons are found throughout the length of the spinal cord, and are concentrated in three main regions: the marginal zone; the lateral reticulated area; and the area surrounding the central canal. These three areas receive direct input from primary afferent ▸ nociceptors, contain many nociceptive neurons, and contribute large numbers of neurons to other ascending nociceptive tracts. It was estimated that the total number of spinothalamic and spinohypothalamic neurons is similar in rats (Burstein et al. 1990a, b), suggesting that the spinohypothalamic tract is large, and it may also make important contributions to nociceptive processing.

The trajectory and termination of the axons of the spinohypothalamic neurons have been examined using antidromic activation (Zhang et al. 1995; Dado et al. 1994a; Kostarczyk et al. 1997) and ▸ anterograde labeling techniques (Cliffer et al. 1991; Newman et al. 1996). The axons cross the midline and ascend in the lateral funiculus on the side contralateral to the cell body (Dado et al. 1994b). In the brainstem, the axons ascend through the lateral reticular formation to the level of the contralateral thalamus (Kostarczyk et al. 1997). Within the thalamus,

brachium inferior colliculus; *CG* central gray; *CN* cuneiform nucleus; *F* fornix; *GP* globus pallidus; *GR* gigantocellular reticular nucleus; *III* oculomotor nucleus; *InC* inferior colliculus; *LL* lateral lemniscus; *MG* medial geniculate nucleus; *ML* medial lemniscus; *MtT* mammillothalamic tract; *NB* nucleus brachium inferior colliculus; *NLL* nucleus of the lateral lemniscus; *OC* optic chiasm; *OT* optic tract; *PN* pontine nucleus; *Po* posterior thalamic nucleus; *PSV* principal sensory nucleus of the trigeminal nerve; *Py* pyramidal tract; *RM* nucleus raphe magnus; *SC* superior colliculus; *SoD* supraoptic decussation; *SoN* supraoptic hypothalamic nucleus; *VII* facial nucleus; *VbC* ventrobasal complex; *VmH* ventromedial hypothalamic nucleus; *ZI* zona incerta

Spinohypothalamic Tract, Anatomical Organization, and Response Properties, Fig. 2 Characterization of a spinohypothalamic neuron. (**a**) Locations of tracks of a stimulating electrode in the hypothalamus, showing thresholds for antidromic activation and a low-threshold point in the supraoptic decussation (*circled*). (**b**) Drawing of lesion marking the recording location. (**c**) Receptive field. (**d**) Oscillographic tracing of responses to brush (*BR*), pressure (*PR*), pinch (*PN*), and squeeze (*SQ*) of the receptive field. (**e**) Responses to noxious heat. Temperature tracings are presented above responses. From Burstein et al. (1991). Abbreviations: *F* fornix; *IC* internal capsule; *MTT* mammillothalamic tract; *OT* optic tract; *VBC* ventrobasal complex; *ZI* zona incerta

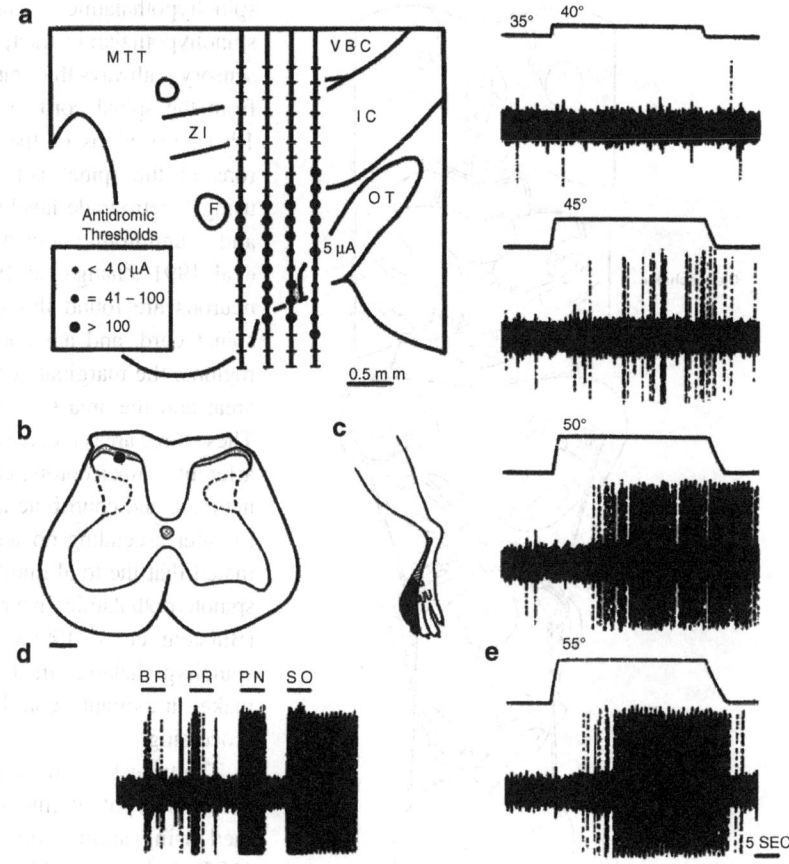

spinohypothalamic axons ascend within the supraoptic decussation, a small area of white matter medially adjacent to the optic tract, to the rostral ventral hypothalamus (Cliffer et al. 1991). In the rostral ventral hypothalamus, the axons cross the midline in the posterior optic chiasm, enter the ipsilateral hypothalamus, turn caudally, and descend along the identical path in which they ascended in the contralateral brain (Zhang et al. 1995). Axons descend and terminate in the ipsilateral hypothalamus, thalamus, midbrain, pons, and some axons project as far as caudally the rostral medulla. The axons of the spinohypothalamic neurons and their collateral branches terminate bilaterally in many nuclei within hypothalamus, thalamus, superior colliculus, and reticular formation of the brainstem. Through this long and complex projection, the axons of the spinohypothalamic neurons are capable of carrying nociceptive

information bilaterally to many areas in the diencephalon and brainstem, which are involved in the production of various responses to noxious stimuli.

Responses of the spinohypothalamic neurons to noxious somatic and visceral stimuli have been described (Burstein et al. 1991; Dado et al. 1994c; Katter et al. 1996; Zhang et al. 2002). The overwhelming majority of the spinohypothalamic neurons responded strongly to noxious mechanical stimuli applied to the skin. Many were also powerfully activated by noxious thermal stimuli. Spinohypothalamic neurons also respond to noxious visceral stimuli such as rectal, vaginal, and bile duct distension. Figure 2 shows that a spinohypothalamic neuron responded to noxious mechanical and thermal stimuli applied to the skin. This spinohypothalamic neuron was located in the marginal zone of the lumbar spinal cord (Fig. 2b), with its axon projected to the contralateral hypothalamus (Fig. 2a). The neuron had a small cutaneous receptive field located in the ipsilateral hindpaw (Fig. 2c). The neuron responded to innocuous brushing of the skin, but responded much more vigorously to noxious stimuli applied to the skin (Fig. 2d). The neuron also responded with increasing intensity to a series of ascending heat stimuli applied to the receptive field (Fig. 2e).

Spinohypothalamic neurons are likely to contribute mainly to ▶ affective responses, although they may also contribute to ▶ sensory discrimination through collateral branches within the thalamus. Through the terminations and collateral branches in the hypothalamus, spinohypothalamic neurons carry information to the hypothalamus that is likely to be involved in pain-related endocrine adjustments or emotional reactions, while the terminations and collateral branches in the brainstem reticular formation and superior colliculus may be involved in pain-related somatic and visceral reflexes. Thus, spinohypothalamic neurons, with their long and complicated axonal projections on both sides of the brain, may directly affect the activity in many areas in the brain that are known to be involved in the production of autonomic, neuroendocrine, and affective responses to noxious stimuli.

References

Burstein, R., Cliffer, K. D., & Giesler, G. J., Jr. (1990a). Cells of origin of the spinohypothalamic tract in the rat. *The Journal of Comparative Neurology, 291*, 329–344.

Burstein, R., Dado, R. J., & Giesler, G. J., Jr. (1990b). The cells of origin of the spinothalamic tract of the rat: A quantitative re-examination. *Brain Research, 511*, 329–337.

Burstein, R., Dado, R. J., Cliffer, K. D., & Giesler, G. J., Jr. (1991). Physiological characterization of spinohypothalamic tract neurons in the lumbar enlargement of rats. *Journal of Neurophysiology, 66*, 261–284.

Cliffer, K. D., Burstein, R., & Giesler, G. J., Jr. (1991). Distributions of spinothalamic, spinohypothalamic and spinotelencephalic fibers revealed by anterograde transport of PHA-L in rats. *Journal of Neuroscience, 11*, 852–868.

Dado, R. J., Katter, J. T., & Giesler, G. J., Jr. (1994a). Spinothalamic and spinohypothalamic tract neurons in the cervical enlargement of rats. I. Locations of antidromically identified axons in the thalamus and hypothalamus. *Journal of Neurophysiology, 71*, 959–980.

Dado, R. J., Katter, J. T., & Giesler, G. J., Jr. (1994b). Spinothalamic and spinohypothalamic tract neurons in the cervical enlargement of rats. III. Locations of antidromically identified axons in the cervical cord white matter. *Journal of Neurophysiology, 71*, 1003–1021.

Dado, R. J., Katter, J. T., & Giesler, G. J., Jr. (1994c). Spinothalamic and spinohypothalamic tract neurons in the cervical enlargement of rats. II. Responses to innocuous and noxious mechanical and thermal stimuli. *Journal of Neurophysiology, 71*, 981–1002.

Giesler, G. J., Jr., Katter, J. T., & Dado, R. J. (1994). Direct spinal pathways to the limbic system for nociceptive information. *Trends in Neurosciences, 17*(6), 244–250.

Katter, J. T., Dado, R. J., Kostarczyk, E., & Giesler, G. J., Jr. (1996). Spinothalamic and spinohypothalamic tract neurons in the sacral spinal cord of rats. II. Responses to cutaneous and visceral stimuli. *Journal of Neurophysiology, 75*, 2606–2628.

Kostarczyk, E., Zhang, X., & Giesler, G. J., Jr. (1997). Spinohypothalamic tract neurons in the cervical enlargement of rats: Locations of antidromically identified ascending axons and their collateral branches in the contralateral brain. *Journal of Neurophysiology, 77*, 435–451.

Newman, H. M., Stevens, R. T., & Apkarian, A. V. (1996). Spinal-suprathalamic projections from the upper cervical and the cervical enlargement in rat and squirrel monkey. *The Journal of Comparative Neurology, 365*, 640–658.

Zhang, X., Kostarczyk, E., & Giesler, G. J., Jr. (1995). Spinohypothalamic tract neurons in the cervical enlargement of rats: Descending axons in the ipsilateral brain. *Journal of Neuroscience, 15*, 8393–8407.

Zhang, X., Gokin, A. P., & Giesler, G. J., Jr. (2002). Responses of spinohypothalamic tract neurons in the thoracic spinal cord of rats to somatic stimuli and to graded distention of the bile duct. *Somatosensory & Motor Research, 19*(1), 5–17.

Spinomedullary Neurons

Definition

Neurons with cell bodies in the spinal cord and axonal projections with sites of termination in the medulla.

Cross-References

▶ Spinal Ascending Pathways, Colon, Urinary Bladder, and Uterus

Spinomesencephalic Neurons

Definition

Neurons with cell bodies in the spinal cord and axonal projections with sites of termination in the midbrain.

Cross-References

▶ Spinal Ascending Pathways, Colon, Urinary Bladder, and Uterus

Spinomesencephalic Tract

Robert P. Yezierski
Comprehensive Center for Pain Research and Department of Neuroscience, The McKnight Brain Institute University of Florida, Gainesville, FL, USA

Synonyms

Spinomesencephalic tract (SMT)

Definition

The spinomesencephalic tract (SMT) with its varied origins, spinal trajectories, and sites of termination is often described as having a prominent role in nociception. This ascending projection is a collection of pathways originating throughout all levels of the spinal cord and terminating in different targets throughout the midbrain. Collateral projections of ascending axons to multiple brainstem targets are not uncommon. The largest percentage of SMT cells is located in the upper cervical spinal cord. The heterogenous population of cells in this region of the cord is involved in the integration of varied inputs from large portions of the body,

including input from muscles, joints, cornea, dura, fore limbs, hind limbs, tail, and/or testicles.

Three prominent pathways represent the spinomesencephalic projection system, including the spinotectal, spinoannular, and spinoparabrachial tracts that project, respectively, to the superior colliculus, periaqueductal, and parabrachial regions. Collateral projections of SMT axons to thalamic, hyopthalamic, and medullary sites make this a pathway with the potential to be involved in a wide variety of functions.

Due to the functional properties of midbrain termination sites it is likely that the SMT is involved in mechanisms of analgesia, fear, anxiety, respiratory and cardiovascular regulation, motor activities, vocalization, mating behavior, and micturition. Cells belonging to this spinal pathway have been shown to express the neurokinin 1 receptor and contain the excitatory amino acid glutamate as well as receive serotonergic input from brainstem locations.

Characteristics

Although there are some variations among species, the majority of cells belonging to the spinomesencephalic tract (SMT) are found in laminae I, III-V, VII, and X at all levels of the cord (Mantyh 1982; Menetrey et al. 1982; Wiberg and Blomqvist 1984). The largest percentage of cells are located in the first four cervical segments. Projections from the cervical enlargement to the parabrachial (PB) area and the periaqueductal gray (PAG) demonstrates a distinct area that projects in different ways to the PB area and PAG (Bernard et al. 1995). Many cells belonging to the SMT have collateral, including bilateral and/or descending, projections to the parabrachial region, thalamus (including intralaminar nuclei), and medullary reticular formation with some having both ascending as well as descending projections (Al-Khater and Todd 2009; Yezierski and Mendez 1991; Pechura and Liu 1986; Liu 1986; Zhang et al. 1990). In the primate, neurons classified as SMT-spinothalamic tract (STT) are located primarily in laminae I, V, VII, and X. In lamina I, fusiform cells project mainly to the contralateral parabrachial nuclei and appear to convey spinal input reflecting visceral input. Pyramidal cells project to the contralateral caudal ventrolateral PAG and represent the ascending branch of the SMT which controls nociceptive transmission in the dorsal horn (Lima and Coimbra 1989).

The number of SMT cells in the monkey was estimated to be approximately 10,000 with 75 % having projections to the contralateral midbrain (Zhang et al. 1990). Cells projecting to both the ventrobasal thalamus and PAG are observed in the cervical and lumbar cord with the greatest number found in the cervical enlargement (Harmann et al. 1988). Chemically, a subpopulation of SMT cells stain for the enzyme glutaminase; thus, the excitatory amino acid glutamate may be a transmitter in this pathway (Yezierski et al. 1993). ▶ Lamina I SMT cells projecting to the parabrachial region stain positive for dynorphin or enkephalin, implicating these peptides as transmitter candidates in this component of the SMT. ▶ Substance P (sub P) receptors are located on this same population of SMT neurons (Todd et al. 2000). The presence of ▶ serotonin (5-HT) immunoreactive contacts suggests that these cells are influenced by descending pathways from the medulla (Hylden et al. 1986a). The spinal trajectory of SMT axons includes the ventral lateral funiculus, although there is evidence that the lamina I component of this pathway ascends in the dorsal part of the lateral funiculus (Hylden et al. 1986b; Swett et al. 1985).

The primary projection targets of the SMT include three prominent regions of the mesencephalon: (1) the caudal midbrain, including the nucleus cuneiformis, parabrachial nucleus, and ▶ periaqueductal gray (PAG); (2) the intercollicular region, including the nucleus cuneiformis, PAG, intercollicular nucleus, and deep layers of the superior colliculus; and (3) the rostral midbrain, including the PAG, the nucleus of Darkschewitsch, and the anterior and posterior pretectal nuclei. The majority of ascending projections terminate contralateral to the site of origin (Wiberg and Blomqvist 1984; Yezierski 1988; Bernard et al. 1995). The

proportion of lamina I neurons projecting to PAG is higher in cervical than lumbar enlargement (Polgar et al. 2010); however, the lamina I projection to PAG has been shown to constitute a limited part of the total projection from the spinal cord to the PAG (Mouton et al. 2001). It is interesting to point out that in the cat there are four times as many lamina I neurons projecting to the parabrachial nuclei and twice as many to the periaqueductal gray as projecting to thalamus (Klop et al. 2005a). Projections to the PAG are topographically organized with distinct subgroups of spinal neurons projecting to specific subregions (Keay et al. 1997). There are also large segmental differences in the number of PAG projecting neurons throughout the length of the spinal cord and different parts of the spinal cord project to specific areas of the PAG (Mouton et al. 1997). The strongest spinal cord projection to the PAG originates from the upper three cervical segments with cells in these segments representing about 30 % of all neurons in the spinal cord projecting to the PAG (Mouton et al. 2005). Cells in the lateral ventral horn of the upper cervical cord not only target the caudal PAG, but terminate abundantly in the intermediate and rostral parts of the PAG (Mouton et al. 2009). It has also been suggested that there exists five separate clusters of spino-PAG neurons with specific segmental and laminar organizations (Mouton and Holstege 2000). Projections from SMT cells in the upper cervical cord are believed to represent the output from a unique integration center for pain arising from deep structures throughout the body (Clement et al. 2000).

As described above, one of the principal targets of the spinomesencephalic projection is the PAG. Originating from cells in all laminae of the gray matter and at all levels of the spinal cord, the spino-PAG component of the SMT is well suited to relay afferent information throughout the rostrocaudal extent of the midbrain. Although the precise functional significance of this input is unknown, the PAG is involved in the control of blood pressure, defense reactions, respiration, vocalization, reproductive behavior, stomach and bladder motility, grooming, and analgesic mechanisms (Depaulis and Bandler 1991; Rossi et al. 1994).

Neurons in the lateral sacral cord project to the PAG suggesting that afferent information regarding micturition and sexual behavior is relayed to the PAG (Klop et al. 2005b). It should also be mentioned that neurons in the lateral two-thirds of the lateral cervical nucleus send fibers to the lateral part of the PAG suggesting the LCN-PAG pathway is involved in alerting animals about the presence of cutaneous stimuli that might represent danger (Mouton et al. 2004).

The heterogeneity of the SMT cell population is exemplified by the responses and receptive field (RF) organization of cells belonging to this pathway. Cells with complex RFs that include the hind limbs, forelimbs, and/or oral-facial structures have been described (Menetrey et al. 1980; Yezierski 1990; Yezierski and Broton 1991; Yezierski and Schwartz 1986). SMT cells in the upper cervical cord are a heterogeneous mix and often have collateral projections to midbrain and thalamus and complex receptive fields involving muscle, cornea, dura, testicles, and skin. In the primate cells projecting to midbrain have complex fields, while those projecting to thalamus and midbrain have restricted excitatory RFs. SMT cells, regardless of segmental location, receive input from large portions of the body. Because of this, it is unlikely that these cells are important in stimulus localization. Since the complex RFs of SMT cells are due, in part, to descending influences (Yezierski 1990), these cells may constitute a component of the SMT involved in a positive feedback loop responsible for autonomic and motor reflexes in the spinal cord. Alternatively, these cells may provide input compatible with known arousal functions of the reticular activating system. Consistent with this are reactions commonly associated with the delivery of noxious stimuli, including changes in motor behavior, respiration and alterations in cardiovascular function, changes in blood pressure, heart rate, and resistance in peripheral vascular beds (Bandler et al. 1991). Lamina 1 cells belonging to the parabrachial projection largely reflect the functional properties of parabrachial neurons suggesting an involvement in the autonomic and emotional/aversive aspects of pain. The spinoparabrachial component of the SMT, which further relays input to the amygdala, hypothalamus, and bed nucleus of the stria terminalis provide afferent input to a major integration site for the autonomic and emotional/aversive aspects of pain.

The response properties of SMT cells have been studied in rat, cat, and monkey (Hylden et al. 1986a, b; Menetrey et al. 1980; Yezierski and Broton 1991; Yezierski and Schwartz 1986; Yezierski et al. 1987; Dougherty et al. 1999). Functional properties include a range of complex and restricted excitatory RFs, often including complex inhibitory components. Consistent with the SMT projection having a prominent role in nociception, cells belonging to this pathway respond to noxious mechanical and thermal stimuli, including inputs from cutaneous and/or deep structures (e.g., joints, muscles, and viscera) (Hylden et al. 1986; Menetrey et al. 1980; Yezierski and Schwartz 1986; Yezierski and Broton 1991; Yezierski et al. 1987; Yezierski 1990). Involvement in chronic pain mechanisms is suggested by the observation that some neurons have the potential for prolonged afterdischarges. These observations, as well as the varied functions associated with SMT projection targets, support a role of the SMT in sensory, motor, and visceral functions. The varied conduction velocities of axons belonging to the SMT reflect the fact that SMT cells vary in size. Cells located in the superficial laminae routinely have slower axons compared with cells in deeper laminae.

Based on responses to mechanical stimuli and input from deep structures (muscles and joints), the response profiles of SMT cells in the lumbosacral and upper cervical spinal cord can be divided into three major categories: (a) wide dynamic range (WDR); (b) high threshold (HT); and (c) deep/tap (D). Responses to peripheral stimuli include both excitatory and inhibitory effects (McMahon and Wall 1985; Yezierski 1990; Yezierski and Broton 1991; Yezierski and Schwartz 1986; Yezierski et al. 1987). Although the functional properties of SMT cells in different laminae have been most frequently studied in the cat (Hylden et al. 1986; Yezierski 1990; Yezierski and Schwartz 1986; Yezierski and Broton 1991), similar cell types are described in the rat (Menetrey et al. 1980) and monkey (Yezierski et al. 1987; Dougherty et al. 1999).

SMT neurons in the monkey projecting to the PAG may evoke components of the systemic response to neurogenic hyperalgesia, as evidenced by responses to intradermal capsaicin. The functional heterogeneity of SMT cells is exemplified by the three patterns of response of primate SMT cells to intradermal capsaicin, which appear to be dependent on the functional regions of the PAG to which they project (Dougherty et al. 1999). The lamina I component of the SMT projecting to the parabrachial region is made up of high-threshold neurons believed to be part of a pathway that contributes to the ▶ motivational-affective aspects of pain. Based on anatomical labeling with Fos protein, it was found that pain arising from visceral versus deep somatic regions influences activity in the vlPAG via distinct spinal pathways (Clement et al. 2000).

Although the responses of SMT cells have been most commonly evaluated with mechanical and thermal stimuli, chemical agents including intramuscular injection of hypertonic saline and intravenous serotonin (i.v. 5-HT) and responses to intradermal capsaicin have also been described (Dougherty et al. 1999; Yezierski and Broton 1991). SMT cells have been proposed to contribute to the spinal mechanism responsible for the development of pain associated with spinal injury. Following injury, cells belonging to this pathway have increased spontaneous activity, expanded RFs, increased responses to mechanical and thermal stimuli, and long afterdischarge responses (Yezierski and Park 1993). These changes in functional state of SMT cells correlate with increased behavioral responses to mechanical and thermal stimuli following injury. The fact that cellular injury is observed throughout the midbrain following peripheral nerve injury suggests that the behavioral disabilities associated with these injuries are related to pathological changes in the spinal projection to different midbrain targets (Mor et al. 2011).

In order to gain a thorough appreciation of the potential functional importance of the SMT, one must acknowledge the varied functions associated with different regions of the midbrain. For example, if one considers the autonomic and somatomotor responses elicited by noxious stimuli and couples this with the control of blood pressure, respiration, and vocalization and defense reactions associated with midbrain regions overlapping the SMT projection, what emerges is a functional relationship between SMT input to the midbrain and behavioral responses to noxious stimuli. Similarly, the relationship between responses of SMT cells to stimulation of the genitals and perineal region and the role of PAG in reproductive behaviors suggests a functional role of the SMT in this behavior. The fact that there exists a specific, monosysnaptic pathway from the lumbosacral spinal cord to estrogen receptor expressing neurons in the PAG suggests that the SMT may represent an important component of afferent systems contributing input to neural substrates directly or indirectly involved in carrying out these functions (Vanderhorst et al. 2002). Ascending projections from PAG also provide insight into the potential functional significance of input to this region. Ascending projections from the PAG terminate in the midline and intralaminar thalamic nuclei, hypothalamus, ventral tegmental area, substantia nigra pars compacta, retrorubral field, preoptic area, diagonal band, lateral division of the bed nucleus (Krout and Loewy 2000; Cameron et al. 1995a). Descending projections include parabrachial nucleus, locus coeruleus, subcoeruleus, gigantocellular nucleus pars alpha, nucleus raphe magnus, pontine reticular formation, pontine and medullary tegmentum, lateral paragigantocellular nucleus, and the rostroventrolateral reticular nucleus (Cameron et al. 1995b; Krout et al. 1998). The fact that the PAG receives input from circumscribed regions of the hypothalamus suggests that the behavioral significance of pain signals is represented in multiple subcortical and brainstem regions (Lumb 2002).

The functional significance of the SMT would not be complete without consideration of the descending control system. To this end, the spinomesencephalic projection overlaps with sites producing analgesia and the inhibition of nociceptive spinal neurons (Hammond 1986; Gerhart et al. 1984). The distribution of midbrain neurons projecting to medullary raphespinal and reticulospinal neurons also overlaps with the terminal domain of the SMT. Consistent with a feedback loop between SMT projection targets and descending control pathways are reports that stimulation at midbrain sites used to antidromically activate SMT cells also inhibits and/or facilitates evoked responses of these cells (Yezierski and Schwartz 1986; Yezierski 1990). Given the distribution of PAG sites producing analgesia and aversive reactions, it is not surprising that excitatory and mixed effects can be obtained with PAG stimulation. Descending excitatory, inhibitory, or mixed effects may be exerted on cells at different levels of the cord for reflex, e.g., autonomic and motor control, or for purposes of stimulus localization. Midbrain control of spinal nociception discriminates between responses evoked by myelinated and unmyelinated heat nociceptors perhaps as part of an active coping strategy in emergency situations (McMullan and Lumb 2006).

The fact that the SMT is involved in the relay of somatosensory input to a region with complex functional significance supports the notion that this pathway is an integral part of a supraspinal control center. Although the emphasis of effects produced by PAG stimulation has been on analgesia and inhibition, it is important to point out that PAG stimulation effects are not limited to inhibition and probably have a functional significance much broader than the analgesic effects that have been described. For example, the excitatory effects of PAG stimulation may facilitate segmental reflex activity. It is also possible that there are different functional compartments within the PAG that exert varied effects on different populations of tract cells and interneurons in different spinal laminae.

Conclusions

The results of anatomical investigations (anterograde and retrograde studies) have provided definitive proof for the existence of a spinal projection to different regions of the midbrain. The spinal distribution of cells responsible for this projection and the response properties of

SMT cells are all consistent with an involvement in conveying information related to the modalities of touch, pressure, pain, and temperature to supraspinal targets. Although the functional relevance of these studies provides the basis for the clinical significance of the SMT projection, any assessment of these functions cannot be made without acknowledging the functional organization of the midbrain. To this end, discussions related to cardiovascular and reproductive function, respiration, vocalization, bladder control, along with analgesic mechanisms must be considered. Somatosensory input to different regions of the midbrain may not alone trigger any of these functions, but in concert with other afferent systems it may be involved in producing a repertoire of behaviorally appropriate responses to different stimulus conditions.

References

Al-Khater, K. M., & Todd, A. J. (2009). Collateral projections of neurons in laminae I, III, and IV of rat spinal cord to thalamus, periaqueductal gray matter, and lateral parabrachial area. *The Journal of Comparative Neurology, 515*, 629–646.

Bandler, R., Carrlve, P., & Zhang, S. P. (1991). Interaction of somatic and autonomic reactions within the midbrain periaqueductal grey: Viscerotopic, somatotopic and functional organization. *Progress in Brain Research, 87*, 269–305.

Bernard, J. F., Dallel, R., Raboisson, P., Villanueva, L., & Le Bars, D. (1995). Organization of the efferent projections from the spinal cervical enlargement to the parabrachial area and periaqueductal gray: A PHA-L study in the rat. *The Journal of Comparative Neurology, 353*, 480–505.

Cameron, A. A., Khan, I. A., Westlund, K. N., Cliffer, K. D., & Willis, W. D. (1995a). The efferent projections of the pariaqueductal gray in the rat: A *Phaseolus vulgaris*-leucoagglutinin study. I. Ascending projections. *The Journal of Comparative Neurology, 351*, 568–584.

Cameron, A. A., Khan, I. A., Westlund, K. N., & Willis, W. D. (1995b). The efferent projections of the pariaqueductal gray in the rat: A *Phaseolus vulgaris*-leucoagglutinin study. II. Descending projections. *The Journal of Comparative Neurology, 351*, 585–601.

Clement, C. L., Keay, K. A., Podzebenko, K., Gordon, B. D., & Bandler, R. (2000). Spinal sources of noxious visceral and noxious deep somatic afferent drive onto the ventrolateral periaqueductal gray of the rat. *The Journal of Comparative Neurology, 425*, 323–344.

Depaulis, A., & Bandler, R. (1991). *The midbrain periaqueductal gray matter: Functional, anatomical and neurochemical organization*. NATO ASI series (Vol. 213). Plenum Press, New York.

Dougherty, P. M., Schwartz, A., & Lenz, F. A. (1999). Responses of primate spinomesencephalic tract cells to intradermal capsaicin. *Neuroscience, 90*, 1377–1392.

Gerhart, K. D., Yezierski, R. P., Wilcox, T. K., et al. (1984). Inhibition of primate spinothalamic tract neurons by stimulation in periaqueductal gray or adjacent midbrain reticular formation. *Journal of Neurophysiology, 51*, 450–466.

Hammond, D. L. (1986). Control systems for nociceptive afferent processing: The descending inhibitory pathways. In T. Yaksh (Ed.), *Spinal afferent processing* (pp. 363–390). New York: Plenum.

Harmann, P. A., Carlton, S. M., & Willis, W. D. (1988). Collaterals of spinothalamic tract cells to the periaqueductal gray: A fluorescent double-labeling study in the rat. *Brain Research, 441*, 87–97.

Hylden, J., Hayashi, H., Ruda, M. A., & Dubner, R. (1986a). Serotonin innervation of physiologically identified lamina I projection neurons. *Brain Research, 370*, 401–404.

Hylden, J., Hayashi, H., & Bennett, G. (1986b). Physiology and morphology of the lamina I spinomesencephalic projection. *The Journal of Comparative Neurology, 247*, 505–515.

Keay, K. A., Feil, K., Gordon, B. D., Herbert, H., & Bandler, R. (1997). Spinal afferents to functionally distinct periaqueductal gray columns in the rat: An anterograde and retrograde tracing study. *The Journal of Comparative Neurology, 385*, 207–229.

Klop, E. M., Mouton, L. J., Hulsebosch, R., & Holstege, G. (2005a). In cat four times as many lamina I neurons project to the parabrachial nuclei and twice as many to the periaqueductal gray as to the thalamus. *Neuroscience, 134*, 189–197.

Klop, E. M., Mouton, L. J., Kuipers, R., & Lolstege, G. (2005b). Neurons in the lateral sacral cord of the cat project to periaqueductal grey, but not to thalamus. *European Journal of Neuroscience, 21*, 2159–2166.

Krout, K. E., Jansen, A. S., & Loewy, A. D. (1998). Periaqueductal gray matter projection to the parabrachial nucleus in rat. *The Journal of Comparative Neurology, 401*, 337–454.

Krout, K. E., & Loewy, A. D. (2000). Periaqueductal gray matter projections to midline and intralaminar thalamic nuclei of the rat. *The Journal of Comparative Neurology, 424*, 111–141.

Lima, D., & Coimbra, A. (1989). Morphological types of spinomesencephalic neurons in the marginal zone (lamina I) of the rat spinal cord, as shown after retrograde labelling with cholera toxin subunit B. *The Journal of Comparative Neurology, 279*, 327–339.

Liu, R. P. (1986). Spinal neuronal collateral to the intralaminar thalamic nuclei and periaqueductal gray. *Brain Research, 365*, 145–150.

Lumb, B. M. (2002). Inescapable and escapable pain is represented in distinct hypothalamic-midbrain circuits: Specific roles for adelta- and C-nociceptors. *Experimental Physiology, 87*, 281–286.

Mantyh, P. W. (1982). The ascending input to the midbrain periaqueductal gray of the primate. *The Journal of Comparative Neurology, 211*, 50–64.

McMahon, S. B., & Wall, P. D. (1985). Electrophysiological mapping of brainstem projections of spinal cord lamina I cells in the rat. *Brain Research, 333*, 19–26.

McMullan, S., & Lumb, B. M. (2006). Midbrain control of spinal nociception discriminates between responses evoked by myelinated and unmyelinated heat nociceptors in the rat. *Pain, 124*, 59–68.

Menetrey, D., Chaouch, A., & Besson, J. M. (1980). Location and properties of dorsal horn neurons at origin of spinoreticular tract in lumbar enlargement of the rat. *Journal of Neurophysiology, 44*, 862–877.

Menetrey, D., Chaouch, A., Binder, D., et al. (1982). The origin of the spinomesencephalic tract in the rat: An anatomical study using the retrograde transport of horseradish peroxidase. *The Journal of Comparative Neurology, 206*, 193–207.

Mor, D., Bembrick, A. L., Austin, P. J., & Keay, K. A. (2011). Evidence for cellular injury in the midbrain of rats following chronic constriction injury of the sciatic nerve. *Journal of Chemical Neuroanatomy, 41*, 158–169.

Mouton, L. J., VanderHorst, V. G., & Holstege, G. (1997). Large segmental differences in the spinal projections to the periaqueductal gray in the cat. *Neuroscience Letters, 238*, 1–4.

Mouton, L. J., & Holstege, G. (2000). Segmental and laminar organization of the spinal neurons projecting to the periaqueductal gray (PAG) in the cat suggests the existence of at least five separate clusters of spino-PAG neurons. *The Journal of Comparative Neurology, 428*, 389–410.

Mouton, L. J., Klop, E., & Holstege, G. (2001). Lamina I-periaqueductal gray (PAG) projections represent only a limited part of the total spinal and caudal medullary input to the PAG in the cat. *Brain Research Bulletin, 54*, 167–174.

Mouton, L. J., Klop, E. M., Broman, J., Zhang, M., & Holstege, G. (2004). Lateral cervical nucleus projections to periaqueductal gray matter in cat. *The Journal of Comparative Neurology, 471*, 434–445.

Mouton, L. J., Klop, E. M., & Holstege, G. (2005). C1-C3 Spinal cord projections to periaqueductal gray and thalamus: A quantitative retrograde tracing study in cat. *Brain Research, 1043*, 87–94.

Mouton, L. J., Eggens-Meijer, E., & Klop, E. M. (2009). The ventrolateral upper cervical cell group in the cat projects to all rostrocaudal levels of the periaqueductal gray matter. *Brain Research, 1300*, 79–96.

Pechura, C. M., & Liu, R. P. (1986). Spinal neurons which project to the periaqueductal gray and the medullary reticular formation via axon collaterals: A double-label fluorescence study in the rat. *Brain Research, 374*, 357–361.

Polgar, E., Wright, L. L., & Todd, A. J. (2010). A quantitative study of brainstem projections from lamina I neurons in the cervical and lumbar enlargement of the rat. *Brain Research, 1308*, 58–67.

Rossi, F., Malone, S., & Berrino, L. (1994). Periaqueductal gray area and cardiovascular function. *Pharmacological Research, 29*, 27–36.

Swett, J. E., McMahon, S. B., & Wall, P. D. (1985). Long ascending projections to the midbrain from cells of lamina I and nucleus of the dorsolateral funiculus of the rat spinal cord. *The Journal of Comparative Neurology, 238*, 401–416.

Todd, A. J., McGill, M. M., & Shehab, S. A. (2000). Neurokinin 1 receptor expression by neurons in laminae I, III and IV of the rat spinal dorsal horn that project to the brainstem. *European Journal of Neuroscience, 12*, 689–700.

Vanderhorst, V. G., Terasawa, E., & Ralston, H. J., III. (2002). Estrogen receptor-alpha immunoreactive neurons in the ventrolateral periaqueductal gray receive monosynaptic input from the lumbosacral cord in the rhesus monkey. *The Journal of Comparative Neurology, 443*, 27–42.

Wiberg, M., & Blomqvist, A. (1984). The spinomesencephalic tract in the cat: Its cells of origin and termination pattern as demonstrated by the intraxonal transport method. *Brain Research, 291*, 1–18.

Willis, W. D., & Coggeshall, R. E. (2004). *Sensory mechanisms of the spinal cord* (Vol. 2). New York: Plenum.

Yezierski, R. P. (1988). The spinomesencephalic tract: Projections from the lumbosacral spinal cord of the rat, cat and monkey. *The Journal of Comparative Neurology, 267*, 131–146.

Yezierski, R. P. (1990). The effects of midbrain and medullary stimulation on spinomesencephalic tract cells in the cat. *Journal of Neurophysiology, 63*, 240–255.

Yezierski, R. P., & Broton, J. G. (1991). Functional properties of spinomesencephalic tract (SMT) cells in the upper cervical spinal cord of the cat. *Pain, 45*, 187–196.

Yezierski, R. P., & Mendez, C. M. (1991). Spinal distribution and collateral projections of rat spinomesencephalic tract cells. *Neuroscience, 44*, 113–130.

Yezierski, R. P., & Park, S. H. (1993). Mechanosensitivity of spinal sensory neurons following intraspinal injections of quisqualic acid. *Neuroscience Letters, 157*, 115–119.

Yezierski, R. P., & Schwartz, R. H. (1986). Response and receptive field properties of spinomesencephalic tract cells in the cat. *Journal of Neurophysiology, 55*, 76–96.

Yezierski, R. P., Sorkin, L. S., & Willis, W. D. (1987). Response properties of spinal neurons projecting to midbrain or midbrain and thalamus in the monkey. *Brain Research, 437*, 165–170.

Yezierski, R. P., Kaneko, T., & Miller, K. E. (1993). Glutaminase-ike immunoreactivity in rat spinomesencephalic tract cells. *Brain Research, 624*, 304–308.

Zhang, D., Carlton, S. M., Sorkin, L. S., et al. (1990). Collaterals of primate spinothalamic tract neurons to the periaqueductal gray. *The Journal of Comparative Neurology, 296*, 277–290.

Spinomesencephalic Tract (SMT)

▶ Spinomesencephalic Tract

Spinoparabrachial Neurons

Definition

Neurons with cell bodies in the spinal cord and axonal projections with sites of termination in the parabrachial nucleus.

Cross-References

▶ Spinal Ascending Pathways, Colon, Urinary Bladder, and Uterus

Spinoparabrachial Pathway

Definition

The projection from the spinal cord to the parabrachial region in the caudal midbrain is known as the spinoparabrachial pathway.

Cross-References

▶ Spinomesencephalic Tract

Spinoparabrachial Tract

Alan R. Light
Department of Anesthesiology, University of
Utah, Salt Lake City, UT, USA

Synonyms

Part of the spinoreticular tract

Definition

Neurons in the spinal cord and their axons that
project and terminate in the parabrachial nuclei of
the brain. Many of these neurons are found in the
most superficial regions of the dorsal horn
(Rexed's lamina 1) and respond best to nociceptive inputs (Andrew 2009; Bester et al. 2000;
Hylden et al. 1985; Light et al. 1993). The
parabrachial nuclei are at the junction between
the midbrain and the pons and are found immediately adjacent to the brachium conjuntivum
(the superior cerebellar peduncle) as it ascends
from the pons, through the midbrain. The function of these neurons in the perception of pain is

unknown (for more information on the function
of this pathway see ▶ parabrachial hypothalamic
and amygdaloid projections and Bernard and
Besson 1990).

Characteristics

Nociceptive neurons (neurons capable of
encoding information discriminating between tissue damaging and non-damaging stimuli) are
concentrated in the most superficial layers of the
spinal cord (laminae 1 and 2) and the neck of
the dorsal horn (lamina 5). Laminae 1 and 5
contain many neurons that project to supraspinal
regions of the brain, including the thalamus and
brain stem (see ▶ Spinohypothalamic Tract,
Anatomical Organization, and Response Properties and other entries on spinothalamic tracts).

One region of the brain stem that receives
many projections from lamina 1 is the region at
the junction of the pons and midbrain adjacent to
the superior cerebellar peduncle called the
parabrachial region (see Fig. 1).

This projection is more prominent
contralaterally than ipsilaterally (Blomqvist
et al. 1989; Cechetto et al. 1985; Craig 1995;
Kitamura et al. 1993; Panneton and Burton

**Spinoparabrachial Tract,
Fig. 1** Diagram of the
spinoparabrachial tract.
Nociceptive primary
afferent axons synapse on
superficial dorsal horn
neurons. Superficial dorsal
horn cell (*bottom panel*)
projects through the
contralateral white matter
to the lateral parabrachial
region to terminate on cells
in this region. Most of the
projections are
contralateral (indicated by
60 % in the diagram), while
some are ipsilateral or
bilateral (indicated by 40 %
in the diagram)

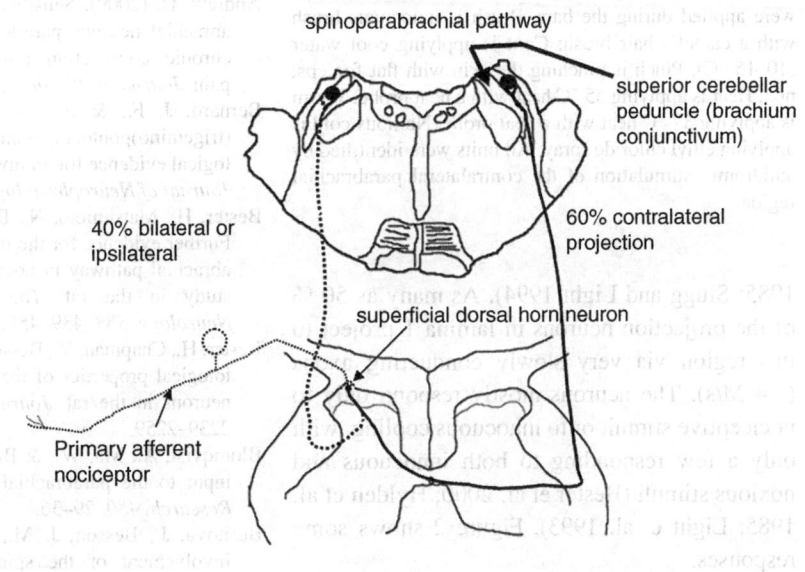

spinoparabrachial pathway

superior cerebellar
peduncle (brachium
conjuntivum)

40% bilateral or
ipsilateral

60% contralateral
projection

superficial dorsal horn neuron

Primary afferent
nociceptor

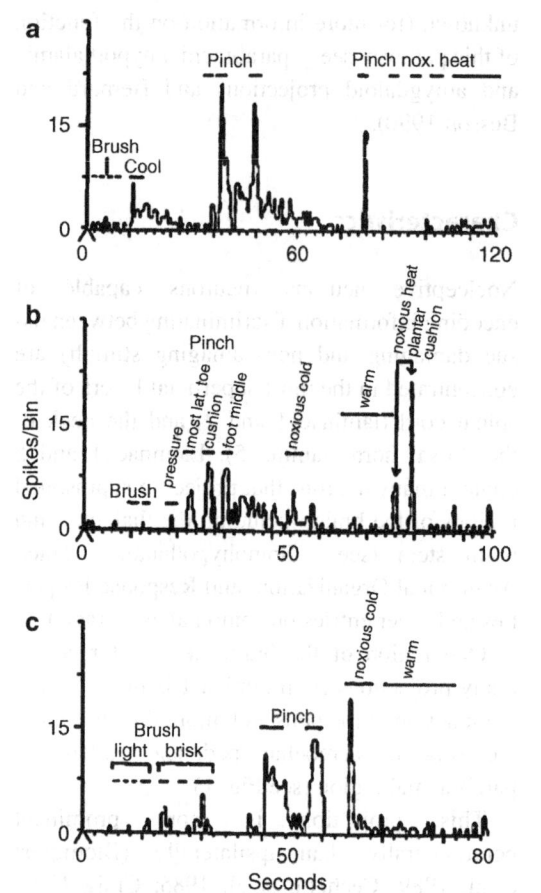

Spinoparabrachial Tract, Fig. 2 Responses of 3 different spinoparabrachial neurons from cat. Action potential rate (action potentials per second) is on the Y-axis, while time in seconds is on the X-axis. The indicated stimuli were applied during the bars. Brush is innocuous brush with a camel's hair brush; Cool is applying cool water (10–15 °C), Pinch is pinching the skin with flat forceps, nox. Heat is applying 55 °C heat with a heat probe. Warm is applying 37 °C heat with a heat probe. Noxious cold is applying ethyl chloride spray. All units were identified by antidromic stimulation of the contralateral parabrachial region

1985; Slugg and Light 1994). As many as 50 % of the projection neurons in lamina 1 project to this region via very slowly conducting axons (<4 M/s). The neurons mostly respond only to nociceptive stimuli or to innocuous cooling, with only a few responding to both innocuous and noxious stimuli (Bester et al. 2000; Hylden et al. 1985; Light et al. 1993). Figure 2 shows some responses.

The pathway has been identified in rats, cats, monkeys, and other species. Its prominence in man is unknown.

The parabrachial region to which the lamina 1 neurons project, itself projects to the amygdala, thalamus, and hypothalamus (e.g., Bernard and Besson 1990), allowing the possibility that information from this pathway may reach higher brain centers, especially those centers involved in emotional and hormonal responses to pain (see entry described above for more information).

The parabrachial region also receives inputs from visceral centers, thus leaving open the possibility that viscero-pain interactions may occur at this level of the CNS (Cechetto et al. 1985, Menetrey and de Pommery 1991; Murphy et al. 2009; Saper 1995; Williams and Ivanusic 2008).

Lesions of the parabrachial region have been associated with increased pain behavior, leading some to postulate that this region may be involved in the descending control of pain (Wall et al. 1988). Still other researchers have stressed the role of the spinoparabrachial pathway in the enhanced pain induced by inflammation and neuropathy (Andrew 2009; Bester et al. 1997; Buritova et al. 1998; Nahin et al. 1992).

References

Andrew, D. (2009). Sensitization of lamina 1 spinoparabrachial neurons parallels heat hyperalgesia in the chronic constriction injury model of neuropathic pain. *Journal of Physiology, 587*, 2005–2017.

Bernard, J. F., & Besson, J. M. (1990). The spino (trigemino)pontoamygdaloid pathway: Electrophysiological evidence for an involvement in pain processes. *Journal of Neurophysiology, 63*, 473–490.

Bester, H., Matsumoto, N., Besson, J. M., et al. (1997). Further evidence for the involvement of the spinoparabrachial pathway in nociceptive processes: A c-Fos study in the rat. *The Journal of Comparative Neurology, 383*, 439–458.

Bester, H., Chapman, V., Besson, J. M., et al. (2000). Physiological properties of the lamina I spinoparabrachial neurons in the rat. *Journal of Neurophysiology, 83*, 2239–2259.

Blomqvist, A., Ma, W., & Berkley, K. J. (1989). Spinal input to the parabrachial nucleus in the cat. *Brain Research, 480*, 29–36.

Buritova, J., Besson, J. M., & Bernard, J. F. (1998). Involvement of the spinoparabrachial pathway in inflammatory nociceptive processes: A c-Fos protein study in the awake rat. *The Journal of Comparative Neurology, 397*, 10–28.

Cechetto, D. F., Standaert, D. G., & Saper, C. B. (1985). Spinal and trigeminal dorsal horn projections to the parabrachial nucleus in the rat. *The Journal of Comparative Neurology, 240*, 153–160.

Craig, A. D. (1995). Distribution of brainstem projections from spinal lamina I neurons in the cat and the monkey. *The Journal of Comparative Neurology, 361*, 225–428.

Hylden, J. L., Hayashi, H., Bennett, G. J., et al. (1985). Spinal lamina I neurons projecting to the parabrachial area of the cat midbrain. *Brain Research, 336*, 195–198.

Kitamura, T., Yamada, J., Sato, H., et al. (1993). Cells of origin of the spinoparabrachial fibers in the rat: A study with fast blue and WGA-HRP. *The Journal of Comparative Neurology, 328*, 449–461.

Light, A. R., Sedivec, M. J., Casale, E. J., et al. (1993). Physiological and morphological characteristics of spinal neurons projecting to the parabrachial region of the cat. *Somatosensory and Motor Research, 10*, 309–325.

Menetrey, D., & De Pommery, J. (1991). Origins of spinal ascending pathways that reach central areas involved in visceroception and visceronociception in the rat. *European Journal of Neuroscience, 3*, 249–259.

Murphy, A. Z., Suckow, S. K., Johns, M., & Traub, R. J. (2009). Sex differences in the activation of the spinoparabrachial circuit by visceral pain. *Physiology & Behavior, 97*, 205–212.

Nahin, R. L., Hylden, J. L., & Humphrey, E. (1992). Demonstration of dynorphin A 1–8 immunoreactive axons contacting spinal cord projection neurons in a rat model of peripheral inflammation and hyperalgesia. *Pain, 51*, 135–143.

Panneton, W. M., & Burton, H. (1985). Projections from the paratrigeminal nucleus and the medullary and spinal dorsal horns to the peribrachial area in the cat. *Neuroscience, 15*, 779–797.

Saper, C. B. (1995). The spinoparabrachial pathway: Shedding new light on an old path. *The Journal of Comparative Neurology, 353*, 477–479.

Slugg, R. M., & Light, A. R. (1994). Spinal cord and trigeminal projections to the pontine parabrachial region in the rat as demonstrated with *Phaseolus vulgaris* leucoagglutinin. *The Journal of Comparative Neurology, 339*, 49–61.

Wall, P. D., Bery, J., & Saade, N. (1988). Effects of lesions to rat spinal cord lamina I cell projection pathways on reactions to acute and chronic noxious stimuli. *Pain, 35*, 327–339.

Williams, M. C., & Ivanusic, J. J. (2008). Evidence for the involvement of the spinoparabrachial pathway, but not the spinothalamic tract or post-synaptic dorsal column, in acute bone nociception. *Neuroscience Letters, 443*, 246–250.

Spinoparabrachial Tract Neurons, Anatomical Features

Andrew J. Todd
Institute of Neuroscience and Psychology,
College of Medical, Veterinary and Life
Sciences, University of Glasgow, Glasgow, UK

Synonyms

Neuronal structure; Spinal cord neurons projecting to the lateral parabrachial area

Definition

The term "morphology," in this context, refers to the shape and size of the cell body of a neuron, and the extent, orientation, and branching pattern of its dendritic tree. The morphology of spinal cord neurons that project to the parabrachial area (spinoparabrachial neurons) has been examined in several studies. It has been suggested that particular morphological types of spinoparabrachial neuron can be recognized and that these may have distinct functional properties.

Characteristics

The Spinoparabrachial Tract

The main spinoparabrachial projection originates from neurons in Rexed's lamina I (also known as the marginal layer) of the spinal dorsal horn and terminates in several nuclei within the lateral parabrachial area (LPb) (Bernard et al. 1995). Although there are some spinoparabrachial cells in deeper laminae, this entry will focus on those in lamina I, since these have been extensively studied. Most studies of this tract have been carried out on the rat, although a similar projection is present in other mammalian species.

The morphology of lamina I projection neurons, including those that belong to the spinoparabrachial tract, has been studied by using retrograde labeling, with tracers injected into the appropriate brain target (in this case, the lateral parabrachial area). Although early retrograde tracing studies generally reported very limited dendritic filling, the use of modern tracers (e.g., cholera toxin B subunit, CTb) and sensitive detection methods has resulted in much more complete labeling. While this does not extend to the extreme distal parts of the dendritic trees, it is often sufficient to allow the morphology of the labeled cells to be examined in some detail. In addition, ~80 % of spinoparabrachial cells in lamina I express the neurokinin 1 receptor (NK1r), and since this receptor is present throughout the somatodendritic plasma membrane, immunocytochemistry with antibodies against NK1r can reveal the entire dendritic trees of these cells (Todd et al. 2002). Recent studies involving retrograde tracing with viral vectors have also been used to provide apparently complete labeling of the dendritic trees of spinoparabrachial neurons (Cordero-Erausquin et al. 2009) (see below).

Lamina I spinoparabrachial neurons belong to a projection system known as the anterolateral tract (ALT). ALT axons generally cross the midline and travel rostrally in the lateral or anterolateral white matter of the spinal cord, giving rise to collateral axons that innervate several brain regions that include the thalamus, periaqueductal gray matter (PAG), LPb, nucleus of the solitary tract, and caudal ventrolateral medulla (Todd 2010). Many of these cells are known to send collateral axons to more than one of these nuclei, and studies in which different tracers have been injected into more than brain region have suggested that virtually all (95 %) of the lamina I projection neurons belonging to the ALT innervate the LPb (Todd 2010). Quantitative studies in the rat indicate that there are around 400 ALT projection neurons in lamina I on each side in the L4 segment, and that these constitute around 5 % of all neurons in this lamina. The number of spinoparabrachial cells is somewhat smaller in the cervical enlargement, since there are only around 215 lamina I ALT projection neurons per side in the C7 segment, although, as in the lumbar enlargement, the vast

majority of these belong to the spinoparabrachial tract (Todd 2010). Studies in which different tracers have been injected into both sides of the brain suggest that around 70 % of lamina I spinoparabrachial neurons project only to the contralateral side, while the remaining 30 % project to the LPb on both sides (Spike et al. 2003).

Morphology of Lamina I Neurons

Neurons in lamina I have been allocated to various morphological classes, based on the shapes of their cell bodies and the branching pattern of their dendritic trees, revealed by Golgi staining (Lima and Coimbra 1986) or retrograde labeling (Todd et al. 2002; Cordero-Erausquin et al. 2009; Spike et al. 2003; Zhang and Craig 1997; Almarestani et al. 2007), or as seen in electrophysiological studies in which individual neurons have been labeled (Han et al. 1998; Prescott and De Koninck 2002). Lima and Coimbra (1986) identified four distinct morphological types: (1) fusiform cells, with dendrites elongated along the rostrocaudal axis; (2) pyramidal cells, which had a triangular soma that typically gave off 3–5 primary dendrites; (3) flattened cells, whose dendrites arborized in both mediolateral and rostrocaudal axes within lamina I; and (4) multipolar cells, with radiating dendrites. The multipolar type was unique since cells belonging to each of the other three types had dendrites that were restricted to lamina I, while those of the multipolar cells were found to penetrate into deeper parts of the dorsal horn. Since this initial classification scheme was based on the results of Golgi staining, it includes both projection neurons and interneurons. Analysis of retrograde tracing experiments has shown that neurons belonging to each of the main types described by Lima and Coimbra, apart from the multipolar cells, are represented among lamina I projection cells (including spinoparabrachial neurons) (Todd et al. 2002; Spike et al. 2003; Zhang and Craig 1997). In other words, the great majority of projection neurons have dendrites that are restricted to lamina I. However, the term "flattened," which was used by Lima and Coimbra to describe the shapes of the cell bodies of these cells, has not been taken up by other authors, and projection cells with dendrites that radiate along both rostrocaudal and mediolateral axes, while remaining in lamina I, have generally been referred to as "multipolar" (Cordero-Erausquin et al. 2009; Spike et al. 2003; Almarestani et al. 2007; Zhang et al. 1996). This terminology will be adhered to in this entry.

There has been considerable interest in the morphology of lamina I projection neurons, including those that belong to the spinoparabrachial tract, since it is unlikely that these cells form a homogeneous group, and in other parts of the CNS, somatodendritic morphology has often been helpful in defining functional populations. Indeed, there have been reports that different morphological types of neuron have specific projection targets (Lima et al. 1991), and also that they have characteristic physiological properties (Almarestani et al. 2007; Han et al. 1998; Prescott and De Koninck 2002).

Morphology and Function of Lamina I Neurons

Lima et al. (1991) concluded on the basis of a series of retrograde-labeling studies that the majority of lamina I neurons belonging to the LPb were of the fusiform type, and that fusiform cells also projected to the caudal ventrolateral medulla. It was also reported that pyramidal cells preferentially targeted the PAG and thalamus, while the flattened cells (described by other authors as multipolar) innervated thalamus and dorsal reticular nucleus (Lima et al. 1991). However, as stated above, it is now known that the vast majority of lamina I ALT projection neurons in both lumbar and cervical enlargement project to the LPb (Todd 2010), and this is not easily reconciled with the suggestion that a distinct morphological class is involved in this projection (Lima et al. 1991). Indeed, two more recent studies have reported that spinoparabrachial lamina I neurons in the rat are morphologically heterogeneous (Spike et al. 2003; Almarestani et al. 2007) (Fig. 1).

Almarestani et al. (2007) found that 39 % of these cells were multipolar, 37 % fusiform, and 23 % pyramidal. In the study of Spike et al. (2003), around one third of the spinoparabrachial cells belonged to each of these morphological classes, and importantly, the proportions in each class did not differ between spinoparabrachial cells and those that were retrogradely labeled from the caudal ventrolateral medulla or PAG. This finding argues against the suggestion that different morphological types of lamina I cell have specific supraspinal targets.

Prescott and De Koninck (2002) carried out a combined electrophysiological and morphological study of lamina I neurons recorded with the whole-cell patch-clamp technique in slices taken from adult rat spinal cord. They reported a strong correlation between morphology of neurons and their firing patterns in response to injection of depolarizing current pulses. However, since projection cells are thought to make up only ~5 % of the neuronal population in lamina I (Todd 2010; Spike et al. 2003), it is likely that many of the cells recorded in this study were interneurons. Further evidence for a relationship between morphology and function came from a study by Han et al. (1998), who obtained intracellular recordings from 38 lamina I neurons in anesthetized cats, of which six could be positively identified as belonging to the spinothalamic tract. Since virtually all spinothalamic lamina I neurons in the rat also project to the LPb, and their sample was likely to be biased toward the larger neurons, most of which are projection cells (Todd 2010), it is likely that at least some of the cells recorded by Han et al. were spinoparabrachial neurons. They reported that there was a clear relationship between morphology and receptive field properties, with pyramidal neurons responding exclusively to innocuous thermal stimuli (cooling of the skin), while fusiform and multipolar cells were activated by noxious mechanical and/or heat stimuli, and in some cases also by noxious cold (Han et al. 1998).

Consistent with the results of Han et al. (1998), Almarestani et al. (2007) found that in the rat, very few lamina I spinoparabrachial neurons with pyramidal morphology expressed the NK1r, which is thought to be restricted to cells that respond to noxious stimuli. However, in contrast, Spike et al. (2003) reported that there was

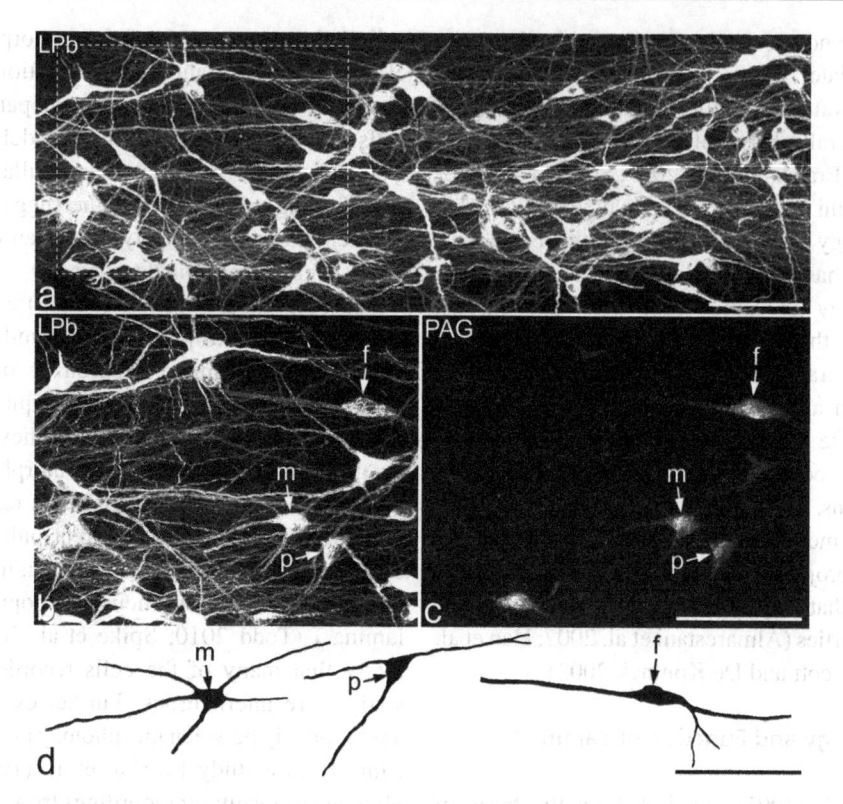

Spinoparabrachial Tract Neurons, Anatomical Features, Fig. 1 Lamina I projection neurons in a horizontal section of rat spinal cord. (**a–c**) confocal images of retrogradely labeled cells in a rat that had received injections of two different tracers: one into the lateral parabrachial area (LPb) and one into the periaqueductal gray matter (PAG). (**a**) Shows the appearance of cells labeled from LPb. Note the high density of neurons and the extensive dendritic filling, which allows morphological types to be recognized. (**b, c**) Show labeling for the two tracers in the boxed area in (**a**). Several cells that were labeled from the PAG are seen in (**c**), and all of these are also visible in (**b**), indicating that they were also labeled from the LPb. (**d**) Drawings of three of the cells that are seen in (**b**) and (**c**), which represent the three main morphological types – multipolar (*m*), pyramidal (*p*), and fusiform (*f*). Scale bars = 100 μm (Reproduced with permission from Spike et al. 2003)

no relationship between morphology and NK1r expression among different morphological types of lamina I projection neuron, with 80 % of the pyramidal cells showing immunoreactivity for the receptor. In addition, the great majority of pyramidal lamina I projection neurons that expressed the NK1r were found to upregulate Fos following noxious stimulation, and cells of this type were also shown to receive a high density of synaptic contacts from substance P-containing (nociceptive) primary afferents (Todd et al. 2002). It is possible that some of the discrepancies between these studies arise from differences in the criteria used by different laboratories to classify the various morphological types, particularly the pyramidal cells (Almarestani et al. 2007). However, there is a more significant problem with the suggestion that pyramidal lamina I spinoparabrachial neurons in the rat respond exclusively to innocuous thermal stimuli. Almarestani et al. (2007) reported that over 20 % of lamina I spinoparabrachial cells in the rat were of the pyramidal type, but two in vivo electrophysiological studies have reported that the vast majority of lamina I spinoparabrachial neurons in the rat are activated by noxious stimuli (Andrew 2009; Bester et al. 2000). Therefore, unless pyramidal neurons were dramatically underrepresented in these two studies, many of them must respond to noxious stimuli.

Although, for the reasons outlined above, there is unlikely to be a simple relationship between morphology and function among lamina I spinoparabrachial neurons in the rat, there is evidence that distinct types can be identified on the basis of their anatomical appearance. Puskár et al. (2001) described a population of very large spinoparabrachial lamina I neurons that either lacked the NK1r or else expressed it at low levels. These neurons, which have been named giant cells (Todd 2010), form a small but easily recognizable group, corresponding to around 3 % of the total population of lamina I spinoparabrachial neurons. They are characterized by a very high density of both excitatory and inhibitory synapses that coat their cell bodies and dendrites, and most of these are thought to involve axons of local interneurons (Todd 2010). In this way, they differ from the NK1r-expressing spinoparabrachial neurons, which receive a high density of synaptic input from substance P-containing primary afferents (Todd 2010). Interestingly, virtually all of the giant cells belong to the multipolar type, although this is not a defining feature, since many of the NK1r-expressing spinoparabrachial cells are also multipolar (Todd et al. 2002; Spike et al. 2003).

Conventional retrograde-labeling studies are not able to reveal the presence of dendritic spines on projection neurons, as it is unlikely that the tracer would diffuse into these in sufficient amounts to reveal them. However, in a recent study, viral vectors that encoded green fluorescent protein (GFP) were used as retrograde tracers (Cordero-Erausquin et al. 2009). This technique resulted in far more complete labeling, due to the extensive synthesis of GFP in the affected neurons, and the authors were able to show that dendritic spines were only infrequently present on the dendrites of spinoparabrachial neurons. In contrast, many interneurons in the superficial dorsal horn are known to have a high density of dendritic spines. Although the significance of this finding is not yet known, it may have implications for our understanding of synaptic plasticity on different classes of dorsal horn neuron, since much of the synaptic input to projection neurons appears to be located on dendritic shafts, rather than spines (Todd et al. 2002).

References

Almarestani, L., Waters, S. M., Krause, J. E., Bennett, G. J., & Ribeiro-da-Silva, A. (2007). Morphological characterization of spinal cord dorsal horn lamina I neurons projecting to the parabrachial nucleus in the rat. *The Journal of Comparative Neurology, 504*, 287–297.

Andrew, D. (2009). Sensitization of lamina I spinoparabrachial neurons parallels heat hyperalgesia in the chronic constriction injury model of neuropathic pain. *The Journal of Physiology, 587*, 2005–2017.

Bernard, J. F., Dallel, R., Raboisson, P., Villanueva, L., & Le Bars, D. (1995). Organization of the efferent projections from the spinal cervical enlargement to the parabrachial area and periaqueductal gray: A PHA-L study in the rat. *The Journal of Comparative Neurology, 353*, 480–505.

Bester, H., Chapman, V., Besson, J. M., & Bernard, J. F. (2000). Physiological properties of the lamina I spinoparabrachial neurons in the rat. *Journal of Neurophysiology, 83*, 2239–2259.

Cordero-Erausquin, M., Allard, S., Dolique, T., Bachand, K., Ribeiro-da-Silva, A., & De Koninck, Y. (2009). Dorsal horn neurons presynaptic to lamina I spinoparabrachial neurons revealed by transynaptic labeling. *The Journal of Comparative Neurology, 517*, 601–615.

Han, Z. S., Zhang, E. T., & Craig, A. D. (1998). Nociceptive and thermoreceptive lamina I neurons are anatomically distinct. *Nature Neuroscience, 1*, 218–225.

Lima, D., & Coimbra, A. (1986). A golgi study of the neuronal population of the marginal zone (lamina I) of the rat spinal cord. *The Journal of Comparative Neurology, 244*, 53–71.

Lima, D., Mendes-Ribeiro, J. A., & Coimbra, A. (1991). The spino-latero-reticular system of the rat: Projections from the superficial dorsal horn and structural characterization of marginal neurons involved. *Neuroscience, 45*, 137–152.

Prescott, S. A., & De Koninck, Y. (2002). Four cell types with distinctive membrane properties and morphologies in lamina I of the spinal dorsal horn of the adult rat. *The Journal of Physiology, 539*, 817–836.

Puskár, Z., Polgár, E., & Todd, A. J. (2001). A population of large lamina I projection neurons with selective inhibitory input in rat spinal cord. *Neuroscience, 102*, 167–176.

Spike, R. C., Puskar, Z., Andrew, D., & Todd, A. J. (2003). A quantitative and morphological study of projection neurons in lamina I of the rat lumbar spinal cord. *European Journal of Neuroscience, 18*, 2433–2448.

Todd, A. J. (2010). Neuronal circuitry for pain processing in the dorsal horn. *Nature Reviews. Neuroscience, 11*, 823–836.

Todd, A. J., Puskar, Z., Spike, R. C., Hughes, C., Watt, C., & Forrest, L. (2002). Projection neurons in lamina I of rat spinal cord with the neurokinin 1 receptor are selectively innervated by substance P-containing afferents and respond to noxious stimulation. *Journal of Neuroscience, 22*, 4103–4113.

Zhang, E. T., & Craig, A. D. (1997). Morphology and distribution of spinothalamic lamina I neurons in the monkey. *Journal of Neuroscience, 17*, 3274–3284.

Zhang, E. T., Han, Z. S., & Craig, A. D. (1996). Morphological classes of spinothalamic lamina I neurons in the cat. *The Journal of Comparative Neurology, 367*, 537–549.

Spinopetal Modulation of Pain

▶ Descending Facilitation and Inhibition in Neuropathic Pain

Spinoreticular Neurons

Definition

Neurons with cell bodies in the spinal cord and axonal projections with sites of termination in supraspinal reticular structures.

Cross-References

▶ Spinal Ascending Pathways, Colon, Urinary Bladder, and Uterus

Spinosolitary Neurons

Definition

Neurons with cell bodies in the spinal cord and axonal projections with sites of termination in the nucleus tractus solitarius.

Cross-References

▶ Spinal Ascending Pathways, Colon, Urinary Bladder, and Uterus

Spinotectal

Definition

The projection from the spinal cord to the superior colliculus, especially the deep layers, is known as the spinotectal pathway.

Cross-References

▶ Spinomesencephalic Tract

Spinothalamic Input: Cells of Origin (Monkey)

William D. Willis[1] and Karin Westlund High[2]
[1]Department of Neuroscience and Cell Biology, University of Texas Medical Branch, Galveston, TX, USA
[2]Department of Physiology, University of Kentucky, Lexington, KY, USA

Synonyms

Spinothalamic neurons; Spinothalamic tract cells

Definition

A spinothalamic tract cell is a neuron whose cell body and dendrites are located in the spinal cord and whose axon projects to and terminates in the thalamus, usually on the side contralateral to the cell body. The spinothalamic tract is a major pathway by which nociceptive information is conveyed to the thalamus and from there to the cerebral cortex, leading to the perception of pain as well as to motivational-affective responses to painful stimuli.

Characteristics

The locations of the cells of origin of the primate spinothalamic tract have been mapped

Spinothalamic Input: Cells of Origin (Monkey), Fig. 1 Distribution of retrogradely labeled spinothalamic tract cells in a monkey. (**a**) shows the site of injection of cholera toxin subunit B in the ventral posterior lateral nucleus of the thalamus. The injection spread into several adjacent nuclei that do not receive spinal projections, but not into the ventral posterior inferior nucleus, posterior complex, or central lateral nuclei. In (**b**) are drawings of representative transverse sections through the spinal cord at different segmental levels from C1–2 through S1–3. The locations of retrogradely labeled spinothalamic tract cells are indicated. The total counts of cell profiles for different levels are indicated on the drawings for the side of the cord contralateral (*left*) and ipsilateral (*right*) to the injection (From Willis et al. 2001)

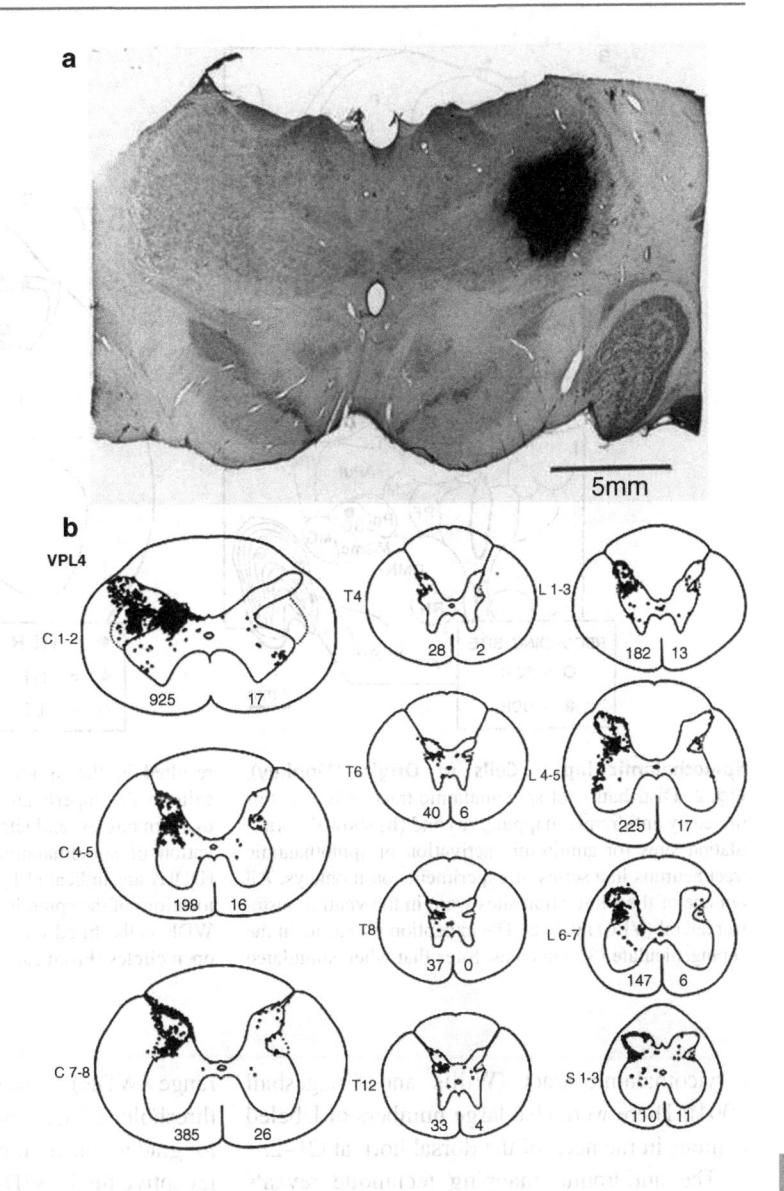

anatomically using the ▶ retrograde tracing technique (Apkarian and Hodge 1989; Willis et al. 1979, 2001) and electrophysiologically by ▶ antidromic microstimulation within the thalamus while recording from spinal cord neurons (Applebaum et al. 1979; Giesler et al. 1981; Zhang et al. 2000).

Figure 1 illustrates the distribution of spinothalamic tract cells that were retrogradely labeled following the injection of cholera toxin subunit B into the ▶ ventral posterior lateral (VPL) nucleus of the thalamus of a monkey

(Willis et al. 2001). The injection site is shown in Fig. 1a. In the cervical and lumbosacral enlargements, numerous labeled neurons were observed in laminae I, V, and VI (Fig. 1b). A few labeled cells were found in laminae II–IV and in the ventral horn. Spinothalamic tract cells in the thoracic and upper lumbar spinal cord had similar locations as in the enlargements, but the density of labeled cells was smaller.

In the uppermost cervical spinal cord, there was a large population of labeled neurons in the ▶ lateral cervical nucleus; these give rise to the

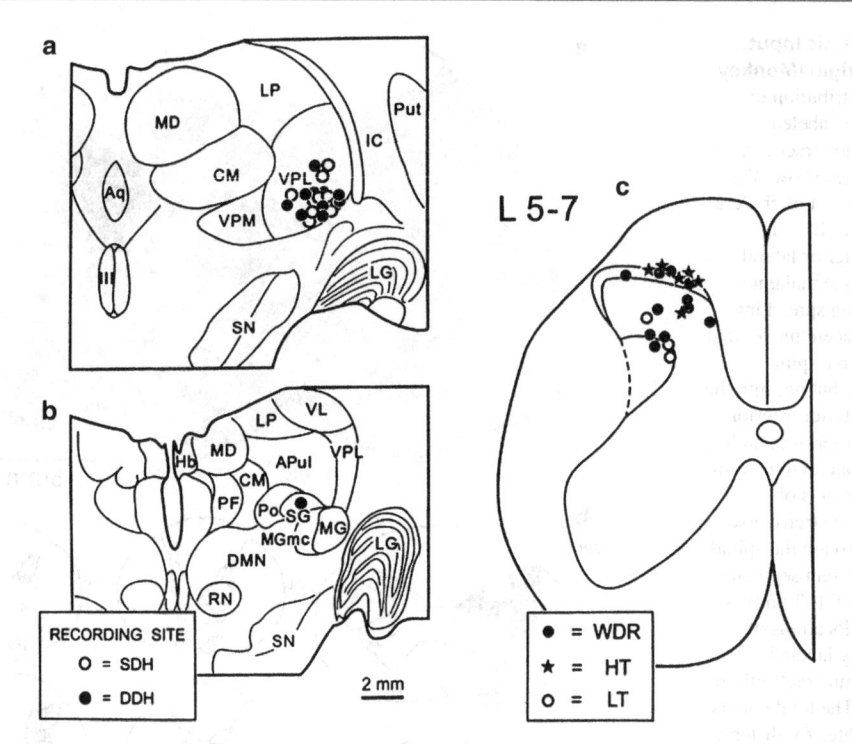

Spinothalamic Input: Cells of Origin (Monkey), Fig. 2 Distribution of spinothalamic tract cells as determined by antidromic mapping. (a) and (b) show the stimulation sites for antidromic activation of spinothalamic tract neurons in a series of experiments on monkeys. All but one of the stimulation sites were in the ventral posterior lateral (VPL) nucleus. The exception was a site in the suprageniculate (SG) nucleus. Sites that when stimulated resulted in the antidromic activation of spinothalamic cells in the superficial dorsal horn (SDH) are indicated by open circles, and sites that resulted in antidromic activation of spinothalamic cells in the deep dorsal horn (DDH) are indicated by the filled circles. (c) shows the locations of the spinothalamic cells. The symbols used are WDR cells, filled circles; HT cells, stars; and LT cells, open circles (From Zhang et al. 2000)

cervicothalamic tract (Willis and Coggeshall 2004). There were also large numbers of labeled neurons in the neck of the dorsal horn at C1–2.

The antidromic mapping technique reveals a comparable distribution of spinothalamic tract cells in the lumbar enlargement of monkeys, as shown in Fig. 2 (Zhang et al. 2000). In these experiments, all of the microstimulation sites were in the ventral posterior lateral thalamic nucleus (Fig. 2a), except for one site that was in the suprageniculate nucleus (Fig. 2b). The antidromically activated spinothalamic tract cells were concentrated in laminae I and IV–V. An advantage of this approach is that the response properties of the recorded spinothalamic cells can be determined. The cells whose locations are shown in Fig. 2c were classed as ► wide dynamic range (WDR), ► high threshold (HT), or ► low threshold (LT), according to their responsiveness to graded mechanical stimuli applied to their receptive fields. WDR cells respond to innocuous and noxious stimuli but best to noxious intensities. HT cells respond essentially only to noxious stimuli. LT cells respond to innocuous and not to noxious stimuli.

The dendrites of spinothalamic tract cells in lamina I are oriented longitudinally, as shown in Fig. 3a in a horizontal section of the spinal cord after retrograde labeling from the thalamus using cholera toxin subunit B (Zhang and Craig 1997). Based on the configuration of the dendrites, lamina I spinothalamic cells could be subdivided into fusiform (47 %), pyramidal (27 %), and multipolar (22 %) types (Zhang and Craig 1997).

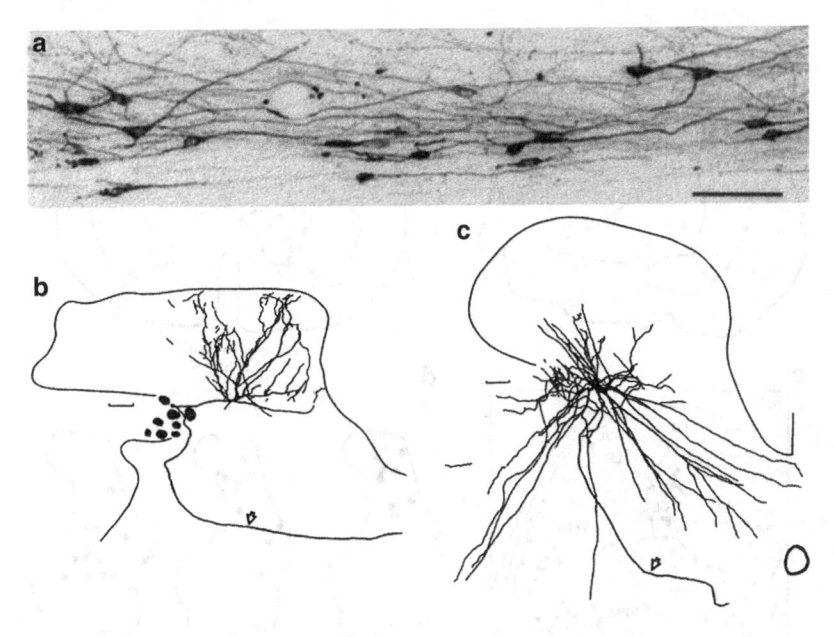

Spinothalamic Input: Cells of Origin (Monkey), Fig. 3 Orientation of the dendrites of spinothalamic tract cells in the monkey. (**a**) shows the cell bodies and dendrites of retrogradely labeled monkey spinothalamic cells following injection of cholera toxin subunit B into the contralateral thalamus. The lumbosacral spinal cord was sectioned horizontally, and the section shown was taken through lamina I. Note that the dendrites are oriented longitudinally. Cells having fusiform, pyramidal, or multipolar configurations can be recognized. (**b**) and (**c**) illustrate the cell bodies and dendrites of two spinothalamic tract cells in the deep dorsal horn of monkey spinal cord. Intracellular recordings were made from these neurons, which were identified by antidromic activation from the ventral posterior lateral nucleus of the thalamus. The responses to graded intensities of mechanical stimuli showed that the neuron in (**b**) was a WDR cell and that in (**c**) a HT cell. After the recordings were completed, the cells were injected with horseradish peroxidase through the microelectrode. The injected cells were later reacted histochemically and then reconstructed from serial sections. The dendrites of the cell in (**b**) extended mostly dorsally and reached lamina I. By contrast, the dendrites of the cell in (**c**) extended mostly ventrally. The *open arrows* indicate the axons of these neurons as they neared the point at which they crossed the midline in the anterior white commissure ((**a**) is from Zhang and Craig 1997); (**b** and **c**) are from Surmeier et al. 1988)

By contrast, the dendrites of spinothalamic tract cells in the deep dorsal horn are oriented transversely, as illustrated in Fig. 3b, c in histological reconstructions of antidromically identified spinothalamic cells that were injected intracellularly with horseradish peroxidase (Surmeier et al. 1988). In some cases, the dendrites extended dorsally as far as to lamina I (Fig. 3b); in other cases the dendrites spread chiefly ventrally (Fig. 3c). The classification of the spinothalamic tract cell in Fig. 3b was WDR and that of the neuron shown in Fig. 3c was HT.

In recordings from 53 lamina I spinothalamic tract cells in monkeys, 69 % were of the WDR type and 29 % of the HT type; only 2 % were LT (Owens et al. 1992). Of 155 spinothalamic tract cells in laminae IV–VI, 59 % were classified as WDR, 22 % as HT, and 20 % as LT cells.

The receptive fields of spinothalamic neurons that terminate in the ventral posterior lateral nucleus of the thalamus can be quite small, or they may be medium sized or even large, but they are usually limited to a single limb (Giesler et al. 1981). Spinothalamic tract cells in lamina I are more likely to have small receptive fields than are spinothalamic cells in laminae IV–VI (Willis and Coggeshall 2004). In addition, the receptive fields of lamina I spinothalamic tract cells have a ▶ somatotopic organization (Willis and Coggeshall 2004). These observations suggest

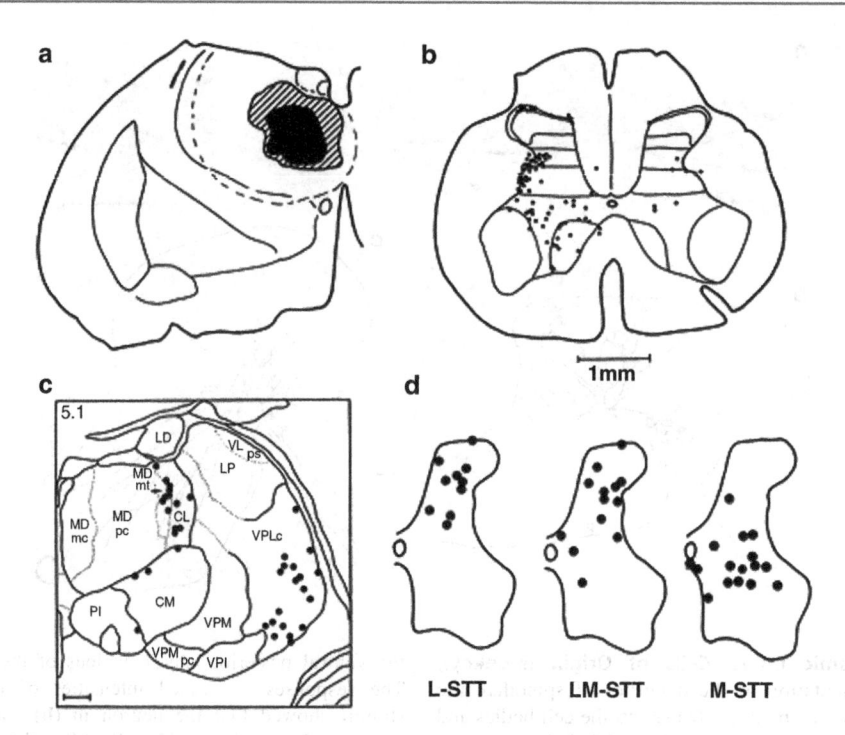

Spinothalamic Input: Cells of Origin (Monkey), Fig. 4 Spinothalamic tract cells that project to the vicinity of the central lateral nucleus in monkeys. (a) shows an injection site in the medial thalamus of a monkey. The retrograde tracer was horseradish peroxidase. In (b) are plotted the retrogradely labeled spinothalamic tract cells in the same animal. Many of the labeled cells are in the ventral horn contralateral to the injection site in the thalamus. Many others are in the deep dorsal horn, and some are in lamina I. (c) shows the stimulation sites for antidromic activation of primate spinothalamic tract cells. Most of the stimulation sites were in the lateral thalamus in the caudal part of the ventral posterior lateral nucleus (VPLc) or in the medial thalamus, especially in the central lateral (CL) nucleus and perhaps in the adjacent part of the medial dorsal (MD) nucleus. (d) shows the locations of spinothalamic tract cells that could be activated antidromically from just the lateral thalamus (L-STT cells), both the lateral and medial thalamus (LM-STT cells), or just the medial thalamus (M-STT cells). Note that spinothalamic tract cells that project just to the medial thalamus are located chiefly in the intermediate region and ventral horn, whereas those that project to the lateral thalamus, with or without collateralizing to the medial thalamus, are mostly in the superficial or deep dorsal horn ((a and b) are from Willis et al. 1979; (c and d) are from Giesler et al. 1981)

that lamina I spinothalamic cells are able to contribute to the accurate localization of painful stimuli.

No studies have been done in which spinothalamic neurons have been retrogradely labeled in a selective way by injections of tracer into the ▶ ventral posterior inferior nucleus or the posterior complex or by antidromic activation from these nuclei. However, the response properties of thalamic neurons in these nuclei have been described in the squirrel monkey (Apkarian and Shi 1994). Half of the neurons sampled in the ventral posterior inferior nucleus and more than a third of those recorded in the posterior complex were nociceptive. Presumably, a major source of the nociceptive input to these thalamic neurons was by way of the spinothalamic tract.

In general, spinothalamic cells that terminate in the ventral posterior lateral nucleus, the ventral posterior inferior nucleus, and the nearby posterior complex appear to be suited to mediate the ▶ sensory-discriminative aspects of pain sensation (Willis and Coggeshall 2004). Cortical projections from the ventral posterior lateral nucleus are to the SI and ▶ SII areas of the somatosensory cortex (Jones 1985). The ventral posterior inferior nucleus and the posterior complex may also project to the SII cortex (Apkarian and Shi 1994; Jones 1985). Functional imaging studies provide evidence that the SI and SII cortices play a major role in sensory discrimination of painful stimuli (Ploner et al. 1999; Rainville et al. 2000).

Some spinothalamic tract cells can be retrogradely labeled (Fig. 4a, b) or antidromically activated from the medial thalamus, in particular from the region of the central lateral nucleus of the intralaminar complex and the adjacent region of the medial dorsal nucleus with projections to limbic cortex (Fig. 4c, d) (Giesler et al. 1981; Apkarian and Hodge 1989). In contrast to spinothalamic tract cells that project only to the ventral posterior lateral nucleus of the lateral thalamus (L-STT cells), some spinothalamic tract cells can be activated both from the lateral thalamus and the medial thalamus (LM-STT cells), while others project only to the medial thalamus (M-STT cells). The locations of L-STT and LM-STT cells (Fig. 4d) are similar to those shown in Figs. 1 and 2. However, the medially projecting spinothalamic cells tend to be located more ventrally, mostly in laminae VII, VIII, and X (Fig. 4d, M-STT). Responses of spontaneously active neurons in the central lateral nucleus of the intralaminar thalamus increase in response to viscerally evoked noxious stimuli, with 69 % excited and 31 % inhibited (Ren et al. 2009). The cells had no cutaneous receptive fields, and responses were significantly reduced by midline dorsal column lesions or spinal morphine.

The response properties of the spinothalamic cells that just project to the ventral posterior lateral nucleus or to both that nucleus and the medial thalamus are similar. However, spinothalamic tract cells that project only to the medial thalamus tend to be of the HT type and often have very large receptive fields that may include the entire surface of the body and head (Giesler et al. 1981). These spinothalamic cells are unlikely to contribute to sensory discrimination. Instead, they presumably participate in the ▶ motivational-affective aspects of pain. They may be especially important for attention and arousal.

References

Apkarian, A. V., & Hodge, C. J. (1989). The primate spinothalamic pathways: I. A quantitative study of the cells of origin of the spinothalamic pathway. *The Journal of Comparative Neurology, 288*, 447–473.

Apkarian, A. V., & Shi, T. (1994). Squirrel monkey lateral thalamus. I. Somatic nociresponsive neurons and their relation to spinothalamic terminals. *Journal of Neuroscience, 14*, 6779–6795.

Applebaum, A. E., Leonard, R. B., Kenshalo, D. R., et al. (1979). Nuclei in which functionally identified spinothalamic tract neurons terminate. *The Journal of Comparative Neurology, 188*, 575–586.

Giesler, G. J., Yezierski, R. P., Gerhart, K. D., et al. (1981). Spinothalamic tract neurons that project to medial and/or lateral thalamic nuclei: Evidence for a physiologically novel population of spinal cord neurons. *Journal of Neurophysiology, 46*, 1285–1308.

Jones, E. G. (1985). *The thalamus.* New York: Plenum Press.

Owens, C. M., Zhang, D., & Willis, W. D. (1992). Changes in the response states of primate spinothalamic tract cells caused by mechanical damage of the skin or activation of descending controls. *Journal of Neurophysiology, 67*, 1509–1527.

Ploner, M., Freund, H. J., & Schnitzler, A. (1999). Pain affect without pain sensation in a patient with a postcentral lesion. *Pain, 81*, 211–214.

Rainville, P., Bushnell, M. C., & Duncan, G. H. (2000). PET studies of the subjective experience of pain. In K. L. Casey & M. C. Bushnell (Eds.), *Pain imaging. Progress in pain research and management* (pp. 123–156). Seattle: IASP Press. vol. 18.

Ren, Y., Zhang, L., Lu, Y., Yang, H., & Westlund, K. N. (2009). Central lateral thalamic neurons receive noxious visceral mechanical and chemical input in rats. *Journal of Neurophysiology, 102*, 244–258.

Surmeier, D. J., Honda, C. N., & Willis, W. D. (1988). Natural groupings of primate spinothalamic neurons based on cutaneous stimulation. Physiological and anatomical features. *Journal of Neurophysiology, 59*, 833–860.

Willis, W. D., & Coggeshall, R. E. (2004). *Sensory mechanisms of the spinal cord* (3rd ed.). New York: Kluwer Academic/Plenum Publishers.

Willis, W. D., Kenshalo, D. R., & Leonard, R. B. (1979). The cells of origin of the primate spinothalamic tract. *Journal of Neurophysiology, 188*, 358–372.

Willis, W. D., Zhang, X., Honda, C. N., et al. (2001). Projections from the marginal zone and deep dorsal horn to the ventrobasal nuclei of the primate thalamus. *Pain, 92*, 267–276.

Zhang, E. T., & Craig, A. D. (1997). Morphology and distribution of spinothalamic lamina I neurons in the monkey. *Journal of Neuroscience, 17*, 3274–3284.

Zhang, X., Honda, C. N., & Giesler, G. J. (2000). Position of spinothalamic tract axons in upper cervical spinal cord of monkeys. *Journal of Neurophysiology, 84*, 1180–1185.

Spinothalamic Neuron

William D. Willis
Department of Neuroscience and Cell Biology,
University of Texas Medical Branch, Galveston,
TX, USA

Definition

Neurons whose cell body and dendrites are located in the spinal cord and whose axons project to and terminate in the thalamus, usually on the side contralateral to the cell body. The axons of spinothalamic neurons form a major nociceptive tract that is responsible for conveying to the brain the information that is required for the perception of the sensory-discriminative components of pain. The spinothalamic tract also contributes to the affective component of pain.

Characteristics

Spinothalamic neurons are a class of sensory projection neurons. They form one of several different sensory pathways that transmit nociceptive signals from the level of the spinal cord to the brain (reviewed in Willis and Coggeshall 2004; Willis and Westlund 1997). The nociceptive signals depend on input to the spinal cord from peripheral ▶ nociceptors in skin, muscle, joints, and viscera. The signals are processed in the spinal cord and then are transmitted to the thalamus and, after a synaptic relay, to the SI and SII somatosensory cortices and other regions of the cerebral cortex (anterior cingulate gyrus, insula). Other pathways, as well as collaterals of spinothalamic tract axons, project to the reticular formation, hypothalamus, limbic system (including the amygdala), and structures usually identified with the motor system, such as the cerebellum and basal ganglia. Different brain structures are thought to be involved in the parallel processing of the nociceptive signals, resulting in different components of the overall pain response, including ▶ sensory

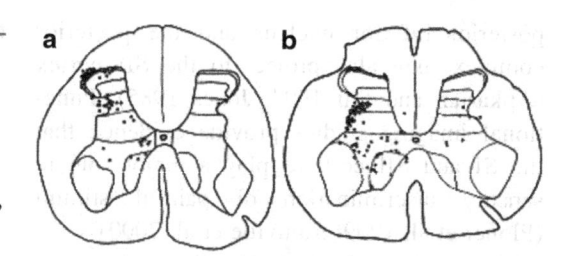

Spinothalamic Neuron, Fig. 1 (a, b) Locations of the cell bodies of spinothalamic tract (STT) neurons in the monkey spinal cord, as determined by retrograde labeling from the thalamus using horseradish peroxidase. (a) Locations of STT neurons that project to the lateral thalamus. (b) Locations of STT neurons that project to the medial thalamus (Modified from Willis et al. 1979)

discrimination (quality, duration, location, and intensity of the pain) and ▶ affective responses, as well as reflex and endocrine adjustments. The role of spinothalamic neurons in pain includes contributions to both sensory discrimination and affective responses (Willis et al. 1974; Giesler et al. 1981). Sensory discrimination is thought to depend on the activation of neurons in the lateral thalamus by the spinothalamic tract, especially neurons in the ventral posterior lateral nucleus, the ventral posterior inferior nucleus, and the posterior complex, whereas terminations of the spinothalamic tract and of a spinoreticulothalamic pathway in the medial thalamus, including the central lateral and parafascicular nuclei of the intralaminar complex and several midline nuclei, are believed to contribute to the affective aspects of pain, as well as to attention and arousal (Willis 1997).

The distribution of the cell bodies of spinothalamic neurons in the spinal cord of monkeys has been mapped using the ▶ antidromic activation (Trevino et al. 1973; Giesler et al. 1981) and the ▶ retrograde labeling techniques (Willis et al. 1979, 2001). Spinothalamic neurons that send projections to the lateral thalamus are concentrated in ▶ Rexed's Laminae I and IV-VI of the dorsal horn (Fig. 1a). Some of those that project to the medial thalamus are found in the same laminae, but others are concentrated in laminae VII and VIII (Fig. 1b). The fact that the cell body of a spinothalamic neuron is located in a particular lamina does not rule out the

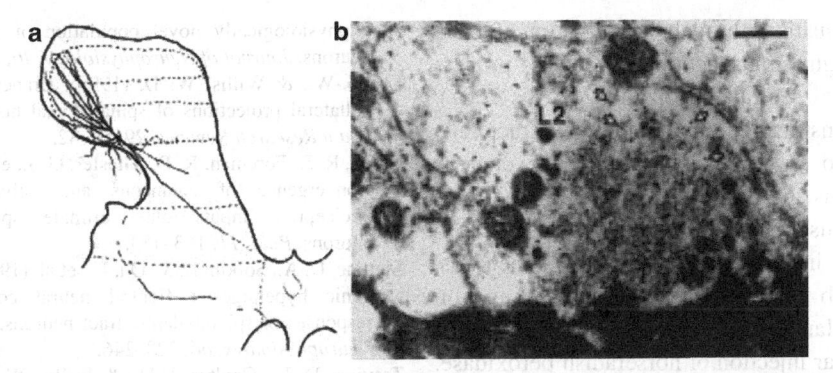

Spinothalamic Neuron, Fig. 2 (a) Drawing of an STT neuron that was identified in an anesthetized monkey by antidromic activation from the ventral posterior lateral nucleus of the thalamus. After the cell was impaled by a microelectrode, horseradish peroxidase was injected intracellularly to label the cell body, dendrites, and axon. The spinal cord tissue containing the cell was processed so that the HRP could be visualized in a light microscope and drawn. (b) Thin sections were made and an immunogold procedure was used to demonstrate the localization of glutamate. Gold particles, some of which are indicated by the *open arrows*, indicate that the presynaptic terminal at the center is glutamatergic. The terminal also contained dense core vesicles, which presumably contain peptides. A part of the labeled STT cell is shown at the bottom and the active zone of an asymmetrical synapse made by the glutamate-containing terminal is shown to be in contact with the STT cell. Scale bar in D = 0.29 μm (Westlund et al. 1992)

possibility of synaptic input to that neuron by synaptic connections formed in other laminae. For example, the dendritic tree of a primate spinothalamic tract neuron is shown in Fig. 2a. This neuron was identified by antidromic activation from the ventral posterior lateral nucleus of the thalamus and then injected intracellularly with horseradish peroxidase. Although the cell body was in lamina IV, some of its dendrites extended dorsally through laminae III and II to reach lamina I. Other dendrites were directed ventrally and reached lamina VI. Thus, this neuron could receive synaptic connections from synaptic terminals in any lamina of the dorsal horn.

Most spinothalamic neurons send their axonal projections to the contralateral thalamus through the spinothalamic tract, which crosses the midline in the anterior white commissure of the spinal cord (Willis et al. 1979; Willis and Coggeshall 2004). However, there are also some uncrossed projections to the ipsilateral thalamus. The axon of the spinothalamic neuron shown in Fig. 2a can be seen to cross just ventral to the central canal. It is instructive that the cell body and the crossing axon are at the same level of the spinal cord. This contrasts with the suggestion in the literature that spinothalamic axons cross several segments rostral to their cell bodies. The spinothalamic tract ascends in the lateral and anterior funiculi of the spinal cord and then through the lateral brain stem to its destinations in the thalamus (Willis and Coggeshall 1991). Collateral branches of spinothalamic axons may terminate in the reticular formation, midbrain periaqueductal gray, hypothalamus, or other brain structures (Lu and Willis 1999). Interruption of the spinothalamic and associated tracts by ▶ anterolateral cordotomy has been used successfully to relieve pain on the side of the body contralateral to the cordotomy (Willis and Coggeshall 1991). It has been especially helpful for cancer pain.

Responses of spinothalamic neurons to noxious stimuli applied to the skin, muscle, joints, and viscera have been described (Willis et al. 1974; Foreman et al. 1979; Milne et al. 1981; Dougherty et al. 1992). Peripheral sensitization of primary afferent nociceptors, including ▶ silent nociceptors, can lead to enhanced responses. Such a change is thought to be responsible for ▶ primary hyperalgesia. The responsiveness of spinothalamic neurons can also be altered by intense noxious stimulation (Simone et al. 1991), after which these neurons may become sensitized through the activation of intra-cellular signal transduction pathways, a process known as ▶ central sensitization (Willis 2001). Central sensitization of spinothalamic tract cells is initiated by synaptic release of excitatory amino acids and peptides, which in turn activate signal transduction pathways in dorsal horn neurons, including spinothalamic neurons. Figure 2b shows a synaptic ending on a spinothalamic tract cell that was labeled by an intracellular injection of horseradish peroxidase. The synaptic terminal was immunolabeled for glutamate, using the immunogold technique. The terminal also contains dense core vesicles that are likely to contain one or more peptides. Presumably, activation of this and other terminals by intense peripheral stimulation would result in the release of glutamate and peptide. With strong synaptic input by such terminals and release of sufficient glutamate and peptide, a state of central sensitization might ensue. It has been suggested that central sensitization of spinothalamic tract neurons helps account for ▶ secondary mechanical allodynia and secondary mechanical hyperalgesia.

Cross-References

▶ Spinal Ascending Pathways, Colon, Urinary Bladder, and Uterus

▶ Spinothalamic Tract Neurons, Descending Control by Brain Stem Neurons

▶ Vagal Input and Descending Modulation

References

Dougherty, P. M., Sluka, K. A., Sorkin, L. S., et al. (1992). Neural changes in acute arthritis in monkeys. I. Parallel enhancement of responses of spinothalamic tract neurons to mechanical stimulation and excitatory amino acids. *Brain Research Reviews, 176*, 1–13.

Foreman, R. D., Schmidt, R., & Willis, W. D. (1979). Effects of mechanical and chemical stimulation of fine muscle afferents upon primate spinothalamic tract cells. *The Journal of Physiology, 286*, 215–231.

Giesler, G. J., Yezierski, R. P., Gerhart, K. D., et al. (1981). Spinothalamic tract neurons that project to medial and/or lateral thalamic nuclei: Evidence for a physiologically novel population of spinal cord neurons. *Journal of Neurophysiology, 46*, 1285–1308.

Lu, G.-W., & Willis, W. D. (1999). Branching and/or collateral projections of spinal dorsal horn neurons. *Brain Research Reviews, 29*, 50–82.

Milne, R. J., Foreman, R. D., Giesler, G. J., et al. (1981). Convergence of cutaneous and pelvic visceral nociceptive inputs onto primate spinothalamic neurons. *Pain, 11*, 163–183.

Simone, D. A., Sorkin, L. S., Oh, U., et al. (1991). Neurogenic hyperalgesia: Central neural correlates in responses of spinothalamic tract neurons. *Journal of Neurophysiology, 66*, 228–246.

Trevino, D. L., Coulter, J. D., & Willis, W. D. (1973). Location of cells of origin of spinothalamic tract in lumbar enlargement of the monkey. *Journal of Neurophysiology, 36*, 750–761.

Westlund, K. N., Carlton, S. M., Zhang, D., et al. (1992). Glutamate-immunoreactive terminals synapse on primate spinothalamic tract cells. *The Journal of Comparative Neurology, 322*, 519–527.

Willis, W. D. (1997). Nociceptive functions of thalamic neurons. In M. Steriade, E. G. Jones, & D. A. McCormick (Eds.), *Thalamus* (Experimental and clinical aspects, Vol. II, pp. 373–424). Amsterdam: Elsevier.

Willis, W. D. (2001). Mechanisms of central sensitization of nociceptive dorsal horn neurons. In M. M. Patterson & J. W. Gray (Eds.), *Spinal cord plasticity: Alterations in reflex function* (pp. 127–183). Boston: Kluwer.

Willis, W. D., & Coggeshall, R. E. (1991). Sensory mechanisms of the Spinal Cord, 2nd edn. Plenum Press, New York.

Willis, W. D., & Westlund, K. N. (1997). Neuroanatomy of the pain system and of the pathways that modulate pain. *Journal of Clinical Neurophysiology, 14*, 2–31.

Willis, W. D., & Coggeshall, R. E. (2004). *Sensory mechanisms of the spinal cord* (3rd ed.). New York: Kluwer /Plenum.

Willis, W. D., Trevino, D. L., Coulter, J. D., et al. (1974). Responses of primate spinothalamic tract neurons to natural stimulation of hindlimb. *Journal of Neurophysiology, 37*, 358–372.

Willis, W. D., Kenshalo, D. R., & Leonard, R. B. (1979). The cells of origin of the primate spinothalamic tract. *The Journal of Comparative Neurology, 188*, 543–574.

Willis, W. D., Zhang, X., Honda, C. N., et al. (2001). Projections from the marginal zone and deep dorsal horn to the ventrobasal nuclei of the primate thalamus. *Pain, 92*, 267–276.

Spinothalamic Neurons

▶ Spinothalamic Input: Cells of Origin (Monkey)

▶ Spinothalamic Tract Neurons, Morphology

Spinothalamic Projections in Rat

Daniel Menétrey
Biomédicale, Paris, France

Definition

The spinal ascending pathway that transmits somesthetic inputs from the level of the spinal cord directly to the thalamus in rat.

Characteristics

The spinothalamic tract (STT) of the rat is a moderately developed pathway comprising 6,000–9,500 spinal neurons according to evaluations based on different tracing techniques (Burstein et al. 1990). This is less than half of the number estimated in macaque monkeys. Spinal segments are unevenly represented (Giesler et al. 1979; Granum 1986; Kemplay and Webster 1986). Half of the neurons lie in the first four cervical segments, the cervical enlargement contains less than 5 %, the lumbar enlargement 33 %, and the thoracic, sacral, and coccygeal segments all together about 10 % (Fig. 1). This indicates that, as for somatosensory information conveyed by dorsal column systems towards the ▶ thalamus, different parts of the body are not represented to the same extent, thus reflecting the relative importance of various body regions in somatic sensibility. The rat's STT development requires a short postnatal maturation period of a few days, during which small-diameter peripheral afferent fiber activities are essential. Its maturation is complete 2 weeks after birth. The STT has a trigeminal counterpart, the trigeminothalamic tract, issuing from neurons lying ahead in the spinal trigeminal nucleus pars caudalis.

The rat's STT is a crossed pathway except for an ipsilateral contingent taking its origin from neurons located deep in the ventral horn of the upper cervical cord. Two main subdivisions have been recognized that differ in the distribution of their cells of origin in the gray matter, the ▶ funiculus of the cord in which their axons ascend and the thalamic nuclei where they project (Giesler et al. 1981). Axons issuing from neurons located within the dorsalmost two thirds of the dorsal horn and the ventral gray matter ascend within the ventrolateral funiculus and project to the lateral thalamus. Axons issuing from neurons lying deep in the dorsal horn and in the intermediate horn ascend within the ventral funiculus and project to the medial thalamus (Giesler et al. 1981). A third subdivision located within the dorsolateral funiculus comprises axons from neurons located in the marginal layer of the dorsal horn (Zemlan et al. 1978). Axon collaterals to extrathalamic targets have been examined but with large discrepancies between results. Three to five percent of cervical STT cells give rise to descending propriospinal collaterals. Rat STT neurons containing neuropeptides have been described for a variety of substances (bombesin, cholecystokinin, dynorphin, enkephalin, galanin, nitric oxide synthase, substance P, somatostatin, vasoactive intestinal polypeptide) that all form very minor subgroups, so that the main neurotransmitter(s) in the rat's STT remain(s) to be identified. The lumbar galanin-containing subgroup is sexually dimorphic with males containing more of these neurons than females. Anatomical evidence has been given in support of rat STT neurons' responsiveness to glutamatergic, GABAergic, serotonergic, thyrotropin-releasing hormone, and substance P inputs.

Terminations of the STT have been reported in a variety of discrete thalamic nuclei sometimes overlapping with those of other somatosensory pathways, the ▶ medial lemniscus (ML), or axon terminals from the ▶ lateral cervical nucleus (LCN). Both main and sparser projection zones have been recognized (Lund and Webster 1967: Mehler 1969; Zemlan et al. 1978). In all, their cells of origin are mostly contralateral. Main targets include the rostral part of the ventroposterolateral nucleus (VPL) where they overlap with those of ML; the intralaminar nuclei, primarily the central lateral nucleus (CL); and the dorsomedial region of the posterior thalamic

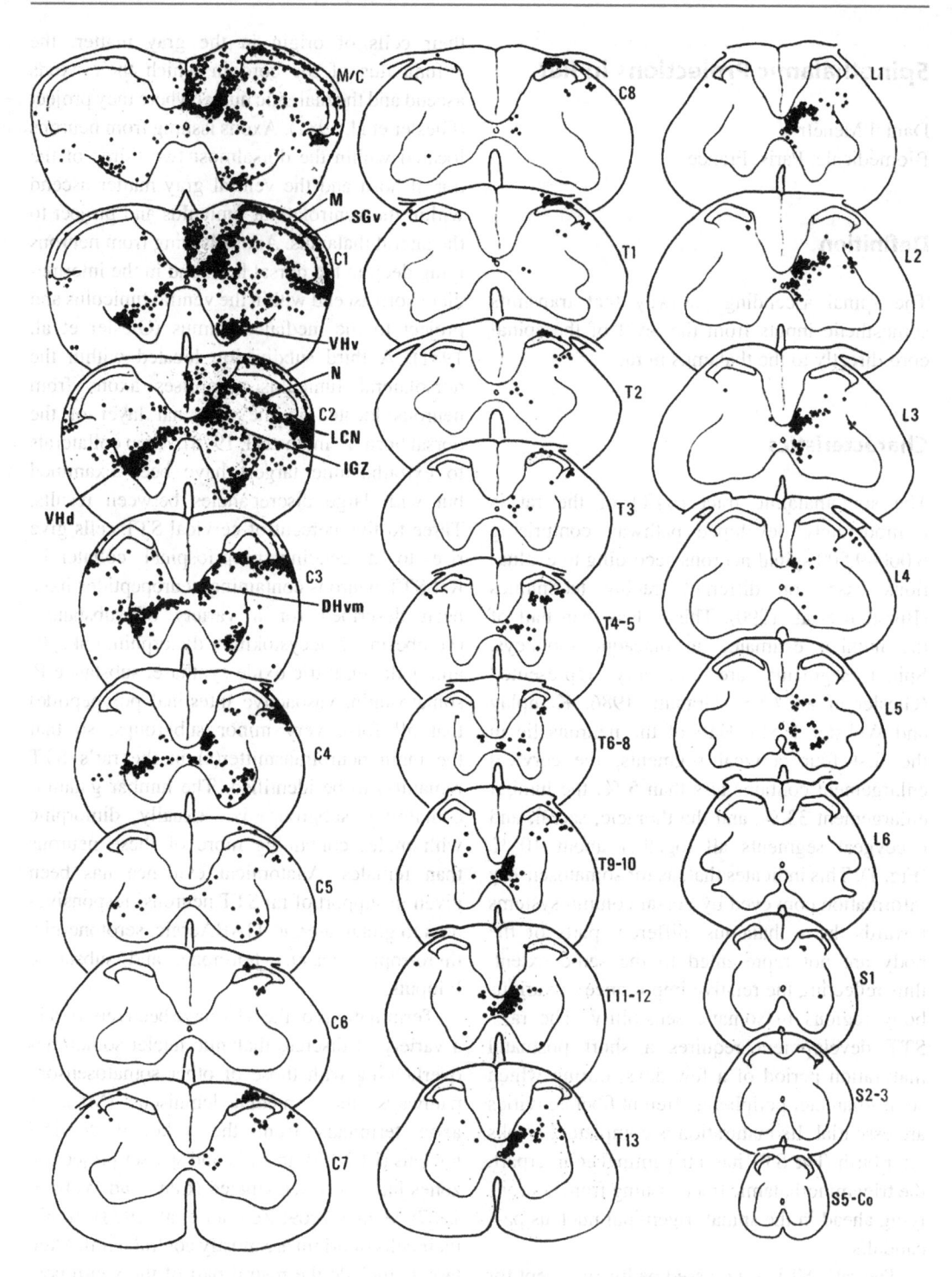

Spinothalamic Projections in Rat, Fig. 1 Cells of origin of the rat's spinothalamic tract. Abbreviations: *Co* coccygeal segments, *C1* to *C8* cervical segments, *DHvm* ventromedial portion of the dorsal horn, *IGZ* intermediate gray zone, *LCN* lateral cervical nucleus, *L1* to *L6* lumbar segments, *M* marginal layer, *M/C* medulla/cord junction, *N* neck of the dorsal horn, *SGv* ventral portion of the substantia gelatinosa, *S1* to *S5* sacral segments, *T1* to *T13* thoracic segments, *VHd and VHv* dorsal and ventral portions of the ventral horn (From Granum (1986))

group (PO). VPL projections are organized in two segregated zones, which are transitional between the ventral lateral nucleus (VL) rostrally and the PO caudally. Terminations in all these main areas share the following features. They have a patchy distribution and terminate in the form of large terminals that contain round synaptic vesicles and make asymmetrical synaptic contacts with large dendrites or dendritic protrusions. STT synaptic profiles cannot be distinguished on a morphological basis from those of ML terminals. However, as ML terminals tend to contact more proximal dendrites as well as cell somata, they probably have greater influence on cell activities (McAllister and Wells 1981; Peschanski et al. 1985; Ma et al. 1987a, b). Since the VPL in rats virtually lacks interneurons, making it simpler than in cat and monkey, its neurons receiving STT synaptic contacts are all ▶ thalamocortical neurons. Sparser targets include the epithalamus, the dorsal portion of the nucleus submedius (gelatinosus), and areas medially bordering the medial geniculate body (Menétrey et al. 1984; Ledoux et al. 1987). This latter area is also a convergent field for projections from the LCN and the inferior colliculus. A role in pairing of acoustic and somesthetic inputs is consequently highly probable in this case.

Cells of origin of the rat's STT have been reported in marginal layer (lamina I), neck of the dorsal horn (nDH, lateral lamina V), ▶ lateral spinal nucleus (LSN), ventromedial portion of the dorsal horn (vmDH, laminae V/VI), and intermediate and ventral gray zones (medial portions of laminae VII/VIII and lamina X). The population of neurons reaching the medial thalamus is distributed preferentially deep in the dorsal horn and intermediate region. With the sole exceptions of the vmDH and LSN groups, the great majority of these cells are confined to the most rostral spinal levels. The vmDH group is a unique feature of the rat's lumbar STT extending from vertebral levels T9 to L4 and projecting to all the identified spinothalamic targets (Fig. 2). Two groups of neurons have been identified electrophysiologically, each constituting a homogeneous population of cells (Menétrey

et al. 1984). Both groups comprise cells supplying the thalamus with integrated hind limb proprioceptive information about limb position in passive postural changes and locomotion, along with the different phases of the ongoing ▶ step cycle (Fig. 3). The LSN constitutes a specific entity in certain mammals including the rat, the guinea pig, the rabbit, and the ferret. Its participation in the STT is moderate compared to other ▶ spinal ascending pathways. LSN neurons have very particular electrophysiological properties. Their axons have slow conduction velocities, sometimes in the unmyelinated range. They do not usually have spontaneous firing or a defined peripheral excitatory receptive field. Receptive fields seem to be located in deep tissues and responses are inconstant. Their exact role in sensory transmission remains to be studied. The LSN and marginal layer receive identical neuropeptidergic afferents of peripheral origin. The LSN and lamina X contain most of the neuropeptidergic ascending tract cells.

Evidence has been given that the rat's STT transmits nociception. It contains dorsal horn cells responding most vigorously to noxious cutaneous stimuli, either marginal layer cells responding solely to noxious stimuli or wide dynamic range cells of nDH responding to both innocuous and noxious stimuli with an increased discharge when the stimulus strength becomes noxious (Giesler et al. 1976). Nociceptive rat STT neurons receive inputs from unmyelinated fibers and develop hypersensitivity in painful pathological conditions (intradermal capsaicin injection, nerve ligature, diabetic neuropathy, spinal cord injury) and are subject to analgesic effects of morphine and brain and vagal stimulation. In spite of these observations, it is however clear that the rat's STT transmits nociception in a way that conflicts with traditional concepts of the STT system as it is regarded from human observations or other animal studies. In fact, anatomical data demonstrate that spinal areas that relay peripheral nociceptive inputs are underrepresented, especially the marginal layer, and exclude some targets such as nucleus submedius, contrary to cat

Spinothalamic Projections in Rat, Fig. 2 Spinal labeling in the lumbar enlargement resulting from discrete horseradish peroxidase injections into five spinothalamic target regions. (**a**) Anterior portion of the VPL as part of the ventrobasal complex (*Vb*), (**b**) nucleus submedius (*Sm*), (**c**) central lateral nucleus (*CL*), (**d**) dorsal portion of the posterior group (*PO*), and (**e**) areas medially bordering the medial geniculate body (*Mg*) (From Menétrey et al. (1984))

and monkey (Menétrey et al. 1984). The role of the rat's STT in transmission of visceral pain has also been questioned. In fact, the rat's STT lacks projections from spinal autonomic columns, as do some other ascending pathways, spinosolitary, lateral spinoreticular, spinoparabrachial, and spinomesencephalic tracts.

Heterogeneity in the functional roles of the rat's STT has thus been established. Since the beginning of the twentieth century, the STT has unequivocally been considered as the main pathway for transmitting information about pain and temperature in all species studied but with a prejudice against the notion that this pathway could assume other putative functions and that

Spinothalamic Projections in Rat, Fig. 3 An example of a vmDH STT cell. This cell had a high ongoing activity directly related to the resting position of the ankle (extension increased it, whereas flexion inhibited it). Innocuous cutaneous stimuli (pressure) excited it, whereas noxious ones (radiant heat) inhibited it. Foot diagrams show the different stages of the application periods for peripheral stimuli, durations of which are indicated by black lines (*lower lines* for the ankle manipulation, upper lines for cutaneous stimulations) (From Menétrey et al. (1984))

other pathways could be involved in pain transmission. Studies in the rat balance this concept.

References

Burstein, R., Dado, R. J., & Giesler, G. J., Jr. (1990). The cells of origin of the spinothalamic tract of the rat: A quantitative reexamination. *Brain Research, 511*, 329–337.

Giesler, G. J., Jr., Menétrey, D., Guilbaud, G., et al. (1976). Lumbar cord neurons at the origin of the spinothalamic tract in the rat. *Brain Research, 118*, 320–324.

Giesler, G. J., Jr., Menétrey, D., & Basbaum, A. I. (1979). Differential origins of spinothalamic tract projections to medial and lateral thalamus in the rat. *The Journal of Comparative Neurology, 184*, 107–126.

Giesler, G. J., Jr., Spiel, H. R., & Willis, W. D. (1981). Organization of spinothalamic tract axons within the rat spinal cord. *The Journal of Comparative Neurology, 195*, 243–252.

Granum, S. L. (1986). The spinothalamic system of the rat. I. Locations of cells of origin. *The Journal of Comparative Neurology, 247*, 159–180.

Kemplay, S. K., & Webster, K. E. (1986). A qualitative and quantitative analysis of the distributions of cells in the spinal cord and spinomedullary junction projecting to the thalamus of the rat. *Neuroscience, 17*, 769–789.

Ledoux, J. E., Ruggiero, D. A., Forest, R., et al. (1987). Topographic organization of convergent projections to the thalamus from the inferior colliculus and spinal cord in the rat. *The Journal of Comparative Neurology, 264*, 123–146.

Lund, R. D., & Webster, K. E. (1967). Thalamic afferents from the spinal cord and trigeminal nuclei. An experimental anatomical study in the rat. *The Journal of Comparative Neurology, 130*, 313–328.

Ma, W., Peschanski, M., & Ralston, H. J., 3rd. (1987a). The differential synaptic organization of the spinal and lemniscal projections to the ventrobasal complex of the rat thalamus. Evidence for convergence of the two systems upon single thalamic neurons. *Neuroscience, 22*, 925–934.

Ma, W., Peschanski, M., & Ralston, H. J., 3rd. (1987b). Fine structure of the spinothalamic projections to the central lateral nucleus of the rat thalamus. *Brain Research, 414*, 187–191.

McAllister, J. P., & Wells, J. (1981). The structural organization of the ventral posterolateral nucleus in the rat. *The Journal of Comparative Neurology, 197*, 271–301.

Mehler, W. R. (1969). Some neurological species differences – A posteriori. *Annals of the New York Academy of Sciences, 167*, 424–468.

Menétrey, D., de Pommery, J., & Roudier, F. (1984). Properties of deep spinothalamic tract cells in the rat, with special reference to ventromedial zone of lumbar dorsal horn. *Journal of Neurophysiology, 52*, 612–624.

Peschanski, M., Roudier, F., Ralston, H. J., 3rd, et al. (1985). Ultrastructural analysis of the terminals of various somatosensory pathways in the ventrobasal complex of the rat thalamus: An electron-microscopic study using wheatgerm agglutinin conjugated to horseradish peroxidase as an axonal tracer. *Somatosensory Research, 3*, 75–87.

Zemlan, F. P., Leonard, C. M., Kow, L. M., et al. (1978). Ascending tracts of the lateral columns of the rat spinal cord: A study using the silver impregnation and horseradish peroxidase techniques. *Experimental Neurology, 62*, 298–334.

Spinothalamic Terminations, Core and Matrix

Edward Jones
Center for Neuroscience, University of California, Davies, CA, USA

Synonyms

Core and matrix; Laminae I/II Inputs to the thalamus; Lateral thalamic nuclei and STT

Definition

The ventral posterior complex can be subdivided into two compartments, based on differential staining with different calcium-binding proteins: a core and a matrix. The spinothalamic terminations from superficial laminae of the spinal cord terminate primarily in the matrix compartment and in multiple nuclei at the caudal pole of the thalamus.

Characteristics

The most direct of the central pain pathways arise in the spinal and medullary dorsal horns and project directly to the thalamus and from thence to the cerebral cortex. The greatest numbers of pain-related fibers in these pathways arise from superficial laminae of the dorsal horn of the spinal and medullary gray matter. For the purposes of this entry, we will not consider to any extent the fibers arising in deep parts of the spinal central gray and we will not consider at all the various brain stem targets of these pathways that may secondarily relay pain information to the thalamus and cerebral cortex.

In monkeys, spinothalamic fibers, whether labeled *en masse* or labeled specifically from the superficial laminae of the spinal dorsal horn, terminate extensively in the thalamus and not exclusively in the ► ventral posterior nucleus, which is the relay for the dorsal column-lemniscal pathway to the primary somatosensory cortex (Rausell et al. 1992; Craig et al. 1994; Graziano and Jones 2004). Thalamic nuclei in which spinal terminations have been identified include the ventral posterior, ► ventral lateral, intralaminar, and ► posterior. Terminations are often identified in the caudal part of the mediodorsal nucleus as well, but this is a matter of nomenclature, for most, if not all, of these are actually in the caudal part of the intralaminar system in a part of the central lateral nucleus often misidentified as the ► densocellular subnucleus of the mediodorsal nucleus (Jones 2007). The terminations in and around the ► ventral posterior nucleus are particularly interesting on account of their relationships to a chemically specified part of the thalamus and their complementarity with the terminations of lemniscal fibers. This is where we come to consider the core and matrix of the primate thalamus.

The *ventral posterior complex* of monkeys and other primates, including humans, is made up of the ► ventral posterior lateral (VPL) and ► ventral posterior medial (VPM) nuclei and two subsidiary nuclei – the ventral posterior inferior nucleus (VPI) and the basal ventral medial nucleus (VMb) (Fig. 1). The basal ventral medial

Spinothalamic Terminations, Core and Matrix, Fig. 1 Schematic view of the background matrix (*left*) of the ventral posterior complex, with its input from the spinothalamic and spinal trigeminothalamic systems and widespread projection to superficial layers of the cerebral cortex, contrasted with the core, restricted to the VPM and VPL nuclei, with its input from the lemniscal system and topographically-organized projection to middle layers of the primary somatosensory cortex (Based on Jones (2007)). Abbreviations: VMb, basal ventral medial nucleus; VPI, ventral posterior inferior nucleus; VPL, ventral posterior lateral nucleus; VPM, ventral posterior medial nucleus

nucleus is a modern name for a nucleus formerly called the parvocellular division of the ▶ VPM. In humans, the ▶ VMb and ▶ VPI together make up the greater part of a nucleus called by Hassler (1959) as ventralis caudalis parvicellularis (v.c.p.c.). Within the four-nucleus complex, two histo- and immunocyto-chemically defined compartments can be identified (Fig. 2) (Rausell and Jones 1991a, b). A *matrix* compartment is found throughout all four nuclei; it stains weakly for ▶ cytochrome oxidase and other oxidative enzymes but contains thalamocortical relay neurons identified by specific immunostaining for the 28 kD calcium binding protein, ▶ calbindin.

A core compartment which stains densely for ▶ cytochrome oxidase and is filled with relay cells that stain specifically for a calcium binding protein, ▶ parvalbumin is imposed on the matrix but only in the VPL and VPM. The core compartment in the VPL and VPM is fragmented, with zones of dense ▶ parvalbumin immunostaining separated by other zones in which the ▶ calbindin immunostained matrix is exposed. The VMb and VPI lack a core compartment and are made up exclusively of the calbindin matrix; fingers of this matrix extend up from the VPI into the VPL and there is a large zone of matrix (the S region) along the medial edge of the VPM.

Spinothalamic Terminations, Core and Matrix, Fig. 2 Photomicrographs of adjacent frontal sections through the thalamus of a macaque monkey, showing the parvalbumin immunoreactive core of the ventral posterior nucleus (a) and the calbindin immunoreactive matrix (b). Bar 500 μm. s, enhanced matrix region of VPM; CL, central lateral nucleus; CM, center médian nucleus; other abbreviations as in Fig. 1. *Arrow* indicates region of parvalbumin and calbindin overlap in medial VPM (From Jones (2007))

The matrix regions of the overall ventral posterior complex extend posteriorly as an enlarged, calbindin-rich, ► cytochrome oxidase-weak and ► parvalbumin-deficient region that is coextensive with nuclei lying posterior and posteromedial to the ventral posterior complex, the ► posterior nucleus and the ► anterior pulvinar nucleus (Jones et al. 2001). The significance of the matrix and core compartments is that they represent the selective termination sites of spino- and spinal trigeminothalamic fibers and of lemniscal fibers respectively (Rausell et al. 1992; Graziano and Jones 2004). Moreover, the cells of the two compartments project differentially upon the cerebral cortex (Fig. 1).

Large regions of the thalamic matrix, such as the S region, VMb, VPI, and the posterior/anterior pulvinar nuclear region, are sufficiently far removed from the core compartment in which lemniscal fibers terminate to represent regions in which spino- and spinal trigeminothalamic fibers exclusively terminate, as emphasized by Craig et al. (1994). In them, there is unlikely to be convergence of the pain and lemniscal systems upon the same thalamocortical relay cells. Within the VPL and VPM nuclei, segregation of the two systems is maintained at a relatively coarse level even within the heart of the VPL and VPM where both the spino- and spinal trigeminothalamic and the lemniscal pathways terminate, but it cannot be ruled out that extension of dendrites of parvalbumin core cells into a calbindin matrix region and vice versa might provide a basis for some convergence. This has been proposed from electron microscopical studies, although terminals of the two systems were not specifically shown to converge on the same cell (Ralston and Ralston 1994).

The parvalbumin immunoreactive cells of the core project in a well-known topographic order on the first somatosensory area of the postcentral gyrus, those receiving proprioceptive inputs ending specifically in area 3a, and those receiving cutaneous inputs in areas 3b and 1 (Fig. 3). In all areas, the terminations of the parvalbumin immunoreactive fibers are concentrated in the middle layers of the cortex (layer IV and deep layer III) (Rausell et al. 1992). The topography in this parvalbumin-defined, dorsal column-lemniscal projection clearly underlies the representational map in the primary somatosensory cortex as commonly defined by multiunit recording of responses to peripheral stimuli. The calbindin-defined cells of the matrix have quite a different pattern of thalamocortical projection. Their fibers

Spinothalamic Terminations, Core and Matrix, Fig. 3 Laminar and areal projections of neurons in the ventral posterior nucleus of monkeys. Core regions (*gray*) receiving inputs from different classes of peripheral mechanoreceptors project to middle layers of specific cortical fields. Matrix regions (*crosses*) receiving inputs from the spinothalamic and spinal trigeminothalamic systems project to superficial layers of all fields (From Jones (2007))

branch more extensively than the fibers of the core and can extend over two or more adjacent cytoarchitectonic areas of the somatosensory cortex and even across the somatosensory and motor areas, without a high degree of topographic order; in these areas, they terminate in superficial layers of the cortex (layers I, II, and upper III) (Fig. 1). There is, thus, a basis for parallel lemniscal and spinothalamic pathways passing through the ventral posterior complex and reaching the cortex of the primary somatosensory cortex in a complementary manner. If, as appears likely, the parvalbumin fibers target not only cortical intrinsic neurons in the middle layers but also the basal and oblique apical side branches of layers III and V pyramidal cells, and the calbindin fibers, the apical tufts of these same cells, then the arrangement seems to form an ideal basis for a coincidence detecting circuit (Jones 2001, 2009). However, the spinothalamic and spinotrigeminothalamic pathway running through the thalamic matrix is more diffuse than that of the core, spreading without topographic order across several areas. Moreover, fibers arising in the matrix, especially that of the VMb and VPI nuclei, project beyond the primary somatosensory cortex into gustatory, visceral, and peri-insular regions of the cortex (Krubitzer and Kaas 1992; Qi et al. 2002).

References

Craig, A. D., Bushnell, M. C., Zhang, E. T., & Blomqvist, A. (1994). A thalamic nucleus specific for pain and temperature sensation. *Nature, 372,* 770–773.

Graziano, A., & Jones, E. G. (2004). Widespread thalamic terminations of fibers arising in the superficial medullary dosal horn of monkeys and their relation to calbindin immunoreactivity. *Journal of Neuroscience, 24,* 248–256.

Hassler, R. (1959). Anatomy of the thalamus. In Schlatenbrand G, Bailey P (eds) Introduction of the stereotaxis with an atlas of the human brain. Stuttgart: Thieme-Verlag, 230–290.

Jones, E. G. (2001). The thalamic matrix and thalamocortical synchrony. *Trends in Neurosciences, 24,* 595–601.

Jones, E. G., Lensky, K. M., & Chan, V. H. (2001). Delineation of thalami nuclei immunoreactive for calcium-binding proteins in and around the posterior pole of the ventral posterior complex. *Thalamus Related Systems, 1,* 213–224.

Jones, E. G. (2007). *The thalamus* (2nd ed.). Cambridge: Cambridge University Press.

Jones, E. G. (2009). Synchrony in the interconnected circuitry of the thalamus and cerebral cortex. *Annals of the New York Academy of Sciences, 1157,* 10–23.

Krubitzer, L. A., & Kaas, J. H. (1992). The somatosensory thalamus of monkeys: Cortical connections and a redefinition of nuclei in marmosets. *The Journal of Comparative Neurology, 319,* 123–140.

Qi, H. X., Lyon, D. C., & Kaas, J. H. (2002). Cortical and thalamic connections of the paricital ventral somatosensory area in marmoset monkeys (*Callithrix jacchus*). *The Journal of Comparative Neurology, 443,* 168–182.

Ralston, H. J. III., & Ralston, D. D. (1994). Medial lemniscal and spinal projections to the macaque thalamus: An electron microscopic study of differing GABAergic circuitry serving thalamic somatosensory mechanisms. *Journal of Neuroscience, 14*, 2485–2502.

Rausell, E., & Jones, E. G. (1991a). Histochemical and immunocytochemical compartments of the thalamic VPM nucleus in monkeys and their relationship to the representational map. *Journal of Neuroscience, 11*, 210–215.

Rausell, E., & Jones, E. G. (1991b). Chemically distinct compartments of the thalamic VPM nucleus in monkeys relay principal and spinal trigeminal pathways to different layers of the somatosensory cortex. *Journal of Neuroscience, 11*, 226–237.

Rausell, E., Bae, C. S., Vinuela, A., Huntley, G. W., & Jones, E. G. (1992). Calbindin and parvalbumin cells in monkey VPL thalamic nucleus: Distribution, laminar cortical projections, and relations to spinothalamic terminations. *Journal of Neuroscience, 12*, 4088–4111.

Spinothalamic Tract Cells

▶ Spinothalamic Input: Cells of Origin (Monkey)

Spinothalamic Tract Neurons, Central Sensitization

Donald A. Simone
Diagnostic and Biological Sciences, University of Minnesota, Minneapolis, MN, USA

Definition

Central sensitization refers to an increase in excitability of dorsal horn neurons following tissue injury and inflammation. The sensitization is typically characterized by an increase in (a) spontaneous activity, (b) responses evoked by mechanical and thermal stimuli applied to the ▶ receptive field, and (c) receptive field area.

Characteristics

Many studies have demonstrated alterations in response properties of spinothalamic tract (STT) neurons following tissue damage. Early studies showed that repeated heating of the skin with heat stimuli from 43 °C to 50 °C produced an increase in spontaneous activity of STT neurons in monkeys and enhanced their responses to heat stimuli applied to the same site (Kenshalo et al. 1979). The sensitization to heat was characterized by a decrease in response threshold and increased response to suprathreshold stimuli. Additional studies showed that repeated noxious heating of the skin decreased the heat response threshold of STT neurons by approximately 5 °C and also increased responses to innocuous cold and mechanical stimuli (Kenshalo et al. 1982; Ferrington et al. 1987). These studies suggested that sensitization of STT neurons may contribute to the ▶ hyperalgesia that occurs at the site of injury (primary hyperalgesia). Other injuries that produce behavioral measures of hyperalgesia, such as joint inflammation and nerve injury, also enhanced the responses of STT neurons. For example, inflammation following intra-articular injection of kaolin and carrageenan-enhanced responses of STT neurons to joint flexion, to mechanical stimuli applied to the skin and to iontophoretic application of excitatory amino acids, suggesting a role for amino acids and their receptors in central sensitization and demonstrating that STT neurons are more sensitive to amino acid neurotransmitters released into the dorsal horn (Dougherty et al. 1992). Similarly, experimental neuropathy produced by spinal nerve ligation in monkeys (Palacek et al. 1992) or by diabetes (streptozotocin model) in rats (Chen and Pan 2002) also resulted in increased responses of STT neurons to mechanical and/or thermal stimuli. Collectively, these studies suggest that sensitization of STT neurons contributes to hyperalgesia produced by inflammation and by nerve injury.

Sensitization of STT neurons that occurs at a site of injury or inflammation may in part be due to sensitization of peripheral nociceptors that project to STT neurons. However, sensitization of STT neurons following a small, localized injury can occur throughout their receptive fields and include areas of the receptive field that are uninjured. Hyperalgesia that extends outside an area of injury is referred to as secondary (Lewis 1942; Hardy et al. 1950) and the contribution of STT neurons to secondary hyperalgesia has been investigated. A role for STT neurons in secondary hyperalgesia was first described by Kenshalo et al. (1982) who reported that responses of STT neurons to innocuous mechanical stimuli, such as brushing the skin, following repeated noxious heating of the skin were enhanced at the stimulation site and also in other areas of the receptive field. To further determine the contribution of STT neurons to secondary hyperalgesia, direct comparisons were made between psychophysical measures of cutaneous hyperalgesia produced by intradermal injection of capsaicin in humans and responses of STT neurons in monkey. In humans, injection of capsaicin (100 μg in 10 μl) produces burning pain, hyperalgesia to heat at the injection site, and secondary mechanical hyperalgesia to stroking the skin with a cotton swab or to poking the skin with punctate stimuli (von Frey monofilaments) within a large area surrounding the injection site (Simone et al. 1989; LaMotte et al. 1991). Injection of capsaicin into the receptive field of STT neurons caused excitation with a time course similar to the capsaicin-evoked pain in humans. In addition, responses of STT to heat stimuli applied to the capsaicin injection site were enhanced and responses to stroking the skin and to stimulation with von Frey monofilaments were increased through the receptive field (Simone et al. 1991a). Responses of STT neurons to iontophoretic application of excitatory amino acids were also increased (Dougherty and Willis 1992).

Sensitization of STT neurons was also found after application of a ▶ pruritic stimulus to the skin (Andrew and Craig 2001). Iontophoretic application of histamine excited a subpopulation of STT neurons in cats and increased their responses to mechanical stimuli. Excitation and sensitization of this unique group of STT neurons may contribute to the sensation of itch as well as to ▶ alloknesis (▶ mechanically evoked itch) that occurs in humans after application of histamine to the skin (Simone et al. 1991b).

Collectively, these data demonstrate that STT neurons develop enhanced responses following tissue injury and inflammation and that the sensitization of STT neurons is not only a reflection of increased input from sensitized nociceptors at the site of injury, but also reflects a general increase in STT neuronal excitability.

References

Andrew, D., & Craig, A. D. (2001). Spinothalamic lamina I neurons selectively sensitive to histamine: A central neural pathway for itch. *Nature Neuroscience, 4*, 72–77.

Chen, S. R., & Pan, H. L. (2002). Hypersensitivity of spinothalamic tract neurons associated with diabetic neuropathic pain in rats. *Journal of Neurophysiology, 87*, 2726–2733.

Dougherty, P. M., & Willis, W. D. (1992). Enhanced responses of spinothalamic tract neurons to excitatory amino acids accompany capsaicin-induced sensitization in the monkey. *Journal of Neuroscience, 12*, 883–894.

Dougherty, P. M., Sluka, K. A., Sorkin, L. S., et al. (1992). Neural changes in acute arthritis in monkeys: I. Parallel enhancement of responses of spinothalamic tract neurons to mechanical stimulation and excitatory amino acids. *Brain Research Reviews, 17*, 1–13.

Ferrington, D. G., Sorkin, L. S., & Willis, W. D. (1987). Responses of spinothalamic tract cells in the superficial dorsal horn of the primate lumbar spinal cord. *Journal of Physiology, 388*, 681–703.

Hardy, J. D., Woolf, H. G., & Goodell, H. (1950). Experimental evidence on the nature of cutaneous hyperalgesia. *The Journal of Clinical Investigation, 29*, 115–140.

Kenshalo, D. R., Leonard, R. B., Chung, J. M., et al. (1979). Responses of primate spinothalamic neurons to graded and to repeated noxious heat stimuli. *Journal of Neurophysiology, 42*, 1370–1389.

Kenshalo, D. R., Leonard, R. B., Chung, J. M., et al. (1982). Facilitation of the responses of primate spinothalamic cells to cold and to tactile stimuli by noxious heating of the skin. *Pain, 12*, 141–152.

LaMotte, R. H., Shain, C. N., Simone, D. A., et al. (1991). Neurogenic hyperalgesia: Psychophysical studies of underlying mechanisms. *Journal of Neurophysiology, 66*, 190–211.

Lewis, T. (1942). *Pain*. London: Macmillan.

Palacek, J., Dougherty, P. M., Kim, S. H., et al. (1992). Responses of spinothalamic tract neurons to mechanical and thermal stimuli in an experimental model of peripheral neuropathy in primates. *Journal of Neurophysiology, 68*, 1951–1966.

Simone, D. A., Baumann, T. K., & LaMotte, R. H. (1989). Dose-dependent pain and mechanical hyperalgesia in humans after intradermal injection of capsaicin. *Pain, 38*, 99–107.

Simone, D. A., Sorkin, L. S., Oh, U., et al. (1991a). Neurogenic hyperalgesia: Central neural correlates in responses of spinothalamic tract neurons. *Journal of Neurophysiology, 66,* 228–246.

Simone, D. A., Alreja, M., & LaMotte, R. H. (1991b). Psychophysical studies of the itch sensation and itchy skin ("alloknesis") produced by intracutaneous injection of histamine. *Somatosensory and Motor Research, 8,* 271–279.

Spinothalamic Tract Neurons, Descending Control by Brain Stem Neurons

William D. Willis
Department of Neuroscience and Cell Biology, University of Texas Medical Branch, Galveston, TX, USA

Synonyms

Centrifugal control of nociceptive processing; Endogenous analgesia system; Supraspinal regulation

Definition

A number of descending pathways to the spinal cord that originate from brain stem neurons have been shown either to inhibit or to excite spinothalamic (STT) and other nociceptive neurons. It is generally accepted that such modulatory actions are likely to affect the processing of nociceptive signals by neural circuits of the spinal cord dorsal horn, as well as of the trigeminal sensory nuclei and hence the experience of pain. Collectively, the descending inhibitory pathways are often referred to as the endogenous analgesia system. By contrast, the descending excitatory pathways are presumed to enhance pain.

Characteristics

Most ► spinothalamic neurons transmit nociceptive signals from relatively restricted cutaneous receptive fields. These cells can usually be activated by volleys in Aαβ, Aδ, and C fibers in the appropriate cutaneous nerve (Fig. 1a, b). The axons of these neurons project directly to the ventral posterior lateral (VPL) nucleus of the contralateral thalamus (Willis and Coggeshall 2004). Some STT cells that project to the contralateral VPL nucleus also convey nociceptive signals from muscles, joints, or viscera (Willis and Coggeshall 2004). The VPL nucleus of the thalamus in turn transmits somatosensory information derived from the spinothalamic tract input to the somatosensory cerebral cortex. Presumably, this information is used for ► sensory discrimination of nociceptive stimuli. Many of these same spinothalamic tract cells also project to the contralateral central lateral nucleus, a part of the thalamic intralaminar complex.

Some nociceptive STT cells have very large cutaneous receptive fields, often including the entire body and face, and project to the medial part of the thalamus, ending in the contralateral central lateral nucleus of the intralaminar complex (Fig. 2) (Giesler et al. 1981). The sensory information conveyed by these STT cells to the intralaminar complex is unlikely to relate to sensory discrimination but instead may play a role in ► arousal, ► attentional mechanisms, or motivational-affective responses to painful stimuli (Giesler et al. 1981; Price 1988).

The nociceptive signals conveyed to the brain by STT cells and other ascending ► nociceptive pathways are necessary for signaling pain. As in other sensory systems, the brain can exert a centrifugal control of the somatosensory information that it receives (Willis 1982). For example, neurons located in the brain stem can modulate nociceptive signals by way of pathways that descend into the spinal cord (Willis 1982; Besson and Chaouch 1987). These brain stem neurons are in turn under the influence of neurons in the cerebral cortex and other areas of the forebrain. In addition, corticospinal projections affect the activity of STT cells. The analgesia that can be produced by stimulation of pathways descending from the brain stem has been termed "stimulation-produced analgesia." The descending neural pathways that cause

Spinothalamic Tract Neurons, Descending Control by Brain Stem Neurons, Fig. 1 Inhibition of the responses of a primate spinothalamic neuron following stimulation in the nucleus raphe magnus (NRM). The peristimulus time histogram in (a) shows the response to stimulation of the sural nerve (arrow) at a strength that activated Aαβ and Aδ fibers. (b) Shows that stimulation in the NRM reduced the response dramatically, especially that to Aδ fibers. In (c) is the response to stimulation of A and C fibers. The inset shows a recording of the C fiber volley from the sural nerve. In (d), stimulation in the NRM eliminated the response to the C-fiber volley. The stimulation site in the NRM is indicated below on a drawing of a midsagittal section through the brain stem. The NRM is outlined, and some of the midline blood vessels are indicated on the drawing (From Gerhart et al. 1981)

Spinothalamic Tract Neurons, Descending Control by Brain Stem Neurons, Fig. 2 An excitatory spino-bulbospinal loop is responsible for the very large receptive fields of primate spinothalamic tract (*STT*) cells that project to the central lateral thalamic nucleus. In (**a**) is shown the excitation of a medially projecting STT cell following stimulation anywhere on the body or face. (**b**) Interruption of the dorsal and lateral funiculi on one side of the spinal cord reduced the responses from most of the receptive field except from one hind limb. (**c**) A lesion of the other dorsal and lateral funiculi restricted the receptive field to one hind limb that which corresponded to the location of the STT cell in the lumbar enlargement. (**d**) Shows sites within the reticular formation that when stimulated caused a powerful excitation of a medially projecting STT cell (From Giesler et al. 1981)

analgesia release endogenous opioid substances, as well as other inhibitory neurotransmitters, such as norepinephrine and serotonin, in the spinal cord dorsal horn (Fields and Besson 1988; Akil and Lewis 1987). The descending pathways that inhibit the processing of nociceptive information are often called the ▶ endogenous analgesia system. The action of morphine on opiate receptors contained in nociceptive neurons explains the ability of morphine and other opiates to produce an effective analgesia. Similarly, other exogenous drugs can affect pain transmission through action on other receptors.

Brain stem regions that have been found to inhibit the activity of spinothalamic neurons include the midbrain ▶ periaqueductal gray matter, the ▶ rostral ventral medulla (which includes the nucleus raphe magnus and adjacent parts of the reticular formation), and the ▶ dorsolateral pons (which includes the locus coeruleus, subcoeruleus, and parabrachial nuclei). Stimulation in the periaqueductal gray or in the nucleus raphe magnus in primates (Fig. 1) can produce a powerful inhibition of spinothalamic neurons (Hayes et al. 1979) (Willis et al. 1977; Gerhart et al. 1981). This occurs in part through an opioid mechanism and in part by the release of serotonin in the spinal cord. Another region that when stimulated causes the inhibition of spinothalamic tract neurons is the dorsolateral pons (Girardot et al. 1987). This inhibition can be attributed to the release of norepinephrine in the spinal cord. Stimulation of forebrain areas, such as the ventrobasal thalamus and the postcentral gyrus, can also result in the inhibition of STT cells (Gerhart et al. 1983; Yezierski et al. 1983).

On the other hand, primate STT cells can be excited following stimulation in the primary motor cortex or the medullary pyramid (Yezierski et al. 1983). Another site that when stimulated can result in the excitation of STT cells is the medullary reticular formation (Haber et al. 1980). However, it is unclear if this excitation is due to activation of reticular formation neurons or of fibers of passage coming from more rostral levels of the brain stem.

The very large cutaneous receptive fields of STT cells that project to the central lateral thalamic nucleus depend on an excitatory loop from the spinal cord to the brain stem and back to the spinal cord, since transection of the spinal cord at a rostral level reduces the receptive fields of such neurons to a restricted area on a single limb (Fig. 2a–c) (Giesler et al. 1981). Stimulation within the pontine reticular formation can produce a long-lasting excitation of such neurons (Fig. 2d) (Giesler et al. 1981).

An excitatory loop through the brain stem is also very important for visceral nociceptive signals (Cervero and Wolstencroft 1984). Although some of the responses of spinal cord neurons to noxious visceral stimuli are mediated by local interneuronal circuits, a major part of these responses depends on the transmission of information to the brain stem and subsequent activation of a descending excitatory projection. This excitatory feedback can be considered a means for amplification of the visceral nociceptive signals.

References

Akil, H., & Lewis, J. W. (1987). *Neurotransmitters and pain control*. Basel: Karger.

Besson, J. M., & Chaouch, A. (1987). Peripheral and spinal mechanisms of nociception. *Physiological Reviews, 67*, 67–186.

Cervero, F., & Wolstencroft, J. H. (1984). A positive feedback loop between spinal cord nociceptive pathways and antinociceptive areas of the cat's brain stem. *Pain, 20*, 125–138.

Fields, H. L., & Besson, J. M. (1988). *Pain modulation. Progress in brain research* (Vol. 77). Amsterdam: Elsevier.

Gerhart, K. D., Wilcox, T. K., Chung, J. M., et al. (1981). Inhibition of nociceptive and nonnociceptive responses of primate spinothalamic cells by stimulation in medial brain stem. *Journal of Neurophysiology, 45*, 121–136.

Gerhart, K. D., Yezierski, R. P., Fang, Z. R., et al. (1983). Inhibition of primate spinothalamic tract neurons by stimulation in ventral posterior lateral (VPL$_c$) thalamic nucleus: Possible mechanisms. *Journal of Neurophysiology, 49*, 406–423.

Giesler, G. J., Yezierski, R. P., Gerhart, K. D., et al. (1981). Spinothalamic tract neurons that project to medial and/or lateral thalamic nuclei: Evidence for a physiologically novel population of spinal cord neurons. *Journal of Neurophysiology, 46*, 1285–1325.

Girardot, M. N., Brennan, T. J., Ammons, W. S., et al. (1987). Effects of stimulating the subcoeruleus-parabrachial region on the non-noxious and noxious responses of T2–T4 spinothalamic tract neurons in the primate. *Brain Research, 409*, 19–30.

Haber, L. H., Martin, R. F., Chung, J. M., et al. (1980). Inhibition and excitation of primate spinothalamic tract neurons by stimulation in region of nucleus reticularis gigantocellularis. *Journal of Neurophysiology, 43*, 1578–1593.

Hayes, R. L., Price, D. D., Ruda, M. A., et al. (1979). Suppression of nociceptive responses in the primate by electrical stimulation of the brain or morphine administration: Behavioral and electrophysiological comparisons. *Brain Research, 167*, 417–421.

Price, D. D. (1988). *Psychological and neural mechanisms of pain*. New York: Raven.

Willis, W. D. (1982). Control of nociceptive transmission in the spinal cord. In D. Ottoson (Ed.), *Progress in sensory physiology 3*. New York: Academic.

Willis, W. D., & Coggeshall, R. E. (2004). *Sensory mechanisms of the spinal cord* (3rd ed.). New York: Kluwer/Plenum.

Willis, W. D., Haber, L. H., & Martin, R. F. (1977). Inhibition of spinothalamic tract cells and interneurons by brain stem stimulation in the monkey. *Journal of Neurophysiology, 40*, 968–981.

Yezierski, R. P., Gerhart, K. D., Schrock, B. J., et al. (1983). A further examination of effects of cortical stimulation in primate spinothalamic tract cells. *Journal of Neurophysiology, 49*, 424–441.

Spinothalamic Tract Neurons, Glutamatergic Input, Fig. 1 Immunogold staining for glutamate (*GLU*) in an "R" type terminal (contains round, clear vesicles) apposing an HRP-labeled STT neuron (From Westlund et al. 1992)

Spinothalamic Tract Neurons, Glutamatergic Input

Susan M. Carlton
Neuroscience and Cell Biology, University of Texas Medical Branch, Galveston, TX, USA

Characteristics

Anatomical Evidence for Glutamatergic Input

Anatomical evidence that ► glutamate-containing terminals synapse on spinothalamic tract (STT) cells is derived from immunohistochemical studies in primate and rat spinal cord. Analysis of high threshold or wide dynamic range primate STT cells demonstrates that glutamate-labeled terminals constitute 46 % of the terminal population apposed to the STT cell body, and 50 % of the profiles apposed to STT dendrites (see Fig. 1; also see ► Spinothalamic Neuron, Fig. 2b) (Westlund et al. 1992). In terms of surface length, this constitutes 54 % of the soma surface length and 50 % of the dendritic surface length. In the rat, type I and II STT cells in laminae I and II of the dorsal horn have been analyzed. Type I STT cells have a pyramidal (triangular) soma with dendrites oriented along the rostro-caudal axis of the cord; type II STT cells have a rounded soma, with numerous primary dendrites extending in all directions (Lima and Coimbra 1988). Glutamatergic input composed 37 % of the terminal population on the 15 STT cells analyzed, and this percentage remains uniform for the soma and dendrites (Lekan and Carlton 1995). There is no significant difference among the 15 STT cells analyzed, suggesting that type I and type II STT cells of the superficial dorsal horn have similar synaptic organizations in relation to glutamatergic input.

A variety of glutamate receptor subunits have been anatomically localized on STT cells including the NMDA receptor subunit NMDAR1 and non-NMDA (AMPA) subunits GluR1 and GluR2/3 (Ye and Westlund 1996). Based on the pharmacological studies discussed below, it is highly likely that metabotropic Glu receptor subunits are also expressed by STT cells.

Pharmacological Evidence for Glutamatergic Input

Several lines of pharmacological evidence indicate that STT cells are modulated by glutamate. Initial studies demonstrated that microiontophoresis of glutamate, or its agonists, in the vicinity of STT cells, excite the cells (Jordan et al. 1978) in a dose-dependent manner (Fig. 2) (Willcockson et al. 1984; Dougherty et al. 1992). Glutamate also participates in the ► sensitization of STT neurons, a phenomenon where responses of STT neurons to the glutamate agonist ► *N*-methyl-D-aspartate (NMDA) are potentiated following a single combined application of

Spinothalamic Tract Neurons, Glutamatergic Input, Fig. 2 Rate histograms of the responses of a representative wide dynamic range STT cell to iontophoretic release of excitatory amino acids (EAAs) before, during, and following dialysis of the spinal cord with CNQX. (a–d) Shows the control responses of the cell to graded iontophoretic current pulses of AMPA, GLUT, NMDA, and aspartate (ASP). (e–h) Shows the responses of the cell to the EAAs following infusion of CNQX. (i–l) Shows the responses to the EAAs 4 h after termination of the CNQX. Bin widths are 100 ms (Dougherty et al. 1992)

NMDA + Substance P. The potentiation usually outlasts the coadministration of drugs by as much as several hours, and is accompanied by an increase in the response of STT cells to mechanical stimulation of the skin. These results provide a cellular model for ▶ secondary hyperalgesia (Dougherty and Willis 1991; Dougherty et al. 1993). Blockade of glutamatergic non-NMDA receptors results in a nearly complete elimination of responses of STT cells to noxious and non-noxious stimuli, while blockade of NMDA receptors attenuates responses to noxious input only (Dougherty et al. 1992). NMDA (Dougherty et al. 1992, 1993) and Group I metabotropic glutamate (mGlu) receptor (Neugebauer and Willis 1999) antagonists, or Group II and III mGlu receptor agonists (Neugebauer et al. 2000), can prevent or reverse STT sensitization that develops after intradermal injection of ▶ capsaicin. Taken together, these results indicate that glutamate is involved in the normal transmission of sensory input from primary afferent fibers to STT cells, and the sensitization of STT cells to peripheral input.

Sources of Glutamate in the Dorsal Horn

There are three possible sources of the glutamate-containing terminals that contact STT cells. Many arise from primary afferent fibers, since it is known that the majority of primary sensory neurons contain glutamate (Battaglia and Rustioni 1988), and that STT cells receive input from primary afferents (Carlton et al. 1989). Spinal interneurons contain glutamate and are likely to synapse onto STT cells. Finally, descending systems originating from the cerebral cortex (Coulter et al. 1974) and reticular formation (Haber et al. 1978) can excite STT neurons, probably through glutamate release.

References

Battaglia, G., & Rustioni, A. (1988). Coexistence of glutamate and substance P in dorsal root ganglion neurons of the rat and monkey. *The Journal of Comparative Neurology, 277*, 302–312.

Carlton, S. M., Lamotte, C. C., Honda, C. N., Surmeier, D. J., Delanerolle, N., & Willis, W. D. (1989). Ultrastructural analysis of axosomatic contacts on functionally identified primate spinothalamic tract neurons. J Comp Neurol. 281(4):555–66.

Coulter, J. D., Maunz, R., & Willis, W. D. (1974). Effects of stimulation of sensorimotor cortex on primate spinothalamic neurons. *Brain Research, 65*, 351–356.

Dougherty, P. M., & Willis, W. D. (1991). Enhancement of spinothalamic neuron responses to chemical and mechanical stimuli following combined micro-iontophoretic application on N-methyl-D-aspartic acid and substance P. *Pain, 47*, 85–93.

Dougherty, P. M., Palecek, J., Paleckova, V., Sorkin, L. S., & Willis, W. D. (1992). The role of NMDA and Non-NMDA excitatory amino acid receptors in the excitation of primate spinothalamic tract neurons by mechanical, chemical, thermal, and electrical stimuli. *Journal of Neuroscience, 12*, 3025–3041.

Dougherty, P. M., Palecek, J., Zorn, S., et al. (1993). Combined application of excitatory amino acids and substance P produces long-lasting changes in responses of primate spinothalamic tract neurons. *Brain Research Reviews, 8*, 227–246.

Haber, L. H., Martin, R. F., Chatt, A. B., et al. (1978). Effects of stimulation in nucleus reticularis gigantocellularis on the activity of spinothalamic tract neurons in the monkey. *Brain Research, 153*, 163–168.

Jordan, L. M., Kenshalo, D., Jr., Martin, R. F., et al. (1978). Depression of primate spinothalamic tract neurons by iontophoretic application of 5-hydroxytryptamine. *Pain, 5*, 135–142.

Lekan, H. A., & Carlton, S. M. (1995). Glutamatergic and GABAergic input to rat spinothalamic tract cells in the superficial dorsal horn. *The Journal of Comparative Neurology, 361*, 417–428.

Lima, D., & Coimbra, A. (1988). The spinothalamic system of the rat: Structural types of retrogradely labeled neurons in the marginal zone (lamina I). *Neuroscience, 27*, 215–230.

Neugebauer, V., & Willis, W. D. (1999). Role of metabotropic glutamate receptor subtype mGluR1 in brief nociception and capsaicin-induced central sensitization of primate STT cells. *Journal of Neurophysiology, 82*, 272–282.

Neugebauer, V., Chen, P.-S., & Willis, W. D. (2000). Groups II and III metabotropic glutamate receptors differentially modulate brief and prolonged nociception in primate STT cells. *Journal of Neurophysiology, 84*, 2998–3009.

Westlund, K. N., Carlton, S. M., Zhang, D., et al. (1992). Glutamate-immunoreactive terminals synapse on primate spinothalamic tract cells. *The Journal of Comparative Neurology, 322*, 519–527.

Willcockson, W. S., Chung, J. M., Hori, Y., et al. (1984). Effects of iontophoretically released amino acids and amines on primate spinothalamic tract cells. *Journal of Neuroscience, 4*, 732–740.

Ye, Z., & Westlund, K. N. (1996). Ultrastructural localization of glutamate receptor subunits (NMDAR1, AMPA GluR1 and GluR2/3) on spinothalamic tract cells. *NeuroReport, 7*, 2581–2585.

Spinothalamic Tract Neurons, in Deep Dorsal Horn

Donald D. Price
Oral Surgery and Neuroscience, McKnight Brain
Institute University of Florida, Gainesville,
FL, USA

Definition

The principal ascending pathways for pain (e.g.,
spinothalamic and trigeminothalamic tracts)
originate partly within deep or most ventral layers
(IV-VI of Rexed) of the dorsal horn of the spinal
cord and medulla, wherein neurons receive synaptic input from primary afferent neurons that
supply ► nociceptors in tissue (Fig. 1). These
second-order neurons within the deep dorsal horn
are mainly ► wide dynamic range (WDR) neurons
and to a lesser extent nociceptive-specific (NS)
neurons, and these two types of neurons have
functional significance for processing both
exteroreceptive and interoreceptive information
associated with pain (Willis 1985; Price and
Dubner 1977; Dubner et al. 1986).

Characteristics

Wide Dynamic Range (WDR) Neurons

WDR neurons are present in both superficial and
deep layers of the dorsal horn (Willis 1985; Price
and Dubner 1977). However, NS neurons occur
in highest percentages in the superficial layers
(layers I–II), and the deep layers contain mostly
WDR neurons. Thus, WDR neurons will be
extensively described in this section. Although
they receive synaptic input from primary
A-delta and C-nociceptive afferent neurons
innervating cutaneous, visceral, muscular, or
other tissues, they are classified on the basis of
their responses to cutaneous stimuli (Willis 1985;
Price and Dubner 1977). WDR neurons have
several distinct and defining characteristics.
First, they respond differentially over a broad
range of stimulus intensity, extending from very
gentle to distinctly painful levels of stimulation;
noxious stimuli delivered to the most sensitive
portion of their receptive fields evoke a higher
impulse frequency than any form of innocuous
stimulation. Second, based on multiple tests,
WDR neurons also receive synaptic input from
A-beta primary afferent neurons that
supply sensitive mechanoreceptors in the skin

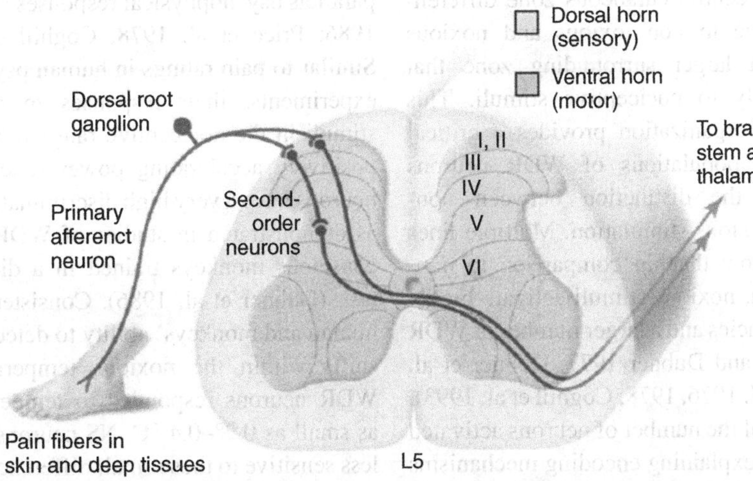

□ Dorsal horn (sensory)

□ Ventral horn (motor)

Dorsal root ganglion

Primary afferent neuron

Second-order neurons

Pain fibers in skin and deep tissues

I, II
III
IV
V
VI

To brain-stem and thalamus

L5

**Spinothalamic Tract Neurons, in Deep Dorsal Horn,
Fig. 1** Schematic of dorsal horn showing superficial
(*I–II*) and deep (*V–VI*) layers. Both regions contain
sensory projection neurons and interneurons and interconnections exist between both regions. Neurons of
the deeper layers tend to be larger than those of
the superficial layers. Superficial neurons (layers *I*
and *II*) and deep dorsal horn neurons (light stippled
zone in layers *V–VI*) tend to be NS and WDR,
respectively

Spinothalamic Tract Neurons, in Deep Dorsal Horn, Fig. 2 Schematics of wide dynamic range (*WDR*) and nociceptive-specific neurons (*NS*). WDR neurons receive synaptic convergence from A-beta mechanoreceptive afferents, A-delta nociceptive afferents, and C-nociceptive afferents, and NS neurons receive synaptic input only from nociceptive afferents. Local inhibitory and excitatory interneurons serve to control the input-output relationships of WDR neurons

(Willis 1985; Price and Dubner 1977; Price et al. 1976, 1978). The demonstration of A-beta input prior to testing with nociceptive stimuli shows that WDR neurons in layer I are not misidentified nociceptive specific, or neurons that respond mainly to heat, pinch, and cold, as has been claimed previously (Craig 2003). WDR neurons have a very distinct receptive field organization that contains a central cutaneous zone differentially responsive to non-noxious and noxious stimuli, and a larger surrounding zone that responds mainly to nociceptive stimuli. This receptive field organization provides a critical basis whereby populations of WDR neurons could encode the distinction between non-noxious and noxious stimulation. Multiple lines of evidence show that, in comparison to non-noxious stimuli, noxious stimuli activate higher impulse frequencies and a larger number of WDR neurons (Price and Dubner 1977; Dubner et al. 1986; Price et al. 1976, 1978; Coghill et al. 1993). Consideration of the number of neurons activated is important in explaining encoding mechanisms because populations are what respond during non-noxious and noxious stimulations (Fig. 2).

Beyond these defining characteristics, several features of WDR neurons are highly consistent with their role in exteroceptive pain. First, they respond in a graded fashion to several forms of noxious stimuli that reflect contact of external objects with the skin. To a major extent, their responses to these stimuli represent characteristics of the objects themselves. Their impulse frequency increases with sharper or hotter objects (above about 45 °C) and in a manner that closely parallels psychophysical responses (Dubner et al. 1986; Price et al. 1978; Coghill et al. 1993). Similar to pain ratings in human psychophysical experiments, their responses to contact heat stimuli in the nociceptive range (45–51 °C) are positively accelerating power functions. WDR neurons have a very high discriminative capacity, as demonstrated in studies of WDR neurons in conscious monkeys trained in a discrimination task (Dubner et al. 1986). Consistent with both human and monkeys' ability to detect very small shifts within the noxious temperature range, WDR neurons responded to temperature shifts as small as 0.2°–0.4 °C. NS neurons were much less sensitive to these small differences, and their discriminative capacity was not sufficient to account for psychophysical discriminations of either humans or monkeys. Deep dorsal horn WDR neurons have this discriminative capacity

(Dubner et al. 1986). This feature of WDR neurons shows that they are a class of neurons that are part of an ascending exteroreceptive pathway that is critical for discrimination of small differences in temperature of external objects. The functional role of such refined discrimination is related to the fact that mammals actually utilize refined nociceptive discrimination to acquire information about properties of external objects. A monkey picking berries out of a thorny bush or an animal stalking prey across hot rocks pursues these tasks in order to eat. The animal must carefully weigh the painfulness of thorns or hot rocks against its need for food. In making such decisions, it must assess whether different objects in the environment or movements in different directions will result in *more* or *less* pain. This high discriminative capacity of the pain system also has an extremely important implication in pain measurement, because it suggests that there are many more discriminative levels of pain than would be implied by simple category scales or numerical rating scales.

A second reason that WDR neurons are involved in exteroreceptive pain is that they are somatotopically organized within the dorsal horn, thereby providing the basis for locating the external source of contact between objects and the body (Price and Dubner 1977; Price et al. 1976, 1978; Yokota and Nishikawa 1980; Yokota 1985). Their ▶ somatotopic organization is a relatively unrecognized feature, and it has recently been claimed that they have only large receptive fields and, therefore, do not display somatotopic organization (Craig 2003). Yet their receptive field sizes range from very small (<2 cm^2), to medium (e.g., a portion of a foot or part of one trigeminal division) to large (larger than the foot or one trigeminal division) in monkeys (Willis 1985; Price and Dubner 1977; Price et al. 1976, 1978), and they are even smaller in cats (Yokota and Nishikawa 1980; Yokota 1985). Moreover, they are somatotopically organized in mediolateral fashion within the trigeminal (Price et al. 1976; Yokota and Nishikawa 1980; Yokota 1985) and spinal dorsal horns (Willis 1985).

A third reason that WDR neurons are involved in exteroreceptive pain is that their responses can distinguish between different forms of externally applied somatic stimulation, including the distinction between tactile stimuli and noxious stimuli. Monkey WDR spinothalamic tract neurons consistently increase their firing frequency (e.g., from below 10 Hz to 15–20 Hz) when a camel-hair brush is dragged gently across the glabrous skin of the foot (Price et al. 1978). They continue to respond at about this frequency for more than 20 s, and as long as 56 s after termination of this stimulus. This phenomenon parallels tactile after-sensations in humans evoked by the same type of stimulus (Melzack and Eisenberg 1968). For example, both human tactile after-sensations and WDR neuron after-responses abruptly terminate when the affected skin is rubbed (Price et al. 1978; Melzack and Eisenberg 1968). Since WDR neurons are the only class of spinal cord neurons that respond with tactile-after-responses, these responses are likely to be sufficient for this form of tactile sensation. Yet the same WDR neurons inevitably respond with a higher impulse frequency to noxious stimulation of the most sensitive portion of their receptive fields (Price and Dubner 1977; Price et al. 1976, 1978; Coghill et al. 1993). Thus, WDR neurons respond differentially to several forms of somatosensory stimulation, particularly when population factors are considered as factors in encoding (Coghill et al. 1993). They encode nociceptive intensity and the distinction between non-noxious and noxious stimuli with exquisite precision (Price and Dubner 1977; Coghill et al. 1993).

Nociceptive-Specific Neurons

Although dorsal horn NS neurons may be somewhat more common within superficial layers of the dorsal horn, they are less commonly found within the deep layers of the dorsal horn (Price and Dubner 1977; Price et al. 1976, 1978; Yokota and Nishikawa 1980; Yokota 1985). NS deep dorsal horn neurons receive synaptic input from the same types of primary nociceptive afferents that contact WDR neurons (Willis 1985;

Price and Dubner 1977; Yokota and Nishikawa 1980). Unlike WDR neurons, however, they do not receive input from A-beta mechanoreceptors, and therefore respond predominantly to nociceptive stimuli. There are at least two types of NS, those receiving exclusive input from A-delta mechanical nociceptive afferents, and those receiving input from different types of A-delta and C-nociceptive afferents. At least some of the latter respond to intense cold and have been termed "heat-pinch-cold" or HPC neurons (Craig 2003). Similar to WDR neurons, they are somatotopically organized within the dorsal horn (Craig 2003). NS appear to encode stimulus location with precision as a result of their small receptive field sizes. The majority of NS neurons have input from multiple types of nociceptive afferent neurons. Thus, the encoding of stimulus submodality within the nociceptive domain (e.g., heat, chemical, and mechanical stimuli) is likely to depend heavily on composition of activated populations of neurons, a composition that includes both WDR and NS neurons. NS neurons may fine-tune the capacity to localize stimulus location as a result of their relatively small receptive fields. A small minority of NS neurons may also fine-tune the ability to detect stimulus submodality as a result of the fact that they respond to input from only one type of primary nociceptive afferent (Willis 1985; Price and Dubner 1977; Price et al. 1976; Craig 2003; Yokota and Nishikawa 1980; Yokota 1985). These functions are exteroreceptive because they deal with detection of the location and physical nature of the object that is in contact with the skin (e.g., pinprick) (Fig. 2).

The Coordination of Superficial and Deep Dorsal Horn Neurons

The physiological characteristics of WDR and NS neurons and the differences between them are found at all levels of the ▶ spinothalamocortical pathway, including the deep dorsal horn where they are intermingled. They are likely to function in concert to encode the sensory discriminative features of noxious stimuli. Indeed, there is a long history of evidence that WDR and NS neurons function together as an integrated system within the dorsal horn (Price and Dubner 1977; Yokota and Nishikawa 1980; Yokota 1985; Khasabov et al. 2002; Suzuki et al. 2002). Recent studies show that destruction of some superficial dorsal horns containing the substance P receptor, some of which are likely to be nociceptive specific, leads to widespread changes throughout the dorsal horn, particularly of WDR neurons (Khasabov et al. 2002; Suzuki et al. 2002). In particular, windup and responses indicative of ▶ central sensitization do not develop in the deep dorsal horn after destruction of neurons in the superficial layers. Thus, after this treatment, WDR neurons of the deep dorsal horn respond normally to innocuous mechanical stimuli and to brief forms of noxious stimulation to some extent, but they no longer display progressive increases in response to repeated C-fiber stimulation, termed windup (Khasabov et al. 2002; Suzuki et al. 2002). Consistent with this change in windup, they also display a large reduction in the capacity to become sensitized (Khasabov et al. 2002; Suzuki et al. 2002). For example, they respond less intensely to intradermal capsaicin and do not show a delayed response to intradermal formalin (i.e., second phase is attenuated). Clearly, superficial and deep dorsal horn neurons are interactive, particularly during persistent pain conditions.

Convergence of Cutaneous, Muscular, and Visceral Primary Afferents on Deep Dorsal Horn Neurons

Primary afferent nociceptive neurons that innervate muscle, skin, and viscera have synaptic terminals on deep dorsal horn nociceptive neurons. These primary afferent neurons from divergent tissue sources often converge on and synaptically activate the same dorsal horn neuron, thereby providing part of the basis of referred pain from these tissues (Willis 1985; Foreman 1977). A common pattern of convergence occurs on these neurons wherein the location of cutaneous, visceral, and muscular "receptive fields" is expected on the basis of known patterns of referred pain. For example, stimulation of cardiac sympathetic nerve axons excites WDR and NS spinothalamic tract neurons that are located at upper thoracic segmental levels of monkeys' spinal cords (Foreman 1977; Milne et al. 1981). These same neurons often have a cutaneous receptive field that extends along the inner aspect of the forearm, the pattern of pain referral in angina pectoralis. Similarly, spinothalamic neurons of upper lumbar segments can be excited both by noxious stimulation of the testicle and by noxious and nonnoxious stimulation of the upper flank and lower abdominal skin areas, areas of pain referral in testicular injury (Willis 1985). Some of these same neurons can also be excited by overdistension of the urinary bladder. The convergence of inputs from different tissue sources onto the same deep dorsal horn neurons provides confirmation of the convergence theory of referred pain, and indicates that deep dorsal horn neurons are involved in interoceptive as well as exteroceptive pain.

Central Projections of Deep Dorsal Horn Neurons

Both WDR and NS neurons of the deep dorsal horn have axons that project to multiple levels of the brain, and the same WDR or NS neuron often has collateral axonal branches projecting to several brain stem sites (Willis 1985; Price et al. 1978). Thus, neurons of the deep dorsal horn receive synaptic input from multiple types of primary nociceptive and non-nociceptive primary afferent neurons, both directly and from superficial neurons. They convey tactile and nociceptive information to multiple brain sites and are related to several functions. Inasmuch as many neurons of ascending spinal pathways origin are WDR neurons, and NS of the deep dorsal horn, these pathways must convey information about a wide variety of somatosensory events, including those related to touch, pain, and possibly innocuous temperature sensibility. This view contrasts sharply with that of "labeled lines" which proposes separate neuron types

encode each form of somatosensation such as touch, warmth, itch, first pain, and second pain (Craig 2003).

References

Coghill, R. C., Mayer, D. J., & Price, D. D. (1993). Spinal cord coding of pain: The role of spatial recruitment and discharge frequency in nociception. *Pain, 53*, 295–309.

Craig, A. D. (2003). Pain mechanisms: Labelled lines versus convergence in central processing. *Annual Review of Neuroscience, 26*, 1–30.

Dubner, R., Bushnell, M. C., & Duncan, G. H. (1986). Sensory-discriminative capacities of nociceptive pathways and their modulation by behavior. In T. L. Yaksh (Ed.), *Spinal afferent processing* (pp. 331–342). New York: Plenum Press.

Foreman, R. D. (1977). Viscerosomatic convergence onto spinal neurons responding to afferent fibers located in the inferior cardiac nerve. *Brain Research, 137*, 164–168.

Khasabov, S. G., Rogers, S. D., Ghilardi, J. R., Peters, C. M., Mantyh, P. W., & Simone, D. (2002). Spinal neurons that possess the substance P receptor are required for the development of central sensitization. *Journal of Neuroscience, 22*, 9086–9098.

Melzack, R., & Eisenberg, H. (1968). Skin sensory afterglows. *Science, 351*, 445–449.

Milne, R. J., Foreman, R. D., Giesler, G. J., & Willis, W. D. (1981). Convergence of cutaneous and pelvic visceral nociceptive inputs onto primate spinothalamic tract neurons. *Pain, 11*, 63–183.

Price, D. D., & Dubner, R. (1977). Neurons that subserve the sensory-discriminative aspects of pain. *Pain, 3*, 307–338.

Price, D. D., Dubner, R., & Hu, J. W. (1976). Trigeminothalamic neurons in nucleus caudalis responsive to tactile, thermal, and nociceptive stimulation of monkey's face. *Journal of Neurophysiology, 39*, 936–953.

Price, D. D., Hayes, R. L., Ruda, M. A., & Dubner, R. (1978). Spatial and temporal transformations of input to spinothalamic tract neurons and their relation to somatic sensation. *Journal of Neurophysiology, 41*, 933–947.

Suzuki, R., Morcuende, S., Webber, M., Hunt, S. P., & Dickenson, A. H. (2002). Superficial NK-[1]expressing neurons control spinal excitability through activation of descending pathways. *Nature Neuroscience, 5*, 1319–1326.

Willis, W. D. (1985). *The pain system*. Basel: Karger.

Yokota, T. (1985). Neural mechanisms of trigeminal pain. In H. L. Fields (Ed.), *Advances in pain research and therapy* (Vol. 9, pp. 211–232). New York: Raven.

Yokota, T., & Nishikawa, N. (1980). Reappraisal of somatotopic tactile representation within trigeminal subnucleus caudalis. *Journal of Neurophysiology, 43*, 700–712.

Spinothalamic Tract Neurons, Morphology

Deolinda Lima[1] and Peter Szucs[2]
[1]Department of Experimental Biology,
Faculdade de Medicina da Universidade do
Porto, Porto, Portugal
[2]Spinal Neuronal Networks, Instituto de Biologia
Molecular e Celular - IBMC, Porto, Portugal

Synonyms

Neuronal Architecture; Neuronal Structure;
Spinal neurons projecting to the thalamus;
Spinothalamic neurons

Definition

Structural characteristics of spinal neurons that
distribute axonal terminals to the thalamus.

Characteristics

The spinothalamic tract differs from several
projecting systems in which the neurons that
contribute ▶ axons to it do not present a unique
and particular morphology. Such diversity cannot
be linearly correlated with the occurrence of
multiple terminal areas in the thalamus, since, in
spite of overall differences, there is a large overlap of the spinal laminae that project to each
thalamic target (see ▶ Spinothalamic Neuron).
The contribution of various spinal laminae to
each spinothalamic ascending pathway is, in itself,
a source of heterogeneity of the participating
neurons. Not only the nature of primary and
modulatory impulses that impinge upon neurons
in each lamina is different, but also several structural aspects, including the size of soma and
dendritic trees, vary considerably from lamina to
lamina. The spinothalamic tract, as well as each
one of its components on its own, is thus likely to
carry an array of various signals that differ as to the
peripheral inputs that generate them as well as the

local and supraspinal actions modulating them.
The spinothalamic system must, therefore,
be considered as particularly fitted to integrate
several kinds of information arriving from the
periphery and filtered in multiple ways in the
spinal cord.

Structural differences between the neuronal
populations at the various laminar sources of
spinothalamic fibers were evaluated taking into
account the size and shape of the soma and the
orientation of the soma and proximal dendritic
arbors (Kobayashi 1998). Neurons in the marginal zone (lamina I) and the lateral spinal
nucleus are smaller than neurons in deeper spinal
gray areas. In the marginal zone, they are oriented
parallel to the dorsal surface of the dorsal horn
when viewed in the transverse plane, whereas in
the lateral spinal nucleus no precise orientation
was depicted. In the deep dorsal horn (laminae
IV–VI), perikarya are larger and either fusiform
or multipolar in shape. However, in what was
called the deep dorsal horn-neck (lamina IV and
dorsal part of lamina V), neurons tend to be
oriented parallel to the laminar borders, whereas
in the deep dorsal horn-base (ventral lamina
V and lamina VI) neurons do not present any
particular orientation. In the internal basilar
group (medial portion of lamina V), perikarya
range in size from small to very large, but in
any case are multipolar in shape and present
dendritic arbors oriented parallel to the lateral
edge of the dorsal funiculus. In the ventral horn
(laminae VII and VIII) large soma prevail. In its
medial portion they are all multipolar and
oriented horizontally, whereas in the lateral
portion they are either multipolar or fusiform,
the latter being oriented vertically to the dorsal
surface of the spinal cord.

A more detailed description of the morphology of the spinothalamic neurons is only liable
for the marginal zone (Lima and Coimbra 1988;
Zhang et al. 1996; Zhang and Craig 1997). In all
the other areas of origin of the spinothalamic
tract, both staining by retrograde tracing and
Golgi impregnations are inconclusive as to
neuronal structural systematization. In the spinal
cord lamina I, based on somatodendritic
morphology, neurons are distributed through

Spinothalamic Tract Neurons, Morphology, Fig. 1 (a, b, e) Photomicrographs of spinal cord lamina I neurons silver-impregnated by the Golgi method or (c, d) retrogradely labeled with CTb from the ventrobasal complex of the thalamus; (a, c) pyramidal neurons from the rat, in horizontal view; (b, d) flattened neurons from the (b) monkey and (d) rat, in horizontal view; (e) fusiform neurons from the cat, in horizontal view

four groups: ▶ pyramidal (Fig. 1a), ▶ flattened (Fig. 1b), ▶ fusiform (Fig. 1e), and ▶ multipolar. These four groups are present in similar proportions in species as different as the pigeon (Galhardo et al. 2000), rat (Lima and Coimbra 1986), cat (Galhardo and Lima 1999), and monkey (Lima et al. 2002), and the fusiform neurons account for almost 50 % of the entire population. They all have dendritic arbors elongated along the rostrocaudal axis, but the mediolateral extent is also considerable in the pyramidal and flattened groups, whereas in the multipolar group ventrally oriented dendrites are characteristically present. In the rat, their soma and dendritic arbors average 25 and 300 μm, respectively, along the rostrocaudal axis (Lima and Coimbra 1986). In the cat (Galhardo and Lima 1999) and monkey (Lima et al. 2002), perikarya and dendritic arbors are slightly larger, averaging respectively, 30 and 320 μm rostrocaudally. There are no differences between the sizes of the various cell groups, but, in all the species studied, each group comprises a few cells (around 6 %) that are twice to three times larger than the remaining. They present nuclei completely devoid of heterochromatin and abundant cytoplasm with particularly extensive rough endoplasmic reticulum (unpublished data). Some of these so-called giant lamina I cells (flattened and pyramidal) appear to receive a particularly important inhibitory GABAergic input (Lima et al. 2002; Puskár et al. 2001). Both small and giant cells participate in the spinothalamic tract (Lima and Coimbra 1988), as well as in other spinofugal pathways. However, lamina I cells "coated" with gephyrin-immunoreactive boutons (Puskár et al. 2001), which make up only 0.2 % of the total neuronal population and ~2.5 % of projection neurons in lamina I, constitute a much higher proportion (~18 %) of the spinothalamic population (Polgár et al. 2008).

Of the four neuronal groups present in lamina I, only two, the pyramidal and the flattened, project to the ventrobasal complex of the thalamus (Lima and Coimbra 1988; Kobayashi 1998). Similar amounts of flattened and pyramidal cells are involved in the lateral spinothalamic tract. As to the medial thalamus, studies based on tracer injections confined to medial thalamic spinal targets did not discriminate the neuronal groups involved. However, in the cat (Zhang et al. 1996) and monkey (Zhang and Craig 1997), retrograde tracing from injections that encompassed both the lateral and medial thalamus revealed labeling of fusiform cells, in addition to pyramidal and flattened cells (the latter misleadingly called multipolar by the authors). It is possible, therefore, that fusiform cells alone, or together with pyramidal and flattened cells, take part in the medial spinothalamic system. In this respect it should be noted that, although species differences between the rat on one side and the cat and monkey on the other have been claimed to

account for the participation of fusiform cells in the spinothalamic tract of the two latter species, the fact that the injection sites are not comparable, but rather include spinal thalamic targets that were not injected in the rat does not allow this sort of conclusion. Moreover, data showing that some spinofugal nociceptive pathways in the pigeon completely match those described in the rat as to the types and relative amounts of lamina I neurons involved (Galhardo et al. 2000), point to a high preservation of the structural characteristics of the ascending nociceptive pathways.

Although distinct in several respects, pyramidal and flattened neurons do share a few characteristics, namely, the mediolateral expansion of the dendritic tree and the scarcity of dendritic spines (Fig. 1a, b) (Lima and Coimbra, 1986, 1988; Galhardo et al. 1999; Lima et al. 2002; Puskár et al. 2001). Pyramidal cells, in addition, have dendritic branches that course inside the white matter overlying the dorsal and dorsolateral surface of lamina I. The dendritic arbor of pyramidal cells ramifies at regular intervals, tracing the direction defined by the primary dendritic trunks (Fig. 1a, c). The dendritic field extends rostrocaudally, mediolaterally, and dorsally as a triangular pyramid protruding into the dorsal funiculus or the dorsolateral fasciculus. In the rat, a few sessile pointed dendritic spines accumulate in proximal dendritic segments, near the soma (Lima and Coimbra, 1986). This feature, however, cannot be observed in the cat and monkey (Galhardo et al. 1999; Lima et al. 2002; Puskár et al. 2001). Axons are of the myelinated type, as revealed by the confinement of silver impregnation to the initial segment.

Flattened cells have disk-shaped soma flattened across the dorsoventral axis, and aspiny, sparsely ramified dendritic arbors that extend as a roughly circular horizontal sheet inside lamina I (Fig. 1b, d) (Lima and Coimbra, 1986, 1988; Galhardo et al. 1999; Lima et al. 2002; Puskár et al. 2001). Axons are of the ▶ myelinated type.

Fusiform cells differ completely from pyramidal and flattened cells as to the strictly longitudinal orientation of the dendritic arbor and the abundance of dendritic spines (Fig. 1e) (Lima and Coimbra, 1986, 1988; Galhardo et al. 1999;

Spinothalamic Tract Neurons, Morphology, Fig. 2 (a, e) Three-dimensional reconstruction of an individually recorded and filled lamina I projection neuron with two lateral collaterals; (b) photomicrographs of the same neuron and of the branchpoint (*arrowheads*) of the first lateral collateral at (c) low and (d) high magnifications; scalebars: (b) 50 μm, (c) 20 μm, (d) 5 μm

Lima et al. 2002; Puskár et al. 2001). The soma is flame-shaped and longitudinally oriented, elongated rostrally and caudally by thick primary dendritic trunks that ramify profusely, but at relatively large distances from the perikarya. Dendritic branches narrow progressively, the more distal ones being extremely thin. Dendritic spines are all of the pedicled type, with round knobs connected to the dendritic shaft by short pedicles of similar length. They are absent from primary dendritic trunks and are particularly abundant in the distal portion of the dendritic arbor. Axons are of the ▶ unmyelinated type. They run longitudinally along with the dendritic arbor, with numerous boutons en passant and a few longitudinal collaterals.

Recent single cell labeling experiments revealed novel features of the axons of lamina I neurons projecting in the contralateral anterolateral tract (ALT), some of which are presumably spinothalamic cells. In ~75 % of the cases, the projecting axon originates from a major dendrite, and this branching point was frequently located after one or two dendritic branches (Szucs et al. 2010). The functional significance of such dendritic origin could be that targeted synaptic input to the intermittent

dendritic region can effectively modulate spike generation.

Local (ipsilateral) axon collaterals (Fig. 2) were frequently encountered in lamina I ALT flattened and pyramidal projection cells (Szucs et al. 2010). The mean diameter of the collaterals was 0.21 ± 0.02 μm, being approximately one-fourth the mean diameter of the main axon (0.79 ± 0.03 μm). Collaterals showed morphological features typical of unmyelinated axons and, based on their trajectory and location, they could be divided into three major categories: dorsal, lateral, and ventral.

Dorsal collaterals branch within the dorsal gray matter (laminae I–IV) of the same segment, occasionally entering the Lissauer tract in the overlying white matter, suggesting that postsynaptic target areas include neighboring lamina I–IV.

Lateral collaterals originate at the lateral edge of the dorsal horn and run parallel with the lateral edge of the dorsal gray matter, with no or only a few (one or two) visible branches along their course (Fig. 2). Their possible roles may include multisegmental integration within lamina I and providing input for neurons in the lateral spinal nucleus, some of which also project to the thalamus.

Ventral collaterals branching from the main axon usually have few (up to five) visible branches and frequently reach laminae V–VII, supporting the hypothesis of sensory information "flow" from superficial to deeper laminae of the dorsal horn (Braz and Basbaum 2009; Wall 1960). The presence of spinal collaterals implies that lamina I spinothalamic neurons may additionally participate in intra- and intersegmental spinal cord processing.

Collaterals of spinothalamic axons were also reported at various supraspinal levels in the rat (Al-Khater and Todd 2009). Virtually all (>95 %) spinothalamic lamina I neurons at both levels C7 and L4 were labeled from the lateral parabrachial area (LPb), and between a third and a half from the periaqueductal gray (PAG). This suggests that a significant number of spinothalamic neurons project to all of these targets.

References

Al-Khater, K. M., & Todd, A. J. (2009). Collateral projections of neurons in laminae I, III and IV of rat spinal cord to thalamus, periaqueductal grey matter and lateral parabrachial area. *The Journal of Comparative Neurology, 515,* 629–646.

Braz, J. M., & Basbaum, A. I. (2009). Triggering genetically-expressed transneuronal tracers by peripheral axotomy reveals convergent and segregated sensory neuron-spinal cord connectivity. *Neuroscience, 163,* 1220–1232.

Galhardo, V., & Lima, D. (1999). Structural characterization of marginal (lamina I) spinal cord neurons in the Cat. A Golgi study. *The Journal of Comparative Neurology, 414,* 315–333.

Galhardo, V., Lima, D., & Necker, R. (2000). Spinomedullary pathways in the pigeon (*Columba livia*): Differential involvement of lamina I cells. *The Journal of Comparative Neurology, 423,* 631–645.

Kobayashi, Y. (1998). Distribution and morphology of spinothalamic tract neurons in the rat. *Anatomy and Embryology, 197,* 51–67.

Lima, D., & Coimbra, A. (1986). A Golgi study of the neuronal population of the marginal zone (lamina I) of the rat spinal cord. *The Journal of Comparative Neurology, 244,* 53–71.

Lima, D., & Coimbra, A. (1988). The spinothalamic system of the rat: structural types of retrogradely labelled neurons in the marginal zone (lamina I). *Neuroscience 27,* 215–230.

Lima, D., Castro, A. R., Castro-Lopes, J. M., & Galhardo, V. (2002). *Differential expression of GABAB receptors in monkey lamina I neuronal types; a golgi and immunocytochemical study.* Society for Neuroscience. Abstract Viewer, 258.4. Washington, DC: Society for Neuroscience.

Polgár, E., Al-Khater, K. M., Shehab, S., Watanabe, M., & Todd, A. J. (2008). Large projection neurons in lamina I of the rat spinal cord that lack the neurokinin 1 receptor are densely innervated by VGLUT2-containing axons and possess GluR4-containing AMPA receptors. *Journal of Neuroscience, 28,* 13150–13160.

Puskár, Z., Polgár, E., & Todd, A. J. (2001). A population of large lamina I projection neurons with selective inhibitory input in rat spinal cord. *Neuroscience, 102,* 167–176.

Szucs, P., Luz, L. L., Lima, D., & Safronov, B. V. (2010). Local axon collaterals of lamina I projection neurons in the spinal cord of young rats. *The Journal of Comparative Neurology, 518*(14), 2645–2665.

Wall, P. D. (1960). Cord cells responding to touch, damage, and temperature of skin. *Journal of Neurophysiology, 23,* 197–210.

Zhang, E.-T., & Craig, A. D. (1997). Morphology and distribution of spinothalamic lamina I neurons in the monkey. *Journal of Neuroscience, 17,* 3274–3284.

Zhang, E.-T., Han, Z.-S., & Craig, A. D. (1996). Morphological classes of spinothalamic lamina I neurons in the Cat. *The Journal of Comparative Neurology, 367,* 537–549.

Spinothalamic Tract Neurons, Peptidergic Input

Virginia S. Seybold
Department of Neuroscience, University of Minnesota, Minneapolis, MN, USA

Synonyms

Co-transmission; Neuromodulators; Neuropeptides

Definition

Peptides are short sequences of amino acids that are clipped from larger precursor proteins during synthesis. Peptides are packaged into large, dense-core vesicles within the cell body and transported to the nerve terminal. The majority

Spinothalamic Tract Neurons, Peptidergic Input, Table 1 Peptides expressed in mammalian primary afferent neurons

Peptide	Normal occurrence	Plasticity
Calcitonin gene-related peptide (CGRP)	Most abundant, occurs primarily in small neurons that give rise to Aδ- and C-fibers	↑ with inflammation ↓ in injured neurons
Substance P	Primarily in small neurons that give rise to C-fibers; extensive coexistence with CGRP	↑with inflammation ↓ in injured neurons
Vasoactive intestinal polypeptide (VIP)	Primarily in visceral afferents	↑ in injured, small diameter somatic afferents
Galanin	Low expression	↑ in injured, small diameter somatic afferents
Neuropeptide Y (NPY)	Low expression	↑ in injured, large diameter somatic afferents
Enkephalin	Primarily in small neurons within superficial dorsal horn	↑ in intrinsic neurons with inflammation
Dynorphin	Primarily in small neurons within superficial dorsal horn	↑ in intrinsic neurons with inflammation

of primary afferent neurons of small and intermediate size contain one or more peptides. Peptides released from the central terminals of primary afferent neurons contribute to ▶ central sensitization and release from the peripheral terminals facilitates the inflammatory response and wound healing. In addition to primary afferent neurons, peptidergic input to spinothalamic neurons arises from neurons within the spinal cord as well as projections from the brain stem.

Characteristics

Peptides occur in primary afferent neurons that innervate somatic and visceral targets (Table 1). Several aspects of peptidergic neurotransmission are important in understanding the contribution of peptides to the processing of nociceptive information by spinothalamic neurons (Hökfelt et al. 2000). (1) Peptides are released with small-molecule neurotransmitters such as glutamate, forming the basis for the concept of "▶ co-transmission." (2) Peptides activate receptors that couple to G-proteins and G-proteins initiate the generation of the intracellular messengers that may ultimately enhance or inhibit excitability of spinothalamic neurons depending on the G-protein. (3) The expression of peptides by primary afferent neurons changes with disease

states. Although a variety of peptides have been localized to primary afferent neurons in many species, the two peptides that have received considerable attention in the study of nociception are ▶ substance P and ▶ calcitonin gene-related peptide (CGRP). Both of these peptides have been localized to ▶ nociceptors (Lawson 2002), both are released in response to intense, persistent, peripheral thermal, mechanical, and chemical stimuli (Duggan et al. 1988) and both have been localized to terminals that make synaptic contacts onto spinal neurons in regions that give rise to spinothalamic neurons (De Koninck et al. 1992). Effects of peptides on nociceptive behaviors and spinal cord physiology have been extensively explored, but less is known about their effects on spinothalamic neurons. Therefore, these two peptides will illustrate potential consequences of peptidergic input to spinothalamic neurons.

Co-transmission

The concept of co-transmission stems from the fact that peptides are released with small-molecule neurotransmitters and sometimes other peptides that may also be contained within the same vesicle (e.g., substance P and CGRP). Peptides are synthesized in the neuronal cell body and stored in large dense-core vesicles in the nerve terminal. In contrast, the small molecule neurotransmitter glutamate is recovered from the synaptic space and stored in small synaptic vesicles. Glutamate is believed to be the universal transmitter of mammalian primary afferent neurons and is the transmitter responsible for activation of spinothalamic neurons in response to acute stimuli. Whereas brief stimuli are sufficient to release glutamate from small synaptic vesicles that cluster at a presynaptic density, persistent stimuli that generate higher firing frequencies of longer duration are required to evoke the release of peptides. The differential release of peptides and glutamate in response to neuronal firing may be due to a number of factors including the greater distance of large dense-core vesicles from the synapse, differences in the vesicle-membrane fusion proteins associated with large dense-core vesicles and small synaptic vesicles, and differences in the intracellular calcium gradients generated by different types of stimuli. Consequently, peptide release encodes the intensity of noxious stimuli. Furthermore, because of differences in signaling mechanisms of receptors for peptides compared to receptors for glutamate, antagonists for peptidergic receptors do not modulate acute nociceptive responses; however, central sensitization and hyperalgesia are decreased with both antagonists and gene deletion of peptidergic receptors (Parsons et al. 1996; Laird et al. 2001).

G-Protein Coupled Receptors
Direct effects of peptides on spinal neurons are mediated by metabotropic receptors that couple to G-proteins. In contrast to ▶ ionotropic receptors for glutamate, in which glutamate-binding sites are part of a cation channel that produces depolarization of the neuron on opening (e.g., AMPA receptors), effects of metabotropic receptors are mediated by intracellular messengers, resulting in a longer latency between receptor activation and physiological effect compared to ionotropic receptors. Metabotropic receptors may also activate multiple intracellular signaling pathways and the downstream effects may or may not include modulation of ion channels. Furthermore, there are families of metabotropic receptors that are activated by peptides as well as glutamate. For example, substance P is

a member of a family of peptides that includes ▶ neurokinin A (which is synthesized in primary afferent neurons as part of the same precursor molecule as substance P) and neurokinin B (synthesized by spinal neurons). These peptides activate a family of receptors, neurokinin (NK) 1, 2, and 3, that are differentiated by the relative potency of the three peptides. All of the neurokinin receptors facilitate nociception, but most is known about the NK1 receptor. Substance P has the greatest potency at NK1 receptors. NK1 receptors occur at the highest density within the dorsal horn of the spinal cord and have been localized to spinothalamic neurons. Neurokinin-1 receptors in the spinal cord couple to activation of phospholipase C resulting in the activation of protein kinase C and release of calcium from intracellular stores (Fig. 1). SP contributes directly to central mechanisms of hyperalgesia by increasing the excitability of spinal neurons (Henry 1976) and indirectly, by facilitating the activation of NMDA receptors (Urban et al. 1994). Neurons in superficial laminae of the spinal cord that express NK1 receptors are required for sensitization of spinal neurons to mechanical and thermal stimuli following treatment with capsaicin (Khasabov et al. 2002). Because of the time required for intracellular events, effects of peptides do not contribute to sensation of acute stimuli. However, they contribute to central sensitization and secondary hyperalgesia.

Peptidergic input to spinal neurons has broader consequences than changing the excitability of the spinothalamic neurons. For example, activation of spinal neurokinin receptors has also been linked to the generation of prostaglandins (Yaksh et al. 1999) and nitric oxide (Linden et al. 1999). Both of these messengers diffuse across cell membranes and can act back on terminals of primary afferent neurons; thus, they are referred to as retrograde messengers. Prostaglandins and nitric oxide contribute to hyperalgesia by facilitating the release of neurotransmitters from presynaptic terminals.

Increasing neuronal excitability and the generation of retrograde messengers are postsynaptic

Spinothalamic Tract Neurons, Peptidergic Input, Fig. 1 Postsynaptic events resulting from the release of glutamate (*glu*) and substance P (*SP*) from primary afferent terminals in the spinal cord. Glutamate activates the ionotropic receptors AMPA and NMDA. Influx of Na^+ through these channels results in depolarization of the neuron within milliseconds of glutamate binding to the receptors. Substance P, however, activates neurokinin (*NK*) receptors, which are metabotropic receptors that couple to the activation of phospholipase C (*PLC*). Activation of PLC results in the generation of intracellular messengers, diacylglycerol (*DAG*) and inositol triphosphate (*IP_3*), as well as release of Ca^{2+} from intracellular stores. These intracellular messengers activate kinases, including protein kinase C (*PKC*), which can phosphorylate AMPA and NMDA receptors to increase their responses to glutamate. Kinases can also phosphorylate K^+ channels, closing them and thereby increasing the excitability of the neuron. Increases in intracellular Ca^{2+} from influx through NMDA receptors and release of Ca^{2+} from intracellular stores can activate cyclooxygenase and nitric oxide synthase, which produce the membrane permeable messengers, nitric oxide (NO) and prostaglandins (*PG*), respectively

consequences of peptides that occur over seconds to minutes. However, the same pathways that mediate these effects also interface with pathways for regulation of gene expression. For example, CGRP has no overt behavioral effect when injected intrathecally and does not depolarize spinal neurons. However, ▶ secondary hyperalgesia is blocked in α-CGRP knockout mice (Zhang et al. 2001). Long-term effects of CGRP on hyperalgesia accompanying peripheral inflammation are likely to include regulation of gene expression (Fig. 2). CGRP receptors on spinal neurons couple to the generation of cyclic-AMP (cAMP), which activates an intracellular pathway that results in phosphorylation and activation of the transcription factor cAMP-response element binding protein (CREB). Several proteins that contribute to central sensitization and are increased in conjunction with peripheral inflammation have CRE-binding sites in the promoter region of their genes (e.g., ▶ NK1 receptors, cyclooxygenase 2, ▶ nitric oxide synthase, dynorphin). Treatment of rat spinal neurons with CGRP increases CRE-dependent gene expression

Spinothalamic Tract Neurons, Peptidergic Input, Fig. 2 CGRP activates metabotropic receptors that couple to the generation of cyclic AMP (*cAMP*). Cyclic AMP activates protein kinase A (*PKA*), which can activate other kinases through phosphorylation. PKA and other kinases can phosphorylate the transcription factor cAMP response element binding protein (*CREB*) within the nucleus. When phosphorylated CREB forms a complex with other proteins at the cAMP response element (*CRE*) site within the promoter region of a gene, transcription is initiated. Several proteins that contribute to central sensitization have CRE-binding sites in the promoter region of their genes. Other peptides, such as vasoactive intestinal polypeptide (*VIP*) may have similar effects because receptors for VIP also couple to the generation of cAMP. Peptides that activate receptors that couple to different G-proteins may regulate gene expression through different transcription factors

and substance P binding to spinal neurons (Seybold et al. 2003). As our understanding of pathways that contribute to the regulation of gene expression expands, it is likely that peptides will assume greater prominence in shaping long-term changes in synaptic plasticity that contribute to chronic pain.

Phenotypic Changes

Just as peptidergic neurotransmission may modulate the expression of proteins in spinal neurons, the expression of peptides by primary afferent neurons and spinal neurons changes with disease states. For example, peripheral inflammation increases levels of mRNA for substance P and CGRP in dorsal root ganglia, the site of cell bodies of primary afferent neurons. Peptide synthesis is increased in neurons that already express the peptides and novel expression is evoked in larger, non-nociceptive neurons that give rise to myelinated fibers (Neumann et al. 1996). The increase in mRNA contributes to increased release of substance P and CGRP from terminals in response to peripheral stimuli; the increased release contributes to enhanced effects of these peptides on central sensitization. Conversely, the expression of substance P and CGRP in small neurons within dorsal root ganglia decreases in models of neuropathic pain, and the expression of other peptides, such as vasoactive intestinal polypeptide (VIP), ▶ galanin, and ▶ neuropeptide Y (NPY), increases (Hökfelt et al. 1994). These peptides activate receptors on primary afferent and spinal neurons that promote hyperalgesia through some of the same intracellular signaling pathways described for substance P and CGRP.

However, VIP, galanin, and NPY will activate their own receptors that are distributed among a unique pattern of spinal neurons.

Other Sources of Peptidergic Input

Descending projections from the brain stem and neurons intrinsic to the spinal cord are additional sources of peptidergic input to spinothalamic tract neurons. ▶ Opioid peptides (met-enkephalin, leu-enkephalin, dynorphin, endomorphin) are noteworthy because they activate opioid receptors (mu, delta, and kappa) that mediate inhibitory effects on spinothalamic neurons. Synaptic contacts of terminals containing these peptides onto spinothalamic neurons have been described (Ruda 1982). Opioid analgesia is effective because it mimics neurotransmission of these endogenous substances.

References

De Koninck, Y., Ribeiro-da-Silva, A., Henry, J. L., et al. (1992). Spinal neurons exhibiting a specific nociceptive response receive abundant substance P-containing synaptic contacts. *Proceedings of National Academic Science USA, 89*, 5073–5077.

Duggan, A. W., Hendry, I. A., Morton, C. R., et al. (1988). Cutaneous stimuli releasing immunoreactive substance P in the dorsal horn of the cat. *Brain Research, 451*, 261–273.

Henry, J. L. (1976). Effects of substance P on functionally identified units in cat spinal cord. *Brain Research, 114*, 439–451.

Hökfelt, T., Zhang, X., & Wiesenfeld-Hallin, Z. (1994). Messenger plasticity in primary sensory neurons following axotomy and its functional implications. *Trends in Neurosciences, 17*, 22–30.

Hökfelt, T., Broberger, C., Xu, Z. Q., et al. (2000). Neuropeptides–an overview. *Neuropharmacology, 39*, 1337–1356.

Khasabov, S. G., Rogers, S. D., Ghilardi, J. R., et al. (2002). Spinal neurons that possess the substance P receptor are required for the development of central sensitization. *Journal of Neuroscience, 22*, 9086–9098.

Laird, J. M., Roza, C., De Felipe, C., et al. (2001). Role of central and peripheral tachykinin NK1 receptors in capsaicin-induced pain and hyperalgesia in mice. *Pain, 90*, 97–103.

Lawson, S. N. (2002). Phenotype and function of somatic primary afferent nociceptive neurones with

C-, Adelta- or alpha/beta-fibres. *Experimental Physiology, 87*, 239–244.

Linden, D. R., Jia, Y. P., & Seybold, V. S. (1999). Spinal neurokinin3 receptors facilitate the nociceptive flexor reflex via a pathway involving nitric oxide. *Pain, 80*, 301–308.

Neumann, S., Doubell, T. P., Leslie, T., et al. (1996). Inflammatory pain hypersensitivity mediated by phenotypic switch in myelinated primary sensory neurons. *Nature, 384*, 360–364.

Parsons, A. M., Honda, C. N., Jia, Y. P., et al. (1996). Spinal NK1 receptors contribute to the increased excitability of the nociceptive flexor reflex during persistent peripheral inflammation. *Brain Research, 739*, 263–275.

Ruda, M. A. (1982). Opiates and pain pathways: Demonstration of enkephalin synapses on dorsal horn projection neurons. *Science, 215*, 1523–1525.

Seybold, V. S., McCarson, K. E., Mermelstein, P. G., et al. (2003). Calcitonin gene-related peptide regulates expression of neurokinin1 receptors by rat spinal neurons. *Journal of Neuroscience, 23*, 1816–1824.

Urban, L., Thompson, S. W., & Dray, A. (1994). Modulation of spinal excitability: co-operation between neurokinin and excitatory amino acid neurotransmitters. *Trends in Neurosciences, 17*, 432–438.

Yaksh, T. L., Hua, X. Y., Kalcheva, I., et al. (1999). The spinal biology in humans and animals of pain states generated by persistent small afferent input. *Proceedings of National Academic Science USA, 96*, 7680–7686.

Zhang, L., Hoff, A. O., Wimalawansa, S. J., et al. (2001). Arthritic calcitonin/alpha calcitonin gene-related peptide knockout mice have reduced nociceptive hypersensitivity. *Pain, 89*, 265–273.

Spinothalamic Tract Neurons, Responses to Chemical Stimulation

Donald A. Simone

Diagnostic and Biological Sciences, University of Minnesota, Minneapolis, MN, USA

Definition

Discharge of spinothalamic tract neurons in response to chemical stimulation of the receptive field. The receptive field may be located in skin or in deep tissues.

Characteristics

Many studies have demonstrated that spinothalamic tract (STT) neurons are excited by noxious chemical stimulation of deep tissues as well as skin. Early studies showed that STT neurons in monkeys were excited by injection of bradykinin into the femoral artery (Levante et al. 1975). Although intra-arterial injection of bradykinin in humans produces pain, which suggests that activation of STT neurons contributed to the bradykinin-evoked pain, it is unclear in which tissues the bradykinin receptors were activated, since several different types of tissue receive blood supply from the injected artery. In subsequent studies, responses of STT neurons were determined following noxious chemical stimulation of identified tissues. For example, Foreman et al. (1979) examined responses of STT neurons produced by intra-arterial injection in preparations in which all nerves of the distal hind limb were denervated except for those innervating the triceps surae muscles. Thus, activation of STT neurons following intra-arterial injection was due to excitation of receptors and primary afferent fibers in muscle. It was found that injection of bradykinin, serotonin, or KCL produced strong excitation of STT neurons. Many of the neurons excited were located in the deep dorsal horn. The time course of response and responses to repeated injections of the same chemical differed for the different chemicals used. For example, excitation by bradykinin had a slower time course than serotonin or KCL, and whereas repeated injections of bradykinin produced similar responses, tachyphylaxis occurred for responses to serotonin.

The effect of chemical stimulation of the heart on responses of STT neurons has also been determined. In initial studies, it was found that occlusion of a coronary artery often produced a delayed excitation of STT neurons, perhaps corresponding to the development of ischemia (and pain) in cardiac muscle (Blair et al. 1984). Since bradykinin excites small afferent fibers that innervate the heart and which may contribute to pain associated with angina, it was determined whether intracardiac injection of bradykinin excited STT neurons (Ammons et al. 1985; Blair et al. 1982, 1984). In one study (Blair et al. 1982), it was found that bradykinin (0.3–3.5 µg) excited approximately 75 % of STT neurons located in the superficial and deep dorsal horn in C8 to T5 spinal segments and the average discharge rate after bradykinin was about twice that of the spontaneous activity prior to injection. In addition, chemical stimulation of cardiac or pericardiac tissue with a chemical mixture containing adenosine, bradykinin, prostaglandin, serotonin, and histamine excited STT neurons in the C1–C2 spinal segments (Chandler et al. 2000). Thus, activation of STT neurons that receive input from small caliber cardiac afferent fibers is likely to contribute to anginal pain.

STT neurons are also excited by chemical stimulation of joints. For example, intra-articular injection of kaolin and carrageenan, which has been used as a model of inflammation, increased spontaneous activity of STT neurons (Dougherty et al. 1992).

Chemical stimulation of the skin also excites STT neurons. Intradermal injection of capsaicin (100 µg) produced a robust excitation of STT neurons located in the superficial and deep dorsal horn in monkeys (Simone et al. 1991a, 2004). The magnitude and time course of excitation of STT neurons was similar to the time course of pain in humans following similar injection of capsaicin (Simone et al. 1989; LaMotte et al. 1991). These STT neurons, which were also sensitive to mechanical stimuli, exhibited weak responses to intradermal injection of histamine (Simone et al. 2004). However in cats, iontophoretic application of histamine produced a greater excitation in STT neurons that were not sensitive to mechanical stimulation of their receptive field (Andrew and Craig 2001). Since histamine evokes a sensation of itch in humans (Magerl et al. 1990; Simone et al. 1991b), these studies indicate that a unique subgroup of mechanically insensitive STT neurons may contribute to the sensation of itch.

References

Ammons, W. S., Girardot, M. N., & Foreman, R. D. (1985). Effects of intracardiac bradykinin on T2-T5 medial spinothalamic cells. *American Journal of Physiology, 249*, R147–R152.

Andrew, D., & Craig, A. D. (2001). Spinothalamic lamina I neurons selectively sensitive to histamine: A central neural pathway for itch. *Nature Neuroscience, 4*, 72–77.

Blair, R. W., Ammons, W. S., & Foreman, R. D. (1984). Responses of thoracic spinothalamic and spinoreticular cells to coronary artery occlusion. *Journal of Neurophysiology, 51*, 636–648.

Blair, R. W., Weber, R. N., & Foreman, R. D. (1982). Responses of thoracic spinothalamic neurons to intracardiac injection of bradykinin in the monkey. *Circulation Research, 51*, 83–94.

Chandler, M. J., Zhang, J., Qin, C., et al. (2000). Intrapericardiac injections of algogenic chemicals excite primate C^1-$C2$ spinothalamic tract neurons. *American Journal of Physiology, 279*, R560–R568.

Dougherty, P. M., Sluka, K. A., Sorkin, L. S., et al. (1992). Neural changes in acute arthritis in monkeys: I. Parallel enhancement of responses of spinothalamic tract neurons to mechanical stimulation and excitatory amino acids. *Brain Research Reviews, 17*, 1–13.

Foreman, R. D., Schmidt, R. F., & Willis, W. D. (1979). Effects of mechanical and chemical stimulation of fine muscle afferents upon primate spinothalamic tract cells. *The Journal of Physiology, 286*, 215–231.

LaMotte, R. H., Shain, C. N., Simone, D. A., et al. (1991). Neurogenic hyperalgesia: Psychophysical studies of underlying mechanisms. *Journal of Neurophysiology, 66*, 190–211.

Levante, A., Lamour, Y., Guilbaud, G., et al. (1975). Spinothalamic cell activity in the monkey during intense nociceptive stimulation: Intra-arterial injection of bradykinin into the limbs. *Brain Research, 88*, 560–564.

Magerl, W., Westerman, R. A., Mohner, B., et al. (1990). Properties of transdermal histamine iontophoresis: Differential effects of season, gender, and body region. *The Journal of Investigative Dermatology, 94*, 347–352.

Simone, D. A., Baumann, T. K., & LaMotte, R. H. (1989). Dose-dependent pain and mechanical hyperalgesia in humans after intradermal injection of capsaicin. *Pain, 38*, 99–107.

Simone, D. A., Sorkin, L. S., Oh, U., et al. (1991a). Neurogenic hyperalgesia: Central neural correlates in responses of spinothalamic tract neurons. *Journal of Neurophysiology, 66*, 228–246.

Simone, D. A., Alreja, M., & LaMotte, R. H. (1991b). Psychophysical studies of the itch sensation and itchy skin ("alloknesis") produced by intracutaneous injection of histamine. *Somatosensory and Motor Research, 8*, 271–279.

Simone, D. A., Zhang, X., Li, J., et al. (2004). Comparison of responses of primate spinothalamic tract neurons to pruritic and algogenic stimuli. *Journal of Neurophysiology, 91*, 213–222.

Spinothalamic Tract Neurons, Role of Nitric Oxide

William D. Willis
Department of Neuroscience and Cell Biology, University of Texas Medical Branch, Galveston, TX, USA

Definition

Nitric oxide is a diffusible gas that may be released in nervous tissue during activity. Nitric oxide plays an important role in the ▶ central sensitization of spinothalamic tract neurons, which occur following strong noxious stimuli. The initiation of central sensitization involves the release of synaptic transmitters, including glutamate and the excitatory neuropeptides, ▶ substance P and ▶ calcitonin gene-related peptide. The actions of these transmitters on glutamate and peptide receptors lead to increased intracellular calcium concentration and the activation of a number of signal transduction cascades. One of these signal transduction cascades involves the activation of ▶ nitric oxide synthase, leading to the synthesis and release of nitric oxide within the spinal cord. Nitric oxide in turn activates guanylyl cyclase, resulting in increased production of cyclic GMP. The cGMP activates protein kinase G, which can then phosphorylate proteins important for central sensitization.

Characteristics

▶ Central sensitization of the responses of nociceptive neurons in the spinal cord, including spinothalamic tract cells, can occur following acute inflammation or as a contributor to other painful conditions, such as neuropathic pain. ▶ Peripheral sensitization of nociceptive afferents may occur, as well, but the demonstration of central sensitization is independent of peripheral sensitization. For example, stimulation of an undamaged or uninflamed area of the body adjacent to the damaged area can produce an enhanced activation of spinal neurons that have undergone central sensitization (Simone et al. 1991; Dougherty and Willis 1992). For this reason, it is thought that central sensitization can help explain the development of ▶ secondary mechanical allodynia and ▶ hyperalgesia in the area of skin that surrounds a site of damage (Hardy et al. 1967).

Central sensitization is a form of activity-dependent plasticity and resembles in many details the process of ▶ long-term potentiation (LTP) that is observed in brain areas, such as the hippocampal formation (Willis 2002). A number of laboratories have described a form of LTP that occurs in the spinal cord (see Willis 2002), so the findings of similarities between central sensitization and LTP are not surprising.

A consistent procedure for evoking central sensitization in primate spinothalamic tract cells is to inject ▶ capsaicin intradermally within the receptive field (Simone et al. 1991) (Dougherty and Willis 1992). This generally leads to a powerful activation of the spinothalamic tract cell and an enhanced responsiveness of the neuron to innocuous and sometimes noxious mechanical stimulation of an area of skin surrounding the injection site. Recordings from primary afferent fibers that innervate the same region of skin show that the responsiveness of the peripheral nerve fibers is unchanged (Baumann et al. 1991; LaMotte et al. 1992). Despite this, the responses of the spinothalamic tract cell can be elevated for a period of hours.

Experiments in which antagonists of various neurotransmitter receptors are introduced into the spinal cord near a spinothalamic tract neuron by ▶ microdialysis have shown that the initiation of central sensitization depends on the activation of several kinds of ▶ neurotransmitter receptors, including non-N-methyl-D-aspartate, N-methyl-D-aspartate, and metabotropic glutamate

Spinothalamic Tract Neurons, Role of Nitric Oxide, Fig. 1 Absorbance peaks for NO^{2-} measured by high-performance liquid chromatography using a UV detector. (NO^{3-} was reduced to NO^{2-} so that the measurements represent the combination of both NO metabolites.) (a) Peak produced by a 25 μM standard; (b) baseline peaks produced by three samples taken from the lumbar spinal cord by microdialysis over a period of 1 h after surgery; (c) peaks from samples taken following intradermal injection of capsaicin; (d) peaks showing recovery from the first capsaicin injection; between (d) and (e), a nitric oxide synthase inhibitor was administered through the microdialysis fiber for 40 min. (e) Peaks showing lack of response to a second injection capsaicin (From Wu et al. 1998)

Spinothalamic Tract Neurons, Role of Nitric Oxide, Fig. 2 Effects of microdialysis administration of a nitric oxide donor (3-morpholinosydnonimine, SIN-1) into the dorsal horn of the spinal cord of a monkey while recordings were made from a spinothalamic tract neuron. (a) Responses are shown to stimulation of different parts of the receptive field using weak (Brush), intermediate strength (Press), and strong (Pinch) mechanical stimuli, as well as a noxious thermal stimulus (Heat). The *horizontal bars* over the left three columns of peristimulus time histograms indicate when the mechanical stimuli were applied. A temperature monitor is shown over the histograms in the rightmost column of histograms of the heat responses. The *top*

receptors, as well as neurokinin-1 receptors (Willis 2002). Calcitonin gene-related peptide receptors (CGRP receptors) are also involved in the initiation of central sensitization in dorsal horn nociceptive neurons, since spinal cord administration of a CGRP receptor antagonist blocks central sensitization of nociceptive dorsal horn neurons by intradermal injection of capsaicin (Sun et al. 2003).

The duration of central sensitization is much longer than would be expected for the normal synaptic actions of glutamate or peptides. This leads to the suggestion that central sensitization depends on the activation of ▶ second messenger cascades. One such cascade that has been investigated involves the synthesis and release of ▶ nitric oxide in the spinal cord following intradermal injection of capsaicin. Figure 1 shows the enhanced level of nitrite (and nitrate that was reduced to nitrite) in dialysate collected from the dorsal horn of a rat's spinal cord by microdialysis following an intradermal injection of capsaicin (Wu et al. 1998). The increase in nitrite, which is a metabolite of nitric oxide, was the result of the action of ▶ nitric oxide synthase, since the increase was prevented by administration of a nitric oxide synthase inhibitor.

Several additional lines of evidence suggest a role of nitric oxide in the central sensitization of spinothalamic tract cells. Administration of a nitric oxide donor causes an enhancement of the responses of spinothalamic tract cells to mechanical stimuli (Fig. 2), and microdialysis administration of a nitric oxide synthase inhibitor prevents central sensitization (Lin et al. 1997). Nitric oxide often has its primary effect by activation of guanylyl cyclase, which results in an increased intracellular synthesis of cyclic GMP and as a consequence the activation of protein kinase G. For this reason, the effects of agents affecting the protein kinase G cascade have

been examined. Microdialysis administration of 8-bromo-cyclic GMP enhances the excitability of spinothalamic tract cells, whereas a guanylyl cyclase inhibitor, ODQ, prevents central sensitization by intradermal capsaicin injection (Lin et al. 1999a, b). Glutamate and aspartate are released into the dorsal horn following intradermal injection of capsaicin, and this release is blocked after administration of a protein kinase G inhibitor, but not after administration of a protein kinase A inhibitor (Sluka and Willis 1998).

Other signal transduction cascades have also been shown to be involved in central sensitization, including the calcium/calmodulin kinase II, protein kinase C, and protein kinase A pathways (Willis 2002).

It is as yet unclear what the action of protein kinase G might be in central sensitization. Protein kinase C and protein kinase A are involved in the phosphorylation of glutamate receptors, including the NR1 subunits of N-methyl-D-aspartate receptors and the GluR1 subunits of AMPA receptors (Willis 2002). It is thought that phosphorylation of these receptors makes them more responsive to the effects of glutamate released at synapses on spinothalamic tract cells. Additionally or alternatively, phosphorylation of glutamate receptors may assist in the insertion of more receptor molecules into the neuronal membrane. In addition, the responses of spinothalamic tract cells to inhibitory amino acids are reduced during central sensitization (Lin et al. 1999b). It is not known if this is the result of phosphorylation of inhibitory amino acid receptors.

Two other roles of nitric oxide have been observed in the studies of central sensitization following intradermal capsaicin administration. One role of nitric oxide is to trigger the expression of ▶ Fos protein in the spinal cord dorsal horn (Wu et al. 2000). Administration of a nitric

---

Spinothalamic Tract Neurons, Role of Nitric Oxide, Fig. 2 (continued) row of histograms shows the control responses. Following these recordings, SIN-1 was infused through the microdialysis fiber. The *lower two rows* are the responses 0.5 h and 2 h after the infusion. (**b**) Grouped

data for 15 spinothalamic tract neurons show the enhancement of the responses to mechanical but not heat stimuli during SIN-1 infusions and for 0.5 h afterward. Background activity (BKG) was also increased (From Lin et al. 1999c)

Spinothalamic Tract Neurons, Role of Nitric Oxide, Fig. 3 Administration of a nitric oxide donor increases the expression of Fos protein in the spinal cord. (**a**) Section of the spinal cord of a rat at the level of placement of a microdialysis fiber. The nitric oxide donor, SIN-1, was infused through the microdialysis fiber for 30 min. Spinal cord tissue was removed at 2 h and immunostained for Fos protein. Labeled neurons are plotted dorsal and ventral to the gap made by the microdialysis fiber. There was very little labeling when artificial cerebrospinal fluid was infused instead of SIN-1 (data not shown). Evidently, SIN-1 triggered the expression of Fos. The increase in Fos labeling spread about 400 μm from the edges of the microdialysis fiber. (**b**) Western blot showing the expression of Fos protein when the spinal cord was removed 30 min or 2 h after infusion of artificial cerebrospinal fluid or SIN-1. Fos expression was greatly increased following SIN-1 administration (From Wu et al. 2000)

oxide donor results in Fos expression (Fig. 3), whereas administration of a nitric oxide synthase inhibitor reduces the expression of Fos following capsaicin injection. The immediate-early gene, c-Fos, is thought to act on the nucleus of neurons through expression of Fos protein, with a consequent regulation of gene expression related to nociception. Another action of nitric oxide is to promote the phosphorylation of ► cyclic adenosine monophosphate-responsive element-binding protein (CREB) in the spinal cord (Wu et al. 2002). CREB is also a transcription factor, and its activation by

phosphorylation also leads to changes in gene expression. Phosphorylation of CREB is seen after either intradermal capsaicin injection or after spinal cord administration of a nitric oxide donor. The phosphorylation is blocked by a nitric oxide synthase inhibitor.

References

Baumann, T. K., Simone, D. A., Shain, C. N., et al. (1991). Neurogenic hyperalgesia: The search for the primary cutaneous afferent fibers that contribute to capsaicin-induced pain and hyperalgesia. *Journal of Neurophysiology, 66*, 212–227.

Dougherty, P. M., & Willis, W. D. (1992). Enhanced responses of spinothalamic tract neurons to excitatory amino acids accompany capsaicin-induced sensitization in the monkey. *Journal of Neuroscience, 12*, 883–894.

Hardy, J. D., Wolff, H. G., & Goodell, H. (1967). *Pain sensations and reactions*. New York: Hafner (reprint of 1952 edn published by Williams and Wilkins).

LaMotte, R. H., Lundberg, L. E. R., & Torebjörk, H. E. (1992). Pain, hyperalgesia and activity in nociceptive C units in humans after intradermal injection of capsaicin. *The Journal of Physiology, 448*, 749–764.

Lin, Q., Peng, Y. B., Wu, J., et al. (1997). Involvement of cGMP in nociceptive processing by and sensitization of spinothalamic neurons in primates. *Journal of Neuroscience, 17*, 3293–3302.

Lin, Q., Palecek, J., Paleckova, V., et al. (1999a). Nitric oxide mediates the central sensitization of primate spinothalamic tract neurons. *Journal of Neurophysiology, 81*, 1075–1085.

Lin, Q., Wu, J., Peng, Y. B., et al. (1999b). Inhibition of primate spinothalamic tract neurons by spinal glycine and GABA is modulated by guanosine 3′,5′-cyclic monophosphate. *Journal of Neurophysiology, 81*, 1095–1103.

Lin, Q., Wu, J., Peng, Y. B., et al. (1999c). Nitric oxide-mediated spinal disinhibition contributes to the sensitization of primate spinothalamic tract neurons. *Journal of Neurophysiology, 81*, 1086–1094.

Simone, D. A., Sorkin, L. S., Oh, U., et al. (1991). Neurogenic hyperalgesia: Central neural correlates in responses of spinothalamic tract neurons. *Journal of Neurophysiology, 66*, 228–246.

Sluka, K. A., & Willis, W. D. (1998). Increased spinal release of excitatory amino acids following intradermal injection of capsaicin is reduced by a protein kinase G inhibitor. *Brain Research, 798*, 281–286.

Sun, R. -Q., Lawand, N. B., Lin, Q., et al. (2003). Capsaicin-induced sensitization of dorsal horn neurons is prevented or reversed by CGRP$_{8-37}$, a specific CGRP 1 receptor antagonist. *Neuroscience Abstracts*.

Willis, W. D. (2002). Long-term potentiation in spinothalamic neurons. *Brain Research. Brain Research Reviews, 40*, 202–214.

Wu, J., Lin, Q., McAdoo, D. J., et al. (1998). Nitric oxide contributes to central sensitization following intradermal injection of capsaicin. *NeuroReport, 9*, 589–592.

Wu, J., Fang, L., Lin, Q., et al. (2000). Fos expression is induced by increased nitric oxide release in rat spinal cord dorsal horn. *Neuroscience, 96*, 351–357.

Wu, J., Fang, L., Lin, Q., et al. (2002). The role of nitric oxide in the phosphorylation of cyclic adenosine monophosphate-responsive element-binding protein in the spinal cord after intradermal injection of capsaicin. *The Journal of Pain, 3*, 190–198.

Spinothalamic Tract Neurons, Visceral Input

Robert D. Foreman
Department of Physiology, College of Medicine, University of Oklahoma Health Sciences Center, Oklahoma City, OK, USA

Synonyms

Angina pectoris; Nociceptive projecting neurons; "Pain" pathway; Visceral pain

Characteristics

Spinothalamic Tract Cells: Spinal Cord Processing

The spinothalamic tract originates from cells in the dorsal gray matter, ascends in the contralateral anterolateral funiculus, and terminates in the thalamus. The dorsal and intermediate gray matter of the spinal dorsal is made up of cells serving as interneurons and as the origin of ascending pathways that transmit visceral information to areas of the brain processing sensory information and participating in pain perception (Fig. 1) (reviewed in Willis and Westlund 1997). Of the ascending pathways, the STT is the most studied system for transmitting visceral afferent information to the brain. Axons of STT cells generally cross over in the white commissure to the

contralateral side within one or two segments and then ascend generally in the anterolateral quadrant. However, some axons remain on the ipsilateral side and some are in the dorsolateral quadrant (Apkarian and Hodge 1989). Visceral information from the upper thoracic segments usually converges with input from somatic structures and ascends to the lateral and medial thalamus (reviewed in Foreman 1997, 1999).

Neurophysiological Mechanisms of Visceral Pain and Spinothalamic Tract Cells

Patients with angina pectoris express three main clinical characteristics to describe their symptoms: (a) pain from the heart is generally referred to somatic structures innervated by the same spinal segments that innervate the heart (Ruch 1961); (b) the pain of angina pectoris is referred to proximal and axial body areas, but generally not to distal limbs (Bonica 1990); and (c) angina pectoris is generally felt as deep and not superficial or cutaneous pain (Lewis 1942). In this section, neurophysiological mechanisms are described to support the patient observations of the referred pain associated with myocardial ischemia and other cardiac diseases. The principles of the mechanisms discussed below can also be used to describe referred pain associated with diseases of the esophagus, gall bladder, stomach, colon and rectum, urinary bladder, and reproductive organs.

Viscerosomatic Convergence

Electrophysiological studies show that electrical stimulation of cardiopulmonary afferent fibers excites STT cells in the T1–T6 segments of the spinal cord (Hobbs et al. 1992). Approximately 80 % of the cells recorded in these segments are strongly activated with cardiopulmonary afferent stimulation, and all these cells receive convergent input from somatic structures. It is interesting that cardiopulmonary afferent stimulation has little effect on the activity of cells in the C7 and C8 segments, where the major innervation is to the distal forelimb and hand. This fits with the clinical observations that anginal pain is usually not referred to the hand and distal forelimb (Harrison and Reeves 1968; Sampson and Cheitlin 1971). Thus, the lack of responsiveness

CHARACTERISTICS OF REFERRED PAIN

1. **Pain of visceral origin referred to somatic regions that are innervated from the same spinal segments as the heart.**

2. **Pain is generally referred to *proximal*, but not *distal* somatic structures**

3. **Referred Pain is experienced as *deep pain***

Spinothalamic Tract Neurons, Visceral Input, Fig. 1 Schematic diagram of the neural mechanisms that explain the characteristics of referred pain resulting from angina pectoris. The *green solid line* represents the spinothalamic tract with the cells located in the spinal gray matter in each segment and the axons projecting to the lateral (caudal ventral posterior lateral nucleus [VPLc] and the medial (central lateral [CL] and central median-parafascicular nucleus [CM-Pf]). The *black* areas on the figurines are representative

somatic fields. The *red line* from the heart to the spinal cord represents the sympathetic afferent fibers from the heart. The *red line* from the T1–T5 spinal segments to the C5–C6 segments is that intraspinal pathway transmitting cardiac afferent information that bypasses the C7–C8 segments and projects to the mid-cervical segments. *NTS* Nucleus Tractus Solitarius, *VPi* Ventral posterior inferior nucleus, *VPM* Ventral Posterior Medial nucleus, *VPMpc* Ventral Posterior Medial parvocellular nucleus)

of these cells agrees with clinical observations. Once past the cervical enlargement, STT cells of the C5–C6 segments are again primarily excited by cardiopulmonary sympathetic afferent and somatic stimulation. Somatic fields for these cells are primarily from the chest. An interesting feature of these segments is that they do not receive cardiac input directly from their dorsal root ganglion cells; in fact, the afferent input is a few segments away. Some evidence suggests that cardiopulmonary afferent fibers may activate cell bodies of a propriospinal path where it directly or indirectly makes synaptic connection with STT cells in the cervical region (Nowicki and Szulczyk 1986). It is also possible that afferent branches of the T2 and T3 sympathetic fibers may travel in the zone of

Lissauer for several segments (Sugiura et al. 1989). Thus, convergence of visceral and somatic input onto a common pool of STT cells provides a substrate for explaining the referral of pain to somatic structures.

Proximal and Axial Somatic Involvement

Neurophysiological mechanisms also provide a basis for the proximal nature of the referred pain of angina pectoris. Electrophysiological studies of the STT show that cardiopulmonary afferent input most commonly excites cells with proximal somatic receptive fields (Hobbs et al. 1992). Cardiopulmonary input strongly excites approximately 80 % of the STT cells with proximal somatic receptor fields but only weakly excites 35 % of the cells with distal somatic input. Thus, the relationship of cells with excitatory visceral input and proximal axial fields is highly significant. These neurophysiological observations support the human studies that angina pectoris is most commonly felt in the proximal and axial regions of the left arm and chest. The frequency distribution of angina pectoris shows that the chest is involved more than 95 % of the time, the pain radiates 30–60 % of the time to the left proximal shoulder, and less involvement occurs down the arm (Sampson and Cheitlin 1971). Proximal and axial pain is common for all forms of visceral pain.

Deep, Dull, Diffuse, Aching Pain

The final characteristic of angina pectoris is the deep, diffuse, dull nature of the symptoms. These sensations are comparable to muscle pain. That is, the pain is typically deep and aching and is often associated with referred muscle hyperalgesia. The similarity of muscle pain and visceral pain was shown in patients who suffer frequently from angina pectoris referred unilaterally to the chest and radiating down the inner aspect of the left arm (Lewis 1942). These patients could not distinguish their anginal pain from pain they experienced when hypertonic saline solution was injected into the interspinus ligament of the left eighth cervical or first thoracic spinal segment. The pain is diffuse, continuous, and difficult to describe, although it can be identified as causing suffering (Lewis 1942).

Visceral Pain and Hyperalgesia

Human studies show that referred muscle hyperalgesia resulting from a diseased visceral organ is a common presentation in the clinic (Giamberardino et al. 1993). Although this section focuses on the results of hyperalgesia from different visceral organs, the possibility of hyperalgesia resulting from angina pectoris should also be considered. Patients suffering from pain caused by calculosis of the upper urinary tract experience muscular hyperalgesia with less involvement of overlying cutaneous structures (Giamberardino et al. 1994; Vecchiet et al. 1989). Experimental studies also show that

stimulation of the ureter results in muscular hyperalgesia and ▶ central sensitization of dorsal horn cells (Giamberardino et al. 1997; Laird et al. 1996). Their results show that muscle and deep structures contribute to referred pain resulting from visceral diseases, while cutaneous pain plays a much smaller role. Evidence exists to suggest that the STT plays a role in processing sensations associated with muscle changes resulting from visceral pain. This evidence is based on studies showing that STT cells responding to nociceptive cardiac information received the most potent somatic inputs from muscle in the proximal structures. In contrast, cells with distal cutaneous input received very little if any noxious cardiac input (Hobbs et al. 1992). Thus, converging input from deep tissue and visceral afferent fibers onto STT cells may provide a basis for explaining why visceral pain, such as that resulting from myocardial ischemia, is felt predominantly as a deep or localized suffering pain, generally in proximal structures, such as muscles, tendons, and ligaments. It also supports the idea that some hyperalgesia may remain after episodes of angina pectoris.

References

Apkarian, A. V., & Hodge, C. J., Jr. (1989). Primate spinothalamic pathways II. The cells of origin of the dorsolateral and ventral spinothalamic pathways. *The Journal of Comparative Neurology, 288*, 474–492.
Bonica, J. J. (1990). *Management of pain* (pp. 133–179). London: Lea & Febiger.
Foreman, R. D. (1997). Organization of visceral input. In T. L. Yaksh et al. (Eds.), *Anesthesia: Biologic foundations* (pp. 663–683). Philadelphia: Lippincott-Raven.
Foreman, R. D. (1999). Mechanisms of cardiac pain. *Annual Review of Physiology, 61*, 143–167.
Giamberardino, M. A., Valente, R., & Vecchiet, L. (1993). Muscular hyperalgesia of renal/ureteral origin. In L. Vecchiet, D. Albe-Fessard, & L. Lindblom (Eds.), *New trends in referred pain and hyperalgesia* (pp. 149–160). New York: Elsevier.
Giamberardino, M. A., de Bigotina, P., Martegiani, C., et al. (1994). Effects of extracorporeal shock-wave lithotripsy on referred hyperalgesia from renal/ureteral calculosis. *Pain, 56*, 77–83.
Giamberardino, M. A., Valente, R., Affaitati, G., et al. (1997). Central neuronal changes in recurrent visceral pain. *International Journal of Clinical Pharmacology Research, 17*, 63–66.

Harrison, T. R., & Reeves, T. J. (1968). Patterns and causes of chest pain. In *Principles and problems of ischemic heart disease* (pp. 97–204). Chicago: Year Book Med.

Hobbs, S. F., Chandler, M. J., Bolser, D. C., et al. (1992). Segmental organization of visceral and somatic input onto C_3–T_6 spinothalamic tract cells of the monkey. *Journal of Neurophysiology, 68,* 1575–1588.

Laird, J. M. A., Roza, C., & Cervero, F. (1996). Spinal dorsal horn neurons responding to noxious distension of the ureter in anesthetized rats. *Journal of Neurophysiology, 76,* 3239–3248.

Lewis, T. (1942). *Pain.* New York: Macmillan.

Nowicki, D., & Szulczyk, P. (1986). Longitudinal distribution of negative cord dorsum potentials following stimulation of afferent fibres in the left inferior cardiac nerve. *Journal of the Autonomic Nervous System, 18,* 185–197.

Ruch, T. C. (1961). Pathophysiology of pain. In T. C. Ruch, H. D. Patton, J. W. Woodbury, et al. (Eds.), *Neurophysiology* (pp. 350–368). Philadelphia: Saunders.

Sampson, J. J., & Cheitlin, M. D. (1971). Pathophysiology and differential diagnosis of cardiac pain. *Progress in Cardiovascular Diseases, 23,* 507–531.

Sugiura, Y., Terul, N., & Hosoya, Y. (1989). Difference in distribution of central terminals between visceral and somatic unmyelinated (C) primary afferent fibers. *Journal of Neurophysiology, 62,* 834–840.

Vecchiet, L., Giamberardino, M. A., Dragani, L., et al. (1989). Pain from renal/ureteral calculosis: Evaluation of sensory thresholds in the lumbar area. *Pain, 36,* 289–295.

Willis, W. D., & Westlund, K. N. (1997). Neuroanatomy of the pain system and of the pathways that modulate pain. *Journal of Clinical Neurophysiology, 14,* 2–31.

Spinothalamic Tract Neurons: Neurotransmitter Inputs and Receptors

Andrew J. Todd
Institute of Neuroscience and Psychology,
College of Medical, Veterinary and Life
Sciences, University of Glasgow, Glasgow, UK

Synonyms

Neurotransmitters used by the axons that form synapses onto these neurons; Receptors for these neurotransmitters that are expressed by the neurons; Spinal cord neurons projecting to the thalamus

Definition

Spinothalamic tract neurons receive numerous synapses from a variety of sources, and these release either excitatory or inhibitory neurotransmitters, which act at receptors located in the postsynaptic membrane. In addition, these neurons are the target for other neurotransmitters that diffuse from axons that may or may not be in synaptic contact, and these act on receptors that are more widely distributed on the surface membrane.

Characteristics

Spinothalamic Tract Neurons

The spinothalamic tract originates from neurons located in several regions of the spinal cord gray matter, including lamina I (the marginal layer) and laminae III-VI of the dorsal horn. Spinothalamic neurons do not form a single, homogeneous group: There are several distinct functional populations that can be recognized on the basis of their differing patterns of synaptic inputs, receptor expression, and responses to natural stimuli. Even within a single location (e.g., lamina I), it is likely that not all of the cells belong to a single population. Spinothalamic tract neurons have been examined in several species, including rats, cats, and monkeys. In anatomical studies, these cells are generally identified by retrograde labeling, following injection of tracers into the thalamus. This method shows the cell bodies of the neurons and may reveal enough of their dendritic trees to allow synaptic inputs and receptor expression patterns to be examined (Naim et al. 1997). In physiological studies, spinothalamic tract neurons can be identified by antidromic activation from the thalamus, and their activity in response to natural stimuli can be monitored with extracellular or intracellular recording electrodes. With intracellular electrodes, dye injection can then be used to reveal the dendritic trees, allowing a subsequent search for synaptic inputs (Westlund et al. 1990).

The axons of spinothalamic cells generally cross the midline and ascend in the anterolateral or lateral white matter of the spinal cord, before entering the thalamus. They form part of a set of ascending projections known as the anterolateral tract (ALT), which also innervates several regions in the brainstem, including the periaqueductal gray matter (PAG), the lateral parabrachial area (LPb), and various nuclei within the medulla. Individual neurons within the ALT may send axon collaterals to several of these structures, and may therefore belong to more than one of the constituent tracts. For example, all of the spinothalamic tract cells in lamina I of the rat also project to the LPb, and some also to the PAG (Al-Khater and Todd 2009).

Spinothalamic tract neurons in lamina I have been studied most extensively, partly because projection neurons are particularly numerous in this region, and partly because it is a major receiving zone for nociceptive primary afferents. Indeed, many of its constituent neurons (including the spinothalamic neurons) respond to noxious stimuli. Another population of spinothalamic tract cells that have been examined in some detail consists of neurons with cell bodies in lamina III or IV and long dorsal dendrites that extend into lamina I. Cells of this type have been identified in both monkey and rat (Westlund et al. 1990; Naim et al. 1997). These two groups of cells will be the main subject of this account, as we know considerably more about their synaptic inputs and receptor expression patterns than is the case for other populations. Although spinothalamic tract neurons account for around 40 % of the projection cells in lamina I in the rat cervical enlargement and are fairly numerous in both enlargements in cat and monkey, they are relatively infrequent in the rat lumbar enlargement, where they constitute only ~4 % of all lamina I projection cells (Todd 2010). Because of the relatively small number of lamina I spinothalamic tract cells in rat lumbar spinal cord, many studies of synaptic circuitry have used retrograde labeling from other regions (e.g., the LPb) to identify projection neurons in this lamina. However, as spinothalamic neurons appear to be a broadly representative subset of the spinoparabrachial population (Al-Khater and Todd 2009), much of what we know about the

properties of lamina I spinoparabrachial neurons can be applied to the spinothalamic tract.

Like all spinal neurons, those belonging to the spinothalamic tract can receive synaptic inputs from primary afferent axons of various types, from other spinal neurons (both excitatory and inhibitory) and from descending axons that originate in the brain. Around 80 % of lamina I spinothalamic neurons in the rat express the NK1 receptor (NK1r) (Todd 2010), the main receptor for the neuropeptide substance P, which is found in many nociceptive primary afferents (Lawson 2002).

Primary Afferent Input

Lamina I projection neurons that express the NK1r are densely innervated by primary afferents that contain a variety of neuropeptides including substance P, and the substance P released by these afferents acts through volume transmission on NK1rs in their somatic and dendritic plasma membranes to cause a slow depolarization. However, these peptidergic afferents also release glutamate (the principal fast excitatory transmitter used by all primary afferents), and this acts at the synapses that they form on the cell bodies and dendrites of these cells (Carlton et al. 1990; Todd et al. 2002). In fact, synapses from neuropeptide-containing primary afferents are so numerous that they account for around 50 % of the glutamatergic (excitatory) synapses on the lamina I projection cells (Todd 2010). In the rat, it has been shown that the spinothalamic neurons with cell bodies in lamina III-IV and long dorsal dendrites also strongly express the NK1r, and as is the case for the lamina I cells, these neurons are densely innervated by substance P-containing (nociceptive) primary afferents, which form multiple synapses on their dendritic trees (Todd 2010) (Fig. 1). Since these cells have dendrites that travel through several dorsal horn laminae, they are potentially able to sample a wider variety of primary afferents than the lamina I neurons (whose dendrites remain within lamina I). It has been shown by using transganglionic tracing that the lamina III/IV spinothalamic cells receive a direct synaptic input from low-threshold mechanoreceptive myelinated

Spinothalamic Tract Neurons: Neurotransmitter Inputs and Receptors, Fig. 1 Substance P-immunoreactive axons in contact with a spinothalamic neuron that expresses the NK1 receptor. These confocal images are taken from a sagittal section through the dorsal horn of a rat that had received an injection of a tracer into the contralateral thalamus. (**a**) Shows staining for the NK1 receptor (*green*), and the cell body and dendrites of a large neuron can be seen. The *boxes* correspond to the locations of the other confocal images (*b–f*). (**b**) The cell body of the neuron contains the tracer (*blue*), indicating that this is a spinothalamic tract neuron. (**c–f**) Show parts of the dendritic tree of the neuron, scanned to reveal the NK1 receptor (*green*) and substance P (*red*). Note that many substance P-containing axons (most of which are primary afferents) form contacts with the dendrites of this cell. Scale bars: **a** = 50 μm; **b** = 20 μm; **c–f** = 10 μm (Reproduced with permission from Naim et al. (1997))

(tactile and/or hair) afferents in lamina III (Naim et al. 1998). However, these synapses are far less numerous than those formed by the substance P-containing afferents, which suggests that the major primary afferent drive to these neurons is from the neuropeptide-containing nociceptors. Not all of the spinothalamic tract neurons in the superficial part of the dorsal horn are densely innervated by nociceptive primary afferents. There is a population of giant lamina I projection neurons that lack the NK1r and appear to receive few synapses from primary afferents (Polgár et al. 2008). Many of these cells belong to the spinothalamic tract, and despite the lack of direct primary afferent input, they do respond to noxious stimuli (Polgár et al. 2008; Todd 2010).

Inputs from Other Spinal Cord Neurons

Spinothalamic neurons in laminae I, III, and IV also receive glutamatergic synapses at which the presynaptic axon is not of primary afferent origin. Many (if not all) of these are derived from local excitatory interneurons, and although we know relatively little about the identity of these cells, they are likely to include a type of interneuron known as vertical (or stalked) cells (Todd 2010). These have cell bodies in the outer half of lamina II and axons that enter lamina I, where their targets include NK1r-expressing projection neurons (Lu and Perl 2005). The giant lamina I spinothalamic neurons receive an extremely dense glutamatergic synaptic input on both their cell bodies and dendritic trees, but the origin of this input is unknown (Polgár et al. 2008). Spinothalamic neurons in these laminae also receive numerous inhibitory synapses from axon terminals that contain GABA and/or glycine. While some of this input may originate from axons that have descended from the brainstem (Antal et al. 1996), it is likely that much of it arises from local inhibitory interneurons, which use one or both of these transmitters (Todd 2010). There is some evidence that specific populations of inhibitory interneurons preferentially innervate particular types of projection neuron. For example, axons that contain GABA and neuropeptide Y, and which are thought to originate from inhibitory interneurons in laminae I-III, give rise to around one third of the inhibitory synapses on the lamina III/IV NK1r-expressing spinothalamic cells (Todd 2010; Polgár et al. 2011).

Descending Inputs

Spinothalamic tract neurons are also innervated by descending axons that originate in the brainstem and contain the monoamine transmitters, norepinephrine (NE) or 5-hydroxytryptamine (5-HT). Synaptic contacts have been identified between NE-immunoreactive axon terminals and the cell bodies and dendrites of monkey spinothalamic neurons located in laminae I, IV, or V (Westlund et al. 1990). In addition, 5-HT-containing axons have been shown to innervate NK1r-expressing projection neurons (presumably including spinothalamic tract cells) in laminae I and III-IV in the rat (Todd and Koerber 2005).

Receptors on Spinothalamic Neurons

Receptors for the main excitatory and inhibitory neurotransmitters (glutamate, GABA, glycine) that act on spinothalamic tract neurons must be located at the appropriate synapses on these cells. However, it is generally difficult to detect synaptic receptors with immunocytochemistry, because the high level of protein cross-linking that occurs at synapses following fixation of the tissue with formaldehyde can severely restrict the access of antibodies to the receptors. This is particularly problematic for glutamatergic synapses, where there is a very high density of protein in the synaptic cleft and postsynaptic density. The main ionotropic receptor for glutamate is the AMPA-type receptor, which has a tetrameric structure, and is made up from four different subunits: GluA1-4. Synaptic AMPA receptors on most spinothalamic neurons in laminae I, III, and IV contain the GluA2, GluA3, and GluA4 subunits, whereas at receptors on interneurons in the superficial dorsal horn, GluA4 is replaced by GluA1 (Todd 2010). The significance of this arrangement is not yet known; however, there is evidence that GluA1 located in synapses on interneurons can undergo phosphorylation, which is likely to underlie some form of excitatory synaptic plasticity (Todd 2010). It is not yet known whether a similar mechanism, involving GluA4 subunits, occurs at synapses on spinothalamic tract neurons. We also know little about the synaptic distribution of the other main type of glutamate receptor (the NMDA receptor) on identified spinal cord neurons (Todd and Koerber 2005). Among the metabotropic receptors for glutamate that are present in the dorsal horn, the main type is mGluR5, which is expressed by many small neurons in laminae I–III, but apparently not by the projection neurons. GABA$_A$ receptors, which are expressed by all of the neurons in the spinal cord, are pentameric structures made up from combinations of α, β, and g subunits. However, we do not know the specific combination(s) of subunits that are present in receptors at the GABAergic synapses on spinothalamic tract neurons. GABA$_B$ receptors are also widely distributed in the spinal cord gray matter, but we know little about their association with spinothalamic neurons. Glycine receptors are also pentameric, and most of them are made up from α1 and β subunits. These are located at glycinergic synapses throughout the spinal cord gray matter, including those on spinothalamic tract neurons.

Many neuropeptides are released by axons in the dorsal horn, and it is therefore not surprising that several neuropeptide receptors are present in this region. As stated above, the NK1r is found on many spinothalamic neurons in laminae I, III, and IV. Several other neuropeptide receptors, including the neurokinin 3 (NK3r), μ opioid, neuropeptide Y Y1, somatostatin 2a (sst2a) receptors have been identified on populations of dorsal horn neurons, although in most cases the expression of these receptors is restricted to interneurons in the superficial laminae (Todd and Koerber 2005). However, the Y1 receptor is expressed by projection cells in laminae I and IV-V (Brumovsky et al. 2006), and some of these may belong to the spinothalamic tract. Although several receptors for the monoamine transmitters have been identified in the dorsal horn, we know little about the expression of these on spinothalamic tract neurons (Todd and Koerber 2005).

References

Al-Khater, K. M., & Todd, A. J. (2009). Collateral projections of neurons in laminae I, III, and IV of rat spinal cord to thalamus, periaqueductal gray matter, and lateral parabrachial area. *The Journal of Comparative Neurology, 515*, 629–646.

Antal, M., Petko, M., Polgar, E., Heizmann, C. W., & Storm-Mathisen, J. (1996). Direct evidence of an extensive GABAergic innervation of the spinal dorsal horn by fibres descending from the rostral ventromedial medulla. *Neuroscience, 73*, 509–518.

Brumovsky, P., Hofstetter, C., Olson, L., Ohning, G., Villar, M., & Hokfelt, T. (2006). The neuropeptide tyrosine Y1R is expressed in interneurons and projection neurons in the dorsal horn and area X of the rat spinal cord. *Neuroscience, 138*, 1361–1376.

Carlton, S. M., Westlund, K. N., Zhang, D. X., Sorkin, L. S., & Willis, W. D. (1990). Calcitonin gene-related peptide containing primary afferent fibers synapse on primate spinothalamic tract cells. *Neuroscience Letters, 109*, 76–81.

Lawson, S. N. (2002). Phenotype and function of somatic primary afferent nociceptive neurones with C-, Adelta- or Aalpha/beta-fibres. *Experimental Physiology, 87*, 239–244.

Lu, Y., & Perl, E. R. (2005). Modular organization of excitatory circuits between neurons of the spinal superficial dorsal horn (laminae I and II). *Journal of Neuroscience, 25*, 3900–3907.

Naim, M., Spike, R. C., Watt, C., Shehab, S. A., & Todd, A. J. (1997). Cells in laminae III and IV of the rat spinal cord that possess the neurokinin-1 receptor and have dorsally directed dendrites receive a major synaptic input from tachykinin-containing primary afferents. *Journal of Neuroscience, 17*, 5536–5548.

Naim, M. M., Shehab, S. A., & Todd, A. J. (1998). Cells in laminae III and IV of the rat spinal cord which possess the neurokinin-1 receptor receive monosynaptic input from myelinated primary afferents. *European Journal of Neuroscience, 10*, 3012–3019.

Polgár, E., Al-Khater, K. M., Shehab, S., Watanabe, M., & Todd, A. J. (2008). Large projection neurons in lamina I of the rat spinal cord that lack the neurokinin 1 receptor are densely innervated by VGLUT2-containing axons and possess GluR4-containing AMPA receptors. *Journal of Neuroscience, 28*, 13150–13160.

Polgár, E., Sardella, T., Watanabe, M., & Todd, A. J. (2011). A quantitative study of NPY-expressing GABAergic neurons and axons in rat spinal dorsal horn. *The Journal of Comparative Neurology, 519*, 1007–1023.

Todd, A. J. (2010). Neuronal circuitry for pain processing in the dorsal horn. *Nature Reviews Neuroscience, 11*, 823–836.

Todd, A. J., & Koerber, H. R. (2005). Neuroanatomical substrates of spinal nociception. In S. McMahon & M. Koltzenburg (Eds.), *Wall and Melzack's textbook of pain* (pp. 73–90). Edinburgh: Elsevier.

Todd, A. J., Puskar, Z., Spike, R. C., Hughes, C., Watt, C., & Forrest, L. (2002). Projection neurons in lamina I of rat spinal cord with the neurokinin 1 receptor are selectively innervated by substance P-containing afferents and respond to noxious stimulation. *Journal of Neuroscience, 22*, 4103–4113.

Westlund, K. N., Carlton, S. M., Zhang, D., & Willis, W. D. (1990). Direct catecholaminergic innervation of primate spinothalamic tract neurons. *The Journal of Comparative Neurology, 299*, 178–186.

Spinothalamocortical Pathway

Definition

The spinothalamocortical pathway is the somatosensory pathway that originates mainly in the dorsal horn and projects to various nuclei within the thalamus that, in turn, project to various regions within the cerebral cortex.

Cross-References

▶ Spinothalamic Tract Neurons, in Deep Dorsal Horn

Spinothalamocortical Projections from SM

Vjekoslav Miletic
School of Medicine and Public Health,
University of Wisconsin, Madison, WI, USA

Synonyms

Nucleus gelatinosus

Definition

The ▶ nucleus submedius (SM) is a small oblong nucleus that is located ventromedial to the internal medullary lamina of the thalamus and is bounded by the nucleus reuniens medioventrally, the rhomboid and central medial nuclei mediodorsally, and the mammillothalamic tract ventrolaterally (Fig. 1). In the cat, the dorsal portion of the SM receives topographically organized projections exclusively from neurons located in ▶ lamina I of both the medullary (trigeminal) and spinal ▶ dorsal horn (Craig and Burton 1981). In the rat, the trigeminal, and especially spinal, afferent projections are substantially smaller than in the cat. Those rat spinal neurons that do send their axons to the SM are located in deeper spinal laminae rather than lamina I (Dado and Giesler 1990). Based on their data, Craig and Burton (1981) proposed that the SM might play an important role in ▶ nociception.

Spinothalamocortical Projections from SM, Fig. 1 Photomicrograph of a cresyl violet-stained section representative of the rat ventromedial thalamus in the coronal plane (50 μm thick section; magnification = 10×). The SM is bounded dorsomedially by the rhomboid (*Rh*) and central medial (*CM*) nuclei, ventromedially by the reuniens nucleus (*Re*), and ventrolaterally by the mammillothalamic tract (*mt*). The dorsal part of the SM (*SM$_d$*) is separated from the ventral part (*SM$_v$*) by a thin fiber lamina extending mediolaterally. Dorsal is up. Scale bar = 200 μm

Characteristics

Coffield et al. (1992) employed the fluorescent tracers Fluoro-Gold and 1,1'-Dioctadecyl-3,3,3,3 -tetramethyl indocarbocyanine perchlorate (DiI) as retrograde markers to examine the organization of reciprocal connections between the SM and the ▶ ventrolateral orbital cortex (VLO) and to investigate the connections of both of these areas with the midbrain in the rat.

VLO Neurons Retrogradely Labeled from the SM

Retrogradely labeled neurons were observed bilaterally throughout the rostrocaudal extent of the VLO, with the majority of labeled cells located ipsilaterally (Fig. 2). Most of the labeled neurons were confined to cortical layers 5 and 6. Densely packed cells were found in layer 5, with more diffuse labeling present in layer 6. In general, when the dye injection filled the whole SM, labeled neurons were concentrated in a narrow band bordering the medial edge of the VLO, curving ventrally toward the olfactory

VLO neurons labeled from SM

4.20   3.20   2.70

Midbrain neurons labeled from SM

-6.72
-7.80
-8.30

SM neurons labeled from VLO

-2.12
-2.56
-3.14

Midbrain neurons labeled from VLO

-6.72
-7.80
-8.30

Spinothalamocortical Projections from SM, Fig. 2 Schematic illustrations of the distribution of retrogradely labeled neurons (*black dots*). The *dots* are meant to indicate only density and distribution and not actual numbers of neurons. The numbers represent rostrocaudal coordinates from the bregma expressed in mm (negative is caudal, positive is rostral). These drawings are slightly modified copies of appropriate sections from the atlas of Paxinos and Watson (1986). Only selected nuclei are labeled. Abbreviations: *A8* dopaminergic cells, *Cg3* cingulate cortex area 3, *CG* central gray, *Cli* caudal linear raphe nucleus, *CM* central medial thalamic nucleus, *DP* dorsal peduncular cortex, *DR* dorsal raphe nucleus, *dtg* dorsal tegmental bundle, *Fr1* frontal cortex, area 1, *Fr2* frontal cortex, area 2, *IAM* interanterodorsal thalamic nucleus, *IC* inferior colliculus, *LO* lateral orbital cortex, *MD* medial dorsal thalamic nucleus, *Me5* mesencephalic trigeminal nucleus, *ml* medial lemniscus, *mlf* medial longitudinal fasciculus, *MnR* median raphe nucleus, *MO* medial orbital cortex, *mt* mammillothalamic tract, *ON* olfactory nuclei, *PAG* periaqueductal gray, *PMR* paramedian raphe nuclei, *PPtg* pedunculopontine

cortex. This was especially evident in the caudal aspect of the VLO. Fewer labeled neurons were noted in layers 5 and 6 of the lateral edge of the VLO and lateral orbital cortex (LO). This distribution of labeled cells was consistent from rat to rat. More diffusely located, labeled neurons were noted in layers 5/6 of the medial prefrontal cortex (medial orbital cortex, cingulate cortex areas 1 and 3, frontal cortex area 2).

Variations in labeling were noted when different regions of the SM were injected. When the injection was centered in the dorsal SM (with little or no dye spread into the ventral SM), the retrogradely labeled neurons extended from the lateral VLO rostrally to the medial VLO caudally. Conversely, when the dye injection included mostly the ventral SM, the greatest number of retrogradely labeled cells extended from the medial VLO rostrally to the lateral VLO caudally. The number of retrogradely labeled neurons in the medial prefrontal cortex increased with the spread of the injection site into the rhomboid and reuniens nuclei. Injection sites centered dorsal to the SM (e.g., in central medial, interanterodorsal, and mediodorsal nuclei) or medial to the SM (rhomboid and reuniens nuclei) resulted in decreased labeling in the VLO and a corresponding increase in labeling in the medial prefrontal cortex.

Retrogradely labeled neurons were also seen in other areas of the prefrontal cortex. These included a limited bilateral focus in area 2 of the frontal cortex and diffuse labeling in the ▶ contralateral VLO, LO, and cingulate cortical areas 1 and 3. Diffusely located, retrogradely labeled neurons were also noted in the somatosensory cortex, the amygdaloid nuclei, and the globus pallidus (not illustrated).

SM Neurons Retrogradely Labeled from the VLO

The greatest number of retrogradely labeled SM neurons was noted when the injection site was

centered around or encompassed the ventrolateral and rostral aspect of the VLO, including the medial aspect of the LO (Fig. 2). The label in the SM was invariably ▶ ipsilateral with few SM neurons labeled ▶ contralateral to the injection site.

Different areas of the SM were labeled differently depending on the placement of the dye in the VLO. When the VLO injection was located medially, retrogradely labeled neurons were noted in the anteroventral, but not dorsal, SM. The dorsal SM was labeled when the injection was situated more laterally and ventrally in cortical layers 1–3. In addition, when the injection site was centered more medially to include the medial and ventral orbital cortices, other thalamic nuclei (mediodorsal, central medial, anteromedial) contained labeled neurons. When the injection spread more ventrally to include the dorsal and medial parts of the rhinal sulcus (layer 1), then the thalamic ventromedial nucleus was labeled. When the injection site was located in the medial and caudal aspect of the VLO, no label was noted in the SM. Injection into the olfactory cortex (just ventral to VLO) or the claustrum (just caudal to VLO) did not lead to labeling in the SM.

Midbrain Neurons Retrogradely Labeled from the VLO

Labeled neurons were found throughout the rostrocaudal extent of the ventral ▶ periaqueductal gray (PAG) (Fig. 2). Most of the labeled neurons were seen in the contralateral caudal 3/4 of the PAG, although some ipsilateral labeling was also noted. No labeled neurons were seen in the dorsal PAG. Many of the labeled neurons were found along the ventral midline of the PAG, in the dorsal raphe nucleus as well as deeper midline nuclei including the caudal linear and median raphe nuclei. A few neurons were seen in the raphe magnus. Labeled neurons were also found in the substantia nigra (mostly contralateral; not illustrated). Occasionally, neurons were found

---

Spinothalamocortical Projections from SM, Fig. 2 (continued) tegmental nucleus, *Re* reuniens thalamic nucleus, *RF* rhinal fissure, *Rh* rhomboid thalamic nucleus, *RMg* raphe magnus nucleus, *RtTg* reticulotegmental nucleus pons, *SM* nucleus submedius or nucleus

gelatinosus, *SuCo* superior colliculus, *VL* ventral lateral thalamic nucleus, *VLO* ventral lateral orbital cortex, *VM* ventral medial thalamic nucleus, *VPL* ventral posterolateral thalamic nucleus, *VPM* ventral posteromedial thalamic nucleus, *VTA*, ventral tegmental area

scattered in the mesencephalic reticular formation and the tegmental nuclei.

Midbrain Neurons Retrogradely Labeled from the SM

The number of midbrain neurons labeled from the SM was smaller than those labeled from the VLO (Fig. 2). The location of neurons in the ventral PAG and dorsal raphe labeled from the SM overlapped with those labeled from the VLO, although they appeared more scattered in their distribution. In experiments in which other midline thalamic nuclei were included in the injection site, increased labeling occurred in the ventral and dorsal PAG and the raphe nuclei.

These data provide anatomical evidence for the existence of a neural circuit between the SM, VLO, and the midbrain PAG. In the prefrontal cortex, input from the SM terminates rostrally within the lateral and ventral areas of the VLO. Conversely, the cortical input to the SM originates from the medial and dorsal parts of the VLO. In addition, neurons from the ventrolateral PAG and the raphe nuclei project to the midline nuclei of the thalamus, including a small projection to the SM. Regions within the ventrolateral PAG and raphe nuclei project to the VLO, and these regions overlap with those that project to the SM.

These data suggest that the SM may provide another link between peripheral noxious input and the limbic, motor, and autonomic systems. The PAG is both the origin and recipient of a diverse array of ascending and descending neural input (Beitz 1982). In particular, the ventral PAG appears strategically situated to transmit or modulate nociceptive inputs between the lower brain stem, thalamus, and cortex. Hence, the connections of the SM with the VLO and the ventral PAG may allow SM neurons to be part of neural circuits involved in both pain and analgesia.

References

Beitz, A. J. (1982). The organization of afferent projections to the midbrain periaqueductal grey of the rat. *Neuroscience, 7,* 133–159.

Coffield, J. A., Bowen, K. K., & Miletic, V. (1992). Retrograde analysis of reciprocal projections between the rat ventrolateral orbital cortex and the nucleus submedius, and their connections to the midbrain. *The Journal of Comparative Neurology, 321,* 488–499.

Craig, A. D., Jr., & Burton, H. (1981). Spinal and medullary lamina I projection to nucleus submedius in medial thalamus: A possible pain center. *Journal of Neurophysiology, 45,* 443–466.

Dado, R. J., & Giesler, G. J., Jr. (1990). Afferent input to nucleus submedius in rats: Retrograde labeling of neurons in the spinal cord and caudal medulla. *Journal of Neuroscience, 10,* 2672–2686.

Paxinos, G., & Watson, C. (1986). *The Rat brain in stereotaxic coordinates* (2nd ed.). New York: Academic.

Spinothalamocortical Projections to Ventromedial and Parafascicular Nuclei

Luis Villanueva and Laurence Bourgeais
INSERM UMR 894, Centre de Psychiatrie et Neurosciences, Paris, France

Definition

Spinoreticulothalamic networks convey nociceptive signals from the whole body to widespread areas of the neocortex, via the ventromedial nuclei and parafascicular thalamic nuclei. This pathway allows any painful stimuli to modify cortical activity in a widespread manner, thus altering behavioral levels of attention and/or the planning of programmed movements.

Characteristics

It has been known for a long time that the majority of ascending axons located in the anterolateral quadrant of the spinal white matter, which contains the pain pathways in mammals, terminate within the medullary ▶ reticular formation (Bowsher 1976; Villanueva et al. 1996). Interestingly, the notion of a receptive center (centrum receptorium or sensorium) within the reticular formation was introduced by Kohnstamm and Quensel (1908) for bulbar reticular areas receiving spinal afferents. In a study of retrograde

a

0.5 mV

50 ms

left cheek

left forepaw   right forepaw

left hindpaw   right hindpaw

base of tail   tip of tail

b

Spinothalamocortical Projections to Ventromedial and Parafascicular Nuclei, Fig. 1 Single sweep recordings showing Aδ- and C-fiber-evoked responses of a VM neuron (*black dot*) following supramaximal percutaneous electrical stimulation (2 ms duration square-wave pulses) of different parts of the body (**a**); Cumulative results showing the magnitudes of the responses of VM neurons to graded mechanical (n = 7) or thermal (n = 16) stimulation of the ipsilateral hind paw (**b**) (Adapted from Monconduit et al. 1999)

cellular reactions in the bulbar reticular formation to high mesencephalic lesions, the same authors demonstrated ascending pathways connecting the centrum receptorium with higher levels of the brain. They postulated that reticulothalamic projections might be part of a polysynaptic path responsible for the conduction of pain and temperature to higher brain levels.

Thus, in addition to spinal pathways that carry nociceptive information directly to the diencephalon, some nociceptive information is relayed to the thalamus via the medullary reticular formation. Widespread areas throughout the brain stem reticular formation contain neurons that are responsive to noxious stimuli. Among these medullary areas, the ▶ subnucleus reticularis dorsalis (SRD) contains neurons that respond and encode selectively noxious stimuli from any part of the body (Villanueva et al. 1996).

Diencephalic projections from the SRD are essentially contralateral and terminate primarily in the lateral half of the ventromedial thalamus (VM), almost throughout its rostro-caudal extent. Moreover, dense terminals are also located in the lateral aspect of the parafascicular nucleus (PF) (Desbois and Villanueva 2001; Villanueva et al. 1998). Reticulo-PF projections also originate from the ▶ gigantocellular reticular nucleus, an area situated rostrally to the SRD, which has also been implicated in nociceptive processing (see refs. in Bowsher 1976).

PF units respond to both cutaneous and visceral noxious stimuli in the rat (Berkley et al. 1995). These neurons are driven from large cutaneous receptive fields and respond to intense, frankly noxious cutaneous and visceral stimuli. Noxious-responding units recorded in the PF thalamic nucleus of alert monkeys also display large receptive fields and some discriminate changes in the intensity of noxious stimuli (Bushnell and Duncan 1989). The lateral PF projects to the dorsolateral part of the caudate-putamen, the lateral subthalamic nucleus, which itself has reciprocal projections with pallidal and nigral motor relays and the sensorimotor cortex (Groenewegen and Berendse 1994). In addition to its striatal afferents, the lateral PF area projects mainly to the lateral agranular field of motor cortex and the rostral parietal cortex. Thus, as

immunohistochemistry. VM cortical efferents terminated as a dense band in layer I of the dorsolateral neocortex (c). They appeared as dense clusters of terminal fibers with small varicosities (d). Afferents to VM appeared as Golgi-like labeled cells (a) that were confined to the dorsal aspect of the medullary subnucleus reticularis dorsalis (SRD). Scale bars: a, b, c = 1 mm, d = 100 μm. Other abbreviations: *DLO* dorsolateral orbital cortex, *LO* lateral orbital cortex, *M2* secondary motor cortex, *MO* medial orbital cortex, *mt* mamillothalamic tract, *NTS* nucleus of the solitary tract, *Pr* prelimbic cortex, *Sp5c* trigeminal subnucleus caudalis, *SRV* subnucleus reticularis ventralis, *VO* ventral orbital cortex, *VPL* thalamic ventroposterolateral nucleus, *VPM* thalamic ventroposteromedial nucleus, *ZI* zona incerta

Spinothalamocortical Projections to Ventromedial and Parafascicular Nuclei, Fig. 2 Bright field images showing an example of the main connections from the nociceptive (lateral) region of the ventromedial thalamus, the VM (b); VM was labeled following an electrophoretic ejection of lysine-fixable dextran of molecular weight 3,000 Da, which is a compound conveyed by both anterograde and retrograde axonal transport and revealed by

most lateral PF projections are related to brain structures involved in motor processing, SRD-PF connections could mediate some motor reactions following noxious stimulation.

VM thalamic neurons are exclusively driven by activities in A∂- and C-cutaneous ▸ polymodal nociceptors from the entire body surface (Monconduit et al. 1999). These neurons present ▸ whole body receptive fields, which are activated by graded stimuli. A linear relationship exists between the evoked firing rate and the intensities of both thermal and mechanical stimuli only within noxious ranges (Fig. 1). In some cases, VM neurons develop residual activity and/or afterdischarges following strong noxious stimulations. Systemic morphine depresses VM neuronal activities evoked by Aδ- and C-fibers and by thermal noxious stimuli, in a dose-dependent and naloxone-reversible fashion (Monconduit et al. 2002). Similar reductions in the C-fiber and noxious thermal evoked activities were obtained with equivalent doses of morphine, suggesting that morphine selectively depresses inputs to the VM that arise from cutaneous polymodal nociceptors. By contrast, these neurons did not respond to the levels of distension that could be considered to be innocuous or noxious of the intact and inflamed colon and rectum. Moreover, following inflammation induced by subcutaneous injections of mustard oil, VM neurons developed responses to both thermal and mechanical innocuous skin stimulation (Monconduit et al. 2003). Nociceptive activity in the VM arises primarily from monosynaptic inputs from the medullary SRD, since a strong reduction in VM responses was obtained following blockade of the contralateral SRD.

VM neurons in turn relay the widespread nociceptive inputs from the SRD to the whole layer I of the dorsolateral neocortex. VM projections are organized as a widespread dense band, covering mainly layer I of the dorsolateral anterior-most aspect of the cortex (Fig. 2; Desbois and Villanueva 2001). This band diminishes progressively as one moves caudally, disappearing completely at 1 mm caudal to bregma level. This reticulo-thalamo-cortical network could allow any painful stimuli to modify cortical activity in a widespread manner, since thalamocortical interactions in layer I are assumed to be a key substrate for the ▸ synchronization of large ensembles of neurons across extensive cortical territories and have been associated with changes in states of consciousness. In this respect, layer I inputs may act as a "mode switch," by activating a spatially restricted low-threshold zone in the apical dendrites of layer V pyramidal neurons and evoking regenerative potentials propagating toward their soma, which in turn could switch layer V neurons into the ▸ burst firing mode (Larkum and Zhu 2002). This hypothesis fits with the facts that painful stimuli can elicit widespread cortical activation in human beings and that increasing stimulus intensity increases the number of brain regions activated, including ventral posterior and medial thalamic regions and prefrontal, premotor, and motor cortices (Porro 2003; Apkarian et al. 2009).

VM neurons cannot be clearly assigned to the clear-cut described ▸ medial pain system or the ▸ lateral pain system. They have fine discriminative properties, as shown by their selective responsiveness to noxious stimuli and their ability to encode precisely different kinds of cutaneous stimuli within noxious ranges. Their activation by innocuous stimuli occurs only under conditions of experimental allodynia. However, they lack topographical discrimination, as illustrated by their whole body receptive fields and their ability to respond to widespread noxious stimuli of cutaneous, muscular, or visceral origins. The VM may constitute an important thalamic nociceptive branch of what was originally termed the "ascending reticular activating system" (Herkenham 1986; Jasper 1961).

References

Apkarian, A. V., Baliki, M. N., & Geha, P. Y. (2009). Towards a theory of chronic pain. *Progress in Neurobiology, 87*, 81–97.

Berkley, K., Benoist, J. M., Gautron, M., et al. (1995). Responses of neurons in the caudal intralaminar thalamic complex of the rat to stimulation of the uterus, vagina, cervix, colon and skin. *Brain Research, 695*, 92–95.

Bowsher, D. (1976). Role of the reticular formation in responses to noxious stimulation. *Pain, 2*, 361–378.

Bushnell, M. C., & Duncan, G. H. (1989). Sensory and affective aspects of pain perception: Is medial thalamus restricted to emotional issues? *Experimental Brain Research, 78*, 415–418.

Desbois, C., & Villanueva, L. (2001). The organization of lateral ventromedial thalamic connections in the rat: A link for the distribution of nociceptive signals to widespread cortical regions. *Neuroscience, 102*, 885–898.

Groenewegen, H. J., & Berendse, H. W. (1994). The specificity of the "non-specific" midline and intralaminar thalamic nuclei. *TINS, 17*, 52–57.

Herkenham, M. (1986). New perspectives on the organization and evolution of nonspecific thalamocortical projections. In E. G. Jones & A. Peters (Eds.), *Cerebral cortex* (Sensory-motor areas and aspects of cortical connectivity, Vol. 5, pp. 403–445). New York: Plenum.

Jasper, H. H. (1961). Thalamic reticular system. In D. E. Sheer (Ed.), *Electrical stimulation of the brain* (pp. 277–287). Austin: University of Texas.

Kohnstamm, O., & Quensel, F. (1908). Das centrum receptorium (sensorium) der formatio reticularis. *Neurologie Zentralbl, 27*, 1046–1047.

Larkum, M. E., & Zhu, J. J. (2002). Signaling of layer 1 and whisker-evoked Ca^{2+} and Na^+ action potentials in distal and terminal dendrites of rat neocortical pyramidal neurons in vitro and in vivo. *Journal of Neuroscience, 22*, 6991–7005.

Monconduit, L., Bourgeais, L., Bernard, J. F., et al. (1999). Ventromedial thalamic neurons convey nociceptive signals from the whole body surface to the dorsolateral neocortex. *Journal of Neuroscience, 19*, 9063–9072.

Monconduit, L., Bourgeais, L., Bernard, J. F., et al. (2002). Systemic morphine selectively depresses a thalamic link of widespread nociceptive inputs in the rat. *European Journal of Pain, 6*, 81–87.

Monconduit, L., Bourgeais, L., Bernard, J. F., et al. (2003). Convergence of cutaneous, muscular and visceral noxious inputs onto ventromedial thalamic neurons in the rat. *Pain, 103*, 83–91.

Porro, C. A. (2003). Functional imaging and pain: Behavior, perception, and modulation. *The Neuroscientist, 9*, 354–369.

Villanueva, L., Bouhassira, D., & Le Bars, D. (1996). The medullary subnucleus reticularis dorsalis (SRD) as a key link in both the transmission and modulation of pain signals. *Pain, 67*, 231–240.

Villanueva, L., Desbois, C., Le Bars, D., et al. (1998). Organization of diencephalic projections from the medullary subnucleus reticularis dorsalis and the adjacent cuneate nucleus: A retrograde and anterograde tracer study in the rat. *The Journal of Comparative Neurology, 390*, 133–160.

Spiral CT

▶ CT Scanning

Spiritual Problems

Definition

Deeper concerns such as life after death, guilt regarding certain life events, the meaning of existence, or the perception of disease as punishment.

Cross-References

▶ Cancer Pain Management, Interface Between Cancer Pain Management and Palliative Care

Splice Variants

Definition

Different forms of a protein that result from the alternative splicing of the mRNA before translation.

Cross-References

▶ Opioid Receptors at Postsynaptic Sites

S

Spondyloarthropathy

Definition

Spondylolisthesis is a defect in the construct of bone between the superior and inferior facets with varying degrees of displacement, so that the vertebra with the defect and the spine above that vertebra are displaced forward in relationship to the vertebrae below. It is usually due to a developmental defect or the result of a fracture.

Cross-References

▶ Chronic Low Back Pain: Definitions and Diagnosis
▶ Sacroiliac Joint Pain

Spondylolisthesis

Definition

Spondylolisthesis is a defect in the construct of bone between the superior and inferior facets with varying degrees of displacement, so that the vertebra with the defect and the spine above that vertebra are displaced forward in relationship to the vertebrae below. It is usually due to a developmental defect or the result of a fracture.

Cross-References

▶ Chronic Low Back Pain: Definitions and Diagnosis
▶ Lower Back Pain, Physical Examination

Spontaneous Aliquorrhea

▶ Headache Due to Low Cerebrospinal Fluid Pressure

Spontaneous Cerebrospinal Fluid (CSF) Leak

Synonyms

CSF Leak

Definition

A syndrome in which a positional headache occurs after a non-lumbar puncture induced CSF leak. Headache will occur out of the blue, normally with no inciting event (although repetitive valsalva maneuvers, e.g., cough, sneeze, and strain, may induce leak). Headache will be better or absent in a supine position and severe in an upright position. Most common location of spontaneous CSF leaks is at the cervical-thoracic junction.

Cross-References

▶ New Daily Persistent Headache
▶ Post-lumbar Puncture Headache
▶ Primary Exertional Headache

Spontaneous Dissection

Definition

There is no obvious trauma involved; no trauma preceding the onset of symptoms, such as a car accident or a heavy hit to the neck.

Cross-References

▶ Headache Due to Dissection

Spontaneous Ectopia

▶ Ectopia, Spontaneous

Spontaneous Intracranial Hypotension

▶ Headache Due to Low Cerebrospinal Fluid Pressure

Spontaneous Onset

Definition

Pain that occurs for no apparent reason and cannot be attributed to a specific cause.

Cross-References

▶ Pain in the Workplace, Risk Factors for Chronicity, Job Demands

Spontaneous Pain

Definition

Pain without any obvious external stimuli, e.g., constant deep aching pain, constant superficial burning pain, and paroxysms of brief lancinating pain. Spontaneous pain is thought to be elicited by epitopic discharge generated along axons of involved neurons.

Cross-References

▶ Antidepressants in Neuropathic Pain
▶ Causalgia: Assessment
▶ Diabetic Neuropathy, Treatment
▶ Dysesthesia, Assessment
▶ Neuropathic Pain Model, Partial Sciatic Nerve Ligation Model
▶ Nocifensive Behaviors, Muscle and Joint

Sports Headache

▶ Primary Exertional Headache

Spouse, Role in Chronic Pain

Alicia A. Heapy and Robert D. Kerns
VA Connecticut Healthcare System, Westhaven, CT, USA

Synonyms

Distracting responding; Negative responding; Pain behaviors; Pain-relevant communication; Significant others; Social support; Solicitousness

Definition

Family systems theories espouse that the patterns of family functioning play a dominant role in determining health and illness among family members. ▶ Social support generally refers to the availability or provision of instrumental and/or emotional assistance from ▶ significant others. The ▶ operant conditioning model of chronic pain hypothesizes that pain and pain-related disability may be maintained by environmental contingencies for overt, observable demonstrations of pain. ▶ Pain-relevant communication involving significant others refers to verbal and nonverbal exchanges in the context of a person's expressions of pain. These expressions of pain are often referred to as ▶ pain behaviors and include distorted ambulation (i.e., use of a prosthesis, limping), facial/audible expressions of pain (e.g., verbal complaints of pain, grimacing), expressions of affective distress (e.g., sighing, crying), and seeking help (e.g., asking for assistance, taking pain medications). ▶ Solicitous responses are defined as a set of positive responses (e.g., expressions of sympathy, taking over of duties and responsibilities) delivered by significant others, including the spouse or family members, contingent on the display of pain behaviors. ▶ Distracting responses refer to cues from significant others intended to encourage alternative, presumably more adaptive, well behaviors (e.g., increased activity, use of distraction to cope with pain) on the part of the person experiencing pain. ▶ Negative or punishing responses (e.g., expressions of irritation, ignoring) represent a third category of responses to the expression of pain. The ▶ cognitive-behavioral transactional model of family functioning emphasizes the role of cognitions and beliefs in the appraisal of pain and pain-related behavioral interactions.

Characteristics

The influence of the social context, and particularly the family environment, on health and illness issues has been widely recognized and studied (Ramsey 1989; Schmaling and Sher 2000). Dominant models for understanding and explaining these influences, and for the development of family-based interventions, include family systems theories (Patterson 1988) and operant behavioral and cognitive-behavioral perspectives (Kerns and Weiss 1994). Within the chronic pain literature, there has been a similar call for attention to the role of the family and social context in studies of patient adjustment and adaptation (Nicassio and Radojevic 1993; Roy 1989; Leonard, Cano and Johanson 2006).

The importance of the family's role can be traced, in part, to the dominance of behavioral and cognitive-behavioral theories in chronic pain–related research and practice (Fordyce 1976). Whereas social support has generally been accepted as having a positive, moderating, or mediating role on health and recovery or adaptation to illness, the operant conditioning model of chronic pain hypothesizes that the specific pattern of pain-relevant communication involving significant others, including family members, may either have positive, or negative, influences on important aspects of the chronic pain experience. According to the operant conditioning model of chronic pain, pain and pain-related disability may be maintained by environmental contingencies for overt expressions of pain, termed pain behaviors, even in the absence of continued nociception (Fordyce 1976). In particular,

positive or solicitous responses from spouses and family members, contingent on pain behaviors, may serve to reinforce these maladaptive behaviors and encourage the development and maintenance of pain and pain-related disability. Examples of solicitous behavior include expressions of sympathy or concern for the spouse, physical assistance or performance of a task, and encouraging rest and discouraging activity.

Reviews of the empirical literature informed by this model concluded that, in general, spouse solicitousness is significantly related to greater pain intensity, greater frequency of pain behavior, higher levels of disability (Kerns and Otis 2003; Leonard et al. 2006; Newton-John 2002), and increased help-seeking behaviors (Kerns & Otis 2003; Newton-John 2002). Interestingly, responses from significant others categorized as distracting responses, although likely intended to cue more adaptive behavior, have been found to be reliably positively related to poor outcomes (Leonard et al. 2006). In contrast to the predictions of operant theory, frequency of negative or punishing responses is not positively associated with increased activity in persons with chronic pain (Leonard et al. 2006; Newton-John 2002). Mood and global marital satisfaction appear to mediate these relationships (Newton-John 2002). Several studies have found that solicitousness only results in poorer pain outcomes in the context of a globally satisfying relationship (Leonard et al. 2006). Higher rates of punishing responses, in the context of well-adjusted marriages, have been found to be associated with lower levels of pain intensity, but negative responses have been associated with greater pain intensity when marital satisfaction is low (Weiss and Kerns 1995). Marital satisfaction has also been identified as a significant predictor of depressive symptoms, and higher rates of punishing responses from spouses have been shown to be associated with higher levels of depression among persons with chronic pain (Turk et al. 1992). Further, one study found that the link between solicitousness and increased physical disability was only present among depressed patients (Romano et al. 1995). Overall, pain-related outcomes and spouse responses most closely reflect those predicted by operant principles in the context of higher levels of marital satisfaction, and depressive symptom severity appears to influence pain-related outcomes (Newton-John 2002).

The inclusion of cognitive factors in theoretical models has been advocated as a way to better understand this pattern of findings. The cognitive-behavioral transactional model of family functioning emphasizes the role of cognitions and beliefs in the appraisal of pain and pain-related behavioral interactions (Kerns 1995; Kerns and Otis 2003). This model also promotes a more sophisticated understanding of the reciprocal influence among spousal interactions, the perceived effectiveness of those exchanges, and adaptation to pain. Another model rooted in the cognitive-behavioral tradition is the communal coping model, which emphasizes the role of pain catastrophizing and its effect on spouse response to pain (Thorn et al. 2003; Leonard et al. 2006). Specifically, catastrophizing, or negative thoughts that concentrate on unlikely but disastrous consequences of chronic pain, may serve the purpose of prompting a spouse to provide pain-related support or attention. An additional implication of this model is that some persons with pain feel that other have a responsibility to act in solicitous ways toward them and provide more support, a phenomenon termed support entitlement. Using the communal coping model, Cano et al. (2009) hypothesized that persons with feelings of support entitlement would be more likely to use catastrophizing to indirectly obtain greater support from a spouse. They further hypothesized these persons would actually receive less support as prior research has found that support providers regard such indirect support seeking negatively. They found that for persons with greater support entitlement, catastrophizing was associated with more negative spousal responses such as punishing and invalidating.

Finally, intimacy models conceptualize some verbal pain behaviors as expressions of distress instead of regarding them solely as descriptions of pain or pain behaviors (Cano and Williams 2010). An intimacy-based model predicts that

emotional validation in response to an expression of pain-related distress would result in increased intimacy, relationship satisfaction, and capacity to address stressful situations. In contrast, operant-based models conceptualize all verbal communication about pain as pain behaviors and validating responses on the part of the spouse as a potential reinforcer. These more sophisticated models, that include cognitive components, may be needed to untangle the complex relationships present among spouse behavior, marital satisfaction, mood, and pain-related outcomes.

In addition to examining how the behavior of the well spouse affects the adaptation of the person experiencing chronic pain, the effect on the well-being of the spouse of living with someone with chronic pain has also been examined. Although estimates vary, high rates of depression have been shown to exist in spouses of persons with chronic pain (Flor et al. 1987; Leonard et al. 2006). However, studies investigating the consequences of living with someone with chronic pain have not demonstrated consistently negative outcomes. Although several studies have found that persons with chronic pain and their spouses report lower levels of marital and sexual satisfaction than spouses with chronic pain, several studies have found normal levels of marital satisfaction (Kerns and Otis 2003; Romano and Schmaling 2001).

Empirical investigations of the efficacy of pain treatments involving the spouse of the person with chronic pain have not demonstrated incremental efficacy beyond treatment of the individual alone (Kerns and Otis 2003). Unfortunately, the number of empirical studies has been small, and many studies have been marked by methodological shortcomings. For example, although controlled trials have examined treatments that included couples treatment components, the specific contribution of this component has not been extensively examined (Kerns and Otis 2003). Despite the wealth of studies that have demonstrated a link between spouse behavior and chronic pain, these findings have not been translated into the development of efficacious and theory-based interventions.

Increased attention to the role of the spouse in treatment for chronic pain and methodologically rigorous studies are needed to advance pain treatment and to fully realize the benefits of the increased knowledge of the role of the spouse in chronic pain outcomes.

Although not a treatment designed to address chronic pain specifically, Cano and Leonard (2006) have proposed the use of integrative behavioral couple therapy (IBCT) as a treatment for couples where one member of the couple has chronic pain and where pain has negatively affected the marital relationship. ICBT (Jacobson et al. 2000) is a behaviorally based intervention that emphasizes reciprocal engagement in behaviors that the spouse has rated as reinforcing, communication training, and emotional acceptance. This treatment has been shown to increase marital satisfaction and decrease psychological distress. In particular, emotional acceptance is hypothesized to be an important treatment ingredient as persons with pain report feeling that others do not understand their pain experiences and respond negatively when pain experiences are discussed (Cano and Leonard 2006). For couples with one member with chronic pain, ICBT may address not only marital dissatisfaction and distress but the emphasis on emotional acceptance may enhance the empathy and intimacy of the relationship. Enhanced empathy and intimacy may alleviate the previously reported perception held by persons with pain that others do not understand their pain and their tendency to conceal pain due to the perception that others will respond negatively (Cano and Leonard 2010). These factors are hypothesized to result in more durable change.

References

Cano, A., Miller, L. R., & Loree, A. (2009). Spouse beliefs about partner chronic pain. *Journal of Pain, 10*, 486–92.

Cano, A. & de C. Williams, A. C., (2010). Social Interaction in Pain: Reinforcing Pain Behaviours or Building Intimacy? *Psychology Faculty Research Publications.* http://digitalcommons.wayne.edu/psychfrp/13.

Cano, A., & Leonard, M. T. (2006). Integrative behavioral couple therapy for chronic pain: Promoting behavior change and emotional acceptance. *Journal of Clinical Psychology: In Session, 62*, 1409–18.

Flor, H., Kerns, R. D., & Turk, D. C. (1987). The role of spouse reinforcement, perceived pain, and activity levels of chronic pain patients. *Journal of Psychosomatic Research, 31*, 251–259.

Fordyce, W. E. (1976). *Behavioral methods in chronic pain and illness.* St. Louis: Mosby.

Jacobson, N. S., Christenson, A., Prince, S. E., Cordova, J. and Eldridge, K. (2000). Integrativebehavioral couple therapy: an acceptance-based, promising new treatment for couple discord. *Journal of Consulting and Clinical Psychology, 68*, 351–355.

Kerns, R. D. (1995). Family assessment and intervention. In P. M. Nicassio & T. W. Smith (Eds.), *Managing chronic illness* (pp. 207–244). Washington, DC: American Psychological Association.

Kerns, R. D., & Otis, J. (2003). Family therapy for persons experiencing pain: Evidence for its effectiveness. *Seminars in Pain Medicine, 1*, 79–89.

Kerns, R. D., & Weiss, L. H. (1994). Family influences on the course of chronic illness: A cognitive-behavioral transactional model. *Annals of Behavioral Medicine, 16*, 116–130.

Leonard, M. T., Cano, A., & Johansen, A. B. (2006). Chronic pain in a couples context: A review and integration of theoretical models and empirical evidence. *The Journal of Pain, 7*, 377–390.

Newton-John, T. R. O. (2002). Solicitousness and chronic pain: A critical review. *Pain Reviews, 9*, 7–27.

Nicassio, P., & Radojevic, V. (1993). Models of family functioning and their contribution to patient outcomes in chronic pain. *Motivation and Emotion, 17*, 349–356.

Patterson, J. M. (1988). Families experiencing stress: The family adjustment and adaptation response model. *Family Systems in Medicine, 5*, 202–237.

Ramsey, C. N. (1989). *Family systems in medicine.* New York: Guilford Press.

Romano, J. M., & Schmaling, K. B. (2001). Assessment of couples and families with chronic pain. In D. C. Turk & R. Melzack (Eds.), *Handbook of pain assessment* (pp. 346–361). New York: Guilford Press.

Romano, J. M., Turner, J. A., Jensen, M. P., et al. (1995). Chronic pain related patient-spouse behavioral interactions predict patient disability. *Pain, 6*, 227–233.

Roy, R. (1989). *Chronic pain and the family: A problem-centered perspective.* New York: Human Sciences Press.

Schmaling, K. B., & Sher, T. G. (2000). *The psychology of couples and illness: Theory, research and practice.* Washington, DC: American Psychological Association.

Thorn, B., Boothy, J., Ward, C., & Chaplin, W. (2003). Emotional vulnerability and catastrophizing in the explanation of sex differences in pain response.

Turk, D. C., Kerns, R. D., & Rosenberg, R. (1992). Effects of marital interaction on chronic pain and disability: Examining the downside of social support. *Rehabilitation Psychology, 37*, 259–274.

Weiss, L. H., & Kerns, R. D. (1995). Patterns of patient-relevant social interactions. *International Journal of Behavioral Medicine, 2*, 157–171.

Sprained Ankle Pain Model

Jin Mo Chung[1] and Sungtae Koo[2]
[1]Department of Neuroscience and Cell Biology, University Chair in Neuroscience, University of Texas Medical Branch, Galveston, TX, USA
[2]Division of Meridian and Structural Medicine, School of Korean Medicine, Pusan National University, Yangsan, Korea

Definition

Ankle sprain commonly occurs during participation in various strenuous sports, as well as during routine locomotive activity, and is a common source of persistent pain and reduction of mobility. The rat model of sprained ankle pain is produced by manually overextending the lateral ligaments, without breaking them, to imitate a lateral ankle sprain in a human. The ankle sprain rat subsequently showed swelling of the ankle and a reduced stepping force of the affected limb for the next several days. The amount of reduction of foot stepping force is used as an index of the degree of pain generated by ankle sprain.

Characteristics

The typical ankle sprain is caused by a combination of forces that tend to adduct, invert, and plantar flex the talus beyond its physiological limit (Connolly 1981). Such motion causes avulsion of lateral ligaments, rupture of the adjacent capsule, and soft tissue hemorrhage and edema, resulting in severe discomfort and talar instability (Cotler 1984; Liu and Nguyen 1999). Examination of the sprained ankle of the rat indicated that there was stretching of ligaments without a macroscopic tear. Such a condition is classified as a mild or grade 1 ankle sprain in humans (Brier 1999; Patel and Warren 1999). Grade 1 can be further subdivided into grade 1 and grade 1+, and rats in the present study displayed behaviors similar to human patients with the latter grade, in that they showed a mild loss of function, which recovered in 2–5 days (Brier 1999).

Procedure for Ankle Sprain Pain Model

Under general anesthesia, ankle sprain is produced by manually overextending the lateral ligaments, without breaking them, to imitate a lateral ankle sprain in a human. As originally described by Koo et al. (2002), the right hind foot was repeatedly bent in the direction of simultaneous inversion and plantar flexion 60 times during a 1-min period with gradually increasing force, so that the foot could eventually be bent to a position of 90° inversion and 90° plantar flexion from the resting position. During the next 1-min period, the foot was further inverted so that it eventually reached 180° inversion (paw facing completely upward). Both of the above-mentioned 1-min procedures were repeated one more time; therefore, the total procedure took 4 min.

Anatomical Observation

Careful observation of the ligaments around the ankle joint in normal rats revealed that the calcaneofibular, calcaneocuboid, and talonavicular ligaments are stretched during plantar flexion. During inversion, on the other hand, the talocalcanean ligament was stretched along with the calcaneofibular and calcaneocuboid ligaments. Therefore, the ankle sprain procedure, which was the combination of the forceful plantar flexion and inversion, most likely induced stretching of four ligaments: calcaneofibular, calcaneocuboid, talonavicular, and talocalcanean ligaments. Careful examination, under a dissecting microscope, of those four ligaments immediately after the sprain procedure indicated that there were no obvious signs of tearing of any ligaments. However, those ligaments were elongated, and thus, gaps between

Sprained Ankle Pain Model, Fig. 1 Changes in weight bearing of the limb and foot volume after ankle sprain. Foot stepping force was measured and expressed as weight bearing ratio to body weight. (**a**) Shows the pattern of the foot stepping force of both the forelimb (*F*) and hind limb (*H*) on one side of a normal rat. This rat weighed 230 g and stepping on the fore-limb and hind limb bore about 45 % and 60 % of the body weight, respectively, when the rat walked at a normal pace. (**b**) Shows the pattern of foot stepping force of the same rat 1 day after ankle sprain. The amount of weight bearing of the hind limb was reduced to about half. (**c**) Shows the average values (± SEM) of peak foot stepping force of the hind limb of a group of nine rats 1 h before (−1 h) and at various times after induction of ankle sprain. (**d**) Shows the average volume (± SEM) of the foot 1 h before (−1 h) and various times after induction of ankle sprain. Foot volume increased rapidly due to swelling of the ankle after the sprain procedure. For both (**c**) and (**d**), post-operative time is expressed as h for hours and d for days. Asterisks indicate values significantly different (*P* < 0.05) from the preoperative value (−1 h) (Koo et al. 2002)

bones were widened. In addition, there was minor hemorrhaging on the retinaculum surrounding the extensor tendons (Koo et al. 2002).

Behavioral Changes

The rat subsequently showed swelling of the ankle and a reduced stepping force of the affected limb for the next several days. Measuring the amount of weight bearing force by the foot (Min et al. 2001; Koo et al. 2002) is employed as a behavioral test method to estimate the level of pain in the sprained ankle. The average weight bearing by the hind limb during normal gait is 60 % of total body weight, which is reduced to less than 10 % after ankle sprain.

Systemically injected morphine improved the stepping force in a dose-dependent manner. The reduced stepping force of the hind limb after the induction of ankle sprain was nearly restored 1 h after an intraperitoneal injection of morphine, at a dose of 5 mg/kg (Koo et al. 2002). The fact that morphine restored the reduced foot stepping force suggests that measuring foot stepping force can be used as an index for ankle sprain pain.

Changes in foot stepping force were measured before and after induction of ankle sprain. Figure 1a shows the pattern of force generated by the stepping of the forelimb, and then the hind limb of one side, during locomotion in a normal rat.

The weight bearing on the forelimb was followed by the hind limb, with the latter being a little stronger, so that it reached about 60 % of the rat's total body weight at the peak point. Immediately after the induction of ankle sprain, the rat started limping on the affected foot. To quantify the limping, the stepping force of the limb was measured. As shown in Fig. 1b, for example, the stepping force was greatly reduced 1 day after induction of ankle sprain. Figure 1c shows average values of stepping force of the hind limb in a group of rats before, and at various times after, ankle sprain.

The stepping force was maximally reduced at the 12th hour after ankle sprain, and then gradually recovered afterward, so that it was fully recovered by the fifth day after the induction of ankle sprain.

Daily measurements of weight bearing of the hind limb on the side contralateral to the sprained ankle revealed that there was a small increase (~5 %) in weight bearing over the next 3 days, as compared to the preoperative control value (Koo et al. 2002). The reduced weight bearing on the foot after induction of ankle sprain is because the rats are trying to avoid bearing weight on the affected foot due to pain associated with the sprained ankle.

The swelling of the ankle roughly paralleled the reduction in weight bearing. As shown in Fig. 1d, the average volume of the foot was just below 2 ml before the induction of ankle sprain.

However, the ankle swelled rapidly after the sprain operation, so that it had almost doubled its volume in 6–12 h. After that, the swelling rapidly subsided so that the foot volume was almost restored to normal volume by 2 days.

There is a certain level of endogenous opioid activity in the ankle sprain model, and blocking it further reduces the foot stepping force. Peritoneal injection of naltrexone, a nonselective long-lasting opioid antagonist (Crabtree 1984), further reduced foot stepping force 1 h after injection (Koo et al. 2002). The effect of naltrexone on the weight bearing suggests that the ankle sprain pain model triggers the release of endogenous opioids, so that the level of pain is maintained at a partially reduced state.

References

Brier, S. R. (1999). *Primary care orthopedics* (pp. 380–386). St. Louis: Mosby.

Connolly, J. F. (1981). *De Palma's the management of fractures and dislocations* (pp. 1806–1808). Philadelphia: W.B. Saunders.

Cotler, J. M. (1984). Lateral ligamentous injuries of the ankle. In W. C. Hamilton (Ed.), *Traumatic disorders of the ankle* (pp. 113–123). New York: Springer.

Crabtree, B. L. (1984). Review of naltrexone, a long-acting opiate antagonist. *Clinical Pharmacy, 3,* 273–280.

Koo, S. T., Park, Y. I., Lim, K. S., Chung, K., & Chung, J. M. (2002). Acupuncture analgesia in a new rat model of ankle sprain pain. *Pain, 99,* 423–431.

Liu, S. H., & Nguyen, T. M. (1999). Ankle sprains and other soft tissue injuries. *Current Opinion in Rheumatology, 11*, 132–137.

Min, S. S., Han, J. S., Kim, Y. I., Na, H. S., Yoon, Y. W., Hong, S. K., & Han, H. C. (2001). A novel method for convenient assessment of arthritic pain in voluntarily walking rats. *Neuroscience Letters, 308*, 95–98.

Patel, D. V., & Warren, R. F. (1999). Ankle sprain: Clinical evaluation and current treatment concepts. In C. S. Ranawat & R. G. Positano (Eds.), *Disorders of the heel, rearfoot, and ankle* (pp. 323–340). Philadelphia: Churchill Livingstone.

Spray and Stretch

Definition

Spray and stretch is a technique used by manual therapists to treat trigger points associated with myofascial pain syndrome. Trigger (tender) points are soft tissue spots, usually in muscle, which on finger tip pressure elicit pain in a related area, but they have so far not been biologically defined. The technique involves stretching local muscles after first producing a superficial anesthesia with a coolant spray.

Cross-References

▶ Stretching

Spreading Depression

Definition

Very brief spreading wave of neuronal and glial depolarization, followed by a long-lasting suppression of neural activity, which is easily evoked in lissencephalic brains by mechanical or chemical stimuli of the brain surface (Leão 1944). As spreading depression invades the cortex at a speed (4–6 mm/min) similar to that of spreading visual aura symptoms and the oligemia found to be associated with them in

functional imaging studies, it is thought to be the pathophysiologic event underlying the migraine aura.

Cross-References

▶ Clinical Migraine Without Aura
▶ Post-seizure Headache

SRD

▶ Subnucleus Reticularis Dorsalis

SSDI

▶ Social Security Disability Insurance

sSNI Model

▶ Neuropathic Pain Model, Spared Nerve Injury

SSR

▶ Sympathetic Skin Response

Stabilizing Exercise

▶ Exercise

Static Allodynia

Definition

See also "Allodynia." A problematic denomination for pain due to tonic pressure, since it is based on the assumption that mechanoreceptors other than nociceptive elements are responsible for this form of hyperalgesia. Experimental evidence speaks more in favor of the assumption that sensitized mechanonociceptors are responsible for increased tonic pressure pain. Therefore the term "static mechanical hyperalgesia" seems to be more appropriate.

Cross-References

▶ Hyperpathia, Assessment
▶ Neuropathic Pain Model, Tail Nerve Transection Model

Statistical Decision Making

Definition

Statistical decision making is the only mathematical model that yields separate measures for discrimination and response bias.

Step Cycle

Definition

The various temporally organized activity phases that give rise to locomotion.

Cross-References

▶ Spinothalamic Projections in Rat

Stepwise Approach

Definition

An approach to psychosocial evaluation that proceeds from assessment of more global indices

of emotional distress and disturbance to more detailed evaluations of specific diagnoses of psychopathology when warranted.

Cross-References

▶ Disability Assessment, Psychological/ Psychiatric Evaluation

Stereological Techniques

Definition

An unbiased way to calculate the number of neurons in a structure, such as the dorsal root ganglion, or in a nucleus in the CNS.

Cross-References

▶ Retrograde Cellular Changes After Nerve Injury

Stereotactic

▶ Pain Treatment, Intracranial Ablative Procedures

Stereotactic Cingulotomy

Definition

Ablation of the anterior cingulum, which is part of the limbic system located in the medial frontal lobe, by use of electrodes heated via radiofrequency energy.

Cross-References

▶ Cancer Pain Management, Neurosurgical Interventions

Stereotactic Medial Thalamotomy

Definition

Precise ablation of the medial thalamus.

Cross-References

▶ Cancer Pain Management, Neurosurgical Interventions

Stereotactic Operation with Thalamic Target

▶ Thalamotomy for Human Pain Relief

Stereotactic Radiosurgery

Definition

Precise delivery of multiple beams of radiation so that they overlap and sum to ablate a target area in the brain.

Cross-References

▶ Cancer Pain Management, Neurosurgical Interventions

Stereotactic Sacroiliac Joint Denervation

▶ Sacroiliac Joint Radiofrequency Denervation

Stereotactic Surgery

Definition

The branch of neurosurgery that involves the use of a precision apparatus to carry out x-ray-guided operations on the nervous system.

Cross-References

▶ Intracranial Ablative Procedures

Stereotaxy

Definition

A system of three dimensional coordinates which can be used to locate a structure within the brain with reference to the anterior posterior, lateral, and vertical axes.

Cross-References

▶ Poststroke Pain Model, Thalamic Pain (Lesion)

Steroid Injections

Nikolai Bogduk
University of Newcastle, Newcastle Bone & Joint Institute, Royal Newcastle Centre, Newcastle, NSW, Australia

Synonyms

Corticosteroid injections; Injections of corticosteroids; Injections of steroids

Definition

Steroid injections are a treatment for pain, in which a corticosteroid is injected into a site of tenderness that is ostensibly the source of pain. Steroid injections differ from intra-articular steroids in that the target tissues are not cavities of synovial joints. Instead, they are typically sites of attachment of tendons, ligaments, or muscles.

Characteristics

It is difficult to obtain the rationale for steroid injections from the literature. They seem to be used as if in the belief that steroids have an undisclosed healing property in areas that are tender and seemingly the source of pain. Although the anti-inflammatory effect of steroids might be invoked as grounds for the injections, there is no evidence that the conditions treated involve inflammation. Indeed, such evidence as is available indicates that degeneration, but not inflammation, is the underlying pathology.

Technique
Steroid injections simply require inserting a needle into the area or point that appears to be tender. Through the needle, a quantity of corticosteroid preparation is injected, typically mixed with a local anesthetic. The volume injected differs between operators and for different sites. It may be a fraction of a milliliter or between 1, 2, and 5 ml. The agents used are based on triamcinolone, dexamethasone, betamethasone, or methylprednisolone, usually as a depot preparation designed for slow release of the active steroid.

Application
In practice, almost any tender site might be a target for steroid injections. For most sites, however, there is little scientific literature. The sites for which the greatest number of controlled trials has been published are the lateral elbow and the low back. These sites have a common feature. The target is a region where over a short distance muscles become tendons, which insert into bone.

Lateral Elbow
Lateral elbow pain is classically known as "tennis elbow" or "lateral epicondylalgia." It is the archetypical condition for which steroid injections have been used. The literature is not consistent as to the exact target point for injection. It ranges from the apex of the lateral epicondyle, to the enthesis of the extensor tendons arising from this eminence, to the terminal tendons themselves, and even the myotendinous junctions of the extensor muscles.

The pathology of lateral elbow pain does not involve inflammation. Biopsy studies have shown that the extensor tendons exhibit partial rupture, hyaline degeneration, fibrofatty change, vascular proliferation, fibroblastic proliferation, glycosaminoglycan infiltration, fibrocartilage formation, calcification, and new bone formation, to various extents (Chard et al. 1994; Regan et al. 1992).

An early controlled study found no difference in outcome between patients treated with steroid injections or saline injections (Saartok and Eriksson 1986). A later study found that steroid injections gave better relief shortly after injection than did lignocaine alone, but by 24 weeks, differences were no longer apparent (Price et al. 1991). Another study found steroid injections to be more effective than physiotherapy for lateral elbow pain (Verhaar et al. 1996). The most recent controlled trial found that steroid injections provided greater relief of pain than either naproxen or placebo at a 4-week follow-up, but thereafter, there was no difference in outcome (Hay et al. 1999). A systematic review, published in 1996 (Assendelft et al. 1996), found that the pooled odds ratio for benefit at less than 6 weeks was 0.15 (95 % CI 0.10–0.23), indicating a statistically significant benefit for corticosteroid injections, but there was no statistically significant benefit at periods in excess of 6 weeks.

Low Back Pain
In patients with low back pain, a site that is commonly tender lies against the iliac crest, supermedial to the posterior superior iliac spine. This site has attracted a variety of injection therapies, mostly with local anesthetic alone but sometimes with steroids. According to a systematic review, the published studies favor an effect of injection therapy, but the small sample sizes used prevent the pooled outcome data from achieving statistical significance. The 95 % confidence intervals of the pooled relative risk cross the critical

value of 1.0 (Nelemans et al. 2001). One controlled study showed that methylprednisolone mixed with lignocaine was more effective than normal saline in securing "improvement" in pain, when assessed at 2 weeks (Sonne et al. 1985).

Other data apply to the injection of steroids at various tender sites in the back. An old study reported that a greater proportion of patients achieve complete relief of pain (odds ratio: 7.7) when corticosteroids are used rather than lignocaine alone (Bourne 1984). This encouraging report, however, has not been replicated.

Discussion

Despite the popularity of steroid injections as a treatment for musculoskeletal pain, the literature offers little support for this intervention. These injections do appear to have an effect in some patients, but the benefit is short-lived.

References

Assendelft, W. J. J., Hay, E. M., Adshead, R., et al. (1996). Corticosteroid injection for lateral epicondylitis: A systematic overview. *British Journal of General Practice, 46*, 209–216.

Bourne, I. H. (1984). Treatment of chronic back pain. Comparing corticosteroid-lignocaine injections with lignocaine alone. *Practitioner, 228*, 333–338.

Chard, M. D., Cawston, T. E., Riley, G. P., et al. (1994). Rotator cuff degeneration and lateral epicondylitis: A comparative histological study. *Annals of the Rheumatic Diseases, 53*, 30–34.

Hay, E. M., Paterson, S. M., Lewis, M., et al. (1999). Pragmatic randomised controlled trial of local corticosteroid injection and naproxen for treatment of lateral epicondylitis of elbow in primary care. *British Medical Journal, 319*, 964–968.

Nelemans, P. J., de Bie, R. A., deVet, H. C. W., et al. (2001). Injection therapy for subacute and chronic benign low back pain. *Spine, 26*, 501–515.

Price, R., Sinclair, H., Heinrich, I., et al. (1991). Local injection treatment of tennis elbow-hydrocortisone, triamcinolone and lignocaine compared. *British Journal of Rheumatology, 30*, 39–44.

Regan, W., Wold, L. E., Coonrad, R., et al. (1992). Microscopic histopathology of chronic refractory lateral epicondylitis. *The American Journal of Sports Medicine, 20*, 746–749.

Saartok, T., & Eriksson, E. (1986). Randomized trial of oral naproxen or local injection of betamethasone in lateral epicondylitis of the humerus. *Orthopedics, 9*, 191–194.

Sonne, M., Christensen, K., Hansen, S. E., et al. (1985). Injection of steroids and local anaesthetics as therapy for low back pain. *Scandinavian Journal of Rheumatology, 14*, 343–345.

Verhaar, J. A., Walenkamp, G. H., van Mameren, H., et al. (1996). Local corticosteroid injection versus cyriax-type physiotherapy for tennis elbow. *Journal of Bone and Joint Surgery, 78B*, 128–132.

Steroids

Definition

Long-lasting therapy with glucocorticosteroids may be complicated by muscle weakness.

Cross-References

▶ Muscle Pain in Systemic Inflammation (Polymyalgia Rheumatica, Giant Cell Arteritis, Rheumatoid Arthritis)

Stimulation Treatments of Central Pain

Antonio Oliviero[1], Vincenzo Di Lazzaro[2], Elisa Lopez-Dolado[3] and Laura Mordillo-Mateos[4]
[1]FENNSI Group, National Hospital of Paraplèjicos, Toledo, Spain
[2]Institute of Neurology, Campus Biomedico University, Rome, Italy
[3]Servicio de Rehabilitacion, National Hospital of Paraplèjicos SESCAM, Toledo, Spain
[4]FENNSI Group, Hospital Nacional de Paraplèjicos, Toledo, Spain

Synonyms

Neuromodulation; Neurostimulation

Definition

Nervous system stimulation is a pain treatment that uses low voltage electrical or magnetic stimulation, delivered to the central or peripheral nervous systems, to inhibit or block the sensation of pain.

Introduction

Despite the discovery of multiple mechanisms causing chronic pain, the handling of ▶ central pain is still insufficient. Pharmacological treatments are widely used but not all the patients improve their quality of life. Stimulation treatments have been proposed to obtain pain relief. An overview of the most commonly used stimulation techniques is provided.

Characteristics

Chronic pain is an important cause of physical and emotional suffering, interference in family and social life, reduced quality of life, disability, work absenteeism, and an increased need for health care. Injury to the nervous system represents a risk for the development of chronic pain, having a major impact on patients as well as society. The IASP (International Association for the Study of Pain) defines "neuropathic pain" as a pain caused by a lesion or disease of the somatosensory nervous system. Peripheral neuropathic pain occurs when the lesion or dysfunction affects the peripheral somatosensory nervous system. Damage to the central nervous system (CNS) can also cause chronic pain. The IASP defines "central pain" as a pain caused by a lesion or disease of the central somatosensory nervous system. The treatment of central pain (CP) is often difficult and it is not known if different central pain conditions respond differently to different treatments. Despite the discovery of multiple mechanisms causing chronic pain, the handling of central pain is still insufficient.

Central Nervous System Stimulation

CNS electrical stimulation has been used to treat different kinds of pain. Investigation has been done into the use of stimulation on well-defined targets including cerebral cortex, sensory thalamus, midbrain periaqueductal or periventricular gray, and spinal cord.

CP is caused by a heterogeneous group of etiologically different diseases (traumatic, vascular, degenerative, inflammatory, tumoral, etc.), of different damage locations (spinal cord, brainstem, thalamus, and cortex), and of both rapidly and slowly developing lesions. It seems that every lesion affecting structures involved in somatic sensibility can produce CP. The physiopathology of central pain is still poorly understood, but the thalamus seems to play a pivotal role in the generation of central pain. Thalamic neurons are hyperactive and cells spontaneously fire (bursting cells) in patients with central pain. In animal studies, it was shown that rats with CP exhibit suppressed activity in the inhibitory nucleus zona incerta (ZI) which leads to increased activity in the posterior thalamus. Intraoperative stimulation of the lateral thalamus in patients with CP (with symptoms of ▶ hyperpathia or ▶ allodynia) elicited a painful sensation, while control patients (patients operated for dyskinesias or tremor) did not. Furthermore, an altered somatotopy and displaced and/or reorganized repetitive fields have been observed in the thalamus of patients with CP. In addition, an unbalanced transmission between sensory motor cortex and the sensory thalamus has also been hypothesized in the generation of CP.

All these considerations suggest that a form of plasticity is present in the thalamus and the cortex, and also probably in other locations of the CNS (Brown and Barbaro 2003). The sensory cortex could be thought of as a primary target for stimulation to obtain pain relief, but the usefulness of the stimulation of this target is still under debate. Stimulation of the sensory cortex in animal models did not produce conclusive evidence of effective pain management. In humans, when the sensory cortex has been stimulated, pain was either exacerbated or remained unchanged. However, contradictory results of sensory cortex stimulation have recently been reported (Bonicalzi and Canavero 2004; Brown and Barbaro 2003). White and

Sweet (1955) performed postcentral gyrectomy and obtained pain relief in only a few (13 %) of the treated patients, whereas the surgical removal of both the post- and the precentral gyri produced long-term relief from pain. These results are suggestive of the importance of motor cortex removal in providing pain relief, due to the fact that precentral gyrus ablation was performed in a second surgical operation, given that the previous postcentral ablation had not been effective. Tsubokawa and his coworkers described their first clinical experience using motor cortex stimulation to treat central pain syndromes in 1991. They reported "excellent or good" pain relief and a long-term efficacy of motor cortex stimulation in most of the patients. Since then many other patients have been treated with motor cortex stimulation with similar, interesting results (Brown and Barbaro 2003). How motor cortex stimulation (MCS) works in producing pain relief is still poorly understood. It is still not clear if MCS acts at cortical level or through the activation of descending axons. An intact corticospinal tract originating from motor cortex would seem to be required to obtain pain relief, while an intact somatosensory system does not appear to be necessary to receive benefit from MCS. Pain relief is achieved at a stimulus intensity below the threshold for muscular movement, and the efficacy of MCS seems to have a somatotopic organization. The corticospinal tract could be activated by motor cortex stimulation below the threshold for muscle movement (Di Lazzaro et al. 1998). The necessity to have an intact corticospinal tract to obtain pain relief, and the possibility to stimulate it with stimulus intensity below the threshold for muscular movement, could suggest that an activation of projecting axons is required. Moreover, MCS seems to activate the thalamus. Positron emission tomography studies during MCS show activation in the ipsilateral thalamus, cingulate gyrus, orbitofrontal cortex, and brainstem; some of these structures could be related to the suffering component of chronic pain, and the removal of this component could be one of the ways in which MCS is effective in providing pain relief (Garcia-Larrea et al. 1999).

Transcranial Magnetic Stimulation

Transcranial magnetic stimulation (TMS) was introduced in 1985, and it permits a noninvasive stimulation of the cerebral cortex. TMS has been used in evaluation of the motor system, the functional study of several cerebral regions, and the physiological study of several neuropsychiatric illnesses. *Repetitive transcranial magnetic stimulation (rTMS)* has become a useful tool for investigating and even modulating human brain function. rTMS of the human motor cortex can produce changes in cortical excitability that outlast the period of stimulation. In addition, it has been postulated that rTMS could be a therapeutic tool in some psychiatric disorders. rTMS of the motor cortex has also been demonstrated as useful in alleviating acute pain (induced by capsaicin) and chronic pain (neuropathic pain of central and peripheral origin). High-frequency rTMS seems effective in alleviating neuropathic pain, and this effect seems capable of predicting the efficacy of invasive MCS (Lefaucheur et al. 2004). How motor cortex rTMS works to provide pain relief is still not clear. rTMS can modulate cortical excitability, promote a short-term change in inhibitory cortical activity, modify corticocortical connections functionally, and produce a net change in synaptic cortical activity. Sensory functions are modulated by rTMS of the sensorimotor cortex. Tactile threshold was increased for a short duration after low-frequency rTMS of the sensorimotor cortex, and thermal perception was reduced after high-frequency rTMS of the same cortical area (unpublished personal observations). rTMS activates a network of primary and secondary cortical motor regions (using a suprathreshold intensity), supplementary motor area, dorsal premotor cortex, cingulate motor area, as well as the putamen and thalamus (using both supra- and subthreshold intensity) (Bestmann et al. 2004). In conclusion, rTMS of the motor cortex and invasive MCS seem to activate overlapping cerebral regions, and activation of the thalamus again seems to be important to obtain pain relief. Moreover, it has recently been observed that invasive MCS and rTMS activate the motor cortex in a similar (though not exactly the same) way (Di Lazzaro et al. 2004). Future studies will clarify the role of rTMS: whether this noninvasive technique could be useful for treatment of CP and predict the efficacy of invasive MCS. A recent meta-analysis suggested that rTMS may be more effective in the treatment of neuropathic pain conditions with a central compared to a peripheral nervous system origin (Leung et al. 2009). Pain relief is sometimes greater when stimulation is applied to the motor cortex region representing an area adjacent to the area in pain (Lefaucheur et al. 2008).

Transcranial Direct Current Stimulation

Transcranial direct current stimulation (tDCS) is another technique of brain stimulation, in which, cerebral cortex is stimulated by a weak DC current in a noninvasive and painless manner. Surface-positive polarization of rat and cat cerebral cortex raises the mean firing rate of neurons recorded in deep cortical layers, whereas surface-negative polarization reduces spontaneous firing. The conditioning effects of DC stimulation on firing rates have been attributed to shifts in the resting membrane potential of cortical neurons. Several studies have shown that this technique modulates cortical excitability in the human motor cortex (Nitsche and Paulus 2000). Anodal stimulation depolarizes (hence excites) and cathodal stimulation hyperpolarizes (hence inhibits) neurons. The efficacy of tDCS depends critically on parameters such as electrode position and current strength. The typical intensities administered are 1 and 2 mA for a maximum of 20 min in a given session. Its effects remains after the end of the stimulation and not only shift activity of cortical areas situated directly under the electrodes, but also distant areas, probably by interconnections of the primary stimulated area with these structures (Lang et al. 2005). The tDCS effects originate at the cortical level (Lang et al. 2011). Motor cortex was the first cortical target that was proved to be efficacious in chronic pain treatment. At present, significant analgesic effects were also shown to occur after the stimulation of other cortical targets (including prefrontal and parietal areas) in acute provoked pain, chronic neuropathic pain,

fibromyalgia, or visceral pain. Therapeutic applications of rTMS in pain syndromes are limited by the short duration of the induced effects, but prolonged pain relief can be obtained by repeating rTMS sessions every day for several weeks. Recent tDCS studies also showed some effects on various types of chronic pain. tDCS could also be applied before rTMS to further enhance the size and the duration of the effects of rTMS itself.

Thalamus has long been known to be involved in the processing of pain in normal and pathological conditions, and it seems to play an important role in the way therapeutic cortical stimulation works. Stimulation of the somatosensory thalamus has been used for many years in the treatment of chronic pain. More recently, the center median-parafascicular complex has also been targeted in order to obtain relief from neuropathic pain. However, despite clinical reports of successful results, little is known about the way this form of analgesia is achieved. Thalamic stimulation activates ipsilateral insula and the anterior cingulate cortex and only produces a slight effect on the somatosensory cortex (Duncan et al. 1998). For these reasons and others, the gate-control theory, which is based on the idea of stimulation of low–threshold somatosensory pathways inhibiting pain (Melzack and Wall 1965), as a basic mechanism of how stimulation works, has been thrown into doubt. An alternative hypothesis is that thalamic stimulation may activate thalamocortical circuits involved in thermal as well as tactile processing, and in this way could provide analgesia. Whichever mechanism is hypothesized, it seems that the cortex and thalamus represent two sites within a pain related circuit, and pain relief can be achieved wherever you stimulate within this circuit.

Spinal cord stimulation (SCS) has been widely used in the management of pain. It emerged as a clinical application of the gate-control theory. Later, the therapeutic properties of SCS were considered to be due to a sympatholytic effect. This effect is considered to be responsible for the effectiveness of SCS in peripheral ischemia, cardiac ischemia, and at least some cases of pain syndromes. The sympatholytic effect has also been considered effective in the treatment of other chronic pain states such as failed back surgery syndrome, diabetic neuropathy, postherpetic neuralgia, and multiple sclerosis. It is surprising that almost 40 years after the first attempt to modulate pain through SCS, only few large-scale trials of its effectiveness have been published, and at present, there is limited evidence that spinal cord stimulators are effective in the management of pain.

Peripheral Nervous System Stimulation

The basic idea of using local application of electricity for pain treatment is ancient, and in ancient Greece, the electrical properties of animals have been used for pain relief in pain conditions of the upper and lower limbs. Transcutaneous electrical nerve stimulation (TENS) is widely believed to be an effective, safe, and relatively noninvasive intervention which can be used to alleviate many different types of pain, including labor pain, surgical pain, back pain, arthritis, neuropathic pain, menstrual pain, migraine, and headache. However, controversy still surrounds TENS and its effectiveness has not been proven conclusively. A Cochrane review recently concluded that it is not possible to provide useful evidence-based information for the use of TENS for CP (Carroll et al. 2004). The gate-control theory could provide a possible means of understanding the way in which TENS works. Peripheral stimulation could activate large myelinated nerve fibers and local inhibitory circuits within the dorsal horn of the spinal cord, regulating the flow of pain information, and thus reducing the perception of pain. Peripheral mechanisms of pain relief have also been hypothesized, but microneurographic studies failed to demonstrate an impulse transmission failure during TENS administration to nerves. TENS could also provide pain relief neurochemically. Contradictory data are present in literature where opioid systems are or are not stimulated by TENS. It seems that low frequency of stimulation is capable of activating opioid endogenous systems, while high-frequency TENS does not. This would suggest that the frequency of stimulation is critical in achieving successful pain management.

Opioid endogenous system activation should work in a similar way to that of acupuncture analgesia.

Summary

Stimulation treatments have been used to treat different kinds of pain. Investigation has been done into the use of stimulation on well-defined targets including cerebral cortex, sensory thalamus, midbrain periaqueductal or periventricular gray, spinal cord, and peripheral nervous system.

References

Bestmann, S., Baudewig, J., Siebner, H. R., et al. (2004). Functional MRI of the immediate impact of transcranial magnetic stimulation on cortical and subcortical motor circuits. *European Journal of Neuroscience, 19*, 1950–1962.

Bonicalzi, V., & Canavero, S. (2004). Motor cortex stimulation for central and neuropathic pain. *Pain, 108*, 199–200.

Brown, J. A., & Barbaro, N. M. (2003). Motor cortex stimulation for central and neuropathic pain: Current status. *Pain, 104*, 431–435.

Carroll, D., Moore, R. A., McQuay, H. J., et al. (2004). Transcutaneous electrical nerve stimulation (TENS) for chronic pain (Cochrane review). In *The Cochrane library* (Issue 2). Chichester: Wiley.

Di Lazzaro, V., Restuccia, D., Oliviero, A., et al. (1998). Effects of voluntary contraction on descending volleys evoked by transcranial stimulation in conscious humans. *The Journal of Physiology, 508*, 625–633.

Di Lazzaro, V., Oliviero, A., Pilato, F., et al. (2004). Comparison of descending volleys evoked by transcranial and epidural motor cortex stimulation in a conscious patient with bulbar pain. *Clinical Neurophysiology, 115*, 834–838.

Duncan, G. H., Kupers, R. C., Marchand, S., et al. (1998). Stimulation of human thalamus for pain relief: Possible modulatory circuits revealed by positron emission tomography. *Journal of Neurophysiology, 80*, 3326–3330.

Garcia-Larrea, L., Peyron, R., & Mertens, P. (1999). Electrical stimulation of motor cortex for pain control: A combined pet-scan and electrophysiological study. *Pain, 83*, 259–273.

Lang, N., Nitsche, M. A., Dileone, M., Mazzone, P., De Andrés-Arés, J., Diaz-Jara, L., Paulus, W., Di Lazzaro, V., & Oliviero, A. (2011). Transcranial direct current stimulation effects on I-wave activity in man. *Journal of Neurophysiology, 105*(6), 2802–2810.

Lang, N., Siebner, H. R., Ward, N. S., Lee, L., Nitsche, M. A., Paulus, W., Rothwell, J. C., Lemon, R. N., & Frackowiak, R. S. (2005). How does transcranial DC stimulation of the primary motor cortex alter regional neuronal activity in the human brain? *European Journal of Neuroscience, 22*(2), 495–504.

Lefaucheur, J. P., Drouot, X., Ménard-Lefaucheur, I., & Nguyen, J. P. (2004). Neuropathic pain controlled for more than a year by monthly sessions of repetitive transcranial magnetic stimulation of the motor cortex. *Neurophysiologie Clinique, 34*(2), 91–95.

Lefaucheur, J. P., Antal, A., Ahdab, R., Ciampi de Andrade, D., Fregni, F., Khedr, E. M., Nitsche, M., & Paulus, W. (2008). The use of repetitive transcranial magnetic stimulation (rTMS) and transcranial direct current stimulation (tDCS) to relieve pain. *Brain Stimulation, 1*(4), 337–344.

Leung, A., Donohue, M., Xu, R., Lee, R., Lefaucheur, J. P., Khedr, E. M., Saitoh, Y., André-Obadia, N., Rollnik, J., Wallace, M., & Chen, R. (2009). rTMS for suppressing neuropathic pain: A meta-analysis. *Journal of Pain, 10*(12), 1205–1216.

Melzack, R., & Wall, P. D. (1965). Pain mechanisms: A new theory. *Science, 150*(3699), 971–979.

Nitsche, M. A., & Paulus, W. (2000). Excitability changes induced in the human motor cortex by weak transcranial direct current stimulation. *Journal of Physiology, 527*(Pt. 3), 633–639.

White, J. C. & Sweet, W. H. (1955). *Pain. Its mechanisms and neurosurgical control.* Springfield: Charles C. Thomas. Massachusetts General Hospital and Harvard Medical School. With Assistance in the Psychiatric Sections from Stanley Cobb, M. D., and Frances J. Bonner, M.D. 10 × 7 in. pp. 736 + xxiv, with 134 illustrations. (Oxford: Blackwell Scientific Publications.) 126 s. *British Journal of Surgery, 43*, 108. doi:10.1002/bjs.18004317730

Stimulation-Produced Analgesia

Gregory W. Terman
Department of Anesthesiology and the Graduate Program in Neurobiology and Behavior, University of Washington, Seattle, WA, USA

Synonyms

Brain stimulation analgesia; Electrical stimulation-induced analgesia

Definition

Although the concept of descending pain modulation goes back to Sherrington in the early 1900s and was an important part of Melzack and Wall's gate control theory of pain (Melzack and Wall 1965), there was little direct evidence until 1970 for ▶ endogenous neural systems descending from brain to spinal cord whose normal function was the inhibition of pain. Reynolds in 1969 reported that focal electrical stimulation of the ▶ periaqueductal gray (PAG) resulted in sufficient analgesia to perform laparotomy in rats. These observations were soon confirmed and extended by Liebeskind and his students (e.g., Mayer et al. 1971) who labeled the phenomenon "▶ stimulation-produced analgesia" (SPA).

Characteristics

These early studies by Liebeskind and associates found SPA to be specifically antinociceptive – producing no generalized sensory, attentional, emotional, or motoric deficits. They also showed that SPA could outlast the period of brain stimulation and that it occurred in a restricted peripheral field such that noxious stimuli applied outside that field elicited normal defensive reactions. They reported that during stimulation, the sensation of light touch was intact, there was no indication of seizure activity, and animals were still capable of eating. Furthermore, rats could be trained to self-administer brain stimulation in the presence of noxious stimuli, and this self-administration did not recur when the noxious stimuli were terminated. The analgesia produced by electrical stimulation of the PAG was of rapid onset and as potent as high doses of morphine, completely inhibiting withdrawal responses to even the most severe noxious stimuli. These findings demonstrated the existence of a potent and selective natural ▶ pain inhibitory system in the PAG (Fig. 1).

SPA's effects depend on descending mechanisms (rather than primarily supraspinal mechanisms) as evidenced by the inhibition of

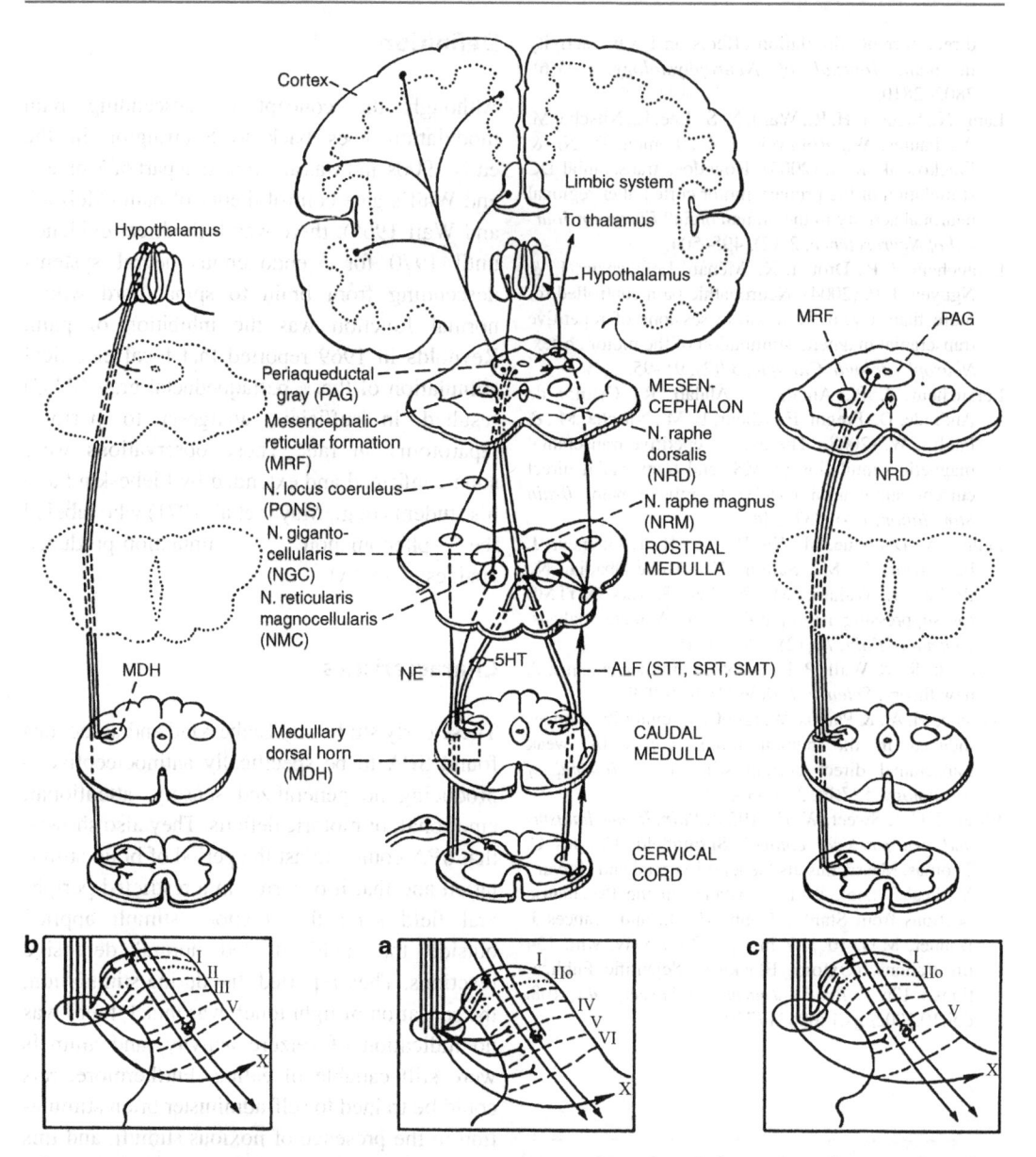

Stimulation-Produced Analgesia, Fig. 1 Descending endogenous pain inhibitory systems activated in stimulation-produced analgesia. (**a**) The periaqueductal *gray* (PAG) receives input from such rostral structures as the frontal and insular cortex and other parts of the cerebrum involved in cognition and from the limbic system, thalamus, and hypothalamus. The PAG sends axons that project to nucleus raphe magnus (NRM) and nucleus reticularis gigantocellularis (NGC). Here, they synapse with neurons that are primarily serotonergic, whose axons project to the medullary dorsal horn or descend in the dorsolateral funiculus to all laminae of the spinal *gray* matter. Noradrenergic fibers from locus coeruleus also descend in the dorsolateral funiculus to the spinal *gray*. (**b**) Direct hypothalamospinal descending control system, which originates in the medial and paraventricular hypothalamic nuclei. The inset shows sites of termination in the spinal cord. (**c**) Direct PAG-spinal projection system, which bypasses the medullary nuclei and projects directly to the medullary dorsal horn or spinal *gray* matter. The inset shows sites of termination in the spinal cord (Reprinted with permission from Loeser et al. 2001)

both spinal nociceptive reflexes and responses of spinal cord dorsal horn nociceptive cells to noxious stimuli (Dubner and Bennett 1983). SPA eliminates behavioral responses to such varied noxious somatic and visceral stimuli as electric shocks applied to the tooth pulp and limbs, heating of the skin, and injection of irritants into the skin and abdominal cavity. Subsequent demonstrations that stimulation of homologous brain regions in humans could relieve chronic pain provided even greater interest in this phenomenon (Hosobuchi 1980). Within 10 years of Reynolds' initial SPA paper, more than 150 reports of stimulation-produced analgesia had been published, including two major reviews (Basbaum and Fields 1978; Mayer and Price 1976) of the anatomical and biochemical substrates of SPA.

Perhaps the most important finding concerning SPA, in these early studies, was that analgesia induced by brain stimulation shared several characteristics with analgesia from opiate drugs. Areas of the brain from which SPA could be elicited were rich in opiate receptors, and microinjection of morphine into these brain regions produced analgesia, indicating that common brain sites support both SPA and opiate analgesia. Like opiate analgesia, ▶ tolerance develops to the analgesic effects of repeated brain stimulation. Moreover, cross-tolerance is observed between opiate analgesia and SPA, suggesting a common underlying mechanism. Finally, opiate antagonists such as naloxone antagonize SPA, a finding which was influential in suggesting the existence of and subsequent sequencing of the endogenous ▶ opioid peptides (Hughes et al. 1975). Ironically, the naloxone sensitivity of SPA has proven controversial, and it is now clear that not all SPA is mediated by opioid peptides. Indeed, differential production of "opioid" and "non-opioid" forms of SPA has been found to rely on the distinct region of the PAG which is stimulated.

Additional work clarifying the circuitry of descending pain inhibitory systems activated by electrical stimulation of the central nervous system (Gebhart 2004) has focused on four tiers of CNS processing.

Cortical and Diencephalic Descending Systems

Stimulation of S1 and S2 cortex inhibits spontaneous activity of wide-dynamic-range spinothalamic tract (STT) neurons, as well as responses to C-fiber volleys in sural nerve and to noxious thermal and mechanical stimuli. Efferents from sensory cortex run with the corticospinal tract, descending contralaterally in the dorsolateral funiculus before terminating in laminae I–VII of the spinal gray matter. Corticospinal fibers in laminae I and II exert direct postsynaptic control on dorsal horn neurons, although the neurotransmitters involved are unknown. Despite these direct connections, the many S1 and S2 cortical projections to striatum, thalamus, and mesencephalon are likely to mediate many of the analgesic effects of somatosensory cortex stimulation.

Similarly, the specific descending influence of limbic structures (such as amygdala and cingulate cortex) on nociception is difficult to identify because of the complex brain circuitry these structures are involved in. Nonetheless, stimulation of limbic structures can cause analgesia, and it is likely that activation of these structures mediate analgesia associated with such diverse phenomena as hypnosis, placebo, learned helplessness, and uncontrollable stress.

Diencephalic structures such as the periventricular gray, the medial and lateral hypothalamus, the medial preoptic/basal forebrain region, and the somatosensory nuclei of the thalamus can all support SPA, albeit by differing mechanisms. Stimulation of the ventroposterolateral and medial nuclei of the thalamus, for example, can inhibit evoked responses of identified spinal nociceptive neurons. There is no evidence for a direct projection of thalamic neurons to the spinal cord, however. It is likely instead that stimulation of the thalamus produces antidromic activation of spinothalamic tract collaterals, which synapse in medullary reticular formation or periaqueductal gray and which, in turn, activate neurons which do project to the spinal cord to exert inhibition.

A direct, predominantly ipsilateral projection to the spinal cord from the medial hypothalamic nuclei and, to a lesser extent, the lateral nuclei of the hypothalamus travels principally in the dorsolateral funiculus and may mediate the antinociceptive effect of hypothalamic stimulation. Although the ▶ spinal cord laminae in which these projections terminate have not been identified, it is known that oxytocin- and vasopressin-containing neurons of the hypothalamus terminate predominantly in laminae I and X and that there are sparse inputs in laminae II, III, and V. Both oxytocin and vasopressin have been reported to produce analgesia at the level of the spinal cord (although the effects of the latter, at least, may be confounded by paralytic motor effects). Another neurochemical possibly involved in such hypothalamic stimulation-produced analgesia is dopamine. Dopaminergic innervation terminates in laminae I and particularly laminae III and IV of the spinal cord, and dopamine has been found to reduce spinal nociception both electrophysiologically and behaviorally. This is one of three major ▶ monoamines thought to mediate some forms of SPA. As in cortical SPA, SPA from the hypothalamus may be mediated by indirect projections to the spinal cord via midbrain and hindbrain relay nuclei.

Mesencephalic Descending Systems

The PAG receives inputs from such rostral structures as the frontal and insular cortex, limbic system, septum, amygdala, and hypothalamus. Input from caudal structures includes those from nucleus cuneiformis, the pontomedullary reticular formation, locus coeruleus, and the spinal cord. Moreover, the PAG has descending connections to the rostral medulla, particularly its ventromedial part (RVM), the medullary reticular nuclei, and the ▶ nucleus raphe magnus (known, as we will see, to be a major source of projections to the medullary and spinal dorsal horns) as well as ascending projections to diencephalon. Thus, the PAG is ideally positioned to integrate information from sensory systems in the brainstem/spinal cord with information from

higher processing in the forebrain and is thought to modulate a number of homeostatic processes including fear, anxiety, cardiovascular tone, and vocalization. The descending nociceptive modulation from PAG activated by SPA appears to be normally under tonic ▶ GABA inhibitory control such that GABA$_A$ antagonists can produce reliable antinociception when microinjected into the PAG in the absence of electrical stimulation. Endogenous and exogenous opiates likely produce analgesia within the PAG via inhibition of inhibitory GABA-containing interneurons and thereby activate (disinhibit) PAG output neurons.

Pontine and Medullary Descending Systems

The ▶ serotonergic-rich raphe nuclei in the brainstem, particularly the nucleus raphe magnus (NRM), were implicated early on in mediating SPA. Raphe-spinal neurons were excited by PAG opiate microinjection or PAG electrical stimulation. Stimulation of the NRM produced analgesia, whereas lesions of this structure inhibited SPA from PAG. Consequently, the NRM appeared to be a relay station that received descending input from the PAG and modulated spinal and medullary dorsal horn.

Subsequent findings suggest other important alternative descending paths are activated by PAG opiates and electrical stimulation. In order to abolish SPA from PAG, lesions of the NRM had to be extended into the adjacent magnocellular reticular field. Electrical stimulation or lesions in the medullary reticular core activate or interfere with, respectively, multiple descending pathways including rubrospinal, tectospinal, pontine reticulospinal, and noradrenergic fibers in the pons – all of which can influence dorsal horn nociceptive activity. In addition to fibers of passage, the reticular formation in the rostral medulla and caudal pons also includes several nuclei which have been implicated in pain modulation (including the reticularis gigantocellularis (NGC) and reticularis paragigantocellularis lateralis (RPGL) in the rat). These nuclei and the NRM have been grouped together by Fields and labeled the ▶ rostroventral medulla (RVM). All of the nuclei in this region receive projections from the PAG, contain cells which can be activated by PAG stimulation, send axons to the spinal cord, and can support SPA.

Microinjection of opiates into RVM, like into the PAG, can produce descending pain inhibition, although via a different mechanism than observed in PAG as detailed by Fields and colleagues (Pan et al. 1997). Nicotinic acetylcholine and endogenous cannabinoid (e.g., anandamide) mechanisms have also been implicated in RVM-mediated pain inhibitory substrates.

Descending Fibers from Supraspinal Nuclei Terminate in Spinal Dorsal Horn and Selectively Inhibit Nociceptive Neurons

In the mid-1970s, Basbaum and associates (Basbaum et al. 1977) lesioned various quadrants of the spinal cord to determine the spinal tracts containing fibers descending from brain stem to dorsal horn which inhibit nociception. They found that lesions of the dorsal part of the lateral funiculus (DLF) markedly reduced or abolished SPA (opiate analgesia was also blocked by these lesions). The neurochemicals underlying this descending inhibition have been much studied but have primarily focused on monoaminergic systems because of the importance of the medullary raphe and reticular nuclei in mediating SPA:

Cell bodies containing ▶ serotonin (5HT) are located in the nucleus raphe magnus (NRM), n. raphe dorsalis (NRD), and several other raphe nuclei in the medulla and pons and are absent in spinal cord. Most NRM-5HT neurons project to spinal cord via the dorsolateral funiculus to terminate predominantly in laminae I, IIo, IV, and V of the spinal dorsal horn and near the central canal. Early studies of SPA from a number of brainstem sites (as well as systemic opiate analgesia) supported a role for 5HT in mediating endogenous pain inhibition (Basbaum and Fields 1978). Inhibition of synthesis of 5HT, destruction of spinal 5HT terminals, and lesions of

5HT-producing cells in the medullary raphe all inhibited SPA. Moreover, an intrathecal serotonin antagonist blocked analgesia produced by microinjection of morphine into the raphe nuclei. Application of 5HT to the spinal cord was most commonly reported to inhibit the response of dorsal horn neurons to noxious stimulation and produced analgesia. Dubner, Ruda, and associates examined sites of interaction between descending 5HT axons and laminae I and II neurons in a series of experiments (Dubner et al. 1984). The majority of spinal 5HT terminals form axosomatic and axodendritic synapses with neurons of the dorsal horn including STT neurons. Stimulation of NRM inhibited all nociceptive-specific (NS) or wide-dynamic-range (WDR) neurons identified, including STT cells, and all of these lamina I neurons exhibited many 5HT immunoreactive contacts. Thus, the 5HT neurons of the nucleus raphe magnus and other nuclei are presumed to provide the serotonergic link in the pain inhibitory controls exerted by SPA. Unfortunately, activation of serotonin-containing neurons also has pro-nociceptive effects (e.g., Calejesan et al. 1998), and serotonin's pro- or antinociceptive effects in the spinal cord may depend on the specific 5HT receptors activated (of which there are many subtypes) and the location of those receptors. Although this does not negate the results of previous SPA studies, it has limited the clinical relevance of, for example, serotonin reuptake inhibitors, where it is difficult to predict the magnitude or even the direction of the effect on pain.

A second neurochemically distinct monoamine descending system involves ▶ noradrenergic neurons whose cell bodies are in A5, A6, and A7 cell groups in the brainstem and whose axons descend in dorsolateral, ventrolateral, and ventral funiculi to end in all of the laminae of the spinal cord. Several studies indicate that a descending norepinephrine (NE) system can mediate analgesia and dorsal horn inhibition and that the NE descending system is critical for opiate-induced analgesia. Adrenergic blockers can attenuate the analgesia produced by systemic morphine administration, microinjection of morphine into the PAG or magnocellular tegmental field, and SPA of magnocellularis. Blockade of PAG SPA by intrathecal α_2-adrenergic antagonists (Budai et al. 1998) emphasizes the importance of noradrenergic systems in descending nociceptive modulation. Intrathecal α_2-adrenergic agonists, of which clonidine is the prototype, produce analgesia in animals, including man, although such analgesia is associated with hemodynamic and sedative side effects. Much ongoing work is attempting to identify α_2-adrenergic receptor subtypes which might mediate analgesia without inducing undesirable side effects to provide targets for development of new analgesics, thus far without success.

The opioid peptide enkephalin and 5-HT coexist in some RVM neurons, and based on the overlap between opiate and SPA mechanisms, it seems reasonable to assume the importance of enkephalins in descending pain inhibitory systems. Nonetheless, there is surprisingly little evidence suggesting the importance of spinal opioids in descending inhibition of nociception. Budai and Fields reported that descending inhibition of heat responses in dorsal horn cells from PAG is antagonized by direct application of mu-antagonists on the spinal cord, although many others have failed to see inhibition of SPA by spinally administered opiate antagonists.

The anatomy and physiology of SPA as revealed in the laboratory have been complementary to the many clinical reports that electrical stimulation of the thalamus, septum, caudate nucleus, medial forebrain bundle, lateral hypothalamus, and internal capsule can all produce analgesia in humans (Richardson 1990). The invasive nature of such human stimulation-produced analgesia and the resultant risks involved in such therapy make such reports of more scientific than therapeutic value. The long-term benefit of SPA in humans has not been demonstrated due to tolerance to SPA, inadequate electrode placement, or innumerable other physiological or clinical reasons. Nonetheless, although the similarities between SPA in humans and laboratory animals do not imply that such stimulation is recommended for either, an understanding of the neurochemical basis of SPA in animals may one day help identify new

therapeutic drugs in man. Moreover, identification of less invasive ways of activating the endogenous pain inhibitory substrates mediating SPA (e.g., by placebo, acupuncture, or exercise) may help us harness these systems for therapeutic benefit in the years to come.

Cross-References

▶ Forebrain Modulation of the Periaqueductal Gray and Its Role in Pain

References

Basbaum, A. I., & Fields, H. L. (1978). Endogenous pain control mechanisms: Review and hypothesis. *Annals of Neurology, 4*, 451–462.

Basbaum, A. I., Marley, N. J., O'Keefe, J., & Clanton, C. H. (1977). Reversal of morphine and stimulus-produced analgesia by subtotal spinal cord lesions. *Pain, 3*, 43–56.

Budai, D., Harasawa, I., & Fields, H. L. (1998). Midbrain periaqueductal gray (PAG) inhibits nociceptive inputs to sacral dorsal horn nociceptive neurons through Alpha2-adrenergic receptors. *Journal of Neurophysiology, 80*, 2244–2254.

Calejesan, A. A., Ch'ang, M. H., & Zhuo, M. (1998). Spinal serotonergic receptors mediate facilitation of a nociceptive reflex by subcutaneous formalin injection into the hindpaw in rats. *Brain Research, 798*, 46–54.

Dubner, R., & Bennett, G. J. (1983). Spinal and trigeminal mechanisms of nociception. *Annual Review of Neuroscience, 6*, 381–418.

Dubner, R., et al. (1984). Neural circuitry mediating nociception in the medullary and spinal dorsal horn. In L. Kruger & J. C. Liebeskind (Eds.), *Advances in pain research and therapy* (Vol. 6, pp. 151–166). New York: Raven Press.

Gebhart, G. F. (2004). Descending modulation of pain. *Neuroscience and Biobehavioral Reviews, 27*, 729–737.

Hosobuchi, Y. (1980). The current status of analgesic brain stimulation. *Acta Neurochirurgica Supplementum (Wien), 30*, 219–227.

Hughes, J., Smith, T. W., Kosterlitz, H. W., Fothergill, L. A., Morgan, B. A., & Morris, H. R. (1975). Identification of two related pentapeptides from the brain with potent opiate agonist activity. *Nature, 258*, 577–580.

Loeser, J. D., Butler, S. F., Chapman, C. R., & Turk, D. C. (2001). *Bonica's management of pain* (3rd ed.). Philadelphia: Lippincott Williams and Wilkins.

Mayer, D. J., & Price, D. D. (1976). Central nervous system mechanisms of analgesia. *Pain, 2*, 379–404.

Mayer, D. J., Wolfle, T. L., Akil, H., Carder, B., Liebeskind, J. C., et al. (1971). Analgesia from electrical stimulation in the brainstem of the rat. *Science, 174*, 1351–1354.

Melzack, R., & Wall, P. D. (1965). Pain mechanisms: A new theory. *Science, 150*, 971–979.

Pan, Z. Z., Tershner, S. A., & Fields, H. L. (1997). Cellular mechanism for anti-analgesic action of agonists of the kappa-opioid receptor. *Nature, 389*, 382–385.

Reynolds, D. V. (1969). Surgery in the rat during electrical analgesia induced by focal brain stimulation. *Science, 164*, 444–445.

Richardson, D. E. (1990). Central stimulation-induced analgesia in humans – Modulation by endogenous opioid peptides. *Critical Reviews in Neurobiology, 6*, 33–37.

Stimulus Quality

Definition

The categorical subjective experience evoked by activation of either primary afferents or central nervous system neurons. Such experiences could be burning, aching, pricking, warm, cool, etc.

Cross-References

▶ Encoding of Noxious Information in the Spinal Cord

Stimulus-dependent Method

▶ Pain Evaluation, Psychophysical Methods

Stimulus-Dependent Pain

Definition

Pain that is evoked by stimulation.

Cross-References

▶ Diagnosis and Assessment of Clinical Characteristics of Central Pain

Stimulus-Evoked Pain

Definition

Pain elicited by external stimulation.

Cross-References

▶ Antidepressants in Neuropathic Pain

Stimulus-Independent Pain

Definition

Pain that occurs spontaneously, not dependent on stimuli applied to the skin.

Cross-References

▶ Diagnosis and Assessment of Clinical Characteristics of Central Pain

Stimulus-Response Functions

Richard H. Gracely
Departments of Medicine-Rheumatology and Neurology, University of Michigan Health System, Ann Arbor, MI, USA

Characteristics

Pain Threshold and Pain Tolerance

The pain threshold is the minimum amount of noxious stimulation that reliably evokes a report of pain. Pain tolerance is similarly defined as the time that a continuous stimulus is endured or the maximally tolerated stimulus intensity. The popularity of threshold and tolerance measures is due in part to their ease of use and a response that is expressed in physical units of stimulus intensity or time, avoiding the problems with developing a subjective scale of pain. Both threshold and tolerance measures are often confounded with time or increasing intensity, and thus consistent data or analgesic effects can be simulated by responding after a specific number of stimuli or after a specific elapsed time. External factors can bias a subject to respond sooner or later or to a lower or higher intensity. Tolerance of a painful stimulus has been shown to be related to a separate endurance factor that is not associated with sensory intensity (Cleeland et al. 1966; Timmermans and Sternbach 1974; Wolff 1971). Although these simple methods may be useful in specific situations, at best they only assess the extremes of the perceptual pain range.

Assessing the Range Between Pain Threshold and Tolerance: Suprathreshold Scaling

Rather than determine pain sensitivity at specific endpoints, suprathreshold scaling methods evaluate the range of pain intensity from low to those deemed tolerable by the subject. Since tolerance is not a measure of pain sensitivity, these scaling methods provide critical information about the pain processing system that are more relevant to the magnitudes of pain experienced in clinical conditions. In addition, these measures are not always associated; in the example of hyperaphia and in other less extreme conditions, the function may rotate, resulting in higher pain thresholds but increased suprathreshold sensitivity. In these cases, the pain threshold is not only irrelevant but actually misleading.

Although there are many types of direct scaling methods, the most common method is to select a set of 5–7 stimulus intensities that (1) cover the range from above pain threshold to that tolerated by the subject and (2) are spaced close enough to be confused with one another. This latter property is important to avoid artifactual reliability (e.g., "that is the third stimulus from the bottom that I call a 5"). Once the stimulus set is chosen, a random sequence is determined to deliver each stimulus at least 2 or 3×; thus, 7 stimuli delivered 3× each would result in 21 trials. After each stimulus, the subject uses a response scale to rate some aspect, usually sensory intensity, of the evoked pain sensation. There is a large and sometimes lively literature about choice of response scale and the appropriate analysis, addressed only briefly below.

The random sequence of stimulus presentation theoretically controls for psychophysical biases. This feature is not usually included in necessarily fast clinical methods because of the enormity of modalities and areas that have to be tested in a short time and because many clinical abnormalities are distinct and adequately captured by these simple methods. Recent empirical evidence supports the theoretical advantage of the random methods. ► Fibromyalgia is diagnosed in part by applying 4 kg of manual pressure to 18 defined tender points and recording a report of pain from at least 11 of these sites. The method can be objectified somewhat by using one of several types of pressure algometer to deliver a pressure stimulus that increases at a constant rate, and the subject indicates the point at which the evoked sensation becomes painful (termed the pain pressure threshold or PPT). Petzke et al. (2003), in a population with both healthy controls and with fibromyalgia patients with a normal spread of pressure sensitivities, observed that the tender point count, and to a lesser extent the PPT, was significantly associated with psychological distress while a random method was not. Giesecke et al. (2003) replicated this study with a cohort of fibromyalgia patients. Although the use of simple clinical measures may be warranted and sufficient in many cases, in this case one of the diagnostic criteria is probably an unknown mix of somatic sensitivity and psychological distress.

The scaling method described above was used extensively before the age of computers and has been adapted to computer administration. However, computers allow for new, interactive methods that take advantage of the ability to adjust the levels of stimulation based on the subject's responses (Duncan et al. 1992; Gracely et al. 1988). These "second-generation" methods are able to overcome a number of problems with conventional direct scaling. For example, in many cases and especially when comparing patient groups to controls, the maximum level of stimulation must be determined for each individual with direct scaling, while newer methods determine this automatically. Also, a successful analgesic intervention may be obvious with direct scaling because all of the evoked sensations seem less intense. An interactive method can increase the stimulus intensities to evoke the same responses (and presumably similar sensation levels) as those produced before the analgesic intervention. Interactive methods express sensitivity in terms of stimulus intensity, providing a common measure across time and subjects.

There are a number of other advanced scaling techniques that have been applied to pain measurement. Stimulus integration methods require subjects to rate some combination of two stimuli or two stimulus characteristics, which are varied independently. Both a nonparametric method termed conjoint measurement and a parametric method termed functional measurement have been applied to pain assessment. In a typical application, subjects are presented with two stimuli and asked to rate the average pain of the two. The two stimuli can be from the same pain modality (Jones 1980), different modalities, a pain modality and an aversive nonpainful modality (Algom et al. 1986), or even a pain modality and words symbolizing pain (Gracely and Wolskee 1983; Heft and Parker 1984). The analysis treats scaling as an experiment that either indicates that subjects can perceive, combine, and reliably rate the concept being measured or indicates a failure at one or more of these three stages. This promising method has only been used in a small number of studies.

Issues Concerning the Choice of Response

As mentioned above, pain scaling can use a number of response measures such as verbal (none, mild, moderate, intense) or numerical (0–10) categories (Brennum et al. 1992; Max and Laska 1991). Probably the most widely used is the visual analog scale (VAS), commonly displayed as a 10 cm horizontal line labeled at the ends with extreme descriptors such as "no pain" and "pain as bad as it could be." Subjects indicate the magnitude of pain by marking the line. This method has been shown to be sensitive (Price 1988)

and is useful in multicenter international trials, avoiding the problems of translating and validating verbal scale in multiple languages. On the other hand, verbal category scales may increase face validity by anchoring judgments to specific levels of pain. The author favors category scales in which the position of descriptors is determined in previous experiments, and different scales display words of a particular pain dimension, such as mild, moderate, and intense for pain intensity and unpleasant, distressing, or very distressing for pain unpleasantness (Gracely et al. 1979). There is some evidence that the use of language may help to facilitate discrimination of different pain dimensions, although it is also likely that sufficient training will result in adequate discrimination with any scale. In addition, the issues about the type of response are dependent on other facets of the experimental situation, such as the goal of the measurement. If the goal is to provide baseline measurement and compare the responses between different groups of subjects, methods with increased face validity, possibly by the use of multiple verbal categories, may be advantageous. If the goal is to address the efficacy of a fast-acting analgesic, sensitivity to change in pain is more important than categorizing initial pain.

If subjects make repeated responses to a number of stimuli, all of the responses and methods above are influenced by biases, such as the desire to spread responses along the entire response continuum. This bias can be controlled by the use of responses with an unlimited range (number use, duration of a button press) or by presenting quantified verbal descriptors in a random sequence, forcing choices based on semantic content rather than the position in a list. A few studies have addressed these biases with scales such as the VAS (Fernandez et al. 1991), but generally these effects have not been investigated in any detail. In many cases, they would tend to diminish analgesic effects, making the many such demonstrations of analgesia more convincing.

The Use of Experimental Pain Scaling in the Assessment of Clinical Pain

In addition to the goals of determining differences in pain sensitivities between groups and assessing the action of interventions, the stimulus-response function can be used to characterize the function of the pain sensory system under normal conditions, under experimental conditions such as slow temporal summation, and under pathophysiological conditions such as nerve injury or inflammation. Indeed, the growth of clinical studies employing quantitative sensory testing (QST) is an excellent example of the successful merging of experimental procedures and clinical evaluation. An additional application is the interesting concept that responses to an experimental pain stimulus can aid the often problematic process of measuring the subjective experience of clinical pain.

One approach to assisting clinical pain measurement is the use of triangular validation in which the vertices represent a scaling method, clinical pain experience, and the experience of experimentally evoked pain. The sides of the triangle represent the three possible comparisons of two vertices. One side is using a scale to clinical pain. Another is using the same scale to establish the stimulus-response function described above. The third is a variant of magnitude matching (Feine et al. 1991) in which subjects perform the novel task of comparing their clinical pain experience directly with an experimental pain experience. In theory, the combination of any two legs predicts the third leg; for example, the combination of stimulus-response functions and clinical-experimental pain matching provides a second measure of clinical pain magnitude that can be compared to the direct scaling of clinical pain experience. Experimental studies using acute dental pain, chronic orofacial pain, and chronic low back pain confirm the theoretical agreement (Gracely 1979; Heft et al. 1980; Price et al. 1984).

Finally, while not using painful stimulation, another approach applies the principles of scaling to clinical pain assessment. Scaling methods collect multiple stimulus-evoked responses in relation to a single subjective standard. A clinical application of scaling turns this process around by using quantified categories to compare a single "clinical stimulus" to multiple subjective standards. The descriptor differential scale (DDS)

uses this approach by presenting 12 quantified descriptors and instructing subjects to rate whether their clinical pain is equal to that implied by the descriptors or how much less or greater on a +10 to −10 scale. The DDS has demonstrated adequate reliability, internal consistency, validity, and sensitivity (Doctor et al. 1995; Gracely and Kwilosz 1988). The use of multiple measures also provides a measure of scaling consistency. Preliminary studies have shown generally high consistency that improves from the first to the second administration (Gracely and Kwilosz 1988). These consistency measures can be used to identify subjects who are obviously not attending to the scaling task. The use of multiple items also allows the construction of alternative forms, each with different descriptor items. Validated alternative forms of the DDS demonstrated accurate memory of acute postsurgical pain over a 1 week period (Kwilosz et al. 1984). The use of alternative forms in such studies increases confidence that the demonstrated recall is of previous pain and not of a previous response. An additional advantage of scaling methods in clinical pain assessment is that the use of multiple items decreases the influence of random scaling errors. The reliability and homogeneity of the overall score of the DDS is improved in comparison to analyses of individual items (Gracely and Kwilosz 1988); an effect consistent with the concept that averages of multiple responses to a single-item pain diary are superior to single responses or to the mean of only a few responses (Jensen and McFarland 1993).

Critics point out that pain evoked by experimental methods cannot duplicate the sensory, emotional, and cognitive dimensions of clinical pain. Perfect duplication of all these features is not a requirement for utility, and scaling methods have provided and continue to provide a wealth of information about the mechanisms of pain and pain control, about normal and abnormal pain processing, about the influence of clinical pain on pain perception, and about clinical pain magnitude. Future studies of pain genetics will be as good as the ability to phenotype individuals accurately and scaling methods that have been shown to be purer measures of factors such as tenderness will probably contribute to such investigations. Despite advances in neuroimaging and human neurophysiology, pain is still defined only by human subjective experience. Clearly, methods that systematically assess this experience are critical for studies that will ultimately lead to improved diagnosis and treatment.

These experiments will probably provide important parts of the puzzle of the multiple mechanisms of pain perception. They will also probably continue to approach one of the most elusive goals: a physiological signature associated with what is otherwise an unobservable and private event.

References

Algom, D., Raphaeli, N., & Cohen-Raz, L. (1986). Integration of noxious stimulation across separate somatosensory communications systems: A functional theory of pain. *Journal of Experimental Psychology: Human Perception and Performance, 12*, 92–102.

Brennum, J., Arendt-Nielsen, L., Secher, N. H., et al. (1992). Quantitative sensory examination during epidural anesthesia in man: Effects of lidocaine. *Pain, 51*, 27–34.

Cleeland, C. S., Nakamura, Y., Howland, E. W., et al. (1966). Effects of oral morphine on cold pressor tolerance time and neuropsychological performance. *Neuropsychopharmacology, 15*, 252–262.

Doctor, J. N., Slater, M. A., & Atkinson, J. H. (1995). The descriptor differential scale of pain intensity, an evaluation of item and scale properties. *Pain, 61*, 251–260.

Duncan, G. H., Miron, D., & Parker, S. R. (1992). Yet another adaptive scheme for tracking threshold. Meeting of the International Society for Psychophysics, Stockholm

Feine, J. S., Bushnell, M. C., Miron, D., et al. (1991). Sex differences in the perception of noxious heat stimuli. *Pain, 44*, 255–262.

Fernandez, E., Nygren, T. E., & Thorn, B. E. (1991). An "open-transformed scale" for correcting ceiling effects and enhancing retest reliability: The example of pain. *Perception & Psychophysics, 49*, 572–578.

Giesecke, T., Gracely, R. H., Harris, R. E., et al. (2003). Factor analysis of various measures of tenderness, clinical pain, and psychological distress, in patients with fibromyalgia. *Arthritis and Rheumatism, 48*, 86 (abstract supplement).

Gracely, R. H. (1979). Psychophysical assessment of human pain. In J. J. Bonica, J. C. Liebeskind, & D. Albe-Fessard (Eds.), *Advances in pain research and therapy* (Vol. 3, pp. 805–824). New York: Raven.

Gracely, R. H., & Kwilosz, D. M. (1988). The descriptor differential scale: Applying psychophysical principles to clinical pain assessment. *Pain, 35*, 279–288.

Gracely, R. H., & Wolskee, P. J. (1983). Semantic functional measurement of pain: Integrating perception and language. *Pain, 15*, 389–398.

Gracely, R. H., Dubner, R., & McGrath, P. A. (1979). Narcotic analgesia: Fentanyl reduces the intensity but not the unpleasantness of painful tooth pulp stimulation. *Science, 203*, 1261–1263.

Gracely, R. H., Lota, L., Walther, D. J., et al. (1988). A multiple random staircase method of psychophysical pain assessment. *Pain, 32*, 55–63.

Heft, M. W., & Parker, S. R. (1984). An experimental basis for revising the graphic rating scale. *Pain, 19*, 153–161.

Heft, M. W., Gracely, R. H., Dubner, R., et al. (1980). A validation model for verbal descriptor scaling of human clinical pain. *Pain, 9*, 363–373.

Jensen, M. P., & McFarland, C. A. (1993). Increasing the reliability and validity of pain intensity measurement in chronic pain patients. *Pain, 55*, 195–203.

Jones, B. (1980). Algebraic models for integration of painful and nonpainful electric shocks. *Perception & Psychophysics, 28*, 572–576.

Kwilosz, D. M., Gracely, R. H., & Torgerson, W. S. (1984). Memory for post-surgical dental pain. *Pain, 2*, 426.

Max, M. B., & Laska, E. M. (1991). Single-dose analgesic comparisons. In M. B. Max, R. K. Portenoy, & E. M. Laska (Eds.), *Advances in pain research and therapy* (The design of analgesic clinical trials, Vol. 18, pp. 55–95). New York: Raven.

Petzke, F., Gracely, R. H., Park, K. M., et al. (2003). What do tender points measure? Influence of distress on 4 measures of tenderness. *Journal of Rheumatology, 30*, 567–574.

Price, D. D. (1988). *Psychological and neural mechanisms of pain*. New York: Raven.

Price, D. D., Rafii, A., Watkins, L. R., et al. (1984). A psychophysical analysis of acupuncture analgesia. *Pain, 19*, 27–42.

Timmermans, G., & Sternbach, R. A. (1974). Factors of human chronic pain: An analysis of personality and pain reaction variables. *Science, 184*, 806–808.

Wolff, B. B. (1971). Factor analysis of human pain responses: Pain endurance as a specific pain factor. *Journal of Abnormal Psychology, 78*, 292–298.

Stomatodynia

▶ Atypical Facial Pain: Etiology, Pathogenesis, and Management

▶ Burning Mouth Syndrome

Stomatopyrosis

▶ Burning Mouth Syndrome

STP Syndromes

▶ Soft Tissue Pain Syndromes

Straight-Leg-Raising Sign

▶ Lower Back Pain, Physical Examination

Strain

Definition

Traumatic injury to the muscle.

Cross-References

▶ Opioids and Muscle Pain

Stratum Corneum

Definition

The stratum corneum is the uppermost layer of the skin and consists of cells containing keratin. Old and dead cells on the surface are continuously shed off and are replaced by new cells formed at the stratum germinativum.

Cross-References

▶ Opioid Therapy in Cancer Pain Management, Route of Administration

Strengthening Exercise

▶ Exercise

Stress

Definition

Any environmental or life event perceived by the individual as threatening to his or her physical or psychological well-being and exceeding his or her capacities to cope. It causes bodily or mental tension and may be a factor in disease causation.

Cross-References

▶ Descending Modulation of Visceral Pain
▶ Pain as a Cause of Psychiatric Illness
▶ Pain in the Workplace, Risk Factors for Chronicity, and Psychosocial Factors
▶ Stress and Pain

Stress and Pain

Jamie L. Rhudy, Emily J. Bartley and
Shreela Palit
Department of Psychology, The University of
Tulsa, Tulsa, OK, USA

Synonyms

Appraisal; Biobehavioral factors; Coping; Physiological pathways; Psychosocial factors

Definition

Homeostasis – The process of maintaining the body's physiology within an optimal range (e.g., temperature, pH, oxygen level), despite constant demands that attempt to alter physiological levels.

Allostasis – A term that has replaced the idea of homeostasis, in that it recognizes that the body's optimal range can be achieved by a number of different strategies and that the optimal range may vary depending on the situation.

Stress – A term that refers to a disturbance of the body's homeostasis. In colloquial terms, it is often used to refer to the negative emotional experiences (e.g., anxiety, fear) that co-occur with disrupted homeostasis.

Stress response – The body's compensatory reaction to a homeostatic challenge, which attempts to restore equilibrium.

Pain – An unpleasant sensory and emotional experience associated with tissue damage or described in terms of such damage, which serves as an alarm system for internal and external sources of threat.

Nociception – The process by which neural signals associated with noxious stimulation are transmitted throughout the nervous system.

Introduction

A Brief Overview of Stress

For a thorough overview of stress, see Sapolsky (2004) or Lovallo (2005). The concept of physiological stress was first introduced by William Cannon in the 1930s. His work elucidated the role of the sympathetic nervous system and the endocrine system in reaction to threat, and demonstrated that when faced with a threat, resources are mobilized (e.g., increased heart rate, respiration, blood flow to large muscles) to help an organism combat or flee the source of the stress (i.e., fight or flight response). However, the notion of stress was later popularized in the 1950s by Hans Seyle who noted that repeated or prolonged physical stressors (e.g., cold, sleep deprivation, infection) caused an organism to move through three stages, that is, alarm, resistance, and exhaustion, a pattern he called the general adaptation syndrome. He found that, regardless of the type of stressor, the physiological consequences of exhaustion were similar and included enlarged adrenal cortex, decreased thymus and lymph glands, and ulceration of the

stomach and duodenum. This illustrated that intense, prolonged, or repeated stressors can have detrimental physiological effects. More recently, the term "allostatic load" has been used by McEwen and others to refer to the physiological costs of chronic exposure to stress and the cumulative wear and tear on the multiple systems that regulate allostasis.

Lazarus and Folkman proposed that the stress response is influenced by the cognitive appraisal of the stressor. If a person believes that he/she has the resources available to handle the threat, then the threat will elicit very little stress. However, if the person believes their resources are not adequate to handle the threat, then the threat will elicit considerable stress. It is now known that a number of different variables can impact stress appraisals and the stress response, such as the predictability or controllability of the stressor (Lovallo 2005). Importantly, when coping resources are inadequate, negative emotions are elicited (e.g., distress, anxiety, fear) that also contribute to the stress response.

The two primary "arms" of the stress response are the hypothalamic-pituitary-adrenocortical (HPA) axis and the hypothalamic-sympathetic-adrenomedullary (HSAM) system. When a stressor activates the HPA axis, this causes the hypothalamus to release corticotropin-releasing hormone (CRH), thus stimulating pituitary to release ACTH, that in turn elicits the release of glucocorticoids from the adrenal cortex. For humans, cortisol is the most important stress-related glucocorticoid. When stress activates the HSAM, this causes the hypothalamus to communicate with the brain stem nucleus of the solitary tract (NST). The NST promotes sympathetic outflow that in turn stimulates the adrenal medulla to release epinephrine (adrenaline) and norepinephrine (noradrenaline). Epinephrine is the more important of the two in the stress response. It is the output of the HSAM that produces responses most associated with fight or flight. The HPA and the HSAM can be activated by sensory channels involved with the detection of external and internal threats (e.g., predator, pain); however, these stress systems are also modulated in a top-down manner from higher brain centers (e.g., prefrontal cortex, limbic system) involved with emotion, evaluation, and planning (Lovallo 2005). Thus, it is through these top-down connections that emotion, appraisals, and coping efforts can moderate the stress response, but also how mental activities, such as rumination, worry, and anticipation, can elicit a stress response in the absence of an immediate threat. While HPA and HSAM represent the two major arms of the stress response, the consequences of stress are wide-ranging, affecting all body systems (e.g., cardiovascular, immune, gastrointestinal, neurological, endocrine, reproductive).

Pain Systems and Connections to the Stress Systems

For a thorough overview of the ascending pain pathway, see other sections in the *Encyclopedia of Pain* (e.g., ▶ Ascending Nociceptive Pathways). Briefly, primary afferent (nociceptors) relay the pain signal to the dorsal horn of the spinal cord where the message is then relayed to supraspinal centers by several ascending tracts, including the spinothalamic, spinohypothalamic, spinoreticular, spinoparabrachial, spinocervicothalamic, postsynaptic dorsal column projection, and spinosolitary. Once the signal reaches the brain it undergoes parallel processing in a number of brain regions, including the thalamus, brain stem (including the nucleus of the solitary tract), anterior cingulate cortex, insula, amygdala, hypothalamus, and somatosensory cortices. It is this supraspinal processing within this circuitry referred to as the pain ▶ neuromatrix that eventually leads to the subjective experience of pain (Melzack 1999). Because of these ascending connections to regions involved with the arms of the stress response (i.e., hypothalamus, nucleus of the solitary tract), pain signaling can directly activate a stress response.

In addition to the ascending pain pathway, there is a descending pain pathway that allows for the inhibition and facilitation of pain signaling (for a thorough overview, see other sections in the *Encyclopedia of Pain*, i.e., ▶ Descending Modulation of Nociceptive Processing, ▶ Descending Circuits in the Forebrain, Imaging,

▶ Descending Facilitatory Systems, ▶ Forebrain Modulation of the Periaqueductal Gray and Its Role in Pain). Some of the supraspinal regions involved in descending modulation include the hypothalamus, amygdala, insula, anterior cingulate cortex, and orbitofrontal cortex. These brain areas have direct or indirect connections with the periaqueductal gray and rostral ventromedial medulla, brain stem regions that play a critical role in brain-to-spinal cord modulation of afferent nociception within the spinal cord dorsal horn. However, the nociceptive signal and pain perception can undergo modulation at the supraspinal level also via supraspinal circuits. Therefore, these supraspinal regions and their connections to the spinal cord provide a means by which stress can modulate nociception and pain experience.

Characteristics

The relationship between stress and pain is bidirectional; therefore, the next sections will cover each direction of influence.

The Influence of Stress on Pain
Stress-Induced Modulation of Pain
The effect of stress on pain is complex because in some circumstances stress can inhibit pain (i.e., hypoalgesia), whereas in other circumstances it can enhance pain (i.e., ▶ hyperalgesia). Based on a review of the literature, Rhudy and Meagher (2001) proposed that the effect of stress on pain may be determined by the type of stressor. Specifically, when a stressor is severe and threat is imminent, an intense emotional state akin to fear is produced which leads to hypoalgesia. A classic example of stress-induced hypoalgesia is the soldier in the midst of battle who does not realize they have suffered a serious injury. Most of the research on stress-induced hypoalgesia comes from animal studies which find that direct exposure to a severe stressor (e.g., electric shock, forced swim in cold water, food restriction, restraint, social isolation/conflict, exposure to a predator) or exposure to stimuli that had been previously paired with a severe stressor (e.g., an

environment in which shock was presented) leads to a reduction in pain-related behavior in the animal (Butler and Finn 2009). However, exposure to severe stressors (e.g., shock, a cue paired with shock, mental arithmetic combined with noise, spider exposure in spider phobics, combat video exposure in persons with posttraumatic stress disorder) also produces hypoalgesia in humans (Butler and Finn 2009). From an evolutionary perspective, stress-induced hypoalgesia is adaptive because active fight or flight is necessary when there is an imminent threat, and tending to a painful injury could interfere with the success of fight or flight behaviors (Bolles and Fanselow 1980).

A number of physiological mechanisms have been implicated in stress-induced hypoalgesia (Butler and Finn 2009). The brain regions of particular importance are the hypothalamus, amygdala, periaqueductal gray, and brain stem (i.e., rostroventral medulla, parabrachial nucleus, nucleus of the solitary tract). As noted in the earlier section on pain modulation, higher forebrain regions have connections with the brain stem, which in turn have projections to the spinal cord that can inhibit afferent nociceptive input at spinal levels. Therefore, these circuits are involved in stress-induced hypoalgesia. Some stress-induced hypoalgesia is mediated by endogenous opioid peptides (Watkins and Mayer 1986). For example, beta-endorphin is released by the pituitary during stress-evoked activation of the HPA axis. However, other forms of stress-induced hypoalgesia are mediated by non-opioid compounds (e.g., endogenous cannabinoids, GABA, oxytocin, adenosine, glycine, vasopressin) and some are dependent on stress hormones (e.g., ACTH) and activation of the HPA axis (Butler and Finn 2009). Monoamines, such as dopamine, serotonin, and norepinephrine, are also likely to play a role in stress-induced hypoalgesia, but may also contribute to some forms of stress-induced hyperalgesia (Butler and Finn 2009).

In contrast to stress-induced hypoalgesia, Rhudy and Meagher (2001) propose that when a stressor is mild or is likely to occur sometime in the future, then a less intense emotional state akin to anxiety is produced, which leads to hyperalgesia. Stress-induced hyperalgesia is also evolutionarily adaptive because in the absence of an imminent threat, hyperalgesia would promote hypervigilance to somatic threats and such hypervigilance may help prevent injuries. Furthermore, if an injury had already occurred, stress-induced hyperalgesia would promote wound tending and other recuperative behaviors (Walters 1994). Stress-induced hyperalgesia has been studied using a variety of mild-to-moderate stressors in animals (e.g., mild electric shocks, changes in environmental temperature, holding by nape of neck, novel environments, vibration) (Imbe et al. 2006), as well as humans (e.g., pictures of threatening scenes, threatening mental imagery, stressful interview, light paired with mild shock, threat of shock, brief noise bursts, unpleasant odors) (Naliboff and Rhudy 2009). The neurocircuitry involved with stress-induced hyperalgesia is similar to that of stress-induced hypoalgesia, with the difference being that different receptor subtypes are involved (Butler and Finn 2009; Imbe et al. 2006).

Given that hyperalgesia and hypoalgesia are subserved by similar neurocircuits, it is not surprising that stress can activate inhibitory and facilitatory mechanisms simultaneously (Crown et al. 2004). This means that the net effect on pain is determined by which process predominates (inhibitory vs. facilitatory). So, if inhibitory mechanisms get depleted or become habituated in some way, then stress-induced facilitatory mechanisms would dominate, thus causing hyperalgesia (Bruehl et al. 2009). Interestingly, evidence suggests that many chronic pain states are associated with depleted inhibitory mechanisms, so there is very little evidence of stress-induced hypoalgesia in chronic pain (Naliboff and Rhudy 2009). As a result, stress-induced hyperalgesia may be one factor that contributes to the initiation and/or maintenance of chronic pain.

Stress-Related Disorders and Pain

Various psychological conditions co-occur with pain disorders and the relationship may be maintained by stress and concomitant emotional factors (Klossika et al. 2006). For example, chronic pain is often comorbid with depression, and depressed patients prone to high anxiety report greater pain, perhaps due to the fact that increased anxiety exacerbates pain experience (Romano and Turner 1985). Zautra et al. (2007) found that rheumatoid arthritis (RA) patients who had experienced multiple depressive episodes had greater pain and exhibited greater stress-induced hyperalgesia. Posttraumatic stress disorder (PTSD) and chronic pain also co-occur and the two may be maintained by the stress caused by both pain and trauma, which in turn leads to attentional biases, anxiety sensitivity, avoidance, anxiety, and cognitive demand (Sharp and Harvey 2001). Comorbidity between borderline personality disorder (BPD) and chronic pain has also been identified, and the link between the two may be due to problems with emotion/stress regulation mechanisms (Sansone et al. 2001).

Early Life Stressors and Pain

Some evidence suggests that early life stress can influence pain processing. For example, animals exposed to prenatal stress or neonate animals exposed to maternal separation are less capable of engaging pain inhibitory mechanisms later in life (Butler and Finn 2009). Further, early life stress is associated with long-term changes in the pain system, which in turn is associated with increased prevalence of human chronic pain such as irritable bowel syndrome and fibromyalgia (Al-Chaer and Weaver 2009). In a prospective longitudinal study, McBeth and colleagues (2007) found that HPA axis dysfunction, specifically high levels of cortisol post-dexamethasone, low levels of morning cortisol, and high levels of evening cortisol, was associated with an increased risk of developing chronic widespread pain. Furthermore, in a large sample of men and women, Anda et al. (2010) found that frequent headaches in adulthood were associated with adverse childhood experiences (e.g., emotional or physical abuse) in a dose-response manner. Therefore, early life stress may have long-lasting effects that promote chronic pain in adulthood.

Stress, Immune Function, and Pain

The immune system plays an important role in protecting the body from invading bacteria, viruses, and foreign substances. Exposure to stressors influences immune function, but the direction of influence depends on timing and the type of stressor. Immediately following an acute stressor, immune activity is generally enhanced. This is adaptive as it allows the immune system to prepare itself for fighting off infection in areas that require it most. However, if the stressor is prolonged or repetitive, the immune system is unable to return to baseline and instead lapses into immunosuppression (Sapolsky 2004). Ultimately, prolonged suppression of the immune system can lead to the development of immunological disease and alterations in pain modulation. In fact, many autoimmune diseases (e.g., lupus erythematosis, rheumatoid arthritis, inflammatory bowel disease) are associated with persistent pain. Some of these effects are due to the release of chemical mediators such as cytokines, substance P, prostaglandins, bradykinin, chemokines, and nerve growth factor. For example, studies have shown that stress increases the production of cytokines (Moons et al. 2010) and that cytokines can increase pain and sickness behaviors by acting at spinal and supraspinal sites (including the hypothalamus) (Watkins et al. 2007). Cortisol release can act as an immunosuppressant by decreasing circulating levels of leukocytes, natural killer cell activity, and cytokines (Zeller et al. 1996). These chemicals are released after the onset of tissue destruction and inflammation and can impact the nervous system both peripherally and centrally. Although they are responsible for preventing and eliminating infection, these chemical mediators can also directly elicit pain through the activation of nociceptors (Rittner et al. 2009) (Fig. 1).

The Influence of Pain on Stress

Pain and Stress

As noted in earlier sections, incoming pain signals can directly activate the HPA and HSAM to evoke stress responses. Thus, pain can directly evoke stress. In chronic pain patients, not only can a lack of proper outlets for coping with

Stress and Pain,
Fig. 1 Diagram of the
routes of communication
between the brain and
immune system, including
the HPA axis, sympathetic
nervous system, and
cytokine feedback to the
brain (from Webster et al.
2002)

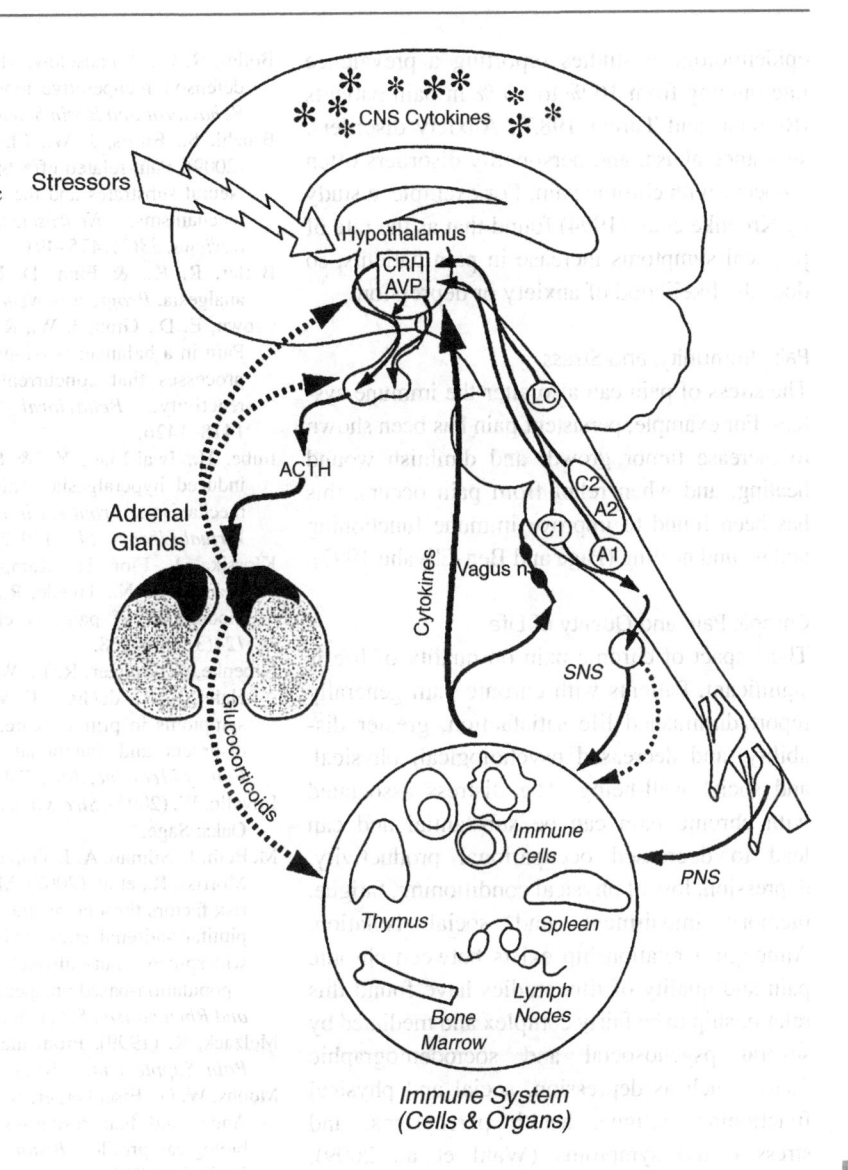

stressful situations lead to the exacerbation of pain, but the consequences of chronic pain (e.g., feelings of threat/loss of social relations, depression, interpersonal conflict, disability, isolation) can amplify stress responses (Zautra et al. 1999). Therefore, pain can affect stress via multiple pathways.

Chronic Pain and Stress-Related Disorders

Chronic pain is associated with increased risk for stress-related disorders. This is not surprising as (1) both nociceptive and affective pathways overlap and (2) neurotransmitters such as norepinephrine, serotonin, and dopamine are associated with pain and emotion. Severity plays an important role in development of stress-related disorders, and as pain transitions from acute to chronic, stress/emotional factors play a more dominant role in its maintenance. For example, Sternbach et al. (1973) examined the psychological profiles of patients with low back pain and found that those with chronic pain demonstrated higher rates of psychopathology when compared to patients with acute pain. As noted earlier, depression is also commonly associated with chronic pain, with epidemiological studies reporting a prevalence rate ranging from 16 % to 50 % in pain patients (Romano and Turner 1985). Anxiety disorders, substance abuse, and personality disorders often co-occur with chronic pain. For example, a study by Kroenke et al. (1994) found that as the rate of physical symptoms increase in pain patients, so does the likelihood of anxiety or depression.

Pain, Immunity, and Stress

The stress of pain can also alter the immune system. For example, persistent pain has been shown to increase tumor growth and diminish wound healing, and when relief from pain occurs, this has been found to improve immune functioning and wound healing (Page and Ben-Eliyahu 1997).

Chronic Pain and Quality of Life

The impact of chronic pain on quality of life is significant. Patients with chronic pain generally report diminished life satisfaction, greater disability, and decreased psychological, physical, and social well-being. The distress associated with chronic pain can be substantial and can lead to decreased occupational productivity, depression, loss of physical conditioning, fatigue, memory impairment, and social isolation. Although a relationship exists between chronic pain and quality of life, studies have found this relationship to be fairly complex and mediated by several psychosocial and sociodemographic factors such as depression, social and physical functioning, fatigue, health perceptions, and stress-related symptoms (Wahl et al. 2009). Therefore, chronic pain may not be directly related to quality of life, but instead may be mediated by multiple factors including stress.

References

Al-Chaer, E. D., & Weaver, S. A. (2009). Early life trauma and chronic pain. In E. A. Mayer & M. C. Bushnell (Eds.), *Functional pain syndromes: Presentation and pathophysiology* (pp. 423–452). Seattle: IASP Press.

Anda, R., Tietjen, G., Schulman, E., Felitti, V., & Croft, J. (2010). Adverse childhood experiences and frequent headaches in adults. *Headache: The Journal of Head and Face Pain, 50*(9), 1473–1481.

Bolles, R. C., & Fanselow, M. S. (1980). A perceptual-defensive-recuperative model of fear and pain. *The Behavioral and Brain Sciences, 3*(2), 291–323.

Bruehl, S., Burns, J. W., Chung, O. Y., & Chont, M. (2009). Pain-related effects of trait anger expression: Neural substrates and the role of endogenous opioid mechanisms. *Neuroscience and Biobehavioral Reviews, 33*(3), 475–491.

Butler, R. K., & Finn, D. P. (2009). Stress-induced analgesia. *Progress in Neurobiology, 88*, 184–202.

Crown, E. D., Grau, J. W., & Meagher, M. W. (2004). Pain in a balance: Noxious events engage opposing processes that concurrently modulate nociceptive reactivity. *Behavioral Neuroscience, 118*(6), 1418–1426.

Imbe, H., Iwai-Liao, Y., & Senba, E. (2006). Stress-induced hyperalgesia: Animal models and putative mechanisms. *Frontiers in Bioscience: A Journal and Virtual Library, 11*, 2179–2192.

Klossika, I., Flor, H., Kamping, S., Bleinchardt, G., Trautmann, N., Treede, R., et al. (2006). Emotional modulation of pain: A clinical perspective. *Pain, 124*(3), 264–268.

Kroenke, K., Spitzer, R. L., Williams, J. B., Linzer, M., Hahn, S. R., deGruy, F. V., et al. (1994). Physical symptoms in primary care. Predictors of psychiatric disorders and functional impairment. *Archives of Family Medicine, 3*(9), 774–779.

Lovallo, W. (2005). *Stress and health* (2nd ed.). Thousand Oaks: Sage.

McBeth, J., Silman, A. J., Gupta, A., Chiu, Y. H., Ray, D., Morriss, R., et al. (2007). Moderation of psychosocial risk factors through dysfunction of the hypothalamic-pituitary-adrenal stress axis in the onset of chronic widespread musculoskeletal pain: Findings of a population-based prospective cohort study. *Arthritis and Rheumatism, 56*(1), 360–371.

Melzack, R. (1999). From the gate to the neuromatrix. *Pain. Supplement, 6*, S121–S126.

Moons, W. G., Eisenberger, N. I., & Taylor, S. E. (2010). Anger and fear responses to stress have different biological profiles. *Brain, Behavior, and Immunity, 24*(2), 215–219.

Naliboff, B., & Rhudy, J. L. (2009). Anxiety and functional pain disorders. In E. A. Mayer & M. C. Bushnell (Eds.), *Functional pain syndromes: Presentation and pathophysiology*. Seattle: IASP Press.

Page, G. G., & Ben-Eliyahu, S. (1997). The immune-suppressive nature of pain. *Seminars in Oncology Nursing, 13*(1), 10–15.

Rhudy, J. L., & Meagher, M. W. (2001). The role of emotion in pain modulation. *Current Opinion in Psychiatry, 14*(3), 241–245.

Rittner, H. L., Machelska, H., & Stein, C. (2009). Immune system, pain and analgesia. In A. I. Basbaum & M. C. Bushnell (Eds.), *Science of pain*. Oxford, UK: Elsevier.

Romano, J. M., & Turner, J. A. (1985). Chronic pain and depression: Does the evidence support a relationship? *Psychological Bulletin, 97*(1), 18–34.

Sansone, R. A., Whitecar, P., Meier, B. P., & Murry, A. (2001). The prevalence of borderline personality among primary care patients with chronic pain. *General Hospital Psychiatry, 23*(4), 193–197.

Sapolsky, R. M. (2004). *Why zebras don't get ulcers.* New York: Henry Holt.

Sharp, T. J., & Harvey, A. G. (2001). Chronic pain and posttraumatic stress disorder: Mutual maintenance? *Clinical Psychology Review, 21*(6), 857–877.

Sternbach, R. A., Wolf, S. R., Murphy, R. W., & Akeson, W. H. (1973). Traits of pain patients: The low-back 'loser'. *Psychosomatics, 14*(4), 226–229.

Wahl, A. K., Rustoen, T., Rokne, B., Lerdal, A., Knudsen, O., Miaskowski, C., et al. (2009). The complexity of the relationship between chronic pain and quality of life: A study of the general Norwegian population. *Quality of Life Research, 18*(8), 971–980.

Walters, E. T. (1994). Injury related behavior and neural plasticity: An evolutionary perspective on sensitization, hyperalgesia, and analgesia. *International Review of Neurobiology, 36,* 325–426.

Watkins, L. R., & Mayer, D. J. (1986). Multiple endogenous opiate and non-opiate analgesia systems: Evidence of their existence and clinical implications. *Annals of the New York Academy of Sciences, 467,* 273–299.

Watkins, L. R., Hutchinson, M. R., Milligan, E. D., & Maier, S. F. (2007). "Listening" and "talking" to neurons: Implications of immune activation for pain control and increasing the efficacy of opioids. *Brain Research Reviews, 56*(1), 148–169.

Webster, J. I., Tonelli, L., & Sternberg, E. M. (2002). Neuroendocrine regulation of immunity. *Annual Review of Immunology, 20,* 125–163.

Zautra, A. J., Hamilton, N. A., & Burke, H. M. (1999). Comparison of stress responses in women with two types of chronic pain: Fibromyalgia and osteoarthritis. *Cognitive Therapy and Research, 23*(2), 209–230.

Zautra, A. J., Parrish, B. P., Van Puymbroeck, C. M., Tennen, H., Davis, M. C., Reich, J. W., et al. (2007). Depression history, stress, and pain in rheumatoid arthritis patients. *Journal of Behavioral Medicine, 30*(3), 187–197.

Zeller, J. M., McCain, N. L., & Swanson, B. (1996). Psychoneuroimmunology: An emerging framework for nursing research. *Journal of Advanced Nursing, 23*(4), 657–664.

Stress Management

▶ Psychological Treatment of Headache
▶ Relaxation Training

Stress Metabolism

▶ Postoperative Pain, Pathophysiological Changes in Metabolism in Response to Acute Pain

Stress Profile

▶ Psychophysiological Assessment of Pain

Stress Response

Definition

The cognitive, behavioral, and/or physiological changes that occur in response to a stressor as the body attempts to maintain allostasis.

Cross-References

▶ Postoperative Pain, Pathophysiological Changes in Neuroendocrine Function in Response to Acute Pain
▶ Stress and Pain

Stress-Induced Analgesia

Definition

Inhibition of pain by environmental stressors.

Cross-References

▶ Pain-Modulatory Systems, History of Discovery

Stressor

Definition

A stimulus of a physical (i.e., frequent lifting) or psychological (i.e., time demands at work) nature that disrupts the body's allostatic balance.

Cross-References

▶ Stress and Pain

Stretching

Robert D. Herbert
Neuroscience Research Australia, Sydney,
Australia

Definition

Stretching involves positioning a joint or joints so that muscles or other structures are placed under passive tension. Stretches may be brief and repetitive (dynamic stretching) or sustained (static stretching). The muscles may remain relaxed as they are stretched (passive stretching) or agonist or antagonist muscles may be made to actively contract (active stretching).

Characteristics

Physiotherapists, chiropractors, and osteopaths (collectively referred to here as manual therapists) often stretch soft tissues. The purpose may be to prevent or treat stiff joints (adaptive shortening of soft tissues that prevents normal joint motion) or to prevent or treat pain. This entry is concerned with the use of soft tissue stretching to prevent or treat pain of presumed musculoskeletal origin.

A cursory inspection of the literature reveals a remarkable diversity in proposed indications for therapeutic stretching of muscles. Equally remarkable is the diversity of theories about mechanisms by which stretching is thought to produce its therapeutic effects. Establishing the effect of muscle stretching is difficult because most research has investigated stretching as part of a combination of therapies, an approach that reflects contemporary clinical practice.

This entry will consider three putative indications for stretching muscles: treatment of musculoskeletal pain, prevention and treatment of malalignment syndromes, and prevention of delayed-onset muscle soreness. These indications were chosen because they provide scenarios where stretching is used as the main or sole component of therapy.

Treatment of Musculoskeletal Pain

A range of conditions is associated with pain of presumed musculoskeletal origin. One school of thought holds that a prevalent cause of pain is the ▶ myofascial pain syndrome, characterized by the presence of trigger points, focal muscle lesions tender to palpation (Travell and Simons 1983). Trigger points are thought to be regions of muscle containing sensitized nerve fibers (Hong and Simons 1998; Travell and Simons 1983).

Trigger point therapists stretch muscles because they believe that stretching "inactivates" trigger points and relieves pain (Travell and Simons 1983). A popular technique (▶ Spray and Stretch) involves first spraying the skin overlying the muscles with a vapocoolant spray to produce a superficial local anesthesia. But soft tissue stretching is not solely the domain of trigger point therapy; manual therapists who do not subscribe to the trigger point theory may also stretch soft tissues to relieve pain.

There have been relatively few randomized investigations of stretching for treatment of pain. Hou and colleagues (2002) randomly allocated 31 subjects with active trigger points in the upper trapezius muscle to receive either spray and stretch or a control condition. Pain levels were measured immediately after the stretch. The effect of spray and stretch procedures was to reduce pain by approximately one-third of initial levels. This study shows that spray and stretch produces an immediate reduction in pain, but it is not clear if the effects are sustained, or whether the effects are due to the spray or the stretch. Khalil et al. (1992) randomly allocated 28 subjects with low back pain to receive four sessions of muscle stretching over a 2-week period or a control condition. All subjects had been diagnosed with myofascial syndrome. After the last stretching session, subjects reported mean pain levels (in cm on a 10 cm VAS scale) of 5.3 (SD 2.0) in the control group and 1.6 (1.4) in the treated group, suggesting that stretching reduced pain by 3.7 cm on a 10 cm VAS scale. Again, it is not clear if this effect was sustained for a clinically useful duration. Hanten and colleagues (2000) randomized 40 subjects with trigger points in the neck or back to receive either self-administered pressure and stretch treatment (10 sessions over 5 days) or a control condition. They measured pain levels 2 days after the last treatment and found modest reductions in pain (VAS pain 1.3 cm (SD 1.6) compared to 2.5 cm (SD 2.1)). Together these studies provide some evidence that stretching can produce short-term reductions in pain of people with myofascial pain syndromes. This conclusion is supported by more recent, larger and well-designed randomised study (Bron et al. 2011)

Malalignment Syndromes

Mainstream medical texts often label conditions such as tendinoses, periostitis, stress fractures, compartment syndromes, impingement syndromes, and even osteoarthritis as diseases of "overuse." Historically there has been little attempt to distinguish why some people develop a particular condition, while others, even those who experience high degrees of use, do not.

A view held by many manual therapists is that the causes of some musculoskeletal pain conditions can be understood in biomechanical terms. According to this view, abnormalities of body alignment place excessive stresses on specific anatomical structures, causing pain. Thus, it is believed that particular pain conditions can be prevented or treated by altering body alignment. Therapists who hold to this view assess body alignment with an aim to identifying ▶ malalignment syndromes. Intervention often involves stretching muscles, provision of orthoses or splints, or specific exercises to correct the malalignment.

One example may suffice. Various lower limb conditions, such as plantar fasciitis, are thought to be due to tightness of the soleus muscle (Pfeffer et al. 1999). Thus, patients with plantar fasciitis may be advised to perform "heel cord stretches" (self-administered calf muscle stretches) or to wear night splints that stretch the ankle into dorsiflexion. The expectation is that stretching of calf muscles will cause the rear foot to be less supinated, causing less stress on painful structures and less pain. Similar theories are used to explain other sorts of foot pain, as well as back and neck pain, shoulder pain, and knee pain. Manual therapists may stretch muscles to treat any of these conditions. If interventions such as muscle stretching are to be generally effective in preventing and treating pain due to malalignment, then there must be a strong causal relationship between body alignment and the development of pain. To what extent is this condition satisfied?

It is difficult to establish if there is a causal relationship between body alignment and the development of pain. The only satisfactory approach involves prospective monitoring of large cohorts, using multivariate statistical techniques to control for potentially confounding factors. The largest studies investigating the relationship between body alignment and development of low back pain (Bigos et al. 1992; Diek et al. 1985) found no association, suggesting that body alignment does not have an important influence on risk of developing back pain. Some relatively small studies have investigated the relationship between lower limb alignment and risk of lower limb overuse injuries and these studies provide inconsistent evidence of a relationship (e.g., Cowan et al. 1996; Wen et al. 1998). It has not been established if the observed relationship between lower limb alignment and subsequent development of painful overuse injuries in the lower limb is causal.

Many studies have investigated the effects of stretching as part of a combination of therapies. For example, stretching has been incorporated into many exercise programs that have been shown to be effective for reducing back pain, neck pain, or pain due to hip or knee osteoarthritis. This makes sense because it is not usual clinical practice to apply stretch as a sole treatment for musculoskeletal pain. Nonetheless it is difficult with such studies to extract effects of stretching from other components of exercise programs.

Muscle Soreness

Unaccustomed exercise can produce muscle soreness that develops 1 or 2 days after the exercise is completed. This type of muscle soreness is called ► delayed-onset muscle soreness. Many people stretch before or after exercise in the belief that this will prevent or minimize delayed-onset muscle soreness. The origin of the belief that stretching of muscles can prevent muscle soreness may lie with physiological studies conducted in the 1960s. De Vries (1966) hypothesized that unaccustomed exercise induced muscle spasm. He believed that spasm impaired blood flow, causing ischemic pain and further spasm (the pain-spasm-pain cycle) and that stretching interrupted this cycle. De Vries' early model of the pain-spasm-pain cycle was probably a precursor of the contemporary theories about mechanisms of trigger points.

The spasm theory of muscle soreness is almost certainly wrong. Moderate isometric muscle contractions are required to prevent blood flow to muscles, but sore muscles do not exhibit this level of spasm and may not exhibit any increase in resting activity (Bobbert et al. 1986). Instead, it is likely that muscle soreness is caused by a series of events beginning with damage to the contractile elements in muscle fibers (Proske and Morgan 2001). This mechanism leaves little room for a role of stretching.

What evidence is there that stretching before or after unaccustomed exercise prevents or reduces delayed-onset muscle soreness? A systematic review of the relevant literature identified randomized studies of the effects of stretching on muscle soreness (Herbert et al. 2011). Eight studies evaluated stretching after exercising, and four evaluated stretching before exercising. As there was no evidence of heterogeneity in the outcomes of the studies, the findings of these studies were combined in a meta-analysis. The pooled estimate of the mean effect of stretching on muscle soreness 24 h after exercising was just 0.93 mm (95 % CI −64.0 mm to 4.2 mm) on a 100 mm scale, where negative values indicate a beneficial effect of stretching. These data suggest indicate that stretching before or after exercise does not produce worthwhile reductions in muscle soreness. A much larger and more recent trial (Jamtvedt et al. 2010; N) showed that stretching before and after exercising produced statistically significant but very small reductions in soreness (mean reduction 3.8 mm; 95 % CI −5.2 mm to −2.4 mm).

Conclusion

A number of theories have been put forward about how stretching could prevent or treat pain. Stretch has been used to reduce pain of presumed musculoskeletal origin, and there is some evidence from several small studies that these techniques are helpful. Many therapists stretch muscles to correct postural malalignment thought to cause overuse injury, but there is little evidence of a causal relationship between abnormal alignment and development of overuse injury and the few relevant trials are equivocal. The practice of stretching immediately before or after exercise has no effect, or only a very small effect, on delayed-onset muscle soreness.

References

Bigos, S. J., Battie, M. C., Spengler, D. M., et al. (1992). A longitudinal, prospective study of industrial back injury reporting. *Clinical Orthopaedics and Related Research, 279,* 21–34.

Bobbert, M. F., Hollander, A. P., & Huijing, P. A. (1986). Factors in delayed onset muscular soreness of man. *Medicine & Science in Sports & Exercise*, *18*, 75–81.

Bron, C., de Gast, A., Dommerholt, J., et al. (2011). Treatment of myofascial trigger points in patients with chronic shoulder pain: a randomized, controlled trial. BMC Medicine, 9, 8.

Crawford, F., & Thomson, C. (2003) Interventions for treating plantar heel pain. In: The Cochrane Library, Issue 4. Chichester: Wiley.

Cowan, D. N., Jones, B. H., Frykman, P. N., et al. (1996). Lower limb morphology and risk of overuse injury among male infantry trainees. *Medicine & Science in Sports & Exercise, 28*, 945–952.

De Vries, H. A. (1966). Quantitative electromyographic investigation of the spasm theory of muscle pain. *American Journal of Physical Medicine, 45*, 119–134.

Dieck, G. S., Kelsey, J. L., Goel, V. K., et al. (1985). An epidemiologic study of the relationship between postural asymmetry in the teen years and subsequent back and neck pain. *Spine, 10*, 872–877.

Hanten, W. P., Olson, S. L., Butts, N. L., et al. (2000). Effectiveness of a home program of ischemic pressure followed by sustained stretch for treatment of myofascial trigger points. *Physical Therapy, 80*, 997–1003.

Herbert, R., de Noronha, M., Kamper, S.K. (2011) Stretching to prevent or reduce muscle soreness after exercise. Cochrane Database of Systematic Reviews, 7.

Hong, C., & Simons, D. G. (1998). Pathophysiological and electrophysiological mechanisms of myofascial trigger points. *Archives of Physical Medicine and Rehabilitation, 79*, 863–872.

Hou, C.-R., Tsai, L.-C., Cheng, K.-F., et al. (2002). Immediate effects of various physical therapeutic modalities on cervical myofascial pain and trigger-point sensitivity. *Archives of Physical Medicine and Rehabilitation, 83*, 1406–1414.

Khalil, T. M., Asfour, S. S., Martinez, L. M., et al. (1992). Stretching in the rehabilitation of low back pain patients. *Spine, 17*, 311–318.

Pfeffer, G., Bacchetti, P., Deland, J., et al. (1999). Comparison of custom and prefabricated orthoses in the initial treatment of proximal plantar fasciitis. *Foot & Ankle International, 20*, 214–221.

Proske, U., & Morgan, D. L. (2001). Muscle damage from eccentric exercise: Mechanism, mechanical signs, adaptation and clinical applications. *The Journal of Physiology, 537*, 333–345.

Travell, J. G., & Simons, D. G. (1983). *Myofascial pain and dysfunction. The trigger point manual*. Baltimore: Williams & Wilkins.

Wen, D. Y., Puffer, J. C., & Schmalzried, T. P. (1998). Injuries in runners: A prospective study of alignment. *Clinical Journal of Sport Medicine, 8*, 187–194.

Stretch-sensitive Receptors

▶ Visceral Mechanoreceptors

Stria Terminalis (ST)

Definition

A curved fiber bundle that accompanies the caudate nucleus and reciprocally connects the amygdala and the medial hypothalamus.

Cross-References

▶ Nociceptive Processing in the Amygdala: Neurophysiology and Neuropharmacology

Stroke

Definition

A sudden disruption of blood flow to the brain, either by a clot or a leak in a blood vessel.

Cross-References

▶ Central Pain, Outcome Measures in Clinical Trials
▶ Headache Due to Arteritis

Structural Lesion

Definition

In relation to low back pain, a structural lesion is an abnormality in the structure of the spine that is thought to cause an individual's back pain. Many physicians assume that the most appropriate strategy for treating low back pain is to identify and correct the structural lesion underlying it.

Cross-References

▶ Compensation, Disability, and Pain in the Workplace
▶ Disability, Effect of Physician Communication

Struggle

Definition

Struggle is a centrally processed reaction (the rat attempts to escape).

Cross-References

▶ Randall-Selitto Paw Pressure Test

Strychnine

Definition

Strychnine is a blocker of inhibitory glycine receptors.

Cross-References

▶ GABA and Glycine in Spinal Nociceptive Processing

Stump Pain

Definition

Stump pain is pain in the portion of the body adjacent to the amputation line. It may be spontaneous or evoked.

Cross-References

▶ Postoperative Pain, Postamputation Pain, Treatment, and Prevention

SUA

▶ Single Unit Activity

Subacute Pain

▶ Postoperative Pain, Persistent Acute Pain

Subarachnoid Drug Administration

▶ Postoperative Pain, Intrathecal Drug Administration

Subarachnoid Space

Definition

The space between the arachnoidea and pia mater membranes surrounding the spinal cord, traversed by delicate fibrous trabeculae and filled with cerebrospinal fluid. This space is especially large around the lumbar level of the human spinal cord.

Cross-References

▶ Cell Therapy in the Treatment of Central Pain

Subarachnoid/Spinal Anesthesia

Definition

A form of regional anesthesia that involves the injection of anesthetic into the cerebrospinal fluid (CSF), at a predetermined space along the spinal canal, to produce anesthesia to all body regions that are supplied by nerves that arise below the anatomic region of the block.

Cross-References

▶ Epidural Infusions in Acute Pain

Subchondral Bone

Definition

A portion of bone that lies beneath the cartilage, typically the ends of bones that form joints.

Cross-References

▶ Arthritis Model, Osteoarthritis

Subclinical Forehead Sweating

Definition

Subclinical forehead sweating refers to increased sweating not evident by observation.

Cross-References

▶ Sunct Syndrome

Subcutaneous Mechanoreception

Definition

An ambiguous term: Mechanonociceptors and sensitive mechanoreceptors are found in the subcutis, e.g., the highly sensitive Pacinian corpuscles (vibration sensors) are situated in the subcutis and in deeper tissues. Muscles, tendons, and periostium contain mechanosensitive nerves. Some of these mechanosensors are nociceptive, but not all of them. The specific muscle spindle afferents are highly mechanosensitive. They do not contribute to mechanoreception, however, but to the position sense of our extremities.

Cross-References

▶ Postsynaptic Dorsal Column Projection, Anatomical Organization

Subdural

Definition

The area between the dura mater and the arachnoid mater.

Cross-Reference

▶ Epidural Infusions in Acute Pain

Subdural Drug Pumps

▶ Pain Treatment, Implantable Pumps for Drug Delivery

Subdural Hematoma: Posttraumatic Subdural Hemorrhage

▶ Headache Due to Intracranial Bleeding

Subfamily V

▶ TRPV1 Receptor, Species Variability

Subject Weights

Definition

Subject weights are the coordinates of an individual in the source space with respect to the dimensions found in the group stimulus space.

Subjective Factors

Definition

Subjective factors are an individual's experiences that make it difficult for them to carry out various activities. Since examiners cannot directly observe subjective factors, they must rely on reports by claimants about them.

Cross-References

▶ Impairment, Pain-Related

Subjective Judgment

Definition

Conclusions drawn about another's pain based on personal feeling or interpretation; not objective. While it is impossible to avoid subjective judgment entirely, it can be minimized through the use of standardized and validated pain assessment tools.

Cross-References

▶ Pain Assessment in Neonates

Subjective Pain Experience

Definition

The multidimensional perceptual qualities that are categorized by people as pain.

Cross-References

▶ McGill Pain Questionnaire

Subjective Pain Measurement

▶ McGill Pain Questionnaire

Sublenticular Extended Amygdala

Synonyms

SLEA

Definition

Neurons that represent an extension of the central and medial nuclei of the amygdala.

Cross-References

▶ Amygdala, Functional Imaging

Sublimaze®

▶ Postoperative Pain, Fentanyl

Subliminal Receptive Fields

Definition

Receptive fields detected after removal of dominant inputs to the neuron.

Cross-References

▶ Thalamic Plasticity and Chronic Pain

Submaximum Effort Tourniquet Technique

▶ Tourniquet Test

Subnucleus Caudalis

Synonyms

Caudalis, Vc; Medullary dorsal horn (MDH); Trigeminal nucleus caudalis; Trigeminal subnucleus caudalis

Definition

The most caudal segment and subnucleus of the trigeminal sensory nuclear complex. The subnucleus caudalis is a laminated structure that closely resembles the dorsal horn of the spinal cord with which it is continuous. In many of its properties, it is functionally and anatomically homologous to the spinal dorsal horn and hence is frequently termed the medullary dorsal horn (MDH). Integrative processing of nociceptive input from facial cutaneous, intraoral, and musculoskeletal orofacial tissues principally occurs in the subnucleus caudalis.

Cross-References

▶ Nociceptors in the Orofacial Region (Temporomandibular Joint and Masseter Muscle)
▶ Trigeminal Brainstem Nuclear Complex, Anatomy

Subnucleus Reticularis Dorsalis

Synonyms

SRD

Definition

These neurons selectively convey nociceptive information from all parts of the body to the thalamocortical system.

Cross-References

▶ Spinothalamocortical Projections to Ventromedial and Parafascicular Nuclei

Subnucleus Reticularis Ventralis

Definition

The ventral portion of the subnucleus reticularis dorsalis (SRD).

Cross-References

▶ Brainstem Subnucleus Reticularis Dorsalis Neuron

Substance Abuse

Definition

The intentional misuse of a medication; either overuse or taken for a purpose not prescribed (i.e., mood alteration).

Cross-References

▶ Cancer Pain, Evaluation of Relevant Comorbidities and Impact
▶ Opioid Therapy in Cancer Patients with Substance Abuse Disorders, Management

Substance P

Definition

Substance P is an 11 amino acid neuropeptide that is a member of the tachykinin neuropeptide family and functions as a neurotransmitter and neuromodulator. It is present in both primary afferent neurons and in neurons within the spinal cord and higher brain levels. The endogenous receptor for substance P is the neurokinin 1 receptor, a G-protein-coupled receptor. The release of substance P from peripheral terminals of DRG neurons is a major mediator of neurogenic inflammation, causing vasodilation of blood vessels, plasma extravasation, and degranulation of mast cells. The release of substance P from central terminals contributes to enhanced function of nociceptive spinal cord neurons, in part by facilitating the actions of glutamate at the N-methyl-D-aspartate (NMDA) receptor.

Cross-References

► Alternative Medicine in Neuropathic Pain
► Fibromyalgia
► Immunocytochemistry of Nociceptors
► Nociceptor, Categorization
► Opioid Modulation of Nociceptive Afferents In Vivo
► Opioids in the Periphery and Analgesia
► Somatic Pain
► Spinomesencephalic Tract

Substance P Regulation in Inflammation

Janet Winter
Novartis Institute for Medical Science,
London, UK

Synonyms

Neurokinin; NK; Substance P tachykinin

Definition

Substance P is an eleven-amino-acid neuropeptide associated with pain transmission and is expressed at high levels in ► dorsal root ganglia and in sensory afferent terminals in the spinal cord, as well as in the sensory innervation of peripheral tissues such as skin and viscera. Electrical stimulation causes release of substance P from spinal cord, but it can also be released from peripheral endings; in fact, the majority of this neuropeptide is transported peripherally, not centrally from the DRG. Peripheral release of substance P is thought to have a proinflammatory role (neurogenic inflammation).

Characteristics

Peptide Family

Substance P is part of the tachykinin family along with neurokinin A and neurokinin B; all share the same carboxy-terminal sequence (Phe-X-Gly-Leu-X-NH_2); see Table 1.

There are three genes encoding tachykinins, PPT-A (preprotachykinin-A), PPT-B, and PPT-C. Mammalian substance P is encoded by the PPT-A gene (Severini et al. 2002). This gene also encodes other tachykinins, including NKA and the two N terminally extended forms of NKA, neuropeptide K (NPK), and neuropeptide γ (NPγ). NKB is derived from a separate gene, PPT-B. The endokinins are encoded by PPT-C. There are four splice variants of PPT-A, (αβγδ) with β PPT-A mRNA containing all seven exons of the corresponding gene. Substance P can be expressed alone (αδ), whereas NKA expression is always accompanied by substance P expression (βγ).

In neurons, substance P is released from its precursor (preprotachykinin) by the action of converting enzymes and is then COOH-terminally amidated, packed into storage vesicles, and transported to terminals for final enzymatic processing.

Substance P is expressed predominantly in the nervous system (although other tissue types have recently been shown to express it, e.g., endothelial cells, inflammatory cells, airway smooth muscle cells). All vertebrate tissues are innervated by networks of tachykinergic fibers. Some of the new members of the tachykinin family, the hemokinins, and endokinins are primarily expressed in nonneuronal tissues.

Substance P Regulation in Inflammation, Table 1 Amino acid sequence of mammalian tachykinins

Tachykinin	Sequence
Substance P	Arg-Pro-Lys-Pro-Gln-Gln-Phe-Phe-Gly-Leu-Met-NH2
Neurokinin A	His-Lys-Thr-Asp-Ser-Phe-Val-Gly-Leu-Met-NH2
Neurokinin B	Asp-Met-His-Asp-Phe-Phe-Val-Gly-Leu-Met-NH2
Hemokinin-1 (m/r)	Arg-Ser-Arg-Thr-Arg-Gln-Phe-Tyr-Gly-Leu-Met-NH2
Hemokinin-1 (h)	Thr-Gly-Lys-Ala-Ser-Gln-Phe-Phe-Gly-Leu-Met-NH2
Endokinin A and B[a]	-Gly-Lys-Ala-Ser-Gln-Phe-Phe-Gly-Leu-Met-NH2
C14TKL-1[b]	Arg-His-Arg-Thr-Pro-Met-Phe-Tyr-Gly-Leu-Met-NH2
Virokinin (viral origin)	Gly-Ile-Pro-Glu-Leu-Ile-His-Tyr-Thr-Arg-Asn-Ser-Thr-Lys-Lys-Phe-Tyr-Gly-Leu-Met-NH2
Tachykinin gene-related peptides	
Endokinin C	Lys-Lys-Ala-Tyr-Gln-Leu-Glu-His-Thr-Phe-Gln-Gly-Leu-Leu-NH2
Endokinin D	Val-Gly-Ala-Tyr-Gln-Leu-Glu-His-Thr-Phe-Gln-Gly-Leu-Leu-NH2

[a]The C-terminal decapeptidic fragment common to endokinin A (47 amino acids) and endokinin B (41 amino acids) is shown
[b]Chromosome 14 tachykinin-like peptide 1. *m/r* mouse/rat, *h* human (Modified from Patacchini et al. (2004))

▶ Neuron restrictive silencer factor (NRSF) is a repressor that is predominantly expressed in nonneuronal cells. NRSF silences neuronal genes in nonneuronal cells by binding to the NRSE (neuron-restrictive silencer element) motif. These neuronal genes include type II sodium channel, synapsin 1, M4 muscarinic acetylcholine receptor, and certain adhesion molecules. The NRSE motif is also found at the major transcriptional start site of the PPT-A gene, and when promoter constructs expressing NRSF are introduced into cultured DRG neurons, there is a marked downregulation of PPT-A transcription (Quinn et al. 2002). This important silencer may therefore be responsible for the inhibition of expression of substance P in nonneuronal cells in normal tissues.

Receptors
Substance P and the other tachykinins act via G protein-coupled, seven transmembrane domain, and neurokinin (NK) receptors (Severini et al. 2002). There are (at least) three of these, NK1, NK2, and NK3, with highest affinities for substance P, NKA, and NKB, respectively, although endogenous tachykinins are not highly selective and can act at all three receptors under certain conditions. A multistate model of GPCR activation has been proposed whereby ligands can specifically stabilize receptor conformations, enabling one receptor to activate one or more G proteins in a ligand-specific manner.

The main substance P receptor, NK1, will be described here. NK1 is a single-copy gene (mapped to human chromosome 2) that has a high degree of sequence homology between species. There is a cAMP/calcium response element in the 5′ untranslated region.

Recent papers suggest that NK1 receptors are expressed both on DRG neurons and terminals and postsynaptically in the spinal cord. Presence on DRG neurons suggests that substance P may regulate its own release. Following substance P application, either exogenously applied to the spinal cord or released from afferent terminals during intense stimulation, there is a dramatic internalization of NK1 receptor expressed on dorsal horn lamina I neurons from the usual cell membrane localization. Substance P sensitizes NMDA receptors in the spinal cord.

However, this entry will focus on the peripheral roles of substance P and NK1 in inflammation.

Neurogenic Inflammation
▶ Neurogenic inflammation is the collective term for the effects produced by neuropeptides released from (mainly capsaicin-sensitive)

sensory neurons and include vasodilatation (flare), plasma protein extravasation (PPE) and leukocyte adhesion to vascular endothelial cell receptors (Patacchini et al. 2004; Severini et al. 2002).

In the skin, some substance P-positive terminals are found near the epidermal basal membrane, but the majority innervate blood vessels. Substance P released from cutaneous sensory nerves binds to NK1 receptors on endothelial cells, resulting in plasma leakage and increased blood flow, dilatation of the arteriole (although CGRP is probably the main vasodilator released), and wheal and flare (Brain and Williams 1988; Holzer 1998; Green et al. 1992). Substance P can also cause itch. These outcomes are due not only to direct action of substance P on blood vessels but also indirectly to mast cell activation. Substance P released by neurogenic inflammation can also promote angiogenesis via NK1 receptors (Seegers et al. 2003). Vascular responses to substance P in the skin are very species dependent, however. These proinflammatory actions of substance P may potentiate nociceptive responses, although neurogenic inflammation alone evoked by antidromic stimulation of primary afferents is not sufficient to sensitize nociceptor terminals in vivo, at least acutely.

Mice with a targeted disruption of either the PPT-A gene or the NK1 receptor gene show a complex pattern of responses to noxious stimuli. PPT-A −/− mice which lack substance P (SP) and neurokinin A (NKA) exhibit reduced behavioral responses to intense somatic stimuli, but the behavioral hypersensitivity following injury is unaltered. From NK1 −/− mice studies, it seems that NK1 receptors have an essential role in mediating central nociceptive and peripheral inflammatory responses to noxious stimuli that provoke neurogenic inflammation (e.g., carrageenan), but no role in non-neurogenic stimuli (such as turpentine or ▶ CFA). A much higher percentage of visceral (80 %) than skin (25 %) afferents express substance P and NK1 −/− mice show profound deficits in spontaneous behavioral reactions to an acute visceral chemical stimulus (intracolonic capsaicin) and fail to develop referred hyperalgesia or tissue edema. However, responses (hyperalgesia and edema) to mustard oil are intact. Two separate hyperalgesia pathways may exist, one that is NK1 receptor dependent and one that is independent (Laird et al. 2000). The lack of neurogenic inflammation in NK1 −/− mice probably does not account for the blunted pain response and lack of hyperalgesia, as NK1 antagonists that do not cross the blood–brain barrier do not affect behavioral nociceptive responses to capsaicin or formalin, but completely block neurogenic inflammation.

In the periphery, density of substance P fibers innervating skeletal muscle increases in inflammation and substance P expression and transport down the sciatic nerve increases following inflammation of the hindpaw, though substance P expression in the paw declines, probably due to decreased storage and increased release. NGF may be an important factor in regulation of substance P in inflammation (Woolf 1996). NGF produced in peripheral targets is retrogradely transported to the cell bodies of sensory neurons innervating that tissue, where it can regulate substance P production. NGF injection into the paw can cause increase in substance P and hyperalgesia. Anti-NGF can block the in vivo inflammation-induced increases in substance P as well as hyperalgesia. Substance P injection or neurogenic inflammation activates NK1 receptors on keratinocytes and causes increased NGF production, which is blocked by an NK1 receptor antagonist (Amann et al. 2000). So substance P-induced NGF upregulation can feedback to further increases in substance P production. After inflammation, there is a phenotypic switch so that some of the large A fibers now also express substance P (Woolf 1996). These may be A fiber nociceptors.

Substance P receptors are expressed on sensory fibers innervating skin (Ruocco et al. 2001) and are also upregulated in inflamed tissues. The proportion of NK[1]-expressing sensory fibers in the ▶ glabrous skin of the hindpaw increases 2 days after CFA inflammation (Coggeshall and Carlton 2003). PGE2 (prostaglandin E2) production increases in inflammation and may influence NK1 expression levels; NK1 receptor levels are upregulated by PGE2 in adult sensory neuron cultures.

There is also a probable role for substance P in inflammation in man, where psoriatic skin shows a large increase in substance P and other neuropeptides and an increase in substance P receptors. In the clinic, levels of substance P in dental pulp tissue increased in patients with irreversible ▶ pulpitis (Bowles et al. 2003).

Antagonists

NK1 antagonists can inhibit PPE and edema in rodent skin. They are also potent anti-hyperalgesics in animal models of inflammation and neuropathic pain (Campbell et al. 2000). In the clinic, an NK1 antagonist was effective for acute pain following dental surgery; however, clinical trials of NK1 antagonists in chronic pain have failed to exhibit efficacy (Hill 2000). No clinical trials have yet been carried out on visceral pain, where NK1 antagonists may be more effective (Laird 2001); antagonists acting at visceral afferents have been effective in the clinic against nausea and vomiting.

References

Amann, R., Egger, T., & Schuligoi, R. (2000). The tachykinin NK(1) receptor antagonist SR140333 prevents the increase of nerve growth factor in rat paw skin induced by substance P or neurogenic inflammation. *Neuroscience, 100*, 611–615.

Bowles, W. R., Withrow, J. C., Lepinski, A. M., et al. (2003). Tissue levels of immunoreactive substance P are increased in patients with irreversible pulpitis. *Journal of Endodontics, 29*, 265–267.

Brain, S. D., & Williams, T. J. (1988). Substance P regulates the vasodilator activity of calcitonin gene-related peptide. *Nature, 335*, 73–75.

Campbell, E. A., Gentry, C., Patel, S., et al. (2000). Oral anti-hyperalgesic and anti-inflammatory activity of NK(1) receptor antagonists in models of inflammatory hyperalgesia of the guinea-pig. *Pain, 87*, 253–263.

Coggeshall, R. E., & Carlton, S. M. (2003). Control of postganglionic sympathetic efferent fibers by neurokinin 1 receptors in rats. *Neuroscience Letters, 353*, 197–200.

Green, P. G., Basbaum, A. I., & Levine, J. D. (1992). Sensory neuropeptide interactions in the production of plasma extravasation in the rat. *Neuroscience, 50*, 745–749.

Hill, R. (2000). NK1 (Substance P) receptor antagonists–why are they not analgesic in humans? *Trends in Pharmacological Sciences, 21*, 244–246.

Holzer, P. (1998). Neurogenic vasodilatation and plasma leakage in the skin. *General Pharmacology, 30*, 5–11.

Laird, J. (2001). Gut feelings about tachykinin NK1 receptor antagonists. *Trends in Pharmacological Sciences, 22*, 169.

Laird, J. M., Olivar, T., Roza, C., et al. (2000). Deficits in visceral pain and hyperalgesia of mice with a disruption of the tachykinin NK1 receptor gene. *Neuroscience, 98*, 345–352.

Patacchini, R., Lecci, A., Holzer, P., et al. (2004). Newly discovered tachykinins raise new questions about their peripheral roles and the tachykinin nomenclature. *Trends in Pharmacological Sciences, 25*, 1–3.

Quinn, J. P., Bubb, V. J., Marshall-Jones, Z. V., et al. (2002). Neuron restrictive silencer factor as a modulator of neuropeptide gene expression. *Regulatory Peptides, 108*, 135–141.

Ruocco, I., Cuello, A. C., Shigemoto, R., et al. (2001). Light and electron microscopic study of the distribution of substance P-immunoreactive fibers and neurokinin-1 receptors in the skin of the rat lower lip. *The Journal of Comparative Neurology, 432*, 466–480.

Seegers, H. C., Hood, V. C., Kidd, B. L., et al. (2003). Enhancement of angiogenesis by endogenous substance P release and neurokinin-1 receptors during neurogenic inflammation. *Journal of Pharmacology and Experimental Therapeutics, 306*, 8–12.

Severini, C., Improta, G., Falconieri-Erspamer, G., et al. (2002). The tachykinin peptide family. *Pharmacological Reviews, 54*, 285–322.

Woolf, C. J. (1996). Phenotypic modification of primary sensory neurons: The role of nerve growth factor in the production of persistent pain. *Philosophical Transactions of the Royal Society of London. Series B, Biological Sciences, 351*, 441–448.

Substance P Tachykinin

▶ Substance P Regulation in Inflammation

Substantia Gelatinosa

Synonyms

Rexed's lamina II

Definition

A narrow, dense, vertical band of gelatinous gray matter forming the dorsal part of the posterior horn of the spinal cord and serving to integrate the sensory stimuli that give rise to the sensations of heat and pain. It is also referred to as Rexed's lamina II. Many afferents synapse on the second-order neuron here.

Cross-References

▶ Nociceptor, Categorization
▶ Opioid Therapy in Cancer Pain Management, Route of Administration
▶ Psychiatric Aspects of Visceral Pain
▶ Somatic Pain
▶ Visceral Nociception and Pain

Substantial Gainful Activity

Definition

Work that involves doing significant and productive physical or mental duties, which is done (or intended) for pay or profit. Work may be substantial even if done on a part-time basis or if the person does less, gets paid less, or has less responsibility than when he or she worked before.

Cross-References

▶ Disability Evaluation in the Social Security Administration

Sucrose

▶ Sweet Solutions for Pain Management

Suddick's Atrophy (Historical Name)

▶ CRPS-1 in Children

Sudeck's Atrophy

▶ Complex Regional Pain Syndrome
▶ Neuropathic Pain Models, CRPS-I Neuropathy Model
▶ Postoperative Pain, Acute Presentation of Complex Regional Pain Syndrome

Sudeck's Atrophy (1901)

▶ CRPS, Evidence-Based Treatment

Sudomotor Function and Dysfunction

Definition

Pertaining to the regulation of sweating in health and disease.

Suffering

Definition

Suffering is the clash arising from an attack to a person's integrity, to its desire of happiness and health, produced or not produced by pain.

Cross-References

▶ Cancer Pain Management, Interface Between Cancer Pain Management and Palliative Care
▶ Ethics of Pain in the Newborn Human

Suggestibility

Definition

The degree to which an individual is likely to respond to suggestion. This may vary within an individual person depending upon context and state of mind. Group pressure may increase suggestibility to conform to the mores of the group. Hypnosis is considered a state of mind that enhances suggestibility.

Cross-References

▶ Therapy of Pain, Hypnosis

Sulci

Definition

Sulci are grooves on the surface of the brain, often used as landmarks to identify areas with a distinct function.

Cross-References

▶ Clinical Migraine with Aura

Sulfur Bath

Definition

Bathing in a sulfur pool heated to 34–37 °C.

Cross-References

▶ Spa Treatment

Sunburn

Definition

Sunburn is a colloquial term for the damage to the skin caused by acute overexposure to ultraviolet light from the sun.

Cross-References

▶ UV Erythema, a Model for Inducing Hyperalgesias
▶ UV-Induced Erythema

Sunburn as a Model of Cutaneous Hyperalgesia

▶ UV-Induced Erythema

SUNCT Status

Definition

Continuous flow of SUNCT attacks for more than 24 h.

Cross-References

▶ Sunct Syndrome

SUNCT Syndrome

Juan A. Pareja
Department of Neurology, Fundación Hospital Alcorcón, Madrid, Spain

Synonyms

Short-lasting unilateral neuralgiform headache attacks with conjunctival injection and tearing

Definition

The acronym SUNCT (short-lasting unilateral neuralgiform headache attacks with conjunctival injection and tearing) (Sjaastad et al. 1989) summarizes the clinical features of this syndrome.

SUNCT (Sjaastad et al. 1989) is a primary disorder characterized by intermittent short-lived (5–240 s), moderate to severe, paroxysms of orbital/periorbital pain, accompanied by ipsilateral, rapidly developing, dramatic, conjunctival injection and lacrimation (International Headache Society 2004). Less frequently, there may also be rhinorrhea or nasal stuffiness. Attacks are usually precipitated from trigeminal and extratrigeminal innervated areas.

Characteristics

SUNCT is a male predominant disorder, with the mean age of onset around 50 years (Matharu et al. 2003; Pareja and Sjaastad 1997). Although available epidemiological data is lacking, the low number of reported cases indicates it is a rare disorder.

Localization
Symptoms and signs are strictly unilateral, generally with the pain persistently confined to the ocular/periocular area (Pareja and Sjaastad 1997; Sjaastad et al. 1989). An occasional spread beyond the midline, or a co-involvement of the opposite side, has been observed in SUNCT (Matharu et al. 2003; Pareja and Sjaastad 1997), with the pain still predominating in the originally symptomatic side. Shifting side attacks have been reported in two patients (e.g., Matharu et al. 2003).

Intensity and Character of Pain (Intensity of Pain)
Most attacks are characterized by moderate to severe pain. Excruciatingly severe pain (▶ excruciating pain) is rarely reported. The pain is frequently described as burning, stabbing, or electric in character (Pareja and Sjaastad 1997). SUNCT attacks start and cease abruptly. The solitary attacks usually have a "plateau-like" pattern (Sjaastad et al. 1989) but also other patterns have been noted (Pareja and Sjaastad 1994): "repetitive" (short-lasting attacks in rapid succession), "sawtooth-like" (and its variant "staccato-like," in which consecutive spikelike paroxysms occur without reaching the pain-free baseline), and "plateau-like plus exacerbations" (admixture of 1–2 s jabs superimposed on top of the conventional plateau-like pattern).

Duration of Attacks
The mean duration of paroxysms is ca. 1 min. Objectively, measurements of 348 attacks in 11 patients rendered a mean duration of 49 s, with a range of 5–250 s (Pareja et al. 1996c). Very rarely, a few patients reported relatively long-lasting attacks, lasting 1–2 h (Pareja et al. 1996b). However, this atypical duration of SUNCT should be taken as a rare variant since, even in these patients, the vast majority of attacks were almost invariably typical.

In between attacks, the patients are completely free of symptoms. Very rarely, however, patients may report very low-grade background pain or discomfort in the symptomatic area during symptomatic periods (Pareja et al. 1996b). This background pain or discomfort could be rather durable, fluctuating, or intermittent.

Temporal Distribution of Attacks
The attacks predominate during daytime. There seems to be a tendency for attacks to occur in the morning and afternoon/evening in a bimodal fashion. Nocturnal attacks are rare but may occur either during the worst periods or as an occasional feature (Pareja et al. 1996c). The reduction in attacks during the night seems to concern patients exhibiting both exclusively spontaneous attacks and those with mostly precipitated attacks. The reduction in precipitating stimuli during sleep does, therefore, not seem critical for the nocturnal break in the flow of attacks.

The temporal pattern is irregular, with symptomatic periods, alternating with remissions in an unpredictable fashion. Symptomatic periods last from a few days to several months but may persist for up to 5 years, in the exceptional case. Remissions may last from 1 week to several years but usually last for several months (Pareja and Sjaastad 1997; Sjaastad et al. 1989). The chronic pattern seems to be much less typical than the remitting form in SUNCT (Matharu et al. 2003). During active periods, the usual frequency of attacks may vary, from a few attacks/day to >30 attacks/h (Sjaastad et al. 1989). Objective assessment of the frequency of 585 attacks in four patients rendered a mean of 16 attacks per day, with a range of 1–86 daily (Pareja et al. 1996c). At the peak of an attack series, a 1–3-day long clinical "status" has been witnessed in several patients (Pareja et al. 1996a) (▶ SUNCT status).

Accompaniments

Attacks are regularly accompanied by prominent, ipsilateral conjunctival injection and lacrimation, both signs appearing in tandem from the onset of symptoms (Pareja and Sjaastad 1997; Sjaastad et al. 1989). Rhinorrhea and/or nasal obstruction is found in approximately 2/3 of the SUNCT cases (Pareja and Sjaastad 1997; Sjaastad et al. 1989). In addition, there is – subclinical – increased forehead sweating (▶ subclinical forehead sweating), increased intraocular pressure, and corneal temperature (Sjaastad et al. 1992). During periods with a marked attack tendency, there may be swelling of the eyelids on the symptomatic side owing to vascular engorgement and eyelid edema, but such ▶ pseudoptosis is not paretic in nature. Otherwise, changes in pupil diameter are not typical of SUNCT (Zhao and Sjaastad 1993).

During SUNCT attacks, there may be increased systemic blood pressure with heart rate decrement. It has been demonstrated during attacks, and less markedly in between attacks, that SUNCT patients hyperventilate (e.g., Pareja and Sjaastad 1997).

Precipitating Mechanisms

In SUNCT, numerous mechanical precipitating factors within the trigeminal and extratrigeminal areas have been described (Pareja and Sjaastad 1997; Sjaastad et al. 1989). Although most SUNCT attacks probably are precipitated, a minority of patients seemingly may exhibit exclusively spontaneous attacks. The apparent spontaneous attacks in genuine SUNCT may have "subclinical triggers." ▶ Refractory periods that are a typical feature of genuine, neuralgic pain seem to be lacking in SUNCT patients (Laín et al. 2000; Matharu et al. 2003; Pareja and Sjaastad 1994, 1997).

Symptomatic SUNCT

In the vast majority of SUNCT cases, etiology and pathogenesis are unknown. Symptomatic SUNCT has been described in some patients with documented intracranial structural lesions (Bussone et al. 1991; Matharu et al. 2003; Pareja and Sjaastad 1997) mostly located in the posterior fossa and in the pituitary region. A relationship between pituitary adenomas and SUNCT has been suggested, and reports of exacerbation by dopamine agonists have been published (Massiou et al. 2002). These findings suggest a functional role for the hypothalamic-pituitary axis in SUNCT. The possibility of symptomatic forms makes neuroimaging exams as mandatory.

Treatment

In SUNCT, there is a lack of persistent, convincingly beneficial effect of drugs or anesthetic blockades generally effective in cluster headache (CH), paroxysmal hemicrania (PH), trigeminal neuralgia, primary stabbing headache (PSH), and other headaches more faintly resembling SUNCT (Matharu et al. 2003; Pareja and Sjaastad 1997; Goadsby et al. 2008). Single reports have claimed that carbamazepine (200–1,600 mg daily), lamotrigine (100–300 mg daily) (D'Andrea et al. 2001), gabapentin (800–2,700 mg daily), topiramate (50–300 mg daily), or surgical procedures may be of help. At present, the drug of choice for SUNCT seems to be lamotrigine. There is no available abortive treatment for the individual attacks. During the worst periods, intravenous lidocaine (1–4 mg/kg/h) may decrease the flow of SUNCT attacks.

Bilateral blockade of the greater occipital nerve (Porta-Etessam et al. 2010) and superior cervical ganglion opioid blockade (Sabatowski et al. 2001) have been reported as temporary/partially effective in one patient each. Botulinum toxin injected around the symptomatic orbit provided sustained relief to one patient (Zabalza 2012). Owing to the scarcity of reports, the results of these interventions should be taken as preliminary (Pareja et al. 2013).

Invasive therapy with interventions directed to the first division of the trigeminal nerve or Gasserian ganglion, with local anesthetics or alcohol, radiofrequency thermocoagulation, microvascular decompression, and gamma-knife neurosurgery, has been tried in the treatment of refractory SUNCT. Some patients seemed to benefit from such interventions, but one should still have a critical attitude to these claims since no convincing results have been obtained as yet. Otherwise, negative results of several surgical procedures such as glycerol rhizotomy, gammaknife radiosurgery, and microvascular decompression of the trigeminal nerve have also been reported. Therefore, at this stage of development, surgical treatment of SUNCT has not been sufficiently validated (Matharu et al. 2003; Pareja et al. 2002). The documented activation of the hypothalamus during SUNCT attacks (May et al. 1999) has prompted the development of deep brain stimulation of the hypothalamus in refractory SUNCT with promising results (Leone et al. 2005).

Differential Diagnosis

SUNCT syndrome should mainly be considered when encountering a case of orbital/periorbital pain with prominent autonomic accompaniments and/or when the orbital pain paroxysms are short lasting. The differential diagnostic possibilities seem to be limited and mainly consist of CH, PH, first branch (V-1) trigeminal neuralgia, and PSH.

Diagnostic criteria of SUNCT syndrome (The International Headache Society Classification, 2nd edn):

(A) At least 20 attacks fulfilling criteria B–D.
(B) Attacks of unilateral orbital, supraorbital, or temporal stabbing or pulsating pain lasting 5–240 s.
(C) Pain is accompanied by ipsilateral conjunctival injection and lacrimation.
(D) Attacks occur with a frequency from 3 to 200 per day.
(E) Not attributed to another disorder.

SUNCT syndrome differs clearly from CH as regards a number of clinical variables such as duration, intensity, frequency, and nocturnal preponderance of attacks. The two syndromes also differ markedly as regards precipitation of attacks, the usual age of onset, and efficacy of various treatment alternatives. Laboratory investigations have disclosed differences as regards presence/absence of Horner-like picture. All in all, these differences seem sufficiently ponderous to make it likely that SUNCT syndrome and CH differ essentially.

SUNCT and PSH are topographically antagonistic in that SUNCT is typically confined to the same orbital/periorbital localization, whereas PSH is characterized by a multidirectional, chaotic, sequence of paroxysms. Such a lack of topographic organization of the symptoms could be the expression of a presumed multi-origin, most probably in the terminal sensitive fibers of the pericranial nerves. However, the pain may be felt within the V-1 territory, frequently in the orbit. Otherwise, PSH is a predominantly female disorder with ultrashort (typically lasting 1 s) spontaneous attacks not accompanied by autonomic features.

Short-lasting attacks of PH may overlap with long-lasting SUNCT attacks. Indeed, PH may rarely be precipitated by spontaneous or provoked neck movements. These features may pose extra difficulties in its differentiation from SUNCT. The absolute responsiveness of PH to indomethacin is diagnostically crucial, but there are other differentiating features (Table 1).

SUNCT syndrome should be differentiated from V-1 trigeminal neuralgia. It is worth noting that only 5 % of trigeminal neuralgia primarily originates in its first division.

V-1 trigeminal neuralgia attacks are much shorter than SUNCT attacks and are regularly followed by a refractory period. V-1 trigeminal neuralgia attacks may be accompanied by local autonomic signs, but contrary to what is the case in SUNCT, autonomic accompaniments in V-I trigeminal neuralgia are not constant, tend to appear after years of headache, and have a modest dimension. In SUNCT, attacks are regularly accompanied by dramatic lacrimation and conjunctival injection, both signs appearing in tandem, whereas the minority of V-1 trigeminal neuralgia patients with autonomic accompaniments of attacks show modest lacrimation without conjunctival injection. Other remarkable differential features are set forth in Table 2.

SUNA (Short-lasting, unilateral, neuralgiform headache attacks with cranial autonomic features) has been proposed as a broader category

SUNCT Syndrome, Table 1 PH and SUNCT syndrome differences

	PH	SUNCT
Sex predominance	Female	Male
Usual age of onset	<40 years	>40 years
Duration of attacks	2–30 min.	10–120 s
Severity of pain	Excruciating	Moderate–severe
Frequency of attacks	Above 5/day (up to 30/day)	16 (1–86)/day (up to 30/h)
Nocturnal attacks	Frequent	Very rare
Mechanical precipitation of attacks	Infrequent	Frequent
Response to indomethacin	Absolute	None

In a minority (12 %) of PH patients, attacks may be precipitated by neck movements (Modified from Pareja et al. 2002)

SUNCT Syndrome, Table 2 First division (V-1) trigeminal neuralgia and SUNCT syndrome differences

	V-1 neuralgia	SUNCT
Sex predominance	Male = Female	Male > female
Usual attack duration	5–10 s	60 s
Usual range of duration	1–30 s	10–120 s
Autonomic phenomena during attacks		
Conjunctival injection	Absent	Frequent/prominent
Lacrimation	Absent or slight	Frequent/prominent
Rhinorrhea/nasal stuffiness	Absent	Frequent
Severity of pain	Excruciating	Moderate–severe
Spreading of pain from V-1 to V-2, V-3	Yes	No
Nocturnal attacks	Rare	Very rare
Refractory periods	Constant	May lack
Carbamazepine effect	Excellent	Partial

Modified from Pareja et al. 2002

of headache that would include SUNCT (Cohen et al. 2006). According to this proposal, SUNA can be diagnosed with the presence of just one cranial autonomic accompaniment. In addition, the range of duration of pain attacks extends up to 10 min.

Diagnostic criteria for SUNA (The International Headache Society Classification, 2nd edn):
(A) At least 20 attacks fulfilling criteria B–E
(B) Attacks of unilateral orbital, supraorbital or temporal stabbing or pulsating pain lasting from 2 seconds to 10 minutes
(C) Pain is accompanied by one of:
1. conjunctival injection and/or lacrimation
2. nasal congestion and/or rhinorrhoea
3. eyelid edema
(D) Attacks occur with a frequency of one or more per day for more than half the time

(E) No refractory period follows attacks triggered from trigger areas
(F) Not attributed to another disorder

SUNA is even more rare than SUNCT and seems to slightly prevail in females. The mean age at onset is quite similar to SUNCT. Otherwise, the clinical phenotypes of SUNCT and SUNA are similar in regards to the cardinal features, such as the temporal and spatial characteristics of the attacks, the precipitating mechanisms, and the frequent absence of refractory periods. Admittedly, SUNCT and SUNA could represent complementary clinical pictures of the same syndrome. Therefore, it is hard to understand that there is a need for two terms for one disorder (Pareja et al. 2013). Further review of the International Classification of Headache Disorders may clarify this cumbersome situation.

Pathophysiology

Pathophysiology of SUNCT is mostly unknown. Based on the clinical features, there seems to be an activation of the V-I trigeminal system, but the "ignition" process is unknown. Since SUNCT attacks are longer, and refractory periods may be lacking, the pain-modulating system may seem to behave differently in SUNCT as compared with, e.g., trigeminal neuralgia. In SUNCT, parasympathetics could be either strongly activated or hypersensitive, bringing about an impressive set of local autonomic signs. The reported activation of the ipsilateral hypothalamic gray during a SUNCT attack (May et al. 1999) has been claimed as pathogenically relevant, as the hypothalamus has connections to the pain-modulating system and the superior salivary nucleus and is, therefore, anatomically positioned to influence both the pain and the autonomic accompaniments (Goadsby and Lipton 1997; May et al. 1999).

Cross-References

▶ Hemicrania Continua (HC)
▶ Paroxysmal Hemicrania
▶ Primary Stabbing Headache

References

Bussone, G., Leone, M., Volta, G. D., et al. (1991). Short-lasting unilateral neuralgiform headache attacks with tearing and conjunctival injection: The first 'symptomatic' case. Cephalalgia, 11, 123–127.

Cohen, A. S., Matharu, M. S., & Goadsby, P. J. (2006). Short-lasting unilateral neuralgiform headache attacks with conjunctival injection and tearing (SUNCT) or cranial autonomic features (SUNA) – a prospective clinical study of SUNCT and SUNA. Brain, 129, 2746–2760.

D'Andrea, G., Granella, F., Ghiotto, N., et al. (2001). Lamotrigine in the treatment of SUNCT syndrome. Neurology, 57, 1723–1725.

Goadsby, P. J., & Lipton, R. B. (1997). A review of paroxysmal hemicranias, SUNCT syndrome and other short-lasting headaches with autonomic features including new cases. Brain, 120, 193–209.

Goadsby, P. J., Cittadini, E., Burns, B., & Cohen, A. S. (2008). Trigeminal autonomic cephalalgias:

Diagnostic and therapeutic developments. Current Opinion in Neurology, 21, 323–330.

International Headache Society. (2004). The international classification of headache disorders, 2nd edition. Cephalalgia, 24(Suppl 1), 1–160.

Laín, A. H., Caminero, A. B., & Pareja, J. A. (2000). SUNCT syndrome; absence of refractory periods and modulation of attack duration by lengthening of the trigger stimuli. Cephalalgia, 20, 671–673.

Leone, M., Franzini, A., D'andrea, G., et al. (2005). Deep brain stimulation to relieve drug-resistant SUNCT. Annals of Neurology, 57, 924–927.

Massiou, H., Launay, J. M., Levy, C., et al. (2002). SUNCT syndrome in two patients with prolactinomas and bromocriptine-induced attacks. Neurology, 58, 1698–1699.

Matharu, M. S., Cohen, A. S., Boes, C. J., et al. (2003). Short-lasting unilateral neuralgiform headache with conjunctival injection and tearing syndrome: A review. Current Pain and Headache Reports, 7, 308–318.

May, A., Bahra, A., Buchel, C., et al. (1999). Functional MRI in spontaneous attacks of SUNCT: Short-lasting neuralgiform headache with conjunctival injection and tearing. Annals of Neurology, 46, 791–793.

Pareja, J. A., & Sjaastad, O. (1994). SUNCT syndrome in the female. Headache, 34, 217–220.

Pareja, J. A., & Sjaastad, O. (1997). SUNCT syndrome: A clinical review. Headache, 37, 195–202.

Pareja, J. A., Caballero, V., & Sjaastad, O. (1996a). SUNCT syndrome. Status-like pattern. Headache, 36, 622–624.

Pareja, J. A., Joubert, J., & Sjaastad, O. (1996b). SUNCT syndrome. Atypical temporal patterns. Headache, 36, 108–110.

Pareja, J. A., Shen, J. M., Kruszewski, P., et al. (1996c). SUNCT syndrome. Duration, frequency, and temporal distribution of attacks. Headache, 36, 161–165.

Pareja, J. A., Caminero, A. B., & Sjaastad, O. (2002). SUNCT syndrome. Diagnosis and treatment. CNS Drugs, 16, 373–383.

Pareja, J. A., Alvarez, M., & Montojo, T. (2013). SUNCT and SUNA: Recognition and Treatment. Current Treatment Options in Neurology, 15, 28–39.

Porta-Etessam, J., Cuadrado, M. L., Galán, L., et al. (2010). Temporal response to bupivacaine great occipital block in a patient with SUNCT syndrome. The Journal of Headache and Pain, 11, 179.

Sabatowski, R., Huber, M., Meuser, T., & Radbruch, L. (2001). SUNCT syndrome: A treatment option with local opioid blockade of the superior cervical ganglion? A case report. Cephalalgia, 21, 154–156.

Sjaastad, O., Saunte, C., Salvesen, R., et al. (1989). Short-lasting, unilateral, neuralgiform headache attacks with conjunctival injection, tearing, sweating, and rhinorrhea. Cephalalgia, 9, 147–156.

Sjaastad, O., Kruszwezki, P., Fostad, K., et al. (1992). SUNCT syndrome. VII. Ocular and related variables. Headache, 32, 489–495.

Zabalza, R. J. (2012). Sustained response to botulinum toxin in SUNCT syndrome. *Cephalalgia, 32*, 869–872.

Zhao, J. M., & Sjaastad, O. (1993). SUNCT syndrome: VIII. Pupillary reaction and corneal sensitivity. *Functional Neurology, 8*, 409–414.

Superconducting Quantum Interference Device

▶ Magnetoencephalography in Assessment of Pain in Humans

Superficial Dorsal Horn

Definition

The spinal cord is arranged in such a way that primary afferent fibers originating in the periphery display specific somatotopic organization upon entry into the cord. Classically, the spinal cord can be divided into the white matter and gray matter, and the latter can be further divided into 10 different laminae, each layer being composed of functionally distinct cells. Laminae I and II make up the superficial dorsal horn, and this area of the spinal cord represents the point at which nociceptive peripheral inputs first synapse.

Cross-References

▶ Opioids in the Spinal Cord and Modulation of Ascending Pathways (N. gracilis)
▶ Spinoparabrachial Tract

Superficial Dry Needling

Definition

The technique of dry needling was developed by western pain physicians. The thin needle is applied subcutaneously but does not penetrate the muscle tissue. "Dry" is the opposite of using injection blocks for pain relief, "wet" needles. The analgesic effect elicited probably resembles that of needle acupuncture.

Cross-References

▶ Dry Needling

Superficial Dyspareunia

Definition

Entry dyspareunia.

Cross-References

▶ Gynecological Pain and Sexual Functioning

Supersensitivity

Definition

Increased response to a drug, presumably due to an increase in the number of receptors activated by the drug (upregulation) or by a functional increase in receptor reactivity.

Cross-References

▶ Opioids and Reflexes

Supplemental Security Income

▶ Disability Evaluation in the Social Security Administration

Supplemental Security Income Disability Program

Definition

Means-tested national program administered by the Social Security Administration that provides monthly payments to needy disabled adults and disabled children (individuals under age 18).

Cross-References

▶ Disability Evaluation in the Social Security Administration

Supportive Care

Definition

Care given to prevent, control, or relieve complications or side effects in general health care and to improve the patient's comfort and quality of life. Can be part of palliative care but does not include disease-modifying therapy with palliative intent.

Cross-References

▶ Cancer Pain Management, Interface Between Cancer Pain Management and Palliative Care

Supportive Rehabilitation

▶ Cancer Pain Management, Rehabilitative Therapies

Suppression

▶ Nociceptor, Fatigue

Supraspinal Antinociceptive Mechanisms

▶ Extrasegmental Antinociceptive Mechanisms

Supraspinal Pain Control Systems

▶ Descending Modulation of Nociceptive Processing

Supraspinal Pain Modulation

▶ Descending Facilitation and Inhibition in Neuropathic Pain

Supraspinal Regulation

▶ GABA Mechanisms and Descending Inhibitory Mechanisms
▶ Spinothalamic Tract Neurons, Descending Control by Brain Stem Neurons

Suprathreshold Stimuli

Definition

These are stimuli whose intensity lies above the subject's pain threshold. It is important to use such stimuli, as opposed to threshold or subthreshold stimuli, to understand true nociceptive processing.

Cross-References

▶ Quantitative Sensory Testing

Sural Nerve Biopsy

Definition

The traditional method of observing and quantitating axons involving partial or total surgical removal of the sural nerve above the ankle and electron microscopic analysis. Complications include numbness of the foot, infection, and persistent pain.

Cross-References

▶ Diabetic Neuropathies

Sural Spared Nerve Injury Model

▶ Neuropathic Pain Model, Spared Nerve Injury

Surgery in the DREZ

Definition

Ablative pain surgery in the dorsal root entry zone.

Cross-References

▶ Brachial Plexus Avulsion and Surgery in the Dorsal Root Entry Zone (DREZ)

Surgical Denervation

Definition

To cut a peripheral nerve surgically.

Cross-References

▶ Cancer Pain Management, Anesthesiologic Interventions, and Neural Blockade

Survivin

Definition

Survivin is a member of the inhibitor of apoptosis protein (IAP) family.

Cross-References

▶ NSAIDs and Cancer

Swaddling

Definition

An infant is wrapped in a light cloth to keep limbs close to the trunk and prevent the child from moving around excessively.

Cross-References

▶ Acute Pain Management in Infants

Swedish Massage

▶ Massage, Basic Considerations

Sweet Solutions for Pain Management

Denise Harrison
Nursing, Children's Hospital of Eastern Ontario (CHEO), University of Ottawa, Ottawa, ON, Canada

Synonyms

Sucrose

Introduction

Sucrose for pain relief

> To cure sometimes, to relieve often, to comfort always.
>
> (Quote attributed to three different sources – Hippocrates, 460 BC–c. 370 BC; Ambroise Paré, 1510–1590 and Edward Livingston Trudeau, 1848–1915)

In 1991, Blass and Hoffmeyer published the first randomized controlled trial (RCT) evaluating analgesic effects of sucrose in newborn infants during painful procedures (Blass and Hoffmeyer 1991). This RCT set the stage for the conduct of almost 160 trials and numerous systematic reviews and meta-analyses evaluating analgesic effects of sweet solutions for infants during painful procedures over the following two decades. With few exceptions, findings of these large numbers of studies consistently showed that sucrose and glucose reduced behavioral responses and composite pain scores during commonly performed painful procedures in diverse populations of preterm infants, term newborn infants, as well as infants up to 12 months of age.

This entry will commence with a brief overview of the proposed mechanisms of sweet taste analgesia. The body of this entry will then address the following frequently asked questions:

- What is the most effective dose?
  - What is the minimum safe effective volume?
  - What is the most effective concentration?
  - What is the most effective delivery method – single dose or given in aliquots throughout the procedure?
- Is sucrose effective in the context of opioid analgesics?
- Are repeated doses over days/weeks/months effective and safe?
- What is the maximum age of effect?

Finally, implications for clinical practice and recommendations for further research will be discussed. The recently published review of studies evaluating analgesic effects of sweet solutions by Harrison et al. (2010a) titled "Analgesic effects of sweet tasting solutions in infants: Do we have equipoise yet?", the Cochrane Systematic review of sucrose for analgesia in newborn infants (Stevens et al. 2010), and a systematic review of glucose in newborn infants (Bueno et al., in press) will serve as key references for this entry, supported by seminal studies by Blass and colleagues (Blass and Ciaramitaro 1994; Blass and Shah 1995; Blass and Smith 1992).

Background

Knowledge that sweet taste induces calming in human infants is not new. Historical references pertaining to the analgesic and calming benefits of sweet substances date back to 632 AD, when Prophet Mohammed recommended giving infants a well-chewed date (Islamic Voice 2002). Centuries later, in 1938, it was recommended that anesthesia for infants during surgery was often not required, and that "a sucker consisting of a sponge dipped in some sugar water will often suffice to calm a baby" (p. 2021) (Thorek 1938). Since these early observations, nearly 160 studies have been published, which consistently show that both sucrose and glucose reduce behavioral responses during single episodes of short-lasting painful procedures in healthy term newborn infants and medically stable preterm infants (Harrison et al. 2010a). Currently (as of June 1, 2012), 159 published studies examining analgesic or calming effects of sucrose in infants have been identified, of which 102 (64 %) have focused on sucrose and 52 (33 %) have studied glucose. Although the exact mechanism of effect remains elusive, the mechanism is considered to be an orally induced endogenous opioid increase, which peaks at around 60–120 s and persists for around 1–5 min depending on the age of the infant (Blass and Ciaramitaro 1994). The effects when used prior to painful procedures are primarily in the domain of behavioral responses (cry, facial expressions, and composite pain scores), with less consistent effects on physiological responses and, although only measured in two studies to date, no effect on cortical indicators (Norman et al. 2008; Slater et al. 2010). Inconsistency between behavioral, physiological, and cortical responses highlights the loose correlation between physiological and behavioral responses to pain (Barr 1998).

Characteristics

What Is the Optimal Dose?

Stevens et al. 2010 and Bueno et al. (in press) concluded that an optimal dose of sucrose or glucose could not be determined. However in terms of the sucrose or glucose concentration, as long as the solutions used are sufficiently sweet, results of RCTs consistently show a reduction in behavioral responses during short-lasting painful procedures. Although sucrose is the sweetest of the four sugars (sucrose > fructose > glucose > lactose) (Blass and Smith 1992), studies using glucose in concentrations from 20 % and stronger have demonstrated analgesic effects during painful procedures. Effects are less consistent with lower concentrations of solutions (10–15 % sucrose or glucose). There are 22 published trials directly comparing the effectiveness of varying concentrations of sweet solutions during painful procedures. Results overall show that the most concentrated solutions are more effective compared to the less sweet solutions. For example, 30 % sucrose was superior to both 30 % glucose and 10 % glucose in reducing crying during heel lance (Isik et al. 2000); 25 % glucose and 24 % sucrose more effectively reduced cry duration during venipuncture than 10 % glucose and 12 % sucrose (Abad et al. 1996; Deshmukh and Udani 2002); and in a stepwise fashion, glucose solutions of increasing concentrations from 10–20 % to 30 % were progressively more effective in reducing pain during venipuncture in healthy preterm and term infants (Dilen and Elseviers 2010). Although there is wide variability of sweet solutions used, the concentrations most frequently studied range from 24 % or 25 % sucrose and 25–30 % glucose.

What Volume Is Effective?

The minimum effective volume of sweet solution and whether this differs by gestational age,

Sweet Solutions for Pain Management, Table 1 Volumes of sweet solutions used in the 159 studies to date

Volumes (mL)	N	(%)
mL/kg doses	6	(4)
Pacifier dip or 0.05 or less	9	(6)
0.1 to <1	29	(18)
1 to <2	32	(20)
2 to <3	67	(42)
5–10	3	(2)
Not specified	13	(8)

postnatal age, or weight has not been established. When all trials of sweet solutions are evaluated, small volumes including doses of 0.05 mL up to 0.2 mL reduce behavioral responses during painful procedures compared to water or no treatment. However, in the only study which directly compared a smaller to larger volume of 30 % glucose in term infants during venipuncture, 0.4 mL was less effective than the larger 2 mL volume. Further research is warranted to establish the minimal effective volume in different populations of hospitalized infants who undergo multiple procedures and may be given multiple doses of sucrose or glucose over prolonged periods. Currently, over two thirds of the published studies (n = 102) have used volumes of more than 1 mL (Table 1). Repeated doses of such volumes could potentially lead to excess amounts of sweet solutions over the course of a hospitalization.

What Is the Most Effective Delivery Method: Single Dose Versus Aliquots Throughout the Procedure?

The most commonly used delivery regimen is a single dose of sucrose or glucose, given over a period of 1 min, 2 min prior to commencement of a procedure, regardless of the age of the infant or duration of the procedure. This regimen is based on one study demonstrating that this 2-min time period most effectively reduced crying duration in term newborn infants undergoing heel lancing (Blass and Shah 1995), although a shorter peak effect time, of around 1 min, in infants beyond the newborn period was reported (Barr et al. 1994). Therefore the question whether a single dose of sucrose or glucose given before a painful procedure is as efficacious as doses given in aliquots throughout the procedure, especially for infants beyond the newborn period and for procedures of prolonged duration needs to be addressed. Only one study has directly compared the effectiveness of single versus repeated doses. Johnston et al. (1999) showed that administrating 24 % sucrose in 0.05 mL aliquots to preterm infants resulted in lower pain scores than a single dose given 2 min prior to a heel lance. Although further research is warranted, administering sweet solutions in aliquots throughout the duration of painful and distressing procedures, especially in infants beyond the newborn period, is recommended to ensure sustained analgesic effects (Harrison et al. 2010).

Is Sucrose Effective for Infants on Opioids?

As the analgesic mechanism of sweet solutions is proposed to be a sweet taste–induced endogenous opioid effect, whether this effect is able to occur in the presence of exogenous opioids is an important question and one which has received little research attention. Marceau et al. (2010) reported that sucrose delivered in small aliquots throughout a heel lance procedure was equally effective in methadone-exposed infants, including eight infants receiving morphine for withdrawal symptoms, as control infants with no antenatal opioid exposure. Yet a study in which sucrose was standard of care, infants receiving opioid analgesics for management of postoperative pain or ventilator-related care had a trend toward higher pain scores during heel lancing compared to infants not receiving opioid analgesics (Harrison et al. 2011a). Conflicting findings from these two studies, both of which acknowledged significant methodological limitations, preclude a definitive answer at this time to the question of effectiveness of sweet solutions during painful procedures for infants receiving opioid analgesics.

Are Repeated Doses Over Days/Weeks/Months Effective and Safe?

Investigations of repeated use of sucrose or glucose over days, weeks, or months of use are limited. Only three studies to date have examined

prolonged use over periods longer than 1 week and all showed that sucrose remained effective (Harrison et al. 2009; Mucignat et al. 2004; Stevens et al. 2005). Two studies reported safety outcomes (Harrison et al. 2009; Stevens et al. 2005) (Harrison; incidence of NEC and Stevens; neurobiological risk scores) and both showed no increased risk over time with repeated doses, or, in Stevens et al. case; no differences in neurobiological risk scores between infants receiving sucrose compared to infants receiving water. However over a 7-day period, more than ten doses of sucrose per day was associated with poorer neurodevelopmental outcomes compared to ten or less doses in preterm infants during the first week of life (Johnston et al. 2007; Johnston et al. 2002). To date, Stevens and Johnston's trials are the only two studies examining neurodevelopmental outcomes following repeated use of sweet solutions for procedural pain management.

Although it has not been reported in studies to date, anecdotally, sweet solutions are used to assist to calm infants in the NICU. This use of sweet solutions for soothing is not recommended due to the short-lasting nature of the endogenous opioid calming effect, potentially leading to multiple doses over the course of a hospitalization for infants who are prone to be unsettled. Further research attention is warranted to establish if this use of sucrose for soothing is indeed occurring, and to what extent.

What Is the Maximum Age of Effect?

Although the majority of studies to date have included newborn infants within the first days, to weeks of life, a systematic review reported that sweet solutions remained effective in infants up to 1 year of life during immunizations (Harrison et al. 2010b). The effects were however reported to be less profound than seen in the newborn period. It is not known if sweet solutions continue to have analgesic effects in young children beyond 1 year of age. Results of only two studies including children in this age group are conflicting (Harrison et al. 2011b) and there is no evidence to support analgesic effects in school-aged children.

Clinical Recommendations

Although there remain a number of uncertainties relating to sweet taste analgesia in infants, the following clinical recommendations are based on available evidence and interpretation of the evidence.

- Use solutions of sufficiently sweet concentrations (at least 24 % sucrose or 25–30 % glucose) in small doses on the top of the tongue, with a pacifier if the infant is able to suck.
- Use small volumes – sufficient for observable effect throughout the procedure being performed. The volume required dose may depend on the gestational and postnatal age of the infant.
- For more painful or prolonged procedures, such as eye examination, use additional strategies as recommended.
- Sucrose or glucose should be considered as adjuvant pain management strategies for more major procedures such circumcision and used in conjunction with topical and regional anesthetics
- Document the administration and effectiveness of solutions
- Use additional strategies as feasible such as nonnutritive sucking and skin-to-skin care. Support parents to breast-feed or skin-to-skin care whenever possible and feasible.

Recommendations for Further Research

As there is abundant evidence of analgesic effects of sucrose and glucose during single episodes of short-lasting painful procedures in medically stable and healthy preterm and term infants up to 1 year of age, further placebo-controlled trials in this population are not ethical. Further research needs to address remaining knowledge gaps.

- Minimum safe effective volume
- Most effective delivery method – single dose compared to doses given in aliquots throughout the procedure
- Effectiveness in the context of opioid analgesics
- Effectiveness in combination with other pain management strategies such as skin-to-skin contact

- Safety and effectiveness of repeated doses over prolonged periods
- Effectiveness in young children

In addition, a key unanswered question is whether consistent use of evidence–based pain management strategies, including sucrose and glucose can ameliorate the adverse effects of pain exposure in preterm and term hospitalized infants.

Summary

In conclusion, abundant evidence consistently demonstrates analgesic effects of sucrose and glucose in infants in the newborn period during single episodes of commonly performed short-lasting painful procedures. In addition, both sugars continue to reduce pain compared to water or no treatment up to 1 year of age, although there is minimal evidence of effectiveness beyond this age. Clinicians are advised to use sweet solutions sparingly as recommended and ensure that their use is well documented to allow systematic studies of use, effectiveness, and outcomes. Further research needs to address remaining knowledge gaps. The ethical care of infants encompasses advocating for the most effective and safe pain management for hospitalized infants. This includes the judicious use of sweet solutions, in conjunction with minimizing painful procedures, and supporting parents to support their infants during painful procedures.

References

Abad, F., Diaz, N. M., Domenech, E., Robayna, M., & Rico, J. (1996). Oral sweet solution reduces pain-related behaviour in preterm infants. *Acta Paediatrica, 85*(7), 854–858.

Barr, R. G. (1998). Reflections on measuring pain in infants: Dissociation in responsive systems and "honest signalling". *Archives of Disease in Childhood. Fetal and Neonatal Edition, 79*(2), F152–F156.

Barr, R. G., Quek, V. S. H., Cousineau, D., Oberlander, T. F., Brian, J. A., & Young, S. N. (1994). Effects of intra-oral sucrose on crying, mouthing, and hand-mouth contact in newborn and six-week old infants. *Developmental Medicine and Child Neurology, 36*(7), 608–618.

Blass, E., & Ciaramitaro, V. (1994). A new look at some old mechanisms in human newborns. *Monographs of the Society for Research in Child Development, 59*(1).

Blass, E. M., & Hoffmeyer, L. B. (1991). Sucrose as an analgesic for newborn infants. *Pediatrics, 87*(2), 215–218.

Blass, E. M., & Shah, A. (1995). Pain-reducing properties of sucrose in human newborns. *Chemical Senses, 20*(1), 29–35.

Blass, E. M., & Smith, B. A. (1992). Differential effects of sucrose, fructose, glucose, and lactose on crying in 1- to 3-day-old human infants: Qualitative and quantitative considerations. *Developmental Psychology, 28*(5), 804–810.

Bueno, M., Yamada, J., Harrison, D., Kahn, S., Adams-Webber, T., Beyene, J., et al. (in press). Pain research and management. A systematic review and meta-analyses of non-sucrose sweet solutions sucrose for pain relief in neonates.

Deshmukh, L. S., & Udani, R. H. (2002). Analgesic effect of oral glucose in preterm infants during venipuncture–a double-blind, randomized, controlled trial. *Journal of Tropical Pediatrics, 48*(3), 138–141.

Dilen, B., & Elseviers, M. (2010). Oral glucose solution as pain relief in newborns: Results of a clinical trial. *Birth, 37*(2), 98–105.

Harrison, D., Loughnan, P., Manias, E., Gordon, I., & Johnston, L. (2009). Repeated doses of sucrose in infants continue to reduce procedural pain during prolonged hospitalizations. *Nursing Research, 58*(6), 427–434.

Harrison, D., Bueno, M., Yamada, J., Adams-Webber, T., & Stevens, B. (2010a). Analgesic effects of sweet tasting solutions in infants: Do we have equipoise yet? *Pediatrics, 126*(5), 894–902.

Harrison, D., Stevens, B., Bueno, M., Yamada, J., Adams-Webber, T., Beyene, J., et al. (2010b). Efficacy of sweet solutions for analgesia in infants between 1 and 12 months of age: A systematic review. *Archives of Disease in Childhood, 95*(6), 406–413.

Harrison, D., Loughnan, P., Manias, E., Smith, K., & Johnston, L. (2011a). Effect of concomitant opioid analgesics and oral sucrose during heel lancing. *Early Human Development, 87*, 147–149.

Harrison, D., Stevens, B., Yamada, J., Adams-Webber, T., Beyene, J., & Ohlsson, A. (2011b). Sweet tasting solutions for reduction of needle-related procedural pain in infants and children aged 1 to 16 years. *Cochrane Database of Systematic Reviews* (10), CD008408.

Isik, U., Ozek, E., Bilgen, H., & Cebeci, D. (2000). Comparison of oral glucose and sucrose solutions on pain response in neonates. *The Journal of Pain, 1*(4), 275–278.

Islamic Voice. (2002). Relief of pain: A medical discovery. Retrieved February 10, 2011, 2004, from http://www.islamicvoice.com/April.2001/quran.htm

Johnston, C. C., Stremler, R., Horton, L., & Friedman, A. (1999). Effect of repeated doses of sucrose during heel stick procedure in preterm neonates. *Biology of the Neonate, 75*(3), 160–166.

Johnston, C. C., Filion, F., Snider, L., Majnemer, A., Limperopoulos, C., Walker, C. D., et al. (2002). Routine sucrose analgesia during the first week of life in neonates younger than 31 weeks' postconceptional age. *Pediatrics, 110*(3), 523–528.

Johnston, C. C., Filion, F., Snider, L., Limperopoulos, C., Majnemer, A., Pelausa, E., et al. (2007). How much sucrose is too much sucrose? *Pediatrics, 119*(1).

Marceau, J. R., Murray, H., & Nanan, R. K. H. (2010). Efficacy of oral sucrose in infants of methadone-maintained mothers. *Neonatology, 97*(1), 67–70.

Mucignat, V., Ducrocq, S., Lebas, F., Mochel, F., Baudon, J. J., & Gold, F. (2004). Analgesic effects of EMLA cream and saccharose solution for subcutaneous injections in preterm newborns: A prospective study of 265 injections. *Archives de Pediatrie, 11*(8), 921–925.

Norman, E., Rosen, I., Vanhatalo, S., Stjernqvist, K., Okland, O., Fellman, V., et al. (2008). Electroencephalographic response to procedural pain in healthy term newborn infants. *Pediatric Research, 64*(4), 429–434.

Slater, R., Cornelissen, L., Fabrizi, L., Patten, D., Yoxen, J., Worley, A., et al. (2010). Oral sucrose as an analgesic drug for procedural pain in newborn infants: A randomised controlled trial. *Lancet, 376*(9748), 1225–1232.

Stevens, B., Yamada, J., Beyene, J., Gibbins, S., Petryshen, P., Stinson, J., et al. (2005). Consistent management of repeated procedural pain with sucrose in preterm neonates: Is it effective and safe for repeated use over time? *The Clinical Journal of Pain, 21*(6), 543–548.

Stevens, B., Yamada, J., & Ohlsson, A. (2010). Sucrose for analgesia in newborn infants undergoing painful procedures. *Cochrane Database of Systematic Reviews* (1), Art. No.: CD001069. doi:10.1002/14651858.CD001069.pub3.

Thorek, M. (1938). *Modern surgical technique* (Vol. 111). Montreal: Lippincott.

Switching

Definition

Pharmacological technique used to improve the balance between analgesia and adverse effects in clinical conditions of poor opioid response. The prior opioid is discontinued and an alternative one is administered at doses substantially lower than those provided by equianalgesic tables.

The rationale is based on the asymmetric tolerance and differences in efficacies among opioids. Equivalence tables are only indicative, as patients who are highly tolerant and receiving high doses of opioids should be carefully monitored, particularly when switching from morphine to methadone, which has a higher potency than expected in patients taking high doses of morphine.

Cross-References

▶ Opioid Responsiveness in Cancer Pain Management

Sylvian Fissure

Synonyms

Lateral sulcus

Definition

Also called lateral sulcus, one of the largest fissures of the brain, the Sylvian fissure separates the temporal lobe from the frontal and parietal lobes. The insula is located deep inside the Sylvian fissure, which is separated from the frontal, parietal, and temporal operculum by the circular sulcus of the insula.

Cross-References

▶ Insular Cortex, Neurophysiology, and Functional Imaging of Nociceptive Processing
▶ Nociceptive Processing in the Secondary Somatosensory Cortex
▶ SII

Symmetric Diabetic Neuropathy

▶ Diabetic Neuropathy, Treatment

Sympathectomy

Definition

Surgical resection or other (e.g., chemical) removal of some portion of the peripheral sympathetic nervous system, e.g., for treatment of sympathetically maintained pain (SMP) or hyperhidrosis (excessive sweating).

Cross-References

▶ Complex Regional Pain Syndrome and the Sympathetic Nervous System
▶ Neurosurgical Treatment of Pain
▶ Sympathetically Maintained Pain in CRPS II, Human Experimentation

Sympathetic Afferents

▶ Angina Pectoris, Neurophysiology and Psychophysics

Sympathetic and Sensory Neurons After Nerve Lesions, Structural Basis for Interactions

Elspeth M. McLachlan
Neuroscience Research Australia, University of New South Wales, Randwick, NSW, Australia
University of Glasgow, Glasgow, UK

Synonyms

Axonal sprouting; Perineuronal terminals; Sympathetic-afferent interaction; Vascular hypertrophy

Definition

After nerve injury, some people develop spontaneous pain and allodynia that is alleviated by sympathetic blockade. Injury-induced influx of macrophages and lymphocytes from the blood into dorsal root ganglia lead to collateral sprouting of sympathetic and afferent axons nearby. These changes are associated with the local production of neurotrophins and cytokines that also raise excitability of sensory neurons. Although some sensory somata express alpha2-adrenoceptors after lesions, there is no clear evidence that noradrenaline released by sympathetic terminals can access these receptors. Indeed hyperexcitability appears before the ingrowing sprouts appear. Structural evidence to support sympathetic-sensory interaction within a neuroma is not strong but, after incomplete lesions, there may be a few sites in the skin where noradrenaline released from sympathetic nerve terminals can access alpha-adrenoceptors expressed *de novo* on nociceptors, generating painful sensations. However, when chronic pain develops, activity in sympathetic neurons may contribute indirectly by generating vasoconstriction, leading to excitation of hyperexcitable afferents by ischemia.

Characteristics

In normal individuals, neurotransmitters such as noradrenaline (NA) released by sympathetic nerve activity have no direct effect on nociceptive terminals in skin although these may be sensitive to changes in the environment mediated by sympathetic effector responses. For example, experimentally induced intense vasoconstriction may lead to ischemic pain after metabolites have accumulated near nociceptive endings. A few sensory receptors are specifically supplied by sympathetic terminals, for example, the neck of hair follicles (Gibbins 1997).

Blockade of alpha-adrenoceptors can alleviate neuropathic pain in a proportion of patients with peripheral nerve injuries, indicating that pain and sympathetic activity are not always independent (see ▶ Sympathetically Maintained Pain, Clinical Pharmacological Tests; ▶ Sympathetically Maintained Pain in CRPS I, Human Experimentation; ▶ Sympathetically Maintained Pain in

CRPS II, Human Experimentation). This has led to the idea that a link can develop after nerve injury, such that NA released during sympathetic activity excites nociceptive neurons or their peripheral terminals. Because sites in normal individuals where sympathetic nerve terminals lie close to nociceptive ones are lacking, the mechanism must involve some form of anatomical or functional plasticity. What is the evidence that afferent neurons are directly or indirectly excited by NA released from sympathetic nerve terminals after injury to peripheral nerves?

What Is the Physical Relationship Between Sympathetic and Sensory Neurons in the Somatic Domain?

Sympathetic postganglionic neurons in the paravertebral chain send their axons in a virtually exclusive bundle (the gray ramus) to join the spinal nerve immediately distal to the dorsal root ganglion (DRG). The axons then run in bundles encased in Schwann cells along the peripheral nerve trunks to supply vessels in skin, muscle and joints, and piloerector muscles and sweat glands in some regions of skin. In several species, including cats and rats, the only sympathetic axons that normally turn toward the DRG are those that innervate small blood vessels in the perineurium or around the nerve roots. Sympathetic terminals are normally not present between or around the somata within the DRG itself.

The small diameter neurons that make up the nociceptive population tend to lie deep in the DRG. They send their unmyelinated axons to the periphery in the same nerve bundles as the sympathetic axons. Within muscle nerves, ~50 % of the unmyelinated axons are sympathetic and ~50 % afferent. Within skin nerves, about 20 % of the unmyelinated axons are sympathetic and 80 % afferent (Baron et al. 1988). Sympathetic axons form varicose terminal plexuses around deep or subdermal arterial vessels, more than 1 mm below the epidermis in humans (Gibbins 1997). In some regions of skin, some peptidergic afferent terminals run along the blood vessel, beside the sympathetic perivascular plexus. However, the majority of afferent axons penetrate the dermis in small bundles and form a plexus below the epidermal-dermal junction from which bare individual axons run at right angles up into the epidermis itself where sympathetic terminals are not found. Deep nociception and pain is subserved by C and Aδ axons with endings in muscle and joints but again close associations with sympathetic terminals are unusual.

Despite the lack of sensation after introducing NA into normal skin (see ▶ Sympathetically Maintained Pain in CRPS II, Human Experimentation; ▶ Sympathetically Maintained Pain, Clinical Pharmacological Tests), use of an alpha1-adrenoceptor antibody appears to have identified the receptor in DRG neurons and some TRPV1+ terminals near the epidermal-dermal junction in normal rat footpad skin (Dawson et al. 2011).

Neuronal Plasticity Distal to a Nerve Injury

When a nerve is cut, the distal axons degenerate. Their Schwann cells become activated and phagocytose myelin; after a few days, hematogenous macrophages enter to complete the clearing of debris (Wallerian degeneration). When a nerve is crushed or partially transected, intact axons survive among this debris and activity in these axons has been attributed to cytokines released by invading macrophages and lymphocytes.

In the target territory in skin, invading immune cells produce neurotrophins that bind to p75 receptors in the dermis, where they attract sprouting of nearby intact axons containing calcitonin gene-related peptide (CGRP) and other peptides (Peleshok and Ribeiro-da-Silva 2012). In the nerve itself, the unmyelinated axons degenerate progressively and only a proportion of them regenerate to reinnervate the skin much later (Wakisaka et al. 1991; McLachlan et al. 1994). The degree to which sympathetic and sensory axons disappear from the skin clearly depends on the extent of damage produced by the lesion. For example, a chronic constriction injury (CCI) may cause substantial loss of sympathetic and

peptidergic axons (Wakisaka et al. 1991; Hu et al. 2006) or may not be detectable, partly because CGRP is expressed in intact larger axons (Zochodne 2012). Occasional sympathetic axons have been identified in nerve bundles within the dermis and these gradually approached the epidermis over several weeks after CCI (Yen et al. 2006). If the injury is far proximal, reinnervation of the original targets is sparse (McLachlan et al. 1994; Tripovic et al. 2011), partly because many of the neurons of origin die (Hu and McLachlan 2003b).

Stimulation of intact sympathetic axons activates nociceptor endings in partially denervated skin, apparently via alpha 2-adrenoceptors (Sato and Perl 1991). Injection of NA into skin of patients rekindles neuropathic pain (Törebjörk et al. 1995; see ▶ Sympathetic-Afferent Coupling in the Afferent Nerve Fiber, Neurophysiological Experiments). Thus, the apparently tenuous anatomical link may be functional in some cases or over time.

Retrograde Changes Proximal to the Injury

Soon after a nerve lesion, a retrograde signal leads to activation of satellite glia around DRG neurons and of microglia and astrocytes in the spinal cord and more centrally. Central glial activation could be the crucial factor in generating activity in nociceptive pathways, but its effects are likely to be triggered or amplified by additional signals from the periphery.

Proximal to the injury, unmyelinated axons retract back along the nerve trunk if they cannot regenerate through the scar. After transection with ligation, very few axons containing NA or peptides are present within the neuroma, although many are associated with the surrounding scar tissue. Thus, it is unlikely that sympathetic-afferent coupling occurs within the neuroma (see ▶ Sympathetic-Afferent Coupling in the Afferent Nerve Fiber, Neurophysiological Experiments).

Within sympathetic ganglia and DRGs, the lesioned cell bodies show typical responses to axotomy, while satellite glia are activated and proliferate around larger neuronal somata.

Within a few days, resident macrophages become activated and lymphocytes and other types of macrophage invade the DRG (Hu and McLachlan 2003a). The neuronal somata downregulate transmitter synthesis and start to express proteins associated with survival and regeneration. However, if regeneration is impeded, ~50 % of the small sensory neurons in the lesioned DRG, and a similar number of sympathetic neurons, die over the ensuing weeks. This neuronal death is not all dependent on axotomy (Hu and McLachlan 2003b).

Over the next few weeks, collateral sprouts from perivascular terminals outside the DRG (McLachlan et al. 1993), from nearby intact sympathetic axons (Xie et al. 2010) and probably from axotomized sympathetic axons themselves (Ramer et al. 1999) invade the DRG, probably triggered by neurotrophins, nerve growth factor (NGF), and neurotrophin-3, released by invading immune cells and concentrated locally by p75 receptors on activated glia (Walsh et al. 1999; Zhou et al. 1999). Inflammation of DRGs by several other means also elicits sympathetic sprouting which, like the associated hyperalgesias, is lessened by anti-inflammatory treatment; this is consistent with activated immune cells being the source of neurotrophins. The signal for changes in the DRG is suggested to be activity in the damaged axons as local anesthetic or tetrodotoxin application to the injured nerve prevents sympathetic sprouting (Xie et al. 2007). It is interesting that most of these changes also occur in trigeminal ganglia after a nerve lesion (Bongenhielm et al. 1999), but these are not invaded by sympathetic sprouts, except the trigeminal ganglia of mice that overexpress NGF in skin (Davis et al. 1994); this suggests that retrogradely transported neurotrophin is also effective in triggering sprouts.

Perineuronal rings or baskets form around large or medium diameter cell bodies of cutaneous afferent neurons, many of which are axotomized by the lesion. The proportion of neurons bearing rings is low (<5 % of DRG neurons after complete sciatic nerve transection) and is maximal many weeks after damage to a major nerve. Not only sympathetic but also trkA-positive peptidergic afferent axons form perineuronal rings within the DRG, to a similar extent (McLachlan and Hu 1998). The peptidergic terminals arise from small diameter DRG somata and contain, notably, CGRP and/or substance P (i.e., nociceptive afferents or larger afferents after injury) and/or galanin and NPY (i.e., either sympathetic or various afferents after injury). Terminals of more than one type of axon sometimes mingle around the same cell or they may form independent rings around adjacent cells. However, most of the varicosities lie among the perineuronal glia (Shinder et al. 1999) and do not make synaptic contacts with neuronal somata.

While these data provide a coherent story for the structural changes in DRGs projecting into damaged nerve trunks, the evidence for excitation of afferent somata by local NA is not particularly convincing. Alpha 2A-adrenoceptors are expressed in medium to large DRG somata after sciatic injury (Birder and Perl 1999) but the upregulated receptors disappear over the period when perineuronal rings develop. Whether alpha1-adrenoceptors, expressed by some normal DRG neurons (Dawson et al. 2011), become translocated to their surface after axon injury remains to be tested.

Direct application of NA has generally been unsuccessful in exciting larger diameter neurons, except when very high concentrations are applied; the latter can depolarize somata in both control and lesioned DRGs via a non-adrenoceptor-mediated mechanism (Lopez de Armentia et al. 2003). Small DRG somata can be excited via alpha2-adrenoceptors (Abdulla and Smith 1997), but these neurons are not associated with sympathetic terminals.

Stimulation of sympathetic sprouts in an isolated in vitro DRG after spinal nerve ligation (Xie et al. 2010) raised the excitability of large diameter neurons. This effect was slow in onset and persisted for many minutes after stimulation ceased; it could be reduced, or spontaneous activity eliminated, by a mixture of antagonists of NA and ATP receptors (not by phentolamine alone), by pretreatment with bretylium, or by cutting the gray ramus prior to ligation of the spinal nerve.

This last intervention also substantially reduced nociceptive behavior, implying that sympathetic sprouts within the DRG are more important than any interactions in the periphery for hyperalgesic responses.

A recent study utilizing mice engineered to express green fluorescent protein in sympathetic postganglionic neurons has reported that DRG somata associated with sympathetic terminals are more often spontaneously active than silent (Xie et al. 2011). This tour de force needs confirmation including an explanation of how activity that is dependent on sympathetic activity can be maintained in vitro. At this stage, whether NA released from ectopic sprouts can activate neuronal alpha-adrenoceptors within a lesioned DRG remains an open question.

Vascular Changes After Nerve Injury

Activation of sympathetic pathways that project to lesioned DRGs leads to increased discharge of spontaneously active muscle afferent neurons with myelinated axons (Häbler et al 2000; Michaelis et al. 2000). However, this results only when high stimulation frequencies produce extreme vasoconstriction within the DRG with markedly reduced blood flow (see ▶ Sympathetic-Afferent Coupling in the Dorsal Root Ganglion, Neurophysiological Experiments). Perhaps the enhanced metabolic activity of macrophages and glia in the lesioned DRG provides an environment where vasoconstriction further compromises the supply of nutrients and oxygen, causing excitation of the hyperexcitable DRG somata.

A structural basis for this observation may be the hypertrophy of some arterial vessels in the lesioned nerve trunks. In the rat sciatic nerve, endoneurial vessels are normally uninnervated capillaries. However, proximal to a ligation and transection, enlarged endoneurial vessels appear over several weeks after the lesion (Zochodne and Nguyen 1997). These vessels develop muscle layers and become densely innervated by varicose sympathetic terminals. Similar enlarged vessels are found around and occasionally within the DRG. As constriction of these arterial vessels during sympathetic activity is likely to restrict endoneurial and ganglionic blood flow, it may lead to ischemic excitation of hyperexcitable afferents (Häbler et al 2000; ▶ Sympathetic-Afferent Coupling in the Dorsal Root Ganglion, Neurophysiological Experiments). Such a scenario could explain why neuropathic pain is sensitive to alpha-adrenoceptor blockade and would be consistent with observations in patients that sympathetic block by antagonists that are not very effective on alpha2-adrenoceptors can reduce sympathetically mediated pain.

Arteries located far from the site of a nerve injury remain denervated for a long period before becoming reinnervated. The regenerated perivascular plexus is much sparser than normal even after many months (Tripovic et al. 2011). However, the peripheral vessels respond to this sparse innervation with exaggerated and prolonged vasoconstriction (Koltzenburg et al. 1995; Tripovic et al. 2011). Such a phenomenon may contribute to raising neuronal excitabiliity by generating ischemia in other conditions.

References

Abdulla, F. A., & Smith, P. A. (1997). Ectopic alpha2-adrenoceptors couple to N-type Ca^{2+} channels in axotomized rat sensory neurons. *Journal of Neuroscience, 17,* 1633–1641.

Baron, R., Jänig, W., & Kollmann, W. (1988). Sympathetic and afferent somata projecting in hindlimb nerves and the anatomical organization of the lumbar sympathetic nervous system of the rat. *The Journal of Comparative Neurology, 272,* 460–468.

Birder, L. A., & Perl, E. R. (1999). Expression of alpha2-adrenergic receptors in rat primary afferent neurones after peripheral nerve injury or inflammation. *The Journal of Physiology, 515,* 533–542.

Bongenhielm, U., Biossonade, F. M., Westermark, A., Robinson, P. P., & Fried, K. (1999). Sympathetic nerve sprouting fails to occur in the trigeminal ganglion after peripheral nerve injury in the rat. *Pain, 82,* 283–288.

Davis, B. M., Albers, K. M., Seroogy, K. B., & Katz, D. M. (1994). Overexpression of nerve growth factor in transgenic mice induces novel sympathetic projections to primary sensory neurons. *The Journal of Comparative Neurology, 349,* 464–474.

Dawson, L. F., Phillips, J. K., Finch, P. M., Inglis, J. J., & Drummond, P. D. (2011). Expression of a_1-adrenoceptors on peripheral nociceptive neurons. *Neuroscience, 175,* 300–314.

Gibbins, I. L. (1997). Autonomic pathways to cutaneous effectors. In J. L. Morris & I. L. Gibbins (Eds.), *Autonomic innervation of the skin* (pp. 1–56). Amsterdam: Harwood Academic Publishers.

Häbler, H. J., Eschenfelder, S., Liu, X. G., & Jänig, W. (2000). Sympathetic-afferent coupling after L5 spinal nerve lesion in the rat and its relation to changes in dorsal root ganglion blood flow. *Pain, 87*, 335–345.

Hu, P., & McLachlan, E. M. (2003a). Distinct functional types of macrophage in dorsal root ganglia and spinal nerve proximal to sciatic and spinal nerve transections in the rat. *Experimental Neurology, 184*, 590–605.

Hu, P., & McLachlan, E. M. (2003b). Selective reactions of cutaneous and muscle afferent neurons to peripheral nerve transection in rats. *Journal of Neuroscience, 23*, 10559–10567.

Hu, P., Bembrick, A. L., Keay, K. A., & McLachlan, E. M. (2006). Immune cell involvement in dorsal root ganglia and spinal cord after chronic contriction or transection of the rat sciatic nerve. *Brain, Behavior, and Immunity, 21*, 599–616.

Koltzenburg, M., Häbler, H. J., & Jänig, W. (1995). Functional reinnervation of the vasculature of the adult cat paw pad by axons originally innervating vessels in hairy skin. *Neuroscience, 67*, 245–252.

Lopez de Armentia, M., Leeson, A. H., Stebbing, M. J., Urban, L., & McLachlan, E. M. (2003). Responses to sympathomimetics in rat sensory neurones after nerve transection. *NeuroReport, 14*, 9–13.

McLachlan, E. M., & Hu, P. (1998). Axonal sprouts containing calcitonin gene-related peptide and substance P form pericellular baskets around large diameter neurones after sciatic nerve transection in the rat. *Neuroscience, 84*, 961–965.

McLachlan, E. M., Jänig, W., Devor, M., & Michaelis, M. (1993). Peripheral nerve injury triggers noradrenergic sprouting within dorsal root ganglia. *Nature, 363*, 543–546.

McLachlan, E. M., Keast, J. R., & Bauer, M. (1994). SP- and CGRP- immunoreactive axons differ in their ability to reinnervate the skin of the rat tail. *Neuroscience Letters, 176*, 147–151.

Michaelis, M., Liu, X., & Jänig, W. (2000). Axotomized and intact muscle afferents but no skin afferents develop ongoing discharges of dorsal root origin after peripheral nerve lesion. *Journal of Neuroscience, 20*, 2742–2748.

Peleshok, J. C., & Ribeiro-da-Silva, A. (2012). Neurotrophic factor changes in the rat thick skin following chronic constriction injury of the sciatic nerve. *Molecular Pain, 8*, 1.

Ramer, M. S., Thompson, S. W. N., & McMahon, S. B. (1999). Causes and consequences of sympathetic basket formation in dorsal root ganglia. *Pain, 6*, S111–S120.

Sato, J., & Perl, E. R. (1991). Adrenergic excitation of cutaneous pain receptors induced by peripheral nerve injury. *Science, 251*, 1608–1610.

Shinder, V., Govrin-Lippmann, R., Cohen, S., Belenky, M., Ilin, P., Fried, K., Wilkinson, H. A., & Devor, M. (1999). Structural basis of sympathetic-sensory coupling in rat and human dorsal root ganglia following peripheral nerve injury. *Journal of Neurocytology, 28*, 743–761.

Törebjörk, E., Wahren, L., Wallin, G., Hallin, R., & Koltzenburg, M. (1995). Noradrenaline-evoked pain in neuralgia. *Pain, 63*, 11–20.

Tripovic, D., Pianova, S., McLachlan, E. M., & Brock, J. A. (2011). Slow and incomplete sympathetic reinnervation of rat tail artery restores the amplitude of nerve-evoked contractions provided a perivascular plexus is present. *American Journal of Physiology. Heart and Circulatory Physiology, 300*, H541–H554.

Wakisaka, S., Kajander, K. C., & Bennett, G. J. (1991). Abnormal skin temperature and abnormal sympathetic vasomotor innervation in an experimental painful peripheral neuropathy. *Pain, 46*, 299–313.

Walsh, G. S., Krol, K. M., & Kawaja, M. D. (1999). Absence of the p75 neurotrophin receptor alters the pattern of sympathosensory sprouting in the trigeminal ganglia of mice overexpressing nerve growth factor. *Journal of Neuroscience, 19*, 258–273.

Xie, W., Strong, J. A., Li, H., & Zhang, J. M. (2007). Sympathetic sprouting near sensory neurons after nerve injury occurs preferentially on spontaneously active cells and is reduced by early nerve block. *Journal of Neurophysiology, 97*, 492–502.

Xie, W., Strong, J. A., & Zhang, J. M. (2010). Increased excitability and spontaneous activity of rat sensory neurons following in vitro stimulation of sympathetic fiber sprouts in the isolated dorsal root ganglion. *Pain, 151*, 447–459.

Xie, W., Strong, J. A., & Zhang, J. M. (2011). Highly localized interactions between sensory neurons and sprouting sympathetic fibers observed in a transgenic tyrosine hydroxylase reporter mouse. *Molecular Pain, 7*, 53.

Yen, L. D., Bennett, G. J., & Ribeiro-da-Silva, A. (2006). *Sympathetic sprouting and changes in nociceptive sensory innervation in the glabrous skin of the rat hind paw following partial peripheral nerve injury., 495*, 679–690.

Zhou, X. F., Deng, Y.-S., Chie, E., Xue, Q., Zhong, J.-H., McLachlan, E. M., Rush, R. A., & Xian, C. (1999). Satellite cell-derived nerve growth factor and neurotrophin-3 are involved in noradrenergic sprouting in the dorsal root ganglia following peripheral nerve injury in the rat. *European Journal of Neuroscience, 11*, 1711–1722.

Zochodne, D. W. (2012). The challenges and beauty of peripheral nerve regrowth. *Journal of the Peripheral Nervous System, 17*, 1–18.

Zochodne, D. W., & Nguyen, C. (1997). Angiogenesis at the site of neuroma formation in transected peripheral nerve. *Journal of Anatomy, 191*, 23–30.

Sympathetic Arousal

Definition

Sympathetic arousal is a physiological process associated with a relative activation of the sympathetic nervous system and mediated by peripheral sympathetic pathways to blood vessels, heart, sweat glands, etc., resulting in responses such as increased heart rate, increased arterial blood pressure, increased activation of the adrenal medulla with increased release of adrenaline, and increased sweating. These autonomic reactions occur in parallel to fast and shallow breathing and increased muscle tension.

Cross-References

▶ Relaxation in the Treatment of Pain

Sympathetic Blockade

Definition

Interruption of impulse transmission through sympathetic ganglia and along pre- and postganglionic nerve fibers by a local anesthetic. This includes criteria for sufficient blocks to avoid false positive and false negative results. As the sympathetic ganglia are largely separated from somatic nerves, it is possible to achieve selective blockade of sympathetic pathways to the limbs or head at these sites without effects on somatic sensory and somatomotor functions. However, most sympathetic blocks interrupt conduction in spinal visceral afferents (e.g., during stellate ganglion block spinal visceral afferents from the heart are blocked, during block of upper lumbar ganglia visceral afferents from pelvic organs and colon are blocked).

Cross-References

▶ Sympathetic Blocks

Sympathetic Blocks

Andrew Jeffreys
Director Department of Anaesthesia & Pain Medicine, Western Health, Melbourne, VIC, Australia

Synonyms

Chemical sympathectomy; Sympathetic nerve block

Definition

Sympathetic blocks are procedures designed to produce temporary or permanent interruption of activity in sympathetic neurons.

Characteristics

Principles

Sympathetic blockade can be produced temporarily by injecting ▶ local anesthetic onto sympathetic nerves. Typically, the nerves are blocked where they form the sympathetic trunks and ganglia, or plexuses, in front of the vertebral column. Sympathetic nerves are most concentrated in these regions (Breivik and Cousins 2009; Burton et al. 2009). The objective of the block may be to interrupt efferent output to a target region, afferent output from that region, or both. The site of the block used depends on the region targeted (Table 1).

Techniques

Blind techniques using anatomical landmarks can be performed, but fluoroscopic or CT guidance allows for safer and more accurate placement of needles. The injection of a test-dose of contrast medium allows confirmation that the agent to be used will spread safely and appropriately.

Stellate Ganglion Block

The patient lies supine with their neck slightly extended and mouth open. Upon palpating the

Sympathetic Blocks, Table 1 The variety of sympathetic blocks and the regions that they target

Block	Target region
Stellate ganglion	Head, neck, and upper limb
Thoracic sympathetic trunk	Upper limb and heart
Coeliac plexus/splanchnic nerve	Upper abdominal viscera
Lumbar sympathetic trunk	Lower limb
Hypogastric plexus	Pelvic viscera
Ganglion impar	Perineum

transverse process of C6, the operator inserts a needle until it contacts the junction of the transverse process and body of C6. Between 10 and 15 ml of local anesthetic is injected slowly. Spread of solution to the midthoracic segments is required to completely block sympathetic efferents to the upper limb.

Thoracic Sympathetic Block
The needle is introduced from behind and is directed to pass ventrally and medially, through the costotransverse ligament, until its tip lies beside the vertebral column at the depth of the sympathetic trunk.

Celiac Plexus Block
A bilateral approach uses two needles, inserted from behind and directed to the anterior aspect of the body of L1 adjacent to the aorta. CT scanning greatly facilitates this block. The insertion points and needle direction can be determined; the pleura and kidneys can be avoided; and the spread of injectate can be ascertained with contrast medium.

Splanchnic Nerve Block
The technique for splanchnic nerve block is similar to that of the celiac plexus block. The main difference is that the needles are aimed at the anterolateral angle of the T12 vertebral body to rest at the same point on the vertebral body as in the lumbar sympathetic block. Pain relief can be achieved with a smaller volume of solution (5–10 ml each side) compared with the celiac plexus block (10–15 ml each side).

Lumbar Sympathetic Block
A needle is introduced from behind and is directed under fluoroscopic guidance through the psoas muscle, until its tip emerges from the psoas fascia immediately adjacent to the vertebral bodies of L2, L3, or L4. A linear spread of 1 ml of contrast medium, with no lateral extension, indicates appropriate positioning. Small volume injections of the active agent can then be made at L2 and L4 or a single injection of a larger volume (15–20 ml) at L3.

Hypogastric Plexus Block
This block is performed in a similar fashion to the lumbar sympathetic block, save that a bilateral approach is used, and the target lies at L5-S1.

Ganglion Impar Block
A needle is passed from below the tip of the coccyx to follow its anterior curvature, until it reaches the anterior surface of the sacrococcygeal junction. An alternative is to pass the needle from behind, through the sacrococcygeal ligament.

Agents Used
Local Anesthetics
For prolonged block, an agent such as bupivacaine or ropivacaine is commonly used. Clinical experience, and some research evidence, suggests that analgesic effects can continue for significantly longer than the described duration of action of these agents (6–12 h) (Price et al. 1998). Volumes in the range of 10–20 ml are used, keeping in mind the total dose being administered to the patient.

Neurolytic Agents
Neurolytic agents have the advantage of producing a prolonged "chemical sympathectomy." The major disadvantage is that unwanted effects on neurological structures, such as somatic nerves, may not recover. The most widely used agents are 50–100 % alcohol or 6–7 % phenol in water.

Validity
The accuracy of a sympathetic block is confirmed by signs of venodilatation, increased capillary blood flow and perfusion, and an increase in

skin surface temperature. A temperature increase of 1–3 °C is indicative of a successful block, but this method is ineffective if skin is warm to begin with (Fisher et al. 1997). Patients with significant vascular disease may not demonstrate these signs.

Complications

The risk of major complications appears to be acceptably low. The published case reports pertain largely to neurolytic blocks.

Systematically, the potential complications of sympathetic blocks include the inadvertent puncture of adjacent structures or injection of agents into them, spread of agents to somatic nerves, and the physiological effects of sympathetic blockade, such as postural hypotension or ejaculatory failure.

Indications and Applications

Rest Pain Due to Chronic Vascular Insufficiency

Chemical lumbar sympathectomy has been used for several decades to treat patients with occlusive vascular disease who suffer ischemic rest pain of the distal lower limbs. Its use has declined significantly in latter years due to the development of successful reconstructive surgical techniques.

Since muscle blood flow is largely regulated by autoregulatory mechanisms, sympathectomy is not effective in the treatment of the pain of claudication (Breivik and Cousins 2009; Gordon et al. 1994).

The role of sympathectomy in the treatment of peripheral vascular disease has come under scrutiny and its long-term effectiveness questioned (Gordon et al. 1994). Nevertheless, its major application remains for patients for whom surgery is not possible or has failed (Alexander 1994).

Chronic Malignant Pain

Neurolysis of visceral nociceptive afferents is an effective treatment for the pain of visceral malignancy. Diagnostic blocks are usually performed to establish efficacy prior to neurolysis.

The most widely used and best studied of these blocks is celiac plexus/splanchnic nerve block, for pain due to malignancy arising from the upper abdomen (Day 2008). Properly performed blocks can provide pain relief lasting for more than 1 month in 70–85 % of patients (Brogan 2010). Best responses occur in those suffering carcinoma of the pancreas.

The most feared complication is damage to spinal cord or somatic nerves. One retrospective questionnaire-based study detected four cases of permanent paraplegia in 2,730 neurolytic celiac plexus blocks (Davies 1993).

Sympathetically Maintained Neuropathic Pain

The prominence of signs suggesting autonomic dysfunction in Complex Regional Pain Syndromes, General Aspects prompted the widespread use of sympathetic blocks in their treatment. In recent years, however, the role and efficacy of sympathetic blocks in these conditions has been questioned (Cepada et al. 2002).

The exact role that the sympathetic nervous system plays in the maintenance of chronic neuropathic pain remains unclear (Fisher et al. 1997). Evidence of sympathetic nervous system dysfunction in these conditions does exist, but a link between this evidence and the appropriate use of sympathetic blocks has not been established (Baron et al. 2002).

Patients who respond to sympathetic blocks are said to have sympathetically maintained pain. This concept has been criticized, due to the variable efficacy of injection techniques in achieving sympathetic blockade and because sympathetic blocks can have affects on pathways other than sympathetic nerves (Fisher et al. 1997).

Clinical trials have suffered from differences in inclusion/exclusion criteria, treatment methods, outcome measurements, and the use of controls (Perez et al. 2001). Convincing evidence confirming the efficacy of sympathetic blocks in the treatment of neuropathic pain has not been produced (Tran et al. 2010). However, some evidence suggests that the use of sympathetic blocks soon after injury may prevent the development of CRPS (Reuben et al. 2000).

Since the placebo effect is powerful in this situation, the use of appropriate controls is vital when conducting studies of this nature. The use of control blocks in routine clinical practice has also been strongly advocated (Bogduk 2002).

Apparent relief of pain with local anesthetic blocks has led to the use of neurolytic blocks, surgical sympathectomy, or radiofrequency denervation in the treatment of CRPS. However, neuro-ablative sympathectomy seems to have, at best, a temporary analgesic effect. A significant risk of these techniques is post-sympathectomy neuralgia, whereby the patient may be left with a worse pain problem than existed at the outset (Anastasios et al. 2003).

Overall, whereas clinical experience suggests that, when used appropriately, sympathetic blocks can be a useful component of a multidisciplinary treatment regimen, questions regarding their use in the treatment of CRPS remain unanswered. The use of control blocks, where possible, in order to properly assess response is to be recommended. Continuing a series of sympathetic blocks in the face of no improvement is futile and potentially harmful.

Cross-References

▶ Epidural Steroid Injections for Chronic Back Pain

References

Alexander, J. P. (1994). Chemical lumbar sympathectomy in patients with severe lower limb ischemia. The Ulster Medical Journal, 63, 137–143.
Anastasios, T., et al. (2003). Characteristics and associated features of persistent post-sympathectomy pain. The Clinical Journal of Pain, 19, 192–199.
Baron, R., et al. (2002). National institutes of health workshop: Reflex sympathetic dystrophy/complex regional pain syndromes – State of the science. Anesthesia and Analgesia, 95, 1812–1816.
Bogduk, N. (2002). Diagnostic nerve blocks in chronic pain. Best Practice & Research. Clinical Anaesthesiology, 16, 565–578.
Breivik, H., & Cousins, M. J. (2009). Sympathetic neural blockade of upper and lower extremity. In M. J. Cousins & P. O. Bridenbaugh (Eds.), Neural blockade in clinical anesthesia and pain medicine (pp. 848–885). Philadelphia: Lippincott Williams & Wilkins.
Brogan, S. E. (2010). Interventional pain therapies. In S. M. Fishman, J. C. Ballantyne, & J. P. Rathmell (Eds.), Bonica's management of pain (pp. 605–617). Philadelphia: Lippincott Williams & Wilkins.
Burton, A. W., Phan, P. C., & Cousins, M. J. (2009). Treatment of cancer pain: Role of neural blockade and neuromodulation. In M. J. Cousins & P. O. Bridenbaugh (Eds.), Neural blockade in clinical anesthesia and pain medicine (pp. 1111–1153). Philadelphia: Lippincott Williams & Wilkins.
Cepada, S. M., et al. (2002). Defining the therapeutic role of local anesthetic sympathetic blockade in complex regional pain syndrome: A narrative and systematic review. The Clinical Journal of Pain, 18, 216–233.
Davies, D. D. (1993). Incidence of major complications of neurolytic coeliac plexus block. Journal of the Royal Society of Medicine, 86, 264–266.
Day, M. (2008). Sympathetic blocks: The evidence. Pain Practice, 8, 98–109.
Fisher, D. M., Abram, S. E., & Hogan, Q. H. (1997). Neural blockade for diagnosis and prognosis: A review. Anesthesiology, 86, 216–241.
Gordon, A., Zechmeister, K., & Collin, J. (1994). The role of sympathectomy in current surgical practice. European Journal of Vascular Surgery, 8, 129–137.
Perez, R. S., et al. (2001). Treatment of reflex sympathetic dystrophy (CRPS type 1): A research synthesis of 21 randomized clinical trials. Journal of Pain and Symptom Management, 21, 511–526.
Price, D. D., Long, S., Wilsey, B., & Rafii, A. (1998). Analysis of peak magnitude and duration of analgesia produced by local anesthetics injected into sympathetic ganglia of complex regional pain syndrome patients. The Clinical Journal of Pain, 14, 216–226.
Reuben, S. S., et al. (2000). Surgery on the affected upper extremity of patients with a history of complex regional pain syndrome: A retrospective study of 100 patients. Journal of hand surgery, 25A, 1147–1151.
Tran, D. Q., Duong, S., Bertini, P., & Finlayson, R. J. (2010). Treatment of complex regional pain syndrome: A review of the evidence. Canadian Journal of Anesthesia, 57, 149–166.

Sympathetic Nerve Block

▶ Sympathetic Blocks

Sympathetic Nervous System

▶ Neurogenic Inflammation and Sympathetic Nervous System

Sympathetic Nervous System and Pain

Wilfrid Jaenig[1] and Ralf Baron[2]
[1]Physiologisches Institut, Christian-Albrechts-Universität zu Kiel, Kiel, Germany
[2]Department of Neurological Pain Research and Therapy, Christian Albrechts University Kiel, Kiel, Germany

Introduction

Research on the relationship between the autonomic nervous system and pain is a developing field with many facets that are interesting from the biological, pathobiological, and clinical point of view. We prefer to use the terms biology and pathobiology, instead of physiology and pathophysiology, because they include physiological, morphological, biochemical, and molecular aspects. Here we will briefly review this field in a broad context and focus on the sympathetic nervous system. We will argue that the peripheral sympathetic noradrenergic neuron may have, in addition to its conventional function to transmit signals generated in the brain to peripheral target cells (e.g., smooth muscle cells, secretory epithelia, heart cells, neurons of the enteric nervous system etc.), quite different functions that are directly or indirectly related to protection of body tissues and pain. Some of these functions have not been studied as extensively as the function to regulate autonomic target cells (Jänig 2006). The parasympathetic nervous system is not involved in the generation of pain, yet may also be important in protection of body tissues (e.g., the gastrointestinal tract). Vagal afferents are involved in integrative aspects of pain, hyperalgesia, and inflammation. This is discussed elsewhere (Jänig 2005; Jänig and Levine 2013).

Table 1 lists different functions of the sympathetic nervous system related to pain and protection of body tissues in biological and pathobiological conditions. The table clearly shows that peripheral as well as central mechanisms have to be considered and that the topic not only encompasses pain and hyperalgesia, but has a broader context including neural and neuroendocrine regulation of inflammation and the immune system. Often biology and pathobiology are not clearly separated.

As shown in Table 1, in this brief review, we will distinguish three aspects of the topic *sympathetic nervous system and pain*:

1. Reactions of the sympathetic nervous system in pain. This addresses the autonomic reactions during pain and includes reflexes to noxious stimulation and adaptive reactions during pain.

2. Peripheral mechanisms underlying coupling (cross talk) from sympathetic (noradrenergic) neurons to afferent neurons in the generation of pain. This consists of two types of coupling: coupling related to excitation of the sympathetic postganglionic neurons and release of noradrenaline; coupling independent of excitation and noradrenaline release.

3. Central aspects of the role of the sympathetic nervous system in the generation of pain. This aspect includes the sympathetic-afferent coupling mentioned before. However, based on observations on patients and animal experimentation, we will discuss whether change of activity in sympathetic noradrenergic neurons or of the sympatho-adrenal system generated in the brain contributes directly or indirectly to hyperalgesia and inflammation.

The role of the sympathetic nervous system and its transmitters in chronic inflammation (such as rheumatoid arthritis) and its interaction with the peptidergic afferent innervation as well as the hypothalamo-pituitary-adrenal system will not be discussed. We are just at the beginning of the experimental investigation of these control mechanisms (Straub and Härle 2005; Straub et al. 2005).

Characteristics

Biology of the Sympathetic Nervous System

The peripheral sympathetic nervous system is by definition an efferent system consisting in the

Sympathetic Nervous System and Pain, Table 1   Sympathetic nervous system and pain

	References
1. Reactions of the sympathetic nervous system in pain • Protective spinal reflexes • Fight, flight, and quiescence organized at the level of the periaqueductal gray • Hyperalgesia and sympathetically mediated changes in referred zones during visceral pain	Bandler and Shipley (1994), Bandler et al. (2000a, b), Jänig (1993, 2006)
2. Role of the sympathetic nervous system in the generation of pain *Sympathetic-afferent coupling in the periphery*: • Coupling after nerve lesion (noradrenaline, α-adrenoceptors) • Coupling via the micromilieu of the nociceptor and the vascular bed • Sensitization of nociceptors mediated by sympathetic terminals independent of excitation and release of noradrenaline • Sensitization of nociceptors initiated by cytokines or nerve growth factor and mediated by sympathetic terminals • Sympatho-adrenal system and nociceptors sensitization	Green et al. (1997), Jänig (2005, 2009a), Jänig and Häbler (2000b), Jänig et al. (2000), Jänig and Levine (2013), Khasar et al. (1998a, b)
3. Sympathetic nervous system and central mechanisms *Control of inflammation and hyperalgesia by sympathetic and neuroendocrine mechanisms* • Sympathetic nervous system and complex regional pain syndromes • Sympathetic nervous system and immune system • Sympathetic nervous system and rheumatic diseases • Sympathetic nervous system and persistent generalizing pain disorders (e.g., fibromyalgia, irritable bowel syndrome, others	Bradesi et al. (2009), Campero et al. (2010), Donello et al. (2011), Harden et al. (2001), Jänig (2009b), Jänig and Baron (2002, 2003, 2004), Jänig and Häbler (2000a), Jänig and Stanton-Hicks (1996), Jänig et al. (2000), Jänig and Levine (2013), Jorum et al. (2007), Nance and Sanders (2007), Pongratz and Straub (2010), Mayer and Bushnell (2009), Staud (2009), Straub and Härle (2005), Straub et al. (2005)

periphery of many functionally distinct pathways that transmit the impulse activity from the spinal cord to the effector cells. These pathways are separate from each other and functionally distinct with respect to the effector cells, as is the impulse transmission in the sympathetic ganglia and at the neuroeffector junctions to the effector cells. Each peripheral sympathetic pathway is connected to distinct neuronal circuits in the spinal cord, brain stem, and hypothalamus that generate the typical discharge patterns in the sympathetic neurons and are responsible for the precise regulation of the target organs (cardiovascular regulation, thermoregulation, regulation of pelvic organs, regulation of gastrointestinal tract, and so forth). The autonomic regulation in which the sympathetic nervous system is involved is represented in these central networks (Jänig 2006; Jänig and McLachlan 2013; Llewellyn-Smith and Verbene 2011). The precise anatomical, histochemical, and functional organization of the sympathetic nervous system, in the periphery and in the brain, often receives insufficient attention in studies of the functioning of this system under pathobiological conditions.

Although primary afferent neurons innervating visceral organs are indispensable to understanding autonomic regulations, they are neither sympathetic nor parasympathetic nor autonomic, but simply *visceral* afferents (specified as spinal or vagal). We lack biological criteria to label these afferents specifically as sympathetic or parasympathetic (Jänig 2006).

Reactions of the Sympathetic Nervous System in Pain

Pain is a complex multidimensional event that cannot be reduced to a nociceptive sensation (sensory-discriminative dimension). Rather, pain includes an affective-motivational and a cognitive-evaluative dimension, all three describing together the secondary affective pain behavior (Price et al. 2009, Fig. 1). Modern imaging and other techniques demonstrate that various centers in spinal cord, brainstem, hypothalamus, and telencephalon are involved in the generation of pain behavior (Bushnell and Apkarian 2006; Casey and Tran 2006; Tracey and Mantyh 2007; Treede and Apkarian 2009) (see left side of Fig. 1) indicating that several cortical areas participate in the different dimensions of pain. These regions are activated by experimental or natural stimulation of peripheral nociceptive afferent neurons. Their integrated pattern of activity has been referred to as a "neural pain matrix" (Melzack 1999) or "cerebral signature of pain" (Tracey and Mantyh 2007).

Paralleling the generation of nociceptive sensations, protective somatic, autonomic, and neuroendocrine motor systems are activated, through circuits that include the spinal cord, brain stem, hypothalamus, amygdala, and other brain centers, in a stereotyped fashion. During the perception of intrusion or threat to the body and during the secondary (cognitive) appraisal, there is a powerful modulation of diverse motor functions involving somatic, autonomic, and neuroendocrine systems. These also involve higher brain centers (see interrupted arrows in Fig. 1). These motor components are integral to the secondary affective pain behavior and are orchestrated by the anterior cingulate cortex and the amygdala (Neugebauer et al. 2009). Thus, the autonomic changes observed under the condition of ongoing pain are dependent on the integrated activity of many centers in the central nervous system (Fig. 1). There is a poor or absent correlation between the magnitude of the changes of the autonomic parameters and the magnitude of pain (Ledowski et al. 2012). In the extreme, higher (cortical) centers may also be activated during imagined painful events where there is

Sympathetic Nervous System and Pain, Fig. 1 The interaction between nociceptive sensations, bodily unpleasant affective sensations, and secondary affective pain behavior and the brain centers possibly involved. Activation of autonomic, endocrine, and somatic motor systems occur on all integrative levels during activation of the centers representing nociception and pain. The mechanisms of activation of these motor systems by higher centers (see *interrupted arrows*) have been little studied. Simplified scheme. *ACC* anterior cingulate cortex, *AMYG* amygdala, *BS* brain stem, *Ento* entorhinal cortex, *Hip* hippocampus, *Hypothal* hypothalamus, *INS* insular cortex, *IPC* inferior parietal cortex, *PAG* periaqueductal gray, *PCC* posterior cingulate cortex, *PFC* prefrontal cortex (*dl* dorsolateral, *m* medial), *PMC* premotor cortex, *SC* spinal cord, *SI, SII* primary and secondary somato-sensory cortex, *SMA* supplementary motor area, *THAL* Thalamus (Modified from Jänig (2012) and designed after Casey and Tran (2006), Price (2000) and Price et al. (2009))

no obvious tissue damage, i.e., without an activation of nociceptors, and these central changes may also be accompanied by autonomic changes (Jänig 2012).

Any acute, but possibly also chronic tissue-damaging stimulus generating pain affects the sympathetic nervous system. Neurons of sympathetic systems exhibit generalized and specific reactions to these stimuli. The generalized reactions probably only occur in certain types of sympathetic system (e.g., muscle vasoconstrictor, visceral vasoconstrictor, sudomotor neurons or sympathetic cardiomotor neurons) but are weak or absent in other systems (e.g., sympathetic systems to pelvic organs). They are organized in spinal cord, brain stem (medulla oblongata, mesencephalon), and hypothalamus and can best be conceptualized by looking at the different patterns of defense behavior, as shown experimentally in animals, including "confrontational defense," "flight," and "quiescence."

Confrontational defense and flight are typical of an active defense strategy and occur when animals encounter threatening stimuli which are potentially injurious for the body and trigger pain. *Confrontational defense* leads potentially to fight or flight and flight forwards avoidance. Both defense patterns are represented in the lateral and dorsolateral periaqueductal gray of the mesencephalon. They are activated or enhanced from the body surface or the cortex and are associated with endogenous non-opioid analgesia, hypertension, and tachycardia. *Quiescence*, by contrast, resembles the natural reactions of mammals when there is serious injury and chronic pain, particularly in deep or visceral body domains. Quiescence is represented in the ventrolateral periaqueductal gray, activated from the deep (somatic and visceral) body domains and is manifested as hyporeactivity of the somatomotor system, hypotension, bradycardia, and an endogenous opioid analgesia. These stereotyped preprogrammed elementary protective behaviors and their association with the endogenous control of analgesia enable the organism to cope with dangerous situations that are usually accompanied by pain or impending pain. The dorsolateral, lateral, and ventrolateral columns of the periaqueductal gray have distinct reciprocal connections with the autonomic centers in the lower brain stem and hypothalamus that differentially regulate the activity in neurons of the autonomic pathways. The different periaqueductal gray centers are under control of the medial and orbital prefrontal cortex and other telencephalic centers (Bandler and Shipley 1994; Bandler et al. 2000a, b; Keay and Bandler 2004; Jänig 2009a).

There are also more localized and distinct reactions of the sympathetic nervous system to noxious stimuli that are organized within the spinal cord and trigeminal nucleus and in the periphery, i.e., somato-sympathetic, viscero-sympathetic, and viscero-visceral reflexes. For example, cutaneous vasoconstrictor neurons exhibit distinct inhibitory reflexes to noxious stimuli of the territories innervated by these neurons. The hypothalamo-mesencephalic and the spinal level of integration are presumably protective under normal biological conditions and are associated with activation of the adreno-cortical system by the hypothalamo-hypophyseal axis (Jänig 2006).

Role of the Sympathetic System in the Generation of Pain

In this section, we will discuss mechanisms by way of which sympathetic noradrenergic neurons may be coupled directly or indirectly to primary afferent neurons in the periphery, leading to their excitation and/or change of their excitability. This sympathetic-afferent coupling is either dependent on the activity in the sympathetic neurons, but not necessarily on an increase of this centrally generated activity, or entirely independent of activity and release of noradrenaline. The latter function of the sympathetic noradrenergic neuron is entirely different from its function to transmit impulses to target cells.

Physiological Conditions

Under physiological conditions, there exists almost no influence of sympathetic activity on sensory neurons projecting to skin and deep somatic tissues in mammals. The effects that have been measured under experimental conditions on receptors with myelinated and unmyelinated axons are weak and can in part be explained by changes of the effector organs (erector pili muscles, blood vessels) induced by the activation of sympathetic neurons. These rather negative results do not rule out that noradrenaline or co-localized substances released by the postganglionic terminals have secondary long-term effects on the excitability of sensory receptors although we have no experimental evidence for this (Jänig 2009a).

Sympathetic-Afferent Coupling Dependent on Activity in Sympathetic Neurons

Pain being dependent on activity in sympathetic neurons is called *sympathetically maintained pain* (SMP; Stanton-Hicks et al. 1995). SMP is a symptom and includes generically spontaneous pain and pain evoked by mechanical or thermal stimuli. It may be present in patients with complex regional pain syndrome (CRPS) type I or type II and in patiens with other neuropathic pain syndromes (Stanton-Hicks et al. 1995). The idea about the involvement of the (efferent) sympathetic nervous system in pain is based on various clinical observations which have been documented in the literature since tens of years (Harden et al. 2001). Representative for these multiple observations on patients with SMP are quantitative experimental investigations (Ali et al. 2000; Baron et al. 2002; Carroll et al. 2009; Meier et al. 2009; Price et al. 1998; Torebjörk et al. 1995; for detailed discussion see Jänig 2009a). These experiments demonstrate that: (1) sympathetic postganglionic neurons can be involved in the generation of pain; (2) blockade of the sympathetic activity can relieve the pain; (3) noradrenaline injected intracutaneously is able to rekindle the pain (▶ Sympathetically Maintained Pain, Clinical Pharmacological Tests; ▶ Sympathetically Maintained Pain in CRPS I, Human Experimentation; ▶ Sympathetically Maintained Pain in CRPS II, Human Experimentation); and (4) α-adrenoceptor blockers or guanethidine (which depletes noradrenaline from its stores) may relieve the pain (Arnér 1991; ▶ Sympathetically Maintained Pain in CRPS II, Human Experimentation).

The interpretation of these data is: Nociceptors are excited and possibly sensitized by noradrenaline released by the sympathetic fibers. Either the nociceptors have expressed adrenoceptors and/or the excitatory effect is generated indirectly, e.g., via changes in blood flow. Sympathetically maintained activity in nociceptive neurons may generate a state of central sensitization/hyperexcitability, leading to spontaneous pain and secondary evoked pain (mechanical and possibly cold allodynia; Baron and Jänig 2010; Jänig 2013) (Fig. 2).

Coupling after Nerve Lesion Via Noradrenaline and α-Adrenoceptors The coupling between sympathetic postganglionic neurons and primary afferent neurons following trauma with nerve lesion which underlies SMP may occur in several ways (Fig. 3a): at the lesion site, along the nerve, and in the dorsal root ganglion. The afferent activation is mediated by noradrenaline released by the postganglionic axons and by α-adrenoceptors expressed by the afferent neurons. Primary afferent nociceptive as well as non-nociceptive neurons may be affected (▶ Sympathetic-Afferent Coupling in the Afferent Nerve fiber, Neurophysiological Experiments). Coupling in the dorsal root ganglion practically only occurs to large diameter afferent cell bodies. The underlying mechanisms are change of blood flow (generating hypoxia?) and possibly a direct effect by release of noradrenaline (▶ Sympathetic-Afferent Coupling in the Dorsal Root Ganglion, Neurophysiological Experiments). It is debated whether this coupling is important in patients with SMP following trauma with nerve lesion. Coupling along the lesioned nerve is still hypothetical. However, in view of the experimental observation showing that peptidergic (afferent) and postganglionic (noradrenergic) neurons undergo continuous sprouting in a lesioned nerve proximal to the nerve injury over <100 days (Barrett and McLachlan 1999), this mechanism is a realistic possibility and needs to be examined. The coupling may also occur indirectly by changes in the micromilieu of the lesioned primary afferent neurons (e.g., by changes of blood flow in the dorsal root ganglion or in the lesioned nerve; ▶ Sympathetic-Afferent Coupling in the Dorsal Root Ganglion, Neurophysiological Experiments).

Experiments supporting these ideas have been performed on human models, on animal behavioral models, and on reduced animal models in vivo and in vitro (▶ Sympathetically Maintained Pain in CRPS I, Human Experimentation; ▶ Sympathetic Nervous System in the Generation of Pain, Animal Behavioral Models; see Baron and Jänig 2010; Harden et al. 2001; Jänig 2009a; Jänig and Baron 2002, 2003; Jänig and Häbler 2000a).

Sympathetic Nervous System and Pain, Fig. 2 Concept of generation of peripheral and central hyperexcitability during inflammatory pain and neuropathic pain. The *upper interrupted arrow* indicates that the central changes are generated (and possibly maintained) (a) by persistent activation of nociceptors with C-fibers (e.g., during chronic inflammation) called here "central sensitization" and (b) after trauma with nerve lesion by ectopic activity and other changes in lesioned afferent neurons called here "central hyperexcitability." The *lower interrupted arrow* indicates the efferent feedback via the sympathetic nervous system and neuroendocrine systems (e.g., the sympatho-adrenal system). Primary afferent nociceptive neurons (in particular those with C-fibers) are sensitized during inflammation. After nerve lesion, *all* lesioned primary afferent neurons (unmyelinated as well as myelinated ones) undergo biochemical, physiological, and morphological changes which become irreversible with time. These peripheral changes entail changes of the central representation (of the somato-sensory system) which become irreversible if no regeneration of primary afferent neurons to their target tissue occurs. Processing of information in the spinal (and trigeminal) dorsal horn is under excitatory and inhibitory control of supraspinal centers. The central changes, induced by persistent activity in afferent nociceptive neurons or after nerve lesions, are also reflected in the efferent feedback systems that may establish positive feedback to the primary afferent neurons (Modified from Jänig (2013))

Coupling via the Micromilieu of the Nociceptor

After trauma without nerve lesion, the sympathetic activity may be mediated indirectly to the afferent neurons (i.e., by way of the vascular bed or other mechanisms). The most convincing experiment supporting this idea was conducted by Baron et al. (2002) on CRPS type I patients with SMP. These patients show that SMP in the skin is enhanced if activity in cutaneous vasoconstrictor neurons is increased (e.g., generated by central cooling). However, the difference in SMP between the states of whole body cooling and whole body warming is significantly smaller than the decrease of SMP following sympathetic block. This important observation argues that the coupling between sympathetic and nociceptive afferent neurons mainly occurs in the deep somatic tissues (Wasner and Baron Essay 37.2). An animal model to study this important way of coupling does not exist and has to be designed.

Sensitization of Nociceptors Mediated by Sympathetic Terminals Independent of Excitation and Release of Noradrenaline

In rat, withdrawal threshold to stimulation of the rat hindpaw with a linearly increasing mechanical stimulus applied to the dorsum of the paw decreases dose-dependently after intradermal injection of the inflammatory mediator bradykinin (an octapeptide cleaved from plasma α_2-globulins, by kallikreins circulating in the

Sympathetic Nervous System and Pain, Fig. 3 Ways hypothesized to couple sympathetic and primary afferent neurons following peripheral nerve lesion (**a**) or during inflammation (**b** to **d**). (**a**) These types of coupling depend on the activity in the sympathetic neurons and on the expression of functional adrenoceptors by the afferent neurons or mediation indirectly via the blood vessels (blood flow). It can occur in the periphery (**1**), in the dorsal root ganglion (**3**), or possibly also in the lesioned nerve (**2**). (**b**) The inflammatory mediator bradykinin acts at B_2 receptors in the membrane of the sympathetic varicosities or in cells upstream of these varicosities, inducing release of prostaglandin E_2 (PGE_2) and sensitization of nociceptors. This way of coupling is probably *not dependent* on activity in the sympathetic neurons. (**c**) Nerve growth factor released during an experimental inflammation reacts with the high-affinity receptor, trkA, in the membrane of the sympathetic varicosities, inducing release of an inflammatory mediator or inflammatory mediators and sensitization of nociceptors. This effect is probably *not* dependent on activity in the sympathetic neurons. (**d**) Activation of the adrenal medulla by sympathetic preganglionic neurons leads to release of adrenaline which generates sensitization of nociceptors. The ? in (**b**) and (**c**) indicates that PGE_2 or other inflammatory mediators may be released by cells other the sympathetic varicosities (Modified from Jänig and Häbler (2000a))

Sympathetic Nervous System and Pain, Fig. 4 Mechanical hyperalgesic behavior and sympathetic innervation. Bradykinin-induced hyperalgesia (decrease of paw-withdrawal threshold) in sham sympathectomized (*circles*; $n = 6$ hindpaws) rats, in surgically sympathectomized rats (*squares*; $n = 13$ hindpaws), and in rats with decentralized lumbar sympathetic chains (preganglionic axons in lumbar sympathetic chain interrupted 8 days before, diamonds; $n = 10$ paws). Paw withdrawal thresholds in sham sympathectomized rats and in rats with sympathetic decentralization were both significantly different from the sympathectomy group ($P < 0.01$) (Modified from Jänig (2012b) and Khasar et al. (1998a))

plasma) (Figs. 3b, 4). Following a single injection of bradykinin this decrease lasts for more than one hour for mechanical stimulation. This type of mechanical hyperalgesic behavior is mediated by the B_2 bradykinin-receptor (Khasar et al. 1995) and is not present when bradykinin is injected subcutaneously (Khasar et al. 1993). Bradykinin-induced hyperalgesic behavior is blocked by the cyclooxygenase inhibitor indomethacin and therefore mediated by a prostaglandin (probably PGE_2) that sensitizes nociceptors for mechanical stimulation (Fig. 4). However, in vagotomized rats in which the bradykinin-induced mechanical hyperalgesia is significantly enhanced (probably due to adrenaline released by the activated adrenal medulla [Khasar et al. 1998a, b]; ▶ Sympatho-adrenal System and Mechanical Hyperalgesic Behavior, Animal Experimentation), indomethacin has almost no effect on bradykinin-induced hyperalgesia. This failure is not related to a switch from B_2- to B_1-receptor subtype because the selective B_1-receptor agonist des-Arg9-BK failed to produce hyperalgesia in vagotomized rats (Khasar et al. 1998a). This shows that brady-kinin-induced hyperalgesia may not be mediated by prostanoids in vagotomized rats. This interest-ing finding needs to be followed experimentally.

The decrease in paw-withdrawal threshold provided by bradykinin is significantly reduced after surgical sympathectomy. This shows that the sympathetic innervation of the skin is involved in the sensitization of nociceptors for mechanical stimulation. Interestingly, decen-tralization of the lumbar paravertebral sympa-thetic ganglia (denervating the postganglionic neurons by cutting the preganglionic sympa-thetic axons) does not abolish bradykinin-induced mechanical hyperalgesic behavior (Fig. 4). This indicates that the sensitizing effect of bradykinin is not dependent on activity in the sympathetic neurons innervating skin and there-fore not on release of noradrenaline (Khasar et al. 1998a; Jänig and Häbler 2000a). It is believed that bradykinin stimulates the release of prostaglandin from the sympathetic terminals (Gonzales et al. 1989). However, this release could also occur from other cells in association with the sympathetic terminals in the skin. Evi-dence for sensitization of cutaneous nociceptors to mechanical stimulation by bradykinin is poor or absent (see Treede et al. 1992). Some sensiti-zation to mechanical stimulation by bradykinin has been demonstrated for afferents from the knee joint (Neugebauer et al. 1989) and from skeletal muscle (Mense and Meyer 1988). Furthermore, injured myelinated and unmyelin-ated cutaneous afferents can be sensitized to mechanical stimuli by the so-called inflamma-tory soup which contains bradykinin (10^{-5} M) in addition to histamine, 5-HT, prostaglandin E_2 (all 10^{-5} M), K^+ (7 mM), and pH = 7 (Grossmann et al. 2009; Michaelis et al. 1998).

Mechanical hyperalgesic behavior generated by intracutaneous injection of bradykinin is an interesting phenomenon. However, the conven-tional explanation that this mechanical hyperalgesia is due to prostanoids sensitizing nociceptors since indomethacin prevents it appears to be too simple. The reasons are (1) the finding that sympathetic fibers mediate this hyperalgesia independent of neural activity and release of noradrenaline; (2) indomethacin does not block this behavior under certain conditions (e.g., when the adrenal medullae are activated after vagotomy; Khasar et al. 1998a); and (3) sen-sitization of nociceptors by bradykinin is weak or absent. Thus, this phenomenon has to be reinvestigated using a rigorous experimental approach.

Sensitization of Nociceptors Initiated by Cytokines or Nerve Growth Factor Possibly Mediated by Sympathetic Terminals

Nerve Growth Factor Systemic injection of nerve growth factor (NGF) is followed by a transient thermal and mechanical hyperalgesia in rats (Lewin et al. 1993, 1994) and humans (Petty et al. 1994). During experimental inflam-mation (evoked by Freund´s adjuvant in the rat hindpaw), NGF increases in the inflamed tissue paralleled by the development of thermal and mechanical hyperalgesia (Donnerer et al. 1992; Woolf et al. 1994). Both are prevented by anti-NGF antibodies (Lewin et al. 1994; Woolf et al. 1994). The mechanisms responsible are sensiti-zation of nociceptors via high-affinity NGF-receptors (trkA receptors) and an induction of increased synthesis of calcitonin-gene related peptide (CGRP) and substance P in the afferent cell bodies by NGF taken up by the afferent terminals and transported to the cell bodies. The NGF-induced sensitization of nociceptors also seems to be mediated indirectly by the sympa-thetic postganglionic terminals. Heat and mechanical hyperalgesic behavior generated by local injection of NGF into the skin is prevented or significantly reduced after chemical or surgical sympathectomy (Andreev et al. 1995; Woolf et al. 1996). These experiments suggest that NGF released during inflammation by inflammatory cells acts on the sympathetic terminals via high-affinity trkA receptors, inducing the release of inflammatory mediators and, subsequently, sensitization of nociceptors for mechanical and heat stimuli (Fig. 3c; McMahon 1996; Woolf 1996; Jänig and Häbler 2000a). It is unclear whether this sensitization of nociceptors mediated by terminal sympathetic nerve fibers (1) is dependent on activity in the sympathetic neurons, release of noradrenaline, and adrenoceptors expressed in the nociceptive afferent neurons, (2) is independent of activity and release of noradrenaline, or (3) is dependent on both.

Proinflammatory cytokines Based on behavioral experiments conducted on rats (studying mechanical and heat hyperalgesia) it has been shown that tissue injury, injection of the bacterial cell wall endotoxin lipopolysaccharide, or injection of carrageenan (a plant polysaccharide) stimulates tissue inflammation and leads to sensitization of nociceptors. Systematic pharmacological interventions using blockers or inhibitors of the various mediators demonstrate that the proinflammatory cytokines, tumor necrosis factor α (TNFα), and interleukin(IL)-1, IL-6, and IL-8 may be involved in this process of sensitization and therefore in the generation of hyperalgesia (Cunha et al. 2007; Verri et al. 2006; Woolf et al. 1996). Pathogenic stimuli lead to activation of resident cells, the release of the inflammatory mediator bradykinin and other mediators. The inflammatory mediators and the pathogenic stimuli themselves activate macrophages, monocytes, and other immune-related cells. These cells release TNFα which generates sensitization of nociceptors by two possible pathways (Fig. 5): (1) It induces production of IL-6 and IL-1β by immune cells, whereby IL-6 enhances the production of IL-1β. These interleukins stimulate cyclooxygenase 2 (COX 2) and the production of prostaglandins (PGE$_2$, PGI$_2$) which in turn react with the nociceptive terminal via E-type prostaglandin receptors. (2) It induces the release of IL-8 from endothelial cells and macrophages. IL-8 reacts with the sympathetic terminals that are supposed to mediate sensitization of

nociceptive afferent terminals by release of noradrenaline to act via β$_2$-adrenoceptors. These two peripheral pathways by which nociceptive afferents can be sensitized involving cytokines are under inhibitory control of circulating glucocorticoids (indicated by asterisks in Fig. 5) and of other, anti-inflammatory, interleukins (e.g., IL-4 and IL-10 indicated by # in Fig. 5).

It is important to emphasize that the mechanisms, involving NGF, proinflammatory cytokines, and noradrenergic sympathetic postganglionic fibers in the sensitization of nociceptive afferents, have been deduced on the basis of behavioral and pharmacological experiments. Proof of such interaction by directly assessing the activity of nociceptors with electrophysiological techniques and of the effect of noradrenergic nerve fibers on the afferent nerve fibers is lacking. These experiments have to be done.

Sympathetic Nervous System and Central Mechanisms in the Control of Hyperalgesia and Inflammation

The question to be addressed here is: Can centrally generated signals channeled through peripheral sympathetic pathways modulate sensitivity of nociceptors and inflammation? The hypothetical mechanisms involved may include and overlap with the peripheral mechanisms discussed in the foregoing section.

Sympatho-Adrenal System and Nociceptor Sensitization

Activation of the sympatho-adrenal system (adrenal medulla; e.g., by release of the central sympathetic circuits from vagal inhibition), but not of the sympatho-neural system, generates mechanical hyperalgesia and enhances bradykinin-induced mechanical hyperalgesia (decrease of paw-withdrawal threshold to mechanical stimulation of the skin). Both develop over 7–14 days following activation of the adrenal medullae and reverse slowly after denervation of the adrenal medullae (Khasar et al. 1998a, b). Application of adrenaline through a minipump over days enhances the slow development of mechanical hyperalgesia. Furthermore, continuous application of a β$_2$-adrenoceptor blocker by way

Stimulus
(pathogenic challenge)

↓

Resident cells

↓ ↓

Bradykinin
other

↓ ↓

*TNFα#

*IL-1#   *IL-8
IL-6

↓   ↓

*COX-2#   Symp. terminal

↓

Eicosanoids

↓

Nociceptive terminal

Cytokines and inflammatory
Hyperalgesia
(Sensitization of nociceptors)

Sympathetic Nervous System and Pain, Fig. 5 Role of cytokines in sensitization of nociceptors during inflammation and the underlying putative mechanisms leading to hyperalgesia. Pathogenic stimuli activate resident cells and lead to release of inflammatory mediators (such as bradykinin). Proinflammatory cytokines are synthesized and released by macrophages and other immune or immune-related cells. Nociceptors are postulated to be sensitized by two pathways involving the cytokines: (1) Tumor necrosis factor α (TNFα) induces synthesis and release of interleukin 1 (IL-1) and IL-6 which, in turn, induce the release of eicosanoids (prostaglandin E_2 and I_2 [PGE_2, PGI_2]) by activating the cyclooxygenase-2 (COX-2). (2) TNFα induces synthesis and release of IL-8. IL-8 activates sympathetic terminals that sensitize nociceptors via β_2-adrenoceptors. Glucocorticoids inhibit the synthesis of the cytokines and the activation of COX-2 (indicated by *asterisks*). Anti-inflammatory cytokines (such as IL-4 and IL-10) that are also synthesized and released by immune cells inhibit the synthesis and release of proinflammatory cytokines (indicated by #). This scheme is fully dependent on behavioral experiments and pharmacological interventions. The different steps will need to be verified experimentally using neurophysiological experiments (Modified from Verri et al. (2006))

of a minipump prevents the enhancement of bradykinin-induced hyperalgesic behavior generated by activation of the adrenal medullae (Khasar et al. 2003) (▶ Sympatho-adrenal

System and Mechanical Hyperalgesic Behavior, Animal Experimentation).

The results of this experiment are interpreted in the following way: Adrenaline released by persistent activation of the adrenal medullae sensitizes cutaneous nociceptors for mechanical stimuli. This sensitization of nociceptors and its reversal are slow and take days to develop. The slow time course implies that the nociceptor sensitization cannot acutely be blocked by an adrenoceptor antagonist given intracutaneously. Adrenaline probably does not act directly on the cutaneous nociceptors but on cells in the microenvironment of the nociceptors inducing slow changes which result in nociceptor sensitization. Candidate cells may be mast cells, macrophages, or keratinocytes which then release substances that generate sensitization (Jänig and Häbler 2000a; Jänig et al. 2000; Khasar et al. 1998b). The change of sensitivity of a population of nociceptors generated by adrenaline released by the adrenal medullae, which is regulated by the brain, would be a novel mechanism of sensitization. This mechanism would be different from mechanisms that lead to activation and/or sensitization of nociceptors by sympathetic-afferent coupling as discussed above (Fig. 3).

This mechanism of pain and hyperalgesia involving the sympatho-adrenal system may operate in ill-defined pain syndromes such as irritable bowel syndrome, functional dyspepsia, fibromyalgia, chronic fatigue syndrome, etc. (Mayer and Bushnell 2009). The conclusions on long-term sensitization of nociceptors for mechanical stimulation by adrenaline are indirect and fully rest on behavioral experiments. Neurophysiological experiments in vivo on primary afferent nociceptive neurons are required to test whether all or only a subpopulation of nociceptors are sensitized (▶ Sympatho-adrenal System and Mechanical Hyperalgesic Behavior, Animal Experimentation).

Sympathetic Nervous System and Complex Regional Pain Syndrome Type I

Complex regional pain syndromes (CRPS) are painful disorders that may develop as a consequence of trauma typically affecting the limbs. Clinically, they are characterized by pain (spontaneous, hyperalgesia, allodynia), active and passive movement disorders including an increased physiological tremor, abnormal regulation of blood flow and sweating, edema of skin and subcutaneous tissues, and trophic changes of skin, appendages of skin and subcutaneous tissues (Baron and Jänig 2013; Harden et al. 2001; Jänig and Baron 2002, 2003; Jänig and Stanton-Hicks 1996; Stanton-Hicks et al. 1995; Wilson et al. 2005). CRPS type I (previously called reflex sympathetic dystrophy) usually develops after minor trauma with a small or no obvious nerve lesion at an extremity (e.g., bone fracture, sprains, bruises or skin lesions, surgeries) and rarely after remote trauma in the visceral domain or after a CNS lesion (e.g., stroke). An important feature of CRPS I is that the severity and combination of clinical symptoms are *disproportionate* to the severity and type of trauma with a tendency to spread in the affected limb. The symptoms are not confined to the innervation zone of an individual nerve. CRPS type II (previously called causalgia) develops after trauma with a mostly large nerve lesion (Harden et al. 2001; Jänig and Stanton-Hicks 1996; Stanton-Hicks et al. 1995; Baron and Jänig 2013).

Recently, we proposed, on the basis of clinical observations and research on humans and animals, that CRPS (in particular type I) is a systemic disorder involving the central and peripheral nervous systems (Fig. 6). Various traumas can trigger variable combinations of clinical phenomena in which the somato-sensory system, the sympathetic nervous system, the somatomotor system, and peripheral (vascular, inflammatory) systems are involved (Baron and Jänig 2013; Harden et al. 2001; Jänig and Baron 2002, 2003). We proposed that the central representations of the sensory, motor, and sympathetic systems are changed. The central changes are reflected in changes of somatic sensations (changes of detection thresholds for mechanical, cold, warm, and heat; Rommel et al. 1999, 2001), of motor performances and autonomically regulated effector systems (vasculature, sweat glands, inflammatory cells, etc.). This idea is fully supported by several observations made on CRPS patients: (1) CRPS patients may exhibit disturbances of the perception of the affected extremity (e.g., foreignness and impaired position sense of the affected extremity) (Lewis et al. 2007, 2010). (2) The subjective body midline in CRPS patients is shifted to the affected extremity (Bultitude and Rafal 2010; Sumitani et al. 2007). (3) Ambiguous visual stimuli (leading to visual illusions) may cause enhanced pain, abnormal autonomic responses, and abnormal motor responses (dystonia) in CRPS I patients (Cohen et al. 2012). (4) Mirror image therapy decreases pain, swelling, and temperature difference between extremities in CRPS patients (McCabe and Blake 2008; Moseley 2004, 2005). (5) The change of processing of tactile stimuli is correlated with a decrease of skin temperature (increase of activity in cutaneous vasoconstrictor neurons) in CRPS patients. These changes depend on the position of the affected extremity in relation to the trunk and not on its somatotopic representation in neural space (Moseley et al. 2009). (6) Imagination of movements can trigger pain and swelling of fingers in CRPS patients (Moseley et al. 2008).

The peripheral changes cannot be seen independently of the central ones; both interact with each other via afferent and efferent signals. Furthermore, the mechanisms that underlie CRPS cannot be reduced to *one* system or to *one* mechanism only (e.g., to sympathetic-afferent coupling, to an adrenoceptor disease, to a peripheral inflammatory disease, to a psychogenic disease) (Harden et al. 2001; Baron and Jänig 2013; Jänig and Baron 2002, 2003).

The centrally generated activity in sympathetic neurons may be involved in various ways in the pathogenesis of CRPS type I, the main argument being that these changes are reversed or aggravated after intervention at the peripheral sympathetic nervous system. Three points that are relevant in the present context of sympathetic nervous system, pain, and body protection need to be emphasized for CRPS type I with SMP:

- In a subgroup of CRPS I patients, pain or a component of pain is obviously dependent on activity in sympathetic neurons and related to activation or sensitization of nociceptors by noradrenaline released by the sympathetic fibers (sympathetically maintained pain, SMP). Either the nociceptors have expressed adrenoceptors (▶ Sympathetic-Afferent Coupling in the Dorsal Root Ganglion, Neurophysiological Experiments; ▶ Sympathetic-Afferent Coupling in the Afferent Nerve Fiber, Neurophysiological Experiments) and/or the excitatory effect is generated indirectly, e.g., by way of changes in blood flow in deep somatic tissues of an extremity (Baron et al. 2002). Sympathetically maintained activity in nociceptive neurons may generate a state of central sensitization/hyperexcitability that is responsible for spontaneous pain and secondary evoked pain (mechanical and cold allodynia).

• Generation of swelling (edema) and inflammation. Sympathetic and peptidergic afferent fibers may interact at the arteriolar site (influencing blood flow) and venular site (influencing plasma extravasation) of the vascular bed. Sympathetic fibers are supposed to influence inflammatory cells (macrophages, mast cells) by way of noradrenaline and adrenoceptors on the inflammatory cells (Fig. 6) (Baron and Jänig 2013; Jänig and Baron 2002, 2003) (▶ Sympathetically Maintained Pain and Inflammation, Human Experimentation).

• Trophic changes in skin, appendages of skin, and subcutaneous tissues (including joints and so forth) are believed to be dependent, at least in part, on the sympathetic innervation. The nature of this neural influence on the tissue structure is unknown (Harden et al. 2001; Jänig and Baron 2003; Jänig and Stanton-Hicks 1996).

These three observations made on patients with CRPS I (who have no nerve lesion and by definition therefore not neuropathic pain) argue that the change of activity in sympathetic neurons entails peripheral changes which in turn may result in secondary activation and/or sensitization of primary afferent nociceptive neurons. The underlying mechanism(s) of this activation/sensitization of nociceptive afferent neurons may involve short- and long-term processes as discussed above.

An important observation, in CRPS I patients with SMP (but possibly also in CRPS II patients with SMP), to be mentioned is that pain relief following conduction block of sympathetic neurons by a local anesthetic applied to the sympathetic chain mostly outlasts the conduction block by at least an order of magnitude (days). This has been measured quantitatively by Price et al. (1998). The long-lasting pain-relieving effect of sympathetic blocks suggests that activity in sympathetic neurons, which is of central origin, maintains a positive feedback circuit via the primary afferent neurons that are probably nociceptive in function. Activity in sympathetic neurons maintains a central state of hyperexcitability (e.g., of neurons in the spinal dorsal horn; Fig. 2), via excitation of afferent neurons started by an intense noxious event. The persistent afferent activity needed to maintain such a state of central hyperexcitability is switched off during temporary block of conduction in the sympathetic chain lasting only a few hours and cannot be immediately switched on again when the block wears off and the activity in the sympathetic postganglionic neurons (and therefore probably also the sympathetically induced activity in afferent neurons) returns. It is hypothesized that the afferent activity has to act over long periods of time to initiate and maintain the central state of hyperexcitability (via the positive feedback). This important and interesting phenomenon needs to be studied experimentally in patients with SMP as well as in animal models that have to be designed.

Sympathetic Nervous System and Immune System

The hypothalamus can influence the immune system by way of the sympathetic nervous system and therefore control protective mechanisms of the body at the cellular level (see Ader 2007; Besedovsky and del Rey 1995; Hori et al. 1995; Madden and Felten 1995; Madden et al. 1995). The parameters of the immune tissues potentially controlled are proliferation, trafficking, and circulation of lymphocytes; functional activity of lymphoid cells (e.g., activity of natural killer cells) and cytokine production; hematopoiesis of bone marrow; mucosal immunity; thymocyte

Sympathetic Nervous System and Pain, Fig. 6 Schematic diagram summarizing the sensory, autonomic, and somatomotor changes in complex regional pain syndromes I (CRPS I) patients. The figure symbolizes the CNS (forebrain, brain stem, and spinal cord). Changes occur in the central representations of the somato-sensory, the somatomotor, and the sympathetic nervous system (which include the spinal circuits) and are reflected in the changes of the sensory painful and non-painful perceptions, of cutaneous blood flow and sweating, and of motor performances. They are triggered and possibly maintained by the nociceptive afferent input from the somatic and visceral body domains. It is unclear whether these central changes are reversible in chronic CRPS I patients. These central changes affect the endogenous control system of nociceptive impulse transmission possibly too. Coupling between the sympathetic neurons and the afferent neurons in the periphery (see *bold closed arrow*) is one component of the pain in CRPS I patients with sympathetically maintained pain (SMP). However, it seems to be unimportant in CRPS I patients without SMP (Modified from Jänig and Baron (2002, 2003))

development, etc. (for details of potential mechanisms see Elenkov et al. 2000). The mechanisms of this influence remain largely unsolved (Ader and Cohen 1993; Besedovsky and del Rey 1995). It is unknown whether there exists a functionally distinct sympathetic system from the brain to the immune system or whether the modulation of the immune system is a general function of the sympathetic nervous system. The more likely hypothesis favored by one of us is that the immune system is modulated by the brain by way of a functionally distinct sympathetic pathway (for discussion, see Jänig and Häbler 2000b; Jänig 2006). In both cases, we hypothesize that activity in the sympathetic neurons supplying the immune system entails indirect modulation of the sensitivity of nociceptors involving various types of immune-related cells and their signaling molecules (e.g., cytokines, see above and Jänig and Levine 2013).

Conclusions

Clinical observations, experimentation on humans, and animal experimentation indicate that the sympathetic nervous system is potentially involved in the generation of pain, hyperalgesia, and inflammation which should conceptually be seen in the frame of protection of body tissues. These functions of the sympathetic nervous system are different from its conventional functions, namely, to transmit impulses from the brain via distinct pathways to peripheral effector cells. They include:

- Sympathetic-afferent coupling following trauma with or without nerve injury
- Sensitization of nociceptors independent of action potential generation and noradrenaline release
- Sensitization of nociceptors by nerve growth factor and cytokines
- Sensitization of nociceptors involving the sympatho-adrenal system
- Neurogenic inflammation (e.g., in the knee joint and its control by neuroendocrine systems)
- Modulation of the immune system by the sympathetic innervation
- Central changes in sympathetic pathways which in turn constitute positive feedback circuits influencing the vasculature, tissue swelling, and trophic structure of tissues

Multiple mechanisms are at the base of these effects of the sympathetic nervous system. Some have been studied in animal models, but have no equivalent in clinical reality yet. The research on the role of the sympathetic nervous system in the generation of pain, hyperalgesia, and inflammation is a developing field. In future, we will gain insight into these new mechanisms that probably underlie various pain diseases.

Acknowledgment This work is supported by the Deutsche Forschungsgemeinschaft, the German Ministry of Research and Education, the German Research Network on Neuropathic Pain, and an unrestricted educational grant from Pfizer, Germany.

References

Ader, A. (Ed.). (2007). *Psychoneuroimmunology*. Amsterdam: Academic Press/Elsevier.

Ader, A., & Cohen, N. (1993). Psychoneuroendocrinology: Conditioning and stress. *Annual Review of Physiology, 44*, 53–85.

Ali, Z., Raja, S. N., Wesselmann, U., Fuchs, P. N., Meyer, R. A., & Campbell, J. N. (2000). Intradermal injection of norepinephrine evokes pain in patients with sympathetically maintained pain. *Pain, 88*, 161–168.

Andreev, N. Y., Dimitrieva, N., Koltzenburg, M., & McMahon, S. B. (1995). Peripheral administration of nerve growth factor in the adult rat produces a thermal hyperalgesia that requires the presence of sympathetic post-ganglionic neurones. *Pain, 63*, 109–115.

Arnér, S. (1991). Intravenous phentolamine test: Diagnostic and prognostic use in reflex sympathetic dystrophy. *Pain, 46*, 17–22.

Bandler, R., & Shipley, M. T. (1994). Columnar organization in the midbrain periaqueductal gray: Modules for emotionel expression? *Trends in Neurosciences, 17*, 379–389.

Bandler, R., Price, J. L., & Keay, K. A. (2000a). Brain mediation of active and passive emotional coping. *Progress in Brain Research, 122*, 333–349.

Bandler, R., Keay, K. A., Floyd, N., & Price, J. (2000b). Central circuits mediating patterned autonomic activity during active vs. passive emotional coping. *Brain Research Bulletin, 53*, 95–104.

Baron, R., & Jänig, W. (2010). Adrenergic and cholinergic targets in pain pharmacology. In P. Beaulieu, D. Lussier, F. Porreca, & A. H. Dickenson (Eds.), *Pharmacology of pain* (pp. 347–381). Seattle: IASP Press.

Baron, R., & Jänig, W. (2013). Mechanisms underlying complex regional pain syndrome. Role of the sympathetic nervous system. In C. J. Mathias & R. Bannister (Eds.), *Autonomic failure* (5th ed., pp. 136–147). Oxford: Oxford University Press.

Baron, R., Schattschneider, J., Binder, A., Siebrecht, D., & Wasner, G. (2002). Relation between sympathetic vasoconstrictor activity and pain and hyperalgesia in complex regional pain syndromes: A case-control study. *Lancet, 359*, 1655–1660.

Barrett, H. L. G., & McLachlan, E. M. (1999). Long-term changes in peripheral nerve trunks following nerve ligation in rats. *Proceedings of the Australian Neuroscience Society, 10*, 116.

Besedovsky, H. O., & del Rey, A. (1995). Immune-neuroendocrine interactions: Facts and hypotheses. *Endocrine Reviews, 17*, 64–102.

Bradesi, S., Mayer, E. A., & Schwetz, I. (2009). Irritable bowel syndrome. In A. L. Basbaum & M. C. Bushnell (Eds.), *Science of pain* (pp. 571–578). San Diego: Academic.

Bultitude, J. H., & Rafal, R. D. (2010). Derangement of body representation in complex regional pain syndrome: Report of a case treated with mirror and prisms. *Experimental Brain Research, 204,* 409–418.

Bushnell, M. C., & Apkarian, A. V. (2006). Representation of pain in the brain. In S. B. McMahon & M. Koltzenburg (Eds.), *Wall and Melzack's textbook of pain* (5th ed., pp. 107–124). Amsterdam/Edinburgh: Elsevier/Churchill Livingstone.

Campero, M., Bostock, H., Baumann, T. K., & Ochoa, J. L. (2010). A search for activation of C nociceptors by sympathetic fibers in complex regional pain syndrome. *Clinical Neurophysiology, 121,* 1072–1079.

Carroll, I., Clark, J. D., & Mackey, S. (2009). Sympathetic block with botulinum toxin to treat complex regional pain syndrome. *Annals of Neurology, 65,* 348–351.

Casey, K. L., Tran, T. D. (2006). Cortical mechanisms mediating acute and chronic pain in humans. In F. Cervero, T. S. Jensen, (Eds.), *Handbook of clinical neurology,* Vol. 81. Aminoff, M. J., Boller, F., Swaab, D. F. (series Eds.) (pp. 159–177). Edinburgh: Elsevier.

Cohen, H. E., Hall, J., Harris, N., McCabe, C. S., Blake, D. R., & Jänig, W. (2012). Enhanced pain and autonomic responses to ambiguous visual stimuli in chronic Complex Regional Pain Syndrome (CRPS) type I. *European Journal of Pain, 16,* 182–195.

Cunha, T. M., Verri, W. A., Jr., Poole, S., Parada, C. A., Cunha, F. Q., & Fereira, S. H. (2007). Pain facilitation of proinflammatory cytokine actions at peripheral nerve terminals. In J. A. DeLeo, L. S. Sorkin, & L. R. Watkins (Eds.), *Immune and glial regulation of pain* (pp. 67–83). Seattle: IASP Press.

Donello, J. E., Guan, Y., Tian, M., Cheevers, C. V., Alcantara, M., Cabrera, S., Raja, S. N., & Gil, D. W. (2011). A peripheral adrenoceptor-mediated sympathetic mechanism can transform stress-induced analgesia into hyperalgesia. *Anesthesiology, 114,* 1403–1416.

Donnerer, J., Schuligoi, R., & Stein, C. (1992). Increased content and transport of substance P and calcitonin gene-related peptide in sensory nerves innervating inflamed tissue: Evidence for a regulatory function of nerve growth factor in vivo. *Neuroscience, 49,* 693–698.

Elenkov, I. J., Wilder, R. L., Chrousos, G. P., & Vizi, E. S. (2000). The sympathetic nerve–an integrative interface between two supersystems: The brain and the immune system. *Pharmacological Reviews, 52,* 595–638.

Gonzales, R., Goldyne, M. E., Taiwo, Y. O., & Levine, J. D. (1989). Production of hyperalgesic prostaglandins by sympathetic postganglionic neurons. *Journal of Neurochemistry, 53,* 1595–1598.

Green, P. G., Jänig, W., & Levine, J. D. (1997). Negative feedback neuroendocrine control of inflammatory

response in the rat is dependent on the sympathetic postganglionic neuron. *The Journal of Neuroscience, 17,* 3234–3238.

Grossmann, L., Gorodetskaya, N., Baron, R., & Jänig, W. (2009). Enhancement of ectopic discharge in regenerating A- and C-fibers by inflammatory mediators. *Journal of Neurophysiology, 101,* 2762–2774.

Harden, R. N., Baron, R., & Jänig, W. (Eds.). (2001). *Complex regional pain syndrome* (Vol. 22). Seattle: IASP Press.

Hori, T., Katafuchi, T., Take, S., Shimizu, N., & Niijima, A. (1995). The autonomic nervous system as a communication channel between the brain and the immune system. *Neuroimmunomodulation, 2,* 203–215.

Jänig, W. (1993). Spinal visceral afferents, sympathetic nervous system and referred pain. In L. Vecchiet, D. Albe-Fessard, U. Lindblom, & M. A. Giamberardino (Eds.), *New trends in referred pain and hyperalgesia* (Pain research and clinical management, Vol. 7, pp. 83–92). Amsterdam: Elsevier Science.

Jänig, W. (2005). Vagal afferents and visceral pain. In B. Undem & D. Weinreich (Eds.), *Advances in vagal afferent neurobiology* (pp. 461–489). Boca Raton: CRC Press.

Jänig, W. (2006). *The integrative action of the autonomic nervous system. Neurobiology of homeostasis.* Cambridge/New York: Cambridge University Press.

Jänig, W. (2009a). Autonomic nervous system and pain. In A. I. Basbaum & M. C. Bushnell (Eds.), *Science of pain* (pp. 193–225). San Diego: Academic.

Jänig, W. (2009b). Autonomic nervous system dysfunction. In E. A. Mayer & M. C. Bushnell (Eds.), *Functional pain syndromes* (pp. 265–300). Seattle: IASP Press.

Jänig, W. (2012). Autonomic reactions in pain. *Pain, 153,* 733–735.

Jänig, W. (2013). Pain and the sympathetic nervous system: pathophysiological mechanisms. In C. J. Mathias & R. Bannister (Eds.), *Autonomic failure* (5th ed., pp. 677–687). New York/Oxford: Oxford University Press.

Jänig, W., & Baron, R. (2002). Complex regional pain syndrome is a disease of the central nervous system. *Clinical Autonomic Research, 12,* 150–164.

Jänig, W., & Baron, R. (2003). Complex regional pain syndrome: Mystery explained? *Lancet Neurology, 2,* 687–697.

Jänig, W., & Baron, R. (2004). Experimental approach to CRPS. *Pain, 108,* 3–7.

Jänig, W., & Häbler, H. J. (2000a). Sympathetic nervous system: Contribution to chronic pain. *Progress in Brain Research, 129,* 451–468.

Jänig, W., & Häbler, H. J. (2000b). Specificity in the organization of the autonomic nervous system: A basis for precise neural regulation of homeostatic and protective body functions. *Progress in Brain Research, 122,* 351–367.

Jänig, W., & Levine, J. D. (2013). Autonomic-neuroendocrine-immune responses in acute and chronic pain. In S. B. McMahon, M. Koltzenburg, I. Tracey & D. C. Turk (Eds.), *Wall & Melzack's textbook of pain* (6th ed.). Edinburgh: Elsevier/Churchill Livinstone.

Jänig, W., & McLachlan, E. M. (2013). Neurobiology of the autonomic nervous system. In C. J. Mathias & R. Bannister (Eds.), *Autonomic failure* (5th ed., pp. 21–34). New York/Oxford: Oxford University Press.

Jänig, W., & Stanton-Hicks, M. (Eds.). (1996). *Reflex sympathetic dystrophy – A reappraisal*. Seattle: IASP Press.

Jänig, W., Khasar, S. G., Levine, J. D., & Miao, F. J. (2000). The role of vagal visceral afferents in the control of nociception. *Progress in Brain Research, 122*, 273–287.

Jorum, E., Orstavik, K., Schmidt, R., Namer, B., Carr, R. W., Kvarstein, G., Hilliges, M., Handwerker, H., Torebjörk, E., & Schmelz, M. (2007). Catecholamine-induced excitation of nociceptors in sympathetically maintained pain. *Pain, 127*, 296–301.

Keay, K. A., & Bandler, R. (2004). Periaqueductal gray. In G. Paxinos (Ed.), *The rat nervous system* (3rd ed., pp. 243–257). San Diego: Academic.

Khasar, S. G., Green, P. G., & Levine, J. D. (1993). Comparison of intradermal and subcutaneous hyperalgesic effects of inflammatory mediators in the rat. *Neuroscience Letters, 153*, 215–218.

Khasar, S. G., Miao, F. J. P., & Levine, J. D. (1995). Inflammation modulates the contribution of receptor-subtypes to bradykinin-induced hyperalgesia in the rat. *Neuroscience, 69*, 685–690.

Khasar, S. G., Miao, F. J. P., Jänig, W., & Levine, J. D. (1998a). Modulation of bradykinin-induced mechanical hyperalgesia in the rat skin by activity in the abdominal vagal afferents. *The European Journal of Neuroscience, 10*, 435–444.

Khasar, S. G., Miao, F. J. P., Jänig, W., & Levine, J. D. (1998b). Vagotomy-induced enhancement of mechanical hyperalgesia in the rat is sympathoadrenal-mediated. *The Journal of Neuroscience, 18*, 3043–3049.

Khasar, S. G., Green, P. G., Miao, F. J., & Levine, J. D. (2003). Vagal modulation of nociception is mediated by adrenomedullary epinephrine in the rat. *The European Journal of Neuroscience, 17*, 909–915.

Ledowski, T., Reimer, M., Chavez, V., Kapoor, V., & Wenk, M. (2012). Effects of acute postoperative pain on catecholamine plasma levels, hemodynamic parameters, and cardiac autonomic control. *Pain, 153*, 759–764.

Lewin, G. R., Ritter, A. M., & Mendell, L. M. (1993). Nerve growth factor-induced hyperalgesia in the neonatal and adult rat. *The Journal of Neuroscience, 13*, 2136–2148.

Lewin, G. R., Rueff, A., & Mendell, L. M. (1994). Peripheral and central mechanisms of NGF-induced hyperalgesia. *The European Journal of Neuroscience, 6*, 1903–1912.

Lewis, J. S., Kersten, P., McCabe, C. S., McPherson, K. M., & Blake, D. R. (2007). Body perception disturbance: A contribution to pain in complex regional pain syndrome (CRPS). *Pain, 133*, 111–119.

Lewis, J. S., Kersten, P., McPherson, K. M., Taylor, G. J., Harris, N., McCabe, C. S., & Blake, D. R. (2010). Wherever is my arm? Impaired upper limb position accuracy in complex regional pain syndrome. *Pain, 149*, 463–469.

Llewellyn-Smith, I. J., & Verbene, A. J. M. (2011). *Central regulation of autonomic functions*. New York: Oxford University Press.

Madden, K. S., & Felten, D. L. (1995). Experimental basis for neural-immune interactions. *Physiological Reviews, 75*, 77–106.

Madden, K. S., Sanders, K., & Felten, D. L. (1995). Catecholamine influences and sympathetic modulation of immune responsiveness. *Review of Pharmacology and Toxicology, 35*, 417–448.

Mayer, E. M., & Bushnell, M. C. (Eds.). (2009). *Functional pain syndromes: presentation and pathophysiology*. Seattle: IASP Press.

McCabe, C. S., & Blake, D. R. (2008). An embarrassment of pain perceptions? Towards an understanding of and explanation for the clinical presentation of CRPS type 1. *Rheumatology (Oxford, England), 47*, 1612–1616.

McMahon, S. B. (1996). NGF as a mediator of inflammatory pain. *Philosophical Transactions of the Royal Society of London. Series B, Biological Sciences, 351*, 431–440.

Meier, P. M., Zurakowski, D., Berde, C. B., & Sethna, N. F. (2009). Lumbar sympathetic blockade in children with complex regional pain syndromes: A double blind placebo-controlled crossover trial. *Anesthesiology, 111*, 372–380.

Melzack R. (1999). From the gate to the neuromatrix. Pain (Suppl 6):S121–S126.

Mense, S., & Meyer, H. (1988). Bradykinin-induced modulation of the response behaviour of different types of feline group III and IV muscle receptors. *The Journal of Physiology, 398*, 49–63.

Michaelis, M., Vogel, C., Blenk, K. H., Arnarson, A., & Jänig, W. (1998). Inflammatory mediators sensitize acutely axotomized nerve fibers to mechanical stimulation in the rat. *The Journal of Neuroscience, 18*, 7581–7587.

Moseley, G. L. (2004). Graded motor imagery is effective for long-standing complex regional pain syndrome: A randomised controlled trial. *Pain, 108*, 192–198.

Moseley, G. L. (2005). Is successful rehabilitation of complex regional pain syndrome due to sustained attention to the affected limb? A randomised clinical trial. *Pain, 114*, 54–61.

Moseley, G. L., Zalucki, N., Birklein, F., Marinus, J., van Hilten, J. J., & Luomajoki, H. (2008). Thinking about movement hurts: The effect of motor imagery on pain and swelling in people with chronic arm pain. *Arthritis and Rheumatism, 59,* 623–631.

Moseley, G. L., Gallace, A., & Spence, C. (2009). Space-based, but not arm-based, shift in tactile processing in complex regional pain syndrome and its relationship to cooling of the affected limb. *Brain, 132,* 3142–3151.

Nance, D. M., & Sanders, V. M. (2007). Autonomic innervation and regulation of the immune system. *Brain, Behavior and Immunity, 21,* 736–745.

Neugebauer, V., Schaible, H. G., & Schmidt, R. F. (1989). Sensitization of articular afferents to mechanical stimuli by bradykinin. *Pflügers Archiv, 415,* 330–335.

Neugebauer, V., Galhardo, V., Maione, S., & Mackey, S. C. (2009). Forebrain pain mechanisms. *Brain Research Reviews, 60,* 226–242.

Petty, B. G., Cornblath, D. R., Adornato, B. T., Chaudhry, V., Flexner, C., Wachsman, M., Sinicropi, D., Burton, L. E., & Peroutka, S. J. (1994). The effect of systemically administered recombinant human nerve growth factor in healthy human subjects. *Annals of Neurology, 36,* 244–246.

Pongratz, G., & Straub, R. H. (2010). The B cell, arthritis, and the sympathetic nervous system. *Brain, Behavior, and Immunity, 24,* 186–192.

Price, D. D. (2000). Psychological and neural mechanisms of the affective dimension of pain. *Science, 288,* 1769–1772.

Price, D. D., Long, S., Wilsey, B., & Rafii, A. (1998). Analysis of peak magnitude and duration of analgesia produced by local anesthetics injected into sympathetic ganglia of complex regional pain syndrome patients. *The Clinical Journal of Pain, 14,* 216–226.

Price, D. D., Hirsh, A., & Robinson, M. E. (2009). Psychological modulation of pain. In A. I. Basbaum & M. C. Bushnell (Eds.), *Science of pain* (pp. 975–1002). San Diego: Academic.

Rommel, O., Gehling, M., Dertwinkel, R., Witscher, K., Zenz, M., Malin, J. P., & Jänig, W. (1999). Hemisensory impairment in patients with complex regional pain syndrome. *Pain, 80,* 95–101.

Rommel, O., Malin, J.-P., Zenz, M., & Jänig, W. (2001). Quantitative sensory testing, neurophysiological and psychological examination in patients with complex regional pain syndrome and hemisensory deficits. *Pain, 93,* 279–293.

Stanton-Hicks, M., Jänig, W., Hassenbusch, S., Haddox, J. D., Boas, R., & Wilson, P. (1995). Reflex sympathetic dystrophy: Changing concepts and taxonomy. *Pain, 63,* 127–133.

Staud, R. (2009). Fibromyalgia. In A. I. Basbaum & M. C. Bushnell (Eds.), *Science of pain* (pp. 775–781). San Diego: Academic.

Straub, R. H., & Härle, P. (2005). Sympathetic neurotransmitters in joint inflammation. *Rheumatic Diseases Clinics of North America, 31,* 43–59. viii.

Straub, R. H., Baerwald, C. G., Wahle, M., & Jänig, W. (2005). Autonomic dysfunction in rheumatic diseases. *Rheumatic Diseases Clinics of North America, 31,* 61–75. viii.

Sumitani, M., Rossetti, Y., Shibata, M., Matsuda, Y., Sakaue, G., Inoue, T., Mashimo, T., & Miyauchi, S. (2007). Prism adaptation to optical deviation alleviates pathologic pain. *Neurology, 68,* 128–133.

Torebjörk, H. E., Wahren, L. K., Wallin, B. G., Hallin, R., & Koltzenburg, M. (1995). Noradrenaline-evoked pain in neuralgia. *Pain, 63,* 11–20.

Tracey, I., & Mantyh, P. W. (2007). The cerebral signature for pain perception and its modulation. *Neuron, 55,* 377–391.

Treede, R. D., & Apkarian, A. V. (2009). Nociceptive processing in the cerebral cortex. In A. I. Basbaum & M. C. Bushnell (Eds.), *Science of pain* (pp. 669–698). San Diego: Academic.

Treede, R. D., Meyer, R. A., Raja, S. N., & Campbell, J. N. (1992). Peripheral and central mechanisms of cutaneous hyperalgesia. *Progress in Neurobiology, 38,* 397–421.

Verri, W. A., Jr., Cunha, T. M., Parada, C. A., Poole, S., Cunha, F. Q., & Ferreira, S. H. (2006). Hypernociceptive role of cytokines and chemokines: Targets for analgesic drug development? *Pharmacology and Therapeutics, 112,* 116–138.

Wilson, P., Stanton-Hicks, M., & Harden, R. N. (Eds.). (2005). *CRPS: Current diagnosis and therapy.* Seattle: IASP Press.

Woolf, C. J. (1996). Phenotypic modification of primary sensory neurons: The role of nerve growth factor in the production of persistent pain. *Philosophical Transactions of the Royal Society of London B, 351,* 441–448.

Woolf, C. J., Safieh-Garabedian, B., Ma, Q.-P., Crilly, P., & Winter, J. (1994). Nerve growth factor contributes to the generation of inflammatory sensory hypersensitivity. *Neuroscience, 62,* 327–331.

Woolf, C. J., Ma, Q.-P., Allchorne, A., & Poole, S. (1996). Peripheral cell types contributing to the hyperalgesic action of nerve growth factor in inflammation. *The Journal of Neuroscience, 16,* 2716–2723.

Sympathetic Nervous System in the Generation of Pain, Animal Behavioral Models

Wilfrid Jaenig[1] and Heinz-Joachim Häbler[2]
[1]Physiologisches Institut, Christian-Albrechts-Universität zu Kiel, Kiel, Germany
[2]FH Bonn-Rhein-Sieg, Rheinbach, Germany

Definition

Behavioral animal models are designed to test quantitatively physiological as well as

Sympathetic Nervous System in the Generation of Pain, Animal Behavioral Models, Fig. 1 Effect of surgical sympathectomy on mechanical allodynic behavior in Sprague-Dawley rats (**a**) and Wistar rats (**b**) following ligation and section of spinal nerves L5 and L6 (**a**) or spinal nerve L5 (**b**). Experiments in **a** from Kim et al. (1993) and experiments in **b** from Ringkamp et al. (1999). Mechanical allodynic behavior was tested by measuring response incidence (**a1–a3**, **b1**) and response threshold (**b2**, **b3**) to stimulation of the plantar skin of the hindpaw ipsilateral to the spinal nerve lesion (*filled symbols*) and of the contralateral hindpaw (*open symbols*) with von Frey hairs. For testing the *incidence of paw withdrawal*, the von Frey hair was applied repetitively six to eight times within a period of 2–3 s (**a**) or eight times (**b1**) with a frequency of about two per second to the plantar test area. Five or ten trials were performed on each hindpaw, and the number of trials with a withdrawal response was noted. The ordinate scale (in **a** and **b1**) shows the percentage of positive responses in groups of rats. The *threshold of paw*

pathological (e.g., neuropathic) pain resulting from defined interventions, in intact animals, or in animals subjected to defined intentional lesions, under conditions that are known to cause pain in humans. As pain cannot be measured directly in animals for obvious reasons, basic behavioral responses that are known normally to accompany pain are taken as an indicator of pain in non-anesthetized animals. Most often, simple motor reflexes, e.g., paw withdrawal, or in some cases more complex motor responses, e.g., guarding behavior of the affected limb, are used to test stimulation-induced pain. To assess spontaneous pain in an animal, spontaneous abnormal motor behavior after an experimental intervention is quantified within a given time interval. To determine the presence of allodynia and/or hyperalgesia, noxious and innocuous stimuli of different modality at defined strengths (e.g., mechanical with calibrated von Frey hairs, graded heat stimuli) are applied to the body region of interest, and paw withdrawal latency or paw withdrawal threshold is determined. However, as motor responses are not specific for painful stimulation, one has to keep in mind that the validity of animal models of pain critically rests on the interpretation that the observed motor behavior is indeed a correlate of the animal experiencing pain.

Several behavioral animal models have been developed of ▶ complex regional pain syndrome (CRPS) which may arise in humans with (CRPS II, formerly causalgia) or without (CRPS I, formerly reflex sympathetic dystrophy) nerve lesion (see below). In a number of CRPS patients, the pain is entirely or in part dependent on the sympathetic nervous system (sympathetically

maintained pain, SMP). The behavioral animal models for CRPS allow the testing of the role of the sympathetic nervous system in these pathological pain states. To this end, pain behavior is assessed before and after interventions that target the sympathetic nervous system. These interventions include surgical sympathectomy, chemical sympathectomy with 6-hydroxy-dopamine, systemic application of adrenoreceptor agonists or antagonists, or systemic application of guanethidine (which is taken up by the noradrenergic terminals and depletes norepinephrine), the most straightforward procedure being surgical sympathectomy.

Characteristics

Models of CRPS I (with No Apparent Nerve Lesion)

As the conditions underlying CRPS I in humans are still poorly understood, there is no animal model available on which the pathophysiological mechanisms and the potential involvement of the sympathetic nervous system could be studied. However, rat models of inflammatory pain and/ or hyperalgesia in the future may prove to be of relevance also for the pathophysiological mechanisms underlying CRPS I. Several rat behavioral models of hyperalgesia have been propagated in which the sympathetic nervous system is supposed to be involved: (a) cutaneous mechanical hyperalgesia elicited by the inflammatory mediator bradykinin, (b) cutaneous hyperalgesia induced by intradermal injection of capsaicin, and (c) cutaneous hyperalgesia generated by nerve growth factor (▶ Sympathetic Nervous

---

Sympathetic Nervous System in the Generation of Pain, Animal Behavioral Models, Fig. 1 (continued) withdrawal (b2, b3) was measured using the up-down testing paradigm (for details see Ringkamp et al. 1999). Surgical sympathectomy (bilateral removal of lumbar paravertebral ganglia L1 to L6 or L2-L4) was performed either after spinal nerve lesion (a1, a2, b1, b2) or before spinal nerve lesion (a3, b3) using a medial or lateral approach. a1, a2, b1, b2. Sympathectomy performed 7 days or 35 days (a2) after spinal nerve lesion. a3, b3. Sympathectomy performed 7 days prior to spinal nerve

lesion. Values are means $\pm$ 1SEM in groups of 13 (a1), 18 (a2), 9 (a3), 10 (b1, b2), or 4 rats (b3). Increase of response incidence and decrease in paw withdrawal threshold following spinal nerve lesion on the ipsilateral side were significant at $p < 0.05$ (a1, a2), $p < 0.001$ (b1, b2), or $p < 0.01$ (b3, Friedman ANOVA). Note that sympathectomy alleviated or prevented the development of the mechanical allodynic behavior in the experiments of Kim et al. (1993) but had no effect in the experiments of Ringkamp et al. (1999)

System and Pain) (Jänig and Baron 2004; Baron and Jänig 2010; Jänig 2009).

Coderre and coworkers showed that an ischemia-reperfusion injury on a rat hindlimb generates a CRPS-I-like picture. A tightly fitting O-ring was placed on one hindlimb proximal to the ankle in anesthetized rats for 3 h. These rats developed hyperemia, edema, plasma extravasation, pinprick, and cold hyperalgesic behavior, as well as mechanical allodynic behavior for up to 4 weeks, including spread of symptoms to the contralateral side (Coderre et al. 2004). Using an ischemia-reperfusion injury by ligation of the femoral artery in rats for 3 h, Ludwig et al. (2007) could not reproduce these results.

Models of CRPS II (After Nerve Lesion)
In several established rat models of neuropathic pain (believed to be models of CRPS II), behavioral experiments have been performed before and after interventions on the sympathetic supply: autotomy model (Wall et al. 1979), partial sciatic nerve lesion (Seltzer et al. 1990), chronic constriction injury (CCI) of the sciatic nerve (Bennett and Xie 1988), and spinal nerve ligation (SNL) (Kim and Chung 1992).

Autotomy Model
Ligation and transection of the sciatic and saphenous nerves in rats leads to autotomy behavior (self-mutilation) which is considered to be due to pain arising in the denervated part of the hindlimb (anesthesia dolorosa). Repeated intraperitoneal application of guanethidine prevented or significantly reduced the autotomy behavior (Wall et al. 1979). The decrease of self-mutilation was interpreted to mean that activity in sympathetic neurons maintains ectopic activity in lesioned afferent neurons which in turn triggers the autotomy behavior. However, the effect of surgical sympathectomy on the autotomy behavior was never tested.

Partial Sciatic Nerve Lesion
Partial ligation (1/3–1/2) of the sciatic nerve leads to signs of spontaneous pain, thermal hyperalgesia, mechanical allodynia, and mechanical hyperalgesia (Seltzer et al. 1990) which partly also extended to the contralateral nonlesioned side. Chemical sympatholysis with repeated intraperitoneal injection of guanethidine prior to nerve injury prevented thermal hyperalgesia from occurring but changed mechanical hyperalgesia only little (Shir and Seltzer 1991). When performed months after nerve lesion, sympatholysis by guanethidine alleviated all sensory disorders (Shir and Seltzer 1991). Consistent results were obtained by Tracey et al. (1995) who found that noradrenaline injected locally into the affected paw exacerbated, while the α_2-adrenoceptor antagonist yohimbine significantly relieved mechanical and thermal hyperalgesia which was abolished by chemical sympathectomy with 6-hydroxy-dopamine. However, thermal and mechanical hyperalgesia were not enhanced by the α_2-adrenoceptor agonist clonidine.

Conflicting evidence came from Kim et al. (1997) who found only small and nonsignificant effects of surgical sympathectomy performed 1 week after partial sciatic nerve lesion on mechanical and cold allodynia and ongoing pain.

Chronic Constriction Injury of the Sciatic Nerve
Chronic constriction injury of the sciatic nerve generates mechanical and thermal hyperalgesic behavior (Bennett and Xie 1988). Guilbaud and collaborators (Desmeules et al. 1995) found that surgical sympathectomy or depletion of sympathetic transmitters by guanethidine both resulted in a reduction of thermal hyperalgesia while mechanical hyperalgesia was little affected. Self-mutilating behavior and spontaneous pain behavior were not changed after guanethidine or surgical sympathectomy. However, only a small nonsignificant reduction of mechanical and cold allodynia was found after surgical sympathectomy in this model by Kim et al. (1997). Rats with cryoneurolysis of the sciatic nerve also develop signs of mechanical allodynia but no thermal hyperalgesic behavior. Surgical sympathectomy performed prior to or after the cryoneurolysis did not influence the development or the maintenance of the mechanical allodynic behavior (Willenbring et al. 1993).

Spinal Nerve Ligation

Ligation and transection of the spinal nerves L5 and L6 or only the spinal nerve L5 is followed by behavioral signs of mechanical hyperalgesia and allodynia, as well as thermal hyperalgesia (Kim and Chung 1992; Blenk et al. 1997). Some authors found that these pain behaviors were permanently reversed by surgical lumbar sympathectomy (Kim et al. 1993, 1997; Kinnman and Levine 1995; Fig. 1a), prevented to develop by surgical sympathectomy prior to spinal nerve lesion (Kim et al. 1993), and temporarily reversed by intraperitoneal injection of the α-adrenoreceptor antagonist phentolamine or guanethidine (Kim et al. 1993). These findings suggested that a coupling between sympathetic noradrenergic neurons and afferent neurons is critical for the development and maintenance of the behavioral changes. However, there were considerable differences between rat strains and substrains in the development of signs of mechanical allodynia and heat hyperalgesia (Yoon et al. 1996) and in the dependence of mechanical allodynic behavior on adrenoceptors.

Completely discrepant observations were made by Ringkamp et al. (1999) (Fig. 1b) who found that surgical sympathectomy performed prior to or after spinal nerve lesion had no effect on mechanical allodynic and hyperalgesic behavior in two different rat strains. Furthermore and most important, Eschenfelder et al. (2000) have shown that transection of the dorsal root L5 alone is followed by the same behavioral signs of allodynia and hyperalgesia over 57 days.

Conclusion

The controversial results with respect to an involvement of the sympathetic nervous system in neuropathic pain, allodynia, and hyperalgesia obtained on the aforementioned animal behavioral models indicate that at present we have no animal model that mimics SMP in patients with CRPS I or CRPS II as defined by the clinical situation (Baron and Jänig 2010).

References

Baron, R., & Jänig, W. (2010). Adrenergic and cholinergic targets in pain pharmacology. In P. Beaulieu, D. Lussier, F. Porreca, & A. H. Dickinson (Eds.), *Pharmacology of pain* (pp. 347–381). Seattle: IASP Press.

Bennett, G. J., & Xie, Y. K. (1988). A peripheral mononeuropathy in rat produces disorders of pain sensation like those seen in man. *Pain, 33,* 87–107.

Blenk, K. H., Häbler, H. J., & Jänig, W. (1997). Neomycin and gadolinium applied to an L5 spinal nerve lesion prevent mechanical allodynia-like behaviour in rats. *Pain, 70,* 155–165.

Coderre, T. J., Xanthos, D. N., Francis, L., & Bennett, G. J. (2004). Chronic postischemia pain (CPIP): A novel animal model of complex regional pain syndrome-type I (CRPS-I; reflex sympathetic dystrophy) produced by prolonged hindpaw ischemia and reperfusion in the rat. *Pain, 112,* 94–105.

Desmeules, J. A., Kayser, V., Weil-Fuggaza, J., Bertrand, A., & Guilbaud, G. (1995). Influence of the sympathetic nervous system in the development of abnormal pain-related behaviours in a rat model of neuropathic pain. *Neuroscience, 67,* 941–951.

Eschenfelder, S., Häbler, H.-J., & Jänig, W. (2000). Dorsal root section elicits signs of neuropathic pain rather than reversing them in rats with L5 spinal nerve injury. *Pain, 87,* 213–219.

Jänig, W. (2009). Autonomic nervous system and pain. In A. I. Basbaum & M. C. Bushnell (Eds.), *Science of pain* (pp. 193–225). San Diego: Academic.

Jänig, W., & Baron, R. (2004). Experimental approach to CRPS. *Pain, 108,* 3–7.

Kim, S. H., & Chung, J. M. (1992). An experimental model for peripheral neuropathy produced by segmental spinal nerve ligation in the rat. *Pain, 50,* 355–363.

Kim, S. H., Na, H. S., Sheen, K., & Chung, J. M. (1993). Effects of sympathectomy on a rat model of peripheral neuropathy. *Pain, 55,* 85–93.

Kim, K. J., Yoon, Y. W., & Chung, J. M. (1997). Comparison of three rodent neuropathic pain models. *Experimental Brain Research, 113,* 200–206.

Kinnman, E., & Levine, J. D. (1995). Sensory and sympathetic contributions to nerve injury-induced sensory abnormalities in the rat. *Neuroscience, 64,* 751–767.

Ludwig, J., Gorodetskaya, N., Schattschneider, J., Jänig, W., & Baron, R. (2007). Behavioral and sensory changes after direct ischemia-reperfusion injury in rats. *European Journal of Pain, 11,* 677–684.

Ringkamp, M., Eschenfelder, S., Grethel, E. J., Häbler, H. J., Meyer, R. A., Jänig, W., & Raja, S. N. (1999). Lumbar sympathectomy failed to reverse mechanical allodynia- and hyperalgesia-like behavior in rats with L5 spinal nerve injury. *Pain, 79,* 142–153.

Seltzer, Z., Dubner, R., & Shir, Y. (1990). A novel behavioral model of neuropathic pain disorders produced in rats by partial sciatic nerve injury. *Pain, 43,* 205–218.

Shir, Y., & Seltzer, Z. (1991). Effects of sympathectomy in a model of causalgiform pain produced by partial sciatic nerve injury in rats. *Pain, 45*, 309–320.

Tracey, D. J., Cunningham, J. E., & Romm, M. A. (1995). Peripheral hyperalgesia in experimental neuropathy: Mediation by alpha 2-adrenoreceptors on postganglionic sympathetic terminals. *Pain, 60*, 317–327.

Wall, P. D., Devor, M., Inbal, R., Scadding, J. W., Schonfeld, D., Seltzer, Z., & Tomkiewicz, M. M. (1979). Autotomy following peripheral nerve lesions: Experimental anaesthesia dolorosa. *Pain, 7*, 103–113.

Willenbring, S., Beauprie, I. G., & DeLeo, J. A. (1993). Sciatic cryoneurolysis in rats: a model of sympathetically independent pain. Part 1: Effects of sympathectomy. *Anesthesia & Analgesia, 81*, 544–548.

Yoon, Y. W., Lee, D. H., Lee, B. H., & Chung, J. M. (1996). Different strains and substrains of rat show different levels of neuropathic pain behaviors. *Experimental Brain Research, 129*, 167–171.

Sympathetic Postganglionic Neurons in Neurogenic Inflammation of the Synovia

Paul G. Green[1] and Wilfrid Jaenig[2]
[1]Oral & Maxillofacial Surgery, University of California San Francisco, San Francisco, CA, USA
[2]Physiologisches Institut, Christian-Albrechts-Universität zu Kiel, Kiel, Germany

Definition

Sympathetic postganglionic neurons (SPGNs) are the final motor neurons of the autonomic nervous system, and it has recently been appreciated that the sympathetic postganglionic nervous system plays a key role in the regulation of the inflammatory response. In particular, the SPGN mediates inflammation produced in the synovia by the potent inflammatory mediator, bradykinin. The synovium may be unique in regards to the dependence on the SPGN in neurogenic inflammation, possibly due to the density of SPGN innervation which accounts for between half and two-thirds of the nerve fibers in the synovium (Hildebrand et al. 1991; Langford and Schmidt 1983).

The role of the SPGN has been extensively studied in the control of plasma extravasation, a primary component of the inflammatory response. This has established the SPGN as a constituent part of a multifaceted integrated system that also includes primary afferent neurons, mast cells, infiltrating leukocytes, and blood vessels.

Characteristics

SPGNs are Involved in Synovial Inflammation

The inflammatory mediator, bradykinin, produces a marked increase in plasma extravasation in the rat knee joint. Evidence over the past 25 years has demonstrated a role for the SPGN in synovial inflammation; specifically, either chemically ablating the sympathetic nervous system with guanethidine or 6-hydroxy-dopamine or surgical excision of postganglionic neurons (Coderre et al. 1989; Levine et al. 1985) attenuates the magnitude of baseline extravasation as well as plasma extravasation produced by bradykinin infused into the knee joint. Bradykinin-induced synovial plasma extravasation is reduced by 60–70 % 7–14 days following surgical sympathectomy (Fig. 1c; Miao et al. 1996a, b). Importantly, at close to physiological levels (30–300 nM), bradykinin-induced increase in plasma extravasation is SPGN-dependent, while at >1 µM the increase in plasma extravasation is independent of the SPGN (Fig. 2).

Mechanism of SPGN Synovial Inflammation

The contribution of the SPGN to synovial inflammation is independent of classical vesicular neurotransmitter release, since bradykinin-induced plasma extravasation is unaffected by decentralization of the lumbar sympathetic chain (ablating preganglionic axons that innervate the postganglionic neurons to the hind limb; closed circle in Fig. 1c), acute interruption of the lumbar sympathetic chain, or perfusion of tetrodotoxin through the knee joint synovial cavity (Miao et al. 1996a). This indicates that mediators released in response to inflammatory agents, such as bradykinin, are not dependent on the excitability of the SPGN, thus not on generation of action potentials and vesicular release of noradrenaline. These inflammatory agents rather are synthesized and released either from the SPGN terminal or from other cells

Sympathetic Postganglionic Neurons in Neurogenic Inflammation of the Synovia, Fig. 1 Bradykinin-induced plasma extravasation in the synovia of the knee joint is largely dependent on the sympathetic innervation but not on activity in the sympathetic neurons. (**a**) The rat knee joint was perfused with saline in the anesthetized rat given at a constant rate of 200 µL/min. Rats were pretreated with Evans blue dye given i.v. (Evans blue bind to albumin and does not normally leave the vascular space). Plasma extravasation of the synovia into the knee joint cavity was determined by measuring Evans blue dye extravasation into the perfusate in 5-min samples. The concentration of Evans blue, determined spectrophotometrically at a wavelength of 620 nm, is proportional to the degree of extravasation (ordinate scale in **c**). Bradykinin (BK, 160 ng/ml, 1.5×10^{-7} M), an inflammatory mediator, was added to the perfusate 15 min after beginning of the perfusion, and the perfusion lasted for

in association with the SPGN terminal. Release of noradrenaline from the SPGN attenuates the increased magnitude of plasma extravasation. Activation of the sympathetic postganglionic neurons by electrical stimulation of the lumbar sympathetic chain reduces both resting and bradykinin-induced plasma extravasation because of a reduction of blood flow through the synovia.

These types of experiments suggest that sympathetic postganglionic neurons innervating the joint capsule and its synovium have two functions: to regulate blood flow (vasoconstrictor function) and to mediate vascular permeability. The first occurs at the precapillary resistance vessels by vesicular release of transmitter(s) which induces vasoconstriction and which is regulated by action potentials in the sympathetic vasoconstrictor neurons. The second function occurs at the postcapillary venules by nonvesicular release of a chemical substance (or more than one substance) (possibly prostaglandin E_2 and/or related substances) that is independent of the electrical activity in the sympathetic neurons. Whether these substance(s) are released by the sympathetic terminals or by other cells in association with these terminals is unknown (Gonzales et al. 1989). This sympathetically mediated component of neurogenic inflammation is an entirely peripheral function of the sympathetic terminals. Whether both functions of the sympathetic postganglionic neurons are represented in the *same* class of neuron or in *distinct* ones remains to be studied.

Primary Afferent Neuron-SPGN Integration in Neurogenic Inflammation

In the rat knee joint stimulation of peptidergic (unmyelinated) primary afferents generates neurogenic inflammation (precapillary vasodilation and

venular plasma extravasation) involving substance P and calcitonin gene-related peptide (CGRP) (Ferrell and Russell 1986; Green et al. 1992). This afferent-induced neurogenic inflammation seems to be independent of the bradykinin-induced synovial plasma extravasation that is mediated by the sympathetic terminals. In fact, chemical destruction of unmyelinated primary afferents (by pretreating the rats on neonatal day 2 with capsaicin) does not affect the SPGN-mediated plasma extravasation (Coderre et al. 1989). However, it is unclear in which way plasma extravasation mediated by the sympathetic terminals and plasma extravasation mediated by peptidergic afferents in the synovia work together. To study this potential interaction requires a controlled selective activation of the afferents supplying the synovia, e.g., by electrical stimulation of the corresponding dorsal roots (Häbler et al. 1997).

Sensory-sympathetic neuron coupling has been well documented in pathophysiogical conditions underlying sympathetically maintained pain (e.g., in complex regional pain syndromes; see ▶ Sympathetic Nervous System and Pain; ▶ Sympathetically Maintained Pain in CRPS I, Human Experimentation; ▶ Sympathetic-afferent Coupling in the Dorsal Root Ganglion, Neurophysiological Experiments; ▶ Sympathetic-afferent Coupling in the Afferent Nerve Fiber, Neurophysiological Experiments).

Endocrine-SPGN Integration in Neurogenic Inflammation

We have demonstrated that the SPGN mediates a negative feedback control of inflammatory response. There are two, independent neuroendocrine circuits which are activated by noxious stimulation. Repetitive transcutaneous electrical

---

Sympathetic Postganglionic Neurons in Neurogenic Inflammation of the Synovia, Fig. 1 (continued) another 30 min. (**b**) Three preparations were used: intact lumbar sympathetic system, decentralized lumbar sympathetic system (preganglionic axons interrupted 7 days before the experiment by sectioning the white rami), and surgical sympathectomy 4 or 14 days before the experiment (removal of the paravertebral ganglia (see Baron et al. 1988)). (**c**) Increase of plasma extravasation induced by BK in control rats (*open circles*, N = 12 knees). This

increase was significantly smaller 14 days after sympathectomy (*closed squares*, N = 12 knees; p<0.01, two-way ANOVA) but was not significantly different from control after decentralization (*closed circles*, N = 12 knees). *Inset*: Baseline plasma extravasation over 15 min preceding infusion of BK was also lower in animals 4 days and 14 days after sympathectomy than in control animals ((**c**) Modified from Miao et al. 1996a; (**a**) from Paul Green)

Sympathetic Postganglionic Neurons in Neurogenic Inflammation of the Synovia, Fig. 2 *Concentration-dependent bradykinin-induced plasma extravasation in control rats (N=8) and in sympathectomized rats (N=8).* For experimental procedure, see legend to Fig. 1. Bradykinin was added cumulatively from 10^{-8} to 10^{-5} M to the perfusate. The bradykinin concentrations which have been measured in inflamed tissue (Hargreaves et al. 1993; Swift et al. 1993) are indicated in bold on the abscissa scale. In sympathectomized rats, the extravasation is significantly reduced at 3×10^{-8} to 3×10^{-7} M bradykinin ($p<0.01$, two-way ANOVA). At higher (pharmacological) concentrations, sympathectomy has no effect because the effect of bradykinin is generated via other than the sympathetic-mediated pathways (Modified from Miao et al. 1996b)

stimulation of unmyelinated afferents of plantar skin activates the hypothalamic-pituitary-adrenal axis to release corticosterone, which inhibits the SPGN-dependent plasma extravasation produced by bradykinin. This inhibitory endocrine circuit appears to specifically affect the SPGN-dependent component of the inflammatory response and does not appear to be a vascular effect since, following surgical sympathectomy, the residual (i.e., SPGN-independent) plasma extravasation produced by bradykinin was *not* inhibited by the noxious stimulation-induced feedback inhibition and neither was the SPGN-independent plasma extravasation produced by platelet activating factor. Corticosterone infused intravenously also depressed bradykinin-induced plasma extravasation at a time course that was similar to that during reflex activation of the hypothalamo-pituitary axis. Also, this inhibitory effect of i.v.-infused corticosterone was no longer present in sympathectomized joints (Green et al. 1995, 1997).

The effect of corticosterone, whether released from the adrenal cortex or injected intravenously, does not appear to be due to a direct action of glucocorticoids on the SPGN since corticosterone administered in the synovium does not itself inhibit bradykinin-induced plasma extravasation. However, the peptide annexin I (lipocortin I) co-infused with bradykinin in the rat knee joint, which is released from many tissues by corticosterone, does inhibit bradykinin-induced plasma extravasation (Green et al. 1998). By the same token is the depression of bradykinin-induced synovial plasma extravasation generated by electrical noxious stimulation or by corticosterone infused i.v. attenuated by annexin I antibody infused into the knee joint (Green et al. 1998). The source for annexin I released by corticosterone has not been determined, although polymorphonuclear leukocytes (PMNs), which are a rich source of annexin I, are required for bradykinin-induced synovial plasma extravasation (Bjerknes et al. 1991).

The other endocrine-SPGN anti-inflammatory circuit is activated by noxious somatic or visceral stimulation with capsaicin (Miao et al. 2000).

In this circuit, the sympatho-adrenal axis mediates the inhibition of bradykinin-induced plasma extravasation since it is abolished following denervation of the adrenal medulla. The depression of bradykinin-induced plasma extravasation in the synovia by adrenaline released by the adrenal medulla is not caused by vasoconstriction in the synovia. Its mechanisms of action are unknown but may involve β-adrenoceptors (Jänig et al. 2000; Miao et al. 2000, 2001).

These results raise the question of the biological significance of two independent anti-inflammatory neuroendocrine pathways. There are differences in the pattern of noxious stimulation that activate these circuits; while capsaicin selectively activates primary afferent C-fiber and a subpopulation of Aδ-afferents, electrical stimulation at the intensities used in these studies *synchronously* activates all C-, Aδ-, and Aβ-fibers. This differential pattern of activation of nociceptive afferents could be responsible for specific encoding of nociceptive signals to selectively activate specific neuroendocrine circuits. The differential activation of either the hypothalamo-pituitary-adrenal system or the sympatho-adrenal system during stimulation of nociceptive afferents by transcutaneous electrical impulses or by capsaicin, respectively, may turn out to be a valuable tool in the experimental investigation of the effects of both neuroendocrine systems on inflammation. However, it has yet to be determined which physiological stimuli (e.g., activation of nociceptive afferents during chronic inflammation or during traumatic injury pain) activate either the hypothalamo-pituitary system or the sympatho-adrenal system.

Conclusions

Experimental inflammation in the rat knee joint generated by the inflammatory mediator bradykinin is largely dependent on the sympathetic innervation. Sympathetic terminals mediate synovial plasma extravasation, possibly by release of a prostanoid either from the sympathetic terminal or from other cells, independent of their excitability and of vesicular release of noradrenaline. This peripheral mechanism is under inhibitory control of the hypothalamo-pituitary-adrenal system (corticosterone) and the sympatho-adrenal system.

The sympathetic terminal mediates the action of corticosterone on bradykinin-induced plasma extravasation, with the intermediary substance probably being the peptide annexin I. These experiments show that the sympathetic terminal nerve fiber is in a strategic position to integrate local inflammatory signals and remote neuroendocrine signals (e.g., corticosterone, adrenaline) in the modulation of venular plasma extravasation. This is a function of the sympathetic terminal that is independent of its excitability and release of noradrenaline.

References

Baron, R., Jänig, W., & Kollmann, W. (1988). Sympathetic and afferent somata projecting in hindlimb nerves and the anatomical organization of the lumbar sympathetic nervous system of the rat. *The Journal of Comparative Neurology, 275*, 460–468.

Bjerknes, L., Coderre, T. J., Green, P. G., et al. (1991). Neutrophils contribute to sympathetic nerve terminal-dependent plasma extravasation in the knee joint of the rat. *Neuroscience, 43*, 679–686.

Coderre, T. J., Basbaum, A. I., & Levine, J. D. (1989). Neural control of vascular permeability; interactions between primary afferents, mast cells, and sympathetic efferents. *Journal of Neurophysiology, 62*, 48–58.

Ferrell, W. R., & Russell, N. J. (1986). Extravasation in the knee induced by antidromic stimulation of articular C fibre afferents of the anaesthetized cat. *The Journal of Physiology, 379*, 407–416.

Gonzales, R., Goldyne, M. E., Taiwo, Y. O., et al. (1989). Production of hyperalgesic prostaglandins by sympathetic postganglionic neurons. *Journal of Neurochemistry, 53*, 1595–1598.

Green, P. G., Basbaum, A. I., & Levine, J. D. (1992). Sensory neuropeptide interactions in the production of plasma extravasation in the rat. *Neuroscience, 50*, 745–749.

Green, P. G., Miao, F. J.-P., Jänig, W., et al. (1995). Negative feedback neuroendocrine control of the inflammatory response in rats. *The Journal of Neuroscience, 15*, 4678–4686.

Green, P. G., Jänig, W., & Levine, J. D. (1997). Negative feedback neuroendocrine control of inflammatory response in the rat is dependent on the sympathetic postganglionic neuron. *The Journal of Neuroscience, 17*, 3234–3238.

Green, P. G., Strausbaugh, H. J., & Levine, J. D. (1998). Annexin I is a local mediator in neural-endocrine feedback control of inflammation. *Journal of Neurophysiology, 80*, 3120–3126.

Häbler, H. J., Wasner, G., & Jänig, W. (1997). Interaction of sympathetic vasoconstriction and antidromic vasodilatation in the control of skin blood flow. *Experimental Brain Research, 113*, 402–410.

Hargreaves, K. M., Roszkowski, M. T., & Swift, J. Q. (1993). Bradykinin and inflammatory pain. *Agents and Actions Supplements, 41*, 65–73.

Hildebrand, C., Oqvist, G., Brax, L., et al. (1991). Anatomy of the rat knee joint and fibre composition of a major articular nerve. *The Anatomical Record, 229*, 545–555.

Jänig, W., Khasar, S. G., Levine, J. D., et al. (2000). The role of vagal visceral afferents in the control of nociception. *Progress in Brain Research, 122*, 271–285.

Langford, L. A., & Schmidt, R. F. (1983). Afferent and efferent axons in the medial and posterior articular nerves of the cat. *The Anatomical Record, 206*, 71–78.

Levine, J. D., Dardick, S. J., Basbaum, A. I., et al. (1985). Reflex neurogenic inflammation. I. Contribution of the peripheral nervous system to spatially remote inflammatory responses that follow injury. *The Journal of Neuroscience, 5*, 1380–1386.

Miao, F. J.-P., Jänig, W., & Levine, J. D. (1996a). Role of sympathetic postganglionic neurons in synovial plasma extravasation induced by bradykinin. *Journal of Neurophysiology, 75*, 715–724.

Miao, F. J.-P., Green, P., Coderre, T. J., et al. (1996b). Sympathetic-dependence in bradykinin-induced synovial plasma extravasation is dose-related. *Neuroscience Letters, 205*, 165–168.

Miao, F. J.-P., Jänig, W., & Levine, J. D. (2000). Nociceptive neuroendocrine negative feedback control of neurogenic inflammation activated by capsaicin in the rat paw: Role of the adrenal medulla. *The Journal of Physiology, 527*, 601–610.

Miao, F. J.-P., Jänig, W., Jasmin, L., et al. (2001). Spino-bulbo-spinal pathway mediating vagal modulation of nociceptive-neuroendocrine control of inflammation in the rat. *The Journal of Physiology, 532*, 811–822.

Swift, J. Q., Garry, M. G., Roszkowski, M. T., et al. (1993). Effect of flurbiprofen on tissue levels of immunoreactive bradykinin and acute postoperative pain. *Journal of Oral and Maxillofacial Surgery, 51*, 112–116.

Sympathetic Skin Response

Synonyms

SSR; Electrodermal activity (EDA); Galvanic skin response (GSR)

Definition

The sympathetic skin response (SSR) is recorded from palmar or plantar skin versus the dorsal skin of hand or foot using AgAgCl-electrodes and standard clinical electrophysiological recording techniques. Activation of eccrine sweat glands by activity in postganglionic sudomotor neurons, generated, e.g., by cutaneous electrical stimulation of afferents, acoustic stimuli, or mental stimulation, is followed by the SSR. The first transient negative potential is ascribed to the synaptic activation of sweat glands. It can be blocked by atropine (Mathias and Bannister 2012).

Cross-References

▶ Congenital Insensitivity to Pain with Anhidrosis

References

Mathias, C. J., & Bannister, R. (Eds.). (2012). *Autonomic failure* (5th ed.). Oxford: Oxford University Press.

Sympathetic-Afferent Coupling in the Afferent Nerve Fiber, Neurophysiological Experiments

Matthias Ringkamp[1] and Heinz-Joachim Häbler[2]
[1]Department of Neurosurgery, School of Medicine Johns Hopkins University, Baltimore, MD, USA
[2]FH Bonn-Rhein-Sieg, Rheinbach, Germany

Synonyms

Coupling of sympathetic postganglionic neurons onto primary afferent nerve fibers; Sympathetic-sensory coupling distal to the dorsal root ganglion (DRG), peripheral sympathetic-afferent coupling; Sympathetic-sensory coupling

Definition

In neuropathic painful conditions, neuronal activity in sympathetic efferent nerve fibers may cause excitation in somatic afferents. This sympathetic-sensory coupling may contribute to the ectopic activity of lesioned and/or unlesioned afferents in animal models of neuropathic pain (models of CRPS II). Ectopic activity is thought to be involved in initiating and maintaining neuropathic pain behavior. Ectopic activity arises from the somata of primary afferent neurons in the dorsal root ganglion (DRG) (see ► Sympathetic-Afferent Coupling in the Dorsal Root Ganglion, Neurophysiological Experiments) and can also be generated at different sites along the axon and receptive terminals. In vivo neurophysiological recordings in anesthetized animals can be used to study coupling between efferent sympathetic postganglionic fibers and nociceptive afferents. Coupling may develop after various types of experimental nerve lesions and could provide a mechanism of SMP in patients. To test for sympathetic-afferent coupling, the sympathetic supply is either activated by natural stimuli (e.g., hypoxia) or directly by electrical stimulation while the activity in primary afferent fibers is measured. In addition, pharmacological tools (application of adrenoceptor agonists and antagonists) are applied to block or to mimic the effects of sympathetic-sensory coupling on activity in nociceptive afferents. Sympathetic-afferent coupling may be direct, i.e., the primary afferent neuron is activated by noradrenaline released from sympathetic axons, and/or indirect via effects, for example, on the vascular bed. Results from these reduced in vivo animal models have to be correlated with the results obtained in behavioral animal models and findings in patients.

Characteristics

Following nerve injury, sympathetic noradrenergic neurons may couple onto afferent neurons at several sites. In addition to coupling at the dorsal root ganglion, coupling may also occur along the afferent axon. Coupling could occur at or close to the injury site as well as proximal to the lesion or distal at the afferent terminal.

Furthermore, it may involve lesioned, intact, or regenerating afferents. That peripheral sympathetic-afferent coupling is a mechanism for SMP is consistent with the finding that intradermal administration of adrenergic agents rekindles pain in patients that underwent a sympathetic block for the treatment and/or diagnosis of SMP (Ali et al. 2000; Torebjork et al. 1995). In microneurographic recordings in human, sympathetic-afferent coupling has been observed by some (Jorum et al. 2007), but not by others (Campero et al. 2010).

Coupling Between Sympathetic Fibers and Afferent Endings in Neuromas Following Nerve Lesion

Sympathetic-afferent coupling in neuromas is *rare*, at least in cats and particularly in old neuromas, taking into account the relatively low proportion of lesioned afferents showing ectopic activity (Blumberg and Jänig 1984). It may be more common in rat neuromas in which the proportion of ectopically active afferents appears to be high in the first 2 weeks after lesion (Devor and Jänig 1981). In the early phase after lesion, coupling to myelinated and unmyelinated fibers occurs in the cat, but mainly to myelinated fibers in rats. In old sciatic nerve neuromas of the rat, sympathetic-afferent coupling occurs mainly in unmyelinated afferents (Jänig 1990). Lesioned afferents with ectopic activity can be activated by epinephrine/norepinephrine and by electrical stimulation of the sympathetic trunk, and this activation is mediated by α-adrenoceptors. However, high stimulation frequencies of ^{3}10 Hz are needed to elicit discharges in afferent neurons from the neuroma (Blumberg and Jänig 1984; Devor and Jänig 1981). Such frequencies never occur in sympathetic neurons under physiological in vivo conditions. Furthermore, there is no evidence that the ongoing ectopic activity in neuroma fibers is generated by ongoing sympathetic activity. Likewise, there is no evidence that postganglionic sympathetic fibers ephaptically couple to afferents from the neuroma. It is possible that the observed sympathetic-afferent coupling in neuromas was mediated indirectly through changes in blood flow.

Sympathetic-Afferent Coupling After Nerve Lesion with Subsequent Regeneration of Afferent and Sympathetic Fibers to the Target Tissue

More than 1 year after an inappropriate cross-union between the proximal stump of either the sural or the superficial peroneal nerve and the distal stump of the tibial nerve of the cat, unmyelinated afferent fibers with low level of ectopic activity could be vigorously excited by low frequency electrical stimulation of the sympathetic chain (frequencies of [3]0.5 Hz) and/or by intravenous injection of epinephrine (Häbler et al. 1987) (Fig. 1). Physiological stimulation of the sympathetic neurons by systemic hypoxia also activated some unmyelinated afferent fibers. The sympathetically and epinephrine-induced discharges were blocked by the a-adrenoceptor antagonist phentolamine indicating an a-adrenoceptor-mediated coupling. These effects were unlikely to be mediated indirectly via the vascular bed in the lesioned nerve, because angiotensin-induced vasoconstriction did not activate afferent neurons. This is yet the only experimental study showing that stimulation of sympathetic neurons at physiological frequencies is able to elicit impulses in unmyelinated, probably nociceptive afferents after nerve lesion.

Coupling Between Unlesioned Postganglionic and Afferent Nerve Terminals Following Partial Nerve Lesion

After partially cutting the great auricular nerve in rabbits, a subset of spared C-fiber polymodal nociceptors and also a few spared Aδ-nociceptors developed sensitivity to electrical stimulation of the cervical sympathetic trunk and to close-arterial injection of norepinephrine or epinephrine. In addition, they were sensitized to heat stimuli by this sympathetic stimulation 4–148 days after lesion. These effects were preferentially mediated by α_2-adrenoceptors (O'Halloran and Perl 1997; Sato and Perl 1991). Indirect

effects mediated via the vascular bed were unlikely, because vasopressin did not affect nociceptors. However, sympathetic-afferent coupling leading to activation and sensitization of nociceptors required electrical stimulation of the sympathetic supply at high, non-physiological frequencies. It was proposed that circulating epinephrine might act via α_2-adrenoceptors that are upregulated on afferent terminals after nerve lesion (Birder and Perl 1999). However, the epinephrine concentrations used to mimic sympathetic-afferent coupling were much higher than plasma concentrations under physiological conditions.

In vitro experiments in nonhuman primates revealed that cutaneous C-fiber afferents spared by the nerve lesion can acquire sensitivity to sympathetic agonists. After partial denervation of the skin resulting from an L6 spinal nerve lesion, uninjured C-fiber nociceptors showed spontaneous activity and displayed a higher incidence of responses to both α_1- and α_2-adrenergic agonists than controls (Ali et al. 1999). However, the rate of spontaneous activity and the response magnitudes were generally low.

In the absence of any direct injury to afferent nerve fibers, a similar sensitivity to norepinephrine was found in C-fiber polymodal nociceptors following surgical sympathectomy (Bossut et al. 1996). This finding was suggested to be a correlate of postsympathectomy neuralgia that is observed in some patients following sympathectomy.

Sympathetic-Afferent Coupling in the Nerve Proximal to the Nerve Lesion

A nerve lesion is also followed by dramatic changes along the nerve proximal to the lesion. Many neurons with unmyelinated axons die with time after nerve lesion (Jänig and McLachlan 1984). Peptidergic and nonpeptidergic afferents and postganglionic fibers start to sprout retro- and anterogradely, and there are signs of inflammation and angiogenesis and changes in the innervation of blood vessels supplying the nerve (see essay McLachlan). In view of these and other changes, it is possible that sympathetic-afferent

Sympathetic-Afferent Coupling in the Afferent Nerve Fiber, Neurophysiological Experiments, Fig. 1 Sympathetic-afferent coupling after nerve lesion with subsequent fiber regeneration. Excitation of unmyelinated afferent units by electrical stimulation of sympathetic fibers following nerve injury. Unmyelinated primary afferents were recorded in cats 11–20 months following a nerve lesion. The central cut stump of a cutaneous nerve innervating hairy skin (sural [SU] or superficial peroneal nerve) had been imperfectly adapted to the distal stump of a transected mixed nerve (tibial nerve). There was a "neuroma-in-continuity" at the site of the lesion and cutaneous nerve fibers had regenerated into skin and deep somatic tissue supplied by the mixed nerve. (a) Experimental setup. preganglionic, preganglionic; *LST* lumbar sympathetic trunk, *TIB* tibial nerve. *Lower record*: The afferent fibers were identified as unmyelinated by electrical stimulation of the neuroma with single impulses. The signal indicated by dot was same as in (**b**); the afferent fiber conducted at 1.3 m/s. (**b**), Record from a single unmyelinated afferent unit. Supramaximal stimulation of the LST with trains of 30 pulses at 1–5 Hz (trains and stimulation artifacts indicated by bars). Note that the afferent unit had some low rate of ongoing activity (impulses before the trains at 1 and 4 Hz). (**c**) Adrenaline (5 μg injected i.v.) activated the fiber. Angiotensin (0.2 μg injected i.v.) generated a large increase of blood pressure (*MAP* mean arterial blood pressure) but did not activate the afferent fiber (Modified from Häbler et al. (1987))

coupling occurs in the nerve proximal to the lesion. However, this possibility has not yet been investigated.

Significance of Sympathetic Coupling onto the Afferent Nerve Fiber

Sympathetic-afferent coupling after nerve lesion can occur in the peripheral nerve at different sites. In neuromas, coupling involves mostly myelinated fibers, and since most of them are probably low-threshold mechanoreceptors, its significance for neuropathic pain is doubtful. This is consistent with clinical experience showing that pain from neuromas is usually independent of the sympathetic nervous system. With the exception of experiments on inappropriate cross-union of two hindlimb nerves, a general problem is that un-physiologically high stimulation frequencies of the sympathetic trunk are necessary to activate afferent fibers. Similar to sympathetic-afferent coupling in the dorsal root ganglion, the exact mechanism of the sympathetic coupling onto afferent nerve fibers remains unclear. While it is an attractive idea that afferent neurons are excited by norepinephrine acting on upregulated, neuronal α_2-adrenoceptors, it has to be reconciled with the well-established finding that intact nociceptors possess presynaptic α_2-adrenoceptors which are inhibitory (e.g., Li and Eisenach 2001). Furthermore, there is evidence indicating that α_2-adrenoceptors remain inhibitory also after partial sciatic nerve ligation (Lavand'homme et al. 2002), and this inhibitory effect may underlie the antihyperalgesic effect of topical clonidine observed in an animal model of neuropathic pain (Li et al. 2007).

Conclusion

The functional significance of sympathetic-afferent coupling at the level of the afferent nerve fiber is still largely unclear. The assumption that it represents an animal model for SMP in patients should be considered with caution.

Acknowledgment This work is supported by the Deutsche Forschungsgemeinschaft.

References

Ali, Z., Ringkamp, M., Hartke, T. V., et al. (1999). Uninjured C-fiber nociceptors develop spontaneous activity and α-adrenergic sensitivity following L₆ spinal nerve ligation in monkey. *Journal of Neurophysiology, 81*, 455–466.

Ali, Z., Raja, S. N., Wesselmann, U., et al. (2000). Intradermal injection of norepinephrine evokes pain in patients with sympathetically maintained pain. *Pain, 88*, 161–168.

Birder, L. A., & Perl, E. R. (1999). Expression of α_2-adrenergic receptors in rat primary afferent neurones after peripheral nerve injury or inflammation. *The Journal of Physiology, 515*, 533–542.

Blumberg, H., & Jänig, W. (1984). Discharge pattern of afferent fibers from a neuroma. *Pain, 20*, 335–353.

Bossut, D. F., Shea, V. K., & Perl, E. R. (1996). Sympathectomy induces adrenergic excitability of cutaneous C-fiber nociceptors. *Journal of Neurophysiology, 75*, 514–517.

Campero, M., Bostock, H., Baumann, T. K., & Ochoa, J. L. (2010). A search for activation of C nociceptors by sympathetic fibers in complex regional pain syndrome. *Clinical Neurophysiology, 121*, 1072–1079.

Devor, M., & Jänig, W. (1981). Activation of myelinated afferents ending in a neuroma by stimulation of the sympathetic supply in the rat. *Neuroscience Letters, 24*, 43–47.

Häbler, H. J., Jänig, W., & Koltzenburg, M. (1987). Activation of unmyelinated afferents in chronically lesioned nerves by adrenaline and excitation of sympathetic efferents in the cat. *Neuroscience Letters, 82*, 35–40.

Jänig, W. (1990). Activation of afferent fibers ending in an old neuroma by sympathetic stimulation in the rat. *Neuroscience Letters, 111*, 309–314.

Jänig, W., & McLachlan, E. M. (1984). On the fate of sympathetic and sensory neurons projecting into a neuroma of the superficial peroneal nerve in the cat. *The Journal of Comparative Neurology, 225*, 302–311.

Jorum, E., Orstavik, K., Schmidt, R., Namer, B., Carr, R. W., Kvarstein, G., Hilliges, M., Handwerker, H., Torebjork, E., & Schmelz, M. (2007). Catecholamine-induced excitation of nociceptors in sympathetically maintained pain. *Pain, 127*, 296–301.

Lavand'homme, P. M., Ma, W., De Kock, M., et al. (2002). Perineural alpha(2A)-adrenoceptor activation inhibits spinal cord neuroplasticity and tactile allodynia after nerve injury. *Anesthesiology, 97*, 972–980.

Li, X., & Eisenach, J. C. (2001). Alpha2A-adrenoceptor stimulation reduces capsaicin-induced glutamate release from spinal cord synaptosomes. *The Journal of Pharmacology and Experimental Therapeutics, 299*, 939–944.

Li, C., Sekiyama, H., Hayashida, M., Takeda, K., Sumida, T., Sawamura, S., Yamada, Y., Arita, H., & Hanaoka, K. (2007). Effects of topical application of clonidine cream on pain behaviors and spinal Fos protein expression in rat models of neuropathic pain, postoperative pain, and inflammatory pain. *Anesthesiology, 107*, 486–494.

O'Halloran, K. D., & Perl, E. R. (1997). Effects of partial nerve injury on the responses of C-fiber polymodal nociceptors to adrenergic agonists. *Brain Research, 759*, 233–240.

Sato, J., & Perl, E. R. (1991). Adrenergic excitation of cutaneous pain receptors induced by peripheral nerve injury. *Science, 251*, 1608–1610.

Torebjork, E., Wahren, L., Wallin, G., et al. (1995). Noradrenaline-evoked pain in neuralgia. *Pain, 63*, 11–20.

Sympathetic-Afferent Coupling in the Dorsal Root Ganglion, Neurophysiological Experiments

Wilfrid Jaenig[1] and Heinz-Joachim Häbler[2]
[1]Physiologisches Institut, Christian-Albrechts-Universität zu Kiel, Kiel, Germany
[2]FH Bonn-Rhein-Sieg, Rheinbach, Germany

Synonyms

Coupling between sympathetic postganglionic neurons and primary afferent neurons (in the dorsal root ganglion); Sympathetic-sensory coupling

Definition

Under physiological conditions, there is no coupling between sympathetic neurons and nociceptive primary afferent neurons at any site in the peripheral nervous system, i.e., activity in sympathetic postganglionic neurons does not activate nociceptors and thus does not elicit pain under any condition. In rat models of neuropathic pain, nerve lesions lead to the generation of ectopic activity in lesioned afferents, part of which arises from the somata of primary afferent neurons in the dorsal root ganglion (DRG). It is believed that the ectopic activity is involved in initiating and maintaining neuropathic pain behavior in these animals. Using neurophysiological recordings in reduced animal models in vivo under general anesthesia or in vitro allows to study whether there is a coupling between efferent sympathetic postganglionic neurons and afferent nociceptive neurons within the DRG which may develop after various types of experimental nerve lesion (i.e., whether this could be a model of SMP). Under these conditions, controlled natural or electrical stimulation of the sympathetic supply and controlled pharmacology (application of adrenoceptor agonists or antagonists) can be applied, and changes of the ectopic activity in axotomized primary afferent neurons can be measured. Furthermore, changes of blood flow resulting from sympathetic stimulation can be measured on the surface of the DRG by laser Doppler flowmetry to investigate the question whether sympathetic-afferent coupling may be mediated by a direct activation of primary afferent neurons by sympathetic transmitters or indirectly mediated by changes in DRG blood flow. Results obtained in these reduced in vivo animal models have to be correlated with the results obtained in behavioral animal models and in patients.

Characteristics

Following transection and/or ligation of the sciatic nerve or the L5/L6 spinal nerve (spinal nerve ligation model) in rats, perivascular catecholamine-containing axons start to invade the DRGs, which contain somata with lesioned axon and form baskets around sensory somata (McLachlan et al. 1993) (see ▶ Sympathetic and Sensory Neurons after Nerve Lesions, Structural Basis for Interactions). Furthermore, following both types of nerve lesion, α_{2A}-adrenoceptors were found to be upregulated, particularly in medium-sized and large-diameter DRG cells (Birder and Perl 1999; Shi et al. 2000). There was evidence that this happened not only in axotomized but also in uninjured neurons (Birder and Perl 1999).

Therefore, the hypothesis has been put forward that the baskets, which form around primary afferent somata in DRGs corresponding to the lesioned nerve and which consist of postganglionic sprouts, may mediate sympathetic-afferent coupling. According to this hypothesis, norepinephrine released from these postganglionic sprouts could then directly bind to α_{2A}-adrenoceptors upregulated on the DRG somata and activate neurons involved in the generation and maintenance of neuropathic pain behavior. Thus, these morphological changes in the DRG would represent the peripheral morphological correlate of sympathetically maintained pain (SMP).

In Vivo Neurophysiological Experiments After Sciatic Nerve Lesion

After tight ligation and section of the sciatic nerve in Wistar-derived Sabra strain rats, many afferent neurons in L4 and L5 DRGs, almost all with myelinated (A-) fibers, develop ongoing ectopic activity originating from within these ganglia (Devor et al. 1994; Michaelis et al. 1996). Electrical stimulation of the sympathetic supply of the L4 and L5 DRGs at unphysiologically high frequencies (≥ 5 Hz), or systemic application of epinephrine, changed the ectopic activity of 55–60 % of the afferent axotomized neurons via coupling within the DRG. Early after lesion (4–22 days), the predominant response was excitation, whereas later on, the predominant response was depression of activity. A few silent afferent A-neurons were recruited by sympathetic stimulation. Only few DRG cells with unmyelinated (C-) fibers were spontaneously active (0.1–8.7 % of neurons with ectopic activity) and most responded to sympathetic stimulation with inhibition. Responses were mediated by α_2-adrenoceptors in 65 % of cases, by α_1-adrenoceptors in 13 %, and by α_1- and α_2-adrenoceptors in 10 % of cases (Chen et al. 1996).

In Vivo Neurophysiological Experiments After Spinal Nerve Lesion

In the spinal nerve ligation (SNL) model, only axotomized neurons with A-fibers develop ectopic activity (Liu et al. 2000). These injured afferent neurons with ectopic activity originally probably supplied skeletal muscle rather than skin (Michaelis et al. 2000).

The incidence of axotomized afferent neurons with ectopic activity, which were responsive to electrical stimulation of the sympathetic chain at high stimulation frequencies of ≥ 5 Hz, was significantly lower (17.6 %) than after sciatic nerve lesion (Häbler et al. 2000). Almost all neurons responding to sympathetic stimulation were excited 3–56 days after lesion, while almost all of the 26.3 % neurons responsive to systemic application of norepinephrine were inhibited.

Registration of blood flow on the surface of the L5 DRG with laser Doppler flowmetry in parallel to recording neural responses revealed that electrical stimulation of the sympathetic chain leads to frequency-dependent vasoconstriction in the DRG (Häbler et al. 2000). When

sympathetic stimulation excited axotomized afferent neurons, the activation occurred once a critical threshold of vascular resistance in the DRG was exceeded and subsided after vasoconstriction had fallen below the critical threshold (Fig. 1). Preconstriction of the vascular bed with the potent vasoconstrictor drug N^G-nitro-L-arginine methyl ester (L-NAME), an unspecific inhibitor of the nitric oxide (NO) synthase, significantly enhanced the incidence and magnitude of excitatory responses in axotomized DRG neurons with ectopic activity induced by sympathetic stimulation (from 17.6 % to 76 % of neurons), while the effective stimulation frequency necessary to excite the afferent neurons decreased significantly. In this way, many axotomized neurons originally unresponsive even to prolonged sympathetic stimulation could be converted to responders after L-NAME. Both the DRG vasoconstriction and the neuronal responses to sympathetic stimulation were antagonized by both α_1- and α_2-adrenoceptor antagonists.

Preconstriction of the DRG vascular bed also increased the incidence of axotomized afferents responding to systemic application of norepinephrine from 23 % to 75 %; all responses were excitatory, and previously inhibitory responses were converted into excitatory ones. Similar to norepinephrine, nonsympathetic vasoconstrictor agents like angiotensin II and vasopressin also induced excitatory responses of axotomized afferent neurons with ectopic activity and a recruitment of some previously silent afferent neurons. Systemic asphyxia was also capable of activating lesioned afferent neurons.

These experiments suggest that sympathetic-afferent coupling in the DRG after spinal nerve lesion and, by analogy, also after sciatic nerve lesion is indirect. It occurs mainly when perfusion of the DRG is impaired, and the effective stimulus, which activates axotomized afferent neurons, is likely to be transient ischemia.

In Vitro Neurophysiological Experiments on Axotomized DRG Neurons

In in vitro experiments, lesioned afferent neurons after chronic constriction or L5 spinal nerve injury showed excitatory responses to

Sympathetic-Afferent Coupling in the Dorsal Root Ganglion, Neurophysiological Experiments, Fig. 1 Activation of an axotomized afferent neuron following lesion of the spinal nerve L5 (performed 35 days prior to the experiment) to electrical stimulation of the lumbar sympathetic trunk (LST, trains of 10 s at 1–50 Hz) in an anesthetized rat. The axon was isolated from the dorsal root L5. Mean arterial blood pressure (MAP), relative blood flow through the dorsal root ganglion (DRG) L5 (measured by laser Doppler flowmetry), and resistance to flow in the DRG were recorded in parallel to the neural activity. The axotomized afferent neuron responded only to the highest frequencies used (20 Hz and 50 Hz). The activation occurred in parallel with increase in vascular resistance exceeding a critical threshold value (broken and *stippled lines*). Fiber activation mirrored the magnitude and time course of vascular resistance above this critical level. When the vascular bed was preconstricted by L-NAME, the axotomized afferent neurons already responded to stimulation frequencies of the LST at ≥ 5 Hz (not shown here) (From Häbler et al. (2000) with permission)

norepinephrine or epinephrine. However, the doses used were relatively high. In similar in vitro experiments on acutely isolated DRG neurons, small depolarizations in both intact and lesioned somata in response to norepinephrine or other sympathomimetics could not be blocked by adrenoceptor antagonists (de Armentia et al. 2003), indicating that the observed effects were nonspecific.

Significance of Sympathetic-Afferent Coupling in the DRG

The critical question is whether sympathetic-afferent coupling in the DRG, as observed after spinal nerve lesion and sciatic nerve lesion, is a neurophysiological correlate of sympathetically maintained pain. This question is directly linked to the question of whether or not the ectopic activity arising in axotomized afferent neurons after nerve lesion is responsible for the

development and/or maintenance of neuropathic pain behavior. This issue is controversially discussed. Earlier experiments in the SNL model indicated that sectioning the corresponding dorsal roots before spinal nerve lesion prevented the development of neuropathic pain behavior, while sectioning the corresponding dorsal roots after spinal nerve lesion abolished neuropathic pain behavior (Kinnman and Levine 1995; Yoon et al. 1996), suggesting that the signal generating neuropathic pain behavior arose from the lesioned segment. However, later studies found that mechanical allodynic and hyperalgesic behavior after L5 spinal nerve lesion was unchanged by an additional L5 dorsal root section after spinal nerve lesion. Furthermore, transection of the dorsal root L5 alone was followed by mechanical allodynic and hyperalgesic behavior, which was undistinguishable from that after spinal nerve lesion

(Eschenfelder et al. 2000; Liu et al. 2000). In view of these conflicting results, it appears doubtful that ectopic activity in lesioned afferents is responsible for the neuropathic pain behavior in the spinal nerve ligation model.

Another major problem is that ectopic activity in the SNL model arises in axotomized afferent A- but not, or almost not, in C-fibers. At present, there is no conclusive explanation how A-fiber activity in the absence of C-fiber discharge can generate neuropathic pain in the early time period after lesion. However, injury of skin or muscle nerves is clearly followed by ectopic ongoing activity in C-fibers up to 15 months after nerve lesion (Gorodetskaya et al. 2009; Jänig et al. 2009; Kirillova et al. 2011). A possible alternative explanation for the more chronic state after lesion might be that nerve lesion entails the death of primary afferent neurons with C-fibers (mostly those to the skin but not those to skeletal muscle, see ▶ Sympathetic and Sensory Neurons after Nerve Lesions, Structural Basis for Interactions and Hu and McLachlan 2003) and that this leads to central reorganization. Now, activity in afferent neurons with A-fibers may be involved in the generation of pain or pain-like behavior.

Finally, according to behavioral experiments, it is at best controversial whether spinal nerve lesion is a model for SMP, first, since surgical sympathectomy alleviated neuropathic pain behavior only in a limited number of rat species, but had no effect in others, and second, since in a given rat strain, the results produced by different laboratories were opposite (see ▶ Sympathetic Nervous System in the Generation of Pain, Animal Behavioral Models).

Conclusion

The generation of neuropathic pain, allodynia, and hyperalgesia by pathophysiological mechanisms acting within the DRG, and a contribution to these symptoms by the sympathetic nervous system via sympathetic-afferent coupling in the DRG, is at present an attractive possibility, but a number of open questions have to be clarified, until firm conclusions can be drawn.

Acknowledgment Supported by the Deutsche Forschungsgemeinschaft and the German Ministry of Research and Education, German Research Network on Neuropathic Pain.

References

Birder, L. A., & Perl, E. R. (1999). Expression of a_2-adrenergic receptors in rat primary afferent neurones after peripheral nerve injury or inflammation. *The Journal of Physiology, 515*, 533–542.

Chen, Y., Michaelis, M., Jänig, W., & Devor, M. (1996). Adrenoceptor subtype mediating sympathetic-sensory coupling in injured sensory neurons. *Journal of Neurophysiology, 76*, 3721–3730.

De Armentia, L. M., Leeson, A. H., Stebbing, M. J., Urban, L., & McLachlan, E. M. (2003). Responses of sympathomimetics in rat sensory neurones after nerve transection. *Neuroreport, 14*, 9–13.

Devor, M., Jänig, W., & Michaelis, M. (1994). Modulation of activity in dorsal root ganglion (DRG) neurons by sympathetic activation in nerve-injured rats. *Journal of Neurophysiology, 71*, 38–47.

Eschenfelder, S., Häbler, H. J., & Jänig, W. (2000). Dorsal root section elicits signs of neuropathic pain rather than reversing them in rats with spinal nerve injury. *Pain, 87*, 213–219.

Gorodetskaya, N., Grossmann, L., Constantin, C., & Jänig, W. (2009). Functional properties of cutaneous A- and C-fibers 1-15 months after a nerve lesion. *Journal of Neurophysiology, 102*, 3129–3141.

Häbler, H. J., Eschenfelder, S., Liu, X. G., & Jänig, W. (2000). Sympathetic-sensory coupling after L5 spinal nerve lesion in the rat and its relation to changes in dorsal root ganglion blood flow. *Pain, 87*, 335–345.

Hu, P., & McLachlan, E. M. (2003). Selective reactions of cutaneous and muscle afferent neurons to peripheral nerve transection in rats. *The Journal of Neuroscience, 23*, 10559–10567.

Jänig, W., Grossmann, L., & Gorodetskaya, N. (2009). Mechano- and thermosensitivity of regenerating cutaneous nerve fibers. *Experimental Brain Research, 196*, 101–114.

Kinnman, E., & Levine, J. D. (1995). Sensory and sympathetic contributions to nerve injury-induced sensory abnormalities in the rat. *Neuroscience, 64*, 751–767.

Kirillova, I., Rausch, V. H., Tode, J., Baron, R., & Jänig, W. (2011). Mechano- and thermosensitivity of injured muscle afferents. *Journal of Neurophysiology, 105*, 2058–2073.

Liu, X. G., Eschenfelder, S., Blenk, K. H., Jänig, W., & Häbler, H. J. (2000). Spontaneous activity of axotomized afferent neurons after L5 spinal nerve injury in the rat. *Pain, 84*, 309–318.

McLachlan, E. M., Jänig, W., Devor, M., & Michaelis, M. (1993). Peripheral nerve injury triggers noradrenergic sprouting within dorsal root ganglia. *Nature, 363*, 543–546.

Michaelis, M., Devor, M., & Jänig, W. (1996). Sympathetic modulation of activity in dorsal root ganglion neurons changes over time following peripheral nerve injury. *Journal of Neurophysiology, 76,* 753–763.

Michaelis, M., Liu, X.-G., & Jänig, W. (2000). Peripheral nerve lesion induces ongoing discharges originating from dorsal root ganglia in axotomized and unlesioned muscle afferents. *The Journal of Neuroscience, 20,* 2742–2748.

Shi, T. J. S., Winzer-Serhan, U., Leslie, F., & Hökfelt, T. (2000). Distribution and regulation of a_2-adrenoceptors in rat dorsal root ganglia. *Pain, 84,* 319–330.

Yoon, Y. W., Na, H. S., & Chung, J. M. (1996). Contributions of injured and intact afferents to neuropathic pain in an experimental rat model. *Pain, 64,* 27–36.

Sympathetic-Afferent Interaction

▶ Sympathetic and Sensory Neurons After Nerve Lesions, Structural Basis for Interactions

Sympathetically Maintained Pain

Synonyms

SMP

Definition

The sympathetic nervous system is normally involved in control of body temperature, cardiovascular system, pelvic organs, etc. However, in certain pathophysiological conditions, this part of the nervous system is involved in the maintenance of pain (sympathetically maintained pain). In such cases, interruption or blockade of the sympathetic nerves with a local anesthetic nerve block or drugs that block transmission from the sympathetic postganglionic neurons can result in alleviation of the pain. A subset of patients with complex regional pain syndrome (CRPS) may have sympathetically maintained pain. The proportion of sympathetically maintained pain may vary between patients or with time after tauma in an individual patient, and the residual pain is termed sympathetically independent pain (SIP).

Cross-References

▶ Complex Regional Pain Syndrome and the Sympathetic Nervous System
▶ Sympathetically Maintained Pain and Inflammation, Human Experimentation
▶ Sympathetically Maintained Pain, Clinical Pharmacological Tests

Sympathetically Maintained Pain (SMP)

▶ Neurosurgical Treatment of Pain

Sympathetically Maintained Pain and Inflammation, Human Experimentation

Frank Birklein[1] and Ralf Baron[2]
[1]Department of Neurology, University of Mainz, Mainz, Germany
[2]Department of Neurological Pain Research and Therapy, Christian Albrechts University Kiel, Kiel, Germany

Synonyms

Sympathetically maintained pain; Sympathonociceptive coupling

Definition

There is strong evidence that some neuropathic pain states are characterized by an inflammatory reaction. In some patients, sympatholytic procedures can ameliorate pain as well as inflammation and edema.

Characteristics

Sympathetic Influence on Neurogenic Inflammation in Human Pain Models

When sympathetic activation is induced by mental stress, experimental pain is suppressed (stress-induced analgesia) (Fechir et al. 2009). When sympathetic activation is modulated by thermoregulation, it does not measurably influence capsaicin-induced pain and neurogenic inflammation (Fig. 1c) (Baron et al. 1999). Accordingly, the sole activation of skin sympathetic vasoconstrictor neurons does not change the capsaicin-induced discharge of single microneurographically recorded cutaneous C-nociceptors (Elam et al. 1999), and sympathetic blocks were ineffective to reduce pain in the capsaicin model (Pedersen et al. 1997; Drummond 2001).

Physiological activity in postganglionic sympathetic efferents or the cutaneous injection of catecholamines into healthy skin does not produce pain. However, immune histochemistry has demonstrated α-adrenoreceptors on primary nociceptive neurons (Dawson et al. 2011). Accordingly, iontophoresis of noradrenalin into healthy skin induces axon reflexes for which a subthreshold-for-pain activation of C-fibers is sufficient (Drummond and Lipnicki 1999). On the contrary, when application of catecholamines is performed by microdialysis without any additional noxious stimulus like the iontophoretic current, even with pharmacological doses of noradrenaline or clonidine, axon reflexes could not be elicited (Fig. 1a, b) (Zahn et al. 2004).

Similar negative results were reported when sympathetic muscle vasoconstrictor activity was altered physiologically by CO_2 stress and the intensity of pain after capsaicin injection into the muscle was quantified. The muscle pain did not change in this experiment (Wasner et al. 2002).

On the other hand, thermal hyperalgesia, which frequently occurs after capsaicin injection or in sun burns, was enhanced if norepinephrine was co-applied by iontophoresis or injection (Drummond 2001). This effect was reversed by combined blockade of a_1- and a_2-adrenoceptors. However, enhancement of capsaicin-induced heat hyperalgesia was also found for other vasoconstrictive agents like vasopressin, angiotensin, or the occlusion of blood flow, and it was antagonized by vasodilators like nitroprusside (Drummond 1999; Drummond and Lipnicki 1999). These findings indicate that at least a part of catecholamine amplification of capsaicin-induced heat sensitization must be mediated by local vasoconstriction, which is induced by catecholamines.

Neurogenic Inflammation in CRPS

In patients with CRPS, axon reflex vasodilatation was significantly increased on the affected side as measured with laser Doppler flowmetry after electrical C-fiber stimulation. Accordingly, systemic CGRP and SP levels were found to be increased in CRPS (Birklein et al. 2001; Schinkel et al. 2006).

In acute untreated CRPS I patients, axon reflex activation was elicited by strong transcutaneous electrical stimulation of peptidergic unmyelinated afferents via intradermal microdialysis capillaries. Protein extravasation that was simultaneously assessed by the microdialysis system was only provoked on the affected extremity as compared with the normal side. The time course of electrically induced protein extravasation in the patients resembled the one observed following application of exogenous substance-P (SP) (Weber et al. 2001).

Plasma extravasation could also be demonstrated with other experimental techniques: Bone scintigraphy demonstrated periarticular tracer uptake in acute CRPS and analysis of joint fluid and synovial biopsies in CRPS patients have shown an increase in protein concentration and synovial hypervascularity. Muscle edema can be visualized on MRI scans (Nishida et al. 2009) and scintigraphic investigations with radiolabeled immunoglobulins further demonstrate extensive plasma extravasation in the affected limb (Baron 2012).

Sympathetically Maintained Pain and Inflammation, Human Experimentation, Fig. 1 (a, b) *Microdialysis* has the advantage that a variety of substances could be applied intradermally without simultaneous C-fiber stimulus. The fibers were inserted and rest in the skin for at least 1 h before catecholamines were perfused (a). Changes of skin blood flow have been monitored by taking repeated pictures covering the whole back of the hand from a laser Doppler perfusion monitor. Even supraphysiological doses of catecholamines (phenylephrine *PE*, clonidine *CL*, norepinephrine *NE*) did not induce axon reflex like vasodilatation (With permission from Zahn et al. 2004). (c) Effects of sympathetic activity on capsaicin-evoked cutaneous pain in 12 individuals. Experimental pain was evoked by topical application of 0.6 % capsaicin at the volar forearm (*arrow*). Pain was measured on a visual analogue scale (VAS) at 5-min intervals. High cutaneous vasoconstrictor activity (*open squares*) was experimentally induced by whole-body cooling, low sympathetic activity (*triangles*) by whole-body warming. The local skin temperature at the application site was kept constant with a feedback-controlled heating device set to 35 °C. There was no difference in capsaicin-evoked pain between high and low activity in cutaneous vasoconstrictor neurons. Mean ± SEM (Modified from Baron et al. 1999)

Immune Cell–Mediated Inflammation and Cytokine Release in CRPS

Skin biopsies in CRPS patients showed a striking increase in the number of Langerhans cells which can release immune cell chemoattractants and proinflammatory cytokines. Accordingly, in the fluid of skin blisters (Huygen et al. 2002) and in punch biopsies from the affected skin (Krämer et al. 2011), significantly higher levels of IL-6 and TNF-alpha were observed in the involved extremity.

The patchy osteoporosis may also be consistent with a regional inflammatory process in deep somatic tissues. Inflammatory cytokines like IL-1, and IL-6 or TNFα cause proliferation and activation of osteoclasts. Changes in hair growth can also be created by proinflammatory cytokines. TNF and IL-1 directly inhibit hair growth. Keratinocyte-derived TNF and IL-6 retard hair growth, induce signs of fibrosis, and in turn induce immune infiltration of the dermis, all present in CRPS patients. On the other hand, SP induces hair growth (Paus et al. 1994).

The Link Between Immune Cell–Mediated Inflammation and Neurogenic Inflammation in CRPS

An extensive release of SP would provide a new hypothesis linking immune cell–mediated and neurogenic inflammation. It has been shown that SP also potently activates immune cells in the human skin. They start to release cytokines, such as TNF-a and IL-1, which, in turn, further activate nociceptive fibers by enhancing sodium

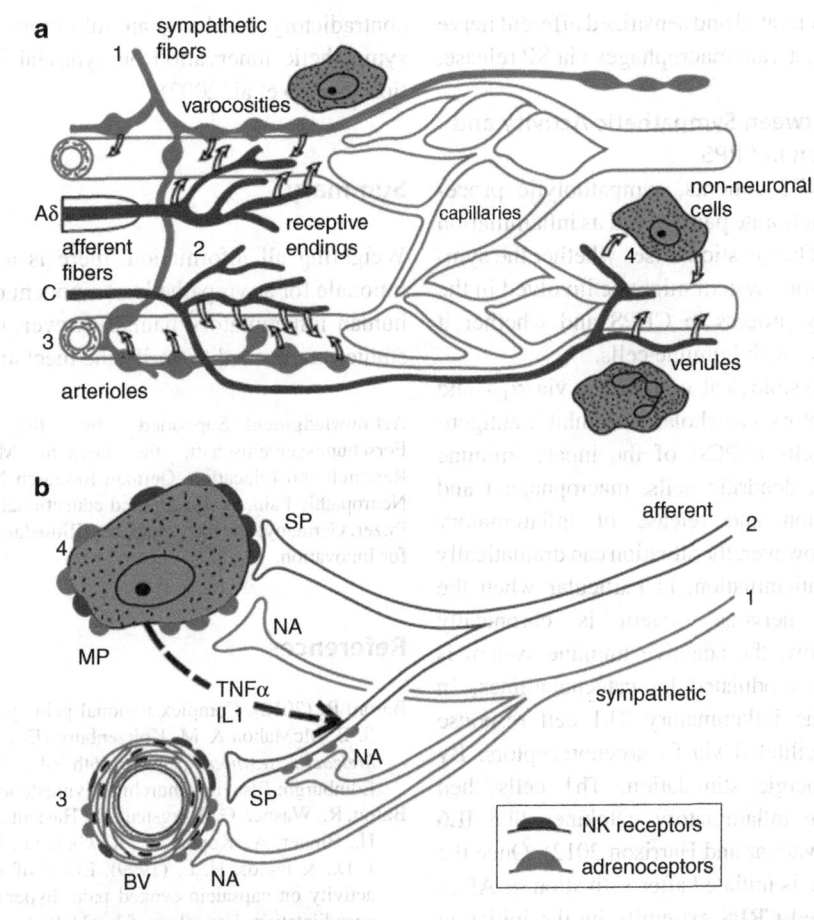

Sympathetically Maintained Pain and Inflammation, Human Experimentation, Fig. 2 (a) *The micromilieu of nociceptors.* The microenvironment of primary afferents is thought to affect the properties of the receptive endings of myelinated (A) and unmyelinated (C) afferent fibers. This has been particularly documented for inflammatory processes, but one may speculate that pathological changes in the direct surroundings of primary afferents may contribute to other pain states as well. The vascular bed consists of arterioles (directly innervated by sympathetic and afferent fibers), capillaries (not innervated and not influenced by nerve fibers), and venules (not directly innervated but influenced by nerve fibers). The micromilieu depends on several interacting components: Neural activity in postganglionic noradrenergic fibers (*1*) supplying blood vessels (*3, BV*) causes release of noradrenaline (*NA*) and possibly other substances and vasoconstriction. Excitation of primary afferents (Ad- and C-fibers) (*2*) causes vasodilation in precapillary arterioles and plasma extravasation in postcapillary venules (C-fibers only) by the release of substance-P (*SP*) and other vasoactive compounds (e.g., calcitonin gene–related peptide, *CGRP*). Some of these effects may be mediated by non-neuronal cells such as mast cells and

macrophages (*4*). Other factors that affect the control of the microcirculation are the myogenic properties of arterioles (*3*) and more global environmental influences such as a change of the temperature and the metabolic state of the tissue. (**b**) *Hypothetical relation between sympathetic noradrenergic nerve fibers (1), peptidergic afferent nerve fibers (2), blood vessels (3), and macrophages (4).* The activated and sensitized afferent nerve fibers activate macrophages (*MP*) possibly via substance-P release. The immune cells start to release cytokines, such as tumor necrosis factor-a (*TNF-a*) and interleukin-1 (*IL-1*) which further activate afferent fibers. Substance-P (and CGRP) released from the afferent nerve fibers reacts with neurokinin-1 (*NK-1*) receptors in the blood vessels (arteriolar vasodilation, venular plasma extravasation; neurogenic inflammation). The sympathetic nerve fibers interact with this system on three levels: (1) via adrenoceptors (mainly alpha) on the blood vessels (vasoconstriction); (2) via different adrenoceptors on macrophages (further release of cytokines), and (3) via adrenoceptors (mainly alpha) on afferents (further sensitization of these fibers) (Modified from Jänig and Baron 2003)

influx. The activated and sensitized afferent nerve fibers again activate macrophages via SP release.

The Link Between Sympathetic Activity and Inflammation in CRPS

In some CRPS patients, sympatholytic procedures can ameliorate pain as well as inflammation and edema. The question arises whether the sympathetic nervous system might be involved in the inflammatory process in CRPS and whether it might interact with immune cells.

Under physiological conditions, via $\alpha_{1/2}$ and β_2-adrenoceptors, catecholamines inhibit antigen-presenting cells (APCs) of the innate immune system (e.g., dendritic cells, macrophages) and the production and release of inflammatory cytokines. However, the situation can dramatically change in inflammation, in particular when the sympathetic nervous system is chronically activated. Now, the adaptive immune system is differentially modulated by catecholamines. In particular, the inflammatory Th1 cell response might be facilitated via β_2 adrenoreceptors. By catecholaminergic stimulation, Th1 cells then secrete more inflammatory cytokines like IL6 and TNFα (Marvar and Harrison 2012). Once the Th1 response is initiated after activation of APCs in the affected CRPS extremity by the initiating trauma, sympathetic activation would be predicted to cause pain and other inflammatory signs via cytokine release.

Summarizing the hypothetical ideas described above, Fig. 2 illustrates the possible interactions between sympathetic fibers, afferent fibers, blood vessels, and nonneural cells related to the immune system (e.g., macrophages), leading theoretically to the inflammatory changes observed in CRPS patients.

Sympathetically Maintained Pain in Other Inflammatory Pain States

There are some reports that sympathetic blocks might relieve pain in inflammatory diseases. In rheumatoid arthritis, a randomized controlled trial (RCT) study of intravenous sympathetic block with guanethidine has been effective in 24 patients (Levine et al. 1986). However, more recent human histological data revealed contradictory results – an inhibitory effect of sympathetic innervation on synovial inflammation (Straub et al. 2002).

Summary

Weighting all information, there is a scientific rationale for a sympathetic component of chronic human inflammatory pain. However, controlled studies are needed to verify the mechanisms.

Acknowledgment Supported by the Deutsche Forschungsgemeinschaft, the German Ministry of Research and Education, German Research Network on Neuropathic Pain, an unrestricted educational grant from Pfizer, Germany, and the Foundation Rhineland Palatinate for Innovation.

References

Baron, R. (2012). Complex regional pain syndromes. In S. B. McMahon & M. Koltzenburg (Eds.), *Wall and Melzacks' textbook of pain* (6th ed.). Amsterdam/Edinburgh: Elsevier Churchill Livingstone, in press.

Baron, R., Wasner, G., Borgstedt, R., Hastedt, E., Schulte, H., Binder, A., Kopper, F., Rowbotham, M., Levine, J. D., & Fields, H. L. (1999). Effect of sympathetic activity on capsaicin-evoked pain, hyperalgesia, and vasodilatation. *Neurology, 52*, 923–932.

Birklein, F., Schmelz, M., Schifter, S., & Weber, M. (2001). The important role of neuropeptides in complex regional pain syndrome. *Neurology, 57*, 2179–2184.

Dawson, L. F., Phillips, J. K., Finch, P. M., Inglis, J. J., & Drummond, P. D. (2011). Expression of alpha1-adrenoceptors on peripheral nociceptive neurons. *Neuroscience, 175*, 300–314.

Drummond, P. D. (1999). Nitroprusside inhibits thermal hyperalgesia induced by noradrenaline in capsaicin-treated skin. *Pain, 80*, 405–412.

Drummond, P. D. (2001). The effect of sympathetic activity on thermal hyperalgesia in capsaicin-treated skin during body cooling and warming. *European Journal of Pain, 5*, 59–67.

Drummond, P. D., & Lipnicki, D. M. (1999). Noradrenaline provokes axon reflex hyperaemia in the skin of the human forearm. *Journal of the Autonomic Nervous System, 77*, 39–44.

Elam, M., Olausson, B., Skarphedinsson, J. O., & Wallin, B. G. (1999). Does sympathetic nerve discharge affect the firing of polymodal C-fibre afferents in humans? *Brain, 122*, 2237–2244.

Fechir, M., Schlereth, T., Kritzmann, S., Balon, S., Pfeifer, N., Geber, C., Breimhorst, M., Eberle, T., Gamer, M., & Birklein, F. (2009). Stress and thermoregulation: Different sympathetic responses and different effects on experimental pain. *European Journal of Pain, 13*, 935–941.

Huygen, F. J., De Bruijn, A. G., De Bruin, M. T., Groeneweg, J. G., Klein, J., & Zijlstra, F. J. (2002). Evidence for local inflammation in complex regional pain syndrome type 1. *Mediators of Inflammation, 11*, 47–51.

Jänig, W., & Baron, R. (2003). Complex regional pain syndrome: Mystery explained? *Lancet Neurology, 2*, 687–697.

Krämer, H. H., Eberle, T., Uceyler, N., Wagner, I., Klonschinsky, T., Muller, L. P., et al. (2011). TNF-alpha in CRPS and "normal" trauma – Significant differences between tissue and serum. *Pain, 152*, 285–290.

Levine, J. D., Fye, K., Heller, P., Basbaum, A. I., & Whiting-O'Keefe, Q. (1986). Clinical response to regional intravenous guanethidine in patients with rheumatoid arthritis. *Journal of Rheumatology, 13*, 1040–1043.

Marvar, P. J., & Harrison, D. K. (2012). Inflammation, immunity and the autonomic nervous system. In D. Robertson, I. Biaggioni, G. Burnstock, P. A. Low, & J. F. R. Paton (Eds.), *Primer on the autonomic nervous system* (pp. 325–329). London: Academic Press.

Nishida, Y., Saito, Y., Yokota, T., Kanda, T., & Mizusawa, H. (2009). Skeletal muscle MRI in complex regional pain syndrome. *Internal Medicine, 48*, 209–212.

Paus, R., Heinzelmann, T., Schultz, K. D., Furkert, J., Fechner, K., & Czarnetzki, B. M. (1994). Hair growth induction by substance P. *Laboratory Investigation, 71*, 134–140.

Pedersen, J. L., Rung, G. W., & Kehlet, H. (1997). Effect of sympathetic nerve block on acute inflammatory pain and hyperalgesia. *Anesthesiology, 86*, 293–301.

Schinkel, C., Gaertner, A., Zaspel, J., Zedler, S., Faist, E., & Schuermann, M. (2006). Inflammatory mediators are altered in the acute phase of posttraumatic complex regional pain syndrome. *The Clinical Journal of Pain, 22*, 235–239.

Straub, R. H., Gunzler, C., Miller, L. E., Cutolo, M., Scholmerich, J., & Schill, S. (2002). Anti-inflammatory cooperativity of corticosteroids and norepinephrine in rheumatoid arthritis synovial tissue in vivo and in vitro. *The FASEB Journal, 16*, 993–1000.

Wasner, G., Brechot, A., Schattschneider, J., Allardt, A., Binder, A., Jensen, T. S., et al. (2002). Effect of sympathetic muscle vasoconstrictor activity on capsaicin-induced muscle pain. *Muscle & Nerve, 26*, 113–121.

Weber, M., Birklein, F., Neundorfer, B., & Schmelz, M. (2001). Facilitated neurogenic inflammation in complex regional pain syndrome. *Pain, 91*, 251–257.

Zahn, S., Leis, S., Schick, C., Schmelz, M., & Birklein, F. (2004). No {alpha}-adrenoreceptor-induced C-fiber activation in healthy human skin. *Journal of Applied Physiology, 96*, 1380–1384.

Sympathetically Maintained Pain in CRPS I, Human Experimentation

Gunnar Wasner[1] and Ralf Baron[2]
[1]Department of Neurological Pain Research and Therapy, Neurological Clinic, Kiel, Germany
[2]Department of Neurological Pain Research and Therapy, Christian Albrechts University Kiel, Kiel, Germany

Synonyms

Algodystrophy; CRPS I (complex regional pain syndrome type I): reflex sympathetic dystrophy; Morbus Sudeck

Definition

▶ Sympathetically maintained pain (SMP) is a symptom of neuropathic pain conditions defined as the pain component that is relieved by specific sympatholytic procedures. If sympatholytic procedures have no influence on the pain, the symptom is called "sympathetically independent pain" (SIP). Controlled studies on therapeutic efficacy of sympatholytic procedures are rare (Price et al. 1998; Straube et al. 2010). SMP is most frequently found in ▶ Complex Regional Pain Syndrome Type I (CRPS I), which develops as a disproportionate consequence of an extremity trauma without any nerve lesion. CRPS I is clinically characterized by spontaneous pain, allodynia, hyperalgesia, autonomic, trophic and motor dysfunction in a distal generalized distribution (see Synopsis Jänig & Baron ▶ Sympathetic Nervous System and Pain). There is no clinical feature allowing a distinction between patients with SMP and with SIP. The only possibility to differentiate between SMP and SIP is the efficacy of a correctly applied sympatholytic intervention. Most animal data for investigation of the pathophysiological mechanisms underlying SMP are derived from nerve lesion models, except a fracture model and a chronic post-ischemia pain model (CPIP) (Coderre and Bennett 2010; Kingery 2010; but see ▶ Sympathetic Nervous System in the Generation of Pain, Animal Behavioral Models and Ludwig et al 2007). Therefore, human experiments for investigation of SMP in neuropathic pain patients, in particular in CRPS I, are essential for the understanding of the pathological interaction between the sympathetic nervous system and nociceptive pathways. In these studies, the influence of increasing activity in sympathetic neurons, of blocking the sympathetic activity, and of externally applied adrenoceptor agonists in experimental pain models and patients were tested.

Characteristics

Physiologically, sympathetic preganglionic neurons that are involved in the regulation of effector cells in somatic tissues, project to the paravertebral ganglia of the sympathetic trunk and synapse with postganglionic neurons that innervate the effector cells. These preganglionic sympathetic neurons are under central control, and the pattern of ongoing and reflex discharges is characteristic for each type of sympathetic pathway, e.g., skin vasoconstrictor, muscle vasoconstrictor, and sudomotor neurons and varies according to the innervated target cells (Jänig 2006). Under normal conditions, sympathetic activity has no influence on nociceptive and other primary afferent neurons leading to pain. However, the clinical picture of CRPS I with autonomic dysfunction including changes in skin blood flow, temperature, and sweating, as well as the symptom of sympathetically maintained pain, suggests that the sympathetic nervous system is involved in the pathophysiology of CRPS (Stanton-Hicks et al. 1995; Wasner 2010).

SMP in CRPS I
Contrasting the results in human experimental pain models, several studies have demonstrated that the sympathetic nervous system can be involved in the generation of pain in CRPS I patients, from which possible underlying pathophysiological mechanisms can be derived. Drummond et al. (Drummond et al. 2001; Drummond and Finch 2004) investigated the effect of sympathetic arousal on pain in CRPS patients. They found that pain ratings increased in over 70 % of CRPS patients during forehead cooling and or after being startled, e.g., by mental arithmetic. This pain-increasing effect of such phasic cutaneous sympathetic activation was most pronounced in patients suffering from cold or punctate hyperalgesia.

Microneurographic recordings in 24 patients with CRPS I and II did not find any strong evidence for activation of nociceptors by sympathetic efferent discharge (Campero et al. 2010). However, in a patient, who clearly responded to sympathetic blockade, pathologically sensitized mechano-insensitive nociceptors were activated by endogenously released or topically applied catecholamines (Jorum et al. 2007). Correspondingly, histological studies demonstrated a close physical association between sympathetic and nociceptive neurons in the upper dermis of human skin (Gibbs et al. 2008; see ▶ Sympathetic and Sensory Neurons after Nerve Lesions, Structural Basis for Interactions).

In another study, the technique of controlled alterations of sympathetic activity by changing environmental temperature that was applied to investigate sympathetic-nociceptive coupling was used in CRPS I patients (Baron et al. 2002): Cutaneous sympathetic vasoconstrictor outflow to the painful limb was activated by whole-body cooling. During the thermal challenge, local skin temperature at the affected region was fixed at 35 °C to avoid thermal effects at the nociceptor level. The intensity of spontaneous pain, the area of mechanical hyperalgesia, and the heat pain thresholds were measured and compared with the results of diagnostic sympathetic blocks that were performed to identify patients with SMP (pain relief >50 % after sympathetic block) and SIP (pain relief <50 %). The results showed that in patients with SMP, the intensity of spontaneous pain significantly increased by 22 %, and the spatial distribution of mechanical dynamic and punctate hyperalgesia increased by 42 % and 27 %, respectively, during high compared with

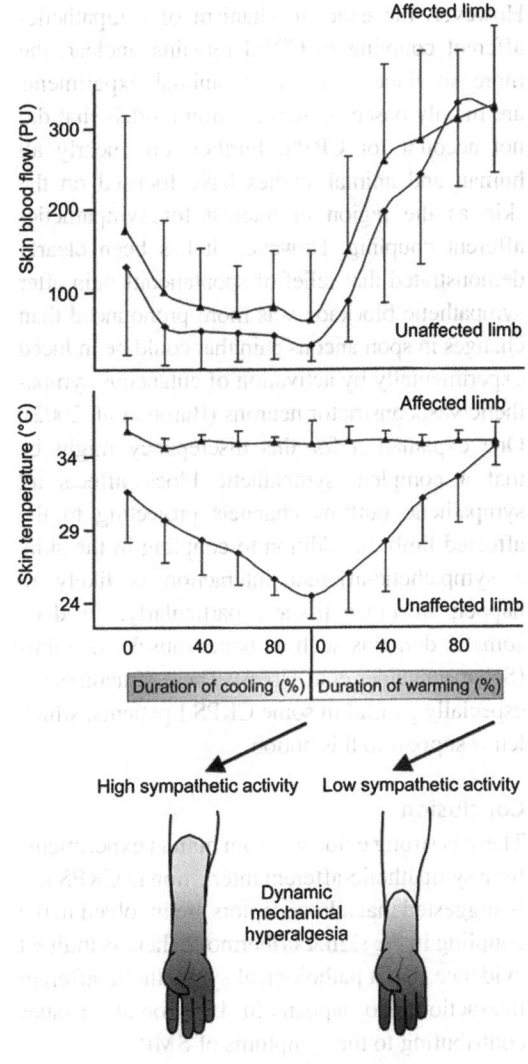

low sympathetic activity (Fig. 1). On the other hand, heat pain thresholds did not change, indicating that specific heat-sensitive primary afferent neurons are not, or are only marginally, affected by sympathetic activity. This observation also lends support to an alternative idea that low threshold Aβ afferent neurons, which are associated with mechanical allodynia, might be implicated in sympathetic-afferent coupling. Interestingly, pain relief after sympathetic blockade correlated with augmentation of spontaneous pain after experimental stimulation of cutaneous vasoconstrictor activity in all patients (both SMP and SIP). This suggests that the sympathetically maintained pain component in CRPS I might form a continuum. Also, in patients classified as SIP (pain relief <50 % after sympathetic block), there might be a pain component that depends on sympathetic activity (Fig. 2).

The study data (Drummond et al. 2001; Baron et al. 2002) gives strong evidence for a pathological coupling between skin sympathetic vasoconstrictor neurons and cutaneous afferents in CRPS. The question arises where this coupling might be located. In CRPS and posttraumatic neuralgias, intracutaneous application of noradrenaline into the symptomatic area rekindled spontaneous pain and mechanical hyperalgesias that had been relieved by sympathetic blockade (Torebjörk et al. 1995). Ali et al. (2000) also found in patients with SMP during local anesthetic sympathetic blockade that intracutaneous injections of noradrenaline in physiologically relevant doses induced pain and mechanical hyperalgesias in the affected region, but not on the contralateral side and not in controls. This effect was blocked in most patients by systemic administration of the α-adrenoceptor-antagonist phentolamine (see ▶ Sympathetically Maintained Pain in CRPS II,

Sympathetically Maintained Pain in CRPS I, Human Experimentation, Fig. 1 Experimental modulation of cutaneous sympathetic vasoconstrictor neurons by physiological thermoregulatory reflex stimuli in 13 CRPS patients. With the help of a thermal suit, whole-body cooling and warming was performed to alter sympathetic skin nerve activity. The subjects were lying in a suit supplied by tubes, in which running water of 12 °C and 50 °C, respectively (inflow temperature), was used to cool or warm the whole body. By these means, sympathetic activity can be switched on and off. *Top*: High sympathetic vasoconstrictor activity during cooling induces considerable drop in skin blood flow on the affected and unaffected extremity (laser Doppler flowmetry). Measurements were taken at 5-min intervals. *Middle*: On the unaffected side, a secondary decrease of skin temperature was documented. On the affected side, the forearm temperature was clamped at 35 °C by a feedback-controlled heat lamp to exclude temperature effects on the sensory

receptor level. Measurements were taken at 5-min intervals. *Bottom:* Effect of cutaneous sympathetic vasoconstrictor activity on dynamic mechanical allodynia in one CRPS patient with sympathetically maintained pain (SMP). Activation of sympathetic neurons (during cooling) leads to a considerable increase of the area of dynamic mechanical allodynia (From Baron et al. (2002) with permission)

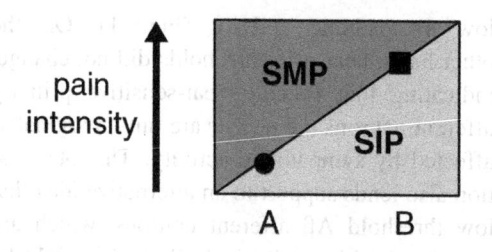

Sympathetically Maintained Pain in CRPS I, Human Experimentation, Fig. 2 A graphic representation of the relative contribution of SMP to overall pain in a given patient. *Point A* represents an individual whose pain is predominantly sympathetically maintained, whereas *point B* represents a situation in which the pain is only slightly responsive to sympatholytic intervention. Points A and B may represent different patients or the same patient at two different times (Adapted from Stanton-Hicks et al. (1995))

Human Experimentation). Accordingly, results of autoradiographic studies showed that the number of alpha-1-adrenoceptors in hyperalgesic skin in CRPS I patients is substantially higher than in the pain-free skin of the patients (Drummond et al. 1996). These findings support a role for adrenoceptors in cutaneous primary afferent neurons in the mechanisms of sympathetically maintained pain. Interestingly, development of adrenergic supersensitivity is suggested to be involved in vasomotor abnormalities in CRPS I (Wasner et al. 2001; Wasner 2010). Evidence from animal experiments favors the idea that sympathetic activity can sensitize nociceptors chemically, via noradrenaline released from sympathetic fibers, acting directly on adrenoceptors expressed on afferent fibers (see ▶ Sympathetic-Afferent Coupling in the Dorsal Root Ganglion, Neurophysiological Experiments, ▶ Sympathetic-Afferent Coupling in the Afferent Nerve Fiber, Neurophysiological Experiments). Furthermore, amplified excitability of nociceptors might be generated indirectly – i.e., via the vascular bed (e.g., change of blood flow), or by other components such as inflammatory cells around nociceptive neurons, which might have a permissive effect on sympathetic to nociceptive coupling, or they might be modulated directly by molecules released from sympathetic postganglionic neurons (▶ Sympathetic nervous system and pain).

However, the exact mechanism of sympathetic-afferent coupling in CRPS remains unclear, the more so since results from animal experiments are mainly based on nerve lesion models that did not account for CRPS. Furthermore, nearly all human and animal studies have focused on the skin as the region of interest for sympathetic-afferent coupling. However, it has been clearly demonstrated that relief of spontaneous pain after sympathetic blockade was more pronounced than changes in spontaneous pain that could be induced experimentally by activation of cutaneous sympathetic vasoconstrictor neurons (Baron et al. 2002). One explanation for this discrepancy might be that a complete sympathetic block affects all sympathetic outflow channels projecting to the affected limb. In addition to coupling in the skin, a sympathetic-afferent interaction is likely to happen in other tissues, particularly, in deep somatic domains such as bone, muscle, or joints (Schattschneider et al. 2006). These structures are especially painful in some CRPS I patients, which lends support to this notion.

Conclusion

There is strong evidence from human experiments for a sympathetic-afferent interaction in CRPS I. It is suggested that adrenoceptors are involved in the coupling in the skin. Furthermore, there is indirect evidence that a pathological sympathetic-afferent interaction also happens in deep somatic tissues contributing to the symptoms of SMP.

Acknowledgment The work was supported by the National Health and Medical Research Council of Australia (NHMRC), the EFIC-Grünenthal-Grant (EGG), and the Ministry of Science, Commerce and Transportation of Schleswig-Holstein, Germany.

References

Ali, Z., Raja, S. N., Wesselmann, U., Fuchs, P. N., Meyer, R. A., & Campbell, J. N. (2000). Intradermal injection of norepinephrine evokes pain in patients with sympathetically maintained pain. *Pain, 88*, 161–168.

Baron, R., Schattschneider, J., Binder, A., Siebrecht, D., & Wasner, G. (2002). Relation between sympathetic vasoconstrictor activity and pain and hyperalgesia in complex regional pain syndromes: A case–control study. *Lancet, 359*, 1655–1660.

Campero, M., Bostock, H., Baumann, T. K., & Ochoa, J. L. (2010). A search for activation of C nociceptors by sympathetic fibers in complex regional pain syndrome. *Clinical Neurophysiology, 121*, 1072–1079.

Coderre, T. J., & Bennett, G. J. (2010). A hypothesis for the cause of complex regional pain syndrome-type I (reflex sympathetic dystrophy): Pain due to deep-tissue microvascular pathology. *Pain Medicine, 11*, 1224–1238.

Drummond, P. D., & Finch, P. M. (2004). Persistence of pain induced by startle and forehead cooling after sympathetic blockade in patients with complex regional pain syndrome. *Journal of Neurology, Neurosurgery, and Psychiatry, 75*, 98–102.

Drummond, P. D., Skipworth, S., & Finch, P. M. (1996). alpha 1-adrenoceptors in normal and hyperalgesic human skin. *Clinical Science (Colch), 91*, 73–77.

Drummond, P. D., Finch, P. M., Skipworth, S., & Blockey, P. (2001). Pain increases during sympathetic arousal in patients with complex regional pain syndrome. *Neurology, 57*, 1296–1303.

Gibbs, G. F., Drummond, P. D., Finch, P. M., & Phillips, J. K. (2008). Unravelling the pathophysiology of complex regional pain syndrome: Focus on sympathetically maintained pain. *Clinical and Experimental Pharmacology and Physiology, 35*, 717–724.

Jänig, W. (2006). *Integrative action of the autonomic nervous system. Neurobiology of homeostasis.* Cambridge, NY: Cambridge University Press.

Jorum, E., Orstavik, K., Schmidt, R., Namer, B., Carr, R. W., Kvarstein, G., Hilliges, M., Handwerker, H., Torebjork, E., & Schmelz, M. (2007). Catecholamine-induced excitation of nociceptors in sympathetically maintained pain. *Pain, 127*, 296–301.

Kingery, W. S. (2010). Role of neuropeptide, cytokine, and growth factor signaling in complex regional pain syndrome. *Pain Medicine, 11*, 1239–1250.

Ludwig, J., Gorodetskaya, N., Schattschneider, J., Jämnig, W., & Baron, R. (2007). Behavioral and sensory changes after direct ischemia-reperfusion injury in rats. *European Journal of Pain, 11*, 677–684.

Price, D. D., Long, S., Wilsey, B., & Rafii, A. (1998). Analysis of peak magnitude and duration of analgesia produced by local anesthetics injected into sympathetic ganglia of complex regional syndrome patients. *The Clinical Journal of Pain, 14*, 216–226.

Schattschneider, J., Binder, A., Siebrecht, D., Wasner, G., & Baron, R. (2006). Complex regional pain syndromes: The influence of cutaneous and deep somatic sympathetic innervation on pain. *The Clinical Journal of Pain, 22*, 240–244.

Stanton-Hicks, M., Jänig, W., Hassenbusch, S., Haddox, J. D., Boas, R., & Wilson, P. (1995). Reflex sympathetic dystrophy: Changing concepts and taxonomy. *Pain, 63*, 127–133.

Straube, S., Derry,.S, Moore, R. A., & McQuay, H. J. (2010). Cervico-thoracic or lumbar sympathectomy for neuropathic pain and complex regional pain syndrome. *Cochrane Database of Systematic Reviews*, (7):CD002918.

Torebjörk, E., Wahren, L., Wallin, G., Hallin, R., & Koltzenburg, M. (1995). Noradrenaline-evoked pain in neuralgia. *Pain, 63*, 11–20.

Wasner, G. (2010). Vasomotor disturbances in complex regional pain syndrome–a review. *Pain Medicine, 11*, 1267–1273.

Wasner, G., Drummond, P., Birklein, F., & Baron, R. (2001). The role of the sympathetic nervous system in autonomic disturbances and "sympathetically maintained pain" in CRPS. In R. N. Harden, R. Baron, & W. Jänig (Eds.), *Complex regional pain syndrome progress in pain research and management* (Vol. 22, pp. 89–118). Seattle: IASP Press.

Sympathetically Maintained Pain in CRPS II, Human Experimentation

George M. Hanna[1], Srinivasa N. Raja[2] and Adam Carinci[1]
[1]Department of Anesthesia, Critical Care and Pain Medicine, Massachusetts General Hospital Harvard Medical School, Boston, MA, USA
[2]Department of Anesthesiology and Critical Care Medicine, Johns Hopkins University, Baltimore, USA

Synonyms

Causalgia; Complex regional pain syndrome II; CRPS II

Definition

Complex regional pain syndromes, ▶ CRPS I and ▶ CRPS II (formerly known as reflex sympathetic dystrophy and causalgia, respectively), are neuropathic pain disorders that develop as a consequence of trauma and can exhibit subsequent dysregulation of the autonomic nervous system. CRPS can be associated *with* (type II) or *without* (type I) evidence of major nerve damage. These syndromes frequently occur in the extremities and are characterized by pain disproportionate to the inciting trauma, ▶ allodynia, ▶ hyperalgesia, ▶ sudomotor, vasomotor, edema, and trophic skin changes, as well as active and passive movement disorders (Jänig and Baron 2003). CRPS II, in contrast to CRPS I, occurs following obvious nerve injury and is commonly caused by high velocity or partial injuries to peripheral nerves such as those occurring in gunshot wounds, stab wounds, or automobile accidents (Harden et al. 2001).

Autonomic dysfunction and sympathetic nervous system involvement have been suggested to explain the pathophysiology in a subset of patients with CRPS, but the exact mechanisms remain unclear. Research studies in humans with CRPS II have focused on two broad areas: (1) studies aimed at understanding the pathophysiologic mechanisms of pain and other associated features and (2) studies to examine the efficacy of treatment strategies. The salient observations related to each of these significant areas with regard to sympathetically maintained pain in CRPS II will be summarized.

Characteristics

Mechanisms

Several studies have explored the role of the sympathetic nervous system (SNS) in CRPS II. The SNS appears to contribute to pain in certain patients after nerve or tissue injury in an extremity. The sympathetic component of pain is often clinically demonstrated by blocking sympathetic activity in the symptomatic limb or by administering an adrenergic receptor agonist or antagonist. Several lines of clinical evidence support the hypothesis that the density of ▶ alpha(α)-adrenoceptors increases in peripheral tissues in ▶ sympathetically maintained pain (SMP), such that the release of noradrenaline (NA) from the sympathetic terminals activates the nociceptors and leads to the sensation of pain (Raja 1998). An example of this is the perineuromal administration of adrenaline or NA that results in increased pain in patients with ▶ neuromas after amputations (Chabal et al. 1992; Katz 1997; Lin et al. 2006). In CRPS II patients with SMP rendered pain-free by a sympathetic block or surgical ▶ sympathectomy, NA injected into the skin

restores the hyperalgesia (Davis et al. 1991; Torebjörk et al. 1995; Wallin et al. 1976). In addition, intravenous phentolamine (an α-adrenoceptor blocker), but not propranolol (a β-adrenoceptor blocker), relieves SMP (Raja et al. 1991).

A potential criticism of the previous studies is that the doses of NA used were much higher than are likely to exist in vivo. However, Ali et al. concluded that NA in physiologically relevant doses resulted in pain in patients with SMP (Ali et al. 2000). Thus, adrenoceptor mechanisms play a critical role in the maintenance of the pain in a subset of patients with chronic pain that is sympathetically maintained.

Other studies have investigated the effect of sympathetic arousal on pain and vasomotor responses in patients with CRPS. One such study examined sensitivity to heat, cold, and mechanical stimulation and concluded that pain decreased during sympathetic arousal in healthy subjects, whereas pain often increased and abnormal sensations sometimes developed in the symptomatic limb of patients with CRPS during the same procedures (Drummond et al. 2001). The authors suggest that a mechanism that normally inhibits pain during sympathetic arousal is compromised in the majority of patients with CRPS.

Wasner and colleagues (2001) have demonstrated a sequential progression from acute, relative sympathetic hypofunction to a chronic state of sympathetic hyperfunction in CRPS patients. This evolution of autonomic dysfunction is concordant with the clinical symptomology exhibited in CRPS – an acute vasodilation (warmth, rubor) in patients of less than 6 months duration followed by a chronic vasoconstriction (cool, blue) (Wasner et al. 2001). The "cold" phase is not accompanied by an increased sympathetic activity but more likely the result of increased density of alpha(1)-adrenoceptors in the skin (Drummond 2009). The intensity and the area of spontaneous pain and cutaneous hyperalgesia increased significantly in patients with sympathetically maintained pain when the central autonomic tone was increased by core body cooling (Baron et al. 2002).

Thermoregulatory testing revealed significant differences in acute and chronic CRPS patients, suggesting a disease progression related to an ultimate adrenergic hypersensitivity in the chronic pathologic state (Marinus et al. 2011).

Treatments

Experimental studies in the treatment of CRPS II have touched on a variety of pharmacological, interventional, and behavioral modalities. However, it is important to note that few of these studies have been placebo-controlled trials.

Muizelaar et al. (1997) explored the use of nifedipine and/or phenoxybenzamine in the treatment of CRPS II and concluded that both are effective first-line treatments for the acute stage of CRPS with a cure rate of 92 %. However, in the chronic stage of CRPS, treatment with these drugs was much less successful, with a documented cure rate of only 40 %, thus underscoring the importance of early recognition and intervention in CRPS II patients.

The alpha 2-adrenergic agonist, clonidine, has been studied because of its ability to reduce sympathetic tone. Clonidine, administered epidurally and transdermally, has been used for the treatment of CRPS, and case series have shown a reduction in hyperalgesia and allodynia (Davis et al. 1991); however, larger systematic reviews have suggested that the overall evidence for a beneficial effect is weak (Kingery 1997).

Calcitonin and bisphosphonates have been studied and may have a role in the treatment of CRPS, possibly secondary to their effect on associated osteopenia in the chronic CRPS sympathetically mediated state (Braga 1994; Robinson et al. 2004). Many of these studies, however, have been performed in patients with CRPS I. Thirty-five subjects with upper extremity CRPS undergoing physical therapy were randomized to an 8-week regimen of nasal calcitonin or oral paracetamol. No differences in pain, edema, or range of motion were observed in each group (Sahin et al. 2006). Bisphosphonates, however, have showed more promise in investigational studies. Robinson et al. (2004) randomized 27 patients to a single dose of an intravenous bisphosphonate or placebo. Three months later,

the patients who had received palmidronate reported lower pain scores and higher functionality. Again, these subjects carried the diagnosis of CRPS I, not CRPS II.

Autonomic dysfunction leading to tissue inflammation in CRPS has led to the study of systemic corticosteroids as a treatment modality. Sixty patients with CRPS following stroke were randomized to receive either a 5-week course of prednisolone or the nonsteroidal anti-inflammatory drug, piroxicam (Kalita et al. 2006). The steroid treatment group reported lower shoulder-hand syndrome scores (pain, edema, range of motion), but no differences were detected in activities of daily living between both groups.

A recent systematic review on the potential benefits of sympathectomy in patients with CRPS has concluded that chemical and surgical sympathectomy should be used cautiously in clinical practice in carefully selected patients (Straube et al. 2010). Based on the comparison of the pain relieving effects of lidocaine/bupivicaine (LA) vs. saline (S) blocks of the sympathetic ganglia, Price et al. (1998) suggest that both magnitude and duration of pain reduction should be closely monitored to provide optimal efficacy in patients undergoing such interventions. A case series of 25 subjects who had 3 stellate ganglion blocks at weekly intervals for upper extremity CRPS reported that 40 % of patients had complete pain relief, 36 % had partial pain relief, and 24 % had no pain relief over a 6-month observation period (Ackerman and Zhang 2006). The addition of botulinum toxin type A to local anesthetic for sympathectomy is thought to prolong the duration of the block by preventing the release of acetylcholine from preganglionic sympathetic nerves. Carroll et al. (2009) compared bupivacaine alone with a combination of bupivacaine-botulinum toxin A in seven patients with CRPS who previously responded to a lumbar sympathetic block. A longer duration of analgesia in the combination group was observed (71 days vs. <10 days, p <.002) Overall, sympathetic blocks are considered as a reasonable treatment option for patients refractory to pharmacologic and other noninterventional treatments, especially when conducted early in the disease course.

Surgical sympathectomy for CRPS II has also been investigated (Singh et al. 2003). In this study, 42 patients with CRPS II of the upper extremity were categorized into one of two groups based upon duration of symptoms, with group I consisting of patients with symptoms less than 3 months and group II consisting of patients with symptoms greater than 3 months. Following sympathectomy, all 42 patients demonstrated improved pain scores (VAS), with the patients in group I showing significantly greater improvement than those in group II. However, long-term results in most cases are disappointing with a high recurrence rate of the pain. Therefore, surgical sympathectomy should not be recommended as the treatment option of choice.

Additionally, spinal cord stimulation (SCS) is increasingly used as a treatment modality for CRPS II in those patients who fail to progress with more conservative measures. In fact, CRPS is the second-largest indication for the use of SCS in the USA (Stanton-Hicks et al. 2002). SCS has been demonstrated to be effective, at least during the first 2 years after implantation, in controlled trials (Kemler et al. 2000; Harke et al. 2005). At 6 months, the reduction in pain in the control and SCS groups were 2.4 and 0.2, respectively, on a 10-cm visual analogue scale. Recent studies indicate that SCS releases substance P, serotonin, noradrenaline, glycine, and GABA in the dorsal horns, resulting in a decrease of neuropathic pain (Linderoth and Meyerson 2002). Animal experts suggest that part of the effects of SCS are dependent on an intact sympathetic efferent system, but human data to confirm this are not available. Nonetheless, for patients with refractory CRPS II, SCS is a promising treatment option with a reasonable success rate in properly selected patients.

Multiple trials have examined the effectiveness of various intrathecal medications in the treatment of CRPS. In particular, intrathecal baclofen may be useful for the treatment of treatment-refractory dystonia associated with CRPS (van Hilten et al. 2000).

Conclusions

CRPS II is a multifaceted and complicated neuropathic pain disorder that develops as a consequence of a traumatic and definable nerve lesion. The mechanisms of, and treatment modalities for, CRPS II have been explored by a multitude of investigators and have shed light on the pathophysiology of this syndrome. However, many aspects of this chronic pain state remain obscure and enigmatic. Future research is needed to elucidate more precise knowledge of this syndrome and to potentially engender more efficacious treatment strategies.

References

Ackerman, W. E., & Zhang, J. M. (2006). Efficacy of stellate ganglion blockade for the management of type 1 complex regional pain syndrome. *Southern Medical Journal, 99*, 1084–1088.

Ali, Z., Raja, S. N., Wesselmann, U., et al. (2000). Intradermal injection of norepinephrine evokes pain in patients with sympathetically maintained pain. *Pain, 88*, 161–168.

Baron, R., Schattschneider, J., Binder, A., Siebrecht, D., & Wasner, G. (2002). Relation between sympathetic vasoconstrictor activity and pain and hyperalgesia in complex regional pain syndromes: A case–control study. *Lancet, 359*(9318), 1655–1660.

Braga, P. C. (1994). Calcitonin and its antinociceptive activity: Animal and human investigations 1975–1992. *Agents and Actions, 41*(3–4), 121–131.

Carroll, I., Clark, J. D., & Mackey, S. (2009). Sympathetic block with botulinum toxin to treat complex regional pain syndrome. *Annals of Neurology, 65*, 348–351.

Chabal, C., Jacobson, L., Russell, L. C., et al. (1992). Pain response to perineuromal injection of normal saline, epinephrine, and lidocaine in humans. *Pain, 49*, 9–12.

Davis, K. D., Treede, R. D., Raja, S. N., et al. (1991). Topical application of clonidine relieves hyperalgesia in patients with sympathetically maintained pain. *Pain, 47*, 309–317.

Drummond, P. D. (2009). Alpha (1)-adrenoceptors augment thermal hyperalgesia in mildly burnt skin. *European Journal of Pain, 13*, 273–279.

Drummond, P. D., Finch, P. M., Skipworth, S., et al. (2001). Pain increases during sympathetic arousal in patients with complex regional pain syndrome. *Neurology, 57*, 1296–1303.

Harden, R. N., Baron, R., & Janig, W. (Eds.). (2001). *Complex regional pain syndrome* (pp. 89–118). Seattle: IASP.

Harke, H., Gretenkort, P., Ladleif, H. U., & Rahman, S. (2005). Spinal cord stimulation in sympathetically maintained complex regional pain syndrome type I with severe disability: A prospective clinical study. *European Journal of Pain, 9*, 363–373.

Jänig, W., & Baron, R. (2003). Complex regional pain syndrome: Mystery explained? *Lancet Neurology, 2*, 687–697.

Kalita, J., Vajpayee, A., & Misra, U. K. (2006). Comparison of prednisolone with piroxicam in complex regional pain syndrome following stroke: A randomized controlled trial. *QJM: An International Journal of Medicine, 99*, 89–95.

Katz, J. (1997). Central nervous system correlates and mechanisms of phantom pain. In R. A. Sherman (Ed.), *Phantom pain* (pp. 89–109). New York: Plenum.

Kemler, M. A., Barendse, G. A., van Kleef, M., de Vet, H. C., Rijks, C. P., Furnee, C. A., & van den Wildenberg, F. A. (2000). Spinal cord stimulation in patients with chronic reflex sympathetic dystrophy. *The New England Journal of Medicine, 343*, 618–624.

Kingery, W. S. (1997). A critical review of controlled clinical trials for peripheral neuropathic pain and complex regional pain syndromes. *Pain, 73*, 123–139.

Lin, E. E., Horasek, S., Agarwal, S., et al. (2006). Local administration of norepinephrine in the stump evokes dose-dependent pain in amputees. *The Clinical Journal of Pain, 22*, 482–486.

Linderoth, B., & Meyerson, B. A. (2002). Spinal cord stimulation: Mechanisms of action. In K. J. Burchiel (Ed.), *Surgical management of pain* (pp. 505–526). New York: Thieme Medical Publishers.

Marinus, J., Moseley, G. L., Birklein, F., Baron, R., Maihöfner, C., Kingery, W. S., & van Hilten, J. J. (2011). Clinical features and pathophysiology of complex regional pain syndrome. *Lancet Neurology, 10*(7), 637–648.

Muizelaar, J. P., Kleyer, M., Hertogs, I. A., et al. (1997). Complex regional pain syndrome (RSD and causalgia): Management with calcium channel blocker nifedipine and/or the alpha-sympathetic blocker phenoxybenzamine in 59 patients. *Clinical Neurology and Neurosurgery, 99*, 26–30.

Price, D. D., Long, S., Wilsey, B., et al. (1998). Analysis of peak magnitude and duration of analgesia produced by local anesthetics injected into sympathetic ganglia of complex regional pain syndrome patients. *The Clinical Journal of Pain, 14*, 216–226.

Raja, S. N. (1998). Peripheral modulatory effects of catecholamines in inflammatory and neuropathic pain. *Advances in Pharmacology, 42*, 567–571.

Raja, S. N., Treede, R. D., Davis, K. D., et al. (1991). Systemic alpha-adrenergic blockade with phentolamine: A diagnostic test for sympathetically maintained pain. *Anesthesiology, 74*, 691–698.

Robinson, J. N., Sandom, J., & Chapman, P. T. (2004). Efficacy of pamidronate in complex regional pain syndrome type I. *Pain Medicine, 5*(3), 276–280.

Sahin, F., Yilmaz, F., et al. (2006). Efficacy of salmon calcitonin in complex regional pain syndrome (type 1) in addition to physical therapy. *Clinical Rheumatology, 25*, 143–148.

Singh, B., Moodley, J., Shaik, A. S., et al. (2003). Sympathectomy for complex regional pain syndrome. *Journal of Vascular Surgery, 37*, 508–511.

Stanton-Hicks, M. D., Burton, A. W., Bruehl, S. P., et al. (2002). An updated interdisciplinary clinical pathway for CRPS: Report of an expert panel. *Pain Practice, 2*, 1–16.

Straube, S., Derry, S., Moore, R. A., & McQuay, H. J. (2010). Cervico-thoracic or lumbar sympathectomy for neuropathic pain and complex regional pain syndrome. Cochrane Database of Systematic Reviews(7), CD002918.

Torebjörk, E., Wahren, L., Wallin, G., et al. (1995). Noradrenaline-evoked pain in neuralgia. *Pain, 63*, 11–20.

van Hilten, B. J., van de Beek, W. J., Hoff, J. I., Voormolen, J. H., & Delhaas, E. M. (2000). Intrathecal baclofen for the treatment of dystonia in patients with reflex sympathetic dystrophy. *The New England Journal of Medicine, 343*, 625–630.

Wallin, B. G., Torebjörk, E., & Hallin, R. G. (1976). Preliminary observations on the pathophysiology of hyperalgesia in the causalgic pain syndrome. In Y. Zotterman (Ed.), *Sensory functions of the skin of primates with special reference to man* (pp. 489–499). Oxford: Pergamon.

Wasner, G., Drummond, P. D., Birklein, F., et al. (2001). The role of the sympathetic nervous system in autonomic disturbances and "sympathetically maintained pain" in CRPS. In R. N. Harden, R. Baron, & W. Jänig (Eds.), *Complex regional pain syndrome* (pp. 89–118). Seattle: IASP.

Sympathetically Maintained Pain, Clinical Pharmacological Tests

Peter D. Drummond
School of Psychology, Murdoch University, Perth, Australia

Synonyms

Complex regional pain syndrome; CRPS; SMP; Sympathetically maintained pain

Definition

Neuropathic pain patients can be divided into groups based on their response to selective sympathetic blockade or antagonism of α-adrenoreceptor mechanisms (Price et al. 1998). The component of spontaneous and evoked pain (hyperalgesia), which is relieved by specific sympatholytic procedures, is termed sympathetically maintained pain (SMP). Hyperalgesia can be defined as a leftward shift of the stimulus response function that relates the magnitude of pain to the intensity of a noxious stimulus. An increased sensitivity to painful heat is termed thermal hyperalgesia. Hyperalgesia can also be evoked by application of different mechanical stimuli; for instance, punctate stimuli (e.g., firm von Frey hairs) can evoke punctate hyperalgesia, whereas gently stroking sensitized skin with a cotton swab can provoke dynamic hyperalgesia (also termed mechanical allodynia).

Intradermal injection or iontophoresis of adrenergic agonists, such as noradrenaline, increases thermal hyperalgesia, but does not normally evoke spontaneous pain or mechanical hyperalgesia. However, in certain neuropathic pain syndromes, intradermal injection of noradrenaline into the affected region exacerbates pain and mechanical hyperalgesia, and can rekindle pain after effective treatment with sympathetic blockade or sympathectomy. These patients are said to have a component of SMP.

Characteristics

Adrenergic Hyperalgesia in a Human Pain Model

Transcutaneous iontophoresis of noradrenaline and drugs that release neural stores of noradrenaline (e.g., tyramine) (Drummond 1998) exacerbate thermal hyperalgesia in skin made sensitive to heat by the topical application of capsaicin. Similarly, intradermal injection of α-adrenergic agonists, such as noradrenaline, phenylephrine and brimonidine, induces thermal but not mechanical hyperalgesia in the skin of normal human subjects (Fuchs et al. 2001). Intradermal injection of the non-adrenergic vasoconstrictors angiotensin II and vasopressin does not increase sensitivity to painful heat, despite producing

comparable decreases in blood flow to adrenergic vasoconstrictors (Fuchs et al. 2001). The adrenergic component of thermal hyperalgesia persists after arterial occlusion, indicating that thermal hyperalgesia is independent of blood flow (Drummond 1999), and is not affected by cyclooxygenase inhibition, indicating that prostaglandins do not account entirely for adrenergic hyperalgesia (Schattschneider et al. 2007).

When administered in capsaicin-treated skin, the α-adrenoceptor antagonist (see adrenergic antagonist) phenoxybenzamine blocks thermal hyperalgesia to the adrenergic releasing agent tyramine (Drummond 1998). In contrast, phenoxybenzamine administered before the topical application of capsaicin does not inhibit thermal hyperalgesia to tyramine. The blood-nerve barrier weakens during inflammation, thus allowing substances to diffuse more readily than usual from the extracellular fluid to the local environment of nerve fascicles. Hence, enhanced access to neural or perineural α-adrenoceptors during inflammation might account for the inhibitory influence of phenoxybenzamine on thermal hyperalgesia to tyramine after capsaicin treatment.

Adrenoreceptor Subtypes

In animal experiments, different adrenoreceptor blockers (e.g., phentolamine: α1 and α2), prazosin (α1) or yohimbine (α2) or adrenoreceptor agonists (e.g., phenylephrine: α1; and clonidine: α2) have been used to examine the role of the sympathetic nervous system in neuropathic pain. In some animal neuropathic pain models, sympathetic-sensory coupling is primarily mediated by α2-adrenoreceptors, whereas α1-adrenoreceptors play a minor role (Chen et al. 1996). However, in other animal models of neuropathic pain (Ali et al. 2001; Kim et al. 2005) and inflammation (Dogrul et al. 2006), nociceptive α1-adrenoreceptor effects predominate.

In humans, topical application of the α2-receptor agonist clonidine relieves hyperalgesia at the site of application in patients with SMP, presumably because activation of presynaptic adrenergic autoreceptors inhibits the release of noradrenaline (Davis et al. 1991). Davis et al. (1991) reported that a subsequent injection of noradrenaline and the α1-receptor agonist phenylephrine in the clonidine treated area rekindled hyperalgesia. Quantitative autoradiographic studies indicate that the number of α1-adrenoreceptors in hyperalgesic skin of patients with complex regional pain syndrome (CRPS) is significantly greater than in the skin of normal subjects (Drummond et al. 1996). In contrast to the data obtained in some animal experiments, it seems that, in humans, sympathetic-sensory coupling is mediated via α1-adrenoreceptors. Pharmacological stimulation of α1-adrenoreceptors with phenylephrine evokes axon reflexes in human skin whereas pharmacological stimulation of α2-adrenoreceptors with clonidine does not (Drummond 2011). Thus, activation of α1-adrenoreceptors on or near primary nociceptive afferents may increase both orthodromic and antidromic neural discharge.

Identifying Patients with a Peripheral Adrenergic Component of Pain

The adrenergic component of pain is often investigated clinically with regional sympathetic blockade. However, there are several difficulties with this approach. For example, pain reduction may be due to placebo effects, parallel decreases in anxiety and pain, inadvertent somatic blockade, or systemic uptake of the local anesthetic agent. The phentolamine test was developed to overcome these difficulties. This procedure typically involves intravenous injection of the α-adrenoceptor antagonist phentolamine or saline in a double-blind manner. A major challenge to the validity of this procedure and, indeed, to the concept of SMP, was mounted by Verdugo and Ochoa (1994). They reported that the frequency of pain reduction did not differ between phentolamine and placebo conditions in CRPS patients. However, adrenergic blockade may not have been complete in this study because only a low dose (35 mg) of phentolamine was infused over a 20-min interval. Furthermore, phentolamine administered by intravenous injection might not block adrenergic receptors protected by

a blood-nerve barrier. Campero et al. (2010) employed microneurography to investigate sympathetically maintained pain in 24 patients diagnosed with CRPS. They found no evidence of nociceptor activation by maneuvers known to increase sympathetic neural discharge (e.g., sudden loud noises or mental stress). Thus, other approaches may be required to identify SMP.

An alternative to microneurography and the phentolamine test is to inject adrenergic agonists intradermally in the affected region. Ali et al. (2000) reported that 1 and 10 µM of noradrenaline, at the threshold for producing vasoconstriction, rekindled pain in the affected limb of CRPS patients whose pain had decreased during regional sympathetic blockade with a local anesthetic agent. Similar doses did not induce pain on the unaffected side or in control subjects, indicating that this test might be useful for identifying patients with an adrenergic component of pain. Nociceptive responses to the iontophoretic application or cutaneous injection of noradrenaline were investigated in 25 patients considered to have SMP on the basis of a positive response to regional sympathetic blockade or sympathectomy, and in 10 others whose pain did not change after sympathetic blockade (Torebjörk et al. 1995). Noradrenaline aggravated pain and hyperalgesia in 7 of 25 patients with SMP but had no effect in the remainder of patients. In another study, intradermal injection of 1 mg phenylephrine in 0.1 ml induced brief stinging pain in normal controls, but evoked pain and mechanical hyperalgesia for 10–60 min both in the non-sympathectomized and sympathectomized limbs in 10 of 17 CRPS patients with signs of SMP (Mailis-Gagnon and Bennett 2004). Pain also developed in the contralateral limb of three patients. Effects of sympathetic activation by whole body cooling on spontaneous pain and mechanical and thermal evoked pain were investigated in 20 patients with median nerve entrapment (Schattschneider et al. 2008). Whole body cooling and iontophoresis of noradrenaline aggravated heat hyperalgesia in ten patients with preexisting heat hyperalgesia, consistent with SMP in this subgroup.

Intravenous regional blockade with guanethidine is sometimes used clinically to identify and treat SMP. Guanethidine occupies adrenergic varicosities in sympathetic nerve fibers, thereby abolishing noradrenaline release and preventing sympathetic vasoconstrictor activity. Controlled trials show that intravenous regional blockade with guanethidine is ineffective for treating the pain of unselected patients with CRPS (Kingery 1997), presumably due to variation of sympathetic involvement in CRPS. However, certain patients may benefit from the guanethidine treatment, particularly those with thermal hyperalgesia and allodynia to cold and vibration (Wahren et al. 1991).

Summary

At present, well-validated procedures to identify SMP are lacking. The ideal test would be free of placebo bias and would localize the source of sensory-sympathetic interaction to particular tissues (e.g., the skin or deeper tissue in the affected limb or within the central nervous system). Promising approaches include intradermal injection of adrenergic agents to identify a cutaneous source of interaction, and procedures that increase sympathetic activity (e.g., whole body cooling) to identify broader influences on pain. A combination of placebo-controlled pharmacological tests, regional sympathetic blockade, and physiological procedures that increase sympathetic activity might ultimately be required to identify the presence and source of SMP.

Acknowledgment Supported by the National Health and Medical Research Council of Australia

References

Ali, Z., Raja, S., Wesselmann, U., et al. (2000). Intradermal injection of norepinephrine evokes pain in patients with sympathetically maintained pain. *Pain, 88,* 161–168.

Ali, Z., Ringkamp, M., Hartke, T. V., et al. (2001). Uninjured C-fiber nociceptors develop spontaneous activity and alpha-adrenergic sensitivity following L6 spinal nerve ligation in monkey. *Journal of Neurophysiology, 81,* 455–466.

Campero, M., Bostock, H., Baumann, T. K., & Ochoa, J. L. (2010). A search for activation of C nociceptors by sympathetic fibers in complex regional pain syndrome. *Clinical Neurophysiology, 121,* 1072–1079.

Chen, Y., Michaelis, M., Jänig, W., et al. (1996). Adrenoceptor subtype mediating sympathetic-sensory coupling in injured sensory neurons. *Journal of Neurophysiology, 76,* 3721–3730.

Davis, K. D., Treede, R. D., Raja, S. N., et al. (1991). Topical application of clonidine relieves hyperalgesia in patients with sympathetically maintained pain. *Pain, 47,* 309–317.

Dogrul, A., Coskun, I., & Uzbay, T. (2006). The contribution of alpha-1 and alpha-2 adrenoceptors in peripheral imidazoline and adrenoceptor agonist-induced nociception. *Anesthesia and Analgesia, 103,* 471–477.

Drummond, P. D. (1998). Enhancement of thermal hyperalgesia by alpha-adrenoreceptors in capsaicin-treated skin. *Journal of the Autonomic Nervous System, 69,* 96–102.

Drummond, P. D. (1999). Nitroprusside inhibits thermal hyperalgesia induced by noradrenaline in capsaicin-treated skin. *Pain, 80,* 405–412.

Drummond, P. D. (2011). Inflammation contributes to axon reflex vasodilatation evoked by iontophoresis of an alpha-1 adrenoceptor agonist. *Autonomic Neuroscience, 159,* 90–97.

Drummond, P. D., Skipworth, S., & Finch, P. M. (1996). Alpha1-Adrenoreceptors in normal and hyperalgesic human skin. *Clinical Science, 91,* 73–77.

Fuchs, P. N., Meyer, R. A., & Raja, S. N. (2001). Heat, but not mechanical hyperalgesia, following adrenergic injections in normal human skin. *Pain, 90,* 15–23.

Kim, S. K., Min, B. I., Kim, J. H., et al. (2005). Effects of alpha1- and alpha2-adrenoreceptor antagonists on cold allodynia in a rat tail model of neuropathic pain. *Brain Research, 1039,* 207–210.

Kingery, W. S. (1997). A critical review of controlled clinical trials for peripheral neuropathic pain and complex regional pain syndromes. *Pain, 73,* 123–139.

Mailis-Gagnon, A., & Bennett, G. J. (2004). Abnormal contralateral pain responses from an intradermal injection of phenylephrine in a subset of patients with complex regional pain syndrome (CRPS). *Pain, 111,* 378–384.

Price, D. D., Long, S., Wilsey, B., et al. (1998). Analysis of peak magnitude and duration of analgesia produced by local anesthetics into sympathetic ganglia of complex regional pain syndrome patients. *The Clinical Journal of Pain, 14,* 216–226.

Schattschneider, J., Zum Buttel, I., Binder, A., Wasner, G., Hedderich, J., & Baron, R. (2007). Mechanisms of adrenosensitivity in capsaicin induced hyperalgesia. *European Journal of Pain, 11,* 756–763.

Schattschneider, J., Scarano, M., Binder, A., Wasner, G., & Baron, R. (2008). Modulation of sensitized C-fibers by adrenergic stimulation in human neuropathic pain. *European Journal of Pain, 12*, 517–524.

Torebjörk, E., Wahren, L., Wallin, G., Hallin, R., & Koltzenburg, M. (1995). Noradrenaline-evoked pain in neuralgia. *Pain, 63*, 11–20.

Verdugo, R. J., & Ochoa, J. L. (1994). Sympathetically maintained pain I. Phentolamine block questions the concept. *Neurology, 44*, 1003–1010.

Wahren, L. K., Torebjörk, E., & Nystrom, B. (1991). Quantitative sensory testing before and after regional guanethidine block in patients with neuralgia in the hand. *Pain, 46*, 23–30.

Sympathetic-Sensory Coupling

▶ Sympathetic-Afferent Coupling in the Afferent Nerve Fiber, Neurophysiological Experiments
▶ Sympathetic-Afferent Coupling in the Dorsal Root Ganglion, Neurophysiological Experiments

---

Sympathetic-Sensory Coupling Distal to the Dorsal Root Ganglion (DRG), Peripheral Sympathetic-Afferent Coupling

▶ Sympathetic-Afferent Coupling in the Afferent Nerve Fiber, Neurophysiological Experiments

---

Sympatho-adrenal System and Mechanical Hyperalgesic Behavior, Animal Experimentation

Wilfrid Jaenig
Physiologisches Institut, Christian-Albrechts-Universität zu Kiel, Kiel, Germany

Definition

The adrenal medullae release both adrenaline and noradrenaline, regulated by activity in preganglionic sympathetic neurons that project from the thoracic spinal cord through the splanchnic nerves to this gland. Histological and immunohistochemical studies show that adrenaline and noradrenaline are synthesized by different cells, which are also innervated by different groups of preganglionic neurons (Edwards et al. 1996). These data suggest that there are two separate sympathetic pathways to the adrenal medulla, one mediating the release of adrenaline and the other noradrenaline (Vollmer 1996). This idea is supported by functional studies implying that the release of adrenaline and noradrenaline from the adrenal medulla can be regulated differentially by the brain (see Folkow 1955 for early studies; Folkow and von Euler 1954; Vollmer 1996). Morrison and Cao (2000) showed in the rat that preganglionic neurons, which innervate adrenergic cells of the adrenal medulla, are not affected by arterial baroreceptor reflexes and respiratory reflexes but are strongly excited by hypoglycemia. In contrast, preganglionic neurons that innervate noradrenergic cells of the adrenal medulla behave like muscle vasoconstrictor neurons. In humans under resting conditions, the concentration of noradrenaline in the blood plasma is 1–1.5 pmol/ml (0.17–0.25 μg/L) and of adrenaline about 0.25 pmol/ml (0.05 μg/L) (Kopin 1989). Furthermore, under almost all physiological conditions, the concentration of circulating noradrenaline is 3–5 × higher than the concentration of adrenaline (Cryer 1980). In humans, only about 2–8 % of the circulating noradrenaline is released by the adrenal medulla, and the rest is released by sympathetic nerve endings (Esler et al. 1990). Under physiological conditions, adrenaline may be involved in the following functions: (1) regulation of metabolism such as catalyzing the mobilization of glucose and lactic acid from glycogen and probably of fatty acids from adipose tissue, (2) modulation of sensitivity of nociceptors, (3) modulation of inflammation (see ▶ Sympathetic Postganglionic Neurons in Neurogenic Inflammation of the Synovia; Miao et al. 2000, 2001), (4) consolidation of memory (via vagal afferents and the nucleus tractus solitarii [NTS]), and (5) probably not support responses of target organs elicited by activation of other sympathetic pathways except under strong emergency conditions (Jänig 2006).

Characteristics

Effect of Subdiaphragmatic Vagotomy on Mechanical Hyperalgesic Behavior

Withdrawal threshold to stimulation of the rat hind paw with a linearly increasing mechanical stimulus applied to the dorsum of the paw decreases dose-dependently after intradermal injection of bradykinin (open circles and closed squares in Fig. 1a; bradykinin-induced mechanical hyperalgesic behavior). If inhibition maintained by activity in vagal afferents acts continuously on the central nociceptive pathway, one would expect, on the basis of studies reported by Gebhart, Randich, and coworkers (see Foreman 1989; Gebhart and Randich 1992; Randich and Gebhart 1992), that subdiaphragmatic vagotomy might enhance the mechanical hyperalgesic behavior, irrespective of the way the nociceptive afferents have been sensitized (e.g., by bradykinin or another hyperalgesic agent), and lower baseline threshold to mechanical stimulation.

Baseline paw-withdrawal threshold (normal rats: 109 ± 2.1 g [mean ± SEM]; sham-vagotomized rats: 107 ± 2.8 g; open and closed circles in Fig. 1a) is significantly decreased 7 days after vagotomy (89 ± 1.7 g; triangles in Fig. 1a). Bradykinin-induced hyperalgesia is significantly enhanced 7 days after subdiaphragmatic vagotomy (triangles in Fig. 1a). There are three important characteristics of the effect of vagotomy on mechanical baseline threshold and on bradykinin-induced decrease of paw-withdrawal threshold to mechanical stimulation:

1. The dramatic enhancement of bradykinin-induced mechanical hyperalgesic behavior occurs also when only the *celiac vagal branches* are interrupted, but not when the gastric and/or hepatic branches of the abdominal vagus nerves are interrupted. Thus, the vagal afferents involved project through the celiac branches of the abdominal vagus nerves, which innervate the small intestine and proximal part of the large intestine, and

not through the hepatic or gastric branches. Surprisingly, the baseline paw-withdrawal threshold to mechanical stimulation does not decrease when only the celiac vagal branches are interrupted (Khasar et al. 1998a).

2. Both vagotomy-induced changes (decrease in baseline paw-withdrawal threshold, bradykinin-induced hyperalgesic behavior) take about 1–2 weeks to reach maximum and remain stable over 5 weeks (Fig. 2; Khasar et al. 1998a, b).

3. Subdiaphragmatic vagotomy does not have a significant effect on cutaneous mechanical hyperalgesic behavior produced by intradermal injection of prostaglandin E_2 (which is supposed to act directly to sensitize nociceptors) (Khasar et al. 1998a).

Thus, the effect of vagotomy is not a general effect of all abdominal vagal afferents and cannot readily be explained by an immediate removal of inhibition from the central nociceptive system (e.g., acting in the dorsal horn) as predicted by the experiments of Foreman, Gebhart, Randich, and coworkers (see above).

Effect of Denervation or Removal of the Sympatho-Adrenal System

Bilateral removal or selective denervation of the adrenal medullae (cutting the sympathetic preganglionic axons) generates a small increase both in baseline paw-withdrawal threshold and paw-withdrawal threshold to intradermal injection of bradykinin compared to the controls (open squares vs. closed circles in Fig. 1b). Under this condition of defunctionalized adrenal medullae (i.e., when the effect of adrenaline described below has been removed), subdiaphragmatic vagotomy is followed by only a small decrease in paw-withdrawal threshold (compare open and closed squares in Fig. 1b) but not by the large changes seen in animals with functioning adrenal medullae. These small changes are significant, with the exception of the change of baseline threshold in animals with denervated adrenal medullae. They can fully be explained by removal of central inhibition of nociceptive impulse transmission occurring probably in the dorsal horn (Gebhart and Randich 1992; Randich and Gebhart 1992).

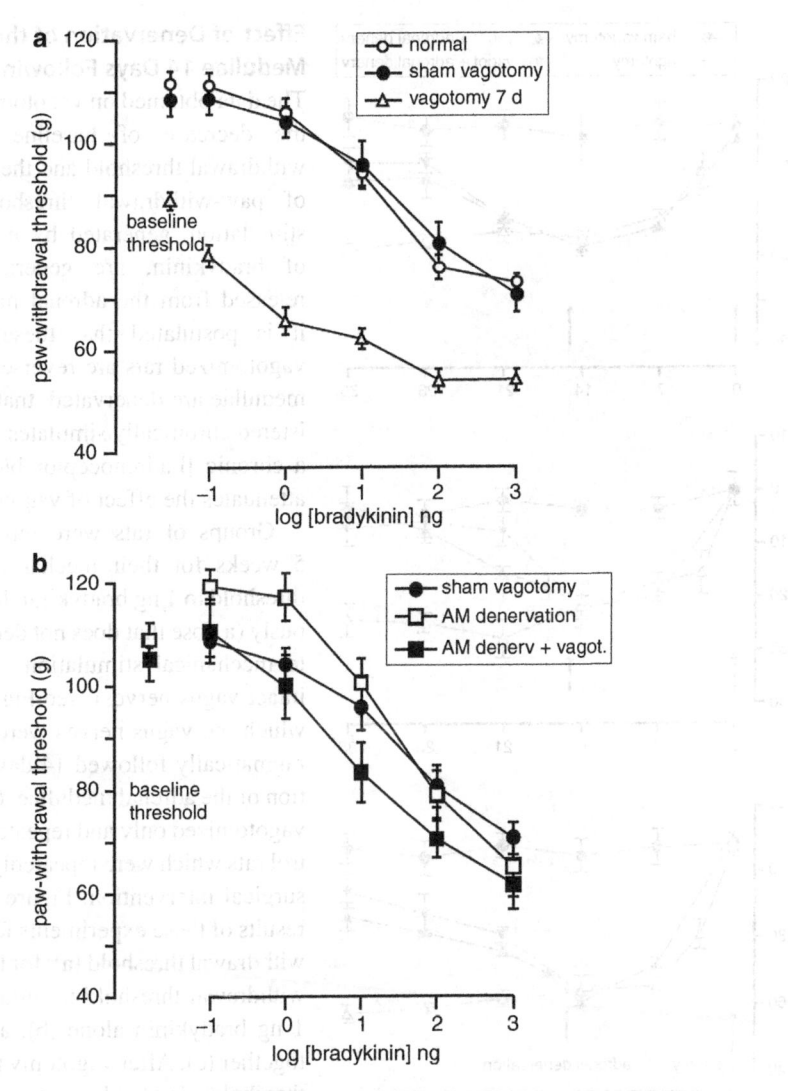

Sympatho-adrenal System and Mechanical Hyperalgesic Behavior, Animal Experimentation, Fig. 1 (a) Decrease of paw-withdrawal threshold to mechanical stimulation of the dorsum of the rat hind paw induced by bradykinin (bradykinin-induced behavioral mechanical hyperalgesia) in normal control (*open circles*, n = 26), vagotomized (*triangles*, n = 16), and sham-vagotomized (*closed circles*, n = 18) rats. Experiments conducted 7 days after vagotomy. Post hoc test shows significant differences between vagotomized and normal ($P < 0.05$) as well as between vagotomized and sham-vagotomized ($P < 0.05$) rats, in response to bradykinin. Cutaneous mechanoreceptors in the hairy skin are stimulated by a linearly increasing mechanical force using a Basile algesimeter (Stoelting, Chicago, IL). Threshold is defined as the minimum force (g) at which the paw is withdrawn by a rat. Ordinate scale expresses paw-withdrawal threshold in grams. The abscissa scale is the log dose of BK (in ng) injected in a volume of 2.5 μL saline into the dermis of the skin of the dorsal aspect of the hind paw. (Data from Khasar et al. (1998a) with permission) (b) Role of denervation of adrenal medullae on bradykinin-induced behavioral mechanical hyperalgesia and its enhancement after subdiaphragmatic vagotomy. Data from experiments in which the adrenal medullae are denervated: rats with denervated adrenal medullae (AM denervation, *open squares*, n = 6); vagotomized rats with denervated adrenal medullae (AM denerv plus vagotomy, *closed squares*; n = 10). Experiments were conducted 7 days after surgery. Paw-withdrawal thresholds of vagotomized rats in which the adrenal medullae were denervated are significantly higher than those of rats which are only vagotomized (see *open triangles* in **a**; $P < 0.05$). Paw-withdrawal thresholds of rats in which the adrenal medullae are denervated are significantly higher than those of rats which were additionally vagotomized ($P < 0.05$; compare *open* with *closed squares*) (Modified from Khasar et al. 1998b)

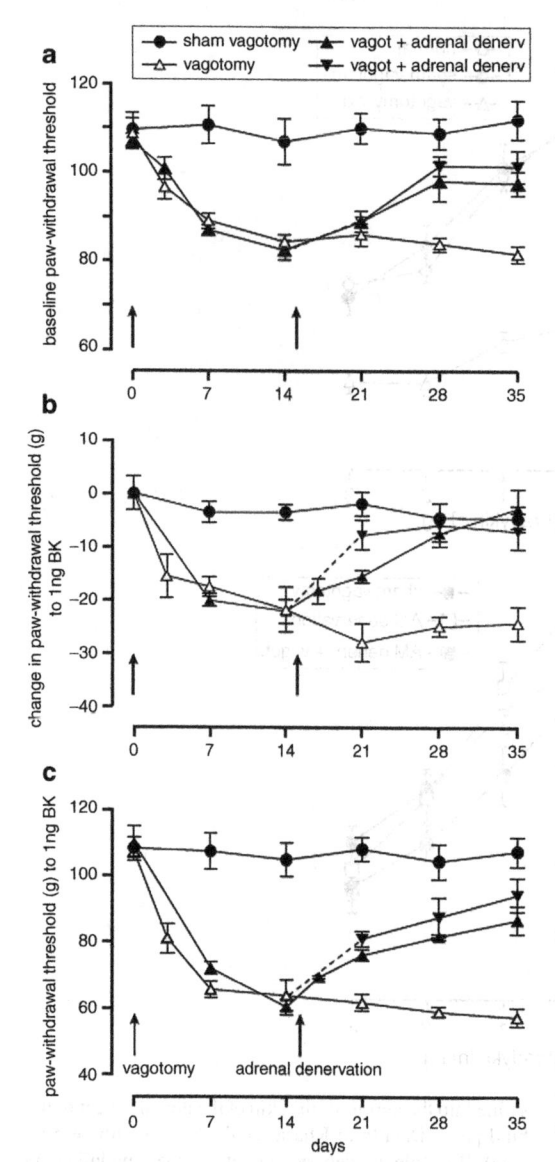

Effect of Denervation of the Adrenal Medullae 14 Days Following Vagotomy

The data obtained on vagotomized rats argue that the decrease of baseline mechanical paw-withdrawal threshold and the enhanced decrease of paw-withdrawal threshold to mechanical stimulation, generated by intradermal injection of bradykinin, are generated by adrenaline released from the adrenal medullae. Therefore, it is postulated that these changes seen in vagotomized rats are reversed when the adrenal medullae are denervated, that adrenaline administered chronically simulates the effects, and that a chronic β-adrenoceptor blockade prevents or attenuates the effect of vagotomy.

Groups of rats were repeatedly tested over 5 weeks for their mechanical paw-withdrawal threshold to 1 ng bradykinin injected intracutaneously (a dose that does not decrease the threshold to mechanical stimulation in normal rats with intact vagus nerves) (see Fig. 1a, b): (1) Rats in which the vagus nerves were severed subdiaphragmatically followed 14 days later by denervation of the adrenal medullae. (2) Rats which were vagotomized only and repeatedly tested. (3) Control rats which were repeatedly tested without any surgical intervention. Figure 2 demonstrates the results of these experiments for the baseline paw-withdrawal threshold (a), for the decrease in paw-withdrawal threshold to intradermal injection of 1 ng bradykinin alone (b), and for both effects together (c). After vagotomy the paw-withdrawal threshold slowly decreases reaching its lowest values after 7–14 days. The reversal of the

Sympatho-adrenal System and Mechanical Hyperalgesic Behavior, Animal Experimentation, Fig. 2 Long-term enhancement of bradykinin-induced behavioral mechanical hyperalgesia after vagotomy and its disappearance after denervation of the adrenal medullae. Baseline paw-withdrawal threshold (a), difference between baseline paw-withdrawal threshold and total paw-withdrawal threshold in response to 1 ng BK injected intradermally (b), and total change of paw-withdrawal threshold in response to intradermal injection of 1 ng bradykinin (c) in rats before and 7–35 days after vagotomy (*open triangles*, $n = 6$), before and 7–35 days after sham-vagotomy (*closed circles*, $n = 8$), and in rats which are first vagotomized and whose adrenal medullae (AM) are denervated 14 days after vagotomy and measurements taken up to 35 days after initial surgery. The latter group

of animals consists of two subgroups: rats which are tested after vagotomy and after additional denervation of the adrenal medullae (*closed normal triangles*, $n = 6$) and rats which are only tested after additional denervation of the adrenal medulla (*closed inverted triangles*, $n = 4$). Ordinate scale is threshold in grams. Data of the sham-vagotomy and the vagotomy group of rats are significantly different 7 days after vagotomy ($P < 0.01$). Data of vagotomized rats with denervated AM and rats that are only vagotomized are significantly different on days 28 and 35 ($P < 0.01$). Data between sham-vagotomized rats and vagotomized rats in which the adrenal medullae are denervated are not significantly different on days 28 and 35 ($P > 0.05$) (Modified from Khasar et al. 1998b)

vagotomy effect after additional denervation of the adrenal medullae has also a similar slow time course (closed triangles in Fig. 2). Repeated testing of sham-vagotomized control rats over the same time period does not reveal a decrease in paw-withdrawal threshold produced by 1 ng bradykinin (closed circles in Fig. 2). The paw-withdrawal thresholds, 14 and 21 days after denervation of the adrenal medullae, are significantly higher than those measured in the animals that are only vagotomized (closed triangles vs open triangles in Fig. 2). The paw-withdrawal thresholds in response to 1 ng bradykinin at 14 and 21 days after denervation of the adrenal medullae in vagotomized animals are not significantly different from those in sham-vagotomized (control) animals (closed triangles vs closed circles in Fig. 2).

Chronic administration of adrenaline (10.8 μg/h; using a subcutaneously implanted microosmotic pump) generates the same effect as vagotomy: The bradykinin-induced paw-withdrawal threshold to mechanical stimulation significantly decreases. This decrease is delayed and reaches its peak effect 14 days after start of epinephrine infusion. After chronic infusion of the β_2-adrenoceptor blocker ICI 118551, the decrease of bradykinin-induced paw-withdrawal threshold following vagotomy is significantly attenuated. The plasma levels of adrenaline following vagotomy significantly increase 3, 7, and 14 days after subdiaphragmatic vagotomy compared to sham-vagotomized animals (Khasar et al. 2003).

Interpretation of the Results

These results suggest that vagotomy does not only remove ongoing inhibition of nociceptive impulses transmission in the dorsal horn (which is small in the rat model of bradykinin-induced mechanical hyperalgesia) but triggers the activation of sympathetic preganglionic neurons innervating the adrenal medullae, probably by removing central inhibition acting at this sympathetic pathway. This leads to an increased release of adrenaline from the adrenal medullae with an increased adrenaline level in the plasma.

Interruption of the sympathetic preganglionic axons innervating the adrenal glands stops the release of adrenaline and therefore prevents or reverses the decrease of baseline mechanical paw-withdrawal threshold and the enhancement of bradykinin-induced decrease of paw-withdrawal threshold to mechanical stimulation. This finding implies that the sensitivity of nociceptors to mechanical stimulation is potentially under control of the sympatho-adrenal system and that nociceptor sensitivity can be regulated from remote body domains by way of the brain and this neuroendocrine pathway.

The novel mechanism has several implications and raises several interesting questions and problems (Jänig et al. 2000):

- The vagal afferents that are involved in modulation of hyperalgesic behavior project through the celiac branches of the abdominal vagus nerves and supply small and large intestines, but probably not liver and stomach. These vagal afferents may monitor toxic and other events at the inner defense line of the body (the "gut associated lymphoid tissue," GALT). The physiological stimuli activating these vagal afferents are unknown (Jänig 2005).
- The changes following vagotomy (decreased mechanical baseline threshold and enhanced bradykinin-induced mechanical hyperalgesic behavior; Khasar et al. 1998a, b) are generated by the *interruption* of vagal afferents. Thus, the vagal afferents involved must be tonically active.
- The mechanism of the slow time course of the changes of paw-withdrawal threshold is unknown. Adrenaline obviously has to act over a long period of time to induce changes in the micromilieu of the nociceptor population which in turn leads to their sensitization (see Khasar et al. 2003). It does most likely act *not directly* on the nociceptors but on other cells (e.g., macrophages, mast cells, keratinocytes) which then release substances that generate the sensitization, this in particular since prostaglandin E_2-induced mechanical hyperalgesic behavior is not changed after vagotomy (Khasar et al. 1998a).

- The change of sensitivity of a population of cutaneous nociceptors generated by adrenaline which is regulated by the brain would be a special mechanism of peripheral sensitization. This mechanism of nociceptor sensitization would be different from mechanisms that lead to activation and/or sensitization of nociceptors by sympathetic-afferent coupling under pathophysiological conditions (see ▶ Sympathetic Nervous System and Pain; ▶ Sympathetic Nervous System in the Generation of Pain, Animal Behavioral Models; ▶ Sympathetic-afferent Coupling in the Dorsal Root Ganglion, Neurophysiological Experiments; ▶ Sympathetic-afferent Coupling in the Afferent Nerve Fiber, Neurophysiological Experiments).

- Which central pathways are involved leading to activation of preganglionic sympathetic neurons that innervate the adrenal medulla after subdiaphragmatic vagotomy? Are only sympathetic neurons that innervate the adrenal medullae activated after vagotomy or also other functional types of sympathetic neuron (Jänig and McLachlan 2012; Jänig 2006)? Experimental investigations performed on rats show that sympathetic preganglionic neurons innervating cells of the adrenal medullae that release adrenaline are connected to distinct neuronal circuits in the spinal cord, brain stem, and hypothalamus that are different from those connected to preganglionic neurons being involved in regulating other autonomic functions (Morrison and Cao 2000; Morrison 2001).

Conclusions

The experiments imply that adrenaline released by the adrenal medullae can sensitize nociceptors for mechanical stimulation. This sensitization is a long-term effect and requires continuous activation of the adrenal medullae. Are all or only subgroups of nociceptors sensitized? Are also nociceptors innervating deep somatic tissues or viscera sensitized?

References

Cryer, P. E. (1980). Physiology and pathophysiology of the human sympathoadrenal neuroendocrine system. *New England Journal of Medicine, 303*, 436–444.

Edwards, S. L., Anderson, C. R., Southwell, B. R., & McAllen, R. M. (1996). Distinct preganglionic neurons innervate noradrenaline and adrenaline cells in the cat adrenal medulla. *Neuroscience, 70*, 825–832.

Esler, M., Jennings, G., Lambert, G., Meredith, I., Horne, M., & Eisenhofer, G. (1990). Overflow of catecholamine neurotransmitters to the circulation: Source fate, and functions. *Physiological Reviews, 70*, 963–985.

Folkow, B. (1955). The control of blood vessels. *Physiological Reviews, 35*, 629–663.

Folkow, B., & Von Euler, U. S. (1954). Selective activation of noradrenaline and adrenaline producing cells in the cat's adrenal gland by hypothalamic stimulation. *Circulation Research, 2*, 191–195.

Foreman, R. D. (1989). Organization of the spinothalamis tract as a relay for cardiopulmonary sympathetic afferent fiber activity. *Progress in Sensory Physiology, 9*, 1–51.

Gebhart, G. F., & Randich, A. (1992). Vagal modulation of nociception. *American Pain Society Journal, 1*, 26–32.

Häbler, H.-J., & Jänig, W. Role of the sympathetic nervous system in the generation of pain: Animal behavioral models (Essay 37.5).

Häbler, H.-J., & Jänig, W. Sympathetic-afferent coupling in the dorsal root ganglion: Neurophysiological experiments (Essay 37.6).

Jänig, W. (2005). Vagal afferents and visceral pain. In B. Undem & D. Weinreich (Eds.), *Advances in vagal afferent neurobiology* (pp. 461–489). Boca Raton: CRC Press.

Jänig, W. (2006). *The Integrative action of the autonomic nervous system: Neurobiology of homeostasis*. Cambridge/New York: Cambridge University Press.

Jänig, W., & Baron, R. Pain and the autonomic nervous system (Synopsis of 37).

Jänig, W., & McLachlan, E. M. (2012). Neurobiology of the autonomic nervous system. In C. J. Mathias & R. Bannister (Eds.), *Autonomic failure* (5th ed., pp. 21–34). Oxford: Oxford University Press.

Jänig, W., Khasar, S. G., Levine, J. D., & Miao, F.-J. P. (2000). The role of vagal visceral afferents in the control of nociception. *Progress in Brain Research, 122*, 273–287.

Khasar, S. G., Miao, F.-J. P., Jänig, W., & Levine, J. D. (1998a). Modulation of bradykinin-induced mechanical hyperalgesia in the rat skin by activity in the abdominal vagal afferents. *European Journal of Neuroscience, 10*, 435–444.

Khasar, S. G., Miao, F.-J. P., Jänig, W., & Levine, J. D. (1998b). Vagotomy-induced enhancement of mechanical hyperalgesia in the rat is sympathoadrenal-mediated. *Journal of Neuroscience, 18*, 3043–3049.

Khasar, S. G., Green, P. G., Miao, F.-J. P., & Levine, J. D. (2003). Vagal modulation of nociception is mediated by adrenomedullary epinephrine in the rat. *European Journal of Neuroscience, 17*, 909–915.

Kopin, I. J. (1989). Plasma levels of catecholamines and dopamine-beta-hydroxylase. In U. Trendelenburg & N. Weiner (Eds.), *Catecholamines II. Handbook of experimental pharmacology* (Vol. 90/II, pp. 211–275). Berlin: Springer.

Miao, F.-J. P., Jänig, W., & Levine, J. D. (2000). Nociceptive-neuroendocrine negative feedback control of neurogenic inflammation activated by capsaicin in the skin: Role of the adrenal medulla. *Journal of Physiology (London), 527*, 601–610.

Miao, F. J.-P., Jänig, W., Jasmin, L., & Levine, J. D. (2001). Spino-bulbo-spinal pathway mediating vagal modulation of nociceptive-neuroendocrine control of inflammation in the rat. *Journal of Physiology (London), 532*, 811–822.

Morrison, S. F. (2001). Differential control of sympathetic outflow. *American Journal of Physiology – Regulatory, Integrative and Comparative Physiology, 281*, R683–R698.

Morrison, S. F., & Cao, W. H. (2000). Different adrenal sympathetic preganglionic neurons regulate epinephrine and norepinephrine secretion. *American Journal of Physiology – Regulatory, Integrative and Comparative Physiology, 279*, R1763–R1775.

Randich, A., & Gebhart, G. F. (1992). Vagal afferent modulation of nociception. *Brain Research Reviews, 17*, 77–99.

Ringkamp, M., & Häbler, H.-J. Sympathetic-afferent coupling in the afferent nerve fiber: Neurophysiological experiments (Essay 37.7).

Vollmer, R. R. (1996). Selective neural regulation of epinephrine and norepinephrine cells in the adrenal medulla – Cardiovascular implications. *Clinical and Experimental Hypertension, 18*, 731–751.

Sympatho-Nociceptive Coupling

▶ Sympathetically Maintained Pain and Inflammation, Human Experimentation

Symphysis Pubis Dysfunction

Definition

Separation of the symphysis pubis is a complication of pregnancy, with an increasing prevalence from 1 in 36 at worst (Owens et al. 2002). In the past, it has largely been ignored as a cause of painful morbidity after delivery. It is characterized by suprapubic pain and tenderness with radiation to the backs of the legs, difficulty in walking, and occasionally bladder dysfunction. A clinical history is usually adequate to make the diagnosis, but confirmation by ultrasound or radiography may be required. It is a condition associated with multiparity, but the underlying etiology has not been fully investigated. Conservative therapy is usual with rest, binders, and mild analgesics, and complete recovery is expected up to 2 months after delivery.

Cross-References

▶ Postpartum Pain

References

Owens K, Pearson A, Mason G. (2002). Symphysis pubis dysfunction-a cause of significant obstetric morbidity. *European Journal of Obstetrics Gynecology and Reproductive Biology, 105*(2), 143–146.

Symptom Magnifier

Definition

A claimant whose verbal or nonverbal communications suggest a degree of symptom severity and incapacitation that appears to be excessive compared to his or her medical findings, or to the communications of other individuals with similar medical findings.

Cross-References

▶ Impairment, Pain-Related

Symptomatic Annular Tear

▶ Discogenic Back Pain

Symptomatic Cramps

Definition

Cramps due to several reasons such as central and peripheral nervous system disorders, muscular diseases, vascular diseases, endocrine-metabolic diseases, electrolyte imbalance, toxic and pharmacological substances, and psychiatric disorders.

Cross-References

► Muscular Cramps

Symptomatic Intracranial Hypotension

► Headache Due to Low Cerebrospinal Fluid Pressure

Symptoms

Definition

Symptoms are sensations that patients experience and interpret as warnings that something is out of balance or wrong within their bodies, and those experiences are communicated to clinicians. The US Social Security Administration holds that no symptom can be the basis for a finding of disability, no matter how genuine the person's complaints may appear to be, unless there is objective medical evidence that demonstrates the existence of a physical or mental impairment(s) that could reasonably be expected to produce the symptom.

Cross-References

► Disability Evaluation in the Social Security Administration
► Hypoalgesia, Assessment
► Hypoesthesia, Assessment

Synapse

Definition

The junction where a signal is transmitted from one nerve cell to another, to a muscle cell, or to a gland cell usually by a neurotransmitter (chemical synapse) but sometimes electrically (electrical synapse).

Cross-References

► Amygdala, Pain Processing and Behavior in Animals

Synaptic Glomerulus

Definition

Synaptic glomerulus is a complex synaptic arrangement representing a central terminal from primary afferents surrounded with many neural elements such as axon terminals and dendrites.

Cross-References

► Trigeminal Brainstem Nuclear Complex, Anatomy

Synaptic Organization

Definition

Synaptic organization is usually used as a morphological characteristic of synaptic contacts. The synapse is composed of different elements, which include a terminal axon, dendrite, dendritic spine, and presynaptic axon. The number of elements varies according to the function of inputs and the output information to secondary neurons.

Cross-References

▶ Morphology, Intraspinal Organization of Visceral Afferents

Synaptic Organization of Afferent Fibers from Viscera in the Spinal Cord

▶ Morphology, Intraspinal Organization of Visceral Afferents

Synchronization

Definition

Large ensembles of neurons firing at the same time.

Cross-References

▶ Spinothalamocortical Projections to Ventromedial and Parafascicular Nuclei

Syndrome X

Definition

Syndrome X is a condition characterized by angina-like chest pain and ST depression on stress ECG in patients that have normal coronary arteries on angiography.

Cross-References

▶ Thalamus and Visceral Pain Processing (Human Imaging)

Synergistic

Definition

The action of two or more substances achieving an effect greater than that expected with doses of the individual components, i.e., the sum is greater than the parts. In medicine, this refers to the potentiation of one drug by one or more additional agents when administered together or by a single drug at two separate sites of biological action.

Cross-References

▶ Opioids and Reflexes
▶ Postoperative Pain, Appropriate Management

Synergy

Definition

Synergy refers to the combined action of two agents that produce an effect that neither agent could produce alone at the given doses or an effect that is greater than the total of those produced by the individual agents.

Cross-References

▶ Drugs with Mixed Action and Combinations: Emphasis on Tramadol

Synovial Joints

Definition

A specialized joint that is characterized by opposing bone surfaces covered with cartilage and enveloped by a synovial membrane, which serves to secrete synovial fluid as a lubricant for the joint. Synovial joints typically allow unrestricted motion within the movement range and are most commonly found in the arms and legs. These joints are the most common joints to be affected by osteoarthritis.

Syntenic

Definition

Syntenic refers to a conserved region of chromosomal DNA in which two or more genes are found adjacent in the chromosomes of two species.

Cross-References

▶ TRPV1 Receptor, Species Variability

Syringomyelia

Definition

A chronic progressive disease of the spinal cord in which cavities form in the gray matter.

Cross-References

▶ Central Pain, Outcome Measures in Clinical Trials
▶ Spinal Cord Injury Pain

Syringomyelia and Pain

▶ Pain in Syringomyelia

Systematic

Definition

Characterized by a series of orderly actions that are planned and deliberate: not haphazard.

Cross-References

▶ Pain Assessment in Neonates

Systematic Review

Definition

A review of the literature, in which there is a clearly formulated question, and systematic and explicit methods are used to identify, select, and critically appraise relevant research and to collect and analyze data from the original studies that are included in the review (Petitti 1994).

Cross-References

▶ Lumbar Traction

References

Petitti, D. B. (1994). Meta-analysis, decision analysis, and cost-effectiveness analysis: Methods for quantitative synthesis in medicine. New York: Oxford University Press.

Systemic Lupus Erythematosus

Synonyms

SLE

Definition

Systemic lupus erythematosus (SLE) is a systemic autoimmune disease with antinuclear antibodies and multilocular manifestations, caused by a thrombotic vasopathy or antibodies interacting with cell membrane function.

Cross-References

▶ Headache Due to Arteritis

Systemic Morphine Effects on Evoked Pain

▶ Opioids, Effects of Systemic Morphine on Evoked Pain

Systems Theory

Definition

Principles, models, and laws applied to the interrelationships and interdependencies of sets of linked components that form a system.

Cross-References

▶ Assessment of Pain Behaviors

caused by ... Microscopic sexuality of subbodies.
Includes ... all cell membrane function.

Cross-References

▶ Headache Due to Arnold's

Systemic Morphine Effects on Evoked Pain

▶ Opioids, Effects of Systemic Morphine on Evoked Pain

Systems Theory

Definition

Principles, models, and laws applied to the interrelationships and interdependence of sets of linked components that form a system

Cross-References

▶ Assessment of Pain Behaviors

Encyclopedia of Pain

Gerald F. Gebhart • Robert F. Schmidt

Editors

Encyclopedia of Pain

Second Edition

Volume 7

T–Z

With 795 Figures and 217 Tables

 Springer Reference

Editors
Gerald F. Gebhart
Department of Anesthesiology
Center for Pain Research
University of Pittsburgh
Pittsburgh, PA, USA

Robert F. Schmidt
Physiologisches Institut der
Universität Würzburg
Würzburg, Germany

ISBN 978-3-642-28752-7   ISBN 978-3-642-28753-4 (eBook)
ISBN 978-3-642-28754-1 (print and electronic bundle)
DOI 10.1007/978-3-642-28753-4
Springer Heidelberg New York Dordrecht London

Library of Congress Control Number: 2013951551

Printed on acid-free paper

Springer is part of Springer Science+Business Media (www.springer.com)

Preface to the Second Edition

As was pointed out in the Preface to the first edition, pain is the principal reason individuals seek medical and dental attention. Fortunately, acute tissue insults associated with pain – and typically inflammation – readily resolve with appropriate treatment and care. Chronic pain states, in contrast, are notoriously difficult to manage and are commonly associated with comorbidities (e.g., depression, cross-organ sensitization) and reduced quality of life. Indeed, chronic pain has come to be considered by some as a disease in its own right. This edition of the *Encyclopedia of Pain* updates the content of the 1st edition and introduces new topics consistent with advances in our knowledge of underlying mechanisms of pain. Thus, new content on channelopathies, expanded and updated material on the role of glia and immune-competent cell interactions with nociceptive neurons, and advances in imaging and changes in brain structure in chronic pain states are included in this edition.

The cardinal objective of research on pain is the translation of knowledge between the laboratory and the clinic. The present prevailing refrain is "bench to bedside," but in fact many ideas for research in pain derive from clinical observations and the absence of knowledge to explain clinical pain states. In the recent past, basic and clinical science research has seen the application of imaging techniques to visualize areas of the brain that are involved in both the sensory and affective aspects of pain, cloning of pain-related channels and receptors, and the application of elegant genetic manipulations that selectively "label" functionally distinct dorsal root ganglion sensory and spinal neurons, two-photon imaging, and optogenetic approaches.

Accordingly, much is new in this second edition of the *Encyclopedia of Pain*. The number of authors who have contributed to this edition is close to 800. In addition to a print edition, the publisher, Springer-Verlag, will produce an electronic version that will make comprehensive searches easy to manage. The electronic version, to be made available on the online publishing platform SpringerReference, can be updated regularly to ensure that the content remains current.

October 2013
Gerald F. Gebhart   Robert F. Schmidt
Pittsburgh, PA, USA   Würzburg, Germany

Preface to the First Edition

As all medical students know, pain is the most common reason for a person to consult a physician. Under ordinary circumstances, acute pain has a useful, protective function. It discourages the individual from activities that aggravate the pain, allowing faster recovery from tissue damage. The physician can often tell from the nature of the pain what its source is. In most cases, treatment of the underlying condition resolves the pain. By contrast, children born with congenital insensitivity to pain suffer repeated physical damage and die young (see Sweet WH (1981) Pain 10:275).

Pain resulting from difficult to treat or untreatable conditions can become persistent. Chronic pain "never has a biologic function but is a malefic force that often imposes severe emotional, physical, economic, and social stresses on the patient and on the family..." (Bonica JJ (1990) The Management of Pain, vol 1, 2nd edn. Lea & Febiger, Philadelphia, p 19). Chronic pain can be considered a disease in its own right.

Pain is a complex phenomenon. It has been defined by the Taxonomy Committee of the International Association for the Study of Pain as "An unpleasant emotional and sensory experience associated with actual or potential tissue damage, or described in terms of such damage" (Merskey H and Bogduk N (1994) Classification of Chronic Pain, 2nd edn. IASP Press, Seattle). It is often ongoing, but in some cases it may be evoked by stimuli. Hyperalgesia occurs when there is an increase in pain intensity in response to stimuli that are normally painful. Allodynia is pain that is evoked by stimuli that are normally non-painful.

Acute pain is generally attributable to the activation of primary efferent neurons called nociceptors (Sherrington CS (1906) The Integrative Action of the Nervous System. Yale University Press, New Haven; 2nd edn, 1947). These sensory nerve fibers have high thresholds and respond to strong stimuli that threaten or cause injury to tissues of the body. Chronic pain may result from continuous or repeated activation of nociceptors, as in some forms of cancer or in chronic inflammatory states, such as arthritis.

However, chronic pain can also be produced by damage to nervous tissue. If peripheral nerves are injured, peripheral neuropathic pain may develop. Damage to certain parts of the central nervous system may result in central neuropathic pain. Examples of conditions that can cause central neuropathic pain include spinal cord injury, cerebrovascular accidents, and multiple sclerosis.

Research on pain in humans has been an important clinical topic for many years. Basic science studies were relatively few in number until experimental work on pain accelerated following detailed descriptions of peripheral nociceptors and central nociceptive neurons that were made in the 1960s and 1970s, by the discovery of the endogenous opioid compounds and the descending pain control systems in the 1970s and the application of modern imaging techniques to visualize areas of the brain that are affected by pain in the 1990s. Accompanying these advances has been the development of a number of animal models of human pain states, with the goal of using these to examine pain mechanisms and also to test analgesic drugs or non-pharmacologic interventions that might prove useful for the treatment of pain in humans. Basic research on pain now emphasizes multidisciplinary approaches, including behavioral testing, electrophysiology, and the application of many of the techniques of modern cell and molecular biology, including the use of transgenic animals.

The *Encyclopedia of Pain* is meant to provide a source of information that spans contemporary basic and clinical research on pain and pain therapy. It should be useful not only to researchers in these fields but also to practicing physicians and other health care professionals and to health care educators and administrators. The work is subdivided into 35 fields, and the Field Editor of each of these describes the areas covered in each in a brief review entry. The topics included in a field are the subject of a series of short essays, accompanied by key words, definitions, illustrations, and a list of significant references. The number of authors who have contributed to the encyclopedia exceeds 550. The plan of the publisher, Springer-Verlag, is to produce both print and electronic versions of this encyclopedia. Numerous links within the electronic version should make comprehensive searches easy to manage. The electronic version will be updated at sufficiently short intervals to ensure that the content remains current.

The editors thank the staff at Springer-Verlag who have provided oversight for this project, including Rolf Lange, Thomas Mager, Claudia Lange, Natasja Sheriff, and Michaela Bilic. Working with these outstanding individuals has been a pleasure.

July 2006
ROBERT F. SCHMIDT   WILLIAM D. WILLIS
Würzburg, Germany   Galveston, Texas, USA

About the Editors

Gerald F. Gebhart received his PhD from The University of Iowa at Iowa City, Iowa, USA, after which he completed a 2-year fellowship at the Université de Montréal, Montréal, Canada, under the direction of Herbert H. Jasper. He began his academic career in the Department of Pharmacology at The University of Iowa in 1973, where he developed a productive research program and led a Pain Interest Group. His career includes a sabbatical year in Heidelberg, Germany (1981–1982), and leadership as head of the Department of Pharmacology from 1996 to 2006. In 2006, Dr. Gebhart was named director of the Center for Pain Research, Department of Anesthesiology, the University of Pittsburgh at Pittsburgh, Pennsylvania, USA. Dr. Gebhart has served the pain community throughout his career. He was elected to the Board of Directors of the American Pain Society (1991–1994) and subsequently elected president of the APS (1997). In 2000, he became the founding editor of *The Journal of Pain*, the official journal of the APS, serving as editor-in-chief until 2010. Dr. Gebhart also served on the Council and later as president-elect, president, and past president of the International Association for the Study of Pain (2005–2012).

Dr. Gebhart is best known for his research contributions in descending modulation of pain and mechanisms of visceral pain, and was designated in 2004 by Thomson ISI as a Highly Cited Researcher (Neuroscience). His scientific contributions have been recognized with several awards, including a 5-year Bristol Myers-Squibb Award for Excellence in Pain Research (1989–1994), a 10-year MERIT Award from the National Institutes of Health (1993–2003),

ix

the Frederick W.L. Kerr Award from the American Pain Society (1994), the Purdue Pharma Prize for Pain Research (2004), the Janssen Award in Gastroenterology (2005), the Founders Award from the American Academy of Pain Medicine (2006), and the Patrick D. Wall Award from the British Pain Society (2012).

Robert F. Schmidt studied medicine at Heidelberg University (Germany). He graduated in 1959 and was promoted to Dr.med. in the same year. After residencies in clinical disciplines, Dr. Schmidt went to the John Curtin School of Medical Research of the Australian National University in Canberra, ACT, where he worked under the head of the Department of Neurophysiology, Sir John C. Eccles. He received his Ph.D. in 1963, the same year Sir John was awarded the Nobel Prize for Physiology and Medicine.

Returning to Heidelberg Medical School, Dr. Schmidt continued his academic career. He completed his habilitation in 1964, becoming a tenured lecturer (Wissenschaftlicher Rat) in 1966 and an associate professor in 1970. After an 8-month sabbatical with Sir John at the State University of New York, Dr. Schmidt became professor and director of the Department of Physiology at the University of Kiel, Germany. In 1982, he moved to the University of Würzburg to take the same position in the Department of Physiology. In 2000, Dr. Schmidt became professor emeritus at this department.

The research work of Dr. Schmidt and his associates in both Kiel and Würzburg focused on the neurophysiology of fine sensory afferents and their connections and functions in the spinal cord. In Würzburg, he concentrated on the processes leading to the changes in response characteristics of nociceptive afferents (i.e., peripheral sensitization) during inflammation of joints, discovering in the process silent nociceptors, which have subsequently been described in all vertebrates studied, including humans. In addition, Dr. Schmidt has edited, coedited, authored, and coauthored numerous textbooks and monographs, most of them having several editions and translations.

In recognition of his research contributions, Dr. Schmidt received many prizes, awards, and honors, including a D.Sc. honoris causa from the University of New South Wales, Sydney, Australia, and an Honorary Professorship from the University of Tübingen, Germany. He was elected to membership in the Akademie der Wissenschaften und der Literatur, Mainz, Germany. Dr. Schmidt is an Honorary Member of the Colombian Association for the Study of Pain, the Japan Physiological Society, the German Society for the Study of Pain, the German Pain Association, the International Association for the Study of Pain (IASP), and the German Physiological Society (in which he served as president in 1979). In 2000, he was awarded the Bundesverdienstkreuz 1st Class (Order of Merit of the German Federal Republic).

In recognition of his research contributions, Dr. Schmidt received many prizes, awards, and honors, including a D.Sc. honoris causa from the University of New South Wales, Sydney, Australia, and an Honorary Professorship from the University of Tübingen, Germany. He was elected to membership in the Akademie der Wissenschaften und der Literatur, Mainz, Germany. Dr. Schmidt is an Honorary Member of the Colombian Association for the Study of Pain, the Japan Physiological Society, the German Society for the Study of Pain (DGSS), and the German Physiological Society, of which he served as president in 1979. In 2000, he was awarded the Bundesverdienstkreuz 1st Class (Order of Merit) of the German Federal Republic.

Section Editors

Ebtesam Ahmed College of Pharmacy and Health Sciences, St. John's University, Department of Pain Medicine and Palliative Care, Beth Israel Medical Center, New York, NY, USA

A. Vania Apkarian Department of Physiology, Northwestern University Feinberg School of Medicine, Chicago, IL, USA

Lars Arendt-Nielsen Center for Sensory-Motor Interaction, Department of Health Science and Technology, School of Medicine, Aalborg University, Aalborg, Denmark

Carlos Belmonte Instituto de Neurociencias, Universidad Miguel Hernandez-CSIC, San Juan de Alicante, Spain

Niels Birbaumer Institute of Medical Psychology and Behavioral Neurobiology, University of Tübingen, Tübingen, Germany

Nikolai Bogduk University of Newcastle, Newcastle Bone & Joint Institute,
Royal Newcastle Centre, Newcastle, NSW, Australia

Jin Mo Chung Department of Neuroscience and Cell Biology, University Chair in Neuroscience, University of Texas Medical Branch, Galveston, TX, USA

Joyce DeLeo Dartmouth Hitchcock Medical Center, Dartmouth Medical School, Hanover, NH, USA

Marshall Devor Department of Cell & Developmental Biology, Institute of Life Sciences and Center for Research on Pain, The Hebrew University of Jerusalem, Jerusalem, Givat Ram, Israel

Hans-Christoph Diener Department of Neurology, University Hospital Essen, University Duisburg-Essen, Essen, Germany

Jonathan O. Dostrovsky Department of Physiology, University of Toronto, Toronto, ON, Canada

Ronald Dubner Department of Neural and Pain Sciences, University of Maryland Dental School, Baltimore, MD, USA

Michel Y. Dubois NYU Medical School, NYU Langone Medical Centers, Center for Research & Treatment of Pain, NYU Ambulatory Care Center, New York, NY, USA

Nanna Brix Finnerup Aarhus University Hospital, Danish Pain Research Center, Aarhus, Denmark

Herta Flor Department of Cognitive and Clinical Neuroscience, Central Institute of Mental Health, Medical Faculty Mannheim, Heidelberg University, Mannheim, Germany

Gerald F. Gebhart Department of Anesthesiology, Center for Pain Research, University of Pittsburgh, Pittsburgh, PA, USA

Maria Adele Giamberardino Pathophysiology of Pain Laboratory and Centre for Fibromyalgia and Musculoskeletal Pain, Department of Medicine and Science of Aging, University of Chieti, Chieti, Italy

Glenn J. Giesler Department of Neuroscience, University of Minnesota, Minneapolis, MN, USA

Peter Goadsby Headache Center, Department of Neurology, University of California, San Francisco, San Francisco, CA, USA

Michael S. Gold Department of Anesthesiology, Center for Pain Research, University of Pittsburgh, Pittsburgh, PA, USA

Hermann O. Handwerker Department of Physiology and Pathophysiology, University of Erlangen/Nürnberg, Erlangen, Germany

Burkhard Hinz Institute of Toxicology and Pharmacology, University of Rostock, Rostock, Germany

Wilfrid Jaenig Physiologisches Institut, Christian-Albrechts-Universität zu Kiel, Kiel, Germany

Michaela Kress Department für Physiologie und Medizinische Physik, Medizinische Universität Innsbruck, Innsbruck, Austria

Frederick A. Lenz Department of Neurosurgery, Johns Hopkins Hospital, Baltimore, MD, USA

Tonya M. Palermo Seattle Children's Hospital Research Institute, CW8-6 - Child Health, Behavior and Development, University of Washington, Seattle, WA, USA

Frank Porreca Department of Pharmacology, University of Arizona Health Sciences Center, Tucson, AZ, USA

Russell K. Portenoy Department of Pain Medicine and Palliative Care, Beth Israel Medical Center, New York, NY, USA

James P. Robinson Department of Anesthesiology & Pain Medicine, University of Washington, Seattle, WA, USA

Hans-Georg Schaible Department of Physiology - Neurophysiology, Friedrich-Schiller-University of Jena, Jena, Germany

Robert F. Schmidt Physiologisches Institut der Universität Würzburg, Würzburg, Germany

Stephan A. Schug School of Medicine and Pharmacology, Royal Perth Hospital, Faculty of Medicine, Dentistry & Health Sciences, The University of Western Australia, UWA Anaesthesiology, Perth, Australia

Barry J. Sessle Faculty of Dentistry, University of Toronto, Toronto, ON, Canada

Bengt H. Sjölund Department of Community Medicine and Rehabilitation Medicine, Umea University, Umea, Sweden

Mark D. Sullivan Psychiatry and Behavioral Sciences, University of Washington, Seattle, WA, USA

Irene Tracey Pain Imaging Neuroscience (PaIN) Group, Department of Physiology, Anatomy and Genetics, Oxford University, Oxford, UK

Dennis C. Turk Department of Anesthesiology & Pain Research, University of Washington, Seattle, WA, USA

Félix Viana Instituto de Neurociencias, Universidad Miguel Hernandez-CSIC, San Juan de Alicante, Alicante, Spain

Katy Vincent Nuffield Department of Obstetrics and Gynaecology, The Women's Centre, John Radcliffe Hospital, University of Oxford, Oxford, UK

Gary A. Walco Department of Anesthesiology and Pain Medicine, University of Washington School of Medicine, Seattle Children's Hospital, Seattle, WA, USA

George L. Wilcox Department of Neuroscience, Pharmacology and Dermatology, University of Minnesota Medical School, Minneapolis, MN, USA

Katja Wiech Nuffield Department of Clinical Neurosciences, John Radcliffe Hospital, Oxford, UK

George L. Wilcox, Department of Neuroscience, Pharmacology and Dermatology, University of Minnesota Medical School, Minneapolis, MN, USA

Katja Wiech, Nuffield Department of Clinical Neuroscience, John Radcliffe Hospital, Oxford, UK.

Contributors

Tipu Aamir The Auckland Regional Pain Service, Auckland Hospital, Auckland, New Zealand

Catherine Abbadie Department of Pharmacology, Merck Research Laboratories, Rahway, NJ, USA

Frances V. Abbott Departments of Psychiatry and Psychology, McGill University, Montreal, QC, Canada

Heather Adams Department of Surgery, McGill University Health Centre, Montreal, QC, Canada

Mary O. Adedoyin Department of Anatomy, University of California San Francisco, San Francisco, CA, USA

Rolf H. Adler University of Berne Medical School, Kehrsatz, Switzerland

Claire D. Advokat Department of Psychology, Louisiana State University, Baton Rouge, LA, USA

Reto M. Agosti Headache Center Hirlsanden, Zurich, Switzerland

Ebtesam Ahmed College of Pharmacy and Health Sciences, St. John's University, Department of Pain Medicine and Palliative Care, Beth Israel Medical Center, New York, NY, USA

Marie-Claire Albanese Department of Psychology, McGill University, Montreal, QC, Canada

Kathryn M. Albers Department of Neurobiology, University of Pittsburgh School of Medicine, Pittsburgh, PA, USA

Phillip J. Albrecht Center for Neuropharmacology and Neuroscience, Albany Medical College, Albany, NY, USA

Elie D. Al-Chaer Center for Pain Research, University of Arkansas for Medical Sciences, Little Rock, AR, USA

Terry Altilio Department of Pain Medicine and Palliative Care, Beth Israel Medical Center, New York, USA

Rainer Amann Medical University Graz, Graz, Austria

Ron Amir Institute of Life Sciences, The Hebrew University of Jerusalem, Jerusalem, Israel

Praveen Anand Peripheral Neuropathy Unit, Imperial College London, London, UK

Karen O. Anderson The University of Texas MD Anderson Cancer Center, Houston, TX, USA

Ole K. Andersen Department of Health Science & Technology, Aalborg University, Aalborg, Denmark

Stanley W. Anderson Department of Neurosurgery, Johns Hopkins Hospital, Johns Hopkins Medical Institutions, Johns Hopkins University, Baltimore, MD, USA

Trevor Anderson Fairview Pain & Palliative Care, University of Minnesota Medical Center, Minneapolis, MN, USA

Karl-Erik Andersson Wake Forest Institute for Regenerative Medicine, Wake Forest University School of Medicine, Winston Salem, USA

Frank Andrasik Department of Psychology, Applied Psychophysiology and Biofeedback, The University of Memphis, Memphis, TN, USA

A. Vania Apkarian Department of Physiology, Northwestern University Feinberg School of Medicine, Chicago, IL, USA

Lars Arendt-Nielsen Center for Sensory-Motor Interaction, Department of Health Science and Technology, School of Medicine, Aalborg University, Aalborg, Denmark

Kirsten Arndt CNS Pharmacology, Boehringer Ingelheim Pharma GmbH & Co. KG, Biberach, Germany

Charles-Daniel Arreto Université Paris Descartes, INSERM UMR 894, France

Erik Askenasy Neurobiology and Anatomy, Health Science Center, University of Texas, Houston, TX, USA

J. H. Atkinson Psychiatry Service, VA San Diego Healthcare System and Department of Psychiatry, University of California, San Diego, CA, USA

Nadine Attal Centre d'Evaluation et de Traitement de la Douleur, Hôpital Ambroise Paré, INSERM U 987 APHP, Boulogne-Billancourt, France

Christoph Aufenberg Functional Neurosurgery, Neurosurgical Clinic University Hospital, Zürich, Switzerland

Beate Averbeck Sanofi-Aventis Deutschland GmbH, Frankfurt, Germany

Qasim Aziz Section of GI Sciences, Hope Hospital University of Manchester, Salford, UK

Cathrine Baastrup Danish Pain Research Center, Aarhus University Aarhus University Hospital, Aarhus, Denmark

F. W. Bach Danish Pain Research Center, Department of Neurology, Aarhus University Hospital, Aarhus, Denmark

S. K. Back Neuroscience Research Institute and Department of Physiology, Korea University, College of Medicine, Seoul, Republic of Korea

Misha-Miroslav Backonja Department of Neurology, University of Wisconsin-Madison, Madison, WI, USA

Banani Banerjee Division of Gastroenterology and Hepatology and Pediatric Gastroenterology, Medical College of Wisconsin, Milwaukee, WI, USA

John D. Banja Department of Rehabilitation Medicine, Center for Ethics Emory University, Atlanta, GA, USA

Carsten Bantel Department of Surgery & Cancer? Anaesthetics Section, Imperial College of Science and Medicine Chelsea and Westminster Campus, London, UK

Andrew Paul Baranowski The Pain Management Centre, National Hospital for Neurology and Neurosurgery, UCL Hospitals NHS Foundation Trust, London, UK

Alex Barling Department of Neurology, Birmingham Muscle and Nerve Centre Queen Elizabeth Hospital, Birmingham, UK

Ralf Baron Department of Neurological Pain Research and Therapy, Christian Albrechts University Kiel, Kiel, Germany

Gordon A. Barr Department of Anesthesiology and Critical Care Medicine, Children's Hospital of Philadelphia University of Pennsylvania, Philadelphia, PA, USA

Emily J. Bartley Department of Psychology, The University of Tulsa, Tulsa, OK, USA

Jeffrey R. Basford Department of Physical Medicine and Rehabilitation, Mayo Clinic and Mayo Foundation, Rochester, NY, USA

Michele C. Battie Department of Physical Therapy, University of Alberta, Edmonton, AB, Canada

Ulf Baumgartner Chair of Neurophysiology, Center for Biomedicine and Medical Technology Mannheim, Heidelberg University, Mannheim, Germany

Nicola Beale Nuffield Department of Anaesthetics, Oxford University Hospitals NHS Trust, UK

Alain Beaudet Montreal Neurological Institute, McGill University, Montreal, QC, Canada

Lino Becerra P.A.I.N. Group, Department of Anesthesiology, Perioperative and Pain Medicine, Boston Children's Hospital Department of Radiology Massachusetts General Hospital Center for Pain and the Brain, Harvard Medicals, Boston, MA, USA

Yaakov Beilin Department of Anesthesiology, and Obstetrics, Mount Sinai School of Medicine, New York, NY, USA

Alvin J. Beitz Department of Veterinary and Biomedical Sciences, University of Minnesota, St. Paul, MN, USA

Miles Belgrade Fairview Pain Management Center, University of Minnesota Medical Center, Minneapolis, MN, USA

Carlo Valerio Bellieni Neonatal Intensive Care Unit, University Hospital, Siena, Italy

Carlos Belmonte Instituto de Neurociencias, Universidad Miguel Hernandez-CSIC, San Juan de Alicante, Spain

Allan J. Belzberg Department of Neurosurgery, Johns Hopkins School of Medicine, Baltimore, MD, USA

Matt Bender Department of Neurosurgery, Johns Hopkins Hospital, Baltimore, MA, USA

Fabrizio Benedetti Department of Neuroscience, Clinical and Applied Physiology Program, University of Turin Medical School, Turin, Italy

David L. H. Bennett Wolfson CARD, Kings College London, London, UK

Nuffield Department of Clinical Neurosciences, John Radclife Hospital, The University of Oxford, Oxford, UK

Robert Bennett Department of Medicine, Oregon Health and Science University, Portland, USA

Edward C. Benzel Department of Neurosurgery, Center for Spine Health Neurological Institute Cleveland Clinic, Cleveland, OH, USA

David A. Bereiter Department of Diagnostic and Biological Sciences, University of Minnesota School of Dentistry, Minneapolis, MN, USA

Elmar Berendes Klinik für Anästhesiologie, Operative Intensivmedizin und Schmerztherapie, Klinikum Krefeld, Krefeld, Germany

Karen J. Berkley Program in Neuroscience, Florida State University, Tallahassee, FL, USA

Peter Berlit Department of Neurology, Alfried Krupp Hospital, Essen, Germany

Jean-François Bernard INSERM UMR-894, UPMC Site Pitié-Salpêtrière, Paris, France

Anne K. Bertelsen Neurology, Aleris Hospital, Bergen, Norway

Jennifer Bickel Integrative Pain Management, Children's Mercy Hospitals and Clinics, University of Missouri-Kansas City School of Medicine, Kansas City, MO, USA

Marcelo E. Bigal Department of Neurology and Montefiore Headache Unit, Albert Einstein College of Medicine, Bronx, NY, USA

Yitzchak M. Binik McGill University, Montreal, QC, Canada

Niels Birbaumer Institute of Medical Psychology and Behavioral Neurobiology, University of Tübingen, Tübingen, Germany

Frank Birklein Department of Neurology, University of Mainz, Mainz, Germany

Kathryn A. Birnie Department of Psychology, Dalhousie University, Halifax, NS, Canada

Thomas Bishop Centre for Neuroscience, King's College London, London, UK

Henning Bliddal The Parker Institute, Frederiksberg Hospital, Copenhagen, Denmark

Nikolai Bogduk University of Newcastle, Newcastle Bone & Joint Institute, Royal Newcastle Centre, Newcastle, NSW, Australia

Jorgen Boivie Department of Neurology, University Hospital, Linköping, Sweden

Michael R. Bond University of Glasgow, Glasgow, UK

Richard L. Boortz-Marx Division of Pain Medicine, Department of Anesthesiology, University of North Carolina, Chapel Hill, North Carolina, USA

James M. Borowczyk Department of Orthopaedics and Musculoskeletal Medicine, University of Otago, Christchurch, Christchurch, New Zealand

David Borsook P.A.I.N. Group, Department of Anesthesiology, Perioperative and Pain Medicine, Boston Children's Hospital Department of Radiology Massachusetts General Hospital Center for Pain and the Brain, Harvard Medicals, Boston, MA, USA

George S. Borszcz Department of Psychology, Wayne State University, Detroit, USA

Jasenka Borzan Department of Anesthesiology and Critical Care Medicine, Johns Hopkins University, Baltimore, USA

Regina M. Botting London, UK

Didier Bouhassira Centre d'Evaluation et de Traitement de la Douleur, Hôpital Ambroise Paré, INSERM U 987 APHP, Boulogne-Billancourt, France

Laurence Bourgeais INSERM UMR 894, Centre de Psychiatrie et Neurosciences, Paris, France

David Bowsher Pain Research Institute, University Hospital Aintree, Liverpool, UK

Emma L. Brandon Anaesthesiology and Pain Medicine, University of Western Australia Royal Perth Hospital, Perth, WA, Australia

Jennifer Brawn P.A.I.N. Group, Center for Pain and the Brain, Boston Children's Hospital, Harvard Medical School, Waltham, MA, USA

Harald Breivik Department of Pain Management and Research, Oslo University Hospital, Rikshospitalet and University of Oslo, Oslo, Norway

Kori L. Brewer Department of Emergency Medicine, Brody School of Medicine East Carolina University, Greenville, NC, USA

Christopher R. Brigham Brigham and Associates, Inc., Portland, USA

Abigail Broderick Nuffield Department of Obstetrics and Gynaecology, University of Oxford, Oxford, UK

Jeffrey A. Brown Department of Neurological Surgery, Wayne State University School of Medicine, Detroit, MI, USA

Stephen C. Brown Department of Anesthesia, The Hospital for Sick Children, Toronto, ON, Canada

Eduardo Bruera Department of Palliative Care and Rehabilitation Medicine, The University of Texas M. D. Anderson Cancer Center, Houston, TX, USA

Pablo Brumovsky Faculty of Biomedical Sciences, Austral University, Buenos Aires, Argentina

Consejo Nacional de Investigaciones Científicas y Técnicas (CONICET), Buenos Aires, Argentina

Kay Brune Department of Experimental and Clinical Pharmacology and Toxicology, Friedrich Alexander University Erlangen-Nürnberg, Erlangen, Germany

Maria Burian Department of General, Visceral and Transplantation Surgery, University of Heidelberg, Heidelberg, Germany

Dan Buskila Rheumatic Disease Unit, Soroka Medical Center, Beer Sheva, Israel

Catherine M. Cahill Anesthesiology & Perioperative Care, University of California, Irvine, Irvine, CA, USA

Brian E. Cairns Faculty of Pharmaceutical Sciences, University of British Columbia, Vancouver, BC, Canada

Nigel A. Calcutt Department of Pathology, University of California San Diego, La Jolla, CA, USA

Margarita Calvo Wolfson CARD, Kings College London, London, UK

James N. Campbell Johns Hopkins University School of Medicine, Baltimore, MD, USA

Lauren Campbell Department of Psychology, York University, Toronto, ON, Canada

Ling Cao Department of Biomedical Sciences, College of Osteopathic Medicine, University of New England, ME, USA

Adam Carinci Department of Anesthesia, Critical Care and Pain Medicine, Massachusetts General Hospital Harvard Medical School, Boston, MA, USA

Giancarlo Carli Department of Physiology, University of Siena, Siena, Italy

Christer P. O. Carlsson Rehabilitation Clinic, Lunds University Hospital, Lund, Sweden

Susan M. Carlton Neuroscience and Cell Biology, University of Texas Medical Branch, Galveston, TX, USA

Richard Carr Clinic for Anaesthesiology, Heidelberg University, Mannheim, Baden-Wuerttemberg, Germany

Benjamin S. Carson Department of Neurosurgery, Johns Hopkins Hospital, Baltimore, MA, USA

Brian Carter University of Missouri at Kansas City Children's Mercy Hospital and Bioethics Center, Kansas City, MO, USA

Kenneth L. Casey Department of Neurology, University of Michigan Medical Center, Ann Arbor, MI, USA

Christine T. Chambers Departments of Pediatrics and Psychology, Dalhousie University, Halifax, NS, Canada

Sandra Chaplan Janssen Research and Development, LLC, San Diego, CA, USA

C. Richard Chapman Department of Anesthesiology, University of Utah School of Medicine, Salt Lake City, UT, USA

Santosh K. Chaturvedi Psychiatry, National Institute of Mental Health & Neurosciences, Bangalore, India

Andrew C. N. Chen Human Brain Mapping and Cortical Imaging Laboratory, Aalborg University, Aalborg, Denmark

Hsinlin Thomas Cheng Department of Neurology, University of Michigan, Ann Arbor, MI, USA

Pavel Cherkas Faculty of Dentistry, University of Toronto, Toronto, ON, Canada

Andrea Cheville Department of Physical Medicine and Rehabilitation, Mayo Clinic, Rochester, MN, USA

Chen Yu Chiang Faculty of Dentistry, University of Toronto, Toronto, ON, Canada

N. L. Chillingworth School of Physiology and Pharmacology, University of Bristol, Bristol, UK

Edward Chow Department of Radiation Oncology, Sunnybrook Health Sciences Centre Odette Cancer Centre (T2-182), Toronto, ON, Canada

Julie A. Christianson Department of Anatomy and Cell Biology, University of Kansas Medical Center, Kansas City, KS, USA

MacDonald J. Christie Discipline of Pharmacology, University of Sydney, Sydney, Australia

Jin Mo Chung Department of Neuroscience and Cell Biology, University Chair in Neuroscience, University of Texas Medical Branch, Galveston, TX, USA

Kyungsoon Chung Department of Neuroscience and Cell Biology, University of Texas Medical Branch, Galveston, TX, USA

Anna K. Clark Wolfson Centre for Age Related Diseases, King's College London, London, UK

W. Crawford Clark College of Physicians and Surgeons, Columbia University, New York, USA

James F. Cleary Pain & Policy Studies Group, UW Carbone Cancer Center, Madison, WI, USA

John De Clercq Department of Rehabilitation Sciences and Physiotherapy, Ghent University Hospital, Ghent, Belgium

Jacqui Clinch Royal National Hospital for Rheumatic Diseases NHS Foundation Trust, Bath, UK

Terence J. Coderre Departments of Anesthesia, Neurology & Neurosurgery and Psychology, McGill University, Montreal, QC, Canada

Robert Coghill Department of Neurobiology and Anatomy, Wake Forest University School of Medicine, Winston-Salem, NC, USA

Lindsey Cohen Psychology, Georgia State University, Atlanta, GA, USA

Alan L. Colledge Utah Labor Commission, International Association of Industrial Accident Boards and Commissions Central Utah Medical Clinic, South Salt Lake City, UT, USA

Beverly J. Collett University Hospitals of Leicester, Leicester, UK

Lesley A. Colvin Department of Anaesthesia, Critical Care and Pain Medicine Western General Hospital, Edinburgh, UK

Kevin C. Conlon Trinity College Dublin, University of Dublin, Dublin, Ireland

Mark Connelly Integrative Pain Management, Children's Mercy Hospitals and Clinics, University of Missouri-Kansas City School of Medicine, Kansas City, MO, USA

Brian Y. Cooper Department of Oral and Maxillofacial Surgery, University of Florida, Gainesville, FL, USA

Christine Cordle University Hospitals of Leicester, Leicester, UK

Laura A. Cousins Psychology, Georgia State University, Atlanta, GA, USA

Michael J. Cousins Department of Anesthesia and Pain Management, Royal North Shore Hospital University of Sydney, Sydney, NSW, Australia

Verne C. Cox Department of Psychology, University of Texas at Arlington, Arlington, TX, USA

George A. Crabill Department of Neurosurgery, Johns Hopkins Hospital, Baltimore, MA, USA

Rebecca M. Craft Psychology, Washington State University, Pullman, WA, USA

Kenneth D. Craig Department of Psychology, University of British Columbia, Vancouver, BC, Canada

Richard Crevenna Medizinische Universität Wien, Wien, Austria

Geert Crombez Department of Experimental Clinical and Health Psychology, Ghent University, Ghent, Belgium

Joan Crook School of Nursing, Faculty of Health Sciences, McMaster University, Hamilton, ON, Canada

Giorgio Cruccu Department of Neurological Sciences, La Sapienza University, Rome, Italy

Michele Curatolo Department of Anesthesiology, Division of Pain Therapy, University Hospital of Bern Inselspital, Bern, Switzerland

Nachum Dafny Neurobiology and Anatomy, Health Science Center, University of Texas, Houston, TX, USA

Jorgen B. Dahl Department of Anaesthesiology, Glostrup University Hospital, Herlev, Denmark

Yi Dai Department of Anatomy and Neuroscience, Hyogo College of Medicine, Hyogo, Japan

Stefaan Van Damme Department of Experimental Clinical and Health Psychology, Ghent University, Ghent, Belgium

Alexandre F. M. DaSilva Biologic & Materials Sciences, University of Michigan Dental School, MI, USA

Steve Davidson Pain Center & Department of Anesthesiology, Washington University School of Medicine, MO, USA

Brian Davis Department of Neurobiology, University of Pittsburgh School of Medicine, Pittsburgh, PA, USA

Karen D. Davis Brain, Imaging and Behaviour – Systems Neuroscience, Toronto Western Research Institute, Toronto, ON, Canada

Department of Surgery, University of Toronto, Toronto, ON, Canada

Richard O. Day Department of Clinical Pharmacology and Toxicology, St Vincent's Hospital, Sydney, NSW, Australia

Isabelle Decosterd Pain Center, University Hospital Center and University of Lausanne, Lausanne, Switzerland

Ruth Defrin Physical Therapy, Faculty of Medicine, Tel-Aviv University, Tel-Aviv, Israel

Antoon F. C. De Laat Department of Oral/Maxillofacial Surgery, Catholic University of Leuven, Leuven, Belgium

Joyce DeLeo Dartmouth Hitchcock Medical Center, Dartmouth Medical School, Hanover, NH, USA

A. Lee Dellon Department of Plastic Surgery and Neurosurgery, Johns Hopkins University, Baltimore, USA

Dellon Institutes for Peripheral Nerve Surgery, Baltimore, MD, USA

Marshall Devor Department of Cell & Developmental Biology, Institute of Life Sciences and Center for Research on Pain, The Hebrew University of Jerusalem, Jerusalem, Givat Ram, Israel

Perry Dhaliwal Department of Neurosurgery, Cleveland Clinic Spine Institute, Cleveland, OH, USA

David Diamant Neurological and Spinal Surgery, Lincoln, NE, USA

Anthony H. Dickenson Department of Pharmacology, University College London, London, UK

Hans-Christoph Diener Department of Neurology, University Hospital Essen, University Duisburg-Essen, Essen, Germany

Natalia Dmitrieva Program in Neuroscience, Florida State University, Tallahassee, FL, USA

L. F. Donaldson School of Life Sciences, University of Nottingham, Nottingham, UK

Henri Doods CNS Pharmacology, Boehringer Ingelheim Pharma GmbH & Co. KG, Biberach, Germany

Victoria Dorf Social Security Administration, Baltimore, MD, USA

Michael J. Dorsi Department of Neurosurgery, Johns Hopkins School of Medicine, Baltimore, MD, USA

Jonathan O. Dostrovsky Department of Physiology, University of Toronto, Toronto, ON, Canada
Faculty of Dentistry, University of Toronto, Toronto, ON, Canada

Robyn Drach Fairview Pain Management Center, Minneapolis, MN, USA

Ron Dressler Utah Labor Commission, International Association of Industrial Accident Boards and Commissions Central Utah Medical Clinic, South Salt Lake City, UT, USA

Paul Dreyfuss Department of Rehabilitation Medicine, University of Washington, Seattle, USA

Peter D. Drummond School of Psychology, Murdoch University, Perth, Australia

Ronald Dubner Department of Neural and Pain Sciences, University of Maryland Dental School, Baltimore, MD, USA

Anne Ducros Headache Emergency Department, INSERM, Paris, France

Gary H. Duncan Département de stomatologie, Faculté de médecine dentaire, Université de Montréal, Montreal, QC, Canada
Department of Neurology and Neurosurgery, McGill University, Montreal, QC, Canada

Mary J. Eaton Research Service (151), Miami VA Health Care System, Miami VAMC, Miami, FL, USA

Andrea Ebersberger Department of Physiology, Friedrich Schiller University of Jena, Jena, Germany

Christopher Eccleston Pain Management Unit, University of Bath, Bath, UK

John Edmeads Sunnybrook and Woman's College Health Sciences Centre, University of Toronto, Toronto, Canada

Edzard Ernst Complementary Medicine, Peninsula Medical School Universities of Exeter and Plymouth, Exeter, UK

Reuben Escorpizo Swiss Paraplegic Research, Nottwil, Switzerland
Department of Health Sciences and Health Policy, University of Lucerne, Lucerne, Switzerland
ICF Research Branch in cooperation with the World Health Organization (WHO) Collaborating Centre for the Family of International Classifications in Germany, Nottwil, Switzerland

Schonstein Eva RehabCo, Fyshwick, ACT, Australia

Joshua C. Eyer Department of Psychology, University of Alabama, Tuscaloosa, AL, USA

Marie Faaborg-Andersen McGill University, Montreal, QC, Canada

Keri L. Fakata College of Pharmacy and Pain Management Center, University of Utah, Salt Lake City, USA

Marie Fallon Department of Oncology, Edinburgh Cancer Center University of Edinburgh, Edinburgh, UK

Melissa Farmer Department of Physiology / Feinberg School of Medicine, Northwestern University, Chicago, IL, USA

Samantha R. Fashler Department of Psychology, University of British Columbia, Vancouver, BC, Canada

Jean-Baptiste Fassier Department of Occupational Health and Medicine, UMRESTTE-Université Claude Bernard Lyon 1, Lyon, France

Charlotte Feinmann Eastman Dental Institute London, University College London, London, UK

Elizabeth Roy Felix The Miami Project to Cure Paralysis, University of Miami Miller School of Medicine, Miami, FL, USA

Bin Feng Center for Pain Research, Department of Anesthesiology, School of Medicine, University of Pittsburgh, Pittsburgh, Pennsylvania, USA

Yi Feng Department of Anesthesiology, Peking University People's Hospital, Peking, China

Andres Fernandez Department of Neurosurgery, Johns Hopkins Hospital, Baltimore, MD, USA

Antonio Vicente Ferrer-Montiel Institute of Molecular and Cell Biology, University Miguel Hernández, Elche, Alicante, Spain

Perry G. Fine Pain Research Center, School of Medicine, University of Utah, Salt Lake City, UT, USA

David J. Fink Department of Neurology, University of Michigan and VA Ann Arbor Healthcare System, Ann Arbor, MI, USA

Nanna Brix Finnerup Aarhus University Hospital, Danish Pain Research Center, Aarhus, Denmark

David A. Fishbain Department of Psychiatry, University of Miami, Miami, FL, USA

Maria Fitzgerald Neuroscience, Physiology & Pharmacology, University College London, London, UK

Per Flisberg Department of Anaesthesia and Intensive Care, Lund University Hospital, Lund, Sweden

Herta Flor Department of Cognitive and Clinical Neuroscience, Central Institute of Mental Health, Medical Faculty Mannheim, Heidelberg University, Mannheim, Germany

Christopher M. Flores Drug Discovery, Johnson and Johnson Pharmaceutical Research and Development, Spring House, PA, USA

Wilbert E. Fordyce Department of Rehabilitation, University of Washington School of Medicine, Seattle, WA, USA

Robert D. Foreman Department of Physiology, College of Medicine, University of Oklahoma Health Sciences Center, Oklahoma City, OK, USA

Gary M. Franklin Department of Environmental and Occupational Health Sciences, University of Washington, Seattle, WA, USA

Department of Health Services, University of Washington, Seattle, WA, USA

Jan Fridén Department of Hand Surgery, Sahlgrenska University Hospital, Gothenborg, Sweden

Justin Friedlander The Arthur Smith Institute for Urology, North Shore-LIJ Healthcare System, Hofstra North Shore LIJ School of Medicine, New Hyde Park, NY, USA

Allan H. Friedman Division of Neurosurgery, Duke University Medical Center, Durham, NC, USA

Perry N. Fuchs Department of Psychology, University of Texas at Arlington, Arlington, TX, USA

Deborah Fulton-Kehoe Department of Environmental and Occupational Health Sciences, University of Washington, Seattle, WA, USA

Andrea D. Furlan Institute for Work & Health, Toronto, ON, Canada

Vasco Galhardo Departamento de Biologia Experimental, Faculdade de Medicina, Universidade do Porto, Portugal

Andreas R. Gantenbein RehaClinic Bad Zurzach, Bad Zurzach, Switzerland

I. Garonzik Department of Neurosurgery, Johns Hopkins Medical Institutions, Baltimore, MD, USA

Robert Gassin Musculoskeletal Medicine Clinic, Frankston, VIC, Australia

Robert J. Gatchel Department of Psychology, The University of Texas, Arlington, TX, USA

Gerald F. Gebhart Department of Anesthesiology, Center for Pain Research, University of Pittsburgh, Pittsburgh, PA, USA

Gerd Geisslinger Department of Clinical Pharmacology, University Frankfurt am Main, Frankfurt, Germany

Robert D. Gerwin Department of Neurology, Johns Hopkins University, Bethesda, MD, USA

Maria Adele Giamberardino Pathophysiology of Pain Laboratory and Centre for Fibromyalgia and Musculoskeletal Pain, Department of Medicine and Science of Aging, University of Chieti, Chieti, Italy

Stephen Gibson National Ageing Research Institute Department of Medicine, University of Melbourne, Parkville, VIC, Australia

Glenn J. Giesler Department of Neuroscience, University of Minnesota, Minneapolis, MN, USA

Philip L. Gildenberg Baylor Medical College, Houston, TX, USA

Myra Glajchen Department of Pain Medicine and Palliative Care, Beth Israel Medical Center Albert Einstein College of Medicine, New York, NY, USA

Hartmut Göbel Kiel Migraine and Headache Center, Kiel Pain Center, Kiel, Germany

Peter Goadsby Headache Center, Department of Neurology, University of California, San Francisco, San Francisco, CA, USA

Philippe Goffaux Faculty of Medicine, University of Sherbrooke, Sherbrooke, QC, Canada

Ana Gomis Instituto de Neurociencias de Alicante, Universidad Miguel Hernández-CSIC, San Juan de Alicante, Spain

Gilbert R. Gonzales Department of Neurology, Memorial Sloan-Kettering Cancer Center, New York, USA

Allan Gordon Wasser Pain Management Centre, Mount Sinai Hospital University of Toronto, Toronto, ON, Canada

Liesbet Goubert Department of Experimental Clinical and Health Psychology, Ghent University, Ghent, Belgium

Jayantilal Govind Pain Clinic, Department of Anaesthesia, University of New South Wales, Sydney, NSW, Australia

Richard H. Gracely Departments of Medicine-Rheumatology and Neurology, University of Michigan Health System, Ann Arbor, MI, USA

Garry G. Graham Department of Pharmacology, School of Medical Sciences, University of New South Wales, Sydney, NSW, Australia

David K. Grandy Department of Physiology and Pharmacology, Oregon Health and Science University, Portland, OR, USA

Thomas Graven-Nielsen Laboratory for Musculoskeletal Pain and Motor Control, Center for Sensory-Motor Interaction, Aalborg University, Aalborg, Denmark

Barry Green Pierce Laboratory, Yale School of Medicine, New Haven, CT, USA

Paul G. Green Oral & Maxillofacial Surgery, University of California San Francisco, San Francisco, CA, USA

Joel D. Greenspan Department of Neural and Pain Sciences, University of Maryland School of Dentistry, Baltimore, MD, USA

John W. Griffin Department of Neurology, Johns Hopkins University School of Medicine, Baltimore, MD, USA

Cornelius Groenwald Department of Anesthesiology and Pain Medicine, Seattle Children's Hospital, University of Washington School of Medicine, Seattle, WA, USA

Douglas P. Gross Department of Physical Therapy, University of Alberta WCB-Alberta Millard Health, Edmonton, AB, Canada

Sabine Grösch Institute of Clinical Pharmacology, Clinical Center of the Goethe University, Frankfurt, Germany

Blair D. Grubb Department of Cell Physiology and Pharmacology, University of Leicester, Leicester, UK

Ruth Eckstein Grunau Pediatrics Dept, University of British Columbia Developmental Neurosciences & Child Health, Child & Family Research Institute, Vancouver, BC, Canada

Christoph Gutenbrunner Department of Physical Medicine and Rehabilitation, Hanover Medical School, Hanover, Germany

Maija Haanpää Department of Neurosurgery, Helsinki University Central Hospital, Helsinki, Finland

Sherri Haas Fairview Pain Management Center, Minneapolis, MN, USA

Ute Habel Department of Psychiatry and Psychotherapy, RWTH Aachen University, Aachen, Germany

Thomas Hachenberg Clinic for Anaesthesiology and Intensive Therapy, Otto-von-Guericke University, Magdeburg, Germany

Nouchine Hadjikhani Martinos Center for Biomedical Imaging, Massachusetts General Hospital Harvard Medical School, Charlestown, MA, USA

Ole Haegg Department of Orthopedic Surgery, Sahlgren University Hospital, Goteborg, Sweden

Bryan C. Hains Department of Neurology, Yale University School of Medicine, West Haven, CT, USA

Else K. B. Hals University of Oslo, Faculty of Dentistry, Oslo, Norway

Darryl T. Hamamoto Diagnostic and Biological Sciences, University of Minnesota, Minneapolis, MN, USA

Donna L. Hammond Department of Anesthesia, University of Iowa, Iowa City, IA, USA

Menachem Hanani Hadassah-Hebrew University Medical Center Hebrew University Medical School, Jerusalem, Israel

Hermann O. Handwerker Department of Physiology and Pathophysiology, University of Erlangen/Nürnberg, Erlangen, Germany

George M. Hanna Department of Anesthesia, Critical Care and Pain Medicine, Massachusetts General Hospital Harvard Medical School, Boston, MA, USA

Jing-Xia Hao Department of Physiology and Pharmacology, Karolinska Institutet, Stockholm, Sweden

Eleni G. Hapidou Department of Psychiatry and Behavioural Neurosciences, McMaster University and Hamilton Health Sciences Chronic Pain Management Unit, Chedoke Hospital, Hamilton, ON, Canada

Geoffrey Harding Sandgate Spinal Medicine Clinic, Sandgate, QLD, Australia

Louise Harding Hospitals NHS Foundation Trust, University College London, London, UK

Kenneth M. Hargreaves Department of Endodontics, University of Texas Health Science Center at San Antonio, San Antonio, USA

D. Paul Harries Pain Treatment Center, Samaritan Hospital Lexington, Lexington, KY, USA

Denise Harrison Nursing, Children's Hospital of Eastern Ontario (CHEO), University of Ottawa, Ottawa, ON, Canada

Steven Harte Anesthesiology and Internal Medicine/Rheumatology, University of Michigan VA Ann Arbor Healthcare System, Ann Arbor, MI, USA

Barbara A. Hastie Community Dentistry & Behavioral Science, University of Florida College of Dentistry, Gainesville, FL, USA

Jennifer A. Haythornthwaite Psychiatry & Behavioral Sciences, Johns Hopkins University, Baltimore, MD, USA

Heinz-Joachim Häbler FH Bonn-Rhein-Sieg, Rheinbach, Germany

John H. Healey Orthopaedic Service, Department of Surgery, Memorial Sloan-Kettering Cancer Center Weill Medical College of Cornell University, New York, NY, USA

Alicia A. Heapy VA Connecticut Healthcare System, Westhaven, CT, USA

Anne M. Heffernan The Adelaide and Meath Hospital Incorporating The National Children's Hospital, Dublin, Ireland

Geert Van der Heijden Julius Center, Department of Epidemiology, University Medical Center Utrecht, Utrecht, The Netherlands

Mary M. Heinricher Departments of Neurological Surgery and Behavioral Neuroscience, Oregon Health & Science University, Portland, USA

Robert D. Helme Department of Medicine, Royal Melbourne Hospital, University of Melbourne, Melbourne, Australia

Kim Helsen Research Group on Health Psychology, KU Leuven Katholieke Universiteit Leuven, Leuven, Belgium

Department of Experimental–Clinical and Health Psychology, Ghent University, Ghent, Belgium

Amin S. Herati The Arthur Smith Institute for Urology, North Shore-LIJ Healthcare System, Hofstra North Shore LIJ School of Medicine, New Hyde Park, NY, USA

Robert D. Herbert Neuroscience Research Australia, Sydney, Australia

Christiane Hermann Department of Clinical Psychology, Justus-Liebig University Giessen, Giessen, Germany

Juan F. Herrero Department of Physiology, Faculty of Medicine University of Alcala, Alcalá de Henares Madrid, Spain

Aki J. Hietaharju Department of Neurology and Rehabilitation, Pain Clinic Tampere University Hospital, Tampere, Finland

Karin Westlund High Department of Physiology, University of Kentucky, Lexington, KY, USA

Marita Hilliges Dalarna University, Halmstad, Sweden

Max J. Hilz University of Erlangen-Nuremberg, Erlangen, Germany

Burkhard Hinz Institute of Toxicology and Pharmacology, University of Rostock, Rostock, Germany

Anthony R. Hobson Section of GI Sciences, Hope Hospital University of Manchester, Salford, UK

Edward Högestätt Department of Laboratory Medicine, Lund University, Division of Clinical Chemistry and Pharmacology, Lund, Sweden

Tomas Hökfelt Department of Neuroscience, Karolinska Institute, Stockholm, Sweden

Sarah V. Holdridge Department of Pharmacology and Toxicology, Queen's University, Kinston, Canada

Kjell Hole University of Bergen, Bergen, Norway

Graham R. Holland Department of Cariology, University of Michigan, Ann Arbor, MI, USA

Chang-Zern Hong Department of Physical Medicine and Rehabilitation, University of California, Irvine, CA, USA

Minoru Hoshiyama Department of Integrative Physiology, National Institute for Physiological Sciences, Okazaki, Japan

Fred M. Howard University of Rochester, Rochester, NY, USA

Sung-Tsang Hsieh Department of Anatomy and Cell Biology, Department of Neurology, National Taiwan University Hospital, National Taiwan University, Taipei, Taiwan

James W. Hu Faculty of Dentistry, University of Toronto, Toronto, Canada

Claire E. Hulsebosch Spinal Cord Injury Research, University of Texas Medical Branch, Galveston, TX, USA

Thomas Hummel Department of Otorhinolaryngology, Technical University of Dresden Medical School, Dresden, Germany

Mark R. Hutchinson School of Medical Sciences, University of Adelaide, Adelaide, SA, Australia

Charles E. Inturrisi Department of Pharmacology, Weill Cornell Medical College, New York, USA

Koji Inui Department of Integrative Physiology, National Institute for Physiological Sciences, Okazaki, Japan

Koichi Iwata Department of Physiology, Nihon University School of Dentistry, Chiyoda-ku, Japan

Suhayl J. Jabbur Anatomy, Cell Biology and Physiology, Faculty of Medicine, American University of Beirut, Beirut, Lebanon

Neuroscience Program, Faculty of Medicine, American University of Beirut, Beirut, Lebanon

Nora Janjan University of Texas M. D. Anderson Cancer Center, Houston, TX, USA

Luc Jasmin Department of Anatomy, University of California San Francisco, San Francisco, CA, USA

Anne Jaumees Royal North Shore Hospital, St Leonards, NSW, Australia

Wilfrid Jaenig Physiologisches Institut, Christian-Albrechts-Universität zu Kiel, Kiel, Germany

Daniel Jeanmonod Functional Neurosurgery, The Center of Ultrasound Functional Neurosurgery, Solothurn, Switzerland

Andrew Jeffreys Director Department of Anaesthesia & Pain Medicine, Western Health, Melbourne, VIC, Australia

Mark P. Jensen Department of Rehabilitation Medicine, University of Washington, Seattle, WA, USA

Troels Staehelin Jensen Danish Pain Research Center, Department of Neurology, Aarhus University Hospital, Aarhus, Denmark

Kurt L. Johnson Department of Rehabilitation Medicine, University of Washington, Seattle, WA, USA

Mark I. Johnson Faculty of Health and Social Sciences, Leeds Pallium Research Group, Leeds Metropolitan University, Leeds, UK

Mark Johnston Hibiscus Coast Highway, Orewa, New Zealand

Edward Jones Center for Neuroscience, University of California, Davies, CA, USA

Jeroen R. De Jong Department of Rehabilitation, University Hospital Maastricht, Maastricht, The Netherlands

Sven-Eric Jordt Department of Pharmacology, Yale University School of Medicine, New Haven, CT, USA

Ellen Jørum The Laboratory of Clinical Neurophysiology, Rikshospitalet University, Oslo, Norway

Stefan Just CNS Pharmacology, Boehringer Ingelheim Pharma GmbH & Co. KG, Biberach, Germany

Timothy K. Y. Kaan Department of Anatomy, University of California San Francisco, San Francisco, CA, USA

Noufissa Kabli Department of Pharmacology and Toxicology, Queen's University, Kinston, Canada

Ryusuke Kakigi Department of Integrative Physiology, National Institute for Physiological Sciences, Okazaki, Japan

Peter C. A. Kam Anaesthesia, Central Clinical School, The University of Sydney, Sydney, Australia

Giresh Kanji The Sports and Pain Clinic, Wellington, New Zealand

Joel Katz Department of Psychology, McGill University, Montreal, QC, Canada

Valerie Kayser NSERM U677 91Bd de l'Hôpital, Paris, France

Francis J. Keefe Pain Prevention and Treatment Research, Department of Psychiatry and Behavioral Sciences, Duke University Medical Center, Durham, NC, USA

Lois J. Kehl Department of Anesthesiology, University of Minnesota, Minneapolis, MN, USA

Maureen Kelley Department of Pediatrics, Bioethics Division, University of Washington School of Medicine, WA, USA

Seattle Children's Hospital, Seattle, WA, USA

Stephen Kennedy Nuffield Department of Obstetrics and Gynaecology, University of Oxford, Oxford, UK

Edmund Keogh Department of Psychology, University of Bath, Bath, UK

Robert D. Kerns VA Connecticut Healthcare System, Westhaven, CT, USA

Sanjay Khanna Department of Physiology, National University of Singapore, Singapore

Kok Eng Khor Department of Pain Management, Prince of Wales Hospital University of New South Wales, Randwick, NSW, Australia

Mina Khoshnejad Department of Neuroscience, University of Montreal, Montreal, QC, Canada

H. J. Kim Department of Life Science, Yonsei University, Wonju, Republic of Korea

Sara King Faculty of Education (School Psychology), Mount Saint Vincent University, Halifax, NS, Canada

Wade King Department of Clinical Research, Royal Newcastle Centre University of Newcastle, Newcastle, NSW, Australia

K. L. Kirsh Behavioral Medicine, The Pain Treatment Center of the Blue grass, Lexington, KY, USA

Henriette Klit Danish Pain Research Center, Aarhus University Hospital, Aarhus C, Denmark

Ricardo J. Komotar Department of Neurological Surgery, Neurological Institute Columbia University, New York, NY, USA

Sungtae Koo Division of Meridian and Structural Medicine, School of Korean Medicine, Pusan National University, Yangsan, Korea

Rhonda K. Kotarinos Oakbrook Terrace, IL, USA

Katalin J. Kovacs Department of Veterinary and Biomedical Sciences, University of Minnesota, St. Paul, MN, USA

Malgorzata Krajnik Chair of Palliative Care, Nicolaus Copernicus University in Torun, Collegium Medicum in Bydgoszcz, Bydgoszcz, Poland

Elliot Krane Anesthesia and Pediatrics, Stanford University School of Medicine Lucile Packard Children's Hospital at Stanford, Stanford, CA, USA

Ajit Krishnaney Department of Neurosurgery, Cleveland Clinic, Cleveland, OH, USA

Greg Krohm International Association of Industrial Accident Boards and Commissions (IAIABC), Utah Valley University, USA

Lawrence Kruger UCLA Geffen School of Medicine, Los Angeles, CA, USA

M. M. ter Kuile Department of Gynecology, Leiden University Medical Center, Leiden, The Netherlands

Promil Kukreja Department of Anesthesiology, University of Alabama at Birmingham, Birmingham, AL, USA

Alice Kvåle Department of Public Health and Primary Health Care, Faculty of Medicine and Dentistry University of Bergen, Bergen, Norway

Gunnvald Kvarstein Department of Pain Management and Pain Research, Oslo University Hospital, Rikshospitalet, Oslo, Norway

Jean-Jacques Labat Neurology and Rehabilitation, Clinique Urologique, Nantes, France

Gaston Labrecque Faculty of Pharmacy, University of Laval, Quebec City, QC, Canada

Marco Lacerenza Pain Medicine Center, Casa di Cura S. Pio X, Opera San Camillo Foundation, Milan, Italy

Uri Ladabaum Gastroenterology & Hepatology, Stanford Cancer Institute, Redwood City, CA, USA

Marie Andree Lahaie McGill University, Montreal, QC, Canada

Miguel J. A. Lainez-Andres Department of Neurology, University of Valencia, Valencia, Spain

Robert H. LaMotte Department of Anesthesiology, Yale School of Medicine, New Haven, CT, USA

James W. Lance Institute of Neurological Sciences, University of New South Wales, Sydney, Australia

Robert G. Large The Auckland Regional Pain Service, Auckland Hospital, Auckland, New Zealand

Alice A. Larson Department of Veterinary and Biomedical Sciences, University of Minnesota, St. Paul, MN, USA

Roberta Lattanzi Department of Physiology and Pharmacology, Sapienza University of Rome, Rome, Italy

Peter Lau Department of Clinical Research, Royal Newcastle Hospital, University of Newcastle, Newcastle, NSW, Australia

Stefan A. Laufer Institute of Pharmacy, Eberhard-Karls University of Tuebingen, Tuebingen, Germany

Stefan Lautenbacher Physiologische Psychologie, Otto-Friedrich-Universität, Bamberg, Germany

Gilles J. Lavigne Facultés de Médecine Dentaire et de Médecine, Université de Montréal, Hopital Sacre Coeur de Montreal. Surgery-trauma dept, Montréal, QC, Canada

Emily Law Center for Child Health, Behavior, and Development, Seattle Children's Research Institute, Seattle, WA, USA

Peter G. Lawlor Our Lady's Hospice, Medical Department, Dublin, Ireland

Michel Lazdunski Institut de Pharmacologie Moléculaire et Cellulaire (IPMC) CNRS/UNS UMR7275, Valbonne-Sophia Antipolis, France

Vincenzo Di Lazzaro Institute of Neurology, Campus Biomedico University, Rome, Italy

Alyssa Lebel Boston Children's Hospital, Boston, MA, USA

Frank Lee Johns Hopkins University, Baltimore, MD, USA

Maaike Leeuw Department of Medical, Clinical and Experimental Psychology, Maastricht University, Maastricht, The Netherlands

Siri Graff Leknes Research Fellow, Psykologisk institutt, Oslo, Norway

Frederick A. Lenz Department of Neurosurgery, Johns Hopkins Hospital, Baltimore, MD, USA

Guillaume Léonard Faculty of Medicine, University of Sherbrooke, Sherbrooke, QC, Canada

Pauline Lesage Department of Pain Medicine and Palliative Care, Beth Israel Medical Center, New York, USA

Isobel Lever Neural Plasticity Unit, Institute of Child Health University College London, London, UK

Khan W. Li Department of Neurosurgery, Johns Hopkins Hospital, Baltimore, MD, USA

Alan R. Light Department of Anesthesiology, University of Utah, Salt Lake City, UT, USA

Michael Lim Department of Neurosurgery and Department of Neurology, Johns Hopkins Hospital, Baltimore, MA, USA

Deolinda Lima Department of Experimental Biology, Faculdade de Medicina da Universidade do Porto, Porto, Portugal

Volker Limmroth Department of Neurology, University of Essen, Essen, Germany

Eric Lingueglia Institut de Pharmacologie Moléculaire et Cellulaire (IPMC) CNRS/UNS UMR7275, Valbonne-Sophia Antipolis, France

Steven J. Linton Department of Occupational and Environmental Medicine, Orebro Medical Center, Orebro, Sweden

Arthur G. Lipman Departments of Pharmscotherpay and Anesthesiology and Pain Management Center, University of Utah, Salt Lake City, UT, USA

Richard B. Lipton Departments of Neurology, Epidemiology and Population Health, Albert Einstein College of Medicine, Bronx, NY, USA

Diana Lisi Department of Psychology, York University, Toronto, ON, Canada

Kenneth M. Little Division of Neurosurgery, Duke University Medical Center, Durham, NC, USA

Edward Littleton Queen Elizabeth Hospital, Birmingham, UK

Spencer S. Liu Virginia Mason Medical Center, Seattle, WA, USA

Anne Elisabeth Ljunggren Section for Physiotherapy Science, Department of Public Health and Primary Health, Faculty of Medicine, University of Bergen, Bergen, Norway

John D. Loeser Department of Neurological Surgery, University of Washington, Seattle, WA, USA

Joern Loetsch Institute for Clinical Pharmacology, Pharmaceutical Center Frankfurt, Johann Wolfgang Goethe University, Frankfurt, Germany

Patrick Loisel Disability Prevention Research and Training Centre, Université de Sherbrooke, Longueuil, Canada

Elisa Lopez-Dolado Servicio de Rehabilitacion, National Hospital of Paraplèjicos SESCAM, Toledo, Spain

Jürgen Lorenz Faculty of Life Science, Competence Center Gesundheit, University of Applied Science Hamburg, Hamburg, Germany

Dayna R. Loyd US Army Institute of Surgical Research, Fort Sam Houston, TX, USA

Daniel Lubelski Cleveland Clinic Lerner College of Medicine, Cleveland, OH, USA

Stephen Lutz Blanchard Valley Regional Cancer Center, Findlay, OH, USA

Bruce Lynn Department of Physiology, University College London, London, UK

Ross MacPherson Department of Anesthesia and Pain Management, Royal North Shore Hospital University of Sydney, Sydney, NSW, Australia

Christophe Maes Manager Innovatie AZ Alma, Belgium

Michel Magnin INSERM, Neurological Hospital, Lyon, France

Bryan Mahony Department of Anesthesiology, Ohio State University Medical Center, Mount Sinai Hospital, New York, NY, USA

Steven F. Maier Department of Psychology and Neuroscience, and The Center for Neuroscience, University of Colorado at Boulder, Boulder, CO, USA

Thorsten J. Maier Institute of Pharmaceutical Chemistry, Goethe University, Frankfurt, Germany

Chris J. Main Keele University, Keele, Staffordshire, UK

Marzia Malcangio Wolfson Centre for Age Related Diseases, King's College London, London, UK

Paolo L. Manfredi Department of Neurology, Memorial Sloan-Kettering Cancer Center, New York, USA

Kaisa Mannerkorpi Department of Rheumatology and Inflammation Research, Institute of Medicine, University of Gothenburg, Göteborg, Sweden

Barton H. Manning Amgen Inc., Thousand Oaks, CA, USA

Patrick W. Mantyh Department of Pharmacology, The University of Arizona, AZ, USA

Jianren Mao Department of Anesthesia and Critical Care, Harvard Medical School, Boston, MA, USA

Serge Marchand Faculty of Medicine, University of Sherbrooke, Sherbrooke, QC, Canada

Paolo Marchettini Pain Medicine Center, Scientific Institute San Raffaele, Milan, Italy

Sarah R. Martin Psychology, Georgia State University, Atlanta, GA, USA

Peggy Mason Department of Neurobiology, Pharmacology and Physiology, University of Chicago, Chicago, IL, USA

Scott Masters Caloundra Spinal and Sports Medicine Centre, Caloundra, QLD, Australia

Laurence E. Mather Department of Anaesthesia and Pain Management, University of Sydney at Royal North Shore Hospital, St. Leonards, Sydney, NSW, Australia

Bill McCarberg Pain Services, Kaiser Permanente, San Diego, USA

Lance McCracken Health Psychology/Psychology Department, King's College London, London, UK

Kathleen McGinn Anesthesiology & Pain Medicine, Seattle Children's Hospital, Seattle, WA, USA

Patricia A. McGrath The Hospital for Sick Children Pain Research Center, Toronto, ON, Canada

Patrick McGrath Departments of Psychology, Pediatrics and Psychiatry, Dalhousie University, Canada

Greg McIntosh Canadian Back Institute, Toronto, ON, Canada

Peter McIntyre Department of Pharmacology, University of Melbourne, Melbourne, Australia

Elspeth M. McLachlan Neuroscience Research Australia, University of New South Wales, Randwick, NSW, Australia

University of Glasgow, Glasgow, UK

Peter A. McNaughton Department of Pharmacology, University of Cambridge, Cambridge, UK

J. Mark Melhorn Section of Orthopaedics, Department of Surgery, University of Kansas School of Medicine, Wichita, KS, USA

Ronald Melzack Department of Psychology, McGill University, Montreal, QC, Canada

Lorne M. Mendell Department of Neurobiology and Behavior, Stony Brook University, Stony Brook, NY, USA

George Mendelson Department of Psychological Medicine, Monash University, Caulfield, VIC, Australia

Siegfried Mense Neuroanatomy, Heidelberg University, Medical Faculty Mannheim, CBTM, Mannheim, Germany

Daniel Menétrey Biomédicale, Paris, France

Sebastiano Mercadante Pain Relief & Palliative Care, La Maddalena Cancer Center University of Palermo, Palermo, Italy

Susan Mercer School of Biomedical Sciences, The University of Queensland, Brisbane, Australia

Karl Messlinger University of Erlangen-Nürnberg, Erlangen, Germany

Richard A. Meyer Department of Neurosurgergy, School of Medicine Johns Hopkins University, Baltimore, MD, USA

Walter J. Meyer III Shriners Hospitals for Children, Shriners Burns Hospital, Galveston, TX, USA

Department of Psychiatry, University of Texas Medical Branch, Galveston, TX, USA

Björn A. Meyerson Department of Clinical Neuroscience, Section of Neurosurgery, Karolinska Institute/University Hospital, Stockholm, Sweden

Martin Michaelis Sanofi-Aventis Deutschland GmbH, Frankurt am Main, Germany

Joseph Mignogna Michael E. DeBakey VA Medical Center, Houston VA HSR&D Center of Excellence (152), Houston, TX, USA

Judith A. Mikacich Department of Obstetrics and Gynecology, University of California Los Angeles Medical Center, Los Angeles, CA, USA

Kensaku Miki Department of Integrative Physiology, National Institute for Physiological Sciences, Okazaki, Japan

Vjekoslav Miletic School of Medicine and Public Health, University of Wisconsin, Madison, WI, USA

Scott R. Miller Memorial Sloan-Kettering Cancer Center, New York, USA

Erin D. Milligan Department of Neurosciences, University of New Mexico, Albuquerque, NM, USA

Michael S. Minett Molecular Nociception Group, Wolfson Institute for Biomedical Research, University College London, London, WC, UK

J. Mocco Department of Neurological Surgery, Neurological Institute Columbia University, New York, NY, USA

Jeffrey S. Mogil Department of Psychology, McGill University, Montreal, QC, Canada

Harvey Moldofsky Sleep Disorders Clinic of the Centre for Sleep and Chronobiology, Toronto, ON, Canada

Robert M. Moldwin The Arthur Smith Institute for Urology, North Shore-LIJ Healthcare System, Hofstra North Shore LIJ School of Medicine, New Hyde Park, NY, USA

Derek C. Molliver Department of Neurobiology, University of Pittsburgh School of Medicine, Pittsburgh, PA, USA

Robert D. Mootz Washington State Department of Labor and Industries, Olympia, WA, USA

Laura Mordillo-Mateos FENNSI Group, Hospital Nacional de Parapléjicos, Toledo, Spain

Anne Morel Laboratory for Functional Neurosurgery, Neurosurgical Clinic University Hospital Zürich, Zürich, Switzerland

Anne Morinville Montreal Neurological Institute, McGill University, Montreal, QC, Canada

Stephen Morley Academic Unit of Psychiatry and Behavioral Sciences, University of Leeds, Leeds, UK

David B. Morris University of Virginia, Charlottesville, VA, USA

G. Lorimer Moseley Sansom Institute for Health Research, University of South Australia Neuroscience Research Australia, Adelaide, Australia

Joachim Mossner Medical Clinic and Policlinic II, University Clinical Center Leipzig AöR, Leipzig, Germany

Eric A. Moulton P.A.I.N. Group, Anesthesiology, Boston Children's Hospital, Harvard Medical School, Waltham, MA, USA

Francis Githae Muriithi Department of Obstetrics and Gynaecology, Aga Khan University, Nairobi, Kenya

Anne Z. Murphy Neuroscience Institute, Georgia State University, Atlanta, GA, USA

Paul M. Murphy Department of Anaesthesia and Pain Management, Royal North Shore Hospital, St. Leonards, NSW, Australia

Alison Murray Foothills Medical Center, Calgary, AB, Canada

H. S. Na Neuroscience Research Institute and Department of Physiology, Korea University, College of Medicine, Seoul, Republic of Korea

Daisuke Naka Department of Integrative Physiology, National Institute for Physiological Sciences, Okazaki, Japan

Yoshio Nakamura Pain Research Center, Department of Anesthesiology University of Utah School of Medicine, Salt Lake City, UT, USA

Barbara Namer Department of Physiology & Pathophysiology, FAU, Erlangen/Nuremberg, Bavaria, Germany

Matti V. O. Närhi Department of Biomedicine/Physiology, Department of Dentistry, University of Eastern Finland, Institute of Medicine, Kuopio, Finland

H. J. W. Nauta University of Texas Medical Branch, Galveston, TX, USA

Lucia Negri Department of Physiology and Pharmacology, Sapienza University of Rome, Rome, Italy

Gary Nesbit Department of Radiology, Oregon Health and Science University, Portland, OR, USA

Timothy Ness Department of Anesthesiology, University of Alabama at Birmingham, Birmingham, AL, USA

Volker Neugebauer Department of Neuroscience & Cell Biology, University of Texas Medical Branch, Galveston, TX, USA

Lawrence C. Newman Albert Einstein College of Medicine, New York, NY, USA

Michael K. Nicholas Royal North Shore Hospital, University of Sydney, St. Leonards, NSW, Australia

Ellen Niederberger Pharmazentrum frankfurt, Institute of Clinical Pharmacology, Goethe-University Hospital, Frankfurt, Germany

Lone Nikolajsen Department of Anaesthesiology, Danish Pain Research Center Aarhus University Hospital, Aarhus, Denmark

Katie Nixdorf Comprehensive Pain Center, Oregon Health and Science University, Portland, OR, USA

Donald R. Nixdorf Department of Diagnostic and Biological Sciences, School of Dentistry University of Minnesota, Minneapolis, MN, USA

Carl E. Noe University of Texas Southwestern Medical Center, Lubbock, TX, USA

Koichi Noguchi Department of Anatomy and Neuroscience, Hyogo College of Medicine, Hyogo, Japan

Richard B. North The Johns Hopkins University, Baltimore, MD, USA

Turo J. Nurmikko Department of Neurological Science, University of Liverpool, and Pain Research Institute, The Walton Centre NHS Foundation Trust, Liverpool, UK

Anne Louise Oaklander Nerve Injury Unit, Departments of Neurology and Pathology, Massachusetts General Hospital Harvard Medical School, Boston, MA, USA

Koichi Obata Department of Anatomy and Neuroscience, Hyogo College of Medicine, Hyogo, Japan

Tim F. Oberlander Division of Developmental Pediatrics, University of British Columbia, Vancouver, BC, Canada

Bruno Oertel Institute for Clinical Pharmacology, Pharmaceutical Center Frankfurt, Johann Wolfgang Goethe University, Frankfurt, Germany

Andrea C. Ogbonna Wolfson Centre for Age Related Diseases, King's College London, London, UK

Alfred T. Ogden Department of Neurological Surgery, Neurological Institute Columbia University, New York, NY, USA

S. Ohara Department of Neurosurgery, Johns Hopkins Hospital, Johns Hopkins Medical Institutions, Johns Hopkins University, Baltimore, MD, USA

Peter T. Ohara Department of Anatomy, University of California San Francisco, San Francisco, CA, USA

Akiko Okifuji Department of Anesthesiology, University of Utah, Salt Lake City, UT, USA

Jean-Louis Oliveras Inserm, Paris, France

Antonio Oliviero FENNSI Group, National Hospital of Parapléjicos, Toledo, Spain

Sean O'Mahony Palliative medicine, Rush University Medical Center, Chicago, IL, USA

Peregrine B. Osborne Dept Anatomy & Neuroscience, University of Melbourne, Melbourne, VIC, Australia

Michael H. Ossipov Department of Pharmacology, University of Arizona College of Medicine, Tucson, AZ, USA

Tonya M. Palermo Seattle Children's Hospital Research Institute, CW8-6 - Child Health, Behavior and Development, University of Washington, Seattle, WA, USA

Shreela Palit Department of Psychology, The University of Tulsa, Tulsa, OK, USA

Juan A. Pareja Department of Neurology, Fundación Hospital Alcorcón, Madrid, Spain

Steven D. Passik Psychiatry and Anesthesiology, Vanderbilt University, Nashville, TN, USA

Gavril W. Pasternak Laboratory of Molecular Neuropharmacology, Memorial Sloan-Kettering Cancer Center, New York, USA

S. H. Patel Department of Neurosurgery, Johns Hopkins Hospital, Baltimore, MD, USA

Gavin Pattullo Department of Anesthesia and Pain Management, Royal North Shore Hospital, University of Sydney, St. Leonards, NSW, Australia

Kevin Pauza Tyler Spine and Joint Hospital, Tyler, TX, USA

Eric M. Pearlman Pediatrics, Mercer University School of Medicine, The Children's Hospital at Memorial University Medical Center, Savannah, USA

Elvira De la Pena Instituto de Neurociencias, Universidad Miguel Hernández-CSIC, Alicante, Spain

Yuan Bo Peng Department of Psychology, University of Texas at Arlington, Arlington, TX, USA

Jose Pereira Foothills Medical Center, Calgary, AB, Canada

Edward Perl Department of Cell and Molecular Physiology, School of Medicine University of North Carolina, Chapel Hill, NC, USA

Tiffany Grace Perry Center for Spine Health, The Cleveland Clinic Foundation, Cleveland, OH, USA

Marie Pertin Pain Center, University Hospital Center and University of Lausanne, Lausanne, Switzerland

Antti Pertovaara Department of Physiology, Institute of Biomedicine University of Helsinki, Helsinki, Finland

Peter Peter Department of Neurobiology and Anatomy, University of Rochester Medical Center, Rochester, NY, USA

Lisa M. Peters Department of Anesthesiology and Pain Medicine, Seattle Children's Hospital, Seattle, WA, USA

Pramit Phal Department of Radiology, Oregon Health and Science University, Portland, OR, USA

Rebecca Pillai Riddell Department of Psychology, York University, Toronto, ON, Canada

Department of Psychiatry, Hospital for Sick Children, Toronto, ON, Canada

Issy Pilowsky University of Adelaide, Adelaide, Australia

Leon Plaghki Institute of Neuroscience (IoNS), Université Catholique de Louvain, Brussels, Belgium

Markus Ploner Department of Neurology, Technische Universität München, Munich, Germany

Esther M. Pogatzki-Zahn Department of Anesthesiology and Intensive Care Medicine, University Muenster, Muenster, Germany

Michael Polydefkis Department of Neurology, Johns Hopkins University School of Medicine, Baltimore, MD, USA

James D. Pomonis Algos Preclinical Services, Roseville, MN, USA

Dieter Pongratz Friedrich Baur Institute, Ludwig Maximilians University, Munich, Germany

Hayley Poole Hadley House, Leicester General Hospital, Leicester, UK

Frank Porreca Department of Pharmacology, University of Arizona Health Sciences Center, Tucson, AZ, USA

Russell K. Portenoy Department of Pain Medicine and Palliative Care, Beth Israel Medical Center, New York, NY, USA

Ian Power Critical Care and Pain Medicine, The University of Edinburgh, Edinburgh, UK

Donald D. Price Oral Surgery and Neuroscience, McKnight Brain Institute University of Florida, Gainesville, FL, USA

Daryl Pullman Medical Ethics, Memorial University of Newfoundland, St. John's, NL, Canada

Yunhai Qiu Department of Integrative Physiology, National Institute for Physiological Sciences, Okazaki, Japan

Michael Quittan Department of Physical Medicine and Rehabilitation, Kaiser-Franz-Joseph Hospital, Vienna, Austria

Raymond M. Quock Department of Pharmaceutical Sciences, Washington State University College of Pharmacy, Pullman, WA, USA

Jennifer Rabbitts Department of Anesthesiology and Pain Medicine, Seattle Children's Hospital, University of Washington School of Medicine, Seattle, WA, USA

Nicole Racine Psychology, York University, Toronto, ON, Canada

Gabor B. Racz Texas Tech University Health Sciences Center, Lubbock, TX, USA

Lukas Radbruch Department of Palliative Medicine, University of Bonn, Bonn, Germany

Robert B. Raffa Department of Pharmaceutical Sciences, Temple University School of Pharmacy, Philadelphia, PA, USA

Vasudeva Raghavendra Algos Therapeutics Inc., Saint Paul, MN, USA

Pierre Rainville Department of Stomatology, Faculty of Dental Medicine, University of Montreal, Montreal, QC, Canada

Srinivasa N. Raja Department of Anesthesiology and Critical Care Medicine, Johns Hopkins University, Baltimore, USA

Alan Randich University of Alabama, Birmingham, AL, USA

Andrea J. Rapkin Department of Obstetrics and Gynecology, University of California Los Angeles Medical Center, Los Angeles, CA, USA

Z. Harry Rappaport Department of Neurosurgery, Rabin Medical Center Tel-Aviv University, Petah Tiqva, Israel

Matthew N. Rasband Department of Molecular and Cellular Biology, Baylor College of Medicine, Farmington, Houston TX, USA

University of Connecticut Health Center, Farmington, CT, USA

Douglas Rasmusson Department of Physiology and Biophysics, Dalhousie University, Halifax, Canada

James P. Rathmell Department of Anesthesiology, University of Vermont College of Medicine, Burlington, USA

K. Ravishankar The Headache and Migraine Clinic, Jaslok Hospital and Research Centre Lilavati Hospital and Research Centre, Mumbai, India

Ke Ren Department of Neural and Pain Sciences, University of Maryland Dental School, Baltimore, MD, USA

Seth Resnick Silver Hill Hospital, New York University School of Medicine, USA

Cielito Reyes-Gibby The University of Texas MD Anderson Cancer Center, Houston, TX, USA

Jamie L. Rhudy Department of Psychology, The University of Tulsa, Tulsa, OK, USA

Alfredo Ribeiro-da-Silva Department of Pharmacology and Therapeutics, McGill University, Montreal, QC, Canada

Frank L. Rice Center for Neuropharmacology and Neuroscience, Albany Medical College, Albany, NY, USA

Matthias Ringkamp Department of Neurosurgergy, School of Medicine Johns Hopkins University, Baltimore, MD, USA

Pierre J. M. Rivière Ferring Research Institute, San Diego, CA, USA

Claude Robert Université Paris Descartes, INSERM UMR 894, France

Roger Robert Neurosurgery, Hotel Dieu, Nantes, CHU, France

James P. Robinson Department of Anesthesiology & Pain Medicine, University of Washington, Seattle, WA, USA

John Robinson Pain Management Centre, Burwood Hospital, Christchurch, New Zealand

Alfonso Romero-Sandoval Department of Pharmaceuticals and Administrative Sciences, Presbyterian College School of Pharmacy, Clinton, SC, USA

David Roselt Aberdovy Clinic, Bundaberg, QLD, Australia

Jackie Ross Early Pregnancy Unit, King's College Hospital Early Pregnancy Unit, London, UK

Shlomo Rotshenker Department of Medical Neurobiology, IMRIC, Faculty of Medicine Hebrew University of Jerusalem, Jerusalem, Israel

Paul C. Rousseau Medical University of South Carolina, Charleston, SC, USA

Ranjan Roy Department of Clinical Health Psychology, Faculty of Medicine, University of Manitoba, Winnipeg, MB, Canada

Carolina Roza Biología de Sistemas (Unidad Fisiología), Universidad de Alcalá, Alcalá de Henares, Madrid, Spain

Todd D. Rozen Department of Neurology, Geisinger Health System, Wilkes-Barre, PA, USA

Roman Rukwied Department of Anaesthesiology and Intensive Care Medicine, Faculty of Clinical Medicine Mannheim University of Heidelberg, Heidelberg, Germany

Irwin Jon Russell Division of Clinical Immunology and Rheumatology, Department of Medicine, The University of Texas Health Science Center, San Antonio, TX, USA

Nayef E. Saadé Anatomy, Cell Biology and Physiology, Faculty of Medicine, American University of Beirut, Beirut, Lebanon

Mary Ann C. Sabino Oral and Maxillofacial Surgery, Hennepin County Medical Center University of Minnesota, Minneapolis, MN, USA

Thomas E. Salt Institute of Ophthalmology, University College London, London, UK

A. Samdani Department of Neurosurgery, Johns Hopkins Medical Institutions, Baltimore, MD, USA

Yasmin Sana King's College Hospital, London, UK

Margarita Sanchez-del-Rio Neurology Department, Hospital Ruber Internacional, Madrid, Spain

Jürgen Sandkühler Department of Neurophysiology, Center for Brain Research Medical University of Vienna, Vienna, Austria

Peter S. Sándor RehaClinic, Cantonal Hospital Baden, Baden, Switzerland

Johannes Sarnthein Functional Neurosurgery, Neurosurgical Clinic University Hospital, Zürich, Switzerland

Susanne K. Sauer Department of Physiology and Pathophysiology, University of Erlangen, Erlangen, Germany

Steven Savvas Biomedical Division, National Ageing Research Institute, Melbourne, VIC, Australia

Jana Sawynok Department of Pharmacology, Dalhousie University, Halifax, Canada

John W. Scadding The National Hospital for Neurology and Neurosurgery, London, UK

Frederieke Geraldine Schaafsma Department of Public and Occupational Health, EMGO+ Institute/VU Medical Center, Amsterdam, MB, The Netherlands

Hans-Georg Schaible Department of Physiology - Neurophysiology, Friedrich-Schiller-University of Jena, Jena, Germany

Steven S. Scherer Department of Neurology, The Perelman School of Medicine at the University of Pennsylvania, Philadelphia, PA, USA

Michael Schäfer Department of Anesthesiology and Intensive Care Medicine, Charité University, Berlin, Germany

Martin Schmelz Department of Anesthesiology in Mannheim, University of Heidelberg, Mannheim, Germany

Robert F. Schmidt Physiologisches Institut der Universität Würzburg, Würzburg, Germany

Suzan Schneeweiss The Hospital for Sick Children, Toronto, ON, Canada

Frank Schneider Department of Psychiatry and Psychotherapy, RWTH Aachen University, Aachen, Germany

Alfons Schnitzler Department of Neurology, Heinrich Heine University, Düsseldorf, Germany

Jean Schoenen Department of Neurology, University of Liège, CHR Citadelle, Belgium

Department of Neuroanatomy, University of Liège, CHR Citadelle, Belgium

Jens Schouenborg Neuronano Research Center, Lund University, Lund, Sweden

Stephan A. Schug School of Medicine and Pharmacology, Royal Perth Hospital, Faculty of Medicine, Dentistry & Health Sciences, The University of Western Australia, UWA Anaesthesiology, Perth, Australia

Patrick Schuss Department of Neurosurgery, University-Hospital Bonn, Bonn, Germany

Anthony C. Schwarzer Faculty of Medicine and Health Sciences, The University of Newcastle, Newcastle, Australia

Whitney Scott Department of Psychology, McGill University, Montreal, QC, Canada

Laura C. Seidman Pediatric Pain Program, Department of Pediatrics David Geffen School of Medicine at UCLA, Los Angeles, CA, USA

Ze'ev Seltzer Faculty of Dentistry, University of Toronto Centre for the Study of Pain, Toronto, ON, Canada

Peter Selwyn Albert Einstein College of Medicine, Bronx, NY, USA

Patrick B. Senatus Department of Neurological Surgery, Neurological Institute Columbia University, New York, NY, USA

Jyoti N. Sengupta Division of Gastroenterology and Hepatology and Pediatric Gastroenterology, Medical College of Wisconsin, Milwaukee, WI, USA

Mariano Serrao Department of Neurology and Otorinolaringoiatry, University of Rome La Sapienza, Rome, Italy

Barry J. Sessle Faculty of Dentistry, University of Toronto, Toronto, ON, Canada

Faculty of Dentistry, University of Toronto, Toronto, ON, Canada

Virginia S. Seybold Department of Neuroscience, University of Minnesota, Minneapolis, MN, USA

T. Shah Royal Perth Hospital and University of Western Australia, Crawley, WA, Australia

Yair Sharav Department of Oral Medicine, School of Dental Medicine, Hebrew University-Hadassah, Jerusalem, Israel

S. Murray Sherman Department of Neurobiology, University of Chicago, Chicago, IL, USA

Yoshio Shigenaga Department of Oral Anatomy and Neurobiology, Osaka University, Osaka, Japan

Susan Shin Neurology, Mount Sinai School of Medicine, New York, NY, USA

Edward A. Shipton Department of Anaesthesia, Christchurch School of Medicine and Health Sciences, University of Otago, Christchurch, New Zealand

Yoram Shir The Alan Edwards Pain Management Unit, McGill University Health Centre, Montreal, QC, Canada

Peter Shrager Department of Neurobiology and Anatomy, University of Rochester Medical Center, Rochester, NY, USA

Marc J. Shulman VA Connecticut Healthcare System, Yale University, New Haven, CT, USA

David M. Sibell Comprehensive Pain Center/Spine Center, Oregon Health and Science University, Portland, OR, USA

Philip J. Siddall Department of Pain Management, Greenwich Hospital, University of Sydney, Sydney, NSW, Australia

Stephen D. Silberstein Jefferson Medical College, Thomas Jefferson University and Jefferson Headache Center, Thomas Jefferson University Hospital, Philadelphia, PA, USA

Donald A. Simone Diagnostic and Biological Sciences, University of Minnesota, Minneapolis, MN, USA

David G. Simons Department of Rehabilitation Medicine, Emory University, Atlanta, GA, USA

David Simpson Clinical Neurophysiology Laboratories, Mount Sinai Medical Center, New York, NY, USA

Marc Sindou Department of Neurosurgery, Hopital Neurologique Pierre Wertheimer, University of Lyon 1, Lyon, France

Soeren H. Sindrup Department of Neurology, Odense University Hospital, Odense, Denmark

Bengt H. Sjölund Department of Community Medicine and Rehabilitation Medicine, Umea University, Umea, Sweden

Carsten Skarke Institute for Translational Medicine and Therapeutics (ITMAT), University of Pennsylvania Perelman School of Medicine, Philadelphia, PA, USA

Paul A. Sloan Department of Anesthesiology, University of Kentucky, Lexington, KY, USA

Kathleen A. Sluka Physical Therapy and Rehabilitation Science Graduate Program, University of Iowa, Iowa City, IA, USA

Seymour Solomon Montefiore Medical Center, Albert Einstein College of Medicine, Bronx, NY, USA

Claudia Sommer Department of Neurology, University of Würzburg, Würzburg, Germany

Linda S. Sorkin Occupational Therapy and Rehabilitation, Anesthesia Research Laboratories, University of California, San Diego, CA, USA

Ingrid Söderback Uppsala University, Uppsala, Sweden

Kari Sørensen Department of Pain Management and Research, Oslo University Hospital, Rikshospitalet, Oslo, Norway

Michael Spaeth Friedrich-Baur-Institute, University of Munich, Munich, Germany

Peregrin O. Spielholz Washington State Department of Labor and Industries, SHARP Program, Olympia, WA, USA

Peter S. Staats Division of Pain Medicine, The Johns Hopkins University, Baltimore, MD, USA

Brett R. Stacey Comprehensive Pain Center, Oregon Health and Science University, Portland, OR, USA

Wendy M. Stein San Diego Hospice and Palliative Care, UCSD, San Diego, CA, USA

Christoph Stein Department of Anesthesiology and Intensive Care Medicine, Freie Universitaet Berlin Charité Campus Benjamin Franklin, Berlin, Germany

Michael P. Steinmetz Department of Neurosurgery, MetroHealth Medical Center, Case Western Reserve University School of Medicine, Cleveland, OH, USA

Jair Stern Functional Neurosurgery, Neurosurgical Clinic University Hospital, Zürich, Switzerland

Bonnie J. Stevens University of Toronto and The Hospital for Sick Children, Toronto, ON, Canada

Carl-Olav Stiller Division of Clinical Pharmacology and Department of Medicine, Karolinska University Hospital, Stockholm, Sweden

Jennifer N. Stinson The Hospital for Sick Children, Toronto, ON, Canada

Henry Stockbridge University of Washington School of Public Health and Community Medicine, Seattle, WA, USA

Christian S. Stohler Baltimore College of Dental Surgery, University of Maryland, Baltimore, MD, USA

Laura Stone Faculty of Dentistry, Alan Edwards Centre for Research on Pain, Montreal, QC, Canada

William Stones Department of Obstetrics and Gynaecology, Aga Khan University, Nairobi, Kenya

Andreas Straube Department of Neurology, Ludwig Maximilians University, Munich, Germany

Jenny Strong University of Queensland, Brisbane, QLD, Australia

Audun Stubhaug Department of Pain Management and Research, Oslo University Hospital, Rikshospitalet and University of Oslo, Oslo, Norway

Gerold Stucki Swiss Paraplegic Research, Nottwil, Switzerland

Department of Health Sciences and Health Policy, University of Lucerne, Lucerne, Switzerland

ICF Research Branch in cooperation with the World Health Organization (WHO) Collaborating Centre for the Family of International Classifications in Germany, Nottwil, Switzerland

Cheryl L. Stucky Department of Cell Biology Neurobiology and Anatomy, Medical College of Wisconsin, Milwaukee, WI, USA

Mathias Sturzenegger Department of Neurology, University Hospital, Inselspital, University of Bern, Bern, Switzerland

Yasuo Sugiura School of Human Care Study, Nagoya University of Arts and Sciences, Nissin City, Aichi, Japan

Mark D. Sullivan Psychiatry and Behavioral Sciences, University of Washington, Seattle, WA, USA

Michael J. L. Sullivan Department of Psychology, McGill University, Montreal, QC, Canada

Rie Suzuki Department of Pharmacology, University College London, London, UK

Kristina B. Svendsen Danish Pain Research Center, Aarhus University Hospital, Aarhus, Denmark

Peter Svensson Clinical Oral Physiology, Aarhus University, Aarhus, Denmark

Peter Szucs Spinal Neuronal Networks, Instituto de Biologia Molecular e Celular - IBMC, Porto, Portugal

A. Taghva Department of Neurosurgery, Johns Hopkins Hospital, Baltimore, MD, USA

Kimberson Cochien Tanco Department of Palliative Care and Rehabilitation Medicine, The University of Texas M. D. Anderson Cancer Center, Houston, TX, USA

Kersi Taraporewalla Royal Brisbane and Womens' Hospital, University of Queensland, Brisbane, QLD, Australia

Irmgard Tegeder Pharmazentrum frankfurt, Institute of Clinical Pharmacology, Goethe-University Hospital, Frankfurt, Germany

Astrid Juhl Terkelsen Danish Pain Research Center, Department of Neurology, Aarhus University Hospital, Aarhus, Denmark

Gregory W. Terman Department of Anesthesiology and the Graduate Program in Neurobiology and Behavior, University of Washington, Seattle, WA, USA

Alison Thomas Lismore, NSW, Australia

Benjamin Thomas Chelsea & Westminster Hospital Foundation Trust, London, UK

Jacob Thomas School of Medical Sciences, University of Adelaide, Adelaide, SA, Australia

Miles Thompson Goldsmiths, University of London, London, UK

Beverly E. Thorn Department of Psychology, University of Alabama, Tuscaloosa, AL, USA

Cindy L. Thurston-Stanfield Department of Biomedical Sciences, University of South Alabama, Mobile, AL, USA

Arne Tjølsen Department of Bomedicine, University of Bergen, Haukeland University Hospital, Bergen, Norway

Andrew J. Todd Institute of Neuroscience and Psychology, College of Medical, Veterinary and Life Sciences, University of Glasgow, Glasgow, UK

Makoto Tominaga Department of Cellular and Molecular Physiology, Mie University School of Medicine, Tsu Mie, Japan

Victor Tortorici Instituto Venezolano de Investigaciones Científicas (IVIC), Caracas, Venezuela

Irene Tracey Pain Imaging Neuroscience (PaIN) Group, Department of Physiology, Anatomy and Genetics, Oxford University, Oxford, UK

Diep Tuan Tran Department of Integrative Physiology, National Institute for Physiological Sciences, Okazaki, Japan

Rolf-Detlef Treede Institute of Physiology and Pathophysiology, Johannes Gutenberg University, Mainz, Germany

Jennie C. I. Tsao Pediatric Pain Program, Department of Pediatrics David Geffen School of Medicine at UCLA, Los Angeles, CA, USA

Lawrence Tsen Department of Anesthesiology, Perioperative and Pain Medicine, Brigham and Women's Hospital Harvard Medical School, Boston, MA, USA

Jeanna Tsenter Department of Physical Medicine and Rehabilitation, Hadassah University Hospital, Jerusalem, Israel

Maurits Van Tulder Institute for Research in Extramural Medicine, Free University Amsterdam, Amsterdam, The Netherlands

Eldon Tunks Hamilton Health Sciences Corporation, Hamilton, ON, Canada

Susan Tupper Pediatrics, University of Saskatchewan, Saskatoon, SK, Canada

Dennis C. Turk Department of Anesthesiology & Pain Research, University of Washington, Seattle, WA, USA

Judith A. Turner Department of Psychiatry and Behavioral Sciences, University of Washington, Seattle, WA, USA

Department of Rehabilitation Medicine, University of Washington, Seattle, WA, USA

Stephen P. Tyrer Royal Victoria Infirmary, University of Newcastle upon Tyne, Newcastle-on-Tyne, UK

Alexander Tzabazis Department of Anesthesia, Stanford University, Stanford, CA, USA

Kene Ugokwe Department of Neurosurgery, St Elizabeth Health Center, Youngstown, OH, USA

Lindsay S. Uman Department of Psychology, IWK Health Centre, Halifax, NS, Canada

Christiane Vahle-Hinz Department of Neurophysiology and Pathophysiology, University Medical Center Hamburg-Eppendorf, Hamburg, Germany

Muthukumar Vaidyaraman Division of Pain Medicine, The Johns Hopkins University, Baltimore, MD, USA

Todd W. Vanderah Department of Pharmacology, College of Medicine University of Arizona, Tucson, AZ, USA

Guy Vanderstraeten Department of Rehabilitation Sciences and Physiotherapy, Ghent University Hospital, Ghent, Belgium

Horacio Vanegas Instituto Venezolano de Investigaciones Cientificas (IVIC), Caracas, Venezuela

Jean-Jacques Vatine Outpatient and Research Division, Reuth Medical Center, Sackler Faculty of Medicine, Tel Aviv University, Tel Aviv, Israel

Leigh M. Vaughan Medical University of South Carolina, Charleston, SC, USA

Linda K. Vaughn Department of Biomedical Sciences, Marquette University, Milwaukee, WI, USA

Enrique Vazquez Instituto Venezolano de Investigaciones Cientificas (IVIC), Caracas, Venezuela

Louis Vera-Portocarrero Department of Pharmacology, College of Medicine University of Arizona, Tucson, AZ, USA

Paul Verrills Metropolitan Spinal Clinic, Prahran, VIC, Australia

Tine Vervoort Department of Experimental Clinical and Health Psychology, Ghent University, Ghent, Belgium

Félix Viana Instituto de Neurociencias, Universidad Miguel Hernández-CSIC, San Juan de Alicante, Alicante, Spain

Charles J. Vierck Jr. Department of Neuroscience and McKnight Brain Institute, University of Florida College of Medicine, Gainesville, FL, USA

Luis Villanueva INSERM UMR 894, Centre de Psychiatrie et Neurosciences, Paris, France

Marcelo Villar Faculty of Biomedical Sciences, Austral University, Buenos Aires, Argentina

Katy Vincent Nuffield Department of Obstetrics and Gynaecology, The Women's Centre, John Radcliffe Hospital, University of Oxford, Oxford, UK

David Vivian Metro Spinal Clinic, South Caulfield, VIC, Australia

Johan W. S. Vlaeyen Research Group Health Psychology, Catholic University of Leuven, Leuven, Belgium

Department of Clinical Psychological Science, Maastricht University, Maastricht, The Netherlands

Stéphanie Volders Research Group on Health Psychology, Leuven, Belgium

Ernest Volinn Pain Research Center, University of Utah, Salt Lake City, UT, USA

Lucy Vulchanova Department of Veterinary and Biomedical Sciences, University of Minnesota, St. Paul, MN, USA

Paul W. Wacnik Neuromodulation Research, Medtronic Inc., Minneapolis, MN, USA

Gordon Waddell University of Glasgow, Glasgow, UK

Gary A. Walco Department of Anesthesiology and Pain Medicine, University of Washington School of Medicine, Seattle Children's Hospital, Seattle, WA, USA

Suellen M. Walker University of Sydney Pain Management and Research Centre, Royal North Shore Hospital, NSW, Australia

Victoria C. J. Wallace Department of Anaesthetics, Pain Medicine and Intensive Care Chelsea and Westminster Hospital Campus, London, UK

Xiaohong Wang Department of Integrative Physiology, National Institute for Physiological Sciences, Okazaki, Japan

Gunnar Wasner Department of Neurological Pain Research and Therapy, Neurological Clinic, Kiel, Germany

Shoko Watanabe Department of Integrative Physiology, National Institute for Physiological Sciences, Okazaki, Japan

Linda R. Watkins Department of Psychology and Neuroscience, and The Center for Neuroscience, University of Colorado at Boulder, Boulder, CO, USA

Peter N. Watson Department of Medicine, University of Toronto, Toronto, ON, Canada

James Watt North Shore Hospital, Auckland, New Zealand

Stephen G. Waxman Department of Neurology, Yale School of Medicine, New Haven, CT, USA

Inge Wegner Julius Center, Department of Epidemiology, University Medical Center Utrecht, Utrecht, The Netherlands

Feng Wei Department of Neural and Pain Sciences, University of Maryland Dental School, Baltimore, MD, USA

Christian Weidner Department of Physiology and Experimental Pathophysiology, University Erlangen, Erlangen, Germany

P. T. M. Weijenborg Department of Gynecology, Leiden University Medical Center, Leiden, The Netherlands

Robin Weir School of Nursing, Faculty of Health Sciences, McMaster University, Hamilton, ON, Canada

Nirit Weiss Department of Neurosurgery, Mount Sinai Medical Center, New York, NY, USA

Ursula Wesselmann Departments of Neurology and Anesthesiology / Division of Pain Management, University of Alabama at Birmingham, Birmingham, AL, USA

Dagmar Westerling Department of Anesthesiology, Central Hospital of Kristianstad, Kristianstad, Sweden

Karin N. Westlund Department of Physiology, University of Kentucky, Lexington, KY, USA

Thomas M. Wickizer Health Services Management and Policy, Ohio State University, Colombus, OH, USA

Eva Widerström-Noga The Miami Project to Cure Paralysis, Department of Neurological Surgery, University of Miami Miller, School of Medicine, Miami, FL, USA

Julie Wieseler Department of Psychology and Neuroscience, and The Center for Neuroscience, University of Colorado at Boulder, Boulder, CO, USA

Zsuzsanna Wiesenfeld-Hallin Karolinska Institute, Stockholm, Sweden

Ann Wigham McCoull Clinic, Prudhoe Hospital, Northumberland, UK

George L. Wilcox Department of Neuroscience, Pharmacology and Dermatology, University of Minnesota Medical School, Minneapolis, MN, USA

Oliver H. G. Wilder-Smith Pain Centre, Department of Anaesthesiology, Radboud University Nijmegen Medical Centre, Nijmegen, The Netherlands

Kjersti Wilhelmsen Section for Physiotherapy Science, Department of Public Health and Primary Health, Faculty of Medicine, University of Bergen, Bergen, Norway

Victor Wilk Brighton Spinal Group, Brighton, VIC, Australia

W. A. Williams Royal Perth Hospital and University of Western Australia, Perth, WA, Australia

Sara E. Williams Division of Pediatric Gastroenterology, Medical College of Wisconsin, Milwaukee, WI, USA

William D. Willis Department of Neuroscience and Cell Biology, University of Texas Medical Branch, Galveston, TX, USA

John B. Winer Department of Neurology, Birmingham Muscle and Nerve Centre Queen Elizabeth Hospital, Birmingham, UK

Christopher J. Winfree Department of Neurological Surgery, Neurological Institute Columbia University, New York, NY, USA

Janet Winter Novartis Institute for Medical Science, London, UK

Alain Woda Faculté de Chirurgie Dentaire, University Clermont-Ferrand, Clermont-Ferrand, France

John N. Wood Molecular Nociception Group, Wolfson Institute for Biomedical Research, University College London, London, WC, UK

Lee Woodson Shriners Hospitals for Children, Shriners Burns Hospital, Galveston, TX, USA

Department of Anesthesia, University of Texas Medical Branch, Galveston, TX, USA

Anava A. Wren Pain Prevention and Treatment Research, Department of Psychiatry and Behavioral Sciences, Duke University Medical Center, Durham, NC, USA

Xiao-Jun Xu Department of Physiology and Pharmacology, Section of Integrative Pain Rsearch Karolinska Institutet, Stockholm, Sweden

Hiroshi Yamasaki Department of Integrative Physiology, National Institute for Physiological Sciences, Okazaki, Japan

Michael Yelland School of Medicine, Logan campus, Griffith University, Logan, Australia

Sriram Yennurajalingam Department of Palliative Care and Rehabilitation Medicine, The University of Texas M. D. Anderson Cancer Center, Houston, TX, USA

David C. Yeomans Department of Anesthesia, Stanford University, Stanford, CA, USA

Robert P. Yezierski Comprehensive Center for Pain Research and Department of Neuroscience, The McKnight Brain Institute University of Florida, Gainesville, FL, USA

Way Yin Department of Anesthesiology, University of Washington, Seattle, WA, USA

Atsushi Yoshida Department of Oral Anatomy and Neurobiology, Osaka University, Osaka, Japan

Joanna M. Zakrzewska Eastman Dental Hospital UCLH NHS Foundation Trust, London, UK

Jenna E. Zauk St. George's University, School of Medicine, Grenada, West Indies

Hanns Ulrich Zeilhofer Institute for Pharmacology and Toxicology, University of Zurich, and Swiss Federal Institute of Technology (ETH) Zurich, Zürich, Switzerland

Lonnie K. Zeltzer Pediatric Pain Program, Department of Pediatrics David Geffen School of Medicine at UCLA, Los Angeles, CA, USA

Giovambattista Zeppetella St. Clare Hospice, Hastingwood, UK

Xijing Zhang Department of Neuroscience, University of Minnesota, Minneapolis, MN, USA

Min Zhuo Department of Physiology, University of Toronto, Toronto, ON, Canada

C. Zöllner Department of Anesthesiology and Intensive Care Medicine, Charité University, Berlin, Germany

Krina Zondervan Genetic and Genomic Epidemiology Unit, Nuffield Dept of Obstetrics and Gynaecology, Wellcome Trust Centre for Human Genetics, Oxford, UK

Peter Zygmunt Department of Laboratory Medicine, Lund University, Division of Clinical Chemistry and Pharmacology, Lund, Sweden

Zbigniew Zylicz Hildegard Hospiz, Basel Switzerland, Basel, Switzerland

David C. Yeomans Department of Anesthesia, Stanford University, Stanford, CA, USA

Robert F. Yezierski Comprehensive Center for Pain Research and Department of Neurosciences, The McKnight Brain Institute University of Florida, Gainesville, FL, USA

Mary Lou Department of Anesthesiology, University of Washington, Seattle, WA, USA

Atsushi Yasuda Department of Oral Anatomy and Neurobiology, Osaka University, Osaka, Japan

Joanna M. Zakrzewska Eastman Dental Hospital UCLH NHS Foundation Trust, London, UK

Jenna K. Zaak St. George's University School of Medicine, Grenada, West Indies

Hans Ulrich Zeilhofer Institute of Pharmacology and Toxicology, University of Zurich and Swiss Federal Institute of Technology (ETH) Zurich, Zurich, Switzerland

Luping Zhou Arthur Fulrow Department of Pediatrics David Geffen School of Medicine at UCLA, Los Angeles, CA, USA

Demetrio Zippolla St. Clare's Hospice, Hastingwood, UK

Xijing Zhang Department of Neuroscience, University of Minnesota, Minneapolis, MN, USA

Min Zhuo Department of Physiology, University of Toronto, Toronto, ON, Canada

Cristina Department of Anesthesiology and Intensive Care, Charité Universität, Berlin, Germany

Selina Zondervan Oxford University Hospitals NHS Trust, Nuffield Department of Anaesthetics and Oxford Centre for Functional Magnetic Resonance Imaging of the Brain, Oxford, UK

Peter Zygmunt Department of Laboratory Medicine, Lund University, Division of Clinical Chemistry and Pharmacology, Lund, Sweden

Magnus Zyber University Hospital, Basel Switzerland, Basel, Switzerland

T

T Cells and HIV Neuropathy

Ling Cao
Department of Biomedical Sciences, College of
Osteopathic Medicine, University of
New England, ME, USA

Introduction

HIV infection is now considered to be a chronic
disease due to more effective viral suppressive
treatments significantly prolonging survival after
infection. At the end of 2007, it was estimated
that 33 million people worldwide (*The 2007
United Nations AIDS Epidemic Update Report*)
and 600,000 residents of 37 states in the United
States (*Centers for Disease Control and Prevention. HIV/AIDS Surveillance Report 2008*) were
positive for HIV. As such, greater attention is
now being focused on the treatment of the numerous HIV infection-related complications, including neurological disorders, opportunistic
infections, and tumors.

HIV-associated peripheral neuropathies are
the most common neurological disorders associated with HIV infection (Phillips et al. 2010).
Several forms of HIV-associated peripheral
neuropathies have been characterized, including
HIV-associated distal symmetrical
polyneuropathy (DSP), mononeuropathy, inflammatory demyelinating polyneuropathy, and
progressive polyradiculopathy. DSP is the most
frequent neurological complication related to

HIV infection and occurs in about one-third of
patients with HIV infection. HIV DSP can be
caused by pathological changes following HIV
infection, selected antiretroviral agents, or the
combination of these two. Thus, HIV DSP does
not always correlate with neurotoxic therapy
(Schifitto et al. 2002; Morgello et al. 2004). HIV
DSP is often diagnosed based on the presentation
of neuropathic signs, including reduced or absent
ankle reflexes, diminished vibration sense, and
altered sensitivity to light tactile or thermal stimuli. Patients with symptoms often experience the
typical symptoms of neuropathic pain, such as
burning sensation, pins and needles sensation,
and numbness (Phillips et al. 2010). It is estimated that HIV-associated neuropathic pain
accounts for 25–50 % of all pain clinic visits in
the United States (Verma et al. 2005). In general,
the symptoms of DSP are more pronounced in the
distal lower extremities and are symmetrically
distributed, while the upper extremities are
often affected later in the course of DSP. In correlation with this clinical course, the neuropathological changes in DSP have been described to
progress in a "dying back" pattern as they usually
affect the distal region of the nerve fibers first and
progress proximally (Pardo et al. 2001;
Polydefkis et al. 2002). The major pathological
changes associated with HIV DSP include
reduced intraepidermal nerve fiber density, damage to primary sensory fibers, including axonal
degeneration and frank nerve fiber loss (mainly
the unmyelinated C-fibers), and macrophage
infiltration into peripheral nerves and the dorsal

G.F. Gebhart, R.F. Schmidt (eds.), *Encyclopedia of Pain*, DOI 10.1007/978-3-642-28753-4,
© Springer-Verlag Berlin Heidelberg 2013

root ganglia (DRG). However, the underlying mechanism of the pathogenesis of HIV DSP is largely unknown. As such, HIV DSP is often underdiagnosed and/or undertreated, and there are no US FDA-approved pharmacological agents specifically designed for the treatment of HIV DSP.

Definition

Neurological disorders, including HIV-1 encephalopathy, HIV peripheral neuropathy, and HIV myelopathy, are the complications most commonly associated with chronic human immunodeficiency virus (HIV)-1 infection and the subsequently developed acquired immunodeficiency syndrome (AIDS). Of these, HIV peripheral neuropathy has the highest prevalence and is often associated with pain. Immune cells (including T lymphocytes) and associated factors have been recognized to play significant roles in the development of neuropathic pain (pain caused by a lesion or disease of the somatosensory system (Jensen et al. 2011)). As the major target of HIV-1, $CD4^+$ T cells can be critical in the development of HIV peripheral neuropathy.

Characteristics

Peripheral $CD4^+$ T Lymphocytes and HIV Neuropathy

HIV-1 is a RNA retrovirus that specifically targets $CD4^+$ cells, which results in a decrease in peripheral $CD4^+$ T lymphocytes, decreased immunity, and the development of the AIDS. It has been shown that DSP is often correlated with low $CD4^+$ T lymphocyte count and high plasma HIV loads prior to the introduction of highly active antiretroviral therapy (HAART). However, the associations between HIV DSP and high viral load and low CD4 cell count have become less clear since the use of HAART has become standard and the recognition that certain types of antiretroviral drugs are neurotoxic (Schifitto et al. 2002; Phillips et al. 2010). One study showed that initiation of HAART at CD4

cell counts $>/=350$ cells/mm^3 reduced the incidence and risk of peripheral neuropathy (Lichtenstein et al. 2008). Whether the presence of sufficient numbers of $CD4^+$ T lymphocytes directly contributes to the reduced risk of HIV peripheral neuropathy is not clear. It is possible that the correlation between low $CD4^+$ T lymphocyte count and the high incidence of HIV DSP is simply due to low $CD4^+$ T lymphocyte count being one of the signs of advanced HIV infection and HIV DSP generally thought being associated with advanced HIV infection.

Infiltrating T Lymphocytes and HIV Neuropathy

It has been extensively reviewed that peripheral T lymphocytes can infiltrate into the spinal cord and contribute to the development and progression of neuropathic pain behaviors in various rodent models (Grace et al. 2011). Animals with no T lymphocytes (nude mice and rats) displayed significantly reduced pain behavior after nerve injury. $CD4^+$ cell-deficient mice exhibit reduced mechanical hypersensitivity during the maintenance (but not the induction) phase of peripheral nerve injury-induced neuropathic pain. It has been proposed that the interaction between central nervous system (CNS)-infiltrating T lymphocytes, particularly $CD4^+$ T lymphocytes, and the subsequent interactions between these infiltrating cells and resident glial cells, are critical in T lymphocyte-mediated behavioral hypersensitivity, a behavioral sign of neuropathic pain. Several observed changes within the CNS following peripheral nerve injury are thought to promote T lymphocyte infiltration and interaction with CNS resident cells: (1) increased permeability of the blood brain barrier (BBB) and blood spinal cord barrier, thus providing more opportunity for T lymphocytes to enter the CNS; (2) recruitment of antigen presenting cells within the CNS, such as activated microglia, that may interact with infiltrating T lymphocytes through surface molecules, including major histocompatibility complex (MHC) II, CD40, and B7; and (3) elevated CNS production of chemokines and proinflammatory factors that can further attract

T lymphocytes to infiltrate into the CNS. Specific subtypes of CD4$^+$ T lymphocytes, including type 1 and type 17 helper T cells (Th1 cells and Th17 cells, respectively), have been implicated in T lymphocyte-mediated neuropathic pain behaviors. Key cytokines produced by these cells (interferon (IFN)-gamma from Th1 cells and interleukin (IL)-17 from Th17 cells) have been found to be critical in the development of peripheral nerve injury-induced behavioral hypersensitivity. Conversely, type 2 helper T cells (Th2 cells) have been shown to reduce peripheral nerve injury-induced behavioral hypersensitivity.

During HIV infection, despite the development of immunodeficiency in the periphery due to the viral infection of CD4$^+$ cells, there is significant immune activation/inflammation in both the CNS and peripheral nerve tissue. Significant macrophage infiltration, activation, and secretion of a wide range of cytokines and other biologically reactive substances have been detected in both the CNS and the peripheral nervous system. Similar phenomena have been observed in several animal models. In murine models of HIV peripheral neuropathy that are induced by injecting the viral envelope protein gp120 either intrathecally (around the spinal cord) or perineurally (around the peripheral nerve), significant activation of spinal cord astrocytes and microglia, a sign of central immune activation, was observed, and this activation was shown to contribute to behavioral hypersensitivity (Milligan et al. 2000; Herzberg and Sagen 2001; Minami et al. 2003; Wallace et al. 2007). In a nonhuman primate model of HIV-sensory neuropathy (simian immunodeficiency virus (SIV)-infected rhesus macaque monkeys with CD8$^+$ T lymphocyte depletion), the severity of the tissue damage in the dorsal root ganglia was found to be positively correlated with the degree of mononuclear cell infiltration (sign of inflammation). Susceptible strains of mice infected with a murine retrovirus, LP-BM5, develop an HIV-like immunodeficiency syndrome (murine AIDS). Long-lasting glial responses have been detected as early as 3 weeks post-LP-BM5 infection, and increased CNS levels of proinflammatory and neurotoxic factors, such as

tumor necrosis factor (TNF)-alpha, that potentially could have been produced by glial cells are also found after LP-BM5 infection (Sei et al. 1998). Further, it has long been known that HIV infection can cause dysregulation and increased permeability of the BBB, which has been thought to contribute to the infiltration of peripheral leukocytes and viral particles into the CNS. An alteration in the BBB has been linked to a number of neurological symptoms associated with HIV infection (Strazza et al. 2011).

To date, most studies have focused on the role of infected, activated macrophages and various circulating HIV proteins (such as Tat and gp120) in immune activation/dysregulation in the CNS and their contribution to HIV-associated neurological diseases, including peripheral neuropathy. Similar to peripheral nerve injury-induced neuropathies, activated CD4$^+$ T lymphocytes may infiltrate into the CNS and interact with CNS resident cells, furthering the development of HIV peripheral neuropathy. CNS-infiltrating CD4$^+$ T lymphocytes have been observed during HIV infection, and this infiltration was suggested as one of the HIV reservoirs within the CNS that contributes to the HIV virus being detected in the cerebral fluid (CSF) (McGee et al. 2006). Increases in CD4$^+$ and CD8$^+$ T lymphocytes and IFN-gamma have been observed in the brains of SIV-infected macaques following HAART treatment (Graham et al. 2011). It is most likely that infiltrating CD4$^+$ T lymphocytes would be involved in the maintenance rather than the initiation of HIV peripheral neuropathy as what is observed in peripheral nerve injury-induced neuropathies. Further investigation of the role of infiltrating T lymphocytes in HIV peripheral neuropathy is particularly important because normal CD4$^+$ T lymphocyte counts have been observed in HIV-infected patients with neurological disorders, including peripheral neuropathy.

Summary

In summary, HIV peripheral neuropathy is the most common complication associated with HIV infection. Prior to the introduction of

HAART, a low CD4$^+$ T lymphocyte count was often correlated with the development of HIV peripheral neuropathy. However, during the era of HAART, it is not uncommon for a patient to have HIV peripheral neuropathy without a low CD4$^+$ T lymphocyte count. HIV infection causes significant immune activation/dysregulation in the nervous system that can contribute to the development of HIV peripheral neuropathy. There is a need for further investigation on whether and how CNS-infiltrating CD4$^+$ T lymphocytes contribute to the development of HIV peripheral neuropathy.

References

Grace, P. M., Rolan, P. E., et al. (2011). Peripheral immune contributions to the maintenance of central glial activation underlying neuropathic pain. *Brain, Behavior, and Immunity, 25*(7), 1322–1332.

Graham, D. R., Gama, L., et al. (2011). Initiation of HAART during acute simian immunodeficiency virus infection rapidly controls virus replication in the CNS by enhancing immune activity and preserving protective immune responses. *Journal of Neurovirology, 17*(1), 120–130.

Herzberg, U., & Sagen, J. (2001). Peripheral nerve exposure to HIV viral envelope protein gp120 induces neuropathic pain and spinal gliosis. *Journal of Neuroimmunology, 116*(1), 29–39.

Jensen, T. S., Baron, R., et al. (2011). A new definition of neuropathic pain. *Pain, 152*(10), 2204–2205.

Lichtenstein, K. A., Armon, C., et al. (2008). Initiation of antiretroviral therapy at CD4 cell counts >/=350 cells/mm^3 does not increase incidence or risk of peripheral neuropathy, anemia, or renal insufficiency. *Journal of Acquired Immune Deficiency Syndromes, 47*(1), 27–35.

McGee, B., Smith, N., et al. (2006). HIV pharmacology: Barriers to the eradication of HIV from the CNS. *HIV Clinical Trials, 7*(3), 142–153.

Milligan, E. D., Mehmert, K. K., et al. (2000). Thermal hyperalgesia and mechanical allodynia produced by intrathecal administration of the human immunodeficiency virus-1 (HIV-1) envelope glycoprotein, gp120. *Brain Research, 861*(1), 105–116.

Minami, T., Matsumura, S., et al. (2003). Functional evidence for interaction between prostaglandin EP3 and [kappa]-opioid receptor pathways in tactile pain induced by human immunodeficiency virus type-1 (HIV-1) glycoprotein gp120. *Neuropharmacology, 45*(1), 96–105.

Morgello, S., Estanislao, L., et al. (2004). HIV-associated distal sensory polyneuropathy in the era of highly active antiretroviral therapy: The Manhattan HIV Brain Bank. *Archives of Neurology, 61*(4), 546–551.

Pardo, C. A., McArthur, J. C., et al. (2001). HIV neuropathy: Insights in the pathology of HIV peripheral nerve disease. *Journal of the Peripheral Nervous System, 6*(1), 21–27.

Phillips, T. J. C., Cherry, C. L., et al. (2010). Painful HIV-associated sensory neuropathy. *The Pain: Clinical Updates, XVIII*(3), P1–8.

Polydefkis, M., Yiannoutsos, C. T., et al. (2002). Reduced intraepidermal nerve fiber density in HIV-associated sensory neuropathy. *Neurology, 58*(1), 115–119.

Schifitto, G., McDermott, M. P., et al. (2002). Incidence of and risk factors for HIV-associated distal sensory polyneuropathy. *Neurology, 58*(12), 1764–1768.

Sei, Y., Kustova, Y., et al. (1998). The encephalopathy associated with murine acquired immunodeficiency syndrome. *Annals of the New York Academy of Sciences, 840*, 822–834.

Strazza, M., Pirrone, V., et al. (2011). Breaking down the barrier: The effects of HIV-1 on the blood–brain barrier. *Brain Research, 1399*, 96–115.

Verma, S., Estanislao, L., et al. (2005). HIV-associated neuropathic pain: Epidemiology, pathophysiology and management. *CNS Drugs, 19*(4), 325–334.

Wallace, V. C., Blackbeard, J., et al. (2007). Pharmacological, behavioural and mechanistic analysis of HIV-1 gp120 induced painful neuropathy. *Pain, 133*(1–3), 47–63.

Tabes Dorsalis

Definition

Tabes dorsalis is a late complication of neurosyphilis. It results in gait impairment, joint deformity, pain, lack of coordination, sensory loss, as well as autonomic dysfunction and ocular symptoms. Pain is felt mostly in the legs and is described as lightning and lancinating. Abdominal colicky pain is also reported.

Cross-References

▶ Central Nervous System Stimulation for Pain

Tachykinin

Definition

Tachykinins are a family of structurally related peptides, widely scattered in vertebrate and invertebrate tissues. Mammalian tachykinins are substance P (SP), neurokinin A (NKA), and neurokinin B (NKB). All mammalian tachykinins share a common C-terminal amino acid sequence, i.e., Phe-x-Gly-Leu- MetNH, which is the minimal structural motif for the activation of tachykinin receptors (NK1, NK2, and NK3). Pharmacologically, they all cause hypotension in mammals, contraction of gut and bladder smooth muscle, and secretion of saliva.

Cross-References

▶ NGF, Regulation During Inflammation
▶ Visceral Nociception and Pain

Tachyphylaxis

▶ Nociceptor, Fatigue

Tactile Allodynia

Definition

See also "Allodynia." A better term may be "touch-evoked pain."

Cross-References

▶ Opioids in the Spinal Cord and Modulation of Ascending Pathways (N. gracilis)

Tactile Allodynia Test

Definition

In human subjects, a typical test consists in stroking the skin with a swab of cotton wool. Though allodynia, or "tactile hyperalgesia," can also be tested with non-moving stimuli, for example,

Von Frey filaments, it is best identified with stimuli slowly moving over hyperalgesic skin.

In animal experiments, the term "allodynia" is generally used more loosely, and no distinction is made between reactions due to sensitized nociceptors (hyperalgesia) and those due to mechanoreceptors (allodynia). Here a typical test procedure is described: The plantar aspects of intact and neuropathic legs of rats are probed with Von Frey hairs of different calibers or strengths. The number of paw withdrawals per 10 trials is counted. In general, a number of 5/10 withdrawals is observed with hairs of strength $\geq$200 mN in rats with intact legs. In rats with mononeuropathy, a hair of 20 mN can produce a score of $\geq$5/10 withdrawals.

Cross-References

▶ Thalamotomy, Pain Behavior in Animals

Tactile Stimuli

Definition

Stimuli of light touch applied to the skin.

Cross-References

▶ Causalgia: Assessment
▶ Dysesthesia, Assessment

Tail Immersion

Definition

Submersion of the tail in hot water may be used as the nociceptive stimulus in the tail-flick test.

Cross-References

▶ Tail-Flick Test

Tail Skin Temperature Recording

Definition

Thermocouples, thermistors, or infrared sensors may be used.

Cross-References

▶ Tail-Flick Test

Tail-Flick Latency

Definition

Tail-flick latency is the time from the start of the noxious stimulus until the animal flicks its tail.

Cross-References

▶ Tail-Flick Test

Tail-Flick Latency Correction

Definition

The tail-flick latency may be corrected for influences of tail skin temperature using a regression analysis or analysis of covariance. Linearity of the data can be assumed only for a limited range of skin temperatures. In some experiments preheating of the tail to a certain temperature may be used.

Cross-References

▶ Tail-Flick Test

Tail-Flick Test

Arne Tjølsen[1] and Kjell Hole[2]
[1]Department of Bomedicine, University of Bergen, Haukeland University Hospital, Bergen, Norway
[2]University of Bergen, Bergen, Norway

Definition

The tail-flick test is a test of nociception used in rats and mice. The noxious stimulus is usually ▶ radiant heat on the tail or ▶ tail immersion in hot water, and the response is a flick of the tail. The outcome is the latency from the stimulus onset to the tail flick.

Introduction

The tail-flick test is an extensively used test of nociception in rats and mice (Le Bars et al. 2001), first described in 1941 (D'Amour and Smith 1941). In the standard method, radiant heat is focused on the tail, and the time it takes until the animal flicks the tail away from the beam is measured. This ▶ tail-flick latency has been regarded a measure of the nociceptive sensitivity of the animal and is prolonged for instance by opioid analgesics. A spinal transection above the lumbar level does not block the tail-flick response. Thus, the test relies on a spinal nociceptive reflex, and pain is not measured directly. Still, this is considered a very useful test of a component of "phasic pain," both in basic pain research and in pharmacological investigations of analgesic drugs. The relevance of the test as a measure of pain has been discussed (Le Bars et al. 2001).

Characteristics

The test stimulus is noxious heat. The stimulus may be radiant heat, e.g., focused light from a light bulb. There are good reasons not to use an incandescent light bulb but instead to employ a laser (Benoist et al. 2008; Pincedé et al. 2012). The stimulus may also be applied by, e.g., direct contact with a heated surface, such as a Peltier element, or by submersion of part of the tail in hot water. The test may be performed in lightly anesthetized rats or mice, as well as in animals that are awake.

Several tail-flick apparatuses are commercially available, and many laboratories have made their own equipment. The main requirements are stable functioning and proper focus of the light beam on the tail. The tail-flick latency may be recorded by means of a photocell that is activated when the animal flicks the tail. When a photocell is used, one should be aware that the reflex response may involve retracting the tail, without immediately removing the tail from the light beam. Thus the rats' behavior should always be observed.

The tail-flick test may be considered a useful test of nociception but only if it is carefully performed and possible sources of error are taken into account. One requirement, particularly in rats, is that the animals are well handled. This may require daily handling for up to a week, including adaptation to the test apparatus. Some researchers confine the animals in a plastic tube during testing. If this is used, it is necessary that the animals are so well handled that they freely walk in and out of the tube. We find it better and faster not to use a tube but to hold the well-adapted animal by hand.

A particular problem with tests that use thermal stimulation is the possible confounding influence of the skin temperature. In electrophysiological studies in animals, it has been reported that changes in the temperature or blood flow of the skin (Duggan et al. 1978) alter the response to cutaneous heat stimulation. More recently, it has been found that the tail skin temperature affects the tail-flick latency as well. This has been described using radiant heat stimulation (Ren and Han 1979; Berge et al. 1988; Roane et al. 1998; Sawamura et al. 2002) as well as with hot water immersion of the tail (Milne and Gamble 1989). For an extensive review, see Le Bars et al. (2001). The psychophysical aspects of the test have been thoroughly investigated both in

rats (Benoist et al. 2008) and in mice (Pincedé et al. 2012). Also in these studies it was clearly shown that spontaneous variations of the temperature of the tail for thermoregulation directly caused variations of the tail-flick latency. However, in many laboratories the tail-flick test is still performed without taking the tail skin temperature into account. This is probably a main confounding factor and therefore needs special consideration in the following.

We have investigated the relationship between skin surface temperature and tail-flick latency in rats in a setup with a radiant heat apparatus, stimulating the distal part of the tail (10–15 mm from the tip), with a stimulated area of 15–20 mm^2 (Fig. 1). Using control latencies of approximately 4 s, we regularly find a clear and reproducible relationship between tail skin temperature and tail-flick latency, with a slope of the regression equation of −0.3 to 0.4 s/°C (Tjølsen et al. 1989). With similar methodology, the same relationship has been found in mice, with a very similar slope (Eide et al. 1988).

The tail is the most important thermoregulatory organ of the rat. The heat loss is regulated by an on-off regulation of blood flow in the tail, which leads to rapid variations in skin temperature (Milne and Gamble 1989; Tjølsen and Hole 1992; Benoist et al. 2008). The amount and duration of vasodilation is partly determined by the relationship between the ambient temperature and the acclimatization temperature. In rats at rest, the ambient temperature where vasodilation occurs is lower after acclimatization to cold, than after acclimatization to a warmer environment. When animals are lightly stressed and activated due to experimental procedures, a considerable increase in tail skin temperature is regularly observed (Tjølsen et al. 1989). Rats restrained in tubes for a short time may show a considerable increase in the temperature of the tail (Tjølsen and Hole 1992), probably due to vasodilation.

The relationship between skin temperature and response latency (Fig. 2) would be expected to vary with different experimental conditions. The most reliable values for the slope are obtained in experiments where data from repeated measures are not pooled but analyzed

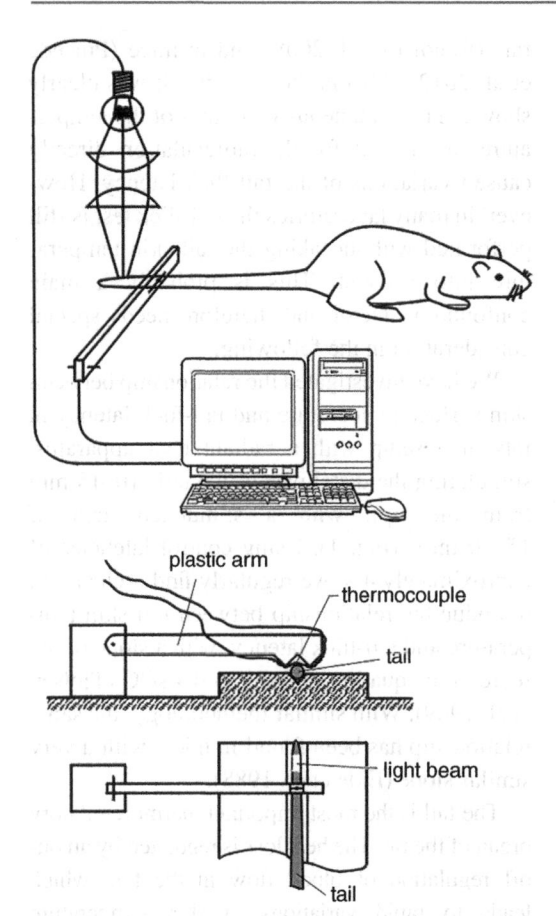

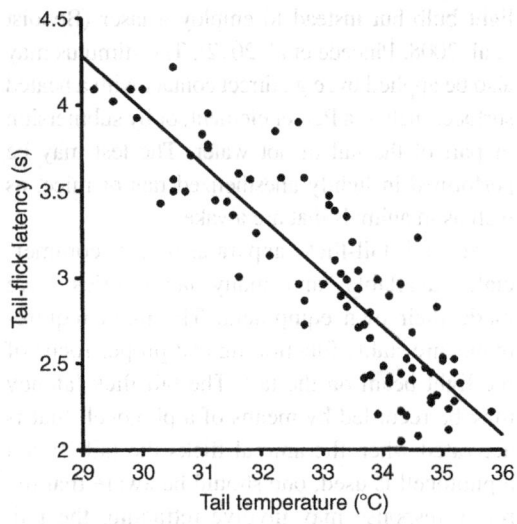

Tail-Flick Test, Fig. 2 The relationship between tail-flick latency and tail skin temperature. Data were obtained from eight measurements in each of 12 rats. Tail skin temperature was controlled by means of a heating blanket (Adapted from Sawamura et al. (2002), with permission)

Tail-Flick Test, Fig. 1 Simple test equipment for concomitant recording of tail skin temperatures and tail-flick latencies. A standard tail-flick apparatus can easily be modified to enable recording of tail skin temperatures. The temperature is measured by means of a small thermocouple mounted on a plastic arm, 65 mm long, which rests on the tail with a force corresponding to approximately 1 g. For a thorough description, see Tjølsen et al. (1989)

separately for each time point (Tjølsen et al. 1989). In fact, if repeated measures on the same animals are pooled in a regression analysis, an error in the calculated slope may be introduced. A possible cause of error is the effect of repeated testing on nociception itself, whether due to stress or to local effects in the skin if the same site is stimulated repeatedly. The time required for heating the tissue to a critical response temperature will depend on the initial skin temperature, which is determined by local blood flow within the limits given by deep body and ambient

temperatures. Measuring subcutaneous tissue temperatures during a radiant heat stimulus, we found that the rate of increase in tissue temperature was independent of initial skin temperature, and the time required to reach a hypothetical threshold temperature was strongly dependent on the initial temperature (Hole and Tjølsen 1993). As a consequence, the tail-flick latency is negatively correlated to the ambient temperature (Berge et al. 1988) and to skin temperature when the heating intensity is kept constant.

The temperature of the tail skin of rats during an experiment may vary as much as 8 °C in untreated animals (Tjølsen and Hole 1992; Benoist et al. 2008). It is reasonable to consider this is the maximal possible difference in skin temperature due to changes in vasodilation. With a change in tail-flick latency of 0.3–0.4 s/°C, it would imply a potential difference in tail-flick latency of up to approximately 3 s. In a group of rats, not all animals would show this degree of vasodilation, and hence, the mean difference in latency would be somewhat smaller. However, this shows that increased vasoconstriction or inhibition of vasodilation may cause differences in tail-flick latency that easily could be misinterpreted as analgesia.

The potential for treatment-induced vasodilation to cause reduction of the tail-flick latencies is approximately the same size. Under circumstances when control animals are relatively vasoconstricted, vasodilation may lead to an increase in tail skin temperature from about ambient temperature to above 30 °C. The effect of vasodilation is particularly important, as smaller changes in the tail-flick latency are required to interpret the results as hyperalgesia than as analgesia. Even a modest increase in the mean tail skin temperature of about 3.5 °C, due to lesioning of descending serotonergic systems, leads to a reduction of the tail-flick latency from 4–4.5 to about 3 s (Hole and Tjølsen 1993). If the change in skin temperature were not taken into consideration, a reduction of the tail-flick latency of this size would have been considered an indication of a hyperalgesic state.

Many experimental treatments affect blood flow and thereby the tail skin temperature. This may by itself influence the tail-flick latency and lead to erroneous conclusions with regard to nociception. An increase in tail skin temperature may shorten tail-flick latencies and may be interpreted as hyperalgesia (Urban and Smith 1994; Roane et al. 1998; Sawamura et al. 2002). Even a reduction in tail skin temperature compared to untreated animals may occur and may be interpreted as analgesia. Desipramine reduced tail skin temperature and increased tail-flick latencies at an ambient temperature of 24–25 °C, while no significant change was observed at 21–22 °C (Hole and Tjølsen 1993). This difference in temperature is well within the variation in ambient temperature between laboratories and even within the range of ambient temperatures that may occur in a laboratory with insufficient control of room temperature. In the experiments at 24–25 °C, desipramine inhibited vasodilation so that the skin temperatures in the drug-treated group were close to the ambient temperature, while control animals showed higher skin temperatures and hence shorter response latencies.

Stress, due to a new environment, handling, or injection procedures, may influence peripheral blood flow and tail temperature. In rats, stress causes motor activation, increased heat production, increased core temperature and an increased frequency and duration of the periods of vasodilation, and increase in skin temperature of the tail. It has been shown (Tjølsen and Hole 1992) that immobilization may cause a considerable increase in core temperature and tail vasodilation, while small doses of morphine (0.5–1 mg/kg) completely abolish the vasodilation.

The importance of the skin temperature for the ordinary use of the tail-flick test has been discussed (Roane et al. 1998). Clearly, when high doses of potent analgesics like opioids are used, the relative influence of the skin temperature may be small. However, when the temperature influence is not known, this will always be an unpredictable confounding factor. As discussed above, this has, in several instances, lead to erroneous conclusions.

Possible Remedies

The temperature of the tail skin should always be considered a possible confounding factor when performing the tail-flick test. A minimal requirement should be that the tail skin temperature is measured before testing, e.g., by means of thermocouples (Fig. 1), thermistors or infrared sensors, and the possible influence of the temperature evaluated.

It is obviously necessary to take the effects of skin temperature into account when investigating factors or using drugs that may influence autonomic activity and thermo- or cardiovascular regulation. Recording the tail skin temperature and correcting the tail-flick latency data for changes in the temperature may reduce the problem. In some cases, a regression analysis or an analysis of covariance may be performed for this purpose. Methods for tail-flick testing with measurement of skin temperature and for correction of tail-flick data (see ► tail-flick latency correction) have been described (Tjølsen et al. 1989; Roane et al. 1998; Sawamura et al. 2002). However, there are some limitations when using this type of statistical analysis on tail-flick data.

In these statistical methods, linearity of the relationship is supposed. It seems to be a reasonable approximation to suppose linearity over a normal, limited range of skin temperatures in untreated animals, e.g., 20–30 °C. In studies where drug administration causes a large increase in tail-flick latency due to changes in nociception, the assumption of linearity may not be correct. Above all, these methods for statistical evaluation cannot adequately handle cut-off values for tail-flick latencies. This should be considered in each experiment, and even when limitations as above are applicable, the temperature of the skin of the tail should be measured and the possible influence on the results should be evaluated.

As a number of factors may possibly influence the relationship between skin temperature and response latency, it seems ideal to adjust data from one experiment according to the regression slope calculated from that experiment. However, this will not always be possible, in that a regression analysis requires an adequate number of measurements to allow calculation of a reliable regression coefficient, and the spread of the independent variable (skin temperature) must be sufficiently large. If these requirements are not fulfilled, the results of the regression analysis will be inconclusive. With an increasing number of measurements in the analysis, there is an increasing probability that a reliable regression coefficient may be calculated. In many cases, it may be a problem to obtain a reliable correction of tail-flick data based on the same experiment, due to a limited number of animals measured. An alternative method for correction of latencies is to establish the relationship between skin temperature and tail-flick latency in an adequate number of animals under similar experimental conditions and subsequently to correct tail-flick latencies according to the calculated regression factor (Ren and Han 1979). This method should be used with caution, because it must be assumed that the experiment is performed under the same conditions as when the correction factor was determined. This is of course an approximation.

Another alternative that has been used is local preheating of the tail to a certain temperature before measuring the tail-flick latency. If the temperatures of the skin and subcutaneous tissue in the stimulated area are constant before the start of stimulation, this may abolish the confounding effect of varying tissue temperatures. In electrophysiological experiments in anesthetized cats, preheating has been used with a heating lamp and a feedback control system, with a thermocouple on the area of skin to be heated (Duggan et al. 1978). This procedure seemed to reduce the confounding effect of differences in blood flow. For tail-flick reflex recordings, this technique may be used in experiments in lightly anesthetized animals for instance (Haws et al. 1990) or when the rat is placed in a restrainer and the tail is fixed (Carstens and Douglass 1995). It may probably be more difficult to use this method in animals that are awake when little restraint of the animal is required to minimize stress.

Summary

The tail-flick test is a potentially useful test of nociception in rodents. The outcome is significantly dependent of the temperature of the skin and the subcutaneous tissue at the point of stimulation, and this has to be taken into account when the test is employed.

References

Benoist, J.-M., Pincedé, I., Ballantyne, K., Plaghki, L., & Le Bars, D. (2008). Peripheral and central determinants of a nociceptive reaction: An approach to psychophysics in the rat. *PLoS One, 3*, e3125.

Berge, O.-G., Garcia-Cabrera, I., & Hole, K. (1988). Response latencies in the tail-flick test depend on tail skin temperature. *Neuroscience Letters, 86*, 284–288.

Carstens, E., & Douglass, D. K. (1995). Midbrain suppression of limb withdrawal and tail-flick reflexes in the rat: Correlates with descending inhibition of sacral spinal neurons. *Journal of Neurophysiology, 73*, 2179–2194.

D'Amour, F. E., & Smith, D. L. (1941). A method for determining loss of pain sensation. *Journal of Pharmacology and Experimental Therapeutics, 72*, 74–79.

Duggan, A. W., Griersmith, B. T., Headley, P. M., & Maher, J. B. (1978). The need to control skin temperature when using radiant heat in tests of analgesia. *Experimental Neurology, 61*, 471–478.

Eide, P. K., Berge, O.-G., Tjølsen, A., & Hole, K. (1988). Apparent hyperalgesia in the mouse tail-flick test due to increased tail skin temperature after lesioning of serotonergic pathways. *Acta Physiologica Scandinavica, 134*, 413–420.

Haws, C. M., Heinricher, M. M., & Fields, H. L. (1990). Î±−adrenergic receptor agonists, but not antagonists, alter the tail-flick latency when microinjected into the rostral ventromedial medulla of the lightly anesthetized rat. *Brain Research, 533*, 192–195.

Hole, K., & Tjølsen, A. (1993). The tail-flick and formalin tests in rodents: Changes in skin temperature as a confounding factor. *Pain, 53*, 247–254.

Le Bars, D., Gozariu, M., & Cadden, S. W. (2001). Animal models of nociception. *Pharmacological Reviews, 53*, 597–652.

Milne, R. J., & Gamble, G. D. (1989). Habituation to sham testing procedures modifies tail-flick latencies: Effects on nociception rather than vasomotor tone. *Pain, 39*, 103–107.

Pincedé, I., Pollin, B., Meert, T., Plaghki, L., & Le Bars, D. (2012). Psychophysics of a nociceptive test in the mouse: Ambient temperature as a key factor for variation. *PLoS One, 7*, e36699.

Ren, M. F., & Han, J. S. (1979). Rat tail-flick acupuncture analgesia model. *Chinese Medical Journal, 92*, 576–582.

Roane, D. S., Bounds, J. K., Ang, C.-Y., & Adloo, A. A. (1998). Quinpirole-induced alterations of tail temperature appear as hyperalgesia in the radiant heat tail-flick test. *Pharmacology Biochemistry and Behavior, 59*, 77–82.

Sawamura, S., Tomioka, T., & Hanaoka, K. (2002). The importance of tail temperature monitoring during tail-flick test in evaluating the antinociceptive action of volatile anesthetics. *Acta Anaesthesiologica Scandinavica, 46*, 451–454.

Tjølsen, A., & Hole, K. (1992). The effect of morphine on core and skin temperature in rats. *NeuroReport, 3*, 512–514.

Tjølsen, A., Lund, A., Berge, O.-G., & Hole, K. (1989). An improved method for tail-flick testing with adjustment for tail-skin temperature. *Journal of Neuroscience Methods, 26*, 259–265.

Urban, M. O., & Smith, D. J. (1994). Nuclei within the rostral ventromedial medulla mediating morphine antinociception from the periaqueductal gray. *Brain Research, 652*, 9–16.

Talairach Coordinates

Definition

Initially developed for a specific stereotactic frame; based on one single brain; frequently used as a common coordinate system; X: left-right, Y: anterior-posterior, Z: superior-inferior; the reference point (0, 0, 0) is the anterior commissure (Talairach and Tournoux 1988).

Cross-References

▶ Nociceptive Processing in the Secondary Somatosensory Cortex

Tampa Scale for Kinesiophobia

Synonyms

TSK

Definition

The Tampa Scale for Kinesiophobia is a questionnaire aimed at the assessment of fear of (re)injury due to movement, consisting of 17 items each scored on 4-point numerical rating scales.

Cross-References

▶ Disability, Fear of Movement

Tapotement

▶ Massage and Pain Relief Prospects

Targeting

▶ Trafficking and Localization of Ion Channels

Task Force on Promotion and Dissemination of Psychological Procedures

Definition

In 1993, Division 12 (Clinical Psychology) of the American Psychological Association appointed a task force, with the goal of identifying and disseminating psychological interventions that could be considered as empirically validated. In its 1995 report, this task force published criteria that allowed the classification of psychological treatments as "well established" and "probably efficacious." In 1999, for its special issue on empirically validated treatments in pediatric psychology, the Journal of Pediatric Psychology defined an additional category of "promising interventions." A treatment was considered to be well established if there were at least two good between-group design experiments (or well-controlled single-case studies) by at least two different investigators that demonstrated the treatment's efficacy over placebo, or at least equal efficacy as compared to an already established treatment. In addition, a treatment manual or a well-defined treatment protocol needed to be available. A treatment was considered as "probably efficacious" if there were at least two experiments showing its superiority to a wait-list control or if there was at least one study (or a small series of single-case designs) that met the well-established treatment criteria. Finally, an intervention was considered "promising" if there was at least one well-controlled and another less well-controlled study by separate investigators, or a small number of single-case experiments, or at least two well-controlled studies by the same investigator.

Cross-References

▶ Modeling, Social Learning in Pain
▶ Psychological Treatment of Pain in Children

Task Force on Vicarious Instigation

Definition

Vicarious instigation describes the phenomenon that, possibly mediated by empathy, mere observation of another person's response to a stimulus or a situation (e.g., a pain response) can induce a similar response in the observer in the absence of any direct experience with the eliciting stimulus or situation. In the context of pain, it is still a matter of debate whether observing another person in pain can induce a pain-like vicarious response in the observer or whether it elicits a more generalized emotional response in the observer.

Cross-References

▶ Modeling, Social Learning in Pain

TAUT

Definition

TAUT is a plasma membrane GABA transporter, which transports taurine with higher affinity than GABA.

Cross-References

▶ GABA and Glycine in Spinal Nociceptive Processing

Taut Band

Definition

A taut band is a string- or cord-like structure in striated muscle that extends the length of the muscle fibers. It consists of a number of fascicles that are most palpable across (at a right angle to) the fiber direction in the region of fiber midpoints, where the myofascial trigger point is located. The taut band is the responsive part of the muscle in a local twitch response.

Cross-References

▶ Myofascial Trigger Points

Taxonomy

Nikolai Bogduk
University of Newcastle, Newcastle Bone & Joint Institute, Royal Newcastle Centre, Newcastle, NSW, Australia

Synonyms

Catalogue; Classification; List of Diagnoses and their Definitions

Definition

A ▶ taxonomy is a catalogue that lists and classifies entities and provides definitions of them. It is like a dictionary restricted to a particular field of scholarship. It is designed to standardize the meaning and use of particular terms. In relation to pain medicine, a taxonomy lists, classifies, and defines terms used to describe pain, and provides criteria for the use of diagnostic labels.

Characteristics

Two taxonomies have been produced for use in pain medicine. One, developed by the International Association for the Study of Pain (IASP), covers pain in general (Merskey and Bogduk

1994). The other, developed by the International Headache Society, relates exclusively to headache (Headache Classification Subcommittee of the International Headache Society 2004). A related taxonomy – the Diagnostic and Statistical Manual of Mental Disorders (▶ DSM, DSM-IV, DSM-IVR), was developed by the American Psychiatric Association and is designed to cover mental disorders, but includes some entries that potentially relate to pain (American Psychiatric Association 2000).

IASP Taxonomy

The taxonomy of the IASP (Merskey and Bogduk 1994) consists of a short, introductory section devoted to the definition of terms used to describe pain, its different forms, (such as ▶ somatic pain, ▶ Visceral Nociception and Pain, ▶ referred pain, and ▶ radicular pain), and its associated clinical features, such as ▶ hyperalgesia, ▶ allodynia, and ▶ hyperpathia). The longer, more substantive section lists various entities that constitute possible diagnoses for patients with ▶ chronic pain. For each entity, criteria for making the ▶ diagnosis are stipulated. The entities are catalogued and listed according to the region of the body that they affect. Conditions that affect the whole body, or which may occur in any region of the body, are described first, followed by conditions that affect the head, the neck and cervical spine, the upper limbs, the thoracic region and thoracic spine, the abdomen and pelvis, the lumbar spine, and the lower limbs.

The IASP Taxonomy was developed because it was recognized that particular terms were being used indiscriminately by practitioners. Different practitioners were using the same term to apply to different conditions, and different terms were used to apply to the same condition. Practitioners were also applying different diagnostic labels to what were essentially the same patients, or were applying labels to patients that were not appropriate. In effect, the use of terms and diagnostic labels was arbitrary. In 1979, Bonica likened the terminology for pain syndromes in use at that time to the "tower of Babel" (Bonica 1979).

Taxonomy, Table 1 Taxonomy

Axis I	Region	Referred to the anatomical region in which the pain was perceived (e.g., head, abdomen, lower limb)
Axis II	System	Referred to the body system that ostensibly was affected by pathology to produce pain (e.g., nervous system, vascular system, musculoskeletal system)
Axis II	Temporal	Described whether the pain was continuous, recurring, paroxysmal, etc
Axis IV	Intensity and Duration	Stated if the pain was mild, medium or severe; and lasted less than 1 month, between 1 and 6 months, or longer than 6 months
Axis V	Aetiology	Stated the nature of the cause of the pain (e.g., infectious, inflammatory, and neuropathic)

The first edition of the Taxonomy of the IASP (Merskey 1986) listed common and rare conditions associated with chronic pain, and provided defining descriptions of each. It allowed each condition to be described along five axes (Table 1).

The axis system, however, pertained mainly to the six-digit alphanumeric code ascribed to each condition. The conditions themselves were classified largely according to Axis I, with only parenthetical mention of pathology, aetiology and other features, if these were known.

The first edition of the Taxonomy was not intended to be, or expected to be, comprehensive or fixed. Indeed, readers were invited to submit revisions (Merskey 1986).

The second edition of the Taxonomy (Merskey and Bogduk 1994) addressed many of the shortcomings of the first edition. Some descriptions were modernized, and involved a name change, e.g., ► reflex sympathetic dystrophy and ► causalgia became ► complex regional pain syndrome Type I and Type II. Some entries were deleted (e.g., prolapsed disc, osteophyte, spondylolysis, arachnoiditis, acute low back

strain, recurrent low back strain, and chronic mechanical low back pain) and were replaced by more generic or alternative entries. Some new entries were added, e.g., cervicogenic headache, xiphoidalgia, carcinoma of the lung, proctalgia fugax, piriformis syndrome, and peroneal muscular atrophy.

The greatest revision pertained to entries on spinal pain. Some 96 new entries replaced all previous entries on neck pain, back pain, and other spinal pain. Moreover, the new entries were systematic and rigorous. They were designed to eliminate the problems of content validity and of former entries.

The new entries covered standard conditions such as spinal pain attributable to tumor, infection, metabolic disease, and arthritis. Radicular pain, due to osteophyte, disc prolapse, cysts, tumours, etc., was strictly distinguished and segregated from spinal pain on the grounds that, although radicular pain might have a spinal aetiology, it was pain perceived in the limbs or trunk wall rather than in the spinal region.

Perhaps the most comprehensive change was the introduction of the rubric – "spinal pain of unknown origin." Users were invited, if not directed, to use this rubric wherever an alternative could not be legitimately, or honestly, applied. Providing this rubric encouraged physicians to avoid other poorly defined, invalid, or arbitrary rubrics, in an effort to reduce confusion and false labelling of patients.

Nevertheless, other rubrics were offered. They covered emerging entities, such as discogenic pain and zygapophysial joint pain, as well as classical entities, such as ligament strain and muscle strain, and allopathic entities such as segmental dysfunction. In providing these entries, however, the Taxonomy stipulated stringent, essential diagnostic criteria, in order to avoid the rubrics being applied on intuitive or presumptive grounds.

Thus, for "ligament strain" the ligament had to be specified, and the diagnosis had to be proven with a test that explicitly showed that the ligament in question was the source of pain. Similar criteria were applied for "muscle strain." For "segmental dysfunction" the essential criteria required clinical tests of proven reliability, and established validity to implicate the specified segment as the source of pain.

These rigorous criteria were stipulated quite deliberately in the full knowledge that the tests required to make the diagnosis did not (yet) exist. In effect, therefore, it was impossible to make the diagnosis in practice; yet it appeared in the Taxonomy. The purpose of this action was to indicate to proponents of specific, but ill-defined, yet perhaps popular, diagnoses, that research was required in order for the entity to satisfy the standards of a responsible Taxonomy, and for the diagnosis to be reliable, valid and, therefore, respectable.

IHS Taxonomy

The taxonomy for headache catalogues the many forms of ▶ headache according to mechanism or cause. Diagnostic criteria are stipulated for each form of headache. These are designed to ensure that practitioners use a particular diagnostic label only in those patients who exhibit the prescribed criteria.

The taxonomy describes and defines those headaches whose mechanism is not known but which have well-defined clinical features, such as ▶ migraine, ▶ cluster headache, and ▶ paroxysmal hemicrania. It continues with descriptions and definitions of headaches associated with particular circumstances (such as the headaches of analgesic abuse, and rebound headache), headaches due to particular causes (such as raised or lowered pressure of cerebrospinal fluid, cerebral tumours, aneurysms, infections and granulomas), and headaches associated with other disorders (such as disorders of the ear, nose, and throat, or the cervical spine) (see essay ▶ Headache).

The headache taxonomy is complemented by a textbook, now in its second edition (Olesen et al. 2000), with a third edition in preparation. The textbook follows the format of the taxonomy, but provides descriptions, in detail, of the entities and their diagnosis and treatment.

Cross-References

▶ Orofacial Pain: Taxonomy/Classification

References

American Psychiatric Association. (2000). *DSM-IV-TR. Diagnostic and statistical manual of mental disorders* (4th edn, text revision). Washington, DC: American Psychiatric Association.

Bonica, J. J. (1979). The need for a taxonomy. *Pain, 6*, 247–252.

Headache Classification Subcommittee of the International Headache Society. (2004). The international classification of headache disorders, 2nd edn. *Cephalalgia, 1*(Suppl. 1), 1–160.

Merskey, H. (Ed.) (1986). Classification of chronic pain. Descriptions of chronic pain syndromes and definition of pain terms. *Pain*, Suppl. 3, S1–S225.

Merskey, H., & Bogduk, N. (1994). *Classification of pain. Descriptions of chronic pain syndromes and definitions of pain terms* (2nd ed., pp. 64–65). Seattle: International Association for the Study of Pain.

Olesen, J., Tfelt-Hansen, P., & Welch, K. M. (2000). *The headaches* (2nd ed.). Philadelphia: Lippincott Williams & Wilkins.

Taxonomy of Pelvic Pain

Andrew Paul Baranowski
The Pain Management Centre, National Hospital for Neurology and Neurosurgery, UCL Hospitals NHS Foundation Trust, London, UK

Introduction

When considering the classification of pelvic pain, it is important to appreciate that the pain may be associated with a specific disease process of an organ or organs (Specific Disease Associated Pelvic Pain) but it may also occur in the absence of classical end-organ pathology (Chronic Pelvic Pain Syndrome) (Fall et al. 2010; Baranowski et al. 2008b). When we are dealing with specific disease associated pelvic pain, there may be certain obvious differences between the male and female as a result of different anatomy and physiology, and this may result in sex-specific pathophysiology (Fillingim et al. 2009). However, when we are dealing with the chronic pain syndromes many of the mechanisms will be similar in both sexes (Giamberardino et al. 2010; Giamberardino et al. 2001).

Characteristics

Classification as a Science

Classification of any condition may be divided into three components: terminology, phenotype, and taxonomy.

Terminology

This is the study of terms and their use. The terms are the words and compound words that we use in a specific context when describing a condition. Over the past 8 years, a working group from the International Association for the Study of Pain (IASP) has been looking at pelvic pain classification, and this is now to be published on the IASP Web site. During the process of developing these guidelines, simply changing some of the terms produced significant problems for certain groups of patients.

Terms usually have a very specific meaning, but unfortunately different groups may use the terms in different ways. For instance, I would suggest the term urinary *urgency* should be used to describe "a sudden, compelling need to urinate for fear of incontinence" and the term urinary *urge* "as the desire to urinate" which is the sensation many patients with bladder pain syndrome/interstitial cystitis may describe (Abrams et al. 2006; van de Merwe et al. 2008). Interestingly, this sensation of urge and the bladder pain syndrome appears more common in women than men, but the mechanisms are probably similar to those of the condition often called prostatitis or male chronic pelvic pain syndrome.

Spurious terms can have a significant negative effect on patient outcome as such terms may lend to inappropriate investigation and treatment (Abrams et al. 2006). It is also well established that certain consultation approaches and term usage can affect prognosis (Selfe et al. 1998; Stones et al. 2006) and certain terms can be associated with catastrophic thought process by the patient (Thompson and Carr 2007).

Certain terms are also important in certain countries for remuneration of medical treatment expenses. For instance, there are remuneration codes available for interstitial cystitis but not bladder pain syndrome.

For a patient to make progress with a chronic disease it is often important for them to have a diagnosis and many patient support groups are diagnosis based. Changing the name of a condition can significantly affect the patient identity and the membership of patient support groups.

So in conclusion, terminology is important. It should not be spurious and needs to represent the disease or the component of the disease being defined. However, changing terms can also have a significant effect.

Phenotype

This is the term used to encapsulate the specific features of a condition, as described by history, examination, and clinical investigation (Baranowski et al. 2008a; Castori et al. 2010; Nickel et al. 2009; Nickel et al. 2010a). As well as descriptors of the pain, when phenotyping pain conditions, there is now a significant wealth of experience to include information relating to mood/emotion (e.g., depression, anxiety), cognition (especially negative cognitive process such as catastrophizing), behavior (especially negative contingent resting), sexual difficulties, and effect on work, social, and relationships. These factors are considered important as they are as important as pain in affecting quality of life as well as also affecting response to pain treatment. This has been confirmed in studies on males (Nickel et al. 2007) as well as for women (Nickel et al. 2010b) and where as there may be subtle differences the data is comparable (Heinberg et al. 2004). The use of assessments of physical and emotional functioning as well as pain scores are recommended for research as well (Dworkin et al. 2008).

Taxonomy

This term finds its roots in the Greek taxis (meaning "order" or "arrangement") and nomos (meaning "law" or "science"). Essentially, this is about the relationship between conditions and classically we are talking about a supertype-subtype relationship or less formally, parent-child relationships. In such an inherent relationship, we are suggesting that two conditions are directly related with subtle differences that are defined. Whereas this term is frequently used in medicine and in the classification of pain, this inherent relationship is often not substantiated and therefore the term should be used with care. Having said that, compartmentalizing conditions associated with pain does help us to focus on areas of management (Fall et al. 2010).

Cause, Predisposition, and Maintenance

When considering the classification of pelvic pain it is important not to confuse cause, predisposing factors, and factors that maintain the pain with pain mechanisms. Data in relation to these factors may form a part of the phenotype.

A simple example in relation to cause, predisposing factors, factors that maintain the pain and pain mechanisms would be to consider a stabbing. The cause of the pain is the stabbing by an instrument, the pain is maintained while the knife is in situ and there is tissue damage, the predisposing factors will be those that led the person to be stabbed. From a classification purpose, the pain is acute tissue trauma with acute pain mechanisms.

When we are dealing with chronic pain there may be many causes that produce the same end result. Often in any one single patient there may be multiple potential causes. If the cause is ongoing (Specific Disease Associated Pain), then management should be aimed at treating the cause with symptomatic treatment of the pain as appropriate. Patients will often focus on cause, if the cause is no longer ongoing then the focus needs to shift to the current mechanisms and treatment aimed at these (Baranowski 2009; Fall et al. 2010; Wesselmann et al. 2009). The patient may need psychological support to help them move forward from a traumatic cause, such as a negative sexual event, to enable treatment of the ongoing pain mechanisms.

There are many potential factors that may predispose patients to developing chronic pain. Genetic predisposition (Stamer and Stuber 2007), trauma, and endometriosis are often sighted (Giamberardino et al. 2010). There is little evidence to support that the majority of patients with pelvic pain have a past history of negative sexual encounter (NSE) (Anda et al. 2006; Raphael 2005). However, patients that have suffered sexual trauma (rape, systematic rape, torture, and abuse) may be at greater risk of real physical pain (Williams et al. 2010). In this situation, interventions aimed at supporting the patient following the NSE are important as well as managing the pain.

When considering maintenance factors, we have to appreciate that the medical system can reinforce pain mechanisms through inappropriate consultations, investigation, and treatment (Abrams et al. 2006; Baskin and Tanagho 1992; Stones et al. 2006). Chronic pain mechanisms are now well described (Baranowski 2009; Giamberardino et al. 2010; Baranowski et al. 2008b) and these may self maintain the pain in their own right, though physical stimuli may also play a role in perpetuating matters (Latremoliere and Woolf 2009; Woolf and Doubell 1994) (e.g., continuing with painful sex) (Pukall et al. 2007).

The Classification of Chronic Pelvic Pain

The basis of the following definitions were published by the chronic pelvic pain working group of the European Association of Urology in 2004 and modified 2010 (Fall et al. 2010). The Pain of Urogenital Origin (PUGO) Taxonomy Working Group have further modified this and incorporated the classification into the International Association for the Study of Pain (IASP) Taxonomy (to be published 2012). There are slight modifications in the classification presented below for clarity, as the document is not presented in its entirety (Table 1).

Chronic pelvic pain is defined as chronic/persistent[$] pain perceived* in structures related to the pelvis of either men or women. It is often associated with negative cognitive, behavioral, sexual, and emotional consequences as well as with symptoms suggestive of lower urinary tract, sexual, bowel, or gynecological dysfunction[D] (Fall et al. 2010).

Please note this definition applies equally to men and women.

Chronic/persistent[$] in the case of documented nociceptive pain mechanisms (such as recurrent infection and local disease process) that becomes chronic/persistent through time, the pain must

Taxonomy of Pelvic Pain, Table 1 The division of chronic pelvic pain into pelvic pain syndromes and non-pelvic pain syndromes

AXIS I Region	Axis II System	Axis III End organ as pain syndrome as identified from Hx, Ex, and Ix	Axis IV Referral characteristics	Axis V Temporal characteristics	Axis VI Characteristics	Axis VII Associated symptoms	Axis VIII Psychological symptoms
Chronic pelvic pain	Chronic pelvic pain syndrome						
	Urological	Bladder pain syndrome — (See ESSIC classification reference)	Suprapubic	ONSET Acute Chronic	Aching Burning Stabbing Electric Other	URINARY Frequency Nocturia Hesitance	Cognitive Behavioral Emotional
		Urethral pain syndrome	Inguinal	ONGOING Sporadic Cyclical Continuous		Poor flow Pis en deux Urge Urgency	
		Prostate pain syndrome — Type A inflammatory	Urethral	TIME Filling Emptying Immediate post		Incontinence Other	
		Type B noninflammatory	Penile/clitoral	Late post		GYNECOLOGICAL Menstrual	
		Scrotal pain syndrome — Testicular pain syndrome	Perineal	PROVOKED		SEXUAL Female dyspareunia	
		Epididymal pain syndrome	Rectal			Impotence	
		Post vasectomy pain syndrome	Back			Anorectal Incontinence	
		Penile pain syndrome	Buttocks			Constipation	
	Gynecological	Vaginal pain syndrome				MUSCULAR Hyperalgesia Dysfunction	
		Vulvar pain syndrome — Generalized vulvar pain syndrome				CUTANEOUS Allodynia	
		Localized vulvar pain syndrome — Vestibular pain syndrome					
		Clitoral pain syndrome					
		Other — Endometriosis-associated pain syndrome					
	Anorectal	Anorectal pain syndrome					
	Neurological	Pudendal pain syndrome					
	Muscular	Pelvic floor muscle pain syndrome					
Specific disease-associated pelvic pain	Neurological	Pudendal neuralgia					
	Urological						

have been continuous or recurrent for at least 6 months. The pain can be cyclical over a 6-month period, such as the cyclical pain of dysmenorrhea. Six months is arbitrary; 6 months was chosen because 3 months was not considered long enough if we include cyclical pain conditions. If non-acute and central sensitization pain mechanisms are inferred from the history, examination, and investigations (Baranowski 2009; Giamberardino et al. 2010; Pukall et al. 2002; Vecchiet et al. 1999), then the pain may be regarded as chronic, irrespective of the time period.

*Perceived** indicates that the patient and clinician, to the best of their ability from the history, examination, and investigations (where appropriate) have localized the pain as being perceived in the specified anatomical pelvic area (Warren et al. 2008).

Dysfunction^D is a term that indicates that there is a mechanistic functional disorder. The term is NOT used to indicate a psychological or psychiatric disorder.

Cyclical pain is included in the classification and hence *dysmenorrhea* needs to be considered as a chronic pain syndrome if it is persistent and associated with negative cognitive, behavioral, sexual, and emotional consequences. The term "dysmenorrhea pain syndrome" could be used.

Chronic pelvic pain may be subdivided into those conditions with well-defined classical pathology (such as infection and cancer) and those where no obvious pathology are found. For the purpose of this classification, the terms *specific disease associated pelvic pain* is proposed for the former and *chronic pelvic pain syndrome* is used for the latter.

Chronic pelvic pain syndrome (CPPS) (Fall et al. 2010) is the occurrence of chronic pelvic pain where there is no proven infection or other obvious local pathology that may account for the pain. It is often associated with negative cognitive, behavioral, sexual, and emotional consequences as well as with symptoms suggestive of lower urinary tract, sexual, bowel, or gynecological dysfunction.

The chronic pelvic pain syndromes are defined by a process of exclusion. In particular, there

should be no evidence of local organ cancer, infection, or inflammation. As a consequence, early investigations by end-organ specialists (e.g., gynecologists, urologists, and proctologists) should be aimed at excluding local end-organ pathology (Fall et al. 2010). As stated above, repeated unnecessary and inappropriate investigations will have a detrimental effect.

Subdivision of the CPPS into another sub-phenotype (Table 1) should only occur if there is adequate evidence to support using that term and sub-phenotype. For instance, in nonspecific, poorly localized pelvic pain without obvious pathology, only the term "chronic pelvic pain syndrome" should be used. If the pain can be well localized to an organ, then a more specific term such as "rectal pain syndrome" may be used. Localizing in this way may assist in the choice of specific end-organ treatments. If the pain is localized to multiple organs, then the syndrome is a regional pain syndrome (see below) and the term "chronic pelvic pain syndrome" should once more be considered (Fall et al. 2010).

When defining a patient, as well as defining the patient by a specific end-organ phenotype, there are multiple other more general descriptors that need to be considered. These are primarily psychological (e.g., cognitive and emotional), sexual, behavioral, and functional, although others exist, axes IV-VII, of Table 1. Psychological and behavioral factors are well established to relate to quality-of-life issues and prognosis (Nickel et al. 2007; Tripp et al. 2009). The other descriptors help to classify the nature of the pain.

Systemic disorders outside the pelvis are very common in those that suffer with chronic pelvic pain (Nickel et al. 2010a; Pasoto et al. 2005; Tietjen et al. 2007; Baranowski et al. 2008b; Warren et al. 2011). In America, a NIH research program, the Multidisciplinary Approach to the Study of Chronic Pelvic Pain (MAPP) program has been devised to investigate the importance of these factors and looks at all types of pelvic pain irrespective of the end organ where the pain is perceived. Moreover, it addresses systemic disorder associations, such as the co-concurrence of fibromyalgia, facial pain, and autoimmune disorders.

The debate in relation to subdividing the pain syndromes is ongoing. As more and more information is collected suggesting that the central nervous system is involved and indeed may be the main cause of many CPP conditions (wherever the pain is perceived – bladder, genitalia, colorectal, or myofascial), there is a general tendency to move away from end-organ nomenclature. Whether this is appropriate or not is something that only time and good research will tell. To enable such research, it is essential to have a framework of classification to work within. Therefore, although Table 1 has its inherent deficiencies, it is considered as a good framework to base management and research from.

Case History

Approximately, one third of the new patients attending the Urogenital/Pelvic Pain Management Center at the National Hospital for Neurology and Neurosurgery, Queen Square fall into the Complex Regional Pelvic Pain Syndrome phenotype. A classical patient is described below.

A lady in her late thirties, early forties with no negative sexual encounter history, but a history of dysmenorrhea and primary dyspareunia. There is a family history of fibromyalgia. At some stage, she has had a laparoscopy and some endometriotic lesions have been identified, but the lesions were small and not greater than those found in pain asymptomatic patients. All of the aforementioned are predisposing factors to developing chronic pelvic pain.

While on holiday, she developed an increase in bladder urge sensations and was treated for a presumed UTI with antibiotics. No cultures were taken but dipsticks demonstrated a trace of blood and protein (remember this can occur with neurogenic edema in the absence of infection (Huygen et al. 2001)). The urge and frequency settled (remember antibiotics can have an analgesic effect in the absence of infection (Toussirot et al. 1997)).

Over a period of time, the urge returned and further antibiotics were given but cultures were negative. She went on to have a cystoscopy and urodynamics. The cystoscopy demonstrated small areas of redness but not frank interstitial cystitis.

The urodynamics demonstrated a sensitive bladder but no other overt abnormality. The lady becomes more anxious as a cause cannot be found and the sensation of urge is now associated with pain if she retains urine. Her voiding frequency is significantly increased. She cannot work and is becoming low of mood. She is unable to have sexual activity as the vulva is now sensitive, but examinations by a dermatologist do not reveal a cause. Local creams do not help. She has always had a tendency to constipation, but now her doctor informs her she has irritable bowel syndrome. All her pains are made worse with menstruation. As a result of her pain, she becomes house bound and because of the pain being exacerbated by sitting she spends a lot of time in bed. Her depression worsens, and this becomes considered as the cause rather than secondary.

Because of her disability and immobility, she develops diffuse widespread pain and she has several other diagnoses tagged on (such as fibromyalgia). She is prone to infections (due to immuno suppression) and may also develop endocrine disorders. She is now dependent and some may say addicted to opioids.

Unfortunately, this is not an uncommon scenario. What was not recognized early on is that the lady had a bladder pain syndrome and not a UTI. Bladder pain syndromes (due to central sensitization processes) can frequently become associated with a vulvar pain syndrome. In the case described, the whole of the pelvic organs have become involved and examination of the muscles (pelvic, abdominal, and back) will probably reveal muscle hyperalgesia. I would therefore consider this lady as having a complex pelvic pain syndrome or my preferred terminology – complex regional pelvic pain syndrome. The word "regional" indicates an overlap with mechanisms of complex regional pain syndrome (Bruehl 2010; Goebel et al. 2010; Janig and Baron 2002; Naleschinski and Baron 2010). The lady has developed a systemic manifestation with psychological and behavioral responses which will significantly affect quality of life and need treatment in their own right (Nickel et al. 2010b; Sutton et al. 2009; Tripp et al. 2009). Management of such a complex patient is probably best within the context of a Multidisciplinary Pain Management Center that includes joint clinic management with a range of specialties that have a primary interest in pelvic pain (Fall et al. 2010).

Classification, Terminology, and Phenotypes as Relevant to Gynecological Practice

(based upon the EAU guidelines and submitted to IASP) (Fall et al. 2010):

Dyspareunia is defined as pain perceived within the pelvis associated with penetrative sex. *It tells us nothing about the mechanism and may be applied to the female or male. It is usually applied to penile penetration, but is often associated with pain during insertion of any object. As well as vaginal intercourse it may apply to anodyspareunia. It is classically subdivided into superficial and deep.*

Perineal pain syndrome is a "neuropathic-type" pain perceived in the distribution of the pudendal nerve that may be associated with symptoms and signs of rectal, urinary tract, or sexual dysfunction. There is no proven obvious pathology. It is often associated with negative cognitive, behavioral, sexual, and emotional consequences as well as with symptoms suggestive of lower urinary tract, sexual, bowel, or gynecological dysfunction. *This should be separated from pudendal neuralgia which is a specific disease associated pelvic pain, because the nerve is damaged.*

Bladder pain syndrome is the occurrence of persistent or recurrent pain perceived in the urinary bladder region, accompanied by at least one other symptom, such as pain worsening with bladder filling and day-time and/or night-time urinary frequency. There is no proven infection or other obvious local pathology. Bladder pain syndrome is often associated with negative cognitive, behavioral, sexual, or emotional consequences as well as with symptoms suggestive of lower urinary tract and sexual dysfunction.

Bladder pain syndrome is believed to represent a heterogeneous spectrum of disorders. There may be specific types of inflammation as a feature in subsets of patients. Localization of the pain can be difficult by examination and as a consequence another localizing symptom is required. Cystoscopy with hydrodistension and biopsy may be indicated to define phenotypes. Recently, ESSIC (van de Merwe et al. 2008) have suggested a standardized scheme of sub-classifications to acknowledge differences and make it easier to compare various studies.

Other terms that have been used include interstitial cystitis, painful bladder syndrome, and PBS/IC or BPS/IC, these terms are no longer recommended by the author.

Urethral pain syndrome is the occurrence of chronic or recurrent episodic pain perceived in the urethra, in the absence of proven infection or other obvious local pathology. Urethral pain syndrome is often associated with negative cognitive, behavioral, sexual, or emotional consequences as well as with symptoms suggestive of lower urinary tract, sexual, bowel, or gynecological dysfunction.

Urethral pain syndrome may occur in men and women.

Vulvar pain syndrome is the occurrence of persistent or recurrent episodic vulvar pain. There is no proven infection or other local obvious pathology. It is often associated with negative cognitive, behavioral, sexual, or emotional consequences as well as with symptoms suggestive of lower urinary tract, sexual, bowel, or gynecological dysfunction.

Although pain perceived in the vulva was subsumed under sexual disorders in the DSM-IV-R manual for classifying psychiatric disorders, there is no scientific basis for this classification and pain perceived in the vulva is best understood as a pain problem that usually has psychological consequences. There is no evidence for its classification as a psychiatric disorder.

Generalized vulvar pain syndrome refers to a vulvar pain syndrome where the pain/burning cannot be consistently and precisely localized by point-pressure "mapping" via probing with a cotton-tipped applicator or similar instrument. Rather, the pain is diffuse and affects all locations of the vulva. The vulval vestibule (part of the vulva which lies between the labia minora into which the urethral meatus and vaginal introitus open) may be involved but the discomfort is not limited to the vestibule. This pain syndrome is often associated with negative cognitive, behavioral, sexual, and emotional consequences. *Previous terms have included dysaesthetic vulvodynia and essential vulvodynia, but are no longer recommended by this author.*

Localized vulvar pain syndrome refers to pain that can be consistently and precisely localized by point-pressure mapping to one or more portions of the vulva. Clinically, the pain usually occurs as a result of provocation (touch, pressure, or friction).

Localized vulvar pain syndrome can be subdivided into *vestibular pain syndrome* and *clitoral pain syndrome.*

Vestibular pain syndrome refers to pain that can be localized by point-pressure mapping to one or more portions of the vulval vestibule. *Previous terms have included vulval/vulvar vestibulitis, vestibulodynia, and localized and focal vulvitis, but are no longer recommended by this author.*

Clitoral pain syndrome refers to pain that can be localized by point-pressure mapping to the clitoris or is well perceived in the area of the clitoris.

Endometriosis-associated pain syndrome is chronic or recurrent pelvic pain in laparoscopically confirmed endometriosis and the term is used when the symptoms persist despite adequate endometriosis treatment. It is often associated with negative cognitive, behavioral, sexual, or emotional consequences as well as with symptoms suggestive of lower urinary tract, sexual, bowel, or gynecological dysfunction.

Many patients have pain above and beyond the endometriotic lesions; this term is used to cover that group of patients. Endometriosis may be an incidental finding, is not always painful, and the degree of disease seen laparoscopically does not correlate with severity of symptoms. As with other patients, they often have more than one end organ involved. It has been suggested that this phenotype should be removed from the classification as the endometriosis may be irrelevant.

Chronic pelvic pain syndrome with cyclical exacerbations is the term that would cover both the non-gynecological organ pains which frequently show cyclical exacerbations (e.g., IBS, bladder pain syndrome) but also pain similar to that associated with endometriosis/adenomyosis but where no pathology is identified. This condition is different from dysmenorrhea where pain is only present with menstruation.

Primary dysmenorrhea is pain with menstruation not associated with a well-defined pathology. Dysmenorrhea needs to be considered as a chronic pain syndrome if it is persistent and associated with negative cognitive, behavioral, sexual, or emotional consequences otherwise it is a symptom.

Pelvic floor muscle pain syndrome is the occurrence of persistent or recurrent, episodic, pelvic floor pain. There is no proven well-defined local pathology. It is often associated with negative cognitive, behavioral, sexual, or emotional consequences as well as with symptoms suggestive of lower urinary tract, sexual, bowel, or gynecological dysfunction.

This syndrome may be associated with overactivity of or trigger points within the pelvic floor muscles. Trigger points may also be found in multiple muscles, such as the abdominal, thigh, and paraspinal muscles and even muscles not directly related to the pelvis.

Irritable bowel syndrome is the occurrence of chronic or recurrent episodic pain perceived in the bowel, in the absence of proven infection or other obvious local pathology. Bowel dysfunction is frequent. Irritable bowel syndrome is often associated with worry and preoccupation about bowel function, negative cognitive, behavioral, sexual, or emotional consequences as well as with symptoms suggestive of lower urinary tract, or gynecological dysfunction.

The above classification is based upon the Rome III Criteria: Three months of continuous or recurring symptoms of abdominal pain or irritation that may be relieved with a bowel movement, may be coupled with a change in frequency, or may be related to a change in the consistency of stools.

Two or more of the following are present at least 25 % (one quarter) of the time: A change in stool frequency (more than three bowel movement per day or fewer than three bowel movements per week), noticeable difference in stool form (hard, loose, and watery stools or poorly formed stools), passage of mucus in stools, bloating or feeling of abdominal distension, altered stool passage (e.g., sensations of incomplete evacuation, straining, or urgency). Extraintestinal symptoms include nausea, fatigue, full sensation after even a small meal, vomiting.

Chronic anal pain syndrome is the occurrence of chronic or recurrent episodic pain perceived in the anus, in the absence of proven infection or other obvious local pathology. Chronic anal pain syndrome is often associated with negative cognitive, behavioral, sexual, or emotional consequences as well as with symptoms suggestive of lower urinary tract, sexual, bowel, or gynecological dysfunction.

Intermittent chronic anal pain syndrome refers to severe, brief, episodic pain that seems to arise in the rectum or anal canal and occurs at irregular intervals. *This is unrelated to the need to or the process of defecation. It may be considered a subgroup of the chronic anal pain syndromes. Previously known as Proctalgia fugax. This term is no longer recommended by the author of this entry.*

Summary

The classification of pelvic pain is complex as the condition may be multidimensional with symptoms in multiple areas. A key point is that spurious terminology should not be used, as this is associated with a negative prognosis. The use of the term "syndrome" indicates the complexity of the situation and allows us to include the full range of physical and psychological symptoms that this group of patients develop. There is a debate about the usefulness of end-organ identification in any pelvic pain classification; however, its usage may help to focus on the use of certain treatments. The downside is that focusing on the end organ where the pain is perceived may distract from the more significant central nervous system mechanisms that are so well described as being involved. Any classification must take into account the negative psychological (emotional and cognitive), behavioral, and sexual symptoms that significantly affect quality of life and treatment outcome.

References

Abrams, P., Baranowski, A., Berger, R. E., Fall, M., Hanno, P., & Wesselmann, U. (2006). A new classification is needed for pelvic pain syndromes – are existing terminologies of spurious diagnostic authority bad for patients? *Journal of Urology, 175*, 1989–1990. United States.

Anda, R. F., Felitti, V. J., Bremner, J. D., Walker, J. D., Whitfield, C., Perry, B. D., Dube, S. R., & Giles, W. H. (2006). The enduring effects of abuse and related adverse experiences in childhood. A convergence of evidence from neurobiology and epidemiology. *European Archives of Psychiatry and Clinical Neuroscience, 256*(3), 174–186. doi:10.1007/s00406-005-0624-4.

Baranowski, A. P. (2009). Chronic pelvic pain. *Best Practice & Research. Clinical Gastroenterology, 23*(4), 593–610. doi:S1521-6918(09)00066-3 [pii] 10.1016/j.bpg.2009.04.013.

Baranowski, A. P., Abrams, P., Berger, R. E., Buffington, C. A., de C Williams, A. C., Hanno, P., Loeser, J. D., Nickel, J. C., & Wesselmann, U. (2008a). Urogenital pain – time to accept a new approach to phenotyping and, as a consequence, management. *European Urology, 53*(1), 33–36. doi:10.1016/j.eururo.2007.10.010.

Baranowski, A. P., Abrams, P., & Fall, M. (Eds.). (2008b). *Urogenital pain in clinical practice.* New York: Informa Healthcare.

Baskin, L. S., & Tanagho, E. A. (1992). Pelvic pain without pelvic organs. *Journal of Urology, 147*(3), 683–686.

Bruehl, S. (2010). An update on the pathophysiology of complex regional pain syndrome. *Anesthesiology, 113*(3), 713–725. doi:10.1097/ALN.0b013e3181e3db38.

Castori, M., Camerota, F., Celletti, C., Danese, C., Santilli, V., Saraceni, V. M., & Grammatico, P. (2010). Natural history and manifestations of the hypermobility type Ehlers-Danlos syndrome: A pilot study on 21 patients. *American Journal of Medical Genetics. Part A, 152A*(3), 556–564. doi:10.1002/ajmg.a.33231.

Curtis Nickel, J., Baranowski, A. P., Pontari, M., Berger, R. E., & Tripp, D. A. (2007). Management of men diagnosed with chronic prostatitis/chronic pelvic pain syndrome who have failed traditional management. *Revista de Urología, 9*(2), 63–72.

Dworkin, R. H., Turk, D. C., Wyrwich, K. W., Beaton, D., Cleeland, C. S., Farrar, J. T., Haythornthwaite, J. A., Jensen, M. P., Kerns, R. D., Ader, D. N., Brandenburg, N., Burke, L. B., Cella, D., Chandler, J.,

Cowan, P., Dimitrova, R., Dionne, R., Hertz, S., Jadad, A. R., Katz, N. P., Kehlet, H., Kramer, L. D., Manning, D. C., McCormick, C., McDermott, M. P., McQuay, H. J., Patel, S., Porter, L., Quessy, S., Rappaport, B. A., Rauschkolb, C., Revicki, D. A., Rothman, M., Schmader, K. E., Stacey, B. R., Stauffer, J. W., von Stein, T., White, R. E., Witter, J., & Zavisic, S. (2008). Interpreting the clinical importance of treatment outcomes in chronic pain clinical trials: IMMPACT recommendations. *The Journal of Pain, 9*, 105–121. United States.

Fall, M., Baranowski, A. P., Elneil, S., Engeler, D., Hughes, J., Messelink, E. J., Oberpenning, F., & de C Williams, A. C. (2010). EAU guidelines on chronic pelvic pain. *European Urology, 57*(1), 35–48. doi:10.1016/j.eururo.2009.08.020.

Fillingim, R. B., King, C. D., Ribeiro-Dasilva, M. C., Rahim-Williams, B., & Riley, J. L., 3rd. (2009). Sex, gender, and pain: A review of recent clinical and experimental findings. *The Journal of Pain, 10*, 447–485. United States.

Giamberardino, M. A., De Laurentis, S., Affaitati, G., Lerza, R., Lapenna, D., & Vecchiet, L. (2001). Modulation of pain and hyperalgesia from the urinary tract by algogenic conditions of the reproductive organs in women. *Neuroscience Letters, 304*, 61–64. Ireland.

Giamberardino, M. A., Costantini, R., Affaitati, G., Fabrizio, A., Lapenna, D., Tafuri, E., & Mezzetti, A. (2010). Viscero-visceral hyperalgesia: Characterization in different clinical models. *Pain, 151*, 307–322. Netherlands: 2010 International Association for the Study of Pain. Published by Elsevier B.V.

Goebel, A., Baranowski, A., Maurer, K., Ghiai, A., McCabe, C., & Ambler, G. (2010). Intravenous immunoglobulin treatment of the complex regional pain syndrome: A randomized trial. *Annals of Internal Medicine, 152*, 152–158. United States.

Heinberg, L. J., Fisher, B. J., Wesselmann, U., Reed, J., & Haythornthwaite, J. A. (2004). Psychological factors in pelvic/urogenital pain: The influence of site of pain versus sex. *Pain, 108*, 88–94. Netherlands.

Huygen, F. J., de Bruijn, A. G., Klein, J., & Zijlstra, F. J. (2001). Neuroimmune alterations in the complex regional pain syndrome. *European Journal of Pharmacology, 429*, 101–113. Netherlands.

Janig, W., & Baron, R. (2002). Complex regional pain syndrome is a disease of the central nervous system. *Clinical Autonomic Research, 12*(3), 150–164.

Latremoliere, A., & Woolf, C. J. (2009). Central sensitization: A generator of pain hypersensitivity by central neural plasticity. *The Journal of Pain, 10*, 895–926. United States.

Naleschinski, D., & Baron, R. (2010). Complex regional pain syndrome type I: Neuropathic or not? *Current Pain and Headache Reports, 14*(3), 196–202. doi:10.1007/s11916-010-0115-9.

Nickel, J. C., Tripp, D., Teal, V., Propert, K. J., Burks, D., Foster, H. E., Hanno, P., Mayer, R., Payne, C. K., Peters, K. M., Kusek, J. W., & Nyberg, L. M. (2007).

Sexual function is a determinant of poor quality of life for women with treatment refractory interstitial cystitis. *Journal of Urology, 177*, 1832–1836. United States.

Nickel, J. C., Shoskes, D., & Irvine-Bird, K. (2009). Clinical phenotyping of women with interstitial cystitis/painful bladder syndrome: A key to classification and potentially improved management. *Journal of Urology, 182*, 155–160. United States.

Nickel, J. C., Tripp, D. A., Pontari, M., Moldwin, R., Mayer, R., Carr, L. K., Doggweiler, R., Yang, C. C., Mishra, N., & Nordling, J. (2010a). Interstitial cystitis/painful bladder syndrome and associated medical conditions with an emphasis on irritable bowel syndrome, fibromyalgia and chronic fatigue syndrome. *Journal of Urology, 184*, 1358–1363. United States: Inc. Published by Elsevier Inc.

Nickel, J. C., Tripp, D. A., Pontari, M., Moldwin, R., Mayer, R., Carr, L. K., Doggweiler, R., Yang, C. C., Mishra, N., & Nordling, J. (2010b). Psychosocial phenotyping in women with interstitial cystitis/painful bladder syndrome: A case control study. *Journal of Urology, 183*, 167–172. United States.

Pasoto, S. G., Abrao, M. S., Viana, V. S., Bueno, C., Leon, E. P., & Bonfa, E. (2005). Endometriosis and systemic lupus erythematosus: A comparative evaluation of clinical manifestations and serological autoimmune phenomena. *American Journal of Reproductive Immunology, 53*, 85–93. Denmark: Blackwell Munksgaard, 2005.

Pukall, C. F., Binik, Y. M., Khalife, S., Amsel, R., & Abbott, F. V. (2002). Vestibular tactile and pain thresholds in women with vulvar vestibulitis syndrome. *Pain, 96*, 163–175. Netherlands.

Pukall, C. F., Smith, K. B., & Chamberlain, S. M. (2007). Provoked vestibulodynia. *Women's Health (London, England), 3*(5), 583–592. doi:10.2217/17455057.3.5.583.

Raphael, K. G. (2005). Childhood abuse and pain in adulthood: More than a modest relationship? *The Clinical Journal of Pain, 21*, 371–373. United States.

Selfe, S. A., Matthews, Z., & Stones, R. W. (1998). Factors influencing outcome in consultations for chronic pelvic pain. *Journal of Women's Health, 7*(8), 1041–1048.

Stamer, U. M., & Stuber, F. (2007). Genetic factors in pain and its treatment. *Current Opinion in Anaesthesiology, 20*, 478–484. United States.

Stones, R. W., Lawrence, W. T., & Selfe, S. A. (2006). Lasting impressions: Influence of the initial hospital consultation for chronic pelvic pain on dimensions of patient satisfaction at follow-up. *Journal of Psychosomatic Research, 60*, 163–167. England.

Sutton, K. S., Pukall, C. F., & Chamberlain, S. (2009). Pain, psychosocial, sexual, and psychophysical characteristics of women with primary vs. secondary provoked vestibulodynia. *The Journal of Sexual Medicine, 6*, 205–214. United States.

Thompson, P., & Carr, E. (2007). Content analysis of general practitioner-requested lumbar spine X-ray reports. *British Journal of Radiology, 80*, 866–871. England.

Tietjen, G. E., Bushnell, C. D., Herial, N. A., Utley, C., White, L., & Hafeez, F. (2007). Endometriosis is associated with prevalence of comorbid conditions in migraine. *Headache, 47*, 1069–1078. United States.

Toussirot, E., Despaux, J., & Wendling, D. (1997). Do minocycline and other tetracyclines have a place in rheumatology? *Revue du Rhumatisme (English Ed.), 64*(7–9), 474–480.

Tripp, D. A., Nickel, J. C., Fitzgerald, M. P., Mayer, R., Stechyson, N., & Hsieh, A. (2009). Sexual functioning, catastrophizing, depression, and pain, as predictors of quality of life in women with interstitial cystitis/painful bladder syndrome. *Urology, 73*, 987–992. United States.

van de Merwe, J. P., Nordling, J., Bouchelouche, P., Bouchelouche, K., Cervigni, M., Daha, L. K., Elneil, S., Fall, M., Hohlbrugger, G., Irwin, P., Mortensen, S., van Ophoven, A., Osborne, J. L., Peeker, R., Richter, B., Riedl, C., Sairanen, J., Tinzl, M., & Wyndaele, J. J. (2008). Diagnostic criteria, classification, and nomenclature for painful bladder syndrome/interstitial cystitis: An ESSIC proposal. *European Urology, 53*, 60–67. Switzerland.

Vecchiet, L., Vecchiet, J., & Giamberardino, M. A. (1999). Referred muscle pain: Clinical and pathophysiologic aspects. *Current Review of Pain, 3*(6), 489–498.

Warren, J. W., Langenberg, P., Greenberg, P., Diggs, C., Jacobs, S., & Wesselmann, U. (2008). Sites of pain from interstitial cystitis/painful bladder syndrome. *Journal of Urology, 180*, 1373–1377. United States.

Warren, J. W., Wesselmann, U., Morozov, V., & Langenberg, P. W. (2011). Numbers and types of nonbladder syndromes as risk factors for interstitial cystitis/painful bladder syndrome. *Urology, 77*, 313–319. United States: 2011 Elsevier Inc.

Wesselmann, U., Baranowski, A. P., Borjesson, M., Curran, N. C., Czakanski, P. P., Giamberardino, M. A., Ness, T. J., Robbins, M. T., & Traub, R. J. (2009). Emerging therapies and novel approaches to visceral pain. *Drug Discovery Today: Therapeutic Strategies, 6*(3), 89–95. doi:10.1016/j.ddstr.2009.05.001.

Williams, A. C., Pena, C. R., & Rice, A. S. (2010). Persistent pain in survivors of torture: A cohort study. *Journal of Pain and Symptom Management, 40*, 715–722. United States: 2010 U.S. Cancer Pain Relief Committee. Published by Elsevier Inc.

Woolf, C. J., & Doubell, T. P. (1994). The pathophysiology of chronic pain–increased sensitivity to low threshold A beta-fibre inputs. *Current Opinion in Neurobiology, 4*(4), 525–534.

TCAs

▶ Tricyclic Antidepressants

TCD

▶ Thalamocortical Dysrhythmia

Team Approach

▶ Physical Medicine and Rehabilitation, Team-Oriented Approach

Technique of Ultrasound Application

Definition

The most common technique is the stroking technique.

Cross-References

▶ Ultrasound Therapy of Pain from the Musculoskeletal System

Tegretol

Synonyms

Generic carbamazepine

Definition

Tegretol (Generic Carbamazepine) is an antiepileptic drug acting at non-voltage-dependent sodium channels.

Cross-References

▶ Trigeminal, Glossopharyngeal, and Geniculate Neuralgias

Telehealth

▶ Remote Management of Pediatric Pain

Telemedicine

▶ Remote Management of Pediatric Pain

Telemetric

Definition

Telemetric means the transmission of data by radio or other means from a remote source.

Cross-References

▶ Opioid Therapy in Cancer Pain Management, Route of Administration

Temperament

▶ Personality and Pain

Temporal Arteritis

Definition

Temporal arteritis is an arterial disease with inflammation of the temporal arteries characterized by fever, anorexia, loss of weight, leukocytosis, and tenderness over the scalp and along the temporal vessels. The giant cell arteritis most often attacks the external arteries in the anterior skull region – branches from the arteria cerebri externa.

Cross-References

▶ Cancer Pain, Assessment in the Cognitively Impaired
▶ Muscle Pain in Systemic Inflammation (Polymyalgia Rheumatica, Giant Cell Arteritis, Rheumatoid Arthritis)

Temporal Association

Definition

Temporal association between two disorders or clinical problems refers to their hypothesized relationship in terms of time of onset, most often inferring a causal or contributory relationship.

Cross-References

▶ Depression and Pain

Temporal Resolution

Definition

The value indicates how reliable the results are in terms of the time period. The higher, the better, and EEG and MEG are much higher than fMRI and PET.

Cross-References

▶ Magnetoencephalography in Assessment of Pain in Humans

Temporal Summation (Windup)

Definition

When synaptic potentials overlap in time, they add together. In this case, repeated administration of the same stimulus, at a given interval of time, produces a progressively increased painful response. Temporal summation is probably the initial part of windup, which is the increased neuronal firing to a train of stimuli recorded in animals.

Cross-References

▶ Encoding of Noxious Information in the Spinal Cord
▶ Exogenous Muscle Pain
▶ Opioids and Muscle Pain
▶ Opioids, Effects of Systemic Morphine on Evoked Pain
▶ Pain in Humans, Electrical Stimulation (Skin, Muscle, and Viscera)

Temporomandibular Disorder

Synonyms

TMD

Definition

A collective term embracing a number of clinical problems that involve the masticatory musculature, the temporomandibular joint and associated structures, or both. Temporomandibular disorders have been identified as a major cause of nondental pain in the orofacial region, and are considered to be a subclassification of musculoskeletal disorders. Also sometimes referred to as craniomandibular or temporomandibular (joint) dysfunction.

Cross-References

▶ Orofacial Pain, Movement Disorders
▶ Orofacial Pain: Taxonomy/Classification
▶ Psychological Aspects of Pain in Women

Temporomandibular Disorders

▶ Orofacial Pain, Movement Disorders

Temporomandibular Disorders (TMDs)

▶ Temporomandibular Joint Disorders

Temporomandibular Joint

Synonyms

TMJ

Definition

The jaw joint. The joint formed between the condylar process of the mandible and the mandibular fossa and articular tubercle of the temporal bone.

Cross-References

▶ Nociceptors in the Orofacial Region (Temporomandibular Joint and Masseter Muscle)
▶ Psychiatric Aspects of Pain and Dentistry
▶ Temporomandibular Joint Disorders

Temporomandibular Joint and Muscle Pain Dysfunction

▶ Temporomandibular Joint Disorders

Temporomandibular Joint Disorders

Christian S. Stohler
Baltimore College of Dental Surgery, University
of Maryland, Baltimore, MD, USA

Synonyms

Craniomandibular disorders (CMDs); Previously
used diagnostic labels; Temporomandibular disorders (TMDs); Temporomandibular joint and
muscle pain dysfunction; Temporomandibular
joint disorders (TMJDs)

Definition

Temporomandibular joint disorders (TMJDs),
often referred to as temporomandibular disorders
(TMDs), comprise a family of musculoskeletal
conditions that involve deep ache or pain in the
area of one or both temporomandibular joints
(TMJs) and/or adjacent tissue structures.
These conditions constitute a major source of
non-dental pain in the craniofacial complex.

Characteristics

Etiology and Pathogenesis

Although the etiology of these musculoskeletal
pain disorders is not established and various pathogenetic constructs have been proposed, these
conditions are believed to develop from the combined action of many genes, risk-conferring
behaviors, and environmental factors. The fact
that pain originates in deep tissue appears to be
relevant to understanding the clinical phenomenon because, unlike superficial pain, deep pain is

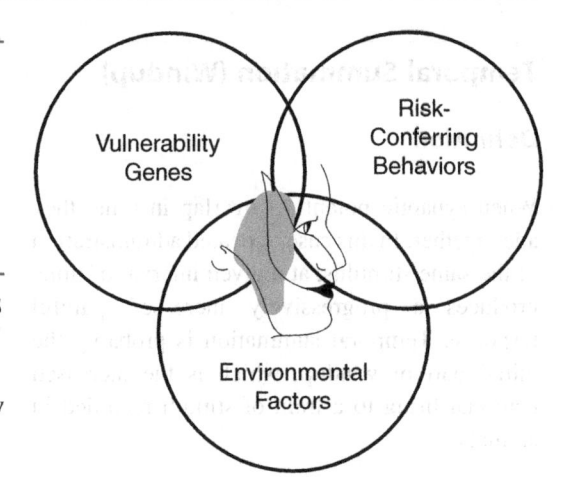

Temporomandibular Joint Disorders, Fig. 1 Etiological construct. Not unlike other complex diseases,
vulnerability is conferred by the interaction of genes,
risk-conferring behaviors, and environmental factors

poorly localized and frequently associated with
pronounced autonomic reactions. Genetic vulnerability is attributed to differences in the genetic
makeup that mediate in part, directly or
indirectly, the vulnerability or conversely the
resiliency to develop these conditions (Fig. 1).

Earlier etiological constructs have placed significant weight on the dental occlusion as a causal
factor in the etiopathogenesis of TMJDs. However, low strengths of association between occlusal features and TMJDs, inconsistent findings
from study to study regarding the role of a given
occlusal attribute, and the absence of any gradient
effect of occlusal factors put these earlier theories
into question.

Case Assignment

Although research in this subject matter has been
intensified in recent years, no biomarkers of
exposure or effect are established for valid and
reliable TMJD case ascertainment. TMJD case
assignment occurs on the basis of clinical
features, which consist of symptoms like pain
and limited range of mandibular motion. Facial
pain reports focus on anatomical regions, such as
the temples, cheeks, pre-auricular area, or inside
the ear, and vary in intensity and spatial distribution, both interindividually and intraindividually,
with time. With respect to corresponding clinical

Temporomandibular Joint Disorders, Fig. 2 Overview of the diagnostic construct adopted by the Research Diagnostic Criteria (RDC-TMD) for temporomandibular disorders (For detail see Dworkin and LeResche 1992)

signs, allodynia in the form of tenderness to palpation is linked to painful topographical sites. Limited range of mandibular motion is often noted and attributed to factors such as the articular disk preventing smooth gliding movement of the mandibular condyle along the articular eminence, constraining mandibular excursion, and/or the recruitment of jaw closing muscles during their function as antagonists, limiting mandibular side-to-side excursions and the capacity to open the jaw fully. However, under no circumstances should a single observable sign be used in isolation to define a TMJD case, because of insufficient diagnostic validity due to high sensitivity and low specificity.

Classification Systems

Important in directing clinical research in the past decades were efforts to produce a dual-axes taxonomy for the major types of TMJDs (Dworkin and LeResche 1992). Focusing on the craniofacial domain, Axis I distinguishes three main diagnostic subsets (Fig. 2):

1. Group I: Masticatory myofascial pain
2. Group II: TMJ internal derangements
3. Group III: TMJ arthritides

Axis II criteria assess pain intensity, pain-related disability, and the presence and severity of depressive and anxiety symptoms. Using this classification scheme, about half of all

TMJD cases are identified as Group I disorders (List and Dworkin 1996).

Due to the overlap with regional myofascial pain, tension-type headache, fibromyalgia, polyarthritides, and possibly connective tissue disorders with impaired collagen makeup, shortcomings of available TMJD taxonomies are increasingly recognized (Fig. 3). The fact that persistent TMJDs are rarely limited to a single topographical domain underscores the need to assess these conditions in the broader context.

Phenomenology

General Characteristics

Poorly localizable ache or pain unrelated to dental pathology constitutes the chief complaint of all major forms of TMJDs. The sensory experience is captured by pain descriptors, such as "aching," "tight," "throbbing," and "tender" (Turp et al. 1997). Besides pain, (a) inability to freely move the jaw due to pain and/or soft or hard tissue interference, (b) sounds originating from the jaw joint, and (c) the disturbing perception of teeth not fitting properly constitute the other shared concerns. With respect to clinically observable signs, pressure allodynia, the experience of pain in response to defined pressure that is rarely identified as painful by subjects without TMJDs, represents the clinical hallmark feature of this family of pain conditions. Inability to

Temporomandibular Joint Disorders, Fig. 3 Overlap of TMJDs with regional and systemic disorders

move the jaw freely is determined by measurements of the mandibular range of motion and
expressed by the clinically observable maximum
mandibular excursions in all directions.
Joint sounds are often linked to mechanical
events between moving articular structures,
such as the temporal component of the TMJ, the
articular disk, and the condyle. With respect to
age, prevalence rates are lower among older subjects, and initial care-seeking in both men and
women is more likely to occur before age 50
than later in life. Differences between the sexes
appear to emerge with pubertal development with
the prevalence of pain reports becoming more
prevalent in females.

Spatial Characteristics
Not only can TMJ arthritides be part of an
existing polyarthritis that affects additional joints
other than the TMJs, those TMJDs that involve
muscle differ in the extent of their bodily involvement as well. Distinction of local and widespread
phenomena is important, because cases with
widespread pain are more likely in pain on

follow-up examination than cases with localized
pain (Raphael et al. 2000).

In contrast to TMJD and muscle pain conditions that involve the face and adjacent head or
neck regions, fibromyalgia (FMS) is understood
as a clinical entity characterized by persistent
widespread pain and tenderness to 4 kg of
pressure at 11 of 18 anatomically defined body
sites (Wolfe et al. 1990). Overlap between
TMJDs and FMS has been demonstrated in
a number of studies (Plesh et al. 1996;
Hedenberg-Magnusson et al. 1997; Korszun
et al. 1998). According to Plesh and coworkers,
75 % of their FMS patients had TMJDs, while, on
the other hand, 18 % of cases with TMJDs met the
diagnostic criteria for FMS. Epidemiological
studies also report high associations between
TMJDs and the two most common types of headache, tension-type headache and migraine headache (Agerberg and Carlsson 1973). In fact,
persistent TMJD pain is associated with comorbid pain in body parts other than the face at much
greater rates than the condition being limited to
the face (Fig. 4) (Turp et al. 1998).

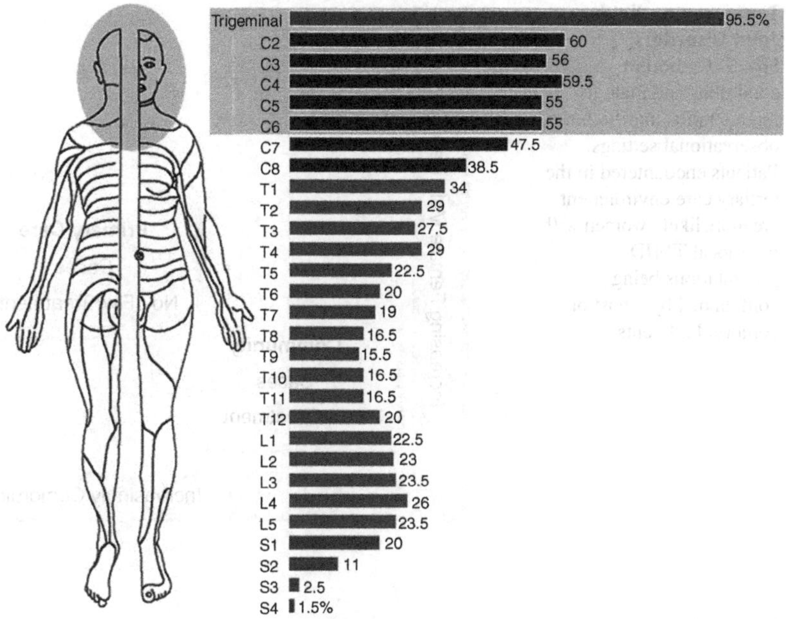

**Temporomandibular
Joint Disorders,**
Fig. 4 Cases (in %)
reporting pain in a given
dermatome. Data were
generated from a case series
of 200 patients, seeking
care for orofacial pain in
a tertiary care clinic
(Adapted from Turp et al.
1998)

Trigeminal — 95.5%
C2 — 60
C3 — 56
C4 — 59.5
C5 — 55
C6 — 55
C7 — 47.5
C8 — 38.5
T1 — 34
T2 — 29
T3 — 27.5
T4 — 29
T5 — 22.5
T6 — 20
T7 — 19
T8 — 16.5
T9 — 15.5
T10 — 16.5
T11 — 16.5
T12 — 20
L1 — 22.5
L2 — 23
L3 — 23.5
L4 — 26
L5 — 23.5
S1 — 20
S2 — 11
S3 — 2.5
S4 — 1.5%

Temporal Characteristics

Complaints of pain range from a local response to simple injury to complaints of persistent widespread bodily involvement without obvious cause. From a phenomenological point of view, it needs to be emphasized that the overwhelming case majority seen in the primary care setting exhibits episodic forms, while cases encountered in the tertiary care environment are more likely affected by persistent conditions. The fact that the personally most devastating and clinically most challenging TMJD presentations occur in females in greater numbers than males results in up to 90 % of tertiary care cases being women (Fig. 5). Among women, prevalence rates are higher for subjects of reproductive age than those in postmenopausal years without hormone replacement therapy based on the formulations used in the 1990s (LeResche et al. 1997).

TMJD pain is characterized as nonprogressive and fluctuating in intensity, which is often translated into "good" and "bad" days in a patient's day-by-day experience. What is applicable to a wide range of pain disorders seems also to be the case for TMJDs. As a generalization, infrequent pains of even high intensity are more likely perceived as a nuisance when compared to persistent pain of lesser intensity. On the other hand, persistent pain disrupts the lifestyle, causing functional limitations and restrictions in daily activities. In this context, it is increasingly understood that time in pain influences the subject's physiological state and response behavior. Initial pain constitutes a warning signal, causing the subject to stop the ongoing activity and to take actions to alleviate the pain. If pain persists, longer-lasting effects on neuronal excitability, such as the upregulation of NMDA-mediated effects and changes in the CNS "hardwiring" occur via a series of events and involve alterations in intermediate and late gene expressions. Binding of c-fos and c-jun to DNA alters the transcription of intermediate and eventually late effector genes, which in turn affect enzymes, growth factors, peptides, and even the phenotype.

Pain Affect

Prolonged and persistent pain can induce significant pain affect, which in itself constitutes an integral part of the TMJDs. Pain affect is captured by pain descriptors, such as "tiring," "exhausting," "frightening," and "fearful." While initial symptoms are linked to seeking advice and possibly care, prolonged, unsuccessfully managed conditions can be emotionally devastating and cause the patient to seek for

Temporomandibular Joint Disorders, Fig. 5 Comorbid conditions and male-to-female ratios in different observational settings. Patients encountered in the tertiary care environment are more likely women with their local TMJD presentations being confounded by a host of comorbid ailments

Temporomandibular Joint Disorders, Fig. 6 Initial symptoms are linked to seeking advice and care. Prolonged, unsuccessfully managed conditions can be emotionally devastating resulting to search for "cures"

"cures" (Fig. 6). Great variations are observed with respect to the degree to which pain affect is expressed from patient to patient, even within a given TMJD subset (Ohrbach and Dworkin 1998). Much of the variability in response to pain and the presence of a stressor are believed to be determined by genetic differences among patients. Consequently, intense research has begun to identify the allelic variants in a host of candidate genes that underlie these response differences. For example, a threefold to fourfold reduction in the enzymatic activity due to a valine-methionine polymorphism of catechol-O-methyltransferase (COMT), an enzyme that catalyzes the O-methylation of compounds with a catechol structure, results in less or greater than normal availability of catecholamines at the site of neurotransmission, which in turn significantly shapes the sensory and affective experience of facial pain, or the pain sensitivity through activation of both beta2 and beta3-adrenergic receptors

(Zubieta et al. 2003; Nackley et al. 2007). Studies examining the effect of global gene expression are under way.

Management

Because the causal sequence of events that leads to pain and dysfunction is not known, therapeutic interventions focus on symptom management rather than on the elimination of the cause. Patients, who seek care for the first time, report symptom relief of TMJD by 65–95 % with a host of interventions that are presumed to exhibit very different modes of action. Treatments include thermal packs, nonsteroidal anti-inflammatory drugs (NSAIDs) and/or muscle relaxants, inter-occlusal appliances, physical therapy, relaxation and stress management, acupuncture, and diet counseling to mention the most common interventions. There are little differences among the various types with respect to symptom relief. Although patients express strong preferences for one treatment over another, properly designed clinical trials show no statistically significant difference in therapeutic efficacy when compared to a credible placebo. Those patients that do not get a satisfactory outcome, which happen to constitute a clear case minority in the primary care setting, are characterized by persistent pain and dysfunction for which all current forms of treatment fall short. Given the questionable superiority of one type of intervention over another, the choice of care is more influenced by unwanted effects attributable to the intervention, and/or the greater cost for care that does not translate into a justifiable improvement of the therapeutic efficacy. Consequently, customary case management is "conservative" and "reversible." In other words, for those cases for whom treatments fall short, primary consideration should emphasize safety rather than efficacy.

References

Agerberg, G., & Carlsson, G. E. (1973). Functional disorders of the masticatory system. II. Symptoms in relation to impaired mobility of the mandible as judged from investigation by questionnaire. *Acta Odontologica Scandinavica, 31*, 337–347.

Dworkin, S. F., & LeResche, L. (1992). Research diagnostic criteria for temporomandibular disorders: Review, criteria, examinations and specifications, critique. *Journal of Craniomandibular Disorders, 6*, 301–355.

Hedenberg-Magnusson, B., Ernberg, M., & Kopp, S. (1997). Symptoms and signs of temporomandibular disorders in patients with fibromyalgia and local myalgia of the temporomandibular system. A comparative study. *Acta Odontologica Scandinavica, 55*, 344–349.

Korszun, A., Papadopoulos, E., Demitrack, M., Engleberg, C., & Crofford, L. (1998). The relationship between temporomandibular disorders and stress-associated syndromes. *Oral Surgery, Oral Medicine, Oral Pathology, Oral Radiology, and Endodontics, 86*, 416–420.

LeResche, L., Saunders, K., Von, K. M., Barlow, W., & Dworkin, S. F. (1997). Use of exogenous hormones and risk of temporomandibular disorder pain. *Pain, 69*, 153–160.

List, T., & Dworkin, S. F. (1996). Comparing TMD diagnoses and clinical findings at Swedish and US TMD centers using research diagnostic criteria for temporomandibular disorders. *Journal of Orofacial Pain, 10*, 240–253.

Nackley, A. G., Tan, K. S., Fecho, K., Flood, P., Diatchenko, L., & Maixner, W. (2007). Catechol-O-methyltransferase inhibition increases pain sensitivity through activation of both beta2- and beta3-adrenergic receptors. *Pain, 128*, 199–208.

Ohrbach, R., & Dworkin, S. F. (1998). Five-year outcomes in TMD: Relationship of changes in pain to changes in physical and psychological variables. *Pain, 74*, 315–326.

Plesh, O., Wolfe, F., & Lane, N. (1996). The relationship between fibromyalgia and temporomandibular disorders: Prevalence and symptom severity. *The Journal of Rheumatology, 23*, 1948–1952.

Raphael, K. G., Marbach, J. J., & Klausner, J. (2000). Myofascial face pain. Clinical characteristics of those with regional vs. widespread pain. *Journal of the American Dental Association, 131*, 161–171.

Turp, J. C., Kowalski, C. J., & Stohler, C. S. (1997). Pain descriptors characteristic of persistent facial pain. *Journal of Orofacial Pain, 11*, 285–290.

Turp, J. C., Kowalski, C. J., O'Leary, T. J., & Stohler, C. S. (1998). Pain maps from facial pain patients indicate a broad pain geography. *Journal of Dental Research, 77*, 1465–1472.

Wolfe, F., Smythe, H. A., Yunus, M. B., Bennett, R. M., Bombardier, C., Goldenberg, D. L., Tugwell, P., Campbell, S. M., Abeles, M., Clark, P., et al. (1990). The American college of rheumatology 1990 criteria for the classification of fibromyalgia. Report of the multicenter criteria committee [see comments]. *Arthritis and Rheumatism, 33*, 160–172.

Zubieta, J. K., Heitzeg, M. M., Smith, Y. R., Bueller, J. A., Xu, K., Xu, Y., Koeppe, R. A., Stohler, C. S., & Goldman, D. (2003). Genotype affects mu-opioid neurotransmitter responses to a pain stressor. *Science, 299*, 1240–1243.

Temporomandibular Joint Disorders (TMJDs)

▶ Temporomandibular Joint Disorders

Temporomandibular Pain

Definition

Acute or chronic pain in the jaw muscles and TM joint, sometimes associated with malocclusion.

Cross-References

▶ Jaw-Muscle Silent Periods (Exteroceptive Suppression)

Tender Points

Michael Spaeth
Friedrich-Baur-Institute, University of Munich,
Munich, Germany

Synonyms

TePs

Definition

If an individual reports local pain when a site is palpated with standardized pressure, this is considered a positive "tender point" (TP).

Characteristics

Anatomy
The anatomic TP sites do not appear to represent a single type of anatomical structure but rather can include ligaments, tendons, skeletal muscles, and bursae. The TP hurts at the site where pressure is applied only, whereas pain induced by pressure at a myofascial pain syndrome (MPS) "trigger point" causes both local pain and pain at a more distant area of reference ("referred pain"). Several efforts have been made to find a primary origin of fibromyalgia (FM) pain at the anatomic sites themselves (Bengtsson et al. 1986; Drewes et al. 1993; Henriksson et al. 1982; Yunus and Kalyan Raman 1989). In fact, most of these investigations studied skeletal muscle exclusively and did not report any findings on other anatomical structures composing the TP regions. Morphological findings in skeletal muscle tissue specimens from FM patients are rather nonspecific and presumably secondary to pain-related reduction of activity. Results of image analysis quantification of substance P immunoreactivity in the trapezius muscle of patients with fibromyalgia and myofascial pain syndrome pointed to a peripheral hyperactivity of the peptidergic nervous system in FM as well as in MPS (De Stefano et al. 2000). Recently reported ultrastructural changes in fibromyalgic muscle fibers may contribute to the induction and/or chronicity of nociceptive transmission from muscle to the central nervous system (Sprott et al. 2004). But these alterations could not be identified as a primary cause of hyperalgesia in FM TP areas.

Clinical Characteristics
Application of pressure on each of the TPs often induces patient's involuntary withdrawal. After the examination of all 18 TPs, patients may report a persisting "deep ache," similar to that of bone pain. Some patients may show the symptoms with either one half (upper or lower) or one side (right or left) of the body preponderating. For standardization and for research purposes, pressure gauges are available. A dolorimeter is commonly used and can further help to standardize the amount of pressure (e.g., 4 kg) applied by the examining finger of each investigator.

Recent Objections
Despite the lack of information on what TPs really do measure, research over the last few years has brought up some interesting findings about these points. There is evidence from different studies that FM patients are tenderer in both TP and non-TP regions than healthy control subjects and that these TP regions represent areas, where anyone is tenderer. TPs were revealed not to be specific to FM (Granges and Littlejohn 1993b; Tunks et al. 1995).

TPs were studied in a random sample from adults in the general population. As a result, tenderness to pressure was found to occur both in people without widespread pain and in people without any pain. The investigators found that TP counts were increased in people who had other symptoms (i.e., poor sleep and/or fatigue), even if they did not complain about pain at all (MacFarlane et al. 1996). Data suggest that TP counts can discriminate between tender and nontender individuals and can therefore be considered as a clinically useful measure of tenderness (Gracely et al. 2003). From another study, it was concluded that the tender point count was associated not only with the extent of rheumatic pain but also independently with the extent of bodily complaints (Schochat and Raspe 2003). Significant correlations were found between TP count and psychological distress, evaluated by analyzing somatic and depressive symptoms (Croft et al. 1994). Another study found that TP pain severity ratings produced higher correlations with symptoms of FM and predicted distress better than TP counts (McCarberg et al. 2003).

Since the TP count seems to be a composite measure of at least tenderness and psychological distress, it is of limited value in research settings but useful in a clinical setting in order to recognize the tenderness-distress nature of FM (Gracely et al. 2003). Furthermore, the TP count does not reflect differences in distress or pressure-pain sensitivity or provide help in subgrouping FM patients (Giesecke et al. 2003). Another study replicated previous findings in population-based samples showing that dolorimeter determinations are less influenced by psychological factors than TP counts (Croft et al. 1994; Granges and Littlejohn 1993a), but there was still an impact of distress even on dolorimetry results (Petzke et al. 2003).

Most of these data-based objections were followed by recommendations: (1) to reconsider the current definition of FM, (2) to be aware of the tenderness-distress nature of both FM and TPs, and (3) to reevaluate chronic widespread pain (CWP) in further population-based studies both to potentially discriminate between CWP and FM and to emphasize FM characteristics.

Cross-References

▶ Muscle Pain, Fibromyalgia Syndrome (Primary, Secondary)

References

Bengtsson, A., Henriksson, K. G., & Larsson, J. (1986). Muscle biopsy in primary fibromyalgia. Light-microscopical and histochemical findings. Scandinavian Journal of Rheumatology, 15, 1–6.

Croft, P., Schollum, J., & Silman, A. (1994). Population study of tender point counts and pain as evidence of fibromyalgia. British Medical Journal, 309, 696–699.

De Stefano, R., Selvi, E., Villanova, M., et al. (2000). Image analysis quantification of substance P immunoreactivity in the trapezius muscle of patients with fibromyalgia and myofascial pain syndrome. Journal of Rheumatology, 27, 2906–2910.

Drewes, A. M., Andreasen, A., Schroder, H. D., et al. (1993). Pathology of skeletal muscle in fibromyalgia: A histo-immuno-chemical and ultrastructural study. British Journal of Rheumatology, 32, 479–483.

Giesecke, T., Williams, D. A., Harris, R. E., et al. (2003). Subgrouping of fibromyalgia patients on the basis of pressure-pain thresholds and psychological factors. Arthritis and Rheumatism, 48, 2916–2922.

Gracely, R. H., Grant, M. A., & Giesecke, T. (2003). Evoked pain measures in fibromyalgia. Best Practice & Research. Clinical Rheumatology, 17, 593–609.

Granges, G., & Littlejohn, G. (1993a). Pressure pain threshold in pain-free subjects, in patients with chronic regional pain syndromes, and in patients with fibromyalgia syndrome. Arthritis and Rheumatism, 36, 642–646.

Granges, G., & Littlejohn, G. O. (1993b). A comparative study of clinical signs in fibromyalgia/fibrositis syndrome, healthy and exercising subjects. Journal of Rheumatology, 20, 344–351.

Henriksson, K. G., Bengtsson, A., Larsson, J., et al. (1982). Muscle biopsy findings of possible diagnostic importance in primary fibromyalgia (fibrositis, myofascial syndrome). Lancet, 2, 1395.

MacFarlane, G. J., Croft, P. R., Schollum, J., et al. (1996). Widespread pain: Is an improved classification possible? Journal of Rheumatology, 23, 1628–1632.

McCarberg, B., Barkin, R. L., Wright, J. A., et al. (2003). Tender points as predictors of distress and the pharmacologic management of fibromyalgia syndrome. American Journal of Therapeutics, 10, 176–192.

Petzke, F., Gracely, R. H., Park, K. M., et al. (2003). What do tender points measure? Influence of distress on 4 measures of tenderness. Journal of Rheumatology, 30, 567–574.

Schochat, T., & Raspe, H. (2003). Elements of fibromyalgia in an open population. Rheumatology (Oxford, England), 42, 829–835.

Sprott, H., Salemi, S., Gay, R. E., et al. (2004). Increased DNA fragmentation and ultrastructural changes in fibromyalgic muscle fibres. Annals of the Rheumatic Diseases, 63, 245–251.

Tunks, E., McCain, G. A., Hart, L. E., et al. (1995). The reliability of examination for tenderness in patients with myofascial pain, chronic fibromyalgia and controls. Journal of Rheumatology, 22, 944–952.

Yunus, M. B., & Kalyan Raman, U. P. (1989). Muscle biopsy findings in primary fibromyalgia and other forms of nonarticular rheumatism. Rheumatic Diseases Clinics of North America, 15, 115–134.

Tenderness

Definition

Tenderness describes a feeling of discomfort or pain caused by light pressure or moving mechanical stimuli that would normally be insufficient to cause such sensations.

Cross-References

▶ Headache, Episodic Tension-Type

Tendinitis

Definition

Tendinitis is a painful tendon, usually resulting from unaccustomed physical activity. It is classified as a localized STP. Fraying and thickening of the tendon may be observed.

Cross-References

▶ Ergonomics Essay
▶ Muscle Pain, Fibromyalgia Syndrome (Primary, Secondary)

Tendon Sheath Inflammation

Definition

Tendon sheaths have synovial lining cells, which are included in the inflammation in rheumatoid arthritis.

Cross-References

▶ Muscle Pain in Systemic Inflammation (Polymyalgia Rheumatica, Giant Cell Arteritis, Rheumatoid Arthritis)

Tenosynovitis

Definition

Tenosynovitis refers to inflammation of the tendon sheaths, through which the tendons slide when the muscle length changes. Excessive fluid accumulation can cause swelling and pain in the affected areas.

Cross-References

▶ Ergonomics Essay

TENS

▶ Transcutaneous Electrical Nerve Stimulation

TENS Outcomes

▶ Transcutaneous Electrical Nerve Stimulation Outcomes

TENS, Mechanisms of Action

Kathleen A. Sluka
Physical Therapy and Rehabilitation Science
Graduate Program, University of Iowa, Iowa City, IA, USA

Synonyms

Electrical stimulation analgesia; PES; Transcutaneous electrical nerve stimulation

Definition

Electrical stimulation applied to the skin for pain relief.

Characteristics

The mechanisms of action of ▶ TENS primarily involve central mechanisms and have been extensively reviewed (see Sluka and Walsh 2009 for more details and references). There are generally two types of TENS applied clinically, low frequency (<10 Hz) and high frequency (>50 Hz). These can be applied at either a sensory intensity that produces a tapping or tingling sensation or at motor intensity that produces an additional motor contraction. The mechanisms of action for TENS appear to be frequency, not intensity, dependent.

High-Frequency (50–100 Hz) TENS

Effects on Behavior and Dorsal Horn Neurons
Early studies utilizing acute pain tests show that high-frequency, motor intensity TENS increases the tail flick latency to heat (i.e., analgesia) and decreases the flexion reflex response to noxious stimuli (reviewed in Sluka and Walsh 2009). Recording from spinothalamic tract cells, stimulation at an intensity activating Aβ fibers (3 × the threshold) has no effect on the spontaneous firing rate. However, increasing the intensity so as to also activate Aδ nociceptors reduces spontaneous activity and responses to noxious heat or pinch (Lee et al. 1985). Similarly, studies by Garrison and Foreman (1997) and by Sjolund (1985) both show that increasing intensity increases inhibition of dorsal horn neurons and the flexion reflex response to noxious stimuli. These data suggest that high- and low-frequency TENS are effective and that increasing intensity increases inhibition.

Utilizing an animal model of joint inflammation reveals that high-frequency, sensory intensity TENS has long-lasting effects on both primary and secondary heat and mechanical ▶ hyperalgesia (reviewed in Sluka and Walsh 2009) (Fig. 1). In fact, these studies show that high-frequency, sensory intensity reverses the primary hyperalgesia and secondary hyperalgesia associated with ▶ carrageenan inflammation for 24 h. Importantly, modulation of frequency (4 Hz vs. 100 Hz), intensity (sensory vs. motor), or pulse duration (100 μs vs. 250 μs) shows a frequency, but pulse duration, dependent effect on primary hyperalgesia to mechanical and heat stimuli in animals with carrageenan paw inflammation. There is a minimum intensity that is needed to produce analgesia with TENS where 50 % motor threshold and below or at sensory perception are not effective, while 90 % motor threshold and above reverse hyperalgesia and increase pain thresholds (Sato et al. 2012; Moran et al. 2011). The increased responsiveness of dorsal horn neurons to innocuous and noxious mechanical stimuli that occurs after inflammation is completely reduced following high-frequency, sensory intensity TENS treatment applied to the inflamed paw (see Sluka and Walsh 2009). Utilizing a ▶ model of neuropathic pain, Somers and Clemente (1998) demonstrated that high-frequency, sensory intensity TENS stimulation over the paraspinal musculature reduced the heat but not the mechanical hyperalgesia that normally occurs in this model. This inhibition of heat hyperalgesia only occurs if TENS was started the first day after injury but not if it was started 3 days after injury.

Pharmacology
In animals that were spinalized to remove descending inhibitory pathways (Fig. 2), inhibition of the tail flick by high-frequency, motor intensity TENS still occurs but is reduced by about 50 % (see Sluka and Walsh 2009 for review). Thus, these studies suggest that both spinal and ▶ descending inhibition are involved in the analgesia produced by high-frequency, motor intensity TENS. Later studies prevented the antihyperalgesia, by blockade of δ-opioid receptors in the rostral ventral medial medulla (RVM), further supporting a role for descending inhibitory systems in the inhibition produced by TENS.

Pharmacologically, opioid peptides mediate the effects of high-frequency TENS. Concentrations of beta-endorphins increase in the bloodstream and cerebrospinal fluid, and methionine-enkephalin increases in the cerebrospinal fluid of human subjects, following

TENS, Mechanisms of Action, Fig. 1 Effects of TENS on primary and secondary and mechanical and heat hyperalgesia induced by carrageenan inflammation. High, but not low, frequency TENS partially reverses primary hyperalgesia to heat and mechanical stimuli induced by carrageenan paw inflammation (*left panels*). In contrast, both high- and low-frequency TENS reverse secondary hyperalgesia induced by carrageenan knee-joint inflammation (*right panels*)

administration of high-frequency, sensory intensity TENS (reviewed in Sluka and Walsh 2009). High-frequency, motor intensity TENS is blocked by systemic blockade of opioid receptors with naloxone and systemic depletion of serotonin (reviewed in Sluka and Walsh 2009). Blockade of δ-opioid receptors in the spinal cord or the rostral ventral medial medulla (RVM) reverses the antihyperalgesia produced by high-frequency sensory intensity TENS in animals with carrageenan knee-joint inflammation (Fig. 3) (Kalra et al. 2001; Sluka et al. 1999). Similarly, spinal δ-opioid receptors are implicated in the antihyperalgesic effects of high-frequency motor intensity TENS, since repeated application of high-frequency TENS produces tolerance (reduced effectiveness) to the antihyperalgesic effects of TENS and at spinal δ-opioid receptors (Chandran and Sluka 2003). Tolerance can be prevented by mixing frequencies at the time of stimulation, alternating

frequencies on different days or increasing intensity daily by 10 % (deSantana et al. 2008; Sato et al. 2012).

Further, blockade of muscarinic receptors (M1 and M3, but not M2) in the spinal cord also partially reverses the antihyperalgesia produced by high-frequency, sensory intensity TENS (Radhakrishnan and Sluka 2003). However, blockade of serotonin or noradrenergic receptors in the spinal cord has no effect on the reversal of hyperalgesia produced by high-frequency, sensory intensity TENS (see Sluka and Walsh (2009) for details) (Fig. 3).

Autonomic and Peripheral Effects

TENS could have effects on autonomic function, blood flow, and peripheral afferent fibers (reviewed in Sluka and Walsh 2009). However, high-frequency, sensory intensity TENS stimulation at intensities just above or below motor threshold does not affect local blood flow.

TENS, Mechanisms of Action, Fig. 2 Schematic drawing demonstrating that *TENS* applied to the periphery at the site of injury activates primary afferent fibers. This information is transmitted to the spinal cord and results in inhibition both locally and from descending inhibitory pathways. Descending inhibition from the rostral ventral medial medulla (*RVM*) involves 5-HT and opioids and can be activated by the periaqueductal gray (*PAG*). Previous studies show that opioid receptors in the spinal cord and RVM and serotoninergic and muscarinic receptors in the spinal cord mediate the reduction in hyperalgesia by *TENS*

In contrast, utilizing laser Doppler imaging, increases in blood flow were observed with high-frequency TENS, at an intensity "that was felt but not painful" (10–15 mA). In human subjects, after application of high-frequency TENS at the threshold for discomfort (strong motor intensity applied to a digit), subjects report numbness and cooling just distal to the stimulation (on the digit). This is associated with decreased temperature and loss of color in the skin suggesting effects on the autonomic nervous system. The primary afferent neuropeptide, substance P, is reduced in dorsal root ganglia neurons and spinal cord dorsal horn by high-frequency, sensory intensity TENS in animals injected with the inflammatory irritant, formalin. Thus, evidence is beginning to emerge that some of the analgesic effects of TENS may be mediated through actions on primary afferent fibers and modulation of autonomic activity.

Low-Frequency (<10 Hz) TENS
Effects on Behavior and Dorsal Horn Neuron Activity
In primates without tissue injury, low-rate burst TENS (3 bursts per second and 7 pulses per burst with an internal frequency of 85 Hz) at an intensity that activates Aβ (presumable sensory intensity, $3 \times$ sensory threshold) fibers has no effect on either the spontaneous activity or responses to noxious stimuli of spinothalamic tract cells. Increasing intensity to activate Aδ fibers reduces spontaneous activity and responses to noxious stimuli of spinothalamic tract cells (Lee et al. 1985). Similarly, low-frequency TENS at an intensity that activates Aδ fibers reduces the ventral root reflex in response to C-fiber stimulation (Sjolund 1985).

Low-frequency, sensory intensity TENS reverses primary and secondary hyperalgesia and mechanical hyperalgesia (reviewed in Sluka and Walsh 2009). Like high-frequency TENS,

TENS, Mechanisms of Action, Fig. 3 Summary bar graph of the effects of blockade of spinal receptors on the antihyperalgesia produced by low- and high-frequency TENS. Approximately 100 % inhibition of hyperalgesia occurs after treatment with either high- or low-frequency TENS (*saline, purple*). Blockade of μ-opioid (*naloxone, dark blue*) and muscarinic (*atropine, red*) receptors prevents the antihyperalgesia produced by high-frequency TENS. Blockade of spinal δ-opioid (*naltrindole, blue*), serotonin (*methysergide, green*), or muscarinic (*atropine, red*) receptors prevents the antihyperalgesia produced by low-frequency TENS. Spinal blockade of α-2 adrenergic (*yohimbine, yellow*) or nicotinic (*mecamylamine, orange*) receptors has no effect on the effects of either high- or low-frequency TENS

analgesic effects depend on intensity with intensities ≥90 % motor threshold producing analgesia and intensities at 50 % motor threshold or at sensory perception threshold or below ineffective (Sato et al. 2012; Moran et al. 2011). The increased responsiveness of dorsal horn

neurons to innocuous and noxious mechanical stimuli that occurs after inflammation is equally and completely reduced following low-frequency, sensory intensity TENS treatment applied to the inflamed paw (Ma and Sluka 2001). Following spinal nerve ligation, TENS reduces the responsiveness to noxious mechanical stimulation of dorsal horn neurons in both normal and neuropathic animals. However, the responsiveness of spinal neurons to innocuous mechanical stimulation is only inhibited by TENS in neuropathic animals (Leem et al. 1995). Behaviorally, low-frequency, motor intensity TENS reduces mechanical hyperalgesia and cold allodynia induced by nerve injury (Nam et al. 2001).

Pharmacology

Low-frequency, sensory intensity TENS antihyperalgesia is prevented by blockade of μ-opioid receptors in the spinal cord or the RVM (Fig. 3) (Kalra et al. 2001; Sluka et al. 1999) or peripherally at the site of application (Sabino et al. 2008). Studies utilizing carrageenan knee-joint inflammation suggest that μ-opioid receptors are also activated by low-frequency TENS since repeated application of TENS produces tolerance (reduced effectiveness) to the antihyperalgesic effects of TENS and of spinal μ-opioid receptors (Chandran and Sluka 2003). Like high-frequency TENS, tolerance can be alleviated with mixing frequencies, alternating frequencies, or increasing intensities (deSantana et al. 2008; Sato et al. 2012). Low-frequency, sensory intensity TENS is also reduced by blockade of serotonin 5-HT2A and 5-HT3 and muscarinic M1 and M3 receptors in the spinal cord (Fig. 3) (Radhakrishnan et al. 2003; Radhakrishnan and Sluka 2003). Taken together, these studies suggest a role of opioid, serotonin, and muscarinic receptors in the spinal cord and supraspinal opioid mechanisms in the action of low-frequency, sensory intensity TENS.

Autonomic Effects of Low-Frequency TENS

The effect of low-frequency, motor intensity TENS on cold allodynia, but not mechanical hyperalgesia, is reduced by systemic phentolamine to block α-adrenergic receptors, suggesting that activation of sympathetic noradrenergic receptors may mediate TENS effects (Nam et al. 2001). However, phentolamine could block central receptors. Transient increases in blood flow with low-frequency, burst-mode (2 Hz) TENS were observed at the area of stimulation, if intensity was 25 % above the motor threshold, but not just below (sensory intensity) or just above motor threshold (Sherry et al. 2001).

References

Chandran, P., & Sluka, K. A. (2003). Development of opioid tolerance with repeated TENS administration. *Pain, 101*, 195–201.

DeSantana, J. M., Santana-Filho, V. J., & Sluka, K. A. (2008). Modulation between high- and low-frequency transcutaneous electric nerve stimulation delays the development of analgesic tolerance in arthritic rats. *Archives of Physical Medicine and Rehabilitation, 89*, 754–780.

Garrison, D. W., & Foreman, R. D. (1997). Effects of prolonged transcutaneous electrical nerve stimulation (TENS) and variation of stimulation variables on dorsal horn cell activity. *European Journal of Physical and Rehabilitation Medicine, 6*, 87–94.

Kalra, A., Urban, M. O., & Sluka, K. A. (2001). Blockade of opioid receptors in rostral ventral medulla prevents antihyperalgesia produced by transcutaneous electrical nerve stimulation (TENS). *Journal of Pharmacology and Experimental Therapeutics, 298*, 257–263.

Lee, K. H., Chung, J. M., & Willis, W. D. (1985). Inhibition of primate spinothalamic tract cells by TENS. *Journal of Neurosurgery, 62*, 276–287.

Leem, J. W., Park, E. S., & Paik, K. S. (1995). Electrophysiological evidence for the antinociceptive effect of transcutaneous electrical nerve stimulation on mechanically evoked responsiveness of dorsal horn neurons in neuropathic rats. *Neuroscience Letters, 192*, 197–200.

Ma, Y. T., & Sluka, K. A. (2001). Reduction in inflammation-induced sensitization of dorsal horn neurons by transcutaneous electrical nerve stimulation in anesthetized rats. *Experimental Brain Research, 137*(1), 94–102.

Moran, F., Leonard, T., Hawthorne, S., Hughes, C. M., McCrum-Gardner, E., Johnson, M. I., Rakel, B. A., Sluka, K. A., & Walsh, D. M. (2011). Hypoalgesia in response to transcutaneous electrical nerve stimulation (TENS) depends on stimulation intensity. *The Journal of Pain, 12*(8), 929–35.

Nam, T. S., Choi, Y., Yeon, D. S., et al. (2001). Differential antinociceptive effect of transcutaneous electrical stimulation on pain behavior sensitive or insensitive to phentolamine in neuropathic rats. *Neuroscience Letters, 301*, 17–20.

Radhakrishnan, R., & Sluka, K. A. (2003). Spinal muscarinic receptors are activated during low or high frequency TENS -induced antihyperalgesia in rats. *Neuropharmacology, 45*, 1111–1119.

Radhakrishnan, R., King, E. W., Dickman, J. K., Herold, C. A., Johnston, N. F., Spurgin, M. L., & Sluka, K. A. (2003). Spinal 5-HT(2) and 5-HT(3) receptors mediate low, but not high, frequency TENS-induced antihyperalgesia in rats. *The Journal of Pain, 105*(1–2), 205–13.

Resende, M. A., Sabino, G. G., Candido, C. R. M., Pereira, L. S. M., & Francischi, J. N. (2004). Transcutaneous electrical stimulation (TENS) effects in experimental inflammatory edema and pain. *European Journal of Pharmacology, 504*, 217–222.

Sabino, G. S., Santos, C. M., Francischi, J. N., de Resende, M. A. (2008). Release of endogenous opioids following transcutaneous electric nerve stimulation in an experimental model of acute inflammatory pain. *The Journal of Pain, 9*, 157–163.

Sato, K. L., Sanada, L. S., Rakel, B. A., & Sluka, K. A. (2012). Increasing intensity of TENS prevents analgesic tolerance in rats. *The Journal of Pain, 13*(9), 884–890.

Sherry, J. E., Oehrlein, K. M., Hegge, K. S., et al. (2001). Effect of burst-mode transcutaneous electrical nerve stimulation on peripheral vascular resistance. *Physical Therapy, 81*, 1183–1191.

Sjolund, B. H. (1985). Peripheral nerve stimulation suppression of C-fiber evoked flexion reflex in rats. Part 1: Parameters of continuous stimulation. *Journal of Neurosurgery, 63*, 612–616.

Sluka, K. A., & Walsh, D. M. (2009). Transcutaneous electrical nerve stimulation and interferential current. In K. A. Sluka (Ed.), *Mechanisms and management of pain for the physical therapist* (pp. 167–190). Seattle: IASP Press.

Sluka, K. A., Deacon, M., Stibal, A., et al. (1999). Spinal blockade of opioid receptors prevents the analgesia produced by TENS in arthritic rats. *Journal of Pharmacology and Experimental Therapeutics, 289*, 840–846.

Somers, D. L., & Clemente, F. R. (1998). High-frequency transcutaneous electrical nerve stimulation alters thermal but not mechanical allodynia following chronic constriction injury of the rat sciatic nerve. *Archives of Physical Medicine and Rehabilitation, 79*, 1370–1376.

Tension Headache

▶ Headache, Episodic Tension-Type

Tension Receptors

▶ Visceral Mechanoreceptors

Tension-Type Headache

Definition

Tension-type headache consists of pain with two of the following characteristics: pressing or tightening quality, mild or moderate intensity, bilateral location, and not aggravated by routine physical activity. Headache should be otherwise featureless. Headache is not attributed to another disorder, an individual's use of analgesics or other acute medication, and is ≤10 days per month.

Cross-References

▶ Headache Due to Arteritis
▶ Headache, Episodic Tension-Type
▶ Sensitization of Muscular and Articular Nociceptors

TePs

▶ Tender Points

Tertiary Gain

Definition

Gains sought or obtained by others from a patient's illness.

Cross-References

▶ Malingering, Primary and Secondary Gain

Testosterone

Definition

Testosterone is normally produced in the testes in men, the ovaries in women, and in the adrenal cortex of both men and women. In men, testosterone is primarily responsible for normal growth and development of male sex and reproductive organs, including the penis, testicles, scrotum, prostate, and seminal vesicles. It facilitates the development of secondary male sex characteristics such as musculature, bone mass, fat distribution, hair patterns, laryngeal enlargement, and vocal cord thickening. In women, testosterone strengthens bone and ensures the nipples and clitoris are sensitive to sexual pleasure. In both men and women, normal testosterone levels maintain energy level, healthy mood, muscle mass, fertility, and sexual desire. Decreased levels of testosterone, induced by aging, disease, surgery, and medications (including opioids), lead to loss of libido and decreased sexual function.

Cross-References

▶ Cancer Pain Management, Opioid Side Effects, Endocrine Changes, and Sexual Dysfunction

Tetrodotoxin

Synonyms

TTX

Definition

Tetrodotoxin, often abbreviated as TTX, is a potent neurotoxin found in the tissues of the puffer fish. It is not proteinaceous and has a complex structure, with the active part of the molecule being a positively charged guanidinium moiety. The molecule blocks action potentials by binding sodium channels from the outside (in contrast to clinically used local anesthetic agents that must first permeate the neuron and then act from the inside). Tetrodotoxin is used by neurophysiologists to categorize various types of sodium channels. In nociceptors, a fraction of the sodium current is resistant to block by TTX.

Cross-References

- ▶ Acute Pain Mechanisms
- ▶ Membrane-Stabilizing Drugs
- ▶ Nociceptors, Action Potentials, and Post-Firing Excitability Changes
- ▶ Tetrodotoxin

Tetrodotoxin (TTX)-Resistant Sodium Channel

Definition

A type of voltage-gated sodium channel that is not inhibited by the highly potent neurotoxin, tetrodotoxin, which is extracted from the puffer fish. Tetrodotoxin (TTX)-resistant sodium channels are found on the membrane of many DRG neurons that have nociceptive response properties and contribute to the excitability of the neurons and generation of action potentials.

Cross-References

- ▶ IB4-Positive Neurons, Role in Inflammatory Pain
- ▶ Nociceptor Generator Potential

TG

- ▶ Trigeminal Ganglion

Thalamatomy, Pain Behaviors in Animals

- ▶ Lateral Thalamic Lesions, Pain Behavior in Animals

Thalamic Bursting Activity

- ▶ Burst Activity in Thalamus and Pain

Thalamic Bursting Activity, Chronic Pain

Frederick A. Lenz[1], A. Taghva[1], Andres Fernandez[1], S. H. Patel[1] and Nirit Weiss[2]
[1]Department of Neurosurgery, Johns Hopkins Hospital, Baltimore, MD, USA
[2]Department of Neurosurgery, Mount Sinai Medical Center, New York, NY, USA

Synonyms

Spike bursting

Definition

Spontaneous thalamic cellular activity is often categorized as either bursting activity (▶ spike bursts, bursting mode) or as tonic firing (tonic mode) (Steriade et al. 1990). Many studies have suggested that increased ▶ spike bursting occurs in the thalamus of patients with chronic pain (Lenz et al. 1994, 1998a; Rinaldi et al. 1991; Jeanmonod et al. 1993).

Characteristics

The Thalamic Region of Vc and Its Importance in Pain Processing

Several lines of evidence demonstrate that the ventral caudal nucleus of the human sensory thalamus (▶ Ventrocaudal Nucleus (Vc)), the human analog of monkey ventral posterior (VP) nucleus (Hirai and Jones 1989), is important in human pain-signaling pathways. Studies of patients at autopsy following lesions of the ▶ spinothalamic tract (STT) show the most dense STT termination in the Vc region including the posterior and inferior subnuclei of Vc, suprageniculate, and posterior subnuclei (Mehler 1962; Walker 1943). In monkeys, STT terminations are found in the Vmpo of Craig, which may, by immunohistochemistry and physiology, have a human analog (Craig et al. 1994).

Is Thalamic Functional Mode Altered in Chronic Pain States?

▶ Spike bursting activity refers to a particular pattern of ▶ interspike intervals (ISI) between action potentials, such that a ▶ spike burst begins after a relatively long ISI and is comprised of a series of action potentials with short ISIs (typically <6 ms) (Lenz et al. 1998a). Thereafter, the ISIs progressively lengthen so that the cell's firing decelerates throughout the spike burst. When the period of the bursting is completed, the cell is said to fire in tonic mode when the firing rate is relatively constant and bursts do not occur.

In patients with spinal transection, the highest rate of bursting occurs in cells that do not have peripheral ▶ receptive fields and that are located in the representation of the anesthetic part of the body. These cells also have the lowest firing rates in the interval between bursts (Lenz et al. 1994). The low firing rates suggest that these cells have decreased tonic excitatory drive and are ▶ hyperpolarized, perhaps due to loss of excitatory input from the STT (Eaton and Salt 1990; Dougherty et al. 1996). Therefore the available evidence suggests that affected thalamic cells in patients with spinal transection were dominated by spike bursting consistent with membrane hyperpolarization (Steriade and Llinas 1988; Lenz et al. 1998a; Davis et al. 1998).

Spike bursting activity is maximal in the region posterior and inferior to the core nucleus of Vc (Table 4 in Lenz et al. 1994). Stimulation in this area evokes the sensation of pain more frequently than does stimulation in the core of Vc (Lenz et al. 1993; Hassler 1970). Thus, increased spike bursting activity may be correlated with some aspects of the abnormal sensations (e.g., ▶ Dysesthesias or pain) that these patients experience. However, in patients with spinal transection, the painful area and the area of sensory loss overlap (Lenz et al. 1994). Thus, the bursting activity might be related to sensory loss, rather than to pain.

These findings about spike bursting activity in spinal patients have been called into question by a recent study in patients with chronic pain (Radhakrishnan et al. 1999). It has been reported that the number of bursting cells per trajectory in patients with movement disorders (controls) is not different from that in patients with chronic pain. However, there are significant differences between the two studies (Lenz et al. 1994; Radhakrishnan et al. 1999) in terms of patient population (spinal cord injury vs. mixed chronic pain), location of cells studied (Vc vs. anterior and posterior to Vc), and analysis methods (incidence of bursting cells vs. bursting parameters). Clearly, the increase in bursting activity demonstrated in the earlier study is more applicable to the region of the principal somatic sensory nucleus of patients with central pain from spinal transection (Lenz et al. 1994).

Further support for increased spike bursts occurring in spinal cord-transected patients is found in thalamic recordings from monkeys with thoracic anterolateral cordotomies (Weng et al. 2000). Some of these animals showed increased responsiveness to electrocutaneous stimuli and thus may represent a model of central pain (Vierck 1991). The most pronounced changes in firing pattern were found in thalamic multi-receptive cells, which respond to both cutaneous brushing and compressive stimuli with activity that is not graded into the noxious range. In comparison with normal controls, multi-receptive cells in the monkeys with cordotomies showed significant increases in the number of bursts occurring spontaneously or in response to brushing or compressive stimuli. The changes in bursting behavior were widespread, occurring in the thalamic representation of upper and lower extremities, both ipsilateral and contralateral to the cordotomy.

Although there is an increase in spike burst activity in chronic pain states, there does not appear to be a direct relationship between spike burst firing and pain. Spike bursts are also found in the thalamic representation of the monkey upper extremity and of the representation of the arm and leg ipsilateral to the ▶ cordotomy (Weng et al. 2000). Pain is not typically experienced in these parts of the body in patients with thoracic spinal cord transection or cordotomy (Beric et al. 1988). Spike bursts are increased in frequency during slow-wave sleep in all mammals studied (Steriade and Llinas 1988) including man (Zirh et al. 1997). However, such bursting could cause pain if stimulation in the vicinity of the bursting cell produced the sensation of pain. This finding has been reported in two recent studies of sensations evoked by stimulation of the region of Vc in patients with chronic pain secondary to neural injury (Davis et al. 1996; Lenz et al. 1998a).

Further evidence that thalamic bursting states help modulate but do not directly cause pain comes from studies in mice. (Cheong et al. 2008) demonstrated that the modulation of burst and tonic firing in the thalamus helps control pain sensory gating in the mouse. Subsequent studies have demonstrated the role of bursting activity and intra-burst intervals in thalamocortical relay neurons, in the response to chronic inflammatory pain.

Thus, there is evidence from both human and animal studies for a correlation between chronic pain states and an altered thalamic neuronal action-potential firing pattern. It appears that there is an increase in spike burst firing in chronic pain conditions. The exact physiological relationships that link the pattern of thalamic firing to the human perception of chronic pain have yet to be elucidated.

References

Beric, A., Dimitrijevic, M. R., & Lindblom, U. (1988). Central dysesthesia syndrome in spinal cord injury patients. *Pain, 34*, 109–116.

Cheong, E., Lee, S., Choi, B. J., et al. (2008). Tuning thalamic firing modes via simultaneous modulation of T- and L-type Ca2+ channels controls pain sensory

gating in the thalamus. *Journal of Neuroscience, 28*(49), 13331–13340.

Craig, A. D., Bushnell, M. C., Zhang, E. T., et al. (1994). A thalamic nucleus specific for pain and temperature sensation. *Nature, 372*, 770–773.

Davis, K. D., Kiss, Z. H. T., Tasker, R. R., et al. (1996). Thalamic stimulation-evoked sensations in chronic pain patients and nonpain (movement disorder) patients. *Journal of Neurophysiology, 75*, 1026–1037.

Davis, K. D., Kiss, Z. H. T., Luo, L., et al. (1998). Phantom sensations generated by thalamic microstimulation. *Nature, 391*, 385–387.

Dougherty, P. M., Li, Y. J., Lenz, F. A., et al. (1996). Evidence that excitatory amino acids mediate afferent input to the primate somatosensory thalamus. *Brain Research, 278*, 267–273.

Eaton, S. A., & Salt, T. E. (1990). Thalamic NMDA receptors and nociceptive sensory synaptic transmission. *Neuroscience Letters, 110*, 297–302.

Hassler, R. (1970). Dichotomy of facial pain conduction in the diencephalon. In A. E. Walker (Ed.), *Trigeminal neuralgia* (pp. 123–138). Philadelphia: Saunders.

Hirai, T., & Jones, E. G. (1989). A new parcellation of the human thalamus on the basis of histochemical staining. *Brain Research Reviews, 14*, 1–34.

Huh, Y., Bhatt, R., Jung, D., H-s, S., & Cho, J. (2012). Interactive responses of a thalamic neuron to formalin induced lasting pain in behaving mice. *PLoS One, 7*(1), e30699. doi:10.1371/journal.pone.0030699.

Jeanmonod, D., Magnin, M., & Morel, A. (1993). Thalamus and neurogenic pain: Physiological, anatomical and clinical data. *Neuroreport, 4*, 475–478.

Lenz, F. A., Seike, M., Richardson, R. T., et al. (1993). Thermal and pain sensations evoked by microstimulation in the area of human ventrocaudal nucleus. *Journal of Neurophysiology, 70*, 200–212.

Lenz, F. A., Kwan, H. C., Martin, R., et al. (1994). Characteristics of somatotopic organization and spontaneous neuronal activity in the region of the thalamic principal sensory nucleus in patients with spinal cord transection. *Journal of Neurophysiology, 72*, 1570–1587.

Lenz, F. A., Gracely, R. H., Baker, F. H., et al. (1998a). Reorganization of sensory modalities evoked by stimulation in the region of the principal sensory nucleus (ventral caudal – Vc) in patients with pain secondary to neural injury. *The Journal of Comparative Neurology, 399*, 125–138.

Lenz, F. A., Zirh, A. T., Garonzik, I. M., et al. (1998b). Neuronal activity in the region of the principle sensory nucleus of human thalamus (ventralis caudalis) in patients with pain following amputations. *Neuroscience, 86*, 1065–1081.

Mehler, W. R. (1962). The anatomy of the so-called "pain tract" in man: An analysis of the course and distribution of the ascending fibers of the fasciculus anterolateralis. In J. D. French & R. W. Porter (Eds.), *Basic research in paraplegia* (pp. 26–55). Springfield: Thomas.

Radhakrishnan, V., Tsoukatos, J., Davis, K. D., et al. (1999). A comparison of the burst activity of lateral thalamic neurons in chronic pain and non-pain patients. *Pain, 80*, 567–575.

Rinaldi, P. C., Young, R. F., Albe-Fessard, D. G., et al. (1991). Spontaneous neuronal hyperactivity in the medial and intralaminar thalamic nuclei in patients with deafferentation pain. *Journal of Neurosurgery, 74*, 415–421.

Steriade, M., & Llinas, R. R. (1988). The functional states of the thalamus and the associated neuronal interplay. *Physiological Reviews, 68*, 649–742.

Steriade, M., Jones, E. G., & Llinas, R. R. (1990). *Thalamic oscillations and signaling*. New York: Wiley.

Vierck, C. J. (1991). Can mechanisms of central pain syndromes be investigated in animal models? In K. L. Casey (Ed.), *Pain and central nervous system disease: The central pain syndromes* (pp. 129–141). New York: Raven.

Walker, A. E. (1943). Central representation of pain. *Research Publications – Association for Research in Nervous and Mental Disease, 23*, 63–85.

Weng, H. R., Lee, J. I., Lenz, F. A., Vierck, C. J., Rowland, L. H., & Dougherty, P. M. (2000). Functional plasticity in primate somatosensory thalamus following chronic lesion of the ventral lateral spinal cord. *Neuroscience, 101*, 393–401.

Zirh, A. T., Lenz, F. A., Reich, S. G., et al. (1997). Patterns of bursting occurring in thalamic cells during Parkinsonian tremor. *Neuroscience, 83*, 107–121.

Thalamic Neurotransmitters and Neurochemical Effector Molecules

▶ Thalamic Neurotransmitters and Neuromodulators

Thalamic Neurotransmitters and Neuromodulators

Karin N. Westlund[1] and Thomas E. Salt[2]

[1]Department of Physiology, University of Kentucky, Lexington, KY, USA

[2]Institute of Ophthalmology, University College London, London, UK

Synonyms

Chemical transmitter; Neuromodulator; Signaling molecules of thalamic regions;

Thalamic neurotransmitters and neurochemical effector molecules; Thalamus, nociceptive neurotransmission

Definition

The ▶ thalamus, a heterogeneous structure located in the dorsal diencephalic region of the brain, processes both sensory and motor input signals prior to transmission to higher cortical areas. Neurotransmitters are chemical messengers that are released from one neural element (e.g. a nerve terminal) to then act upon a receptor located on or in another neural element (e.g. a dendrite), including in the thalamus. This transfer of information is neurotransmission, and this contribution describes neurotransmitter mechanisms which mediate the transmission and integration of nociceptive information in the thalamus. ▶ Neuromodulators are chemical effector molecules that influence the neural activity enabling this process. While the microcircuitry and content of neuromodulators in the thalamus can be partially detailed at this point in time, the precise role of transmitters of the sensory thalamus in the transmission and generation of what is perceived as "pain" is still largely undefined. More recent observations suggest that neuromodulation occurs throughout the excitatory and the inhibitory neurocircuitry of the thalamus. It is clear that the chemical neuromodulation occurring in the thalamus permits a significantly more important role for the thalamus than simply as a "relay station". Rather, the thalamus is a site of integration where filtering and consolidation of sensory information occurs through chemical neurotransmission.

Characteristics

The integrative role of the thalamus in the processing of nociceptive information is complex and diverse. A variety of different neurotransmitters and an array of receptors take part in this process, and it has become clear that the nature of these processes is pivotal to the function of the

Thalamic Neurotransmitters and Neuromodulators, Table 1 Nociceptive processing in the thalamus

Thalamic transmitters and neuromodulators	
Thalamic input	
Spinothalamic tract and medial lemniscus	GLU, CCK
Corticothalamic tract	GLU
Central tegmental field and reticular activation systems	ACH (ChAT), 5-HT, NA, (TH, DBH), CGRP
Input to intralaminar and ventral thalamic nuclei (from hypothalamus, basal telencephalon, and lateral midbrain)	ENK, SOM, CCK, SP, NPY
Intrinsic interneurons and reticular nucleus	GABA, ACH, SOM
Thalamic output	
Thalamocortical projections	GLU
Thalamoamygdalar projections	CGRP

thalamo-cortico-thalamic circuitry (McCormick 1992; Broman 1994; Salt and Eaton 1996; Millan 1999). The majority of work carried out in the field of thalamic neurotransmitters has been in the so-called relay nuclei, of which the ventrobasal complex (ventroposterolateral and ventroposteromedial nuclei) is the somatosensory representative. Some of this function pertains to nociception, but it is important to remember that the ventrobasal complex (VB) has an important role in the processing of non-nociceptive somatosensory information and that many other thalamic nuclei (whose detailed transmitter functions are much less well studied) also participate in nociceptive functions, including medial and intralaminar nuclei (Giesler et al. 1981).

Neuromodulation in the sensory thalamus can best be described in terms of whether the neurochemicals are the content of input fibers, intrinsic interneurons or output components (Table 1; Fig. 1). The primary integrative sites in the thalamus for nociception are located medially and laterally in the posterior thalamus. The ventral posterior lateral thalamus is chiefly responsible for determining the intensity and location of the painful stimulus before relaying that information to the appropriate higher cortical regions of the brain. The medial thalamus is involved with emotional responses to nociceptive input.

Ascending Sensory Input to Thalamus

The primary sources of sensory information provided to the thalamus are the incoming spinothalamic, spinotrigeminal and medial lemniscal pathways. The corticothalamic tract, from cortical areas SI and SII, also provides a large contribution of incoming axonal fibers to the thalamus. There is overwhelming neuroanatomical evidence, at both the light-microscopical and ultrastructural levels, to favour a neurotransmitter role for glutamate in ascending afferent fibres in several mammalian species including rodents and primates (Broman 1994). Glutamate, the classical excitatory amino acid, is believed to be the primary transmitter of these fiber tracts, based on physiological and anatomical data. These fibres impinge upon ionotropic ▶ glutamate receptors of the ▶ NMDA and ▶ AMPA varieties located upon proximal dendrites of VB thalamic relay neurones (Broman 1994; Liu 1997). Ascending afferents to other thalamic nuclei that may be important in nociception are probably also glutamatergic (Broman 1994). Electrophysiological studies confirm a functional role for these receptor types in somatosensory transmission to the VB in rodents and primates (Salt and Eaton 1991; Dougherty et al. 1996) and it appears probable that, as in many other central synapses, the initial synaptic response is mediated via AMPA receptors with a following longer duration NMDA receptor mediated component which may become more prominent upon repetitive stimulation (Salt and Eaton 1991).

In other brain regions, increased release of glutamate and activation of neurons by glutamate is typically enhanced by the modulatory effects of ▶ neuropeptides, such as substance P (SP), CGRP and vasoactive intestinal polypeptide (VIP). Neuromodulators such as SP are found in abundance and probably function in a similar manner in the sensory thalamus (Fig. 2). The neuropeptide content of neuronal terminals in

Thalamic Neurotransmitters and Neuromodulators, Fig. 1 Sensory input to the thalamus is transmitted by glutamatergic (*GLU*) spinothalamic (*STT*), spinal trigeminal (*SpV*), dorsal column nucleus (*DCN*) and reticular formation (*RF*) neurons. The information, modulated by neuropeptides, is relayed to the medial and lateral thalamus (*Thal*) where it is influenced by intrinsic GABAergic (*GABA*) interneurons. The integrated information is routed by GLU thalamic projection neurons to sites including the cerebral cortex (*CTh*) and amygdala. Reciprocal input back to the thalamus from the cortex is also GLU. Abbreviations: *ACh* acetylcholine, *CCK* cholecystokinin, *CRF* corticotrophin releasing factor, *Enk* leu- or met-enkephalin, *Glu* glutamate, *5-HT* serotonin, *NA* norepinephrine, *NPY* neuropeptide Y, *SOM* somatostatin, *SP* substance P, *VIP* vasoactive intestinal polypeptide

the thalamus is believed to be partially from thalamic input sources (Jones 1988).

Spinothalamic Tract

Glutamate has been observed in spinothalamic projection neurons by electron microscopy (Westlund et al. 1992) (Fig. 1). Most spinothalamic neurons are also believed to contain at least one neuropeptide, such as SP, cholecystokinin (CCK), bombesin, dynorphin, enkephalin, galanin, corticotrophin releasing hormone (CRH) (Fig. 3) and/or vasoactive intestinal polypeptide (VIP) (Coffield and Miletic 1987; Leah et al. 1988; Nahin 1988). In fact, some have been shown to contain multiple neuropeptides (Ju et al. 1987). Lamina X STT cells have been shown to predominantly contain bombesin, enkephalin, CCK, somatostatin (SOM), CRF,

neuropeptide Y and SP (Battaglia et al. 1988; Leah et al. 1988). SP seems to function as a slow intermediary for transmission of noxious stimuli as well as a modulator of fast excitatory neurotransmission (Battaglia et al. 1988). Most of the terminals staining for SP are located in the medial rather than the lateral thalamus (Fig. 2). Enkephalin has also been identified in spinotrigeminal neurons (Nahin 1988). These findings are supported by both tract tracing/immunocytochemical studies and negative staining after spinal hemisection.

Medial Lemniscus

Input to the thalamus from the dorsal column nuclei is less well studied, but overlaps substantially with the spinothalamic terminations (Ralston, 1983). Neuropeptide CCK content in

Thalamic Neurotransmitters and Neuromodulators, Fig. 2 Schematic diagram illustrating the distribution of substance P immunoreactive fibers (*dots*) and cell somata (*stars*) in the monkey diencephalon (Jones 1988)

the ventral posterolateral (VPL) nucleus is, however, reduced upon lesion of the contralateral dorsal column nucleus, suggesting that this modulator is involved in the relay of sensory information by the medial lemniscus (Jones 1988).

Spinoreticular Pathways

Innervation by brainstem cholinergic, serotonergic and noradrenergic neurons is also present in the thalamus (Jones 1988; Westlund et al. 1990; Vogt et al. 2008) (Fig. 4). Sparse innervation by ▶ serotonergic fibers from the raphe nucleus and adjacent periaqueductal gray region has been demonstrated by dual labeling in monkeys. Similarly, in the same studies, norepinephrine was found sparsely innervating the VPL.

However, innervation by serotonergic, noradrenergic and ▶ acetylcholine terminals is found more prominently in the reticular and dorsal lateral geniculate nuclei (as reviewed in McCormick 1992). The purpose of these classical neurotransmitters in the thalamus remains unclear, though iontophoretic application of acetylcholine results in marked inhibition of the activity of reticular nucleus neurons, which, if it persists, can shift reticular nucleus cells into bursting activity. Iontophoresis of norepinephrine or serotonin markedly activates the reticular nucleus neurons through alpha$_1$ and 5-HT$_2$ (and possibly, 5-HT$_{1C}$) receptors, due to the decrease in resting potassium conductance. The net result is facilitation of single spike activity

Thalamic Neurotransmitters and Neuromodulators, Fig. 3 After fluorescent tracer is injected into the ventral thalamus, spinothalamic tract cells containing peptides such as corticotrophin releasing factor (*CRF*) can be identified in the sacral spinal cord, lamina V. The panel labeled VPL illustrates two cells retrogradely labeled after injection of fluorescent tracer into ventral thalamus. One of these cells also stained for CRF. Scale bar = 15 μm (Westlund et al. unpublished)

and marked inhibition of the rhythmic bursting activity normally promoted by the slow release of norepinephrine in this nucleus. It is assumed that these processes are part of the general arousal system throughout the neuraxis in which these transmitters participate. The highest levels of 5-HT$_7$ receptors are found on the intralaminar and midline thalamic neurons. Stimulation of the raphe nuclei can alter the responses of these neurons to nociceptive input (Goaillard and Vincent 2002).

Corticothalamic Input

Corticothalamic input to the thalamus is also glutamatergic since there is dense staining of both glutamate and aspartate in cells in layers IV and III of SI sensory cortex and many of these cortical cells can be double labeled with tract tracers injected into the thalamus (Guiffrida and Rustioni 1988; Rustioni et al. 1988) Only a small population of neurons is labeled with both glutamate and aspartate. Corticofugal fibers utilizing glutamate as the primary neurotransmitter terminate as small endings containing round vesicles primarily contacting fine caliber (distal) dendrites of thalamic neurons and are particularly dense in the reticular nucleus (Guiffrida and Rustioni 1988; McCormick 1992).

The cortical inputs to thalamic relay neurones have been a focus of much study and speculation over many years (Sherman and Guillery 2000). Electrophysiological studies of these pathways have focused on the role of NMDA receptors and, latterly, ► metabotropic glutamate (mGlu) receptors (Salt 2002). A particular focus has been the function of mGlu1 receptors, as these have been localised postsynaptically beneath terminals of cortico-thalamic fibres (Martin et al. 1992). However, it is also evident that there is a contribution from AMPA receptors to cortico-thalamic transmission (Golshani et al. 2001), a finding supported by ultrastructural evidence which indicates that there are AMPA receptor subunits, predominantly GluR2/3 and GluR4, located postsynaptically at cortico-thalamic synapses in VB (Golshani et al. 2001). More recently a low level of ► kainate receptor subunits (GluR5/6/7) has been found postsynaptically beneath corticothalamic synapses in VB, although a synaptic role for these receptors has not been detected at this location.

Intrinsic Interneurons

The inhibitory neurotransmitter, gamma amino butyric acid (► GABA), has been well characterized in the thalamus, particularly in the reticular nucleus of the thalamus where it is the predominant feature (Fig. 5). ► GABAergic inhibitory interneurones are a prominent feature of thalamic relay nuclei, and it is well known that ► GABAA and ► GABAB receptors play a prominent part in synaptic processing at both the pre- and post-synaptic level (Crunelli and Leresche 1991). There are two major groups of GABAergic neurones in thalamic relay nuclei: intrinsic

Thalamic Neurotransmitters and Neuromodulators, Fig. 4 Transverse sections of the brainstem of a monkey showing retrograde labeling of neurons (*FB*) after injection of Fast Blue in the ipsilateral ventral posterolateral nucleus of the thalamus, and the distribution of choline acetyltransferase (*ChAT*)-, tyrosine hydroxylase (*TH*)- and serotonin (*5HT*)- immunoreactive cells (Jones 1988)

Golgi II type interneurones, and neurones of the ▶ thalamic reticular nucleus (TRN) that exert their influence via their projection into relay nuclei (Ralston 1983). In rodents, only the latter population appears to be present (Ralston 1983) and performs a profound gating function upon thalamic transmission (Sherman and Guillery 2000). These GABAergic mechanisms may play an important part in the processing of sensory information in both acute and chronic nociception (Roberts et al. 1992). Intriguingy, the GABAergic output from TRN is itself modulated by metabotropic ▶ glutamate receptors (Salt 2002) and ▶ kainate receptors

(Binns et al. 2003). Such mechanisms indicate that sensory transmission through VB is not only dependent upon excitatory transmission, but that reduction of inhibitory transmission (i.e., functional disinhibition) could also have a significant potentiating influence on transmission. It has been determined that GABA is present in interneurons of the ventral posterior thalamus only in carnivorous animals (i.e. present in primates and cats, but not in rats) (Rustioni et al. 1988). Intrinsic ▶ substance P interneurons have also been described in the thalamus in regions receiving spinothalamic tract input as demonstrated by anterograde tracing (Battaglia et al. 1988).

Thalamic Neurotransmitters and Neuromodulators, Fig. 5 (**a**) Immunocytochemical staining of GABAergic neurons in the reticular nucleus (R) of a cat, identified using antiserum to glutamic acid decarboxylase (GAD). Fine dots in the dorsal thalamus are GAD-positive somata of interneurons. (**b**) Higher-power photomicrograph of GAD-positive cells in reticular nucleus (R) and ventral posterior (VP) nucleus of a cat. Bars = 1 mm (**a**), 250 μm (**b**) (Jones 1988)

Glia

The concept that glial cells or astrocytes may be active participants in brain function is supported by several findings, including that astrocytes possess ion channels and neurotransmitter receptors for a variety of neurotransmitters, and that astrocytes contain and can release excitatory amino acids such as glutamate and homocysteate (Haydon 2001). It is known that, in the VB thalamus, activity in astrocytes can evoke NMDA-receptor mediated responses in thalamic relay neurones in vitro (Parri et al. 2001), and that homocysteate can be released from thalamus in vivo and activate NMDA receptors (Do et al. 2004). This raises the possibility that astrocytes play a key role in the responses of thalamic neurones to sensory stimuli.

Thalamic Output

Information regarding the sensory-discriminative aspect of pain is sent to parts of the cortex involved in somatosensory processing, such as precisely localized areas I (SI), II (SII) and area 4, for further site specific association, in order that pain-appropriate response instructions can be dispatched via motor systems. The thalamocortical efferent pathways (Fig. 1) were identified as glutamate positive neurons with collaterals to the reticular nucleus (Guiffrida and Rustioni 1988; as reviewed in McCormick 1992). Thalamic output to the cortex is activated by glutamate NMDA and non-NMDA receptors, since specific antagonists applied to the thalamus can block transmission of thalamic outflow. The glutamatergic input to the thalamocortical neurons probably arises from the input sources detailed above.

There is also an affective component to pain perception, which contributes to the emotional responses generated in response to the stimulus. This aids in memory development for future avoidance of the painful situation. These functions are associated with regions of the thalamus with inputs to the limbic cortex, including the mediodorsal (ventrocaudal) (MDvc) and intralaminar and midline nuclei (central lateral, central dorsal, central medial, paracentral, reuniens, and paraventricular nuclei, and the ventral medial and ventral anterior nuclei) (Thompson and Robertson 1987; Price, 2000). The intralaminar (centromedian, centrolateral, paraventricular) thalamic nuclei receive input mainly from laminae I and X of the spinal cord, including information about noxious visceral stimulation (Wang et al., 1999; Ren et al. 2009). The ventral posterior inferior nuclei send signals directly to the somatosensory area of the parietal cortex (SII) and indirectly to the insula (Millan 1999). Both may be joined by additional ipsilateral input. A thin layer of CGRP-containing neurons is found just ventral to the thalamus near to the meso-diencephalic junction (Kruger et al. 1988) and is associated with the thalamus and the somatosensory pathways, as are all CGRP components of the nervous system. These CGRP-containing neurons have been shown to provide a thalamo-amygdalar projection which may also be important in generating the emotional responses to painful stimuli.

Calcium calmodulin-dependent protein kinase (CaM kinase II) has been found in thalamic dendrites apposed by glutamate terminals, suggesting that this major post-synaptic activity regulator may assist in modulating synaptic strength, transmitter release and/ or vesicle movement in thalamic neurons (Liu and Jones 1996). The CaM kinase II terminals in the cerebral cortex are thought to be terminals of thalamocortical neurons.

Neurotransmitters and Thalamic Integrative Function in Nociception

A role for NMDA receptors in the signalling of acute thermal and mechanical nociceptive responses in the VB thalamus at the single-neurone and behavioural level is now well established (Salt and Eaton 1996; Millan 1999). However, it is important to note that transmission of non-nociceptive sensory information to the thalamus can also show substantial NMDA receptor involvement (Salt and Eaton 1996). In addition, *both* Group I mGlu (mGlu1 and mGlu5) receptors also participate in the signalling of

acute nociceptive information but not in the signalling of non-nociceptive mechanoreceptor input to the VB thalamus (Salt and Binns 2000). This functional distinction is intriguing, as there appears to be remarkable anatomical and neurochemical similarity between lemniscal (which carries non-nociceptive information) and spinothalamic (which carries nociceptive and convergent multimodal somatosensory information) inputs to the VB thalamus (Ralston 1983; Ma et al. 1987; Liu 1997). In view of this, it is conceivable that the recruitment of additional neural circuitry during noxious stimuli underlies the Group I mGlu receptor involvement in nociceptive responses. A possible source of this additional input could be the dense cortico-thalamic projection (Eaton and Salt 1995), which is known to be glutamatergic and which impinges upon mGlu receptors, particularly mGlu1 (see above). This is a particularly attractive hypothesis in the case of mGlu1 receptors, as these are restricted to corticothalamic synapses and because NMDA-receptor mediated responses have been shown to be modulated by activation of Group I (i.e., mGlu1/mGlu5) receptors in several brain areas, as has modulation of AMPA-receptor mediated responses. In the VB, activation of mGlu1 receptors potentiates responses mediated via either AMPA or NMDA receptors in vivo (Salt and Binns 2000). It is probable that this is due to the direct effects of mGlu1 activation on neuronal membrane potential and resistance rather than a specific interaction at the receptor level, or that the potentiation that is seen is a combination of these factors (Salt and Binns 2000). Thus, the cortico-thalamic input would be able to exert a profound influence on ionotropic receptor mediated responses if the sensory stimulus was appropriate to recruit activity in the cortico-thalamic output. This may well be the case for nociceptive stimuli (Eaton and Salt 1995; Millan 1999). A further enabling factor could be the removal of the inhibitory influence arising from the TRN (see above) (Roberts et al. 1992), and in this respect it is interesting to note that TRN neurones are inhibited by noxious peripheral stimulation (Peschanski et al. 1980).

Modulatory Systems

Transmission through the thalamic relay nuclei, including those serving somatosensation and nociception, can be modulated by amine neurotransmitters such as serotonin, noradrenaline or acetylcholine (McCormick 1992). These systems appear to be associated with activating systems that govern states of wakefulness and arousal, and it is unclear to what extent these specifically affect nociceptive processing at the thalamic level. In addition, the nitric oxide (NO) system is associated with some of these activating systems (Vincent 2000), and it has been shown that NO can modulate somatosensory transmission through the thalamus (Shaw and Salt 1997). Similarly, a number of ► neuropeptides have been located in thalamic nuclei and afferents, but their function remains unclear (Sherman and Guillery 2000). Of particular interest to nociceptive processing is the finding that ► cannabinoid receptors modulate acute nociceptive responses of VB neurones in the rat (Martin et al. 1996). However, the precise mechanisms of this action remain to be elucidated.

Thalamic Transmitter Mechanisms and Adaptive Changes

It is known that thalamic neurones change their response and firing characteristics in conditions of chronic pain or chronic pain models, and a role for mGlu receptors and NMDA receptors in models of synaptic plasticity has been known for some time. Thus it is conceivable that changes in thalamic neurone responses may be due to changes in glutamate receptor function and may even be a consequence of activation of these receptors as has been suggested for the spinal cord (Willis 2002). Indeed there is already evidence to suggest that NMDA receptors in the thalamus are involved in inflammation-produced hyperalgesia in the rat. In arthritic rats, decreases in thalamic expression (including in VB) of mRNA for mGlu1, mGlu4 and mGlu7 receptors have been observed (Neto et al. 2000). Interestingly, in these same animals, mGlu3 mRNA expression is elevated in the TRN and this expression appears to be both in presumed GABAergic neurones and in glial cells (Neto et al. 2000).

Furthermore, injection into TRN of an antagonist for this receptor was found to be anti-hyperalgesic in such rats (Neto et al. 2000). Thus it may be that thalamic mGlu receptor mechanisms are important in both the induction of hyperalgesia and in the expression of hyperalgesic behaviours. It is noteworthy, however, that changes in other thalamic transmitter systems may also occur in chronic pain conditions, for example in the serotonergic system (Goettl et al. 2002). It is therefore important to note that transmitter systems should not be regarded in isolation.

Summary

It is evident that glutamate receptors are of fundamental importance in the transmission of nociceptive and other sensory information through the thalamus. Activation of NMDA receptors and certain mGlu receptors may be particularly important in the signalling and processing of nociceptive information, as well as in the induction of longer-term plastic changes in response to chronic noxious stimulation or injury. Molecular intervention at some of these sites may have considerable therapeutic potential.

References

Battaglia, G., Spreafico, R., & Rustioni, A. (1988). Substance P-immunoreactive fibers in the thalamus from ascending somatosensory pathways. In M. Bentivoglio & R. Spreafico (Eds.), *Cellular thalamic mechanisms*. Amsterdam: Elsevier Science Publishers.

Binns, K. E., Turner, J. P., & Salt, T. E. (2003). Kainate receptor (GluR5)-mediated disinhibition of responses in rat ventrobasal thalamus allows a novel sensory processing mechanism. *Journal of Physiology (London), 551,* 525–537.

Broman, J. (1994). Neurotransmitters in subcortical somatosensory pathways. *Anatomy and Embryology, 189,* 181–214.

Coffield, J. A., & Miletic, V. (1987). Immunoreactive enkephalin is contained within some trigeminal and spinal neurons projecting to the rat medial thalamus. *Brain Research, 425,* 380–383.

Crunelli, V., & Leresche, N. (1991). A role for GABAB receptors in excitation and inhibition of thalamocortical cells. *Trends in Neurosciences, 14,* 16–21.

Do, K. Q., Benz, B., Binns, K. E., Eaton, S. A., & Salt, T. E. (2004). Release of homocysteic acid from rat thalamus following stimulation of somatosensory afferents in vivo: Feasability of glial participation in synaptic transmission. *Neuroscience, 124*, 387–393.

Dougherty, P. M., Li, Y. J., Lenz, F. A., Rowland, L., & Mittman, S. (1996). Evidence that excitatory amino acids mediate afferent input to the primate somatosensory thalamus. *Brain Research, 728*, 267–273.

Eaton, S. A., & Salt, T. E. (1995). The role of excitatory amino acid receptors in thalamic nociception. In J. M. Besson, G. Guilbaud, & H. Ollat (Eds.), *Forebrain areas involved in pain processing* (pp. 131–141). Paris: John Libbey Eurotext.

Giesler, G. J., Yezierski, R. P., Gerhart, K. D., & Willis, W. D. (1981). Spinothalamic tract neurons that project to medial and/or lateral thalamic nuclei: Evidence for a physiologically novel population of spinal cord neurons. *Journal of Neurophysiology, 46*, 1285–1308.

Goaillard, J. M., & Vincent, P. (2002). Serotonin suppresses the slow afterhyperpolarization in rat intralaminar and midline thalamic neurones by activating 5-HT(7) receptors. *The Journal of Physiology, 541*, 453–465.

Goettl, V. M., Huang, Y., Hackshaw, K. V., et al. (2002). Reduced basal release of serotonin from the ventrobasal thalamus of the rat in a model of neuropathic pain. *Pain, 99*, 359–366.

Golshani, P., Liu, X. B., & Jones, E. G. (2001). Differences in quantal amplitude reflect glur4-subunit number at corticothalamic synapses on two populations of thalamic neurons. *Proceedings of the National Academy of Sciences of the United States of America, 98*, 4172–4177.

Guiffrida, R., & Rustioni, A. (1988). Glutamate and aspartate immunoreactivity in corticothalamic neurons of rats. In M. Bentivoglio & R. Spreafico (Eds.), *Cellular thalamic mechanisms* (pp. 311–320). Amsterdam: Elsevier Science Publishers.

Haydon, P. G. (2001). Glia: Listening and talking to the synapse. *Nature Reviews Neuroscience, 2*, 185–193.

Jones, E. G. (1988). Modern views of cellular thalamic mechanisms. In M. Bentivoglio & R. Spreafico (Eds.), *Cellular thalamic mechanisms* (pp. 1–22). Elsevier Science Publishers, New York.

Ju, G., Melander, T., Ceccatelli, S., et al. (1987). Immunohistochemical evidence for a spinothalamic pathway co-containing cholecystokinin- and galanin-like immunoreactivities in the rat. *Neuroscience, 20*, 439–456.

Kruger, L., Mantyh, P. W., Sternini, C., Brecha, N. C., & Mantyh, C. R. (1988). Calcitonin gene-related peptide (CGRP) in the rat central nervous system: patterns of immunoreactivity and receptor binding sites. *Brain Research, 463*, 223–244.

Leah, J., Menetrey, D., & de Pommery, J. (1988). Neuropeptides in long ascending spinal tract cells in the rat: Evidence for parallel processing of ascending information. *Neuroscience, 24*, 195–207.

Liu, X. B. (1997). Subcellular distribution of AMPA and NMDA receptor subunit immunoreactivity in ventral posterior and reticular nuclei of rat and cat thalamus. *The Journal of Comparative Neurology, 388*, 587–602.

Liu, X. B., & Jones, E. G. (1996). Localization of alpha type II calcium calmodulin-dependent protein kinase at glutamatergic but not gamma-aminobutyric acid (GABAergic) synapses in thalamus and cerebral cortex. *Proceedings of the National Academy of Science, 93*, 7332–7336.

Ma, W., Peschanski, M., & Ralston, H. J., III. (1987). The differential synaptic organization of the spinal and lemniscal projections to the ventrobasal complex of the rat thalamus. Evidence for convergence of the two systems upon single thalamic neurons. *Neuroscience, 22*, 925–934.

Martin, L. J., Blackstone, C. D., Huganir, R. L., et al. (1992). Cellular localization of a metabotropic glutamate receptor in rat brain. *Neuron, 9*, 259–270.

Martin, W. J., Hohmann, A. G., & Walker, J. M. (1996). Suppression of noxious stimulus-evoked activity in the ventral posterolateral nucleus of the thalamus by a cannabinoid agonist: Correlation between electrophysiological and antinociceptive effects. *Journal of Neuroscience, 16*, 6601–6611.

McCormick, D. A. (1992). Neurotransmitter actions in the thalamus and cerebral cortex and their role in neuromodulation of thalamocortical activity. *Progress in Neurobiology, 39*, 337–388.

Millan, M. J. (1999). The induction of pain: An integrative review. *Progress in Neurobiology, 57*, 1–164.

Nahin, R. L. (1988). Immunocytochemical identification of long ascending, peptidergic lumbar spinal neurons terminating in either the medial or lateral thalamus in the rat. *Brain Research, 443*, 345–349.

Neto, F. L., Schadrack, J., Platzer, S., et al. (2000). Expression of metabotropic glutamate receptors mRNA in the thalamus and brainstem of monoarthritic rats. *Molecular Brain Research, 81*, 140–154.

Parri, H. R., Gould, T. M., & Crunelli, V. (2001). Spontaneous astrocytic Ca2+ oscillations in situ drive NMDA- mediated neuronal excitation. *Nature Neuroscience, 4*, 803–812.

Peschanski, M., Guilbaud, G., & Gautron, M. (1980). Neuronal responses to cutaneous electrical and noxious mechanical stimuli in the nucleus reticularis thalami of the rat. *Neuroscience Letters, 20*, 165–170.

Price, D. D. (2000). Psychological and neural mechanisms of the affective dimension of pain. *Science, 288*(5472), 1769–1772.

Ralston, H. J., III. (1983). The synaptic organization of the ventrobasal thalamus in the rat, cat and monkey. In G. Macchi, A. Rustioni, & R. Spreafico (Eds.), *Somatosensory integration in the thalamus* (pp. 241–250). Amsterdam: Elsevier Science Publishers.

Ren, Y., Zhang, L., Lu, Y., Yang, H., & Westlund, K. N. (2009). Central lateral thalamic neurons receive noxious visceral mechanical and chemical input in rats. *Journal of Neurophysiology, 102*, 244–258.

Roberts, W. A., Eaton, S. A., & Salt, T. E. (1992). Widely distributed GABA-mediated afferent inhibition processes within the ventrobasal thalamus of rat and their possible relevance to pathological pain states and somatotopic plasticity. *Experimental Brain Research, 89*, 363–372.

Rustioni, A., Battaglia, G., De Biasi, S., et al. (1988). Neuromediators in somatosensory thalamus: An immunocytochemical overview. In M. Bentivoglio & R. Spreafico (Eds.), *Cellular thalamic mechanisms* (pp. 271–296). Amsterdam: Elsevier Science Publishers.

Salt, T. E. (2002). Glutamate receptor functions in sensory relay in the thalamus. *Philosophical Transactions of the Royal Society of London. Series B, Biological Sciences, 357*, 1759–1766.

Salt, T. E., & Binns, K. E. (2000). Contributions of mGlu1 and mGlu5 receptors to interactions with N-methyl-D-aspartate receptor-mediated responses and nociceptive sensory responses of rat thalamic neurones. *Neuroscience, 100*, 375–380.

Salt, T. E., & Eaton, S. A. (1991). Sensory excitatory postsynaptic potentials mediated by NMDA and non-NMDA receptors in the thalamus in vivo. *European Journal of Neuroscience, 3*, 296–300.

Salt, T. E., & Eaton, S. A. (1996). Functions of ionotropic and metabotropic glutamate receptors in sensory transmission in the mammalian thalamus. *Progress in Neurobiology, 48*, 55–72.

Shaw, P. J., & Salt, T. E. (1997). Modulation of sensory and excitatory amino acid responses by nitric oxide donors and glutathione in the ventrobasal thalamus of the rat. *European Journal of Neuroscience, 9*, 1507–1513.

Sherman, S. M., & Guillary, R. W. (2000). *Exploring the thalamus.* New York: Academic Press.

Thompson, S. M., & Robertson, R. T. (1987). Organization of subcortical pathways for sensory projections to the limbic cortex. I. Subcortical projections to the medial limbic cortex in the rat. *The Journal of Comparative Neurology, 265*, 175–188.

Vincent, S. R. (2000). The ascending reticular activating system – from aminergic neurons to nitric oxide. *Journal of Chemical Neuroanatomy, 18*, 23–30.

Vogt, B. A., Hof, P. R., Friedman, D. P., Sikes, R. W., & Vogt, L. J. (2008). Norepinephrinergic afferents and cytology of the macaque monkey midline, mediodorsal, and intralaminar thalamic nuclei. *Brain Structure & Function, 212*, 465–479.

Wang, C. C., Willis, W. D., & Westlund, K. N. (1999). Ascending projections from the central, visceral processing region of the spinal cord: A PHA-L study in rats. *Journal of Comparative Neurology, 415*, 341–367.

Westlund, K. N., Carlton, S. M., Zhang, D., et al. (1992). Glutamate-immunoreactive terminals synapse on primate spinothalamic tract cells. *The Journal of Comparative Neurology, 322*, 519–527.

Westlund, K. N., Sorkin, L. S., Ferrington, D. G., Carlton, S. M., Willcockson, H. H., & Willis, W. D. (1990). Serotonergic and noradrenergic projections to the ventral posterolateral nucleus of monkey thalamus. *The Journal of Comparative Neurology, 295*, 197–207.

Willis, W. D. (2002). Long-term potentiation in spinothalamic neurons. *Brain Research Reviews, 40*, 202–214.

Thalamic Nociceptive System

▶ Thalamic Nuclei Involved in Pain: Cat and Rat

Thalamic Nuclei

Definition

The thalamus, which is located in the center of the cerebral cortex, comprises many nuclei. The thalamus relays information to and from the cerebral cortex and also plays a part in modulating sensory information.

Cross-References

▶ Central Pain
▶ Diagnosis

Thalamic Nuclei Involved in Pain, Human and Monkey

Edward Jones
Center for Neuroscience, University of California, Davies, CA, USA

Synonyms

Lateral thalamus encodes pain; Nociceptive coding in lateral thalamus

Characteristics

Description
Based on anatomical considerations, afferent inputs, cortical projections and electrophysiological evidence, a number of nuclei in the posterior regions of the thalamus as well as the core of the ventral posterior nuclear complex must be involved in transmitting nociceptive information to the cortex.

It is likely that many thalamic nuclei and therefore their projection areas in the cerebral cortex are "involved" in pain. However, despite many attempts to discredit it, it remains apparent that the ventral posterior nuclear complex is that part of the thalamus that possesses the appropriate input connections and the relevant cellular machinery for relaying the qualitative features and peripheral localization of a painful stimulus to the cortex. A necessary corollary of this is that the cortical projection target of the ventral posterior complex, the post central gyrus, is a central element in the onward pathways to the perception of pain.

The essay ▶ Spinothalamic Terminations, Core and Matrix summarizes the newer connection tracing studies that demonstrate not only the widespread terminations of the ▶ central pain pathways in the primate thalamus but also their relationships to cells within the heart of the ventral posterior nucleus, the ▶ ventral posterior medial (VPM) and ▶ ventral posterior lateral (VPL) nuclei. Here they appear to terminate preferentially on the *matrix* cells of the two nuclei, extending these terminations beyond the confines of ▶ VPM and ▶ VPL into and surrounding nuclei that are also characterized by the presence of ▶ calbindin-immunoreactive matrix cells.

Single unit recordings both within and around the perimeter of the ventral posterior nucleus of the thalamus in monkeys and humans reveal the presence of neurons with both nociceptive-specific and ▶ wide dynamic range properties akin to those found in dorsal horn neurons (e.g., Bushnell et al. 1993; Craig et al. 1994; Lee et al. 1999; Ohara and Lenz 2003; Willis and Coggeshall 2004). Unfortunately, these have never been satisfactorily identified as exclusively within the matrix, but it is clear that many of these recordings have come not just from the matrix regions surrounding VPM and VPL but also from within the heart of these nuclei themselves. From what we have described in the essay on spino- and spinal trigeminothalamic terminations mentioned above, it is unlikely (although some have considered it controversial) that noci- or thermo-specific neurons of spinal lamina I project only to neurons outside the confines of the ventral posterior nucleus, although their projections may be concentrated around the posterior pole of the VPM nucleus in the matrix domain in monkeys and humans (and in comparable regions in cats). Of special note is the presence of neurons with nociceptive specific properties within the heart of the ventral posterior nucleus. This carries the implication that these neurons project to the primary somatosensory cortex and that has indeed been directly demonstrated in a small number of cases (Kenshalo et al. 1980) (Fig. 1).

The presence of neurons with properties appropriate for relaying the details of the modality, location and intensity of a painful stimulus within the heart of the thalamic nucleus that projects to the primary somatosensory cortex should not be taken to imply that these neurons are found only therein. They are indeed found within all or most of the other sites of termination of the spinothalamic tracts as a whole and within those sites outside the ventral posterior nucleus to which neurons specifically located in the superficial dorsal horn project (reviewed in Willis et al. 2002; Lenz et al. 2009). The cortical targets of these other parts of the thalamus, which include the ▶ ventral posterior inferior and ▶ basal ventral medial nuclei of the ventral posterior complex, the adjacent matrix-filled domains of the ▶ posterior nucleus and ▶ anterior pulvinar nucleus and the caudal intralaminar nuclei, are diverse and include the second somatosensory area, other peri-insular areas and parts of the prefrontal and possibly anterior cingulate cortex. The broad implication is the classical one that the multi-dimensional character of that sensation that we call pain is reflected in inputs to regions of cortex whose known or conjectured functional connotations are appropriate for each of these

Thalamic Nuclei Involved in Pain, Human and Monkey,
Fig. 1 Schematic view of the distribution of spinothalamic tract fibers in the thalamus of the macaque monkey and the cortical projections of the receiving nuclei. *CL* central lateral nucleus, *Po* posterior nucleus, *Pla* anterior pulvinar nucleus, *SG-L* limitans-suprageniculate nucleus, *VLp* posterior ventral lateral nucleus, *VMb* basal ventral medial nucleus, *VPI* ventral posterior inferior nucleus, *VPL* ventral posterior lateral nucleus, *VPM* ventral posterior medial nucleus, *ZI* zona incerta

Frontal, cingulate, parietal cortex

CL,VPL,VPI,VMb, ZI

Primary somatosensory cortex

VPL, VPI, VMb, Pla, CL, VLp

Peri-insular cortex

Po, SG-L, VPI

dimensions. But there is still no reason to rule out the belief that the primary somatosensory cortex is the primary route to the centers for perception of the quality, location and intensity of a painful stimulus. This is not the place to consider all the data that imply the dissociation of the sensory discriminative and affective-motivational dimensions of pain in relation to different cortical areas but there is a growing literature (often misinterpreted or misrepresented) in this field that is covered in other essays in this encyclopedia.

Functional imaging studies in humans support the view of widespread thalamocortical projections in the central pain system. Most of the areas referred to in the preceding paragraph show functional activation in response to application of painful stimuli to the skin or to stimuli engendered in deeper tissues. Areas of the cortex reportedly activated during the appreciation of a painful stimulus include the first and second somatosensory areas, the insula and the anterior

cingulate gyrus (e.g., Coghill et al. 1994; Davis et al. 1998; Gelnar et al. 1999; Craig et al. 2000; Lenz et al. 2009). Trying to correlate these observations with electrophysiological reports from monkeys and the effects of stimulation or lesions in humans presents a confusing picture from which workers committed to any point of view about the cortical representation of pain, however eccentric, can derive sustenance.

When we look at the results of recording or imaging of the postcentral gyrus after presentation of painful stimuli to human subjects, it is difficult to escape the conclusion that the primary somatosensory cortex is at the heart of the territory involved in nociception but more lateral areas are likely to be involved as well (Treede et al. 2000; Ohara et al. 2004a, b). These imaging results, moreover, tend to be supported by the results of single unit studies of areas outside the primary somatosensory area in monkeys (Robinson and Burton 1980; Kenshalo et al. 2000). The role of the lateral areas has been adduced mainly from lesion studies in humans, but these have been variously interpreted.

My own view is that lesions involving the parietal operculum (which might mean involvement of the second somatosensory area alone but could equally undercut fibers approaching the postcentral gyrus) seem to alter the sensory-discriminative aspects of pain appreciation, while those affecting regions located in insular and postinsular regions tend to alter affective and motivational aspects and can result in pain asymbolia (see, for example, Greenspan et al. 1999; Ostrowsky et al. 2002). This still leaves out, however, the prefrontal and cingulate areas. Perhaps it is their activation in pain states that represents the inputs from thalamic nuclei such as the caudal intralaminar?

References

Bushnell, M. C., Duncan, D. H., & Tremblay, N. (1993). Thalamic VPM nucleus in the behaving monkey. I. Multimodal and discriminative properties of thermosensitive neurons. *Journal of Neurophysiology, 69*, 739–752.

Coghill, R. C., Talbot, J. D., Evans, A. C., et al. (1994). Distributed processing of pain and vibration by the human brain. *Journal of Neuroscience, 14*, 4095–4108.

Craig, A. D., Bushnell, M. C., Zhang, E.-T., et al. (1994). A thalamic nucleus specific for pain and temperature sensation. *Nature, 372*, 770–773.

Craig, A. D., Chen, K., Bandy, D., & Reiman, E. M. (2000). Thermosensory activation of nsular cortex. *Nature Neuroscience, 3*, 184–190.

Davis, K. D., Kwan, C. L., Crawley, A. P., et al. (1998). Functional MRI study of thalamic and cortical activations evoked by cutaneous heat, cold, and tactile stimuli. *Journal of Neurophysiology, 80*, 1533–1546.

Gelnar, P. A., Krauss, B. R., Sheehe, P. R., et al. (1999). A comparative fMRI study of cortical representations for thermal painful, vibrotactile, and motor performance tasks. *NeuroImage, 10*, 460–482.

Greenspan, J. D., Lee, R. R., & Lenz, F. A. (1999). Pain sensitivity alterations as a function of lesion location in the parasylvian cortex. *Pain, 81*, 273–282.

Kenshalo, D. R., Jr., Giesler, G. J., Jr., Leonard, R. B., et al. (1980). Responses of neurons in primate ventral posterior lateral nucleus to noxious stimuli. *Journal of Neurophysiology, 43*, 1594–1614.

Kenshalo, D. R., Iwata, K., Sholas, M., & Thomas, D. A. (2000). Response properties and organization of nociceptive neurons in area 1 of monkey primary somatosensory cortex. *Journal of Neurophysiology, 84*, 719–729.

Lee, J.-I., Dougherty, P. M., Antezana, D., et al. (1999). Responses of neurons in the region of human thalamic principal somatic sensory nucleus to mechanical and thermal stimuli graded into the painful range. *The Journal of Comparative Neurology, 410*, 541–555.

Lenz, F. A., Casey, K. L., Jones, E. G., & Willis, W. D. (2009). *The human pain system. Experimental and clinical perspectives*. Cambridge: Cambridge University Press.

Ohara, S., & Lenz, F. A. (2003). Medial lateral extent of thermal and pain sensations evoked by microstimulation in somatic sensory nuclei of human thalamus. *Journal of Neurophysiology, 90*, 2367–2377.

Ohara, S., Crone, N. E., Weiss, N., et al. (2004a). Cutaneous painful laser stimuli evoke responses recorded directly from primary somatosensory cortex in awake humans. *Journal of Neurophysiology, 91*, 2734–2746.

Ohara, S., Crone, N. E., Weiss, N., et al. (2004b). Attention to pain is processed at multiple cortical sites in man. *Experimental Brain Research, 156*, 513–517.

Ostrowsky, K., Magnin, M., Ryvlin, P., et al. (2002). Representation of pain and somatic sensation in the human insula: A study of responses to direct electrical cortical stimulation. *Cerebral Cortex, 12*, 376–385.

Robinson, C. J., & Burton, H. (1980). Somatic submodality distribution within the second somatosensory (SII), 7b, retroinsular, postauditory and granular insular cortical areas of *M. fascicularis*. *The Journal of Comparative Neurology, 192*, 93–108.

Treede, R. D., Apkarian, A. V., Bromm, B., et al. (2000). Cortical representation of pain: functional characterization of nociceptive areas near the lateral sulcus. *Pain, 87*, 113–119.

Willis, J., Morello, A., Davie, A., Rice, J., Bennett, J. (2002). Forced use treatment of childhod hemiparesis. *Pediatrics, 110*, 1 I, 94–96.

Willis, W. D., Coggeshall, R. E. (2004). Sensory mechanisms of the spinal cord. Ed 3. New York: Kluwer Academic/Plenum.

Thalamic Nuclei Involved in Pain: Cat and Rat

Christiane Vahle-Hinz
Department of Neurophysiology and Pathophysiology, University Medical Center Hamburg-Eppendorf, Hamburg, Germany

Synonyms

Lateral and medial thalamic nociceptive system; Thalamic nociceptive system

Definition

Nuclei of the dorsal thalamus whose neurons receive inputs from ascending nociceptive pathways and project to cortical nociceptive areas. The neurons respond to stimulation of ▶ nociceptors in skin, muscles, joints, and viscera. The different thalamic regions process and transmit information subserving the sensory-discriminative and the ▶ motivational-affective components of pain.

Characteristics

A distinction is generally made in all species between a lateral and a medial thalamic system, the lateral with neurons encoding stimulus quality, duration, intensity, and location on the body and thus subserving the ▶ sensory-discriminative component of pain and the medial system with neurons activated from large areas of the body and/or internal organs and involved in the motivational-affective component of pain. The different response properties reflect ascending inputs dominated by the spinothalamic and spinal trigeminothalamic tracts (STT) and projection to the SI and SII somatosensory cortices (lateral system) versus STT and spinoreticulothalamic tract and widespread cortical projections including limbic areas (medial system).

In contrast to other sensory nuclei in the thalamus, nociceptive "nuclei" cannot be delineated histologically. Rather, nociceptive neurons are found within nuclei primarily subserving other functions like the mediodorsal or intralaminar nuclei or the ventral posterior complex (VP). Thus, "thalamic nuclei involved in pain" have been identified by neuroanatomical tracing of terminations of ascending tracts carrying nociceptive signals or by ▶ electrophysiological mapping of neurons responsive to noxious stimuli. The latter is hindered further by the problems of applying painful stimuli in awake animals or the inherent ▶ antinociceptive effects of general anesthetics when studies are done under ▶ anesthesia (Vahle-Hinz and Detsch 2002;

Vahle-Hinz et al. 2002). At the spinal cord level, the majority of nociceptive somatic and visceral neurons in addition have low-threshold somatic receptive fields (RFs), which may be preserved in the thalamus while their nociceptive components are abolished by the anesthetic. These difficulties are reflected in the heterogeneous picture of thalamic nociception in the literature (for review, see Willis 1997), and it still awaits clarification whether species differences originate from principally different organizations or a unifying picture may emerge.

Lateral Thalamus

Nociceptive neurons are found in regions of terminations of the STT that are patchy and dispersed in and around the VP and the posterior complex (PO) in the rat, while they are confined to the PO and the margins of the VP in the cat. These regions are further characterized by small neurons and may correspond to the small-celled matrix regions in and around the VP of monkeys. Thus, although the location of nociceptive neurons may differ between species (outside versus inside VP), this may result from a different degree of invasion of this small-celled matrix into the VP.

Region of the Ventral Posterior Complex (VP)

The VP is characterized by input from the medial and trigeminal lemnisci carrying a somatotopic representation of low-threshold mechanoreceptors of the contralateral half of the body surface including the furry buccal pad (rat) or a bilateral representation of intraoral structures (cat) (Vahle-Hinz and Gottschaldt 1983). The cartoon in Fig. 1b shows the approximate proportions in the cat's VP in a frontal plane, while in fact the complex extends caudomedially to rostrolaterally with the hind limb area rostral to the face in both rat and cat. The subnuclei of the VP holding the representations of the different parts of the body stand out from adjacent regions histologically by denser packing and larger somata of neurons (Fig. 1a). Interspersed are smaller ▶ GABAergic cells with local axonal arbors (interneurons, about 25 % of the neurons) in the cat; these are virtually absent in the rat. The projection neurons

Thalamic Nuclei Involved in Pain: Cat and Rat, Fig. 1 Locations of nociceptive neurons in the lateral thalamus of the cat. (**a, b**) Subnuclei of the ventral posterior complex and adjacent regions as determined by histology (**a**), Nissl stain, coronal section, medial to the left, *scale bar*: 1 mm) and electrophysiological mapping of neuronal receptive fields (**b**). (**c**) Examples of neurons from the VP$_{vp}$ and the VP (camera lucida drawings from Golgi-stained tissue, scale bar: 50 μm). (**d**) Recording sites in VP$_p$ of nociceptive neurons with cutaneous receptive fields. Nociceptive-specific (*dots*) and multireceptive neurons (*triangles*). (**e**) Recording sites of visceroceptive neurons in VP$_p$ and PO determined by electrical stimulation of the pelvic nerve. *Eml* external medullary lamina, *POl, m* medial and lateral parts of the posterior complex, *R* thalamic reticular nucleus, *VMb* basal ventral medial nucleus (gustatory relay), *VP* ventral posterior complex, *VP$_p$* periphery of VP, *VP$_{vp}$* ventral periphery of VP, *VPMm, l* medial (intraoral RFs) and lateral (face RFs) parts of the ventral posterior medial nucleus, *VPL* ventral posterior lateral nucleus (postcranial body RFs), *ZI* zona incerta; scale bar: 1 mm (Modified from Vahle-Hinz et al. 1995 (**b**), Kniffki et al. 1993 (**c**), Kniffki and Vahle-Hinz 1987 (**e**), Brüggemann et al. 1994 (**e**))

of the VP have large round somata with dense radial dendritic arbors (Fig. 1c) (Kniffki et al. 1993).

Nociceptive neurons occur in the periphery of the VP (VP$_p$) in the cat (Kniffki and Vahle-Hinz 1987; Vahle-Hinz et al. 1987). This peripheral region is characterized by scattered neurons with small, spindle-shaped somata and few primary dendrites, giving rise to sparse dendritic arbors oriented within the confines of the region, i.e., mediolaterally along its borders (Fig. 1c)

(Kniffki et al. 1993). The narrow sheath of cells wraps around the VP dorsally and laterally, extending towards the external medullary lamina on its ventral border, thus widening to a sizeable area ventral to the ventral posterior medial nucleus (VPM). This part resembles the ventral posterior inferior nucleus (VPI) of the monkey. Most of the neurons in the VPM$_p$ have somatic RFs, and the majority are located on the head (Fig. 1d). The lateral part of the VP$_p$ (VPL$_p$) is a narrow band surrounding the ventral posterior lateral nucleus (VPL) ventrally, laterally, and dorsally. It merges with the lateral and medial parts of the PO, and its RFs are located on the fore- and hind limbs as well as on visceral organs (Fig. 1e). Thus, with respect to somatic RFs, a coarse somatotopy is present running parallel to that of the VP proper.

While no nociceptive neurons are found inside the VP of the cat, they may invade the laminae between subnuclei of the VP as shown in Fig. 1b. In the rat, clustering of VP cells with similar RFs is even more pronounced, and thus, laminae with nociceptive cells are more numerously interspersed, contributing to the appearance of a mixed representation of mechanoreceptive and nociceptive neurons described in the literature (for review, see Willis 1997). Most of the somatic nociceptive neurons in the rat, however, are found in the PO.

Posterior Complex (PO)

The somatosensory parts of PO adjoin VPL laterally and dorsally and VPM dorsally (POm). The nuclear and subnuclear boundaries are not discernible histologically. PO receives inputs from the STT, the lateral cervical nucleus, the dorsal column, the nucleus of the solitary tract, and the parabrachial area. These afferents carry nociceptive somatic and visceral signals. In the cat, the VPL_p and adjoining PO were found to hold visceroceptive neurons with inputs from thoracic and pelvic visceral organs (Brüggemann et al. 1993, 1994; Horn et al. 1997, 1999; Vahle-Hinz et al. 1995). Also in the rat, a concentration of visceroceptive neurons in the region immediately surrounding the VP was found (Berkley et al. 1993). No viscerotropic organization is discernible, but the response properties indicate that this region may be involved in the encoding, localization, and referral of visceral pain.

The somatic low-threshold RFs of PO neurons are often large and bilateral and include deep structures like muscles and joints. In the rat, somatic nociceptive neurons with small RFs and stimulus-encoding properties similar to those of the cat's VP_p are found in the POm.

Medial Thalamus

Nociceptive somatic and visceral neurons are found in a number of nuclei of the medial thalamus, including the central lateral and parafascicular nucleus of the intralaminar complex, the mediodorsal nucleus, and the nucleus submedius (for review, see Willis 1997). The RFs are usually large, often bilateral, with convergent input from skin, muscles, joints, and viscera. Most medial and intralaminar neurons respond to convergent noxious somatic and visceral stimuli (Berkley et al. 1995); in contrast, visceroceptive-specific neurons have been described in the central lateral nucleus (Ren et al. 2009).

There is no somatotopic or viscerotopic organization in medial thalamic nuclei; thus, these regions may not be involved in the spatial localization of a painful stimulus. However, intensity is encoded in graded responses, a property important for the affective component of pain. This has been corroborated in studies on chronically implanted awake rats where nocifensive behavior and neuronal responses to noxious laser stimulation were compared (Zhang et al. 2011). The nociceptive responses are particularly sensitive to the kind and depth of anesthesia. NS, WDR, and multireceptive neurons occur in the mediodorsal and intralaminar nuclei. The nucleus submedius receives a dense projection from lamina I of the spinal cord in cats, and nociceptive-specific somatic neurons here are more abundant than WDR neurons in both rats and cats. Recently, behavioral studies in rats have identified a differential involvement of medial thalamic nuclei in endogenous descending control of discriminative aspects of nociception (MD, VM; You et al. 2013) and suppression of pain affect (parafascicular nucleus; Harte et al. 2011).

References

Berkley, K. J., Guilbaud, G., Benoist, J.-M., & Gautron, M. (1993). Responses of neurons in and near the thalamic ventrobasal complex of the rat to stimulation of uterus, vagina, colon, and skin. *Journal of Neurophysiology, 69,* 557–568.

Berkley, K. J., Benoist, J.-M., Gautron, M., & Guilbaud, G. (1995). Responses of neurons in the caudal intralaminar thalamic complex of the rat to stimulation of the uterus, vagina, cervix, colon, and skin. *Brain Research, 695*, 92–95.

Brüggemann, J., Vahle-Hinz, C., & Kniffki, K.-D. (1993). Representation of the urinary bladder in the lateral thalamus of the cat. *Journal of Neurophysiology, 70*, 482–491.

Brüggemann, J., Vahle-Hinz, C., & Kniffki, K.-D. (1994). Projections from the pelvic nerve to the periphery of the cat's thalamic ventral posterolateral nucleus and adjacent regions of the posterior complex. *Journal of Neurophysiology, 72*, 2237–2245.

Harte, S. E., Spuz, C. A., & Borszcz, G. S. (2011). Functional interaction between medial thalamus and rostral anterior cingulate cortex in the suppression of pain affect. *Neuroscience, 172*, 460–473.

Horn, A. C., Vahle-Hinz, C., Petersen, M., et al. (1997). Projections from the renal nerve to the cat's lateral somatosensory thalamus. *Brain Research, 736*, 47–55.

Horn, A. C., Vahle-Hinz, C., Brüggemann, J., et al. (1999). Responses of neurons in the lateral thalamus of the cat to stimulation of urinary bladder, colon, esophagus, and skin. *Brain Research, 851*, 164–174.

Kniffki, K.-D., & Vahle-Hinz, C. (1987). The periphery of the cat's ventroposteromedial nucleus (VPM$_p$): Nociceptive neurons. In J.-M. Besson, G. Guilbaud, & M. Peschanski (Eds.), *Thalamus and pain* (pp. 245–257). Amsterdam: Elsevier.

Kniffki, K.-D., Pawlak, M., & Vahle-Hinz, C. (1993). Scaling behavior of the dendritic branches of thalamic neurons. *Fractals, 1*, 171–178.

Ren, Y., Zhang, L., Lu, Y., Yang, H., & Westlund, K. N. (2009). Central lateral thalamic neurons receive noxious visceral mechanical and chemical inputs in rats. *Journal of Neurophysiology, 102*, 244–258.

Vahle-Hinz, C., & Detsch, O. (2002). What can in vivo electrophysiology in animal models tell us about mechanisms of anesthesia? *British Journal of Anaesthesia, 89*, 123–142.

Vahle-Hinz, C., & Gottschaldt, K.-M. (1983). Principal differences in the organization of the thalamic face representation in rodents and felids. In G. Macchi, A. Rustioni, & R. Spreafico (Eds.), *Somatosensory integration in the thalamus* (pp. 125–145). Amsterdam: Elsevier.

Vahle-Hinz, C., Freund, I., & Kniffki, K.-D. (1987). Nociceptive neurons in the ventral periphery of the cat thalamic ventroposteromedial nucleus. In R. F. Schmidt, H.-G. Schaible, & C. Vahle-Hinz (Eds.), *Fine afferent nerve fibers and pain* (pp. 440–450). Weinheim: VCH Verlagsgesellschaft.

Vahle-Hinz, C., Brüggemann, J., & Kniffki, K.-D. (1995). Thalamic processing of visceral pain. In B. Bromm & J. Desmedt (Eds.), *Pain and the brain. From nociception to cognition. Advances in pain research and therapy* (Vol. 22, pp. 125–141). New York: Raven.

Vahle-Hinz, C., Reeker, W., Detsch, O., et al. (2002). Antinociceptive effects of anesthetics in vivo: Neuronal responses and cellular mechanisms. In B. W. Urban & M. Barann (Eds.), *Molecular and basic mechanisms of anesthesia* (pp. 516–524). Lengerich: Pabst Science.

Willis, W. D. (1997). Nociceptive functions of thalamic neurons. In M. Steriade, E. G. Jones, & D. A. McCormick (Eds.), *Thalamus (Experimental and clinical aspects*, Vol. II, pp. 373–424). Amsterdam: Elsevier.

You, H.-J., Lei, J., Niu, N., Yang, L., Fan, X.-L., Tjolsen, A., et al. (2013). Specific thalamic nuclei function as novel 'nociceptive discriminators' in the endogenous control of nociception in rats. *The Journal of Neuroscience, 232*, 53–63.

Zhang, Y., Wang, N., Wang, J.-Y., Chang, J.-Y., Woodward, D. J., & Luo, F. (2011). Ensemble encoding of nociceptive stimulus intensity in the rat medial and lateral pain systems. *Molecular Pain, 7*, 64.

Thalamic Pain

▶ Central Pain, Diagnosis
▶ Central Poststroke Pain

Thalamic Plasticity

▶ Thalamus, Dynamics of Nociception

Thalamic Plasticity and Chronic Pain

Jonathan O. Dostrovsky[1,2] and
Douglas Rasmusson[3]
[1]Department of Physiology, University of Toronto, Toronto, ON, Canada
[2]Faculty of Dentistry, University of Toronto, Toronto, ON, Canada
[3]Department of Physiology and Biophysics, Dalhousie University, Halifax, Canada

Synonyms

Chronic pain, thalamic plasticity; Functional changes in thalamus

Definition

Long-term changes in processing of somatosensory inputs in thalamus as a result of peripheral nerve damage, ▶ deafferentation, spinal cord and brain damage, and/or chronic pain.

Characteristics

Most of the information regarding ▶ plasticity in the somatosensory system at supraspinal levels concerns the processing and representation of non-nociceptive tactile inputs. This is mainly because it is difficult to determine a clear map of the body representation of neurons responding to nociceptive inputs. However, it is likely that there are at least some common mechanisms between plasticity of innocuous information and nociceptive information processing, and thus, the first section will deal with findings regarding plasticity in processing of tactile information.

In contrast to the hundreds of papers that have described reorganization of the primary somatosensory cortex after deafferentation, relatively few have studied changes at the thalamic level. However, the variables that affect thalamic and cortical reorganization are the same: the age of the animal at the time of the damage; the extent and location of the damage; and the time that has expired since the damage. Neonatal damage, or even trimming of whiskers from birth to decrease sensory experience, can produce dramatic changes in the somatotopic map within the rodent thalamus (e.g., Nicolelis et al. 1997; Simons and Land 1994). These developmental changes are reflected in morphological changes in the structural representation of whiskers in the rodent thalamus, the "▶ barreloids." In particular, the separation between affected barreloids is lost, suggesting a profound interference with the migration of thalamic neurons. The critical period for the thalamus is intermediate between the earlier developing brainstem and the later developing cortex.

In the adult animal, immediate reorganization of the ▶ ventral posterior lateral nucleus (VPL) thalamus, by reversibly blocking afferent inputs from the skin following local anesthetic injections, has been demonstrated in several species including man (Nicolelis et al. 1993; see Dostrovsky 1999 for other references). This results in the appearance of new and/or enlargement of the RFs of individual neurons. Of particular interest is a study that found that this unmasking was greatly reduced if the somatosensory cortex was inactivated at the time of peripheral deafferentation, suggesting that ▶ cortical feedback plays an important role in these immediate changes (Krupa et al. 1999).

Longer-term plasticity of the thalamus (requiring several months) has been demonstrated after a variety of different types of peripheral nerve damage. These include transection of median and ulnar nerves in squirrel monkeys (Garraghty and Kaas 1991), digit amputation in raccoons (Rasmusson 1996), dorsal rhizotomies in primates (Jones and Pons 1998), as well as central lesions affecting the ascending branch of peripheral afferents in the dorsal columns (Pollin and Albe-Fessard 1979). Thalamic reorganization also results from destruction of the dorsal column nuclei, which provide the major, direct input to VPL and to the spinothalamic tract. Lesioning the gracile nucleus in rats results in expansion of upper limb and shoulder representations (Parker et al. 1998; Wall and Egger 1971). In all these cases, the major effect is an expansion of intact adjacently represented regions into the region that has been deafferented. In addition, spinothalamic tract lesions in subhuman primates have been shown to increase the spontaneous firing rate and responses to mechanical stimulation of the skin, and increase the degree of ▶ bursting activity in low threshold slowly adapting neurons (Weng et al. 2003).

Recordings and stimulation in human thalamus of chronic pain patients suffering spinal cord damage or amputation reveal an expansion of the intact regions into deafferented regions, and the existence of neurons firing spontaneously in bursts. Stimulation in such regions usually evokes sensations referred to the deafferented region or phantom limb. Some of these alterations may also be involved in mediating the patients' pain (see references in Dostrovsky 1999; Weng et al. 2003).

Few studies have examined the mechanisms responsible for thalamic plasticity. Electron microscopic evidence for synaptogenesis, presumably resulting from sprouting of surviving axons, has been found in the rat after dorsal column nuclei lesions (Wells and Tripp 1987). Of potential interest is the decrease in ▶ GABAA receptors that follows dorsal ▶ rhizotomy in primates (Rausell et al. 1992). In addition, other studies have revealed changes in the morphology of ▶ GABA terminals following dorsal column lesions. Decreased inhibitory control could contribute directly or indirectly to the plasticity mechanisms responsible for thalamic reorganization, and may also result in increased RF sizes and lowered thresholds of nociceptive neurons in thalamus (see references in Weng et al. 2003). There is also electrophysiological evidence that many cells in VPL have ▶ subliminal receptive fields, and these may provide an important substrate for the expansion of ▶ receptive fields following deafferentation (Dostrovsky 1999).

There have been several studies in animals showing greatly increased responses, receptive field sizes, and decreased thresholds of nociceptive neurons in medial and lateral (VPL) thalamus and alterations in sodium channel expression in VPL following peripheral damage, inflammation, and/or central damage. Although some of these alterations were due to changes at the thalamic level (see Zhao et al. 2006; Hains and Waxman 2007), it is difficult to assess to what extent the changes were due also to neuroplastic changes occurring at spinal and trigeminal levels (e.g., Dostrovsky and Guilbaud 1990; see references in Weng et al. 2003).

Several studies have reported that following deafferentation and ▶ CNS damage leading to chronic pain, there is an increase in bursting activity of neurons in and near the deafferented region in human thalamus, and have suggested that this may be related to the development of central pain. Bursting cells have also been observed in chronic pain patients (see ▶ Thalamic Bursting Activity and references in Weng et al. 2003).

Of considerable interest are observations in humans on the incidence of stimulation-evoked pain. In non-pain patients, the incidence of evoking pain by stimulation within ▶ ventral basal nucleus (VB) is very low. However, in some types of pain patients, and in particular in poststroke central pain patients, the incidence is much higher, suggesting that neuroplastic changes have occurred in the thalamocortical system in these cases that are causing pain to be perceived by thalamic stimuli that normally elicit only nonpainful paresthesia (see references in Dostrovsky 1999; Weng et al. 2003).

Cross-References

▶ Thalamus, Dynamics of Nociception

References

Dostrovsky, J. O. (1999). Immediate and long-term plasticity in human somatosensory thalamus and its involvement in phantom limbs. *Pain, 6,* 37–43.

Dostrovsky, J. O., & Guilbaud, G. (1990). Nociceptive responses in medial thalamus of the normal and arthritic rat. *Pain, 40,* 93–104.

Garraghty, P., & Kaas, J. (1991). Functional reorganization in adult monkey thalamus after peripheral nerve injury. *Neuroreport, 2,* 747–750.

Hains, B. C., & Waxman, S. G. (2007). Sodium channel expression and the molecular pathophysiology of pain after SCI. *Progress in Brain Research, 161,* 195–203.

Jones, E. G., & Pons, T. P. (1998). Thalamic and brainstem contributions to large-scale plasticity of primate somatosensory cortex. *Science, 282,* 1121–1125.

Krupa, D. J., Ghazanfar, A. A., & Nicolelis, M. A. (1999). Immediate thalamic sensory plasticity depends on corticothalamic feedback. *Proceedings of the National Academy of Sciences of the United States of America, 96,* 8200–8205.

Nicolelis, M. A. L., Lin, R. C. S., Woodward, D. J., et al. (1993). Induction of immediate spatiotemporal changes in thalamic networks by peripheral block of ascending cutaneous information. *Nature, 361,* 533–536.

Nicolelis, M. A., Lin, R. C. S., & Chapin, J. K. (1997). Neonatal whisker removal reduces the discrimination of tactile stimuli by thalamic ensembles in adult rats. *Journal of Neurophysiology, 78,* 1691–1706.

Parker, J. L., Wood, M. L., & Dostrovsky, J. O. (1998). A focal zone of thalamic plasticity. *Journal of Neuroscience, 18,* 548–558.

Pollin, L., & Albe-Fessard, D. (1979). Organization of somatic thalamus in monkeys with and without section of dorsal spinal tracts. *Brain Research, 173,* 431–449.

Rasmusson, D. D. (1996). Changes in the organization of the ventroposterior lateral thalamic nucleus after digit removal in adult raccoon. *The Journal of Comparative Neurology, 364*, 92–103.

Rausell, E., Cusick, C. G., Taub, E., et al. (1992). Chronic deafferentation in monkeys differentially affects nociceptive and nonnociceptive pathways distinguished by specific calcium-binding proteins and down-regulates Î³-aminobutyric acid type A receptors at thalamic levels. *Proceedings of the National Academy of Sciences of the United States of America, 89*, 2571–2575.

Simons, D. J., & Land, P. W. (1994). Neonatal whisker trimming produces greater effects in nondeprived than deprived thalamic barreloids. *Journal of Neurophysiology, 72*, 1434–1447.

Wall, P. D., & Egger, M. D. (1971). Formation of new connexions in adult rat brains after partial deafferentation. *Nature, 232*, 542–545.

Wells, J., & Tripp, L. N. (1987). Time course of reactive synaptogenesis in the subcortical somatosensory system. *The Journal of Comparative Neurology, 255*, 466–475.

Weng, H. R., Lenz, F. A., Vierck, C., et al. (2003). Physiological changes in primate somatosensory thalamus induced by deafferentation are dependent on the spinal funiculi that are sectioned and time following injury. *Neuroscience, 116*, 1149–1160.

Zhao, P., Waxman, S. G., & Hains, B. C. (2006). Sodium channel expression in the ventral posterolateral nucleus of the thalamus after peripheral nerve injury. *Molecular Pain, 2*, 27.

Thalamic Projections

▶ Corticothalamic and Thalamocortical Interactions

Thalamic Reorganization

▶ Thalamus, Dynamics of Nociception

Thalamic Response to Experimental Pain in Humans

▶ Human Thalamic Response to Experimental Pain (Neuroimaging)

Thalamic Reticular Nucleus

Synonyms

TRN

Definition

A nucleus that surrounds the thalamus. It receives inputs from the cortex and thalamus and projects back to the thalamus with inhibitory connections. Neurons in this thalamic nucleus are inhibitory GABAergic neurons that provide inhibition onto thalamic relay neurons. TRN neurons do not project to the cerebral cortex.

Cross-References

▶ Thalamus, Nociceptive Neurotransmission

Thalamo-Amygdala Interactions and Pain

George S. Borszcz[1] and Steven Harte[2]
[1]Department of Psychology, Wayne State University, Detroit, USA
[2]Anesthesiology and Internal Medicine/ Rheumatology, University of Michigan VA Ann Arbor Healthcare System, Ann Arbor, MI, USA

Synonyms

Affective analgesia; Affective pain processing; Limbic forebrain matrix; Pain asymbolia

Definition

Processing the affective-motivational dimension of the pain experience. A limbic forebrain matrix underlies generation of the affective-motivational dimension of pain (Borszcz and Spuz 2009), and suppression of nociceptive transmission through this limbic forebrain matrix produces selective inhibition of the affective response to pain (affective analgesia). Interaction between limbic forebrain sites and the midbrain periaqueductal gray is one mechanism through which morphine acts to reduce the affective-motivational dimension of pain.

Characteristics

W. H. Sweet (1980), the eminent neurosurgeon and pain researcher, observed the critical need to evaluate the neural mechanisms that generate and suppress the affective-motivational dimension of pain. In reviewing the extant human neurosurgical data, he noted that ablation of several limbic forebrain sites proved effective in reducing the "agonizing" pain associated with advanced cancer. These areas include the anterior cingulate cortex (ACC), medial thalamic nuclei (centromedian and parafascicular nuclei), amygdala, and prefrontal cortex. Damage of these structures produced no loss of the sensory-discriminative dimension of pain but was efficacious in alleviating its affective-motivational dimension. Sweet pointed out that the mechanisms responsible for this dissociative analgesia were poorly understood, and called for the development of animal models and systematic research programs that could provide insights into the neurobiology of the affective dimension of pain. Sweet's seminal call to action has led to development of an understanding of the neurobiological, psychological, and social underpinnings of pain affect (Lumley et al. 2011).

Subsequent neurobiological research provided convergent evidence in support of the involvement of the limbic forebrain in processing the affective-motivational dimension of human pain. Additional human neurosurgical observations confirmed the involvement of the ▶ medial thalamus and ACC in pain perception. Ablations of these areas remain an effective treatment in alleviating chronic intractable pain (Hassenbusch et al. 1990; Romanelli et al. 2004; Whittle and Jenkinson 1995). The chronic pain state appears to be mediated in part by sustained abnormal neural activity within the limbic forebrain. High-frequency spontaneous neural activity was recorded from medial thalamic nuclei in patients with deafferentation pain (Rinaldi et al. 1991). High-frequency stimulation of the medial thalamus produced reports of intense pain in human subjects, which was suppressed by treatment with the μ-opiate agonist fentanyl (Velasco et al. 1998). Electrophysiological recordings from the human medial temporal lobe also demonstrated noxious evoked activity within the central nucleus of the amygdala (Liu et al. 2010). Neuroimaging studies of the human pain, experimental and clinical, also revealed a limbic forebrain matrix that underlies the processing of the affective-motivational dimension of pain (Derbyshire 2000; Kulkarni et al. 2007). These studies consistently report activation of the amygdala, hypothalamus, ACC, insula, prefrontal cortex, and medial thalamus.

Following the ▶ parallel pain processing model of Melzack and Casey (1968), the results of these neurosurgical and neuroimaging studies are interpreted as reflecting the processing of noxious stimulation by the ▶ medial pain system (Kulkarni et al. 2005; Vogt and Sikes 2000). The medial pain system is proposed to process noxious stimulation transmitted by medially projecting spinothalamic pathways (i.e., spinoreticulothalamic tract). This projection system terminates in medial thalamic nuclei that project to limbic forebrain structures that underlie processing of the affective-motivational dimension of pain. The ACC, insula, and prefrontal cortex are principal forebrain targets of medial thalamic projections (Vogt and Sikes 2000). The medial thalamus also projects to the amygdala which is reciprocally interconnected with the medial thalamus, ACC, and insula. The amygdala and hypothalamus also receive direct nociceptive afferents from the spinal dorsal horn via spinoamygdaloid and spinohypothalamic pathways (Newman et al. 1996), as well as nociceptive afferents relayed via the parabrachial nucleus (Bernard and Besson 1990; Bester et al. 1995; Bourgeais et al. 2001). The excitatory amino acid glutamate, acting at NMDA and non-NMDA receptors, in the central nucleus of the amygdala and ACC is the principal neuro-transmitter of nociceptive afferents to these sites (Lee et al. 2007; Spuz and Borszcz 2012; Yang et al. 2006). The processing of noxious stimulation by this limbic forebrain matrix generates the emotional response to pain, coordinates relevant motor activity, and supports the development of fear conditioning and avoidance responding (Derbyshire 2000). For example, stimulation of the dorsomedial subdivision of the ventromedial hypothalamus in rats, that receives nociceptive afferents, generates affective pain-like behaviors and supports fear conditioning, and subthreshold stimulation of this site enhances affective responses to noxious stimulation and fear conditioning supported by noxious stimulation (Borszcz 2006).

Alternately, the ▶ lateral pain system processes noxious stimulation transmitted by laterally projecting spinothalamic and trigeminothalamic pathways that terminate in lateral thalamic nuclei (i.e., VPL and VPM). Projections of the lateral thalamic nuclei to the primary and secondary somatosensory cortices are proposed to underlie the processing of the sensory-discriminative dimension of the pain experience. This dimension of pain signals the location, intensity, and physical properties of a noxious stimulus.

Systemic administration of morphine preferentially suppresses the affective response of humans and animals to noxious stimulation (Borszcz et al. 1994; Price et al. 1985). This effect is mediated by the inhibition of pain transmission through limbic forebrain structures that process the affective dimension of pain (Casey et al. 2000). The ventrolateral division of the midbrain periaqueductal gray (vPAG) is a major site through which morphine acts to suppress pain transmission. Through its descending projections to the rostral ventromedial medulla, the vPAG inhibits pain transmission at the level of the spinal dorsal horn. The vPAG also contributes ascending projections that suppress pain

transmission directly within the limbic forebrain (Borszcz 1999). The vPAG and adjacent dorsal raphé nucleus provide ascending serotonergic projections to the medial thalamus (parafascicular nucleus) and amygdala, and the microinjection of morphine into the vPAG increases the release and metabolism of serotonin in these sites (Munn and Borszcz 2002; Tao and Auerbach 1995). Moreover, the suppression of rats' affective reaction to noxious stimulation following injection of morphine into vPAG is reversed by bilateral administration of serotonin antagonists into either the parafascicular nucleus or amygdala (Borszcz 1999).

The parafascicular nucleus and amygdala appear to interact in the production of affective analgesia following the injection of morphine into the vPAG. The unilateral administration of serotonin antagonists into the parafascicular nucleus or amygdala failed to alter the antinociceptive action of vPAG-administered morphine. However, the combined unilateral administration of serotonin antagonists into the parafascicular nucleus and amygdala was as effective as their bilateral administration into either site in reversing the antinociceptive action of morphine injected into the vPAG (Borszcz and Streltsov 2000). These findings suggest that a functional interaction exists between the medial thalamus, amygdala, and vPAG in mediating affective analgesia. For example, suppression of rats' affective response to noxious stimulation that follows administration of morphine into the parafascicular nucleus is blocked by inhibition of the vPAG via microinjection of the GABAA agonist muscimol (Munn et al. 2009) and by microinjection of glutamate receptor antagonists (NMDA and non-NMDA) into the rostral ACC (Harte et al. 2011). The rostral ACC projects to the vPAG, and stimulation of the rostral ACC supports analgesia through its activation of the vPAG (Hardy 1985; Hardy and Haigler 1985; LaBuda and Fuchs 2005). Evidence also exists for the interaction between the vPAG and other limbic forebrain sites (nucleus accumbens, habenula) in mediating the antinociceptive action of vPAG-administered morphine (Ma and Han 1991).

References

Bernard, J. F., & Besson, J. M. (1990). The spino (trigemino)pontoamygdaloid pathway: Electrophysiological evidence for an involvement in pain processes. *Journal of Neurophysiology, 63*(3), 473–490.

Bester, H., Menendez, L., Besson, J. M., & Bernard, J. F. (1995). Spino (trigemino) parabrachiohypothalamic pathway: Electrophysiological evidence for an involvement in pain processes. *Journal of Neurophysiology, 73*(2), 568–585.

Borszcz, G. S. (1999). Differential contributions of medullary, thalamic, and amygdaloid serotonin to the antinociceptive action of morphine administered into the periaqueductal gray: A model of morphine analgesia. *Behavioral Neuroscience, 113*(3), 612–631.

Borszcz, G. S. (2006). Contribution of the ventromedial hypothalamus to generation of the affective dimension of pain. *Pain, 123*(1–2), 155–168.

Borszcz, G. S., & Spuz, C. A. (2009). Hypothalamic control of pain vocalization and affective dimension of pain signaling. In S. M. Brudzynski (Ed.), *Handbook of mammalian vocalization* (pp. 281–292). Oxford: Academic.

Borszcz, G. S., & Streltsov, N. G. (2000). Amygdaloid-thalamic interactions mediate the antinociceptive action of morphine microinjected into the periaqueductal gray. *Behavioral Neuroscience, 114*(3), 574–584.

Borszcz, G. S., Johnson, C. P., & Fahey, K. A. (1994). Comparison of motor reflex and vocalization thresholds following systemically administered morphine, fentanyl, and diazepam in the rat: Assessment of sensory and performance variables. *Pharmacology, Biochemistry, and Behavior, 49*(4), 827–834.

Bourgeais, L., Monconduit, L., Villanueva, L., & Bernard, J. F. (2001). Parabrachial internal lateral neurons convey nociceptive messages from the deep laminas of the dorsal horn to the intralaminar thalamus. *Journal of Neuroscience, 21*(6), 2159–2165.

Casey, K. L., Svensson, P., Morrow, T. J., Raz, J., Jone, C., & Minoshima, S. (2000). Selective opiate modulation of nociceptive processing in the human brain. *Journal of Neurophysiology, 84*(1), 525–533.

Derbyshire, S. W. (2000). Exploring the pain "neuromatrix". *Current Review of Pain, 4*(6), 467–477.

Hardy, S. G. (1985). Analgesia elicited by prefrontal stimulation. *Brain Research, 339*(2), 281–284.

Hardy, S. G., & Haigler, H. J. (1985). Prefrontal influences upon the midbrain: A possible route for pain modulation. *Brain Research, 339*(2), 285–293.

Harte, S. E., Spuz, C. A., & Borszcz, G. S. (2011). Functional interaction between medial thalamus and rostral anterior cingulate cortex in the suppression of pain affect. *Neuroscience, 172*, 460–473.

Hassenbusch, S. J., Pillay, P. K., & Barnett, G. H. (1990). Radiofrequency cingulotomy for intractable cancer pain using stereotaxis guided by magnetic resonance imaging. *Neurosurgery, 27*(2), 220–223.

Kulkarni, B., Bentley, D. E., Elliott, R., Youell, P., Watson, A., Derbyshire, S. W., et al. (2005). Attention to pain localization and unpleasantness discriminates the functions of the medial and lateral pain systems. *European Journal of Neuroscience, 21*(11), 3133–3142.

Kulkarni, B., Bentley, D. E., Elliott, R., Julyan, P. J., Boger, E., Watson, A., et al. (2007). Arthritic pain is processed in brain areas concerned with emotions and fear. *Arthritis and Rheumatism, 56*(4), 1345–1354.

LaBuda, C. J., & Fuchs, P. N. (2005). Attenuation of negative pain affect produced by unilateral spinal nerve injury in the rat following anterior cingulate cortex activation. *Neuroscience, 136*(1), 311–322.

Lee, C. M., Chang, W. C., Chang, K. B., & Shyu, B. C. (2007). Synaptic organization and input-specific short-term plasticity in anterior cingulate cortical neurons with intact thalamic inputs. *European Journal of Neuroscience, 25*(9), 2847–2861.

Liu, C. C., Ohara, S., Franaszczuk, P., Zagzoog, N., Gallagher, M., & Lenz, F. A. (2010). Painful stimuli evoke potentials recorded from the medial temporal lobe in humans. *Neuroscience, 165*(4), 1402–1411.

Lumley, M. A., Cohen, J. L., Borszcz, G. S., Cano, A., Radcliffe, A. M., Porter, L. S., et al. (2011). Pain and emotion: A biopsychosocial review of recent research. *Journal of Clinical Psychology, 67*(9), 942–968.

Ma, Q. P., & Han, J. S. (1991). Neurochemical studies on the mesolimbic circuitry of antinociception. *Brain Research, 566*(1–2), 95–102.

Melzack, R., & Casey, K. L. (1968). Sensory, motivational, and central control determinants of pain. In D. Kenshalo (Ed.), *The skin senses* (pp. 423–439). Springfield, IL: Thomas.

Munn, E. M., & Borszcz, G. S. (2002). Increases in the release and metabolism of serotonin in nucleus parafascicularis thalami following systemically administered morphine in the rat. *Neuroscience Letters, 332*(3), 151–154.

Munn, E. M., Harte, S. E., Lagman, A., & Borszcz, G. S. (2009). Contribution of the periaqueductal gray to the suppression of pain affect produced by administration of morphine into the intralaminar thalamus of rat. *The Journal of Pain, 10*(4), 426–435.

Newman, H. M., Stevens, R. T., & Apkarian, A. V. (1996). Direct spinal projections to limbic and striatal areas: Anterograde transport studies from the upper cervical spinal cord and the cervical enlargement in squirrel monkey and rat. *The Journal of Comparative Neurology, 365*(4), 640–658.

Price, D. D., Von der Gruen, A., Miller, J., Rafii, A., & Price, C. (1985). A psychophysical analysis of morphine analgesia. *Pain, 22*(3), 261–269.

Rinaldi, P. C., Young, R. F., Albe-Fessard, D., & Chodakiewitz, J. (1991). Spontaneous neuronal hyperactivity in the medial and intralaminar thalamic nuclei of patients with deafferentation pain. *Journal of Neurosurgery, 74*(3), 415–421.

Romanelli, P., Esposito, V., & Adler, J. (2004). Ablative procedures for chronic pain. *Neurosurgery Clinics of North America, 15*(3), 335–342.

Spuz, C. A., & Borszcz, G. S. (2012). NMDA or non-NMDA receptor antagonism within the amygdaloid central nucleus suppresses the affective dimension of pain in rats: Evidence for hemispheric synergy. *The Journal of Pain, 13*(4), 328–337.

Sweet, M. H. (1980). Mechanisms of chronic pain (neuralgias and certain other neurogenic pain). In J. J. Bonica (Ed.), *Pain* (pp. 287–303). New York: Raven.

Tao, R., & Auerbach, S. B. (1995). Involvement of the dorsal raphe but not median raphe nucleus in morphine-induced increases in serotonin release in the rat forebrain. *Neuroscience, 68*(2), 553–561.

Velasco, M., Brito, F., Jimenez, F., Gallegos, M., Velasco, A. L., & Velasco, F. (1998). Effect of fentanyl and naloxone on a thalamic induced painful response in intractable epileptic patients. *Stereotactic and Functional Neurosurgery, 71*(2), 90–102.

Vogt, B. A., & Sikes, R. W. (2000). The medial pain system, cingulate cortex, and parallel processing of nociceptive information. *Progress in Brain Research, 122*, 223–235.

Whittle, I. R., & Jenkinson, J. L. (1995). CT-guided stereotactic antero-medial pulvinotomy and centromedian-parafascicular thalamotomy for intractable malignant pain. *British Journal of Neurosurgery, 9*(2), 195–200.

Yang, J. W., Shih, H. C., & Shyu, B. C. (2006). Intracortical circuits in rat anterior cingulate cortex are activated by nociceptive inputs mediated by medial thalamus. *Journal of Neurophysiology, 96*(6), 3409–3422.

Thalamocortical and Corticothalamic Interactions

▶ Corticothalamic and Thalamocortical Interactions

Thalamocortical Dysrhythmia

Synonyms

TCD

Definition

A pathophysiological chain reaction at the origin of neurogenic pain. It consists of (1) a reduction of excitatory inputs onto thalamic cells, which results in cell membrane hyperpolarization; (2) the production of low-threshold calcium spike bursts by deinactivation of calcium T-channels, discharging at low (theta) frequency; (3) a progressive increase of the number of thalamocortical modules discharging at theta frequency; and (4) a cortical high-frequency activation through asymmetric corticocortical inhibition. These events have been documented by thalamic and cortical recordings in patients suffering from peripheral and central neurogenic pain.

Cross-References

▶ Thalamotomy for Human Pain Relief
▶ Thalamus, Dynamics of Nociception

Thalamocortical Fibers

Definition

Axons with cell bodies located in the thalamus and terminations in the cortex.

Cross-References

▶ Corticothalamic and Thalamocortical Interactions

Thalamocortical Module

Definition

An anatomofunctional entity comprised of thalamic cells and their cortical partners, interconnected by thalamocortical and corticothalamic projections and sustaining perceptual, motor, and cognitive hemispheric functions. The thalamocortical loop is accompanied by a shorter thalamoreticulothalamic loop. Each module may be subdivided into a specific or content subpart, providing the substrate for the integration of a given function, and a nonspecific or context subpart, dealing with the interactions between functional domains.

Cross-References

▶ Thalamotomy for Human Pain Relief

Thalamocortical Neurons

Definition

Neurons located within the thalamus and projecting directly to the cerebral cortex.

Cross-References

▶ Spinothalamic Projections in Rat

Thalamotomy

Definition

A neurosurgical procedure in which a therapeutic lesion is made in a specific subnucleus of the thalamus.

Cross-References

▶ Pain Treatment, Intracranial Ablative Procedures
▶ Thalamotomy for Human Pain Relief
▶ Thalamotomy, Pain Behavior in Animals
▶ Thalamus, Receptive Fields, Projected Fields, Human

Thalamotomy for Human Pain Relief

Daniel Jeanmonod[1], Johannes Sarnthein[2], Jair Stern[2], Michel Magnin[3], Christoph Aufenberg[2] and Anne Morel[4]
[1]Functional Neurosurgery, The Center of Ultrasound Functional Neurosurgery, Solothurn, Switzerland
[2]Functional Neurosurgery, Neurosurgical Clinic University Hospital, Zürich, Switzerland
[3]INSERM, Neurological Hospital, Lyon, France
[4]Laboratory for Functional Neurosurgery, Neurosurgical Clinic University Hospital Zürich, Zürich, Switzerland

Synonyms

First-order subtype: medial thalamotomy; Second-order subtype: central lateral thalamotomy; Stereotactic operation with thalamic target

Definition

Neurophysiological studies at the cellular level (▶ single unit activity and ▶ local field potentials or LFPs) as well as electroencephalographic (EEG) and ▶ magnetoencephalographic (MEG) recordings provide converging evidence for a thalamocortical dysregulation at the origin of chronic ▶ neurogenic pain of both peripheral and central origin. These data suggest an increase of low frequency thalamocortical rhythmicity originating in disfacilitation of thalamic relay neurons, followed by cortical activation due to asymmetries of corticocortical inhibition. The process, called ▶ thalamocortical dysrhythmia (TCD), may become self-sustained and thus chronic, due to recurrent thalamoreticulothalamic and corticoreticulothalamic feedback inhibition. The surgical approach presented here is centered on a reestablishment of a normal thalamocortical oscillatory activity using a strategically placed medial thalamic lesion, which reduces or abolishes the TCD via low frequency desamplification and provides long-term therapeutic efficiency coupled with sparing of the specific ▶ thalamocortical modules. This physiopathological framework underscores the risks run by any surgical procedure aiming at further reducing the activation of specific thalamic relay cells and thus increasing thalamic disfacilitation and dysrhythmic pain mechanisms. This entry is thus focused on ▶ medial thalamotomies, more specifically on central lateral thalamotomy (CLT).

In 1989, Lenz and collaborators (1989) provided the first evidence for the presence of ▶ low-threshold calcium spike (LTS) bursts in the thalamus of patients suffering from neurogenic pain. They found them in the somatosensory ▶ ventral posterior (VP) complex, localized in and around the portion of VP representing the deafferented and, thus, painful body part (Lenz et al. 1994). The presence of the same activities was also described widely spread in and around the posterior part of the ▶ central lateral nucleus (CL) of the medial thalamus (Jeanmonod et al. 1993, 1994, 1996). An example of such LTS bursting activity in CL is shown in Fig. 1a–c, with the typical progressive increase in duration of each successive interspike interval (ISI) and the inverse relationship between the duration of the first interspike interval and the number of spikes in a burst. Furthermore, it showed the following: (1) half of the recorded neurons presented LTS bursting activity, (2) only a minority (less than 1 %) had somatosensory receptive fields, (3) LTS bursts displayed a theta rhythmicity, with a mean interburst discharge rate of 4 Hz, and (4) there were no significant differences between recordings performed in patients suffering from peripheral and central neurogenic pain.

A recent analysis (Sarnthein et al. 2003) of thalamic LFP recordings in CL showed the presence of a high theta (4–8 Hz) power (Fig. 1d), correlating closely with the theta rhythmicity displayed by LTS bursts. In addition, an increase in the cortical theta power was recorded by EEG (Fig. 1d) and a high coherence between EEG and thalamic theta activities was found (Fig. 1e). This underscores the expected high level of functional coupling between thalamus and cortex. In Fig. 1f,

EEG power spectra of patients have been averaged and compared with those of controls, confirming the existence of a disease-related increase in theta power.

Based on converging evidence from experimental and clinical data over the last 20 years (Llinás and Jahnsen 1982; Steriade et al. 1990; Jeanmonod et al. 1993, 1996, 2001; Steriade et al. 1997; Llinás et al. 1999, 2001), a thalamocortical concept of chronic neurogenic pain was initially proposed at the thalamic level (Jeanmonod et al. 1993), and extended to the cortical level, with the denomination of thalamocortical dysrhythmia (Llinás et al. 1999). It is characterized by the following sequential set of events.

1. Bottom-up or top-down disfacilitation by deafferentation of excitatory inputs onto thalamic relay cells through a somatosensory lesion, either peripheral or central, is at the source of the neurogenic pain syndrome. This results in cell membrane hyperpolarization.

2. In this hyperpolarized state, deinactivation of calcium T-channels causes thalamic relay neurons to fire LTS bursts at theta frequency.

3. These thalamic neurons impose a theta rhythmicity to the thalamocortical modules they are part of, as demonstrated by theta power increases in both LFP and EEG recordings. The tight functional coupling between thalamus and cortex is confirmed by the high theta coherence between the two. This coupling is not only sustained by thalamocorticothalamic, but also by thalamoreticulothalamic and corticoreticulothalamic recurrent projections.

4. Divergent thalamocortical, corticothalamic, and reticulothalamic projections provide the anatomical basis for the coherent diffusion of low frequencies to an increasing number of neighboring thalamocortical modules. This phenomenon may explain the frequently observed delay between the occurrence of the causal insult and the beginning of pain.

5. The final step consists in the activation of high frequency (beta and gamma) cortical domains in the vicinity of low frequency theta areas: constraining corticocortical GABAergic

Thalamotomy for Human Pain Relief, Fig. 1 (a, b, c) LTS bursts. (a) LTS bursting cell recorded in the posterior part of CL. Note the progressive increase in duration of each successive interspike interval (ISI). (b) Progressive increase of the ISI within the burst. (c) Logarithmic decrease of the length of the first ISI as a function of the number of spikes in the burst. (d, e, f) Spectral analysis of LFP and EEG. (d) LFP and EEG power spectra of one neurogenic pain patient (power units: μV^2/Hz). (e) Coherence between EEG and LFP in the same patient. (f) Power spectra of scalp EEG recordings in 11 patients (mean age 59 +/−13 years) and 12 healthy controls (mean age 56 +/−10 years). Spectra were averaged over subjects

inhibitory interneurons to theta rhythmicity may indeed reduce lateral inhibitory drive. This leads to disinhibition and thus activation of neighboring thalamocortical modules (edge effect), with production of pain sensation. MEG (Llinás et al. 1999) and LFP power correlation studies as well as LFP bicoherence data (Sarnthein et al. 2003) provide evidence for such a phenomenon, showing an increased interfrequency coherence between theta and

beta domains and thus indicating a coupling of low and high frequency activities.

Characteristics

Exploration of the human medial thalamus allowed the identification of a zone located in the posterior part of the CL, which harbored a large majority of the recorded LTS bursts

Thalamotomy for Human Pain Relief, Fig. 2 (a) Sagittal atlas section (Morel et al. 1997) 7.2 mm distant from the ventricular border. The horizontal scale line is aligned to the intercommissural plane, with a cross indicating the level of the posterior commissure. Scale: 1 mm between graduations. The *shaded area* displays the position and extent of CLT, centered on the penetration track (*dashed line*). *AV* anteroventral, *CL* central lateral, *CM* centre médian, *LD* lateral dorsal, *Li* limitans, *MDpc* parvocellular part of the mediodorsal nucleus, *Pf* parafascicular, *PuM* medial pulvinar, *VAmc* magnocellular part of the ventral anterior (VA) nucleus, *VLpd* and *VLpv* dorsal and ventral lateral divisions of the ventral lateral posterior nucleus, *VM* ventral medial, *VPMpc* parvocellular part of the ventral posterior medial nucleus, *ZI* zona incerta. (**b**) Histogram displaying the clinical results of the CL thalamotomy for 96 patients with a mean follow-up of 3 years and 9 months

(Jeanmonod et al. 1993, 1994). Considering in addition the evidence for low frequency recruitment by CL stimulation (Morison and Dempsey 1942) and surgical experience in the medial thalamus, a medial thalamic target was redefined, centered in the posterior part of the CL (for review, see Jeanmonod et al. 2001). This target aims at rendering the low frequency thalamocortical power increase subcritical by reducing theta overamplification and oversynchronization, without reducing the specific and remaining unaffected nonspecific thalamocortical loops. A magnetic resonance- and microelectrode-guided stereotactic thalamotomy in the posterior part of the CL (central lateral thalamotomy or CLT, Fig. 2a) was implemented in 96 patients (Jeanmonod et al. 2001) suffering from chronic therapy-resistant peripheral or central neurogenic pain (mean age 56 years; mean pain duration before surgery 7.5 years). At a mean follow-up of 3 years 9 months, 53 % of the patients benefited from a relief of more than 50 % (Fig. 2b). Patients with continuous pain showed only a mean relief of 20 % in contrast to the 66 % obtained for patients

with phasic or intermittent pain manifestations. Allodynia was suppressed in 57 % of the patients. There was only a trend for better relief in patients with peripheral neurogenic pain. An increase of continuous pain relief from 14 % to 49 % was observed when CLT was performed bilaterally instead of only contralaterally. The CLT procedure produced no somatosensory deficits and entailed no risk of pain increase, but only the usual vascular risk related to the stereotactic procedure. The localization and expanse of the posterior part of the CL, away from neurologically eloquent areas, allow the placing of the therapeutic lesion with minimal risks for neurological functions. Extension of bleeding into adjacent structures, such as the centre médian-parafascicular complex, posterior complex or medial pulvinar, does not correlate with observable deficits and could even contribute to pain relief. The medial location of the CL target guarantees a good distance away from the VP and lateral spinothalamic tract and its anteroposterior coordinate is posterior enough to avoid a significant encroachment on the mediodorsal nucleus. At the ventral limit of the target, 1–2 mm may be left intact in the area of the limitans nucleus to avoid any intrusion into the pretectum below.

From a review of the literature (Jeanmonod et al. 2001), relief percentages between 50 % and 100 % were obtained in averages of 53 %, 47 %, and 53 % after dorsal column stimulation, VP stimulation, and motor cortex stimulation, respectively. These results thus come close to those obtained after CLT. In our study, all patients selected clinically as having a neurogenic pain diagnosis are included in the results. To compare with stimulation studies, it is necessary to include in the failure percentages of these studies those patients who received only a temporary unsatisfactory stimulation test, thus showing an immediate resistance to stimulation. In many studies, quoted success rates are based only on the patients who underwent permanent implantation. A correction integrating all patients with and without permanent implantation reduces pain relief percentages very significantly.

The fact that CLT does not produce clinically relevant deficits suggests that the posterior part of the CL is no longer normally functional, that is, only serves as a generator of low frequencies. This is supported by the unresponsiveness of 99 % of the recorded cells (see above) and implies a redistribution of its functions to other thalamic nuclei. Such a transfer may be all the more complete in our patients who have been operated on after years of suffering. CLT being restricted to the medial thalamic area harboring LTS activities and exhibiting a widespread functional block leaves other medial thalamic nuclei, all potential candidates for the redistribution of CL functions, intact. We are thus facing the paradoxical and conceptually intriguing situation where a lesion not only produces no or only clinically irrelevant deficits but also results in beneficial effects by the suppression of a dysfunctional area (Welsh 1998).

The therapeutic possibilities of other medial thalamic targets, particularly the posterior complex, the medial pulvinar and the centre médian-parafascicular complex seem, in our experience, to be inferior to those of the CLT. We have, however, not explored these areas sufficiently to make a definitive statement.

References

Jeanmonod, D., Magnin, M., & Morel, A. (1993). Thalamus and neurogenic pain: Physiological, anatomical and clinical data. *Neuroreport, 4*, 475–478.

Jeanmonod, D., Magnin, M., & Morel, A. (1994). A thalamic concept of neurogenic pain. In G. F. Gebhart, D. L. Hammond, & T. S. Jensen (Eds.), *Progress in pain research and management* (pp. 767–787). Seattle: IASP Press.

Jeanmonod, D., Magnin, M., & Morel, A. (1996). Low-threshold calcium spike bursts in the human thalamus. Common physiopathology for sensory, motor and limbic positive symptoms. *Brain, 119*, 363–375.

Jeanmonod, D., Magnin, M., Morel, A., et al. (2001). Surgical control of the human thalamocortical dysrhythmia: I. Central lateral thalamotomy in neurogenic pain. *Thalamus and Related Systems, 1*, 71–79.

Lenz, F. A., Kwan, H. C., Dostrovsky, J. O., et al. (1989). Characteristics of the bursting pattern of action potentials that occurs in the thalamus of patients with central pain. *Brain Research, 496*, 357–360.

Lenz, F. A., Kwan, H. C., Martin, R., et al. (1994). Characteristics of somatotopic organization and spontaneous neuronal activity in the region of the thalamic principal sensory nucleus in patients with spinal cord transection. *Journal of Neurophysiology, 72*, 1570–1587.

Llinás, R., & Jahnsen, H. (1982). Electrophysiology of mammalian thalamic neurones in vitro. *Nature, 297*, 406–408.

Llinás, R. R., Ribary, U., Jeanmonod, D., et al. (1999). Thalamocortical dysrhythmia: A neurological and neuropsychiatric syndrome characterized by magnetoencephalography. *Proceedings of the National Academy of USA, 96*, 15222–15227.

Llinás, R., Ribary, U., Jeanmonod, D., et al. (2001). Thalamocortical dysrhythmia I. Functional and imaging aspects. *Thalamus and Related Systems, 1*, 237–244.

Morel, A., Magnin, M., & Jeanmonod, D. (1997). Multiarchitectonic and stereotactic atlas of the human thalamus. *The Journal of Comparative Neurology, 387*, 588–630.

Morison, R., & Dempsey, E. (1942). A study of thalamocortical relations. *American Journal of Physiology, 135*, 281–292.

Sarnthein, J., Morel, A., von Stein, A., et al. (2003). Thalamic theta field potentials and EEG: High thalamocortical coherence in patients with neurogenic pain, epilepsy and movement disorders. *Thalamus and Related Systems, 2*, 231–238.

Steriade, M., Jones, E. G., & Llinas, R. (1990). *Thalamic oscillations and signalling*. New York: Wiley.

Steriade, M., Jones, E. G., & McCormick, D. A. (1997). *Thalamus: Organisation and function*. Oxford: Elsevier.

Welsh, J. P. (1998). Systemic harmaline blocks associative and motor learning by the actions of the inferior olive. *European Journal of Neuroscience, 10*, 3307–3320.

Thalamotomy, Pain Behavior in Animals

Nayef E. Saadé and Suhayl J. Jabbur
Anatomy, Cell Biology and Physiology, Faculty of Medicine, American University of Beirut, Beirut, Lebanon

Definition

Textbooks of neurology continue to regard the thalamus as an obligatory relay station for all sensory pathways (except olfaction) on their way to the cerebral cortex. Moreover, the thalamus has been considered to be a most important brain center for the perception of pain (Head and Holmes 1911). Over the last century, several clinical reports described disturbed sensations and spontaneous pain produced by thalamic lesion, a pathology labeled "thalamic syndrome" (Dejerine and Roussy 1906). Therefore, clinical and experimental investigations have been devoted to the understanding of the mechanisms underlying sensory disturbances following thalamic lesions.

Characteristics

The work of Roussy (Roussy 1907) was among the first attempts to reproduce the "thalamic syndrome" in animals by controlled thalamic lesions (thalamotomies). The stereotaxic method and the techniques used for the placement of small or large experimental lesions in the thalamus are explained in ▶ Post-stroke pain model, thalamic pain (lesion).

Placement of a controlled lesion in the human thalamus cannot be performed for experimental purposes. Therefore, studies involving partial or total thalamotomies can only be performed on experimental animals under controlled conditions and with strict adherence to ethical guidelines for pain experimentation on animals. The experimental protocol should take into consideration the effects of lesions placed in different thalamic nuclear groups known to be involved in the processing of various aspects of pain. These areas can be classified under two major headings, the lateral and the medial nuclear groups. The lateral system, consisting of the ventrobasal complex, was assigned the sensory-discriminative role of pain; while the medial system, made of the midline and intralaminar nuclear groups, was assigned the motivational affective role. The protocol should also be based on special animal models simulating different pain conditions (including acute and chronic pain) and using appropriate tests that allow the assessment of the different qualities of nociception (such as mechanonociception or thermonociception etc.).

Despite abundant clinical literature on the thalamic syndrome (about 2,717 titles, listed in a recent Pub Med search), few studies have been devoted to the investigation of the effects of various thalamic lesions on nociceptive behavior in animals. Thalamic lesion, involving about 50 % of the medial thalamic nuclear groups and the adjacent area in cats, can block the escape reaction to grid electrical shock stimulation (Mitchell and Kaelber 1967). Another study showed that lesions of centromedian and parafascicular nuclei of the thalamus did not interfere with the escape reaction to noxious stimuli, but suppressed learning and avoidance conditioning (Delacour and Borst 1972). More recently, Casey and Morrow (1983) showed that bilateral medial thalamic lesions increased nociceptive responses in cats. This finding received further confirmation by the reported effects of lesions of the nucleus submedius in rats (Roberts and Dong 1994). A more recent study (Norrsell and Craig 1999) however, reported a mild thermosensory deficiency in cats subjected to ▶ electrolytic lesions, placed in the various medial nuclear groups (nucleus submedius, posterior medial nucleus, basal ventral medial nucleus). It is important to note that the electrolytic lesions used can involve neuronal cell bodies in addition to passing fibers, which can have a functional role different from that of the lesioned centers.

Recent work from our laboratory has aimed at studying the effects of lesions of various sizes and

Thalamotomy, Pain Behavior in Animals, Fig. 1 Temporal evolution of the effects of lesions placed in the lateral thalamus of rats. Mechanical (paw pressure) and thermal (hot plate) nociceptive thresholds. The latencies of the nociceptive tests elicit a significant decrease (hyperalgesia) after the lesion (time 0) when compared to the latencies of the same test observed in intact rats (control)

Thalamotomy, Pain Behavior in Animals, Fig. 2 Summary of the effects of different types of thalamic lesions on the formalin tests performed on three groups of rats subjected to thalamic lesions as compared to a fourth intact group (sham). (*top*) A composite drawing showing a microscopic view of a transverse section of a rat brain (*right*) and schematic drawing (*left*) illustrating the placement of lesions in the thalamus. (*bottom*) Each bar represents the average nociceptive score measured during a period of 12 min in phase I (early phase 3–15 min), or phase II (late or tonic phase 30–42 min) in the various groups of rats

locations in the thalamus on the nociceptive behavior in rats. These lesions were performed by injecting either electrical current (electrolytic) or excitotoxic substances (for selective lesion of cell bodies). Their effects were assessed on acute nociceptive reactions and on a rat model for mononeuropathy (Saadé et al. 1999). Chronic unilateral lesions, which were either subtotal or placed in the lateral or the medial thalamus produced significant and persistent decreases in the mechanical and thermal nociceptive thresholds as assessed by the ▶ paw pressure test (PP) and ▶ hot plate (HP) tests, respectively (Fig. 1). No significant differences were noticed between the

effects of electrolytic and excitotoxic lesions. Furthermore, the increased nociception was bilateral (Fig. 1) and more pronounced on supraspinally coordinated nociceptive tests such as the HP. Thalamic lesions also exerted differential effects on the aversive behavior induced by intraplantar injection of 0.05 ml of formalin 2.5 %, known as the ▶ formalin test (Dubuisson and Dennis 1977). The most pronounced effect, observed as increases in the nociceptive scores, was obtained with subtotal thalamic lesion (Fig. 2).

Thalamotomy, Pain Behavior in Animals, Fig. 3 Transient attenuation of neuropathic manifestations by electrolytic lesions in the lateral or the medialthalamic nuclei. Each lesion was performed at day zero on a group of rats subjected to mononeuropathy. Tactile allodynia (assessed by von frey hair filaments {VF} of two calibers) and heat hyperalgesia (assessed by the duration of paw withdrawal {PWD}to heating spot) were decreased during the first 2 weeks, after which they recovered to their pre-lesion levels. Medial lesion produced more sustained effects. Cold allodynia (assessed by the duration of reaction to a drop of acetone apllied on the paw) was mildly affected (Saadé et al. 2006, 2007)

The effects of either lateral or medial thalamic lesions on the neuropathic-like behavior observed in a rat animal model for mononeuropathy were also examined. Neuropathy was induced by selective lesion of two components (peroneal and tibial nerves) of the sciatic nerve supplying the hind paw and sparing the third component (the sural). This model is characterized by a persistent ▶ allodynia (nociceptive reaction induced by non-noxious stimulus) and ▶ hyperalgesia (increased reactivity to a noxious stimulus). Both electrolytic or excitotoxic lesions placed either in the lateral or medial thalamic sensory nuclei produced transient decreases in neuropathic manifestations, which recovered their levels before thalamotomy within 1 or 2 weeks (Fig. 3). However, when mononeuropathy was induced 1 or 2 weeks after thalamic lesions, allodynia and hyperalgesia developed without any significant differences (in intensity or temporal evolution) from those observed in animals with an intact thalamus (Saadé et al. 2006, 2007). In a more recent study, using other rat models for pain, lesion of the medial thalamus did not elicit significant alteration neither in the mechanical paw withdrawal thresholds nor in the escape/avoidance behavior (Wilson et al. 2008). This result appears to be in discordance with the traditional view attributing a key role of the medial thalamus in the affective dimension of pain.

Although it is difficult to produce exact simulation of clinical syndromes such as spontaneous pain by animal models, the batteries of tests employed appear to reflect significant changes in nociceptive reactivity in animals subjected to thalamic lesions. Starting with the earliest animal experiments by Roussy in 1907, the outcome of lesions in sensory thalamic nuclei often resulted in either a decrease or no change in the nociceptive thresholds. It becomes necessary, therefore, to explain the resulting paradox as to how a function of a center can be either unaffected or exaggerated when that center is ablated? Possible answers to this question include the fact that a lesion to a nervous center can either affect, in varying proportions, inhibitory and excitatory mechanisms or lead to plastic changes that adjust and compensate for the insult or injury.

Cross-References

▶ Lateral Thalamic Lesions, Pain Behavior in Animals

References

Casey, K. L., & Morrow, T. J. (1983). Supraspinal pain mechanisms in the cat. In K. L. Kitchell & H. H. Erickson (Eds.), *Animal pain perception and alleviation* (pp. 63–82). Bethesda: American Physiological Society.
Dejerine, J., & Roussy, G. (1906). Le syndrome thalamique. *Revista de Neurologia, 14*, 521–532.
Delacour, J., & Borst, A. (1972). Failure to find homology in rat, cat, and monkey for functions of a subcortical structure in avoidance conditioning. *Journal of Comparative and Physiological Psychology, 80*, 458–468.
Dubuisson, D., & Dennis, S. G. (1977). The formalin test: A quantitative study of the analgesic effects of morphine, meperidine and brain stem stimulation in rats and cats. *Pain, 4*, 161–174.
Head, H., & Holmes, G. (1911). Sensory disturbances from cerebral lesions. *Brain, 34*, 102–254.
Mitchell, C. L., & Kaelber, W. W. (1967). Unilateral vs bilateral medial thalamic lesions and reactivity to noxious stimuli. *Archives of Neurology, 17*, 653–660.
Norrsell, U., & Craig, A. D. (1999). Behavioral thermosensitivity after lesions of thalamic target areas of a thermosensory spinothalamic pathway in the cat. *Journal of Neurophysiology, 82*, 611–625.
Roberts, V. J., & Dong, W. K. (1994). The effect of thalamic nucleus submedius lesions on nociceptive responding in rats. *Pain, 57*, 341–349.
Roussy, G. (1907). Le syndrome thalamique. In G. Steinheil (Ed.), *La couche optique*. Paris: G. Steinheil.
Saadé, N. E., Al Amin, H., AbdelBaki, S., Safieh-Garabedian, B., Atweh, S. F., & Jabbur, S. J. (2006). Transient attenuation of neuropathic manifestations in rats following lesion or reversible block of the lateral thalamic somatosensory nuclei. *Experimental Neurology, 197*, 157–166.
Saadé, N. E., Al Amin, H., Abdel Baki, S., Chalouhi, S., Jabbur, S. J., & Atweh, S. F. (2007). Reversible attenuation of neuropathic-like manifestations in rats by lesions or local blocks of the intralaminar or the medial thalamic nuclei. *Experimental Neurology, 204*, 205–219.
Saadé, N. E., Kafrouni, A. I., Saab, C. Y., et al. (1999). Chronic thalamotomy increases pain-related behavior in rats. *Pain, 83*, 401–409.
Wilson, H. D., Uhelski, M. L., & Fuchs, P. N. (2008). Examining the role of the medial thalamus in modulating the affective dimension of pain. *Brain Research, 1229*, 90–99.

Thalamus

Definition

The thalamus is derived from the Greek word "Thálamos" (bedroom, chamber). It represents the biggest structure of the diencephalon. Its two hemispheres are located in the center of the brain next to the third ventricle. It has reciprocal connections with the cerebral cortex and relays sensory signals from all senses except that of olfaction to the cerebral cortex and is also involved in motor, arousal, and mood functions.

Cross-References

▶ Angina Pectoris, Neurophysiology and Psychophysics
▶ Deep Brain Stimulation
▶ Lateral Thalamic Pain-Related Cells in Humans

▶ Pain Treatment, Intracranial Ablative
Procedures
▶ Prefrontal Cortex, Effects on Pain-Related
Behavior
▶ Spinothalamic Projections in Rat

Thalamus and Cardiac Pain

▶ Thalamus, Clinical Visceral Pain, Human
Imaging

Thalamus and Gastrointestinal Pain

▶ Thalamus, Clinical Visceral Pain, Human
Imaging

Thalamus and Pain

Definition

Thalamic structures (including intralaminar
nuclei, principal sensory nucleus, nuclei posterior
to it including supra-geniculate, posterior, ventral
medial-posterior nucleus) that have pain-related
activity, as identified by anatomic and physiologic studies in primates.

Cross-References

▶ Pain Treatment, Motor Cortex Stimulation

Thalamus and Visceral Pain (Positron Emission Tomography or Functional Magnetic Resonance Imaging)

▶ Thalamus, Clinical Visceral Pain, Human
Imaging

Thalamus and Visceral Pain Processing (Human Imaging)

Anthony R. Hobson and Qasim Aziz
Section of GI Sciences, Hope Hospital University
of Manchester, Salford, UK

Definition

While there is ample evidence from animal
studies to support the role of the thalamus in
visceral pain processing (Al-Chaer et al. 1998;
Willis and Westlund 1997), studies in humans
have, until recently, been limited to intraoperative observations from patients with
implanted deep brain–stimulating electrodes
(Lenz et al. 1994). However, the availability of
▶ functional brain imaging techniques (FBI)
such as ▶ positron emission tomography (PET)
and ▶ functional magnetic resonance imaging
(fMRI) has allowed visceral pain researchers to
confirm that the thalamus is not only an important
relay station for visceral pain transmission, but
may also be pivotal in the generation of symptoms in several visceral pain conditions.

Characteristics

To date, information regarding the brain
processing of visceral sensations and pain has
been obtained following stimulation of the esophagus, stomach, rectum, bladder, and vagina, in
addition to that acquired during ▶ dobutamine-
induced chest pain (Aziz et al. 1997; Hobday
et al. 2001; Ladabaum et al. 2001; Matsuura
et al. 2002; Rosen et al. 1996; Whipple and
Komisaruk 2002). In healthy subjects, thalamic
activity has only been reported in approximately
50 % of studies, and this has been predominantly
in response to noxious rather than innocuous
visceral stimulation.

Strigo et al. have compared the processing of
visceral and cutaneous pain in the human brain
using fMRI (Strigo et al. 2003). In this study,
esophageal distension and contact heat

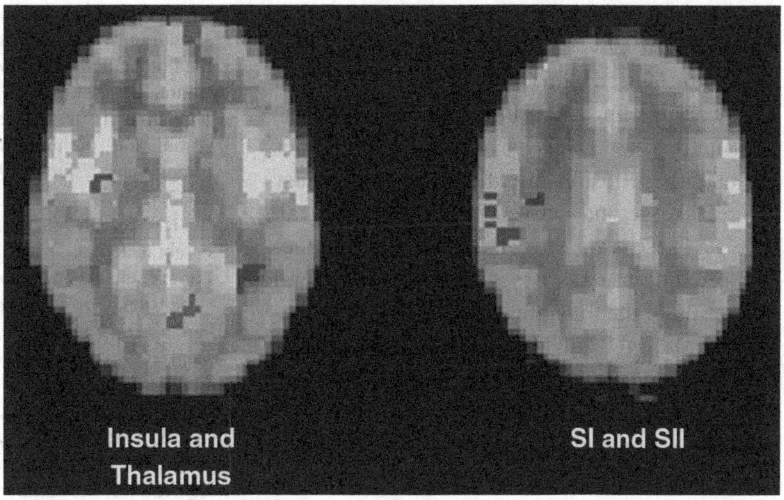

Thalamus and Visceral Pain Processing (Human Imaging),
Fig. 1 This image shows bilateral activation of the thalamus following painful esophageal balloon distension. In addition, activity is also seen in the primary/secondary somatosensory cortex and insula. These four regions are robustly activated following noxious visceral stimulation

Insula and Thalamus

SI and SII

stimulation of the anterior chest wall were matched for subjective intensity and applied to healthy subjects in a counterbalanced order. Analysis revealed that while differences were seen in cortical regions such as the anterior insula, a common pain neural network was activated encompassing the second somatosensory cortex, posterior parietal cortex, basal ganglia, and thalamus. The authors concluded that the similar activations seen within these regions implicate their function in the identification of a stimulus as painful, rather than in the differentiation between the nature of painful stimuli (Strigo et al. 2003).

In a PET study that examined the effect of increasing intensity of gastric distension on cortical and subcortical activation, Ladabaum et al. reported significant thalamic activity only at the highest distension volumes, which produced a noxious stimulus. Peak activations were noted bilaterally in the ventral posterolateral (VPL) nuclei and in the left dorsomedial nuclei (Ladabaum et al. 2001), leading the authors to conclude that noxious gastric stimulation activates both the lateral and medial pain systems.

The ability to noninvasively map the neuroanatomy of the visceral pain matrix has led to several researchers embarking on clinical studies in patients with ▶ functional gastrointestinal disorders, predominantly in patients with ▶ irritable bowel syndrome (IBS). Patients with IBS commonly report heightened perception of rectal distension (▶ Visceral Hypersensitivity), and the aim of these studies has been to identify objective neural correlates of visceral hypersensitivity.

While the heterogeneity of the IBS population has meant that results from group imaging data have been inconsistent, several groups have shown increased thalamic activation in response to noxious rectal distension (Mertz et al. 2000; Verne et al. 2003; Yuan et al. 2003). The thalamic regions of interest identified in these studies encompassed the VPL and dorsomedial nuclei. The limited spatial resolution of the fMRI imaging techniques used in these studies meant that it was not possible to comment on activation of specific thalamic nuclei, and thus, no comment could be made on the specific contribution of the sensory and affective dimensions of pain. Future clinical studies using high-resolution imaging of the thalamus may be able to provide more information regarding the contribution of different thalamic regions to aberrant visceral pain processing in IBS.

Perhaps the most dramatic evidence to support the role of the thalamus in aberrant visceral pain processing comes from studies by Rosen et al. in patients with various forms of ▶ angina pectoris (Rosen et al. 1996, 2002). A previous case report from 1994 had reported that microstimulation of the VPL induced angina-like symptoms in a single patient, which was not coincident with changes in cardiovascular function, strongly implicating the VPL in visceral and referred pain (Lenz et al. 1994). Rosen followed this up with a PET study in which he measured brain activity during intravenous dobutamine infusion in patients with coronary artery disease and active angina (Rosen et al. 1994). These data revealed that dobutamine-induced angina was associated with bilateral activation of the thalamus, in addition to activation of a number of higher cortical structures. Following the cessation of dobutamine infusion, symptoms ceased and no cortical activity was seen; however, thalamic activity was still evident. The authors concluded that the thalamus acted as a gateway for afferent visceral pain signals; however, cortical activity was needed to bring this to a level of conscious perception. Further evidence to support this has been provided by additional studies in patients with silent ischemia and ▶ syndrome X (Rosen et al. 1996, 2002) (Fig. 1).

In summary, painful visceral stimuli produce bilateral activation of the thalamus incorporating the regions of the VPL and dorsomedial nuclei. Heightened activation of these regions has been associated with aberrant pain processing in patients with gastrointestinal and cardiac disease. As the spatial resolution of FBI techniques improves, it will be possible to study the thalamus in greater detail, shedding further light on its role in visceral pain processing.

References

Al-Chaer, E. D., Feng, Y., & Willis, W. D. (1998). A role for the dorsal column in nociceptive visceral input into the thalamus of primates. *Journal of Neurophysiology, 79*, 3143–3150.

Aziz, Q., Andersson, J. L., Valind, S., et al. (1997). Identification of human brain loci processing esophageal sensation using positron emission tomography. *Gastroenterology, 113*, 50–59.

Hobday, D. I., Aziz, Q., Thacker, N., et al. (2001). A study of the cortical processing of ano-rectal sensation using functional MRI. *Brain, 124*, 361–368.

Ladabaum, U., Minoshima, S., Hasler, W. L., et al. (2001). Gastric distention correlates with activation of multiple cortical and subcortical regions. *Gastroenterology, 120*, 369–376.

Lenz, F. A., Gracely, R. H., Hope, E. J., et al. (1994). The sensation of angina can be evoked by stimulation of the human thalamus. *Pain, 59*, 119–125.

Matsuura, S., Kakizaki, H., Mitsui, T., et al. (2002). Human brain region response to distention or cold stimulation of the bladder: A positron emission tomography study. *Journal of Urology, 168*, 2035–2039.

Mertz, H., Morgan, V., Tanner, G., et al. (2000). Regional cerebral activation in irritable bowel syndrome and control subjects with painful and non-painful rectal distension. *Gastroenterology, 118*, 842–848.

Rosen, S. D., Paulesu, E., Frith, C. D., et al. (1994). Central nervous pathways mediating angina pectoris. *Lancet, 344*, 147–150.

Rosen, S. D., Paulesu, E., Nihoyannopoulos, P., et al. (1996). Silent ischemia as a central problem: Regional brain activation compared in silent and painful myocardial ischemia. *Annals of Internal Medicine, 124*, 939–949.

Rosen, S. D., Paulesu, E., Wise, R. J., et al. (2002). Central neural contribution to the perception of chest pain in cardiac syndrome X. *Heart, 87*, 513–519.

Strigo, I. A., Duncan, G. H., Boivin, M., et al. (2003). Differentiation of visceral and cutaneous pain in the human brain. *Journal of Neurophysiology, 89*, 3294–3303.

Verne, G. N., Himes, N. C., Robinson, M. E., et al. (2003). Central representation of visceral and cutaneous hypersensitivity in the irritable bowel syndrome. *Pain, 103*, 99–110.

Whipple, B., & Komisaruk, B. R. (2002). Brain (PET) responses to vaginal-cervical self-stimulation in women with complete spinal cord injury: Preliminary findings. *Journal of Sex & Marital Therapy, 28*, 79–86.

Willis, W. D., & Westlund, K. N. (1997). Neuroanatomy of the pain system and of the pathways that modulate pain. *Journal of Clinical Neurophysiology, 14*, 2–31.

Yuan, Y. Z., Tao, R. J., Xu, B., et al. (2003). Functional brain imaging in irritable bowel syndrome with rectal balloon-distention by using fMRI. *World Journal of Gastroenterology, 9*, 1356–1360.

Thalamus Lesion

▶ Lateral Thalamic Lesions, Pain Behavior in Animals

Thalamus Metabotropic Glutamate Receptors

▶ Metabotropic Glutamate Receptors in the Thalamus

Thalamus, Clinical Pain, Human Imaging

A. Vania Apkarian
Department of Physiology, Northwestern
University Feinberg School of Medicine,
Chicago, IL, USA

Synonyms

BOLD activity; Chronic pain; Gray matter density; Voxel-based morphometry

Definition

Noninvasive human brain imaging provides the opportunity to examine central processes that may be critically involved in the induction and/or maintenance of clinical chronic pain conditions. Such studies point to the notion that the thalamus shows reduced signaling and reduced gray matter, suggesting that the region is at least an active player in chronic pain conditions.

Characteristics

The advent of noninvasive brain imaging technologies affords a unique opportunity for unraveling brain processes that may be critical in induction and/or maintenance of clinical chronic pain conditions. Such studies have the potential for replacing speculations, psychosocial interpretations, and accusations of patients by dubbing them malingerers and other such labels, by physiological parameters that characterize these conditions and then hopefully lead to new, more science based development of therapies. Here, we briefly discuss the current understanding of the role of the thalamus in clinical pain states based on human brain imaging studies.

Brain activity as determined by ▶ PET or ▶ fMRI has established a reproducible pattern of cortical activity associated with acute or experimental painful conditions. A recent meta-analysis, of such studies over the last 15 years, estimated that the incidence of reporting of thalamic activity in experimental pain conditions is 84 % (16/19 studies) in PET studies and 81 % (13/16 studies) in fMRI studies (Apkarian et al. 2005) (incidence here is the ratio of the number of studies where the area was investigated in contrast to the number of studies where the area was reported to be activated). In contrast to this value, when the incidence of thalamic activity is examined in brain imaging studies in clinical pain conditions, the incidence is 59 % (16/27 PET, ▶ SPECT and fMRI studies combined). The conditions included in the clinical cases are cancer pain, cluster headache, migraine, cardiac pain, irritable bowel syndrome, fibromyalgia, CRPS, phantom pain, and mono- or poly-neuropathies (Apkarian et al. 2005). Contrasting the incidence of thalamic activity in normal subjects to clinical pain conditions indicates a borderline significant decrease in clinical conditions ($p = 0.09$, Fisher's exact test). The result seems paradoxical since one would naturally associate decreased thalamic activity with decreased pain and not the other way. We argue below that this observation is consistent with the notion that chronic pain conditions are more emotional states and hence less sensory; as a result, they may involve less spinothalamic activation and enhanced activations through pathways more directly accessing emotional regions of the brain.

The first clinical pain study was done in five patients with chronic cancer pain, where brain activity as determined by PET was compared between them and normal subjects before and after high cervical cordotomies that resulted in significant pain relief (Di Piero et al. 1991). The main outcome of the study was the observation that thalamic activity was low in the patients during chronic pain and normalized after the cordotomy. At least another five studies report that chronic clinical pain conditions are associated with decreased baseline activity or decreased stimulus-related activity in the thalamus. A SPECT blood flow study (Fukumoto et al. 1999) has shown a strong relationship between time of onset of CRPS symptoms and thalamic activity. The ratio between contralateral and ipsilateral thalamic perfusion was larger than 1.0, indicating hyperperfusion, for patients with symptoms for only 3–7 months and smaller than 1.0, indicating hypoperfusion, for patients with longer-term symptoms (24–36 months), with a correlation coefficient of 0.97 (normal subjects had a thalamic perfusion ratio of about 1.0). These results strongly imply that the thalamus undergoes adaptive changes in the course of CRPS. Thus, it can be asserted that thalamic activity for pain in chronic clinical conditions is different from that for acute painful stimuli in normal subjects.

Proving long-term reorganization of the CNS is hard with functional imaging, since one cannot disentangle reorganization from modulation of responses due to the presence of the condition, for example, in chronic pain. On the other hand, examination of brain chemistry by ▶ MRS indicates changes in various metabolites, most of which are not affected by the current cognitive state of the person. Thus, such measures document long-term changes more readily. The limitation of the technique regards the specific chemicals that can be detected (the specific functions of many of which remain unclear), and the need for regional imaging which limits the spatial extent of the measurement (Salibi and Brown 1998). One such study examined thalamic metabolites in chronic back pain patients and observed no changes in comparison to healthy subjects, although there were decreased measures for multiple metabolites in the prefrontal cortex (Grachev et al. 2000). Another similar study examined thalamic metabolites (Pattany et al. 2002) in patients with chronic neuropathic pain after spinal cord injury and did see decreased metabolites in the thalamus and observed a negative correlation between pain intensity and concentration in the thalamus. Since a decreased concentration of N-acetyl-aspartate has been reported in most neurodegenerative conditions, it is reasonable to conclude that the observation of decreased N-acetyl-aspartate in the chronic spinal cord injury patients suggests that the condition is associated with a neurodegenerative process in the thalamus. This idea was tested directly in a morphometric study. The gray matter density of chronic back pain patients was contrasted with age- and gender-matched normal subjects using high-resolution anatomical MRI. The results indicate that the thalamus gray matter density, mainly on the right, is significantly lower in density in the chronic back pain patients (Apkarian et al. 2004). The mechanisms inducing this atrophy remain to be determined, as well as the extent of its reversibility and its impact on information processing in the thalamus. However, the observation clearly implies that the thalamus is undergoing long-term reorganization due to the presence of the pain condition. It is also possible that some of this reorganization has a genetic component that may predispose these subjects to develop chronic pain. Thus, there are more questions raised than answered. Fortunately, these questions can be answered in future studies and should provide a more accurate understanding of the role of the thalamus in chronic pain conditions.

Decreased brain responses to painful stimuli in clinical pain conditions have been observed in other meta-analyses of the literature (Derbyshire 1999; Peyron et al. 2000), and interpreted as evidence for a generalized decrease in brain activity in such patients. However, when the incidence of prefrontal cortical activity is examined, one observes increased incidence of activating this region in chronic pain patients in contrast to normal subjects. In normal subjects, prefrontal cortex is activated in 55 % (23/42) of functional imaging studies for pain, while in pain patients, this rate increases to 81 % (21/26, $p = 0.04$, Fisher's exact statistics). Thus, in chronic pain patients, decreased thalamic activity seems to be accompanied by increased prefrontal cortical activity. This implies a switch of nociceptive inputs away from spinothalamic afferents and enhanced nociceptive inputs through brainstem prefrontal cortical regions. This has been suggested in animal models of neuropathic pain (Hunt and Mantyh 2001), based on the response changes observed in spinal cord neurons following neuropathic injury. The shift suggests that chronic clinical pain conditions are more emotional/cognitive than sensory. Whether the thalamic atrophy observed in back pain patients directly contributes to this shift remains to be determined. It is at least consistent with the model.

Overall, multiple lines of investigations regarding the involvement of the thalamus in clinical pain conditions imply that the region actively reorganizes in such conditions. It should be emphasized that most probably, the details of this reorganization are specific to different types of clinical pains. The extent to which this reorganization contributes to such conditions remains unclear. More importantly, these observations strongly suggest that at least part of this reorganization may be a consequence of neuronal death and hence irreversible, pointing to the urgency for further elaborating these details and for more effective therapeutic approaches for these conditions.

References

Apkarian, A. V., Sosa, Y., Sonty, S., et al. (2004). Chronic back pain is associated with decreased prefrontal and thalamic gray matter density. *Journal of Neuroscience, 24,* 10410–10415.

Apkarian, A. V., Bushnell, M. C., Treede, R. D., et al. (2005). Human brain mechanisms of pain perception and regulation in health and disease. *European Journal of Pain, 9,* 463–484.

Derbyshire, S. W. (1999). Meta-analysis of thirty-four independent samples studied using PET reveals a significantly attenuated central response to noxious stimulation in clinical pain patients. *Current Review of Pain, 3,* 265–280.

Di Piero, V., Jones, A. K., Iannotti, F., et al. (1991). Chronic pain: A PET study of the central effects of percutaneous high cervical cordotomy. *Pain, 46,* 9–12.

Fukumoto, M., Ushida, T., Zinchuk, V. S., et al. (1999). Contralateral thalamic perfusion in patients with reflex sympathetic dystrophy syndrome. *Lancet, 354,* 1790–1791.

Grachev, I. D., Fredrickson, B. E., & Apkarian, A. V. (2000). Abnormal brain chemistry in chronic back pain: An in vivo proton magnetic resonance spectroscopy study. *Pain, 89,* 7–18.

Hunt, S. P., & Mantyh, P. W. (2001). The molecular dynamics of pain control. *Nature Reviews Neuroscience, 2,* 83–91.

Pattany, P. M., Yezierski, R. P., Widerstrom-Noga, E. G., et al. (2002). Proton magnetic resonance spectroscopy of the thalamus in patients with chronic neuropathic pain after spinal cord injury. *AJNR American Journal of Neuroradiology, 23,* 901–905.

Peyron, R., Laurent, B., & Garcia-Larrea, L. (2000). Functional imaging of brain responses to pain. A review and meta-analysis (2000). *Neurophysiologie Clinique, 30,* 263–288.

Salibi, N., & Brown, M. A. (1998). *Clinical MR spectroscopy.* New York: Wiley-Liss.

Thalamus, Clinical Visceral Pain, Human Imaging

Uri Ladabaum
Gastroenterology & Hepatology, Stanford Cancer Institute, Redwood City, CA, USA

Synonyms

Thalamus and cardiac pain; Thalamus and gastrointestinal pain; Thalamus and visceral pain (positron emission tomography or functional magnetic resonance imaging)

Definition

Visceral pain arises from the internal organs, such as the heart and the gastrointestinal tract. In contrast, somatic pain arises from the skin and deeper tissues, including muscle. The central nervous system regions activated by visceral pain in humans, including the thalamus, have been studied noninvasively with ▶ positron emission tomography (PET) and ▶ functional magnetic resonance imaging (fMRI), functional imaging techniques that measure increased regional cerebral blood flow as a marker of neuronal activation.

Characteristics

Thalamic activation (defined as a statistically significant increase in regional blood flow during the condition of interest) has been reported in many but not all functional cerebral imaging studies of visceral pain in humans. The cerebral representation of visceral pain was first studied by inducing angina in humans and later by distending gastrointestinal viscera to induce pain. A recent systematic review that included inspection of published images by the author in addition to the results explicitly reported by investigators found evidence for thalamic activation with angina, noxious esophageal stimulation, gastric distension, and noxious lower gastrointestinal distension in healthy volunteers and patients with ▶ irritable bowel syndrome (IBS) (Derbyshire 2003).

Cardiac Ischemia and Angina

A pioneering ▶ PET study reported the central nervous system pathways mediating ▶ dobutamine-induced angina in patients with coronary artery disease (Rosen et al. 1994). Compared to the resting state, regional blood flow increased during angina in a number of cerebral structures, including increases of 2.7 % in the left thalamus and 3.7 % in the right thalamus. After the resolution of angina, thalamic activity remained significantly increased, whereas the cortical activity associated with angina was no longer detected. The authors suggested that the thalamus receives input from the heart during angina and continues to receive such input even when angina is no longer felt, with this less intense signal not being transmitted to the cerebral cortex and therefore not associated with conscious perception.

Cerebral PET imaging was then used to gain insight into the problem of ▶ silent myocardial ischemia (Rosen et al. 1996). Cerebral activation patterns were compared between patients with stress-induced angina and patients with stress-induced myocardial ischemia but no angina. During myocardial ischemia, significant left thalamic activation was detected in patients with angina, and bilateral thalamic activation was detected in patients with silent ischemia. However, much more extensive cortical activation was seen in those with angina. Because thalamic activation was seen in both groups of patients, the authors concluded that ▶ silent myocardial ischemia cannot be explained by peripheral nerve dysfunction. They proposed that abnormal central processing of afferent pain signals (possibly abnormal gating at the level of the thalamus) may contribute to the pathophysiology of silent myocardial ischemia.

Esophageal Stimulation

An early PET study of the cerebral regions involved in esophageal sensation found bilateral activations along the central sulcus, insulae, and frontal and parietal operculum during non-painful esophageal distension and more intense activations in these regions as well as additional activation in the right anterior insular cortex and the anterior cingulate gyrus during painful distension (Aziz et al. 1997). Thalamic activation was not detected.

Subsequently, an fMRI study examined the cerebral cortical response to esophageal distension or acid perfusion (Kern et al. 1998). Acidification and distension generally resulted in activations of Brodmann's areas 7, 23, 30, 32, insula, operculum, and anterior cingulate cortex. Although the activated regions were similar, the temporal characteristics of the activation were different (slower for acidification). Significant thalamic activation was again not detected.

The cortical processing of distal and proximal esophageal sensation has been compared using fMRI (Aziz et al. 2000). Among other differences, proximal distension was localized precisely to the upper chest and activated the left primary somatosensory cortex, whereas distal distension was perceived diffusely over the lower chest and activated the junction of the primary and secondary somatosensory cortices bilaterally. Significant thalamic activation was not reported.

The cerebral processing of visceral and cutaneous pains was compared using distal esophageal distension and application of heat to the chest during cerebral fMRI scanning (Strigo et al. 2003). Painful esophageal distension and painful heat stimulation both induced statistically significant thalamic activation on the left and nonsignificant activation on the right. Overall, a similar neural network was activated with visceral and somatic stimulation, but notable differences were also apparent that probably relate to the differences in the experience of visceral and cutaneous pain.

The different results among these studies probably relate to differences in experimental design, study population, and chance. Regional cerebral blood flow changes with "activation" are generally small (a few percent). Because the thalamus is a major relay and processing station between the periphery and the cortex and basal ganglia and because thalamic activation has been reported in a substantial proportion of all available visceral pain studies in humans, it is likely that the lack of significant thalamic activation in the first three studies reflects the limitations of the available techniques.

Gastric Stimulation

Cerebral PET imaging during increasing levels of gastric distension detected progressive increases in activation in multiple cortical and subcortical regions, including the thalami, insulae, anterior cingulate cortex, periaqueductal gray matter, and cerebellum (Ladabaum et al. 2000). Statistically significant activation centered in the right ventral posterolateral (VPL) nucleus of the thalamus was detected with distension producing threshold pain as well as distension producing moderate pain. Statistically significant activations centered in the VPL and dorsomedial (DM) nuclei of the left thalamus were detected with distension producing moderate pain. The VPL nucleus is a key component of the lateral pain system, which is believed to subserve the sensory-discriminative component of pain, and the DM nucleus is part of the medial pain system, which is believed to mediate the affective-motivational component of pain. The identification of precise thalamic nuclei as the regions of peak activation must be interpreted with caution given the spatial resolution limits of PET.

Rectosigmoid Stimulation

A pioneering PET study of visceral sensation in healthy volunteers and patients with IBS found thalamic activation of borderline significance during rectal distension in healthy volunteers (Silverman et al. 1997). An early fMRI study detected activation in numerous regions including the anterior cingulate, prefrontal, insular, and sensorimotor cortices during rectal distension in healthy volunteers, but no thalamic activation was detected (Baciu et al. 1999).

A larger study of healthy volunteers (16) and IBS patients (18) using fMRI during rectal

distension reported increased activity in the anterior cingulate, prefrontal, and insular cortices, as well as increased thalamic activity in nearly all subjects (Mertz et al. 2000). In contrast to earlier results (Silverman et al. 1997), this study found comparable patterns of activation in healthy volunteers and IBS patients, but greater activation of the anterior cingulate cortex during painful compared to non-painful stimulation only in IBS patients.

An intriguing study examined cerebral activation during subliminal visceral stimulation caused by rectal distension (Kern and Shaker 2002). Cerebral activation during subliminal distension was generally bilateral in regions including the sensory/motor, parieto-occipital, anterior cingulate, prefrontal, and insular cortices. Distension to liminal (threshold sensation) and supraliminal (above threshold sensation) levels activated regions similar to those activated with subliminal stimulation, but the volume of cortical activity increased with the stimulus intensity. In contrast to studies of myocardial ischemia in which thalamic activity was detected even after resolution of angina (Rosen et al. 1994; Rosen et al. 1996), thalamic activation was not detected in this study of subliminal rectal sensation.

Some subsequent studies involving rectal distension have reported significant thalamic activation (Verne et al. 2003; Yuan et al. 2003), including greater activation in the thalamus and multiple other regions in IBS patients compared to healthy volunteers during both rectal and somatic stimulation (Verne et al. 2003). However, thalamic activation has not been detected in all subsequent studies (Berman et al. 2002). As with the esophageal stimulation studies that failed to detect thalamic activation, it seems likely that the lack of significant thalamic activation in some rectal distension studies reflects the limitations of the available technology.

Summary

The thalamus is a major relay and processing station between the peripheral nervous system and higher centers in the central nervous system. It would be anticipated that cerebral functional imaging studies using PET or fMRI in humans would detect thalamic activation during visceral pain.

Many such studies have indeed detected thalamic activation. The results include thalamic activation during myocardial ischemia with or without angina and after the resolution of angina and thalamic activation during distal esophageal distension, gastric distension, and rectal distension. However, not all studies of functional cerebral imaging during gastrointestinal visceral distension have detected thalamic activation. These seemingly inconsistent results are probably due to limitations in the sensitivity of the available technology.

References

Aziz, Q., Andersson, J. L., Valind, S., et al. (1997). Identification of human brain loci processing esophageal sensation using positron emission tomography. *Gastroenterology, 113*, 50–59.

Aziz, Q., Thompson, D. G., Ng, V. W., et al. (2000). Cortical processing of human somatic and visceral sensation. *Journal of Neuroscience, 20*, 2657–2663.

Baciu, M. V., Bonaz, B. L., Papillon, E., et al. (1999). Central processing of rectal pain: A functional MR imaging study. *AJNR. American Journal of Neuroradiology, 20*, 1920–1924.

Berman, S. M., Chang, L., Suyenobu, B., et al. (2002). Condition-specific deactivation of brain regions by 5-HT3 receptor antagonist alosetron. *Gastroenterology, 123*, 969–977.

Derbyshire, S. W. (2003). A systematic review of neuroimaging data during visceral stimulation. *American Journal of Gastroenterology, 98*, 12–20.

Kern, M. K., & Shaker, R. (2002). Cerebral cortical registration of subliminal visceral stimulation. *Gastroenterology, 122*, 290–298.

Kern, M. K., Birn, R. M., Jaradeh, S., et al. (1998). Identification and characterization of cerebral cortical response to esophageal mucosal acid exposure and distention. *Gastroenterology, 115*, 1353–1362.

Ladabaum, U., Minoshima, S., Hasler, W., et al. (2000). Gastric distention correlates with activation of multiple cortical and subcortical regions. *Gastroenterology, 120*, 369–376.

Mertz, H., Morgan, V., Tanner, G., et al. (2000). Regional cerebral activation in irritable bowel syndrome and control subjects with painful and nonpainful rectal distention. *Gastroenterology, 118*, 842–848.

Rosen, S. D., Paulesu, E., Frith, C. D., et al. (1994). Central nervous pathways mediating angina pectoris. *Lancet, 344*, 147–150.

Rosen, S. D., Paulesu, E., Nihoyannopoulos, P., et al. (1996). Silent ischemia as a central problem: Regional brain activation compared in silent and painful myocardial ischemia. *Annals of Internal Medicine, 124*, 939–949.

Silverman, D. H., Munakata, J. A., Ennes, H., et al. (1997). Regional cerebral activity in normal and pathological perception of visceral pain. *Gastroenterology, 112*, 64–72.

Strigo, I. A., Duncan, G. H., Boivin, M., et al. (2003). Differentiation of visceral and cutaneous pain in the human brain. *Journal of Neurophysiology, 89*, 3294–3303.

Verne, G. N., Himes, N. C., Robinson, M. E., et al. (2003). Central representation of visceral and cutaneous hypersensitivity in the irritable bowel syndrome. *Pain, 103*, 99–110.

Yuan, Y. Z., Tao, R. J., Xu, B., et al. (2003). Functional brain imaging in irritable bowel syndrome with rectal balloon-distention by using fMRI. *World Journal of Gastroenterology, 9*, 1356–1360.

Thalamus, Dynamics of Nociception

Vasco Galhardo
Departamento de Biologia Experimental,
Faculdade de Medicina, Universidade do Porto,
Portugal

Synonyms

Sensitization; Thalamic plasticity; Thalamic reorganization

Definition

The dynamics of the thalamic responses to noxious stimuli are the signature of neural mechanisms that make the responses not rigid and immutable, since they reflect both the nature of the incoming afferent signal and the internal state of the neuronal populations that process nociceptive signals.

Characteristics

Somatosensory noxious information is processed by the thalamus in a complex manner that is still little understood. Far from being a simple relay station for painful stimuli on their way from the periphery to the cortex, thalamic neurons integrate and modulate the pain signals in a dynamic way. This implies that the modulation itself is the result of the instantaneous status of the thalamic neuronal network. By addressing the topic of the dynamics of thalamic responses we acknowledge the fact that the arriving nociceptive information will not always be processed by the thalamic neural networks in exactly the same manner, since the functional properties of the processing neural networks will change over time, reflecting the history of the incoming signals.

The factors that affect the status of the processing neural networks may be *extrinsic* to the thalamus – meaning that they are caused by the qualities of the past and immediate afferent information – or *intrinsic* – meaning that they are the result of the status of the thalamus, namely, the occurrence of population ▶ oscillations.

This brief entry will review four key aspects of nociceptive information processing dynamics within the thalamus.

Sensitization of the Thalamus

The most basic aspect of thalamic somatosensory dynamics is the change in the response properties of thalamic neurons induced by states of peripheral persistent pain. Several studies have shown that in conditions of ▶ hyperalgesia or ▶ allodynia, the neurons in the ventrobasal complex of the lateral thalamus have enhanced responsiveness, i.e., they have lower activation thresholds for both thermal and mechanical stimuli and larger peripheral receptive fields and continue to discharge spontaneously for long periods after cessation of the noxious stimulation (Guilbaud et al. 1990; Sherman et al. 1997). Ultimately, it is difficult to determine to what extent this functional plasticity corresponds to a change within the thalamic networks or if the enhanced thalamic activity is instead simply a consequence of the enhanced somatosensory information generated at spinal levels during chronic pain. One strong piece of evidence for the occurrence of plasticity at the thalamic level is that painful sensations are evoked more frequently by microstimulation of the lateral thalamus in patients with chronic neuropathic pain than in

patients with non-painful movement disorders (Davis et al. 1996). Further microstimulation studies suggest that in neuropathic pain patients, thalamic areas usually signaling thermal non-painful discrimination are now evoking painful sensations when stimulated (Lenz et al. 1998).

It is very tempting to assume that the enhanced sensory activity corresponds to a progressive change of thalamic nociceptive-specific neurons into the wide-dynamic range category, as was implied in some studies (Guilbaud et al. 1990). However, multielectrode recordings lasting several hours in the rat somatosensory thalamus (Brueggemann et al. 2001) showed that immediately after a nerve lesion, no rigid pattern of change between cell types could be found (Fig. 1). Although the results from multielectrode studies show immediate fluctuations in response properties, they do not show a clear net increase in nociceptive responses in the affected portion of the lateral thalamus.

Deafferentation-Induced Plasticity

Paradoxically, just as the enhanced somatosensory information leads to a sensitization of thalamic networks, the decrease in somatosensory information arriving at the thalamus due to peripheral nerve lesions also leads to higher levels of pain sensitivity. This is probably due to the fact that both the functional loss of afferents by peripheral lesions and the functional gain of afferents by peripheral ▶ hyperalgesia lead to similar net effects; some regions of the thalamus are suddenly more active than their neighbors and this spatial imbalance causes a regional peak of activity that leads to somatosensory hypersensitivity. ▶ Deafferentation studies also reveal a crosstalk between somatosensory modalities in which both the disruption of thick myelinated non-nociceptive fibers leads to altered pain perception and the disruption of unmyelinated high-threshold ▶ C-fibers leads to altered tactile perception. A recent multielectrode study in the rat lateral thalamus (Katz et al. 1999) showed that the silencing of C-fibers after perioral capsaicin injections changed the pattern of sensitivity to tactile deflection of the whiskers, resulting in "unmasking" of new tactile responses.

Thalamus, Dynamics of Nociception, Fig. 1 Changes
in classification of lateral thalamic neurons at the beginning (Before, before sciatic nerve ligation) and at the end
of the experiments (After, after nerve ligation). *Horizontal
lines* indicate no change; *diagonal lines* indicate changes
in neuronal class. Abbreviations: Ø unresponsive neurons,
LT low-threshold neurons, *WDR* wide-dynamic range neurons, *NS* nociceptive-specific neurons (Figure taken from
Brueggemann et al. 2001)

Spatiotemporal Dynamics in the Thalamus

Several studies have shown novel properties in
the thalamic integration of nociceptive signals, in
both space and time domains. Finding spatiotemporal patterns of coherence in the activity of
neuronal populations in the thalamus during the
processing of pain will help unravel the intrinsic
organization of the processing units within the
lateral thalamus. The use of multielectrode arrays
with precise spatial arrangements showed that the
functional connectivity between pairs of neurons
depends on the distance separating them; neighboring nociceptive neurons tend to be positively
correlated, while most negative correlations in

evoked spike timings occur between neurons
separated by more than 50 μm (Apkarian et al.
2000). Interestingly, the spatial maps of evoked
spike coherence between pairs of neurons were
different for noxious and non-noxious stimulations,
with tactile stimuli inducing almost no inhibitory
spatial effects. The results of that study also suggest
that in the primate thalamus, nociceptive and
non-nociceptive neurons are clustered separately
in a lattice manner. This spatial arrangement of
the lateral thalamus would have profound implications for the spatiotemporal dynamics of thalamic
function, since nociceptive clusters would receive
spinothalamic afferents while non-nociceptive
clusters would receive mainly dorsal column afferents and both have distinct velocity and synchrony
characteristics that would result in complex spatiotemporal patterns, difficult to interpret and predict.

By the same token, it has been conclusively
shown that for fine tactile discrimination also, the
spatiotemporal maps are complex to determine
and reorganize after peripheral ▶ deafferentation
and that distributed coherent activity in the thalamus plays a key role in the encoding of somatosensory signals (Nicolelis 1997).

Thalamic Oscillations and Pain Processing

The thalamus is known to have two modes of firing:
tonic mode, in which neurons fire single action
potentials and bursting mode, in which neurons
fire in rhythmic rapid bursts of several action potentials. Traditionally, thalamic bursts were associated
with non-aware states, while tonic mode was
associated with alert states, although this simple
scheme has been questioned in recent studies
(Nicolelis and Fanselow 2002). High-frequency
spike bursts are one of the most characteristic
features observed in the thalamus following
▶ deafferentation by lesion of peripheral nerves,
dorsal roots, or the spinal cord (Albe-Fessard
et al. 1985; Guilbaud et al. 1990). In the thalamus,
spike bursts have been described with very specific
membrane physiology and have been shown
to function normally to synchronize activity of
neurons both within and between thalamic nuclei
as well as between the thalamus and cortex. In
human thalamic recordings performed during
surgery (Lenz et al. 1989), a characteristic series

Thalamus, Dynamics of Nociception, Fig. 2 Brush stimuli evoke oscillatory activity in the lateral thalamus of animals with complete sectioning of the anterolateral spinal quadrant (Figure taken from Weng et al. 2000)

of spontaneous short high-frequency spike trains (3–8 action potentials occurring at 60–160 Hz) or spike bursts was found in long-term para- and tetraplegics. These bursts repeat themselves in low-frequency ▶ oscillations in pain patients and the clinical condition has been termed ▶ thalamocortical dysrhythmia (Llinas et al. 1999).

Significance has been assigned to changes in the temporal patterns of activity in thalamic neurons following deafferentation because spike bursts in pain patients are especially concentrated in regions of the lateral thalamus representing the painful part of the body. In Fig. 2, we present an example of such a burst activity. The figure is taken from Weng et al. (2000) where the authors report that thalamic wide-dynamic range neurons (but not the low-threshold neurons) of monkeys partially deafferented by spinal cord lesions showed enhanced activity compared with the same types of cells in thalamus with intact innervation; both the spontaneous and evoked discharges of the cells were altered, so that there was an increased incidence of spike bursts in cells of deafferented thalamus.

Recently, the importance of this abnormal thalamic firing for the genesis of chronic central pain has been challenged by the finding of no differences between the number of oscillating neurons in the thalamus of allodynic versus non-allodynic rats with spinal cord injury (Gerke et al. 2003), suggesting that ▶ thalamic dysrhythmia is linked to cord injury but not to the presence of ▶ allodynia. However, this study was performed

in anesthetized animals, raising the question of its applicability to awake thalamic processing. By the same token, Radhakrishnan et al. (1999) showed that the incidence of thalamic bursting was similar between painful and non-painful neurological injuries. Hence, the role of the internal oscillatory states of the thalamus in pain processing is still to be clarified.

References

Albe-Fessard, D., Berkley, K. J., Kruger, L., et al. (1985). Diencephalic mechanisms of pain sensation. *Brain Research, 356*, 217–296.

Apkarian, A. V., Shi, T., Brueggemann, J., et al. (2000). Segregation of nociceptive and non-nociceptive networks in the squirrel monkey somatosensory thalamus. *Journal of Neurophysiology, 84*, 484–494.

Brueggemann, J., Galhardo, V., & Apkarian, A. V. (2001). Immediate reorganization of the rat somatosensory thalamus following peripheral partial nerve ligation. *The Journal of Pain, 2*, 220–228.

Davis, K. D., Kiss, Z. H. T., Tasker, R. R., et al. (1996). Thalamic stimulation-evoked sensations in chronic pain patients and non-pain (movement disorder) patients. *Journal of Neurophysiology, 75*, 1026–1037.

Gerke, M., Duggan, A., Xu, L., et al. (2003). Thalamic neuronal activity in rats with mechanical allodynia following contusive spinal cord injury. *Neuroscience, 117*, 715–722.

Guilbaud, G., Benoist, J. M., Jazat, F., et al. (1990). Neuronal responsiveness in the ventrobasal thalamic complex of rats with an experimental peripheral mononeuropathy. *Journal of Neurophysiology, 64*, 1537–1554.

Katz, D. B., Simon, S. A., Moody, A., et al. (1999). Simultaneous reorganization in thalamocortical ensembles evolves over several hours after perioral capsaicin injections. *Journal of Neurophysiology, 82*, 963–977.

Lenz, F. A., Kwan, H. C., Dostrovsky, J. O., et al. (1989). Characteristics of the bursting pattern of action potentials that occurs in the thalamus of patients with central pain. *Brain Research, 496*, 357–360.

Lenz, F. A., Gracely, R. H., Baker, F. H., et al. (1998). Reorganization of sensory modalities evoked by microstimulation in region of the thalamic principal sensory nucleus in patients with pain due to nervous system injury. *The Journal of Comparative Neurology, 399*, 125–138.

Llinas, R. R., Ribary, U., Jeanmonod, D., et al. (1999). Thalamocortical dysrhythmia: A neurological and neuropsychiatric syndrome characterized by magnetoencephalography. *Proceedings of the National Academy of Sciences USA, 96*, 15222–15227.

Nicolelis, M. A. L. (1997). Dynamic and distributed somatosensory representations as the substrate for cortical and subcortical plasticity. *Seminars in Neuroscience, 9*, 24–33.

Nicolelis, M. A. L., & Fanselow, E. E. (2002). Thalamocortical optimization of tactile processing according to behavioral state. *Nature Neuroscience, 5*, 517–523.

Radhakrishnan, V., Tsoukatos, J., Davis, K. D., et al. (1999). A comparison of the burst activity of lateral thalamic neurons in chronic pain and non-pain patients. *Pain, 80*, 567–575.

Sherman, S. E., Luo, L., & Dostrovsky, J. O. (1997). Altered receptive fields and sensory modalities of rat VPL thalamic neurons during spinal strychnine-induced allodynia. *Journal of Neurophysiology, 78*, 2296–2308.

Weng, H. R., Lee, J. I., Lenz, F. A., et al. (2000). Functional plasticity in primate somatosensory thalamus following chronic lesion of the ventral lateral spinal cord. *Neuroscience, 101*, 393–401.

Thalamus, Nociceptive Cells in VPI, Cat and Rat

Christiane Vahle-Hinz
Department of Neurophysiology and Pathophysiology, University Medical Center Hamburg-Eppendorf, Hamburg, Germany

Synonyms

High-threshold neurons; Multireceptive neurons (MR); Nociceptive-specific neurons (NS); Wide-dynamic range neurons (WDR)

Definition

Neurons of the lateral thalamus encoding stimulus quality, duration, intensity, and location on the body and thus subserving the ▶ sensory-discriminative component of pain.

The neurons respond with an increase or decrease of discharge activity to noxious stimuli exclusively (▶ nociceptive-specific neurons, NS) or with a higher discharge to noxious mechanical than to innocuous mechanical stimuli (▶ Wide-dynamic range neurons (WDR)) or to both noxious and innocuous mechanical as well as thermal stimuli (multireceptive neurons, MR). As defined in "Thalamic Nuclei Involved in Pain, Cat and Rat," the region termed ▶ ventral posterior inferior nucleus (VPI) in monkeys may correspond to the small celled region of the ventral periphery of the ventral posterior medial nucleus (VPMvp) in cats, the extensions of which surround the ventral, lateral, and dorsal periphery of the ▶ ventral posterior complex (VP). In rats, nociceptive neurons are also found within and around the VP, but most are concentrated in the adjoining ▶ medial part of the posterior complex (POm). Since these have similar ▶ receptive field (RF) and response properties, it appears that these regions in the lateral thalamus of rats and cats are involved in processing and transmission of nociceptive signals in a similar way.

Characteristics

Nociceptive neurons in the VPp have a less precise somatotopic organization than neurons of the VP proper, largely due to numerous neurons with larger and complex ▶ RFs consisting of discontinuous areas on the body surface. However, a coarse mediolateral sequence of head, forelimb, and hind limb RFs is present, running parallel to that of the VP proper, thus forming a second representation of the body. Due to a lower ongoing activity and a sparser packing of neurons, even low impedance electrodes hardly pick up any background activity in contrast to the noisy "hash" characteristic of the VP (Kniffki and Vahle-Hinz 1987; Vahle-Hinz et al. 1987).

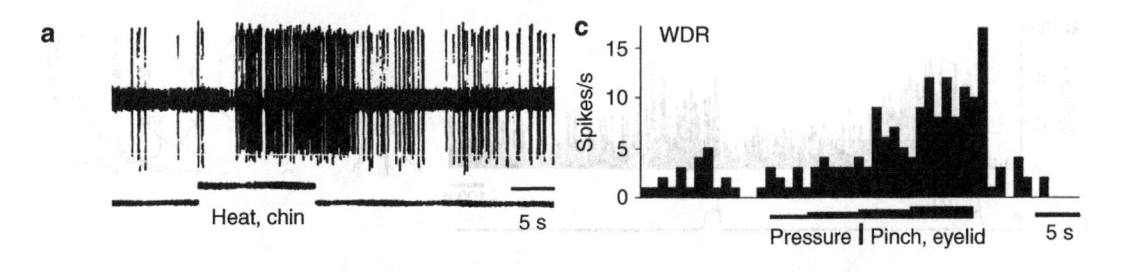

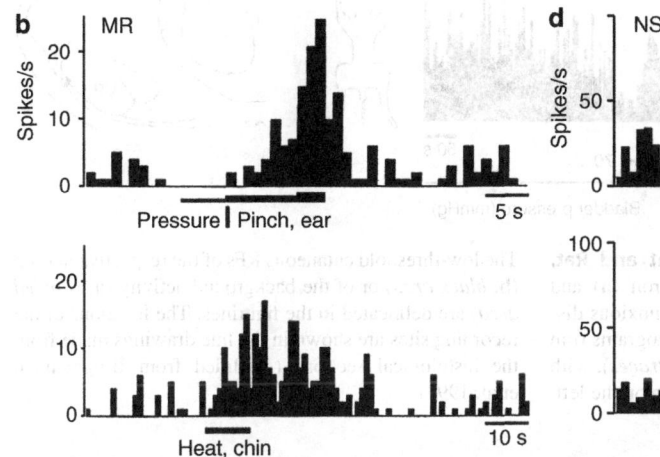

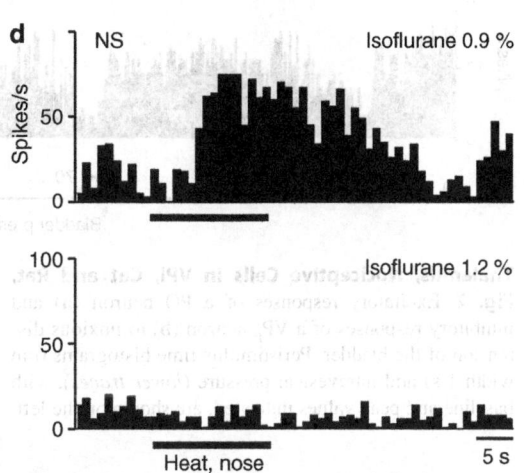

Thalamus, Nociceptive Cells in VPI, Cat and Rat,
Fig. 1 Response characteristics of nociceptive neurons
of the cat's VP_p (**a**, **b**, **c**) and the rat's POm (**d**). (**a**) Spike
record of a heat-evoked response. The elevated activity
representing stimulus duration is followed by a lower
afterdischarge for about 30 s before return to prestimulus
activity. (**b**) Multireceptive neuron responding to noxious
pressure and heat stimuli in different parts of the receptive
field. (**c**) Wide-dynamic range neuron with graded

responses to innocuous pressure and two intensities of
noxious pressure. (**d**) The response elicited by a radiant
heat stimulus is abolished by an increase of isoflurane
concentration from 0.9 % to 1.2 %. Spike histograms,
bin width 1 s; the bar below each histogram represents
the duration of the stimulus (Modified from Vahle-Hinz
et al. 1987 (**a**), Vahle-Hinz et al. 2002 (**d**), Kniffki and
Vahle-Hinz 1987 (**c**))

In both cats and rats, the neurons respond to
graded noxious mechanical and/or heat stimuli
(for review, see Willis 1997). Responses to noxious heat stimuli characteristically occur after
long latencies (several seconds), which may
result from receptor activation time and transmission by C fibers. The discharge increase in many
neurons encodes stimulus duration, often
followed by a lower but still increased
afterdischarge for several tens of seconds
(Fig. 1a, b, d). A threshold temperature and a stimulus–response function resembling psychophysical heat pain in humans are found. Similar
correlation between neuronal responses to noxious laser stimulation and nocifensive behavior

has been reported in studies on chronically
implanted awake rats (Zhang et al. 2011).
All types of nociceptive neurons are present,
nociceptive-specific (Fig. 1a, d), multireceptive
(Fig. 1b), and WDR neurons (Fig. 1c). NS and
WDR/MR neurons occur in about equal proportions in the cat, while the majority in the rat is of
the NS type. Thalamic nociceptive neurons are
sensitive to anesthetics; the response to a noxious
stimulus may be abolished by a slight increase of
anesthetic dose without a marked suppression of
the neuron's ongoing discharge activity (Fig. 1d).
Different classes of anesthetic agents appear to
block nociceptive signal transmission at different
subthalamic/thalamic sites within the ascending

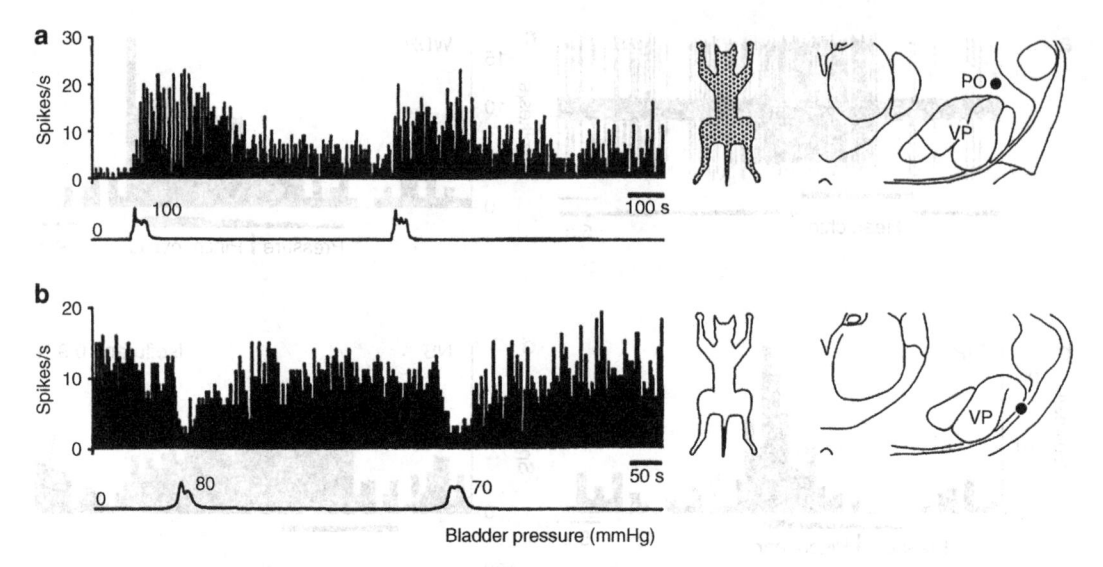

Thalamus, Nociceptive Cells in VPI, Cat and Rat, Fig. 2 Excitatory responses of a PO neuron (**a**) and inhibitory responses of a VP_p neuron (**b**) to noxious distension of the bladder. Peristimulus-time histograms (bin width 1 s) and intravesical pressure (*lower traces*), with baseline and peak values indicated, are shown on the left. The low-threshold cutaneous RFs of the respective neuron (**b**, *black area*) or of the background activity (**a**, *stippled area*) are delineated in the figurines. The locations of the recording sites are shown in the line drawings made from the histological sections (Modified from Brüggemann et al. 1993)

pathways. The low-threshold mechanoreceptive responses in contrast are more robust (Vahle-Hinz and Detsch 2002; Vahle-Hinz et al. 2002).

As with noxious somatic stimuli and also with visceral stimulation, both excitation and inhibition of thalamic neuronal discharges can be elicited (Fig. 2). In contrast to the spinal cord level, where the majority of visceroceptive neurons have convergent somatic inputs of the WDR type, visceroceptive-specific neurons are found in the thalamus of both rats and cats. Some of the additional somatic RFs of viscero-somatic convergent neurons in the rat are nociceptive (Berkley et al. 1993), while in the cat most are of the low-threshold type. The majority of these somatic low-threshold RFs are located in areas including dermatomes to which pain is referred from the respective visceral organs (Brüggemann et al. 1993; Horn et al. 1999; Vahle-Hinz et al. 1995).

The neuronal responses in the lateral thalamus indicate that distinct response properties are segregated (e.g., somatic and visceral nociceptive specific), thus increasing the diversity of response characteristics contained in the population of nociceptive neurons. Excitatory and inhibitory interactions between inputs from different somatic sources (mechanoreceptive/nociceptive) or visceral organs may help to focus the activity and hence the sensation to a certain stimulus or organ.

References

Berkley, K. J., Guilbaud, G., Benoist, J.-M., et al. (1993). Responses of neurons in and near the thalamic ventrobasal complex of the rat to stimulation of uterus, vagina, colon, and skin. *Journal of Neurophysiology, 69,* 557–568.

Brüggemann, J., Vahle-Hinz, C., & Kniffki, K.-D. (1993). Representation of the urinary bladder in the lateral thalamus of the cat. *Journal of Neurophysiology, 70,* 482–491.

Horn, A. C., Vahle-Hinz, C., Brüggemann, J., et al. (1999). Responses of neurons in the lateral thalamus of the cat to stimulation of urinary bladder, colon, esophagus, and skin. *Brain Research, 851,* 164–174.

Kniffki, K.-D., & Vahle-Hinz, C. (1987). The periphery of the cat's ventroposteromedial nucleus (VPM_p): Nociceptive neurones. In J.-M. Besson, G. Guilbaud, & M. Peschanski (Eds.), *Thalamus and pain* (pp. 245–257). Amsterdam: Elsevier.

Vahle-Hinz, C., & Detsch, O. (2002). What can *in vivo* electrophysiology in animal models tell us about mechanisms of anesthesia? *British Journal of Anaesthesia, 89*, 123–142.

Vahle-Hinz, C., Freund, I., & Kniffki, K.-D. (1987). Nociceptive neurons in the ventral periphery of the cat thalamic ventroposteromedial nucleus. In R. F. Schmidt, H.-G. Schaible, & C. Vahle-Hinz (Eds.), *Fine afferent nerve fibers and pain* (pp. 440–450). VCH Verlagsgesellschaft: Weinheim.

Vahle-Hinz, C., Brüggemann, J., & Kniffki, K.-D. (1995). Thalamic processing of visceral pain. In B. Bromm & J. Desmedt (Eds.), *Pain and the brain. From nociception to cognition. Advances in pain research and therapy* (Vol. 22, pp. 125–141). New York: Raven.

Vahle-Hinz, C., Reeker, W., Detsch, O., et al. (2002). Antinociceptive effects of anesthetics *in vivo*: Neuronal responses and cellular mechanisms. In B. W. Urban & M. Barann (Eds.), *Molecular and basic mechanisms of anesthesia* (pp. 516–524). Lengerich: Pabst Sci Publ.

Willis, W. D. (1997). Nociceptive functions of thalamic neurons. In M. Steriade, E. G. Jones, & D. A. McCormick (Eds.), *Thalamus, Vol. II, experimental and clinical aspects* (pp. 373–424). Amsterdam: Elsevier.

Zhang, Y., Wang, N., Wang, J.-Y., Chang, J.-Y., Woodward, D. J., & Luo, F. (2011). Ensemble encoding of nociceptive stimulus intensity in the rat medial and lateral pain systems. *Molecular Pain, 7*, 64.

Thalamus, Nociceptive Inputs in the Rat (Spinal)

Jean-François Bernard
INSERM UMR-894, UPMC Site Pitié-Salpétrière, Paris, France

Definition

Layers of the Spinal Cord

The gray matter of the spinal cord is divided, from dorsal to ventral, into ten laminae on the basis of cytoarchitectonic criteria. The dorsal horn includes the laminae I to VI; the lamina VII is an intermediate area; the ventral horn includes the laminae VIII and IX (motoneurons); and the region around the central canal corresponds to the lamina X.

Thalamus

The caudal portion of the thalamus is an important brain center for somatosensory and nociceptive processing. This caudal region is often divided into the lateral and the medial thalamus. (1) The lateral thalamus is primarily a relay that includes chiefly, for somatosensory functions, the ventral posterolateral (VPL), the ventral posteromedial (VPM), the posterior group (Po), and the triangular posterior group (PoT) thalamic nuclei. (2) The medial thalamus might be regarded as a more integrative region that includes intralaminar (chiefly the central lateral (CL), paracentral (PC), central medial (CM), and parafascicular (Pf) nuclei), paralaminar (notably the ventral medial (VM) nucleus), and median thalamic nuclei.

Characteristics

The Spinal Relay

Noxious messages are conveyed from the periphery (trunk and limbs) to the thalamus via a primary relay, the dorsal horn of the spinal cord. Two individualized regions of the dorsal horn, the superficial and the deep laminae, have key roles in the processing of nociceptive messages.

The superficial laminae (I and II) have a major role in nociceptive processing because this region is the main recipient for peripheral nociceptive inputs conveyed by Aδ- and C-fibers. Whereas both the laminae I and II receive nociceptive inputs, only the lamina I neurons project to supraspinal centers. These neurons are the main output of this superficial region. The lamina I neurons are primarily nociceptive, a majority of them being nociceptive specific (Christensen and Perl 1970). In addition, a lower proportion of lamina I neurons can encode specifically innocuous thermal stimuli (Light et al. 1993).

The second area of the dorsal horn, which processes nociceptive messages, includes the deep laminae V and VI and the adjacent portion of the lamina VII. The involvement of deep laminae in nociceptive processing was chiefly demonstrated by electrophysiology; this region contains numerous wide dynamic range neurons that have a great ability to encode noxious stimuli but from a clearly innocuous range (Besson and Chaouch 1987). The anatomical link of this region

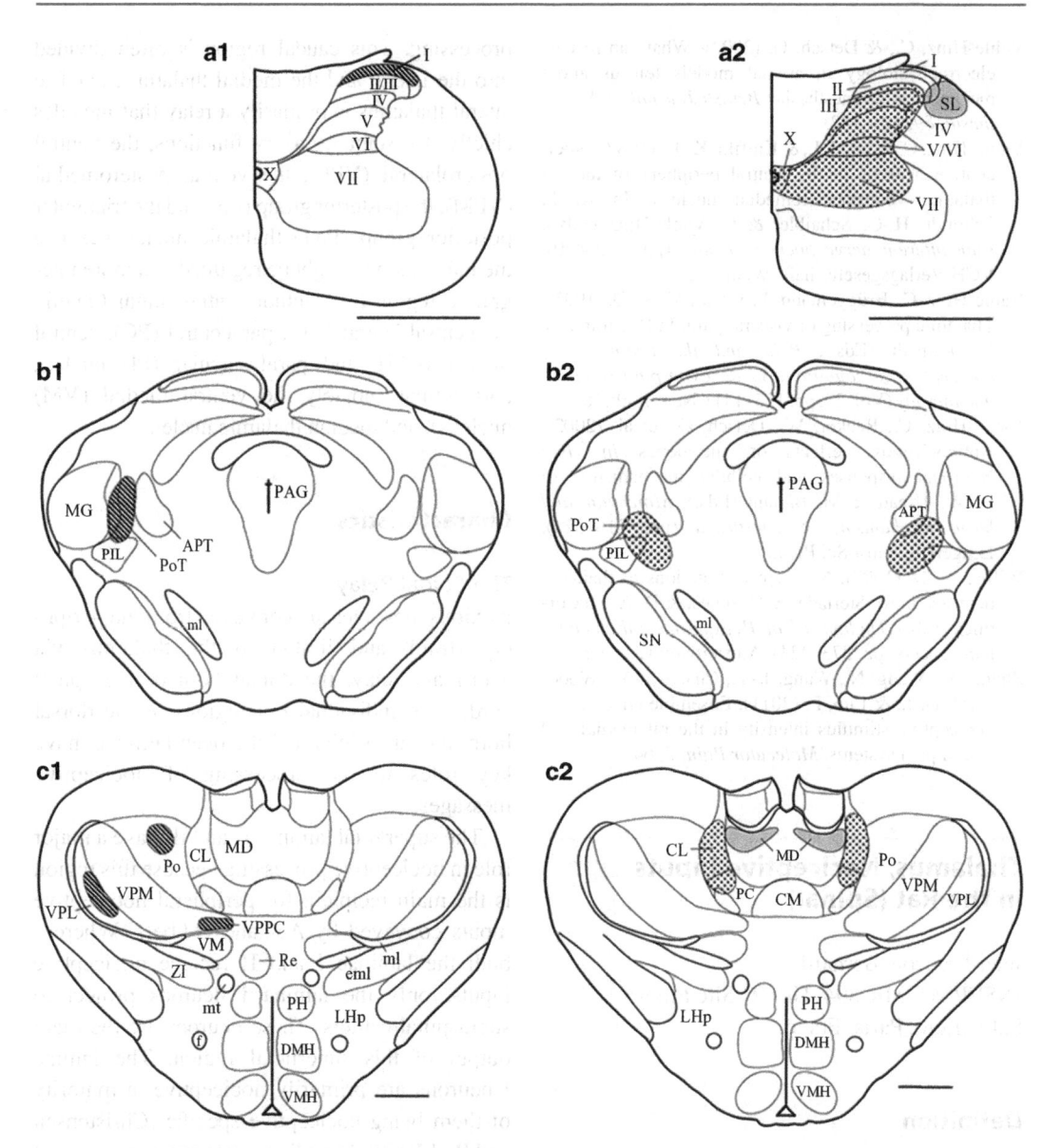

Thalamus, Nociceptive Inputs in the Rat (Spinal), Fig. 1 Summary diagram of spinal projections to the thalamus. (**a1**) Lamina I (*hatching*) projecting area in the cervical enlargement of the spinal cord. (**b1**) Projection in the PoT (*hatching*) from lamina I. (**c1**) Projection to VPL, Po, and VPPC thalamic nuclei (*hatching*) from lamina I. (**a2**) Deep laminae (*black points*) and SL nucleus (*gray*) projecting area in the cervical enlargement. (**b2**) Projection to PoT thalamic nucleus (*points*) from the deep laminae. (**c2**) Projection to CL (*points*) and MD (*gray*) thalamic nuclei from deep laminae and SL nucleus, respectively. Scale bars = 1 mm. Abbreviations: *I–X* laminae I–X of the spinal cord, *APT* anterior pretectal nucleus, *CL* central lateral thalamic nucleus, *CM* central medial thalamic nucleus, *DMH* dorsomedial hypothalamic nucleus, *eml* external medullary lamina, *f* fornix, *LHp* lateral hypothalamus posterior, *MD* mediodorsal nucleus, *MG* medial geniculate nucleus, *ml* medial lemniscus, *mt* mammillothalamic tract, *PAG* periaqueductal gray matter, *PC* paracentral thalamic nucleus, *PH* posterior hypothalamus, *PIL* posterior intralaminar thalamic nucleus, *Po* posterior thalamic group, *PoT* posterior thalamic group, triangular part, *Re* reuniens nucleus, *SL* spinal lateral nucleus, *SN* substantia nigra, *VM* ventromedial thalamic nucleus, *VMH* ventromedial hypothalamic nucleus, *VPL* ventral posterolateral thalamic nucleus, *VPM* ventral posteromedial thalamic nucleus, *VPPC* ventral posterior parvocellular thalamic nucleus, *ZI* zona incerta

Thalamus, Nociceptive Inputs in the Rat (Spinal),
Fig. 2 Photomicrographs of the projection from lamina I to the thalamus. (**a**) PHA-L injection site in laminae I/II of the cervical enlargement. (**b**) Extensive terminal labeling with large varicosities in the VPL thalamic nucleus resulting from the injection in (**a**). Scale bars = 500 μm in (**a**), 100 μm in (**b**). Abbreviations: *I–X* laminae I–X of the spinal cord, *VPL* ventral posterolateral thalamic nucleus

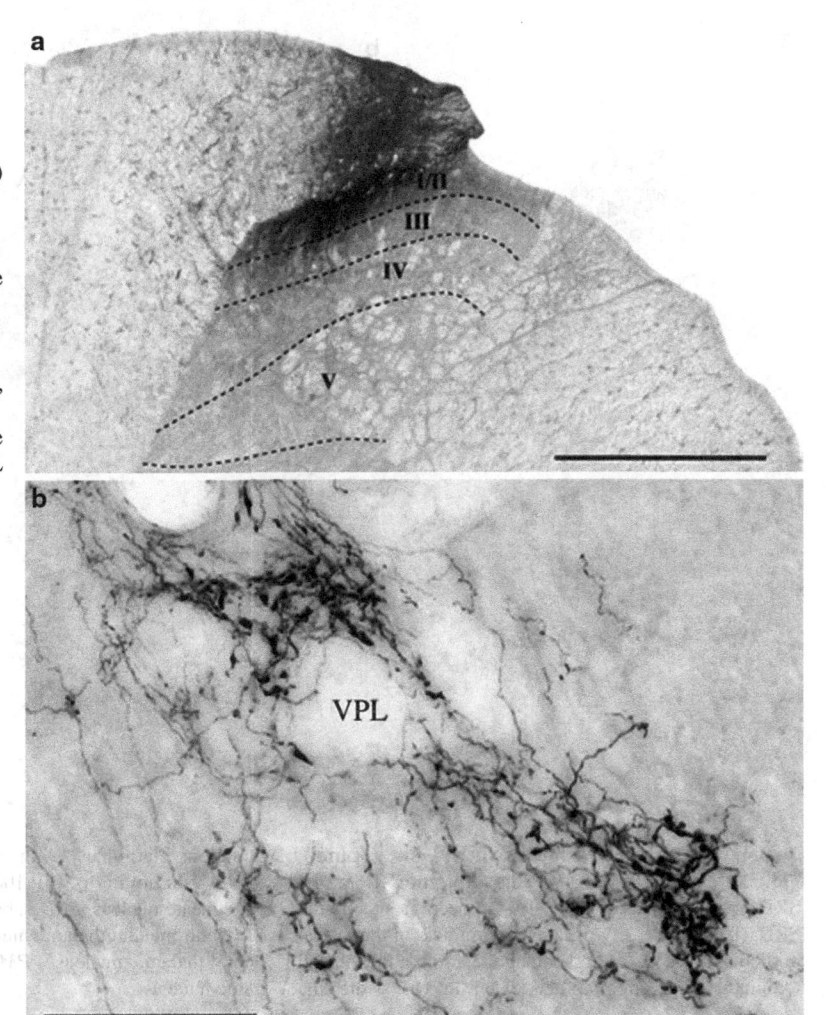

with peripheral Aδ- and C-nociceptive fibers is less clear; deep laminae neurons receive some collateral projections from Aδ- and C-fibers, but the main nociceptive input might be conveyed indirectly via superficial laminae.

The caudal thalamus is a primary brain center for pain processing. Nociceptive information is conveyed to the thalamus from nociceptive neurons in lamina I and deep laminae of the spinal cord (1) directly via the spinothalamic tract and (2) indirectly via relay nuclei in the brainstem.

Spinal Nociceptive Inputs to the Thalamus

Retrograde tracing studies show that neurons projecting to the thalamus are located in lamina I and in deep laminae of the spinal cord (Burstein et al. 1990). Thalamic projecting superficial spinal neurons are clearly concentrated in lamina I of the dorsal horn, only very few are located in lamina II. Lamina I neurons are evenly distributed along the spinal cord, with higher concentrations at the levels of cervical and lumbar enlargements. On the other hand, thalamic projecting deep laminae neurons are scattered in deep laminae of the spinal cord (III to VIII and X), although they are more numerous around the laminae IV–VI. These thalamic projecting deep laminae neurons are not evenly distributed in the spinal cord, since almost half of them are concentrated in the two first spinal segments (C1, C2). In the cervical enlargement (C5–C8), which relays nociceptive messages from the

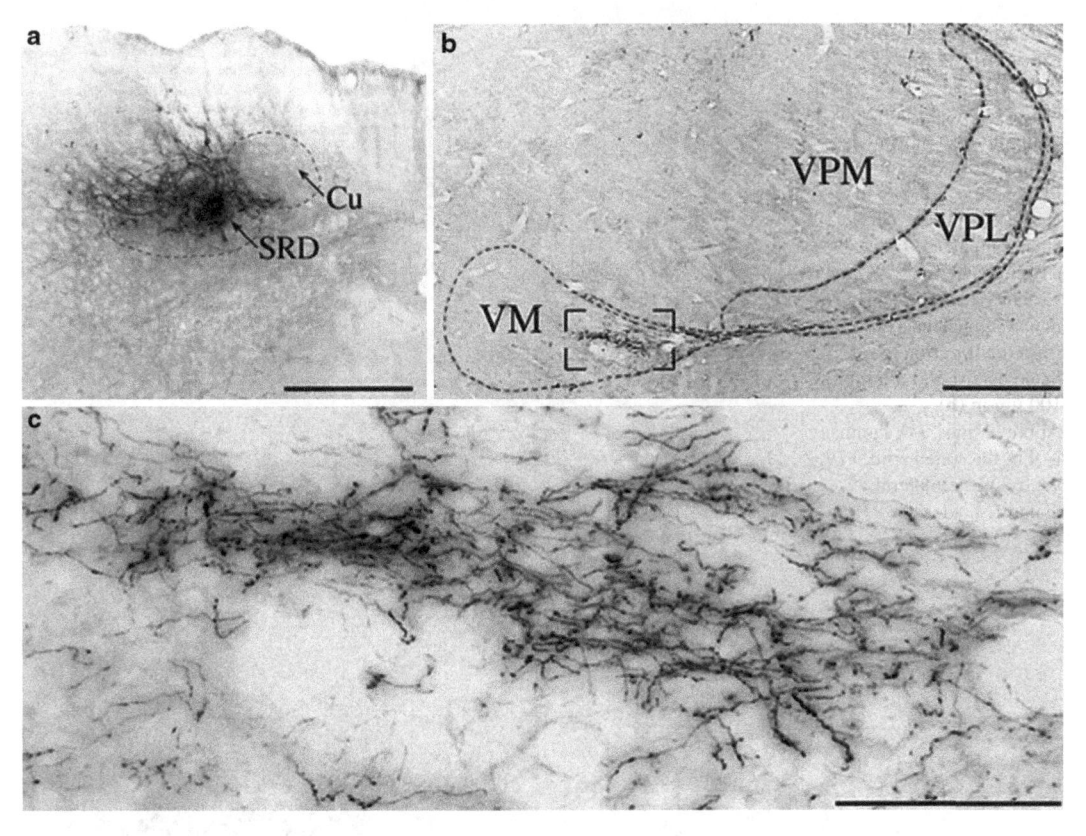

Thalamus, Nociceptive Inputs in the Rat (Spinal), Fig. 3 Photomicrographs of the projection from the SRD to the thalamus. (a) PHA-L injection site in the SRD. (b) High density of labeled terminals in the lateral portion of the VM. (c) Higher magnification of the labeled terminals in the region delineated in (b). Note the numerous terminals with small varicosities. Scale bars = 500 μm in (a), (b), 100 μm in (c). Abbreviations: *Cu* cuneate nucleus, *SRD* subnucleus reticularis dorsalis, *VM* ventromedial thalamic nucleus, *VPL* ventral posterolateral thalamic nucleus, *VPM* ventral posteromedial thalamic nucleus

forelimb, one spinothalamic neuron in lamina I and one in the deep lamina were counted on average per 50 μm section (Burstein et al. 1990).

Anterograde tracing studies, using a high-resolution tracer such as the *Phaseolus vulgaris* leucoagglutinin (PHA-L), allowed the demonstration of separate projections from lamina I and deep laminae of the spinal cord upon the thalamus (Gauriau and Bernard 2003). The projections from lamina I are markedly different from those of deep laminae (Fig. 1).

Lamina I neurons project primarily to restricted portions of the VPL and rostral Po (Fig. 1a1–c1). In these areas, axonal endings have large varicosities (Fig. 2). Other substantial projections with small varicosities were observed in the PoT and the ventral posterior parvocellular nucleus (VPPC). Only a moderate number of projections are observed in a few additional thalamic targets such as the periventricular, the subparafascicular, the reuniens, and the mediodorsal nuclei; no projection is observed in the submedius nucleus. Lamina I projections to the thalamus are almost exclusively contralateral. Because the VPL/Po system is also known to be the thalamic relay conveying somatotopically tactile messages from gracile and cuneate nuclei to somatosensory SI and SII cortex, it appears likely that the lamina I (VPL/Po system) is devoted to somatosensory discriminative aspects of nociception. The role of the lamina I (PoT/VPPC system) is less clear; it could participate to the recognition of the "painful" nature of nociceptive stimuli.

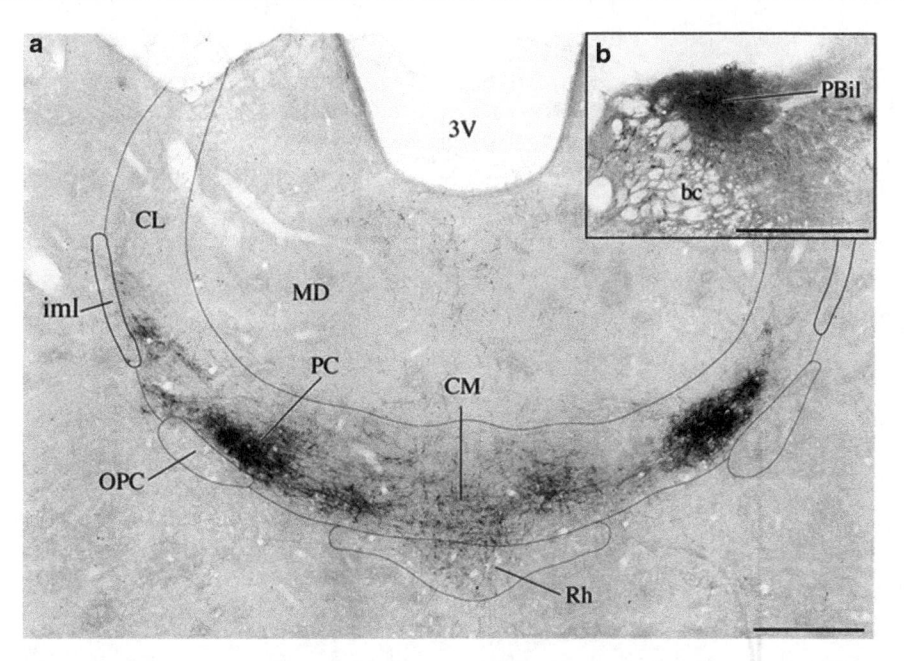

Thalamus, Nociceptive Inputs in the Rat (Spinal),
Fig. 4 Photomicrographs of the projection from the
PBil to the thalamus. Note the high density of labeled
terminals concentrated in the PC (*black area on both
sides*) with a lower density of labeling in the CM (**a**)
resulting from the PHA-L injection covering the PBil
(**b**). Scale bars = 500 μm. Abbreviations: *3 V* third
ventricle, *bc* brachium conjunctivum, *CL* central lateral
thalamic nucleus, *CM* central medial thalamic nucleus,
iml internal medullary lamina, *MD* mediodorsal thalamic
nucleus, *OPC* oval paracentral thalamic nucleus, *PBil*
internal lateral parabrachial nucleus, *PC* paracentral thalamic nucleus, *Rh* rhomboid thalamic nucleus

Deep laminae neurons project chiefly to the
central lateral intralaminar and the PoT thalamic
nuclei on both sides (Fig. 1a2–c2). Very few
projections are observed in other thalamic nuclei.
In fact, it appears that deep laminae neurons
project as much to the thalamus as to extrathalamic
targets, such as the substantia innominata, the
globus pallidus, the posterior and lateral hypothalamus, and the central amygdaloid nucleus. These
recent findings clearly question the projections of
deep laminae to the VPL, suggesting strongly that
the main nociceptive inputs to the VPL originate
primarily from the lamina I neurons, in the rat.

A last notable projection is from the spinal
lateral neurons to the caudal portion of the
mediodorsal nucleus (Fig. 1a2–c2).

Brainstem Nociceptive Inputs
to the Thalamus
The spinothalamic tract is often considered to be
the primary pathway conveying nociceptive
messages from the spinal cord to the thalamus.

In fact, the number of spinothalamic neurons
(9,500 is the higher estimation in the rat; Burstein
et al. 1990) is substantial but represents only
a very small proportion of dorsal horn neurons.
Furthermore, anterograde studies in primate
(Mehler et al. 1960) as well as in the rat (Gauriau
and Bernard 2003) indicate that spinal projections are clearly more extensive upon the
brainstem than to the thalamus. These data
strongly suggest that the brainstem should have
an important role in conveying nociceptive input
from the spinal cord to the thalamus and the other
brain centers.

The gigantocellular reticular (Gi) nucleus,
located in the center of the medulla, was the
first candidate to fulfill this role (Casey 1971;
Bowsher 1976). This nucleus receives an
extensive projection from deep laminae of the
spinal cord, it contains numerous nociceptive
neurons, and its stimulation produces aversive
reactions. However, the portion of the Gi that
receives the densest spinal projection projects

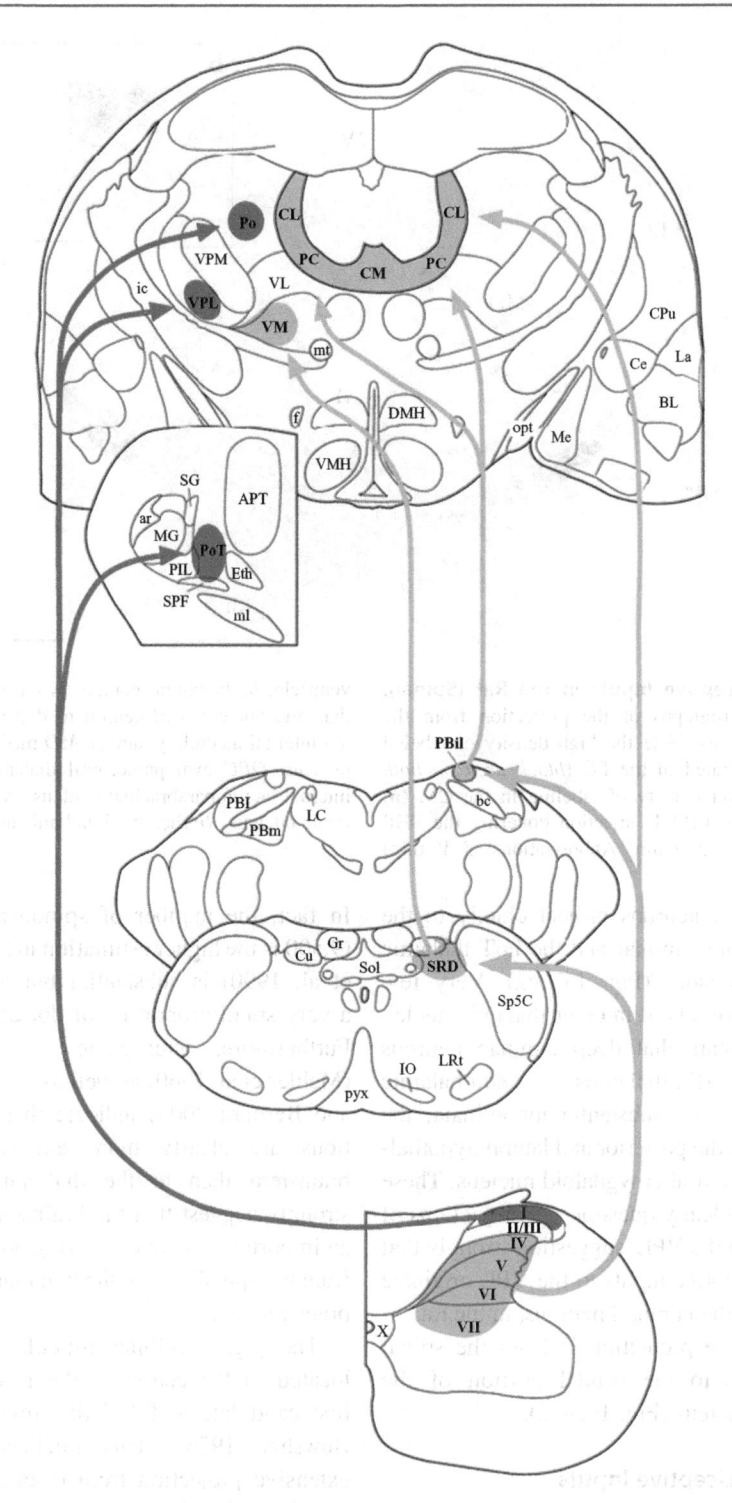

Thalamus, Nociceptive Inputs in the Rat (Spinal),
Fig. 5 Schematic representation of the main nociceptive
inputs to the thalamus, in the rat. *Red*: Main inputs to the
lateral thalamus from the lamina I of the spinal cord. *Blue*:

Main inputs to the medial thalamus originating from
the deep laminae of the spinal cord, directly via the
spinothalamic tract, and indirectly via the SRD and
the PBil nuclei. Abbreviations: *I–X* laminae I–X of

primarily to the locus coeruleus and the spinal cord and only weakly to the thalamus.

The subnucleus reticularis dorsalis (SRD), a very caudal reticular area of the medulla, located just ventral to the cuneate nucleus, now appears to be the best candidate to convey nociceptive messages from the spinal cord to the thalamus. Indeed, the SRD receives an extensive projection from deep laminae of the spinal cord (Raboisson et al. 1996). Electrophysiological studies demonstrate the involvement of this reticular region in nociceptive processing. Indeed, most of the SRD neurons are strongly excited by noxious stimuli from a low spontaneous activity and do not respond to multisensory (visual and auditory) stimuli. SRD neurons encode the intensity of thermal, mechanical, and visceral noxious stimuli. They respond exclusively to the activation of peripheral Aδ- or Aδ- and C-fibers. Such responses are depressed by intravenous morphine in a dose-dependent and naloxone-reversible fashion (Villanueva et al. 1996). The receptive fields of SRD neurons are very large; they often include the whole body. The main thalamic targets of the SRD are the lateral portions of the VM nucleus, (Fig. 3) and, to a lesser extent, of the Pf nucleus (Villanueva et al. 1998).

The internal lateral parabrachial (PBil) nucleus, located in a dorsal position at the pontomesencephalic junction, is another candidate for conveying nociceptive messages from the spinal cord to the thalamus. The PBil should not be mixed up with the lateral parabrachial area, which receives an extensive projection from lamina I and does not project substantially to the thalamus. In fact, the PBil specifically receives a dense projection from

deep laminae of the spinal cord (especially from the reticular portion of laminae IV and V) (Bernard et al. 1995). The PBil neurons project primarily to the PC thalamic nucleus (Fig. 4) and, to a lesser extent, to the CM and the Pf thalamic nuclei (Bester et al. 1999). Electrophysiological studies demonstrate the involvement of thalamic projecting PBil neurons in nociceptive processing. Indeed, most of them respond to thermal and noxious stimuli, with a maximum response in the mid-nociceptive scale (48 °C and 16 N/cm^2). The PBil neurons exhibit strong "windup" and long lasting afterdischarge in response to noxious stimuli (Bourgeais et al. 2001).

With regard to the deep laminae system, the brainstem clearly has an important complementary role to the spinothalamic tract. Indeed, deep laminae neurons send nociceptive messages to the medial thalamus directly via the spinothalamic tract, as well as indirectly via SRD and PBil neurons. Thus, this system could deal with alertness and emotional and motor aspects of pain through a general arousal of the prefrontal and frontal (motor) cortices (the cortical targets of the medial thalamus).

Synthesis

Nociceptive inputs to the thalamus can be classified in the two different systems summarized in Fig. 5 as follows:
1. The lamina I – (lateral) thalamic system, which could be chiefly responsible for sensory discrimination of nociceptive stimuli via thalamic projections to somatosensory cortices (SI, SII, Insular).

Thalamus, Nociceptive Inputs in the Rat (Spinal), Fig. 5 (continued) the spinal cord, APT anterior pretectal nucleus, ar acoustic radiation, bc brachium conjunctivum, BL basolateral amygdaloid nucleus, Ce central amygdaloid nucleus, CL central lateral thalamic nucleus, CM central medial thalamic nucleus, CPu caudate putamen (striatum), Cu cuneate nucleus, DMH dorsomedial hypothalamic nucleus, Eth ethmoid thalamic nucleus, f fornix, Gr gracilis nucleus, ic internal capsule, IO inferior olive, La lateral amygdaloid nucleus, LC locus coeruleus, LRt lateral reticular nucleus, Me medial amygdaloid nucleus, MG medial geniculate nucleus, ml medial lemniscus, mt mammillothalamic tract, opt optic tract, PBil internal

lateral parabrachial nucleus, PBl lateral parabrachial nucleus, PBm medial parabrachial nucleus, PC paracentral thalamic nucleus, PIL posterior intralaminar thalamic nucleus, Po posterior thalamic group, PoT posterior thalamic group, triangular part, pyx pyramidal decussation, SG suprageniculate thalamic nucleus, Sol solitary tract nucleus, Sp5C spinal trigeminal nucleus caudal part, SPF subparafascicular nucleus, SRD subnucleus reticularis dorsalis, VL ventrolateral thalamic nucleus, VM ventromedial thalamic nucleus, VMH ventromedial hypothalamic nucleus, VPL ventral posterolateral thalamic nucleus, VPM ventral posteromedial thalamic nucleus

2. The deep laminae – (medial) thalamic system, which includes two subsystems: a direct deep laminae (medial thalamic pathway) and an indirect deep laminae (SRD/PBil – medial thalamic pathway). This system could be involved in motor and alertness/arousal emotional features of pain via the medial thalamic projections to the frontal motor and the prefrontal medial (cingulate) cortices.

The spinal ascending axons of all these systems are located around the same region of the spinal lateral/ventrolateral quadrant in the rat. Thus, it appears clear that the strikingly acute effectiveness of ventrolateral cordotomy is due to the interruption of both a direct spinothalamic tract and a strong spino-brainstem-thalamic pathway.

An additional nociceptive pathway to the thalamus (not illustrated) was proposed by the group of Willis. The visceral (colorectal) nociceptive information, after a relay in laminae X and VII of the lumbosacral spinal cord, would be conveyed via the medial portion of the dorsal column and the gracile nucleus to the VPL thalamic nucleus. The key point in the demonstration of this pathway is the strong effect of a medial commissurotomy (in the middle of dorsal column fasciculus) upon the response of VPL neurons to visceral noxious stimuli and visceral pain (Willis et al. 1999). However, the existence of such a pathway remains difficult to reconcile with previous clinical and experimental data.

References

Bernard, J. F., Dallel, R., Raboisson, P., et al. (1995). Organization of the efferent projections from the spinal cervical enlargement to the parabrachial area and periaqueductal gray: A PHA-L study in the rat. *The Journal of Comparative Neurology, 353*, 480–505.

Besson, J. M., & Chaouch, A. (1987). Peripheral and spinal mechanisms of nociception. *Physiological Reviews, 67*, 67–186.

Bester, H., Bourgeais, L., Villanueva, L., et al. (1999). Differential projections to the intralaminar and gustatory thalamus from the parabrachial area: A PHA-L study in the rat. *The Journal of Comparative Neurology, 405*, 421–449.

Bourgeais, L., Monconduit, L., Villanueva, L., et al. (2001). Parabrachial internal lateral neurons convey nociceptive messages from the deep laminas of the dorsal horn to the intralaminar thalamus. *Journal of Neuroscience, 21*, 2159–2165.

Bowsher, D. (1976). Role of the reticular formation in responses to noxious stimulation. *Pain, 2*, 361–378.

Burstein, R., Dado, R. J., & Giesler, G. J., Jr. (1990). The cells of origin of the spinothalamic tract of the rat: A quantitative reexamination. *Brain Research, 511*, 329–337.

Casey, K. L. (1971). Somatosensory responses of bulboreticular units in awake cat: Relation to escape-producing stimuli. *Science, 173*, 77–80.

Christensen, B. N., & Perl, E. R. (1970). Spinal neurons specifically excited by noxious or thermal stimuli: Marginal zone of the dorsal horn. *Journal of Neurophysiology, 33*, 293–307.

Gauriau, C., & Bernard, J. F. (2003). A comparative reappraisal of projections from the superficial laminae of the dorsal horn in the rat: The forebrain. *The Journal of Comparative Neurology, 468*, 24–56.

Light, A. R., Sedivec, M. J., Casale, E. J., et al. (1993). Physiological and morphological characteristics of spinal neurons projecting to the parabrachial region of the cat. *Somatosensory and Motor Research, 10*, 309–325.

Mehler, W. R., Feferman, M. E., & Nauta, W. (1960). Ascending axon degeneration following anterolateral cordotomy. An experimental study in the monkey. *Brain, 83*, 718–751.

Raboisson, P., Dallel, R., Bernard, J. F., et al. (1996). Organization of efferent projections from the spinal cervical enlargement to the medullary subnucleus reticularis dorsalis and the adjacent cuneate nucleus: A PHA-L study in the rat. *The Journal of Comparative Neurology, 367*, 503–517.

Villanueva, L., Bouhassira, D., & Le Bars, D. (1996). The medullary subnucleus reticularis dorsalis (SRD) as a key link in both the transmission and modulation of pain signals. *Pain, 67*, 231–240.

Villanueva, L., Desbois, C., Le Bars, D., et al. (1998). Organization of diencephalic projections from the medullary subnucleus reticularis dorsalis and the adjacent cuneate nucleus: A retrograde and anterograde tracer study in the rat. *The Journal of Comparative Neurology, 390*, 133–160.

Willis, W. D., Al-Chaer, E. D., Quast, M. J., et al. (1999). A visceral pain pathway in the dorsal column of the spinal cord. *Proceedings of the National Academy of Sciences of the United States of America, 96*, 7675–7679.

Thalamus, Nociceptive Neurotransmission

▶ Thalamic Neurotransmitters and Neuromodulators

Thalamus, Receptive Fields, Projected Fields, Human

Nirit Weiss[1], Stanley W. Anderson[2] and Frederick A. Lenz[3]
[1]Department of Neurosurgery, Mount Sinai Medical Center, New York, NY, USA
[2]Department of Neurosurgery, Johns Hopkins Hospital, Johns Hopkins Medical Institutions, Johns Hopkins University, Baltimore, MD, USA
[3]Department of Neurosurgery, Johns Hopkins Hospital, Baltimore, MD, USA

Synonyms

Changes in thalamic physiology occurring in patients with chronic pain

Definition

Physiological changes in the thalamus seen in patients with chronic pain syndromes. These changes are reflected in alterations in the properties of neurons and assemblies of neurons recorded in patients who suffer from chronic pain.

Characteristics

Studies of plasticity of the somatosensory system in non-human primates have focused on maps of cortical function determined by examining neuronal receptive fields (RFs) (Kaas 1991). Studies in humans can explore both maps determined from RFs (RF maps) and maps of the locations and quality of sensations evoked by stimulation of the brain (projected field or PF maps) (Lenz et al. 1994). While RF maps are a reflection of the organization of inputs to the central nervous system, PF maps give an indication of the image of the body contained in the thalamus and cortex. There are numerous studies of cortical plasticity secondary to injuries of the nervous system (Kaas 1991). Although much of the reorganization of cortical maps may reflect underlying changes in

thalamic organization (Pons et al. 1991), there are relatively few direct studies of thalamic plasticity (Rasmusson 1996b).

Thalamic Neuronal Activity in Primates injuries of the Nervous System

Medically intractable chronic pain due to nervous system injury or medically intractable movement disorders may be surgically treated by implantation of deep brain stimulating electrodes in the principal sensory nucleus of the thalamus (ventral caudal – Vc) (Hosobuchi 1986) or by thalamotomy (Tasker et al. 1988; Zirh et al. 1999; Cardoso et al. 1995). During such operations, microelectrode recordings may be used to confirm the target predicted by the radiological studies. The physiological studies of Vc in patients with chronic pain syndromes are compared with the recordings of Vc in patients with movement disorders, and thus, plasticity of pain-related neuronal activity in the human thalamus can be studied directly.

The region of the principal sensory nucleus of thalamus (Vc) was explored during stereotactic surgical procedures for the treatment of patients with pain following spinal cord transaction and compared to results of patients with movement disorders (Lenz et al. 1994). Many cells in the expected representation of the anesthetic part of the body did not have RFs. However, in the borderzone/anesthetic zone between the anesthetic and innervated areas of the thalamus, there was frequently a mismatch between the location of RFs and PFs. Often, RFs were located on the chest and abdominal wall above the level of the spinal transection, while PFs were located in the lower extremity below the spinal transection.

Borderzone/anesthetic areas of thalamus often exhibited increased representations of the border of the anesthetic part of the body. Two sites were said to have a consistent RF if the RF of both sites included the same part of the body. The length of a trajectory with consistent RFs is the distance along the trajectory where each RF continues to include the same part of the body (Lenz et al. 1988b). The maximal distance along a trajectory over which the RF or the PF stays consistent is longer for body parts with larger representations (Lenz et al. 1994). Lengths of trajectory with a consistent RF in a particular part of the body were significantly longer in borderzone/anesthetic zones than in control areas with apparently normal input.

Neurons with RFs adjacent to the area of sensory loss in amputation patients ($n = 3$) occupied a larger part of the thalamic homunculus (Lenz et al. 1998b) than found for the same part of the body in patients with movement disorders (Lenz et al. 1988a, 1994). This result is consistent with somatotopic reorganization of afferent inputs from the limb. Similarly, the large area over which PFs include the stump (Lenz et al. 1993, 1994) suggest that there has been reorganization of the perceptual image of the limb in the central nervous system (Jensen and Rasmussen 1994). It has also been reported that phantom sensations can be evoked by stimulation of the region where stump RFs are located in the region of Vc (Davis et al. 1998). Thus, in the case of amputations and spinal cord injury, the alteration in the image of the body in the thalamus is less than that of the inputs from that part of the body.

In primates, there are well documented alterations in thalamic anatomy and physiology after peripheral nerve injury. The distributions of thalamic regions with characteristic histology are altered (Rausell et al. 1992) in monkeys with a C2-T4 dorsal rhizotomy (Sweet 1980; Levitt 1985). In the affected arm area of rhizotomized animals, there is a reduction in the density of large cells, and of parvalbumin and CO staining, all characteristic of the terminal zone for dorsal column inputs. There is corresponding increase in the calbindin staining in the arm area, characteristic of the terminal zone for STT inputs.

After cervical dorsal rhizotomy, large numbers of cells without RFs are encountered (Albe-Fessard and Lombard 1983; Lombard et al. 1979) in the forelimb region of the monkey VP. Following adult digit amputation (Rasmusson 1996a, b), increased representation of the stump is found with large RFs including adjacent digits (Rasmusson 1996a, b). The thalamic representation of the border of the anesthetic part is increased in monkeys with nerve sections (Garraghty and Kaas 1991).

Thalamic Microstimulation in Humans with Chronic Pain

Studies have shown that stimulation of the somatic sensory thalamus is more likely to evoke pain in patients with chronic pain after nervous system injury than in patients without somatic sensory abnormalities (patients with movement disorders) (Lenz et al. 1998a; Davis et al. 1996). The region of Vc was divided on the basis of projected fields into areas representing the part of the body where the patients experienced chronic pain (pain affected) or did not experience chronic pain (pain unaffected) and into a control area located in the thalamus of patients with movement disorders and no experience of chronic pain.

In both the core and posterior inferior regions of the thalamic sensory nucleus, the proportion of sites where threshold microstimulation evoked pain was larger in pain affected and unaffected areas than in control areas. The number of sites where thermal (warm or cold) sensations were evoked was correspondingly smaller, so that the total of pain plus thermal sites was not significantly different across all areas (Lenz et al. 1998a). Therefore, sites where stimulation-evoked pain in patients with neuropathic pain may correspond to sites where thermal sensations were evoked by stimulation in patients without somatic sensory abnormality. In the posterior inferior region, the number of sites where cold was evoked by stimulation decreased significantly while the number of sites where pain was evoked increased significantly.

These results suggest that pain is evoked in patients with neuropathic pain by stimulation at sites where thermal sensations would normally be evoked. Therefore, the present data suggest that the STT or elements to which the STT projects signal pain rather than thermal sensations in patients with neuropathic pain. This is consistent with the finding that stimulation of the STT evokes pain in patients with neuropathic pain but evokes non-painful thermal sensations in patients who do not have neuropathic pain (Tasker 1988). Cordotomy relieves pain in a much greater proportion of patients with somatic pain than it does in patients with neuropathic pain (Sweet et al. 1994; Tasker et al. 1980). The failure of cordotomy to relieve neuropathic pain might be anticipated from the occurrence of central pain in patients with impaired function of the STT (Cassinari and Pagni 1969; Boivie et al. 1989; Andersen et al. 1995). These results suggest that the generator for pain in patients with central pain is the terminus of the STT.

In patients with central pain, anatomical evidence of damage to STT is a common finding (Cassinari and Pagni 1969) and loss of STT function, indicated by impaired thermal and pain sensibility, is a uniform finding (Boivie 1994; Boivie et al. 1989; Andersen et al. 1995). In patients with central pain, pain is more common than in controls while threshold microstimulation-evoked cold sensations are correspondingly less common. These findings suggest that there has been a reorganization so that cold modalities are relabeled to signal pain in the thalamus of patients with central pain. (Ralston et al. 1996). The relationship between thalamic stimulation-evoked cold and pain in patients with central pain may explain the perception of cold as pain (cold hyperalgesia) which can occur in these patients (Boivie 1994).

Similar changes are observed after central nervous system injury. In patients with spinal transection, the numbers of cellular RFs representing the border of the anesthetic part of the body are increased (Lenz et al. 1994). Following transection of the dorsal columns at the T3-T5 level, activity in simian VP (Pollin and Albe-Fessard 1979) shows a decrease in the percentage of cells with hindlimb RFs and an increase in the percentage of cells with forelimb RFs. Many cells have large RFs and respond to high threshold inputs, consistent with inputs from STT. Therefore, loss of input due to peripheral or central nervous system injury leads to significant reorganization of the thalamic representation of inputs from different parts of the body.

Studies of the thalamus in monkeys and humans with nervous system injury demonstrate large changes in the perceptual representation of the body in the thalamus, as revealed by somatotopic maps of projected fields (Lenz et al. 1994, 1998b). Patients with amputations show increases in the thalamic area from which stimulation evokes sensations in the part of a limb that has become the stump (Lenz et al. 1998b). In patients with spinal transections, the incidence of mismatches between neuronal RFs and threshold microstimulation-evoked PFs at the same sites is much higher than in patients with movement disorders (Lenz et al. 1994). In patients with spinal transection, these mismatches occur because sensations in the anesthetic part are evoked by stimulation at sites where cellular RFs represent parts of the body proximal to the anesthetic part. In these patients, stimulation in the thalamic area where the representation of the anesthetic part of the body is usually located evokes sensations in the anesthetic part (Lenz et al. 1994). These results suggest that the perceptual representation of the body in the thalamus reorganizes less than the representation of inputs to the thalamus. Although the central image of the body is relatively constant in the face of altered input, our studies show that changes in the modality organization of this image can change dramatically (Lenz et al. 1998b). This plasticity of modality organization may contribute to the development of chronic pain in patients with nervous system injury (Flor et al. 1998).

A recent study studied plasticity by analysis of pain perception in subjects with forebrain strokes leading to neglect syndromes for visual or tactile modalities (Liu et al. 2010). These syndromes are defined as the failure of a subject "to report, respond, or orient to meaningful stimuli (of particular modalities) contralateral to the (cortical) brain lesion" (Vallar and Perani 1986; Heilman et al. 1993). Spatial attention for painful stimuli was assessed by tests of extinction, in which stimuli are perceived on the left when they are presented on the left alone, but not when both sides are stimulated simultaneously.

The results demonstrate that thermal pain extinction of hot and cold pain stimuli occurs in a proportion of subjects with hemi-neglect (Liu et al. 2010). In the subjects with visual spatial hemi-neglect but without thermal pain extinction, the sensation of the thermal pain stimulus on the affected (left) side was not extinguished but was often localized to the unaffected (right) side, and the submodality of the stimulus (cold or hot) was often misidentified. These abnormalities were significantly larger on the left side of subjects with visual spatial neglect than in healthy controls or in controls with stroke but without hemineglect. Therefore, plasticity of the human pain system can result from lesions damaging pain pathways or producing cognitive deficits or combinations of the two.

References

Albe-Fessard, D. G., & Lombard, M. C. (1983). Use of an animal model to evaluate the origin of and protection against deafferentation pain. *Advances in Pain Research and Therapy, 5*, 691–700.

Andersen, P., Vestergaard, K., Ingeman-Nielsen, M., & Jensen, T. S. (1995). Incidence of post-stroke central pain. *Pain, 61*, 187–193.

Boivie, J. (1994). Central pain. In P. D. Wall & R. Melzack (Eds.), *Textbook of pain* (pp. 871–902). Edinburgh: Churchill Livingston.

Boivie, J., Leijon, G., & Johansson, I. (1989). Central post-stroke pain-a study of the mechanisms through analyses of the sensory abnormalities. *Pain, 37*, 173–185.

Cardoso, F., Jankovic, J., Grossman, R. G., & Hamilton, W. (1995). Outcome after stereotactic thalamotomy for dystonia and hemiballismus. *Neurosurgery, 36*, 501–508.

Cassinari, V., & Pagni, C. A. (1969). *Central pain. A neurosurgical survey.* Cambridge, MA: Harvard University Press.

Davis, K. D., Kiss, Z. H. T., Luo, L., et al. (1998). Phantom sensations generated by thalamic microstimulation. *Nature, 391*, 385–387.

Davis, K. D., Kiss, Z. H. T., Tasker, R. R., & Dostrovsky, J. O. (1996). Thalamic stimulation-evoked sensations in chronic pain patients and nonpain (movement disorder) patients. *Journal of Neurophysiology, 75*, 1026–1037.

Flor, H., Elbert, T., Muhlnickel, W., Pantev, C., Wienbruch, C., & Taub, E. (1998). Cortical reorganization and phantom phenomena in congenital and traumatic upper-extremity amputees. *Experimental Brain Research, 119*, 205–212.

Garraghty P. E., & Kaas K. H. (1991). Functional reorganization in adult monkey thalamus after peripheral nerve injury. *Neuroreport , 2*(12), 747–750.

Heilman, K. M., Watson, R. T., & Valenstein, E. (1993). Neglect and related disorders. In K. M. Heilman & E. Valenstein (Eds.), *Clinical neuropsychology* (pp. 279–336). New York: Oxford.

Hosobuchi, Y. (1986). Subcortical electrical stimulation for control of intractable pain in humans. *Journal of Neurosurgery, 64*, 543–553.

Jensen, T. S., & Rasmussen, P. (1994). Phantom pain and related phenomena after amputation. In P. D. Wall & R. Melzack (Eds.), *Textbook of pain* (pp. 651–665). New York: Churchill Livingstone.

Kaas, J. H. (1991). Plasticity of sensory and motor maps in adult mammals. *Annual Review of Neuroscience, 14*, 137–167.

Lenz, F. A., Dostrovsky, J. O., Tasker, R. R., et al. (1988a). Single-unit analysis of the human ventral thalamic nuclear group: somatosensory responses. *Journal of Neurophysiology, 59*, 299–316.

Lenz, F. A., Tasker, R. R., Kwan, H. C., et al. (1988b). Single unit analysis of the human ventral thalamic nuclear group: correlation of thalamic "tremor cells" with the 3–6 Hz component of parkinsonian tremor. *Journal of Neuroscience, 8*, 754–764.

Lenz, F. A., Seike, M., Lin, Y. C., et al. (1993). Thermal and pain sensations evoked by microstimulation in the area of the human ventrocaudal nucleus (Vc). *Journal of Neurophysiology, 70*, 200–212.

Lenz, F. A., Kwan, H. C., Martin, R., et al. (1994). Characteristics of somatotopic organization and spontaneous neuronal activity in the region of the thalamic principal sensory nucleus in patients with spinal cord transection. *Journal of Neurophysiology, 72*, 1570–1587.

Lenz, F. A., Gracely, R. H., Baker, F. H., et al. (1998a). Reorganization of sensory modalities evoked by stimulation in the region of the principal sensory nucleus (ventral caudal – Vc) in patients with pain secondary to neural injury. *The Journal of Comparative Neurology, 399*, 125–138.

Lenz, F. A., Zirh, A. T., Garonzik, I. M., et al. (1998b). Neuronal activity in the region of the principal sensory nucleus of human thalamus (ventralis caudalis) in patients with pain following amputations. *Neurosci, 86*, 1065–1081.

Liu, C. C., Veldhuijzen, D. S., Ohara, S., Winberry, J., Greenspan, J. D., & Lenz, F. A. (2010). Spatial attention to thermal pain stimuli in subjects with visual spatial hemi-neglect: extinction, mislocalization and misidentification of stimulus modality. *Pain, 152*, 498–506.

Levitt, M. (1985). Dysesthesias and self-mutilation in humans and subhumans: a review of clinical and experimental studies. *Brain Research, 357*, 247–290.

Lombard, M. C., Nashold, B. S., & Pelissier, T. (1979). Thalamic recordings in rats with hyperalgesia. *Advances in Pain Research and Therapy, 3*, 767–772.

Pollin, B., & Albe-Fessard, D. G. (1979). Organization of somatic thalamus in monkeys with and without section of dorsal spinal tracts. *Brain Research, 173*, 431–449.

Pons, T. P., Garraghty, P. E., Ommaya, A. K., et al. (1991). Massive cortical reorganization after sensory deafferentation in adult macaques. *Science, 252*, 1857–1860.

Ralston, H. J., Ohara, P. T., Meng, X. W., et al. (1996). Transneuronal changes in the inhibitory circuitry of the macaque somatosensory thalamus following lesions of the dorsal column nuclei. *The Journal of Comparative Neurology, 371*, 325–335.

Rasmusson, D. (1996a). Changes in the organization of ventroposterior lateral thalamic nucleus after digit removal in the adult raccoon. *The Journal of Comparative Neurology, 364*, 92–103.

Rasmusson, D. D. (1996b). Changes in the response properties of neurons in the ventroposterior lateral thalamic nucleus of the raccoon after peripheral deafferentation. *Journal of Neurophysiology, 75*, 2441–2450.

Rausell, E., Cusick, C. G., Taub, E., et al. (1992). Chronic Deafferentation in monkeys differentially affects nociceptive and non-nociceptive pathway distinguished by specific calcium-binding protiens and down-regulates gamma-aminobutyric acid type A receptors at thalamic levels. *Proceedings of National Academy of Science USA, 89*, 2571–2575.

Sweet, W. H. (1980). Central mechanisms of chronic pain (neuralgias and certain other neurogenic pain). Res Publ Assoc Res Nerve Ment Dis, *58*, 287–303.

Sweet, W. H., Poletti, C. E., & Gybels, G. M. (1994). Operations in the brainstem and spinal canal with an appendix on the relationship of open and percutaneous cordotomy. In P. D. Wall & R. Melzack (Eds.), *Textbook of pain* (pp. 1113–1136). New York: Churchill and Livingston.

Tasker, R. R. (1988). Percutaneous cordotomy: the lateral high cervical technique. In H. H. Schmidek & W. H. Sweet (Eds.), *Operative neurosurgical techniques indications, methods, and results* (pp. 1191–1205). Philadelphia: Saunders, W.B.

Tasker, R. R., Doorly, T., & Yamashiro, K. (1988). Thalamotomy in generalized dystonia. *Advances in Neurology, 50*, 615–631.

Tasker, R. R., Organ, L. W., & Hawrylyshyn, P. (1980). Deafferentation and causalgia. In J. J. Bonica (Ed.), *Pain* (pp. 305–329). New York: Raven.

Vallar, G., & Perani, D. (1986). The anatomy of unilateral neglect after right-hemisphere stroke lesions. A clinical/CT-scan correlation study in man. *Neuropsychologia, 24*, 609–622.

Zirh, A., Reich, S. G., Dougherty, P. M., & Lenz, F. A. (1999). Stereotactic thalamotomy in the treatment of essential tremor of the upper extremity: reassessment including a blinded measure of outcome. *Journal of Neurology, Neurosurgery & Psychiatry, 66*, 772–775.

Thalamus, Visceral Representation

A. Vania Apkarian
Department of Physiology, Northwestern University Feinberg School of Medicine, Chicago, IL, USA

Synonyms

Interoception; Referred pain; Visceral modulation; Visceral representation

Definition

Stimulation of visceral organs distinctly activates different populations of neurons in the thalamus. Neurons located in ▶ parvocellular VP (VPpc) are described as visceral-specific, while neurons located in other medial and lateral thalamic nuclei respond to multiple viscera and to somatic inputs convergently. Other thalamic nuclei do not receive visceral inputs.

Characteristics

In contrast to somatic stimuli, visceral stimuli may or may not be accompanied by perception, and the perceptions show distinct properties. For example, baro- and chemoreceptor activation do not give rise to sensations/perceptions, although they are represented in visceral-specific portions of the thalamus and in insular cortex. On the other hand, distension of most hollow viscera is associated with ill-defined sensations. Visceral nociceptive stimuli in such organs are generally characterized by a sense of malaise or discomfort, with no clear ability to localize the source of the stimulus. Moreover, visceral pains are usually associated with pain referred on the skin, the location of which is characteristic for different viscera, an association used since antiquity to pinpoint specific viscera as the source of injury or inflammation. Thus, the contrast between visceral and somatic representation and associated differences in perception provides an opportunity for dissecting conscious perceptions from unconscious modulation, in relationship to thalamocortical connectivity. Unfortunately, very little effort has been invested in this direction. It should also be added that visceral stimuli that do not evoke any conscious sensations and are represented at the level of the thalamus and cortex most probably play a modulatory role in emotional responses to the environment and exert emotionally driven modulatory control over somatic and visceral responses. This notion, however, remains mainly a speculation, since direct studies on the topic are minimal (Gebhart 1995). Here, we concentrate on thalamic neuronal response properties from the viewpoint of coding visceral stimuli and being modulated by visceral stimuli. All the available data are from anesthetized preparations; thus, they should be regarded as examples of responses undoubtedly stunted due to anesthesia.

The earliest evidence for visceral inputs to the thalamus and cortex used electrical stimulation of the vagus (Bailey and Bremer 1938; Dell and Olson 1951) and demonstrated a relay from the lateral thalamus to the ▶ insular cortex from visceral organs. More recent anatomic studies have elucidated that the visceral inputs through vagal afferents terminate in the solitary tract, second-order neurons then project ipsilaterally to the parabrachial nucleus, and third-order neurons travel contralaterally to the ventroposterior parvocellular nucleus of the thalamus (VPpc) and then project to the insula. A separate projection from parabrachial regions directly accesses the insula too, see chapter by Cechetto in Gebhart (1995). ▶ Calcitonin gene-related peptide (CGRP) seems to label this pathway in rats as well as in man. Using a combination of physiological recordings from insular cortex and tracing techniques, it was demonstrated that the most medial portion of the VPpc projects to the insular region with gustatory responses, while a more lateral portion projects to part of the insula with gastric mechanoreceptor

responsive neurons and the most lateral portion projects to an area with cardiopulmonary responses; parabrachial inputs to the thalamus seem to have a parallel organization as well (Cechetto and Saper 1987; Gebhart 1995). More recent studies in the rat confirm the parabrachial inputs and also show inputs from spinal cord and trigeminal lamina I (Bester et al. 1999; Gauriau and Bernard 2004). Thus, the VPpc portion of the thalamus seems to be viscerotopically organized, where gustatory responses have been shown physiologically (Ganchrow and Erickson 1972).

Inputs from pelvic viscera to the cat lateral thalamus have been mapped with electrical stimulation and natural stimulation of the urinary bladder (Brüggemann et al. 1993, 1994b). Identified neurons were all found located outside the ▶ VPL, either just in its periphery or in the adjacent ▶ posterior nucleus (PO). All visceral responsive neurons also responded to low-threshold stimuli applied to the skin. A similar study examined renal nerve stimulus responses to map kidney representation in and around the ▶ VPL in the cat (Horn et al. 1997). Most responsive units were located in the periphery of the VPL, dorsal PO, or lateral PO; none were found within the VPL. Response latency suggested large and small myelinated afferent fibers mediating kidney inputs. The somatic receptive fields of the cells with renal nerve inputs showed minimal correspondence with the dermatomes of the renal nerve. A subsequent study mapped the same region of the thalamus in the cat for bladder, colon, and esophagus inputs and again found the majority of cells localized to either the periphery of the VPL or the ▶ PO and no indication of segregation of neurons based on responses to any particular viscus (Ganchrow and Erickson 1972; Horn et al. 1999). Visceral responsive cells have been studied in the medial thalamus of the cat (1995) and localized within the ▶ mediodorsal (MD), ▶ central medial (CM), ▶ central lateral (CL), periphery of the VPL and ▶ ventral posterior medial (VPM) nuclei, and in the ▶ zona incerta (ZI). More than 23 % responded to esophagus (of 120 neurons examined), 8 % to bladder, and 6 % to colon distension. In contrast to the lateral thalamic neurons, all medial thalamic visceral responsive cells had nociceptive inputs from the skin.

In the squirrel monkey, lateral thalamus responses to distending the urinary bladder, distal colon and lower esophagus, and to noxious and innocuous somatic stimuli have been mapped (Brüggemann et al. 1994a). Eighty-five percent (of 106 neurons studied) responded to at least one of the viscera. Most visceroceptive cells had somatic low-threshold responses and convergent multivisceral responses. Figure 1 is an example of a neuron located in the VPL; it responds to all three viscera by excitation and has a somatic tactile receptive field on foot digit 5. The visceral responsive cells showed increased or decreased firing for distensions in the noxious range, and some also coded distensions in the innocuous and noxious ranges. Figure 2 shows visceral and somatic properties of 20 VPL neurons identified in a single electrode penetration. The extent of unpredictability between adjacent neurons, regarding specific visceral inputs and their excitatory or inhibitory responses, is evident. Modulatory effects of visceral stimulation have been examined in squirrel monkey VPL (Brüggemann et al. 1998) and indicate that noxious distensions of urinary bladder, distal colon, or lower esophagus decrease responses to somatic stimuli by about 50 %. The high incidence of visceral responsive cells in the VPL implies that this information must be transmitted to the cortex and such visceral responsive neurons have in fact been reported in the monkey primary somatosensory cortex (Brüggemann et al. 1997).

Neuronal responses to uterus, cervix, vagina, colon, and skin have been examined in the rat thalamus (Berkley et al. 1993, 1995; Guilbaud et al. 1993). In the lateral thalamus, most neurons responded to multiple viscera, and most were located in and around the border of ventrobasal complex (VB). Like results seen in the cat medial thalamus, most neurons in the caudal intralaminar thalamic nuclei (IL) with visceral

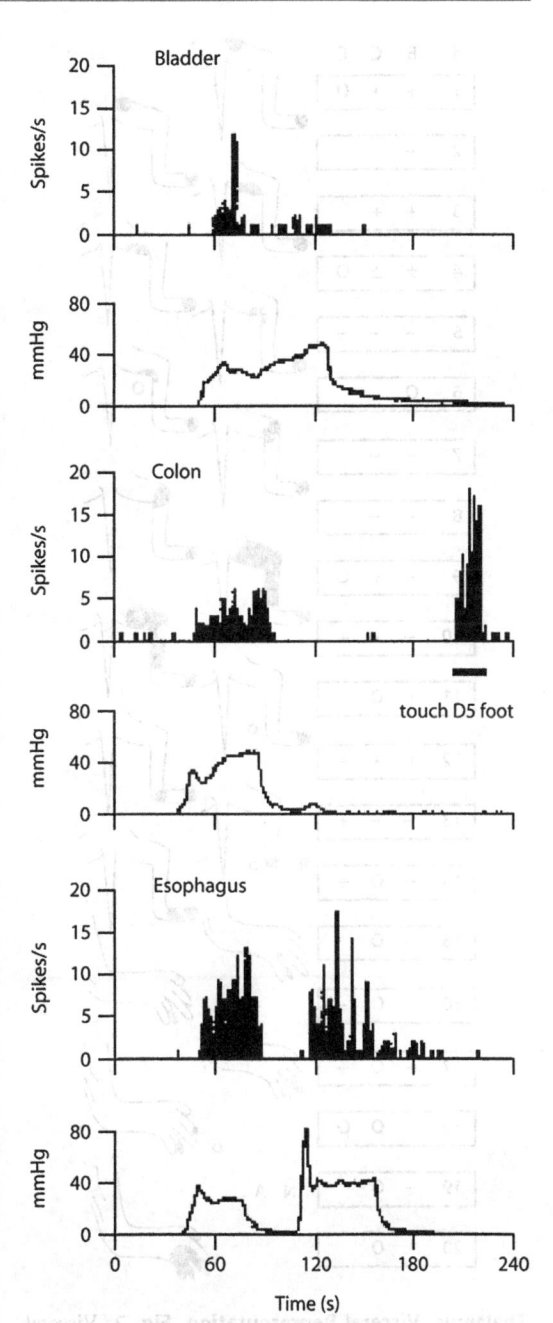

Thalamus, Visceral Representation, Fig. 1 Visceral responses of a ▶ VPL neuron with low-threshold somatic response (touching digit 5, *D5* on the foot contralateral to the recording) in the anesthetized squirrel monkey. Intraluminal pressures are shown for each viscus distended. The neuron responds by increased activity to noxious distension of the bladder, colon, and esophagus (Brüggemann et al. 1994a)

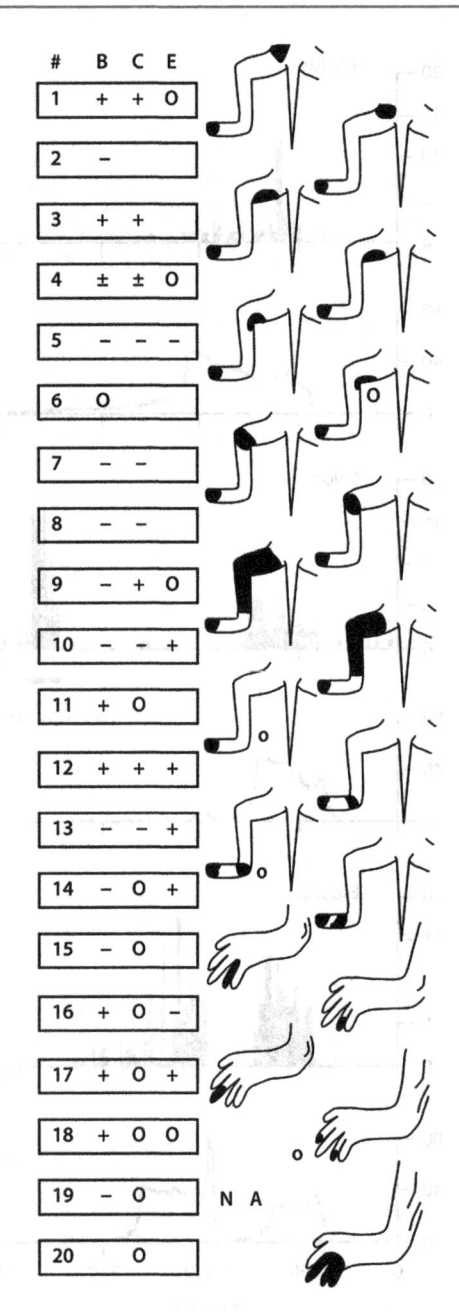

#	B	C	E
1	+	+	O
2	–		
3	+	+	
4	±	±	O
5	–	–	–
6	O		
7	–	–	
8	–	–	
9	–	+	O
10	–	–	+
11	+	O	
12	+	+	+
13	–	–	+
14	–	O	+
15	–	O	
16	+	O	–
17	+	O	+
18	+	O	O
19	–	O	N A
20	O		

Thalamus, Visceral Representation, Fig. 2 Visceral and somatic responses of 20 neurons identified in a single track by an electrode traversing ▶ VPL dorsoventrally. Cell 1 (#) is located most dorsally. Somatic receptive fields are either low-threshold tactile (*unmarked*) or wide-dynamic range type marked next to the receptive fields (*open circle*). Visceral responses are as follows: 0, no response; –, inhibition; +, excitation; ±, mixed; due to distension of the urinary bladder (*B*), colon (*C*), or lower esophagus (*E*) (missing symbols, not tested; *N.A.* somatic receptive field not found) (From Brüggemann et al. 1994a)

responses also responded to noxious somatic stimuli, and most had inputs from multiple viscera (Berkley et al. 1995). The lateral portion of ventral medial nucleus (VM_l) has recently been established as a region important in signaling nociceptive information from the brainstem to the cortex (see chapter in this encyclopedia by Villanueva). Cells in this region respond to noxious mechanical and thermal stimuli applied anywhere on the body, yet surprisingly these neurons do not respond to visceral stimuli (Monconduit et al. 2003).

Overall, thalamic nuclei can be differentiated into three kinds: nuclei with visceral-specific inputs, nuclei with visceral convergent inputs, and nuclei with no visceral inputs. Surprisingly, this differentiation does not follow the organization of the thalamus regarding somatic nociceptive and non-nociceptive responses, which was a primary hypothesis for most of the studies in the topic. The region of the thalamus presumed to be organized viscerotopically has not been adequately studied, especially physiologically, and, although it is commonly referred as the visceral-specific part of the thalamus, its detailed response properties remain unknown. However, it is possible that the observations in the cat VPL periphery and PO are actually part of the same visceral-specific nucleus (VPpc) extending laterally and dorsally. There are important species differences in the visceral organization of the thalamus. In the cat, there is good reproducible evidence that the VPL proper and most likely also the VPM proper do not receive visceral inputs, while the VPL and VPM in rat and monkey do. In fact, in the squirrel monkey there is good evidence that all neurons in the VPL and VPM receive visceral inputs, non-viscerotopically. There is also good agreement between studies and across species for medial visceroceptive cells having convergent somatic nociceptive inputs, while lateral visceroceptive cells have mainly innocuous somatic inputs. Given the widely convergent viscero-visceral inputs on somatic responsive cells in the monkey lateral thalamus, it has been proposed that the visceral inputs through dorsal column inputs provide the signal for poorly localized visceral perceptions and that the spinothalamic inputs (presumed to be more specific and more dominantly somatic) may then lead to the perception of referred pain on the skin; see chapter by Apkarian in (Gebhart 1995). Undoubtedly, visceral inputs to the medial thalamus would complement this model by providing modulation of affect. The distinct visceral organizational rules of the thalamus, coupled with unique perceptions, should provide the opportunity for studying the brain circuitry regarding different types of perceptions, yet current research in this field remains limited.

References

Bailey, P., & Bremer, F. (1938). A sensory cortical representation of the vagus nerve with a note on the effects of the low blood pressure on the cortical electrogram. *Journal of Neurophysiology, 1,* 405–412.

Berkley, K. J., Guilbaud, G., Benoist, J. M., et al. (1993). Responses of neurons in and near the thalamic ventrobasal complex of the rat to stimulation of uterus, cervix, vagina, colon, and skin. *Journal of Neurophysiology, 69,* 557–568.

Berkley, K. J., Benoist, J. M., Gautron, M., et al. (1995). Responses of neurons in the caudal intralaminar thalamic complex of the rat to stimulation of the uterus, vagina, cervix, colon and skin. *Brain Research, 695,* 92–95.

Bester, H., Bourgeais, L., Villanueva, L., et al. (1999). Differential projections to the intralaminar and gustatory thalamus from the parabrachial area: A PHA-L study in the rat. *The Journal of Comparative Neurology, 405,* 421–449.

Brüggemann, J., Vahle-Hinz, C., & Kniffki, K. D. (1993). Representation of the urinary bladder in the lateral thalamus of the cat. *Journal of Neurophysiology, 70,* 482–491.

Brüggemann, J., Shi, T., & Apkarian, A. V. (1994a). Squirrel monkey lateral thalamus. II. Viscerosomatic convergent representation of urinary bladder, colon, and esophagus. *Journal of Neuroscience, 14,* 6796–6814.

Brüggemann, J., Vahle-Hinz, C., & Kniffki, K. D. (1994b). Projections from the pelvic nerve to the periphery of the cat's thalamic ventral posterolateral nucleus and adjacent regions of the posterior complex. *Journal of Neurophysiology, 72,* 2237–2245.

Brüggemann, J., Shi, T., & Apkarian, A. V. (1997). Viscero-somatic neurons in the primary somatosensory cortex (SI) of the squirrel monkey. *Brain Research, 756,* 297–300.

Brüggemann, J., Shi, T., & Apkarian, A. V. (1998). Viscerosomatic interactions in the thalamic ventral posterolateral nucleus (VPL) of the squirrel monkey. *Brain Research, 787,* 269–276.

Cechetto, D. F., & Saper, C. B. (1987). Evidence for a viscerotopic sensory representation in the cortex and thalamus in the rat. *The Journal of Comparative Neurology, 262,* 27–45.

Dell, P., & Olson, R. (1951). Projections thalamiques, corticales et cerebelleuses des afferences viscerales vagales. *Society of Biology (Paris), 145,* 1084–1088.

Ganchrow, D., & Erickson, R. P. (1972). Thalamocortical relations in gustation. *Brain Research, 36,* 298–305.

Gauriau, C., & Bernard, J. F. (2004). A comparative reappraisal of projections from the superficial laminae of the dorsal horn in the rat: The forebrain. *The Journal of Comparative Neurology, 468,* 24–56.

Gebhart, G. F. (1995). *Visceral pain.* Seattle: IASP Press.

Guilbaud, G., Berkley, K. J., Benoist, J. M., et al. (1993). Responses of neurons in thalamic ventrobasal complex of rats to graded distension of uterus and vagina and to uterine suprafusion with bradykinin and prostaglandin F2 alpha. *Brain Research, 614,* 285–290.

Horn, A. C., Vahle-Hinz, C., Petersen, M., et al. (1997). Projections from the renal nerve to the cat's lateral somatosensory thalamus. *Brain Research, 763,* 47–55.

Horn, A. C., Vahle-Hinz, C., Bruggemann, J., et al. (1999). Responses of neurons in the lateral thalamus of the cat to stimulation of urinary bladder, colon, esophagus, and skin. *Brain Research, 851,* 164–174.

Monconduit, L., Bourgeais, L., Bernard, J. F., et al. (2003). Convergence of cutaneous, muscular and visceral noxious inputs onto ventromedial thalamic neurons in the rat. *Pain, 103,* 83–91.

Thalassotherapy

Definition

Bathing in the sea as a form of therapy.

Cross-References

▶ Spa Treatment

Therapeutic Acupuncture

Definition

Therapeutic acupuncture refers to the clinical use of acupuncture for slowly arising, long-term relief of different symptoms (weeks-months) after a course of treatments, usually with mild acupuncture stimulation.

Cross-References

▶ Acupuncture Mechanisms

Therapeutic Alliance

▶ Chronic Pain, Patient-Therapist Interaction

Therapeutic Block

▶ Pain Treatment: Spinal Nerve Blocks

Therapeutic Blocks

▶ Peripheral Nerve Blocks

Therapeutic Cold

▶ Therapeutic Heat, Microwaves, and Cold

Therapeutic Drug Monitoring

Definition

The use of serum drug concentrations to guide dosing of the drug in order to obtain optimum treatment effect and avoid toxicity.

Cross-References

▶ Antidepressants in Neuropathic Pain

Therapeutic Exercise

Definition

Exercise as part of a treatment program designed to improve an individual's ability to move, balance and coordination, endurance, flexibility, muscle tone, posture, and strength.

Cross-References

▶ Chronic Pain in Children, Physical Medicine and Rehabilitation

Therapeutic Gain

Definition

Therapeutic gain is the difference between the therapeutic response to the verum and the placebo in a randomized controlled trial, in migraine prophylaxis, between the percentage of "responders," i.e., of patients with a 50 % reduction of attack frequency, to the active drug and the percentage of responders to placebo.

Cross-References

▶ Clinical Migraine Without Aura

Therapeutic Gene Transfer

▶ Opioids and Gene Therapy

Therapeutic Heat

▶ Therapeutic Heat, Microwaves, and Cold

Therapeutic Heat, Microwaves, and Cold

Jeffrey R. Basford
Department of Physical Medicine and
Rehabilitation, Mayo Clinic and Mayo
Foundation, Rochester, NY, USA

Synonyms

Cold therapy; Diathermy; Microwaves; Modality; Physical agent; Therapeutic cold; Therapeutic heat

Definition

The use of heat and cold to lessen pain, promote healing, or obtain other therapeutic goals.

Characteristics

The body is sensitive to its environment. Temperatures above 42 °C or below 13 °C produce discomfort and those only a few degrees higher or lower may actually injure it. Furthermore, temperature changes easily obtained in the clinic alter enzymatic activity, nerve conduction, and tissue viscosity (Oosterveld and Rasker 1994). While heat and cold may be used to alleviate the cause of pain or to produce analgesia, the latter is far more common.

Thermal agents achieve their effects as the result of changing the temperature of a tissue. Specifically, most heat-based treatments attempt to warm tissues to by 3–8 °C and most cold therapies attempt to reduce tissue temperatures by a similar amount. As a result, the overlap between their indications and contraindications is surprisingly broad (Tables 1 and 2). It should be remembered that many of the contraindications of heat and cold are relative, i.e., localized heating might be acceptable in a patient with cardiac disease, while systemic heating

Therapeutic Heat, Microwaves, and Cold, Table 1 Indications and precautions for the use of therapeutic heat

Indications	Precautions and contraindications
Pain	Acute inflammation, trauma, or hemorrhage
Muscle spasm	Bleeding dyscrasias
Contractures	Insensitivity
Fibromyalgia	Inability to communicate or respond to pain
Hyperemia	Poor thermal regulation (systemic heating situation)
Acceleration of metabolic processes	Malignancy
Chronic Hematoma	Edema
Bursitis/Tenosynovitis	Ischemia
Superficial thrombophlebitis	Atrophic skin/scar tissue
	Unstable angina or blood pressure
	Decompensated heart failure/recent myocardial infarction

Therapeutic Heat, Microwaves, and Cold, Table 2 Indications and precautions for the use of therapeutic cold

Indications	Precautions and contraindications
Acute musculoskeletal trauma	Ischemia
Pain	Cold intolerance\uticaria
Muscle spasm	Raynaud's phenomenon and disease
Spasticity	Severe cold pressor responses
Reduction of metabolic activity	Inability to communicate or respond to pain
	Poor thermal regulation
	Insensitivity

might place unacceptable demands on his cardiac function (Keast and Adamo 2000).

Mechanism of Action

Tissue can be heated or cooled by conduction, convection, or the conversion of a different form of energy into heat. Hot packs epitomize conductive heating, hydrotherapy convection, and ultrasound conversion, due to its reliance on the conversion of sound into heat.

Superficial Heat

The physiological effects of the superficial heating agents differ little and choice depends on the situation and patient/therapist preference.

Hot Packs

Hot packs are typically constructed of porous bags filled with a hygroscopic material. They are either kept in 70–80 °C water baths or warmed in a microwave before being covered with an absorbent wrap and placed on the body. Treatments may last for 30 min with the packs slowly cooling. Alternatives such as electrically heated pads that do not cool spontaneously are convenient but may increase the risk of burns.

Heat Lamps

Heat lamp use is declining but is still common. Skin temperatures are controlled by adjusting the distance between the lamp and the patient. The precautions in Table 1 apply to these agents but it should also be noted that chronic use of superficial heat can produce a permanent mottling of the skin (erythema *ab igne*).

Hydrotherapy

Hydrotherapy uses a fluid medium to produce heating, cooling, massage, and debridement. Neither agitation nor water is necessary and solutes in the medium (e.g., NaCl for wound care) may be a significant factor in treatment. Hydrotherapy may also be performed with substances such as finely ground solid materials suspended by jets of hot air. The benefits of this dry heat "fluidotherapy" approach over conventional water-based hydrotherapy remain unclear.

Whirlpool Baths

Whirlpool baths range in size from those intended to treat a single extremity to others in which the entire body is immersed. Temperatures of 33–36 °C are usually well tolerated, although for a healthy patient, elevations to 43 °C are possible on limited portions of the body. Treatment with stationary mediums may be beneficial. Sitz baths, for example, are beneficial in the treatment of anorectal pain and research demonstrates that bathing at 40–50 °C is not only comfortable but also lessens anal tone (Pinho et al. 1993). Agitation increases the efficiency of thermal transfer, and it should be remembered that a temperature that is comfortable with a motionless medium may become painfully hot or cold with agitation.

Contrast Bathing

Contrast bathing involves the patient shifting their treated extremities alternatively between a warm (38–40 °C) and a cool (13–16 °C) bath about ten times. These baths have often been used in the treatment of complex regional pain syndrome with benefits thought to result from reflex hyperemia and desensitization.

Paraffin Baths

Paraffin baths consist of a basin filled with a 1:7 mixture of mineral oil and paraffin. Temperatures (45–54 °C) are higher than those of water-based hydrotherapy and are tolerated due to the lack of agitation, the insulation provided by wax as it solidifies on the treated area and the low heat capacity of the medium. Treatment is usually performed by dipping the involved extremity in the bath about ten times, covering it and then allowing it to cool slowly in an insulated wrap. An alternative approach in which the treated extremity is dipped once and then kept in the bath permits more vigorous heating. Thirty minutes treatments may produce increases in the intramuscular temperature of superficial muscles of about 3 °C (Abramson et al. 1964).

Diathermy

▶ Diathermy can be performed with short waves (SWD), microwaves (MWD), or ultrasound (USD). USD is the most frequently used agent (Lindsay et al. 1995) but as it is discussed in another chapter will be only summarized here. Discussion here will emphasize SWD which is still in relatively widespread use and to a lesser extent, MWD whose medical use is now quite restricted.

Ultrasound

Ultrasound has both thermal and ▶ nonthermal effects. The production and effects of heat are well understood. The benefits of nonthermal processes, which include such phenomena as cavitation (the production and destruction of small bubbles), tissue micro-streaming, and mechanical deformation, remain to be established. Treatment usually involves stroking an USD applicator over the treated tissue for 5–7 min. Intensities range from mWs to 1.5 + W per cm^2. Continuous waves are employed when the goal is heating and pulsed treatments are chosen to emphasize nonthermal effects. Temperature elevations of 5 °C are easily possible and may be particularly large at bone/soft tissue interfaces. USD phonophoresis is also used to introduce topical medication (e.g., lidocaine) through the skin.

USD used as a treatment for conditions ranging from contractures, sprains, muscle strains, wounds, tendinitis and non-healing fractures to carpal tunnel syndrome. Benefits are controversial and in some situations may be no more effective than placebos or anti-inflammatories (Basford 1998). Precautions include those for heat in general, as well as avoidance of fluid-filled cavities, the gravid or mensurating uterus, the heart, brain, cervical ganglia, tumors, laminectomy sites, and acutely inflamed joints.

Short-Wave Diathermy

The use of short wave diathermy as is true for many modalities is declining. Nevertheless, it remains in use and will benefit from a review. In summary SWD uses radio waves to heat tissue. Use is restricted to a limited range of frequencies (27.12 MHz, 13.56 MHz, and 40.68 MHz in the USA). In one approach, the body acts as an antenna and the SWD machine induces eddy currents that produce heat as they flow through the body. This method delivers the most energy to water-rich conductive tissues such as muscles. In the second approach, the body serves as the dielectric of a capacitor that is charged by the SWD machine. In this case, the tissues are in series and heating may be most marked in water-poor, high-resistance tissues, such as fat and ligaments. Applicators range from inductive

pads that are placed on the patient to the flat plates of a simple capacitor. Although now rarely done, specialized heating can be performed by wrapping coils around a patient's limb or with rectal or vaginal probes. Continuous waveforms are used when the goal is heating, while pulsed waves are used when nonthermal effects are desired. SWD can increase subcutaneous fat temperatures by 15 °C and 3–5 cm deep intramuscular temperatures by 4–6 °C (Draper et al. 1999). A SWD device is, in effect, a radio transmitter that is used to produce heat in two ways. As is true for USD, the benefits of SWD nonthermal phenomena (such as possible frequency-dependent effects on cell function) remain to be established.

Microwaves

Microwaves (915 MHz and 2,456 MHz) do not penetrate tissue as deeply as SWD. In fact, their absorption is so rapid that fat overlying a site of interest will absorb a significant portion of the beam. Thus, microwave diathermy (MWD) may increase subcutaneous fat temperatures by 10–12 °C, while the underlying muscles will be warmed only a third as much (Basford 1998). Microwaves have been replaced in therapy by other modalites. Today MSD use in medicine appears to be restricted to the production of local hyperthermia and the potentiation of cancer chemotherapy and radiation treatment.

SW and MW diathermy are both subject to the general precautions for heat outlined in Table 1. However, both are electromagnetic in nature and metal implants/devices, pacemakers, stimulators, contact lenses, and the menstruating or pregnant uterus should be avoided. Risks are real; diathermy treatment of the jaw resulted in severe brain damage in a man with a deep brain stimulator (Nutt et al. 2001).

Cold Therapy

Cold decreases metabolic activity, slows nerve conduction, produces analgesia, lessens muscle tone, inhibits spasticity, and increases gastrointestinal motility (Basford 1998; Denys 1991). The application of ice to the body decreases skin temperature by 20 °C in about 10 min. Subcutaneous temperatures decrease 3–5 °C over the same period. If cooling is continued, forearm intramuscular temperature may decrease by 6–16 °C and muscle blood flow by as much as 30 % (Oosterveld and Rasker 1994). Although chemical and refrigerated agents may have temperatures below 0 °C and can produce frostbite, ice treatments of healthy people for periods of less than 30 min do not seem to cause injury.

Technique

Ice has a high heat capacity; ice packs, massage, compression wraps, and slushes all cool tissues rapidly. Treatments tend to last 10–20 min and a slightly damp, thin towel may be placed between the ice and the skin. Iced whirlpools provide particularly rapid cooling. However, they are poorly accepted by most people. Insulated foot coverings or fabric socks and gloves may lessen discomfort and increase acceptance. Ice massage produces rapid cooling over a limited area. Treatment involves rubbing ice (often prefrozen in a paper cup) over the painful area. Analgesia can be achieved in 7–10 min.

There are a number of other cooling agents. Vapocoolant sprays can reduce skin temperature by 20 °C (Oosterveld and Rasker 1994) and are used for local skin analgesia and the "spray and stretch" techniques. Chemical ice packs cool via the production of an endothermic reaction and while convenient tend to be expensive. Refrigerated and pressurized water pressure cuffs are also available. Frozen juice and vegetable packages are convenient for home use.

The application of superficial cold may lessen hypoxic damage, edema, and compartmental pressures after injury (Bert et al. 1991), but the magnitude of its benefits in the ultimate recovery remain debatable despite its ability to lessen metabolic activity and blood flow (Ho et al. 1995). In practice, ice is typically used in conjunction with rest, compression, and elevation (RICE) in the treatment of many musculoskeletal injuries. A common regimen, such as for an ankle sprain, consists of using ice acutely for about 20 min every 2 h for 6–24 h. Although icing is almost the automatic response to acute soft tissue injury,

it may not be a panacea in that studies of the postsurgical knee (Daniel et al. 1994) and cesarean sections (Amin-Hanjani et al. 1992) may not show benefit.

After the first day or two, the choice between ice-based therapy and heat appears to depend on personal choice. Many find heat more comfortable and use it unless there is a worsening of edema or pain. Others find a combination of icing and active exercising a more effective way to speed recovery. In any case, long-term recovery depends on mobilization and exercise. Heat and cold are only important as the agents used to assist in gaining it. Patients with long-standing trochanteric bursitis and lateral epicondylitis may find ice in combination with friction massage extraordinarily effective. Ice massage and transcutaneous electrical nerve stimulation (TENS) may be equally beneficial in low back pain (Basford 1998). The precautions of Table 2 should be heeded. Elevation of blood pressure as well as the effects of cold-induced vasoconstriction in people with ischemia and Raynaud's phenomenon may be important considerations.

In summary, there is little evidence that heat and cold alone are of much benefit except for the temporary reduction of pain. As a result, they are usually prescribed in conjunction with a program of education, activity modification, and exercise. Preference plays a role in agent choice but there are some guidelines that are appropriate. Thus, acute musculoskeletal conditions are usually initially treated with ice. In addition, hydrotherapy or SWD are used to treat large areas of the body, while USD is used for more focal conditions. Deep heat may appear physiologically appealing, but at times the comforting effects of superficial heat may prove as beneficial.

References

Abramson, D. I., Tuck, S., Chu, L. S. W., et al. (1964). Effect of paraffin bath and hot fomentations on local tissue temperatures. *Archives of Physical Medicine and Rehabilitation, 45,* 87–94.

Amin-Hanjani, S., Corcoran, J., & Chatwani, A. (1992). Cold therapy in the management of postoperative cesarean section pain. *Journal of Obstetrics and Gynecology, 167,* 108–109.

Basford, J. R. (1998). *Physical agents. Rehabilitation medicine: Principles and practice* (2nd ed., pp. 483 – 504). Philadelphia: Lippincott-Raven.

Beenakker, E. A., Oparina, T. I., Teelken, A., et al. (2001). Cooling garment treatment in MS: Clinical improvement and decrease in leukocyte NO production. *Neurology, 57*, 892–894.

Bert, J. M., Stark, J. G., Maschka, K., et al. (1991). The effect of cold therapy on morbidity subsequent to arthroscopic lateral retinacular release. *Orthopedic Reviews, 20*, 755–758.

Daniel, D. M., Stone, M. L., & Arendt, D. L. (1994). The effect of cold therapy on pain, swelling, and range of motion after anterior cruciate ligament reconstructive surgery. *Arthroscopy, 10*, 530–533.

Denys, E. H. (1991). AAEM minimonograph #14: The influence of temperature in clinical neurophysiology. *Muscle & Nerve, 14*, 795–811.

Draper, D. O., Knight, K., Fujiwara, T., et al. (1999). Temperature change in human muscle during and after pulsed short-wave diathermy. *The Journal of Orthopaedic and Sports Physical Therapy, 29*, 13–18. discussion 19-22.

Ho, S. S. W., Illgen, R. L., Meyer, R. W., et al. (1995). Comparison of various icing times in decreasing bone metabolism and blood flow in the knee. *The American Journal of Sports Medicine, 23*, 74–76.

Keast, M. L., & Adamo, K. B. (2000). The Finnish sauna bath and its use in patients with cardiovascular disease. *Journal of Cardiopulmonary Rehabilitation, 20*, 225–230.

Lindsay, D. M., Dearness, J., & McGinley, C. C. (1995). Electrotherapy usage trends in private physiotherapy practice in Alberta. *Physiotherapy Canada, 47*, 30–34.

Nutt, J. G., Anderson, V. C., Peacock, J. H., et al. (2001). DBS and diathermy interaction induces severe CNS damage. *Neurology, 56*, 1384–1386.

Oosterveld, F. G., & Rasker, J. J. (1994). Effects of local heat and cold treatment of surface and articular temperature of arthritic knees. *Arthritis and Rheumatism, 31*, 1578–1582.

Pinho, M., Correa, J. C. O., Furtado, A., et al. (1993). Do hot baths promote anal sphincter relaxation? *Diseases of the Colon and Rectum, 36*, 273–274.

Therapeutic Relationships

▶ Chronic Gynecological Pain, Doctor-Patient Interaction

Therapeutic Ultrasound

▶ Modalities

Therapy of Pain, Hypnosis

Robert G. Large
The Auckland Regional Pain Service, Auckland Hospital, Auckland, New Zealand

Synonyms

Hypnotherapy; Hypnotism; Mesmerism

Definition

Hypnosis refers to the "state" of consciousness associated with the phenomenon in question.

Hypnotism refers to the science and art of inducing and utilizing this phenomenon. This term is now less used in current research literature.

Mesmerism is a term associated with the controversial work of Franz Anton Mesmer (1734–1815) and replaced his concept of "animal magnetism." Modern accounts occasionally use the term to describe the nonverbal "mesmeric passes" used in some contexts.

Hypnotherapy is a term used to describe various forms of psychotherapy utilizing hypnosis as the major ingredient.

Hypnosis has been notoriously difficult to define. The British Medical Association (1955) introduced the following operational definition:

Hypnosis is a temporary condition of altered perception in the subject which may be induced by another person and in which a variety of phenomena may appear spontaneously or in response to verbal or other stimuli. These phenomena include alterations in consciousness and memory, increased susceptibility to suggestion, and the production in the subject of responses and ideas unfamiliar to him in his normal state of mind. Further phenomena such as anaesthesia, paralysis and the rigidity of muscles, and vasomotor changes can be produced and removed in the hypnotic state.

Most investigators emphasize one or more of four characteristics: expectations and the hypnotist-subject interaction, ▶ suggestibility, a cognitive dimension related to relaxation and/ or ▶ imagery, and ▶ dissociation (Evans 2001).

Characteristics

Of the wide range of phenomena associated with hypnosis, hypnotic analgesia is obviously the most useful in treating pain. Clinicians have typically used one or more of the following approaches: direct suggestion of pain reduction or insensitivity, suggestions aimed at altering the experience of pain, or suggestions directing attention away from pain and its source.

Social role theorists have proposed that hypnotic analgesia is simply a consequence of compliance with the ▶ demand characteristics of the experimental or clinical situation. In other words, subjects respond in the way they expect the hypnotist wishes them to respond. However, hypnotic analgesia shows a moderately strong correlation of 0.50 with measured hypnotic susceptibility, and there is a marked difference in the subjective experience of subjects simulating hypnosis compared with authentic hypnosis (Hilgard and Hilgard 1975). Many studies have shown that pain reduction in response to suggestion can occur without any apparent hypnotic induction. Nevertheless, accumulating evidence suggests that a hypnotic induction at least facilitates more profound analgesia. Price (1999) proposes that the sense of ease, absorption of attention, and lack of monitoring and censoring that are characteristic of hypnosis lays the foundation for an increased responsiveness to suggestion. Modern brain imaging techniques show differences in brain activity between normal waking and hypnotic states. Furthermore, specific analgesia suggestions are accompanied by specific brain changes; e.g., suggestions to enhance or reduce pain unpleasantness, with no change in pain sensation, are accompanied by changes in the anterior cingulate cortex and not in the primary sensory cortex. Similarly,

suggestions of sensory reduction produce parallel changes in subjective ratings and in activity in the primary somatosensory cortex. There is now some evidence that hypnotic analgesia also involves descending brain-to-spinal-cord inhibitory mechanisms and can inhibit spinal nociceptive reflexes. Our emerging understanding of the nature of hypnotic analgesia is that it relates to a wide range of cognitive variables, such as placebo/nocebo, attention, distraction, and emotional tone which modulate pain through changes in neural activity in many brain structures involved in nociception and pain (Large et al. 2003).

Treatment for Acute Pain

Clinicians over many years have reported remarkable success in using hypnosis to manage acute pain. Mesmer's explorations of "animal magnetism" seemed often to precipitate pain in his patients as an indication of the healing process (Bloch 1980). His followers, such as John Elliotson (1791–1868), reported surgical operations performed under "mesmeric sleep." James Esdaile (1808–1859) described 345 operations performed in India with mesmerism as the sole anesthetic. Modern day accounts of major surgery performed under hypnotic analgesia continue to be collected (Hilgard and Hilgard 1975). Rapid induction techniques have been developed for use in managing the acute pain of trauma and the imposed pain of procedures (Barber 1982).

Despite its long history and the experience of many clinicians and patients, good outcome research has been somewhat sparse. A recent meta-review concluded that hypnosis had been shown to be effective in controlling the pain of procedures in children with cancer. It is also effective in reducing the acute pain of burns and childbirth (Hawkins 2001).

Treatment for Chronic Pain

The persistent and relapsing nature of chronic pain presents a challenge to the therapist to develop strategies that will endure beyond the laboratory or consultation room. Hypnotic analgesia in the therapy session may offer some respite from pain but not a long-term cure. Clinicians have therefore turned to developing self-management techniques and the teaching of self-hypnosis (Eimer 2002). A typical clinical approach is to work with the patient in exploring hypnotic responsiveness, emphasizing what is possible rather than what is not possible. Techniques are explored and developed in collaboration between therapist and patient. These may range from simple ► relaxation to ► "special place" imagery, to inducing hypnotic analgesia in a hand (► Glove Anesthesia) and transferring this to the site of pain, and to using ► posthypnotic suggestion with or without specific cues. The patient's own sense of control is enhanced and encouraged.

A review of clinical trials of hypnosis for chronic pain, where there has been some attempt at systemization and control, suggests that hypnosis is effective in a variety of pain conditions. It has not been shown to be consistently superior to relaxation or other psychological interventions, however.

Hypnosis proved superior to propranolol in children with migraine, but few other comparisons with drug treatments have been made. In studies with adults, hypnosis and ► autogenic training appear to be equally efficacious for migraine.

Studies in irritable bowel syndrome have shown not only a reduction in pain and distension but also changes in rectal sensitivity assessed by balloon manometry. This finding hints at some basic psychophysiological change induced by hypnosis. However, a similar effect on tender points in fibromyalgia was looked for but not found. Work adherence was also improved in a hypnosis group treated for irritable bowel syndrome, suggesting an important shift in social and occupational functioning. Large and James (1988) found that patients trained in self-hypnosis changed their ► self-constructs as they gained mastery over their pain. From viewing themselves as physically ill people, they moved closer to their construction of their ideal self. This and other studies suggest

that one of the major gains that can be made through hypnosis is the enhancement of the patient's sense of control and ► self-efficacy. Price (1999) points out that unlike placebo suggestion, which implies an external authoritative agent, hypnotic suggestion refers to a more innate, internal, and self-directed capacity to alter experience.

Overall, the outcome research on hypnosis is promising, but there is a difficulty in relating current clinical approaches to the results of randomized controlled trials. Rigorous experimental design favors standardized approaches to hypnosis, with the use of scripts and prepared tape recordings. In contrast, many clinicians use sophisticated and individualized techniques that are difficult to replicate across subjects (Large et al. 2003). For example, some anecdotal reports have described using hypnosis as a means of returning to a presumed crucial traumatic event, so-called ► age regression. These reports suggest that resolution of the trauma can lead to a dramatic resolution of a chronic pain problem (e.g., Gainer 1992). It is difficult to evaluate the importance of such reports in the absence of reports of failures, which are seldom, if ever, published.

As noted, ► hypnotizability is correlated with hypnotic analgesia in laboratory studies. A number of studies using hypnosis in chronic pain management have found better results in higher hypnotizable subjects. However, it would seem that high hypnotizable subjects are also more likely to respond to relaxation techniques in general. This suggests that there are some commonalities in the various forms of psychological interventions that utilize relaxation. It is possible that highly hypnotizable subjects begin to access the hypnotic state in response to relaxation training in general. Hilgard's research in the laboratory has suggested that the simple induction of hypnosis without analgesic suggestions does not induce hypnotic analgesia. Specific analgesic suggestions are required. Many relaxation scripts, however, do include suggestions for comfort and pain reduction so that there is considerable crossover between strategies labeled "hypnosis" and those not labeled "hypnosis"!

Conclusion

Hypnosis has waxed and waned in popularity since the days of Mesmer. Systematic research in cognitive psychology and physiological psychology has improved our understanding of the nature of hypnotic responding and hypnotic analgesia. The introduction of modern neuroimaging techniques has begun to validate the construct of hypnosis as a phenomenon that has both subjective and objective reality. Clinical research continues to encourage the development of effective strategies utilizing hypnosis. Hypnosis continues to pose important questions in our understanding of pain, consciousness, and cognitive influences on brain processes, as well as challenging us to refine therapeutic strategies utilizing the full potential of hypnotic analgesia.

References

Barber, J. (1982). Managing acute pain. In J. Barber & C. Adrian (Eds.), *Psychological approaches to the management of pain* (pp. 168–185). New York: Brunner/Mazel.

Bloch, G. (1980). *Mesmerism. A translation of the original scientific and medical writings of F.A. Mesmer.* Los Altos: William Kaufmann.

British Medical Association Report. (1955). Medical use of hypnotism. *BMJ, 1*(Supplement), 190.

Eimer, B. N. (2002). *Hypnotize yourself out of pain now!* Oakland: New Harbinger Publications.

Evans, F. J. (2001). Hypnosis in chronic pain management. In G. D. Burrows, R. O. Stanley, & P. B. Bloom (Eds.), *International handbook of clinical hypnosis* (pp. 247–260). Chichester: Wiley.

Gainer, M. J. (1992). Hypnotherapy for reflex sympathetic dystrophy. *American Journal of Clinical Hypnosis, 34,* 227–232.

Hawkins, R. M. F. (2001). A systemic meta-review of hypnosis as an empirically supported treatment for pain. *Pain Review, 8,* 47–73.

Hilgard, E. R., & Hilgard, J. R. (1975). *Hypnosis in the relief of pain.* Los Altos: William Kaufmann.

Large, R. G., & James, F. R. (1988). Personalised evaluation of self-hypnosis as a treatment of chronic pain: A repertory grid analysis. *Pain, 35,* 155–169.

Large, R. G., Price, D. D., Hawkins, R. (2003). Hypnotic analgesia and its applications in pain management.

In J. O. Dostrovsky, D. B. Carr, M. Koltzenburg (Eds.) *Proceedings of the 10th world congress on pain.* Progress in pain research and management, Vol. 24. IASP, Seattle, pp. 839–851.

Price, D. D. (1999). Mechanisms of hypnotic analgesia. In D. D. Price (Ed.), *Psychological mechanisms of pain and analgesia* (Progress in pain research and management, Vol. 15, pp. 183–204). Seattle: IASP.

Thermal Allodynia

Definition

See also "▶ Allodynia." A problematic term: The underlying hypothesis is that this thermal pain is induced by another class of sensors than thermal pain in healthy tissue, since "allo-" means "other." However, it has been never proven that, e.g., heat hyperalgesia is induced by another class of afferent fibers than sensitized nociceptors.

Cross-References

▶ Cognitive-Behavior Treatment of Pain
▶ Neuropathic Pain Model, Tail Nerve Transection Model
▶ Spinal Cord Injury Pain Model, Contusion Injury Model

Thermal Effects of Ultrasound

Definition

The thermal effects on the target tissue result in an increased local metabolism, circulation, extensibility of connective tissue, tissue regeneration, and bone growth.

Cross-References

▶ Ultrasound Therapy of Pain from the Musculoskeletal System

Thermal Grill Illusion of Pain

Jonathan O. Dostrovsky
Department of Physiology, University of
Toronto, Toronto, ON, Canada

Definition

The painful and/or unpleasant sensation frequently induced by the simultaneous application to the skin of spatially interlaced innocuous warm and cool stimuli.

Introduction

The application to the skin of an alternating pattern of warm and cool elements, usually metallic bars, evokes in most people an unpleasant and burning thermal sensation. Since the individual bars are set at innocuous temperatures, the resulting paradoxical unpleasant sensation is generally termed the thermal grill illusion (TGI) of pain. Although the detailed mechanisms underlying this illusion are not fully understood, it is believed to result from central integration between the thermoreceptive and nociceptive pathways. There is increasing interest in utilizing the thermal grill in studies of nociception and in patients with various painful conditions in order to gain further insights into central pain modulatory mechanisms and the pathophysiology underlying some types of chronic pain.

Characteristics

The thermal grill illusion was first described by Thundberg in 1876. In 1994, Craig and Bushnell published the first modern-day study on the TGI (Craig and Bushnell 1994). They provided quantitative data from psychophysical studies documenting the painful qualities of the TGI and also presented data from animal studies on the effects of TG stimulation on responses of lamina I thermoreceptive cold and nociceptive spinal cord dorsal horn neurons. They proposed that the illusion resulted from altered central integration of ascending pain and temperature sensory channels and more specifically from a reduction in cold inhibition of central pain processing. In a subsequent PET study, they demonstrated that stimulation of the skin with the thermal grill but not with the component warm and cool stimuli produced activation in the cingulate cortex similar to that evoked by noxious thermal stimuli (Craig et al. 1996). Since their pioneering studies, there have been in recent years several other studies by other investigators which have confirmed and extended the original psychophysical findings (e.g., Leung et al. 2005; Bouhassira et al. 2005).

Studies by Bouhassira and colleagues (Kern et al. 2008a, b) have shown that the TGI can be modulated pharmacologically. They demonstrated that both the NMDA glutamate receptor antagonist ketamine and the analgesic morphine reduced the TGI and the latter effect could be reversed by the opioid antagonist naloxone. It has been suggested that some types of neuropathic pain (e.g., central post stroke pain) may be caused at least in part by a disruption of the central integration of the thermoreceptive and pain pathways (Craig 1998) and in such patients the TGI would be absent. Thus, the thermal grill may be a useful diagnostic tool for identifying the underlying mechanism of neuropathic pain (Craig 2008). This hypothesis was examined in one multiple sclerosis patient with central pain (Morin et al. 2002), and indeed, this patient's thermal grill stimulation was not painful. Further clinical studies utilizing a thermal grill stimulator in various types of painful disorders are necessary to determine whether the TGI is a useful measure for improving our understanding of underlying pathophysiological mechanisms, diagnosis, and treatment.

References

Bouhassira, D., Kern, D., Rouaud, J., Pelle-Lancien, E., & Morain, F. (2005). Investigation of the paradoxical painful sensation ('illusion of pain') produced by a thermal grill. *Pain, 114*, 160–167.
Craig, A.D. (1998). A new version of the thalamic disinhibition hypothesis of central pain. *Pain Forum, 7*, 1–14.

Craig, A. D. (2008). Can the basis for central neuropathic pain be identified by using a thermal grill? *Pain, 135*, 215–216.

Craig, A. D., & Bushnell, M. C. (1994). The thermal grill illusion: Unmasking the burn of cold pain. *Science, 265*, 252–255.

Craig, A. D., Reiman, E. M., Evans, A., & Bushnell, M. C. (1996). Functional imaging of an illusion of pain. *Nature, 384*, 258–260.

Kern, D., Pelle-Lancien, E., Luce, V., & Bouhassira, D. (2008a). Pharmacological dissection of the paradoxical pain induced by a thermal grill. *Pain, 135*, 291–299.

Kern, D., Plantevin, F., & Bouhassira, D. (2008b). Effects of morphine on the experimental illusion of pain produced by a thermal grill. *Pain, 139*, 653–659.

Leung, A., Wallace, M. S., Schulteis, G., & Yaksh, T. L. (2005). Qualitative and quantitative characterization of the thermal grill. *Pain, 116*, 26–32.

Morin, C., Bushnell, M. C., Luskin, M. B., & Craig, A. D. (2002). Disruption of thermal perception in a multiple sclerosis patient with central pain. *The Clinical Journal of Pain, 18*, 191–195.

Thermal Hyperalgesia

Definition

Thermal hyperalgesia is a condition of altered perception of temperature. Describes heightened sensitivity to noxious heat or cold. This includes also the perception of stimuli as painful which are normally perceived as just "warm" or "cool."

Cross-References

▶ Neuropathic Pain Model, Chronic Constriction Injury
▶ Opioids in the Spinal Cord and Modulation of Ascending Pathways (N. gracilis)
▶ TRPV1, Regulation by Nerve Growth Factor
▶ TRPV1, Regulation by Protons

Thermal Hyperalgesia Test

▶ Thermal Nociception Test

Thermal Neuroablation

▶ Radiofrequency Neurotomy, Electrophysiological Principles

Thermal Nociception Test

Kenneth M. Hargreaves[1] and
Christopher M. Flores[2]
[1]Department of Endodontics, University of Texas Health Science Center at San Antonio, San Antonio, USA
[2]Drug Discovery, Johnson and Johnson Pharmaceutical Research and Development, Spring House, PA, USA

Synonyms

Hargreaves test; Plantar test; Radiant heat test; Thermal hyperalgesia test

Definition

The thermal hyperalgesia test permits highly reproducible evaluation of paw withdrawal thresholds of animals to a beam of radiant heat applied to the plantar surface of the paw. The endpoint is detected automatically, thereby removing an important potential source of observer bias. The test can be used to measure normal nociceptive thresholds and to quantify hyperalgesia and allodynia in models of inflammatory or neuropathic pain (Hargreaves et al. 1988).

Characteristics

Overview
Our understanding of the chemical and anatomical substrates that underlie hyperalgesia and allodynia has benefited tremendously from numerous important animal models.

Consequently, we have a more profound appreciation for the processes that contribute to the development and maintenance of certain behavioral responses to noxious stimulation and tissue injury. Nonetheless, such models are only as valuable as are the methods employed to detect and quantitate these behavioral responses. Thus, validity, reliability, sensitivity, assay reproducibility, and flexibility are all critical determinants of any paradigm by which to assess nociception and its physiological sequelae.

Attempts to assess pain quantitatively in humans and animals dates back to the first half of this century, and there are numerous reports of "pain" measurements in animals responding to thermal, chemical, mechanical, or electrical stimuli. However, these methods all suffered from a variety of drawbacks including, but not limited to, a lack of correlation between human clinical and experimental animal studies, a lack of assay sensitivity to pharmacological manipulation, especially with regard to ▶ NSAIDs, Survey (NSAIDS), the inability to perform within-subject controls, and a virtually uniform reliance on noncomplex, reflex behaviors (often involving the uniquely scaly, non-glabrous skin of the rat tail) that are difficult if not impossible to equate with the human sensation of pain.

The plantar test avoids many of these limitations since it measures complex, nociceptive behaviors following thermal stimulation of cutaneous tissues, using a method allowing for not only the determination of thermal nociceptive thresholds but also the quantitation of ▶ primary hyperalgesia, ▶ secondary hyperalgesia, and ▶ allodynia. The test is performed in awake, freely moving, and unrestrained animals, thereby avoiding the potential generation of stress responses. In addition, the device is flexible enough to permit the independent testing of selected, individual receptive fields (e.g., independent assessment of both hind paws), permitting the use of within-subject controls, sophisticated enough to yield a quantifiable, automated endpoint, thereby removing most of the potentially confounding observer interaction and sensitive enough to detect relatively small changes in response to relevant perturbations,

such as the induction of inflammation and hyperalgesia and/or the administration of drugs. In 1988, after 2 years of design, development, and validation, the plantar test method paper was published. According to an ISI Web of Science™ analysis of papers published through May 2005, this paper has been cited in more than 1,000 publications, making it one of the most heavily cited articles ever published in the journal *Pain* (Terajima and Aneman 2003).

Procedure

The device (for detailed description, see Hargreaves et al. 1988) essentially consists of a raised glass floor beneath which is placed a movable case containing a radiant heat source and photoelectric cell that are aimed through an aperture at the top of the case. The method involves placing an animal on top of the glass floor and enclosing it in a small, clear plastic cage. Depending on the size of the floor and the enclosures employed, several animals may be positioned simultaneously in individual cages. Following an acclimation period, at which point the animal has come to rest, the aperture is positioned directly beneath the plantar surface of the animal's hind paw. The trial is commenced by a switch that simultaneously activates the heat source and an electronic timer. Upon withdrawal of the paw, the resulting interruption of reflected light is detected by the photoelectric cell, thereby signaling the lamp and timer to turn off and a tone to be emitted. The electronic clock circuit is wired to a microcomputer and LED readout. The paw withdrawal latency (PWL), measured as the time to the nearest 0.1 s between points at which the switch is turned on and the beam of light is interrupted, is taken as a measure of the thermal nociceptive threshold of the animal and is displayed on the readout. The test may then be repeated on the same or the other paw. Owing to the fact that the procedure is carried out on unrestrained, freely moving animals, this method also allows for the concurrent measurement of complex, organized behaviors in addition to the PWL, including the duration and rapidity of withdrawal as well as various locomotor activities such as licking, each a separate measure of

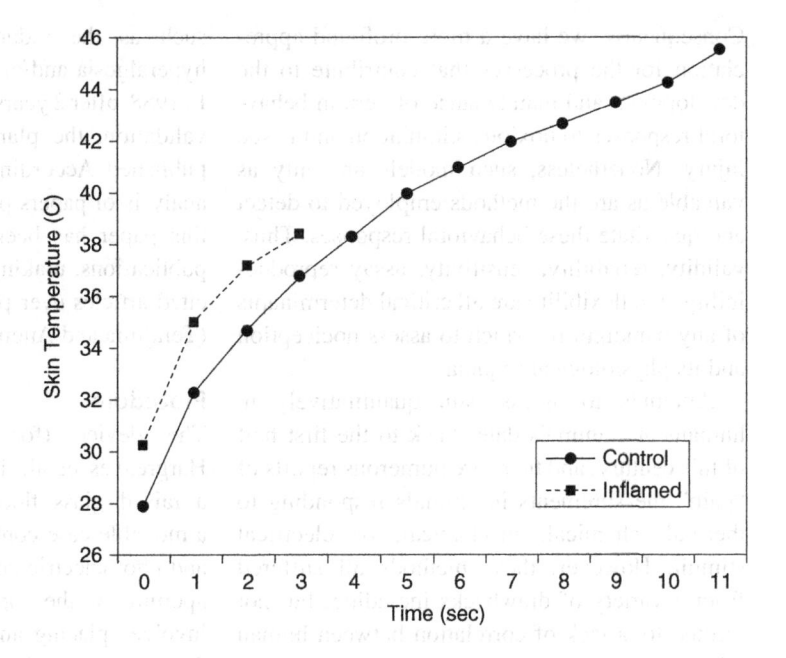

Thermal Nociception Test, Fig. 1 Effects of radiant heat on the cutaneous temperature of rat hind paws (Reproduced with permission from: Hargreaves et al. 1988, Elsevier©)

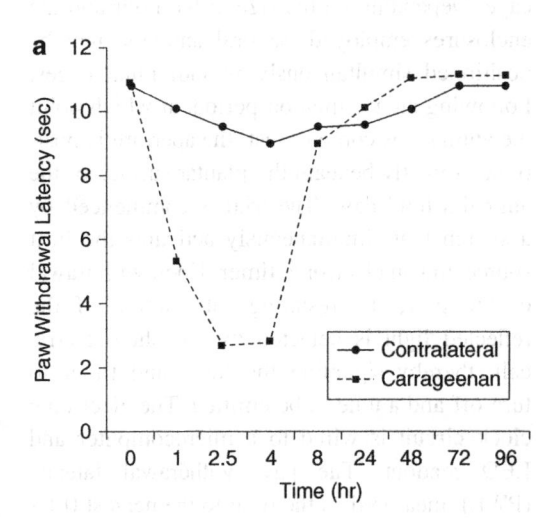

Thermal Nociception Test, Fig. 2 (a) Effects of carrageenan on the latency for hind paw withdrawal following injection of carrageenan into one hind paw with measurement of both the ipsilateral injected hind paw and the contralateral control hind paw. (b) Effects of carrageenan on the poststimulus duration of hind paw elevation above the glass floor in the same animals as (a) (Reproduced with permission from: Hargreaves et al. 1988, Elsevier©)

hyperalgesia (defined as an exaggerated response to a noxious stimulus). To measure duration of paw withdrawal, a stopwatch is started when the device emits the tone and is stopped when the paw is returned to the glass floor. Similarly, the duration of licking or other complex behavior can be measured. Changes in latency correspond to thermal allodynia, since latency is directly correlated with cutaneous temperature (Fig. 1). In contrast, changes in duration of hind paw elevation following the stimulus may reflect aspects of hyperalgesia (Fig. 2).

Advantages and Disadvantages

Many of the advantages and limitations of this radiant heat method compared with previously existing methods have been summarized (Table 1). Perhaps of greatest importance is that this paradigm conforms to the guidelines for the ethical and responsible use of animals as experimental subjects established by the International Association for the Study of Pain and the Society for Neuroscience. In this regard, it is of profound importance that the noxious stimulus applied during testing of the animals is readily escapable. Thus, the animal itself terminates the experimental session after a brief stimulus and before any thermal injury can occur, even after repeated application of the test. In addition, the capacity to perform the test on an awake, freely moving animal that appears ostensibly calm and comfortable during the acclimation and testing period markedly reduces the concern that any stress-related neuroendocrine circuits will be engaged. Another advantage to the freedom of movement afforded by this method is that a number of more complex behaviors related to an induced hyperalgesic state and occurring secondary to or as a result of the testing can be measured. The ability to quantitate such an array of nociceptive behaviors gives rise to a "hyperalgesia profile" for a given animal that may be compared to other individuals under a variety of experimental conditions. This setup also readily permits the administration of test articles or other perturbations, on a repeated basis if necessary, in between PWL measurements.

As already alluded to, this method, in addition to measuring thermal nociceptive thresholds in normal animals/hind paws, may also be used to quantitate hyperalgesia, operationally defined as an increased responsiveness to noxious stimulation. It is also theoretically possible to alter the device to deliver a subthreshold stimulus as a means of measuring allodynia. In the context of this discussion then, it may be appreciated that this method is amenable to measuring hyperalgesia that develops due to inflammation or nerve injury induced by a variety of perturbations. Thus, the method has been applied to several models, including the intraplantar

Thermal Nociception Test, Table 1 Methodological papers evaluating the radiant heat device

Modification of parameter evaluated	Authors
Methods paper: thermal hyperalgesia in a model of inflammation	Hargreaves et al. (1988)
Thermal hyperalgesia in a model of peripheral neuropathy	Bennett and Xie (1988)
Glass surface acts as a heat sink	Hirata et al. (1990)
To avoid heat sink effect, only test a paw when it is in contact with the glass surface	Bennett and Hargreaves (1990)
A nylon mesh over the glass surface reduces the heat sink effect	Murphy et al. (1991)
Animals restrained during thermal testing	Kerns et al. (1991)
Testing thermal thresholds in normal animals	Yamamoto and Yaksh (1991)
Heated glass surface using feedback control to avoid a heat sink effect	Yamamoto and Yaksh (1991)
Correlation of paw temperature and escape latencies are consistent with measuring nociception	Yeomans et al. (1992)
Thermal hyperalgesia in a model of peripheral neuropathy	Kim and Chung (1992)
Thermal hyperalgesia following i.t. NMDA	Malmberg and Yaksh (1995)
Heated glass floor, joystick control of radiant heat source, illustration of a circuit diagram	Galbraith et al. (1993)
Radiant heat device more sensitive than tail flick for detecting thermal hyperalgesia during tolerance	Mao et al. (1994)
Frogs as test subject	Willenbring and Stevens (1996)
Radiant heat applied to the dorsum of the hindpaw	
Gerbils as test subject	Rupniak et al. (1995)
Evaluated two intensities of radiant heat in the same subjects	O'Connor and Abram (1995)

injection of carrageenan (Hargreaves et al. 1988), complete Freund's adjuvant (Iadarola et al. 1988), yeast (Iadarola et al. 1988), nerve growth factor (Andreev et al. 1995), or the induction of a peripheral neuropathy by nerve constriction (Bennett and Xie 1988; Kim and Chung 1992; Bennett 1999).

One of the most important advantages of this method compared with previous methods is the ability to have within-subject controls. This may be represented statistically as left versus right or baseline versus post-perturbation hind paw difference scores. The utility of this transformation of the data is highlighted by the demonstration of Bennett and Xie that 140 normal rats exhibited difference scores that fit a Gaussian distribution with a mean and standard deviation of 0.04 ± 0.66 s (i.e., a right-left difference score of approximately zero) (Bennett and Xie 1988). This feature of the paradigm has been exploited in myriad, within-animal studies evaluating the mechanisms of hyperalgesia and inflammation and the effects of drugs or other therapeutic interventions used to counteract these processes. Thus, one could design a study to determine whether a given inflammatory agent results in a localized versus systemic hyperalgesia. In addition, one could distinguish between peripheral and central mechanisms of action of antinociceptive agents (Jackson et al. 1995), in that PWL measurements can be made in a given paw following injection of a potentially antinociceptive test substance directly into that paw or alternatively into the contralateral paw or some other remote location, usually at a dose that is not sufficient to reach pharmacologically relevant systemic concentrations. Another application of this method is the ability to correlate somatotopically localized biochemical, cellular, and molecular alterations with behavior in the affected limb. Thus, several interesting studies were performed in which carrageenan- or CFA-induced hyperalgesia was correlated with somatotopically appropriate increases in opioid peptide gene expression in the dorsal spinal cord on the side ipsilateral to the inflammation (Iadarola et al. 1988). In addition, in vivo microdialysis has been used to measure and correlate the time course of the local production and/or release of inflammatory mediators with nociceptive behavior in parallel groups of animals (Hargreaves and Costello 1990).

Ultimately, the true measure of any method rests on its sensitivity, in other words, the ability to detect changes in the dependent measurement when in fact they occur. In this respect, it is significant that in the periphery, the sensitivity to detect hyperalgesia is greater for thermal versus mechanical stimuli in several tests (Handwerker et al. 1987). Consistent with this concept, the radiant heat device was shown to be more sensitive than the Randall and Selitto mechanical test in detecting hyperalgesia in response to a subcutaneous injection of carrageenan, which was manifested as an increase in signal-to-noise ratios, as indicated by improved ANOVA F ratios and increased carrageenan dose-response curve slopes (Hargreaves et al. 1988). Moreover, the radiant heat device has been used to evaluate several hundred compounds, representing at least 20 distinct classes for their ability to alter nociception and nociceptive behaviors, highlighting its utility as a tool not only for investigating the neurochemistry of pain but also for the development of novel analgesic therapeutics.

The most obvious limitation of the current method is that it does not measure mechanical nociceptive thresholds, allodynia, or hyperalgesia. Moreover, the extent to which thermal hypersensitivity, as measured by this method, accurately predicts the efficacy of clinically useful drugs (based on type 1 and type 2 errors) or reliably mimics one or more than one clinically relevant pain condition/symptom remains to be determined (although the test reliably demonstrates the efficacy of a variety of currently marketed NSAID and opioid analgesics with reasonably appropriate rank order potency). Notwithstanding these caveats, the combination of the thermal method with mechanical tests, such as the use of von Frey filaments, offers a powerful approach to the study of a variety of nociception-modifying perturbations as well as a means for discriminating potential sensory modality differences between them. The other major disadvantage of the radiant heat method is that the glass surface may act as a heat sink, leading to artifactually short paw withdrawal latencies and a subsequent increase in assay variability (Bennett and Hargreaves 1990; Hirata et al. 1990). This factor will vary with the specific conductive properties of the floor, the ambient temperature in the experimental procedure room, and the pressure exerted by the rat on the paw being tested, in which latter will be affected under conditions of induced inflammation or neuropathy (Bennett and Hargreaves 1990; Murphy et al. 1991). However, several suggestions have been made to counteract this effect, the most important of these being that the skin being irradiated should be in contact with the glass floor (Table 1). Modifications to the device itself include placing a nylon mesh screen between the rat and the glass (Murphy et al. 1991) or heating the floor to maintain it at a constant temperature, either by implanting a heating element within the glass or applying warm air to its underside (Galbraith et al. 1993). The minor inconveniences created by the rats urinating or defecating on the glass floor are easily remedied with a spray bottle of cleanser and a roll of paper towels.

In conclusion, the radiant heat test described here offers a simple, highly reproducible, and useful approach to the study of thermal nociception in naïve and injured/inflamed rodents. Accordingly, it may be considered one of several powerful behavioral tools available to pain scientists in the investigation of nociceptive mechanisms and in the search for better and safer analgesic drugs. It is hoped that further improvements in this method, including those mentioned herein, and its use in combination with complementary behavioral assays will lead to a better understanding of the neurophysiological underpinnings of pain in humans and ultimately better treatment options.

References

Andreev, N. Y., Dimitrieva, N., Koltzenburg, M., et al. (1995). Peripheral administration of nerve growth-factor in the adult-rat produces a thermal hyperalgesia that requires the presence of sympathetic postganglionic neurons. *Pain, 63*, 109–115.

Bennett, G. J. (1999). Does a neuroimmune interaction contribute to the genesis of painful peripheral neuropathies? *Proceedings of the National Academy of Sciences of the United States of America, 96*, 7737–7738.

Bennett, G. J., & Hargreaves, K. M. (1990). A model of peripheral mononeuropathy in the rat – Reply. *Pain, 42*, 255.

Bennett, G. J., & Xie, Y.-K. (1988). A peripheral mononeuropathy in rat that produces disorders of pain sensation like those seen in man. *Pain, 33*, 87–107.

Galbraith, J. A., Mrosko, B. J., & Myers, R. R. (1993). A system to measure thermal nociception. *Journal of Neuroscience Methods, 49*, 63–68.

Handwerker, H. O., Anton, F., Kocher, L., et al. (1987). Nociceptor functions in intact skin and in neurogenic or non-neurogenic inflammation. *Acta Physiologica Hungarica, 69*, 333–342.

Hargreaves, K. M., & Costello, A. (1990). Glucocorticoids suppress levels of immunoreactive bradykinin in inflamed tissue as evaluated by microdialysis probes. *Clinical Pharmacology and Therapeutics, 48*, 168–178.

Hargreaves, K. M., Dubner, R., Brown, F., et al. (1988). A new and sensitive method for measuring thermal nociception in cutaneous hyperalgesia. *Pain, 32*, 77–88.

Hirata, H., Pataky, A., Kajander, K., et al. (1990). A model of peripheral mononeuropathy in the rat. *Pain, 42*, 253–254.

Iadarola, M. J., Brady, L. S., Draisci, G., et al. (1988). Enhancement of dynorphin gene-expression in spinal-cord following experimental inflammation – Stimulus specificity, behavioral parameters and opioid receptor-binding. *Pain, 35*, 313–326.

Jackson, D. L., Graff, C. B., Richardson, J. D., et al. (1995). Glutamate participates in the peripheral modulation of thermal hyperalgesia in rats. *European Journal of Pharmacology, 284*, 321–325.

Kerns, J. M., Fakhouri, A. J., Weinrib, H. P., & Freeman, J. A. (1991). Electrical stimulation of nerve regeneration in the rat: The early effects evaluated by a vibrating probe and electron microscopy. *Neuroscience, 40*, 93–107.

Kim, S. H., & Chung, J. M. (1992). An experimental-model for peripheral neuropathy produced by segmental spinal nerve ligation in the rat. *Pain, 50*, 355–363.

Malmberg, A. B., & Yaksh, T. L. (1995 March). The effect of morphine on formalin-evoked behaviour and spinal release of excitatory amino acids and prostaglandin E2 using microdialysis in conscious rats. *British Journal of Pharmacology, 114*(5), 1069–1075.

Mao, J., Price, D. D., & Mayer, D. J. (1994). Thermal hyperalgesia in association with the development of morphine tolerance in rats: Roles of excitatory amino acid receptors and protein kinase C. *Journal of Neuroscience, 14*, 2301–2312.

Murphy, L. G., Alexander, G. M., & Schwartzman, R. J. (1991). Improvement to the Hargreaves paw flick method. *Pain, 46*, 347.

O' Connor, T. C., & Abram, S. E. (1995). Inhibition of nociception-induced spinal sensitization by anesthetic agents. *Anesthesiology, 82,* 259–266.

Rupniak, N. M., Webb, J. K., Williams, A. R., Carlson, E., Boyce, S., & Hill, R. G. (1995). Antinociceptive activity of the tachykinin NK1 receptor antagonist, CP-99,994, in conscious gerbils. *British Journal of Pharmacology, 116,* 1937–1943.

Terajima, K., & Aneman, A. (2003). Citation classics in anaesthesia and pain journals: A literature review in the era of the internet. *Acta Anaesthesiologica Scandinavica, 47,* 655–663.

Willenbring, S., & Stevens, C. W. (1996). Thermal, mechanical and chemical peripheral sensation in amphibians: Opioid and adrenergic effects. *Life Sciences, 58*(2), 125–133.

Yamamoto, T., & Yaksh, T. L. (1991). Stereospecific effects of a nonpeptidic NK-1 selective antagonist, CP96345: Antinociception in the absence of motor dysfunction. *Life Sciences, 49,* 1955–1963.

Yeomans, F. E., Selzer, M. A., & Clarkin, J. F. (1992). *Treating the borderline patient: A contract-based approach.* New York, NY: Basic Books.

Thermal Receptors

Definition

Sensory receptors sensitive to thermal stimuli. The skin of vertebrates contains separate populations of receptors for warming and for cooling. Thermal receptors are also found in the interior of our body, e.g., intraperitoneally. There are also neurons in the central nervous system which are thermosensors.

Cross-References

► Hyperaesthesia, Assessment
► Hypoesthesia, Assessment

Thermal Stimulation (Skin, Muscle, Viscera)

► Pain in Humans, Thermal Stimulation (Skin, Muscle, Viscera), Laser, Peltier, Cold (Cold Pressure), Radiant, Contact

Thermal Stimulation (Stimuli)

Definition

Thermal stimulation (warm and cold stimuli) consists of transferring (adding or subtracting) calorific energy between the skin (or mucosa, muscle, viscera) and its surroundings.

Cross-References

► Causalgia: Assessment
► Dysesthesia, Assessment
► Pain in Humans, Thermal Stimulation (Skin, Muscle, Viscera), Laser, Peltier, Cold (Cold Pressure), Radiant, Contact

Thermal Therapy

► Spa Treatment

Thermal Transduction

► Polymodal Nociceptors, Heat Transduction

Thermocoagulation

Definition

Destroying tissue (e.g., nerves) by heating.

Cross-References

► Cancer Pain Management, Anesthesiologic Interventions, and Neural Blockade
► Radiofrequency Neurotomy, Electrophysiological Principles

Thermode

Definition

A thermode is a device used to apply controlled temperature to the skin of subjects or animals. One surface of the thermode is placed in contact with the skin. The temperature at that surface is controlled with the use of heating elements, Peltier elements, and/or circulating water. The surface temperature is measured with a thermocouple or thermistor, which is used in a feedback manner to regulate the thermode's surface temperature.

Cross-References

▶ Threshold Determination Protocols

Thermography

Definition

Devices which are able to detect infrared radiation. If these devices are scanners, we speak of "thermographic" apparatus. Thermography has been used to visualize the distribution of skin surface temperature changes.

Cross-References

▶ Postoperative Pain, Acute Presentation of Complex Regional Pain Syndrome

Thermoreception

▶ Lateral Thalamic Pain-Related Cells in Humans

Thermoreceptor

Definition

See "Thermal Receptors."

Cross-References

▶ Nociceptors, Cold Thermotransduction

Thermoregulation

Definition

Rodents regulate body temperature mainly by regulating blood flow in the tail by an on-off mechanism, suddenly changing the tail blood flow and, thus, skin temperature by up to 8–10°C.

Cross-References

▶ Tail-Flick Test

Theta Rhythm

Definition

Frequency domain of oscillatory hemispheric activity between 4 and 8 Hz. It has been associated with different functional brain states, e.g., somnolence, cognitive activations, altered states of consciousness like meditation, and, relevant here, dysfunctional brain states like neurogenic pain and tinnitus, abnormal movements, epilepsy, and neuropsychiatric disorders (see ▶ Thalamocortical Dysrhythmia).

Cross-References

▶ Thalamotomy for Human Pain Relief

Thoracic Epidural Analgesia

Definition

Pain relief obtained by drugs acting directly on the spinal cord, for example, morphine, local anesthetics, adrenaline (epinephrine), clonidine, and neostigmine.

Cross-References

▶ Postoperative Pain, Acute Pain Management Principles
▶ Postoperative Pain, Acute Pain Team

Thoracic Epidural Anesthesia

Definition

For thoracic epidural anesthesia the catheters are inserted depending on the surgical procedure, between level T4 and T12 by a median or paramedian approach and usually with the loss of resistance technique. Postoperatively, the patient has perfect analgesia and can be mobilized.

Cross-References

▶ Postoperative Pain, Thoracic and Cardiac Surgery

Thoracic Medial Branch Blocks and Intra-Articular Blocks

Paul Verrills
Metropolitan Spinal Clinic, Prahran, VIC, Australia

Synonyms

Intra-articular blocks and thoracic medial branch blocks

Definition

Thoracic medial branch blocks and intra-articular blocks are both diagnostic procedures designed to test if the patient's pain arises from a thoracic zygapophysial joint. They involve anesthetizing the joint or its nerve supply with injections of small volumes of local anesthetic.

Characteristics

The thoracic zygapophysial joints may be a cause of posterior thoracic spinal pain. Studies in normal volunteers (Dreyfuss et al. 1994a) and in patients (Fukui et al. 1997) have shown that noxious stimulation of these joints produces local and ▶ somatic referred pain in a segmental pattern along the posterior chest wall (Fig. 1).

There are no means clinically, or by medical imaging, by which pain from these joints can be diagnosed. The only means of determining if they are a source of pain is to anesthetize putatively symptomatic joints.

The thoracic zygapophysial joints can be anesthetized either by injecting local anesthetic into the joint cavity or by anesthetizing the medial branches of the dorsal rami that innervate them. Respectively, these procedures are called thoracic intra-articular zygapophysial joint blocks and thoracic medial branch blocks.

Intra-Articular Block

The thoracic zygapophysial joints are not directly evident on anteroposterior views of the thoracic spine, because their joint cavities are orientated in a coronal plane. Nevertheless, the location of each joint can be inferred. The joint lies opposite the intervertebral disk of the segment, between the pedicles of the two vertebrae that contribute to the joint.

Once the target joint has been identified, a puncture point on the skin is selected, about three-quarters of a segment caudal to the joint, on a caudal extension of the sagittal bisector of the pedicle below the target joint. A spinal needle is introduced at around 60° caudal to the perpendicular and is advanced towards the lower quarter

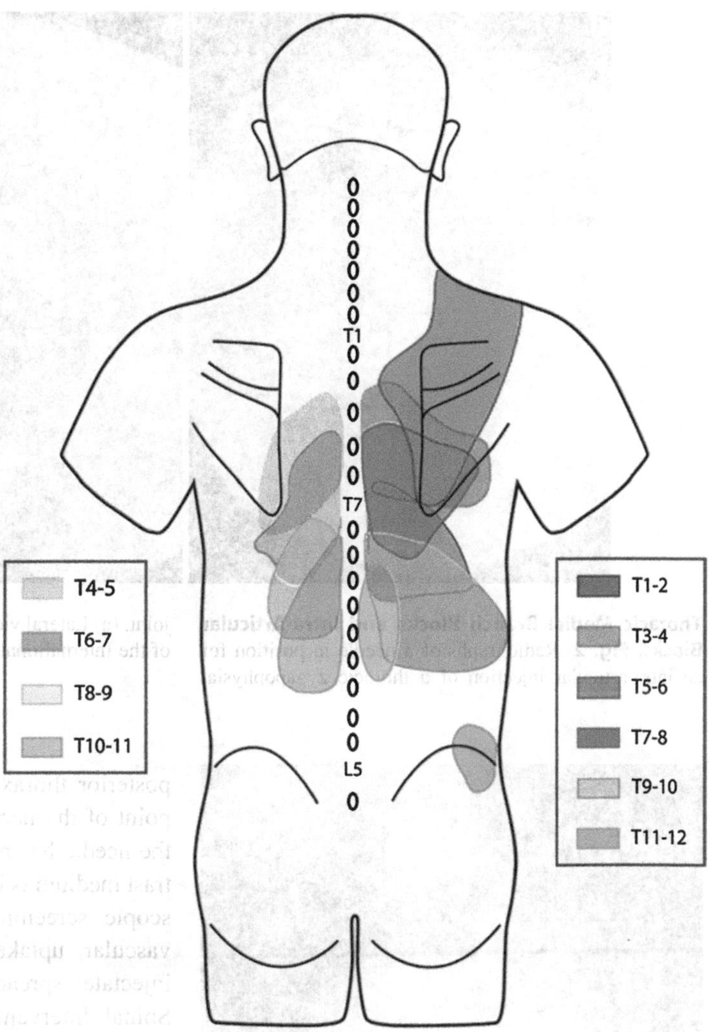

**Thoracic Medial Branch
Blocks and Intra-
Articular Blocks,**
Fig. 1 The distribution of
referred pain from the
thoracic zygapophysial
joints (Based on Dreyfuss
et al. (1994) and Fukui et al.
(1997))

of the silhouette of the pedicle below the joint
(Dreyfuss et al. 1994b; International Spinal Intervention Society 2004a). A contralateral oblique
view can then be used to carefully advance the
needle into the inferior margin of the joint
(Fig. 2a). Once the needle has entered the joint,
an anteroposterior view is used in order to check
that the needle has not strayed medially or laterally (Fig. 2b).

Intra-articular placement is verified by
injecting 0.1 ml of contrast medium, in order to
obtain an arthrogram (Dreyfuss et al. 1994b;
International Spinal Intervention Society 2004a)
(Fig. 3). Subsequently, the joint is anesthetized
by injecting 0.75 ml of local anesthetic.

Medial Branch Block

The thoracic medial branches cross the thoracic
transverse processes obliquely, passing from the
region of the superior lateral corner of the process to its inferomedial corner. At segmental
levels T1-4 and T9 and T10, the nerves lie on
the posterior surface of the transverse process,
slightly below and just medial to the superior
lateral corner of the transverse process. This
point constitutes the target point for medial
branch blocks at these levels (Chua and Bogduk
1995; International Spinal Intervention Society
2004b).

Under posteroanterior fluoroscopic screening,
a spinal needle is inserted through the skin of the

Thoracic Medial Branch Blocks and Intra-Articular Blocks, Fig. 2 Radiographs of a needle in position for an intra-articular injection of a thoracic zygapophysial joint. (a) Lateral view. (b) AP view (Reproduced courtesy of the International Spinal Intervention Society (2004a))

Thoracic Medial Branch Blocks and Intra-Articular Blocks, Fig. 3 A lateral radiograph of an arthrogram of a thoracic zygapophysial joint. The *arrows point* to contrast medium inside the joint (Reproduced courtesy of the International Spinal Intervention Society (2004a))

posterior thorax and passed towards the target point of the nerve to be blocked (Fig. 4). Once the needle has reached bone, a test-dose of contrast medium is injected under continuous fluoroscopic screening to ensure that there is no vascular uptake and to determine that the injectate spreads appropriately (International Spinal Intervention Society 2004b). To block the nerve, 0.3 ml of local anesthetic is injected. To anesthetize a given zygapophysial joint, both the nerves that innervate it are blocked.

At the T4–T8 levels, the medial branches do not run on bone. From the intertransverse space, they are suspended dorsal to the depth of the transverse process (Chua and Bogduk 1995). In order to block the nerve, a needle is delivered into the intertransverse space, slightly cranial to the tip of the transverse process (International Spinal Intervention Society 2004b). In this location, a small volume of contrast medium is injected, in order to confirm appropriate spread of injectate. Thereupon, local anesthetic can be injected to anesthetize the nerve.

The T11 and T12 medial branches assume a course similar to that of the lumbar

Thoracic Medial Branch Blocks and Intra-Articular Blocks, Fig. 4 An AP radiograph showing a needle in place for a thoracic medial branch block and an injection of contrast medium to show spread of injectate over the superior lateral corner of the transverse process. The inferior border of the transverse process (*TP*) is labeled (Reproduced courtesy of the International Spinal Intervention Society (2004a))

medial branches. The target points and technique for these nerves are like those for ▶ lumbar medial branch blocks.

Evaluation

The optimal means of reducing error and securing reliable diagnostic information is real time assessment. The response to the diagnostic block is evaluated immediately after the block and, for some time afterwards, at the clinic at which the block was performed, and by an independent observer using validated and objective instruments or tools (Dreyfuss et al. 1994b; International Spinal Intervention Society 2004b).

Visual analogue scores are recorded before the block, immediately afterwards, at 30 min, and then hourly. The patient is instructed to monitor the extent and duration of any relief that ensues. Further, if relief occurs, the patient should carefully attempt movements and activities that are usually restricted by pain to assess their response during the anesthetic phase.

If the response to the block is negative, then zygapophysial joint pain can effectively be ruled

out at the level tested. Adjacent levels may then be blocked based on clinical indication.

If the response to the block is positive, then a control block is undertaken. If the patient has a concordant response to controlled blocks, then the putative diagnosis of zygapophysial joint pain is confirmed.

Validity

Single diagnostic blocks of thoracic medial branches carry a false-positive rate of 58 % (Manchikanti et al. 2002). Control blocks are, therefore, mandatory (International Spinal Intervention Society 2004b; Manchikanti et al. 2002).

Although placebo-controlled blocks could be used, they require three injections and are not readily implemented in clinical practice. Comparative local anesthetic blocks, using lignocaine 2 %, as a short-acting agent, and bupivacaine 0.5 % as a long-acting agent, are a valid alternative (International Spinal Intervention Society 2004b; Manchikanti et al. 2002; Barnsley et al. 1993; Lord et al. 1995). A concordant response would be complete relief of pain for a shorter duration following the short-acting agent and a longer duration of relief following the long-acting agent (see ▶ Peripheral Nerve Blocks). Complete relief of pain on each of the two occasions, regardless of duration of relief, increases the positive yield of blocks but increases the risk of false-positive responses.

Similar controls have not been evaluated for intra-articular blocks. Their validity has not been established.

Application

Only one study has described the application of thoracic medial branch blocks in clinical practice (Manchikanti et al. 2002). It reported that 48 % of patients had their pain relieved with the medial branch blocks. The study thereby established that thoracic zygapophysial joint blocks have a considerable diagnostic utility in clinical practice. However, it did not indicate which segmental levels were most commonly the source of pain.

By analogy, with lumbar and cervical spine, one would expect the therapeutic utility of thoracic medial branch blocks would be to select patients for medial branch ▶ radiofrequency neurotomy. However, there are no publications to describe a valid technique or to establish the efficacy of such treatment in the thoracic spine. Therefore, the therapeutic utility of thoracic zygapophysial joint blocks is only potential.

Although intra-articular blocks constitute an alternative means of diagnosing thoracic zygapophysial joint pain, no studies have described any results from their use. Some operators administer intra-articular injections of corticosteroids as a treatment for thoracic zygapophysial joint pain, but no studies have reported the efficacy of this treatment.

Patient Selection

Thoracic zygapophysial joint blocks are not indicated in acute pain. They are relevant only for patients with persistent thoracic spinal pain, in whom a diagnosis is required. A prerequisite is that serious possible causes of pain such as infection, tumors, vascular disease, and metabolic disease have been excluded by careful and thorough history and examination, laboratory tests, and medical imaging if necessary. The working diagnosis should be that of thoracic spinal pain of unknown origin.

The minimum criterion should be that the patient has pain in a distribution that resembles that shown to emanate from thoracic zygapophysial joint in normal volunteers (Dreyfuss et al. 1994a; Fukui et al. 1997).

Contraindications

Absolute contraindications include bacterial infection, either systemic or localized in the region that blocks are to be performed, bleeding diatheses, or possible pregnancy. Relative contraindications include allergy to contrast media or local anesthetics.

References

Barnsley, L., Lord, S., & Bogduk, N. (1993). Comparative local anaesthetic blocks in the diagnosis of cervical zygapophysial joint pain. *Pain, 55*, 99–106.

Chua, W. H., & Bogduk, N. (1995). The surgical anatomy of thoracic facet denervation. *Acta Neurochirurgica, 136*, 140–144.

Dreyfuss, P., Tibiletti, C., & Dreyer, S. J. (1994a). Thoracic zygapophyseal joint pain patterns. A study in normal volunteers. *Spine, 19*, 807–811.

Dreyfuss, P., Tibiletti, C., Dreyer, S., & Sobel, J. (1994b). Thoracic zygapophyseal joint pain: A review and description of an intra-articular block technique. *Pain Digest, 4*, 46–64.

Fukui, S., Ohseto, K., & Shiotani, M. (1997). Patterns of pain induced by distending the thoracic zygapophyseal joints. *Regional Anesthesia, 22*, 332–336.

International Spinal Intervention Society. (2004a). Thoracic intra-articular zygapophyseal joint blocks. In N. Bogduk (Ed.), *Practice guidelines for spinal diagnostic and treatment procedures*. San Francisco: International Spinal Intervention Society.

International Spinal Intervention Society. (2004b). Thoracic medial branch blocks. In N. Bogduk (Ed.), *Practice guidelines for spinal diagnostic and treatment procedures*. San Francisco: International Spinal Intervention Society.

Lord, S. M., Barnsley, L., & Bogduk, N. (1995). The utility of comparative local anaesthetic blocks versus placebo-controlled blocks for the diagnosis of cervical zygapophysial joint pain. *The Clinical Journal of Pain, 11*, 208–213.

Manchikanti, L., Singh, V., Pampati, V., Beyer, C. D., & Damron, K. S. (2002). Evaluation of the prevalence of facet joint pain in chronic thoracic pain. *Pain Physician, 5*, 354–359.

Thoracic Outlet Syndrome

A. Lee Dellon
Department of Plastic Surgery and Neurosurgery, Johns Hopkins University, Baltimore, USA
Dellon, Institutes for Peripheral Nerve Surgery, Baltimore, MD, USA

Synonyms

Brachial plexus compression; Costoclavicular syndrome; Scalenus anticus syndrome

Definition

Thoracic outlet syndrome is a misnomer for the symptoms caused by compression of the brachial plexus in the thoracic inlet. The symptoms have a wide range that can encompass any manifestation related to the motor or sensory functions of the cervical nerve roots C5, C6, C7, C8, and T1. There can be secondary symptoms related to the cervical plexus and shoulder dysfunction. The most common symptoms are aching in the shoulder and numbness in the fingers, with the symptoms being aggravated or initiated by elevating the arm above the shoulder level. The syndrome most commonly occurs in the setting of neck or shoulder trauma and is often related to the presence of anatomical congenital anomalies. Due to the inability of traditional electrodiagnostic testing to identify this syndrome, except for the isolated problem with the lower trunk of the brachial plexus, diagnosis and treatment for brachial plexus compression in the thoracic inlet remains a controversial source of neck, shoulder, and upper extremity pain.

Characteristics

The subclavian artery and subclavian vein exit the thorax and enter the upper extremity by a path that takes them into the supraclavicular region and then between the clavicle and the first rib and into the axilla. The anterior scalene muscle is usually located between these two large blood vessels as it inserts into the first rib. Posttraumatic tightness or anatomic anomalies in this region can cause vascular symptoms such as purplish color and swelling due to venous obstruction, or coldness and digital necrosis due to arterial obstruction and emboli. These vascular forms of the "thoracic outlet syndrome" are responsible for about 5 % of the patients with symptoms in this region, and the diagnosis is made by radiologic imaging of these vessels with the arm at rest and the arm elevated. These vessels actually leave the thorax and pass across the thoracic inlet to reach the axilla (Fig. 1). Relief of obstruction is surgical, requiring resection of the anterior scalene muscle or excision of the first rib. Different anatomic approaches, supraclavicular or transaxillary, are currently in use. These vascular syndromes do not have neurologic symptoms but can be associated with compression of the brachial plexus in this same region (Mackinnon and Dellon 1988; Leffert 1992). The classic physical examination maneuvers related to the disappearance of the radial pulse when the hand is elevated or the head is turned are best related to the arterial form of compression in this region but can be present in 33 % of the normal population.

When the lower trunk of the brachial plexus, formed from the C8 nerve root and T1 nerve root, is compressed against the posterior border of the first rib by a tumor (Pancoast syndrome, upper pulmonary lobe bronchogenic carcinoma, or benign neural tumor like schwannoma or neurofibroma), trauma, or congenital anomaly, the patient experiences symptoms of coldness in the arm, apart from subclavian artery compression (because the sympathetic input from the stellate ganglion enters the lower trunk). They also experience numbness of the little and ring finger, as well as the upper inner arm, and intrinsic muscle weakness, because these are the skin territories innervated by these nerve roots, and all the intrinsic muscles of the hand are innervated by these nerve roots. Note that the compression site is at the thoracic inlet (the thoracic outlet is the diaphragm). Therefore, this is the one form of thoracic outlet syndrome that can be identified by electrodiagnostic testing (EDT) (Gilliatt et al. 1970). The EDT will demonstrate abnormal electromyography of both median- and ulnar-nerve-innervated intrinsic muscles (typically the abductor digiti minimi, the first dorsal interosseous, and the opponens pollicis and the abductor pollicis brevis), with a decreased sensory amplitude for the little finger, while the sensory amplitude remains normal for the index finger. This "true neurogenic" thoracic outlet syndrome is rare, accounting for less than 1 % of patients with thoracic outlet syndrome.

The vast majority of patients (94 %) have complaints that are often global in nature, which has led some to doubt the existence of this syndrome (Cherington and Cherington 1992; Roos and Wilbourn 1990). Headaches, shoulder pain, diffuse weakness in the upper extremity, numbness in the little and ring finger, but also in all fingers, and these symptoms being made

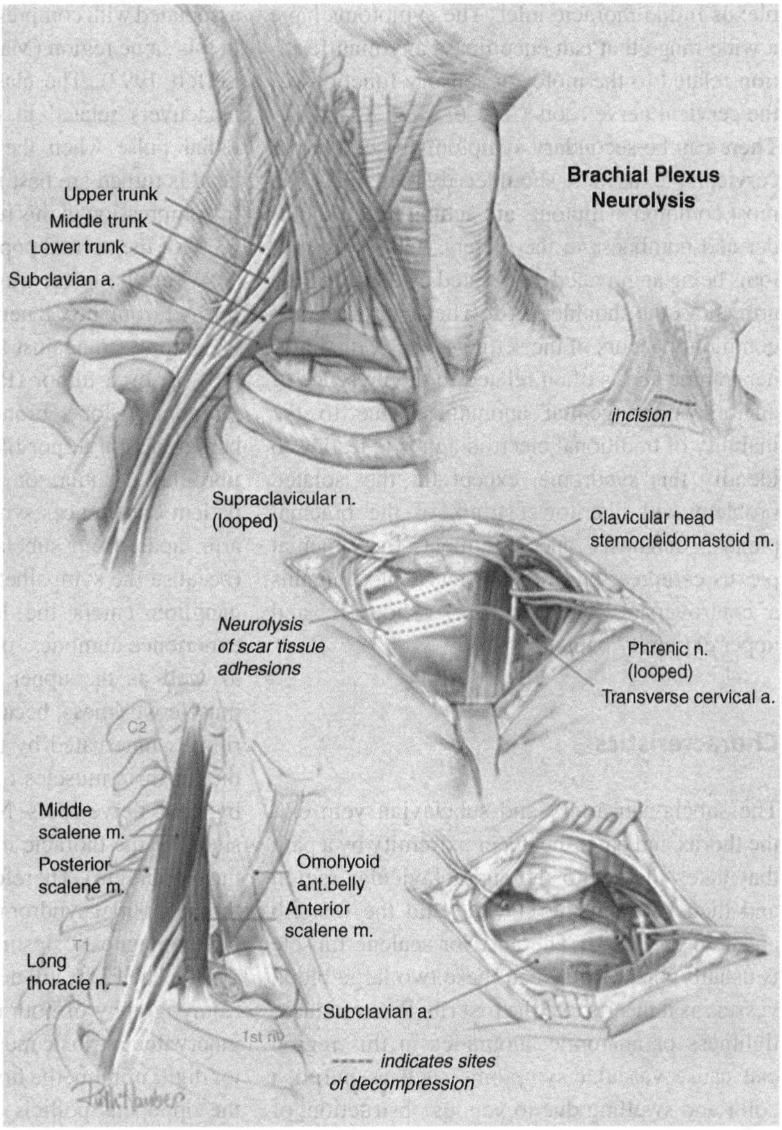

**Thoracic Outlet
Syndrome,**
Fig. 1 Brachial plexus
neurolysis

worse by any activity that requires elevation of
the shoulder. In some women there is breast pain.
In some patients there is upper back or scapular
pain. In some patients there is pain in the temporomandibular joint, requiring evaluation by
a dentist or oral surgeon. In some there is neck
pain. It is critical to appreciate that the brachial
plexus does take a course across the thoracic inlet
to enter the axilla between the clavicle (which is
concave at this location) and the second rib; it
does not travel between the clavicle and the first

rib. The upper trunk of the brachial plexus does
travel beneath the anterior scalene muscle. There
is a spectrum of congenital anomalies that course
across or through or under the brachial plexus that
are responsible for the symptoms of the nerve
compression, and depending upon the location
of the compression sites, the symptoms will
vary. These include the following:
• Cervical rib
• Fibrous bands from C7 transverse process
• Extra origins for scalene muscles

- Pre- or postfixed brachial plexus
- Intraplexus anomalous connections
- Elevated position of subclavian artery
- Muscle of albinus (scalenus minimus)
- Fibrous edges of scalene muscles
- Anomalous vessels crossing plexus
- Sibson's fascia crossing T1 nerve root
- Proximal junction of T1 to C8

Secondary compression of the cervical plexus by a tight or spasmodic anterior scalene muscle is the cause of the facial pain, and this muscle pulling upon the occiput is the source of the headache (Mackinnon and Dellon 1988). A winging component to the scapula is due to compression of the long thoracic nerve as it exits between the medial and posterior scalene muscles, and this can be corrected by including a neurolysis of this branch of the brachial plexus in the surgical approach (Disa et al. 2001).

Diagnosis of brachial plexus compression in the thoracic inlet has been a challenge. Traditional EDT cannot identify this compression because the site is too close to the spinal nerve roots, because the variation in thickness of the chest wall invalidates amplitude measurements, and because the inability to measure distance variable invalidates nerve conduction measurements. The H wave is not reliable because as the impulse travels from the finger to the spinal cord and back to the hand, multiple potential sites of entrapment are encountered at the wrist, forearm, and elbow (Wilbourn and Urschel 1984). Since the symptoms are produced or aggravated when the hands are elevated above the head, provocation of the plexus in such a manner has been used to identify patients with thoracic outlet syndrome, a positive Roos sign (Roos 1966). Palpation over the brachial plexus in the thoracic inlet will produce a distally radiating sign when the plexus is compressed beneath anatomic structures in this location, positive Tinel sign or pressure provocative test (Novak et al. 1995). Based upon the assumption that measurement of the cutaneous pressure threshold would change for the index finger (representing the upper trunk) and the little finger (representing the lower trunk), when the threshold was compared between the at rest and the provoked measurement, the Pressure-

Specified Sensory Device™ was introduced for diagnosis (Lee et al. 2000). This approach was refined to include measurements of one- and two-point static and moving touch, plus pinch and grip strength (Howard et al. 2003). A sensitivity of 100 % and a specificity of 88 % were obtained in the diagnosis of severe brachial plexus compression using this approach.

Treatment of brachial plexus compression should include stretching of the anterior scalene and strengthening of the upper trapezius, rhomboids, and serratus anterior muscles (Novak et al. 1995). This approach can relieve symptoms in 90 % of patients. In those patients who fail to respond to nonoperative methods, an exhaustive diagnostic approach must be undertaken to rule out causes that may give similar pain, prior to undertaking a surgical decompression of the plexus. Additional diagnostic testing should include chest X-ray, cervical and shoulder MRI, and evaluation of peripheral nerve entrapments such as carpal and cubital tunnel syndrome, since these entrapments, cervical disc disease, and intrinsic shoulder pathology are the commonest causes of these types of pain (Campbell et al. 1991; Levin and Dellon 1992).

The surgical treatment of brachial plexus compression must accomplish a decompression of the brachial plexus from whichever structure or structures are causing the compression. Historically, just the scalenus anterior was resected (Mackinnon and Dellon 1988). In 1966, the transaxillary first rib resection was introduced, with the concept that the plexus was compressed between the first rib and the clavicle (Roos 1966). This is a difficult approach, requiring wide retraction from the axilla in order to reach the posterior border of the first rib. Complications include burning pain in the distribution of the second or third intercostobrachial nerve due to stretch/traction injury and injury to the subclavian artery, pneumothorax, and injury to the C8 or T1 nerve roots, as they cross the posterior border of the first rib. Due to persistent upper trunk symptoms following transaxillary first rib resection, an anterior scalene resection often had to be added as a secondary procedure. This gave rise to the supraclavicular surgical approach through which the anterior (and medial) scalene muscles could be resected, a neurolysis of the brachial plexus could be done under direct vision, the major vessels could be identified and protected, and a cervical or first rib could still be resected if necessary (Dellon 1993; Hempel et al. 1996; Sanders 1991). The results reported by Hempel et al. in 1996 are worth noting: During a 28-year period, 637 patients underwent 770 supraclavicular first rib resections and scalenectomies for thoracic outlet syndrome. Following surgery, an excellent response was achieved in 59 %, good in 27 %, fair in 13 %, and poor in 1 %. One lymphatic leakage and no brachial plexus injuries resulted. Postoperative causalgia requiring subsequent sympathectomy developed in two cases. No vascular or permanent phrenic nerve injuries occurred, and 12 patients (2 %) required operative intervention for recurrence. Sanders, in 1991, reported no difference in the outcome whether he resected the first rib or did not resect it. The author's preferred approach is a supraclavicular anterior scalenectomy and neurolysis of the brachial plexus, preserving the first rib. During 15 years with this approach, there has not been one pneumothorax, or injury to the brachial plexus or major vessels. There have been two patients with transient phrenic nerve palsy.

References

Campbell, J. N., Naff, N., & Dellon, A. L. (1991). Thoracic outlet syndrome: A neurosurgical perspective. *Neurosurgery Clinics of North America, 2*, 227–234.

Cherington, M., & Cherington, C. (1992). Thoracic outlet syndrome reimbursement patterns and patient profiles. *Neurology, 42*, 492–495.

Dellon, A. L. (1993). The results of supraclavicular brachial plexus neurolysis (without first rib resection) in management of post-traumatic "thoracic outlet syndrome". *Journal of Reconstructive Microsurgery, 9*, 11–17.

Disa, J., Wang, B., & Dellon, A. L. (2001). Correction of scapular winging by neurolysis of the long thoracic nerve. *Journal of Reconstructive Microsurgery, 17*, 79–84.

Gilliatt, R. W., Le Quesne, P. M., Logue, V., et al. (1970). Wasting of the hand associated with a cervical rib or band. *Journal of Neurology, Neurosurgery & Psychiatry, 33*, 615–619.

Hempel, G. K., Shutze, W. P., Anderson, J. F., et al. (1996). 770 consecutive supraclavicular first rib resections for thoracic outlet syndrome. *Annals of Vascular Surgery, 10*, 456–463.

Howard, M., Lee, C., & Dellon, A. L. (2003). Documentation of brachial plexus compression in the thoracic inlet utilizing provocation with neurosensory and motor testing. *Journal of Reconstructive Microsurgery, 19*, 303–312.

Lee, G. W., Massry, D. R., Kupfer, D. M., et al. (2000). Documentation of brachial plexus compression in the thoracic inlet with quantitative sensory testing. *Journal of Reconstructive Microsurgery, 16*, 15–20.

Leffert, R. D. (1992). Thoracic outlet syndromes. *Hand Clinics, 8*, 285–291.

Levin, L. S., & Dellon, A. L. (1992). Pathology of the shoulder as it relates to the differential diagnosis of thoracic outlet compression. *Journal of Reconstructive Microsurgery, 8*, 313–317.

Mackinnon, S. E., & Dellon, A. L. (1988). *Surgery of the peripheral nerve* (1st ed., pp. 175–191). New York: Thieme.

Novak, C. B., Collins, E. D., & Mackinnon, S. (1995). Outcome following conservative management of thoracic outlet syndrome. *Journal of Hand Surgery, 20A*, 542–548.

Roos, D. (1966). Transaxillary approach to the first rib to relieve thoracic outlet syndrome. *Annals of Surgery, 163*, 354–358.

Roos, D., & Wilbourn, A. J. (1990). Thoracic outlet syndrome is underrated/overdiagnosed. *Archives of Neurology, 47*, 228–230.

Sanders, R. J. (1991). *Thoracic outlet syndrome: A common sequelae of neck injuries*. Philadelphia: JB Lippincott.

Wilbourn, A., & Urschel, H. C. (1984). Evidence for conduction delay in thoracic outlet syndrome is challenged. *New England Journal of Medicine, 310*, 1052–1053.

Thoracic Surgery

Definition

Thoracic surgery includes all thoracic surgical procedures performed with lateral thoracotomy, median sternotomy, or minimally invasive techniques.

Cross-References

▶ Postoperative Pain, Thoracic and Cardiac Surgery

Thought Suppression

Definition

The intentional act of excluding certain thoughts from consciousness.

Cross-References

▶ Psychology of Pain, Assessment of Cognitive Variables

Threat Appraisal

Definition

The judgment that a stressful event, such as pain, is threatening.

Cross-References

▶ Psychological Aspects of Pain in Women

Three-Way Scaling Models

Definition

Three-way scaling models provide, in addition to the group stimulus space, a subject weight space that provides coordinates for the dimensions most salient to each individual.

Threshold

Definition

The threshold is the endpoint for the appearance of a given reaction or behavior.

Cross-References

▶ Randall-Selitto Paw Pressure Test

Threshold Determination Protocols

Joel D. Greenspan
Department of Neural and Pain Sciences, University of Maryland School of Dentistry, Baltimore, MD, USA

Definition

Threshold determination protocols are those psychophysical procedures used to estimate perceptual threshold. These protocols (of which there are several variations) are defined by the rules by which stimuli are applied, the ways that responses are measured, and how the entire sequence of events is orchestrated.

Introduction

There are several protocols in the literature that estimate pain threshold. To a large extent, these protocols are the same as those used to estimate the threshold of other sensory systems. These protocols have been developed and modified over the last 150 years, resulting in many variations (Boring 1942).

Characteristics

One important distinction is between those protocols that depend upon the subject's reaction time (RT) and those that do not (Yarnitsky 1997). The most commonly used RT-dependent protocol is referred to as the *method of limits*. Within this protocol, the stimulus intensity (i.e., temperature with a thermal probe or force with a pressure ▶ algometer) is gradually increased until the point at which the subject indicates he feels pain, typically by pushing

Threshold Determination Protocols, Fig. 1 Representation of a reaction-time-dependent (*top*) and a reaction-time-independent (*bottom*) protocol. *Top*: stimulus intensity increases until the subject indicates (s) he feels pain (or another specified percept), at which time the stimulus is returned to its baseline level. The stimulus intensity at the time of the report is taken as an estimate of threshold. This process is repeated, and the threshold estimates averaged across trials. *Bottom*: a prescribed stimulus is presented, and the subject is instructed to provide a response at the end of the stimulus (i.e., "painful" or "not painful"). Depending upon the response, subsequent stimuli are of either greater or lesser intensity. See text for further details

a button. The stimulus level at the time of that report is taken as the pain threshold (Fig. 1, top). One can also use this same approach and have the subject indicate the limits of his tolerance.

A major reason for its wide use is the simplicity of this protocol, particularly in the clinical environment. However, it has some distinct disadvantages, including the obvious confound of a motor response embedded in the measure. One must consider, for instance, in comparing two groups of people, whether a group difference in motor reaction time could be large enough to produce different thresholds with this measure. Alternatively, and perhaps more likely, is whether slower decision making in one group versus another (i.e., aged or patient groups) could account for slower responses, thus leading to higher thresholds.

One of the common variants of this protocol uses a contact ▶ thermode and runs a series of trials to estimate the threshold for innocuous thermal sensation (warm and cool) and then cold pain and heat pain. In the example provided in Fig. 2 (Greenspan 2001), the thermode was placed on the test site (right foot) at an

Threshold Determination Protocols, Fig. 2 An example of a threshold data set gathered using the method of limits protocol. The *bars* represent the temperature reached for each trial at the time the subject pressed a button to indicate she felt a particular perception. The

particular sensation for any given set of trials is indicated at the *bottom* of the graph. The threshold for each sensory percept is calculated as the average of each trial and indicated at the *top* of the graph (Adapted from data presented in Greenspan 2001)

adapting temperature of 32 °C. Then, some seconds after asking the subject to pay attention, the temperature was decreased (cool or cold pain trials) or increased (warm or heat pain trials) at a fixed rate of temperature change (1.0 °C/s). The subject depressed a button when she felt the particular sensation for which she was instructed to attend. These trials are repeated typically three to four times for each sensory modality, and an average value is calculated for threshold.

Almost all other pain threshold protocols use fixed duration stimuli of a prescribed intensity and do not rely on the subject's reaction time (see Fig. 1, bottom). The most common protocol of this type is a *staircase* protocol, or one of its variants, including the *method of levels*. In this type of method, a series of stimulus intensities is prescribed with a few simple rules. After a single stimulus is applied, the subject reports whether it was painful or not. If it was not painful, the next stimulus is more intense. If it was painful, the next stimulus is less intense. Stimuli are applied according to these rules until a predefined endpoint is reached. That endpoint can be

defined in terms of the number of "response reversals" (i.e., going from "painful" to "nonpainful," or vice versa, from one trial to the next). Variations of this protocol include the way in which stimulus intensity is changed from trial to trial. For instance, it is common to start the protocol with relatively large changes (or "steps") in stimulus intensity from trial to trial (such as 2–4 °C steps to derive heat or cold pain) and reduce step size as the session progresses. The initial temperature is typically chosen to be well below pain threshold, so subsequent stimuli will be of larger intensity, until the subject reports that one of the stimuli is painful. At that point, the next stimulus is of a lower intensity, but the change in temperature will be less (for instance, half of the step size used to that point). Once a stimulus intensity is reached that is not painful, the next stimulus will be of a greater intensity but will be yet a smaller step size. In this manner, the stimuli are "titrated" to reach the intensity closest to the painful/nonpainful level, ultimately at the smallest step size (resolution) desired. For the method of levels, the step sizes are gradually reduced, and the protocol is complete when the minimally desired step size is achieved. In this case, the threshold is the average of the temperature at the last two response reversals. For the staircase protocol, the step sizes reduce quickly to a minimal value, and the protocol continues using that step size for several response reversals. In this case, the threshold is calculated as an average of the last few (typically 4–6) response reversal temperatures.

A simpler variant of this type of protocol is to use a series of stimuli in ascending intensity, until a stimulus is given that the subject reports as painful. This protocol is something like the method of limits (in terms of increasing stimulus intensity) but uses discrete stimulus trials, thus making it a reaction-time-independent protocol. The threshold would be estimated as the stimulus intensity between the first painful stimulus and the immediately preceding one. This stepwise method of limits would be repeated several times, and threshold estimates averaged. A protocol of this type has been successfully used for mechanical pain thresholds (Greenspan and McGillis 1991) and heat pain thresholds (Dyck et al. 1996).

One problem with these protocols is the subjects' predictability of the stimulus sequence, particularly after some experience with it. One way to thwart such predictability is to include some "out of sequence" stimulus intensities at random times, which are irrelevant to the course of the staircase. Another way is to use a multiple staircase protocol. For this protocol, one starts two or more staircases, each operating by the same rules described above. However, from trial to trial, the stimulus may be drawn from any of the staircases, determined randomly. Thus, at the end of the protocol, one has two or more estimates of pain threshold – one derived from each staircase – and the stimulus sequence was unpredictable on a trial by trail basis. Another variation of the multiple staircase protocol allows one to derive thresholds for more than just the pain threshold. If the subject is instructed to report from a list of qualitative descriptors (i.e., "warm," "hot," "slight pain," "moderate pain," etc.), one can prescribe a staircase to titrate

to any one of the qualitative percepts. Gracely et al. (1988) used this protocol to derive thresholds for "slight pain," "moderate pain," and "intense pain" in a single test session, with heat stimuli. Taylor et al. (1993) and Greenspan et al. (1993) used a multiple staircase protocol to derive thresholds for "cool," "slight heat pain," and "moderate heat pain" (see Fig. 3). However, the Greenspan et al. report revealed that mixing of cool and heat stimuli increased the incidence of paradoxical heat sensations, rendering cool threshold estimates problematic in some instances. Thus, it may be preferable to restrict multiple staircases (or any other threshold protocol) to either detection or pain thresholds, within a single modality, for any single block of trials.

In principle, one should be able to compare directly threshold data gathered at different times, and by different investigators, using the same protocol. Such comparisons are tenuous, due to the fact that subtle variations in execution of the protocol may systematically alter the resulting threshold values. Differences in stimulus area, rate of stimulus change, interstimulus interval, and precise body site tested can significantly alter pain thresholds. Furthermore, situational factors such as the subject's general comfort, or the instructions to the subject, can affect threshold values. It is for these reasons that comparisons of threshold values are better made within a given study, rather than across studies (Shy et al. 2003).

However, a large-scale, concerted effort has been made by a group of scientists and clinicians in Germany to develop a sensory testing protocol in sufficient detail to be used across groups and institutions (Rolke et al. 2006a, b). The protocols developed by the German Research Network on Neuropathic Pain (DFNS) encompass more than just pain thresholds. They include innocuous tactile and thermal thresholds and ratings of suprathreshold stimuli in the painful range. The availability of their protocols and normative values collected from healthy volunteers makes it possible for other groups to evaluate patients without having to generate a separate set of normative results for comparison.

Threshold Determination Protocols, Fig. 3 Data from two multiple staircase sessions using thermal stimuli. Three different staircases were prescribed: one for cool sensation, one for slight heat pain sensation, and one for moderate heat pain sensation. Each *filled symbol* represents a single temperature presentation (a trial). After each stimulus was presented, the subject chose a descriptor that best represented his sensation from a list of thermal and pain terms. For each staircase, successive stimuli were more intense until the criterion response was elicited. Then, subsequent stimuli were progressively less intense, until a response other than the criterion response was elicited. Thus, the stimulus at which the subject's response changed with respect to the criterion response produced a reversal of the staircase. The trials producing such staircase reversals are marked by the *open symbols* surrounding the *filled symbols*. Each staircase was continued until six reversals occurred. The *dotted lines* denote the calculated threshold for a given staircase, based on the average of its last four reversals. On any given trial, a random process selected one of the three staircase algorithms to determine the temperature, thus preventing any prediction of the stimulus sequence. The adapting temperature was 34.0 °C (Reprinted with permission from Taylor et al. 1993; http://www.tandf.co.uk/journals)

Summary

Several different threshold determination protocols have been described in the literature, although they may be considered as falling into one of two categories: response-dependent and response-independent protocols. The most commonly used response-dependent type of protocol is referred to as the method of limits. It consists of gradually increasing stimulus intensity, until the subject provides a response indicating his experience of pain. The stimulus level at that point is the threshold. Another response-dependent protocol consists of maintaining a fixed stimulus and using the duration of exposure until perceiving pain as a threshold measure. Response-independent protocols consist of applying a fixed duration stimulus and having the subject report whether it is painful or not at the end of the stimulation period. The most common version of this type of protocol is the staircase method. For either type of protocol, parameters of stimulus delivery will influence the result, making it invalid to compare threshold results across studies using different parameters of stimulation, even if using the same protocol.

References

Boring, E. G. (1942). *Sensation and perception in the history of experimental psychology*. New York: Appleton.

Dyck, P. J., Zimmerman, I. R., Johnson, D. M., et al. (1996). A standard test of heat-pain responses using CASE IV. *Journal of Neurological Sciences, 136,* 54–63.

Gracely, R. H., Lota, L., Walter, D. J., et al. (1988). A multiple random staircase method of psychophysical pain assessment. *Pain, 32,* 55–63.

Greenspan, J. D. (2001). Quantitative assessment of neuropathic pain. *Current Pain and Headache Reports, 5,* 107–113.

Greenspan, J. D., & McGillis, S. L. B. (1991). Stimulus features relevant to the perception of sharpness and mechanically evoked cutaneous pain. *Somatosensory and Motor Research, 8,* 137–147.

Greenspan, J. D., Taylor, D. J., & McGillis, S. L. B. (1993). Body site variation of cool perception thresholds, with observations on paradoxical heat. *Somatosensory and Motor Research, 10,* 467–474.

Rolke, R., Baron, R., Maier, C., Tolle, T. R., Treede, R. D., Beyer, A., Binder, A., Birbaumer, N., Birklein, F., & Botefur, I. C. (2006a). Quantitative sensory testing in the German research network on neuropathic pain (DFNS): Standardized protocol and reference values. *Pain, 123*(3), 231–243.

Rolke, R., Magerl, W., Campbell, K. A., Schalber, C., Caspari, S., Birklein, F., & Treede, R. D. (2006b). Quantitative sensory testing: A comprehensive protocol for clinical trials. *European Journal of Pain, 10*(1), 77–88. PM:16291301.

Shy, M. E., Frohman, E. M., So, Y. T., et al. (2003). Quantitative sensory testing: Report of the therapeutics and technology assessment subcommittee of the American academy of neurology. *Neurology, 60,* 898–904.

Taylor, D. J., McGillis, S. L. B., & Greenspan, J. D. (1993). Body site variation of heat pain sensitivity. *Somatosensory and Motor Research, 10,* 455–466.

Yarnitsky, D. (1997). Quantitative sensory testing. *Muscle & Nerve, 20,* 198–204.

Tic Doloreux

▶ Trigeminal and Cranial Nerve Neuralgias

Tic Douloureux

▶ Trigeminal and Cranial Nerve Neuralgias
▶ Trigeminal Neuralgia
▶ Trigeminal Neuralgia: Diagnosis and Treatment

Time Constant

Definition

For a variable that is an exponential function of time $x(t) = x_0 e^{\pm \frac{t}{\tau}}$, the time constant ($\tau$) is the time after which the variable (x) has decreased by $1/e$ or increased by e.

Cross-References

▶ Mechano-Insensitive C-Fibers, Biophysics

Time Loss Payment

▶ Worker Compensation Benefits

Time-Contingent Medication

▶ Time-Locked Medication

Time-Locked Medication

Synonyms

Time-contingent medication

Definition

Time-locked medication refers to medication intake not on an as needed (prn) schedule but at fixed points in time to avoid maladaptive learning processes.

Cross-References

▶ Operant Treatment of Chronic Pain

Timing

Definition

In the context of preemptive analgesia, an analgesic treatment can be initiated (or "timed" to start) before or after the surgical injury.

Cross-References

▶ Postoperative Pain, Preemptive or Preventive Analgesia

Timolol

Definition

A beta adrenoceptor-blocker.

Cross-References

▶ Preventive Migraine Therapy

Tinel Sign

Turo J. Nurmikko
Department of Neurological Science, University of Liverpool, and Pain Research Institute, The Walton Centre NHS Foundation Trust, Liverpool, UK

Synonyms

Hoffmann-Tinel sign; Sign of formication; Tinel's sign

Definition

Tingling or sensation of pins and needles elicited by an examiner tapping over a nerve. Tingling must be felt in the territory supplied by the nerve.

Introduction

The Tinel sign is a simple clinical test to identify local nerve damage or map the progress of nerve regeneration following nerve injury. It was championed in the 1910s by two neuroscientists independently of each other, Dr. Jules Tinel (1879–1952), a French neurologist, and Paul Hoffmann (1884–1962), a German neurophysiologist, but there are earlier descriptions of the sign in medical literature. The phenomenon was probably well known to the medical community before their reports. Tinel and

Hoffmann are, however, credited for highlighting the clinical potential of the sign and giving it a measure of scientific validity (Wilkins and Brody 1971).

Characteristics

The Tinel sign is a tingling perceived by a patient with nerve injury when the examiner taps over the trunk of the nerve. Tingling is felt in the skin supplied by the nerve, does not change on repetition, and outlasts the stimulus by no more than a few seconds. Tinel and Hoffmann considered this to signal the presence of regenerating ▶ axons, which start to sprout from the proximal part of the nerve after injury. The further the recovery advances, the further away from the site of injury the Tinel sign can be elicited. In this way, the successful progress of nerve recovery can be tracked.

In its classical form, the test is carried out by applying repeated percussions by one's extended finger over the presumed course of the nerve, initially well below the expected target, and continued upward until the first tingling sensation is elicited. The site where this happens is the maximum distance over which the nerve fibers have regenerated. In order to prevent false results from vibrations induced on the skin, the examiner places his free hand between the percussing finger and the site of injury.

In the early years of its popularity, the sign was used to estimate the rate of recovery of the nerve following injury. Large case series were published, mostly involving patients with gunshot wounds who had undergone surgery for their injury. At the time these papers provided the best estimates of recovery times of peripheral nerves, ranging from 1 to 3 mm per day (Sunderland 1968).

Not long after Tinel and Hoffmann published their first papers, there were reports by several surgeons that the sign failed to indicate accurately whether or not recovery of the injured nerve was taking place. Reports of patients with a positive Tinel sign, who at operation were found to have total interruption of the nerve

with no chance of recovery, were published. It appears that the essential requirement for the sign to be positive – that it was progressing – was overlooked by many. Nevertheless, more criticism toward the sign was generated by observations of good neural recovery in patients in whom the Tinel sign was missing (Napier 1949; David and Chung 2004), and the test fell into disfavor. Later, a few investigators published relatively large case series and concluded that it is, when used critically, of moderate use as a rough guide of nerve recovery (Napier 1949; Nathan and Rennie 1946). There are today other more advanced methods for assessment of such recovery, and the sign is to be regarded more as a curiosity than a valid clinical tool. Despite this, it is always introduced to medical students and regularly appears in surgical and neurological textbooks.

The basis for the test lies in increased excitability of the recovering nerve fibers. Following injury, nerve fibers begin to grow sprouts from the proximal stump. These sprouts are far more sensitive to mechanical stimuli than are mature nerve fibers in intact nerves. As the sprouts go longer, the most distant point of mechanical sensitivity moves further away from the injury. ▶ Myelin breakage and restoration, and recovery of support structures of the nerve trunk, also have an impact on the general excitability of the recovering nerve. In the part of the nerve in which full or near-full recovery of fibers has taken place, the nerve trunk loses its sensitivity to percussion (Yarnitsky and Ochoa 1991).

Tinel, and later advocates of his method, stressed that tapping should produce tingling, not pain in the reference area (Moldaver 1978). Tinel did acknowledge that at times mechanical compression of the site of injury could cause pain, prominent on the site of compression, with less intense pain referred to the territory of the nerve. To Tinel, the natural explanation for a stationary sign was that nerve degeneration did not take place and that the regenerating nerves grew aimlessly into a bundle of sprouts, forming a ▶ neuroma. Neuromas are known for their excessive mechanical sensitivity.

Inevitably, the sign was soon adopted for diagnosing neuromas, with no evidence that it provides more information than careful manual palpation or compression. It has been suggested that a neuroma can be distinguished from other tender structures at the operation site by using a series of light weight percussion tools (Tucker and Nancarrow 2000), but this method is not in common use. There are sophisticated electrophysiological and neuroimaging methods that can be used instead.

In the 1950s, Phalen and coworkers reported that they could frequently elicit the Tinel sign in patients with ▶ carpal tunnel syndrome and added the test to their other diagnostic criteria (Phalen et al. 1950). In carpal tunnel syndrome, the median nerve is under constant compression, imposing increasing structural damage on nerve fibers. Axonal degeneration and regeneration take place concurrently, and with varying levels of demyelination and remyelination present, the nerve becomes excessively sensitive to mechanical stimuli (Dellon 1984). However, several studies have been published that show convincingly that the Tinel sign has little to offer for differential diagnosis. It is too crude a test for accurate differentiation between patients with carpal tunnel syndrome from those with other peripheral nerve diseases, or even healthy people (D'Arcy and McGee 2000).

The significance of the Tinel sign has faded with time, and these days it remains more a curiosity than a serious clinical tool. It is still used occasionally in the clinical setting as a complementary test for carpal tunnel syndrome (Davis and Chang 2004). In pain research it may be seen used out of its original context – more as a "neuroma sign," rather than the sign of a recovering nerve than it was used for in the early part of the twentieth century.

Cross-References

- ▶ Neuroma Pain
- ▶ Peripheral Neuropathic Pain
- ▶ Ulceration, Prevention by Nerve Decompression

References

D'Arcy, C. A., & McGee, S. (2000). Does this patient have Carpal Tunnel Syndrome? *Journal of the American Medical Association, 283*, 3110–3117.

David, E. N., & Chung, K. C. (2004). The Tinel sign: A historical perspective. *Plastic and Reconstructive Surgery, 114*, 494–499.

Davis, E. N., & Chang, K. C. (2004). The Tinel sign: A historical perspective. *Plastic and Reconstructive Surgery, 114*, 494–499.

Dellon, A. L. (1984). Tinel or not Tinel. *Journal of Hand Surgery, 9*, 216.

Moldaver, J. (1978). Tinel's sign. *Journal of Bone & Joint Surgery, 60-A*, 412–413.

Napier, J. R. (1949). The significance of Tinel's sign in peripheral nerve injuries. *Brain, 72*, 63–82.

Nathan, P. W., & Rennie, A. M. (1946). Value of Tinel's sign. *Lancet, I*, 610–611.

Phalen, G. S., Gardner, W. J., & LaLonde, A. A. (1950). Neuropathy of the median nerve due to compression beneath the transverse Carpal Ligament. *Journal of Bone & Joint Surgery, 32A*, 109–112.

Sunderland, S. (1968). *Nerves and nerve injuries.* Edinburgh: E&S Livingstone.

Tucker, S. C., & Nancarrow, J. D. (2000). Objective assessment of post-traumatic nerve repairs and neuromas. *British Journal of Plastic Surgery, 53*, 694–696.

Wilkins, R. H., & Brody, I. A. (1971). Neurological classics XXXIV. *Tinel's Sign, 24*, 573–575.

Yarnitsky, D., & Ochoa, J. L. (1991). The sign of Tinel can be mediated either by myelinated or unmyelinated primary afferents. *Muscle & Nerve, 14*, 379–380.

Tinel's Sign

▶ Tinel Sign

Tinnitus

Definition

A ringing sensation in one or both ears.

Cross-References

- ▶ NSAIDs, Adverse Effects

Tissue Fatigue

Definition

Tissue fatigue refers to a decline in the load-bearing capacity of tissue in response to an applied load. This phenomenon is reversible provided that there is sufficient time for recovery.

Cross-References

▶ Ergonomic Counseling

TLIF

Synonyms

Transforaminal interbody fusion

Definition

Graft/cages placed between the vertebral bodies by posterior approach through the neural foramina without retracting the thecal sac.

Cross-References

▶ Spinal Fusion for Chronic Back Pain

TMD

▶ Temporomandibular Disorder

TMJ

▶ Temporomandibular Joint

TMS

▶ Transcranial Magnetic Stimulation

TNF alpha(α)

▶ Tumor Necrosis Factor Alpha(α)

Toddler

▶ Pain in Children

Tolerance

Definition

Tolerance refers to diminishing effects of a drug with continued use. The phenomenon of tolerance is characteristic of opioid analgesics. The development of tolerance to the analgesic effects of the opioids means that with continued opioid use, progressively higher doses may be required to maintain the same analgesic effect. Tolerance also develops to the nonanalgesic side effects of opioids, including respiratory depression, nausea, and sedation. The development of tolerance to these effects is desirable as it permits dose titration.

Cross-References

▶ CRPS, Evidence-Based Treatment
▶ Opiates During Development
▶ Opioid Receptors
▶ Opioid Responsiveness in Cancer Pain Management
▶ Opioid Therapy in Cancer Patients with Substance Abuse Disorders, Management
▶ Opioids and Inflammatory Pain
▶ Opioids in the Periphery and Analgesia

- Postoperative Pain, Opioids
- Purine Receptor Targets in the Treatment of Neuropathic Pain
- Stimulation-Produced Analgesia

Tolerance Thresholds

- Pain in Humans, Thresholds

Tolosa-Hunt Syndrome

Definition

Also called "painful ophthalmoplegia" a variable combination of periorbital pain, ipsilateral oculomotor nerve palsies, oculosympathetic palsy, and trigeminal sensory loss.

Cross-References

- Headache due to Dissection

Tolosa-Hunt Syndrome (Painful Ophthalmoplegia)

- Headache Due to Dissection

Tonic Firing Mode

Definition

A pattern of spontaneous action potential firing demonstrated by thalamic neurons, in which no clear pattern of interspike intervals exists.

Cross-References

- Central Pain, Human Studies of Physiology
- Thalamic Bursting Activity, Chronic Pain

Tonic Rebalance

Definition

Tonic rebalance is a spontaneous, physiological process by which the system tries to restore symmetry at the level of the vestibular nuclei in the acute phase after a vestibular crisis. The process occurs independently of the patient's activities.

Cross-References

- Coordination Exercises in the Treatment of Cervical Dizziness

Toothache

- Dental Pain, Etiology, Pathogenesis, and Management

Top-Down Control of Pain

Definition

Neural pathways descending from higher brain structures to lower brain structures modulate ascending nociceptive transmission.

Cross-References

- Descending Modulation and Persistent Pain

Topiramate

Definition

Originally an anticonvulsant medication widely used, including indications for seizure prevention and migraine.

Topographical

Definition

Topographical refers to the arrangement or reference to regions of the body or of a body part, especially the regions of a definite and limited area of the surface.

Cross-References

▶ Postsynaptic Dorsal Column Projection, Anatomical Organization

Topography

▶ Magnetoencephalography in Assessment of Pain in Humans

Touch-Evoked Pain

▶ Allodynia (Clinical, Experimental)

Tourniquet Ischemia

▶ Tourniquet Test

Tourniquet Pain Ratio

Definition

It is calculated as follows: Ongoing Clinical Pain Match Time/Tolerance Time × 100 where the Ongoing Clinical Pain Match Time is the time point, since the beginning of the tourniquet test, at which the patient reports a pain intensity that matches the clinical pain intensity, and the Tolerance Time is the time the patient can tolerate the ischemia in the tourniquet test.

Cross-References

▶ Tourniquet Test

Tourniquet Test

Giancarlo Carli
Department of Physiology, University of Siena, Siena, Italy

Synonyms

Acute ischemia test; Forearm ischemia procedure; Forearm occlusion pain; Ischemic pain; Ischemic test; Submaximum effort tourniquet technique; Tourniquet ischemia

Definition

Ischemic ▶ pain is elicited by having the subject squeeze a handspring exercises 20 times after a tourniquet is inflated around his upper arm. The quality of sensation is dull-aching or stinging muscular pain, which closely resembles most types of pathologic pain, but increases progressively after cessation of squeezing. Test performance is measured in terms of elapsed time between cessation of squeezing and report of slight (threshold) and unbearable (tolerance) pain. Muscular pain from ischemic contractions, which is due to transient stimulation of peripheral ▶ nociceptors (LaMotte and Campbell 1978), is based on the ▶ algogenic actions of protons (Issberner et al. 1996). The muscle pain is not specific since skin, periosteum, and other tissues contribute to the overall pain perception. In addition, the contact of the tourniquet with the skin activates low threshold non-nociceptive afferents that might exert an inhibitory influence on pain mechanisms. Noxious forearm ischemia also evokes substantial elevations in arterial blood pressure (Maixner et al. 1990) and activation of endogenous opioid systems (Goldstein and Grevert 1978), thus representing a ► diffuse noxious inhibitory controls (DNIC) paradigm (Willer et al. 1989). A pneumatic tourniquet cuff connected to a computer controlled air compressor has been used recently to produce pressure pain (Polianskis et al. 2002). Computerized cuff algometry may continuously control stimulus or adjust the pressure to maintain the programmed pain level. This set up, which elicits an increasing pain immediately after tourniquet cuff application around the gastrocnemius-soleus muscle, represents a highly configurable tool for assessment of pain sensitivity by pressure, but does not involve the ischemic mechanisms as does the classical tourniquet test described in this entry.

Characteristics

The need for a dependable experimental method to test new analgesics in man was one of the main motivations to modify the earlier techniques of producing ► Ischemic Pain/Test (Lewis et al. 1931; Hewer and Keele 1948), by changing from maximum to submaximum squeezing effort.

Originally, the ► submaximum tourniquet test (Smith et al. 1966, 1968) was suggested with the following order (see Fig. 1a–c).

The subject, reclining on a bed, is asked to comply with the following procedure:

1. To extend his nondominant arm toward the ceiling and, with the arm raised, an Esmarch bandage is wrapped from the fingers to the elbow to drain the arm of venous blood (see Fig. 1a).
2. To apply a tourniquet bandage around the upper arm and inflate it to obtain a pressure of 250 mm Hg, to abolish arterial supply, and to render the arm hypoxic.
3. To lower the arm, remove the band and, after a pause of 60 s, to squeeze a hand spring exerciser 20 times, while his arm is rested. Each squeeze is timed to last 2 s followed by a 2 s rest, so that the whole exercise time lasts 80 s. The schedule is presented to the subject by means of tape-recorded signals.

4. After completing the squeezes, to remain in a reclining position and rest his arm at his side with the tourniquet still inflated, not to move his arm until the pressure has been released (see Fig. 1c).
5. To rate, when asked at irregularly spaced intervals, the incoming sensations of pain according to a predetermined scale, which includes the following categories of experience: 0 = none; 1 = slight; 2 = moderately distressing; 3 = very distressing; 4 = unbearable. Irregular intervals should be used to minimize the cues to the subject regarding the time he had tolerated the pain.

► Pain threshold (slight pain) occurs after about 3–6 min, while unbearable pain (tolerance) varies from 7 to 53 min (Smith et al. 1966, 1968). The latency of unbearable pain is usually called ► pain tolerance, but formally it represents ischemia tolerance, i.e., the measurement of time elapsed since exercise termination (beginning of ischemia) and not since pain onset. When the cuff is released, pain intensity and numbness sensations sharply decrease, but their time course, as not included in the formal test, have not been systematically investigated.

Several authors have modified the test, but the main changes that have improved the test have been proposed by Moore et al. (1979). According to new suggestions, the individual maximal grip strength of the nondominant arm has to be assessed using an isometric exerciser equipped with a force gauge (Fig. 1b). After lowering the arm, the squeeze exercise must start immediately after the cuff has been inflated to 200 mm Hg, instead of 250, avoiding a pause of 60 s as suggested by Smith et al. (1966). The individual grip strength should be kept at the fixed percentage (50 %) of the maximum grip strength, in order to avoid fatigue and to reduce intersubject differences. The measurement of elapsed time should begin at the time of blood pressure cuff inflation, i.e., at the beginning of the ischemia, rather than at exercise termination. In addition, the rating must be in a visual analog scale from 0 (no pain) to 100 (pain so severe that you would commit suicide if you had to bear it for more than a few minutes).

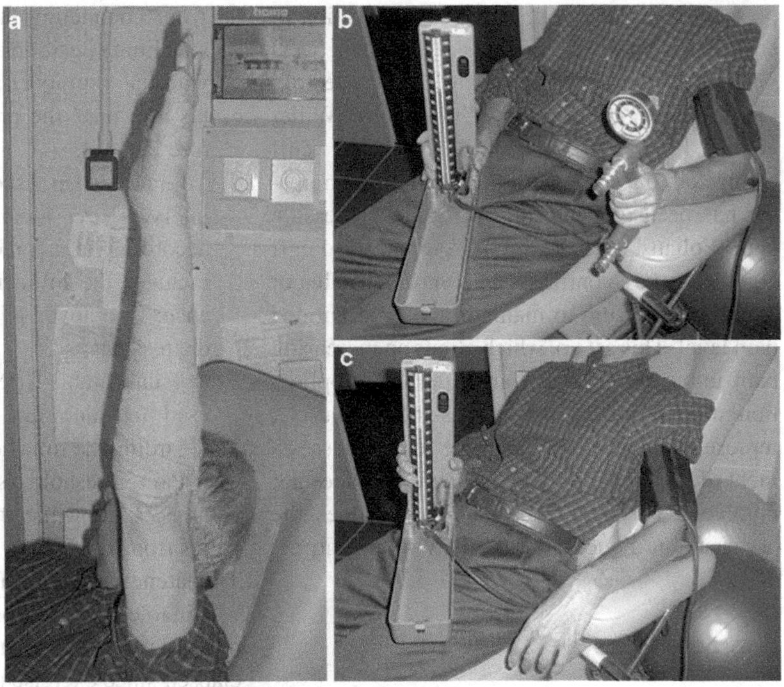

Tourniquet Test,
Fig. 1 Main phases of the
tourniquet test.
(**a**) elevation of the
bandaged forearm;
(**b**) squeezes of isometric
exerciser, while the
pneumatic tourniquet
remains inflated to block
forearm circulation; (**c**) arm
at rest, while the tourniquet
is still inflated

In the same research (Moore et al. 1979), it was observed that the amount of exercise used during the submaximal effort tourniquet test significantly affected the level of reported pain intensity versus time elapsed: in fact, by increasing the duration of each contraction from 0.6 s. to 2 s, the threshold pain intensity was reached much more rapidly.

Tourniquet Pain Ratio: A Method to Evaluate Clinical Pain

Sternback et al. (1974) originally suggested that the experimental pain generated by ischemic exercise could be used to evaluate the severity of the ongoing clinical pain. This method, called ▶ Tourniquet Pain Ratio, requires the patients to report when their arm pain intensity matches the clinical pain intensity: this time point is then used to measure the subjective severity of patient's clinical pain and is calculated as follows: (Ongoing Clinical Pain Match Time/Tolerance Time) × 100. This method has several limitations, mainly because the Tourniquet Pain Matching Score is significantly lower than the patient's own pain estimate (Moore et al. 1979). As for the relationship between pain intensity and elapsed time,

the data indicate that the pain ratings produced by the test are not a linear function of elapsed time but rather a sigmoid shaped curve. Nonetheless, the Tourniquet Pain Ratio is still used to provide an additional index of clinical pain modulation (Sigurdsson and Maixner 1994).

Tourniquet Test in the Diagnosis of Altered Nociceptive Transmission

Both threshold and tolerance to ischemic pain has been repetitively tested in clinical syndromes such as localized musculoskeletal pain, ▶ fibromyalgia, peripheral neuralgias, bulimia, hypertension, and other conditions in which a dysfunction in the excitability of the nociceptive system is expected.

Tourniquet Test and DNIC

The efficacy of DNIC, the neurophysiological mechanism that underlines the long established clinical phenomenon of ▶ counterirritation (Wand-Tetley 1956), has been successfully tested in healthy subjects following the procedure of the submaximum tourniquet test, in electrical threshold to evoke muscle pain (Pantaleo et al. 1988), cutaneous warming pain threshold

(Pertovara et al. 1982), and in other sensory modalities. In patients with ongoing pain, it has been established that pain by clinical origin does not stimulate DNIC on ischemic or thermal pain perception (Ekblom and Hansson 1987; Hansson et al. 1988). On the contrary, forearm ischemia elicited by the procedure of the submaximum tourniquet test elicits a generalized non-segmental inhibition of tooth pain, resulting from acute irreversible pulpitis (Sigurdsson and Maixner 1994). The latter effects remain for at least 5 min after removal of the tourniquet, while the arm is free from pain. It has to be underlined that the mechanisms of DNIC are not univocal since, for instance, the tourniquet test elicits inhibition of static mechano-allodynia sensations, triggered by pressure stimuli, but has no effect on dynamic mechano-allodynia sensations, elicited by brushing (Bouhassira et al. 2003). In fibromyalgia and painful osteoarthritis patients, the procedure of the submaximum tourniquet test does not elicit modulation of pressure pain, as opposed to controls (Kosek et al. 1996), while the DNIC is still effective in patients suffering from long-term trapezius myalgia.

Tourniquet Test and Drug Screening

It has been repeatedly established that placebo and suggestions of analgesia, both in subjects who are awake and during hypnosis, can reduce pain elicited by the tourniquet test. Similarly, small doses of opiates (morphine, dipipanone, codeine) and the NMDA-receptor antagonist ketamine display analgesic effects. On the contrary, the NMDA-receptor antagonist dextromethorphan, diazepam, and anti-inflammatory drugs such as aspirin or indomethacin do not affect ischemic pain. Finally, adenosine, which mediates ischemic pain in humans, at low doses exerts analgesic effects on the tourniquet pain by membrane-bound peripheral adenosine receptors (Eriksson et al. 2000).

Concluding Remarks

In conclusion, the tourniquet test can provide useful indications about individual reactivity and tolerance to deep ischemic stimuli, both in healthy subjects and in patients. There are, however, some limitations to be underlined. First of all, in patients suffering from deep pressure pain or other deep pains in the forearms, unbearable pain can occur following the application and/or inflation of the bandage. Moreover, the tourniquet procedure has shown a considerable difference between subjects and between session variability (Sigurdsson and Maixner 1994), and the sensitivity displayed may be inadequate to assess the analgesic effects of some pharmacological agents (Sternbach et al. 1977).

References

Bouhassira, D., Danziger, N., Atta, N., et al. (2003). Comparison of the pain suppressive effects of clinical and experimental painful conditioning stimuli. *Brain, 126*, 1068–1078.

Ekblom, A., & Hansson, P. (1987). Thermal sensitivity is not changed by acute pain or afferent stimulation. *Journal of Neurology, Neurosurgery, and Psychiatry, 50*, 1216–1220.

Eriksson, B. E., Sadig, B., Svedenhag, J., et al. (2000). Analgesic effects of adenosine in syndrome X are counteracted by theophylline: A double-blind placebo-controlled study. *Clinical Science, 98*, 15–20.

Goldstein, A., & Grevert, P. (1978). Placebo analgesia, endorphins and naloxone. *Lancet, 2*, 1385.

Hansson, P., Ekbloom, A., Lindblom, U., et al. (1988). Does acute intraoral pain alter cutaneous sensibility? *Journal of Neurology, Neurosurgery, and Psychiatry, 51*, 10032–11036.

Hewer, A. J., & Keele, C. A. (1948). A method of testing analgesics in man. *Lancet, 2*, 683–688.

Issberner, U., Rehe, P. W., & Steen, K. H. (1996). Pain due to tissue acidosis: A mechanism for inflammatory and ischemic myalgia? *Neuroscience Letters, 208*, 191–194.

Kosek, E., Ekholm, J., & Hansson, P. (1996). Sensory dysfunction in fibromyalgia patients with implications for pathogenic mechanisms. *Pain, 68*, 375–383.

LaMotte, R. H., & Campbell, J. N. (1978). Comparison of responses of warm and nociceptive C-fiber afferents in monkey with human judgements of thermal pain. *Journal of Neurophysiology, 41*, 509–528.

Lewis, D., Pikering, G. W., & Rothschild, P. (1931). Observations upon muscular pain in intermittent claudication. *Heart, 15*, 359–383.

Maixner, W., Gracely, R. H., Zuniga, J. R., et al. (1990). Cardiovascular and sensory responses to forearm ischemia and dynamic hand exercise. *American Journal of Physiology, 259*, R1156–R1163.

Moore, P. A., Duncan, G. H., Scott, D. S., et al. (1979). The submaximal effort tourniquet Test: Its use in evaluating experimental and chronic pain. *Pain, 6*, 375–382.

Pantaleo, T., Duranti, R., & Bellini, F. (1988). Effects of heterotopic ischemic pain on muscular pain threshold and blink reflex in humans. *Neuroscience Letter, 85*, 56–60.

Pertovara, A., Kemppainen, P., Johannson, G., et al. (1982). Ischemic pain nonsegmentally produces a predominant reduction of pain and thermal sensitivity in man: A selective role for endogenous opioids. *Brain Research, 251*, 82–93.

Polianskis, R., Graven Nielsen, T., & Arendt-Nielsen, L. (2002). Spatial and temporal aspects of deep tissue pain assessed by cuff algometry. *Pain, 100*, 19–26.

Sigurdsson, A., & Maixner, W. (1994). Effects of experimental and clinical noxious counterirritants on pain perception. *Pain, 57*, 265–275.

Smith, G. M., Egbert, L. D., Markowitz, R. A., et al. (1966). An experimental pain method sensitive to morphine in man: The submaximum effort tourniquet technique. *Journal of Pharmacology and Experimental Therapeutics, 154*, 324–332.

Smith, G. M., Lowenstein, E., Hubbard, J. H., et al. (1968). Experimental pain produced by the submaximum effort tourniquet technique: Further evidence and validity. *Journal of Pharmacology and Experimental Therapeutics, 163*, 468–474.

Sternbach, R. A., Deems, L. M., Timmermans, G., et al. (1977). On the sensitivity of the tourniquet pain test. *Pain, 3*, 105–110.

Wand-Tetley, J. I. (1956). Historical methods of counter-irritation. *Annals of Physical Medicine, 3*, 90–98.

Willer, J. C., DeBroucker, T., & LeBars, D. (1989). Encoding of nociceptive thermal stimuli by diffuse noxious inhibitory controls in humans. *Journal of Neurophysiology, 62*, 1028–1038.

Toxic Neuropathies

Sung-Tsang Hsieh
Department of Anatomy and Cell Biology,
Department of Neurology, National Taiwan
University Hospital, National Taiwan University,
Taipei, Taiwan

Definition

Toxic neuropathies are peripheral nerve disorders due to acute or chronic adverse effects of chemicals or medications which occur in individuals or as an outbreak in a certain population after occupational exposure or environmental contamination.

Introduction

Toxic neuropathies are a group of peripheral nerve disorders caused by various chemicals ranging from environmental contaminants to therapeutic medications. The symptoms are diverse including motor, sensory, and autonomic manifestations and their combinations. Mechanisms underlying nerve injury after toxic insults include axonal degeneration, demyelination, and vascular lesions. New research tools, particularly the imaging of nerve terminals, help to elucidate the pathology of toxic neuropathies. This entry will summarize current understandings of the spectrum, pathology, and mechanisms of toxic neuropathies.

Characteristics

Neurons, Axons, and Neuropathy

Neuropathies are diseases of the peripheral nervous system from various etiologies, including metabolic disorders, such as diabetic neuropathy due to diabetes mellitus (Said 2007), and toxic effects of chemotherapy for cancer. Peripheral nerves, composed of different categories of nerve fibers, are responsible for conducting impulses between the central nervous system and target organs (Chen et al. 1999). Motor nerves and proprioceptive nerves, which convey sensory information from bones, muscles, and joints, are large myelinated nerves (large-diameter nerves or large fibers). Nociceptive, thermal, and autonomic nerves belong to small myelinated nerves or unmyelinated nerves (small-diameter nerves or small fibers). All of these nerve fibers have different cytoskeletal organizations, origins, and terminals (Chen et al. 1999). Axons, the technical term for nerve fibers, are an extension of the neuronal cell body, the perikaryon. In the peripheral nervous system, the length of axons is longer than the diameter of the neuronal cell body. This results in a much larger volume of the entire axon compared to the volume of the neuronal cell body, which is of an order of one to several magnitudes. As an example, take a motor neuron of an adult human. The diameter of a motor neuron in the lumbar spinal cord is up to 100 μm; in contrast, the length of a motor axon from this motor neuron innervating the foot muscles is up to 1 m in length. This greatly differs from other cells in the body, such as fibroblasts, which have a cellular extension of only a limited length.

To maintain such unusual structural requirements, axons possess a rather complicated cytoskeletal system (Perrot et al. 2008; Chevalier-Larsen and Holzbaur 2011; Luo and O'Leary 2005). Components of the cytoskeleton include microtubules (>24 nm), intermediate filaments (10 nm), and microfilaments (6~8 nm), according to their diameters and composition. These cytoskeletal proteins interact with each other through associated proteins, such as microtubule-associated proteins, to form a three-dimensional interlacing structure. In addition, the metabolic demands of axons heavily depend on the neuronal cell body and an axoplasmic transport system has developed to transport materials both to and from neuronal cell bodies. Some cytoskeleton-associated proteins, such as the kinesin superfamily, are also responsible for transporting organelles and transmitters (Hirokawa and Noda 2008). Peripheral nerves, particularly those in nerve terminal regions, are distributed throughout the entire body and, thus, are the most vulnerable part of the nervous system during toxin exposure and metabolic derangements. Consequences of these insults, including dysregulated cytoskeletal maintenance or blockade of axonal transport, impair the functioning of axons, and eventually lead to axonal degeneration, the fundamental pathologic basis of clinical neuropathies.

Structural Organization of Peripheral Nerves and Pathology of Toxic Neuropathies

The structural organization of peripheral nerves includes two major components: neurons with cytoplasmic extensions and axons, and ensheathing cells (Schwann cells) with myelin sheaths, modified membranous insulating materials. Nerve injury in toxic neuropathy occurs at different levels along the neural axis. Neuronal degeneration results in subsequent nerve fiber degeneration. Alternately, the major target site is at axons, either proximally or distally along the nerve fibers, while neuronal cell bodies largely remain intact. The common outcome of neuronal injury or direct nerve injury is axonal degeneration, or Wallerian-like degeneration, and secondary demyelination. At the very early stage, axonal degeneration in toxic neuropathy mainly occurs at the most distal terminals, such as with acrylamide. Only rare experimental toxins, such as β,β'-iminodipropionitrile, act on proximal axons (Coleman and Freeman 2010; Day and Thompson 2010; Jortner 2000). If a toxin selectively damages Schwann cells or impairs myelin organization, neuropathies of primary demyelination will develop. Because of conduction failure in primary demyelination, the major electrophysiological abnormality of demyelinating neuropathy is the marked slowing of nerve conduction. This type of toxic neuropathy is relatively rare compared to neuropathies of the primarily axonal degeneration type. An example of a primarily demyelinating neuropathy is tellurium-induced toxic neuropathy (Toews et al. 1997). In addition, the accumulation of polymeric chemical components in the vascular wall contributes to axonal degeneration, for example, familial amyloid polyneuropathy due to mutations of transthyretin, a major etiology of painful neuropathy clinically (Yang et al. 2010; Ando et al. 2005).

Clinical Applications

Peripheral neuropathies are common neurological disorders of the community with an annual incidence of 100–200 per 100,000 people. Toxic neuropathies account for a minor proportion of etiology (−5 %) depending on economical and medical factors and geographic distribution. In the 1950s–1970s, most toxic neuropathies have been related to outbreaks of industrial chemicals, such as acrylamide and organic solvents (hexacarbons). The list of these etiologies depends on the industrial development of different countries. In recent years, major

Toxic Neuropathies, Fig. 1 Structural organization of peripheral nerves. Peripheral nerves consist of: (1) motor nerves from ventral horn motor neurons of the spinal cord (*black*), (2) proprioceptive nerves from large-diameter sensory neurons of dorsal root ganglia which terminate in skeletal muscles, tendons, and joints (*olive*), and (3) nociceptive and thermal nerves from small-diameter sensory neurons of dorsal root ganglia which terminate in the skin (*pink*) and autonomic nerves from autonomic ganglia innervating smooth muscles and glands (*blue*)

outbreaks of toxic neuropathy have decreased to a great extent due to stringent regulation of chemicals used in industry.

In contrast, most reported toxic neuropathies have been due to side effects of medications or from accidental exposure. Common chemotherapeutic agents causing toxic neuropathies include cis-platinum, taxol, vincristine, and thalidomide. Thus, toxic neuropathies are relatively uncommon compared to other etiologies, such as diabetes mellitus, genetic neuropathies, and inflammatory disorders. Nevertheless, toxic neuropathies have great impact on environmental and industrial regulation, and serve as important models for investigating mechanisms of nerve injury and the correlation between pathology and clinical manifestations.

Clinical Manifestations of Toxic Neuropathies
Clinical presentations of toxic neuropathies depend on the involvement of peripheral nerve components: motor, sensory, and autonomic nerve fibers (Fig. 1). The severity of neurological deficits depends on several factors, including the vulnerability of each nerve fiber type, the duration of toxin exposure, the total dose of toxins,

and the extent of nerve degeneration. A major factor in determining clinical presentations is the differential susceptibility of neurons to toxins or pathologic processes, a characteristic of neurological diseases: for example, motor neurons in amyotrophic lateral sclerosis (Lou Gehrig's disease), cortical neurons in Alzheimer's disease, and striatal neurons in Parkinson's disease. Each disease has a distinct pattern of clinical presentations because certain parts of the nervous system are selectively damaged.

Clinical manifestations of neuropathies, therefore, depend on the damage to different fiber types (Pascuzzi 2009). When motor nerve fibers are injured, neurological deficits range from mild weakness of distal limb muscles to marked paralysis of all four limbs. When sensory nerves responsible for joint movements are injured, unsteadiness during walking may develop. This type of neuropathy is sometimes called large-fiber neuropathy. On the other end, in patients with damaged sensory nerves, neuropathic pain, such as tingling or electric-shock sensations, can be experienced. Various autonomic complaints, including orthostatic hypotension, tachycardia, constipation, diarrhea, or anhidrosis, can be presenting symptoms in patients with damaged autonomic nerves. Sensory neuropathy with painful features and loss of nociceptive functions and autonomic neuropathy are termed as small-fiber neuropathy. Often, neurological presentations are combined effects due to deficits of different fibers types: For example, a patient with motor, sensory, and autonomic neuropathy may have manifestations of weakness, sensory disturbances, and dysautonomia. Various etiologies of toxic neuropathies with their pathologic mechanisms and manifestations are summarized in Table. It is recommended that the Table is not intended to be a comprehensive catalog. Readers can refer to various sources for detailed information (Spencer and Schaumburg 2000).

The terminals of the longest nerves are the most susceptible part during toxin exposure, and clinical symptoms usually develop in the corresponding innervated areas of these damaged nerves in the early phase of toxic neuropathy. In addition, clinical symptoms and signs are usually more severe in regions with early-onset symptoms than in regions with late-onset symptoms. For example, numbness and neuropathic pain may appear in the toes and feet earlier than in the legs and thighs. Similarly, symptoms in the fingers and hands are earlier than those in the forearms. Neurological deficits are usually more severe in the lower limbs than in the upper limbs. This "glove-stocking" type of distribution is characteristic of length-dependent neuropathies, particularly for the majority of toxic neuropathies although exceptions may exist in certain types of toxic neuropathies.

Evolution of Pathology in Toxic Neuropathies
The typical pathology of toxic neuropathies of large-diameter motor and sensory nerves has been demonstrated in both humans and experimental animals over the past few decades: for example, neuropathies due to exposure to hexacarbons, particularly the active components, 2,5-hexanediol, acrylamide, and carbon disulfide. The longest and largest nerves are affected earlier with the major pathology in the terminal parts of axons, i.e., distal axonopathy. Intoxication results from absorption of organic solvents in poorly

ventilated environments. Figure 2 illustrates the scenario of pathology in neuromuscular junctions after hexacarbon and acrylamide intoxication (Ko et al. 1999). In animals intoxicated with such compounds, for example, acrylamide, there is no visible weakness or ataxia during the early phase. At that stage, however, motor nerve terminals have already begun to swell up. Obvious weakness and ataxia of hind limbs gradually develops. The weakness and unsteadiness progress at variable speeds, and eventually, the forelimbs are affected, resulting in quadriparesis at the late stage. As intoxication proceeds, axonal swelling extends from junctional folds into the intramuscular nerves, which results in Wallerian-like degeneration of motor nerves and denervation of neuromuscular junctions (Fig. 2a). Similar changes also develop in the central terminals of long axons, such as terminals of the posterior column (nucleus gracilis and nucleus cuneatus) at the brainstem, the central axons of dorsal root ganglia, thus the term, central-peripheral distal axonopathy, or dying-back neuropathy (LoPachin and Gavin 2008).

Therapeutic Consequences
Nociceptive Nerves in Toxic Neuropathies
Nociceptive nerves subserving thermal sensations are Aδ- or C-fibers based on physiological classification with diameters in the range of 1~5 μm. These axons, with free nerve endings, are peripheral processes of small neurons in dorsal root ganglia, and terminate in the most superficial layer of the skin, the epidermis (Fig. 1). Their central processes end in the dorsal horn and synapse with spinothalamic neurons. Because of their small size, these nerve fibers traditionally have been studied with high-resolution electron microscopy. The application of sensitive immunohistochemistry with various neuronal markers has enabled the evaluation of skin innervation at a global scale (Hsieh et al. 2000). Among these neuronal proteins, protein gene product 9.5, an ubiquitin C-terminal hydrolase, is particularly useful for demonstrating the rich innervation of the epidermis (Fig. 2b). Based on technical improvements, it is now possible to address the issue of neuropathy of small-diameter nerves due to toxin exposure.

Toxic Neuropathies, Fig. 2 Diagram of the progression of toxic neuropathy. The diagram illustrates degeneration of nerve terminals in neuromuscular junctions (**a**) and in the epidermis of the skin (**b**) during the evolution of toxic neuropathy with acrylamide-induced neuropathy as an example. Axons and nerve terminals were immunostained with protein gene product 9.5 (*dark brown*), and neuromuscular junctions with substrates of cholinesterase (*blue*). (**a**) In normal muscles, motor nerve terminals travel into every fold of the neuromuscular junction (*NMJ*) with quite uniform thickness. In the early phase of degeneration, focal swelling of nerve terminals in the neuromuscular junctions develops, and axonal swelling extends into the motor nerve trunks. Finally, in the late stage, neuromuscular junctions are denervated, and motor nerves become degenerated with a segmented appearance. (**b**) Bundles of nerve fibers form subepidermal nerve plexuses (*SNP*) in the dermis (*DERM*) below the epidermis (*EPI*) in normal skin. Epidermal nerves arise from the subepidermal nerve plexuses, and ascend vertically in the epidermis with a varicose appearance. In the early phase of degeneration, epidermal nerves become swollen, and branching increases. During the late phase, epidermal nerves disappear, and the epidermis becomes denervated. Subepidermal nerve plexuses have a fragmented appearance, indicating that they are undergoing degeneration

Capsaicin is an active compound from hot peppers, which is known to activate the transient receptor potential vallinoid type-1 (TRPV1) channel. A major use of capsaicin ointment is to treat neuropathic pain; its efficacy probably occurs through damage to sensory nerve endings. Systemic injection of capsaicin in neonatal rats abolishes primary afferent terminals in the dorsal horn of the spinal cord. In human studies, capsaicin causes degeneration of nerve terminals in the epidermis after local application with corresponding loss of thermal sensations in the application area (Simone et al. 1998). A certain degree of epidermal nerve regeneration can later be observed, suggesting that the major target of capsaicin-induced nerve damage by local treatment is at the terminal portion instead of at the neuronal cell body.

Traditionally, acrylamide and cisplatin have been considered to cause neuropathy of large-diameter nerves. The development of staining for cutaneous nerve terminals allowed the influence of acrylamide on small-diameter sensory nerves in the skin to be assessed (Fig. 2b). At the initial stage of intoxication, epidermal nerves show two major changes: terminal swelling and increased branching (Ko et al. 2002). There is a progressive reduction in epidermal nerve density thereafter. At the late stage, there is significant dermal nerve degeneration with ultrastructural demonstration of vacuolar changes. These findings have established the pathological consequences of acrylamide neurotoxicity in cutaneous sensory nerves for studying the "dying-back" pathology of nociceptive nerves. These phenomena are quite similar to the pathology of cutaneous nerves in human skin, including epidermal nerve swelling, increased branching points of epidermal nerves, and fragmentation of dermal nerve fibers.

In addition to structural changes, toxic neuropathies, in particular, chemotherapy-induced peripheral neuropathy, serve as excellent system to explore neuropathic pain mechanisms because patients treating with chemotherapy frequently suffer from neuropathic pain (Balayssac et al. 2011; Authier et al. 2009). After chemotherapy-induced peripheral neuropathy by paclitaxel, there was significant increase in spontaneous discharges of A-fibers and C-fibers in sural nerves (Xiao and Bennett 2008) and increased spontaneous discharges of wide dynamic neurons in the dorsal horn (Cata et al. 2006). The mechanisms of neuropathic pain include the dysregulations of protein expression, for example, upregulation of P2X3 and ASIC3 channels at the dorsal root ganglia level (Hori et al. 2010) and decreased glutamate transporter at the dorsal horn (Cata et al. 2006). Functional changes in T-type calcium channel without alterations of protein expression also account for mechanisms of neuropathic pain in paclitaxel neuropathy (Okubo et al. 2011). These changes lead to ectopic sensory neuronal hyperexcitability (Xiao and Bennett 2012; Okubo et al. 2011; Xiao and Bennett 2012).

Our knowledge of toxic neuropathies has broadly expanded because many new investigative techniques have been developed, particularly toxic neuropathies of nociceptive nerves, whose dysfunctions cause diverse manifestations of neuropathic pain (Hsieh 2010).

Classification of Toxic Neuropathies

Sources
- Environmental contaminants: lead
- Occupational outbreaks: n-hexane
- Adverse effects of medications: cisplatin
- Accidental exposure or ingestion: lead
- Experimental uses: e.g., hexacarbons, 2,5-hexanediol, acrylamide, capsaicin
- Tissue deposition of polymeric compounds: amyloid (transthyretin)

Functional Components
- Motor neuropathy: lead
- Sensory neuropathies
  - Proprioceptive type: cis-platinum, vincristine, isoniazid, amyloid
  - Nociceptive type: capsaicin, amyloid
- Autonomic neuropathy: acrylamide, alcoholic, arsenic (inorganic), diphtheritic toxin, amyloid

Clinical Presentations
- Weakness
- Sensory disturbances: neuropathic pain, hypesthesia
- Dysautonomia: orthostatic hypotension, diarrhea, constipation, anhidrosis

Structural Components
- Large myelinated nerves
  - Motor nerves: acrylamide
  - Sensory nerves (proprioception): n-hexane
- Small myelinated nerves
  - Sensory nerves (thermal sense, nociception): capsaicin
  - Autonomic nerves: acrylamide, alcoholic, arsenic (inorganic), diphtheritic toxin
- Unmyelinated nerves
  - Sensory nerves (thermal senses, nociception): capsaicin
  - Autonomic nerves: acrylamide, alcoholic, arsenic (inorganic), diphtheritic toxin

Sites of Actions and Mechanisms
- Proximal axons: filamentous swelling
  - β,β'-iminodipropionitrile
- Terminal portion: swelling and degeneration
  - Neuromuscular junctions: 2,5-hexanediol, acrylamide
  - Pacinian corpuscles: 2,5-hexanediol
  - Epidermal nerves of the skin: acrylamide, cisplatin
  - Vasculature: amyloid

Pathology
- Primarily axonal degeneration: hexacarbon
- Primarily demyelination: tellurium
- Primarily vascular deposition: amyloid

Summary

The patterns of toxic neuropathies have changed, i.e., the incidence and spectrum. Among these trends, medication-related toxic neuropathies emerge as an important issue, particularly chemotherapy-induced peripheral neuropathy. A further disease entity is toxic products due to polymeric compounds of genetic mutations, for example, family amyloid polyneuropathy due to transthyretin mutations, a major etiology of painful neuropathy. These new types of toxic neuropathies provide new insights to explore mechanisms of nerve injury and novel therapeutic strategies.

Acknowledgment The works were supported by grants from Nation Science Council and National Health Research Institute, Taiwan.

References

Ando, Y., Nakamura, M., & Araki, S. (2005). Transthyretin-related familial amyloidotic polyneuropathy. *Archives of Neurology, 62*, 1057–1062.

Authier, N., Balayssac, D., Marchand, F., et al. (2009). Animal models of chemotherapy-evoked painful peripheral neuropathies. *Neurotherapeutics, 6*, 620–629.

Balayssac, D., Ferrier, J., Descoeur, J., et al. (2011). Chemotherapy-induced peripheral neuropathies: From clinical relevance to preclinical evidence. *Expert Opinion on Drug Safety, 10*, 407–417.

Cata, J. P., Weng, H. R., Chen, J. H., & Dougherty, P. M. (2006). Altered discharges of spinal wide dynamic range neurons and down-regulation of glutamate transporter expression in rats with paclitaxel-induced hyperalgesia. *Neuroscience, 138*, 329–338.

Chen, W. P., Chang, Y. C., & Hsieh, S. T. (1999). Trophic interactions between sensory nerves and the targets. *Journal of Biomedical Science, 6*, 79–85.

Chevalier-Larsen, E., & Holzbaur, E. L. F. (2011). Axonal transport and neurodegenerative disease. *Biochimica et Biophysica Acta, 1762*, 1094–1108.

Coleman, M. P., & Freeman, M. R. (2010). Wallerian degeneration, WldS, and Nmnat. *Annual Review of Neuroscience, 33*, 245–267.

Day, I. N. M., & Thompson, R. J. (2010). UCHL1 (PGP 9.5): Neuronal biomarker and ubiquitin system protein. *Progress in Neurobiology, 90*, 327–362.

Hirokawa, N., & Noda, Y. (2008). Intracellular transport and kinesin superfamily proteins, KIFs: Structure, function, and dynamics. *Physiological Reviews, 88*, 1089–1118.

Hori, K., Ozaki, N., Suzuki, S., & Sugiura, Y. (2010). Upregulations of P2X3 and ASIC3 involve in hyperalgesia induced by cisplatin administration in rats. *Pain, 149*, 393–405.

Hsieh, S. T. (2010). Pathology and functional diagnosis of small-fiber painful neuropathy. *Acta Neurologica Taiwanica, 19*, 82–89.

Hsieh, S. T., Chiang, H. Y., & Lin, W. M. (2000). Pathology of nerve terminal degeneration in the skin. *Journal of Neuropathology and Experimental Neurology, 59*, 297–307.

Jortner, B. S. (2000). Mechanisms of toxic injury in the peripheral nervous system: Neuropathologic considerations. *Toxicologic Pathology, 28*, 54–69.

Ko, M. H., Chen, W. P., Lin-Shiau, S. Y., & Hsieh, S. T. (1999). Age-dependent acrylamide neurotoxicity in mice: Morphology, physiology, and function. *Experimental Neurology, 158*, 37–46.

Ko, M. H., Chen, W. P., & Hsieh, S. T. (2002). Neuropathology of skin denervation in acrylamide-induced neuropathy. *Neurobiology of Disease, 11*, 155–165.

LoPachin, R. M., & Gavin, T. (2008). Acrylamide-induced nerve terminal damage: Relevance to neurotoxic and neurodegenerative mechanisms. *Journal of Agricultural and Food Chemistry, 56*, 5994–6003.

Luo, L., & O'Leary, D. D. M. (2005). Axon retraction and degeneration in development and disease. *Annual Review of Neuroscience, 28*, 127–156.

Okubo, K., Takahashi, T., Sekiguchi, F., et al. (2011). Inhibition of T-type calcium channels and hydrogen sulfide-forming enzyme reverses paclitaxel-evoked neuropathic hyperalgesia in rats. *Neuroscience, 188*, 148–156.

Pascuzzi, R. M. (2009). Peripheral neuropathy. *The Medical Clinics of North America, 93*, 317–342.

Perrot, R., Berges, R., Bocquet, A., & Eyer, J. (2008). Review of the multiple aspects of neurofilament functions, and their possible contribution to neurodegeneration. *Molecular Neurobiology, 38*, 27–65.

Said, G. (2007). Diabetic neuropathy-a review. *Nature Clinical Practice Neurology, 3*, 331–340.

Simone, D. A., Nolano, M., Johnson, T., Wendelschafer-Crabb, G., & Kennedy, W. R. (1998). Intradermal injection of capsaicin in humans produces degeneration and subsequent reinnervation of epidermal nerve fibers: Correlation with sensory function. *The Journal of Neuroscience, 18*, 8947–8959.

Spencer, P. S., & Schaumburg, H. H. (2000). *Experimental and clinical neurotoxicology*. New York: Oxford University Press.

Toews, A. D., Hostettler, J., Barrett, C., & Morell, P. (1997). Alterations in gene expression associated with primary demyelination and remyelination in the peripheral nervous system. *Neurochemical Research, 22*, 1271–1280.

Xiao, W. H., & Bennett, G. J. (2008). Chemotherapy-evoked neuropathic pain: Abnormal spontaneous discharge in A-fiber and C-fiber primary afferent neurons and its suppression by acetyl-L-carnitine. *Pain, 135*, 262–270.

Xiao, W. H., & Bennett, G. J. (2012). Effects of mitochondrial poisons on the neuropathic pain produced by the chemotherapeutic agents, paclitaxel and oxaliplatin. *Pain, 153*(3), 704–709.

Yang, N. C., Lee, M. J., Chao, C. C., et al. (2010). Clinical presentations and skin denervation in amyloid neuropathy due to transthyretin Ala97Ser. *Neurology, 75*, 532–538.

Traction

Susan Mercer
School of Biomedical Sciences, The University of Queensland, Brisbane, Australia

Synonyms

Autotraction; Gravity-assisted traction; Intermittent; Manual or continuous traction; Progressive adjustive; Self-traction; Unloading

Definition

Traction is a treatment for spinal pain. It involves applying a pulling force to the lower limbs or to the head, in order to separate the vertebrae of the spine and/or to stretch the surrounding muscles and ligaments. The traction force may be delivered manually or via weights, pulleys, or mechanical devices as a continuous, sustained, intermittent, or intermittent pulsed force.

Characteristics

Mechanism
No mechanism has been established whereby traction might relieve pain. Nevertheless, proponents of traction believe that it works and have speculated on its mechanism of effect (Krause et al. 2000).

There is clear evidence that spinal traction does separate vertebral bodies. However, in the lumbar spine much of the separation observed arises from flattening of the lumbar lordosis (Twomey 1985). In the cervical spine, 30 lb of traction achieves only fractions of a millimeter separation between vertebral bodies, amounting to 2 mm total elongation anteriorly and 6 mm posteriorly between C2 and T1 (Colachis and Strohm 1969). The purported benefit of this separation is, however, ambiguous.

When traction is used to treat radicular pain, the implicit mechanism is decompression of the affected spinal nerve. However, separation of vertebrae increases the longitudinal dimension of the intervertebral foramina, but longitudinal compression of spinal nerves is an uncommon phenomenon. Most commonly, spinal nerves are affected in the sagittal dimension, anteriorly by disc herniations or osteophytes or posteriorly by osteophytes of the zygapophysial joints. Longitudinal separation does not relieve encroachment in the sagittal dimension. Moreover, upon the patient resuming the upright posture, any benefit of traction is immediately lost, as gravity restores the compression load on the spine. Indeed, it has been shown that, without rising, after simply resting on the traction table for 20 min, the effects of cervical traction are all but lost (Colachis and Strohm 1969).

Another conjecture is that traction reduces disc herniations. The available data, however, are limited and conflicting. In one small study, although traction did reduce disc herniations in two of three patients, the protrusions reappeared within 14 min after release of the force (Matthews 1968).

In the absence of firm evidence for a mechanical effect of traction, some authorities have pursued alternative rationales, such as silencing ectopic impulse generators and normalization of conduction in spinal nerves (Krause et al. 2000). These speculations, however, nevertheless presuppose that traction reverses compression of the spinal nerve by separating the vertebral bodies. They also require that relatively brief traction somehow achieves lasting reversal of the pathophysiology that causes pain.

There is even less of a physiological rationale for traction when it is applied for spinal pain, as opposed to radicular pain. In the first instance, nerve root irritation causes pain in the limbs, not in the back or in the neck. The rationale for traction for spinal pain, therefore, cannot involve decompression of spinal nerves. Instead, it has been proposed that perhaps spinal pain might be relieved by "increasing non-nociceptive input and recruitment of descending inhibition" (Krause et al. 2000). While perhaps attractive as a conjecture, such a statement falls short of actually constituting evidence of how traction might relieve spinal pain. Another proposition is that traction serves to stretch spinal tissues (Krause et al. 2000), but in that event it is questionable whether elaborate and passive traction offers any advantage over simple stretching exercises that the patients can undertake themselves. The proposition that traction reduces intervertebral disc pressures is confounded not only by the lack of experimental data that demonstrate this effect but also by the lack of a cogent theory as to how raised disc pressure causes back pain and why that pain should stay relieved once the traction is released and the patient resumes an upright posture.

Applications
Traction has been used to treat neck pain, cervical radicular pain and radiculopathy, and back pain and lumbar radicular pain and radiculopathy. Some practitioners use it to treat thoracic spinal pain, but there is no literature on its efficacy for this condition.

Traction can be applied manually by the therapist (manual traction), by a motorized pulley (motorized lumbar or cervical traction), by the patients themselves providing the pulling force (autotraction), or by suspension from a device (gravitational traction) (Twomey 1985).

Efficacy
Systematic reviews have been confounded by the irregular and inconsistent diagnostic criteria used in the studies reviewed. It has not always been evident if the patients had somatic pain, somatic referred pain, or radicular pain. Although some earlier reviews offered open or encouraging conclusions, these have been supplanted by subsequent studies and reviews.

No randomized controlled trials have shown if traction is effective for neck pain (Aker et al. 1996; Harms-Ringdahl and Nachemson 2000). The available studies indicated that it is not effective for acute neck pain (Bogduk and McGuirk 2006). For chronic neck pain, traction has not been subjected to scientific studies. For cervical radicular pain, studies have shown that traction is no more effective than sham traction or placebo treatment (British Association of Physical Medicine 1966; Goldie and Landquist 1970; Klaber et al. 1990). The same conclusions apply for the treatment of lumbar radicular pain (Coxhead et al. 1981; Matthews and Hickling 1975; Pal et al. 1986). For nonspecific low back pain, traction is no more effective than a placebo treatment (Beurskens et al. 1997).

Indications
In the face of the available scientific evidence, there are no legitimate indications for traction in the modern era. It is a treatment, steeped in tradition but devoid of evidence.

Cross-References
▶ Lumbar Traction

References
Aker, P. D., Gross, A. R., Goldsmith, C. H., et al. (1996). Conservative management of mechanical neck pain: Systematic overview and meta-analysis. *British Medical Journal, 313*, 1291–1296.

Beurskens, A. J., de Vet, H. C., Koke, A. J., et al. (1997). Efficacy of traction for non-specific low back pain: 12-week and 6-month results of a randomized clinical trial. *Spine, 22*, 2756–2762.

Bogduk, N., & McGuirk, B. (2006). *Medical management of acute and chronic neck pain. An evidence-based approach*. Amsterdam: Elsevier.

British Association of Physical Medicine. (1966). Pain in the neck and arm: A multicentre trial of the effects of physiotherapy. *British Medical Journal, 1*, 253–258.

Colachis, S. C., & Strohm, B. R. (1969). Effects of intermittent traction on separation of lumbar vertebrae. *Archives of Physical Medicine and Rehabilitation, 44*, 251–258.

Coxhead, C. E., Inskip, H., Meade, T. W., et al. (1981). Multicentre trial of physiotherapy in the management of sciatic symptoms. *Lancet, 1*, 1065–1068.

Goldie, I., & Landquist, A. (1970). Evaluation of the effects of different forms of physiotherapy in cervical pain. *Scandinavian Journal of Rehabilitation Medicine, 2–3*, 117–121.

Harms-Ringdahl, K., & Nachemson, A. (2000). Acute and subacute neck pain: Nonsurgical treatment. In A. Nachemson & E. Jonsson (Eds.), *Neck and back pain: The scientific evidence of causes, diagnosis, and treatment* (pp. 327–338). Philadelphia: Lippincott Williams & Wilkins.

Klaber, M. J. A., Hughes, G. I., & Griffiths, P. (1990). An investigation of the effects of cervical traction. Part 1: Clinical effectiveness. *Clinical Rehabilitation, 4*, 205–211.

Krause, M., Refshauge, K. M., Dessen, M., et al. (2000). Lumbar spine traction: Evaluation of effects and recommended application for treatment. *Manual Therapy, 5*, 72–81.

Matthews, J. A. (1968). Dynamic discography: A study of lumbar traction. *Annals of Physical Medicine, 9*, 275–279.

Matthews, J. A., & Hickling, J. (1975). Lumbar traction: A double-blind controlled study for sciatica. *Rheumatology and Rehabilitation, 14*, 222–225.

Pal, B., Mangion, P., Hossain, M. A., et al. (1986). A controlled trial of continuous lumbar traction in the treatment of back pain and sciatica. *British Journal of Rheumatology, 25*, 181–183.

Twomey, L. T. (1985). Sustained lumbar traction. An experimental study of long spine segments. *Spine, 10*, 146–149.

Tractus Trigeminothalamicus

▶ Trigeminothalamic Tract Projections

Traditional Pharmacological Pain Relief

Definition

Pain relieved by oral and subcutaneous injections of analgesic drugs when the patient demands pain relief. Often ineffective in that doses are too low and dosing intervals are too long.

Cross-References

▶ Postoperative Pain, Acute Pain Team

Trafficking

Definition

Trafficking refers to synthesis and targeting of proteins to specific locations within the cell.

Cross-References

▶ Opioid Receptor Trafficking in Pain States
▶ Trafficking and Localization of Ion Channels

Trafficking and Localization of Ion Channels

Matthew N. Rasband[1,2] and Peter Peter[3]
[1]Department of Molecular and Cellular Biology, Baylor College of Medicine, Farmington, Houston TX, USA
[2]University of Connecticut Health Center, Farmington, CT, USA
[3]Department of Neurobiology and Anatomy, University of Rochester Medical Center, Rochester, NY, USA

Synonyms

Clustering; Ion channel trafficking; Targeting; Voltage-dependent pores

Definition

Ion channels are pore-forming proteins in the surface membrane through which electrical charges move to establish and alter the membrane potential. These channels must be synthesized and inserted into the membrane in the proper locations within the neuron in order for signaling to occur normally.

Characteristics

Neurons are highly polarized cells, and during development, individual neurites become either dendrites or axons. An essential element in this polarization is the proper ▶ trafficking of proteins to these processes and, in particular, the targeting of ion channels to specific locations. In dendrites, ligand-gated channels must accumulate at post-synaptic regions. In axons, Na^+ channels are clustered at the initial segment, which is the site of integration of postsynaptic potentials and of initiation of the action potential. Na^+ and K^+ channels are distributed at low density throughout the remainder of the fiber, and if the axon is myelinated, they are also clustered within specific zones in the region of the node of Ranvier. Ectopic localization of these channels can lead to repetitive or spontaneous firing and can thus contribute to pain. Localization may be controlled by domains within the protein that direct transport or insertion to specific sites, or by differential rates of endocytosis. For example, a protein may be axonal because its endocytotic removal rate is much higher in dendrites. There has been much progress recently in uncovering these mechanisms.

Multiple targeting domains have been discovered that direct K^+ channels to different sites within the neuron. The differential targeting of Kv2.1 and Kv2.2 to alternative dendritic regions is governed, at least in part, by a 26-amino-acid ▶ motif on the cytoplasmic tail of Kv2.1. Manganas et al. (2001) have found that specific residues within the pore region of Kv1. × control surface expression of those channels. Voltage-dependent ion channels are typically heteromultimers consisting of a pore-forming alpha subunit and one or more beta subunits that modulate gating. Kv channel beta subunits are bound at the cytoplasmic surface of the channel. It was recently found that beta subunits also participate in localization. Axonal targeting of Kv1.2 in hippocampal neurons is driven by the tetramerization domain on the N-terminus, at a site that binds Kvbeta2 (Gu et al. 2003). Kvbeta subunits have a binding pocket for NADP+, and mutation of this site eliminates their axonal

targeting capability (Campomanes et al. 2002). Nodal regions have three major zones: the nodal gap, paranodes in which axoglial junctions link the paranodal loops to the axolemma, and the juxtaparanodes, zones that flank the paranodes, begin the internode, and have no special morphological characteristics. At nodal regions KCNQ2 and KCNQ3, slowly activating channels that can be modulated by neurotransmitters are found within the nodal gap (Devaux et al. 2004). Kv1.1 and Kv1.2 are clustered in the juxtaparanodes in both the PNS and CNS. The localization of Kv1. × is dependent on the integrity of the axoglial paranodal junctions. When the latter are disrupted, these K^+ channels are found in the paranodes and diffusely throughout the internode.

At both axon initial segments and ▶ nodes of Ranvier, voltage-dependent ▶ Na^+ channels are linked to the spectrin-actin cytoskeleton by the adapter protein ankyrinG. AnkyrinG also serves to link Na^+ channels in a large molecular complex that includes the L1-family proteins NrCAM and neurofascin. The beta subunits of Na^+ channels are membrane proteins, and they, along with NrCAM, neurofascin, and contactin (a GPI-anchored protein), are all members of the immunoglobulin superfamily, implicating them in intercellular (perhaps neuron-glial) signaling. The link to ankyrinG also appears to be important in targeting. In spinal motor neurons in culture, Na^+ channels are normally synthesized in the soma but are inserted in the surface membrane only at the axon initial segment (Alessandri-Haber et al. 1999). Nav1.2 has a C-terminal motif that is involved in axonal targeting. It has also recently been demonstrated that the intracellular linker between domains II and III serves as a localization signal to the initial segment (Garrido et al. 2003). Of particular interest, Lemaillet et al. (2003) showed that this latter loop contains an ankyrinG-binding motif. Thus, establishment of these regions of high Na^+ channel density appears to involve targeting to the axon and insertion in the surface membrane, followed by cytoskeletal immobilization via ankyrinG. In the case of nodes of Ranvier, there is an additional requirement for glial modeling of the axonal surface. Myelinating Schwann cells in the PNS direct the clustering of Na^+ channels, perhaps by reorganizing the low density of these channels present throughout the axon prior to myelin formation (Dugandzija-Novakovic et al. 1995). The neuron-glial communication involved in this trafficking may involve the other members of the Na^+ channel complex mentioned above (Custer et al. 2003). In the CNS, some early aspects of clustering may involve soluble factors released by oligodendrocytes, but final node formation seems to be contact dependent, as in the PNS (Kaplan et al. 2001). Finally, there is a developmentally regulated progression in Na^+ channel subtype in axons. Nav1.2 is expressed early and is later replaced by Nav1.6 (Boiko et al. 2001). The physiological consequences of this shift are not known, but Nav1.6 has a unique resurgent current (Raman and Bean 1997), which could render axons more susceptible to reentry excitation that may contribute to pain, if other regulatory zones of nodes are compromised. For example, loss of juxtaparanodal Kv1. × can lead to instabilities.

Turnover rates of ion channels in myelinated axons have not been measured directly, but some data are available. Nodal Na^+ channels are particularly stable, and clusters can be detected up to 9 days after myelin disruption (Custer et al. 2003). This is undoubtedly due to the cytoskeletal link via ankyrinG. The stabilization of these channels seems to occur gradually after their initial clustering and may involve the formation of large complexes through the multiple binding sites of ankyrinG (Custer et al. 2003). Juxtaparanodal Kv1. × channel clusters are significantly less stable and break up within 1–2 days after initiating demyelination (Rasband et al. 1998).

There is considerable evidence that altered ion channel trafficking, localization, and turnover are involved in painful neuropathies, although the precise mechanisms remain to be determined. Painful neuromas develop within peripheral nerves after axonal injury, and the spontaneous discharge at these sites is thought to result from an accumulation of Na^+ channels (Devor et al. 1989). Further, there is evidence that specific subtypes of Na^+ channels are involved, including PN1 and PN3. The Na^+ channels accumulating at neuromas are colocalized with ankyrinG (Kretschmer et al. 2002), and the increased levels of both proteins may reflect dysregulation of trafficking and/or turnover rates. As more information becomes available on the neurobiological mechanisms regulating ion channel biosynthesis, trafficking, and targeting, it will be interesting to apply these principles to models of neuropathic pain. For example, if some forms of neuropathic pain are a consequence of altered Na^+ channel trafficking, then therapeutic strategies designed to disrupt or perturb the trafficking of these proteins may prove useful in the treatment of these disorders.

References

Alessandri-Haber, N., Paillart, C., Arsac, C., Gola, M., Couraud, F., & Crest, M. (1999). Specific distribution of sodium channels in axons of rat embryo spinal motoneurones. *The Journal of Physiology, 518*, 203–214.

Boiko, T., Rasband, M. N., Levinson, S. R., Caldwell, J. H., Mandel, G., Trimmer, J. S., & Matthews, G. (2001). Compact myelin dictates the differential targeting of two sodium channel isoforms in the same axon. *Neuron, 30*, 91–104.

Campomanes, C. R., Carroll, K. I., Manganas, L. N., Hershberger, M. E., Gong, B., Antonucci, D. E., Rhodes, K. J., & Trimmer, J. S. (2002). Kv beta subunit oxidoreductase activity and Kv1 potassium channel trafficking. *The Journal of Biological Chemistry, 277*, 8298–8305.

Custer, A. W., Kazarinova-Noyes, K., Sakurai, T., Xu, X., Simon, W., Grumet, M., & Shrager, P. (2003). The role of the ankyrin-binding protein NrCAM in node of ranvier formation. *The Journal of Neuroscience, 23*, 10032–10039.

Devaux, J. J., Kleopa, K. A., Cooper, E. C., & Scherer, S. S. (2004). KCNQ2 is a nodal K^+ channel. *The Journal of Neuroscience, 24*, 1236–1244.

Devor, M., Keller, C. H., Deerinck, T. J., Levinson, S. R., & Ellisman, M. H. (1989). Na^+ channel accumulation on axolemma of afferent endings in nerve end neuromas in apteronotus. *Neuroscience Letters, 102*, 149–154.

Dugandzija-Novakovic, S., Koszowski, A. G., Levinson, S. R., & Shrager, P. (1995). Clustering of Na channels and node of ranvier formation in remyelinating axons. *The Journal of Neuroscience, 15*, 492–502.

Garrido, J. J., Giraud, P., Carlier, E., Fernandes, F., Moussif, A., Fache, M. P., Debanne, D., & Dargent, B. (2003). A targeting motif involved in sodium channel clustering at the axonal initial segment. *Science, 300*, 2091–2094.

Gu, C., Jan, Y. N., & Jan, L. Y. (2003). A conserved domain in axonal targeting of Kv1 (Shaker) voltage-gated potassium channels. *Science, 301*, 646–649.

Kaplan, M. R., Cho, M. H., Ullian, E. M., Isom, L. L., Levinson, S. R., & Barres, B. A. (2001). Differential control of clustering of the sodium channels Na(v)1.2 and Na(v)1.6 at developing CNS nodes of ranvier. *Neuron, 30*, 105–119.

Kretschmer, T., England, J. D., Happel, L. T., Liu, Z. P., Thouron, C. L., Nguyen, D. H., Beuerman, R. W., & Kline, D. G. (2002). Ankyrin G and voltage gated sodium channels co-localize in human neuroma-key proteins of membrane remodeling after axonal injury. *Neuroscience Letters, 323*, 151–155.

Lemaillet, G., Walker, B., & Lambert, S. (2003). Identification of a conserved ankyrin-binding motif in the family of sodium channel alpha subunits. *The Journal of Biological Chemistry, 278*, 27333–27339.

Manganas, L. N., Wang, Q., Scannevin, R. H., Antonucci, D. E., Rhodes, K. J., & Trimmer, J. S. (2001). Identification of a trafficking determinant localized to the Kv1 potassium channel pore. *Proceedings of the National Academy of Sciences of the United States of America, 98*, 14055–14059.

Raman, I. M., & Bean, B. P. (1997). Resurgent sodium current and action potential formation in dissociated cerebellar purkinje neurons. *The Journal of Neuroscience, 17*, 4517–4526.

Rasband, M. N., Trimmer, J. S., Schwarz, T. L., Levinson, S. R., Ellisman, M. H., Schachner, M., & Shrager, P. (1998). Potassium channel distribution, clustering, and function in remyelinating rat axons. *The Journal of Neuroscience, 18*, 36–47.

Trafficking of Proteins

Definition

Proteins are synthesized on ribosomes in the cell body, packaged into vesicles in the Golgi apparatus, and then transported along cytoskeletal structures to their final location in the cell. The process of targeted transport is trafficking.

Cross-References

▶ Peripheral Neuropathic Pain
▶ Trafficking and Localization of Ion Channels

Training by Quotas

Wilbert E. Fordyce
Department of Rehabilitation, University of Washington School of Medicine, Seattle, WA, USA

Synonyms

Contingency management; Exercise; Quotas

Definition

Training by quota describes methods for increasing exercise and activity level by use of the quota system, a form of contingency management based on behavioral science. The methods make use of sensitivity to our environments to improve our state. Review of the theoretical and conceptual background can be found in texts by Fordyce (1976, 1990). "How" to do it will be described. *Pinpoint* the behavior to be changed (i.e., specify in ▶ movement cycles, walking steps, exercise repetitions), *record* the pre-intervention rate of the behavior, and *consequate*, i.e., spell out the rate of reinforcement and by whom, always with the patient's informed consent.

Characteristics

The term chronic pain encompasses many conditions. Increasing exercise and/or activity level is not indicated for all of them. Evaluation of a problem of chronic pain (ideally a multidisciplinary, multimodal process) indicating increased exercise or activity is clearly to the patient's benefit points to using these methods. Quota methods help patients increase exercise or activity levels. They are not methods for treating pain. Increases in activity level help restore patient access to activities usually engaged in. When target levels of activity or exercise are reached, the program can be faded to maintenance levels or stopped altogether (Lindström et al. 1992). Reentry into sustaining activities should be achieved before halting or reducing exercise levels.

▶ Reinforcers are usually simple to define and apply in the practical case, using the Premack principle (Premack 1959). This states that high strength behaviors reinforce low strength behaviors; paraphrased as, what a person does a lot of can be used to reinforce activities or movements targeted to be increased. Observing the consequences of what the person does a lot can usually readily identify effective reinforcers. To a restless person frequently on the move, activity, a high strength behavior, is likely to be an effective reinforcer. If programmed or scheduled carefully, it can serve to enhance or reinforce exercise. To an inactive person, one who moves relatively little, activity is unlikely to be reinforcing but rest or "time out" from activity is. In the context of pain management, rest or "time out" from activity and attention or encouragement of important others around the person (e.g., the therapist) usually suffice as reinforcers, when programmed appropriately.

Baseline

Baseline is the starting point or rate (i.e., number of repetitions) at which the target behavior is occurring, e.g., number of steps walked before weakness, pain, or fatigue causes stopping. If the target behavior is not in the person's present repertoire, establishing it by teaching or "shaping" is indicated (Fordyce 1976).

Walking, performing selected exercises, and engaging in activities intended to diminish reclining time and increasing activity are common targets for behavior change. Talking about pain, grimacing, asking for and/or taking analgesics are less frequently targets for change and should become so only with careful consideration of appropriateness to the patient's condition and with full informed consent.

Quota

Once baseline performance is identified, set quotas to be applied for each trial. The initial quota should be approximately 2/3 or 3/4 of baseline, e.g., laps walked = 7, initial quota = 4 or 5, baseline of an exercise repetition = 5, and initial quota = 2 or 3.

Initial quota level should be determined by confidence that it is an amount well within the patient's current repertoire, ensuring successes at early trials. It is better to have too low an initial quota and be sure of success than too high and a risk of failure. Quotas are then incremented at a rate driven mainly by an ascending performance curve confidently expected to provide several early succeeding trials before initial baseline is reached. In short, "success breeds success."

Quota Increment Rates

Most commonly, incrementing by 1 repetition per trial or session is appropriate. In cases of marginally adequate ability to increment, a slower rate (e.g., 1 × each 2 or 3 trials) may suffice. An amount easily performed may indicate increments of 2 or more with each trial, but do that only if confident that several successful sessions are achieved prior to reaching baseline. In all cases, full informed consent prior to performance is indicated. Quota increment rates should be spelled out to the patient following their determination, including the expected endpoint.

Quota Endpoints

Within physiological limits, quota endpoints or targets are defined mainly by practical considerations. Target levels must consider time to permit achieving current quota levels, as well as overall fatigue considerations if there is an array of exercises to be performed. Quota increments and endpoints are also moderated by patient access to monitoring. For example, for a patient for whom 3 weeks of daily trials is the logistical limit, quota endpoints may be limited by the time available to reach them using the quota increments selected, unless surrogate monitoring is available, e.g., family.

Examples

Walking
Pinpoint
Lay out the course to be walked (e.g., a 25 m lap in a hospital corridor or distance in meters from the front door of home to a selected landmark).

Baseline

The initial baseline should be within expected patient tolerance. The patient is asked to walk laps (start-end-return to start) until "pain, weakness or fatigue cause you to want to stop. You decide when to stop." If the patient is unable to do at least one lap, shorten the lap distance until two or more repetitions are within starting tolerance. Theoretically, baseline is defined by a stabilizing performance curve. Usually, however, in the exercise context, 2–3 baseline trials suffice. Each trial is spaced to provide ample rest between trials.

Record

Record laps walked each trial. A graph on the wall at the bedside or in a folder carried by the patient is adequate. By making recorded results "visible," an element of social support or reinforcement is also provided.

Quota

One lap increments per trial usually suffices.

Consequate

Reinforcement from rest (i.e., time out from the activity) and social approval of contextual personnel usually suffice. Resort to some form of tangible reinforcement is rarely needed. See Fordyce (1976, 1990) for more on this.

Exercise (e.g., Hip Abduction, Deep Knee Bends, Arm Extension)

Pinpoint

The patient is asked to perform the assigned exercise (always with full informed consent) and instructed as to how to do it.

Baseline

The patient is asked to perform the assigned exercise "until pain, weakness or fatigue cause you to want to stop. You decide when to stop." Amounts performed across 2–3 trials, spaced appropriately, are recorded.

Quota

Usually one repetition per trial suffices. Increment rates of >1 repetition per trial could be designated if there is ample confidence that

performance will succeed, particularly if access to monitoring is limited by the calendar.

Consequate

Rest or time out from the exercise plus therapist attention and regard suffice.

Quota Failures

The options re failure divide into (a) problems of patient status and access and (b) performance (see also Fordyce 1976, 1990).

Status

If opportunity to perform was not the problem, ask the patient why failure? If change in health or status of pain is the problem, clarify with the attending physician whether to proceed with quotas, change exercises, etc. Scheduling problems or access to equipment with which to perform can be dealt with directly.

Performance

Your value as a source of reinforcement should not be expended carelessly. Admonitions or urging better performance lets your attention become contingent on failure. Instead, convey that you hope and expect he/she will be able to resume meeting quotas and leave it at that. If on subsequent trials success is achieved, be quick to praise.

If the patient is not invested in getting better, the patient should choose whether to proceed or to halt the program. Discussions with significant others to the patient may also help.

Acknowledgement Professor Wilbert "Bill" Evans Fordyce passed away October 15, 2009, at age 86. We thank Prof. Dr. med. Bengt H Sjölund for the update of his entry.

References

Fordyce, W. E. (1976). *Behavioral methods in chronic pain and illness* (p. 236). St. Louis: C.V. Mosby.
Fordyce, W. E. (1990). Contingency management. In J. J. Bonica (Ed.), *Management of pain* (2nd ed., Vol. 2, pp. 1702–1710). Philadelphia: Lea & Febiger (also in

Bonica, J. J. (2001). *Management of pain* (3rd ed., pp. 1745–1750). Philadelphia: Lippincott, Williams & Wilkins).

Lindström, I., Ohlund, C., Eek, C., Wallin, L., Peterson, L. E., Fordyce, W. E., & Nachemson, A. L. (1992). The effect of graded activity on patients with subacute low back pain: A randomized prospective clinical study with an operant-conditioning behavioral approach. *Physical Therapy, 72*(4), 279–290.

Premack, D. (1959). Toward empirical behavior laws: I Positive reinforcement. *Psychological Review, 66*, 219–233.

Trait

▶ Personality and Pain

Trajectory

Definition

A developmental trajectory describes the course of a behavior over age or time. As related to pain, that may include the behavioral expression or the subjective experience as predictably changing as a function of time.

Cross-References

▶ Pain Assessment in Neonates

Tramadol

Definition

Tramadol is a synthetic, centrally acting analgesic agent that is structurally related to codeine and morphine; it is a racemic mixture, and the two enantiomers function in a complementary manner to enhance analgesic actions. It also acts as a weak opioid agonist and inhibits the serotonin release and reuptake of norepinephrine and serotonin. The analgesic effects of oral tramadol (100 mg) peak at 1–2 h and last for 3–6 h after drug administration.

Cross-References

▶ Acute Pain in Children, Postoperative
▶ Drugs with Mixed Action and Combinations: Emphasis on Tramadol
▶ Postoperative Pain, Tramadol
▶ Tramadol Hydrochloride

Tramadol Hydrochloride

Jayantilal Govind
Pain Clinic, Department of Anaesthesia,
University of New South Wales, Sydney,
NSW, Australia

Synonyms

Tramadol; Tramal; Zydol

Definition

An analogue of codeine, tramadol hydrochloride is a centrally acting analgesic. It is not derived from natural sources and structurally is not related to opioids but does exhibit certain opioid characteristics.

Characteristics

Mechanism

Animal and in vitro studies suggest that in addition to its mu-opioid effect, tramadol synergistically inhibits the reuptake of norepinephrine (NE) and serotonin and simultaneously stimulates the presynaptic release of serotonin (Raffa and Fridericks 1996; Bamigbade et al. 1997). It has a weak affinity for opioid receptors and is less potent than morphine. The analgesic effect is apportioned between the opioid and monoaminergic components (Desmoules et al. 1996). The response is dose dependent but the relationship between analgesic effect and serum concentration varies considerably between individuals.

Pharmacokinetics

After oral administration, tramadol is rapidly and almost completely absorbed with a mean bioavailability of 68–72 %. The drug is widely distributed and 20 % is bound to plasma proteins. The peak serum level is reached within 2 h (range 1–3 h) (Tramal 2004).

Tramadol is metabolized, via the CY2D6 isoenzyme of cytochrome P450, to 11 metabolites of which only O-desmethyltramadol (M1) is active. M1 has a greater affinity for opioid receptors and exerts greater analgesia than tramadol. About 7–8 % of Caucasians lack this isoenzyme. Tramadol has a half-life of about 6–7 h and is principally excreted via the kidneys. At least 30 % remains unchanged (Tramal 2004).

Routes of Administration

Tramadol can be administered orally, rectally, and parenterally. For formulations and dosing schedules, readers should consult the product information pertinent to their respective jurisdictions.

Applications

The agent is indicated for the treatment of mild to moderate pain. It is less effective than morphine for severe pain (Kaye 2004; Pang et al. 1999; Osipova et al. 1991).

Efficacy

When administered to postsurgical and posttraumatic patients, 100 mg injectable tramadol is equivalent to 5–10 mg of morphine. As an oral dose, 100 mg tramadol is as effective as 1,000 mg paracetamol (Moore and McQuay 1997). For achieving 50 % reduction of pain, tramadol has an NNT of 4.6 (Moore and McQuay 1997). For 150 mg the NNT is 2.9 (Drugdex Drug Evaluations. Micromedex® Healthcare series).

Tramadol has been shown to be effective for the relief of neuropathic pain (Duhmke et al. 2004). Patients with chronic painful neuropathy reported relief of their pain, paresthesia, and touch-evoked allodynia by 2 median points on a 10-point scale (Sindrup et al. 1999). To achieve a 50 % reduction in pain level, the ▶ NNT is between 3 and 6, using a daily dose of 200–400 mg. Studies investigating the efficacy of tramadol for the treatment of pain of diabetes neuropathy (Harati et al. 1998) and ▶ postherpetic neuralgia have been inconclusive (Wareham 2004).

The safety and efficacy of tramadol in patients under the age of 16 years have not been established (Drugdex Drug Evaluations. Micromedex® Healthcare series).

Contraindications

Tramadol is contraindicated in patients with porphyria; with known hypersensitivity to tramadol, opioids, or any excipients; with acute intoxication with alcohol, opioids, hypnotics, analgesics, or psychotropic drugs; and who use monoamine oxidase inhibitors or have used them in the last 14 days. Tramadol must not be used for opioid dependency, addiction, or for opioid withdrawal treatment (Kaye 2004; Pang et al. 1999; Drugdex Drug Evaluations. Micromedex® Healthcare series).

Side Effects and Complications

Adverse reactions are common and patients must be given guidance about appropriate action. The potential for serious side effects including anaphylaxis should not be underestimated. Either the monoaminergic or opioid effects may predominate. The adverse effects of tramadol may be difficult to distinguish or recognize in patients taking multiple medications. Even with therapeutic doses, opioid side effects can occur, except respiratory depression and inhibition of smooth muscle, which are usually less pronounced (Kaye 2004; Drugdex Drug Evaluations. Micromedex® Healthcare series).

Central effects include seizures, hallucinations, euphoria, ataxia, and suicidal ideation. Coadministration with alcohol, general anesthetics, or other respiratory depressants can precipitate profound respiratory decompensation. Dizziness can cause at-risk patients to fall.

Gastrointestinal side effects are very common, including nausea, vomiting, and constipation.

Dyspepsia, diarrhea, and increased flatulence have been reported (Kaye 2004).

Although uncommon, cardiovascular side effects such as hypertension, palpitations, tachycardia, bradycardia, other ECG abnormalities, and orthostatic dysregulation do occur and can be serious.

Additional reported side effects include urinary retention, erythema, hypertonia, hyponatremia, itch, rhabdomyolysis, and diaphoresis.

Drug Interactions

Tramadol is known to interact with at least 72 other drugs, some with serious consequences (Kaye 2004; Drugdex Drug Evaluations. Micromedex® Healthcare series). Carbamazepine reduces tramadol's analgesic effect by increasing its metabolism. Drugs that inhibit CY2D6 activity (e.g., Selective Serotonin Reuptake Inhibitors [SSRI] or quinidine) will prevent conversion to its active metabolite, M1.

Serotonin syndrome is a potentially serious toxic state caused mainly by excess serotonin within the central nervous system (Hall and Buckley 2003). It manifests as a dose-related range of toxic symptoms due to a variety of mental, autonomic, and neuromuscular changes. The clinical features are highly variable, and the onset could be insidious or dramatic. It is nearly always caused by a drug interaction involving two or more "serotonergic" drugs including SSRIs, MAOI, pethidine, tricyclic antidepressants, St John's wort, and lithium. Severe hyperthermia, rhabdomyolysis, disseminated intravascular coagulation, and adult respiratory distress syndrome are potentially life-threatening (Hall and Buckley 2003).

Reproduction

Tramadol is embryotoxic and fetotoxic but not teratogenic in animal studies (Drugdex Drug Evaluations. Micromedex® Healthcare series).

Precautions

Tramadol is best avoided in patients with epilepsy and in patients who are at risk of developing respiratory depression. Because of miosis and its central effects, it is inadvisable to prescribe tramadol for patients with a head injury or raised intracranial pressure. Elderly patients and patients with renal or hepatic decompensation, myxedema, hypothyroidism, and hypoadrenalism are particularly vulnerable to severe side effects and complications. Its administration may complicate the clinical assessment of patients with acute abdominal conditions (Kaye 2004; Drugdex Drug Evaluations. Micromedex® Healthcare series).

Pregnancy and Lactation

Tramadol has been detected in breast milk. It is best avoided during pregnancy and while breast-feeding (Kaye 2004).

Overdose

Treatment is symptomatic. The effects are not completely reversed by naloxone (Kaye 2004), and the latter may increase the risk of seizure (Drugdex Drug Evaluations. Micromedex® Healthcare series).

Acknowledgement Dr. Jayantilal Govind passed away on June 20, 2009, at age 86. We thank Prof. Nik Bogduk for the update of his entry.

References

Bamigbade, T. A., Davidson, C., Langdord, R. M., & Stamford, J. A. (1997). Actions of tramadol: Its enantiomers and principle metabolite, O-desmethyltramadol on serotonin (5-HT) efflux and uptake in the rat dorsal raphe nucleus. *British Journal of Anaesthesia, 79,* 352–356.

Desmoules, J. B., Piquet, R., Collart, L., & Dayer, P. (1996). Combination of monoaminergic modulation to the analgesic effect of tramadol. *British Journal of Clinical Pharmacology, 41,* 7–12.

Drugdex Drug Evaluations. Micromedex® Healthcare series. http://micromedex.hen.net.au/mde-3261/display.exeCTL=apache/products/micromede. Accessed 13 Oct 2004.

Duhmke, R. M., Cornblath, D. D., Hollingshead, J. R. (2004). Tramadol for neuropathic pain. *Cochrane Database Systematic Review,* (2): CD003726.

Hall, M., & Buckley, N. (2003). Serotonin syndrome. *Australian Prescriber, 26,* 62–63.

Harati, Y., Gooch, C., Swenson, M., et al. (1998). Double blind randomized trial of tramadol for the treatment of the pain of diabetic neuropathy. *Neurology, 50,* 1842–1846.

T

Kaye, K. (2004). Trouble with tramadol. *Australian Prescriber, 27,* 26–27 (editorial).

Moore, R. A., & McQuay, H. J. (1997). Single-patient data meta-analysis of 3453 postoperative patients: Oral tramadol versus placebo, codeine and combination analgesics. *Pain, 69,* 287–294.

Osipova, N. A., Novikov, G. A., Beresnev, V. A., et al. (1991). Analgesic effect of tramadol in cancer patients with chronic pain: A comparison with prolonged action morphine sulfate. *Current Therapeutic Research, 50,* 812–821.

Pang, W. W., Mok, M. S., Lin, C. H., et al. (1999). Comparison of patient controlled analgesia (PCA) with tramadol or morphine. *Canadian Journal of Anesthesia, 46,* 1030–1035.

Raffa, R. B., & Fridericks, E. (1996). The basic science aspect of tramadol hydrochloride. *Pain Reviews, 3,* 249–271.

Sindrup, S. H., Andersen, G., Madsen, C., Smith, T., Brosen, K., & Jensen, T. S. (1999). Tramadol relieves pain and allodynia in polyneuropathy: A randomised double blind controlled trial. *Pain, 83,* 85–90.

Tramal (2004). Product information. CSL Limited. Parkville, VIC, Australia.

Wareham, D. W. (2004). Postherpetic neuralgia. In: F. Godlee (Ed.), *Clinical evidence.* edn.

Tramal

▶ Tramadol Hydrochloride

Tramal®

▶ Postoperative Pain, Tramadol

Transcranial Magnetic Stimulation

Synonyms

TMS

Definition

Transcranial Magnetic Stimulation refers to stimulation of the cerebral cortex by an electromagnetic field, generated by a magnetic stimulator

placed over the scalp without the need for surgery or external electrodes. Magnetic stimuli can also activate the peripheral nervous system.

Cross-References

▶ Clinical Migraine Without Aura
▶ Motor Cortex, Effect on Pain-Related Behavior
▶ Stimulation Treatments of Central Pain

Transcription Factor

Definition

A protein, product of gene transcription and translation, which enters the nucleus where it binds to a nucleotide sequence in the regulatory regions of responsive genes and has the effect of either enhancing or repressing the expression of one or more genes downstream of the binding site. Expression of one gene is normally controlled by several transcription factors.

Cross-References

▶ Central Changes after Peripheral Nerve Injury
▶ COX-1 and COX-2 in Pain
▶ Cytokines, Effects on Nociceptors
▶ NSAIDs and Cancer

Transcutaneous Electrical Nerve Stimulation

James Watt
North Shore Hospital, Auckland, New Zealand

Synonyms

AL-TENS (acupuncture-like TENS); Electrical therapy; TENS

Definition

Transcutaneous electrical nerve stimulation (TENS) is a means of relieving pain. It entails delivering an electrical stimulus through electrodes to the skin overlying or near the region in which pain is perceived. The stimulus is delivered from a battery-driven generator.

Characteristics

TENS is a commonly used noninvasive modality that provides an alternative to medication for pain relief. It has been used for more than 30 years, but its effectiveness remains controversial.

Mechanism

TENS was developed on the basis of the gate control theory of pain proposed by Melzack and Wall (1965). This theory predicted that stimulation of large-diameter primary afferent fibers (A fibers) would have an inhibitory effect on transmission from the small-diameter, unmyelinated afferent fibers (C fibers). Accordingly, pain should be relieved if cutaneous afferents from a region of pain could be artificially stimulated using an electrical current.

A battery-operated, transistorized unit generates the electrical stimulus. There are usually three separate controls. The first varies the amplitude or intensity and allows for administration of a range from low (2–4 Hz) to high (100–250 Hz) frequency stimulation. The second varies pulse width, usually from 0.04 to 0.1 ms, and the third controls the mode to select continuous or pulsed ± ramped, ± random stimulus. The stimulus is delivered through pads attached to the skin surface. The electrical field generated has to be of sufficient magnitude to excite the adjacent afferent nerve fibers without damaging the local skin, producing dysesthesia or producing painful muscle contractions.

The mechanism of action of TENS has not been explicitly or directly demonstrated. However, it is believed that low frequency stimulation releases mu opioids (β-endorphins), while high frequency stimulation releases delta opioids (met-enkephalin and leu-enkephalin) in the central nervous system. The mu receptor antagonist – naloxone – inhibits the effects of low frequency stimulation but not those of high frequency stimulation (Sjolund and Eriksson 1979; Freeman et al. 1983). Rats develop opioid tolerance after 4 days when treated with TENS for 20 min. a day (Chandran and Sluka 2003).

Due to the development of tolerance, treatment is more effective when TENS is used intermittently. Daily administration lessens the analgesic effect. The short duration of benefit (260 min) (Cheing et al. 2003), however, encourages frequent application in chronic conditions; but this in turn reduces pain relief.

Applications

TENS is used for the control of pain of various types in various regions of the body. The portability, safety, and low cost of the transistorized impulse generator have all contributed to its wide and varied use in the field of pain medicine.

A number of factors need to be considered when offering TENS as a treatment option. The aim is to produce the optimal tolerable stimulus. Application is critical to the success of the treatment. The site of stimulation should be chosen to produce maximal input in the segment where the pain originates and should be proximal to it. It is often not possible to predict the optimal placement or type of electrodes or the stimulus that will produce maximal pain relief. Consequently, with guidance, the patient should be encouraged to experiment with amplitude, frequency, and pulse width, as well as electrode placement and duration of stimulus, in order to maximize their relief of pain. Most choose frequencies from 40 to 70 Hz; with a pulse width of 0.1–0.5 ms. Stimulation at lower frequencies requires higher intensity of stimulus, which tends to produce painful muscle contractions.

Maximal comfortable stimulus is more effective in relieving pain than stimulation that is barely detectable. However, the theoretical argument of providing a sufficiently powerful stimulus to activate both small myelinated A-delta fibers and large A-beta fibers is negated by the difficulty patients have in tolerating continuous high-intensity stimulation.

Efficacy

Two Cochrane reviews, one of chronic pain (Carroll et al. 2002) and the other of chronic low back pain (Milne et al. 2002), demonstrate lack of effectiveness. There were wide variations in the parameters of type, site, and frequency of application, treatment duration, and intensity and frequency of the electrical impulse in both reviews.

Many studies show lack of evidence of benefit. There was no evidence that TENS was any more effective than other conservative treatments for acute and chronic nonspecific low back pain (van Tulder et al. 1997) or for chronic pain (McQuay et al. 1997). TENS was found not to be effective in relieving labor pain (Carroll et al. 1997).

In contrast, a Cochrane review by Proctor (Proctor et al. 2002) showed high frequency TENS to be more effective than placebo in relieving primary dysmenorrhea, while low frequency TENS showed no difference. A review by Osiri (Osiri et al. 2001) on knee osteoarthritis showed TENS provided significantly better pain relief and improved movement, compared with placebo. The effect had previously been shown in three randomized controlled trials reviewed in a paper by Puett and Griffin (1994). However, Osiri's review showed that TENS only differed from placebo when treatment continued for more than 4 weeks in duration.

The efficacy of TENS in acute situations is attested in a meta-analysis of studies of reduction of analgesic use in postoperative patients. Bjordal (Bjordal et al. 2003) reviewed 21 randomized placebo-controlled trials with a total of 1,350 patients and demonstrated a mean reduction in analgesic consumption after TENS/AL-TENS of 26.5 % (range −6 % to +51 %). Eleven of these trials, covering 964 patients, reported that a strong sub-noxious electrical impulse of adequate stimulus frequency was administered. These trials demonstrated a reduction in analgesic consumption of 35.5 % (range 1–51 %). In comparison, the nine trials that did not confirm sufficient current intensity or adequate frequency demonstrated a reduction of only 4.1 % (range −10 % to +29 %).

The short-term analgesic efficacy of TENS was demonstrated in a trial using distension arthrography (a moderately painful procedure) for frozen shoulder. There was a 50 % reduction using high-intensity TENS and 38 % reduction using low intensity TENS, compared to controls (Morgan et al. 1996).

Thus, the evidence is mixed. This is due, in part, to the poor quality of studies available and the lack of clear definition of the treatment given. The effect has been shown to differ depending on a number of variables in the treatment application. These are frequency, intensity, and waveform of the stimulus, site and duration of its application, and conduction medium. All need to be recorded clearly for every subject in a trial of the efficacy of TENS.

Side Effects

Opioid tolerance can develop with repeated daily use. There are few other side effects from the administration of TENS, probably due to the application generally being under the patient's control. There is the theoretical risk that too high a stimulus could damage the skin by electrical burn, but it is unusual for patients to achieve such damage. The range of machine settings and the noxious stimulus produced by painful muscle contraction at the higher stimulus range affords protection. Side effects really only occur in the presence of excessive zeal.

Cross-References

► Acupuncture Mechanisms
► Chronic Pain in Children, Physical Medicine and Rehabilitation
► Complex Chronic Pain in Children, Interdisciplinary Treatment
► McGill Pain Questionnaire
► Pain in Humans, Electrical Stimulation (Skin, Muscle, and Viscera)
► Postoperative Pain, Appropriate Management
► TENS, Mechanisms of Action
► Transcutaneous Electrical Nerve Stimulation Outcomes
► Transcutaneous Electrical Nerve Stimulation (TENS) in Treatment of Muscle Pain

References

Bjordal, J. M., Johnson, M. I., & Ljunggreen, A. E. (2003). Transcutaneous Electrical Nerve Stimulation (TENS) can reduce postoperative analgesic consumption. A meta-analysis with assessment of optimal treatment parameters for postoperative pain. *European Journal of Pain, 7*, 181–188.

Carroll, D., Moore, R. A., Tramer, M. R., & McQuay, H. J. (1997). Transcutaneous electrical nerve stimulation does not relieve labour pain: Updated systematic review. *Contemporary Reviews in Obstetrics and Gynaecology*, 195–205.

Carroll, D., Moore, R. A., McQuay, H. J., Fairman, F., Tramer, M., & Leijon, G. (2002). Transcutaneous Electrical Nerve Stimulation (TENS) for chronic pain (Cochrane Review). *The Cochrane Library*, (2).

Chandran, P., & Sluka, K. A. (2003). Development of opioid tolerance with repeated transcutaneous electrical nerve stimulation administration. *Pain, 102*(1–2), 195–201.

Cheing, G. L., Tsui, A. Y., Lo, S. K., & Hui-Chan, C. W. (2003). Optimal stimulation duration of TENS in the management of osteoarthritic knee pain. *Journal of Rehabilitation Medicine, 35*(2), 62–68.

Freeman, T. B., Campbell, J. N., & Long, D. M. (1983). Naloxone does not affect pain relief induced by electrical stimulation in man. *Pain, 17*, 189–195.

McQuay, H. J., Moore, R. A., Eccleston, C., Morley, S., & DeC Williams, A. C. (1997). Systematic review of outpatient services for chronic pain control. *Health Technology Assessment, 1*(6), 1–137.

Melzack, R., & Wall, P. D. (1965). Pain mechanisms: A new theory. *Science, 15*, 971–979.

Milne, S., Welch, V., Brosseau, L., Saginur, M., Shea, B., Tugwell, P., & Wells, G. (2002). Transcutaneous Electrical Nerve Stimulation (TENS) for chronic low back pain (Cochrane Review). *The Cochrane Library*, (1).

Morgan, B., Jones, A. R., Mulcahy, K. A., Finlay, D. B., & Collett, B. (1996). Transcutaneous Electrical Nerve Stimulation (TENS) during distension shoulder arthrography: A controlled trial. *Pain, 64*, 265–267.

Osiri, M., Welch, V., Brosseau, L., Shea, B., McGowan, J., Tugwell, P., & Wells, G. (2001). Transcutaneous electrical nerve stimulation for knee osteoarthritis (Cochrane Review). *The Cochrane Library*, (1).

Proctor, M. L., Smith, C. A., Farquhar, C. M., & Stones, R. W. (2002). Transcutaneous electrical nerve stimulation and acupuncture for primary dysmenorrhoea (Cochrane Review). *The Cochrane Library*, (1).

Puett, D. W., & Griffin, M. R. (1994). Published trials of non-medicinal and non-invasive therapies for hip and knee osteoarthritis. *Annals of Internal Medicine, 121*, 133–140.

Sjolund, B. H., & Eriksson, M. B. E. (1979). The influence of naloxone on analgesia produced by peripheral conditioning stimulation. *Brain Research, 173*, 295–301.

van Tulder, M. W., Koes, B. W., & Bouter, L. M. (1997). Conservative treatment of acute and chronic non-specific low back pain: A systematic review of randomised controlled trials of the most common interventions. *Spine, 22*, 2128–2156.

Transcutaneous Electrical Nerve Stimulation (TENS) in Treatment of Muscle Pain

Mark I. Johnson
Faculty of Health and Social Sciences, Leeds
Pallium Research Group, Leeds Metropolitan
University, Leeds, UK

Synonyms

Electrical stimulation therapy (EST); Electroanalgesia; Percutaneous electrical nerve stimulation; Transcutaneous electrical stimulation (TES); Transcutaneous nerve stimulation (TNS)

Definition

▶ Transcutaneous electrical nerve stimulation (TENS) is a noninvasive analgesic intervention, which delivers electrical currents across the intact surface of the skin to stimulate the underlying nerves (see Fig. 1). TENS is used extensively for the symptomatic relief of all types of pain, including pains of musculoskeletal origin. The purpose of TENS is to activate selectively those populations of nerve fibers that are concerned with segmental and ▶ extrasegmental antinociceptive mechanisms. The two main TENS techniques are ▶ conventional TENS and ▶ acupuncture-like TENS (AL-TENS).

Characteristics

TENS is popular because patients can administer TENS themselves and can titrate the dosage of treatment as required. TENS effects are often rapid in onset, and there are few side effects and no potential for toxicity or overdose. The technical specifications of TENS devices vary according to manufacturer, although most utilize biphasic pulsed currents that may be either symmetrical or nonsymmetrical square waves. Pulse durations lie between 50 and 1,000 μs, pulse frequencies between 1 and 200 pulses per second (pps), and pulse amplitudes between 1 and 60 mA. Most devices offer continuous, burst (used for AL-TENS), and modulated pulse patterns. Factors that influence the success of TENS include the patient, the condition, and the appropriateness of the TENS technique employed (see Fig. 2).

Transcutaneous Electrical Nerve Stimulation (TENS) in Treatment of Muscle Pain, Fig. 1 An electrical pulse generator delivers currents via conducting electrodes attached to the intact surface of the skin. Traditionally, carbon rubber electrodes smeared with conducting gel and attached to the skin using self-adhesive tape were used to deliver the electrical currents. Nowadays, self-adhesive electrodes are used

TENS is commonly used to treat chronic pains, including those of musculoskeletal origin. Systematic reviews on TENS and chronic pain have been inconclusive due to the low methodological quality of RCTs (Bennett et al. 2011; Johnson and Bjordal 2011; Nnoaham and Kumbang 2008). A meta-analysis on TENS for chronic musculoskeletal pain found a significant decrease in pain during TENS suggesting that TENS is beneficial (Johnson and Martinson 2007). Meta-analyses have also found that that TENS is beneficial for post-operative pain (Bjordal et al. 2003) and rheumatoid arthritis in the hand (Brosseau et al. 2003). Meta-analysis on TENS for non-specific low back pain are conflicting (Dubinsky and Miyasaki 2010) with the magnitude of effects similar to that seen with other analgesic treatments (Machado et al. 2009). Underdosing of TENS and the use of inappropriate TENS techniques in some RCTs have influenced the findings of systematic reviews (Bennett et al. 2011).

There have been no systematic reviews on TENS and muscle pain because of a lack of randomized controlled clinical trials. TENS is used extensively for reducing localized muscle pain, especially in the neck and shoulder, arising from muscle tension. TENS is also used to treat acute traumatic muscle pain that results from physical injury from sport and minor accidents and to treat postexertional muscle pain. Reports

Transcutaneous Electrical Nerve Stimulation (TENS) in Treatment of Muscle Pain, Fig. 2 The output characteristics of a typical TENS device (topographic view). The amplitude, frequency, duration, and pattern of electrical pulses can be controlled by the end user (*I* intensity, *F* frequency, *D* duration, *C* continuous, *B* burst, *M* modulation, *pps* pulses per second)

Transcutaneous Electrical Nerve Stimulation (TENS) in Treatment of Muscle Pain, Fig. 3 The purpose of conventional TENS is to activate non-nociceptive cutaneous afferents (Aβ) without concurrent activation of nociceptive (Aδ/C) or muscle efferents and afferents. *Arrows* indicate impulses traveling towards the central nervous system

suggest that TENS is helpful for myofascial pain syndrome (Graff-Radford et al. 1989; Hou et al. 2002) and localized pain in fibromyalgia (Offenbacher and Stucki 2000), although in practice, TENS reduces pain in some of these patients and aggravates it in others. It is unlikely that TENS will be helpful in widespread pain in fibromyalgia, because it is difficult to direct TENS currents into the painful area.

In clinical practice, conventional TENS is used in the first instance for most pains, including those of musculoskeletal origin. The purpose of conventional TENS is to activate selectively large-diameter non-nociceptive cutaneous afferents (Aβ) without concurrently activating small-diameter nociceptive afferents (Aδ and C), which would cause pain, or muscle efferents, which would cause muscle contractions. TENS-induced Aβ activity has been shown to inhibit ongoing activity in second-order nociceptive neurons in the dorsal horn of the spinal cord with much of the physiological and pharmacological detail known (DeSantana et al. 2008). In practice, large-diameter non-nociceptive cutaneous afferent activity is recognized by a "strong but comfortable" nonpainful electrical paresthesia beneath the electrodes, and patients are trained to titrate current amplitude to achieve this outcome (see Fig. 3).

During conventional TENS, electrodes are applied to healthy skin, at the site of pain, to stimulate large-diameter non-nociceptive cutaneous afferents which enter the same spinal segment as the nociceptive fibers associated with the origin of the pain. When it is not possible to apply electrodes at the site of pain, for example, when the skin is damaged and/or sensitive to touch, electrodes can be applied proximally over the main nerve trunk that innervates the skin at the site of pain. Alternatively, electrodes can be placed over the spinal cord at a level segmentally related to the site of origin of the pain, or at a site which is contralateral (mirror image) to the site of pain. Dual-channel devices using four electrodes can be used for pains covering large areas such as in a gluteal region or lower limb and for multiple pains such as low back pain and sciatica (see Fig. 4).

For conventional TENS, maximum pain relief is achieved when the TENS device is switched on. Therefore, patients should be encouraged to use conventional TENS whenever the pain is present, although it is wise to instruct the patient to monitor their skin condition under the electrodes on a regular basis and perhaps take regular (although short) breaks from stimulation. Dosing regimens of 20 min at daily, weekly, or monthly intervals are likely to be ineffective for conventional TENS. Some patients report poststimulation analgesia following conventional TENS, although reports of the duration of this effect varies widely and may reflect natural fluctuations in symptoms rather than specific TENS-induced effects. The relationship between the pulse frequency, duration, and pattern of TENS and the magnitude of analgesia for different pain conditions has not been fully confirmed in the clinical setting as much available evidence is conflicting, inconclusive, or methodologically flawed. Encouraging patients to experiment with all TENS settings while they maintain a strong but comfortable electrical paresthesia within the site of pain may be the most effective approach for conventional TENS (Johnson and Bjordal 2011).

AL-TENS is a variant of conventional TENS and has been successfully used for pains of

Transcutaneous Electrical Nerve Stimulation (TENS) in Treatment of Muscle Pain, Fig. 4 (a) Electrode positions for common pain conditions – anterior view. (b) Electrode positions for common pain conditions – posterior view

a

Trigeminal neuralgia

Thalamic pain
Where pain is most pronounced

Anterior shoulder

Angina

Rib metastasis

Postherpetic neuralgia
Above affected dermatome
Across affected dermatome

Postoperative pain
Large electrodes if appropriate

Phantom limb pain
Over main nerve bundle arising from phantom

Phantom limb pain
Contralateral site-median nerve

Dysmenorrhoea
(women)

Postoperative pain
Large electrodes if appropriate

Knee pain (Osteoarthritis)
Dual channel if appropriate

Ankle pain

b

Neck pain
Bilateral
Unilateral

Shoulder pain

Postherpetic neuralgia
Above affected dermatome
Across affected dermatome

Low back pain or
Dysmenorrhoea

Stump pain

Hip pain

Sciatica

Peripheral vascular disease

Stump and phantom limb

Tendonitis

musculoskeletal origin. The purpose of AL-TENS is to stimulate axons of muscle efferent neurons (α motor axons) to generate a forceful but nonpainful phasic muscle twitch. This muscle twitch generates activity in small-diameter group III muscle afferent neurons, which trigger extrasegmental antinociceptive mechanisms, which lead to the release of opioid

Electrodes TENS Currents

Cathode | Anode
Motor point | Proximal

Aβ

MUSCLE ← ← ← GI

CONTRACTION

MUSCLE → → → GIII activity

Transcutaneous Electrical Nerve Stimulation (TENS) in Treatment of Muscle Pain, Fig. 5 The purpose of AL-TENS is to elicit a nonpainful muscle twitch by activating large-diameter motor efferents. The muscle twitch generates activity in ergoreceptors and small-diameter group III (GIII) muscle afferents which initiates extrasegmental antinociceptive mechanisms. Aβ afferents may also become active as currents pass through the skin. *Arrows* indicate direction of relevant impulse information

peptides in a manner similar to that suggested for acupuncture. AL-TENS is helpful for radiating neurogenic pain and for patients who have decreased skin sensitivity from damage to cutaneous afferents in the painful region. AL-TENS has been used successfully for muscle pain by stimulating the painful muscles and trigger points. However, some patients report that this aggravates their pain. In such circumstances, AL-TENS can be administered on the contralateral myotome (Sjölund et al. 1990; Johnson 1998) (see Fig. 5).

AL-TENS is administered over muscles and motor points using low-frequency burst patterns of pulse delivery. Currents are delivered at high but nonpainful intensities, to generate a forceful but nonpainful phasic muscle twitch. Currents delivered during AL-TENS will also activate Aβ fibers during their passage through the skin leading to segmental analgesia. AL-TENS is administered intermittently for 20–30 min at a time to reduce excessive muscle fatigue. The general impression of users is that post-TENS analgesia is longer for AL-TENS than conventional TENS, and this is supported by initial findings in experimental studies. For this reason, AL-TENS is useful for patients who obtain brief aftereffects from conventional TENS, or who are resistant to conventional TENS.

TENS should not be administered to patients fitted with cardiac pacemakers or women in the first trimester of pregnancy. TENS should not be used while operating vehicles or potentially hazardous equipment, and electrodes should not be positioned over the anterior part of the neck, over areas of broken skin, or directly over a pregnant uterus (although it is safe when applied to the lower back to treat labor pains). Patients should be tested for normal skin sensation prior to using TENS. Patients should be warned not to use TENS in the shower or bath and to keep TENS appliances out of the reach of children.

References

Bennett, M. I., Hughes, N., & Johnson, M. I. (2011). Methodological quality in randomised controlled trials of transcutaneous electric nerve stimulation for pain: Low fidelity may explain negative findings. *Pain, 152*(6), 1226–1232. Epub 2011 Mar 23. Review.

Bjordal, J. M., Johnson, M. I., & Ljunggreen, A. E. (2003). Transcutaneous electrical nerve stimulation (TENS) can reduce postoperative analgesic consumption. A meta-analysis with assessment of optimal treatment parameters for postoperative pain. *European Journal of Pain, 7*(2), 181–188.

Brosseau, L., Milne, S., Robinson, V., et al. (2002). Efficacy of the transcutaneous electrical nerve stimulation for the treatment of chronic low back pain: A meta-analysis. *Spine, 27*, 596–603.

Brosseau, L., Yonge, K. A., Robinson, V., et al. (2003). Transcutaneous electrical nerve stimulation (TENS) for the treatment of rheumatoid arthritis in the hand (Cochrane Review). *The Cochrane Library Oxford: Update Software Issue, 3*, 1–19.

Carroll, D., Moore, R. A., McQuay, H. J., et al. (2003). Transcutaneous electrical nerve stimulation (TENS) for chronic pain. In *The Cochrane library*, Issue 3. Update Software, Oxford.

DeSantana, J. M., Walsh, D. M., Vance, C., Rakel, B. A., Sluka, K. A. (2008). Effectiveness of transcutaneous electrical nerve stimulation for treatment of hyperalgesia and pain. *Current Rheumatology Reports, 10*(6), 492–499. Review.

Dubinsky, R. M., Miyasaki, J. (2010). Assessment: efficacy of transcutaneous electric nerve stimulation in the treatment of pain in neurologic disorders (an evidence-based review): report of the Therapeutics and Technology Assessment Subcommittee of the American Academy of Neurology. *Neurology, 74*(2):173–176. Epub 2009 Dec 30. Review.

Flowerdew, M., & Gadsby, G. (1997). A review of the treatment of chronic low back pain with acupuncture-like transcutaneous electrical nerve stimulation and transcutaneous electrical nerve stimulation. *Complementary Therapies in Medicine, 5*, 193–201.

Garrison, D. W., & Foreman, R. D. (1994). Decreased activity of spontaneous and noxiously evoked dorsal horn cells during transcutaneous electrical nerve stimulation (TENS). *Pain, 58*, 309–315.

Graff-Radford, S. B., Reeves, J. L., Baker, R. L., et al. (1989). Effects of transcutaneous electrical nerve stimulation on myofascial pain and trigger point sensitivity. *Pain, 37*, 1–5.

Hou, C. R., Tsai, L. C., Cheng, K. F., et al. (2002). Immediate effects of various physical therapeutic modalities on cervical myofascial pain and trigger-point sensitivity. *Archives of Physical Medicine and Rehabilitation, 83*, 1406–1414.

Johnson, M. (1998). The analgesic effects and clinical use of acupuncture-like TENS (AL-TENS). *Physical Therapy Reviews, 3*, 73–93.

Johnson, M. I. (2001). Transcutaneous electrical nerve stimulation. In S. Kitchen (Ed.), *Electrotherapy: Evidence-based practice* (pp. 259–286). Edinburgh: Churchill Livingstone.

Johnson, M. I., & Walsh, D. M. (2010). Pain: Continued uncertainty of TENS' effectiveness for pain relief. *Nature Reviews. Rheumatology, 6*(6), 314–316.

Johnson, M. I., & Bjordal, J. M. (2011). Transcutaneous electrical nerve stimulation for the management of painful conditions: Focus on neuropathic pain. *Expert Review of Neurotherapeutics, 11*(5), 735–753. Review.

Johnson, M., & Martinson, M. (2007). Efficacy of electrical nerve stimulation for chronic musculoskeletal pain: A meta-analysis of randomized controlled trials. *Pain, 130*(1–2), 157–165. Epub 2007 Mar 23.

Machado, L. A., Kamper, S. J., Herbert, R. D., Maher, C. G., & McAuley, J. H. (2009). Analgesic effects of treatments for non-specific low back pain: A meta-analysis of placebo-controlled randomized trials. *Rheumatology (Oxford), 48*(5), 520–527. Epub 2008 Dec 24. Review.

McQuay, H., & Moore, A. (1998). TENS in chronic pain. In H. McQuay & A. Moore (Eds.), *An evidence-based resource for pain relief* (pp. 207–211). Oxford: Oxford University Press.

Nnoaham, K. E., & Kumbang, J. (2008). Transcutaneous electrical nerve stimulation (TENS) for chronic pain. *Cochrane Database Systematic Reviews*, (3), CD003222. Review.

Offenbacher, M., & Stucki, G. (2000). Physical therapy in the treatment of fibromyalgia. *Scandinavian Journal of Rheumatology, 113*, 78–85.

Osiri, M., Welch, V., Brosseau, L., et al. (2003). Transcutaneous electrical nerve stimulation for knee osteoarthritis. In *The Cochrane library*, Issue 3. Update Software, Oxford.

Reeve, J., Menon, D., & Corabian, P. (1996). Transcutaneous electrical nerve stimulation (TENS): A technology assessment. *International Journal of Technology Assessment in Health Care, 12*, 299–324.

Sjölund, B., Eriksson, M., & Loeser, J. (1990). Transcutaneous and implanted electric stimulation of peripheral nerves. In J. J. Bonica (Ed.), *The management of pain* (Vol. 2, pp. 1852–1861). Philadelphia: Lea & Febiger.

Transcutaneous Electrical Nerve Stimulation Outcomes

Mark I. Johnson
Faculty of Health and Social Sciences, Leeds
Pallium Research Group, Leeds Metropolitan
University, Leeds, UK

Synonyms

Electrical stimulation therapy (EST); Electroanalgesia; Percutaneous electrical nerve stimulation; TENS outcomes; Transcutaneous electrical stimulation (TES); Transcutaneous nerve stimulation (TNS)

Introduction

Clinical experience supports a role for TENS in the management of acute and chronic pain. Clinical research on TENS also reports beneficial effects for a wide range of chronic pain conditions including back pain, neck pain, headache, osteoarthritis, rib fracture, orofacial pain, postherpetic neuralgia, trigeminal neuralgia, poststroke pain, phantom limb and stump pain, brachial plexus avulsion, complex regional pain syndrome, angina pectoris, myalgia, postoperative pain, labor pain, dental pain, and cancer pain (Johnson and Bjordal 2011). However, many clinical trials lack appropriate controls and/or randomization leading to overestimation of treatment effects (Carroll et al. 1996). In recent years, systematic reviews and meta-analyses have produced inconsistent findings on the effectiveness of TENS, and this may explained in part by poor implementation fidelity including TENS being administered using inappropriate techniques and suboptimal dosing (Bennett et al. 2011; Johnson and Walsh 2010).

Definition

▶ Transcutaneous electrical nerve stimulation (TENS) is a noninvasive analgesic intervention that delivers electrical currents across the intact surface of the skin to stimulate the underlying nerves. TENS is used extensively for the symptomatic relief of all types of pain including pains of nociceptive, neuropathic, and musculoskeletal origin. The purpose of TENS is to activate selectively those populations of nerve fibers that are concerned with segmental and extrasegmental antinociceptive mechanisms. The two main TENS techniques are ▶ conventional TENS and ▶ acupuncture-like TENS (AL-TENS).

Characteristics

Efficacy of TENS for Chronic Pain

The most recent Cochrane review of TENS for chronic pain (25 RCTs, 1,281 participants) found that TENS was superior to inactive TENS controls in 13/22 studies (Nnoaham and Kumbang 2008). It was concluded that evidence was inconclusive.

The largest meta-analysis of TENS studies to date (32 studies on TENS, six studies on percutaneous electric nerve stimulation, 1,227 participants) found that TENS was superior to inactive TENS controls for reducing chronic musculoskeletal pain (Johnson and Martinson 2007). A Cochrane review of TENS for osteoarthritic knee pain (18 RCTs, 813 participants) found evidence to be inconclusive. The standard mean difference between active and placebo TENS was approximately 20 mm on a 100 mm Visual Analogue Scale (VAS) (Rutjes et al. 2009) which was similar to an earlier meta-analysis of seven RCTs delivering TENS at optimal doses (22.2 mm (95 % CI: 18.1–26.3)) that found that TENS reduced pain to a greater extent than

placebo TENS (Bjordal et al. 2007). Cochrane reviews of TENS for rheumatoid arthritis of the hand (three small RCTs) were inconclusive.

A Cochrane review of TENS for chronic low back pain (3 RCTs, 110 participants receiving TENS and 87 receiving placebo TENS) was inconclusive (Khadilkar et al. 2008). The National Institute of Health and Clinical Excellence (NICE) in the UK found insufficient evidence to judge effectiveness for persistent nonspecific low back pain (3 RCTs, 331 participants received TENS and 168 received placebo TENS). In contrast, a systematic review by the North American Spine Society found that TENS reduced pain intensity in the short term (6 RCTs, 375 participants receiving TENS and 192 receiving placebo TENS) (Poitras and Brosseau 2008). Recently, an assessment by the Therapeutics and Technology Assessment Subcommittee of the American Academy of Neurology found that TENS was not effective for chronic low back pain (2 RCTs, 114 participants receiving TENS and 87 receiving placebo) (Dubinsky and Miyasaki 2010). The authors claimed that this judgement was based on level A evidence (i.e., good-quality RCTs), although one of the RCTs was criticized at the time of publication for significant clinical heterogeneity, the use of a suboptimal TENS technique, and the concurrent use of hot packs, which could have masked the effects of TENS. The other RCT used participants with multiple sclerosis with participants in the placebo TENS group taking additional analgesics. The effect size for TENS is of a similar magnitude to other therapies used to manage nonspecific low back pain including NSAIDs and muscle relaxants (Machado et al. 2009).

An evaluation by the European Federation of Neurological Societies (EFNS) Task Force for neurostimulation therapy for neuropathic pain found TENS to be superior to placebo (9 controlled trials, 200 participants) and suggested that TENS should be given as an add-on therapy (Cruccu et al. 2007). Systematic reviewers claim that TENS is superior to placebo (no current) TENS at reducing diabetic peripheral neuropathy pain (3 RCTs, 78 participants

Transcutaneous Electrical Nerve Stimulation Outcomes, Table 1 Physical medicine, TENS outcomes – chronic pain

Condition	Ref	Sample	Reviewers' conclusion
Chronic pain			
Chronic pain (Update of Carroll et al. 2001)	Nnoaham and Kumbang (2008)	25 RCTs (1,281 participants) on TENS	Evidence inconclusive
Chronic musculoskeletal pain	Johnson and Martinson (2007)	38 RCTs (1,227 participants) of any electrical nerve stimulation with 32 RCTs on TENS	Evidence of effect
Whiplash and mechanical neck disorders	Kroeling et al. (2009)	18 RCTs (1,043 participants) of any electrotherapy with 7 RCTs (88 participants) on TENS	Evidence of effect
Low back pain	Dubinsky and Miyasaki (2010)	2 RCTs (201 participants) on TENS	Evidence of no effect
Low back pain	Khadilkar et al. (2008)	3 RCTs (197 participants) on TENS	Evidence inconclusive
Low back pain	Poitras and Brosseau (2008)	6 RCTs (375 participants) on TENS	Evidence of effect
Osteoarthritic knee pain (Update of Osiri et al. 2003)	Rutjes et al. (2009)	18 RCTs (275 participants) on TENS	Evidence inconclusive
Osteoarthritic knee pain	Bjordal et al. (2007)	36 RCTs (2,434 participants) of physical agents with 7 RCTs (414 participants) on TENS using adequate technique	Evidence of effect
Rheumatoid arthritis of the hand	Brosseau et al. (2003)	3 RCT (78 participants) on TENS	Evidence of effect
Neuropathic pain	Cruccu et al. (2007)	9 RCTs (200 participants) on TENS	Evidence of effect
Painful diabetic neuropathy	Jin et al. (2010)	3 RCTs (78 participants) on TENS	Evidence of effect
Painful diabetic neuropathy	Dubinsky and Miyasaki (2010)	2 RCTs (55 participants) on TENS	Evidence of effect
Phantom limb and stump pain	Mulvey et al. (2010)	0 RCTs	No evidence available
Poststroke shoulder pain	Price and Pandyan (2001)	4 RCTs (170 participants) of any surface electrical stimulation with 2 RCTs on TENS	Evidence inconclusive
Chronic recurrent headache	Bronfort et al. (2004)	22 RCTs (2,628 participants) of physical agents but no RCTs on TENS	Lack of available evidence
Cancer pain	Hurlow et al. (2012)	3 RCTs (88 participants) on TENS	Evidence inconclusive

(Jin et al. 2010); 3 RCTs, 31 participants receiving TENS and 24 receiving placebo TENS (Dubinsky and Miyasaki 2010)), although studies did not use standard TENS devices and had insufficient participants. Cochrane reviews on TENS for poststroke shoulder pain (Price and Pandyan 2001), amputee pain (Mulvey et al. 2010), and cancer pain (Hurlow et al. 2012) found insufficient evidence to judge effectiveness (Table 1).

Efficacy of TENS for Acute Pain

Cochrane reviews have found that evidence is inconclusive for the use of TENS to relieve acute pain (Walsh et al. 2009) and the pain of establish labor (Dowswell et al. 2009). There was weak evidence that TENS reduced pain associated with dysmenorrhea (Proctor et al. 2003). A systematic review in 1996 concluded that TENS was not effective at relieving postoperative pain (Carroll et al. 1996). However, pain relief scores were compromised in some of the included RCTs because patients had free access to analgesic drugs, so they could titrate analgesic consumption to achieve similar levels of pain relief in sham and active TENS groups. Other confounding factors include the difficulty of dichotomizing multiple outcome measures, heterogeneous baseline pain measures, and sample sizes with insufficient statistical power to detect potential differences between groups. A meta-analysis of 1,350 patients (21 RCTs) accounted for some of these issues and found that TENS reduced analgesic consumption when compared to sham TENS (Bjordal et al. 2003). A subgroup analysis of 964 patients (11 RCTs) that used optimal TENS dosage (i.e., a strong, subnoxious electrical stimulation) found a significant improvement in outcome suggesting that adequate TENS technique is necessary in order to achieve an effect. Recently, a meta-analysis found that TENS was effective in combination with analgesics for relieving moderate post-thoracotomy pain and effective as a stand-alone therapy for mild post-thoracotomy pain (Freynet and Falcoz 2010) (Table 2).

The Challenges of TENS Research

The low methodological quality of RCTs has created uncertainty in the clinical research evidence for TENS. Often the appropriateness of TENS technique is not accounted for in methodological quality rating scales and/or inclusion criteria used in systematic reviews and meta-analyses (Bennett et al. 2011). Underdosing of TENS has occurred in many trials using short duration, single, or infrequent TENS interventions. TENS effects are maximal when the device is switched on, and in practice users of conventional TENS keep the device switched on whenever they need pain relief. However, many RCTs record pain outcome before and after TENS rather than during stimulation. Conventional TENS and AL-TENS are ill defined in published reports and are often categorized according to the electrical characteristics of TENS rather than the users' intention to stimulate particular types of

Transcutaneous Electrical Nerve Stimulation Outcomes, Table 2 Physical medicine, TENS outcomes – acute pain

Condition	Ref	Sample	Reviewers' conclusion
Acute pain			
Acute pain	Walsh et al. (2009)	12 RCTs (919 participants) on TENS	Evidence inconclusive
Post-thoracotomy pain	Freynet and Falcoz (2010)	9 RCTs (645 participants) on TENS	Evidence of effect as an adjuvant
Postoperative analgesic consumption	Bjordal et al. (2003)	21 RCTs (964 participants) on TENS	Evidence of effect
Postoperative pain	Carroll et al. (1996)	17 RCTs (786 participants) on TENS	Evidence of no effect
Labor pain	Mello et al. (2011)	9 RCTs (1,076 women) on TENS	Evidence inconclusive
Labor pain	Dowswell et al. (2009)	19 RCTs (1,671 women) on TENS	Evidence inconclusive
Labor pain	Carroll et al. (1997)	10 RCTs (877 women) on TENS	Evidence of no effect
Primary dysmenorrhea	Proctor et al. (2003)	7 RCTs, (213 participants) on TENS	Evidence of effect

nerve fiber and whether or not this was achieved during the trial. Analysis of adequate stimulation technique is absent from the majority of published RCT reports. In future, RCTs on TENS must take these factors into account because they have been shown to alter TENS outcomes.

TENS is a technique-based intervention, so outcome is dictated by the appropriateness of TENS technique. The potential number of TENS protocols is vast, as users can alter the characteristics of the electrical currents (i.e., the output characteristics), the application procedure (i.e., electrode type and location), and the dosing regimen. Attempts to improve clinical effectiveness by searching for optimal TENS settings have largely been unsuccessful. Nevertheless, an increasing number of nonstandard TENS-like devices have appeared on the market (e.g., interferential therapy (IFT), microcurrent electrical therapy (MET), transcranial electrical stimulation, and transcutaneous spinal electroanalgesia (TSE)). Manufacturers overstate the potential effects of these TENS-like devices, and often similar levels of pain relief can be achieved using a standard TENS device (Johnson 2003).

Health-care professionals should not dismiss the use of TENS for any condition until the issues in clinical trial design and review methodology have been resolved.

Summary

Clinical experience suggests that TENS relieves acute and chronic pain. Meta-analyses of RCTs suggest positive outcomes for musculoskeletal and postoperative pain. However, many systematic reviews are inconclusive, and this may be because investigators use inappropriate TENS technique and small study sample sizes. The National Institute of Health and Clinical Excellence (NICE) in the UK evaluates the cost-effectiveness of treatments for use by the National Health Service (NHS) and recommends that TENS should be offered for rheumatoid arthritis, osteoarthritis, and musculoskeletal pain associated with multiple sclerosis. The National Institute of Health and Clinical Excellence recommends that TENS should not be offered for nonspecific low back pain, labor pain, or stable angina because evidence is inconclusive. TENS is inexpensive, safe, and popular with patients so it seems sensible to offer patients TENS.

References

Bennett, M. I., Hughes, N., & Johnson, M. I. (2011). Methodological quality in randomised controlled trials of TENS for pain: Low fidelity may explain negative findings. *Pain, 152*(6), 1226–1232.

Bjordal, J. M., Johnson, M. I., & Ljunggreen, A. E. (2003). Transcutaneous electrical nerve stimulation (TENS) can reduce postoperative analgesic consumption. A meta-analysis with assessment of optimal treatment parameters for postoperative pain. *European Journal of Pain, 7*, 181–188.

Bjordal, J. M., Johnson, M. I., Lopes-Martins, R. A., et al. (2007). Short-term efficacy of physical interventions in osteoarthritic knee pain. A systematic review and meta-analysis of randomised placebo-controlled trials. *BMC Musculoskeletal Disorders, 8*, 51.

Bronfort, G., Nilsson, N., Haas, M., et al. (2004). Non-invasive physical treatments for chronic/recurrent headache. *Cochrane Database of Systematic Reviews*, (3), 1–68, (Art. No.: CD001878).

Brosseau, L., Yonge, K. A., Robinson, V., et al. (2003). Transcutaneous electrical nerve stimulation (TENS) for the treatment of rheumatoid arthritis in the hand. *Cochrane Database of Systematic Reviews*, (3), 1–18, Update Software, Oxford.

Carroll, D., Tramer, M., McQuay, H., et al. (1996). Randomization is important in studies with pain outcomes: Systematic review of transcutaneous electrical nerve stimulation in acute postoperative pain. *British Journal of Anaesthesia, 77*, 798–803.

Carroll, D., Moore, A., Tramer, M., et al. (1997). Transcutaneous electrical nerve stimulation does not relieve in labour pain: updated systematic review. *Contemporary Reviews in Obstetrics and Gynecology*, 195–205.

Carroll, D., Moore, R. A., McQuay, H. J., Fairman, F., Tramèr, M., & Leijon, G. (2001) Transcutaneous electrical nerve stimulation (TENS) for chronic pain. *Cochrane Database of Systematic Reviews*, (3), CD003222.

Carroll, D., Moore, R. A., Tramèr, M. R., & McQuay, H. J. (2007). Transcutaneous electrical nerve stimulation does not relieve labour pain: Updated systematic review. *Contemporary Reviews in Obstetrics and Gynecology, September*, 195–205.

Cruccu, G., Aziz, T. Z., Garcia-Larrea, L., et al. (2007). EFNS guidelines on neurostimulation therapy for neuropathic pain. *European Journal of Neurology, 14*(9), 952–970.

Dowswell, T., Bedwell, C., Lavender, T., et al. (2009). Transcutaneous electrical nerve stimulation (TENS) for pain relief in labour. *Cochrane Database of Systematic Reviews*, (2), 1–79, (Art. No.: CD007214).

Dubinsky, R. M., & Miyasaki, J. (2010). Assessment: Efficacy of transcutaneous electric nerve stimulation in the treatment of pain in neurologic disorders (an evidence-based review): Report of the therapeutics and technology assessment subcommittee of the American academy of neurology. *Neurology, 74*(2), 173–176.

Freynet, A., & Falcoz, P. E. (2010). Is transcutaneous electrical nerve stimulation effective in relieving postoperative pain after thoracotomy? *Interactive Cardiovascular and Thoracic Surgery, 10*(2), 283–288.

Hurlow, A., Bennett, M. I., Robb, K. A., Johnson, M. I., Simpson, K. H., & Oxberry, S. G. (2012). Transcutaneous electric nerve stimulation (TENS) for cancer pain in adults. *Cochrane Database of Systematic Reviews*, (3), CD006276.

Jin, D. M., Xu, Y., Geng, D. F., et al. (2010). Effect of transcutaneous electrical nerve stimulation on symptomatic diabetic peripheral neuropathy: A meta-analysis of randomized controlled trials. *Diabetes Research and Clinical Practice, 89*(1), 10–15.

Johnson, M. (2003). Transcutaneous electrical nerve stimulation (TENS) and TENS-like devices. Do they provide pain relief? *Pain Reviews, 8*, 121–128.

Johnson, M. I., & Bjordal, J. M. (2011). Transcutaneous electrical nerve stimulation for the management of painful conditions: Focus on neuropathic pain. *Expert Reviews in Neurotherapeutics, 11*(5), 735–753.

Johnson, M., & Martinson, M. (2007). Efficacy of electrical nerve stimulation for chronic musculoskeletal pain: A meta-analysis of randomized controlled trials. *Pain, 130*, 157–165.

Johnson, M. I., & Walsh, D. M. (2010). Continued uncertainty of TENS effectiveness for pain relief. *Nature Reviews. Rheumatology, 6*(6), 314–316.

Khadilkar, A., Odebiyi, D. O., Brosseau, L., et al. (2008). Transcutaneous electrical nerve stimulation (TENS) versus placebo for chronic low-back pain. *Cochrane Database of Systematic Reviews*, (4), 1–58, (Art. No.: CD003008).

Kroeling, P., Gross, A., & Goldsmith, C. H. (2009). Electrotherapy for neck pain. *Cochrane Database of Systematic Reviews*, (4), 1–76, (Art. No.: CD004251).

Machado, L. A., Kamper, S. J., Herbert, R. D., et al. (2009). Analgesic effects of treatments for non-specific low back pain: A meta-analysis of placebo-controlled randomized trials. *Rheumatology, 48*(5), 520–527.

Mello, L. F., Nobrega, L. F., & Lemos, A. (2011). Transcutaneous electrical stimulation for pain relief during labor: A systematic review and meta analysis. *Revista Brasileira de Fisioterapia, 15*(3), 175–184.

Mulvey, M. R., Bagnall, A. M., Johnson, M. I., et al. (2010). Transcutaneous electrical nerve stimulation (TENS) for phantom pain and stump pain following amputation in adults. *Cochrane Database of Systematic Reviews*, (5), 1–15, (Art. No.: CD007264).

Nnoaham, K. E., & Kumbang, J. (2008). Transcutaneous electrical nerve stimulation (TENS) for chronic pain. *Cochrane Database of Systematic Reviews*, (3), 1–61, (Art. No.: CD003222).

Osiri, M., Welch, V., Brosseau, L., et al. (2003). Transcutaneous electrical nerve stimulation for knee osteoarthritis. In: The Cochrane Library, Issue 3, 1–26, Update Software, Oxford.

Poitras, S., & Brosseau, L. (2008). Evidence-informed management of chronic low back pain with transcutaneous electrical nerve stimulation, interferential current, electrical muscle stimulation, ultrasound, and thermotherapy. *The Spine Journal, 8*(1), 226–233.

Price, C. I., & Pandyan, A. D. (2001). Electrical stimulation for preventing and treating post-stroke shoulder pain: A systematic Cochrane review. *Clin Rehab, 15*, 5–19.

Proctor, M. L., Smith, C. A., Farquhar, C. M., et al. (2003). Transcutaneous electrical nerve stimulation and acupuncture for primary dysmenorrhoea. (*Cochrane Review*). The Cochrane Library, Issue 3. Update Software, Oxford.

Rutjes, A. W., Nuesch, E., Sterchi, R., et al. (2009). Transcutaneous electrostimulation for osteoarthritis of the knee. *Cochrane Database of Systematic Reviews*, (4), 1–79, (Art. No.: CD002823).

Walsh, D. M., Howe, T. E., Johnson, M. I., et al. (2009). Transcutaneous electrical nerve stimulation for acute pain. *Cochrane Database of Systematic Reviews*, (2), 1–72, (Art. No.: CD006142).

Transcutaneous Electrical Stimulation (TES)

▶ Transcutaneous Electrical Nerve Stimulation Outcomes

▶ Transcutaneous Electrical Nerve Stimulation (TENS) in Treatment of Muscle Pain

Transcutaneous Nerve Stimulation (TNS)

▶ Transcutaneous Electrical Nerve Stimulation Outcomes

▶ Transcutaneous Electrical Nerve Stimulation (TENS) in Treatment of Muscle Pain

Transduction

Definition

Transduction is the conversion of one form of signal to another, for example, in sensory endings, the conversion of the stimulus (e.g., pressure or cold) into an electrical signal.

Cross-References

▶ Somatic Pain
▶ Species Differences in Skin Nociception
▶ Visceral Nociception and Pain

Transduction and Encoding of Noxious Stimuli

Carlos Belmonte and Félix Viana
Instituto de Neurociencias, Universidad
Miguel Hernandez-CSIC, San Juan de Alicante,
Spain

Characteristics

Sensory Transduction

Sensory transduction is the process by which external physical or chemical changes are transformed into internal biochemical and/or electrical signals that are propagated and processed through different levels of the central nervous system to elicit a sensation. Despite the specialized design of sensory receptors for the different sensory modalities, the cellular and molecular mechanisms involved in sensory transduction have certain common basic principles that can be outlined into a unified scheme (Fig. 1) (Block 1992; Belmonte 1996). This process involves sequential detection, amplification, and filtering of the incoming signal. In nociceptor neurons, as in other types of receptor neurons, the transformation of physical and chemical stimuli into electrical signals normally takes place at the peripheral nerve terminals where the transduction machinery is located. This machinery is formed by a variety of specialized proteins called receptors. Activation of these receptors by the different stimuli leads to a conformational change in ion channels, pore-forming membrane proteins that regulate the flow of ions.

Stimulating Energy

Nociceptors were defined by Charles Sherrington as sensory receptors activated by stimuli that potentially lead to tissue injury (noxious stimuli). Nociceptors, as is the case with other sensory receptors, are directly activated by a limited number of the various forms of energy that are continuously impinging upon the external surface and internal organs of the

Transduction and Encoding of Noxious Stimuli, Fig. 1 Schematic diagram for sensory transduction steps (Adapted from Belmonte (1996) and Block (1992))

body and only within a narrow range of intensities (noxious stimuli). Effective stimuli include mechanical forces as well as temperature and a relatively large number of chemical substances. The largest part of the electromagnetical spectrum goes undetected by nociceptors. Only when the action of these otherwise unnoticed forces leads to cell damage, indirect stimulation of nociceptors may follow due to release of chemical mediators by injured cells located nearby. This is the case, for instance, of ultraviolet light or nuclear radiation exposure. What distinguishes nociceptors from low-threshold sensory receptors is not the physical nature of the adequate stimuli but the required threshold intensity for their activation (Belmonte 1996).

Perireceptor Elements

Nociceptors are naked, also called free, nerve terminals embedded in an intercellular matrix that contains collagen fibrils and proteoglycans and devoid of specialized structures (perireceptor elements) that in other receptors act as filters for the transmission of forces to the transduction sites. Receptive areas in the nociceptor membrane appear to be discontinuous patches of bare axolemma, covered only by the basal lamina of the nerve fiber and thus exposed to the direct action of stimuli. However, macromolecules present in the intercellular matrix surrounding nociceptive nerve endings may play a role in the filtering of noxious stimuli reaching the receptive membrane. The chemical and physical characteristics, including mechanical properties, of the cellular microenvironment can strongly influence sensory transduction. Moreover, mechanical, thermal, and chemical cues from the extracellular matrix sensed and transduced by nociceptors may regulate gene expression thereby producing long-term effects in primary nociceptive neurons.

Transduction Molecules

The diversity of forces that selectively activate nociceptor endings suggests that nociceptive neurons are equipped with separate transduction mechanisms for each of the stimulating energies. Nevertheless, some of the transducer molecules are multimodal and can be activated by more than one class of stimulus.

The detection of stimuli by nociceptor neurons is based on membrane signaling molecules which convert the stimulus energy into an allosteric molecular change, leading ultimately to the gating of membrane ion channels, influx of cations, and depolarization of the nerve terminal. In nociceptors, most transduction molecules are ion channels directly gated by the stimulus. Several classes of ion channels have been associated to the transduction of the various forms of energy and the production of generator potentials at nociceptor nerve terminals (Fig. 2). The list of channels located at nociceptor nerve terminals continues to grow, and their functional role is under intense scrutiny. In addition, stimuli can influence other ion channels in the terminal and receptor molecules that are part of G-protein signaling cascades that will exert a modulatory role in the genesis and propagation of impulse discharges.

Transduction Channels Directly Gated by the Stimulus

The **TRP superfamily of cation channels** is composed of several subfamilies (TRPC, TRPV, TRPM, TRPP, TRPN, TRPML, and TRPA). Many TRPs have emerged as important cellular sensors in a variety of sensory modalities including taste, olfaction, phototransduction, osmosensation, touch, hearing, thermal sensations, and pain (Clapham 2003; Damann et al. 2008). Members of the TRPC, TRPV, and TRPM families and TRPA1 (the single member of the TRPA subfamily) are expressed in mammalian primary sensory neurons and participate in the sensory transduction of heat, cold, chemical stimuli, and mechanical forces.

The degenerin/epithelium sodium channel superfamily (DEG/ENaC) includes five major subfamilies of channels (Kellenberger and Schild 2002). Mutations in some of the members, named degenerins, make constitutively active channels that lead to neurodegeneration. DEG/ENaC channels are gated by gentle mechanical forces and mediate touch in nematodes and flies. In vertebrates, the homologs of degenerins include various subunits of Na^+-selective, ▶ acid-sensing ion channels that are blocked by amiloride and were called ASICs (Waldmann and Lazdunski 1998). These subunits are encoded by 4 separate genes and are processed in different splice variants. Some subunits, like ASIC1b and ASIC3, are expressed almost exclusively in the peripheral nervous system. Others, like ASIC2a, are expressed in specialized mechanosensory structures. The crystal structure of ASIC1 has been solved. These channels have two transmembrane domains separated by a large extracellular cysteine-rich region and ensemble into homomeric and heteromeric combinations of 3 subunits (i.e., trimers). They are gated by protons and give rise to transient inward currents with fast-activating and variable desensitizing kinetics. A recent study identified novel gating mechanisms of human ASIC3 channels during alkalization (Delaunay et al. 2012). In addition to activation by low pH, some of these channels are activated by mechanical forces, temperature, and animal toxins and

Transduction Channels

Purinergic	Chloride	2P Domain K⁺	TRP	PIEZO	DEG/ENaCs
P2X2	ANO1	TASK-1	TRPV1	PIEZO2	ASIC1a
P2X3		TASK-3	TRPV2		ASIC1b
P2Y1		TREK-1	TRPV3		ASIC2a
		TREK-2	TRPV4		ASIC2b
		TRAAK	TRPM3		ASIC3
		TRESK	TRPM5		
			TRPM8		
			TRPA1		
			TRPC3		
			TRPC5		
			TRPC6		

Voltage Gated Channels

Hyperpolarisation Activated	Na⁺$_v$		Ca²⁺$_v$		K⁺$_v$	
	TTX-S	TTX-R	High Threshold	Low Threshold	K$_v$1.1	K$_v$7.2
HCN1	Na$_v$1.3	Na$_v$1.8	Ca$_v$2.1	Ca$_v$3.1	K$_v$1.2	K$_v$7.3
HCN2	Na$_v$1.6	Na$_v$1.9	Ca$_v$2.2	Ca$_v$3.2	K$_v$1.4	K$_v$7.5
	Na$_v$1.7			Ca$_v$3.3		

Modulation

Ligand-gated channel	Peptides, aminoacids, purines,
G-protein coupled receptorss	5-HT, histamine, growth factors
Tyrosine kinase receptors	AA metabolites, opioids
Guanylyl cyclase receptors	

Transduction and Encoding of Noxious Stimuli, Fig. 2 Summary of transduction channels, modulatory molecules, and voltage-gated channels involved in noxious stimulus transduction and modulation and encoding of propagated nerve impulses by the various classes of nociceptors

could play important roles in nociception (Waldmann and Lazdunski 1998; Bohlen et al. 2011).

The two-pore domain superfamily of K⁺ channels (KCNK) is a class of voltage-insensitive, K⁺-selective channels that are open at rest. For this property they are also known as background or leak K⁺ channels. They control neuronal excitability by changing the resting membrane potential and membrane resistance (Talley et al. 2003). Activity of these channels is regulated by a variety of signaling molecules, including protons, polyunsaturated fatty acids, and phospholipids (Patel et al. 2001). They are also modulated by physical variables like mechanical stretch and temperature, making them candidates for sensory transduction channels. Various

KCNK subunits have been identified in primary sensory neurons. However, very little information is available regarding the expression pattern in specific subpopulations of somatosensory afferents. TREK-1, TREK-2, and TRAAK may play a role in mechanosensory and thermal transduction (Patel et al. 2001; Kang et al. 2005). The TWIK-related spinal cord potassium channel (TRESK) is highly expressed in DRG and TG. Moreover, a mutation with loss of TRESK channel function is linked to familial migraine with aura (Lafreniere et al. 2010). Some of these channels, like members of the TASK subfamily, are closed by extracellular acidic solutions, leading to depolarization, and could be involved in pH sensing by nociceptors (Talley et al. 2003).

Ligand-gated ion channels are also present in nociceptor neurons and may play a role in the transduction of stimuli. These channels open in response to a variety of chemical substances, some of which are released locally following tissue injury. A prominent example is offered by the family of ionotropic purinergic receptors (P2X). These are ATP-gated cation-selective ion channels. Of the seven P2X channels identified so far, all except $P2X_7$ are expressed in sensory neurons (Burnstock 2000). Only the $P2X_3$ subunit is selectively expressed in nociceptors (Chen et al. 1995). Also, some members of the **nicotinic acetylcholine receptor** superfamily of ion channels are present in nociceptive terminals. These include the **cation-permeable nicotinic ACh receptors** and the 5-HT3 serotonin receptors that are nonselective cation channels, gated, respectively, by acetylcholine and serotonin, substances that are released in inflamed tissues. Recently, ionotropic receptors for excitatory amino acids have been reported in nociceptive terminals associated to their excitation during injury.

Remote Sensors and Modulators of the Stimulus

Activation of metabotropic membrane receptor proteins by endogenous substances released locally in the environment of nociceptive nerve endings plays an important role in modulating their activity and responsiveness following injury. These substances include peptides, kinins, purines, and excitatory amino acids. In this case, the effects on nociceptor membrane excitability are not exerted directly on an ion channel. Rather, the presence of an agonist substance causes the activation of second-messenger cascades that will finally lead to the opening or closure of ion channels. The effects can be immediate or be the consequence of long-term changes in gene expression that include variations in the number of ion channels and, consecutively, sustained modifications of neuronal excitability.

Receptors of this type include the superfamily of G-protein-coupled receptors (**GPCR**); a large number of G-protein-coupled receptors have

been shown to be present in sensory nerve endings and up- and down-modulate their excitability following application of the specific ligand. Very often sensitization is specific to a certain type of stimulus (mechanical versus thermal). The list of pro-algesic substances includes acetylcholine, bradykinin (BK), noradrenaline, histamine, serotonin, PGE2, and nerve growth factor (NGF).

Among the GPCRs capable of acutely sensitizing nociceptors, we count B1 and B2 BK receptors and alpha-1, alpha-2, and beta-2 adrenergic receptors. Histamine 1 receptors also sensitize nociceptors. The sensitization of nociceptors produced by bradykinin through the activation of B2 and B1 receptors involves the phospholipase C and protein kinase C signaling pathways that lead ultimately to the modulation of the TRPV1 (capsaicin) ion channel (Chuang et al. 2001; Vellani et al. 2001).

In contrast, M2 muscarinic receptors participate in the depression of nociceptive responsiveness to heat and mechanical stimulation produced by the acetylcholine liberated during injury by skin keratinocytes and other local cells. Activation of somatostatin type 2 receptors, present in about 10 % of cutaneous afferent terminals, also has antinociceptive effects. Intradermal injection of metabotropic glutamate receptors type 5 antagonists produces full reversal of thermal hyperalgesia in animal models of neuropathic pain. Opiates also exert a peripheral antinociceptive activity, and these unconventional effects of opioids appear to be mediated by opening of K^+ ATP channels following activation of the arginine/NO/cGMP/pathway by protein kinase G.

Tyrosine kinase receptors, another superfamily of membrane receptor proteins, also participate in the modulation of nociceptor excitation. This group includes TrkA receptors that are activated by nerve growth factor (NGF) released by injury in the environment of nociceptive terminals. The effect of NGF appears to be mediated in part by the same mechanism that is activated by BK, namely, PLC stimulation with subsequent hydrolysis of membrane phosphatidylinositol-4,5-bisphosphate (PIP2) yielding IP3 and diacylglycerol. One final target of this cascade is the TRPV1 channel that is modulated by both PKC-dependent and PKC-independent mechanisms (Vellani et al. 2001; Chuang et al. 2001).

▶ Proteinase-activated receptors (PARs) are also members of the superfamily of G-protein-coupled receptors. They initiate intracellular signaling by the proteolytic activity of extracellular serine proteases that cleave the N terminus of the receptor. PAR-1 and PAR-2 are expressed in many peptidergic sensory neurons, and their activation induces neurogenic inflammation and mediates peripheral sensitization of nociceptors (Vergnolle et al. 2003).

Membrane lipids are emerging as a new class of ion channel modulators, with important implications in nociceptor activation. A growing list of ion channels is modulated directly by interactions with phosphatidylinositol-4,5-bisphosphate (**PIP2**) (Suh and Hille 2008). As already mentioned, a number of agonists exert their actions by their ability to activate the phospholipase C (PLC) signaling pathway, which stimulates phosphoinositide hydrolysis. The end result is a reduction in the concentrations of plasma membrane phosphoinositides to produce inositol 1,4,5-trisphosphate (IP3). For most sensory channels, exogenous application of PIP2 leads to channel activation (Rohacs and Nilius 2007), and elevated PIP2 levels in lumbar DRGs lead to thermal and mechanical hypersensitivity (Sowa et al. 2010). In the case of TRPV1, the regulation of activity by membrane PIP2 levels is complex; some studies indicate channel activation downstream of PLC activation by NGF or BK (Chuang et al. 2001).

Transduction Mechanisms for the Different Stimuli

Ion channels and receptor proteins of the different superfamilies exhibit a variable sensitivity to mechanical forces, chemical substances, heat, or cold. The characteristic expression patterns of these proteins in the various subtypes of nociceptor neurons confer them their specific transduction capabilities for the different forms of stimulating energy.

Mechanotransduction

Propagated impulse responses to mechanical forces are prominent in the peripheral endings of nociceptor neurons exclusively activated by mechanical stimuli (▶ mechano-nociceptors); they also appear in ▶ polymodal nociceptor neurons that are additionally excited by chemical and thermal stimuli. Mechanically evoked nerve impulse discharges are absent in ▶ "silent" nociceptor nerve fibers, but they are obtained following local tissue inflammation. Ion channels directly gated by mechanical stimuli were first recognized in the 1980s, but the identification of the molecular entities involved in the transduction of mechanical forces remained elusive until recently.

Mechanical distortion, like stretch or pressure, produces the opening of mechanically sensitive channels (mechanosensitive channels, MSCs). MSCs are a heterogeneous population of channels with differences in sensitivity, type of response and biophysical properties like ionic selectivity and adaptation, and pharmacology. They are present in a great number of cells and, in addition to stimulus detection, participate in a variety of other cell functions, like volume regulation, cell movement, cell division, osmosensation, and contraction. The effect of mechanical force on a channel can be direct, leading to gating by tension exerted directly on the channel proteins, or indirect, involving second messengers controlled by mechanosensitive enzymes. However, the precise mechanism coupling the supply of energy provided by the mechanical stimulus to the gating of the channels is still unresolved.

The molecular identity of ion channels involved in the transduction of low-intensity mechanical forces by specific mechanosensory cells has been partly elucidated in ciliated mechanosensory cells of invertebrates and in the hair cells of the auditory, vestibular, and lateral line organs of vertebrates (reviewed by Corey 2003). With the application of genetic screens, DEG/ENaC channels and TRP channels have been identified as essential for mechanosensation in different types of mechanosensors in flies and worms. In most cases, however, it remains uncertain whether they are the transduction channel themselves and what the specific mechanism of activation is. The fact that mechanosensation requires the concerted function of several proteins acting in an ensemble (transduction apparatus) makes these studies especially difficult.

In the nematode *C. elegans*, gentle touch mechanosensation depends on members of the DEG/ENaC family. In particular, mutations in the channel subunits MEC-4 and MEC-10 led to loss of responses to touch and abolished mechanotransduction currents. OSM-9 and OCR-2, two related TRPV channels, are also required for touch sensitivity in the nose region in *C. elegans*. In flies TRPN1 (previously known as NompC) is essential for mechanotransduction in sensory bristles. This channel is also at the core of mechanotransduction in zebra fish hair cells. Surprisingly, no TRPN-like genes could be found in higher vertebrates. Also in flies, a TRPV-like channel protein, NAN, is expressed selectively in chordotonal neurons and is essential for hearing. No orthologues of *nanchung* have been identified in vertebrates, suggesting that, in this case, hearing is mediated by different transducers. Genetic screens in *Drosophila* larvae for mutations that alter responses to noxious mechanical and thermal stimuli identified *Painless*, a TRPA channel (Tracey, Jr. et al. 2003). The localization of TRPA1 channels at the apical hair bundles (i.e., the mechanotransduction site) of vertebrate hair cells and the effects of TRPA1 silencing on mechanosensitive currents generated by the movement of the cilia suggested an important in role in hearing (Corey et al. 2004). However, TRPA1-deficient mice have unimpaired auditory function (Bautista et al. 2006; Kwan et al. 2006).

Until recently, the nature of the channels directly involved in the transduction of mechanical stimuli by mammalian somatosensory endings was largely ignored (Nilius and Honore 2012). Moreover, the differences in threshold and adaptation characteristics found between low- and high-threshold mechanoreceptor neurons are still unexplained at the cellular level. These differences may simply lie in the density of mechanosensory channels in the transducing areas, due to the presence of different types of mechanosensory transduction channels or to other factors like variations in the arrangement or composition of the mechanotransduction apparatus, cytoskeleton organization, and/or second-messenger pathways.

A major breakthrough in the mechanosensory field was the recent identification of two novel, evolutionary conserved, cation channels, piezo1 and piezo2, with pronounced mechanosensitivity (Coste et al. 2010). Piezo2 is expressed in primary sensory neurons and knockdown of the protein reduces rapidly adapting mechanosensitive currents. Moreover, in *Drosophila*, knocking down the piezo protein in sensory neurons impairs responses to noxious mechanical stimuli (Kim et al. 2012). Several other channels with apparent mechanosensitivity have been identified in primary sensory neurons of mammals, but evidence for their direct mechanical activation is slim in all cases. Candidate transducer molecules for mechanotransduction include TRP channels and members of the acid-sensing ion channel (ASIC) family. Modulatory roles have been assigned to other channels like 2P-domain K^+ channels and purinergic receptors.

Within the TRP superfamily of ion channels, TRPV4 appears to be sensitive to membrane stretch. TRPV4 opens in response to osmotic cell swelling when expressed in mammalian cultured cells. TRPV4 is a multifunctional channel that is also activated by warm temperatures and lipoxygenase metabolites (Patapoutian et al. 2003). Perception of noxious pressure is reduced in TRPV4-/- mice, while gentle touch detection is unimpaired. However, TRPV4 does not appear to be directly mechanosensitive; rather, its activation appears to depend on a second-messenger cascade. TRPA1 channels are present in a large proportion of mammalian DRG neurons. Mechanosensitivity is reduced in specific subpopulations of mechanosensory afferents in TRPA1-deficient mice (Vilceanu and Stucky 2010; Brierley et al. 2011; Kwan et al. 2009). Moreover, mechanical hypersensitivity following inflammation or tissue injury is severely reduced in TRPA1 KO mice. The role of canonical TRP proteins in somatosensory mechanosensitivity is still controversial (Gottlieb et al. 2008). Nevertheless, TRPC5 is gated by mechanical stimuli and shows a clear sensitivity to osmotic stimuli in primary sensory neurons (Gomis et al. 2008). The combined deletion of TRPC3 and TRPC6 causes deficits in light touch, hearing, and balance (Quick et al. 2012). These deficits were not observed in single mutants, suggesting redundancy in the function of these channels.

Among channels of the DEG/ENaCs superfamily, ASIC1, ASIC2, and ASIC3 subunits are expressed in somas and peripheral terminals of primary sensory neurons (reviewed by Kellenberger and Schild 2002). Nonetheless, assigning a functional role to specific subunits is complicated by the fact that, in most cases, functional channels in vivo are heteromultimeric. Somewhat predictably, knockout of individual ASIC subunits in mice only leads to modest deficits in touch sensitivity. Furthermore, the existence of numerous touch receptors with overlapping mechanical sensitivities and intermixed receptive fields limits the possibility of marked functional deficits in case of single-gene disruptions. In addition, properties of human subunits are not identical to those displayed by homologs in other species. So far, there is no definitive evidence for a major role of a particular ASIC subunit in somatosensory mechanotransduction. Early studies suggested that loss of ASIC2a reduces the sensitivity of low-threshold rapidly adapting Ab mechanoreceptors to touch. Deletion of ASIC3, also known as DRASIC, increased the sensitivity of mechanoreceptors detecting light touch but reduced the sensitivity of an $A\delta$ nociceptors responding to noxious pinch. However, other studies indicate that lack of ASIC1, ASIC2, or ASIC3 does not impair cutaneous or visceral mechanosensation nor does it reduce mechanogated currents in the soma of DRG neurons (Roza et al. 2004).

The role of the lipid-sensitive, mechano-gated K^+ channels TREK-1, TREK-2, and TRAAK of the 2P-domain K^+ channel family in nociceptor mechanotransduction is still open to discussion (Patel et al. 2001).

P2 purinergic receptors also seem to participate indirectly in mechanotransduction, possibly modulating the opening probability of other classes of mechanosensory channels. Expression of $P2Y_1$ receptors in oocytes determines the development of electrical responses to the application of light mechanical stimuli. These stimuli appear to release intracellular ATP that in turn activates $P2Y_1$ receptors (Nakamura and Strittmatter 1996). Similarly, release of ATP from damaged cells following mechanical injury may stimulate P2X receptors present in small unmyelinated fibers and may be involved in mechanical sensitivity of nociceptors (Cook and McCleskey 2002).

Finally, mechanical stretching modifies the activity of voltage-dependent Ca^{2+} channels but not of Na^+ and K^+ channels or K^+ leakage channels. The mechanosensitivity of Ca^{2+} channels may contribute to modulate the neuron response to mechanical stimuli through changes in intracellular Ca^{2+}.

Extracellular matrix attachments have been proposed as the general mechanism involved in transmitting external mechanical forces to the neuron surface and, subsequently, to MSCs. In turn, MSCs are tethered to the internal cytoskeleton. Relative displacement of these structures would transmit tension to the gate of the mechanosensory channel. Altogether they form a functional unity called the mechanosensory apparatus (Fig. 3). It has been suggested that transmembrane integrins act as a molecular linker between the extracellular mechanical signal and the cytoskeleton, because they bind actin-associated proteins and therefore link physically the extracellular matrix with the microfilaments. Other cellular elements, like certain enzymes, can be directly sensitive to stress and act additionally as mechanotransducers. However, the low sensitivity and slow time course of the evoked responses make their direct responsibility in sensory mechanotransduction dubious.

Chemotransduction

A great variety of molecules, either exogenous or endogenous, act on nociceptive terminals, producing a change in membrane potential and eventually a discharge of nerve impulses (Belmonte 1996; Basbaum et al. 2009). Sensitivity to

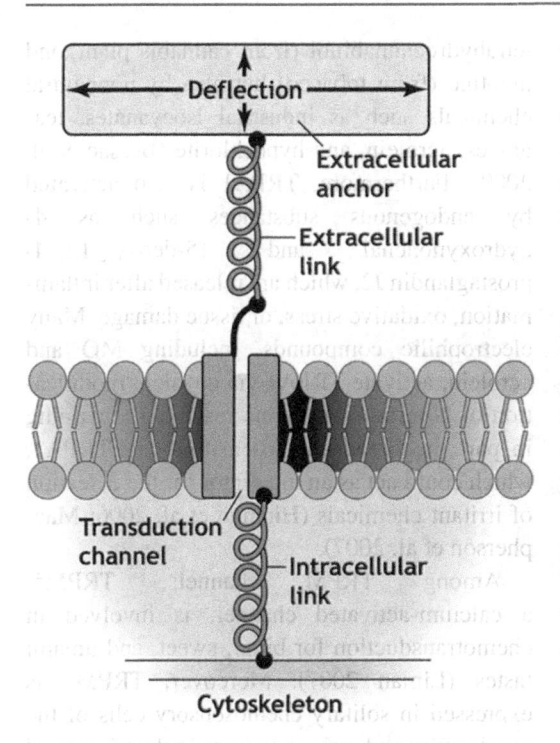

Transduction and Encoding of Noxious Stimuli, Fig. 3 Hypothetical arrangement for the different elements of the mechanotransduction apparatus. External mechanical forces are transmitted by extracellular matrix molecules and cytoskeleton proteins to mechanotransduction channels, causing an ion current flow through the open channel

chemicals is present to a variable degree in all subpopulations of nociceptors (Viana 2011). In some cases (polymodal nociceptors, silent nociceptors), chemical mediators activate directly the nerve terminals. In others (mechano-nociceptors), chemical mediators may alter the excitability of nerve terminals without producing a propagated discharge but changing their responsiveness to further stimuli.

Endogenous compounds that influence the excitability of nociceptors include protons, low oxygen (i.e., hypoxia), arachidonic acid and their metabolites (e.g., prostaglandins), kinins, amines like serotonin and histamine, cytokines (e.g., tumor necrosis factor-a, IL-1b, and IL-8), acetylcholine, amino acids, NO, opioids, ATP, adenosine, endocannabinoids, and other neuropeptides (e.g., endothelin-1). Many of these substances are released as part of the

injury/inflammatory response caused by noxious stimuli and will trigger or modulate the transduction process (Petho and Reeh 2012). In addition, a number of growth factors and inflammatory mediators influence nociceptor excitability by regulating gene transcription, resulting in altered ion channel expression. These types of substances include tumor necrosis factor-a and interleukin-1a (Hatano et al. 2012), nerve growth factor (NGF), brain-derived neurotrophic factor (BDNF), and glial cell-derived neurotrophic factor (GDNF) and related factors.

Considering the diverse physicochemical properties of exogenous sensory irritants, it is doubtful that all of them act through a specific receptor molecule. Some of the compounds that stimulate nociceptive nerve endings may partition in the membrane, according to their liposolubility, altering membrane and cellular properties, including surface charge, gating of ion channels, and metabolic state of the cell. As detailed below, some nociceptor channels (e.g., TRPA1) can be gated by reversible covalent modification. The net result of these actions is a depolarization of nerve terminals and a discharge of nerve impulses, whose firing frequency may be proportional to the concentration of the irritant substance within certain limits. In contrast, many other exogenous and endogenous chemicals and protons gate directly or indirectly members of the various superfamilies of ion channels, leading to changes in membrane potential (Julius and Basbaum 2001).

TRPV1 (originally named VR1) is the founding member of the TRPV subfamily and a receptor for capsaicin, the pungent compound found in hot peppers. TRPV1 appears to play a central role in the detection and integration of chemical and thermal noxious stimuli leading to sensitization of nociceptors (Caterina et al. 1997). This channel is activated by many endogenous substances and also by protons and noxious heat (over 42 °C). TRPV1 is a nonselective cation channel with a very high permeability to Ca^{2+}. Sensory neurons expressing native TRPV1 receptors, and oocytes or mammalian cells transfected with TRPV1, exhibit robust membrane currents in response to capsaicin that are desensitized by repeated exposures to the agonist and blocked by the competitive vanilloid receptor antagonist capsazepine and by the noncompetitive antagonists ruthenium red and peptoids. The channel is also activated by various endogenous lipids, like anandamide, a ligand for cannabinoid receptor 1. Moreover, the endogenous inflammatory mediator bradykinin (BK) that is known to sensitize polymodal nociceptor endings through activation of the G-protein-coupled receptor BK2 also enhances capsaicin currents. A similar effect is produced by nerve growth factor (NGF) that also sensitizes polymodal nerve endings to noxious stimuli by activation of the tyrosine-kinase receptor TrkA. In addition, GDNF upregulates B1 receptor expression in small non-peptidergic nociceptive neurons. Furthermore, activation of B1 receptor causes a marked increase in the amplitude of the heat-activated current in these neurons. BK2 and TrkA receptors act through the stimulation of phospholipase C (PLC) that enhances both basal TRPV1 activity and capsaicin-evoked responses. PLC catalyzes the hydrolysis of membrane PIP2 to form inositol trisphosphate (IP3) and diacylglycerol. PIP2 inhibits TRPV1 channels directly and its hydrolysis results in TRPV1 potentiation (Chuang et al. 2001). Of note, others have found a clear potentiation of TRPV1 by PIP2 (Klein et al. 2008). B2 and $P2Y_2$ receptors appear to be mediated by protein kinase C (PKC) activation.

Other members of the TRPV family exhibit chemical sensitivity relevant for nociception. TRPV4 channels are activated by the endocannabinoid anandamide and its metabolite arachidonic acid, as well by compounds obtained from traditional medicinal plants and used as anti-inflammatory remedies like bisandrographolide A (Vriens et al. 2007).

TRPA1 was initially identified as a putative noxious cold sensor (Story et al. 2003), but more recent studies evidenced that this channel acts as a quite promiscuous chemosensor, as it can be activated by a broad variety of natural compounds such as allyl isothiocyanate (the pungent compound in mustard oil (MO) and wasabi), allicin (from garlic), cinnamaldehyde (from cinnamon), menthol (from mint),

tetrahydrocannabinol (from cannabis plant),and nicotine (from tobacco) but also by nonnatural chemicals such as industrial isocyanates, tear gasses, acrolein, and hypochlorite (Bessac et al. 2009). Furthermore, TRPA1 is also activated by endogenous substances such as 4-hydroxynonenal and 15-deoxy-_12,14-prostaglandin J2, which are released after inflammation, oxidative stress, or tissue damage. Many electrophilic compounds, including MO and acrolein, activate TRPA1 via covalent modification of N-terminal cysteine residues, explaining in part the broad chemosensitivity of TRPA1, which could act as an integrator for the detection of irritant chemicals (Hinman et al. 2006; Macpherson et al. 2007).

Among TRPM channels, TRPM5, a calcium-activated channel, is involved in chemotransduction for bitter, sweet, and umami tastes (Liman 2007). Moreover, TRPM5 is expressed in solitary chemosensory cells of the nasal cavity and may participate in the trigeminal response to airborne irritants (Lin et al. 2008). Recently, TRPM3, a channel activated by pregnenolone sulfate, was found to be expressed in many nociceptors and involved in noxious heat detection (Vriens et al. 2011).

Evidence for the role of purinergic signaling in pain is based on multiple observations (reviewed by Burnstock 2009). ATP elicits pain when applied to human skin blisters. Application of ATP or a,β-methylene ATP excites capsaicin-sensitive trigeminal lingual afferents. Impulse responses to ATP have been evoked in the soma and neurites of tooth pulp cultured nociceptive neurons labelled retrogradely (Cook et al. 1997). It has been claimed that ATP released by injured cells excites nearby nociceptive terminals. Moreover, in mammalian cells expressing cloned TRPV1 channels and in DRG neurons, ATP enhances the response to capsaicin, heat, and protons (Moriyama et al. 2003). This effect seems to be secondary to activation of the metabotropic purinergic receptor $P2Y_2$, which activates phospholipase C through a G?protein $(G_{q/11})$, leading to the production of IP3 and diacylglycerol. Primary sensory neurons and nerve terminals express all subtypes of ionotropic purinergic receptors (P2X). Of those, P2X3 ionotropic channels mediate the above-described effects.

Among 2-pore domain K^+ channels, TREK-1, TREK-2, and TRAAK are gated by different biolipids like arachidonic acid and lysophosphatidic acid (Patel et al. 2001). Alkylamides are a unique class of compounds that elicit a distinct tingling sensation when applied to the tongue. Hydroxy-a-sanshool, a natural alkylamide found in Szechuan pepper, inhibits several pH-sensitive, 2-pore domain K^+ channels (TASK-1 or KCNK3, TASK-3 or KCNK9, and TRESK or KCNK18), depolarizing capsaicin-sensitive nociceptors and D-type mechanoreceptors (Bautista et al. 2008). Other studies have established excitatory effects of sanshool on TRPV1 and TRPA1 channels as well.

Acid Sensing

Intradermal injection of acidic solutions induces pain. Acidosis activates and sensitizes polymodal nerve terminals (Belmonte et al. 1991). Furthermore, tissue acidosis is a common occurrence following inflammation, ischemia, and tissue injury. Protons could also play a role during postoperative pain. In particular, localized low tissue pH following a surgical incision correlated with the time course of pain behaviors in animal models (Woo et al. 2004). Most acid-sensing neurons have small diameters, typical for nociceptors. The excitatory effects of protons could be mediated by a variety of ion channels including TRPV1, ASICs, and TASK channels that are sensitive to pH changes. In TRPV1-transfected cells, protons evoke distinct inward currents and enhance the response to capsaicin up to five times over control values (Caterina et al. 1997). Single-channel activity in outside-out membrane patches of TRPV1-transfected cells also increases when external pH is reduced.

The acid-sensing ion channels (ASICs), members of the DEG/ENaC superfamily, are also activated by extracellular protons and are overexpressed during inflammation. The contribution of different DEG/ENaC channel subunits

to pH-sensitive currents in DRG neurons has been investigated with targeted disruption of the various subunits. Deletion of any one subunit did not abolish proton-gated currents suggesting that two or more ASIC subunits coassemble as heteromultimers. As mentioned above, ASIC3 is present in presumed primary nociceptive neurons of the mouse and possibly participates in the detection of strong local pH reductions accompanying ischemia, inflammation, and postsurgical incisions. This channel is likely involved in the sensation of cardiac pain (Sutherland et al. 2001). The ASIC3 knockout mouse exhibits a lowered sensitivity to intramuscular injection of acid, and their acid-sensitive DRG neurons respond less to low-pH solutions. Pharmacological inhibition of ASIC3 channels or in vivo knockdown of ASIC3 subunit by small interfering RNA led to a significant reduction of postoperative spontaneous, thermal, and postural pain behaviors evoked by skin incisions, suggesting an important role of this channel in prolonged pain evoked by skin and muscle incisions (Deval et al. 2011). In fact, psychophysical and pharmacological data in humans suggest that ASIC channels have a more important role than TRPV1 channels in the sensation of moderate cutaneous acid-induced pain.

Among the 2P-domain K^+ channels, TASK subunits (TASK-1, TASK-2, TASK-3) represent background outward rectifiers that are constitutively active at all voltages and are inhibited by extracellular acidic pH values. Closure of these channels depolarizes the membrane potential. TASK channels have been found in some sensory neurons with nociceptive properties, but their role in pH-induced pain has not been firmly established yet.

Thermal Transduction

Extremely low or high temperatures evoke distinct pain sensations that are mediated by activation of subpopulations of sensory afferents. Heat activates polymodal nociceptor fibers, which respond to temperatures overpassing 41–42 °C. For the sensation of pain evoked by cold, it was suggested that a specific group of "cold" peripheral sensory fibers may exist.

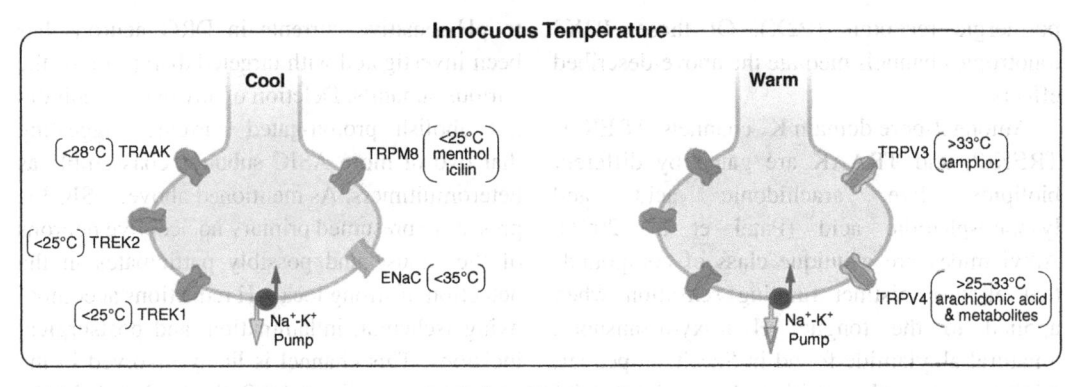

Transduction and Encoding of Noxious Stimuli, Fig. 4 Diagram of ion channels hypothetically involved in transduction of innocuous and noxious temperatures by peripheral sensory receptor terminals

However, there is also evidence favoring the view that the sensation of cold pain rather results from the concomitant activation by low temperatures of nonnociceptive cold thermoreceptors and a fraction of polymodal nociceptors (Campero et al. 1996). The transduction mechanisms for heat and cold in primary sensory neurons appear to be different at the cellular and molecular level (Fig. 4). Several receptor molecules have been identified recently that seem to be involved in the detection of temperature changes overpassing normal values (Patapoutian et al. 2003). In contrast, the cellular and molecular mechanisms involved in the detection of innocuous versus noxious temperature decreases are still incompletely understood.

Heat

Temperature elevations influence ionic pumps and conductances in all cell types including nociceptive terminals. However, polymodal nociceptor nerve fibers, unlike thermal sensory fibers that detect innocuous warming, begin to discharge nerve impulses when the tissue temperature increases over 40 °C. Moreover, they become sensitized by repeated thermal stimulation in the noxious range. Based on threshold, latency, and peak discharges to controlled heat pulses, two types of impulse responses were identified in separate groups of Ad nociceptors, one with a high threshold (over 53 °C), slow buildup, and latencies measured in seconds and a second type characterized by a lower threshold (46 °C), latencies measured in milliseconds, and rapid peak discharges. In turn, C nociceptors begin to respond at around 41 °C, giving a peak discharge near the stimulus onset. Overall, these observations suggest that various transduction mechanisms may contribute to the final activation of nociceptors by heat. Changes in responsiveness of polymodal nociceptive terminals to heat often go in parallel with a modified chemosensitivity. Thus, thermo- and chemosensitivities were simultaneously altered by capsaicin and exhibited cross-sensitization (Belmonte et al. 1991). These observations suggested that in polymodal neurons the specific transduction mechanisms for heat were closely associated with chemosensitivity to capsaicin.

Cesare and McNaughton (Cesare and McNaughton 1996) first described an inward current in a subpopulation of small dorsal root ganglion neurons that was activated by noxious heat, sensitized by bradykinin, and presumably mediated by a nonselective cation channel. Identification and cloning of the "capsaicin receptor" TRPV1 (Caterina et al. 1997) proved that heat is a stimulus for this channel as well as for other channels of the TRPV family. Heat-induced currents in membrane patches of TRPV1-transfected cells showed the same outwardly rectifying current–voltage relations as those evoked with capsaicin. Both were blocked by capsazepine and displayed analogous ionic selectivity. However, in transgenic mice where the TRPV1 gene was disrupted, sensitivity to capsaicin was lost, while responses to noxious heat still persisted. Also, a fraction of cultured primary sensory neurons respond to heat but not to capsaicin, indicating the existence of additional heat sensors. A candidate molecule to the additional sensitivity to noxious heat is another vanilloid-receptor-like channel, named TRPV2, that is 50 % identical to TRPV1 and responds to heat over 52 °C and is insensitive to capsaicin, being inhibited by the noncompetitive antagonist ruthenium red (Caterina et al. 1999). However, some nociceptors lacking TRPV1 and TRPV2 have normal heat responses (Woodbury et al. 2004), and TRPV2 KO mice display normal thermal and mechanical nociception (Park et al. 2011). Recently, TRPM3, a channel activated by pregnenolone sulfate, was found to be expressed in many nociceptors and shown to be involved in noxious heat detection and heat hyperalgesia (Vriens et al. 2011). Of note, simultaneously blocking TRPM3 and TRPV1 does not abrogate noxious thermosensation.

Sensitivity to rising temperatures has been shown in two more members of the TRPV family. TRPV3 is structurally analogous to TRPV1 and is expressed in the skin keratinocytes, the brain, and other tissues. When transfected to mammalian cells, this channel responds to temperature with a threshold around 33 °C but not to capsaicin or pH changes, showing an activation profile closely similar to TRPV1 (reviewed by Patapoutian et al. 2003). TRPV3 is also activated by camphor. Although no immunoreactivity to this channel could be initially found in the neurons of peripheral sensory ganglia, TRPV3 null mice have strong deficits in responses to innocuous and noxious heat indicating that it participates in thermosensation, perhaps responding to ATP released by keratinocytes during heating (Moqrich et al. 2005). TRPV4, a cation channel that has been implicated in hypo-osmolality sensing by neurons of the anterior hypothalamus, is also temperature sensitive when expressed in oocytes and HEK 293 cells, with an activation threshold of 25 °C and maximal responses around 40 °C (Guler et al. 2002). Thus, transduction of heat by primary sensory neurons appears to be mediated by several ion channel proteins that cover different temperature ranges and may explain the differences in heat threshold among nociceptors. Formation of heteromeric channel assemblies by the members of the TRPV subfamily in nociceptive neurons is an additional possibility in explaining the presence of a range of receptors with a wide spectrum of responses to temperature. However, recent evidence in mice deficient of TRPV3 and TRPV4 suggests that these channels play a minimal role in the detection of innocuous or noxious high temperatures and thermal sensation (Huang et al. 2011).

Anoctamin 1 (ANO1) also known as TMEM16A is a Ca^{2+}-activated chloride channel expressed by small primary sensory neurons where it co-localizes with nociceptor markers. ANO1 activity increases steeply at temperatures above 44 °C, and mice twice-reduced ANO1 levels showed impaired responses to heat, suggesting a role for this protein in the transduction of noxious heat signals (Cho et al. 2012).

All together, these studies emphasize the complexity and apparent redundancy of molecular mechanisms involved in sensing noxious heat.

Cold

During the past decade, great progress has been made in the identification of the cellular mechanisms involved in the transduction of cold stimuli by low- and high-threshold cold thermal receptors. Low-threshold (innocuous) cold sensory fibers increase their firing frequency with changes of skin temperature as low as 0.01 °C (Gallar et al. 2003). On the other hand, high-threshold cold fibers responding specifically to cutaneous cold below 27 °C have been reported (reviewed by Reid 2005). Finally, a fraction of polymodal nociceptor afferents discharge when they are exposed to very low temperatures (Campero et al. 1996; Acosta et al. 2001).

The unifying hypothesis originally proposed to explain sensitivity to low temperatures of primary sensory neurons was that the electrogenic sodium-potassium pump was highly temperature-dependent so that temperature decreases cause a reduction in pump activity, leading to depolarization. The differential temperature sensitivity of sodium and potassium ion channels would additionally contribute to a depolarized membrane potential during cooling. However, blockers of the Na^+-K^+ pump do not eliminate sensitivity to temperature decreases of cold sensory fibers.

In 2002, TRPM8, an ion channel of the TRPM family, originally named CMR1 because of its cold and menthol sensitivity, was cloned and characterized (McKemy et al. 2002; Peier et al. 2002). TRPM8 is a tetramer, each subunit consisting of 6 transmembrane domains (S1–S6) and intracellular C- and N-terminal. TRPM8 is a sensor of moderate temperature decreases and is also activated by compounds producing cooling sensations such as menthol, icilin, and eucalyptol. TRPM8 is found in about 10–15 % of primary sensory neurons, being expressed in low-threshold cold thermoreceptor neurons but also in a fraction of nociceptive neurons. Experiments in genetically modified mice lacking functional expression of TRPM8 strongly suggest that this ion channel is the principal receptor for sensing cold in low-threshold cold thermoreceptor neurons (reviewed by Daniels

and McKemy 2007). In heterologous expression systems, TRPM8 channels activate at temperatures below 25 °C, a temperature notably lower than the activity threshold of cold thermoreceptors (de la Pena et al. 2005). In sensory neurons activated by cold, thermal sensitivity is largely determined by the expression level of TRPM8 and a voltage-gated potassium current (IK_D) that acts as an excitability break (Madrid et al. 2009). Low concentrations of 4-AP or a-dendrotoxin selectively block this current and render some cold-insensitive neurons responsive to cold (Viana et al. 2002; Madrid et al. 2009). Cold also closes a background K^+ current in the soma of cultured cold-sensitive neurons (Viana et al. 2002; Reid 2005), producing a net inward current (I_{cold}) that leads to depolarization and impulse firing. This effect is partly counteracted by the prominent inwardly rectifying I_h current present in these cells that tend to depolarize the neuron at the basal membrane potential but closes progressively with depolarization as well as a consequence of the direct action of cold (Orio et al. 2009).

Noxious Cold

Reduction of cutaneous surface temperature below 15 °C elicits a sensation of cold pain. The discovery of TRPA1, a TRP channel expressed in a fraction of primary sensory neurons, some of which also harbor TRPV1 channels, and the fact that TRPA1 is activated by temperatures below 18 °C made it a good candidate for the transduction of noxious cold (Story et al. 2003). However, as said above, TRPA1 is also gated by mechanical forces and a number of pungent compounds like cinnamaldehyde, mustard oil, and allicin that, when applied to the skin, evoke a burning rather than a cold sensation (Namer et al. 2005). These findings and the disputed activation of TRPA1 by low temperature leave unsettled the specific role of TRPA1 in the transduction of physiological cold pain. In neuropathic pain models, the time course of cold hyperalgesia matches the expansion of a population of TRPA1-expressing neurons but only in the uninjured ganglion (Obata et al. 2005). Moreover, mild cooling strongly potentiates the response of TRPA1 to reactive chemical agonists (del Camino et al. 2010). These results, combined with the absence of cold hypersensitivity in inflammatory and neuropathic pain models in TRPA1 KO mice, suggest a key role of this channel in cold hypersensitivity under pathological conditions.

There is growing experimental evidence that TRPM8-expressing neurons could also participate in sensing not only innocuous but also unpleasant or painful cold (Knowlton et al. 2011; Belmonte et al. 2009). Both functional and immunohistological observations suggest that in a subpopulation of somatosensory neurons, TRPM8 is co-expressed with classical nociceptor markers, such as TRPV1 channels, calcitonin gene-related peptide (CGRP), and substance P. Cultured primary sensory neurons responding to cold have a wide range of temperature thresholds from 32 °C to 18 °C but exhibit homogeneous membrane properties and firing characteristics, and the great majority are sensitive to the TRPM8 activator, menthol. About 50 % of them are also activated by capsaicin (Viana et al. 2002). In this context, an important finding is the discovery of a-dendrotoxin-sensitive voltage-gated potassium currents in TRPM8-expressing neurons that strongly modulate their firing threshold (Madrid et al. 2009). As a matter of fact, intradermal injection of a-dendrotoxin increased nocifensive responses to cold in mice (Viana et al. 2002). Therefore, the possibility that noxious cold is signaled by a fraction of cold-sensitive neurons biophysically analogous to those responding to innocuous cold but with different central connections cannot be excluded. In that respect, it is interesting that topical application of menthol, the specific agonist of TRPM8 channels, can also induce sensations of irritation and pain (Acosta et al. 2001; Wasner et al. 2004). Nonetheless, there is experimental evidence that noxious cooling also activates identified polymodal nociceptor fibers in animals and humans. Thus, it is also possible that noxious cold sensations are evoked through the parallel excitation by very low temperatures of polymodal nociceptor and innocuous cold receptor fibers.

Generation and Encoding of Nerve Impulses

In nociceptors, the change of local membrane conductance caused by physical and chemical stimuli is expected to produce a **generator potential**, whose magnitude and duration would reflect the intensity, duration, and time course of the stimulus. These generator potentials associated to gating of transduction channels in nociceptive terminals have never been recorded due to technical limitations imposed by the small size of nociceptive endings. The generator potential will in turn propagate electrotonically and initiate at the closest point of the axonal membrane endowed with regenerative properties, a discharge of propagated nerve impulses whose frequency of discharge is also proportional to the amplitude of the stimulus. The various ion channels associated with the stimulus transduction and the generation of electrical activity in primary nociceptive neurons appear to be distributed unevenly among the peripheral terminal arborizations, the parent axon, and the cell body, thus conferring different electrical properties to the different portions of the neuron.

Voltage-Gated Channels and the Encoding of Noxious Stimuli

Primary sensory neurons also express ion channels that gate in response to changes in membrane potential and whose primary role in terms of stimulus detection is not transduction but propagation and modulation of the impulse discharge. They form the gene superfamily of **voltage-gated channels** that includes a large variety of K^+, Na^+, and Ca^{2+} channels, Cl^- channels, and nonselective cation channels like the HCN channels (hyperpolarization-activated cyclic nucleotide-gated K^+ channels) (Hille 2001). Nociceptors have characteristic electrophysiological properties (Koerber and Mendell 1992), and many different voltage-gated ion channels contribute to their cellular excitability.

Abnormal excitability of injured neurons has been linked to alterations in the expression and functional characteristics of many types of voltage-gated ion channels (Cummins et al. 2000).

Given the diversity of ion channels in nerve terminals and their opposite effects on cell firing, the net change in excitability may not always correspond to that predicted by disturbances in a single class of voltage-gated ion channel.

Na$^+$ Channels

Pharmacologically, ▶ voltage-gated Na + channels have been broadly separated into two groups: one class is blocked by nanomolar concentrations of the natural marine toxin tetrodotoxin, while the other group of Na$^+$ channels is resistant to tetrodotoxin. The results of various molecular approaches indicate that nine of the ten alpha subtypes of sodium channels are present in sensory neurons. Some of these channels are found in virtually all sensory neurons, while others, specially the TTXr subtypes, show a more restricted expression pattern. The expression profile also changes during development.

Propagation of action potentials along all sensory axons is mediated by TTXs channels. In contrast, some of the TTXr Na$^+$ channels subtypes (Na$_v$1.8 and Na$_v$1.9) are selectively found in small diameter, primarily nociceptive neurons (Akopian et al. 1996; McCleskey and Gold 1999). Na$_v$1.8 channels are particularly abundant in polymodal nociceptive terminals, where they can sustain propagated action potentials that may contribute to local release of ▶ neuropeptides (Brock et al. 1998).

Expression of Na$^+$ channels is up- or downregulated following local inflammation and injury of peripheral sensory axons (Lai et al. 2004). Functional properties are also affected by injury. These factors possibly contribute to the changes in sensitivity and the ectopic activity found in damaged sensory nerves which may give rise to ▶ neuropathic pain. The role of Na$^+$ channels in hyperalgesia is further substantiated by the effectiveness of use-dependent sodium channel blockers for the treatment of various types of chronic pain. It must be noted that abnormal activity in uninjured primary afferents may also be critical for the observed hypersensitivity to sensory input in animal pain models (Lai et al. 2004).

K$^+$ Channels

Potassium channels form a diverse superfamily of ion channels. On the basis of their structure and functional properties, they can be separated into voltage-gated (Kv), calcium-activated (KCa), inward rectifier (Kir), and two-pore domain (KCNK) channels. They are involved in a variety of neuronal functions including maintenance of membrane potential, action potential repolarization, and regulation of firing frequency and are critical regulators of neuronal excitability (Hille 2001). Because K$^+$ ions have a negative equilibrium potential across the plasma membrane, activation of these channels tends to dampen excitation.

Voltage-Gated K$^+$ Channels (Kv) A large number of Kv channels are expressed in nociceptors (McCleskey and Gold 1999). A variable combination of rapidly inactivating A-type (I_A) and slowly or noninactivating (I_K) K$^+$ currents is observed in the various types of nociceptor neurons. These functional currents can be further subdivided into several types with distinct kinetic and pharmacological components. Pharmacological, functional, and genetic studies confirm that K$^+$ channels play a role in nociceptor excitability. Furthermore, several Kv channel genes are downregulated in sensory neurons following axotomy or chronic constriction of peripheral nerves, suggesting that, together with Na$_v$ channels, they participate in the alteration of axonal excitability that accompanies peripheral nerve injury (Rasband et al. 2001). Substances like PGE$_2$ can sensitize nociceptive sensory neurons by reducing activity in Kv channels.

K$_v$1.4, an A-type K$^+$ channel, is expressed in sensory neurons of small size that co-express the capsaicin channel TRPV1 and the TTX-R Na$^+$ channel Na$_v$1.8 and are presumably polymodal nociceptor neurons (Rasband et al. 2001). Kv1.1 and Kv1.2 channels are also expressed in small DRG neurons, and genetic studies have linked the lack of Kv1.1 channels with thermal hyperalgesia. KCNQ channels, responsible for M-type K$^+$ currents, are also present in nociceptive sensory neurons, and pharmacological activation of these channels inhibits responses to algesic substances in animal models of pain (Passmore et al. 2003). Finally, activation of small conductance calcium-activated K^+ channels and ATP-sensitive K^+ channels has been also implicated in antinociception. Kir3.2 knockout and Kir3.3 knockout mice display hyperalgesia to elevated temperatures (>50 °C) suggesting an implication G-protein-gated K^+ channels in thermal nociception. These modulatory effects appear to be occurring at the level of the spinal cord rather than at the periphery.

Two-Pore Domain K^+ Channels (KCNK) In addition to their role as cellular sensors, these channels, responsible for leak K^+ currents, are key elements in the regulation of background excitability. As such, they are thought to play an important role in triggering ectopic discharges associated with chronic pain conditions.

Ca^{2+} Channels

Voltage-activated Ca^{2+} channels are channels with steeply voltage-dependent gates that open in response to membrane depolarization. They have the special role of translating electrical signals into chemical signals through their control of the flow of Ca^{2+} ions into the cytoplasm, thereby regulating a variety of Ca^{2+}-dependent intracellular events (Hille 2001). Functional and pharmacological studies, followed by cloning and genetic analysis, have led to the identification of a large number of voltage-gated Ca^{2+} channels. Subpopulations of primary sensory neurons exhibit differences in their functional type of Ca^{2+} channels, and some subunits are selectively expressed in nociceptors (Bell et al. 2004). The properties of voltage-gated Ca^{2+} channels are modulated directly and indirectly by a number of endogenous mediators released during injury, thereby changing the excitability and responsiveness of nociceptive neurons. The expression of voltage-gated Ca^{2+} channel subunits is also altered in conditions like peripheral nerve injury, contributing to abnormal nerve activity and neuropathic pain. In addition, Ca^{2+} channels also play a key role in the function of central neurons involved in nociceptive

processing. The end result of these findings is that N-type, T-type, and P-type calcium channels have been validated as useful targets for the treatment of pain (reviewed by Bourinet and Zamponi 2005).

HCN Channels The currents carried by hyperpolarization-activated, cyclic nucleotide-gated channels (HCN) have been termed If, Ih, or Iq and define poorly selective K^+ currents activated by membrane hyperpolarization. The current reverses at membrane potentials of about −25 mV and is inwardly directing at rest. Ih currents contribute to the resting membrane potential, input conductance, and subthreshold membrane oscillations in many types of neurons (Robinson and Siegelbaum 2003). HCN channels, particularly HCN1 and HCN2, are abundantly expressed in sensory neurons. Inflammatory mediators, such as serotonin, raise intracellular cAMP levels, which, in turn, increase I_h current by binding to the HCN channel, shifting its voltage-dependent activation to less negative potentials (reviewed by Robinson and Siegelbaum 2003). Likewise, the amplitude of I_h currents in DRG neurons increases markedly following spinal nerve ligation or chronic compression of the ganglion (Yao et al. 2003). Genetic deletion of HCN2 abolishes action potential firing caused by an elevation of cAMP in nociceptors. In these mice, inflammatory or neuropathic hypersensitivity to thermal or mechanical stimuli was abolished (Emery et al. 2011). Furthermore, pharmacological blockade of HCN activity with the specific inhibitor ZD7288 reduces the spontaneous ectopic activity secondary to nerve injury and reverses abnormal hypersensitivity to light touch. These results suggest that HCN channels play a critical role in the initiation and expression of neuropathic pain.

Electrical Activity in Peripheral Terminal Arborizations

Parent sensory axons of nociceptors branch extensively in the peripheral territory. The distribution of the various transduction and voltage-gated channels is presumably nonhomogeneous

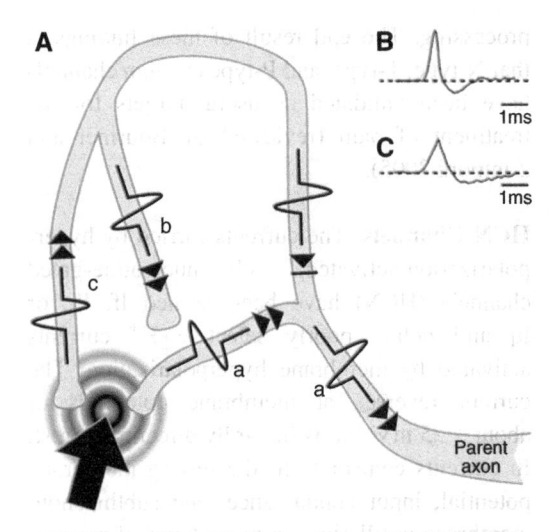

Transduction and Encoding of Noxious Stimuli, Fig. 5 (a) Propagation of nerve impulses at the peripheral branches of nociceptor fibers. The stimulus extends to various branches where it generates propagated action potentials that travel centripetally. These action potentials may collide at the branching points and this determines the final firing frequency in the parent axon (Modified from Weidner et al. 2003). (b, c) Nerve terminal impulses recorded from nociceptors endings in the cornea, before (b) and after (c) local application of lidocaine. The local anesthetic blocks the regenerative nerve impulse that normally occurs in the terminal (transducing) region of the branch (b), and only impulses passively propagated from neighbor endings are recorded (c)

among the terminal branches of the various subclasses of nociceptor fibers and perhaps even between branches of a single neuron. It is conceivable that in nociceptors, each terminal branch of an axonal arborization acts as an independent site for transduction and generation of action potentials. Action potentials of nociceptor branches travel centripetally at a conduction velocity that is below that of the parent axon (Weidner et al. 2003). When arriving at a branching point, impulses of the fastest branches will usually travel antidromically and invade the slower branch, changing transiently its excitability or even occluding the generation of action potentials (Fig. 5). In polymodal nociceptors antidromic action potentials invade the terminals due to the high density of TTXr Na$^+$ channels present in their membrane (Brock et al. 1998).

Impulse Firing in Parent Axons and Soma

Impulses generated at the endings progress centrally through terminal branches of different length, conduction velocity, and possibly duration of their refractory and supernormal periods becoming finally integrated in the parent axon. This integration will determine the frequency and firing pattern of the discharge of propagated nerve impulses travelling to the CNS along the complete axonal path of the sensory ganglion neuron.

The differences in membrane properties between subclasses of nociceptive neurons are reflected in the conduction velocity and the shape of propagated action potentials at the parent axon and soma, far away from the peripheral endings. C nociceptive axons have longer duration APs and AP undershoot durations than low-threshold mechanosensory axons (Koerber and Mendell 1992). Further differences in axonal spike duration and post-spike excitability are noticed between classes of nociceptors (Lawson 2002), possibly influencing the pattern and frequency of their impulse discharge. The soma of the various classes of nociceptive neurons also possesses a number of specific membrane characteristics. Neurons tentatively identified as polymodal have slow somatic action potentials with a hump in the falling phase and do not present inward rectification in response to hyperpolarizing pulses, while low-threshold mechanosensory neurons produce short-lasting action potentials and have a prominent inward rectification (Lawson 2002). These data further suggest that nociceptive neurons possess a specific set of voltage-activated Na$^+$, K$^+$, and Ca^{2+} channels that confer them distinct electrical properties and ultimately determine their excitability and pattern of impulse firing in response to peripheral stimuli.

It is worth noticing that the electrical behavior of nociceptor neurons is profoundly influenced by previous history. Repeated stimulation can transiently modify their excitability, through changes in some of the ionic mechanisms associated with the codification and propagation of nerve impulses, contributing to fatigue and/or sensitization (Serra et al. 1999). Similar, changes have been observed after nerve damage. This plasticity is important in understanding the altered excitability of nociceptor neurons following injury.

Summary

Electrophysiological, pharmacological, and molecular evidence suggests that nociceptor neurons possess multiple mechanisms for detection, amplification, and encoding of input signals. These mechanisms will mediate transduction of the various forms of stimulating energy and also modulate the input signal at the successive steps of the transduction and encoding processes. Different types of stimuli may interact at threshold or subthreshold levels to produce propagated responses. Short- and long-term modulatory mechanisms will further modify the final response composed by a discharge of nerve impulses of a given frequency, time course, and firing pattern. The interplay of transduction and modulatory mechanisms also define other characteristics of the impulse response such as ongoing activity, post-discharge, sensitization, fatigue, and inactivation.

From the perspective that several molecular mechanisms may coexist in nociceptive endings for the transduction and amplification of various forms of stimulating energy, a sharp categorization of nociceptors based only on the dominant presence of a given transduction molecule appears to be somewhat simplistic. Many of the "specific" transduction molecules for a particular form of energy are also present in sensory neurons that respond preferentially to other modalities of stimulus, as is the case of TRPV channels in cold receptor neurons. Nonetheless, it is also incomplete to distinguish nociceptors accordingly only with the type of stimulus that evokes a propagated impulse response. Nociceptor functional subtypes categorized with this criterion may in fact represent cases where the transducer mechanism for a given form of energy is prevalent over the others so that under normal circumstances only this form of stimulus will elicit a propagated response. When changes in

excitability and/or summation of the effects of other subthreshold stimuli take place, as is probably the case following inflammation and axonal injury, propagated responses evoked by other forms of energy may become apparent, as occurs, for example, with "silent" nociceptors. In spite of their biophysical specialization for certain stimuli, most nociceptor neurons appear to be potentially polymodal and endowed with a high degree of plasticity to modify their response according to the characteristics of the stimulus and its temporal course.

Acknowledgments The authors are grateful to Stuart Ingham for the help with the illustrations. Supported by grants SAF2010-14990 to F.V. and BFU2008-04425 and CONSOLIDER-INGENIO 2010 CSD2007-00023 to C.B.

References

Acosta, M. C., Belmonte, C., & Gallar, J. (2001). Sensory experiences in humans and single-unit activity in cats evoked by polymodal stimulation of the cornea. *The Journal of Physiology, 534*, 511–525.

Akopian, A. N., Sivilotti, L., & Wood, J. N. (1996). A tetrodotoxin-resistant voltage-gated sodium channel expressed by sensory neurons. *Nature, 379*, 257–262.

Basbaum, A. I., Bautista, D. M., Scherrer, G., & Julius, D. (2009). Cellular and molecular mechanisms of pain. *Cell, 139*, 267–284.

Bautista, D. M., Jordt, S. E., Nikai, T., Tsuruda, P. R., Read, A. J., Poblete, J., et al. (2006). TRPA1 mediates the inflammatory actions of environmental irritants and proalgesic agents. *Cell, 124*, 1269–1282.

Bautista, D. M., Sigal, Y. M., Milstein, A. D., Garrison, J. L., Zorn, J. A., Tsuruda, P. R., et al. (2008). Pungent agents from Szechuan peppers excite sensory neurons by inhibiting two-pore potassium channels. *Nature Neuroscience, 11*, 772–779.

Bell, T. J., Thaler, C., Castiglioni, A. J., Helton, T. D., & Lipscombe, D. (2004). Cell-specific alternative splicing increases calcium channel current density in the pain pathway. *Neuron, 41*, 127–138.

Belmonte, C. (1996). Signal transduction in nociceptors: General principles. In C. Belmonte & F. Cervero (Eds.), *Neurobiology of nociceptors* (pp. 243–257). New York: Oxford University Press.

Belmonte, C., Brock, J. A., & Viana, F. (2009). Converting cold into pain. *Experimental Brain Research, 196*, 13–30.

Belmonte, C., Gallar, J., Pozo, M. A., & Rebollo, I. (1991). Excitation by irritant chemical substances of sensory afferent units in the cat's cornea. *The Journal of Physiology, 437*, 709–725.

Bessac, B. F., Sivula, M., von Hehn, C. A., Caceres, A. I., Escalera, J., & Jordt, S. E. (2009). Transient receptor potential ankyrin 1 antagonists block the noxious effects of toxic industrial isocyanates and tear gases. *The FASEB Journal, 23*, 1102–1114.

Block, S. (1992). Biophysical principles of sensory transduction. In D. Corey & S. D. Roper (Eds.), *Sensory transduction* (pp. 1–17). New York: The Rockefeller University Press.

Bohlen, C. J., Chesler, A. T., Sharif-Naeini, R., Medzihradszky, K. F., Zhou, S., King, D., et al. (2011). A heteromeric Texas coral snake toxin targets acid-sensing ion channels to produce pain. *Nature, 479*, 410–414.

Bourinet, E., & Zamponi, G. W. (2005). Voltage gated calcium channels as targets for analgesics. *Current Topics in Medicinal Chemistry, 5*, 539–546.

Brierley, S. M., Castro, J., Harrington, A. M., Hughes, P. A., Page, A. J., Rychkov, G. Y., et al. (2011). TRPA1 contributes to specific mechanically activated currents and sensory neuron mechanical hypersensitivity. *The Journal of Physiology, 589*, 3575–3593.

Brock, J. A., McLachlan, E. M., & Belmonte, C. (1998). Tetrodotoxin-resistant impulses in single nociceptor nerve terminals in guinea-pig cornea. *The Journal of Physiology, 512*(Pt 1), 211–217.

Burnstock, G. (2000). P2X receptors in sensory neurones. *British Journal of Anaesthesia, 84*, 476–488.

Burnstock, G. (2009). Purinergic receptors and pain. *Current Pharmaceutical Design, 15*, 1717–1735.

Campero, M., Serra, J., & Ochoa, J. L. (1996). C-polymodal nociceptors activated by noxious low temperature in human skin. *The Journal of Physiology, 497*(Pt 2), 565–572.

Caterina, M. J., Rosen, T. A., Tominaga, M., Brake, A. J., & Julius, D. (1999). A capsaicin-receptor homologue with a high threshold for noxious heat. *Nature, 398*, 436–441.

Caterina, M. J., Schumacher, M. A., Tominaga, M., Rosen, T. A., Levine, J. D., & Julius, D. (1997). The capsaicin receptor: A heat-activated ion channel in the pain pathway. *Nature, 389*, 816–824.

Cesare, P., & McNaughton, P. (1996). A novel heat-activated current in nociceptive neurons and its sensitization by bradykinin. *Proceedings of the National Academy of Sciences of the United States of America, 93*, 15435–15439.

Chen, C. C., Akopian, A. N., Sivilotti, L., Colquhoun, D., Burnstock, G., & Wood, J. N. (1995). A P2X purinoceptor expressed by a subset of sensory neurons. *Nature, 377*, 428–431.

Cho, H., Yang, Y. D., Lee, J., Lee, B., Kim, T., Jang, Y., et al. (2012). The calcium-activated chloride channel anoctamin 1 acts as a heat sensor in nociceptive neurons. *Nature Neuroscience, 15*, 1015–1021.

Chuang, H. H., Prescott, E. D., Kong, H., Shields, S., Jordt, S. E., Basbaum, A. I., et al. (2001). Bradykinin and nerve growth factor release the capsaicin receptor from PtdIns(4,5)P2-mediated inhibition. *Nature, 411*, 957–962.

Clapham, D. E. (2003). TRP channels as cellular sensors. *Nature, 426*, 517–524.

Cook, S. P., & McCleskey, E. W. (2002). Cell damage excites nociceptors through release of cytosolic ATP. *Pain, 95*, 41–47.

Cook, S. P., Vulchanova, L., Hargreaves, K. M., Elde, R., & McCleskey, E. W. (1997). Distinct ATP receptors on pain-sensing and stretch-sensing neurons. *Nature, 387*, 505–508.

Corey, D. P. (2003). New TRP channels in hearing and mechanosensation. *Neuron, 39*, 585–588.

Corey, D. P., Garcia-Anoveros, J., Holt, J. R., Kwan, K. Y., Lin, S. Y., Vollrath, M. A., et al. (2004). TRPA1 is a candidate for the mechanosensitive transduction channel of vertebrate hair cells. *Nature, 432*, 723–730.

Coste, B., Mathur, J., Schmidt, M., Earley, T. J., Ranade, S., Petrus, M. J., et al. (2010). Piezo1 and Piezo2 are essential components of distinct mechanically activated cation channels. *Science, 330*, 55–60.

Cummins, T. R., Dib-Hajj, S. D., Black, J. A., & Waxman, S. G. (2000). Sodium channels and the molecular pathophysiology of pain. *Progress in Brain Research, 129*, 3–19.

Damann, N., Voets, T., & Nilius, B. (2008). TRPs in our senses. *Current Biology, 18*, R880–R889.

Daniels, R. L., & McKemy, D. D. (2007). Mice left out in the cold: Commentary on the phenotype of TRPM8-nulls. *Molecular Pain, 3*, 23.

de la Pena, E., Malkia, A., Cabedo, H., Belmonte, C., & Viana, F. (2005). The contribution of TRPM8 channels to cold sensing in mammalian neurones. *The Journal of Physiology, 567*, 415–426.

del Camino, D., Murphy, S., Heiry, M., Barrett, L. B., Earley, T. J., Cook, C. A., et al. (2010). TRPA1 contributes to cold hypersensitivity. *Journal of Neuroscience, 30*, 15165–15174.

Delaunay, A., Gasull, X., Salinas, M., Noel, J., Friend, V., Lingueglia, E., et al. (2012). Human ASIC3 channel dynamically adapts its activity to sense the extracellular pH in both acidic and alkaline directions. *Proceedings of the National Academy of Sciences of the United States of America, 109*, 13124–13129.

Deval, E., Noel, J., Gasull, X., Delaunay, A., Alloui, A., Friend, V., et al. (2011). Acid-sensing ion channels in postoperative pain. *Journal of Neuroscience, 31*, 6059–6066.

Emery, E. C., Young, G. T., Berrocoso, E. M., Chen, L., & McNaughton, P. A. (2011). HCN2 ion channels play a central role in inflammatory and neuropathic pain. *Science, 333*, 1462–1466.

Gallar, J., Acosta, M. C., & Belmonte, C. (2003). Activation of scleral cold thermoreceptors by temperature and blood flow changes. *Investigative Ophthalmology & Visual Science, 44*, 697–705.

Gomis, A., Soriano, S., Belmonte, C., & Viana, F. (2008). Hypoosmotic- and pressure-induced membrane stretch activate TRPC5 channels. *The Journal of Physiology, 586*, 5633–5649.

Gottlieb, P., Folgering, J., Maroto, R., Raso, A., Wood, T. G., Kurosky, A., et al. (2008). Revisiting TRPC1 and TRPC6 mechanosensitivity. *Pflügers Archiv, 455*, 1097–1103.

Guler, A. D., Lee, H., Iida, T., Shimizu, I., Tominaga, M., & Caterina, M. (2002). Heat-evoked activation of the ion channel, TRPV4. *Journal of Neuroscience, 22*, 6408–6414.

Hatano, N., Itoh, Y., Suzuki, H., Muraki, Y., Hayashi, H., Onozaki, K., et al. (2012). Hypoxia-inducible factor-1alpha (HIF1alpha) switches on transient receptor potential ankyrin repeat 1 (TRPA1) gene expression via a hypoxia response element-like motif to modulate cytokine release. *Journal of Biological Chemistry, 287*, 31962–31972.

Hille, B. (2001). *Ion channels of excitable membranes.* Sunderlad, MA: Sinauer Associates.

Hinman, A., Chuang, H. H., Bautista, D. M., & Julius, D. (2006). TRP channel activation by reversible covalent modification. *Proceedings of the National Academy of Sciences of the United States of America, 103*, 19564–19568.

Huang, S. M., Li, X., Yu, Y., Wang, J., & Caterina, M. J. (2011). TRPV3 and TRPV4 ion channels are not major contributors to mouse heat sensation. *Molecular Pain, 7*, 37.

Julius, D., & Basbaum, A. I. (2001). Molecular mechanisms of nociception. *Nature, 413*, 203–210.

Kang, D., Choe, C., & Kim, D. (2005). Thermosensitivity of the two-pore domain K+ channels TREK-2 and TRAAK. *The Journal of Physiology, 564*, 103–116.

Kellenberger, S., & Schild, L. (2002). Epithelial sodium channel/degenerin family of ion channels: A variety of functions for a shared structure. *Physiological Reviews, 82*, 735–767.

Kim, S. E., Coste, B., Chadha, A., Cook, B., & Patapoutian, A. (2012). The role of drosophila piezo in mechanical nociception. *Nature, 483*, 209–212.

Klein, R. M., Ufret-Vincenty, C. A., Hua, L., & Gordon, S. E. (2008). Determinants of molecular specificity in phosphoinositide regulation. Phosphatidylinositol (4,5)-bisphosphate (PI(4,5)P2) is the endogenous lipid regulating TRPV1. *Journal of Biological Chemistry, 283*, 26208–26216.

Knowlton, W. M., Daniels, R. L., Palkar, R., McCoy, D. D., & McKemy, D. D. (2011). Pharmacological blockade of TRPM8 ion channels alters cold and cold pain responses in mice. *PLoS One, 6*, e25894.

Koerber, H. R., & Mendell, L. M. (1992). Functional heterogeneity of dorsal root ganglion cells. In S. A. Scott (Ed.), *Sensory neurons diversity, development, and plasticity* (pp. 77–96). New York: Oxford University Press.

Kwan, K. Y., Allchorne, A. J., Vollrath, M. A., Christensen, A. P., Zhang, D. S., Woolf, C. J., et al. (2006). TRPA1 Contributes to cold, mechanical, and chemical nociception but is not essential for hair-cell transduction. *Neuron, 50*, 277–289.

Kwan, K. Y., Glazer, J. M., Corey, D. P., Rice, F. L., & Stucky, C. L. (2009). TRPA1 modulates mechanotransduction in cutaneous sensory neurons. *Journal of Neuroscience, 29*, 4808–4819.

Lafreniere, R. G., Cader, M. Z., Poulin, J. F., Andres-Enguix, I., Simoneau, M., Gupta, N., et al. (2010). A dominant-negative mutation in the TRESK potassium channel is linked to familial migraine with aura. *Nature Medicine, 16*, 1157–1160.

Lai, J., Porreca, F., Hunter, J. C., & Gold, M. S. (2004). Voltage-gated sodium channels and hyperalgesia. *Annual Review of Pharmacology and Toxicology, 44*, 371–397.

Lawson, S. N. (2002). Phenotype and function of somatic primary afferent nociceptive neurones with C-, Adelta- or Aalpha/beta-fibres. *Experimental Physiology, 87*, 239–244.

Liman, E. R. (2007). TRPM5 and taste transduction. *Handbook of Experimental Pharmacology, 179*, 287–298.

Lin, W., Ogura, T., Margolskee, R. F., Finger, T. E., & Restrepo, D. (2008). TRPM5-expressing solitary chemosensory cells respond to odorous irritants. *Journal of Neurophysiology, 99*, 1451–1460.

Macpherson, L. J., Dubin, A. E., Evans, M. J., Marr, F., Schultz, P. G., Cravatt, B. F., et al. (2007). Noxious compounds activate TRPA1 ion channels through covalent modification of cysteines. *Nature, 445*, 541–545.

Madrid, R., de la PE, Donovan-Rodriguez, T., Belmonte, C., & Viana, F. (2009). Variable threshold of trigeminal cold-thermosensitive neurons is determined by a balance between TRPM8 and Kv1 potassium channels. *Journal of Neuroscience, 29*(29), 3120–3131.

McCleskey, E. W., & Gold, M. S. (1999). Ion channels of nociception. *Annual Review of Physiology, 61*, 835–856.

McKemy, D. D., Neuhausser, W. M., & Julius, D. (2002). Identification of a cold receptor reveals a general role for TRP channels in thermosensation. *Nature, 416*, 52–58.

Moqrich, A., Hwang, S. W., Earley, T. J., Petrus, M. J., Murray, A. N., Spencer, K. S., et al. (2005). Impaired thermosensation in mice lacking TRPV3, a heat and camphor sensor in the skin. *Science, 307*, 1468–1472.

Moriyama, T., Iida, T., Kobayashi, K., Higashi, T., Fukuoka, T., Tsumura, H., et al. (2003). Possible involvement of P2Y2 metabotropic receptors in ATP-induced transient receptor potential vanilloid receptor 1-mediated thermal hypersensitivity. *Journal of Neuroscience, 23*, 6058–6062.

Nakamura, F., & Strittmatter, S. M. (1996). P2Y1 Purinergic receptors in sensory neurons: Contribution to touch-induced impulse generation. *Proceedings of the National Academy of Sciences of the United States of America, 93*, 10465–10470.

Namer, B., Seifert, F., Handwerker, H. O., & Maihofner, C. (2005). TRPA1 and TRPM8 activation in humans: Effects of cinnamaldehyde and menthol. *Neuroreport, 16*, 955–959.

Nilius, B., & Honore, E. (2012). Sensing pressure with ion channels. *Trends in Neurosciences, 35,* 477–486.

Obata, K., Katsura, H., Mizushima, T., Yamanaka, H., Kobayashi, K., Dai, Y., Fukuoka, T., Tokunaga, A., Tominaga, M., & Noguchi, K. (2005). TRPA1 induced in sensory neurons contributes to cold hyperalgesia after inflammation and nerve injury. *Journal of Clinincal Investigation, 115,* 2393–2401.

Orio, P., Madrid, R., de la Peña, E., Parra, A., Meseguer, V., Bayliss, D. A., Belmonte, C., & Viana, F. (2009). Characteristics and physiological role of hyperpolarization activated currents in mouse cold thermoreceptors. *The Journal of Physiology, 587*(Pt 9), 1961–1976.

Park, U., Vastani, N., Guan, Y., Raja, S. N., Koltzenburg, M., & Caterina, M. J. (2011). TRP vanilloid 2 knockout mice are susceptible to perinatal lethality but display normal thermal and mechanical nociception. *Journal of Neuroscience, 31,* 11425–11436.

Passmore, G. M., Selyanko, A. A., Mistry, M., Al Qatari, M., Marsh, S. J., Matthews, E. A., et al. (2003). KCNQ/ M currents in sensory neurons: Significance for pain therapy. *Journal of Neuroscience, 23,* 7227–7236.

Patapoutian, A., Peier, A. M., Story, G. M., & Viswanath, V. (2003). ThermoTRP channels and beyond: Mechanisms of temperature sensation. *Nature Reviews Neuroscience, 4,* 529–539.

Patel, A. J., Lazdunski, M., & Honore, E. (2001). Lipid and mechano-gated 2P domain K(+) channels. *Current Opinion in Cell Biology, 13,* 422–428.

Peier, A. M., Moqrich, A., Hergarden, A. C., Reeve, A. J., Andersson, D. A., Story, G. M., et al. (2002). A TRP channel that senses cold stimuli and menthol. *Cell, 108,* 705–715.

Petho, G., & Reeh, P. W. (2012). Sensory and signaling mechanisms of bradykinin, eicosanoids, platelet-activating factor, and nitric oxide in peripheral nociceptors. *Physiological Reviews, 92,* 1699–1775.

Quick, K., Zhao, J., Eijkelkamp, N., Linley, J. E., Rugiero, F., Cox, J. J., et al. (2012). TRPC3 And TRPC6 are essential for normal mechanotransduction in subsets of sensory neurons and cochlear hair cells. *Open Biology, 2,* 120068.

Rasband, M. N., Park, E. W., Vanderah, T. W., Lai, J., Porreca, F., & Trimmer, J. S. (2001). Distinct potassium channels on pain-sensing neurons. *Proceedings of the National Academy of Sciences of the United States of America, 98,* 13373–13378.

Reid, G. (2005). ThermoTRP channels and cold sensing: What are they really up to? *Pflügers Archiv, 451,* 250–263.

Robinson, R. B., & Siegelbaum, S. A. (2003). Hyperpolarization-activated cation currents: From molecules to physiological function. *Annual Review of Physiology, 65,* 453–480.

Rohacs, T., & Nilius, B. (2007). Regulation of transient receptor potential (TRP) channels by phosphoinositides. *Pflügers Archiv, 455,* 157–168.

Roza, C., Puel, J. L., Kress, M., Baron, A., Diochot, S., Lazdunski, M., et al. (2004). Knockout of the ASIC2 channel in mice does not impair cutaneous mechanosensation, visceral mechanonociception and hearing. *Journal of Physiology (London), 558,* 659–669.

Serra, J., Campero, M., Ochoa, J., & Bostock, H. (1999). Activity-dependent slowing of conduction differentiates functional subtypes of C fibres innervating human skin. *The Journal of Physiology, 515*(Pt 3), 799–811.

Sowa, N. A., Street, S. E., Vihko, P., & Zylka, M. J. (2010). Prostatic acid phosphatase reduces thermal sensitivity and chronic pain sensitization by depleting phosphatidylinositol 4,5-bisphosphate. *Journal of Neuroscience, 30,* 10282–10293.

Story, G. M., Peier, A. M., Reeve, A. J., Eid, S. R., Mosbacher, J., Hricik, T. R., et al. (2003). ANKTM1, a TRP-like channel expressed in nociceptive neurons, is activated by cold temperatures. *Cell, 112,* 819–829.

Suh, B. C., & Hille, B. (2008). PIP2 is a necessary cofactor for ion channel function: how and why? *Annual Review of Biophysics, 37,* 175–195.

Sutherland, S. P., Benson, C. J., Adelman, J. P., & McCleskey, E. W. (2001). Acid-sensing ion channel 3 matches the acid-gated current in cardiac ischemia-sensing neurons. *Proceedings of the National Academy of Sciences of the United States of America, 98,* 711–716.

Talley, E. M., Sirois, J. E., Lei, Q., & Bayliss, D. A. (2003). Two-pore-domain (KCNK) potassium channels: Dynamic roles in neuronal function. *The Neuroscientist, 9,* 46–56.

Tracey, W. D., Jr., Wilson, R. I., Laurent, G., & Benzer, S. (2003). Painless, a drosophila gene essential for nociception. *Cell, 113,* 261–273.

Vellani, V., Mapplebeck, S., Moriondo, A., Davis, J. B., & McNaughton, P. A. (2001). Protein kinase C activation potentiates gating of the vanilloid receptor VR1 by capsaicin, protons, heat and anandamide. *The Journal of Physiology, 534,* 813–825.

Vergnolle, N., Ferazzini, M., D'Andrea, M. R., Buddenkotte, J., & Steinhoff, M. (2003). Proteinase-activated receptors: Novel signals for peripheral nerves. *Trends in Neurosciences, 26,* 496–500.

Viana, F. (2011). Chemosensory properties of the trigeminal system. *ACS Chemical Neuroscience, 2,* 38–50.

Viana, F., de la Pena, E., & Belmonte, C. (2002). Specificity of cold thermotransduction is determined by differential ionic channel expression. *Nature Neuroscience, 5,* 254–260.

Vilceanu, D., & Stucky, C. L. (2010). TRPA1 mediates mechanical currents in the plasma membrane of mouse sensory neurons. *PLoS One, 5,* e12177.

Vriens, J., Owsianik, G., Hofmann, T., Philipp, S. E., Stab, J., Chen, X., et al. (2011). TRPM3 is a nociceptor channel involved in the detection of noxious heat. *Neuron, 70,* 482–494.

Vriens, J., Owsianik, G., Janssens, A., Voets, T., & Nilius, B. (2007). Determinants of 4 alpha-phorbol sensitivity in transmembrane domains 3 and 4 of the cation channel TRPV4. *Journal of Biological Chemistry, 282,* 12796–12803.

Waldmann, R., & Lazdunski, M. (1998). H(+)-gated cation channels: Neuronal acid sensors in the NaC/DEG
family of ion channels. *Current Opinion in Neurobiology, 8,* 418–424.
Wasner, G., Schattschneider, J., Binder, A., & Baron, R.
(2004). Topical menthol–a human model for cold pain
by activation and sensitization of C nociceptors. *Brain,
127,* 1159–1171.
Weidner, C., Schmidt, R., Schmelz, M., Torebjork, H. E., &
Handwerker, H. O. (2003). Action potential conduction
in the terminal arborisation of nociceptive C-fibre afferents. *Journal of Physiology (London), 547,* 931–940.
Woo, Y. C., Park, S. S., Subieta, A. R., & Brennan, T. J.
(2004). Changes in tissue pH and temperature after
incision indicate acidosis may contribute to postoperative pain. *Anesthesiology, 101,* 468–475.
Woodbury, C. J., Zwick, M., Wang, S., Lawson, J. J.,
Caterina, M. J., Koltzenburg, M., et al. (2004).
Nociceptors lacking TRPV1 and TRPV2 have normal
heat responses. *Journal of Neuroscience, 24,* 6410–6415.
Yao, H., Donnelly, D. F., Ma, C., & LaMotte, R. H. (2003).
Upregulation of the hyperpolarization-activated cation
current after chronic compression of the dorsal root
ganglion. *Journal of Neuroscience, 23,* 2069–2074.

Transduction Channel

Definition

Ionic channel gated (i.e., open or closed) by
a physical or chemical stimulus applied to the
membrane of a peripheral sensory cell.

Cross-References

▶ Nociceptor Generator Potential

Transduction Sites

Definition

Membrane patches of sensory receptor terminals
where sensory transduction takes place.

Cross-References

▶ Nociceptor Generator Potential

Transfected Cells

Definition

Transfection refers to the introduction of exogenous genetic material encoding a gene of interest
into cells. Transfected cells are used as model
systems to study functional and structural properties of the transfected protein.

Cross-References

▶ Purine Receptor Targets in the Treatment of
Neuropathic Pain

Transfer and Generalization

Herta Flor
Department of Cognitive and Clinical
Neuroscience, Central Institute of Mental
Health/Medical Faculty Mannheim, Heidelberg
University, Mannheim, Germany

Definition

Transfer refers to the maintenance of learned
behavior change in the patient's environment
outside the treatment context; generalization
refers to the extension of learned behaviors to
similar problems.

Cross-References

▶ Operant Treatment of Chronic Pain

Transforaminal Epidural Steroid Injection

▶ Epidural Steroid Injections for Chronic Back
Pain

Transforaminal Injection of Steroids

▶ Epidural Steroid Injections

Transforaminal Interbody Fusion

▶ TLIF

Transforaminal Steroids

▶ Lumbar Transforaminal Injection of Steroids

Transformed Migraine

▶ Chronic Daily Headache in Children

Transformed Migraine Headache

Definition

Defunct terminology, see chronic migraine.

Cross-References

▶ Chronic Daily Headache in Children

Transganglionic Changes after Peripheral Nerve Injury

▶ Central Changes After Peripheral Nerve Injury

Transganglionic Transport

Definition

One of the tract-tracing methods, e.g., a neural tracer taken up by a peripheral axon or branches of primary afferents is transported centripetally and then centrifugally to their central terminals.

Cross-References

▶ Trigeminal Brainstem Nuclear Complex, Anatomy

Transgene

Definition

Any gene inserted into a vector that is expressed in cells in vitro or animals in vivo following vector-mediated transduction.

Cross-References

▶ Opioids and Gene Therapy

Transgenic Knockout Mice

Definition

Genetically engineered null mutant mice. Via blastocyst microinjection of transgenic (i.e., possessing an altered DNA sequence) embryonic stem cells, mice can be created that lack all expression of a particular gene. These mice can be evaluated for their altered pain sensitivity compared to "wild-type" mice and the function of the missing gene/protein inferred.

Cross-References

▶ Nerve Growth Factor Overexpressing Mice as Models of Inflammatory Pain

Transgenic Mice

▶ Nerve Growth Factor Overexpressing Mice as Models of Inflammatory Pain

Transient Headache and CSF Lymphocytosis

Hans-Christoph Diener
Department of Neurology, University Hospital
Essen, University Duisburg-Essen, Essen,
Germany

Synonyms

Migraine with pleocytosis; Pseudomigraine with lymphocytic pleocytosis

Definition

The International Headache Society proposed the following diagnostic criteria for HaNDL (IHS 2004):
1. Moderate or severe ▶ headache.
2. Cerebrospinal fluid pleocytosis with lymphocytic predominance (greater than 15 cells/µl) and normal neuroimaging, CSF culture, and other tests for etiology.
3. Headache occurs at time of CSF pleocytosis.
4. Episodes of headache and neurological deficits last <3 months.

Characteristics

HaNDL is a transient syndrome characterized by severe headache, focal neurological symptoms, and CSF lymphocytosis (IHS 2003; for review, see Pascual and Valle 2003).

The HaNDL syndrome was first described by Bartleson et al. (Bartleson et al. 1981). Until now about 100 cases of HaNDL have been published. Fifty patients have been reported by Gomez-Aranda et al. (1997). Berg and Williams presented a series of seven patients, as well as a systemic review of the literature (Berg and Williams 1995).

Age of onset of HaNDL ranges between 7 and 50 years, with a mean of 27 years. HaNDL is more frequent in men, with a male to female ratio of 3:1 (Berg and Williams 1995; Gomez-Aranda et al. 1997).

Approximately 25–30 % of patients report a viral-like illness 2–3 weeks before the onset of HaNDL. Typical symptoms are cough, rhinitis, diarrhea, and fatigue (Gomez-Aranda et al. 1997).

The timing between the headache and neurologic deficit is variable. In some patients the initial symptom is headache, in others neurologic deficits occur first. In some patients neurologic deficits with CSF lymphocytosis can occur without a severe headache (Oldani et al. 1998).

Headache is moderate to severe, pulsating or throbbing, and mostly bilateral. In some patients the headache is unilateral and localized contralateral to the neurologic symptoms. Duration of headache ranges between 1 h and 1 week, on average 19 h (Gomez-Aranda et al. 1997). Accompanying symptoms are nausea and vomiting and photo- and phonophobia.

Neurologic signs usually consist of hemiparesis, hemisensory symptoms, or aphasia. Sensory symptoms are more frequent than motor signs. Pure motor aphasia is the most frequent speech disorder followed by global aphasia and pure sensory aphasia (Gomez-Aranda et al. 1997). Visual symptoms are rare. The focal neurologic symptoms usually evolve progressively, in some cases, however, suddenly. The deficits are transient, lasting from 5 min to hours and at most 3 days. Atypical cases with longer duration up to a week have been described. The recovery is always complete.

Routine blood and immunological tests are usually normal. Two exceptional cases associated with a cytomegalovirus infection have been reported (Ferrari et al. 1983; Richert et al. 1987). The examination of cerebrospinal fluid reveals a lymphocytic pleocytosis (range 10–760/µl) and increased total protein. In 50 % of the cases, an elevated CSF pressure has been measured (Berg and Williams 1995; Gomez-Aranda et al. 1997). Immunological, bacterial, and viral studies produce normal results.

Electroencephalography reveals a unilateral focal slowing during the acute HaNDL phase in up to 70 % of the patients (Berg and Williams 1995; Gomez-Aranda et al. 1997). In 10 % ▶ EEG slowing was observed bilaterally (Gomez-Aranda et al. 1997). The changes normalized again after the symptomatic period.

Cranial computed tomography, as well as cranial magnetic resonance imaging, is usually normal. In a few patients nonspecific small areas of high signal in T2-weighted images have been observed (Berg and Williams 1995; Gomez-Aranda et al. 1997). In one patient, a diffusion-weighted MRI could be performed during acute HaNDL, which did not reveal diffusion changes (Gekeler et al. 2002). In another case diffusion restriction in the splenium of the corpus callosum was observed (Raets 2012).

Single photon emission computed tomography has been suggested to be the most informative neuroimaging technique. A focus of decreased tracer uptake has been detected during the acute phase, which becomes normal several days after recovery (Caminero et al. 1997; Fuentes et al. 1998).

Cerebral angiography is usually not informative. In the vast majority of patients studied, no changes could be found. In some patients, inflammation-like processes in the wall of the opercular arteries have been detected, and in some patients new episodes of HaNDL were triggered by the angiography; therefore, it has to be avoided.

The pathophysiology of HaNDL is still unclear. There are some obvious similarities to ▶ migraine with aura. However, only few patients with HaNDL report a history of migraine. The duration of the focal neurological symptoms in HaNDL is longer than in migraine with aura. The most important difference between the HaNDL and migraine is the CSF pleocytosis. HaNDL has also been separated from other diseases such as familial hemiplegic migraine or progressive cerebellar ataxia.

There are several infectious diseases that can cause headache, focal neurologic symptoms, and CSF pleocytosis. Mollaret's meningitis is characterized by recurrent episodes of aseptic meningitis with headache and symptoms of meningeal irritation. However, focal neurological symptoms are not typical for Mollaret's meningitis. Furthermore, the pleocytosis in Mollaret's meningitis is a mixture of lymphocytes, polymorphonuclear, and large "Mollaret's" cells.

HaNDL has to be separated from further infectious conditions such as Lyme disease (Pal et al. 1987), neurosyphilis (Berg and Williams 1995), neurobrucellosis (Roldan-Montaud et al. 1991), mycoplasma infection (Dalton and Newton 1991), granulomatous meningitis (Mayer et al. 1993), and secondary cytomegalovirus encephalitis associated with human immunodeficiency virus (HIV) infection (Richert et al. 1987).

HaNDL is always a self-limiting disorder. Therefore, the therapy is restricted to symptomatic treatment with analgesics and antiemetics. No preventive treatment is needed.

References

Bartleson, J. D., Swanson, J. W., & Whisnant, J. P. (1981). A migrainous syndrome with cerebrospinal fluid pleocytosis. *Neurology, 31*, 1257–1262.

Berg, M., & Williams, L. (1995). The transient syndrome of headache with neurologic deficits and CFS lymphocytosis. *Neurology, 45*, 1648–1954.

Caminero, A. B., Pareja, J. A., Arpa, J., et al. (1997). Migrainous syndrome with CSF pleocytosis. SPECT findings. *Headache, 37*, 511–515.

Dalton, M., & Newton, R. W. (1991). Aseptic meningitis. *Developmental Medicine and Child Neurology, 33*, 446–451.

Ferrari, M. D., Buruma, O. J., van Laar-Ramaker, M., et al. (1983). A migrainous syndrome with pleocytosis. *Neurology, 33*, 813.

Fuentes, B., Diez Tejedor, E., Pascual, J., et al. (1998). Cerebral blood flow changes in pseudomigraine with pleocytosis analyzed by single photon emission computed tomography. A spreading depression mechanism? *Cephalalgia, 18*, 570–573, discussion 531.

Gekeler, F., Holtmannspotter, M., Straube, A., et al. (2002). Diffusion-weighted magnetic resonance imaging during the aura of pseudomigraine with temporary neurologic symptoms and lymphocytic pleocytosis. *Headache, 42*, 294–296.

Gomez-Aranda, F., Canadillas, F., Marti-Masso, J. F., et al. (1997). Pseudomigraine with temporary neurological symptoms and lymphocytic pleocytosis. A report of 50 cases. *Brain, 120*, 1105–1113.

IHS. (2004). Headache Classification Committee of the International Headache Society. Classification and diagnostic criteria for headache disorders, cranial neuralgias and facial pain. *Cephalalgia, 24*(Suppl 1), 9–160.

Mayer, S. A., Yim, G. K., Onesti, S. T., et al. (1993). Biopsy-proven isolated sarcoid meningitis. Case report. *Journal of Neurosurgery, 78*, 994–996.

Oldani, A., Marcone, A., Zamboni, M., et al. (1998). The transient syndrome of headache with neurologic deficits and CSF lymphocytosis. Report of a case without severe headache. *Headache, 38*, 135–137.

Pal, G. S., Baker, J. T., & Humphrey, P. R. (1987). Lyme disease presenting as recurrent acute meningitis. *British Medical Journal (Clinical Research Edition), 295*, 367.

Pascual, J., & Valle, N. (2003). Pseudomigraine with lymphocytic pleocytosis. *Current Pain and Headache Reports, 7*, 224–228.

Raets, I. (2012). Diffusion restriction in the splenium of the corpus callosum in a patient with the syndrome of transient headache with neurological deficits and CSF lymphocytosis (HaNDL): A challenge to the diagnostic criteria? *Acta Neurologica Belgica, 112*(1), 67–69.

Richert, J. R., Potolicchio, S., Jr., Garagusi, V. F., et al. (1987). Cytomegalovirus encephalitis associated with episodic neurologic deficits and OKT-8+ pleocytosis. *Neurology, 37*, 149–152.

Roldan-Montaud, A., Jimenez-Jimenez, F. J., Zancada, F., et al. (1991). Neurobrucellosis mimicking migraine. *European Neurology, 31*, 30–32.

Transient Receptor Potential

Definition

Transient change in electrical potential across the cell membrane, for example, induced by activation of a TRP receptor.

Cross-References

▶ TRPV1 Receptor, Species Variability

Transient Receptor Potential Cation Channel

▶ TRPV1 Receptor, Species Variability

Transient Receptor Potential Family of Ion Channels

Definition

The mammalian transient receptor potential (TRP) ion channels are named after the role of these channels in *Drosophila* phototransduction. They are encoded by at least 21 different channel genes. The TRP channel primary structures predict six transmembrane domains, with a pore domain between the fifth and sixth segments, and both the C- and N-termini are located intracellularly. The mammalian TRP channel family is comprised of three subfamilies, including TRPC, TRPV, and TRPM. The family members are at least 25 % homologous within their amino acid sequences. Most of the channels are nonselective to cations, allowing sodium and calcium to flow in and depolarize neurons. The most well-characterized TRP channels in DRG neurons are the vanilloid family of TRPV1–4 channels, which are activated by a range of heat and/or warm temperatures. TRPV1 is the prototype vanilloid channel and is activated by noxious heat, acidic pH, and the alkaloid irritant capsaicin. Additional TRP channels in DRG neurons are TRPM8, which is activated by cool temperatures and menthol, and ANKTM1, which is activated by mustard oil derivatives and may be activated by cold.

Cross-References

▶ Immunocytochemistry of Nociceptors
▶ TRPV1 Receptor, Species Variability

Transient Receptor Potential Vanilloid Type 1

▶ TRPV1

Transition from Acute to Chronic Pain

Stephan A. Schug
School of Medicine and Pharmacology, Royal
Perth Hospital, Faculty of Medicine, Dentistry &
Health Sciences, The University of Western
Australia, UWA Anaesthesiology, Perth,
Australia

Synonyms

Central sensitization; Pain chronification; Pain
progression

Definition

It is now well established that chronic pain can
develop as the consequence of repeated or severe
episodes of acute pain, e.g., in the context of
trauma, surgery, or acute painful illness. This
progression is most likely to be the consequence
of central nervous processes, commonly called
central ▶ neuroplasticity or sensitization.

Characteristics

It is well recognized that chronic pain states often
follow an acutely painful stimulus such as surgery or trauma. A large study of over 5,000
patients referred to chronic pain clinics in the
UK revealed that 22.5 % of these patients had
developed their pain after surgery and 18.7 %
after trauma (Crombie et al. 1998). Similar observations have been made when following patients
after surgery or trauma, and severe early pain
after thoracotomy (Katz et al. 1996) and orthopedic trauma (Gehling et al. 1999) predicts development of chronic pain states.

Risk Factors for Transition to Chronic Pain States

In 2000, Perkins and Kehlet published a ▶ meta-
analysis of the predictive factors of chronic

pain after surgery (Perkins and Kehlet 2000).
They identified a number of significant risk factors for this transition, which include type of
surgery and preoperative, intraoperative, and
postoperative factors.

Type of Surgery

After lower limb amputation, phantom limb pain
occurs in 30–81 % of patients and stump pain in
>50 %. Post-thoracotomy pain syndrome (PTPS)
occurs in >50 % of patients. Breast surgery can
give rise to chest wall, scar, breast, or shoulder
pain in 11–57 % of patients and phantom breast
pain in 13–24 %; postmastectomy pain syndrome
(PMPS) has an overall incidence of 50 % at 1 year
(Kwekkeboom 1996). Gallbladder surgery
carries a risk between 3 % and 56 % of postchole-
cystectomy syndrome (PCS).

Preoperative Factors

Intense preoperative pain increases the incidence
of phantom limb pain (from 33 % to 72 %)
(Nikolajsen et al. 1997; Krane and Heller 1995).
Preoperative risk factors for PCS include "psychological vulnerability," female gender, and
long-standing preoperative symptoms. Other factors are repeat surgery and issues of compensation. Preoperative epidural pain control may
decrease the risk of chronic pain after amputation
(Schug 2004; Bach et al. 1988).

Intraoperative Factors

Technical issues of the surgery, including the
surgical approach and the risk of nerve injury,
are important here. Video-assisted thoracoscopic
lung surgery reduces the risk of PTPS compared
to open lung resection. The position of the incision for thoracotomy may influence the incidence
of PTPS. Intraoperative epidural analgesia seems
to reduce the risk of PTPS. The incidence of
PMPS may be increased by breast-conserving
surgery, immediate insertion of implants, extent
of axillary dissection, and damage to the intercostobrachial nerve.

Early Postoperative Factors

Stump pain at one-week post-amputation correlates with the risk of phantom pain. The intensity

of acute postoperative pain is an independent predictor for PTPS and PMPS. Intercostal nerve dysfunction (loss of the superficial abdominal reflex) is associated with more acute, subacute, and chronic pain. Adjuvant postoperative radiotherapy to the breast increases the risk of development of PMPS.

Late Postoperative Factors

Long-term stump pain predicts long-term phantom limb pain (Nikolajsen et al. 1997). More severe or prolonged acute pain in the postoperative period as well as postoperative complications, commonly leading to increased ▶ nociception, significantly predicts the development of chronic pain after surgery (Katz et al. 1996; Gehling et al. 1999). Other relevant factors are radiotherapy and neurotoxic chemotherapy but also psychological factors such as anxiety, depression, neuroticism, and psychological vulnerability.

Mechanisms for the Transition from Acute to Chronic Pain

The development of chronic pain is based on the phenomenon of central neuroplasticity. Disruption of the normal specialization of the somatosensory system leads to increasing mismatch between stimulus and response. The underlying mechanisms are not fully elucidated yet; however, the physiological principles have been reviewed in detail (Pockett 1995). The implications for the development of chronic pain as a separate disease entity have been summarized recently by Siddall and Cousins (Siddall and Cousins 2004). Some of these mechanisms are presented in the following.

Wind-Up

A progressive increase in the number of action potentials elicited per stimulus occurs in the dorsal and ventral horn neurons when the stimulus exceeds 0.5 Hz. Above this frequency, the postsynaptic depolarizing responses summate to produce a cumulative ▶ depolarization, resulting in a burst of action potentials, instead of a single action potential, in response to each stimulus. It is mediated via N-methyl-D-aspartic acid (NMDA) glutamate receptors and therefore blocked and reversed by NMDA antagonists.

▶ Wind-up lasts as long as ventral horn cell depolarization, i.e., about 60 s.

Long-Term Potentiation

Repeated episodes of wind-up may trigger ▶ long-term potentiation (LTP). It was first studied in the hippocampus and is now known to occur in visual, sensorimotor, and prefrontal cortex, as well as in the spinal cord. Its mechanism is complex, but in essence high-frequency presynaptic activity causes a presynaptic glutamate release, which activates α-amino-3-hydroxy-5-methyl-4-isoxazole propionic acid (AMPA) receptors. AMPA receptor activation opens ion channels allowing postsynaptic depolarization. If the depolarization reaches a certain threshold, a magnesium-dependent block of NMDA receptors is released, and these then open their associated ion channels. There is an overall influx of calcium ions which triggers additional calcium release from intracellular calcium stores. The intracellular calcium rise triggers a complex chain of events, which includes the release of one or more retrograde factors by the postsynaptic cell. By diffusing back to the presynaptic membrane, these cause an increased transmitter release in response to each presynaptic action potential. Subsequently, calcium-dependent enzymes are activated, such as protein kinases A and C (PKA, PKC) and calcium/calmodulin kinase, leading to phosphorylation of membrane proteins including receptors and ion channels. This makes the postsynaptic cell more excitable, and upregulation of AMPA receptors and growth of dendrites/spines on the postsynaptic cell occurs.

The overall result is LTP, which can last from one hour to several months. It can be slowed or prevented from occurring in vitro by NMDA antagonists, early cooling, or PKC inhibitors, but in contrast to wind-up cannot be reversed.

Recruitment

Chronic inflammation and nerve injury have an effect on the presence and distribution of voltage-gated sodium channels, which can become concentrated in areas of injury and produce ectopic discharges. Studies have shown that neuron-specific sodium channels become concentrated in neurons proximal to a site of nerve injury and play a role in the ▶ hyperalgesia and ▶ allodynia of chronic pain states. Not all sensory neurons are active all the time, and this ▶ peripheral sensitization will "recruit" dormant nociceptors, thus increasing the receptive fields of dorsal horn neurons and increasing the intensity and area of pain (Mannion and Woolf 2000).

Immediate Early Gene Expression

Immediate early genes are a family of genes (e.g., c-fos, c-jun) that share the characteristic of having their expression rapidly and transiently induced upon stimulation of neuronal and nonneuronal cells (Caputto and Guido 2000). Damaged sensory neurons may undergo altered gene expression, such that they release a different type of neurotransmitter. The release of neurotransmitters usually associated with noxious stimuli, such as substance P, may contribute to ▶ central sensitization.

A change in gene expression can also lead to up- or downregulation of ion channels leading to changes in cellular excitability.

Excitotoxicity

Excitotoxicity is a phenomenon that was first described by Olney in the 1970s (Olney et al. 1972). It involves the activation of glutamate receptors in the central nervous system (CNS). Glutamate, an excitatory amino acid, activates different types of ion channel-forming receptors to develop their essential role in the functional activity of the brain. However, high concentrations of glutamate or neurotoxins acting at the same receptors cause cell death by apoptosis through the excessive activation of these receptors.

The physiological role of the NMDA receptor seems to be related to synaptic plasticity and learning. In addition, working together with G-protein-coupled glutamate receptors, it ensures the establishment of the long-term potentiation phenomenon (LTP) described above. Research into the phenomenon has focused on finding clinically useful NMDA receptor antagonists, for use

in both chronic pain conditions and neurodegenerative disorders in which excitotoxicity plays a part, such as ▶ Parkinson's disease and Alzheimer's disease (Sureda 2000).

Conclusion

These mechanisms imply that chronic pathological pain may persist long after the initial noxious insult has ceased and tissue damage has healed. The process of synaptic plasticity and learning begins early and is difficult to reverse. It seems that untreated acute pain persisting for long periods of time can imprint memory-like processes into the central nervous system (Schug 2004). Currently the recommendation is that an extended, balanced, multimodal approach to pain management should begin in the preoperative period and continue postoperatively.

References

Bach, S., Noreng, M. F., & Tjellden, N. U. (1988). Phantom limb pain in amputees during the first 12 months following limb amputation, after preoperative lumbar epidural blockade. *Pain, 33,* 297–301.

Caputto, B. L., & Guido, M. E. (2000). Immediate early gene expression within the visual system: Light and circadian regulation in the retina and the suprachiasmatic nucleus. *Neurochemical Research, 25,* 153–162.

Crombie, I. K., Davies, H. T., & Macrae, W. A. (1998). Cut and thrust: Antecedent surgery and trauma amongst patients attending a chronic pain clinic. *Pain, 76,* 167–171.

Gehling, M., Scheidt, C.-E., Neibergall, H., et al. (1999). Persistent pain after elective trauma surgery. *Acute Pain, 2,* 110–114.

Katz, J., Jackson, M., Kavanaugh, B., & Sandler, A. (1996). Acute pain after thoracic surgery predicts long-term post-thoracotomy pain. *The Clinical Journal of Pain, 12,* 50–55.

Krane, E. J., & Heller, L. B. (1995). The prevalence of phantom limb sensation and pain in pediatric amputees. *Journal Pain Symptom Management, 10,* 21–29.

Kwekkeboom, K. (1996). Postmastectomy pain syndromes. *Cancer Nursing, 19,* 37–43.

Mannion, R. J., & Woolf, C. J. (2000). Pain mechanisms and management: A central perspective. *The Clinical Journal of Pain, 16*(Suppl. 3), S144–156.

Nikolajsen, L., Ilkjaer, S., Kroner, K., et al. (1997). The influence of preamputation pain on postamputation stump and phantom pain. *Pain, 72,* 393–405.

Olney, J. W., Sharpe, L. G., & Feigin, R. D. (1972). Glutamate-induced brain damage in infant primates. *Journal of Neuropathology and Experimental Neurology, 31,* 464–488.

Perkins, F. M., & Kehlet, H. (2000). Chronic pain as an outcome of surgery – a review of predictive factors. *Anesthesiology, 93,* 1123–1133.

Pockett, S. (1995). Spinal cord synaptic plasticity and chronic pain. *Anesthesia and Analgesia, 80,* 173–179.

Schug, S. A. (2004). Acute pain management – its role in the prevention of chronic pain states. *ASEAN Journal Anaesthesiology, 5,* 166–169.

Siddall, P. J., & Cousins, M. J. (2004). Persistent pain as a disease entity: Implications for clinical management. *Anesthesia and Analgesia, 99,* 510–520.

Sureda F.X. (2000). Excitotoxicity and the NMDA receptor. From EUROSIVA meeting Vienna. www.eurosiva.org/Archive//Vienna/abstracts/Speakers.

Transition from Parenteral to Oral Analgesic Drugs

▶ Postoperative Pain, Transition from Parenteral to Oral Drugs

Transition to Chronicity

▶ Central Changes After Peripheral Nerve Injury

Translaminar Epidural Steroid Injection

▶ Epidural Steroid Injections for Chronic Back Pain

Transmitters in the Descending Circuitry

▶ Descending Circuitry, Transmitters and Receptors

Transmucosal

Definition

Absorption across the buccal mucosa.

Cross-References

▶ Pain Control in Children with Burns

Transmucosal Fentanyl

Definition

Transmucosal fentanyl is also available as a lollipop preparation (transmucosal preparation) that has been successfully used in children.

Cross-References

▶ Postoperative Pain, Fentanyl

Transmural

Definition

Passing through a wall, as of the body or of a cyst or any hollow structure.

Cross-References

▶ Animal Models of Inflammatory Bowel Disease

Traumatic Angiospasm

▶ CRPS, Evidence-Based Treatment

Traumatic Nerve Endbulb Pain

▶ Neuroma Pain

Traumatic Peripheral Neuropathy

▶ Causalgia

Treating Pediatric Burns

▶ Pain Control in Children with Burns

Treatment Adherence

Definition

The extent to which patients or clients follow treatment recommendations made by the clinician or treatment goals negotiated with the clinician.

Cross-References

▶ Chronic Pain, Patient-Therapist Interaction

Treatment Alliance

Definition

The extent to which patients and clinicians agree on treatment goals and hold each other in positive regard.

Cross-References

▶ Chronic Pain, Patient-Therapist Interaction

Treatment Matching

Definition

Prescribing treatments based on specific features of patients that are believed to be important to outcomes. Treatments may be matched on the basis of physical, behavioral, or any unique individual differences associated with pain and disability.

Cross-References

▶ Psychological [Behavioral Health] Assessment of Pain

Treatment of Neuropathic Pain

▶ Phantom Limb Pain, Treatment

Treatment of Phantom Pain

▶ Phantom Limb Pain, Treatment

Treatment Outcome

▶ Psychological Treatment of Chronic Pain, Prediction of Outcome

Treatment Outcome Research

▶ Psychology of Pain, Efficacy of Cognitive-Behavioral Treatment

Treatment-Related Neutropenia

Definition

Low white blood counts that develop after chemotherapy or radiation therapy.

Cross-References

▶ Cancer Pain Management, Orthopedic Surgery

Tremor

▶ Orofacial Pain, Movement Disorders

Tricyclic Antidepressants

Mark Johnston
Hibiscus Coast Highway, Orewa, New Zealand

Synonyms

TCAs; Tricyclics

Definition

Tricyclic antidepressants are a group of drugs developed and prescribed primarily as antidepressants. It was discovered in the 1960s that they had an analgesic effect, which was separate from their antidepressant effect. They have since been used increasingly as an analgesic in managing chronic pain.

TCAs consist of a 3-ring central core (Fig. 1). Their properties are related to the degree of saturation of the terminal amine. Tertiary amines available at present include amitriptyline, clomipramine, doxepin, imipramine, and trimipramine. Some of these compounds are metabolized to active secondary amine tricyclic

Tricyclic Antidepressants, Fig. 1 The core chemical structure of the tricyclic antidepressants, showing the three-ring structure. Individual agents differ in their substitution of carbon or nitrogen in the central ring and in the radical on the amine chain

compounds, e.g., imipramine to desipramine and nortriptyline from amitriptyline.

Characteristics

Mechanisms

Whereas their antidepressant action takes about 2–3 weeks to develop, the analgesic effect of TCAs occurs in 3–7 days. For many years, the most appealing hypothesis was that the analgesic, and antidepressant effects are related to the block of reuptake of norepinephrine and serotonin (5-HT) at spinal dorsal horn synapses. Presynaptically, TCAs inhibit the reuptake of serotonin, norepinephrine and, to a lesser degree, dopamine. Postsynaptic activity is variable. Amitriptyline blocks cholinergic, histamine, alpha adrenergic, muscarinic, N-methyl-D-aspartate (NMDA), substance P, and various subsets of serotonergic receptors.

Recent evidence from animal studies has shown that other mechanisms may also be having antinociceptive effects. These include modulation of the sympathetic nervous system, blockade of sodium channels, some anti-inflammatory effects, and mechanisms involving GABA or opioid receptors (Cohen and Abdi 2001).

Routes of Administration

TCAs are only administered orally. However, intravenous preparations can be obtained.

Applications

TCAs have been recommended and used in many pain conditions, but this is not supported by the evidence. Evidence supports their use only for:
• Diabetic neuropathy
• Postherpetic neuralgia

- Tension-type headache
- Prevention of migraine and tension-type headache

There is no evidence, or no more than dubious evidence, for the use of TCA in acute or chronic low back pain, "rheumatologic pain" (including fibromyalgia), and chronic facial pain.

Combining TCAs with other analgesics has been recommended in patients with chronic pain, but there are no good studies to support this practice.

Side Effects

Cardiovascular

The most serious side effects of TCAs are on the cardiovascular system, where they can cause heart block, arrhythmias, and postural hypotension.

TCAs slow conduction through the heart. The effect is more marked in patients with preexisting cardiac conduction disease, particularly bundle branch block. Ischemic heart disease is frequently associated with conduction defects. Some 18 % of depressed patients with ischemic heart disease, treated with nortriptyline, developed potentially dangerous sinus tachycardia or complex arrhythmias, compared with 2 % in those treated with paroxetine (Roose et al. 1998). Treating the elderly is a problem, as ischemic heart disease is common, and as many episodes may be silent, it may not be obvious from the history that the patient is suffering from ischemic heart disease.

Postural hypotension is a result of α-adrenergic receptor blockade. Syncope can occur at any age, but it is more serious in the elderly. The elderly are more prone to syncope, as part of the normal aging process. Other diseases (congestive heart failure) and medication (vasodilators, diuretics, etc.) may intensify the normal postural blood pressure drop. The consequences of falls are also more serious in the elderly.

Sedation

Sedation can be a useful side effect if the patient has insomnia. Sedation is caused by blockade of H1 receptors. Different TCAs produce different degrees of sedation. The sedative effects may diminish after several weeks.

Sexual Dysfunction

TCAs have all been reported to delay or prevent orgasm in both sexes.

Psychiatric

Manic episodes may be triggered in patients with bipolar disease. A greater concern is that of a possible suicide attempt using TCAs, in someone who has severe depression. In large doses the cardiovascular effects can be lethal.

Withdrawal Syndromes

Sudden or gradual discontinuation of TCAs may cause a number of symptoms. These include nausea, vomiting, headache, malaise, sleep disturbance, akathisia, or paradoxical behavioral activation resulting in hypomanic symptoms. These effects may start within 24–48 h of the last dose and last up to 1 month. To reduce withdrawal symptoms, the dose should be tapered gradually; if symptoms are upsetting, the dose may even be increased a little.

Serotonin Syndrome

Combining TCAs with selective serotonin reuptake inhibitors or monoamine reuptake inhibitors may precipitate a "serotonin overload syndrome," characterized by myoclonus, hyperreflexia, tremor, increased muscle tone, fever, shivering, sweating, diarrhea, delirium, coma, or death (Bodner et al. 1995). The condition is usually reversible if the drugs involved are stopped.

Efficacy

For the treatment of chronic low back pain, a systematic review (van Tulder et al. 1997) found moderate evidence that TCAs are not effective for chronic LBP. There is no evidence of efficacy in acute low back pain.

For ▶ neuropathic pain, TCAs are considered "adjuvant analgesics." A systematic review (McQuay et al. 1996) concluded that TCA are effective for the treatment of neuropathic pain: "of 100 patients . . . 30 will obtain more than 50 %

pain relief." Evidence is strongest for pain relief with amitriptyline and desipramine.

For ▶ fibromyalgia, TCAs provide pain relief and improve sleep (O'Malley et al. 2000). The ▶ NNT was 4 for the outcome "to obtain significant benefit."

A meta-analysis (Tomkins et al. 2001) of the outcomes from TCAs for the treatment of chronic headache showed an NNT of 3.2, for "improvement in headaches." The effect was the same for TCA and serotonin antagonists and for both tension-like headaches and migraines.

Selection

When deciding which TCA to use, the broader spectrum ones, such as amitriptyline, imipramine, and nortriptyline, have greater efficacy than the selective reuptake blockers. Nortriptyline may be the best as it has less sedation, less postural hypotension, and less anticholinergic effects.

Cross-References

- ▶ Antidepressant Analgesics in Pain Management
- ▶ Antidepressants in Neuropathic Pain
- ▶ Drugs Targeting Voltage-Gated Sodium and Calcium Channels
- ▶ Drugs with Mixed Action and Combinations: Emphasis on Tramadol
- ▶ Postoperative Pain, Postamputation Pain, Treatment, and Prevention

References

Bodner, R. A., Lynch, T., Lewis, L., & Kahn, D. (1995). Serotonin syndrome. *Neurology, 45*, 219–223.

Cohen, S. P., & Abdi, S. (2001). New developments in the use of tricyclic antidepressants for the management of pain. *Current Opinion in Anaesthesiology, 14*, 505–511.

McQuay, H. J., Tramer, M., Nye, B. A., Carroll, D., Wiffen, J., & Moore, R. A. (1996). A systematic review of antidepressants in neuropathic pain. *Pain, 68*, 217–227.

O'Malley, P. G., Balden, E., Tompkins, G., Santoro, J., Kroenke, K., & Jackson, J. L. (2000). Treatment of fibromyalgia with antidepressants: A meta-analysis. *Journal of General Internal Medicine, 15*, 659–666.

Roose, S. P., Laghrissi-Thode, F., Kennedy, J. S., Nelson, J. C., Bigger, J. T., Pollock, B. G., Gaffney, A., Narayan, M., Finkel, M. S., McCafferty, J., & Gergel, I. (1998). Comparison of paroxetine and nortriptyline in depressed patients with ischemic heart disease. *Journal of the American Medical Association, 279*, 287–291.

Tomkins, G. E., Jackson, J. L., O'Malley, P. G., Balden, E., & Santoro, J. E. (2001). Treatment of chronic headache with antidepressants: A meta-analysis. *The American Journal of Medicine, 111*, 54–63.

van Tulder, M. W., Koes, B. W., & Bouter, L. M. (1997). Conservative treatment of acute and chronic nonspecific low back pain. *Spine, 22*, 2128–2156.

Tricyclics

▶ Tricyclic Antidepressants

Trigeminal and Cranial Nerve Neuralgias

Z. Harry Rappaport[1] and Marshall Devor[2]
[1]Department of Neurosurgery, Rabin Medical Center, Tel-Aviv University, Petah Tiqva, Israel
[2]Department of Cell & Developmental Biology, Institute of Life Sciences and Center for Research on Pain, The Hebrew University of Jerusalem, Jerusalem, Givat Ram, Israel

Synonyms

Neuralgia of cranial nerve V: fifth nerve neuralgia; Tic doloreux; Tic douloureux; Trigeminal neuralgia

Other cranial nerve neuralgias: neuralgia of cranial nerve IX with or without cranial nerve X; Glossopharyngeal neuralgia; Vagoglossopharyngeal neuralgia; Neuralgia of cranial nerve VII; Geniculate neuralgia

Definition

Cranial nerve neuralgia refers to a clinical pain symptom complex that consists of recurrent, intermittent, often paroxysmal pain felt in the head, in the distribution of a specific cranial nerve. By far the most commonly encountered cranial nerve neuralgia is that of the trigeminal nerve and root. This is trigeminal neuralgia (tic douloureux). Neuralgias of the vago-glossopharyngeal nerves and the nervus intermedius branch of the facial nerve are much rarer. Cranial nerve neuralgias are severely painful, but non-life-threatening conditions.

Characteristics

Clinical Phenomenology

Trigeminal neuralgia is unique among cephalic-facial pain syndromes both for its dramatic symptomatology as for its susceptibility to both drug and interventional therapy. The incidence is approximately four new cases per 100,000 population/year. It is much more common in the elderly than in young patients. There is a small female to male preponderance. Intense, brief, intermittent paroxysms of pain in the distribution of the trigeminal branches are followed by variable pain-free periods. The pain is unilateral and does not cross the midline of the face. There is usually a distinct trigger point, most commonly in the perinasal or perioral area, but sometimes within the oral cavity. When the trigger point is stimulated, after a short latency period, the painful paroxysm begins (Fig. 1). Pain may radiate from the trigger area along the appropriate trigeminal division, sometimes crossing into adjacent divisions. Trigger stimuli are usually quite weak, such as light touch, chewing, talking movement, or a puff of wind. Paroxysms may also arise spontaneously. Noxious stimulation of the trigger points is usually ineffective (Rasmussen 1990; Kugelberg and Lindblom 1959).

The pain onset is usually sudden, and has an electric shock–like quality. A series of attacks, building to a crescendo, is common. Patients describe the pain as "shooting," "stabbing," "electric shock–like," or "cutting" in quality. While the paroxysms tend to be short, in the order of seconds or occasionally minutes, they can run into each other giving the impression of

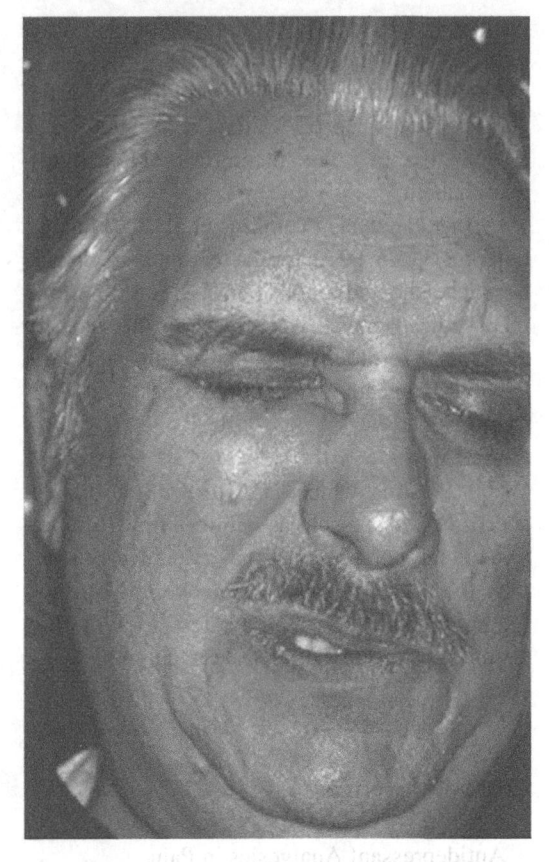

Trigeminal and Cranial Nerve Neuralgias, Fig. 1 A patient winces during a right-sided trigeminal neuralgia attack

a prolonged attack over hours. Typically, pain attacks come in clusters separated by remissions of days to months, and sometimes even years (Rasmussen 1990). Following an attack, the trigger area remains relatively refractory to further pain inducing stimuli for a few minutes (Kugelberg and Lindblom 1959). Attacks may be less frequent during sleep, perhaps because of reduced chances of triggering, but they do occur and frequently awaken sufferers from their sleep (Devor et al. 2008). Some patients additionally report an unremitting background pain, usually of burning quality. Routine sensory examination of the face usually does not reveal any changes, but quantitative sensory testing may reveal elevated tactile and thermal detection thresholds in the area surrounding and including the trigger point (Nurmikko 1991).

Trigeminal neuralgia is usually idiopathic. Only rarely is there an obvious precipitating injury or disease. When there is, it can usually be detected by clinical examination and imaging. For example, prominent hypoesthesia in the presence of typical trigeminal neuralgia symptoms implies the presence of a paratrigeminal space occupying lesion or multiple sclerosis ("symptomatic" trigeminal neuralgia). In a new classification of facial pain (Burchiel 2003), type 1 trigeminal neuralgia refers to those conditions where greater than 50 % of the pain is paroxysmal and episodic, while in type 2 trigeminal neuralgia, greater than 50 % of the pain has a constant, burning quality. This classification is of importance when estimating the prognosis of therapy, as the episodic pain is much more responsive to therapeutic intervention than the constant pain.

Trigeminal neuralgia must be distinguished from trigeminal neuropathic pain that can occur after injuries to the trigeminal nerve and from atypical facial pain, which is often considered a somatoform pain disorder. Both tend to be resistant to most therapies. An initial response to anticonvulsant medication is often useful in distinguishing TN from the various other types of chronic orofacial pain.

The symptomatology of glossopharyngeal neuralgia is similar to that of trigeminal neuralgia, but pain is felt in the distributions of the ninth (glossopharyngeal) and/or tenth (vagus) cranial nerves (Rushton et al. 1981). Pain paroxysms are usually triggered during swallowing, talking, or chewing, with the trigger point located unilaterally at the base of the tongue, tonsillar region, or in the ear. Concomitant vagal autonomic involvement probably accounts for episodes of syncope and bradycardia that may accompany pain attacks. In more severe cases, prolonged asystole may be life threatening. The disease is equally distributed between the sexes and is most frequently seen between the fifth and seventh decades of life, with an incidence of 0.8 cases/100,000/year. It is occasionally bilateral in presentation. Pain tends to be less severe than trigeminal neuralgia, and more often steady and burning in quality (Katusic et al. 1991).

Geniculate neuralgia involves the sensory distribution of the seventh cranial nerve (facial nerve) via the nervus intermedius (of Wrisberg). The painful attacks, which are sharp and stabbing, are centered within the ear canal (Pulec 2002). A dull background pain may persist for several hours after an attack. Affected patients tend to be younger than in trigeminal and glossopharyngeal neuralgia. When seen together with hemifacial spasm the term "tic convulsive" has been used (Yeh and Tew 1984).

Pathophysiology

Medical and surgical experience in treating tic, combined with pathological findings and laboratory experiments using animal models, sheds light on the etiology of pain paroxysms in the cranial nerve neuralgias. During exploration of the cerebellopontine angle on the side of the pain, vascular compression of the appropriate nerve root is usually observed (Fig. 2). Following decompression, the patients are relieved of their pain in a high percentage of cases (Yeh and Tew 1984; Patel et al. 2002; Sindou et al. 2002). Ultrastructural examination of postmortem specimens and cranial nerve biopsies taken during surgical procedures show extensive pathological change in the cranial nerve root and ganglion (Devor et al. 2002a, b). Disordering and loss of myelin is seen in the area compressed by the vessel loop, as are signs of axonal sprouting and regions of close membrane contact between adjacent denuded axons. The later observation is a likely anatomical substrate of ephaptic (electrical, nonsynaptic) coupling between axons. These types of structural change can give rise to ectopic pacemaker sites and the generation of abnormal impulse discharge. Moreover, they facilitate neuron-to-neuron spread of discharge through the injured cranial nerve root and ganglion by means of two distinct neurophysiological mechanisms, ephaptic cross talk, and non-synaptic chemically mediated crossed afterdischarge (Rappaport and Devor 1994). The generation of self-sustaining discharge in the trigeminal root and/or ganglion following peripheral triggering, and the subsequent rapid spread of this activity to neighboring trigeminal afferents, is thought to

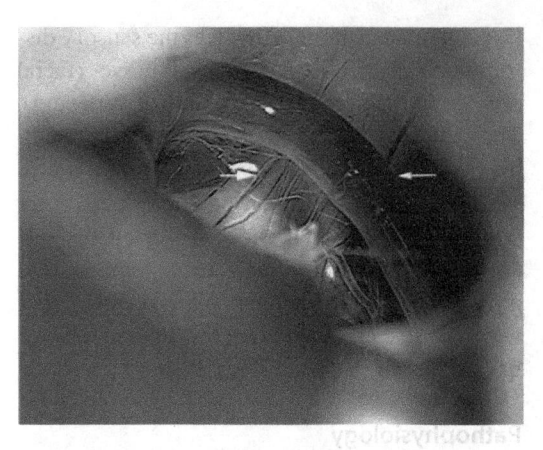

Trigeminal and Cranial Nerve Neuralgias, Fig. 2 A photomicrograph of a compressed trigeminal root (*short arrow*) and the compressing blood vessel (*long arrow*), taken during microvascular decompression surgery in trigeminal neuralgia. Note the discoloration of the root at the compression site

cause paroxysms of pain. This is the "ignition hypothesis" of trigeminal neuralgia (Rappaport and Devor 1994; Burchiel and Baumann 2004). Other properties of cranial nerve neuralgias, such as post-paroxysm refractoriness and pain suppression by membrane stabilizing anticonvulsants, are consistent with this mechanism (Rappaport and Devor 1994). Central changes in signal processing within the trigeminal brain stem have also been proposed as causes of tic pain (Fromm and Sessle 1991).

Treatment

The hallmark of the cranial nerve neuralgias is their response to specific anticonvulsant medications. Notable among them is carbamazepine, but gabapentin, pregabalin, oxcarbazepine, lamotrigine, sodium valproate, and to a lesser extent phenytoin and baclofen may also be effective. Barbiturates are not effective, presumably because they act synaptically rather than by membrane stabilization and reduction of abnormal discharge. The effectiveness of drug therapy declines over time in nearly one half of cases. When drug therapy no longer controls the pain paroxysms, or its side effects impair quality of life, interventional procedures may be considered (Rappaport 1996). Peripheral nerve blocks or neurectomies have only a short-term effect.

Many elderly patients undergo percutaneous partial damage to the trigeminal root and ganglion using radiofrequency coagulation (Kanpolat et al. 2001), retro-Gasserian glycerol injection (Pollock 2005), or balloon compression (Correa and Teixeira 1997). The initial success rate is in the 80 % range, with recurrence rates of up to 30 % at 1 year and up to 45 % at 5 years. These procedures entail various degrees of facial hypoesthesia. Their efficacy is probably due to lessened triggering and spread of electrical activity within the trigeminal root and ganglion. This, in turn, is due to a reduction in the size of trigeminal afferent populations, and hence reduction of the critical amount of activity available to reach the threshold required for pain ignition. Radiofrequency rhizolysis of the glossopharyngeal nerve is also effective for ninth nerve neuralgia. However, the procedure is technically more demanding and, as a result, is less commonly performed. Another method of partially damaging trigeminal pain generators that has recently become popular is stereotactic radiosurgery (gamma knife treatment) (Régis et al. 2006). This procedure is especially useful in patients under anti-coagulation therapy or with failures or contraindications to the other lesioning procedures. Pain cessation is achieved after a somewhat longer latency (at least 2 weeks) and the long-term success rate is lower than in other invasive procedures (Dhople et al. 2009; Linskey et al. 2008). Persistent anesthesia dolorosa (deafferentation pain) is a potential complication.

For patients who can tolerate the surgery, open exposure of the posterior fossa via a retromastoid craniotomy, and microvascular decompression (MVD) of the neural complex, is the preferred interventional procedure for all cranial nerve neuralgias (Barker et al. 1996; Resnick et al. 1995). Preoperative MRI scanning using high-resolution 3-D time of flight protocols can accurately visualize the blood vessel relationship to the trigeminal root (Miller et al. 2008). Microvascular decompression has the advantage of treating the root cause of the symptoms without destroying neural tissue and superimposing hypoesthesia. The low morbidity (less than 4 %) and the very rare mortality associated with this procedure, its impressive efficacy at relieving pain with only modest rates of recurrence (80 % still with significant pain relief at 5 year follow-up), and the fact that it is followed by little if any sensory loss make MVD the surgical treatment of choice, especially in younger patients (Tyler-Kabara et al. 2002).

Acknowledgment We wish to thank Yair Sharav of the Hebrew University School of Dentistry for providing us with the photograph of a patient during a trigeminal neuralgia paroxysm used in Fig. 1, and Irene Wood of the Trigeminal Neuralgia Association of Australia for helpful comments on the manuscript.

References

Barker, F. G., II, Jannetta, P. J., Bissonette, D. J., Larkins, M. V.,& Jho, H. D. (1996). The long-term outcome of microvascular decompression for trigeminal neuralgia. *New England Journal of Medicine, 334,* 1077–1083.

Burchiel, K. J. (2003). A new classification for facial pain. *Neurosurgery, 53,* 1164–1167.

Burchiel, K. J., & Baumann, T. K. (2004). Pathophysiology of trigeminal neuralgia: New evidence from a trigeminal ganglion intraoperative microneurographic recording. Case report. *Journal of Neurosurgery, 101,* 872–873.

Correa, C. F., & Teixeira, M. F. (1997). Balloon compression of the gasserian ganglion for the treatment of trigeminal neuralgia. *Stereotactic and Functional Neurosurgery, 71,* 83–89.

Devor, M., Govrin-Lippmann, R., & Rappaport, Z. H. (2002a). Mechanism of trigeminal neuralgia: An ultrastructural analysis of trigeminal root specimens obtained during microvascular decompression surgery. *Journal of Neurosurgery, 96,* 532–543.

Devor, M., Govrin-Lippmann, R., Rappaport, Z. H., Tasker, R. R., & Dostrovsky, J. O. (2002b). Cranial root injury in glossopharyngeal neuralgia: Electron microscopic observations. *Journal of Neurosurgery, 96,* 603–606.

Devor, M., Wood, I., Sharav, Y., & Zakrzewska, J. M. (2008). Trigeminal neuralgia during sleep. *Pain Practice, 8,* 263–268.

Dhople, A. A., Adams, J. R., Maggio, W. W., Naqvi, S. A., Regine, W. F., & Kwok, Y. (2009). Long-term outcomes of gamma knife radiosurgery for classic trigeminal neuralgia: Implications of treatment and critical review of the literature. *Journal of Neurosurgery, 111,* 351–358.

Fromm, G. H., & Sessle, B. J. (Eds.). (1991). *Trigeminal neuralgia: Current concepts regarding pathogenesis and treatment* (pp. 1–230). Boston: Butterworth-Heinemann.

Kanpolat, Y., Savas, A., Bekar, A., & Berk, C. (2001). Percutaneous controlled radiofrequency trigeminal rhizotomy for the treatment of idiopathic trigeminal neuralgia: 25-year experience with 1,600 patients. *Neurosurgery, 48,* 524–532.

Katusic, S., Williams, D. B., Beard, C. M., Bergstralh, E. J., & Kurland, L. T. (1991). Epidemiology and clinical features of idiopathic trigeminal neuralgia and glossopharyngeal neuralgia similarities and differences, Rochester, Minnesota, 1945–1984. *Neuroepidemiology, 10,* 276–281.

Kugelberg, E., & Lindblom, U. (1959). The mechanism of the pain in trigeminal neuralgia. *Journal of Neurology, Neurosurgery & Psychiatry, 22,* 36–43.

Linskey, M. E., Ratanatharathorn, V., & Peñagaricano, J. (2008). A prospective cohort study of microvascular decompression and gamma knife surgery in patients with trigeminal neuralgia. *Journal of Neurosurgery, 109*(Suppl.), 160–172.

Miller, J., Acar, F., Hamilton, B., & Burchiel, K. J. (2008). Preoperative visualization of neurovascular anatomy in trigeminal neuralgia. *Journal of Neurosurgery, 108,* 477–482.

Nurmikko, T. J. (1991). Altered cutaneous sensation in trigeminal neuralgia. *Archives of Neurology, 48,* 523–527.

Patel, A., Kassam, A., Horowitz, M., & Chang, Y. F. (2002). Microvascular decompression in the management of glossopharyngeal neuralgia: Analysis of 217 cases. *Neurosurgery, 50,* 705–710.

Pollock, B. E. (2005). Percutaneous retrogasserian glycerol rhizotomy for patients with idiopathic trigeminal neuralgia: A prospective analysis of factors related to pain relief. *Journal of Neurosurgery, 102,* 223–228.

Pulec, J. L. (2002). Geniculate neuralgia: Long-term results of surgical treatment. *Ear, Nose, & Throat Journal, 81,* 30–33.

Rappaport, Z. H. (1996). The choice of therapy in medically intractable trigeminal neuralgia. *Israel Journal of Medical Sciences, 32,* 1232–1234.

Rappaport, Z. H., & Devor, M. (1994). Trigeminal neuralgia: The role of self-sustaining discharge in the trigeminal ganglion. *Pain, 56,* 127–138.

Rasmussen, P. (1990). Facial pain: A prospective study of 1052 patients with a view of: Character of the attacks, onset, course, and character of pain. *Acta Neurochirurgica, 107,* 121–128.

Régis, J., Metellus, P., Hayashi, M., Roussel, P., Donnet, A., & Bille-Turc, F. (2006). Prospective controlled trial of gamma knife surgery for essential trigeminal neuralgia. *Journal of Neurosurgery, 104,* 913–924.

Resnick, D. K., Jannetta, P. J., Bissonnette, D., Jho, H. D., & Lanzino, G. (1995). Microvascular decompression for glossopharyngeal neuralgia. *Journal of Neurosurgery, 36,* 64–68.

Rushton, J. G., Stevens, J. C., & Miller, R. H. (1981). Glossopharyngeal (vagoglossopharyngeal) neuralgia: A study of 217 cases. *Archives of Neurology, 38,* 201–205.

Sindou, M., Howeidy, T., & Acevedo, G. (2002). Anatomical observations during microvascular decompression for idiopathic trigeminal neuralgia (with correlations between topography of pain and site of the neurovascular conflict). Prospective study in a series of 579 patients. *Acta Neurochirurgica, 144*, 1–12.

Tyler-Kabara, E. C., Kassam, A. B., Horowitz, M. H., Urgo, L.,Hadjipanayis, C., Levy, E. I., & Chang, Y. F. (2002). Predictors of outcome in surgically managed patients with typical and atypical trigeminal neuralgia: Comparison of results following microvascular decompression. *Journal of Neurosurgery, 96*, 527–531.

Yeh, H. S., & Tew, J. M., Jr. (1984). Tic convulsive, thecombination of geniculate neuralgia and hemifacial spasm relieved by vascular decompression. *Neurology, 34*, 682–683.

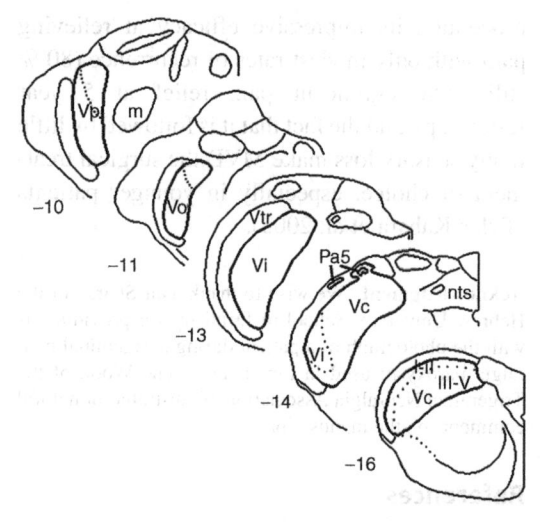

Trigeminal Brain Stem Nuclear Complex, Immunohistochemistry, and Neurochemistry, Fig. 1 Trigeminal brain stem nuclear complex of the rat. Abbreviations: I–II, laminae I–II; III–V, laminae III–V; m, trigeminal motor n.; nts, n. tractus solitarii; Pa5, paratrigeminal islands; Vc, subnucleus caudalis; Vi, subnucleus interpolaris; ▶ Vi/Vc, interpolaris/caudalis transition region; Vo, subnucleus oralis; Vp, principal sensory nucleus; Vtr, spinal trigeminal tract. Numbers below each outline indicate approximate distance in mm caudal to bregma, a skull surface landmark

Trigeminal Brain Stem Nuclear Complex, Immunohistochemistry, and Neurochemistry

David A. Bereiter
Department of Diagnostic and Biological Sciences, University of Minnesota School of Dentistry, Minneapolis, MN, USA

Definition

The trigeminal brain stem nuclear complex (TBNC) is comprised of the principal sensory nucleus and spinal trigeminal nucleus. The spinal trigeminal nucleus is further subdivided, from rostral to caudal, into subnucleus oralis, subnucleus interpolaris, and subnucleus caudalis (Vc) which is often referred to as the medullary dorsal horn since it displays several features, such as a laminated organization, similar to the spinal dorsal horn (Fig. 1).

Characteristics

The role of the different subnuclei of the TBNC in mediating various aspects of pain is confounded by several factors. Craniofacial tissues are represented somatotopically at multiple, and interconnected, levels of the TBNC which allows injurious stimuli to excite neurons throughout the nuclear complex resulting in neuroplasticity (see Sessle 2000; Bereiter et al. 2009). Although the Vc is associated with nearly all forms of craniofacial pain, the rostral subnuclei (i.e., Vp and Vo) provide additional processing critical for intraoral and dental pain (Sessle 2000; Bereiter et al. 2009). This organizational feature may help explain the distribution of anatomical markers within rostral portions of the TBNC in relation to nociceptive processing. Immunohistochemical studies identify two broad classes of small-diameter myelinated or unmyelinated nerve fibers: (1) those containing the neuropeptides substance P, calcitonin gene-related peptide, and neurotrophins and (2) those that express the cell surface marker, isolectin B4, and are often negative for neuropeptides. Table 1 summarizes the pattern of distribution within the TBNC for selected neurochemical markers known to contribute to craniofacial pain transmission. Note that for

Trigeminal Brain Stem Nuclear Complex, Immunohistochemistry, and Neurochemistry, Table 1 Distribution of selected neurochemical markers associated with nociceptive processing in different portions of the trigeminal brain stem nuclear complex

	Vp		Vo		Vi	dPa5	Vi/Vc$_{vl}$	Vc		
	dm	vl	dm	vl				I–II	III–IV	V
IB4	+	−	+	−	+	++	−	+++	−	−
TRPV1	+	−	+	−	+	+++	++	+++	−	−
TRPA1	++	−	+	−	−	+	+	+++	−	−
SP	++	−	++	−	+	+++	++	+++	−	+
NK1	+	−	+	−	+	+	+	+++	++	+
CGRP	++	−	++	−	+	+++	++	+++	−	+
P2X$_2$	+	+	+	++	++	+	++	++	+	+
P2X$_3$	+	+	+	+	+	++	++	+++	−	−
trkA	++	+	++	−	+	++	++	+++	+	+
BDNF	+	−	+	−	++	++	++	+++	−	+
trkB	++	++	++	++	++	++	++	+++	+	+
ChAT	+	−	+	−	+	−	−	++	−	−
nNOS	++	+	++	+	+	+	+	+++	+	+
NR1	++	+	++	+	++	+++	++	+++	+	+
GABA$_A$a1R	+	+	+	+	++	+	++	++	+	+
GABA$_B$b2/3R	+	+	+	+	+	++	++	+++	+	+
AR$_{a/b}$	+	−	+	−	−	−	−	++	−	−
5HT	+	+	+	+	+	+++	++	+++	+	+
5HT$_{1B/1D}$	++	−	++	−	+	++	+	+++	+	+
Endomor2	−	−	+	−	+	++	+	+++	−	−
MOR1	−	−	+	−	+	++	+	+++	−	−
ER$_a$	−	−	+	−	−	+	+	+++	+	+

Symbols and abbreviations: − = very weak or no staining; +,++,+++ = weak, moderate, and dense staining; AR$_{a/b}$, adrenergic receptor subtypes; BDNF, brain-derived neurotrophic factor; CGRP, calcitonin gene-related peptide; ChAT, choline acetyltransferase; Endomor2, endomorphin 2; ERa, estrogen receptor alpha subtype; GABA$_A$a1R, gamma aminobutyric acid A receptor subtype, alpha1 subunit; GABAbR, GABA$_B$ receptor subtype, beta2/3 subunit, 5HT, serotonin; 5HT$_{1B/1D}$, serotonin receptors; IB4, isolectin B4; MOR1, mu opioid receptor; nNOS, neuronal nitric oxide synthase; NK1, neurokinin1 receptor; NR1, NMDA receptor subunit; P2X$_2$ and P2X$_3$ ATP receptor subtypes; SP, substance P; trkA, tyrosine kinase A receptor; trkB, tyrosine kinase B receptor; TRPA1, transient receptor potential ankyrin 1; TRPV1, transient receptor potential vanilloid 1

both IB4 and sensory neuropeptides there is dense terminal labeling in laminae I–II of Vc and the paratrigeminal islands as well as significant labeling in the dorsomedial portions of Vp and Vo (Connor et al. 1997; Sugimoto et al. 1997a, b). Similarly, TRPV1 (Bae et al. 2004) and TRPA1 (Kim et al. 2010) receptors, strongly associated with neuropeptide-containing nociceptive processing, are densely distributed in superficial laminae of Vc and significant staining in dorsomedial Vp and Vo. The distribution of NK1, the substance P receptor, is similar to that for substance P in dorsomedial Vp and

Vo and deeper laminae of Vc; however, in superficial laminae of Vc, dense labeling for substance P is seen in laminae I and II, while NK1 staining occurs mainly in lamina I (Nakaya et al. 1994). Markers associated with opioid analgesia, such as the endogenous opioid peptide, endomorphin 2 (Martin-Schild et al. 1999), and the mu opioid receptor subtype, MOR1 (Bereiter and Bereiter 2000; Ding et al. 1996), display moderate-to-dense immunoreactivity in laminae I–II of Vc, paratrigeminal islands, and the Vi/Vc$_{vl}$ transition region, while deeper laminae of Vc and more rostral regions of the TBNC display weak staining. This suggests that opioid modulation of dental pain occurs outside the rostral portions of the TBNC. Physiological studies reveal that morphine differentially affects ocular-responsive neurons at the Vi/Vc$_{vl}$ transition and caudal Vc regions, and all caudal Vc neurons are inhibited, whereas many Vi/Vc$_{vl}$ transition neurons are enhanced by morphine suggesting a role in the recruitment of endogenous controls in ocular pain (see Bereiter et al. 2009). Receptors for mono-amine transmitters (norepinephrine and serotonin) display diffuse staining throughout the TBNC; however, the superficial laminae of Vc have moderate-to-dense levels of immunoreactivity (Day et al. 1997; Thor et al. 1992; Talley et al. 1996; Potrebic et al. 2003; Harding et al. 2004). Staining for choline acetyltransferase (ChAT), the biosynthesis enzyme for acetylcholine, is weak in most rostral regions of TBNC, although the superficial laminae of Vc display moderate levels (Pioro and Cuello 1990; Sobreviela et al. 1994). Receptors for neurotrophic factors (trkA and trkB) are widely distributed in TBNC with the highest densities being located in superficial laminae of Vc (Pioro and Cuello 1990; Connor et al. 1997; Yan et al. 1997). Neurons that express receptor subtypes for GABA, the inhibitory amino acid transmitter (i.e., ligand-gated subtype A or G-protein-coupled subtype B) or nitric oxide synathase are found in all regions of the TBNC with the highest density in laminae I–II of Vc (Rodrigo et al. 1994; Fritschy and Mohler 1995; Margeta-Mitrovic et al. 1999). Immunoreactivity for the purinergic receptors, P2X$_2$ and P2X$_3$, ATP-gated ion channels often associated with unmyelinated nerve fibers, displays a high density in laminae I–II of Vc as well as significant expression in more rostral subnuclei (Kanjhan et al. 1999; Kim et al. 2008). The NMDA receptor subunit, NR1, is widely distributed throughout the TBNC (Petralia et al. 1994) consistent with evidence that glutamatergic mechanisms contribute to neuroplasticity that likely accompanies all forms of craniofacial pain. Many craniofacial pain conditions (e.g., migraine headache, temporomandibular joint disorders) display a significant female prevalence. Estrogen receptor subtype alpha, ERa, displays a high density of staining in neurons of Vc, mainly in lamina II, with only weak staining in other regions of the TBNC (Bereiter et al. 2005; VanderHorst et al. 2005). This suggests that the Vc may play a significant role in mediating sex-related differences in craniofacial pain.

Summary

The neurochemical organization of the TBNC is consistent with the notion that laminae I–II of Vc and the paratrigeminal islands receive direct input from small-diameter nerve fibers and play an essential role in the processing and modulation of trigeminal pain from all craniofacial tissues. Although rostral regions of the TNBC receive weaker direct input from small-diameter nerve fibers than Vc, the distribution of selected markers for nociceptive processing in the dorsomedial portions of Vp and Vo suggests that these regions likely contribute significantly to the integration of input from specific craniofacial tissues, such as teeth and oral mucosa.

Acknowledgment The author thanks Dr. Keiichiro Okamoto for helpful comments. The author was supported by NIH Grants DE12758 and EY021447.

References

Bae, Y. C., Oh, J. M., Hwang, S. J., Shigenaga, Y., & Valtschanoff, J. G. (2004). Expression of vanilloid receptor TRPV1 in the rat trigeminal sensory nuclei. *The Journal of Comparative Neurology, 478*, 62–71.

Bereiter, D. A., & Bereiter, D. F. (2000). Morphine and NMDA receptor antagonism reduce c-*fos* expression in spinal trigeminal nucleus produced by acute injury to the TMJ region. *Pain, 85*, 65–77.

Bereiter, D. A., Cioffi, J. L., & Bereiter, D. F. (2005). Oestrogen receptor-immunoreactive neurons in the trigeminal sensory system of male and cycling female rats. *Archives of Oral Biology, 50*, 971–979.

Bereiter, D. A., Hargreaves, K. M., & Hu, J. W. (2009). Trigeminal mechanisms of nociception: Peripheral and brain stem organization. In M. C. Bushnell & A. I. Basbaum (Eds.), *Handbook of the senses* (Science of pain, Vol. 5, pp. 435–460). San Diego: Elsevier.

Connor, J. M., Lauterborn, J. C., Yan, Q., Gall, C. M., & Varon, S. (1997). Distribution of brain-derived neurotrophic factor (BDNF) protein and mRNA in the normal adult rat CNS: Evidence for anterograde axonal transport. *Journal of Neuroscience, 17*, 2295–2313.

Day, H. E., Campeau, S., Watson, S. J., & Akil, H. (1997). Distribution of a1a, a1b, and a1d-adrenergic receptor mRNA in the rat brain and spinal cord. *Journal of Chemical Neuroanatomy, 13*, 115–139.

Ding, Y.-Q., Kaneko, T., Nomura, S., & Mizuno, N. (1996). Immunohistochemical localization of μ-opioid receptors in the central nervous system of the rat. *The Journal of Comparative Neurology, 367*, 375–402.

Fritschy, J.-M., & Mohler, H. (1995). GABAA-receptor heterogeneity in the adult rat brain: Differential regional and cellular distribution of seven major subunits. *The Journal of Comparative Neurology, 359*, 154–194.

Harding, A., Paxinos, G., & Halliday, G. (2004). The serotonin and tachykinin systems. In G. Paxinos (Ed.), *The rat nervous system* (pp. 1205–1256). New York: Elsevier.

Kanjhan, R., Housley, G. D., Burton, L. D., Christie, D. L., Kippenberger, A., Thorne, P. R., Luo, L., & Ryan, A. F. (1999). Distribution of the P2X$_2$ receptor subunit of the ATP-gated ion channels in the rat central nervous system. *The Journal of Comparative Neurology, 407*, 11–32.

Kim, Y. S., Paik, S. K., Cho, Y. S., Shin, H. S., Bae, J. Y., Moritani, M., et al. (2008). Expression of P2X3 receptor in the trigeminal sensory nuclei of the rat. *The Journal of Comparative Neurology, 506*, 627–639.

Kim, Y. S., Son, J. Y., Kim, T. H., Paik, S. K., Dai, Y., Noguchi, K., et al. (2010). Expression of transient receptor potential ankyrin 1 (TRPA1) in the rat trigeminal sensory afferents and spinal dorsal horn. *The Journal of Comparative Neurology, 518*, 687–698.

Margeta-Mitrovic, M., Mitrovic, I., Riley, R. C., Jan, L. Y., & Basbaum, A. I. (1999). Immunohistochemical localization of GABAB receptors in the rat central nervous system. *The Journal of Comparative Neurology, 405*, 299–321.

Martin-Schild, S., Gerall, A. A., Kastin, A. J., & Zadina, J. E. (1999). Differential distribution of endomorphin1- and endomorphin2-like immunoreactivities in the CNS of the rodent. *The Journal of Comparative Neurology, 405*, 450–471.

Nakaya, Y., Kaneko, T., Shigemoto, R., Nakanishi, S., & Mizuno, N. (1994). Immunohistochemical localization of substance P receptor in the central nervous system of the adult rat. *The Journal of Comparative Neurology, 347*, 249–274.

Petralia, R. S., Yokotani, N., & Wenthold, R. J. (1994). Light and electron microscope distribution of the NMDA receptor subunit NMDAR1 in the rat nervous system using a selective anti-peptide antibody. *Journal of Neuroscience, 14*, 667–696.

Pioro, E. P., & Cuello, A. C. (1990). Distribution of nerve growth factor receptor-like immunoreactivity in the adult rat central nervous system. Effect of colchicine and correlation with the cholinergic system–II. Brainstem, cerebellum and spinal cord. *Neuroscience, 34*, 89–110.

Potrebic, S., Ahn, A. H., Skinner, K., Fields, H. L., & Basbaum, A. I. (2003). Peptidergic nociceptors of both trigeminal and dorsal root ganglia express serotonin 1D receptors: Implications for the selective antimigraine action of triptans. *Journal of Neuroscience, 23*, 10988–10997.

Rodrigo, J., Springall, D. R., Uttenthal, O., Bentura, M. L., Abadia-Molina, F., Riveros-Moreno, V., et al. (1994). Localization of nitric oxide synthase in the adult rat brain. *Philosophical Transactions of the Royal Society of London. Series B, Biological Sciences, 345*, 175–221.

Sessle, B. J. (2000). Acute and chronic craniofacial pain: Brainstem mechanisms of nociceptive transmission and neuroplasticity, and their clinical correlates. *Critical Reviews in Oral Biology and Medicine, 11*, 57–91.

Sobreviela, T., Clary, D. O., Reichardt, L. F., Brandabur, M. M., Kordower, J. H., & Mufson, E. J. (1994). TrkA-immunoreactive profiles in the central nervous system: Colocalization with neurons containing p75 nerve growth factor receptor, choline acetyltransferase, and serotonin. *The Journal of Comparative Neurology, 350*, 587–611.

Sugimoto, T., Fujiyoshi, Y., He, Y.-F., Xiao, C., & Ichikawa, H. (1997a). Trigeminal primary projection to the rat brain stem sensory trigeminal nuclear complex and surrounding structures revealed by anterograde transport of cholera toxin B subunit-conjugated and *Bandeiraea simplicifolia* isolectin B4-conjugated horseradish peroxidase. *Neuroscience Research, 28*, 361–371.

Sugimoto, T., Fujiyoshi, Y., Xiao, C., He, Y.-F., & Ichikawa, H. (1997b). Central projection of calcitonin gene-related peptide (CGRP)- and substance P (SP)-immunoreactive trigeminal primary neurons in the rat. *The Journal of Comparative Neurology, 378*, 425–442.

Talley, E. M., Rosen, D. I., Lee, A., Guyenet, P. G., & Lynch, K. R. (1996). Distribution of a2A-adrenergic receptor-like immunoreactivity in the rat central nervous system. *The Journal of Comparative Neurology, 372*, 111–134.

Thor, K. B., Blitz-Siebert, A., & Helke, C. J. (1992). Autoradiographic localization of 5HT1 binding sites in the medulla oblongata of the rat. *Synapse, 10*, 185–205.

Vanderhorst, V. G., Gustafsson, J. A., & Ulfhake, B. (2005). Estrogen receptor-alpha and -beta immunoreactive neurons in the brainstem and spinal cord of male and female mice: Relationships to monoaminergic, cholinergic, and spinal projection systems. *The Journal of Comparative Neurology, 488*, 152–179.

Yan, Q., Radeke, M. J., Matheson, C. R., Talvenheimo, J., Welcher, A. A., & Feinstein, S. C. (1997). Immunocytochemical localization of TrkB in the central nervous system of the adult rat. *The Journal of Comparative Neurology, 378*, 135–157.

Trigeminal Brainstem Nuclear Complex, Anatomy

Yoshio Shigenaga and Atsushi Yoshida
Department of Oral Anatomy and Neurobiology,
Osaka University, Osaka, Japan

Definition

The trigeminal sensory nuclear complex in mammals is generally recognized to include the nucleus principalis (Vp), the spinal tract nucleus (Vsp; sometimes also termed the spinal nuclear complex), and the ▶ mesencephalic nucleus (Vmes). It extends through the pons and medulla from the C2 spinal segment and upward through the mesencephalon. In a caudorostral direction, these nuclei are the Vsp, Vp, and Vmes (Fig. 1). Three further divisions of the Vsp are differentiated (Olszewski 1950): the subnuclei caudalis (Vc) (Fig. 1e), interpolaris (Vi) (Fig. 1d), and oralis (Vo) (Fig. 1b, c). The Vmes differs from the other nuclei in that it contains cell bodies of primary afferents innervating jaw muscle spindles or periodontal ligaments. The Vmes neurons send their axon collaterals mainly to the supratrigeminal nucleus (Vsup), trigeminal motor nucleus (Vmo), and the reticular formation intercalated between the Vp and Vmo, named the intertrigeminal region (Vint), and adjacent to the Vo or Vi, named the juxtatrigeminal region (Vjuxt) (Fig. 1). In contrast to cell bodies in the trigeminal ganglion (TG), those in the Vmes receive many synaptic inputs (Honma et al. 2001). Although the Vp and Vsp, except for Vc, receive input from Vmes neurons, the main input is supplied by primary afferents whose cell bodies are located in the TG. Note that primary afferents of the periodontal ligament have their cell bodies in both the Vmes and the TG. Therefore, the trigeminal brainstem nuclear complex consists of the trigeminal sensory nuclear complex and the nuclei or the regions that receive a direct projection from primary afferents with their cell bodies in the Vmes.

Characteristics

The Vp and Vsp integrate the trigeminal afferent system that is organized to serve the ▶ exteroceptive, interoceptive, and proprioceptive sensory functions of the oral and craniofacial structures, but also receive projections from the facial, glossopharyngeal, and vagus nerves and upper cervical nerves. The two nuclei receive both the noxious and innocuous inputs from their primary afferents, except for Vmes afferents.

Intra-axonal labeling shows that stained axons (central processes) of primary afferents in the trigeminal spinal tract (Vtr) ascend and descend (bifurcating fibers), or descend without bifurcation (descending non-bifurcating fibers) (Tsuru et al. 1989; Hayashi 1985). The ascending fiber and the descending fiber give off axon collaterals mainly in the Vp and the Vsp, respectively.

The Vtr is somatotopically organized, with fibers of the ophthalmic, maxillary, and mandibular divisions lying successively more dorsally. This dorsoventral arrangement is also applicable to the facial region of the Vp and Vsp, while the intraoral representation is organized in a complex manner (see below) (Shigenaga et al. 1986a, 1986b, 1989b).

Primary afferent projections are not identical at each level, as the density of terminals varies along different nuclei or in different parts of the same nucleus (Tsuru et al. 1989). The individual nuclei are connected by ascending and descending internuclear pathways (Nasution and Shigenaga 1987).

Nucleus Principalis (Vp)

This is divided into a subnucleus dorsomedialis (Vpd) and a subnucleus ventrolateralis (Vpv; Fig. 1a). The Vpv extends further caudally than the Vpd, to the rostral pole of the facial nucleus. The caudal limit of Vpd corresponds to the caudal pole of the Vmo. The neurons are densely packed and have a uniform appearance, with small- and medium-sized, round or oval cell bodies. The Vpd is distinguished by the compact aggregation of its cells, although caudally it becomes loosely arranged (Shigenaga et al. 1986b).

Trigeminal Brainstem Nuclear Complex, Anatomy, Fig. 1 Schematic drawings of nociceptive afferent projections from the upper and lower tooth pulps, jaw muscles, and cornea in the trigeminal sensory nuclear complex. (a–e) are arranged rostrocaudaly. Note the projections of jaw muscle and tooth pulp afferents overlap each other in the Vpd and laminae I and V. BC, brachium conjunctivum; SO, superior olivary nucleus; Vi, subnucleus interpolaris; Vinst, interstitial trigeminal nucleus (d, l, and v indicate dorsal, lateral, and ventral, respectively); Vint, intertrigeminal region; Vjuxt, juxtatrigeminal region; Vmes, mesencephalic trigeminal nucleus; Vmo, trigeminal motor nucleus, Vo.c, caudal part of subnucleus oralis (Vo); Vo.dm, dorsomedial part of Vo; Vo.r, rostrodorsomedial part of Vo; Vpd, dorsomedial subnucleus of nucleus principalis (Vp); Vpv, ventrolateral subnucleus of Vp; Vsup, supratrigeminal nucleus; Vtr, spinal trigeminal tract; VII, facial nucleus; 5N, trigeminal motor nerve; 7N, facial nerve

The central projections of primary afferents have been examined using the techniques of ▶ transganglionic transport and intra-axonal labeling of horseradish peroxidase (HRP). The Vpd receives a projection from trigeminal primary afferents innervating intraoral structures, but both the intraoral and facial afferents project to the Vpv (Shigenaga et al. 1986b). Tooth pulp afferents terminate in the Vpd in the cat (Shigenaga et al. 1989b; Westrum et al. 1981) and in the Vp in the rat (Marfurt and Turner 1984). In cats, jaw muscle afferents with the cell bodies in the TG terminate in the Vpd (Fig.1a) (Shigenaga et al. 1988). The fact that these afferents are nociceptors has been demonstrated by an intra-axonal study, which shows that masseter muscle noxious mechanoreceptic afferents (group III) make synapses with neurons in the dorsomedial portion of Vp in rats (Dessem et al. 2007). Corneal afferents with unmyelinated nociceptors project to the ventralmost portion of the Vpv (Fig.1a) in cats (Shigenaga et al. 1989a, b, c). The dorsomedial portion of rat Vp (Vpd in cats) contains axons and terminals immunoreactive for transient receptor potential vanilloid 1 (TRPV1) (Bae et al. 2004), a marker of nociceptive primary afferents that are activated by capsaicin and noxious heat (>43 °C) (Caterina et al. 1997), and for TRP ankyrin 1 (TRPA1) (Kim et al. 2010), responding to noxious cold and pungent compounds (Bautista et al. 2005).

At the electron microscopic level (Bae et al. 2003), all pulpal afferent boutons in rats contain clear, round vesicles (S-type) with few dense-cored vesicles and make synaptic contact with non-primary dendrites with ▶ asymmetric junctions. Approximately one-third of the boutons show characteristic glomeruli, in which synaptic contact is made with small dendrites as well as with other axon terminals. The ▶ synaptic glomerulus has an important role to activate multiple second-order neurons at the same time. The presynaptic axon terminals contain ▶ pleomorphic vesicles and form symmetric contacts with the postsynaptic membrane. The presynaptic boutons are immunoreactive for GABA. These ultrastructural features are also common in axon terminals from low-threshold mechanoreceptive afferents.

The distribution of cell bodies of trigeminothalamic neurons has been mapped in cats with the ▶ retrograde labeling technique (Shigenaga et al. 1983). It was found that most Vpv neurons project to the contralateral thalamus, the nucleus ventralis posteromedialis (VPM), via the crossed ventral tract, while most Vpd neurons project to the ipsilateral VPM via the uncrossed dorsal tract. Thus, sensory information from the intraoral structures is mediated through both crossed (via Vpv) and uncrossed (via Vpd) pathways, whereas facial sensory information is mediated via Vpv by a single crossed pathway, although in the rat most Vp neurons project into the crossed pathway. The axons of the crossed pathway give off collaterals in the central lateral nucleus (CL) of the intralaminar complex (Shigenaga et al. 1983). An ▶ intracellular labeling study in cats (Yoshida et al. 1998) has shown that the Vp also contains ▶ local-circuit cells whose axon collaterals terminate in the jaw-closing region of the Vmo and the lateral reticular formation (Vint, Vjuxt).

Subnucleus Oralis (Vo)

This extends from the caudal tip of the facial nucleus to the level a little posterior to the caudal end of Vmo, and thus, the rostral half of the Vo (termed Vo.r) is situated in the region dorsomedial to the Vpv (Fig. 1b). Its caudal limit corresponds to the rostral pole of the facial nucleus. The Vo.r is characterized by large ▶ multipolar cells and merges medially with the lateral reticular formation (Vjuxt). More caudally, two regions represent subgroupings, a ventrolateral region (Vo.c) and a dorsomedial region (Vo.dm, Fig. 1c). The Vo.dm is composed mainly of small, compactly arranged cells and merges with the ventrolateral border of the solitary nucleus. In the cat, this subdivision is present between the levels corresponding to the facial nucleus (VII), whereas in the rat, it continues without change of structure caudally into the Vi. The Vo.c is composed of oval- or spindle-shaped small cells, triangular or fusiform medium-sized cells, and large multipolar cells. The large cells are sparsely scattered throughout this subdivision (Shigenaga et al. 1986b). The Vo.r and Vo.dm receive projections from primary afferents innervating intraoral structures as well as from mesencephalic primary (Vmes) afferents (Shigenaga et al. 1989a), while facial afferents project to the Vo.c. A dorsoventral organization is not apparent in the Vo.r and Vo.dm. In cats, pulpal afferents terminate in the Vo.r and Vo.dm (Fig. 1c, d), where the upper and lower teeth are represented in a mediolateral sequence, and projections from the anterior to posterior teeth are organized in a ventrolateral to dorsomedial sequence, with an extensive overlap in projections from adjoining teeth (Shigenaga et al. 1989b). In rat, masseter muscle noxious afferents terminate in the dorsal portion of Vo (Dessem et al. 2007).

At the electron microscopic level (Bae et al. 2003), synaptic organization of rat pulpal afferent boutons in the Vo differs from that in Vp, in that the number of postsynaptic elements per bouton and the frequency of axoaxonic contacts are lower in the Vo, indicating less frequent synaptic glomeruli. These ultrastructural features are common to those of low-threshold mechanoreceptive afferent terminals in the Vp and Vo in cats.

Vo neurons, especially in Vo.r and Vo.dm, send few axons to the VPM in cats (Shigenaga et al. 1983), while a considerable number of Vo.r neurons project to the CL of the intralaminar complex.

There are, however, species differences, and a significant projection to the VPM might exist in the rat (see ▶ Trigeminal Brainstem Nuclear Complex, Physiology). In addition, the Vo.r and Vo.dm contain premotoneurons projecting to either jaw-closing or jaw-opening region of the Vmo (Yoshida et al. 1994). Although many neurons respond to light mechanical stimulation of intraoral structures, some are activated by noxious stimuli. These nociceptive neurons send their axon collaterals to the jaw-closing region of the Vmo, suggesting that they may be involved in a reflex circuit that modulates jaw-closing alpha-motoneurons.

Subnucleus Interpolaris (Vi)

This lies between the Vo and Vc and ends a little caudal to the obex and is composed of three neural populations with small, medium, and large cells. Its dorsomedial margin, however, contains ovoid- or spindle-shaped small- or medium-sized cells, where pulpal afferents terminate partly in cats (Shigenaga et al. 1986b).

The projection pattern of primary afferents in the Vi is organized in a similar fashion to that in the Vpv, with the exception of lingual and pulpal afferents, which do not terminate in the Vpv in the cat (Shigenaga et al. 1986b, 1989b).

The pulpal projections from the cat's upper and lower teeth are represented in the medial Vi (Fig. 1d) with a topographic fashion, similar to that found in the Vo.r and Vo.dm (Fig. 1b, c) (Shigenaga et al. 1989b). In addition, the Vi receives projections of primary afferents, with the cell bodies in the TG, from jaw-closing and jaw-opening muscles (Shigenaga et al. 1988). Their projection sites are confined to the caudal levels of Vi. They are in the most lateral part of the nucleus, with an extensive overlap in projections, save for the deep temporal nerve, which projects to the dorsal portion of the interstitial nucleus of Cajal (Gobel and Purvis 1972) or paratrigeminal nucleus in rats (Chan-Paley 1978). The interstitial nucleus of Vtr (Vinst) is considered as a rostral extension of cell islands of lamina I of the Vc, which is seen intercalated in the dorsal (dVinst), ventral (vVinst), and lateral (lVinst) parts at the caudal levels of Vi in cats (Fig. 1d) (Shigenaga et al. 1986b). The dVinst and vVinst that receive projections from jaw muscle and corneal afferents, respectively, may function in nociception. In fact, the paratrigeminal nucleus and the lateral margin of Vi, bordering the Vtr, contain axons and terminals immunoreactive for TRPA1 (Kim et al. 2010) and for TRPV1 (Bae et al. 2004) in rats. Neurons in the dVinst and lamina I in Vc send their axons into the medial parabrachial nucleus (Nasution and Shigenaga 1987).

Similar to the Vpv, the Vi neurons project to the contralateral VPM, but not to the CL (Shigenaga et al. 1983). The Vi neurons in the ▶ Vi/Vc transition zone in rats project to the ▶ nucleus submedius (Sm) of the thalamus (Yoshida et al. 1991).

Trigeminal Brainstem Nuclear Complex, Anatomy, Fig. 2 Schematic drawings of the oral and facial representation in laminae I and II of the medullary and upper cervical dorsal horns. A series of concentric bands (a–e) depicting the "onionskin" representation is shown in the drawing of a cat's face and mouth. In a three-dimensional diagram of the medullary and upper cervical dorsal horns, intraoral divisions are marked by O, and numbers indicated on the right side show distance in millimeters rostral or caudal (−) to the obex. Each terminal zone of the trigeminal afferent branches examined is filled in the right column, where b and m indicate presumed terminal zones of the facial branch of the buccal nerve and the most anterior branch of the mental nerve, respectively. The ophthalmic (V1), maxillary (V2), and mandibular (V3) divisions of the trigeminal nerve are illustrated by straight lines, oblique lines, and black dots, respectively. ai, inferior alveolar nerve; as, superior alveolar nerve; at, auriculotemporal nerve; b, buccal nerve; c, cornea; f, frontal nerve; io, infraorbital nerve; l, lingual nerve; m, mental nerve; my, mylohyoid nerve; p, palatine nerve; z, zygomatic nerve (Adapted from Fig. 13 in Shigenaga et al. (1986a))

Subnucleus Caudalis (Vc)

This extends from the obex to the level of the pyramidal decussation or the C2 segment. The structure resembles closely the spinal dorsal horn; thus, it is often termed the medullary dorsal horn. Olszewski (1950) divided the Vc into three laminar zones: marginalis, gelatinosus, and magnocellularis. The Rexed lamination scheme of the spinal dorsal horn is also applicable to the Vc, and the marginalis, gelatinosus, and magnocellularis correspond to lamina I, lamina II, and laminae III/IV, respectively (Fig. 1e). The detailed morphology of cells in the different laminae has been reported by several studies (Gobel et al. 1981; Renehan et al. 1986). In cats, laminae I, outer II (IIo) and V (the lateral parts of medullary reticular formation) receive inputs from small A-delta fibers and C-fibers, which convey information about pain or thermal sensation. Laminae I and IIo that receive projections from tooth pulp, cornea, and jaw muscle afferent projections express immunoreactivity of TRPA1 and TRPV1 (for references, see above). However, the tooth pulp and muscle afferents terminate in lamina V also. Pulpal afferent boutons in the superficial layers in rats can be classified into two types, those with clear, round vesicles (nonpeptidasic) and those with dense-cored vesicles (peptidasic) (Bae et al. 2003). Unmyelinated pulpal afferents have been suggested to bear terminal varicosities that contain a large number of dense-cored vesicles with few clear, round vesicles (Bae et al. 2003). The occurrence of synaptic glomeruli and of axoaxonic contacts is much less frequent in the Vc than in the Vp.

The projection patterns of primary afferents in the Vc have also been examined in cats using the technique of transganglionic transport of HRP (Shigenaga et al. 1986a, 1988, 1989b). All three trigeminal divisions terminate throughout laminae I–V. However, the mediolateral arrangements and caudal extensions differ between the different nerves or branches.

The intraoral and facial structures are arranged as a series of concentric semicircular rings that are centered around the midline of the most anterior face and mouth, and are represented, especially in laminae I/II, in a consecutive order. In this way, the midline of the mouth and the most anterior face are represented most rostrally, while more lateral or posterior structures are represented at successively more caudal levels in the medullary and upper cervical dorsal horns (Fig. 2). The fact that neurons in laminae I, IIo, and V receive inputs from nociceptive or thermoreceptive afferents, and that lamina

I neurons project to the VPM (Shigenaga et al. 1983), Sm (Yoshida et al. 1991) and in monkey, the posterior ventral medial nucleus (VMpo) (Beggs et al. 2003), support the concept that the onionskin-like organization of pain and thermal sensations is defined by the arrangement of the sensory projections to laminae I and IIo. An anatomical substrate for referred pain phenomena may be provided by the extensive mediolateral overlap in projections from different nerve branches. Neurons in laminae III and IV project to the rostral nuclei (Nasution and Shigenaga 1987) and medullary reticular formation.

References

Bae, Y. C., Kim, J. P., Choi, B. J., et al. (2003). Synaptic organization of tooth pulp afferent terminals in the Rat trigeminal sensory nuclei. *The Journal of Comparative Neurology, 463,* 13–24.

Bae, Y. C., Oh, J. M., Hwang, S. J., et al. (2004). Expression of vanilloid receptor (TRPV1) in the rat trigeminal sensory nuclei. *The Journal of Comparative Neurology, 478,* 62–71.

Bautista, D. M., Movahed, P., Hinman, A., et al. (2005). Pungent products from garlic activate the sensory Ion channel TRPA1. *Proceedings of the National Academy of Sciences, 102,* 12248–12252.

Beggs, J., Jordan, S., Ericson, A. C., et al. (2003). Synaptology of trigemino- and spinothalamic lamina I terminations in the posterior ventral medial nucleus of the macaque. *The Journal of Comparative Neurology, 459,* 334–354.

Caterina, M. J., Schumacher, M. A., Tominaga, M., et al. (1997). The capsaicin receptor: A heat-activated Ion channel in the pain pathway. *Nature, 389,* 816–824.

Chan-Palay, V. (1978). The paratrigeminal nucleus: I. Neurons and synaptic organization. *Journal of Neurocytology, 7,* 405–418.

Dessem, D., Moritani, M., & Ambalavanar, R. (2007). Nociceptive craniofacial muscle primary afferent neurons synapses in both the rostral and caudal brain stem. *Journal of Neurophysiology, 98,* 214–223.

Gobel, S., & Purvis, M. B. (1972). Anatomical studies of the organization of the spinal V nucleus: The deep bundles and spinal V tract. *Brain Research, 48,* 27–44.

Gobel, S., Hockfield, S., & Ruda, M. A. (1981). An anatomical analysis of the similarities between medullary and spinal dorsal horns. In Y. Kawamura & R. Dubner (Eds.), *Oral-facial sensory and motor functions* (pp. 211–223). Tokyo: Quintessence.

Hayashi, H. (1985). Morphology of terminations of small and large myelinated trigeminal primary afferent fibers in the cat. *The Journal of Comparative Neurology, 240,* 71–89.

Honma, S., Moritani, M., Zhang, L. F., et al. (2001). Quantitative ultrastructure of synapses on functionally identified primary afferent neurons in the cat trigeminal mesencephalic nucleus. *Experimental Brain Research, 137,* 150–162.

Kim, Y. S., Son, J. Y., Kim, T. H., et al. (2010). Expression of transient receptor potential Ankyrin1 (TRPA1) in the rat trigeminal sensory afferents and spinal dorsal horn. *The Journal of Comparative Neurology, 518,* 687–698.

Marfurt, C. F., & Turner, D. F. (1984). The central projections of tooth pulp afferent neurons in the Rat as demonstrated by the transganglionic transport of horseradish peroxidase. *The Journal of Comparative Neurology, 223,* 535–547.

Nasution, I. D., & Shigenaga, Y. (1987). Ascending and descending internuclear projections within the trigeminal sensory nuclear complex. *Brain Research, 425,* 234–247.

Olszewski, J. (1950). On the anatomical and functional organization of the spinal trigeminal nucleus. *The Journal of Comparative Neurology, 92,* 402–413.

Renehan, W. E., Jacquin, M. F., Mooney, R. D., & Rhoades, R. W. (1986). Structure-function relationships in rat medullary and cervical dorsal horns. II. Medullary dorsal horn cells. *Journal of Neurophysiology, 55,* 1187–1201.

Shigenaga, Y., Nakatani, Z., Nishimori, T., et al. (1983). The cells of origin of cat trigeminothalamic projections: Especially in the caudal medulla. *Brain Research, 277,* 201–222.

Shigenaga, Y., Chen, I. C., Suemune, S., et al. (1986a). Oral and facial representation within the medullary and upper cervical dorsal horns in the cat. *The Journal of Comparative Neurology, 243,* 388–408.

Shigenaga, Y., Okamoto, T., Nishimori, T., et al. (1986b). Oral and facial representation in the trigeminal principal and rostral spinal nuclei of the Cat. *The Journal of Comparative Neurology, 244,* 1–18.

Shigenaga, Y., Suemune, S., Nishimura, T., et al. (1986c). Topographic representation of lower and upper teeth within the trigeminal sensory nuclei od adult cat as demonstrated by the transganglionic transport of horseradish peroxidase. *The Journal of Comparative Neurology, 251,* 299–316.

Shigenaga, Y., Sera, M., Nishimori, T., et al. (1988). The central projection of masticatory afferent fibers to the trigeminal sensory nuclear complex and upper cervical spinal cord. *The Journal of Comparative Neurology, 268,* 489–507.

Shigenaga, Y., Doe, K., Suemune, S., et al. (1989a). Physiological and morphological characteristics of periodontal mesencephalic trigeminal neurons - intraaxonal staining with HRP. *Brain Research, 505,* 91–110.

Shigenaga, Y., Nishimura, M., Suemune, S., et al. (1989b). Somatotopic organization of tooth pulp primary afferent neurons in the cat. *Brain Research, 477,* 66–89.

Tsuru, K., Otani, K., Kajiyama, K., et al. (1989). Central terminations of periodontal mechanoreceptive and tooth pulp afferents in the trigeminal principal and oral nuclei of the cat. *Brain Research, 485,* 29–61.

Westrum, L. E., Canfield, R. C., & O'Conner, T. A. (1981). Each canine tooth projects to all brain stem trigeminal nuclei in cat. *Experimental Neurology, 74,* 787–799.

Yoshida, A., Dostrovsky, J. O., Sessle, B. J., et al. (1991). Trigeminal projections to the nucleus submedius of the thalamus in the rat. *The Journal of Comparative Neurology, 307,* 609–625.

Yoshida, A., Yasuda, K., Dostrovsky, J. O., et al. (1994). Two major types of premotoneurons in the feline trigeminal nucleus oralis as demonstrated by intracellular staining with horseradish peroxidase. *The Journal of Comparative Neurology, 347,* 495–514.

Yoshida, A., Hiraga, T., Moritani, M., et al. (1998). Morphologic characteristics of physiologically defined neurons in the cat trigeminal nucleus principalis. *The Journal of Comparative Neurology, 401,* 308–328.

Trigeminal Brainstem Nuclear Complex, Physiology

James W. Hu[1] and Alain Woda[2]
[1]Faculty of Dentistry, University of Toronto, Toronto, Canada
[2]Faculté de Chirurgie Dentaire, University Clermont-Ferrand, Clermont-Ferrand, France

Synonyms

Trigeminal sensory nucleus

Definition

The trigeminal (V) brainstem nuclear complex (VBNC, Fig. 1) is the site of the first synapse of most V sensory primary afferents and of considerable sensory processing and modulation. Its anatomy and physiology, although generally similar to the spinal sensory system, display several differences. These differences, and the existence of specialized organs and tissues, are important features that have prompted the study of orofacial nociception and may be manifested clinically in

Trigeminal Brainstem Nuclear Complex, Physiology, Fig. 1 Schema of nociceptive somatosensory organization of the orofacial area. The trigeminal nerve is made up of three divisions, ophthalmic (V1), maxillary (V2), and mandibular (V3), which supply a wide variety of tissues; note that V2 and V3 as well as V1 also supply part of the meninges. The primary afferent neurons mediating nociceptive messages (A-delta and C) project to the spinal tract nucleus. Ascending pathways arise from spinal tract nucleus to reach suprasegmental levels such as the thalamus and cerebral cortex. There are also important pathways descending from the suprasegmental centers to modulate the afferent messages, such as the cerebral cortex (Modified from Dallel et al. 2003)

the unique trigeminal pain conditions, such as toothache, migraine headache, V neuralgia, temporomandibular disorders, and some other idiopathic orofacial pains.

Characteristics

Trigeminal Brainstem Nuclear Complex (VBNC) Organization

Most V primary afferents have their cell bodies in the V ganglion and project centrally to the VBNC. The large diameter myelinated (A-β) fibers ascend to the main (or principal, Vp) sensory nucleus, whereas both large and small diameter fibers descend to the spinal tract nucleus, which includes subnuclei oralis (Vo), ▶ interpolaris (Vi) and ▶ caudalis (Vc); see article by Shigenaga and Yoshida (Fig. 1). The most caudal, Vc, is a laminated structure that extends caudally into the dorsal horn of the cervical spinal cord with which it is homologous and continuous. For this reason, it is also called the medullary dorsal horn (MDH; Gobel et al. 1981). Almost all nociceptive ▶ C Fiber afferents from the V nerve have their central terminals distributed in the Vc and ▶ Vi/Vc transition zone. Terminals of the A-δ fiber afferents are also found in Vc; however, some, mostly from the oral and perioral area, are found rostral to Vc in Vi and Vo; in particular, A-δ fibers from the dental pulp have been described to be in the rostral divisions, Vp, Vo, and Vi. The VBNC is somatotopically organized.

General Features of Nociceptive Processing in Subnucleus Caudalis

In 1938, Sjöqvist reported that ▶ trigeminal tractotomy, which interrupts the afferent inputs to the Vc by severing the V spinal tract at the upper level of Vc, was an efficient treatment for ▶ trigeminal neuralgia and also produces facial analgesia. The effect of Vc deafferentation by tractotomy has repeatedly been confirmed by neurosurgeons and animal researchers, and it has therefore long been considered that Vc is the crucial brainstem center for orofacial nociceptive processing. However, this procedure has little effect on tactile sensation and intraoral pain. These results indicate the involvement of the rostral nuclei, such as the Vo, in toothache or other pain sensations from the oral area (Young 1982). Thus, neurons in V_0, as well as Vc, are involved in ascending pathways contributing to orofacial pain (also see below). In the case of autonomic responses to noxious stimuli, such as cardiovascular and secretory reflexes (e.g., lacrimation and salivation), the trigeminal brainstem regions that mediate these evoked responses include the Vi/Vc transition zone and medial regions of the VBNC that border the reticular formation. The ▶ Vi/Vc transition zone is also involved in endocrine and muscle responses (Bereiter et al. 2000; also see below).

The fact that Vc is critical for V nociception is supported by eight pieces of evidence:

1. The facial analgesia produced by Vc deafferentation (see above).
2. Vc is a laminated structure with a substantia gelatinosa (lamina II) region, which is known to be associated with pain processing and a high concentration of neuropeptide and other neurochemical markers for nociception.

Trigeminal Brainstem Nuclear Complex, Physiology, Table 1 Convergence of nociceptive afferent inputs to nociceptive neurons in trigeminal subnucleus caudalis of cats

Facial skin/oral mucosa	100 %
Tooth pulp	66 %
Muscle (jaw or tongue)	55 %
Temporomandibular joint (TMJ)	35 %
Laryngeal mucosa (viscera)	55 %
Upper cervical nerves	50 %

Based upon a sample of approximately 100 WDR and NS neurons with cutaneous or intraoral mechanoreceptive fields (From Sessle et al. 1986)

3. Vc receives A-δ and C nociceptive afferents that terminate in its laminae I and II.
4. Vc contains nociceptive neurons, either wide dynamic range (WDR) or nociceptive specific (NS). These nociceptive neurons predominate in the laminae I, V, and VI and display positive stimulus-response functions to various afferent inputs and can therefore code the intensity of different types of noxious stimuli, including those arising from specialized orofacial tissues such as tooth pulp, cornea, and meninges (Price et al. 1976; Sessle 2000). In addition, these Vc nociceptive neurons also receive a wide range of inputs (see Table 1). This provides the basis for pain referral and spread in orofacial region.
5. Some Vc nociceptive neurons send their axons into ascending nociceptive pathways that reach the thalamus (Price et al. 1976).
6. Some Vc neurons serve as reflex interneurons in reflex responses to noxious orofacial stimuli (see Sessle 2000).
7. Vc nociceptive neurons are subject to several afferent and descending antinociceptive modulatory influences, as well as pharmacological antinociceptive modulations (Sessle 2000; Dubner and Ren 2004).
8. In certain conditions (e.g., inflammation or nerve injury), these nociceptive neurons manifest neuroplastic changes, such as increase in spontaneous activity, reduction of activation threshold, increase in response magnitude to noxious stimuli, and receptive field

Trigeminal Brainstem Nuclear Complex, Physiology, Fig. 2 Rat caudalis nociceptive-specific (NS) neuron showing neuroplastic changes in neuronal properties after small-fiber excitant and inflammatory irritant mustard oil (MO) application to the maxillary *right* molar pulp. In (**a**), histologically retrieved recording site and the *arrow* indicate the neuronal recording site in caudalis and, in (**b**), mechanoreceptive field (RF) sizes before (0 min.) and 10 min. after MO application (i.e., solid area represents pinch RF, and "red" area, with a thin red *arrow*, represents a newly defined tactile RF that also responded to pinch stimulus after MO application). In (**b**), blackened area indicates the RF location and size, which only noxious pinch stimulation activated, before MO application. The RF expansion 10 min after MO application and a small stripped area that appeared only after the application of MO, indicated by a thin *arrow*, that a brush (low-threshold mechanical) stimulus can also activate this NS neuron that previously was unable to respond

to this brush stimulus. In (**c**), MO induced a brief burst of discharges followed by higher firing rate than the baseline level (0 min). This response represents an input from the maxillary tooth pulp. In (**d**), neuronal responses to mechanical and thermal stimuli applied to the cutaneous RF. *Top* trace: marker of brush (Br), pressure (Pr), pinch (Pi), and radiant heat (RH). *Middle* trace: neuronal responses in control conditions (i.e., Pre-MO before MO application). *Bottom* trace: neuronal responses to same stimuli 20 min after MO application (i.e., Post-MO after MO application). In (**e**), neuronal responses to graded mechanical stimuli (50, 100, and 200 g). Each stimulus lasts for 3 s, and the y-axis scale (omitted) is the same for *middle* and *bottom* traces in a and b. Binwidth is 1 s. Note that after MO application, this NS neuron became responsive to light brushing (*thin arrow*) and radiant heating (*thick arrow*) of the cutaneous RF in (d) and strongly responsive to graded pinch stimuli in (e) (Modified with permission from Chiang et al. (1998))

expansion (Hu et al. 1992), (Fig. 2). These neuroplastic changes reflect a central sensitization of the nociceptive neurons, and glutamatergic and purinergic processes, as well as glial cells, are especially involved (Chiang et al. 2011; Dubner and Ren 2004; Sessle 2000, 2011). These neuroplastic changes correlate well with the clinical observation of allodynia and hyperalgesia associated with inflammatory or nerve injury-induced neuropathic (Dubner and Ren 2004;

Iwata et al. 2011; Sessle 2000, 2011; see ▶ Glial cells in orofacial pathological pain mechanisms) pain conditions.

Special Features of Trigeminal Brainstem Nociceptive Processing

1. Nociceptive processing in VBNC is not restricted to Vc: Nociceptive neurons activated from the orofacial area are also observed rostrally in Vo and Vi (Dallel et al. 1990; Sessle 2000; Dubner and Ren 2004) and

Trigeminal Brainstem Nuclear Complex, Physiology, Fig. 3 Schematic dorsal view of the spinal tract nucleus of VBNC that illustrates the Vo-Vc experimental model. The Vo is activated only indirectly by primary afferent C-fibers. A relay in the substantia gelatinosa in Vc is mandatory, since there is no direct C-fiber projection to Vo. This allows for recordings from nociceptive neurons while injecting chemicals or cold block into the substantia gelatinosa

manifest many of the properties of other neurons of the deep spinal dorsal horn or of deep layers of Vc (Dallel et al. 1990; Hu et al. 1992; Dubner and Ren 2004), including plasticity (Sessle 2000; see ▶ Glial cells in orofacial pathological pain mechanisms). These neurons may play a role in reflex responses to noxious orofacial stimulation, but results of V tractotomy (see above) indicate that oral and perioral pain sensation also depends on Vo. Another indication of a sensory role for Vo as well as Vc is the presence of a similar density of anatomical projections from these two subnuclei to higher levels of the brain. In addition, nociceptive thalamic neurons can still be recorded after stimulation of oral and perioral areas, in spite of the deafferentation of Vc by trigeminal tractotomy (Raboisson et al. 1989). In Vo, the C-fiber primary afferent endings are scarce and a substantia gelatinosa is lacking.

However, C-fiber-evoked activities recorded in Vo relay in the substantia gelatinosa of Vc (Dallel et al. 1998). The influence on the properties of Vo neurons of the substantia gelatinosa of Vc constitutes a unique feature, which can be used as an experimental model (Fig. 3), since it offers the possibility of recording from deep layer nociceptive neurons, i.e., in Vo, while injecting neurochemicals in the Vc substantia gelatinosa. Indeed, it has been shown that plastic changes observed in nociceptive neurons of Vo can be modulated, either facilitated or inhibited, by neurochemicals injected into the Vc substantia gelatinosa (Dallel et al. 1998; Woda et al. 2001; Hu et al. 2002). Nociceptive neurons in the ▶ Vi/Vc transition zone also play a special role in autonomic, endocrine, and muscle responses to noxious orofacial stimuli (Dubner and Ren 2004; Bereiter et al. 2000).

2. Some specialized orofacial tissues receive innervation patterns unique to the trigeminal system. Unlike the skin, specialized V structures, such as the tooth pulp, meninges, and cornea, are innervated mainly by A-δ and C nociceptive afferents, since very few or no A-β afferents can be found. Single-unit recording studies have shown that the VBNC nociceptive neurons activated from these specialized tissues often receive convergent inputs from a wide range of structures including facial skin. The distribution of V neurons that encode nociceptive information from these specialized orofacial tissues displays a complex pattern, which is characterized by their occurrence at different rostrocaudal levels of the VBNC.

3. The VBNC also offers the possibility of extending the knowledge base relevant to spinal neuronal properties from anesthetized to awake conditions. Taking advantage of the unique location of the medullary dorsal horn (Vc), Dubner and his colleagues recorded Vc single neurons in conscious monkeys while performing behavioral tasks relevant to nociception (Dubner et al. 1981). While the properties of WDR and NS neurons in Vc

observed in anesthetized animals during recordings were confirmed in the conscious monkeys, these experiments also showed that many of these Vc neurons could be modulated through attentional and motivational factors.

References

Bereiter, D. A., Hirata, H., & Hu, J. W. (2000). Trigeminal subnucleus caudalis: Beyond homologies with the spinal dorsal horn. *Pain, 88,* 221–224.

Chiang, C. Y., Park, S. J., Kwan, C. L., Hu, J. W., & Sessle, B. J. (1998). NMDA receptor mechanisms contribute to the trigeminal nociceptive neuronal plasticity induced by mustard oil application to the rat molar tooth pulp. *Journal of Neurophysiology, 80,* 2621–2631.

Chiang, C. Y., Dostrovsky, J. O., Iwata, K., & Sessle, B. J. (2011). Role of glia in orofacial pain. *The Neuroscientist, 17,* 303–320.

Dallel, R., Raboisson, P., Woda, A., & Sessle, B. J. (1990). Properties of nociceptive and non-nociceptive neurons in trigeminal subnucleus oralis of the rat. *Brain Research, 521,* 95–106.

Dallel, D., Duale, C., & Molat, J. L. (1998). Morphine administered in the substantia gelatinosa of the spinal trigeminal nucleus caudalis inhibits nociceptive activities in the spinal trigeminal nucleus oralis. *Journal of Neuroscience, 18,* 3529–3536.

Dallel, R., Villanueva, L., Woda, A., & Voisin, D. (2003). Neurobiology of trigeminal pain (in French). *Medecine Sciences, 10,* 567–574.

Dubner, R., & Ren, K. (2004). Brainstem mechanisms of persistent pain following injury. *Journal of Orofacial Pain, 18,* 299–305.

Dubner, R., Hoffman, D. S., & Hayes, R. L. (1981). Neuronal activity in medullary dorsal horn of awake monkeys trained in a thermal discrimination task. III. Task-related responses and their functional role. *Journal of Neurophysiology, 46,* 444–464.

Gobel, S., Hockfield, S., & Ruda, M. A. (1981). Anatomical similarities between medullary and spinal dorsal horns. In Y. Kawamura & R. Dubner (Eds.), *Oral-facial sensory and motor functions* (pp. 211–223). Tokyo/Berlin: Quintessence.

Hu, J. W., Sessle, B. J., Raboisson, P., Dallel, R., & Woda, A. (1992). Stimulation of craniofacial muscle afferents induces prolonged facilitatory effects in trigeminal nociceptive brain-stem neurons. *Pain, 48,* 53–60.

Hu, B., Chiang, C. Y., Hu, J. W., Dostrovsky, J. O., & Sessle, B. J. (2002). P2X receptors in trigeminal subnucleus caudalis modulate central sensitization in trigeminal subnucleus oralis. *Journal of Neurophysiology, 88,* 1614–1624.

Iwata, K., Imamura, Y., Honda, K., & Shinoda, M. (2011). Physiological mechanisms of neuropathic pain: The orofacial region. *International Review of Neurobiology, 97,* 227–250.

Price, D. D., Dubner, R., & Hu, J. W. (1976). Trigeminothalamic neurons in nucleus caudalis responsive to tactile, thermal, and nociceptive stimulation of monkey's face. *Journal of Neurophysiology, 39,* 936–953.

Raboisson, P., Dallel, R., & Woda, A. (1989). Responses of neurons in the ventrobasal complex of the thalamus to orofacial noxious stimulation after large trigeminal tractotomy. *Experimental Brain Research, 77,* 569–576.

Sessle, B. J. (2000). Acute and chronic craniofacial pain: Brainstem mechanisms of nociceptive transmission and neuroplasticity, and their clinical correlates. *Critical Reviews in Oral Biology and Medicine, 11,* 57–91.

Sessle, B. J. (2011). Peripheral and central mechanisms of orofacial inflammatory pain. *International Review of Neurobiology, 97,* 179–206.

Sessle, B. J., Hu, J. W., Amano, N., & Zhong, G. (1986). Convergence of cutaneous, tooth pulp, visceral, neck and muscle afferents onto nociceptive and non-nociceptive neurones in trigeminal subnucleus caudalis (medullary dorsal horn) and its implications for referred pain. *Pain, 27,* 219–235.

Woda, A., Molat, J. L., & Luccarini, P. (2001). Low doses of N-methyl-D-aspartate antagonists in superficial laminae of medulla oblongata facilitate wind-up of convergent neurones. *Neuroscience, 107,* 317–327.

Young, R. F. (1982). Effect of trigeminal tractotomy on dental sensation in humans. *Journal of Neurosurgery, 56,* 812–818.

Trigeminal Ganglion

Synonyms

Gasserian ganglion, TG

Definition

A nerve ganglion where the cell bodies of most of the primary afferent neurons innervating the orofacial area are located.

Cross-References

▶ Nociceptors in the Dental Pulp

Trigeminal Lemniscus

Definition

Ascending projection principally from the trigeminal brainstem nuclear complex to the thalamus.

Cross-References

▶ Parafascicular Nucleus, Pain Modulation

Trigeminal Motor Nucleus

Synonyms

Masticatory nucleus

Definition

Also called the masticatory nucleus, it is located in the dorsal mid-pons, close to the trigeminal principal sensory nucleus, and contains motoneurons that supply most of the muscles involved in jaw movements.

Cross-References

▶ Jaw-Muscle Silent Periods (Exteroceptive Suppression)

Trigeminal Nerve

Definition

The trigeminal nerve is the largest cranial nerve and innervates much of the craniofacial region, including the muscles of mastication. The three major divisions of the trigeminal nerve are ophthalmic, maxillary, and mandibular nerves. All three branches provide sensory innervation to these craniofacial tissues, and the mandibular branch also carries motor fibers that innervate the jaw muscles.

Cross-References

▶ Amygdala, Pain Processing and Behavior in Animals
▶ Nociception in Nose and Oral Mucosa
▶ Trigeminothalamic Tract Projections

Trigeminal Neuralgia

Synonyms

Tic douloureux

Definition

Trigeminal neuralgia is an idiopathic, episodic severe pain condition of the orofacial region, often described as electric-shock-like sensations that can be triggered by innocuous stimuli, associated with injury or dysfunction of the fifth cranial nerve (trigeminal nerve) or its ganglion and felt in the distribution of this nerve. It is considered as a prototype of neuropathic pain. Trigeminal neuralgia (tic douloureux) may have no apparent cause or be associated with neurovascular interaction between the trigeminal root and an anomalous vascular loop (classical trigeminal neuralgia) or be secondary to benign tumors of the cerebellopontine angle or multiple sclerosis.

Cross-References

▶ Central Pain in Multiple Sclerosis
▶ Demyelination
▶ Jaw-Muscle Silent Periods (Exteroceptive Suppression)
▶ Neuralgia, Assessment

▶ Neuralgia, Diagnosis

▶ Pain Paroxysms

▶ Paroxysmal Hemicrania

▶ Primary Stabbing Headache

▶ Trigeminal and Cranial Nerve Neuralgias

▶ Trigeminal Brainstem Nuclear Complex, Physiology

Trigeminal Neuralgia, Aims of Surgical Management

Definition

Surgical management of trigeminal neuralgia to provide complete or partial pain relief for a period that ranges from a year to 10 years or so.

Cross-References

▶ Postherpetic Neuralgia, Etiology, Pathogenesis, and Management

Trigeminal Neuralgia, Diagnostic Method

Definition

Diagnosis of trigeminal neuralgia is made principally on history as there are no diagnostic tests to validate the diagnosis.

Trigeminal Neuralgia, Ignition Theory

Definition

A prevailing concept of trigeminal neuralgia is that it is caused by chronic irritation of the trigeminal nerve which leads to ectopic hyperexcitability of trigeminal sensory fibers that prompts the

paroxysm of pain, and this hyperexcitability forms the basis of the ignition hypothesis.

Cross-References

▶ Postherpetic Neuralgia, Etiology, Pathogenesis, and Management

Trigeminal Neuralgia, Medical Management

Definition

Medical management of trigeminal neuralgia is principally with the use of anticonvulsant drugs, few of which have undergone evaluation under randomized controlled trial conditions.

Cross-References

▶ Postherpetic Neuralgia, Etiology, Pathogenesis, and Management

Trigeminal Neuralgia, Microvascular Decompression

Definition

Microvascular decompression is a major neurosurgical procedure stemming from reports that trigeminal nerve (e.g., sensory root) compression may be of etiological importance in this condition. It involves gaining entry into the posterior fossa of the skull and identifying and decompressing the trigeminal nerve in order to provide pain relief without sensory loss.

Cross-References

▶ Postherpetic Neuralgia, Etiology, Pathogenesis, and Management

Trigeminal Neuralgia, Patient Information

Definition

There is a considerable amount of information, both written and electronic, on trigeminal neuralgia for both healthcare workers and patients, which would aid in management.

Trigeminal Neuralgia, Types of Ablative Surgery

Definition

Ablative surgery for trigeminal neuralgia involves selective destruction of the ▶ trigeminal nerve, and many of these procedures can be done in elderly medically unfit patients, but most result in facial sensory loss. These procedures include radiofrequency thermocoagulation (electrically heating the Gasserian ganglion), percutaneous glycerol injection (injecting glycerol round the Gasserian ganglion), balloon microcompression (applying pressure on the Gasserian ganglion for a few seconds), and recently also Gamma Knife radiation. Other ablative procedures that have been used include peripheral neurectomy and alcohol or other neurotoxic injections into trigeminal nerve branches.

Trigeminal Neuralgia: Diagnosis and Treatment

Joanna M. Zakrzewska[1] and Donald R. Nixdorf[2]
[1]Eastman Dental Hospital UCLH NHS Foundation Trust, London, UK
[2]Department of Diagnostic and Biological Sciences, School of Dentistry University of Minnesota, Minneapolis, MN, USA

Synonyms

Tic douloureux

Definition

The International Association for the Study of Pain (IASP) defines trigeminal neuralgia (TN) as "A sudden, usually unilateral, severe, brief, stabbing, recurrent pain in the distribution of one or more branches of the fifth cranial nerve (Merskey and Bogduk 1994)." A more specific definition put forward by the International Headache Society's classification subcommittee is given in Anon (2004) (Table 1).

When the symptoms of pain fit the characteristics as defined above, but also have a persistent and/or continuous pain in the same distribution or region of the paroxysmal pain, classification becomes less clear. The majority of experts refer to such a presentation as atypical trigeminal neuralgia or type 2 trigeminal neuralgia. This is reasonable, since a large portion of people presenting with such symptomology seem to be similar to those with "classical trigeminal neuralgia." Unfortunately, there can be many reasons for the co-presentation of the symptoms of pain in the region, such as dental disease in people not able to perform adequate oral hygiene, sinusitis, or any comorbid chronic pain disorder, such as temporomandibular disorders. For this reason, the authors prefer to refer to it as trigeminal neuralgia with concomitant pain, so as to clearly identify the duality of possible disorders presenting, and therefore, the need to further work through the diagnosis process for this "other" pain as opposed to assuming it as part and parcel with trigeminal neuralgia.

Characteristics

Epidemiology
The incidence of trigeminal neuralgia (TN) is 4.3/100,000 (2.96 male, 3.47 female/100,000 based on American data). The point prevalence is 0.1 %. The peak incidence is in age group 60–69, and it is rare in patients under the age of 40. There is a strong link between multiple sclerosis and TN. A recent population-based study suggests that hypertension increases the

Trigeminal Neuralgia: Diagnosis and Treatment, Table 1 Diagnostic criteria for trigeminal neuralgia

A. Paroxysmal attacks of pain lasting from a fraction of a second to 2 min, affecting one or more of the divisions of the trigeminal nerve and fulfilling criteria B and C
B. Pain has at least one of the following characteristics: Intense, sharp, superficial, or stabbing Precipitation from trigger areas or by trigger factors
C. Attacks are stereotyped in the individual patient
D. There is no clinically evident neurological deficit
E. Not attributed to another disorder

risk for trigeminal neuralgia and that having it increases the risk of stroke (Pan et al. 2011a, b). There is little data on the natural history and prognostic features of people with TN, but data suggests that it does not affect survival, although attacks get more severe with time (Zakrzewska and Hamlyn 1999). More recently, with the use of electronic primary care databases, research suggests a higher incidence rate and at a younger age of first onset. In the United Kingdom (Hall et al. 2008) and The Netherlands (Dieleman et al. 2008), the overall yearly incidence is 26.8 and 28.9 per 100,000 patient years, respectively. However, it is likely that some of these cases were other pain provoking disorders misclassified as TN, such as dental pain, sinusitis, or temporomandibular disorders. Koopman et al. (2009) showed that the rate of misdiagnosis among general practitioners was as high as 48 %. After diagnosis by experts, the overall incidence rate was 12.6 per 100,000 person years with a mean age of 51.5 years and a 66 % female predominance (Koopman et al. 2009).

Etiology and Pathogenesis

The etiology and pathogenesis of TN are not known, and no satisfactory animal model for trigeminal neuralgia exists. The ignition hypothesis, proposed by Devor et al. (2002), is accepted by many, although direct support is very limited. According to this hypothesis, chronic irritation of the trigeminal nerve leads to focal demyelination, which results in the generation of ectopic action potentials and impaired segmental inhibition.

This is thought to lead to hyperexcitability of the afferent neurons that then give rise to pain paroxysms as a result of synchronized after-discharge activity (Devor et al. 2002). This hypothesis is supported by clinical observations that patients with TN in the majority of cases are found to have blood vessels compressing the trigeminal nerve, either at the nerve root entry zone, or less commonly, the brain stem. However, more recent studies utilizing more sophisticated imaging techniques are showing that arterial contact with the trigeminal nerve can occur in people who do not have TN. Electron microscopic examination of nerve roots taken from patients with such compressions has revealed focal demyelination in the region of the compression with close apposition of demyelinated axons and an absence of intervening glial processes. A process of re-myelination occurs and this could be responsible for the spontaneous remission of the neuralgia. Increasingly, there is evidence suggesting that central facilitation and resulting hyperexcitability of the trigeminal system involves both peripheral central mechanisms. Obermann et al. (2007) have found that using nociceptive blink reflexes and pain-related evoked potentials, the atypical forms of TN exhibit a higher degree of central sensory transmission over-activation than typical TN (Obermann et al. 2007). Furthermore, fMRI studies demonstrate deficits in descending inhibition only in patients with atypical TN, supporting the concept that there is sustained sensitization of the trigeminal system in this subgroup of people (Leonard et al. 2009; Moisset et al. 2011).

Clinical History

It is essential to take a very careful history, as this is the only reliable method of making the diagnosis. It is especially important to elucidate the sharpness and paroxysmal quality of this pain, which are the chief characteristics that differentiate them from most other disorders presenting facial pain. Each bout of pain is very brief (seconds) but patients may get many of these in quick succession, so the pain may seem to last longer. The involved nerve eventually becomes refractory, and there is a period when the patient is pain-free. Classically, there are also periods of complete pain remission, which may last for weeks or months. Clinical observations and patient reports suggest that these periods of pain remission gradually get shorter and shorter over time.

Neuroanatomically, the most common divisions of the trigeminal nerve to be involved are the second and third divisions, while it is rare for someone to report that only the first division is involved. TN is characterized as a unilateral pain disorder, but about 3 % of patients with TN report a bilateral presentation. In these patients, it is unusual for both sides to be active at the same time, meaning episodic bouts of pain seem to occur on one side or the other. The characteristic pain of TN can be triggered by light touch or other A-beta nerve fiber mediated sensations, but it is also common for them to occur spontaneously. A recent fMRI study suggests that the mechanism between the two may be different, with spontaneous pain being correlated with patients' fear of pain (Moisset et al. 2011). This research finding is consistent with the clinical observation that spontaneous bouts of pain seem to occur when there are periods of increased emotion, such as during periods of elevated anxiety or loss of a loved one. Therefore, it is important to determine if there are any contributing factors present, and address them and their effect on TN before coming to the conclusion that the patient's TN is progressing.

Other features that need to be assessed include the quality of life and the level of anxiety and depression of the patient. Most patients with severe trigeminal neuralgia find it impossible to socialize because of fear of developing an attack of pain while eating. These patients will often lose weight and become depressed. Ideally, patients should be evaluated with standard assessment measures, such as the McGill Pain Questionnaire, or some form of anxiety or depression scale, such as the Hospital Anxiety and Depression Scale, and a quality of life assessment, such as Brief Pain Inventory, or SF36.

Examination

In most patients, a clinical neurological examination will be negative but some subtle sensory changes may be present. In some cases, this could indicate that there is a secondary cause for trigeminal neuralgia but this is not pathognomonic (Cruccu et al. 2008; Gronseth et al. 2008). It is important to examine the oral cavity to exclude any dental causes for pain, especially if the patient reports intraoral pain. Assessment of hearing may be prudent if patients wish to undergo an operation that carries the risk of causing hearing loss.

Investigation

There are no diagnostic tests specifically for trigeminal neuralgia; rather, imaging is routinely employed to rule out the secondary reasons for the presentation of TN. Both CT and MRI scans can be used to assess for such lesions; however, MRI scans using contrast are useful in demonstrating whether there is a compression of the nerve root by neighboring blood vessels. Even so, the specificity and sensitivity in being able to predict operative findings still remains under investigation, as was found in an evaluation of MRI-based diagnostic methods in recent TN guidelines (Cruccu et al. 2008; Gronseth et al. 2008). Increasingly sophisticated MRIs are now being used, and these are confirming many of the laboratory and operative findings.

Differential Diagnosis

The differential diagnosis for unilateral facial pain in the distribution of the trigeminal nerve is extensive. The initial clinical setting in which the patient is seen can influence the diagnosis and then subsequent management. If there are any autonomic features then conditions such as the trigeminal autonomic cephalalgias need to be considered (Goadsby et al. 2007). Some disease entities that need to be considered are listed in Table 2.

Management

Trigeminal neuralgia is a rare condition, and although it can be treated in its initial phases within primary care, most of these patients

Trigeminal Neuralgia: Diagnosis and Treatment, Table 2 List of diagnoses that can be misclassified as trigeminal neuralgia

Dental	Reversible or irreversible pulpitis (inflammation of the tooth pulp, such as with a cracked tooth)
	Symptomatic apical periodontitis (inflammation in the supporting structures of the tooth, such as with necrosis of the pulp)
Headaches	SUNCT (short-lasting, unilateral, neuralgiform headache attacks with conjunctival injection and tearing)
	SUNA (short-lasting, unilateral, neuralgiform headache attacks with cranial autonomic symptoms)
	Paroxysmal hemicrania
TMD	Myofascial pain of the masticatory muscles (primarily the masseter)
	TMJ arthralgia

benefit from referral to the secondary care sector, to clinicians who specialize in the management of this condition. Patients themselves should become well informed, as compliance and satisfaction are shown to improve in those patients who have a better understanding of their condition and are given an opportunity to express their wishes in terms of treatment. The expert patient is also able to give truly informed consent to treatment. Systematic reviews of the use of antiepileptic drugs (AEDs) and non-AED medication surgical interventions (Zakrzewska and Akram 2011) have been published as Cochrane reviews (Wiffen et al. 2011a, b; Yang et al. 2011), and there are regularly updated reviews in Clinical Evidence (Zakrzewska and Linskey 2009) and international guidelines on TN management (Cruccu et al. 2008; Gronseth et al. 2008). With that said, most patients will initially be managed medically, and then surgery will be offered as a secondary procedure. There remains considerable debate as to the best timing of surgery.

Pharmacological Approaches

The mainstay of treatment for this condition is the use of AEDs, and carbamazepine is the only drug that has been the subject of three randomized, controlled trials, and therefore, remains the gold standard against which other drugs are evaluated. Table 3 gives an indication of the drugs that are currently in use. Although only published in abstract form, oxcarbazepine, which is a drug chemically related to carbamazepine, demonstrated equal clinical efficacy with fewer side effects. Gabapentin, combined with weekly ropivacaine injections into the area of trigger pain, was assessed in a small randomized controlled trials (RCT) in newly diagnosed TN patients, and the combination was found to be more effective than either of these drugs alone (Lemos et al. 2008). In 53 patients receiving pregabalin prospectively followed for 1 year, "good" pain control was achieved, especially in those with classical TN (Obermann et al. 2008).

Surgical Approaches

Table 3 provides data on different surgical procedures from recent recommendations. However, there is insufficient evidence to provide strict guidelines as to the optimum timing of surgery (Cruccu et al. 2008; Gronseth et al. 2008). A recent Cochrane systematic review of surgical interventions for TN has shown a paucity of RCTs (Zakrzewska and Akram 2011), and cohort studies using independently assessed outcome measures are hard to find (Tatli et al. 2008). Also, the quality of data reporting outcomes after surgical treatments is relatively poor (Zakrzewska and Lopez 2003).

Surgical treatment is usually reserved for patients who fail first-line drug therapy due to uncontrollable pain or intolerable side effects. For classical TN, the target for surgery is the peripheral nerve, with either nerve ablation or decompression as the goal. There are only three randomized controlled trials that investigated the effects of ablative surgical treatment options; thus, the advantages and disadvantages of the various procedures are based on case series, many of which are retrospective and lack independent assessment of outcome (Zakrzewska and Akram 2011). Historically, peripheral neurectomy was

Trigeminal Neuralgia: Diagnosis and Treatment, Table 3 Drugs that have been evaluated either in randomized controlled trials or case series for the management of trigeminal neuralgia. Most drugs need to be escalated and withdrawn slowly

Drug	Daily dosage range	Outcome Number needed to treat NNT (95 % CI)	Side effects, Number needed to harm NNH (95 % CI)	Comments
Drugs evaluated in randomized controlled trials				
Baclofen	50–80 mg	NNT 1.4 (1–2.6) only 10 patients, possibly effective	Ataxia, lethargy, fatigue, nausea	Useful as add on therapy
Carbamazepine	300–1,200 mg	NNT 2.6 (2–4), effective	Ataxia, dizziness, diplopia, lethargy NNH 3.4 (2.5–5.2) for side effects, NNH for withdrawal 24 (13–110)	Reduced white cell count, hyponatremia with higher doses
Gabapentin with Ropivacaine injections	1,200–3,600 mg	May be effective	Ataxia, dizziness, drowsiness, nausea, headache, edema	Small study in new patients
Lamotrigine	200–400 mg	NNT 2.1 (1.3–6.1) as add on medication	Dizziness, drowsiness, constipation, ataxia, diplopia, irritability	Rapid dose escalation increases incidence of rashes
Oxcarbazepine	300–1,200 mg	Effective	Ataxia, dizziness, diplopia, lethargy, which may be related to hyponatremia and is dose-dependent	Better tolerated than with carbamazepine
Pimozide	4–12 mg	NNT 2 (2–3)	Extrapyramidal, for example, tremor, rigidity NNH 2.9 (2–4)	Side effects too severe to recommend routine use
Proparacaine	2 drops of 0.5 % solution	Not effective	Toxic keratopathy with long-term use	Short lasting even if given repeatedly
Tizanidine	6–18 mg	May be effective	None reported	Effect is short lasting
Tocainide	60 mg/kg	Not effective	Nausea, paresthesia, rash	Risk of aplastic anemia precludes its routine use
Drugs evaluated in case reports only				
Capsaicin topical on skin	3 g for 21–28 days	Little benefit	Burning sensation	Temporary relief, avoid contact with eye
Clonazepam	2–8 mg	May be effective	Lethargy in 60 %, fatigue, dizziness, personality change	Thrombocytopenia can occur
Phenytoin	200–300 mg	Effective	Ataxia, lethargy, nausea, headache, behavioral changes, folate deficiency with prolonged use, gingival hypertrophy	Small margin for dose escalation
Pregabalin	300–600 mg	Effective	Ataxia, dizziness, drowsiness, nausea, headache, edema	Prospective 1 year cohort study
Valproic acid	600–2,000 mg	May be effective	Irritability, restlessness, tremor, confusion, nausea, rash, weight gain	

the preferred procedure for trigeminal neuralgia, using either mechanical or chemical means. This approach has largely been replaced by other surgical options that result in reduced sensory deficits and higher success. Dental treatments, which result in peripheral nerve ablation, can result in a short-term remission, but pain usually reoccurs in 3–6 weeks.

Gasserian or trigeminal ganglion rhizotomy is a neuroablative procedure performed by inserting a needle within the trigeminal cistern. Several neurolytic methods have been developed and are routinely employed, each of them having relative advantages and disadvantages. They all provide immediate pain relief in more than 90 % of patients. Radiofrequency thermocoagulation can be directed to a specific division of the trigeminal nerve, and therefore has the advantage of providing a limited and well-controlled lesion. Correct positioning of the needle tip can be verified by electrical stimulation of the awakened patient prior to lesioning. Disadvantages of this technique include a potentially higher risk for anesthesia dolorosa and corneal anesthesia. In order to reduce these complications, pulsed radiofrequency has been used in an RCT, but showed to be ineffective for pain control (Erdine et al. 2007). Also, other techniques may be superior and more practical when large areas of the trigeminal nerve are involved, since multiple lesions would be required. The injection of glycerol as a neurolytic agent is another possible approach. Glycerol is thought to affect unmyelinated nerve fibers selectively, thereby sparing proprioception, touch and motor functions of the trigeminal nerve. Altering the patient's head position and adjusting the volume of glycerol injected achieved anatomic variations in nerve ablation. The major disadvantages are the relatively imprecise technique, with the potential of glycerol spreading into the brain, as well as the moderate risk for anesthesia dolorosa and corneal anesthesia. Balloon compression is the newest technique, and has the lowest reported complications of anesthesia dolorosa and corneal anesthesia. This procedure has the advantage of not requiring patient participation, but it has the highest incidence of muscle weakness and bradycardia, requiring pacing during balloon inflation. Overall, these three techniques provide comparable pain relief and differ in their potential advantages and disadvantages.

Stereotactic radiosurgery, or gamma knife, is a procedure performed using radiation to cause nerve injury. The target of treatment is along the trigeminal nerve about 2–4 mm after it exits the brain stem at the level of the pons. Advantages of this technique are that pain relief percentages are comparable to rhizotomy, which has a low risk for anesthesia dolorosa and corneal anesthesia. An RCT comparing the use of one or two isocenters was stopped early because of development of severe dysesthesias following use of two isocenters (Flickinger et al. 2001). Disadvantages include a potential for delayed onset of pain relief and high rates of pain recurrence (Cruccu et al. 2008; Gronseth et al. 2008).

Microvascular decompression, also known by the acronym MVD, is an open surgical procedure of the brain. The goal is the alleviation of existing vascular pressure on the trigeminal nerve root. It does offer the longest pain-free period, and may be potentially curative, but this procedure has initial success rates in pain elimination comparable to those of rhizotomy, but less pain recurrence, and a much lower chance of anesthesia dolorosa, corneal anesthesia, and muscle weakness. A potential disadvantage is the mortality rate which is less than 0.2–0.5 %. Other possible complications include stroke, meningitis, cerebrospinal fluid leak, and various ipsilateral cranial nerve deficits, such as permanent hearing loss in 2 % (Gronseth et al. 2008). As surgical microvascular decompression techniques become less invasive, such as with endoscopic assistance, it is likely that it will become more commonly practiced, and with favorable research published, it may even become a first-line treatment option. However, there are no RCTs or even prospective cohort studies published at present (Table 4).

Other Treatment Approaches
Patients with TN have traditionally been managed using a biomedical model with little emphasis on psychological support. However, narrative accounts (Zakrzewska 2006) obtained from patients show that fear; loneliness; and depression, including suicidal ideation, anxiety, anger, and cognitive impairment, are all encountered and need to be managed. Recurrences of pain, especially after major surgical procedures, will often result in loss of confidence and the return of fear and despair. Support from pain psychologists

Trigeminal Neuralgia: Diagnosis and Treatment, Table 4 Surgical management of trigeminal neuralgia

Procedure	%Probability of being pain-free	Mortality	Morbidity
Peripheral neurectomy, cryotherapy, alcohol, injection, acupuncture	2 years – 22	Nil	Low, sensory loss, transient hematoma, edema
Radiofrequency thermorhizotomy (RFT)*	2 years – 68 5 years – 48	Low	28 % complications mainly relating to trigeminal nerve, dysesthesia, anesthesia dolorosa, eye problems, masseteric problems
Percutaneous glycerol rhizotomy*	2 years – 63 5 years – 45	Low	25 % complications as for RFT
Balloon microcompression*	2 years – 79	Low	10 % complications as for RFT
Microvascular decompression (MVD)	2 years – 81 5 years – 76 10 years – 71	0.5 %	Overall 75 % no complications, 14 % peri-operative complications, 5 % transient cranial nerve fourth, sixth, eighth dysfunction, 2 % permanent deafness
Gamma knife surgery	5 years – 58	Nil	Late onset of relief, may only be partial, 8 % sensory loss up to 2 years post treatment

*Sensory loss common in all Gasserian ganglion procedures

or psychiatrists may be important to improve mood, physical functioning, and quality of life.

Patient Information and Support

Due to the varied presentation of trigeminal neuralgia and the reported relapse of positive treatment effects, long-term follow-up and management is essential in providing the best possible patient care. There is now a considerable amount of literature on trigeminal neuralgia for patients in the English language. The information is in the form of printed articles as well as books (Weigel and Casey 2000), and there is a considerable body of data on the Internet produced by patient support groups (www.tna-uk.org.uk; www.tna-support.org). These national trigeminal neuralgia associations also provide access to other sufferers, organize conferences with participation of healthcare practitioners, and have been shown to result in numerous benefits (Zakrzewska et al. 2009).

References

Anonymous. (2004). The international classification of headache disorders: 2nd edition. *Cephalalgia, 9* (Suppl. 1).

Cruccu, G., Gronseth, G., Alksne, J., Argoff, C., Brainin, M., Burchiel, K., Nurmikko, T., & Zakrzewska, J. M. (2008). AAN-EFNS guidelines on trigeminal neuralgia management. *European Journal of Neurology, 15*, 1013–1028.

Devor, M., Amir, R., & Rappaport, Z. H. (2002). Pathophysiology of trigeminal neuralgia: The ignition hypothesis. *The Clinical Journal of Pain, 18*, 4–13.

Dieleman, J. P., Kerklaan, J., Huygen, F. J., Bouma, P. A., & Sturkenboom, M. C. (2008). Incidence rates and treatment of neuropathic pain conditions in the general population. *Pain, 137*, 681–688.

Erdine, S., Ozyalcin, N. S., Cimen, A., Celik, M., Talu, G. K., & Disci, R. (2007). Comparison of pulsed radiofrequency with conventional radiofrequency in the treatment of idiopathic trigeminal neuralgia. *European Journal of Pain, 11*, 309–313.

Flickinger, J. C., Pollock, B. E., Kondziolka, D., Phuong, L. K., Foote, R. L., Stafford, S. L., & Lunsford, L. D. (2001). Does increased nerve length within the treatment volume improve trigeminal neuralgia radiosurgery? A prospective double-blind, randomized study. *International Journal of Radiation Oncology, Biology, and Physics, 51*, 449–454.

Goadsby, P. J., Cohen, A. S., & Matharu, M. S. (2007). Trigeminal autonomic cephalalgias: Diagnosis and treatment. *Current Neurology and Neuroscience Reports, 7*, 117–125.

Gronseth, G., Cruccu, G., Alksne, J., Argoff, C., Brainin, M., Burchiel, K., Nurmikko, T., & Zakrzewska, J. M. (2008). Practice parameter: The diagnostic evaluation and treatment of trigeminal neuralgia (an evidence-based review): Report of the Quality Standards Subcommittee of the American Academy of Neurology and the European Federation of Neurological Societies. *Neurology, 71*, 1183–1190.

Hall, G. C., Carroll, D., & McQuay, H. J. (2008). Primary care incidence and treatment of four neuropathic pain conditions: A descriptive study, 2002-2005. *BMC Family Practice, 9*, 26.

Koopman, J. S., Dieleman, J. P., Huygen, F. J., De Mos, M., Martin, C. G., & Sturkenboom, M. C. (2009). Incidence of facial pain in the general population. *Pain, 147*, 122–127.

Lemos, L., Flores, S., Oliveira, P., & Almeida, A. (2008). Gabapentin supplemented with ropivacain block of trigger points improves pain control and quality of life in trigeminal neuralgia patients when compared with gabapentin alone. *The Clinical Journal of Pain, 24*, 64–75.

Leonard, G., Goffaux, P., Mathieu, D., Blanchard, J., Kenny, B., & Marchand, S. (2009). Evidence of descending inhibition deficits in atypical but not classical trigeminal neuralgia. *Pain, 147*, 217–223.

Merskey, H., & Bogduk, N. (1994). *Classification of chronic pain. Descriptors of chronic pain syndromes and definitions of pain terms.* Seattle: IASP Press.

Moisset, X., Villain, N., Ducreux, D., Serrie, A., Cunin, G., Valade, D., Calvino, B., & Bouhassira, D. (2011). Functional brain imaging of trigeminal neuralgia. *European Journal of Pain, 15*, 124–131.

Obermann, M., Yoon, M. S., Ese, D., Maschke, M., Kaube, H., Diener, H. C., & Katsarava, Z. (2007). Impaired trigeminal nociceptive processing in patients with trigeminal neuralgia. *Neurology, 69*, 835–841.

Obermann, M., Yoon, M. S., Sensen, K., Maschke, M., Diener, H. C., & Katsarava, Z. (2008). Efficacy of pregabalin in the treatment of trigeminal neuralgia. *Cephalalgia, 28*, 174–181.

Pan, S. L., Yen, M. F., Chiu, Y. H., Chen, L. S., & Chen, H. H. (2011a). Increased risk of trigeminal neuralgia after hypertension: A population-based study. *Neurology, 77*(17), 1605–1610.

Pan, S. L., Chen, L. S., Yen, M. F., Chiu, Y. H., & Chen, H. H. (2011b). Increased risk of stroke after trigeminal neuralgia–a population-based follow-up study. *Cephalalgia, 8*, 937–942.

Tatli, M., Satici, O., Kanpolat, Y., & Sindou, M. (2008). Various surgical modalities for trigeminal neuralgia: Literature study of respective long-term outcomes. *Acta Neurochirurgica, 150*, 243–255.

Weigel, G., & Casey, K. F. (2000). *Striking back. The trigeminal neuralgia handbook.* Barnegat Light: The Trigeminal Neuralgia Association.

Wiffen, P. J., Derry, S., & Moore, R. A. (2011a). Lamotrigine for acute and chronic pain. *Cochrane Database of Systematic Reviews*, (2), CD006044. DOI: 10.1002/14651858.CD006044.pub3

Wiffen, P. J., Derry, S., Moore, R. A., & McQuay, H. J. (2011b). Carbamazepine for acute and chronic pain in adults. *Cochrane Database of Systematic Reviews*, (1), CD005451.

Yang, M., Zhou, M., He, L., Chen, N., & Zakrzewska, J. M. (2011). Non-antiepileptic drugs for trigeminal neuralgia. *Cochrane Database of Systematic Reviews*, (1), CD004029.

Zakrzewska, J. M. (2006). *Insights: Facts and stories behind trigeminal neuralgia.* Gainesville: Trigeminal neuralgia association.

Zakrzewska, J. M., & Akram, H. (2011). Neurosurgical interventions in classical trigeminal neuralgia. *Cochrane Database of Systematic Reviews*, (9), CD007312. DOI: 10.1002/14651858.CD007312.pub2.

Zakrzewska, J. M., & Hamlyn, P. J. (1999). Facial pain. In I. K. Crombie, S. J. Linton, L. LeResche, & M. Von Korff (Eds.), *Epidemiology of pain* (pp. 171–202). Seattle: IASP.

Zakrzewska, J. M., & Linskey, M. E. (2009). Trigeminal neuralgia. *Clin Evid* (Online) 1207, 2009 PMCID: PMC2907816.

Zakrzewska, J. M., & Lopez, B. C. (2003). Quality of reporting in evaluations of surgical treatment of trigeminal neuralgia: Recommendations for future reports. *Neurosurgery, 53*, 110–120.

Zakrzewska, J. M., Jorns, T. P., & Spatz, A. (2009). Patient led conferences–who attends, are their expectations met and do they vary in three different countries? *European Journal of Pain, 13*, 486–491.

Trigeminal Neuralgia: *Tic Douloureux*, Fothergill's Disease, Epileptiform Neuralgia

▶ Trigeminal, Glossopharyngeal, and Geniculate Neuralgias

Trigeminal Nucleus Caudalis

▶ Subnucleus Caudalis

Trigeminal Sensory Nucleus

▶ Trigeminal Brainstem Nuclear Complex, Physiology

Trigeminal Subnucleus Caudalis

Definition

Trigeminal subnucleus caudalis is the most caudal component of the trigeminal brainstem sensory nuclear complex, continuous with the dorsal
horn of the spinal cord (See ▶ Subnucleus
Caudalis, ▶ Trigeminal Brainstem Nuclear Complex, Anatomy, ▶ Trigeminal Brainstem Nuclear
Complex, Physiology).

Cross-References

▶ Dental Pain, Etiology, Pathogenesis, and
  Management
▶ Subnucleus Caudalis

Trigeminal Subnucleus Interpolaris, Vi, Trigeminal Nucleus Interpolaris

▶ Interpolaris

Trigeminal Tractotomy

Definition

A surgical procedure to sever the trigeminal
spinal tract. The lesion also involves some lateral part of ▶ subnucleus caudalis. When
performed in humans, the purpose is to deprive
subnucleus caudalis of the inputs from the
primary afferents, and therefore, to provide
relief to patients from the excruciating pain of
▶ trigeminal neuralgia. Not in use much anymore, following the introduction of medical
management and other ablative procedures for
this condition.

Cross-References

▶ Trigeminal Brainstem Nuclear Complex,
  Physiology

Trigeminal Transition Zone

Synonyms

Vi/Vc transition zone

Definition

A region of the Trigeminal Brain stem Nuclear
Complex (VBNC) at the obex level where
the subnuclei interpolaris (Vi) and caudalis (Vc)
of the spinal trigeminal nucleus converge.
The ventral portion of the laminated Vc is pushed
dorsomedially by Vi (see Yoshida et al. 1991;
Ren and Dubner 2011). The substantia gelatinosa
is still identifiable but often interrupted. Neurons
in the trigeminal transition zone project to the
thalamus and are involved in processing of nociceptive information from the orofacial regions
and have some unique physiological properties
(see Ren and Dubner 2011).

Cross-References

▶ Trigeminothalamic Tract Projections
▶ Vi/Vc Transition Zone

References

Ren, K., & Dubner, R. (2011). The role of trigeminal
  interpolaris-caudalis transition zone in persistent
  orofacial pain. *International Review of Neurobiology,*
  97, 207–225.
Yoshida, A., Dostrovsky, J. O., Sessle, B. J., & Chiang, C. Y.
  (1991). Trigeminal projections to the nucleus submedius
  of the thalamus in the rat. *The Journal of Comparative*
  Neurology, 307, 609–625.

Trigeminal, Glossopharyngeal, and Geniculate Neuralgias

Matt Bender[1], George A. Crabill[1],
Benjamin S. Carson[1] and Michael Lim[2]
[1]Department of Neurosurgery, Johns Hopkins
Hospital, Baltimore, MA, USA
[2]Department of Neurosurgery and Department of
Neurology, Johns Hopkins Hospital, Baltimore,
MA, USA

Synonyms

Geniculate neuralgia: *nervus intermedius*,
primary otalgia; Glossopharyngeal neuralgia:
vagoglossopharyngeal neuralgia; Trigeminal
neuralgia: *tic douloureux*, Fothergill's disease,
epileptiform neuralgia

Definition

Cranial neuralgias are idiopathic chronic pain
conditions characterized by sudden, often
short-lived, episodes of pain that may result
from non-noxious cutaneous stimulation and
that involve the distributions of cranial nerves.

Characteristics

Demographics

Cranial neuralgias are syndromes of usually
paroxysmal pain in the distribution of individual
cranial nerves unilaterally. They include trigeminal
neuralgia, glossopharyngeal neuralgia, and geniculate neuralgia. Among these, trigeminal neuralgia
(TN) is easily the most common cranial neuralgia.
Demographic studies in Minnesota report a TN
incidence of approximately 4/100,000 population,
with higher rates in women than men (Katusic et al.
1990). The incidence of glossopharyngeal
neuralgia (Henson et al. 2005) is approximately
1 % of that of TN, and geniculate neuralgia
(GenN) is even more rare (Teixeira et al. 2008).

Cranial neuralgias increase in prevalence
with age. The median age of diagnosis in the
Minnesota series was 67 (Katusic et al. 1990),
although the range is wide and younger patients
are common, especially those with multiple
sclerosis (MS) affected by trigeminal neuralgia.
There are case series describing familial
occurrences of TN. No specific genetic defects
have been identified, and there have not been
linkage studies performed on these families. An
aggregation of the case reports shows that family
members tend to have a younger age of onset, to
have an increased prevalence of bilateral TN,
and to be afflicted on the same side of the face.
There is an autosomal dominant mode of transmission with reduced penetrance (Fleetwood
et al. 2001).

Clinical Features

Classical trigeminal neuralgia is characterized
by paroxysms of electric shock-like pain within
one or more divisions of the trigeminal nerve.
The pain can occur spontaneously or be elicited
by non-noxious stimuli such as touch, talking,
eating, or wind. The pain is acute in onset and
termination and may show periods of remission.
Pain is usually unilateral, but bilateral pain is
present ~5 % of patients. Trigeminal neuralgia
is said to be atypical when the pain is constant or
when there is a component of numbness. In
either case, neurologic deficits generally are
absent.

Most cases of TN are caused by compression
of the nerve, usually by vascular structures.
A recent operative series of microvascular
decompression procedures showed that 90 % of
first operations revealed compression of the nerve
root by either an artery or vein, while in the rest of
the cases compression of the root was caused by
arachnoid thickening (Ishikawa et al. 2002).
Approximately 70 % of this series had arterial
compression, with the most common compressive vessel being the superior cerebellar artery
(SCA). Compression results in axonal demyelination and the destruction of glial fibers that normally separate axons, resulting in close anatomic
apposition of these axons. This arrangement

facilitates ephaptic transmission, the spontaneous transmission of impulses between closely apposed nerve fibers, which in the case of cranial neuralgias originate from non-noxious stimuli (Love and Coakham 2001).

Among other causes of TN, multiple sclerosis is the most common, with TN-MS patients representing 2–8 % of patients with TN (Chakravorty 1966). TN is found in 2 % of MS patients and usually occurs more than 10 years after MS is diagnosed but can be the first MS symptom in up to 15 % of patients (Hooge and Redekop 1995). Unlike TN, there is no increased prevalence of glossopharyngeal or geniculate neuralgia in multiple sclerosis (Rushton et al. 1981).

Glossopharyngeal neuralgia is paroxysmal pain felt in the root of the tongue and pharynx that radiates to the throat and deep ear structures. GN is often brought on by eating, swallowing, yawning, or sneezing. Pain is bilateral in as many as 12 % of patients (Rushton et al. 1981). Patients can also experience vagally mediated symptoms, including syncope and hemodynamic instability, due to the concomitant involvement of cranial nerve X in addition to CN IX. Relief with the application of a local anesthetic to the oropharynx can be both diagnostic and therapeutic. The differential diagnosis of GN includes oropharyngeal malignancy and peritonsillar infection, which can both be identified on MRI. MRI may also reveal a vascular loop compressing cranial nerve IX, which, like TN, is the presumed cause in most cases of GN (Rozen 2004).

Geniculate neuralgia involves intractable deep ear pain and prosopalgia, or pain in the deep structures of the face, including the posterior orbital, nasal, and palatal regions. Like TN and GN, the pain is often triggered by innocuous stimuli such as cold, noise, swallowing, and gentle pressure on the tragus. Unlike TN and GN, however, the pain of GenN is often constant in addition to being paroxysmal (Lovely and Jannetta 1997). An extensive differential diagnosis must be ruled out before the diagnosis of geniculate neuralgia can be made. This includes acute otitis media, otitis media externa, temporomandibular joint disease, carcinoma of the nasopharynx, compacted upper molars, vascular lesions, referred pain from the larynx or thyroid, and tumors of the cerebellopontine angle.

Identifying the nerve responsible for otalgia can be very challenging diagnostically. The sensory innervation of the ear is complex, with contributions from the fifth, seventh, ninth, and tenth cranial nerves. Among these, sensory fibers of the facial nerve, which are carried along with preganglionic parasympathetic fibers in its nervus intermedius branch, have been most commonly implicated. Historical information on pain character, localization, and triggers is of little predictive value. Local nerve blocks are imprecise because of sensory overlap. Nerve stimulation under regional anesthesia at the time of surgery can be similarly equivocal and is a harrowing experience for patients (Rupa et al. 1991). Because of the complex innervation of the external ear, surgery for geniculate neuralgia typically requires an exploration of multiple cranial nerves and may require treatment at multiple sites.

Pharmacological Treatment

Patients with TN are initially medically managed, and the first-line drug therapy is carbamazepine, with typical maintenance doses of 1,500–2,000 mg/day (Sindrup and Jensen 2002). Carbamazepine was shown to be effective at reducing pain severity and frequency in approximately 75 % of patients. The most common side effects include sedation, rash, and hyponatremia, with agranulocytosis occurring very rarely. Many clinicians have switched patients to oxcarbamazepine, a keto-derivative of carbamazepine with a less burdensome side-effect profile and fewer associated hematologic problems.

Phenytoin is a commonly used second-line agent, but few controlled trials have been performed with this drug; typical maintenance doses are 300–400 mg/day (Sindrup and Jensen 2002). Other agents that are commonly attempted include baclofen, lamotrigine either alone or adjuvantly, gabapentin, and Lyrica.

The relative roles of pharmacotherapy and surgery for TN are in a state of flux, although certainly medically refractory patients tend to progress to surgery more quickly. Medical management is frequently unsatisfactory for patients with TN-MS. First-line agents like carbamazepine are poorly tolerated in MS patients because side effects may enhance MS symptomatology, including weakness, dizziness, and ataxia (Hooge and Redekop 1995) or the liver toxicity associated with some immunomodulatory medications used in MS.

Glossopharyngeal neuralgia is treated with a similar pharmacologic armamentarium to TN. Carbamazepine, phenytoin, baclofen, and gabapentin are the leading agents. Patients with GN typically become medically refractory earlier in their disease history than patients with TN (Sampson et al. 2004). Geniculate neuralgia is more resistant and typically does not respond as well to medications.

Surgical Treatment
The spectrum of non-pharmacologic treatments for TN includes microvascular decompression (MVD), percutaneous ablative procedures, and radiotherapeutic options. The literature on the outcomes of TN surgery is predominantly retrospective with few comparisons of the different procedures (Lopez et al. 2004).

Microvascular decompression is a definitive, etiologic treatment of TN that is especially indicated in young patients and those without medical comorbidities with acceptable surgical and anesthetic risks. In MVD, a suboccipital craniectomy is employed to access the trigeminal nerve near its point of entry into the brain stem, where it is examined under the microscope for compressing vessels. Compressive arteries may be padded away with muscle or Teflon; compressive veins may be coagulated and divided. MVD offers the most durable outcomes of any treatment for TN. In a study of 1,185 patients, immediate complete relief was present in 82 %, and 70 % still had complete relief 10 years after surgery (Barker et al. 1996). For patients with TN secondary to MS, MVD is not as effective. In the largest study of MVD for TN-MS, mean recurrence time was 13.5 months in 35 TN-MS patients, and 30 % experienced sustained relief beyond 7 years.

Percutaneous ablative procedures are a less invasive treatment option than MVD for patients with TN-MS as well as elderly patients with significant comorbidities or those with high surgical or anesthetic risk, patients who refuse open surgery, and those who have failed MVD. Percutaneous ablative procedures include glycerol rhizotomy (GR), radiofrequency thermocoagulation, and balloon compression (BC), all of which work by causing controlled injury of the trigeminal nerve root. Each of these procedures is performed through a needle inserted in the cheek and directed toward the trigeminal cistern.

Glycerol injected into the trigeminal cistern selectively ablates myelinated axons and can provide significant relief of TN pain with minimal sensory deficits (Hakanson 1981). Glycerol rhizotomy achieves significant initial pain relief in 70–90 % of patients with results that last 2–3 years on average (Pollock 2005). Because of this finite duration of effect, patients often require multiple interventions over time. Glycerol rhizotomy can be repeated without a decrement in pain relief or durability or an increased complication rate (Bender et al. 2012). In contrast with glycerol rhizotomy, radiofrequency ablation, which relies on thermal destruction of the nerve root, is less technique-based, and results are less user-dependent. Initial pain relief is invariably greater than 90 %, although durability is comparable to GR, and rates of sensory loss are higher (Broggi et al. 1990; Kanpolat et al. 2001). Percutaneous balloon compression is less frequently practiced but has shown comparable results to RF and GR (Fraioli et al. 1989). Side effects associated with percutaneous ablative procedures include herpes simplex virus reactivation, dysesthesias, and corneal numbness.

Stereotactic radiosurgery is the least invasive treatment option for TN but takes longer to deliver pain relief and as a result is reserved for patients with the most severe comorbidities, bleeding diathesis, those who have failed multiple other modalities, and those who request it. In a study of 220 patients undergoing SRS as a treatment for TN, more than 60 % of whom had undergone at least one prior TN treatment with a different surgical modality, rates of complete pain relief were 65 %, 70 %, and 75 % at 6, 12, and 33 months, respectively (Maesawa et al. 2001). SRS delivers partial but incomplete pain relief in many cases. Another study of 117 patients undergoing SRS showed 55 % complete pain relief with or without medications 3 years after surgery (Pollock et al. 2002). Although it is a newer treatment with fewer studies in the literature, the overall durability of SRS appears comparable to percutaneous ablative procedures in the treatment of TN.

There is a more limited surgical armamentarium for GN, with the two main options being nerve sectioning and neurovascular decompression. The largest series of patients undergoing nerve sectioning was reported by Rushton et al. from the Mayo Clinic. Between 1922 and 1977, they described 129 patient who underwent intracranial sectioning of the glossopharyngeal nerve with or without sectioning of the upper two to three rootlets of the vagus nerve through a suboccipital craniectomy. Pain relief was "good" in 85 % of patients, but severe postoperative swallowing difficulties were present in 19 %. After the success of MVD for TN, evidence of neurovascular compression in GN was sought and found. MVD has since become the mainstay procedure for GN. The vessels most commonly found to be compressing the ninth and tenth cranial nerves are the posterior inferior cerebellar artery and the vertebral artery. In a series of 217 patients undergoing MVD, initial complete pain relief was achieved in greater than 90 % of patients, with long-term complete relief in 58 % and partial relief in 18 % of patients over 4 years (Patel et al. 2002). In contrast with nerve sectioning, cranial nerve palsies are observed in less than 3 % of patients after undergoing MVD.

Because of the complex innervation of the external ear, surgery for geniculate neuralgia typically requires the exploration of multiple cranial nerves and may require treatment at multiple sites. The mainstay procedure is sectioning of the nervus intermedius with neurovascular decompression or sectioning of other nerves used adjunctively. The nervus intermedius is approached through a retromastoid craniectomy. This allows the surgeon to visualize the trigeminal root entry zone and the CN VII and VIII complex and inspect for compressive vascular loops. Microsurgical dissection between the facial and vestibulocochlear nerves allows identification of the nervus intermedius, which can then be sectioned. Because of the rarity of the diagnosis, outcome data for GenN is limited. In a series of 18 cases, 14 of which were treated with nervus intermedius sectioning, often in combination with other procedures, 72 % of patients achieved pain relief over a mean follow-up period of 3 years (Rupa et al. 1991). Other series have shown similar initial pain relief achieved by neurovascular decompression in combination with sectioning of the nervus intermedius, with 71 % of patients experiencing complete relief initially, but poor durability, with just 30 % continuing to have complete relief 1 year after the procedure. It is the prosopalgic component, the deep orbital and nasal pain, that tends to recur rather than the otalgia, which is more durably relieved (Lovely and Jannetta 1997).

Summary

Cranial neuralgias are syndromes of paroxysmal pain in the distribution of individual cranial nerves unilaterally. Among these, the most common by far is trigeminal neuralgia, followed by glossopharyngeal and geniculate neuralgia. Patients with TN are initially treated medically with carbamazepine or other antiepileptics. Surgical treatments include microvascular decompression, which attempts to pad a compressing vessel away from the trigeminal nerve, and ablative therapies such as glycerol or radiofrequency rhizotomy. Glossopharyngeal and geniculate neuralgia are less amenable to medical management and are most often treated with neurovascular decompression or nerve sectioning.

References

Barker, F. G., 2nd, Jannetta, P. J., Bissonette, D. J., Larkins, M. V., & Jho, H. D. (1996). The long-term outcome of microvascular decompression for trigeminal neuralgia. *The New England Journal of Medicine, 334*(17), 1077–1083. doi:10.1056/NEJM199604253341701.

Bender, M. T., Pradilla, G., Batra, S., See, A. P., James, C., Pardo, C. A., & Lim, M. (2012). Glycerol rhizotomy and radiofrequency thermocoagulation for trigeminal neuralgia in multiple sclerosis. *Journal of Neurosurgery.* doi:10.3171/2012.9.jns1226.

Broggi, G., Franzini, A., Lasio, G., Giorgi, C., & Servello, D. (1990). Long-term results of percutaneous retrogasserian thermorhizotomy for "essential" trigeminal neuralgia: Considerations in 1000 consecutive patients. *Neurosurgery, 26*(5), 783–786; discussion 786–787.

Chakravorty, B. G. (1966). Association of trigeminal neuralgia with multiple sclerosis. *Archives of Neurology, 14*(1), 95–99.

Fleetwood, I. G., Innes, A. M., Hansen, S. R., & Steinberg, G. K. (2001). Familial trigeminal neuralgia. Case report and review of the literature. *Journal of Neurosurgery, 95*(3), 513–517. doi:10.3171/jns.2001.95.3.0513.

Fraioli, B., Esposito, V., Guidetti, B., Cruccu, G., & Manfredi, M. (1989). Treatment of trigeminal neuralgia by thermocoagulation, glycerolization, and percutaneous compression of the gasserian ganglion and/or retrogasserian rootlets: Long-term results and therapeutic protocol. *Neurosurgery, 24*(2), 239–245.

Hakanson, S. (1981). Trigeminal neuralgia treated by the injection of glycerol into the trigeminal cistern. *Neurosurgery, 9*(6), 638–646.

Henson, C. F., Goldman, H. W., Rosenwasser, R. H., Downes, M. B., Bednarz, G., Pequignot, E. C., & Andrews, D. W. (2005). Glycerol rhizotomy versus gamma knife radiosurgery for the treatment of trigeminal neuralgia: An analysis of patients treated at one institution. *International Journal of Radiation Oncology, Biology, and Physics, 63*(1), 82–90. doi:S0360-3016(05)00150-1 [pii].

Hooge, J. P., & Redekop, W. K. (1995). Trigeminal neuralgia in multiple sclerosis. *Neurology, 45*(7), 1294–1296.

Ishikawa, M., Nishi, S., Aoki, T., Takase, T., Wada, E., Ohwaki, H., & Fukuda, H. (2002). Operative findings in cases of trigeminal neuralgia without vascular compression: Proposal of a different mechanism. *Journal of Clinical Neuroscience, 9*(2), 200–204. doi:10.1054/jocn.2001.0922.

Kanpolat, Y., Savas, A., Bekar, A., & Berk, C. (2001). Percutaneous controlled radiofrequency trigeminal rhizotomy for the treatment of idiopathic trigeminal neuralgia: 25-year experience with 1,600 patients. *Neurosurgery, 48*(3), 524–532; discussion 532–524.

Katusic, S., Beard, C. M., Bergstralh, E., & Kurland, L. T. (1990). Incidence and clinical features of trigeminal neuralgia, Rochester, Minnesota, 1945–1984. *Annals of Neurology, 27*(1), 89–95. doi:10.1002/ana.410270114.

Lopez, B. C., Hamlyn, P. J., & Zakrzewska, J. M. (2004). Systematic review of ablative neurosurgical techniques for the treatment of trigeminal neuralgia. *Neurosurgery, 54*(4), 973–982; discussion 982–973.

Love, S., & Coakham, H. B. (2001). Trigeminal neuralgia: Pathology and pathogenesis. *Brain, 124*(Pt 12), 2347–2360.

Lovely, T. J., & Jannetta, P. J. (1997). Surgical management of geniculate neuralgia. *The American Journal of Otology, 18*(4), 512–517.

Maesawa, S., Salame, C., Flickinger, J. C., Pirris, S., Kondziolka, D., & Lunsford, L. D. (2001). Clinical outcomes after stereotactic radiosurgery for idiopathic trigeminal neuralgia. *Journal of Neurosurgery, 94*(1), 14–20. doi:10.3171/jns.2001.94.1.0014.

Patel, A., Kassam, A., Horowitz, M., & Chang, Y. F. (2002). Microvascular decompression in the management of glossopharyngeal neuralgia: Analysis of 217 cases. *Neurosurgery, 50*(4), 705–710; discussion 710–701.

Pollock, B. E. (2005). Percutaneous retrogasserian glycerol rhizotomy for patients with idiopathic trigeminal neuralgia: A prospective analysis of factors related to pain relief. *Journal of Neurosurgery, 102*(2), 223–228. doi:10.3171/jns.2005.102.2.0223.

Pollock, B. E., Phuong, L. K., Gorman, D. A., Foote, R. L., & Stafford, S. L. (2002). Stereotactic radiosurgery for idiopathic trigeminal neuralgia. *Journal of Neurosurgery, 97*(2), 347–353. doi:10.3171/jns.2002.97.2.0347.

Rozen, T. D. (2004). Trigeminal neuralgia and glossopharyngeal neuralgia. *Neurologic Clinics, 22*(1), 185–206. doi:10.1016/s0733-8619(03)00094-x.

Rupa, V., Saunders, R. L., & Weider, D. J. (1991). Geniculate neuralgia: The surgical management of primary otalgia. *Journal of Neurosurgery, 75*(4), 505–511. doi:10.3171/jns.1991.75.4.0505.

Rushton, J. G., Stevens, J. C., & Miller, R. H. (1981). Glossopharyngeal (vagoglossopharyngeal) neuralgia: A study of 217 cases. *Archives of Neurology, 38*(4), 201–205.

Sampson, J. H., Grossi, P. M., Asaoka, K., & Fukushima, T. (2004). Microvascular decompression for glossopharyngeal neuralgia: Long-term effectiveness and complication avoidance. *Neurosurgery, 54*(4), 884–889; discussion 889–890.

Sindrup, S. H., & Jensen, T. S. (2002). Pharmacotherapy of trigeminal neuralgia. *The Clinical Journal of Pain, 18*(1), 22–27.

Teixeira, M. J., de Siqueira, S. R., & Bor-Seng-Shu, E. (2008). Glossopharyngeal neuralgia: Neurosurgical treatment and differential diagnosis. *Acta Neurochirurgica (Wien), 150*(5), 471–475. doi:10.1007/s00701-007-1493-6; discussion 475.

Trigeminocervical Complex

Definition

Neurons in the ► trigeminal subnucleus caudalis and dorsal horn of spinal cord segments C_1 and C_2, which act together as a functional relay for pain input from intracranial structures, and overlap with input from the front and back of the head, and from the mouth.

Cross-References

► Migraine, Pathophysiology

Trigeminohypothalamic Tract

Xijing Zhang
Department of Neuroscience, University of
Minnesota, Minneapolis, MN, USA

Definition

The trigeminohypothalamic tract is a bundle of nerve fibers originating from all of the subnuclei of the trigeminal brainstem nuclear complex and the gray matter of upper cervical spinal cord segments (C1-2), and projecting to or through the hypothalamus. The trigeminohypothalamic tract is responsible for conveying sensory information, especially nociceptive information, from facial skin, cornea, oral mucosa, and intracranial dura to the hypothalamus, a brain area that regulates homeostasis and other hormonal responses required for survival of the organism.

Characteristics

Anatomical retrograde labeling and physiological antidromic activation techniques have identified the locations of the trigeminohypothalamic neurons in the trigeminal brainstem nuclear complex and the gray matter of upper cervical spinal cord segments (Malick and Buratein 1998; Malick et al. 2000). After the retrograde tracer Fluoro-Gold injection into the hypothalamus, trigeminohypothalamic neurons are found throughout all of the subnuclei of the trigeminal brainstem nuclear complex (56 %) and upper cervical spinal cord segment (44 %) (Malick and Buratein 1998). Within the trigeminal brainstem nuclear complex, over 75 % of the neurons were distributed caudal to the obex; most of the neurons were located in the nucleus caudalis. Over 90 % of the neurons in the nucleus caudalis were distributed in laminae I, II, and V. The nucleus caudalis is the subnucleus in the trigeminal brainstem nuclear complex that processes nociceptive information arising in oral and facial organs (Sessle 1987; Jacquin et al. 1986). In cervical cord segments 1 and 2, approximately 85 % of the neurons were found in laminae I, II, and V. These areas receive direct input from trigeminal primary afferent fibers and contain neurons that respond to mechanical or thermal stimulation of the cornea, oral mucosa, temporomandibular joint, facial skin, and intracranial dura (Lu et al. 1993; Broton et al. 1988, Burstein et al. 1998). Trigeminohypothalamic neurons were also recorded in laminae I, II and IV, V of the nucleus caudalis, and upper cervical (C1) spinal cord by physiological antidromic activation studies (Malick et al. 2000). Most of the trigeminohypothalamic neurons that were recorded responded maximally or exclusively to noxious mechanical or thermal stimulation to the head and orofacial receptive fields innervated by the trigeminal nerve.

The trajectory and termination of the trigeminohypothalamic tract have been examined using antidromic activation (Malick et al. 2000). The trigeminohypothalamic axons cross the midline and ascend on the contralateral side of the brainstem to the level of the contralateral thalamus. Within the thalamus, trigeminohypothalamic axons shift through the optic tract, internal capsule, and supraoptic decussation and reach the rostral ventral hypothalamus. In the rostral ventral hypothalamus, more than half of the axons cross the midline again at the posterior optic chiasm to the ipsilateral hypothalamus, turn caudally, and descend along the identical path in

Trigeminohypothalamic Tract, Fig. 1 Response property of a high-threshold trigeminohypothalamic neuron. (**a**) Location of low-threshold point for antidromic activation in the hypothalamus. (**b**) Recording site in dorsomedial lamina V. (**c**) Receptive field. (**d**) Peristimulus histogram and record of window discriminator output (below histogram) illustrating the response to mechanical stimuli applied to the receptive field. Numbers in parentheses depict mean response (spikes/sec) to each stimulus. (From Malick et al. 2000). Hyp, hypothalamus; IC, internal capsule; VBC, ventrobasal complex

which they ascended in the contralateral hypothalamus (Malick et al. 2000). The axons of the trigeminohypothalamic neurons and their collateral branches terminate bilaterally in many nuclei within hypothalamus, such as the lateral, perifornical, dorsomedial, suprachiasmatic, and supraoptic hypothalamic nuclei (Malick et al. 2000). Through this complex projection, the axons of the trigeminohypothalamic neurons are capable of carrying nociceptive information bilaterally to many nuclei in the hypothalamus that are involved in the production of various responses to noxious stimuli.

Responses of the trigeminohypothalamic neurons to noxious stimuli have been described (Malick et al. 2000). The overwhelming majority of the trigeminohypothalamic neurons responded strongly to noxious mechanical and thermal stimuli applied to the skin. The receptive fields of the trigeminohypothalamic neurons included the skin of the head and neck, as well as orofacial organs

such as the oral mucosa, tongue, lips, cornea, and intracranial dura. Laminae I-II neurons generally exhibited small to medium receptive fields, while those located in deeper layers had medium to large receptive fields. Over 80 % of the recorded trigeminohypothalamic neurons were nociceptive. Among them, 42 % were wide dynamic range, and 38 % were high-threshold neurons. The remaining 20 % were low-threshold neurons (Malick et al. 2000). Figure 1 shows that a high-threshold trigeminohypothalamic neuron responded to noxious mechanical stimulus applied to its receptive field. The neuron was located in dorsomedial lamina V of the nucleus caudalis (b), and its axon projected to the contralateral hypothalamus (a). The neuron had a small cutaneous receptive field located around the ipsilateral side of the mouth (c) and only responded to noxious mechanical stimuli applied to the skin (d).

The majority of nociceptive neurons responded primarily to noxious intensities of thermal stimulation. In contrast, all low-threshold neurons responded to both heat and cold stimuli at innocuous and noxious intensities, a phenomenon that has not been reported for other populations of spinal or trigeminal low-threshold neurons (Malick et al. 2000). Thus, trigeminohypothalamic neurons can transmit innocuous and primarily noxious mechanical and thermal information from facial skin directly to the hypothalamus.

Many trigeminohypothalamic neurons were also powerfully activated by electrical and mechanical stimuli of the oral mucosa, tongue, lips, cornea, and intracranial dura mater (Malick et al. 2000). Most (85 %) oral-sensitive trigeminohypothalamic neurons encoded the intensity of the noxious mechanical and thermal stimuli. Since pain is the only sensation that can be evoked by stimulating the cornea and intracranial dura and sinuses, regardless of whether the stimulus is electrical, mechanical, or chemical (Lele and Weddell 1959, Ray and Wolff 1940), the cornea- and dura-sensitive neurons were considered nociceptive. Therefore, trigeminohypothalamic neurons can transmit nociceptive information from these specific organs in the head to the hypothalamus.

The hypothalamus plays an important role in regulating body temperature, food and water intake, sleep and circadian rhythms, endocrine adjustments, and a wide range of behavior (Swanson 1987). Through the terminations and collateral branches in the hypothalamus, the trigeminohypothalamic tract is likely to bring noxious and innocuous sensory signals from orofacial skin and organs to the hypothalamus, and is involved in pain-related autonomic responses, endocrine adjustments, or emotion reactions.

References

Broton, J. G., Hu, J. W., & Sessle, B. J. (1988). Effects of temporomandibular joint stimulation on nociceptive and Non-nociceptive neurons of the cat's trigeminal subnucleus caudalis (medullary dorsal horn). *Journal of Neurophysiology, 59*, 1575–1589.

Burstein, R., Yamamura, H., Malick, A., & Strassman, A. M. (1998). Chemical stimulation of the intracranial dura induces enhanced responses to facial stimulation in brainstem trigeminal neurons. *Journal of Neurophysiology, 79*, 964–982.

Jacquin, M. F., Renehan, W. E., Mooney, R. D., & Rhoades, R. W. (1986). Structure-function relationships in rat medullary and cervical dorsal horns. I. Trigeminal primary afferents. *Journal of Neurophysiology, 55*, 1153–1186.

Lele, P. P., & Weddell, G. (1959). Sensory nerves of the cornea and cutaneous sensibility. *Experimental Neurology, 1*, 334–359.

Lu, J., Hathaway, C. B., & Bereiter, D. A. (1993). Adrenalectomy enhances Fos-like immunoreactivity within the spinal trigeminal nucleus induced by noxious thermal stimulation of the cornea. *Neuroscience, 54*, 809–818.

Malick, A., & Buratein, R. (1998). Cells of origin of the trigeminohypothalamic tract in the rat. *The Journal of Comparative Neurology, 400*, 125–144.

Malick, A., Strassman, A. M., & Burstein, R. (2000). Trigeminohypothalamic and reticulohypothalamic tract neurons in the upper cervical spinal cord and caudal medulla of the rat. *Journal of Neurophysiology, 84*, 2078–2112.

Ray, B. S., & Wolff, H. G. (1940). Experimental studies on headache. Pain-sensitive structures of the head and their significance in headache. *Archives of Surgery, 41*, 813–856.

Sessle, B. J. (1987). The neurobiology of facial and dental pain: Present knowledge, future directions. *Journal of Dental Research, 66*, 617–626.

Swanson, L. W. (1987). The hypothalamus. In T. Hökfelt & L. W. Swanson (Eds.), *Handbook of chemical neuroanatomy* (Integrated systems of the CNS, part I, Vol. 5, pp. 1–124). Amsterdam: Elsevier.

Trigeminothalamic Tract Projections

Ke Ren
Department of Neural and Pain Sciences,
University of Maryland Dental School,
Baltimore, MD, USA

Synonyms

Lemniscus trigeminalis; Tractus trigeminothalamicus

Definition

The axon bundles of the secondary sensory neurons in the spinal trigeminal nucleus and principal sensory nucleus of the trigeminal nerve that carry somatosensory information from the head and face and terminate in the ventral posterior part of the opposite thalamus. As the counterpart of the spinothalamic tract, the trigeminothalamic axons from the spinal trigeminal nucleus, mainly the subnucleus caudalis, convey nociceptive input. The trigeminothalamic axons from the principal sensory nucleus, equivalent to the medial lemniscal fibers, transmit discriminative tactile as well as proprioceptive impulses.

Characteristics

The somesthetic information from the orofacial regions is conveyed via the fifth cranial nerve, the trigeminal nerve, to the two major trigeminal sensory nuclei, the spinal trigeminal, and the principal sensory nuclei. Populations of neurons in the trigeminal sensory nuclei then send their axons to the thalamus via trigeminothalamic projections. The pattern and functions of trigeminothalamic projections are specifically related to the individual subnuclei of the spinal trigeminal nucleus and the thalamus (Fukushima and Kerr 1979; Rausell and Jones 1991). For the purpose of completeness, the following discussion will go beyond the projection to the ventrobasal thalamus to also include trigeminal projections to other nuclei of the thalamus that are related to pain processing.

Spinal Trigeminal Nucleus

The spinal trigeminal nucleus is further divided rostrocaudally into the subnuclei oralis, interpolaris, and caudalis (Fig. 1). Cytoarchitecturally, the general classification of the subnuclei for humans is applicable to lower animals such as the rat.

The subnucleus caudalis appears at approximately the level of the obex and continues caudally to merge with the upper cervical spinal dorsal horn. The caudal portion of the subnucleus caudalis is analogous to the spinal dorsal horn both histologically and physiologically. The Rexed classification of the dorsal horn can be

Trigeminothalamic Tract Projections, Fig. 1 Major thalamic projections from the trigeminal sensory nuclei. Sensory inputs from the head and face are relayed in the trigeminal brain stem nuclear complex, including the principal sensory nucleus (*PrV*) and spinal trigeminal nucleus (*SpV*). The spinal trigeminal nucleus is further divided into subnuclei oralis (*Vo*), interpolaris (*Vi*), and caudalis (*Vc*). The most caudal subnucleus interpolaris and rostral caudalis merge to form a transition zone (*Vi/Vc*) at the obex level. Large-diameter myelinated primary afferents mainly synapse in the principal sensory nucleus, and small-diameter A-delta and C-fibers travel caudally to terminate in the subnucleus caudalis. Neurons in the spinal trigeminal nucleus, primarily the subnucleus caudalis, project to contralateral medial ventroposterior nucleus of the thalamus (*VPM*), posterior thalamic nucleus (not shown), as well as bilateral intralaminar nuclei (*ILN*) and nucleus submedius (*Sm*). The trigeminal projection to the submedius is predominantly contralateral. The principal sensory nucleus and the subnucleus caudalis project to the barreloid and non-barreloid regions (*open circles* in VPM) of the VPM, respectively. Note that discriminative and affective components of pain may be relayed by distinct pathways. *Dashed lines* indicate transmission of discriminative tactile information from the head and face

applied to its outer layers. Thus, the subnucleus caudalis is regarded as a laminated structure, sometimes referred to as the medullary dorsal horn. After entering the central nervous system at the middle of the pons, small-diameter primary afferent (mostly nociceptive) fibers in the trigeminal nerve turn caudally, then descend in the spinal trigeminal tract, and finally terminate in the subnucleus caudalis. Similar to the spinal dorsal horn, nociceptive neurons, both nociceptive-specific and wide dynamic range types, can be recorded from the subnucleus caudalis. The thalamic-projecting axons of subnucleus caudalis neurons carry pain and temperature information to the thalamus.

The subnucleus interpolaris appears gradually from a transition region at the level of the obex where the subnuclei caudalis and interpolaris coexist in the same transverse plane for a small distance (the trigeminal Vi/Vc transition zone). The subnucleus interpolaris neurons have a well-defined and extensive projection to the medial ventroposterior thalamic nucleus (VPM, the medial part of the ventrobasal complex) (Phelan and Falls 1991). Nociceptive thalamic-projecting neurons have been identified in subnucleus interpolaris. However, most thalamic-projecting subnucleus interpolaris neurons respond to deflection of mystacial vibrissae and are low-threshold mechanoreceptive (Jacquin et al. 1986; Hayashi et al. 1984). Two types of thalamic ascending fibers have been identified in the subnucleus interpolaris: Fast-conducting thick fibers terminate in the posterior nucleus of the thalamus and slow-conducting thin fibers project to VPM (Veinante et al. 2000).

The subnucleus oralis is the most rostral subnucleus of the spinal trigeminal nucleus. Fukushima and Kerr (1979) initially found that the subnucleus oralis had no projections to the thalamus in the rat. Using fluorescent dye tracers, however, subnucleus oralis neurons are shown to project to the thalamus in rodents (Bruce et al. 1987). The neurons of the subnucleus oralis may project to the contralateral thalamus via a relay in the juxtatrigeminal nucleus and principal sensory nucleus (Zhang and Yang 1999).

Principal Sensory Nucleus of the Trigeminal Nerve: Trigeminal Nucleus Principalis

The principal sensory nucleus (main sensory nucleus) can be traced caudally to the level of the rostral tip of the facial nucleus and contiguous obliquely with subnucleus oralis. The principal sensory nucleus mainly receives large-diameter primary afferents and contributes to discriminative sensations. In the rat, the principal sensory nucleus thalamic-projecting axons arise from small- and medium-sized neurons; and large neurons (20–42 mm) of the principal sensory nucleus do not project to the thalamus (Fukushima and Kerr 1979). The ascending fibers from the ventral and dorsal principal sensory nucleus project to the contralateral and ipsilateral VPM, respectively.

Subnuclei of the Thalamus

The ascending trigeminal projections terminate in several thalamic subnuclei including the VPM, posterior thalamic nucleus, nucleus submedius, and intralaminar nuclei centralis medialis and lateralis and parafascicular nucleus. Calbindin D-28k and parvalbumin belong to calcium-binding protein families and can be used as selective neural markers of central neurons. The principal sensory nucleus and spinal trigeminal nucleus have complementary projection foci in the thalamus (Rausell and Jones 1991; Williams et al. 1994). The principal sensory nucleus input enters the barreloid portion of VPM that contains parvalbumin-immunoreactive cells. The ascending fibers from spinal trigeminal nucleus terminate in non-barreloid VPM that contains another calcium-binding protein, calbindin-D28k (but see Craig 2004). The trigeminal inputs to the posterior thalamic nucleus are mainly from the principal sensory nucleus and subnucleus interpolaris (Waite and Tracey 1995). The parafascicular nucleus of the thalamus receives input from subnuclei caudalis and interpolaris, but not from principal sensory nucleus and subnucleus oralis (Krout et al. 2002). The nucleus submedius receives a major input from the Vi/Vc transition zone as well as from the subnucleus caudalis (Yoshida et al. 1991; Craig and Dostrovsky 2001; Ikeda et al. 2003).

Most trigeminothalamic projections are contralateral, except that the subnucleus caudalis projects to bilateral intralaminar thalamic nuclei (Peschanski 1984) and principal sensory nucleus has an ipsilateral projection to VPM (Fukushima and Kerr 1979) (Fig. 1). Although the major trigeminothalamic projections are the inputs to VPM and posterior thalamic nucleus from the principal sensory nucleus and subnucleus interpolaris (Waite and Tracey 1995), they are generally not concerned with nociceptive transmission, or at least do not play a major role. Most VPM neurons relay precisely organized tactile information from the face and mouth. Neurons in the posterior nucleus and medial thalamus including nucleus submedius, intralaminar nuclei, and parafascicular nucleus are less somatotopically organized in relaying sensory information.

Trigeminothalamic Projection and Pain

Trigeminothalamic neurons are involved in transmitting pain information from the craniofacial region. The trigeminothalamic projection involved in pain is mainly from the subnucleus caudalis where small-diameter trigeminal primary afferents terminate. Specifically, neurons in the marginal layer (lamina I) of the subnucleus caudalis project to VPM and posterior thalamic nucleus (Shigenaga et al. 1979). Subnucleus caudalis nociceptive neurons are antidromically activated by stimulation of the posterior nucleus of the thalamus in rats (Hirata et al. 1999). Neurons in the subnuclei interpolaris and oralis are also involved in orofacial pain process, although limited information is available on their ascending thalamic projections related to pain.

One characteristic of nociceptive trigeminothalamic projection is that a relatively small number of nociceptive neurons in the subnucleus caudalis project to the VPM. This is in sharp contrast with the principal sensory nucleus where a vast majority of cells are connected to VPM. However, nociceptive trigeminothalamic neurons also project to the posterior thalamic nucleus, nucleus submedius, and intralaminar thalamic nuclei. In general, the trigeminal nociceptive input to the VPM is further relayed to the somatosensory cerebral cortex and related to discriminative information of pain. In contrast, the information of affective and emotional aspects of pain from the head and face is likely related through the thalamic nucleus submedius and intralaminar nuclei (Fig. 1).

Most lamina I trigeminothalamic nociceptive neurons in the subnucleus caudalis project to nucleus submedius, but not VPM. In contrast, most thermoreceptive trigeminothalamic cells project to VPM in cats (Craig and Dostrovsky 2001). In the primate, the axons of spinal and trigeminal lamina I neurons mainly terminate in the ventral caudal part of the medial dorsal nucleus (MDvc) and posterior part of the ventral medial nucleus (VMpo), which further relay sensory input to the anterior cingulate cortex and the dorsal posterior insula, respectively (Craig 2004). In the rat, thalamic-projecting lamina I neurons in the subnucleus caudalis also send axon collaterals to the parabrachial nucleus, as shown by dual retrograde tracing strategy (Li et al. 2006). The trigeminal projections to the parabrachial nucleus, the anterior cingulate cortex, and the insula likely transmit nociceptive information related to pain affect.

Using combined retrograde tracing and nuclear Fos protein expression technique, it has been shown that a population of nucleus submedius-projecting neurons in the Vi/Vc transition zone are activated by inflammation of the masseter muscle (Ikeda et al. 2003). In contrast, corneal-responsive units in the Vi/Vc transition zone do not appear to project to the nucleus submedius (Hirata et al. 1999). Thus, the Vi/Vc submedius projection may be selectively activated by injury of deep orofacial structures.

Trigeminothalamic Projection and Discriminative Sensations

Information related to discriminative sensations from the head and face is relayed in the VPM through trigeminothalamic projections. Neurons in the principal sensory nucleus receive synaptic contact from large-diameter primary afferent terminals of the trigeminal nerve and send axons ascending to VPM. This pathway corresponds to the spinal dorsal column/medial lemniscal system and is highly somatotopically organized. In rodents, distinct clusters of VPM cells receive inputs from designated whiskers mediated by corresponding aggregates of neurons in the principal sensory nucleus. Neurons in the principal sensory nucleus, VPM, and the somatosensory cortex are grouped into rows (rods, cellular cylinders) to give topographical representation of the whisker pad. This specific array of cells is named "barrels" in cortex, "barreloids" in VPM, and "barrelets" in the principal sensory nucleus (Waite and Tracey 1995; Hendry and Hsiao 2003). Information related to discriminative pain from the head and face is relayed in the non-barreloid thalamic regions of the VPM complementary to those mediating tactile sensations, or the barreloid areas (Rausell and Jones 1991; Williams et al. 1994).

References

Bruce, L. L., McHaffie, J. G., & Stein, B. E. (1987). The organization of trigeminotectal and trigeminothalamic neurons in rodents: A double-labeling study with fluorescent dyes. *The Journal of Comparative Neurology, 262*, 315–330.

Craig, A. D. (2004). Distribution of trigeminothalamic and spinothalamic lamina I terminations in the macaque monkey. *The Journal of Comparative Neurology, 477*, 119–148.

Craig, A. D., & Dostrovsky, J. O. (2001). Differential projections of thermoreceptive and nociceptive lamina I trigeminothalamic and spinothalamic neurons in the cat. *Journal of Neurophysiology, 86*, 856–870.

Fukushima, T., & Kerr, F. W. (1979). Organization of trigeminothalamic tracts and other thalamic afferent systems of the brainstem in the rat: Presence of gelatinosa neurons with thalamic connections. *The Journal of Comparative Neurology, 183*, 169–184.

Hayashi, H., Sumino, R., & Sessle, B. J. (1984). Functional organization of trigeminal subnucleus interpolaris: Nociceptive and innocuous afferent inputs, projections to thalamus, cerebellum, and spinal cord, and descending modulation from periaqueductal gray. *Journal of Neurophysiology, 51*, 890–905.

Hendry, S. H., & Hsiao, S. S. (2003). The somatosensory system. In L. R. Squire, F. E. Bloom, S. K. McConnell, J. L. Roberts, N. C. Spitzer, & M. J. Zigmond (Eds.), *Fundamental neuroscience* (2nd ed., pp. 667–697). London: Academic.

Hirata, H., Hu, J. W., & Bereiter, D. A. (1999). Responses of medullary dorsal horn neurons to corneal stimulation by CO(2) pulses in the rat. *Journal of Neurophysiology, 82*, 2092–2107.

Ikeda, T., Terayama, R., Jue, S.-S., Sugiyo, S., Dubner, R., & Ren, K. (2003). Differential rostral projections of caudal brainstem neurons receiving trigeminal input after masseter inflammation. *The Journal of Comparative Neurology, 465*, 220–233.

Jacquin, M. F., Mooney, R. D., & Rhoades, R. W. (1986). Morphology, response properties, and collateral projections of trigeminothalamic neurons in brainstem subnucleus interpolaris of rat. *Experimental Brain Research, 61*, 457–468.

Krout, K. E., Belzer, R. E., & Loewy, A. D. (2002). Brainstem projections to midline and intralaminar thalamic nuclei of the rat. *The Journal of Comparative Neurology, 448*, 53–101.

Li, J., Xiong, K., Pang, Y., Dong, Y., Kaneko, T., & Mizuno, N. (2006). Medullary dorsal horn neurons providing axons to both the parabrachial nucleus and thalamus. *The Journal of Comparative Neurology, 498*, 539–551.

Peschanski, M. (1984). Trigeminal afferents to the diencephalon in the rat. *Neuroscience, 12*, 465–487.

Phelan, K. D., & Falls, W. M. (1991). A comparison of the distribution and morphology of thalamic, cerebellar and spinal projection neurons in rat trigeminal nucleus interpolaris. *Neuroscience, 40*, 497–511.

Rausell, E., & Jones, E. G. (1991). Chemically distinct compartments of the thalamic VPM nucleus in monkeys relay principal and spinal trigeminal pathways to different layers of the somatosensory cortex. *Journal of Neuroscience, 11*, 226–237.

Shigenaga, Y., Takabatake, M., Sugimoto, T., & Sakai, A. (1979). Neurons in marginal layer of trigeminal nucleus caudalis projecting to ventrobasal complex (VB) and posterior nuclear group (PO) demonstrated by retrograde labeling with horseradish peroxidase. *Brain Research, 166*, 391–396.

Veinante, P., Jacquin, M. F., & Deschenes, M. (2000). Thalamic projections from the whisker-sensitive regions of the spinal trigeminal complex in the rat. *The Journal of Comparative Neurology, 420*, 233–243.

Waite, P. M. E., & Tracey, D. J. (1995). Trigeminal sensory system. In G. Paxinos (Ed.), *The rat nervous system* (2nd ed., pp. 705–724). San Diego: Academic.

Williams, M. N., Zahm, D. S., & Jacquin, M. F. (1994). Differential foci and synaptic organization of the principal and spinal trigeminal projections to the thalamus in the rat. *European Journal of Neuroscience, 6*, 429–453.

Yoshida, A. K., Dostrovsky, J. O., Sessle, B. J., & Chiang, C. Y. (1991). Trigeminal projections to the nucleus submedius of the thalamus in the rat. *The Journal of Comparative Neurology, 307*, 609–625.

Zhang, J. D., & Yang, X. L. (1999). Projections from subnucleus oralis of the spinal trigeminal nucleus to contralateral thalamus via the relay of juxtatrigeminal nucleus and dorsomedial part of the principal sensory trigeminal nucleus in the rat. *Journal für Hirnforschung, 39*, 301–310.

Trigeminovascular System

Definition

The trigeminovascular system comprises the primary afferent neurons innervating the cerebral blood vessels and surrounding meninges. Their cell bodies are located in the ophthalmic division of the trigeminal ganglion, and their central terminals project primarily to the trigeminal subnucleus caudalis and dorsal horn of the rostral cervical spinal cord. Vessels posterior to the rostral basilar artery are innervated by afferents whose cell bodies are in the C_2 and C_3 dorsal root ganglia.

Cross-References

▶ Clinical Migraine Without Aura
▶ Migraine, Pathophysiology

Trigger Point

Definition

A trigger point is a spot of localized tenderness in a palpable taut band of muscle. Pressure stimulation or palpation on trigger points evokes a characteristic pattern of referred pain and typically a local twitch response. There are also scar, skin, and connective tissue sensitive spots that, when stimulated, mechanically refer to pain and are sometimes called trigger points.

Cross-References

▶ Chronic Pelvic Pain, Musculoskeletal Syndromes
▶ Muscle Pain, Fibromyalgia Syndrome (Primary, Secondary)
▶ Myofascial Trigger Points
▶ Psychophysiological Assessment of Pain
▶ Referred Muscle Pain, Assessment
▶ Stretching

Trigger Point Pain

▶ Myofascial Pain

Trigger Point Pressure Release

Definition

Trigger point pressure release is a simple, often effective, manual technique for the treatment of myofascial trigger points. Slowly increasing, gentle, digital pressure is applied to the trigger point until a barrier of tissue resistance is encountered. Constant pressure is maintained until the resistance of the barrier tension decreases. Then pressure is slowly increased to reach a new barrier. The term "ischemic compression" has previously been used to describe this treatment approach, but a different concept of the treatment mechanism was attributed to it, and the procedure was unnecessarily painful.

Cross-References

▶ Myofascial Trigger Points

Triple Response

Definition

The term has been introduced by Thomas Lewis who experimented with intracutaneous histamine injections. In these experiments, he observed three prominent reactions with different time courses. The *first* reaction is a local erythema occurring within a few seconds. We know now that this reaction is due to a vasodilation following the binding of histamine to H1 receptors in terminal arteriols. *Secondly*, a local swelling (the wheal reaction) is observed, which is due to plasma-extravastion from capillaries and venols which also have H1 receptors. At the same time histamine activates a particular subgroup of unmyelinated C-fibers at the injection site. They are responsible for the *third* response, an extended erythema, the "flare," which requires more time to develop and reaches its final extension only after several minutes. This flare is due to the axon reflex (see the respective definition) by which these histamine-sensitive nociceptors release vasoactive Calcitonin-gene-related peptide (CGRP) from their axon collaterals.

Only the flare response is mediated by nerves, the C-fiber pruriceptors. It is therefore diminished, or abolished if these nerve fibers are damaged (e.g., in diabetic neuropathy).

Cross-References

► Congenital Insensitivity to Pain with Anhidrosis

Triptans

Definition

Serotonin, 5-HT$_{1B/1D}$, receptor agonists, i.e., compounds that activate or turn on both these receptor subtypes. These compounds are highly effective in the treatment of acute migraine and cluster headache.

Cross-References

► Clinical Migraine Without Aura
► Migraine, Pathophysiology

trkA

► Tyrosine Kinase A

TrkA Receptor(s)

Definition

TrkA is a member of the trk family of tyrosine kinase receptors and is the high-affinity receptor for nerve growth factor found on NGF-dependent sensory neurons. When trkA binds NGF, the receptor autophosphorylates on tyrosine residues, leading to activation of multiple downstream effector proteins, including the Ras-MAP kinase signaling cascade and PI$_3$-kinase activation.

Cross-References

► Congenital Insensitivity to Pain with Anhidrosis
► IB4-Positive Neurons, Role in Inflammatory Pain
► Immunocytochemistry of Nociceptors
► Nerve Growth Factor, Sensitizing Action on Nociceptors
► TRPV1, Regulation by Nerve Growth Factor

TrkA-IgG

Definition

NGF-sequestering molecule.

Cross-References

► Spinal Cord Nociception, Neurotrophins

TRN

► Thalamic Reticular Nucleus

Trochanteric Bursitis

Definition

Inflammation of the bursa near the greater trochanter of the hip. This condition can produce pain in the thigh.

Cross-References

▶ Sciatica

Trophic Factors

Definition

Molecules that cause growth/regeneration of various parts of a cell/neuron; more recently some trophic factors have also been shown to have transmitter-like actions.

Cross-References

▶ Retrograde Cellular Changes after Nerve Injury

Tropism

Definition

Movement or growth of an organism in response to an external stimulus.

Cross-References

▶ Hansen's Disease

TRP Channels

Definition

Denominates ion channels of the transient receptor potential (TRP) family. The first member of this ion channel family was identified by localization of a gene that caused deficiencies in signaling in the visual system of the model organism *Drosophila melanogaster* (fruit fly). Mammals have more than 14 TRP channel genes. TRP channels play important roles in the regulation of neural excitability and in sensory systems, such as vision, olfaction, and pheromone sensation. TRP channels serve as receptors for temperature and irritant chemicals (capsaicin, mustard oil) in the pain pathway and play important functional roles in neurogenic inflammation and hyperalgesia.

Cross-References

▶ Nociceptors, Cold Thermotransduction
▶ TRPV1, Regulation by Nerve Growth Factor
▶ TRPV1, Regulation by Protons

TRPA1

Definition

Abbreviation for Transient Receptor Potential Ankyrin 1, a nonselective cation channel selectively expressed in a subpopulation of peptidergic nociceptors, together with TRPV1. The recombinant channel is activated by temperatures below 17 °C, and it was proposed to function as a cold sensor in nociceptors, a controversial issue. The channel is activated by many natural pungent compounds like cinnamon and mustard oil and chemical irritants by an electrophilic mechanism. It is also activated by the proalgesic peptide bradykinin. It plays an important role in the mechanisms of mechanical and cold hyperalgesia following inflammatory and neuropathic injury of nociceptors.

Cross-References

▶ Nociceptors, Cold Thermotransduction

TRPA1 Channel

Peter Zygmunt and Edward Högestätt
Department of Laboratory Medicine, Lund University, Division of Clinical Chemistry and Pharmacology, Lund, Sweden

Synonyms

ANKTM1

Definition

The transient receptor potential ankyrin subtype-1 (TRPA1) is a nonselective cation channel permeable to Ca^{2+}, Na^+, and K^+. In primary sensory neurons, Ca^{2+} and Na^+ flowing through TRPA1 into the cell cause neuronal excitability and neurotransmitter release. This triggers nociceptive signaling, including ▶ pain and the release of inflammatory neuropeptides in the periphery.

Introduction

In 1999, Jaquemar and coworkers described an mRNA transcript encoding a TRP-like protein with 18 putative N-terminal ankyrin repeats in human fibroblasts (Jaquemar et al. 1999). Four years later, Story and colleagues showed that the mouse orthologue of this gene is expressed in ▶ sensory ganglia and that the gene product ANKTM1, later termed TRPA1, is an ion channel

responding to noxious cold temperatures (Story et al. 2003). In the meantime, other researchers used an expression-cloning strategy to identify rat and human ANKTM1 (TRPA1) as chemosensors activated by mustard oil, which had been widely used to trigger ▶ nociception in animal pain models, and cannabinoids (Jordt et al. 2004). Of further importance, TRPA1 was proposed to be activated by known pro-inflammatory agents via the phospholipase C (PLC) signaling pathway, and cytosolic Ca^{2+} was identified as a regulator of channel gating (Bandell et al. 2004; Jordt et al. 2004). Altogether, these findings linked TRPA1 to noxious thermal and chemical transduction, highlighting TRPA1 as a novel target for ▶ analgesics.

Characteristics

The mammalian TRPA1 is a nonselective cation channel with six putative transmembrane segments (S1–S6), intracellular N- and C-termini, and a pore loop between S5 and S6 (Fig. 1). The channel belongs to the transient receptor potential (TRP) channel superfamily that consists of 28 different ion channels divided into six subgroups, of which TRPA1 constitutes its own subgroup. The N-terminus contains between 14 and 18 ankyrin repeats and has several cysteine residues that are intimately involved in channel regulation. The C-terminus is considerably shorter than the N-terminus and has binding sites for Ca^{2+} and Zn^{2+}, both of which are important regulators of TRPA1 activity (Fig. 1). TRPA1 has a high Ca^{2+} permeability compared to other TRP ion channels and a unitary conductance of approximately 60 and 100 pS in the inward and outward directions, respectively, under physiological conditions (Nilius et al. 2012). However, TRPA1 may undergo pore dilatation, as shown in the presence of the activator mustard oil, increasing the permeability and reducing the relative cation selectivity of the channel (Nilius et al. 2012).

In sensory neurons from trigeminal, dorsal root, and nodose ganglia, TRPA1 is expressed

TRPA1 Channel, Fig. 1 The functional ion channel TRPA1 is a putative homotetrameric protein where each monomer contains six transmembrane segments (*S1–S6*) with cytoplasmic N- and C-termini. Between S5 and S6 is the pore region, through which Ca^{2+} and Na^{2+} enter the cell and K^+ can flow in the outward direction at physiological membrane potentials. The pore loop (*P*) contains Asp918 (*D*) that is supposed to be critical for Ca^{2+} selectivity. The N-terminus with its many ankyrin repeats contains cysteine (*C*) and lysine (*K*) residues that are targets for electrophiles and oxidants. Cysteines outside the N-terminus, such as the oxidation-sensitive Cys856 (*C, S4*), as well as other aminoacid residues (*S and I*) in S6 may also contribute to the complex regulation of TRPA1 gating. Hydroxylation of Pro394 (*P*) in the N-terminus reduces TRPA1 activity at normoxia. A gain-of-function mutation at Asn855 (*N*) in S4 of TRPA1 causes familial episodic pain syndrome. Menthol is a non-electrophilic compound that activates TRPA1 by binding to the amino acid residues Ser873 and Thr874 (*S and T*) in S5. Phosphatidylinositol-4,5-bisphosphate (PIP_2), which is cleaved by phospholipase C into diacylglycerol (DAG) and inositol 1,4,5-trisphosphate (IP_3), may regulate TRPA1 activity by interacting with charged residues in the C-terminus, as suggested for TRPV1. Cysteine (*C*) and histidine (*H*) residues on the C-termini are binding sites for Zn^{2+}, which activates TRPA1. The N- and C-termini contain putative binding sites for Ca^{2+} (*D and E*), which can both sensitize and desensitize TRPA1

in both peptidergic and non-peptidergic neurons classified as C- and Aδ-fiber ▶ primary afferents (Andrade et al. 2012; Kim et al. 2010). TRPA1 is mostly found in a subpopulation of ▶ TRPV1-positive neurons, but non-TRPV1-containing neurons expressing TRPA1 exist (La et al. 2011). Interestingly, a small population large myelinated Aß-fibers, which are activated by innocuous mechanical force, expresses TRPA1 (Kim et al. 2010). Outside sensory neurons, TRPA1 is found in epithelial cells, vascular endothelium, keratinocytes, fibroblasts, odontoblasts, and in enterochromaffin cells

(Andrade et al. 2012; Baraldi et al. 2010; Earley 2012; Nilius et al. 2012). Many of these cells have sensory properties and communicate with nearby ▶ nociceptors (Lumpkin and Caterina 2007).

TRPA1: A Chemosensor

The proposal of an ionotropic ▶ cannabinoid receptor belonging to the TRP ion channel family (Zygmunt et al. 2002) and the frequent use of mustard oil in animal models of acute pain and ▶ hyperalgesia prompted a search for Δ^9-tetrahydrocannabinol- and mustard oil-sensitive clones upon expression of a rat trigeminal ganglia cDNA library (Jordt et al. 2004). This led to the identification of the rat and human TRPA1 as ionotropic cannabinoid receptors and targets for mustard oil (Jordt et al. 2004). Another study identified mouse TRPA1 also as a target for both electrophilic and non-electrophilic compounds (Bandell et al. 2004). Shortly after, it was shown that certain thiol-reactive garlic-derived compounds activated TRPA1 (Bautista et al. 2005; Macpherson et al. 2005). Together, these studies revealed TRPA1 as a detector of thiol reactive electrophiles and oxidants. Today, a large number of endogenous and exogenous irritants are known to activate mammalian TRPA1, mainly by acting on cysteine residues in the N-terminus (Andrade et al. 2012; Baraldi et al. 2010; Holzer 2011; Nilius et al. 2012). However, other amino acids, such as lysine in the N-terminus and cysteines outside the N-terminus, are potential targets for electrophiles and oxidants and may contribute to the regulation of TRPA1 (Macpherson et al. 2007; Nilius et al. 2012; Wang et al. 2012). Furthermore, the chemical nature of the covalent interaction differs between TRPA1 activators (Bessac and Jordt 2008; Nilius et al. 2012), which may contribute to ligand-specific pharmacological properties in terms of sensitization, desensitization, channel trafficking, pungency, and duration of action. Whereas the sensitivity of TRPA1 to potentially harmful electrophiles and oxidants is in line with its proposed role as a nocisensor, the ability of TRPA1 to respond to many non-electrophilic compounds of diverse chemical structure is intriguing (Andrade et al. 2012; Baraldi et al. 2010; Holzer 2011; Nilius et al. 2012). Except for menthol, which binds to S5, the binding sites for other non-electrophilic compounds on TRPA1 remain to be identified. Whether endogenous ligands sharing binding sites with these non-electrophiles exist and regulate TRPA1 is not known.

Agents such as ► Complete Freund's adjuvant (CFA) and ► carrageenan are widely used to evoke ► inflammation and related pain in animals by the production of numerous inflammatory mediators, some of which may act directly or indirectly on TRPA1 (Andrade et al. 2012; Mogil 2009; Moilanen et al. 2012). Many of these mediators bind to G protein-coupled receptors that regulate TRPA1 via the PLC signaling pathway as well through protein kinase A-mediated phosphorylation (Andrade et al. 2012; Bautista et al. 2012; Bessac and Jordt 2008; Nilius et al. 2012). The nociceptive responses triggered by intraplantar injection of CFA and carrageenan are sensitive to TRPA1 antagonists (Andrade et al. 2012; Gregus et al. 2012), and TRPA1 plays a key role in the development of carrageenan-induced inflammation (Moilanen et al. 2012). The rat and mouse formalin test is also a widely used pain model for evaluating analgesic compounds on acute pain (first phase) and inflammatory pain (second phase). Interestingly, this biphasic nociceptive response is driven by a direct action of formaldehyde on TRPA1 (Moran et al. 2011).

TRPA1 displays a U-shaped intracellular pH sensitivity and the N-terminal plays a key role in the activation induced by both ► acidosis and alkalosis, conditions that are potentially tissue damaging (Nilius et al. 2012). TRPA1 activity is also influenced by hydroxylation of a proline residue (Pro394) in the N-terminus, and ► hypoxia indirectly activates TRPA1 by preventing this hydroxylation of TRPA1 (Nilius et al. 2012). This further underscores the importance of the N-terminus in controlling TRPA1 activity. However, the C-terminal domain should not be neglected, as it mediates Ca^{2+} and Zn^{2+} sensitivities and modulates the response to electrophiles and the voltage dependence of TRPA1 (Andrade et al. 2012; Nilius et al. 2012).

TRPA1: A Cold Sensor

TRPA1 has been suggested to respond to noxious cold temperatures (Story et al. 2003). However, the mechanism behind cold activation of TRPA1 and its role in noxious cold sensation is debated (Andrade et al. 2012; Belmonte et al. 2009). Methodological differences as well as regional differences in neuronal phenotype may explain some of the conflicting findings (Belmonte et al. 2009), but whether or not TRPA1 is gated directly by cold remains to be demonstrated and may require studies on the purified ion channel. Nevertheless, several studies have reported that acute noxious cold sensation is impaired in TRPA1-deficient mice and, more importantly, that TRPA1 is involved in cold hypersensitivity induced by nerve injury and inflammation in animals (Andrade et al. 2012; Belmonte et al. 2009; Moran et al. 2011; Nilius et al. 2012). Interestingly, an allosteric gating model has been proposed to explain how cold and inflammatory agents may synergistically regulate TRPA1 (Salazar et al. 2011).

TRPA1: A Mechanosensor

Shortly after the publication of TRPA1 as a sensor of noxious cold and chemical irritants (Story et al. 2003; Bandell et al. 2004; Jordt et al. 2004), TRPA1 was proposed as a mechanosensor in auditory hair cells (Andrade et al. 2012; Baraldi et al. 2010; Holzer 2011; Nilius et al. 2012). Although the role of TRPA1 as a mechanosensor was later discarded in studies of TRPA1-deficient mice (Andrade et al. 2012; Nilius et al. 2012), such mice clearly display an increased threshold for noxious mechanical stimuli (Andersson et al. 2011). Interestingly, it has been speculated that TRPA1 expressed in Aß-fibers contributes to mechanical hypersensitivity in the setting of inflammation and nerve injury (Kim et al. 2010). In the intestine, TRPA1 seems to be involved in both normal and pathophysiological mechanosensation (Holzer 2011; Nilius et al. 2012). However, the exact role of TRPA1 in mechanosensation remains to be elucidated (Andrade et al. 2012; Baraldi et al. 2010; Holzer 2011; Nilius et al. 2012).

TRPA1: A Pharmacological Target for Treatment of Pain

Animal studies have shown that TRPA1 is involved in cold and mechanical sensitization induced by inflammation and nerve injury (Andrade et al. 2012; Holzer 2011). ► Cold allodynia and ► mechanical allodynia are common symptoms in patients with ► neuropathic pain. TRPA1 may also drive the development of ► diabetic neuropathy and the associated

mechanical sensitization (Moran et al. 2011). Methylglyoxal, a potent TRPA1 activator produced under hyperglycemic conditions, has been implicated as a mediator of diabetic neuropathy, although other endogenous electrophilic compounds may also contribute (Koivisto et al. 2012). Thermal and mechanical allodynia are common adverse effects in patients undergoing ► cancer chemotherapy. TRPA1 together with ► TRPM8 and TRPV1 may contribute to these adverse effects, as shown in animals exposed to chemotherapeutic agents (Belmonte et al. 2009; Nilius et al. 2012). It is interesting that TRPA1 activators, including metabolites of ► paracetamol and ► prostaglandins, can induce desensitization of the channel and/or the whole neuron thereby inhibiting nociceptive signaling (Andersson et al. 2011; Weng et al. 2012). Thus, both antagonists and agonists of TRPA1 may be useful to treat various pain conditions, as shown for TRPV1 (Andersson et al. 2011; Mallet et al. 2010; Moran et al. 2011; Patapoutian et al. 2009). It is encouraging that the TRPA1 antagonist A-967079 is able to attenuate cold allodynia in animals without affecting noxious heat sensation and body temperature, as such unwanted effects have hampered the development of TRPV1 antagonists as analgesics (Moran et al. 2011). Taken together, TRPA1 is a promising drug target, but more information about the consequences of interfering with TRPA1 in non-neuronal cells is needed to develop safe therapies.

Summary

TRPA1 is a promiscuous chemical nocisensor and possibly also a detector of noxious cold and mechanical stimuli. It is present mainly in a subpopulation of small- and medium-sized nociceptive primary sensory neurons as well as in non-neuronal cells, such as epithelial cells. TRPA1 is therefore an interesting drug target for novel treatments of pain and other symptoms or diseases associated with enhanced sensory signaling (Andrade et al. 2012; Baraldi et al. 2010; Bautista et al. 2012;

Bessac and Jordt 2008; Earley 2012; Holzer 2011; Moran et al. 2011; Nilius et al. 2012).

Acknowledgements The authors acknowledge the Swedish Research Council (2010-3347 and 2010-5787) and Lund University for the fundings, and Lavanya Moparthi for drawing of the figure.

References

Andersson, D. A., Gentry, C., Alenmyr, L., Killander, D., Lewis, S. E., Andersson, A., et al. (2011). TRPA1 mediates spinal antinociception induced by acetaminophen and the cannabinoid Delta(9)-tetrahydrocannabiorcol. *Nature Communications, 2*, 551.

Andrade, E. L., Meotti, F. C., & Calixto, J. B. (2012). TRPA1 antagonists as potential analgesic drugs. *Pharmacology and Therapeutics, 133*(2), 189–204.

Bandell, M., Story, G. M., Hwang, S. W., Viswanath, V., Eid, S. R., Petrus, M. J., et al. (2004). Noxious cold ion channel TRPA1 is activated by pungent compounds and bradykinin. *Neuron, 41*(6), 849–857.

Baraldi, P. G., Preti, D., Materazzi, S., & Geppetti, P. (2010). Transient receptor potential ankyrin 1 (TRPA1) channel as emerging target for novel analgesics and anti-inflammatory agents. *Journal of Medicinal Chemistry, 53*(14), 5085–5107.

Bautista, D. M., Movahed, P., Hinman, A., Axelsson, H. E., Sterner, O., Högestätt, E. D., et al. (2005). Pungent products from garlic activate the sensory ion channel TRPA1. *Proceedings of the National Academy of Sciences of the United States of America, 102*(34), 12248–12252.

Bautista, D. M., Pellegrino, M., & Tsunozaki, M. (2013). TRPA1: A gatekeeper for Inflammation. *Annual Review of Physiology, 75,* 181–200.

Belmonte, C., Brock, J. A., & Viana, F. (2009). Converting cold into pain. *Experimental Brain Research, 196*(1), 13–30.

Bessac, B. F., & Jordt, S. E. (2008). Breathtaking TRP channels: TRPA1 and TRPV1 in airway chemosensation and reflex control. *Physiology (Bethesda), 23,* 360–370.

Earley, S. (2012). TRPA1 channels in the vasculature. *British Journal of Pharmacology, 167*(1), 13–22.

Gregus, A. M., Doolen, S., Dumlao, D. S., Buczynski, M. W., Takasusuki, T., Fitzsimmons, B. L., et al. (2012). Spinal 12-lipoxygenase-derived hepoxilin A3 contributes to inflammatory hyperalgesia via activation of TRPV1 and TRPA1 receptors. *Proceedings of the National Academy of Sciences of the United States of America, 109*(17), 6721–6726.

Holzer, P. (2011). Transient receptor potential (TRP) channels as drug targets for diseases of the digestive system. *Pharmacology and Therapeutics, 131*(1), 142–170.

Jaquemar, D., Schenker, T., & Trueb, B. (1999). An ankyrin-like protein with transmembrane domains is specifically lost after oncogenic transformation of human fibroblasts. *Journal of Biological Chemistry, 274*(11), 7325–7333.

Jordt, S. E., Bautista, D. M., Chuang, H. H., McKemy, D. D., Zygmunt, P. M., Högestätt, E. D., et al. (2004). Mustard oils and cannabinoids excite sensory nerve fibres through the TRP channel ANKTM1. *Nature, 427*(15), 260–265.

Kim, Y. S., Son, J. Y., Kim, T. H., Paik, S. K., Dai, Y., Noguchi, K., et al. (2010). Expression of transient receptor potential ankyrin 1 (TRPA1) in the rat trigeminal sensory afferents and spinal dorsal horn. *The Journal of Comparative Neurology, 518*(5), 687–698.

Koivisto, A., Hukkanen, M., Saarnilehto, M., Chapman, H., Kuokkanen, K., Wei, H., et al. (2012). Inhibiting TRPA1 ion channel reduces loss of cutaneous nerve fiber function in diabetic animals: sustained activation of the TRPA1 channel contributes to the pathogenesis of peripheral diabetic neuropathy. *Pharmacological Research, 65*(1), 149–158.

La, J. H., Schwartz, E. S., & Gebhart, G. F. (2011). Differences in the expression of transient receptor potential channel V1, transient receptor potential channel A1 and mechanosensitive two pore-domain K+ channels between the lumbar splanchnic and pelvic nerve innervations of mouse urinary bladder and colon. *Neuroscience, 186,* 179–187.

Lumpkin, E. A., & Caterina, M. J. (2007). Mechanisms of sensory transduction in the skin. *Nature, 445*(7130), 858–865.

Macpherson, L. J., Geierstanger, B. H., Viswanath, V., Bandell, M., Eid, S. R., Hwang, S., et al. (2005). The pungency of garlic: Activation of TRPA1 and TRPV1 in response to allicin. *Current Biology, 15*(10), 929–934.

Macpherson, L. J., Dubin, A. E., Evans, M. J., Marr, F., Schultz, P. G., Cravatt, B. F., et al. (2007). Noxious compounds activate TRPA1 ion channels through covalent modification of cysteines. *Nature, 445*(7127), 541–545.

Mallet, C., Barriere, D. A., Ermund, A., Jönsson, B. A., Eschalier, A., Zygmunt, P. M., et al. (2010). TRPV(1) in brain is involved in acetaminophen-induced antinociception. *PLoS One, 5*(9), e12748.

Mogil, J. S. (2009). Animal models of pain: Progress and challenges. *Nature Reviews Neuroscience, 10*(4), 283–294.

Moilanen, L. J., Laavola, M., Kukkonen, M., Korhonen, R., Leppanen, T., Hogestatt, E. D., et al. (2012). TRPA1 contributes to the acute inflammatory response and mediates carrageenan-induced paw edema in the mouse. *Sciences Report, 2,* 380.

Moran, M. M., McAlexander, M. A., Biro, T., & Szallasi, A. (2011). Transient receptor potential channels as therapeutic targets. *Nature Reviews Drug Discovery, 10*(8), 601–620.

Nilius, B., Appendino, G., & Owsianik, G. (2012). The transient receptor potential channel TRPA1: from gene to pathophysiology. *Pflügers Archives, 464*(5), 425–458.

Patapoutian, A., Tate, S., & Woolf, C. J. (2009). Transient receptor potential channels: Targeting pain at the source. *Nature Reviews Drug Discovery, 8*(1), 55–68.

Salazar, M., Moldenhauer, H., & Baez-Nieto, D. (2011). Could an allosteric gating model explain the role of TRPA1 in cold hypersensitivity? *Journal of Neuroscience, 31*(15), 5554–5556.

Story, G. M., Peier, A. M., Reeve, A. J., Eid, S. R., Mosbacher, J., Hricik, T. R., et al. (2003). ANKTM1, a TRP-like channel expressed in nociceptive neurons, is activated by cold temperatures. *Cell, 112*(6), 819–829.

Wang, L., Cvetkov, T. L., Chance, M. R., & Moiseenkova-Bell, V. Y. (2012). Identification of in vivo disulfide conformation of TRPA1 ion channel. *Journal of Biological Chemistry, 287*(9), 6169–6176.

Weng, Y., Batista, P. A., Barabas, M. E., Harris, E. Q., Dinsmore, T. B., Kossyreva, E. A., et al. (2012). Prostaglandin metabolite induces inhibition of TRPA1 and channel-dependent nociception. *Molecular Pain, 8*(1), 75.

Zygmunt, P. M., Andersson, D. A., & Högestätt, E. D. (2002). Δ⁹-Tetrahydrocannabinol and cannabinol activate capsaicin-sensitive sensory nerves via a CB1 and CB2 cannabinoid receptor-independent mechanism. *Journal of Neuroscience, 22*(11), 4720–4727.

TRPM8 Channel

Definition

Acronym for Transient Receptor Potential Melastatin type 8, a nonselective cation channel of the TRP family gated by moderate cool temperatures and natural cooling compounds, like menthol. TRPM8 is expressed in a subpopulation of small diameter primary sensory neurons and is a fundamental molecular sensor for cold temperature detection in peripheral cold thermoreceptors. Pharmacological and behavioral results suggest that TRPM8 may also play a role in cold discomfort and noxious cold pain.

Cross-References

▶ Nociceptors, Cold Thermotransduction

TRPV1

Synonyms

Transient receptor potential vanilloid type 1

Definition

The TRPV1 receptor is a member of the transient receptor vanilloid subfamily. It is sensitive to capsaicin and has a well-defined temperature threshold of about 43 °C, similar to the threshold for thermal nociception. The response of this receptor can be sensitized to NGF and other proinflammatory mediators providing a basis for thermal hyperalgesia. Other TRP subfamily members are not sensitive to capsaicin and have different temperature thresholds (e.g., TRPV2 – 52 °C; TRPV3 – 35 °C). Still other members of this subfamily appear to respond to other stimuli, e.g., TRPV4 – osmoreceptors. TRPV1 is the charter member of a family of TRP channels identified in mammalian sensory neurons that are activated by temperatures ranging from cold to warm to intense heat.

Cross-References

▶ Capsaicin Receptor
▶ ERK Regulation in Sensory Neurons during Inflammation
▶ IB4-Positive Neurons, Role in Inflammatory Pain
▶ Muscle Pain Model, Ischemia-Induced and Hypertonic Saline-Induced
▶ Nerve Growth Factor, Sensitizing Action on Nociceptors
▶ NGF, Regulation During Inflammation
▶ Nociceptor, Categorization
▶ Species Differences in Skin Nociception
▶ TRPV1 Modulation by p2Y Receptors
▶ TRPV1 Receptor, Species Variability
▶ TRPV1, Regulation by Nerve Growth Factor
▶ TRPV1, Regulation by Protons
▶ Visceral Pain Model; Esophageal Pain

TRPV1 Modulation by P2Y Receptors

Makoto Tominaga
Department of Cellular and Molecular
Physiology, Mie University School of Medicine,
Tsu Mie, Japan

Synonyms

Modulation by P2Y receptors; TRPV1 receptor

Definition

▶ Capsaicin receptor ▶ TRPV1 is a nonselective cation channel expressed in a subset of sensory neurons, ▶ nociceptors. ▶ P2Y receptor activation potentiates or sensitizes TRPV1 activity through a ▶ PKC-dependent pathway. In the presence of ▶ ATP released in the context of tissue damage, the temperature threshold for TRPV1 activation is reduced to less than 35 °C, so that body temperature is capable of activating TRPV1, leading to the sensation of pain.

Characteristics

Pain is initiated when noxious thermal, mechanical, or chemical stimuli excite the peripheral terminals of specialized primary afferent neurons called nociceptors. Many different kinds of ionotropic and metabotropic receptors are known to be involved in this process (McCleskey and Gold 1999). Vanilloid receptors are nociceptor-specific cation channels that serve as the molecular target of capsaicin, the pungent ingredient in hot chili peppers (Szallasi and Blumberg 1999). A gene encoding a capsaicin receptor was isolated using an expression-cloning method, and the receptor protein was found to be an ion channel with six transmembrane domains having high Ca^{2+} permeability (Caterina et al. 1997). When expressed in heterologous systems, the cloned capsaicin receptor TRPV1 can also be activated by noxious heat

(with a thermal threshold >43 °C) or protons (acidification), both of which cause pain in vivo (Caterina et al. 1997; Tominaga et al. 1998; Caterina and Julius 2001). Furthermore, analyses of mice lacking TRPV1 have shown that TRPV1 is essential for selective modalities of pain sensation and for tissue injury-induced thermal hyperalgesia (Caterina et al. 2000; Davis et al. 2000).

Tissue damage associated with infection, inflammation, or ischemia produces an array of chemical mediators that activate or sensitize nociceptor terminals to elicit pain at the site of injury. One important component of this proalgesic response is adenosine triphosphate (ATP) released from different cell types (North and Barnard 1997; Burnstock and Williams 2000). Extracellular ATP excites the nociceptive endings of nearby sensory nerves, evoking a sensation of pain. In these neurons, the most widely studied targets of extracellular ATP have been ionotropic ATP ▶ (P2X) receptors. Indeed, several P2X receptor subtypes have been identified in sensory neurons, including one ($P2X_3$) whose expression is largely confined to these cells (North and Barnard 1997). The importance of widely distributed metabotropic ATP (P2Y) receptors in nociception has been recently reported (Molliver et al. 2002; Zimmermann et al. 2002). In particular, the functional interaction between P2Y receptors and TRPV1 has received much attention because TRPV1 is one of the key molecules detecting nociceptive stimuli and because the signaling pathway downstream of P2Y receptor activation can be applicable for other G_q-protein-coupled receptor activation.

Pretreatment with 100 μM extracellular ATP (an attainable concentration in the context of tissue damage) causes more than six times the potentiation of capsaicin (low doses)-evoked current responses in HEK293 cells heterologously expressing TRPV1 channels (Fig. 1a) (Tominaga et al. 2001). A similar potentiating effect of extracellular ATP is observed on proton-evoked activation of TRPV1. The dose–response curves for capsaicin in the presence or absence of ATP demonstrate that ATP enhances capsaicin and

TRPV1 Modulation by P2Y Receptors, Fig. 1
(a) Extracellular ATP (100 µM) potentiates capsaicin (CAP)-activated currents in HEK293 cells expressing TRPV1. Whole-cell patch-clamp recordings were carried out with holding potential of -60 mV. Cells were perfused for 2 min with solution containing ATP before reexposure to capsaicin. (b) Capsaicin dose–response curves for TRPV1 in the absence (λ) and presence (μ) of 100 µM extracellular ATP. Currents were normalized to the currents maximally activated by 1 µM capsaicin in the absence of ATP. Figure shows averaged data fitted with the Hill equation. $EC_{50} = 114.7$ nM and Hill coefficient = 1.48 in the absence of ATP. $EC_{50} = 49.3$ nM and Hill coefficient = 1.56 in the presence of ATP. Values represent the mean $\pm$ SEM. (c) Reduction of the threshold temperature for TRPV1 activation by extracellular ATP. Representative temperature-response profiles in the absence and presence of 100 µM extracellular ATP. *Dashed lines* show the threshold temperature for heat activation of VR1. Temperature threshold for activation of TRPV1 in the presence of ATP (35.3 ± 0.7 °C) was significantly lower than that in the absence of ATP (41.7 ± 1.1 °C). *, $p < 0.001$ (Tominaga et al. 2001)

proton action on TRPV1 by lowering EC50 values without altering maximal responses (Fig. 1b). Extracellular ATP also lowers the threshold temperature for TRPV1 activation significantly (from about 43 °C to about 35 °C) (Fig. 1c). Thus, in the presence of ATP, normally non-painful thermal stimuli (even body temperature) are capable of activating TRPV1. These data show that TRPV1 currents evoked by any of three different stimuli (capsaicin, proton, or heat) are potentiated or sensitized by extracellular ATP. Activation of protein kinase C (PKC) by diacylglycerol (DAG) downstream of P2Y$_1$ receptors is found to be a mechanism for

TRPV1 Modulation by P2Y Receptors, Fig. 2
(a) Regulation mechanisms of TRPV1 by P2Y receptors. G_q-coupled P2Y receptor activation leads to production of IP_3 and DAG through PLCβ. PKC activation by DAG causes phosphorylation of TRPV1, leading to functional potentiation. *PLCβ* phospholipase Cβ, *DAG* diacylglycerol, *IP_3* inositol 1,4,5-trisphosphate, *P* phosphorylation, *o* and *I* outside and inside of cell, respectively. (b) TRPV1 is essential for the development of ATP-induced thermal hypersensitivity in vivo. Wild-type (λ) or TRPV1−/− mice (μ) were injected intraplantarly with ATP (100 nmol), and the response latency to radiant heating of the hind paw was measured at various time points after injection. * $p < 0.05$ and ** $p < 0.01$ vs. wild-type mice (Moriyama et al. 2003)

the ATP-induced potentiation from various pharmacological analyses in HEK293 cells (Fig. 2a), consistent with the report that PKC-ε is specifically involved in sensitization of heat-activated channels by ► bradykinin in dorsal-root ganglion (DRG) neurons.

The interaction between ATP and TRPV1 in the context of ATP-induced hyperalgesia in vivo is confirmed by a behavioral analysis using wild-type mice and TRPV1-deficient mice (Moriyama et al. 2003). A significant reduction in paw withdrawal latency to radiant paw heating is observed for 5–30 min following ATP injection in wild-type mice. On the other hand, TRPV1-deficient mice develop no such thermal hypersensitivity in response to ATP injection, suggesting a functional interaction between ATP and TRPV1 (Fig. 2b). Mice lacking P2Y₁ receptors, a subtype proved to be involved in ATP-induced potentiation of TRPV1 currents in HEK293 cells, showed similar thermal hyperalgesia to wild-type mice, indicating that other P2Y subtypes are involved in ATP-induced thermal hyperalgesia in mice.

Electrophysiological and pharmacological analyses using DRG neurons of mice revealed that P2Y₂ is a subtype involved in ATP-induced potentiation of TRPV1 currents and ATP-induced thermal hyperalgesia in mice (Moriyama et al. 2003). UTP, a potent P2Y₂ receptor agonist, potentiates the capsaicin-evoked current responses and causes thermal hyperalgesia to a similar extent as ATP in mice, further confirming the involvement of $P2Y_2$ receptors. $P2Y_2$ receptor mRNA but not $P2Y_1$ mRNA is found to be co-expressed with TRPV1 mRNA in the rat lumbar DRG using double in situ hybridization, suggesting that $P2Y_2$, not $P2Y_1$ receptors, can functionally interact with TRPV1 in DRG neurons.

The data described above suggest that direct phosphorylation of TRPV1 or a closely associated protein by PKC changes the agonist sensitivity of this ion channel. The in vivo phosphorylation of TRPV1 by PKC is confirmed in HEK293 cells expressing TRPV1 (Numazaki et al. 2002). Furthermore, two serine residues as substrates for PKC-dependent phosphorylation, S502 in the first intracellular loop and S800 in the C-terminus, were identified using an in vitro kinase assay (Numazaki et al. 2002). A mutant lacking substrates for PKC-dependent phosphorylation (S502A/S800A) exhibited almost no potentiation effects of capsaicin- or proton-evoked current responses by PMA (a direct activator of PKC). Furthermore, the double mutant showed no reduction in temperature threshold for TRPV1 activation, suggesting that the two serine residues are the major substrates for PKC-dependent phosphorylation.

Inflammatory pain is initiated by tissue damage/inflammation and is characterized by hypersensitivity both at the site of damage and in adjacent tissue. One mechanism underlying these phenomena is the modulation (sensitization) of ion channels, such as TRPV1, that detect noxious stimuli at the nociceptor terminal. Sensitization is triggered by extracellular inflammatory mediators that are released in vivo from surrounding damaged or inflamed tissue and from nociceptive neurons themselves. Among the mediators, ATP not only potentiates capsaicin- or proton-evoked currents but also lowers the temperature threshold for heat activation of TRPV1, such that normally non-painful thermal stimuli (i.e., normal body temperature) are capable of activating TRPV1, making ATP act like a direct activator of TRPV1.

This represents a novel mechanism through which extracellular ATP may cause inflammatory pain. Most attention in the pain field has focused on the role of ionotropic ATP receptors in ATP-evoked nociception. The above findings suggest that $P2Y_2$ is also involved in this process and may represent a fruitful target for the development of drugs that blunt nociceptive signaling through capsaicin receptors. $P2Y_2$ receptors confer responsiveness to uridine triphosphate (UTP) and ATP to a similar extent, suggesting a possible role for UTP as an important component of pro-algesic response in the context of tissue injury. UTP was indeed found to potentiate capsaicin-activated currents and cause thermal hyperalgesia in mice (Moriyama et al. 2003).

These results suggest that activation of similar PKC-dependent events might underlie certain nociceptive effects of other G_q-coupled metabotropic receptors. Indeed, bradykinin is found to potentiate or sensitize TRPV1 in a similar PKC-dependent pathway through the activation of B_2 receptors (Sugiura et al. 2002).

References

Burnstock, G., & Williams, M. (2000). P2 purinergic receptors: Modulation of cell function and therapeutic potential. *Journal of Pharmacology and Experimental Therapeutics, 295*, 862–869.

Caterina, M. J., & Julius, D. (2001). The vanilloid receptor: A molecular gateway to the pain pathway. *Annual Review of Neuroscience, 24*, 487–517.

Caterina, M. J., Schumacher, M. A., Tominaga, M., et al. (1997). The capsaicin receptor: A heat-activated ion channel in the pain pathway. *Nature, 389*, 816–824.

Caterina, M. J., Leffler, A., Malmberg, A. B., et al. (2000). Impaired nociception and pain sensation in mice lacking the capsaicin receptor. *Science, 288*, 306–313.

Davis, J. B., Gray, J., Gunthorpe, M. J., et al. (2000). Vanilloid receptor-1 is essential for inflammatory thermal hyperalgesia. *Nature, 405*, 183–187.

McCleskey, E. W., & Gold, M. S. (1999). Ion channels of nociception. *Annual Review of Physiology, 61*, 835–856.

Molliver, D. C., Cook, S. P., Carlsten, J. A., et al. (2002). ATP and UTP excite sensory neurons and induce CREB phosphorylation through the metabotropic receptor, P2Y2. *European Journal of Neuroscience, 16*, 1850–1860.

Moriyama, T., Iida, T., Kobayashi, K., et al. (2003). Possible involvement of $P2Y_2$ metabotropic receptors in

ATP-induced transient receptor potential vanilloid receptor 1–mediated thermal hypersensitivity. *Journal of Neuroscience, 23*, 6058–6062.

North, A. N., & Barnard, E. A. (1997). Nucleotide receptors. *Current Opinion in Neurobiology, 7*, 346–357.

Numazaki, M., Tominaga, T., Toyooka, H., et al. (2002). Direct phosphorylation of capsaicin receptor VR1 by PKCÎμ and identification of two target serine residues. *Journal of Biological Chemistry, 277*, 13375–13378.

Sugiura, T., Tominaga, M., Katsuya, H., et al. (2002). Bradykinin lowers the threshold temperature for heat activation of vanilloid receptor 1. *Journal of Neurophysiology, 88*, 544–548.

Szallasi, A., & Blumberg, P. M. (1999). Vanilloid (capsaicin) receptors and mechanisms. *Pharmacological Reviews, 51*, 159–211.

Tominaga, M., Caterina, M. J., Malmberg, A. B., et al. (1998). The cloned capsaicin receptor integrates multiple pain-producing stimuli. *Neuron, 21*, 531–543.

Tominaga, M., Wada, M., & Masu, M. (2001). Potentiation of capsaicin receptor activity by metabotropic ATP receptors as a possible mechanism for ATP-evoked pain and hyperalgesia. *Proceedings of National Academy of Sciences USA, 98*, 6951–6956.

Zimmermann, K., Reeh, P. W., & Averbeck, B. (2002). ATP can enhance the proton-induced CGRP release through P2Y receptors and secondary PGE(2) release in isolated rat dura mater. *Pain, 97*, 259–265.

TRPV1 Modulation by PKC

Peter A. McNaughton
Department of Pharmacology, University of Cambridge, Cambridge, UK

Synonyms

PKC; Protein kinase C; VR1

Definition

TRPV1 is an ion channel located in the surface membrane of pain-sensitive nerve terminals or nociceptors, which is activated by capsaicin, heat, and protons. TRPV1 when open conducts electric charge, in the form of cations (mainly Na^+ and Ca^{2+}), and therefore makes the interior of the nociceptor more positive, leading to the generation of action potentials. The temperature threshold of TRPV1 is lowered when inflammatory mediators such as bradykinin or ATP interact with surface membrane receptors on nociceptive nerve terminals, and so the nerve terminal becomes hyperalgesic (hypersensitive to heat pain). Many inflammatory mediators act by stimulating an intracellular enzyme known as protein kinase C (a member of the large kinase family of enzymes), which attach phosphate groups to specific locations on the portion of the TRPV1 protein exposed to the intracellular milieu of the nociceptor terminal.

Characteristics

A role for PKC in nociceptor sensitization was first suspected in studies of the potent sensitizing agent bradykinin, a pro-inflammatory nonapeptide which is released from a larger precursor protein by proteolytic cleavage following tissue damage (Dray and Perkins 1993). Bradykinin is one of the most potent pain-producing substances known, and causing pain directly, it acts as a sensitizing agent which lowers the temperature threshold for the activation of heat pain in vivo (Mizumura and Kumazawa 1996). In isolated nociceptive neurons, action potentials are elicited by heat stimuli (Fig. 1a), just as they are in vivo, and the inward membrane current responsible can be recorded using the whole-cell voltage-clamp technique (Fig. 1b). The properties of the heat-gated ion current, such as its threshold and dependence on temperature, closely correspond to those of heat pain in vivo (Cesare and McNaughton 1996). Bradykinin is a potent agonist at the G-protein-coupled B2 receptor, leading to activation of G_q and phospholipase C β (PLCβ), followed by metabolism of phosphatidylinositol bisphosphate (PIP$_2$) and release of diacylglycerol (DAG) and inositol trisphosphate (IP$_3$). These two products have different actions: DAG activates protein kinase C (PKC), while IP$_3$ releases calcium from intracellular organelles. The critical member of this cascade which is responsible for causing sensitization of the heat-gated membrane current is PKC because the sensitizing effects of

a

b

TRPV1 Modulation by PKC, Fig. 1 Action potentials and generator currents elicited in a nociceptive neuron by a noxious heat stimulus. (**a**) Application of a 49 °C heat stimulus depolarizes a nociceptive neuron to threshold and elicits a train of action potentials. This temperature did not damage the neuron as repeated application of the stimulus gave similar results. (**b**) Dependence of the membrane current on temperature in a heat-sensitive neuron before and after activation of PKC by phorbol myristate acetate (PMA), a PKC-specific activator. After activation of PKC, the activation threshold shifts to lower temperatures and the magnitude of the current increases

bradykinin are mimicked by direct PKC activation (Fig. 1b), are antagonized by PKC inhibitors such as staurosporine, and are promoted by phosphatase inhibitors such as calyculin A (Cesare and McNaughton 1996, 1997).

TRPV1 is activated by heat stimuli and is the molecule responsible for the heat hyperalgesia caused by inflammation, because no heat hyperalgesia is seen in animals from which TRPV1 has been genetically deleted (Davis et al. 2000; Caterina et al. 2000). Two main models have been proposed for the molecular basis of this hyperalgesia. In experiments on isolated neurons (see above), PKC has been identified as a critical mediator of heat hyperalgesia. PKC acts by phosphorylating serine or threonine residues on its target proteins, and phosphorylation sites relevant for the process of sensitization have been identified by the use of site-directed mutagenesis of individual serine or threonine residues, followed by expression of the mutant TRPV1 in a heterologous expression system (Numazaki et al. 2002). Mutation of two serines to alanine, which cannot be phosphorylated, abolished sensitization following PKC activation (Fig. 2). The isoform of PKC responsible has also been identified (Cesare et al. 1999). Of the 11 known isoforms of PKC, only 5, namely, PKCβI, βII, δ, ε, and ζ, are expressed to any significant extent in sensory neurons, and of these, only

PKCε is translocated to the membrane following exposure to bradykinin, suggesting that it is this isoform which is responsible for sensitization of TRPV1. A central role for PKCε in sensitization was confirmed by showing that constitutively active PKCε incorporated into the cell was indeed capable of sensitizing the heat-gated current and that the incorporation of a specific PKCε inhibitor into the cell largely abolished sensitization (Cesare et al. 1999). A specific role for PKCε in nociceptor sensitization is also suggested by the reduced heat hyperalgesia seen in studies using PKCε knockout mice (Khasar et al. 1999). The model of specific phosphorylation of TRPV1 by PKCε is shown in outline in Fig. 2.

In an alternative model (Chuang et al. 2001), TRPV1 is proposed to be tonically inhibited by PIP_2 in the neuronal cell membrane. Removal of PIP_2, when bradykinin activates PLCβ, is then proposed to release TRPV1 from inhibition. This model is supported by experiments in which application of PLCβ to isolated membrane patches in order to break down PIP_2, or removal of PIP_2 with an antibody, potentiated gating of TRPV1 (Chuang et al. 2001). The proposal also draws on work carried out on other ion channels, particularly on inward rectifier potassium channels, where a modulation of ion channel function by PIP_2 is well established (Hilgemann 2003). It has yet to be clearly established whether this

TRPV1 Modulation by PKC, Fig. 2 Pathways leading to heat hyperalgesia. Binding of bradykinin to its receptor activates PLCβ, which releases DAG from PIP₂, leading to activation of PKCε, which in turn phosphorylates TRPV1 at two serine residues, leading to the lowering of heat threshold shown in Fig. 1b

mechanism plays an important role in TRPV1 sensitization, but the observations cited above suggest that it plays at most a minor role by comparison with the PKCε pathway summarized in Fig. 2.

References

Caterina, M. J., Leffler, A., Malmberget, A. B., et al. (2000). Impaired nociception and pain sensation in mice lacking the capsaicin receptor. *Science, 288*, 306–313.

Cesare, P., & McNaughton, P. A. (1996). A novel heat-activated current in nociceptive neurons, and its sensitization by bradykinin. *Proceedings of National Academy of Sciences USA, 93*, 15435–15439.

Cesare, P., & McNaughton, P. A. (1997). Peripheral pain mechanisms. *Current Opinion in Neurobiology, 7*, 493–499.

Cesare, P., Dekker, L. V., Sardini, A., et al. (1999). Specific involvement of PKC-epsilon in sensitization of the neuronal response to painful heat. *Neuron, 23*, 617–624.

Chuang, H. H., Prescott, E. D., Kong, H., et al. (2001). Bradykinin and nerve growth factor release the capsaicin receptor from PtdIns(4,5)P2-mediated inhibition. *Nature, 411*, 957–962.

Davis, J. B., Gray, J., Gunthorpe, M. J., et al. (2000). Vanilloid receptor-1 is essential for inflammatory thermal hyperalgesia. *Nature, 405*, 183–187.

Dray, A., & Perkins, M. (1993). Bradykinin and inflammatory pain. *Trends in Neurosciences, 16*, 99–104.

Hilgemann, D. W. (2003). Getting ready for the decade of the lipids. *Annual Review of Physiology, 65*, 697–700.

Khasar, S. G., Lin, Y.-H., Martin, A., et al. (1999). A novel nociceptor signaling pathway revealed in protein kinase c epsilon mutant mice. *Neuron, 24*, 253–260.

Mizumura, K., & Kumazawa, T. (1996). Modification of nociceptor responses by inflammatory mediators and second messengers implicated in their action – A study in canine testicular polymodal receptors. *Progress in Brain Research, 113*, 115–141.

Numazaki, M., Tominaga, T., Toyooka, H., et al. (2002). Direct phosphorylation of capsaicin receptor VR1 by protein kinase Cepsilon and identification of two target serine residues. *Journal of Biological Chemistry, 277*, 13375–13378.

TRPV1 Receptor

▶ TRPV1 Modulation by P2Y Receptors

TRPV1 Receptor, Species Variability

Peter McIntyre
Department of Pharmacology, University of Melbourne, Melbourne, Australia

Synonyms

Capsaicin receptor; Member 1; Subfamily V; Transient receptor potential cation channel; TRPV1; Vanilloid receptor subtype 1; VR1

Definition

TRPV1 is a nonspecific cation channel of the ▶ transient receptor potential protein family, which is expressed in ▶ polymodal sensory neurons in the peripheral sensory nervous system and is activated by noxious heat (>43 °C), pH below 6.5, capsaicin, the pungent ingredient of chili peppers, resiniferatoxin and a number of other vanilloid, phorbol-related compounds as well as the endogenous activators anandamide and some products of lipoxygenase action on arachidonic acid.

Introduction

TRPV1 is the ion channel target of capsaicin and a range of nociceptive stimuli outlined in the definition above. It is a key marker of polymodal nociceptive neurons and is considered to be an excellent target for new drug therapies for the treatment of chronic pain. Understanding the structure and function of ion channel necessitates investigating the species-differences of the channel, since there are species selectivities in the action of agonists, like capsaicin and in the action of antagonists, like capsazepine.

Characteristics

Relatives of TRPV1

TRPV1 is a nonselective cation channel with high calcium permeability. It has a long cytoplasmic N-terminal tail with 6 ankyrin repeats and 6 hydrophobic domains (here called S1-S6) which are transmembrane domains and a putative pore-forming region between S5 and S6. It is a member of the superfamily of ion channels with six hydrophobic transmembrane domains and a putative pore-forming loop between the last two hydrophobic domains. This superfamily includes the voltage-activated potassium channels, hyperpolarization and cyclic nucleotide–gated channels (HCN), and the transient receptor potential (TRP) ion channels, which are involved in many sensory processes.

Splice Variants

TRPV1 was first isolated by expression cloning from rat sensory neurons (Caterina et al. 1997) and has subsequently been isolated from human (McIntyre et al. 2001; Cortright et al. 2001; Hayes et al. 2000), guinea pig (Savidge et al. 2002), mouse (Accession number AJ620495), rabbit (Accession number AY487342), and chicken (Jordt and Julius 2002). There are 58 sequences of vertebrate TRPV1 in the NCBI Refseq database. Of these, 11 are cDNAs rather than predicted sequences from genomic sequencing (see Table 1). Cloning and sequencing of the human and mouse genes has shown that the trpv1 gene has 16 exons in mouse and human predicted by Ensembl with 15 coding exons and that there are four splice variants of mRNA. The TRPV1 gene is found at position 17p13.2 in the human genome and in a highly ▶ syntenic region on chromosome 10 in the rat genome (Xue et al. 2001) and chromosome 11 in the mouse genome (Caterina et al. 2000). The predicted protein size is 839 amino acids in human, guinea pig, and mouse (~95 kD); 838 amino acids in rat; 842 in the rabbit; and 843 (96.5 kD) in chicken. TRPV1 is a glycoprotein and the rat protein is glycosylated at asparagine 604, migrates with a higher apparent molecular mass in its glycosylated form, and can form tetramers that are likely to be the functional form of the channel (Kedei et al. 2001).

An N-terminal splice variant, with most of the ankyrin domains and the cytoplasmic tail missing, has been observed (Schumacher et al. 2000b). The transcript termed VR.5'sv is not functional, but the protein is expressed in rat kidney and this splice variant is expressed at low levels in dorsal root ganglia compared to TRPV1 (Sanchez et al. 2001). The importance of this transcript, if any, is still not understood, but it appears that an intact N-terminus is essential for TRPV1 to be functional. An earlier report of a TRPV1 transcript termed stretch-inhibited channel (SIC) with differences at both C- and N-termini (Schumacher et al. 2000a) seems to be derived from two independent genes and is unlikely to be a functional TRPV1 transcript (Xue et al. 2001). Another human TRPV1 splice

TRPV1 Receptor, Species Variability,
Table 1 TRPV1 normalized amino acid substitutions calculated using Clustal W

TRPV1 Receptor, Species Variability,
Table 2 TRPV1% amino acid identity (top triangle) and divergence (lower triangle) calculated using Clustal W

	rat	mouse	rabbit	human	guinea pig	chicken
rat	***	94.9	86.9	85.7	88.0	65.8
mouse	5.1	***	88.0	86.3	88.5	66.0
rabbit	14.2	13.0	***	88.0	87.5	65.2
human	15.4	14.5	13.0	***	86.5	64.0
guinea pig	12.6	12.3	13.6	14.3	***	66.0
chicken	41.0	40.9	41.3	43.1	40.7	***

variant termed TRPV1b is missing coding exon 6 resulting in a 60 amino acid deletion immediately proximal to S1 domain. This variant responds to heat (47 °C) but not to protons or capsaicin (Lu et al. 2005).

Protein Sequence Alignment

Alignment of the proteins predicted from the cDNAs using Clustal W shows that the highest variation occurs at the N- and C-terminal regions of the protein and within the putative pore region. Overall sequence relatedness is shown in Table 1, and sequence identity is shown in Table 2. A limited alignment is shown in Fig. 1. The alignment shows that rat and mouse sequences have most identity (94.9 %) and that apart from these closely related sequences, the mammalian proteins determined so far share between 85.7 % and 88.5 % identity. The mammalian proteins have between 64 % and 66 % identity to the chicken protein.

Pharmacological Differences

Agonists The most extreme example of variation in pharmacological characteristics of TRPV1 is between the rat and chicken ▶ orthologues. Primary sensory neurons from chicks are responsive to heat (~45 °C) and low pH (pH 4) but not capsaicin (Marin-Burgin et al. 2000). The cloned chicken TRPV1 has these properties when expressed in *Xenopus* oocytes or mammalian cells (Jordt and Julius 2002). Jordt and Julius

showed, using chimeric receptors, that the molecular determinants for this difference lay in sequence within and between the S3 and S4 regions; specifically residues responsible for capsaicin sensitivity are tyrosine 511 and serine 512, whereas arginine 491 was shown to modulate the ratio of capsaicin-evoked current to pH-evoked current in electrophysiological experiments.

Differences in agonist sensitivity have been found within TRPV1 from mammalian species. The rabbit TRPV1 is unresponsive to capsaicin, and substitution of threonine 553 for isoleucine is largely responsible for this phenotype (Gavva et al. 2003). Rat TRPV1 responds well to the agonist phorbol 12-phenylacetate 13-acetate 20-homovanillate (PPAHV), but the human (McIntyre et al. 2001) and guinea pig (Savidge et al. 2002) channels do not. This difference has been mapped to a single conservative amino acid change. Mutation of leucine 547 to methionine enables the human TRPV1 to respond to PPAHV, and introducing the equivalent mutation into guinea pig TRPV1 (L549M) increased the agonist potency of PPAHV by more than tenfold in an intracellular calcium assay and increased the amplitude of the evoked current. The rat M547L mutation reduced the affinity of RTX binding (Phillips et al. 2004). Thus, amino acids within the S2-S4 region are important sites of agonist and antagonist interaction with TRPV1.

Antagonists There is a pharmacological difference in the ability of capsazepine to antagonize

```
rat          -------MEQRASLDSEES----ESPPQENSCLDPPDRDPNCKPPPVKPHIFTT-RSRTR
mouse        -------MEKWASLDSDES----EPPAQENSCPDPPDRDPNSKPPPAKPHIFAT-RSRTR
guinea pig   -------MKKRASVDSKES----EDPPQEDYSLDTLDVDANSKTPPAKPHTFSVSKSRNR
rabbit       -------MKRWVSLDSGES----EDPLPEDTCPDLLDGDSNAKPPAKPHIFSTAKSRSR
human        -------MKKWSSTDLGAA----ADPLQKDTCPDPLDGDPNSRPPPAKP-QLSTAKSRTR
chicken      MSSILEKMKKFGSSDIEESEVTDEHTDGEDSALETADNLQGTFSNKVQPSKSNIFARRGR
                   *::    *  *      :         .  ::  .  :  *   .   . .:*       * *

rat          LFGKGDSEEASPLDCPYEEGGLASCPIITVSSVLTIQRPGDGPASVRPSSQDSVSAG--E
mouse        LFGKGDSEEASPMDCPYEEGGLASCPIITVSSVVTLQRSVDGPTCLRQTSQDSVSTG-VE
guinea pig   LFGKSDLEESSPIDCSFREGEAASCPTITVSSVVTSPRPADGPTSTRQLTQDSIPTS-AE
rabbit       LFGKGDSEETSPMDCSYEEGELAPCPAITVSSVIIVQRSGDGPTCARQLSQDSVAAAGAE
human        LFGKGDSEEAFPVDCPHEEGELDSCPTITVSPVITIQRPGDGPTGARLLSQDSVAAS-TE
chicken      FVMGDCDKDMAPMDSFYQMD------HLMAPSVIKFHANMERGKLHKLLSTDSITGC-SE
                :..  ..  ::  *:.*  ...     :....*:  .:  :  :  **:.    *

rat          KPPRLYDRRSIFDAVAQSNCQELESLLPFLQRSKKRLTDSEFKDPETGKTCLLKAMLNLH
mouse        TPPRLYDRRSIFDAVAQSNCQELESLLSFLQKSKKRLTDSEFKDPETGKTCLLKAMLNLH
guinea pig   KPPLKFYDRRSIFDAVAQNNCQDLDSLLPFLQKSKKRLTDTEFKDPETGKTCLLKAMLNLH
rabbit       KPPLKLYDRRRIFEAVAQNNCQELESLLCFLQRSKKRLTDSEFKDPETGKTCLLKAMLNLH
human        KTLRLYDRRSIFEAVAQNNCQDLESLLLFLQKSKKHLTDNEFKDPETGKTCLLKAMLNLH
chicken      KAFKFYDRRRIFDAVARGSTKDLDDLLLYLNRTLKHLTDDEFKEPETGKTCLLKAMLNLH
                ..  ::****  **:.***:..  ::*:..**  :*:::.  *:***  ***:***********

rat          NGQNDTIALLLDVARKTDSLKQFVNASYTDSYYKGQTALHIAIERRNMTLVTLLVENGAD
mouse        NGQNDTIALLLDIARKTDSLKQFVNASYTDSYYKGQTALHIAIERRNMALVTLLVENGAD
guinea pig   NGQNDTISLLLDIARQTNSLKEFVNASYTDSYYRGQTALHIAIERRNMVLVTLLVENGAD
rabbit       SGQNDTIPLLLEIARQTDSLKEFVNASYTDSYYKGQTALHIAIERRNMALVTLLVENGAD
human        DGQNTTIPLLLEIARQTDSLKELVNASYTDSYYKGQTALHIAIERRNMALVTLLVENGAD
chicken      DGKNDTIPLLLDIAKKTGTLKEFVNAEYTDNYYKGQTALHIAIERRNMYLVKLLVQNGAD
                .*:*  **.***:::*:.:*  .:**:.***  .***.  **.:*************  **.***:****

rat          VQAAANGDFFKKTKGRPGFYFGELPLSLAACTNQLAIVKFLLQNSWQPADISARDSVGNT
mouse        VQAAANGDFFKKTKGRPGFYFGELPLSLAACTNQLAIVKFLLQNSWQPADISARDSVGNT
guinea pig   VQAAANGDFFKKTKGRPGFYFGELPLSLAACTNQLAIVKFLLQNSWQPADISARDSVGNT
rabbit       VQAAANGDFFKKTKGRPGFYFGELPLSLAACTNQLAIVKFLLQNSWQPADISARDSVGNT
human        VQAAAHGDFFKKTKGRPGFYFGELPLSLAACTNQLGIVKFLLQNSWQTADISARDSVGNT
chicken      VHARACGEFFRKIKGKPGFYFGELPLSLAACTNQLCIVKFLLENPYQAADIAAEDSMGNM
                *:*  *  *:**:*  *:.**:*  ********  ******.*  .:*.***:*.**:**

rat          VLHALVEVADNTVDNTKFVTSMYNEILILGAKLHPTLKLEEITNRKGLTPLALAASSGKI
mouse        VLHALVEVADNTADNTKFVTNMYNEILILGAKLHPTLKLEELTNKKGLTPLALAASSGKI
guinea pig   VLHALVEVADNTKFVTSMYNEILILGAKLYPTLKLEELTNKKGFTPLALAASSGKI
rabbit       VLHALVEVADNTPDNTKFVTSMYNEILILGAKLHPTLKLEELINKKGLTPLALAAGSGKI
human        VLHALVEVADNTADNTKFVTSMYNEILILGAKLHPTLKLEELTNKKGMMPLALAAGTGKI
chicken      VLHTLVEIADNTKDNTKFVTKMYNNILILGAKINPILKLEELTNKKGLTPLTLAAKTGKI
                ***:***:****  *******.  ***.*******:  *  *****:  *  *****:  *:**  **:***   :***

rat          GVLAYILQREIHEPECRHLSRKFTEWAYGPVHSSLYDLSCIDTCEKNSVLEVIAYSSSET
mouse        GVLAYILQREIHEPECRHLSRKFTEWAYGPVHSSLYDLSCIDTCEKNSXLEVIAYSSSET
guinea pig   GVLAYILQREIPEPECRHLSRKFTEWAYGPVHSSLYDLSCIDTCEKNSVLEVIAYSSSET
rabbit       GVLAYILQREILEPECRHLSRKFTEWAYGPVHSSLYDLSCIDTCERNSVLEVIAYSSSET
human        GVLAYILQREIQBPECRHLSRKFTEWAYGPVHSSLYDLSCIDTCEKNSVLEVIAYSSSET
chicken      GIFAYILRREIKDPECRHLSRKFTEWAYGPVHSSLYDLSCIDTCEKNSVLEIIAYSS-ET
                *:.****.***   :.****************************:**  **.*****  **

rat          PNRHDMLLVEPLNRLLQDKWDRFVKRIFYFNFFVYCLYMIIFTAAAYYRPVEG--LPPYK
mouse        PNRHDMLLVEPLNRLLQDKWDRFVKRIFYFNFFVYCLYMIIFTAAAYYRPVEG--LPPYK
guinea pig   PNRHDMLLVEPLNRLLQDKWDRFVKRIFYFNFFIYCLYMIIFTMAAYYRPVDG--LPPYK
rabbit       PNRHDMLLVEPLNRLLQDKWDRVVKRIFYFNFFVYCLYMIIFTTAAYYRPVDG--LPPYK
human        PNRHDMLLVEPLNRLLQDKWDRFVKRIFYFNFLVYCLYMIIFTMAAYYRPVDG--LPPFK
chicken      PNRHEMLLVEPLNRLLQDKWDRFVKHLFYFNFFVYAIHISILTTAAYYRPVQKGDKPPFA
                ****:*************************..*:****:.*.::  *.*  *******:    **:
```
                                                                       S1

TRPV1 Receptor, Species Variability, Fig. 1 (continued)

```
rat          LKNTVGDYFRVTGEILSVSGGVYFFFRGIQYFLQRRPSLKSLFVDSYSEILFFVQSLFML
mouse        LNNTVGDYFRVTGEILSVSGGVYFFFRGIQYFLQRRPSLKSLFVDSYSEILFFVQSLFML
guinea pig   MKNTVGDYFRVTGEILSVIGGFHFFFRGIQYFLQRRPSVKTLFVDSYSEILFFVQSLFLL
rabbit       LRNLPGDYFRVTGEILSVSGGVYFFFRGIQYFLQRRPSMKALFVDSYSEMLFFVQALFML
human        MEKT-GDYFRVTGEILSVLGGVYFFFRGIQYFLQKRPSMKTLFVDSYSEMLFFLQSLFML
chicken      FGHSTGEYFRVTGEILSVLGGLYFFFRGIQYFVQRRPSLKTLIVDSYSEVLFFVHSLLLL
             : :   *:**********  **.:*********:*:.***:*:*:******:***::*::*
                       S2                                    S3
```

```
rat          VSVVLYFSQRKEYVASMVFSLAMGWTNMLYYTRGFQQMGIYAVMIEKMILRDLCRFMFVY
mouse        VSVVLYFSHRKEYVASMVFSLAMGWTNMLYYTRGFQQMGIYAVMIEKMILRDLCRFMFVY
guinea pig   ASVVLYFSHRKEYVACMVFSLALGWTNMLYYTRGFQQMGIYAVMIEKMILRDLCRFMFVY
rabbit       ATVVLYFSHCKEYVATMVFSLALGWINMLYYTRGFQQMGIYAVMIEKMILRDLCRFMFVY
human        ATVVLYFSHLKEYVASMVFSLALGWTNMLYYTRGFQQMGIYAVMIEKMILRDLCRFMFVY
chicken      SSVVLYFCGQELYVASMVFSLALGWANMLYYTRGFQQMGIYSVMIAKMILRDLCRFMFVY
             :*****.  :  *** ******:**  ****************:***  *************
                         S4
```

```
rat          LVFLFGFSTAVVTLIEDGKNNSLPMESTPHKCRGSACKPG-NSYNSLYSTCLELFKFTIG
mouse        LVFLFGFSTAVVTLIEDGKNNSLPVESPPHKCRGSACRPG-NSYNSLYSTCLELFKFTIG
guinea pig   LVFLFGFSTAVVTLIEDGKNESLSAE--PHRWRGPGCRSAKNSYNSLYSTCLELFKFTIG
rabbit       LVFLFGFSTAVVTLIEDGKNSSTSAESTSHRWRGFGCRSSDSSYNSLYSTCLELFKFTIG
human        IVFLFGFSTAVVTLIEDGKNDSLPSESTSHRWRGPACRPPDSSYNSLYSTCLELFKFTIG
chicken      LVFLLGFSTAVVTLIED--DNEGQDTNSSEYARCSHTKRGRTSYNSLYYTCLELFKFTIG
             :***:************   :..        ..   *    :    .****** **********
                  S5
```

```
rat          MGDLEFTENYDFKAVFIILLLAYVILTYILLLNMLIALMGETVNKIAQESKNIWKLQRAI
mouse        MGDLEFTENYDFKAVFIILLLAYVILTYILLLNMLIALMGETVNKIAQESKNIWKLQRAI
guinea pig   MGDLEFTENYDFKAVFIILLLAYVILTYILLLNMLIALMGETVNKIAQESKNIWKLQRAI
rabbit       MGDLEFTENYDFKAVFIILLLAYVILTYILLLNMLIALMGETVNKIAQESKSIWKLQRAI
human        MGDLEFTENYDFKAVFIILLLAYVILTYILLLNMLIALMGETVNKIAQESKNIWKLQRAI
chicken      MGDLEFTENYRFKSVFVILLVLYVILTYILLLNMLIALMGETVSKIAQESKSIWKLQRAI
             **********  **.:**:.***:  ***********************.*******.********
                                           S6
```

```
rat          TILDTEKSFLKCMRKAFRSGKLLQVGFTPDGKDDYRWCFRVDEVNWTTWNTNVGIINEDP
mouse        TILDTEKSFLKCMRKAFRSGKLLQVGFTPDGKDDFRWCFRVDEVNWTTWNTNVGIINEDP
guinea pig   TILDTEKSFLKCMRKAFRSGKLLQVGYTPDGKDDYRWCFRVDEVNWTTWNTNVGIINEDP
rabbit       TILDTEKGFLKCMRKAFRSGKLLQVGYTPDGKDDCRWCFRVDEVNWTTWNTNVGIINEDP
human        TILDTEKSFLKCMRKAFRSGKLLQVGYTPDGKDDYRWCFRVDEVNWTTWNTNVGIINEDP
chicken      TILDIENSYLNCLRRSFRSGKRVLVGITPDGQDDYRWCFRVDEVNWSTWNTNLGIINEDP
             **** *:.:*:*.*:*:*::***** :  ** ****:** **********:*****:*******
```

```
rat          GNCEGVKRTLSFSLRSGRVSGRNWKNFALVPLLRDASTRDRHATQQEEVQLKHYTG-SLK
mouse        GNCEGVKRTLSFSLRSGRVSXRNWKNFALVPLLRDASTRDRHSTQPEEVQLKHYTG-SLK
guinea pig   GNCEGVKRTLSFSLRSGRVSGRNWKNFALVPLLRDASTRDRHSAQPEEVHLKHFSG-SLK
rabbit       GNCEGVKRTLSFSLRSGRVSGRNWKNFALVPLLRDASTRDRHPXPPEDVHLRPFVG-SLK
human        GNCEGVKRTLSFSLRSSRVSGRHWKNFALVPLLREASARDRQSAQPEEVYLRQFSG-SLK
chicken      GCSGDLKRNPSYCIKPGRVSGKNWK--TLVPLLRDGSRREETPKLPEEIKLKPILEPYYE
             *  . ..:**. *:.::.  *** ::**  :******:.* *:. .  *:: *:       :
```

```
rat          PEDAEVFKDSMVPGEK
mouse        PEDAEVFKDSMAPGEK
guinea pig   PEDAEVFKDSAVPGEK
rabbit       PGDAELFKDSVAAAEK
human        PEDAEVFKSPAASGEK
chicken      PEDCETLKESLAKSV-
             * *.*  :*... . .
```

TRPV1 Receptor, Species Variability, Fig. 1 Alignment of the available predicted full-length proteins using Clustal W (*DNAstar*). Predicted ankyrin repeat regions are shaded *gray*, and hydrophobic transmembrane domains are *underlined* and marked *S1–S6*. The most conserved regions are the hydrophobic domains which are thought to be buried in the membrane and the least conserved regions are the N-terminal and C-terminal regions and the proximal part of the putative pore region between S5 and S6

the low pH and heat activation of the rat TRPV1 (McIntyre et al. 2001). Whereas it blocks these modalities well in human and guinea pig TRPV1, it has little or no effect on these modalities in rat TRPV1. The residues responsible for this change were mutated to the human equivalents and restored the phenotype, and thus, the difference was mapped to amino acids isoleucine 514, valine 518, and methionine 547 in the rat TRPV1 (Phillips et al. 2004).

In summary, several groups have shown that residues between hydrophobic domains S2 and S4 are important in recognition of agonists and antagonists and that small variations in sequence can have significant effects on the pharmacology of this ion channel.

References

Caterina, M. J., Schumacher, M. A., Tominaga, M., et al. (1997). The capsaicin receptor: A heat-activated ion channel in the pain pathway. *Nature, 389*, 816–824.

Caterina, M. J., Leffler, A., Malmberg, A. B., et al. (2000). Impaired nociception and pain sensation in mice lacking the capsaicin receptor. *Science, 288*, 306–313.

Cortright, D. N., Crandall, M., Sanchez, J. F., et al. (2001). The tissue distribution and functional characterization of human VR1. *Biochemical and Biophysical Research Communications, 281*, 1183–1189.

Gavva, N. R., Klionsky, L., et al. (2003). Molecular determinants of vanilloid sensitivity in TRPV1. *Journal of Biological Chemistry, 279*(19), 20283–20295.

Hayes, P., Meadows, H. J., Gunthorpe, M. J., et al. (2000). Cloning and functional expression of a human orthologue of rat vanilloid receptor-1. *Pain, 88*, 205–215.

Jordt, S. E., & Julius, D. (2002). Molecular basis for species-specific sensitivity to "hot" chili peppers. *Cell, 108*, 421–430.

Kedei, N., Szabo, T., Lile, J. D., et al. (2001). Analysis of the native quaternary structure of vanilloid receptor 1. *Journal of Biological Chemistry, 276*, 28613–28619.

Lu, G., Henderson, D., Liu, L., Reinhart, P. H., Simon, S. A. (2005). Cloning and functional characterization of TRPV1b, a new human vanilloid receptor. *Molecular Pharmacology, 67*, 1119–1127.

Marin-Burgin, A., Reppenhagen, S., Klusch, A., et al. (2000). Low-threshold heat response antagonized by capsazepine in chick sensory neurons, which are capsaicin-insensitive. *European Journal of Neuroscience, 12*, 3560–3566.

McIntyre, P., McLatchie, L. M., Chambers, A., et al. (2001). Pharmacological differences between the human and rat vanilloid receptor 1 (VR1). *British Journal of Pharmacology, 132*, 1084–1094.

Phillips, E., Reeve, A., et al. (2004). Identification of species-specific determinants of the action of the antagonist capsazepine and the agonist PPAHV on TRPV1. *Journal of Biological Chemistry, 279*(17), 17165–17172.

Sanchez, J. F., Krause, J. E., & Cortright, D. N. (2001). The distribution and regulation of vanilloid receptor VR1 and VR1 5' splice variant RNA expression in rat. *Neuroscience, 107*, 373–381.

Savidge, J., Davis, C., Shah, K., et al. (2002). Cloning and functional characterization of the guinea pig vanilloid receptor 1. *Neuropharmacology, 43*, 450–456.

Schumacher, M. A., Jong, B. E., Frey, S. L., et al. (2000a). The stretch-inactivated channel, a vanilloid receptor variant, is expressed in small-diameter sensory neurons in the rat. *Neuroscience Letters, 287*, 215–218.

Schumacher, M. A., Moff, I., Sudanagunta, S. P., et al. (2000b). Molecular cloning of an N-terminal splice variant of the capsaicin receptor. Loss of N-terminal domain suggests functional divergence among capsaicin receptor subtypes. *Journal of Biological Chemistry, 275*, 2756–2762.

Xue, Q., Yu, Y., Trilk, S. L., et al. (2001). The genomic organization of the gene encoding the vanilloid receptor: Evidence for multiple splice variants. *Genomics, 76*, 14–20.

TRPV1, Regulation by Acid

▶ TRPV1, Regulation by Protons

TRPV1, Regulation by Nerve Growth Factor

Sven-Eric Jordt
Department of Pharmacology, Yale University
School of Medicine, New Haven, CT, USA

Synonyms

Capsaicin receptor, regulation by NGF; TRPV1, regulation by NGF; Vanilloid receptor, regulation by NGF; VR1, regulation by NGF

Definition

▶ TRPV1, the receptor for ▶ capsaicin, the pungent ingredient in chili peppers, is a polymodal receptor for physical (heat) and chemical painful stimuli in sensory neurons. The neuropeptide ▶ nerve growth factor (NGF), which is generated during injury or inflammation, causes ▶ thermal hyperalgesia, an increased sensitivity to thermal stimuli. NGF binds to its receptor, TrkA, which activates intracellular signaling pathways that shift the thermal dependence of TRPV1 activation to lower temperatures.

Characteristics

Tissue damage produces a variety of mediators that activate or sensitize nociceptor terminals and elicit pain. These mediators include peptides such as bradykinin or nerve growth factor (NGF) that bind to their receptors on the nociceptor membrane and activate intracellular signaling pathways that lead to neural activation or sensitization.

NGF is the prototypical member of the neurotrophin family. During mammalian development, NGF is essential for the survival of sensory neurons and contributes to the maintenance of their phenotypes within the first 2 weeks after birth. In the adult, NGF is not required for cell survival.

The NGF peptide consists of three subunits: alpha, beta, and gamma. The active neurotrophic peptide is the beta subunit that is processed by proteases and contains 118 amino acids. NGF binds to a specific high-affinity receptor, ▶ TrkA, on the cell membrane (Kaplan and Miller 2000). TrkA belongs to the receptor tyrosine kinase family of membrane receptors. This peptide receptor protein family has >50 members and includes the receptors for other neurotrophins and epidermal growth factor (EGF), platelet-derived growth factor (PDGF), and insulin. The TrkA protein consists of an extracellular ligand-binding domain, a single transmembrane domain that transmits the extracellular signal and an intracellular domain that is responsible for intracellular signaling and for the formation of a signaling complex with other proteins.

While the role of NGF in the developing nervous system is firmly established, the concept of NGF as a nociceptive signaling molecule is relatively new. NGF has been shown to contribute to inflammation and inflammatory pain in numerous studies. During injury and inflammation, the concentration of NGF in the affected tissue is increased (Constantinou et al. 1994; Donnerer et al. 1992; Woolf et al. 1994). For example, inflammatory interleukins that are produced by macrophages and mast cells have been found to trigger the release of NGF from keratinocytes in the skin. NGF release is also activated by tumor necrosis factor (TNF alpha), a potent inflammatory peptide. When NGF is injected into rats, the animals develop reduced paw withdrawal thresholds for mechanical and thermal stimuli (Lewin et al. 1993). After injection of NGF, mechanical and thermal hyperalgesia develop with different time courses. Thermal hyperalgesia can be observed within minutes and is thought to be mediated through a peripheral mechanism, whereas mechanical hyperalgesia has a longer onset and is believed to be caused by ▶ central sensitization on the level of the spinal cord (Lewin et al. 1993, 1994).

The capsaicin receptor, TRPV1, is an ion channel that is activated by heat and by painful chemical stimuli such as acid (Caterina et al. 1997). Activation of TRPV1 by capsaicin does not only induce acute pain but it also causes thermal hyperalgesia. Heat hyperalgesia is absent in mice deficient in TRPV1 (Caterina et al. 2000; Davis et al. 2000). In addition, injection of NGF does not result in thermal hyperalgesia in these mice, indicating that TRPV1 might be the downstream target of NGF signaling in nociceptors. Indeed, when TrkA and TRPV1 were coexpressed in a heterologous expression system, activation of TrkA by NGF resulted in a pronounced sensitization of TRPV1 currents towards temperature, with channels already active below body temperature (Chuang et al. 2001).

Recent studies revealed the signaling mechanisms that are involved in the sensitization of TRPV1 by NGF. The initial signaling event after NGF binding is the dimerization of TrkA, followed by autocatalytic phosphorylation of tyrosine residues within the intracellular domain of the receptor (Kaplan and Miller 2000). This results in the activation of phospholipase C (PLC)

TRPV1, Regulation by Nerve Growth Factor, Fig. 1 Sensitization of TRPV1 by nerve growth factor – activated signaling pathways. The capsaicin receptor, TRPV1, is coexpressed with the receptor for nerve growth factor (NGF), TrkA, in the plasma membrane of many sensory neurons. Binding of NGF to TrkA leads to TrkA dimerization and subsequent activation of intracellular signaling. TrkA activates phospholipase C (PLC) and phosphoinositide-3-kinase (PI3K). Both enzymes catalyze reactions that reduce the concentration of the phospholipid phosphoinositol-(4,5)-bisphosphate (PIP2) in the plasma membrane. Whereas PLC hydrolyses PIP2, yielding inositol triphosphate (IP3) and diacylglycerol (DAG), PI3K phosphorylates PIP2, thereby generating PIP3. PIP2 inhibits TRPV1 currents by direct interaction with the receptor protein. PIP2 removal leads to sensitization of TRPV1, resulting in thermal hyperalgesia

that catalyzes the hydrolysis of the membrane phospholipid phosphoinositol-4,5-bisphosphate (PIP2), yielding inositol-1,4,5-trisphosphate (IP3) and diacylglycerol (DAG). PIP2 has been shown to inhibit TRPV1 through interaction with a C-terminal domain in the protein (Chuang et al. 2001; Prescott and Julius 2003). Hydrolysis of PIP2 relieves this inhibition and leads to sensitization of TRPV1 (Chuang et al. 2001). Biochemical studies revealed that TrkA, TRPV1, and PLC proteins interact in vitro and may form a signaling complex. If this interaction occurs, in vivo remains to be determined (Chuang et al. 2001) (Fig. 1).

In addition to PIP2 hydrolysis by PLC, TrkA activates intracellular kinase pathways that play important roles in the sensitization of TRPV1 in sensory neurons. One of the kinases involved is phosphoinositide-3-kinase (PI3K), a kinase that phosphorylates membrane phospholipids. PI3K contributes to TRPV1 sensitization by reducing PIP2 levels in the membrane. In addition, PI3K activates downstream kinases such as ▶ ERK that may induce long-term sensitization of the neuron (Bonnington and McNaughton 2003; Zhuang et al. 2004).

References

Bonnington, J. K., & McNaughton, P. A. (2003). Signalling pathways involved in the sensitisation of mouse nociceptive neurones by nerve growth factor. *The Journal of Physiology, 551*, 433–446.

Caterina, M. J., Schumacher, M. A., Tominaga, M., et al. (1997). The capsaicin receptor: A heat-activated ion channel in the pain pathway. *Nature, 389*, 816–824.

Caterina, M. J., Leffler, A., Malmberg, A. B., et al. (2000). Impaired nociception and pain sensation in mice lacking the capsaicin receptor. *Science, 288*, 306–313.

Chuang, H. H., Prescott, E. D., Kong, H., et al. (2001). Bradykinin and nerve growth factor release the capsaicin receptor from PtdIns(4,5)P2-mediated inhibition. *Nature, 411*, 957–962.

Constantinou, J., Reynolds, M. L., Woolf, C. J., et al. (1994). Nerve growth factor levels in developing rat skin: Upregulation following skin wounding. *Neuroreport, 5*, 2281–2284.

Davis, J. B., Gray, J., Gunthorpe, M. J., et al. (2000). Vanilloid receptor-1 is essential for inflammatory thermal hyperalgesia. *Nature, 405*, 183–187.

Donnerer, J., Schuligoi, R., & Stein, C. (1992). Increased content and transport of substance P and calcitonin gene-related peptide in sensory nerves innervating inflamed tissue: Evidence for a regulatory function of nerve growth factor in vivo. *Neuroscience, 49*, 693–698.

Kaplan, D. R., & Miller, F. D. (2000). Neurotrophin signal transduction in the nervous system. *Current Opinion in Neurobiology, 10*, 381–391.

Lewin, G. R., Ritter, A. M., & Mendell, L. M. (1993). Nerve growth factor-induced hyperalgesia in the neonatal and adult rat. *The Journal of Neuroscience, 13*, 2136–2148.

Lewin, G. R., Rueff, A., & Mendell, L. M. (1994). Peripheral and central mechanisms of NGF-induced hyperalgesia. *The European Journal of Neuroscience, 6*, 1903–1912.

Prescott, E. D., & Julius, D. (2003). A modular PIP2 binding site as a determinant of capsaicin receptor sensitivity. *Science, 300*, 1284–1288.

Woolf, C. J., Safieh-Garabedian, B., Ma, Q. P., et al. (1994). Nerve growth factor contributes to the generation of inflammatory sensory hypersensitivity. *Neuroscience, 62*, 327–331.

Zhuang, Z. Y., Xu, H., Clapham, D. E., et al. (2004). Phosphatidylinositol 3-kinase activates ERK in primary sensory neurons and mediates inflammatory heat hyperalgesia through TRPV1 sensitization. *The Journal of Neuroscience, 24*, 8300–8309.

TRPV1, Regulation by NGF

▶ TRPV1, Regulation by Nerve Growth Factor

TRPV1, Regulation by Protons

Sven-Eric Jordt
Department of Pharmacology, Yale University
School of Medicine, New Haven, CT, USA

Synonyms

Capsaicin receptor, regulation by protons; TRPV1, regulation by acid; Vanilloid receptor, regulation by protons; VR1, regulation by protons

Definition

▶ TRPV1, the receptor for ▶ capsaicin, the pungent ingredient in chili peppers, is a polymodal receptor for physical (heat) and chemical painful stimuli in sensory neurons. Extracellular acidification, caused by inflammation or tissue injury, increases TRPV1 activation through interaction with glutamate residues in the channel protein.

Characteristics

Tissue damage produces a variety of chemical mediators that activate or sensitize nociceptor terminals to elicit pain. An important component of this proalgesic response is local ▶ acidosis, namely, a reduction in extracellular pH to levels below the physiological norm of ~7.4 (Reeh and Steen 1996). Tissue acidosis has been observed in many painful clinical disorders, which include inflammation, skeletal muscle and cardiac ▶ ischemia, ▶ arthritis, hematoma, and bone cancer. In chronic cough and asthma, acidification is thought to contribute to the induction of cough through sensitization of sensory neurons. Tissue acidification is perceived as painful by humans. Test subjects report a significant correlation of the perceived intensity of pain with decreasing pH when acidic solution is perfused into the forearm muscle (Issberner et al. 1996).

Protons cause excitation or sensitization of sensory neurons by activating ionic currents across the neural membrane. Two different proton-activated inward cationic currents have been described (Bevan and Yeats 1991). One current is sustained and non-desensitizing. The other current is a transient, rapidly activating and desensitizing current. In concert with these currents, a proton-dependent reduction in background potassium conductances has been observed in some neurons that may contribute to an increase in neural excitability.

Protons are capable of modulating the activity of a number of cloned receptors and ion channels expressed by primary afferent nociceptors. Pharmacological, electrophysiological, and genetic

TRPV1, Regulation by Protons, Fig. 1 Activation and modulation or TRPV1 by protons (*left*): TRPV1 currents recorded from a TRPV1-expressing Xenopus oocyte are shown, recorded at a membrane voltage of −40 mV. TRPV1 channels were first activated for 20 s by acidic solution (pH 4.0) at room temperature, inducing a rapidly activating inward current. To record heat-activated currents, the bath temperature was then elevated from room temperature to 47 °C within 10 s (*red* bar). This procedure was repeated seven times in 2 min. intervals. During the fifth heat application, the bath pH was decreased from 7.6 to 6.3. Heat-activated currents are potentiated greater than twofold by the drop in pH. Localization of proton-sensing residues in the TRPV1 channel (*right*): TRPV1 has a transmembrane moiety with six putative transmembrane domains and a P-loop structure between transmembrane domains 5 and 6 that contributes to the ion permeation pathway. The N- and C-termini are localized intracellularly. Two glutamate residues (E) are essential for regulation by protons (E600) and activation by protons (E648). Both residues are in or near the putative pore domain of the channel protein

evidence suggest that the capsaicin receptor, TRPV1, is underlying the sustained proton-activated current, whereas different combinations of ► ASICs (acid-sensitive ion channels of the degenerin family) may give rise to transient acid-sensitive currents (Krishtal 2003). In addition, ATP-gated channels (P$_2$X receptors) and background potassium channels (TASK channels) have been shown to be modulated by protons.

The capsaicin receptor, TRPV1, is a nociceptor-specific cation channel that serves as the molecular target for capsaicin, the main pungent ingredient in "hot" chili peppers (Caterina et al. 1997). TRPV1 is a member of the transient receptor potential (TRP) ion channel gene family that includes several other channels involved in sensory transduction, such as ► TRPM8, the receptor for cold temperature and ► menthol, and TRPV2, a channel activated at high temperatures (Jordt et al. 2003). In addition to being sensitive to capsaicin and protons,

TRPV1 is also activated by noxious heat and is essential for ► bradykinin- or ► nerve growth factor-dependent ► thermal hyperalgesia (Caterina and Julius 2001) (Fig. 1).

TRPV1 is predominantly expressed in small-diameter unmyelinated neurons. In these neurons, measurements of capsaicin-activated currents correlate with the presence of the sustained proton-activated current (Bevan and Geppetti 1994; Petersen and LaMotte 1993). The sustained current can be reduced by the TRPV1 antagonists capsazepine (Liu and Simon 1994) and iodo-resiniferatoxin, indicating that capsaicin and protons use the same target to activate neural excitation. The effectiveness of TRPV1 antagonists, especially ► capsazepine, is strongly dependent on the species used in the individual studies. Whereas proton-induced fiber responses in guinea pigs are strongly reduced by capsazepine, responses in rodents are less affected. However, new high-affinity TRPV1 antagonists such as BCTC

(N-(4-tertiarybutylphenyl)-4-(3-chloropyridin-2-yl)-tetrahydropyrazine-1(2H)-carboxamide) are very effective at blocking proton-induced fiber responses in rats at nanomolar concentrations. Capsaicin-activated currents in sensory neurons show sensitivity to extracellular acidification. For example, a decrease in extracellular pH from 7.3 to 6.3 leads to a sevenfold potentiation of the capsaicin-activated current (300 nM) in dissociated rat sensory neurons (Petersen and LaMotte 1993). Single-ion channel recordings from sensory neurons show that acidification leads to a dramatic increase in the open probability of capsaicin-activated channels, whereas the single channel conductance is slightly diminished (Baumann and Martenson 2000).

The analysis of mice with a targeted deletion in the gene encoding for TRPV1 provided further evidence for an important role of TRPV1 in proton-dependent signaling in sensory neurons. In these mice, proton-activated sustained currents in DRG neurons were absent and proton-activated Ca^{2+} uptake was greatly diminished (Caterina et al. 2000). In addition, recordings in the skin-nerve preparation showed a dramatic reduction in the prevalence of proton-sensitive C-fibers.

In heterologous systems, TRPV1 channels are activated by extracellular protons in the absence of other activating stimuli, such as capsaicin or heat. When expressed in cultured mammalian cells, protons activate TRPV1 starting at pH around 6.5, with a half maximal activation at pH 5.4 (Tominaga et al. 1998). These currents are inhibited by TRPV1 antagonists. Like the situation in the native neuron, the degree of inhibition depends on the individual antagonist and differs between TRPV1 species homologs. For example, proton-activated human and guinea pig TRPV1 channels are more strongly inhibited by capsazepine than rat TRPV1 channels.

The potentiation of capsaicin-activated currents by protons in sensory neurons can be recapitulated with cloned TRPV1 channels. More importantly, it was found that protons also potentiate heat-activated TRPV1 currents, effectively lowering the temperature activation threshold of TRPV1 channels during persistent activation and increasing current amplitudes (Tominaga et al. 1998). This indicates that at body temperature and especially in inflammatory situations with elevated concentrations of other sensitizing factors (e.g., bradykinin or ▶ nerve growth factor), even moderate acidification can lead to strong activation of TRPV1.

The capsaicin receptor is a nonselective cation channel with a central ion pore. The channel is formed by four homomeric protein subunits. Each monomer has a transmembrane moiety with six predicted transmembrane domains and extended cytosolic N-terminal and C-terminal domains. The transmembrane moiety contains a region, the P-loop between the fifth and sixth transmembrane domains, which contributes to the ion conduction pathway of the central pore. Extensive site-directed mutagenesis studies pinpointed putative proton interaction sites to regions near or within the pore domain of the channel (Jordt et al. 2000; Welch et al. 2000). In particular, two glutamate residues, at positions E600 and E648 in the rat protein, were identified as essential for the regulation and activation of TRPV1 by protons. Negatively charged glutamate residues can serve as acceptors for protons. In addition, glutamate residues may interact with permeating cations near or in the conduction pathway. These mutagenesis studies also provided further evidence that protons play a dual role in the activation of TRPV1. Mutations introduced at position E600 dramatically shifted or eliminated the pH dependence of heat- and capsaicin-activated TRPV1 currents. On the other hand, mutations at position E648 led to a significant reduction in proton-activated currents, whereas heat- and capsaicin-activated currents maintained their pH dependence. These results indicated that the modulation of TRPV1 currents by protons and direct channel gating by protons can be separated at the structural level.

References

Baumann, T. K., & Martenson, M. E. (2000). Extracellular protons both increase the activity and reduce the conductance of capsaicin-gated channels. *The Journal of Neuroscience, 20*, RC80.

Bevan, S., & Geppetti, P. (1994). Protons: Small stimulants of capsaicin-sensitive sensory nerves. *Trends in Neurosciences, 17*, 509–512.

Bevan, S., & Yeats, J. (1991). Protons activate a cation conductance in a sub-population of rat dorsal root ganglion neurones. *Journal of Physiology (London), 433*, 145–161.

Caterina, M. J., & Julius, D. (2001). The vanilloid receptor: A molecular gateway to the pain pathway. *Annual Review of Neuroscience, 24*, 487–517.

Caterina, M. J., Schumacher, M. A., Tominaga, M., et al. (1997). The capsaicin receptor: A heat-activated ion channel in the pain pathway. *Nature, 389*, 816–824.

Caterina, M. J., Leffler, A., Malmberg, A. B., et al. (2000). Impaired nociception and pain sensation in mice lacking the capsaicin receptor. *Science, 288*, 306–313.

Issberner, U., Reeh, P. W., & Steen, K. H. (1996). Pain due to tissue acidosis: A mechanism for inflammatory and ischemic Myalgia? *Neuroscience Letters, 208*, 191–194.

Jordt, S. E., Tominaga, M., & Julius, D. (2000). Acid potentiation of the capsaicin receptor determined by a key extracellular site. *Proceedings of the National Academy of Sciences USA, 97*, 8134–8139.

Jordt, S. E., McKemy, D. D., & Julius, D. (2003). Lessons from peppers and peppermint: The molecular logic of thermosensation. *Current Opinion in Neurobiology, 13*, 487–492.

Krishtal, O. (2003). The ASICs: Signaling molecules? Modulators? *Trends in Neurosciences, 26*, 477–483.

Liu, L., & Simon, S. A. (1994). A rapid capsaicin-activated current in rat trigeminal ganglion neurons. *Proceedings of the National Academy of Sciences USA, 91*, 738–741.

Petersen, M., & LaMotte, R. H. (1993). Effect of protons on the inward current evoked by capsaicin in isolated dorsal root ganglion cells. *Pain, 54*, 37–42.

Reeh, P. W., & Steen, K. H. (1996). Tissue acidosis in nociception and pain. *Progress in Brain Research, 113*, 143–151.

Tominaga, M., Caterina, M. J., Malmberg, A. B., et al. (1998). The cloned capsaicin receptor integrates multiple pain-producing stimuli. *Neuron, 21*, 531–543.

Welch, J. M., Simon, S. A., & Reinhart, P. H. (2000). The activation mechanism of rat vanilloid receptor 1 by capsaicin involves the pore domain and differs from the activation by either acid or heat. *Proceedings of the National Academy of Sciences USA, 97*, 13889–13894.

TRPV1-Null Mice

Definition

Mice that have a disruption of the TRPV1 gene. These mice lack expression of the TRPV1 protein and have no response to capsaicin.

Cross-References

▶ IB4-Positive Neurons, Role in Inflammatory Pain

TSK

▶ Tampa Scale for Kinesiophobia

TTX

▶ Tetrodotoxin

Tuberculoid Leprosy

Definition

The relatively benign and least infectious type of leprosy, which is characterized by early severe damage to the nerves and by the presence of one to a few sharply defined, anesthetic, hypopigmented skin lesions.

Cross-References

▶ Hansen's Disease

Tubo-ovarian Complex

▶ Chronic Pelvic Pain, Pelvic Inflammatory Disease and Adhesions

Tumor (National Cancer Institute Terminology: Neoplasm)

Definition

An abnormal mass of tissue that results when cells divide more than they should or do not die when they should. Tumors may be benign (not cancerous) or malignant (cancerous).

Cross-References

▶ Cancer Pain Management, Treatment of Neuropathic Components

Tumor Necrosis Factor Alpha(α)

Synonyms

TNF alpha(α)

Definition

TNFα is a pro-inflammatory cytokine and member of the "TNF-superfamily" with algesic actions. It participates in inflammation, wound healing, and remodeling of tissue.

Cross-References

▶ Cytokines as Targets in the Treatment of Neuropathic Pain
▶ Inflammatory Neuritis
▶ Wallerian Degeneration

Twin Studies

Definition

The comparison of traits among pairs of monozygotic (MZ, identical) versus dizygotic (DZ, fraternal) twins. MZ twins are clones, sharing 100 % of their DNA sequence, whereas DZ twins share only 50 % (no more than any other siblings). If a painful pathology occurs more often in both individuals in an MZ twin pair than a DZ twin pair, it can be said to be *heritable*. A caveat is that this analysis is dependent on the assumption of equal environments of MZ and DZ twins, which may not be realistic.

Cross-References

▶ Heritable

Twitch-Obtained Intramuscular Stimulation

▶ Dry Needling

Twitch-Obtaining Intramuscular Stimulation

Definition

This is a technique of dry needling developed by Chu, using an EMG needle to perform dry needling.

Cross-References

▶ Dry Needling

Two Pore Domain K⁺ Channels

Definition

A large, structurally related, class of K^+ channels that provide a large fraction of the background conductance in many neurons. As their name implies, these channels are unique from most other channels in that they appear to have two pore or ion channel domains. They are modulated by a variety of physical (temperature, mechanical) and chemical stimuli (pH, anesthetics, lipids).

Cross-References

▶ Nociceptors, Cold Thermotransduction

Two-Way Scaling Models

Definition

Two-way scaling models yield only the group stimulus space; individual differences are lost.

Type 1 Reaction (Leprosy)

Synonyms

Reversal reaction

Definition

A leprosy reaction usually occurring during chemotherapy in borderline leprosy, representing a delayed hypersensitivity reaction with upgrading of cell-mediated immunity to *Mycobacterium leprae*. It is characterized by erythema, edema, and tenderness of preexisting skin lesions, neuritis with nerve damage, and fever.

Cross-References

▶ Hansen's Disease

Type 2 Reaction (Leprosy)

Synonyms

Erythema nodosum leprosum

Definition

A leprosy reaction resembling an Arthus reaction and representing humoral hypersensitivity. It is due to an antigen-antibody reaction with the formation of immune complexes at the site of antigen deposits in various tissues and gives rise to acute inflammatory foci. It is characterized by painful red nodules or plaques and distributes especially on the face and limbs. It may be associated with severe systemic and visceral symptoms.

Cross-References

▶ Hansen's Disease

Type-II Receptors

Definition

Type II receptors are G-protein coupled receptors with seven transmembrane domains.

Cross-References

▶ Opioid Modulation of Nociceptive Afferents In Vivo

Tyrosine Kinase A

Synonyms

trkA

Definition

The biological actions of neurotrophins are mediated by specific neurotrophin receptor tyrosine kinases (trks). The proto-oncogene trkA is now recognized as the primary high affinity receptor for nerve growth factor (NGF), while two related tyrosine kinase receptors, trkB and trkC, mediate the biological functions of brain-derived neurotrophic factor (BDNF) and neurotrophin-3 (NT-3), respectively.

Cross-References

▶ ERK Regulation in Sensory Neurons During Inflammation

U

U-1 Type Units

Definition

Small population of mechano-sensitive units identified in the guinea-pig's ureter (in vitro preparation), which are low-threshold mechanoreceptors concerned with the regulation of ureteric motility [all sensitive to peristaltic contractions of the ureter, responding to imposed distensions with a short latency and a low threshold (mean: 8 mmHg), and not displaying spontaneous activity or afterdischarges)] (Cervero and Sann 1989, *Journal of Physiology, 412*, 245).

Cross-References

▶ Visceral Pain Model, Kidney Stone Pain

References

Cervero, F., & Sann, H. (1989). Mechanically evoked responses of afferent fibers innervating the guinea-pig's ureter: an in vitro study. *Journal of Physiology, 412*, 245–66.

U-2 Type Units

Definition

Large population of mechano-sensitive units identified in the guinea-pig's ureter that have the characteristics of nociceptors (not sensitive to peristaltic contractions of the ureter, all responding to imposed distensions with a latency greater than 3 s and a high threshold (mean: 34 mmHg), showing spontaneous activity and afterdischarges to mechanical stimuli lasting up to several minutes). These units are also sensitive to strong local distensions imposed with an artificial kidney stone, to hypoxia of the ureteric mucosa, and to intraluminal application of bradykinin, capsaicin, and potassium (Cervero and Sann 1989, *Journal of Physiology, 412*, 245). The proportion of U-2 units is virtually identical to the proportion of visceral afferent fibers from the guinea-pig's ureter that contains substance P, calcitonin gene-related peptide (CGRP), or both peptides (Semenenko and Cervero 1992, *Neuroscience, 47*, 1197).

Cross-References

▶ Ureteral Nociceptors
▶ Visceral Pain Model, Kidney Stone Pain

References

Cervero, F., & Sann, H. (1989). Mechanically evoked responses of afferent fibers innervating the guinea-pig's ureter: an in vitro study. *Journal of Physiology, 412*, 245–66.
Semenenko, F. M., & Cervero, F. (1992). Afferent fibres from the guinea-pig ureter: size and peptide content of the dorsal root ganglion cells of origin. *Neuroscience, 47*, 197–201.

G.F. Gebhart, R.F. Schmidt (eds.), *Encyclopedia of Pain*, DOI 10.1007/978-3-642-28753-4,
© Springer-Verlag Berlin Heidelberg 2013

UAB Pain Behavior Scale

Definition

The UAB Pain Behavior Scale is an instrument where pain behavior is assessed by a trained clinician, who observes a series of behaviors demonstrated by the patient. It consists of 10 target behaviors, each of which contributes equally to a total score.

Cross-References

▶ Pain Inventories

Ulceration

Definition

Ulceration refers to a circumscribed inflammatory and often suppurating lesion on the skin or an internal mucous surface resulting in necrosis of tissue.

Cross-References

▶ NSAIDs and Cancer

Ulceration, Prevention by Nerve Decompression

A. Lee Dellon
Department of Plastic Surgery and Neurosurgery, Johns Hopkins University, Baltimore, USA
Dellon Institutes for Peripheral Nerve Surgery, Baltimore, MD, USA

Synonyms

Chemotherapy-induced neuropathy; Diabetic neuropathy; Large-fiber neuropathy; Sensibility

Definition

Ulceration in the lower extremities occurs in the presence of neuropathy, which usually implies loss of protective sensation and absence of pain. The magnitude of the problem may be envisioned by considering ▶ diabetic neuropathy; there are about 20 million diabetics in the United States, and 60 % will develop neuropathy, and one in six diabetics with neuropathy will develop an ulcer. The neuropathy that produces the loss of sensibility, however, can be associated with significant pain. While the mechanism of pain in neuropathy is not well understood, strategies are now available to identify the presence of neuropathy at an early stage, document progression of the neuropathy, and document improvement in sensibility after treatment. For those neuropathies that have an increased susceptibility to chronic nerve compression, such as the diffuse, distal, symmetrical form of diabetic neuropathy, or ▶ chemotherapy-induced neuropathy, decompression of the peroneal and tibial nerves at known sites of anatomic narrowing can relieve pain, restore sensibility, and, therefore, prevent ulceration and amputation.

Characteristics

Ulceration in the foot is due to pressure applied to the skin between some external object and the underlying foot skeleton. Painful small-fiber neuropathy is not associated with ulceration, because the normal functioning large-fiber populations of the peripheral nerve permit perception of pressure against the skin. Small-fiber neuropathy is best documented by cutaneous thermal threshold testing and skin biopsy, with histologic measurement of intraepidermal nerve fiber density (Boulton and Malik 1998). Electrodiagnostic testing (EDT) is normal in the patient with small-fiber neuropathy, because EDT measures impulse generation in the fastest conducting (large-fiber) population.

Neuropathy that involves both the small-fiber and large-fiber nerve populations creates the clinical conditions in which a patient can have both pain and the large-fiber symptoms of numbness, with associated loss of the perception of touch and vibration. The large-fiber population generates impulses due to ischemia, which can themselves be perceived as painful. Such patients typically complain of burning pain in their feet, shooting hot or sometimes cold flashes, while having an underlying heaviness and tightness in their feet. The disturbing buzzing and tingling sensations are termed paresthesias. The most common cause of this type of problem is the neuropathy in diabetics that is termed diffuse, distal, or symmetrical and affects primarily sensory, but can also affect motor, function (Boulton and Malik 1998). More recently, with the use of cisplatin and taxol for the treatment of cancer, a similar chemotherapy-induced neuropathy is being increasingly seen (Mollman 1990; Rowinsky et al. 1993). Documentation of large-fiber neuropathy can be obtained with EDT, or the measurement of the cutaneous vibratory or pressure threshold. Recently, with the addition of thalidomide to the chemotherapy regimen, a new chemotherapy-induced neuropathy has been documented with the Pressure-Specified Sensory Device, a painless method for evaluation of the large-fiber population for touch (Marx et al. 2001).

The approach to treatment involves neuropathic pain medication, such as tramadol and gabapentin, since the underlying cause of the neuropathy still cannot be treated (Harati et al. 1998; Backonja et al. 1998). Continued investigation of nonoperative modalities continue, such as that recently reported for low-intensity laser light (Bril et al. 2003), but this too failed to offer hope for clinical relief. In contrast, a source of optimism was first reported more than 15 years ago, related to the concept that in some neuropathies, like diabetes, the increased intraneural water content (Jakobsen 1978) and the decrease in the slow component of anterograde axoplasmic flow (Jakobsen and Sidenius 1980) render the nerve susceptible to chronic compression (Dellon et al. 1988): the streptozotocin-induced diabetic rat was more likely to develop sciatic nerve compression to a more profound degree than the banded nondiabetic control group of rats. Decompression of the tarsal tunnel in streptozotocin-induced

diabetic rats prevented them from developing the typical neuropathic walking track pattern observed in diabetic rats (blood sugar of 400) with an intact tarsal tunnel. This normal walking pattern in the diabetic rats without a site of anatomic compression continued for the 1 year of the study, half the rat's expected lifetime (Dellon et al. 1994). This research approach was then applied to a model of cisplatin chemotherapy-induced neuropathy, and the study demonstrated that the established neuropathy could be reversed by decompression of the tibial nerve in the tarsal tunnel (Tassler et al. 2000). This research has recently been confirmed for the diabetic rat model, with the documentation that adding an internal neurolysis to the sciatic nerve relieves the increased intraneural pressure and further improves peripheral nerve function, as demonstrated by walking track analysis (Kale et al. 2003).

In 1992, Dellon reported the first clinical series of patients with diabetic neuropathy who had decompression of multiple peripheral nerves, to treat the symptoms of stocking- and glove-type sensory symptoms; there was an overall 0 % improvement in function, not only for the median and ulnar nerves in the upper extremity, but also for the peroneal and tibial nerves in the lower extremity. Motor function was also improved. In the lower extremity, this approach incorporated the anatomic knowledge that the tarsal tunnel was not the homologue of the carpal tunnel but rather represented the forearm, and therefore, the traditional tarsal tunnel decompression did not address the site of anatomic narrowing that was responsible for the pressure upon the tibial nerve branches. Instead, Dellon demonstrated that the medial plantar tunnel was the homologue of the carpal tunnel. Dellon's approach decompresses four medial ankle tunnels. In 1995, independent confirmation of Dellon's approach to lower extremity nerve decompression was reported for the relief of painful neuropathy in the feet of diabetics by Wieman and Patel (see Table 1). The most recent review of this subject demonstrates that relief of pain may be expected in 0 % of patients, improvement in sensibility may be expected in 0 % of patients, existing ulcerations will heal, and future ulcerations and

Ulceration, Prevention by Nerve Decompression, Table 1 Results of peripheral nerve decompression in diabetic neuropathy

Study	Number of Nerves	Patients	Preoperative Ulcers	Amput	Results Improved	Recurrent ulceration
Dellon (1992)	31	22	0	0	Pain 5 %	0 %
					2PD 2 %	
Wieman and Patel (1995)	33	26	13	0	Pain 2 %	7 %
					2PD 2 %	
					Ulcer 3 %	
Chafee (2000)	58	36	11	6	Pain 6 %	0 %
					Touch 0 %	
					Ulcer n/a	
Aszmann et al. (2000)	16	12	0	0	2PD 9 %	0 %
Wood and Wood (2003)	33	33	0	0	Pain 0 %	0 %
					2PD 7 %	
Biddinger and Amend (2004)	15	22	0	0	Pain 0 %	0 %
					2PD 0 %	

Posterior tibial nerve: tarsal tunnel syndrome

amputations will be prevented by this approach (Dellon 2002). The most up-to-date information on this subject is available at dellonipns.com and neuropathyregistry.com, public information websites that track reports of peripheral nerve decompressions for relief of pain, restoration of sensation, and prevention of ulceration and amputation. This surgery is now being done in 40 states within the USA and 12 other countries.

Selection of the patient with symptoms of neuropathy for surgical decompression is based upon the following: (1) failure to improve with medical treatment of the cause of the neuropathy, e.g., achieve euglycemia for a diabetic and achieve euthyroid for hypothyroid; (2) failure to achieve pain relief with neuropathic medication either due to ineffectiveness or complications of the medications; (3) documented large-fiber neuropathy with axonal loss either with electrodiagnostic testing or testing with the Pressure-Specified Sensory Device; and (4) presence of a positive ▶ Tinel sign over the site of known anatomic narrowing, e.g., the median nerve at the wrist and the tibial nerve at the medial ankle. Patients with neuropathy of unknown etiology have the same possibility of relief of pain and recovery of sensibility as those with known etiology, like diabetic or chemotherapy-induced neuropathy, if there is a positive Tinel sign at the site of the proposed nerve decompression.

The most recent publications in this field, from a prospective, multicenter study offer Level I evidence that a positive Tinel sign correlates with reduction of pain and improvement in sensation (Dellon et al. 2012a) and that decompression of the tibial and peroneal nerves in the lower extremity prevents ulceration and amputation, and reduces admission to hospital for foot infection compared to traditional approaches (Dellon et al. 2012b), representing a huge health-care cost savings.

References

Aszmann, O. A., Kress, K. M., & Dellon, A. L. (2000). Results of decompression of peripheral nerves in diabetics: A prospective, blinded study. *Plastic and Reconstructive Surgery, 106*, 816–821.

Backonja, M., Beydoun, A., Edwards, K. R., & Gabapentin Diabetic Neuropathy Study Group. (1998). Gabapentin for symptomatic treatment of painful neuropathy in patients with diabetes mellitus: A randomized controlled trial. *Journal of the American Medical Association, 380*, 1831–1836.

Biddinger, K., & Amend, M. A. (2004). The role of surgical decompression for diabetic neuropathy. *Foot & Ankle Clinics North America, 9*, 239–254.

Boulton, A. J., & Malik, R. A. (1998). Diabetic neuropathy. *Medical Clinics of North America, 82*, 909–929.

Bril, V., Ngo, M., Ng, E., New, P., Gogov, S., & Skandarajah, S. (2003). The results of low-intensity laser therapy for painful diabetic sensorimotor polyneuropathy. *Journal of the Peripheral Nervous System, 8,* 1–78.

Chafee, H. (2000). Decompression of peripheral nerves for diabetic neuropathy. *Plastic and Reconstructive Surgery, 106,* 813–815.

Dellon, A. L. (1992). Treatment of symptoms of diabetic neuropathy by peripheral nerve decompression. *Plastic and Reconstructive Surgery, 89,* 689–697.

Dellon, A. L. (2002). Prevention of foot ulceration and amputation by decompression of peripheral nerves in patients with diabetic neuropathy. *Ostomy/Wound Management, 48,* 36–45.

Dellon, A. L., Mackinnon, S. E., & Seiler, W. A. (1988). Susceptibility of the diabetic nerve to chronic compression. *Annals of Plastic Surgery, 20,* 117–119.

Dellon, A. L., Dellon, E. S., & Seiler, W. A. (1994). Effect of tarsal tunnel decompression in the streptozotocin-induced diabetic rat. *Microsurgery, 15,* 265–268.

Dellon, A. L., Muse, V. L., & Nickerson, D. S. (2012a). A positive Tinel sign as predictor of pain relief or sensory recovery after decompression of chronic tibial nerve compression in patients with diabetic neuropathy. *Journal of Reconstructive Microsurgery, 28,* 235–240.

Dellon, A. L., Muse, V. L., & Nickerson, D. S. (2012b). Prevention of ulceration, amputation and reduction of hospitalization: Outcomes of a prospective multi-center trial of tibial neurolysis in patients with diabetic neuropathy. *Journal of Reconstructive Microsurgery, 28,* 241–246.

Harati, Y., Gooch, C., Swenson, M., Edelman, S., Greene, D., Raskin, P., Donofrio, P., Cornblath, D., Sachdeo, R., Siu, C. O., & Kamin, M. (1998). Double-blind randomized trial of tramadol for the treatment of the pain of diabetic neuropathy. *Neurology, 50,* 1842–1846.

Jakobsen, J. (1978). Peripheral nerves in early experimental diabetes. Expansion of the endoneurial space as a cause of increased water content. *Diabetologia, 14,* 113–119.

Jakobsen, J., & Sidenius, P. (1980). Decreased axonal transport of structural proteins in streptozotocin diabetic rats. *The Journal of Clinical Investigation, 66,* 292–296.

Kale, B., Yuksel, F., Celikoz, B., Sirvanci, S., Ergun, O., & Arbak, S. (2003). Effect of various nerve decompression procedures on the functions of distal limbs in streptozotocin-induced diabetic rats: Further optimism in diabetic neuropathy. *Plastic and Reconstructive Surgery, 111,* 2265–2272.

Marx, G. M., Pavlakis, N., McCowatt, S., Boyle, F. M., Levi, J. A., Bell, D. R., Cook, R., Biggs, M., Little, N., & Wheeler, H. R. (2001). Phase II study of thalidomide in the treatment of recurrent glioblastoma multiforme. *Journal of Neuro-Oncology, 54,* 31–38.

Mollman, J. E. (1990). Cisplatin neurotoxicity. *The New England Journal of Medicine, 322,* 126–129.

Rowinsky, E. K., Eisenhauer, E. A., Chaudhry, V., Griffin, J., & Cornblath, D. (1993). Clinical toxicities encountered with paclitaxel (Taxol). *Seminars in Onclology, 20,* 1–14.

Tassler, P. L., Dellon, A. L., Lesser, G., & Grossman, S. (2000). Utility of decompressive surgery in the prophylaxis and treatment of cisplatin neuropathy in adult rats. *Journal of Reconstructive Surgery, 16,* 457–463.

Wieman, T. J., & Patel, V. G. (1995). Treatment of hyperesthetic neuropathic pain in diabetics; decompression of the tarsal tunnel. *Annals of Surgery, 221,* 660–665.

Wood, W. A., & Wood, M. A. (2003). Decompression of peripheral nerve for diabetic neuropathy in the lower extremity. *The Journal of Foot and Ankle Surgery, 42,* 268–275.

Ulcerative Colitis

▶ Animal Models of Inflammatory Bowel Disease

Ultrasonography

▶ Ultrasound

Ultrasound

Peter Lau
Department of Clinical Research, Royal Newcastle Hospital, University of Newcastle, Newcastle, NSW, Australia

Synonyms

Sonography; Ultrasonography

Definition

Diagnostic ultrasound is a means of obtaining images of the internal structure of the body by recording the pattern of echoes obtained when

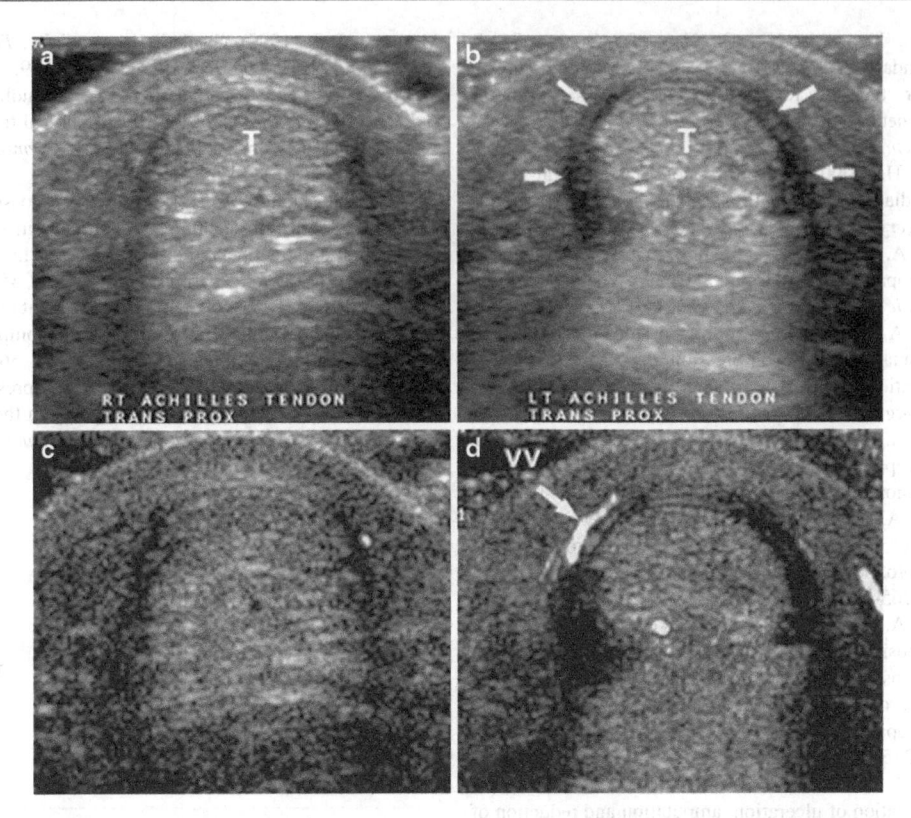

Ultrasound, Fig. 1 Ultrasonography of peritendinitis of an Achilles tendon. (**a**) Conventional transverse scan of a normal Achilles tendon. (**b**) Conventional transverse scan showing edema of the peritendinous tissues (*arrows*). (**c**) Color Doppler vascular scan of the normal tendon shown in (**a**). (**d**) Color Doppler vascular scan of the peritendinitis shown in (**b**). The dilated peritendinous vessels (*vv*) appear *pink* (Images provided by courtesy of Dr. J. Linklater, Sonic Health, Australia)

high-frequency acoustic waves are beamed at a selected region of the body.

Characteristics

Principles

Ultrasound does not depict tissues. Rather, it depicts the boundaries between tissues of different density. The identity of the tissues depicted, however, can be inferred from a knowledge of the anatomy of the region.

Ultrasound is particularly useful for demonstrating hollow organs or cystic pathology because there is a major difference in density between the wall of the organ or lesion and its lumen. Ultrasound also depicts well-laminated structures, such as muscles and tendons arranged in parallel layers. In such structures, it can reveal swelling, displacement, and discontinuities. It has, therefore, been applied in the pursuit of tendonitis (Figs. 1 and 2), tendinopathy and tears of tendons, muscle tears (Fig. 3), plantar fasciitis (Fig. 4), and Morton's neuroma (Fig. 5).

Advances in technology, such as higher-frequency transducers and color and power Doppler capability, allow modern ultrasound scanners to provide high-tech facilities such as 3-D and 4-D scans. Using the power Doppler facility, increased vascularity due to inflammation can be directly visualized (Figs. 1 and 2). 3-D ultrasound allows for more rapid acquisition of reproducible scans, particularly when ultrasound is used as a screening test (Blohmer et al. 1996). Under harmonic scanning, the ultrasound machine scans images at twice the frequency transmitted. This technique

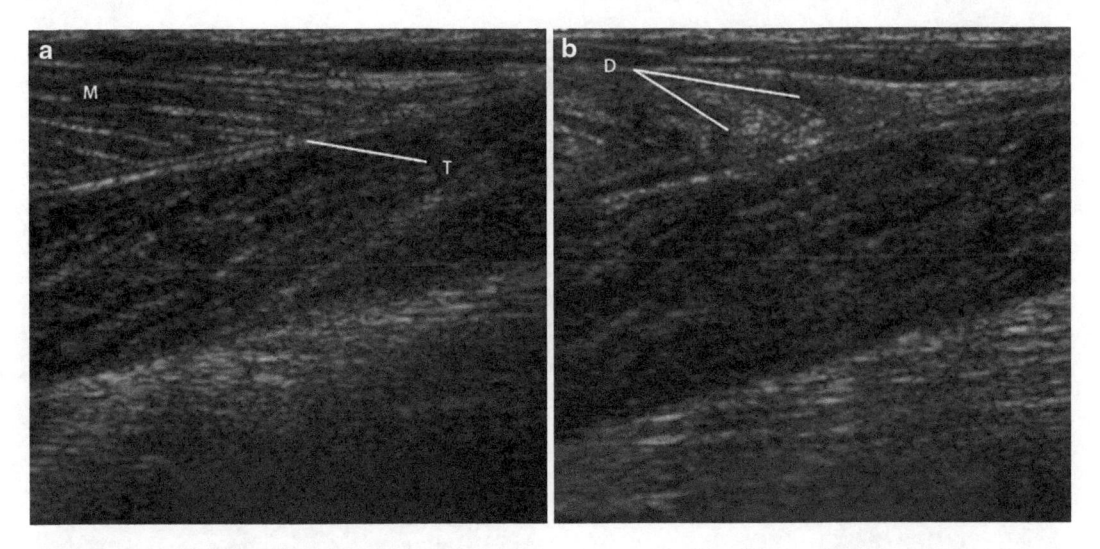

Ultrasound, Fig. 2 Ultrasound images of a tear of the myotendinous junction of the gastrocnemius. (**a**) Normal muscle (*M*) inserting into the intramuscular tendon (*T*). (**b**) Muscle sprain, showing disruption (*D*) of the muscle where it approaches its tendon (Images provided by courtesy of Dr. J. Linklater, Sonic Health, Australia)

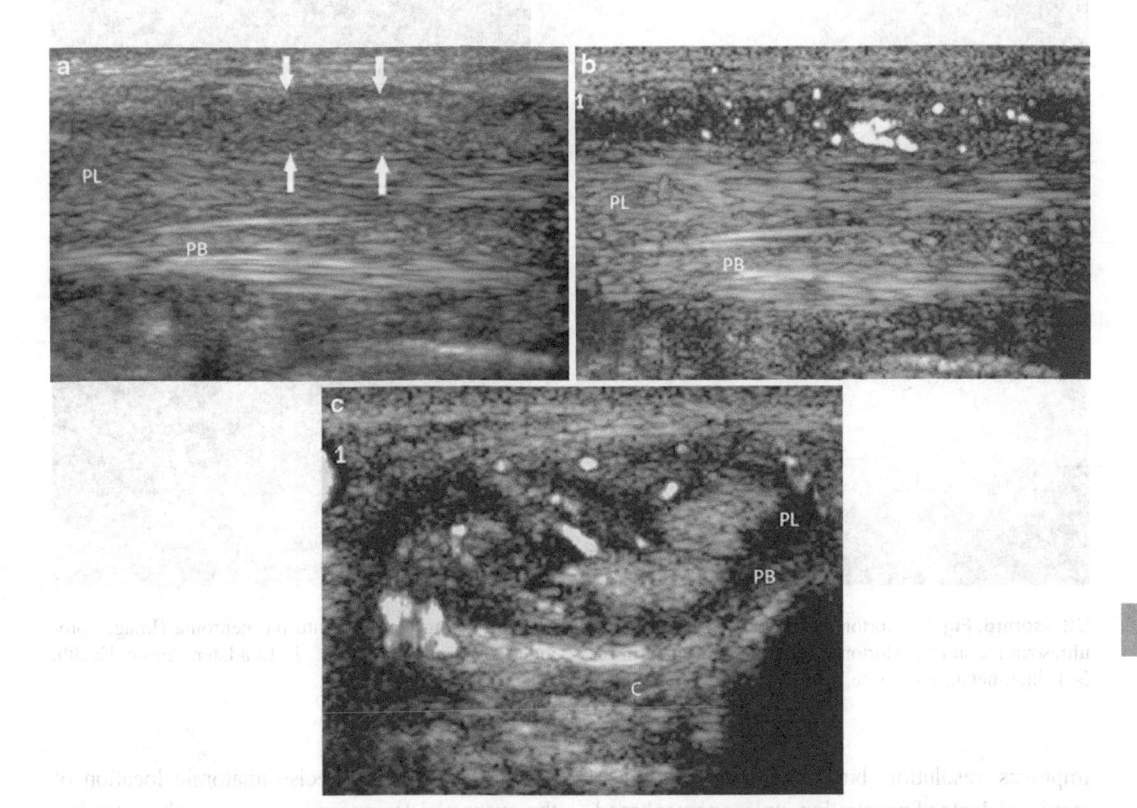

Ultrasound, Fig. 3 Ultrasound images of peritendinitis of the peroneal tendons. (**a**) Conventional longitudinal scan of peroneus longus (*PL*) and peroneus brevis (*PB*) with swollen peritendinous tissues (*arrows*). (**b**) Color Doppler vascular scan showing dilated vessels in the swollen peritendinous tissues. (**c**) Transverse vascular scan showing dilated vessels around the peroneus longus (*PL*) and peroneus brevis (*PB*) lateral to the calcaneus (*C*) (Images provided by courtesy of Dr J Linklater, Sonic Health, Australia)

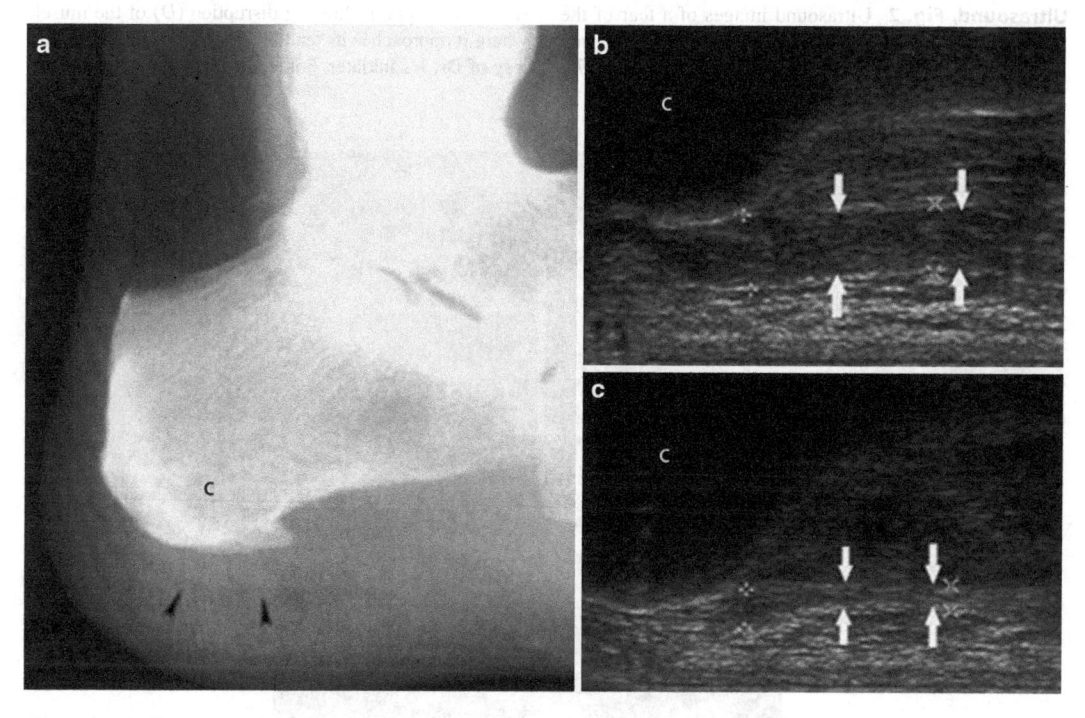

Ultrasound, Fig. 4 Plantar fasciitis. (**a**) A lateral plain radiograph showing the swollen plantar fascia as a thickened soft tissue density (*arrowheads*) beneath the calcaneus (*C*). (**b**) Longitudinal ultrasound scan showing the thickened plantar fascia (*arrows*). (**c**) The normal plantar fascia from the asymptomatic side (*arrows*) (Images provided by courtesy of Dr. J. Linklater, Sonic Health, Australia)

Ultrasound, Fig. 5 Morton's neuroma. (**a**) Longitudinal ultrasound scan of a Morton's neuroma and bursa in the 3–4 intermetatarsal space. (**b**) Longitudinal scan of a needle being directed onto the neuroma (Images provided by courtesy of Dr. J. Linklater, Sonic Health, Australia)

improves resolution but has the limitation of restricted depth of penetration, unless newer broad-band harmonic imaging techniques are used.

A particular advantage of ultrasonography is that it allows the examiner to interact with the patient. Scanning can be directed by and correlated with the precise anatomic location of the suspected lesion and features such as tenderness on palpation and aggravation of pain upon movement. Since ultrasonography is an innocuous procedure, scanning of the opposite side of the body can readily be undertaken in order to provide a comparison between symptomatic and asymptomatic sides (Figs. 1 and 3). This allows abnormalities on the symptomatic side to be checked for normal variants or age changes that might otherwise constitute false-positive findings.

The main disadvantages of ultrasonography are questionable reliability and validity. Technical artifacts are easily produced by ultrasound and can be misconstrued as lesions. Consequently there is a steep learning curve. For this reason, authorities recommend that ultrasound be performed only by experienced operators (Tyson 1995). However, no one has defined what the criteria for "experienced" are, and while this may secure reliability, it does not guarantee validity.

Validity

In the evaluation of chronic shoulder pain, when compared with arthrography and operative findings, the sensitivity and specificity of ultrasound for the detection of rotator cuff tears ranges from 60 % to 100 % (Stiles and Otte 1993). Missing, however, are the data that such tears are the cause of pain or that repairing such tears guarantees relief of pain. In orthopedic circles, it has been assumed that tears in the rotator cuff seen on ultrasound must be the cause of pain. The validity of this assumption is fatally challenged when ultrasound demonstrates the same findings in the contralateral, but asymptomatic, shoulder. Tears of the rotator cuff occur in totally asymptomatic individuals and with increasing frequency with age (Needell et al. 1996; Sher et al. 1995). Tears are not a surrogate for shoulder pain. For the detection of acute and chronic inflammation, such as in rheumatoid arthritis, ultrasound can detect humeroscapular erosions, synovitis, and bursitis better than plain radiography but is less effective than MRI.

In about 1995, there was a wave of enthusiasm in spinal sonography for assessing soft tissue trauma, spinal canal stenosis, and nerve root pathology. Claims were made that ultrasound measurements could demonstrate reductions in the size of the vertebral canal in patients with back pain and significant reductions in patients with neurogenic claudication and nerve root symptoms. Lesions were supposedly confirmed at surgery in over 80 % of cases. Consequently, ultrasound was promoted as a safe, non-invasive, atraumatic method of spinal imaging. Further claims were made that ultrasound showed a correlation of at least 90 % with MRI, plain radiography, and orthopedic and neurological examination. Alarmed by the unbridled application of ultrasound that these claims invited, the American Institute of Ultrasound Medicine (AIUM) issued the following official statement:

> There is insufficient evidence in the peer-reviewed medical literature establishing the value of diagnostic spinal ultrasound. Therefore, the AIUM states that, at this time, the use of diagnostic spinal ultrasound (for study of facet joints and capsules, nerve and fascial oedema and other subtle paraspinous abnormalities) for diagnostic evaluation, for evaluation of pain or radiculopathy syndromes and monitoring of therapy has no proven clinical utility.
>
> Diagnostic spinal ultrasound should be considered investigational. The AIUM urges investigators to perform proper double-blind research projects to evaluate the efficacy of these diagnostic spinal ultrasound examinations.

In other regions of the body, ultrasound is increasingly used to demonstrate inflamed or swollen tendons in patients with soft tissue pain. While this is attractive, it is arguably superfluous to do so. Tendinopathy is readily diagnosed on the basis of local tenderness on clinical examination and does not require ultrasound confirmation. Exercise may be prescribed and local anesthetic or corticosteroids can be injected, without recourse to ultrasound. Nevertheless, ultrasound can be used to guide injection therapy in a variety of sports injuries, such as "tennis elbow," "jumper's knee," and Achilles tendinopathy. Ultrasound is used to better visualize the abnormal tendons and to direct the needle accurately to the lesion (Lin et al. 1999).

Similarly, groin pain and hip pain are amendable to ultrasound evaluation. The common causes are adductor tendon lesions, osteitis pubis, and inguinal canal lesions, such as sports hernia or inguinal canal posterior wall deficiency. Ultrasound has previously been shown to be of value in diagnosing adductor tendon pathology. The ability to identify pathology of the abdominal muscular wall is of value too in directing further clinical management. However, data are still lacking that these lesions are the cause of pain. There is only a weak correlation between abdominal wall deficiency and groin pain, and a temporal relationship between the lesion and symptoms has not been established (Orchard et al. 1998).

Ultrasound has been used to assess carpal tunnel syndrome. One study has claimed that it is a more reliable method than nerve conduction testing (El Miedany et al. 2004), but nerve conduction itself may not be a valid criterion standard. A large proportion of the asymptomatic population exhibits abnormal conduction velocities through the carpal tunnel (Atroshi et al. 1999).

Ultrasound is particularly useful in the evaluation of visceral disorders. Accordingly it has a particular application in the assessment of acute pelvic pain in women. It has an established reputation for detecting gynecological disease, but it is also a useful modality for assessing non-gynecological pains due to diverticulitis, urinary tract calculi, appendicitis, or incarcerated hernia.

Summary

Ultrasound is an attractive imaging modality because it does not involve ionizing radiation, it allows for real-time assessment, and it is relatively portable. It has an established reputation for the assessment of visceral disorders and is promoted as the investigation of choice for soft tissue musculoskeletal disorders. Sorely lacking, however, are data that establish the validity of ultrasound as a means of establishing sources and causes of musculoskeletal pain or show that such establishment makes a significant difference to management.

Cross-References

▶ Ultrasound Therapy of Pain from the Musculoskeletal System

References

Atroshi, I., Gummesson, C., Johnsson, R., et al. (1999). Prevalence of carpal tunnel syndrome in a general population. *Journal of the American Medical Association, 282*, 153–158.

Blohmer, J. U., Bollmann, R., Heinrich, G., et al. (1996). Three-dimensional ultrasound study of the female breast. *Geburtshilfe und Frauenheilkunde, 56*, 161–165.

El Miedany, Y. M., Aty, S. A., & Ashour, S. (2004). Ultrasonography versus nerve conduction study in patients with carpal tunnel syndrome: Substantive or complementary tests? *Rheumatology, 43*, 887–895.

Lin, E. C., Middleton, W. D., & Teefey, S. A. (1999). Extended field of view sonography in musculoskeletal imaging. *Journal of Ultrasound in Medicine, 18*, 147–152.

Needell, S. D., Zlatkin, M. B., Sher, J. S., et al. (1996). MR imaging of the rotator cuff: Peritendinous and bone abnormalities in an asymptomatic population. *AJR. American Journal of Roentgenology, 166*, 863–867.

Orchard, J. W., Read, J., Neophyton, J., et al. (1998). Groin pain associated with ultrasound findings of inguinal canal posterior wall deficiency in Australian rules footballers. *Br J Sport Med, 32*, 134–139.

Sher, J. S., Uribe, J. W., Posada, A., et al. (1995). Abnormal findings on magnetic resonance images of asymptomatic shoulders. *J Bone Joint Surg, 77A*, 10–15.

Stiles, R. G., & Otte, M. T. (1993). Imaging of the shoulder. *Radiology, 188*, 603–613.

Tyson, L. L. (1995). Imaging of the painful shoulder. *Current Problems in Diagnostic Radiology, 24*, 110–140.

Ultrasound Delivery

Definition

Ultrasound delivery may be continuous or pulsed. Pulsed delivery results in less heating than continuous wave.

Cross-References

▶ Ultrasound Therapy of Pain from the Musculoskeletal System

Ultrasound Therapy

▶ Ultrasound Therapy of Pain from the Musculoskeletal System

Ultrasound Therapy of Pain from the Musculoskeletal System

Richard Crevenna
Medizinische Universität Wien, Wien, Austria

Synonyms

Ultrasound; Ultrasound therapy; Ultrasound treatment

Definition

► Ultrasound (US) is defined as acoustic vibration with frequencies above the audible range (i.e., greater than 20,000 Hz). Medical uses of US can be diagnostic or therapeutic. Therapeutic US involves the use of high-frequency acoustic energy to produce thermal and nonthermal effects in tissue.

Ultrasound selectively heats interfaces between tissues of different acoustic impedance because of reflection, formation of shear waves, and the high selective absorption in the superficial layers of tissue with a high coefficient of absorption (Table 1). It is characteristic of ultrasound that a selective increase in temperature may occur at the interface between tissues of different acoustic impedance (Fig. 1) (Lehmann et al. 1968).

Characteristics

Parameters for therapeutic US are:
- Frequency
- Power
- Effective radiating area
- Intensity
- Duration
- Additional parameters for pulsed ultrasound ("time on" and "time off")
- Pulse repetition frequency
- Duty factor (Table 2)

Frequency (► Frequency of Ultrasound Treatment): the most commonly used frequencies

Ultrasound Therapy of Pain from the Musculoskeletal System, Table 1 Ultrasound penetration into tissues

Half value depth of ultrasound	
Muscle	3.0 cm
Fatty tissue	8.0 cm
Bones	0.5 cm

are in the range of 0.8–1.1 MHz, although frequencies around 3.0 MHz are also fairly common. Power (► Power of Ultrasound) is total energy per unit time. Intensity (► Intensity of Ultrasound) is power per unit area. The World Health Organization and the International Electrical Commission both recommend limiting spatial average intensity to 3 W/cm^2. Most clinically used intensities of therapeutic US are in the 0.5–2.0 W/cm^2 range. Temperatures of up to 46 °C (114.8 °F) in deep tissues (e.g., bone-muscle interface) are easily achieved with US. Duration (► Duration of Ultrasound Treatment): 1–15 min, usually the duration should be at least 3 min. ► Ultrasound delivery may be continuous or pulsed. Pulsed delivery results in less heating than continuous wave.

Coupling media (► Coupling Media of Ultrasound) like mineral oil and several commercially available coupling gels have similar transmissivities to the reference standard of distilled degassed water (Warren et al. 1976). Predictable and controlled dosages of ultrasound in water can be reliably calculated using dosage correction factors. Those factors enable the user to compensate for applying ultrasound at different skin-to-applicator distances. However, dosage correction is not recommended at distances above 3 cm because gas bubble formation during treatment could seriously impede energy transmission. The dosage correction factors are easily remembered as 0 cm = 0 change, 1 cm = increase by 30 %, 2 cm = increase by 55 %, and 3 cm = increase by 80 %.

The most common ► technique of ultrasound application is the stroking technique. The applicator is moved slowly over an area of approximately 25 cm^2 (4 in.^{2}) in a circular or longitudinal manner. The applicator size (usually 5–10 cm^2)

Ultrasound Therapy of Pain from the Musculoskeletal System, Fig. 1 Temperature distribution 2 cm proximal to the joint space (Lehmann et al. 1968)

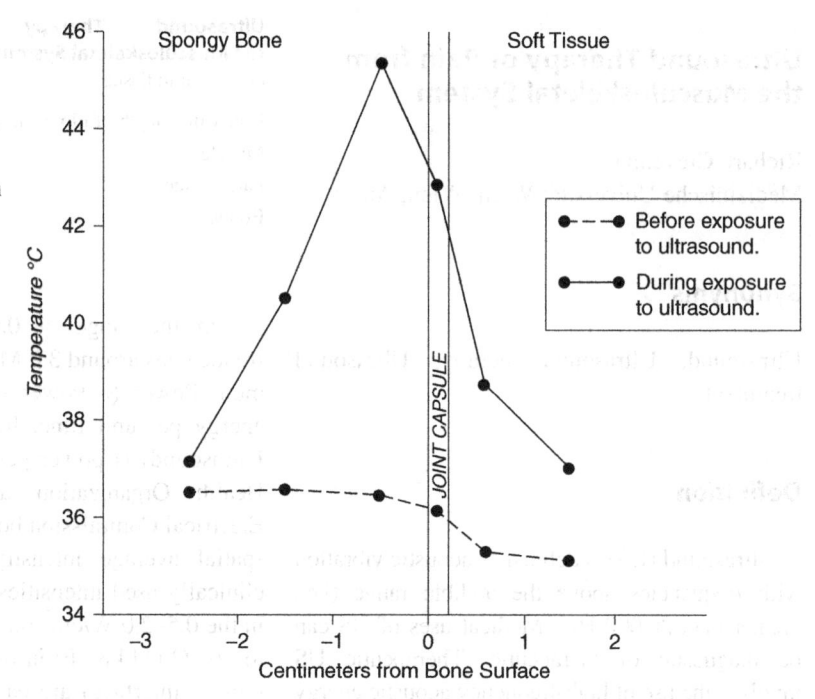

Ultrasound Therapy of Pain from the Musculoskeletal System, Table 2 Parameters for therapeutic ultrasound dependent on the target tissue and on the stage of the disease

Depth of the target from the surface	Superficial	⇒	Deep
Frequency	3 MHz		0.75 MHz
Intensity	0.1 W/cm^2		3 W/cm^2
Size of the target area to be treated	Small	⇒	Large
Time	3 min		15 min
Condition	Acute	⇒	Chronic
Intensity	0.1 W/cm^2		3 W/cm^2
Time	3 min		15 min
Pattern	Pulsed		Continuous
Heating required	No heat	⇒	Heat
Intensity	0.1 W/cm^2		3 W/cm^2
Time	3 min		15 min
Pattern	Pulsed		Continuous

limits the size of the area that can be treated. The stationary technique should generally be avoided because of the potential for standing waves and the production of hot spots.

Indications

Thermal effects (▶ Thermal Effects of Ultrasound) and mechanical effects (▶ Mechanical Effects of Ultrasound) on the target tissue result in an increased local metabolism, circulation, extensibility of connective tissue, tissue regeneration, and bone growth. The nonthermal effects of ultrasound include cavitation and microstreaming. Beneficial effects for the patient with musculoskeletal disorders should include improvements in pain, swelling, and range of motion.

The indication for ultrasound therapy is wide and covers many chronic and acute disorders. Ultrasound therapy with an intensity ranging from 0.5 to 2.0 W per square centimeter of body-surface area is widely used for the treatment of painful musculoskeletal disorders. However, the clinical efficacy of this approach for most such applications has not been confirmed (Gam and Johannsen 1995). Weak evidence for effectiveness of ultrasound was found for lateral epicondylitis (van der Windt et al. 1999). Gam and Johannsen (1995) hypothesized that ultrasound therapy might augment the effect of exercise therapy. In patients with symptomatic calcific tendinitis of the shoulder, ultrasound treatment helped resolve calcifications and was associated with short-term clinical improvement (Ebenbichler et al. 1999). Patients received 24 15-min sessions of either pulsed ultrasound (frequency 0.89 MHz, intensity 2.5 W per square centimeter, pulsed mode 1:4) or sham treatment to the area over the calcification. The first 15 treatments were given daily and the remainder were given three times a week for 3 weeks. Criteria for the adequate use of ultrasound have recently been proposed by Roebroeck et al. (1998); ultrasound should be used primarily (1) for soft tissue injuries, in particular of the shoulder and elbow; (2) in recent injuries and in the first phases of treatment; (3) to treat signs of inflammation (pain, swelling, restricted range of motion); and (4) in combination with other forms of therapy, especially exercise therapy.

Experiments on the stimulation of nerve regeneration (Hong et al. 1988) and on nerve conduction by ultrasound treatment (Currier et al. 1978; Kramer 1989) and findings of an anti-inflammatory effect of such treatment (El Hag et al. 1985) support the concept that ultrasound treatment might facilitate recovery from nerve compression (Hong et al. 1988). However, few studies report a benefit of ultrasound treatment in carpal tunnel syndrome under clinical conditions (Edel and Bergmann 1970). Results suggest there are satisfying short- to medium-term effects due to ultrasound treatment in patients with mild to moderate idiopathic carpal tunnel syndrome. In the study by Ebenbichler et al. (1998), 20 sessions of ultrasound treatment showed good short- and medium-term efficacy in patients with mild to moderate forms of idiopathic carpal tunnel syndrome. Serial ratings by patients of overall improvement suggest that ultrasound treatment would be best administered every day. Frequent treatment is time consuming, but ultrasound treatment could be performed by compliant patients at home. Findings need to be confirmed and ultrasound treatment will have to be compared with standard conservative and invasive treatment options.

Ultrasound Precautions (See General Heat Precautions)

Deep heating over an open epiphysis could result in either increased growth (from hyperemia) or decreased growth (from thermal injury). Avoiding US near pacemakers is reasonable because of potential thermal or mechanical injury to the pacemaker. Ultrasound over laminectomy sites could theoretically result in spinal cord heating (McLeod and Fowlow 1989; Oakley 1978). Metal per se should not be a contraindication to US. However, Lehmann (1990) cautions against the use of US near methyl methacrylate or high-density polyethylene because of their high coefficients of absorption.

General Heat Precautions

- Near brain, eyes, reproductive organs
- Pregnant or menstruating uterus
- Near pacemaker
- Near spine, laminectomy sites
- Malignancy
- Skeletal immaturity
- Arthroplasties
- Methyl methacrylate or high-density polyethylene

References

Currier, D. P., Greathouse, D., & Swift, T. (1978). Sensory nerve conduction: Effect of ultrasound. *Archives of Physical Medicine and Rehabilitation, 59*, 181–185.

Ebenbichler, G. R., Resch, K. L., Nicolakis, P., et al. (1998). Ultrasound treatment for treating the carpal tunnel syndrome: Randomised "sham" controlled trial. *BMJ, 316*, 731–735.

Ebenbichler, G. R., Erdogmus, C. B., Resch, K. L., et al. (1999). Ultrasound therapy for calcific tendinitis of the shoulder. *The New England Journal of Medicine, 340*, 1533–1538.

Edel, H., & Bergmann, P. (1970). Studies on the effect of ultrasonics in different dosage on the neural conduction velocity in man. *Archives Physical Therapy Leipz, 22*, 255–259.

El Hag, M., Coghlan, K., Christmas, P., et al. (1985). The anti-inflammatory effects of Dexamethasone and therapeutic ultrasound in oral surgery. *British Journal of Oral & Maxillofacial Surgery, 23*, 17–23.

Gam, A. N., & Johannsen, F. (1995). Ultrasound therapy in musculoskeletal disorders: A meta-analysis. *Pain, 63*, 85–91.

Hong, C. Z., Liu, H. H., & Yu, J. (1988). Ultrasound thermotherapy effect on the recovery of nerve conduction in experimental compression neuropathy. *Archives of Physical Medicine and Rehabilitation, 69*, 410–414.

Kramer, J. F. (1989). Effect of therapeutic ultrasound intensity on subcutaneous tissue temperature and ulnar nerve conduction velocity. *American Journal of Physical Medicine, 64*, 1–9.

Lehmann, J. F. (1990). *Therapeutic heat and cold* (4th ed.). Baltimore: Williams & Wilkins.

Lehmann, J. F., de Lateur, B. J., Warren, C. G., et al. (1968). Heating of joint structures by ultrasound. *Archives of Physical Medicine and Rehabilitation, 49*, 28–30.

McLeod, D. R., & Fowlow, S. B. (1989). Multiple malformation and exposure to therapeutic ultrasound during organogeneses. *American Journal of Medical Genetics, 34*, 317–319.

Oakley, E. M. (1978). Dangers and contraindication of therapeutic ultrasound. *Physiotherapy, 64*, 173–174.

Roebroeck, M. E., Dekker, J., & Oostendorp, R. A. B. (1998). The use of therapeutic ultrasound in physical therapy: Practice patterns in Dutch primary health care. *Physical Therapy, 78*, 470–478.

Van der Windt, D. A. W. M., van der Heijden, G. J. M. G., van den Berg, S. G. M., et al. (1999). Ultrasound therapy for musculoskeletal disorders: A systematic review. *Pain, 81*, 257–271.

Warren, C. G., Koblanski, J. N., & Sigelmann, R. A. (1976). Ultrasound coupling media: Their relative transmissivity. *Archives of Physical Medicine and Rehabilitation, 57*, 218–222.

Ultrasound Treatment

▶ Ultrasound Therapy of Pain from the Musculoskeletal System

Ultrastructure

Definition

Ultrastructure refers to a fine structure; structures seen using an electron microscope.

Cross-References

▶ Opioid Receptor Trafficking in Pain States

Ultraviolet Light

Synonyms

UV light

Definition

Ultraviolet (UV) radiation is classified in wave bands of UV-A (320–400 nm), UV-B (280–320 nm), and UV-C (below 280 nm). UV-A is of low energy and penetrates the skin up to the dermis, UV-B is absorbed by the dermis, and UV-C is of the highest energy and is completely absorbed by the epidermis. UV-C should not reach the earth's surface as it is fully abolished by the ozone layer.

Cross-References

▶ UV Erythema, a Model for Inducing Hyperalgesias

Ultraviolet Radiation

▶ UV Erythema, a Model for Inducing Hyperalgesias

Unconventional Medicine

▶ Alternative Medicine in Neuropathic Pain

Underdiagnosing of Pain in the Elderly

Definition

Refers to the tendency of clinicians to overlook pain as a significant problem. This can occur for several reasons: accepting the pain as a natural consequence of old age, holding the misconception that treatment options are less effective with increasing age, and misunderstanding the clinical relevance of laboratory studies about elevated pain thresholds in the elderly.

Cross-References

▶ Psychological Treatment of Pain in Older Populations

Underreporting of Pain

Definition

The elderly have a tendency to underreport pain, frequently for fear of having their negative expectations verified by further diagnostic procedures, or for fear of the procedures themselves, or just for the reason of accepting pain as a concomitant of old age. Cancer patients may also under report pain because they believe it might distract treatment providers from focusing on treatment of the disease itself.

Cross-References

▶ Psychological Treatment of Pain in Older Populations

Understanding the Neural Basis of Pain in Human Infants

▶ Infant Pain Mechanisms

Undertreated Pain in Children

▶ Evolution of Pediatric Pain Treatment

Unidimensional

Definition

A pain assessment tool that includes measurement of a single indicator of newborn pain behavioral or physiological or biochemical (hormonal).

Cross-References

▶ Pain Assessment in Neonates

Unilateral

Definition

On either the right or left side, but not crossing the midline.

Cross-References

▶ Hemicrania Continua

Unilateral Headache

▶ Hemicrania Continua (HC)

Unloading

▶ Traction

Unmyelinated Axon

Definition

Small diameter axon that is just enveloped by glial cytoplasm. Axons of this type conduct at low velocities and frequently present boutons en passant and numerous collaterals near the dendritic tree of the neuron of origin. Unmyelinated nerves are bundles of unmyelinated nerve fibers

Cross-References

▶ Spinothalamic Tract Neurons, Morphology
▶ Toxic Neuropathies

Unpleasant

▶ Dysesthesia, Assessment

Unpleasant Sensation

Definition

A sensation that is not agreeable.

Cross-References

▶ Dysesthesia, Assessment

(Un)provoked Vestibulodynia

▶ Gynecological Pain and Sexual Functioning

Unrecognized Pain in Children

▶ Evolution of Pediatric Pain Treatment

Upregulated

Definition

A state whereby a physiologic feedback loop causes a substance to increase the production or action of another substance.

Cross-References

▶ Cancer Pain Management, Orthopedic Surgery

Upregulation

Definition

Upregulation refers to an increase in the amount of mRNA and/or protein of a certain gene in a cell or tissue. It may be mediated by an increase of transcription or decrease of metabolism.

Cross-References

▶ COX-1 and COX-2 in Pain
▶ NSAIDs, Adverse Effects

Ureteral Nociceptors

Synonyms

U-2 type units

Definition

Ureteral units identified in the guinea pig's ureter that are sensitive to high-threshold mechanical stimuli, hypoxia of the ureteric mucosa, and intraluminal application of bradykinin, capsaicin, and potassium.

Cross-References

▶ U-2 Type Units
▶ Visceral Pain Model, Kidney Stone Pain

Urethral Dyssynergia

Definition

Contraction of the bladder against a contracting rather than relaxing internal sphincter.

Cross-References

▶ Clitoral Pain

Urinary Bladder

Definition

A hollow smooth muscle organ consisting of the bladder body and the bladder base. Its inside is lined by a mucous membrane and its outside is partly covered by peritoneal serosa and partly by fascia.

Cross-References

▶ Opioids and Bladder Pain/Function

Urinary Bladder Nocifensive Behaviors

▶ Nocifensive Behaviors of the Urinary Bladder

Urogenital Pain

▶ Dyspareunia and Vaginismus

Urticaria

Definition

Commonly "hives," characterized by sudden eruption of widespread wheals connected with itch sensations. Acute urticaria is often allergic in origin. Chronic urticaria has also other kinds of reasons.

Cross-References

▶ NSAIDs, Adverse Effects

UV Erythema

▶ UV-Induced Erythema

UV Erythema, a Model for Inducing Hyperalgesias

Roman Rukwied
Department of Anaesthesiology and Intensive Care Medicine, Faculty of Clinical Medicine Mannheim University of Heidelberg, Heidelberg, Germany

Synonyms

Sunburn; Ultraviolet Radiation; UV-B erythema

Definition

Dependent on the wavelength, solar ▶ ultraviolet (UV) light is classified into UV-A (320–400 nm), UV-B (280–320 nm), and UV-C (below 280 nm) wave bands. Of the UV band, the long wave UV-A is of lowest energy and penetrates the skin up to the dermis. UV-B is of higher energy, penetrates the epidermis, and is completely absorbed by the superficial layers of the dermis. UV-C radiation is completely absorbed by the epidermis. Noteworthy, both UV-B and UV-C are potentially detrimental to living cells and effective promoters of skin cancer, probably due to their damaging effect on the DNA.

Sunburned skin shows tenderness to heat and mechanical stimuli. Increased suprathreshold responses and reduced thresholds correspond to mechanical and heat hyperalgesias. Hyperalgesias developing in the damaged area, the primary zone, are termed ▶ primary hyperalgesias; those occurring in the area surrounding the injured site – the secondary zone – are called ▶ secondary hyperalgesia. The nature of the eliciting stimulus differentiates ▶ thermal hyperalgesia from ▶ mechanical hyperalgesia. The former develops in the primary hyperalgesic zone only, whereas the latter may occur in both primary and secondary hyperalgesic areas. Mechanical hyperalgesia to non-painful stimuli, for instance touch or light stroking, is termed "▶ allodynia" and differentiated from "pinprick hyperalgesia."

UV-A irradiation of the skin may cause pronounced long-lasting tanning of the skin. At higher energy levels, UV-A irradiation leads to increasing heat pain and accompanying dermal changes, including vascular damage, blisters, endothelial cell enlargement, extravasation of red blood cells, and extravascular fibrin deposition. However, apart from the consequences of possible heating trauma, no specific mechanical or thermal hyperalgesia occurs after UV-A irradiation. In contrast, exposure of the skin to UV-B induces a localized skin reddening – an ▶ erythema – and subacute primary hyperalgesias to mechanical and thermal stimuli. Sensitization of nociceptive afferents at the UV-treated site is most likely the underlying mechanism leading to primary hyperalgesia. If large skin areas are irradiated, even sensitization processes of dorsal horn neurons of the spinal cord may develop – probably due to an ongoing activity of C-nociceptors that induce and maintain secondary hyperalgesia. In addition, UV-B irradiation causes the release of various endogenous inflammatory mediators, many of which can promote or facilitate neuronal sensitization.

Characteristics

A plethora of experiments have been performed over the last decades to study the release of inflammatory mediators following UV radiation of the skin. It has been postulated that these mediators are considerably involved in the development of UV-B-induced erythema and primary hyperalgesia, respectively. Increased levels of cytokines, such as interleukins (IL) IL-1, IL-6, IL-8, and TNF-α, have been analyzed during the first few hours after UV-B treatment. Moreover, the levels of histamine and prostaglandins (PG), particularly PGD-2, PGE-2, and PGF-2alpha, increased 4–8 h and with maximum levels 18–24 h after UV exposure. Onset of the UV-B erythema paralleled the mediator production; administration of acetylsalicylic acid or indomethacin reduced the erythema development during 24 h by about 50 %. These findings suggest an enhanced cyclooxygenase activity after UV-B irradiation and an augmented arachidonic acid turnover with elevated prostanoid release, which is supported by the finding of an upregulation of inducible cyclooxygenase (COX-2) after UV-B irradiation. However, PGE-2 levels returned to baseline concentrations within 48 h after UV-B treatment, even though the erythema reaction was still 1.5 fold higher than at control sites (Hoffmann and Schmelz 1999). Apparently, prostanoids are not crucially involved in the late phase of a UV-B erythema; however, systemic (oral) treatment with nonsteroidal anti-inflammatory drugs (NSAIDs) significantly attenuated the erythema and primary hyperalgesias to heat and pinprick after UV-B skin irradiation (Bickel et al. 1998).

Obviously, there is an interaction between the mediators released upon UV-B irradiation that induce the erythema reaction and neuronal sensitization. Inflammatory mediators like PGE-2 or bradykinin (BK) are known to sensitize nociceptors, particularly to thermal stimuli (Schmelz et al. 2003). Moreover, prostaglandins sensitize bradykinin-B2-receptor responses and cause an increase in BK-induced pain. In UV-B-irradiated skin, administration of both des-Arg^{10}-kallidin (B^1 receptor agonist) and BK

(B2-receptor agonist) induced an increased pain sensation, augmented erythema, and enhanced axon reflex vasodilatation (Eisenbarth et al. 2004). Higher pain ratings and enlarged axon reflexes indicate a stronger activation of primary afferent nociceptors. This reaction might be promoted by the sensitization of nociceptors, either due to UV-B-induced release of prostaglandins with subsequently altered BK-responses or due to an enhanced BK-receptor expression and stimulation.

In recent years, the controlled exposure of the skin to UV light has been investigated as a novel method for the induction and investigation of hyperalgesias. Several comparative experiments have been performed to reveal the dose-dependency and time course of UV-induced erythema and hyperalgesias in humans (Bickel et al. 1998; Hoffmann and Schmelz 1999; Harrison et al. 2004; Gustorff et al. 2004a). The investigators found that UV-B-exposed skin areas reveal an increased blood flow that peaks at 12–24 h after irradiation and lasts at least 48 h. At irradiated skin sites, pronounced primary hyperalgesias to mechanical and thermal stimuli develop dose dependently after UV-B irradiation. Thermal hyperalgesia occurs within 6–24 h and remains persistent even 96 h following UV-B exposure (Bickel et al. 1998; Hoffmann and Schmelz 1999). A distinct primary hyperalgesia to mechanical impact stimuli develops 6–48 h after UV-B irradiation, which gradually declines within 5 days post UV-B treatment. Compared to the development of UV-B-induced primary hyperalgesias, no secondary hyperalgesia to mechanical stimuli was determined in the vicinity of UV-B-treated skin (Harrison et al. 2004). However, a recent investigation demonstrated stable secondary hyperalgesia to pinprick that continued over 10 h (Gustorff et al. 2004a). Notably, Gustorff et al. irradiated skin areas covering 5 cm in diameter, whereas UV-B irradiation of a skin patch either 1.5 cm in diameter (Hoffmann and Schmelz 1999) or a ring 6 cm in diameter and a 2-cm un-irradiated zone (Harrison et al. 2004) does not provoke any secondary hyperalgesia to mechanical (pinprick) stimuli. Apparently, induction of UV-B-evoked secondary pinprick hyperalgesia requires widespread

irradiation of rather large skin areas in order to induce a continuous ongoing activity of nociceptive afferent units to facilitate a ▶ central sensitization. It is well accepted that secondary (mechanical) hyperalgesia is based on sensitization processes of dorsal horn neurons of the spinal cord, promoting enhanced efficacy of synaptic transmission in the central nervous system. Capsaicin-insensitive A-fiber nociceptors (A-delta fibers) play a major role in the conveyance of pain and secondary hyperalgesia to pinprick (Magerl et al. 2001). However, no evidence could be provided to demonstrate the involvement of peripheral sensitization, e.g., of "sleeping" capsaicin-sensitive mechano-insensitive C-nociceptors in the secondary zone (Magerl et al. 2001), even though some data might suggest a peripheral component of secondary hyperalgesia (Serra et al. 2004). Employing the UV-B irradiation model, Gustorff and colleagues demonstrated that systemic (intravenous) administration of the μ-opioid receptor agonist remifentanil reduced the area of secondary pinprick hyperalgesia (Gustorff et al. 2004b). This study suggests an opioid-sensitive pathway of central sensitization in the mediation of secondary pinprick hyperalgesia. However, given that irradiation of large skin areas with UV-B comprises apparently both central sensitization processes and peripheral neuronal components to induce secondary hyperalgesia, antihyperalgesic effects of the investigated drug are difficult to attribute to either a peripheral or a central mode of action of the pharmaceutical.

The UV-B erythema model has been used to assess the peripheral antihyperalgesic and analgesic effects of drugs. It was demonstrated that systemic (oral) administration of a kappa-opioid agonist had no impact on thermal and mechanical hyperalgesia that had been induced experimentally in UV-B-irradiated skin spots (Bickel et al. 1998). In contrast, regionally administered morphine (IV regional anesthesia; "Bier block") significantly increased heat pain thresholds in UV-B skin, whereas pain ratings after mechanical impact stimulation were not affected (Koppert et al. 1999). These studies suggest that μ-opioid and delta-opioid receptors may play a predominant role in thermal hyperalgesia, which corroborates the fact that once inflammation is induced, opioid receptors are newly expressed, producing an upregulation in the nerve ending by their axonal transport to the peripheral terminal. A recent experiment demonstrated that an intravenous bolus injection of the local anesthetic lidocaine significantly reduced repetitive mechanical hyperalgesia at the site of UV-B-irradiated skin (Koppert et al. 2004). The authors concluded that systemically administered lidocaine exerts its antihyperalgesic effect on mechanical stimuli via a peripheral mechanism. Central blockade of spinal NMDA-receptors, on the other hand, had no impact on UV-B-induced mechanical hyperalgesia (Bishop et al., 2010).

UV-B irradiation does not evoke uniform increases in sensitivity of all peripheral nociceptor classes. It was shown recently after UV-B irradiation that A-delta nociceptors respond diminished to suprathreshold mechanical stimulation (Bishop et al. 2010). Similarly, "polymodal" heat-sensitive CMH-nociceptors did not change their firing pattern to suprathreshold mechanical stimuli, but revealed increased discharges upon thermal stimuli, whereas heat-insensitive CM-nociceptors exhibited enhanced mechanically evoked responses (Bishop et al. 2010). Apparently, primary mechanical hyperalgesia is differentially mediated from heat hyperalgesia, and UV-B inflammation alters the coding properties of C-nociceptors.

Few experimental models have been developed that reliably induce hyperalgesias in humans. Available models include electrically, chemically, and thermally induced hyperalgesias. The administration of chemicals, for instance, the injection of capsaicin, the pungent ingredient of chili peppers, or the topical application of mustard oil, induces an instant burning pain. Subsequently, an acute primary hyperalgesia to thermal or mechanical stimuli occurs and, with a delay of a few minutes, a secondary mechanical hyperalgesia to punctate pressure (pinprick hyperalgesia) and light touch (allodynia) develops. Chemically as well as electrically induced hyperalgesias represent acute nociceptive responses that gradually decline with time, usually within a few hours. Thus, they are inadequate models to assess the time-dependent dose response of analgesic drugs, which require rather persistent areas of hyperalgesia. Only recently,

a heat/capsaicin sensitization model has been established that synergistically combines repetitive thermal and chemical stimuli to induce stable areas of secondary hyperalgesias for 4 h and a persistent erythema lasting up to 12 h (Petersen and Rowbotham 1999). Thermal stimuli on their own may induce sustained and subacute ("chronic") hyperalgesias, as demonstrated in a freeze lesion model (Kilo et al. 1994) and after burn injury (Moiniche et al. 1993). These models induce a clear primary hyperalgesia to thermal and mechanical stimuli; burn injury, in addition, evokes a secondary hyperalgesia to pinprick and touch that lasts up to 72 h. However, burn injuries cause acute pain and provoke considerable tissue damage, whereas freeze lesions induce a persistent pigmentation of the skin. Given these limitations, only the UV erythema model meets the requirements of a reliable experimental method that induces a sustained and subacute primary hyperalgesia. Even better, UV-B irradiation is non-painful and may induce a secondary hyperalgesia, provided a large skin area had been irradiated. Therefore, the UV-B erythema represents a model of first choice to explore the efficacy of newly developed antihyperalgesic drugs.

References

Bickel, A., Dorfs, S., Schmelz, M., et al. (1998). Effects of antihyperalgesic drugs on experimentally induced hyperalgesia in man. *Pain, 76,* 317–325.

Bishop, T., Marchand, F., Young, A. R., Lewin, G. R., & McMahon, S. B. (2010). Ultraviolet-B-induced mechanical hyperalgesia: A role for peripheral sensitisation. *Pain, 150*(1), 141–152.

Eisenbarth, H., Rukwied, R., Petersen, M., et al. (2004). Sensitization to bradykinin B1 and B2 receptor activation in UV-B irradiated human skin. *Pain, 110,* 197–204.

Gustorff, B., Anzenhofer, S., Sycha, T., et al. (2004a). The sunburn pain model: the stability of primary and secondary hyperalgesia over 10 hours in a crossover setting. *Anesthesia and Analgesia, 98,* 173–177.

Gustorff, B., Hoechtl, K., Sycha, T., et al. (2004b). The effects of remifentanil and gabapentin on hyperalgesia in a new extended inflammatory skin pain model in healthy volunteers. *Anesthesia and Analgesia, 98,* 401–407.

Harrison, G. I., Young, A. R., & McMahon, S. B. (2004). Ultraviolet radiation-induced inflammation as a model for cutaneous hyperalgesia. *The Journal of Investigative Dermatology, 122*, 183–189.

Hoffmann, R. T., & Schmelz, M. (1999). Time course of UVA- and UVB-induced inflammation and hyperalgesia in human skin. *European Journal of Pain, 3*, 131–139.

Kilo, S., Schmelz, M., Koltzenburg, M., & Handwerker, H. O. (1994). Different patterns of hyperalgesia induced by experimental inflammation in human skin. *Brain, 117*, 385–396.

Koppert, W., Likar, R., Geisslinger, G., et al. (1999). Peripheral antihyperalgesic effect of morphine to heat, but not mechanical, stimulation in healthy volunteers after ultraviolet-B irradiation. *Anesthesia and Analgesia, 88*, 117–122.

Koppert, W., Brueckl, V., Weidner, C., et al. (2004). Mechanically induced axon reflex and hyperalgesia in human UV-B burn are reduced by systemic lidocaine. *European Journal of Pain, 8*, 237–244.

Magerl, W., Fuchs, P. N., Meyer, R. A., et al. (2001). Roles of capsaicin-insensitive nociceptors in cutaneous pain and secondary hyperalgesia. *Brain, 124*, 1754–1764.

Moiniche, S., Dahl, J. B., & Kehlet, H. (1993). Time course of primary and secondary hyperalgesia after heat injury to the skin. *British Journal of Anaesthesia, 71*, 201–205.

Petersen, K. L., & Rowbotham, M. C. (1999). A new human experimental pain model: the heat/capsaicin sensitization model. *NeuroReport, 10*, 1511–1516.

Schmelz, M., Schmidt, R., Weidner, C., et al. (2003). Chemical response pattern of different classes of C-nociceptors to pruritogens and algogens. *Journal of Neurophysiology, 89*, 2441–2448.

Serra, J., Campero, M., Bostock, H., et al. (2004). Two types of C nociceptors in human skin and their behavior in areas of capsaicin-induced secondary hyperalgesia. *Journal of Neurophysiology, 91*, 2770–2781.

UV Light

▶ Ultraviolet Light

UV-B Erythema

▶ UV Erythema, a Model for Inducing Hyperalgesias

UV-Induced Erythema

Thomas Bishop
Centre for Neuroscience, King's College London, London, UK

Synonyms

Sunburn as a model of cutaneous hyperalgesia; UV erythema; UV-induced erythema as a model of inflammatory pain

Definition

▶ Sunburn is the commonly experienced cutaneous inflammation that occurs in response to acute overexposure of the skin to ultraviolet radiation (UVR). Along with the classical characteristics of inflammation (▶ erythema, edema, and increased tissue temperature), very marked increased sensitivities to both thermal and mechanical stimuli emerge in response to ultraviolet radiation within the burn and are typically present for several days. Several characteristics of these sensory changes make sunburn an interesting and useful model for the study of pain.

Characteristics

UVR-Induced Inflammation in Humans

Virtually all Caucasians are acquainted with the effects of excessive cutaneous exposure to UV radiation resulting in sunburn. The associated sensory changes are readily appreciable to anyone who has taken a hot shower with a sunburnt back.

Many different UV sources have been used in the study of pain, including UVA, UVB, and solar simulated radiation (SSR) sources. All appear to share similar gross patterns of inflammation and sensory changes (Benrath et al. 2001; Harrison et al. 2004; Gustorff et al. 2004a), providing appropriate doses that are used.

This inflammatory response is dependent on the wavelength of the ultraviolet radiation.

In humans, UVB (280–320 nm) is two to four orders of magnitude more effective at inducing erythema compared to UVA (320–400 nm) (Young et al. 1998). In SSR-induced inflammation, the UVB component of the spectrum accounts for approximately 90 % of the erythema.

Individual responses to UV radiation vary greatly and, as a result, are largely expressed and normalized in terms of the biological effect generated. Doses are given in terms of the erythema that is generated (minimal erythemal dose, MED; the minimal dose required to produce a clearly demarcated area of erythema assessed at 24 h) and are tailored to the individual subject. Doses of 1–3 MEDs of SSR are obtainable clinically and are equivalent to exposure of an hour or less of summer sun at temperate latitudes.

Well-demarcated erythema (with no surrounding flare) emerges soon after exposure to a SSR source and peaks at between 24 and 48 h. It has been demonstrated that during this time, the blood flow to the skin is increased (Benrath et al. 2001; Harrison et al. 2004). The blood flow to the skin remains significantly raised for 72 h but is gradually falling from its peak (24–48 h) and has returned to normal within a week to 10 days.

Human Models of UVR-Induced Pain

Doses of 2 MEDs produce significant ▶ hyperalgesia to both thermal and mechanical stimuli in skin exposed to the UVR (Benrath et al. 2001; Harrison et al. 2004; Gustorff et al. 2004a) though the irradiation is painless. The sensory changes are dose dependent with the magnitude of hyperalgesia positively correlating with the increasing biologically active dose (MED, level of inflammation) (Benrath et al. 2001; Harrison et al. 2004; Gustorff et al. 2004a).

The hyperalgesia develops in parallel with the erythema and also peaks at approximately 24 h. The magnitude of hyperalgesia produced by UVR is surprising, with modest clinically achievable doses of UVR producing large reductions in thermal thresholds and mechanical pain thresholds to levels where brushing the burnt skin can be painful.

The sensory changes return to normal over several days at a rate dependent on the initial dose. Though the erythema may be biphasic, the sensory changes appear to be monophasic.

Burns created by these clinically relevant doses of UVR (1–3 MEDs) do not produce blistering or breakdown of the skin and are not associated with ongoing spontaneous pain.

Primary vs. Secondary Sensory Changes

Many conventional models of inflammatory pain produce an area of ▶ primary hyperalgesia associated with the area of noxious stimuli and an adjacent area of ▶ secondary hyperalgesia. The neural correlates of these sensory changes are thought to differ between the areas of primary and secondary hyperalgesia.

▶ Sensitization of the receptive terminals of ▶ nociceptors is thought to underpin a large proportion of primary hyperalgesia. Activity-dependent changes in the central processing of sensory stimuli are thought to account for the secondary hyperalgesia. Peripheral neuronal inputs converge on the topographical map in the spinal cord, resulting in these central changes affecting uninjured areas adjacent to the primary area of insult.

One of the key points of debate surrounding the UV model is that despite clear evidence of changes in primary mechanical sensitivity, there are contrasting reports concerning the emergence of secondary mechanical hyperalgesia surrounding UV irradiated skin. One group elegantly investigated the distribution of primary and secondary changes in mechanical pain thresholds and observed that increased mechanical sensitivity was restricted to the exact area of UVR inflammation without any alterations in secondary hyperalgesia (see Fig. 1) (Harrison et al. 2004).

A second group using different mechanical stimulation protocols demonstrated significant and stable secondary mechanical hyperalgesia (Gustorff et al. 2004a). Further work is needed to clarify these findings, though a working model with decreased primary mechanical pain thresholds superimposed on an area of secondary mechanical hyperalgesia (mediated by changes in the central responses) may provide a starting point. It must be noted that neither group could

UV-Induced Erythema,
Fig. 1 UVR-induced
changes in sensitivity to
mechanical pain are
localized. The investigators
exposed an annulus-shaped
region of skin to 3 MEDs
worth of UVR, which
produced a ring of burnt
skin surrounding a circle of
unburnt skin (*A*).
Mechanical pain thresholds
were assessed 24 h after
irradiation inside (*A*),
within (*B, C*), and at
varying distances outside
(*D–F*) the ring of burnt
skin. The thresholds are
shown for each point as
mean +/− SEM (Taken
from Harrison et al. (2004))

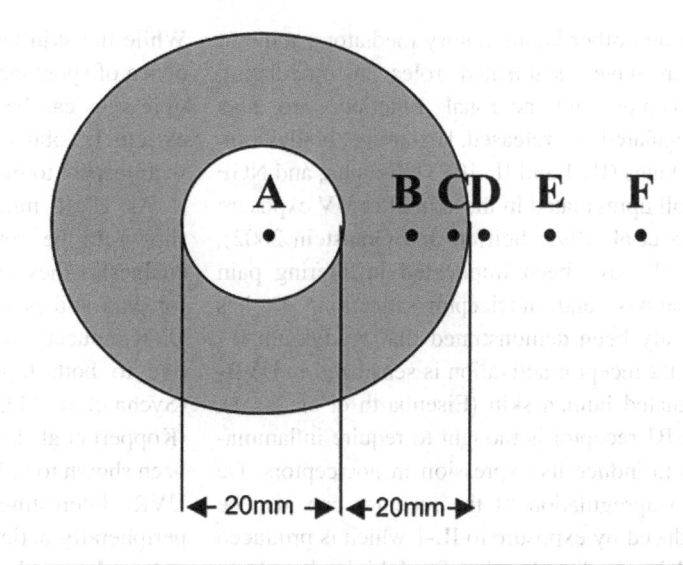

Site	A	B	C	D	E	F
Distance from centre (mm)	0	20	28.5	31.5	40	60

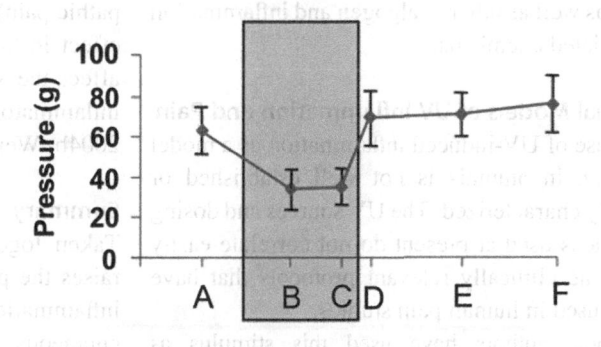

demonstrate secondary brush evoked pain surrounding irradiated skin.

The power of this model may reside in the opportunities to study the mechanistic basis of mechanical hyperalgesia, which is a feature of many clinical states. Whereas many substances have been shown to sensitize the peripheral terminals of nociceptors to thermal stimuli, very few substances or conditions produce sensitization of mechanical response profiles.

Mediators Raised in the Skin

Very few studies have looked at the mechanisms that subserve the altered sensation in this model. There is however a large body of work looking at the inflammatory mediators that are produced in the skin by UVR, some of which are algogenic.

There is some evidence that there is a degree of neurogenic inflammation which results from and causes the release of substance P and calcitonin gene-related peptide. Both of these neuropeptides are reported to contribute to the hyperalgesia (Seiffert and Granstein 2002).

The cyclooxygenase (COX) pathway may yield mediators which contribute to the development of the hyperalgesia. COX inhibitors applied topically to irradiated skin can reduce the hyperalgesia (Bayerl et al. 1998). Moreover several members of the prostaglandin family are produced during the time course of the erythema and hyperalgesia in sunburnt skin (Rhodes et al. 2001). PGE2, one of the prostaglandins produced, is known to sensitize peripheral nociceptors to thermal stimuli.

Many other inflammatory mediators, some of which have established roles in modulating nociceptor and neuronal function, are also upregulated and released. Histamine, bradykinin, cytokines (IL-1 and IL-10) TNF-alpha, and NGF are all upregulated in the skin after UV exposure (Barr et al. 1999; Seiffert and Granstein 2002), and all have been implicated in altering pain sensations and nociceptor function. It has recently been demonstrated that bradykinin B1 and B2 receptor activation is sensitized in UVB-irradiated human skin (Eisenbarth et al. 2004). The B1 receptor is thought to require inflammation to induce its expression in nociceptors. De novo upregulation of the receptor can also be produced by exposure to IL-1 which is produced in UV irradiated skin. Bradykinin has been strongly implicated in the sensitization of several properties of nociceptors, increasing responses to heat as well as other ▶ algogen and inflammation associated chemicals.

Animal Models of UV Inflammation and Pain

The use of UV-induced inflammation as a model of pain in animals is not well established or clearly characterized. The UV sources and dosing regimens used at present do not correlate easily with the clinically relevant protocols that have been used in human pain studies.

Some authors have used this stimulus as a surrogate for chronic inflammatory pain states by producing frank tissue damage. However, it is hard to see the benefit of these protocols employing very large doses of UVR, over conventional models produced by other inflammatory stimuli.

Recently two groups have characterized rat UVR-induced models of pain using dosing regimens which are more analogous to the human protocols (Davies et al. 2005; Themistocleous et al. 2006). The sensory phenotypes produced are equivalent to the changes seen in the established human models.

Analgesic Interventions in Human Models

There are very few experimental models of sub-acute inflammatory pain. The UVR-based models offer a quantifiable stimulus which provokes little discomfort during or after the irradiation. While the skin becomes tender, there is no evidence of spontaneous tissue injury or pain. Multiple sites can be exposed. Sensory changes are extremely robust and subjects are easily blinded with respect to treatments.

As UVR inflammatory models have been shown to be sensitive to several established analgesics they are emerging as a valuable tool for pain and pharmaceutical researchers. Acute UVR-induced cutaneous inflammation is sensitive to both topical and systemic ▶ NSAIDs (Sycha et al. 2003). Morphine and other opioids (Koppert et al. 1999; Gustorff et al. 2004b) have been shown to reduce the hyperalgesia caused by UVR. Interestingly it has been reported that peripherally acting morphine may differentially reduce thermal hyperalgesia pointing to different effector pathways in UV-inflamed skin (Koppert et al. 1999). Gabapentin (a treatment for neuropathic pain) does not seem to have any analgesic effect in this model though it has been seen to affect the secondary component of some other inflammatory pain models (Gustorff et al. 2004b; Werner et al. 2001).

Summary

Taken together, work in humans and animals raises the prospect of "sunburn" (UVR-induced inflammation) becoming a translational model of cutaneous inflammatory pain. Moreover, the changes in responses to mechanical stimuli (magnitude of reduction in thresholds and significant primary hypersensitivity component) suggest that skin exposure to clinically relevant doses of UVR represents a powerful model of mechanical sensitization of nociceptors.

References

Barr, R. M., Walker, S. L., Tsang, W., et al. (1999). Suppressed alloantigen presentation, increased TNF-alpha, IL-1, IL-1Ra, IL-10, and modulation of TNF-R in UV-irradiated human skin. *The Journal of Investigative Dermatology, 112*, 692–698.

Bayerl, C., Pagung, R., & Jung, E. G. (1998). Meloxicam in acute UV dermatitis – A pilot study. *Photodermatology, Photoimmunology and Photomedicine, 14*, 167–169.

Benrath, J., Gillardon, F., & Zimmermann, M. (2001). Differential time courses of skin blood flow and hyperalgesia in the human sunburn reaction following ultraviolet irradiation of the skin. *European Journal of Pain, 5*, 155–167.

Davies, S. L., Siao, C., & Bennett, G. J. (2005). Characterization of a model of cutaneous inflammatory pain produced by an ultraviolet irradiation-evoked sterile injury in the rat. *Journal of Neuroscience Methods, 148*, 161–166.

Eisenbarth, H., Rukwied, R., Petersen, M., et al. (2004). Sensitization to bradykinin B1 and B2 receptor activation in UVB irradiated human skin. *Pain, 110*, 197–204.

Gustorff, B., Anzenhofer, S., Sycha, T., et al. (2004a). The sunburn pain model: The stability of primary and secondary hyperalgesia over 10 hours in a crossover setting. *Anesthesia and Analgesia, 98*, 173–177.

Gustorff, B., Hoechtl, K., Sycha, T., et al. (2004b). The effects of remifentanil and gabapentin on hyperalgesia in a new extended inflammatory skin pain model in healthy volunteers. *Anesthesia and Analgesia, 98*, 401–407.

Harrison, G. I., Young, A. R., & McMahon, S. B. (2004). Ultraviolet radiation-induced inflammation as a model for cutaneous hyperalgesia. *The Journal of Investigative Dermatology, 122*, 183–189.

Koppert, W., Likar, R., Geisslinger, G., et al. (1999). Peripheral antihyperalgesic effect of morphine to heat, but not mechanical, stimulation in healthy volunteers after ultraviolet-B irradiation. *Anesthesia and Analgesia, 88*, 117–122.

Rhodes, L. E., Belgi, G., Parslew, R., et al. (2001). Ultraviolet-B-induced erythema is mediated by nitric oxide and prostaglandin E2 in combination. *The Journal of Investigative Dermatology, 117*, 880–885.

Seiffert, K., & Granstein, R. D. (2002). Neuropeptides and neuroendocrine hormones in ultraviolet radiation-induced immunosuppression. *Methods, 28*, 97–103.

Sycha, T., Gustorff, B., Lehr, S., et al. (2003). A simple pain model for the evaluation of analgesic effects of NSAIDs in healthy subjects. *British Journal of Clinical Pharmacology, 56*, 165–172.

Themistocleous, A., Fick, L., Plessis, I. D., et al. (2006). Exposure of the rat tail to ultraviolet A light produces sustained hyperalgesia to noxious thermal and mechanical challenges. *Journal of Neuroscience Methods, 152*, 267–273.

Werner, M. U., Perkins, F. M., Holte, K., et al. (2001). Effects of gabapentin in acute inflammatory pain in humans. *Regional Anesthesia and Pain Medicine, 26*, 322–328.

Young, A. R., Chadwick, C. A., Harrison, G. I., et al. (1998). The similarity of action spectra for thymine dimers in human epidermis and erythema suggests that DNA is the chromophore for erythema. *The Journal of Investigative Dermatology, 111*, 982–988.

UV-induced Erythema as a Model of Inflammatory Pain

▶ UV-Induced Erythema

UV-induced Erythema as a Model of Inflammatory Pain

UV-Induced Erythema

V

24-h Variation

▶ Diurnal Variations of Pain in Humans

Vagal Input and Descending Modulation

Alan Randich[1] and Cindy L. Thurston-Stanfield[2]
[1]University of Alabama, Birmingham, AL, USA
[2]Department of Biomedical Sciences, University of South Alabama, Mobile, AL, USA

Synonyms

Vagal stimulation-produced antinociception; Vagally mediated analgesia; Vagally mediated hyperalgesia

Definition

Input from vagal afferents to neurons in the brain control spinal nociceptive transmission. The vagus nerve (cranial nerve X) contains both afferent and efferent fibers that innervate viscera throughout the body. Activation of certain vagal afferents produces ▶ antinociception, whereas activation of others produces ▶ hyperalgesia. Both the antinociception and hyperalgesia result from altering neural activity in certain areas of the brain. These brain regions, through a descending pathway, then alter communication between primary afferent neurons and ▶ spinothalamic neurons.

Characteristics

Activation of a number of regions in the brainstem produces antinociception, as evidenced by either the attenuation/absence of behavioral responses to noxious stimuli or decreases in the responses of spinal/trigeminal dorsal horn neurons to noxious stimuli. Studies that are more recent indicate that some of these same regions can also produce hyperalgesia, when activated by either different types of stimuli or different intensities of electrical stimulation. These descending inhibitory and facilitatory pain-modulating systems are affected by a variety of input, including signals generated by visceral stimuli and transmitted to the central nervous system by vagal afferents. Vagal afferents transmit signals from the esophagus, lower airways, heart, gastrointestinal tract, liver, gallbladder, and pancreas to the ▶ nucleus tractus solitarius in the medulla oblongata.

Electrical, pharmacological, and physiological activation of either cervical or subdiaphragmatic vagal afferents alters nociception, by activating these central descending pain modulatory systems. Analogous to central stimulation, activation of vagal afferents can either facilitate or inhibit nociception. For example, electrical stimulation of vagal

G.F. Gebhart, R.F. Schmidt (eds.), *Encyclopedia of Pain*, DOI 10.1007/978-3-642-28753-4,
© Springer-Verlag Berlin Heidelberg 2013

afferents has been shown to both facilitate and attenuate nociception, depending on the intensity of stimulation.

In rats, low-intensity electrical stimulation of cervical vagal afferents facilitates the nociceptive tail flick reflex and enhances noxious heat-evoked responses in spinal cord dorsal horn neurons (Randich and Gebhart 1992). High-intensity electrical stimulation of cervical or subdiaphragmatic vagal afferents has the opposite effect, inhibiting the tail flick reflex, decreasing formalin-induced nociceptive behavior, attenuating heat-evoked responses in spinal cord dorsal horn neurons, and decreasing formalin-induced Fos expression in trigeminal nuclei (Bohotin et al. 2003; Randich and Gebhart 1992; Ren et al. 1993; Tanimoto et al. 2002; Thurston and Randich 1992).

Human studies of vagal stimulation are limited and most often have been performed using experimental pain assessments in individuals implanted with vagal stimulators for the treatment of either epilepsy or depression. For instance, electrical stimulation of the cervical vagus in epileptic or depressed humans has been shown to decrease thermal pain, thermal pain threshold, and thermal pain tolerance (Borckardt et al. 2005; Lange et al. 2011; Ness et al. 2000) and decrease the intensity of tonic pressure pain perception (Kirchner et al. 2000). In an uncontrolled proof of concept pilot study in fibromyalgia (FM) patients refractory to treatment, Lange et al. (2011) found substantial improvement in efficacy measures over time, e.g., reduction in tender point thresholds, and surprisingly found a loss of caseness (using diagnostic criteria for the 1990 ACR FM case definition) in two individuals. This therapeutic effect appeared to increase as a function of time of stimulation during an 11-month trial. These results notwithstanding, many patients experience undesirable side effects of vagal activation, e.g., dyspnea, neck/throat pain, and nausea, that may interfere with widespread use of the technique. While parameter manipulation may reduce some of these effects, alternatives include more selective activation of vagal afferent fiber branches or use of noninvasive auricular vagal afferent stimulation. The latter has been shown to produce a trend for reducing both evoked pain intensity and temporal summation and significantly reduced anxiety in chronic pelvic pain patients resulting from endometriosis (Napadow et al. 2012).

Other possible alternatives include more selective pharmacological or physiological activation of vagal afferents. This could be accomplished with administration of serotonin (e.g., acute intraluminal administration; Zhang et al. (2011)) or low-dose opioids, which also inhibits spinal nociceptive reflexes and the response of spinothalamic neurons to noxious stimuli (Meller et al. 1992; Randich et al. 1993). The opioids that support a vagally induced antinociception include morphine, (D-Ala$_2$)-methionine enkephalinamide, and DAGO, all mu agonists (Randich et al. 1993). Certain stimuli that reproduce physiological activation of vagal afferents produce either antinociception or hyperalgesia. Volume expansion produces a vagally mediated attenuation of nociceptive reflexes and inhibition of trigeminal neuronal responses to a noxious stimulus (Takeda et al. 2002). Morgan and Fields (1993) showed that the volume-expander Ficoll increased the activity of OFF cells (associated with pain inhibition) and decreased the activity of ON cells (associated with facilitation of pain) in the rostral ventral medulla and simultaneously inhibited the tail flick reflex in these rats although these changes could not be shown to be a sufficient condition for the observed behavioral outcomes. In a model of inflammation, bradykinin-induced plasma extravasation and hyperalgesia are enhanced by subdiaphragmatic vagotomy, indicating that subdiaphragmatic vagal afferent activity normally suppresses these inflammatory responses (Miao et al. 2001). In these same studies, subdiaphragmatic vagotomy decreased paw withdrawal thresholds in untreated rats, also indicating an attenuation of nociception by tonic vagal activity. In terms of hyperalgesia, both lipopolysaccharide-induced illness and fasting produce a hyperalgesia that is reversed by subdiaphragmatic vagotomy (Khasar et al. 2003; Watkins et al. 1995). The illness-induced hyperalgesia can also

Vagal Input and Descending Modulation, Fig. 1 Vagal input to descending pain inhibitory systems in the nucleus raphe magnus and locus coeruleus. Ascending pathways are shown on the *left*. Vagal afferents terminate in the nucleus tractus solitarius (*NTS*). Pathways from the NTS to the nucleus raphe magnus (*NRM*) and locus coeruleus (*LC*) are unknown, but both regions are necessary for vagally mediated antinociception. Descending pathways are shown on the *right*. Activation of vagal afferents results in the spinal release of serotonin (5-HT), norepinephrine (*NE*), and opioids, which inhibits transmission between the primary afferent and spinothalamic neurons (not shown)

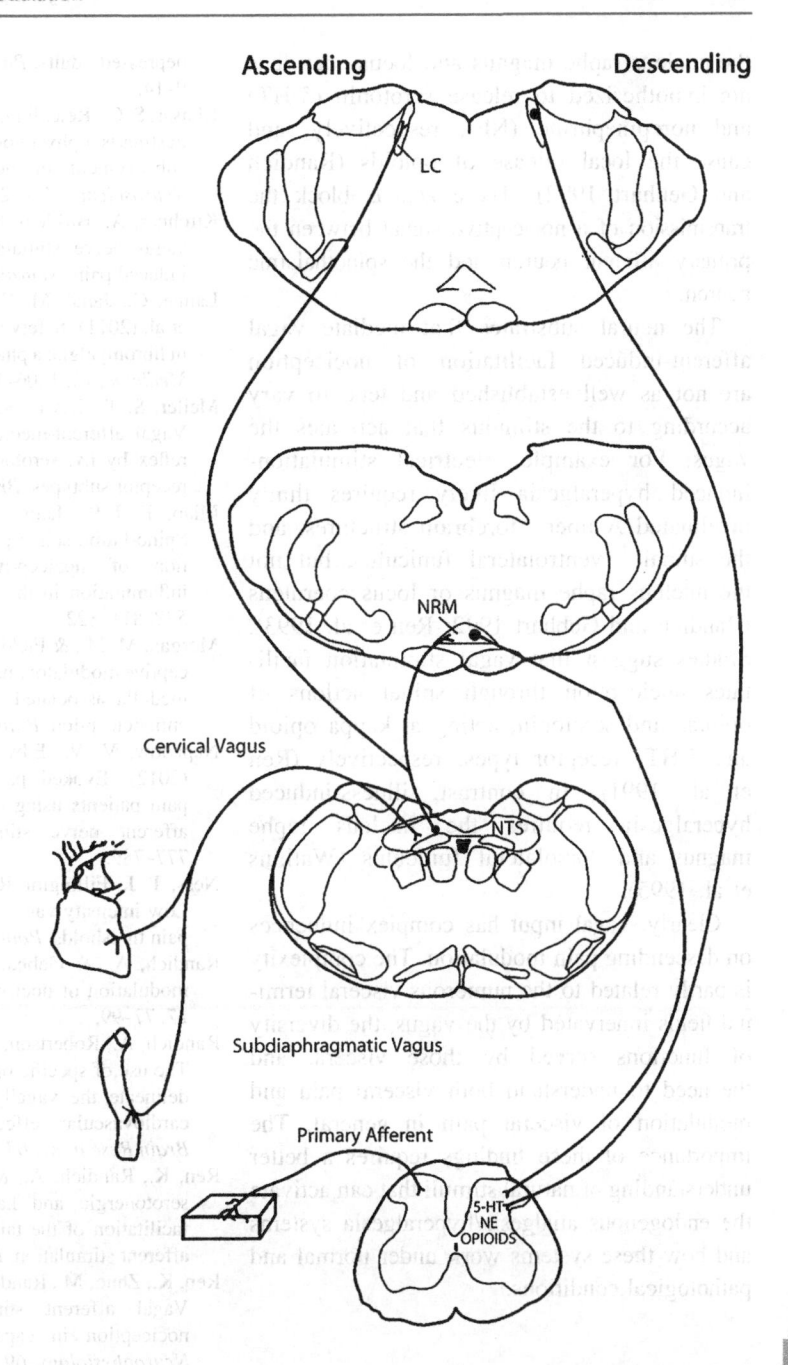

Ascending

Descending

LC

NRM

Cervical Vagus

NTS

Subdiaphragmatic Vagus

Primary Afferent

5-HT
NE
OPIOIDS

be attenuated by the more selective resection of the hepatic branch of the vagus (Watkins et al. 1995).

Much of the circuitry, whereby activation of vagal afferents attenuates nociception, has been established (see Fig. 1). Attenuation of nociception requires activation of vagal ▶ C fibers, possibly associated with ▶ visceral nociception and pain (Ren et al. 1993). Some of the neural substrates mediating vagally induced attenuation of nociception include the nucleus tractus solitarius (NTS), the ▶ nucleus raphe magnus (NRM), the ▶ locus coeruleus (LC), and a descending spinal pathway in the ▶ dorsolateral funiculus (Miao et al. 2001; Randich and Gebhart 1992; Thurston and Randich 1992; Thurston-Stanfield et al. 1999). Neurons descending from the nucleus raphe magnus and locus coeruleus are hypothesized to release serotonin (5-HT) and norepinephrine (NE), respectively, and cause the local release of opioids (Randich and Gebhart 1992). These actions block the transmission of a nociceptive signal between the primary afferent neuron and the spinothalamic neuron.

The neural substrates that mediate vagal afferent-induced facilitation of nociception are not as well established and tend to vary according to the stimulus that activates the vagus. For example, electrical stimulation-induced hyperalgesia likely requires thinly myelinated A fibers, forebrain structures, and the spinal ▶ ventrolateral funiculus, but not the nucleus raphe magnus or locus coeruleus (Randich and Gebhart 1992; Ren et al. 1993). Studies suggest that vagal stimulation facilitates nociception through spinal actions of opioids and serotonin, acting at kappa opioid and 5-HT$_1$ receptor types, respectively (Ren et al. 1991). In contrast, illness-induced hyperalgesia requires the nucleus raphe magnus and dorsolateral funiculus (Watkins et al. 1995).

Clearly, vagal input has complex influences on descending pain modulation. The complexity is partly related to the numerous visceral terminal fields innervated by the vagus, the diversity of functions served by those viscera, and the need to understand both visceral pain and modulation of visceral pain in general. The importance of these findings requires a better understanding of natural stimuli that can activate the endogenous analgesia/hyperalgesia systems and how these systems work under normal and pathological conditions.

References

Bohotin, C., Scholsem, M., Multon, S., et al. (2003). Vagus nerve stimulation in awake rats reduces formalin-induced nociceptive behavior and fos-immunoreactivity in trigeminal nucleus caudalis. *Pain, 101*, 3–12.

Borckardt, J. J., Kozel, F. A., Anderson, B., et al. (2005). Vagus nerve stimulation affects pain perception in depressed adults. *Pain Research & Management, 10*, 9–14.

Khasar, S. G., Reichling, D. B., Green, P. G., et al. (2003). Fasting is a physiological stimulus of vagus-mediated enhancement of nociception in the female rat. *Neuroscience, 119*, 215–221.

Kirchner, A., Birklein, F., Stefan, H., et al. (2000). Left vagus nerve stimulation suppresses experimentally induced pain. *Neurology, 55*, 1167–1171.

Lange, G., Janal, M. N., Manikier, A., Fitzgibbons, J., et al. (2011). Safety and efficacy of vagus stimulation in fibromyalgia: a phase i/ii proof of concept trial. *Pain Medicine, 12*, 1406–1413.

Meller, S. T., Lewis, S. J., Brody, M. J., et al. (1992). Vagal afferent-mediated inhibition of a nociceptive reflex by i.v. serotonin in the rat. II. Role of 5-HT receptor subtypes. *Brain Research, 585*, 71–86.

Miao, F. J.-P., Janig, W., Jasmin, L., et al. (2001). Spino-bulbo-spinal pathway mediating vagal modulation of nociceptive-neuroendocrine control of inflammation in the rat. *The Journal of Physiology, 532*, 811–822.

Morgan, M. M., & Fields, H. L. (1993). Activity of nociceptive modulatory neurons in the rostral ventromedial medulla associated with volume expansion-induced antinociception. *Pain, 52*(1), 1–9.

Napadow, V. V., Edwards, R. R., Cahalan, C., et al. (2012). Evoked pain analgesia in chronic pelvic pain patients using respiratory-gated auricular vagal afferent nerve stimulation. *Pain Medicine, 13*, 777–789.

Ness, T. J., Fillingim, R. B., Randich, A., et al. (2000). Low intensity vagal nerve stimulation lowers human pain thresholds. *Pain, 86*, 81–85.

Randich, A., & Gebhart, G. F. (1992). Vagal afferent modulation of nociception. *Brain Research Reviews, 17*, 77–99.

Randich, A., Robertson, J. D., & Willingham, T. (1993). The use of specific opioid agonists and antagonists to delineate the vagally mediated antinociceptive and cardiovascular effects of intravenous morphine. *Brain Research, 603*, 186–200.

Ren, K., Randich, A., & Gebhart, G. F. (1991). Spinal serotonergic and kappa opioid receptors mediate facilitation of the tail flick reflex produced by vagal afferent stimulation. *Pain, 45*, 321–329.

Ren, K., Zhuo, M., Randich, A., & Gebhart, G. F. (1993). Vagal afferent stimulation-produced effects on nociception in capsaicin-treated rats. *Journal of Neurophysiology, 69*, 1530–1540.

Takeda, M., Tanimoto, T., Nashikawa, T., et al. (2002). Volume expansion suppresses the tooth-pulp evoked jaw-opening reflex related activity of trigeminal neurons in rats. *Brain Research Bulletin, 58*, 83–89.

Tanimoto, T., Takeda, M., & Matsumoto, S. (2002). Suppressive effect of vagal afferents on cervical dorsal horn neurons responding to tooth pulp electrical stimulation in the rat. *Experimental Brain Research, 145*, 468–479.

Thurston, C. L., & Randich, A. (1992). Electrical stimulation of the subdiaphragmatic vagus in rats: Inhibition of heat-evoked responses of spinal dorsal horn neurons and central substrates mediating inhibition of the nociceptive tail flick reflex. *Pain, 51,* 349–365.

Thurston-Stanfield, C. L., Ranieri, J. T., Vallabhapurapu, R., et al. (1999). Role of vagal afferents and the rostral ventral medulla in intravenous serotonin-induced changes in nociception and arterial blood pressure. *Physiology and Behavior, 67,* 753–767.

Watkins, L. R., Maier, S. F., & Goehler, L. E. (1995). Immune activation: The role of pro-inflammatory cytokines in inflammation, illness responses and pathological pain states. *Pain, 63,* 289–302.

Zhang, L. Y., Dong, X., Liu, Z. L., et al. (2011). Luminal serotonin time-dependently modulates vagal afferent driven antinociception in response to colorectal distention in rats. *Neurogastroenterology and Motility, 23,* 62–69.

Vagal Stimulation-produced Antinociception

▶ Vagal Input and Descending Modulation

Vagally Mediated Analgesia

▶ Vagal Input and Descending Modulation

Vagally Mediated Hyperalgesia

▶ Vagal Input and Descending Modulation

Vaginal Hyperalgesia Model

▶ Visceral Pain Models, Female Reproductive Organ Pain

Vagoglossopharyngeal Neuralgia

▶ Trigeminal and Cranial Nerve Neuralgias

Valdecoxib

Definition

(Trade name: Bextra) Valdecoxib is a COX-2 inhibitor that is more selective for COX-2 than celecoxib. In addition to the risk of myocardial infarction, it carries the risk of serious skin reactions and its sales have been suspended.

Cross-References

▶ Cyclooxygenases in Biology and Disease
▶ Pharmacology of Cyclooxygenase Inhibitors

Validity

David Vivian
Metro Spinal Clinic, South Caulfield, VIC, Australia

Definition

▶ Validity is the measure of the extent to which a diagnostic test actually does what it is supposed to do, i.e., how well it actually detects the condition it is supposed to detect. The term is derived from the Latin – *validus* – meaning strong or robust.

Characteristics

Validity is determined by correlating the results of a test with those of a reference standard, known as the criterion standard, or formerly known as the "gold standard." The criterion standard differs according to the test whose validity is being sought. The essential feature of a criterion standard is that, by consensus, it provides a more direct observation of the abnormality in question than the test in question and is, therefore, less susceptible to error. Moreover, the criterion standard must use methods independent of those used by the test in question. For clinical tests, such as those that rely on palpation, the criterion standard may be a radiographic finding, a postmortem finding, or an observation at surgery, for these standards are less susceptible to error than palpation through skin and muscle, and rely on vision rather than touch.

Sometimes, when a criterion standard is not available, investigators have used a consensus view of a panel of experts, who determine whether or not the index condition is present, as the criterion standard. However, if those experts establish their diagnosis using the same or similar tests as the one in question, they are not assessing validity. Rather, theirs is a test of ▶ reliability, for they are measuring the extent to which two observers using the same test agree on the results.

Validity is determined by subjecting the same sample of patients to the test in question and to the criterion standard. The results of this exercise can be summarized in a 2 × 2 contingency table, from which numerical indices of validity can be calculated (Table 1). Four cells are generated, in which "a" is the number of cases in which the test and the criterion standard are both positive (i.e., both tests agree that the condition was present); "b" is the number of cases in which the test was positive but the criterion standard was negative (i.e., the condition was not present despite the test being positive); "c" is the number of cases in which the test was negative but the criterion standard was positive (i.e., the condition was present but the test missed it); and "d" is the number of cases in which the test and the criterion standard were both negative (i.e., both tests agreed that the condition was absent).

Clearly, the "a" and "d" cases are correct results of the test, but the "b" and "c" cases are mistakes. In the "b" cases, the test was positive but should not have been. Those results are false-positive. In the "c" cases, the test was negative, but should not have been (false-negative). A good test is one that carries few, if any, false-positive and false-negative results. From the contingency table, two fundamental values can be derived. These stem from the columns under criterion standard.

The ▶ sensitivity of the test measures how well it detects the condition when the condition is present (Sackett et al. 1991). It is derived down the

Validity, Table 1 The structure of a 2 × 2 contingency table comparing the results of a diagnostic test with those of a criterion standard

Result of Test	Criterion standard	
	Positive	Negative
Positive	a	b
Negative	c	d

first column. Numerically, it is the ratio: $a/(a + c)$. Idiomatically, it is the ratio between the number of true-positive results and all positive cases. This ratio constitutes the true-positive rate.

The ▶ specificity of the test measures how well it excludes the condition when the condition is absent (Sackett et al. 1991). It is derived up the second column. Numerically, it is the ratio: $d/(b + d)$. Idiomatically, it is the ratio between the number of true negative results and all negative cases, and constitutes the true-negative rate. Its complement, $b/(b + d)$, i.e., [1 − specificity], is the false-negative rate.

A valid test is one with high sensitivity and high specificity. The validity of the test is compromised if either of these indices is low.

From the sensitivity and the specificity, a single measure of the validity of a test can be derived. It is the positive likelihood ratio, which indicates how often positive results of the test are truly positive. Idiomatically, the positive likelihood ratio is the true-positive rate discounted by the false-positive rate, and is defined as (true-positive rate) / (false-positive rate). Numerically, it is (sensitivity) / (1 − specificity) (Sackett et al. 1991).

The virtue of the positive likelihood ratio is that it reveals the extent to which confidence in diagnosis is increased by a positive result of the diagnostic test. This can be quantified from the relationship:

$$(\text{prevalence odds}) \times (\text{positive likelihood ratio}) = (\text{diagnostic confidence odds})$$

where prevalence odds are the pretest likelihood that the condition is present, expressed as odds, and diagnostic confidence odds is the posttest likelihood that the condition is present, expressed as odds (Sackett et al. 1991).

Odds, instead of percentages, are used in order for the equation to work mathematically (Bogduk 1999). If the prevalence of a condition is 40 %, the odds that it is present are 40:60, i.e., 40: (100–40). If the diagnostic confidence odds are 8:2, the diagnostic confidence is 80 %, i.e., 8/(8 + 2).

If a diagnostic test has a positive likelihood ratio of 1.0, the test has no validity. From the equation, it can be seen that applying the test does not improve diagnostic confidence. The odds after the test are the same as before the test.

For a test to be valid, the positive likelihood ratio must be substantially greater than 1.0. Exactly by how much greater it has to be depends on the prevalence of the condition and the diagnostic confidence required. As a rule, if the condition is common, the positive likelihood ratio needs only to be modestly large. If the condition is rare, the positive likelihood ratio must be extremely large. If greater diagnostic confidence is required or desired, the test must have a proportionately larger positive likelihood ratio.

If the natural prevalence of a condition is, say, 40 %, and one wants to be 80 % confident in the diagnosis, the value required of the positive likelihood ratio (PLR) can be derived:

$$40 : 60 \times PLR = 80 : 20$$
$$PLR = 6$$

Thus, for a condition with a prevalence of 40 %, if the desired diagnostic confidence is 80 %, a test is sufficiently valid if its positive likelihood ratio is 6 or more. A test whose positive likelihood ratio is less than 6 lacks sufficient validity to serve the purpose required.

If the natural prevalence of a condition is, say 10 %, and one wants to be 80 % confident in the diagnosis, the value required of the positive likelihood ratio is different:

$$PLR = 36$$

Thus, in this example, a much larger positive likelihood ratio is required. This arises because of the lower prevalence of the condition being sought.

When the condition in question is uncommon, patients who do not have the condition, but who might nevertheless generate a positive result, outnumber those patients who do have the condition. This increases the chances of a false-positive result. To overcome this false-positive rate, a large positive likelihood ratio is required.

As a general rule, tests that have a positive likelihood ratio with a magnitude of the order of 1.0 or 2.0 have little or no useful validity. Few clinical tests in pain medicine exceed these values. Consequently, few tests are valid.

Cross-References

▶ Disability Assessment, Psychological/Psychiatric Evaluation
▶ Multiaxial Assessment of Pain
▶ Oswestry Disability Index
▶ Pain Assessment in Children
▶ Pain Assessment in Neonates
▶ Pain Inventories
▶ Whiplash

References

Bogduk, N. (1999). Truth in musculoskeletal medicine. Truth in diagnosis – validity. *Australasian Musculoskeletal Medicine, 4,* 32–39.
Sackett, D. L., Haynes, R. B., Guyatt, G. H., & Tugwell, P. (1991). *Clinical epidemiology. A basic science for clinical medicine* (2nd ed., pp. 119–139). Boston: Little, Brown & Co.

Valsalva Maneuver

Definition

The Valsalva Maneuver consists of a forced expiration against a closed glottis and nasal airways. This obstructs venous return to the heart and increases intrathoracic and intracranial pressure (Mathias and Bannister 2012).

Cross-References

▶ Primary Cough Headache
▶ Primary Exertional Headache

References

Mathias, C. J., & Bannister, R. (Eds.). (2012). *Autonomic failure* (5th ed.). Oxford: Oxford University Press.

Valsalva Maneuver Headache

▶ Primary Cough Headache

Vanilloid Receptor Subtype 1

▶ TRPV1 Receptor, Species Variability

Vanilloid Receptor Subunit 1

▶ Capsaicin Receptor

Vanilloid Receptor, Regulation by NGF

▶ TRPV1, Regulation by Nerve Growth Factor

Vanilloid Receptor, Regulation by Protons

▶ TRPV1, Regulation by Protons

Vanilloids

Definition

Vanilloids are a group of compounds containing a vanillyl residue. Most data exists about capsaicin, the pungent substance in hot pepper varieties. Vanilloids act by stimulating certain transient receptor potentials (TRPV1) on nociceptive afferent neurons, cation channels that are also opened by noxious heat or acidic pH.

VAP

▶ Ventral Amygdaloid Pathway

Varicosities

Definition

The terminals of unmyelinated axons bear a number of swellings called varicosities that contain accumulations of mitochondria, small and large synaptic vesicles, and other organelles. In autonomic axons, the small vesicles contain neurotransmitter agents (noradrenaline in most sympathetic postganglionic axons, acetylcholine in all parasympathetic postganglionic axons and in all preganglionic axons), while the large vesicles also contain neuropeptides. The varicosities are the sites of transmitter release, although the probability of release of the contents of one vesicle from a single varicosity is normally very low ($p < 0.01$) (Jänig 2006).

Cross-References

▶ Sympathetic and Sensory Neurons After Nerve Lesions, Structural Basis for Interactions

References

Jänig, W. (2006). *The integrative action of the autonomic nervous system: Neurobiology of homeostasis.* Cambridge/New York: Cambridge University Press.

VAS

▶ Visual Analog Scale

Vasa Nervorum

Definition

Blood vessels supplying, and located within, a nerve trunk. Endoneural blood vessels are normally not innervated but may become innervated chronically after nerve injury.

Cross-References

▶ Sympathetic and Sensory Neurons After Nerve Lesions, Structural Basis for Interactions

Vascular Compression Syndromes

Definition

Vascular compression of the cranial nerves leading to syndromes characterized by increased neuronal activity, such as spontaneous pain (this entry) or muscle spasm, that is, hemifacial spasm or twisting neck movements known as torticollis.

Cross-References

▶ Trigeminal, Glossopharyngeal, and Geniculate Neuralgias

Vascular Hypertrophy

▶ Sympathetic and Sensory Neurons After Nerve Lesions, Structural Basis for Interactions

Vascular Neuropathies

Claudia Sommer
Department of Neurology, University of Würzburg, Würzburg, Germany

Synonyms

Ischemic neuropathies; Vasculitic neuropathies

Definition

Neuropathies caused by malfunction of nerve blood vessels.

Characteristics

Neuropathies in Peripheral Arterial Occlusive Disease (PAOD)

Peripheral nerve function depends on an adequate blood supply. Since peripheral nerves have a dual blood supply (an intrinsic system consisting of longitudinal microvessels within the endoneurium and an extrinsic system of regional arteries, arterioles, venules, and epineurial vessels, in addition, extensive anastomosing), peripheral nerves are relatively protected from ischemic injury (Nukada et al. 1993). However, acute ischemia leads to axonal nerve damage (Wilbourn et al. 1983). In chronic endoneurial ischemia, histological studies of nerves show signs of axonal pathology, and signs of demyelination have been reported (Chalk and Dyck 1993). Furthermore, there is a correlation between the severity of nerve damage and the stage of the vascular insufficiency (Laghi Pasini et al. 1996). Diabetes being the most common cause of small vessel disease in industrialized societies, PAOD is often caused by or associated with diabetic complications. Diabetes leads to a multifactorial neuropathy by itself, which may be enhanced by ischemia due to small vessel occlusion.

By definition, activity-induced or resting pain is a prevalent symptom in patients with PAOD.

Few studies have tried to distinguish between ischemic and neuropathic pain in these patients. In one study, paresthesias, indicating neuropathic pain, were present in about 50 % of patients with PAOD. Since the patients' symptoms are usually dominated by activity-dependent pain and restrictions in activity are associated with vascular claudication, the neuropathic component receives less attention and often does not require separate treatment (Weber and Ziegler 2002). In a study with 19 patients with PAOD, equally non-neuropathic and local causes of pain (superficial skin ulceration, osteomyelitis, claudication) dominated the patient's concern and attention. Pain symptoms attributed to the neuropathy were rest pain in the toes or foot (58 %), numbness (58 %), burning (42 %), and paresthesias (37 %) (Weinberg et al. 2001). Trophic skin changes may be caused by both the arterial disease itself and by the neuropathy. Treatment is aimed at improvement of blood flow. In a recent pilot study, autologous transplantation of bone marrow mononuclear cells improved nerve function and reduced pain (Arima et al. 2010). Spinal cord stimulation has been recommended for cases in which pain persists in spite of other treatment directed at the ischemia, for example, endovascular interventions (Lang et al. 2009; Pedrini and Magnoni 2007).

Vasculitic Neuropathies

Vasculitic neuropathies are immune-mediated diseases of the peripheral nervous system, in which inflammation of the blood vessels causes damage to the nerves. Vasculitis may be part of a systemic disease in primary systemic vasculitis or in connective tissue diseases, or may be an isolated phenomenon in peripheral nerves, called isolated (or nonsystemic) peripheral nerve vasculitis (NSVN). The typical clinical picture consists of an asymmetric or multifocal, painful sensorimotor neuropathy with an acute, subacute, or chronic course and acute relapses. However, vasculitis can also cause a rather unspecific syndrome of distal symmetric, sensorimotor neuropathy.

Systemic vasculitis has an incidence of 4/100,000 per year and, untreated, has a poor prognosis, which is greatly improved by the use of immunosuppressive treatment. The prognosis of NSVN is generally better, although many patients need long-term immunosuppression (Collins et al. 2010).

Primary systemic vasculitis and secondary vasculitis in connective tissue diseases has been classified in different ways, the latest generally accepted classification being the one of the Chapel Hill Consensus Conference in 1994 (Jennette et al. 1994), which groups diseases according to the size of the affected arteries. Neuropathies are a frequent manifestation in polyarteritis nodosa, in Wegener's granulomatosis and Churg-Strauss-syndrome, in Sjögren's syndrome, and in rheumatoid arthritis. Stage II and III borreliosis may also lead to vasculitic neuropathy. Hepatitis B and C may be associated with cryoglobulinemic vasculitic neuropathy. Malignant tumors may rarely cause vasculitic neuropathy as a paraneoplastic syndrome. One third of the patients coming to attention with vasculitic neuropathy have isolated peripheral nerve vasculitis. A diagnosis of definite vasculitis can be made with evidence of vasculitis in a biopsy specimen (Fig. 1); histologic criteria for this diagnosis have been defined (Collins et al. 2000).

The neuropathy in vasculitis is partially caused by ischemia, partially due to direct inflammatory damage of the nerves. Endothelial cells have antigen-presenting capacities, they express adhesion molecules which then recruit inflammatory cells. Immunoglobulin complex deposits in the vessel walls may trigger the activation of the complement system and generate chemotactic factors attracting granulocytes. These further produce cytokines, prostaglandins, metalloproteases, and perforin, leading to tissue damage and sensitization of nerve fibers (Satoi et al. 1998; Kissel 2001; Lindenlaub and Sommer 2003). Nerve growth factor (NGF) has also been implied as a pain-inducing factor in vasculitic neuropathies (Yamamoto et al. 2003). DNA microarray analysis showed upregulation of the chemokines CXCL9 and CCR2 (Steck et al. 2011). C-fibers are affected in vasculitic neuropathy (Lee et al. 2005), and the skin itself may be affected by the vasculitis (Üçeyler et al. 2010), thus providing an inflammatory environment for the diseased nociceptors.

Vascular Neuropathies, Fig. 1 (a) Frozen section from a sural nerve biopsy specimen from a patient with isolated peripheral nerve vasculitis, immunostained for T-cells. Note dense infiltration of the vessel wall. (b) Paraffin section from a sural nerve biopsy specimen from a patient with isolated peripheral nerve vasculitis, elastica-van-Gieson stain. Note narrowing of vessel lumen and perivascular infiltrates. Bar = 10 µm

Thus, in the vasculitic neuropathies, pain is thought to be induced by a combination of axonal injury and inflammatory changes. Changes induced directly through axonal injury are, for example, the altered distribution of sodium channels. In addition, the release of inflammatory mediators in the vicinity of the nerve fibers is increased. Inflammatory mediators, in particular in combination ("inflammatory soup"), are known to increase the excitability of nociceptors, and even more so of those that have suffered an injury (Schäfers et al. 2003; Michaelis et al. 1998). This may be a reason why pain and paresthesias are usually the first symptoms to respond to treatment with corticosteroids in patients with vasculitic neuropathies. However, many patients still need symptomatic treatment for pain, at least temporarily.

Treatment of the underlying disorders consists in immunosuppression, to which most patients respond. Depending on the severity of the disorder and on the general organ involvement, treatment may consist of a combination of corticosteroids with cyclophosphamide or methotrexate. After achievement of remission, in general after 6–12 months, cyclophosphamide is substituted by azathioprine or methotrexate (Jayne 2001). Nonsystemic isolated peripheral nerve vasculitis can sometimes be treated with corticosteroids alone. If the neuropathy progresses, when these are tapered, azathioprine may be necessary, or even temporarily cyclophosphamide (Collins et al. 2010). Pain in isolated peripheral nerve vasculitis is treated according to the general rules of treatment of neuropathic pain.

References

Arima, K., Katsuda, Y., Takeshita, Y., Saito, Y., Toyama, Y., Katsuki, Y., Ootsuka, M., Koiwaya, H., Sasaki, K., Kai, H., & Imaizumi, T. (2010). Autologous transplantation of bone marrow mononuclear cells improved ischemic peripheral neuropathy in humans. *Journal of the American College of Cardiology, 56,* 238–239.

Chalk, C. H., & Dyck, P. J. (1993). Ischemic neuropathy. In P. J. Dyck, P. K. Thomas, J. W. Griffin, P. A. Low, & J. F. Poduslo (Eds.), *Peripheral neuropathy* (pp. 980–989). Philadelphia: WB Saunders.

Collins, M. P., Mendell, J. R., Periquet, M. I., Sahenk, Z., Amato, A. A., Gronseth, G. S., Barohn, R. J., Jackson, C. E., & Kissel, J. T. (2000). Superficial peroneal nerve/peroneus brevis muscle biopsy in vasculitic neuropathy. *Neurology, 55,* 636–643.

Collins, M. P., Dyck, P. J., Gronseth, G. S., Guillevin, L., Hadden, R. D., Heuss, D., Leger, J. M., Notermans, N. C., Pollard, J. D., Said, G., Sobue, G., Vrancken, A. F., & Kissel, J. T. (2010). Peripheral nerve society guideline on the classification, diagnosis, investigation, and immunosuppressive therapy of non-systemic vasculitic neuropathy: Executive summary. *Journal of the Peripheral Nervous System, 15,* 176–184.

Jayne, D. (2001). Update on the European vasculitis study group trials. *Current Opinion in Rheumatology, 13,* 48–55.

Jennette, J. C., Falk, R. J., Andrassy, K., Bacon, P. A., Churg, J., Gross, W. L., Hagen, E. C., Hoffman, G. S., Hunder, G. G., Kallenberg, C. G., et al. (1994).

Nomenclature of systemic vasculitides. Proposal of an international consensus conference. *Arthritis and Rheumatism, 37*, 187–192.

Kissel, J. (2001). Vasculitic neuropathies. In S. Gilman (Ed.), *MedLink neurology*. San Diego: MedLink Corporaation.

Laghi Pasini, F., Pastorelli, M., Beermann, U., de Candia, S., Gallo, S., Blardi, P., & Di Perri, T. (1996). Peripheral neuropathy associated with ischemic vascular disease of the lower limbs. *Angiology, 47*, 569–577.

Lang, P. M., Vock, G., Schober, G. M., Kramer, S., Abahji, T., Crispin, A., Irnich, D., & Hoffmann, U. (2009). Impact of endovascular intervention on pain and sensory thresholds in nondiabetic patients with intermittent claudication: A pilot study. *The Journal of Pain, 10*, 264–273.

Lee, J. E., Shun, C. T., Hsieh, S. C., & Hsieh, S. T. (2005). Skin denervation in vasculitic neuropathy. *Archives of Neurology, 62*, 1570–1573.

Lindenlaub, T., & Sommer, C. (2003). Cytokines in sural nerve biopsies from inflammatory and non-inflammatory neuropathies. *Acta Neuropathologica, 105*, 593–602.

Michaelis, M., Vogel, C., Blenk, K. H., Arnarson, A., & Janig, W. (1998). Inflammatory mediators sensitize acutely axotomized nerve fibers to mechanical stimulation in the rat. *The Journal of Neuroscience, 18*, 7581–7587.

Nukada, H., Powell, H. C., & Myers, R. R. (1993). Spatial distribution of nerve injury after occlusion of individual major vessels in rat sciatic nerves. *Journal of Neuropathology and Experimental Neurology, 52*, 452–459.

Pedrini, L., & Magnoni, F. (2007). Spinal cord stimulation for lower limb ischemic pain treatment. *Interactive Cardiovascular and Thoracic Surgery, 6*, 495–500.

Satoi, H., Oka, N., Kawasaki, T., Miyamoto, K., Akiguchi, I., & Kimura, J. (1998). Mechanisms of tissue injury in vasculitic neuropathies. *Neurology, 50*, 492–496.

Schäfers, M., Lee, D. H., Brors, D., Yaksh, T. L., & Sorkin, L. S. (2003). Increased sensitivity of injured and adjacent uninjured rat primary sensory neurons to exogenous tumor necrosis factor-alpha after spinal nerve ligation. *The Journal of Neuroscience, 23*, 3028–3038.

Steck, A. J., Kinter, J., & Renaud, S. (2011). Differential gene expression in nerve biopsies of inflammatory neuropathies. *Journal of the Peripheral Nervous System, 16*(Suppl. 1), 30–33.

Üçeyler, N., Devigili, G., Toyka, K. V., & Sommer, C. (2010). Skin biopsy as an additional diagnostic tool in non-systemic vasculitic neuropathy. *Acta Neuropathologica, 120*, 109–116.

Weber, F., & Ziegler, A. (2002). Axonal neuropathy in chronic peripheral arterial occlusive disease. *Muscle & Nerve, 26*, 471–476.

Weinberg, D. H., Simovic, D., Isner, J., & Ropper, A. H. (2001). Chronic ischemic monomelic neuropathy from critical limb ischemia. *Neurology, 57*, 1008–1012.

Wilbourn, A. J., Furlan, A. J., Hulley, W., & Ruschhaupt, W. (1983). Ischemic monomelic neuropathy. *Neurology, 33*, 447–451.

Yamamoto, M., Ito, Y., Mitsuma, N., Hattori, N., & Sobue, G. (2003). Pain-related differential expression of NGF, GDNF, IL-6, and their receptors in human vasculitic neuropathies. *Internal Medicine, 42*, 1100–1103.

Vascular Orofacial Pain

Synonyms

Vascular-type craniofacial pain; VOP

Definition

Vascular orofacial pain (VOP) shares many of the signs and symptoms of other vascular-type craniofacial pains. It is confined to the oral cavity, the perioral region, and the lower part of the face. When affecting a tooth, it may mimic pulpitis.

Cross-References

▶ Atypical Facial Pain: Etiology, Pathogenesis, and Management
▶ Atypical Odontalgia

Vascular-Type Craniofacial Pain

Definition

This is a unilateral, periodic, pulsatile, severe pain in the craniofacial region, which is accompanied by autonomic phenomena. This includes pain syndromes such as migraine, cluster headache, and paroxysmal hemicrania.

Cross-References

▶ Atypical Facial Pain: Etiology, Pathogenesis, and Management
▶ Atypical Odontalgia
▶ Vascular Orofacial Pain

Vasculitic

Definition

Pertaining to vasculitis.

Cross-References

▶ Hansen's Disease

Vasculitic Neuropathies

▶ Vascular Neuropathies

Vasculitic Neuropathy

▶ Diabetic Neuropathies

Vasculitis

▶ Headache Due to Arteritis

Vasoactive Intestinal Polypeptide

Synonyms

VIP

Definition

Vasoactive intestinal polypeptide (VIP), a 28-amino acid peptide, was isolated as a vasodilator peptide from lung and intestine and was later also localized to nervous tissue. VIP acts as an anti-inflammatory peptide but is itself also produced by immune cells. In the nervous system, it modulates pain sensation as well as other neuronal processes. Apart from its involvement in brain metabolism, it relaxes smooth muscle and stimulates secretory glands. In postganglionic fibers of the autonomic nervous system, it is a co-transmitter of acetylcholine.

Cross-References

▶ Peptides in Neuropathic Pain States
▶ Postsynaptic Dorsal Column Projection, Anatomical Organization

VDCCs

▶ Calcium Channels in the Spinal Processing of Nociceptive Input

Vector

Definition

Gene delivery is accomplished through the use of "vectors." These may be derived from viruses (viral vectors) or constructed using nonviral elements (nonviral vectors). In ex vivo applications, cells removed from the body are cultured in a dish, transduced with the vector of choice, and, after determining that the transgene is expressed appropriately, are transplanted back into the body to achieve the desired therapeutic effect. In vivo applications of gene therapy rely on the use of vectors to deliver the therapeutic transgene to an intact organism by injection into a blood vessel or directly into a target tissue.

Cross-References

▶ Opioids and Gene Therapy

Vein Thrombosis

▶ Postoperative Pain, Venous Thromboembolism

Venereal Arthritis

▶ Reiter's Syndrome

Venous Sinus Thrombosis

▶ Headache Due to Venous Sinus Thrombosis

Ventral Amygdaloid Pathway

Synonyms

VAP

Definition

A collection of fibers that reciprocally connect the amygdala with the lateral hypothalamus and brain stem areas, such as the parabrachial area.

Cross-References

▶ Nociceptive Processing in the Amygdala: Neurophysiology and Neuropharmacology

Ventral Basal Complex of the Lateral Thalamus

Definition

Human nomenclature (VC), VP or VB in monkeys or cats, includes VPL and VPM.

Cross-References

▶ Thalamus, Dynamics of Nociception

Ventral Basal Nucleus (VB)

Synonyms

Ventrocaudal nucleus (Vc)

Definition

The main somatosensory relay nucleus in the thalamus and consisting of the ventroposterolateral (VPL) and ventroposteromedial (VPM) nuclei. Orofacial inputs are relayed from the trigeminal main sensory nucleus to VPM and from the rest of the body are relayed from the dorsal column nuclei to VPL.

Cross-References

▶ Thalamic Plasticity and Chronic Pain

Ventral Caudal Nucleus (Vc)

Definition

Human nomenclature for VP or VB in monkeys or cats includes VPL and VPM.

Cross-References

▶ Burst Activity in Thalamus and Pain
▶ Human Thalamic Nociceptive Neurons
▶ Thalamic Bursting Activity, Chronic Pain
▶ Thalamus, Receptive Fields, Projected Fields, Human

Ventral Lateral Nucleus

Definition

It is the ventral nucleus anterior to VP. The region is heavily connected with the motor cortex.

Cross-References

▶ Core and Matrix

Ventral Medial Nucleus

Synonyms

Vmpo

Definition

The posterior part (Vmpo) of the ventral medial nucleus of the thalamus. The region is claimed to be the main spinothalamic termination site for lamina I neurons in humans and in the monkey.

Cross-References

▶ Human Thalamic Nociceptive Neurons

Ventral Posterior Complex (VP)

Definition

A group of nuclei in the ventral tier of the thalamus, receiving somatosensory inputs from the whole body. The complex is organized in a topological way, the head being placed medially (ventral posterior medial nucleus), and the foot laterally close to the internal capsule (ventral

posterior lateral nucleus). Nociceptive cells have been found in different parts of the complex, but more specifically in the ventral posterior inferior nucleus, where a pain homunculus was described. The complex projects mainly to cortical areas SI, SII, and the insula.

Cross-References

▶ Thalamotomy for Human Pain Relief

Ventral Posterior Inferior Nucleus

Synonyms

VPI

Definition

The VPI nucleus is part of the somatosensory thalamus. It is located just ventral to the ventral posterior lateral and medial nuclei. Many of the neurons in the VPI nucleus are nociceptive. One source of input to the VPI nucleus is the spinothalamic tract. The thalamocortical output probably includes SII.

Cross-References

▶ Spinothalamic Input: Cells of Origin (Monkey)
▶ Spinothalamic Terminations, Core and Matrix
▶ Thalamic Nuclei Involved in Pain, Human and Monkey
▶ Thalamus, Nociceptive Cells in VPI, Cat and Rat

Ventral Posterior Lateral Nucleus

Synonyms

VPL

Definition

The VPL nucleus is a part of the somatosensory thalamus. As the name implies, this nucleus is located in the ventral part of the posterior thalamus. It is lateral to another part of the somatosensory thalamus, the ventral posterior medial (VPM) nucleus. The VPL nucleus is concerned with somatosensory input from the body, whereas the VPM nucleus receives information concerning the head. In primates, the VPL nucleus receives synaptic input from both the dorsal column-medial lemniscus pathway, spinocervicothalamic pathway, and the spinothalamic tract. Some VPL neurons are nociceptive, although most are mechanoreceptive. The VPL nucleus projects to, and thus conveys somatosensory information to, the SI and SII regions of the cerebral cortex.

Cross-References

▶ Burst Activity in Thalamus and Pain
▶ Lateral Thalamic Lesions, Pain Behavior in Animals
▶ Spinothalamic Input: Cells of Origin (Monkey)
▶ Spinothalamic Terminations, Core and Matrix
▶ Thalamic Nuclei Involved in Pain, Human and Monkey
▶ Thalamic Plasticity and Chronic Pain
▶ Thalamus, Visceral Representation

Ventral Posterior Medial

Synonyms

VPM

Definition

Medial portion of VP. Neurons in this nucleus respond mainly to touch and proprioception applied to the head.

Cross-References

▶ Burst Activity in Thalamus and Pain
▶ Human Thalamic Response to Experimental Pain (Neuroimaging)
▶ Spinothalamic Terminations, Core and Matrix
▶ Thalamic Nuclei Involved in Pain, Human and Monkey
▶ Thalamus, Visceral Representation

Ventral Posterior Nuclear Group of the Thalamus

Definition

Ventral Posterior Nuclear Group of the Thalamus is a summary term for the specific somatosensory nuclei of the thalamus. The most commonly used taxonomy is for monkey thalamus – VPL (ventro-postero-lateral), VPM (ventro-postero-medial), and VPI (ventro-postero-inferior) – and in humans Vc (ventro-caudal).

Cross-References

▶ Nociceptive Processing in the Secondary Somatosensory Cortex

Ventral Posterior Nucleus (Human Ventral Caudal)

Definition

The principal somatic sensory nucleus of thalamus; it receives projections from the spinothalamic tract. Other nuclei receiving such input are posterior and inferior to VP including the posterior nucleus, the ventral posterior inferior nucleus, and the ventral medial nucleus – posterior part (VMpo).

Cross-References

▶ Angina Pectoris, Neurophysiology and Psychophysics
▶ Lateral Thalamic Pain-Related Cells in Humans

Ventral Posterior Nucleus of Thalamus

▶ Lateral Thalamic Pain-Related Cells in Humans

Ventral Tegmental Area

Definition

Region of the midbrain containing dopaminergic neurons that project to cortical and limbic structures.

Cross-References

▶ Nociceptive Processing in the Nucleus Accumbens, Neurophysiology and Behavioral Studies

Ventricular Collapse

▶ Headache Due to Low Cerebrospinal Fluid Pressure

Ventrocaudal Nucleus (Vc)

▶ Ventral Basal Nucleus (VB)

Ventrolateral Funiculus

Definition

White matter of the ventral and lateral aspects of the spinal cord containing ascending and descending tracts.

Cross-References

▶ Vagal Input and Descending Modulation

Ventrolateral Orbital Cortex

Synonyms

VLO

Definition

The VLO is a region of the frontal cortex that envelops the rhinal fissure at its medial limits, and is respectively bounded by the lateral and medial orbital regions. Its location in the so-called limbic (emotional) areas of the cortex, and its interconnections with the SM and PAG, suggest that VLO neurons may process motivational-affective aspects of nociception.

Cross-References

▶ Spinothalamocortical Projections from SM

Ventrolateral System, Ascending Spinal Projection, Development

▶ Development of Spinothalamic Tract Neurons

Ventromedial Nucleus

Definition

The most medial portion of the ventrobasal or ventral posterior nucleus.

Cross-References

▶ Brainstem Subnucleus Reticularis Dorsalis Neuron
▶ Spinothalamocortical Projections to Ventromedial and Parafascicular Nuclei

Ventroposterior Lateral Nucleus

Synonyms

VPL

Definition

Lateral portion of VP. Neurons in this nucleus respond mainly to touch and proprioception applied to the lower body and lower limbs.

Cross-References

▶ Human Thalamic Response to Experimental Pain (Neuroimaging)

Verapamil

Definition

Calcium channel blocker.

Cross-References

▶ Preventive Migraine Therapy

Vertebrobasilar Arterial Disease

Definition

This is a condition involving insufficient delivery of blood flow through the vertebral or basilar arteries, or both, to the brain. The vertebrobasilar system supplies the hindbrain and brain stem with blood.

Cross-References

▶ Primary Cough Headache

Vertigo

Definition

An illusory sensation of motion (rotational, translational, or tilting of the visual environment) of either the self or the surrounding.

Cross-References

▶ Coordination Exercises in the Treatment of Cervical Dizziness

Vesical

Definition

Of the bladder.

Cross-References

▶ Nocifensive Behaviors of the Urinary Bladder
▶ Visceral Pain Model, Urinary Bladder Pain (Irritants or Distension)

Vesical Pain Models

▶ Visceral Pain Model, Urinary Bladder Pain (Irritants or Distension)

Vesicular Inhibitory AminoAcid Transporter

Synonyms

VIAAT

Definition

A transporter protein that transports GABA and glycine into presynaptic vesicles.

Cross-References

▶ GABA and Glycine in Spinal Nociceptive Processing

Vestibular Adaptation Exercises

▶ Coordination Exercises in the Treatment of Cervical Dizziness

Vestibular Compensation Exercises

▶ Coordination Exercises in the Treatment of Cervical Dizziness

Vestibular Habituation Exercises

▶ Coordination Exercises in the Treatment of Cervical Dizziness

Vestibular Rehabilitation

▶ Coordination Exercises in the Treatment of Cervical Dizziness

Vestibulectomy

Definition

A vestibulectomy is a minor surgical procedure involving the excision of a portion of the vestibular area in women with vulvodynia with the aim of improving vulvar pain.

Cross-References

▶ Dyspareunia and Vaginismus
▶ Vulvodynia

Vestibulodynia

Definition

A localized form of vulvodynia (pain is localized to the vulvar vestibule); this definition has been introduced to replace the term "vulvar vestibulitis," since it was felt that the suffix "-itis" in vestibulitis might incorrectly suggest an inflammatory etiology. However, the term "vestibular pain syndrome" is more appropriate.

Cross-References

▶ Vulvodynia

VGCCs

▶ Calcium Channels in the Spinal Processing of Nociceptive Input

VGSC

▶ Nociception: Role of Voltage-Gated Sodium Channels

VH1

▶ Phosphatases and Glia

VH6

▶ Phosphatases and Glia

Vi/Vc Transition Zone

Synonyms

Trigeminal transition zone

Definition

A region of the Trigeminal Brainstem Nuclear Complex (VBNC) at the obex level where the subnuclei interpolaris (Vi) and caudalis (Vc) of the spinal trigeminal nucleus converge. The ventral portion of the laminated Vc is pushed dorsomedially by Vi (see Yoshida et al. 1991). The substantia gelatinosa is still identifiable but often interrupted. Neurons in the trigeminal transition zone project to the thalamus, and are involved in processing of nociceptive information from the orofacial regions and have some unique physiological properties (see Ren and Dubner 2011).

References

Ren, K., & Dubner, R. (2011). The role of trigeminal interpolaris-caudalis transition zone in persistent orofacial pain. *International Review of Neurobiology, 97*, 207–225.

Yoshida, A., Dostrovsky, J. O., Sessle, B. J., & Chiang, C. Y. (1991). Trigeminal projections to the nucleus submedius of the thalamus in the rat. *The Journal of Comparative Neurology, 307*, 609–625.

VIAAT

▶ Vesicular Inhibitory AminoAcid Transporter

Viagra

▶ Sildenafil

Vibration

Definition

Oscillating movements of body parts or of the whole body relative to a midposition. These movements are generated by constantly alternating forces or moments at a certain frequency, usually produced by an external motor-driven vibrator. Biologically, vibration-sensitive receptors, usually innervated by coarse myelinated nerve afferents, would be driven and be expected to produce an analgesic effect analogous to that of transcutaneous electrical nerve stimulation. Insufficient clinical data for effect evaluation are at hand.

Cross-References

▶ Ergonomic Counseling
▶ Massage and Pain Relief Prospects

Vicarious Experience

Definition

Observing another individual enacting a particular behavior or responding to particular environmental conditions leads to a similar experience in the observer.

Cross-References

▶ Psychology of Pain, Self-Efficacy

Vicarious Instigation

Definition

Vicarious instigation describes the phenomenon that mere observation of another person's response to a stimulus or a situation (e.g., a pain response), possibly mediated by empathy, can induce a similar response in the observer, in the absence of any direct experience with the eliciting stimulus or situation. In the context of pain, it is still a matter of debate whether observing another person in pain can induce pain-like vicarious responses in the observer, or whether it elicits a more generalized emotional response in the observer.

Cross-References

▶ Modeling, Social Learning in Pain

Vicarious Learning

▶ Modeling, Social Learning in Pain

VIP

▶ Vasoactive Intestinal Polypeptide

Viral Meningitis

▶ Headache in Aseptic Meningitis

Viral Neuropathies

Susan Shin[1] and David Simpson[2]
[1]Neurology, Mount Sinai School of Medicine, New York, NY, USA
[2]Clinical Neurophysiology Laboratories, Mount Sinai Medical Center, New York, NY, USA

Synonyms

Nerve viral infection; Virus-induced nerve dysfunction

Definition

Neuropathies that are associated with viral infection.

Introduction

Virus-induced neuropathy is an important cause of painful peripheral neuropathies. Pain is often the most prominent symptom after a virus causes destruction to a nerve. The mechanisms by which viruses can induce pain include direct infection of neurons or nerve fibers; they may also cause indirect nerve destruction by activating an immune-mediated inflammatory response. In this chapter, we will focus on a few important viruses that may cause painful peripheral neuropathies.

Characteristics

Human Immunodeficiency Virus (HIV)

Epidemiology

Distal sensory polyneuropathy (DSP) is a common neurologic complication and major cause of morbidity in HIV-infected patients (Gonzalez-Duarte et al. 2008). The estimated incidence of DSP is 30 %, with a prevalence ranging from 25 % to 53 % (Morgello et al. 2004). Risk factors that are associated with a higher occurrence of DSP include older age, low nadir CD4 count, use of neurotoxic drugs, and history of opioid abuse or dependence (Cornblath and McArthur 1988).

Pathogenesis

Although the exact means by which HIV causes DSP is unclear, both in vitro and animal studies suggest that multiple mechanisms exist to produce axonal injury associated with HIV-DSP. HIV virus is not thought to directly invade the axon or cells of the peripheral nervous system, e.g., Schwann cells. Its destructive mechanism is thought to be closely linked to an HIV-associated protein gp120. It is postulated that gp120 may induce immune-mediated damage in several ways. These include induction of proinflammatory cytokines (IL-1 and TNF-alpha), expression of the capsaicin receptor VR1/TRKV1 in gp120 responsive neurons, ligation of chemokine receptors on glial cells and neurons, or by acting on chemokine receptors related to Schwann cell-neuron interactions. The consequence of this immune-mediated destruction has been seen on pathologic examination as axonal loss in myelinated fibers, distal axonal degeneration affecting small unmyelinated fibers, and occasionally demyelination and remyelination. Proximal changes in the dorsal root ganglion have also been identified; these include activation of macrophages, infiltrates of inflammatory lymphocytes, and production of proinflammatory cytokines (TNF-alpha, IL-6, and Nitric Oxide) (Gonzalez-Duarte et al. 2008).

Clinical Features

Patients affected with DSP often complain of a multitude of symptoms with pain being the most debilitating symptom. Painful dysesthesias; sensation of burning, pins, and needles; numbness; and allodynia are some common symptoms. It is usually bilateral and often starts in the soles of the feet, but as the condition progresses, it can affect the legs and the hands. Muscle weakness is usually minimal or absent. On examination, patients may have sensory deficits in a "stocking and glove distribution" and may have absent or depressed tendon reflexes in the ankles. Hyperactive knee reflexes together with normal or depressed ankle reflexes may be seen in patients with combined disease of the peripheral and central nervous systems.

Diagnostic Studies

Although the diagnosis of DSP is usually made on clinical grounds, some studies can aid in the diagnosis or exclusion of the condition. Routine blood work to screen for other causes of peripheral neuropathy is recommended. These conditions include diabetes mellitus, thyroid dysfunction, vitamin B12 deficiency, renal or liver impairment, and syphilis. The presence of a low CD4 count or history or current use of neurotoxic antiretroviral drugs (ddC, ddI, or d4T) also suggests an increased risk of developing DSP (Gonzalez-Duarte et al. 2008).

Findings on electrodiagnostic tests such as nerve conduction studies (NCS) and electromyography (EMG) may show a length-dependent axonal, predominantly sensory polyneuropathy and chronic partial denervation and reinnervation in the muscles of the foot and distal leg (Gonzalez-Duarte et al. 2008). However, EMG and NCS studies may be normal since DSP tends to effect small myelinated and unmyelinated fibers (Luciano et al. 2003). A validated tool in the evaluation of small fiber neuropathy, including DSP, is skin biopsy with quantitative analysis of intraepidermal nerve fiber densities (Holland et al. 1997). A low epidermal nerve density correlates with increased severity of painful symptoms (McArthur et al. 2000). In a 6- to 12-month follow-up study of asymptomatic patients, a low epidermal nerve fiber density of less than 11 fibers/mm increased the chance of the development of symptoms. Therefore, the skin biopsy technique is not only helpful in establishing a diagnosis, but it can also help monitor disease progression over time and help predict the likelihood of symptoms developing in asymptomatic patients (Simpson et al. 2006).

In clinical practice, nerve biopsy should be reserved only for patients that present with atypical clinical features or when the diagnosis of DSP is being questioned. For example, when vasculitis or CMV infection is being considered, sural nerve biopsy may be informative.

Nerve biopsy specimens do not consistently reveal the virus or related viral proteins, likely due to the fact that the mechanism of disease is immune-mediated rather than direct viral infection.

Treatment

While treatment of DSP is challenging, there are several treatment options for the symptomatic relief of painful neuropathy due to DSP. Common agents that are used include antidepressants, antiepileptic drugs, capsaicin, topical anesthetics, NSAIDs, and opioids.

The two commonly prescribed classes of antidepressants include tricyclic antidepressants (amitriptyline and nortriptyline) and selective serotonin-norepinephrine reuptake inhibitors (duloxetine and venlafaxine).

Gabapentin and pregabalin are commonly used antiepileptic drugs in the treatment of HIV-DSP. They have similar mechanisms of action, by inhibiting presynaptic voltage-gated calcium channels. Gabapentin has been widely used in clinical settings and has been shown to be effective in reducing pain related to HIV-DSP in a small placebo-controlled trial (Hahn et al. 2004). However, a more recent randomized double-blind, placebo-control trial failed to show that pregabalin was superior to placebo in the treatment of painful HIV neuropathy (Simpson et al. 2010). Due to potential methodological limitations in that study, another placebo-controlled trial of pregabalin in HIV-DSP is underway.

More recently, the use of high-concentration capsaicin patches has been shown to provide up to 12 weeks of pain reduction in patients with postherpetic neuralgia and HIV-DSP. The desensitization of cutaneous nociceptors by capsaicin is thought to result in reduced pain (Simpson et al. 2008). The high-concentration capsaicin patch has been approved in the European Union for the treatment of peripheral neuropathic pain, with the exception of diabetics, and in the USA for postherpetic neuralgia.

Successful pain control may require a combination of therapies, targeting a variety of mechanisms that lead to neuropathic pain. This approach is termed rational polypharmacy.

Varicella-Zoster Virus (VZV)

Epidemiology

Herpes zoster is a reactivation of the varicella-zoster virus (VZV), which causes the disease shingles. Pain can be a prominent feature of the disease, at various phases of the illness. It can also persist once the acute illness has subsided in a condition known as postherpetic neuralgia. Postherpetic neuralgia (PHN) is defined as pain that persists more than 4 months after rash onset (Dworkin and Portenoy 1996), and it can afflict up to 27–73 % of individuals older than 55 years of age (Kost and Straus 1997).

Pathogenesis

After primary infection, VZV lays dormant in sensory ganglia. The virus then reactivates and spreads from the sensory ganglia to the corresponding dermatome; this results in shingles. At a pathologic level, the virus causes hemorrhagic necrosis in the peripheral nerve and dorsal root ganglion and is accompanied by neuronal loss.

Clinical Features

Up to 70-80 % of patients experience a prodrome of dermatomal pain that precedes the appearance of the characteristic rash by several days (Dworkin and Portenoy 1996). Patients may describe painful dysesthesias with adjectives that describe the pain as sharp, shooting, tender, burning, or throbbing.

Diagnostic Studies

There are no specific tests to diagnose PHN. The diagnosis is usually made on clinical grounds, as in patients who experience persistent pain after the eruption of a characteristic rash of herpes zoster.

Treatment

The approach to treating PHN is similar to most neuropathic pain syndromes. A combination of antidepressants, anticonvulsants, topical anesthetics, capsaicin, and opiates can be used to control painful symptoms. One randomized trial found some benefit in intrathecal methylprednisolone for intractable postherpetic neuralgia (Kotani et al. 2000). Systematic reviews of multiple trials only showed marginal evidence that treatment with acyclovir reduces the incidence of PHN.

Cytomegalovirus (CMV)

Epidemiology

Cytomegalovirus infection can cause a progressive polyradiculopathy (PP) as well as a mononeuropathy multiplex (MM) in patients with advanced AIDS. Progressive polyradiculopathy usually occurs when CD4 lymphocytes counts are below 50 cells/mm^3. Early diagnosis is crucial as it can progress to a rapidly devastating condition, with irreversible lower extremity paralysis and incontinence, but can respond to early antiviral therapy (Wulff et al. 2000).

Pathogenesis

Immunocytochemical studies have documented the presence of CMV inclusions in neural tissue. Marked inflammation and extensive necrosis of ventral and dorsal nerve roots have been seen in autopsy studies. Focal myelitis has also been seen in the area of the inflamed nerve roots. Areas of severe necrosis appear to have infiltrates of both polymorphonuclear and mononuclear cells, leading to vascular congestion and edema (Verma 2001).

Clinical Features

Rapidly progressive lower extremity weakness is the most common clinical presentation of PP. Radiating pain and parasthesias in the cauda equina distribution, mild sensory loss, and sphincter dysfunction are frequent accompanying symptoms. On examination, lower extremity flaccid paraparesis and areflexia are present, and a thoracic sensory level may be found in some patients (Wulff et al. 2000).

Diagnostic Studies

CSF profile usually shows a polymorphonuclear pleocytosis, elevated protein (above 50 mg/ml), and low glucose. Polymerase chain reaction (PCR) of spinal fluid for CMV is the hallmark of diagnosis. Electromyography (EMG) studies show a pattern of polyradiculopathy, with lower extremity axonal loss. Radiologic imaging of the spinal cord should be performed to exclude focal compressive lesions of the cauda equina (Wulff et al. 2000). In PP, lumbar MRI may reveal enhancement of nerve roots.

Treatment

CMV-related progressive polyradiculopathy is a neurologic emergency. Patients should be started on antiviral treatment for CMV infection immediately if the clinical suspicion is high, regardless of pending CSF studies (Verma 2001) Delayed treatment can result in irreversible neurologic deficits due to nerve root necrosis. Ganciclovir is an antiviral agent effective against CMV. Other antiviral medications that can be used include foscarnet, cidofovir, valganciclovir, and fomivirsen (Wulff et al. 2000). While controlled studies have not been performed, combination antiviral therapy may provide greater efficacy.

Hepatitis C Virus (HCV)

Epidemiology

Hepatitis C infection is commonly associated with peripheral neuropathy. The presence of parasthesias, numbness, and lancinating pain can occur in approximately 50 % of patients (Gorwon et al. 2007).

Pathogenesis

Although the pathogenesis of HCV-associated neuropathy remains unclear, several theories discussing the mechanism of action have been reported. These include peripheral nerve damage through direct viral infection, immune-mediated necrotizing vasculitis via presence of cryoglobulins, and damage from interferon treatment.

Clinical Features

Patients with HCV-associated peripheral neuropathy may complain of pain, cramping, numbness, parasthesias, and fatigue in all four limbs; physical exertion may exacerbate these symptoms (Gorwon et al. 2007). The triad of palpable purpura, joint pains, and weakness can be seen in HCV-related mixed cryoglobulinemia (Iannuzzella et al. 2010). This condition can also cause a more widespread vasculitis in some patients leading to end-organ damage: membranoproliferative glomerulonephritis, bronchiectasis, hepatomegaly, splenomegaly, and cerebrovascular events.

Diagnostic Studies

Electrophysiological abnormalities can be detected in patients who present with symptoms. These include decreased sensory or motor conduction velocities, reduced amplitude or absence of sensory nerve action potential (SNAP), and low amplitude compound muscle action potential (CMAP). Sensory nerves of the lower limbs are more commonly affected (Ammendola et al. 2005). Nerve biopsy can reveal epineural vasculitis, fascicular loss of axons, and signs of both demyelination and axonal degeneration (Nemni et al. 2003). Serum should also be tested for the presence of cryoglobulins.

Treatment

Treatment of HCV-associated neuropathy is aimed at treating the hepatitis C infection and cryoglobulinemia. Although interferons are used to treat the active infection, the treatment itself can sometimes cause a painful neuropathy. The use of steroids, plasma exchange, and immune-modulating agents such as cyclophosphamide and rituximab has been used with varying success. Other agents used to control neuropathic pain can be reasonably used in clinical practice to help reduce painful symptoms, although controlled studies have not been done.

Summary

Viral neuropathies are an important cause of neuropathic pain. Careful clinical assessment along with relevant diagnostic studies can help identify the particular type of neuropathy. A multidimensional approach that targets various mechanisms should be implemented to provide the patient with optimal symptomatic relief.

References

Ammendola, A., Sampaolo, S., Ambrosone, L., et al. (2005). Peripheral neuropathy in hepatitis related mixed cryoglobulinemia: Electrophysiologic follow up study. *Muscle & Nerve, 31*, 382–385.

Cornblath, D. R., & McArthur, J. C. (1988). Predominantly sensory neuropathy in patients with AIDS and AIDS-related complex. *Neurology, 38*, 794–796.

Dworkin, R. H., & Portenoy, R. K. (1996). Pain and its persistence in herpes zoster. *Pain, 67*(2–3), 241–251.

Gonzalez-Duarte, A., Robinson-Papp, J., & Simpson, D. M. (2008). Diagnosis and management of HIV-associated neuropathy. *Neurologic Clinics, 26*(3), 821–832.

Gorwon, K. C., Herrmann, D. N., et al. (2007). Non-length dependent small fiber neuropathy/ganglionopathy. *Journal of Neurology, Neurosurgery, and Psychiatry, 79*, 163–169.

Hahn, K., Arendt, G., Braun, J. S., et al. (2004). A placebo-controlled trial of gabapentin for painful HIV-associated sensory neuropathie. *Journal of Neurology, 251*, 1260–1266.

Holland, N., Stocks, A., Hauer, P., et al. (1997). Intraepidermal nerve fiber density in patients with painful sensory neuropathy. *Neurology, 48*, 708–711.

Iannuzzella, F., Vaglio, A., & Garini, G. (2010). Management of hepatitis C virus-related mixed cryoglobulinemia. *American Journal of Medicine, 123*(5), 400–408.

Kost, R. G., & Straus, S. E. (1997). Postherpetic neuralgia. Predicting and preventing risk. *Archives of Internal Medicine, 157*(11), 1166–1167.

Kotani, N., Kushikata, T., Hashimoto, H., Kimura, F., Muraoka, M., Yodono, M., et al. (2000). Intrathecal methylprednisolone for intractable postherpetic neuralgia. *The New England Journal of Medicine, 343*, 1514–1519.

Luciano, C. A., Pardo, C. A., & McArthur, J. C. (2003). Recent developments in the HIV neuropathies. *Current Opinion in Neurology, 16*, 403–409.

McArthur, J. C., Yiannoutsos, C., Simpson, D. M., & ACTG Clinical Trials Group Team 291. (2000). A phase II trial of recombinant human nerve growth factor for sensory neuropathy associated with HIV infection. *Neurology, 54*, 1080–1088.

Morgello, S., Estanislao, L., Simpson, D., et al. (2004). HIV-associated distal sensory polyneuropathy in the era of highly active antiretroviral therapy: the Manhattan HIV brain bank. *Archives of Neurology, 61*, 546–551.

Nemni, R., Sanvito, L., et al. (2003). Peripheral neuropathy in hepatitis C virus infection with and without cryoglobulinemia. *Journal of Neurology, Neurosurgery, and Psychiatry, 74*, 1267–1271.

Simpson, D. M., Kitch, D., Evans, S. R., et al. (2006). HIV neuropathy natural history cohort study: Assessment measures and risk factors. *Neurology, 66*, 1679–1687.

Simpson, D. M., Brown, S., Tobias, J., et al. (2008). Controlled trial of high-concentration capsaicin patch for treatment of HIV neuropathy. *Neurology, 70*(24), 2305–2313.

Simpson, D. M., Schifitto, G., Clifford, D. B., et al. (2010). Pregabalin for painful HIV neuropathy: A randomized, double-blind, placebo-controlled trial. *Neurology, 74*(5), 413–420.

Verma, Ashok. (2001). Epidemiology and clinical features of HIV-1 associated neuropathies. *Journal of the Peripheral Nervous System, 6,* 8–13.

Wulff, E. A., Wang, A. K., & Simpson, D. M. (2000). HIV associated peripheral neuropathy. *Drugs, 59*(6), 1251–1260.

Viral Vectors

▶ Opioids and Gene Therapy

Virus-induced Nerve Dysfunction

▶ Viral Neuropathies

Visceral

Definition

Pertaining to any organ within the great cavities of the body, mainly describing the soft internal abdominal organs, particularly the intestines.

Cross-References

▶ Functional Abdominal Pain in Children

Visceral Afferent

Definition

Afferent fibers from the viscera to the central nervous system. These fibers originate from dorsal root ganglion cells and vagal (nodose) ganglion cells.

Cross-References

▶ Morphology, Intraspinal Organization of Visceral Afferents

Visceral Hyperalgesia

▶ Descending Modulation of Visceral Pain

Visceral Hypersensitivity

Definition

Heightened perception of visceral events, both physiological and experimental.

Cross-References

▶ Descending Modulation of Visceral Pain
▶ Thalamus and Visceral Pain Processing (Human Imaging)
▶ Visceral Pain Model, Irritable Bowel Syndrome Model

Visceral Inflammation

▶ Visceral Pain Model, Pancreatic Pain

Visceral Mechanoreceptors

Bin Feng[1] and Gerald F. Gebhart[2]
[1]Center for Pain Research, Department of Anesthesiology, School of Medicine, University of Pittsburgh, Pittsburgh, Pennsylvania, USA
[2]Department of Anesthesiology, Center for Pain Research, University of Pittsburgh, Pittsburgh, PA, USA

Synonyms

Mechanonociceptors; Mechanoreceptive endings; Mucosal receptors; Stretch-sensitive receptors; Tension receptors

Definition

Stimuli adequate for activation of visceral nociceptors include distension of hollow organs, traction on the mesentery, ischemia, and chemicals typically associated with organ inflammation. As experimental hollow organ distension (e.g., with a balloon) reproduces in human subjects the quality and localization of pathophysiological visceral disorders (see Ness and Gebhart 1990 for review), hollow organ distension and other mechanical stimuli (e.g., probing, stretch) have been most widely employed to examine the characteristics of visceral nociceptors. Powley and co-workers (2011) recently reported the presence of three vagal afferent specializations innervating the mucosa of the proximal GI tract: villus afferents, crypt afferents, and antral gland afferents. Their location suggests roles in chemosensation (response to nutrients, etc.); whether that includes chemo-nociception remains to be determined.

Characteristics

Visceral mechanonociceptors in hollow organs are assumed to be associated with muscle layers, and to respond to tension developed in muscle produced, for example, by distension or stretch and are, accordingly, also referred to as stretch-sensitive or stretch-responsive. The functional properties of visceral mechanoreceptive nociceptors (▶ mechanoreceptive/mechanosensitive visceral receptors) have been characterized using both in vivo and in vitro preparations to record activity from axons innervating various organs. Two populations of stretch-sensitive visceral afferents have been characterized: those with low thresholds for activation and those with high thresholds for activation, the latter of which presumably fulfills the role of providing information about acute visceral nociception. Approximately 75–80 % of stretch/tension-sensitive visceral afferents recorded in a variety of animal species, and innervating a variety of organs (e.g., stomach, gall bladder, urinary bladder, colon, uterus), have low thresholds for response within what is considered the physiological range (i.e., < 5 mmHg). The remaining 20–25 % have thresholds for response in the noxious range (i.e., ≥ 20 mmHg in mouse colon and 30 mmHg in rat colon), an intensity of hollow organ distension typically associated with ▶ nocifensive behaviors when applied in behaving animals (▶ Nocifensive Behaviors, Gastrointestinal Tract). Both low- and high-threshold stretch/tension-sensitive visceral afferents are slowly adapting and both encode stimulus intensity into the noxious range. Thus, unlike cutaneous low-threshold mechanoreceptors, which do not respond to noxious intensities of mechanical stimulation of the skin, low-threshold stretch/tension-sensitive visceral afferents continue to encode into the noxious range (e.g., see Sengupta and Gebhart 1995 for review). To further distinguish low-threshold stretch/tension-sensitive visceral receptors from low-threshold cutaneous mechanoreceptors, most low-threshold stretch/tension-sensitive visceral receptors sensitize after experimental exposure of receptive endings to an inflammogen/inflammatory soup or after organ insult (see ▶ Sensitization of Visceral Nociceptors). Accordingly, because both low- and high-threshold stretch/tension-sensitive visceral receptors (1) encode into the noxious range and (2) sensitize, they are functionally distinct from low- and high-threshold mechanoreceptors in the cutaneous realm, and thus both are likely able to contribute to nociception after organ insult.

In vitro organ-nerve preparations have led to broader functional characterization of mechanosensitive endings in the viscera. In all organs studied to date, including the esophagus, stomach, urinary bladder, and colon, three receptors – mucosal, tension (muscular), and serosal – are present. These classes of receptive endings are based on responses to application of different mechanical stimuli and are not yet verified to also represent histological location. In addition, receptive endings that respond to both mucosal stimulation and stretch (muscular-mucosal receptors) have been found in the stomach, colon, and urinary bladder, and mesenteric receptors have been identified in the lumbar splanchnic nerve innervation of the mouse colorectum. These in vitro preparations are also convenient for application of electrical stimulation as a search strategy, which reveals the presence of mechanically insensitive, silent

Visceral Mechanoreceptors, Fig. 1 Mouse colorectal pelvic nerve (*top*) and lumbar splanchnic nerve (*bottom*) afferents identified by electrical stimulation (e-stim; *left-most column, red arrow* indicates stimulus artifact) and classified based on responses to mechanical stimuli. All mechanosensitive afferents respond to probing (0.4–1.0 g; *horizontal lines* under records); muscular afferents are also activated by circumferential stretch (0–170 mN, 5 mN/s), mucosal afferents also by stroking (10 mg; *horizontal dashes* under records), and muscular-mucosal afferents also by stretch and stroking. Serosal afferents are activated only by probing. Mechanically insensitive afferents (*MIA*) do not respond to mechanical stimuli. Muscular-mucosal afferents are unique to the pelvic pathway; mesenteric afferents are unique to the splanchnic pathway (not shown) (Adapted from Feng and Gebhart 2011)

afferents (MIAs) in the visceral innervation (Feng and Gebhart 2011). Figure 1 shows examples of mechanosensitive and mechano*in*sensitive colorectal afferents recorded from the two nerves innervating the mouse colorectum: the pelvic nerve (PN) and lumbar splanchnic nerve (LSN).

The nature of these in vitro organ-nerve preparations allows for localization of receptive endings in the organ, which is not easily achieved using in vivo preparations, and also permits direct application of putative sensitizing mediators to the receptive ending in the organ (e.g., ▶ Sensitization of Visceral Nociceptors). When the distribution of receptive endings, which are typically 1–4 mm^2 in area, are plotted on cartoons of the mouse colorectum, significant differences between the pelvic and lumbar splanchnic innervations of the organ become readily apparent, supporting the notion that the nerves innervating the same organ have some overlapping but also clearly different functions. Figure 2 illustrates the distribution of mechanosensitive and mechano*in*sensitive afferent endings in the two nerves innervating the mouse colorectum and Fig. 3 illustrates the differences in proportions of afferent classes between the two nerves.

Not only does the topographical distribution of endings differ between the two nerves innervating

Visceral Mechanoreceptors, Fig. 2 Topographical distributions of colorectal afferent receptive endings in the pelvic nerve (*PN*; *left*) and the lumbar splanchnic nerve (*LSN*; *right*) innervations of the mouse colorectum (Data are derived and summarized from Brierley et al. 2004 and Feng and Gebhart 2011). The distribution of receptive endings in the PN are located in the caudal ~2 cm of mouse colorectum, whereas the distribution of receptive endings in the LSN are along the mesenteric side of the colorectum and include a significant proportion of endings localized along the vasculature of the mesenteric attachment (unique to this innervation). Similarly, the distribution of mechanosensitive endings in the mouse urinary bladder differs between the PN and LSN (Xu and Gebhart 2008)

the mouse colorectum, their proportions also differ significantly in the two nerves (Fig. 3).

Stretch-sensitive muscular and muscular-mucosal afferents represent about one third of the pelvic nerve afferent innervation of the colorectum, whereas stretch-sensitive muscular afferents are not prominent in the lumbar splanchnic innervation. This is consistent with reports that behavioral responses to balloon distension of the mouse colorectum are unaffected by transection of the lumbar splanchnic nerve but absent after pelvic nerve transection (Kyloh et al. 2011). There are also very few afferent endings in the lumbar splanchnic innervation that respond to mucosal stroking, whereas about one-quarter of pelvic nerve afferent endings respond to this stimulus. There are a significant proportion of mesenteric afferents in the lumbar splanchnic innervation, but none in the pelvic innervation. Serosal and MIA proportions are similar between the two colorectal innervations, although the MIA endings likely differ functionally. Seventy percent of MIAs in the pelvic nerve colorectal innervation acquired mechanosensitivity when exposed to a cocktail of inflammatory mediators (inflammatory soup), whereas less than 25 % of MIAs in the colorectal lumbar splanchnic innervation did so (Feng and Gebhart 2011). The extent to which MIA endings that did not acutely acquire mechanosensitivity after exposure to an inflammatory soup represent endings that are selectively chemo- or thermo-sensitive or represent efferent fibers remains to be determined.

Visceral Mechanoreceptors, Fig. 3 Proportions of colorectal afferent receptive endings in the pelvic nerve (*left*) and the lumbar splanchnic nerve (*right*) innervations of the mouse colorectum (Data are derived and summarized from Brierley et al. 2004 and Feng and Gebhart 2011)

References

Brierley, S. M., Jones, R. C. W., 3rd, Gebhart, G. F., et al. (2004). Splanchnic and pelvic mechanosensory afferents signal different qualities of colonic stimuli in mice. *Gastroenterology, 127*, 166–178.

Feng, B., & Gebhart, G. F. (2011). Characterization of silent afferents in the pelvic and splanchnic innervations of the mouse colorectum. *The American Journal of Physiology, 300*, G170–G180.

Kyloh, M., Nicholas, S., Zagorodnyuk, V. P., Brookes, S. J., & Spencer, N. J. (2011). Identification of the visceral pain pathway activated by noxious colorectal distension in mice. *Frontiers in Neuroscience, 5*, 16.

Ness, T. J., & Gebhart, G. F. (1990). Visceral pain: A review of experimental studies. *Pain, 41*, 167–234.

Powley, T. L., Spaulding, R. A., & Haglof, S. A. (2011). Vagal afferent innervation of the proximal gastrointestinal tract mucosa: Chemoreceptor and mechanoreceptor architecture. *The Journal of Comparative Neurology, 519*, 644–660.

Sengupta, J. N., & Gebhart, G. F. (1995). Mechanosensitive afferent fibers in the gastrointestinal and lower urinary tracts. In *Progress in pain research and management. Visceral pain* (Vol. 5, pp. 75–98). Seattle: IASP Press.

Xu, L., & Gebhart, G. F. (2008). Mouse lumbar splanchnic and pelvic nerve mechanosensory afferents encode different qualities of urinary bladder stimuli. *Journal of Neurophysiology, 99*, 244–253.

Visceral Modulation

▶ Thalamus, Visceral Representation

Visceral Nociception and Pain

Paul M. Murphy
Department of Anaesthesia and Pain Management, Royal North Shore Hospital, St. Leonards, NSW, Australia

Definition

▶ Visceral nociception and pain originates from body organs. Visceral ▶ nociceptors are located within body organs and internal cavities. The relative scarcity of nociceptors in these

Visceral Nociception and Pain, Table 1 Comparison of somatic versus visceral pain

	Somatic	Visceral
Stimuli	Mechanical	Ischemia
	Thermal	Distension
	Inflammatory	Inflammatory
Localization	Precise	Poor,
		Referred to somatic structures
Autonomic symptoms	Yes	No

areas results in a pain that is often of a vague cramping/aching quality, diffuse, poorly localized, and of a longer duration than somatic pain (Table 1).

Characteristics

To date, most fundamental basic scientific research has focused on ▶ somatic pain rather than visceral pain states. In part, this is due to the relative ease with which somatic pain can be evoked in both animal and human volunteer models. Despite this, visceral pain is an important clinical problem, which differs from somatic nociception in many fundamental respects.

Phylogenetically, somatic and visceral nociceptive processes subserve different biological functions and thus it is unsurprising that the systems respond to different stimuli. Mechanical and thermal stimuli, which have been demonstrated to evoke somatic nociceptive responses when applied to the skin, often fail to produce pain when applied experimentally to the viscera. Indeed, application of thermal and mechanical stimuli to healthy viscera often evokes no sensory response. Alternative stimuli such as inflammation, ischemia, or distension of hollow organs are however demonstrated to evoke pain.

Visceral Nociception

It is a subject of controversy as to whether specific visceral nociceptors or an alteration in electrophysiological parameters, such as signal pattern or intensity, in nonspecific afferent receptors normally involved in regulatory visceral reflexes ("pattern" theory) is responsible for the ▶ transduction of visceral pain. Recent evidence suggests that in hollow viscera, such as the urinary bladder, a homogeneous group of visceral afferents respond in a graded manner to distensile stimulation. Increasing distension beyond the biological range results in altered neuronal firing patterns and the perception of visceral pain. Animal studies (Sengupta and Gebhart 1995) have demonstrated the presence of both low threshold "intensity encoding" and high threshold afferent neurons in the cat colon. Low threshold neurons respond to colonic distension across both the biological and noxious range, while high threshold neurons respond only to noxious distension. It appears that both afferent fiber types play an important role in visceral pain encoding. High threshold afferents have been demonstrated in the gastrointestinal system (colon, small bowel, and biliary tree), the genitourinary system (ureter, bladder), and in the uterus. Intensity encoding afferents have been demonstrated in the heart, esophagus, colon, bladder, and testes. A third subtype of nociceptor, the "silent receptor" has been described. This receptor appears to be "sensitized" by alterations, such as hypoxia, acidosis, or hyperkalemia, in the local tissue environment that are associated with tissue damage. Initial studies estimated that some 80–90 % of all visceral nociceptors were of this "silent" type, although this has been subsequently revised to approximately 40–45 % (Cervero 1994; McMahon et al. 1995).

Spinal Processing

The dorsal horn of the spinal cord is the first relay point for conveying sensory information from the periphery to the brain. Somatic nociceptive inputs terminate on projection neurons and inhibitory/excitatory interneurons in ▶ Rexed's Laminae I (marginal zone) and II (▶ substantia gelatinosa). Some fibers also project more deeply into lamina V. Visceral nociceptive inputs terminate in laminae I and V, where they converge with somatic inputs, although they are less well somatotopically represented. Somatovisceral afferents also end in lamina X. Visceral inputs are predominantly of the peptidergic subtype

(substance P, CGRP). NK-1 receptor (Sub P) knockout mice fail to develop primary ▶ hyperalgesia or ▶ referred pain after visceral injury. Blockade of ▶ tachykinin (NK-2) receptors in animal models of colonic distension does not alter the pain response; however, if the animal colon is inflamed prior to distension, NK-2 blockade inhibits visceral pain transmission. It has been suggested that this response may be important in conditions such as irritable bowel syndrome. Unlike somatic spinal neurons, "▶ wind-up" is not identified in visceral neurons. Interaction between the viscerosomatic neurons and supraspinal centers may result in enhanced autonomic and motor responsiveness characteristic of visceral pain (Gebhart 1995; Beaulieu and Rice 2003; Laird et al. 2000, 2001).

Referred Pain

Three theories exist regarding the phenomenon of referred visceral pain.

1. Axon divergence: A bifurcating primary afferent from visceral and cutaneous structures terminates on a spinal projection neuron.
2. Axon convergence:
   (a) Projection: Somatic and visceral afferents terminate on the same spinal projection neuron. Electrophysiological activity in the projection neuron is perceived as somatic pain.
   (b) Facilitation: Visceral afferent input to the spinal projection neuron reduces the threshold for stimulation from somatic afferent input.
3. Thalamic: The presence of specific visceral pain projection fibers in the thalamus supports this theory although it cannot account for referred hyperalgesia (McMahon 1994).

Supraspinal Visceral Pain Pathways

Three separate visceral pain pathways have been identified:

1. Spinohypothalamic
2. Spino-amygdaloid
3. Dorsal column

Traditionally, the dorsal column-medial lemniscus system has not been considered to be involved in pain perception. Recent clinical and experimental data suggest that this pathway may play an important role in relaying visceral nociceptive information (Houghton et al. 2001). Clinical studies indicate that small ▶ dorsal column lesions relieve pain and decrease analgesic requirements in the setting of visceral cancer pain (Al-Chaer et al. 1998).

Role of Supraspinal Structures

Various imaging and electrophysiological techniques have been utilized to study the perception of visceral pain. Thalamic stimulation has been demonstrated to evoke visceral pain sensations analogous to angina or labor pain in patients, often years after the original pain. This suggests a potential integrative role for the thalamus in visceral "pain memory."

The cortical representation of visceral pain of enteric origin has been assessed by means of positron emission tomography (PET). Rectal stimulation in normal subjects is associated with activation of the anterior cingulate cortex, a region associated with affective components of the pain experience. The activated areas are quite different to the cingulate cortical areas activated by somatic pain and correspond to areas typically involved in visceromotor reactions. In patients suffering from irritable bowel syndrome, the same stimulus did not activate the anterior cingulate cortex but rather activation of the dorsolateral prefrontal cortex was identified *in anticipation* of the stimulus. This appears to indicate that persistent visceral pain states are associated with ▶ hypervigilance to visceral events. Functional magnetic resonance imaging (fMRI) studies on distension of the proximal (somatic) and distal (visceral) esophagus demonstrate differential cortical activation patterns (Aziz et al. 2000). Differences in perception of somatic and visceral pain are thus partly dependent upon differential cortical representation (Treede et al. 1999).

Cross-References

▶ Animal Models and Experimental Tests to Study Nociception and Pain

▶ Chronic Pelvic Pain, Musculoskeletal Syndromes
▶ Postsynaptic Dorsal Column Projection, Functional Characteristics
▶ Rest and Movement Pain
▶ Spinothalamic Tract Neurons, Visceral Input
▶ Vagal Input and Descending Modulation
▶ Visceral Pain and Nociception
▶ Visceral Pain Model, Pancreatic Pain
▶ Visceral Referred Pain

References

Al-Chaer, E. D., Feng, Y., & Willis, W. D. (1998). A role for the dorsal column in nociceptive visceral input into the thalamus of primates. *Journal of Neurophysiology, 79*, 3143–3150.

Aziz, Q., Thompson, D. G., Ng, V. W. K., et al. (2000). Cortical processing of human somatic and visceral sensation. *Journal of Neuroscience, 20*, 2657–2663.

Beaulieu, P., & Rice, A. S. C. (2003). Applied physiology of nociception. In D. J. Rowbotham & P. E. Macintyre (Eds.), *Clinical pain management: Acute pain*. New York: Oxford University Press.

Cervero, F. (1994). Sensory innervation of the viscera: Peripheral basis of visceral pain. *Physiological Reviews, 75*, 95–134.

Gebhart, G. F. (1995). Visceral pain. In G. F. Gebhard (Ed.), *Progress in pain research and management* (Vol. 5). Seattle: IASP Press.

Houghton, A. K., Wang, C. C., & Westlund, K. N. (2001). Do nociceptive signals from the pancreas travel in the dorsal column? *Pain, 89*, 207–220.

Laird, J. M. A., Olivar, T., Lopez-Garcia, J. A., et al. (2001). Responses of rat spinal neurons to distension of inflamed colon: Role of tachykinin NK2 receptors. *Neuropharmacology, 40*, 696–701.

Laird, J. M. A., Olivar, T., Roza, C., et al. (2000). Deficits of visceral pain and hyperalgesia of mice with a disruption of the tachykinin NK1 receptor gene. *Neuroscience, 98*, 345–352.

McMahon, S. B. (1994). Mechanisms of cutaneous deep and visceral pain. In P. D. Wall & R. Melzack (Eds.), *Textbook of pain* (pp. 129–151). Edinburgh: Churchill Livingstone.

McMahon, S. B., Dmitrieva, N., & Koltzenberg, M. (1995). Visceral pain. *British Journal of Anaesthesia, 75*, 132–144.

Sengupta, J. N., & Gebhart, G. F. (1995). Mechanosensitive afferent fibers in the gastrointestinal and lower urinary tracts in Visceral pain. In: G. F. Gebhart (Ed.), *Progress in Pain Research and Management* (Vol. 5, pp 75–98). Seattle, IASP Press.

Treede, R.-D., Kenshalo, D. R., Gracely, R. H., et al. (1999). The cortical representation of pain. *Pain, 79*, 105–111.

Visceral Nociceptive Tracts

▶ Postsynaptic Dorsal Column Neurons, Responses to Visceral Input
▶ Spinal Dorsal Horn Pathways, Dorsal Column (Visceral)

Visceral Nociceptors

▶ Sensitization of Visceral Nociceptors

Visceral Pain

▶ Spinothalamic Tract Neurons, Visceral Input
▶ Visceral Pain Model, Pancreatic Pain
▶ Visceral Referred Pain

Visceral Pain and Nociception

James M. Borowczyk
Department of Orthopaedics and
Musculoskeletal Medicine, University of Otago,
Christchurch, Christchurch, New Zealand

Definition

Visceral pain is pain that is caused by nociceptive information arising from the organs (viscera) of the body, e.g., the heart, stomach, intestine, bladder, and uterus.

Characteristics

Of all the different pain types that have undergone increasingly intensive study in the past 10–20 years, visceral pain is the poor cousin (Cervero and Laird 1999). Due to recent developments in imaging (Hobson and Aziz 2003) and the development of improved animal models with a focus on gastrointestinal tract pain (Al-Chaer et al. 2000), this deficiency is being addressed.

Lewis distinguished two types of visceral pain: true visceral pain arising from a viscus or part of a viscus and pain arising from the parietal peritoneum or pleura (Lewis 1942). More recently, Cervero and Laird (1999) ascribed five cardinal characteristics to visceral pain:

1. Visceral pain may not be evoked from all organs, either because they are not innervated or because the stimulus is inappropriate.
2. Visceral pain is not always linked to injury.
3. Visceral pain is referred to the body wall.
4. Visceral pain is diffuse and poorly localized.
5. Visceral pain is often accompanied by accentuated motor and autonomic reflexes (Cervero and Laird 1999).

Mechanisms
In terms of mechanism, visceral pain can be caused by (Procacci et al. 1991):

- Spasm of smooth muscle in the hollow viscera
- Contraction of smooth muscle in the gastrointestinal or genitourinary tracts against obstruction
- Sudden, abnormal stretching, distension, or tearing
- Rapid stretching of the capsule of solid viscera
- Rapidly developing ischemia
- Inflammation of the lining of hollow viscera
- Mechanical or chemical stimulation of inflamed mucous membranes
- Traction compression or twisting of the mesentery, organ ligaments, or their blood vessels and necrosis of viscera such as the pancreas or myocardium
- Stimuli that tend not to produce visceral pain, when applied, include cutting, pressure, and burning

The viscera are innervated by either or both two sets of primary afferent fibers that project to separate regions of the neuraxis (Al-Chaer and Traub 2002) and which constitute approximately 10 % of all afferent input into the spinal cord. Firstly, the heart, lungs, and the gastrointestinal tract, from the level of the esophagus to the transverse colon, are innervated by the vagus nerve, with central projection to the nucleus of the solitary tract. The remainder of the colon and the pelvic viscera is supplied by pelvic splanchnic nerves, whose afferent fibers project centrally to the sacral spinal cord. Secondly, all the viscera of the chest, abdomen, and pelvis are innervated by splanchnic nerves from the thoracolumbar, sympathetic trunk (Berkley et al. 1993). Their afferents terminate in the dorsal gray column between T1 and L2.

These two separate systems seem to serve a different purpose. Vagal nerves do not directly convey nociceptive signals, but vagal activity does modify nociceptive processing in the spinal cord (Ness et al. 2000). Nociceptive afferents are conveyed by the sympathetic splanchnic nerves. Centrally, they terminate in the dorsal gray column between T1 and L2.

Visceral sensory afferent fibers innervate all layers of a given viscus. They are exclusively small diameter, myelinated A-delta, or unmyelinated C fibers. They have no end organs or any evidence of morphological specialization. They exhibit chemosensitivity, thermosensitivity, and/or mechanosensitivity.

Evidence suggests that there are two physiological populations of these receptors (Sengupta and Gebhart 1994). The first are "high-threshold" receptors that respond to stimuli only within the noxious range and are found predominantly in organs in which pain is the dominant sensation (heart, ureter). The second are "low-threshold" receptors, which respond to the intensity of a stimulus by increasing their frequency of discharge from the innocuous into the noxious range. They are found predominantly in viscera such as bowel and bladder, which respond to both innocuous and noxious stimuli such as filling and distension. In addition, there is evidence to support a third population of "silent" nociceptive afferents that are sensitized by ischemia and inflammation (McMahon et al. 1995; Gebhart 2000).

As in the somatic nociceptive system, the visceral nociceptive system exhibits both peripheral and ▶ central sensitization. The mechanisms of peripheral sensitization contain elements in common with that of the somatic system. Low- and high-threshold receptors signal acute pain. So-called silent nociceptors are activated by the pathophysiological processes as described above, and inflammatory mediators released at the site of tissue damage lower receptor firing thresholds, therefore facilitating the central transmission of noxious stimuli (Cervero and Laird 1999). In the gut, it is thought that this is mediated by increased activity in a particular subpopulation of sodium channels found in all colon and bladder afferents (Al-Chaer and Traub 2002). Furthermore, as compared with the somatic system, visceral afferents contain a greater percentage of neuropeptides, particularly substance P, but also including CGRP, somatostatin, and vasoactive intestinal peptide.

It is likely that the mechanism of central sensitization in the visceral system differs from that in the somatic system. Cardinal to this difference is the observation that NMDA receptor antagonists inhibit primary afferent and spinal responses to both innocuous and acute noxious colonic stimuli (McRoberts et al. 2001), suggesting that NMDA receptors have a role in the transmission of both types of stimuli. This is in stark contrast to the somatic system, where NMDA receptors are thought to contribute mainly to the development of central sensitization and do not operate under acute or innocuous situations.

In the visceral system, it is possible that excessive NMDA receptor activity in the absence of noxious stimuli or inflammation could explain phenomena such as irritable bowel syndrome (Ji and Traub 2001). With respect to neurokinin receptors, NK-1 receptors have been shown to signal neurogenic inflammation, but not nonneurogenic inflammation (Laird et al. 2000), whereas a different inflammagen is thought to work through NK-3 and NMDA receptors (Kamp et al. 2001). All of this points to a probable different pathophysiology for the two systems.

Future Developments

Combinations of traditional therapeutic agents remain the mainstay of treatment for visceral pain management. More recent studies focusing on the role of kappa opioid receptors, sodium channels, and neurokinin and NMDA receptors promise new interventions in pain management. This, combined with improved models of study, and improved imaging of pain pathways are likely to provide the discipline with exciting research tools and novel pharmacological approaches to the treatment of visceral pain.

Cross-References

▶ Animal Models and Experimental Tests to Study Nociception and Pain
▶ Chronic Pelvic Pain, Musculoskeletal Syndromes
▶ Postsynaptic Dorsal Column Projection, Functional Characteristics
▶ Rest and Movement Pain
▶ Spinothalamic Tract Neurons, Visceral Input
▶ Vagal Input and Descending Modulation
▶ Visceral Nociception and Pain
▶ Visceral Pain Model, Pancreatic Pain
▶ Visceral Referred Pain

References

Al-Chaer, E. D., & Traub, R. J. (2002). Biological basis of visceral pain: Recent developments. *Pain, 96*, 221–225.

Al-Chaer, E. D., Kawasaki, M., & Pasricha, P. J. (2000). A new model of chronic visceral hypersensitivity in adult rats induced by colon irritation during postnatal development. *Gastroenterology, 119*, 1276–1285.

Berkley, K. J., Hubscher, C. H., & Wall, P. D. (1993). Neuronal responses to stimulation of the cervix, uterus, colon, and skin in the rat spinal cord. *Journal of Neurophysiology, 69*, 545–556.

Cervero, F., & Laird, J. M. (1999). Visceral Pain. *Lancet, 353*, 2145–2148.

Gebhart, G. F. (2000). Physiology, pathophysiology, and pharmacology of visceral pain. *Regional Anesthesia and Pain Medicine, 25*, 632–638.

Hobson, A. R., & Aziz, Q. (2003). Central nervous system processing of human visceral pain in health and disease. *News in Physiological Sciences, 18*, 109–114.

Ji, Y., & Traub, R. J. (2001). Spinal NMDA receptors contribute to neuronal processing of acute noxious and non-noxious colorectal

stimulation in the rat. *Journal of Neurophysiology, 86*, 1783–1791.

Kamp, E. H., Beck, D. R., & Gebhart, G. F. (2001). Combinations of neurokinin receptor antagonists reduce visceral hyperalgesia. *Journal of Pharmacology and Experimental Therapeutics, 299*, 105–113.

Laird, J. M., Olivar, T., Roza, C., De Felipe, C., Hunt, S. P., & Cervero, F. (2000). Deficits in visceral pain and hyperalgesia of mice with a disruption of the tachykinin NK1 receptor gene. *Neuroscience, 98*, 345–352.

Lewis, T. (1942). *Pain*. New York: Macmillan.

McMahon, S. B., Dmitrieva, N., & Koltzenburg, M. (1995). Visceral pain. *British Journal of Anaesthesia, 75*, 132–144.

McRoberts, J. A., Coutinho, S. V., Marvizon, J. C., Grady, E. F., Tognetto, M., Sengupta, J. N., Ennes, H. S., Chaban, V. V., Amadesi, S., Creminon, C., Lanthorn, T., Geppetti, P., Bunnett, N. W., & Mayer, E. A. (2001). Role of peripheral N-methyl-D-aspartate (NMDA) receptors in visceral nociception in rats. *Gastroenterology, 120*, 1737–1748.

Ness, T. J., Fillingim, R. B., Randich, A., Backensto, E. M., & Faught, E. (2000). Low intensity vagal nerve stimulation lowers human thermal pain thresholds. *Pain, 86*, 81–85.

Procacci, P., Maresca, M., & Cersosimo, R. M. (1991). Visceral pain: Pathophysiology and clinical aspects. *Advances in Experimental Medicine and Biology, 298*, 175–181.

Sengupta, J. N., & Gebhart, G. F. (1994). Characterization of mechanosensitive pelvic nerve afferent fibers innervating the colon of the rat. *Journal of Neurophysiology, 71*, 2046–2060.

Visceral Pain from the Bladder

Definition

Visceral pain from the bladder is evoked from distension, chemicals, inflammation, or trauma. It is often diffuse and poorly located. It may be felt in the midline above the symphysis, and can also be referred to the inner thighs, the groin, or even to the knees. Bladder pain may be accompanied by sweating, nausea, or dizziness.

Cross-References

▶ Opioids and Bladder Pain/Function

Visceral Pain Model, Angina Pain

Robert D. Foreman
Department of Physiology, College of Medicine,
University of Oklahoma Health Sciences Center,
Oklahoma City, OK, USA

Synonyms

Angina pectoris; Chest pain; Coronary artery disease; Coronary insufficiency; Heart pain; Ischemic heart disease; Myocardial ischemia

Characteristics

Adequate Stimulus for Angina Pectoris

▶ Mechanosensitive sympathetic afferent fibers and ▶ chemosensitive sympathetic afferent fibers innervate the heart. Both classes of cardiac afferent fibers respond to ▶ bradykinin; however, only chemosensitive endings are sensitized by prostaglandins, especially PGE_1 (Baker et al. 1980; Nerdrum et al. 1986). After sensitization, bradykinin increases the magnitude and duration of the afferent impulses being transmitted in the chemosensitive endings, but does not affect mechanosensitive endings. It appears, however, that the chemical effects of these neurons are far more dramatic than are the mechanical effects for producing pain of ▶ angina pectoris. Thus, chemosensitive endings are better candidates for carrying information that leads to angina pectoris resulting from ▶ ischemic heart disease. Often the environment around receptors leads to sensitization because the release of chemicals changes the responsiveness of the afferent endings.

Debates about mechanical and chemical mechanisms to activate sensory receptors generating cardiac pain symptoms have persisted for years. The mechanical hypothesis centered on the idea that distension of the ventricular wall resulted in pain (Colbeck 1903). Clinical studies showing that painful and painless episodes of transient ischemia are produced by similar patterns of ventricular dilation do not support this hypothesis (Davies et al. 1988). In addition, acute ventricular failure, valvuloplasty, and myocardial biopsy all produce non-painful ventricular dilation. Thus, the general sense is that mechanical stimulation does not play a major role in the pain associated with myocardial ischemia (Davies et al. 1988).

Chemical Stimulation of Sensory Endings

Unmyelinated sympathetic afferent fibers respond vigorously to chemical substances such as bradykinin, potassium, adenosine, acids, and veratridine when applied to the epicardial surface or injected directly into coronary arteries (Foreman 1999). Several chemicals have been examined for their role in activating afferent fibers during myocardial ischemia; however, bradykinin has long been used as a chemical of choice to produce nociceptive responses. Although some disagreements exist, bradykinin is increased in the effluent of the coronary sinus following coronary artery occlusion (Hashimoto et al. 1977; Kimura et al. 1973), and intracoronary infusions of bradykinin lead to angina-like pain (Gaspardone et al. 1999). An important observation about angina pectoris and ischemic heart disease is that patients may experience significant ischemic episodes, as demonstrated by ECG changes, but no pain is experienced. These episodes are called silent ischemia. Thus, the variability of the pain sensation and its lack of direct relationship with myocardial ischemia shows that a complex system may be operating in a way not yet discovered. This lack of understanding is due to the fact that little information is available about the characteristics of sensory receptors in the heart, and how the interstitial environment surrounding the receptors influences their processing of information.

Adenosine may also be an important component of the chemicals contributing to cardiac pain. Intracoronary infusions of adenosine cause angina-like pain in healthy volunteers (Sylvén et al. 1986) and in patients who suffer from ischemic heart disease and experience angina pectoris (Sylvén et al. 1988; Crea et al. 1990) without expressing any electrocardiographic changes. This pain most likely originates from the sympathetic afferent fibers innervating the heart, because similar doses of adenosine infused into the right atrium did not evoke any pain, and afferent activity is increased.

Methods of Chemical Stimulation of Cardiac Receptors

Intracardiac Injections (Blair et al. 1982; Ammons et al. 1985): In cats and monkeys, the left thoracic cavity is exposed by removing the second through the fifth ribs and the anterior lobe of the lung. A small hemostat is used to grip the tip of the left atrial appendage. A small incision is made into the tip and then a cannula (PE-90) with a small flanged end is inserted into the left atrial appendage. A purse string is used to clasp the atrial tissue tightly against the cannula. This ligation prevents leakage of blood chemicals into the thoracic cavity. Commonly, another cannula (PE-90) is inserted into the left femoral artery and is moved up the aorta, until the tip is placed just above the coronary ostia. Chemical injections are made through this cannula to serve for control measurements.

Intracoronary Injections: In dog studies, several routes of anesthesia administration and techniques for intracoronary implantation can be used. The anesthesia and techniques described here will serve as a prototype for administering intracoronary injections of chemicals. Healthy mongrel dogs, generally weighing between 20 and 30 kg, are tranquilized with morphine sulfate (2 mg/kg I.P) and anesthetized with pentobarbital sodium (30 mg/kg), intubated, and placed on a Harvard apparatus room air ventilator. During the course of the experiment, additional doses of anesthesia are administered as needed (5 mg/kg I.V.). The appropriate level of anesthesia is checked by making sure the withdrawal reflex is absent when the forepaw is pinched with a small hemostat and the jaw tension is low. Access to the right femoral artery is obtained by inserting an 8F sheath. An 8F hockey stick guide catheter is then passed via the femoral sheath and positioned in the ostium of the left coronary artery. The chemicals can then be injected directly into the coronary artery.

Intrapericardial Injections (Euchner-Wamser et al. 1994; Qin et al. 2001; commonly used procedure in rats): A midline thoracotomy is used to place a silicone catheter (0.020 ID, 0.037 OD, 14–16 cm length) into the pericardial sac over the left ventricle. Individual chemicals such as bradykinin, capsaicin, or adenosine, or a mixture of algesic compounds can be injected into the pericardial sac to stimulate both vagal and sympathetic afferent endings. Solutions are injected and withdrawn via a 1-ml syringe connected to the catheter. The protocol is to inject 0.2 ml of warm saline into the sac and then withdraw the saline after 60 s. This is followed by the injection of 0.2 ml of the algesic mixture, which is also withdrawn after 60 s. The pericardial sac is then flushed twice or three times with warm saline. It is important to make the injection in at least 15-min intervals to prevent tachyphylaxis.

Coronary artery occlusion (Blair et al. 1984): To expose the left coronary artery, the chest is opened at the fourth or fifth intercostal space on the left side of the chest. The ribs are retracted and the lungs are reflected to expose the heart. For cats and monkeys, tissue surrounding the left anterior descending branch of the left main coronary artery is dissected, while carefully avoiding damage to the pericoronary nerves. The hearts of cats and monkeys are large enough to place occluders around both the left circumflex and left anterior descending branch of the left coronary artery. For rats, the heart is exteriorized by pressing gently on the right chest wall. For all species, a silk ligature (2–0 to 4–0) is placed around the left coronary artery, approximately 2–3 mm from its origin. The ends of the sutures are slipped through a small polyethylene tube. For the occlusion, the tubing is moved until it touches the artery and then the suture strings are pulled to collapse the vessel. The ECG is monitored via Lead II on an oscilloscope and stored on a data acquisition system. We have shown that repeated occlusions can be done without diminishing the cellular responses. This technique can be adapted for the rat, cat, monkey, and dog. Another approach for the dog is to introduce balloon catheters in closed chest preparations.

References

Ammons, W. S., Girardot, M.-N., & Foreman, R. D. (1985). Effects of Intracardiac Bradykinin on T2-T5 Medial Spinothalamic Cells. *American Journal of Physiology, 249*, R147–R152.

Baker, D. G., Coleridge, H. M., Coleridge, J. C. G., & Nerdrum, T. (1980). Search for a cardiac nociceptor: Stimulation by bradykinin of sympathetic afferent nerve endings in the heart of cat. *The Journal of Physiology, 306*, 519–536.

Blair, R. W., Weber, R. N., & Foreman, R. D. (1982). Responses of thoracic spinothalamic neurons to intracardiac injection of bradykinin in the monkey. *Circulation Research, 51*, 83–94.

Blair, R. W., Ammons, W. S., & Foreman, R. D. (1984). Responses of thoracic spinothalamic and spinoreticular cells to coronary artery occlusion. *Journal of Neurophysiology, 51*, 636–648.

Colbeck, E. H. (1903). Angina pectoris: A criticism and a hypothesis. *Lancet, 1*, 793–795.

Crea, F., Pupita, G., Galassi, A. R., El-Tammi, H., Kaski, J. C., et al. (1990). Role of adenosine in pathogenesis of anginal pain. *Circulation, 81*, 164–167.

Davies, G., Bencivelli, W., Chierchia, S., Fragasso, G., Crea, F., et al. (1988). Sequence and magnitude of ventricular volume changes in painful and painless myocardial ischemia. *Circulation, 78*, 310–319.

Euchner-Wamser, I., Meller, S. T., & Gebhart, G. F. (1994). A model of cardiac nociception in chronically instrumented rats: Behavioral and electrophysiological effects of pericardial administration of algogenic substances. *Pain, 58*, 117–128.

Foreman, R. D. (1999). Mechanisms of cardiac pain. *Annual Review of Physiology, 61*, 143–167.

Gaspardone, A., Crea, F., Tomai, F., Versaci, F., Pellegrino, A., Chiariello, L., & Gioffre, P. A. (1999). Effect of acetylsalicyclate on cardiac and muscular pain induced by intracoronary and intra-arterial infusion of bradykinin in humans. *Journal of the American College of Cardiology, 34*, 216–222.

Hashimoto, K., Hirose, M., Furukawa, S., Hayakawa, H., & Kimura, E. (1977). Changes in hemodynamics and bradykinin concentration in coronary sinus blood in experimental coronary artery occlusion. *Japanese Heart Journal, 18*, 679–689.

Kimura, E., Hashimoto, K., Furukawa, S., & Hayakawa, H. (1973). Changes in bradykinin level in coronary sinus blood after the experimental occlusion of coronary artery. *American Heart Journal, 85*, 635–647.

Nerdrum, T., Baker, D. G., Coleridge, H. M., & Coleridge, J. C. G. (1986). Interaction of bradykinin and prostaglandin E1 on cardiac pressor reflex and sympathetic afferents. *American Journal of Physiology, 250*, R815–R822.

Qin, C., Chandler, M. J., Miller, K. E., & Foreman, R. D. (2001). Responses and afferent pathways of superficial and deeper C_1-C_2 spinal cells to intrapericardial algogenic chemicals in rats. *Journal of Neurophysiology, 85*, 1522–1532.

Sylvén, C., Beermann, B., Jonzon, B., & Brandt, R. (1986). Angina pectoris-like pain provoked by intravenous adenosine in healthy volunteers. *British Medical Journal, 293*, 227–230.

Sylvén, C., Beermann, B., Edlund, A., Lewander, R., Jonzon, B., & Mogensen, L. (1988). Provocation of chest pain in patients with coronary insufficiency using the vasodilator adenosine. *European Heart Journal, 9*, 6–10.

Visceral Pain Model, Irritable Bowel Syndrome Model

Elie D. Al-Chaer
Center for Pain Research, University of Arkansas for Medical Sciences, Little Rock, AR, USA

Synonyms

Chronic neural sensitization; Functional abdominal pain; IBS; Irritable bowel syndrome model; Neonatal inflammation; Neonatal pain; Nonstructural disorders; Visceral hypersensitivity

Definition

▶ Irritable bowel syndrome (IBS) is a functional disorder of the large and small intestines, which causes abdominal pain or discomfort in the absence of identifiable physical pathology such as inflammation or tumors. The pain occurs along with constipation or diarrhea. Other common symptoms are bloating, passing mucus in the stools, or a sense that the bowels have not been completely emptied. The exact cause of irritable bowel syndrome is not known. However, a number of clinical studies have suggested a link between childhood pain or abuse and adult irritable bowel syndrome. This hypothesis was tested in an animal model of neonatal colon pain or inflammation. The results indicate that adult rats, exposed to neonatal colon pain or inflammation, exhibit symptoms of visceral hypersensitivity in the absence of colon inflammation. This hypersensitivity is often associated with altered fecal output, symptoms typical of irritable bowel syndrome.

Introduction

Recurrent abdominal pain in children places them at high risk of functional disorders as adults (Blanchard and Scharff 2002). The causes of neonatal pain vary, from the often necessary but painful medical procedures, to unavoidable medical conditions (e.g., sepsis and inflammation), to readily preventable abuse. The newborn gut may well be exposed to a variety of factors resulting in mucosal inflammation and tissue irritation. These factors include inflammation resulting from food allergies, viral infections, and acid reflux, all of which are common in early life. Given the immature neonatal nervous system (e.g., underdeveloped endogenous pain inhibition systems), this may result in permanent neuroplastic alterations, manifesting as an altered pain response to visceral stimuli in adults.

Characteristics

Irritable bowel syndrome comprises a group of ▶ functional bowel disorders, in which abdominal discomfort or pain is associated with defecation or a change in bowel habits and with features of disordered defecation. The diagnostic criteria for irritable bowel syndrome are abnormal discomfort or pain for at least 12 weeks or more, which need not be consecutive, in the preceding 12 months and that has 2 out of 3 features: (1) relieved with defecation, (2) onset associated with a change in frequency of stool, and/or (3) onset associated with a change in form (appearance) of stool. These symptoms occur in the absence of structural or metabolic abnormalities to explain them (Thompson et al. 2000).

The cause of irritable bowel syndrome is not known. Even though currently used symptom criteria implicitly assume a homogenous disorder of "IBS," it is likely that the nonspecific symptom complex of abdominal pain and discomfort associated with alterations in bowel habits reflects a heterogeneous group of disorders with different etiologies, pathophysiologies, and clinical presentations. For example, a number of etiologies have been advanced to explain symptom generation in irritable bowel syndrome; these include recurrent abdominal pain in children (Hyams et al. 1999), a patient history of sexual and physical abuse (Drossman et al. 1995), and disturbances in motor activity, psychosocial factors, or dietary factors. On the other hand, strong evidence exists for an altered sensitivity of gastrointestinal afferent pathways as a necessary component of any etiological theory of functional bowel disorders.

Animal Models

These can be evaluated based on their face validity (how well the model resembles one or several clinical or pathophysiological characteristics of the clinical condition), construct validity (how well the model is consistent with a particular theoretical rationale), and predictive validity (how well the model predicts treatment responses to specific drugs or non-pharmacologic interventions). Even though high face validity of an animal model for the entire syndrome is desirable, it is not essential, given some of the unique clinical features of patients with irritable bowel syndrome (cognitive and behavioral). However, an animal model that adequately models specific aspects of the syndrome, such as chronic visceral hypersensitivity aggravated by stress or motility changes in the absence of intestinal inflammation, would be of significant benefit, particularly if such a model showed good predictive validity.

Neonatal Pain/Neonatal Inflammation Model

In an animal model of neonatal colon injury, Al-Chaer et al. reported that exposing newborn rat pups (days 8–21) to painful ▶ colorectal distension (CRD, using angioplasty balloon, 2×60 s distensions separated by 30 min) – a painful stimulus but arguably a reproducible experimental form of physical abuse – or colon inflammation (daily intracolonic injection of mustard oil) during preadolescence can cause long-term visceral and somatic hypersensitivity,

Long-term Consequences of Neonatal Injury

Visceral Pain Model, Irritable Bowel Syndrome Model, Fig. 1 Long-term consequences of neonatal colon injury. Experimental injury to the colon in neonate rats (<22 days), caused by repeated colorectal distension (CRD) or colon inflammation with mustard oil (MO), can cause long-term behavioral and physiological changes measured in the adults (>5 weeks). These changes include visceral hypersensitivity, central sensitization to visceral and somatic stimuli in the absence of colon inflammation (Al-Chaer et al. 2000), and peripheral sensitization in both pelvic and hypogastric afferents (Lin & Al-Chaer 2003). It can also cause alternating episodes of constipation and diarrhea in a subgroup of adult rats (Ma et al. 2002). These symptoms are aggravated by water avoidance stress (Hinze et al. 2002)

which lingers beyond the age of 4 months in the adult rat and long after the initial injury has resolved (Al-Chaer et al. 2000) (Fig. 1). In fact, minimal neonatal colon irritation (CI) in rats given at postnatal days 8, 10, and 12 can yield a variety of sensory symptoms in adults including changes in somatic and visceral sensitivity, in addition to gastrointestinal symptoms including changes in stool consistency and frequency (Wang et al. 2008).

Behavioral pain responses to CRD are significantly increased in the neonatally irritated rats when assessed at postnatal week 5 and persist up to postnatal week 12 (Fig. 2), while colon histology is not different from controls (Al-Chaer et al. 2000).

A hallmark of these sensory changes is visceral hypersensitivity. It is associated with central and peripheral neural sensitization (Al-Chaer et al. 2000; Lin and Al-Chaer 2003; Wang et al. 2008) and functional changes in pain processing pathways (Saab et al. 2004). Electrophysiologic recordings of the responses of viscerosensitive primary afferent fibers converging onto the L6–S1 or T13-L2 segments of the spinal cord, and of individual viscerosensitive dorsal horn neurons at L6-S1 or T13-L2 spinal segments, show increased baseline firing rates, as well as increased responses, to graded CRD and somatic stimuli applied to the same dermatome (Lin and Al-Chaer 2003; Wang et al. 2008). Evidence of central (spinal) sensitization is further demonstrated by an increased expression of c-Fos in spinal segments receiving colorectal input.

Adult rats exposed to neonatal colon irritation show disturbances in colon motility in the form of altered fecal output (Ma et al. 2002), symptoms commonly seen in patients with IBS. Furthermore, exposure to neonatal colon pain can cause adult rats of both sexes to reduce their exploratory activity and confine themselves to a limited area of an open field. The decrease in exploratory activity is aggravated by stress (Hinze et al. 2002; Wang et al. 2008). Similar observations were made in the mouse where neonatal colon insult was shown to alter growth factor expression and TRPA1 responses in adults (Christianson 2010).

Sensitization of EMG Responses to CRD

EMG responses were recorded from the external oblique muscle.
The responses were larger in CI rats than in controls

Visceral Pain Model, Irritable Bowel Syndrome Model, Fig. 2 Sensitization of EMG responses to CRD. *Left Panel*: waveforms (traces 2 and 4) showing the responses to CRD (80 mmHg) recorded in the external oblique muscles of a control rat and a rat with neonatal colon irritation. The upper traces (1 and 3) show the rectified signal used to measure the area under the curve. *Right Panel*: bar graphs showing the average EMG responses (+/− SEM), quantified by measuring the area under the curve, to graded colorectal distension measured in control rats (dark) or rats with neonatal CI (light)

Other Animal Models

Other animal models include the following: (1) Neonatal maternal separation stress model, in which the normal mother-infant interaction is compromised. Daily separation of the litter from their mothers for 180 min each day during postnatal days 4–18 results in an alteration of maternal behavior, with significantly reduced times of the licking/grooming behavior. Adult maternally separated rats show evidence of alterations in stress-induced visceral and somatic pain sensitivity, consistent with compromised stress-induced engagement of opioid systems. Additionally, colonic motor function in response to stress is also enhanced in these animals (Coutinho et al. 2001). (2) Post-traumatic stress disorder model, which uses daily sessions of brief electric footshock, or other interventions such as repeated injections of psychostimulants or corticotropin-releasing factor (CRF), to induce a sensitized state in the adult rat. Evidence suggests that such animals also show increased visceral afferent responses to colonic distension (Stam et al. 2000), in addition to postinfectious or post-inflammatory models (Collins et al. 1996).

Summary

Adverse experiences in the neonatal period can lead to chronic intractable disorders that may occur in the absence of identifiable pathology or structural problems and which can sometimes be very painful; examples include fibromyalgia and functional abdominal pain. Animal studies have shown that an individual's early life history is a source of variability in adult pain experiences and one that can affect sensory processing, pain sensitivity, and other outcomes. Adult rats and mice that receive a neonatal colon insult exhibit a cluster of symptoms as diverse as those observed in patients diagnosed with irritable bowel syndrome. These symptoms occur in the absence of colon inflammation and include visceral hypersensitivity with referred somatic pain, altered fecal output, and reduced exploratory behavior. This is commensurate with the diversity of symptoms in functional GI disorders and the ambiguity of its pathophysiology, making these models highly valid tools to study the mechanisms of complex functional disorders and target them with novel therapeutic approaches.

References

Al-Chaer, E. D., Kawasaki, M., & Pasricha, P. J. (2000). A New model of chronic visceral hypersensitivity in adult rats induced by colon irritation during postnatal development. *Gastroenterology, 119*, 1276–1285.

Blanchard, E. B., & Scharff, L. (2002). Psychosocial aspects of assessment and treatment of irritable bowel syndrome in adults and recurrent abdominal pain in children. *Journal of Consulting and Clinical Psychology, 70*, 725–738.

Collins, S. M., McHugh, K., Jacobson, K., Khan, I., Riddell, R., Murase, K., & Weingarten, H. P. (1996). Previous inflammation alters the response of the Rat colon to stress. *Gastroenterology, 111*, 1509–1515.

Coutinho, S., Plotsky, P., Sablad, M., Miller, J., Zhou, H., Bayati, A., McRoberts, J. A., & Mayer, E. A. (2001). Neonatal maternal separation alters stress-induced responses to viscerosomatic nociceptive stimuli in the adult rat. *American Journal of Physiology – Gastrointestinal and Liver Physiology, 282*, G307–316.

Drossman, D. A., Talley, N. J., Leserman, J., Olden, K. J., & Barreiro, M. (1995). Sexual and physical abuse and gastrointestinal illness: Review and recommendations. *Annals of Internal Medicine, 123*, 782–794.

Hinze, C. L., Lin, C., Al-Chaer, E. D. (2002). Estrous cycle and stress related variations of open field activity in adult female rats with neonatal colon irritation (CI). Program No. 155.14. 2002. Abstract Viewer/Itinerary Planner. Society for Neuroscience, Washington, DC, CD-ROM.

Hyams, J. S., Hyman, P. E., & Rasquin-Weber, A. (1999). Childhood recurrent abdominal pain and subsequent adult irritable bowel syndrome. *Journal of Developmental and Behavioral Pediatrics, 20*, 318–319.

Lin, C., & Al-Chaer, E. D. (2003). Long-term sensitization of primary afferents in adult rats exposed to neonatal colon pain. *Brain Research, 971*, 73–82.

Ma, H., Park, Y., & Al-Chaer, E. D. (2002). Functional outcomes of neonatal colon pain measured in adult rats. *Journal of Pain, 3*(Suppl. 1), 27, #707. American Pain Society.

Saab, C. Y., Arai, Y.-C. P., & Al-Chaer, E. D. (2004). Modulation of visceral nociceptive processing in the lumbar spinal cord following thalamic stimulation or inactivation and after dorsal column lesion in rats with neonatal colon irritation. *Brain Research, 1008*, 186–192.

Stam, R., Bruijnzeel, A. W., & Wiegant, V. M. (2000). Long-lasting stress sensitization. *European Journal of Pharmacology, 405*, 217–224.

Thompson, W. G., Longstreth, G., Drossman, D. A., Heaton, K., Irvine, E. J., & Muller-Lissner, S. (2000). Functional bowel disorders and functional abdominal pain. In D. A. Drossman, E. Corazziari, N. J. Talley, W. G. Thompson, & W. E. Whitehead (Eds.), *Rome II: The functional gastrointestinal disorders: Diagnosis, pathophysiology and treatment: A multinational consensus* (2nd ed., pp. 351–375). USA: Degnon Associates, McLean VA.

Additional References

Christianson, J. A., Bielefeldt, K., Malin, S. A., & Davis, B. M. (2010). Neonatal colon insult alters growth factor expression and TRPA1 responses in adult mice. *Pain, 151*, 540–549. doi:10.1016/j.pain.2010.08.029.

Wang, J., Gu, C., & Al-Chaer, E. D. (2008). Altered behavior and digestive outcomes in adult male rats primed with minimal colon pain as neonates. *Behavioral and Brain Functions, 4*, 28. doi:10.1186/1744-9081-4-28.

Visceral Pain Model, Kidney Stone Pain

Maria Adele Giamberardino
Pathophysiology of Pain Laboratory and Centre for Fibromyalgia and Musculoskeletal Pain, Department of Medicine and Science of Aging, University of Chieti, Chieti, Italy

Definition

Animal model (rat) of urinary colic and referred lumbar muscle hyperalgesia from an artificial ureteral calculosis.

Characteristics

Renal/ureteral colic, evoked by the passage of kidney stones, is intensely painful in humans (Bihl and Meyers 2001). Calculosis of one upper urinary tract typically produces pain in the ipsilateral lumbar region (L1), which radiates downward on the anterior flank toward the groin, and ▶ referred hyperalgesia in the ipsilateral oblique musculature. The hyperalgesia is already detectable after a few colic episodes, is accentuated in extent by the repetition of the colic, and is long-lasting, i.e., it outlasts not only the spontaneous pain (it is detectable in the pain-free interval) but often also the presence of the stone in the urinary tract (it can persist long after the stone is expelled through the urine) (Giamberardino et al. 1994; Vecchiet et al. 1989).

Rats with an artificial stone in one upper ureter present characteristics closely mimicking the clinical condition. They display both repeated

Visceral Pain Model, Kidney Stone Pain, Fig. 1 Sites of ureteral stone formation (*left*) and referred lumbar muscle hyperalgesia (*right*) in the rat model of artificial ureteral calculosis

"crises" of abnormal behavior (direct visceral pain/colic) and hypersensitivity of the ipsilateral oblique musculature (referred hyperalgesia), whose extent and temporal evolution are in parallel with those of patients (see below). Spontaneous behavior is monitored, post-stone formation, through nonstop videotape recordings, and referred hyperalgesia is measured as a decrease in the ▶ vocalization threshold to electrical muscle stimulation in the post-stone period, with respect to the pre-stone formation period (Giamberardino et al. 1990, 1995). This model has been developed to investigate pathophysiological mechanisms of visceral pain/referred phenomena from the urinary tract in standardized conditions.

Operation

Under general anesthesia (pentobarbital or halothane) (Giamberardino et al. 1990; Roza et al. 1998), the obliquus externus muscle (L1) of both sides is implanted with bipolar wire electrodes (insulated nickelchrome wires), 1 cm apart. The wires are passed under the skin to reach the skull, where connectors are fixed into the parietal bones via small screws and dental cement, to form a small device. This permits testing of muscle sensitivity in the awake, freely moving rats, for many days after implantation. On the third day after electrode implantation, a second general anesthesia is induced for ureteral stone formation. The left ureter is approached via a suprapubic vertical incision, and 0.02 ml of dental lamell resin cement is injected, while still fluid, in the upper third of the lumen, using an insulin syringe with a 0.4-mm-diameter needle. The cement hardens quickly and forms a small stone (Fig. 1, left) (Giamberardino et al. 1995).

Behavior in Stone Rats

Ureteral "crises": Ninety-eight percent of stone rats show spontaneous nocifensive behavior over several days postoperatively, i.e., repeated "crises" similar in nature to the writhing behavior typical with noxious visceral stimulation (Le Bars et al. 2001). These crises, which never appear after non-noxious ureteral procedures (sham intervention, i.e., injection of saline in the lumen, ligature of the lowest ureter) or interventions on somatic structures (e.g., muscle implantation), are significantly and dose-dependently reduced in number and duration by chronic treatment with morphine, tramadol, or metamizol (Affaitati et al. 2002; Giamberardino et al. 1995; Laird et al. 1998). All these observations strongly support their being an index of perceived visceral pain and thus the equivalent of urinary colics in humans.

During ▶ continuous videotape recordings, the typical movements of a ureteral "crisis" are:
1. Hump-backed position
2. Licking of lower abdomen and/or ipsilateral flank

3. Repeated waves of contraction of flank muscles with inward movements of ipsilateral hind limb
4. Body stretching
5. Squashing of lower abdomen against the floor
6. Supine position with ipsilateral hind limb adducted and compressed against the abdomen

A "crisis" is a sequence of at least three of the above movements for a minimum duration of 2 min. A lesser number of movements (1 or 2) and/or duration of the sequence (<2 min) is termed "subcrisis." Complexity of crises is estimated via a 4-point arbitrary scale: a sequence of three movements is scored 1, and sequences of four, five, or six movements are scored 2, 3, or 4, respectively.

From one rat to another, the crises vary from very few to a very high number ($\sim$ 60), last a few min to over 45 min, and are of variable complexity. In each rat, they predominate in the dark phase of the day; their number and duration decrease progressively with time after stone formation, so that they are concentrated mostly during the first 3-4 days (Giamberardino et al. 1995). Number and duration of crises are significantly higher in female than male rats. In female rats, the percentage of time spent in crisis is significantly higher during the metestrus/diestrus than during the proestrus/estrus phases of the estrus cycle (Affaitati et al. 2002), in parallel with the clinical evidence of urinary colics prevailing during the premenstrual/menstrual phases in fertile women (Giamberardino et al. 2001).

Oblique Muscle Hyperalgesia
Calculosis rats develop hypersensitivity of the left oblique musculature (Fig. 1, right), revealed by vocalization to slight digital pressure and pinching of the muscle. The hypersensitivity is quantified in terms of a decrease in the vocalization threshold to electrical muscle stimulation, with respect to values preceding stone formation (thresholds are measured daily for 3 days before and 7–10 days after stone implantation). The decrease starts from the first post-stone implantation day, is maximal (peak hyperalgesia) on the

second and third day, and lasts over a week, i.e., it outlasts the presence of ureteral crises (which cease after 4 days) and even the presence of the stone in the urinary tract. In fact, it is still detectable in rats that, at autopsy, appear to have spontaneously eliminated the stone. The muscle threshold lowering is significantly and directly correlated to the number and duration of "crises" displayed over the first 4 days (Giamberardino et al. 1995). Referred muscle hyperalgesia in calculosis rats has, therefore, the same characteristics as in humans, i.e., early phenomenon, accentuated in extent by the repetition of visceral "crises," and long-lasting (Giamberardino et al. 1994). It is also significantly reduced in extent and duration by chronic administration of NSAIDs (ketoprofen), spasmolytics (hyoscine-N-butylbromide), or both compounds in combination, similar to what observed in the clinical setting (Affaitati et al. 2002).

Destiny of the Stone
The destiny of the stone in rats is variable, just as it is in patients. Autopsy shows different patterns: stone spontaneously eliminated through the urine or stone still present in the ureter, either in the original position or lower, with or without signs of urinary tract occlusion (proximal ureter dilated, kidney enlarged) (Affaitati et al. 2002; Giamberardino et al. 1990, 1995).

Generation of Ureteral Pain
Pain from urinary calculosis is thought to derive mainly from acute pelvis dilatation, due to stone impact in the ureteral lumen. When a stone moves into the renal collecting system, the resulting increase in intraluminal pressure stretches nerve endings in the mucosa. Ureteral obstruction furthermore causes increased synthesis and release of prostaglandins, which in turn both increase glomerular filtration and renal pelvic pressure and sensitize nociceptors locally. An important algogenic role is also played by ureteral wall contractility, i.e., hypermotility taking place above the obstacle (Bihl and Meyers 2001).

Abnormal ureteral peristalsis has been demonstrated in calculosis rats (Laird et al. 1997), which may contribute to the pain behavior, although the mechanisms producing the change in peristalsis are unknown. One possibility is that damage and inflammation of the ureteral wall was caused by the stone-induced cyclooxygenase activity (endogenous prostaglandins have been shown to enhance the activity of the pacemaker in the renal pelvis in vitro). Another possibility is that stimulation of ureteral nociceptors by the stone provokes axon reflexes, with consequent release of neuropeptides such as tachykinins (affecting ureteral motility) from the nociceptive terminals (Laird et al. 1997).

The ▶ ureteral nociceptors stimulated in the calculosis rat model are presumably ▶ U-2 type units. These are a large population of high threshold units originally identified in the guinea pig's ureter. Unlike the ▶ U-1 type units, they respond to intense distensions but not to peristalsis (Cervero 1996).

Generation of Referred Hyperalgesia
Electrophysiology in the calculosis model has shown changes in activity and excitability of spinal sensory neurons with convergent input from the ureter and the somatic area of referral (Giamberardino et al. 1996; Laird et al. 1998; Roza et al. 1998). Central mechanisms, i.e., sensitization of convergent neurons, are thus likely to be involved in the increased pain response to stimulation of the referred area.

Based on the clinical observation that the hyperalgesic muscle is often contracted, a reflex arc mechanism has also been hypothesized; the afferent visceral barrage would trigger activation of somatic efferents to the muscle, generating sustained contraction, which, in turn, would be responsible for sensitization of nociceptors locally (Vecchiet et al. 1989). Experimental studies in the calculosis rat model provide some experimental support to this theory. In specimens of the ipsilateral (hyperalgesic) versus contralateral (non-hyperalgesic) oblique musculature, significant changes were found in several parameters indicative of skeletal muscle contraction. These changes were directly proportional to the number of ureteral crises, which in turn were proportional to the degree of the hyperalgesia (Giamberardino et al. 2003).

Pros and Cons of the Ureteral Stone Model
Ideal requisites for pain modeling (Gebhart et al. 1993; Le Bars et al. 2001) are natural, minimally invasive, reproducible and reliable stimulus, quantifiable responses, and unanesthetized animals. In the calculosis model, the mechanical stimulus exerted by the stone can be regarded as "natural," being similar to that occurring spontaneously in humans with urinary stones. Though the procedure for stone formation is invasive, the stimulus per se is scarcely so, as the stone can even be eliminated spontaneously. The behavioral responses are quantifiable in terms of number/duration/complexity of ureteral crises and changes in vocalization thresholds to muscle stimulation, with a mathematical relationship between the amount of ureteral behavior and the extent of muscle hyperalgesia. All responses are observed in the unanesthetized animal. The requisite "reproducible and reliable stimulus" is only satisfied in part. Though the technique for stone formation is standardized and easily reproducible, and the stimulus is reliably "nociceptive" (nocifensive behavior observed in almost 100 % of the cases), the outcome in terms of extent of noxious stimulation of the ureter is not predictable. The amount of pain behavior varies considerably from rat to rat, probably due to the different destiny of the stone. This variability renders the experiments with calculosis rats more elaborate, with high numbers of animals often being required for reliable statistical evaluations. However, an identical variability is seen in patients with urinary stones (different number/duration of colics for calculosis of similar characteristics). Due to the close similarity to the clinical condition, the calculosis model offers more advantages than ▶ alternative rat models of ureteral nociceptive stimulation in vivo.

References

Affaitati, G., Giamberardino, M. A., Lerza, R., Lapenna, D., De Laurentis, S., & Vecchiet, L. (2002). Effects of tramadol on behavioural indicators of colic pain in a rat model of ureteral calculosis. *Fundamental and Clinical Pharmacology, 16*, 23–30.

Bihl, G., & Meyers, A. (2001). Recurrent renal stone disease – Advances in pathogenesis and clinical management. *Lancet, 358*, 651–656.

Cervero, F. (1996). Visceral Nociceptors. In C. Belmonte & F. Cervero (Eds.), *Neurobiology of nociceptors* (pp. 220–240). Oxford/New York/Tokyo: Oxford University Press.

Gebhart, G. F., Meller, S. T., Euchner-Wamser, I., & Sengupta, J. N. (1993). Modeling visceral pain. In L. Vecchiet, D. Albe-Fessard, & U. Lindblom (Eds.), *New trends in referred pain and hyperalgesia* (pp. 129–148). Amsterdam: Elsevier.

Giamberardino, M. A., Affaitati, G., Lerza, R., Fanò, G., Fulle, S., Belia, S., Lapenna, D., & Vecchiet, L. (2003). Evaluation of indices of skeletal muscle contraction in areas of referred hyperalgesia from an artificial ureteric stone in rats. *Neuroscience Letters, 338*, 213–216.

Giamberardino, M. A., Dalal, A., Valente, R., & Vecchiet, L. (1996). Changes in activity of spinal cells with muscular input in rats with referred muscular hyperalgesia from ureteral calculosis. *Neuroscience Letters, 203*, 89–92.

Giamberardino, M. A., De Laurentis, S., Affaitati, G., Lerza, R., Lapenna, D., & Vecchiet, L. (2001). Modulation of pain and hyperalgesia from the urinary tract by algogenic conditions of the reproductive organs in women. *Neuroscience Letters, 304*, 61–64.

Giamberardino, M. A., de Bigontina, P., Martegiani, C., & Vecchiet, L. (1994). Effects of extracorporeal shock-wave lithotripsy on referred hyperalgesia from renal/ureteral calculosis. *Pain, 56*, 77–83.

Giamberardino, M. A., Valente, R., de Bigontina, P., & Vecchiet, L. (1995). Artificial ureteral calculosis in rats: Behavioural characterization of visceral pain episodes and their relationship with referred lumbar muscle hyperalgesia. *Pain, 61*, 459–469.

Giamberardino, M. A., Vecchiet, L., & Albe-Fessard, D. (1990). Comparison of the effects of ureteral calculosis and occlusion on muscular sensitivity in rats. *Pain, 43*, 227–234.

Laird, J. M. A., Roza, C., & Cervero, F. (1997). Effects of artificial calculosis on rat ureter motility: Peripheral contribution to the pain of ureteric colic. *American Journal of Physiology, 272*, 1409–1416.

Laird, J. M. A., Roza, C., & Olivar, T. (1998). Antinociceptive activity of metamizol in rats with experimental ureteric calculosis: Central and peripheral components. *Inflammation Research, 47*, 389–395.

Le Bars, D., Gozariu, M., & Cadden, S. W. (2001). Animal models of nociception. *Pharmacological Reviews, 53*, 597–652.

Roza, C., Laird, J. M. A., & Cervero, F. (1998). Spinal mechanisms underlying persistent pain and referred hyperalgesia in rats with an experimental ureteric stone. *Journal of Neurophysiology, 79*, 1603–1612.

Vecchiet, L., Giamberardino, M. A., Dragani, L., & Albe-Fessard, D. (1989). Pain from renal/ureteral calculosis: Evaluation of sensory thresholds in the lumbar area. *Pain, 36*, 289–295.

Visceral Pain Model, Lower Gastrointestinal Tract Pain

Gerald F. Gebhart
Department of Anesthesiology, Center for Pain Research, University of Pittsburgh, Pittsburgh, PA, USA

Synonyms

Lower gastrointestinal tract pain models

Definition

Animal models of lower gastrointestinal tract pain include graded ▶ colorectal distention, experimental ▶ colitis produced by intracolonic treatments (e.g., trinitrobenzene sulfonic acid) or ingestion of DSS (dextran sodium sulfate).

Characteristics

Methods

Distension of the Lower Gastrointestinal Tract

Distension of hollow organs has a long history in human medicine (e.g., see Ness and Gebhart 1990) and is rational because balloon distension reproduces the character and localization of pathological pain. Accordingly, graded colorectal distension (CRD) has been used to model in animals the study of lower gastrointestinal pain and hypersensitivity (Ness and Gebhart, 1987, 1988a, b; Traub et al. 1993; Al-Chaer et al. 1996, 1998). A device for controlling the pressure of intracolonic balloon distension in nonhuman animals has been described (Anderson et al. 1987). Humans report distinct sensations of pressure and of pain with different intensities of CRD, although CRD is not commonly used in humans. Rather, studies in humans typically employ barostat-controlled *rectal* distension. Repeated CRD in human beings results in progressively greater reports of pain intensity and increased areas of pain referral (i.e., sensitization; Ness et al. 1990).

CRD in nonhuman animals has proven useful because it produces robust, reliable, and reproducible neuronal and autonomic/motor responses. In addition, the colorectum is easily accessible and the stimulus readily controllable.

Construction of Balloon for Colorectal Distension in Rats and Mice

For colorectal distension in rats, an inflatable balloon is inserted through the anus into the rectum and lower colon for a distance of about 7 cm. The balloon can be made from a finger from a latex surgical glove or a condom attached to tygon tubing. The resistance to inflation of the balloon material is reduced by overinflating the balloon and leaving it inflated overnight. Before inserting the balloon, the tubing is connected by way of a T-connector to a manual pump and to a pressure transducer. Balloons of smaller sizes (e.g., 2–5 cm in length) have also been used, but it should be kept in mind that spatial summation is important in an organ like the colon and thus, a longer balloon is preferable. Distensions that result in pressures ranging from 20 to 80 mmHg are used. Pressures above 30–40 mmHg are aversive to rats, as shown by passive avoidance behavior (Ness et al. 1991). Care needs to be taken not to apply pressures in excess of 100 mmHg, since the colon is ruptured by pressures of 120 mmHg or greater (Ness and Gebhart 1987). Even pressures of 80 mmHg produce some damage to the colon (Traub et al. 1993).

For CRD in mice, balloons are made from polyethylene sheets, are usually 1.5–2-cm long, and distending pressures range from 15 to 75 mmHg. Methods of balloon manufacture, distension, and data analysis are described in Christianson and Gebhart (2007).

Methods for Determining the Effects of Colorectal Distension in Rats and Mice

Experiments using a passive avoidance paradigm in rats have provided evidence that colorectal distension to 80 mmHg for 20 s is aversive (Ness et al. 1991). A more convenient behavioral response that can be used to demonstrate the effects of colorectal distension in rats is the recording of ► visceromotor reflexes from the external abdominal oblique muscle, using electromyography (Ness and Gebhart 1988a). In rats, there usually is no or minimal response when the distension pressure is 20 mmHg or less, but graded responses are evoked by pressures of 40, 60, and 80 mmHg (Ness and Gebhart 1988a; Palecek et al. 2002). Other pseudoaffective responses to CRD that have been quantified include changes in heart rate or blood pressure (Ness and Gebhart 1988a). In mice, visceromotor responses quantified by electomyography are graded as in rats between distending pressures of 14 and 75 mmHg.

In addition, responses to CRD in rats have been assessed by electrophysiological recordings from viscero-responsive neurons in the spinal cord or brain (Ness and Gebhart 1987, 1988b, 1989; Al-Chaer et al. 1996, 1998), Fos expression in spinal cord (Traub et al. 1993), and internalization of substance P receptors (NK-1 receptors) in spinal cord neurons following colorectal distension (Honore et al. 2002).

Methods for Inducing Colorectal Hypersensitivity

Acutely, irritant chemicals such as mustard oil, capsaicin, or formaldehyde have been instilled intracolonically or injected into the colon wall to produce relatively short-lived hypersensitivity. Longer lasting hypersensitivity (i.e., inflammatory colonic insult; colitis) can be produced by intracolonic instillation of trinitrobenzene sulfonic acid (TNBS), zymosan (e.g., Coutinho et al. 1996), deoxycholic acid (e.g., Traub et al. 2008) or by oral ingestion over time of dextran sodium sulfate (DSS). Most of these strategies have been used in both mice and rats and, generally, the results are qualitatively similar (i.e., they produce colorectal hypersensitivity), although onset, magnitude, and duration of colorectal hypersensitivity vary based on animal species and strain, concentration of chemical, etc. Assessment of hypersensitivity includes quantification of the visceromotor response by electromyography (as above) or by monitoring open field activity, which is typically decreased in hypersensitive animals.

References

Al-Chaer, E. D., Feng, Y., & Willis, W. D. (1998). A role for the dorsal column in nociceptive visceral input into the thalamus of primates. *Journal of Neurophysiology, 79*, 3143–3150.

Al-Chaer, E. D., Lawand, N. B., Westlund, K. N., et al. (1996). Visceral nociceptive input into the ventral posterolateral nucleus of the thalamus: A new function for the dorsal column pathway. *Journal of Neurophysiology, 76*, 2661–2674.

Anderson, R., Ness, T. J., & Gebhart, G. F. (1987). A distension control device useful for studies of hollow organ sensation. *Physiology and Behavior, 41*, 635–638.

Christianson, J. C., & Gebhart, G. F. (2007). Assessment of colon sensitivity by luminal distension in mice. *Nature Protocols, 2*, 2624–2631.

Coutinho, S. V., Meller, S. T., & Gebhart, G. F. (1996). Intracolonic zymosan produces visceral hyperalgesia in the rat that is mediated by spinal NMDA and non-NMDA receptors. *Brain Research, 736*, 7–15.

Honore, P., Kamp, E. H., Rogers, S. D., et al. (2002). Activation of lamina I spinal cord neurons that express the substance P receptor in visceral nociception and hyperalgesia. *The Journal of Pain, 3*, 3–11.

Ness, T. J., & Gebhart, G. F. (1987). Characterization of neuronal responses to noxious visceral and somatic stimuli in the medial lumbosacral spinal cord of the rat. *Journal of Neurophysiology, 57*, 1867–1892.

Ness, T. J., & Gebhart, G. F. (1988a). Colorectal distension as a noxious visceral stimulus: Physiologic and pharmacologic characterization of pseudoaffective reflexes in the rat. *Brain Research, 450*, 153–169.

Ness, T. J., & Gebhart, G. F. (1988b). Characterization of neurons responsive to noxious colorectal distension in the T13-L2 spinal cord of the rat. *Journal of Neurophysiology, 60*, 1419–1438.

Ness, T. J., & Gebhart, G. F. (1989). Characterization of superficial dorsal horn neurons encoding for colorectal distension in the rat: Comparison with neurons in deep laminae. *Brain Research, 486*, 301–309.

Ness, T. J., & Gebhart, G. F. (1990). Visceral pain: A review of experimental studies. *Pain, 41*, 167–234.

Ness, T.J., Randich, A. & Gebhart, G.F. (1991) Further evidence that colorectal distension is a noxious visceral stimulus in rats. *Neuroscience Letters, 131*, 113–116.

Palecek, J., Paleckova, V., & Willis, W. D. (2002). The roles of pathways in the spinal cord lateral and dorsal funiculi in signaling nociceptive somatic and visceral stimuli in rats. *Pain, 96*, 297–307.

Traub, R. J., Pechman, P., Iadarola, M. J., et al. (1993). Fos-like proteins in the lumbosacral spinal cord following noxious and non-noxious colorectal distention in the rat. *Pain, 49*, 393–403.

Traub, R. J., Tang, B., Ji, Y., Pandya, S., Yfantis, H., & Sun, Y. (2008). A rat model of chronic postinflammatory visceral pain induced by deoxycholic acid. *Gastroenterology, 135*, 2075–2083.

Visceral Pain Model, Pancreatic Pain

Louis Vera-Portocarrero[1] and
Karin Westlund High[2]
[1]Department of Pharmacology, College of Medicine University of Arizona, Tucson, AZ, USA
[2]Department of Physiology, University of Kentucky, Lexington, KY, USA

Synonyms

Acinar cell injury; Alcohol-induced pancreatitis; Epigastric pain; Pancreatalgia; Pancreatitis pain; Visceral inflammation; Visceral pain

Definition

Pancreatic pain (▶ pancreatalgia) is a clinical indicator of an inflammatory or malignant condition involving the pancreas. Pancreatitis is the inflammation of the pancreas. Clinically ▶ pancreatitis is defined by a sudden severe abdominal pain, nausea, fever, and leukocytosis accompanied by high levels of pancreatic enzymes, such as α-amylase and lipase, in blood serum.

Characteristics

Pancreatitis and Pancreatic Pain

The pancreas is a roughly triangular organ in the ▶ retroperitoneal area of the upper abdomen. Pain is the only sensation that can be elicited from the pancreas (Cervero 1994), and ▶ visceralgia is the major complaint bringing patients with pancreatitis to the clinic. Pancreatitis pain is usually ▶ epigastric in nature, is felt in the right and/or left upper quadrant, and may radiate to the back. The etiology of this referred pain can be from several diseases affecting the pancreas, including pancreatitis in its acute and chronic forms. While some patients are afflicted with an idiopathic pancreatitis such as with blockage of the biliary duct, most pancreatitis is the result of chronic alcohol abuse and/or industrial chemical, paint, or solvent exposures. Inflammation of the pancreas is defined clinically by a sudden severe abdominal pain, nausea, fever, and leukocytosis. Pancreatitis is characterized by severe histopathological changes, such as the presence of inflammatory mediators, acinar atrophy, fatty necrosis, intraductal hemorrhage, periductal fibrosis and stromal proliferation (Schmidt et al. 1995), and elevated levels of α-amylase and lipase in the serum. Acute pancreatitis ranges from a mild edematous condition that is self-limited to severe hemorrhagic necrotizing inflammation that is often fatal over a period of days as patients succumb to abdominal sepsis and multi-organ failure. Persistent pancreatitis is typically characterized by unremitting pain and progressive destructive fibrosis. The ► visceral nociception and pain can be severe and become intractable, resistant even to morphine. In patient surveys, 32 % of patients with chronic pancreatic pain report being willing to try any new therapy for relief and some may resort to suicide for this intractable pain state.

Animal Models of Pancreatitis Pain

Anatomy and Function of Pancreatic Innervation
Much of the present knowledge about pancreatic pain has come from clinical work. For example, the effectiveness of blocks or sectioning of the sympathetic innervation, but not the vagus, indicates that primary afferent fibers traveling with the sympathetic nerves are involved in transmitting nociceptive information from the pancreas. Experimental anatomical studies, however, have been useful in showing that pancreatic afferent fibers reach the spinal cord through Lissauer's tract where they may terminate in the spinal segment in which they enter the spinal cord or within one or two segments rostrally or caudally (Sugiura et al. 1989). Small-caliber visceral afferent fibers have a wide, diffuse termination pattern extending over more than two segments with some terminating bilaterally in laminae V and X. This diffuse termination in contrast to clearly defined terminal fields for somatic afferent fibers may contribute to the diffuse and difficult to localize

Visceral Pain Model, Pancreatic Pain, Fig. 1 The visceral nociceptive pathway arises from postsynaptic dorsal column projection neurons located around the central canal. The pathway was visualized after small injections of anterograde transporter, *Phaseolus vulgaris leucoagglutinin*, were made in the gray matter near the central canal (Wang et al. 1999). The visceral pathway arising from the lumbosacral cord travels in the dorsal column midline. The ascending fibers from thoracic levels travel between the gracile and cuneate fasciculi. Many of the postsynaptic dorsal column neurons send their axons through the dorsal column to relay their information at the dorsal column nuclei. The information then travels with the lemniscal system to the thalamus. Axons from the region around the spinal cord are also seen crossing to travel in the ventral medial white matter (Reprinted with permission from Westlund 2000)

nature of this visceral pain state. Direct stimulation of the pancreas with bradykinin in anesthetized rats induces FOS expression most abundantly in segments T7-T10 of the spinal cord, indicating that nociceptive information from the pancreas is mainly processed in those segments (Wang and Westlund, unpublished). With bradykinin-induced noxious stimulation

Visceral Pain Model, Pancreatic Pain, Fig. 2 The summary diagram depicts the two routes taken by axons ascending as separate pathways in the dorsal column and the ventrolateral spinal white matter depending on whether they are transmitting somatic or visceral pain. Two separate pathways to higher brain centers are taken by cutaneous and visceral pain information. It is well known that cutaneous inputs are relayed in the spinal cord onto spinothalamic tract cells, sending their axons up the ventrolateral white matter on the opposite side. Clinical and basic studies have shown that visceral pain is transmitted through the dorsal spinal white matter (Reprinted with permission from Westlund 2000)

Ventroposterolateral thalamus

Nucleus gracilis

Cervical cord

Thoracic cord

Skin

Lumbar cord

Viscera

Sacral cord

Visceral Pain Model, Pancreatic Pain, Fig. 3 Rate histograms showing the responses of a thalamic cell to bradykinin (10^{-5} M, 20 s) application to the pancreas both before (**a**) and (**b**) after DC lesions (Reprinted with permission from Westlund 2000)

Visceral Pain Model, Pancreatic Pain, Fig. 4 (a) The dark area delineates the cutaneous receptive field of a thalamic cell on the shoulder of a rat. (b) The cell was responsive to bradykinin activation of the pancreas and was located at the border between the ventral posterolateral (*VPL*) and ventromedial (*VPM*) thalamus. Responses ceased after the dorsal column was lesioned. (c) Camera lucida reconstruction from several spinal cord sections located the lesion as shown in the blackened areas in the dorsal columns (Reprinted with permission from Westlund 2000)

of the pancreas, evidence is provided that spinal cord cells are activated in the region around the central canal, which is known to be involved in visceral processing. The visceral (pancreatic) nociceptive information travels in the dorsal

column pathway with relays in the dorsal column nuclei and the thalamus (Figs. 1 and 2) (Wang et al. 1999; Houghton et al. 2001; Wang and Westlund 2001; Westlund 2000). Rate histograms show decreased activity after dorsal column lesions (Figs. 3 and 4). Thus, animal models have shown that much of the visceral pain is relayed to the thalamus through a dorsal column pathway separate from that relaying information about somatosensation (Fig. 2).

Animal models of pancreatic pain have also been useful in determining the functional aspects of pancreatitis as an example of visceral pain. All of the animal models described below have been shown histologically to be confined to the pancreas with limited or no involvement of the liver or other organs. Direct intraductal stimulation (distension) of the pancreas with bradykinin (0.5 ml of 10-5 M in lactated Ringer's solution) has been used, as has intraductal distension with another chemical irritant, glycodeoxycholic acid, combined with intraperitoneal caerulean (Schmidt et al. 1992). With these distension/irritation models, pancreatic hypersensitization behavior has been observed acutely in conscious animals (Houghton et al. 2001; Lu et al. 2003; Zhang et al. 2004; Smiley et al. 2004). Decreases in cage crossing and rearing exploratory behavior are noted in open field testing. Hind limb extensions, a specific pain-related behavior, are observed in this model. Restoration of behavior is noted after treatment with specific neurotransmitter receptor blockers and modulators as described below.

A model of more persistent pancreatic pain develops after a tail vein injection of dibutyltin dichloride (DBTC) in commercially available Lewis rats (Vera-Portocarrero et al. 2003). This model is clinically relevant in that both histological and immunological responses resemble human pancreatitis and involvement is limited to the pancreas without significant involvement of the liver. The symptoms persist for up to 3 weeks and are maximal at 1 week. The method was originally described by Merkord and colleagues (1989) and is described as chronic when applied to their highly inbred strain of rats. The method involves a tail vein injection of

Visceral Pain Model, Pancreatic Pain, Fig. 5 Time course of nociceptive behaviors. (**a**) Secondary mechanical hypersensitivity. The y-axis indicates the average number of reflexive withdrawals from von Frey stimuli applied to the abdominal surface of rats (AWE). Rats with an injection of dibutyltin dichloride, DBTC, demonstrate a greater sensitivity to von Frey stimuli (204.1 mN) compared to naïve animals and vehicle-injected animals at time points 3 and 7 days after dibutyltin dichloride injection. + Denotes $p < 0.05$ when comparing the respective animal group behavior to its baseline measure at time point 0. * denotes $p < 0.05$ when compared to corresponding controls (naïve and vehicle-injected groups). Error bars denote standard error of the mean. (**b**) Secondary heat hyperalgesia. Time course for the withdrawal latencies of rats injected with DBTC. The rats receiving the injection of DBTC demonstrated a significantly more rapid reflexive withdrawal to abdominal stimulation with a radiant heat source on days 1–7 after injection compared with the vehicle-injected and naïve groups *$p < 0.05$ compared to vehicle control; + $p < 0.05$ compared to baseline (time point 0). Error bars denote standard error of the mean (Reprinted with permission from Vera-Portocarrero et al. 2003)

dibutyltin dichloride (8 mg/kg body weight). Dibutyltin dichloride is an active component of some paint thinners, a stabilizer in plastics manufacturing, and is used as a biocide in agriculture. The damage induced after tail vein injection in rats occurs slowly in the pancreas over the subsequent 2 weeks. Extreme care must be taken to ensure that the compound remains in the vessel, as it is most caustic upon leakage. Animals are fed with a low soy content diet, such as Harlan TekLad Diet 8626, and provided with 10 % grain ethanol in their drinking water. Serum parameters that serve as markers of pancreatitis remain elevated during the first week and histopathological changes occur throughout the observation period of 21 days. Nociceptive characterization of this animal model of persistent visceral pain has determined that the animals have ▶ central sensitization. Secondary mechanical allodynia and ▶ thermal hyperalgesia present along the abdomen are persistent during the first week after tail vein injection in rats (Fig. 5) (Vera-Portocarrero et al. 2003; Westlund 2000).

Pharmacological Characterization of Pancreatitis Pain in Animal Models

Behavioral studies using these pancreatitis models in Lewis rats have examined the role of specific receptors in the maintenance of a nociceptive state in this model (Vera-Portocarrero et al. 2003; Vera-Portocarrero and Westlund 2004; Lu et al. 2003; Zhang et al. 2004; Smiley et al. 2004). Thus far, studies have shown that blockers of neurokinin NK-1, opiate, and NMDA receptors (CP99,994, morphine, and AP-5, respectively) can significantly reduce the pain-related behaviors in these pancreatitis rats. Block of the $\alpha_2\delta_1$ subunit of N-type voltage activated calcium channels with gabapentin (Neurontin), or protein kinase C inhibition (GF 109302X) is also highly effective. The restoration of open field behavior and reduction of pain related behaviors observed in these pancreatitis models after treatment with the NMDA antagonist or gabapentin can be potentiated with subtherapeutic doses of morphine (Fig. 6) (Lu et al. 2003; Smiley et al. 2004).

Visceral Pain Model, Pancreatic Pain, Fig. 6 Bar chart illustrating the antinociceptive effects of intrathecal co-administration of a single ineffective dose of gabapentin (300 μg) when combined with various low doses of morphine (0.1, 0.2, 0.5, 1.0 μg) (*white bar*) during bradykinin infusion of the pancreatic duct. Measures examined included cage crossing, rearing, and hind limb extension activities in rats with acute pancreatitis (*n* = 5 per dose) or for morphine alone (*n* = 6 per dose). Values are represented as the mean ± SEM for the initial 10 min of activity in a novel cage experience. Surgical baseline behaviors were tested for each of the animals in the drug studies for comparison to its own behavior after a single drug treatment (total *n* =20 for the combined morphine plus gabapentin groups and total *n* = 24 for the morphine alone doses). The 0 μg dose (*hatched bar*) represents the aCSF vehicle-treated control group in animals with intrathecal and intraductal catheters after bradykinin infusion (*n* = 6) used as controls combined for both studies. The *dotted line* illustrates the values for gabapentin alone (300 μg) from Fig. 1A (9.5 ±1.45 for crossing, 7.0 ±1.79 for rearing, and 6.33 ±1.94 for hind limb extension). *Black bars* indicate treatment with morphine alone for comparisons. Comparisons were made between surgical baseline and combined drug groups using the Wilcoxon test (# $p < 0.05$; ## $p < 0.01$). Comparisons were made to the combined gabapentin and morphine treatment groups using Mann-Whitney U tests versus the 0 μg aCSF vehicle-treated control group (*$p < 0.05$; **$p < 0.01$), versus morphine alone (+ $p < 0.05$; ++$p < 0.01$), and versus gabapentin alone (^ $p < 0.05$; ^^$p < 0.01$). *MS* morphine sulfate, *GBP* gabapentin (Reprinted with permission from Smiley et al. 2004)

References

Cervero, F. (1994). Sensory innervation of the viscera: Peripheral basis of visceral pain. *Physiological Reviews, 74*, 95–138.

Houghton, A. K., Wang, C.-C., & Westlund, K. N. (2001). Do nociceptive signals from the pancreas travel in the dorsal column? *Pain, 89*, 207–220.

Lu, Y., Vera-Portocarrero, L. P., & Westlund, K. N. (2003). Intrathecal co-administration of D-APV and morphine is maximally effective in a rat experimental pancreatitis model. *Anesthesiology, 98*, 734–740.

Merkord, J., & Hennighausen, G. (1989). Acute pancreatitis and bile duct lesions in rat induced by dibutyltin dichloride. *Experimental Pathology, 36*, 59–62.

Schmidt, J., Compton, C. C., Rattner, D. W., et al. (1995). Late histopathological changes and healing in an improved rodent model of acute necrotizing pancreatitis. *Digestion, 56*, 246–252.

Schmidt, J., Rattner, D. W., Lewandrowski, K., et al. (1992). A better model of acute pancreatitis for evaluating therapy. *Annals of Surgery, 215*, 44–56.

Smiley, M. M., Lu, Y., Vera-Portocarrero, L. P., et al. (2004). Intrathecal gabapentin enhances the analgesic effects of subtherapeutic-dose morphine in a rat experimental pancreatitis model. *Anesthesiology, 101*, 759–765.

Sugiura, Y., Terui, N., & Hosoya, Y. (1989). Difference in distribution of central terminal between visceral and somatic unmyelinated primary afferent fibers. *Journal of Neurophysiology, 62*, 834–840.

Vera-Portocarrero, L. P., Lu, Y., & Westlund, K. N. (2003). Nociception in persistent pancreatitis in rats: The effects of morphine and neuropeptide alterations. *Anesthesiology, 98*, 474–84.

Vera-Portocarrero, L. P., & Westlund, K. N. (2004). Attenuation of nociception in a model of acute pancreatitis by an NK-1 antagonist. *Pharmacology, Biochemistry, and Behavior, 77*, 631–640.

Wang, C.-C., & Westlund, K. N. (2001). Responses of rat dorsal column neurons to pancreatic nociceptive stimulation. *NeuroReport, 12*, 2527–2530.

Wang, C.-C., Willis, W. D., & Westlund, K. N. (1999). Ascending projections from the area around the spinal cord central canal: A *PHA-L* study in rats. *The Journal of Comparative Neurology, 415*, 341–367.

Westlund, K. N. (2000). Visceral nociception. *Current Review of Pain, 4*, 478–487.

Zhang, L., Zhang, X., & Westlund, K. N. (2004). Restoration of spontaneous exploratory behaviors with an intrathecal NMDA receptor antagonist or a PKC inhibitor in rats with acute pancreatitis. *Pharmacology, Biochemistry, and Behavior, 77*, 145–153.

Visceral Pain Model, Small Intestinal Bowel Distension Pain

Yi Feng
Department of Anesthesiology, Peking
University People's Hospital, Peking, China

Definition

The small intestinal distension pain model is an abdominal mechanical visceral pain model induced by intermittent small intestinal distention, utilizing a balloon catheter that is implanted in the small intestine.

Characteristics

Clinical studies in patients and healthy volunteers have consistently documented that human experimental pain evoked by distention of the gastrointestinal tract is similar to the sensation of pathologic pain. Inflation of a balloon inserted into the small intestine produces an intense pain, and the sites of pain referral are predominantly epigastric and periumbilical. In rats, responses elicited by distension of the small intestine are considered to be indicative of visceral nociception, as they are blocked by morphine (Colburn et al. 1989) and abolished by capsaicin (Lembeck and Skofitsch 1982). Small intestinal distension in animals induces a range of nociceptive responses, including writhing-like responses and pseudo-affective reflexes, which have been used as indexes of visceral nociception (Lembeck and Skofitsch 1982; Colburn et al. 1989; Feng et al. 1998; McLean et al. 1998; Timar-Peregrin et al. 2001). The proximal part of the small intestine, the duodenum or jejunum, is often chosen because gastric or pyloric distention and/or obstruction is frequently fatal to the rats. The duodenum is found to be more sensitive to distention than more distal sections of the small intestine (Jänig and Morrison 1986). Here, we are going to put more emphasis on the duodenal distention model. This model was first created by Dr. Colburn and his colleagues in 1988 and was later modified by several investigators.

This model has some advantages over other visceral pain models. The stimulus (distention) is reversible and repeatable. A quantifiable nociceptive behavior can be induced and used as a bioassay intended for repetitive use. This repeatable visceral pain model has many of the study design advantages found in somatic pain models, such as serial tail-flick or hot plate tests. It allows for paired observations in small animal populations with a minimum of discomfort to the animals. The model can reproduce duplicate visceral symptoms without serious tissue injury, and permit repetitive testing of potential antinociceptive agents in awake or anesthetized rats.

The model can be used to study the pharmacology and mechanisms of visceral pain including anatomical pathways and physiological processing of nociceptive signals.

Experimental Model Preparation

As experimental animals, adult Sprague-Dawley rats, weighing 200–350 g, are acceptable. Under halothane inhalation anesthesia (2–2.5 %), the rats are positioned supine, the abdomen is shaved, and prepped with iodine. A 1.5-cm laparotomy is made at the midline of the upper abdomen (see Fig. 1) and the gastroduodenal junction is exposed. A purse string is accomplished with 4.0 silk on the greater curvature of the stomach, at least 10 mm from the pylorus. The stomach wall is punctured with an 18-gauge needle and a latex rubber balloon catheter (15-cm long epidural or PE50 catheter with an attached 8-mm-long balloon distensible to a 2-ml fluid volume) is introduced and advanced through the pylorus into the first portion of the duodenum (see Fig. 2, area marked "2" and Fig. 3, area marked "4"). The purse string is tied properly affecting a gastrostomy closure around the catheter. A 5-mm-long silicone rubber sleeve (the inner diameter is slightly larger than the outer diameter of the catheter) is advanced over the protruding gastrostomy catheter to the closure purse string (see Fig. 2, area marked "5" and

Visceral Pain Model, Small Intestinal Bowel Distension Pain, Fig. 1 The rat is placed in a supine position. The solid line represents the incision (1–1.5 cm)

Visceral Pain Model, Small Intestinal Bowel Distension Pain, Fig. 2 Schematic drawing of surgical implantation of the balloon catheter. *1*, the distal end of catheter; *2*, the balloon; *3*, the duodenum; *4*, the greater curvature of the stomach; *5*, the internal silicon sleeve at the gastrostomy site

Fig. 3, area marked "6"). The free end of the purse string is tied to anchor the silicone sleeve and the enclosed catheter. A tangential tunnel is made through the abdominal musculature in the left upper quadrant, and the distal end of the catheter is drawn through using an 18-gauge Tuohy epidural needle and tunneled subcutaneously to the base of the skull, externalized,

Visceral Pain Model, Small Intestinal Bowel Distension Pain, Fig. 3 Schematic drawing of implanted balloon catheter. *1*, the silicone sleeve at the exit site; *2*, the balloon catheter; *3*, the stomach; *4*, the balloon; *5*, the first portion of the duodenum; *6*, the silicone sleeve at the gastrostomy site

and anchored to the dermis with another silicone sleeve and super glue (see Fig. 3, area marked "1"). The distal external catheter length is trimmed to 2–3 cm (see Fig. 3, area marked "2"). The wound is closed in two layers, and the animals are allowed 5–7 days to recover before any experimental manipulation.

Postimplantation Care

The intraduodenal balloon catheter implantation will not affect the rats' normal diet or behavior. The weight of the rats usually decreases 5–20 g during the initial 3–5 days after implantation surgery and then gradually increases. Except for slight acute inflammation, there is no severe mucosal injury 7–12 days after the balloon catheter implantation. The balloon catheter may be lost in some rats, either because it backs into the stomach or passes through the rectum. This lost catheter problem can be solved by careful application of silastic sleeves and ligatures. The incidence of intestinal rupture will increase if the intra-balloon volume is greater than 0.75 ml. Duodenal dilation at the site of the balloon may occur in some rats for several weeks (>4 weeks) postimplantation.

Nociceptive Behavioral Response

Rats will show unconditioned responses to noxious duodenal distension, which are characterized by repeated contractions of the abdominal muscles accompanied by extension of the hind limbs. The intraduodenal balloon can be distended by saline using a 1-ml syringe for 1–2 min. at a 15-min. interval. Upon increasing the intra-balloon volume, the rats display uniform behavioral responses within a discrete volume range. Behavioral responses can be scored on a 0–4 scale:

- 0: Normal body position and exploratory behavior
- 1: Halt in activity, "wet dog" shaking, excessive facial grooming, and teeth chattering
- 2: Hunching, abdominal nipping, hind paw biting, and immobility of hind limbs
- 3: Stretching of the hind limbs, arching, and dorsiflexion of the hind paws
- 4: Stretching of the body, extension of the hind limbs, squatting of the body to the floor of the test box, or rotation of the pelvis sideways

The most exaggerated level is characterized by contraction of abdominal muscles followed by a stretch of the body with hind limb extension (the classic writhing behavior). The greater the volume of balloon inflation, the higher the score that is seen. The threshold volume to elicit classic writhing behavior is around 0.5–0.8 ml, depending on the weight of the rat.

References

Colburn, R. W., Coombs, D. W., Degnan, C. C., & Rogers, L. (1989). Mechanical visceral pain model: Chronic intermittent intestinal distention in the Rat. *Physiology & Behavior, 45*, 191–197.

Feng, Y., Cui, M., Al-Chaer, E. D., & Willis, W. D. (1998). Epigastric antinociception by cervical dorsal column lesions in rats. *Anesthesiology, 89*, 411–420.

Jänig, W., & Morrison, J. F. (1986). Functional properties of spinal visceral afferents supplying abdominal and pelvic organs with special emphasis on visceral nociception. In F. Cervero & J. F. B. Morrison (Eds.), *Progress in brain research* (pp. 87–144). Amsterdam: Elsevier.

Lembeck, F., & Skofitsch, G. (1982). Visceral pain reflex after pretreatment with capsaicin and morphine. *Naunyn-Schmiedeberg's Archives of Pharmacology, 321*, 116–122.

McLean, P. G., Garcia-Villar, R., Fioramonti, J., & Bueno, L. (1998). Effects of tachykinin receptor antagonists on the rat jejunal distension pain response. *European Journal of Pharmacology, 26*, 247–252.

Timar-Peregrin, A., Kumano, K., Khalil, Z., Sanger, G. J., & Furness, J. B. (2001). The relationship between propagated contractions and pseudoaffective changes in blood pressure in response to intestinal distension. *Neurogastroenterology and Motility, 13*, 575–584.

Visceral Pain Model, Urinary Bladder Pain (Irritants or Distension)

Timothy Ness and Promil Kukreja
Department of Anesthesiology, University of Alabama at Birmingham, Birmingham, AL, USA

Synonyms

Cystitis models; Vesical pain models

Definition

The urinary bladder is a common site of visceral pain generation and/or localization. Numerous animal models of pain exist utilizing a bladder stimulus to evoke responses, which are summarized in Fig. 1. These models can be broadly stratified into those that use mechanical stimuli to activate afferents arising in the bladder (typically by air or fluid distension using a catheter) and those that chemically activate/sensitize afferents arising in the bladder using irritants. These irritants may be administered directly into the bladder using a catheter or indirectly by the renal excretion of irritating urinary constituents.

Characteristics

Adequate, Appropriate, and Useful Stimuli

There are many considerations related to the development of models for the study of visceral pain (Cervero and Laird 1999; Al-Chaer and Traub 2002; Ness and Gebhart 2001; Fry et al. 2010), but the most important considerations must be whether the stimulus is an "adequate" stimulus and whether the models using the stimulus are appropriate to the organ system and useful. By Sherrington's definition, an adequate noxious stimulus is one which damages or threatens to damage the organism. Whereas this definition is valid for stimulating skin, mechanical, thermal, and chemical stimuli that produce pain when applied to skin may not produce pain in humans when applied to their viscera. Mechanical distension and the inflammation of hollow viscera are stimuli that are reliably associated with pain in humans and with behaviors in nonhuman animals consistent with nociception, and so these stimuli have been viewed as adequate.

To determine whether models of visceral pain are appropriate and useful in the study of visceral pain mechanisms, additional important criteria need to be considered. To the extent possible, the stimulus should reproduce a natural stimulus for the tissue or organ. Distension and inflammation of the urinary bladder meet this criterion. Ideally, the stimulus should be reproducible in terms of onset, intensity, and duration and the responses produced in animals should be quantifiable, reliable, and ideally reproducible. Models which meet these criteria are presented here.

It should be noted that models of bladder pain may or may not be models of a specific disease state. In this particular case, models of bladder pain are often viewed as models of interstitial cystitis, a chronic painful bladder syndrome (Westropp and Buffington 2002). In some cases there may be some validity to this extrapolation, particularly when an attempt to mimic pathophysiological events is undertaken (e.g., Chuang et al. 2003), but in general, data gathered from animal studies must be narrowly interpreted as indicative of a type of pain processing and not necessarily representative of a specific pathophysiological process.

Urinary Bladder Distension (UBD)

A limited number of studies have formally utilized the stimulus of UBD in models of nociception. One might expect more pain-related studies, since classic ▶ pseudoaffective responses to UBD (vigorous cardiovascular and visceromotor reflexes) have been demonstrated in numerous species including

Visceral Pain Model, Urinary Bladder Pain (Irritants or Distension), Fig. 1 Examples of responses to urinary bladder stimulation using mechanical or chemical stimuli. Urinary bladder distension (*UBD*) can evoke pseudaffective reflexes (*top left* – data adapted from Ness et al. 2001) and neuronal responses (*top right* – data adapted from Ness and Castroman 2001) in addition to cystometrograms (*middle right* – data adapted from

rats, mice, rabbits, cats, dogs, and monkeys. Instead, most existent studies have focused upon the evocation of micturition reflexes and their modulation by pharmacological and surgical manipulations. The intensity of the distending stimulus and its rate of delivery are often chosen based upon the planned measure: for micturition-related studies, a cystometrogram can be produced by the slow infusion of fluid into the bladder with a simultaneous measure of the ▶ intravesical pressure; studies related to nociception have more commonly used a much more rapid UBD stimulus through the use of a rapid infusion of air or saline. In anesthetized female rats, vigorous neuronal responses can be evoked (e.g., Ness and Castroman 2001) which are inhibited by opioids, local anesthetics, and NMDA receptor antagonists. Reliable and reproducible pressor responses and contractions of the abdominal musculature (a visceromotor response measured by electromyogram) can also be evoked by UBDs (Ness et al. 2001). Gender differences and hormonal influences are apparent. Genetic models of painful bladder disorders such as the feline interstitial cystitis model have also noted increased sympathoneural activation associated with bladder stimulation (Westropp and Buffington 2002). Neonatal bladder inflammation led to accentuated responses to UBD when tested as adults, suggesting early developmental events may permanently sensitize visceral sensory systems (Randich et al. 2006b). Intravesical treatment with sensitizing chemicals such as those associated with inflammation leads to more vigorous responses to UBD, particularly at low intensities of stimulation (Dmitrieva and McMahon 1996). This phenomenon will be discussed further in the next section.

Direct Instillation of Inflammatory Compounds

Even without an evocative stimulus such as distension, inflammation of the bladder commonly produces reports of pain and urgency in patients suffering from a urinary tract infection. Experiments in nonhuman animals have artificially inflamed or sensitized the bladder with the intravesical administration of numerous irritants. Such irritation leads to altered micturition reflexes, alterations in the spontaneous activity of primary afferent neurons (e.g., Dmitrieva and McMahon 1996), and spontaneous behavioral responses. Most reflex studies have been performed in rats, although behavioral and neurophysiological studies have been performed in primates and cats.

Guerios et al. (2008) monitored expression of nerve growth factor (NGF) and tyrosine kinase receptors (trk) in cystitis model in female rats induced by one or three (at 72-h intervals) 400-microl intravesical instillations of 1mM acrolein. McMahon and Abel (1987) performed an extensive characterization of visceromotor and altered micturition reflexes in chronically decerebrate rats following inflammation of the bladder with 25 % turpentine, 2.5 % mustard oil, or 2 % croton oil administered directly into the bladder via a urethral cannula. Subsequently, the bladder was slowly filled with saline through the urethral cannula and the intravesical pressures were measured,

---

Visceral Pain Model, Urinary Bladder Pain (Irritants or Distension), Fig. 1 (continued) Ness et al. 2001). Pseudaffective reflexes are typically visceromotor responses that can be measured by direct visualization or by electromyographically (VM-EMG) or autonomic responses such as alterations in heart rate (HR in beats per minute, *bpm*) and alterations in arterial pressure (AP measured in mm Hg). Neuronal responses have been demonstrated to be graded in response to graded UBD (40 or 60 mm Hg in this example) and can be quantified as activity per unit time (peristimulus time histograms with one second bins are indicated above oscillographic tracings demonstrating neuronal action potentials). Bladder irritants have been demonstrated to induce c-fos in central nervous system structures (Lanteri-Minet et al. 1995; Vizzard 2000) (stylized spinal cord slice of immunohistochemical localization – *middle left*), to evoke behaviors such as abdominal licking and head turns (Data adapted from Abelli et al. 1989), and to produce immobility with characteristic postures (Lanteri-Minet et al. 1995; Vizzard et al. 1996) (*top-middle*) or secondary hyperalgesia of the hind limbs (indicated as a decrease in the thermal threshold/latency for hind limb withdrawal [Mean Δ Threshold]) (Data adapted from Jaggar et al. 1999). In all of these models, pharmacological manipulation with analgesic drugs such as morphine leads to reduced responses (Total figure is adapted from Ness and Gebhart 2001)

thereby generatinga cystometrogram. Following administration of an irritant, increased responses to urinary bladder filling correlated with measures of inflammation such as tissue edema, plasma extravasation, and leukocyte infiltration. Rats became hypersensitive to noxious stimuli applied to the lower abdomen, perineum, and tail as measured by the number of "kicks" evoked by a given stimulus. This model has also been modified for use in both anesthetized and unanesthetized rat preparations (e.g., Jaggar et al. 1999). In those models, the focus of study has been novel mechanisms related to visceral hyperreflexia (altered cystometrograms) or secondary hyperalgesia (decreased thresholds to heat stimuli in hind limbs). Intact female rats are anesthetized, and a 25 % solution of turpentine is instilled into the bladder for 1 h using a transurethral angiocatheter. Subsequently, repeat cystometrograms or thermal/mechanical testing of the hind paws is performed 1–24 h after the intravesical treatment in awake or anesthetized preparations. Modulatory effects of glutamate receptor antagonists, nitric oxide synthase inhibitors, and bradykinin receptor antagonists have all been noted.

Direct Application of Irritants

Another model of bladder pain using intravesical irritant administration is that described by Abelli et al. (1989), which was modified by Craft et al. (1993). This model examines immediate effects of the irritants on animal behaviors. In this preparation a midline laparotomy is performed 24 h prior to testing, and a 1 mm diameter polyethylene cannula is inserted through the dome of the bladder. The cannula is held in place by a purse string suture and the opposite end is tunneled subcutaneously to the midline upper back, where it is externalized and anchored by a skin button. On the day of testing, rats receive an injection through the cannula of 0.3 cc of xylene (30 % in silicone oil vehicle), capsaicin, or its related compound resiniferatoxin (0.1 or 3.0 nmoles). Immediate behavioral responses are evoked which consist of abdominal/perineal licking, head turns, hind limb hyperextension, head grooming, biting, vocalization, defecation, scratching, and salivation. The time to onset, incidence, and number of individual behavioral

responses is recorded for 15 or more minutes following intravesical drug administration. Baclofen; mu, kappa, and delta opioid receptor agonists; and intravesical tetracaine have all been demonstrated to inhibit these behavioral responses. Pretreatment with systemic capsaicin as adults or neonates abolishes the licking of the lower abdomen/perineum but does not block hind limb hyperextension. Bladder denervation abolishes all behavioral responses. This model of acute urinary bladder pain is methodologically simple and only requires a minimally invasive surgical preparation a day prior to testing. The noxious stimulus is normally intermediate in duration (15 min) but is not escapable. There is no parallel human literature related to intravesical xylene application, but there is an increasing anecdotal literature related to the painful nature of intravesical capsaicin instillation in humans. Numerous analgesic manipulations have been demonstrated to be inhibitory in this model and no nonspecific effects of non-analgesics have been noted. A recent addition to the bladder irritant models is that of Chuang et al. (2003) who have demonstrated that the intravesical administration of protamine, which disrupts the urothelial barrier, makes the bladder easily irritated by physiological concentrations of urinary potassium. This model has a pathophysiological correlate in that defects in the urothelial lining of the bladder have been observed in patients with interstitial cystitis. Intravesical zymosan, a yeast cell-wall component, was used to study bladder-related nociceptive responses by Randich et al. (2006).

Cyclophosphamide-Induced Cystitis

An additional, distinctly different, model of subacute, irritant-induced urinary bladder pain mimics clinically noted pain that occurs due to cystitis that is secondary to antineoplastic treatments. In this model in rats, the cancer chemotherapeutic agent cyclophosphamide (CP) is administered intraperitoneally and subsequently metabolized and excreted as urinary acrolein, a potent irritant of the bladder (Lanteri-Minet et al. 1995). Beginning approximately one hour after the systemic administration of CP and continuing for approximately four hours, unanesthetized rats demonstrate decreased locomotion coupled with altered postures and "crises" (sudden alterations in postures). These behaviors are scored and a cumulative sum for relevant time periods assigned. Measures of bladder inflammation correlate with some behaviors and with the induction of c-fos and Krox-24 proteins within the spinal cord. There is the ethical concern that the stimulus is inescapable and of 4 h duration. Chronic (2-week) treatments with CP have also been described with numerous physiological and biochemical alterations (e.g., Vizzard et al. 1996). Olivar and Laird (1999) have characterized responses to acutely administered CP in mice, noting a 53 % reduction in spontaneous locomotion and intermittent crises, similar to those observed in rats. Subsequent use of the CP model in transgenic mice has suggested a role for NK1 receptor activation and altered Nav1.8 sodium channel activity. Recently CP-induced cystitis model has suggested that nitric oxide may have an antinociceptive function and modulate bladder hyperactivity induced by pathological conditions (Yu and deGroat 2013).

Comparison of Models

As noted above, numerous models of urinary bladder pain have been developed because of the clinical importance of this type of pain. Species and technique differences have occasionally resulted in apparently contradictory results related to the efficacy of novel analgesics. However, to date, all traditional analgesics (i.e., opioids) have been demonstrated to have efficacy in the different model systems. Subsequent translation of laboratory data into clinical trials will determine which model systems are best predictive of the human condition and therefore the most useful.

References

Abelli, L., Conte, B., Somme, V., Maggi, C. A., Girliani, S., & Meli, A. (1989). A method for studying pain arising from the urinary bladder in conscious, freely-moving rats. *Journal of Urology, 141*, 148–151.

Al-Chaer, E. D., & Traub, R. J. (2002). Biological basis of visceral pain: Recent developments. *Pain, 96*, 221–225.

Cervero, F., & Laird, J. M. (1999). Visceral pain. *Lancet, 353*, 2145–2148.

Chuang, Y. C., Chancellor, M. B., Seki, S., Yoshimura, N., Tyagi, P., Huang, L., Lavelle, J. P., DeGroat, W. C., & Fraser, M. O. (2003). Intravesical protamine sulfate and potassium chloride as a model for bladder hyperactivity. *Urology, 61*, 664–670.

Craft, R. M., Carlisi, V. J., Mattia, A., Herman, R. M., & Porreca, F. (1993). Behavioral characterization of the excitatory and desensitizing effects of intravesical capsaicin and resiniferatoxin in the rat. *Pain, 55*, 205–215.

Dmitrieva, N., & McMahon, S. B. (1996). Sensitisation of visceral afferents by nerve growth factor in the adult rat. *Pain, 66*, 87–97.

Fry, C. H., Daneshgari, F., Thor, K., Drake, M., Eccles, R., Kanai, A. J., & Birder, L. A. (2010). Animal models and their use in understanding lower urinary tract dysfunction. *Neurourology and Urodynamics, 29*(4), 603–608.

Guerios, S. D., Wang, Z. Y., Boldon, K., Bushman, W., & Bjorling, D. E. (2008). Blockade of NGF and trk receptors inhibit increased peripheral mechanical sensitivity accompanying cystitis in rats. *American Journal of Physiology. Regulatory, Integrative and Comparative Physiology, 295*(1), R111–R122.

Jaggar, S. I., Scott, H. C. F., & Rice, A. S. C. (1999). Inflammation of the Rat urinary bladder is associated with a referred hyperalgesia which is NGF dependent. *British Journal of Anaesthesia, 83*, 442–448.

Lanteri-Minet, M., Bon, K., de Pommery, J., Michiels, J. F., & Menetrey, D. (1995). Cyclophosphamide cystitis as a model of visceral pain in rats: Model elaboration and spinal structures involved as revealed by the expression of c-fos and Krox-24 proteins. *Experimental Brain Research, 105*, 220–232.

McMahon, S. B., & Abel, C. (1987). A model for the study of visceral pain states: Chronic inflammation of the chronic decerebrate rat urinary bladder by irritant chemicals. *Pain, 28*, 109–127.

Ness, T. J., & Castroman, P. J. (2001). Evidence for two populations of Rat spinal dorsal horn neurons excited by urinary bladder distension. *Brain Research, 932*, 147–156.

Ness, T. J., & Gebhart, G. F. (2001). Methods in visceral pain research. In L. Kruger (Ed.), *Methods in pain research* (pp. 93–108). New York: CRC Press.

Ness, T. J., Lewis-Sides, A., & Castroman, P. J. (2001). Characterization of pseudaffective responses to urinary bladder distension in the rat: Sources of variability and effect of analgesics. *Journal of Urology, 165*, 968–974.

Olivar, T., & Laird, J. M. (1999). Cyclophosphamide cystitis in mice: Behavioral characterization and correlation with bladder inflammation. *European Journal of Pain, 3*, 141–149.

Randich, A., Uzzell, T., Cannon, R., & Ness, T. J. (2006a). Inflammation and enhanced nociceptive responses to bladder distention produced by intravesical zymosan in the rat. *BMC Urology, 6*, 2.

Randich, A., Uzzell, T., DeBerry, J. J., & Ness, T. J. (2006b). Neonatal urinary bladder inflammation produces adult bladder hypersentivity. *The Journal of Pain, 7*(7), 469–479.

Vizzard, M. A. (2000). Alterations in spinal cord Fos protein expression induced by bladder stimulation following cystitis. *American Journal of Physiology. Regulatory, Integrative and Comparative Physiology, 278*(4), R1027–R1039.

Vizzard, M. A., Erdman, S. L., & de Groat, W. C. (1996). Increased expression of neuronal nitric oxide synthetase in bladder afferent pathways following chronic bladder irritation. *The Journal of Comparative Neurology, 370*, 191–202.

Westropp, J. L., & Buffington, C. A. T. (2002). In vivo models of interstitial cystitis. *Journal of Urology, 167*, 694–702.

Yu, Y., & deGroat, W. C. (2013). Nitric oxide modulates bladder afferent nerve activity in the in vitro urinary bladder-pelvic nerve preparation from rats with cyclophosphamide induced cystitis. *Brain Research, 1490*, 83–94.

Visceral Pain Model; Esophageal Pain

Jyoti N. Sengupta and Banani Banerjee
Division of Gastroenterology and Hepatology
and Pediatric Gastroenterology, Medical College
of Wisconsin, Milwaukee, WI, USA

Synonyms

Central hyperexcitability; Esophageal pain, Noncardiac chest pain; Gastroesophageal reflux disease (GERD)

Definition

Esophageal pain is a painful sensation that originates from the esophagus due to frequent acid reflux, dysmotility, and inflammation of the organ.

Introduction

The typical nature of esophageal chest pain is recurrent retrosternal pain, which closely resembles the pain that occurs due to coronary artery disease. Patients having such unexplained chest pain are excluded from coronary artery disease during an angiogram (Katz and Castell 2000; Richter et al. 1989). These patients represent a major economic burden because of frequent visits to an emergency room for recurring chest pain (Richter et al. 1989). Noncardiac chest pain (NCCP) is confused with that of cardiac angina, since there is a common convergence of cardiac and esophageal sensory afferents onto thoracic and cervical spinal dorsal horn neurons (Euchner-Wamser et al. 1994; Garrison et al. 1992). The etiologies of chest pain of esophageal origin may include frequent gastric acid reflux into the esophagus, abnormal esophageal motility, thoracic musculoskeletal pain, and even psychiatric disorders such as panic attack-induced prolonged esophageal spasm. In the United States, the annual prevalence of NCCP is approximately 25 %. The natural history and the underlying pathophysiology of NCCP still remains poorly understood. Most of the available clinical information has been derived from patients who seek treatment only after the development of symptoms. As a result, the inciting events can only be evaluated retrospectively, making it difficult to establish a direct cause-and-effect relationship. Several esophageal abnormalities have been identified in patients suffering from NCCP, including gastroesophageal reflux disease (GERD), esophageal dysmotility, esophageal hypersensitivity, altered cerebral processing, autonomic dysregulation, and abnormal mechanophysical properties of the esophagus (Fass and Achem 2011). Among these abnormalities, it is thought that frequent gastroesophageal reflux is the major contributor in the pathogenesis of NCCP. Ambulatory intraesophageal pH monitoring has demonstrated that approximately 50 % of patients with recurrent NCCP have abnormal esophageal acid exposure (Hewson et al. 1991; DeMeester et al. 1982; Hershcovici et al. 2012). Further confirmation regarding the role of acid in the genesis of NCCP arises from a clinical study where acid perfusion and recording symptoms of pain during 24-h pH monitoring revealed that chest pain closely correlated with the presence of acid in the esophagus (Richter et al. 1991). In addition to acid-induced esophageal pain, these patients also experience greater chest pain at lower volumes of intraesophageal balloon distension (Balaban et al. 1999; Richter et al. 1986; Hobson et al. 2006; Paterson et al. 1995). Several studies have shown that a prolonged (~15 min) acid infusion into the esophagus of healthy human subject produces hyperalgesia to mechanical distension (Richter et al. 1998; Wayne et al. 2000; Sarkar et al. 2000, 2001; Hobson et al. 2004; Willert et al. 2004; Mehta et al. 1995; Hu et al. 2000; Shaker 2007). Esophageal chest pain can also occur due to prolonged esophageal contractions (Balaban et al. 1999; Mittal 2005; Richter 1998). Studies using intraluminal ultrasonography suggest that muscle contractions producing thickening of the esophageal wall are associated with unexplained chest pain (Rao et al. 1996; Dogan et al. 2007). Healthy volunteers develop mechanical and heat hyperalgesia in the esophagus after prolonged acid exposure (Pedersen et al. 2004). Similarly, NCCP patients hyperalgesic to balloon distension exhibit chemical hyperalgesia to acid perfusion (Barish et al. 1986). These findings suggest that in addition to epithelial injury, there are significant alterations in pain transmission pathways following frequent acid exposure. A multimodal pain assessment model has been developed in humans using a probe that can deliver mechanical, electrical, and thermal stimuli (Drewes et al. 2002). Healthy subjects reported increased pain intensity and expansion of the referred pain area with increased stimulus intensity. However, the type of pain sensation varied with different modes of stimulation, suggesting the recruitment of different types of sensory neurons and pathways to different stimuli (Drewes et al. 2002). A subsequent study reported that 15 min of acid exposure in the esophagus produces mechanical and thermal hyperalgesia and a significant expansion of areas of referred pain (Drewes et al. 2003). Because the altered sensory perception in NCCP patients is generally not associated with tissue pathology or motility disorders, it has been speculated that the etiology of NCCP involves either hypersensitivity of primary sensory neurons (peripheral sensitization) and/or the central nervous system (central sensitization).

In patients with NCCP, acid infusion into the lower segment of the esophagus induces enhanced pain sensitivity in nonacid-exposed upper esophagus (secondary hyperalgesia), suggesting that central sensitization contributes to pain sensitivity in functional esophageal disorders (Sarkar et al. 2000). Decreased pain thresholds to esophageal electrical stimulation after the infusion of acid into the duodenum, a sign of overlapping pain, have also been reported (Hobson et al. 2004). The involvement of cortical regions of the brain in central sensitization also has been suggested (Hollerbech et al. 2000). Several studies suggest that esophageal acid activates a number of cerebral cortical regions, including cingulate gyrus, the prefrontal cortex, and the insula (Hobson et al. 2005; Drewes et al. 2005, 2006; Kern et al. 1998, 2004a, b; Lawal et al. 2008). Furthermore, recent functional magnetic resonance imaging (fMRI) demonstrates distinctly different cerebral activation patterns among subgroups of GERD patients following esophageal acid stimulation and psychological anticipation (Wang et al. 2011).

Characteristics

There are several underlying mechanisms involved in esophageal pain. The following sections will describe acid reflux-induced cellular structure of the esophagus, inflammation, neural responses, and molecular changes following acid reflux.

Influence of Acid Reflux on Cellular Structure and Inflammation

The mucosal layer of the esophagus is the major defensive barrier that protects the esophagus from acid-induced injury. However, frequent acid reflux may result in increased intracellular spaces, a marker of esophageal injury and increased epithelial permeability (Barlow and Orlando 2005; Tobey et al. 1996; Ravelli et al. 2006). The dilated intracellular space (DIS) is thought to alter esophageal mucosal barrier function and allow H^+ to permeate deeper into the esophageal wall, resulting in activation of ion-sensitive channels (TRPV1, ASIC) of primary sensory afferents (Ang et al. 2008). This could be one of the mechanisms for esophageal hypersensitivity observed in nonerosive reflux disease (NERD). Changes in intracellular spacing can have consequences such as irreversible and long-term sensitization of primary afferents responsible for esophageal pain. In patients having normal acid exposure times but positive symptom correlation with gastroesophageal reflux events, the sensory threshold to mechanical distension is reduced (Trimble et al. 1995), which may be related to DIS of the mucosal barrier.

Several studies suggest the contribution of proton-gated ion channels such as transient receptor potential vanilloid 1 (TRPV1) and acid-sensing ion channels (ASIC) in visceral hyperalgesia (Page et al. 2007; Peles et al. 2009). A close association between TRPV1 and reflux-related esophageal inflammation has been reported lately. Mice lacking TRPV1 channels (TRPV1-/-) develop significantly less inflammation to fundus ligation-induced acid reflux compared with wild-type mice, and treating wild-type mice with a TRPV1 receptor antagonist prevents esophageal inflammation (Fujino et al. 2006). Similarly, in humans with GERD or NERD, an overexpression of TRPV1 channels in the esophageal mucosa has been reported (Matthews et al. 2004; Bhat and Bielefeldt 2006). Activation of TRPV1 channels by acid causes release of neuropeptides including substance P (SP) and calcitonin gene-related peptide (CGRP) from the peripheral terminals of primary sensory afferents in the esophagus and platelet-activating factor (PAF) from epithelial cells (Cheng et al. 2009). These neuropeptides contribute to neurogenic inflammation, producing vasodilation and plasma extravasation in esophageal tissue (Geppetti and Trevisani 2004). Increased content of CGRP and SP in the esophageal mucosa correlate negatively with perception thresholds in NERD patients (Xu et al. 2013). Upregulation of TRPV1 channels and co-expression with SP in sensory neurons has been reported in an experimental model of chronic reflux-induced esophagitis in rats (Banerjee et al. 2007). Other peripheral receptors involved in mechanical and chemical stimulation

of the esophagus include TRPA1, galanin, ghrelin, bradykinin, mGluR, iGluR, and 5-HT (Miwa et al. 2010). A recent study indicates the involvement of mast cell activation in esophageal distension-induced mechano-excitability and phosphorylation of ERK1/2 in esophageal vagal afferent fibers (Gao et al. 2011). Reflux of proteases such as trypsin and pepsin is associated with the occurrence of GERD, and a recent study has documented an increase of mRNA for protease-activated receptor 2 (PAR2) in patients with NERD (Yoshida et al. 2012). This study also demonstrates a positive correlation between SP and reflux symptoms in reflux patients. In esophageal epithelial cells, stimulation of PAR2 by trypsin resulted in higher expression of proinflammatory cytokines such as interleukin-8 (IL-8), whereas stimulation of PAR2 on nerve fibers triggered the sensation of pain via SP release and other neuropeptides (Yoshida et al. 2007; Bunnett 2006).

Neural Mechanisms

The majority of human studies evaluating esophageal pain have been carried out by scoring pain perception and/or by imaging cortical brain activity in response to different stimuli. Unfortunately, no behavioral model has been developed to evaluate esophageal pain perception in experimental animals. Animal studies evaluating esophageal pain have been generally restricted to recordings of neuronal activity from primary sensory afferents, spinal cord, and/or brainstem neurons in an acutely anesthetized condition. The esophagus is primarily innervated by vagal and spinal sensory neurons, which are pseudounipolar cells with their cell bodies located either in the nodose ganglia (for vagal afferents) or in thoracic and cervical dorsal root ganglia (DRG) for spinal afferents. Sensory neurons in vagal and spinal pathways primarily innervate the mucosa and muscles of the esophagus. The vast majority of neuron somata in both pathways are associated with unmyelinated C-fibers or thinly myelinated Ad-fibers. Both mucosal and muscle afferent fibers have been classified on the basis of stimulus modality to which they respond: mechano-, chemo-, or

Visceral Pain Model; Esophageal Pain, Fig. 1 Illustrates the effect of capsaicin on esophageal distension-sensitive afferent fibers from the rat before and after the application of the TRPV1 receptor selective antagonist AMG9810. In each panel, the *top trace* represents the frequency histogram (1s binwidth) and the *lower trace* action potentials. (**a**) responses to intravenous injection of capsaicin were dose-dependent. Capsaicin (Cap) was injected at a 30-min interval to avoid desensitization. (**b**) responses to capsaicin-induced excitation were blocked by AMG9810 (10 μmol/kg, iv.). AMG9810 was injected 2 min before the first injection of capsaicin. *Arrows* indicate the time of capsaicin injection (Sengupta JN, unpublished data)

thermoreceptors. However, many vagal afferents respond to multiple modes of stimuli (i.e., are polymodal). Figure 1 illustrates the chemosensitive property of an esophageal distension-sensitive vagal afferent fiber that responded to capsaicin in a dose-dependent manner. Capsaicin-induced excitation was significantly blocked by the selective TRPV1 receptor antagonist AMG9810 (Sengupta, unpublished data). Almost all vagal afferent fibers have low thresholds for response to changes in intraluminal pressure, suggesting their role in the regulation of peristalsis. Spinal afferents, on the other hand, have either low (<5 mmHg, 63 %) or high thresholds (>40 mmHg, 37 %) for response to esophageal distension (Sengupta and Gebhart 1994). Their encoding characteristics to graded distension differ from vagal afferent fibers. Most vagal afferent fibers exhibit a steep increase in firing within a narrow range of distending pressures

(5–40 mmHg); with increasing distending pressures >40 mmHg, firing frequency typically plateaus. Unlike vagal afferents, spinal afferents continue to exhibit a monotonic increase in firing in response to increasing distending pressures (Sengupta and Gebhart 1994). The encoding properties of spinal afferents thus suggest a role in nociceptive signaling to the CNS. Vagal sensory afferents convey non-nociceptive information and modulate motor function. Interestingly, a recent study documents that a proportion of vagal sensory neurons supplying the thoracic esophagus project to the upper cervical (C1-C2) spinal cord (Qin et al. 2004). It is not known whether this vagal pathway to the spinal cord is involved in modulation of esophageal pain.

Existing data regarding mucosal afferents are exclusively from the vagus nerve. Mucosal afferent endings are mostly located in the mucosa

Visceral Pain Model; Esophageal Pain, Fig. 2 Illustrates the chemosensitive properties of a vagal mucosal afferent innervating the rat esophagus. (a) At rest, this afferent was spontaneously active and did not respond to intraluminal saline (pH 5.6) infusion. (b and c) The fiber was excited following the intraluminal application of capsaicin or acid (0.1N HCl, pH 2.1). (d) Action potential waveform of the fiber (Sengupta JN, unpublished data)

and are sensitive to light touch of the mucosal surface and generally insensitive to peristaltic contractions. Unlike muscle afferents, mucosal afferents do not always exhibit spontaneous firing. Mucosal afferents are, however, pH sensitive and many of them also exhibit a response to other chemicals including capsaicin (Page et al. 2000). Figure 2 illustrates responses of a vagal mucosal afferent fiber to intraesophageal infusion of saline, 0.1N HCl acid, and capsaicin (Sengupta, unpublished data). Both acid and capsaicin, but not saline, produced marked excitation of the fiber. It has also been shown that chronic acid exposure alters the sensitivity of esophageal mucosal afferents to chemical stimuli. For example, afferent fibers that were initially insensitive or minimally sensitive to chemicals, (e.g., α,β-methylene ATP, capsaicin, and HCl)

became highly sensitive after chronic acid exposure (Page et al. 2000; Peles et al. 2009).

Although the thermosensitivity of primary afferents in the esophagus has not been examined in electrophysiology experiments, it is very likely that vagal and spinal afferent fibers are capable of sensing temperature. A recent study has documented the existence of thermosensitive transient receptor potential channels (TRPV1-4, TRPN1, and TRPM8) in nodose ganglion somata (Ward et al. 2003; Patterson et al. 2003; Zhang et al. 2004). In addition, human studies have documented reports of pain following application of hot (50 °C) and cold (7 °C) stimuli in the esophagus (Drewes et al. 2002, 2003).

Animal studies involving electrophysiological recordings from the spinal cord complement many of the human psychophysical studies.

For example, spinal neurons receiving input from the esophagus have been shown to exhibit sensitization to esophageal distension and expansion of the area of somatic referral following acid infusion into the esophagus (Garrison et al. 1992). Acid exposure of the distal esophagus produces a secondary hyperalgesia distant from the area of acid exposure, suggesting sensitization of spinal neurons that may receive projections from distant areas (Sarkar et al. 2001). In the rat, approximately 26 % of thoracic (T3-T4) dorsal horn neurons respond to cervical and thoracic esophageal distension (Qin et al. 2003), which suggests that viscero-visceral convergence plays a major role in the development of secondary visceral hyperalgesia in the esophagus. In addition, it has been reported that 42 % of thoracic spinal dorsal horn neurons in the rat respond to esophageal distension and pericardial administration of the algogenic substance bradykinin (Euchner-Wamser et al. 1994). As a result of this viscero-visceral convergence from the heart and esophagus, pain originating from the esophagus often mimics that of cardiac angina described above.

Prolonged esophageal acid exposure leads to sensitization of primary afferents and, subsequently, central neurons. It has been shown that frequent acid exposure initiates overexpression of many ion channels (Page et al. 2000). ASIC and TRPV1 channels may play major roles in sensitization of primary afferent fibers. Activation of these channels by acid leads to Na^+ and Ca^{++} influx and excitation of the neuron, including peripheral and central release of glutamate and neuropeptides SP and CGRP. These (and other mediators) contribute to sensitization of primary afferents and thus central neurons. In recent study, it has been shown that the reflux-induced chronic esophagitis in rats causes hypersensitivity of esophageal vagal afferents to mechanical distension, 0.1N HCl, and capsaicin (Peles et al. 2009). This result was also supported by immunohistochemical studies, which indicate a significant increase in expressions of TRPV1 and $P2X_3$ channels in the nodose ganglia (NG) and DRGs following chronic esophagitis, suggesting that the sensitization of esophageal

afferents to chemical stimuli may occur due to changes in the expression of ion channels in sensory afferents (Banerjee et al. 2007, 2009). There is evidence that activation of glutamate receptors, especially N-methyl D-aspartic acid (NMDA) receptors, contributes to long-term sensitization of spinal dorsal horn neurons. In the acute setting, preemptive administration of the NMDA channel blocker ketamine can prevent esophageal acid-induced hypersensitivity in humans, whereas ketamine administration following acid-induced hypersensitivity reverses the hypersensitivity (Willert et al. 2004).

Molecular Mechanisms

In recent studies, attempts have been made to investigate the underlying molecular mechanism(s) of acid-induced long-term cortical sensitization and its contribution to esophageal hypersensitivity. An emerging notion is that the neonatal period is a critical time for development of nociceptive neural pathways, and an episode of inflammation during this period may be sufficient to alter the development and physiology of the maturing brain. Among various mechanisms of neuronal transmission, excitatory neurotransmission via NMDA receptors play a crucial role in neuronal development and plasticity, and a variety of other brain functions including memory formation to chronic pain (Hobson et al. 2005; Collingridge and Bliss 1995; Zhao et al. 2005). The enhancement of NMDA receptor-mediated synaptic transmission in rostral cingulate cortex neurons contributes to visceral hyperalgesia and allodynia (Cao et al. 2008). Long-term molecular alteration in cortical NMDA receptor subunit composition and postsynaptic membrane protein expression has been reported in adult rats that had received esophageal acid exposure during early stage of development from postnatal (P) days P7 through P14 (Banerjee et al. 2011). Interestingly, acute esophageal acid exposure in adult rats results in a transient increase in NR1 and NR2A subunit expression in the rostral cingulate cortex, whereas neonatally acid-exposed rats receiving acute rechallenge in adulthood (P60) exhibited the highest level of NR1 and NR2A expression,

V

a

b

Visceral Pain Model; Esophageal Pain, Fig. 3 Illustrates phosphorylation of NMDA receptor subunits in the rostroventral medulla following early life (P7-P14) esophageal bile exposure in rats (unpublished data Sengupta et al.). A significant increase in Tyr1336 phosphorylation of the NR2B subunit was observed in bile-exposed animals compared with a catheter-only group, whereas no significant difference in the expression of Tyr1471NR2B was observed between the bile-treated and catheter-only groups (Banerjee and Sengupta, unpublished data)

indicating their enhanced susceptibility to molecular changes following acid exposure. Such molecular changes induced during development could be associated with the symptoms of heartburn and esophageal pain observed in NERD and NCCP patients in the absence of esophageal inflammation. The study also emphasizes the synaptic strengthening and upregulation of postsynaptic density protein 95 (PSD-95) in cortical neurons in rats receiving neonatal acid exposure (Banerjee et al. 2011). The long-term effect of neonatal treatment on PSD-95 expression in cingulate cortex neurons following esophageal acid exposure indicates effective surface expression and clustering of NMDA receptors at synapses that eventually may play an important role in NMDA receptor-dependent esophageal hypersensitivity. Phosphorylation of NMDA receptor subunits regulate many cellular processes including surface expression and protein activity, resulting in changes in synaptic strength and neuronal plasticity. Several kinases such as protein kinase C (PKC), calcium calmodulin kinase II (CaMKII), and tyrosine kinases are reported to phosphorylate various serine/threonine and tyrosine residues at the C-terminal ends of NMDA receptor subunits. Figure 3 illustrates phosphorylation of NMDA receptor subunits in the rostroventral medulla (RVM) following early life (P7-P14) esophageal bile exposure in rats (unpublished data Sengupta et al.). A significant increase in Tyr1336 phosphorylation of the NR2B

subunit was observed in bile-exposed animals compared to the catheter-only group. A detailed study of the downstream signaling mechanism in esophageal acid-induced cortical sensitization suggests the involvement of NMDA receptor subunit NR2B phosphorylation at Ser1303 in cingulate cortex neurons (Banerjee et al. 2011). In addition, microinjection of the CaMKII inhibitor KN-93 into the rostral cingulate cortex of rats prevented acid-induced phosphorylation of NR2B in cortical neurons. Moreover, phosphorylation of Ser1303 NR2B by CaMKII promotes slow dissociation of preformed CaMKII-NR2B complexes and stabilizes the receptor-enzyme in the membrane (Strack et al. 2000). This stabilization of NMDA receptor-CaMKII complex in the postsynaptic density may in turn activate multiple proteins and enzymes, such as Ca^{2+}-ATPase, tyrosine hydroxylase, and transcription factor cAMP-responsive-element-binding protein (CREB), and further initiate a cascade of transcriptional and translational processes.

Summary

Human psychophysical studies have shown that frequent acid reflux can cause chest pain of esophageal origin. Although significant advances have been made in recent years in understanding the pathophysiology, neurophysiology, and molecular mechanisms of esophageal pain, several questions still remain to be answered, especially regarding central processing of esophageal pain transmission. Due to lack of reliable animal models to measure pain originating from the esophagus, it is difficult to understand the complexity of reflux-related esophageal pain. The etiologies of GERD, NERD, NCCP, and functional heartburn could be entirely different. Current animal models, including frequent acid infusion and fundus-ligated reflux promoting esophagitis, can only approximate these disease conditions, and may not be reliable prototypes. Therefore, there are always limitations in extrapolating animal experimental data to human conditions. In spite of such limitations, recent molecular studies in the CNS complement many of the findings of human fMRI experiments. With further development of small animal fMRI in conjunction with pharmacological intervention, there will be rapid progress in the understanding of esophageal pain mechanisms.

Acknowledgement We acknowledge the research support provided by National Institutes of Health award R56DK089493 to Jyoti N. Sengupta and Banani Banerjee.

References

Ang, D., Sifrim, D., & Tack, J. (2008). Mechanisms of heartburn. *Nature Clinical Practice. Gastroenterology & Hepatology, 5*, 383–392.

Balaban, D. H., Yamamoto, Y., Liu, J., Pehlivanov, N., Wisniewski, R., DeSilvey, D., & Mittal, R. K. (1999). Sustained esophageal contraction: A marker of esophageal chest pain identified by intraluminal ultrasonography. *Gastroenterology, 116*, 29–37.

Banerjee, B., Medda, B. K., Lazarova, Z., Bansal, N., Shaker, R., & Sengupta, J. N. (2007). Effect of reflux-induced inflammation on transient receptor potential vanilloid one (TRPV1) expression in primary sensory neurons innervating the oesophagus of rats. *Neurogastroenterology and Motility, 19*, 681–691.

Banerjee, B., Medda, B. K., Schmidt, J., Zheng, Y., Zhang, Z., Shaker, R., & Sengupta, J. N. (2009). Altered expression of P2X3 in vagal and spinal afferents following esophagitis in rats. *Histochemistry and Cell Biology, 132*, 585–597.

Banerjee, B., Medda, B. K., Schmidt, J., Lang, I. M., Sengupta, J. N., & Shaker, R. (2011). Neuronal plasticity in the cingulate cortex of rats following esophageal acid exposure in early life. *Gastroenterology, 141*, 544–552.

Barish, C. F., Castell, D. O., & Richter, J. E. (1986). Graded esophageal balloon distention. A new provocative test for noncardiac chest pain. *Digestive Diseases and Sciences, 31*, 1292–1298.

Barlow, W. J., & Orlando, R. C. (2005). The pathogenesis of heartburn in nonerosive reflux disease: A unifying hypothesis. *Gastroenterology, 128*, 771–778.

Bhat, Y. M., & Bielefeldt, K. (2006). Capsaicin receptor (TRPV1) and non-erosive reflux disease. *European Journal of Gastroenterology & Hepatology, 18*, 263–270.

Bunnett, N. W. (2006). Protease-activated receptors: How proteases signal to cells to cause inflammation and pain. *Seminars in Thrombosis and Hemostasis, 32* (Suppl 1), 39–48.

Cao, Z., Wu, X., Chen, S., Fan, J., Zhang, R., Owyang, C., & Li, Y. (2008). Anterior cingulate cortex modulates visceral pain as measured by visceromotor responses in viscerally hypersensitive rats. *Gastroenterology, 134*, 535–543.

Cheng, L., de la Monte, S., Ma, J., Hong, J., Tong, M., Cao, W., Behar, J., Biancani, P., & Harnett, K. M. (2009). HCl-activated neural and epithelial vanilloid receptors (TRPV1) in cat esophageal mucosa. *American Journal of Physiology. Gastrointestinal and Liver Physiology, 297*, G135–G143.

Collingridge, G. L., & Bliss, T. V. (1995). Memories of NMDA receptors and LTP. *Trends in Neurosciences, 18*, 54–56.

DeMeester, T. R., O'Sullivan, G. C., Bermudez, G., Midell, A. I., Cimochowski, G. E., & O'Drobinak, J. (1982). Esophageal function in patients with angina-type chest pain and normal coronary angiograms. *Annals of Surgery, 196*, 488–498.

Dogan, I., Puckett, J. L., Padda, B. S., & Mittal, R. K. (2007). Prevalence of increased esophageal muscle thickness in patients with esophageal symptoms. *American Journal of Gastroenterology, 102*, 137–145.

Drewes, A. M., Reddy, H., Staahl, C., Pedersen, J., Funch-Jensen, P., Arendt-Nielsen, L., & Gregersen, H. (2005). Sensory-motor responses to mechanical stimulation of the esophagus after sensitization with acid. *World Journal of Gastroenterology, 11*, 4367–4374.

Drewes, A. M., Reddy, H., Pedersen, J., Funch-Jensen, P., Gregersen, H., Arendt-Nielsen, L. (2006). Multimodal pain stimulations in patients with grade B oesophagitis. *Gut, 55*, 926–932.

Drewes, A. M., Schipper, K.-P., Dimcevski, G., Petersen, P., Andersen, O. K., Gregersen, H., & Nielsen, L. A. (2002). Multimodal assessment of pain in the esophagus: A new experimental model. *American Journal of Physiology, 283*, G95–G103.

Drewes, A. M., Schipper, K.-P., Dimcevski, G., Petersen, P., Andersen, O. K., Gregersen, H., & Nielsen, L. A. (2003). Multi-modal induction and assessment of allodynia and hyperalgesia in the human esophagus. *European Journal of Pain, 7*, 539–549.

Euchner-Wamser, I., Meller, S. T., & Gebhart, G. F. (1994). A model of cardiac nociception in chronically instrumented rats: Behavioral and electrophysiological effects of pericardial administration of algogenic substances. *Pain, 58*, 117–128.

Fass, R., & Achem, S. R. (2011). Noncardiac chest pain: Epidemiology, natural course and pathogenesis. *Journal of Neurogastroenterology and Motility, 17*, 110–123.

Fujino, K., de la Fuente, S. G., Takami, Y., Takahashi, T., & Mantyh, C. R. (2006). Attenuation of acid induced oesophagitis in VR-1 deficient mice. *Gut, 55*, 34–40.

Gao, G., Ouyang, A., Kaufman, M. P., & Yu, S. (2011). ERK1/2 signaling pathway in mast cell activation-induced sensitization of esophageal nodose C-fiber neurons. *Diseases of the Esophagus, 24*, 194–203.

Garrison, D. W., Chandler, M. J., & Foreman, R. D. (1992). Viscerosomatic convergence onto feline neurons from esophagus, heart and somatic fields: Effects of inflammation. *Pain, 49*, 373–382.

Geppetti, P., & Trevisani, M. (2004). Activation and sensitisation of the vanilloid receptor: Role in gastrointestinal inflammation and function. *British Journal of Pharmacology, 141*, 1313–1320.

Hershcovici, T., Achem, S. R., Jha, L. K., & Fass, R. (2012). Systematic review: The treatment of noncardiac chest pain. *Alimentary Pharmacology and Therapeutics, 35*, 5–14.

Hewson, E. G., Sinclair, J. W., Dalton, C. B., & Richter, J. E. (1991). Twenty-four-hour esophageal pH monitoring: The most useful test for evaluating noncardiac chest pain. *American Journal of Medicine, 90*, 576–583.

Hobson, A. R., Khan, R. W., Sarkar, S., Furlong, P. L., Aziz, Q. (2004). Development of esophageal hypersensitivity following experimental duodenal acidification. *American Journal of Gastroenterology, 99*, 813–820.

Hobson, A. R., Furlong, P. L., Worthen, S. F., Hillebrand, A., Barnes G. R., Singh, K. D., Aziz, Q. (2005). Real-time imaging of human cortical activity evoked by painful esophageal stimulation. *Gastroenterology, 128*, 610–619.

Hobson, A. R., Furlong, P. L., Sarkar, S., Matthews, P. J., Willert, R. P., Worthen, S. F., Unsworth, B. J., & Aziz, Q. (2006). Neurophysiologic assessment of esophageal sensory processing in noncardiac chest pain. *Gastroenterology, 130*, 80–88.

Hollerbach, S., Bulat, R., May, A., Kamath, M. V., Upton, A. R., Fallen, E. L., & Tougas, G. (2000). Abnormal cerebral processing of oesophageal stimuli in patients with noncardiac chest pain (NCCP). *Neurogastroenterology and Motility, 12*, 555–565.

Hu, W. H., Martin, C. J., & Talley, N. J. (2000). Intraesophageal acid perfusion sensitizes the esophagus to mechanical distension: A Barostat study. *The American Journal of Gastroenterology, 95*, 2189–2194.

Katz, P. O., & Castell, D. O. (2000). Approach to the patient with unexplained chest pain. *American Journal of Gastroenterology, 95*, S4–S8.

Kern, M. K., Birn, R. M., Jaradeh, S., Jesmanowicz, A., Cox, R. W., Hyde, J. S., & Shaker, R. (1998). Identification and characterization of cerebral cortical response to esophageal mucosal acid exposure and distention. *Gastroenterology, 115*, 1353–1362.

Kern, M., Hofmann, C., Hyde, J., & Shaker, R. (2004a). Characterization of the cerebral cortical representation of heartburn in GERD patients. *American Journal of Physiology. Gastrointestinal and Liver Physiology, 286*, G174–G181.

Kern, M. K., Arndorfer, R. C., Hyde, J. S., & Shaker, R. (2004b). Cerebral cortical representation of external anal sphincter contraction: Effect of effort. *American Journal of Physiology. Gastrointestinal and Liver Physiology, 286*, G304–G311.

Lawal, A., Kern, M., Sanjeevi, A., Antonik, S., Mepani, R., Rittmann, T., Hussaini, S., Hofmann, C., Tatro, L., Jesmanowicz, A., Verber, M., & Shaker, R. (2008). Neurocognitive processing of esophageal central sensitization in the insula and cingulate gyrus. *American Journal of Physiology. Gastrointestinal and Liver Physiology, 294*, G787–G794.

Matthews, P. J., Aziz, Q., Facer, P., Davis, J. B., Thompson, D. G., & Anand, P. (2004). Increased capsaicin receptor TRPV1 nerve fibres in the inflamed human oesophagus. *European Journal of Gastroenterology & Hepatology, 16*, 897–902.

Mehta, A. J., De Caestecker, J. S., Camm, A. J., & Northfield, T. C. (1995). Sensitization to painful distention and abnormal sensory perception in the esophagus. *Gastroenterology, 108*, 311–319.

Mittal, R. K. (2005). Motor and sensory function of the esophagus: Revelations through ultrasound imaging. *Journal of Clinical Gastroenterology, 39*, S42–S48.

Miwa, H., Kondo, T., Oshima, T., Fukui, H., Tomita, T., & Watari, J. (2010). Esophageal sensation and esophageal hypersensitivity – Overview from bench to bedside. *Journal of Neurogastroenterology and Motility, 16*, 353–362.

Page, A. J., O'Donnell, T. A., & Blackshaw, L. A. (2000). P2X purinoceptor-induced sensitization of ferret vagal mechanoreceptors in esophageal inflammation. *The Journal of Physiology, 523*, 403–411.

Page, A. J., Brierley, S. M., Martin, C. M., Hughes, P. A., & Blackshaw, L. A. (2007). Acid sensing ion channels 2 and 3 are required for inhibition of visceral nociceptors by benzamil. *Pain, 133*, 150–160.

Paterson, W. G., Wang, H., & Vanner, S. J. (1995). Increasing pain sensation to repeated esophageal balloon distension in patients with chest pain of undetermined etiology. *Digestive Diseases and Sciences, 40*, 1325–1331.

Patterson, L. M., Zheng, H., Ward, S. M., & Berthoud, H. R. (2003). Vanilloid receptor (VR1) expression in vagal afferent neurons innervating the gastrointestinal tract. *Cell and Tissue Research, 311*, 277–287.

Pedersen, J., Reddy, H., Funch-Jensen, P., Arendt-Nielsen, L., Gregersen, H., & Drewes, A. M. (2004). Cold and heat pain assessment of the human oesophagus after experimental sensitization with acid. *Pain, 110*, 393–399.

Peles, S., Medda, B. K., Zhang, Z., Banerjee, B., Lehmann, A., Shaker, R., & Sengupta, J. N. (2009). Differential effects of transient receptor vanilloid one (TRPV1) antagonists in acid-induced excitation of esophageal vagal afferent fibers of rats. *Neuroscience, 161*, 515–525.

Qin, C., Chandler, M. J., & Foreman, R. D. (2003). Afferent pathways and responses of T3-T4 spinal neurons to cervical and thoracic esophageal distensions in rats. *Autonomic Neuroscience, 109*, 10–20.

Qin, C., Chandler, M. J., Jou, C. J., & Foreman, R. D. (2004). Response and afferent pathways of C1-C2 spinal neurons to cervical and thoracic esophageal stimulation in rats. *Journal of Neurophysiology, 91*, 2227–2235.

Rao, S. S., Gregersen, H., Hayek, B., Summers, R. W., & Christensen, J. (1996). Unexplained chest pain: The hypersensitive, hyperreactive, and poorly compliant esophagus. *Annals of Internal Medicine, 124*, 950–958.

Ravelli, A. M., Villanacci, V., Ruzzenenti, N., Grigolato, P., Tobanelli, P., Klersy, C., & Rindi, G. (2006). Dilated intercellular spaces: A major morphological feature of esophagitis. *Journal of Pediatric Gastroenterology and Nutrition, 42*, 510–515.

Richter, J. E. (1998). Dysphagia, odynophagia, heartburn and other esophageal symptoms. In M. Feldman, B. F. Scharschmidt, M. Sleisenger, & R. Zorab (Eds.), *Gastrointestinal and liver disease* (pp. 97–105). Philadelphia: W.B. Saunders.

Richter, J. E., Barish, C. F., & Castell, D. O. (1986). Abnormal sensory perception in patients with esophageal chest pain. *Gastroenterology, 91*, 845–852.

Richter, J. E., Bradley, L. A., & Castell, D. O. (1989). Esophageal chest pain: Current ontroversies in pathogenesis, diagnosis, and therapy. *Annals of Internal Medicine, 110*, 66–78.

Richter, J. E., Hewson, E. G., Sinclair, J. W., & Dalton, C. B. (1991). Acid perfusion test and 24-h esophageal pH monitoring with symptom index. Comparison of tests for esophageal acid sensitivity. *Digestive Diseases and Sciences, 36*, 565–571.

Sarkar, S., Aziz, Q., Woolf, C. J., Hobson, A. R., & Thompson, D. G. (2000). Contribution of central sensitisation to the development of non-cardiac chest pain. *Lancet, 356*, 1154–1159.

Sarkar, S., Hobson, A. R., Furlong, P. L., Woolf, C. J., Tomson, D. G., & Aziz, Q. (2001). Central neural mechanisms mediating human visceral hypersensitivity. *American Journal of Physiology, 281*, G1196–G1202.

Sengupta, J. N., & Gebhart, G. F. (1994). Gastrointestinal afferents fibers and sensation. In L. R. Johnson (Ed.), *Physiology of the gastrointestinal tract* (pp. 483–519). New York: Raven.

Shaker, R. (2007). Gastroesophageal reflux disease: Beyond mucosal injury. *Journal of Clinical Gastroenterology, 41*(Suppl 2), S160–S162.

Strack, S., McNeill, R. B., & Colbran, R. J. (2000). Mechanism and regulation of calcium/calmodulin-dependent protein kinase II targeting to the NR2B subunit of the *N*-methyl-D-aspartate receptor. *Journal of Biological Chemistry, 275*, 23798–23806.

Tobey, N. A., Carson, J. L., Alkiek, R. A., & Orlando, R. C. (1996). Dilated intercellular spaces: A morphological feature of acid reflux–damaged human esophageal epithelium. *Gastroenterology, 111*, 1200–1205.

Trimble, K. C., Pryde, A., & Heading, R. C. (1995). Lowered oesophageal sensory thresholds in patients with symptomatic but not excess gastro-oesophageal reflux: Evidence for a spectrum of visceral sensitivity in GORD. *Gut, 37*, 7–12.

Wang, K., Duan, L. P., Zeng, X. Z., Liu, J. Y., & Xu-Chu, W. (2011). Differences in cerebral response to esophageal acid stimuli and psychological anticipation in GERD subtypes – an fMRI study. *BMC Gastroenterology, 11*, 28.

Ward, S. M., Bayguinov, J., Won, K. J., Grundy, D., & Berthoud, H. R. (2003). Distribution of the vanilloid receptor (VR1) in the gastrointestinal tract. *The Journal of Comparative Neurology, 465*, 121–235.

Wayne, H. C., Hu Martin, C. J., & Talley, N. J. (2000). Intraesophageal acid perfusion sensitizes the esophagus to mechanical distension: A barostat study. *The American Journal of Gastroenterology, 95*, 2189–2194.

Willert, P. R., Woolf, C. J., Hobson, A. R., Delaney, C., Thomson, D., & Aziz, Q. (2004). The development and maintenance of human visceral pain hypersensitivity is dependent on the *N*-methyl-D-aspartate receptor. *Gastroenterology, 126*, 683–692.

Xu, X., Li, Z., Zou, D., Yang, M., Liu, Z., Wang, X. (2013). High expression of calcitonin gene-related peptide and substance P in esophageal mucosa of patients with non-erosive reflux disease. *Digestive Diseases and Sciences, 58*, 53–60.

Yoshida, N., Katada, K., Handa, O., Takagi, T., Kokura, S., Naito, Y., Mukaida, N., Soma, T., Shimada, Y., Yoshikawa, T., & Okanoue, T. (2007). Interleukin-8 production via protease-activated receptor 2 in human esophageal epithelial cells. *International Journal of Molecular Medicine, 19*, 335–340.

Yoshida, N., Kuroda, M., Suzuki, T., Kamada, K., Uchiyama, K., Handa, O., et al. (2012). Role of nociceptors/neuropeptides in the pathogenesis of visceral hypersensitivity of nonerosive reflux disease. *Digestive Diseases and Sciences, 10*, 1–7.

Zhang, L., Jones, S., Brody, K., Costa, M., & Brookes, S. J. H. (2004). Thermosensitive transient receptor potential channels in vagal afferent neurons of the mouse. *American Journal of Physiology, 286*, G983–G991.

Zhao, M. G., Toyoda, H., Lee, Y. S., Wu, L. J., Ko, S. W., Zhang, X. H., Jia, Y., Shum, F., Xu, H., Li, B. M., Kaang, B. K., & Zhuo, M. (2005). Roles of NMDA NR2B subtype receptor in prefrontal long-term potentiation and contextual fear memory. *Neuron, 47*, 859–872.

Visceral Pain Models, Female Reproductive Organ Pain

Karen J. Berkley and Natalia Dmitrieva
Program in Neuroscience, Florida State
University, Tallahassee, FL, USA

Synonyms

Adenomyosis model; Chronic pelvic pain model; Dyspareunia model; Endometriosis model; Endometritis model; Female reproductive organ pain model; Gynecological pain model; Labor pain model; Referred abdominal muscle pain model; Vaginal hyperalgesia model; Vulvodynia model

Definition

An "animal model" of pain refers to a nonhuman animal treated in a manner to create a condition that mimics a painful human syndrome or disease. The model can then be used both to improve knowledge of mechanisms that might underlie the condition/disease and its relation to pain and to develop better treatments.

Introduction

Models of visceral pain usually refer to nonhuman animals that currently have or have had some type of pathophysiology of an internal organ. For female reproductive organs, the models are created in ways that mimic pain associated with female reproductive organs, i.e., gynecological pain. Importantly, however, as discussed below, such pain can sometimes be derived from pathophysiology of *non*reproductive organs. Of utmost importance in interpreting results from studies using animal models is recognition of limitations inherent in translating results from nonhuman animals to humans.

Characteristics

In general, there are six procedural steps for developing and using models of painful conditions/diseases. The first step is to mimic in animals the pathophysiological conditions and/or signs/symptoms associated with a particular painful condition/disease in humans. The second is to develop the means for reliable assessment of nociception in the model. The third is to validate the model and understand its interpretative limitations. The fourth is to use the model to study mechanisms underlying the pain or changes in nociceptive sensitivity associated with it. The fifth is to use the model to assess potential treatments and their efficacy. The sixth is to improve translation in both directions, i.e., from the model to women as well as from women to the model. Of particular importance for gynecological pain models in all steps is to include considerations of the model's and the woman's reproductive and hormonal status.

There are a large number of painful gynecological conditions/diseases/syndromes, and new models continue to be developed, most using rodents. Some models currently in use include those for:

1. *Labor pain*, assessed by distending the cervix and measuring the reflex abdominal muscle contractions and neuronal activation patterns (Sandner-Kiesling et al. 2002; Tong et al. 2003) (see below for more details).
2. *Uterine inflammation-associated pain* (e.g., endometritis), modeled by inflaming the uterus and assessed by videotaping spontaneous "pain behaviors" (Wesselmann et al. 1998) and neurogenic plasma extravasation in the skin (as an indirect measure of referred nociception; Wesselmann and Lai 1997) (see below for more details).
3. *Adenomyosis*, modeled by oral dosing of neonatal mice with tamoxifen (daily, 2–5 days after birth) and assessed in adult mice via histological examination of the uterus and both immunohistochemical visualization and molecular measurements of nerve growth factor and its receptors trkA, trkA mRNA, and p75NTR in

Visceral Pain Models, Female Reproductive Organ Pain, Fig. 1 (a) Diagram of apparatus designed to distend the rat cervix as a model for labor pain (Modified from Sandner-Kiesling et al. 2002). (b) Diagram of abnormal body positions used to quantify pain behaviors in a rat model of uterine inflammation (Modified from Wesselmann et al. 1998)

the uterus and T13-L2 dorsal root ganglia. They found that as the disease became more severe as the rats matured, the levels of NGF increased in the uterus, while trkA mRNA increased in the dorsal root ganglia (Li et al. 2011).

4. *Endometriosis-induced dyspareunia*, modeled by surgical transplantation of uterine tissue in the abdomen and assessed with behavioral methods to measure vaginal nociception (e.g., Cason et al. 2003; McAllister et al. 2012) (see below for more details).

5. *Endometriosis-associated referred abdominal muscle pain*, modeled surgically as in 4 and assessed indirectly by telemetered electrical activity responses from abdominal muscles (visceromotor reflex, VMR) elicited by vaginal distention (Dmitrieva et al. 2012) (see below for explanation and more details).

6. *Postmenopausal-induced dyspareunia*, modeled and assessed in senescent rats with behavioral methods to measure vaginal nociception (Bradshaw and Berkley 2002; Berkley et al. 2007) and neuronal activation patterns (Bradshaw and Berkley 2003) (see below for more details).

7. *Vulvodynia*, modeled in rats by localized exposure of the vulva to *Candida albicans* infection or zymosan and assessed behaviorally by mechanical (von Frey) sensitivity testing. The treatments induce mechanical allodynia that may be due to treatment-induced vulvar hyperinnervation by peptidergic sensory and sympathetic fibers (Farmer et al. 2011).

Although none of these models has yet been studied using all six procedural steps, results so far are beginning to provide insights into mechanisms of gynecological pain and some potential clinical applications. Here are some examples:

Labor Pain (Cervix Distention)
In an attempt to mimic the cervical distention that occurs during parturition, Gintzler and colleagues and later Eisenach and colleagues developed similar methods that allowed them to distend the rat's uterine cervix mechanically (UCD) with known force (Fig. 1a).

Gintzler and Komisaruk (1991) used the technique in adult, nonpregnant, nulliparous rats of unspecified estrous stage to study the analgesic effects of cervix stimulation. They found that UCD activated neurons in the deep dorsal horn of spinal segments T12-L2 and produced an increase in tail-flick latency (analgesia), which was abolished by pelvic, but not hypogastric, neurectomy. Unlike UCD-induced analgesia in nonpregnant rats, analgesia produced by pregnancy was significantly reduced by hypogastric nerve transection (Gintzler et al. 1983).

Eisenach and his colleagues (Sandner-Kiesling et al. 2002; Shin and Eisenach 2003; Shin et al. 2003; Tong et al. 2003; 2006; Yan et al. 2007; Liu et al. 2008) applied different forces of cervical distention and studied UCD-induced force-dependent hypogastric nerve activity and evoked abdominal muscle contractions. When studied in ovariectomized rats with (OVX + E2) or without (OVX) estradiol replacement, they found that estradiol enhanced the cervix-induced activity and contractions. Such enhancement of the activity also occurred in rats during the late stages of pregnancy. They also studied how systemic opioids and systemic (intravenous) or intrathecal cyclooxygenase inhibitors (ketorolac) or a TRPV1 antagonist (capsaizepine) influenced these effects. They found that systemic delivery of mu and kappa opioids and systemic, but not intrathecal, delivery of ketorolac reduced the inhibition of responses to UCD. Estrogen reduced the efficacy of the mu opioid inhibition, but not of the kappa opioid or intrathecal ketorolac. They also found that TRPV1 became relevant only during estrogen-induced sensitization. The authors concluded that whereas morphine acts at estrogen-influenced spinal and supraspinal sites, the cyclooxygenase inhibitor acts at an estrogen-independent, non-spinal site(s). They also concluded that cervix-related pain during labor or cancer, induced by input to spinal cord from cervix afferents, might be enhanced by estradiol and reduced by TRPV1 antagonists.

Neither of these two groups or others has yet used their model to study parturient rats, so although interesting with respect to pain associated with cervix stimulation, it is yet to be determined how these results relate to labor pain itself.

Uterine Inflammation

To mimic the several painful clinical conditions in women that involve uterine inflammation, Wesselmann and her colleagues (Wesselmann et al. 1998; Wesselmann and Lai 1997) developed a model in which they inflamed with a mustard oil the left uterine horn of adult, nonpregnant, nulliparous rats in unspecified estrous stages. To validate the model, they used methods developed by Giamberardino et al. (1995) for assessing pain behaviors produced by a ureteral stone ("ureteral" pain behaviors) to videotape and quantify "uterine" pain behaviors for 7 days after the inflammation (Fig. 1b). To assess referred hyperalgesia, they examined whether this manipulation produced plasma extravasation of Evans blue (EB) dye in the skin and increased vocalization thresholds to stimulation of the left flank musculature. They found that in most, but not all, rats (72 %), the inflammation resulted in clear-cut and quantifiable pain behaviors that persisted for 2–4 days posttreatment. The treatment also reduced muscle nociceptive vocalization thresholds and induced dye extravasation in flank skin, indicating activation of nociceptors distant from the originally inflamed tissue and the presence of referred hyperalgesia. The authors concluded that their behavioral model closely resembles pain exhibited by women who suffer from inflammatory uterine conditions and therefore will be of value for future studies of mechanisms and potential treatment.

Using a different approach, our group created a method for behavioral assessment of uterine nociception in rats (Berkley et al. 1995). The method is identical to that used for behavioral assessment of vaginal nociception (see Fig. 2 in Berkley and Dmitrieva, Pain Mechanisms in Endometriosis, and Fig. 2a, b, d here), except that the distending balloon is permanently implanted in the left uterine horn (Fig. 2a, c, e). We found that ~70 % of rats would learn to perform operant escape responses to distention of one mid-uterine horn and that the probability of escape increased as distention volume (and pressure) increased. The responses depend on estrous stage (rats exhibit fewer escape responses during periovulatory stages than in other stages, Fig. 2e). These responses are eliminated by hypogastric neurectomy (a procedure similar to presacral neurectomy in women; Temple et al. 1999). However, the neurectomy is ineffective if the uterine horn exhibits signs of inflammation (or if inflammation is induced by prostaglandin 2a-containing tablets). As discussed in Temple et al. (1999), these results support clinical evidence indicating the limited efficacy of presacral neurectomy for severe dysmenorrhea in women.

Postmenopausal Dyspareunia

Using behavioral methods to assess vaginal nociception described in another entry (Fig. 2a, b, d here and Fig. 2 in Berkley and Dmitrieva, Endometriosis essay), our group examined whether surgical menopause (ovariectomy, OVX) increased escape responding, which would indicate OVX-induced vaginal hyperalgesia, and thus serve as a model for postmenopausal dyspareunia in women (Bradshaw and Berkley 2002). We found that OVX produced vaginal hyperalgesia in all rats. Furthermore, in most, but not all, rats (87 %), the vaginal hyperalgesia was alleviated by estradiol replacement, a situation similar to the individual differences in efficacy of estrogen replacement on postmenopausal dyspareunia in women.

Because these results indicated that this model could help improve our understanding of mechanisms of postmenopausal dyspareunia, we examined the influence of OVX and OVX + E2 on neural responses of neurons in the gracile nucleus in the brainstem to stimulation of the vaginal canal and other pelvic organs and of the skin of the hindquarters (Bradshaw and Berkley 2003). We found that OVX reduced responses to skin stimulation of the perineum, hip, and tail (but not foot or leg) and that these responses were partially restored by estradiol replacement. Importantly moreover, OVX enhanced excitatory responses to vaginal distention, and estradiol replacement reversed this effect. Thus, it seems likely that the activity of gracile nucleus neurons contributes to the effects of surgically induced reproductive senescence and estrogen replacement on vaginal nociception.

Endometriosis-Induced Dyspareunia

Endometriosis is defined by the presence of ectopic uterine tissue. The condition is often associated with dyspareunia and vaginal hyperalgesia and other pains (Stratton and Berkley 2011). To improve understanding of mechanisms underlying this hyperalgesia, a rat model of endometriosis (surgically induced) that had originally been developed in 1985 to assess mechanisms of endometriosis-associated reduced fertility/fecundity (Vernon and Wilson 1985) was replicated in our laboratory and adapted to study endometriosis-associated pains. As described in detail in another entry (Figs. 1 and 2 in Berkley and Dmitrieva, Endometriosis essay), this model has been used extensively and translationally to reveal a number of peripheral and central neural mechanisms, particularly sprouting of sensory and sympathetic fibers into the abnormal endometrial growths that can trigger central changes.

Endometriosis-Induced, Abdominal Muscle-Referred Vaginal Hyperalgesia

One of the symptoms associated with endometriosis in women is abdominal muscle pain (spontaneous pain, muscle hyperalgesia/tenderness, or muscle cramping), some of which could be referred from pelvic gynecological organs. Such referral to abdominal muscles also occurs with pathophysiology in other pelvic organs such as the colon or bladder (Giamberardino and Vecchiet 1995). Ness and colleagues (Ness and Gebhart 1988; Ness et al. 2001) developed an indirect means to assess visceral pain referred to muscles from the colon by measuring the reflex responses of abdominal muscles (the "VMR") to distention of the colon. Later studies extended use of the VMR to the bladder and to conditions of bladder or colon pathophysiology (by Al Chaer, Visceral Pain Model, Irritable Bowel Syndrome Model, and by Ness, Visceral Pain Model, Urinary Bladder Pain [irritants or Distension]).

Recently, we extended this model further to assess the effects of endometriosis on abdominal muscle reflexes induced by vaginal distention. We measured the VMR to vaginal distention in rats with endometriosis surgery (ENDO)

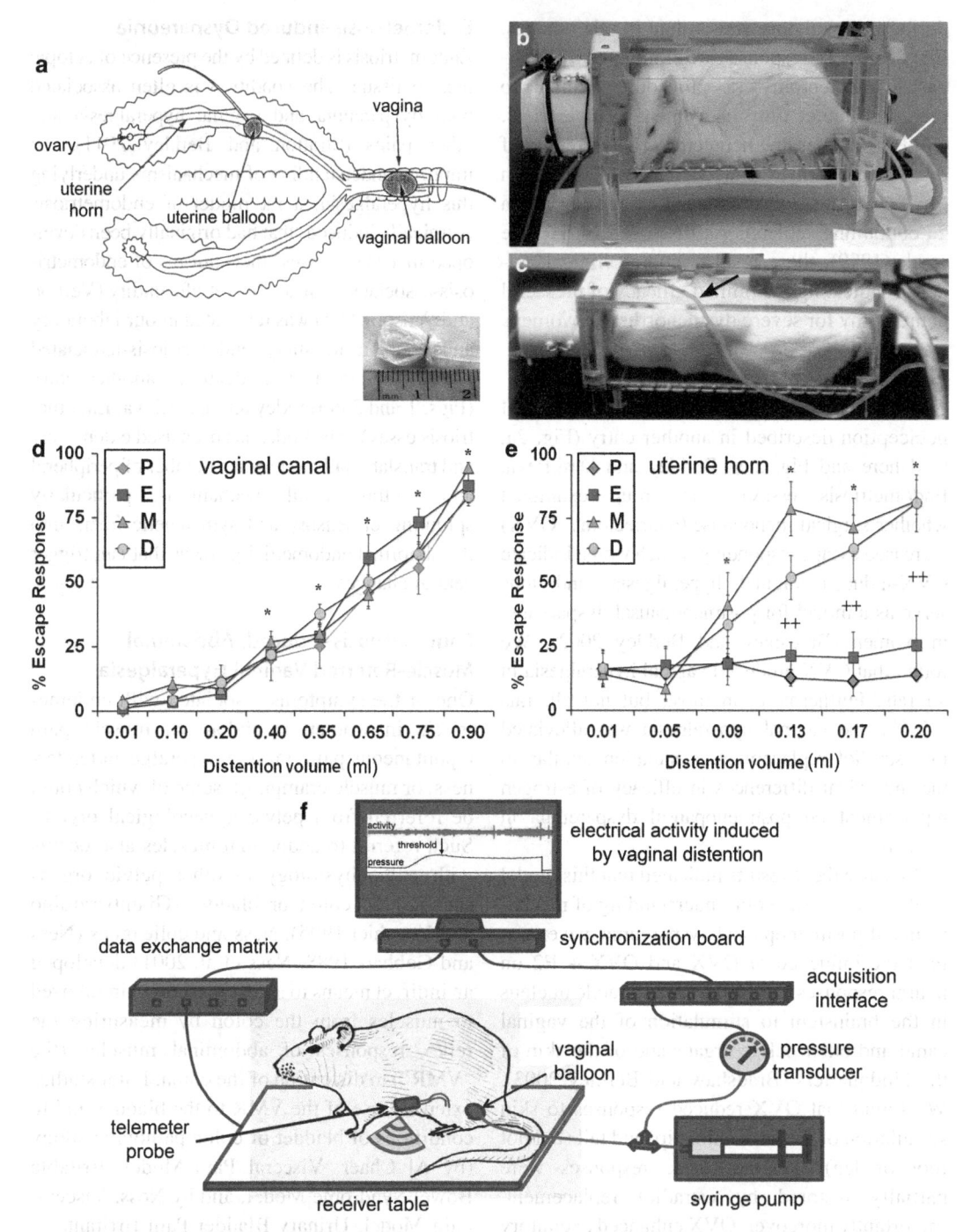

Visceral Pain Models, Female Reproductive Organ Pain, Fig. 2 (**a**) Diagram showing location of distensible balloons placed in either the rat uterus or vagina. The inset is a close-up photograph of the vaginal balloon. (**b**, **c**) Photographs showing rat in testing chamber. To escape a stimulus, the rat must extend its head or nose into the white tube at the front of the chamber. Vaginal or uterine nociception is quantified by measuring the probability of an escape response at different volumes of distention (**d**, **e**), respectively. (**b**) This photo shows a rat in the chamber not making an escape response. This rat has an uninflated balloon in its vaginal canal (note *blue* catheter exiting her vaginal canal, *white arrow*). (**c**) This photo shows a rat making an escape response to distention of

compared with rats with a sham surgery (shamENDO) and with naïve rats. Initially, we found, in lightly anesthetized rats, that ENDO reduced VMR thresholds to vaginal distention compared to the other groups and that this effect was greater in proestrus (estradiol levels high) than estrus (estradiol levels low) (Nagabukuro and Berkley 2007; Dmitrieva et al. 2012). We then found, in unanesthetized, but tethered, rats, that the ENDO-induced effects on the vaginally triggered VMR were alleviated by treatment with a cannabinoid receptor agonist (Dmitrieva et al. 2010). We then improved the method further by developing a technique to measure the abdominal muscle activity in untethered rats via telemetry (Fig. 2f) and showing as proof of principle that the abdominal muscle-referred vaginal hyperalgesia assessed in this manner in ENDO rats could be reduced by indomethacin (Dmitrieva et al. 2012).

Cross System, Viscero-visceral Interactions (Also Called "Cross Talk")

One clinical feature of common gynecological pains, e.g., severe dysmenorrhea or dyspareunia, is that they often co-occur with chronic painful disorders associated with other functional systems such as the urinary tract (interstitial cystitis, ureteral/kidney calculosis), gastrointestinal (irritable bowel disorder), musculoskeletal (fibromyalgia), orofacial (migraine headaches), and more. Mechanisms underlying such comorbidity are poorly understood. Despite the high prevalence of dysmenorrhea (Berkley and McAllister 2011), and while studies on pelvic organ cross talk in general are

increasing, few of the "cross talk" models include gynecological pain.

In one attempt to address this lack, our group (Winnard et al. 2006) inflamed either the colon, the bladder, or one uterine horn with turpentine in anesthetized rats and assessed the effects of this inflammation in one organ on signs of neurogenic inflammation (EB dye extravasation; Fig. 3c) in the otherwise uninflamed other two organs. We studied this outcome in different estrous stages. We found that indirect "cross organ" effects occurred in all three organs. For example, colon inflammation with turpentine gave rise to increased extravasation in both the otherwise healthy bladder and mid-uterine horn, but only during the estrous stage when estradiol levels were high (Fig. 3a, b). Furthermore, the cross system effects were eliminated by prior hypogastric neurectomy. From the perspective of gynecological pain, this result suggests that the "source" of some gynecological pains (such as the uterus) could arise from the influence of pathophysiology in other pelvic organs (e.g., the colon), particularly when estradiol levels are high, and that mechanisms most likely involve central (CNS) interactions.

In another study, two models were combined: the endometriosis model discussed above with a second rat model of ureteral calculosis (stone) (Fig. 3d-f). In this second model, rats are videotaped for seven days after implanting an artificial stone into the left ureter, and their pain behaviors quantified (Giamberardino et al. 1995). The combined model involved several steps: First the endometriosis surgery (or the control surgery) was performed and the rat's videotaped

---

Visceral Pain Models, Female Reproductive Organ Pain, Fig. 2 (continued) a balloon that had been implanted into one uterine horn (note *blue* catheter that had been threaded from the uterus under the skin to exit at the nape of neck, *black arrow*) (**a–c** are modified from Berkley et al. 1995; Bradshaw and Berkley 2002; **d** and **f** are modified from Bradshaw et al. 1999). (**f**) Diagram of apparatus for telemetric measurement of abdominal muscle activity in response to vaginal distention (Modified from Dmitrieva et al. 2012). The vaginal balloon is distended by an infusion pump. The pressure

generated by this infusion is recorded by a pressure transducer. The abdominal muscle electrical activity in response to this distention is sent via electrode wires surgically-implanted in the muscle to the telemetric probe that is implanted subcutaneously. The muscle electrical activity is converted into radio signals by the probe and transmitted to a receiver connected to the acquisition module and synchronized with the incoming signal from the pressure transducer. The equipment used is the Ponemah Telemetric System (DSI, St. Paul, MN, USA)

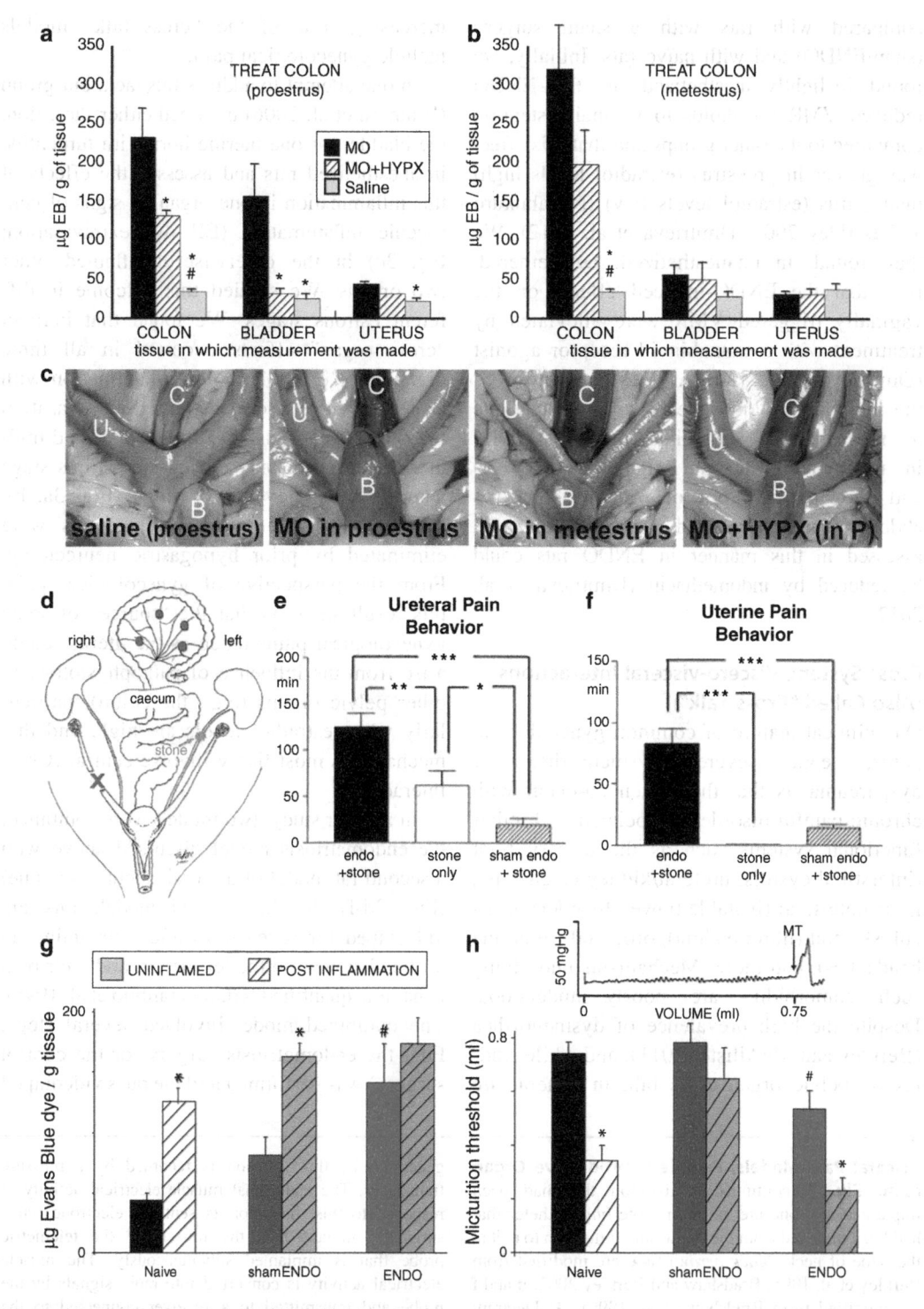

Visceral Pain Models, Female Reproductive Organ Pain, Fig. 3 Models of "cross-talk" involving gynecological organs. (**a–c**) These panels illustrate cross-talk between colon, bladder and uterus (The figures are

modified and adapted from Winnard et al. 2006). (**a**) This graph shows that, when the rat is in proestrus (high estradiol levels), treatment of the colon with mustard oil (*MO*; *solid bars*), when compared to treatment of the

pain behaviors (if any) were quantified. About 3 weeks later, a ureteral stone was implanted and the rat's videotaped pain behaviors quantified again. The results were then compared with pain behaviors after only a stone had been implanted. Vocalization thresholds to stimulation of the oblique musculature (L1 dermatome) were also measured. The results showed that whereas endometriosis increased pain behaviors and referred muscle hyperalgesia associated with a ureteral stone, the control surgical procedures *reduced* them. We called this latter effect "silent stones."

Another approach was developed to assess gynecological and urinary tract cross talk. Here, the impact of endometriosis on bladder contractility and neurogenic inflammation (Evans blue dye extravasation) in otherwise healthy bladder as well as the inflamed bladder (turpentine) was studied in three groups of anesthetized rats, ENDO, shamENDO, and Naïve (Morrison et al. 2006) (Fig. 3f–g). The endometriosis surgery

increased in dye extravasation (Fig. 3f) and produced bladder hyperactivity (Fig. 3g) in the otherwise healthy bladder. Although bladder inflammation increased extravasation in the bladders of all three groups (Fig. 3f), it only increased bladder contractility in the Naïve and endometriosis groups (Fig. 3g). Thus, once again, the sham surgery somehow *prevented* the occurrence of inflammation-induced bladder hyperactivity. We called this effect "silent bladder inflammation."

As discussed in both articles, the findings indicate the existence of cross system interactions within the CNS that can produce powerful effects on symptoms, not only to exacerbate them but also, importantly, to reduce them. This situation has obvious implications for diagnosis as well as therapy and warrants further study not only in animal models but also in the clinic, i.e., translational research in both directions: not only from animal to clinic but also from clinic to animal.

Visceral Pain Models, Female Reproductive Organ Pain, Fig. 3 (continued) colon with saline (*grey bars*), increases EB dye extravasation in the otherwise untreated bladder and uterus. Hypogastric neurectomy (*hatched bars*) reduces this effect in the bladder and eliminates it in the uterus. (**b**) This graph shows that this cross-talk does not occur in metestrus, when estradiol levels are low. (**c**) These photos illustrate examples of the effects in the graphs. *C* colon, *B* bladder, *U* uterine horn. (**d–f**) These panels illustrate the cross-talk effects of combining endometriosis and ureteral stone surgeries on "ureteral" and "uterine" pain behaviors. Ureteral pain behaviors are readily distinguishable from uterine behaviors. Uterine pain behaviors are illustrated in Fig. 1b (The figures are adapted from Giamberardino et al. 2002). (**d**) This diagram shows the surgical model. (**e** and **f**) These graphs show that the amount of time that the rats engage in either ureteral or uterine pain behaviors is increased in rat with both endometriosis surgery and a stone (*black bar*) compared with rats that have only a ureteral stone (*clear bar*). In contrast, when the sham endometriosis surgery is combined with the stone (*hatched bars*), ureteral pain behaviors are reduced and uterine pain behaviors are eliminated compared with stone only. (**g** and **h**) These graphs and the inset illustrate the cross-talk-like effects that occur when the influence of endometriosis surgery on

reflexes and EB dye extravasation in the otherwise healthy or turpentine-inflamed bladder is assessed (in proestrus). (**g**) By comparing the *solid* (uninflamed) and *hatched bars* (inflamed), this graph shows that inflaming the bladder increases dye extravasation in all three groups: Naïve, sham endometriosis surgery (shamENDO) and endometriosis surgery (ENDO). The graph also shows that EB dye extravasation is increased in the otherwise healthy bladder of ENDO rats (*solid red bar*) compared with bladders in Naïve or shamENDO rats (*solid black, blue bars*, respectively). (**h**) The *inset* is an example measurement of a bladder reflex. The bladder is infused via the urethra with water, and the volume at which the bladder contracts (in this case, 0.75 ml), is designated as the micturition threshold (*MT*). Reduction in the MT indicates hyperreflexivity. The graph shows that ENDO, but not shamENDO induced bladder hyper-reflexivity (compare *solid bars*). In contrast, Whereas MTs were reduced in the inflamed bladders of Naïve and ENDO rats, this reduction failed to occur in shamENDO rats. Together sections (**d–f**) and (**g, h**) show that whereas ENDO can increase pain or indications of pathophysiology associated with urinary tract organs (ureter, bladder), the shamENDO surgery somehow prevents or reduces the impact of pathophysiology on those organs. These phenomena were referred to as "silent stones" and "silent bladder inflammation"

Summary

Challenges for the Future

Although a number of models of gynecological pain have been developed, and some have had an impact clinically and improved our understanding of mechanisms (e.g., see Stratton and Berkley 2011), more models of gynecological pains are needed, particularly for the most common of clinical problems such as dysmenorrhea, ovarian cysts, gynecological cancers, mastitis (painful inflammation of the breast), and postmastectomy pain. Of value would be concurrent development of visceral pains unique to males (e.g., prostatitis, testicular pain, etc.) because comparisons between females and males could produce meaningful insights. Although the situation is improving, also needed are studies that incorporate into their design analyses of how reproductive and hormonal status can influence signs/symptoms. Of clear value will be more studies aimed at understanding how interactions across organ systems can influence pain symptoms and therapy. And finally, of perhaps greatest importance is that there are currently no models of clinical conditions that, surprisingly, *fail to produce pain* when in fact the pain could save lives, e.g., early stage gynecological cancers.

Acknowledgments Studies cited from the authors' laboratory were supported by NIH grant RO1 NS 11892, with the exception of Dmitrieva et al. 2012, which was supported by Bayer HealthCare Pharmaceuticals.

References

Berkley, K. J., & McAllister, S. L. (2011). Don't dismiss dysmenorrhea! *Pain, 152*, 1940–1941.

Berkley, K. J., Wood, E., Scofield, S. L., & Little, M. (1995). Behavioral responses to uterine or vaginal distension in the rat. *Pain, 61*, 121–131.

Berkley, K. J., McAllister, S. L., Accius, B. E., & Winnard, K. P. (2007). Endometriosis-induced vaginal hyperalgesia in the rat: Effect of estropause, ovariectomy, and estradiol replacement. *Pain, 132*(Suppl 1), S150–S159.

Bradshaw, H. B., & Berkley, K. J. (2002). Estrogen replacement reverses ovariectomy-induced vaginal hyperalgesia in the rat. *Maturitas, 41*, 157–165.

Bradshaw, H. B., & Berkley, K. J. (2003). The influence of ovariectomy with or without estrogen replacement on responses of rat gracile nucleus neurons to stimulation of hindquarter skin and pelvic viscera. *Brain Research, 986*, 82–90.

Bradshaw, H. B., Temple, J. L., Wood, E., & Berkley, K. J. (1999). Estrous variations in behavioral responses to vaginal and uterine distention in the rat. *Pain, 82*, 187–197.

Cason, A. M., Samuelsen, C. L., & Berkley, K. J. (2003). Estrous changes in vaginal nociception in a rat model of endometriosis. *Hormones and Behavior, 44*, 123–131.

Dmitrieva, N., Nagabukuro, H., Resuehr, D., Zhang, G., McAllister, S. L., McGinty, K. A., Mackie, K., & Berkley, K. J. (2010). Endocannabinoid involvement in endometriosis. *Pain, 151*, 703–710.

Dmitrieva, N., Faircloth, E. K., Pyatok, S., Sacher, F., & Patchev, V. (2012). Telemetric assessment of referred vaginal hyperalgesia and the effect of indomethacin in a rat model of endometriosis. *Frontiers in Pharmacology, 3*, 158.

Farmer, M. A., Taylor, A. M., Bailey, A. L., Tuttle, A. H., MacIntyre, L. C., Milagrosa, Z. E., Crissman, H. P., Bennett, G. J., Ribeiro-da-Silva, A., Binik, Y. M., & Mogil, J. S. (2011). Repeated vulvovaginal fungal infections cause persistent pain in a mouse model of vulvodynia. *Science Translational Medicine, 3*, 101ra91.

Giamberardino, M. A., & Vecchiet, L. (1995). Visceral pain, referred hyperalgesia and outcome: New concepts. *European Journal of Anaesthesiology, Supplement, 10*, 61–66.

Giamberardino, M. A., Valente, R., de Bigontina, P., & Vecchiet, L. (1995). Artificial ureteral calculosis in rats: Behavioural characterization of visceral pain episodes and their relationship with referred lumbar muscle hyperalgesia. *Pain, 61*, 459–469.

Giamberardino, M. A., Berkley, K. J., Affaitati, G., Lerza, R., Centurione, L., Lapenna, D., & Vecchiet, L. (2002). Influence of endometriosis on pain behaviors and muscle hyperalgesia induced by a ureteral calculosis in female rats. *Pain, 95*, 247–257.

Gintzler, A. R., & Komisaruk, B. R. (1991). Analgesia is produced by uterocervical mechanostimulation in rats: Roles of afferent nerves and implications for analgesia of pregnancy and parturition. *Brain Research, 566*, 299–302.

Gintzler, A. R., Peters, L. C., & Komisaruk, B. R. (1983). Attenuation of pregnancy-induced analgesia by hypogastric neurectomy in rats. *Brain Research, 277*, 186–188.

Li, B., Tong, C., & Eisenach, J. C. (2008). Pregnancy increases excitability of mechanosensitive afferents innervating the uterine cervix. *Anesthesiology, 108*, 1087–1092.

Li, Y., Zhang, S. F., Zou, S. E., Xia, X., & Bao, L. (2011). Accumulation of nerve growth factor and its receptors in the uterus and dorsal root ganglia in a mouse model of adenomyosis. *Reproductive Biology and Endocrinology, 9*, 30.

McAllister, S. L., Dmitrieva, N., & Berkley, K. J. (2012). Sprouted innervation into uterine transplants contributes to the development of hyperalgesia in a rat model of endometriosis. *PLoS One, 7*, e31758.

Morrison, T. C., Dmitrieva, N., Winnard, K. P., & Berkley, K. J. (2006). Opposing viscerovisceral effects of surgically induced endometriosis and a control abdominal surgery on the rat bladder. *Fertility and Sterility, 86*(4 Suppl), 1067–1073.

Nagabukuro, H., & Berkley, K. J. (2007). Influence of endometriosis on visceromotor and cardiovascular responses induced by vaginal distention in the rat. *Pain, 132*(Suppl 1), S96–103.

Ness, T. J., & Gebhart, G. F. (1988). Colorectal distension as a noxious visceral stimulus: Physiologic and pharmacologic characterization of pseudaffective reflexes in the rat. *Brain Research, 450*, 153–169.

Ness, T. J., Lewis-Sides, A., & Castroman, P. (2001). Characterization of pressor and visceromotor reflex responses to bladder distention in rats: Sources of variability and effect of analgesics. *Journal of Urology, 165*, 968–974.

Sandner-Kiesling, A., Pan, H. L., Chen, S. R., James, R. L., DeHaven-Hudkins, D. L., Dewan, D. M., & Eisenach, J. C. (2002). Effect of kappa opioid agonists on visceral nociception induced by uterine cervical distension in rats. *Pain, 96*, 13–22.

Shin, S. W., & Eisenach, J. C. (2003). Intrathecal morphine reduces the visceromotor response to acute uterine cervical distension in an estrogen-independent manner. *Anesthesiology, 98*, 1467–1471.

Shin, S. W., Sandner-Kiesling, A., & Eisenach, J. C. (2003). Systemic, but not intrathecal ketorolac is antinociceptive to uterine cervical distension in rats. *Pain, 105*, 109–114.

Stratton, P., & Berkley, K. J. (2011). Chronic pelvic pain and endometriosis: Translational evidence of the relationship and implications. *Human Reproduction Update, 17*, 327–346.

Temple, J., Bradshaw, H. B., Wood, E., & Berkley, K. J. (1999). Effects of hypogastric neurectomy on escape responses to uterine distention in the rat. *Pain, 6*, S13–20.

Tong, C., Ma, W., Shin, S. W., James, R. L., & Eisenach, J. C. (2003). Uterine cervical distension induces cFos expression in deep dorsal horn neurons of the rat spinal cord. *Anesthesiology, 99*, 205–211.

Tong, C., Conklin, D., Clyne, B. B., Stanislaus, J. D., & Eisenach, J. C. (2006). Uterine cervical afferents in thoracolumbar dorsal root ganglia express transient receptor potential vanilloid type 1 channel and calcitonin gene-related peptide, but not P2X3 receptor and somatostatin. *Anesthesiology, 104*, 651–657.

Vernon, M. W., & Wilson, E. A. (1985). Studies on the surgical induction of endometriosis in the rat. *Fertility and Sterility, 44*, 684–694.

Wesselmann, U., & Lai, J. (1997). Mechanisms of referred visceral pain: Uterine inflammation in the adult virgin rat results in neurogenic plasma extravasation in the skin. *Pain, 73*, 309–317.

Wesselmann, U., Czakanski, P. P., Affaitati, G., & Giamberardino, M. A. (1998). Uterine inflammation

as a noxious visceral stimulus: Behavioral characterization in the rat. *Neuroscience Letters, 246*, 73–76.

Winnard, K. P., Dmitrieva, N., & Berkley, K. J. (2006). Cross-organ interactions between reproductive, gastrointestinal, and urinary tracts: Modulation by estrous stage and involvement of the hypogastric nerve. *American Journal of Physiology - Regulatory, Integrative and Comparative Physiology, 291*, R1592–R1601.

Yan, T., Liu, B., Du, D., Eisenach, J. C., & Tong, C. (2007). Estrogen amplifies pain responses to uterine cervical distension in rats by altering transient receptor potential-1 function. *Anesthesia and Analgesia, 104*, 1246–1250.

Visceral Pain Pathway

▶ Postsynaptic Dorsal Column Projection, Functional Characteristics

Visceral Referred Pain

James M. Borowczyk
Department of Orthopaedics and
Musculoskeletal Medicine, University of Otago,
Christchurch, Christchurch, New Zealand

Synonyms

Referred pain; Visceral pain

Definition

Referred pain is pain perceived in a region innervated by nerves other than those that innervate the source of the pain (Merskey and Bogduk 1994).
▶ Visceral referred pain is explicitly ▶ visceral nociception and pain that becomes referred.

Characteristics

Visceral pain is not perceived locally. Viscera are diffusely innervated, and the central pathways of visceral pain are poorly organized somatotopically. This means that the nervous system is not "wired" in a manner to localize visceral pain accurately. Consequently, patients with visceral pain are not aware of its precise source. Instead, they experience a sensation whose location is attributed within and throughout the territory subtended by the same spinal cord segment, or segments, that innervate the source of pain. Typically, the pain is perceived in the body wall surrounding the viscus or in other territories with the same segmental innervation. As this perception is in an area remote from the viscus, it has been interpreted as an example of ▶ referred pain.

The viscera are innervated by the thoracolumbar outflow of the sympathetic nervous system, between spinal cord segments T1 and L2. Their segmental innervation is established early during development, before the viscera are fully differentiated and before they assume their final topographic positions. In this state, the systematic pattern of innervation is apparent (Fig. 1). Within this pattern, the progenitors of proximal viscera receive their innervation from proximal segments, and progressively more caudal viscera are innervated by progressively more caudal segments. The gastrointestinal tract is innervated by segments T1 to L2. The heart is innervated by T1–4, and the urinary system by T11 to L2.

As the viscera grow and the gastrointestinal tract rotates and migrates, the viscera draw their nerve supply with them. The final location of viscera bears little resemblance to their original, embryonic location. Consequently, the segmental innervation of the viscera is dissonant with the innervation of those parts of the chest and abdominal wall under which they lie (Fig. 2). For example, in the upper abdomen lie viscera innervated by T5, T6, T6–9, and T11–12.

The site to which visceral pain is referred is dictated by its segmental innervation. It will be perceived in those segments of the body wall that have the same segmental innervation as the viscus (Fig. 3). Thus, gastric pain is perceived in the epigastrium, not because the stomach lies in the epigastrium but because the abdominal wall of the epigastrium is innervated by the same

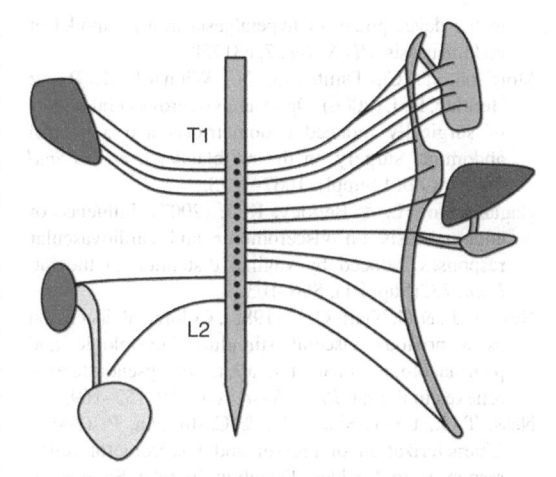

Visceral Referred Pain, Fig. 1 A sketch of the systematic pattern of innervation of the viscera during early embryonic development. From proximal to distal sites, the gastrointestinal tract progressively receives its innervation from segments T1 to L2

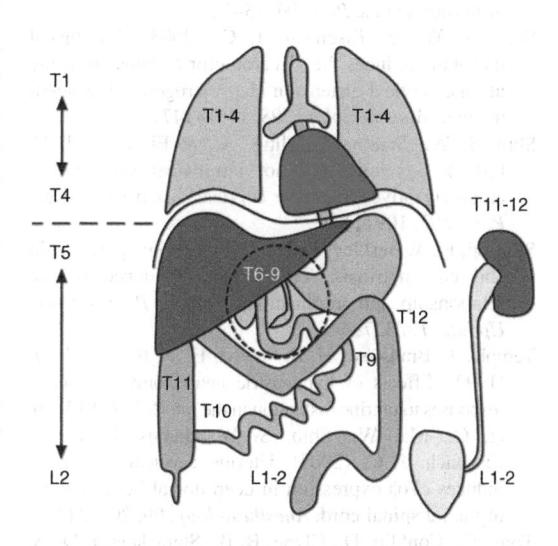

Visceral Referred Pain, Fig. 2 A sketch of the relative topographic disposition of the thoracic and abdominal viscera, with the segmental innervation of each organ labeled. Above the diaphragm, all the thoracic viscera are innervated by T1–4. Below the diaphragm, the abdominal viscera are serially innervated from T5 to L2, along the length of the gastrointestinal tract

segments (T5, T6) that innervate the stomach. The transverse colon may lie in the epigastrium, but its segmental innervation is T11. Its pain is therefore perceived in the lower abdominal wall. Pain from the urinary tract is felt in the loin and

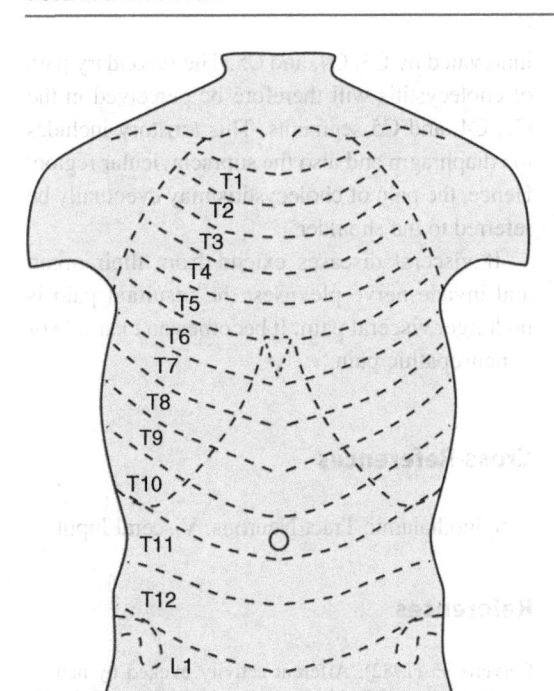

Visceral Referred Pain, Fig. 3 The segmental pattern of innervation of the chest and abdominal wall. Pain will be referred from viscera with the same segmental innervation to these areas

groin, because the abdominal wall spanning these regions is innervated by T12 and L1. Pain from the appendix is perceived in the umbilical region, because this region is innervated by T10. Cardiac pain is perceived in the central chest wall, because the central chest is innervated by T1–4.

Referral to the anterior chest wall, and to the anterior abdominal wall, is the classical pattern of behavior of visceral pain with which most physicians will be familiar, but other patterns of referral may occur. For example, viscera may refer pain posteriorly, where it is perceived in regions of the spine with the same segmental innervation of the viscus. However, although such patterns of referral are known by anecdote and experience, no studies have formally addressed its incidence. Nevertheless, posterior referral of visceral pain is an important differential diagnosis of spinal pain. Cardinal examples include uterine pain and the pain of abdominal aortic aneurysm presenting as low back pain.

At a neurological level, the mechanism underlying pain referral is generally held to be due to convergence. Afferents from viscera and from the body wall converge onto common neurons in the dorsal horn. Evidence for this was first presented in 1968 (Pomeranz et al. 1968), and further evidence was presented in the 1980s (Cervero 1983). This convergence occurs mainly at the spinal level, but also occurs at higher levels, such as the thalamus and even in the cortex. Some 50 % of all neurons at the termination of the spinothalamic tract in the thalamus can be excited by both cutaneous and noxious visceral input.

Due to convergence, the brain cannot determine the exact peripheral origin of information arriving to it from the spinal cord. It cannot determine if the information was initiated in an organ or in that part of the body wall with the same segmental innervation as the organ. Being more accustomed to stimuli from the body wall, the brain attributes the origin of visceral pain to the body wall.

Convergence also allows for referred hyperalgesia. If dorsal horn neurons are facilitated by ongoing nociceptive input from an organ, areas of the skin supplied by the same spinal segment may exhibit hyperalgesia or hyperesthesia. This explains certain phenomena that, in the surgery literature, were classically described as "oddities" among the features of visceral disease. A good example is "Boas" sign, where superficial cutaneous hyperalgesia may be found on the posteroinferior chest wall and flank in acute cholecystitis (Cervero 1982).

Reflex Muscle Contraction

An accompanying feature of visceral pain is reflex muscle contraction. In response to pain in an abdominal organ, muscles of the anterolateral abdominal wall contract. The particular muscles involved depend on both the spinal segments supplying the viscera and the intensity of the stimulus. The degree of muscle spasm ranges from mild to intense and if prolonged can cause muscle pain and prolonged muscle tenderness (McLellan and Goodell 1943). If the noxious stimulus is extreme, such as in pancreatitis, muscles above and below the initially involved spinal segments may also be recruited. This muscle spasm also prolongs and aggravates the initial condition and may lead to ongoing or chronic pain that confounds further diagnosis and therapy (Zimmerman and Mayer-Rivera 1979). Associated phenomena include deep hyperalgesia and tenderness that may outlast the actual muscle spasm.

Distinctions

Visceral pain and its referral must be distinguished from other types of pain that may be associated with disorders of viscera. Some disorders may initially affect the organ intrinsically, but subsequently they extend to involve tissues adjacent to the organ. That secondary involvement of adjacent structures can become a second source of pain. Typical examples include the spread of infection or inflammation to involve the parietal peritoneum of the abdomen, inflammation or hemorrhage over the posterior abdominal wall, and invasion of visceral nerve plexuses by spreading tumors.

These secondary sources of pain have their own mechanisms and patterns of behavior, distinct from those of intrinsic visceral pain. Inflammation may be felt locally or may itself become referred. This underlies classical examples of so-called visceral pain, but becomes examples of primary and secondary visceral pain, when properly understood.

Intrinsic pain from the appendix will be perceived in the central abdominal wall, because the appendix and the umbilical region are both innervated by T10. If, and once, the inflammation spreads to involve the overlying parietal peritoneum, a new (second and additional) pain occurs. It is felt locally in the iliac region of the abdominal wall, under which the appendix lies. Technically, this latter pain is not pain of the appendix, but pain caused by the inflammation of the appendix spreading to involve the abdominal wall. Similarly, an inflamed retrocaecal appendix may irritate the posterior abdominal wall and present with pain in the back.

The inflammation of cholecystitis may spread to involve the parietal peritoneum of the diaphragm. In this condition, the intrinsic pain of the organ will be perceived in the epigastric region, because it is innervated by the same segments (T6–8) that innervate the gall bladder (T6–9), but the diaphragm is

innervated by C3, C4, and C5. The secondary pain of cholecystitis will therefore be perceived in the C3, C4, and C5 segments. This territory includes the diaphragm and also the supraclavicular region. Hence, the pain of cholecystitis may eventually be referred to the shoulder.

If visceral diseases extend from their organ and invade nerve plexuses, the resultant pain is no longer visceral pain. It becomes an example of ▶ neuropathic pain.

Cross-References

▶ Spinothalamic Tract Neurons, Visceral Input

References

Cervero, F. (1982). Afferent activity evoked by natural stimulation of the biliary system in the ferret. *Pain, 13*, 137–151.

Cervero, F. (1983). Mechanisms of visceral pain. In S. Lipton (Ed.), *Persistent pain* (pp. 1–19). New York: Grune and Stratton.

McLellan, A., & Goodell, H. (1943). Pain from the bladder, ureter and kidney pelvis. *Proceedings Association for Research in Nervous and Mental Disease, 23*, 252–262.

Merskey, H., & Bogduk, N. (1994). *Classification of chronic pain. Descriptions of chronic pain syndromes and definitions of pain terms* (2nd ed., pp. 11–16). Seattle: IASP Press.

Pomeranz, B., Wall, P. D., & Weber, W. V. (1968). Cord cells responding to fine myelinated afferents from viscera, muscle and skin. *Journal de Physiologie, 199*, 511–532.

Zimmerman, M., & Mayer-Rivera, F. (1979). Peripheral and central nervous mechanisms of nociception, pain and pain therapy: Facts and hypotheses. In J. J. Bonica (Ed.), *Advances in pain research and therapy* (pp. 3–32). New York: Raven.

Visceral Representation

▶ Thalamus, Visceral Representation

Visceral Sensation

▶ Angina Pectoris, Neurophysiology and Psychophysics

Visceral Spinohypothalamic Tract (See Spinohypothalamic Neurons)

▶ Spinal Ascending Pathways, Colon, Urinary Bladder, and Uterus

Visceral Spinomedullary Tract (See Spinomedullary Neurons)

▶ Spinal Ascending Pathways, Colon, Urinary Bladder, and Uterus

Visceral Spinomesencephalic Tract (See Spinomesencephalic Neurons)

▶ Spinal Ascending Pathways, Colon, Urinary Bladder, and Uterus

Visceral Spinoreticular Tract

▶ Spinal Ascending Pathways, Colon, Urinary Bladder, and Uterus

Visceral Spinothalamic Tract

▶ Spinal Ascending Pathways, Colon, Urinary Bladder, and Uterus

Visceral Sympathetic Blocks

Definition

Blocking conduction in nerves to and from internal organs, commonly by injection into a pre-vertebral ganglion (e.g., celiac ganglion for cancer pain management). Time-limited blocks are produced by local anesthetic injections, irreversible blocks by injection of phenol, alcohol, etc. to destroy the nerves. The term "sympathetic" relates to the anatomical association of visceral afferent (sensory) fibers with the efferent sympathetic (motor) innervation.

Cross-References

▶ Cancer Pain Management, Anesthesiologic Interventions, and Neural Blockade

Visceralgia

Definition

Pain may arise under certain conditions from the viscera, the organs of splanchnic origin.

Cross-References

▶ Animal Models and Experimental Tests to Study Nociception and Pain
▶ Visceral Pain Model, Pancreatic Pain

Visceromotor Reflex

Definition

Visceromotor reflex is a reflex response to noxious stimulation of a visceral organ. An example is the reflex contraction of the external abdominal oblique muscle in response to colorectal distention. In the case of colon or bladder distension, the abdominal muscles will contract. The visceromotor response requires a spino-bulbar-spinal loop, distinguishing it from a simple nociceptive withdrawal reflex organized within the spinal cord. The reflex may be recorded by electromyography.

Cross-References

- ► Hypothalamus and Nociceptive Pathways
- ► Nocifensive Behaviors, Gastrointestinal Tract
- ► Nocifensive Behaviors of the Urinary Bladder
- ► Spinal Dorsal Horn Pathways, Dorsal Column (Visceral)
- ► Visceral Pain Model, Lower Gastrointestinal Tract Pain
- ► Visceral Pain Model, Urinary Bladder Pain (Irritants or Distension)

Visceroreception

Definition

Nociceptive awareness of visceral structures may be referred to as visceroreception. Visceral afferents are normally silent, but sensitization of afferent input can lead to heightened awareness or increased responsivity to visceral afferent input.

Cross-References

- ► Postsynaptic Dorsal Column Projection, Anatomical Organization

Viscerosensation

Definition

Sensory input derived from the internal organs or viscera.

Cross-References

- ► Amygdala, Pain Processing and Behavior in Animals

Viscerosomatic Pain Syndrome

Definition

A syndrome in which related visceral and body wall pain interact with each other, as interstitial cystitis is related to suprapubic body wall pain.

Cross-References

- ► Chronic Pelvic Pain, Musculoskeletal Syndromes
- ► Myalgia

Viscerosomatic Pain Syndromes

- ► Chronic Pelvic Pain, Musculoskeletal Syndromes

Viscerotome

Definition

Representation on the body surface of visceral innervation, first described by Sir Henry Head (1893). Unlike dermatomes, which are associated with sensory innervation of skin contained in one dorsal root (or trigeminal counterpart) and named by spinal segment (e.g., T1-12, L1-5), viscertomes are associated with sensory innervation of viscera typically contained in multiple dorsal roots. For example, the T-9 viscerotome was represented by Head (Head 1893) over the T-11 and T-12 dermatomes and the T-10 viscerotome over L-1, L-2, and L-3 dermatomes. Viscerotomes have recently regained significance in relation to traditional acupuncture back Shu points.

Cross-References

- ► Central Nervous System Stimulation for Pain

References

Head, H. (1893). On disturbances of sensation with especial reference to the pain of visceral disease. *Brain, 16*, 1–133.

Viscero-Visceral Hyperalgesia

Synonyms

VVH

Definition

Phenomenon of visceral pain symptom enhancement due to the interaction between algogenic conditions from internal organs that share at least part of their central sensory projection, e.g., heart and gallbladder (T5), uterus and colon (T10-L1), and uterus and upper urinary tract (T10-L1).

Cross-References

▶ Muscle Pain, Referred Pain

Viscoelastic Elements

Definition

Molecules that exhibit properties combining the elastic behavior of a gel and the viscous behavior of a solution.

Cross-References

▶ Perireceptor Elements

Viscosupplementation

▶ Hyaluronan

Visual Analog Scale

Synonyms

VAS

Definition

The visual analog scale is often used for ratings of pain intensity in experimental studies. It is a scale that is comprised of a horizontal or vertical line and anchored at both ends by words indicative of extremes of magnitude, such as "no pain" to "most intense pain sensation imaginable." For children, a 10-cm vertical VAS is typically used to represent a continuum from no pain at all (at the bottom) to the worst pain imaginable (at the top of the VAS scale). To enhance comprehension in children, the VAS may include visual cues. For example, the VAS may employ color cues, graded from white at the bottom to dark red at the top, as well as a neutral face at the bottom and a negative facial expression at the top.

Cross-References

▶ Central Pain, Outcome Measures in Clinical Trials
▶ Experimental Pain in Children
▶ Magnetoencephalography in Assessment of Pain in Humans
▶ Pain in Humans, Psychophysical Law
▶ Quantitative Sensory Testing

Visual Evoked Potentials

Definition

The retina is activated by visual stimuli such as stroboscopic flashes or an alternating checkered board pattern. A signal is then obtained using electroencephalogram electrodes over the occipital region.

Cross-References

▶ Metabolic and Nutritional Neuropathies

Vitamin Neuropathy

▶ Metabolic and Nutritional Neuropathies

VLO

▶ Ventrolateral Orbital Cortex

VM, VMb

▶ Basal Ventral Medial Nuclei

VMpo

▶ Posterior Part of the Ventromedial Nucleus
▶ Ventral Medial Nucleus

Vocalization

Definition

Vocalization is a centrally processed reaction.

Cross-References

▶ Randall-Selitto Paw Pressure Test

Vocalization Threshold to Electrical Muscle Stimulation

Definition

The minimal current intensity (in mA) corresponding to the first vocalization of the animal upon electrical stimulation of a skeletal muscle. Vocalization, being a highly integrated test, is regarded as a reliable index of perceived pain in animals (Le Bars et al. 2001); vocalization thresholds can thus be considered as an equivalent of pain thresholds in humans. In normal rats, electrical muscle thresholds are stable and reproducible over long periods of time (Giamberardino et al. 1990).

In rats with ureteric stones, the vocalization test is adopted to assess changes in sensitivity of the oblique muscles (L1 level) chronically implanted with nickel chrome wires. The stimuli employed are 250 ms trains of 1 ms square waves, frequency 200/s, delivered automatically every 2 s by a constant current electrical stimulator. The intensity of the current is gradually increased in 0.3 mA steps/2 s. Thresholds are measured by the method of the limits. The intensity is increased in 0.3 mA steps until vocalization occurs, then decreased in 0.1 mA steps until it disappears, and then re-increased at the latter rate until vocalization returns, noting the corresponding values in mA. The threshold is calculated as the mean of these values.

Cross-References

▶ Referred Hyperalgesia
▶ Visceral Pain Model, Kidney Stone Pain

References

Le Bars, D., Gozariu, M., & Cadden, S. W. (2001). Animal models of nociception. *Pharmacol Rev, 53*, 597–652.
Giamberardino, M. A., Vecchiet, L., & Albe-Fessard, D. (1990). Comparison of the effects of ureteral calculosis and occlusion on muscular sensitivity to painful stimulation in rats. *Pain, 43*, 227–234.

Vocational Appraisal

▶ Vocational Assessment in Chronic Pain

Vocational Aptitudes

Definition

Vocational aptitudes refer to a client's natural ability, other characteristics, and necessary skills. Aptitude tests assess client's abilities (general learning ability, verbal, numerical, spatial, form perception, clerical perception, motor coordination, finger, and manual dexterity) and are designed to reveal whether the client is suitable for a particular type or course of training or occupation.

Cross-References

▶ Vocational Counseling

Vocational Assessment

▶ Disability, Functional Capacity Evaluations

Vocational Assessment in Chronic Pain

Kurt L. Johnson
Department of Rehabilitation Medicine,
University of Washington, Seattle, WA, USA

Synonyms

Career assessment; Employment assessment;
Vocational appraisal; Vocational evaluation

Definition

Vocational assessment is a dynamic, interactive process of data collection and analysis to assist individuals in determining their preferences, strengths, limitations, and barriers with respect to education and employment.

Characteristics

Among people 45 years of age or younger, back pain is the most common, and most expensive, cause of work-related disability (Deyo and Weinstein 2001). About 50 % of working-age people surveyed report back pain symptoms each year (Sternbach 1986). About 2 % of the work force in the USA may be expected to report a compensable back injury during any given year (Snook 1987; Andersson 1999). Of those, approximately 10 % are limited in their ability to work at 3 months, and these few workers, whose recovery takes longest and suffer the most, use up to 80 % of the total costs of back pain care (Bigos et al. 1986), or 88 % of wage compensation and medical costs (Volinn et al. 1991). For people with chronic low back pain, and those with other chronic musculoskeletal pain without history of traumatic injury or clear objective findings, return to work after extended pain related disability can be very difficult.

Individuals with chronic pain seeking to return to employment face significant obstacles related to their own deconditioning and self-perception of competence and capacity for employment, often in the context of an adversarial relationship with their worker's compensation or other disability insurance carrier, and psychosocial variables such as changes in family roles. Many individuals may need to change careers due to their functional limitations or changes in the labor market. Vocational assessment is a process that provides a framework to assist individuals in understanding their preferences for various aspects of work, strengths and aptitudes, limitations, and barriers they may confront as they seek to reenter employment.

Vocational assessment is a global term describing the appraisal of an individual's entire educational and employment background, and characteristics of temperament, medical issues, critical behaviors, skills, and abilities necessary to obtain and retain employment. A more narrow appraisal of specific occupational interests, specific job skills, would be termed a vocational evaluation, which is now considered one element of vocational assessment (Wesolek and McFarlene 1992). It is not sufficient simply to assess the individual with chronic pain. One must also have a mechanism to evaluate the characteristics of various workplaces. The Minnesota Theory of Work Adjustment (MTWA) (Dawes 1987) provides a scaffold that includes both the individual and the work environment. The MTWA has served as the basis for the way the U.S. Dept. of Labor codes the characteristics of workers and jobs.

Under the MTWA, successful work adjustment is determined by two factors: job satisfaction on the part of the employee, and the satisfactoriness of job performance as perceived by the employer. A person who is able to satisfy his or her needs, in terms of temperament, work values, compensation, and preferred activities, is likely to continue employment. Similarly, if the employer perceives the performance of the employee as satisfactory, and if the economy supports the position, the employment is likely to be continued. The MTWA does not include a focus on factors that may moderate satisfaction and satisfactoriness, nor does it include consideration of cognitive variables, such as perceived competence and self-efficacy, with respect to various elements of employment (Tinsley 1993).

Satisfaction includes a variety of factors that must be assessed. These include preferences for reinforcers including ability utilization, achievement, activity, advancement, authority, company policies and practices, compensation, coworkers, creativity, independence, moral values, recognition, responsibility, security, social service, social status, supervision-human relations, supervision-technical, variety, and working conditions. There are a variety of paper and pencil inventories that can be used to assess subgroups of these elements.

Holland's theory of occupational choice is currently the most widely used, to help people consider the relationship between their career interests and various occupations (Holland 1997). Under Holland's theory, career choice is an extension of one's personality, and vocational personality can be categorized by the degree to which one shares characteristics of six behavioral styles: realistic, investigative, artistic, social service, enterprising, and conventional. The most widely used inventory for career decision making, the Strong Interest Inventory, is based on Holland's model. People who have taken the Strong Interest Inventory can see from the results the pattern of their interests (e.g., their dominant behavioral style or styles), how their interests compare with men and women in general, and the degree to which their interests conform to those in large criterion groups in a variety of occupations. Holland contends that when people share interest patterns or vocational personality styles with others in the occupation, they are more likely to be satisfied and successful. The Strong is primarily geared toward respondents with some college education, but other similar instruments, such as the Career Assessment Inventory, are appropriate for respondents with high school or vocational technical training.

Job Satisfactoriness involves those attributes of the person with chronic pain that are necessary for general employability and specific employability with reference to the performance requirements of various occupations. General employability may encompass characteristics common to many jobs such as personal hygiene, grooming, punctuality, stamina, tolerance of frustration, ability to get along with coworkers, tolerance for supervision, etc. These factors are evaluated by interview of the individual with chronic pain, review of employment records, interviews with past employers or coworkers, and/or observation.

Specific employability refers to those characteristics necessary to perform various specific occupations such as intelligence, aptitude, temperament, physical capacity, etc. These may be evaluated through the use of psychometric testing using instruments such as the Wechsler Adult Intelligence Scale III, Minnesota Multiphasic Personality Inventory 2, General Aptitude Test Battery, Wide Range Achievement Test, etc. More frequently, they are evaluated less formally through interview, observation, and record review. Physical capacity is often estimated using a physical capacities evaluation, where the subject is asked to perform a number of tasks such as lifting, reaching, kneeling, etc. From a practical perspective, simply asking the person with chronic pain his or her capacity in each area is likely to give a valid indicator of physical capacity, since it is the individual's perception of his or her capacity that is relevant.

In the United States, it is important to frame vocational evaluations in the context of the American's with Disabilities Act. Many people with chronic pain will be qualified as individuals with disabilities under the Act, so the question becomes: "Can this individual do the essential functions of this job with or without accommodation."

Who Conducts Vocational Assessment

Typically, vocational assessment would be conducted by someone with at least a master's degree in rehabilitation counseling or counseling, who has had training at graduate level in career assessment, functional assessment, and job placement. Here, the assessor will be referred to as the "counselor."

Strategies for Vocational Assessment

Although some of the psychometric instruments described above are often used diagnostically, the vocational assessment process is decidedly not diagnostic in the traditional sense. Rather, the vocational assessment involves collaboration between the assessor and the subject. The counselor works with the individual with chronic pain to guide or facilitate an empirical process, where the individual gathers data about him or her around the factors described above, and gathers data about the characteristics and demands of various jobs. To have a successful outcome (e.g., reentering employment), it is critical that the subject invests in the process and takes responsibility for the results. Vocational assessment is also dynamic. Hypotheses are developed, tested, and refined through experience.

Most frequently, the vocational assessment may include interviews with the subject and various key informants (e.g., previous employer, coworker, or spouse), a review of history including education and employment, and a review of relevant medical information. In some cases, the addition of paper and pencil tests may be helpful. For example, it may be important to establish a literacy level, or a subject may not have a clear understanding of his or her ability to profit from higher education, or what his or her interests are. In these cases, using interest inventories, achievement tests, intelligence tests, personality inventories, or other tools may help the subject gain a better understanding of these variables. As part of the assessment, the counselor will identify with the subject a set of transferable skills, that is, skills that the subject has acquired or demonstrated in the past that may be applied to other employment areas. This also allows for planning for education and training, which may be necessary to expand the skill set so that more occupational choices are available.

Situational assessment can also play a key role for some people with chronic pain. Situational assessment involves the subject performing real or simulated work. For example, a 48-year-old man has previously worked at finishing concrete and cannot return to that job. He has a high school education and reads at the tenth grade level. He reports that he enjoys working with people and "helping" people but is not aware of what sorts of jobs are available and how they would suit him. He also is not certain he has the stamina for an 8 h day. For situational assessment, the counselor could set up an opportunity for him to work 4 h a day, 3 days a week, for 2 weeks at a patient reception counter in a medical center. At the end of the assessment, the counselor and the man could discuss various aspects of that job, his endurance, and could get feedback on performance from the site supervisor. These data would then become part of the career counseling process.

Job analysis is another key component of vocational assessment. In a job analysis, the counselor evaluates the various demands of a particular job. It is important to evaluate not only the physical demands but also the social and cognitive demands. The U.S. Department of Labor has cataloged job analyses of representative jobs in the national economy. While these generic job analyses are useful for career counseling, they rarely describe adequately the requirements of a specific job. Job analyses can be conducted by interviewing workers or supervisors but are best conducted by also conducting observations in the workplace.

The outcome of the vocational assessment is a synthesis of the variables related to satisfaction and satisfactoriness, an exposition of the subject's strengths, limitations, barriers, and strategies to enhance success, including potential accommodations. There should also be examples of potential jobs that would be a good fit, and issues that should be considered in job placement.

References

Andersson, G. B. J. (1999). Epidemiological features of chronic low-back pain. *Lancet, 354*, 581–585.

Bigos, S., Spengler, D., Martin, N., et al. (1986). Back injuries in industry: A retrospective study II. Injury factors. *Spine, 11*, 121–130.

Dawes, R. (1987). A theory of work adjustment. In B. Bolton (Ed.), *A handbook on the measurement and evaluation in rehabilitation* (2nd ed.). Baltimore: Paul Brooks.

Deyo, R., & Weinstein, J. N. (2001). Low back pain. *The New England Journal of Medicine, 344*, 363–370.

Holland, J. (1997). *Making vocational choices: A theory of vocational personalities and work environments.* Odessa, Fl: Psychological Assessment Resources.

Snook, S. H. (1987). The costs of back pain in industry. In R. A. Deyo (Ed.), *Occupational back pain: State of the Art reviews* (pp. 1–5). Philadelphia: Hanley and Belfus.

Sternbach, R. (1986). Survey of pain in the united states: The nuprin pain report. *The Clinical Journal of Pain, 2*, 49–53.

Tinsley, D. (1993). Extensions, elaboration, and content validation of the TWA. *Journal of Vocational Behavior, 43*, 67–74.

Volinn, E., Van Koevering, D., & Loeser, J. (1991). Back sprain in industry: The role of socioeconomic factors in chronicity. *Spine, 16*, 542–548.

Wesolek, J., & McFarlene, F. (1992). Vocational assessment and evaluation: Some observations for the past and anticipation for the future. *Vocational Evaluation and Work Adjustment Bulletin, 25*, 51–54.

Vocational Counseling

Ingrid Söderback
Uppsala University, Uppsala, Sweden

Definition

Vocational counseling is one among several interventions such as work conditioning, workplace modifications, job coaching, and work supervision applied in multimodal vocational rehabilitation. It is aimed at assisting the client to understand vocational assets and liabilities for a suitable occupation. Vocational counseling contains the process of giving professional education, advice, and determining viable and reasonable employment goals to people with musculoskeletal pain impairments regarding their needs for retaining or resumption of a particular job. It is a highly individualized measure whereas intensive interactions between the team members of physical rehabilitation medicine, health care, employment office, insurance office, and social services are initiated with focuses on client's interests and suitability for vocational options available in the open or adapted labor market (Chamberlain et al. 2009; Skrida 2009).

Characteristics

The prevalence rate of long-term pain impairments varies according to nation-wide population studies between 11 % and 66 %, with predominance among people with brief education and among women aged 50–58 years. Activity limitations and participation restrictions (World Health Organisation 2006) (prevalence 26 %; incidence 0.07), e.g., inability to perform employed work (93.1 %) due to long-term/recurrent pain were self-reported by a Swedish population (n = 6,419) (Müllersdorf and Söderback 2000). Factors indicating need for multidisciplinary rehabilitative measures, e.g., vocational counseling, were feelings of insufficiency such as irresolution and gnawing or searing pain. More precisely, indicators for need of occupational therapy interventions were (1) pain behavior that prevented engagement in activities; (2) lack of knowledge about pain mechanisms; (3) occupational imbalance in work; (4) emotional stress and depression due to pain; and (5) physical or environmental strain resulting in limitations in occupational performance (Skjutar et al. 2010). People living with long-term (>3 months) pain due to diseases in the musculoskeletal system and connective tissues and have work disability or instability, i.e., provided by a long-term incapacity or disability pension, constitute about half a million of the population in Austria and Sweden (Chamberlain et al. 2009).

Theoretical Perspectives in the Vocational Counseling Process

The professional work of counseling in general is applied on several theoretical approaches, e.g., the client centered, whereas a growth-promoting climate is created; the holistic health (bio-psycho-social) that emphasizes supplying of clients' rehabilitative needs, the cognitive-behavioral therapy encouraging the positive and solution-focused approaches to the client's present and future situation (Steven (2010)). Vocational counseling aimed for people with long-term pain disabilities seems to be lacking a general theoretical base but encourages the idea of ▶ work fitness. It comprises measures for maximizing the client's ▶ vocational aptitudes and skills and/or functional capacity that promote return to work. Recently, a scientific discussion has originated about what theoretical perspectives are effective for contributing to the counseling and rehabilitative process that results in clients' return to work. Among available perspectives Antonovsky's salutogenic model, Kielhofner's Model of Human Occupation, and Sceff's sociological theory of "shame and pride" are judged possible to be used for understanding clients' self-experience and thus indicates client's perception of work ability or disability that may contribute to return-to-work interventions (Svensson and Björklund 2010).

Goals and Assessment Methods of Vocational Counseling

The goals of vocational counseling are (a) to facilitate a client's knowledge of available occupational opportunities and resources, (b) to identify a client's present vocational aptitudes, (c) to identify environmental occupational barriers (World Health Organisation 2001), (d) to identify a vocational goal consistent with identified individual vocational aptitudes and preferences, and (e) to outline a rehabilitative plan for vocational assessment and vocational rehabilitation measures (Leahy 1995).

Vocational counseling comprises vocational assessment and vocational rehabilitative interventions.

Through vocational assessment, a client's vocational strengths, values, suitability, readiness, and potential for an optimal return to work outcome are identified. Moreover, a client's weaknesses or lack of skills needed to be able to perform a work is identified and constitutes the rehabilitation plan.

Vocational assessment is a comprehensive, interdisciplinary, and systematic process where real or simulated occupational tasks are used. The results reveal the contents of a client's job tasks and his/her individual characteristics (i.e., physical, mental, and emotional aptitudes/limitations, education and work tolerance). The assessment serves as the basis for planning vocational rehabilitation measures. The goal of return to work implies full- or part-time work at the previous workplace or another, performing job tasks adjusted to the rehabilitee's aptitude and skill and using adaptive equipment.

Vocational counseling assessment methods include (a) an evaluation of vocational capacity, (b) a ▶ functional capacity evaluation, and (c) a ▶ vocational assessment. However, the client's return to work potential is determined by the results of a vocational assessment. Vocational assessment is a multi-professional investigation. It is performed for several days and based on the main principle of using ▶ criterion-referenced assessments, meaning that the vocational skills required of the worker in a specific work environment are compared with the personal characteristics, personal factors, and vocational aptitudes the worker brings to a job. These influence optimal healthy performance and work productivity (Spenner 1990). Assessors systematically use multidimensional assessment tools and methods, e.g., interviews, self-reports, and behavioral observation. A vocational assessment contains an intake interview, a general medical examination, a ► job analysis, and an evaluation of the client's characteristics.

Methods for Performing a Vocational Assessment

A vocational assessment may be based on the following methods (Stein et al. 2006):

- Analyzing the actual job: Information about the ► job requirements of a client's actual occupation is available from electronic sources (US Department of Labor 1996) which combine information from the ► Dictionary of Occupational Titles (US Department of Labor 1991a), the Worker Traits Data Book (US Department of Labor 1994) on Occupational Information Network (2012) (O*NET) database (US Department of Labor 2006). These databases contain variables that represent descriptors of work and worker characteristics, including skill requirements. Selecting from the electronic sources the occupation that is the most comparable to the client's present circumstances is determined by (a) structured interviews with the worker, workmates, and manager and (b) a ► situational assessment or an ► on the job site evaluation. These observations are video-recorded and systematically analyzed according to the Revised Handbook of Analyzing Jobs (US Department of Labor 1991b), and ergonomics. The key characteristic of an occupation (job substantive complexity, physical demands, and circumstances of physical environmental factors) is numerically determined using an index.

- Assessing the client's occupational performance. Worker traits, i.e., the client's vocational aptitudes and skills that contribute to performance of his/her actual occupation are determined by (a) observing the client's

► occupational behavior (b) observing the client's performance of real and/or ► simulated job tasks, (c) ► work samples, e.g., the VAL PAR component work samples (VCWS) is an example of a commonly used work sample (Christopherson and Hayes 1992), and/or (d) *functional capacity assessments*, e.g., use of the progressive isoinertial lifting evaluation (PILE Test), Isernhagen's functional capacity evaluation, and the ERGOS work simulator are commonly used in Europe (Chamberlain et al. 2009) following the process applied to the Functional Capacity Evaluation aimed at assessing work activity limitations (Gibson 2008/2009).

The job analysis constitutes the base for the choice of a, b, c, or d as being the most appropriate occupational tasks for assessing the client's behavior. The client's results when performing at least two VCWS samples give a worker qualification profile (WQP) (VAL PAR 1998), which demonstrates whether the client's achievement corresponds to (a) the required DOT variables and (b) the standard criterion of the methods-time measurement (MTM) score. Self-reported assessments of the client's capability to perform daily occupations may contribute valuable information (Schult 2002). The client's performance of the chosen job tasks is systematically observed and video recorded. The analysis focuses on the client's biomechanical and ergonomic work technique with the aim of preventing adverse health.

- Assessing the client's personal behavior. The client's personal behavior as affected by internal strengths, perception of present life values, and coping style regarding the workmates and manager and by pain-associated problems, such as fear, avoidance behavior, depression, and negative orientation to life are assessed using client reported, reliable, and valid assessment questionnaires (Schult 2002).

- Analyzing and assessing the work environment. The physical work environment, e.g., exposure/no exposure to weather, noise intensity level and vibration, is determined by systematic observation of the real work conditions according to the Worker Traits Data Book. The ergonomic and biomechanical conditions at the workplace are observed and video recorded. The psychological work environment that has an impact on the worker's satisfaction is related to organizational risk factors for job stress, psychological work demands, employee control over the work process, on the job social supports, uncertainly on the job and on the job conflict. These factors are estimated with client-reported assessment instruments. The Demand-Control-Support Questionnaire (Theorell et al. 1990) estimates how the actual work organization, which may be characterized by, e.g., high levels of psychological demands, a low level of decision latitude or poor social support, affects the worker's health adverse. The Job Diagnostic Survey estimates the worker's need for workplace redesign to increase the positive personal outcome of high internal motivation, work satisfaction, quality performance, and low absenteeism (Schult and Söderback 2000).

The Validity of Vocational Assessments

The criterion-referenced multidimensional job-related vocational assessment (CMVA) model seeks to assess the client's ▶ work capacity. It addresses the variables; "Work demands" derived from the DOT variables, i.e., physical demands, substantive complexity, and physical environmental factors, "Occupational performance," i.e., the individual's results on two VCWS, "Personal," i.e., individual perception of capacity for occupational performance of work tasks and the difference between desired values of work and attained values and "Environment," i.e., the individual's perception of how much social support he or she receives. With the CMVA model it is possible to predict a long-term pain sufferer's capacity for work, as demonstrated by the multivariate linear regression model with the forward selection of variables. The result explains 78 % of the variance. Further studies may demonstrate the value of the CMVA for preventing people from becoming sick listed or out of work. Among the variables in the

CMVA model, "occupational performance" (self-perceived ability and satisfaction with the performance of a daily occupation) is the strongest predictor. This is shown by the fact that the results of logistic regression analysis correctly classified 88 % of participants who were out of work due to long-term pain impairments (Söderback et al. 2000; Stein et al. 2006).

Predictors for Returning to Work

The main goals of medical rehabilitation for people with long-term musculoskeletal pain impairments are to attain work fitness and to return to work. Variables of the bio-psycho-social model (Hanson and Gerber 1990) significantly predict return to work and therefore partly constitute the components of a vocational assessment. These variables are:

The *disability factor* (World Health Organisation 2006), which consists of a low or average impairment sensation of pain (<5.4 score on a 10° VAS scale) and the *medical factor*, e.g., the prognoses of the disease.

The contextual factors (World Health Organisation 2006), which consist of:

- ▶ *Personal factors* (World Health Organisation 2006): possessing (a) vocational training, (b) a strong intention to come back to and remain at one's previous work, (c) a positive perception of one's previous employment status, a moderate or high level of self-perceived capacity for work, general satisfaction with present life value and job and a feeling that one's job tasks are meaningful
- *Environmental factors* (World Health Organisation 2006): (a) availability of suitable job tasks that are (b) comparable to one's vocational aptitude, (c) early referral to vocational counseling, and (d) less than 1 month elapsing between completed vocational rehabilitation and returning to work (Fishbain et al. 1997; Schult 2002; van der Giezen et al. 2000)

Predictors of nonreturn to work are older age, language barriers, and the emotional impairment of depression (>16 scores on the Beck depression scale).

Variables that do not significantly predict return to work are the client's medical history, physical capacity status, range of motion, and findings from biomechanical tests (Hunt et al. 2002).

Moreover, rehabilitation actors' *recourses, effectiveness, and the cooperation* between institutes involved in the counseling process as well as *application of laws* and *principles of social welfare* are other factors that strongly influence the vocational outcome (Chamberlain et al. 2009).

In conclusion, psychosocial aspects of health and work in combination with *economic aspects and present status of the labor market* have a significantly larger impact on return to work than medical and physical aspects. People work because they need to earn money to survive or to establish a career and/or because the work itself gives them satisfaction and membership in a social group that may increase their meaning of life.

Effectiveness

The primarily required outcome variables of vocational counseling and vocational rehabilitation are proposed to be work resumption or days of sick leave. These parameters are relatively rarely used (n = 13) in vocational rehabilitation. However, among these available randomized controlled studies, the conclusion is convincing results showing more people resume and retain to work provided that the intervention includes multimodal vocational rehabilitation (Chamberlain et al. 2009). Furthermore, two studies mentioned vocational counseling specifically as parts of the multimodal vocational rehabilitation intervention. The German randomized controlled study including people (n = 307) with long-term chronic back pain showed decreased number of sick days, which was interpreted partly as results of intense and multilevel counseling of work-related problems (Dibbelt et al. 2006). Apart from this study, Magnusson et al. (2007) investigated the effects of an intervention including education, reassurance, motivational interviewing (Shannon 2008/2009), and vocational counseling among disabled pensioners (n = 89) with low back pain. At 1 year

follow-up a nonsignificant result showed higher participation in work-related activities, but no decrease in pension payment was proved.

It might be without any doubt that several more randomized control studies among people with musculoskeletal pain and no-working are required, especially when vocational counseling solely constitutes the intervention. For these studies, an action theory perspective is proposed that is based on understanding of human experience as containing a goal-directed, integrative approach including continuity of action, project, and career, as well as goals, functional steps, and behavioral elements (Young and Valach 2009).

Summary

- There seems to be a need for vocational counseling assessment, especially among people with long-term gnawing or searing pain and those who perceive themselves as incapable of performing daily occupations.
- Vocational counseling assessment methods include evaluation of vocational capacity, evaluation of functional capacity and vocational assessment. Of these, vocational assessment is judged the most valid for assessing capacity for work among people with long-term pain. This is because these assessment results demonstrate the client's capacity for work compared to the job requirements.
- Research is required about the effectiveness of vocational counseling that is based on specific theoretical perspectives that contribute to effective interventions among people with long-term pain and long sick leave.

References

Chamberlain, M. A., Moser, V. F., Schüldt Ekholm, K., O'Connor, J., Herceg, M., & Ekholm, J. (2009). Vocational rehabilitation in an educational review. *Journal of Rehabilitation Medicine, 41*, 856–869.

Christopherson, B. B., & Hayes, P. D. (1992). *VAL PAR Component Work Samples uses in allied health.* Tucson: VAL PAR International Corporation.

Dibbelt, S., Greitemann, B., & Büschel, C. (2006). Long-term efficiency of orthopedic rehabilitation in chronic back pain – The integrative orthopedic psychosomatic concept (IopKol). *Rehabilitaion (Stuttg.), 45*(6), 324–325.

Fishbain, D. A., Cutler, R. B., Rosomoff, H. L., et al. (1997). Impact of chronic pain patient's job perception variables or actual return to work. *The Clinical Journal of Pain, 13*, 197–206.

Gibson, L. (2008/2009). Functional capacity evaluation: An integrated approach to assessing work activity limitation. In Söderback, I. (Ed.), *International handbook of occupational therapy interventions* (pp. 497–514). New York: Springer.

Hanson, R., & Gerber, K. (1990). *Coping with chronic pain. A guide to patient self-management.* New York: The Guilford Press.

Hunt, D. G., Zuberbier, O. A., Kozlowski, A. J., et al. (2002). Are components of a comprehensive medical assessment predictive of work disability after an episode of occupational low back trouble? *Spine, 27*(23), 2715–2719.

Leahy, M. J. (1995). Assessment of vocational interests and aptitudes in rehabilitation setting. In M. J. Scherer (Ed.), *Psychological assessment in medical rehabilitation* (pp. 299–324). Easton: American Psychological Association.

Magnusson, L., Stand, L. I., Skouen, J. S., & Eriksen, H. R. (2007). Motivation disability pensioners with back pain to return to work – A randomized controlled trail. *Journal of Rehabilitation Medicine, 39*, 81–87.

Müllersdorf, M., & Söderback, I. (2000). Assessing health care needs. The actual state of self-perceived activity limitation and participation restrictions due to pain in a national-wide Swedish population. *International Journal of Rehabilitation Research, 23*, 201–207.

Occupational Information network (O*NetOnLine). (2012). The O*NET database. http://www.online.onetcenter.org. 10 March 2006.

PAR Component Val Work Samples. (1998). *Worker Qualification Profile. Computer program: S 2000.* Tucson: VAL PAR International Corporation.

Schult, M.- L. (2002). Multidimensional assessment of people with chronic pain. A critical appraisal of the person, environment, occupation model. Acta Dissertations of Uppsala University, Uppsala, Sweden.

Schult, M.-L., & Söderback, I. (2000). A method for 'diagnosing' jobs before re-design in chronic-pain patients. Preliminary findings. *Journal of Occupational Rehabilitation, 10*, 295–307.

Shannon, R. J. (2008/2009). Motivational interviewing: Enhancing patient motivation for behavior change. In Söderback, I. (Ed.), *International handbook of occupational therapy interventions* (pp. 515–523). New York: Springer.

Skjutar, A., Schult, M. L., Christensson, K., & Müllersdorf, M. (2010). Indicators of need for occupational therapy in patients with chronic pain:

Occupational therapists' focus groups. *Occupational Therapy International, 17*(2), 92–103.

Skrida, R.J. & Associates Inc. (2009). Forensic and vocational rehabilitation & employability specialits. http://www.rjskirda.com/services.html. Retrieved September, 2010.

Söderback, I., Schult, M.-L., & Jacobs, K. (2000). A criterion-referenced multidimensional job-related model predicting capability to perform occupations among persons with chronic pain. *Work, 15*, 25–39.

Spenner, K. (1990). Skill. Meanings, methods and measurements. *Work and Occupations, 17*, 399–421.

Stein, F., Söderback, I., Cutler, S. K., & Larson, B. (2006). *Occupational therapy and ergonomics. Applying ergonomics principles to everyday occupation in the home and at the work* (1st ed.). London/Philadelphia: Whurr Publisher/Wiley.

Steven, C. (2010). Basic-counseling-skills – Introduction to theories. Laurentian University's counseling and support program. www.basic.counseling.skills.com. Retrieved September, 2010.

Svensson, T., & Björklund, A. (2010). Focus on health, motivation and pride: A discussion of three theoretical perspectives on the rehabilitation of sick-listed people. Work. *A Journal of Prevention, Assessment and Rehabilitation, 36*(3), 237–282.

Theorell, T., Ahlberg-Hultén, G., & Härenstam, A. (1990). A psychosocial and biomedical comparison between men in six contrasting service occupations. *Work and Stress, 4*, 51–53.

US Department of Labor. (1991a). *Dictionary of occupational titles.* Indianapolis: JIST Works.

US Department of Labor. (1991b). *The revised handbook for analyzing jobs.* Indianapolis: JIST Works.

US Department of Labor. (1994). *Worker traits data book.* Indianapolis: JIST Works.

US Department of Labor. (1996). *Jist's Electronic Enhanced Dictionary of Occupational Titles (Jist's DOT). Computer program.* Indianapolis: JIST Works.

van der Giezen, A. M., Bouter, L. M., & Nijuhuis, F. J. (2000). Prediction of return-to-work of low back pain patients sick-listed for 3–4 months. *Pain, 87*, 285–294.

World Health Organisation. (2001). International Classification of Functioning, Disability and Health (ICF). http://www.who.int/classifications/icf/en. Retrieved September 14, 2012.

Young, R. A., & Valach, S. (2009). Evaluating the processes and outcomes of vocational counseling: An action theory perspective. *L'orientation scolaire et professionnelle* [En ligne]. Consulté le 13 septembre 2010. http://osp.revues.org/index1944.html. Retrieved September, 2010.

Vocational Evaluation

▶ Vocational Assessment in Chronic Pain

Voltage-dependent Calcium Channels

▶ Calcium Channels in the Spinal Processing of Nociceptive Input

Voltage-dependent Pores

▶ Trafficking and Localization of Ion Channels

Voltage-Gated Calcium Channel

Definition

Voltage-gated calcium channels represent a range of membrane ion channels that are selectively permeable to calcium ions. Permeation of calcium into nerve terminals and cell bodies through these channels is gated by membrane depolarization to produce calcium entry and excitation. The major protein classes are N- (Ca_v 2.2), P/Q (Ca_v 2.1), R (Ca_v 2.3), T (Ca_v 3.X), and L (Ca_v 1.X).

Cross-References

▶ Opioid Electrophysiology in PAG

Voltage-Gated Channel

Definition

A pore in the cell membrane that allows passage of a substance dependent upon the voltage across the cell membrane.

Cross-References

▶ Drugs Targeting Voltage-Gated Sodium and Calcium Channels
▶ Migraine, Pathophysiology

Voltage-Gated Channels

▶ Drugs Targeting Voltage-Gated Sodium and Calcium Channels

Voltage-Gated Potassium Channels

Definition

Voltage-gated K_V channels represent a large family of membrane ion channels that are selectively permeable to potassium ions. Activation of these channels by changes in membrane voltage produces membrane hyperpolarization to control action potential repolarization, frequency, and firing patterns.

Cross-References

▶ Opioid Electrophysiology in PAG

Voltage-Gated Sodium Channel

Definition

A pore in the neuronal cell membrane that permits the influx of sodium ions, which is opened by a change in membrane potential.

Cross-References

▶ Molecular Contributions to the Mechanism of Central Pain

Voltage-sensitive Calcium Channels

▶ Calcium Channels in the Spinal Processing of Nociceptive Input

Volume Conduction

▶ Cross Excitation

Volume of Distribution (Vd)

Definition

The volume of distribution (Vd) is a hypothetical volume. It is the volume that would be required to dissolve the administered amount of drug (A) at the measured drug concentration (C):

$$Vd = \frac{A}{C}$$

Cross-References

▶ NSAIDs, Pharmacokinetics
▶ Pharmacology of Cyclooxygenase Inhibitors

Volume Transmission

Definition

The term "volume transmission" is used to describe a mechanism whereby a neurotransmitter diffuses through the neuropil to reach the receptors on which it acts rather than merely crossing a synaptic cleft. Both neuropeptides and monoamines are thought to act through volume transmission.

Cross-References

▶ Nociceptive Circuitry in the Spinal Cord

Voluntary Control

▶ Biofeedback in the Treatment of Pain

Von Frey Hair

Definition

Von Frey hairs (named after the German physiologist Max von Frey, 1852–1932) have been originally produced from animal and human hairs of different diameters. Nowadays they are nylon monofilaments of different diameters, each of them mounted at right angles to the end of a plastic handle. The diameter determines the resistance of the monofilament to bending and thereby determines the amount of force applied by each filament. A filament is placed perpendicularly to the skin with slowly increasing force until it bends. It stays bent for 1–2 s and is then removed with slowly decreasing strength. Each of the filaments exerts a stimulus of a distinct force which is often expressed in grams (not correctly, since in physics "gram" is a mass unit), or more correctly in milli-Newton (in physics a measure of force). Using a series of von Frey hairs with ascending stiffness can be employed, for example, to determine sensory or neuronal thresholds.

Cross-References

▶ Cognitive-Behavior Treatment of Pain
▶ Hyperaesthesia, Assessment
▶ Hypoesthesia, Assessment
▶ Muscle Pain Model, Inflammatory Agents-Induced
▶ Neuropathic Pain Model, Tail Nerve Transection Model
▶ Species Differences in Skin Nociception
▶ Spinal Cord Injury Pain Model, Contusion Injury Model

VOP

▶ Vascular Orofacial Pain

Voxel-Based Morphometry

Definition

A computational approach to neuroanatomy, which measures differences in local concentrations of brain tissue through a voxel-wise comparison of multiple brain images.

Cross-References

▶ Thalamus, Clinical Pain, Human Imaging

VPI

▶ Ventral Posterior Inferior Nucleus

VPL

▶ Ventral Posterior Lateral Nucleus
▶ Ventroposterior Lateral Nucleus

VPM

▶ Ventral Posterior Medial

VPpc

▶ Parvocellular VP

VR1

▶ TRPV1 Modulation by PKC
▶ TRPV1 Receptor, Species Variability

VR1, Regulation by NGF

▶ TRPV1, Regulation by Nerve Growth Factor

VR1, Regulation by Protons

▶ TRPV1, Regulation by Protons

VR1, VRL1, TRPV1, TRPV2

Definition

The transient receptor potential (TRP) family of proteins assembles to form a variety of ligand and thermally gated ion channels. Some of these were originally designated vanilloid receptors (VR). TRPV1 and TRPV2 assemble to form heat-sensitive ion channels with thresholds near 45 and 50°, respectively. TRPV1 is also proton sensitive and can also be gated by a number of pro-inflammatory lipids.

Cross-References

▶ Nociceptors in the Orofacial Region (Skin/Mucosa)

VSCCs

▶ Calcium Channels in the Spinal Processing of Nociceptive Input

Vulvar Dysesthesia

▶ Vulvodynia

Vulvar Vestibulitis

▶ Vulvodynia

Vulvar Vestibulitis Syndrome

▶ Gynecological Pain and Sexual Functioning

Vulvodynia

Ursula Wesselmann[1] and Melissa Farmer[2]
[1]Departments of Neurology and Anesthesiology /
Division of Pain Management, University of
Alabama at Birmingham, Birmingham, AL, USA
[2]Department of Physiology / Feinberg School
of Medicine, Northwestern University, Chicago,
IL, USA

Synonyms

Dysesthetic vulvodynia; Essential vulvodynia; Idiopathic vulvar pain; Localized provoked vulvodynia; Vestibulodynia; Vulvar dysesthesia; Vulvar vestibulitis

Definition

The term vulvodynia (Latin: vulva; Greek: -odynia, pain) is a modern word for an age-old pain condition. The first recorded reference to vulvar pain is embedded in the ancient Egyptian Ramesseum papyrus (Farmer et al. 2008). The phenomenon was not described in medical texts until the nineteenth century, where it was variably called vulvar hyperesthesia, vulvar "supersensitiveness," and heightened vulvar "sensitivity" (Amalraj et al. 2008). The first attempt to develop formal terminology for vulvar pain occurred in 1976, when the International Society for the Study of Vulvovaginal Disease (ISSVD) introduced the terminology "burning vulva syndrome" based on women's

reports of a hot, burning vulvar sensation, and soon thereafter, this term was replaced with "vulvodynia." During the 1980s, the early gynecological literature differentiated two subtypes of vulvodynia based on the perceived location of vulvar pain. Localized vulvar pain, or vulvar vestibulitis, was exclusively elicited by touch or pressure to the vulvar vestibule (Friedrich 1987), whereas a more diffuse or "generalized" vulvar pain was termed dysesthetic (or "essential") vulvodynia (McKay 1989). Despite multiple revisions of the vulvodynia nomenclature, the ISSVD's 2003 vulvar pain terminology has essentially preserved these two subtypes as localized vulvodynia and generalized vulvodynia (Moyal-Barracco and Lynch 2004). Epidemiological studies have confirmed that localized provoked vulvodynia (LPV) is the most common vulvodynia subtype in premenopausal women (Harlow and Stewart 2003; Harlow et al. 2001), and recent evidence suggests LPV may also explain a majority of postmenopausal dyspareunia (pain during sex; Kao et al. 2012).

Vulvodynia is an umbrella term that can refer to pain experienced at any location on the vulva (external genital organs), including the clitoris, labia majora, labia minora, Bartholin's glands, and the vaginal entrance (vulvar vestibule). Pain can be highly localized (e.g., LPV) or more diffuse or radiating (generalized vulvodynia). Clinically, two different groups of patients with LPV have been described: Primary provoked vulvodynia is defined as dyspareunia from the first attempt of sexual intercourse (and is often associated with a history of pain with tampon use), whereas in secondary provoked vulvodynia, the dyspareunia appears after a period of pain-free sexual intercourse. Different subtypes of vulvodynia are often characterized by unique pain descriptors. For instance, LPV (also known as provoked vestibulodynia, formerly vulvar vestibulitis syndrome) is described as "burning" or "cutting" pain, whereas dysesthetic vulvodynia elicits "tingling" and "shooting" sensations. "Sore" and "aching" sensations may refer to pain originating in the vulvar tissue or underlying muscle, potentially due to pelvic floor tension secondary to vulvar pain. Provoked vulvar pain is initiated with mechanical stimulation (from clothes, sitting, or sexual activity) and may be of brief or prolonged duration, depending on the intensity of stimulation and the degree of vulvar allodynia. In contrast, unprovoked vulvar pain is unpredictable in onset and duration, although mechanical provocation can at times elicit this spontaneous pain as well. These vulvar pain characteristics are not associated with any evident pathology, meaning that vulvodynia is a diagnosis of exclusion rather than a unitary condition.

In a broader view, vulvodynia can be grouped with the chronic nonmalignant syndromes of urogenital origin occurring in both sexes (Wesselmann et al. 1997). In addition to vulvodynia, these pain syndromes include urethral syndrome, coccygodynia, generalized perineal pain, orchialgia, prostatodynia (chronic pelvic pain syndrome in men), chronic penile pain, and interstitial cystitis/painful bladder syndrome.

Characteristics

The current understanding of vulvodynia as a multifactorial chronic pain condition can be traced to small, independent literatures about vulvar pain from a variety of disciplines, including gynecology, neurology, dermatology, physical therapy, and psychiatry. These literatures originally consisted of case studies or uncontrolled, cross-sectional research that conceptualized vulvar pain using the vocabulary of the respective discipline. Accordingly, gynecologists theorized that vulvodynia may result from anatomical abnormalities, hormonal imbalances, urogenital disease or infection, inflammatory conditions, or bladder dysfunction (interstitial cystitis, descended bladder). Neurologists suspected peripheral nerve injuries, such as pudendal neuralgia or abnormalities in peripheral pain transmission. Dermatologists investigated lichen sclerosis, lichen simplex chronicus, lichen planus, allergic contact dermatitis, and poor vulvar hygiene, whereas physical therapists targeted pelvic floor abnormalities, including heightened muscle tension, weakness,

and instability. When no medical cause of vulvar pain could be identified, women were referred to psychiatrists who often interpreted the pain as a confabulation of female neuroses. This fragmented history of the study of vulvodynia provides some key caveats for the interpretation of past research. First, the term "vulvodynia" likely refers to a pain with multiple etiological pathways, suggesting the existence of unidentified vulvodynia subtypes. Secondly, approaching vulvar pain from any single perspective will limit an understanding of its multidimensional nature. Finally, successful treatment of the condition that originally caused the vulvar pain may not address the mechanisms that have chronically maintained the pain. In summary, vulvodynia likely consists of multiple subtypes, each with unique etiologies, and therapeutic interventions that target these initiating factors may fail to address the pathological mechanisms underlying the maintenance of chronic vulvar pain.

Quantitative sensory testing has confirmed that vulvar allodynia and hyperalgesia can be replicated in the laboratory. Women with LPV have reduced touch detection and mechanical pain thresholds compared to age-matched controls (Pukall et al. 2002), as well as heightened pain sensitivity at nongenital body sites (Giesecke et al. 2004; Pukall et al. 2002). Thus, in some women, vulvar pain may not simply be a pelvic phenomenon; rather, its presence may signal a systemic condition driven by central changes in sensory processing. It is unclear whether heightened body sensitivity is a predisposing factor to vulvar pain or whether it is the result. Notably, some evidence suggests symptomatic overlap between vulvodynia and systemic pain disorders like fibromyalgia (Pukall et al. 2006).

Peripheral abnormalities may indicate one mechanism by which localized pain perception is enhanced. In women with localized provoked vulvodynia, vulvar biopsy tissue exhibits increased innervation of nerve fibers, including nociceptors expressing the vanilloid receptor VR1 (Bohm-Starke 2010). Increased vulvar innervation could conceivably enhance transmission of painful sensation, particularly in the presence of local inflammation that can sensitize nerve endings, and contribute to peripheral sensitization of vulvar nociceptors. However, the presence of increased vulvar innervation does not point to a specific etiology. Furthermore, an animal model examining persistent vulvar pain following repeated and prolonged *Candida albicans* infections has demonstrated that active inflammation is not necessary to evoke substantial vulvar innervation changes, following previous periods of inflammation (Farmer et al. 2011).

Systemic factors that facilitate abnormal inflammatory processes may also explain increased vulvar pain perception. In vitro and in vivo evidence of enhanced vulvar and systemic inflammatory responses to pathogens and allergens suggests that vulvar pain may be symptomatic of an abnormal inflammatory response. Similarly, increasing evidence of genetic polymorphisms that are found more often in women with LPV (including genes for mannose-binding lectin, cytokines, interleukin-1 receptor antagonist, and interleukin-1 beta) indicates that in some women, the over- or underexpression of certain pro-inflammatory cytokines could underlie abnormal physiological mechanisms driving vulvar pain (Gerber et al. 2008). In such cases, the magnitude or duration of the inflammatory response may be altered in response to otherwise innocuous levels of stimulation. However, equivocal findings of occasional redness and inflammatory infiltrate in vulvar biopsy tissue of women with provoked vulvar pain have called this inflammatory hypothesis into question (Bohm-Starke 2010; Gerber et al. 2008). It is therefore feasible that a subgroup of women experience vulvodynia of an inflammatory origin, yet the existence of such a subgroup has not been confirmed or refuted by current clinical phenotyping efforts.

Evidence of central alterations in women with vulvodynia is mixed. Central abnormalities in pain processing were first suspected when women with provoked vulvar pain were found to have altered mechanical and thermal sensitivity in unaffected dermatomes (Bohm-Starke 2010). A lack of an abnormal diffuse noxious inhibitory control response in women with LPV indicates that such changes do not reflect descending control (Johannesson et al. 2007).

During painful vulvar stimulation, women with provoked vestibulodynia exhibit brain activity that is typical of acute pain (Pukall et al. 2005), but it is unclear whether this functional pattern differs from pain-related brain activity in healthy controls, as this control was not examined. Increased regional grey matter volume in the parahippocampal gyrus/hippocampus and basal ganglia in the brains of women with LPV suggests that the experience of vulvar pain may alter brain structure (Schweinhardt et al. 2008). These brain regions participate in the processing of a broad range of cognitive and sensorimotor information that far surpass the experience of vulvar pain.

Because the pathophysiological mechanisms of vulvodynia are not yet known, targeted therapy is not available at present (Haefner et al. 2005). Treatment approaches are arbitrarily chosen and target presumed etiologies that cannot be clinically confirmed, and they may not address the underlying mechanisms that maintain vulvar pain once it has become chronic. Proper vulvar hygiene is emphasized as well as the use of lubricant. Treatments for localized vulvar pain include topical applications (estrogen, lidocaine, corticosteroid, or capsaicin creams), subcutaneous or intramuscular injections (of botulinum toxin, bupivacaine, inteferon alpha), or surgical excision of painful vestibular tissue (vestibulectomy or perineoplasty). In cases of presumed infectious origins (e.g., vulvovaginal candidiasis, bacterial vaginosis), oral antifungal or antibiotic treatments are given. Of these interventions, vestibulectomy has received strong empirical support from randomized controlled trials, with more modest empirical support showing that EMG biofeedback reduces LPV as well (Bergeron et al. 2001; Weijmar Schultz et al. 2005). At a 2.5-year follow-up of Bergeron's study, reductions in vulvar pain following group cognitive behavioral therapy rivaled pain reductions seen with the vestibulectomy group.

The rapidly evolving field of vulvodynia research is increasingly focusing on multidisciplinary, concurrent assessment and treatment for vulvar pain, combining biomedical efforts with pelvic floor physical therapy and cognitive pain management approaches. Vulvar pain is increasingly studied in relation to the surrounding structures (pelvic floor musculature, adjacent visceral organs), in relation to the mind that must react and adapt to the repeated presence of pain, and in relation to intimate partners who are affected by the psychological and behavioral consequences of vulvar pain.

Acknowledgement U. Wesselmann is supported by NIH grants: R01 DK066641, R01 HD39699 and the Office of Research for Women's Health. M. Farmer is supported by an NIH National Research Service Award (F31NS062611) from the NINDS.

Cross-References

▶ Dyspareunia and Vaginismus
▶ Gynecological Pain and Sexual Functioning
▶ Gynecological Pain, Neural Mechanisms

References

Amalraj, P., Kelly, S., & Bachmann, G. A. (2008). Historical perspective of vulvodynia. In A. T. Goldstein, C. F. Pukall, & I. Goldstein (Eds.), *Female sexual pain disorders: Evaluation and management* (pp. 1–3). New York: Wiley-Blackwell.

Bergeron, S., Binik, Y. M., Khalife, S., et al. (2001). A randomized comparison of group cognitive-behavioral therapy, surface electromyographic biofeedback, and vestibulectomy in the treatment of dyspareunia resulting from vulvar vestibulitis. *Pain, 91*, 297–306.

Bohm-Starke, N. (2010). Medical and physical predictors of localized provoked vulvodynia. *Acta Obstetricia et Gynecologica, 89*, 1504–1510.

Farmer, M. A., Kukkonen, T., & Binik, Y. M. (2008). Female genital pain and its treatment. In D. L. Rowland & L. Inrocci (Eds.), *Handbook of sexual and identity disorders* (pp. 220–250). Hoboken: Wiley.

Farmer, M. A., Taylor, A. M., Bailey, A. L., et al. (2011). Repeated vulvovaginal fungal infections cause persistent pain in a mouse model of vulvodynia. *Science Translational Medicine, 3*(101), 101ra91.

Friedrich, E. G. (1987). Vulvar vestibulitis syndrome. *The Journal of Reproductive Medicine, 32*, 110–114.

Gerber, S., Witkin, S. S., & Stucki, D. (2008). Immunological and genetic characterization of women with vulvodynia. *Journal of Medical Life, 1*, 432–438.

Giesecke, J., Reed, B., Haefner, H. K., et al. (2004). Quantitative sensory testing in vulvodynia patients and increased peripheral pressure pain sensitivity. *Obstetrics and Gynecology, 104*, 126–133.

Haefner, H. K., Collins, M. E., Davis, G. D., et al. (2005). The vulvodynia guideline. *Journal of Lower Genital Tract Disease, 9*, 40–51.

Harlow, B. L., & Stewart, E. G. (2003). A Population-based assessment of chronic unexplained vulvar pain: Have we underestimated the prevalence of vulvodynia? *Journal of the American Medical Women's Association, 58*, 82–88.

Harlow, B. L., Wise, L. A., & Stewart, E. G. (2001). Prevalence and predictors of chronic lower genital tract discomfort. *American Journal of Obstetrics and Gynecology, 185*, 545–550.

Johannesson, U., de Boussard, C. N., Brodda Jansen, G., et al. (2007). Evidence of diffuse noxious inhibitory controls (DNIC) elicited by cold noxious stimulation in patients with provoked vestibulodynia. *Pain, 130*, 31–39.

Kao, A., Binik, Y. M., Amsel, R., et al. (2012). Challenging atrophied perspectives on postmenopausal dyspareunia: A systematic description and synthesis of clinical pain characteristics. *Journal of Sex & Marital Therapy, 38*(2), 128–150.

McKay, M. (1989). Vulvodynia: A multifactoral clinical problem. *Archives of Dermatology, 125*, 256–262.

Moyal-Barracco, M., & Lynch, P. J. (2004). 2003 ISSVD terminology and classification of vulvodynia: A historical perspective. *The Journal of Reproductive Medicine, 49*, 772–777.

Pukall, C. F., Binik, Y. M., Khalife, S., et al. (2002). Vestibular tactile and pain thresholds in women with vulvar vestibulitis syndrome. *Pain, 96*, 163–175.

Pukall, C. F., Strigo, I. A., Binik, Y. M., et al. (2005). Neural correlates of painful genital touch in women with vulvar vestibulitis syndrome. *Pain, 115*, 118–127.

Pukall, C. F., Baron, M., Amsel, R., et al. (2006). Tender point examination in women with vulvar vestibulitis syndrome. *The Clinical Journal of Pain, 22*, 601–609.

Schweinhardt, P., Kuchinad, A., Pukall, C. F., et al. (2008). Increased grey matter density in young women with chronic vulvar pain. *Pain, 140*, 411–419.

Weijmar Schultz, W., Basson, R., Binik, Y. M., et al. (2005). Women's sexual pain and its management. *The Journal of Sexual Medicine, 2*, 301–316.

Wesselmann, U., Burnett, A. L., & Heinberg, L. J. (1997). The urogenital and rectal pain syndromes. *Pain, 73*, 269–294.

Vulvodynia Model

▶ Visceral Pain Models, Female Reproductive Organ Pain

Vulvo-Vaginal Atrophy

Definition

Vulvo-vaginal atrophy is a manifestation of tissue aging and the cytological and chemical transformations within the internal and external genitalia, resulting from declining levels of endogenously produced estrogens.

Cross-References

▶ Dyspareunia and Vaginismus

VVH

▶ Viscero-Visceral Hyperalgesia

Vulvo-Vaginal Atrophy

Definition

Vulvo vaginal atrophy is a manifestation of tissue aging and the cytological and chemical transformations within the internal and external genitalia, resulting from declining levels of endogenously produced estrogens.

Cross-References

▶ Dyspareunia and Vaginismus

VVA

▶ Vulvo-Vaginal Atrophy

W

Waddell Signs

▶ Nonorganic Symptoms and Signs

Wage Replacement

Definition

The wage replacement benefit is the ratio of the expected disability benefit to the preinjury wage.

Cross-References

▶ Pain in the Workplace, Risk Factors for Chronicity, and Workplace Factors

Wage Replacement Benefit

▶ Worker Compensation Benefits

Walking Epidural

Definition

"Walking epidural" is a term used to describe an epidural technique whereby a woman in labor has analgesia but is also able to ambulate.

The medication used to confer epidural analgesia typically causes lower extremity motor weakness. With the "walking epidural" technique, a small concentration of local anesthetic with an opioid is used to achieve analgesia while maintaining lower extremity motor function. There is no scientific evidence demonstrating a benefit with respect to obstetric outcome, but women are more satisfied when they can move their legs during labor.

Cross-References

▶ Analgesia During Labor and Delivery

Wallerian Degeneration

Shlomo Rotshenker
Department of Medical Neurobiology, IMRIC,
Faculty of Medicine Hebrew University of
Jerusalem, Jerusalem, Israel

Synonyms

Cytokines, upregulation in inflammation neuropathic pain model, chronic constriction injury; Dorsal root ganglionectomy and dorsal rhizotomy; Neuropathic pain model, diabetic neuropathy model; Painless neuropathies

G.F. Gebhart, R.F. Schmidt (eds.), *Encyclopedia of Pain*, DOI 10.1007/978-3-642-28753-4,
© Springer-Verlag Berlin Heidelberg 2013

Definition

▶ Wallerian degeneration, named after Waller (1850), defines the array of cellular and molecular events that follow a traumatic injury to PNS (peripheral nervous system) ▶ axons. Wallerian degeneration takes place throughout the nerve segment situated distal to a lesion site: anterograde degeneration.

Introduction

Wallerian degeneration can be viewed as the inflammatory, innate immune response of the PNS to traumatic injury. The production of cytokines, the mediator molecules of inflammation; the involvement of monocytes/macrophages, which are inflammatory/innate immune cells; and the phagocytosis of degenerated myelin, which is an innate immune function, indicate this.

The term Wallerian degeneration is sometimes used in reference to a traumatic injury to CNS (central nervous system) axons even though cell types, cellular events, and outcomes differ substantially. The term Wallerian degeneration is also occasionally used to define events that develop during PNS neuropathies without trauma but those that differ from injury-induced Wallerian degeneration. The term Wallerian degeneration that is used here refers to injury-induced Wallerian degeneration in the PNS.

Characteristics

Cellular Characteristics of Wallerian Degeneration

In intact PNS (Fig. 1a), ▶ Schwann cells surround axons and form myelin sheaths around the larger diameter sensory and motor axons. In between nerve fibers, ▶ fibroblasts and a few mast cells and macrophages are scattered. Endothelial cells are present within walls of capillaries that nourish the PNS tissue. Traumatic injury produces immediate tissue damage at the lesion site where physical impact occurred (Fig. 1b). Monocytes/macrophages that originate

from ruptured vasculature rapidly accumulate at the lesion site. Then, the nerve stump that is located distal to the lesion site undergoes Wallerian degeneration even though it did not encounter the physical trauma directly. Among others, axons break down, Schwann cells reject the myelin portion of their membranes and proliferate, fibroblasts proliferate, and bone-marrow-derived monocytes/macrophages are recruited from the circulation. Recruitment begins on the second day after the injury, but macrophages reach significant numbers as of the fourth day after the injury. Activated macrophages and Schwann cells complete the removal of degenerated myelin by phagocytosis within 8–12 days. The clearance of the degenerated myelin is essential for successful regeneration and restoration of function since myelin contains molecules which inhibit the growth of severed adult axons.

The simplistic view that Wallerian degeneration emanates from the loss of metabolic support of axons due to disconnection from their cell bodies is at most partial since in mutant Wld^s mice axon/myelin degeneration, monocytes/macrophages recruitment, and myelin clearance are delayed for many days after the injury. The finding of the aberrant molecule that is composed of the N-terminal 70 amino acid of the multiubiquitination factor Ube4b fused to NAD$^+$-synthesizing enzyme Nmnat1 in Wld^s mice led to the notion that isoform(s) of Nmnat, which are produced in neuronal cell bodies and transported anterogradely, protects axons by inhibiting a self-destructing mechanism (Gilley and Coleman 2010; Avery et al. 2009; Sasaki et al. 2009).

Taken together, Wld^s mice display abnormal slow progression of Wallerian degeneration compared to the normal faster progression of Wallerian degeneration that wild-type mice display as described above (Brown et al. 1991; Reichert et al. 1994). Wallerian degeneration in wild-type mice will be referred to as Wallerian degeneration or "normal Wallerian degeneration," and Wallerian degeneration in Wld^s mice will be referred to as "slow Wallerian degeneration."

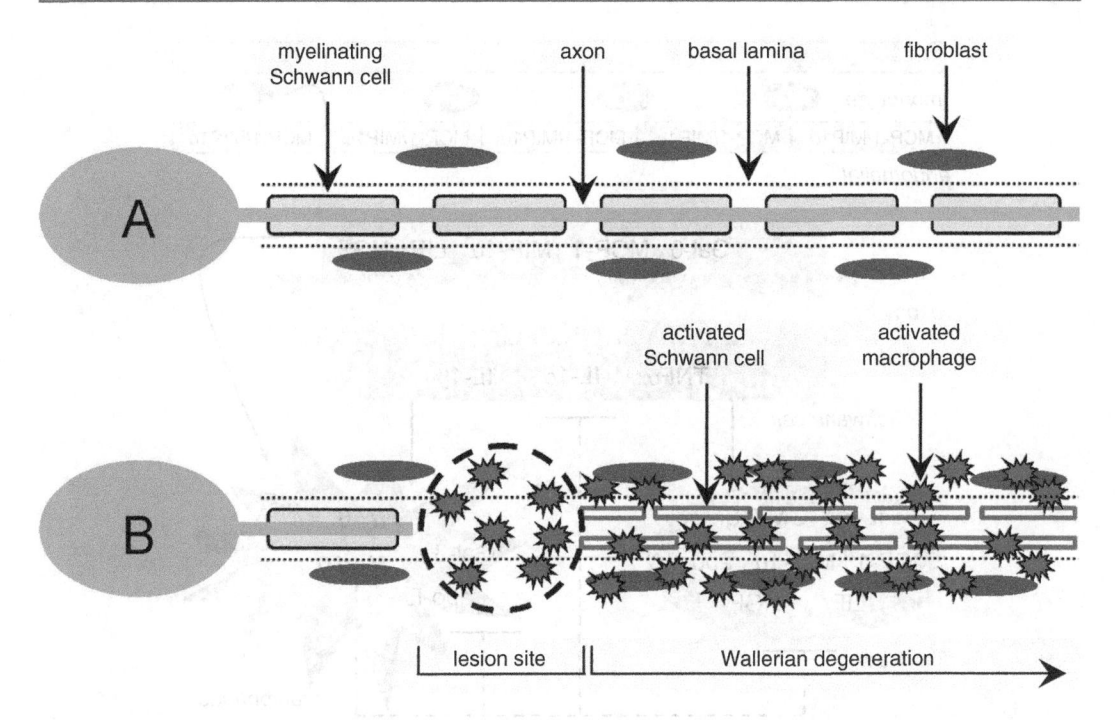

Wallerian Degeneration, Fig. 1 *Intact (A) and injured (B) PNS*. (A) Intact axons are surrounded by myelin-forming Schwann cells, and fibroblasts are scattered between nerve fibers. (B) Traumatic injury produces immediate tissue damage at lesion sites where macrophages accumulate. Normal Wallerian degeneration then develops throughout distal to lesion sites. Among others, axons degenerate, Schwann cells reject their myelin, macrophages are recruited, and macrophages and Schwann cells are activated to phagocytose myelin. In complete PNS injury, all axons undergo normal Wallerian degeneration. In partial PNS injury, lesioned axons and nonneuronal cells participating in Wallerian degeneration are situated next to intact axons and their associated nonneuronal cells and receptors (envisage axons A and B next to each other)

The Cytokine Network of Wallerian Degeneration

Detailed examinations of cytokine-mRNA expression and cytokine-protein synthesis and secretion in wild-type mice that display normal Wallerian degeneration indicate that cytokine production is orchestrated in time and magnitude, thereby forming the cytokine network of Wallerian degeneration (Fig. 2) (Rotshenker et al. 1992; Reichert et al. 1996; Saada et al. 1996; Be'eri et al. 1998; Shamash et al. 2002; Mirski et al. 2003; Rotshenker 2011). The producing nonneuronal cell types, their spatial distribution in the PNS tissue, and the timing of monocyte/macrophage recruitment determine timing and magnitude.

Schwann cells are the first to respond rapidly to axotomy by producing the inflammatory cytokines TNFα followed by IL-1α. The rapid response of Schwann cells is possible since (1) they form intimate contact with axons and are thus the first among the nonneuronal cells to "sense" axonal injury, (2) they normally express TNFα and IL-1α mRNAs, and (3) they normally contain low levels of TNFα protein. Fibroblasts follow by producing IL-6 and GM-CSF within 2 and 4 h after the injury, respectively. IL-6 and GM-CSF production can be induced by diffusible TNFα and IL-1α synthesized and secreted by Schwann cells. IL-1β, whose onset of production by Schwann cells is delayed by 5–10 h after the injury, can further contribute to IL-6 and GM-CSF production.

Schwann cell-derived TNFα, IL-1α, and IL-1β advance monocyte/macrophage recruitment directly and indirectly by inducing the production of MCP-1 (chemoattractant protein-1, known also as CCL2, (C-C motif ligand 2) and

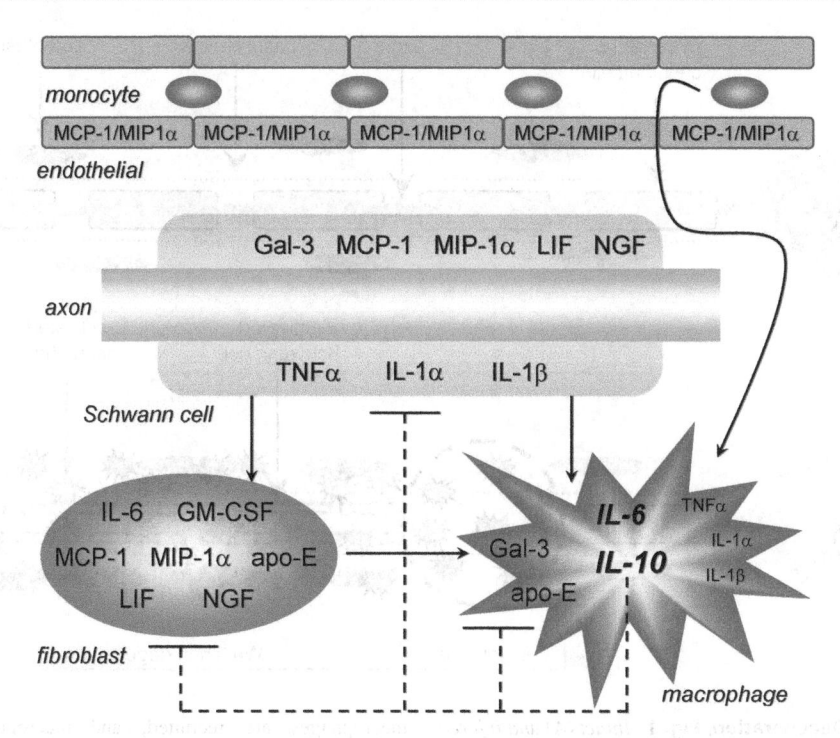

Wallerian Degeneration, Fig. 2 *The cytokine network of Wallerian degeneration.* The cellular elements depicted are a resident Schwann cell surrounding an axon, a resident fibroblast, endothelial cells forming the wall of a capillary containing circulating monocytes, and a recruited macrophage that acquired the M2 phenotype. *Solid* and *dotted lines,* respectively, represent up- and downregulation of cytokines' protein production, and the *curved line* represents the transmigration of a recruited monocyte/macrophage. Axotomy induces the production of TNFα and IL-1α in resident Schwann cells first. Sequentially thereafter follow IL-6 and GM-CSF production in resident fibroblasts and IL-1 β production in resident Schwann cells. Schwann cells, fibroblasts, and endothelial cells further produce chemokines MCP-1α and MIP-1α which advance monocyte/macrophage recruitment. Apo-E (apolipoprotein-E) and Gal-3 (Galectin-3) drive recruited monocytes/macrophages to differentiate into the M2 phenotype which produces high levels of IL-10 and IL-6 but low levels of TNFα, IL-1α, and IL-1 β. The anti-inflammatory cytokine IL-10 downregulates the production of all inflammatory cytokines and itself in all nonneuronal cells. Schwann cells and fibroblasts also produce the neurotrophic factors NGF and LIF. Not shown are the breakdown of the axon and the myelin, the phagocytosis of the degenerated myelin, and the additional cellular and molecular interactions that are detailed in text (After Shamash et al. 2002; Rotshenker 2011)

MIP-1α (macrophage inflammatory protein-1α, known also as CCL3) in Schwann cells, fibroblasts, mast cells, and endothelial cells. MCP-1/CCL2 and MIP-1α/CCL3 then promote the transmigration of circulating monocytes across the endothelial cell wall of blood vessels (Subang and Richardson 2001; Perrin et al. 2005). Concomitantly, Schwann cell-derived TNFα, IL-1α, and IL-1β induce recruited monocytes/macrophages to synthesize high levels of IL-6 but low levels of TNFα, IL-1α, and IL-1β.

Schwann cell-derived TNFα, IL-1α, and IL-1β further induce the production of the anti-inflammatory cytokine IL-10 in fibroblasts and the recruited monocytes/macrophages (Be'eri et al. 1998; Rotshenker 2011). Indeed, the onset of IL-10 production by fibroblasts is rapid, but levels of production are low and insignificant. High levels of IL-10 are produced by, and therefore concomitant with, monocyte/macrophage recruitment from the fourth day of Wallerian degeneration. IL-10 then downregulates the production of the inflammatory cytokines and itself, thereby downregulating the inflammatory aspects of Wallerian degeneration. IL-6, which macrophages also produce, contributes to the downregulation of TNFα production. Remarkably, all cytokines augment myelin phagocytosis by macrophages.

The inflammatory nature of Wallerian degeneration and the role of cytokines are supported by observations of deficient cytokine production during slow Wallerian degeneration in mutant Wld^s mice. Notably, TNFα and IL-1α protein production fails in slow Wallerian degeneration in Wld^s mice even though their mRNAs are expressed, which suggests differential regulation between mRNA expression and protein synthesis. It is likely, therefore, that Schwann cell-derived TNFα and IL-1α play a critical role in setting the normal cytokine network and normal Wallerian degeneration in motion, and the failure of their production results in an abnormal deficient cytokine network and slow Wallerian degeneration.

Of note, production levels of TNFα, IL-1α, and IL-1β peak on the first day following the injury prior to monocyte/macrophage recruitment, and, furthermore, production levels decrease as recruited monocytes/macrophages increase in number (Shamash et al. 2002; Rotshenker 2011). Taken that recruited macrophages produce high levels of the anti-inflammatory cytokine IL-10 but low levels of the inflammatory cytokines TNFα, IL-1α, and IL-1β strongly suggests that these macrophages are of the M2 "wound healing" phenotype (Auffray et al. 2009; Benoit et al. 2008). This is in contrast to high-level production of inflammatory but low-level production of anti-inflammatory cytokines by macrophages of the M1 phenotype. Nonetheless, both M1 and M2 macrophages function as phagocytes.

Immune Inhibitory Receptors in Wallerian Degeneration

Innate immune functions are regulated by the interplay between activating and inhibitory signals; neither act in an "all or none" fashion. Inhibition is produced by a family of immune inhibitory receptors. SIRPα

(signal-regulatory-protein-α, known also as CD172α and SHPS1) is a member of this family (Barclay and Brown 2006; Matozaki et al. 2009). SIRPα is expressed on the surface of macrophages and microglia. Its ligand is the cell membrane protein receptor CD47 (known also as IAP – integrin-associated protein) that many cells express on their surface. Cells that express CD47 can downregulate their own phagocytosis by macrophages after CD47 binds and activates SIRPα on phagocytes. CD47 functions, therefore, as a marker of "self" that protects cells from activated autologous macrophages by sending a "do not eat me" signal. CD47 is expressed on myelin and the myelin-forming Schwann cells and oligodendrocytes, and, furthermore, myelin downregulates its own phagocytosis by macrophages and microglia through CD47-SIRPα interactions (Gitik et al. 2011). CD47 can function, therefore, as a marker of "self" that protects intact myelin and Schwann cells from activated macrophages in the PNS. This mechanism may be useful under normal conditions and while combating invading pathogens since it protects bystander intact myelin and myelin-forming cells from macrophages that are activated to scavenge and kill pathogens. However, the very same mechanism may turn harmful when faster removal of degenerated myelin is useful as after traumatic PNS injury. Therefore, it is most likely that normal Wallerian degeneration does not display the fastest possible rate of in vivo myelin clearance.

Neurotrophic Factors in Wallerian Degeneration

▶ Neurotrophic factors are an additional class of molecules whose expression is altered after PNS injury (reviewed in Terenghi 1999). For example, NGF (▶ Nerve Growth Factor), BDNF (brain-derived neurotrophic factor), NT-4 (neurotrophin-4), GDNF (glial-derived neurotrophic factor), and LIF (leukemia inhibitory factor) expressions are upregulated in normal Wallerian degeneration. In contrast, CNTF (ciliary neurotrophic factor) and NT-3 expressions are downregulated in normal Wallerian degeneration.

Cytokines and neurotrophic factors are closely associated. For example, NGF production upregulation is integrated into the cytokine network of Wallerian degeneration. NGF mRNA expression and protein synthesis is efficient in normal Wallerian degeneration but deficient in slow Wallerian degeneration (Brown et al. 1991). This discrepancy can be partially explained by the following: (1) the efficient versus deficient TNFα, IL-1α, and IL-1β production in normal and slow Wallerian degeneration, respectively (see above), and (2) the induction of NGF production in fibroblasts by these cytokines (Hattori et al. 1994). Furthermore, some molecules display both neurotrophic factor and cytokine properties, for example, IL-6, LIF, and CNTF (Patterson 1994; Stahl and Yancopoulos 1994).

Neuropathic Pain and Wallerian Degeneration

Delayed and reduced neuropathic pain in Wld^s mice that display slow Wallerian degeneration (Myers et al. 1996) and the ability to provoke neuropathic pain by inducing inflammation without axonal injury (Safieh-Garabedian et al. 1995; Woolf et al. 1997; Eliav et al. 2001) suggest that the molecular events associated with Wallerian degeneration play a major role in the development of neuropathic inflammatory pain (e.g., IL-1β, TNFα, and NGF). There are several potential sites of action for molecules produced in Wallerian degeneration. First, secreted/diffusible molecules may act upon the producing and neighboring nonneuronal cells in an autocrine/paracrine fashion. For example, TNFα and IL-1α secreted from Schwann cells may induce productions, among others, of IL-6, GM-CSF, MCP-1/CCL-2, and NGF in nearby fibroblasts (Fig. 2). Second, in instances of partial PNS injury, some axons are cut but others remain intact (Fig. 1; envisage axons A and B next to each other). Molecules (e.g., TNFα and NGF) secreted from the nonneuronal cells that participate in Wallerian degeneration may affect the neighboring intact axons, myelinated and non-myelinated; their surrounding Schwann cells and sensory

receptors/endings to alter electrical properties and thresholds; and the mechanisms suggested to be instrumental in the pathophysiology of neuropathic pain (Wu et al. 2002). Third, at the neuroma site, the region immediately proximal to the injury site (<1 mm in length), nonneuronal cells may produce cytokines and neurotrophic factors (e.g., documented for IL-1α, IL-6, IL-10, and NGF), which, in turn, may affect properties of lesioned axons and surrounding nonneuronal cells (Zimmermann 2001). Fourth, secreted molecules (e.g., NGF, BDNF, NT-3/4, CNTF, and LIF; Curtis et al. 1998) can be taken up by lesioned axons and reach sensory and motor nerve cell bodies by the retrograde axonal transport, where they may affect cell survival and phenotype. It is unclear to what extent this mechanism replaces target-derived retrograde transport of neurotrophic factors occurring under normal conditions.

Summary

Wallerian degeneration is the inflammatory innate immune response of the PNS to traumatic injury since inflammatory innate immune cells, functions, and molecules are involved. The characteristics of normal Wallerian degeneration are rapid onset and final conclusion, which is based on the orchestrated interplay between Schwann cells, fibroblasts, macrophages, endothelial cells, and the molecules they produce. The innate immune response of Wallerian degeneration helps turning the peripheral nerve tissue into an environment that supports axonal regeneration by removing inhibitory myelin and by upregulating neurotrophic properties. Therefore, normal Wallerian degeneration serves as a prelude for successful repair. The inflammatory innate immune response of Wallerian degeneration further controls some aspects of neuropathic pain. Taken together, the inflammatory innate immune mechanisms of Wallerian degeneration may be targeted to ensure successful functional recovery from trauma and, furthermore, to control neuropathic pain.

References

Auffray, C., Sieweke, M. H., & Geissmann, F. (2009). Blood monocytes: Development, heterogeneity, and relationship with dendritic cells. *Annual Review of Immunology, 27,* 669–692.

Avery, M. A., Sheehan, A. E., Kerr, K. S., Wang, J., & Freeman, M. R. (2009). WldS requires Nmnat1 enzymatic activity and N16-VCP interactions to suppress Wallerian degeneration. *The Journal of Cell Biology, 184,* 501–513.

Barclay, A. N., & Brown, M. H. (2006). The SIRP family of receptors and immune regulation. *Nature Reviews Immunology, 6,* 457–464.

Be'eri, H., Reichert, F., Saada, A., & Rotshenker, S. (1998). The cytokine network of Wallerian degeneration: IL-10 and GM-CSF. *European Journal of Neuroscience, 10,* 2707–2713.

Benoit, M., Desnues, B., & Mege, J. L. (2008). Macrophage polarization in bacterial infections. *Journal of Immunology, 181,* 3733–3739.

Brown, M. C., Perry, V. H., Lunn, E. R., Gordon, S., & Heumann, R. (1991). Macrophage dependence of peripheral sensory nerve regeneration: Possible involvement of nerve growth factor. *Neuron, 6,* 359–370.

Curtis, R., Tonra, J. R., Stark, J. L., Adryan, K. M., Park, J. S., Cliffer, K. D., Lindsay, R. M., & DiStefano, P. S. (1998). Neuronal injury increases retrograde axonal transport of the neurotrophins to spinal sensory neurons and motor neurons via multiple receptor mechanisms. *Molecular and Cellular Neuroscience, 12,* 105–118.

Eliav, E., Benoliel, R., & Tal, M. (2001). Inflammation with no axonal damage of the rat saphenous nerve trunk induces ectopic discharge and mechanosensitivity in myelinated axons. *Neuroscience Letters, 311,* 49–52.

Gilley, J., & Coleman, M. P. (2010). Endogenous Nmnat2 is an essential survival factor for maintenance of healthy axons. *PLoS Biology, 8,* e1000300.

Gitik, M., Liraz, Z. S., Oldenborg, P. A., Reichert, F., & Rotshenker, S. (2011). Myelin down-regulates myelin phagocytosis by microglia and macrophages through interactions between CD47 on myelin and SIRPalpha (signal regulatory protein-alpha) on phagocytes. *Journal of Neuroinflammation, 8,* 24.

Hattori, A., Iwasaki, S., Murase, K., Tsujimoto, M., Sato, M., Hayashi, K., & Kohno, M. (1994). Tumor necrosis factor is markedly synergistic with interleukin 1 and interferon-gamma in stimulating the production of nerve growth factor in fibroblasts. *FEBS Letters, 340,* 177–180.

Matozaki, T., Murata, Y., Okazawa, H., & Ohnishi, H. (2009). Functions and molecular mechanisms of the CD47-SIRPalpha signalling pathway. *Trends in Cell Biology, 19,* 72–80.

Mirski, R., Reichert, F., Klar, A., & Rotshenker, S. (2003). Granulocyte macrophage colony stimulating factor (GM-CSF) activity is regulated by a GM-CSF binding molecule in Wallerian degeneration following injury to peripheral nerve axons. *Journal of Neuroimmunology, 140,* 88–96.

Myers, R. R., Heckman, H. M., & Rodriguez, M. (1996). Reduced hyperalgesia in nerve-injured WLD mice: Relationship to nerve fiber phagocytosis, axonal degeneration, and regeneration in normal mice. *Experimental Neurology, 141,* 94–101.

Patterson, P. H. (1994). Leukemia inhibitory factor, a cytokine at the interface between neurobiology and immunology. *Proceedings of the National Academy of Sciences of the United States of America, 91,* 7833–7835.

Perrin, F. E., Lacroix, S., Aviles-Trigueros, M., & David, S. (2005). Involvement of monocyte chemoattractant protein-1, macrophage inflammatory protein-1alpha and interleukin-1beta in Wallerian degeneration. *Brain, 128,* 854–866.

Reichert, F., Saada, A., & Rotshenker, S. (1994). Peripheral nerve injury induces Schwann cells to express two macrophage phenotypes: Phagocytosis and the galactose-specific lectin MAC-2. *The Journal of Neuroscience, 14,* 3231–3245.

Reichert, F., Levitzky, R., & Rotshenker, S. (1996). Interleukin 6 in intact and injured mouse peripheral nerves. *European Journal of Neuroscience, 8,* 530–535.

Rotshenker, S. (2011). Wallerian degeneration: The innate-immune response to traumatic nerve injury. *Journal of Neuroinflammation, 8,* 109.

Rotshenker, S., Aamar, S., & Barak, V. (1992). Interleukin-1 activity in lesioned peripheral nerve. *Journal of Neuroimmunology, 39,* 75–80.

Saada, A., Reichert, F., & Rotshenker, S. (1996). Granulocyte macrophage colony stimulating factor produced in lesioned peripheral nerves induces the up-regulation of cell surface expression of MAC-2 by macrophages and Schwann cells. *The Journal of Cell Biology, 133,* 159–167.

Safieh-Garabedian, B., Poole, S., Allchorne, A., Winter, J., & Woolf, C. J. (1995). Contribution of interleukin-1 beta to the inflammation-induced increase in nerve growth factor levels and inflammatory hyperalgesia. *British Journal of Pharmacology, 115,* 1265–1275.

Sasaki, Y., Vohra, B. P. S., Baloh, R. H., & Milbrandt, J. (2009). Transgenic mice expressing the Nmnat1 protein manifest robust delay in axonal degeneration in vivo. *The Journal of Neuroscience, 29,* 6526–6534.

Shamash, S., Reichert, F., & Rotshenker, S. (2002). The cytokine network of Wallerian degeneration: Tumor necrosis factor- alpha, interleukin-1alpha, and interleukin-1beta. *The Journal of Neuroscience, 22,* 3052–3060.

Stahl, N., & Yancopoulos, G. D. (1994). The tripartite CNTF receptor complex: Activation and signaling involves components shared with other cytokines. *Journal of Neurobiology, 25,* 1454–1466.

Subang, M. C., & Richardson, P. M. (2001). Influence of injury and cytokines on synthesis of monocyte chemoattractant protein-1 mRNA in peripheral nervous tissue. *European Journal of Neuroscience, 13*, 521–528.

Terenghi, G. (1999). Peripheral nerve regeneration and neurotrophic factors. *Journal of Anatomy, 194*(Pt 1), 1–14.

Waller, A. (1850). Experiments on the section of the glossopharyngeal and hypoglossal nerves of the frog, observations of the alterations produced thereby in the structure of their primitive fibers. *Philosophical Transactions of Royal Society of London, 140*, 423–429.

Woolf, C. J., Allchorne, A., Safieh-Garabedian, B., & Poole, S. (1997). Cytokines, nerve growth factor and inflammatory hyperalgesia: The contribution of tumour necrosis factor alpha. *British Journal of Pharmacology, 121*, 417–424.

Wu, G., Ringkamp, M., Murinson, B. B., Pogatzki, E. M., Hartke, T. V., Weerahandi, H. M., Campbell, J. N., Griffin, J. W., & Meyer, R. A. (2002). Degeneration of myelinated efferent fibers induces spontaneous activity in uninjured C-fiber afferents. *The Journal of Neuroscience, 22*, 7746–7753.

Zimmermann, M. (2001). Pathobiology of neuropathic pain. *European Journal of Pharmacology, 429*, 23–37.

Warm Allodynia

Definition

See also "allodynia." Heat pain sensations to stimuli which normally are regarded as warm but not painful. This expression should be avoided since it is unclear if the pain is really evoked by sensitive warm receptors ("allo" means "other," i.e., another sensory channel but nociceptors) or rather by sensitized heat nociceptors.

Therefore, a better term is "heat hyperalgesia."

Cross-References

▶ Neuropathic Pain Model, Tail Nerve Transection Model

WDR Neuron

▶ Wide Dynamic Range Neuron

Wegener's Granulomatosis (WG)

Definition

A rare systemic necrotizing vasculitis associated with antineutrophil cytoplasm antibodies (c-ANCA) leading to affections of the lung and kidney.

Cross-References

▶ Headache Due to Arteritis

Weighted Scores Technique

Definition

The weighted scores technique identifies distinct (usually four) categories (0 = injected paw is not favored, 1 = injected paw has little or no weight on it, 2 = injected paw is elevated and not in contact with any surface, 3 = injected paw is licked, bitten, or shaken). A weighted pain score (which ranges from 0 to 3) is calculated by multiplying the time spent in each category by the category weight, summing these products and dividing by the total time in a given time interval.

Cross-References

▶ Formalin Test

Weight-Lifter's Headache

▶ Primary Exertional Headache

Wernicke-Korsakoff Syndrome

Definition

This consists of two distinct syndromes caused by the same pathology caused by thiamine deficiency often secondary to alcohol abuse. Wernicke syndrome describes ophthalmoparesis, ataxia, and global confusion. Korsakoff syndrome affects cognitive function and is dominated by both anterograde and retrograde amnesia often associated with confabulation and impaired insight.

Cross-References

▶ Metabolic and Nutritional Neuropathies

West Haven-Yale Multidimensional Pain Inventory

Synonyms

WHYMPI

Definition

The West Haven-Yale Multidimensional Pain Inventory (WHYMPI) is a multidimensional self-report instrument that consists of 52 questions and is designed to measure psychosocial and behavioral aspects of chronic pain across a variety of clinical populations.

Cross-References

▶ Pain Inventories

WGA-HRP

Definition

Wheat germ agglutinin coupled with horseradish peroxidase. This protein is extracted from wheat and coupled to an enzyme that has a high affinity for the soma and axon of a neuron. It migrates in both retrograde (antidromic) and anterograde

(orthodromic) directions. It labels somata of afferent projections and, less effectively than PHA-L, efferent axonal projections.

Cross-References

▶ Parabrachial Hypothalamic and Amygdaloid Projections

Whiplash

Michele Curatolo[1] and Nikolai Bogduk[2]
[1]Department of Anesthesiology, Division of Pain Therapy, University Hospital of Bern Inselspital, Bern, Switzerland
[2]University of Newcastle, Newcastle Bone & Joint Institute, Royal Newcastle Centre, Newcastle, NSW, Australia

Synonyms

Acceleration-deceleration injury; Whiplash-associated disorders; Whiplash syndrome

Definition

The term "▶ whiplash" is used to describe both an event and an injury. The whiplash event is the movement experienced by the head and neck during a motor vehicle accident in which no force is directly exerted onto the head. The whiplash injury is the injury that may occur to the neck as a result of this movement. The symptoms that arise from either the event or the injury are described as a whiplash-associated disorder (WAD).

Characteristics

Etiology

Classically, whiplash occurs during a rear-end collision. However, only about 45 % of cases of

WAD arise from rear-end collisions. The remainder occur in front-end, side-impact, and combined collisions (Bogduk 2000).

Clinical Features

Not all patients involved in a whiplash event suffer, or seek treatment for, symptoms. In those that do, the symptoms typically arise within 24 h of the event. Patients who do develop symptoms in this period are three times more likely to suffer chronic neck pain than members of the general community or individuals who do not develop symptoms after a motor vehicle accident (Berglund et al. 2000).

In the acute phase, the most frequent symptoms are neck pain and headache. Other features that can occur in a minority of patients include anxiety, sleep disturbances, back pain, visual disturbances, dizziness, inability to concentrate, and other cognitive disturbances (Barnsley et al. 2002). In patients who do not recover, these same symptoms persist, but with greater prevalence in affected individuals, together with psychological distress (Radanov et al. 1995). In the chronic phase, patients display ▶ hyperalgesia to stimuli applied to both painful and non-painful areas of the body (Banic et al. 2004).

Natural History and Prognostic Factors

Most patients who suffer a whiplash-associated disorder recover. In the acute phase, possible predictors of non-recovery are pain and/or disability levels, neck range of movement, cold and mechanical hyperalgesia, psychological factors of recovery beliefs/expectations, post-traumatic stress symptoms, depression, and pain catastrophizing, but the evidence is still limited (Sterling et al. 2011). The chance of spontaneous recovery after two years is minimal.

Mechanisms

During the whiplash event, the torso is thrust upward and forward and compresses the neck from below. Initially, the cervical spine undergoes a sigmoid deformation, but subsequently the head drops into extension, and later rebounds into flexion (Bogduk and Yoganandan 2001). At no time does the head and neck exceed normal physiologic range of motion. The offending insult is the initial compression and sigmoid deformation of the cervical spine. During this motion, the lower vertebrae undergo an abnormal pattern of extension, in which anteriorly, the intervertebral discs are strained, and posteriorly, the zygapophysial joints are impacted (Bogduk and Yoganandan 2001). Injury to these structures may or may not occur, depending on the severity of impact and the morphology and strength of the vertebrae and their joints and ligaments.

Most patients suffer no substantive or lasting injury, but a minority does. Tears of the ▶ annulus fibrosus and contusion or small fractures of the zygapophysial joints have been demonstrated in postmortem studies, but these defy resolution by current medical imaging techniques (Yoganandan et al. 2001), giving the false impression that no injury is present.

Thus, there is evidence supporting a lesion-based model in whiplash-associated disorders (Curatolo et al. 2011). Additionally, psychosocial factors such as beliefs, expectations, coping, and depression have been shown to influence the outcome of whiplash (Carroll 2011). Hypersensitivity of central nociceptive pathways in chronic whiplash patients has been demonstrated (Banic et al. 2004). It may amplify the nociceptive signal arising from areas of minimal and undetectable tissue damage, which could partly explain the discrepancy between objective clinical findings and extent of pain and disability that occurs in these patients. There is some evidence for an influence of sensory hypersensitivity on the outcome of whiplash (Sterling et al. 2003).

Physical Examination

Both in the acute phase and in patients with chronic pain, physical examination may detect areas of tenderness, impaired range of movement of the neck, or additional functional disturbances. However, the validity of these signs to make a patho-anatomic diagnosis is uncertain.

Neurological examination is usually normal, but a small proportion of patients may suffer cervical disc prolapse and nerve root compression as a result of whiplash. These patients will exhibit signs of ▶ radiculopathy.

Investigations

Medical imaging is not indicated for patients with just neck pain after whiplash. According to the Canadian C-spine rule, radiography is not justified in a patient who has been in a motor-vehicle collision and can rotate their neck at least 45° to the left and right (Stiell et al. 2001) Magnetic resonance imaging is of no diagnostic value for most patients (Anderson et al. 2012). Its main indication is the presence of neurological signs.

In patients with chronic neck pain after whiplash, the zygapophysial joints are the single most common source of pain, the prevalence being about 50 % (Bogduk 2011). Zygapophysial joint pain cannot be diagnosed by physical examination or medical imaging; it requires controlled diagnostic blocks performed under fluoroscopic guidance (Curatolo and Bogduk 2010).

Treatment

Given the favorable prognosis of acute pain after whiplash, the first imperative is for the physician to reassure the patient of the high likelihood of complete recovery. Passive therapy is not indicated and, indeed, the evidence shows that conventional interventions do not work (Jull et al. 2011). There is even evidence that too much health care and rehabilitation too early after the injury can be associated with delayed recovery and the development of chronic pain and disability (Cote and Soklaridis 2011). It is therefore imperative to reassure the patients, encourage them to resume normal activities, and to keep the neck moving through simple exercises that the patient can do regularly, themselves, at home. Analgesics may be prescribed to reduce pain, but no evidence supports their efficacy in preventing persistent pain.

For patients with acute neurological symptoms, one study has shown that high-dose intravenous ▶ methylprednisolone was effective in reducing the number of sick days (Pettersson and Toolanen 1998), but this treatment may not apply for patients who do not have radiculopathy.

For patients with chronic neck pain after whiplash, few treatments have been vindicated by controlled studies. Inconclusive evidence exists for the effectiveness of physiotherapy (Rushton et al. 2011). A systematic review could neither support nor refute the effectiveness of conservative treatments: Individual studies demonstrated effectiveness of one treatment over another, but the comparisons were varied and the results inconsistent (Verhagen et al. 2007). Intra-articular injections of corticosteroids for cervical zygapophysial joint pain offer no long-term and little short-term benefit (Barnsley et al. 1995).

Multidisciplinary biopsychosocial rehabilitation programs are widely used. However, these treatment programs are expensive and evidence of effectiveness is equivocal. Prospective uncontrolled studies show benefits (Angst et al. 2010), but different controlled trials did not find evidence of efficacy (Cassidy et al. 2007). Early rehabilitation may even be detrimental (Pape et al. 2009; Cote and Soklaridis 2011). There is some evidence that the presence of widespread hyperalgesia is a predictor of poor outcome of rehabilitation programs (Jull et al. 2007). Thus, it is still challenging to identify patients who are more likely to benefit from these programs. Randomized controlled trials that evaluate their effectiveness are urgently needed.

A further option for pain relief may be the long-term use of analgesics. Unfortunately, however, there is no evidence of their efficacy. Opioids may be required for severe, persistent pain, but these need to be prescribed and monitored carefully.

Radiofrequency denervation of painful zygapophysial joints has proven effective in a randomized, double-blind, placebo-controlled trial (Lord et al. 1996), and subsequent, long-term follow-up studies (Bogduk 2011). This procedure is indicated only in those patients in whom controlled diagnostic blocks have identified the joint responsible for the patient's pain (Bogduk 2011).

Discussion

Pain after whiplash injury is a temporary experience for most patients: spontaneous healing is almost the rule. Patients with persistent symptoms, however, constitute an important medical and social problem.

As patients with persisting pain lack conventional, objective signs of organic lesions, they are often treated in a pejorative manner. Resources, such as zygapophysial joint blocks, by which the source of pain can be determined, are not readily available or are not implemented. Conventional therapies have been found to have little or no effect, and few practitioners are able to implement those few interventions that have been shown to be effective. As a result of these factors, patients continue to suffer pain and disability. Beyond the misfortune of getting an illness that changes their life, patients are confronted with the new experience of having to fight insurance companies, employers, and doctors, to convince them that they really "do have something wrong with their neck." The resulting psychological distress is an obvious consequence of this state of affairs and likely affects the magnitude of symptoms.

Patients with acute neck pain after whiplash have the advantage of a favorable prognosis. Being reassured and resuming normal activities is all that they may require for recovery. For patients with persisting pain after whiplash, two main problems apply. First is the lack of reliable diagnostic tools to identify the source of pain. The most honest approach by the physician is to declare and explain this deficiency, as opposed to insisting that there is nothing wrong. The second problem is the lack of proven treatments. This too should be explained, while making the best use of the available evidence.

References

Anderson, S. E., Boesch, C., Zimmermann, H., Busato, A., Hodler, J., Bingisser, R., Ulbrich, E. J., Nidecker, A., Buitrago-Tellez, C. H., Bonel, H. M., Heini, P., Schaeren, S., & Sturzenegger, M. (2012). Are there cervical spine findings at MR imaging that are specific to acute symptomatic whiplash injury? A prospective controlled study with four experienced blinded readers. *Radiology, 262*, 567–575.

Angst, F., Francoise, G., Verra, M., Lehmann, S., Jenni, W., & Aeschlimann, A. (2010). Interdisciplinary rehabilitation after whiplash injury: An observational prospective outcome study. *Journal of Rehabilitation Medicine, 42*, 350–356.

Banic, B., Petersen-Felix, S., Andersen, O. K., Radanov, B. P., Villiger, P. M., Arendt-Nielsen, L., & Curatolo, M. (2004). Evidence for spinal cord hypersensitivity in

chronic pain after whiplash injury and in fibromyalgia. *Pain, 107*, 7–15.

Barnsley, L., Lord, S. M., Wallis, B. J., & Bogduk, N. (1995). The prevalence of chronic cervical zygapophysial joint pain after whiplash. *Spine, 20*, 20–25.

Barnsley, L., Lord, S. M., & Bogduk, N. (2002). The pathophysiology of whiplash. In G. A. Malanga & S. F. Nadler (Eds.), *Whiplash* (pp. 41–77). Philadelphia: Hanley & Belfus.

Berglund, A., Alfredsson, L., Cassidy, J. D., Jensen, I., & Nygren, A. (2000). The association between exposure to a rear-end collision and future neck or shoulder pain: A cohort study. *Journal of Clinical Epidemiology, 53*, 1089–1094.

Bogduk, N. (2000). An overview of whiplash. In N. Yoganandan & F. A. Pintar (Eds.), *Frontiers in whiplash trauma* (pp. 3–9). Amsterdam: IOS Press.

Bogduk, N. (2011). On cervical zygapophysial joint pain after whiplash. *Spine, 36*, S194–S199.

Bogduk, N., Yoganandan N. (2001). Biomechanics of the cervical spine Part 3: minor injuries. *Clin Biomech (Bristol, Avon), 16*, 267–275.

Carroll, L. J. (2011). Beliefs and expectations for recovery, coping, and depression in whiplash-associated disorders: Lessening the transition to chronicity. *Spine, 36*, S250–S256.

Cassidy, J. D., Carroll, L. J., Cote, P., & Frank, J. (2007). Does multidisciplinary rehabilitation benefit whiplash recovery?: Results of a population-based incidence cohort study. *Spine, 32*, 126–131.

Cote, P., & Soklaridis, S. (2011). Does early management of whiplash-associated disorders assist or impede recovery? *Spine, 36*, S275–S279.

Curatolo, M., & Bogduk, N. (2010). Diagnostic and therapeutic nerve blocks. In S. M. Fishman, J. C. Ballantyne, & J. P. Rathmell (Eds.), *Bonica's manament of pain* (pp. 1401–1423). Philadelphia: Lippincott William & Wilkins.

Curatolo, M., Bogduk, N., Ivancic, P. C., McLean, S. A., Siegmund, G. P., & Winkelstein, B. A. (2011). The role of tissue damage in whiplash-associated disorders: Discussion paper 1. *Spine, 36*, S309–S315.

Jull, G., Sterling, M., Kenardy, J., & Beller, E. (2007). Does the presence of sensory hypersensitivity influence outcomes of physical rehabilitation for chronic whiplash?–A preliminary RCT. *Pain, 129*, 28–34.

Jull, G. A., Soderlund, A., Stemper, B. D., Kenardy, J., Gross, A. R., Cote, P., Treleaven, J., Bogduk, N., Sterling, M., & Curatolo, M. (2011). Toward optimal early management after whiplash injury to lessen the rate of transition to chronicity: Discussion paper 5. *Spine, 36*, S335–S342.

Lord, S. M., Barnsley, L., Wallis, B. J., McDonald, G. J., & Bogduk, N. (1996). Percutaneous radio-frequency neurotomy for chronic cervical zygapophyseal-joint pain. *The New England Journal of Medicine, 335*, 1721–1726.

Pape E., Hagen K. B., Brox J. I., Natvig B., & Schirmer H. (2009). Early multidisciplinary evaluation and advice was ineffective for whiplash-associated disorders. *Eur J Pain 13*, 1068–1075.

Pettersson, K., & Toolanen, G. (1998). High-dose methylprednisolone prevents extensive sick leave after whiplash injury. A prospective, randomized, double-blind study. *Spine, 23*, 984–989.

Radanov, B. P., Sturzenegger, M., & Di Stefano, G. (1995). Long-term outcome after whiplash injury. A 2-year follow-up considering features of injury mechanism and somatic, radiologic, and psychosocial findings. *Medicine, 74*, 281–297.

Rushton, A., Wright, C., Heneghan, N., Eveleigh, G., Calvert, M., & Freemantle N. (2011). Physiotherapy rehabilitation for whiplash associated disorder II: a systematic review and meta-analysis of randomised controlled trials. *BMJ Open, 1*, e000265.

Sterling, M., Jull, G., Vicenzino, B., & Kenardy, J. (2003). Sensory hypersensitivity occurs soon after whiplash injury and is associated with poor recovery. *Pain, 104*, 509–517.

Sterling, M., Carroll, L. J., Kasch, H., Kamper, S. J., & Stemper, B. (2011). Prognosis after whiplash injury: where to from here? Discussion paper 4. *Spine, 36*, S330–334.

Stiell, I. G., Wells, G. A., Vandemheen, K. L., et al. (2001). The Canadian C-spine rule for radiography in alert and stable trauma patients. *Journal of the American Medical Association, 286*, 1841–1848.

Verhagen, A. P., Scholten-Peeters, G. G., van Wijngaarden, S., de Bie, R. A., & Bierma-Zeinstra S.M. (2007). Conservative treatments for whiplash. *Cochrane Database Syst Rev, CD003338*.

Yoganandan, N., Cusick, J. F., Pintar, F. A., Rao, R. D (2001). Whiplash injury determination with conventional spine imaging and cryomicrotomy. *Spine, 26*, 2443–2448.

Whiplash Syndrome

▶ Whiplash

Whiplash-associated Disorders

▶ Whiplash

WHO Analgesic Ladder

▶ Cancer Pain Management, Principles of Opioid Therapy, Drug Selection

WHO International Classification of Functioning

▶ WHO System on Impairment and Disability

WHO System on Impairment and Disability

Jenny Strong
University of Queensland, Brisbane, QLD, Australia

Synonyms

Disability and health (ICF) and ICF core sets; WHO International classification of functioning

Definition

The World Health Organization's (WHO) International Classification of Functioning, Disability and Health (ICF) (WHO 2001) provides a framework for health practitioners to evaluate the impact of health conditions upon particular individuals. Against the background of contextual, personal, and environmental factors, the impact of the health condition upon the person is considered in terms of the interaction between body structures and functions, activities, and participation (Gibson and Strong 2003; WHO 2001). The ICF model conceptualizes the functional aspects of health, moving from an impairment-focused or deficit-focused framework to a participative framework. The WHO ICF model is illustrated in Fig. 1. Instead of focusing on impairments, which are problems in body function, such as significant deviation or loss, attention is given to not only the difficulties the person may have in executing activities, but also to the problems the person has in participating in life situations.

Health condition
(disorder or disease)

**WHO System on Impairment and Disability,
Fig. 1** Interaction of Concepts ICF 2001 (WHO 2001),
reproduced with permission

Characteristics

The ICF (http://www.who.int) was developed
to provide a unified and universally understood
language to describe the outcomes of all
health conditions (Gray and Hendershot 2000).
It provides a useful framework for considering
the impacts of chronic pain upon the individual
client or patient. It is also suggested as
a useful framework for considering the outcomes
of *treatments of pain*. The ICF particularly directs
the health practitioner to be clear about the level
at which they are measuring the outcomes,
namely, are they confining their outcome assessment to body structures and impairment-based
systems (e.g., active and passive range of
motion, allodynia, and presence of vasomotor
and sudomotor changes); or are they considering
outcomes in terms of activity limitations or participation restrictions (e.g., ADLs, and life roles)
(Unsworth 2000). Furthermore, are the individual
patient's particular personal and environmental
factors being considered in such outcome
measurements (e.g., financial pressures upon the
family unit, and fear of failure).

Personal factors to be considered may include
sex, age, comorbidities, usual coping style, social
background and education, job, past experiences,
and type of personality. Patients with chronic
pain are embedded within contexts that can facilitate or hinder their rehabilitation. Environmental

factors may be familial, work-related, cultural,
political, societal, or of the built environment
(WHO 2001). The ICF classification is, however,
exhaustive, containing more than 1,400 categories, thereby rendering it impractical for routine
use (Stucki et al. 2008). To facilitate operationalization of the ICF and its use in practice, ICF
Core Sets were developed for pain conditions
ranging from widespread chronic pain, low back
pain, and rheumatoid arthritis (Cieza et al. 2004).
The Core Sets provide a smaller number of the
most salient categories for a given condition.
Consequently, the Core Sets have been proposed
as a tool to guide the development, documentation, and monitoring of patient-generated
rehabilitation goals (Lohmann et al. 2011; Muller
et al. 2011; Soberg et al. 2008).

Considerable attention has been given to
mapping the existing outcome measures to the
ICF, with results suggesting good coverage of
body functions and structure, activities, and
participation, but less satisfactory coverage of
environmental and personal factors (Alviar et al.
2011; Weigl et al. 2006).

Use of the ICF framework can assist in the
measurement of outcomes of relevance for the
individual client/patient, the health care provider,
and the insurer or employer. The patient will most
probably be concerned with the elimination of
pain, an increase in function, and a return to
work. The health care professional will be
concerned with pain control, assisting the patient
to gain an increase in function, and a decrease in
suffering. The insurer or employer will primarily
be concerned with return to work. The value of
the ICF framework is illustrated by considering
the case of a 42-year-old man who sustained
a soft tissue injury at work, and developed complex regional pain syndrome Type I (CRPSI).
The patient was a miner and sole provider for
the family. He was married with two children.
Assessment using a body structure and impairment framework revealed he had no active range
of motion of his right hand (guarding), a reduced
passive range of motion of his right hand,
allodynia, and marked vasomotor and sudomotor
changes.

WHO System on Impairment and Disability, Table 1 Assessment matrix, based upon the WHO ICF 2001

Body function and structures	Activities	Participation
Muscle testing	ADL assessment	Home assessment
Range of Motion testing	Functional Capacity Evaluation	Work-place assessment[a]
NB Psychosocial assessment of personal factors		

[a]Ergonomic assessment of environmental factors

Considering the patient's ability to engage in activities, he reported limitations in all functional activities that required bilateral upper limb involvement, including personal activities of daily living, for example, pushing, lifting, using controls (driving), handling, and carrying tasks. He was fearful of increased pain, and therefore, self-limited many of his instrumental activities, such as walking near other people, or going out of the house to shop, or socialize. He reported reduced participation in all home-based activities, and was currently off work. There were significant financial pressures upon his family. This was combined with self-esteem problems he faced, having been "transformed" from a strong, independent, successful miner to a weak, disabled person. His employer and fellow workers did not understand his pain problem, and suspected he was malingering. His current problems with body function would preclude him from working with heavy machinery in an underground mine. The ICF provides a useful framework for evaluating such patients with chronic pain, and highlights the multidimensional nature of assessment, enabling a matrix of outcome measurements to be identified for use in a pain rehabilitation program (see Table 1).

By using an assessment matrix such as the one in Table 1, which is based upon the WHO ICF, it is easier for health practitioners working with patients with pain, and pain clinics in general, to ensure that the focus of assessment is not only focused upon the impairment, or body function and structures level. Good outcomes for pain treatments should be focused upon the activities

and participation levels as indicated by the WHO ICF classification. It is the outcome at these levels that are of most salience to patients and to insurers, and surely to health practitioners.

References

Alviar, M. J., Olver, J., Brand, C., Hale, T., & Khan, F. (2011). Do patient-reported outcome measures used in assessing outcomes in rehabilitation after hip and knee arthroplasty capture issues relevant to patients? Results of a systematic review and linking process. *Journal of Rehabilitation Medicine, 43*, 374–381.

Cieza, A., Ewart, T., Ustun, B., Chatterji, S., Kostanjsek, N., & Stucki, G. (2004). Development of ICF core sets for patients with chronic conditions. *Journal of Rehabilitation Medicine, 44*, 9–11.

Gibson, L., & Strong, J. (2003). A conceptual framework of functional capacity evaluation for occupational therapy in work rehabilitation. *Australian Occupational Therapy Journal, 50*, 64–71.

Gray, D. B., & Hendershot, G. E. (2000). The ICIDH-2: Developments for a new era of outcomes research. *Archives of Physical Medicine and Rehabilitation, 81* (Suppl 2), S10–S14.

Lohmann, S., Decker, J., Muller, M., Strobl, R., & Grill, E. (2011). The ICF forms a useful framework for classifying individual patient goals in post-acute rehabilitation. *Journal of Rehabilitation Medicine, 43*, 151–155.

Muller, M., Grill, E., & Strobl, R. (2011). Goals of patients with rehabilitation needs in acute hospitals: Goal achievement is an indicator for improved functioning. *Journal of Rehabilitation Medicine, 43*, 145–150.

Soberg, H. L., Finset, A., Roise, O., & Bautz-Holter, E. (2008). Identification and comparison of rehabilitation goals after multiple injuries: An ICF analysis of the patients', physiotherapists' and other allied professionals' reported goals. *Journal of Rehabilitation Medicine, 40*, 340–346.

Stucki, G., Kostanjsek, N., Ustun, B., & Cieza, A. (2008). ICF-based classification and measurement of functioning. *European Journal of Physical and Rehabilitation Medicine, 44*, 315–328.

Unsworth, C. (2000). Measuring of outcome of occupational therapy: Tools and resources. *Australian Occupational Therapy Journal, 47*, 147–158.

Weigl, M., Cieza, A., Kostanjsek, N., Kirschneck, M., & Stucki, G. (2006). The ICF comprehensively covers the spectrum of health problems encountered by health professionals in patients with musculoskeletal conditions. *Rheumatology, 45*, 1247–1254.

World Health Organisation (2001). International classification of functioning, disability and health. Geneva: World Health Organisation.

W

Whole Body Receptive Fields

Hans-Georg Schaible
Department of Physiology - Neurophysiology,
Friedrich-Schiller-University of Jena, Jena,
Germany

Definition

Stimulation anywhere on the body activates these neurons.

Cross-References

▶ Spinothalamocortical Projections to Ventromedial and Parafascicular Nuclei

Whole Cell Patch Clamp Recordings

Definition

A variant of the patch clamp electrophysiological technique that allows the characterization of the electrical behavior of the entire cell. Applying brief, gentle suction to a micropipette patched onto the cell membrane, the investigator ruptures this membrane, creating an opening that allows low resistance electrical access to the interior of the cell, and the ability to measure transmembrane ionic currents.

Cross-References

▶ Amygdala, Pain Processing and Behavior in Animals

Whole-Brain/Partial-Brain Coverage

Definition

The most common forms of functional imaging employ whole-brain coverage. This means that signal changes are measured throughout the whole brain. However, to achieve greater spatial resolution, e.g., for a study of differences between hippocampus and entorhinal cortex processing (or if faster acquisition of data is required), one option is to acquire a signal from a smaller section of the brain only (partial-brain coverage).

Cross-References

▶ Hippocampus and Entorhinal Complex: Functional Imaging

WHYMPI

▶ West Haven-Yale Multidimensional Pain Inventory

Wide Dynamic Range Neuron

Synonyms

WDR neuron

Definition

Wide dynamic range (WDR) neurons are nociceptive neurons that respond with small responses to innocuous stimuli and show stronger and graded responses to noxious mechanical stimulation of peripheral tissues. They often also respond to noxious, thermal and chemical stimuli. This type of neuron is defined physiologically by its response properties and can be found at spinal, thalamic, and cortical sites important in the processing of noxious information.

Cross-References

▶ Alternative Medicine in Neuropathic Pain
▶ Arthritis Model, Kaolin-Carrageenan-Induced Arthritis (Knee)

Wide-Dynamic Range Neurons (WDR)

Wide-Field Radiation

Definition

Radiation of a large area.

Cross-References

Willingness

Wind-up

Wind-Up (Phenomenon)

Definition

Repetitive stimulation of unmyelinated C-fibers
can result in prolonged discharge of dorsal horn
cells, termed "wind-up." It is characterized by
a progressive increase in the number of action
potentials elicited, per stimulus, which occurs in
dorsal horn neurons. "Wind-up" is a short-lived
process; however, repetitive episodes may precipitate long-term potentiation (LTP), which involves
a long-lasting increase in the efficacy of synaptic
transmission and thus alters synaptic plasticity.
Both "wind-up" and LTP are believed to be important components of central sensitization.

Cross-References

Windup of Spinal Cord Neurons

Juan F. Herrero
Department of Physiology, Faculty of
Medicine University of Alcala, Alcalá de
Henares Madrid, Spain

Synonyms

Frequency-dependent nociceptive facilitation

Definition

Progressive and frequency-dependent facilitation of the responses of a spinal cord neuron observed on the application of constant and high-intensity repetitive electrical stimuli. It is a phenomenon that shares some common mechanisms with central sensitization and is mediated by ▶ NMDA receptors and ▶ NK1 receptors, although cyclooxygenases, nitric oxide, TRH, adenosine, and other systems are also involved in its generation or maintenance.

Characteristics

Windup is a phenomenon described in terms of neuronal responses to repetitive electrical stimulation. It can be defined as a progressive and frequency-dependent facilitation of the responses of a spinal cord neuron observed on the application of constant and high-intensity repetitive electrical stimuli. It was first described by Lorne Mendell (Mendell and Wall 1965) as a frequency-dependent facilitation of spinal cord neuronal responses mediated by afferent C-fibers, suggesting that it might be due to a reverberatory activity evoked by C-fibers in interneurons of the spinal cord. Systematic studies of the in vivo activity of spinal cord neurons (Schouenborg and Sjölund 1983) established that under normal conditions, wide dynamic range or class 2 neurons showed the most pronounced windup, whereas classes 1 and 3, equivalent to low-threshold and high-threshold neurons, showed a very weak response. Windup is also observed in nociceptive withdrawal reflexes recorded as single motor units (SMU), spinal cord field potentials, isolated in vitro spinal recordings from immature rats (see for review Herrero et al. 2000), or even dorsal horn neurons in cell culture (Vikman et al. 2001). The generation of windup depends critically on the frequency of stimulation of afferent C-fibers. The greatest windup is seen at frequencies of around 1–2 Hz. It is not observed below frequencies of 0.2–0.3 Hz. Above frequencies of 20 Hz, rather than observing windup, the usual observation is a habituation of the response or wind-down. This is due to a progressive slowing of the conduction velocity of afferent C-fibers when stimulated at high frequency, resulting in a conduction blockade and a reduction in the number of impulses reaching the spinal cord with sufficiently high frequencies (Fig. 1).

The generation of windup does not only depend on the frequency of stimulation; it also depends on other parameters such as the duration and the intensity of the stimuli used. The level of excitability of spinal cord neurons and the integrity of spinal cord connections to the brain are also crucial in the generation of windup. Prolonged noxious stimulation, peripheral injury, and inflammation evoke an enhancement of the excitability of spinal cord neurons and an increase in the degree of windup as well as a reduction in the threshold for the generation of windup, which has been interpreted as a result of loss of inhibitory modulation.

In hyperalgesic states induced by peripheral injury or inflammation, windup is also evoked by stimulation of Aβ-fibers. This new phenomenon has been described either in isolated preparations of the spinal cord from newborn rats (Thompson et al. 1995) or in adult in vivo preparations (Herrero and Cervero 1996). Reflex windup facilitation and a novel A-fiber-mediated windup evoked in arthritic animals were, however, only observed in intact, not in spinalized, preparations (Herrero and Cervero 1996). The enhancement of the reflex windup observed during hyperalgesia, therefore, required an intact spinal cord and

**Windup of Spinal Cord
Neurons, Fig. 1** Windup
of a single motor unit
recorded in an in vivo
preparation. Sixteen stimuli
of 2 ms of duration, at an
intensity of 6 mA, twice the
threshold for C-fiber
activation, were applied at
1 Hz frequency in the
cutaneous receptive field.
Note how the number of
spikes recorded increases
progressively despite
a constant stimulus
intensity

Windup of Spinal Cord Neurons, Fig. 1 Windup of a single motor unit recorded in an in vivo preparation. Sixteen stimuli of 2 ms of duration, at an intensity of 6 mA, twice the threshold for C-fiber activation, were applied at 1 Hz frequency in the cutaneous receptive field. Note how the number of spikes recorded increases progressively despite a constant stimulus intensity

might be explained as the consequence of a direct descending excitatory influence on spinal cord neurons. Supraspinal modulation is also evidenced by the reduction of windup in deep dorsal horn class 2 neurons observed after applying a remote noxious mechanical stimulation to the nose of the rat (Schouenborg and Dickenson 1985), showing that windup is also affected by diffuse noxious inhibitory controls (DNIC).

The mechanisms underlying the generation of windup are still unresolved. Multiple factors may contribute to windup: network properties, presynaptic mechanisms, postsynaptic receptors, and postsynaptic membrane properties. Glutamate NMDA receptors are intimately involved in the generation of windup, as demonstrated by experiments showing that the administration of NMDA antagonists, such as ketamine, D-AP5 (2-amino-5-phosphonopentanoate), or kynurenic acid, abolished windup (Davies and Lodge 1987, Dickenson and Sullivan 1987). At normal resting membrane potential, NMDA receptors are

blocked by Mg^{2+} ions. During repetitive stimulation, each stimulus leaves the neuron at a more depolarized membrane potential, which in turn may contribute to the removal of the Mg^{2+} block from the NMDA receptors. The progressive release of the Mg^{2+} block would act as an amplifier mechanism, boosting the summation of EPSPs and hence potentiating subsequent afferent input. Furthermore, as a consequence of NMDA receptor overactivity, an increase in intracellular Ca^{2+} may activate protein kinase C and this, in turn, would increase the probability of NMDA receptor channel opening and reduce the voltage dependence of the Mg^{2+} block of the receptor (Chen and Huang 1992).

Blockade of windup by NMDA receptor antagonists is, however, only partial, and based on this observation, other modulators have been proposed to contribute to the long latency depolarization and to the hyperactivity observed during windup. Substance P, a member of the tachykinin family of peptide neurotransmitters, was proposed to be involved in windup, and this was confirmed in NK1 knockout mice by De Felipe et al. (1998). Some ▶ cyclooxygenase inhibitors, especially nitric oxide derivatives such as nitroparacetamol, are very effective inhibitors of windup (for review see Herrero et al. 2000); two complementary mechanisms of action have been proposed to explain this effect. On the one hand, nitric oxide seems to inhibit the reuptake of some monoamines like serotonin. The enhancement of the amount of monoamines in the synaptic space would reduce the release of glutamate (Kiss and Vizi 2001). On the other hand, the cyclooxygenase inhibition would reduce the level of circulating prostaglandins, which have been shown to enhance the release of glutamate and aspartate in the spinal cord (Nishihara et al. 1995).

Windup results from a synchronous electrical (and therefore artificial) stimulation of a peripheral nerve, producing a pattern of input distinct from the asynchronous discharge evoked by natural stimulation. Thus, it is not clear if this represents a genuine physiological response or is simply an artificially generated phenomenon. However, since nociceptive neurons respond in a very characteristic way to this specific pattern of artificial stimulation, it seems probable that there is a physiological correlate. Further, repetitive natural stimuli in human subjects can induce a temporal summation of pain sensation very similar, if not identical, to that induced by repetitive electrical stimuli.

Windup is also similar in many respects to the process that evokes short-term potentiation, i.e., an increase in hippocampal excitability that lasts for a few seconds up to an hour. Windup may underlie the continuation of pain sensation in response to prolonged or repetitive stimuli, despite a reduction in the number of action potentials in afferent C-fibers, and may represent an amplification mechanism or even a "pain memory" in spinal cord neurons. In addition, windup-inducing stimuli evoke some of the manifestations of ▶ central sensitization in spinal neurons, such as enlarged receptive fields and enhanced responses to input from afferent C-fibers. NMDA receptors are involved in the generation of both phenomena.

However, it has been shown that while stimuli that evoke windup may be sufficient to induce central sensitization, windup is not essential for the induction of central sensitization (Woolf 1996). There is no doubt that windup is a useful tool in the study of spinal cord processing of somatosensory information and provides an experimental correlate of spinal cord hyperexcitability, sharing some common mechanisms with central sensitization or hyperalgesia.

References

Chen, L., & Huang, L. Y. M. (1992). Protein kinase C reduces Mg block of NMDA-receptor channels as a mechanism of modulation. *Nature, 356*, 521–523.

Davies, S. N., & Lodge, D. (1987). Evidence for the involvement of N-methylaspartate receptors in 'wind-up' of class 2 neurons in the dorsal horn of the rat. *Brain Research, 424*, 402–406.

De Felipe, C., Herrero, J. F., O'Brien, J. A., et al. (1998). Altered nociception, analgesia and aggression in the mice lacking the substance P receptor. *Nature, 392*, 394–397.

Dickenson, A. H., & Sullivan, A. F. (1987). Evidence for a role of the NMDA receptor in the frequency dependent potentiation of deep rat dorsal horn nociceptive neurons following C fiber stimulation. *Neuropharmacology, 26*, 1235–1238.

Herrero, J. F., & Cervero, F. (1996). Supraspinal influences on the facilitation of rat nociceptive reflexes induced by carrageenan monoarthritis. *Neuroscience Letters, 209*, 21–24.

Herrero, J. F., Laird, J. M. A., & Lopez-Garcia, J. A. (2000). Wind-up of spinal cord neurons and pain sensation: Much ado about something? *Progress in Neurobiology, 61*, 169–203.

Kiss, J. P., & Vizi, E. S. (2001). Nitric oxide: A novel link between synaptic and nonsynaptic transmission. *Trends in Neurosciences, 24*, 211–215.

Mendell, L. M., & Wall, P. D. (1965). Responses of single dorsal cells to peripheral cutaneous unmyelinated fibers. *Nature, 206*, 97–99.

Nishihara, I., Minami, T., Watanabe, Y., et al. (1995). Prostaglandin E2 stimulates glutamate release from synaptosomes of rat spinal cord. *Neuroscience Letters, 196*, 57–60.

Schouenborg, J., & Dickenson, A. H. (1985). The effects of a distant noxious stimulation on A and C fiber evoked flexion reflexes and neuronal activity in the dorsal horn of the rat. *Brain Research, 328*, 23–32.

Schouenborg, J., & Sjölund, B. H. (1983). Activity evoked by A- and C-afferent fibers in rat dorsal horn neurons and its relation to a flexion reflex. *Journal of Neurophysiology, 50*, 1108–1121.

Thompson, S. W. N., Dray, A., McCarson, K. E., et al. (1995). Nerve growth factor induces mechanical allodynia associated with novel A fiber-evoked spinal reflex activity and enhanced neurokinin-1 receptor activation in the rat. *Pain, 62*, 219–231.

Vikman, K. S., Kristensson, K., & Hill, R. H. (2001). Sensitization of dorsal horn neurons in a two-compartment cell culture model: Wind-up and long-term potentiation-like responses. *Journal of Neuroscience, 21*, RC169.

Woolf, C. J. (1996). Windup and central sensitization are not equivalent. *Pain, 66*, 105–108.

Winging Scapula

Definition

Winging of the scapula is a protrusion of the scapula away from the thorax when the patient attempts to raise the arm above the shoulder level. The serratus anterior muscle keeps the scapula close to the ribs while stabilizing the scapula for shoulder abduction, overhead use of the hand, and push-up-type work. Any injury or compression of the long thoracic nerve, which innervates this muscle, will cause scapular winging. A neurolysis of the long thoracic nerve in the thoracic inlet can resolve this problem.

Cross-References

▶ Thoracic Outlet Syndrome

Withdrawal

Definition

Withdrawal is a reflex reaction (e.g., the rat withdrawing its limb or tail) from a noxious stimulus.

Cross-References

▶ Randall-Selitto Paw Pressure Test

Withdrawal Symptoms

Definition

Withdrawal symptoms are a result of abrupt discontinuation of a drug, and in the case of opioid analgesics can include diarrhea, sweating, insomnia, and goose bumps.

Cross-References

▶ Opioids, Clinical Opioid Tolerance

Work Capacity

Definition

Work capacity is the individual's ability to experience, understand, and learn to perform job tasks/occupations resulting in production that is at least comparable with the industrial standard for that product or service. In choosing a suitable occupation, work capacity is understood as the individual's valuable and useful quality and skills, indicating that he/she is specially matched for a certain type of work.

Cross-References

▶ Vocational Counseling

Work Capacity Evaluation

▶ Disability, Functional Capacity Evaluations

Work Conditioning

▶ Physical Conditioning Programs

Work Duties

▶ Pain in the Workplace, Risk Factors for Chronicity, Job Demands

Work Environment

▶ Pain in the Workplace, Risk Factors for Chronicity, and Workplace Factors

Work Fitness

Definition

Work fitness is a capacity to successfully meet the present and potential challenges of work requirements with vigor, and to demonstrate the traits and capacities that will prevent occupational injuries.

Cross-References

▶ Vocational Counseling

Work Fitness Test

▶ Disability, Functional Capacity Evaluations

Work Hardening

▶ Physical Conditioning Programs

Work Intolerance

▶ Back Pain in the Workplace

Work Performance Evaluation

▶ Disability, Functional Capacity Evaluations

Work Sample

▶ Disability, Functional Capacity Evaluations

Work Samples

Definition

Work samples are designed to appraise whether an individual's ability to perform actual work matches the job description and skills required for that job. The individual's vocational aptitudes (physical and mental), skills, and occupational behavior are exposed. In a therapeutic environment, the individual performs well-defined simulated work activities, involving tasks, materials, and tools resembling those in actual job tasks, occupations, or cluster of occupations. A work sample simulates factors that are required in thousands of specific jobs.

One example is the VALPAR Component Work Sample (VCWS), which covers 21 individual criterion-referenced work samples. Each VCWS job requirement corresponds to the classification and analysis system of the Dictionary of Occupational Titles (DOT) and the Worker Qualification Profile (WQP). Ability to perform a cluster of VCWS demonstrates whether the individual has met the WQP. The WQP is rated in a time-standard known as Methods-Time Measurement (MTM) scores. MTMs represent the standards of time a well-trained worker would be expected to perform tasks like those of the VALPAR's VCWS within a typical industrial setting during an 8 h working day.

Cross-References

▶ Disability, Functional Capacity Evaluations
▶ Vocational Counseling

Worker Compensation Benefits

Synonyms

Disability payment; Time loss payment; Wage replacement benefit

Definition

Benefits paid to a worker who is unable to work because of an injury that occurred during the conduct of his or her job. They are based on a proportion of the worker's pre-injury salary.

Cross-References

▶ Impairment Rating, Ambiguity
▶ Impairment Rating, Ambiguity, IAIABC System
▶ Pain in the Workplace, Risk Factors for Chronicity, and Workplace Factors

Workers' Compensation

▶ Impairment Rating, Ambiguity
▶ Impairment Rating, Ambiguity, IAIABC System
▶ Rating Impairment Due to Pain in a Workers' Compensation System

Workplace

▶ Pain in the Workplace, Risk Factors for Chronicity, and Workplace Factors

Workplace Design

▶ Ergonomics Essay

Workplace Factors

▶ Pain in the Workplace, Risk Factors for Chronicity, and Workplace Factors

Work-Related Assessment

▶ Disability, Functional Capacity Evaluations

Work-Related Musculoskeletal Disorder

Synonyms

WRMD

Definition

This collective term is used to define the whole of muscle, bone, tendon, nerve, and fascia disorders which are presumed to be work related.

Cross-References

▶ Ergonomic Counseling

Work-Related Musculoskeletal Disorders

▶ Ergonomic Counseling

Work-related Upper-Extremity Disorders

▶ Disability, Upper Extremity

World Health Organization (WHO) Analgesic Ladder

Definition

The World Health Organization Analgesic Ladder is a structured approach to cancer pain management, proposed by an expert committee convened by the Cancer and Palliative Care Unit of the World Health Organization, which focuses on selecting analgesics on the basis of pain intensity.

Cross-References

▶ Cancer Pain Management, Nonopioid Analgesics
▶ Cancer Pain Management, Principles of Opioid Therapy, Drug Selection

WRMD

▶ Work-Related Musculoskeletal Disorder

WRUEDs

▶ Disability, Upper Extremity

X

Xanthochromia

Definition

Yellowish color of the cerebrospinal fluid due to intracranial bleeding or blood–brain-barrier defects.

Cross-References

▶ Headache Due to Low Cerebrospinal Fluid Pressure

G.F. Gebhart, R.F. Schmidt (eds.), *Encyclopedia of Pain*, DOI 10.1007/978-3-642-28753-4,
© Springer-Verlag Berlin Heidelberg 2013

Xanthochromia

Definition

Yellowish color of the cerebrospinal fluid due to antecedent bleeding, open blood-brain-barrier, etc.

Cross-References

▶ Headache - Due to Low Cerebrospinal Fluid Pressure

F. Gerhan, R.F. Schmidt (eds.), *Encyclopedia of Pain*, DOI 10.1007/978-3-642-28753-4,
© Springer-Verlag Berlin Heidelberg 2013

Y

Yellow Flags

Chris J. Main
Keele University, Keele, Staffordshire, UK

Synonyms

Psychosocial obstacles to recovery; Psychosocial risk factors

Definition

Yellow Flags are psychosocial factors associated with risk of the development of chronicity in patients/employees with low back pain. The increasing costs of chronic low back pain, despite advances in technological medicine, stimulated the search for new solutions to the problem of low back disability. They have more recently been subdivided into *Yellow Flags* (clinical factors), *Blue Flags* (occupational factors), *and Black Flags* (contextual factors), although there is a degree of overlap among them.

Characteristics

In New Zealand, the increasing costs of chronic nonspecific low back pain had become an unmanageable burden. This fuelled a new initiative (Kendall et al. 1997) designed to complement an available set of acute back pain

management guidelines, with a psychosocial assessment system designed systematically to address the psychosocial risk factors that had been shown in the scientific literature to be predictive of chronicity. The term "Yellow flags" was coined to encompass psychological and social/environmental risk factors for prolonged disability and failure to return to work as a consequence of musculoskeletal symptoms.

The stated purposes of assessment of Yellow flags were to
- Provide a method for screening for psychosocial factors
- Provide a systematic approach to the assessment of psychosocial factors
- Suggest better strategies for better management for those with back pain who appear at a higher risk of chronicity

The focus was on a number of key psychological factors:
- Belief that back pain is harmful or severely disabling
- Fear-avoidance behavior patterns with reduced activity levels
- Tendency to low mood and withdrawal from social interaction
- Expectation that passive treatments rather than active participation will help

The Yellow flags were integrated into an assessment system, which included a screening questionnaire, interview guidelines, and recommendations for early behavioral management in individuals with low back pain. They consisted of both psychological and socio-occupational

G.F. Gebhart, R.F. Schmidt (eds.), *Encyclopedia of Pain*, DOI 10.1007/978-3-642-28753-4,
© Springer-Verlag Berlin Heidelberg 2013

risk factors. The main categories of the Yellow flags are as follows:

- Attitudes and beliefs about back pain
- Behaviors
- Compensation issues
- Diagnostic and treatment issues
- Emotions
- Family
- Work

They also included a number of specific guidelines for behavioral management which are as follows:

1. Provide a positive expectation that the individual will return to work.
2. Be directive in scheduling regular reviews of progress.
3. Keep the individual active and at work.
4. Acknowledge difficulties of daily living.
5. Help maintain positive cooperation.
6. Communicate that having more time off work reduces the likelihood of successful return.
7. Beware of expectations of "total cure" or expectation of simple "techno-fixes."
8. Promote self-management and self-responsibility.
9. Be prepared to say "I don't know."
10. Avoid confusing the report of symptoms with the presence of emotional distress.
11. Discourage working at home.
12. Encourage people to recognize that pain can be controlled.
13. If barriers are too complex, arrange multidisciplinary referral (Kendall et al. 1997).

In addition to current management advice, there are clear preventative components within the guidelines.

The monograph provided a guide to the assessment of Yellow flags that included a clinical interview and a psychosocial screening questionnaire. This approach assumed that individuals at risk of poor outcomes could be identified on the basis of either a small cluster of highly salient factors, or the cumulative combination of several factors. Since many of these are potentially modifiable, the monograph also contained additional advice on how to incorporate cognitive-behavioral change principles into early management. The concept of Yellow flags has subsequently sparked much attention and debate, and has been adopted in several guidelines on the early management of work-related low back injuries.

Nicholas et al. (2011) offer a recent review of the evidence base on the utility of yellow flags in screening and targeting psychosocial risk factors, while Hill et al. (2011), in a prospective RCT of secondary prevention of LBP, demonstrated the superiority of stratified care, using a screening and targeting approach, over best practice usual care.

More recently, a distinction has been drawn between psychological risk factors that might be considered essentially "normal" but unhelpful psychological reactions to musculoskeletal symptoms (e.g., the belief that pain necessarily implies damage) and clearly "abnormal" psychological/psychiatric factors or disorders (e.g., Post-Traumatic Stress Disorder or Major Depression) suggestive of diagnosable psychopathology. It has been suggested that the "normal" but unhelpful psychological reactions should be described as "Yellow flags," while those meeting criteria for psychopathology should be termed "Orange flags" (Main et al. 2005), and trigger referral for mental health evaluation. The primary significance of this distinction is to differentiate Yellow flag factors which might be amenable to change by suitably trained health-care providers, such as general medical practitioners and physiotherapists, from Orange flag factors that require specialist mental health referral.

Beyond Yellow and Orange Flags: The Blue Flags and Black Flags

Traditionally, clinical rehabilitation had focused primarily on clinical outcomes, with relatively little attention directed specifically at work. Although designed originally for a case-management system, the original Yellow Flags focused primarily on perceptions of health. Initially, the prime reason for development of the flags was to offer a complementary analysis to the ▶ Red Flags (based exclusively on a medical model of illness). Initially, it was decided that the collective importance of psychosocial factors

Yellow Flags, Fig. 1 Psychologically informed practice (From Main and George 2011)

was such that it made sense to integrate them under a single heading, thereby distinguishing them from biomedical factors. It became clear, however, that implicit within the Yellow Flag initiative was the possibility of a range of different solutions, involving both health-care providers and occupational personnel. Burton and Main (2002) decided that insufficient attention had been given to specific occupational factors, and recommended that the term "Yellow flags" should be reserved for more overtly psychological risk factors, such as fears and unhelpful beliefs, and these should be distinguished from workplace and other contextual factors. It was, therefore, decided to subdivide the yellow flags into clinical Yellow Flags and occupationally focused Blue Flags. A more recent overview is presented in Main et al. (2008).

These perceived features of work, generally associated with higher rates of symptoms, ill health, and work loss have been termed Blue flags. In the context of injury, these factors may delay recovery or constitute a major obstacle to it; and for those at work there may be major contributory factors to suboptimal performance or "Presenteeism," a term recently reviewed by Johns (2010). They are characterized by features, such as high demand/low control, unhelpful management style, poor social support from colleagues, perceived time pressure, and lack of job satisfaction. Individual workers may differ in their perception of the same working environment.

Blue flags incorporate not only issues related to the perception of job characteristics, such as job demand, but also perception of social interactions (whether with management or fellow-workers). From a work retention or work rehabilitation perspective, however, a further distinction has been made between two types of occupational risk factors: those concerning the perception of work (Blue Flags) and organizational obstacles to recovery, including objective work characteristics and conditions of employment (Black flags), recently also described in Main et al. (2008). The distinction is similar to the distinction between intrinsic and extrinsic factors by Deci and Ryan (1974), but with a focus on obstacles to recovery (i.e., potential targets for some sort of biopsychosocial intervention).

Flag identification and management is an integral part of *Psychologically Informed Practice* (Main and George 2011). The distinction is depicted in Fig. 1 with the color coding referring to the type of flag.

In 2007, a major review of the Flags initiative was undertaken as part of a *Decade of the* Flags consensus meeting of experts (with associated conference), hosted by Keele University.

As a result of the meeting, publications on the nature, and utility of the Yellow Flags (Nicholas et al. 2011) and Blue Flags (Shaw et al. 2009) have been produced as well as a practical monograph on flag management in clinical and occupational settings (Kendall et al. 2009).

▶ Black Flags are not a matter of perception, and affect all workers equally. Originally their primary focus included both nationally established policy concerning conditions of employment and sickness policy; and working conditions specific to a particular organization. Recently, the distinction more helpfully has been drawn between specific occupational factors, which can be either worker-centered or workplace-centered, and the much wider set of contextual factors influencing the health-work interface, including economic, social, and cultural factors (Waddell and Aylward 2010).

Some jobs require a higher level of working capability than others for successful work retention. After certain types of injury, such jobs may be specifically contraindicated and therefore, constitute an absolute obstacle to return to work. It might be hoped that many such risk factors could be "designed" out of the working environment, but such factors certainly need to be evaluated in the context of work retention or rehabilitation. Specifically, in terms of prevention they are a matter primarily for legislation, establishment of satisfactory job design, and adherence to recommended work practices, in negotiation, where appropriate, between employers and employee organizations.

In summary, the Yellow Flag initiative has been important in two major ways:

Firstly, in promulgation of a "systems approach," involving all key stakeholders, in addressing issues of clinical and occupational rehabilitation, with emphasis on the need for better clinical-occupational interfaces, and secondly, in attempting to identify and reduce risk factors for chronic incapacity, whether in health care or in occupational settings.

Arguably, the strength of the Yellow Flags initiative has been the fact that it has offered a conceptual alternative to the biomedical model of low back pain disability. The development of the Blue and Black Flags has offered a more detailed conceptual framework within which a range of initiatives may be linked. At the core of the "flag" construct is a conceptual shift from risks (many of which may be immutable), to obstacles, to recovery. This may be individual clinical factors or perceptions of work or organization, which in turn require a range of solutions from individual clinical interventions and work reintegration strategies to fundamental redesign of the individual-organization interface.

The conceptual framework of the "Flags" approach now needs to be supported by further empirical research in five key areas:

- The statistical construction and validation of screening strategies
- Refinement of cognitive-behavioral strategies appropriate for different contexts
- Identification of appropriate competencies for the delivery of psychosocial interventions
- Integration of sickness/disability management strategies with the development and enhancement of positive/adaptive coping strategies
- Developing an evidence base of context-specific interventions, including benchmarking of successful initiatives

In conclusion, the Yellow Flags initiative has offered a way of understanding and addressing psychosocial obstacles to recovery (or to optimal function). In its more recent conceptual development (incorporating both Blue and Black Flags), it may also offer an integrative framework within which to consider opportunities for change.

References

Burton, A. K., & Main, C. J. (2002). Obstacles to recovery from work-related musculoskeletal disorders. In W. Karwowski (Ed.), *International encyclopedia of ergonomics and H factors* (pp. 1542–1544). London: Taylor and Francis.

Deci, E. L., & Ryan R. M. (2012). Motivation, Personality and Development within embedded social contexts: an overview of self-determination theory. In R.M. Ryan (Ed.), *the Oxford Handbook of Human Motivation* (pp. 85–110). Oxford: Oxford University Press.

Hill, J. C., Whitehurst, D. G. T., Lewis, M., Bryan, S., Dunn, K. M., Foster, N. E., Konstantinou, K., Main, C. J., Mason, E., Somerville, S., Sowden, S.,

Vohora, K., & Hay, E. M. (2011). Comparison of stratified primary care management for low back pain with current best practice (STarT Back): A randomised controlled trial. *Lancet, 378*, 1560–1571.

Johns, G. (2010). Presenteeism in the workplace: A review and research agenda. *Journal of Organizational Behavior, 31*, 519–542. doi:10.1002/job.630.

Kendall, N., Linton, S. L., & Main, C. J. (1997). *Guide to assessing psychological yellow flags in acute low back pain: Risk factors for long term disability and work loss*. Wellington: Accident Rehabilitation and Compensation Insurance Corporation of New Zealand and the National Health Committee.

Kendall, N. A. S., Burton, A. K., Main, C. J., & Watson, P. J. (2009). *Tackling musculoskeletal problems. The psychosocial flags framework: A guide for clinic and workplace*. London: The Stationary Office.

Main, C. J., & George, S. Z. (2011). Psychologically informed practice for management of low back pain: Future directions in practice and research. *Physical Therapy, 91*, 820–827.

Main, C. J., Phillips, C. J., & Watson, P. J. (2005). Secondary prevention in health-care and occupational settings in musculoskeletal conditions (focusing on low back pain), Chap. 21. In I. Z. Schultz & R. J. Gatchel (Eds.), *Handbook of complex occupational disability claims: Early risk identification, intervention and prevention* (pp. 387–404). New York: Springer Science and Business Media.

Main, C. J., Sullivan, M. J. L., & Watson, P. J. (2008). Psychological perspectives on work, Chap 13. In *Pain Management: Practical applications of the biopsychosocial perspective in clinical and occupational settings* (pp. 343–353). Edinburgh: Churchill-Livingstone.

Nicholas, M. K., Watson, P. J., Linton, S. J., & Main, C. J. (2011). The identification and management of psychosocial risk factors (Yellow Flags) in patients with low back pain. *Physical Therapy, 91*, 737–753.

Shaw, W. S., van der Windt, D. A., Main, C. J., Loisel, P., & Linton, S. J. (2009). Now tell me about your work: The feasibility of early screening and intervention to address occupational factors ("Blue Flags") in back disability. *Journal of Occupational Rehabilitation, 19*, 64–80.

Waddell, G., & Aylward, M. (2010). *Models of sickness and disability: Applied to common health problems*. London: The Royal Society of Medicine Press.

Z

Z Joint

▶ Zygapophyseal Joint

Z Joint Blocks

▶ Cervical Medial Branch Blocks

Zidovudine (AZT)-induced Headache

▶ Pain in Human Immunodeficiency Virus Infection and Acquired Immune Deficiency Syndrome

Zona Incerta

Definition

A region just ventral to the thalamus and dorsal hypothalamus. Receives inputs from the sensory motor areas, ventral lateral geniculate, trigeminal complex, and the spinal cord. Recent animal studies suggest that the zona incerta (ZI) is involved in modulating nociception via its projection to thalamus.

Cross-References

▶ Brainstem Subnucleus Reticularis Dorsalis Neuron
▶ Thalamus, Visceral Representation

Zoster-associated Pain

▶ Postherpetic Neuralgia, Etiology, Pathogenesis, and Management

Zydol

▶ Tramadol Hydrochloride

Zygapophyseal Joint

Synonyms

Apophyseal joint; Posterior intervertebral joint; Z joint

G.F. Gebhart, R.F. Schmidt (eds.), *Encyclopedia of Pain*, DOI 10.1007/978-3-642-28753-4,
© Springer-Verlag Berlin Heidelberg 2013

Definition

The facet joint bridging the vertebrae behind the vertebral foramina. May also be known as the apophyseal joint, posterior intervertebral joint, or Z joint.

Cross-References

▶ Cervical Transforaminal Injection of Steroids
▶ Chronic Low Back Pain: Definitions and Diagnosis

Zygapophyseal Joint Injection

▶ Facet Joint Procedures for Chronic Back Pain

Zygapophysial Joint Injection

▶ Facet Joint Procedures for Chronic Back Pain

Zygapophysial Joint Pain

▶ Facet Joint Pain

Zygapophysis

Definition

One of the articular processes of a vertebra, of which there are usually four, two anterior and two posterior.

Cross-References

▶ Facet Joint Pain

Zymosan

Definition

Zymosan is an insoluble carbohydrate from the cell wall of yeast that leads to immune activation after peri-sciatic application. It is frequently used in animal studies to produce a painful inflammation and hyperalgesia.

Cross-References

▶ Amygdala, Pain Processing and Behavior in Animals

List of Entries

G.F. Gebhart, R.F. Schmidt (eds.), *Encyclopedia of Pain*, DOI 10.1007/978-3-642-28753-4,
© Springer-Verlag Berlin Heidelberg 2013